Molecular Biology of
THE CELL
Seventh Edition

Molecular Biology of THE CELL

Seventh Edition

Bruce Alberts

Rebecca Heald

Alexander Johnson

David Morgan

Martin Raff

Keith Roberts

Peter Walter

With problems by

John Wilson

Tim Hunt

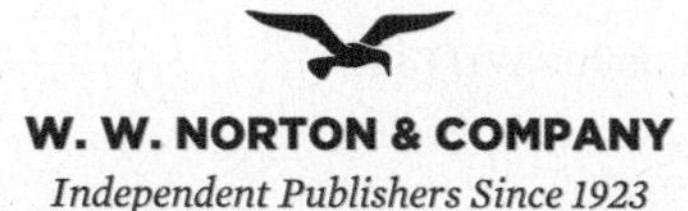

W. W. NORTON & COMPANY

Independent Publishers Since 1923

W. W. Norton & Company has been independent since its founding in 1923, when William Warder Norton and Mary D. Herter Norton first published lectures delivered at the People's Institute, the adult education division of New York City's Cooper Union. The firm soon expanded its program beyond the Institute, publishing books by celebrated academics from America and abroad. By midcentury, the two major pillars of Norton's publishing program—trade books and college texts—were firmly established. In the 1950s, the Norton family transferred control of the company to its employees, and today—with a staff of five hundred and hundreds of trade, college, and professional titles published each year—W. W. Norton & Company stands as the largest and oldest publishing house owned wholly by its employees.

Printed in Canada

Editor: Betsy Twitchell
Editorial Advisor: Denise Schanck
Senior Associate Managing Editor, College: Carla L. Talmadge
Assistant Editor: Danny Vargo
Director of College Production: Jane Searle
Copyeditor: Christopher Curioli
Proofreaders: Julie Henderson, Susan McColl
Managing Editor, College: Marian Johnson
Media Editor: Todd Pearson
Smartwork Editor: Christopher Rapp
Media Project Editor: Jesse Newkirk
Associate Media Editor: Jasmine N. Ribeaux
Media Assistant Editor: Lindsey Heale
Ebook Production Manager: Kate Barnes
Managing Editor, College Digital Media: Kim Yi
Marketing Manager, Biology: Ruth Bolster
Director of College Permissions: Megan Schindel
Photo Editor: Thomas Persano
Permissions Associate: Patricia Wong
Design: Juan Paolo Francisco
Illustrator: Nigel Orme
Composition: Graphic World, Inc.
Manufacturing: Transcontinental—Beauceville

Library of Congress Cataloging-in-Publication Data

Names: Alberts, Bruce, author.
Title: Molecular biology of the cell / Bruce Alberts, Rebecca Heald, Alexander Johnson, David Morgan, Martin Raff, Keith Roberts, Peter Walter.
Description: Seventh edition. | New York : W. W. Norton & Company, [2022] | Includes bibliographical references and index.
Identifiers: LCCN 2021049376 | **ISBN 9780393884821 (hardcover)** | ISBN 9780393884630 (epub)
Subjects: MESH: Cells | Molecular Biology
Classification: LCC QH581.2 | NLM QU 300 | DDC 572.8—dc23/eng/20211015
LC record available at https://lccn.loc.gov/2021049376

W. W. Norton & Company, Inc., 500 Fifth Avenue, New York, NY 10110

wwnorton.com

W. W. Norton & Company Ltd., 15 Carlisle Street, London W1D 3BS

3 4 5 6 7 8 9 0

Preface

Why a cell biology textbook? What is its value in a world of online resources so vast that any information you might want about cells is, in principle, freely available a few taps away?

The answer is that a textbook provides what open-ended Internet searches cannot—a curation of knowledge and an expert, accurate guide to the beauty and complexities of cells. Our book provides a narrative that leads the reader logically and progressively through the key concepts, components, and experiments in such a way that readers can build for themselves a memorable, conceptual framework for cell biology—a framework that will allow them to understand and critically evaluate the exciting rush of new discoveries. That is what we have tried to do in *Molecular Biology of the Cell* for each of its seven editions.

This edition was completed during the COVID-19 pandemic. Many of the questions that this global crisis generated are cell biological questions—including how the virus gets into our cells, how it replicates, how our immune system responds, how vaccines are developed, and how scientists produce the molecular details of virus structure. Required for the rapid development of safe and effective COVID-19 vaccines, answers to all of these questions can be found in this textbook. To make room for them, as well as for many other major recent advances in our knowledge, much previous content had to be removed.

Understanding the inner workings of cells requires more than words. Our book contains more than 1500 illustrations that create a parallel narrative, closely interwoven with the text. Each figure has been designed to highlight a key concept. The unique clarity, simplicity, and consistency of the figures across chapters, achieved by use of a common set of icon designs and colors (for example, DNA *red* and proteins *green*), enables students to scan them as chapter overviews. In this edition, important protein structures are depicted and their Protein Data Bank (PDB) codes provided; these codes link to tools on the RCSB PDB website (www.rcsb.org), where students can more fully explore the proteins that lie at the core of cell biology. In addition, more than 180 narrated movies have been produced for the book, each linked to the text to provide additional insights.

John Wilson and Tim Hunt have again contributed their distinctive and imaginative problems to help students gain a more active understanding of the text. The end-of-chapter problems emphasize experiments and quantitative approaches in order to encourage critical thinking. Their Digital Problems Book in Smartwork greatly expands on these self-assessment problems and includes data analysis and video review questions that are based on the movie links in the textbook.

Many millions of scientific papers are relevant to cell biology, and many important new ones are published daily. The challenge for textbook writers is to sort through this overwhelming wealth of information to produce a clear and accurate conceptual platform for understanding how cells work. We have aimed high, seeking primarily to support the education of cell biology students, including the next generation of bioscientists, but also to support active scientists pursuing new fundamental research and the search for practical advances to improve the human condition.

So, why read a textbook? We live in a world that presents humanity with many challenging problems related to cell biology, including declining biodiversity, climate change, food insecurity, environmental degradation, resource depletion, and animal and plant diseases. We hope that this new edition of our textbook will help the reader to better understand these problems and—for many—to contribute to solving them.

Note to the Reader

What's New in the Seventh Edition?

Every chapter in the Seventh Edition has been significantly updated with information on new discoveries in the field of cell biology. Examples of this new content include:

- Updated information on the continuing impact of human genome research, including what has been learned from sequencing hundreds of thousands of human genomes (Chapter 4), and updated coverage of tumor genomes (Chapter 20).
- New research on pathogens, diseases, and methods of combating them, including discussion of COVID-19 (Chapters 1, 5, and 23) and mRNA vaccines (Chapter 24).
- Updated research on cellular organization, including new information on biomolecular condensates (Chapters 3, 6, 7, 12, and 14) and on chromosome organization by DNA loop extrusion (Chapters 4, 7, and 17).
- Expanded coverage of new microscope technologies, including superresolution light microscopy and atomic resolution electron microscopy (Chapter 9), and new research breakthroughs from cryo-electron microscopy, such as stretch-activated Piezo channels (Chapter 11).
- New coverage of evolution, including a new discussion on the diversity of life (Chapter 1), plus updates on both human (Chapter 4) and HIV (Chapter 23) evolution.

In addition, a quarter of the book's illustrations are either completely new or significantly updated for accuracy, clarity, and visual appeal.

Finally, we are thrilled to offer online assessment, for the first time, with the Digital Problems Book in Smartwork—reimagining the classic companion text, *The Problems Book*, for twenty-first century instructors and students.

Structure of the Book

Although the chapters of this book can be read independently of one another, they are arranged in a logical sequence of five parts. The first three chapters of Part I cover elementary principles and basic biochemistry. They can serve either as an introduction for those who have not studied biochemistry or as a refresher course for those who have. Part II deals with the storage, expression, and transmission of genetic information. Part III presents the principles of the main experimental methods for investigating and analyzing cells; here, a section titled "Mathematical Analysis of Cell Function" in Chapter 8 provides an extra dimension in our understanding of cell regulation and function. Part IV describes the internal organization of the cell. Part V follows the behavior of cells in multicellular systems, starting with how cells become attached to each other and concluding with chapters on pathogens and infection and on the innate and adaptive immune systems.

End-of-Chapter Problems

A selection of problems, written by John Wilson and Tim Hunt, appears in the text at the end of each chapter. Solutions to these problems are available on the Norton Teaching Tools site.

References

A concise list of selected references is included at the end of each chapter. These are arranged in alphabetical order by author surname under the main chapter section headings. These references often include the original papers in which the most critical discoveries were first reported. The ebook also includes the DOI identifier for the references, making it easy for students to access the articles.

Glossary Terms

Throughout the book, boldface type has been used to highlight key terms at the point in a chapter where the main discussion occurs. Italic type is used to set off important terms with a lesser degree of emphasis. At the end of the book is an expanded glossary, covering all the major terms common to cell biology; it should be the first resort for a reader who encounters an unfamiliar technical word.

Website for Students

Resources for students are available at **digital.wwnorton.com/mboc7**. The complete glossary as well as a set of flashcards are available on this student website.

Nomenclature for Genes and Proteins

Each species has its own conventions for naming genes; the only common feature is that they are always set in italics. In some species (such as humans), gene names are spelled out all in capital letters; in other species (such as zebrafish), all in lowercase; in yet others (most mouse genes), with the first letter in uppercase and the rest in lowercase; or (as in *Drosophila*) with different combinations of uppercase and lowercase, according to whether the first mutant allele to be discovered produced a dominant or recessive phenotype. Conventions for naming protein products are equally varied.

This typographical chaos drives everyone crazy. Moreover, there are many occasions, especially in a book such as this, where we need to refer to a gene generically—without specifying the mouse version, the human version, the chick version, or the hippopotamus version—because the gene variants across species are all equivalent for the purposes of our discussion. What convention then should we use?

We have decided in this book to follow a uniform rule. We write all gene names with the first letter in uppercase and the rest in lowercase, and all in italics, thus: *Bazooka, Cdc2, Dishevelled, Egl1*. The corresponding protein, where it is named after the gene, will be written in the same way, but in roman rather than italic letters: Bazooka, Cdc2, Dishevelled, Egl1. When it is necessary to specify the organism, this can be done with a prefix to the gene name.

For completeness, we list a few further details of naming rules that we shall follow. In some instances, an added letter in the gene name is traditionally used to distinguish between genes that are related by function or evolution; for those genes, we put that letter in uppercase if it is usual to do so (*LacZ, RecA, HoxA4*). Proteins are more of a problem. Many of them have names in their own right, assigned to them before the gene was named. Such protein names take many forms, although most of them traditionally begin with a lowercase letter (actin, hemoglobin, catalase); others are acronyms (such as GFP, for green fluorescent protein, or BMP4, for bone morphogenetic protein 4). To force all such protein names into a uniform style would do too much violence to established usages, and we shall simply write them in the traditional way. For the corresponding gene names in all these cases, we shall nevertheless follow our standard rule: *Actin, Hemoglobin, Catalase, Bmp4, Gfp*.

For those who wish to know them, the table shows some of the official conventions for individual species—conventions that we shall mostly violate in this book, in the manner shown.

	Species-specific convention		Unified convention used in this book	
Organism	**Gene**	**Protein**	**Gene**	**Protein**
Mouse	*Hoxa4*	Hoxa4	*HoxA4*	HoxA4
	Bmp4	BMP4	*Bmp4*	BMP4
	integrin α-1, Itgα1	integrin α1	*Integrin α1, Itgα1*	integrin α1
Human	*HOXA4*	HOXA4	*HoxA4*	HoxA4
Zebrafish	*cyclops, cyc*	Cyclops, Cyc	*Cyclops, Cyc*	Cyclops, Cyc
Caenorhabditis	*unc-6*	UNC-6	*Unc6*	Unc6
Drosophila	*sevenless, sev* (named after recessive phenotype)	Sevenless, SEV	*Sevenless, Sev*	Sevenless, Sev
	Deformed, Dfd (named after dominant mutant phenotype)	Deformed, DFD	*Deformed, Dfd*	Deformed, Dfd
Yeast				
Saccharomyces cerevisiae (budding yeast)	*CDC28*	Cdc28, Cdc28p	*Cdc28*	Cdc28
Schizosaccharomyces pombe (fission yeast)	*Cdc2*	Cdc2, Cdc2p	*Cdc2*	Cdc2
Arabidopsis	*GAI*	GAI	*Gai*	GAI
Escherichia coli	*uvrA*	UvrA	*UvrA*	UvrA

Resources for Instructors

digital.wwnorton.com/mboc7

Designed to enrich the classroom experience, Instructor Resources are available at **digital.wwnorton.com/mboc7**. Adopting instructors can obtain access to the site from their sales representative, who can be identified by visiting wwnorton .com/educator and clicking the "Find My Rep" button.

The Digital Problems Book in Smartwork

For the first time, the popular print supplement *Molecular Biology of the Cell: The Problems Book* is now available in Smartwork. Easier for instructors to assign and more helpful to students because of each question's pedagogical scaffolding, the Digital Problems Book in Smartwork features the questions authored by Tim Hunt and John Wilson adapted for digital delivery. An enormous library of almost 3500 questions that include critical thinking questions, data analysis questions, and animation and video questions, allows instructors to deliver the exact type of assessment that their students need. The Digital Problems Book in Smartwork comes at no additional cost with all new copies of *Molecular Biology of the Cell.*

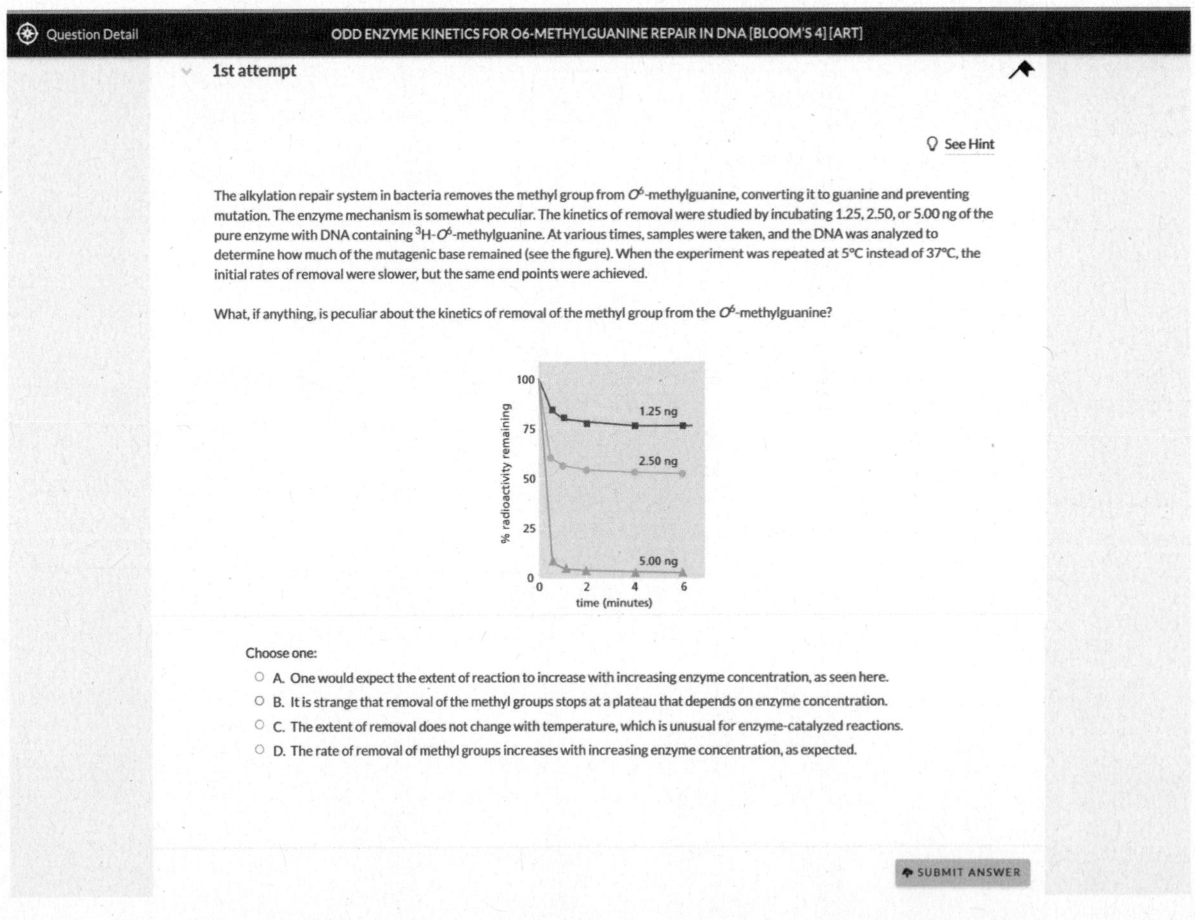

Norton Teaching Tools

The Norton Teaching Tools site for *Molecular Biology of the Cell* provides creative and engaging resources to refresh a syllabus or to design a new one. Dynamic, experienced instructors have created primary literature suggestions, active learning activities, lecture PowerPoint files, descriptions of all of the animations and videos, and much more. All of the teaching tools are aligned with chapter topics and organized by activity type, making it easily sortable. The site also features tips for assigning Norton's digital learning tools and addressing the most common course challenges.

Animations and Videos

Under the authorial direction of Michele M. McDonough and Thomas A. Volpe, both of Northwestern University, the animations and video library has been thoroughly updated and expanded. The more than 180 animations and videos are integrated into the ebook and also available to students and instructors at digital.wwnorton.com/mboc7. Instructors can view descriptions of each on the Norton Teaching Tools site.

Norton Ebook

The purchase of any new print copy of the Seventh Edition of *Molecular Biology of the Cell* includes access to the Norton Ebook version of the text at no additional cost. The Norton Ebook can be purchased as an affordable stand-alone option that provides an active reading experience, enabling students to take notes, bookmark, search, highlight, and read offline. All of the videos and animations appear directly in the ebook, and instructors can add notes that students can see as they are reading the text.

Art of *Molecular Biology of the Cell*, Seventh Edition

The images from the book are available in two convenient formats: PowerPoint and JPEG, and in both labeled and unlabeled versions.

Figure-integrated Lecture Outlines

The section headings, concept headings, and figures from the text have been integrated into PowerPoint presentations and can be customized. For example, the content of these presentations can be combined with videos, questions from the book, or activities in the Norton Teaching Tools site, in order to create unique lectures that facilitate interactive learning.

Test Bank

Updated for the Seventh Edition, the test bank includes a variety of question formats: multiple choice, short answer, fill-in-the-blank, true–false, and matching. The test bank was created with the philosophy that a good exam should require students to reflect upon and integrate information as a part of a sound understanding. Questions are classified by section and difficulty, making it easy to construct tests and quizzes. The test bank question library includes about 70 questions per chapter, ensuring instructors can find the right questions for their exams. It will be delivered through Norton Testmaker, which brings the high-quality questions in the test bank online. Create assessments for your course without downloading files or installing specialized software, customize test bank questions, and easily export your tests to Microsoft Word or Common Cartridge files for your learning management system (LMS).

About the Authors

Bruce Alberts received his PhD from Harvard University and is the Chancellor's Leadership Chair in Biochemistry and Biophysics for Science and Education, University of California, San Francisco. He was the editor-in-chief of *Science* magazine from 2008 until 2013, and for twelve years he served as president of the U.S. National Academy of Sciences (1993–2005).

Rebecca Heald received her PhD from Harvard University and is professor of molecular and cell biology at the University of California, Berkeley. She also serves as the co-chair of that department.

Alexander Johnson received his PhD from Harvard University and is professor of microbiology and immunology at the University of California, San Francisco. He is also director of the Program in Biological Sciences (PIBS) at UCSF.

David Morgan received his PhD from the University of California, San Francisco, and is professor of the Department of Physiology there as well as vice dean for research in the School of Medicine.

Martin Raff received his MD from McGill University and is an emeritus professor of biology and an affiliated member of the Medical Research Council Laboratory for Molecular Cell Biology at University College London.

Keith Roberts received his PhD from the University of Cambridge and was deputy director of the John Innes Centre, Norwich. He is emeritus professor at the University of East Anglia.

Peter Walter received his PhD from the Rockefeller University in New York and is professor of the Department of Biochemistry and Biophysics at the University of California, San Francisco, and an investigator at the Howard Hughes Medical Institute.

John Wilson received his PhD from the California Institute of Technology. He is Distinguished Emeritus Professor of Biochemistry and Molecular Biology at Baylor College of Medicine in Houston.

Tim Hunt received his PhD from the University of Cambridge, where he taught biochemistry and cell biology for more than 20 years. He worked at Cancer Research UK from 1990 to 2010. He shared the 2001 Nobel Prize in Physiology or Medicine with Lee Hartwell and Paul Nurse. He moved to Okinawa in 2016.

Acknowledgments

We once again thank our spouses, partners, families, friends, and colleagues for their continuing patience and support, without which the writing of this new edition of our textbook would not have been possible. As always, we are also indebted to a large number of scientists whose generous help has been essential for making the text as clear, up-to-date, and accurate as possible.

Deserving special credit are the following five outstanding scientists who accepted the task of re-drafting chapters in their areas of expertise: Chapters 12 and 13, Ramanujan Hegde (MRC Laboratory of Molecular Biology and Cambridge University, United Kingdom); Chapter 14, Jared Rutter (University of Utah); Chapter 21, David Bilder (University of California, Berkeley); Chapter 22, Yukiko Yamashita (Whitehead Institute, Massachusetts Institute of Technology); Chapter 23, Matthew Welch (University of California, Berkeley).

In what follows, we acknowledge and thank all of the scientists whose suggestions have helped us to prepare this edition. (A combined list of those who helped with our first, second, third, fourth, fifth, and sixth editions is also provided.)

General: Joseph Ahlander (Northeastern State University), Buzz Baum (Molecular Research Institute, United Kingdom), Michael Burns (Loyola University Chicago), Silvia C. Finnemann (Fordham University), Nora Goosen (Leiden University, The Netherlands), Harold Hoops (State University of New York, Buffalo), Joanna Norris (University of Rhode Island), Mark V. Reedy (Creighton University), Jeff Singer (Portland State University), Amy Springer (University of Massachusetts), Andreas Wodarz (University of Cologne, Germany). In addition, Tiago Barros produced new molecular models for the Seventh Edition.

Chapter 1: Sage Arbor (Marian University Indianapolis), Stephen E. Asmus (Centre College), Jill Banfield (University of California, Berkeley), Zoë Burke (University of Bath, United Kingdom), Elizabeth Good (University of Illinois, Urbana-Champaign), Julian Guttman (Simon Fraser University, Canada), Sudhir Kuman (Temple University), Sue Hum-Musser (Western Illinois University), Brad Mehrtens (University of Illinois, Urbana-Champaign), Inaki Ruiz-Trillo (University of Barcelona, Spain), David Stern (Janelia Research Campus), Andrew Wood (Southern Illinois University), Lidan You (University of Toronto, Canada).

Chapter 2: Sage Arbor (Marian University Indianapolis), Stephen E. Asmus (Centre College), Monica Brinchmann (Nord University, Norway), Michael Cardinal-Aucoin (Lakehead University, Orillia, Canada), Venugopalan Cheriyath (Texas A&M University, Commerce), Mark Grimes (University of Montana), Sue Hum-Musser (Western Illinois University), Brad Mehrtens (University of Illinois, Urbana-Champaign), Richard Roy (McGill University, Canada), Seth Rubin (University of California, Santa Cruz), Laura Serbus (Florida International University), Pei-Lan Tsou (Grand Valley State University), Andrew Wood (Southern Illinois University).

Chapter 3: Sage Arbor (Marian University Indianapolis), Jennifer Armstrong (Scripps College), David Baker (University of Washington, Seattle), Anne Bertolotti (MRC Laboratory of Molecular Biology, United Kingdom), Ashok Bidwai (West Virginia University), Douglas Briant (University of Victoria, Canada), Ken Dill (State University of New York, Stony Brook), David S. Eisenberg (University of California, Los Angeles), James Fraser (University of California, San Francisco), Elizabeth Good (University of Illinois, Urbana-Champaign), Ramanujan Hegde (MRC Laboratory of Molecular Biology, United Kingdom), Jerry E. Honts (Drake University), Sue Hum-Musser (Western Illinois University), Tony Hyman (Max Planck Institute of Molecular Cell Biology and Genetics, Germany), Albert Lee (Macquarie University, Australia), Susan Marqusee (University of California, Berkeley), Brad Mehrtens (University of Illinois, Urbana-Champaign), Jose Rodriguez (University of California, Los Angeles), Michael Rosen (University of Texas Southwestern), Seth Rubin (University of California, Santa Cruz), Jonathan Weissman (University of California, San Francisco), Andrew Wood (Southern Illinois University).

Chapter 4: Blake Bextine (University of Texas, Tyler), Wendy Bickmore (University of Edinburgh, United Kingdom), Zoë Burke (University of Bath, United Kingdom), Greg Cooper (HudsonAlpha Institute for Biotechnology, Huntsville), Caroline Dean (John Innes Centre, United Kingdom), Job Dekker (University of Massachusetts Medical School), Barbara Ehlting (University of Victoria, Canada), Lisa Farmer (University of Houston), Shiv Grewal (National Institutes of Health),

Paul Himes (University of Louisville), Sue Hum-Musser (Western Illinois University), Tom Misteli (National Institutes of Health), Rick Myers (HudsonAlpha Institute for Biotechnology, Huntsville), Geeta Nalikar (University of California, San Francisco), Craig Peterson (University of Massachusetts Medical School), Allison Piovesan (University of Bologna, Italy), Saumya Ramanathan (Fisk University), Irina Solovei (Ludwig Maximilians University, Germany), Chisato Ushida (Hirosaki University, Japan), Ken Zaret (University of Pennsylvania).
Chapter 5: Stephen E. Asmus (Centre College), Blake Bextine (University of Texas, Tyler), Stephen P. Bell (Massachusetts Institute of Technology), Michael Cardinal-Aucoin (Lakehead University, Orillia, Canada), Julie Cooper (National Institutes of Health), John Diffley (The Francis Crick Institute, United Kingdom), Barbara Ehlting (University of Victoria, Canada), Lisa Farmer (University of Houston), Elizabeth Good (University of Illinois, Urbana-Champaign), James Haber (Brandeis University), Paul Himes (University of Louisville), Sue Hum-Musser (Western Illinois University), Neil Hunter (University of California, Davis), Keiko Kono (Okinawa Institute of Science and Technology, Japan), Karim Labib (University of Dundee, United Kingdom), Joachim Li (University of California, San Francisco), Rob Martienssen (Cold Spring Harbor Laboratory), Luiza Nogaj (Mount St. Mary's University, Los Angeles), Bruce Stillman (Cold Spring Harbor Laboratory), Johannes Walter (Harvard Medical School), Stephen C. West (The Francis Crick Institute, United Kingdom), Richard D. Wood (MD Anderson Cancer Center).
Chapter 6: Katsura Asano (Kansas State University), Ahmed Badran (Broad Institute, Massachusetts Institute of Technology), Matthew Christians (Grand Valley State University), Patrick Cramer (Max Planck Institute for Biophysical Chemistry, Germany), Daniel J. Finley (Harvard Medical School), Stephen Floor (University of California, San Francisco), Elizabeth Good (University of Illinois, Urbana-Champaign), Michael R. Green (University of Massachusetts Medical School), Arthur L. Horwich (Yale School of Medicine), Hiten Madhani (University of California, San Francisco), Saumya Ramanathan (Fisk University), German Rosas-Acosta (University of Texas, El Paso), Mahito Sadaie (Tokyo University of Science, Japan), Eric Spana (Duke University), David Tollervey (University of Edinburgh, United Kingdom), Chisato Ushida (Hirosaki University, Japan), Alex Varshavsky (California Institute of Technology), Max Wilkinson (MRC Laboratory of Molecular Biology, United Kingdom).
Chapter 7: Katsura Asano (Kansas State University), David Auble (University of Virginia), David Bartel (Massachusetts Institute of Technology), Adrian Bird (University of Edinburgh, United Kingdom), Matthew Christians (Grand Valley State University), Kathleen Collins (University of California, Berkeley), Elizabeth Good (University of Illinois, Urbana-Champaign), Gordon Hagar (National Institutes of Health), Edith Heard (European Molecular Biology Laboratory, Heidelberg), Agnese Loda (European Molecular Biology Laboratory, Heidelberg), Sue Hum-Musser (Western Illinois University), Tracy Nissan (University of Sussex), Sofia Origanti (Saint Louis University), German Rosas-Acosta (University of Texas, El Paso), Schraga Schwartz (Weizmann Institute of Science), Phillip A. Sharp (Massachusetts Institute of Technology), Kevin Struhl (Harvard Medical School), Igor Ulitsky (Weizmann Institute of Science), Ken Zaret (University of Pennsylvania, Perelman School of Medicine).
Chapter 8: Eric Chow (University of California, San Francisco), Barbara Ehlting (University of Victoria, Canada), Lisa Farmer (University of Houston), Daniel Frigo (University of Texas MD Anderson Cancer Center), Sue Hum-Musser (Western Illinois University), Véronique Moulin (Université Laval, Canada), John Steele (Humboldt State University), Venkat Ventkataraman (Rowan University).
Chapter 9: Eric Betzig (Howard Hughes Medical Institute), Stefano Guido (Università degli Studi di Napoli Federico II, Italy), Julian Guttman (Simon Fraser University, Canada), Richard Henderson (MRC Laboratory of Molecular Biology, United Kingdom), Sue Hum-Musser (Western Illinois University), Lee Ligon (Rensselaer Polytechnic Institute), Jennifer Lippincott-Schwartz (Janelia Research Campus), Eva Nogales (University of California, Berkeley), Helen Saibil (Birkbeck College, University of London, United Kingdom), Peter Shaw (John Innes Centre, United Kingdom), Phoebe Stavride (University of Nicosia, Cyprus), John Steele (Humboldt State University), Paul Tillberg (Janelia Research Campus).
Chapter 10: Stephen E. Asmus (Centre College), Francis Barr (Oxford University, United Kingdom), Adam Frost (University of California, San Francisco), Mark Grimes (University of Montana), Ramanujan Hegde (MRC Laboratory of Molecular Biology and Cambridge University, United Kingdom), Gunnar von Heinje (Stockholm University, Sweden), Sue Hum-Musser (Western Illinois University), Michael Jonz (University of Ottawa, Canada), Keiko Kono (Okinawa Institute of Science and Technology, Japan), Werner Kühlbrandt (Max Planck Institute of Biophysics, Germany), Herman Lehman (Hamilton College), Satyajit Mayor (National Centre for Biological Sciences, India), Richard Posner (Northern Arizona University), Richard Roy (McGill University, Canada), Gregory Schmaltz (University of the Fraser Valley, Canada), Andrew Swan (University of Windsor, Canada), Tobias Walther (Harvard Medical School), Graham Warren (University College London, United Kingdom).
Chapter 11: Mauro Costa-Mattioli (Baylor College of Medicine), Robert Edwards (University of California, San Francisco), Zheng Fan (The University of Tennessee Health Science Center), Mark Grimes (University of Montana), Ramanujan Hegde (MRC Laboratory of Molecular Biology and Cambridge University, United Kingdom), Bertil Hille (University of Washington), Olivia S. Long (University of Pittsburgh, Greensburg), Werner Kühlbrandt (Max Planck Institute of Biophysics, Germany), Liqun Luo (Stanford University), Poul Nissen (Aarhus University, Denmark), Richard Posner (Northern Arizona University), Benoît Roux (University of Chicago), Gregory Schmaltz (University of the Fraser Valley, Canada), Susan Spencer (Saint Louis University), Robert Stroud (University of California, San Francisco), Christine Suetterlin (University of California, Irvine).
Chapter 12: Major contributor Ramanujan Hegde (MRC Laboratory of Molecular Biology and Cambridge University, United Kingdom). Cedric Asensio (University of Denver), Buzz Baum (MRC Laboratory of Molecular Biology, United

Kingdom) Mike Chao (California State University, San Bernardino), Edward B. Cluett (Ithaca College), Elizabeth Good (University of Illinois, Urbana-Champaign), Yun Hyun Huh (Gwangju Institute of Science and Technology, South Korea), Tim Levine (University College London, United Kingdom), Sue Hum-Musser (Western Illinois University), Wendy Innis-Whitehouse (University of Texas, Rio Grande Valley), Mack Ivey (University of Arkansas), Stephen Rogers (University of North Carolina), Michael Silverman (Simon Fraser University, Canada), Gina Voeltz (University of Colorado, Boulder), Graham Warren (University College London, United Kingdom), Weiliang Xia (Shanghai Jiao Tong University, China), Ge Yang (Carnegie Mellon University).

Chapter 13: Major contributor Ramanujan Hegde (MRC Laboratory of Molecular Biology and Cambridge University, United Kingdom). Swapna Bhat (University of North Georgia, Gainesville), Pete Cullen (University of Bristol, United Kingdom), Zheng Fan (The University of Tennessee Health Science Center), Adam Frost (University of California, San Francisco), Wendy Innis-Whitehouse (University of Texas, Rio Grande Valley), Reinhard Jahn (Max Planck Institute for Biophysical Chemistry, Germany), Elizabeth Miller (MRC Laboratory of Molecular Biology, United Kingdom), Sean Munro (MRC Laboratory of Molecular Biology, United Kingdom), Michael Silverman (Simon Fraser University, Canada), Susan Spencer (Saint Louis University), Christine Suetterlin (University of California, Irvine), Graham Warren (University College London, United Kingdom).

Chapter 14: Major contributor Jared Rutter (University of Utah). Cedric Asensio (University of Denver), Alice Barkan (University of Oregon), Tessa Burch-Smith (University of Tennessee, Knoxville), Navdeep Chandel (Northwestern University), Elizabeth Good (University of Illinois, Urbana-Champaign), Mark Grimes (University of Montana), Yun Hyun Huh (Gwangju Institute of Science and Technology, South Korea), Sue Hum-Musser (Western Illinois University), Mack Ivey (University of Arkansas), Werner Kühlbrandt (Max Planck Institute of Biophysics, Germany), Dario Leister (University of Munich, Germany), Song-Tao Liu (University of Toledo), Amit Singh (University of Dayton), Jonathan Snow (Barnard College), Amy Sprowles (Humboldt State University), Dennis Winge (University of Utah).

Chapter 15: Josefino Castillo (University of Santo Tomas, Philippines), Elizabeth Good (University of Illinois, Urbana-Champaign), Mark Grimes (University of Montana), Sue Hum-Musser (Western Illinois University), Natalia Jura (University of California, San Francisco), In Hye Lee (Ewha Womans University, South Korea), Song-Tao Liu (University of Toledo), Bruce Mayer (University of Connecticut), Alexandra Newton (University of California, San Diego), Roeland Nusse (Stanford University), Amit Singh (University of Dayton), Jonathan Snow (Barnard College), Amy Sprowles (Humboldt State University).

Chapter 16: Anna Akhmanova (Utrecht University, The Netherlands), Alexander Bershadsky (National University of Singapore, Singapore), David Bilder (University of California, Berkeley), Josefino Castillo (University of Santo Tomas, Philippines), Ann Cavanaugh (Creighton University), Andrew Carter (Medical Research Council, United Kingdom), Roger Craig (University of Massachusetts Medical School), Lillian Fritz-Laylin (University of Massachusetts), Elizabeth Good (University of Illinois, Urbana-Champaign), Julian Guttman (Simon Fraser University, Canada), Sue Hum-Musser (Western Illinois University), Yulia Komarova (University of Illinois, Chicago), Arash Komeili (University of California, Berkeley), Chao Liu (Stanford University), Richard McIntosh (University of Colorado), Dyche Mullins (University of California, San Francisco), Maxence Nachury (University of California, San Francisco), Daniela Nicastro (University of Texas Southwestern), Sam Reck-Peterson (University of California, San Diego), Luke Rice (University of Texas Southwestern), Stephen Rogers (University of North Carolina), Trina Schroer (Johns Hopkins University), Michael Sixt (IST, Austria), Thomas Surrey (The Francis Crick Institute, United Kingdom), Tatyana Svitkina (University of Pennsylvania), Katherine Warpeha (University of Illinois, Chicago), Matthew Welch (University of California, Berkeley), Lidan You (University of Toronto, Canada).

Chapter 17: Michael W. Black (California Polytechnic State University), Ann Cavanaugh (Creighton University), Yu Chen (University of Massachusetts), Frederick Dick (University of Western Ontario, Canada), Bruce Edgar (University of Utah), Michael Glotzer (University of Chicago), Elizabeth Good (University of Illinois, Urbana-Champaign), Greg Hermann (Lewis and Clark College), Sue Hum-Musser (Western Illinois University), Peter Mayinger (Oregon Health and Science University), Mindy McCarville (Dalhousie University, Canada), Jennifer McDonough (Kent State University), Yuko Miyamoto (Elon University), Jonathon Pines (Institute of Cancer Research, United Kingdom), Stephen Rogers (University of North Carolina), Frank Uhlmann (The Francis Crick Institute, United Kingdom), Johannes Walter (Harvard Medical School), Katherine Warpeha (University of Illinois, Chicago), Christina Zito (University of New Haven).

Chapter 18: Stephen E. Asmus (Centre College), Yu Chen (University of Massachusetts), Elizabeth Good (University of Illinois, Urbana-Champaign), Douglas R. Green (St. Jude's Children's Research Hospital), Mindy McCarville (Dalhousie University, Canada), Sue Hum-Musser (Western Illinois University), Yuko Miyamoto (Elon University), Mark Running (University of Louisville), Christina Zito (University of New Haven).

Chapter 19: Stephen E. Asmus (Centre College), Zoë Burke (University of Bath, United Kingdom), John Couchman (University of Copenhagen, Denmark), Elizabeth Good (University of Illinois, Urbana-Champaign), Tony J.C. Harris (University of Toronto), Sue Hum-Musser (Western Illinois University), Yulia Komarova (University of Illinois, Chicago), Zhengchang Liu (University of New Orleans), Mark Running (University of Louisville), Kara Sawarynski (Oakland University), Masatoshi Takeichi (RIKEN Center for Developmental Biology, Japan), Zena Werb (formerly of University of California, San Francisco), Alpha Yap (University of Queensland, Australia), Christina Zito (University of New Haven).

Chapter 20: Alan Ashworth (University of California, San Francisco), Anton Berns (Netherlands Cancer Institute, The Netherlands), David Bilder (University of California, Berkeley), Fred Bunz (Johns Hopkins University), Venugopalan Cheriyath (Texas A&M University,

Commerce), Michel DuPage (University of California, Berkeley), Paul Edwards (University of Cambridge, United Kingdom), Elizabeth Good (University of Illinois, Urbana-Champaign), Jeff Hadwiger (Oklahoma State University), Monika Haoui (University of California, Berkeley), Lin He (University of California, Berkeley), Dirk Hockemeyer (University of California, Berkeley), Sue Hum-Musser (Western Illinois University), Doug Kellogg (University of California, Santa Cruz), Maria Krasilnikova (Pennsylvania State University), Christopher Lassiter (Roanoke College), Kunxin Luo (University of California, Berkeley), Adam G. Matthews (Wellesley College), Mike Misamore (Texas Christian University), Alexander K. Murashov (East Carolina University), Kristina Pazehoski (Grove City College), David Pellman (Dana Farber Cancer Institute), Jody Rosenblatt (University of Utah), Peter Savage (University of Chicago), Kara Sawarynski (Oakland University), Timothy Shannon (Francis Marion University), Frederick Sproull (LaRoche University), Robert A. Weinberg (Massachusetts Institute of Technology), Harold Varmus (Weill Cornell Medicine), Kenneth Yip (University of Toronto, Canada), Christina Zito (University of New Haven).

Chapter 21: Major contributor David Bilder (University of California, Berkeley). Paul Cullen (State University of New York, Buffalo), Claude Desplan (New York University), Bruce Edgar (University of Utah), Stephen Jesch (Cornell University), Christopher Lassiter (Roanoke College), Laura Mathies (Virginia Commonwealth University), Mike Misamore (Texas Christian University), Alexander K. Murashov (East Carolina University), Nipam Patel (Marine Biological Laboratory), Ruth Phillips (Syracuse University), Olivier Pourquié (Harvard University), Joshua Sanes (Harvard University), Ramaswamy Sharma (University of Texas Health Science Center at San Antonio), Jan Skotheim (Stanford University), David Somers (The Ohio State University), Didier Stanier (Max Planck Institute for Heart and Lung Research, Germany), Shannon Stevenson (University of Minnesota, Duluth), Nancy Toffolo (University of Florida), Jean-Paul Vincent (The Francis Crick Institute, United Kingdom), John Wallingford (University of Texas, Austin), Neal Williams (University of California, Davis).

Chapter 22: Major contributor Yukiko Yamashita (Whitehead Institute, Massachusetts Institute of Technology). Stefano Biressi (University of Trento, Italy), Steven Clark (University of Michigan), Elizabeth Good (University of Illinois, Urbana-Champaign), Konrad Hocheldinger (Harvard University), Shagun Khera (University of Plymouth, United Kingdom), Carien Niessen (University of Cologne, Germany), Jenean O'Brien (The College of St. Scholastica), Peter Reddien (Massachusetts Institute of Technology), Edward Seto (George Washington University), Arianna Smith (Kenyon College), Elly Tanaka (Imperial College London, United Kingdom)

Chapter 23: Major contributor Matthew Welch (University of California, Berkeley). Steven Clark (University of Michigan), Ding Jeak Ling (National University of Singapore, Singapore), Stefan Fischer (Deggendorf Institute of Technology, Germany), Elizabeth Good (University of Illinois, Urbana-Champaign), Karen Guillemin (University of Oregon), Julian Guttman (Simon Fraser University, Canada), Jenean O'Brien (The College of St. Scholastica), Daniel A. Portnoy (University of California, Berkeley), David Sibley (Washington University in Saint Louis), Michael Way (The Francis Crick Institute, United Kingdom)

Chapter 24: Steven Clark (University of Michigan), Elizabeth Good (University of Illinois, Urbana-Champaign), Jenean O'Brien (The College of St. Scholastica), Peter Parham (Stanford University), Shamith Samarajiwa (University of Cambridge, United Kingdom), Phoebe Stavride (University of Nicosia, Cyprus), Hariharan Subramanian (Michigan State University), Colin Watts (University of Dundee, United Kingdom), Donald Very (Duquesne University), Debby Walser-Kuntz (Carleton College).

REVIEWERS OF PREVIOUS EDITIONS

Jerry Adams (The Walter and Eliza Hall Institute of Medical Research, Australia), Ralf Adams (London Research Institute, United Kingdom), Markus Affolter (University of Basel, Switzerland), David Agard (University of California, San Francisco), Julie Ahringer (The Gurdon Institute, United Kingdom), John Aitchison (Institute for System Biology, Seattle), Michael Akam (University of Cambridge, United Kingdom), Anna Akhmanova (Utrecht University, The Netherlands), David Allis (The Rockefeller University), Wolfhard Almers (Oregon Health and Science University), Fred Alt (CBR Institute for Biomedical Research, Boston), Victor Ambros (University of Massachusetts, Worcester), Linda Amos (MRC Laboratory of Molecular Biology, United Kingdom), Raul Andino (University of California, San Francisco), Oscar Aparicio (University of Southern California), Clay Armstrong (University of Pennsylvania), Martha Arnaud (University of California, San Francisco), Najla Arshad (Indian Institute of Science, India), Spyros Artavanis-Tsakonas (Harvard Medical School), Michael Ashburner (University of Cambridge, United Kingdom), Jonathan Ashmore (University College London, United Kingdom), Laura Attardi (Stanford University), Tayna Awabdy (University of California, San Francisco), Jeffrey Axelrod (Stanford University Medical Center), Peter Baker (deceased), David Baldwin (Stanford University), Michael Banda (University of California, San Francisco), Cornelia Bargmann (The Rockefeller University), Ben Barres (Stanford University), David Bartel (Massachusetts Institute of Technology), Konrad Basler (University of Zurich, Switzerland), Wolfgang Baumeister (Max Planck Institute of Biochemistry, Germany), Michael Bennett (Albert Einstein College of Medicine), Darwin Berg (University of California, San Diego), Anton Berns (Netherlands Cancer Institute, The Netherlands), Bradley E. Bernstein (Harvard Medical School), Merton Bernfield (Harvard Medical School), Michael Berridge (The Babraham Institute, United Kingdom), Wendy Bickmore (MRC Human Genetics Unit, United Kingdom), Walter Birchmeier (Max Delbrück Center for Molecular Medicine, Germany), Adrian Bird (Wellcome Trust Centre, United Kingdom), David Birk (UMDNJ—Robert Wood Johnson Medical School), Michael Bishop (University of California, San Francisco), Trever Bivona (University of California, San Francisco), Elizabeth Blackburn (University of California, San Francisco), Tim Bliss (National Institute for Medical Research, United Kingdom), Hans Bode (University of California, Irvine), Piet Borst (Jan Swammerdam Institute, University of

Amsterdam, The Netherlands), Henry Bourne (University of California, San Francisco), Alan Boyde (University College London, United Kingdom), Donita Brady (Duke University), Martin Brand (University of Cambridge, United Kingdom), Carl Branden (deceased), Andre Brandli (Swiss Federal Institute of Technology, Zurich, Switzerland), Dennis Bray (University of Cambridge, United Kingdom), Mark Bretscher (MRC Laboratory of Molecular Biology, United Kingdom), Douglas J. Briant (University of Victoria, Canada), Jason Brickner (Northwestern University), James Briscoe (National Institute for Medical Research, United Kingdom), Neil Brockdorff (University of Oxford, United Kingdom), Marianne Bronner-Fraser (California Institute of Technology), Robert Brooks (King's College London, United Kingdom), Barry Brown (King's College London, United Kingdom), Donald Brown (Carnegie Institution for Science, Baltimore), Michael Brown (University of Oxford, United Kingdom), Michael Bulger (University of Rochester Medical Center), Fred Bunz (Johns Hopkins University), Steve Burden (New York University School of Medicine), Max Burger (University of Basel, Switzerland), Stephen Burley (SGX Pharmaceuticals), Briana Burton (Harvard University), Keith Burridge (University of North Carolina, Chapel Hill), John Cairns (Radcliffe Infirmary, United Kingdom), Patricia Calarco (University of California, San Francisco), Zacheus Cande (University of California, Berkeley), Lewis Cantley (Harvard Medical School), Charles Cantor (Columbia University), Roderick Capaldi (University of Oregon), Mario Capecchi (University of Utah), Michael Carey (University of California, Los Angeles), Adelaide Carpenter (University of California, San Diego), John Carroll (University College London, United Kingdom), Tom Cavalier-Smith (King's College London, United Kingdom), Pierre Chambon (University of Strasbourg, France), Moses Chao (New York University School of Medicine), Venice Chiueh (University of California, Berkeley), Hans Clevers (Hubrecht Institute, The Netherlands), Enrico Coen (John Innes Institute, United Kingdom), Philip Cohen (University of Dundee, United Kingdom), Robert Cohen (University of California, San Francisco), Stephen Cohen (EMBL Heidelberg, Germany), Roger Cooke (University of California, San Francisco), Steven Cook (Imperial College London, United Kingdom), John Cooper (Washington University School of Medicine, St. Louis), Julie P. Cooper (National Cancer Institute), John Couchman (University of Copenhagen, Denmark), Jose A. Costoya (Universidade de Santiago de Compostela, Spain), Michael Cox (University of Wisconsin, Madison), Nancy Craig (Johns Hopkins University), Emily D. Crawford (University of California, San Francisco), James Crow (University of Wisconsin, Madison), Stuart Cull-Candy (University College London, United Kingdom), Leslie Dale (University College London, United Kingdom), Caroline Damsky (University of California, San Francisco), Graeme Davis (University of California, San Francisco), Caroline Dean (John Innes Centre, United Kingdom), Johann De Bono (The Institute of Cancer Research, United Kingdom), Anthony DeFranco (University of California, San Francisco), Abby Dernburg (University of California, Berkeley), Johan de Rooij (The Hubrecht Institute, The Netherlands), Arshad Desai (University of California, San Diego), Michael Dexter (The Wellcome Trust, United Kingdom), John Dick (University of Toronto, Canada), Christopher Dobson (University of Cambridge, United Kingdom), Chris Doe (University of Oregon, Eugene), Russell Doolittle (University of California, San Diego), W. Ford Doolittle (Dalhousie University, Canada), Julian Downward (Cancer Research UK, United Kingdom), Uwe Drescher (King's College London, United Kingdom), Keith Dudley (King's College London, United Kingdom), Graham Dunn (MRC Cell Biophysics Unit, United Kingdom), Jim Dunwell (John Innes Institute, United Kingdom), Susan K. Dutcher (Washington University in St. Louis), Richard H. Ebright (Rutgers University), Bruce Edgar (Fred Hutchinson Cancer Research Center), Paul Edwards (University of Cambridge, United Kingdom), Robert Edwards (University of California, San Francisco), David Eisenberg (University of California, Los Angeles), Sarah Elgin (Washington University in St. Louis), Ruth Ellman (Institute of Cancer Research, United Kingdom), Michael Elowitz (California Institute of Technology), Hana El-Samad (University of California, San Francisco), Beverly Emerson (The Salk Institute), Charles Emerson (University of Virginia), Scott D. Emr (Cornell University), Sharyn Endow (Duke University), Amber English (University of Colorado at Boulder), Lynn Enquist (Princeton University), Tariq Enver (Institute of Cancer Research, United Kingdom), David Epel (Stanford University), Ralf Erdmann (Ruhr University of Bochum, Germany), Gerard Evan (University of California, Comprehensive Cancer Center), Ray Evert (University of Wisconsin, Madison), Matthias Falk (Lehigh University), Stanley Falkow (Stanford University), Douglas Fearon (University of Cambridge, United Kingdom), Gary Felsenfeld (National Institutes of Health), Stuart Ferguson (University of Oxford, United Kingdom), James Ferrell (Stanford University), Christine Field (Harvard Medical School), Daniel Finley (Harvard University), Gary Firestone (University of California, Berkeley), Gerald Fischbach (Columbia University), Gordon Fishell (New York University School of Medicine), Robert Fletterick (University of California, San Francisco), Harvey Florman (Tufts University), Judah Folkman (Harvard Medical School), Larry Fowke (University of Saskatchewan, Canada), Velia Fowler (The Scripps Research Institute), Thomas D. Fox (Cornell University), Jennifer Frazier (Exploratorium, San Francisco), Matthew Freeman (Laboratory of Molecular Biology, United Kingdom), Daniel Friend (University of California, San Francisco), Elaine Fuchs (University of Chicago), Joseph Gall (Carnegie Institution of Washington), Richard Gardner (University of Oxford, United Kingdom), Anthony Gardner-Medwin (University College London, United Kingdom), Peter Garland (Institute of Cancer Research, United Kingdom), David Garrod (University of Manchester, United Kingdom), Susan M. Gasser (University of Basel, Switzerland), Walter Gehring (Biozentrum, University of Basel, Switzerland), Benny Geiger (Weizmann Institute of Science, Rehovot, Israel), Vladimir Gelfand (Northwestern University), Larry Gerace (The Scripps Research Institute), Holger Gerhardt (London Research Institute, United Kingdom), John Gerhart (University of California, Berkeley), Günther Gerisch (Max Planck Institute of Biochemistry, Germany), Frank Gertler (Massachusetts Institute of Technology), Sankar Ghosh (Yale University School of Medicine), Alfred Gilman (The University of Texas Southwestern Medical Center), Andrew P. Gilmore (University of Manchester, United Kingdom), Reid Gilmore (University of

Massachusetts, Amherst), Bernie Gilula (deceased), Charles Gilvarg (Princeton University), Benjamin S. Glick (University of Chicago), Michael Glotzer (University of Chicago), Robert Goldman (Northwestern University), Larry Goldstein (University of California, San Diego), Bastien Gomperts (University College Hospital Medical School, United Kingdom), Daniel Goodenough (Harvard Medical School), Jim Goodrich (University of Colorado, Boulder), Jeffrey Gordon (Washington University in St. Louis), Peter Gould (Middlesex Hospital Medical School, United Kingdom), Alan Grafen (University of Oxford, United Kingdom), Walter Gratzer (King's College London, United Kingdom), Michael Gray (Dalhousie University, Canada), Douglas Green (St. Jude Children's Hospital), Howard Green (Harvard University), Michael R. Green (University of Massachusetts Medical School), Shiv Grewal (National Cancer Institute), Leslie Grivell (University of Amsterdam, The Netherlands), Carol Gross (University of California, San Francisco), Frank Grosveld (Erasmus Universiteit, The Netherlands), Michael Grunstein (University of California, Los Angeles), Barry Gumbiner (Memorial Sloan Kettering Cancer Center), Brian Gunning (Australian National University, Australia), Christine Guthrie (University of California, San Francisco), James Haber (Brandeis University), Ernst Hafen (University of Zurich, Switzerland), David Haig (Harvard University), Andrew Halestrap (University of Bristol, United Kingdom), Alan Hall (Memorial Sloan Kettering Cancer Center), Jeffrey Hall (Brandeis University), John Hall (University of Southampton, United Kingdom), Zach Hall (University of California, San Francisco), Douglas Hanahan (University of California, San Francisco), David Hanke (University of Cambridge, United Kingdom), Gregory Hannon (Cold Spring Harbor Laboratory), Nicholas Harberd (University of Oxford, United Kingdom), Graham Hardie (University of Dundee, United Kingdom), Richard Harland (University of California, Berkeley), Adrian Harris (Cancer Research UK, United Kingdom), John Harris (University of Otago, New Zealand), Tony Harris (University of Toronto, Canada), Stephen Harrison (Harvard University), F. Ulrich Hartl (Max Planck Institute of Biochemistry, Germany), Leland Hartwell (University of Washington), Adrian Harwood (MRC Laboratory for Molecular Cell Biology and Cell Biology Unit, United Kingdom), Scott Hawley (Stowers Institute for Medical Research), John Heath (University of Birmingham, United Kingdom), Ramanujan Hegde (National Institutes of Health), Gunnar von Heijne (Stockholm University, Sweden), Carl-Henrik Heldin (Uppsala University, Sweden), Ari Helenius (Swiss Federal Institute of Technology, Switzerland), Richard Henderson (MRC Laboratory of Molecular Biology, United Kingdom), Glenn Herrick (University of Utah), Ira Herskowitz (deceased), Martin W. Hetzer (The Salk Institute), Bertil Hille (University of Washington), Lindsay Hinck (University of California, Santa Cruz), Alan Hinnebusch (National Institutes of Health), Brigid Hogan (Duke University), Nancy Hollingsworth (State University of New York, Stony Brook), Frank Holstege (University Medical Center, The Netherlands), Leroy Hood (Institute for Systems Biology), Karen Hopkin, John Hopfield (Princeton University), Robert Horvitz (Massachusetts Institute of Technology), Art Horwich (Yale University School of Medicine), Alan Rick Horwitz (University of Virginia), David Housman (Massachusetts Institute of Technology), Joe Howard (Max Planck Institute of Molecular Cell Biology and Genetics, Germany), Jonathan Howard (University of Washington), James Hudspeth (The Rockefeller University), Simon Hughes (King's College London, United Kingdom), Martin Humphries (University of Manchester, United Kingdom), Tim Hunt (Cancer Research UK, United Kingdom), Neil Hunter (University of California, Davis), Laurence Hurst (University of Bath, United Kingdom), Quyen Huynh (University of Toronto, Canada), Jeremy Hyams (University College London, United Kingdom), Tony Hyman (Max Planck Institute of Molecular Cell Biology and Genetics, Germany), Richard Hynes (Massachusetts Institute of Technology), Philip Ingham (University of Sheffield, United Kingdom), Kenneth Irvine (Rutgers University), Robin Irvine (University of Cambridge, United Kingdom), Norman Iscove (Ontario Cancer Institute, Canada), David Ish-Horowicz (Cancer Research UK, United Kingdom), Rudolf Jaenisch (Massachusetts Institute of Technology), Reinhard Jahn (Max Planck Institute for Biophysical Chemistry, Germany), Lily Jan (University of California, San Francisco), Charles Janeway, Jr. (deceased), Tom Jessell (Columbia University), Arthur Johnson (Texas A&M University), Louise Johnson (deceased), Andy Johnston (John Innes Institute, United Kingdom), Laura Johnston (Columbia University), E.G. Jordan (Queen Elizabeth College, United Kingdom), Ron Kaback (University of California, Los Angeles), Michael Karin (University of California, San Diego), Gary Karpen (University of California, Berkeley), Eric Karsenti (European Molecular Biology Laboratory, Germany), David Kashatus (University of Virginia), Stefan Kanzok (Loyola University Chicago) Ken Keegstra (Michigan State University), Doug Kellogg (University of California, Santa Cruz), Ray Keller (University of California, Berkeley), Douglas Kellogg (University of California, Santa Cruz), Regis Kelly (University of California, San Francisco), John Kendrick-Jones (MRC Laboratory of Molecular Biology, United Kingdom), Cynthia Kenyon (University of California, San Francisco), Roger Keynes (University of Cambridge, United Kingdom), Judith Kimble (University of Wisconsin, Madison), David Kimelman (University of Washington), David Kingsley (Stanford University), Robert Kingston (Massachusetts General Hospital), Marc Kirschner (Harvard University), Richard Klausner (National Institutes of Health), Nancy Kleckner (Harvard University), Mike Klymkowsky (University of Colorado, Boulder), Kelly Komachi (University of California, San Francisco), Rachel Kooistra (Loyola University Chicago), Eugene Koonin (National Institutes of Health), Juan Korenbrot (University of California, San Francisco), Roger Kornberg (Stanford University), Tom Kornberg (University of California, San Francisco), Stuart Kornfeld (Washington University in St. Louis), Daniel Koshland (University of California, Berkeley), Douglas Koshland (Carnegie Institution of Washington, Baltimore), Marilyn Kozak (University of Pittsburgh), Maria Krasilnikova (Pennsylvania State University), Mark Krasnow (Stanford University), Arnold Kriegstein (University of California, San Francisco), Werner Kühlbrandt (Max Planck Institute for Biophysics, Germany), John Kuriyan (University of California, Berkeley), Robert Kypta (MRC Laboratory for Molecular Cell Biology, United Kingdom), Karim Labib (University of Manchester, United Kingdom), Peter Lachmann (MRC Centre, United Kingdom), Ulrich

Laemmli (University of Geneva, Switzerland), Trevor Lamb (University of Cambridge, United Kingdom), Hartmut Land (Cancer Research UK, United Kingdom), David Lane (University of Dundee, United Kingdom), Jane Langdale (University of Oxford, United Kingdom), Lewis Lanier (University of California, San Francisco), Nils-Göran Larsson (Max Planck Institute for Biology of Aging, Germany), Jay Lash (University of Pennsylvania), Peter Lawrence (MRC Laboratory of Molecular Biology, United Kingdom), Paul Lazarow (Mount Sinai School of Medicine), Jeannie Lee (Harvard Medical School), Robert J. Lefkowitz (Duke University), Michael Levine (University of California, Berkeley), Warren Levinson (University of California, San Francisco), Alex Levitzki (Hebrew University, Israel), Wes Lewis (University of Alabama), Ottoline Leyser (University of York, United Kingdom), Joachim Li (University of California, San Francisco), Jeffrey Lichtman (Harvard University), H. Lill (VU University Amsterdam, The Netherlands), Tomas Lindahl (Cancer Research UK, United Kingdom), Vishu Lingappa (University of California, San Francisco), Jennifer Lippincott-Schwartz (National Institutes of Health), Joseph Lipsick (Stanford University School of Medicine), Dan Littman (New York University School of Medicine), Clive Lloyd (John Innes Institute, United Kingdom), Richard Locksley (University of California, San Francisco), Richard Losick (Harvard University), Daniel Louvard (Institut Curie, France), Robin Lovell-Badge (National Institute for Medical Research, United Kingdom), Scott Lowe (Cold Spring Harbor Laboratory), Shirley Lowe (University of California, San Francisco), Reinhard Lührman (Max Planck Institute of Biophysical Chemistry, Germany), Michael Lynch (Indiana University), Laura Machesky (University of Birmingham, United Kingdom), Hiten Madhani (University of California, San Francisco), James Maller (University of Colorado Medical School), Tom Maniatis (Harvard University), Richard Mann (Columbia University), Colin Manoil (Harvard Medical School), Elliott Margulies (National Institutes of Health), Philippa Marrack (National Jewish Medical and Research Center, Denver), Mark Marsh (Institute of Cancer Research, United Kingdom), Wallace Marshall (University of California, San Francisco), Gail Martin (University of California, San Francisco), Paul Martin (University College London, United Kingdom), Joan Massagué (Memorial Sloan Kettering Cancer Center), Christopher Mathews (Oregon State University), Satyajit Mayor (National Centre for Biological Sciences, India), Brian McCarthy (University of California, Irvine), Richard McCarty (Cornell University), Melanie McGill (University of Toronto, Canada), William McGinnis (University of California, San Diego), Anne McLaren (Wellcome/Cancer Research Campaign Institute, United Kingdom), Frank McNally (University of California, Davis), James A. McNew (Rice University), J. Richard McIntosh (University of Colorado, Boulder), Andy McMahon (University of Southern California), Frederick Meins (Friedrich Miescher Institute, Switzerland), Stephanie Mel (University of California, San Diego), Ira Mellman (Genentech), Doug Melton (Harvard University), Barbara Meyer (University of California, Berkeley), Elliot Meyerowitz (California Institute of Technology), Chris Miller (Brandeis University), Robert Mishell (University of Birmingham, United Kingdom), Tom Misteli (National Cancer Institute), Avrion Mitchison (University College London, United Kingdom), N.A. Mitchison (University College London, United Kingdom), Timothy Mitchison (Harvard Medical School), Quinn Mitrovich (University of California, San Francisco), Marek Mlodzik (Mount Sinai Hospital, New York), Alex Mogilner (University of California, Davis), Peter Mombaerts (The Rockefeller University), Mark Mooseker (Yale University), David Morgan (University of California, San Francisco), Michelle Moritz (University of California, San Francisco), Richard Morris (John Innes Centre, United Kingdom), Montrose Moses (Duke University), Keith Mostov (University of California, San Francisco), Anne Mudge (University College London, United Kingdom), Hans Müller-Eberhard (Scripps Clinic and Research Institute), Annette Müller-Taubenberger (Ludwig Maximilians University, Germany), Alan Munro (University of Cambridge, United Kingdom), J. Murdoch Mitchison (Harvard University), Richard Myers (Stanford University), Diana Myles (University of California, Davis), Andrew Murray (Harvard University), Maxence Nachury (Stanford School of Medicine), Shigekazu Nagata (Kyoto University, Japan), Eric Nam (University of Toronto, Canada), Geeta Narlikar (University of California, San Francisco), Kim Nasmyth (University of Oxford, United Kingdom), Mark E. Nelson (University of Illinois, Urbana-Champaign), Michael Neuberger (deceased), Walter Neupert (University of Munich, Germany), David Nicholls (University of Dundee, United Kingdom), Roger Nicoll (University of California, San Francisco), Poul Nissen (Aarhus University, Denmark), Suzanne Noble (University of California, San Francisco), Eva Nogales (University of California, Berkeley), Harry Noller (University of California, Santa Cruz), Peter Novick (University of California, San Diego), Susan Ferro-Novick (University of California, San Diego), Jodi Nunnari (University of California, Davis), Paul Nurse (The Francis Crick Institute, United Kingdom), Roel Nusse (Stanford University), Michael Nussenzweig (The Rockefeller University), Duncan O'Dell (deceased), Duncan Odom (Cancer Research UK, United Kingdom), Patrick O'Farrell (University of California, San Francisco), Bjorn Olsen (Harvard Medical School), Maynard Olson (University of Washington), Stuart Orkin (Harvard University), Terry Orr-Weaver (Massachusetts Institute of Technology), Erin O'Shea (Harvard University), Dieter Osterhelt (Max Planck Institute of Biochemistry, Germany), William Otto (Cancer Research UK, United Kingdom), John Owen (University of Birmingham, United Kingdom), Dale Oxender (University of Michigan), George Palade (deceased), Albert Pan (Georgia Regents University), Duojia Pan (Johns Hopkins Medical School), Barbara Panning (University of California, San Francisco), Roy Parker (University of Arizona, Tucson), William W. Parson (University of Washington), Terence Partridge (MRC Clinical Sciences Centre, United Kingdom), William E. Paul (National Institutes of Health), Tony Pawson (deceased), Hugh Pelham (Medical Research Council, United Kingdom), Robert Perry (Institute for Cancer Research, Philadelphia), Gordon Peters (Cancer Research UK, United Kingdom), Greg Petsko (Brandeis University), Nikolaus Pfanner (University of Freiburg, Germany), David Phillips (The Rockefeller University), Jeremy Pickett-Heaps (The University of Melbourne, Australia), Jonathan Pines (The Gurdon Institute, United Kingdom), Julie Pitcher (University College London, United

Kingdom), Jeffrey Pollard (Albert Einstein College of Medicine), Tom Pollard (Yale University), Bruce Ponder (University of Cambridge, United Kingdom), Prasanth Potluri (The Children's Hospital of Philadelphia Research Institute), Daniel Portnoy (University of California, Berkeley), Olivier Pourquie (Harvard Medical School), James Priess (University of Washington), Darwin Prockop (Tulane University), Andreas Prokop (University of Manchester, United Kingdom), Mark Ptashne (Memorial Sloan Kettering Cancer Center), Dale Purves (Duke University), Efraim Racker (Cornell University), Jordan Raff (University of Oxford, United Kingdom), Klaus Rajewsky (Max Delbrück Center for Molecular Medicine, Germany), George Ratcliffe (University of Oxford, United Kingdom), Elio Raviola (Harvard Medical School), Erez Raz (University of Münster, Germany), Martin Rechsteiner (University of Utah, Salt Lake City), Samara Reck-Peterson (Harvard Medical School), David Rees (National Institute for Medical Research, United Kingdom), Thomas A. Reh (University of Washington), Peter Rehling (University of Göttingen, Germany), Louis Reichardt (University of California, San Francisco), Renee Reijo (University of California, San Francisco), Caetano Reis e Sousa (Cancer Research UK, United Kingdom), Fred Richards (Yale University), Conly Rieder (Wadsworth Center, Albany), Phillips Robbins (Massachusetts Institute of Technology), Elizabeth Robertson (The Wellcome Trust Centre for Human Genetics, United Kingdom), Elaine Robson (University of Reading, United Kingdom), Robert Roeder (The Rockefeller University), Michael Rout (The Rockefeller University), Joel Rosenbaum (Yale University), Janet Rossant (Mount Sinai Hospital, Canada), Jesse Roth (National Institutes of Health), Jim Rothman (Memorial Sloan Kettering Cancer Center), Rodney Rothstein (Columbia University), Erkki Ruoslahti (La Jolla Cancer Research Foundation), Chris Rushlow (New York University), Gary Ruvkun (Massachusetts General Hospital), Vladislav Ryvkin (Stony Brook University), David Sabatini (New York University), Alan Sachs (University of California, Berkeley), Edward Salmon (University of North Carolina, Chapel Hill), Laasya Samhita (Indian Institute of Science, India), Aziz Sancar (University of North Carolina, Chapel Hill), Joshua Sanes (Harvard University), Peter Sarnow (Stanford University), Lisa Satterwhite (Duke University Medical School), Robert Sauer (Massachusetts Institute of Technology), Ken Sawin (The Wellcome Trust Centre for Cell Biology, United Kingdom), Howard Schachman (University of California, Berkeley), Gerald Schatten (Pittsburgh Development Center), Gottfried Schatz (Biozentrum, University of Basel, Switzerland), Randy Schekman (University of California, Berkeley), Richard Scheller (Stanford University), Stephen Scherer (University of Toronto, Canada), Giampietro Schiavo (Cancer Research UK, United Kingdom), Ueli Schibler (University of Geneva, Switzerland), Alex Schier (Harvard University), Joseph Schlessinger (New York University Medical Center), Danny J. Schnell (University of Massachusetts, Amherst), Sandra Schmid (University of Texas Southwestern), Michael Schramm (Hebrew University, Israel), Robert Schreiber (Washington University School of Medicine), Sebastian Schuck (University of Heidelberg, Germany), James Schwartz (Columbia University), Ronald Schwartz (National Institutes of Health), François Schweisguth (Institut Pasteur, France), John Scott (University of Manchester, United Kingdom), John Sedat (University of California, San Francisco), Peter Selby (Cancer Research UK, United Kingdom), Zvi Sellinger (Hebrew University, Israel), Gregg Semenza (Johns Hopkins University), John Senderak (Jefferson Medical College), Philippe Sengel (University of Grenoble, France), Peter Shaw (John Innes Institute, United Kingdom), Michael Sheetz (Columbia University), Morgan Sheng (Massachusetts Institute of Technology), Charles Sherr (St. Jude Children's Hospital), David Shima (Cancer Research UK, United Kingdom), Marc Shuman (University of California, San Francisco), David Sibley (Washington University in St. Louis), Samuel Silverstein (Columbia University), Melvin I. Simon (California Institute of Technology), Kai Simons (Max Planck Institute of Molecular Cell Biology and Genetics, Germany), Phillipa Simons (Imperial College, United Kingdom), Robert H. Singer (Albert Einstein School of Medicine), Jonathan Slack (Cancer Research UK, United Kingdom), Stephen Small (New York University), Alison Smith (John Innes Institute, United Kingdom), Austin Smith (University of Edinburgh, United Kingdom), Jim Smith (The Gurdon Institute, United Kingdom), John Maynard Smith (University of Sussex, United Kingdom), Mitchell Sogin (Woods Hole Oceanographic Institution), Frank Solomon (Massachusetts Institute of Technology), Michael Solursh (University of Iowa), Bruce Spiegelman (Harvard Medical School), Timothy Springer (Harvard Medical School), Mathias Sprinzl (University of Bayreuth, Germany), Scott Stachel (University of California, Berkeley), Andrew Staehelin (University of Colorado, Boulder), David Standring (University of California, San Francisco), Margaret Stanley (University of Cambridge, United Kingdom), Martha Stark (University of California, San Francisco), Wilfred Stein (Hebrew University, Israel), Malcolm Steinberg (Princeton University), Ralph Steinman (deceased), Len Stephens (The Babraham Institute, United Kingdom), Paul Sternberg (California Institute of Technology), Rolf Sternglanz (Stony Brook University), Chuck Stevens (The Salk Institute), Alastair Stewart (The Victor Chang Cardiac Research Institute, Australia), Murray Stewart (MRC Laboratory of Molecular Biology, United Kingdom), Bruce Stillman (Cold Spring Harbor Laboratory), Mike Stratton (Wellcome Trust Sanger Institute, United Kingdom), Charles Streuli (University of Manchester, United Kingdom), Monroe Strickberger (University of Missouri, St. Louis), Daniela Stock (The Victor Chang Cardiac Research Institute, Australia), Charles Streuli (University of Manchester, United Kingdom), Robert Stroud (University of California, San Francisco), Kevin Struhl (Harvard Medical School), Michael Stryker (University of California, San Francisco), Suresh Subramani (University of California, San Diego), William Sullivan (University of California, Santa Cruz), Azim Surani (The Gurdon Institute, United Kingdom), Jesper Svejstrup (Cancer Research UK, United Kingdom), Karel Svoboda (Howard Hughes Medical Institute), Daniel Szollosi (Institut National de la Recherche Agronomique, France), Jack Szostak (Harvard Medical School), Clifford Tabin (Harvard Medical School), Masatoshi Takeichi (RIKEN Center for Developmental Biology, Japan), Robert Tampé (Goethe University, Germany), Nicolas Tapon (London Research Institute, United Kingdom), Diethard Tautz (University of Cologne, Germany), Marc Tessier-Lavigne

(The Rockefeller University), Julie Theriot (Stanford University), Roger Thomas (University of Bristol, United Kingdom), Barry Thompson (Cancer Research UK, United Kingdom), Craig Thompson (Memorial Sloan Kettering Cancer Center), Kurt Thorn (University of California, San Francisco), Janet Thornton (European Bioinformatics Institute, United Kingdom), Vernon Thornton (King's College London, United Kingdom), Cheryll Tickle (University of Dundee, United Kingdom), Jim Till (Ontario Cancer Institute, Canada), Lewis Tilney (University of Pennsylvania), David Tollervey (University of Edinburgh, United Kingdom), Ian Tomlinson (Cancer Research United Kingdom), Nick Tonks (Cold Spring Harbor Laboratory), Alain Townsend (Institute of Molecular Medicine, John Radcliffe Hospital, United Kingdom), Paul Travers (Scottish Institute for Regeneration Medicine, United Kingdom), Robert Trelstad (UMDNJ—Robert Wood Johnson Medical School), Anthony Trewavas (Edinburgh University, United Kingdom), Nigel Unwin (MRC Laboratory of Molecular Biology, United Kingdom), Victor Vacquier (University of California, San Diego), Ronald D. Vale (University of California, San Francisco), Richard B. Vallee (Columbia University), Tom Vanaman (University of Kentucky), Harry van der Westen (Wageningen, The Netherlands), Harold Varmus (National Cancer Institute), Alexander J. Varshavsky (California Institute of Technology), Danielle Vidaurre (University of Toronto, Canada), Anna Constance Vind (University of Copenhagen, Denmark), Gia Voeltz (University of Colorado, Boulder), Donald Voet (University of Pennsylvania), Harald von Boehmer (Harvard Medical School), Amy Wagers (Harvard University), Madhu Wahi (University of California, San Francisco), Virginia Walbot (Stanford University), Frank Walsh (GlaxoSmithKline, United Kingdom), D. Eric Walters (Chicago Medical School), Tobias Walther (Harvard University), Trevor Wang (John Innes Institute, United Kingdom), Xiaodong Wang (The University of Texas Southwestern Medical School), Yu-Lie Wang (Worcester Foundation for Biomedical Research), Gary Ward (University of Vermont), Anne Warner (University College London, United Kingdom), Carmen Warren (University of California, Los Angeles), Graham Warren (Yale University School of Medicine), Paul Wassarman (Mount Sinai School of Medicine), Clare Waterman-Storer (The Scripps Research Institute), Fiona Watt (Cancer Research UK, United Kingdom), John Watts (John Innes Institute, United Kingdom), Michael Way (Cancer Research UK, United Kingdom), Klaus Weber (Max Planck Institute for Biophysical Chemistry, Germany), Martin Weigert (Institute for Cancer Research, Philadelphia), Robert Weinberg (Massachusetts Institute of Technology), Orion Weiner (University of California, San Francisco), Harold Weintraub (deceased), Karsten Weis (Swiss Federal Institute of Technology, Switzerland), Irving Weissman (Stanford University), Jonathan Weissman (University of California, San Francisco), Matthew Welch (University of California, Berkeley), Steve Wellard (Pennsylvania State University), Jim Wells (University of California, San Francisco), Susan R. Wente (Vanderbilt University School of Medicine), Norman Wessells (University of Oregon, Eugene), Stephen West (Cancer Research UK, United Kingdom), Judy White (University of Virginia), Evan Whitehead (University of California, Berkeley), William Wickner (Dartmouth College), Carrie Wilczewski (Loyola University Chicago), Michael Wilcox (deceased), Lewis T. Williams (Chiron Corporation), Patrick Williamson (University of Massachusetts, Amherst), Keith Willison (Chester Beatty Laboratories, United Kingdom), John Wilson (Baylor College of Medicine), Anna Wing (Pennsylvania State University), Douglas J. Winton (Cancer Research UK, United Kingdom), Alan Wolffe (deceased), Richard Wolfenden (University of North Carolina, Chapel Hill), Sandra Wolin (Yale University School of Medicine), Lewis Wolpert (University College London, United Kingdom), Richard D. Wood (University of Pittsburgh Cancer Institute), Chris L. Woodcock (University of Massachusetts, Amherst), Ian Woods (Ithaca College), Abraham Worcel (University of Rochester), John Wright (University of Alabama), Nick Wright (Cancer Research UK, United Kingdom), John Wyke (Beatson Institute for Cancer Research, United Kingdom), Johanna Wysocka and lab members (Stanford School of Medicine), Michael P. Yaffe (California Institute for Regenerative Medicine), Kenneth M. Yamada (National Institutes of Health), Keith Yamamoto (University of California, San Francisco), Shinya Yamanaka (Kyoto University, Japan), Alpha Yap (The University of Queensland, Australia), Charles Yocum (University of Michigan, Ann Arbor), Peter Yurchenco (UMDNJ—Robert Wood Johnson Medical School), Rosalind Zalin (University College London, United Kingdom), Patricia Zambryski (University of California, Berkeley), Marino Zerial (Max Planck Institute of Molecular Cell Biology and Genetics, Germany).

OUR PUBLISHER

We thank the dozens of people whose dedication brought this project to fruition. Special thanks go to our editor, Betsy Twitchell, whose talents and personality kept everyone focused. As for previous editions, Denise Schanck directed operations, displaying unflagging patience, insight, tact, and energy throughout the entire journey. Once again, we are in debt to our talented artist, Nigel Orme, who—with author Keith Roberts—produced the captivating art and design that grace the covers, as well as the hundreds of distinctive, highly informative figures throughout the text. Carla Talmadge, our tireless project editor, juggled the many moving parts of this textbook and ushered it gracefully into the book you hold in your hands today. In addition, we are immensely grateful to our copyeditor, Christopher Curioli, who carefully read every page. We also thank Danny Vargo, our assistant editor who tracked the countless emails, chapter drafts, and Excel files that constantly zipped back and forth between editors, authors, and reviewers. And production manager Jane Searle adeptly managed the process of translating our raw material into the polished final product.

An absolutely tireless team at Norton created the print and digital supplementary resources for our book. Media editor Todd Pearson, associate editor Jasmine Ribeaux, and media assistant editor Lindsey Heale worked on every element of the package as a team. Chris Rapp's dedicated and focused efforts on the Smartwork course have produced an invaluable resource for students and instructors. Megan Schindel, Patricia Wong, and Tommy Persano handled the permissions for this edition, and Kim Yi's media project editorial group, specifically Jesse Newkirk, skillfully shepherded the content through its many stages of development.

Marketing manager Ruth Bolster's expertise in direct marketing has helped ensure that this book makes it into the hands of as many instructors and students as possible. We thank her and everyone involved in Norton's sales, marketing, and management teams for their unflagging support of our book, including Erik Fahlgren, Michael Wright, Ann Shin, and Julia Reidhead.

Brief Contents

Special Features

Contents

INTRODUCTION TO THE CELL

CHAPTER 1

Cells, Genomes, and the Diversity of Life

IN THIS CHAPTER

The surface of our planet is populated by living things—*organisms*—curious, intricately organized chemical factories that take in matter from their surroundings and use these raw materials to generate copies of themselves. These organisms appear extraordinarily diverse. What could be more different than a tiger and a piece of seaweed or a butterfly and a tree? Yet our ancestors, knowing nothing of cells or DNA, saw that all these things had something in common. They called that something "life," marveled at it, struggled to define it, and despaired of explaining what it was or how it worked in terms that relate to nonliving matter.

The remarkable discoveries of the past 100 years or so have not diminished the marvel—quite the contrary. But they have removed the central mystery regarding the nature of life. We can now see that all living things are made of cells: small, membrane-enclosed units filled with a concentrated aqueous solution of chemicals and endowed with the extraordinary ability to create copies of themselves by growing and then dividing in two.

Because cells are the fundamental units of life, it is to *cell biology*—the study of the structure, function, and behavior of cells—that we must look for answers to the questions of what life is and how it works. With a deeper understanding of cells and their evolution, we can begin to tackle the grand historical problems of life on Earth: its mysterious origins, its stunning diversity, and its invasion of every conceivable habitat. Indeed, as emphasized long ago by the pioneering cell biologist E. B. Wilson, "the key to every biological problem must finally be sought in the cell; for every living organism is, or at some time has been, a cell."

Despite their apparent diversity, living things are fundamentally similar inside. The whole of biology is thus a counterpoint between two themes: astonishing variety in individual particulars and astonishing constancy in fundamental mechanisms. In this chapter, we begin by outlining the universal features common to all life on our planet, along with some of the fundamental properties of their cells. We then discuss how an analysis of DNA *genomes* allows scientists to position the wide variety of organisms in an evolutionary "tree of life." This approach, which quantifies how closely organisms are related to one another, allows us to identify the three major branches of life on Earth, *eukaryotes, bacteria,* and *archaea*—each with unique qualities. We shall see that the familiar world of plants and animals—the focus of scientists for many centuries—makes up only a small slice of the complete diversity of life, the vast majority of which is invisible to the unaided human eye.

After exploring some of the ways that genomes change over evolutionary times, we highlight the handful of *model organisms* that biologists have chosen to focus on to dissect the molecular mechanisms underlying life. A few specific viruses, including SARS-CoV-2, pose grave threats to humans, so they too have become objects of intensive study. For this reason, this section also includes an introduction to viruses, the ubiquitous parasites that have evolved to feed on cells. Viruses are now recognized to be the most abundant biological entities on the planet.

THE UNIVERSAL FEATURES OF LIFE ON EARTH

There are more than 2 million described species living on Earth today, but many, many more are yet to be discovered. Each species is different, and each reproduces itself faithfully, yielding progeny that are unique to that species. Thus, the parent organism hands down information specifying, in extraordinary detail, the characteristics that the offspring will have. This phenomenon of *heredity* is central to the definition of life: it distinguishes life from other processes, such as the growth of a crystal, or the burning of a candle, or the formation of waves on water, in which structures are generated without the same type of link between the peculiarities of parents and offspring. A living organism must consume free energy to exist, as does a candle flame. But life employs this *free energy* to drive a very complex system of chemical reactions that create and maintain the intricate organization of its cells, all as specified by the hereditary information in those cells.

Most living organisms are single cells. Others, such as us, are like vast multicellular cities in which groups of cells perform specialized functions that are linked by intricate systems of intercellular communication. But even for the aggregate of more than 10^{13} cells that makes up a human body, the whole organism has been generated by cell divisions from a single cell. The single cell therefore contains all of the hereditary information that defines a species (**Figure 1–1**). The cell must also contain all of the machinery needed to gather raw materials from the environment and to construct from them a new cell in its own image, complete with a new copy of the hereditary information of its parent. Every cell on Earth is truly amazing.

All Cells Store Their Hereditary Information in the Form of Double-Strand DNA Molecules

Computers have made us familiar with the concept of information as a measurable quantity—10^6 bytes to record a few hundred pages of text or an image from a digital camera, 10^9 bytes for a 60-minute video streamed from the Internet, and so on. Computers have also made us well aware that the same information can be recorded in many different physical forms: the discs and tapes that we used 25 years ago for our electronic archives have become unreadable on present-day machines. Living cells, like computers, store information, and it is estimated that they have been evolving and diversifying for more than 3.5 billion years. One might not expect that they would all store their information in the same form or that the hereditary information carried by one type of cell should be readable by the information-handling machinery of another. And yet it is so. This fact provides compelling evidence that all living things on Earth have inherited the form of their genetic instructions, as well as how to use them, from a *universal common ancestral cell.* This ancestor is thought to have existed roughly 3.5–3.8 billion years ago.

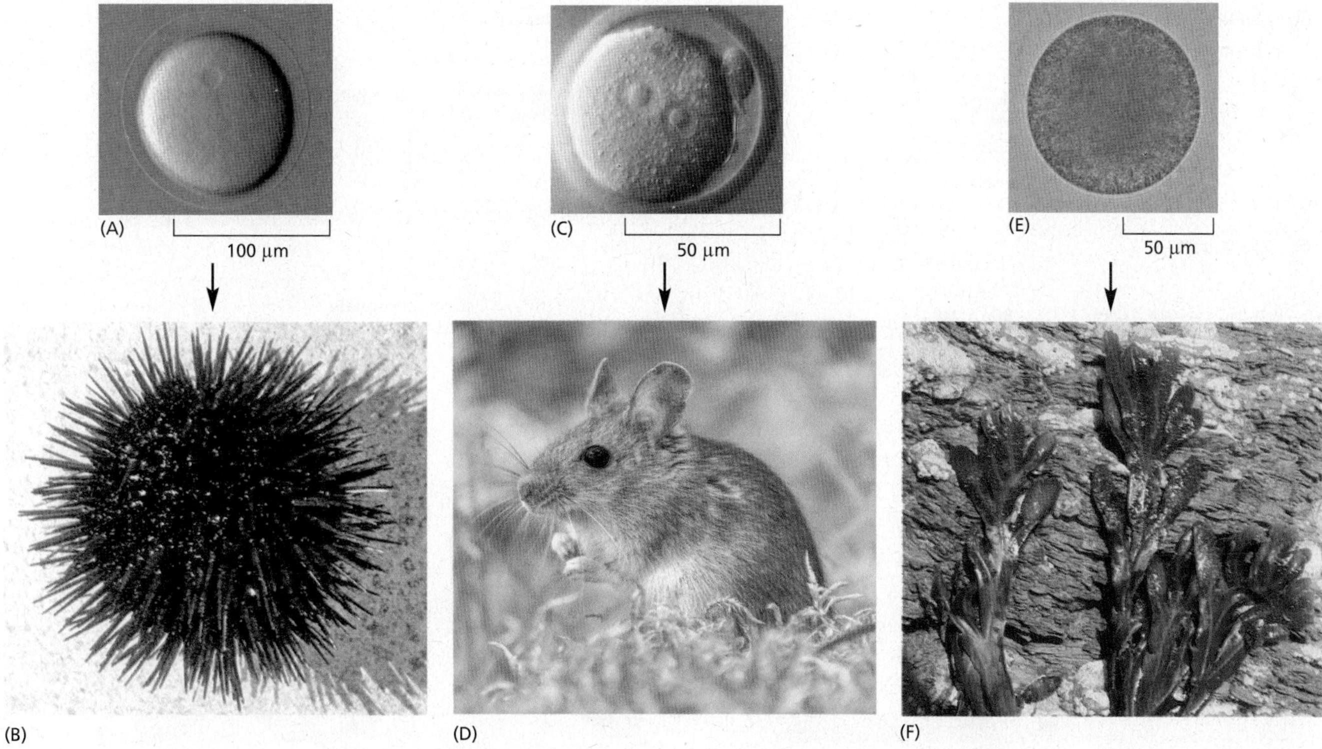

Figure 1–1 The hereditary information in the fertilized egg cell determines the nature of the whole multicellular organism that will develop from it. As indicated, although their starting cells look superficially similar, the egg of a sea urchin gives rise to a sea urchin (A and B), the egg of a mouse gives rise to a mouse (C and D), and the egg of the seaweed *Fucus* gives rise to a *Fucus* seaweed (E and F). (A, courtesy of David McClay; B, courtesy of Tim Hunt; C, courtesy of Patricia Calarco, from G. Martin, *Science* 209:768–776, 1980. With permission from AAAS; D, Rudmer Zwerver/Alamy Stock Photo; E and F, courtesy of Colin Brownlee.)

All cells on Earth today store their hereditary information in the form of double-strand molecules of DNA—long, unbranched, paired *polymer* chains, which are always composed of the same four types of *monomers*. These monomers, chemical compounds known as *nucleotides*, have nicknames drawn from a four-letter alphabet—A, T, C, G—and they are strung together in a long linear sequence that encodes the hereditary information, just as the sequence of 1's and 0's encodes the information in a computer file. We can take a piece of DNA from a human cell and insert it into a bacterium or a piece of bacterial DNA and insert it into a human cell, and, with only a few minor modifications, the information will be successfully read, interpreted, and copied. As we describe in Chapter 8, scientists can now rapidly read out the sequence of nucleotides in any DNA molecule and thereby determine the complete DNA sequence of any cell's **genome**—the totality of its hereditary information embodied in the linear sequence of nucleotides in its DNA. As a result, we now know the complete genome sequences for tens of thousands of species, ranging from the smallest bacterium to the largest plants and animals on Earth.

All Cells Replicate Their Hereditary Information by Templated Polymerization

The mechanisms that make life possible depend on the structure of the double-strand DNA molecule. We discuss this remarkable molecule in detail in Chapters 4 and 5; here we provide only an overview of its structure and means of reproduction. Each monomer in a single DNA strand—that is, each **nucleotide**—consists of two parts: a sugar (deoxyribose) with a phosphate group attached to it, and a *base*, which may be either adenine (A), guanine (G),

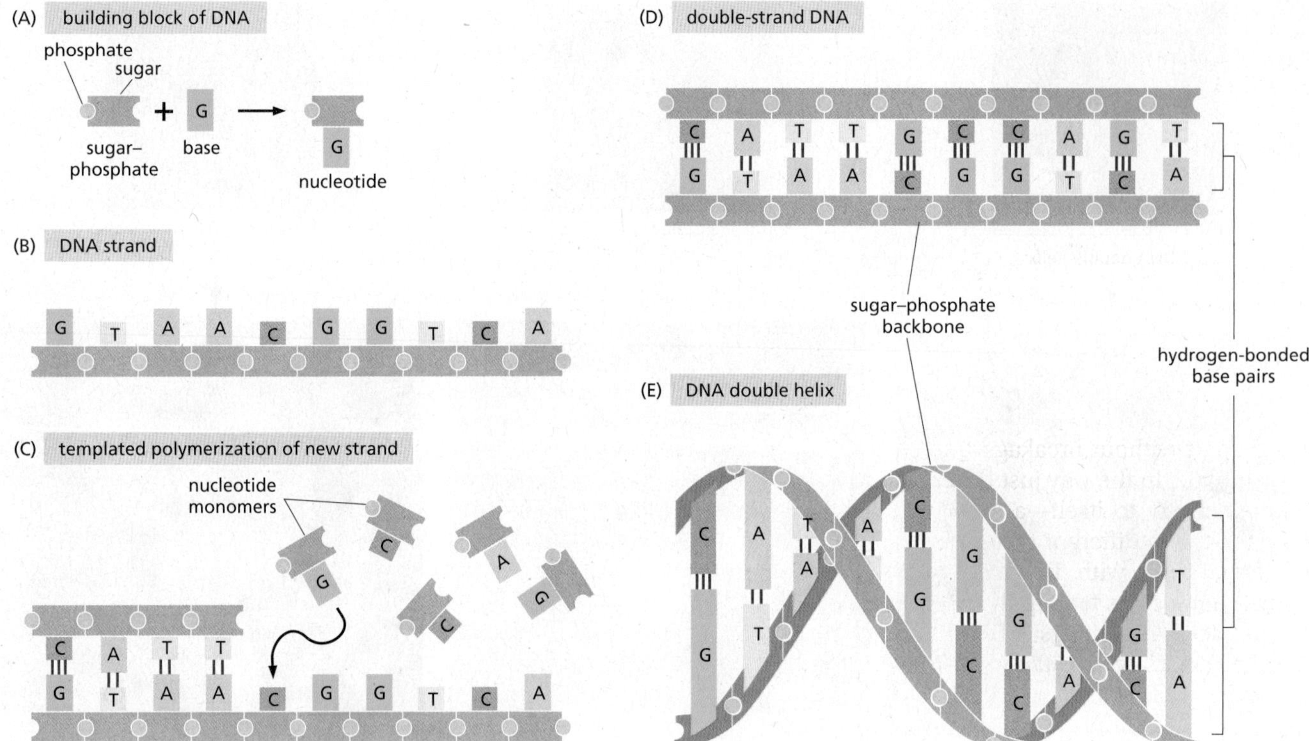

Figure 1–2 **DNA and its building blocks.** (A) DNA is made from simple subunits, called nucleotides. Each nucleotide consists of a specific arrangement of about 35 covalently linked atoms, forming a sugar–phosphate molecule with a nitrogen-containing side group, or base, attached to it. The bases are of four types (adenine, guanine, cytosine, and thymine), corresponding to four distinct nucleotides, labeled A, G, C, and T. (B) A single strand of DNA consists of nucleotides joined together by sugar–phosphate linkages. Note that the individual sugar–phosphate units are asymmetric, giving the backbone of the strand a definite directionality, or polarity. This directionality guides the molecular processes by which the information in DNA is both interpreted and copied (*replicated*) in cells: the information is always "read" in a consistent order, just as written English text is read from left to right. (C) Through templated polymerization, the sequence of nucleotides in an existing DNA strand controls the sequence in which nucleotides are joined together in a new DNA strand; T in one strand pairs with A in the other, and G in one strand with C in the other. The new strand therefore has a nucleotide sequence *complementary* to that of the old strand and a backbone with opposite directionality: thus, GTAA... in the original strand, is ...TTAC in the new strand. (D) A normal DNA molecule consists of two such complementary strands. The nucleotides within each strand are linked by strong (covalent) chemical bonds; the complementary nucleotides on opposite strands are held together more weakly, by hydrogen bonds. (E) The two strands twist around each other to form a double helix—a robust structure that can accommodate any sequence of nucleotides without altering its basic double-helical structure (see Movie 4.1).

cytosine (C), or thymine (T) (**Figure 1–2**). Each sugar is linked to the next via the phosphate group, creating a polymer chain composed of a repetitive sugar-phosphate backbone with a series of bases protruding from it. The DNA polymer is extended by adding monomers at one end. For a single isolated strand, these monomers can, in principle, be added in any order, because each one links to the next in the same way, through the part of the molecule that is the same for all of them. In the living cell, however, DNA is not synthesized as a free strand in isolation, but on a template formed by a preexisting DNA strand. The bases that protrude from this template can bind to bases of the strand being synthesized, according to a strict rule defined by the complementary structures of the bases: A binds to T, and C binds to G. This *base-pairing* holds fresh monomers in place and thereby controls the selection of which one of the four monomers will next be added to a growing strand. In this way, a double-strand structure is created, consisting of two exactly complementary sequences of A's, C's, T's, and G's. These two strands twist around each other, forming a DNA double helix (see Figure 1–2E).

Compared with the covalent sugar-phosphate bonds, the hydrogen bonds between the base pairs are weak, which allows the two DNA strands to be

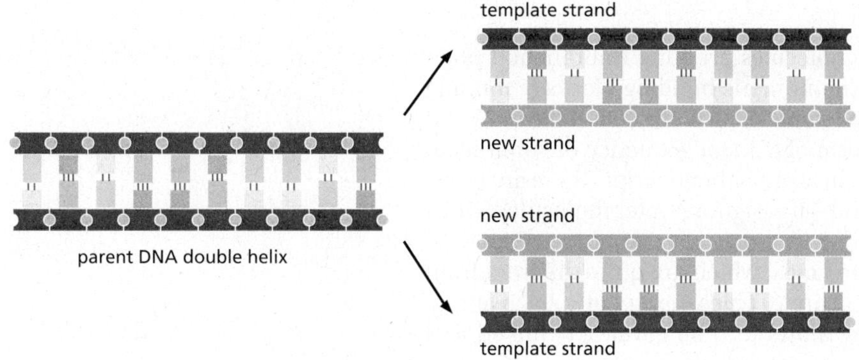

Figure 1–3 The copying of genetic information by DNA replication. In this process, the two strands of a DNA double helix are pulled apart, and each serves as a template for the synthesis of a new complementary strand. The end result is two daughter DNA double helices that are identical in sequence to the parent double helix.

pulled apart without breakage of their backbones. Each strand then can serve as a template, in the way just described, for the synthesis of a fresh DNA strand complementary to itself—a fresh copy, that is, of the hereditary information (Figure 1–3). In different types of cells, this process of **DNA replication** occurs at different rates, with different controls to start it or stop it, and with different auxiliary molecules to help the process along (discussed in Chapters 5 and 17). But the basics are universal: DNA is the information store for heredity, and *templated polymerization* is the way in which this information is copied throughout the living world.

All Cells Transcribe Portions of Their DNA into RNA Molecules

To carry out its information-bearing function, DNA must do more than copy itself. It must also *express* its information, by letting the information guide the synthesis of other molecules in the cell. This expression occurs by a mechanism that is the same in all living organisms, leading first and foremost to the production of two other crucial classes of biological polymers: RNA molecules and protein molecules. The process begins with a templated polymerization called **transcription**, in which segments of the DNA sequence are used as templates for the synthesis of shorter molecules of the closely related polymer **ribonucleic acid, or RNA**. Subsequently, in a process called **translation**, many of these RNA molecules direct the synthesis of polymers of a radically different chemical class—the *proteins* (Figure 1–4). The detailed chemical reactions involved are presented in Chapter 6; here they will only be briefly outlined.

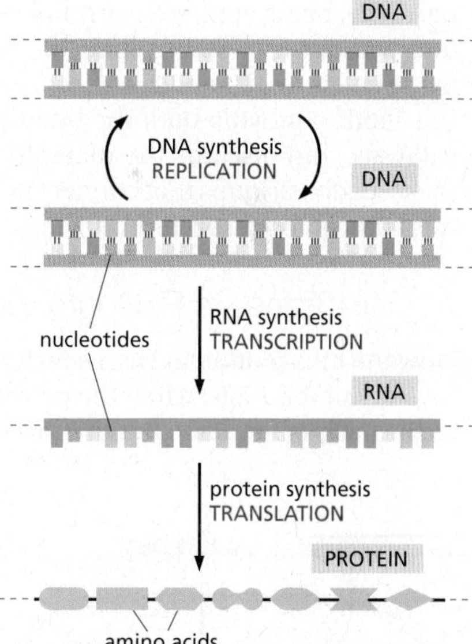

Figure 1–4 From DNA to protein. In addition to DNA *replication* (shown at the top of the figure), genetic information is read out and put to use through a two-step process: First, in *transcription*, segments of the DNA sequence are used to guide the synthesis of molecules of RNA. Then, in *translation*, RNA molecules are used to guide the synthesis of proteins, which are polymers made of amino acid subunits (discussed shortly).

The backbone of an RNA molecule is formed by a slightly different sugar from that in DNA—ribose instead of deoxyribose; in addition, one of the four bases is slightly different—uracil (U) replaces thymine (T). Most important, however, the other three bases—A, C, and G—are identical to those in DNA, and all four bases will pair with their complementary counterparts in DNA—the A, U, C, and G of RNA with the T, A, G, and C of DNA, respectively. During transcription, this pairing allows the RNA monomers to be lined up and selected for polymerization on a template strand of DNA, just as DNA monomers are selected during replication. The outcome is a single-strand polymer molecule whose sequence of nucleotides faithfully represents a portion of the cell's genetic information, even though it is written in a slightly different alphabet—consisting of the four RNA monomers instead of the four DNA monomers.

The same segment of DNA can be used repeatedly to guide the synthesis of many identical RNA molecules. Thus, whereas the cell's archive of genetic information in the form of DNA is fixed and sacrosanct, *RNA transcripts* are mass-produced and disposable. Most of these transcripts function as intermediates in the transfer of genetic information by serving as *messenger RNA* (*mRNA*) molecules that guide the synthesis of proteins according to the genetic instructions stored in the DNA. But as we discuss in Chapter 6, some RNA transcripts do not serve as information carriers; instead, they function directly in the cell to carry out a variety of other functions.

All Cells Use Proteins as Catalysts

Like DNA and RNA molecules, **protein** molecules are long unbranched polymer chains, formed by stringing together monomeric building blocks (subunits) drawn from a standard repertoire that is the same for all living cells. Like DNA and RNA, proteins carry information in the form of a linear sequence of subunits in the same way as a human message written in an alphabetic script. There are many different protein molecules in each cell, and—if we ignore water molecules—they form the major portion of the cell's mass.

The subunits of proteins are the **amino acids**, which are quite different from the nucleotides of DNA and RNA, and there are 20 types instead of 4. Each amino acid is built around a core structure that allows it to be covalently linked in a standard way to any other amino acid in the set; attached to this core is a side group of atoms that gives each amino acid a distinctive chemical character. Each protein molecule is a *polypeptide* chain that is created by joining its amino acids in a particular sequence; this sequence determines how the polypeptide folds up, giving the protein its unique three-dimensional structure. Through several billion years of evolution, these sequences have been selected to give each protein a useful function.

By folding into a precise structure that binds with high specificity to other molecules, each protein performs a specific function according to its genetically specified sequence of amino acids. Proteins form and maintain diverse cell and extracellular structures, generate movements, sense signals, and so on. Many have reactive sites on their surface, allowing them to act as *enzymes* that catalyze reactions that make or break specific covalent bonds. Proteins, above all, are the main molecules that put the cell's genetic information into action. Thus, polynucleotides (DNA and mRNAs) specify the amino acid sequences of proteins. Proteins, in turn, serve as *catalysts* to cause many different chemical reactions to occur, including those that synthesize new DNA and RNA molecules.

In everyday speech, a catalyst refers to "any agent that provokes or speeds significant change or action." But in chemistry, the term **catalyst** is defined more narrowly, being applied to any molecule that speeds up a specific chemical reaction without itself being changed. From the most fundamental point of view, a living cell is a self-replicating collection of catalysts that takes in food, processes this food to provide both the building blocks and energy needed to make more catalysts, and discards the materials left over as waste (**Figure 1–5A**). Together, these feedback loops that connect proteins and polynucleotides form the basis for this autocatalytic, self-reproducing behavior of all living organisms (**Figure 1–5B**).

All Cells Translate RNA into Protein in the Same Way

How the information in DNA specifies the production of proteins was a complete mystery in the 1950s when the double-strand structure of DNA was first revealed as the basis of heredity. But in subsequent years, scientists discovered the elegant

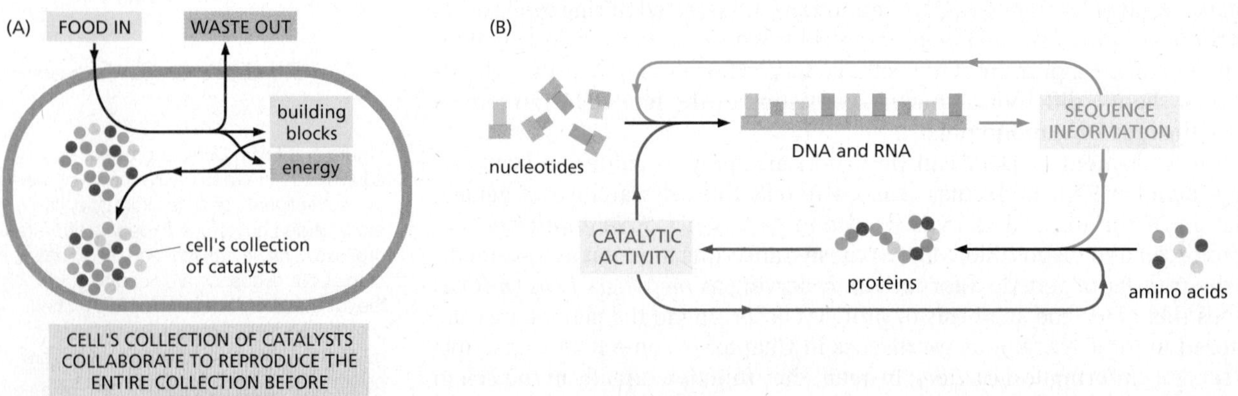

Figure 1–5 **Life as an autocatalytic process.** (A) The living cell is a self-replicating collection of catalysts. (B) Life can be viewed as an autocatalytic process. DNA and RNA molecules provide the nucleotide sequence information *(green arrows)* that is used both to produce proteins and to copy themselves. Proteins, in turn, provide the catalytic activity *(red arrows)* needed to synthesize DNA, RNA, and proteins themselves. Together, these feedback loops create the self-replicating system that endows cells with the ability to reproduce. Although the great majority of the catalysts in the cell are proteins (known as enzymes), a few RNA molecules (known as ribozymes) also have this property, as we will see in Chapter 6.

mechanisms involved. The translation of genetic information from the 4-letter alphabet of polynucleotides into the 20-letter alphabet of proteins is a complex process. The rules of this translation seem in some respects neat and rational but in other respects strangely arbitrary, given that they are (with minor exceptions) identical in all living things. These arbitrary features, it is thought, reflect frozen accidents in the early history of life. They stem from the chance properties of the earliest organisms that were passed on by heredity and have become so deeply embedded in the constitution of all living cells that they cannot be changed without disastrous consequences.

It turns out that the information in the sequence of a messenger RNA (mRNA) molecule is read out in groups of three nucleotides at a time: each triplet of nucleotides, or *codon*, specifies (codes for, or encodes) a single amino acid in a corresponding protein. Because the number of distinct triplets that can be formed from four nucleotides is 4^3, there are 64 possible codons, all of which occur in nature. However, there are only 20 naturally occurring amino acids, which means there are necessarily many cases in which several codons correspond to the same amino acid. This *genetic code* is read out by a special class of small RNA molecules, called *transfer RNAs* (*tRNAs*). Each type of tRNA becomes attached at one end to a specific amino acid and displays at its other end a specific sequence of three nucleotides—an *anticodon*—that enables it to recognize, through base-pairing, a particular codon or subset of codons in mRNA. The intricate chemistry that enables these tRNAs to translate a specific sequence of A, C, G, and U nucleotides in an mRNA molecule into a specific sequence of amino acids in a protein molecule occurs on a *ribosome*, a large multimolecular machine composed of both protein and *ribosomal RNA*. All of these processes will be described in detail in Chapter 6.

Each Protein Is Encoded by a Specific Gene

DNA molecules as a rule are very large, containing the specifications for thousands of proteins and RNA molecules. Special sequences in the DNA serve as punctuation, defining where the information for each RNA and protein begins and ends. And individual segments of the long DNA sequence are transcribed into separate mRNA molecules, coding for different proteins. Each such DNA segment represents one gene. As previously mentioned, some DNA segments—a smaller number—are transcribed into RNA molecules that are not translated into protein but have other functions in the cell; such DNA segments also count as genes. A **gene** therefore is defined as the segment of DNA sequence corresponding either to a single protein (but sometimes to a set of closely related, alternative protein variants) or to a single catalytic, regulatory, or structural RNA molecule.

In all cells, the *expression* of individual genes is regulated: instead of manufacturing its full repertoire of possible proteins and RNAs at full tilt all the time, the cell adjusts the rate of transcription and translation of different genes independently, according to need. As we shall see in Chapter 7, stretches of *regulatory DNA* are interspersed among the segments that code for protein, and these noncoding regions bind to special protein molecules that control the rate of transcription of individual genes. The organization of this regulatory DNA varies widely from one class of organisms to another, but the basic strategy is universal. In this way, the genome of the cell dictates not only the nature of the cell's proteins but also when and where they are to be made.

Life Requires a Continual Input of Free Energy

A living cell is a dynamic chemical system, operating far from chemical equilibrium. For a cell to grow or to make a new cell in its own image, it must take in *free energy* from the environment, as well as raw materials, to drive the necessary synthetic reactions. This consumption of free energy is fundamental to life. When this energy is not available, a cell decays toward chemical equilibrium and soon dies.

As one example, free energy is required for the propagation of genetic information. Picture the molecules in a cell as a swarm of objects endowed with thermal energy, moving around violently at random, buffeted by collisions with one another. To copy genetic information—in the form of a DNA sequence, for example—nucleotide molecules from this wild crowd must be captured, arranged in a specific order defined by a preexisting template, and linked together in a fixed relationship. The bonds that hold the nucleotides in their proper places on the template and join them together must be strong enough to resist the disordering effect of thermal motion, which we describe shortly. The joining process is driven forward by a consumption of free energy, which is needed to ensure that the correct bonds are made, and made robustly. As an analogy, the molecules might be compared with spring-loaded traps, ready to snap into a more stable, lower-energy attached state when they meet their proper partners. As they snap together into the bonded arrangement, their available stored energy—their free energy—like the energy of the spring in the trap, is released and dissipated as heat. In a cell, the chemical processes underlying information transfer are more complex, but the same basic principle applies: free energy must be spent for the creation of order.

To replicate its genetic information faithfully, and indeed to make all its complex molecules according to the correct specifications, the cell therefore requires free energy, which has to be imported somehow from the surroundings. As we will discuss in detail in Chapter 2, the free energy required by animal cells is derived from chemical bonds in food molecules that the animals eat, whereas plants get their free energy from sunlight.

All Cells Function as Biochemical Factories

Because all cells make DNA, RNA, and protein, they all have to contain and manipulate a similar collection of small organic (carbon-containing) molecules, including simple sugars, nucleotides, and amino acids, as well as other substances that are universally required. All cells, for example, require the phosphorylated nucleotide *ATP* (*adenosine triphosphate*), not only as a building block for the synthesis of DNA and RNA but also as a carrier of the free energy that is needed to drive a huge number of chemical reactions in the cell.

Although all cells function as biochemical factories of a broadly similar type, many of the details of their small-molecule transactions differ. Plants, for example, require only the simplest of nutrients because they harness the energy of sunlight to make all their own small organic molecules. Other organisms, such as animals and some bacteria, feed on living (or once living) organisms and must obtain many of their organic molecules ready-made. We return to this point later in the chapter.

All Cells Are Enclosed in a Plasma Membrane Across Which Nutrients and Waste Materials Must Pass

Each living cell is enclosed by a membrane—the **plasma membrane**. This membrane acts as a selective barrier that enables the cell to concentrate nutrients gathered from its environment and retain the products it synthesizes for its own use, while excreting its waste products. Without a plasma membrane, the cell could not maintain its integrity as a coordinated chemical system.

The molecules that form cell membranes have the simple physicochemical property of being *amphiphilic*; that is, they consist of one part that is hydrophilic (water-soluble) and another part that is hydrophobic (water-insoluble). Such molecules placed in water aggregate spontaneously, arranging their hydrophobic portions to be as much in contact with one another as possible to hide them from the water, while keeping their hydrophilic portions exposed. Amphiphilic molecules of appropriate shape, such as the *phospholipid* molecules that compose most of the molecules of the plasma membrane, spontaneously aggregate in water to create a *bilayer* that forms small closed vesicles (**Figure 1–6**).

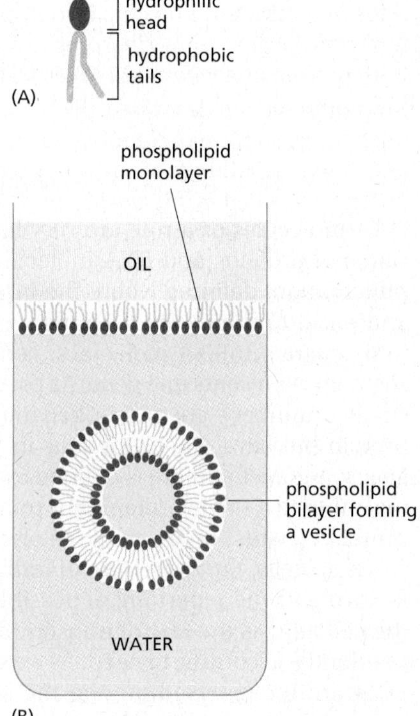

Figure 1–6 Behavior of phospholipid molecules in water. (A) A phospholipid molecule is amphiphilic, having a hydrophilic (water-loving) phosphate head group and a hydrophobic (water-avoiding) hydrocarbon tail. (B) At an interface between oil and water, phospholipids arrange themselves as a single sheet (a monolayer), with their head groups facing the water and their tail groups facing the oil. When immersed in water, however, phospholipids aggregate to form lipid bilayers that fold in on themselves to form sealed aqueous compartments known as vesicles.

Although the chemical details vary, the hydrophobic tails of the predominant lipid molecules in all cells are hydrocarbon polymers ($-CH_2-CH_2-CH_2-$), and their spontaneous assembly into a lipid bilayer is but one of many examples of an important general principle: cells produce molecules whose chemical properties cause them to *self-assemble* into the structures that a cell needs.

The cell boundary cannot be totally impermeable. If a cell is to grow and reproduce, it must be able to import raw materials and export waste across its plasma membrane. All cells therefore have specialized proteins embedded in their plasma membrane that transport specific molecules from one side to the other. Some of these *membrane transport proteins*, like some of the proteins that catalyze the fundamental small-molecule reactions inside the cell, have been so well conserved over the course of evolution that we can recognize the family resemblances between them when even the most distantly related organisms are compared.

The transport proteins in the plasma membrane largely determine which molecules enter the cell, while the catalytic proteins (enzymes) inside the cell determine the reactions that the entering molecules undergo. Thus, by specifying the RNAs and proteins that the cell produces, the genetic information recorded in the DNA sequence dictates the entire chemistry of the cell—in fact, not only its chemistry but also its form and its behavior, for these too are chiefly determined and controlled by the cell's proteins.

Cells Operate at a Microscopic Scale Dominated by Random Thermal Motion

Thus far we have described the cell as a self-replicating, membrane-bound bag of chemicals and macromolecules; but, as the unit of life, the cell is much more than the sum of its parts. Although not obvious from microscopy images, even the simplest cell is highly ordered internally: its individual components must self-assemble and become highly organized for the cell to function. And the cell contents are in perpetual motion. The most obvious movements are catalyzed by *motor proteins*, enzymes that use the energy of ATP hydrolysis for a wide variety of purposes; these include pumping ions across the plasma membrane, translocating large assemblies from one intracellular site to another, and propelling the cell through its environment. In addition, and as previously mentioned, random thermal motions of molecules (including water) are prominent at the scale of cells—whose dimensions can be as small as a micrometer (10^{-6} meters) in diameter. This type of spontaneous movement, called *thermal* or **Brownian motion**, was first observed by Robert Brown in 1827, while looking through a microscope at pollen grains immersed in water. Caused by random molecular collisions, the constant fluctuating movement has important repercussions. Brownian motion drives a process called *diffusion*, and it determines the rates of biochemical reactions as molecules collide with one another within the interior of a cell (described in Chapter 2; see Movie 2.4).

Even though random, the cell can harness Brownian motion for its own advantage. For example, during one step in the crawling migration of animal cells, the plasma membrane at the leading edge extends forward (see Chapter 16). This movement does not involve motor proteins. Instead, a cytoskeletal filament (an actin polymer) polymerizes adjacent to the inner membrane surface. When the membrane fluctuates in the forward direction, actin quickly fills in the gap so that the membrane cannot slip back to its original position. This phenomenon, in which random thermal motions are harnessed in a directed way, creates a *Brownian ratchet* (**Figure 1–7**).

Because an object at the micrometer scale is constantly buffeted by water molecules, its movement requires overcoming high viscous drag forces. As a result, the directed movement of a complex of molecules inside the cell (by a motor protein, for example) will stop immediately when the motor disengages, leaving the complex to be randomly buffeted by thermal motion. There is no "gliding" inside the cell.

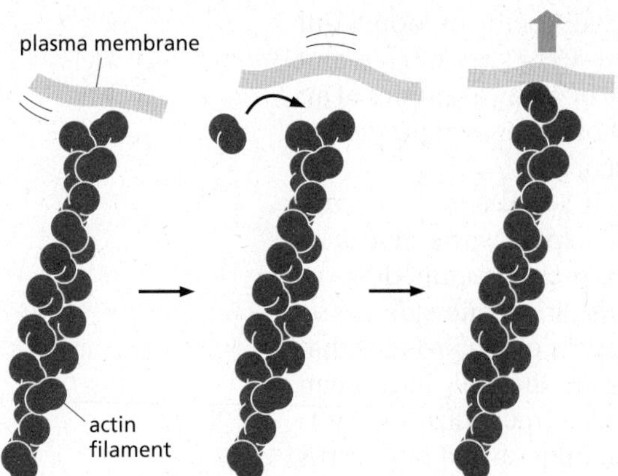

Figure 1–7 How membrane protrusion is driven by a simple Brownian ratchet. A single actin filament is shown abutting the plasma membrane, which is fluctuating back and forth because of random thermal motions. When the membrane happens to move away from the end of the filament, it creates sufficient space for an additional subunit, which quickly adds on. The slightly longer filament acts as a ratchet and prevents the membrane from moving back to its original position. In a migrating animal cell, this Brownian ratchet process drives protrusion of the membrane and contributes to forward movement of the cell, as described in Chapter 16 (see pp. 956–957).

A Living Cell Can Exist with 500 Genes

We have seen how genomes carry the information for all the proteins and RNA molecules of a cell, and how, through catalysis, all the other building blocks of the cell are made. But how complex are real living cells? In particular, what are the minimum requirements of a living cell? One measure of complexity is based on the total number of genes in an organism's genome. A species that has one of the smallest known genomes is the bacterium *Mycoplasma genitalium*, which causes a common, sexually transmitted, human disease (**Figure 1–8**). This organism lives as a parasite in mammals, where the environment provides it with many of the small molecules it needs ready-made. Nevertheless, it still has to make all the large molecules—DNA, RNAs, and proteins. It has 525 genes, most of which are essential. Its genome of 580,070 nucleotide pairs represents 145,018 bytes of information—about as much as it takes to record the text of one chapter of this book. Cell biology may be complicated, but it is not unimaginably so.

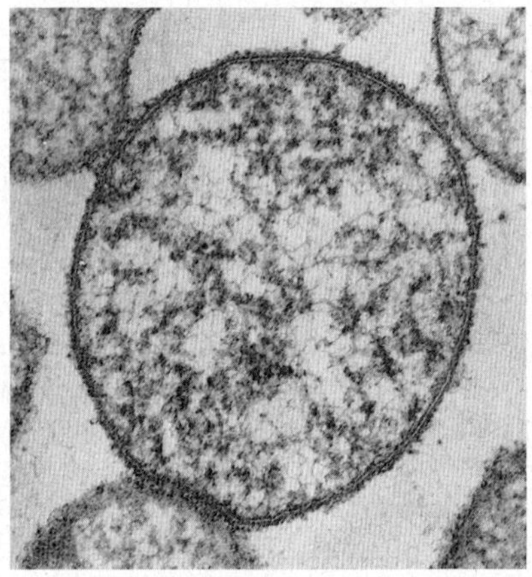

Figure 1–8 The small bacterium *Mycoplasma genitalium*. It is viewed here in cross section in an *electron microscope*, which uses a beam of electrons instead of light to create an image with a resolution that is many times higher than that of an image viewed in a conventional light microscope. Of the 525 genes this bacterium contains, 43 code for transfer RNAs, ribosomal RNAs, and other nonprotein-coding RNAs. Of the 482 protein-coding genes, 154 are involved in replication, transcription, translation, and related processes involving DNA, RNA, and protein; 98 are involved in the membrane and surface structures of the cell; 46 are involved in the transport of nutrients and other molecules across the plasma membrane; and 71 are involved in energy conversion and the synthesis and degradation of small molecules. (Courtesy of Roger Cole, in Medical Microbiology, 4th ed. [S. Baron, ed.]. Galveston: University of Texas Medical Branch, 1996.)

Summary

The individual cell is the minimal self-reproducing unit of life. A cell consists of a self-replicating collection of catalysts, enclosed in a plasma membrane. All cells operate as biochemical factories, driven by the free energy released in a complicated network of chemical reactions. Central to a cell's ability to reproduce is the transmission of its genetic information to its progeny cells when it divides. All cells store their genetic information in double-strand DNA, and the complete sequence of DNA nucleotides for each organism is known as its genome. The cell replicates this information by separating the paired DNA strands and using each as a template for polymerization to make a new DNA strand with a complementary sequence of nucleotide subunits. The same strategy of templated polymerization is used in the transcription of portions of the DNA into molecules of the closely related polynucleotide polymer, RNA. Most of these RNA molecules are mRNAs that in turn guide the synthesis of protein molecules by the process of translation. Proteins are polymers of amino acid subunits and are the catalysts for almost all the cell's chemical reactions. They are also responsible for the selective import and export of molecules across the plasma membrane that surrounds each cell. The specific shape and function of each protein depend on its amino acid sequence, which is specified by the nucleotide sequence of a corresponding segment of the DNA—the gene that codes for that protein. In this way, the DNA of the cell determines the cell's chemistry, which is fundamentally similar in all cells, reflecting their ultimate origin from a common ancestor cell that existed on Earth more than 3.5 billion years ago.

GENOME DIVERSIFICATION AND THE TREE OF LIFE

The success of living organisms based on DNA, RNA, and protein has been spectacular. Through its billions of years of proliferation, life has populated the oceans, covered the land, penetrated deep into Earth's crust, and molded the surface of our planet. Our oxygen-rich atmosphere, the deposits of coal and oil, the layers of iron ores, the cliffs of chalk and limestone and marble—all these are products, directly or indirectly, of past biological activity on Earth.

Living things are not confined to the familiar temperate realm of land, water, and sunlight inhabited by plants and animals. They are found in the darkest depths of the ocean, in hot volcanic mud, in pools beneath the frozen surface of the Antarctic, and buried kilometers deep in Earth's crust. The creatures that live in these extreme environments are generally unfamiliar, not only because they are inaccessible, but also because they are mostly microscopic and cannot be maintained in a laboratory. Even in more familiar habitats, most organisms are too small for us to see without special equipment: they tend to go unnoticed, unless they cause a disease or rot the timbers of our houses. Yet such *microorganisms* (microbes) are by far the most numerous living organisms on our planet. Only recently, through new methods of molecular analysis including rapid *DNA sequencing*, have we begun to get a picture of life on Earth that is not grossly distorted by our biased perspective as large animals living on dry land.

In this section, we consider—in very broad terms—the diversity of organisms on our planet and the relationships among them. Because the genetic information for every organism is written in the universal language of DNA sequences, and because the DNA sequence of any organism's genome can be readily determined, it is now possible to characterize, catalog, and compare any set of living organisms with reference to these sequences. From such comparisons we can specify the place of each organism in the family tree of living species—the "tree of life."

The Tree of Life Has Three Major Domains: Eukaryotes, Bacteria, and Archaea

The classification of living things has traditionally depended on comparisons of their outward appearances: we can see that a fish has eyes, jaws, backbone, brain, and so on, just as humans do, and that a worm does not—just as we can see that a rosebush is more similar to an apple tree than to grass. As Darwin showed, we can readily interpret such close family resemblances in terms of an evolution from common ancestors, and we can find the remains of many of these ancestors preserved in the fossil record. In this way, it became possible to draw a family tree of living organisms, showing the various lines of descent, as well as branch points in evolutionary history, where the ancestors of one group of species became different from those of another.

When the disparities between organisms become very great, however, these methods begin to fail. How do we decide whether a fungus is more closely related to a plant or to an animal? When it comes to microscopic organisms such as bacteria, the task becomes harder still: one tiny rod or sphere can look much like another. Moreover, much of our knowledge of the microbial world was traditionally restricted to those species that can be isolated and cultured in the laboratory. But direct DNA sequencing of populations of microbes in their natural habitats—such as soil, ocean water, or even the human mouth—has taught us that the vast majority of microbes cannot be easily cultured in the laboratory. Often, they thrive in the wild as components of complex ecosystems and—when separated from their natural surroundings—cannot survive. Until modern DNA sequencing was developed, these organisms were largely unknown to us, especially those that inhabit extreme environments such as the deep Earth's crust or seawater miles below the ocean surface.

Genome analysis has now provided us with a simple, direct, and powerful way to determine evolutionary relationships. The complete DNA sequence of an organism defines its nature with almost perfect precision and in exhaustive detail. Moreover, this specification is in a digital form—a string of letters—that can be entered into a computer and compared with the corresponding information for any other organism. Because DNA is subject to random changes that accumulate over long periods of time (as we will see shortly), the number of differences between the DNA sequences of two organisms can provide a direct, objective, and quantitative indication of the evolutionary distance between them.

For constructing a comprehensive tree of life, it is necessary to begin with a segment of DNA that is easily recognized in the genomes of all organisms. We discussed earlier how all cells use the same fundamental mechanism to translate a nucleotide sequence into a protein sequence, and we saw that the ribosome is the "decoding machine" that carries this out. Ribosomes are fundamentally similar in all organisms, and an especially well-conserved component of them is the RNA molecules that make up their core. Although the exact sequence of these *ribosomal RNAs* (*rRNAs*) differs across organisms, they are similar enough to use them as a ruler to judge how closely two species are related: the more similar the ribosomal RNA sequences, the more recently the two species diverged from a common ancestor and the more related they must be. Once a rough approximation of the tree of life has been obtained in this way, many additional DNA sequences in genomes—those that might not be identifiable in all organisms—can accurately determine relationships among more closely related species.

This approach has revealed that the living world consists of three major divisions, or *domains*: eukaryotes, bacteria, and archaea, as illustrated in Figure 1–9; in the following paragraphs, we briefly introduce each in turn.

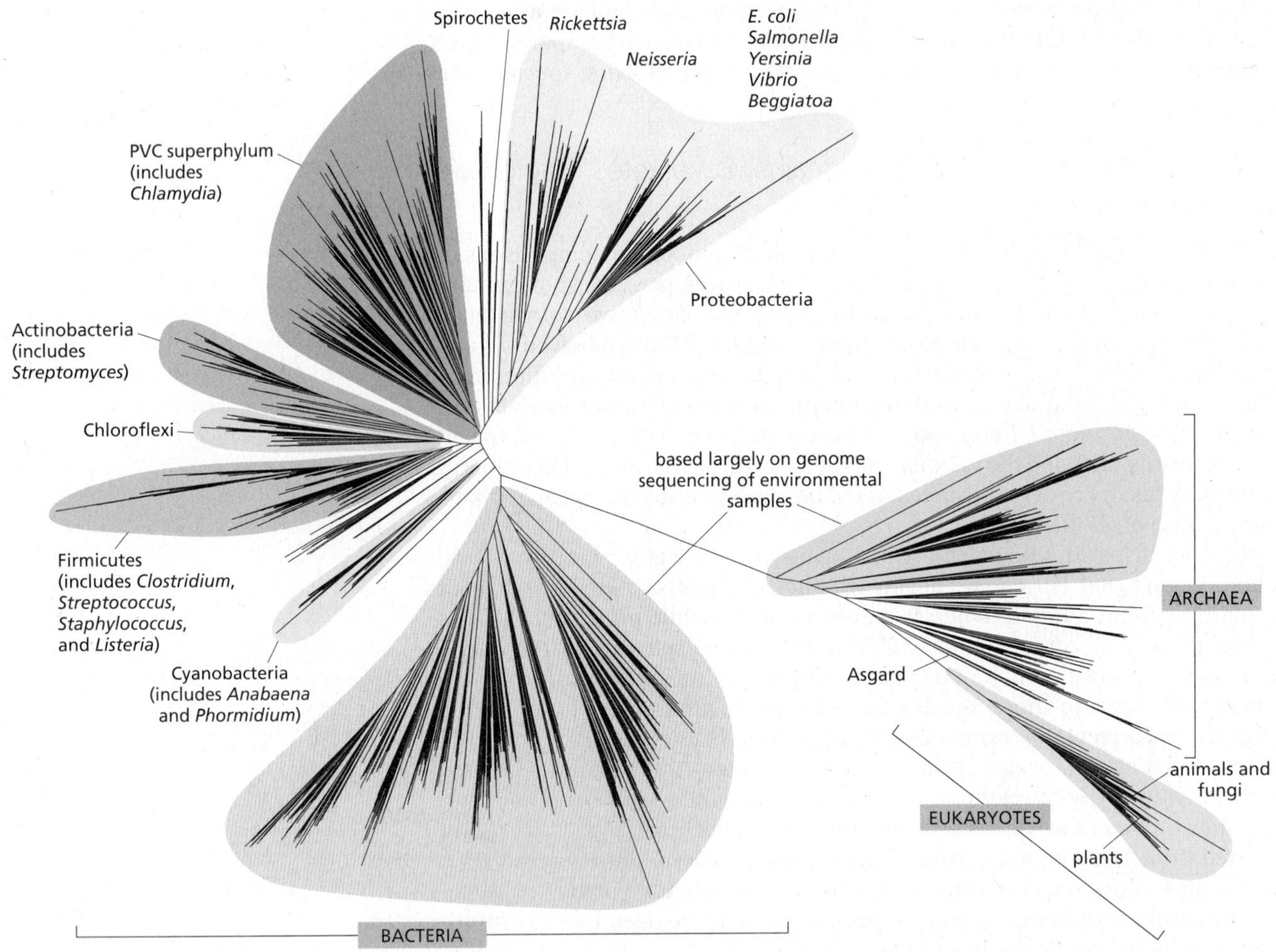

Figure 1–9 **A global tree of life, based on genome comparisons, shows the three major divisions (domains) of the living world.** The lengths of the branches are proportional to differences among genomes using common genes that can be recognized and compared across many different species. Some of the organisms discussed in this and later chapters are indicated. Of the three domains of life (bacteria, archaea, and eukaryotes), bacteria encompass by far the greatest diversity, commensurate with their ability to colonize nearly every ecological niche on the planet. So many new bacterial species are currently being identified through DNA sequencing of environmental samples that simply naming them has become a challenge. Although eukaryotes (and especially animals) are the main focus of this book, they comprise only a small slice of the global diversity. An expanded eukaryotic tree is shown in Figure 1–35, and an expanded tree of mammals is given in Figure 4–67. (Adapted from C.J. Castelle and J.F. Banfield, *Cell* 172:1181–1197, 2018.)

Eukaryotes Make Up the Domain of Life That Is Most Familiar to Us

The great variety of living creatures that we see around us are eukaryotes. The name is from the Greek, meaning "truly nucleated" (from the words *eu*, "well" or "truly," and *karyon*, "kernel" or "nucleus"), reflecting the fact that the cells of these organisms have their DNA enclosed in a membrane-bound organelle called the *nucleus*. Visible by simple light microscopy, this feature was used in the early twentieth century to classify living organisms as either **eukaryotes** (those with a nucleus) or **prokaryotes** (those without a nucleus). We now know that prokaryotes comprise two of the three major domains of life, the bacteria and archaea. Eukaryotic cells are typically much larger than those of bacteria and archaea; in addition to a nucleus, they typically contain a variety of membrane-bound organelles that are also lacking in the prokaryotes. The genomes of eukaryotes also tend to run much larger—containing more than 20,000 genes for humans and corals, for example, compared with 4000–6000 genes for the typical bacteria or archaea.

In addition to plants and animals, the eukaryotes include fungi (such as mushrooms or the yeasts used in beer- and bread-making), as well as an astonishing variety of single-celled, microscopic forms of life. Most of this book is focused on the cell biology of eukaryotic organisms (especially animals); in the final sections of this chapter, we shall return to eukaryotes and focus on the variety within this group.

On the Basis of Genome Analysis, Bacteria Are the Most Diverse Group of Organisms on the Planet

When modern trees of life were constructed using genome information, one of the big surprises was how much more evolutionarily diverse the bacterial world is compared with the eukaryotes; we now know that this great diversity reflects the much earlier appearance of bacteria in the evolutionary history of the planet. Bacteria are usually very small (and invisible to the unaided eye), and they generally live as independent individuals or in loosely organized communities, rather than as multicellular organisms. They are typically spherical or rod-shaped and measure a few micrometers (μm) in linear dimension (**Figure 1–10**). They often have a tough protective coat, called a cell wall, beneath which a plasma membrane encloses a single cytoplasmic compartment—the *cytoplasm*—containing DNA, RNA, proteins, and the many small molecules needed for life (**Figure 1–11**). Although difficult to discern in the light microscope, the interior of a bacterium is nevertheless highly organized, a topic we discuss in Chapter 16.

Commensurate with the diversity of their genomes, bacteria live in an enormous variety of ecological niches, and they are astonishingly varied in their

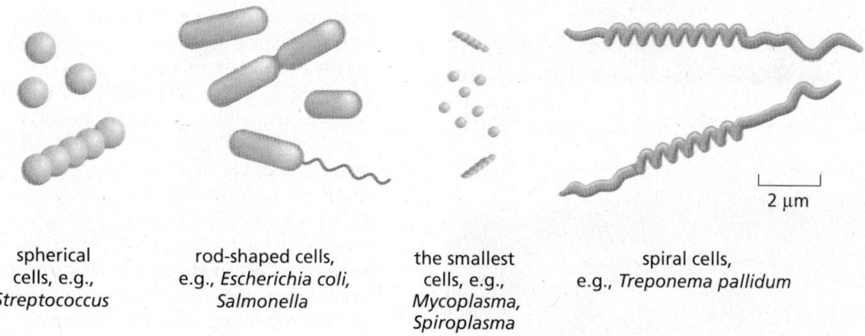

Figure 1–10 Shapes and sizes of some bacteria. Although most are small, as shown, measuring a few micrometers in linear dimension, there are also some giant species. An extreme example is the cigar-shaped bacterium *Epulopiscium fishelsoni*, which lives in the gut of a surgeonfish and can be up to 600 μm long (not shown).

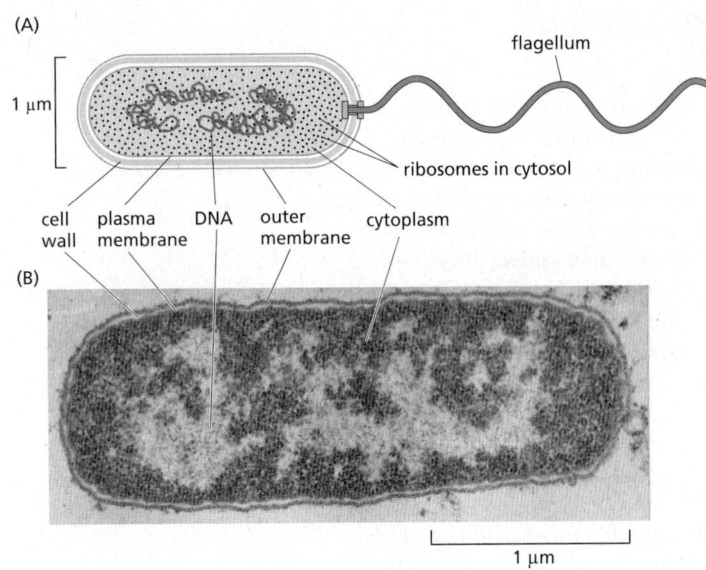

Figure 1–11 Bacterial structure. (A) A drawing of the bacterium *Vibrio cholerae*, showing its simple internal organization. This species can infect the human small intestine to cause cholera; the severe diarrhea that accompanies this disease kills more than 100,000 people a year worldwide. Like many other bacteria, *Vibrio* has a helical appendage at one end—a flagellum—that rotates as a propeller to drive the cell forward. (B) An electron micrograph of a longitudinal section through the widely studied bacterium *Escherichia coli* (*E. coli*). *E. coli* is part of our normal intestinal *microbiota*, the complete collection of microbes in our gut. It has many flagella distributed over its surface, but they are not visible in this section. Both of the bacteria shown here are Gram negative, having both an outer and an inner (plasma) membrane. However, many bacterial species lack the outer membrane; these are classified as Gram positive. (B, courtesy of E. Kellenberger.)

biochemical capabilities. There exist species that can utilize virtually any type of organic molecule as food, ranging from sugars and amino acids to hydrocarbons, including the simplest hydrocarbon, methane gas (CH_4). Other species (**Figure 1–12**) harvest light energy in a variety of ways; some, like plants, carry out photosynthesis and generate oxygen as a by-product. Still others can feed on a plain diet of inorganic nutrients, getting their carbon from CO_2, and relying on a host of other chemicals that occur in the environment to fuel their energy needs—including H_2, Fe^{2+}, H_2S, and elemental sulfur (**Figure 1–13**).

A wide range of bacteria directly affect human health. The bubonic plague of the Middle Ages (estimated to have killed half the population of Europe) and the current tuberculosis pandemic (more than a million deaths a year) are each due to a specific species of bacteria. And thousands of different bacterial species reside in our gut and on our skin, where they are often beneficial to us. We shall discuss bacteria throughout the book, as it is the study of these relatively simple cells that led to much of our understanding of basic biological processes—including DNA replication, transcription, and translation. We focus again on bacteria in Chapter 24 when we examine the cell biology of infectious disease. Finally, genetic

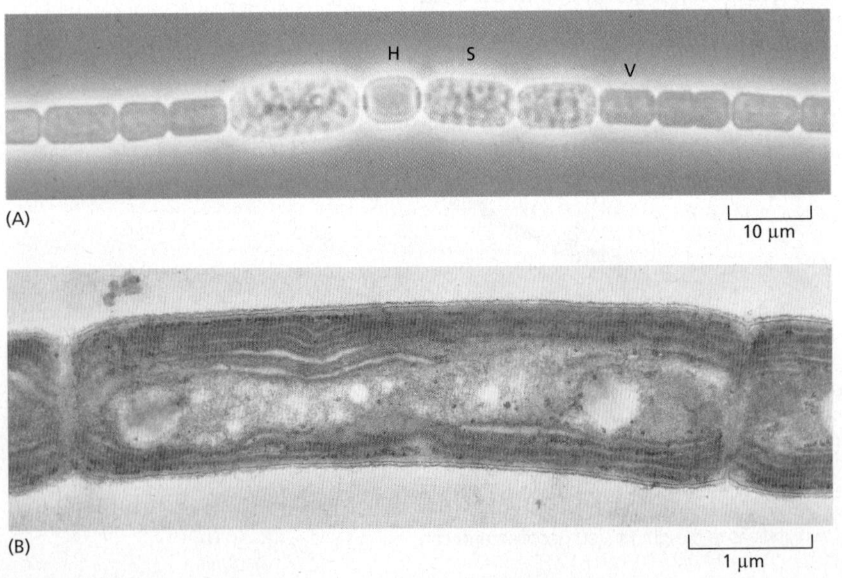

Figure 1–12 Photosynthetic bacteria. (A) A light micrograph of the bacterium *Anabaena cylindrica*. Its cells form long chains, in which most of the cells (labeled V) perform photosynthesis (and thereby capture CO_2 and incorporate C into organic compounds); others (labeled H) become specialized for fixing N from N_2; and still others (labeled S) develop into spores, which can resist unfavorable conditions. (B) An electron micrograph of a related photosynthetic bacterium, *Phormidium laminosum*, which shows the intracellular membranes where photosynthesis occurs. As shown in these micrographs, some prokaryotes have intracellular membranes and form colonies that resemble simple multicellular organisms. (A, courtesy of David Adams; B, courtesy of D.P. Hill and C.J. Howe.)

engineering techniques allow bacteria to be put to use as small "factories" to produce human pharmaceuticals, biofuels, and other high-value chemical products, as we discuss in Chapter 8.

Archaea: The Most Mysterious Domain of Life

Of the three domains of life, archaea remains the most poorly understood. Most of its members have been identified only by DNA sequencing of samples from the environment, and relatively few have been cultured and studied up close in the laboratory. Like bacteria, the archaea we know most about are small and lack the internal, membrane-bound organelles that distinguish the eukaryotes. But they differ from bacteria in many ways, including the chemistry of their cell walls, the kinds of lipids that make up their membrane, and the range of biochemical reactions that they can carry out. Another surprising conclusion came from genome comparisons: although archaea resemble bacteria in their outward appearances, their genomes are much more closely related to eukaryotes than to bacteria (see Figure 1–9). It has even been proposed that the tree of life should be considered to have only two principal domains, with the archaea and eukaryotes making up one domain and bacteria constituting the other. The close relationship of archaea and eukaryotes has also changed our views on how the earliest eukaryotic cell evolved, a topic addressed later in this chapter.

At first it was thought that archaea occupied only extreme environments such as volcanoes, salt lakes, acid hot springs, and the stomachs of cattle, but they are now recognized to be present also in more congenial surroundings such as soils, seawater, and our skin. Commensurate with the wide variety of ecological niches in which they have been found, different species of archaea have highly diverse chemistries. They are believed to be the predominant life-form in soil and seawater, and they play major roles in recycling nitrogen and carbon, two of the most important elements for all cells.

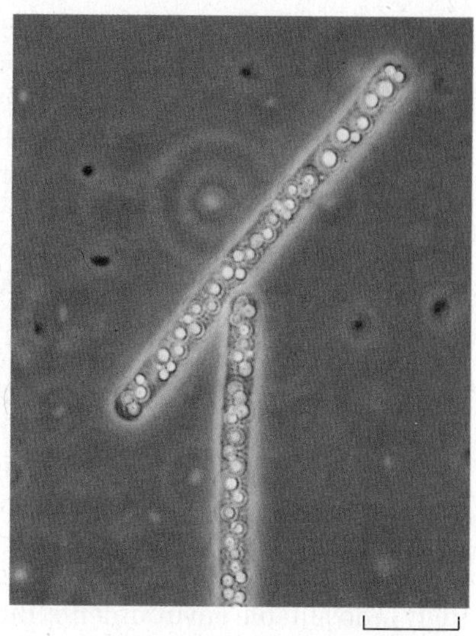

Figure 1–13 The bacterium *Beggiatoa*. It lives in sulfurous environments (for example, see Figure 1–15) and gets its energy by oxidizing H_2S; it can fix carbon even in the dark. Note the yellow deposits of sulfur inside the cells. (Courtesy of Ralph S. Wolfe.)

Organisms Occupy Most of Our Planet

To understand life on Earth, we need to understand more than its diversity; we also need to know where life is found on our planet and how various living species are distributed. Organisms inhabit nearly all of the planet, and we continue to discover new habitats. Amazingly, some bacteria and archaea even live miles down in Earth's deep crust and in the deepest and most hostile parts of the oceans.

How are the main groups of organisms distributed among different environments? DNA sequencing and other advanced technologies have been used recently to address this question. The total biomass on Earth is estimated to contain ~550 gigatons (10^{15} grams) of carbon, of which 450 gigatons of carbon (Gt C) is plants, 70 Gt C is bacteria, 7 Gt C is archaea, and 2 Gt C is animals (**Figure 1–14**). The plants are mainly terrestrial; the bacteria and archaea are mainly in the soil and Earth's crust. Total terrestrial biomass is 100 times greater than that in the oceans, although most of the animal mass is found in the oceans. The human biomass is 10 times greater than that of all measurable wild animals together, and—while human biomass continues to increase—that of wild animals is falling, largely as a result of human activities.

Although humans and other animals make up a small fraction of Earth's biomass, their existence depends completely on other forms of life. In the next section, we shall see some of the ways that these different life-forms work together to capture and recycle energy from Earth's inanimate features.

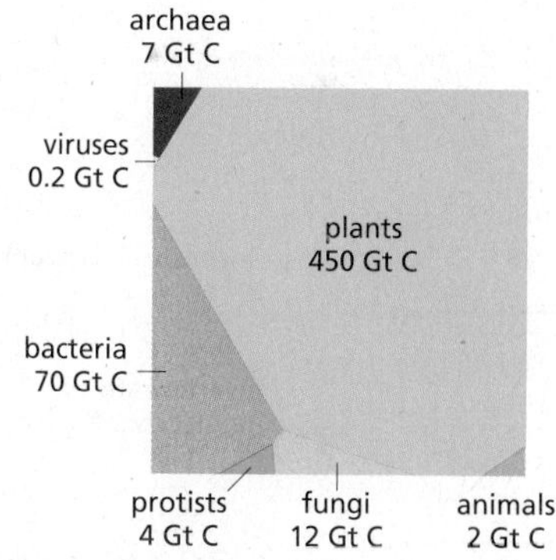

Figure 1–14 The distribution of living biomass on Earth. The total biomass on Earth expressed as gigatons of carbon (Gt C) is estimated to be ~550 Gt C. In the graph shown, the area of each taxon represented is proportional to the taxon's global biomass, so plants account for about 80% (450/550) of the total biomass, whereas animals account for 0.4% (2/550). These recent estimates are based on various advanced techniques, including DNA sequencing and remote sensing. (Adapted from Y.M. Bar-On et al., *Proc. Natl. Acad. Sci. USA* 115:6506–6511, 2018. With permission from the authors.)

Cells Can Be Powered by a Wide Variety of Free-Energy Sources

Organisms obtain the free energy needed for life in different ways. Some—such as animals, fungi, and the many different bacteria that live in the human gut—get it by feeding on other living things or the organic chemicals they produce; such organisms are called *organotrophic* (from the Greek word *trophe*, meaning "food"). Others derive their free energy directly from the nonliving world. These primary energy converters fall into two classes: those that harvest the energy of sunlight, and those that capture their energy from energy-rich systems of inorganic chemicals in the environment (chemical systems that are far from chemical equilibrium). Organisms of the former class are called *phototrophic* (feeding on sunlight); those of the latter are called *lithotrophic* (feeding on rock). The organotrophic organisms like ourselves could not exist without these primary energy converters, which are the most plentiful form of life.

The phototrophic organisms include many types of bacteria, as well as algae and plants, on which we—and virtually all the living things that we ordinarily see around us—depend. Phototrophic organisms have changed the whole chemistry of our environment: as a prime example, the oxygen in Earth's atmosphere is a by-product of their biosynthetic activities.

Lithotrophic organisms are not such an obvious feature of our world, because they are microscopic and mostly live in habitats that humans do not frequent—deep in the ocean, buried in Earth's crust, or in various other seemingly inhospitable environments. But they are a major part of the living world, and they are especially important in any consideration of the history of life on Earth.

Some lithotrophs get energy from *aerobic* reactions, which use molecular oxygen from the environment; because atmospheric O_2 is ultimately the product of living phototrophic organisms, these aerobic lithotrophs are, in a sense, feeding on the products of past life. There are, however, many other lithotrophs that live *anaerobically*, in places where little or no molecular oxygen is present; these are circumstances similar to those that existed in the early days of life on Earth, before oxygen had accumulated.

The most dramatic of the anaerobic sites are the hot *hydrothermal vents* on the floor of the Pacific and Atlantic Oceans. They are located where the ocean floor is spreading as new portions of Earth's crust form by a gradual upwelling of material from Earth's interior (Figure 1–15). Downward-percolating seawater is heated and driven back upward as a submarine geyser, carrying with it a current of chemicals from the hot rocks below. A typical cocktail might include H_2S, H_2, CO, Mn^{2+}, Fe^{2+}, Ni^{2+}, CH_4, NH_4^+, and phosphorus-containing compounds.

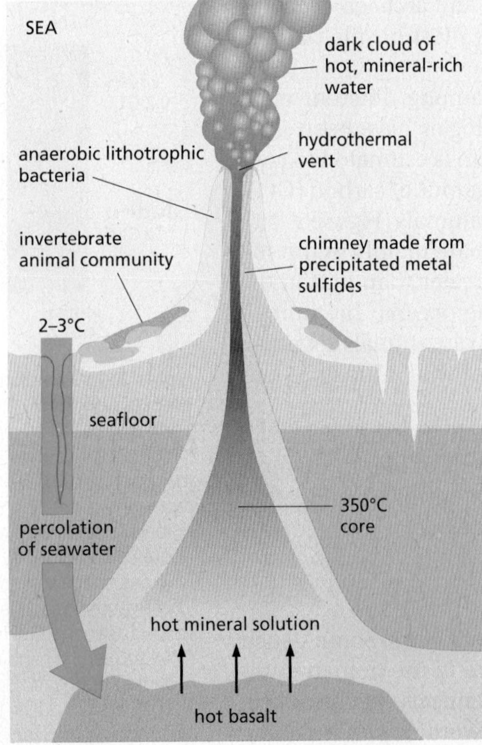

Figure 1–15 The geology of a hot hydrothermal vent in the ocean floor. As indicated, seawater percolates down toward the hot, molten, volcanic rock upwelling (basalt) from Earth's interior and is heated and driven back upward, carrying a mixture of minerals leached from the hot rock. A temperature gradient is set up, from more than 350°C near the core of the vent, down to 2–3°C in the surrounding ocean. Minerals precipitate from the water as it cools, forming a chimney. Different classes of organisms, thriving at different temperatures, live in different neighborhoods of the chimney. A typical chimney might be a few meters tall, spewing out hot, mineral-rich water. The locations of lithotrophic bacteria and the invertebrate marine animals that depend on them are also shown (see Figure 1–16).

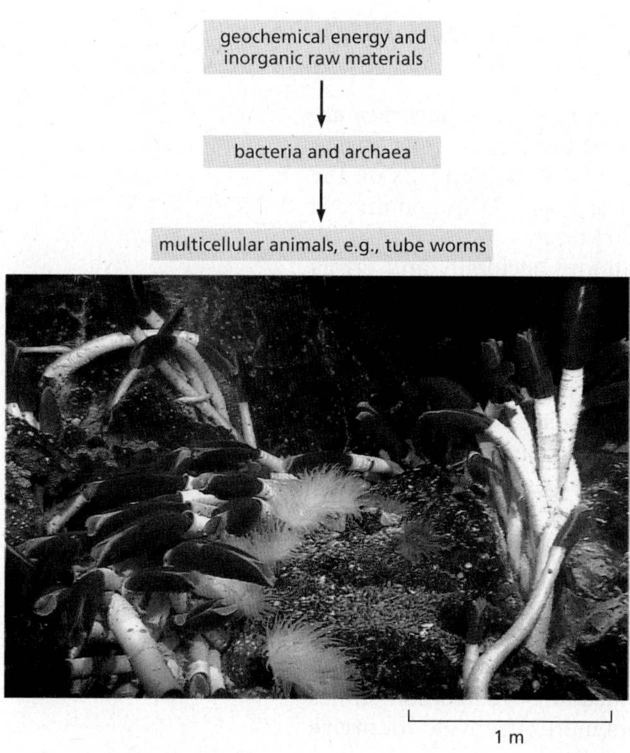

Figure 1–16 Organisms living at a depth of 2500 meters near a vent in the ocean floor. Close to the vent, at temperatures up to about 120°C, various lithotrophic species of bacteria and archaea live, directly fueled by geochemical energy. A little further away, where the temperature is lower, various invertebrate animals live by feeding on these microorganisms. Most remarkable are the giant (2-meter-long) tube worms, *Riftia pachyptila*, which are shown in the photograph. Rather than feed on the lithotrophic microbes, these worms live in symbiosis with them: specialized organs in the worms harbor huge numbers of symbiotic sulfur-oxidizing bacteria, which harness geochemical energy and supply nourishment to their hosts, which have no mouth, gut, or anus. The tube worms are thought to have evolved from more conventional animals and to have become secondarily adapted to life at hydrothermal vents. (Science History Images/Alamy Stock Photo.)

A dense population of microorganisms lives in the neighborhood of the vent, thriving on this austere diet and harvesting free energy from reactions between the available chemicals. Various invertebrate marine animals—clams, mussels, and giant marine worms—in turn, live off the microbes at the vent, forming an entire ecosystem analogous to the world of plants and animals that we belong to, but one powered by geochemical energy instead of light (**Figure 1–16**).

Some Cells Fix Nitrogen and Carbon Dioxide for Other Cells

To make a living cell requires matter, as well as free energy. DNA, RNA, and protein are composed of just six elements: hydrogen, carbon, nitrogen, oxygen, sulfur, and phosphorus. These are all plentiful in the nonliving environment, in Earth's rocks, water, and atmosphere. But they are not present in chemical forms that allow easy incorporation into biological molecules. Atmospheric N_2 and CO_2, in particular, are extremely unreactive. A large amount of free energy is required to drive the reactions that use these inorganic molecules to make the organic compounds needed for further biosynthesis; that is, to *fix* nitrogen and carbon dioxide, so as to make N and C available to living organisms. Many types of cells lack the biochemical machinery to achieve this fixation; they instead rely on other classes of cells to do the job for them. We animals depend on plants, directly or indirectly, for our supplies of carbon- and nitrogen-containing organic compounds. Plants in turn, although they can fix carbon dioxide from the atmosphere, lack the ability to fix atmospheric nitrogen; they depend in part on nitrogen-fixing bacteria to supply their need for nitrogen-containing organic compounds. Plants of the pea family, for example, harbor symbiotic nitrogen-fixing bacteria in nodules in their roots.

Because living cells can differ widely in some of the most basic aspects of their biochemistry, cells with complementary needs and capabilities have frequently developed close associations. Some of these *symbiotic* associations, as we will see later, have evolved to the point where the partners have lost their separate identities altogether: they have joined forces to form a single composite cell—an *endosymbiotic* association, as opposed to an *ectosymbiotic* one between separate organisms.

Genomes Diversify Over Evolutionary Time, Producing New Types of Organisms

Having discussed our current views on the diversity of life-forms, how they are distributed across Earth, and how they depend on one another, we now turn to the question of how this great diversity was generated. All life depends on the storage of genetic information in the form of each organism's DNA genome, so our focus is on how genomes change over evolutionary time.

In storing and copying genetic information, random accidents and errors occur, altering the nucleotide sequence; that is, creating **mutations**. Therefore, when a cell divides, the genomes of its two daughters are often not quite identical to each other or to that of the parent cell. On rare occasions, the error may represent a change for the better; more probably, it will cause no significant difference in the cell's prospects. But in some cases, the error will cause serious damage; for example, by disrupting the coding sequence for a key protein or RNA molecule. Changes due to mistakes of the first type will tend to be perpetuated, because the altered cell has an increased likelihood of surviving and reproducing itself. Changes due to mistakes of the second type—*neutral* changes—may be perpetuated or not: in the competition for limited resources, it is a matter of chance whether the altered cell or its cousins will succeed. But changes that cause serious damage lead nowhere: the cell that suffers them dies, leaving no progeny. Through endless repetition of this cycle of error and trial—of *mutation* and *natural selection*—organisms evolve: their genetic specifications change, sometimes giving organisms new ways to exploit the environment more effectively, to survive in competition with others, and to reproduce successfully.

Some parts of the genome will change more readily than others in the course of evolution. A segment of DNA that does not code for protein or RNA and has no significant regulatory role is free to change at a rate limited only by the frequency of random errors. In contrast, a gene that codes for a highly optimized, essential protein or RNA molecule cannot alter so easily: when mistakes occur, the faulty cells are almost always disabled and eliminated. Genes of this latter sort are therefore *highly conserved*. Through 3.5 billion years or more of evolutionary history, many DNA sequences have changed beyond all recognition, but the most highly conserved genes remain perfectly recognizable in all living species.

These latter genes are the ones we must examine if we wish to trace family relationships between the most distantly related organisms in the tree of life. We discussed an example of one such gene—that for ribosomal RNA—when we introduced the classification of the living world into the three domains of eukaryotes, bacteria, and archaea. Because the production of proteins is fundamental to all living cells, this component of the ribosome has been highly conserved since early in the history of life on Earth (**Figure 1–17**).

The ribosomal RNA genes are exceptional in being so well conserved, whereas most parts of genomes have diversified much more dramatically over evolutionary time. A complete DNA sequence for an organism—its genome sequence—reveals all the genes that an organism possesses, as well as those it lacks. When we

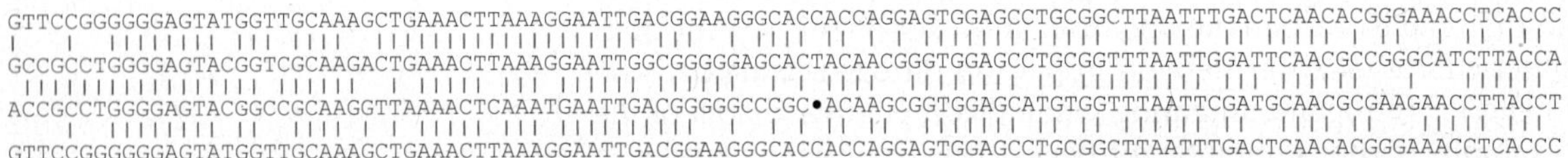

Figure 1–17 Genetic information conserved since the days of the last universal common ancestor of all living things. A part of the gene that codes for the smaller of the two main ribosomal RNA (rRNA) molecules in the ribosome is shown. (The complete molecule is about 1500–1900 nucleotides long, depending on the species.) Corresponding segments of nucleotide sequences from an archaeon (*Methanococcus jannaschii*), a bacterium (*Escherichia coli*), and a eukaryote (*Homo sapiens*) are aligned. The *red vertical lines* indicate sites where the nucleotides are identical between the species; the human sequence is repeated at the bottom of the alignment so that all three two-way comparisons can be seen. The *black dot* halfway along the *E. coli* sequence denotes a site where a nucleotide has been either deleted from the bacterial lineage in the course of evolution or inserted in the other two lineages. Note that the sequences from these three organisms, representative of the three domains of the living world, still retain unmistakable similarities.

compare the three domains of the living world, we can begin to see which genes are common to all of them—and must therefore have been present in the last universal common ancestral cell that was the founder of all present-day living things. We can also identify those genes that are peculiar to a single branch in the tree of life. To explain such findings, we need to consider how new genes arise and, more generally, how genomes evolve.

New Genes Are Generated from Preexisting Genes

The raw material of evolution is the DNA sequence that already exists: there is no natural mechanism for making long stretches of new, random, DNA sequence. In this sense, no gene is ever entirely new. Innovation can, however, occur in several ways (**Figure 1–18**):

1. *Intragenic mutation*: an existing gene can be randomly modified by changes in its DNA sequence, through various types of errors that occur in the process of DNA replication and DNA repair.
2. *Gene duplication*: an existing gene can be accidentally duplicated, creating a pair of initially identical genes within a single cell; these two genes may then diverge in the course of evolution.
3. *DNA segment shuffling*: two or more existing genes can break and rejoin to make a hybrid gene consisting of DNA segments that originally belonged to separate genes.
4. *Horizontal (intercellular) DNA transfer*: a piece of DNA can be transferred from the genome of one cell to that of another—including between species. This process contrasts with the usual *vertical transfer* of genetic information from parent to progeny.

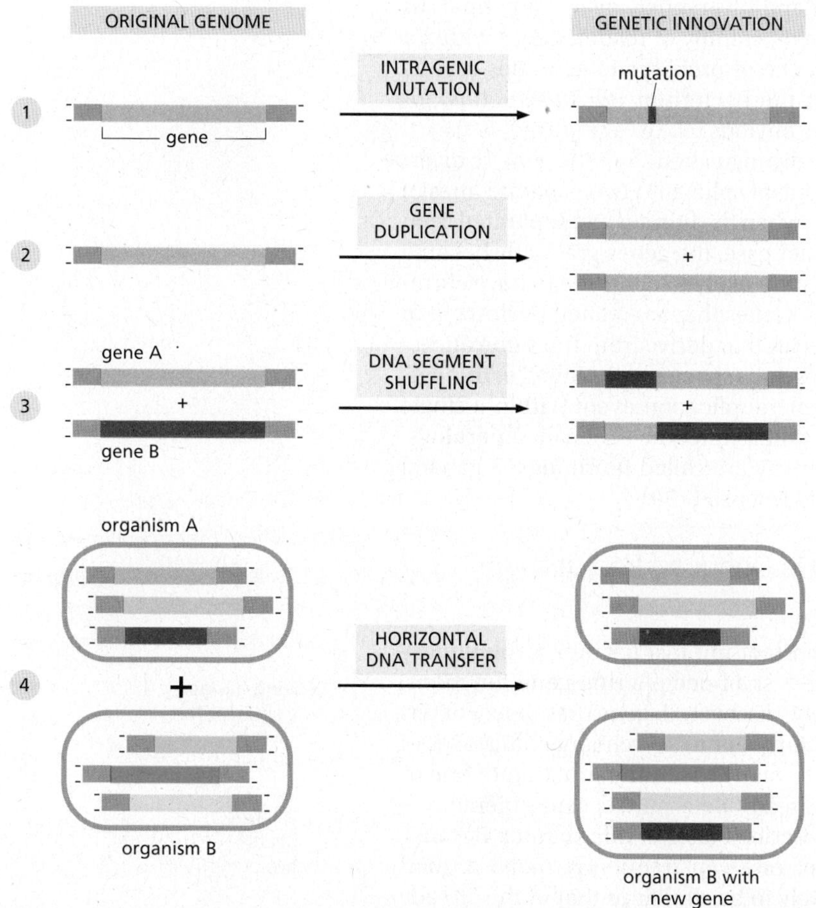

Figure 1–18 Four modes of genetic innovation and their effects on the DNA sequence of an organism. A special form of horizontal transfer occurs when cells of two different species enter into a permanent symbiotic association; genes from one of the cells may subsequently be transferred to the genome of the other, as we will see later when we discuss the likely evolutionary origins of mitochondria and chloroplasts.

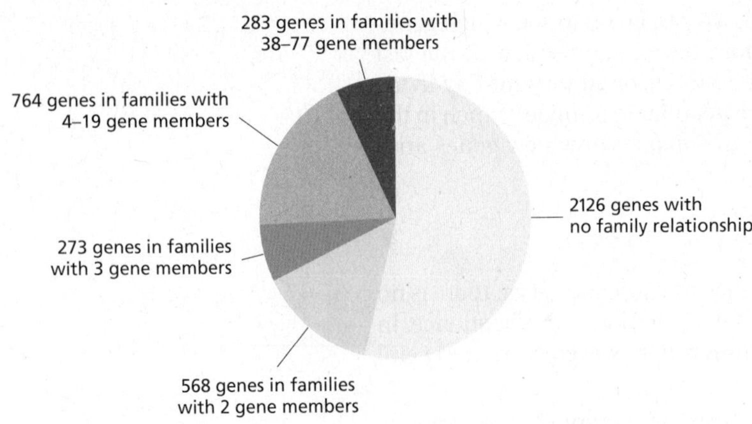

Figure 1–19 **Families of evolutionarily related genes in the genome of *Bacillus subtilis*.** The largest gene family in this bacterium consists of 77 genes coding for varieties of a class of membrane transport proteins called ABC transporters, which are found in all three domains of the living world. (Adapted from F. Kunst et al., *Nature* 390:249–256, 1997.)

Each of these types of change leaves a characteristic trace in the DNA sequence of the organism, and there is clear evidence that all four processes have occurred frequently during evolution. In Chapters 4 and 5, we discuss the mechanisms underlying these changes, but for the present we focus on the consequences.

Gene Duplications Give Rise to Families of Related Genes Within a Single Genome

A cell duplicates its entire genome each time it divides into two daughter cells. However, accidents occasionally result in the inappropriate duplication of just part of the genome, with retention of both the original and duplicate segments in a single cell. Once a gene has been duplicated in this way (see mode 2 in Figure 1–18), the two gene copies can acquire mutations and become specialized to perform different functions within the same cell and its descendants. Repeated rounds of this process of gene duplication and divergence, over many millions of years, have enabled one gene to give rise to a family of related genes within a single genome. Analysis of the DNA sequence of prokaryotic genomes reveals many examples of such **gene families**: in the bacterium *Bacillus subtilis*, for example, 47% of the genes have one or more obvious relatives (**Figure 1–19**).

The above evolutionary process must be distinguished from the genetic divergence that occurs when one species of organism splits into two separate lines of descent at a branch point in the family tree—when the human line separated from that of chimpanzees, for example. In the latter case, the genes gradually become different in the course of evolution, but they are likely to continue to have corresponding functions in the two sister species. Genes that are related by descent in this way—that is, genes in two separate species that derive from the same ancestral gene in the last common ancestor of those two species—are called **orthologs**. Related genes that have resulted from a gene duplication event within a single genome—and are likely to have diverged in their function—are called **paralogs**. Genes that are related by descent in either way are called **homologs**, a general term used to cover both types of relationship (**Figure 1–20**).

The Function of a Gene Can Often Be Deduced from Its Nucleotide Sequence

Family relationships among genes are important not just for their evolutionary interest, but also because they simplify the task of deciphering gene functions. Once the nucleotide sequence of a newly discovered gene has been determined, a scientist can tap a few keys on a computer to search large databases of known gene sequences for gene relatives. In many cases, the function of one or more of these homologs will have been already determined experimentally—generally in one of the model organisms described later in this chapter. Because gene sequence determines gene function, one can frequently make a good guess at the new gene's function, as it is likely to be similar to that of the already

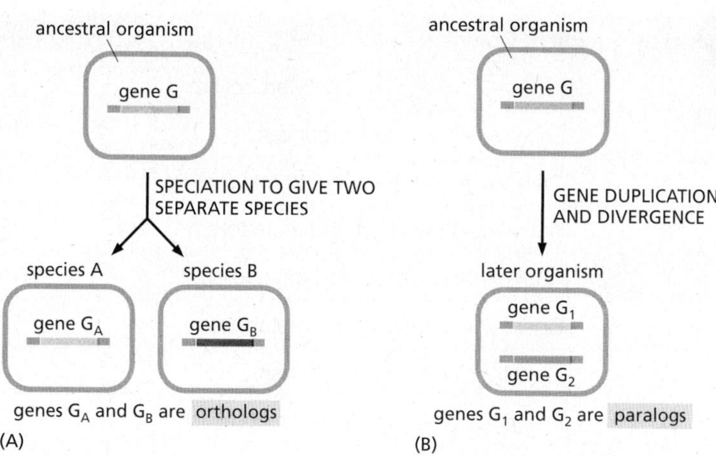

Figure 1–20 Two types of gene homology based on different evolutionary pathways. (A) Orthologs. (B) Paralogs. Genes related by either mechanism are called homologs.

known homologs. In this way, it is possible to decipher a great deal about the biology of an organism simply by analyzing the DNA sequence of its genome.

More Than 200 Gene Families Are Common to All Three Domains of Life

Given the complete genome sequences of representative organisms from all three domains of life—eukaryotes, bacteria, and archaea—we can search systematically for homologies that span this enormous evolutionary divide. In this way, we can begin to take stock of the common inheritance of all living things. There are considerable difficulties in this enterprise. For example, individual species have often lost some of the ancestral genes, and other genes have almost certainly been acquired by horizontal transfer from another species and therefore are not truly ancestral. In fact, genome comparisons strongly suggest that both lineage-specific gene loss and horizontal gene transfer, in some cases between evolutionarily distant species, have been major factors in evolution, at least among bacteria and archaea. As an additional difficulty, in the course of 2 or 3 billion years, some genes that were initially shared will have changed beyond recognition through mutation.

Because of all these vagaries of the evolutionary process, it is difficult, if not impossible, to determine the ancestral gene set that diversified into the present-day variety of life. A crude approximation can be obtained by tallying the gene families that have representatives in multiple—but not necessarily all—species from the three major domains of life. One such analysis revealed 264 ancient conserved families, each of which could be assigned a function on the basis of the best-characterized family member. As shown in **Table 1–1**, the largest number of shared gene families were involved in translation and in amino acid metabolism and transport. However, it must be emphasized that this set of highly conserved gene families represents only a very rough sketch of the common inheritance of all modern life.

Summary

For most of human history, the living world around us was classified by what we could see. Genome sequencing has radically changed our view of life on the planet, and we now realize that living things fall into three broad domains: bacteria, archaea, and eukaryotes. The organisms in the first two domains are largely invisible to our naked eye, and many of them cannot yet be grown in a laboratory—being known only by their DNA sequences. But they make up the vast majority of life's evolutionary diversity, including species that can obtain all their energy and nutrients from inorganic chemical sources—such as the reactive mixtures of minerals released at hydrothermal vents on the ocean floor—the sort of diet that may

TABLE 1–1 The Number of Gene Families, Classified by Function, Common to All Three Domains of the Living World

Information processing		Metabolism	
Translation	63	Energy production and conversion	19
Transcription	7	Carbohydrate transport and metabolism	16
DNA replication, recombination, and repair	13	Amino acid transport and metabolism	43
Cellular processes and signaling		Nucleotide transport and metabolism	15
Cell-cycle control, mitosis, and meiosis	2	Coenzyme transport and metabolism	22
Defense mechanisms	3	Lipid transport and metabolism	9
Signal-transduction mechanisms	1	Inorganic ion transport and metabolism	8
Cell wall/membrane biogenesis	2	Secondary metabolite biosynthesis, transport, and catabolism	5
Intracellular trafficking and secretion	4	**Poorly characterized**	
Post-translational modification, protein turnover, chaperones	8	General biochemical function predicted; specific biological role unknown	24

For the purpose of this analysis, gene families are defined as "universal" if they are represented in the genomes of at least two diverse archaea (*Archaeoglobus fulgidus* and *Aeropyrum pernix*), two evolutionarily distant bacteria (*Escherichia coli* and *Bacillus subtilis*), and one eukaryote (yeast, *Saccharomyces cerevisiae*). (Data from R.L. Tatusov et al., *Science* 278:631–637, 1997; R.L. Tatusov et al., *BMC Bioinformatics* 4:41, 2003; and the COGs database at the US National Library of Medicine.)

have nourished the first living cells more than 3.5 billion years ago. The eukaryotes (whose cells are larger and contain a variety of membrane-bound organelles) evolved later in evolutionary history and are consequently less diverse as a group than either the bacteria or archaea. Eukaryotes, which include all plants and animals, are the organisms most familiar to us, and they are the main focus of this textbook.

Many of the genes within a single organism or species show strong family resemblances in their DNA sequences, implying that they originated from the same ancestral gene through gene duplication and divergence. Family resemblances (homologies) are also clear when gene sequences are compared between different species, and more than 200 gene families have been so highly conserved that they can be recognized as common to most species from all three domains of the living world, suggesting they were present in the ancestral cell from which all life evolved. Given the DNA sequence of a newly discovered gene in any organism, it is therefore often possible to deduce the gene's function from the known function of a homologous gene in a better-studied organism.

EUKARYOTES AND THE ORIGIN OF THE EUKARYOTIC CELL

Eukaryotic cells, in general, are bigger and more elaborate than bacterial and archaeal cells, and their genomes are bigger and more elaborate, too. The greater cell size is accompanied by radical differences in cell structure and function: in particular, eukaryotes contain a diverse set of intracellular **organelles**—discrete membrane-enclosed subcompartments and large membraneless macromolecular assemblies—each with a distinct composition and function. Some eukaryotic cells live independent lives as single-cell organisms. Others live in multicellular assemblies—indeed, all of the more complex multicellular organisms on Earth, including plants, animals, and fungi, are formed from eukaryotic cells.

We begin by discussing how eukaryotic cells are organized and how they might have evolved from more ancient prokaryotes. We then briefly consider how eukaryotic genomes differ from those of prokaryotes, as well as how the cells in multicellular organisms become differently specialized as an embryo develops, so as to contribute to the welfare of the organism as a whole.

Eukaryotic Cells Contain a Variety of Organelles

By definition, eukaryotic cells keep almost all their DNA in a membrane-enclosed internal compartment—the *nucleus,* which is usually the most conspicuous organelle (**Figure 1–21**). The long DNA polymers in the nucleus are packaged with proteins to form *chromosomes,* which only become visible in a light microscope when they condense in preparation for cell division. The *nuclear envelope,* a double layer of membrane, surrounds the nucleus and separates the nuclear DNA from the *cytoplasm,* which, in a eukaryotic cell, includes everything between the plasma membrane and the nucleus. As shown in the figure, the nuclear envelope is perforated by *nuclear pores,* which are channels formed by protein complexes that mediate the two-way traffic of large molecules between the nucleus and the cytoplasm.

Eukaryotic cells have many other features that set them apart from bacterial and archaeal cells. They are typically 10–30 times bigger in linear dimension and 1000–10,000 times larger in volume than a typical prokaryotic cell. They have an elaborate *cytoskeleton* in the cytoplasm, consisting of several types of protein filaments (see Figure 1–21) that, together with the many proteins that attach to them, form a network of girders, ropes, and motors that gives the cell mechanical strength and performs various other functions: when the cell divides, for example, the cytoskeleton reorganizes and pulls the replicated chromosomes apart and distributes them equally to the two daughter cells. In the case of animal cells and some free-living, single-cell eukaryotes, the cytoskeleton controls cell shape

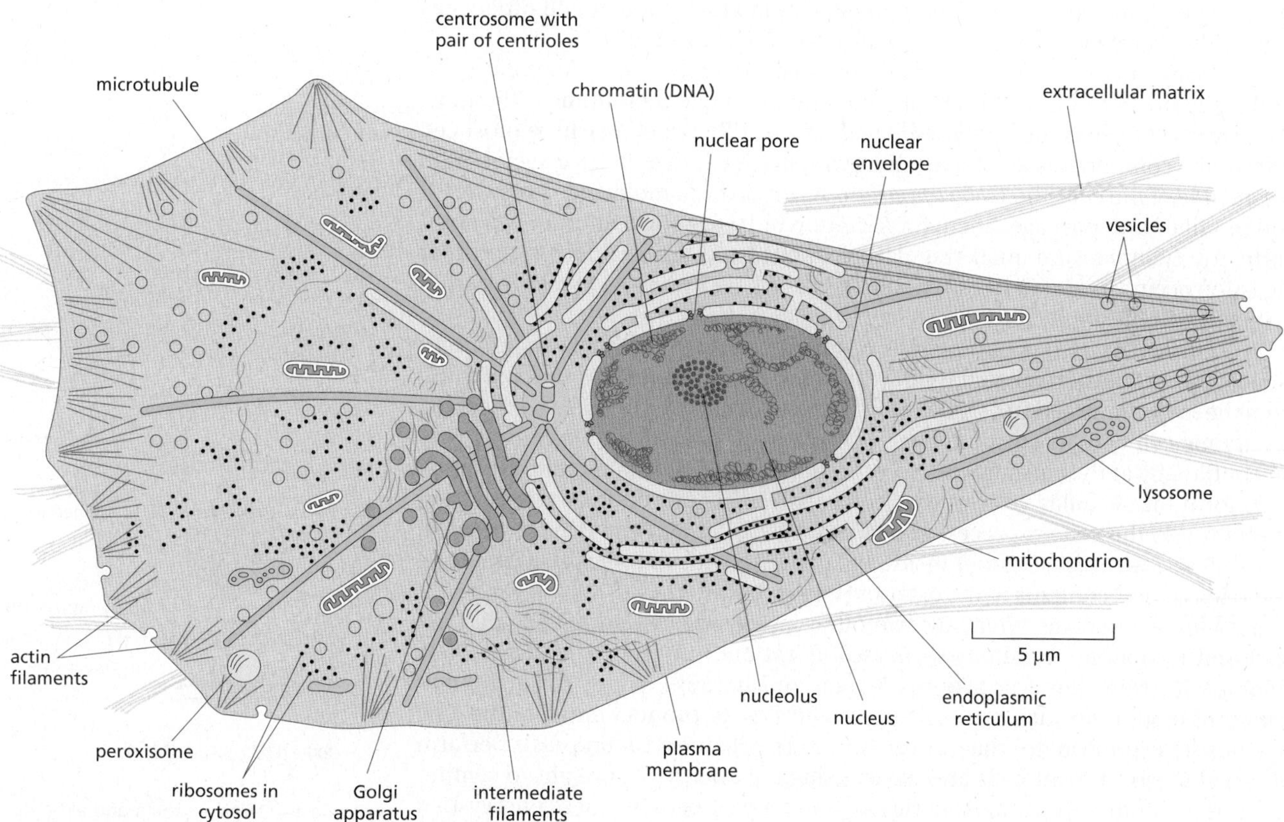

Figure 1–21 **The major features of eukaryotic cells.** The drawing depicts the major contents of a typical animal cell seen in cross section, but almost all the same components are found in plant cells and fungi, as well as in single-cell eukaryotes. The cytoskeleton (discussed in Chapter 16) consists of three types of protein filaments: actin filaments *(red)*, microtubules *(green)*, and intermediate filaments *(blue)*. Plant cells (not shown) contain chloroplasts in addition to the components shown here; they also have a rigid external *cell wall* that contains cellulose surrounding their plasma membrane, which means they are largely immobile. The interior of cells is, in reality, much more crowded than depicted in this simplified diagram.

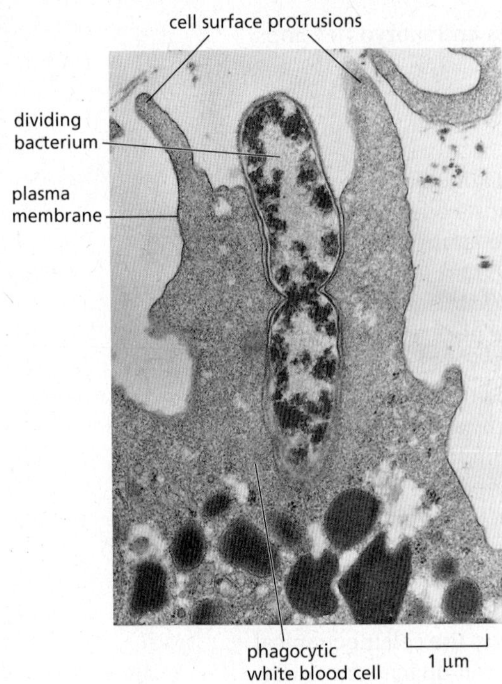

Figure 1–22 Phagocytosis. An electron micrograph of a mammalian phagocytic white blood cell (a neutrophil) ingesting a bacterium that is in the process of dividing. Only the part of the cell that is extending surface protrusions to engulf the bacterium is shown. (Courtesy of Dorothy Bainton.)

and drives and guides cell movements (**Movie 1.1**). Lacking the kind of tough cell wall characteristic of bacteria and archaea, these eukaryotic cells can change their shape rapidly, in some cases enabling them to move and engulf other cells and small objects by a process called *phagocytosis* (**Figure 1–22**).

There are many other membrane-enclosed organelles in eukaryotic cells. Unlike the nucleus, most of them are enclosed by single membranes. The most extensive organelle is the *endoplasmic reticulum* (*ER*), which is where most cell membrane components are made, along with materials destined for secretion to the outside of the cell. The *Golgi apparatus* receives these molecules from the ER and modifies and packages them for secretion or transport to another cell compartment. *Lysosomes* are small irregularly shaped organelles in which intracellular digestion occurs. *Peroxisomes* are small vesicles where hydrogen peroxide is used to inactivate toxic molecules.

A continual exchange of materials occurs between these single-membrane-enclosed organelles, mediated mainly by small *transport vesicles* that pinch off from the membrane of one organelle and fuse with that of another. To connect the eukaryotic cell with its surroundings, a similar vesicle-mediated exchange goes on continually at the cell surface. Here, portions of the plasma membrane pinch in to form intracellular vesicles that carry material captured from the external medium into the cell—a process called *endocytosis*; and in the reverse process, called *exocytosis*, vesicles from inside the cell fuse with the plasma membrane and release their contents to the exterior (**Figure 1–23**).

Besides the nucleus, there are two other eukaryotic cell organelles that are enclosed in double membranes—*mitochondria* and, in plant cells and algae, *chloroplasts*. Mitochondria take up oxygen and harness energy from the oxidation of food molecules, such as sugars and fats, to produce most of the ATP (adenosine triphosphate) that powers the cell's activities. *Chloroplasts* perform photosynthesis in plant cells and algae, using the energy of sunlight to synthesize carbohydrates from atmospheric CO_2 and water, delivering these energy-rich products to the host cell as food. In many eukaryotic cells, roughly half of the cytoplasm is occupied by membrane-enclosed organelles. The surrounding fluid is called the *cytosol*. It contains *ribosomes*, which translate RNAs into proteins, and it is also where most of the cell's other metabolic reactions take place.

In addition to the membrane-enclosed organelles just described, eukaryotic cells contain a variety of smaller organelles that lack membranes. Instead,

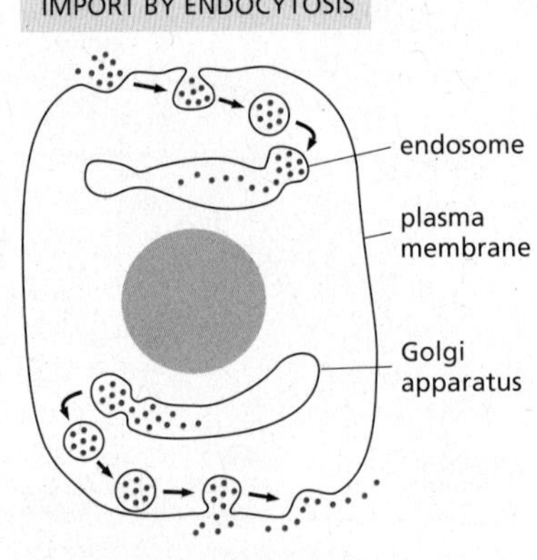

Figure 1–23 Endocytosis and exocytosis across the plasma membrane. Eukaryotic cells import extracellular materials by endocytosis and secrete intracellular materials by exocytosis. The endocytosed material is first delivered to single-membrane-enclosed organelles called *endosomes*, discussed in Chapter 12.

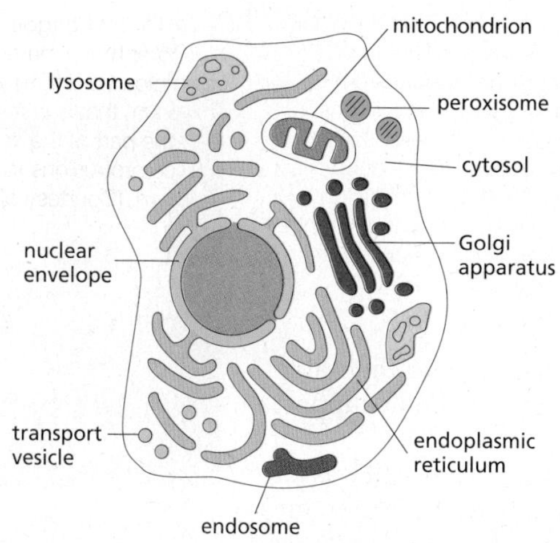

Figure 1–24 Membrane-enclosed organelles are distributed throughout the eukaryotic cell cytoplasm. The membrane-enclosed organelles, shown in different colors, are each specialized to perform a different function. The cytoplasm that fills the space outside of these organelles is called the cytosol.

multiple components that work together are held in close proximity in **biomolecular condensates**, which will be described in Chapters 3, 6, and 12. An example is the nucleolus, where ribosome assembly takes place (see Figure 1–21). Figure 1–24 summarizes the membrane-enclosed organelles in eukaryotic cells, each of which will be discussed in detail in later chapters.

Mitochondria Evolved from a Symbiotic Bacterium Captured by an Ancient Archaeon

A fundamental question in both evolution and cell biology is how did the first eukaryotic cell arise? The evidence suggests that it happened when an archaeal and a bacterial cell merged about 2 billion years ago, in a world that had contained only prokaryotes for more than 1.5 billion years.

All eukaryotic cells contain (or at one time did contain) mitochondria (Figure 1–25). Mitochondria are similar in size to small bacteria, and both reproduce by dividing. The mitochondria contain their own DNA, with genes that resemble bacterial genes; they also contain their own ribosomes and translation factors that resemble those in bacteria. These and other similarities between mitochondria and present-day bacteria provide strong evidence that mitochondria evolved from an aerobic bacterium (one that harvested energy by combining electrons derived from foodstuffs with oxygen gas) that was captured

(A)

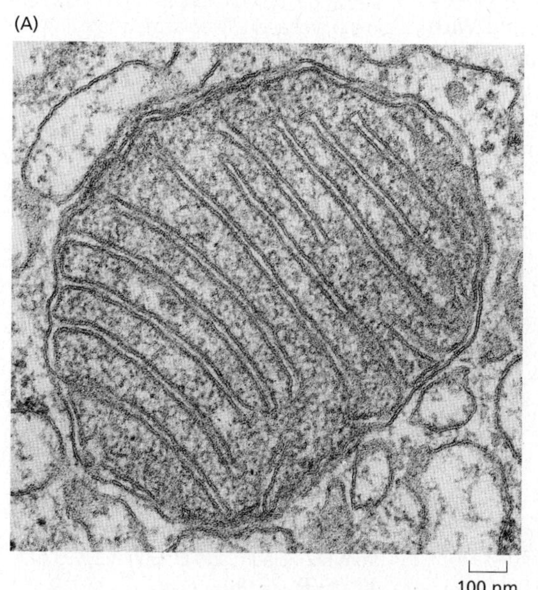

(B)

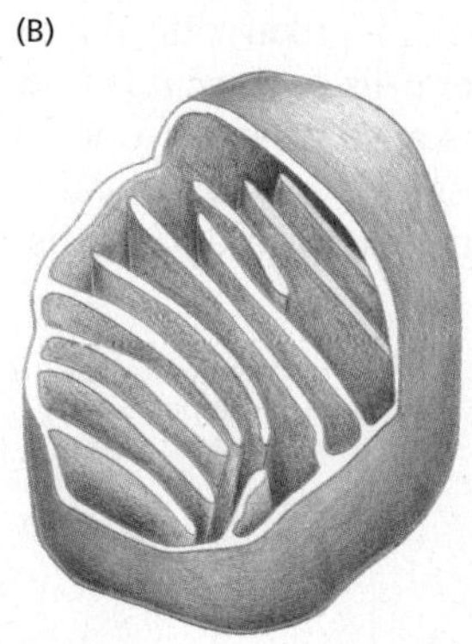

Figure 1–25 A mitochondrion. (A) An electron micrograph of the organelle seen in cross section. (B) A drawing of a mitochondrion with part of it cut away to show the three-dimensional structure (Movie 1.2). Note the smooth outer membrane and the convoluted inner membrane, which houses the proteins that generate ATP from the oxidation of food molecules. (A, courtesy of Daniel S. Friend and by permission of E.L. Bearer.)

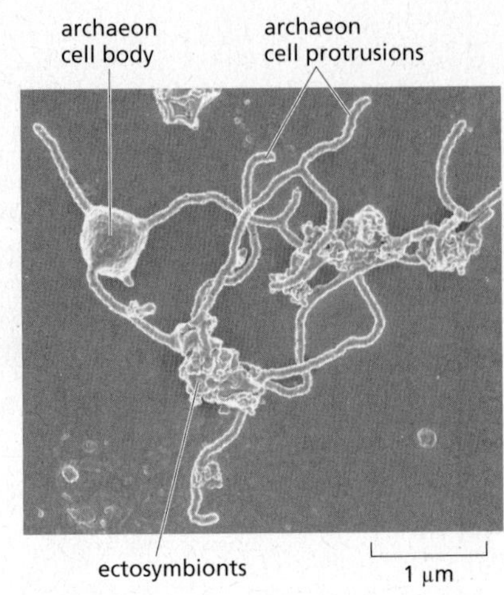

Figure 1–26 **A scanning electron micrograph of an Asgard archaeon in culture.** This anaerobic cell proliferates very slowly, doubling only about every 20 days (compared to every half hour or so for the bacterium *E. coli*). It can be seen to extend elaborate membranous protrusions from its surface—including "blebs" and unique branched and unbranched structures. These protrusions are intimately associated with two other species—one bacterial, one archaeal—that were isolated with the Asgard strain as ectosymbionts, as indicated. The scientists had maintained the deep marine sediment under anaerobic conditions in a bioreactor for more than 2000 days, mimicking conditions of the seabed, and they attempted to culture samples from this bioreactor under a range of different conditions. Only after many years and repeated subculturing were they able to isolate this archaeon with its ectosymbionts. (From H. Imachi et al., *Nature* 577:519–525, 2020.)

by an anaerobic cell. These two cells (and their descendants) were then able to evolve an endosymbiotic relationship, providing mutual metabolic support within a common cytoplasm.

There are also good reasons to believe that the ancestral capturing cell was an archaeon. As we have seen, the genomes of present-day archaea encode many proteins that are characteristic of present-day eukaryotic cells. Those archaea with the most eukaryotic-like genes belong to the Asgard lineage, first identified by sequencing DNA fragments obtained from the seabed. But no living example had been seen or cultivated—until very recently, when, after a heroic 12-year-long isolation procedure, the first Asgard archaeon was propagated in culture. This remarkable, living, anaerobic archaeon looked like no other prokaryote—having long branching protrusions—and it seemed to live in an ectosymbiotic relationship with another bacterium and another archaeon, both of which were isolated with it (**Figure 1–26**). The discovery of the strange Asgard archaeon provides a glimpse of how an ancient archaeon might eventually have captured an aerobic bacterium to initiate the eukaryotic lineage, with a hypothetical pathway being illustrated in **Figure 1–27** and discussed in detail in Chapter 12.

Chloroplasts Evolved from a Symbiotic Photosynthetic Bacterium Engulfed by an Ancient Eukaryotic Cell

Chloroplasts (**Figure 1–28**) perform photosynthesis in plant cells and algae, using the energy of sunlight to synthesize their own "food" (in the form of carbohydrates) from atmospheric CO_2 and water. Like mitochondria, they are enclosed in double membranes, have their own "circular" genomes, and reproduce by dividing. They almost certainly evolved from a symbiotic photosynthetic bacterium that was captured by an ancient eukaryotic cell that already possessed mitochondria. This bacterium may have been captured by phagocytosis, a frequent process in eukaryotes (see Figure 1–22).

Whereas some single-cell eukaryotes are hunters that live by capturing other cells and eating them (see Figure 1–34), a eukaryotic cell equipped with chloroplasts has no need to chase after other cells as prey; it is nourished by the captive chloroplasts it has inherited from its ancient eukaryotic ancestors. Correspondingly, plant cells, although they possess the cytoskeletal equipment for movement, have lost the ability to change shape rapidly and to engulf other cells

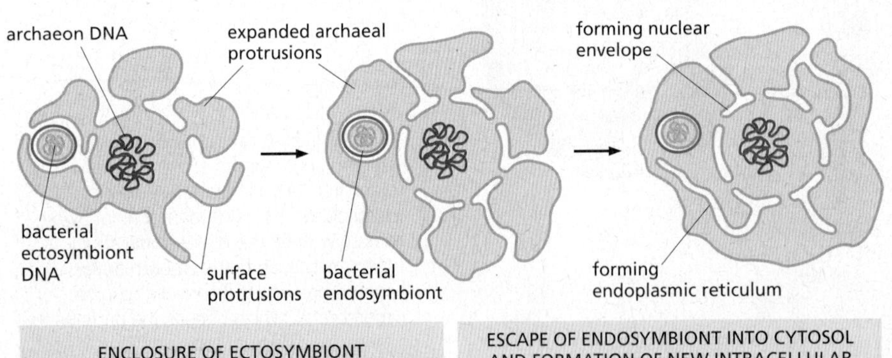

Figure 1–27 **A possible model for some early steps in eukaryotic cell evolution.** In this model, the surface protrusions of an ancient Asgard archaeon expanded and surrounded an ectosymbiotic aerobic bacterium to create a symbiotic relationship between the two types of cells. Eventually, the protrusions fused with one another, trapping the bacterium as an *endosymbiont* in the archaeon cytoplasm, where it was initially enclosed by an internal membrane derived from the archaeon's plasma membrane (the bacterium itself retaining its own membranes). At some point, the endosymbiont escaped from the enclosing archaeon-derived membrane and entered the cytosol, where it eventually evolved into a mitochondrion—with both its DNA and membranes derived from the engulfed bacterium. As shown, it is postulated that the internal archaeon membranes generated by this mechanism of protrusion expansion and fusion progressively formed both the nucleus and single-membrane-enclosed organelles, such as the endoplasmic reticulum. The evidence for this general type of model of eukaryotic-cell evolution is discussed further in Chapter 12. (Adapted from H. Imachi et al., *Nature* 577:519–525, 2020, and from D.A. Baum and B. Baum, *BMC Biol.* 12:76–92, 2014.)

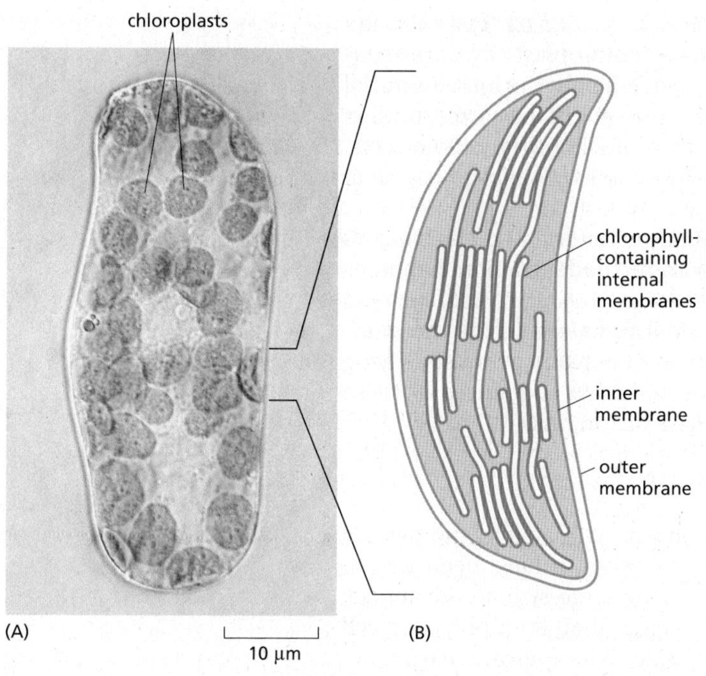

Figure 1–28 **Chloroplasts.** In plant cells and single-celled photosynthetic eukaryotes, these organelles capture the energy of sunlight. (A) A light micrograph of a single cell isolated from a leaf of a flowering plant, showing the green chloroplasts (**Movie 1.3** and see Movie 14.10). (B) A drawing of one chloroplast, showing the highly folded system of internal membranes containing the chlorophyll molecules that absorb light. (A, courtesy of Preeti Dahiya.)

by phagocytosis. Instead, they wrap themselves in a tough, protective cell wall. If some eukaryotic cells can be viewed as hunters, then one might view plant cells as having given up hunting for farming.

Fungi represent yet another eukaryotic way of life. Fungal cells, like animal cells, possess mitochondria but not chloroplasts; they have a tough outer wall that limits their ability to move rapidly or to take up other cells. Fungi, it seems, have turned from hunters into scavengers. Other cells secrete nutrient molecules or release them after death, and fungi feed on these leavings—often performing whatever digestion is necessary extracellularly, by secreting digestive enzymes to the exterior.

Eukaryotes Have Hybrid Genomes

As just discussed, the genetic information of eukaryotic cells has a hybrid origin—from an ancestral anaerobic archaeon and from the bacteria it adopted as endosymbionts (**Figure 1–29**). Most of this genetic information (DNA) is stored in the nucleus, although a small amount remains inside the organelles that evolved

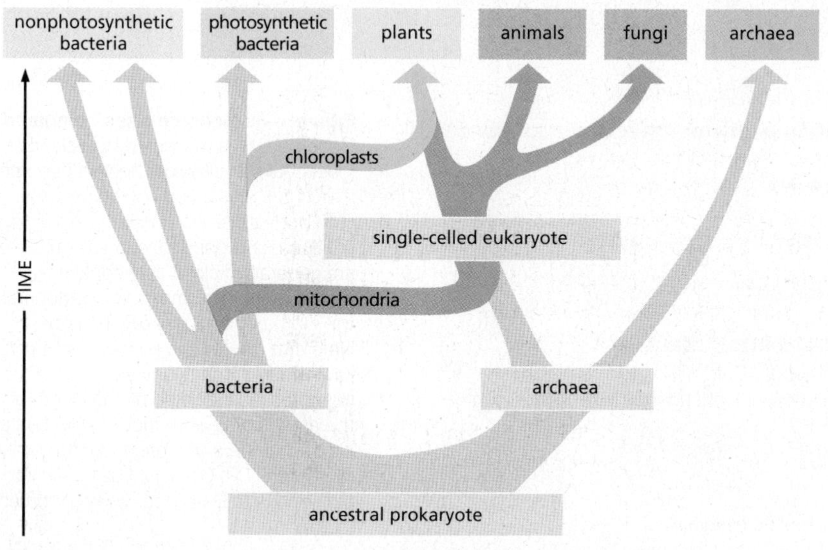

Figure 1–29 **A model for the evolution of eukaryotic cells in the tree of life.** All living cells are thought to have evolved from an ancestral prokaryotic cell (the last universal common ancestor) between 3.5 and 3.8 billion years ago). Many millions of years later, it seems an anaerobic archaeon acquired an aerobic bacterial symbiont, which evolved into mitochondria (see Figure 1–27). Later still, a mitochondria-containing eukaryotic cell acquired a photosynthetic bacterium, which evolved into chloroplasts. Mitochondria are essentially the same in plants, animals, and fungi, indicating that they were acquired before these three lineages diverged about 1.5 billion years ago.

from the captured bacteria—the mitochondria and, in plant and algal cells, the chloroplasts. When the mitochondrial DNA and the chloroplast DNA are separated from the nuclear DNA and individually sequenced, both the mitochondrial and chloroplast genomes are found to be cut-down versions of the corresponding bacterial genomes: in a human cell, for example, the mitochondrial genome consists of only 16,569 nucleotide pairs, and it codes for only 13 proteins plus a set of 24 RNAs involved in protein synthesis.

Many of the genes that are missing from the mitochondria and chloroplasts have not been lost; instead, they have moved from the endosymbiont genomes into the DNA of the host-cell nucleus. Thus the nuclear DNA of animals contains many genes coding for proteins that serve essential functions inside the mitochondria; in plants and algae, the nuclear DNA contains many genes specifying proteins required in chloroplasts. In both cases, the DNA sequences of these nuclear genes still show clear evidence of their bacterial origins.

Eukaryotic Genomes Are Big

Natural selection has evidently favored mitochondria with small genomes. By contrast, the nuclear genomes of most eukaryotes seem to have been free to enlarge. Perhaps the eukaryotic way of life has made large size an advantage: predatory cells, for example, typically need to be bigger than their prey, and cell size generally increases in proportion to genome size. Whatever the reason, the genomes of most eukaryotes have become hundreds of times larger than those of bacteria and archaea (**Figure 1–30**).

The freedom to be extravagant with DNA has had profound implications. Eukaryotes not only have more genes than prokaryotes; they also have vastly more DNA that does not code for protein or RNA. The human genome contains about 700 times as many nucleotide pairs as the genome of a typical bacterium such as *E. coli*, but it contains only about 4.5 times as many protein-coding genes because a much greater proportion of the human genome does not code for protein (~98.5% compared to 11% in *E. coli*). The estimated genome sizes and gene numbers for a few selected eukaryotes are compared with the bacterium *E. coli* in **Table 1–2**; we will discuss shortly how each of these organisms serves as a model organism.

Eukaryotic Genomes Are Rich in Regulatory DNA

As discussed in Chapter 4, much of our nonprotein-coding DNA is almost certainly dispensable "junk," retained during evolution like a mass of old papers because, when there is little pressure to keep an archive small, it is easier to retain

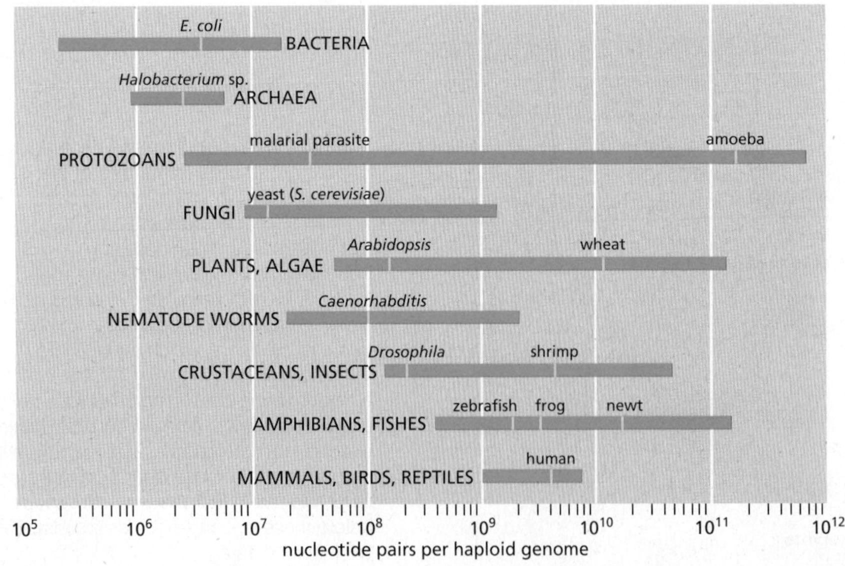

Figure 1–30 **Genome sizes compared.** Genome size is measured in nucleotide (base) pairs of DNA per haploid genome, that is, per single copy of the genome. (The body cells of sexually reproducing, multicellular organisms such as ourselves are generally diploid: they contain two copies of the genome, one inherited from the mother, the other from the father.) Note that closely related organisms can vary widely in the quantity of DNA in their genomes (as indicated by the length of the *green bars*), even though they contain similar numbers of protein-coding genes. (Data from T.R. Gregory, 2021. Animal Genome Size Database: www.genomesize.com.)

TABLE 1–2 Some Model Organisms and Their Genomes

Organism	Approximate genome size* (nucleotide pairs)	Approximate number of protein-coding genes**
Escherichia coli (bacterium)	4.6×10^6	4300
Saccharomyces cerevisiae (yeast)	12.5×10^6	6600
Caenorhabditis elegans (roundworm)	100×10^6	20,000
Arabidopsis thaliana (plant)	135×10^6	27,000
Drosophila melanogaster (fruit fly)	180×10^6	14,000
Danio rerio (zebrafish)	1400×10^6	26,000
Mus musculus (mouse)	2800×10^6	20,000
Homo sapiens (human)	3100×10^6	20,000

*Genome size includes an estimate for the amount of highly repeated, noncoding DNA sequence, which does not appear in genome databases.
**There are also genes that code for functional RNA molecules that do not code for proteins.

everything than to sort out the valuable information and discard the rest. Certain exceptional eukaryotic species, such as the puffer fish, bear witness to the profligacy of their relatives; they have somehow managed to rid themselves of large quantities of nonprotein-coding DNA, and, yet, they appear similar in structure, behavior, and fitness to related species that have vastly more such DNA.

Even in compact eukaryotic genomes such as that of the puffer fish, there is more nonprotein-coding DNA than protein-coding DNA. As in all eukaryotic organisms, at least some of the noncoding DNA certainly has important functions. In particular, it regulates the expression of genes. With this *regulatory DNA*, eukaryotes have evolved distinctive, highly sophisticated ways of controlling when and where a gene is brought into play. Elaborate mechanisms for gene regulation are especially crucial for the formation and function of complex multicellular organisms, which have many different cell types, each with different functions, as we now discuss.

Eukaryotic Genomes Define the Program of Multicellular Development

The cells in an individual animal or plant are extraordinarily varied. Blood cells, skin cells, bone cells, nerve cells—they seem as dissimilar as any cells could be (**Figure 1–31**). Yet all these cell types are the descendants of a single fertilized egg cell, and all (with very minor exceptions) contain identical copies of the genome of the species.

The differences result from the way in which the cells make selective use of their genetic instructions according to their developmental history and the cues they receive from their surroundings in a developing embryo. The DNA is not just a shopping list specifying the molecules that every cell must have, and the cell is not an assembly of all the items on the list. Rather, the cell behaves as a multipurpose machine, with sensors to receive environmental signals and with highly developed abilities to call different sets of genes into action according to signals it receives. The genome in each cell is big enough to accommodate the information that specifies an entire multicellular organism, but in any individual cell only part of that information is used.

Many genes in the eukaryotic genome code for proteins that regulate the activities of other genes, a topic discussed in detail in Chapter 7. Most of these regulatory genes encode *transcription regulators* that act by binding, directly or

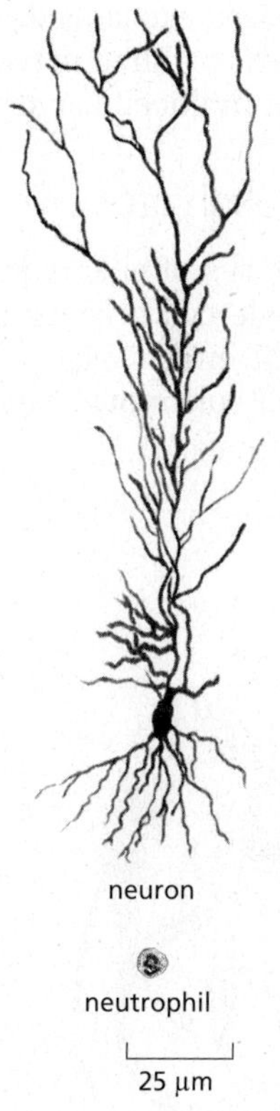

Figure 1–31 **Cell types can vary enormously in size and shape.** A human nerve cell is compared here with a human neutrophil, a type of white blood cell. Both are drawn to scale.

Figure 1–32 **Genetic control of the program of multicellular development.** The role of a regulatory gene is demonstrated in the snapdragon *Antirrhinum*. In this example, a mutation in a single gene coding for a regulatory protein causes leafy shoots to develop in place of flowers: because the regulator protein has been changed, the cells adopt characters that would be appropriate to a different location in the normal plant. The mutant is on the left, the normal plant on the right. (Courtesy of Enrico Coen and Rosemary Carpenter.)

indirectly, to the many different DNA sequences that control which genes are to be expressed, and at what levels. Eukaryotic genomes also produce many noncoding RNA molecules; as the name implies, they are not translated into protein but control gene expression in a variety of ways. The expanded genome of eukaryotes therefore not only specifies the hardware of the cell but also stores the software that controls how that hardware is used (**Figure 1–32**).

Cells do not just passively receive signals; rather, they actively exchange signals with their neighbors. Thus, in a developing multicellular organism, an internal control system governs each cell that has different consequences depending on the messages exchanged. The outcome, astonishingly, is a precisely patterned array of cells of different types, each displaying a character appropriate to its position in the multicellular structure.

Many Eukaryotes Live as Solitary Cells

Many species of eukaryotic cells lead a solitary life. As we have seen, some of these single-cell organisms are hunters, some are photosynthesizers, still others are scavengers. **Figure 1–33** conveys something of the astonishing variety of single-cell eukaryotes, whose anatomy can be remarkably elaborate, including

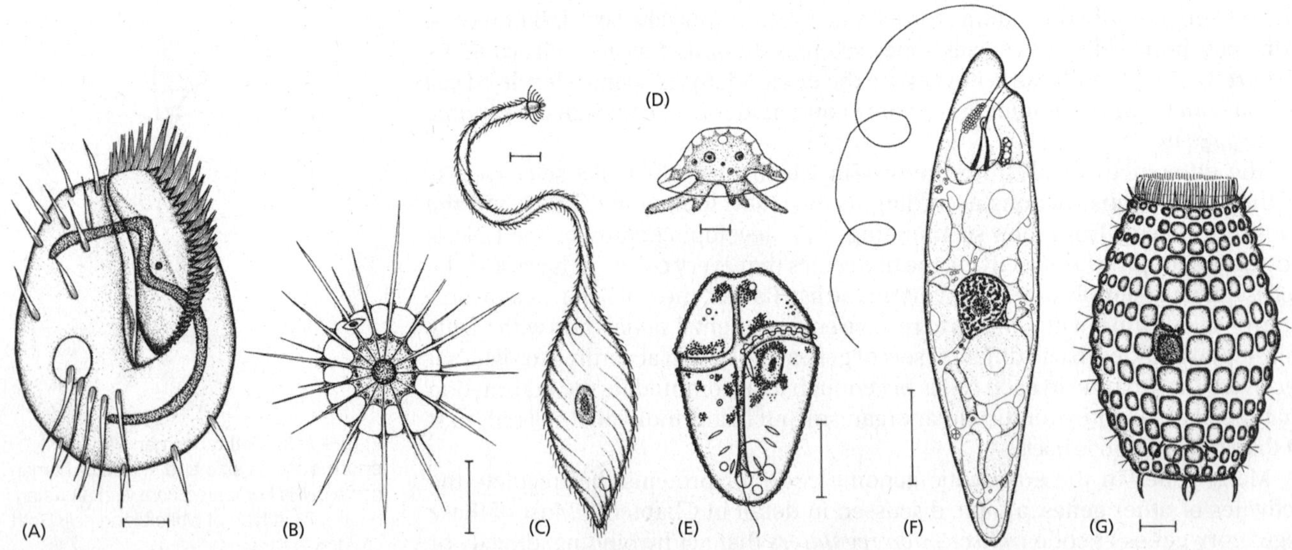

Figure 1–33 **An assortment of single-cell eukaryotes.** The drawings are done to different scales, but in each case the scale bar represents 10 μm. The organisms in (A), (C), and (G) are ciliates; (B) is a heliozoan; (D) is an amoeba; (E) is a dinoflagellate; and (F) is a euglenoid. (Courtesy of Michael Sleigh, from M.A. Sleigh, Biology of Protozoa. Edinburgh: Edward Arnold, 1973.)

(A)

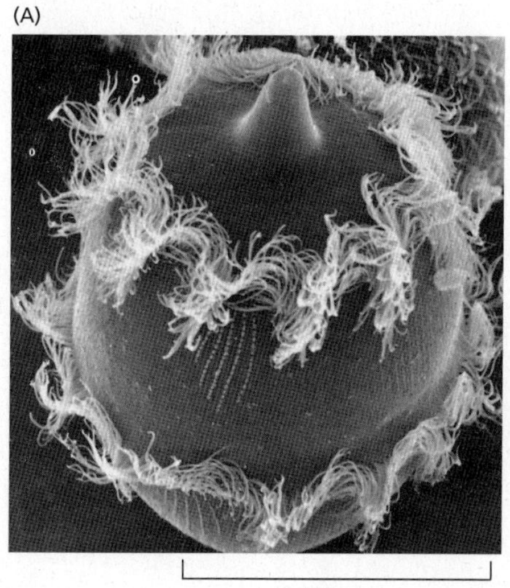

(B)

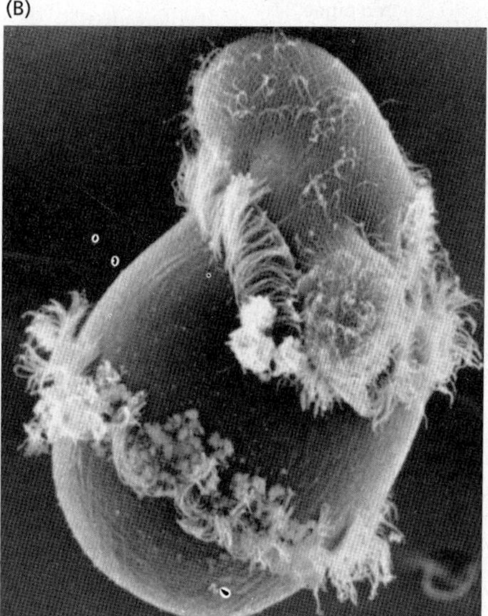

Figure 1–34 A single-cell eukaryote that eats other cells. (A) *Didinium* is a carnivorous protist, belonging to the group of *ciliates*. (A protist is defined as a free-living, single-cell, mobile eukaryote.) *Didinium* has a globular body, about 150 μm in diameter, encircled by two fringes of cilia—sinuous, whiplike appendages that beat continually; its front end is flattened except for a single protrusion, rather like a snout. (B) A *Didinium* engulfing its prey. It normally swims around in the water at high speed by means of the synchronous beating of its cilia. When it encounters a suitable prey, usually another type of protist, it releases numerous small paralyzing darts from its snout region. Then, it attaches to and devours its prey (artificially colored *yellow*) by phagocytosis, inverting like a hollow ball to engulf its victim, which can be almost as large as itself. (Courtesy of D. Barlow.)

such structures as sensory bristles, photoreceptors, sinuously beating cilia, leglike appendages, mouth parts, stinging darts, and muscle-like contractile bundles. Although they are single cells, they can be as intricate, as versatile, and as complex in their behavior as many multicellular organisms (**Figure 1–34**, **Movie 1.4**, and **Movie 1.5**).

Humans tend to focus on plants and animals, while neglecting single-cell eukaryotes because (along with bacteria and archaea) they are microscopic. But thanks to DNA comparisons, we now know that the genomes of single-cell eukaryotes are far more evolutionarily diverse than those of multicellular animals and plants, meaning that animals and plants arose relatively late in the complex and fascinating eukaryotic pedigree. With genome data, we can position the many different single-celled eukaryotes in the tree of life and identify our closest relatives (**Figure 1–35**). Scientists are using this information to probe the origins of multicellularity, with a focus on what these strange creatures can tell us about our own evolutionary past.

Summary

Eukaryotic cells, by definition, keep most of their DNA in a separate membrane-enclosed compartment—the nucleus. They have, in addition, an elaborate set of other organelles, each carrying out different functions, such as the oxidation of food-derived molecules and production of ATP in mitochondria. Eukaryotic cells also contain a cytoskeleton for structural support and movements. There is compelling evidence that mitochondria and, in plants and algae, chloroplasts evolved from captured symbiotic bacteria, which explains why these organelles contain their own DNA and ribosomes.

Eukaryotic cells typically have 3–8 times as many protein-coding genes as bacteria and archaea and often a thousand times more noncoding DNA. Although much of this DNA is probably unimportant, some of it allows for great complexity in the regulation of gene expression, as required for the construction of complex multicellular organisms containing many different cell types.

MODEL ORGANISMS

Because all cells appear to have descended from a common ancestor, whose fundamental properties have been conserved through evolution, the knowledge gained from the study of one organism contributes to our understanding of all others, including ourselves. It turns out that certain organisms are much more accessible than others for study in the laboratory. Some reproduce rapidly and

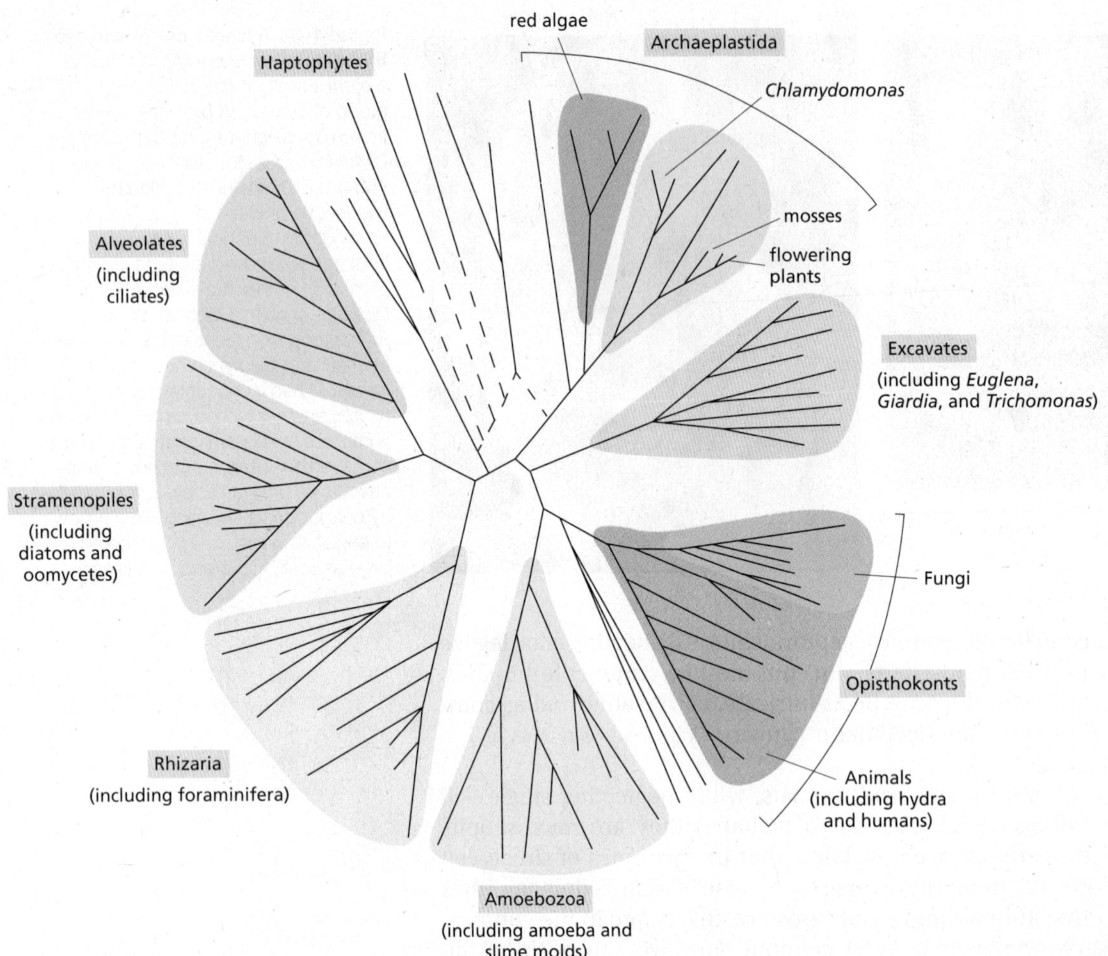

Figure 1–35 **A eukaryotic tree of life based on genome comparisons.** The lengths of the lines are proportional to the extents of genome diversity, with broken lines indicating uncertain relationships. Note that animals and plants are separated by many single-cell species (including the Excavates), suggesting that multicellularity arose independently several times during eukaryotic evolution. (Adapted from F. Burki, *Cold Spring Harb. Perspect. Biol.* 6:a016147, 2014.)

are easily manipulated using powerful genetic techniques. Others are transparent and readily develop in the laboratory from a fertilized egg to a multicellular organism, so that one can readily trace how their cells behave to produce internal tissues and organs.

Over time, different groups of biologists have focused on studying a few chosen species, which allows their knowledge and research tools to be pooled to gain a deeper understanding than could be achieved if their efforts were spread over many different organisms. Although the list of these representative, model organisms is continually expanding, a few stand out in terms of the breadth and depth of information that has been accumulated about them over the years—knowledge that has been essential for our understanding of how all cells work. In this section, we examine some of these organisms and review the benefits that each offers to the study of cell biology and, in many cases, to the promotion of human health. We begin with a discussion of some especially powerful strategies that scientists have developed to understand the cell, and we shall see how these approaches dictated the choice of model organisms.

Mutations Reveal the Functions of Genes

Without additional information, no amount of gazing at genome sequences will reveal the functions of genes. We may recognize that gene B is like gene A, but how do we discover the function of gene A in the first place? And even if we know the function of gene A, how do we test whether the function of gene B is truly the same as the sequence similarity suggests? How do we connect the world of abstract genetic information that was introduced in the previous sections with the world of living cells and organisms?

The analysis of gene functions depends on two highly complementary approaches: biochemistry and genetics. Biochemistry directly examines the functions of purified molecules, such as the protein and RNA produced from a specific gene: first we obtain that molecule from an organism and then study its chemical activities in detail. In contrast, genetics starts with the study of mutants: we either find or make an organism in which the specific gene is altered, and we then examine the effects on the mutant organism's structure and performance (**Figure 1–36**). When combined with biochemistry, careful studies of an organism (and its isolated cells) mutated for a particular protein or RNA molecule can reveal the biological role of that molecule.

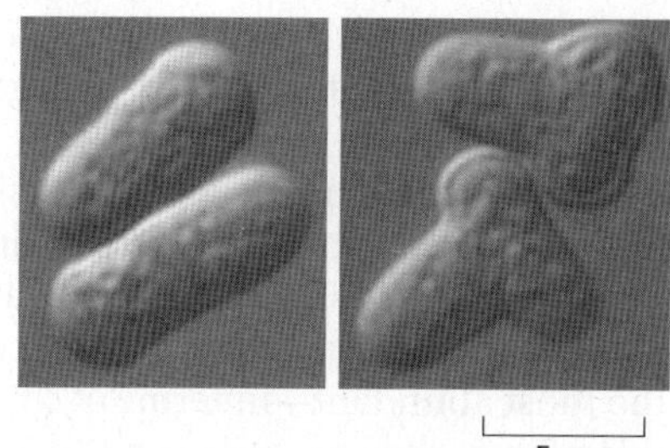

Figure 1-36 **An alteration in organism shape resulting from a gene mutation.** Scanning electron micrographs (discussed in Chapter 9) of a normal yeast (of the species *Schizosaccharomyces pombe*) compared with a mutant yeast, where a change in a single gene has converted the cell from a cigar shape *(left)* to a T shape *(right)*. The mutant gene therefore has a function controlling cell shape. But how, in molecular terms, does the gene product perform that function? That is a harder question, and it needs biochemical analysis to answer it. (Courtesy of Kenneth Sawin and Paul Nurse.)

Biochemistry and genetics, used in combination with cell biology, provide a powerful way to connect genes and molecules directly to cell and organism structure and function. In recent years, DNA sequence information and the powerful tools of molecular biology have greatly accelerated progress in this endeavor. From sequence comparisons, we can often identify particular subregions within a gene that have been conserved nearly unchanged over the course of evolution. These subregions are often the most important parts of the gene in terms of function. We can test their individual contributions to the gene's function by creating in the laboratory mutations of specific sites within the subregion or by constructing artificial hybrid genes that combine part of one gene with parts of another. Organisms can be engineered to make either the RNA or protein specified by the gene in large quantities to facilitate biochemical analysis. Specialists in molecular structure can determine the three-dimensional conformation of the gene product, revealing the exact position of every atom in it. Biochemists can determine how each of the parts of the genetically specified molecule contributes to its chemical behavior and function in a test tube. Cell biologists determine the many other molecules that interact with the molecule of interest and where all these molecules are located within a cell. And they also analyze the behavior of cells that are engineered to express a mutant version of the gene.

There is, however, no one simple universal recipe for discovering a gene's function. We may discover, for example, that the product of a given gene is an enzyme that catalyzes a certain chemical reaction, and yet have no idea how or why that reaction is important to the organism. The functional characterization of each new gene product or family of gene products, unlike the description of the gene sequences, presents a fresh challenge to the biologist's ingenuity. Moreover, we will never fully understand the function of a gene until we learn its role in the life of the organism, which means studying whole organisms, not just isolated molecules or cells.

Molecular Biology Began with a Spotlight on One Bacterium and Its Viruses

Because living organisms are so complex, the more we learn about any particular species, the more attractive it becomes as an object for further study. Each discovery in such a chosen organism raises new questions and provides new tools with which to tackle general biological questions. For this reason, large communities of biologists have become dedicated to studying different aspects of the same **model organism.**

In the early days of molecular biology, the chosen model was the bacterium *Escherichia coli* (*E. coli*—see Figure 1–11B). This small, rod-shaped cell normally lives in the gut of humans and other vertebrates, but it can be grown easily in a simple nutrient broth in a culture bottle or dish, where under favorable conditions, it can reproduce every 20 minutes or so. It adapts to variable chemical conditions and can evolve by mutation and selection at a remarkable speed.

Also of special early interest were a few of the viruses that infect this bacterium—inasmuch as their much smaller genomes made them even easier to analyze in detail.

Viruses are small packets of genetic material that have evolved as parasites that depend on the reproductive and biosynthetic machinery of the host cells they infect. Viruses are not strictly alive, because they depend on the machinery of their host cells for their reproduction. Although we now know that viruses are the most abundant—in terms of sheer numbers—of all the biological entities on this planet, they are too small to be seen in the light microscope. For this reason, they were completely missed until the end of the nineteenth century, when a few viruses were identified as infectious agents that pass through filters that trap bacteria, but are retained by the even-finer filters that allow large molecules to pass. Only with the invention of the electron microscope could viruses finally be visualized as tiny particles with defined shapes and sizes. We now know that viruses consist of many families, with different families having distinct structures and modes of replication (discussed in Chapters 6 and 23).

Those viruses that infect bacteria are called *bacteriophages*, and two that infect *E. coli* have played critical roles as model organisms that advanced our understanding of molecular cell biology. Detailed genetic analyses of these two viruses, bacteriophage lambda and bacteriophage T4, came first, followed by biochemistry that used the analysis of mutant genes to identify and characterize specific proteins of interest. Geneticists, for example, generated and then characterized more than a hundred different mutant genes in bacteriophage T4, a large virus with a double-strand DNA genome (**Figure 1–37**). Sets of T4 genes that encode components of the head and the tail of the bacteriophage were identified, allowing biochemical studies to reveal important principles of biological assembly processes. Similarly, a set of T4 genes that geneticists showed were essential for T4

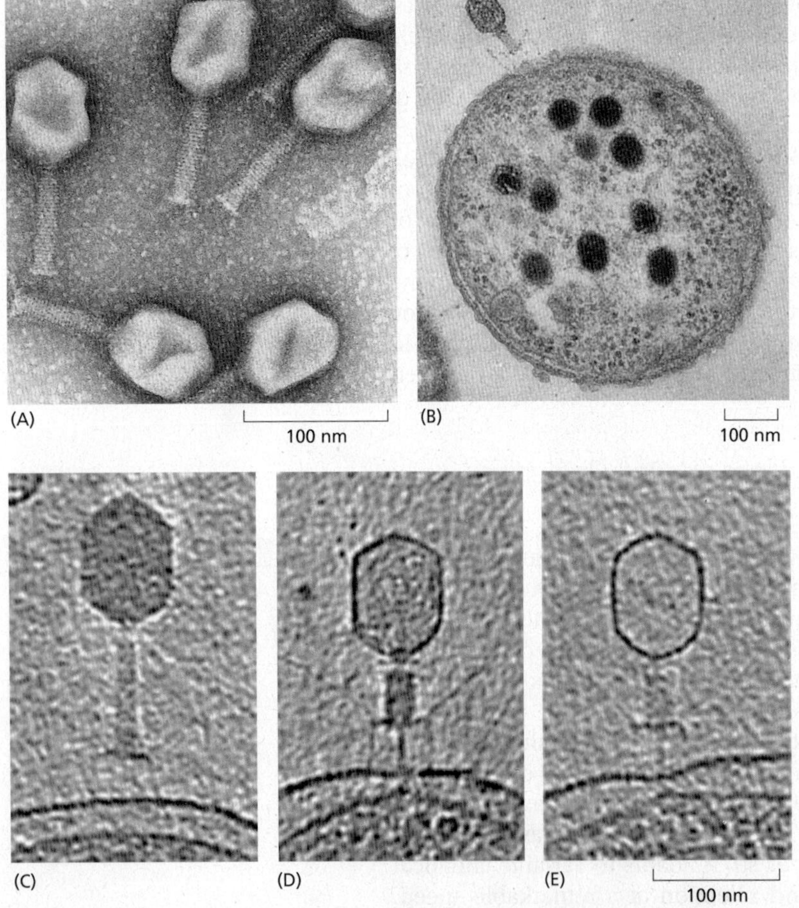

Figure 1–37 **The T4 bacteriophage.** (A) An electron micrograph of particles of the T4 bacteriophage, a virus that infects *E. coli* bacteria. The hexagonal head of the virus contains the viral DNA; the tail contains the apparatus for injecting the DNA into a host bacterium. (B) A cross section of an *E. coli* bacterium with a T4 bacteriophage attached to its surface. The large dark objects inside the bacterium are the assembling heads of new T4 particles. When the particles are mature, the bacterium will burst open and release them. (C–E) The process of DNA injection into the bacterium, as visualized in unstained, frozen samples by cryo-electron microscopy (discussed in Chapter 9). (C) Attachment begins. (D) Attached state during DNA injection. (E) Virus head has emptied its entire DNA into the bacterium. (A, courtesy of James Paulson; B, courtesy of Jonathan King and Erika Hartwig from G. Karp, Cell and Molecular Biology, 2nd ed. New York: John Wiley & Sons, 1999. With permission from John Wiley & Sons; C–E, courtesy of Ian Molineux, University of Texas at Austin, and Jun Liu.)

DNA replication allowed those proteins to be purified, so that biochemists could decipher the central mechanisms of DNA replication in a test tube. In the same way, it was extensive studies of bacteriophage lambda that led to our early understanding of transcription regulators and gene regulatory networks (see Panel 7–1 on pp. 404–405 and Figure 7–43).

We now know that these two bacteriophages have many close relatives distributed throughout the biosphere. Relatives of bacteriophage T4, for example, are abundant in the ocean, where they infect the ubiquitous marine cyanobacteria. As a whole, ocean viruses are present in enormous numbers, estimated at 10^{30}. If lined up end to end, they would extend beyond our nearest galaxies; they kill approximately 20% of the total ocean microbial biomass per day. Because these viruses have such a huge role in nutrient recycling, they profoundly affect Earth's ecology.

Although not themselves living cells, viruses often serve as vectors for gene transfer between cells. A virus will replicate in one cell, emerge from it with a protective wrapping, and then enter and infect another cell, which may be of the same or different species. Often, the infected host cell is killed by the massive proliferation of virus particles inside it, but sometimes the viral DNA, instead of directly generating new virus particles, may persist in its host for many cell generations as a relatively innocuous passenger—either as a separate intracellular fragment of DNA, known as a *plasmid*, or as a DNA sequence inserted into the cell's own genome. In their travels, viruses can accidentally pick up fragments of DNA from the genome of one host cell and ferry them into another cell. Such transfers of genetic material are very common in prokaryotes.

Many bacterial and archaeal species have a remarkable capacity to take up even nonviral DNA molecules from their surroundings and thereby capture the genetic information these molecules carry. By this route or by virus-mediated gene transfer, bacteria and archaea in the wild can acquire genes from neighboring cells relatively easily. Genes that confer resistance to an antibiotic or an ability to produce a toxin, for example, can be transferred from species to species and provide the recipient bacterium with a selective advantage, greatly enhancing its rate of spread. In this way, new and sometimes dangerous strains of antibiotic-resistant bacteria have been observed to evolve in the bacterial ecosystems that inhabit hospitals or various niches in the human body. On a longer time scale, the results can be even more profound; it has been estimated that at least 18% of all the genes in the present-day genome of *E. coli* have been acquired by horizontal transfer from another species within the past 100 million years.

The Focus on *E. coli* as a Model Organism Has Accelerated Many Subsequent Discoveries

The standard laboratory strain *E. coli* K-12 has a genome of approximately 4.6 million nucleotide pairs contained in a single circular molecule of DNA that codes for about 4300 different kinds of proteins (**Figure 1–38**). In molecular terms, we probably have a more complete understanding of *E. coli* than of any other living organism. Most of our understanding of the fundamental mechanisms of life—for example, how cells replicate their DNA or how they decode the instructions represented in the DNA to direct the synthesis of specific RNAs and proteins—initially came from studies of *E. coli* and its viruses. This is because the basic genetic mechanisms have turned out to be highly conserved throughout evolution and are essentially the same in our own cells as in *E. coli*.

It should be noted that, as with other bacteria, different strains of *E. coli*, though classified as members of a single species, differ genetically to a much greater degree than do different varieties of an organism such as a plant or animal. One *E. coli* strain may possess many hundreds of genes that are absent from another, and the two strains could have as little as 50% of their genes in common. These differences are largely the result of rampant horizontal gene transfer, characteristic of this and many other bacterial and archaeal species.

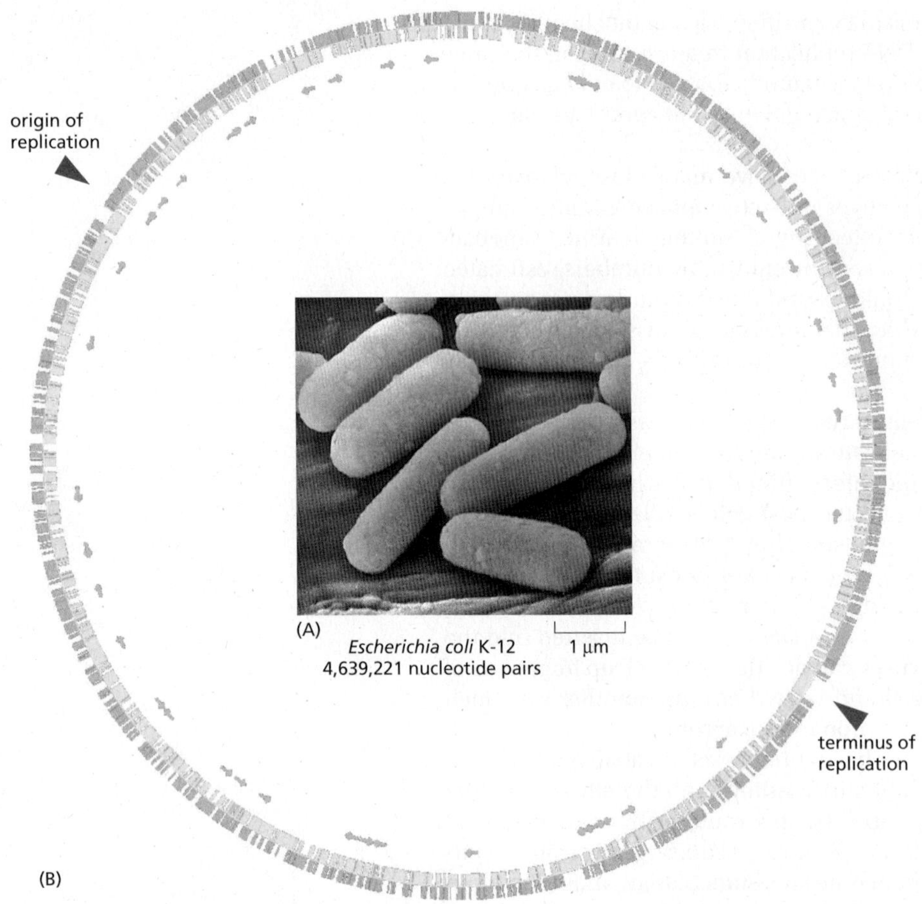

Figure 1–38 The genome of *E. coli*. (A) A cluster of *E. coli* cells viewed in a scanning electron microscope. (B) A diagram of the genome of *E. coli* strain K-12. The diagram is circular because the DNA of *E. coli*, like that of most other bacteria, forms a single, closed loop. Protein-coding genes are shown as *yellow* or *orange bars*, depending on the DNA strand from which they are transcribed; RNA molecules produced from nonprotein-coding genes are indicated by *green arrows*, with the arrowheads indicating their direction of transcription. Some genes are transcribed from one strand of the DNA double helix (in a clockwise direction in this diagram), others from the other strand (counterclockwise). The origin and terminus of DNA replication are marked with *red arrowheads*. (A, Dr. Tony Brain & David Parker/Science Source; B, adapted from F.R. Blattner et al., *Science* 277:1453–1462, 1997.)

A Yeast Serves as a Minimal Model Eukaryote

The molecular and genetic complexity of eukaryotes is daunting, and biologists need to concentrate their limited resources on a small number of selected model organisms to unravel this complexity.

To analyze the internal workings of the eukaryotic cell without the additional problems of multicellular development, it makes sense to use a single-cell species that is as simple as possible. The popular choice for this role of minimal model eukaryote has been the yeast *Saccharomyces cerevisiae* (**Figure 1–39**)—the same species that is used by brewers of beer and bakers of bread.

S. cerevisiae is a small, single-cell member of the kingdom of fungi and, in terms of its genome sequence, much more closely related to animals than to plants (see Figure 1–35). It is robust and easy to grow in a simple nutrient medium. Like other fungi, it has a tough cell wall, is relatively immobile, and possesses mitochondria but not chloroplasts. When nutrients are plentiful, it grows and divides about every hundred minutes. It can reproduce either vegetatively (that is, by ordinary cell division, or mitosis), or sexually: two yeast cells that are *haploid* (possessing a single copy of each chromosome, $n = 1$) can fuse to create a cell that is *diploid* (containing two copies of each chromosome, $n = 2$), and the diploid cell can undergo *meiosis* (a reduction division) to produce cells that are once again

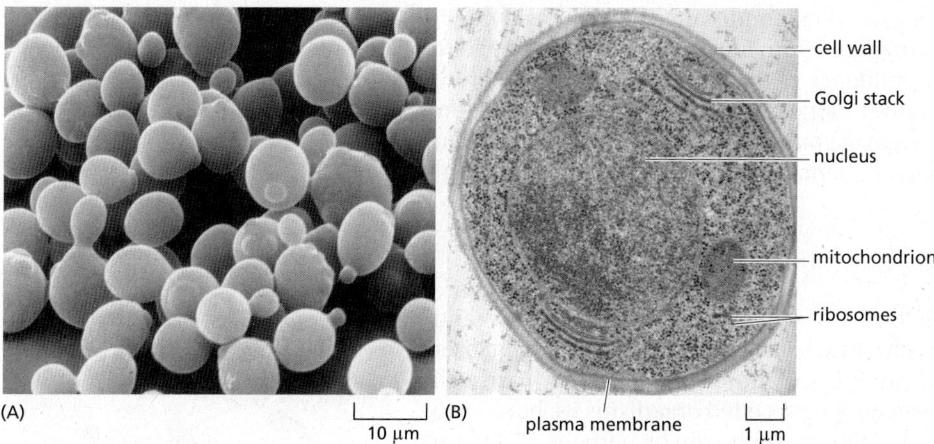

Figure 1–39 The yeast *Saccharomyces cerevisiae*. (A) A scanning electron micrograph of a cluster of yeast cells. This species is also known as budding yeast, because it proliferates by forming a protrusion, or bud, that enlarges and then separates from the mother cell. Many cells with buds are visible in this micrograph. (B) An electron micrograph of a cross section of a yeast cell, showing its plasma membrane and thick cell wall, as well as some of its intracellular organelles. (A, courtesy of Ira Herskowitz and Eric Schabtach; B, courtesy of Andrew Staehelin.)

haploid (**Figure 1–40**). In contrast to most animals, this yeast can therefore proliferate either sexually or asexually, a choice that an experimenter can make simply by changing the growth conditions.

In addition to these features, the yeast has a further property that makes it a convenient organism for genetic studies: its genome, by eukaryotic standards, is exceptionally small (see Table 1–2), yet it suffices for all the basic tasks that every eukaryotic cell must perform. Mutants are available for every gene, and thus the consequence of missing each gene—one by one—can be observed under any environmental condition using the high-throughput procedures described in Chapter 8. Over the past 50 years, extensive studies of yeast cells carried out by many laboratories have provided keys to crucial "eukaryotic-only" processes. These include the cell-division cycle (the critical chain of events by which the nucleus and all the other components of a cell are duplicated and parceled out to create two daughter cells from one) and meiosis (the process through which an organism's reproductive cells are formed). In addition, important insights into eukaryotic chromosome structure, the organization of the nucleus, the mechanisms of gene expression, the formation of organelles, and the ways that proteins are secreted from cells have come out of the work on yeasts. Many of these fundamental processes are so similar between yeasts and humans that a human homolog of a yeast protein will often faithfully carry out its functions when artificially expressed in yeast cells.

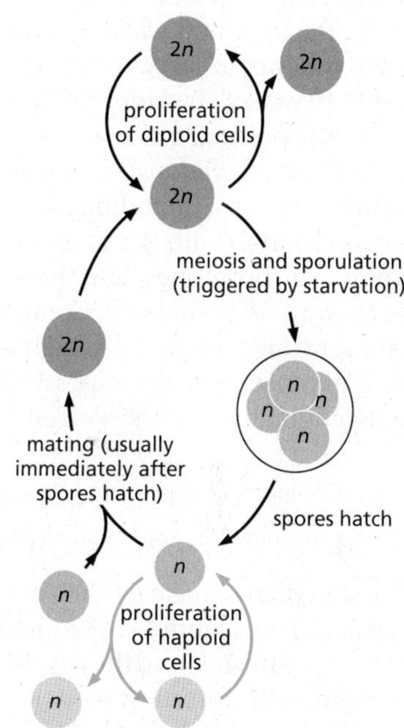

Figure 1–40 The reproductive cycles of the yeast *S. cerevisiae*. Depending on environmental conditions and on details of the genotype, cells of this species can exist in either a diploid (2*n*) state, with a double chromosome set, or a haploid (*n*) state, with a single chromosome set. The diploid form can either proliferate by ordinary cell-division cycles (mitosis) or undergo meiosis to produce haploid cells. The haploid form can either proliferate by ordinary cell-division cycles or undergo sexual fusion with another haploid cell to become diploid. Meiosis is triggered by starvation, and it gives rise to spores—haploid cells in a dormant state, resistant to harsh environmental conditions.

The Expression Levels of All the Genes of an Organism Can Be Determined

The complete genome sequence of *S. cerevisiae* consists of approximately 12,500,000 nucleotide pairs, including the small contribution (about 78,500 nucleotide pairs) of the mitochondrial DNA. This total is only about 2.7 times as much DNA as there is in *E. coli*, and it codes for only about 1.5 times as many distinct proteins (see Table 1–2). The way of life of *S. cerevisiae* is similar in many ways to that of a bacterium, and it seems that this yeast has likewise been subject to selection pressures (for rapid proliferation, for example) that have kept its genome compact.

Knowledge of the complete genome sequence of any organism—be it a yeast or a human—opens up new perspectives on the workings of the cell: many things that once seemed impossibly complex now seem to be within our grasp. Using techniques described in Chapter 8, it is possible, for example, to monitor simultaneously, the amount of mRNA produced from every gene in the yeast genome under any environmental condition. It is also possible to determine in real time how the pattern of gene activity changes when conditions change. This type of analysis can be repeated with mRNA prepared from mutant cells lacking any gene we care to test, and, in this way, the influence of that gene on the expression of all other genes can be observed. Although pioneered in yeast, this approach now provides a way to reveal the entire system of controls that govern gene expression in any organism, as long as its genome sequence is known and it can be manipulated genetically.

Arabidopsis Has Been Chosen as a Model Plant

The large multicellular organisms that we see around us—the plants and animals—seem fantastically varied, but, as we have seen, they are much closer to one another in their evolutionary origins, and more similar in their basic cell biology, than the great host of microscopic single-celled organisms we have been discussing. Thus, while bacteria and archaea are separated by perhaps 3.5 billion years of evolution, vertebrates and insects are separated by about 700 million years, fish and mammals by about 450 million years, and the different species of flowering plants by only about 150 million years (see Figure 1–35).

Because of the close evolutionary relationship between all flowering plants (see Figure 1–35), we can, once again, gain insight into the cell and molecular biology of this whole class of organisms by focusing on just one or a few species for detailed analysis. Out of the nearly 400,000 known species of flowering plants, molecular biologists have chosen to concentrate their efforts on a small weed in the cabbage family, the common wall cress *Arabidopsis thaliana* (**Figure 1–41**), which can be grown indoors in large numbers and produces thousands of offspring per plant after 8–10 weeks. *Arabidopsis* has a total genome size of approximately 135 million nucleotide pairs, about 10 times the size of the yeast genome (see Table 1–2).

Work on *Arabidopsis* has provided a deep understanding of numerous key features of plants, including the mechanisms that cause flower development and its coordination with the seasons, the ability to grow toward sunlight, cell-to-cell signaling by hormones, and the special type of innate immune system that plants use to ward off pathogens. Comparison of the developmental programs between plants and animals has also highlighted some common principles, thereby allowing a glimpse into the basic logic through which large, highly differentiated, multicellular organisms evolved from single-cell ancestors.

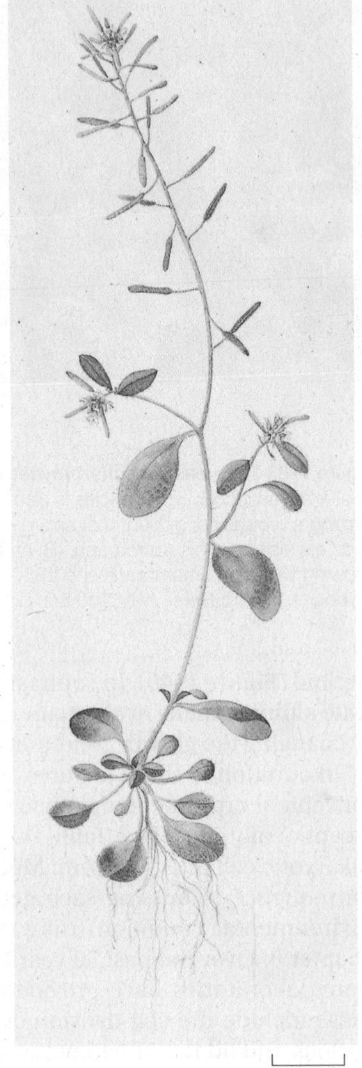

Figure 1–41 ***Arabidopsis thaliana*, the plant chosen as the primary model for studying plant molecular genetics.** (Courtesy of Toni Hayden, FLS; and the John Innes Foundation.)

The World of Animal Cells Is Mainly Represented by a Worm, a Fly, a Fish, a Mouse, and a Human

Although plants make up 80% of the biomass on Earth and animals make up less than 0.4% (see Figure 1–14), animals account for the majority of all named species of living organisms, and they are by far the most intensely studied. Five species have emerged as the foremost model organisms for molecular, cell, and developmental biological studies. In order of increasing body size, they are the nematode worm *Caenorhabditis elegans*, the fly *Drosophila melanogaster*, the zebrafish *Danio rerio*, the mouse *Mus musculus*, and the human, *Homo sapiens*. Genome sequences from many different individuals within each species have been determined.

Caenorhabditis elegans (**Figure 1–42**) is a small, harmless relative of the eelworm that attacks crops. With a life cycle of only a few days, an ability to survive in a freezer indefinitely in a state of suspended animation, a simple body

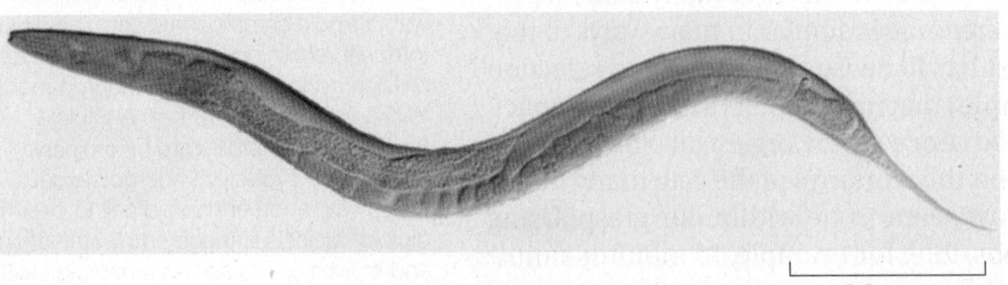

Figure 1–42 ***Caenorhabditis elegans*, the first multicellular organism to have its complete genome sequence determined.** This nematode is only about 1 mm long and normally lives in the soil. Most individuals are hermaphrodites, producing both eggs and sperm. (Courtesy of Maria Gallegos, University of Wisconsin, Madison.)

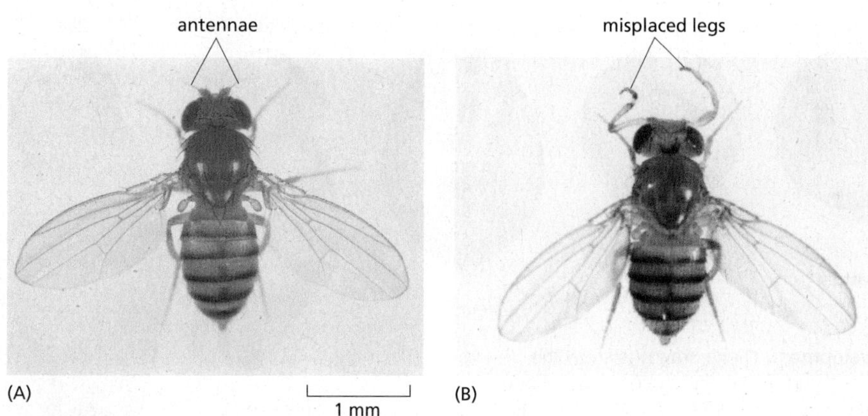

Figure 1–43 *Drosophila melanogaster.* (A) A normal adult *Drosophila* fruit fly. Molecular genetic studies of this fly have provided the main key to understanding how all animals develop from a fertilized egg into an adult. (B) A mutant adult fly, in which a mutation in a regulatory DNA sequence has caused genes for leg formation to be abnormally activated in positions normally reserved for antennae; as a result, legs have developed where antennae should be. (A, Edward B. Lewis, Courtesy of the Archives, California Institute of Technology; B, courtesy of Matthew Scott.)

plan, and an unusual life cycle that is well suited for genetic studies, it is an attractive model animal. *C. elegans* develops with clockwork precision from a fertilized egg cell into an adult worm with exactly 959 body cells (plus a variable number of egg and sperm cells)—an unusual degree of regularity for animal development. We now have a minutely detailed description of the sequence of events by which this development occurs, as the cells divide, move, and change their character according to strict and predictable rules (see Figure 21–42). The genome of about 100 million nucleotide pairs codes for about 20,000 proteins, and many mutants and other tools are available for testing gene functions. Although the worm has a body plan very different from our own, the conservation of biological mechanisms has been sufficient for the worm to be a model for many of the developmental and cell-biological processes that occur in the human body. Thus, for example, studies of the worm have been critical for understanding the molecular mechanisms that mediate and regulate the many cell deaths that help control animal-cell numbers, both in normal development and during human cancer growth. This crucial process, called *programmed cell death* or *apoptosis*, is the subject of Chapter 18. In addition, studies in *C. elegans* first revealed many fascinating features of RNA interference (discussed in Chapters 7 and 8). They have also provided key insights into the ways neurons make their proper connections (discussed in Chapter 21) and informed many additional areas of cell biology.

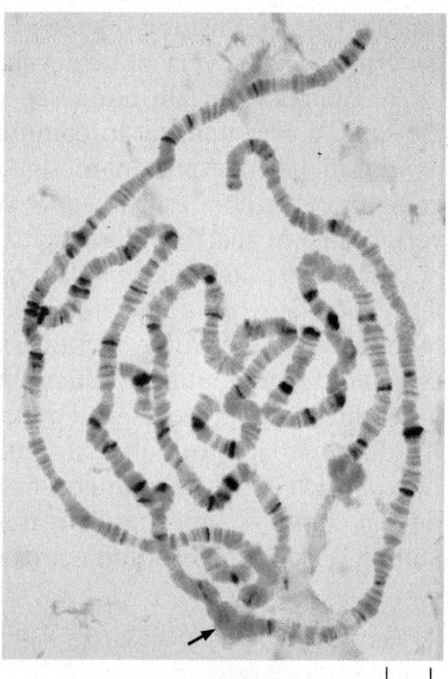

Figure 1–44 Giant chromosomes from salivary gland cells of *Drosophila*. Because many rounds of DNA replication have occurred without an intervening cell division, each of the chromosomes in these unusual cells contains more than 1000 identical double-strand DNA molecules, all aligned in register. This makes them easy to see in the light microscope, where they display a characteristic and reproducible pattern of bands. Specific bands can be identified as the locations of specific genes: a mutant fly with a region of the banding pattern missing or altered shows a phenotype reflecting loss of the genes in that region (not shown). Genes that are being transcribed at a high rate correspond to bands with a "puffed" appearance *(black arrow)*. The bands stained *dark brown* in the micrograph are sites where a particular regulatory protein is bound to the DNA; the regulatory protein is identified by the binding of a specific antibody. (From R. Paro, *Trends Genet.* 6:416–421, 1990. With permission from Elsevier.)

Studies in the Fruit Fly *Drosophila* Provide a Key to Vertebrate Development

The fruit fly *Drosophila melanogaster* (**Figure 1–43**) has been used as a model for animal genetic studies for longer than any other organism; in fact, the foundations of classical genetics were built to a large extent on studies of this insect. Nearly 100 years ago, for example, the fly provided definitive proof that genes—the abstract units of hereditary information at the time—are carried on chromosomes, whose behavior had been closely followed with the light microscope during eukaryotic cell division but whose function was at first unknown. The proof depended on one of the many features that make *Drosophila* especially convenient for molecular genetic studies—the giant chromosomes, which have a characteristic banded appearance that is visible in some of its cells (**Figure 1–44**). Specific changes in the hereditary information, manifest in families of mutant flies, were found to correlate exactly with the loss or alteration of specific bands in the giant chromosomes.

In more recent times, *Drosophila*, more than any other organism, has shown us how to trace the chain of cause and effect from the genetic instructions encoded in the chromosomal DNA to the structure of the adult multicellular body. *Drosophila* mutants with body parts strangely misplaced (Figure 1–43) or mispatterned provided the key to the identification and characterization of the genes required to make a properly structured body, with gut, limbs, eyes, and all the other parts in their correct places. Once these *Drosophila* genes were identified, scientists could identify homologous genes in vertebrates, and then test their functions

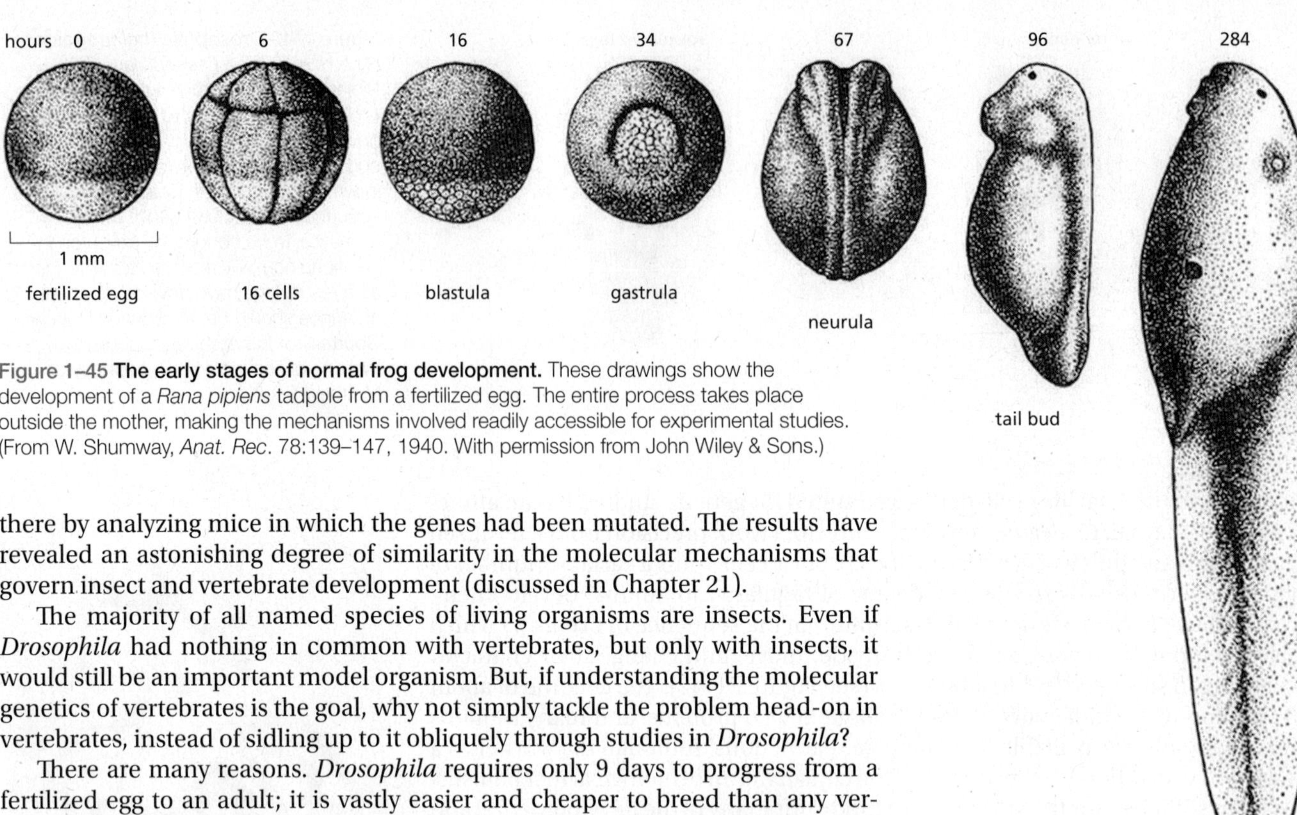

Figure 1–45 **The early stages of normal frog development.** These drawings show the development of a *Rana pipiens* tadpole from a fertilized egg. The entire process takes place outside the mother, making the mechanisms involved readily accessible for experimental studies. (From W. Shumway, *Anat. Rec.* 78:139–147, 1940. With permission from John Wiley & Sons.)

there by analyzing mice in which the genes had been mutated. The results have revealed an astonishing degree of similarity in the molecular mechanisms that govern insect and vertebrate development (discussed in Chapter 21).

The majority of all named species of living organisms are insects. Even if *Drosophila* had nothing in common with vertebrates, but only with insects, it would still be an important model organism. But, if understanding the molecular genetics of vertebrates is the goal, why not simply tackle the problem head-on in vertebrates, instead of sidling up to it obliquely through studies in *Drosophila*?

There are many reasons. *Drosophila* requires only 9 days to progress from a fertilized egg to an adult; it is vastly easier and cheaper to breed than any vertebrate, and its genome is much smaller—about 180 million nucleotide pairs, compared with about 3.1 billion for a human (see Table 1–2). Its genome codes for about 14,000 proteins, and mutants are now available for essentially any gene. In addition to its foundational contributions to animal development, research on *Drosophila* continues to uncover many other insights into biology, ranging from deeply conserved mechanisms that neutralize pathogens to ways that external stimuli from the environment are processed in the brain.

The Frog and the Zebrafish Provide Highly Accessible Vertebrate Models

Frogs have long been used to study the early steps of embryonic development in vertebrates. Because their eggs are big, easy to manipulate, and fertilized outside of the animal, the subsequent development of the early embryo can be easily followed (**Figure 1–45**). *Xenopus laevis,* the African clawed frog, continues to be an important model organism (**Movie 1.6** and see Movie 21.1). Although the species is poorly suited for genetic analysis, cytoplasm isolated from unfertilized *Xenopus* eggs has the remarkable ability to recapitulate the formation of cellular structures and organelles in a test tube. These *egg extracts* allow powerful biochemical approaches to study such fundamental processes as the cell division cycle, described in Chapter 17.

The zebrafish *Danio rerio,* in contrast, is well suited for genetic analysis. Its genome is compact—only half as big as that of a mouse or a human (see Table 1–2)—and it has a generation time of only about 3 months, which is much

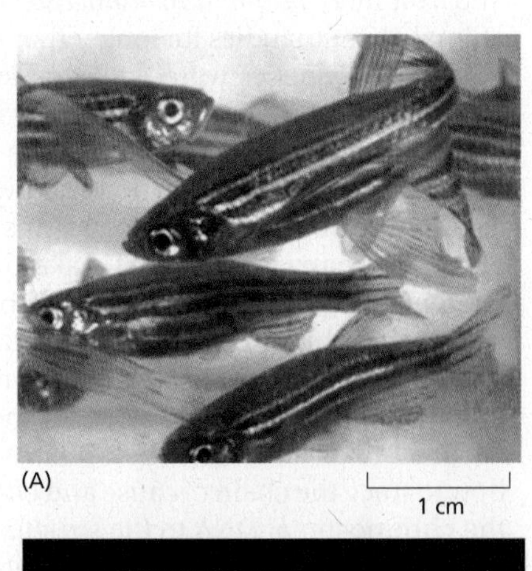

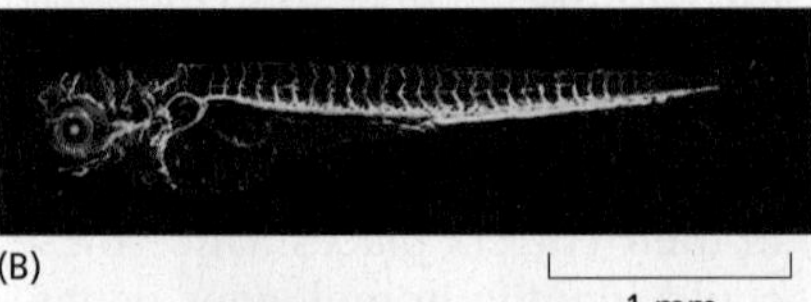

Figure 1–46 **Zebrafish as a model for studies of vertebrate development.** (A) These small, hardy tropical fish are found in many home aquaria and are convenient for laboratory genetic studies. They are ideal for developmental studies as their embryos develop outside of the mother and are transparent, so one can observe cells and internal structures in the living organism throughout its development from an egg to an adult. (B) In this fluorescence image of a 2-day-old embryo, a red fluorescent protein marks the developing blood vessels, and a green fluorescent protein marks the developing lymphatic vessels; regions where the two markers coincide appear *yellow*. (A, courtesy of Steve Baskauf; B, from H.M. Jung et al., *Development* 144:2070–2081, 2017. With permission from The Company of Biologists.)

shorter than that of *Xenopus laevis*. Many mutants are available, and genetic manipulation is relatively simple. The zebrafish has the added virtue that it is transparent for the first 2 weeks of its life, so behavior of specific tissues and individual cells can easily be followed in the living organism as it develops (**Figure 1–46**; see Movie 21.2). All this has made it an increasingly important model vertebrate, one that has been especially crucial for understanding the development of the heart and the circulatory system, as discussed in Chapter 22.

The Mouse Is the Predominant Mammalian Model Organism

In terms of genome size and function, cell biology, and molecular mechanisms, mammals are a highly uniform group of organisms. Even anatomically, the differences among mammals are chiefly a matter of size and proportions; it is hard to think of a human body part that does not have a counterpart in elephants and mice, and vice versa. Evolution plays freely with quantitative features, but it does not readily change the logic of the basic structure.

Mammals have typically about 1.5 times as many protein-coding genes as *Drosophila*, a genome that is about 16 times larger, and an adult body made up of millions or billions of times as many cells. For an exact measure of how closely mammalian species resemble one another genetically, we can compare the nucleotide sequences of corresponding (orthologous) genes or the amino acid sequences of the proteins that these genes encode. The results for individual genes and proteins vary widely. But typically, if we line up the amino acid sequence of a human protein with that of the orthologous protein from, say, an elephant, more than 80% of the amino acids are identical. A similar comparison between human and bird shows an amino acid identity of about 70%—because the bird and mammalian lineages have had longer to diverge than those of the elephant and the human, they have accumulated more differences (**Figure 1–47**).

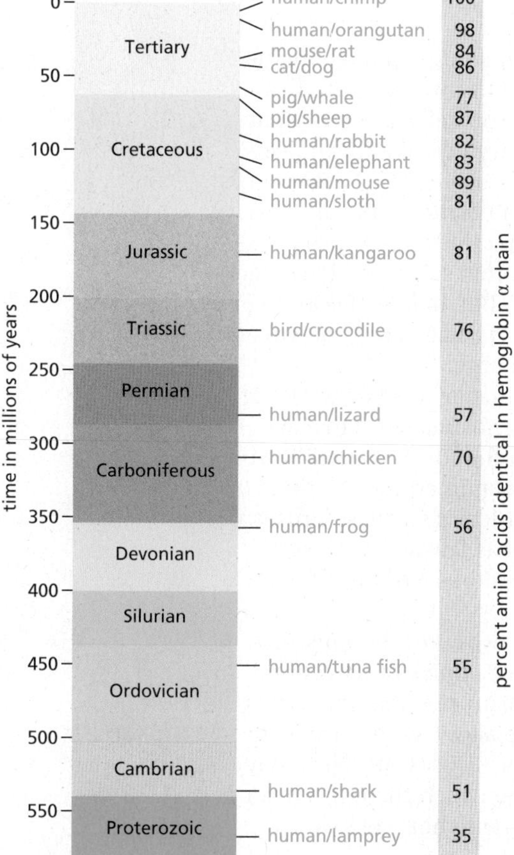

Figure 1–47 **Times of divergence of different vertebrates.** The scale on the left shows the estimated date and geological era of the last common ancestor of each specified pair of animals. Each time estimate is based on comparisons of the amino acid sequences of orthologous proteins; the longer the animals of a pair have had to evolve independently, the smaller the percentage of amino acids that remain identical. The time scale has been calibrated to match the fossil evidence showing that the last common ancestor of mammals and birds lived about 320 million years ago.

The figures on the right show the amino acid sequence divergence for one particular protein—the α chain of hemoglobin. Note that although there is a clear general trend of increasing divergence with increasing time for this protein, there are irregularities that are thought to reflect the action of natural selection causing especially rapid changes in hemoglobin sequence when the organisms experienced special physiological demands. Some proteins that are subject to stricter functional constraints evolve much more slowly than hemoglobin, whereas others evolve as much as five times faster. (Adapted from S. Kumar and S.B. Hedges, *Nature* 392:917–920, 1998.)

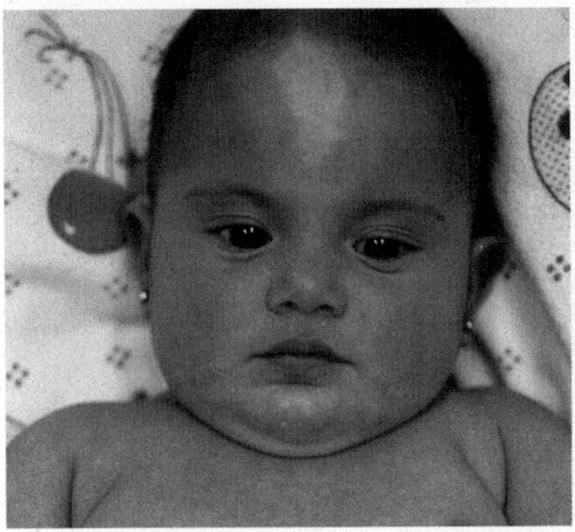

Figure 1–48 Similar mutations produce the same effect in human and mouse. The human baby and the mouse shown here have remarkably similar abnormal white patches on their foreheads as a result of a mutation in the same gene (called *Kit*), which is required for the normal development, migration, and survival of some skin pigment cells. (Courtesy of R.A. Fleischman, from R.A. Fleischman et al., *Proc. Natl. Acad. Sci. USA* 88:10885–10889, 1991.)

The mouse, being small, hardy, and a rapid breeder, has become the foremost model organism for experimental studies of mammalian molecular cell biology. Many naturally occurring mutations are known, often mimicking the effects of corresponding mutations in humans to a remarkable extent (**Figure 1–48**). Moreover, methods have been developed to test the function of any chosen mouse gene or of any noncoding portion of the mouse genome by artificially creating mutations in the relevant part of the gene or genome, as we explain in Chapter 8.

Just one made-to-order mutant mouse can provide a wealth of information for the cell biologist. It reveals the effects of the chosen mutation in various contexts, simultaneously testing the action of the gene in the many different types of cells in the body that could in principle be affected. Studies of the mouse are so fundamental to understanding mammalian biology that we will encounter them in nearly every chapter of this book.

The COVID-19 Pandemic Has Focused Scientists on the SARS-CoV-2 Coronavirus

Having discussed several of the most prominent and well-studied model organisms—which are based on the cell as their fundamental unit—we now turn to an intensively studied virus. Viruses, which in essence feed on cells, are prevalent in all three domains of life: bacteria, archaea, and eukaryotes. We introduced them earlier in this chapter when we discussed several *E. coli* viruses that served as critical experimental systems for the initial development of molecular biology. Here, we focus on one prominent virus, SARS-CoV-2, that infects our own cells and has, due to the widespread attention it has received from scientists, become a model system for understanding eukaryotic viruses. But before discussing this virus in detail, we consider how viruses—genomes packaged in protective shells—first came to be, and how they have evolved over time.

As described in Chapter 6, cells are believed to have first evolved in an "RNA world," before there were proteins or DNA molecules. Scientists suspect that even at that time, parasitic genetic elements were present, in the form of small RNA molecules that took advantage of more advanced replicating entities to proliferate. These are believed to have been the ancestors of today's smallest viruses, which contain single-strand RNA genomes composed of as few as 3000 nucleotides. Thus, virus-like entities have probably been a ubiquitous feature of life on Earth for more than 3 billion years.

At a minimum, a virus requires a genome that encodes two core functions: first, a nucleic acid replication process that produces multiple copies of its genome once inside its host cell, and second, a genome-packaging process that surrounds these new genomes with a protective protein coat, while allowing the viruses to exit the host cell and subsequently enter others. But the viruses present today have evolved through billions of infectious cycles, during which there has been a constant war between host organisms and the viruses—with host cells evolving

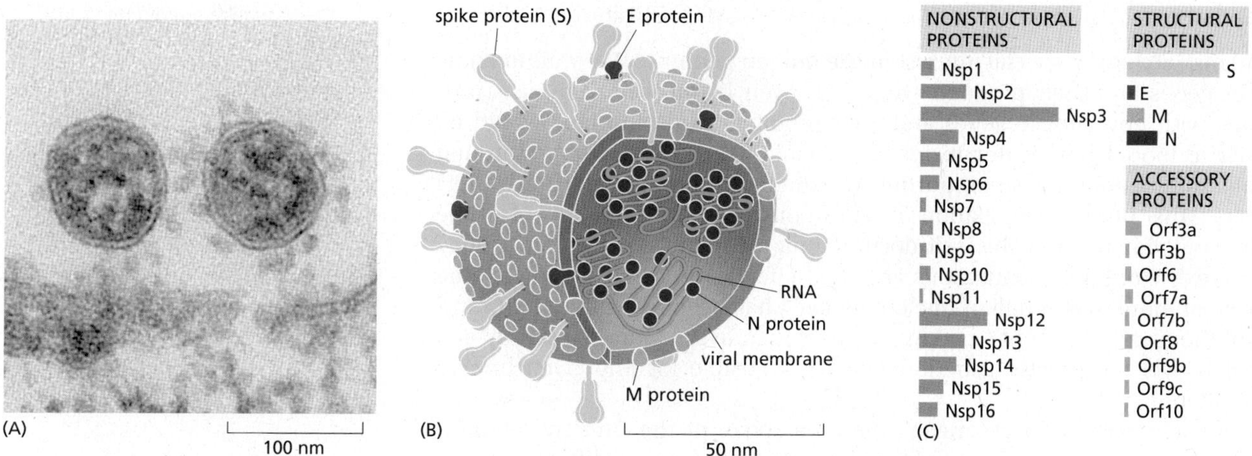

Figure 1–49 The coronavirus. (A) Electron micrograph of SARS-CoV-2 virus particles attached to the surface of a cultured monkey cell. (B) A cut-away drawing of the virus that highlights its protruding spike protein molecules plus a few other major proteins. The spike protein is the major target for vaccines that are designed to block infections, because it attaches the virus to the outside of host cells and then catalyzes transfer of the viral genome into the cell interior. As indicated, the RNA genome is packaged unevenly inside the enveloped virus particle. (C) The 29 proteins produced by SARS-CoV-2, grouped into three different categories. The locations of the structural proteins S, M, E, and N in the virus are indicated in panel B. Each of the proteins listed in the "accessory" category has a role in protecting the virus from host antiviral responses. The functions of the nonstructural proteins include binding to ribosomes to block host protein synthesis (Nsp1), forming a double-membrane "replication organelle" from host-cell membranes (Nsp 3, 4, and 6), and forming the RNA-dependent RNA polymerase (Nsp 7, 8, and 12). The way in which the virus reproduces itself, once inside a host cell, is shown in Figure 5–62. (A, from M. Laue et al. *Sci. Rep.* 11:3515, 2021. With permission from Cold Spring Harbor Press.)

multiple antivirus defenses and viruses evolving various ways to overcome these defenses. As a result, through cycles of random mutation followed by natural selection over long evolutionary times, most virus genomes have grown much larger than needed for their two core functions, with many of the additional genes encoding proteins that help the viruses to circumvent their host-cells' defenses.

Coronavirus genomes are large, single-strand RNA molecules, about 30,000 nucleotides long. This RNA is packaged in a protein coat that is covered with a lipid bilayer envelope, from which protein spikes protrude (**Figure 1–49A and B**). Many coronavirus strains circulate in animal species, including pigs, birds, and bats. Some strains also circulate among humans; these so-called "endemic" strains cause only mild symptoms and are responsible for about one in four common colds. But on rare occasions, a bat coronavirus mutates in a way that allows it to infect humans, where it can cause very severe, even fatal, disease. It is thought that the COVID-19 pandemic of 2020 originated in this way.

The virus that causes COVID-19, SARS-CoV-2, produces 29 proteins (**Figure 1–49C**). Some are structural proteins that package the virus's RNA genome into the virus particle. The nonstructural proteins are critical for replicating the viral genome inside of the host cell, as well as for ensuring that the viral genes are appropriately translated into proteins, including the viral RNA polymerase complex. And, as one would expect, other proteins help the virus to avoid the host's immune defenses, which are described in Chapter 24.

The SARS-CoV-2 virus is closely related to the coronaviruses that cause colds, as well as to the SARS-CoV virus that emerged from bats in 2002 and killed nearly 1 in 10 of the humans it infected. We still do not understand what makes SARS-CoV and SARS-CoV-2 infections so much more dangerous to humans than the infections caused by their close relatives that cause only a mild cold. But, given the thousands of research laboratories currently focused on understanding the cell biology of SARS-CoV-2 with the aim of ameliorating the COVID-19 pandemic, we should know the answers to these questions in the near future. These studies are certain to make us much better prepared to deal with the next virus that emerges to threaten us.

Humans Are Unique in Reporting on Their Own Peculiarities

As humans, we have a special interest in the human genome. We want to know how our genes and their products work. But, even if you were a mouse, preoccupied with how mouse genes and their products work, humans would be attractive as model genetic organisms because of one special property: through medical examinations and self-reporting, we catalog our own genetic (and other) disorders. The human population is enormous, consisting today of some 8 billion individuals, and this self-documenting property means that a huge database exists of human mutations and their effects. And the human genome sequence of more than 3 billion nucleotide pairs has been determined for hundreds of thousands of people, making it easier than ever before to identify at a molecular level the precise genetic change responsible for any given human mutant phenotype.

But what precisely do we mean when we speak of the human genome? Whose genome? On average, any two people taken at random will differ at roughly 4 million different sites in their DNA sequence (see Table 4–3, p. 247). Thus, the human genome is very complex, embracing the entire pool of variant genes found in the human population. As described in Chapter 4, knowledge of this variation is helping us to understand human biology; for example, why some people are prone to one disease, others to another, and why some respond well to a drug, but others badly. It is also providing clues to our history, including population movements, interbreeding among our ancestors, the infections they suffered, and the diets they ate. All these things have left traces in the variant forms of genes that survive today in the human communities that populate our planet, and by exploiting this fact, scientists have been discovering fascinating aspects of our past.

By drawing together the insights from humans, mice, fish, flies, worms, yeasts, plants, and bacteria—using DNA sequence similarities to map out the correspondences between one model organism and another—we are greatly enriching our understanding of them all.

To Understand Cells and Organisms Will Require Mathematics, Computers, and Quantitative Information

Empowered by knowledge of complete genome sequences, we can list the genes, proteins, and RNA molecules in a cell, and we have powerful methods to analyze the complex web of interactions between them. But how are we to use all this information to understand how cells work? Even for a single cell type belonging to a single species of organism, the current deluge of data seems overwhelming. The informal reasoning that biologists usually rely on seems increasingly inadequate in the face of such complexity.

The difficulty is more than just a matter of information overload. Biological systems are, for example, full of feedback loops, and the behavior of even the simplest of systems with feedback is remarkably difficult to predict by intuition alone (**Figure 1–50**); small changes in parameters can cause radical changes in outcome. To go from a circuit diagram to a prediction of the behavior of the

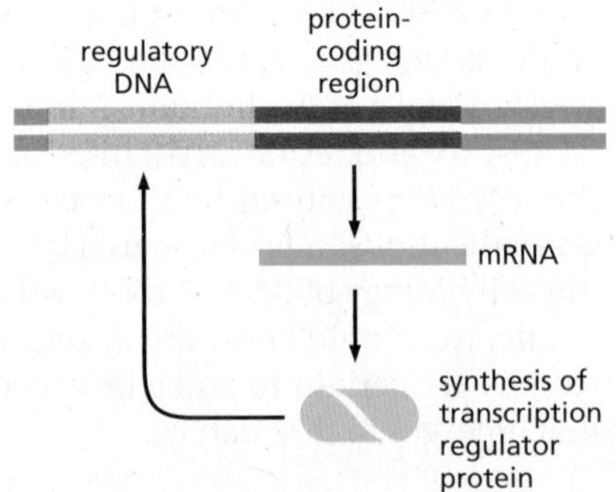

Figure 1–50 **A very simple gene regulatory circuit.** A single gene regulates its own expression because its protein product is a transcription regulator that binds to the regulatory DNA of its own gene. Simple schematic diagrams such as this are found throughout this book. They are often used to summarize what we know, but they leave many questions unanswered. When the protein binds, does it inhibit or stimulate transcription from the gene? How steeply does the transcription rate depend on the protein concentration? How long, on average, does a molecule of the protein remain bound to the DNA? How long does it take to make each molecule of mRNA or protein, and how quickly does each type of molecule get degraded? As explained in Chapter 8, mathematical modeling shows that we need quantitative answers to all these and other questions—obtained by direct observations and experiments—before we can predict the behavior of even this simple circuit. For different parameter values, the system may settle to a unique steady state; or it may behave as a switch, capable of existing in one or another of a set of alternative states; or it may oscillate; or it may even show large random fluctuations.

system, we need detailed quantitative information, and to draw deductions from that information we need mathematics and computers.

Such tools for quantitative reasoning are essential, but they are not all-powerful. You might think that, knowing how each protein in a cell influences each other protein, and how the expression of each gene is regulated by the products of other genes, we should soon be able to calculate how the cell as a whole will behave, just as an astronomer can calculate the orbits of the planets or a chemical engineer can calculate the flows through a chemical plant. But any attempt to perform this feat for anything close to an entire living cell rapidly reveals the limits of our present knowledge. The information we have, plentiful as it is, is full of gaps and uncertainties, and it is largely qualitative rather than quantitative. Most often, cell biologists studying a cell's control systems sum up their knowledge in simple schematic diagrams—this book is full of them—rather than in numbers, graphs, and differential equations.

To progress from qualitative descriptions and intuitive reasoning to quantitative descriptions and mathematical deduction is one of the biggest challenges for contemporary cell biology. So far, the challenge has been met for only a few very simple fragments of the machinery of living cells—subsystems involving a handful of different proteins, or two or three genes that regulate one another, where theory and experiment go closely hand in hand. We discuss some of these examples later in the book and devote much of Chapter 8 to some new approaches designed to answer the increasingly complex questions that arise in biology.

Knowledge and understanding bring the power to intervene—with humans, to prevent and treat disease; with plants, to create better crops; with bacteria, archaea, and fungi, to control them for our own benefit. All these biological enterprises are linked, because the genetic information of all living organisms is written in the same language. The recent ability of molecular biologists to read and decipher this language has already begun to transform our relationship to the living world. The account of cell biology in the subsequent chapters will, we hope, equip the reader to understand, and possibly to contribute to, the great biosciences adventure that we can anticipate through the rest of this century.

Summary

Powerful new technologies, including rapid and cheap genome sequencing, are enabling rapid advances in our knowledge of human biology, with implications for understanding and treating human disease. But living systems are incredibly complex, and simpler model organisms have played a critical part in revealing universal genetic and molecular cell biological mechanisms. Thus, for example, early research on the bacterium E. coli and its viruses provided the foundations needed to decipher the fundamental genetic mechanisms in all cells. And research on the unicellular yeast Saccharomyces cerevisiae, which continues to serve as a simple model organism for eukaryotic cell biology, has revealed the molecular basis for many critical processes that have been strikingly conserved during more than a billion years of eukaryotic evolution. Biologists have also chosen a small number of multicellular organisms for intensive study: a worm, a fly, a fish, the mouse, and humans serve as model organisms for animals, and a small member of the cabbage family serves as a model for plant biology. Even today, research that focuses on these and other model organisms remains crucial for understanding ourselves, as well as for driving scientific and medical advances.

PROBLEMS

Which statements are true? Explain why or why not.

1–1 DNA and RNA use the same four-letter alphabet.

1–2 Each member of the human hemoglobin gene family, which consists of seven genes arranged in two clusters on different chromosomes, is an ortholog to all of the other members.

1–3 Most of the DNA sequences in a bacterial genome code for proteins, whereas most of the DNA sequences in the human genome do not.

1–4 Without additional information, no amount of gazing at genome sequences will reveal the functions of genes.

Discuss the following problems.

1–5 "Life" is easy to recognize but difficult to define. Dictionaries commonly define life as "The state or quality that distinguishes living beings or organisms from dead ones and from inorganic matter, characterized chiefly by metabolism, growth, the ability to reproduce, and the ability to respond to stimuli." Score a car, a cactus, and yourself with respect to these characteristics.

1–6 Since it was deciphered more than five decades ago, some have claimed that the genetic code must be a frozen accident, while others have argued that it was shaped by natural selection. A striking feature of the genetic code is its inherent resistance to the effects of mutation. For example, a change in the third position of a codon often specifies the same amino acid or one with similar chemical properties. The natural code resists mutation more effectively (is less susceptible to error) than most other possible versions, as illustrated in Figure Q1–1. Only one in a million computer-generated "random" codes is more error-resistant than the natural genetic code. Does the extraordinary mutation resistance of the genetic code argue in favor of its origin as a frozen accident or as a result of natural selection? Explain your reasoning.

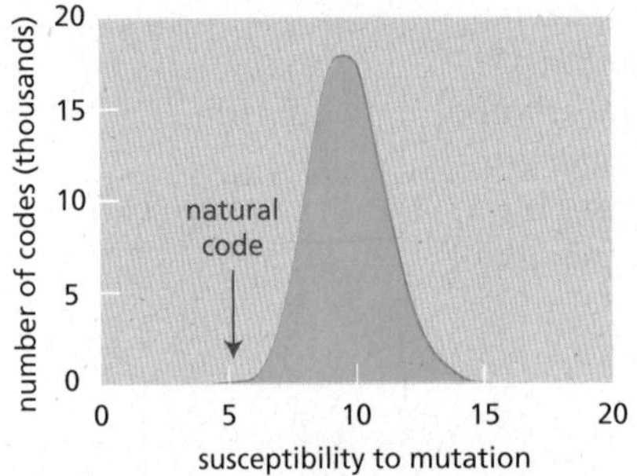

Figure Q1–1 Susceptibility to mutation of the natural code shown relative to that of millions of computer-generated alternative genetic codes (Problem 1–6). Susceptibility measures the average change in amino acid properties caused by random mutations in a genetic code. A small value indicates that mutations tend to cause minor changes. (Data courtesy of Steve Freeland.)

1–7 You have begun to characterize a sample obtained from the depths of the oceans on Europa, one of Jupiter's moons. Much to your surprise, the sample contains a life-form that grows well in a rich broth. Your preliminary analysis shows that it is cellular and contains DNA, RNA, and protein. When you show your results to a colleague, she suggests that your sample was contaminated with an organism from Earth. What approaches might you try to distinguish between contamination and a novel cellular life-form that is based on DNA, RNA, and protein?

1–8 It is not so difficult to imagine what it means to feed on the organic molecules that living things produce. That is, after all, what we do. But what does it mean to "feed" on sunlight, as phototrophs do? Or, even stranger, to "feed" on rocks, as lithotrophs do? Where is the "food," for example, in the mixture of chemicals (H_2S, H_2, CO, Mn^+, Fe^{2+}, Ni^{2+}, CH_4, and NH_4^+) that spews from a hydrothermal vent?

1–9 How many possible different trees (branching patterns) can be drawn to display the evolution of bacteria, archaea, and eukaryotes, assuming that they all arose from a common ancestor?

1–10 The genes for ribosomal RNA are highly conserved (relatively few sequence changes) in all organisms on Earth; thus, they have evolved very slowly over time. Were ribosomal RNA genes "born" perfect?

1–11 Rates of evolution appear to vary in different lineages. For example, the rate of evolution in the rat lineage is significantly higher than in the human lineage. These rate differences are apparent whether one looks at changes in nucleotide sequences that encode proteins and are subject to selective pressure or at changes in noncoding nucleotide sequences, which are not under obvious selection pressure. Can you offer one or more possible explanations for the slower rate of evolutionary change in the human lineage versus the rat lineage?

1–12 Genes participating in informational processes such as replication, transcription, and translation undergo horizontal gene transfer between species much less often than do genes involved in metabolism. The basis for this inequality is unclear at present, but one suggestion is that it relates to the underlying complexity of the two types of processes. Informational processes tend to involve large aggregates of different gene products, whereas metabolic reactions are usually catalyzed by enzymes composed of a single protein. Why would the complexity of the underlying process—informational or metabolic—have any effect on the rate of horizontal gene transfer?

1–13 Animal cells have neither cell walls nor chloroplasts, whereas plant cells have both. Fungal cells are somewhere in between; they have cell walls but lack chloroplasts. Are fungal cells more likely to be animal cells that gained the ability to make cell walls or to be plant cells that lost their chloroplasts? This question represented a difficult issue for early investigators who sought to assign evolutionary relationships solely on the basis of cell characteristics and morphology. How do you suppose that this question was eventually decided?

1–14 *Giardia lamblia* is a fascinating eukaryotic parasite; it contains a nucleus but no mitochondria and no discernible endoplasmic reticulum or Golgi apparatus—one of the very rare examples of such a cellular organization among eukaryotes. This cell organization might have arisen because *Giardia* is an ancient lineage that separated from the rest of the eukaryotes before mitochondria were acquired and internal membranes were developed. Or it might be a stripped-down version of a more standard eukaryote that has lost these structures because they are not necessary for its parasitic lifestyle. How might you use nucleotide sequence comparisons to distinguish between these alternatives?

1–15 When plant hemoglobin genes were first discovered in legumes, it was so surprising to find a gene typical of animal blood that it was hypothesized that the plant gene arose by horizontal transfer from an animal. Many more hemoglobin genes have now been sequenced, and a phylogenetic tree based on some of these sequences is shown in Figure Q1–2.

A. Does this tree support or refute the hypothesis that the plant hemoglobins arose by horizontal gene transfer?

B. Supposing that the plant hemoglobin genes were originally derived from a parasitic nematode, for example, what would you expect the phylogenetic tree to look like?

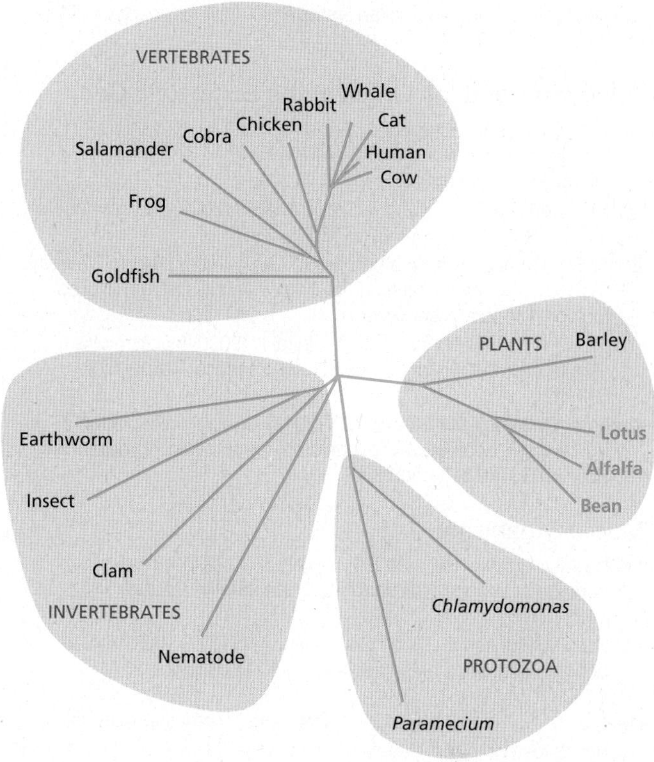

Figure Q1–2 Phylogenetic tree for hemoglobin genes from a variety of species (Problem 1–15). The legumes are highlighted in *green*. The lengths of lines that connect the present-day species represent the evolutionary distances that separate them.

REFERENCES

General

Alberts B, Hopkin K, Johnson A, et al. (2019) Essential Cell Biology, 5th ed. New York: Norton.

Barton NH, Briggs DEG, Eisen JA, et al. (2007) Evolution. Cold Spring Harbor, NY: Cold Spring Harbor Laboratory Press.

Darwin C (1859) On the Origin of Species. London: Murray.

Hall BK & Hallgrímsson B (2014) Strickberger's Evolution, 5th ed. Burlington, MA: Jones & Bartlett.

Lynch M (2007) The Origins of Genome Architecture. Oxford: Oxford University Press.

Madigan MT, Bender KS, Buckley DH et al. (2018) Brock Biology of Microorganisms, 15th ed. London: Pearson.

Margulis L & Chapman MJ (2009) Kingdoms and Domains: An Illustrated Guide to the Phyla of Life on Earth. San Diego: Academic Press.

Moore JA (1993) Science as a Way of Knowing. Cambridge, MA: Harvard University Press.

The Universal Features of Life on Earth

Blain JC & Szostak JW (2014) Progress toward synthetic cells. *Annu. Rev. Biochem*. 83, 615–640.

Brenner S, Jacob F & Meselson M (1961) An unstable intermediate carrying information from genes to ribosomes for protein synthesis. *Nature* 190, 576–581.

Gibson DG, Benders GA, Andrews-Pfannkoch C . . . Smith HO (2008) Complete chemical synthesis, assembly, and cloning of a *Mycoplasma genitalium* genome. *Science* 319, 1215–1220.

Koonin EV (2005) Orthologs, paralogs, and evolutionary genomics. *Annu. Rev. Genet.* 39, 309–338.

Noller H (2005) RNA structure: reading the ribosome. *Science* 309, 1508–1514.

Watson JD & Crick FHC (1953) Molecular structure of nucleic acids: a structure for deoxyribose nucleic acid. *Nature* 171, 737–738.

Genome Diversification and the Tree of Life

Baker BJ, De Anda V, Seitz KW . . . Lloyd KG (2020) Diversity, ecology and evolution of Archaea. *Nat. Microbiol.* 5, 887–900.

Doolittle WF & Brunet TDP (2016) What is the tree of life. *PLoS Genet.* 12(4), e1005912.

Eme L, Spang A, Lombard J . . . Thijs JG (2017) Archaea and the origin of eukaryotes. *Nat. Rev. Microbiol.* 15(12), 711–723.

Hug LA, Baker BJ, Anantharaman K . . . Banfield JF (2016) A new view of the tree of life. *Nat. Microbiol.* 1, 16048.

Kerr RA (1997) Life goes to extremes in the deep earth—and elsewhere? *Science* 276, 703–704.

Woese C (1998) The universal ancestor. *Proc. Natl. Acad. Sci. USA* 95, 6854–6859.

Eukaryotes and the Origin of the Eukaryotic Cell

Andersson SG, Zomorodipour A, Andersson JO . . . Kurland CG (1998) The genome sequence of *Rickettsia prowazekii* and the origin of mitochondria. *Nature* 396, 133–140.

Burki F, Roger AJ, Brown MW & Simpson AGB (2020) The new tree of eukaryotes. *Trends Ecol. Evol.* 35(1), 43–55.

Carroll SB, Grenier JK & Weatherbee SD (2005) From DNA to Diversity: Molecular Genetics and the Evolution of Animal Design, 2nd ed. Maldon, MA: Blackwell Science.

Imachi H, Nobu MK, Nakahara N . . . Takai K (2020) Isolation of an archaeon at the prokaryote–eukaryote interface. *Nature* 577, 519–525.

Spang A, Caceres EF & Ettema TJG (2017) Genomic exploration of the diversity, ecology, and evolution of the archaeal domain of life. *Science* 357(6351), eaaf3883.

Model Organisms

Adams MD, Celniker SE, Holt RA . . . Venter JC (2000) The genome sequence of *Drosophila melanogaster. Science* 287, 2185–2195.

Blattner FR, Plunkett G, Bloch CA . . . Shao Y (1997) The complete genome sequence of *Escherichia coli* K-12. *Science* 277, 1453–1474.

Goffeau A, Barrell BG, Bussey H . . . Oliver SG (1996) Life with 6000 genes. *Science* 274, 546–567.

International Human Genome Sequencing Consortium (2001) Initial sequencing and analysis of the human genome. *Nature* 409, 860–921.

Krupovic M, Dolja VV & Koonin EV (2019) Origin of viruses: primordial replicators recruiting capsids from hosts. *Nat. Rev. Microbiol.* 17(7), 449–458.

Lander ES (2011) Initial impact of the sequencing of the human genome. *Nature* 470, 187–197.

Lynch M & Conery JS (2000) The evolutionary fate and consequences of duplicate genes. *Science* 290, 1151–1155.

Masters PS (2006) The molecular biology of coronaviruses. *Adv. Virus Res.* 66, 193–292.

Prangishvili D, Bamford DH, Forterre P, . . . Krupovic M (2017) The enigmatic archaeal virosphere. *Nat. Rev. Microbiol.* 15(12), 724–739.

Reed FA & Tishkoff SA (2006) African human diversity, origins and migrations. *Curr. Opin. Genet. Dev.* 16, 597–605.

The *Arabidopsis* Initiative (2000) Analysis of the genome sequence of the flowering plant *Arabidopsis thaliana. Nature* 408, 796–815.

The *C. elegans* Sequencing Consortium (1998) Genome sequence of the nematode *C. elegans*: a platform for investigating biology. *Science* 282, 2012–2018.

Tinsley RC & Kobel HR (eds.) (1996) The Biology of *Xenopus*. Oxford: Clarendon Press.

Weiss SR (2020) Forty years with coronaviruses. *J. Exp. Med.* 217(5), e20200537.

Cell Chemistry and Bioenergetics

CHAPTER

2

IN THIS CHAPTER

It is at first sight difficult to accept the idea that living creatures are merely chemical systems. Their incredible diversity of form, their seemingly purposeful behavior, and their ability to grow and reproduce all seem to set them apart from the world of solids, liquids, and gases that chemistry normally describes. Indeed, until the late nineteenth century, animals were generally believed to contain a Vital Force—an "animus"—that was responsible for their distinctive properties.

We now know that there is nothing in living organisms that disobeys chemical or physical laws. However, the chemistry of life is indeed special. First, life depends on chemical reactions that take place in aqueous solution, and it is based overwhelmingly on carbon compounds, the study of which is known as *organic chemistry*. Second, although cells contain a variety of small carbon-containing molecules, most of the carbon atoms present are incorporated into enormous polymeric molecules—chains of chemical subunits linked end-to-end. It is the unique properties of these *macromolecules* that enable cells and organisms to grow and reproduce—and to do all the other things that are characteristic of life. Third, and most important, cell chemistry is enormously complex: even the simplest cell is vastly more complicated in its chemistry than any other chemical system known. In fact, we now recognize that the many interlinked networks of chemical reactions in cells can give rise to so-called *emergent properties*, which will require the development of new experimental and computational methods to understand.

Much of the information in this chapter is summarized—and in some cases further elaborated—in the nine two-page Panels with which the chapter ends (Panels 2–1 to 2–9). Although the Panels will be cited at appropriate places in the text, they should also be useful for refreshing background knowledge when reading later chapters.

THE CHEMICAL COMPONENTS OF A CELL

Living organisms are made of only a small selection of the 92 naturally occurring elements, four of which—carbon (C), hydrogen (H), nitrogen (N), and oxygen (O)—make up 96.5% of an organism's weight (**Figure 2–1**). The *atoms* of these

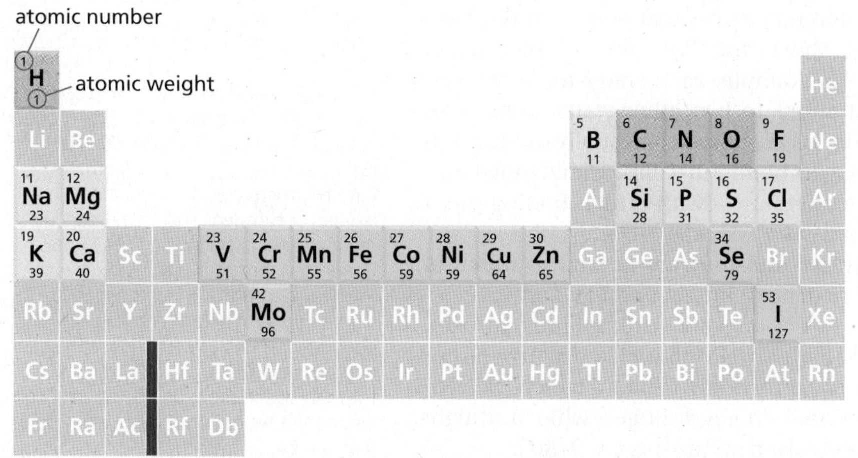

Figure 2–1 The main elements in cells, highlighted in the periodic table. When ordered by their atomic number and arranged in this manner, elements fall into vertical columns that show similar properties.

The four elements highlighted in *red* constitute 99% of the total number of atoms present in the human body (and 96.5% of its weight). An additional seven elements, highlighted in *blue*, together represent about 0.9% of the total atoms in our bodies. The elements shown in *green* are required in trace amounts by humans. It remains unclear whether those elements shown in *yellow* are essential in humans.

The chemistry of life is therefore predominantly the chemistry of lighter elements. The atomic weights shown here are those of the most common isotope of each element. The vertical *red* line marks a break in the periodic table where a group of large atoms with similar chemical properties is omitted.

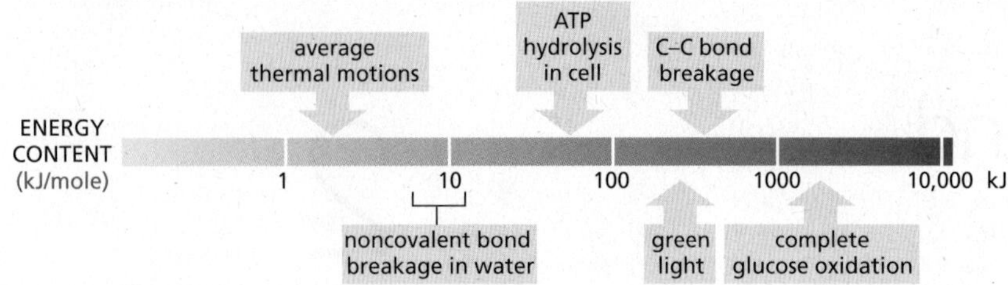

Figure 2–2 **Some energies important for cells.** A crucial property of any bond—covalent or noncovalent—is its strength. *Bond strength* is measured by the amount of energy that must be supplied to break it, expressed in units of either kilojoules per mole (kJ/mole) or kilocalories per mole (kcal/mole). Thus if 100 kJ of energy must be supplied to break 6×10^{23} bonds of a specific type (that is, 1 mole of these bonds), then the strength of that bond is 100 kJ/mole. Note that, in this diagram, energies are compared on a logarithmic scale. (Typical strengths and lengths of the main classes of chemical bonds are given in Table 2–1, later in text.)

One joule (J) is the amount of energy required to move an object a distance of 1 meter (m) against a force of 1 newton (N). This measure of energy is derived from the SI units (Système International d'Unités) universally employed by physical scientists. A second unit of energy, often used by cell biologists, is the kilocalorie (kcal):1 calorie (cal) is the amount of energy needed to raise the temperature of 1 gram (g) of water by 1°C. One kilojoule (kJ) is equal to 0.239 kcal.

elements are linked together by **covalent bonds** to form *molecules* (see **Panel 2–1**, pp. 94–95). Because covalent bonds are typically 100 times stronger than the thermal energies within a cell, they resist being pulled apart by thermal motions, and they are normally broken only during biologically catalyzed chemical reactions that are of use to the cell. *Noncovalent bonds* are much weaker (**Figure 2–2**), but sets of them allow molecules to recognize each other and reversibly associate, which is critical for the vast majority of biological functions.

Water Is Held Together by Hydrogen Bonds

Because 70% of the weight of a cell is water, the reactions that make life possible occur in an aqueous environment. Life on Earth is thought to have begun in shallow bodies of water that had concentrated essential molecules, and the conditions in that primeval environment have left a permanent stamp on the chemistry of all living things.

The chemical properties of water are reviewed in **Panel 2–2** (pp. 96–97). In a water molecule (H_2O), the two H atoms are linked to the O atom by covalent bonds that are highly polar, inasmuch as the O atom attracts electrons more strongly than does the H atom. Consequently, there is a preponderance of positive charge on the two H atoms and of negative charge on the O atom. When a positively charged region of one water molecule (that is, one of its H atoms) approaches a negatively charged region (that is, the O atom) of a second water molecule, the electrical attraction between them can result in a **hydrogen bond** (**Figure 2–3A**). These bonds are much weaker than covalent bonds and are easily broken by the random thermal motions that reflect the heat energy of the molecules. Thus, each bond lasts only a very short time. But the combined effect of many weak bonds can be profound. For example, each water molecule can form hydrogen bonds through its two H atoms to two other water molecules, producing a network in which hydrogen bonds are being continually broken and formed. It is only because of these hydrogen bonds that link water molecules together that water is a liquid at room temperature—with a high boiling point and high surface tension—rather than a gas.

Hydrogen bonds are not limited to water, and they are central to much of biology. This bond represents a special form of polar interaction in which an electropositive hydrogen atom is shared by two electronegative atoms. The hydrogen in this bond can be viewed as a proton that has partially dissociated from a donor atom, allowing it to be shared by a second, acceptor atom. Unlike a typical electrostatic interaction, this bond is highly directional—being strongest when a straight line can be drawn between all three of the involved atoms (**Figure 2–3B**).

(A)

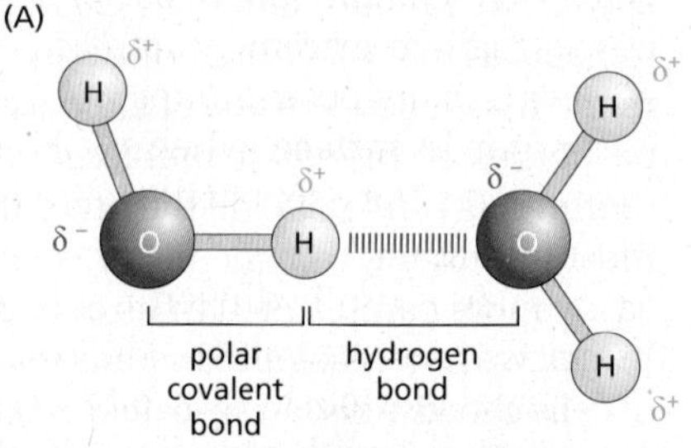

(B)

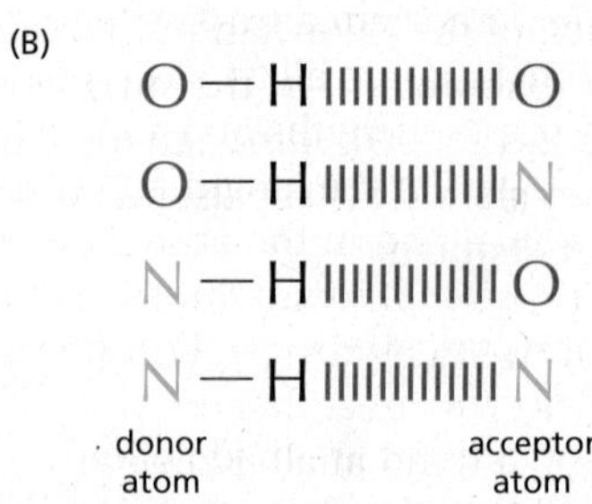

Figure 2–3 **The noncovalent hydrogen bond.** (A) A hydrogen bond forms between two water molecules. The slight positive charge associated with the hydrogen atom is electrically attracted to the slight negative charge of the oxygen atom. This causes water to exist as a large hydrogen-bonded network (see Panel 2–2, pp. 96–97). (B) In cells, hydrogen bonds commonly form between molecules that contain an oxygen or nitrogen. The atom bearing the hydrogen is considered the H-bond donor, and the atom that interacts with the hydrogen is the H-bond acceptor. This type of dipole–dipole interaction is of critical importance in biology. For this reason, and because it is highly directional, the hydrogen bond receives special attention among the set of noncovalent attractions that we discuss next.

Molecules, such as alcohols, that contain polar bonds and that can form hydrogen bonds with water dissolve readily in water. Molecules carrying charges (ions) likewise interact favorably with water. Such molecules are termed **hydrophilic**, meaning that they are water-loving. Many of the molecules in the aqueous environment of a cell necessarily fall into this category, including sugars, DNA, RNA, and most proteins. **Hydrophobic** (water-fearing) molecules, by contrast, are uncharged and form few or no hydrogen bonds, and so do not dissolve in water. Hydrocarbons are an important example. In these molecules, all of the H atoms are covalently linked to C atoms by a largely nonpolar bond; thus, they cannot form effective hydrogen bonds to other molecules (see Panel 2–1, pp. 94–95). This makes the hydrocarbon as a whole hydrophobic—a property that is exploited in cells, whose membranes are constructed from molecules that have long hydrocarbon tails, as we shall see in Chapter 10.

Four Types of Noncovalent Attractions Help Bring Molecules Together in Cells

Much of biology depends on the specific binding between different molecules caused by three types of **noncovalent bonds**—hydrogen bonds, electrostatic attractions (ionic bonds), and van der Waals attractions—combined with a fourth factor that can push molecules together: the hydrophobic force.

Electrostatic attractions are strongest when the atoms involved are fully charged, or ionized. But a weaker electrostatic attraction occurs between molecules that contain polar covalent bonds. Like hydrogen bonds, electrostatic attractions are extremely important in biology. For example, any large molecule with many polar groups will have a pattern of partial positive and negative charges on its surface. When such a molecule encounters a second molecule with a complementary set of charges, the two will be drawn to each other by electrostatic attraction.

In addition to hydrogen bonds and electrostatic attractions, a third type of noncovalent bond, called a **van der Waals attraction**, comes into play when any two atoms approach each other closely. These weak, nonspecific interactions are due to fluctuations in the distribution of electrons in every atom, which can generate a transient attraction when the atoms are in very close proximity. These attractions occur in all types of molecules, even those that are nonpolar.

The relative lengths and strengths of these three types of noncovalent bonds are compared to the length and strength of covalent bonds in **Table 2–1**, both in the presence and in the absence of water. Note that, because water forms competing interactions with the involved molecules, the strength of both electrostatic attractions and hydrogen bonds is greatly weakened inside of the cell.

The fourth effect that often brings molecules together in water is not, strictly speaking, a bond at all. However, a very important **hydrophobic force** is caused by a pushing of nonpolar surfaces out of the hydrogen-bonded water network, where they would otherwise physically interfere with the highly favorable

TABLE 2–1 Covalent and Noncovalent Chemical Bonds

Bond type		Length (nm)	Strength (kJ/mole**)	
			In vacuum	In water
Covalent		0.10	377 (90)	377 (90)
Noncovalent	Ionic*	0.25	335 (80)	12.6 (3)
	Hydrogen	0.17	16.7 (4)	4.2 (1)
	van der Waals attraction (per atom)	0.35	0.4 (0.1)	0.4 (0.1)

*An ionic bond is an electrostatic attraction between two fully charged atoms.
**Values in parentheses are kcal/mole. 1 kJ = 0.239 kcal and 1 kcal = 4.18 kJ.

interactions between water molecules. Bringing any two nonpolar surfaces together reduces their contact with water, and in this sense, the force is nonspecific. Nevertheless, we shall see in Chapter 3 that hydrophobic forces are central to the proper folding of protein molecules.

The properties of the four types of noncovalent attractions are presented in Panel 2–3 (pp. 98–99). Although each individual noncovalent attraction would be much too weak to be effective in the face of thermal motions, the energies of these noncovalent attractions can sum to create a strong force between two separate molecules. Thus, it is an entire set of noncovalent attractions that enables the complementary surfaces of two macromolecules to hold the two macromolecules together (Figure 2–4).

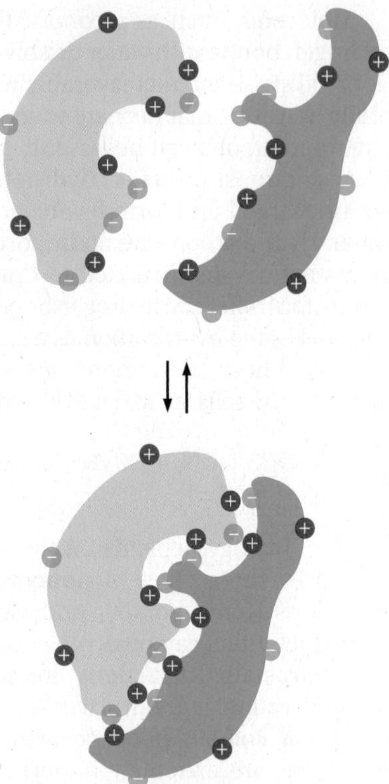

Figure 2–4 **Schematic indicating how two macromolecules with complementary surfaces can bind tightly to one another through noncovalent interactions.** Noncovalent chemical bonds have less than 1/20 the strength of a covalent bond. They are able to produce tight binding only when many of them are formed simultaneously. Although only electrostatic attractions are illustrated here, in reality all four noncovalent forces often contribute to holding two macromolecules together (Movie 2.1).

Some Polar Molecules Form Acids and Bases in Water

One of the simplest kinds of chemical reaction, and one that has a considerable significance for cells, takes place when a molecule containing a highly polar covalent bond between a hydrogen and another atom dissolves in water. The hydrogen atom in such a molecule has given up its electron almost entirely to the companion atom, and so exists as an almost naked positively charged hydrogen nucleus; in other words, a **proton (H^+)**. When this polar molecule becomes surrounded by water molecules, the proton will be attracted to the partial negative charge on the O atom of an adjacent water molecule. The proton can easily dissociate from its original partner and associate instead with the oxygen atom of the water molecule, generating a **hydronium ion (H_3O^+)** (Figure 2–5A). The reverse reaction also takes place very readily, so in an aqueous solution protons are constantly flitting to and fro between one molecule and another.

Substances that release protons when they dissolve in water, thus forming H_3O^+, are termed **acids**. The higher the concentration of H_3O^+, the more acidic the solution. H_3O^+ is present even in pure water, at a concentration of 10^{-7} M, as a result of the movement of protons from one water molecule to another (Figure 2–5B). By convention, the H_3O^+ concentration is usually referred to as the H^+ concentration, even though most protons in an aqueous solution are present as H_3O^+. As explained in Panel 2–2, to avoid the use of unwieldy numbers the concentration of H_3O^+ is expressed using a logarithmic scale called the **pH scale**. Pure water has a pH of 7.0 and is said to be neutral; that is, neither acidic ($pH < 7$) nor basic ($pH > 7$).

Acids are characterized as being strong or weak, depending on how readily they give up their protons to water. Strong acids, such as hydrochloric acid (HCl), easily lose their protons. Acetic acid, on the other hand, is a weak acid because it holds on to its proton more tightly when dissolved in water. Many of the acids important in the cell—such as molecules containing a carboxyl (COOH) group—are weak acids.

Because the proton of a hydronium ion can be passed readily to many types of molecules in cells, altering their character, the concentration of H_3O^+ inside a cell (the acidity) must be closely regulated. Acids—especially weak acids—will give

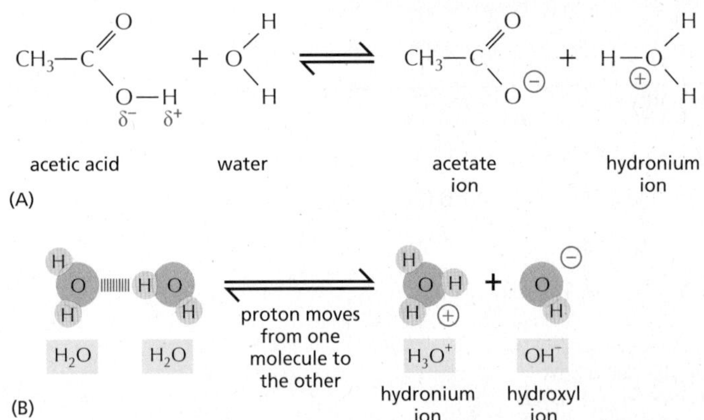

Figure 2–5 **How protons readily move in aqueous solutions.** (A) The reaction that takes place when a molecule of acetic acid dissolves in water. At pH 7, nearly all of the acetic acid is present as acetate ion. (B) Water molecules continually exchange protons with each other to form hydronium and hydroxyl ions. These ions in turn rapidly recombine to form water molecules.

up their protons more readily if the concentration of H_3O^+ in solution is low and will tend to receive them back if the concentration in solution is high.

The opposite of an acid is a **base**. Any molecule capable of accepting a proton from a water molecule is called a base. Sodium hydroxide (NaOH) is basic (the term *alkaline* is also used) because it dissociates readily in aqueous solution to form Na^+ ions and OH^- ions. Because of this property, NaOH is called a strong base. More important in living cells, however, are the weak bases—those that have a weak tendency to reversibly accept a proton from water. Many biologically important molecules contain an amino (NH_2) group. This group is a weak base that can generate OH^- by taking a proton from water: $-NH_2 + H_2O \rightarrow -NH_3^+ + OH^-$ (see Panel 2–2, pp. 96–97).

Because an OH^- ion combines with an H_3O^+ ion to form two water molecules, any increase in the OH^- concentration forces a decrease in the concentration of H_3O^+, and vice versa. Thus the product of the two values, $[OH^-] \times [H_3O^+]$, is always 10^{-14} $(\text{moles/liter})^2$. A pure solution of water contains an equal concentration (10^{-7} M) of both ions, rendering it neutral. The interior of a cell is also kept close to neutrality by the presence of **buffers**: weak acids and bases that can release or take up protons near pH 7, keeping the environment of the cell relatively constant under a variety of conditions.

A Cell Is Formed from Carbon Compounds

Having briefly reviewed the ways that atoms combine into molecules and how these molecules behave in an aqueous environment, we now examine the main classes of small molecules found in cells. We shall see that a few categories of molecules, formed from a handful of different elements, give rise to all the extraordinary richness of form and behavior shown by living things.

If we disregard water and inorganic ions such as potassium, nearly all the molecules in a cell are based on carbon. Carbon is outstanding among all the elements in its ability to form large molecules; silicon is a poor second. Because carbon is small and has four electrons and four vacancies in its outermost shell, a carbon atom can form four covalent bonds with other atoms. Most important, one carbon atom can join to other carbon atoms through highly stable covalent C–C bonds to form chains and rings and hence generate large and complex molecules with no obvious upper limit to their size. The carbon compounds made by cells are called *organic molecules*. In contrast, all other molecules, including water, are said to be *inorganic*.

Certain combinations of atoms, such as the methyl ($-CH_3$), hydroxyl ($-OH$), carboxyl ($-COOH$), carbonyl ($-C{=}O$), phosphate ($-PO_3^{2-}$), sulfhydryl ($-SH$), and amino ($-NH_2$) groups, occur repeatedly in the molecules made by cells. Each such **chemical group** has distinct chemical and physical properties that influence the behavior of the molecule in which the group occurs. The most common chemical groups and some of their properties are summarized in Panel 2–1 (pp. 94–95).

Cells Contain Four Major Families of Small Organic Molecules

The small organic molecules of the cell are carbon-based compounds that have masses in the range of 100–1000 daltons and contain up to 30 or so carbon atoms. They are usually found free in solution and have many different fates. Some are used as *monomer* subunits to construct the giant polymeric *macromolecules* that make up most of the mass of the cell—proteins, nucleic acids, and large polysaccharides. Others act as energy sources and are broken down and transformed into other small molecules in a maze of intracellular metabolic pathways. Many small molecules have more than one role in the cell; for example, acting both as a potential subunit for a macromolecule and as an energy source. Small organic molecules account for only about one-tenth of the total mass of organic matter in a cell, but they are very diverse. Nearly 4000 different kinds of small organic molecules have been detected in the well-studied bacterium, *Escherichia coli*.

All organic molecules are synthesized from and are broken down into the same set of simple compounds. As a consequence, the compounds in a cell are

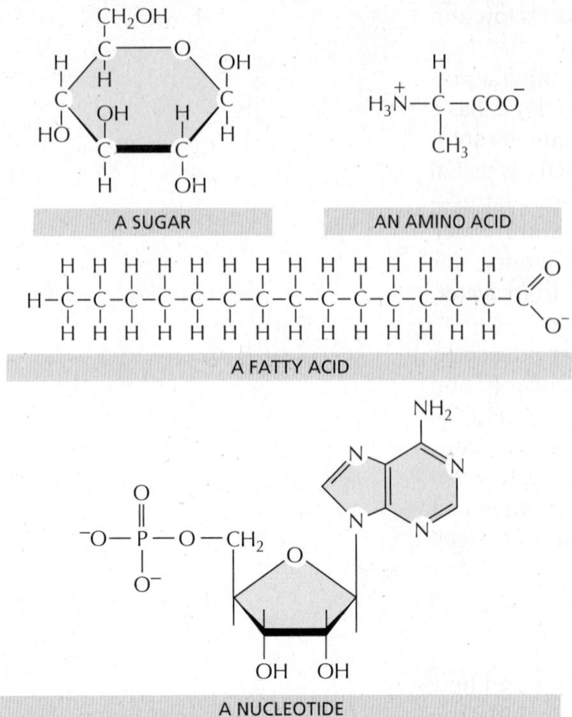

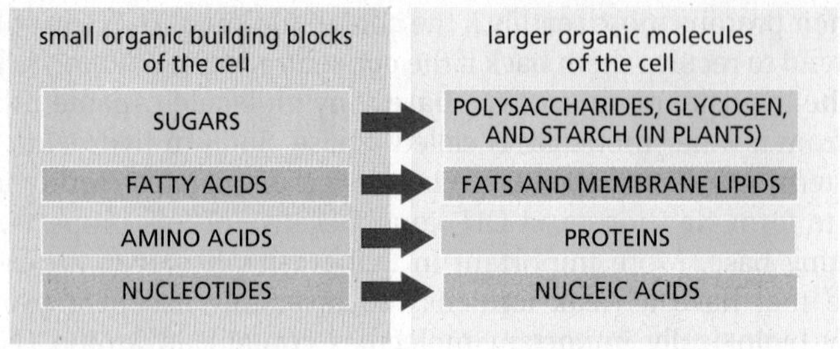

Figure 2–6 The four main families of small organic molecules in cells. These small molecules form the monomeric building blocks, or subunits, for most of the macromolecules and other assemblies of the cell. Some, such as the sugars and the fatty acids, are also energy sources. Their structures are outlined here and shown in more detail in the Panels at the end of this chapter and in Chapter 3.

chemically related and most can be classified into a few distinct families. Broadly speaking, cells contain four major families of small organic molecules: the *sugars*, the *fatty acids*, the *nucleotides*, and the *amino acids* (**Figure 2–6**). Although many compounds present in cells do not fit into these categories, these four families of small organic molecules, together with the macromolecules made by linking them into long chains, account for a large fraction of the cell mass.

Amino acids and the proteins that they form will be the subject of Chapter 3. A summary of the structures and properties of the remaining three families—sugars, fatty acids, and nucleotides—is presented in **Panels 2–4**, **2–5**, and **2–6**, respectively (see pp. 100–105).

The Chemistry of Cells Is Dominated by Macromolecules with Remarkable Properties

By weight, **macromolecules** are the most abundant carbon-containing molecules in a living cell (**Figure 2–7**). They are the principal components from which a cell is constructed, and they also determine the most distinctive properties of living organisms. The macromolecules in cells are polymers that are constructed

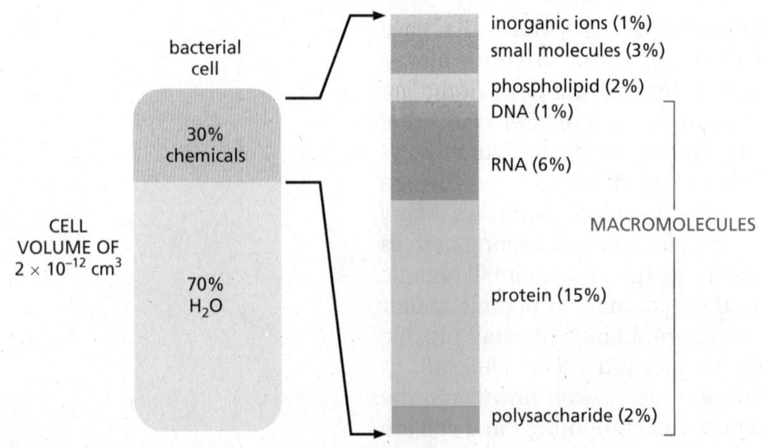

Figure 2–7 The distribution of molecules in cells. The approximate composition of a bacterial cell is shown by weight. The composition of an animal cell is similar, even though its volume is roughly 1000 times greater. Note that macromolecules dominate. The major inorganic ions include Na^+, K^+, Mg^{2+}, Ca^{2+}, and Cl^-.

by covalently linking small organic molecules (called *monomers*) into long chains (**Figure 2–8**). They have remarkable properties that could not have been predicted from their simple constituents.

Proteins are abundant and spectacularly versatile, performing thousands of distinct functions in cells. Many proteins serve as *enzymes*, the catalysts that facilitate the many covalent bond-making and bond-breaking reactions that the cell needs. Enzymes catalyze all of the reactions in which cells extract energy from food molecules, for example. Other proteins are used to build structural components, such as tubulin, a protein that self-assembles to make the cell's long microtubules, or histones, proteins that compact the DNA in chromosomes. Many proteins serve as signaling devices, producing networks that control cell functions. Yet other proteins act as molecular motors to produce force and movement, as for myosin in muscle. We shall describe the remarkable chemistry that underlies these diverse roles throughout this book.

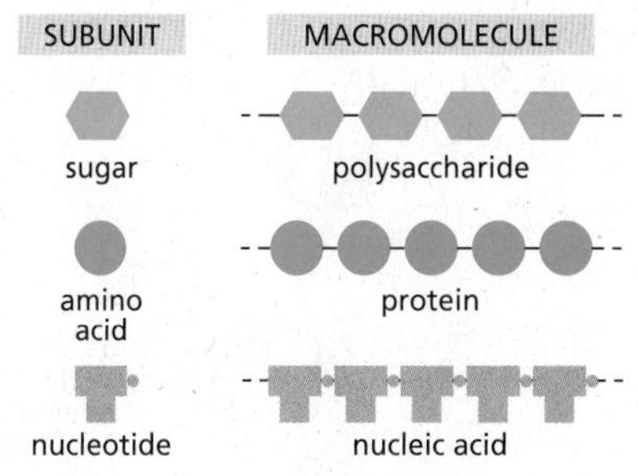

Figure 2–8 **Three families of macromolecules.** Each is a polymer formed from small molecules (called monomers) linked together by covalent bonds. There are two types of nucleic acid: RNA and DNA.

Although the chemical reactions that add subunits to each polymer are different in detail for proteins, nucleic acids, and polysaccharides, they share important features. Each polymer grows by the addition of a monomer onto the end of a growing chain in a *condensation reaction*, in which one molecule of water is lost with each subunit added (**Figure 2–9**). The stepwise polymerization of monomers into a long chain is a simple way to manufacture a large, complex molecule, because the subunits are added by the same reaction performed over and over again by the same set of enzymes. Apart from some of the polysaccharides, most macromolecules are made from a limited set of monomers that are slightly different from one another; for example, the 20 different amino acids from which proteins are made. It is critical to life that the polymer chain is not assembled at random from these subunits; instead, the subunits are added in a precise order, or *sequence*. The elaborate mechanisms that allow enzymes to accomplish this task are described in detail in Chapters 5 and 6.

Noncovalent Bonds Specify Both the Precise Shape of a Macromolecule and Its Binding to Other Molecules

Most of the covalent bonds in a macromolecule allow rotation of the atoms they join, giving the polymer chain great flexibility. In principle, this allows a macromolecule to adopt an almost unlimited number of shapes, or *conformations*, as random thermal energy causes the polymer chain to writhe and rotate. However, the shapes of most biological macromolecules are highly constrained because of the many weak *noncovalent bonds* that form between different parts of the same molecule. If these noncovalent bonds are formed in sufficient numbers, the polymer chain can strongly prefer one particular conformation, determined by the linear sequence of monomers in its chain. Most protein molecules and many of the small RNA molecules found in cells fold tightly into a highly preferred conformation in this way (**Figure 2–10**).

The four types of noncovalent interactions important in biological molecules were presented earlier (see also Panel 2–3, pp. 98–99). In addition to folding biological macromolecules into unique shapes, they can also add up to create a strong attraction between two different molecules (see Figure 2–4). This form of

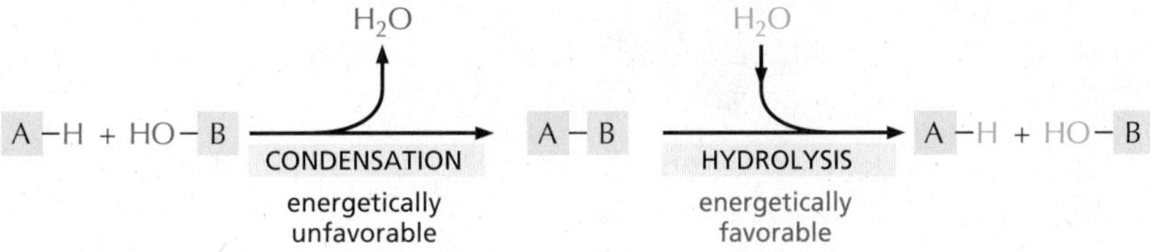

Figure 2–9 **Condensation and hydrolysis as opposite reactions.** The macromolecules of the cell are polymers that are formed from subunits (or monomers) by a condensation reaction, and they are broken down by hydrolysis. The condensation reactions are all energetically unfavorable; thus, polymer formation requires an energy input, as will be described in the text.

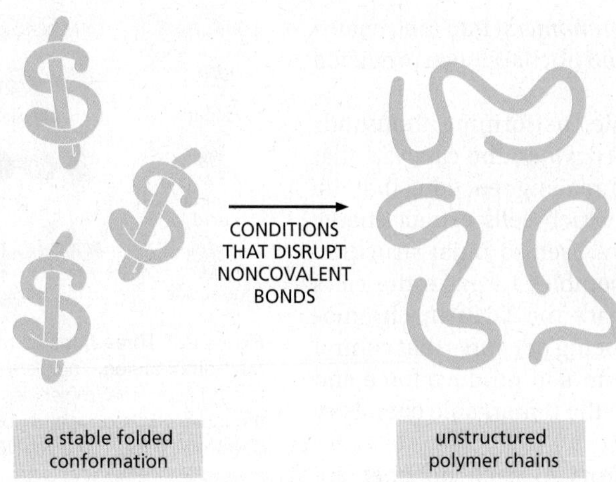

Figure 2–10 **Proteins and RNA molecules are folded into a particularly stable three-dimensional shape, or conformation.** If the noncovalent bonds maintaining the stable conformation are disrupted, the molecule becomes a flexible chain that loses its biological activity.

molecular interaction provides for great specificity, inasmuch as the close multipoint contacts required for strong binding make it possible for a macromolecule to select out—through binding—just one of the many thousands of types of molecules present inside a cell. Moreover, because the strength of the binding depends on the number of noncovalent bonds that are formed, interactions of almost any affinity are possible—allowing rapid dissociation where appropriate.

As we discuss next, binding of this type underlies all biological catalysis, making it possible for proteins to function as enzymes. In addition, noncovalent interactions allow macromolecules to be used to build larger structures, thereby forming intricate machines with multiple moving parts that perform such complex tasks as DNA replication and protein synthesis (**Figure 2–11**).

Summary

Living organisms are autonomous, self-propagating chemical systems. They are formed from a distinctive and restricted set of small carbon-based molecules that are essentially the same for every living species. Each of these small molecules is composed of a set of atoms linked to each other in a precise configuration through covalent bonds. The main categories are sugars, fatty acids, amino acids, and nucleotides.

Most of the dry mass of a cell consists of macromolecules that have been produced as linear polymers of amino acids (proteins) or nucleotides (DNA and RNA), covalently linked to each other in an exact order. Most of the protein molecules and many of the RNAs fold into a particular conformation that is determined

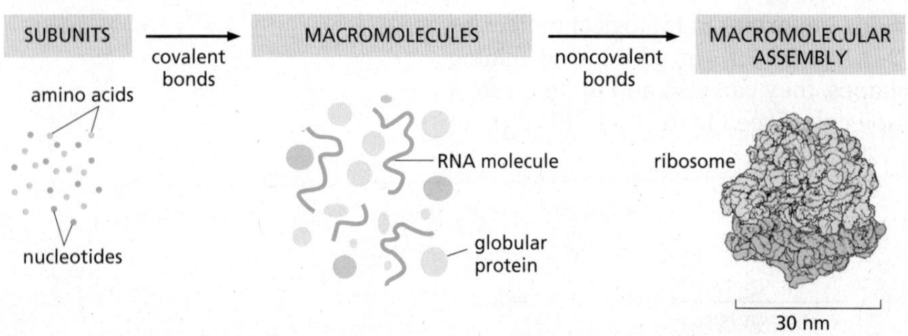

Figure 2–11 **Small molecules are covalently linked to form macromolecules, which in turn can assemble through noncovalent interactions to form large complexes.** Small molecules, proteins, and a ribosome are drawn approximately to scale. Ribosomes are a central part of the machinery that the cell uses to make proteins: each ribosome is formed as a complex of about 90 macromolecules (protein and RNA molecules).

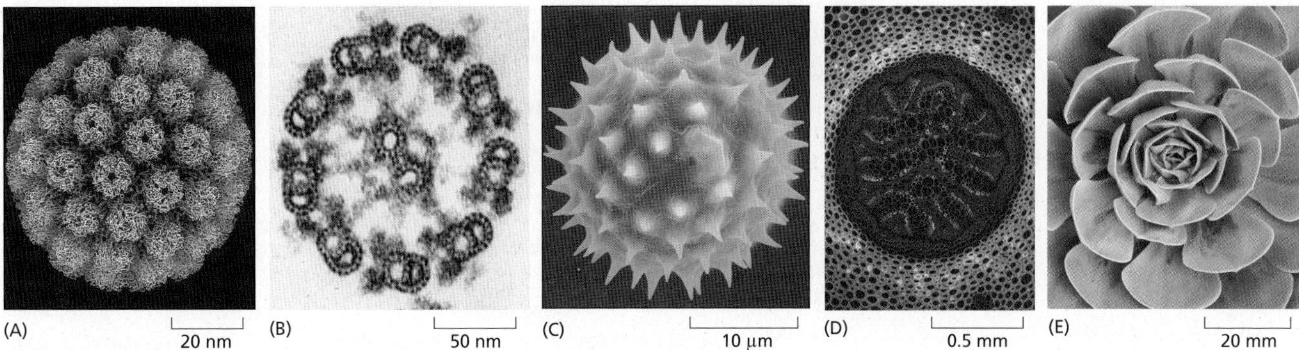

Figure 2–12 **Biological structures are highly ordered.** Well-defined, ornate, and beautiful spatial patterns can be found at every level of organization in living organisms. In order of increasing size: (A) protein molecules in the coat of a virus (a parasite that, although not technically alive, contains the same types of molecules as those found in living cells); (B) the regular array of microtubules seen in a cross section of a sperm tail; (C) surface contours of a pollen grain (a single cell); (D) cross section of a fern stem, showing the patterned arrangement of cells; and (E) a spiral arrangement of leaves in a succulent plant. (A, courtesy of Robert Grant, Stéphane Crainic, and James M. Hogle; B, courtesy of Lewis Tilney; C, courtesy of Colin MacFarlane and Chris Jeffree; D, courtesy of Jim Haseloff; E, courtesy of Aron van de Selenib.)

by their sequence of subunits. This folding process creates unique surfaces, and it depends on a large set of weak attractions produced by noncovalent forces between atoms. These forces are of four types: electrostatic attractions, hydrogen bonds, van der Waals attractions, and an attraction between nonpolar groups caused by their hydrophobic expulsion from water. The same set of weak forces governs the specific binding of a macromolecule to both small molecules and other macromolecules, producing the myriad associations between biological molecules that generate the structure and the chemistry of a cell.

CATALYSIS AND THE USE OF ENERGY BY CELLS

One property of living things above all makes them seem almost miraculously different from nonliving matter: they create and maintain order in a universe that is tending always to greater disorder (**Figure 2–12**). To create this order, the cells in a living organism must perform a never-ending stream of chemical reactions. In some of these reactions, small organic molecules—amino acids, sugars, nucleotides, and lipids—are being taken apart or modified to supply the many other small molecules that the cell requires. In other reactions, small molecules are being used to construct an enormously diverse range of proteins, nucleic acids, and other macromolecules that endow living systems with all of their most distinctive properties. Each cell can be viewed as a tiny chemical factory, performing many millions of reactions every second.

Cell Metabolism Is Organized by Enzymes

The chemical reactions that a cell carries out would normally proceed at an appreciable rate only at much higher temperatures than those existing inside cells. For this reason, each reaction requires a specific boost in chemical reactivity. This requirement is crucial, because it allows the cell to control its chemistry. The control is exerted through specialized biological *catalysts.* These are almost always proteins called *enzymes*, although RNA catalysts also exist, called *ribozymes*. Each enzyme accelerates, or *catalyzes*, just one of the many possible kinds of reactions that a particular molecule might undergo. Enzyme-catalyzed reactions are connected in series, so that the product of one reaction becomes the starting material, or *substrate*, for the next (**Figure 2–13**). Long linear reaction pathways are in turn linked to one another, forming a maze of interconnected reactions that enable the cell to survive, grow, and reproduce.

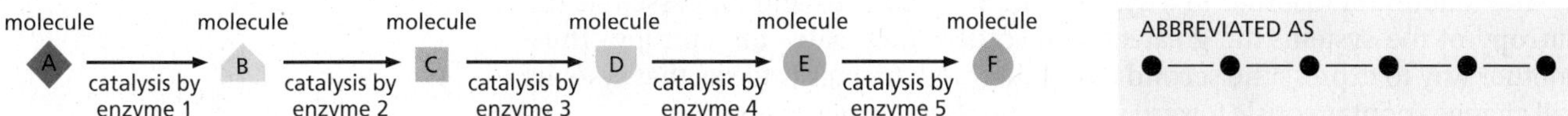

Figure 2–13 **How a set of enzyme-catalyzed reactions generates a metabolic pathway.** Each enzyme catalyzes a particular chemical reaction, leaving the enzyme unchanged. In this example, a set of enzymes acting in series converts molecule A to molecule F, forming a metabolic pathway. (For a diagram of many of the reactions in a human cell, abbreviated as shown, see Figure 2–62.)

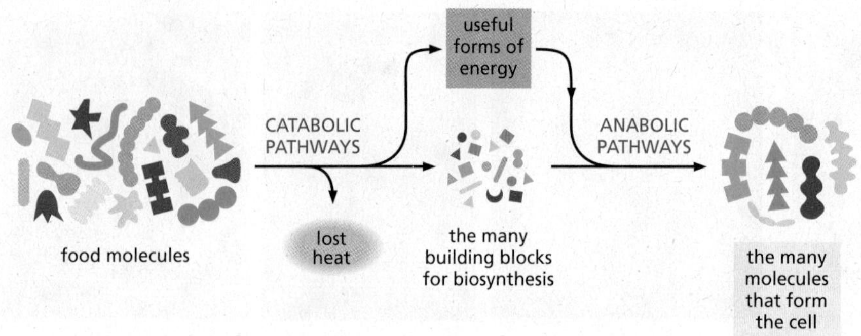

Figure 2–14 **Schematic representation of the relationship between catabolic and anabolic pathways in metabolism.** Catabolism produces both the building blocks and the energy required for biosynthesis. As indicated, a major portion of the energy stored in the chemical bonds of food molecules is dissipated as heat. As also suggested in this diagram, the mass of food required by any organism that derives all of its energy from catabolism is much greater than the mass of the molecules that it can produce by anabolism.

Two opposing streams of chemical reactions occur in cells: (1) the *catabolic* pathways break down foodstuffs into smaller molecules, thereby generating both a useful form of energy for the cell and some of the small molecules that the cell needs as building blocks; (2) the *anabolic*, or *biosynthetic*, pathways use the small molecules plus the energy harnessed by catabolism to drive the synthesis of the many other molecules that form the cell. Together these two sets of reactions constitute the **metabolism** of the cell (Figure 2–14).

The many details of cell metabolism form the traditional subject of *biochemistry*. Most of these details need not concern us here. But the general principles by which cells obtain energy from their environment and use it to create order are central to cell biology. We begin with a discussion of why a constant input of energy is needed to sustain all living things.

Biological Order Is Made Possible by the Release of Heat Energy from Cells

The universal tendency of things to become disordered is a fundamental law of physics—the *second law of thermodynamics*—which states that in the universe or in any isolated system (a collection of matter that is completely isolated from the rest of the universe), the degree of disorder always increases. This law has such profound implications for life that we will restate it in several ways.

For example, we can present the second law in terms of probability by stating that systems will change spontaneously toward those arrangements that have the greatest probability. If we consider a box of 100 coins all lying heads-up, a series of accidents that disturbs the box will tend to move the arrangement toward a mixture of 50 heads and 50 tails. The reason is simple: there is a huge number of possible arrangements of the individual coins in the mixture that can achieve the 50–50 result, but only one possible arrangement that keeps all of the coins oriented heads-up. Because the 50–50 mixture is therefore the most probable, we say that it is more "disordered." For the same reason, it is a common experience that one's living space will become increasingly disordered without intentional effort: the movement toward disorder is a *spontaneous process*, requiring a periodic effort to reverse it (Figure 2–15).

The amount of disorder in a system can be quantified and expressed as the **entropy** of the system: the greater the disorder, the greater the entropy. Thus, another way to express the second law of thermodynamics is to say that systems will change spontaneously toward arrangements with greater entropy.

Living cells—by surviving, growing, and forming complex organisms—are generating order and thus might appear to defy the second law of thermodynamics. How is this possible? The answer is that a cell is not an isolated system: it takes in energy from its environment in the form of the chemical bonds in food

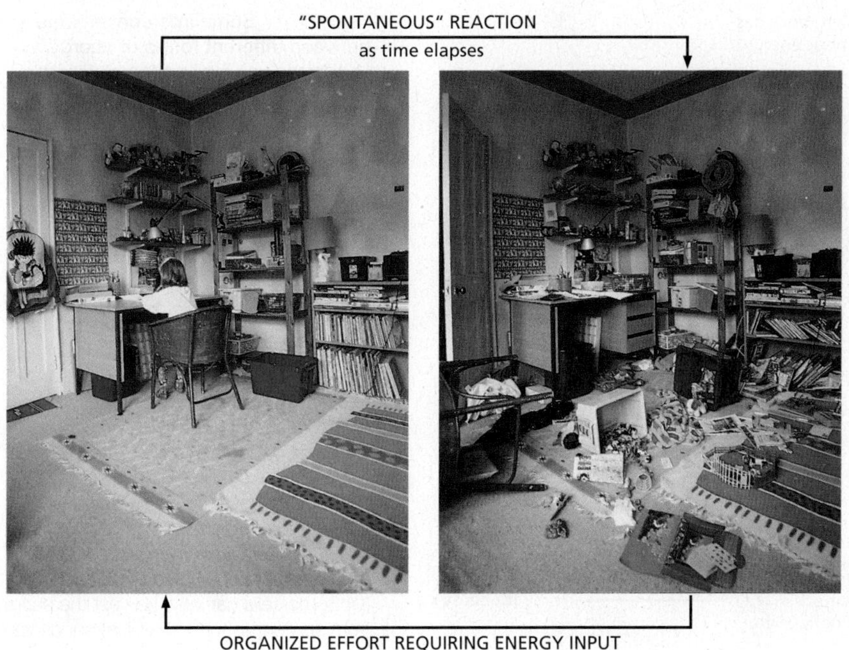

Figure 2–15 **An everyday illustration of the spontaneous drive toward disorder.** Reversing this tendency toward disorder requires an intentional effort and an input of energy: it is not spontaneous. In fact, from the second law of thermodynamics, we can be certain that the human intervention required will release enough heat to the environment to more than compensate for the reordering of the items in this room.

or as photons from the Sun (or even, as in some chemosynthetic bacteria, from inorganic molecules alone). It then uses this energy to generate order within itself. Critically, during the chemical reactions that generate order, the cell converts part of the energy it uses into heat. The heat is discharged into the cell's environment and disorders the surroundings. As a result, the total entropy—that of the cell plus its surroundings—increases, as demanded by the second law of thermodynamics.

To understand the principles governing these energy conversions, think of a cell surrounded by a sea of matter representing the rest of the universe. As the cell lives and grows, it creates internal order. But it constantly releases heat energy as it synthesizes molecules and assembles them into cell structures. Heat is energy in its most disordered form—the random jostling of molecules. When the cell releases heat to the sea, it increases the intensity of molecular motions there (thermal motion)—thereby increasing the randomness, or disorder, of the sea. The second law of thermodynamics is satisfied because the increase in the amount of order inside the cell is always more than compensated for by an even greater decrease in order (increase in entropy) in the surrounding sea of matter (Figure 2–16).

Where does the heat that the cell releases come from? Here we encounter another important law of thermodynamics. The *first law of thermodynamics*

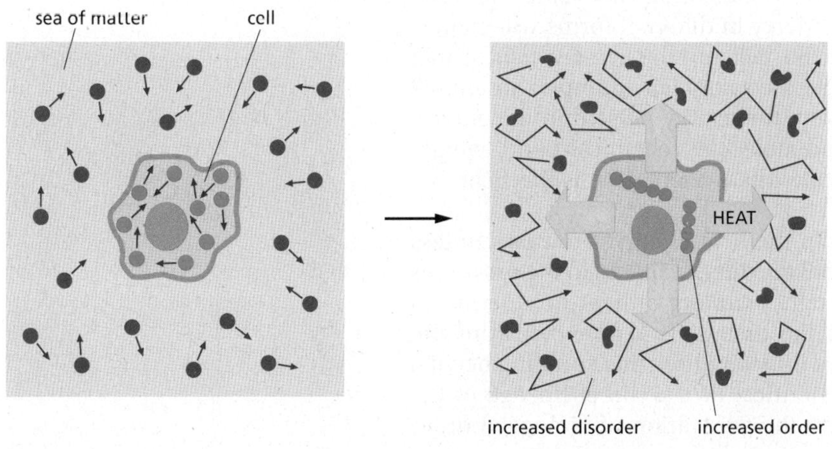

Figure 2–16 **A simple thermodynamic analysis of a living cell.** In the diagram on the left, the molecules of both the cell and the rest of the universe (the sea of matter) are depicted in a relatively disordered state. In the diagram on the right, the cell has taken in energy from food molecules and released heat through reactions that order the molecules the cell contains. The heat released increases the disorder in the environment around the cell (depicted by *jagged arrows* and distorted molecules, indicating increased molecular motions caused by heat). As a result, if enough heat is released, the second law of thermodynamics—which states that the amount of disorder in the universe must always increase—is satisfied as the cell grows and divides. For a detailed discussion, see Panel 2–7 (pp. 106–107).

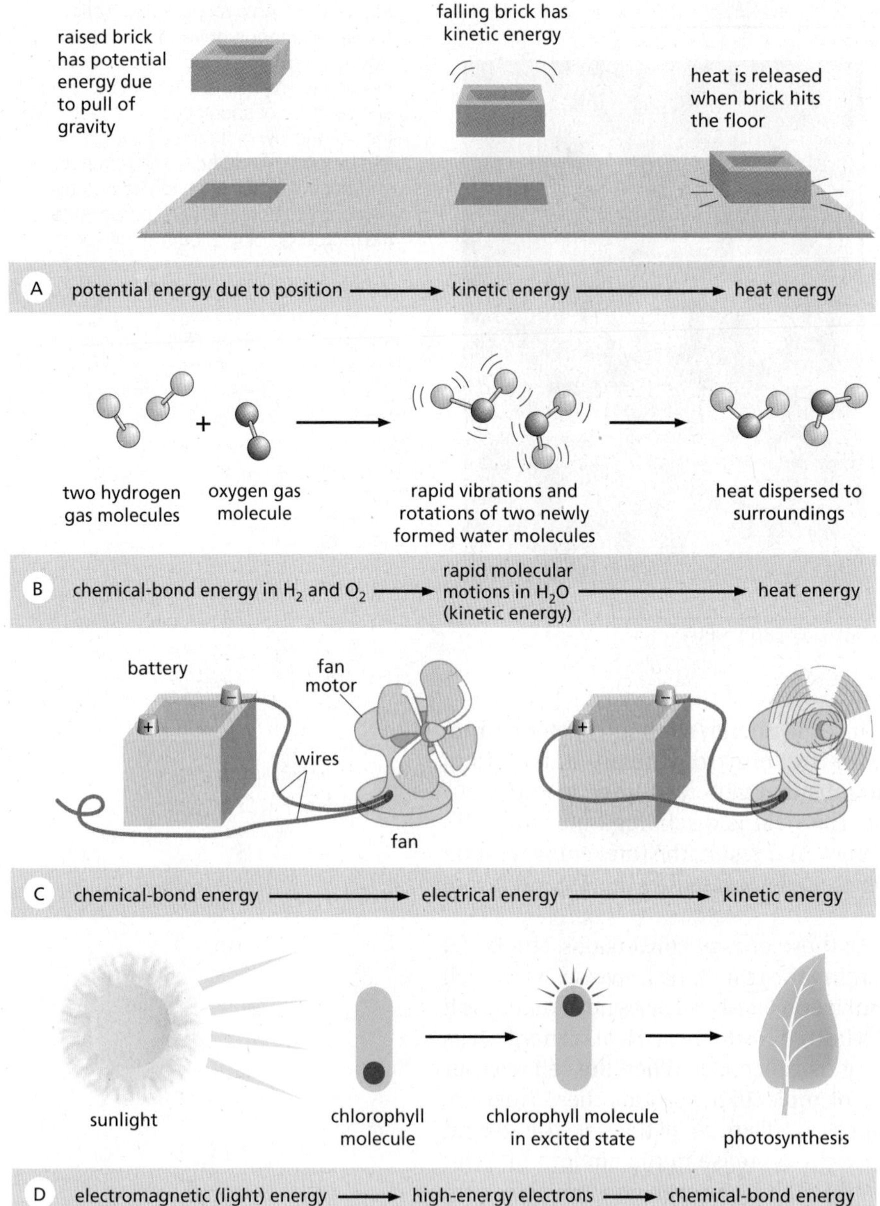

Figure 2–17 **Some interconversions between different forms of energy.** (A) We can use the height and weight of the brick to predict exactly how much heat will be released when it hits the floor. (B) The large amount of chemical-bond energy released when water (H_2O) is formed from H_2 and O_2 is initially converted to very rapid thermal motions in the two new H_2O molecules; however, collisions with other H_2O molecules almost instantaneously spread this kinetic energy evenly throughout the surroundings (heat transfer), making the new H_2O molecules indistinguishable from all the rest. (C) Through coupled processes to be described later, cells can convert chemical-bond energy into kinetic energy to drive, for example, molecular motor proteins; however, this occurs without the intermediate conversion of chemical energy to electrical energy that a man-made appliance such as this fan requires. (D) Some cells can also harvest the energy from sunlight to form chemical bonds via photosynthesis.

states that energy can be converted from one form to another, but that it cannot be created or destroyed. Figure 2–17 illustrates some interconversions between different forms of energy. The amount of energy in different forms will change as a result of the chemical reactions inside the cell, but the first law tells us that the total amount of energy must always be the same. For example, an animal cell takes in foodstuffs and converts some of the energy present in the chemical bonds between the atoms of these food molecules (chemical-bond energy) into the random thermal motion of molecules (heat energy)—it is this heat that keeps our bodies warm.

The cell cannot derive any benefit from the heat energy it releases unless the heat-generating reactions inside the cell are directly linked to the processes that generate molecular order. It is the tight *coupling* of heat production to an increase in order that distinguishes the metabolism of a cell from the wasteful burning of fuel in a fire. Later, we illustrate how this coupling occurs. For now, it is sufficient to recognize this critical fact: a direct linkage of the "controlled burning" of food molecules to the generation of biological order

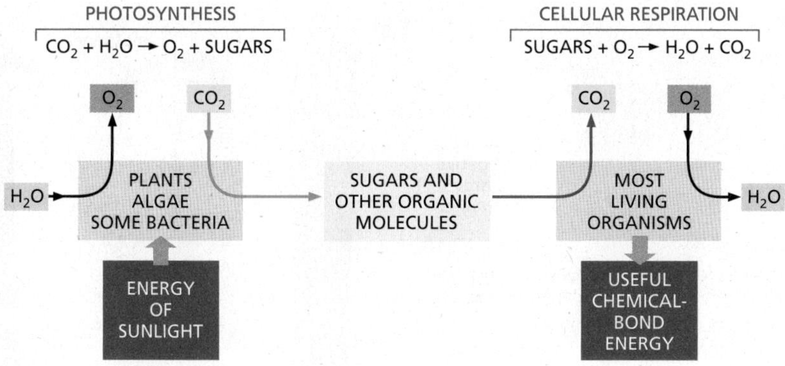

Figure 2–18 Photosynthesis and respiration as complementary processes in the living world. Photosynthesis converts the electromagnetic energy in sunlight into chemical-bond energy in sugars and other organic molecules. Plants, algae, and cyanobacteria obtain the carbon atoms that they need for this purpose from atmospheric CO_2 and the hydrogen from water, producing sugars and releasing O_2 gas as a by-product. The organic molecules produced by photosynthesis in turn serve as food for other organisms. Many of these organisms carry out aerobic respiration, a process that uses O_2 to form CO_2 from the same carbon atoms that had been taken up as CO_2 and converted into sugars by photosynthesis. In the process, the organisms that respire obtain the chemical-bond energy that they need to survive.

The first cells on Earth are thought to have been capable of neither photosynthesis nor respiration (discussed in Chapter 14). However, photosynthesis must have preceded respiration on Earth, because there is strong evidence that billions of years of photosynthesis were required before O_2 had been released in sufficient quantity to create an atmosphere rich in this gas. (Earth's atmosphere currently contains 21% O_2.)

is required for cells to create and maintain an island of order in a universe tending toward chaos.

Cells Obtain Energy by the Oxidation of Organic Molecules

All animal and plant cells are powered by energy stored in the chemical bonds of organic molecules, whether they are sugars that a plant has photosynthesized as food for itself or the mixture of large and small molecules that an animal has eaten. Organisms must extract this energy in usable form to live, grow, and reproduce. In both plants and animals, energy is extracted from food molecules by a process of gradual oxidation, or controlled burning.

Earth's atmosphere contains a great deal of oxygen, and in the presence of oxygen the most energetically stable form of carbon is CO_2 and that of hydrogen is H_2O. A cell is therefore able to obtain energy from sugars or other organic molecules by allowing their carbon and hydrogen atoms to combine with oxygen to produce CO_2 and H_2O, respectively—a process called **aerobic respiration**.

Photosynthesis (discussed in detail in Chapter 14) and respiration are complementary processes (**Figure 2–18**). This means that the transactions between plants and animals are not all one way. Plants, animals, and microorganisms have existed together on this planet for so long that many of them have become an essential part of the others' environments. The oxygen released by photosynthesis is consumed in the combustion of organic molecules during aerobic respiration. And some of the CO_2 molecules that are fixed today into organic molecules by photosynthesis in a green leaf were yesterday released into the atmosphere by the respiration of an animal—or by the respiration of a fungus or bacterium decomposing dead organic matter. We therefore see that carbon utilization forms a huge cycle that involves the *biosphere* (all of the living organisms on Earth) as a whole (**Figure 2–19**). Similarly, atoms of nitrogen, phosphorus, and sulfur move between the living and nonliving worlds in cycles that involve plants, algae, animals, fungi, and bacteria.

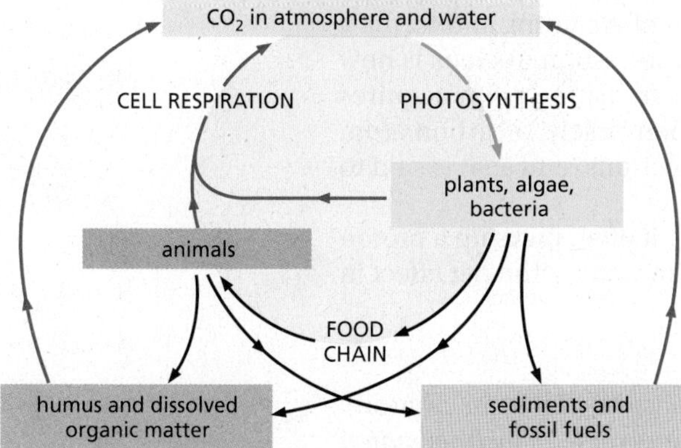

Figure 2–19 How carbon atoms cycle through the biosphere. Individual carbon atoms are incorporated into organic molecules of the living world by the photosynthetic activity of bacteria, algae, and plants. They pass to animals, microorganisms, and organic material in soil and oceans in cyclic paths. CO_2 is restored to the atmosphere when organic molecules are oxidized by cells during respiration or burned by humans as fossil fuels. In this diagram, the *green arrow* denotes an uptake of CO_2, whereas a *red arrow* indicates CO_2 release.

As indicated in Chapter 1, the total biomass on Earth is estimated to contain ~550 gigatons (10^{15} grams) of carbon (Gt C), of which 450 Gt C are plants, 70 are bacteria, 7 are archaea, and 2 are animals (see Figure 1–14).

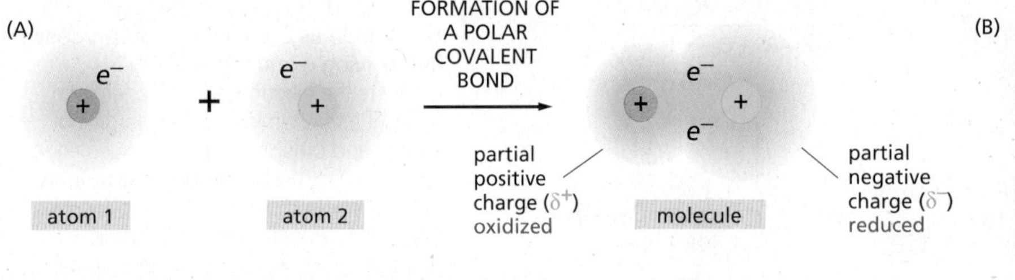

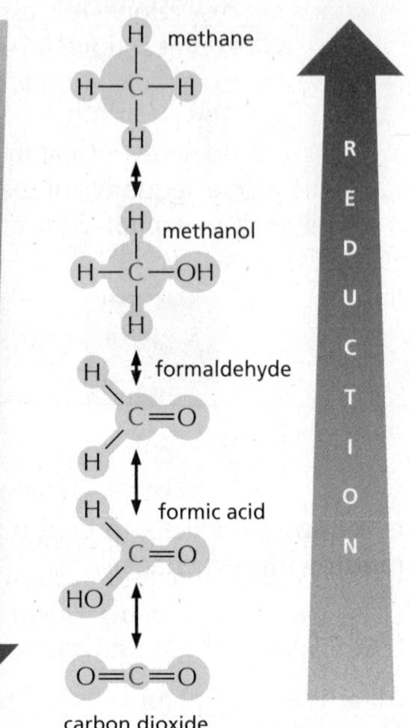

Figure 2–20 Oxidation and reduction. (A) When two atoms form a *polar* covalent bond, the atom ending up with a greater share of electrons is said to be reduced, while the other atom acquires a lesser share of electrons and is said to be oxidized. The reduced atom has acquired a partial negative charge (δ^-) as the positive charge on the atomic nucleus is now more than equaled by the total charge of the electrons surrounding it, and conversely, the oxidized atom has acquired a partial positive charge (δ^+). (B) The single carbon atom of methane can be converted to that of carbon dioxide by the successive replacement of its covalently bonded hydrogen atoms with oxygen atoms. With each step, electrons are shifted away from the carbon (as indicated by the changes in the amount of *blue* shading), and the carbon atom becomes progressively more oxidized. Each of these steps is energetically favorable under the conditions present inside a cell.

Oxidation and Reduction Involve Electron Transfers

The cell does not oxidize organic molecules in one step, as occurs when organic material is burned in a fire. Through the use of enzyme catalysts, metabolism takes these molecules through a large number of reactions that only rarely involve the direct addition of oxygen. Before we consider some of these reactions and their purpose, we discuss what is meant by the process of oxidation.

Oxidation refers to more than the addition of oxygen atoms; the term applies more generally to any reaction in which electrons are transferred from one atom to another. Oxidation in this sense refers to the removal of electrons, and **reduction**—the converse of oxidation—means the addition of electrons. Thus, Fe^{2+} is oxidized if it loses an electron to become Fe^{3+}, and a chlorine atom is reduced if it gains an electron to become Cl^-. Because the number of electrons is conserved (no loss or gain) in a chemical reaction, oxidation and reduction always occur simultaneously; that is, if one molecule gains an electron in a reaction (reduction), a second molecule loses the electron (oxidation). When a sugar molecule is oxidized to CO_2 and H_2O, for example, the O_2 molecules involved in forming H_2O gain electrons and thus are said to have been reduced.

Why is a "gain" of electrons referred to as a "reduction"? The term arose before anything was known about the movement of electrons. Originally, reduction reactions involved a liberation of oxygen—for example, when metals are extracted from ores by heating—which caused the samples to become lighter; in other words, "reduced" in mass.

It is important to recognize that the terms "oxidation" and "reduction" apply even when there is only a partial shift of electrons between atoms linked by a covalent bond (Figure 2–20). When a carbon atom becomes covalently bonded to an atom with a strong affinity for electrons, such as oxygen, chlorine, or sulfur, for example, it gives up more than its equal share of electrons and forms a *polar* covalent bond. Because the positive charge of the carbon nucleus is now somewhat greater than the negative charge of its electrons, the atom acquires a partial positive charge and is said to be oxidized. Conversely, a carbon atom in a C–H linkage has slightly more than its share of electrons, and so it is said to be reduced.

When a molecule in a cell picks up an electron (e^-), it often picks up a proton (H^+) at the same time (protons being freely available in water). The net effect in this case is to add a hydrogen atom to the molecule.

$$A + e^- + H^+ \rightarrow AH$$

Even though a proton plus an electron is involved (instead of just an electron), such *hydrogenation* reactions are reductions, and the reverse *dehydrogenation* reactions are oxidations. It is especially easy to tell whether an organic molecule is being oxidized or reduced: reduction is occurring if its number of C–H bonds increases, whereas oxidation is occurring if its number of C–H bonds decreases (see Figure 2–20B).

Cells use enzymes to catalyze the oxidation of organic molecules in small steps, through a sequence of reactions that allows useful energy to be harvested. We now need to explain how enzymes work and some of the constraints under which they operate.

Enzymes Lower the Activation-Energy Barriers That Block Chemical Reactions

Consider the reaction

$$\text{paper} + O_2 \rightarrow \text{smoke} + \text{ashes} + \text{heat} + CO_2 + H_2O$$

Once ignited, the paper burns readily, releasing to the atmosphere both energy as heat and water and carbon dioxide as gases. The reaction is irreversible, as the smoke and ashes never spontaneously retrieve these entities from the heated atmosphere and reconstitute themselves into paper. When the paper burns, its chemical energy is dissipated as heat—not lost from the universe, as energy can never be created or destroyed, but irretrievably dispersed in the chaotic random thermal motions of molecules. At the same time, the atoms and molecules of the paper become dispersed and disordered. In the language of thermodynamics, there has been a loss of *free energy*; that is, of energy that can be harnessed to do work or drive chemical reactions. This loss reflects a reduction of orderliness in the way the energy and molecules were stored in the paper.

We shall discuss free energy in more detail shortly, but the general principle is clear enough intuitively: chemical reactions proceed spontaneously only in the direction that leads to a loss of free energy. In other words, the spontaneous direction for any reaction is the direction that goes "downhill," where a "downhill" reaction is one that is *energetically favorable*.

Although the most energetically favorable form of carbon under ordinary conditions is CO_2, and that of hydrogen is H_2O, a living organism does not disappear in a puff of smoke, and the paper book in your hands does not burst into flames. This is because the molecules both in the living organism and in the book are in a relatively stable state, and they cannot be changed to a state of lower energy without an input of energy; in other words, a molecule requires **activation energy**—a kick over an energy barrier—before it can undergo a chemical reaction that leaves it in a more stable state (**Figure 2–21**). In the case of a burning book, the activation energy can be provided by the heat of a lighted match. For the molecules in the watery solution inside a cell, the kick is delivered by an unusually energetic random collision with surrounding molecules.

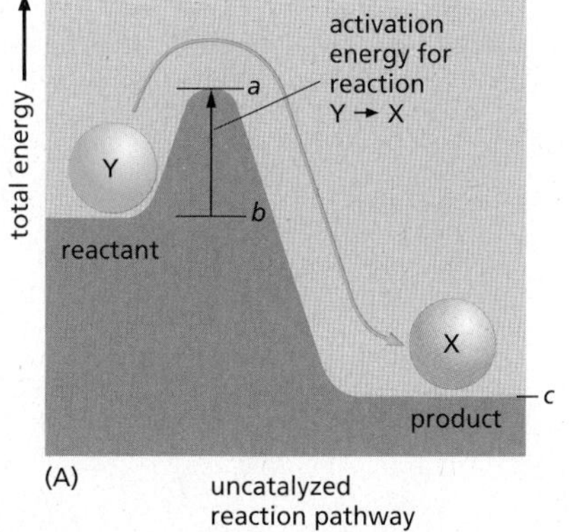

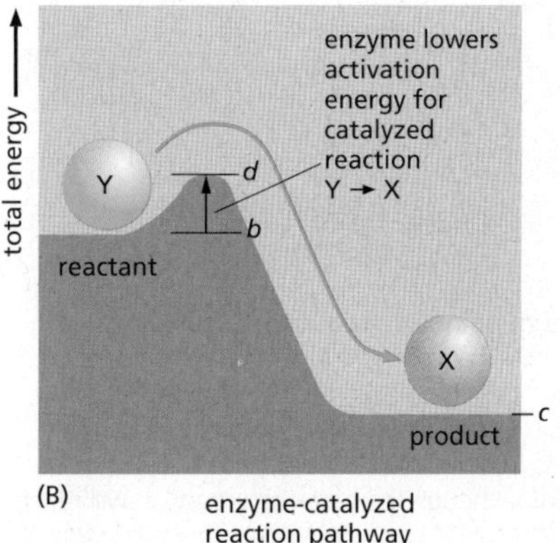

Figure 2–21 **The important principle of activation energy.** (A) Compound Y (a reactant) is in a relatively stable state, and energy is required to convert it to compound X (a product), even though X is at a lower overall energy level than Y. This conversion will not take place, therefore, unless compound Y can acquire enough activation energy (*energy a minus energy b*) from its surroundings to undergo the reaction that converts it into compound X. This energy may be provided by means of an unusually energetic collision with other molecules. For the reverse reaction, X → Y, the activation energy will be much larger (*energy a minus energy c*); this reaction will therefore occur much more rarely. Activation energies are always positive; note, however, that the total energy change for the energetically favorable reaction Y → X is *energy c minus energy b*, a negative number. (B) Energy barriers for specific reactions can be lowered by catalysts, as indicated by the line marked *d*. Enzymes are particularly effective catalysts because they greatly reduce the activation energy for the reactions they perform.

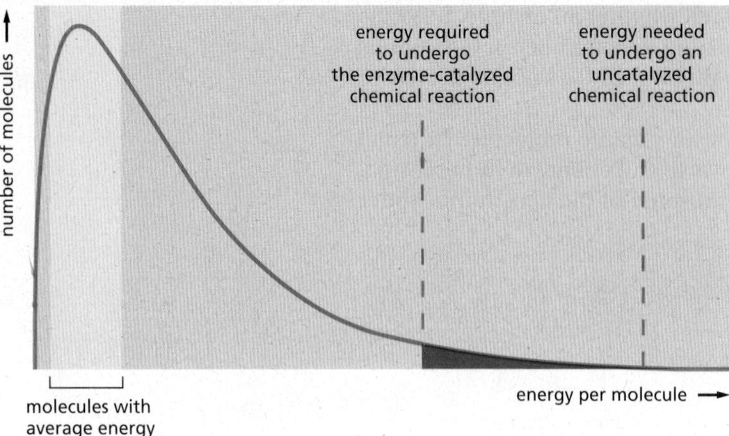

Figure 2–22 **Lowering the activation energy greatly increases the probability of a reaction.** At any given instant, a population of identical substrate molecules will have a range of energies, distributed as shown on the graph. The varying energies come from collisions with surrounding molecules, which make the substrate molecules jiggle, vibrate, and spin. For a molecule to undergo a chemical reaction, the energy of the molecule must exceed the activation-energy barrier for that reaction *(dashed lines)*. For most biological reactions, this almost never happens without enzyme catalysis. Even with enzyme catalysis, the substrate molecules must experience a particularly energetic collision to react *(red shaded area)*. Raising the temperature will also increase the number of molecules with sufficient energy to overcome the activation energy needed for a reaction; but in marked contrast to enzyme catalysis, this effect is nonselective, speeding up all reactions (**Movie 2.2**).

The chemistry in a living cell is tightly controlled, because the kick over energy barriers is greatly aided by a specialized class of proteins—the **enzymes**. Each enzyme binds tightly to one or more molecules, called **substrates**, and holds them in a way that greatly reduces the activation energy of a particular chemical reaction that the bound substrates can undergo. A substance that can lower the activation energy of a reaction is termed a **catalyst**; catalysts increase the rate of chemical reactions because they allow a much larger proportion of the random collisions with surrounding molecules to kick the substrates over the energy barrier, as illustrated in **Figure 2–22**. Enzymes are among the most effective catalysts known: some are capable of speeding up reactions by factors of 10^{14} or more. Enzymes thereby allow reactions that would not otherwise occur to proceed rapidly at normal temperatures.

Enzymes Can Drive Substrate Molecules Along Specific Reaction Pathways

An enzyme cannot change the equilibrium point for a reaction. The reason is simple: when an enzyme (or any catalyst) lowers the activation energy for the reaction $Y \rightarrow X$, of necessity it also lowers the activation energy for the reverse reaction $X \rightarrow Y$ by exactly the same amount (see Figure 2–21). The forward and backward reactions will therefore be accelerated by the same factor by an enzyme, and the equilibrium point for the reaction will be unchanged (**Figure 2–23**).

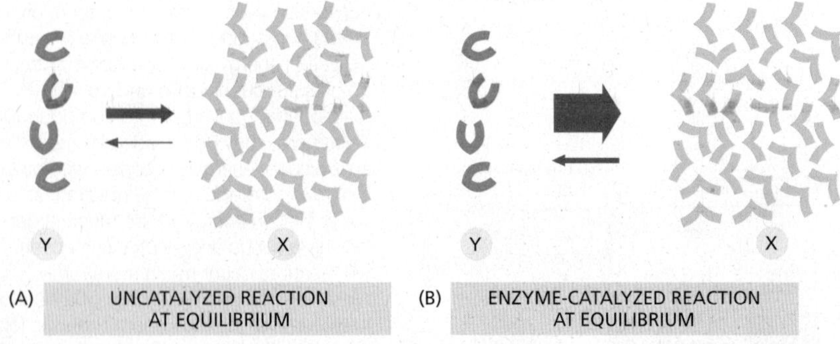

Figure 2–23 **Enzymes cannot change the equilibrium point for reactions.** Enzymes, like all catalysts, speed up the forward and backward rates of a reaction by the same factor. Therefore, for both the catalyzed and the uncatalyzed reactions shown here, the number of molecules undergoing the transition $X \rightarrow Y$ is equal to the number of molecules undergoing the transition $Y \rightarrow X$ when the ratio of X molecules to Y molecules is 7 to 1. In other words, the two reactions will eventually reach exactly the same equilibrium point, although the catalyzed reaction will reach equilibrium much faster.

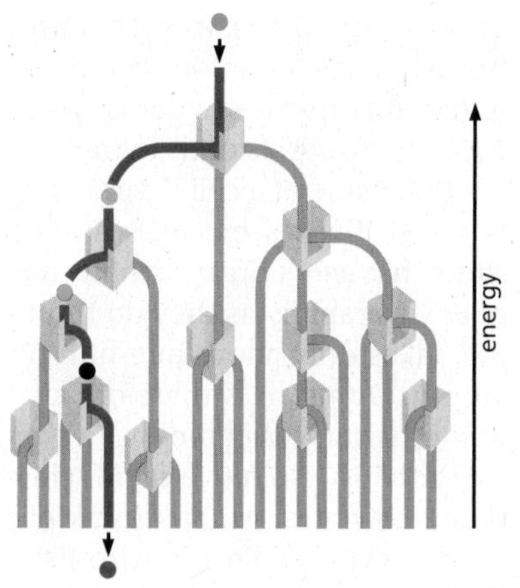

Figure 2–24 **Directing substrate molecules through a specific reaction pathway by enzyme catalysis.** A substrate molecule in a cell *(green ball)* is converted into a different molecule *(red ball)* by means of a series of enzyme-catalyzed reactions. As indicated *(yellow box),* several reactions are energetically favorable at each step, but only one is catalyzed by each enzyme. Sets of enzymes thereby determine the exact reaction pathway that is followed by each molecule inside the cell.

Thus no matter how much an enzyme speeds up a reaction, it cannot change its direction.

Despite the above limitation, enzymes steer all of the reactions in cells through specific reaction paths. This is because enzymes are both highly selective and very precise, usually catalyzing only one particular reaction. In other words, each enzyme selectively lowers the activation energy of only one of the several possible chemical reactions that its bound substrate molecules could undergo. In this way, sets of enzymes can direct each of the many different molecules in a cell along a particular reaction pathway (**Figure 2–24**).

The success of living organisms is attributable to a cell's ability to make enzymes of many types, each with precisely specified properties. Each enzyme has a unique shape containing an *active site,* a pocket or groove in the enzyme into which only particular substrates will fit (**Figure 2–25**). Like all other catalysts, enzyme molecules themselves remain unchanged after participating in a reaction and therefore can function over and over again. In Chapter 3, we discuss further how enzymes work.

How Enzymes Find Their Substrates: The Enormous Rapidity of Molecular Motions

An enzyme will often catalyze the reaction of thousands of substrate molecules every second. This means the enzyme must be able to bind a new substrate molecule in a fraction of a millisecond. But both enzymes and their substrates are present in relatively small numbers in a cell. How do they find each other so fast? Rapid binding is possible because the motions caused by heat energy are enormously fast at the molecular level. These molecular motions can be classified broadly into three kinds: (1) the movement of a molecule from one place to another (*translational motion*), (2) the rapid back-and-forth movement of covalently linked atoms with respect to one another (*vibrations*), and (3) *rotations.* All three of these motions help to bring the surfaces of interacting molecules together.

The rates of molecular motions can be measured by a variety of spectroscopic techniques. A large globular protein is constantly tumbling, rotating about its axis approximately a million times per second. Molecules are also in constant translational motion, which causes them to explore the space inside the cell very efficiently by wandering through it—a process called **diffusion**. In this way, every molecule in a cell collides with a huge number of other molecules each second. As the molecules in a liquid collide and bounce off one another, an individual molecule moves first one way and then another, its path constituting a *random*

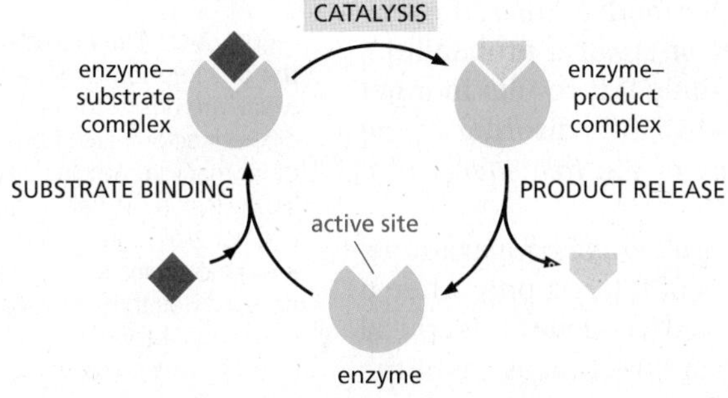

Figure 2–25 **How enzymes work.** Each enzyme has an active site to which one or more *substrate* molecules bind, forming an enzyme–substrate complex. A reaction occurs at the active site, producing an enzyme–product complex. The *product* is then released, allowing the enzyme to bind further substrate molecules.

walk (Figure 2–26). In such a walk, the average net distance that each molecule travels (as the "crow flies") from its starting point is proportional to the square root of the time involved; that is, if it takes a molecule 1 second on average to travel 1 μm, it takes 4 seconds to travel 2 μm, 100 seconds to travel 10 μm, and so on.

The inside of a cell is very crowded (Figure 2–27). Nevertheless, experiments in which fluorescent dyes and other labeled molecules are injected into cells show that small organic molecules diffuse through the watery gel of the cytosol nearly as rapidly as they do through water. A small organic molecule, for example, takes only about one-fifth of a second on average to diffuse a distance of 10 μm. Diffusion is therefore an efficient way for small molecules to move the limited distances in the cell (a typical animal cell is 15 μm in diameter).

Proteins also move rapidly in cells. But because enzymes move more slowly than substrates, we can think of them as sitting still. The rate of encounter of each enzyme molecule with its substrate will depend on the concentration of the substrate molecule. For example, some abundant substrates are present at a concentration of 0.5 mM. As pure water is 55.5 M, there is only about one such substrate molecule in the cell for every 10^5 water molecules. Nevertheless, the active site on an enzyme molecule that binds this substrate will be bombarded by about 500,000 random collisions with the substrate molecule per second. (For a substrate concentration tenfold lower, the number of collisions drops to 50,000 per second, and so on.) A random collision between the active site of an enzyme and the matching surface of its substrate molecule often leads immediately to the formation of an enzyme–substrate complex. A reaction in which a covalent bond is broken or formed can then occur extremely rapidly. When one appreciates how quickly molecules move and react, the observed rates of enzymatic catalysis do not seem so amazing.

Two molecules that are held together by noncovalent bonds can also dissociate. The multiple weak noncovalent bonds that they form with each other will persist until random thermal motion causes the two molecules to separate. In general, the stronger the binding of the enzyme and substrate, the slower their rate of dissociation. In contrast, whenever two colliding molecules have poorly matching surfaces, they form few noncovalent bonds and the total energy of association will be negligible compared with that of thermal motion. In this case, the two molecules dissociate as rapidly as they come together, preventing incorrect and unwanted associations between mismatched molecules, such as between an enzyme and the wrong substrate.

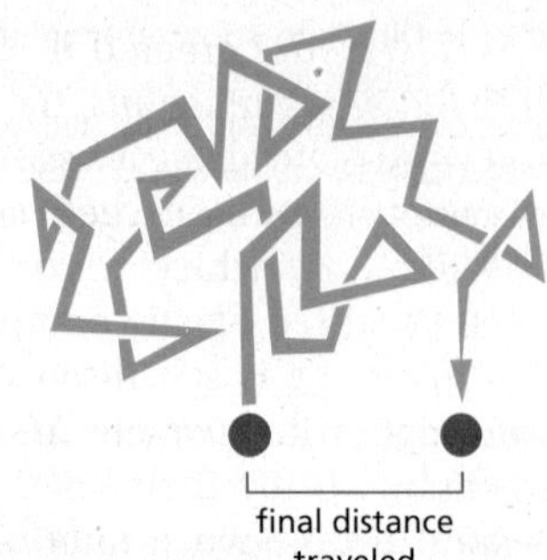

Figure 2–26 **A random walk.** Molecules in solution move in a random fashion as a result of the continual buffeting they receive in collisions with other molecules. This movement allows small molecules to diffuse rapidly from one part of the cell to another, as described in the text (Movie 2.3).

The Free-Energy Change for a Reaction, ΔG, Determines Whether It Can Occur Spontaneously

Although enzymes speed up reactions, they cannot by themselves force energetically unfavorable reactions to occur. In terms of a water analogy, enzymes by themselves cannot make water run uphill. Cells, however, must do just that in order to grow and divide: they must build highly ordered and energy-rich molecules from small and simple ones. We shall see that this is done through enzymes that directly *couple* energetically favorable reactions, which release energy and produce heat, to energetically unfavorable reactions, which produce biological order.

What do cell biologists mean by the term "energetically favorable," and how can this be quantified? According to the second law of thermodynamics, the universe tends toward maximum disorder (largest *entropy* or greatest probability). Thus, a chemical reaction can proceed spontaneously only if it results in a net increase in the disorder of the universe (see Figure 2–16). This disorder of the universe can be expressed most conveniently in terms of the *free energy* of a system, a concept we touched on earlier.

Free energy, *G*, is an expression of the *energy available to do work*; for example, the work of driving chemical reactions. The value of *G* is of interest only when a system undergoes a *change*. The **free-energy change**, denoted **Δ*G*** (delta *G*), is critical because, as explained in Panel 2–7 (pp. 106–107), it is a direct measure of the

Figure 2–27 **The crowded structure of the cell interior.** Only the macromolecules, which are drawn to scale and displayed in different colors, are shown. Enzymes and other macromolecules diffuse relatively slowly inside the cell, in part because they interact with many other macromolecules; small molecules, by contrast, diffuse nearly as rapidly as they do in water (Movie 2.4). (From S.R. McGuffee and A.H. Elcock, *PLoS Comput. Biol.* 6(3):e1000694, 2010.)

amount of disorder created in the universe when a reaction takes place. *Energetically favorable reactions*, by definition, are those that decrease free energy; in other words, they have a *negative* ΔG and disorder the universe (**Figure 2–28**).

An example of an energetically favorable reaction on a macroscopic scale is the "reaction" by which a compressed spring relaxes to an expanded state, releasing its stored elastic energy as heat to its surroundings; an example on a microscopic scale is salt dissolving in water. Conversely, *energetically unfavorable reactions* with a *positive* ΔG—such as the joining of two amino acids to form a peptide bond—by themselves create order in the universe. Therefore, these reactions can take place only if they are coupled to a second reaction with a *negative* ΔG so large that the ΔG of the overall process is negative (**Figure 2–29**).

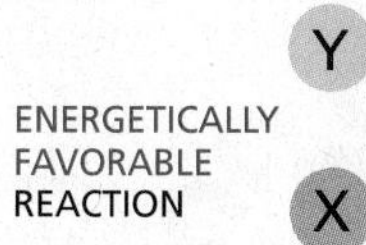
Y
X
ENERGETICALLY FAVORABLE REACTION
The free energy of Y is greater than the free energy of X. Therefore $\Delta G < 0$, and the disorder of the universe increases during the reaction Y → X.
this reaction can occur spontaneously

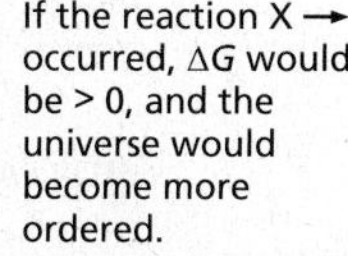
Y
X
ENERGETICALLY UNFAVORABLE REACTION
If the reaction X → Y occurred, ΔG would be > 0, and the universe would become more ordered.
this reaction can occur only if it is coupled to a second, energetically favorable reaction

Figure 2–28 **The distinction between energetically favorable and energetically unfavorable reactions, and how they relate to ΔG.**

The Concentration of Reactants Influences the Free-Energy Change and a Reaction's Direction

As we have just described, a reaction Y ↔ X will go in the direction Y → X when the associated free-energy change, ΔG, is negative, just as a tensed spring left to itself will relax and lose its stored energy to its surroundings as heat. For a chemical reaction, however, ΔG depends not only on the energy stored in each individual molecule but also on the concentrations of the molecules in the reaction mixture. Remember that ΔG reflects the degree to which a reaction creates a more disordered—in other words, a more probable—state of the universe. Recalling our coin analogy, it is very likely that a coin will flip from a head to a tail orientation if a jiggling box contains 90 heads and 10 tails, but this is a less probable event if the box has 10 heads and 90 tails.

The same is true for a chemical reaction. For a reversible reaction Y ↔ X, a large excess of Y over X will tend to drive the reaction in the direction Y → X. Therefore, as the ratio of Y to X increases, the ΔG becomes more negative for the transition Y → X (and more positive for the transition X → Y).

The amount of concentration difference that is needed to compensate for a given decrease in chemical-bond energy (and accompanying heat release) is not intuitively obvious. In the late nineteenth century, the relationship was determined through a thermodynamic analysis that makes it possible to separate the concentration-dependent and the concentration-independent parts of the free-energy change, as we describe next.

The Standard Free-Energy Change, $\Delta G°$, Makes It Possible to Compare the Energetics of Different Reactions

Because ΔG depends on the concentrations of the molecules in the reaction mixture at any given time, it is not a particularly useful value for comparing the relative energies of different types of reactions. To consider reactions on a comparable basis, we need to turn to the **standard free-energy change** of a reaction, $\boldsymbol{\Delta G°}$. The $\Delta G°$ is the change in free energy under a standard condition, defined as that where the concentrations of all the reactants are set to the same fixed value of 1 mole/liter. Defined in this way, $\Delta G°$ depends only on the intrinsic characters of the reacting molecules.

For the simple reaction Y → X at 37°C, $\Delta G°$ is related to ΔG as follows:

$$\Delta G = \Delta G° + RT \ln \frac{[X]}{[Y]}$$

where ΔG is in kilojoules per mole, [Y] and [X] denote the concentrations of Y and X in moles/liter, ln is the *natural logarithm*, and RT is the product of the gas constant, R, and the absolute temperature, T. At 37°C, $RT = 2.58$ kJ mole^{-1}. (A mole is 6×10^{23} molecules of a substance.)

A large body of thermodynamic data has been collected that has made it possible to determine the standard free-energy change, $\Delta G°$, for the important metabolic reactions of a cell. Given these $\Delta G°$ values, combined with additional information about metabolite concentrations and reaction pathways, it is possible to quantitatively predict the course of most biological reactions.

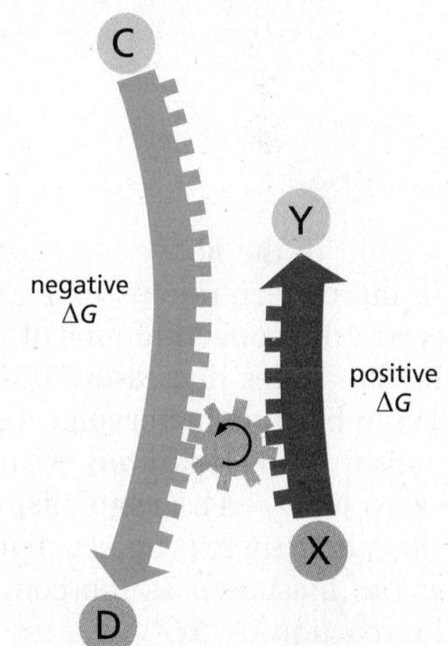
C
D
Y
X
negative ΔG
positive ΔG
the energetically unfavorable reaction X → Y is driven by the energetically favorable reaction C → D, because the net free-energy change for the pair of **coupled** reactions is less than zero

Figure 2–29 **How reaction coupling is used to drive energetically unfavorable reactions.**

FOR THE ENERGETICALLY FAVORABLE REACTION Y → X,

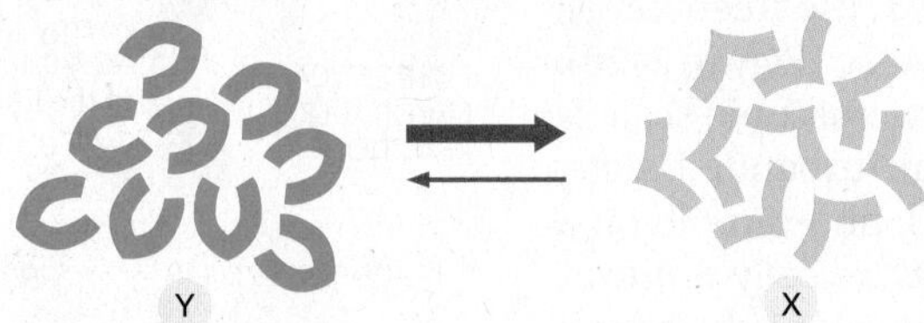

if X and Y are at equal concentrations, [Y] = [X], the formation of X is energetically favored. In other words, the ΔG of Y → X is negative and the ΔG of X → Y is positive. Nevertheless because of thermal bombardments, there will always be some X converting to Y.

FOR EACH INDIVIDUAL MOLECULE,

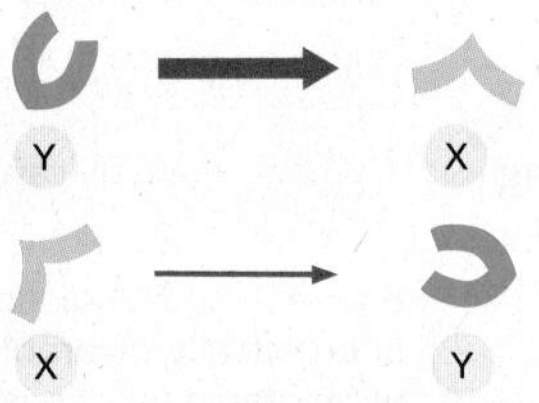

conversion of Y to X will occur often.

Conversion of X to Y will occur less often than the transition Y → X, because it requires a more energetic collision.

Therefore, if one starts with an equal mixture, the ratio of X to Y molecules will increase

EVENTUALLY, there will be a large enough excess of X over Y to just compensate for the slow rate of X → Y, such that the number of X molecules being converted to Y molecules each second is exactly equal to the number of Y molecules being converted to X molecules each second. At this point, the reaction will be at equilibrium.

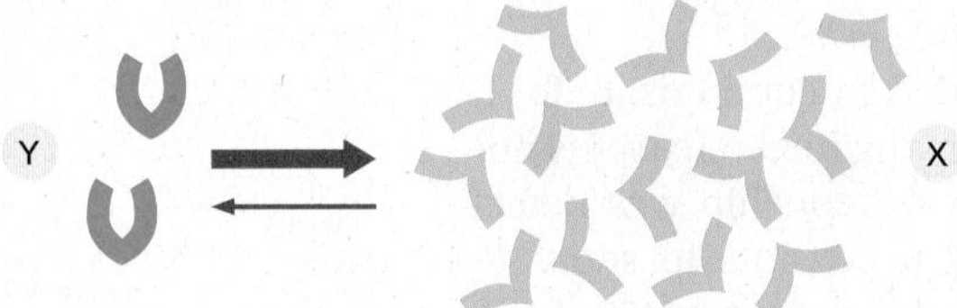

AT EQUILIBRIUM, there is no net change in the ratio of Y to X, and the ΔG for both forward and backward reactions is zero.

Figure 2–30 Chemical equilibrium. When a reaction reaches equilibrium, the forward and backward fluxes of reacting molecules are equal and opposite. In these diagrams, the widths of the *red* arrows indicate the relative rates at which an *individual* molecule reacts.

The Equilibrium Constant and $\Delta G°$ Are Readily Derived from Each Other

Inspection of the above equation reveals that the ΔG equals the value of $\Delta G°$ when the concentrations of Y and X are equal. But as any favorable reaction proceeds, the concentrations of the products will increase as the concentration of the substrates decreases. This change in relative concentrations will cause [X]/[Y] to become increasingly large, making the initially favorable ΔG less and less negative (the logarithm of a number x is positive for $x > 1$, negative for $x < 1$, and zero for $x = 1$). Eventually, when $\Delta G = 0$, a chemical **equilibrium** will be attained; here there is no net change in free energy to drive the reaction in either direction, inasmuch as the concentration effect just balances the push given to the reaction by $\Delta G°$. As a result, the ratio of product to substrate reaches a constant value at chemical equilibrium (**Figure 2–30**).

We can define the **equilibrium constant, *K*,** for the reaction Y → X as

$$K = \frac{[X]}{[Y]}$$

where [X] is the concentration of the product and [Y] is the concentration of the reactant at equilibrium. Remembering that $\Delta G = \Delta G° + RT \ln [X]/[Y]$, and that $\Delta G = 0$ at equilibrium, we see that

$$\Delta G° = -RT \ln \frac{[X]}{[Y]} = -RT \ln K$$

At 37°C, where $RT = 2.58$, the equilibrium equation is therefore:

$$\Delta G° = -2.58 \ln K$$

Converting this equation from the natural logarithm (ln) to the more commonly used base 10 logarithm (log), we get

$$\Delta G° = -5.94 \log K$$

The above equation reveals how the equilibrium ratio of X to Y (expressed as the equilibrium constant, K) depends on the intrinsic character of the molecules (as expressed in the value of $\Delta G°$ in kilojoules per mole). The more energetically favorable a reaction, the more product will accumulate when the reaction proceeds to equilibrium. More precisely, for every 5.94 kJ/mole difference in free energy at 37°C, the equilibrium constant changes by a factor of 10 (Table 2–2). Note that this amount of free-energy difference is roughly equivalent to the free energy available from a single hydrogen bond.

More generally, for a reaction that has multiple reactants and products, such as A + B → C + D,

$$K = \frac{[C][D]}{[A][B]}$$

The concentrations of the two reactants and the two products are multiplied because the rate of the forward reaction depends on the collision of A and B and the rate of the backward reaction depends on the collision of C and D. Thus, at 37°C,

$$\Delta G° = -5.94 \log \frac{[C][D]}{[A][B]}$$

where $\Delta G°$ is in kilojoules per mole, and [A], [B], [C], and [D] denote the concentrations of the reactants and products in moles/liter.

TABLE 2–2 Relationship Between the Standard Free-Energy Change, $\Delta G°$, and the Equilibrium Constant

Equilibrium constant $\frac{[X]}{[Y]} = K$	Free energy of X minus free energy of Y [kJ/mole (kcal/mole)]
10^5	–29.7 (–7.1)
10^4	–23.8 (–5.7)
10^3	–17.8 (–4.3)
10^2	–11.9 (–2.8)
10^1	–5.9 (–1.4)
1	0 (0)
10^{-1}	5.9 (1.4)
10^{-2}	11.9 (2.8)
10^{-3}	17.8 (4.3)
10^{-4}	23.8 (5.7)
10^{-5}	29.7 (7.1)

Values of the equilibrium constant were calculated for the simple chemical reaction Y ↔ X using the equation given in the text. The $\Delta G°$ given here is in kilojoules per mole at 37°C, with kilocalories per mole in parentheses. One kilojoule (kJ) is equal to 0.239 kilocalories (kcal) (1 kcal = 4.18 kJ). As explained in the text, $\Delta G°$ represents the free-energy difference under standard conditions (where all components are present at a concentration of 1.0 mole/liter).

From this table, we see that if there is a favorable standard free-energy change ($\Delta G°$) of –17.8 kJ/mole (–4.3 kcal/mole) for the transition Y → X, there will be 1000 times more molecules in state X than in state Y at equilibrium (K = 1000).

The Free-Energy Changes of Coupled Reactions Are Additive

We have pointed out that unfavorable reactions can be coupled to favorable ones to drive the unfavorable ones forward (see Figure 2–29). In thermodynamic terms, this is possible because the overall free-energy change for a set of coupled reactions is the sum of the free-energy changes in each of its component steps. Consider, as a simple example, two sequential reactions

$$X \rightarrow Y \text{ and } Y \rightarrow Z$$

whose $\Delta G°$ values are +5 and –13 kJ/mole, respectively. If these two reactions occur sequentially, the $\Delta G°$ for the coupled reaction will be –8 kJ/mole. This means that, with appropriate conditions, the unfavorable reaction X → Y can be driven by the favorable reaction Y → Z, provided that this second reaction follows the first. For example, several of the reactions in the long pathway that converts sugars into CO_2 and H_2O have positive $\Delta G°$ values. But the pathway nevertheless proceeds because the total $\Delta G°$ for the series of sequential reactions has a large negative value.

Forming a sequential pathway is not adequate for many purposes. Often the desired pathway is simply X → Y, without further conversion of Y to some other product. Fortunately, there are other more general ways of using enzymes to couple reactions together. In order to explain how this is possible, we need to introduce the activated carrier molecules that we discuss next.

Activated Carrier Molecules Are Essential for Biosynthesis

To make life possible, the energy released by the oxidation of food molecules must be stored temporarily before it can be channeled into the construction of the many other molecules needed by the cell. In most cases, the energy is stored as chemical-bond energy in a small set of activated "carrier molecules," which contain one or more energy-rich covalent bonds. These molecules diffuse rapidly throughout the cell and thereby carry their bond energy from sites of energy

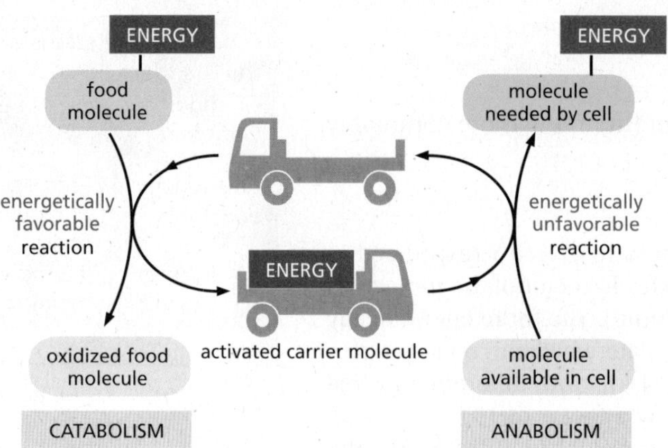

Figure 2–31 **Energy transfer and the role of activated carriers in metabolism.** By serving as energy shuttles, activated carrier molecules perform their function as go-betweens that link the breakdown of food molecules and the release of energy (*catabolism*) to the energy-requiring biosynthesis of small and large organic molecules (*anabolism*).

generation to the sites where the energy will be used for biosynthesis and other cell activities (**Figure 2–31**).

The **activated carriers** store energy in an easily exchangeable form, either as a readily transferable chemical group or as electrons held at a high energy level, and they can serve a dual role as a source of both energy and chemical groups in biosynthetic reactions. For historical reasons, these molecules are also sometimes referred to as *coenzymes*. The most important of the activated carrier molecules are ATP and two molecules that are closely related to each other, NADH and NADPH. As we discuss next, coupled reactions allow cells both to generate such activated carrier molecules and to use them like money to pay for reactions that otherwise could not take place.

The Formation of an Activated Carrier Is Coupled to an Energetically Favorable Reaction

Coupling mechanisms require enzymes and are fundamental to all the energy transactions of the cell. The nature of a **coupled reaction** is illustrated by a mechanical analogy in **Figure 2–32**, in which an energetically favorable chemical reaction is represented by rocks falling from a cliff. The energy of falling rocks would normally be entirely wasted in the form of heat generated by friction when the rocks hit the ground (see the falling-brick diagram in Figure 2–17). By careful design, however, part of this energy could be used instead to drive a paddle wheel

Figure 2–32 **A mechanical model illustrating the principle of coupled chemical reactions.** The spontaneous reaction shown in (A) could serve as an analogy for the direct oxidation of glucose to CO_2 and H_2O, which produces heat only. In (B), the same reaction is coupled to a second reaction; this second reaction is analogous to the synthesis of activated carrier molecules. The energy produced in (B) is in a more useful form than in (A) and can be used to drive a variety of otherwise energetically unfavorable reactions (C).

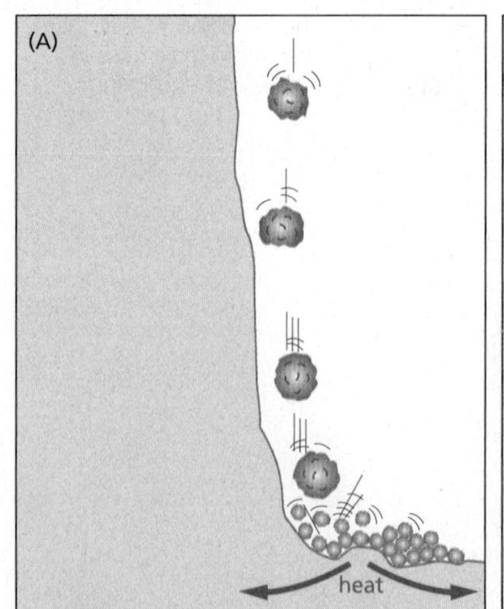

kinetic energy of falling rocks is transformed into heat energy only

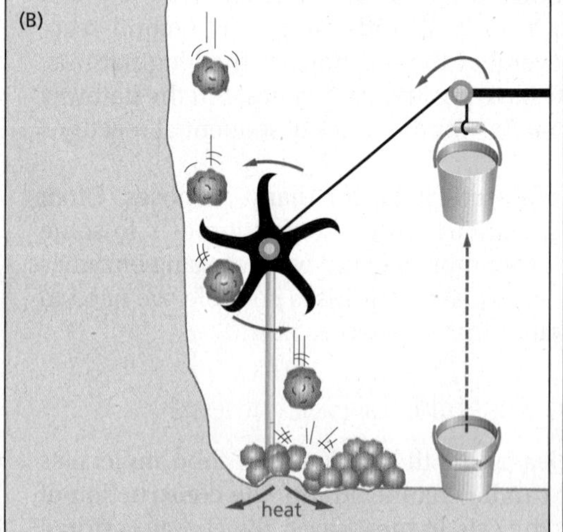

part of the kinetic energy is used to lift a bucket of water, and a correspondingly smaller amount is transformed into heat

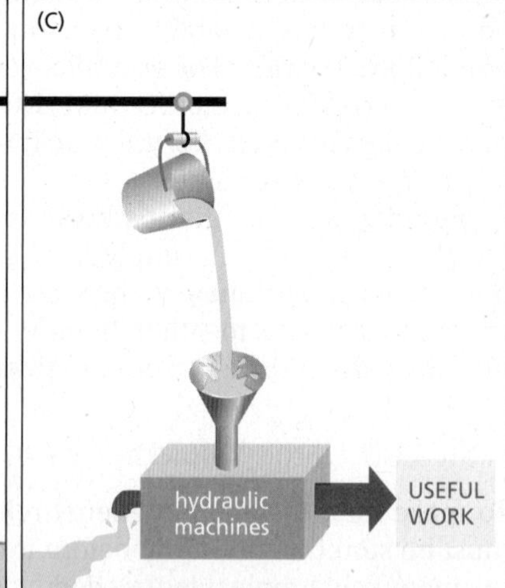

the potential kinetic energy stored in the raised bucket of water can be used to drive hydraulic machines that carry out a variety of useful tasks

that lifts a bucket of water (Figure 2–32B). Because the rocks can now reach the ground only after moving the paddle wheel, we say that the energetically favorable reaction of rock falling has been directly *coupled* to the energetically unfavorable reaction of lifting the bucket of water. Note that because part of the energy is used to do work in Figure 2–32B, the rocks hit the ground with less velocity than in Figure 2–32A, and correspondingly less energy is dissipated as heat.

Similar processes occur in cells, where enzymes play the role of the paddle wheel. By mechanisms that we discuss later in this chapter, enzymes couple an energetically favorable reaction, such as the oxidation of foodstuffs, to an energetically unfavorable reaction, such as the generation of an activated carrier molecule. As in this example, the amount of heat released by the oxidation reaction is reduced by exactly the amount of energy stored in the energy-rich covalent bonds of the activated carrier molecule. And the activated carrier molecule picks up a packet of energy of a size sufficient to power a chemical reaction elsewhere in the cell.

ATP Is the Most Widely Used Activated Carrier Molecule

The most important and versatile of the activated carriers in cells is **ATP** (adenosine triphosphate). Just as the energy stored in the raised bucket of water in Figure 2–32B can drive a wide variety of hydraulic machines, ATP is a convenient and versatile store, or currency, of energy used to drive a huge variety of chemical reactions in cells. ATP is synthesized by coupling a reaction that is highly energetically favorable to an energetically unfavorable phosphorylation reaction in which a phosphate group is added to **ADP** (adenosine diphosphate). When required, ATP gives up its energy packet through its energetically favorable hydrolysis to ADP and inorganic phosphate (**Figure 2–33**). The regenerated ADP is then available to be used for another round of the phosphorylation reaction that forms ATP.

The two outermost phosphate groups in ATP are said to be held to the rest of the molecule by "high-energy" covalent bonds, because each of these phosphoanhydride linkages releases a great deal of free energy when hydrolyzed. The unusually large negative free-energy change for these hydrolysis reactions arises from a number of factors. The release of the terminal phosphate group when ATP forms ADP removes an unfavorable repulsion between adjacent negative charges; in addition, the inorganic phosphate ion released is stabilized both by resonance and by favorable hydrogen-bond formation with water.

The energetically favorable reaction of ATP hydrolysis is coupled to many otherwise unfavorable reactions through which needed molecules are synthesized.

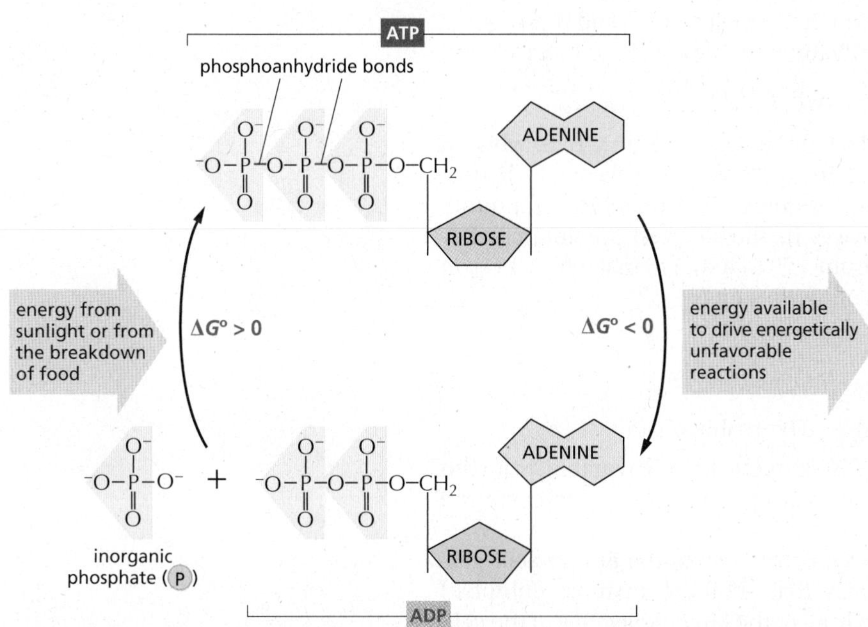

Figure 2–33 The interconversion of ATP and ADP occurs in a cycle. The two outermost phosphates in ATP are held to the rest of the molecule by "high-energy" phosphoanhydride bonds and are readily hydrolyzed or transferred to other molecules. Water can be added to ATP to form ADP and inorganic phosphate. This hydrolysis of the terminal phosphate of ATP yields between 46 and 54 kJ/mole of usable energy, depending on the intracellular conditions. (Although the $\Delta G°$ of this reaction is –30.5 kJ/mole, its ΔG inside cells is much more negative, because the ratio of ATP to the products ADP and phosphate is kept so high.)

The formation of ATP from ADP and phosphate reverses the hydrolysis reaction; because this condensation reaction is energetically unfavorable, it must be coupled to a highly energetically favorable reaction to occur.

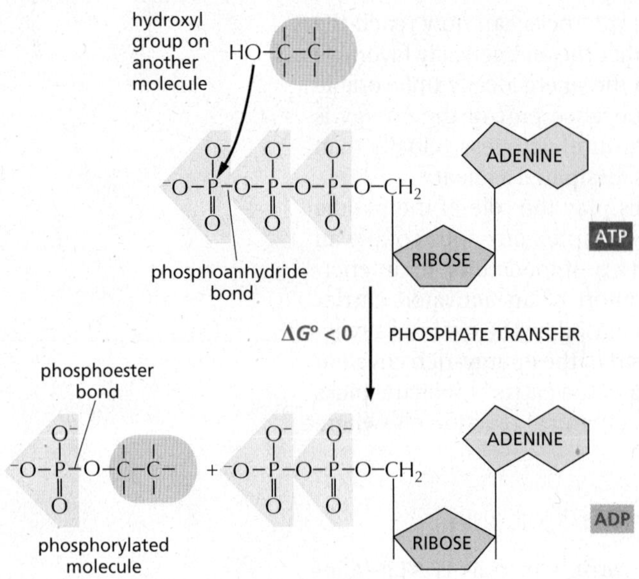

Figure 2–34 **An example of a phosphate transfer reaction.** Because an energy-rich phosphoanhydride bond in ATP is converted to a phosphoester bond, this reaction is energetically favorable, having a large negative ΔG. Reactions of this type are involved in the synthesis of phospholipids and in the initial steps of reactions that catabolize sugars, as well as in many other metabolic pathways.

Many of these coupled reactions involve the transfer of the terminal phosphate in ATP to another molecule, as illustrated by the phosphorylation reaction in **Figure 2–34**.

As the most abundant activated carrier in cells, ATP is the principal energy currency. To give just two examples, it supplies energy for many of the pumps that transport substances into and out of the cell (discussed in Chapter 11), and it powers the molecular motors that enable muscle cells to contract and nerve cells to transport materials from one end of their long axons to another (discussed in Chapter 16).

Energy Stored in ATP Is Often Harnessed to Join Two Molecules Together

We have previously discussed one way in which an energetically favorable reaction can be coupled to an energetically unfavorable reaction, X → Y, so as to enable it to occur. In that scheme, a second enzyme catalyzes the energetically favorable reaction Y → Z, pulling all of the X to Y in the process. But when the required product is Y and not Z, this mechanism is not useful.

A typical biosynthetic reaction is one in which two molecules, A and B, are joined together to produce A–B in the energetically unfavorable *condensation* reaction

$$\text{B–H} + \text{A–OH} \rightarrow \text{A–B} + H_2O$$

There is an indirect pathway that allows B–H and A–OH to form A–B, in which a coupling to ATP hydrolysis makes the reaction go. Here, energy from ATP hydrolysis is first used to convert A–OH to a higher-energy intermediate compound, which then reacts directly with B–H to give A–B. The simplest possible mechanism involves the transfer of a phosphate from ATP to A–OH to make A–O–PO_3, in which case the reaction pathway contains only two steps:

1. A–OH + ATP → A–O–PO_3 + ADP

2. B–H + A–O–PO_3 → A–B + phosphate

Net result: A–OH + ATP + B–H → A–B + ADP + phosphate

The condensation reaction, which by itself is energetically unfavorable, is forced to occur by being directly coupled to ATP hydrolysis in an enzyme-catalyzed reaction pathway (**Figure 2–35A**).

A biosynthetic reaction of exactly this type synthesizes the amino acid glutamine (**Figure 2–35B**). We will see shortly that similar (but more complex) mechanisms are also used to produce nearly all of the large molecules of the cell.

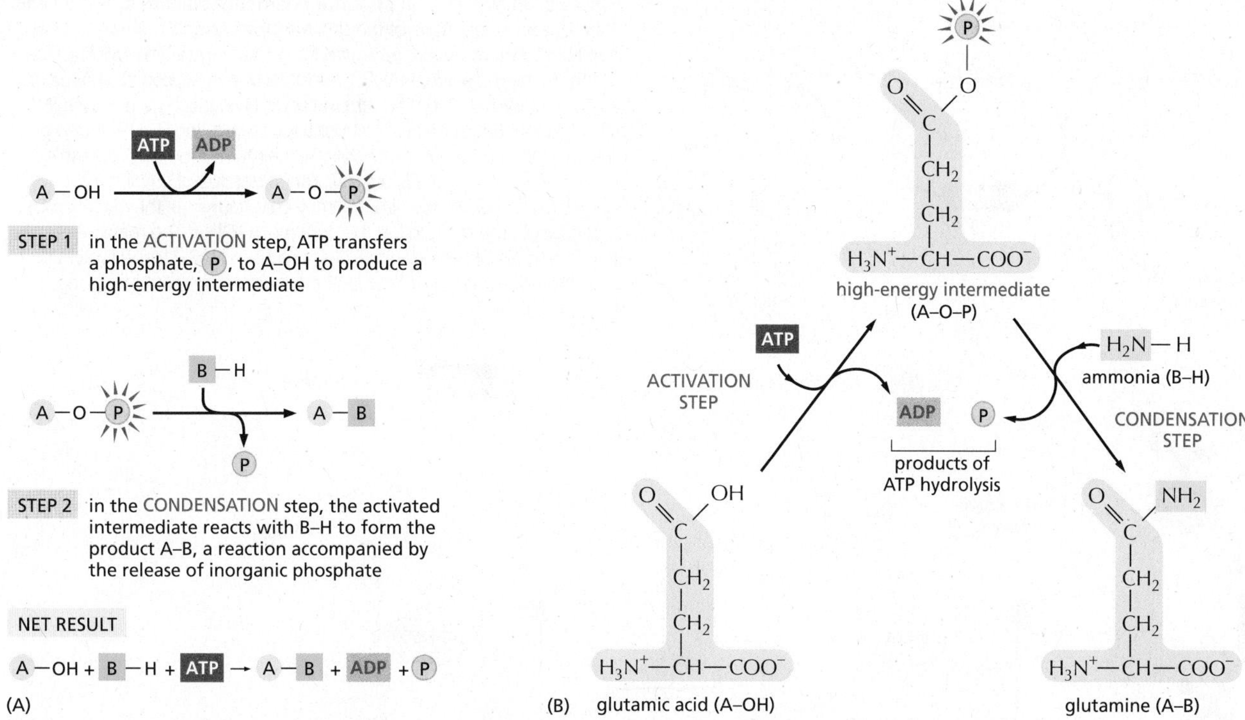

ATP ADP
A—OH → A—O—P
STEP 1 in the ACTIVATION step, ATP transfers a phosphate, P, to A–OH to produce a high-energy intermediate
B—H
A—O—P → A—B
P
STEP 2 in the CONDENSATION step, the activated intermediate reacts with B–H to form the product A–B, a reaction accompanied by the release of inorganic phosphate
NET RESULT
A—OH + B—H + ATP → A—B + ADP + P
(A)
high-energy intermediate (A–O–P)
ACTIVATION STEP
ATP
ADP P
products of ATP hydrolysis
H₂N—H
ammonia (B–H)
CONDENSATION STEP
(B) glutamic acid (A–OH)
glutamine (A–B)

Figure 2–35 **An example of an energetically unfavorable biosynthetic reaction driven by ATP hydrolysis.** (A) Schematic illustration of the condensation reaction described in the text. In this set of reactions, a phosphate group is first donated by ATP to form a high-energy intermediate, A–O–PO_3, which then reacts with the other substrate, B–H, to form the product A–B. (B) Reaction showing the biosynthesis of the amino acid glutamine from glutamic acid. Glutamic acid, which corresponds to the A–OH shown in (A), is first converted to a high-energy phosphorylated intermediate, which corresponds to A–O–PO_3. This intermediate then reacts with ammonia (which corresponds to B–H) to form glutamine. In this example, both steps occur on the surface of the same enzyme, glutamine synthetase (not shown). ATP hydrolysis can drive this energetically unfavorable reaction because it produces a favorable free-energy change ($\Delta G°$ of –30.5 kJ/mole) that is larger in magnitude than the energy required for the synthesis of glutamine from glutamic acid plus NH_3 ($\Delta G°$ of +14.2 kJ/mole). For clarity, the glutamic acid side chain is shown in its uncharged form.

NADH and NADPH Are Important Electron Carriers

Other important activated carrier molecules participate in oxidation–reduction reactions and are commonly part of coupled reactions in cells. These activated carriers are specialized to carry electrons held at a high energy level (sometimes called "high-energy" electrons) and hydrogen atoms. The most important of these electron carriers are **NAD$^+$** (nicotinamide adenine dinucleotide) and the closely related molecule **NADP$^+$** (nicotinamide adenine dinucleotide phosphate). As part of an enzyme-catalyzed reaction in which a substrate molecule is oxidized, each picks up a "packet of energy" corresponding to two electrons plus a proton (H^+), and they are thereby converted to **NADH** (*reduced* nicotinamide adenine dinucleotide) and **NADPH** (*reduced* nicotinamide adenine dinucleotide phosphate), respectively (Figure 2–36). These molecules can therefore be regarded as carriers of hydride ions (the H^+ plus two electrons, or H^-).

Like ATP, NADPH is an activated carrier that participates in many important biosynthetic reactions that would otherwise be energetically unfavorable. The NADPH is produced according to the general scheme shown in Figure 2–36A. During a special set of energy-yielding catabolic reactions, two hydrogen atoms are removed from a substrate molecule. Both electrons but just one hydrogen atom (that is, a hydride ion, H^-) are added to the nicotinamide ring of NADP$^+$ to form NADPH; the second hydrogen atom is released as a proton (H^+) into solution. This is a typical oxidation–reduction reaction, in which the substrate is oxidized and NADP$^+$ is reduced.

NADPH readily gives up the hydride ion it carries in a subsequent oxidation–reduction reaction, because the nicotinamide ring can achieve a more stable

(A)

H—C—OH
C=O
NADP+
NADPH
+ H+
H—C—
H—C—
C—
C—
oxidation of molecule 1
reduction of molecule 2

Figure 2–36 NADPH, an important carrier of electrons. (A) NADPH is produced in reactions of the general type shown on the left, in which two hydrogen atoms are removed from a substrate. The oxidized form of the carrier molecule, $NADP^+$, receives one hydrogen atom plus an electron (a hydride ion); the other H atom is released as a proton (H^+) into solution. Because NADPH holds its hydride ion in a high-energy linkage, the hydride ion can easily be transferred to other molecules, as shown on the right. (B) and (C) The structures of $NADP^+$ and NADPH. The part of the $NADP^+$ molecule known as the nicotinamide ring accepts the hydride ion, H^-, forming NADPH. The molecules NAD^+ and NADH are identical in structure to $NADP^+$ and NADPH, respectively, except that they lack the indicated phosphate group.

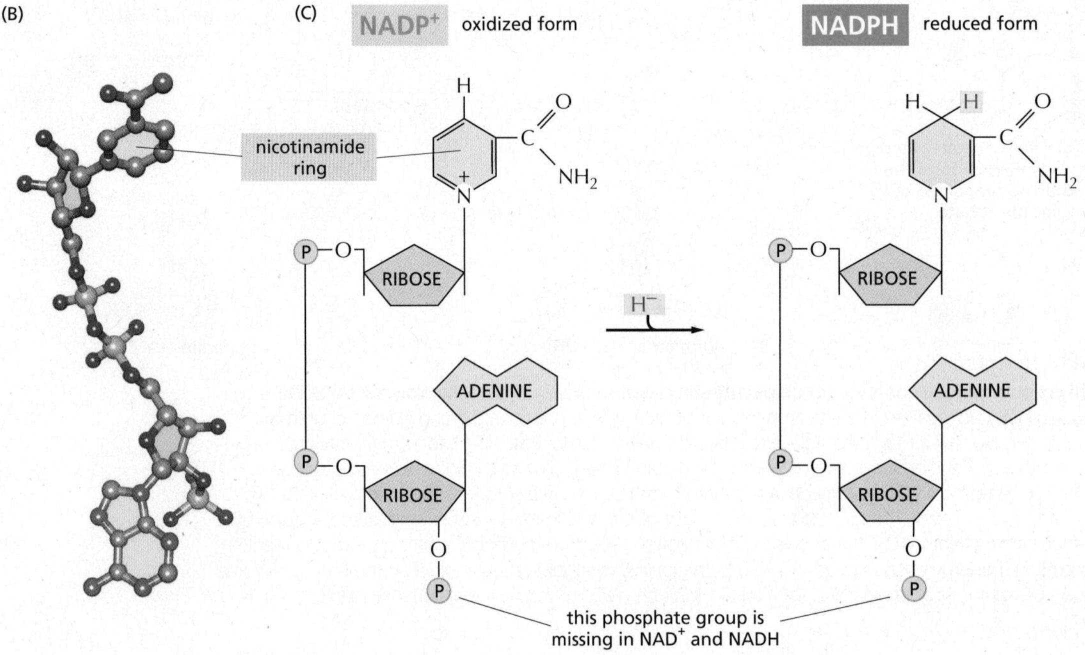

arrangement of electrons without it. In this subsequent reaction, which regenerates $NADP^+$, it is the NADPH that is oxidized and the substrate that is reduced. The NADPH is an effective donor of its hydride ion to other molecules for the same reason that ATP readily transfers a phosphate: in both cases the transfer is accompanied by a large negative free-energy change. One example of the use of NADPH in biosynthesis is shown in **Figure 2–37**.

The extra phosphate group on NADPH has no effect on the electron-transfer properties of NADPH compared with NADH, being far away from the region involved in electron transfer (see Figure 2–36C). It does, however, give a molecule of NADPH a slightly different shape from that of NADH, making it possible for NADPH and NADH to bind as substrates to completely different sets of enzymes. Thus, the two types of carriers are used to transfer electrons (or hydride ions) between two different sets of molecules.

Why should there be this division of labor? The answer lies in the need to regulate two sets of electron-transfer reactions independently. NADPH operates chiefly with enzymes that catalyze anabolic reactions, supplying the high-energy electrons needed to synthesize energy-rich biological molecules. NADH, by contrast, has a special role as an intermediate in the catabolic system of reactions that generate ATP through the oxidation of food molecules, as we will discuss shortly. The geneses of NADH from NAD^+ and of NADPH from $NADP^+$ occur by different pathways and are independently regulated so that the cell can adjust the supply of electrons for these two contrasting purposes. Inside the cell the ratio of NAD^+ to NADH is kept high, whereas the ratio of $NADP^+$ to NADPH is kept low. This

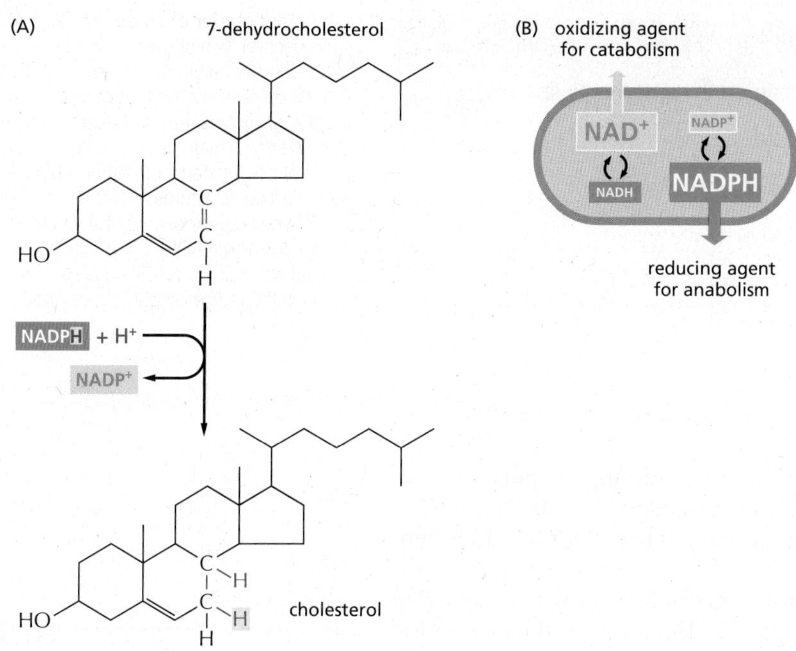

Figure 2–37 **NADPH as a reducing agent.** (A) The final stage in a biosynthetic route leading to cholesterol. As in many other biosynthetic reactions, the reduction of the C=C bond is achieved by the transfer of a hydride ion from the carrier molecule NADPH, plus a proton (H^+) from the solution. (B) Keeping NADPH levels high and NADH levels low alters the redox potential for each of the two redox pairs—NADPH:$NADP^+$ and NADH:NAD^+ (see Panel 14–1, p. 825). It causes NADPH to be a much stronger electron donor (reducing agent) than NADH, and it causes NAD^+ to be a much better electron acceptor (oxidizing agent) than $NADP^+$, as indicated.

provides plenty of NAD^+ to act as an oxidizing agent and plenty of NADPH to act as a reducing agent (Figure 2–37B)—as required for their special roles in catabolism and anabolism, respectively.

There Are Many Other Activated Carrier Molecules in Cells

Other activated carriers also pick up and carry a chemical group in an easily transferred, high-energy linkage. For example, coenzyme A carries a readily transferable acetyl group in a thioester linkage and in this activated form is known as **acetyl CoA** (acetyl coenzyme A). Acetyl CoA (**Figure 2–38**) is used to add two carbon units in the biosynthesis of larger molecules.

In acetyl CoA, as in other carrier molecules, the transferable group makes up only a small part of the molecule. The rest consists of a large organic portion that serves as a convenient "handle," facilitating the recognition of the carrier molecule by specific enzymes. As with acetyl CoA, this handle portion very often contains a nucleotide derivative (usually adenosine diphosphate), a curious fact that may be a relic from an early stage of evolution. It is currently thought that the main catalysts for early life-forms—before DNA or proteins—were RNA molecules (or

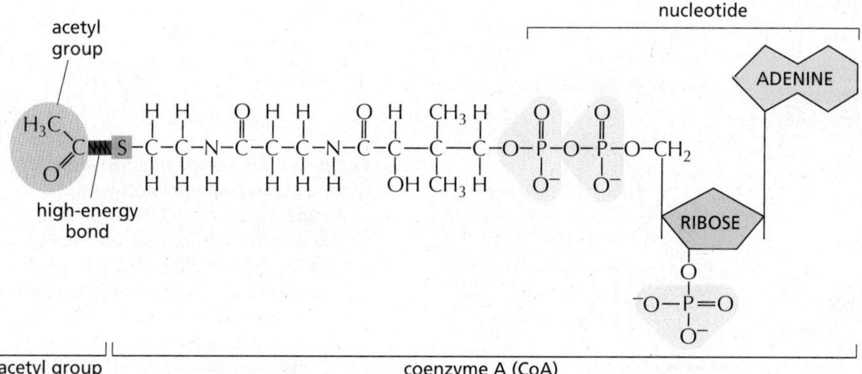

Figure 2–38 **The structure of the important activated carrier molecule acetyl CoA.** The sulfur atom *(orange)* forms a thioester bond to acetate. Because this is a high-energy linkage, releasing a large amount of free energy when it is hydrolyzed, the acetate molecule can be readily transferred to other molecules.

TABLE 2–3 Some Activated Carrier Molecules Widely Used in Metabolism

Activated carrier	Group carried in high-energy linkage
ATP	Phosphate
NADH, NADPH, $FADH_2$	Electrons and hydrogens
Acetyl CoA	Acetyl group
Carboxylated biotin	Carboxyl group
S-Adenosylmethionine	Methyl group
Uridine diphosphate glucose	Glucose

their close relatives), as described in Chapter 6. It is tempting to speculate that many of the carrier molecules that we find today originated in this earlier RNA world, where their nucleotide portions could have been useful for binding them to RNA enzymes (ribozymes).

Thus, ATP transfers phosphate, NADPH transfers electrons and hydrogen, and acetyl CoA transfers two-carbon acetyl groups. **$FADH_2$** (reduced flavin adenine dinucleotide) is used like NADH in electron and proton transfers (**Figure 2–39**). The reactions of other activated carrier molecules involve the transfer of a methyl, carboxyl, or glucose group for biosyntheses (**Table 2–3**). These activated carriers are generated in reactions that are coupled to ATP hydrolysis, as in the example in **Figure 2–40**. Therefore, the energy that enables their groups to be used for biosynthesis ultimately comes from the catabolic reactions that generate ATP. Similar processes occur in the synthesis of the very large molecules of the cell—the nucleic acids, proteins, and polysaccharides—that we discuss next.

Figure 2–39 **$FADH_2$ is a carrier of hydrogens and high-energy electrons, like NADH and NADPH.** (A) Structure of $FADH_2$, with its hydrogen-carrying atoms highlighted in *yellow.* (B) The formation of $FADH_2$ from FAD (flavin adenine dinucleotide).

The Synthesis of Biological Polymers Is Driven by ATP Hydrolysis

As discussed previously, the macromolecules of the cell constitute most of its dry mass (see Figure 2–7). These molecules are made from subunits (or monomers) that are linked together in a *condensation* reaction, in which the constituents of a

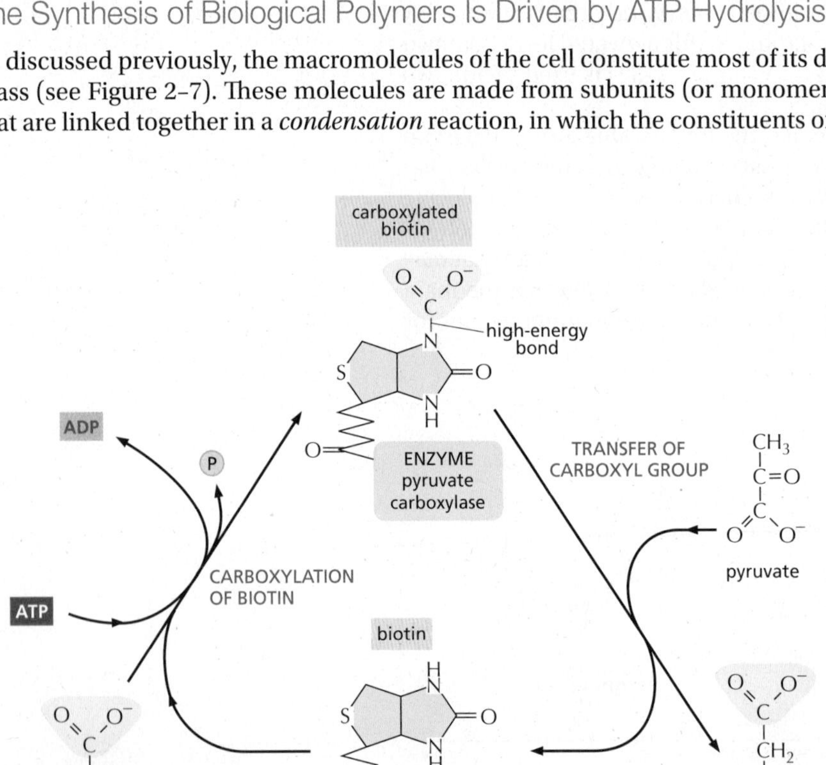

Figure 2–40 **A carboxyl group–transfer reaction using an activated carrier molecule.** Carboxylated biotin is used by the enzyme *pyruvate carboxylase* to transfer a carboxyl group in the production of oxaloacetate, a molecule needed for the citric acid cycle. The acceptor molecule for this group-transfer reaction is pyruvate. Other enzymes use biotin, a B-complex vitamin, to transfer carboxyl groups to other acceptor molecules. Note that synthesis of carboxylated biotin requires energy that is derived from ATP—a general feature of many activated carriers.

water molecule (OH plus H) are removed from the two reactants. Consequently, the reverse reaction—the breakdown of all three types of polymers—occurs by the enzyme-catalyzed addition of water (*hydrolysis*). This hydrolysis reaction is energetically favorable, whereas the biosynthetic reactions require an energy input (see Figure 2–9).

The nucleic acids (DNA and RNA), proteins, and polysaccharides are all polymers that are produced by the repeated addition of a monomer onto one end of a growing chain. The synthesis reactions for these three types of macromolecules are outlined in **Figure 2–41**. As indicated, the condensation step in each case depends on energy from nucleoside triphosphate hydrolysis. And yet, except for the nucleic acids, there are no phosphate groups left in the final product molecules. How are the reactions that release the energy of ATP hydrolysis coupled to polymer synthesis?

For each type of macromolecule, an enzyme-catalyzed pathway exists, which resembles that discussed previously for the synthesis of the amino acid glutamine (see Figure 2–35). The principle is exactly the same, in that the -OH group that will be removed in the condensation reaction is first activated by becoming involved in a high-energy linkage to a second molecule. However, the actual mechanisms used to link ATP hydrolysis to the synthesis of proteins and polysaccharides are more complex than that used for glutamine synthesis, inasmuch as more than

Figure 2–41 **How energy is used to synthesize macromolecules.** Synthesis of a portion of (A) a polysaccharide, (B) a nucleic acid, and (C) a protein is shown here. In each case, synthesis involves a condensation reaction in which water is lost; the atoms involved are shaded in *pink*. Not shown is the consumption of high-energy nucleoside triphosphates that is required to activate each subunit prior to its addition. In contrast, the reverse reaction—the breakdown of all three types of polymers—occurs through the simple addition of water, or hydrolysis (not shown), which is energetically favorable and does not require an energy carrier.

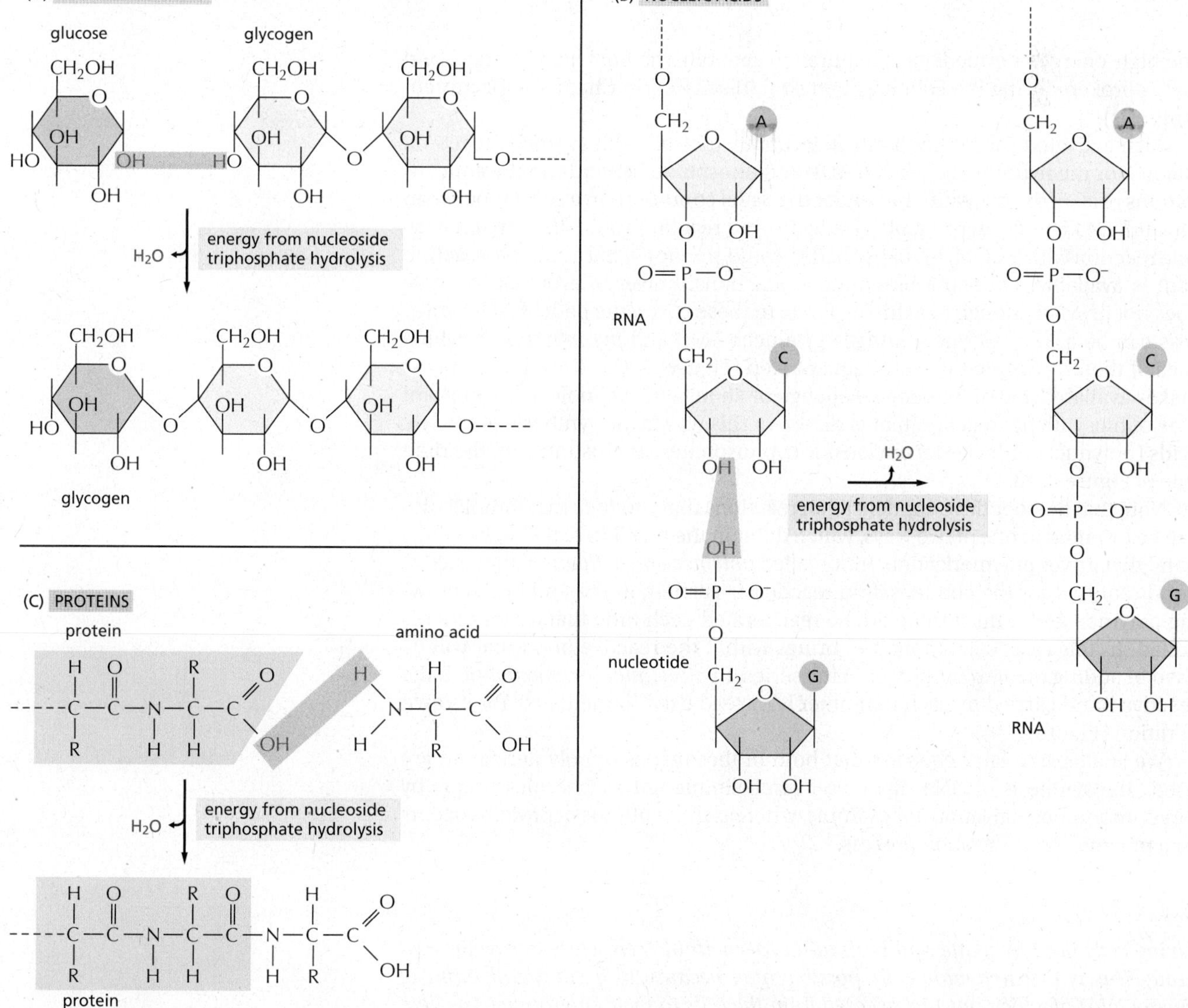

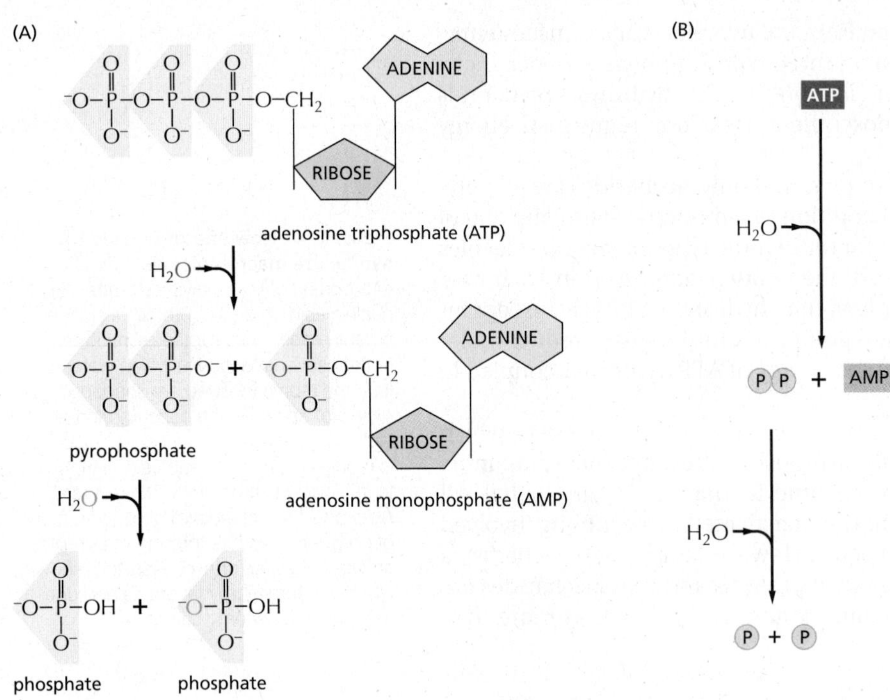

Figure 2–42 An alternative pathway of ATP hydrolysis, in which pyrophosphate is first formed and then hydrolyzed. This route releases about twice as much free energy (approximately −100 kJ/mole) as the reaction shown earlier in Figure 2–33, and it forms AMP instead of ADP. (A) In the two successive hydrolysis reactions, oxygen atoms from the participating water molecules are retained in the products, as indicated, whereas the hydrogen atoms dissociate to form free hydrogen ions (H^+, not shown). (B) Summary of overall reaction.

one high-energy intermediate is required to generate the final high-energy bond that is broken during the condensation step (discussed in Chapter 6 for protein synthesis).

Each activated carrier has limits in its ability to drive a biosynthetic reaction. The ΔG for the hydrolysis of ATP to ADP and phosphate depends on the concentrations of all of the reactants, but under the usual conditions in a cell it is between −46 and −54 kJ/mole. In principle, this hydrolysis reaction could drive an unfavorable reaction with a ΔG of, perhaps, +40 kJ/mole, provided that a suitable reaction path is available. For some biosynthetic reactions, however, even −50 kJ/mole does not provide enough of a driving force. In these cases, the path of ATP hydrolysis can be altered so that it initially produces AMP and *pyrophosphate*, which is itself then hydrolyzed in a subsequent step (**Figure 2–42**). The whole process makes available a total free-energy change of about −100 kJ/mole. An important type of biosynthetic reaction that is driven in this way is the synthesis of nucleic acids (polynucleotides) from nucleoside triphosphates, as illustrated on the right side of **Figure 2–43**.

Note that the repetitive condensation reactions that produce macromolecules can be oriented in one of two ways, which differ in the position of the high-energy bond that drives polymerization. In so-called *polymer-end activation*, the reactive bond required for the condensation reaction is carried on the end of the growing polymer, and it must therefore be regenerated each time that a monomer is added. In this case, each monomer brings with it the reactive bond that will be used in adding the *next* monomer in the series. In *direct-monomer activation*, the reactive bond carried by each monomer is instead used immediately for its own addition (**Figure 2–44**).

We shall see in later chapters that both of these types of polymerization are used. The synthesis of DNA, RNA, and some simple polysaccharides occurs by direct-monomer activation, for example, whereas the synthesis of proteins occurs by a polymer-end activation process.

Summary

Living cells need to create and maintain order within themselves to survive and grow. This is thermodynamically possible only because of a continual input of energy, part of which must be released from the cells to their environment as heat

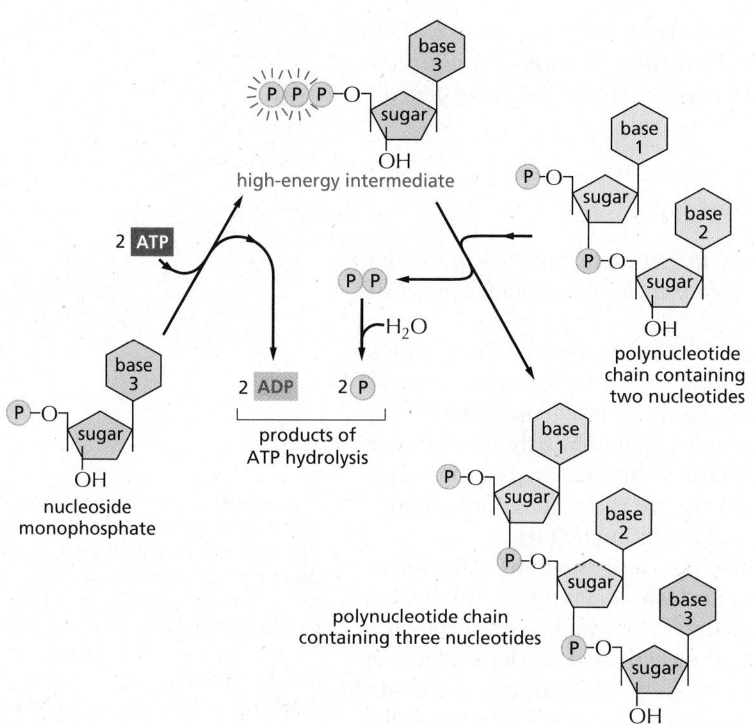

Figure 2–43 **Synthesis of a polynucleotide, RNA or DNA, is a multistep process driven by ATP hydrolysis.** In the first step, a nucleoside monophosphate (a nucleotide) is activated by the sequential transfer of the terminal phosphate groups from two ATP molecules. The high-energy intermediate formed—a nucleoside triphosphate—exists free in solution until it reacts with the growing end of an RNA or a DNA chain, which leads to release of pyrophosphate. Hydrolysis of the latter to inorganic phosphate is highly favorable and helps to drive the overall reaction in the direction of polynucleotide synthesis. For details, see Chapter 5.

that disorders the surroundings. The only chemical reactions possible are those that increase the total amount of disorder in the universe. The free-energy change for a reaction, ΔG, measures this disorder, and it must be less than zero for a reaction to proceed spontaneously. This ΔG depends both on the intrinsic properties of the reacting molecules and their concentrations, and it can be calculated from these concentrations if either the equilibrium constant (K) for the reaction or its standard free-energy change, ΔG°, is known.

The energy needed for life comes ultimately from the electromagnetic radiation of the Sun, which drives the formation of organic molecules in photosynthetic organisms such as green plants. Animals obtain their energy by eating organic molecules and oxidizing them in a series of enzyme-catalyzed reactions that are coupled to the formation of ATP—a common currency of energy in all cells.

To make possible the continual generation of order in cells, energetically favorable reactions, such as the hydrolysis of ATP, are coupled to energetically unfavorable reactions. In the biosynthesis of macromolecules, ATP is used to form reactive phosphorylated intermediates. Because the energetically unfavorable reaction of biosynthesis now becomes energetically favorable, ATP hydrolysis is said to drive the reaction. Polymeric molecules such as proteins, nucleic acids, and polysaccharides are assembled from small activated precursor molecules by repetitive condensation reactions that are driven in this way. Other reactive

Figure 2–44 **Two active-intermediate orientations are utilized for the repetitive condensation reactions that form biological polymers.** A polymer-end activation mechanism for polymer synthesis is compared with its alternative, direct-monomer activation. As indicated, these two mechanisms are used to produce different types of biological macromolecules.

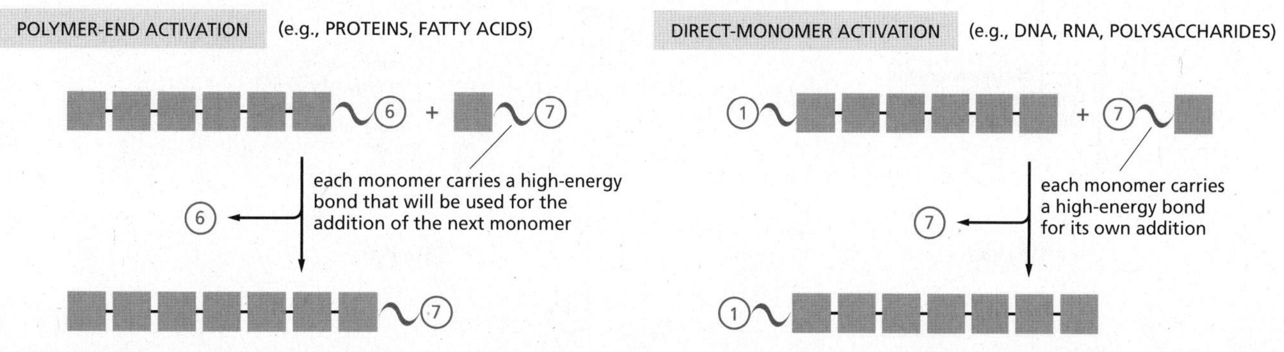

molecules, called either activated carriers or coenzymes, transfer other chemical groups in the course of biosynthesis: NADPH transfers hydrogen as a proton plus two electrons (a hydride ion), for example, whereas acetyl CoA transfers an acetyl group.

HOW CELLS OBTAIN ENERGY FROM FOOD

The constant supply of energy that cells need to generate and maintain the biological order that keeps them alive comes from the chemical-bond energy in food molecules.

The proteins, lipids, and polysaccharides that make up most of the food we eat must be broken down into smaller molecules before our cells can use them—either as a source of energy or as building blocks for other molecules. Enzymatic digestion breaks down the large polymeric molecules in food into their monomer subunits—proteins into amino acids, polysaccharides into sugars, and fats into fatty acids and glycerol. After digestion, the small organic molecules derived from food enter the cytosol of cells, where their gradual oxidation begins.

Sugars are particularly important fuel molecules, and they are oxidized in small controlled steps to carbon dioxide (CO_2) and water (**Figure 2–45**). In this section, we trace the major steps in the breakdown, or catabolism, of sugars and show how they produce ATP, NADH, and other activated carrier molecules in animal cells. A very similar pathway also operates in plants, fungi, and many bacteria. As we shall see, the oxidation of fatty acids is equally important for cells. Other molecules, such as proteins, can also serve as energy sources when they are funneled through appropriate enzymatic pathways.

Glycolysis Is a Central ATP-producing Pathway

The major process for oxidizing sugars is the sequence of reactions known as **glycolysis**, which produces ATP without the involvement of molecular oxygen (O_2 gas). It occurs in the cytosol of most cells, including many anaerobic microorganisms. Glycolysis probably evolved early in the history of life, before photosynthetic organisms introduced oxygen into the atmosphere. During glycolysis, a glucose molecule with six carbon atoms is converted into two molecules of *pyruvate*, each of which contains three carbon atoms. For each glucose molecule, two molecules of ATP are hydrolyzed to provide energy to drive the early steps, but four

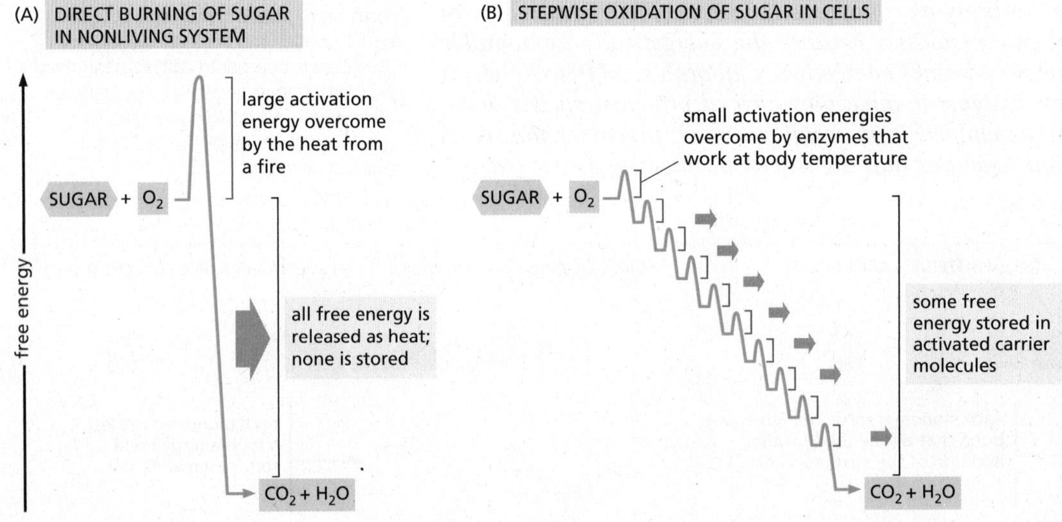

Figure 2–45 Schematic representation of the controlled stepwise oxidation of sugar in a cell, compared with ordinary burning. (A) If the sugar were oxidized to CO_2 and H_2O in a single step, it would release an amount of energy much larger than could be captured for useful purposes. (B) In the cell, enzymes catalyze oxidation via a series of small steps in which free energy is transferred in conveniently sized packets to carrier molecules—most often ATP and NADH. Heat is also released, enabling the universe to be disordered sufficiently to make the entire pathway energetically favorable (ΔG is negative). Each step is catalyzed by an enzyme that lowers the activation-energy barrier that must be surmounted by the random collision of molecules to allow the reaction to occur at the temperature inside cells (see Figure 2–24). Note that the total free energy released by the complete oxidative breakdown of glucose to CO_2 and H_2O—2880 kJ/mole—is exactly the same in (A) and (B).

molecules of ATP are produced in the later steps. At the end of glycolysis, there is consequently a net gain of two molecules of ATP for each glucose molecule broken down.

Piecing together the complete glycolytic pathway in the 1930s was a major triumph of biochemistry, as the pathway consists of a sequence of 10 separate reactions, each producing a different sugar intermediate and each catalyzed by a different enzyme. These enzymes, like most enzymes, all have names ending in "-ase"—like isomerase and dehydrogenase—which specify the type of reaction they catalyze. The sugar oxidation occurs when electrons are removed by NAD^+ (producing NADH) from some of the carbons derived from the glucose molecule. Some of the energy released by this oxidation drives the direct synthesis of ATP molecules from ADP and phosphate, and some remains with the electrons in the activated electron carrier NADH. The pathway is outlined in Figure 2–46 and shown in detail in Panel 2–8 (pp. 108–109) and Movie 2.5.

Two molecules of NADH are formed per molecule of glucose in the course of glycolysis. As we shall see, in aerobic organisms these NADH molecules donate their electrons to the electron-transport chain described in Chapter 14, which enables the NAD^+ formed by NADH oxidization to be reused for glycolysis.

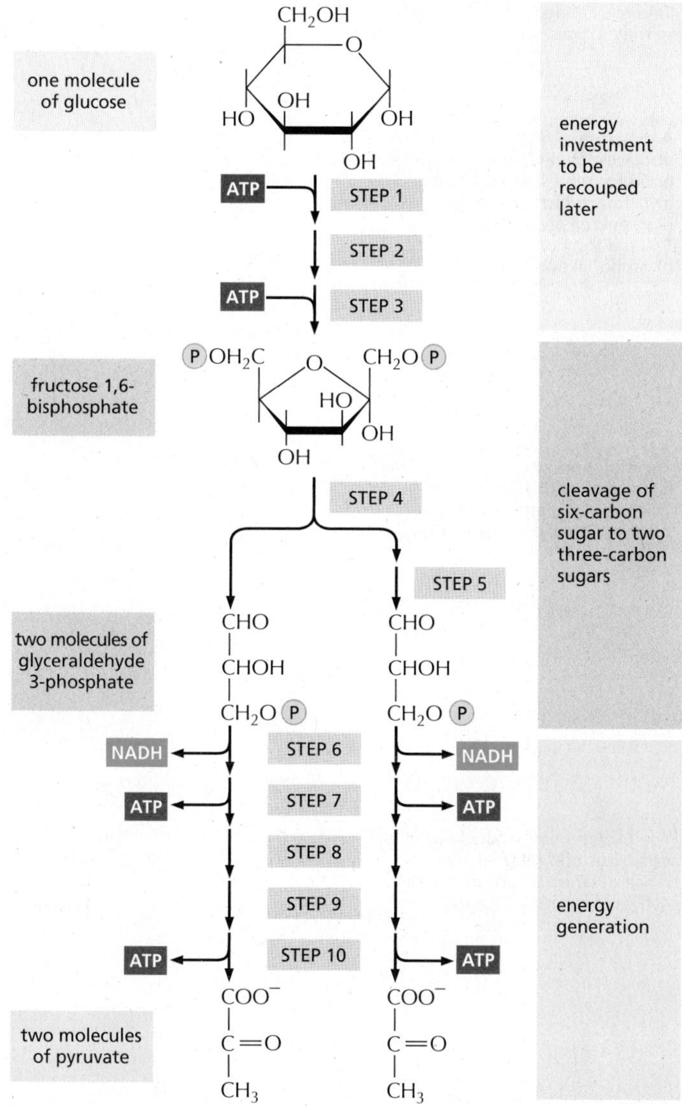

Figure 2–46 **An outline of glycolysis.** Each of the 10 steps shown is catalyzed by a different enzyme. Note that step 4 cleaves a six-carbon sugar into two three-carbon sugars, so that the number of molecules at every stage after this doubles. Note also that one of the two products of step 4 is modified (isomerized) in step 5 to convert it into a second molecule of glyceraldehyde 3-phosphate, the other product of step 4. As indicated, step 6 begins the energy-generation phase of glycolysis. Because two molecules of ATP are hydrolyzed in the early, energy-investment phase, glycolysis results in the net synthesis of 2 ATP and 2 NADH molecules per molecule of glucose. Glycolysis is also referred to as the Embden–Meyerhof pathway, named for the biochemists who first described it. All the steps of glycolysis are reviewed in Panel 2–8 and Movie 2.5.

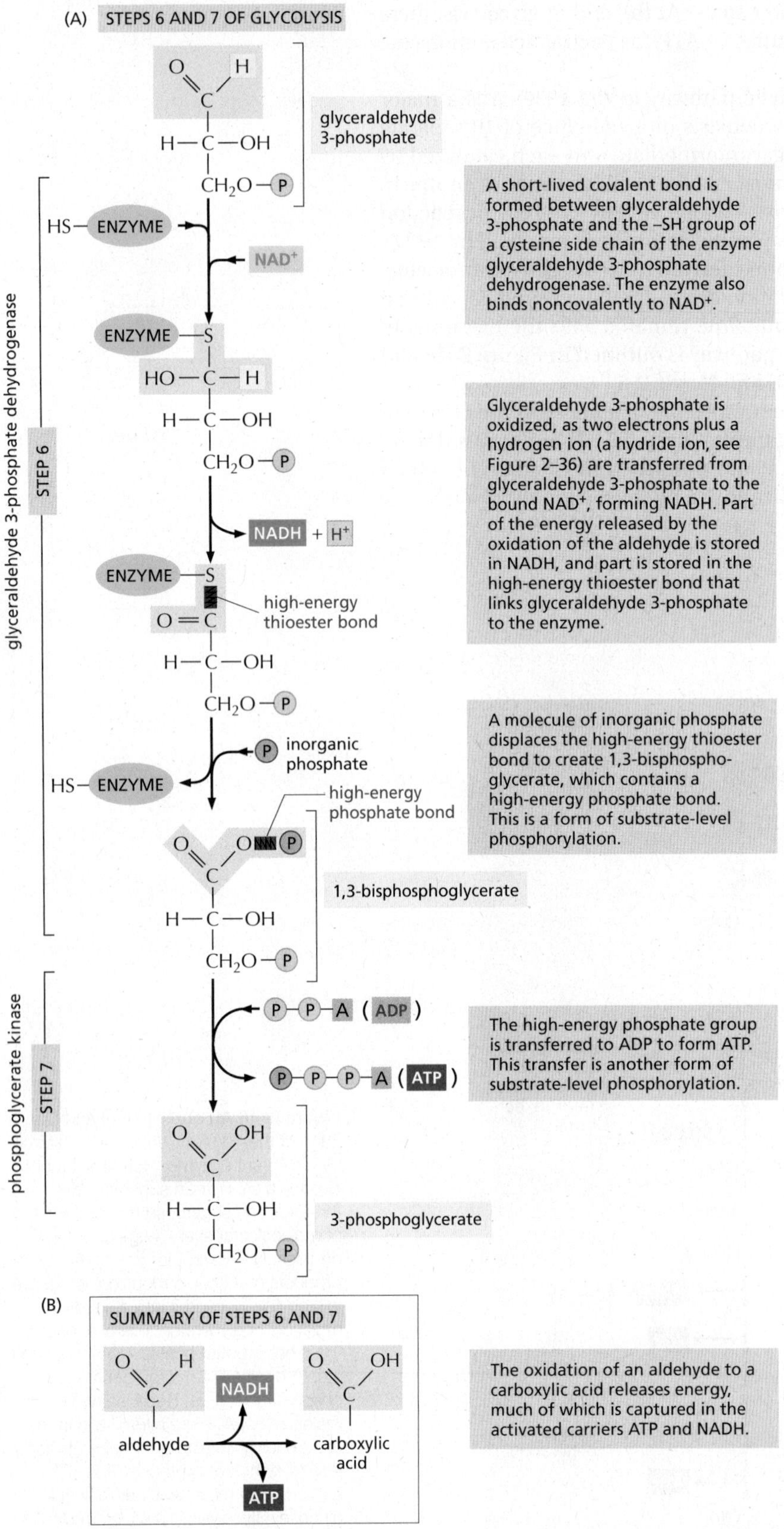

(A) STEPS 6 AND 7 OF GLYCOLYSIS

glyceraldehyde 3-phosphate

HS— ENZYME

NAD^+

ENZYME —S

HO—C—H

H—C—OH

CH_2O—(P)

NADH + H^+

high-energy thioester bond

inorganic phosphate

high-energy phosphate bond

1,3-bisphosphoglycerate

3-phosphoglycerate

glyceraldehyde 3-phosphate dehydrogenase

STEP 6

phosphoglycerate kinase

STEP 7

(P)—(P)—A (ADP)

(P)—(P)—(P)—A (ATP)

A short-lived covalent bond is formed between glyceraldehyde 3-phosphate and the –SH group of a cysteine side chain of the enzyme glyceraldehyde 3-phosphate dehydrogenase. The enzyme also binds noncovalently to NAD^+.

Glyceraldehyde 3-phosphate is oxidized, as two electrons plus a hydrogen ion (a hydride ion, see Figure 2–36) are transferred from glyceraldehyde 3-phosphate to the bound NAD^+, forming NADH. Part of the energy released by the oxidation of the aldehyde is stored in NADH, and part is stored in the high-energy thioester bond that links glyceraldehyde 3-phosphate to the enzyme.

A molecule of inorganic phosphate displaces the high-energy thioester bond to create 1,3-bisphosphoglycerate, which contains a high-energy phosphate bond. This is a form of substrate-level phosphorylation.

The high-energy phosphate group is transferred to ADP to form ATP. This transfer is another form of substrate-level phosphorylation.

(B) SUMMARY OF STEPS 6 AND 7

aldehyde

NADH

ATP

carboxylic acid

The oxidation of an aldehyde to a carboxylic acid releases energy, much of which is captured in the activated carriers ATP and NADH.

Figure 2–47 How the oxidation of glyceraldehyde 3-phosphate is coupled to the formation of ATP and NADH in steps 6 and 7 of glycolysis. (A) In step 6, the enzyme glyceraldehyde 3-phosphate dehydrogenase couples the energetically favorable oxidation of an aldehyde to the energetically unfavorable formation of a high-energy phosphate bond. At the same time, this oxidation enables energy to be stored in NADH. In step 7, the newly formed high-energy phosphate bond in 1,3-bisphosphoglycerate is transferred to ADP by the enzyme phosphoglycerate kinase, forming a molecule of ATP and leaving a free carboxylic acid group on the oxidized sugar. The part of the molecule that undergoes a change is shaded in *blue*; the rest of the molecule remains unchanged throughout all these reactions. (B) Summary of the overall chemical change produced by the reactions of steps 6 and 7.

Glycolysis Illustrates How Enzymes Couple Oxidation to Energy Storage

The formation of ATP during glycolysis provides a particularly clear demonstration of how enzymes couple energetically unfavorable reactions with favorable ones, thereby driving the many chemical reactions that make life possible. Two central reactions in glycolysis (steps 6 and 7) convert the three-carbon sugar intermediate glyceraldehyde 3-phosphate (an aldehyde) into 3-phosphoglycerate (a carboxylic acid; see Panel 2–8, pp. 108–109), thus oxidizing an aldehyde group to a carboxylic acid group. This sugar oxidation process releases enough free energy to convert a molecule of ADP to ATP and to transfer two electrons (and a proton) from the aldehyde to NAD^+ to form NADH, while still liberating enough heat to the environment to make the overall reaction energetically favorable ($\Delta G°$ for the overall reaction is –12.5 kJ/mole).

Figure 2–47 details this remarkable feat of energy harvesting. The chemical reactions are precisely guided by two enzymes to which the sugar intermediates are tightly bound. The first enzyme (glyceraldehyde 3-phosphate dehydrogenase) forms a covalent bond to the sugar aldehyde group through a reactive –SH group on the enzyme and then catalyzes the oxidation of this aldehyde group to a carboxylic acid. This creates a highly reactive enzyme–substrate bond that is displaced by an inorganic phosphate ion to produce a high-energy sugar phosphate intermediate. This intermediate (1,3-bisphosphoglycerate) then binds to the second enzyme (phosphoglycerate kinase), which catalyzes an energetically favorable transfer of its high-energy phosphate to ADP, forming ATP and completing the process of oxidizing a sugar aldehyde to a carboxylic acid.

We have shown this particular oxidation process in some detail because it provides a clear example of enzyme-mediated energy storage through coupled reactions (**Figure 2–48**). Steps 6 and 7 are the only reactions in glycolysis that create a high-energy phosphate linkage directly from inorganic phosphate. As such, they account for the net yield of two ATP molecules and two NADH molecules per molecule of glucose (see Panel 2–8, pp. 108–109).

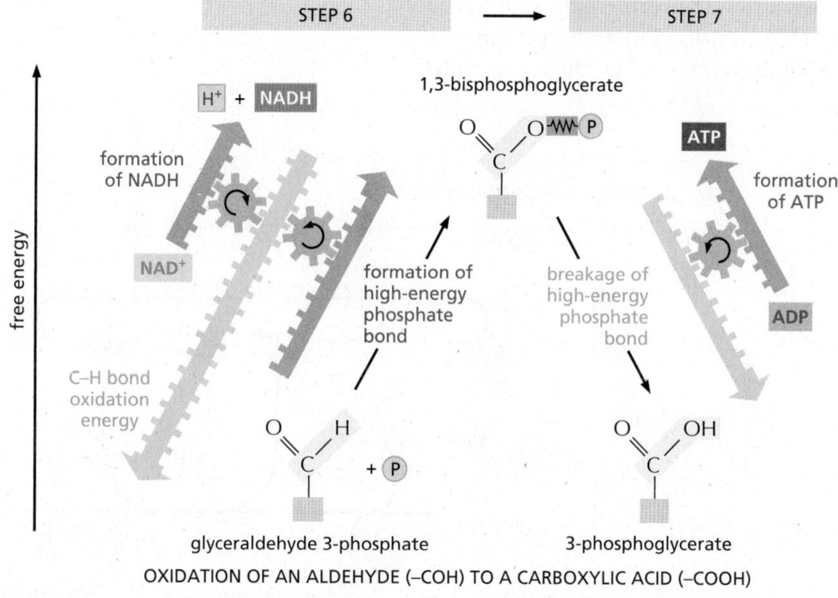

Figure 2–48 How a pair of coupled reactions drives the energetically unfavorable formation of NADH and ATP in steps 6 and 7 of glycolysis. In this diagram, energetically favorable reactions are represented by *blue arrows* and energetically costly reactions by *red arrows*. In step 6, the energy released by the energetically favorable oxidation of a C–H bond in glyceraldehyde 3-phosphate *(blue arrow)* is large enough to drive two energetically costly reactions: the formation of both NADH and a high-energy phosphate bond in 1,3-bisphosphoglycerate *(red arrows)*. The subsequent energetically favorable breakage of that high-energy phosphate bond in step 7 then drives the formation of ATP.

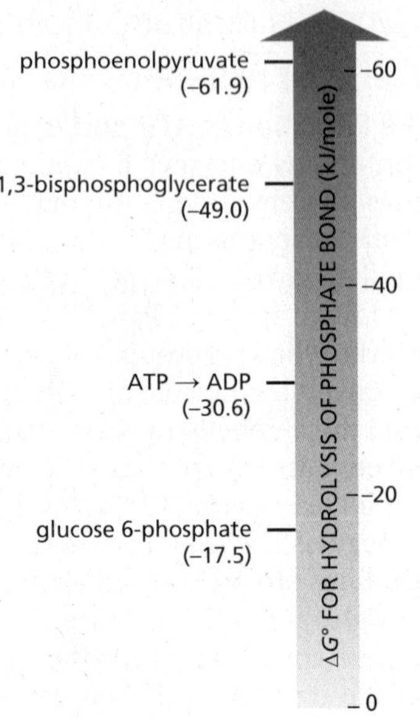

Figure 2–49 **Phosphate bonds with different hydrolysis energies.** Examples of molecules formed during glycolysis that contain different types of phosphate bonds are shown, along with the standard free-energy change for hydrolysis of those bonds in kJ/mole. The transfer of a phosphate group from one molecule to another is energetically favorable if the standard free-energy change ($\Delta G°$) for hydrolysis of the phosphate bond is more negative for the donor molecule than for the acceptor. (The hydrolysis reactions can be thought of as the transfer of the phosphate group to water.) Thus, a phosphate group is readily transferred from 1,3-bisphosphoglycerate to ADP to form ATP. Transfer reactions involving the phosphate groups in these molecules are detailed in Panel 2–8 (pp. 108–109). Note that standard conditions often do *not* pertain to living cells, where the relative concentrations of reactants and products will influence the actual change in free energy (see p. 67).

As we have just seen, ATP can be formed readily from ADP when a reaction intermediate is formed with a phosphate bond of higher energy than that of the terminal phosphate bond in ATP. Phosphate bonds can be ordered in energy by comparing the standard free-energy change ($\Delta G°$) for the breakage of each bond by hydrolysis. Figure 2–49 compares the high-energy phosphoanhydride bonds in ATP with the energy of other types of phosphate bonds that are generated during glycolysis.

Fermentations Produce ATP in the Absence of Oxygen

For most animal and plant cells, glycolysis is only a prelude to the final stage of the breakdown of food molecules. In these cells, the pyruvate formed by glycolysis is rapidly transported into the mitochondria, where it is converted into CO_2 plus acetyl CoA, whose acetyl group is then completely oxidized to CO_2 and H_2O.

In contrast, for many anaerobic organisms—which do not utilize molecular oxygen and can grow and divide without it—glycolysis is the principal source of the cell's ATP. Certain animal tissues, such as skeletal muscle, can also continue to function when molecular oxygen is limited. In these anaerobic conditions, the pyruvate and the NADH electrons stay in the cytosol. The pyruvate is converted into products excreted from the cell; for example, into ethanol and CO_2 in the yeasts used in brewing and breadmaking or into lactic acid (lactate) in muscle. In this process, the NADH gives up its electrons and is converted back into NAD^+. This is crucial, because a regeneration of NAD^+ is required to maintain the reactions of glycolysis (Figure 2–50). Energy-yielding pathways like these are called **fermentations**.

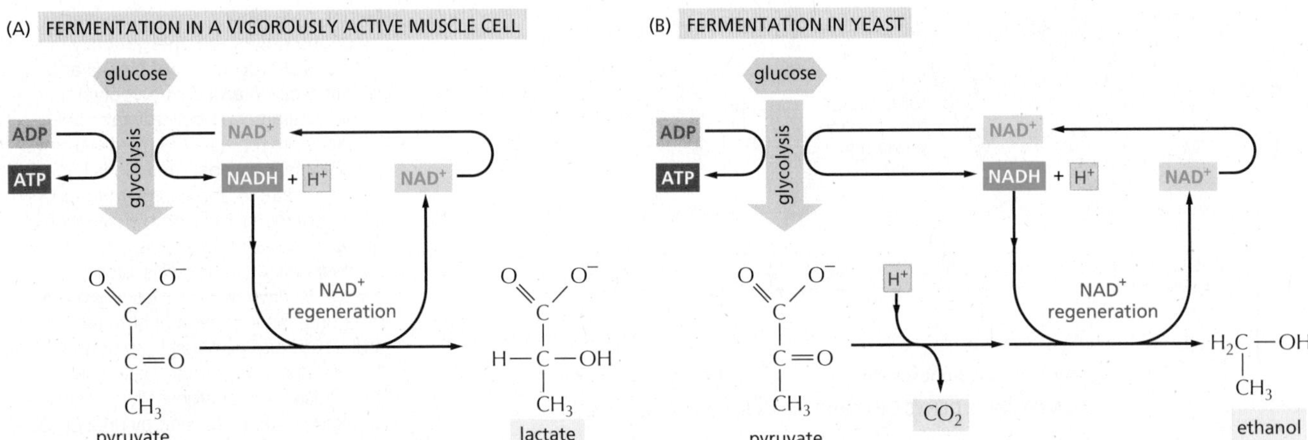

Figure 2–50 **Two pathways for the anaerobic breakdown of pyruvate.** (A) When there is inadequate oxygen in a muscle cell undergoing vigorous contraction, the pyruvate produced by glycolysis is converted to lactic acid (lactate) as shown. This reaction regenerates the NAD^+ required in step 6 of glycolysis, but the whole pathway yields much less energy overall than complete oxidation does. (B) In some organisms that can grow anaerobically, such as yeasts, pyruvate is converted via acetaldehyde into carbon dioxide and ethanol. Again, this pathway regenerates NAD^+ from NADH, as needed for glycolysis to continue. Both (A) and (B) are examples of *fermentations*.

Organisms Store Food Molecules in Special Reservoirs

All organisms need to maintain a high ATP/ADP ratio to maintain biological order in their cells. Yet animals have only periodic access to food, and plants need to survive overnight without sunlight, when they are unable to produce sugar from photosynthesis. For this reason, both plants and animals convert sugars and fats to special forms for storage.

To compensate for long periods of fasting, animals store fatty acids as **fat** droplets composed of water-insoluble *triacylglycerols* (also called triglycerides). The triacylglycerols in animals are mostly stored in the cytoplasm of specialized fat cells called adipocytes (Figure 2–51). For shorter-term storage, sugar is stored as glucose subunits in the large branched polysaccharide **glycogen**, which is present as small granules in the cytoplasm of many cells, with the largest stores in liver and muscle.

The synthesis and degradation of glycogen are rapidly regulated according to need. When cells need more ATP than they can generate from the food molecules taken in from the bloodstream, they break down glycogen in a reaction that produces glucose 1-phosphate, which is rapidly converted to glucose 6-phosphate for glycolysis (Figure 2–52). During fasting, liver cells release glucose derived from breakdown of their glycogen stores into the bloodstream for use by other cells, while muscle cells hoard their supplies for their own use.

Quantitatively, fat is far more important than glycogen as an energy store for animals, because it provides for more efficient storage. The oxidation of a gram of fat releases about twice as much energy as the oxidation of a gram of glycogen does. Moreover, glycogen differs from fat in binding a great deal of water, producing a sixfold difference in the actual mass of glycogen required to store the same amount of energy as fat. An average adult human stores enough glycogen for only about a day of normal activities but enough fat to last for nearly a month.

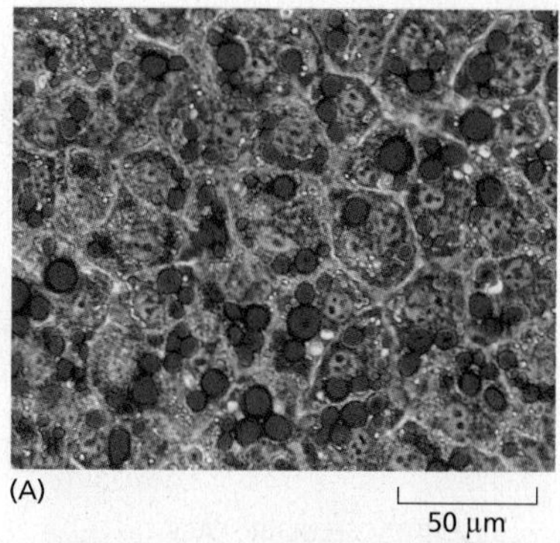

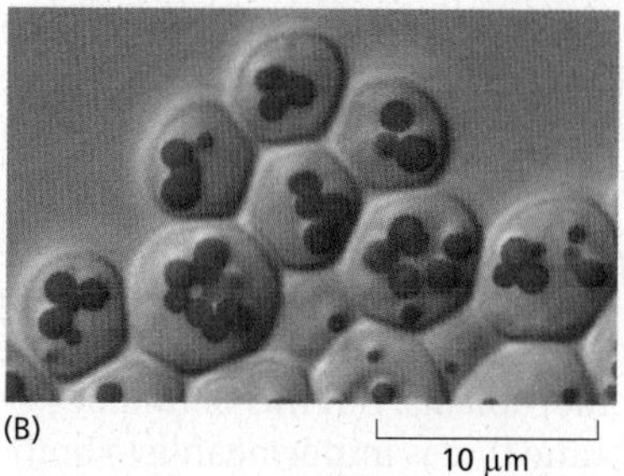

Figure 2–51 **Fats stored in the form of lipid droplets in cells.** (A) Fat droplets (stained *red*) in the cytoplasm of developing adipocytes. (B) Lipid droplets *(red)* in yeast cells, which also use them as a reservoir of energy and as building blocks for membrane lipid biosynthesis. (A, courtesy of Peter Tontonoz and Ronald M. Evans; B, courtesy of Sepp D. Kohlwein.)

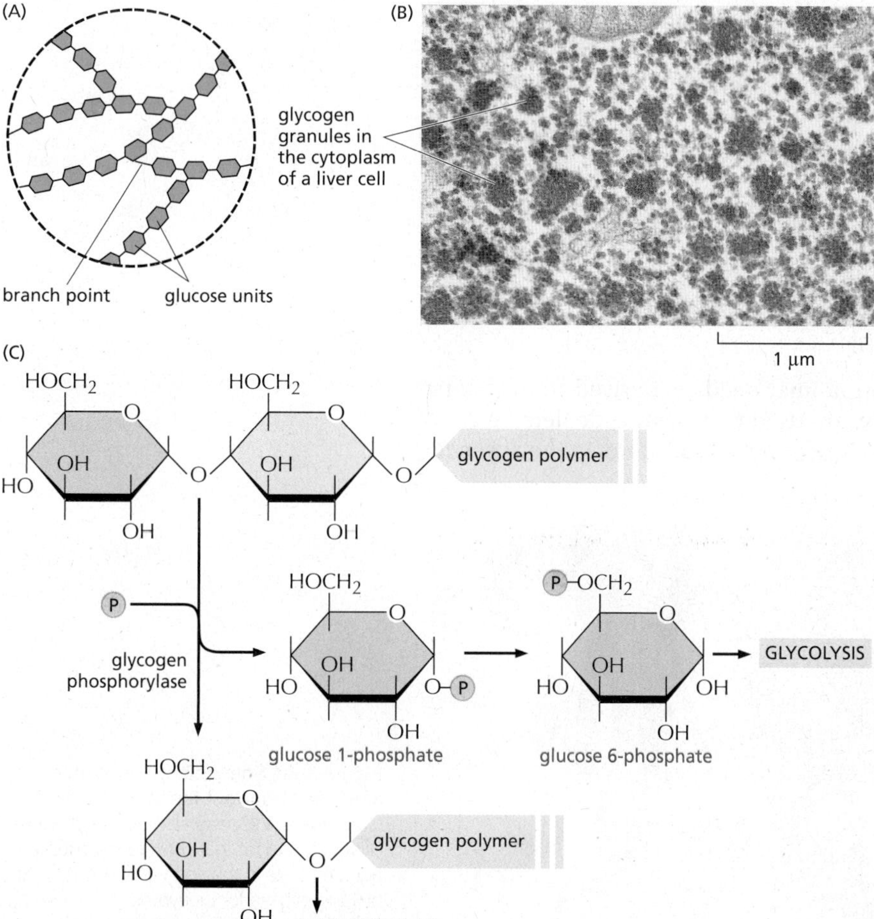

Figure 2–52 **How animal cells store glucose in the form of glycogen to provide energy in times of need.** (A) The structure of glycogen; starch in plants is a very similar branched polymer of glucose but has many fewer branch points. (B) An electron micrograph showing glycogen granules in the cytoplasm of a liver cell; each granule contains both glycogen and the enzymes required for glycogen synthesis and breakdown. (C) The enzyme glycogen phosphorylase breaks down glycogen when cells need more glucose. (B, courtesy of Robert Fletterick and Daniel S. Friend, by permission of E.L. Bearer.)

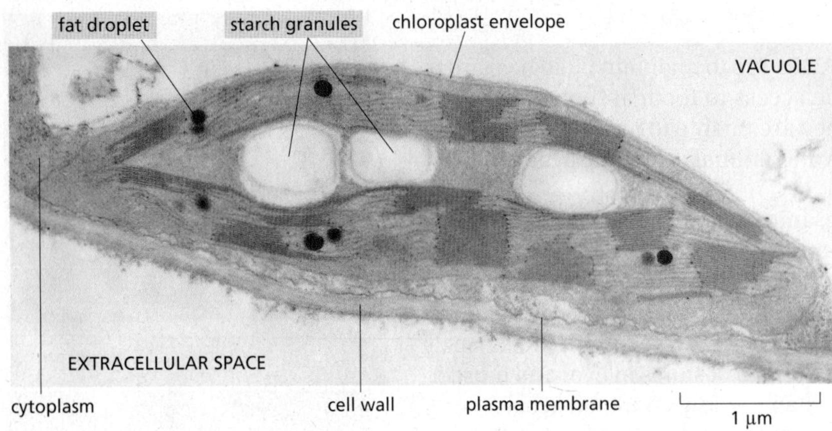

Figure 2–53 **Plant cells store both starch and fats in their chloroplasts.** An electron micrograph of a single chloroplast in a plant cell shows the starch granules and fat droplets that have been synthesized in the organelle. (Courtesy of K. Plaskitt.)

If our main fuel reservoir had to be carried as glycogen instead of fat, body weight would increase by an average of about 60 pounds.

How do plants store sugars and fats? Plants produce abundant amounts of both ATP and NADPH by the photosynthesis that is carried out in their chloroplasts. But this organelle is isolated from the rest of its plant cell by a membrane that is impermeable to both types of activated carrier molecules. Moreover, the plant contains many cells—such as those in the roots—that lack chloroplasts and therefore cannot produce their own sugars. Thus, sugars are exported from chloroplasts to the mitochondria present in all cells of the plant. Most of the ATP needed for general plant cell metabolism is synthesized in these mitochondria, using exactly the same pathways for the oxidative breakdown of sugars as in nonphotosynthetic organisms; this ATP is then passed to the rest of the cell.

During periods of excess photosynthetic capacity during the day, chloroplasts convert some of the sugars that they make into fats and into **starch**, a polymer of glucose analogous to the glycogen of animals. Both are stored inside the chloroplast until needed for energy-yielding oxidation during periods of darkness (**Figure 2–53**).

The embryos inside plant seeds must live on stored sources of energy for a prolonged period until they germinate and produce leaves that can harvest the energy in sunlight. For this reason, plant seeds often contain especially large amounts of fats and starch—which makes them a major food source for animals, including ourselves (**Figure 2–54**).

Between Meals, Most Animal Cells Derive Their Energy from Fatty Acids Obtained from Fat

After a meal, most of the energy that an animal needs is derived from sugars obtained from food. Excess sugars, if any, are used to replenish depleted glycogen stores or to synthesize fats as a stable longer-term food store. In the absence

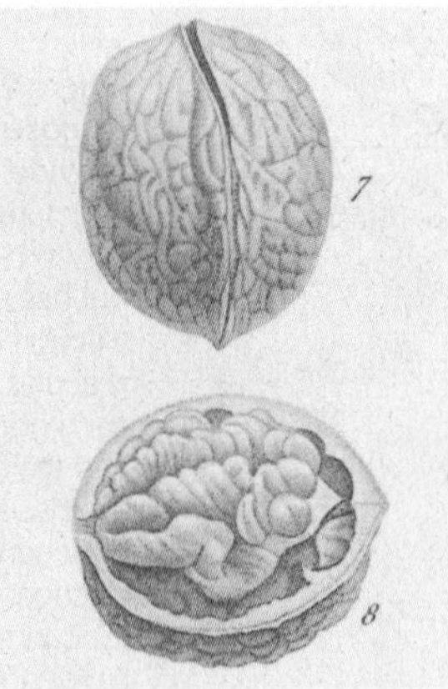

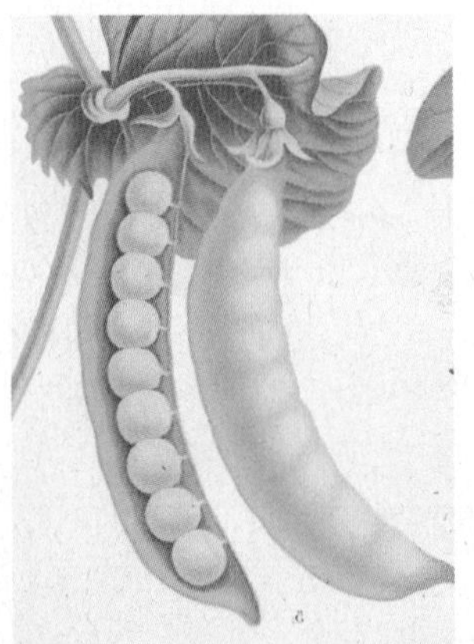

Figure 2–54 **Some plant seeds that serve as important foods for humans.** Corn, nuts, and peas all contain rich stores of starch and fat that provide the young plant embryo in the seed with energy and building blocks for biosynthesis. (Courtesy of the John Innes Foundation.)

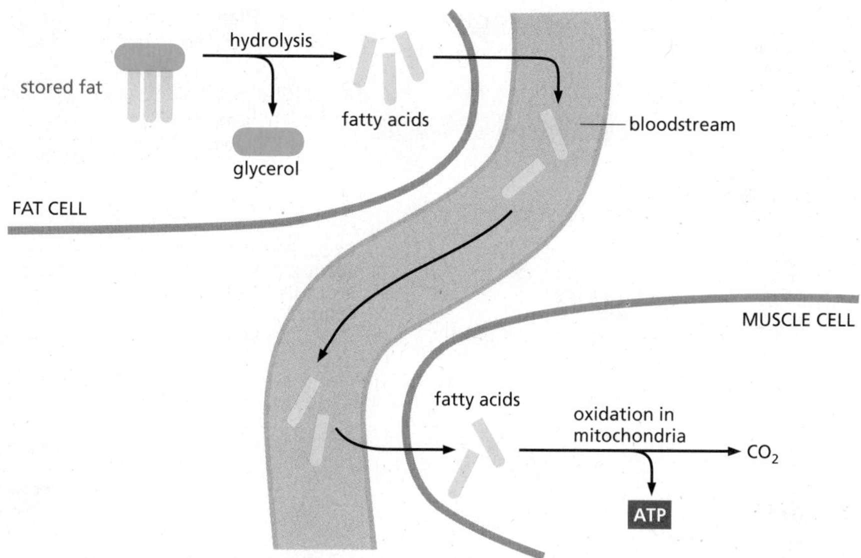

Figure 2–55 **How stored fats are mobilized for energy production in animals.** Low glucose levels in the blood trigger the hydrolysis of the triacylglycerol molecules in fat droplets to free fatty acids and glycerol. These fatty acids enter the bloodstream, where they bind to the abundant blood protein, serum albumin. Special fatty acid transporters in the plasma membrane of cells that oxidize fatty acids, such as muscle cells, then pass these fatty acids into the cytosol, from which they are moved into mitochondria for energy production.

of another meal, the liver begins to release glucose from its store of glycogen to maintain circulating glucose levels, and the fat stored in adipose tissue is also called into play. By the morning after an overnight fast, the glycogen reserves are exhausted, and fatty acid oxidation generates most of the ATP we need.

Low glucose levels in the blood trigger the breakdown of fats for energy production. As illustrated in Figure 2–55, the triacylglycerols stored in fat droplets in adipocytes are hydrolyzed to produce fatty acids and glycerol, and the fatty acids released are transferred to cells in the body through the bloodstream. Notably, the brain must rely on circulating glucose—or *ketone bodies* when available—because fatty acids are poorly utilized by the brain.

What are ketone bodies? During prolonged periods of fasting, when the circulating supply of glucose is mostly maintained by its synthesis from amino acids derived from the breakdown of proteins in muscle, the liver assumes its role as a central metabolic hub to convert fatty acids to the energy-rich molecules acetoacetate and β-hydroxybutyrate. These ketone bodies are released into the bloodstream to serve as an alternative fuel in heart and brain cells, where they are oxidized through the citric acid cycle to generate ATP. By thereby meeting most of the energy needs of the brain and heart, this process partially spares protein breakdown. Consumption of a diet very low in carbohydrate (a *ketogenic diet*) leads to the production of ketone bodies and can enable weight loss in most individuals. A spontaneous breakdown of acetoacetate to acetone generates the bad breath often associated both with such diets and with prolonged fasting.

Sugars and Fats Are Both Degraded to Acetyl CoA in Mitochondria

The fatty acids imported from the bloodstream are moved into mitochondria, where all of their oxidation takes place (Figure 2–56). Each molecule of fatty acid (as the activated molecule *fatty acyl CoA*) is broken down completely by a cycle

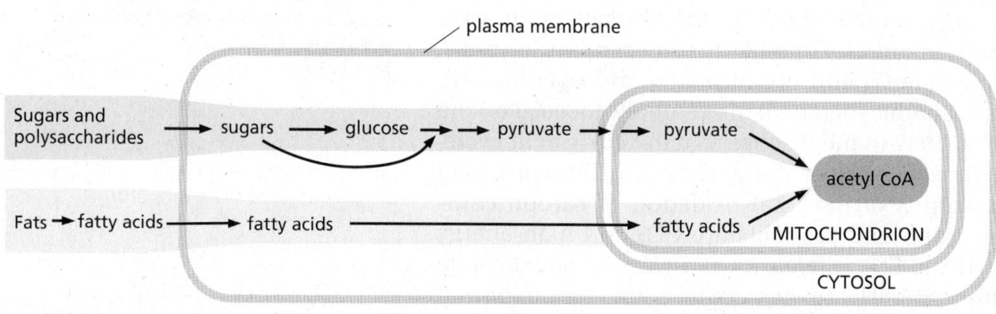

Figure 2–56 **Pathways for the production of acetyl CoA from sugars and fats.** The mitochondrion in eukaryotic cells is where acetyl CoA is produced from both types of food molecules. It is therefore the place where most of the cell's oxidation reactions occur and where most of its ATP is made. Amino acids (not shown) can also enter mitochondria to be converted there into acetyl CoA or another intermediate of the citric acid cycle. The structure and function of mitochondria are discussed in detail in Chapter 14.

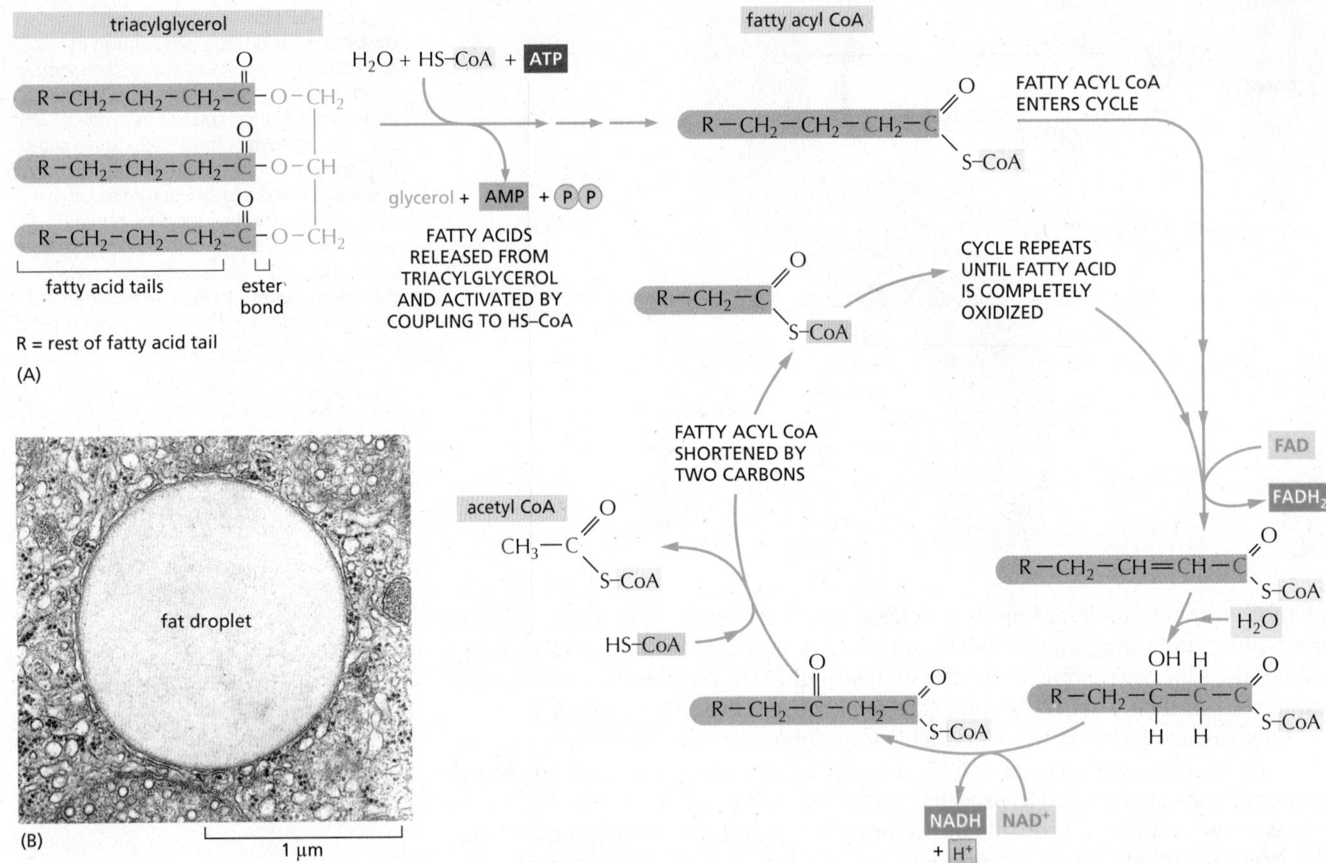

of reactions that trims two carbons at a time from its carboxyl end, generating one molecule of acetyl CoA for each turn of the cycle. A molecule of NADH and a molecule of $FADH_2$ are also produced in this process (**Figure 2–57**).

As was shown in Figure 2–56, in eukaryotes the pyruvate that was produced by glycolysis from sugars in the cytosol is likewise transported into mitochondria, where it is rapidly decarboxylated by a giant complex of three enzymes, called the *pyruvate dehydrogenase complex*. The products are a molecule of CO_2 (a waste product), a molecule of NADH, and acetyl CoA (see Panel 2–9, pp. 110–111).

Sugars and fats are the major energy sources for most nonphotosynthetic organisms, including humans. Most of the useful energy that can be extracted from their oxidation remains stored in the acetyl CoA molecules that are produced from both sources. In the *citric acid cycle* of reactions, the acetyl group ($-COCH_3$) in acetyl CoA is completely oxidized to CO_2 and H_2O. This cycle, to be described next, therefore plays a critical, central role in the energy metabolism of aerobic organisms.

Figure 2–57 **How the oxidation of fatty acids produces acetyl CoA.** (A) Fats are stored in the form of triacylglycerol, the glycerol portion of which is shown in *blue*. Three fatty acid chains (shaded in *red*) are linked to this glycerol through ester bonds that can be hydrolyzed by enzymes called lipases to allow the fatty acids to enter the bloodstream (see Figure 2–55). Once the fatty acids enter mitochondria (see Figure 2–56), they are coupled to coenzyme A in a reaction requiring ATP. These activated fatty acids (fatty acyl CoA molecules) are then oxidized in a cycle containing four enzymes, which are not shown. As indicated, each turn of the cycle shortens the fatty acyl CoA molecule by two carbons *(red)* and generates one molecule each of the energy-rich molecules $FADH_2$, NADH, and acetyl CoA. (B) Fats are insoluble in water and spontaneously form large lipid droplets. This electron micrograph shows a lipid droplet in the cytoplasm of a specialized fat cell, an adipocyte. (B, courtesy of Daniel S. Friend and by permission of E.L. Bearer.)

The Citric Acid Cycle Generates NADH by Oxidizing Acetyl Groups to CO_2

In the nineteenth century, biologists noticed that in the absence of air, cells produce lactic acid (for example, in muscle) or ethanol (for example, in yeast), while in its presence they consume O_2 and produce CO_2 and H_2O instead. Efforts to define the pathways of aerobic metabolism eventually focused on the oxidation of pyruvate and led in 1937 to the discovery of the **citric acid cycle**, also known as the *tricarboxylic acid cycle* or the *Krebs cycle*. The citric acid cycle accounts for about two-thirds of the total oxidation of carbon compounds in most cells, and its major end products are CO_2 and high-energy electrons in the form of NADH. The CO_2 is released as a waste product, while the high-energy electrons from NADH are passed to a membrane-bound

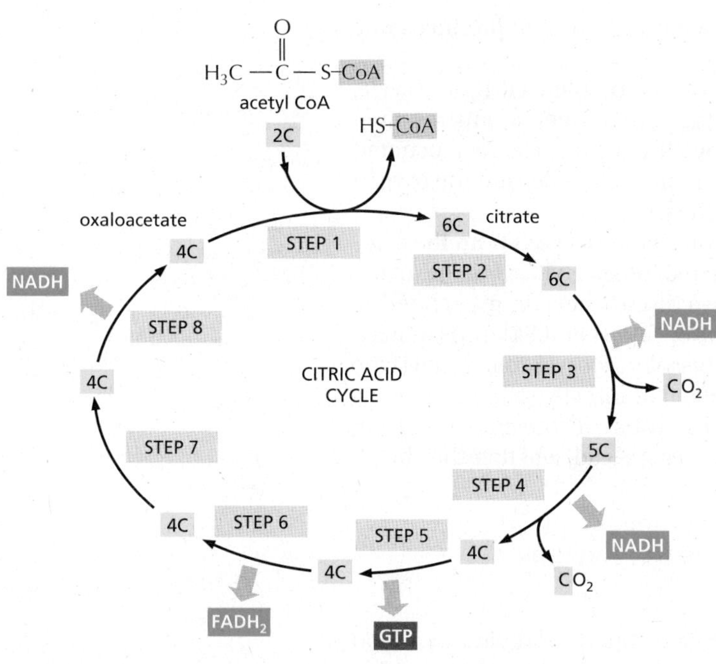

NET RESULT: ONE TURN OF THE CYCLE PRODUCES THREE NADH, ONE GTP, AND ONE $FADH_2$, AND RELEASES TWO MOLECULES OF CO_2

Figure 2–58 **Simple overview of the citric acid cycle.** The reaction of acetyl CoA with oxaloacetate starts the cycle by producing citrate (citric acid). In each turn of the cycle, two molecules of CO_2 are produced as waste products, plus three molecules of NADH, one molecule of GTP, and one molecule of $FADH_2$. The number of carbon atoms in each intermediate is shown in a *yellow box*. For details, see Panel 2–9 (pp. 110–111).

electron-transport chain that will be discussed in Chapter 14, eventually combining with O_2 to produce H_2O.

The citric acid cycle itself does not use gaseous O_2 (it uses oxygen atoms from H_2O). But the cycle does require O_2 in subsequent reactions to keep it going. This is because there is no other efficient way for the NADH to get rid of its electrons and thus regenerate the NAD^+ that the cycle requires.

The citric acid cycle takes place inside mitochondria in eukaryotic cells. The process begins when the acetyl group is transferred from acetyl CoA to a four-carbon molecule, *oxaloacetate*, to form the six-carbon tricarboxylic acid, *citric acid*, for which the subsequent cycle of reactions is named. This citric acid molecule is then gradually oxidized, allowing the energy of oxidation to be harnessed to produce activated carrier molecules. The chain of eight reactions forms a cycle because at the end the oxaloacetate is regenerated to enter a new turn of the cycle, as shown in outline in **Figure 2–58.**

We have thus far highlighted only one of the three types of activated carrier molecules that are produced by the citric acid cycle: NADH, the reduced form of the NAD^+/NADH electron carrier system (see Figure 2–36). In addition to three molecules of NADH, each turn of the cycle also produces one molecule of $FADH_2$ from FAD (see Figure 2–39) and one molecule of GTP, guanosine triphosphate, from GDP. The structure of **GTP** is illustrated in **Figure 2–59**. GTP is a close relative

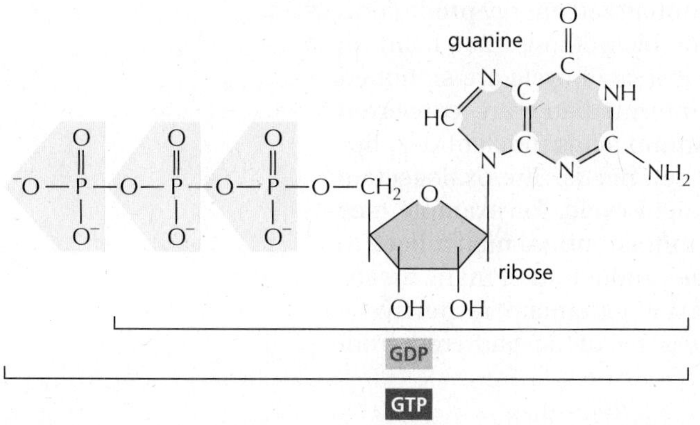

Figure 2–59 **The structure of GTP, guanosine triphosphate.** GTP and GDP are close relatives of ATP and ADP, respectively.

of ATP, and the transfer of its terminal phosphate group to ADP produces one ATP molecule in each cycle.

Panel 2–9 (pp. 110–111) and **Movie 2.6** present the complete citric acid cycle. Water, rather than molecular oxygen, supplies the extra oxygen atoms required to make CO_2 from the acetyl groups entering the citric acid cycle. As illustrated in the Panel, three molecules of water are taken up in each cycle, and the oxygen atoms of some of them are ultimately used to make CO_2.

In addition to pyruvate and fatty acids, some amino acids pass from the cytosol into mitochondria, where they are also converted into acetyl CoA or one of the other intermediates of the citric acid cycle. As we discuss next, in mitochondria the large amount of energy stored in the electrons of NADH and $FADH_2$ is utilized for ATP production through the process of *oxidative phosphorylation*, which is the only step in the oxidative catabolism of foodstuffs that directly requires gaseous oxygen (O_2) from the atmosphere. Thus, in the eukaryotic cell, the mitochondrion is the center toward which all energy-yielding processes lead, whether they begin with sugars, fats, or proteins.

Electron Transport Drives the Synthesis of the Majority of the ATP in Most Cells

Most chemical energy is released in the last stage in the degradation of a food molecule. This process begins when NADH and $FADH_2$ transfer the electrons that they obtained by oxidizing food-derived organic molecules to an **electron-transport chain** embedded in the inner membrane of the mitochondrion. As the electrons pass along this long chain of specialized electron acceptor and donor molecules, they fall to successively lower energy states, being finally passed to molecules of oxygen gas (O_2) that have diffused into the mitochondrion, reducing the oxygen to produce water. The electrons have now reached a very low energy level, and all the available energy has been extracted from the oxidized food molecule.

The energy released by this chain of electron transfers is used to pump H^+ ions (protons) across the membrane—from the innermost mitochondrial compartment (the matrix) to the intermembrane space (and then to the cytosol). The resulting electrochemical proton gradient across the inner mitochondrial membrane serves as a major source of energy for cells, being tapped like a battery to drive a variety of energy-requiring reactions. The most prominent of these reactions is the generation of ATP by the phosphorylation of ADP, as part of a process known as **oxidative phosphorylation** (see Figure 14–12). This harnessing of chemical energy through membrane-based electron transfers is one of the most remarkable achievements of cell evolution—as such, we shall devote an entire chapter to it (Chapter 14).

Many Biosynthetic Pathways Begin with Glycolysis or the Citric Acid Cycle

Catabolic reactions, such as those of glycolysis and the citric acid cycle, produce both energy for the cell and the building blocks from which many other organic molecules are made. Thus far, we have emphasized energy production rather than the provision of starting materials for biosynthesis. But many of the intermediates formed in glycolysis and the citric acid cycle are siphoned off by such anabolic pathways, in which the intermediates are converted by a series of enzyme-catalyzed reactions into amino acids, nucleotides, lipids, and other small organic molecules that the cell needs. The oxaloacetate and α-ketoglutarate produced during the citric acid cycle, for example (see Panel 2–9, pp. 110–111), are transferred from the mitochondrial matrix back to the cytosol, where they serve as precursors for the production of many essential molecules, such as the amino acids aspartate and glutamate, respectively. An idea of the extent of these anabolic pathways can be gathered from

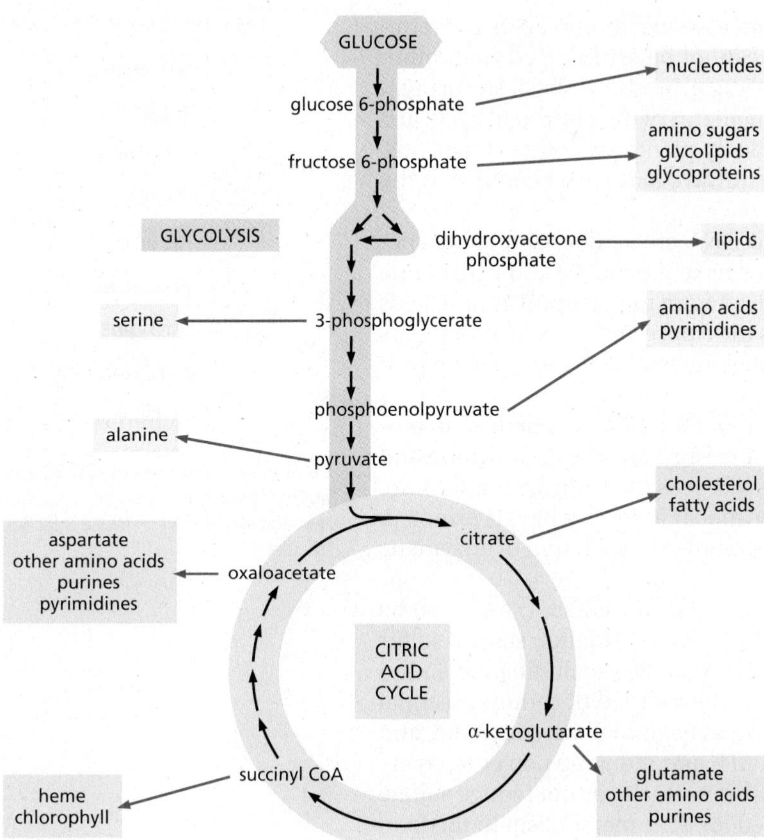

Figure 2–60 **Glycolysis and the citric acid cycle provide the precursors needed to synthesize many important biological molecules.** The amino acids, nucleotides, lipids, sugars, and other molecules—shown here as products—in turn serve as the precursors for the many macromolecules of the cell. Each *black arrow* in this diagram denotes a single enzyme-catalyzed reaction; the *red arrows* generally represent pathways with many steps that are required to produce the indicated products.

Figure 2–60, which illustrates some of the branches leading from the central catabolic reactions to biosyntheses.

Animals Must Obtain All the Nitrogen and Sulfur They Need from Food

So far we have concentrated mainly on carbohydrate metabolism and have not yet considered the metabolism of nitrogen or sulfur. These two elements are important constituents of biological macromolecules. Nitrogen and sulfur atoms pass from compound to compound and between organisms and their environment in a series of reversible cycles.

Although molecular nitrogen constitutes nearly 80% of Earth's atmosphere, nitrogen is chemically unreactive as a gas. Only a few living species are able to incorporate it into organic molecules, a process called **nitrogen fixation**. Nitrogen fixation occurs both in certain microorganisms and by some geophysical processes, such as lightning discharge. It is essential to the biosphere as a whole, for without it life could not exist on this planet. Only a small fraction of the nitrogenous compounds in today's organisms, however, is due to fresh products of nitrogen fixation from the atmosphere. Most organic nitrogen has been in circulation for some time, passing from one living organism to another. Thus, present-day nitrogen-fixing reactions can be said to perform a "topping-up" function for the total nitrogen supply.

Vertebrates receive virtually all of their nitrogen from their dietary intake of proteins and nucleic acids. In the body, these macromolecules are broken down to amino acids and the components of nucleotides, and the nitrogen they contain is used to produce new proteins and nucleic acids—or other molecules. About half of the 20 amino acids found in proteins are essential amino acids for vertebrates (Figure 2–61), which means that they cannot be synthesized from other ingredients of the diet. The other amino acids can be so synthesized, using a variety of raw materials that include the intermediates of the citric acid cycle just described. The essential amino acids are made by plants and other organisms, usually by long and energetically expensive pathways that have been lost in the course of vertebrate evolution.

The nucleotides needed to make RNA and DNA are synthesized using specialized biosynthetic pathways. All of the nitrogens in the purine and pyrimidine bases (as well as some of the carbons) are derived from the plentiful amino acids glutamine, aspartic acid, and glycine, whereas the ribose and deoxyribose sugars are derived from glucose. There are no "essential nucleotides" that must be provided in the diet.

As we have seen, the amino acids derived from food that are not used in biosynthesis can be oxidized to generate metabolic energy. Most of their carbon and hydrogen atoms eventually form CO_2 or H_2O, whereas their nitrogen atoms are shuttled through various forms and eventually appear as urea, which is excreted. Each amino acid is processed differently, and a whole constellation of enzymatic reactions exists for their catabolism.

Sulfur is abundant on Earth in its most oxidized form, sulfate (SO_4^{2-}). To be useful for life, sulfate must be reduced to sulfide (S^{2-}), the oxidation state of sulfur required for the synthesis of essential biological molecules, including the amino acids methionine and cysteine, coenzyme A, and the iron–sulfur centers essential for electron transport. The sulfur-reduction process begins in bacteria, fungi, and plants, where a special group of enzymes use ATP and reducing power to create a sulfate assimilation pathway. Humans and other animals cannot reduce sulfate and must therefore acquire the sulfur they need for their metabolism in the food that they eat.

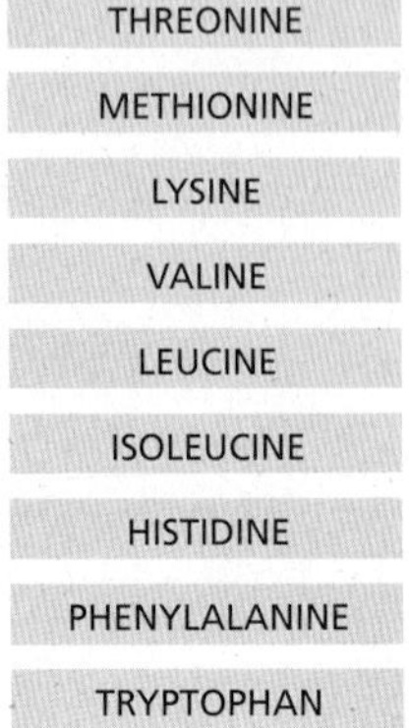

Figure 2–61 **The nine essential amino acids.** These cannot be synthesized by human cells and so must be supplied in the diet.

Metabolism Is Highly Organized and Regulated

One can get a sense of the intricacy of a cell as a chemical machine from the relation of glycolysis and the citric acid cycle to the other metabolic pathways sketched out in Figure 2–62. This diagram represents only some of the enzymatic pathways in a human cell. It is obvious that our discussion of cell metabolism has dealt with only a tiny fraction of the broad field of cell chemistry.

All these reactions occur in a cell that is less than 0.1 mm in diameter, and each requires a different enzyme. The same molecule can often be part of many different pathways. Pyruvate, for example, is a substrate for half a dozen or more different enzymes, each of which modifies it chemically in a different way. One enzyme converts pyruvate to acetyl CoA, another to oxaloacetate; a third enzyme changes pyruvate to the amino acid alanine, a fourth to lactate, and so on. All of these different pathways compete for the same pyruvate molecule, and similar competitions for thousands of other small molecules go on at the same time.

The situation is further complicated in a multicellular organism. Different cell types require somewhat different sets of enzymes. And different tissues make distinct contributions to the chemistry of the organism as a whole. All types of cells have their distinctive metabolic traits, and they cooperate extensively in the normal state, as well as in response to stress and starvation. One might think that the whole system would need to be so finely balanced that any minor upset, such as a temporary change in dietary intake, would be disastrous.

In fact, the metabolic balance of a cell is amazingly stable. Whenever the balance is perturbed, the cell reacts so as to restore the initial state. The cell can adapt and continue to function during starvation or disease. Mutations of many kinds can damage or even eliminate particular reaction pathways, and yet—provided that certain minimum requirements are met—the cell survives. It does so because an elaborate network of *control mechanisms* regulates and coordinates the rates of all of its reactions. These controls rest, ultimately, on the remarkable abilities

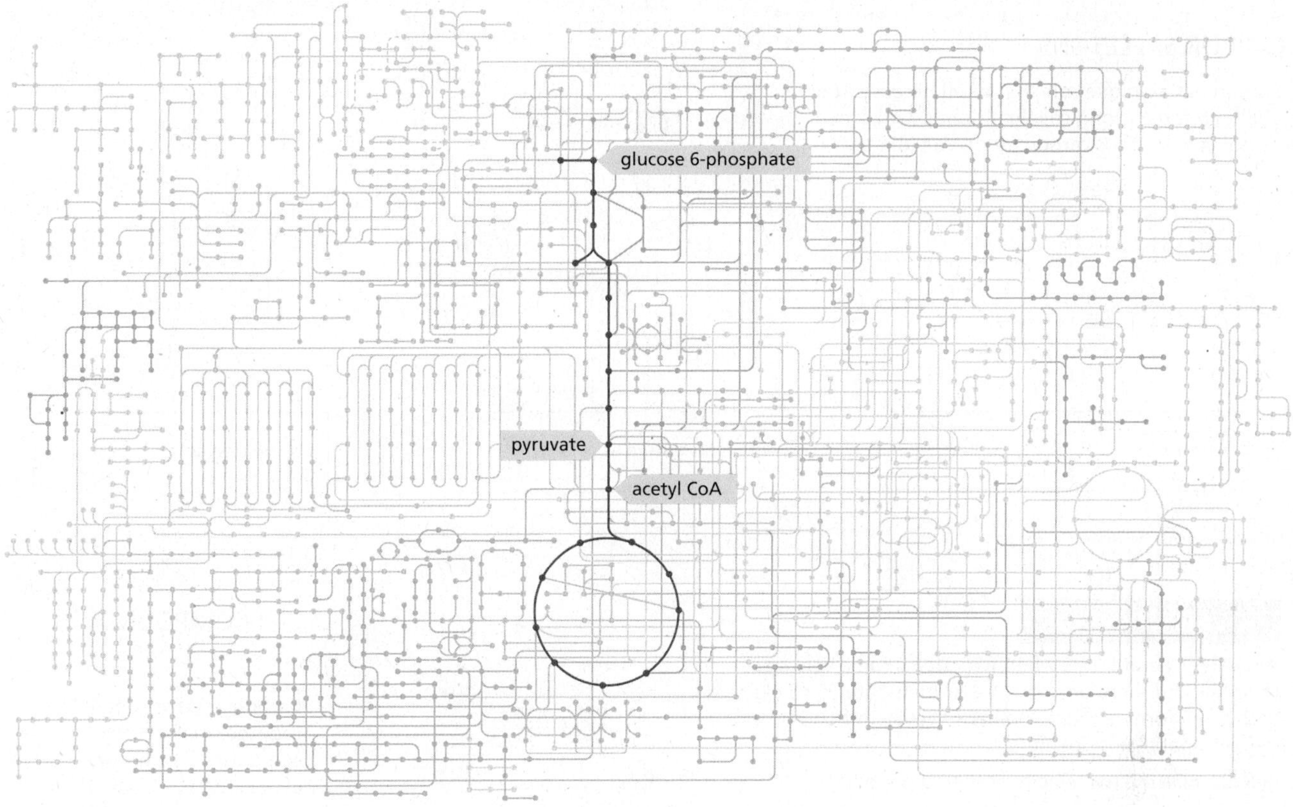

Figure 2–62 **Glycolysis and the citric acid cycle are at the center of an elaborate set of metabolic pathways in human cells.** Some 2000 metabolic reactions are shown schematically with the reactions of glycolysis and the citric acid cycle in *red.* Many other reactions either lead into these two central pathways—delivering small molecules to be catabolized with production of energy—or they lead outward as in Figure 2–60 to supply carbon compounds for the purpose of biosynthesis. (Adapted from KEGG Database. With permission from Kanehisa Laboratories.)

of proteins to change their shape and their chemistry in response to changes in their immediate environment.

The principles that underlie how proteins are built and the chemistry behind their regulation are clearly central to all of biology. And it is proteins that will be our next concern.

Summary

Food molecules are broken down by controlled stepwise oxidation to provide chemical energy in the form of ATP and NADH. Following the breakdown of large food molecules to their simple subunits, three main sets of reactions act in series, the products of each being the starting material for the next: glycolysis (which occurs in the cytosol), the citric acid cycle (in the mitochondrial matrix), and oxidative phosphorylation (on the inner mitochondrial membrane). The intermediate products of glycolysis and the citric acid cycle are used both as sources of metabolic energy and to produce many of the small molecules used as the raw materials for biosynthesis. Cells store sugar molecules as glycogen in animals and starch in plants; both plants and animals also use fats extensively as a food store. These storage materials in turn serve as a major source of food for humans, along with the proteins that comprise the majority of the dry mass of most of the cells in the foods we eat.

CARBON SKELETONS

Carbon has a unique role in the cell because of its ability to form strong covalent bonds with other carbon atoms. Thus carbon atoms can join to form:

chains

also written as

branched trees

also written as

rings

also written as

COVALENT BONDS

A covalent bond forms when two atoms come very close together and share one or more of their outer-shell electrons. Each atom forms a fixed number of covalent bonds in a defined spatial arrangement.

SINGLE BONDS: two electrons shared per bond

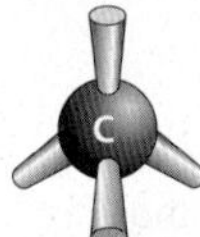

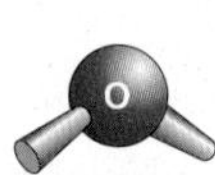

DOUBLE BONDS: four electrons shared per bond

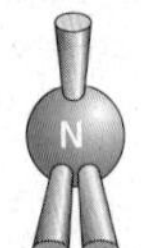

The precise spatial arrangement of covalent bonds influences the three-dimensional structure and chemistry of molecules. In this review Panel, we see how covalent bonds are used in a variety of biological molecules.

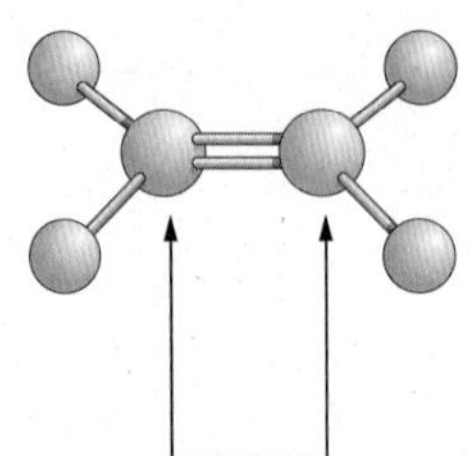

Atoms joined by two or more covalent bonds cannot rotate freely around the bond axis. This restriction has a major influence on the three-dimensional shape of many macromolecules.

C–H COMPOUNDS

Carbon and hydrogen together make stable compounds (or groups) called hydrocarbons. These are nonpolar, do not form hydrogen bonds, and are generally insoluble in water.

methane

methyl group

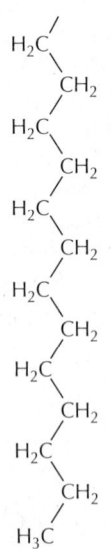

part of the hydrocarbon "tail" of a fatty acid molecule

ALTERNATING DOUBLE BONDS

A carbon chain can include double bonds. If these are on alternate carbon atoms, the bonding electrons move within the molecule, stabilizing the structure by a phenomenon called *resonance*.

the truth is somewhere between these two structures

Alternating double bonds in a ring can generate a very stable structure.

benzene

often written as

C–O COMPOUNDS

Many biological compounds contain a carbon covalently bonded to an oxygen. For example,

alcohol

The –OH is called a hydroxyl group.

aldehyde

ketone

The C=O is called a carbonyl group.

carboxylic acid

The –COOH is called a carboxyl group. In water, this loses an H^+ ion to become $–COO^-$.

esters — Esters are formed by combining an acid and an alcohol.

acid + alcohol → ester + H_2O

C–N COMPOUNDS

Amines and amides are two important examples of compounds containing a carbon linked to a nitrogen.

Amines in water combine with an H^+ ion to become positively charged.

Amides are formed by combining an acid and an amine. Unlike amines, amides are uncharged in water. An example is the peptide bond that joins amino acids in a protein.

acid + amine → amide + H_2O

Nitrogen also occurs in several ring compounds, including important constituents of nucleic acids: purines and pyrimidines.

cytosine (a pyrimidine)

SULFHYDRYL GROUP

The –C–SH is called a sulfhydryl group. In the amino acid cysteine, the sulfhydryl group may exist in the reduced form –C–SH or more rarely in an oxidized, cross-bridging form –C–S–S–C–

PHOSPHATES

Inorganic phosphate is a stable ion formed from phosphoric acid, H_3PO_4. It is also written as Ⓟ.

Phosphate esters can form between a phosphate and a free hydroxyl group. Phosphate groups are often covalently attached to proteins in this way.

also written as –C–O–Ⓟ

The combination of a phosphate and a carboxyl group, or two or more phosphate groups, produces an acid anhydride. Because compounds of this type release a large amount of free energy when the bond is broken by hydrolysis in the cell, they are often said to contain a *high-energy* bond.

high-energy acyl phosphate bond (carboxylic–phosphoric acid anhydride) found in some metabolites

also written as –C(=O)–O–Ⓟ

high-energy phosphoanhydride bond found in molecules such as ATP

also written as –O–Ⓟ–Ⓟ

HYDROGEN BONDS

Because they are polarized, two adjacent H_2O molecules can form a noncovalent linkage known as a hydrogen bond. Hydrogen bonds have only about 1/20 the strength of a covalent bond.

Hydrogen bonds are strongest when the three atoms lie in a straight line.

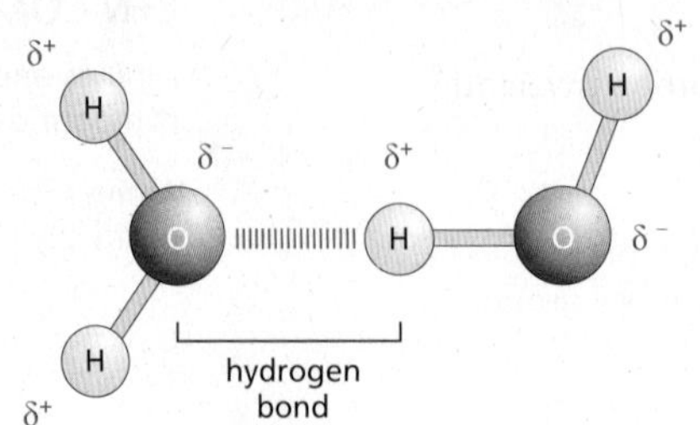

bond lengths

hydrogen bond 0.17 nm

0.10 nm covalent bond

WATER

Two atoms connected by a covalent bond may exert different attractions for the electrons of the bond. In such cases, the bond is polar, with one end slightly negatively charged (δ−) and the other slightly positively charged (δ+).

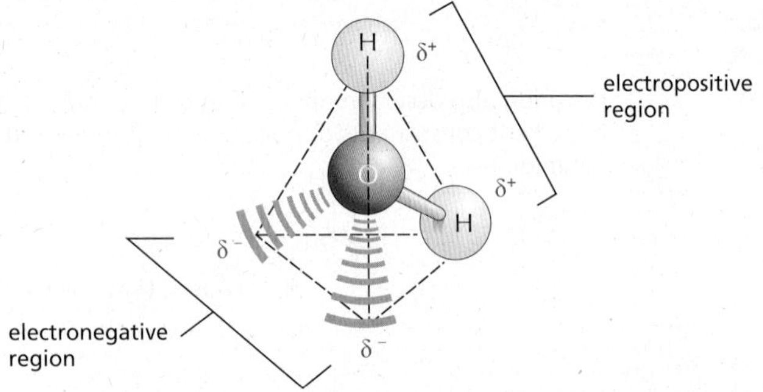

Although a water molecule has an overall neutral charge (having the same number of electrons and protons), the electrons are asymmetrically distributed, making the molecule polar. The oxygen nucleus draws electrons away from the hydrogen nuclei, leaving the hydrogen nuclei with a small net positive charge. The excess of electron density on the oxygen atom creates weakly negative regions at the other two corners of an imaginary tetrahedron. On these pages, we review the chemical properties of water and see how water influences the behavior of biological molecules.

WATER STRUCTURE

Molecules of water join together transiently in a hydrogen-bonded lattice. Even at 37°C, 15% of the water molecules are joined to four others in a short-lived assembly known as a *flickering cluster*.

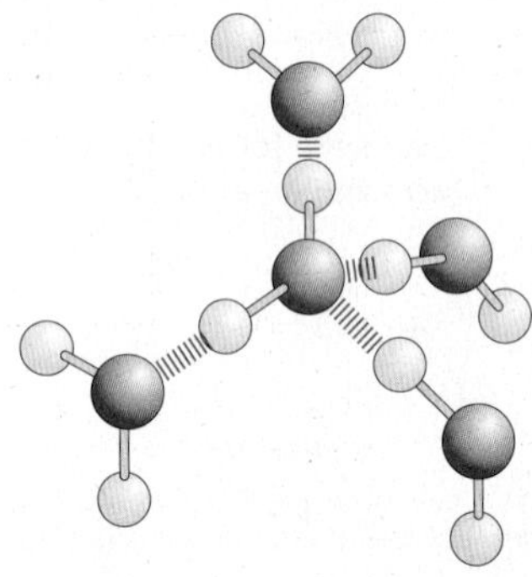

The cohesive nature of water is responsible for many of its unusual properties, such as high surface tension, high specific heat capacity, and high heat of vaporization.

HYDROPHILIC MOLECULES

Substances that dissolve readily in water are termed hydrophilic. They include ions and polar molecules that attract water molecules through electrical charge effects. Water molecules surround each ion or polar molecule and carry it into solution.

Ionic substances such as sodium chloride dissolve because water molecules are attracted to the positive (Na^+) or negative (Cl^-) charge of each ion.

Polar substances such as urea dissolve because their molecules form hydrogen bonds with the surrounding water molecules.

HYDROPHOBIC MOLECULES

Substances that contain a preponderance of nonpolar bonds are usually insoluble in water and are termed hydrophobic. Water molecules are not attracted to such hydrophobic molecules and so have little tendency to surround them and bring them into solution.

Hydrocarbons, which contain many C–H bonds, are especially hydrophobic.

WATER AS A SOLVENT

Many substances, such as household sugar (sucrose), dissolve in water. That is, their molecules separate from each other, each becoming surrounded by water molecules.

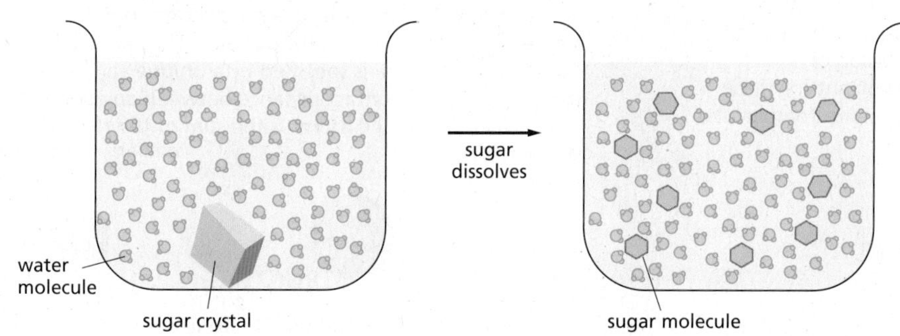

When a substance dissolves in a liquid, the mixture is termed a solution. The dissolved substance (in this case sugar) is the solute, and the liquid that does the dissolving (in this case water) is the solvent. Water is an excellent solvent for hydrophilic substances because of its polar bonds.

ACIDS

Substances that release hydrogen ions (protons) into solution are called acids.

$$\mathrm{HCl} \longrightarrow \mathrm{H^+} + \mathrm{Cl^-}$$

hydrochloric acid (strong acid) → hydrogen ion + chloride ion

Many of the acids important in the cell are not completely dissociated, and they are therefore weak acids; for example, the carboxyl group (–COOH), which dissociates to give a hydrogen ion in solution.

$$-\mathrm{COOH} \rightleftharpoons \mathrm{H^+} + -\mathrm{COO^-}$$

carboxyl group (weak acid)

Note that this is a reversible reaction.

HYDROGEN ION EXCHANGE

Positively charged hydrogen ions (H^+) can spontaneously move from one water molecule to another, thereby creating two ionic species.

$$\mathrm{H_2O \cdots H{-}OH} \rightleftharpoons \mathrm{H_3O^+} + \mathrm{OH^-}$$

hydronium ion, hydroxyl ion

often written as: $H_2O \rightleftharpoons H^+ + OH^-$

hydrogen ion, hydroxyl ion

Because the process is rapidly reversible, hydrogen ions are continually shuttling between water molecules. Pure water contains equal concentrations of hydronium ions and hydroxyl ions (both 10^{-7} M).

pH

The acidity of a solution is defined by the concentration (conc.) of hydronium ions (H_3O^+) it possesses, generally abbreviated as H^+. For convenience, we use the pH scale, where

$$\mathrm{pH} = -\log_{10}[\mathrm{H^+}]$$

For pure water

$[H^+] = 10^{-7}$ moles/liter

pH = 7.0

	H^+ conc. moles/liter	pH
ACIDIC	1	0
	10^{-1}	1
	10^{-2}	2
	10^{-3}	3
	10^{-4}	4
	10^{-5}	5
	10^{-6}	6
	10^{-7}	7
BASIC	10^{-8}	8
	10^{-9}	9
	10^{-10}	10
	10^{-11}	11
	10^{-12}	12
	10^{-13}	13
	10^{-14}	14

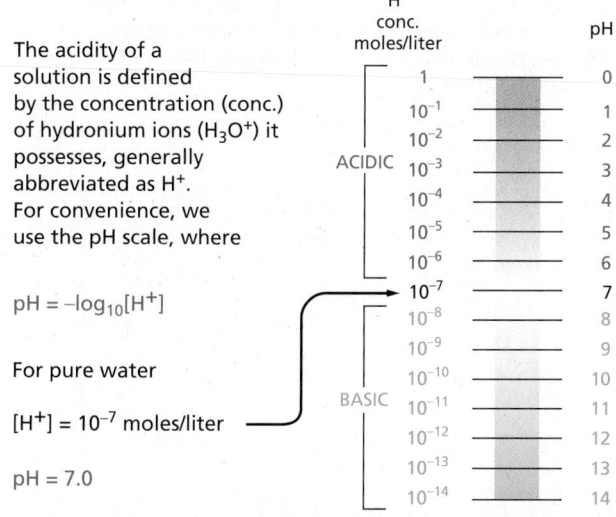

BASES

Substances that reduce the number of hydrogen ions in solution are called bases. Some bases, such as ammonia, combine directly with hydrogen ions.

$$\mathrm{NH_3} + \mathrm{H^+} \longrightarrow \mathrm{NH_4^+}$$

ammonia + hydrogen ion → ammonium ion

Other bases, such as sodium hydroxide, reduce the number of H^+ ions indirectly, by producing OH^- ions that then combine directly with H^+ ions to make H_2O.

$$\mathrm{NaOH} \longrightarrow \mathrm{Na^+} + \mathrm{OH^-}$$

sodium hydroxide (strong base) → sodium ion + hydroxyl ion

Many bases found in cells are partially associated with H^+ ions and are termed weak bases. This is true of compounds that contain an amino group ($-NH_2$), which has a weak tendency to reversibly accept an H^+ ion from water, thereby increasing the concentration of free OH^- ions.

$$-\mathrm{NH_2} + \mathrm{H^+} \rightleftharpoons -\mathrm{NH_3^+}$$

WEAK NONCOVALENT CHEMICAL BONDS

Organic molecules can interact with other molecules through three types of short-range attractive forces known as *noncovalent bonds*: van der Waals attractions, electrostatic attractions, and hydrogen bonds. The repulsion of hydrophobic groups from water is also important for these interactions and for the folding of biological macromolecules.

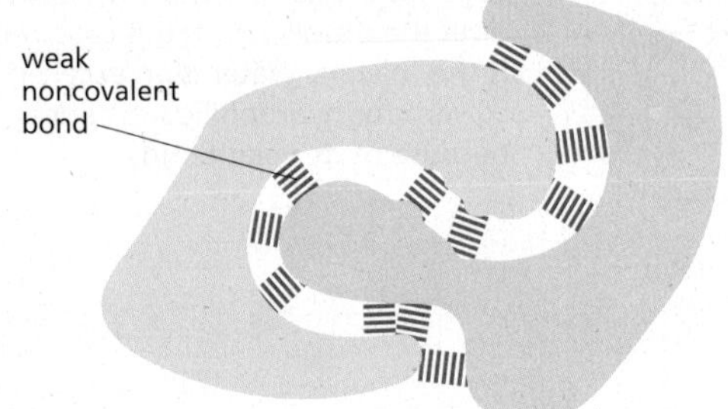

Weak noncovalent bonds have less than 1/20 the strength of a strong covalent bond. They are strong enough to provide tight binding only when many of them are formed simultaneously.

VAN DER WAALS ATTRACTIONS

If two atoms are too close together, they repel each other very strongly. For this reason, an atom can often be treated as a sphere with a fixed radius. The characteristic "size" for each atom is specified by a unique van der Waals radius. The contact distance between any two noncovalently bonded atoms is the sum of their van der Waals radii.

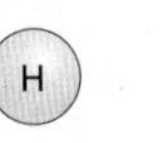
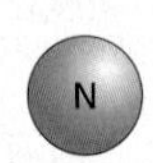

H	C	N	O
0.12 nm radius	0.2 nm radius	0.15 nm radius	0.14 nm radius

At very short distances, any two atoms show a weak bonding interaction due to their fluctuating electrical charges. The two atoms will be attracted to each other in this way until the distance between their nuclei is approximately equal to the sum of their van der Waals radii. Although they are individually very weak, such van der Waals attractions can become important when two macromolecular surfaces fit together very closely, because many atoms are involved.

Note that when two atoms form a covalent bond, the centers of the two atoms (the two atomic nuclei) are much closer together than the sum of the two van der Waals radii. Thus,

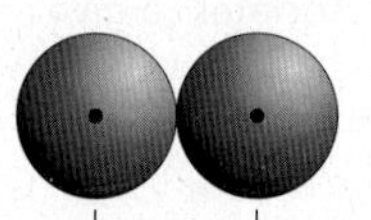
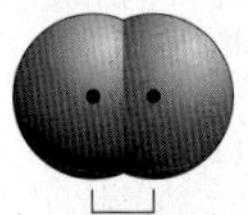
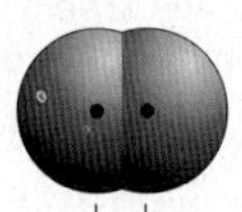

0.4 nm	0.15 nm	0.13 nm
two nonbonded carbon atoms	two carbon atoms held by a single covalent bond	two carbon atoms held by a double covalent bond

HYDROGEN BONDS

As already described for water (see Panel 2–2, pp. 96–97), hydrogen bonds form when a hydrogen atom is "sandwiched" between two electron-attracting atoms (usually oxygen or nitrogen).

Hydrogen bonds are strongest when the three atoms are in a straight line:

O—H ||||||||| O   N—H ||||||||| O

Examples in macromolecules:

Amino acids in a polypeptide chain can be hydrogen-bonded together in a folded protein.

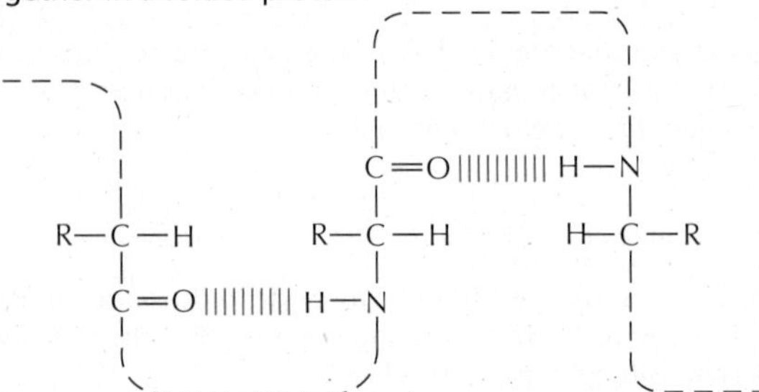

Two bases, G and C, are hydrogen bonded in a DNA double helix.

C   G

HYDROGEN BONDS IN WATER

Any two atoms that can form hydrogen bonds to each other can alternatively form hydrogen bonds to water molecules. Because of this competition with water molecules, the hydrogen bonds formed in water between two peptide bonds, for example, are relatively weak.

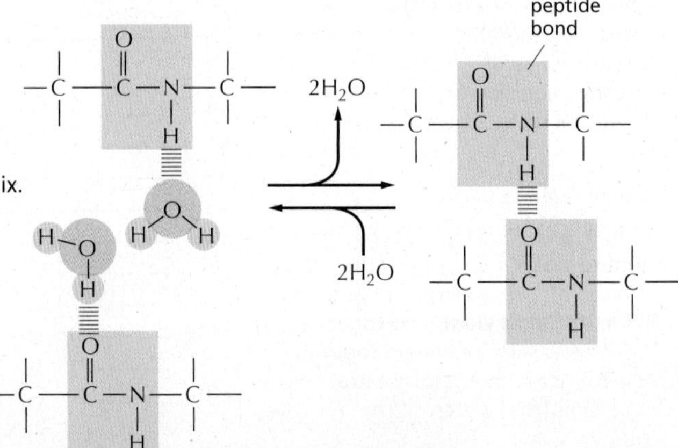

ELECTROSTATIC ATTRACTIONS

Electrostatic attractions occur both between fully charged groups (ionic bond) and between partially charged groups on polar molecules.

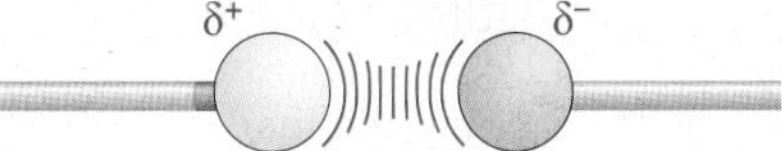

The force of attraction between the two partial charges, δ^+ and δ^-, falls off rapidly as the distance between the charges increases.

In the absence of water, ionic bonds are very strong. They are responsible for the strength of such minerals as marble and agate and for crystal formation in common table salt, NaCl.

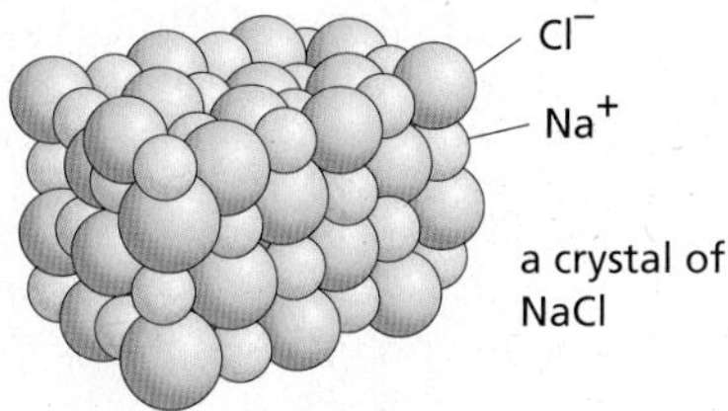

CATION–π INTERACTIONS

The structure of a typical protein reveals that it contains several cation–π (pi) interactions, these electrostatic attractions being about half as abundant as the electrostatic attractions between positively and negatively charged amino-acid side chains. In this energetically favorable interaction, a cation is paired with the aromatic (π) electrons of either a tryptophan, tyrosine, or phenylalanine side chain. As shown, the cation can either be an inorganic ion or a positively charged lysine or arginine side chain. A tryptophan–arginine pair is the most common, as illustrated here.

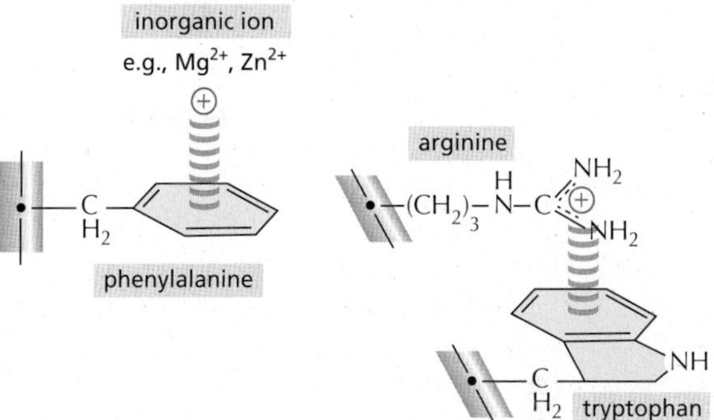

HYDROPHOBIC FORCES

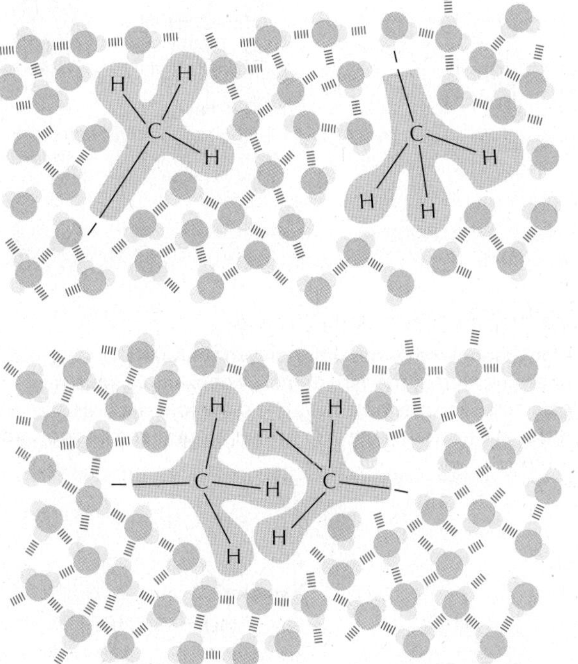

Water forces hydrophobic groups together in order to minimize their disruptive effects on the water network formed by the hydrogen bonds between water molecules. Hydrophobic groups held together in this way are sometimes said to be held together by "hydrophobic bonds," even though the attraction is actually caused by a repulsion from water.

ELECTROSTATIC ATTRACTIONS IN WATER

Charged groups are shielded by their interactions with water molecules. Electrostatic attractions are therefore quite weak in water.

Inorganic ions in solution can also cluster around charged groups and further weaken these electrostatic attractions.

Despite being weakened by water and inorganic ions, electrostatic attractions are very important in biological systems.

MONOSACCHARIDES

Monosaccharides usually have the general formula $(CH_2O)_n$, where *n* can be 3, 4, 5, 6, 7, or 8, and have two or more hydroxyl groups. They contain either an aldehyde group ($-C{=}O$, H) and are called aldoses or a ketone group ($>C{=}O$) and are called ketoses.

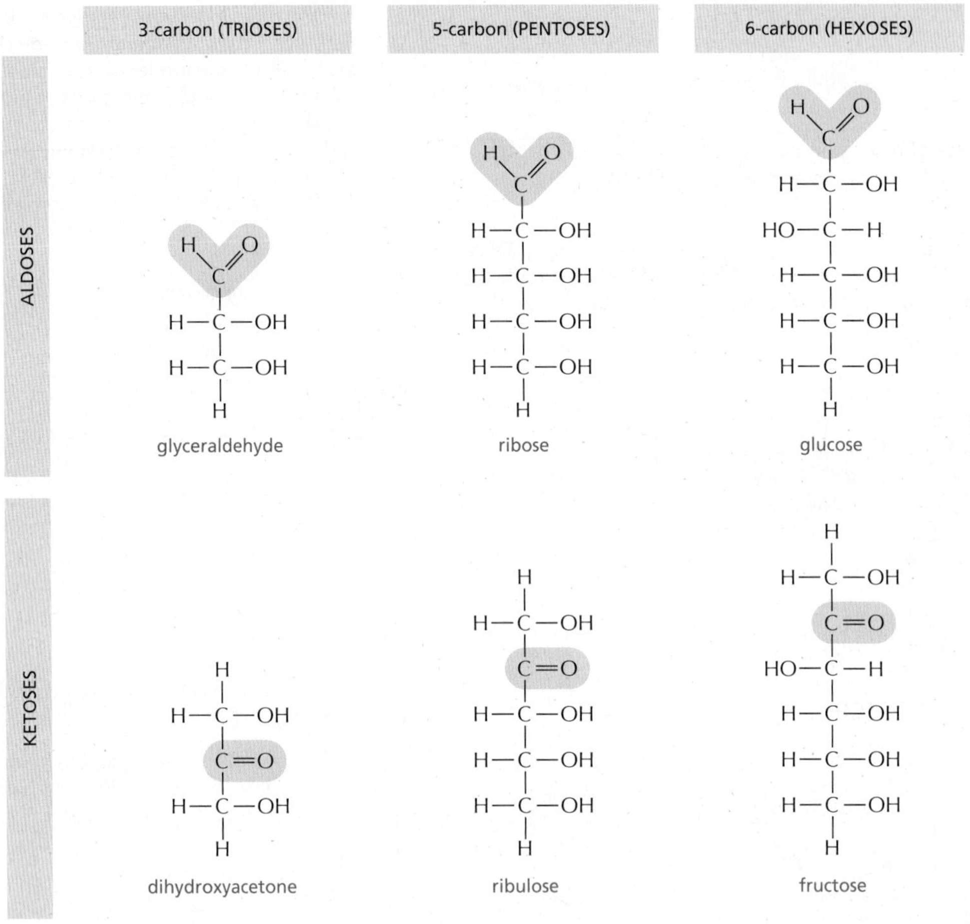

	3-carbon (TRIOSES)	5-carbon (PENTOSES)	6-carbon (HEXOSES)
ALDOSES	glyceraldehyde	ribose	glucose
KETOSES	dihydroxyacetone	ribulose	fructose

RING FORMATION

In aqueous solution, the aldehyde or ketone group of a sugar molecule tends to react with a hydroxyl group of the same molecule, thereby closing the molecule into a ring.

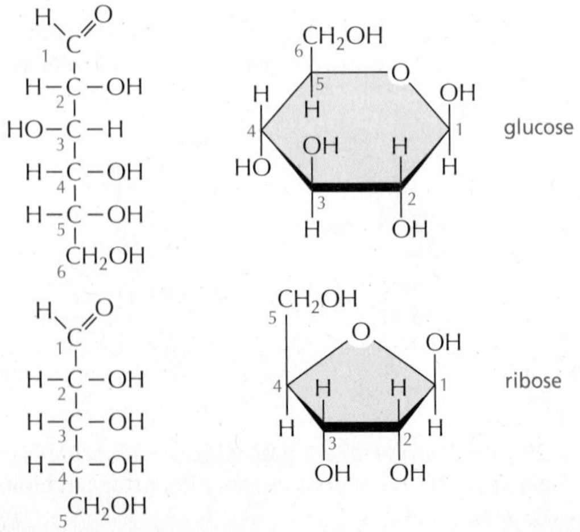

glucose

ribose

Note that each carbon atom has a number.

ISOMERS

Many monosaccharides differ only in the spatial arrangement of atoms; that is, they are isomers. For example, glucose, galactose, and mannose have the same formula ($C_6H_{12}O_6$) but differ in the arrangement of groups around one or two carbon atoms.

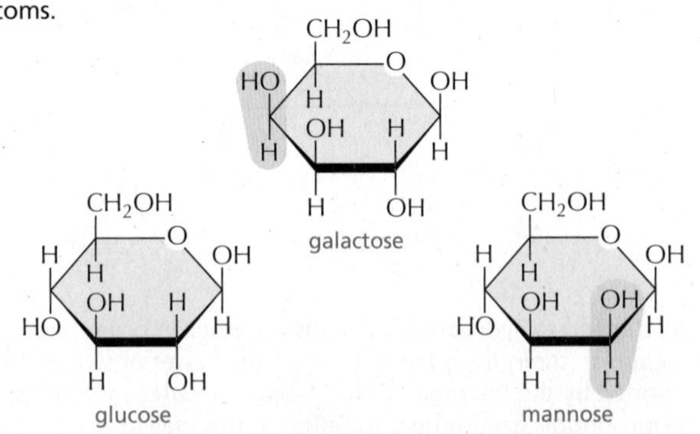

galactose

glucose

mannose

These small differences make only minor changes in the chemical properties of the sugars. But the differences are recognized by enzymes and other proteins and therefore can have major biological effects.

α AND β LINKS

The hydroxyl group on the carbon that carries the aldehyde or ketone can rapidly change from one position to the other. These two positions are called α and β.

OH ⇌ OH

β hydroxyl α hydroxyl

As soon as one sugar is linked to another, the α or β form is frozen.

SUGAR DERIVATIVES

The hydroxyl groups of a simple monosaccharide, such as glucose, can be replaced by other groups.

glucuronic acid

glucosamine (NH_2)

N-acetylglucosamine (NH–C=O–CH_3)

DISACCHARIDES

The carbon that carries the aldehyde or the ketone can react with any hydroxyl group on a second sugar molecule to form a disaccharide. Three common disaccharides are

maltose (glucose + glucose)
lactose (galactose + glucose)
sucrose (glucose + fructose)

The reaction forming sucrose is shown here.

α glucose + β fructose → sucrose + H_2O

OLIGOSACCHARIDES AND POLYSACCHARIDES

Large linear and branched molecules can be made from simple repeating sugar subunits. Short chains are called oligosaccharides, and long chains are called polysaccharides. Glycogen, for example, is a polysaccharide made entirely of glucose subunits joined together.

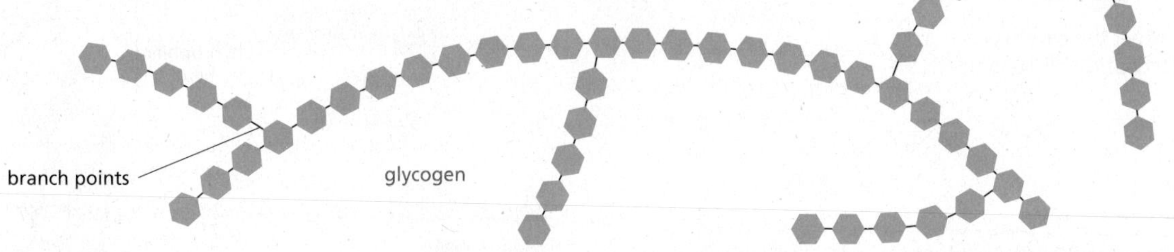

COMPLEX OLIGOSACCHARIDES

In many cases, a sugar sequence is nonrepetitive. Many different molecules are possible. Such complex oligosaccharides are usually linked to proteins or to lipids, as is this oligosaccharide, which is part of a cell-surface molecule that defines a particular blood group.

FATTY ACIDS

All fatty acids have a carboxyl group at one end and a long hydrocarbon tail at the other.

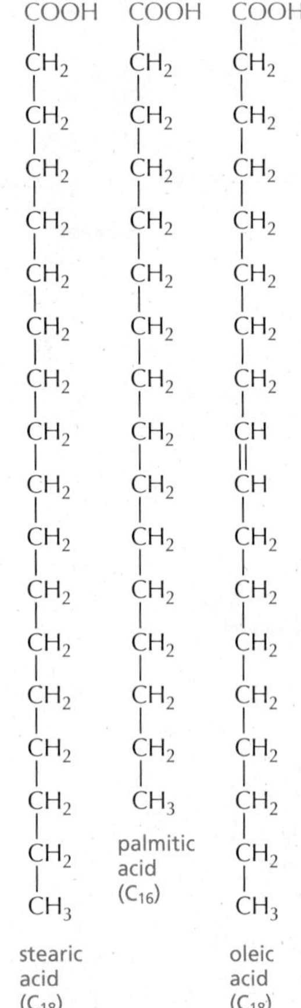

Hundreds of different kinds of fatty acids exist. Some have one or more double bonds in their hydrocarbon tail and are said to be unsaturated. Fatty acids with no double bonds are saturated.

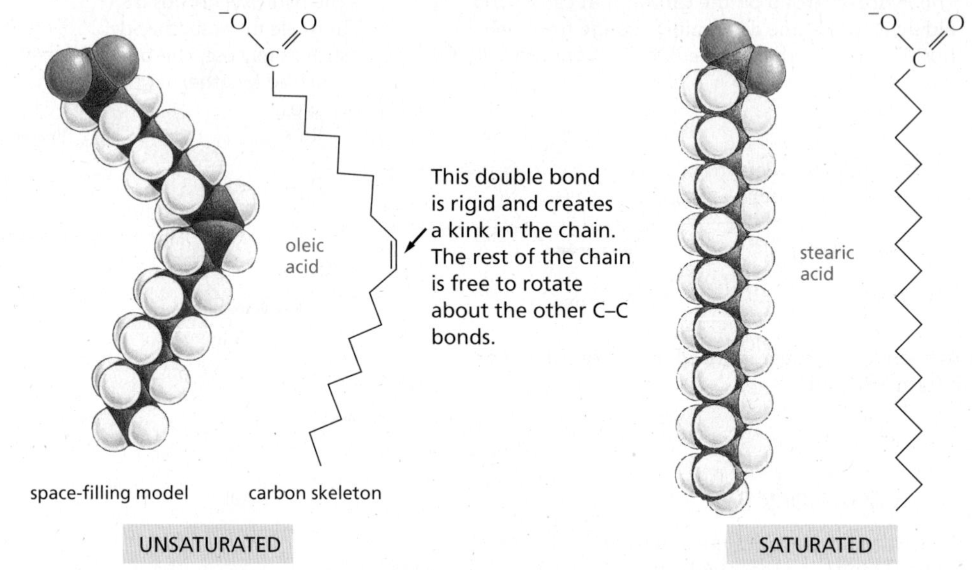

TRIACYLGLYCEROLS

Fatty acids are stored in cells as an energy reserve (fats and oils) through an ester linkage to glycerol to form triacylglycerols, also known as triglycerides.

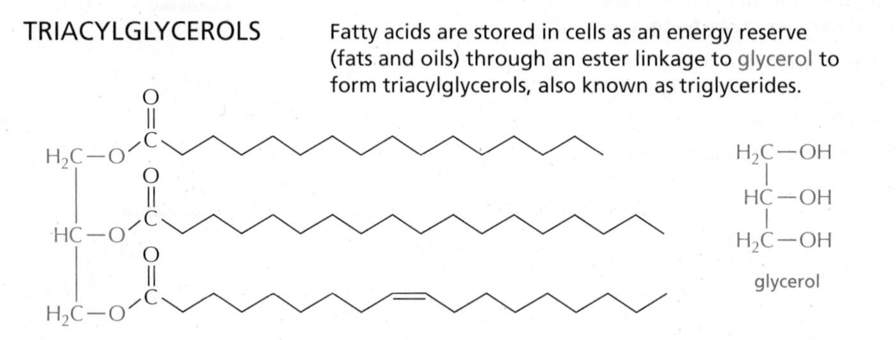

CARBOXYL GROUP

If free, the carboxyl group of a fatty acid will be ionized.

But more often it is linked to other groups to form either esters

or amides.

PHOSPHOLIPIDS

Phospholipids are the major constituents of cell membranes.

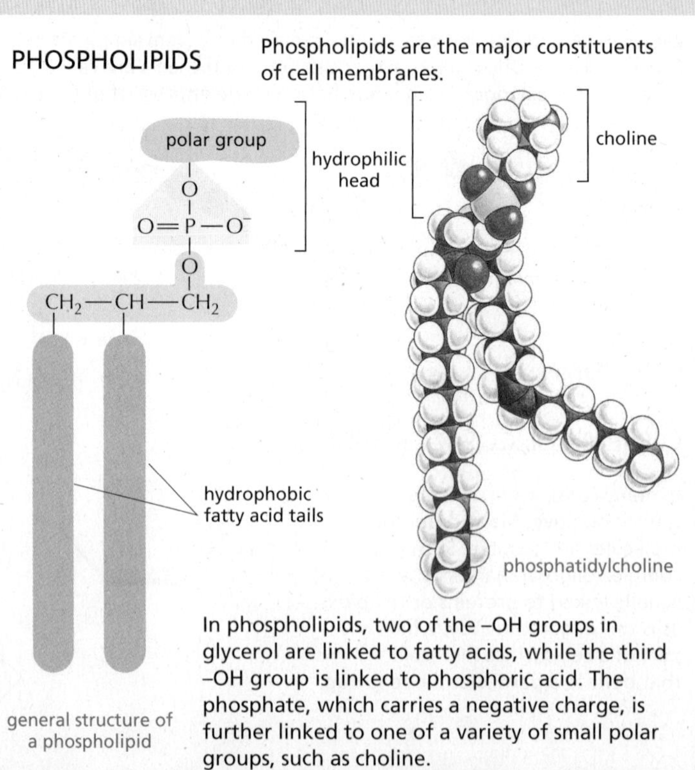

In phospholipids, two of the –OH groups in glycerol are linked to fatty acids, while the third –OH group is linked to phosphoric acid. The phosphate, which carries a negative charge, is further linked to one of a variety of small polar groups, such as choline.

LIPID AGGREGATES

Fatty acids have a hydrophilic head and a hydrophobic tail.

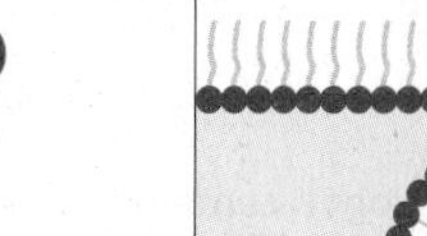

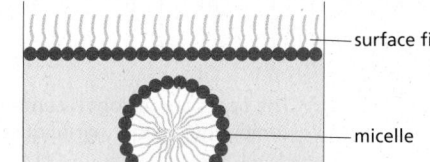

surface film

micelle

In water, they can form either a surface film or small, spherical micelles.

Their derivatives can form larger aggregates held together by hydrophobic forces:

Triacylglycerols form large, spherical fat droplets in the cell cytoplasm.

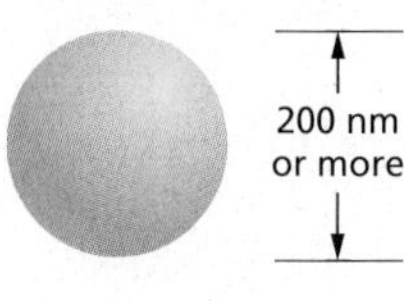

Phospholipids and glycolipids form self-sealing *lipid bilayers*, which are the basis for all cell membranes.

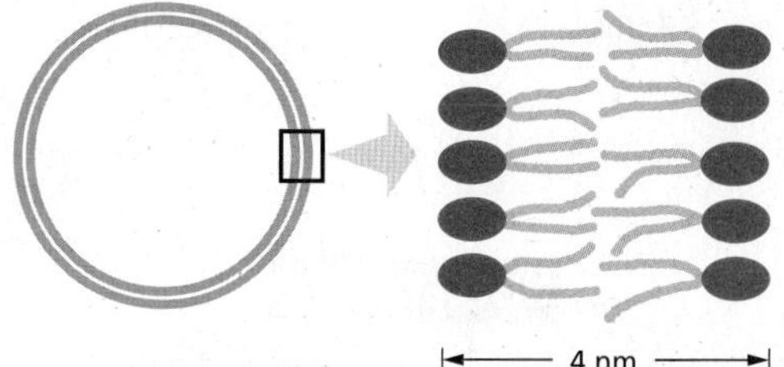

OTHER LIPIDS

Lipids are defined as water-insoluble molecules that are soluble in organic solvents. Two other common types of lipids are steroids and polyisoprenoids. Both are made from isoprene units.

CH_3—C(=CH_2)—CH=CH_2 isoprene

STEROIDS

Steroids have a common multiple-ring structure.

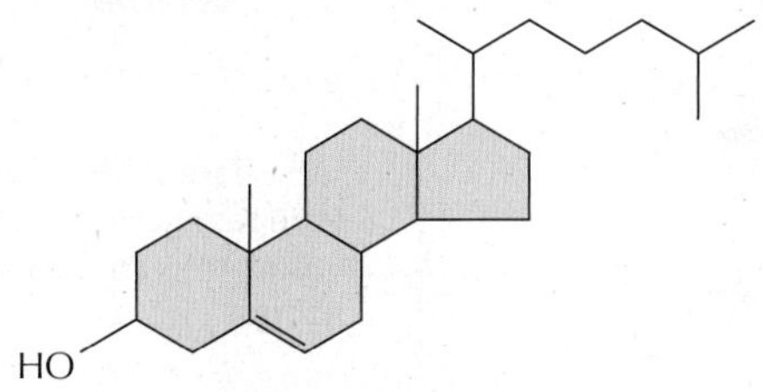

cholesterol—found in many cell membranes

testosterone—male sex hormone

GLYCOLIPIDS

Like phospholipids, these compounds are composed of a hydrophobic region, containing two long hydrocarbon tails, and a polar region, which contains one or more sugars. Unlike phospholipids, there is no phosphate.

galactose

sugar

a simple glycolipid

POLYISOPRENOIDS

Long-chain polymers of isoprene

dolichol phosphate—used to carry activated sugars in the membrane-associated synthesis of glycoproteins and some polysaccharides

BASES

The bases are nitrogen-containing ring compounds, either pyrimidines or purines.

uracil

cytosine

thymine

PYRIMIDINE

PURINE

adenine

guanine

PHOSPHATES

The phosphates are normally joined to the C5 hydroxyl of the ribose or deoxyribose sugar (designated 5′). Mono-, di-, and triphosphates are common.

$^{-}O-P(=O)(O^{-})-O-CH_2$ as in AMP

$^{-}O-P(=O)(O^{-})-O-P(=O)(O^{-})-O-CH_2$ as in ADP

$^{-}O-P(=O)(O^{-})-O-P(=O)(O^{-})-O-P(=O)(O^{-})-O-CH_2$ as in ATP

The phosphate makes a nucleotide negatively charged.

NUCLEOTIDES

A nucleotide consists of a nitrogen-containing base, a five-carbon sugar, and a phosphate group.

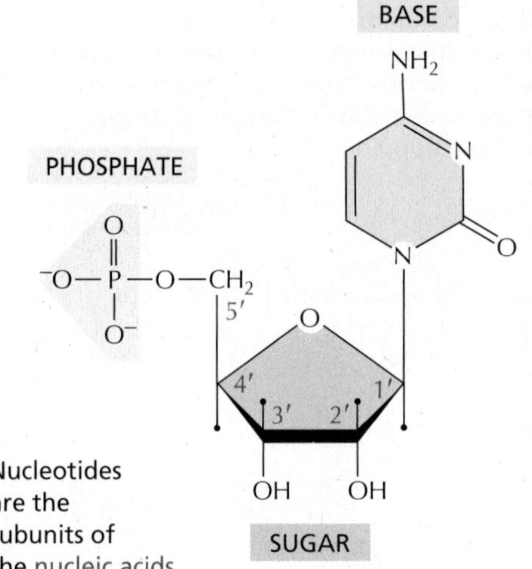

Nucleotides are the subunits of the nucleic acids.

BASE–SUGAR LINKAGE

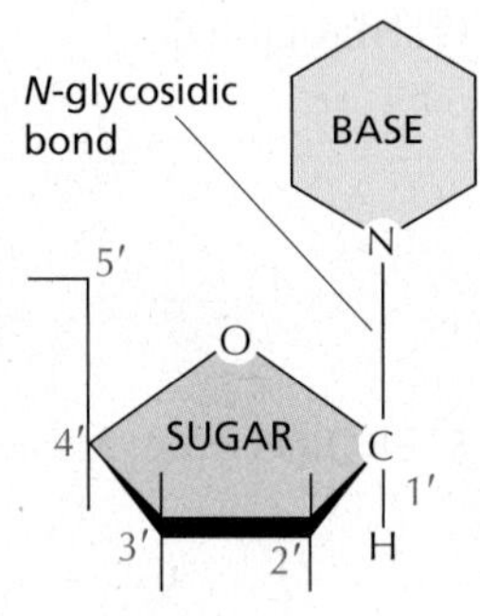

The base is linked to the same carbon (C1) used in sugar–sugar bonds.

SUGARS

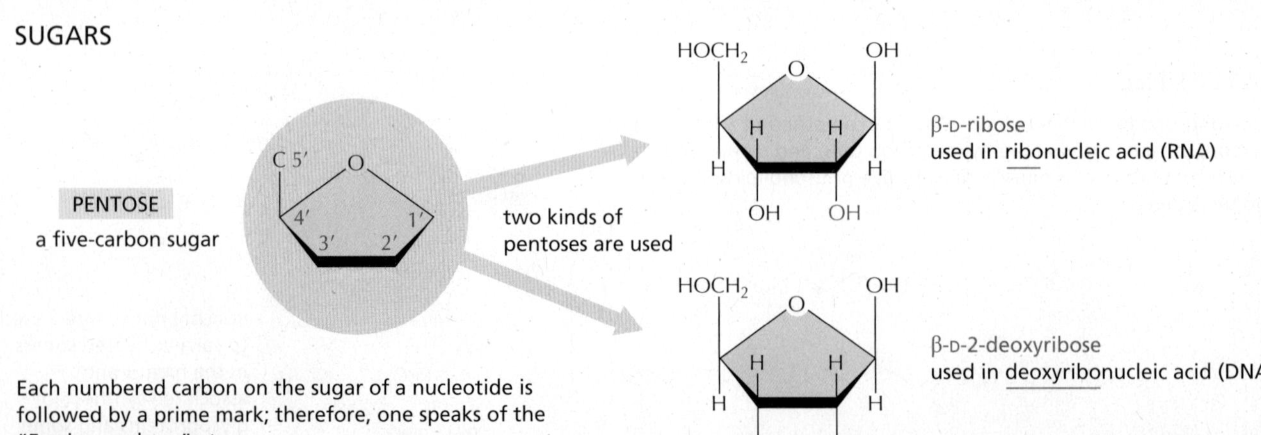

Each numbered carbon on the sugar of a nucleotide is followed by a prime mark; therefore, one speaks of the "5-prime carbon," etc.

NOMENCLATURE

A nucleoside or nucleotide is named according to its nitrogenous base.

BASE	NUCLEOSIDE	ABBR.
adenine	adenosine	A
guanine	guanosine	G
cytosine	cytidine	C
uracil	uridine	U
thymine	thymidine	T

Single-letter abbreviations are used variously as shorthand for (1) the base alone, (2) the nucleoside, or (3) the whole nucleotide—the context will usually make clear which of the three entities is meant. When the context is not sufficient, we will add the terms "base," "nucleoside," "nucleotide," or—as in the examples below—use the full 3-letter nucleotide code.

AMP = adenosine monophosphate
dAMP = deoxyadenosine monophosphate
UDP = uridine diphosphate
ATP = adenosine triphosphate

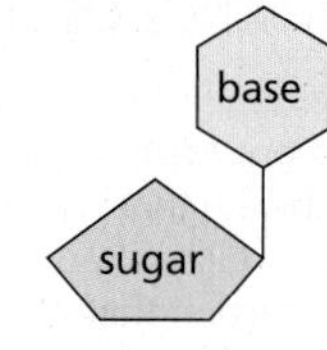

BASE + SUGAR = NUCLEOSIDE

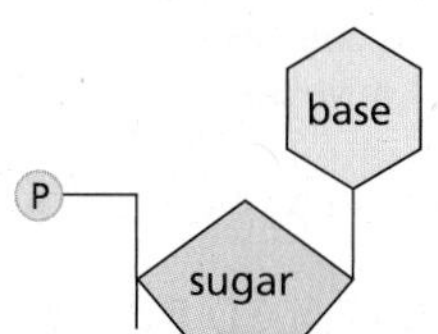

BASE + SUGAR + PHOSPHATE = NUCLEOTIDE

NUCLEIC ACIDS

To form nucleic acid polymers, nucleotides are joined together by phosphodiester bonds between the 5′ and 3′ carbon atoms of adjacent sugar rings. The linear sequence of nucleotides in a nucleic acid chain is abbreviated using a one-letter code, such as AGCTT, starting with the 5′ end of the chain.

phosphodiester bond

5′ end of chain

3′ end of chain

example: DNA

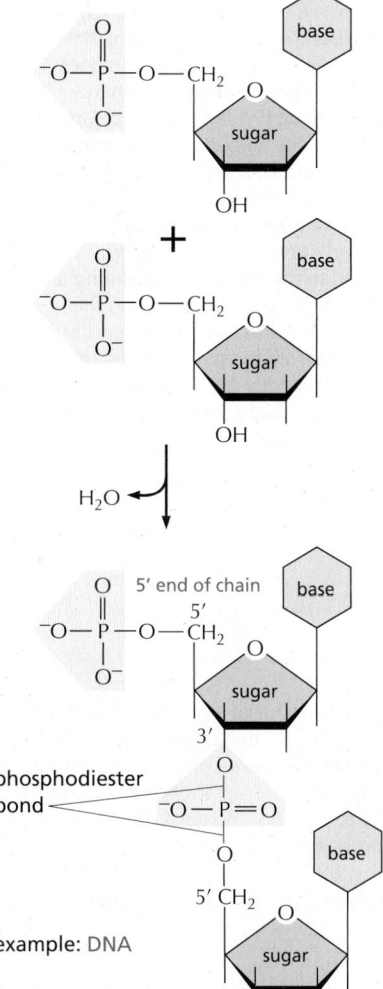

NUCLEOTIDES AND THEIR DERIVATIVES HAVE MANY OTHER FUNCTIONS

1. As nucleoside di- and triphosphates, they carry chemical energy in their easily hydrolyzed phosphoanhydride bonds.

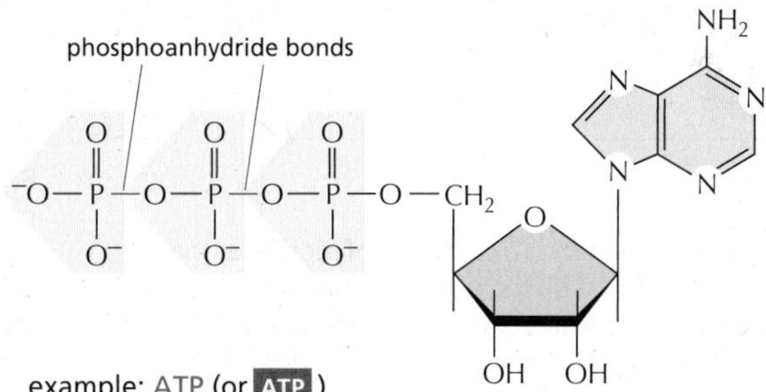

example: ATP (or ATP)

2. They combine with other groups to form coenzymes.

example: coenzyme A (CoA)

3. They are used as small intracellular signaling molecules in the cell.

example: cyclic AMP (cAMP)

THE IMPORTANCE OF FREE ENERGY FOR CELLS

Life is possible because of the complex network of interacting chemical reactions occurring in every cell. In viewing the metabolic pathways that comprise this network, one might suspect that the cell has had the ability to evolve an enzyme to carry out any reaction that it needs. But this is not so. Although enzymes are powerful catalysts, they can promote only those reactions that are thermodynamically possible; other reactions proceed in cells only because they are *coupled* to very favorable reactions that drive them.

The question of whether a reaction can occur spontaneously, or instead needs to be coupled to another reaction, is central to cell biology. The answer is obtained by reference to a quantity called the *free energy*. In this Panel, we shall explain some of the fundamental ideas—derived from a special branch of chemistry and physics called *thermodynamics*—that are required for understanding what free energy is and why it is so important to cells.

ENERGY RELEASED BY CHANGES IN CHEMICAL BONDING IS CONVERTED INTO HEAT

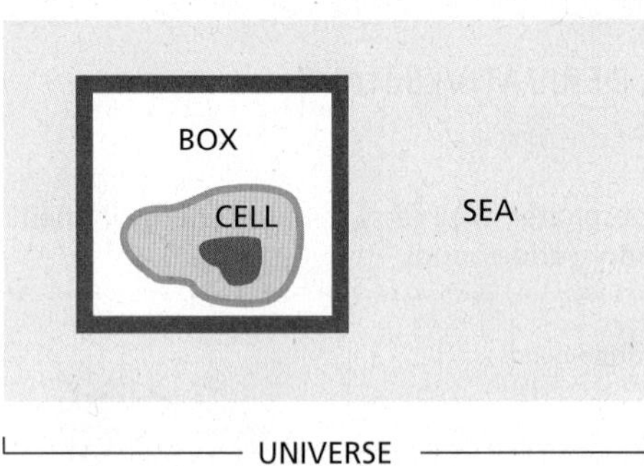

An *enclosed system* is defined as a collection of molecules that does not exchange matter with the rest of the universe (e.g., the "cell in a box" shown above). Any such system will contain molecules with a total energy E. This energy will be distributed in a variety of ways: some as the translational energy of the molecules, some as their vibrational and rotational energies, but most as the bonding energies between the individual atoms that make up the molecules. Suppose that a reaction occurs in the system. The first law of thermodynamics places a constraint on what types of reactions are possible: it states that "in any process, the total energy of the universe remains constant." For example, suppose that reaction A→B occurs somewhere in the box and releases a great deal of chemical-bond energy. This energy will initially increase the intensity of molecular motions (translational, vibrational, and rotational) in the system, which is equivalent to raising its temperature. However, these increased motions will soon be transferred out of the system by a series of molecular collisions that heat up first the walls of the box and then the outside world (represented by the sea in our example). In the end, the system returns to its initial temperature, by which time all the chemical-bond energy released in the box has been converted into heat energy and transferred out of the box to the surroundings. According to the first law, the change in the energy in the box (ΔE_{box}, which we shall denote as ΔE) must be equal and opposite to the amount of heat energy transferred, which we shall designate as h; that is, $\Delta E = -h$. Thus, the energy in the box (E) decreases when heat leaves the system.

E also can change during a reaction as a result of work being done on the outside world. For example, suppose that there is a small increase in the volume (ΔV) of the box during a reaction. Because the walls of the box must push against the constant pressure (P) in the surroundings in order to expand, this does work on the outside world and requires energy. The energy used is $P(\Delta V)$, which according to the first law must decrease the energy in the box (E) by the same amount. In most reactions, chemical-bond energy is converted into both work and heat. *Enthalpy* (H) is a composite function that includes both of these ($H = E + PV$). To be rigorous, it is the change in enthalpy (ΔH) in an enclosed system, and not the change in energy, that is equal to the heat transferred to the outside world during a reaction. Reactions in which H decreases release heat to the surroundings and are said to be "exothermic," while reactions in which H increases absorb heat from the surroundings and are said to be "endothermic." Thus, $-h = \Delta H$. However, the volume change is negligible in most biological reactions, so to a good approximation

$$-h = \Delta H \approx \Delta E$$

THE SECOND LAW OF THERMODYNAMICS

Consider a container in which 1000 coins are all lying heads-up. If the container is shaken vigorously, subjecting the coins to the types of random motions that all molecules experience due to their frequent collisions with other molecules, one will end up with about half the coins oriented heads-down. The reason for this reorientation is that there is only a single way in which the original orderly state of the coins can be reinstated (every coin must lie heads-up), whereas there are many different ways (about 10^{298}) to achieve a disorderly state in which there is an equal mixture of heads and tails; in fact, there are more ways to achieve a 50–50 state than to achieve any other state. Each state has a probability of occurrence that is proportional to the number of ways it can be realized. The second law of thermodynamics states that "systems will change spontaneously from states of lower probability to states of higher probability." Because states of lower probability are more ordered than states of higher probability, the second law can be restated: "the universe constantly changes so as to become more disordered."

THE ENTROPY, S

The second law (but not the first law) allows one to predict the *direction* of a particular reaction. But to make it useful for this purpose, one needs a convenient measure of the probability or, equivalently, the degree of disorder of a state. The entropy (S) is such a measure. It is a logarithmic function of the probability. Thus the *change in entropy* (ΔS) that occurs when the reaction A → B converts 1 mole of A into 1 mole of B is

$$\Delta S = R \ln p_B/p_A$$

where p_A and p_B are the probabilities of the two states A and B, R is the gas constant (8.31 J K^{-1} $mole^{-1}$), and ΔS is measured in entropy units (eu).

For 1000 coins, the relative probability of all heads (state A) versus half heads and half tails (state B) is equal to the ratio of the number of different ways that the two results can be obtained. One can calculate that $p_A = 1$ and $p_B = 1000!(500! \times 500!) = 10^{298}$. Therefore, the entropy change for the reorientation of the coins when their container is vigorously shaken and an equal mixture of heads and tails is obtained is $R \ln (10^{298})$, or about 1370 eu per mole of such containers (6×10^{23} containers). Because ΔS defined above is positive for the transition from state A to state B ($p_B/p_A > 1$), reactions with an *increase* in S (i.e., for which $\Delta S > 0$) are favored and will occur spontaneously.

Heat energy causes the random commotion of molecules. Because the transfer of heat from an enclosed system to its surroundings increases the number of different arrangements that the molecules in the outside world can have, it increases their entropy. It can be shown that the release of a fixed quantity of heat energy has a greater disordering effect at low temperature than at high temperature, and that the value of ΔS for the surroundings, as described above (ΔS_{sea}), is precisely equal to h, the amount of heat transferred to the surroundings from the system, divided by the absolute temperature (T):

$$\Delta S_{sea} = h/T$$

THE GIBBS FREE ENERGY, G

When dealing with an enclosed biological system, one would like to have a simple way of predicting whether a given reaction will or will not occur spontaneously in the system. We have seen that the crucial question is whether the entropy change for the universe is positive or negative when that reaction occurs. In our idealized system, the cell in a box, there are two separate components to the entropy change of the universe—the entropy change for the system enclosed in the box and the entropy change for the surrounding "sea"—and both must be added together before any prediction can be made. For example, it is possible for a reaction to absorb heat and thereby decrease the entropy of the sea ($\Delta S_{sea} < 0$) and at the same time cause such a large degree of disordering inside the box ($\Delta S_{box} > 0$) that the total $\Delta S_{universe} = \Delta S_{sea} + \Delta S_{box}$ is greater than zero. In this case, the reaction will occur spontaneously, even though the sea gives up heat to the box during the reaction. An example of such a reaction is the dissolving of sodium chloride in a beaker containing water (the "box"), which is a spontaneous process even though the temperature of the water drops as the salt goes into solution.

Chemists have found it useful to define a number of new "composite functions" that describe *combinations* of physical properties of a system. The properties that can be combined include the temperature (T), pressure (P), volume (V), energy (E), and entropy (S). The enthalpy (H) is one such composite function. But by far the most useful composite function for biologists is the *Gibbs free energy, G*. It serves as an accounting device that allows one to deduce the entropy change of the universe resulting from a chemical reaction in the box, while avoiding any separate consideration of the entropy change in the sea. The definition of G is

$$G = H - TS$$

where, for a box of volume V, H is the enthalpy described earlier ($E + PV$), T is the absolute temperature, and S is the entropy. Each of these quantities applies to the inside of the box only. The change in free energy during a reaction in the box (the G of the products minus the G of the starting materials) is denoted as ΔG and, as we shall now demonstrate, it is a direct measure of the amount of disorder that is created in the universe when the reaction occurs.

At constant temperature the change in free energy (ΔG) during a reaction equals $\Delta H - T\Delta S$. Remembering that $\Delta H = -h$, the heat absorbed from the sea, we have

$$-\Delta G = -\Delta H + T\Delta S$$

$$-\Delta G = h + T\Delta S, \text{ so } -\Delta G/T = h/T + \Delta S$$

But h/T is equal to the entropy change of the sea (ΔS_{sea}), and the ΔS in the above equation is ΔS_{box}. Therefore

$$-\Delta G/T = \Delta S_{sea} + \Delta S_{box} = \Delta S_{universe}$$

We conclude that the free-energy change is a direct measure of the entropy change of the universe. A reaction will proceed in the direction that causes the change in the free energy (ΔG) to be less than zero, because in this case there will be a positive entropy change in the universe when the reaction occurs.

For a complex set of coupled chemical reactions involving many different molecules, the total free-energy change can be computed simply by adding up the free energies of all the different molecular species after the reaction and comparing this value with the sum of free energies before the reaction. For common substances, the required free-energy values can be found from published tables. In this way, one can predict the direction of a reaction and thereby readily check the feasibility of any proposed mechanism. (Thus, for example, from the magnitude of the electrochemical proton gradient across the inner mitochondrial membrane and the ΔG for ATP hydrolysis inside the mitochondrion, one can be certain that the ATP synthase enzyme requires that more than one proton pass through it for each molecule of ATP that it synthesizes.)

The value of ΔG for a reaction is a direct measure of how far the reaction is from equilibrium. The large negative value for ATP hydrolysis in a cell merely reflects the fact that cells keep the ATP hydrolysis reaction as much as 10 orders of magnitude away from equilibrium. If a reaction reaches equilibrium, $\Delta G = 0$, that reaction then proceeds at precisely equal rates in the forward and backward directions. For ATP hydrolysis, equilibrium is reached when the vast majority of the ATP has been hydrolyzed, as occurs in a dead cell.

For each step, the part of the molecule that undergoes a change is shadowed in *blue*, and the name of the enzyme that catalyzes the reaction is in a *yellow* box. Reactions represented by double arrows (⇄) are readily reversible, whereas those represented by single arrows (→) are effectively irreversible. To watch a video of the reactions of glycolysis, see Movie 2.5.

STEP 1

Glucose is phosphorylated by ATP to form a sugar phosphate. The negative charge of the phosphate prevents passage of the sugar phosphate through the plasma membrane, trapping glucose inside the cell.

glucose + ATP —hexokinase→ glucose 6-phosphate + ADP + H^+

STEP 2

A readily reversible rearrangement of the chemical structure (isomerization) moves the carbonyl oxygen from carbon 1 to carbon 2, forming a ketose from an aldose sugar. (See Panel 2–4, pp. 100–101.)

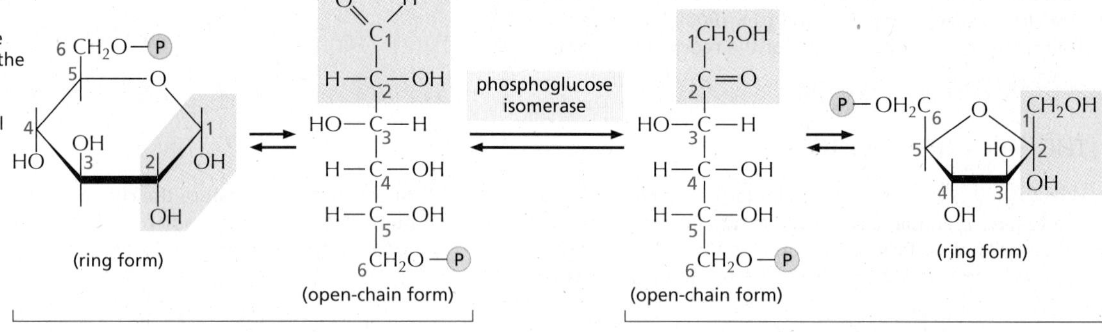

STEP 3

The new hydroxyl group on carbon 1 is phosphorylated by ATP, in preparation for the formation of two three-carbon sugar phosphates. The entry of sugars into glycolysis is controlled at this step, through regulation of the enzyme *phosphofructokinase*.

fructose 6-phosphate + ATP —phosphofructokinase→ fructose 1,6-bisphosphate + ADP + H^+

STEP 4

The six-carbon sugar is cleaved to produce two three-carbon molecules. Only the glyceraldehyde 3-phosphate can proceed immediately through glycolysis.

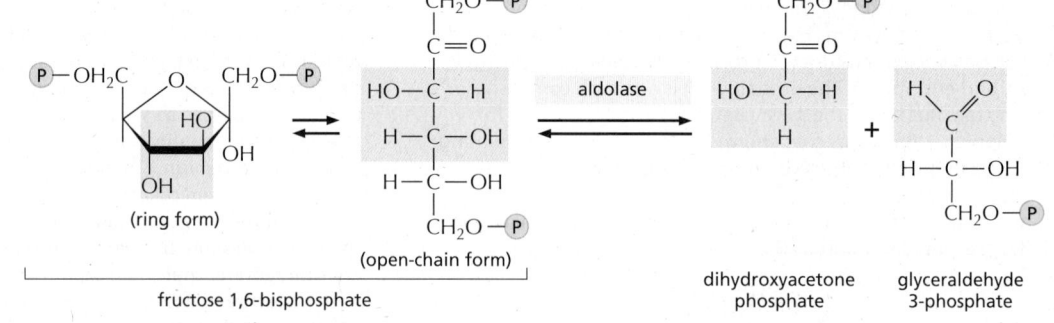

fructose 1,6-bisphosphate (ring form) ⇄ (open-chain form) —aldolase⇄ dihydroxyacetone phosphate + glyceraldehyde 3-phosphate

STEP 5

The other product of step 4, dihydroxyacetone phosphate, is isomerized to form a second molecule of glyceraldehyde 3-phosphate.

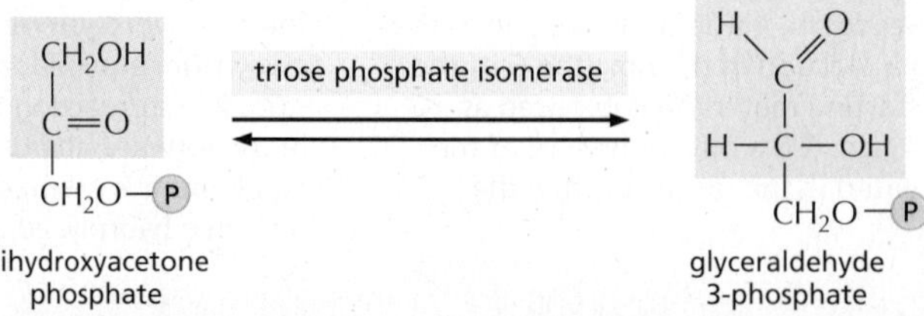

STEP 6

The two molecules of glyceraldehyde 3-phosphate produced in steps 4 and 5 are oxidized. The energy-generation phase of glycolysis begins, as NADH and a new high-energy anhydride linkage to phosphate are formed (see Figure 2–47).

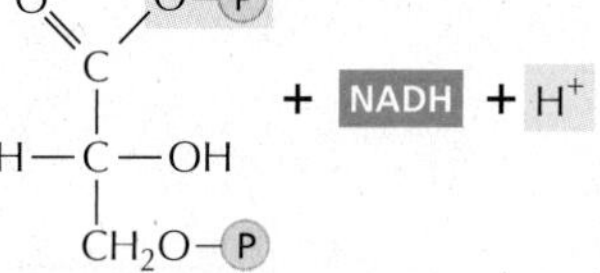

STEP 7

The transfer to ADP of the high-energy phosphate group that was generated in step 6 forms ATP.

STEP 8

The remaining phosphate ester linkage in 3-phosphoglycerate, which has a relatively low free energy of hydrolysis, is moved from carbon 3 to carbon 2 to form 2-phosphoglycerate.

STEP 9

The removal of water from 2-phosphoglycerate creates a high-energy enol phosphate linkage.

STEP 10

The transfer to ADP of the high-energy phosphate group that was generated in step 9 forms ATP, completing glycolysis.

NET RESULT OF GLYCOLYSIS

In addition to the pyruvate, the net products of glycolysis are two molecules of ATP and two molecules of NADH.

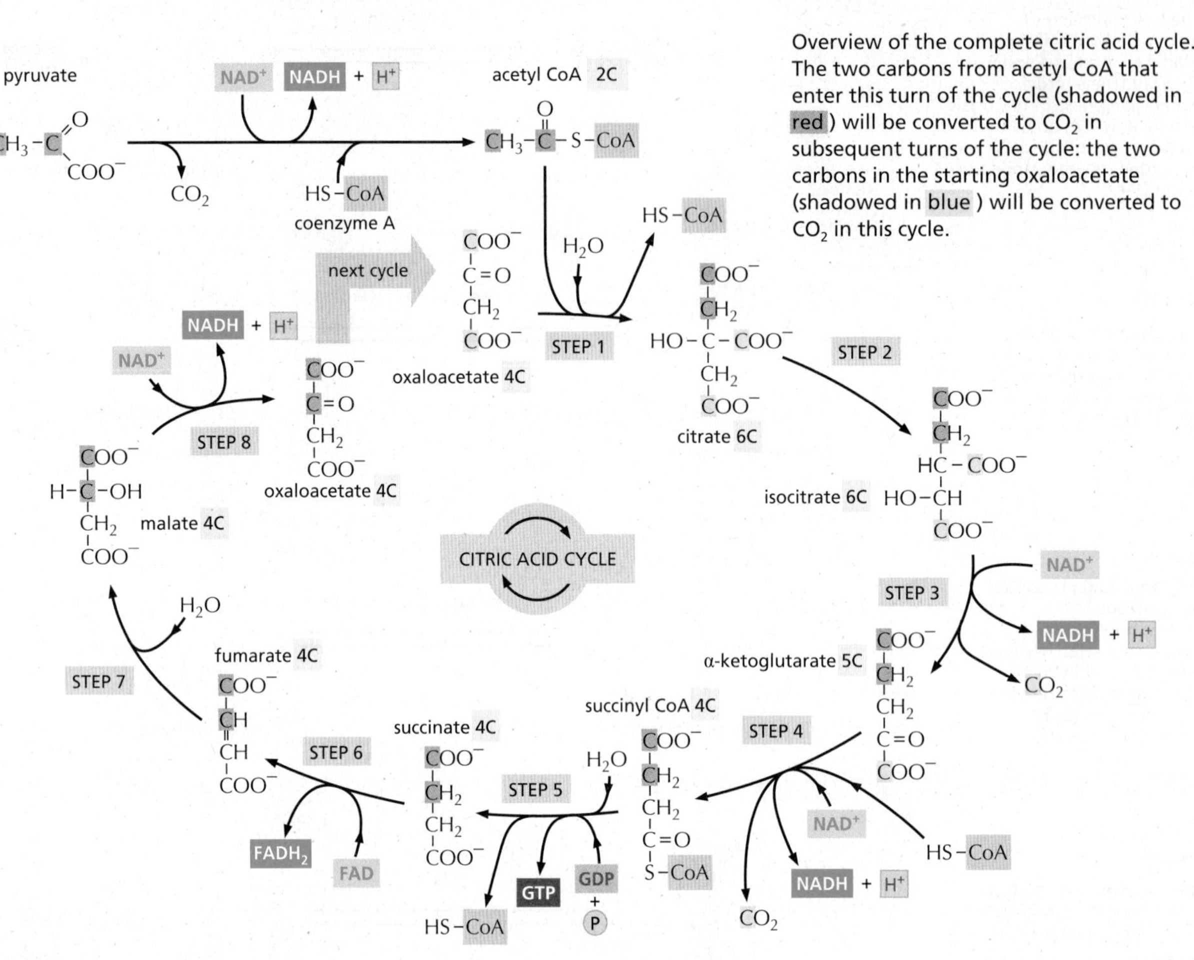

Overview of the complete citric acid cycle. The two carbons from acetyl CoA that enter this turn of the cycle (shadowed in red) will be converted to CO_2 in subsequent turns of the cycle: the two carbons in the starting oxaloacetate (shadowed in blue) will be converted to CO_2 in this cycle.

Details of these eight steps are shown below. In this part of the Panel, for each step, the part of the molecule that undergoes a change is shadowed in blue, and the name of the enzyme that catalyzes the reaction is in a yellow box. To watch a video of the reactions of the citric acid cycle, see Movie 2.6.

STEP 1

After the enzyme removes a proton from the CH_3 group on acetyl CoA, the negatively charged CH_2^- forms a bond to a carbonyl carbon of oxaloacetate. The subsequent loss by hydrolysis of the coenzyme A (HS–CoA) drives the reaction strongly forward.

acetyl CoA + oxaloacetate ⇌ (citrate synthase) [S-citryl-CoA intermediate] → (H_2O) citrate + HS–CoA + H^+

- acetyl CoA: O=C(–S–CoA)–CH_3
- oxaloacetate: COO^-–C(=O)–CH_2–COO^-
- S-citryl-CoA intermediate: O=C(–S–CoA)–CH_2–C(OH)(COO^-)–CH_2–COO^-
- citrate: COO^-–CH_2–C(OH)(COO^-)–CH_2–COO^-

STEP 2

An isomerization reaction, in which water is first removed and then added back, moves the hydroxyl group from one carbon atom to its neighbor.

citrate ⇌ (aconitase; H_2O out) [*cis*-aconitate intermediate] ⇌ (H_2O in) isocitrate

- citrate: COO^-–C(H)(H)–C(OH)(COO^-)–C(H)(H)–COO^-
- *cis*-aconitate intermediate: COO^-–C(H)(H)–C(COO^-)=C(H)–COO^-
- isocitrate: COO^-–C(H)(H)–C(H)(COO^-)–C(OH)(H)–COO^-

STEP 3

In the first of four oxidation steps in the cycle, the carbon carrying the hydroxyl group is converted to a carbonyl group. The immediate product is unstable, losing CO_2 while still bound to the enzyme.

isocitrate → (isocitrate dehydrogenase; NAD^+ → NADH + H^+) → [oxalosuccinate intermediate] → (H^+ in, CO_2 out) → α-ketoglutarate

STEP 4

The α-*ketoglutarate dehydrogenase complex* closely resembles the large enzyme complex that converts pyruvate to acetyl CoA, the pyruvate dehydrogenase complex. It likewise catalyzes an oxidation that produces NADH, CO_2, and a high-energy thioester bond to coenzyme A (CoA).

α-ketoglutarate + HS–CoA → (α-ketoglutarate dehydrogenase complex; NAD^+ → NADH + H^+; CO_2 released) → succinyl CoA

STEP 5

An inorganic phosphate displaces the CoA, forming a high-energy phosphate linkage to succinate. This phosphate is then passed to GDP to form GTP. (In bacteria and plants, ATP is formed instead.)

succinyl CoA → (succinyl CoA synthetase; H_2O, P, GDP → GTP) → succinate + HS–CoA

STEP 6

In the third oxidation step of the cycle, FAD accepts two hydrogen atoms from succinate.

succinate → (succinate dehydrogenase; FAD → $FADH_2$) → fumarate

STEP 7

The addition of water to fumarate places a hydroxyl group next to a carbonyl carbon.

fumarate → (fumarase; H_2O) → malate

STEP 8

In the last of four oxidation steps in the cycle, the carbon carrying the hydroxyl group is converted to a carbonyl group, regenerating the oxaloacetate needed for step 1.

malate → (malate dehydrogenase; NAD^+ → NADH + H^+) → oxaloacetate

PROBLEMS

Which statements are true? Explain why or why not.

2–1 A 10^{-8} M solution of HCl has a pH of 8.

2–2 Most of the interactions between macromolecules could be mediated just as well by covalent bonds as by noncovalent bonds.

2–3 Animals and plants use oxidation to extract energy from food molecules.

2–4 If an oxidation occurs in a reaction, it must be accompanied by a reduction.

2–5 Linking the energetically unfavorable reaction A → B to a second, favorable reaction B → C will shift the equilibrium constant for the first reaction.

2–6 The criterion for whether a reaction proceeds spontaneously is ΔG, not $\Delta G°$, because ΔG takes into account the concentrations of the substrates and products.

2–7 The oxygen consumed during the oxidation of glucose in animal cells is returned as CO_2 to the atmosphere.

Discuss the following problems.

2-8 During an all-out sprint, muscles produce a high concentration of lactic acid, which lowers the pH of the blood and of the cytosol. The lower pH inside the cell reduces the rate of ATP production and contributes to the fatigue that sprinters experience well before their fuel reserves are exhausted. The main blood buffer against pH changes is the bicarbonate/CO_2 system.

$$\underset{\text{(gas)}}{CO_2} \rightleftharpoons \underset{\text{(dissolved)}}{CO_2} \overset{pK_1=2.3}{\rightleftharpoons} H_2CO_3 \overset{pK_2=3.8}{\rightleftharpoons} H^+ + HCO_3^- \overset{pK_3=10.3}{\rightleftharpoons} H^+ + CO_3^{2-}$$

To improve their performance, would you advise sprinters to hold their breath or to breathe rapidly just before the race? Why?

A. Breathe rapidly to decrease the pH of the blood.

B. Breathe rapidly to increase the pH of the blood.

C. Hold their breath to decrease the pH of the blood.

D. Hold their breath to increase the pH of the blood.

2–9 The three molecules in Figure Q2–1 contain the seven most common functional groups in biology. Most molecules in the cell are built from these functional groups. Indicate and name the functional groups in these molecules.

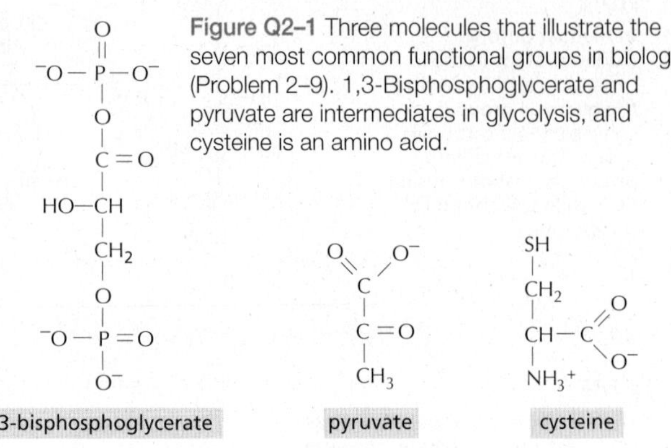

Figure Q2–1 Three molecules that illustrate the seven most common functional groups in biology (Problem 2–9). 1,3-Bisphosphoglycerate and pyruvate are intermediates in glycolysis, and cysteine is an amino acid.

2–10 The molecular weight of ethanol (CH_3CH_2OH) is 46 and its density is 0.789 g/cm^3.

A. What is the molarity of ethanol in beer that is 5% ethanol by volume? [Alcohol content of beer varies from about 4% (lite beer) to 8% (stout beer).]

B. The legal limit for a driver's blood alcohol content varies, but 80 mg of ethanol per 100 mL of blood (usually referred to as a blood alcohol level of 0.08) is typical. What is the molarity of ethanol in a person at this legal limit?

C. How many 12-oz (355-mL) bottles of 5% beer could a 70-kg person drink and remain under the legal limit? A 70-kg person contains about 40 liters of water. Ignore the metabolism of ethanol and assume that the water content of the person remains constant and that ethanol distributes evenly in that volume.

2–11 A histidine side chain is known to play an important role in the catalytic mechanism of an enzyme; however, it is not clear whether histidine is required in its protonated (charged) or unprotonated (uncharged) state. To answer this question, you measure enzyme activity over a range of pH, with the results shown in Figure Q2–2. Which form of histidine is required for enzyme activity?

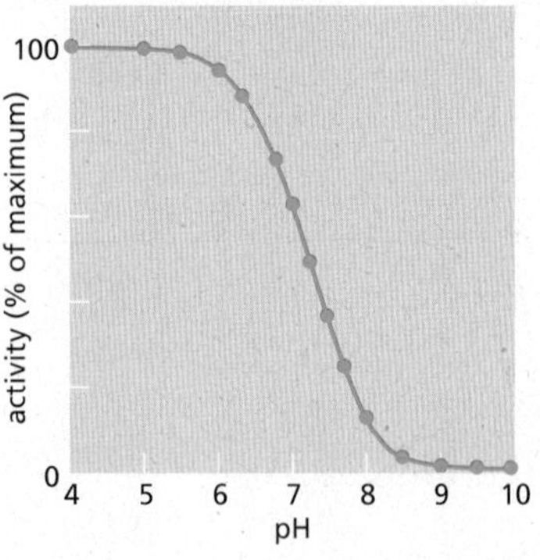

Figure Q2–2 Enzyme activity as a function of pH (Problem 2–11).

2–12 The organic chemistry of living cells is said to be special for two reasons: it occurs in an aqueous environment, and it accomplishes some very complex reactions. But do you suppose it is really all that much different from the organic chemistry carried out in the top laboratories in the world? Why or why not?

2–13 Polymerization of tubulin subunits into microtubules occurs with an increase in the orderliness of the subunits. Yet tubulin polymerization occurs with an increase in entropy (decrease in order). How can that be?

2–14 "Diffusion" sounds slow—and over everyday distances it is—but on the scale of a cell it is very fast. The average instantaneous velocity of a particle in solution—that is, the velocity between the very frequent collisions—is

$$v = (kT/m)^{1/2}$$

where $k = 1.38 \times 10^{-16}$ g cm^2/K sec^2, T = temperature in K (37°C is 310 K), and m = mass in g/molecule.

Calculate the instantaneous velocity of a water molecule (molecular mass = 18 daltons), a glucose molecule (molecular mass = 180 daltons), and a myoglobin molecule (molecular mass = 15,000 daltons) at 37°C.

2–15 Each phosphoanhydride bond between the phosphate groups in ATP is a high-energy linkage with a $\Delta G°$ value of –30.5 kJ/mole. Hydrolysis of this bond in cells normally liberates usable energy in the range of 45 to 55 kJ/mole. Why do you think a range of values for released energy is given for ΔG, rather than a precise number, as for $\Delta G°$?

A. ΔG cannot be accurately measured and can only be estimated as a range of values.

B. Differences in the concentrations of ATP, ADP, and phosphate can significantly change ΔG.

C. The specific enzyme catalyzing hydrolysis determines how much energy is released.

D. The temperature fluctuations occurring inside the cell can significantly change ΔG.

2–16 A 70-kg adult human (154 lb) could meet his or her entire energy needs for one day by eating 3 moles of glucose (540 g). (We do not recommend this.) Each molecule of glucose generates 30 molecules of ATP when it is oxidized to CO_2. The concentration of ATP is maintained in cells at about 2 mM, and a 70-kg adult has about 25 liters of intracellular fluid. Given that the ATP concentration remains constant in cells, calculate how many times per day, on average, each ATP molecule in the body is hydrolyzed and resynthesized.

2–17 What is the "packet of energy" that NADH and NADPH carry?

A. A hydride ion (two electrons and one proton)

B. A hydrogen atom (one electron and one proton)

C. A hydrogen ion (one proton and no electrons)

D. A hydronium ion (a protonated water molecule)

2–18 Cancer cells can increase the rate of glycolysis up to 200-fold relative to normal differentiated cells. This effect, known as the Warburg effect after its discoverer, is exploited in an imaging technique that is commonly used to detect tumors. An individual is dosed with a molecule labeled with a radioactive isotope of fluorine (^{18}F). The labeled molecule is preferentially taken up by cancer cells and is detected by positron emission tomography (PET) scanning. Which one of the following molecules would you label with ^{18}F to detect tumors via the Warburg effect?

A. Acetyl CoA

B. Glucose

C. Lactate

D. Pyruvate

2–19 Does a Snickers candy bar (65 g, 1360 kJ) provide enough energy to climb from Zermatt (elevation 1660 m) to the top of the Matterhorn (4478 m; Figure Q2–3), or might you need to stop at Hörnli Hut (3260 m) to eat another one? Imagine that you and your gear have a mass of 75 kg and that all of your work is done against gravity (that is, you are just climbing straight up). Remember from your introductory physics course that

$$\text{work (J)} = \text{mass (kg)} \times g\,(\text{m/sec}^2) \times \text{height gained (m)}$$

where g is acceleration due to gravity (9.8 m/sec^2). One joule is 1 kg m^2/sec^2. What assumptions made here will greatly underestimate how much candy you need?

Figure Q2–3 The Matterhorn (Problem 2–19). (Earth Trotter Photos/Shutterstock.)

2–20 Assuming that there are 5×10^{13} cells in the human body and that ATP is turning over at a rate of 10^9 ATP molecules per minute in each cell, how many watts is the human body consuming? (A watt is a joule per second.) Assume that hydrolysis of ATP yields 50 kJ/mole.

REFERENCES

General

Berg JM, Tymoczko JL, Gatto GJ & Stryer L (2021) Biochemistry, 9th ed. New York: WH Freeman.

Miesfeld RL & McEvoy MM (2021) Biochemistry, 2nd ed. New York: Norton.

Moran LA, Horton HR, Scrimgeour G & Perry M (2012) Principles of Biochemistry, 5th ed. Upper Saddle River, NJ: Prentice Hall.

Nelson DL & Cox MM (2021) Lehninger Principles of Biochemistry, 8th ed. New York: Macmillan.

van Holde KE, Johnson WC & Ho PS (2005) Principles of Physical Biochemistry, 2nd ed. Upper Saddle River, NJ: Prentice Hall.

Voet D, Voet JG & Pratt CM (2016) Fundamentals of Biochemistry, 5th ed. New York: Wiley.

The Chemical Components of a Cell

Atkins PW (2003) Molecules, 2nd ed. New York: WH Freeman.

Baldwin RL (2014) Dynamic hydration shell restores Kauzmann's 1959 explanation of how the hydrophobic factor drives protein folding. *Proc. Natl. Acad. Sci. USA* 111, 13052–13056.

Ball P (2017) Water is an active matrix of life for cell and molecular biology. *Proc. Natl. Acad. Sci. USA* 114, 13327–13335.

Bloomfield VA, Crothers DM & Tinoco I (2000) Nucleic Acids: Structures, Properties, and Functions. Sausalito, CA: University Science Books.

Branden C & Tooze J (1999) Introduction to Protein Structure, 2nd ed. New York: Garland Science.

de Duve C (2005) Singularities: Landmarks on the Pathways of Life. Cambridge, UK: Cambridge University Press.

Dowhan W (1997) Molecular basis for membrane phospholipid diversity: why are there so many lipids? *Annu. Rev. Biochem.* 66, 199–232.

Franks F (1993) Water. Cambridge, UK: Royal Society of Chemistry.

Henderson LJ (1927) The Fitness of the Environment, 1958 ed. Boston: Beacon.

Laage D, Elsaesser T & Hynes, JT (2017) Water dynamics in the hydration shells of biomolecules. *Chem. Rev.* 117, 10694–10725.

Milo R & Phillips R (2009) A feeling for the numbers in biology. *Proc. Natl. Acad. Sci. USA* 106, 21465–21471.

Phillips R & Milo R (2015) Cell Biology by the Numbers. New York: Garland Science.

Taylor ME & Drickamer K (2011) Introduction to Glycobiology, 3rd ed. New York: Oxford University Press.

van Meer G, Voelker DR & Feigenson GW (2008) Membrane lipids: where they are and how they behave. *Nat. Rev. Mol. Cell Biol.* 9, 112–124.

Varki A, Cummings RD, Esko JD, Freeze HH, Stanley P, Bertozzi CR et al., eds. (2009) Essentials of Glycobiology. Cold Spring Harbor, NY: Cold Spring Harbor Laboratory Press.

Catalysis and the Use of Energy by Cells

Alon U (2020) An Introduction to Systems Biology: Design Principles of Biological Circuits, 2nd ed. London: Chapman & Hall.

Atkins PW (1994) The Second Law: Energy, Chaos and Form. New York: Scientific American Books.

Baldwin JE & Krebs H (1981) The evolution of metabolic cycles. *Nature* 291, 381–382.

Berg HC (1983) Random Walks in Biology. Princeton, NJ: Princeton University Press.

Dill KA & Bromberg S (2010) Molecular Driving Forces: Statistical Thermodynamics in Biology, Chemistry, Physics, and Nanoscience, 2nd ed. New York: Garland Science.

Einstein A (1956) Investigations on the Theory of the Brownian Movement. New York: Dover.

Fruton JS (1999) Proteins, Enzymes, Genes: The Interplay of Chemistry and Biology. New Haven, CT: Yale University Press.

Hohmann-Marriott MF & Blankenship RE (2011) Evolution of photosynthesis. *Annu. Rev. Plant Biol.* 62, 515–548.

Kauzmann W (1967) Thermal Properties of Matter, Vol. 2. Thermodynamics and Statistics: With Applications to Gases. New York: WA Benjamin.

Kornberg A (1989) For the Love of Enzymes. Cambridge, MA: Harvard University Press.

Kuriyan J, Konforti B & Wemmer D (2012) The Molecules of Life: Physical and Chemical Principles. New York: Norton.

Lipmann F (1941) Metabolic generation and utilization of phosphate bond energy. *Adv. Enzymol.* 1, 99–162.

Lipmann F (1971) Wanderings of a Biochemist. New York: Wiley.

Nisbet EG & Sleep NH (2001) The habitat and nature of early life. *Nature* 409, 1083–1091.

Racker E (1980) From Pasteur to Mitchell: a hundred years of bioenergetics. *Fed. Proc.* 39, 210–215.

Sasselov DD, Grotzinger JP, & Sutherland JD (2020) The origin of life as a planetary phenomenon. *Sci. Adv.* 6, eaax3419.

Schrödinger E (1944 & 1958) What Is Life? The Physical Aspect of the Living Cell and Mind and Matter, 1992 combined ed. Cambridge, UK: Cambridge University Press.

Spang A, Saw JH, Jørgensen SL . . . Ettema TJG (2015) Complex archaea that bridge the gap between prokaryotes and eukaryotes. *Nature* 52, 173–179.

Tinoco I, Sauer K, Wang JC, Puglisi JD, Harbison G & Rovnyak D (2013) Physical Chemistry: Principles and Applications in Biological Sciences, 5th ed. Glenview, IL: Pearson Education.

van Meer G, Voelker DR & Feigenson GW (2008) Membrane lipids: where they are and how they behave. *Nat. Rev. Mol. Cell Biol.* 9, 112–124.

Walsh C (2001) Enabling the chemistry of life. *Nature* 409, 226–231.

Westheimer FH (1987) Why nature chose phosphates. *Science* 235, 1173–1178.

How Cells Obtain Energy from Food

Caetano-Anollés D, Kim KM, Mittenthal JE & Caetano-Anollés G (2011) Proteome evolution and the metabolic origins of translation and cellular life. *J. Mol. Evol.* 72, 14–33.

Dismukes GC, Klimov VV, Baranov SV, Kozlov YN, DasGupta J & Tyryshkin A (2001) The origin of atmospheric oxygen on Earth: the innovation of oxygenic photosynthesis. *Proc. Natl. Acad. Sci. USA* 98, 2170–2175.

Fothergill-Gilmore LA (1986) The evolution of the glycolytic pathway. *Trends Biochem. Sci.* 11, 47–51.

Friedmann HC (2004) From *Butyribacterium* to *E. coli*: an essay on unity in biochemistry. *Perspect. Biol. Med.* 47, 47–66.

Huynen MA, Dandekar T & Bork P (1999) Variation and evolution of the citric-acid cycle: a genomic perspective. *Trends Microbiol.* 7, 281–291.

Koonin EV (2014) The origins of cellular life. *Antonie van Leeuwenhoek* 106, 27–41.

Kornberg HL (2000) Krebs and his trinity of cycles. *Nat. Rev. Mol. Cell Biol.* 1, 225–228.

Krebs HA (1970) The history of the tricarboxylic acid cycle. *Perspect. Biol. Med.* 14, 154–170.

Lane N & Martin WF (2012) The origin of membrane bioenergetics. *Cell* 151, 1406–1416.

Martijn J & Ettema TJG (2013) From archaeon to eukaryote: the evolutionary dark ages of the eukaryotic cell. *Biochem. Soc. Trans.* 41, 451–457.

Mulkidjanian AY, Bychkov AY, Dibrova DV, Galperin MY & Koonin EV (2012) Open questions on the origin of life at anoxic geothermal fields. *Orig. Life Evol. Biosph.* 42, 507–516.

Zimmer C (2009) On the origin of eukaryotes. *Science* 325, 666–668.

CHAPTER

3

Proteins

When we look at a cell through a microscope or analyze its electrical or biochemical activity, we are, in essence, observing proteins. Proteins constitute most of a cell's dry mass. They are not only the cell's building blocks; they also execute the majority of the cell's functions. Proteins that are enzymes provide the intricate molecular surfaces inside a cell that catalyze its many chemical reactions. Proteins embedded in the plasma membrane form channels and pumps that control the passage of small molecules into and out of the cell. Other proteins carry messages from one cell to another or act as signal integrators that relay sets of signals inward from the plasma membrane to the cell nucleus. Yet others serve as tiny molecular machines with moving parts: kinesin, for example, propels organelles through the cytoplasm; topoisomerase can untangle knotted DNA molecules. Other specialized proteins act as antibodies, toxins, hormones, antifreeze molecules, elastic fibers, ropes, or sources of luminescence. Before we can hope to understand how genes work, how muscles contract, how nerves conduct electricity, how embryos develop, or how our bodies function, we must attain a deep understanding of proteins.

IN THIS CHAPTER

THE ATOMIC STRUCTURE OF PROTEINS

From a chemical point of view, proteins are by far the most structurally complex and functionally sophisticated molecules known. This is perhaps not surprising, once we realize that the structure and chemistry of each protein have been developed and fine-tuned over billions of years of evolutionary history. The theoretical calculations of population geneticists reveal that, over evolutionary time periods, a surprisingly small selective advantage is enough to cause a randomly altered protein sequence to spread through a population of organisms. Yet, even to experts, the remarkable versatility of proteins can seem truly amazing.

In this section, we consider how the location of each amino acid in a protein's long string of amino acids determines its three-dimensional shape. Later in the chapter, we use this understanding of protein structure at the atomic level to describe how the precise shape of each protein molecule determines its function in a cell.

The Structure of a Protein Is Specified by Its Amino Acid Sequence

There are 20 different types of amino acids in proteins that are encoded directly in an organism's DNA, each with different chemical properties. Every **protein** molecule consists of a long unbranched chain of these amino acids, each linked to its neighbor through a covalent *peptide bond* (**Figure 3–1A**). Proteins are therefore also known as *polypeptides*. Each type of protein has a unique sequence of amino acids, and there are many thousands of different proteins in a cell.

The repeating sequence of atoms along the core of the polypeptide chain is referred to as the **polypeptide backbone**. Attached to this repetitive backbone are those portions of the amino acids that are not involved in making

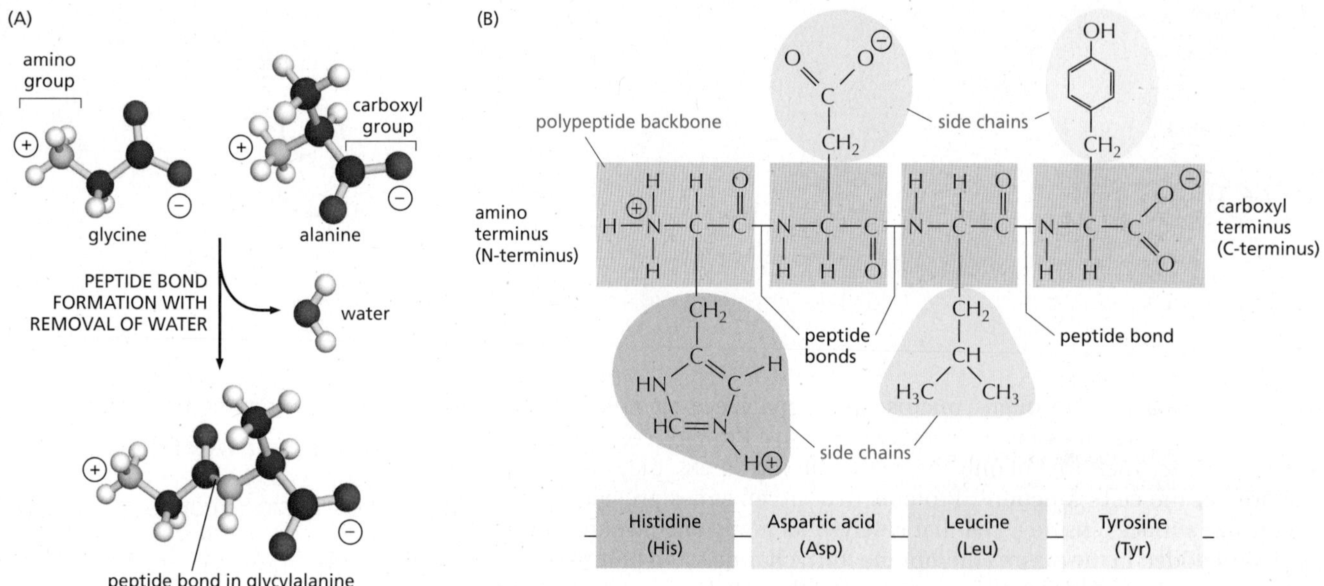

Figure 3–1 **The components of a protein.** (A) Formation of a peptide bond. This covalent bond forms when the carbon atom of the carboxyl group of one amino acid (such as glycine) shares electrons with the nitrogen atom from the amino group of a second amino acid (such as alanine). As indicated, a molecule of water is eliminated in this condensation reaction (see Figure 2–9). In this model, carbon atoms are *black*, nitrogen *blue*, oxygen *red*, and hydrogen *white*. (B) A two-dimensional representation of a short section of polypeptide backbone with its attached side chains. Each type of protein differs in its sequence and number of amino acids; it is the sequence of the chemically different side chains that makes each protein distinct. The two ends of a polypeptide chain are chemically different: the end carrying the free amino group (NH_2, which takes up a proton at neutral pH to become NH_3^+) is the amino terminus, or N-terminus, and the end carrying the free carboxyl group (COOH, which loses a proton at neutral pH to become COO^-) is the carboxyl terminus, or C-terminus. Note that, for simplicity, in many figures in this textbook, NH_2 and COOH are used to denote these termini, instead of their actual ionized forms. The amino acid sequence of a protein is always presented in the N-to-C direction, reading from left to right.

a peptide bond; these are the 20 different amino acid **side chains** that give each amino acid its unique properties (Figure 3–1B). Some of these side chains are nonpolar and hydrophobic ("water-fearing"), others are negatively or positively charged, some can readily form covalent bonds, and so on. Panel 3–1 (pp. 118–119) shows their atomic structures, and Figure 3–2 lists their abbreviations.

AMINO ACID			SIDE CHAIN
Aspartic acid	Asp	D	acidic (negative charge)
Glutamic acid	Glu	E	acidic (negative charge)
Arginine	Arg	R	basic (positive charge)
Lysine	Lys	K	basic (positive charge)
Histidine	His	H	basic (positive charge)
Asparagine	Asn	N	uncharged polar
Glutamine	Gln	Q	uncharged polar
Serine	Ser	S	uncharged polar
Threonine	Thr	T	uncharged polar
Tyrosine	Tyr	Y	uncharged polar

POLAR AMINO ACIDS

AMINO ACID			SIDE CHAIN
Alanine	Ala	A	nonpolar
Glycine	Gly	G	nonpolar
Valine	Val	V	nonpolar
Leucine	Leu	L	nonpolar
Isoleucine	Ile	I	nonpolar
Proline	Pro	P	nonpolar
Phenylalanine	Phe	F	nonpolar
Methionine	Met	M	nonpolar
Tryptophan	Trp	W	nonpolar
Cysteine	Cys	C	nonpolar

NONPOLAR AMINO ACIDS

Figure 3–2 **The 20 amino acids commonly found in proteins.** Each amino acid has a three-letter and a one-letter abbreviation. There are equal numbers of polar and nonpolar side chains; however, some side chains listed here as polar are large enough to have some nonpolar properties (for example, Thr, Tyr, Arg, Lys). For atomic structures, see Panel 3–1 (pp. 118–119).

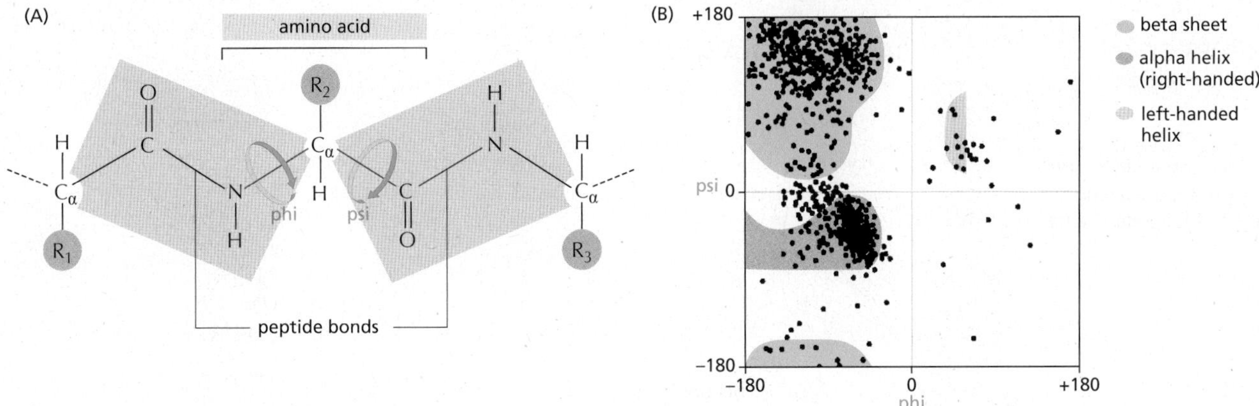

Figure 3–3 **Steric limitations on the bond angles in a polypeptide chain.** (A) Each amino acid contributes three bonds *(red)* to the backbone of the chain. Because it has a partial double-bond character, the peptide bond is planar *(gray shading)* and does not permit free rotation. By contrast, rotation can occur about the C_α–C bond, whose angle of rotation is called psi (Ψ), and about the N–C_α bond, whose angle of rotation is called phi (ϕ). By convention, an R group is often used to denote an amino acid side chain *(purple circles)*. (B) The conformation of the main-chain atoms in a protein is determined by one pair of ϕ and Ψ angles for each amino acid; because of steric restrictions, most of the possible pairs of ϕ and Ψ angles do not occur. In this so-called Ramachandran plot, each dot represents an observed pair of angles in a protein. The three differently shaded clusters of dots reflect three different *secondary structures* repeatedly found in proteins. Most prominent are the alpha helix and the beta sheet, as will be described in the text. (B, from J. Richardson, *Adv. Prot. Chem.* 34:174–175, 1981. With permission from Elsevier.)

As discussed in Chapter 2, atoms behave almost as if they were hard spheres with a definite radius (their *van der Waals radius*). Other constraints limit the possible bond angles in a polypeptide chain, and this—plus the requirement that no two atoms overlap—severely restricts the possible three-dimensional arrangements (or *conformations*) of proteins. As illustrated in Figure 3–3, these steric restrictions (which include a delocalization of electrons in the peptide bond that makes that linkage planar) confine the energy minima for the bond angles in polypeptides to a narrow range. But a long flexible chain such as a protein can still fold in an enormous number of different ways.

The folding of a protein chain is determined by many different sets of weak *noncovalent bonds* that form between one part of the chain and another. These involve atoms in the polypeptide backbone, as well as atoms in the amino acid side chains. There are three types of these weak bonds: *hydrogen bonds, electrostatic attractions,* and *van der Waals attractions,* as explained in Chapter 2 (see p. 51). Individual noncovalent bonds are 30–300 times weaker than the typical covalent bonds that create biological molecules. But many weak bonds acting in parallel can hold two regions of a polypeptide chain tightly together. It is the combined strength of large numbers of these noncovalent bonds that stabilizes each protein's folded shape (Figure 3–4).

A fourth weak force—a hydrophobic clustering force—also has a central role in determining the shape of a protein. As described in Chapter 2, hydrophobic molecules, including the nonpolar side chains of particular amino acids, tend to be forced together in an aqueous environment in order to minimize their disruptive effect on the hydrogen-bonded network of water molecules (see Panel 2–2, pp. 96–97). Therefore, an important factor governing the folding of any protein is the distribution of its polar and nonpolar amino acids. The nonpolar (hydrophobic) side chains in a protein—belonging to such amino acids as phenylalanine, leucine, valine, and tryptophan—tend to cluster in the interior of the molecule (just as hydrophobic oil droplets coalesce in water to form one large droplet). This enables these side chains to avoid contact with the water that surrounds them inside a cell. In contrast, polar groups—such as those belonging to arginine, glutamine, and histidine—tend to arrange themselves near the outside of the molecule, where they can form hydrogen bonds with water and with other polar molecules (Figure 3–5). Any polar amino acids that are left buried within the protein are usually hydrogen-bonded to other polar amino acids or to the polypeptide backbone.

FAMILIES OF AMINO ACIDS

The common amino acids are grouped according to whether their side chains are

acidic
basic
uncharged polar
nonpolar

These 20 amino acids are given both three-letter and one-letter abbreviations.

Thus: alanine = Ala = A

BASIC SIDE CHAINS

lysine (Lys, or K)

This group is very basic because its positive charge is stabilized by resonance (see Panel 2–1).

arginine (Arg, or R)

histidine (His, or H)

These nitrogens have a relatively weak affinity for an H^+ and are only partly positive at neutral pH.

THE AMINO ACID

The general formula of an amino acid is

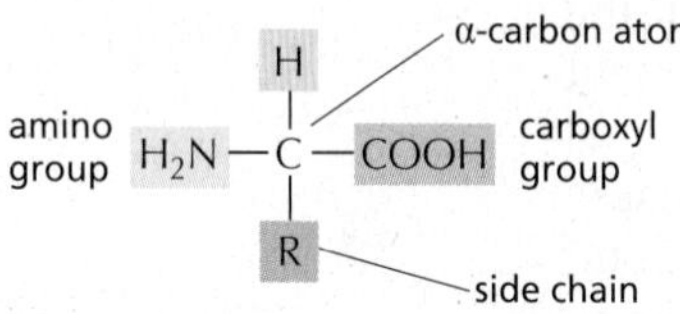

R is commonly one of 20 different side chains. At pH 7, both the amino and carboxyl groups are ionized.

OPTICAL ISOMERS

The α-carbon atom is asymmetric, allowing for two mirror-image (or stereo-) isomers, L and D.

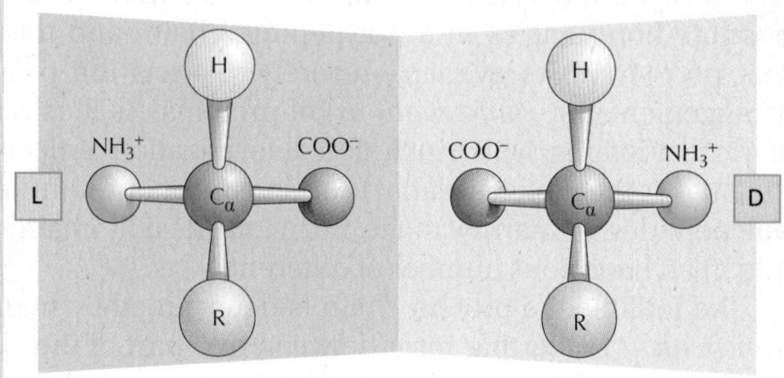

Proteins contain exclusively L-amino acids.

PEPTIDE BONDS

In proteins, amino acids are joined together by an amide linkage, called a peptide bond.

The four atoms involved in each peptide bond form a rigid planar unit *(red box)*. There is no rotation around the C–N bond.

H_2O

peptide bond

Proteins are long polymers of amino acids linked by peptide bonds, and they are always written with the N-terminus toward the left. Peptides are shorter, usually fewer than 50 amino acids long. The sequence of this tripeptide is histidine–cysteine–valine.

amino terminus, or N-terminus

carboxyl terminus, or C-terminus

These two single bonds allow rapid rotation, so that long chains of amino acids are very flexible.

ACIDIC SIDE CHAINS

aspartic acid

(Asp, or D)

glutamic acid

(Glu, or E)

UNCHARGED POLAR SIDE CHAINS

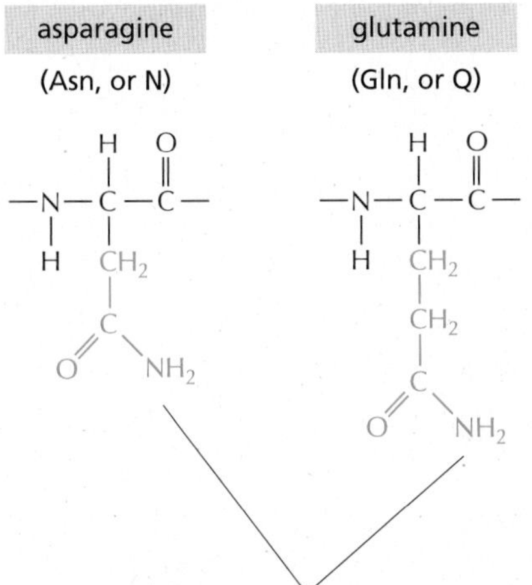

Although the amide N is not charged at neutral pH, it is polar.

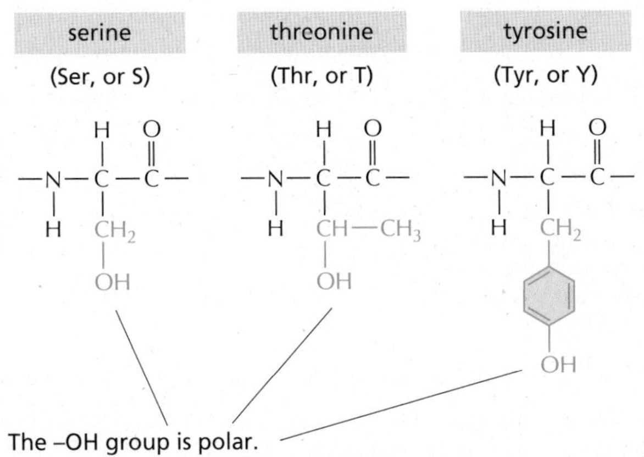

NONPOLAR SIDE CHAINS

alanine

(Ala, or A)

valine

(Val, or V)

leucine

(Leu, or L)

isoleucine

(Ile, or I)

proline

(Pro, or P)

(actually an imino acid)

phenylalanine

(Phe, or F)

methionine

(Met, or M)

tryptophan

(Trp, or W)

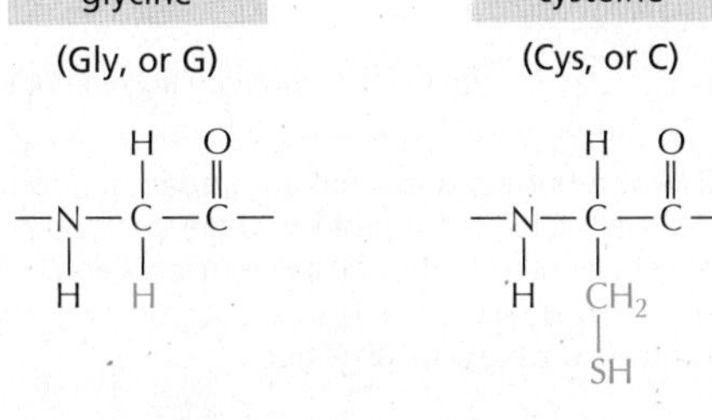

A disulfide bond *(red)* can form between two cysteine side chains in proteins.

$- - CH_2 - S - S - CH_2 - -$

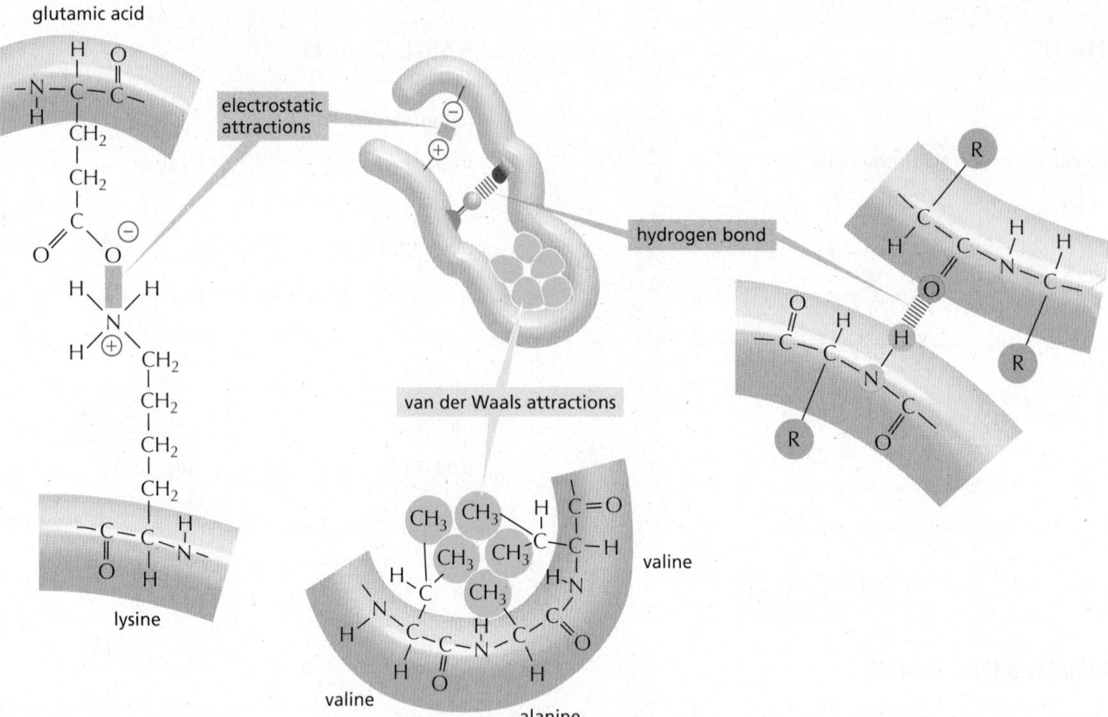

Figure 3–4 **Three types of noncovalent bonds help proteins fold.** Although a single one of these bonds is quite weak, many of them often act together to create a strong bonding arrangement, as in the example shown. As in the previous figure, R is used as a general designation for an amino acid side chain.

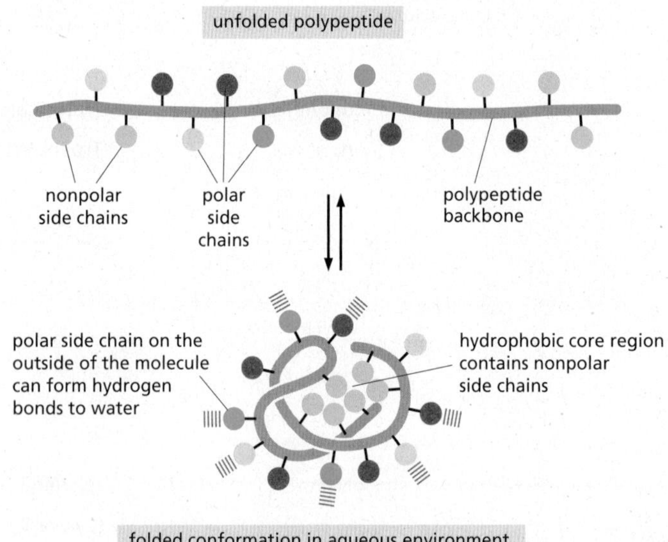

Figure 3–5 **How a protein folds into a compact conformation.** The polar amino acid side chains tend to lie on the outside of the protein, where they can interact with water; the nonpolar amino acid side chains are buried on the inside forming a tightly packed hydrophobic core of atoms that are hidden from water. In this highly schematic drawing, the protein contains only 17 amino acids; actual proteins are generally much larger.

Proteins Fold into a Conformation of Lowest Energy

As a result of all of these interactions, most proteins have a particular three-dimensional structure, which is determined by the order of the amino acids in a protein's chain. The final folded structure, or **conformation**, of any polypeptide chain is generally the one that minimizes its free energy. Biologists have studied protein folding in a test tube using highly purified proteins. Treatment with certain solvents, which disrupt the noncovalent interactions holding the folded chain together, unfolds, or *denatures*, a protein. This treatment converts the protein into a flexible polypeptide chain that has lost its natural shape. When the denaturing solvent is removed, the protein often refolds spontaneously, or *renatures*, into its original conformation. This indicates that the amino acid sequence contains all of the information needed for specifying the three-dimensional shape of a protein, a critical point for understanding cell biology.

Most proteins fold up into a single stable conformation. However, this conformation is very dynamic, experiencing constant fluctuations caused by thermal energy. In addition, a protein's conformation can change when the protein interacts with other molecules in the cell. This change in shape is often crucial to the function of the protein, as we explain in detail later.

Although a protein chain can fold into its correct conformation without outside help, special proteins called *molecular chaperones* often assist in protein folding (see Chapter 6). Molecular chaperones bind to partly folded polypeptide chains and help them progress along the most energetically favorable folding pathway. In the crowded conditions of the cytoplasm, chaperones are required to prevent the temporarily exposed hydrophobic regions in newly synthesized protein chains from associating with each other to form protein aggregates. However, the final three-dimensional shape of the protein is still specified by its amino acid sequence: chaperones simply make reaching the folded state more reliable.

The α Helix and the β Sheet Are Common Folding Motifs

When we compare the three-dimensional structures of many different protein molecules, it becomes clear that, although the overall conformation of each protein is unique, two regular folding patterns are often found within them. Both patterns were discovered 70 years ago from studies of hair and silk. The first folding pattern to be described, called the **α helix**, was found in the protein *α-keratin*, which forms the filaments in hair. Within a year of the discovery of the α helix, a second folded structure, called a **β sheet**, was found in the protein *fibroin*, the major constituent of silk. These two patterns are common because they result from hydrogen-bonding between the N—H and C═O groups in the polypeptide backbone, without involving the side chains of the amino acids. Thus, although incompatible with some amino acid side chains, many different amino acid sequences can form them. In each case, the protein chain adopts a regular, repeating conformation. Figure 3–6 illustrates the detailed structures of these two important conformations, which in ribbon models of proteins are represented by a helical ribbon and by a set of aligned arrows, respectively.

The cores of many proteins contain extensive regions of β sheet. As shown in Figure 3–7, these β sheets can form either from neighboring segments of the polypeptide backbone that run in the same orientation (parallel chains) or from a polypeptide backbone that folds back and forth upon itself, with each section of the chain running in the direction opposite to that of its immediate neighbors (antiparallel chains). Both types of β sheet produce a very rigid structure, held together by hydrogen bonds that connect the peptide bonds in neighboring chains (see Figure 3–6C).

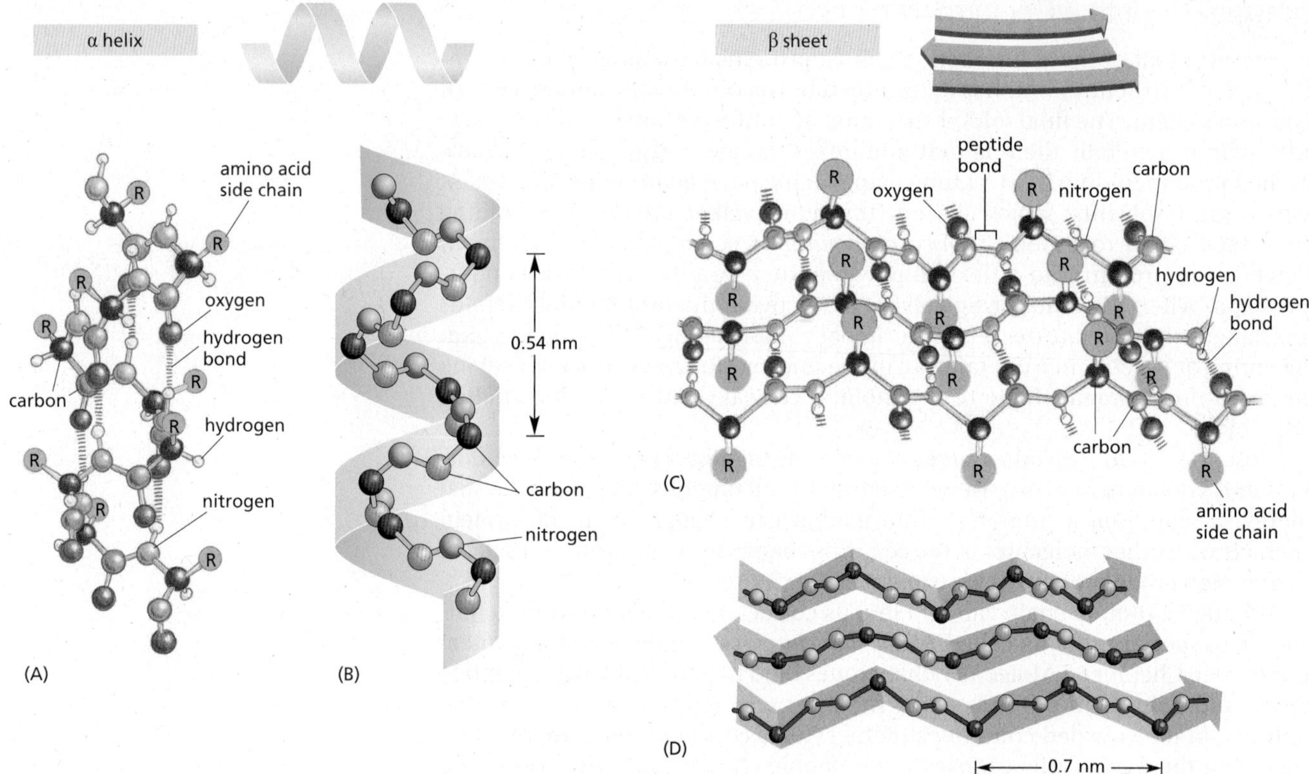

Figure 3–6 **The regular conformation of the polypeptide backbone in the α helix and the β sheet.** The α helix (alpha helix) is shown in (A) and (B). The N—H of every peptide bond is hydrogen-bonded to the C═O of a neighboring peptide bond located four peptide bonds away in the same chain. Note that all of the N—H groups point up in this diagram and that all of the C═O groups point down (toward the C-terminus); this gives a polarity to the helix, with the C-terminus having a partial negative and the N-terminus a partial positive charge (Movie 3.1). The β sheet (beta sheet) is shown in (C) and (D). In this example, adjacent peptide chains run in opposite (antiparallel) directions. Hydrogen-bonding between peptide bonds in different strands holds the individual polypeptide chains (strands) together in a β sheet, and the amino acid side chains in each strand alternately project above and below the plane of the sheet. By convention, when arrows are used to represent a β sheet, the arrowheads point toward the C-terminus (Movie 3.2). (A) and (C) show all the atoms in the polypeptide backbone, but the amino acid side chains are truncated and denoted by R. (It has long been a convention to use R in this way.) In contrast, (B) and (D) show only the carbon and nitrogen backbone atoms.

An α helix is generated when a single polypeptide chain twists around on itself to form a rigid cylinder. A hydrogen bond forms between every fourth peptide bond, linking the C═O of one peptide bond to the N—H of another (see Figure 3–6A). This gives rise to a regular helix with a complete turn every 3.6 amino acids.

Regions of α helix are abundant in proteins located in cell membranes, such as transport proteins and receptors. As we discuss in Chapter 10, those portions of a transmembrane protein that cross the lipid bilayer usually cross as α helices composed largely of amino acids with nonpolar side chains. The polypeptide backbone, which is hydrophilic, is hydrogen-bonded to itself in the α helix and shielded from the hydrophobic lipid environment of the membrane by its protruding nonpolar side chains (see Figure 10–19).

In other proteins, α helices can wrap around each other to form a particularly stable structure, known as a **coiled-coil**. This structure can form when the two (or in some cases, three or four) α helices have most of their nonpolar (hydrophobic) side chains on one side, so that they can twist around each other with these side chains facing inward (Figure 3–8). Long rodlike coiled-coils provide the structural framework for many elongated proteins. Examples are α-keratin, which forms the intracellular fibers that reinforce the outer layer of the skin and its appendages, and the myosin molecules responsible for muscle contraction.

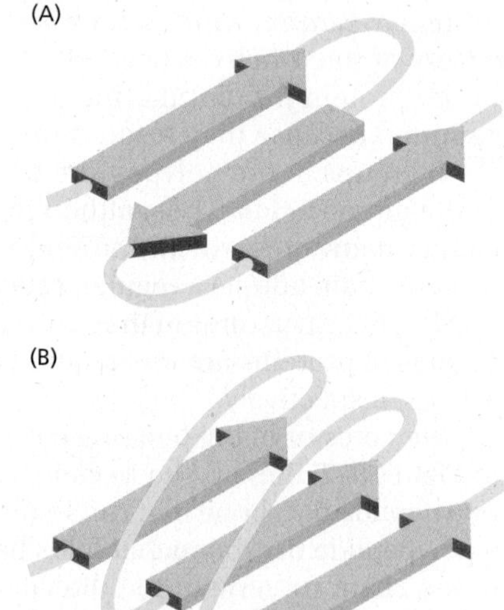

Figure 3–7 **Two types of β sheet structures.** (A) An antiparallel β sheet (see Figure 3–6C). (B) A parallel β sheet. Both of these structures are common in proteins.

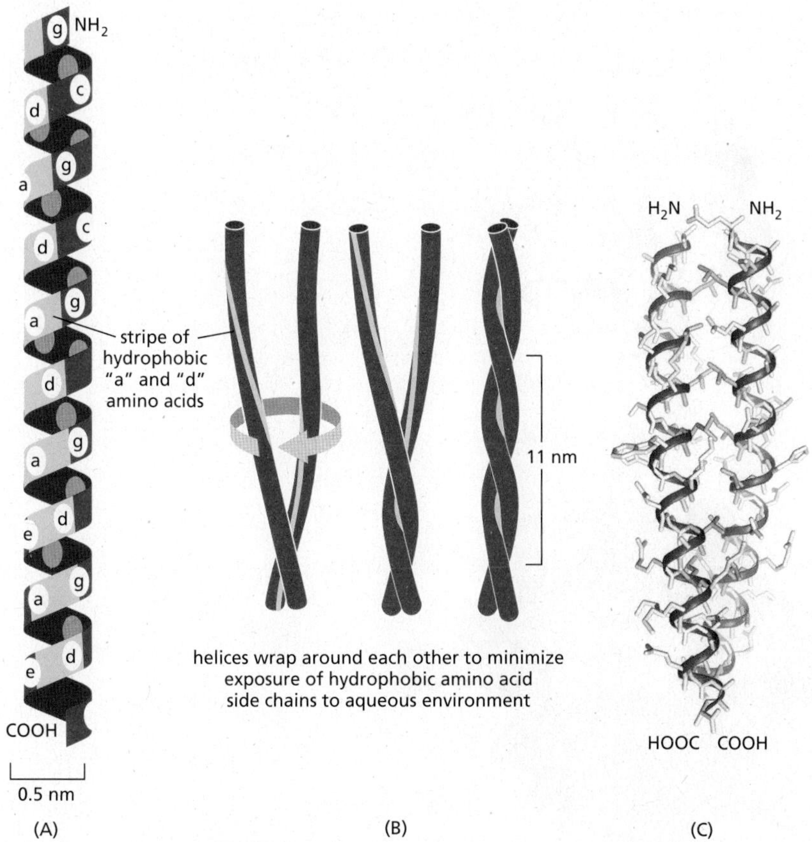

Figure 3–8 **A coiled-coil.** (A) A single α helix, with successive amino acid side chains labeled in a sevenfold sequence, "abcdefg" *(from top to bottom)*. Amino acids "a" and "d" in such a sequence lie close together on the cylinder surface, forming a "stripe" *(green)* that winds slowly around the α helix. Proteins that form coiled-coils typically have nonpolar amino acids at positions "a" and "d." Consequently, as shown in (B), the two α helices can wrap around each other with the nonpolar side chains of one α helix interacting with the nonpolar side chains of the other. (C) The atomic structure of a coiled-coil determined by x-ray crystallography. The α-helical backbone is shown in *red* and the nonpolar side chains in *green*, while the more hydrophilic amino acid side chains, shown in *gray*, are left exposed to the aqueous environment (Movie 3.3). Coiled-coils can also form from three α helices. (PDB code: 3NMD.)

Four Levels of Organization Are Considered to Contribute to Protein Structure

Scientists have found it useful to define four levels of organization that successively generate the structure of a protein. The first level is the protein's amino acid sequence, which is known as its **primary structure**; this sequence is unique for each protein, as determined by the gene that encodes that protein. At the next level, those stretches of the polypeptide chain that form α helices and β sheets constitute the protein's **secondary structure**. The full three-dimensional organization of a polypeptide chain—including its α helices, β sheets, and the many twists and turns that form between its N- and C-termini—is referred to as the protein's **tertiary structure**. And finally, if a protein molecule is formed as a complex of more than one polypeptide chain, its complete conformation is designated as its **quaternary structure**.

Because even a small protein molecule is built from thousands of atoms linked together by precisely oriented covalent and noncovalent bonds, biologists are aided in visualizing these extremely complicated structures by computer-based three-dimensional displays. The student resource site that accompanies this book contains computer-generated images of selected proteins, which can be displayed and rotated on the screen in a variety of formats (Movie 3.4).

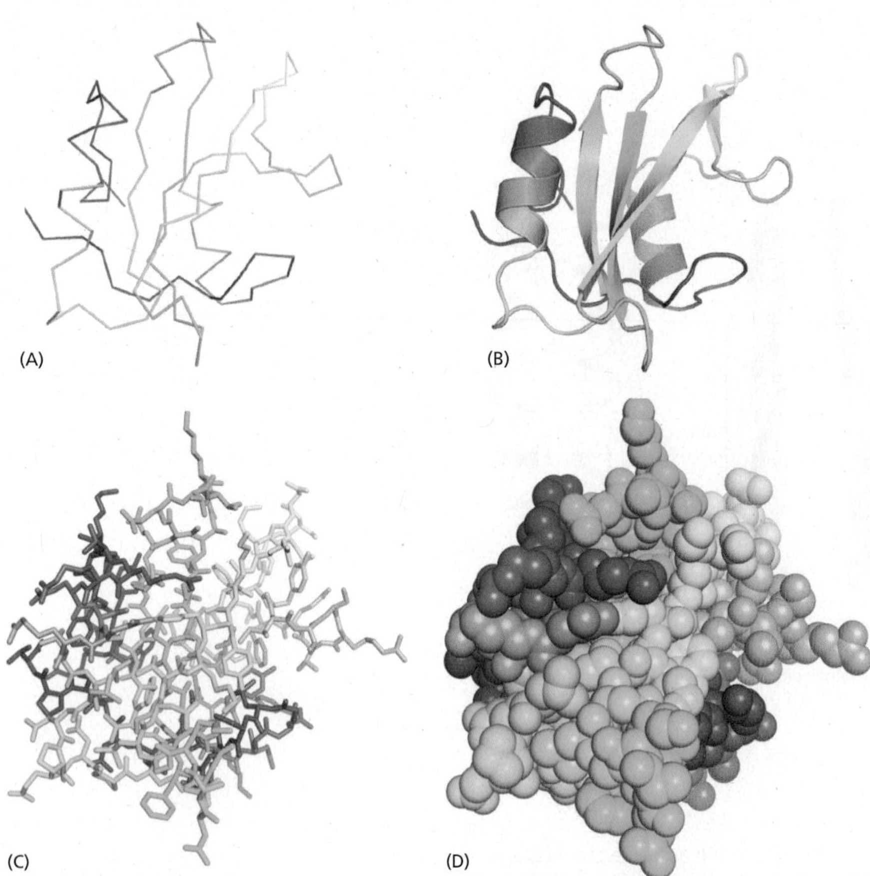

Figure 3–9 **Four representations that are commonly used to describe the structure of a protein.** Constructed from a string of 100 amino acids, the SH2 domain is part of many different proteins. Here, its structure is displayed as (A) a polypeptide backbone model, (B) a ribbon model, (C) a wire model that includes the amino acid side chains, and (D) a space-filling model (**Movie 3.4**). Each image is colored in a way that allows the polypeptide chain to be followed from its N-terminus *(purple)* to its C-terminus *(red)*. (PDB code: 1SHA.)

Protein Domains Are the Modular Units from Which Larger Proteins Are Built

Proteins come in a wide variety of shapes, and most are between 50 and 2000 amino acids long. Large proteins usually consist of a set of smaller *protein domains* that are joined together. A **domain** is a structural unit that folds more or less independently, being formed from perhaps 40 to 350 contiguous amino acids, and it is a modular unit from which larger proteins are constructed.

To display a protein structure in three dimensions, several different representations are conventionally used, each of which emphasizes distinct features. As an example, **Figure 3–9** presents four representations of an important protein structure called the *SH2 domain*. The SH2 domain is present in many different proteins in eukaryotic cells, where it responds to cell signals to cause selected protein molecules to bind to each other, thereby altering cell behavior (see Chapter 15). Contributing to the tertiary structure of this domain are two α helices and a three-stranded, antiparallel β sheet, which are its critical secondary structure elements (see Figure 3–9B).

Figure 3–10 presents ribbon models of three differently organized protein domains. As these examples illustrate, the central core of a domain can be constructed from α helices, from β sheets, or from various combinations of these two fundamental folding elements.

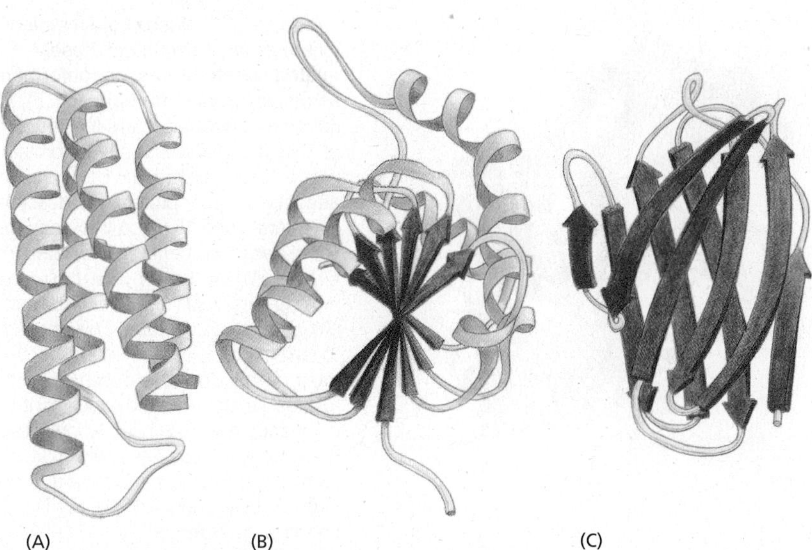

Figure 3–10 **Ribbon models of three different protein domains.** (A) Cytochrome b_{562}, a single-domain protein involved in electron transport in mitochondria. This protein is composed almost entirely of α helices. (B) The NAD-binding domain of the enzyme lactate dehydrogenase, which is composed of a mixture of α helices and parallel β sheets. (C) The variable domain of an immunoglobulin (antibody) light chain, composed of a sandwich of two antiparallel β sheets. In these examples, the α helices are shown in *green*, while strands organized as β sheets are denoted by *red arrows*. Note how the polypeptide chain generally traverses back and forth across the entire domain, making sharp turns (**Movie 3.5**) only at the protein surface. It is the protruding loop regions *(yellow)* that often form the binding sites for other molecules.

The different domains of a protein are often associated with different functions. Figure 3–11 shows an example—the Src protein kinase, which functions in signaling pathways inside vertebrate cells (Src is pronounced "sarc"). This protein is considered to have three domains: its SH2 and SH3 domains have regulatory roles—responding to signals that turn the kinase on and off—while its C-terminal domain is responsible for the kinase catalytic activity. Later in the chapter, we shall return to this protein to explain how proteins can form molecular switches that transmit information throughout cells.

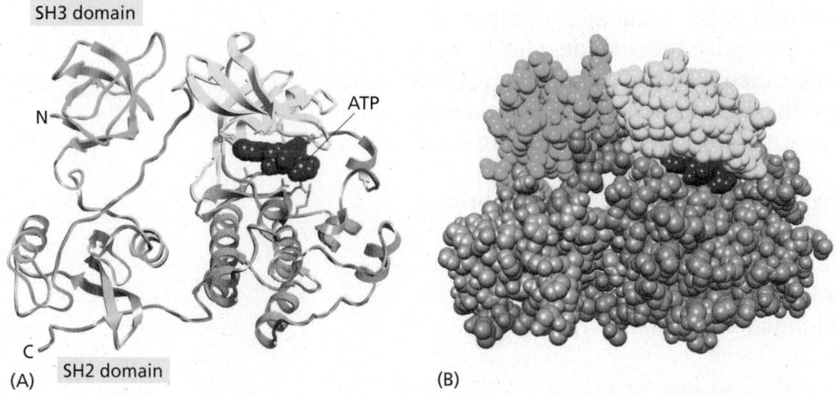

Figure 3–11 **A protein formed from multiple domains.** In the Src protein shown, a C-terminal domain with two lobes (*yellow* and *orange*) forms the core protein kinase enzyme, while its SH2 and SH3 domains perform regulatory functions. Note that both the SH2 and SH3 domains derive their names from this protein, being abbreviations for "Src homology 2" and "Src homology 3," respectively. (A) A ribbon model, with ATP substrate in *red*. (B) A space-filling model, with ATP substrate in *red*. Note that the site that binds ATP is positioned at the interface of the two lobes that form the kinase domain. The human genome encodes about 300 different SH3 domains and 120 SH2 domains. The structure of the SH2 domain was illustrated in Figure 3–9. (PDB code: 2SRC.)

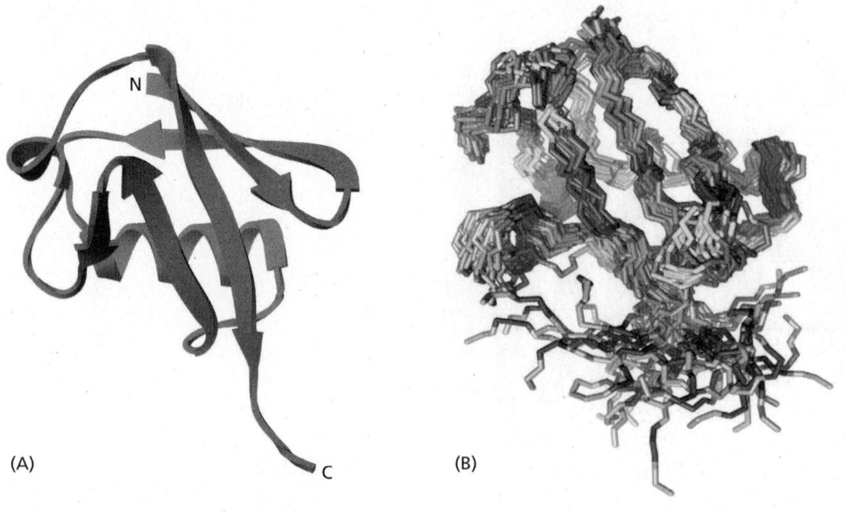

Figure 3–12 **A folded protein molecule exists as an ensemble of closely related substructures, or conformers, as displayed here for ubiquitin.** (A) A ribbon model that displays the structure of ubiquitin. Ubiquitin is a small protein widely used in cells, often being covalently attached to larger proteins, as described in Chapters 6 and 15. (B) In this diagram, a set of backbone conformations determined for ubiquitin has been overlaid to reveal regions that rapidly transition between different substructures. Superimposed on these structures are the rates of motion of the protein's atoms, as observed in NMR residual dipolar coupling experiments. A color code has been used to indicate the magnitude of these rates, which are largest for *red*, with *orange* and *yellow* also being high. (A, PDB code 1UBI; B, from O.F. Lange et al., *Science* 320:1471–1475, 2008. With permission from AAAS.)

Proteins Also Contain Unstructured Regions

The smallest protein molecules contain only a single domain, whereas larger proteins can contain several dozen domains, often connected to each other by short, relatively unstructured lengths of polypeptide chain that can act as flexible hinges between domains. The ubiquity of such intrinsically disordered sequences, which continually bend and flex due to thermal buffeting, became appreciated only after bioinformatics methods were developed that could recognize them from their amino acid sequences. Current estimates suggest that a third of all eukaryotic proteins also possess longer, *intrinsically disordered regions* (*IDRs*)—greater than 30 amino acids in length—in their polypeptide chains. These intrinsically disordered regions can be very long, and they have important functions in cells, as discussed later in this chapter.

All Protein Structures Are Dynamic, Interconverting Rapidly Between an Ensemble of Closely Related Conformations Because of Thermal Energy

Even though a protein has folded into a conformation of lowest free energy, this conformation is always being subjected to thermal bombardment from the Brownian motions of the many molecules that constantly collide with it. Thus the atoms in the protein are always moving, which causes neighboring regions of the protein to oscillate in concerted ways. These motions can now be precisely traced using special NMR techniques, as illustrated in **Figure 3–12** for the small protein ubiquitin.

From recent studies combining many types of analyses, we know that protein function exploits these rapid fluctuations—as when a loop on the surface of a protein flips out to expose a binding site for a second molecule. In fact, the function of a protein is generally dependent on that protein's dynamic character, as we explain later when we discuss protein function in detail.

Function Has Selected for a Tiny Fraction of the Many Possible Polypeptide Chains

Because each of the 20 amino acids is chemically distinct and each can, in principle, occur at any position in a protein chain, there are $20 \times 20 \times 20 \times 20 =$ 160,000 different possible polypeptide chains four amino acids long, or 20^n different possible polypeptide chains n amino acids long. For a typical protein length of about 300 amino acids, a cell could theoretically make more than 10^{390} (20^{300}) different polypeptide chains. This is such an enormous number that to produce just one molecule of each kind would require many more atoms than exist in the universe.

Only a very small fraction of this vast set of conceivable polypeptide chains would adopt a stable three-dimensional conformation—by some estimates, less than one in a billion. And yet the majority of proteins present in cells do adopt unique and stable conformations. How is this possible? The answer lies in natural selection. A protein with an unpredictably variable structure and biochemical activity is unlikely to help the survival of a cell that contains it. Such proteins would therefore have been eliminated by natural selection through the enormously long trial-and-error process that underlies biological evolution.

Because evolution has selected for protein function in living organisms, present-day proteins have chemical properties that enable the protein to perform a particular catalytic or structural function in the cell. Proteins are so precisely built that the change of even a few atoms in one amino acid can sometimes disrupt the structure of the whole molecule so severely that all function is lost. And, as discussed later in this chapter, when certain rare protein misfolding accidents occur, the results can be disastrous for the organisms that contain them.

Proteins Can Be Classified into Many Families

Once a protein had evolved that folded up into a stable conformation with useful properties, its structure was often modified during evolution to enable it to perform new functions. As we will discuss in Chapter 4, this process has been greatly accelerated by genetic mechanisms that duplicate genes accidentally, which allows gene copies to evolve independently to perform new functions. Because this type of event occurred frequently in the past, present-day proteins can be grouped into protein families, each family member having an amino acid sequence and a three-dimensional conformation that resemble those of the other family members.

Consider, for example, the *serine proteases*, a large family of protein-cleaving (proteolytic) enzymes that includes the digestive enzymes chymotrypsin, trypsin, and elastase, as well as several proteases involved in blood clotting. When the protease portions of any two of these enzymes are compared, parts of their amino acid sequences are found to match. The similarity of their three-dimensional conformations is even more striking: most of the detailed twists and turns in their polypeptide chains, which are several hundred amino acids long, are virtually identical (**Figure 3–13**). The many different serine proteases nevertheless have distinct enzymatic activities, each cleaving different proteins or the peptide bonds between different types of amino acids. Each therefore performs a distinct function in an organism.

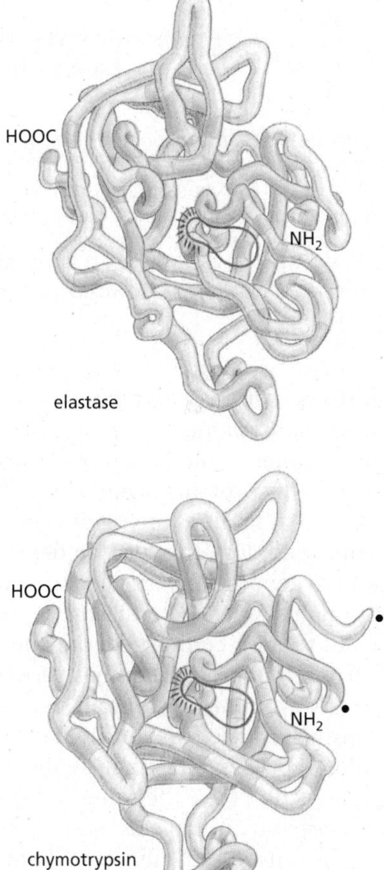

Figure 3–13 A comparison of the conformations of two serine proteases. The backbone conformations of elastase and chymotrypsin. Although only those amino acids in the polypeptide chain shaded in *green* are the same in the two proteins, the two conformations are very similar nearly everywhere. The active site of each enzyme is circled in *red*; this is where the peptide bonds of the proteins that serve as substrates are bound and cleaved by hydrolysis. The serine proteases derive their name from the amino acid serine, whose side chain is part of the active site of each enzyme and directly participates in the cleavage reaction. The two dots on the right side of the chymotrypsin molecule mark the new ends created when this enzyme cuts its own backbone.

The story we have told for the serine proteases could be repeated for hundreds of other protein families. In general, the structure of the different members of a protein family has been more highly conserved than has the amino acid sequence. In many cases, the amino acid sequences have diverged so far that we cannot be certain of a family relationship between two proteins without determining their three-dimensional structures. The yeast α2 protein and the *Drosophila* engrailed protein, for example, are both transcription regulatory proteins in the homeodomain family (discussed in Chapter 7). Because they are identical in only 17 of the 60 amino acids of their homeodomain, their relationship became certain only by comparing their three-dimensional structures (**Figure 3–14**). Many similar examples show that two proteins with more than 25% identity in their amino acid sequences usually share the same overall structure.

The various members of a large protein family often have distinct functions. Mutation is a random process. Some of the amino acid changes that make family members different were selected in the course of evolution because they resulted in useful changes in biological activity; these give the individual family members the different functional properties they have today. Other amino acid changes were effectively "neutral," having neither a beneficial nor a damaging effect on

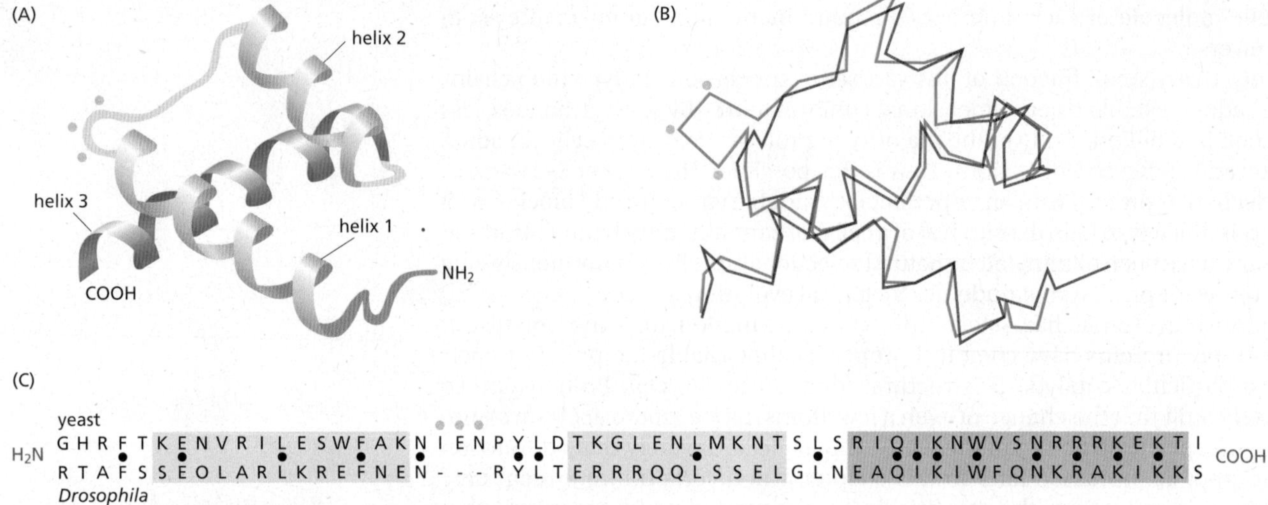

Figure 3–14 **A comparison of a class of DNA-binding domains, called homeodomains, in a pair of proteins from two organisms separated by more than a billion years of evolution.** (A) A ribbon model of the structure common to both proteins. (B) A trace of the α-carbon positions. The three-dimensional structures shown were determined by x-ray crystallography for the yeast α2 protein *(green)* and the *Drosophila* engrailed protein *(red)*. (C) A comparison of amino acid sequences for the region of the proteins shown in A and B. *Black dots* mark sites with identical amino acids. *Green shading* has been used to mark the three α helices shown in A. *Orange dots* indicate the position of a three-amino-acid insert in the α2 protein. (Adapted from C. Wolberger et al., *Cell* 67:517–528, 1991.)

the basic structure and function of the protein. In addition, because mutation is random, there must also have been many deleterious changes that altered the three-dimensional structure of these proteins sufficiently to make them useless. Such faulty proteins would have been readily lost during evolution.

Protein families are readily recognized when the genome of any organism is sequenced; for example, the determination of the DNA sequence for the entire human genome has revealed that we contain about 20,000 protein-coding genes. Through sequence comparisons, we can assign the products of more than half of our protein-coding genes to known protein structures belonging to more than 500 different protein families. Most of the proteins in each family have evolved to perform somewhat different functions, as for the enzymes elastase and chymotrypsin illustrated previously in Figure 3–13. These family members are sometimes called *paralogs* to distinguish them from *orthologs*—those evolutionarily related proteins that have the same function in different organisms (such as the mouse elastase and human elastase enzymes).

The current database of known protein sequences contains more than 100 million entries, and it is growing very rapidly as more and more genomes are sequenced—revealing huge numbers of new genes that encode proteins. The encoded polypeptides range widely in size, from 6 amino acids to a gigantic protein of 34,000 amino acids (titin, a structural protein in muscle).

As described in Chapters 8 and 9, because of the powerful techniques of x-ray crystallography, nuclear magnetic resonance (NMR), and cryo-electron microscopy, we now know the three-dimensional shapes, or conformations, of more than 100,000 of these proteins. By carefully comparing the conformations of these proteins, structural biologists (that is, experts on the structure of biological molecules) have concluded that there are a limited number of ways in which protein domains usually fold up in nature—estimated to be about 2000, if we consider all organisms. For most of these so-called *protein folds*, representative structures have been determined.

Protein comparisons are important because related structures often imply related functions. Many years of experimentation can be saved by discovering that a new protein has an amino acid sequence similarity with a protein of known function. Such sequence relationships, for example, first indicated that certain genes that cause mammalian cells to become cancerous encode protein kinases (discussed in Chapter 20).

Some Protein Domains Are Found in Many Different Proteins

As previously stated, most proteins are composed of a series of protein domains in which different regions of the polypeptide chain fold independently to form compact structures. Such multidomain proteins are believed to have originated from the accidental joining of the DNA sequences that encode each domain, creating a new gene. In an evolutionary process called *domain shuffling*, many large proteins have evolved through the joining of preexisting domains in new combinations (Figure 3–15). Novel binding surfaces have often been created at the juxtaposition of domains, and many of the functional sites where proteins bind to small molecules are found to be located there.

A subset of protein domains has been especially mobile during evolution; these seem to have particularly versatile structures and are sometimes referred to as *protein modules*. The structure of one such module, the SH2 domain, was featured in Figure 3–9. Three other abundant protein domains are illustrated in Figure 3–16.

Each of these three domains has a stable core structure formed from strands of β sheets, from which less-ordered loops of polypeptide chain protrude. The loops are ideally situated to form binding sites for other molecules, as most clearly demonstrated for the immunoglobulin fold, which forms the basis for antibody molecules. Such β sheet–based domains may have achieved their evolutionary success because they provide a convenient framework for the generation of new binding sites for ligands, requiring only small changes to their protruding loops (see Figure 3–40).

A second feature of these protein domains that explains their utility is the ease with which they can be integrated into other proteins. Two of the three domains illustrated in Figure 3–16 have their N- and C-terminal ends at opposite poles of the domain. When the DNA encoding such a domain undergoes tandem duplication, which is not unusual in the evolution of genomes (discussed in Chapter 4), the duplicated domains with this *in-line* arrangement can be readily linked in series to form extended structures—either with themselves or with

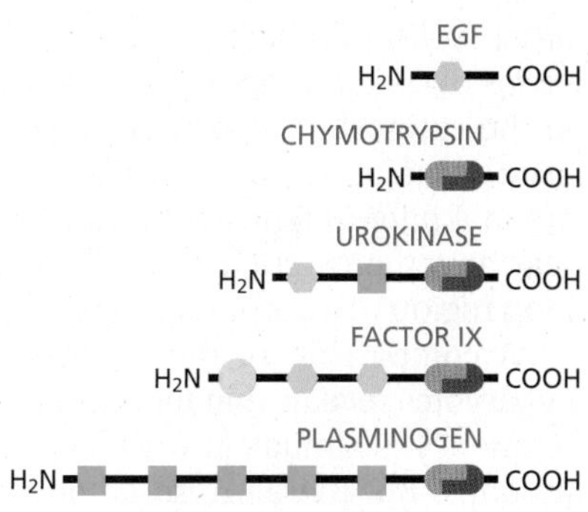

Figure 3–15 **Domain shuffling.** An extensive shuffling of blocks of protein sequence (protein domains) has occurred during protein evolution. Those portions of a protein denoted by the same shape and color in this diagram are evolutionarily related. Serine proteases such as chymotrypsin are formed from two domains *(brown)*. In the three other proteases shown, which are highly regulated and more specialized, these two protease domains are connected to one or more domains that are similar to domains found in epidermal growth factor (EGF; *green*), to a calcium-binding protein *(yellow)*, or to a *kringle* domain *(blue)*. Chymotrypsin is illustrated in Figure 3–13.

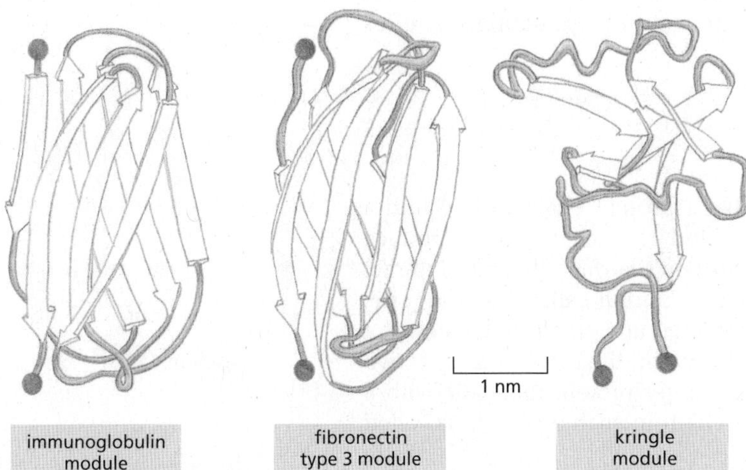

Figure 3–16 **The three-dimensional structures of three commonly used protein domains.** In these ribbon diagrams, β-sheet strands are shown as *arrows*, and the N- and C-termini are indicated by *red spheres*. Many more such "protein modules" exist in nature. (Adapted from D.J. Leahy et al., *Science* 258:987–991, 1992. With permission from AAAS.)

other in-line domains (**Figure 3–17**). Stiff extended structures composed of a series of domains are especially common in extracellular matrix molecules and in the extracellular portions of cell-surface receptor proteins. Other frequently used domains, including the SH2 domain and the kringle domain in Figure 3–16, are of a *plug-in* type, with their N- and C-termini close together. After genomic rearrangements, such domains are usually accommodated as an insertion into a loop region of a second protein.

A comparison of the relative frequency of domain utilization in different eukaryotes reveals that for many common domains, such as protein kinases, this frequency is similar in organisms as diverse as yeast, plants, worms, flies, and humans. But there are some notable exceptions, such as the major histocompatibility complex (MHC) antigen-recognition domain (see Figure 24–36) that is present in 57 copies in humans, but absent in the other four organisms just mentioned. Domains such as these have specialized functions that are not shared with the other eukaryotes; they are assumed to have been strongly selected for during recent evolution to produce the multiple copies observed.

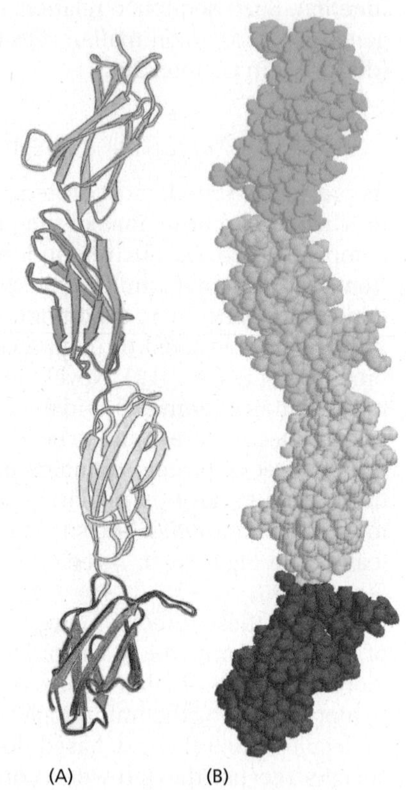

Figure 3–17 An extended structure formed from a series of protein domains. Four fibronectin type 3 domains (see Figure 3–16) from the extracellular matrix molecule fibronectin are illustrated in (A) ribbon and (B) space-filling models. (Adapted from D.J. Leahy et al., *Cell* 84:155–164, 1996.)

The Human Genome Encodes a Complex Set of Proteins, Revealing That Much Remains Unknown

The result of sequencing the human genome has been surprising, because it reveals that our chromosomes contain only about 20,000 protein-coding genes. On the basis of this number alone, we would appear to be no more complex than the tiny mustard weed, *Arabidopsis*, and only about 1.3-fold more complex than a nematode worm. The genome sequences also reveal that vertebrates have inherited nearly all of their protein domains from invertebrates—with only 7% of identified human domains being vertebrate specific.

Each of our proteins is on average more complicated, however (**Figure 3–18**). Domain shuffling during vertebrate evolution has given rise to many novel combinations of protein domains, with the result that there are nearly twice as many combinations of domains found in human proteins as in a worm or a fly. This extra variety in our proteins greatly increases the range of protein-protein interactions possible, but how it contributes to making us human is not known.

The complexity of living organisms is staggering, and it is quite sobering to note that we currently lack even the tiniest hint of what the function might be for more than 10,000 of the proteins that have been identified through examining the human genome. There are certainly enormous challenges ahead for the next generation of cell biologists, with no shortage of fascinating mysteries to solve.

Protein Molecules Often Contain More Than One Polypeptide Chain

The same weak noncovalent bonds that enable a protein chain to fold into a specific conformation also allow proteins to bind to each other to produce larger structures in the cell. Any region of a protein's surface that can interact with another molecule through sets of noncovalent bonds is called a **binding site**. A protein can contain binding sites for various large and small molecules. If a binding site recognizes the surface of a second protein, the tight binding of two folded polypeptide chains at this site creates a larger protein molecule with a precisely defined geometry. Each polypeptide chain in such a protein is called

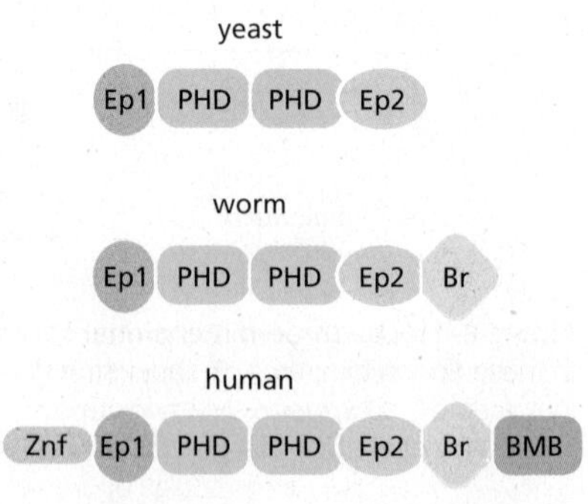

Figure 3–18 Domains in a group of evolutionarily related proteins that have a similar function. In general, there is a tendency for the proteins in more complex organisms, such as humans, to contain additional domains compared to a less complex organism such as yeast—as is the case for the DNA-binding protein compared here.

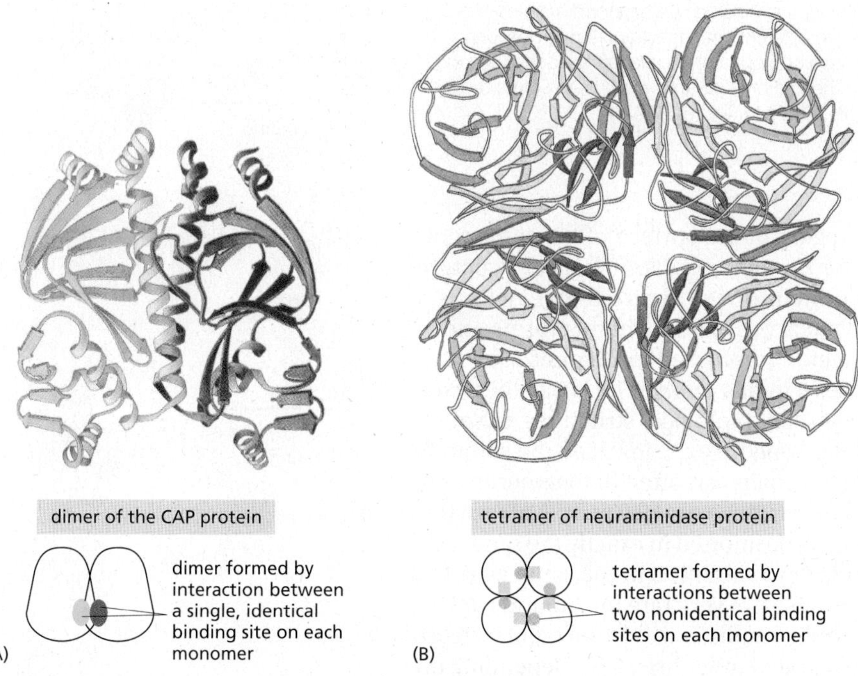

Figure 3–19 **Many protein molecules contain multiple copies of the same protein subunit.** (A) A symmetrical dimer. The CAP protein, a bacterial transcription regulatory protein, is a complex of two identical polypeptide chains. (B) A symmetrical homotetramer. The enzyme neuraminidase exists as a ring of four identical polypeptide chains. For both A and B, a small schematic below the structure emphasizes how the repeated use of the same binding interaction forms the structure. In A, the use of the same binding site on each monomer (represented by *brown* and *green ovals*) causes the formation of a symmetrical dimer. In B, a pair of nonidentical binding sites (represented by *orange circles* and *blue squares*) causes the formation of a symmetrical tetramer.

a **protein subunit**. And the precise way that these subunits are arranged creates the protein's *quaternary structure*—as introduced previously.

In the simplest case, two identical, folded polypeptide chains form a symmetrical complex of two protein subunits (called a *dimer*) that is held together by interactions between two identical binding sites. (Figure 3–19A). Symmetrical protein complexes that are formed from more than two copies of the same polypeptide chain are also commonly found in cells (Figure 3–19B).

Many other proteins contain two or more types of polypeptide chains. *Hemoglobin*, the protein that carries oxygen in red blood cells, contains two identical α-globin subunits and two identical β-globin subunits, symmetrically arranged (Figure 3–20). Such multisubunit proteins can be very large (Movie 3.6).

Some Globular Proteins Form Long Helical Filaments

The proteins that we have discussed so far are *globular proteins*, in which the polypeptide chain folds up into a compact shape like a ball with an irregular surface. Some of these protein molecules can nevertheless assemble to form filaments that may span the entire length of a cell. Most simply, a long chain of identical protein molecules can be constructed if each molecule has a

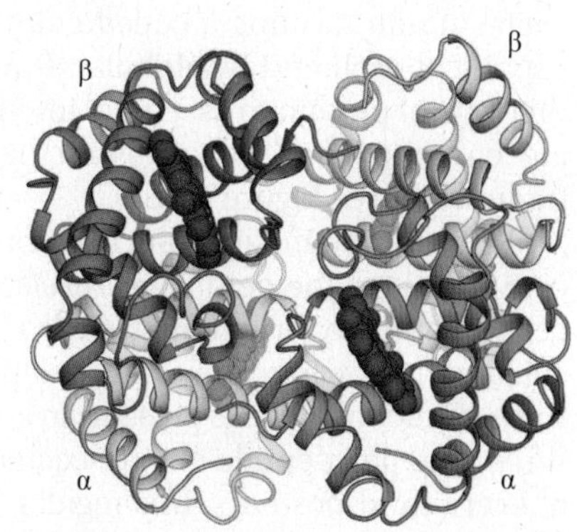

Figure 3–20 **Hemoglobin is a protein formed as a symmetrical assembly using two each of two different subunits.** This abundant, oxygen-carrying protein in red blood cells contains two copies of α-globin *(green)* and two copies of β-globin *(blue)*. Each of these four polypeptide chains contains a heme molecule *(red)*, which is the site that binds oxygen (O_2). Thus, each molecule of hemoglobin carries four molecules of oxygen. (PDB code: 2DHB.)

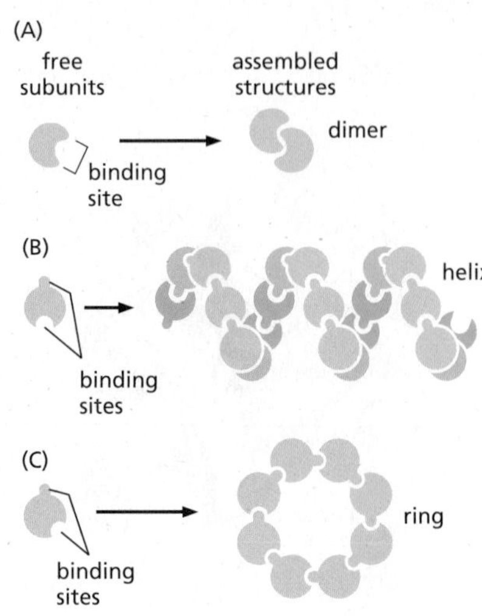

Figure 3–21 **Protein assemblies.** (A) A protein with just one binding site can form a dimer with another identical protein. (B) Identical proteins with two different binding sites often form a long helical filament. (C) If the two binding sites are disposed appropriately in relation to each other, the protein subunits may form a closed ring instead of a helix. (For an example of A, see Figure 3–19A; for an example of B, see Figure 3–22; for an example of C, see Figure 14–32.)

binding site complementary to another region of the surface of the same molecule (Figure 3–21). An actin filament, for example, is a long helical structure produced from many molecules of the protein *actin* (Figure 3–22). Actin is a globular protein that is very abundant in eukaryotic cells, where it forms one of the major filament systems of the cytoskeleton (discussed in Chapter 16).

We will encounter many helical structures in this book. Why is a helix such a common structure in biology? As we have seen, biological structures are often formed by linking similar subunits into long, repetitive chains. If all the subunits are identical, the neighboring subunits in the chain can often fit together in only one way, adjusting their relative positions to minimize the free energy of the contact between them. As a result, each subunit is positioned in exactly the same way in relation to the next, so that subunit 3 fits onto subunit 2 in the same way that subunit 2 fits onto subunit 1, and so on. Because it is very rare for subunits to join up in a straight line, this arrangement generally results in a helix—a regular structure that resembles a spiral staircase, as illustrated in Figure 3–23. Depending on the twist of the staircase, a helix is said to be either right-handed or left-handed (see Figure 3–23E). Handedness is not affected by turning the helix upside down, but it is reversed if the helix is reflected in the mirror.

The observation that helices occur commonly in biological structures holds true whether the subunits are small molecules linked together by covalent bonds (for example, the amino acids in an α helix) or large protein molecules that are linked by noncovalent forces (for example, the actin molecules in actin filaments). This is not surprising. A helix is an unexceptional structure, and it is generated simply by placing many similar subunits next to each other, each in the same strictly repeated relationship to the one before; that is, with a fixed rotation followed by a fixed translation along the helix axis.

Protein Molecules Can Have Elongated, Fibrous Shapes

Enzymes tend to be globular proteins: even though many are large and complicated, with multiple subunits, most have an overall rounded shape. In Figure 3–22, we saw that a globular protein can associate to form long filaments. But some functions require that an individual protein molecule span a large distance. These *fibrous proteins* generally have a relatively simple, elongated three-dimensional structure.

One large family of intracellular fibrous proteins consists of α-keratin, introduced when we described the α helix. Keratin filaments are extremely stable and are the main component in long-lived structures such as hair, horn, and nails. An α-keratin molecule is a dimer of two identical subunits, with the long α helices of each subunit forming a coiled-coil (see Figure 3–8). The coiled-coil regions are capped at each end by globular domains containing binding sites. This enables this type of protein to assemble into ropelike *intermediate filaments*—an important component of the cytoskeleton that creates the cell's internal structural framework (see Figure 16–62).

Fibrous proteins are especially abundant outside the cell, where they are a main component of the gel-like *extracellular matrix* that helps to bind collections of cells together to form tissues. Cells secrete extracellular matrix proteins into their surroundings, where they often assemble into sheets or long fibrils. *Collagen* is the most abundant of these proteins in animal tissues. A collagen molecule consists of three long polypeptide chains, each containing the nonpolar amino acid glycine at every third position. This regular structure allows the chains to wind around

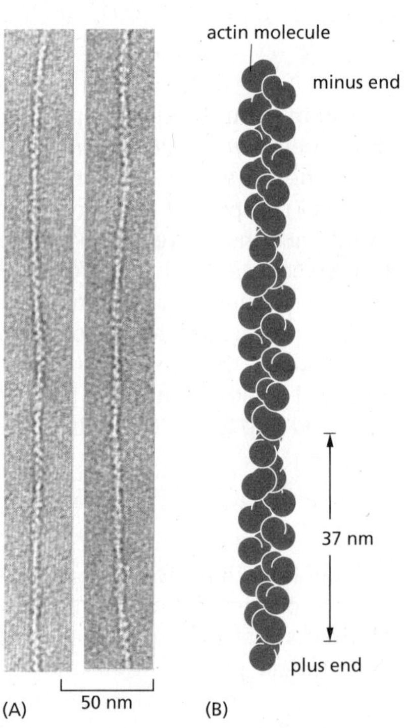

Figure 3–22 **Globular actin monomers assemble to produce an actin filament.** (A) Transmission electron micrographs of negatively stained actin filaments. (B) The helical arrangement of actin molecules in an actin filament. (A, courtesy of Roger Craig.)

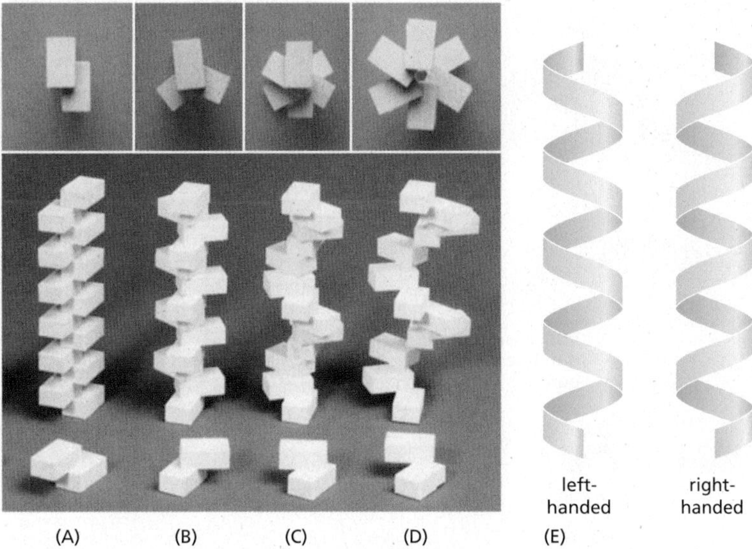

Figure 3–23 **Some properties of a helix.** (A–D) A helix forms when a series of subunits (here represented by rectangular bricks) bind to each other in a regular way. At the top, each of these helices is viewed from directly above the helix and seen to have two (A), three (B), and six (C and D) subunits per helical turn. Note that the helix in D has a wider path than that in C but the same number of subunits per turn. (E) As discussed in the text, a helix can be either right-handed or left-handed. As a reference, it is useful to remember that standard metal screws, which insert when turned clockwise, are right-handed. Note that a helix retains the same handedness when it is turned upside down.

one another to generate a long, regular triple helix (**Figure 3–24**). Many collagen molecules then bind to one another side-by-side and end-to-end to create long overlapping arrays—thereby generating the extremely tough collagen fibrils that give connective tissues their tensile strength, as described in Chapter 19.

Covalent Cross-Linkages Stabilize Extracellular Proteins

Many protein molecules are either attached to the outside of a cell's plasma membrane or secreted to form part of the extracellular matrix. All such proteins are directly exposed to extracellular conditions. To help maintain their structures, the polypeptide chains in such proteins are often stabilized by covalent cross-linkages. These linkages can either tie together two amino acids in the same protein or join together many polypeptide chains in a large protein complex—as for the collagen fibrils just described.

A variety of such cross-links exist, but the most common are covalent sulfur-sulfur bonds. These *disulfide bonds* (also called *S–S bonds*) form as cells prepare newly synthesized proteins for export. As described in Chapter 12, their formation is catalyzed in the endoplasmic reticulum by an enzyme that links together

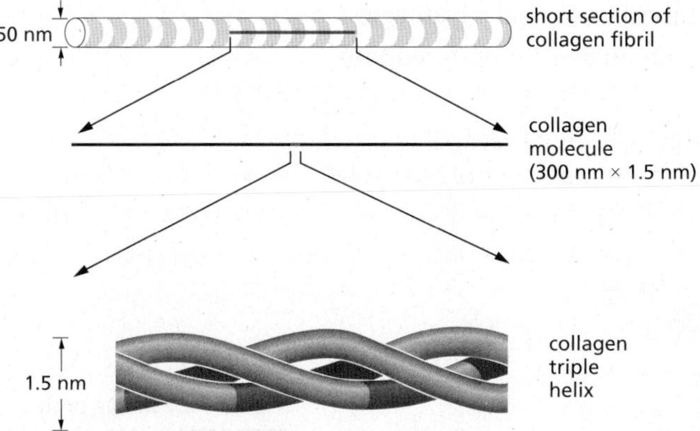

Figure 3–24 **The fibrous protein collagen.** The collagen molecule is a triple helix formed by three extended protein chains that wrap around one another *(bottom)*. In the extracellular space, many rodlike collagen molecules become covalently linked together through their lysine side chains to form collagen fibrils *(top)* that have the tensile strength of steel. The striping on the collagen fibril is caused by the regular repeating arrangement of the collagen molecules within the fibril.

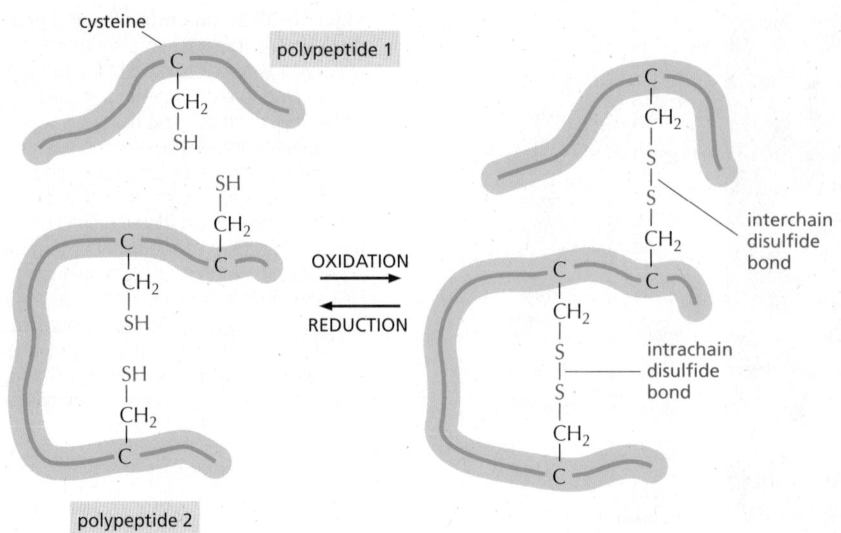

Figure 3–25 **Disulfide bonds.** Covalent disulfide bonds form between adjacent cysteine side chains. These cross-linkages can join either two parts of the same polypeptide chain or two different polypeptide chains. Because the energy required to break one covalent bond is much larger than the energy required to break even a whole set of noncovalent bonds (see Table 2–1, p. 51), a disulfide bond can have a major stabilizing effect on a protein (**Movie 3.7**).

the -SH groups of two cysteine side chains that are adjacent in the folded protein (**Figure 3–25**). Disulfide bonds do not change the conformation of a protein but instead act as atomic staples to reinforce its most favored conformation. For example, lysozyme—an enzyme in tears that dissolves bacterial cell walls—retains its antibacterial activity for a long time because it is stabilized by such cross-linkages.

Disulfide bonds generally fail to form in the cytosol, where a high concentration of reducing agents converts S-S bonds back to cysteine -SH groups. Apparently, proteins do not require this type of reinforcement in the relatively mild environment inside the cell.

Protein Molecules Often Serve as Subunits for the Assembly of Large Structures

The same principles that enable a protein molecule to associate with itself to form rings or a long filament also operate to generate structures that are formed from a set of different macromolecules, such as enzyme complexes, ribosomes, viruses, and membranes. These much larger objects are not made as single, giant, covalently linked molecules. Instead they are formed by the noncovalent assembly of many separately manufactured molecules, which serve as the *subunits* of the final structure.

The use of smaller subunits to build larger structures has several advantages:

1. A large structure built from one or a few repeating smaller subunits requires only a small amount of genetic information.
2. Both assembly and disassembly can be readily controlled reversible processes, because the subunits associate through multiple bonds of relatively low energy.

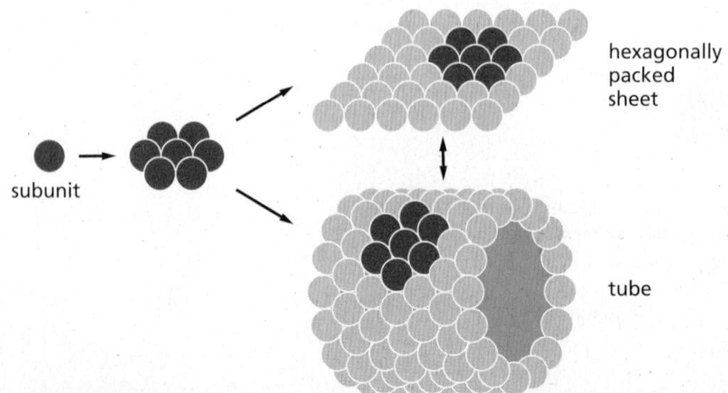

Figure 3–26 **Single protein subunits form protein assemblies that feature multiple protein–protein contacts.** Hexagonally packed globular protein subunits are shown here forming either flat sheets or tubes. Such large structures are not considered to be single "molecules." Instead, like the actin filament described previously, they are viewed as assemblies formed of many different molecules.

3. Errors in the synthesis of the structure can be more easily avoided, because correction mechanisms can operate during the course of assembly to exclude malformed subunits.

To focus on a well-studied example, we can consider how a virus forms from a mixture of proteins and nucleic acids. Some protein subunits are found to assemble into flat sheets in which the subunits are arranged in hexagonal patterns, but with a slight change in the geometry of the individual subunits, a hexagonal sheet can be converted into a tube (**Figure 3–26**) or, with more changes, into a hollow sphere. Protein tubes and spheres that bind specific RNA and DNA molecules in their interior form the coats of viruses.

The formation of closed structures, such as rings, tubes, or spheres, provides additional stability because it increases the number of noncovalent bonds between the protein subunits. Moreover, because such a structure is created by mutually dependent, cooperative interactions between subunits, a relatively small change that affects each subunit individually can cause the structure to assemble or disassemble. These principles are dramatically illustrated in the protein coat, or *capsid*, of many simple viruses, which takes the form of a hollow sphere based on an icosahedron (**Figure 3–27**). Capsids are often made of hundreds of identical protein subunits that enclose and protect the viral nucleic acid (**Figure 3–28**). The protein in such a capsid must have a particularly adaptable

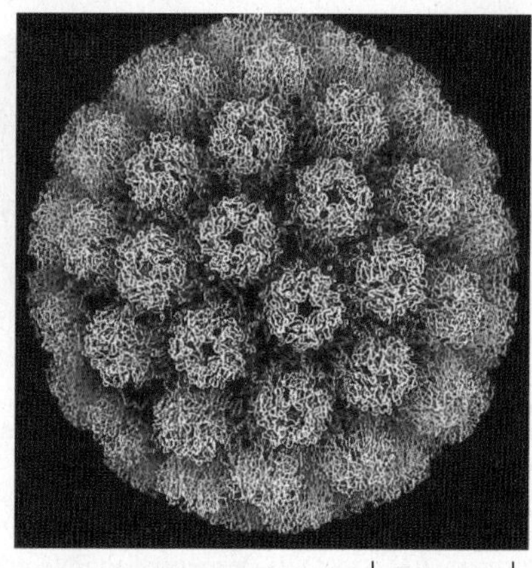

Figure 3–27 **The protein capsid of a virus.** The structure of the simian virus SV40 capsid has been determined by x-ray crystallography and, as for the capsids of many other viruses, it is known in atomic detail. (Courtesy of Robert Grant, Stephan Crainic, and James M. Hogle.)

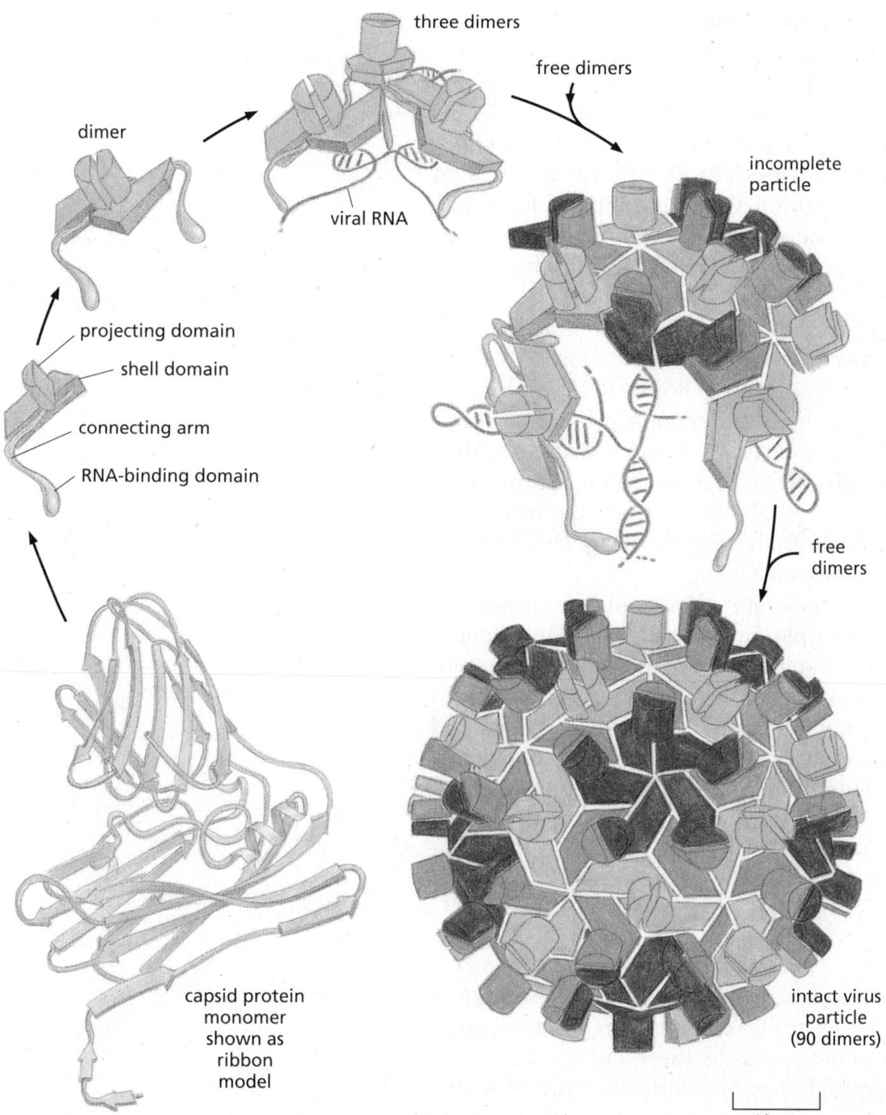

Figure 3–28 **The structure of a spherical virus.** In viruses, many copies of a single protein subunit often pack together to create a spherical shell (a capsid). This capsid encloses the viral genome, composed of either RNA or DNA. For geometric reasons, no more than 60 identical subunits can pack together in a precisely symmetrical way. If slight irregularities are allowed, however, more subunits can be used to produce a larger capsid that retains icosahedral symmetry. The tomato bushy stunt virus (TBSV) shown here, for example, is a spherical virus about 33 nm in diameter formed from 180 identical copies of a 386-amino-acid capsid protein (90 dimers) plus an RNA genome of 4500 nucleotides. To construct such a large capsid, the protein must be able to fit into three somewhat different environments. This requires three slightly different conformations, each of which is differently colored in the virus particle shown here. The postulated pathway of assembly is shown; the precise three-dimensional structure has been determined by x-ray diffraction. (Courtesy of Steve Harrison.)

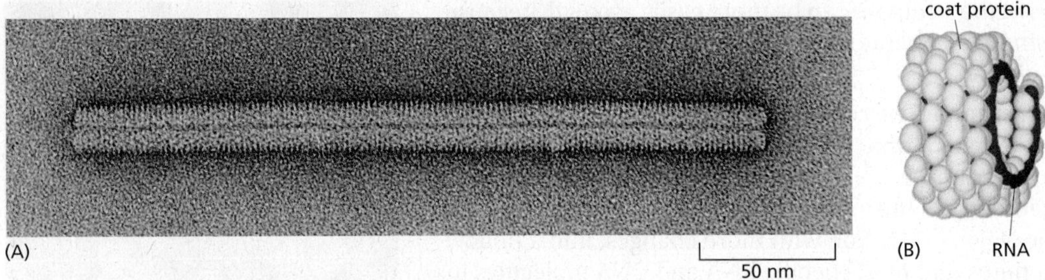

Figure 3–29 The structure of tobacco mosaic virus (TMV). (A) An electron micrograph of the viral particle, which consists of a single long RNA molecule enclosed in a cylindrical protein coat composed of identical protein subunits. (B) A model showing part of the structure of TMV. An RNA molecule of 6395 nucleotides, present as a single strand, is packaged in a helical coat constructed from 2130 copies of a coat protein 158 amino acids long. Fully infective viral particles can self-assemble in a test tube from purified RNA and protein molecules. (A, courtesy of Robley Williams; B, courtesy of Richard J. Feldmann.)

structure: not only must it make several different kinds of contacts to create the sphere, it must also change this arrangement to let the nucleic acid out to initiate viral replication once the virus has entered a cell.

Many Structures in Cells Are Capable of Self-Assembly

The information for forming many of the complex assemblies of macromolecules in cells must be contained in the subunits themselves, because purified subunits can spontaneously assemble into the final structure under the appropriate conditions. The first large macromolecular aggregate shown to be capable of self-assembly from its component parts was *tobacco mosaic virus* (*TMV*). This virus is a long rod in which a cylinder of protein is arranged around a helical RNA core, which constitutes the viral genome (**Figure 3–29**). If the dissociated RNA and protein subunits are mixed together in solution, they recombine to form fully active viral particles. The assembly process is unexpectedly complex and includes the formation of double rings of protein, which serve as intermediates that add to the growing viral coat.

Another complex macromolecular aggregate that can reassemble from its component parts is the bacterial ribosome. This structure is composed of about 55 different protein molecules and 3 different ribosomal RNA (rRNA) molecules. Incubating a mixture of the individual components under appropriate conditions in a test tube causes them to spontaneously re-form the original structure. Most important, such reconstituted ribosomes are able to catalyze protein synthesis. As might be expected, the reassembly of ribosomes follows a specific pathway: after certain proteins have bound to the RNA, this complex is then recognized by other proteins, and so on, until the structure is complete.

It is still not clear how some of the more elaborate self-assembly processes are regulated. Many structures in the cell, for example, have a precisely defined length that appears to be many times greater than that of their component macromolecules. How such length determination is achieved is in many cases a mystery. In the simplest case, a long core protein or other macromolecule provides a scaffold that determines the extent of the final assembly. This is the mechanism that determines the length of the TMV particle, where the RNA chain provides the core. Similarly, a core protein interacting with actin is thought to determine the length of the thin filaments in muscle.

Assembly Factors Often Aid the Formation of Complex Biological Structures

Not all cellular structures held together by noncovalent bonds self-assemble. A cilium, or a myofibril of a muscle cell, for example, cannot form spontaneously from a solution of its component macromolecules. In these cases, part of the assembly information is provided by special enzymes and other proteins that

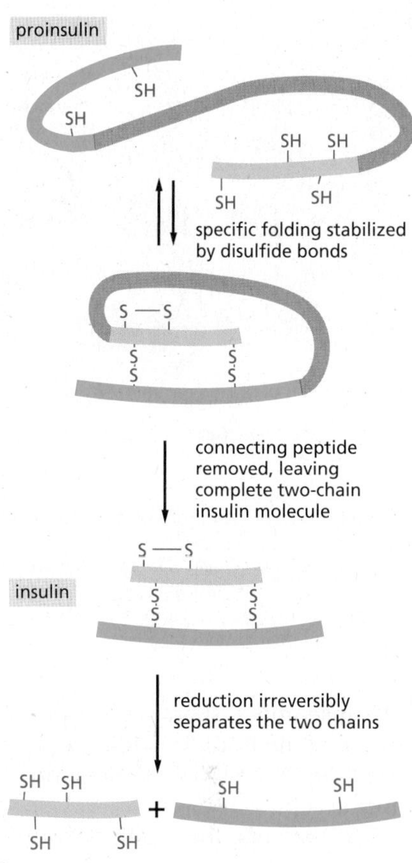

Figure 3–30 **Proteolytic cleavage in insulin assembly.** The polypeptide hormone insulin cannot spontaneously re-form efficiently if its disulfide bonds are disrupted. It is synthesized as a larger protein (*proinsulin*) that is cleaved by a proteolytic enzyme after the protein chain has folded into a specific shape. Excision of part of the proinsulin polypeptide chain removes some of the information needed for the protein to fold spontaneously into its normal conformation. For this reason, once insulin has been denatured and its two polypeptide chains have separated, its ability to reassemble is lost.

perform the function of templates, serving as *assembly factors* that guide construction but take no part in the final assembled structure.

Even relatively simple structures may lack some of the ingredients necessary for their own assembly. In the formation of certain bacterial viruses, for example, the head, which is composed of many copies of a single protein subunit, is assembled on a temporary scaffold composed of a second protein that is produced by the virus. Because the second protein is absent from the final viral particle, the head structure cannot spontaneously reassemble once it has been taken apart. Other examples are known in which proteolytic cleavage is an essential and irreversible step in the normal assembly process. This is even the case for some small protein assemblies, including the structural protein collagen and the hormone insulin (**Figure 3–30**). From these relatively simple examples, it seems certain that the assembly of a structure as complex as a cilium will involve a temporal and spatial ordering that is imparted by numerous other components.

When Assembly Processes Go Wrong: The Case of Amyloid Fibrils

A special class of protein structure, utilized for some normal cell functions, can also contribute to human diseases when not controlled. These are self-propagating, very stable β-sheet aggregates called **amyloid fibrils**. These fibrils are built from a series of identical polypeptide chains that become layered one over the other to create a continuous stack of β strands, with each of the β strands oriented perpendicular to a fibril axis (**Figure 3–31**). In a fibril, two of these stacks of β strands are paired with each other to form a long *cross-beta filament*, with many hundreds of monomers producing an unbranched fibrous structure that can be several micrometers long and 5–15 nm in width (**Figure 3–32**). A surprisingly large fraction of proteins have the potential to adopt such structures,

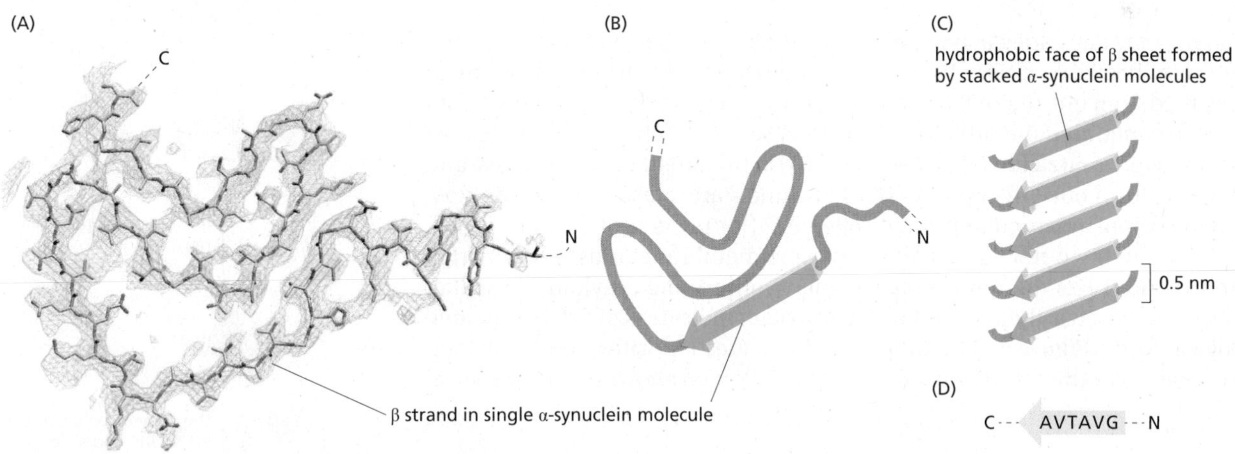

Figure 3–31 **How an amyloid fibril forms from a protein associated with Parkinson's disease.** Illustrated here is the structure of one-half of an amyloid fibril that is formed by the protein α-synuclein, whose abnormal aggregates contribute to Parkinson's disease. The conformation of the α-synuclein monomer is shown as an atomic model in (A) and schematically in (B), with the β strand that will form the cross-beta spine of the filament colored *blue* (only 57 of α-synuclein's 140 amino acids are shown). (C) How the monomer associates to form a long sheet of stacked β strands. As illustrated in Figure 3–32, a second, identical sheet of β strands pairs with this one to form a two-sheet motif that runs the entire length of the fibril. (D) The amino acid sequence that creates a hydrophobic zipper joining the two sheets, forming the cross-beta spine of the fibril. (From R. Guerrero-Ferreira et al., *eLife* 7:e36402, 2018.)

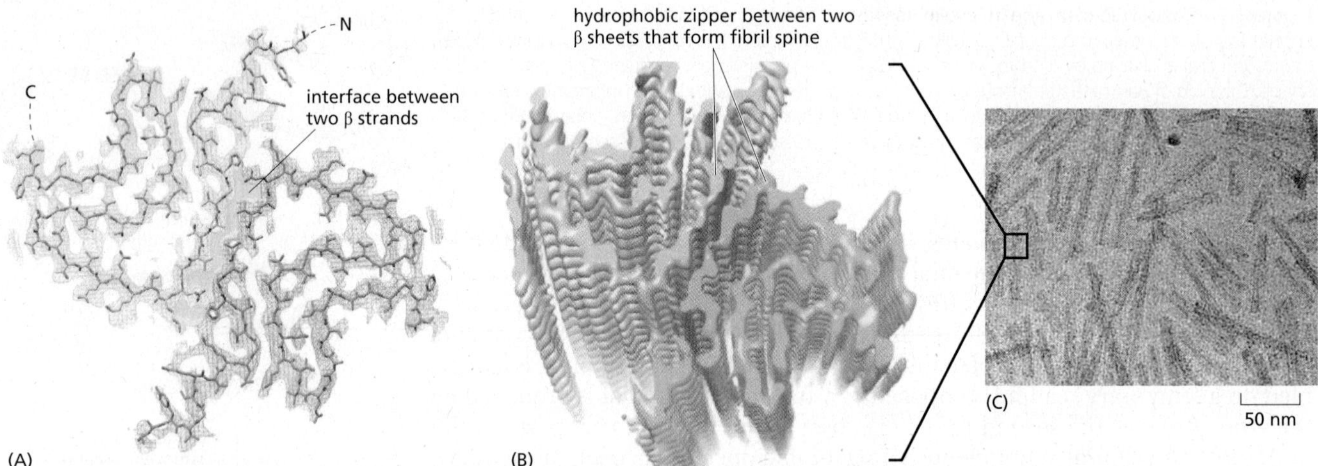

Figure 3–32 The structure of an amyloid fibril. (A) How two monomers of α-synuclein pair to create an amyloid fibril. (B) A three-dimensional rendering of a section of the complete fibril, as determined by cryo-electron microscopy. (C) Electron micrograph of α-synuclein amyloid fibrils. The α-synuclein protein, like some other amyloid-forming proteins, can form several different variants of amyloid fibrils from the same polypeptide chain—only one of which is illustrated here. (From R. Guerrero-Ferreira et al., *eLife* 7:e36402, 2018. This article is distributed under a Creative Commons Attribution 4.0 International license.)

because only a short segment of the polypeptide chain is needed to form the spine of the fibril; in addition, the spine can accommodate a variety of amino acid sequences. Nevertheless, very few proteins will actually form this structure inside cells.

In humans, the quality-control mechanisms governing proteins gradually decline with age, occasionally permitting normal proteins to form pathological aggregates. In extreme cases, the accumulation of such amyloid fibrils in the cell interior can kill the cells and damage tissues. Because the brain is composed of a highly organized collection of nerve cells that cannot regenerate, the brain is especially vulnerable to this sort of cumulative damage. Thus, although amyloid fibrils may form in different tissues and are known to cause pathologies in several sites in the body, the most severe amyloid pathologies are neurodegenerative diseases. For example, an abnormal formation of amyloid fibrils is thought to play a central causative role in both Alzheimer's and Parkinson's diseases.

Prion diseases are a special type of these pathologies. They have attained special notoriety because, unlike Parkinson's or Alzheimer's, prion diseases can readily spread from one organism to another, providing that the second organism eats a tissue containing the protein aggregate. A set of closely related diseases—scrapie in sheep, Creutzfeldt–Jakob disease (CJD) in humans, kuru in humans, and bovine spongiform encephalopathy (BSE) in cattle—are caused by a misfolded, aggregated form of a particular protein called PrP (for prion protein). PrP is normally located on the outer surface of the plasma membrane, most prominently in neurons, and it has the unfortunate property of forming amyloid fibrils that are "infectious" because they convert normally folded molecules of PrP to the same pathological form (**Figure 3–33**). This property creates a positive feedback loop that propagates the abnormal form of PrP, called PrP*, and allows the pathological

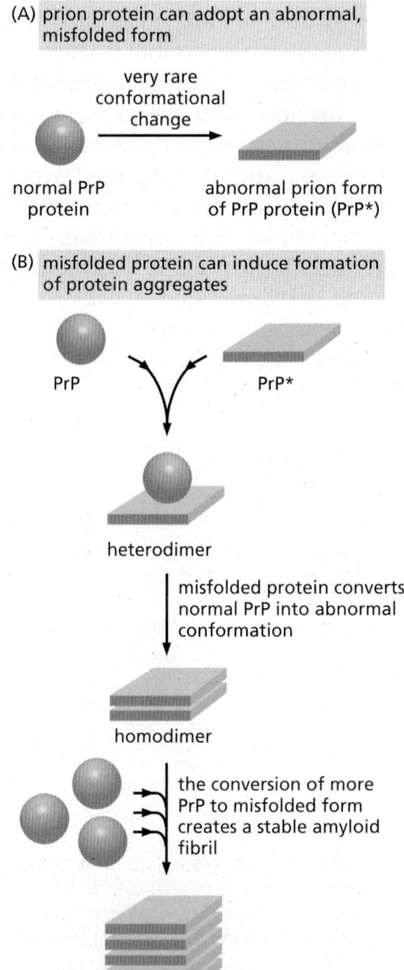

Figure 3–33 Prion diseases are caused by proteins whose misfolding is infectious. (A) Schematic illustration of the type of conformational change in the prion protein (PrP) that produces material for an amyloid fibril. (B) The self-infectious nature of the protein aggregation that is central to prion diseases. The misfolded version of the protein, called PrP*, induces the normal PrP protein it contacts to change its conformation, as shown. PrP* is extremely stable, and if eaten, it can produce amyloid fibrils that disrupt brain-cell function, causing a deadly neurodegenerative disorder. Some of the abnormal amyloid fibrils that form in common noninfectious neurodegenerative disorders, including Parkinson's and Alzheimer's diseases, appear to propagate from cell to cell within the brain in a similar way.

conformation to spread rapidly from cell to cell in the brain, eventually causing death. It can be dangerous to eat the tissues of animals that contain PrP*, as witnessed by the spread of BSE (commonly referred to as "mad cow disease") from cattle to humans. Fortunately, in the absence of PrP*, PrP is extraordinarily difficult to convert to its abnormal form.

A closely related *protein-only inheritance* has been observed in yeast cells. The ability to study infectious proteins in yeast has clarified another remarkable feature of prions. These protein molecules can form several distinctively different types of amyloid fibrils from the same polypeptide chain. Moreover, each type of aggregate can be infectious, forcing normal protein molecules to adopt the same type of abnormal structure. Thus, several different "strains" of infectious particles can arise from the same polypeptide chain.

Recent data suggest that at least some of the abnormal amyloids that form in common human neurological diseases promote the disease by spreading from cell to cell in the brain in a "prion-like" manner, with the abnormally folded form of the protein being taken up by neighboring cells to seed a more widespread formation of the same abnormal structures (for example, α-synuclein in Parkinson's disease, tau protein in Alzheimer's disease). Drugs and antibody treatments are currently being designed in attempts to block these spreading events—and thereby reduce the terrible human toll created by these widespread, common diseases.

Amyloid Structures Can Also Perform Useful Functions in Cells

Amyloid fibrils were initially studied because they cause disease. But the same type of structure is now known to be exploited by cells for useful functions. Eukaryotic cells, for example, store many different peptide and protein hormones that they will secrete in specialized *secretory vesicles*, which package a high concentration of their cargo in dense cores with a regular structure (see Figure 13–43). We now know that these structured cores consist of amyloid fibrils, which in this case have a structure that causes them to dissolve to release soluble cargo after being secreted by exocytosis to the cell exterior (**Figure 3–34A**). Many bacteria use the amyloid structure in a very different way, secreting proteins that form long amyloid fibrils that project from the cell exterior to help bind bacterial neighbors into biofilms (**Figure 3–34B**). Because these biofilms help bacteria to survive in adverse environments (including in humans treated with antibiotics), new drugs that specifically disrupt the fibrous networks formed by bacterial amyloids have promise for treating human infections.

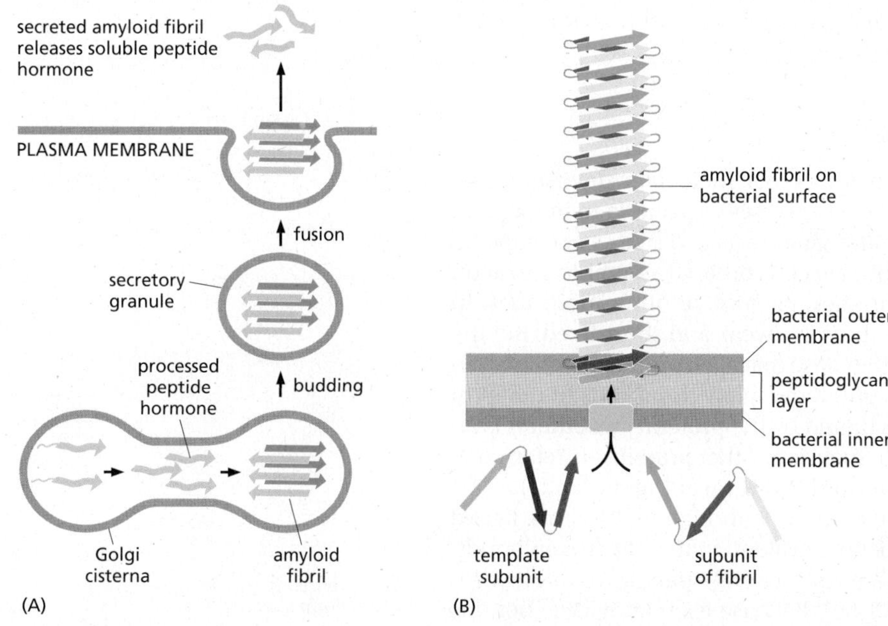

Figure 3–34 **Two normal functions for amyloid fibrils.** (A) In eukaryotic cells, protein cargo can be packed very densely in secretory vesicles and stored until signals cause a release of this cargo by exocytosis. For example, proteins and peptide hormones of the endocrine system, such as glucagon and calcitonin, are efficiently stored as short amyloid fibrils, which dissociate when they reach the cell exterior. (B) Bacteria produce amyloid fibrils on their surface by secreting their precursor proteins; these fibrils then create biofilms that link together, and help to protect, large numbers of individual bacteria.

Summary

A protein molecule's amino acid sequence determines its three-dimensional conformation. Large numbers of noncovalent attractions between different parts of the polypeptide chain stabilize its folded structure. For example, amino acids with hydrophobic side chains tend to cluster in the interior of the molecule, and local hydrogen-bond interactions between neighboring peptide bonds give rise to α helices and β sheets.

Regions of contiguous amino acid sequence fold into globular protein domains. These domains generally contain 40–350 amino acids, and they are the modular units from which larger proteins are constructed. Small proteins typically consist of only a single domain, while large proteins are formed from multiple domains linked together by various lengths of relatively disordered polypeptide chain. As organisms have evolved, the DNA sequences that encode these domains have duplicated, mutated, and been combined with other domains to construct large numbers of new proteins.

Proteins are brought together into larger structures by the same noncovalent attractions that determine protein folding. Proteins with binding sites for their own surface can assemble into dimers, closed rings, spherical shells, or helical polymers. The amyloid fibril is a long unbranched structure assembled through a repeating aggregate of β sheets.

Some mixtures of proteins and nucleic acids can assemble spontaneously into complex structures in a test tube. But not all structures in the cell are capable of spontaneous reassembly after they have been dissociated into their component parts, because many biological assembly processes involve assembly factors that have been removed from the final structure.

PROTEIN FUNCTION

We have seen that each type of protein consists of a precise sequence of amino acids that allows it to fold up into a particular three-dimensional shape, or conformation. These proteins can also have moving parts whose mechanical actions are coupled to chemical events. This coupling of chemistry and movement helps to give proteins the extraordinary capabilities that underlie the dynamic processes in living cells.

In this section, we explain how proteins bind to other selected molecules and how a protein's activity depends on such binding. We will use selected examples to demonstrate how their ability to bind to other molecules enables proteins to act as catalysts, signal receptors, switches, motors, or tiny pumps. These examples by no means exhaust the vast functional repertoire of proteins. You will encounter the specialized functions of many other proteins elsewhere in this book, based on similar principles.

All Proteins Bind to Other Molecules

A protein molecule's physical interaction with other molecules determines its biological properties. Thus, antibodies attach to viruses or bacteria to mark them for destruction, the enzyme hexokinase binds glucose and ATP so as to catalyze a reaction between them, actin molecules bind to each other to assemble into actin filaments, and so on. Indeed, all proteins stick, or *bind*, to other molecules. In some cases, this binding is very tight; in others it is weak and short-lived. But the binding always shows great *specificity*, in the sense that each protein molecule can usually bind just one or a few molecules out of the many thousands of different types it encounters. The substance that is bound by the protein—whether it is an ion, a small molecule, or a macromolecule such as another protein—is referred to as a **ligand** for that protein (from the Latin word *ligare*, meaning "to bind").

The ability of a protein to bind selectively and with high affinity to a ligand depends on the formation of a set of weak noncovalent bonds—hydrogen bonds, electrostatic attractions, and van der Waals attractions—plus favorable hydrophobic interactions (see Panel 2–3, pp. 98–99). Because each individual bond is

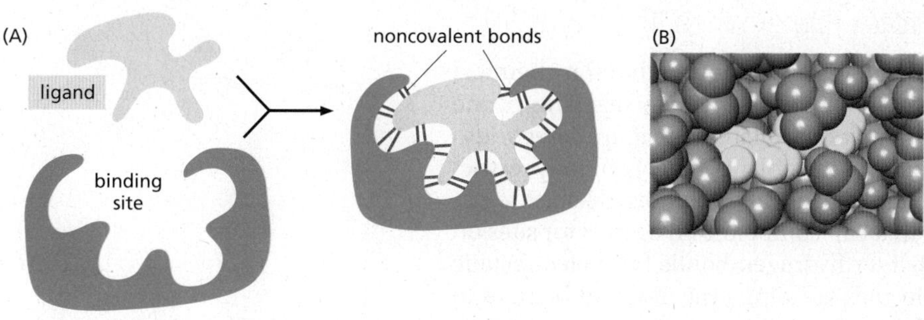

Figure 3–35 **The selective binding of a protein to another molecule.** Many weak bonds are needed to enable a protein to bind tightly to a second molecule, or *ligand*. A ligand must therefore fit precisely into a protein's binding site, like a hand into a glove, so that a large number of noncovalent bonds form between the protein and the ligand. (A) Schematic; (B) space-filling model. (PDB code: 1G6N.)

weak, effective binding occurs only when many of these bonds form simultaneously. Such binding is possible only if the surface contours of the ligand molecule fit very closely to the protein, matching it like a hand in a glove (**Figure 3–35**).

The region of a protein that associates with a ligand, known as the ligand's *binding site*, usually consists of a cavity in the protein surface formed by a particular arrangement of amino acids. These amino acids can belong to different portions of the polypeptide chain that are brought together when the protein folds (**Figure 3–36**). Separate regions of the protein surface generally provide binding sites for different ligands, allowing the protein's activity to be regulated, as we shall see later. And other parts of the protein act as a handle to position the protein in the cell—an example is the SH2 domain discussed previously, which often moves a protein containing it to particular intracellular sites in response to signals.

Although the atoms buried in the interior of the protein have no direct contact with the ligand, they form the framework that gives the surface its contours and its chemical and mechanical properties. Even small changes to the amino acids in the interior of a protein molecule can change its three-dimensional shape enough to destroy a binding site on the surface.

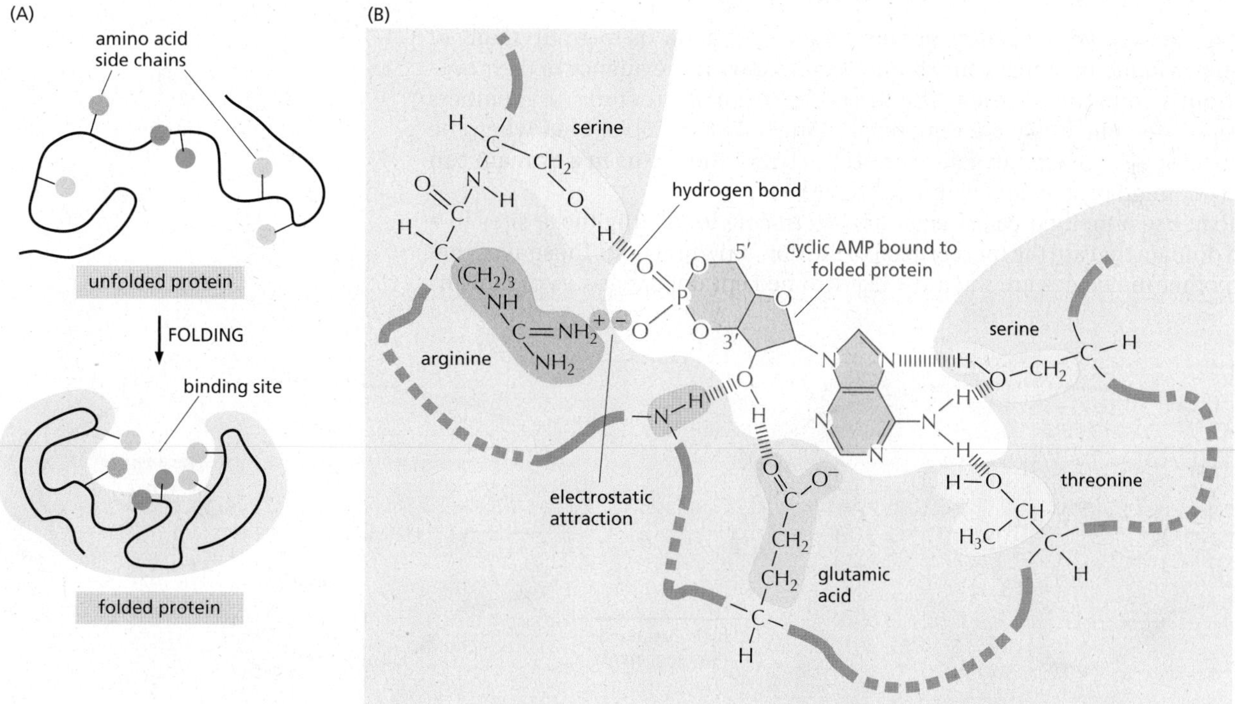

Figure 3–36 **The binding site of a protein.** (A) The folding of the polypeptide chain typically creates a crevice or cavity on the protein surface. This crevice contains a set of amino acid side chains disposed in such a way that they can form noncovalent bonds only with certain ligands. (B) A close-up of an actual binding site, showing the hydrogen bonds and electrostatic interactions formed between a protein and its ligand. In this example, a molecule of cyclic AMP is the bound ligand, shown in *dark yellow*.

The Surface Conformation of a Protein Determines Its Chemistry

The impressive chemical capabilities of proteins often require that the chemical groups on their surface interact in ways that enhance the chemical reactivity of one or more amino acid side chains. These interactions fall into two main categories.

First, the interaction of neighboring parts of the polypeptide chain may restrict the access of water molecules to that protein's ligand-binding sites. Because water molecules readily form hydrogen bonds that can compete with ligands for sites on the protein surface, a ligand will form tighter hydrogen bonds (and electrostatic interactions) with a protein if water molecules are kept away. It might be hard to imagine a mechanism that would exclude a molecule as small as water from a protein surface without affecting the access of the ligand itself. However, because of the strong tendency of water molecules to form water–water hydrogen bonds, water molecules exist in a large hydrogen-bonded network (see Panel 2–2, pp. 96–97). In effect, a protein can keep a ligand-binding site dry, increasing that site's reactivity, because it is energetically unfavorable for individual water molecules to break away from this network—as they must do to reach into a crevice on a protein's surface.

Second, the clustering of neighboring polar amino acid side chains can alter their reactivity. If protein folding brings together a number of negatively charged side chains against their mutual repulsion, for example, the affinity of the site for a positively charged ion is greatly increased. In addition, when amino acid side chains interact with one another through hydrogen bonds, normally unreactive groups (such as the -CH_2OH on the serine shown in **Figure 3–37**) can become reactive, enabling them to be used to make or break selected covalent bonds.

The surface of each protein molecule therefore has a unique chemical reactivity that depends not only on which amino acid side chains are exposed, but also on their exact orientation relative to one another. For this reason, two slightly different conformations of the same protein molecule can differ greatly in their chemistry.

Sequence Comparisons Between Protein Family Members Highlight Crucial Ligand-binding Sites

As we have described previously, genome sequences allow us to group many of the domains found in proteins into families that show clear evidence of their evolution from a common ancestor. The three-dimensional structures of members of the same domain family are remarkably similar. For example, even when the amino acid sequence identity falls to 25%, the backbone atoms in a domain can follow a common protein fold within 0.2 nm (2 Å).

We can use a method called *evolutionary tracing* to identify those sites in a protein domain that are the most crucial to the domain's function. Those sites that bind to other molecules are the most likely to be kept unchanged as organisms

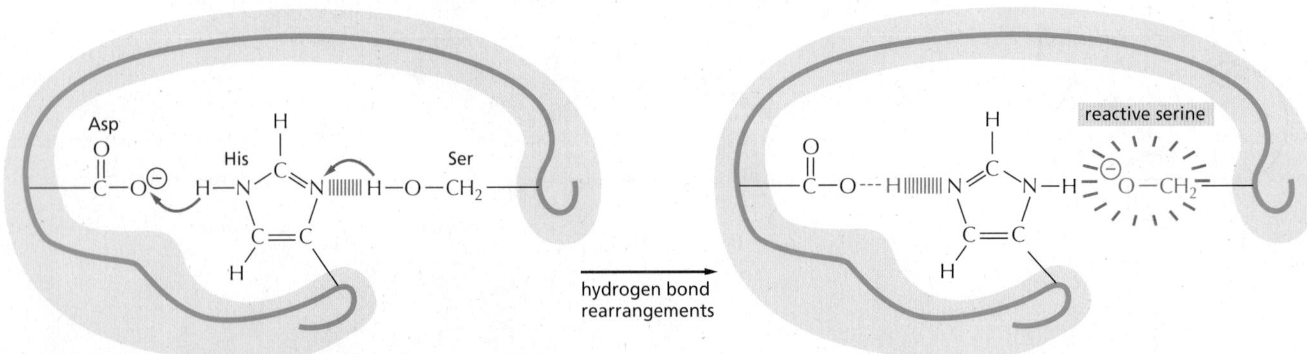

Figure 3–37 An unusually reactive amino acid at the active site of an enzyme. This example is the *catalytic triad* Asp-His-Ser found in chymotrypsin, elastase, and other serine proteases (see Figure 3–13). The aspartic acid side chain (Asp) induces the histidine (His) to remove the proton from a particular serine (Ser). This activates the serine and enables it to form a covalent bond with the enzyme's substrate, hydrolyzing a peptide bond. The many convolutions of the polypeptide chain are omitted here.

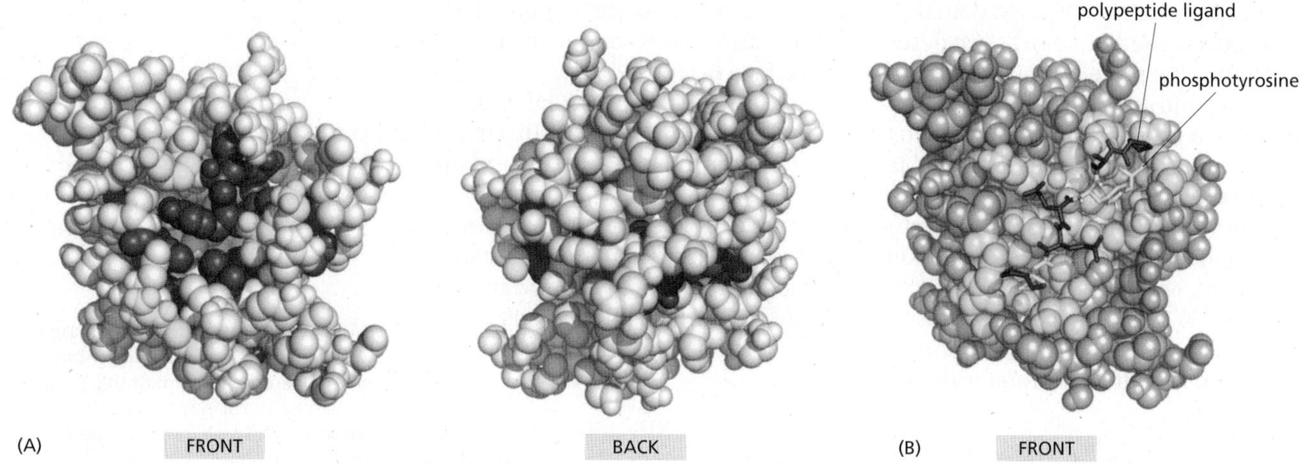

Figure 3–38 **The evolutionary trace method applied to a protein domain.** (A) Front and back views of a space-filling model of the SH2 domain, with evolutionarily conserved amino acids on the protein surface colored *yellow*, and those more toward the protein interior colored *red*. (B) The structure of one specific SH2 domain with its bound polypeptide. Here, those amino acids located within 0.4 nm of the bound ligand are colored *blue*. The two key amino acids of the ligand are *yellow*, and the others are *purple*. Note the high degree of correspondence between A and B. (Adapted from O. Lichtarge et al., *J. Mol. Biol.* 257:342–358, 1996. PDB codes: 1SPR, 1SPS.)

evolve. Thus, in this method, those amino acids that are the same, or nearly so, in all of the known protein family members are mapped onto a model of the three-dimensional structure of a single family member. When this is done, the most invariant positions often form one or more clusters on the protein surface, as illustrated in **Figure 3–38A** for the SH2 domain described previously (see Figure 3–9). These clusters generally correspond to ligand-binding sites.

The SH2 domain functions to link two proteins together. It binds the protein containing it to a second protein that contains a phosphorylated tyrosine side chain in a specific amino acid sequence context, as shown in **Figure 3–38B**. The amino acids located at the binding site for the phosphorylated polypeptide have been the slowest to change during the long evolutionary process that produced the large SH2 family of peptide recognition domains. Mutation is a random process; survival is not. Thus, natural selection (random mutation followed by nonrandom survival) produces the sequence conservation by preferentially eliminating organisms whose SH2 domains have become altered in a way that inactivates the SH2 binding site, destroying SH2 function.

Genome sequencing has revealed huge numbers of proteins whose functions are unknown. Once a three-dimensional structure has been determined for one member of a protein family, evolutionary tracing allows biologists to determine binding sites for the members of that family, and this provides a useful start in deciphering protein function.

Proteins Bind to Other Proteins Through Several Types of Interfaces

Proteins can bind to other proteins in multiple ways. In many cases, a portion of the surface of one protein contacts an extended loop of polypeptide chain (a *string*) on a second protein (**Figure 3–39A**). Such a surface-string interaction,

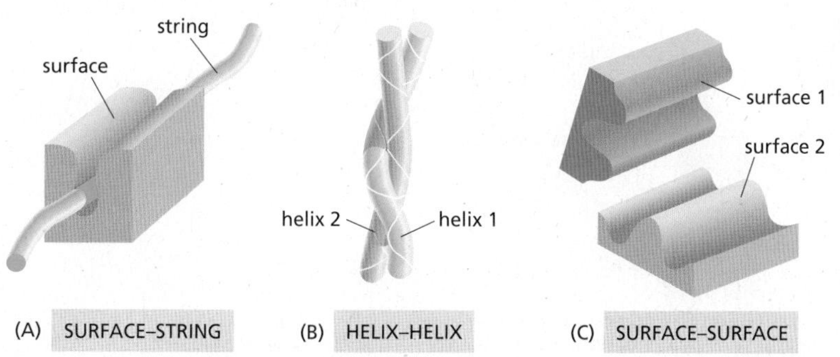

Figure 3–39 **Three ways in which two proteins can bind to each other.** Only the interacting parts of the two proteins are shown. (A) A rigid surface on one protein can bind to an extended loop of polypeptide chain (a *string*) on a second protein. (B) Two α helices can bind together to form a coiled-coil. (C) Two complementary rigid surfaces often link two proteins together. Binding interactions can also involve the pairing of β strands (see, for example, Figure 3–19B).

for example, allows the SH2 domain to recognize a phosphorylated polypeptide loop on a second protein, as just described, and it also enables a protein kinase to recognize the proteins that it will phosphorylate (see below).

A second type of protein–protein interface forms when two α helices, one from each protein, pair together to form a coiled-coil (**Figure 3–39B**). This type of protein interface is found in several families of transcription regulatory proteins, as discussed in Chapter 7.

Another common way for proteins to interact is by the precise matching of one rigid surface with that of another (**Figure 3–39C**). Such interactions can be very tight, because a large number of weak bonds can form between two surfaces that match well. For the same reason, such surface–surface interactions can be extremely specific, enabling a protein to select just one partner from the many thousands of different proteins found in a cell.

Antibody Binding Sites Are Especially Versatile

All proteins must bind to particular ligands to carry out their various functions, and the antibody family is notable for its capacity for tight, highly selective binding (see Chapter 24).

Antibodies, or **immunoglobulins**, are proteins produced by the immune system in response to foreign molecules, such as those on the surface of an invading microorganism. Each antibody binds tightly to a particular target molecule, thereby either inactivating the target molecule directly or marking it for destruction. An antibody recognizes its target (called an **antigen**) with remarkable specificity. Because there are potentially billions of different antigens that humans might encounter, we need to be able to produce billions of different antibodies.

Antibodies are Y-shaped molecules with two identical binding sites that are complementary to a small portion of the surface of the antigen molecule. A detailed examination of the antigen-binding sites of antibodies reveals that they are formed from several loops of polypeptide chain that protrude from the ends of a pair of closely juxtaposed protein domains (**Figure 3–40**). The genes that encode different antibodies generate an enormous diversity of antigen-binding sites by changing only the length and amino acid sequence of these loops, without altering the basic protein structure.

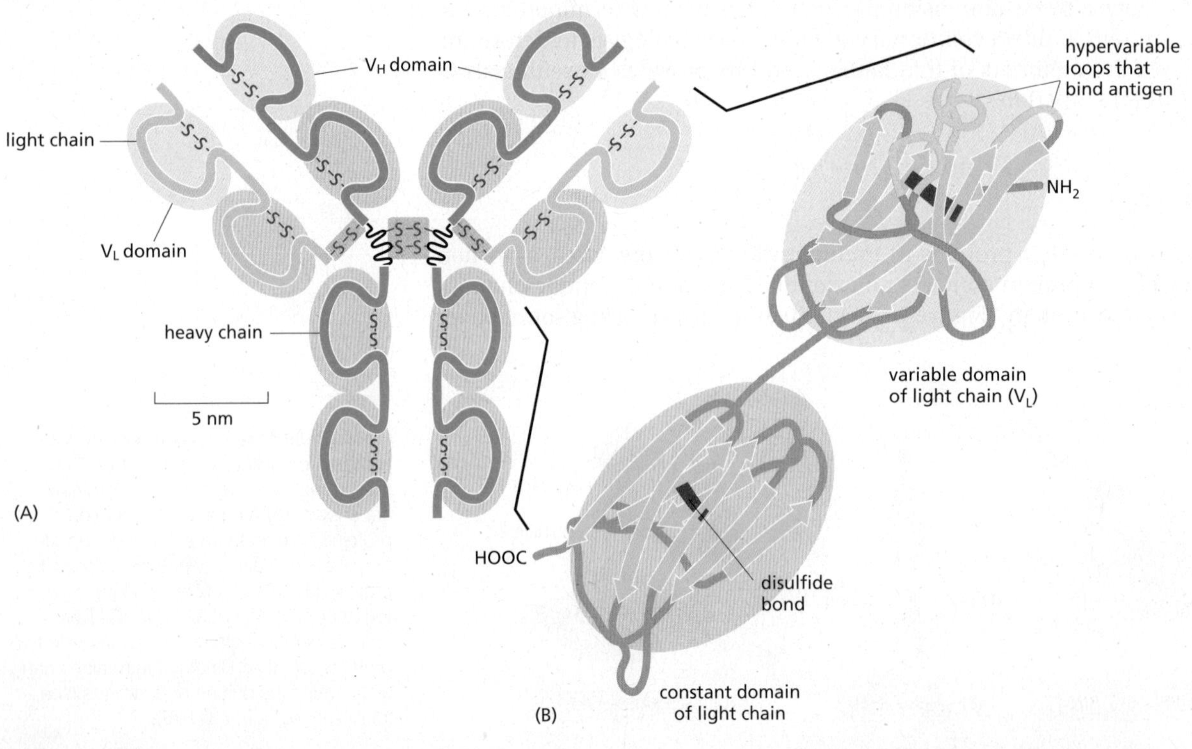

Figure 3–40 An antibody is Y-shaped and has two identical antigen-binding sites, one on each arm of the Y. (A) Schematic drawing of a typical antibody molecule. The protein is composed of four polypeptide chains (two identical heavy chains and two identical, smaller light chains), stabilized and held together by disulfide bonds *(red)*. Each chain is made up of several similar domains, here shaded with *blue*, for the variable domains, or *gray*, for the constant domains. The antigen-binding site is formed where a heavy-chain variable domain (V_H) and a light-chain variable domain (V_L) come close together. These are the domains that differ most in their amino acid sequence in different antibodies—hence their name. (B) Ribbon drawing of a single light chain showing that the most variable parts of the polypeptide chain *(orange)* extend as loops at one end of the variable domain (V_L) to form half of one antigen-binding site of the antibody molecule shown in A. Note that both the constant and variable domains are composed of a sandwich of two antiparallel β sheets connected by a disulfide bond *(red)*. (See Movie 24.5.)

Loops of this kind are ideal for grasping other molecules. They allow a large number of chemical groups to surround a ligand so that the protein can link to it with many weak bonds. For this reason, loops often form the ligand-binding sites in proteins.

The Equilibrium Constant Measures Binding Strength

Molecules in the cell encounter each other very frequently because of their continual random thermal movements. Colliding molecules with poorly matching surfaces form few noncovalent bonds with one another, and the two molecules dissociate as rapidly as they come together. At the other extreme, when many noncovalent bonds form between two colliding molecules, the association can persist for a very long time (**Figure 3–41**). Such strong interactions occur in cells whenever a biological function requires that molecules remain associated; for example, when a group of RNA and protein molecules come together to make a subcellular structure such as a ribosome.

We can measure the strength with which any two molecules bind to each other. As an example, consider a population of identical antibody molecules that suddenly encounters a population of ligands diffusing in the fluid surrounding them. At frequent intervals, one of the ligand molecules will bump into the binding site of an antibody and form an antibody–ligand complex. The population of antibody–ligand complexes will therefore increase, but not without limit: over time, a second process, in which individual complexes break apart because of thermally induced motion, will become increasingly important. Eventually, any population of antibody molecules and ligands will reach a steady state, or **equilibrium**, in which the number of binding (association) events per second is precisely equal to the number of "unbinding" (dissociation) events (see Figure 2–30).

From the concentrations of the ligand, antibody, and antibody–ligand complex at equilibrium, we can calculate a convenient measure of the strength of binding—the **equilibrium constant** (*K*; **Figure 3–42A**). This constant was described in detail in Chapter 2, where its connection to free-energy differences was derived (see pp. 68–69). The equilibrium constant for a reaction in which two molecules (A and B) bind to each other to form a complex (AB) has units of liters/mole, and half of the binding sites will be occupied by ligand when that ligand's concentration (in moles/liter) reaches a value that is equal to $1/K$. This equilibrium constant is larger the greater the binding strength, and it is a direct measure of the free-energy difference between the bound and free states (**Figure 3–42B**). Even a change of a few noncovalent bonds can have a striking effect on a binding interaction,

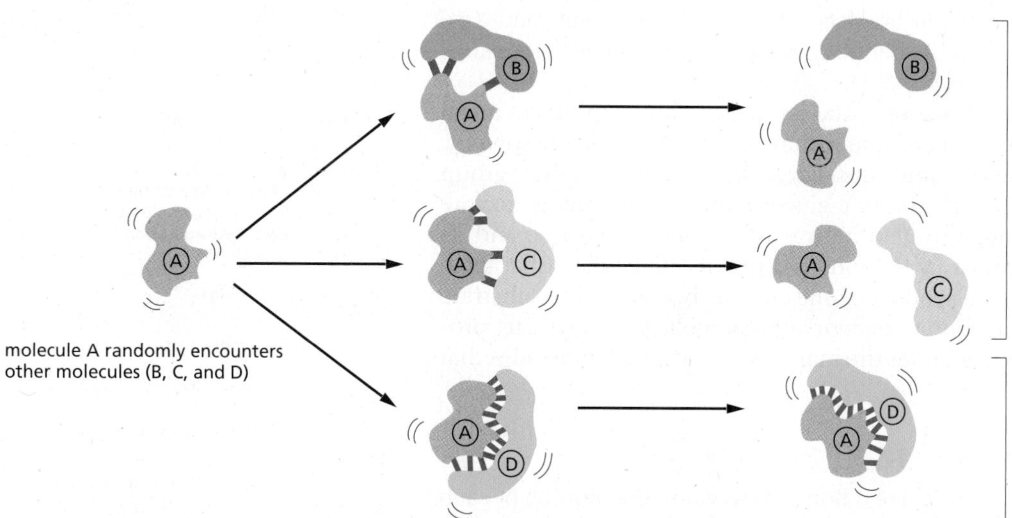
molecule A randomly encounters other molecules (B, C, and D)
A
B
C
D
the surfaces of molecules A and B, and A and C, are a poor match and are capable of forming only a few weak bonds; thermal motion rapidly breaks them apart
the surfaces of molecules A and D match well and therefore can form enough weak bonds to withstand thermal jolting; they therefore stay bound to each other

Figure 3–41 How noncovalent bonds mediate interactions between macromolecules (see Movie 2.1).

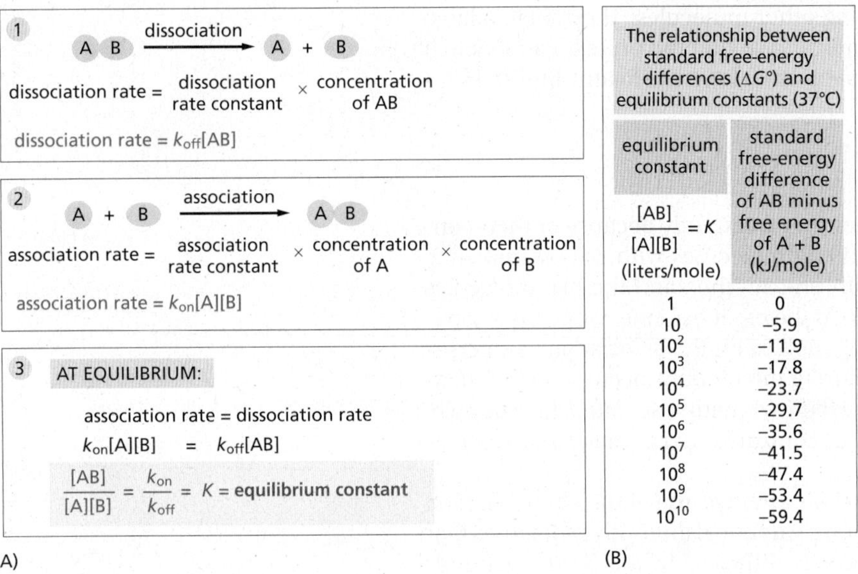

The relationship between standard free-energy differences ($\Delta G°$) and equilibrium constants (37°C)

equilibrium constant $\frac{[AB]}{[A][B]} = K$ (liters/mole)	standard free-energy difference of AB minus free energy of A + B (kJ/mole)
1	0
10	–5.9
10^2	–11.9
10^3	–17.8
10^4	–23.7
10^5	–29.7
10^6	–35.6
10^7	–41.5
10^8	–47.4
10^9	–53.4
10^{10}	–59.4

Figure 3–42 Relating standard free-energy difference ($\Delta G°$) to the equilibrium constant (K). (A) The equilibrium between molecules A and B and the complex AB is maintained by a balance between the two opposing reactions shown in panels 1 and 2. Molecules A and B must collide if they are to react, and the association rate is therefore proportional to the product of their individual concentrations [A] × [B]. (Square brackets indicate concentration.) As shown in panel 3, the ratio of the rate constants for the association and the dissociation reactions is equal to the equilibrium constant (K) for the reaction. (B) The equilibrium constant in panel 3 is that for the reaction A + B ⇌ AB, and the larger its value, the stronger the binding between A and B. Note that for every 5.91 kJ/mole decrease in standard free energy, the equilibrium constant increases by a factor of 10 at 37°C.

The equilibrium constant here has units of liters/mole; for simple binding interactions it is also called the *affinity constant* or *association constant*, denoted K_a. The reciprocal of K_a is called the *dissociation constant*, K_d (in units of moles/liter).

as shown by the example in Figure 3–43. (Note that the equilibrium constant, as defined here, is the **association** or **affinity constant,** K_a; the reciprocal of K_a is the **dissociation constant,** K_d, which is also widely used.)

We have used the case of an antibody binding to its ligand to illustrate the effect of binding strength on the equilibrium state, but the same principles apply to any molecule and its ligand. Many proteins are enzymes, which, as we now discuss, first bind to their ligands and then catalyze the breakage or formation of covalent bonds in these molecules.

Enzymes Are Powerful and Highly Specific Catalysts

Many proteins can perform their function simply by binding to another molecule. An actin molecule, for example, need only associate with other actin molecules to form a filament. There are other proteins, however, for which ligand binding is only a necessary first step in their function. This is the case for the large and very important class of proteins called **enzymes.** As described in Chapter 2, enzymes are remarkable molecules that cause the chemical transformations that make and break covalent bonds in cells. They bind to one or more ligands, called **substrates,** and convert them into one or more chemically modified *products,* doing this over and over again with amazing rapidity. Enzymes speed up reactions, often by a factor of a million or more, without themselves being changed; that is, they act as **catalysts** that permit cells to make or break covalent bonds in a controlled way. It is the catalysis of organized sets of chemical reactions by enzymes that creates and maintains the cell, making life possible.

We can group enzymes into functional classes that perform similar chemical reactions (Table 3–1). Each type of enzyme within such a class is highly specific, catalyzing only a single type of reaction. Thus, *hexokinase* adds a phosphate group to D-glucose but ignores its optical isomer L-glucose; the blood-clotting enzyme *thrombin* cuts one type of blood protein between a particular arginine and its adjacent glycine and nowhere else, and so on. As discussed in detail in Chapter 2, enzymes work in teams, with the product of one enzyme becoming the substrate for the next. The result is an elaborate network of metabolic pathways that provides the cell with energy and generates the many large and small molecules that the cell needs (see Figure 2–62).

Consider 1000 molecules of A and 1000 molecules of B in a eukaryotic cell. The concentration of both will be about 10^{-9} M.

If the equilibrium constant (K) for A + B ⇌ AB is 10^{10}, then one can calculate that at equilibrium there will be

270	270	730
A molecules	B molecules	AB molecules

If the equilibrium constant is a little weaker at 10^8, which represents a loss of 11.9 kJ/mole of binding energy from the example above, or 2–3 fewer hydrogen bonds, then there will be

915	915	85
A molecules	B molecules	AB molecules

Figure 3–43 Small changes in the number of weak bonds can have drastic effects on a binding interaction. This example illustrates the dramatic effect of the presence or absence of a few weak noncovalent bonds in a biological context.

Substrate Binding Is the First Step in Enzyme Catalysis

For a protein that catalyzes a chemical reaction (an enzyme), the binding of each substrate molecule to the protein is an essential prelude. In the simplest case, if we denote the enzyme by E, the substrate by S, and the product by P, the basic reaction

TABLE 3–1 Some Common Types of Enzymes

Enzyme	Reaction catalyzed
Hydrolases	General term for enzymes that catalyze a hydrolytic cleavage reaction; *nucleases* and *proteases* are more specific names for subclasses of these enzymes
Nucleases	Break down nucleic acids by hydrolyzing bonds between nucleotides. *Endonucleases* and *exonucleases* cleave nucleic acids *within* and *from the ends of* the polynucleotide chains, respectively
Proteases	Break down proteins by hydrolyzing bonds between amino acids
Synthases	Synthesize molecules in anabolic reactions by condensing two smaller molecules together
Ligases	Join together (ligate) two molecules in an energy-dependent process. DNA ligase, for example, joins two DNA molecules together end-to-end through phosphodiester bonds
Isomerases	Catalyze the rearrangement of bonds within a single molecule
Polymerases	Catalyze polymerization reactions such as the synthesis of DNA and RNA
Kinases	Catalyze the addition of a phosphate group to a molecule. Protein kinases are an important group of kinases that attach phosphate groups to proteins
Phosphatases	Catalyze the hydrolytic removal of a phosphate group from a molecule
Oxido-reductases	General name for enzymes that catalyze reactions in which one molecule is oxidized while the other is reduced. Enzymes of this type are often more specifically named *oxidases, reductases,* or *dehydrogenases*
ATPases	Hydrolyze ATP. Many proteins with a wide range of roles have an energy-harnessing ATPase activity as part of their function; for example, motor proteins such as *myosin* and membrane transport proteins such as the *sodium–potassium pump*
GTPases	Hydrolyze GTP. A large family of GTP-binding proteins are GTPases with central roles in the regulation of cell processes

Enzyme names typically end in "-ase," with the exception of some enzymes, such as pepsin, trypsin, thrombin, and lysozyme, that were discovered and named before the convention became generally accepted at the end of the nineteenth century. The common name of an enzyme usually indicates the substrate or product and the nature of the reaction catalyzed. For example, citrate synthase catalyzes the synthesis of citrate by a reaction between acetyl CoA and oxaloacetate.

path is E + S → ES → EP → E + P. As illustrated in **Figure 3–44**, there is a limit to the amount of substrate that a single enzyme molecule can process in a given time. Although an increase in the concentration of substrate increases the rate at which product is formed, this rate eventually reaches a maximum value. At that point the enzyme molecule is saturated with substrate, and the rate of reaction (V_{max})

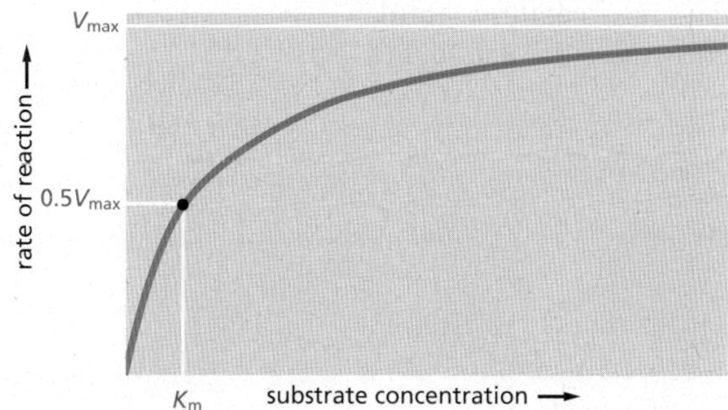

Figure 3–44 **Enzyme kinetics.** The rate of an enzyme reaction (V) increases as the substrate concentration increases until a maximum value (V_{max}) is reached. At this point all substrate-binding sites on the enzyme molecules are fully occupied, and the rate of reaction is limited by the rate of the catalytic process on the enzyme surface. For most enzymes, the concentration of substrate at which the reaction rate is half-maximal (K_m) is a measure of how tightly the substrate is bound, with a large value of K_m corresponding to weak binding (K_m approximates the dissociation constant, K_d, for substrate binding).

depends only on how rapidly the enzyme can process the substrate molecule. This maximum rate divided by the enzyme concentration is called the *turnover number*. Turnover numbers are often about 1000 substrate molecules processed per second per enzyme molecule, although turnover numbers between 1 and 10,000 are known.

The other kinetic parameter frequently used to characterize an enzyme is its K_m, the concentration of substrate that allows the reaction to proceed at one-half its maximum rate ($0.5V_{max}$) (see Figure 3–44). A *low* K_m value means that the enzyme reaches its maximum catalytic rate at a *low concentration* of substrate and generally indicates that the enzyme binds to its substrate very tightly, whereas a *high* K_m value corresponds to weak binding. The methods used to characterize enzymes in this way are explained in Panel 3–2 (pp. 150–151).

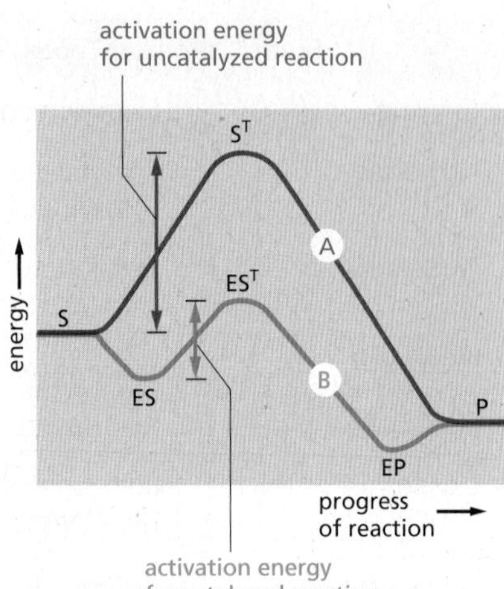

Figure 3–45 **Enzymes accelerate chemical reactions by decreasing the activation energy.** There is a single transition state in this example. However, often both the uncatalyzed reaction (A) and the enzyme-catalyzed reaction (B) go through a series of transition states. In that case, it is the transition state with the highest energy (S^T and ES^T) that determines the activation energy and limits the rate of the reaction. (S = substrate; P = product of the reaction; ES = enzyme–substrate complex; EP = enzyme–product complex.)

Enzymes Speed Reactions by Selectively Stabilizing Transition States

Enzymes achieve extremely high rates of chemical reaction—rates that are far higher than for any synthetic catalysts. There are several reasons for this efficiency. First, when two molecules need to react, the enzyme greatly increases the local concentration of both of these substrate molecules at the catalytic site, holding them in the correct orientation for the reaction that is to follow. More important, however, some of the binding energy contributes directly to the catalysis. Substrate molecules must pass through a series of intermediate states of altered geometry and electron distribution before they form the ultimate products of the reaction. The free energy required to attain the most unstable intermediate state, called the **transition state**, is known as the *activation energy* for the reaction, and it is the major determinant of the reaction rate. Enzymes have a much higher affinity for the transition state of the substrate than they have for the stable form. Because this tight binding greatly lowers the energy of the transition state, the enzyme greatly accelerates a particular reaction by lowering the activation energy that is required (**Figure 3–45**; see also p. 63).

Enzymes Can Use Simultaneous Acid and Base Catalysis

Figure 3–46 compares the spontaneous reaction rates and the corresponding enzyme-catalyzed rates for five enzymes. Rate accelerations range from 10^9 to 10^{23}. This is possible because enzymes not only bind tightly to a transition state, they also contain precisely positioned atoms that alter the electron distributions in the atoms that participate directly in the making and breaking of covalent bonds. Peptide bonds, for example, can be hydrolyzed in the absence of an enzyme by exposing a polypeptide to either a strong acid or a strong base. Enzymes are unique, however, in being able to use acid and base catalysis simultaneously, because the rigid framework of the protein constrains the acidic and

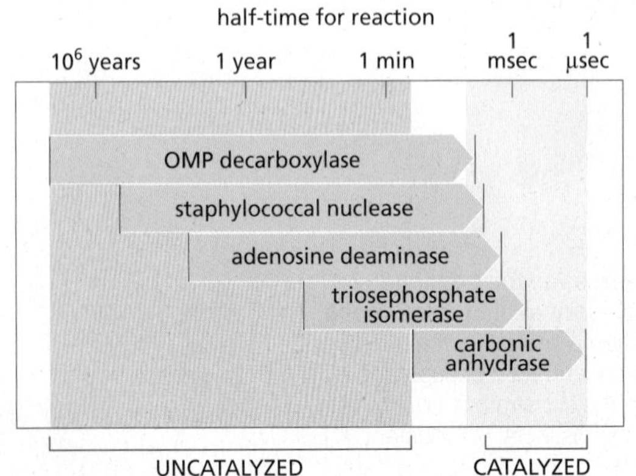

Figure 3–46 **The rate accelerations caused by five different enzymes.** (Adapted from A. Radzicka and R. Wolfenden, *Science* 267:90–93, 1995.)

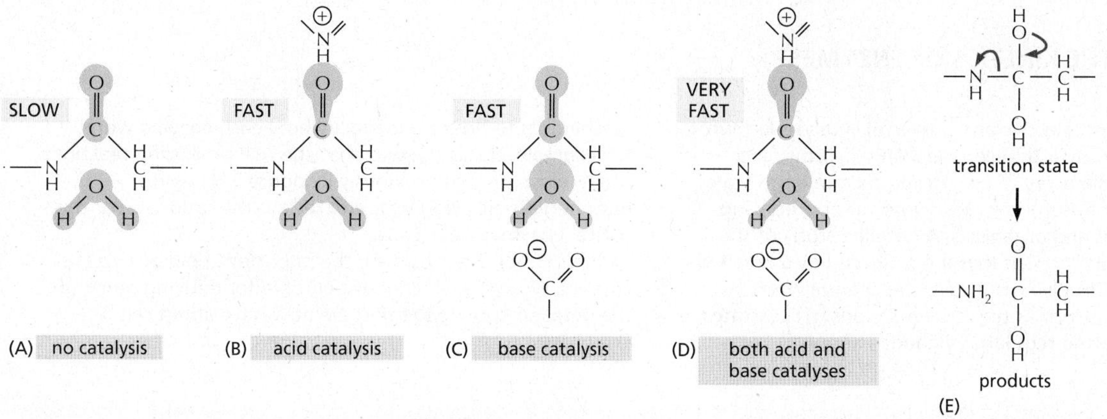

Figure 3–47 **Simultaneous acid catalysis and base catalysis by an enzyme.** (A) The start of the uncatalyzed reaction that hydrolyzes a peptide bond, with *blue* shading used to indicate electron distribution in the water and carbonyl bonds. (B) An acid likes to donate a proton (H^+) to other atoms. By pairing with the carbonyl oxygen, an acid causes electrons to move away from the carbonyl carbon, making this atom much more attractive to the electronegative oxygen of an attacking water molecule. (C) A base likes to take up H^+. By pairing with a hydrogen of the attacking water molecule, a base causes electrons to move toward the water oxygen, making it a better attacking group for the carbonyl carbon. (D) By having appropriately positioned atoms on its surface, an enzyme can perform both acid catalysis and base catalysis at the same time. (E) A tetrahedral intermediate is formed by the attack of the water oxygen atom on the carbonyl carbon atom, and this intermediate rapidly decays to hydrolysis products. The *red arrows* denote the electron shifts associated with product formation.

basic residues and prevents them from combining with each other, as they would do in solution (**Figure 3–47**).

The fit between an enzyme and its substrate needs to be precise. A small change introduced by genetic engineering in the active site of an enzyme can therefore have a profound effect. Replacing a glutamic acid with an aspartic acid in one enzyme, for example, shifts the position of the catalytic carboxylate ion by only 1 Å (about the radius of a hydrogen atom), yet this is enough to decrease the activity of the enzyme a thousandfold.

Lysozyme Illustrates How an Enzyme Works

To demonstrate how enzymes catalyze chemical reactions, we examine an enzyme that acts as a natural antibiotic in egg white, saliva, tears, and other secretions. **Lysozyme** catalyzes the cutting of polysaccharide chains in the cell walls of bacteria. The bacterial cell is under pressure from osmotic forces, and cutting even a small number of these chains causes the cell wall to rupture and the cell to burst. A relatively small and stable protein that can be easily isolated in large quantities, lysozyme was the first enzyme to have its structure worked out in atomic detail by x-ray crystallography (in the mid-1960s).

The reaction that lysozyme catalyzes is a hydrolysis: it adds a molecule of water to a single bond between two adjacent sugar groups in the polysaccharide chain, thereby causing the bond to break (see Figure 2–9). The reaction is energetically favorable because the free energy of the severed polysaccharide chain is lower than the free energy of the intact chain. However, there is an energy barrier for the reaction (its activation energy). In particular, a colliding water molecule can break a bond linking two sugars only if the polysaccharide molecule is distorted into a particular shape—the transition state—in which the atoms around the bond have an altered geometry and electron distribution. Because of this requirement, random collisions must supply a very large activation energy for the reaction to take place. In an aqueous solution at room temperature, the energy of collisions almost never exceeds the activation energy. The pure polysaccharide can therefore remain for years in water without being hydrolyzed to any detectable degree.

This situation changes drastically when the polysaccharide binds to lysozyme. The active site of lysozyme, because its substrate is a polymer, is a long groove that holds six linked sugars at the same time. As soon as the polysaccharide

WHY ANALYZE THE KINETICS OF ENZYMES?

Enzymes are the most selective and powerful catalysts known. An understanding of their detailed mechanisms provides a critical tool for the discovery of new drugs, for the large-scale industrial synthesis of useful chemicals, and for appreciating the chemistry of cells and organisms. A detailed study of the rates of the chemical reactions that are catalyzed by a purified enzyme—more specifically how these rates change with changes in conditions such as the concentrations of substrates, products, inhibitors, and regulatory ligands—allows biochemists to figure out exactly how each enzyme works. For example, this is the way that the ATP-producing reactions of glycolysis, shown previously in Figure 2–47, were deciphered—allowing us to appreciate the rationale for this critical enzymatic pathway.

In this Panel, we introduce the important field of enzyme kinetics, which has been indispensable for deriving much of the detailed knowledge that we now have about cell chemistry.

STEADY-STATE ENZYME KINETICS

Many enzymes have only one substrate, which they bind and then process to produce products according to the scheme outlined in Figure 3–48A. In this case, the reaction is written as

$$E + S \underset{k_{-1}}{\overset{k_1}{\rightleftharpoons}} ES \xrightarrow{k_{cat}} E + P$$

Here we have assumed that the reverse reaction, in which E + P recombine to form EP and then ES, occurs so rarely that we can ignore it. In this case, EP need not be represented, and we can express the rate of the reaction, known as its velocity, V, as

$$V = k_{cat}[ES]$$

where [ES] is the concentration of the enzyme–substrate complex, and k_{cat} is the turnover number, a rate constant that has a value equal to the number of substrate molecules processed per enzyme molecule each second.

But how does the value of [ES] relate to the concentrations that we know directly, which are the total concentration of the enzyme, $[E_o]$, and the concentration of the substrate, [S]? When enzyme and substrate are first mixed, the concentration [ES] will rise rapidly from zero to a so-called steady-state level, as illustrated below.

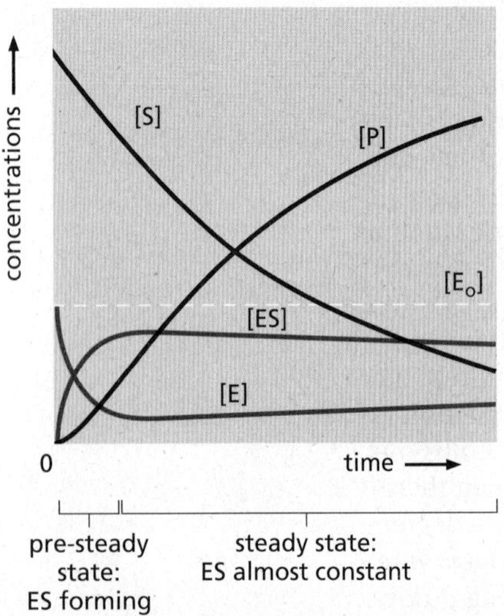

At this steady state, [ES] is nearly constant, so that

$$\underset{k_{-1}[ES] + k_{cat}[ES]}{\text{rate of ES breakdown}} = \underset{k_1[E][S]}{\text{rate of ES formation}}$$

or, because the concentration of the free enzyme, [E], is equal to $[E_o] - [ES]$,

$$[ES] = \left(\frac{k_1}{k_{-1} + k_{cat}}\right)[E][S] = \left(\frac{k_1}{k_{-1} + k_{cat}}\right)\left([E_o] - [ES]\right)[S]$$

Rearranging, and defining the constant K_m as

$$\frac{k_{-1} + k_{cat}}{k_1}$$

we get

$$[ES] = \frac{[E_o][S]}{K_m + [S]}$$

or, remembering that $V = k_{cat}$ [ES], we obtain the famous Michaelis–Menten equation

$$V = \frac{k_{cat}[E_o][S]}{K_m + [S]}$$

As [S] is increased to higher and higher levels, essentially all of the enzyme will be bound to substrate at steady state; at this point, a maximum rate of reaction, V_{max}, will be reached where $V = V_{max} = k_{cat}[E_o]$. Thus, it is convenient to rewrite the Michaelis–Menten equation as

$$V = \frac{V_{max}[S]}{K_m + [S]}$$

THE DOUBLE-RECIPROCAL PLOT

A typical plot of V versus [S] for an enzyme that follows Michaelis–Menten kinetics is shown below. From this plot, neither the value of V_{max} nor of K_m is immediately clear.

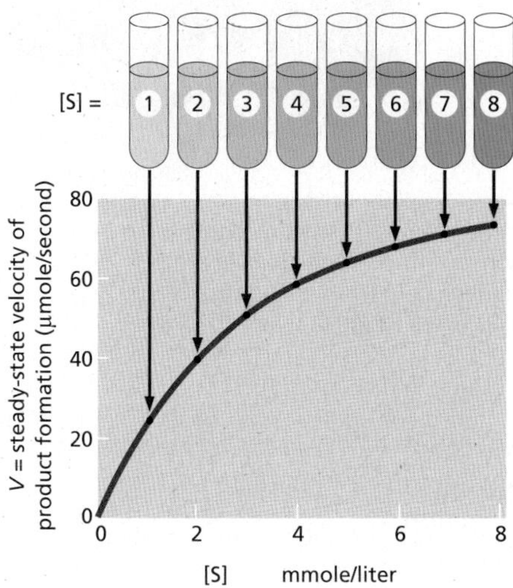

To obtain V_{max} and K_m from such data, a double-reciprocal plot is often used, in which the Michaelis–Menten equation has merely been rearranged, so that $1/V$ can be plotted versus $1/[S]$.

$$1/V = \left(\frac{K_m}{V_{max}}\right)\left(\frac{1}{[S]}\right) + 1/V_{max}$$

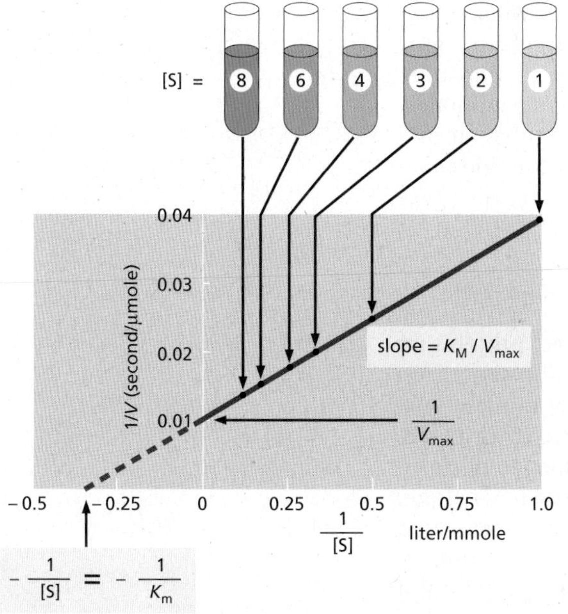

THE SIGNIFICANCE OF K_m, k_{cat}, and k_{cat}/K_m

As described in the text, K_m is an approximate measure of substrate affinity for the enzyme: it is numerically equal to the concentration of [S] at $V = 0.5V_{max}$. In general, a lower value of K_m means tighter substrate binding. In fact, for those cases where k_{cat} is much smaller than k_{-1}, the K_m will be equal to K_d, the dissociation constant for substrate binding to the enzyme ($K_d = 1/K_a$).

We have seen that k_{cat} is the turnover number for the enzyme. At very low substrate concentrations, where $[S] << K_m$, most of the enzyme is free. Thus we can think of $[E] = [E_o]$, so that the Michaelis–Menten equation can be simplified as $V = k_{cat}/K_m[E][S]$. Thus, the ratio k_{cat}/K_m is equivalent to the rate constant for the reaction between free enzyme and free substrate.

A comparison of k_{cat}/K_m for the same enzyme with different substrates, or for two enzymes with their different substrates, is widely used as a measure of enzyme effectiveness.

For simplicity, in this Panel we have discussed enzymes that have only one substrate, such as the lysozyme enzyme described in the text (see p. 152). Most enzymes have two substrates, one of which is often an active carrier molecule—such as NADH or ATP.

A similar, but more complex, analysis is used to determine the kinetics of such enzymes—allowing the order of substrate binding and the presence of covalent intermediates along the pathway to be revealed.

SOME ENZYMES ARE DIFFUSION LIMITED

The values of k_{cat}, K_m, and k_{cat}/K_m for some selected enzymes are given below:

enzyme	substrate	k_{cat} (sec^{-1})	K_m (M)	k_{cat}/K_m (sec^{-1} M^{-1})
acetylcholinesterase	acetylcholine	1.4×10^4	9×10^{-5}	1.6×10^8
catalase	H_2O_2	4×10^7	1	4×10^7
fumarase	fumarate	8×10^2	5×10^{-6}	1.6×10^8

Because an enzyme and its substrate must collide before they can react, k_{cat}/K_m has a maximum possible value that is limited by collision rates. If every collision forms an enzyme–substrate complex, one can calculate from diffusion theory that k_{cat}/K_m will be between 10^8 and 10^9 sec^{-1} M^{-1}, in the case where all subsequent steps proceed immediately. Thus, it is claimed that enzymes like acetylcholinesterase and fumarase are "perfect enzymes," each enzyme having evolved to the point where nearly every collision with its substrate converts the substrate to a product.

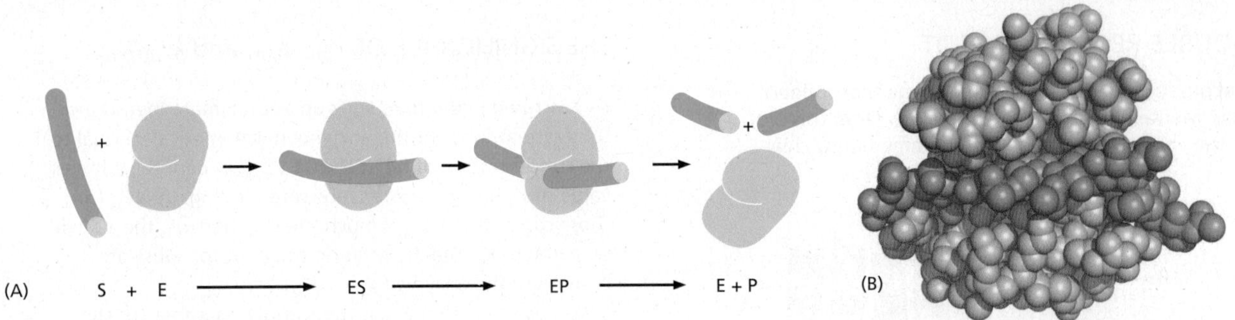

Figure 3–48 **The overall reaction catalyzed by lysozyme.** (A) The enzyme lysozyme (E) catalyzes the cutting of a polysaccharide chain, which is its substrate (S). The enzyme first binds to the chain to form an enzyme–substrate complex (ES) and then catalyzes the cleavage of a specific covalent bond in the backbone of the polysaccharide, forming an enzyme–product complex (EP) that rapidly dissociates. Release of the severed chain (the products P) leaves the enzyme free to act on another substrate molecule. (B) A space-filling model of the lysozyme molecule bound to a short length of polysaccharide chain before cleavage (**Movie 3.8**). (PDB code: 3AB6.)

binds to form an enzyme–substrate complex, the enzyme cuts the polysaccharide by adding a water molecule across one of its sugar–sugar bonds. The product chains are then quickly released, freeing the enzyme for further cycles of reaction (**Figure 3–48**).

An impressive increase in hydrolysis rate is possible because conditions are created in the microenvironment of the lysozyme active site that greatly reduce the activation energy necessary for the hydrolysis to take place. In particular, lysozyme distorts one of the two sugars connected by the bond to be broken from its normal, most stable conformation. The bond to be broken is also held close to two amino acids with acidic side chains (a glutamic acid and an aspartic acid) that participate directly in the reaction. **Figure 3–49** highlights the three central steps in this enzymatically catalyzed reaction, which occurs millions of times faster than uncatalyzed hydrolysis.

Other enzymes use similar mechanisms to lower activation energies and speed up the reactions they catalyze. In reactions involving two or more reactants, the active site also acts like a template, or mold, that brings the substrates together in the proper orientation for a reaction to occur between them (**Figure 3–50A**). As we saw for lysozyme, the active site of an enzyme contains precisely positioned atoms that speed up a reaction by using charged groups to alter the distribution of electrons in the substrates (**Figure 3–50B**). And as we have also seen, when a substrate binds to an enzyme, bonds in the substrate are often distorted, changing the substrate shape. These changes drive a substrate toward a particular transition state (**Figure 3–50C**). Finally, like lysozyme, many enzymes participate intimately in the reaction by transiently forming a covalent bond between the substrate and a side chain of the enzyme. Subsequent steps in the reaction restore the side chain to its original state, so that the enzyme remains unchanged after the reaction (see also Figure 2–47).

Tightly Bound Small Molecules Add Extra Functions to Proteins

Although we have emphasized the versatility of enzymes—and proteins in general—as chains of amino acids that perform remarkable functions, there are many instances in which the amino acids by themselves are not enough. Just as humans employ tools to enhance and extend the capabilities of their hands, enzymes and other proteins often use small nonprotein molecules to perform functions that would be difficult or impossible to do with amino acids alone. Thus, enzymes frequently have a small molecule or metal atom tightly associated with their active site that assists with their catalytic function. *Carboxypeptidase*, for example, an enzyme that cuts polypeptide chains, carries a tightly bound zinc ion in its active site. During the cleavage of a peptide bond by carboxypeptidase, the zinc ion forms a transient bond with one of the substrate atoms,

SUBSTRATE

This substrate is an oligosaccharide of six sugars, labeled A through F. Only sugars D and E are shown in detail.

side chain on sugar E

STEP 1: SUBSTRATE BINDING

PRODUCTS

The final products are an oligosaccharide of four sugars *(left)* and a disaccharide *(right)*, produced by hydrolysis.

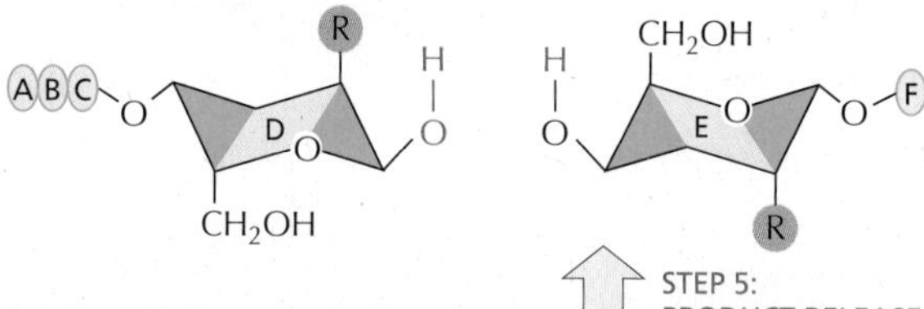

STEP 5: PRODUCT RELEASE

Glu35

C1 carbon

Asp52

H—O—H

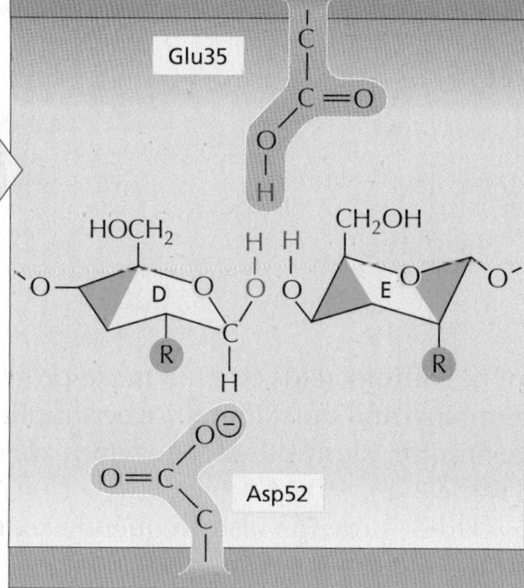

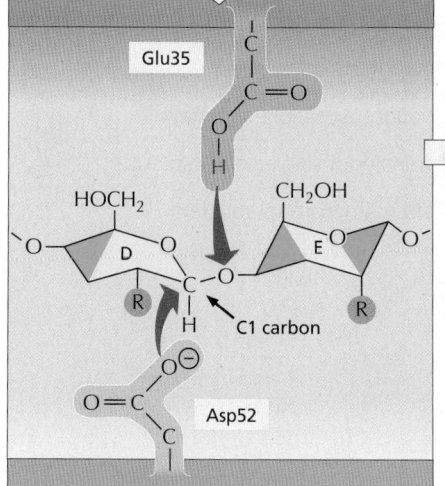

STEP 2: FORMATION OF ES

In the enzyme–substrate complex (ES), the lysozyme forces sugar D into a strained conformation. The Glu35 in the active site is positioned to serve as an acid that attacks the adjacent sugar–sugar bond by donating a proton (H^+) to sugar E; Asp52 is poised to attack the C1 carbon atom of sugar D.

STEP 3: TRANSITION STATE

The Asp52 has formed a covalent bond between the enzyme and the C1 carbon atom of sugar D. The Glu35 then polarizes a water molecule *(red),* so that its oxygen can readily attack the C1 carbon atom of sugar D and displace Asp52.

STEP 4: FORMATION OF EP

The water molecule splits: its –OH group attaches to sugar D and its remaining proton replaces the proton donated by Glu35 in step 2. This completes the hydrolysis and returns the enzyme to its initial state, forming the final enzyme–product complex (EP).

Figure 3–49 **Events at the active site of lysozyme.** The *top left* and *top right* drawings show the free substrate and the free products, respectively. The other three drawings show the sequential events at the enzyme active site, where a sugar–sugar covalent bond is bent and then broken. Note the change in the conformation of sugar D in the enzyme–substrate complex compared with its conformation in the free substrate. This changed conformation favors the formation of the transition state shown in the middle panel, greatly lowering the activation energy that is required for the reaction. This reaction, and the structure of lysozyme bound to its product, are shown in Movie 3.8 and Movie 3.9. (Based on D.J. Vocadlo et al., *Nature* 412:835–838, 2001.)

thereby assisting the hydrolysis reaction. In other enzymes, a small organic molecule serves a similar purpose. Such organic molecules are often referred to as **coenzymes.** An example is *biotin,* which is found in enzymes that transfer a carboxylate group ($-COO^-$) from one molecule to another (see Figure 2–40). Biotin participates in these reactions by forming a transient covalent bond to the $-COO^-$ group to be transferred, being better suited to this function than any

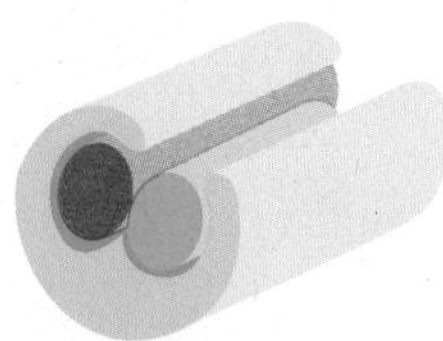

(A) enzyme binds to two substrate molecules and orients them precisely to encourage a reaction to occur between them

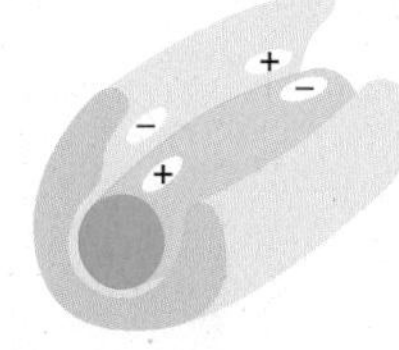

(B) binding of substrate to enzyme rearranges electrons in the substrate, creating partial negative and positive charges that favor a reaction

(C) enzyme strains the bound substrate molecule, forcing it toward a transition state that favors a reaction

Figure 3–50 **Some general strategies used for enzyme catalysis.** (A) Holding substrates together in a precise alignment. (B) Charge stabilization of reaction intermediates. (C) Applying forces that distort bonds in the substrate to increase the rate of a particular reaction.

TABLE 3–2 Many Vitamin Derivatives Are Critical Coenzymes for Human Cells

Vitamin	Coenzyme	Enzyme-catalyzed reactions requiring these coenzymes
Thiamine (vitamin B_1)	Thiamine pyrophosphate	Activation and transfer of aldehydes
Riboflavin (vitamin B_2)	FADH	Oxidation–reduction
Niacin	NADH, NADPH	Oxidation–reduction
Pantothenic acid	Coenzyme A	Acyl group activation and transfer
Pyridoxine	Pyridoxal phosphate	Amino acid activation; also glycogen phosphorylase
Biotin	Biotin	CO_2 activation and transfer
Lipoic acid	Lipoamide	Acyl group activation; oxidation–reduction
Folic acid	Tetrahydrofolate	Activation and transfer of single carbon groups
Vitamin B_{12}	Cobalamin coenzymes	Isomerization and methyl group transfers

of the amino acids used to make proteins. Because it cannot be synthesized by humans, and must therefore be supplied in small quantities in our diet, biotin is a *vitamin*. Many other coenzymes are either vitamins or derivatives of vitamins (Table 3–2).

Other proteins also frequently require specific small-molecule adjuncts to function properly. Thus, the signal receptor protein *rhodopsin*, which is made by the photoreceptor cells in the retina, detects light by means of a small molecule, *retinal*, embedded in the protein (Figure 3–51A). Retinal, which is derived from vitamin A, changes its shape when it absorbs a photon of light, and this change causes the protein to trigger a cascade of enzymatic reactions that eventually lead to an electrical signal being carried to the brain.

Another example of a protein with a nonprotein portion is hemoglobin (see Figure 3–20). Each molecule of hemoglobin carries four *heme* groups, ring-shaped molecules each with a single central iron atom (Figure 3–51B). Heme

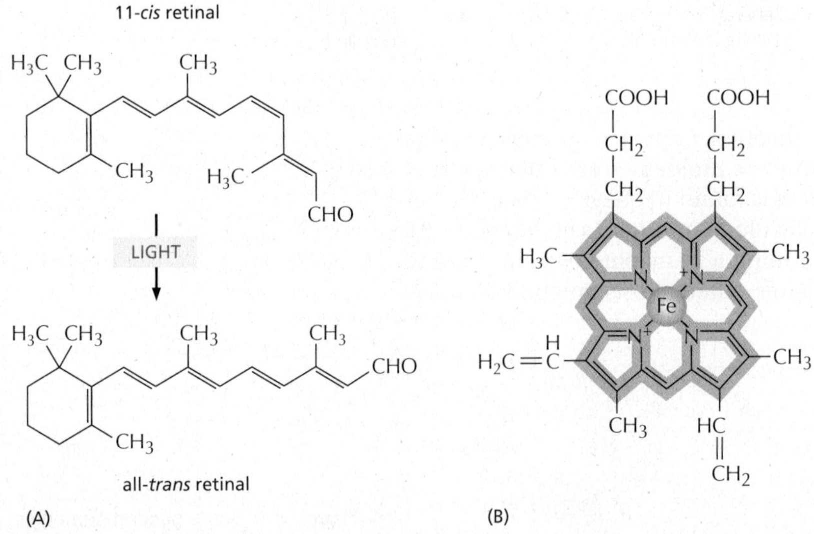

Figure 3–51 Retinal and heme. (A) The structure of retinal, the light-sensitive molecule attached to rhodopsin in the eye, showing its isomerization when it absorbs light. (B) The structure of a heme group. The carbon-containing heme ring is *red* and the iron atom at its center is *orange*. A heme group is tightly bound to each of the four polypeptide chains in hemoglobin, the oxygen-carrying protein whose structure is shown in Figure 3–20.

gives hemoglobin (and blood) its red color. By binding reversibly to oxygen gas through its iron atom, heme enables hemoglobin to pick up oxygen in the lungs and release it in the tissues.

Sometimes these small molecules are attached covalently and permanently to their protein, thereby becoming an integral part of the protein molecule itself. We shall see in Chapter 10 that proteins are often anchored to cell membranes through covalently attached lipid molecules. And membrane proteins exposed on the surface of the cell, as well as proteins secreted outside the cell, are often modified by the covalent addition of sugars and oligosaccharides.

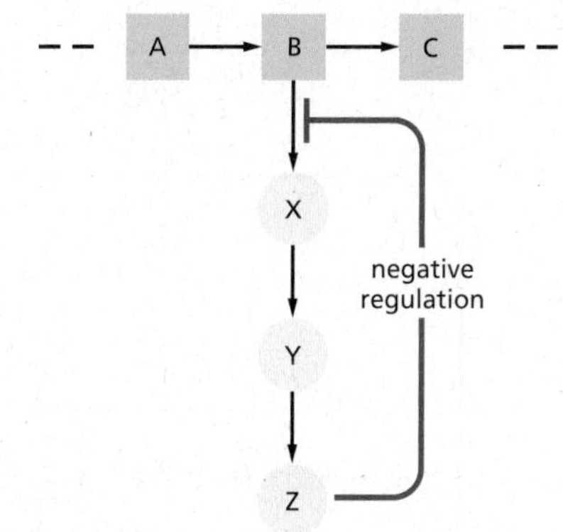

Figure 3–52 **Feedback inhibition of a single biosynthetic pathway.** The end product Z inhibits the first enzyme that is unique to its synthesis and thereby controls its own level in the cell. This is an example of *negative regulation*.

The Cell Regulates the Catalytic Activities of Its Enzymes

A living cell contains thousands of enzymes, many of which operate at the same time and in the same small volume of the cytosol. By their catalytic action, these enzymes generate a complex web of metabolic pathways, each composed of chains of chemical reactions in which the product of one enzyme becomes the substrate of the next. In this maze of pathways, there are many branch points (nodes) where different enzymes compete for the same substrate. The system is complex (see Figure 2–62), and elaborate controls are required to regulate when and how rapidly each reaction occurs.

Regulation occurs at many levels. At one level, the cell controls how many molecules of each enzyme it makes by regulating the expression of the gene that encodes that enzyme (discussed in Chapter 7). The cell also controls enzymatic activities by confining sets of enzymes to particular subcellular compartments (discussed in Chapters 12 and 14) or by concentrating them on protein scaffolds (see pp. 170–173). As will be explained later in this chapter, enzymes are also covalently modified to control their activity. The rate of protein destruction by targeted proteolysis represents yet another important regulatory mechanism (see Figure 6–89). But the most general process that adjusts reaction rates operates through a direct, reversible change in the activity of an enzyme in response to the specific small molecules that it binds.

The most common type of control occurs when an enzyme binds a molecule that is not a substrate to a special regulatory site outside the active site, thereby altering the rate at which the enzyme converts its substrates to products. For example, in **feedback inhibition**, a product produced late in a reaction pathway inhibits an enzyme that acts earlier in the pathway. Thus, whenever large quantities of the final product begin to accumulate, this product binds to the enzyme and slows down its catalytic action, thereby limiting the further entry of substrates into that reaction pathway (**Figure 3–52**). Where pathways branch or intersect, there are usually multiple points of control by different final products, each of which works to regulate its own synthesis (**Figure 3–53**). Feedback inhibition can work almost instantaneously, and it is rapidly reversed when the level of the product falls.

Feedback inhibition is *negative regulation*: it prevents an enzyme from acting. Enzymes can also be subject to *positive regulation*, in which a regulatory molecule stimulates the enzyme's activity rather than shutting the enzyme down. Positive regulation occurs when a product in one branch of the metabolic network stimulates the activity of an enzyme in another pathway. As one example, the accumulation of ADP activates several enzymes involved in the oxidation of sugar molecules, thereby stimulating the cell to convert more ADP to ATP.

Allosteric Enzymes Have Two or More Binding Sites That Interact

A striking feature of both positive and negative feedback regulation is that the regulatory molecule often has a shape totally different from the shape of the substrate of the enzyme. This is why the effect on a protein is termed **allostery** (from the Greek words *allos*, meaning "other," and *stereos*, meaning "solid" or "three-dimensional"). As biologists learned more about feedback regulation,

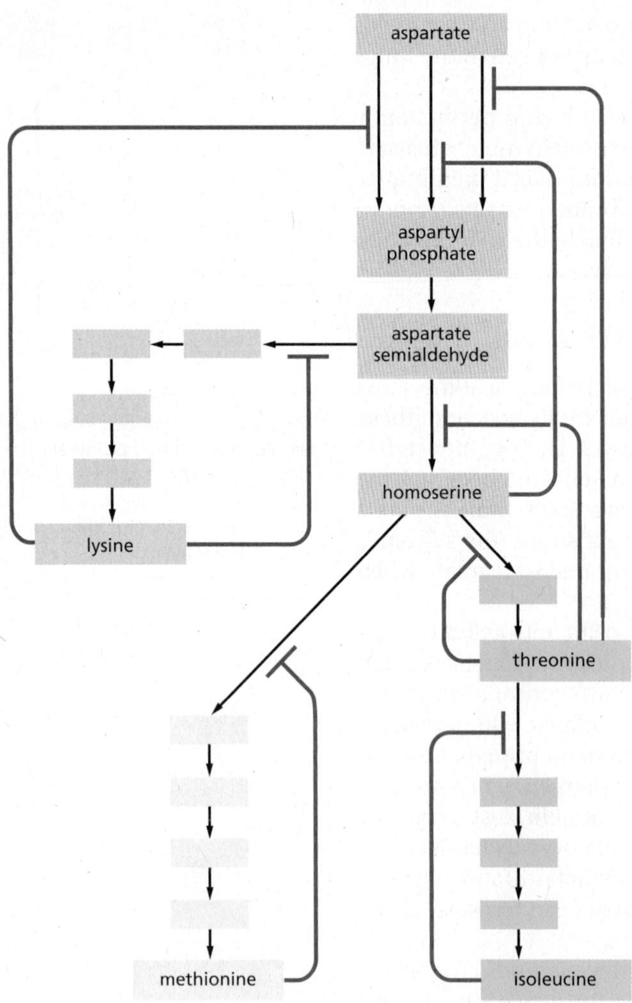

Figure 3–53 Multiple feedback inhibition. In this example, which shows the biosynthetic pathways for four different amino acids in bacteria, the *red lines* indicate positions at which products feed back to inhibit enzymes. Each amino acid controls the first enzyme specific to its own synthesis, thereby controlling its own levels and avoiding a wasteful or even dangerous buildup of intermediates. The products can also separately inhibit the initial set of reactions common to all the syntheses; in this case, three different enzymes catalyze the initial reaction, each inhibited by a different product.

they recognized that the enzymes involved must have at least two different binding sites on their surface—an **active site** that recognizes the substrates, and a **regulatory site** that recognizes a regulatory molecule. These two sites must somehow communicate, so that the catalytic events at the active site can be influenced by the binding of the regulatory molecule at its separate site on the protein's surface.

The interaction between separated sites on a protein molecule is now known to depend on a *conformational change* in the protein: binding at one of the sites causes a shift from one folded shape to a slightly different folded shape. During feedback inhibition, for example, the binding of an inhibitor at one site on the protein causes the protein to shift to a conformation that incapacitates its active site located elsewhere in the protein.

It is thought that most protein molecules are allosteric. They can adopt many slightly different conformations, and a shift from one to another caused by the binding of a ligand can alter their activity. This is true not only for enzymes but also for many other proteins, including receptors, structural proteins, and motor proteins. In all instances of allosteric regulation, each conformation of the protein has somewhat different surface contours, and the protein's binding sites for ligands are altered when the protein changes shape. Importantly, as we discuss next, each ligand will stabilize the conformation that it binds to most strongly, and thus—at high enough concentrations—will tend to "switch" the protein toward the conformation that has a high affinity for that ligand.

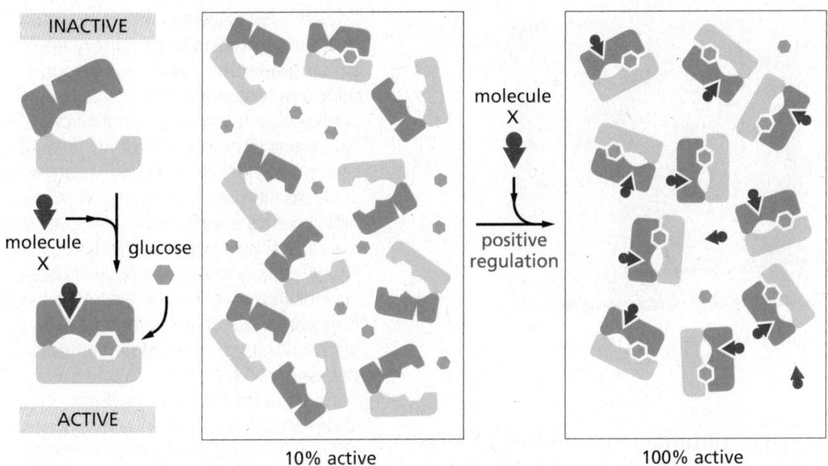

Figure 3–54 Positive regulation caused by conformational coupling between two separate binding sites. In this example, both glucose and molecule X bind best to the *closed* conformation of a protein with two domains. Because both glucose and molecule X drive the protein toward its closed conformation, each ligand helps the other to bind. Glucose and molecule X are therefore said to bind *cooperatively* to the protein.

Two Ligands Whose Binding Sites Are Coupled Must Reciprocally Affect Each Other's Binding

The effects of ligand binding on a protein follow from a fundamental chemical principle known as **linkage**. Suppose, for example, that a protein that binds glucose also binds another molecule, X, at a distant site on the protein's surface. If the binding site for X changes shape as part of the conformational change in the protein induced by glucose binding, the binding sites for X and for glucose are said to be *coupled*. Whenever two ligands prefer to bind to the *same* conformation of an allosteric protein, it follows from basic thermodynamic principles that each ligand must increase the affinity of the protein for the other. For example, if the shift of a protein to a conformation that binds glucose best also causes the binding site for X to fit X better, then the protein will bind glucose more tightly when X is present than when X is absent. In other words, X will positively regulate the protein's binding of glucose (**Figure 3–54**).

Conversely, linkage operates in a negative way if two ligands prefer to bind to *different* conformations of the same protein. In this case, the binding of the first ligand discourages the binding of the second ligand. Thus, if a shape change caused by glucose binding decreases the affinity of a protein for molecule X, the binding of X must also decrease the protein's affinity for glucose (**Figure 3–55**). The linkage relationship is quantitatively reciprocal, so that, for example, if glucose has a very large effect on the binding of X, X has a very large effect on the binding of glucose.

The relationships shown in Figures 3–54 and 3–55 apply to all proteins, and they underlie all of cell biology. The principle seems so obvious in retrospect

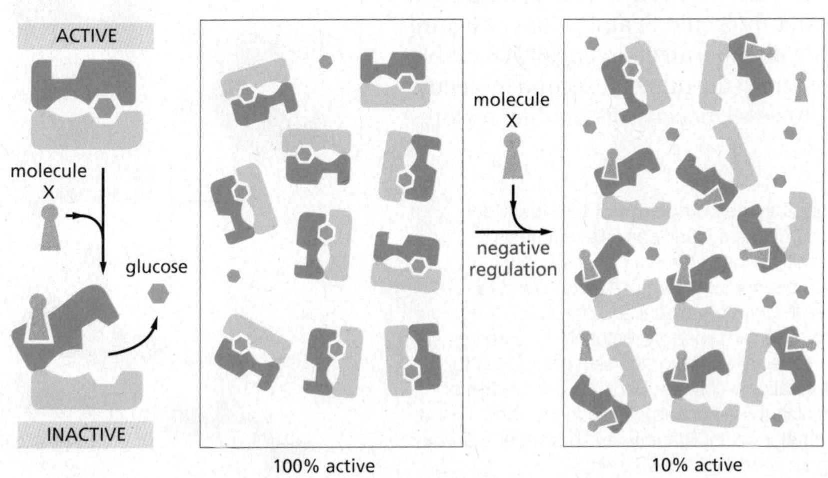

Figure 3–55 Negative regulation caused by conformational coupling between two separate binding sites. The scheme here resembles that in the previous figure, but here molecule X prefers the *open* conformation, while glucose prefers the *closed* conformation. Because glucose and molecule X drive the protein toward opposite conformations (closed and open, respectively), the presence of either ligand interferes with the binding of the other.

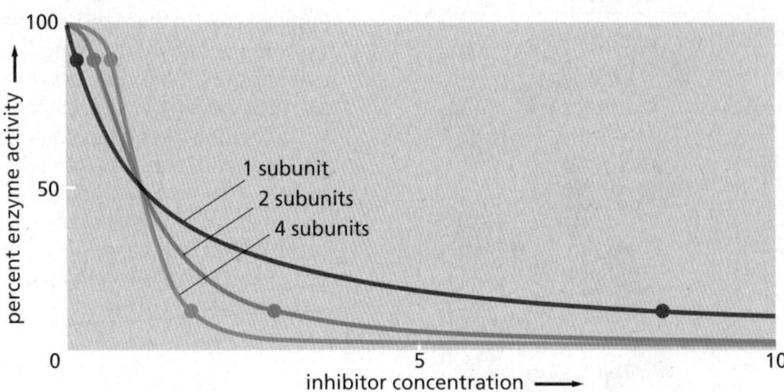

Figure 3–56 **Enzyme activity versus the concentration of inhibitory ligand for single-subunit and multisubunit allosteric enzymes.** For an enzyme with a single subunit *(red line)*, a drop from 90% enzyme activity to 10% activity (indicated by the two dots on the curve) requires a 100-fold increase in the concentration of inhibitor. The enzyme activity is calculated from the simple equilibrium relationship $K = [IP]/[I][P]$, where P is active protein, I is inhibitor, and IP is the inactive protein bound to inhibitor. An identical curve applies to any simple binding interaction between two molecules, A and B. In contrast, a multisubunit allosteric enzyme can respond in a switchlike manner to a change in ligand concentration: the steep response is caused by a cooperative binding of the ligand molecules, as explained in Figure 3–57. Here, the *green line* represents the idealized result expected for the cooperative binding of two inhibitory ligand molecules to an allosteric enzyme with two subunits, and the *blue line* shows the idealized response of an enzyme with four subunits. As indicated by the two dots on each of these curves, the more complex enzymes drop from 90% to 10% activity over a much narrower range of inhibitor concentration than does the enzyme composed of a single subunit.

that we now take it for granted. But the discovery of linkage in studies of a few enzymes in the 1950s, followed by an extensive analysis of allosteric mechanisms in proteins in the early 1960s, had a revolutionary effect on our understanding of biology. Because molecule X in these examples binds at a site on the enzyme that is distinct from the site where catalysis occurs, it need not have any chemical relationship to the substrate that binds at the active site. Moreover, as we have just seen, for enzymes that are regulated in this way, molecule X can either turn the enzyme on (positive regulation) or turn it off (negative regulation). By such a mechanism, **allosteric proteins** serve as general switches that, in principle, can allow one molecule in a cell to affect the fate of any other.

Symmetrical Protein Assemblies Produce Cooperative Allosteric Transitions

A single-subunit enzyme that is regulated by negative feedback can at most decrease from 90% to about 10% activity in response to a 100-fold increase in the concentration of an inhibitory ligand that it binds (**Figure 3–56**, *red line*). Responses of this type are apparently not sharp enough for optimal cell regulation, and most enzymes that are turned on or off by ligand binding consist of symmetrical assemblies of identical subunits. With this arrangement, the binding of a molecule of ligand to a single site on one subunit can promote an allosteric change in the entire assembly that helps the neighboring subunits bind the same ligand. As a result, a *cooperative allosteric transition* occurs (Figure 3–56, *blue line*), allowing a relatively small change in ligand concentration in the cell to switch the whole assembly from an almost fully active to an almost fully inactive conformation (or vice versa).

The principles involved in a cooperative "all-or-none" transition are the same for all proteins, whether or not they are enzymes. Thus, for example, they are critical for the efficient uptake and release of O_2 by hemoglobin in our blood. But they are perhaps easiest to visualize for an enzyme that forms a symmetrical dimer. In the example shown in **Figure 3–57**, the first molecule of an inhibitory ligand binds with great difficulty because its binding disrupts an energetically favorable interaction between the two identical monomers in the dimer. A second molecule of inhibitory ligand now binds more easily, however, because its binding restores

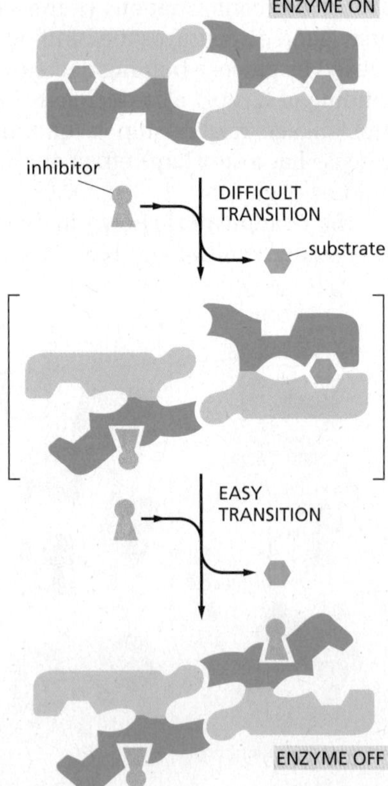

Figure 3–57 **A cooperative allosteric transition in an enzyme composed of two identical subunits.** This diagram illustrates how the conformation of one subunit can influence that of its neighbor. The binding of a single molecule of an inhibitory ligand *(orange)* to one subunit of the enzyme occurs with difficulty because it changes the conformation of this subunit and thereby disrupts the energetically favorable interactions in the symmetrical enzyme. Once this conformational change has occurred, however, the free energy gained by restoring the symmetrical pairing interaction between the two subunits makes it especially easy for the second subunit to bind the inhibitory ligand and undergo the same conformational change. Because the binding of the first molecule of ligand increases the affinity with which the other subunit binds the same ligand, the response of the enzyme to changes in the concentration of the ligand is much steeper than the response of an enzyme with only one subunit (see Figure 3–56 and **Movie 3.10**).

the energetically favorable monomer–monomer contacts of a symmetrical dimer (this also completely inactivates the enzyme).

As an alternative to this *induced fit* model for a cooperative allosteric transition, we can view such a symmetrical enzyme as having only two possible conformations, corresponding to the "enzyme on" and "enzyme off" structures in Figure 3–57. In this view, ligand binding perturbs an all-or-none equilibrium between these two states, thereby changing the proportion of active molecules. Both models represent true and useful concepts.

Many Changes in Proteins Are Driven by Protein Phosphorylation

Proteins are regulated by more than the reversible binding of other molecules. A second method that eukaryotic cells use extensively to regulate a protein's function is the covalent addition of a smaller molecule to one or more of its amino acid side chains. The most common such regulatory modification in higher eukaryotes is the addition of a phosphate group. We shall therefore use *protein phosphorylation* to illustrate some of the general principles involved in the control of protein function through the covalent modification of amino acid side chains.

A phosphorylation event (by a kinase) can affect the protein that is modified in three important ways. First, because each phosphate group carries two negative charges, the enzyme-catalyzed addition of a phosphate group to a protein can cause a major conformational change in the protein by, for example, attracting a cluster of positively charged amino acid side chains. This can, in turn, affect the binding of ligands elsewhere on the protein surface, dramatically changing the protein's activity. When a second enzyme (called a phosphatase) removes the phosphate group, the protein returns to its original conformation and restores its initial activity.

Second, an attached phosphate group can form part of a structure that the binding sites of other proteins recognize. As previously discussed, the SH2 domain binds to a short peptide sequence containing a phosphorylated tyrosine side chain (see Figure 3–38B). More than 10 other common domains provide binding sites for attaching their protein to phosphorylated peptides in other protein molecules, each recognizing a phosphorylated amino acid side chain in a different protein context. Third, the addition of a phosphate group can mask a binding site that otherwise holds two proteins together, and thereby disrupt protein–protein interactions. As a result of the last two effects, protein phosphorylation and dephosphorylation very often drive the regulated assembly and disassembly of protein complexes.

Reversible protein phosphorylation controls the activity, structure, and cellular localization of enzymes and many other types of proteins in eukaryotic cells. In fact, this regulation is so extensive that more than one-third of the 10,000 or so proteins in a typical mammalian cell are thought to be phosphorylated at any given time—many with more than one phosphate.

As might be expected, the addition and removal of phosphate groups from specific proteins often occur in response to signals that specify some change in a cell's state. For example, the complicated series of events that takes place as a eukaryotic cell divides is largely timed in this way (discussed in Chapter 17), and many of the signals mediating cell–cell interactions are relayed from the plasma membrane to the nucleus by a cascade of protein phosphorylation events (discussed in Chapter 15).

A Eukaryotic Cell Contains a Large Collection of Protein Kinases and Protein Phosphatases

Protein phosphorylation involves the enzyme-catalyzed transfer of the terminal phosphate group of an ATP molecule to the hydroxyl group on a serine, threonine, or tyrosine side chain of the protein (Figure 3–58). A **protein kinase** catalyzes this reaction, and the reaction is essentially unidirectional because of the large amount of free energy released when the phosphate–phosphate bond in

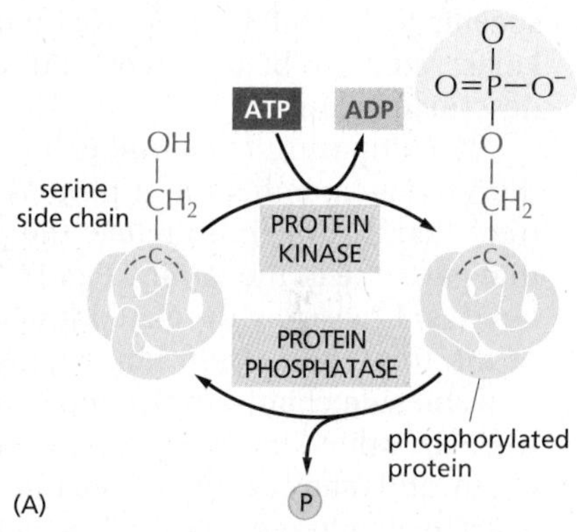

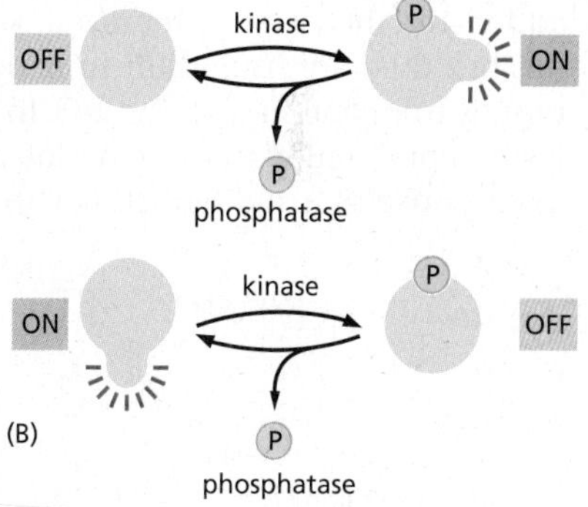

Figure 3–58 **Protein phosphorylation.** Many thousands of proteins in a typical eukaryotic cell are modified by the covalent addition of a phosphate group. (A) The general reaction transfers a phosphate group from ATP to an amino acid side chain of the target protein, catalyzed by a protein kinase. Removal of the phosphate group is catalyzed by a second enzyme, a protein phosphatase. In this example, the phosphate is added to a serine side chain; in other cases, the phosphate is instead linked to the –OH group of a threonine or a tyrosine in the protein. (B) The phosphorylation of a protein by a protein kinase can either increase or decrease the protein's activity, depending on the site of phosphorylation and the structure of the protein.

ATP is broken to produce ADP (discussed in Chapter 2). A **protein phosphatase** catalyzes the reverse reaction of phosphate removal, or *dephosphorylation*. Cells contain hundreds of different protein kinases, each responsible for phosphorylating a different protein or set of proteins. There are also many different protein phosphatases; some are highly specific and remove phosphate groups from only one or a few proteins, whereas others act on a broad range of proteins and are targeted to specific substrates by regulatory subunits. The state of phosphorylation of a protein at any moment, and thus its activity, depends on the relative activities of the protein kinases and phosphatases that modify it.

The protein kinases that phosphorylate proteins in eukaryotic cells belong to a very large family of enzymes that share a catalytic (kinase) sequence of about 290 amino acids. The various family members contain different amino acid sequences on either end of the kinase sequence (for example, see Figure 3–11) and often have short amino acid sequences inserted into loops within it. Some of these additional amino acid sequences enable each kinase to recognize the specific set of proteins it phosphorylates or to bind to structures that localize it in specific regions of the cell. Other parts of the protein regulate the activity of each kinase, so it can be turned on and off in response to different specific signals, as described below.

By comparing the number of amino acid sequence differences between the various members of a protein family, we can construct an "evolutionary tree" that is thought to reflect the pattern of gene duplication and divergence that gave rise to the family. **Figure 3–59** shows an evolutionary tree for protein kinases. Kinases with related functions are often located on nearby branches of the tree: the protein kinases involved in cell signaling that phosphorylate tyrosine side chains, for example, are all clustered in the top left corner of the tree. The other kinases shown phosphorylate either a serine or a threonine side chain, and many are organized into clusters that seem to reflect their function—in transmembrane signal transduction, intracellular signal amplification, cell-cycle control, and so on.

As a result of the combined activities of protein kinases and protein phosphatases, the phosphate groups on proteins are continually turning over—being added and then rapidly removed. Such phosphorylation cycles may seem wasteful, but they are important in allowing the phosphorylated proteins to switch rapidly from one state to another. In fact, the more rapid this cycle is "turning," the faster a population of protein molecules can change its state of phosphorylation in response to a sudden change in its phosphorylation rate (see Figure 15–15).

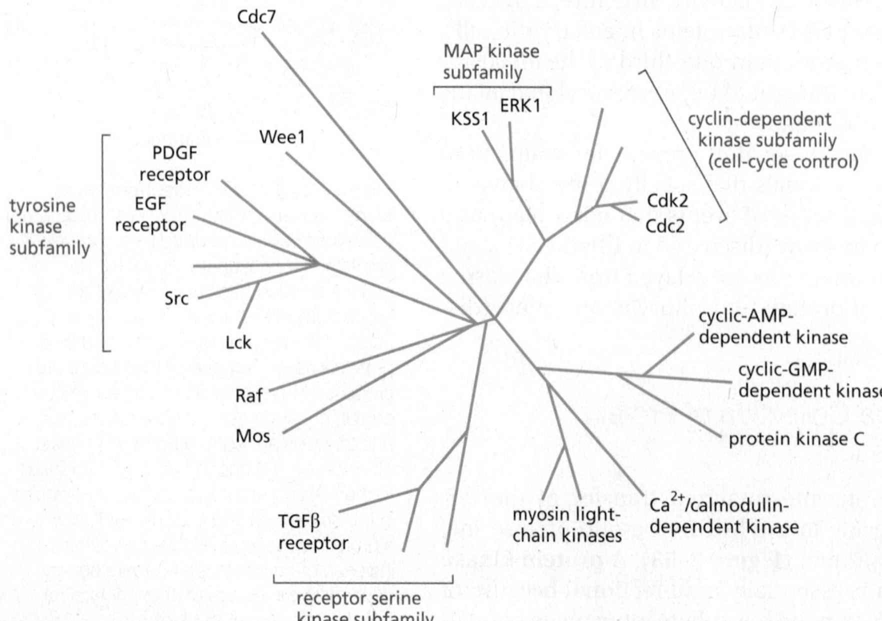

Figure 3–59 An evolutionary tree of selected protein kinases. A higher eukaryotic cell contains hundreds of such enzymes, and the human genome codes for more than 500. Note that only some of these, those discussed in this book, are shown.

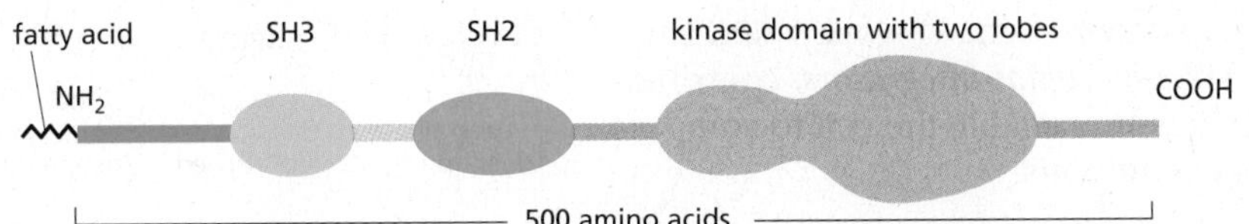

Figure 3–60 The domain structure of the Src family of protein kinases, mapped along the amino acid sequence. For the three-dimensional structure of Src, see Figure 3–11.

The energy required to drive this phosphorylation cycle is derived from the free energy of ATP hydrolysis, one molecule of which is consumed for each phosphorylation event.

The Regulation of the Src Protein Kinase Reveals How a Protein Can Function as a Microprocessor

The hundreds of different protein kinases in a eukaryotic cell are organized into complex networks of signaling pathways that help to coordinate the cell's activities, drive the cell cycle, and relay signals into the cell from the cell's environment. Many of the extracellular signals involved need to be both integrated and amplified by the cell. Individual protein kinases (and other signaling proteins) serve as input-output devices, or "microprocessors," in the integration process. An important part of the input to these signal-processing proteins comes from the control that is exerted by phosphates added and removed from them by protein kinases and protein phosphatases, respectively.

The Src family of protein kinases (see Figure 3–11) exhibits such behavior. The *Src protein* (pronounced "sarc" and named for the type of tumor, a sarcoma, that its deregulation can cause) was the first tyrosine kinase to be discovered. It is now known to be part of a subfamily of nine very similar protein kinases, which are found only in multicellular animals. As indicated by the evolutionary tree in Figure 3–59, sequence comparisons suggest that tyrosine kinases as a group were a relatively late innovation that branched off from the serine/threonine kinases, with the Src subfamily being only one subgroup of the tyrosine kinases created in this way.

The Src protein and its relatives contain a short N-terminal region that becomes covalently linked to a strongly hydrophobic fatty acid, which anchors the kinase at the cytoplasmic face of the plasma membrane. Next along the linear sequence of amino acids come two peptide-binding domains, a Src homology 3 (SH3) domain and an SH2 domain, followed by the kinase catalytic domain (**Figure 3–60**). These kinases normally exist in an inactive conformation, in which a phosphorylated tyrosine near the C-terminus is bound to the SH2 domain, and the SH3 domain is bound to an internal peptide in a way that distorts the active site of the enzyme and helps to render it inactive.

As shown in **Figure 3–61**, turning the kinase on involves at least two specific inputs: removal of the C-terminal phosphate and the binding of the SH3 domain by a specific activating protein. In this way, the activation of the Src kinase signals

Figure 3–61 The activation of a Src-type protein kinase by two sequential events. As described in the text, the requirement for multiple upstream events to trigger these processes allows the kinase to serve as a signal integrator (**Movie 3.11**). (Adapted from S.C. Harrison, *Cell* 112: 737–740, 2003.)

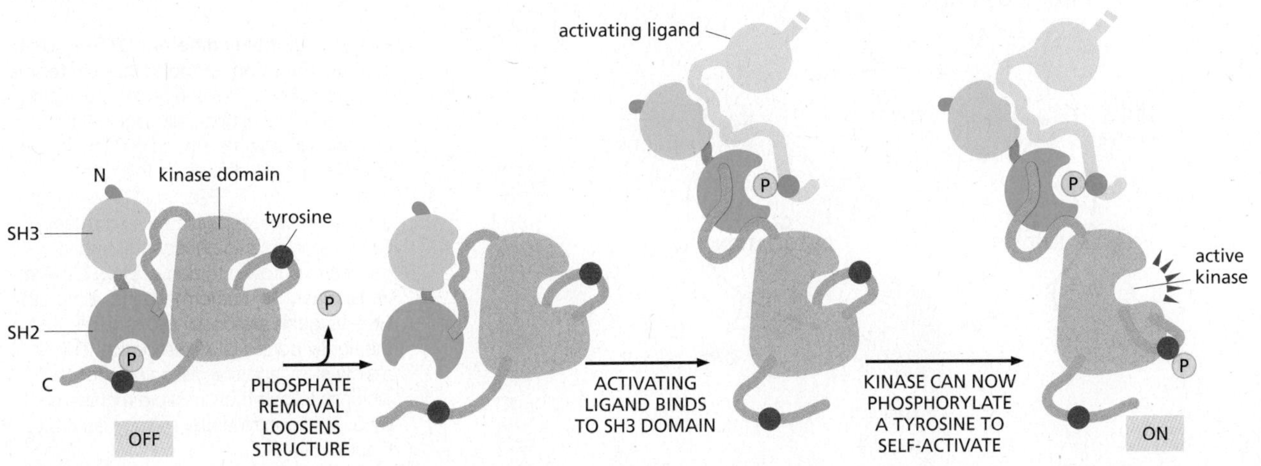

the completion of a particular set of separate upstream events (Figure 3–62). Thus, the Src family of protein kinases serves as specific *signal integrators*, contributing to the web of information-processing events that enable the cell to compute useful responses to a complex set of different conditions.

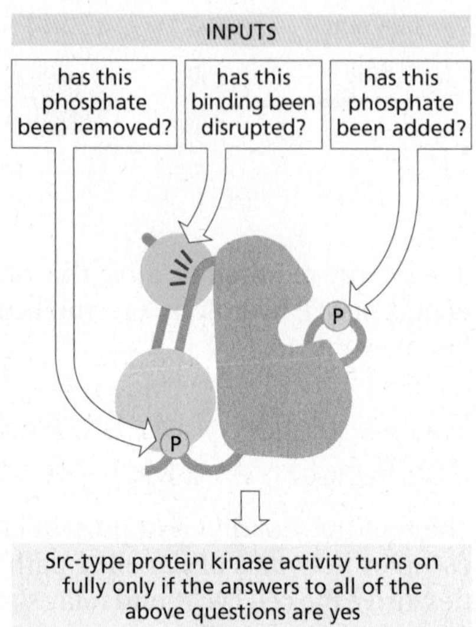

Figure 3–62 How a Src-type protein kinase acts as a signal-integrating device. A disruption of the inhibitory interaction illustrated for the SH3 domain *(green)* occurs when its binding to the indicated *orange* linker region is replaced with its higher-affinity binding to an activating ligand.

Regulatory GTP-binding Proteins Are Switched On and Off by the Gain and Loss of a Phosphate Group

Eukaryotic cells have a second way to regulate protein activity by phosphate addition and removal. In this case, however, the phosphate is not enzymatically transferred from ATP to the protein. Instead, the phosphate is part of a guanine nucleotide—guanosine triphosphate (GTP)—that binds tightly to various types of **GTP-binding proteins**. These proteins, also called **GTPases**, bind to other proteins to regulate their activities. They serve as molecular switches: GTP-binding proteins are in their "on" conformation when GTP is bound, but they can hydrolyze this GTP to GDP—which releases a phosphate and flips the protein to its "off" conformation. As with protein phosphorylation, this process is reversible: the active conformation is regained by dissociation of the GDP, followed by the rapid binding of a fresh molecule of GTP (Figure 3–63).

Hundreds of different GTP-binding proteins function as such molecular switches in cells. They all contain variations of the same globular domain that undergoes a conformational change when its tightly bound GTP is hydrolyzed to GDP. The three-dimensional structure of a prototypical member of this family, the *monomeric GTPase* called **Ras** that plays important roles in cell signaling, is shown in Figure 3–64.

The crucial role that GTP-binding proteins play in intracellular signaling pathways is discussed in detail in Chapter 15.

Proteins Can Be Regulated by the Covalent Addition of Other Proteins

Cells contain a special family of small proteins whose members are covalently attached to many other proteins to determine the activity or fate of the second protein. In each case, the carboxyl end of the small protein becomes linked to the amino group of a lysine side chain of a target protein through an isopeptide bond. The first such protein discovered, and the most abundantly used, is **ubiquitin** (Figure 3–65A). Ubiquitin can be covalently attached to target proteins in a variety of ways, each of which has a different meaning for cells. The major form of ubiquitin addition produces *polyubiquitin* chains in which—once the first ubiquitin molecule is attached to the target—each subsequent ubiquitin molecule links to Lys48 of the previous ubiquitin, creating a chain of Lys48-linked ubiquitins that are attached to a single lysine side chain of the target protein. This form

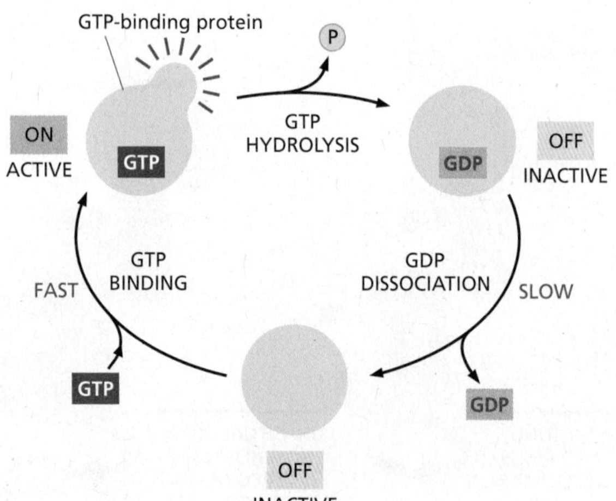

Figure 3–63 Many different GTP-binding proteins function as molecular switches. The activity of a GTP-binding protein (also called a GTPase) generally requires the presence of a tightly bound GTP molecule (switch "on"). Hydrolysis of this GTP molecule by the GTP-binding protein—at a rate that can be regulated—produces GDP and inorganic phosphate, and it causes the protein to convert to a different, usually inactive, conformation (switch "off"). Resetting the switch to "on" requires that the tightly bound GDP dissociate. This is a slow step, and the dissociation of GDP, which is followed by its rapid replacement by GTP, is controlled by cell signals (see Figure 15–8).

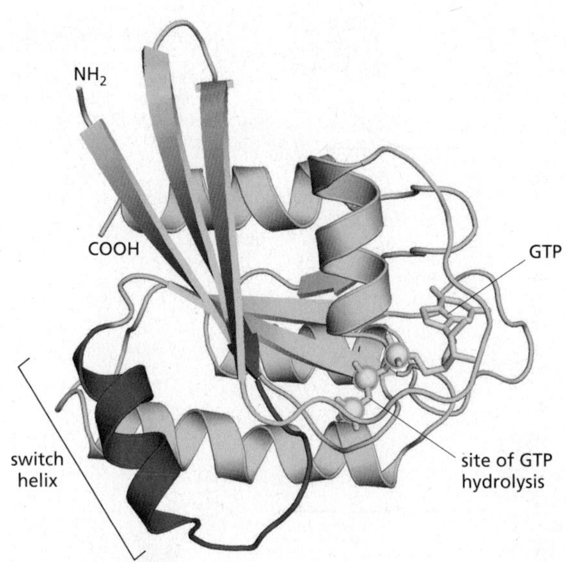

Figure 3–64 The structure of the Ras protein in its GTP-bound form. This monomeric GTPase illustrates the structure of a GTP-binding domain, which is present in a large family of GTP-binding proteins. The *red* regions change their conformation when the GTP molecule is hydrolyzed to GDP and inorganic phosphate by the protein; the GDP remains bound to the protein, while the inorganic phosphate is released. The special role of the *switch helix* in proteins related to Ras is explained in the text (see Figure 3–68 and Movie 15.7). (PDB code: 121P.)

of polyubiquitin directs the target protein to the interior of a proteasome, where it is digested to small peptides (see Figure 6–87). In other circumstances, only single molecules of ubiquitin are added to proteins. In addition, some target proteins are modified with a different type of polyubiquitin chain. These modifications have different functional consequences for the protein that is targeted (**Figure 3–65B**).

Related structures are created when a different member of the ubiquitin family, such as SUMO (small ubiquitin-related modifier), is covalently attached to a lysine side chain of target proteins. Not surprisingly, all such modifications are reversible. Cells contain sets of ubiquitylating and deubiquitylating (and sumoylating and desumoylating) enzymes that manipulate these covalent adducts, thereby playing roles analogous to the protein kinases and protein phosphatases that add and remove phosphates from protein side chains.

An Elaborate Ubiquitin-conjugating System Is Used to Mark Proteins

How do cells select target proteins for ubiquitin addition? As an initial step, the carboxyl end of ubiquitin needs to be activated. This activation is accomplished when a protein called a *ubiquitin-activating enzyme* (E1) uses ATP

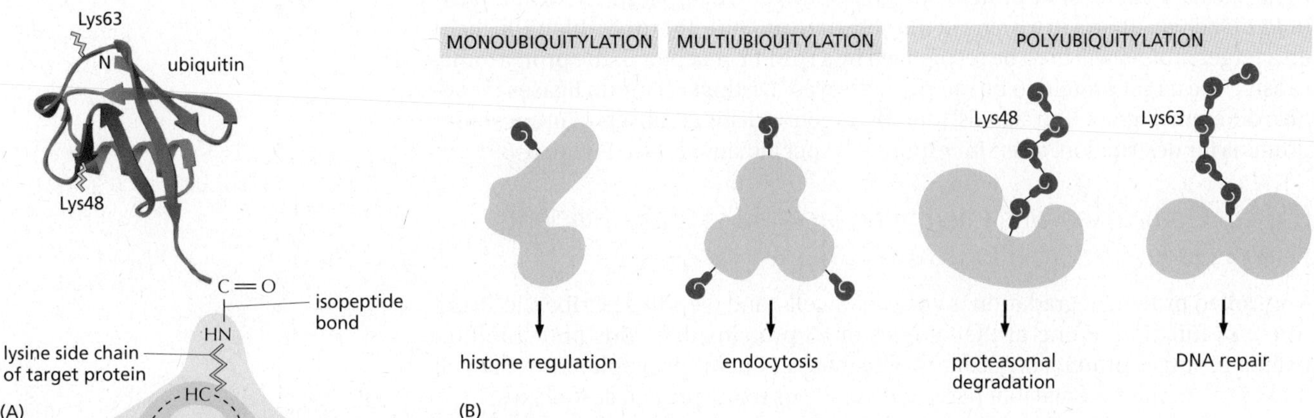

Figure 3–65 The marking of proteins by ubiquitin. (A) The three-dimensional structure of ubiquitin, a small protein of 76 amino acids. A family of special enzymes couples its carboxyl end to the amino group of a lysine side chain in a target protein molecule, forming an isopeptide bond. (B) Some modification patterns that have specific meanings to the cell. Note that the two types of polyubiquitylation differ in the way the ubiquitin molecules are linked together. Linkage through Lys48 signifies degradation by the proteasome (see Figure 6–87), whereas that through Lys63 has other meanings. Ubiquitin markings are "read" by proteins that specifically recognize each type of modification.

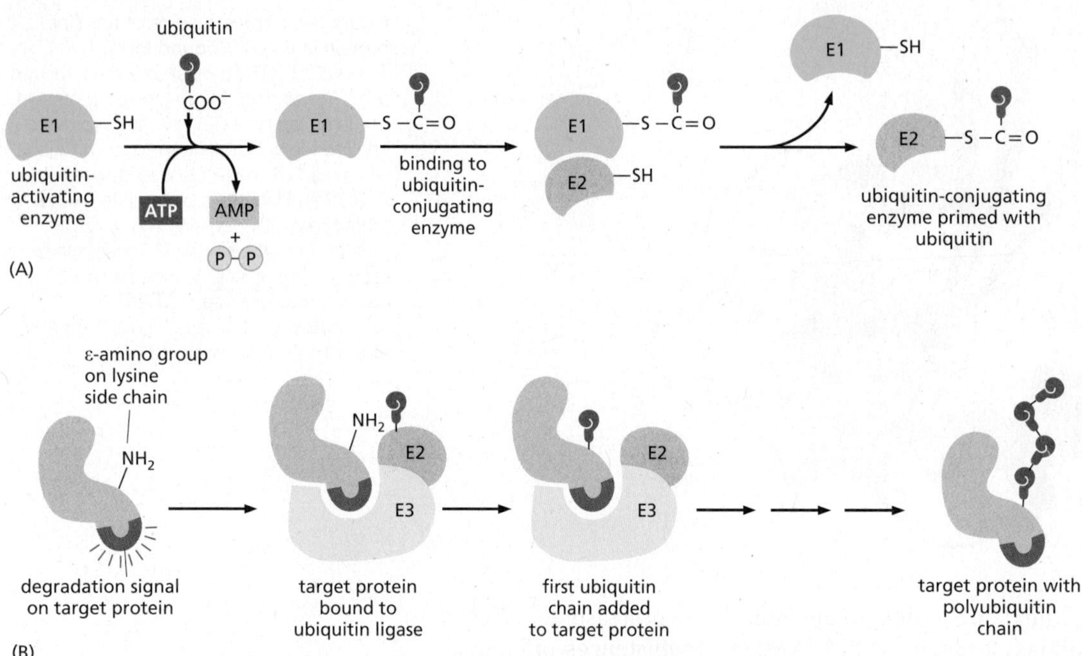

Figure 3–66 How ubiquitin is added to proteins. (A) Ubiquitin activations. The C-terminus of ubiquitin is initially activated by being linked via a high-energy thioester bond to a cysteine side chain on the E1 protein. This reaction requires ATP, and it proceeds via a covalent AMP–ubiquitin intermediate. The activated ubiquitin on E1, also known as the ubiquitin-activating enzyme, is then transferred to the cysteine on an E2 molecule. (B) The addition of a polyubiquitin chain to a target protein. In a mammalian cell, there are several hundred distinct E2–E3 complexes. The E2s are called ubiquitin-conjugating enzymes. The E3s are referred to as ubiquitin ligases. (Adapted from D.R. Knighton et al., *Science* 253:407–414, 1991.)

hydrolysis energy to attach ubiquitin to itself through a high-energy covalent bond (a thioester). E1 then passes this activated ubiquitin to one of a set of *ubiquitin-conjugating* (E2) enzymes, each of which acts in conjunction with a set of accessory (E3) proteins called **ubiquitin ligases** that select the target proteins to be modified. There are roughly 30 structurally similar but distinct E2 enzymes in mammals and hundreds of different E3 proteins that form complexes with specific E2 enzymes.

Figure 3–66 illustrates the process used to mark proteins for proteasomal degradation. [Similar mechanisms are used to attach ubiquitin (and SUMO) to other types of target proteins.] Here, the ubiquitin ligase binds to specific degradation signals, called *degrons*, in protein substrates, thereby helping E2 to form a *polyubiquitin* chain linked to a lysine of the substrate protein. This polyubiquitin chain on a target protein will then be recognized by a specific receptor in the proteasome, causing the target protein to be rapidly destroyed. Distinct ubiquitin ligases recognize different degradation signals, thereby targeting distinct subsets of intracellular proteins for destruction, often in response to specific signals (see Figure 6–89).

Protein Complexes with Interchangeable Parts Make Efficient Use of Genetic Information

Controlled protein degradation is critical for cells, and we will describe the structure and function of one of the families of E3 proteins that adds polyubiquitin chains to target proteins in order to illustrate a general principle: how the cell makes use of interchangeable parts to diversify its many protein complexes.

The *SCF ubiquitin ligase* is a C-shaped structure that is formed from five protein subunits, the largest of which serves as a scaffold on which the rest of the complex is built. The structure underlies a remarkable mechanism (**Figure 3–67**). At one end of the C is an E2 ubiquitin-conjugating enzyme. At the other end is a substrate-binding arm, a subunit known as an *F-box protein*. These two subunits are separated by a gap of about 5 nm. When this protein complex is activated, the

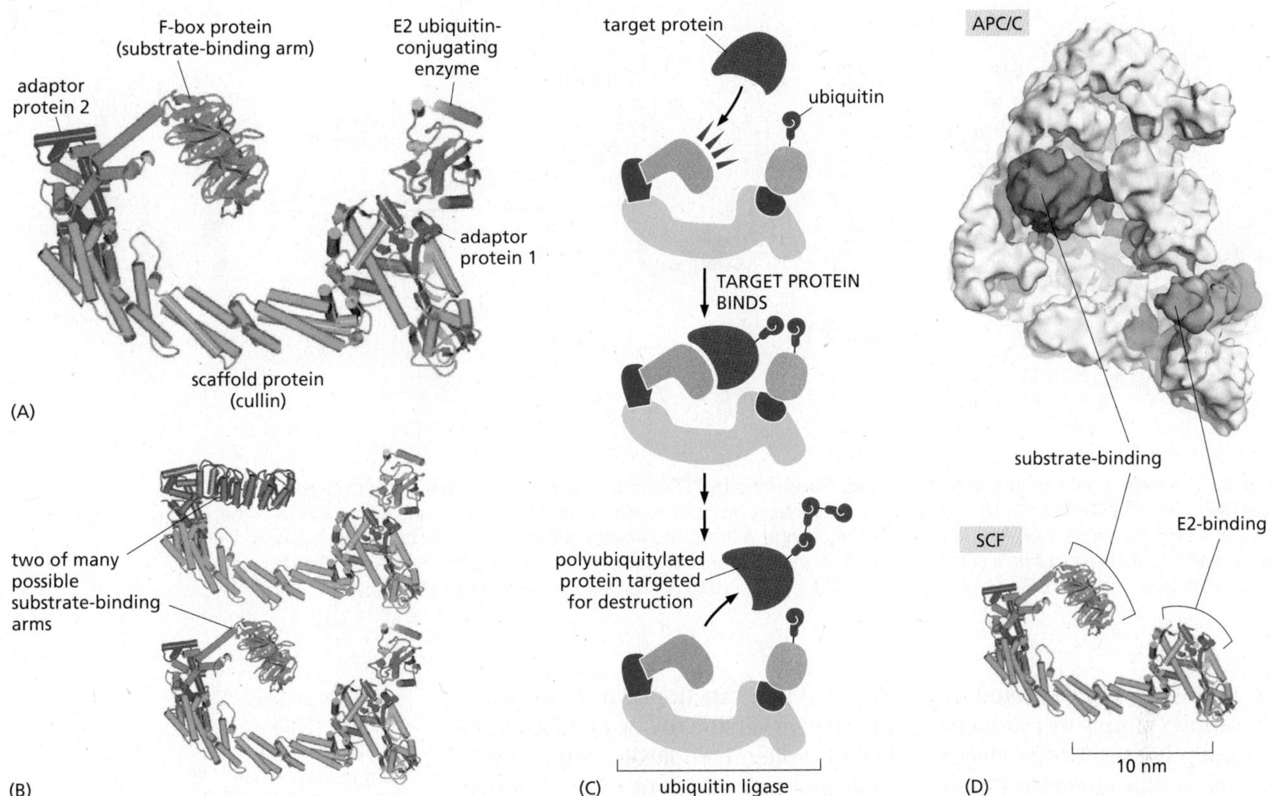

Figure 3–67 The structure and mode of action of a ubiquitin ligase. (A) The structure of the five-protein SCF ubiquitin ligase complex that includes an E2 ubiquitin-conjugating enzyme. Four proteins form the E3 portion. The protein denoted here as adaptor protein 1 is the Rbx1/Hrt1 protein, adaptor protein 2 is the Skp1 protein, and cullin is the Cul1 protein. One of the many different F-box proteins completes the complex. (B) Comparison of the same complex with two different substrate-binding arms, the F-box proteins Skp2 *(top)* and β-trCP1 *(bottom)*, respectively. (C) The binding and ubiquitylation of a target protein by the SCF ubiquitin ligase. If, as indicated, a chain of ubiquitin molecules is added to the same lysine of the target protein, that protein is marked for rapid destruction by the proteasome. (D) Comparison of SCF *(bottom)* with a low-resolution electron microscopy structure of a ubiquitin ligase called the anaphase-promoting complex (APC/C; *top*) at the same scale. The APC/C is a large, 15-protein complex. As discussed in Chapter 17, its ubiquitylations control the late stages of mitosis. It is distantly related to SCF and contains a cullin subunit *(green)* that lies along the side of the complex at right, only partly visible in this view. E2 proteins are not shown here, but their binding sites are indicated in *orange,* along with substrate-binding sites in *purple.* (A and B, adapted from G. Wu et al., *Mol. Cell* 11:1445–1456, 2003. D, adapted from P. da Fonseca et al., *Nature* 470:274–278, 2011.)

F-box protein binds to a specific site on a target protein, positioning the protein in the gap so that some of its lysine side chains contact the ubiquitin-conjugating enzyme. The enzyme can then catalyze repeated additions of ubiquitin to these lysines (see Figure 3–67C), producing the polyubiquitin chains that mark its target proteins for destruction in a proteasome.

In this manner, specific proteins are targeted for rapid destruction in response to specific signals, thereby helping to drive the cell cycle (discussed in Chapter 17). The timing of the destruction often involves creating a specific pattern of phosphorylation on the target protein that is required for its recognition by the F-box subunit. It also requires the activation of an SCF-like ubiquitin ligase that carries the appropriate substrate-binding arm. Many of these arms (the F-box subunits) are interchangeable in the protein complex (see Figure 3–67B), and there are more than 70 human genes that encode them.

As emphasized previously, once a successful protein has evolved, its genetic information tends to be duplicated to produce a family of related proteins. Thus, for example, not only are there many F-box proteins—making possible the recognition of different sets of target proteins—but there is also a family of scaffolds (known as cullins) that give rise to a family of SCF-like ubiquitin ligases.

A protein machine like the SCF ubiquitin ligase, with its interchangeable parts, makes economical use of the genetic information in cells. It also creates

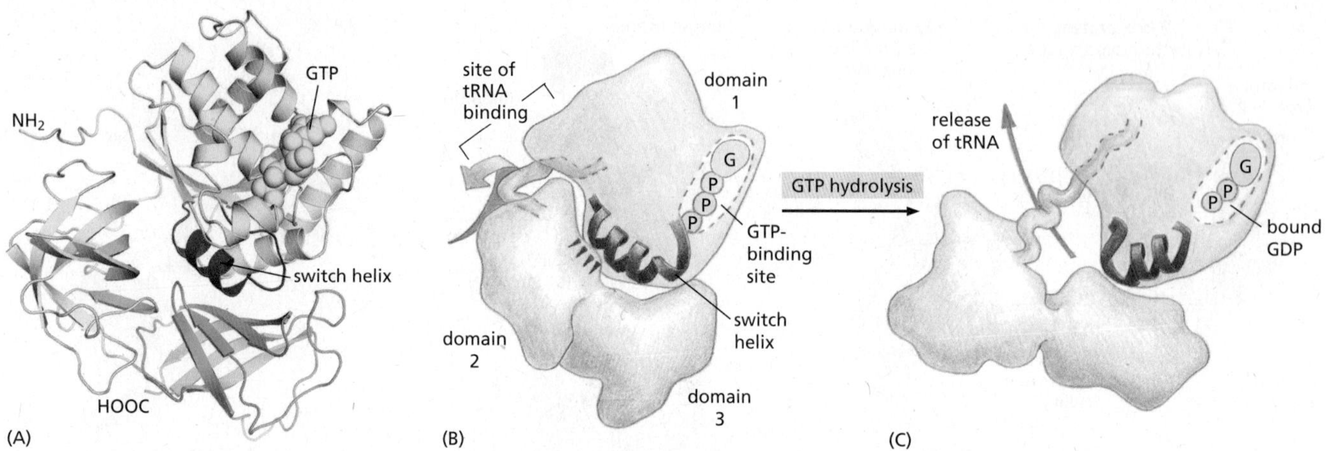

Figure 3–68 The large conformational change in EF-Tu caused by GTP hydrolysis. (A and B) The three-dimensional structure of EF-Tu with GTP bound. The domain at the top has a structure similar to the Ras protein, and its red α helix is the switch helix, which moves after GTP hydrolysis. (C) The change in the conformation of the switch helix in domain 1 allows domains 2 and 3 to rotate as a single unit by about 90 degrees toward the viewer, which releases the tRNA that was bound to this structure (see also Figure 3–69). (A, PDB code: 1EFT; B, courtesy of Mathias Sprinzl and Rolf Hilgenfeld.)

opportunities for "rapid" evolution, inasmuch as new functions can evolve for the entire complex simply by producing an alternative version of one of its subunits.

Ubiquitin ligases form a diverse family of protein complexes. Some of these complexes are far larger and more complicated than SCF, but their underlying enzymatic function remains the same (see Figure 3–67D).

A GTP-binding Protein Shows How Large Protein Movements Can Be Generated from Small Ones

Detailed structures obtained for one of the GTP-binding protein family members, the *EF-Tu protein*, provide a good example of how allosteric changes in protein conformations can produce large movements by amplifying a small, local conformational change. As will be discussed in Chapter 6, EF-Tu is an abundant molecule that serves as an elongation factor (hence the EF) in protein synthesis, loading each aminoacyl-tRNA molecule onto the ribosome. EF-Tu contains a Ras-like domain (see Figure 3–64), and the tRNA molecule forms a tight complex with its GTP-bound form. For the tRNA molecule to transfer its amino acid to the growing polypeptide chain requires that the GTP bound to EF-Tu be hydrolyzed, dissociating the EF-Tu from the tRNA. Because this GTP hydrolysis is triggered by a proper fit of the tRNA to the mRNA molecule on the ribosome, the EF-Tu serves to discriminate between correct and incorrect mRNA–tRNA pairings (see Figure 6–69).

By comparing the three-dimensional structure of EF-Tu in its GTP-bound and GDP-bound forms, we can see how the repositioning of the tRNA occurs. The dissociation of the inorganic phosphate group, which follows the reaction GTP → GDP + phosphate, causes a shift of a few tenths of a nanometer at the GTP-binding site, just as it does in the Ras protein. This tiny movement, equivalent to a few times the diameter of a hydrogen atom, causes a conformational change to propagate along a crucial piece of α helix, called the *switch helix*, in the Ras-like domain of the protein. The switch helix seems to serve as a latch that adheres to a specific site in another domain of the molecule, holding the protein in a "shut" conformation. The conformational change triggered by GTP hydrolysis causes the switch helix to detach, allowing separate domains of the protein to swing apart, through a distance of about 4 nm (**Figure 3–68**). This releases the tRNA, allowing its attached amino acid to be used for protein synthesis (**Figure 3–69**).

Notice in this example how cells have exploited a simple chemical change that occurs on the surface of a small protein domain to create a movement 50 times

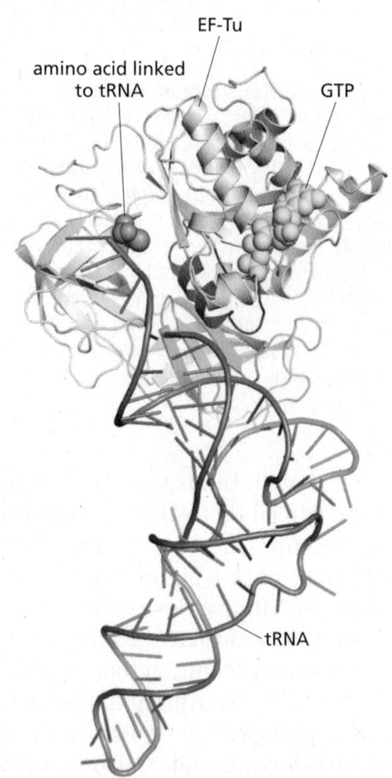

Figure 3–69 An aminoacyl tRNA molecule bound to EF-Tu. Note how the bound protein blocks the use of the tRNA-linked amino acid *(dark green)* for protein synthesis until GTP hydrolysis triggers the conformational changes shown in Figure 3–68C, dissociating the protein–tRNA complex. EF-Tu is a bacterial protein; however, a very similar protein exists in eukaryotes, where it is called EF-1 (**Movie 3.12**). (PDB code: 1B23.)

larger. Dramatic shape changes of this type also cause the very large movements that occur in motor proteins, as we discuss next.

Motor Proteins Produce Directional Movement in Cells

We have seen that conformational changes in proteins have a central role in enzyme regulation and cell signaling. We now discuss proteins whose major function is to move other molecules. These *motor proteins* generate the forces responsible for muscle contraction and the crawling and swimming of cells. Motor proteins also power smaller-scale intracellular movements: they help to move chromosomes to opposite ends of the cell during mitosis (discussed in Chapter 17), to move organelles along molecular tracks within the cell (discussed in Chapter 16), and to move enzymes along a DNA strand during the synthesis of a new DNA molecule (discussed in Chapter 5). All these fundamental processes depend on proteins with moving parts that operate as force-generating machines.

How do these machines work? It is a challenge for cells to use shape changes in proteins to generate persistent movements in a single direction. If, for example, a protein is required to walk along a long cytoskeletal filament, it can do this by undergoing a series of conformational changes, such as those shown in **Figure 3–70**. But with nothing to drive these changes in an orderly sequence, they are perfectly reversible, and the protein can only wander randomly back and forth along the thread. We can look at this situation in another way. Because the directional movement of a protein does work, the laws of thermodynamics (discussed in Chapter 2) demand that such movement use free energy from some other source (otherwise the protein could be used to make a perpetual motion machine). Therefore, without an input of energy, the protein molecule can only wander aimlessly.

How can the cell make such a series of conformational changes unidirectional? To force the entire cycle to proceed in one direction, it is enough to make any one of the changes in shape irreversible. Most proteins that are able to walk in one direction for long distances achieve this motion by coupling one of the conformational changes to the hydrolysis of an ATP molecule that is tightly bound to the protein. The mechanism is similar to the one discussed earlier that drives allosteric protein shape changes by GTP hydrolysis. Because ATP (or GTP) hydrolysis releases a great deal of free energy, it is very unlikely that the nucleotide-binding protein will undergo the reverse shape change needed for moving backward—as this would require that it also reverse the ATP hydrolysis by adding a phosphate molecule to ADP to form ATP.

In the model shown in **Figure 3–71A**, ATP binding shifts a motor protein from conformation 1 to conformation 2. The bound ATP is then hydrolyzed to produce ADP and inorganic phosphate, causing a change from conformation 2 to conformation 3. Finally, the release of the bound ADP and phosphate drives the protein back to conformation 1. Because the energy provided by ATP hydrolysis drives the transition 2 → 3, this series of conformational changes is effectively irreversible. Thus, the entire cycle goes in only one direction, causing the protein molecule to walk continuously to the right in this example.

Many motor proteins generate directional movement through the use of a similar unidirectional ratchet, including the muscle motor protein *myosin*, which walks along actin filaments (**Figure 3–71B**), and the *kinesin* proteins that walk along microtubules (both discussed in Chapter 16). These movements can be rapid: some of the motor proteins involved in DNA replication (the DNA helicases) propel themselves along a DNA strand at rates as high as 1000 nucleotides per second.

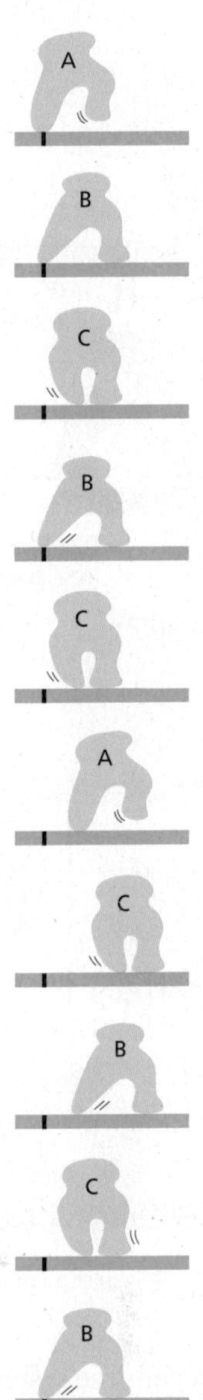

Figure 3–70 Changes in conformation can cause a protein to "walk" along a cytoskeletal filament, driven by its constant collisions with other molecules (thermal energy). This protein cycles between three different conformations (A, B, and C) as it moves along the filament. But, without an input of energy to drive its movement in a single direction, the protein can only wander randomly back and forth, ultimately getting nowhere.

Proteins Often Form Large Complexes That Function as Protein Machines

Large proteins formed from many domains are able to perform more elaborate functions than small, single-domain proteins. But large protein assemblies formed from many protein molecules linked together by noncovalent bonds perform the most impressive tasks. Now that it is possible to reconstruct most biological

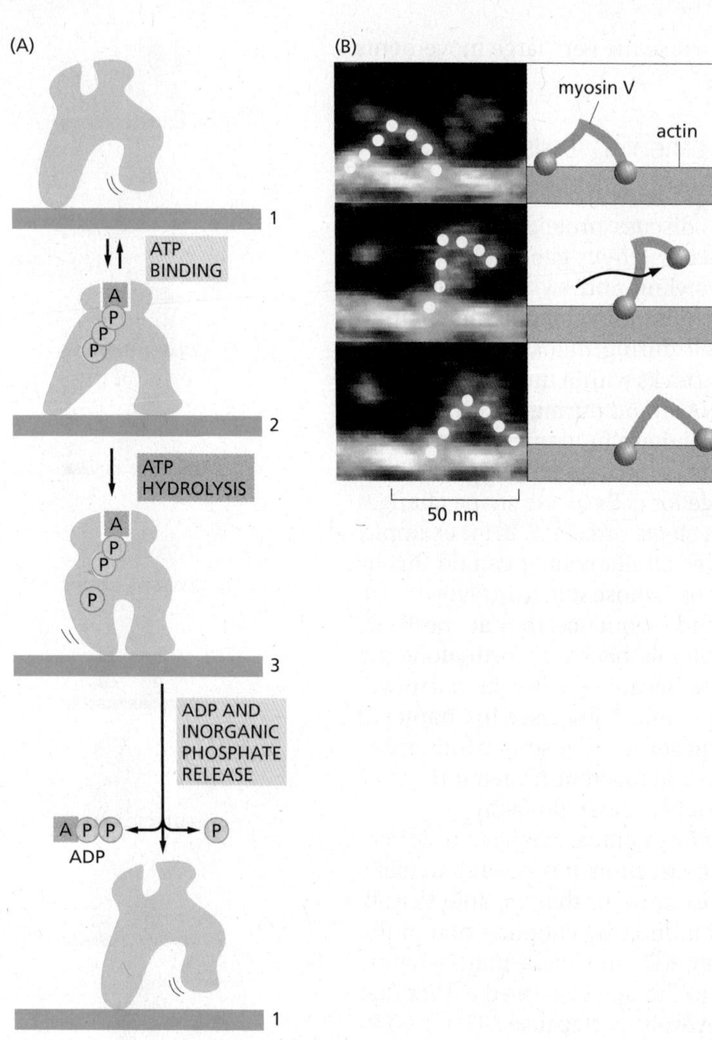

Figure 3–71 How a protein can walk in one direction. (A) An allosteric motor protein driven by ATP hydrolysis. The transition between three different conformations includes a step driven by the hydrolysis of a tightly bound ATP molecule, creating a "unidirectional ratchet" that makes the entire cycle essentially irreversible. By repeated cycles, the protein therefore moves continuously to the right along the thread. (B) Direct visualization of a walking myosin motor protein by high-speed atomic force microscopy; the elapsed time between steps was less than 0.5 sec (see Movie 16.3). (B, adapted from N. Kodera et al., *Nature* 468:72–76, published 2010 by Macmillan Publishers Ltd. Reproduced with permission of SNCSC.)

processes in cell-free systems in the laboratory, it is clear that each of the central processes in a cell—such as DNA replication, protein synthesis, vesicle budding, or transmembrane signaling—is catalyzed by a highly coordinated, linked set of 10 or more proteins. In most such *protein machines,* energetically favorable reactions such as the hydrolysis of bound nucleoside triphosphates (ATP or GTP) drive an ordered series of conformational changes in one or more of the individual protein subunits, enabling the ensemble of proteins to move in a coordinated way. As a result, each enzyme can be moved directly into position as the machine catalyzes successive reactions in a series (**Figure 3–72**). This is what occurs, for example, in protein synthesis on a ribosome (an RNA–protein, or *macromolecular machine,* discussed in Chapter 6)—or in DNA replication, where a large multiprotein complex moves rapidly along the DNA (discussed in Chapter 5).

Cells have evolved protein machines for the same reason that humans have invented mechanical and electronic machines. For accomplishing almost any task, manipulations that are spatially and temporally coordinated through linked processes are much more efficient than the use of many separate tools.

The Disordered Regions in Proteins Are Critical for a Set of Different Functions

Scientists have discovered that proteins contain a surprisingly large amount of intrinsically disordered polypeptide chain. Thus, as previously mentioned, it is estimated that about a third of all eukaryotic proteins contain unstructured

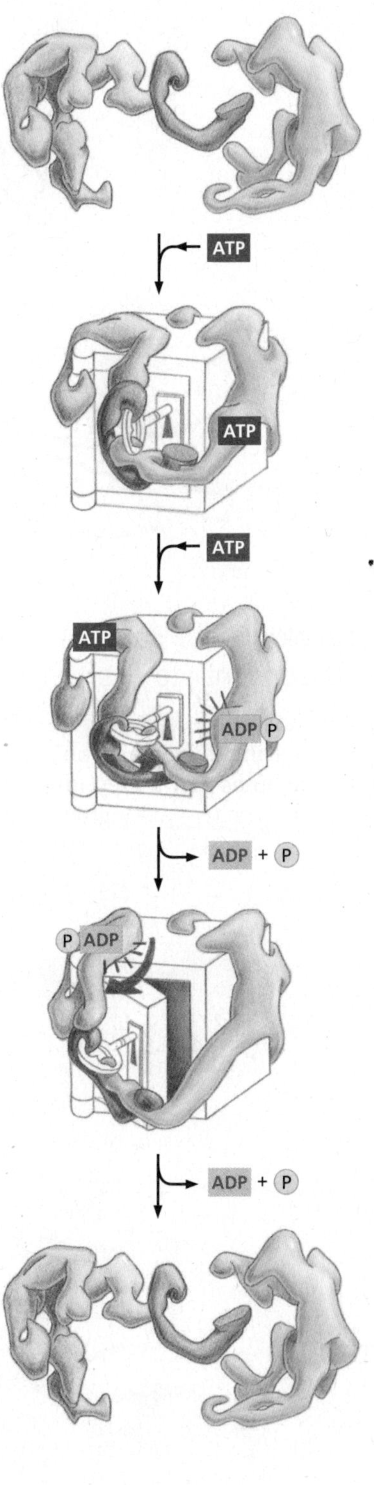

Figure 3–72 Schematic example showing how protein machines can carry out complex functions. These machines are made of individual proteins that collaborate to perform a specific task (Movie 3.13). As in this example, the movement of these proteins is often coordinated by the hydrolysis of a bound nucleotide such as ATP or GTP. Directional allosteric conformational changes of proteins driven in this way often occur in a large protein assembly, thereby allowing directed movements within the complex to coordinate the activities of its individual molecules. (See also Movie 5.5.)

regions greater than 30 amino acids in length. Some of these regions are formed from only a limited subset of the 20 amino acids and are therefore designated as *low-complexity domains.* Because many unstructured regions have been conserved in a particular protein over long periods of evolutionary time, their presence must benefit the organisms that contain them. What do these disordered regions do?

Intrinsically disordered regions of proteins often form specific binding sites for other proteins that are of high specificity, as illustrated in Figure 3–73A. In addition, this type of binding interaction is easily controlled. Most protein phosphorylation sites are in intrinsically disordered regions, not in globular domains, and these regions are central to regulatory mechanisms. As one example, the eukaryotic RNA polymerase enzyme that produces mRNAs contains an unstructured C-terminal tail of 200 amino acids that is covalently modified as the RNA polymerase proceeds, thereby attracting specific other proteins to the transcription complex at different times (see Figure 6–23). Disordered regions tend to evolve rapidly, and the type of binding diagrammed in Figure 3–73B facilitates the fine-tuning and evolution of cell signaling networks (see Chapter 15).

A very different type of function is exemplified by *elastin,* an abundant protein in the extracellular matrix that is formed as a highly disordered polypeptide. Elastin's relatively loose and unstructured polypeptide chains are covalently cross-linked to produce an elastic meshwork that can be stretched like a rubber band, as illustrated in Figure 3–74. The elastic fibers that result enable skin and other tissues, such as arteries and lungs, to stretch and recoil without tearing.

Perhaps most uniquely, intrinsically disordered regions are widely used as tethers to concentrate reactants and thereby accelerate the reactions needed by a cell. For example, within large multienzyme complexes, unstructured regions of polypeptide chain can allow substrates to be carried sequentially between different active sites (Figure 3–75).

In their most general tethering role, unstructured regions allow large *scaffold proteins* with multiple binding sites to concentrate sets of interacting RNA and/or protein molecules at a particular site in a cell, as we discuss next.

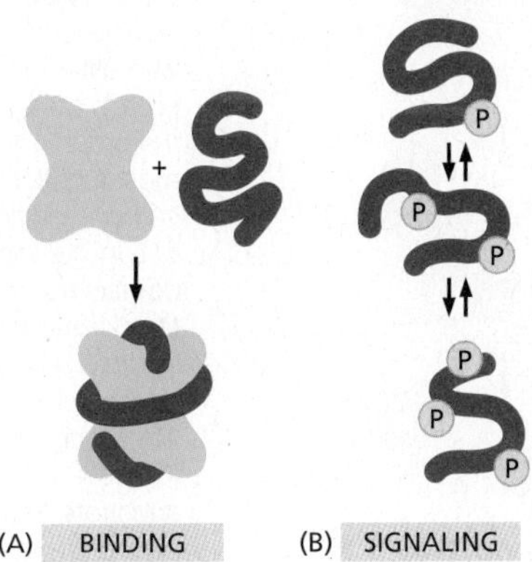

Figure 3–73 Intrinsically disordered protein sequences provide versatile binding sites. (A) Unstructured regions of polypeptide chain often form binding sites for other proteins. Although these binding events are of high specificity, they are often of low affinity because of the free-energy cost of folding the normally unfolded partner (and they are thus readily reversible). (B) Unstructured regions can be easily modified covalently to change their binding preferences, and they are therefore frequently involved in cell signaling processes. In this schematic, multiple sites of protein phosphorylation are indicated.

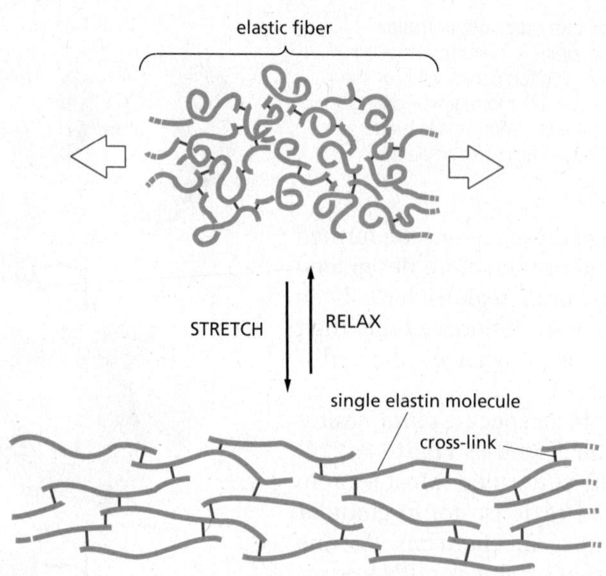

Figure 3–74 Intrinsically disordered protein chains are used to produce elastic structures. The polypeptide chains of the protein *elastin* are cross-linked together in the extracellular space to form rubberlike, elastic fibers. Each elastin molecule uncoils into a more extended conformation when the fiber is stretched, and it recoils spontaneously as soon as the stretching force is relaxed.

Scaffolds Bring Sets of Interacting Macromolecules Together and Concentrate Them in Selected Regions of a Cell

As scientists have learned more of the details of cell biology, they have recognized an increasing degree of sophistication in cell chemistry. We now know that protein machines play a predominant role and that all of their activities—like those of other proteins—are highly regulated. In addition, it has also become clear that

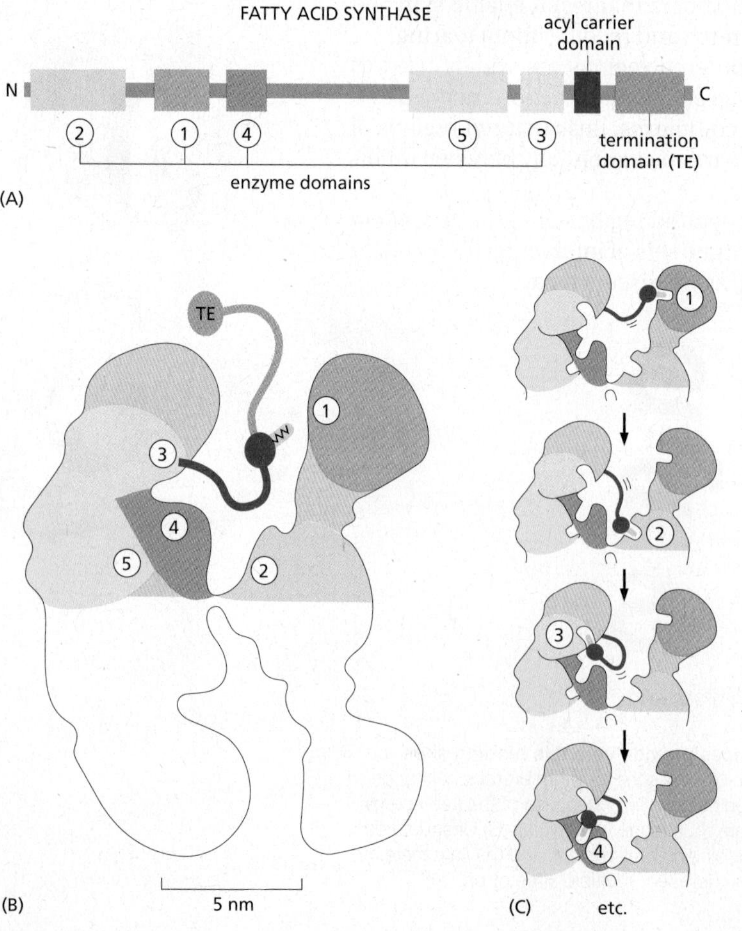

Figure 3–75 How unstructured regions of polypeptide chain can serve as tethers to allow reaction intermediates to be passed from one active site to another in a large multienzyme complex, the fatty acid synthase in mammals. (A) The locations of seven protein domains with different activities in this 270-kilodalton protein are shown. The numbers refer to the order in which each enzyme domain must function to complete each two-carbon addition step. After multiple cycles of two-carbon addition, the termination domain releases the final product once the desired length of fatty acid has been synthesized. (B) The structure of the dimeric enzyme, with the location of the five active sites in one monomer indicated. (C) How a flexible tether allows the substrate that remains linked to the acyl carrier domain *(red)* to be passed from one active site to another in each monomer, sequentially elongating and modifying the bound fatty acid intermediate *(yellow)*. The five steps are repeated until the final length of fatty acid chain has been synthesized. (Only steps 1 through 4 are illustrated here.) (Adapted from T. Maier et al., *Q. Rev. Biophys.* 43:373–422, 2010.)

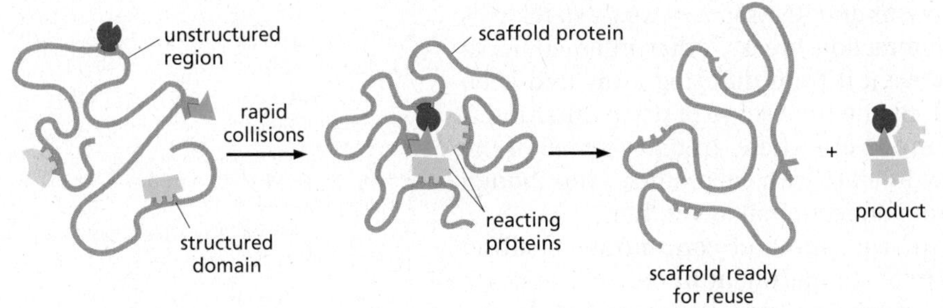

Figure 3–76 **How the proximity created by scaffold proteins can greatly speed reactions in a cell.** In this example, long unstructured regions of polypeptide chain in a large scaffold protein connect a series of structured domains that bind a set of reacting proteins or RNA molecules. The unstructured regions serve as flexible tethers that greatly speed reaction rates by causing a rapid, random collision of all of the molecules that are bound to the scaffold. (For specific examples of protein tethering, see Figure 3–75 and Figure 16–14; for scaffold RNA molecules, see Figure 7–82.)

these machines are often localized to specific sites in the cell, being assembled and activated only where and when they are needed. As one example, when extracellular signaling molecules bind to receptor proteins in the plasma membrane, the activated receptors often recruit a set of other proteins to the inside surface of the plasma membrane to form a large protein complex that passes the signal on (illustrated and discussed in Chapter 15).

The mechanisms generally involve **scaffold proteins** that have binding sites for multiple other proteins and/or RNA molecules. Such scaffolds serve both to link together specific sets of interacting macromolecules and to position them at specific locations inside a cell. At one extreme are rigid scaffolds, such as the cullin in SCF ubiquitin ligase (see Figure 3–67). At the other extreme are large, flexible scaffold proteins that create special regions inside the cell that have a unique biochemistry. Networks of such large scaffolds often underlie regions of specialized plasma membrane. For example, the *Discs-large protein* (Dlg) of about 900 amino acids is concentrated in special regions beneath the plasma membrane in epithelial cells and at synapses. Dlg contains binding sites for at least seven other proteins interspersed with regions of more flexible polypeptide chain. An ancient protein, conserved in organisms as diverse as sponges, worms, flies, and humans, Dlg derives its name from the mutant phenotype of the organism in which it was first discovered. In a *Drosophila* embryo with a mutation in the *Dlg* gene, the imaginal disc cells fail to stop proliferating when they should, and they produce unusually large discs whose epithelial cells can form tumors.

Dlg and a large number of similar scaffold proteins are thought to function like the protein that is schematically illustrated in **Figure 3–76**. By binding a specific set of interacting proteins and/or RNA molecules, these scaffolds can enhance the rate of critical reactions, while also confining them to the particular region of the cell that contains the scaffold. For similar reasons, cells also make extensive use of *scaffold RNA molecules*, as discussed in Chapter 7.

Macromolecules Can Self-assemble to Form Biomolecular Condensates

The macromolecular assemblies and protein machines that we have discussed so far are defined by physical interactions that organize individual proteins and nucleic acids at defined positions relative to each other. Each copy of a macromolecular machine generally is built from the same parts and assembled into the same three-dimensional structure. For example, the bacterial ribosome responsible for synthesizing new proteins is built from 55 proteins and three RNA molecules arranged in an invariant complex (see Figure 6–65). Even in the case of protein complexes containing flexible scaffolds (see Figure 3–75), the macromolecular assembly has a characteristic (albeit flexible) conformation.

In contrast, **biomolecular condensates** are a different type of cellular structure built from proteins (and often RNA) held together by a large number of weak and constantly changing interactions among them. Each condensate is created by at least one scaffold macromolecule (a protein or RNA molecule) that is capable of making multiple independent interactions with either itself or with other macromolecules, which themselves often make multiple interactions. These types of macromolecules are said to be *multivalent*. Typically, each of the individual interactions among these multivalent proteins and RNAs is very weak, so it forms and breaks frequently. When any one interaction breaks, other interactions at different sites in that macromolecule prevent it from diffusing away and keep the macromolecule locally concentrated. By the time some of these other interactions break, new ones have already formed elsewhere. In this way, all of the proteins within a condensate continually interact with each other, even though the specific set of interactions changes from one moment to another.

Formation of a condensate serves to segregate and concentrate a subset of the cell's macromolecules into a separate compartment in the cell. In some cases, these macromolecules perform specialized biochemistry within the condensate—forming a biochemical "factory" that efficiently produces a specific product, as for the ribosomes that are produced by the nucleolus. In other cases, sequestration into a condensate can serve as a temporary storage depot for a set of macromolecules while blocking their activity, as for the stress granules that can form when a cell is perturbed.

The disordered, low-complexity domains of proteins are often found to mediate the fluctuating, weak binding interactions that form a condensate, frequently making a major contribution to their formation. In addition, other types of binding can also drive condensate formation (**Figure 3–77A**). The dynamic, fluctuating interactions within a condensate cause it to behave like a liquid: all of the participating molecules within it jostle around and rapidly exchange their relative positions; in addition, they often exchange rapidly with their equivalents outside the condensate (**Figure 3–77B**). Because the condensate remains intact and

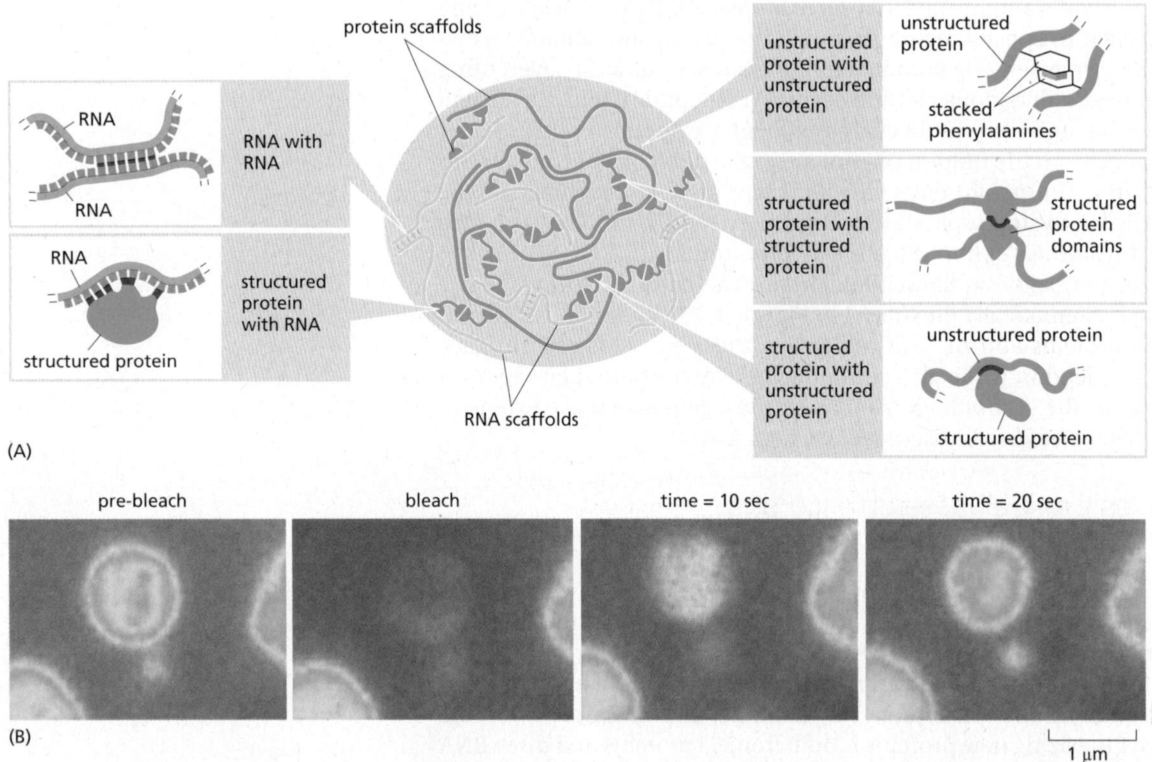

Figure 3–77 The multivalent interactions between scaffold macromolecules that drive the formation of biomolecular condensates. (A) Schematic diagram of a biomolecular condensate that contains both RNA and proteins; illustrated are some types of weak binding interactions frequently involved. Note that the low-complexity domains of scaffold proteins are often critical for forming these condensates, and that several different types of binding interactions are known to cause these unstructured regions to adhere to each other. In addition to the stacking of aromatic side chains (phenylalanine is shown), these include ionic attractions, cation–pi interactions, and the formation of kinked cross-beta structures that resemble amyloids. (B) A fluorescence recovery after photobleaching (FRAP) experiment reveals that the protein molecules inside a condensate are mobile. Here the multiple nucleoli in a mammalian cell have been fluorescently labeled by fusing GFP to the scaffold protein fibrillin, and this fibrillin in one of the nucleoli has been bleached with a flash from a focused laser beam. A rapid recovery of fluorescence demonstrates that the fibrillin in this condensate is continually exchanging with the fibrillin molecules in its surroundings. (B, from R.D. Phair and T. Misteli, *Nature* 404:604–609, 2000. Reproduced with permission of SNCSC.)

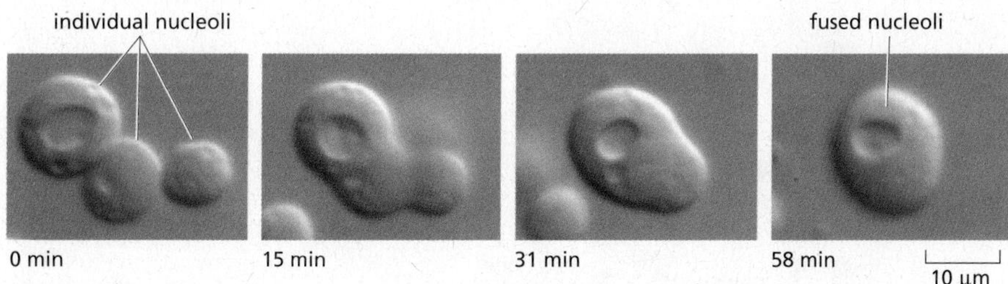

Figure 3–78 **Spherical, liquid-droplet-like nucleoli can be seen to fuse in the light microscope.** In these experiments, the nucleoli are present inside a nucleus that has been dissected from *Xenopus* oocytes and placed under oil on a microscope slide. Here, three nucleoli are seen fusing to form a larger biomolecular condensate. A very similar process occurs after each round of division, when small nucleoli initially form on multiple chromosomes but then coalesce to form a single, large nucleolus (see Figure 6–47). (From C.P. Brangwynne et al., *Proc. Natl. Acad. Sci. USA* 108:4334–4339, 2011. With permission from National Academy of Sciences.)

distinct from the surrounding liquid, the process of condensate formation is commonly termed **liquid–liquid phase separation** or liquid–liquid demixing.

A characteristic feature of biomolecular condensates that reflects their dynamic nature is the readily reversible assembly and disassembly of many of these structures. Thus, for example, the nucleolus disappears during mitosis, and it reforms in early interphase by fusion of the initially separate droplets that form on different chromosomes at the start of each interphase (**Figure 3–78**). Likewise, the DNA repair, DNA replication, and DNA transcription factories in the nucleus appear only where and when each of these processes occurs (**Figure 3–79**; see also Figure 6–51C).

Classical Studies of Phase Separation Have Relevance for Biomolecular Condensates

A familiar phase-separation process is that between oil and water, which occurs in some salad dressings. A phase separation occurs whenever forming two phases instead of one minimizes the free energy of a mixture, and it requires overcoming the large unfavorable free-energy change caused by the entropic cost of demixing. Thus, in the oil and water example, there are many more ways of distributing the small oil molecules in between water molecules than there are ways of condensing the oil molecules all together. The completely mixed state is by far the most probable, and the act of demixing therefore involves a large unfavorable (negative) entropy change that produces a large unfavorable (positive) change in the ΔG for phase separation (remembering that $\Delta G = \Delta H - T\Delta S$). But because of an even larger, favorable ΔG derived from preventing the oil molecules from disrupting the hydrogen-bonded network of water molecules, the oil and water separate into distinct phases (see Panel 2–2, pp. 96–97).

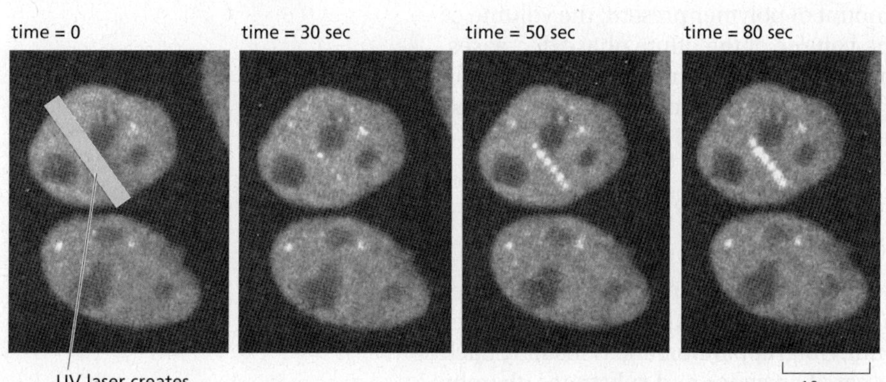

Figure 3–79 **The formation of a biomolecular condensate in response to DNA damage.** Here, a brief irradiation flash from a UV laser has been used to create a narrow line of DNA damage in the interphase nucleus of a mammalian cell. Because the FUS scaffold protein has been fluorescently labeled with GFP, the formation of the liquid-droplet-like DNA repair factories that this scaffold helps to generate can be followed in a living cell. (Adapted from Movie S1 in A. Patel et al., *Cell* 162:1066–1077, 2015. With permission from Elsevier.)

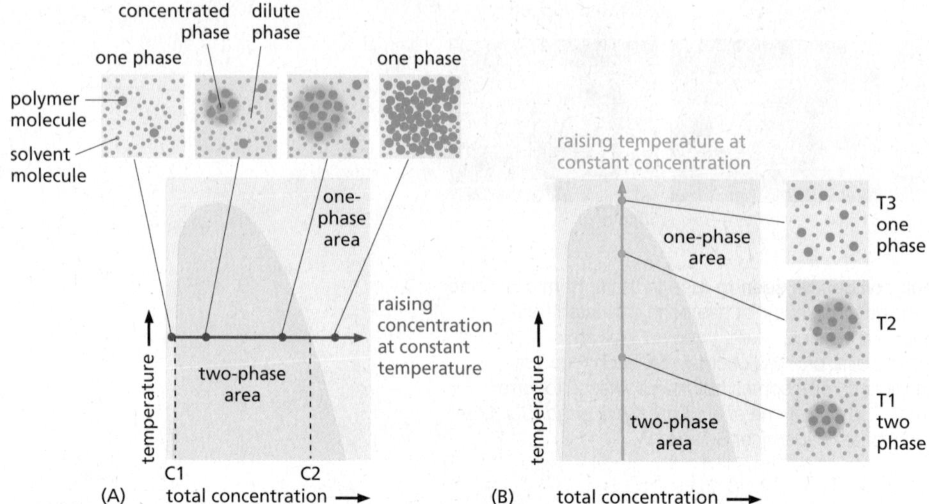

Figure 3–80 How phase diagrams are used to describe phase separations. (A) The effect of increasing the polymer concentration at constant temperature. At a low total concentration of a polymer (Ct), only a single dilute phase is observed. But as the polymer concentration is increased *(red arrow)*, phase separation begins for Ct > C1, and a new concentrated phase now forms with the polymer at concentration C2 in equilibrium with a dilute phase at polymer concentration C1. As Ct is further increased, the phase with polymer concentration C2 increases in volume, while remaining in equilibrium with the polymer in the dilute phase at an unchanging concentration C1. Finally, for Ct > C2, there is only a single phase with concentration Ct. In the example illustrated, C2 is more than 10-fold greater than C1. (B) The effect of increasing temperature (T) at a constant total polymer concentration. As the temperature is raised in a solution that contains a phase-separated polymer from T1 to T2 *(blue arrow)*, the concentration of polymer in its dilute phase (C1) increases and the concentration of polymer in its concentrated phase (C2) decreases. At a higher critical temperature C1 = C2, and the two phases become one. This occurs because the unfavorable entropy change for demixing (ΔS) makes an increasingly large, unfavorable contribution to the net free-energy change at higher temperatures (via the $-T\Delta S$ term in the equation for ΔG), eventually preventing any separation of phases.

For large polymers, which include proteins and nucleic acids, the entropic cost of demixing is considerably less than that for an equivalent mass of small molecules. This is because the monomeric subunits of a polymer are already greatly constrained in their possible arrangements through their covalent attachment to other subunits. As a result, a set of relatively weak attractions between the polymer molecules can often provide a large enough favorable free-energy change to drive phase separation—overcoming the unfavorable free-energy change of demixing.

Chemists have developed *phase diagrams* to describe what happens when chemically synthesized polymers phase-separate (**Figure 3–80**). As illustrated, when a threshold concentration of a polymer is reached, the solution separates into two distinct phases, one dilute and the other considerably more concentrated. The most important feature to notice is that, as more polymer is added at a fixed temperature (Figure 3–80A), its concentration in each phase remains the same. To accommodate the increased amount of polymer present, the volume of the concentrated phase increases and the volume of the dilute phase decreases. These and other features of phase separation are relevant when considering biomolecular condensates, even though the latter are generally composed of mixtures of more complex biological polymers (proteins and RNA molecules).

A Comparison of Three Important Types of Large Biological Assemblies

It has long been recognized that eukaryotic cells contain many membrane-enclosed compartments central to cell biology. These take the form of organelles such as the nucleus, endoplasmic reticulum, Golgi apparatus, and lysosome. Each such organelle concentrates a particular set of enzymes and substrates, thereby

TABLE 3–3 Macromolecular Machines Compared to Biomolecular Condensates and Membrane-enclosed Compartments

	Comparison of three types of large biochemical assemblies		
	Macromolecular machine	Biomolecular condensate	Membrane-enclosed compartment
Properties	Fixed macromolecular composition, with a defined stoichiometry and spatial organization of constituents Formed from a specific set of protein molecules or from protein and RNA molecules Assembles spontaneously and can form *de novo* Nevertheless, in many cases assembly is regulated to occur at specific sites, as needed	Dynamic, often liquidlike or gel-like organization, in which RNAs and low-complexity domains of proteins form specific, but transient, interactions Readily permeable to small molecules Larger than most macromolecular machines Macromolecule composition is selective, but stoichiometry is usually not fixed Can assemble *de novo* and be disassembled in response to changing conditions or cellular need	Creates a distinct chemical and protein environment that is maintained by active transport across the enclosing membrane Interior contains a variable stoichiometry of macromolecules in solution, as determined by the above transport processes Not permeable to most small molecules Formation usually requires a preexisting membrane-enclosed compartment of a special kind, different for each compartment
Examples	SCF ubiquitin ligase DNA replication protein machine Ribosome Nuclear pore	Nucleolus Centrosome Stress granule Neuronal RNA transport granule	Endoplasmic reticulum Mitochondrion Transport vesicle Lysosome

creating a specialized biochemistry in its interior. Those compartments will be the subject of Chapter 12, where we will also discuss biomolecular condensates in more detail. In **Table 3–3**, we compare the properties of the protein machines and biomolecular condensates introduced in this chapter, both with each other and with membrane-enclosed compartments.

Many Proteins Are Controlled by Covalent Modifications That Direct Them to Specific Sites Inside the Cell

In this chapter, we have thus far described only a few ways in which proteins are post-translationally modified. A large number of other such modifications also occur, more than 200 distinct types being known. To give a sense of the variety, **Table 3–4** presents a few of the modifying groups with known regulatory roles.

TABLE 3–4 Some Molecules Covalently Attached to Proteins That Regulate Protein Function

Modifying group	Some prominent functions
Phosphate on Ser, Thr, or Tyr	Drives the assembly of a protein into larger complexes (see Figure 15–11)
Methyl on Lys	Helps to create distinct regions in chromatin by forming either monomethyl, dimethyl, or trimethyl lysine in histones (see Figure 4–34)
Acetyl on Lys	Helps to activate genes in chromatin by modifying histones (see Figure 4–34)
Palmityl group on Cys	This fatty acid addition drives protein association with membranes (see Figure 10–18)
N-Acetylglucosamine on Ser or Thr	Controls enzyme activity and gene expression in glucose homeostasis
Ubiquitin on Lys	Monoubiquitin addition regulates the transport of membrane proteins in vesicles (see Figure 13–59)
	A polyubiquitin chain targets a protein for degradation (see Figure 3–66)
	(Ubiquitin is a 76-amino-acid polypeptide; there are at least 10 other ubiquitin-related proteins in mammalian cells.)

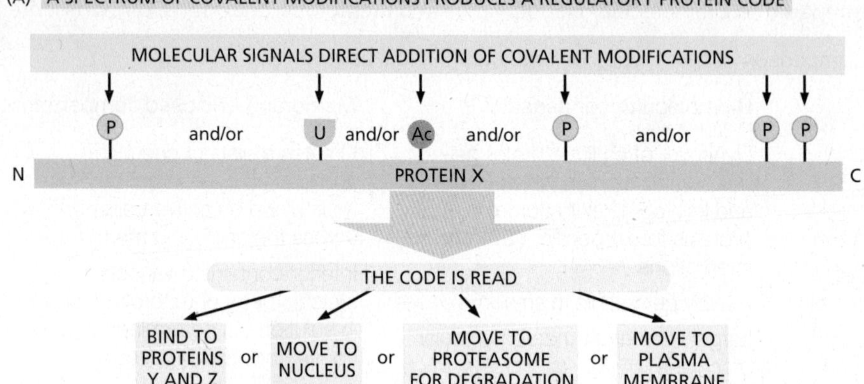

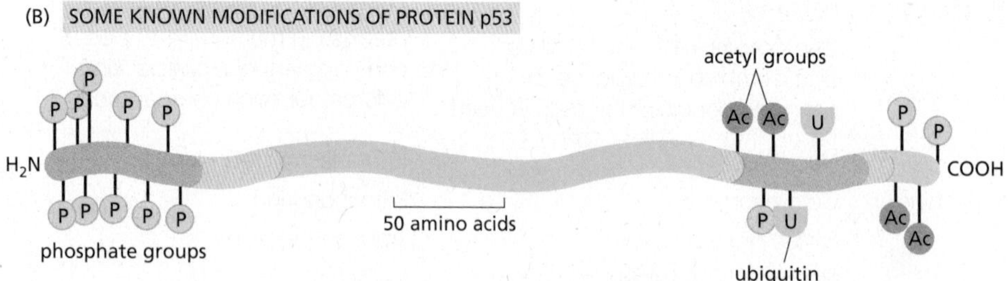

Figure 3–81 Multisite protein modification and its effects. (A) A protein that carries a post-translational addition to more than one of its amino acid side chains can be considered to carry a combinatorial regulatory code. Multisite modifications are added to (and removed from) a protein through signaling networks, and the resulting combinatorial regulatory code on the protein is read to alter its behavior in the cell. (B) The pattern of some covalent modifications to the protein p53.

Like the phosphate and ubiquitin additions described previously, these groups are added and then removed from proteins according to the needs of the cell.

A large number of proteins are modified on more than one amino acid side chain, with different regulatory events producing a different pattern of such modifications. A striking example is the protein p53, which plays a central part in controlling a cell's response to adverse circumstances (see Figure 17–60). Through one of four different types of molecular additions, this protein can be modified at 20 different sites. Because an enormous number of different combinations of these 20 modifications are possible, the protein's behavior can in principle be altered in a huge number of ways. Such modifications will often create a site on the modified protein that binds it to a scaffold protein in a specific region of the cell, thereby connecting it—via the scaffold—to the other proteins required for a reaction at that site. The effects can include moving the modified protein either into or out of a specific biomolecular condensate.

One can view each protein's set of covalent modifications as a *combinatorial regulatory code.* Specific modifying groups are added to or removed from a protein in response to signals, and the code then alters protein behavior—changing the activity or stability of the protein, its binding partners, and/or its specific location within the cell (**Figure 3–81**). As a result, the cell is able to respond rapidly and with great versatility to changes in its condition or environment.

A Complex Network of Protein Interactions Underlies Cell Function

There are many challenges facing cell biologists in this information-rich era when a huge number of complete genome sequences are known. One is the need to dissect each one of the thousands of protein machines that exist in an organism such as ourselves. To understand these remarkable protein complexes, each will need to be reconstituted from its purified protein parts, so that we can study its detailed mode of operation under controlled conditions in a test tube, free from all other cell components. This alone is a massive task. But we now know that each of these subcomponents of a cell also interacts with other sets of macromolecules, creating a large network of protein-protein and protein-nucleic acid interactions throughout the cell. To understand the cell, therefore, we will need to analyze most of these other interactions as well.

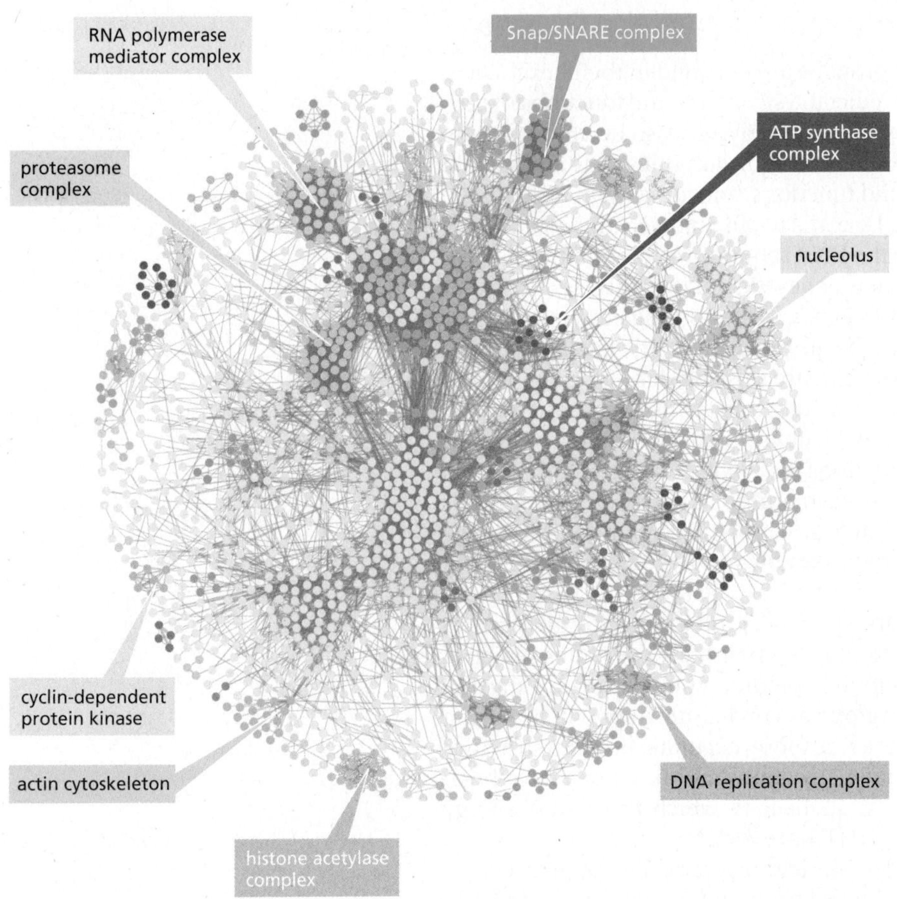

Figure 3–82 **A network of protein-binding interactions in the cells of the fruit fly, *Drosophila*.** Each line connecting a pair of dots (proteins) indicates a protein–protein interaction. Labels are used to denote a few of the highly interactive groups of proteins whose functions are described in this textbook. (From K.G. Guruharsha et al., *Cell* 147:690–703, 2011. With permission from Elsevier.)

We can begin to gain a sense of the nature of intracellular protein networks from a particularly well-studied example described in Chapter 16: the many dozens of proteins in the actin cytoskeleton that interact to control actin filament behavior (see Panel 16–3, p. 965). Biochemists and structural biologists are, in principle, able to purify all of these different actin-accessory proteins to study their effects on actin filaments individually and in combination, and to determine all of their protein–protein interactions and their atomic structures. But to truly understand the actin cytoskeleton will require that we also learn how to use this data to compute how any particular mixture of these components present in an individual cell creates that cell's observed set of three-dimensional networks of actin structures—a goal that currently seems out of reach.

Of course, understanding the cell will require much more than understanding actin. In recent years, as described in Chapter 8, robotics has been harnessed to a set of powerful technologies to produce enormous protein interaction maps (**Figure 3–82**). The data obtained suggest that each of the roughly 10,000 different proteins in a human cell interacts with 5–10 different partners, illustrating the challenges that face scientists working to understand the complexity of cell chemistry.

What does the future hold? Despite the enormous progress made in recent years, we cannot yet claim to understand even the simplest known cells, such as the small *Mycoplasma* bacterium formed from only about 500 gene products (see Figure 1–8). How then can we hope to understand a human? Clearly, a great deal of new biochemistry will be essential, in which each protein in a particular interacting set is purified so that its chemistry and interactions can be dissected in a test tube. But in addition, more powerful ways of analyzing networks will be needed using mathematical and computational tools not yet invented. Clearly, there are many wonderful challenges that remain for future generations of cell biologists.

Protein Structures Can Be Predicted and New Proteins Designed

Because the structures and functions of proteins are encoded in their amino acid sequences, in principle it is possible to predict the structures and functions of proteins directly from their amino acid sequences. We should also be able to create proteins with entirely new structures and functions by designing new amino acid sequences to produce these structures and functions, encoding them in synthetic genes. Success in the first endeavor would transform our ability to understand how the biology of an organism is encoded in the DNA sequence of its genome. Success in the second endeavor could lead to a new generation of designed proteins that address some of the twenty-first-century challenges confronting humanity.

There are major challenges in both of the above areas. A first challenge is the very large number of potential structures that are possible for any given amino acid sequence. Because, as we have seen, a protein folds to its lowest free-energy state, one needs to use physics to compute the energy of each protein conformation. But the number of possible conformations for even a relatively short protein of 100 amino acids is of the order 3^{100}, as each amino acid has on average 3 or more rotatable bonds. Success in predicting protein structure and in designing new proteins thus requires computational methods for very efficient searching through huge numbers of structures.

Progress has been made in recent years. For small proteins or for proteins from very large families to help constrain the problem, large-scale computer searches for the lowest energy state can often accurately predict protein structure starting from amino acid sequence. Recently developed deep learning approaches using artificial intelligence (AI) can produce even more accurate protein structure predictions. Conversely, many new protein structures and functions have been created from scratch by designing new sequences in which the lowest energy state has the desired structure and function (**Figure 3–83**).

While this progress suggests that the protein-folding problem is not intractable, huge challenges remain. Predicting function from structure is even more difficult: while in some cases function can be predicted from structure by analogy to other proteins with similar structures and already known functions, this can be problematic because even a few amino acid changes can considerably change function; for example, the identity of the substrate that an enzyme acts upon. On the design side, while it has been possible to design new proteins with new structures and binding

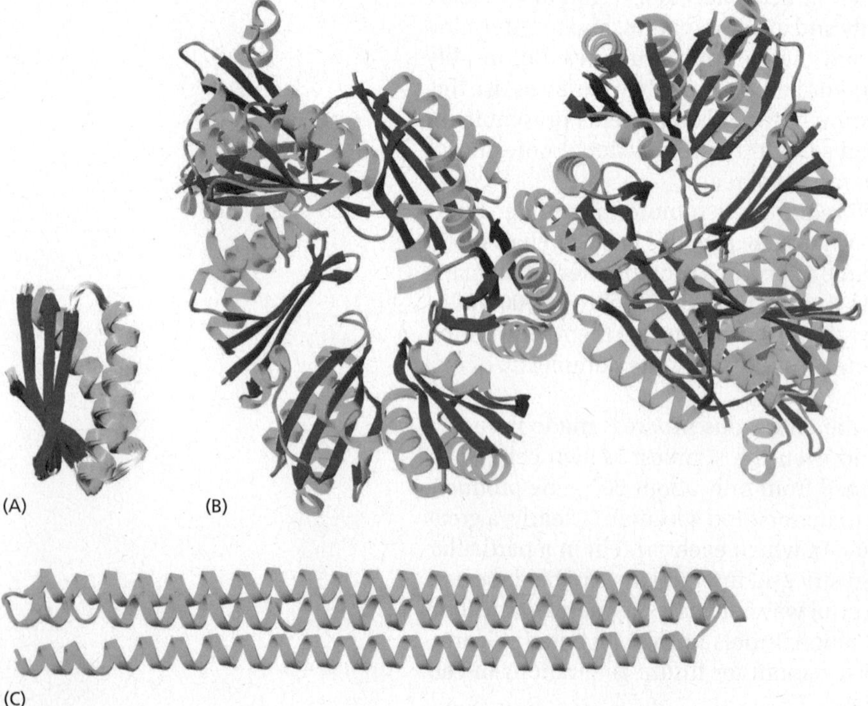

Figure 3–83 **Some examples of designed proteins.** The amino acid sequences of the three proteins illustrated were determined computationally, being selected so that each protein would adopt a specifically designed three-dimensional conformation in its lowest energy state. After each protein was produced in a bacterium using genetic engineering techniques, its actual structure was then determined and shown to be the same as that intended by the designer. (A) A small protein of 122 amino acids. (B) A protein that creates an octahedral shell formed from 24 identical subunits, only 8 of which are shown. (C) A protein that consists of an antiparallel three-helix bundle. (A, PDB code 2N76; B, PDB code 3VCD; C, PDB code 4TQL).

activities, it remains a big challenge to match the remarkable activities of natural enzymes and the sophisticated information integration and force generation of natural molecular machines.

Summary

The function of a protein largely depends on the detailed chemical properties of its surface. Enzymes are catalytic proteins that greatly accelerate the rates of covalent bond making and breaking. They do this by binding the high-energy transition state for a specific reaction path, lowering that reaction's activation energy. The rates of enzyme-catalyzed reactions are often so fast that they are limited only by diffusion.

Proteins can reversibly change their shape when ligands bind to their surface. The allosteric changes in protein conformation produced by one ligand affect the binding of a second ligand, and this linkage between two ligand-binding sites provides a crucial mechanism for regulating cell processes. Metabolic pathways, for example, are controlled by feedback regulation: some small molecules inhibit and other small molecules activate enzymes early in a pathway. Enzymes controlled in this way generally form symmetrical assemblies, allowing cooperative conformational changes to create a steep response to changes in the concentrations of the ligands that regulate them.

The expenditure of chemical energy can drive unidirectional changes in protein shape. By coupling allosteric shape changes to the hydrolysis of a tightly bound ATP molecule, for example, proteins can do useful work, such as generating a mechanical force or moving for long distances in a single direction. The three-dimensional structures of proteins have revealed how a small local change caused by nucleoside triphosphate hydrolysis is amplified to create major changes elsewhere in the protein. Highly efficient protein machines are formed by incorporating many different protein molecules into larger assemblies that coordinate the allosteric movements of the individual components. Machines of this type perform most of the important reactions in cells. They and other specific macromolecules can be brought together in large, liquid-like assemblies known as biomolecular condensates, which are created by weak, fluctuating interactions between multivalent protein and RNA scaffolds.

Proteins are subjected to many reversible, post-translational modifications, such as the covalent addition of a phosphate or an acetyl group to a specific amino acid side chain. The addition of these modifying groups is used to regulate the activity of a protein, changing its conformation, its binding to other proteins, and its location inside the cell. A typical protein in a cell will interact with more than five different partners. Understanding the large protein networks inside cells will require biochemistry, through which small sets of interacting proteins can be purified and their chemistry dissected in detail. In addition, new computational approaches will be required to make sense of the enormous complexity of these networks.

PROBLEMS

Which statements are true? Explain why or why not.

3–1 Each strand in a β sheet is a helix with two amino acids per turn.

3–2 Loops of polypeptide that protrude from the surface of a protein often form the binding sites for other molecules.

3–3 An enzyme reaches a maximum rate at high substrate concentration because it has a fixed number of active sites where substrate binds.

3–4 Higher concentrations of enzyme give rise to a higher turnover number for that enzyme.

3–5 Enzymes that undergo cooperative allosteric transitions invariably consist of symmetrical assemblies of multiple subunits.

3–6 Continual addition and removal of phosphates by protein kinases and protein phosphatases is wasteful of energy—because their combined action consumes ATP—but it is a necessary consequence of effective regulation by phosphorylation.

Discuss the following problems.

3–7 Titin, which has a mass of about 3×10^6 daltons, is the largest polypeptide yet described. Titin molecules extend from muscle thick filaments to the Z disc; they are thought to act as springs to keep the thick filaments centered in the sarcomere. Titin is composed of a large number of repeated immunoglobulin (Ig) sequences of 89 amino acids, each of which is folded into a domain about 4 nm in length (**Figure Q3–1A**).

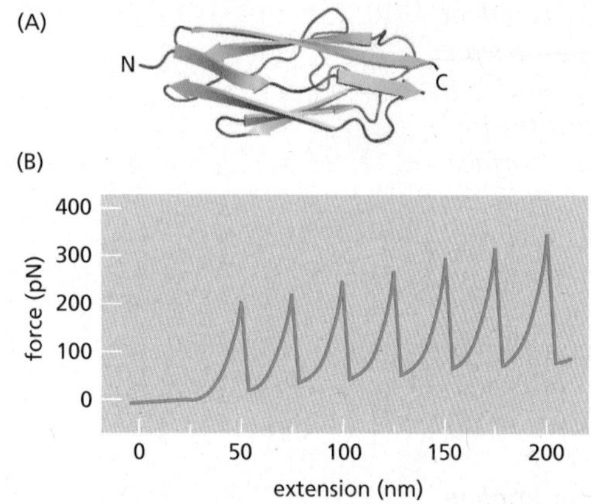

Figure Q3–1 Springlike behavior of titin (Problem 3–7). (A) The structure of an individual Ig domain. (B) Force in piconewtons versus extension in nanometers obtained by atomic force microscopy.

You suspect that the springlike behavior of titin is caused by the sequential unfolding (and refolding) of individual Ig domains. You test this hypothesis using an atomic force microscope, which allows you to pick up one end of a protein molecule and pull with an accurately measured force. For a fragment of titin containing seven repeats of the Ig domain, this experiment gives the sawtooth force-versus-extension curve shown in **Figure Q3–1B**. If the experiment is repeated in a solution of 8 M urea (a protein denaturant), the peaks disappear and the measured extension becomes much longer for a given force. If the experiment is repeated after the protein has been cross-linked by treatment with glutaraldehyde, once again the peaks disappear but the extension becomes much smaller for a given force.

A. Are the data consistent with your hypothesis that titin's springlike behavior is due to the sequential unfolding of individual Ig domains? Explain your reasoning.

B. Is the extension for each putative domain-unfolding event the magnitude you would expect? (In an extended polypeptide chain, amino acids are spaced at intervals of 0.34 nm.)

C. Why does the force collapse so abruptly after each peak?

3–8 Consider the following statement. "To produce one molecule of each possible kind of polypeptide chain, 300 amino acids in length, would require more atoms than exist in the universe." Given the size of the universe, do you suppose this statement could possibly be correct? Because counting atoms is a tricky business, consider the problem from the standpoint of mass. The observable universe is estimated to contain 10^{80} protons plus neutrons, which have a total mass of about 1.7×10^{56} grams. Assuming that the average mass of an amino acid is 110 daltons, what would be the total mass of one molecule of each possible kind of polypeptide chain 300 amino acids in length? Is this greater than the mass of the universe?

3–9 The so-called kelch motif consists of a four-stranded β sheet shaped like the blade of a propeller. It is usually found to be repeated four to seven times, forming a β propeller, or kelch repeat domain, in a multidomain protein. One such kelch repeat domain is shown in **Figure Q3–2**. Would you classify this domain as an *in-line* or *plug-in* type domain?

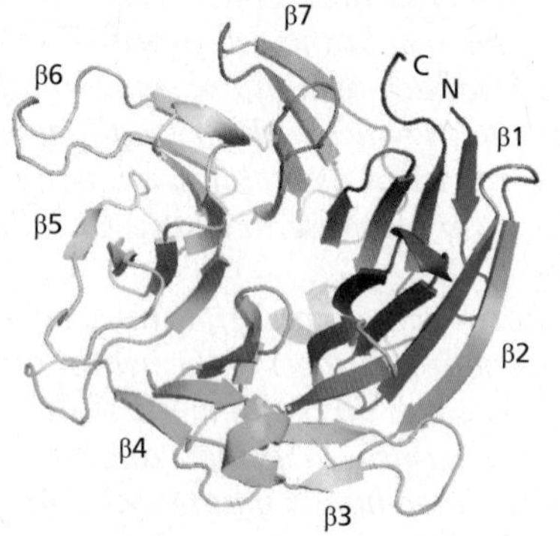

Figure Q3–2 The kelch repeat domain of galactose oxidase from *D. dendroides* (Problem 3–9). The seven individual blades of the β propeller are *color coded* and labeled. The N- and C-termini are indicated by N and C.

3–10 In principle, dimers of identical proteins could be arranged either "head-to-tail" (same as "tail-to-head") or "head-to-head" (equivalent to "tail-to-tail"), as illustrated schematically in **Figure Q3–3**. Do you suppose that one type of dimer is significantly more common than the other? Why or why not?

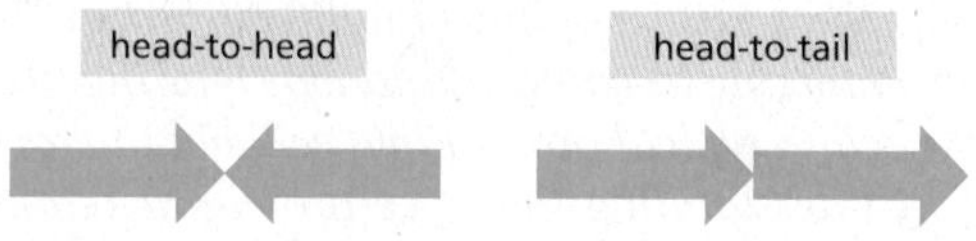

Figure Q3–3 Head-to-head and tail-to-tail dimers (Problem 3–10).

3–11 An antibody binds to another protein with an equilibrium constant, *K*, of 5×10^9 M^{-1}. When it binds to a second, related protein, it forms three fewer hydrogen bonds, reducing its binding affinity by 11.9 kJ/mole. What is the *K* for its binding to the second protein? [Free-energy change is related to the equilibrium constant by the equation $\Delta G° = -2.3\ RT \log K$, where *R* is 8.3×10^{-3} kJ/(mole K) and *T* is 310 K.]

3–12 In bacteria, the protein SmpB binds to a special species of tRNA, tmRNA, to eliminate the incomplete proteins made from truncated mRNAs. If the binding of SmpB to tmRNA is plotted as fraction tmRNA bound versus SmpB concentration, one obtains a symmetrical S-shaped curve as shown in **Figure Q3–4**. This curve is a visual display of a very useful relationship between K_d and concentration, which has broad applicability. The general expression for fraction of ligand (L) bound to a protein (Pr) is derived from the equation for K_d (K_d = [Pr][L]/[Pr–L]) by substituting ([L]$_{TOT}$ – [L]) for [Pr–L] and rearranging. Because the total concentration of ligand ($[L]_{TOT}$) is equal to the free ligand ([L]) plus bound ligand ([Pr–L]),

$$\text{fraction bound} = [\text{Pr–L}]/[\text{L}]_{TOT} = [\text{Pr}]/([\text{Pr}] + K_d)$$

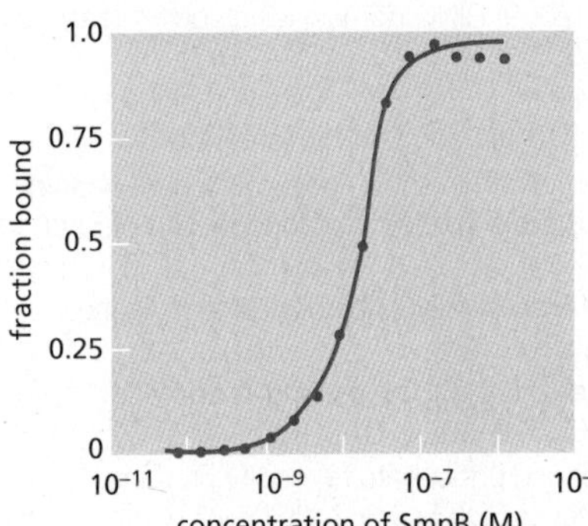

Figure Q3–4 Fraction of tmRNA bound versus SmpB concentration (Problem 3–12). (From A.W. Karzai, M.M. Susskind, and R.T. Sauer, *EMBO J* 18:3793–3799, 1999. With permission from European Molecular Biology Organization.)

For SmpB and tmRNA, the fraction bound = [SmpB–tmRNA]/$[\text{tmRNA}]_{TOT}$ = [SmpB]/([SmpB] + K_d). Using this relationship, calculate the fraction of tmRNA bound for SmpB concentrations equal to $10^4\,K_d$, $10^3\,K_d$, $10^2\,K_d$, $10^1\,K_d$, K_d, $10^{-1}\,K_d$, $10^{-2}\,K_d$, $10^{-3}\,K_d$, and $10^{-4}\,K_d$.

3–13 The enzyme hexokinase adds a phosphate to D-glucose but ignores its mirror image, L-glucose. Suppose that you were able to synthesize hexokinase entirely from D-amino acids, which are the mirror image of the normal L-amino acids.

A. Assuming that the "D" enzyme would fold to a stable conformation, what relationship would you expect it to have to the normal "L" enzyme?

B. Do you suppose the "D" enzyme would add a phosphate to L-glucose and ignore D-glucose?

3–14 Many enzymes obey simple Michaelis–Menten kinetics, which are summarized by the equation

$$\text{rate} = V_{max}[S]/([S] + K_m)$$

where V_{max} = maximum velocity, [S] = concentration of substrate, and K_m = the Michaelis constant.

It is instructive to plug a few values of [S] into the equation to see how rate is affected. What are the rates for [S] equal to zero, equal to K_m, and equal to infinite concentration?

3–15 Synthesis of the purine nucleotides AMP and GMP proceeds by a branched pathway starting with ribose 5-phosphate (R5P), as shown schematically in **Figure Q3–5**. Using the principles of feedback inhibition, propose a regulatory strategy for this pathway that ensures an adequate supply of both AMP and GMP and minimizes the buildup of the intermediates (*A–I*) when supplies of AMP and GMP are adequate.

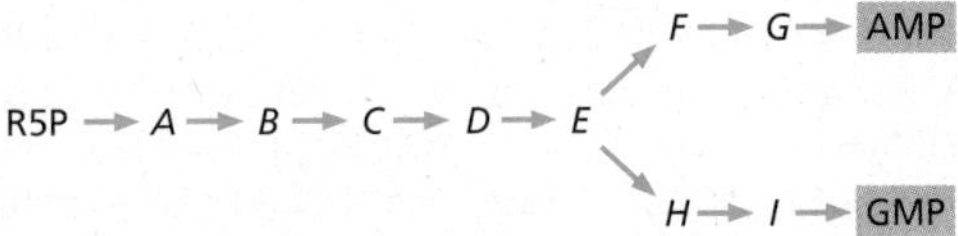

Figure Q3–5 Schematic diagram of the metabolic pathway for synthesis of AMP and GMP from R5P (Problem 3–15).

3–16 How do you suppose that a molecule of hemoglobin is able to bind oxygen efficiently in the lungs and yet release it efficiently in the tissues?

3–17 Rous sarcoma virus (RSV) carries an oncogene called *Src*, which encodes a continually active protein tyrosine kinase that leads to unchecked cell proliferation. Normally, Src carries an attached fatty acid (myristoylate) group that allows it to bind to the cytoplasmic side of the plasma membrane. A mutant version of Src that does not allow attachment of myristoylate does not bind to the membrane. Infection of cells with RSV encoding either the normal or the mutant form of Src leads to the same high level of protein tyrosine kinase activity, but the mutant Src does not cause cell proliferation.

A. Assuming that the normal Src is all bound to the plasma membrane and that the mutant Src is distributed throughout the cytoplasm, calculate their relative concentrations in the neighborhood of the plasma membrane. For the purposes of this calculation, assume that the cell is a sphere with a radius (r) of 10 μm and that the mutant Src is distributed throughout the cell, whereas the normal Src is confined to a 4-nm-thick layer immediately beneath the membrane. [For this problem, assume that the membrane has no thickness. The volume of a sphere is $(4/3)\pi r^3$.]

B. The target (X) for phosphorylation by Src resides in the membrane. Explain why the mutant Src does not cause cell proliferation.

REFERENCES

General

Berg JM, Tymoczko JL, GJ Gatto & Stryer L (2021) Biochemistry, 9th ed. New York: W.H. Freeman.

Branden C & Tooze J (1999) Introduction to Protein Structure, 2nd ed. New York: Garland Science.

Dickerson RE (2005) Present at the Flood: How Structural Molecular Biology Came About. Sunderland, MA: Sinauer.

Kuriyan J, Konforti B & Wemmer D (2013) The Molecules of Life: Physical and Chemical Principles. New York: Norton.

Petsko GA & Ringe D (2004) Protein Structure and Function. London: New Science Press.

Steven AC, Baumeister W, Johnson LN & Perham RN (2016) Molecular Biology of Assemblies and Machines. New York: Garland Science.

The Atomic Structure of Proteins

Anfinsen CB (1973) Principles that govern the folding of protein chains. *Science* 181, 223–230.

Caspar DLD & Klug A (1962) Physical principles in the construction of regular viruses. *Cold Spring Harb. Symp. Quant. Biol.* 27, 1–24.

Dill KA & MacCallum JL (2012) The protein-folding problem, 50 years on. *Science* 338, 1042–1046.

Eisenberg D (2003) The discovery of the alpha-helix and beta-sheet, the principal structural features of proteins. *Proc. Natl. Acad. Sci. USA* 100, 11207–11210.

Fraenkel-Conrat H & Williams RC (1955) Reconstitution of active tobacco mosaic virus from its inactive protein and nucleic acid components. *Proc. Natl. Acad. Sci. USA* 41, 690–698.

Goodsell DS & Olson AJ (2000) Structural symmetry and protein function. *Annu. Rev. Biophys. Biomol. Struct.* 29, 105–153.

Greenwald J & Riek R (2010) Biology of amyloid: structure, function, and regulation. *Structure* 18, 1244–1260.

Harrison SC (1992) Viruses. *Curr. Opin. Struct. Biol.* 2, 293–299.

Hudder A, Nathanson L & Deutscher MP (2003) Organization of mammalian cytoplasm. *Mol. Cell. Biol.* 23, 9318–9326.

Ke PC, Zhou R, Serpell LC . . . Mezzenga R (2020) Half a century of amyloids: past, present and future. *Chem. Soc. Rev*. 49(15), 5473–5509.

Li P, Banjade S, Cheng H-C . . . Rosen MK (2012) Phase transitions in the assembly of multivalent signalling proteins. *Nature* 483, 336–340.

Lindquist SL & Kelly JW (2011) Chemical and biological approaches for adapting proteostasis to ameliorate protein misfolding and aggregation diseases—progress and prognosis. *Cold Spring Harb. Perspect. Biol.* 3, a004507.

Maji SK, Perrin MH, Sawaya MR . . . Riek R (2009) Functional amyloids as natural storage of peptide hormones in pituitary secretory granules. *Science* 325, 328–332.

Nelson R, Sawaya MR, Balbirnie M . . . Eisenberg D (2005) Structure of the cross-beta spine of amyloid-like fibrils. *Nature* 435, 773–778.

Nomura M (1973) Assembly of bacterial ribosomes. *Science* 179, 864–873.

Oldfield CJ & Dunker AK (2014) Intrinsically disordered proteins and intrinsically disordered protein regions. *Annu. Rev. Biochem.* 83, 553–584.

Orengo CA & Thornton JM (2005) Protein families and their evolution—a structural perspective. *Annu. Rev. Biochem.* 74, 867–900.

Pauling L & Corey RB (1951) Configurations of polypeptide chains with favored orientations around single bonds: two new pleated sheets. *Proc. Natl. Acad. Sci. USA* 37, 729–740.

Pauling L, Corey RB & Branson HR (1951) The structure of proteins: two hydrogen-bonded helical configurations of the polypeptide chain. *Proc. Natl. Acad. Sci. USA* 37, 205–211.

Prusiner SB (1998) Prions. *Proc. Natl. Acad. Sci. USA* 95, 13363–13383.

Toyama BH & Weissman JS (2011) Amyloid structure: conformational diversity and consequences. *Annu. Rev. Biochem.* 80, 557–585.

Weiss MC, Sousa FL, Mrnjavac N . . . Martin WF (2006) The physiology and habitat of the last universal common ancestor. *Nat. Microbiol.* 1, 16116.

Wright PE & Dyson HJ (2015) Intrinsically disordered proteins in cellular signalling and regulation. *Nat. Rev. Mol. Cell Biol*. 16(1), 18–29.

Protein Function

Alberts B (1998) The cell as a collection of protein machines: preparing the next generation of molecular biologists. *Cell* 92, 291–294.

Bahar I, Jernigan RL & Dill KA (2017) Protein Actions: Principles and Modeling. New York: Garland Science.

Banani SF, Lee HO, Hyman AA, & Rosen MK (2017) Biomolecular condensates: organizers of cellular biochemistry. *Nat. Rev. Mol. Cell Biol.* 18, 285–298.

Benkovic SJ (1992) Catalytic antibodies. *Annu. Rev. Biochem.* 61, 29–54.

Berg OG & von Hippel PH (1985) Diffusion-controlled macromolecular interactions. *Annu. Rev. Biophys. Biophys. Chem.* 14, 131–160.

Bergeron-Sandoval LP, Safaee N & Michnick SW (2016) Mechanisms and consequences of macromolecular phase separation. *Cell* 165, 1067–1079.

Bourne HR (1995) GTPases: a family of molecular switches and clocks. *Philos. Trans. R. Soc. Lond. B Biol. Sci.* 349, 283–289.

Case LB, Ditlev JA & Rosen MK (2019) Regulation of transmembrane signaling by phase separation. *Annu. Rev. Biophys.* 48, 465–494.

Costanzo M, Baryshnikova A, Bellay J . . . Boone C (2010) The genetic landscape of a cell. *Science* 327, 425–431.

Deshaies RJ & Joazeiro CAP (2009) RING domain E3 ubiquitin ligases. *Annu. Rev. Biochem.* 78, 399–434.

Dickerson RE & Geis I (1983) Hemoglobin: Structure, Function, Evolution, and Pathology. Menlo Park, CA: Benjamin Cummings.

Fersht AR (2017) Structure and Mechanism in Protein Science: A Guide to Enzyme Catalysis and Protein Folding. Cambridge, UK: Kaissa Publications.

Hua Z & Vierstra RD (2011) The cullin-RING ubiquitin-protein ligases. *Annu. Rev. Plant Biol.* 62, 299–334.

Huang PS, Boyken SE & Baker D (2016) The coming of age of de novo protein design. *Nature* 537, 320–327.

Hunter T (2012) Why nature chose phosphate to modify proteins. *Philos. Trans. R. Soc. Lond. B Biol. Sci.* 367, 2513–2516.

Hyman AA, Weber CA & Jülicher F (2014) Liquid–liquid phase separation in biology. *Annu. Rev. Cell Dev. Biol.* 30, 39–58.

Johnson LN & Lewis RJ (2001) Structural basis for control by phosphorylation. *Chem. Rev.* 101, 2209–2242.

Kantrowitz ER & Lipscomb WN (1988) *Escherichia coli* aspartate transcarbamylase: the relation between structure and function. *Science* 241, 669–674.

Kerscher O, Felberbaum R & Hochstrasser M (2006) Modification of proteins by ubiquitin and ubiquitin-like proteins. *Annu. Rev. Cell Dev. Biol.* 22, 159–180.

Koshland DE, Jr (1984) Control of enzyme activity and metabolic pathways. *Trends Biochem. Sci.* 9, 155–159.

Krogan NJ, Cagney G, Yu H . . . Greenblatt JF (2006) Global landscape of protein complexes in the yeast *Saccharomyces cerevisiae*. *Nature* 440, 637–643.

Lichtarge O, Bourne HR & Cohen FE (1996) An evolutionary trace method defines binding surfaces common to protein families. *J. Mol. Biol.* 257, 342–358.

Lu J, Cao Q, Hughes MP . . . Eisenberg DS (2020) CryoEM structure of the low-complexity domain of hnRNPA2 and its conversion to pathogenic amyloid. *Nat. Commun.* 11, 4090.

Maier T, Leibundgut M & Ban N (2008) The crystal structure of a mammalian fatty acid synthase. *Science* 321, 1315–1322.

Monod J, Changeux JP & Jacob F (1963) Allosteric proteins and cellular control systems. *J. Mol. Biol.* 6, 306–329.

Perutz M (1990) Mechanisms of Cooperativity and Allosteric Regulation in Proteins. Cambridge, UK: Cambridge University Press.

Radzicka A & Wolfenden R (1995) A proficient enzyme. *Science* 267, 90–93.

Riback JA, Zhu L, Ferrolino MC . . . Brangwynne CP (2020) Composition-dependent thermodynamics of intracellular phase separation. *Nature* 581, 209–214.

Schramm VL (2011) Enzymatic transition states, transition-state analogs, dynamics, thermodynamics, and lifetimes. *Annu. Rev. Biochem.* 80, 703–732.

Scott JD & Pawson T (2009) Cell signaling in space and time: where proteins come together and when they're apart. *Science* 326, 1220–1224.

Taylor SS, Keshwani MM, Steichen JM & Kornev AP (2012) Evolution of the eukaryotic protein kinases as dynamic molecular switches. *Philos. Trans. R. Soc. Lond. B Biol. Sci.* 367, 2517–2528.

Tunyasuvunakool K, Adler J, Wu Z . . . Hassabis D (2021) Highly accurate protein structure prediction for the human proteome. *Nature* 596, 590–596.

Vale RD & Milligan RA (2000) The way things move: looking under the hood of molecular motor proteins. *Science* 288, 88–95.

BASIC GENETIC MECHANISMS

CHAPTER 4

DNA, Chromosomes, and Genomes

Life depends on the ability of cells to store, retrieve, and translate the genetic instructions required to make and maintain a living organism. This *hereditary information* is passed on from a cell to its daughter cells at cell division and from one generation of an organism to the next through the organism's reproductive cells. The instructions are stored in every living cell's genes, the information-containing units in its genome that determine the characteristics of a species as a whole and of the individuals within it.

As soon as genetics emerged as a science at the beginning of the twentieth century, scientists became intrigued by the nature of the hereditary information. They knew that it was copied and transmitted from cell to daughter cell millions of times to produce the many generations of a multicellular organism, and that it survives the process essentially unchanged. What form of molecule could be capable of such accurate and almost unlimited replication and also be able to exert precise control, directing multicellular development as well as the daily life of every cell? What kind of instructions does the hereditary information contain? And how can the enormous amount of information required for the development and maintenance of an organism fit within the tiny space of a cell?

The answers to some of these questions began to emerge in the 1940s. At this time researchers discovered, from studies in simple fungi, that genes consist largely of instructions for making proteins. Proteins are phenomenally versatile macromolecules that perform most cell functions. As we saw in Chapter 3, they serve as building blocks for cell structures and form the enzymes that catalyze most of the cell's chemical reactions. They also regulate gene expression (Chapter 7), and they enable cells to communicate with each other (Chapter 15) and to move (Chapter 16). The properties and functions of cells and organisms are determined to a great extent by the proteins that they are able to make, and these are determined by their genes.

Painstaking observations of cells and embryos in the late nineteenth century set the stage for experiments that led to the realization that the hereditary information is carried on *chromosomes*—threadlike structures in the nucleus of a eukaryotic cell that become visible by light microscopy as the cell begins to

IN THIS CHAPTER

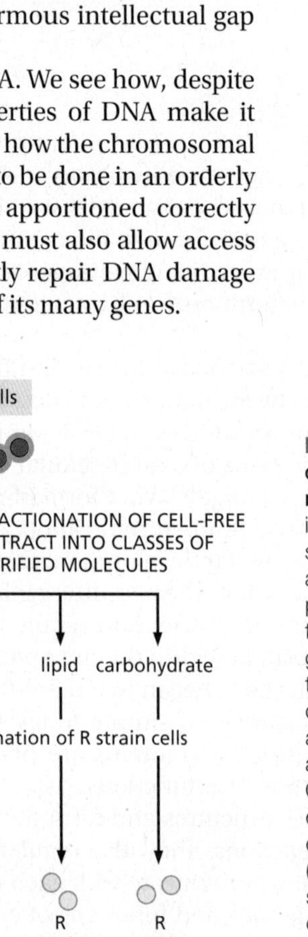

Figure 4–1 Chromosomes in cells. (A) Two adjacent plant cells photographed through a light microscope. The DNA has been stained with a fluorescent dye (DAPI) that binds to it. The DNA is present in chromosomes, which become visible as distinct structures in the light microscope only when they become compact, sausage-shaped structures just before cell division. This can be easily seen in the cell on the *left*, where for clarity a single chromosome has been shaded *dark brown*. The cell on the *right* contains identical chromosomes, but they cannot be clearly distinguished at this phase in the cell's life cycle, because they are in a more extended conformation. (B) Schematic diagram of the outlines of the two cells along with their chromosomes. (A, courtesy of Peter Shaw.)

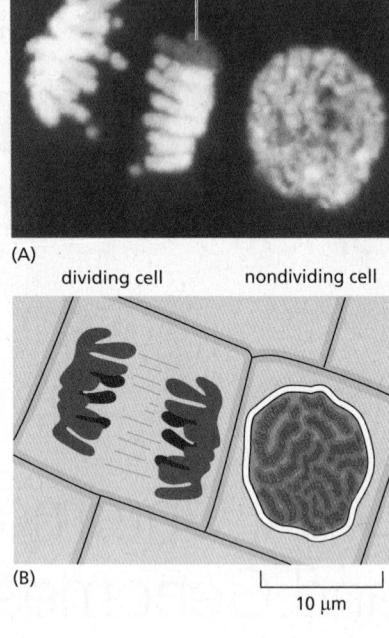

divide (**Figure 4–1**). Later, when biochemical analysis became possible, chromosomes were found to consist of deoxyribonucleic acid (DNA) and protein, with both being present in roughly the same amounts. For many decades, the DNA was thought to be merely a structural element. However, the other crucial advance made in the 1940s was the identification of DNA as the likely carrier of hereditary information. This breakthrough in our understanding of cells came from studies of inheritance in bacteria (**Figure 4–2**). But still, as the 1950s began, both how proteins could be specified by instructions in the DNA and how this information could be copied for transmission from cell to cell seemed completely mysterious. The puzzle was suddenly solved in 1953, when James Watson and Francis Crick derived the answers from their model for the structure of a DNA molecule. As outlined in Chapter 1, the determination of the double-helical structure of DNA immediately solved the problem of how the information in this molecule might be copied, or *replicated*. It also provided the first clues as to how a molecule of DNA might use the sequence of its subunits to encode the instructions for making proteins. Today, the fact that DNA is the genetic material is so fundamental to biological thought that it is difficult to appreciate the enormous intellectual gap that was filled by this sudden breakthrough discovery.

We begin this chapter by describing the structure of DNA. We see how, despite its chemical simplicity, the structure and chemical properties of DNA make it ideally suited as the raw material of genes. We then consider how the chromosomal proteins arrange and package this DNA. The packaging has to be done in an orderly fashion so that the chromosomes can be replicated and apportioned correctly between the two daughter cells at each cell division. And it must also allow access to chromosomal DNA, both for the enzymes that constantly repair DNA damage and for the specialized proteins that direct the expression of its many genes.

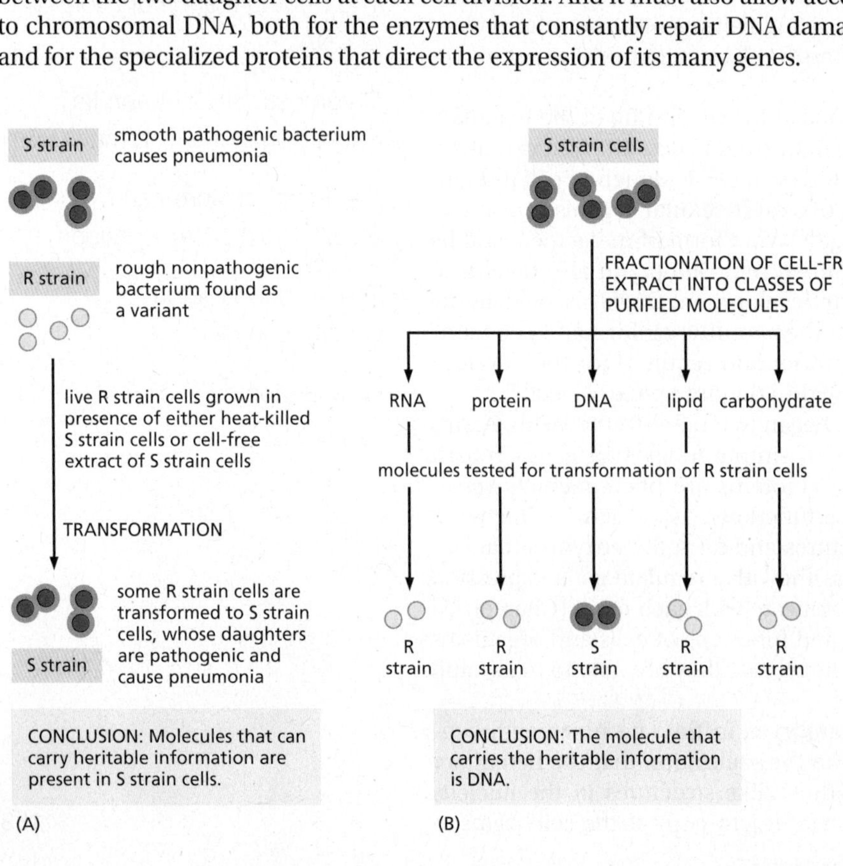

Figure 4–2 The first experimental demonstration that DNA is the genetic material. These experiments, carried out in the 1920s (A) and 1940s (B), eventually showed that adding purified DNA to a bacterium changed the bacterium's properties and that this change was faithfully passed on to subsequent generations. Two closely related strains of the bacterium *Streptococcus pneumoniae* differ from each other in both the appearance of their colonies grown on an agar surface and their pathogenicity. One strain appears smooth (S) and causes death when injected into mice, and the other appears rough (R) and is nonlethal. (A) An initial experiment shows that some substance present in the S strain can change (or transform) the R strain into the S strain and that this change is inherited by subsequent generations of bacteria. (B) This later experiment, in which the R strain has been incubated with various types of biological molecules that were purified from the S strain, identifies the active substance as DNA.

In the past few decades, there has been a revolution in our ability to determine the exact order of subunits in a DNA molecule, as described in Chapter 8. As a result, we now know the sequence of the entire *human genome*—the 3.1 billion nucleotide pairs that provide the information for producing a human adult from a fertilized egg. We also have the DNA sequences for many thousands of other organisms. Detailed analyses of these sequences are providing exciting insights into the process of genome evolution, and it is with this subject that the chapter ends.

This is the first of four chapters that deal with basic genetic mechanisms—the ways in which the cell maintains, replicates, and expresses the hereditary information carried in its DNA. In the next chapter (Chapter 5), we shall discuss the mechanisms by which the cell accurately replicates and repairs DNA; we also describe how DNA sequences can be rearranged through the process of genetic recombination. Gene expression—the process through which the information encoded in DNA is interpreted by the cell to guide the synthesis of proteins—is the main topic of Chapter 6. In Chapter 7, we describe how this gene expression is controlled by the cell to ensure that each of the many thousands of proteins and RNA molecules encrypted in its DNA is manufactured at the proper time and place in the life of the cell.

THE STRUCTURE AND FUNCTION OF DNA

Biologists in the 1940s had difficulty in conceiving how DNA could be the genetic material. The molecule seemed too simple: a long polymer composed of only four types of nucleotide subunits, which resemble one another chemically. In the early 1950s, DNA was examined by x-ray diffraction analysis, a technique for determining the three-dimensional atomic structure of a molecule (discussed in Chapter 8). The x-ray diffraction results indicated that DNA was composed of two strands of the polymer wound into a helix. The observation that DNA was double-stranded provided one of the major clues that led to the Watson–Crick model for DNA structure that, as soon as it was proposed in 1953, made DNA's potential for replication and information storage immediately apparent.

A DNA Molecule Consists of Two Complementary Chains of Nucleotides

A **deoxyribonucleic acid (DNA)** molecule consists of two long polynucleotide chains composed of four types of nucleotide subunits. Each of these chains is known as a *DNA strand*. The two strands run antiparallel to each other, and *hydrogen bonds* between the base portions of the nucleotides hold the two strands together (**Figure 4–3**). As we saw in Chapter 2 (Panel 2–6, pp. 104–105), a nucleotide is composed of a five-carbon sugar to which a phosphate group and a nitrogen-containing base are attached. In the case of the nucleotides in DNA, the sugar is deoxyribose attached to a phosphate group (hence the name deoxyribonucleic acid), and the base may be either *adenine* (*A*), *cytosine* (*C*), *guanine* (*G*), or *thymine* (*T*). The nucleotides are covalently linked together in a chain through the sugars and phosphates, which thus form a "backbone" of alternating sugar-phosphate-sugar-phosphate (**Figure 4–4**). Because only the base differs in each of the four types of nucleotide subunit, each polynucleotide chain in DNA is analogous to a sugar-phosphate necklace (the backbone), from which hang four types of beads (the bases A, C, G, and T). These same symbols (A, C, G, and T) are commonly used to denote either the four bases or the four entire nucleotides; that is, the bases with their attached sugar and phosphate groups.

The way in which the nucleotides are linked together gives a DNA strand a chemical polarity. If we think of each sugar as a block with a protruding knob (the 5′ phosphate) on one side and a hole (the 3′ hydroxyl) on the other (see Figure 4–3), each completed chain, formed by interlocking knobs with holes, will have all of its subunits lined up in the same orientation. Moreover, the two ends of the chain will be easily distinguishable, as one has a hole (the 3′ hydroxyl) and

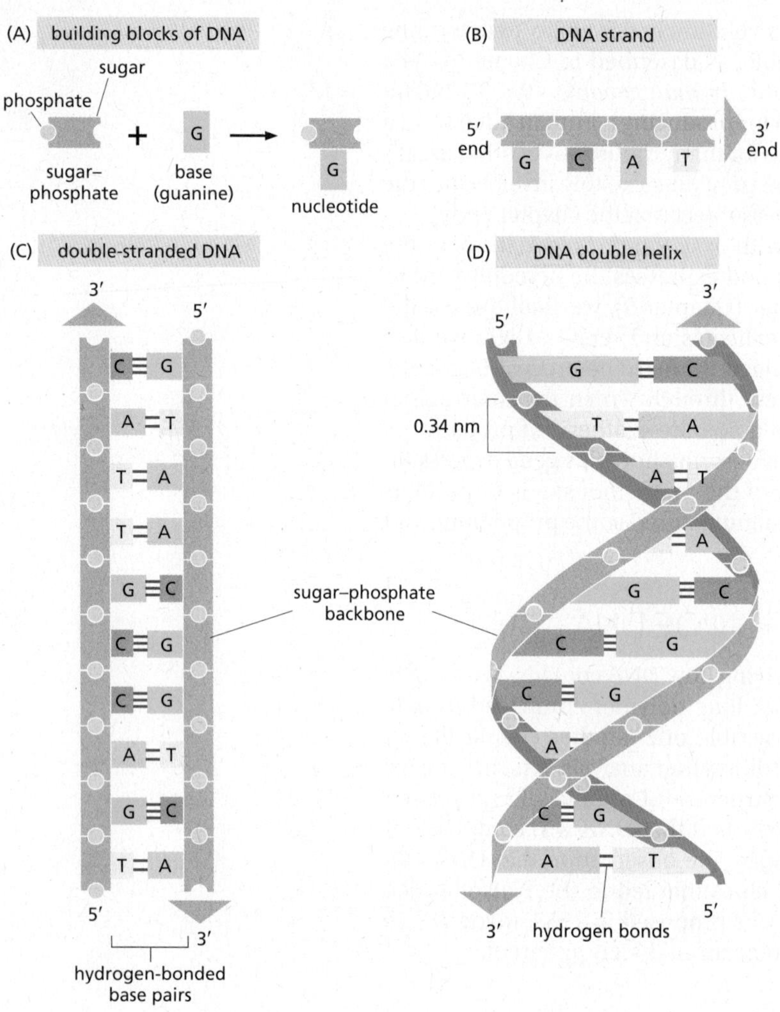

Figure 4–3 **DNA and its four nucleotide building blocks.** (A) Each nucleotide is composed of a sugar–phosphate covalently linked to a base—guanine (G) in this figure. (B) The nucleotides are covalently linked together into polynucleotide chains, with a sugar–phosphate backbone from which the bases—adenine, cytosine, guanine, and thymine (A, C, G, and T)—extend. (C) A DNA molecule is composed of two polynucleotide chains (complementary DNA strands) held together by hydrogen bonds between the paired bases—denoted here by either two (A-T) or three (G-C) red lines. The arrows on the DNA strands indicate the polarities of the two strands, which run antiparallel to each other (with opposite chemical polarities) in the DNA molecule. (D) Although the DNA molecule is shown straightened out in panel C, in reality it is wound into a double helix, as shown here. For details, see the figures that follow.

the other a knob (the 5′ phosphate) at its terminus. This polarity of a DNA strand is indicated by referring to one end as the *3′ end* (pronounced "3 prime end") and the other as the *5′ end* (pronounced "5 prime end"), names derived from the orientation of the deoxyribose sugar (see Figure 4–4). With respect to DNA's information-carrying capacity, the chain of nucleotides in a DNA strand, being both directional and linear, can be read in much the same way as the letters on this page.

The three-dimensional structure of DNA—the DNA **double helix**—arises from the chemical and structural features of its two polynucleotide chains. Because the DNA strands are held together by hydrogen-bonding between the bases on the two different strands, all the bases are on the inside of the double helix, and the sugar–phosphate backbones are on the outside (see Figure 4–3). In each case, a bulkier two-ring base (a purine; see Panel 2–6, pp. 104–105) is paired with a single-ring base (a pyrimidine): A always pairs with T, and G with C (**Figure 4–5A**). This *complementary base-pairing* enables the **base pairs** to be packed in the energetically most favorable arrangement in the interior of the double helix. In this arrangement, each base pair is of similar width, thus holding the sugar–phosphate backbones a constant distance apart along the DNA molecule. To maximize the efficiency of base-pair packing, the two sugar–phosphate backbones wind around each other to form a right-handed double helix, with one complete turn every 10.4 base pairs (**Figure 4–5B** and **Figure 4–6**).

The members of each base pair can fit together within the double helix only if the two strands of the helix are **antiparallel**; that is, only if the polarity of one strand is oriented opposite to that of the other strand (see Figures 4–3 and 4–5). As a consequence of DNA's structure and base-pairing requirements, each strand of a

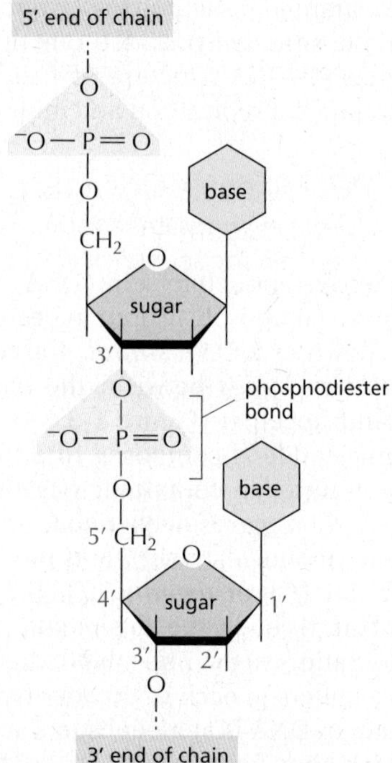

Figure 4–4 **The nucleotide subunits within a DNA strand are held together by a phosphodiester bond.** This bond connects one sugar to the next. The chemical differences in the ester linkages—between the 5′ carbon of one sugar and the 3′ carbon of the other—give rise to the polarity of the resulting DNA strand, as indicated. For simplicity, only two nucleotides are shown here.

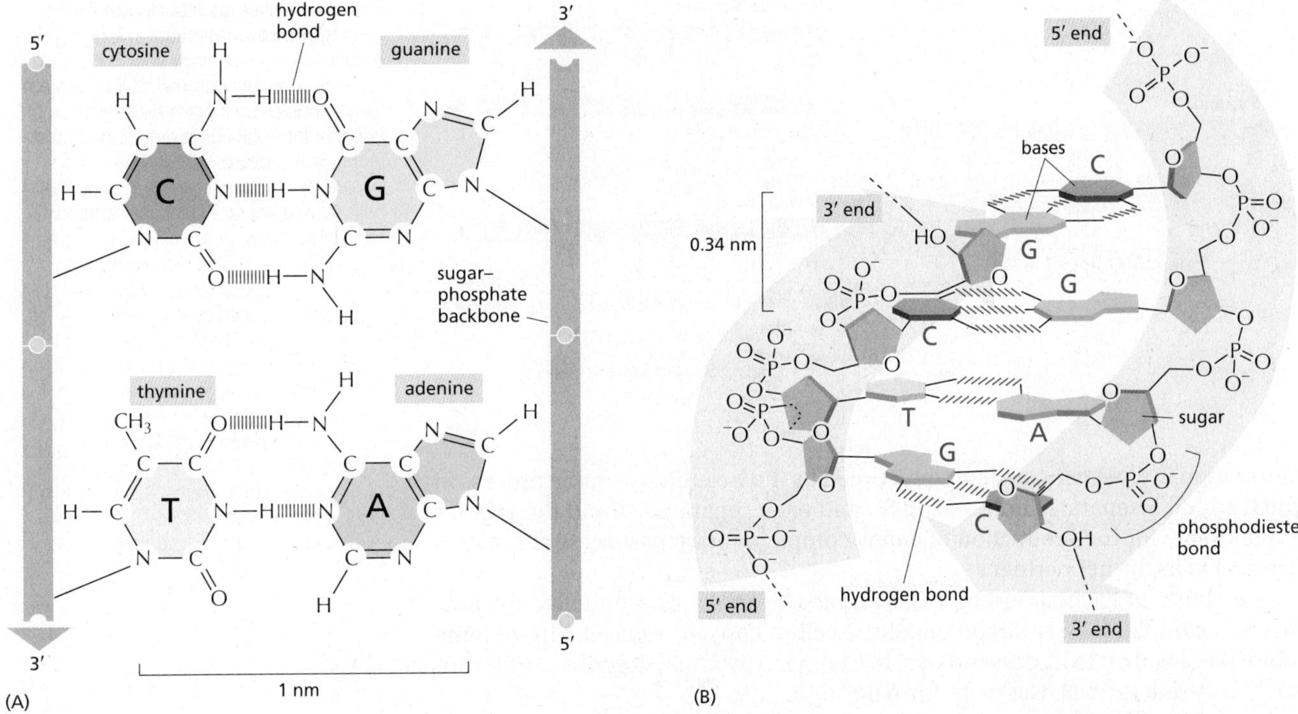

Figure 4–5 **How the two strands of the DNA double helix are held together by hydrogen bonds between complementary base pairs.** (A) Schematic illustration showing how the shapes and chemical structures of the bases allow hydrogen bonds to form efficiently only between A and T and between G and C. As shown, two hydrogen bonds form between A and T, whereas three form between G and C. The bases can pair in this way only if the two polynucleotide chains that contain them are antiparallel; that is, oriented in opposite directions. (B) A short section of the double helix viewed from its side. Four base pairs are illustrated; note that they lie perpendicular to the axis of the helix, unlike those shown in the schematic of panel A. As shown in Figure 4–4, the nucleotides are linked together covalently by a *phosphodiester bond* that connects the 3′-hydroxyl (–OH) group of one sugar and the 5′ phosphate ($-PO_3$) attached to the next (see Figure 4–4 to review how the carbon atoms in the sugar ring are numbered). This linkage gives each polynucleotide strand a chemical polarity; that is, its two ends are chemically different. The 3′ end carries an unlinked –OH group attached to the 3′ position on the sugar ring; the 5′ end carries a free phosphate group attached to the 5′ position on the sugar ring.

DNA molecule contains a sequence of nucleotides that is exactly **complementary** to the nucleotide sequence of its partner strand.

The Structure of DNA Provides a Mechanism for Heredity

The discovery of the structure of DNA immediately suggested answers to the two most fundamental questions about heredity. First, how could the information to specify an organism be carried in a chemical form? And second, how could this information be duplicated and copied from generation to generation?

The answer to the first question came from the realization that DNA is a linear polymer formed from four different kinds of monomers, strung out in a very long, defined sequence like the letters of a document written in an alphabetic script.

The answer to the second question came from the double-stranded nature of the structure: because each strand of DNA contains a sequence of nucleotides that is exactly complementary to the nucleotide sequence of its partner strand, each strand can act as a **template**, or mold, for the synthesis of a new complementary strand. In other words, if we designate the two DNA strands as S and S′, strand S can serve as a template for making a new strand S′, while strand S′ can serve as a template for making a new strand S (**Figure 4–7**). Thus, the genetic

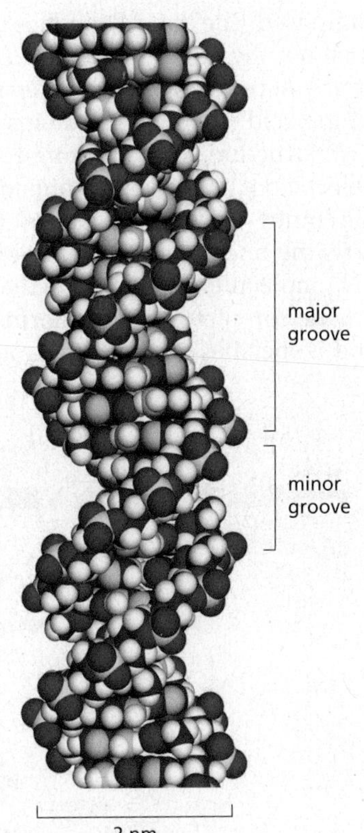

Figure 4–6 **A space-filling model of the DNA double helix.** The two strands wind around each other to form a right-handed helix (see Figure 3–23). Each turn of this helix contains 10.4 nucleotide pairs, and the center-to-center distance between adjacent nucleotide pairs is 0.34 nm. The coiling of the two strands around each other creates two grooves in the double helix: the wider groove is called the major groove, and the narrower groove is called the minor groove, as indicated. The colors of the atoms are N, *blue*; O, *red*; P, *yellow*; H, *white*; and C, *black*. (See **Movie 4.1**.)

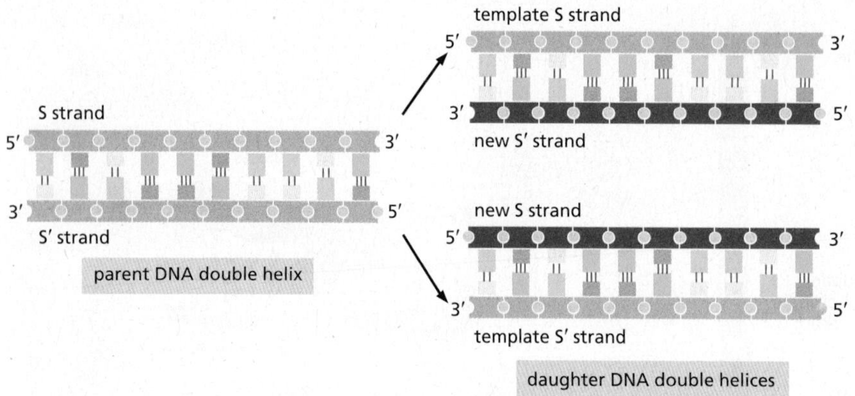

Figure 4–7 **DNA as a template for its own duplication.** Because the nucleotide A successfully pairs only with T, and G pairs with C, each strand of DNA can act as a template to specify the sequence of nucleotides in its complementary strand. In this way, double-helical DNA can be copied precisely, with each parent DNA helix producing two identical daughter DNA helices.

information in DNA can be accurately copied by the beautifully simple process in which strand S separates from strand S′, and each separated strand then serves as a template for the production of a new complementary partner strand that is identical to its former partner.

The ability of each strand of a DNA molecule to act as a template for producing a complementary strand enables a cell to copy, or *replicate*, its genome before passing it on to its descendants. In Chapter 5, we shall describe the elegant machinery that the cell uses to perform this task.

Organisms differ from one another because their respective DNA molecules have different nucleotide sequences and, consequently, carry different biological messages. But how is the nucleotide alphabet used to make messages, and what do they spell out?

As discussed earlier, it was known well before the structure of DNA was determined that genes contain the instructions for producing proteins. If genes are made of DNA, the DNA must therefore somehow encode proteins. As discussed in Chapter 3, the properties of a protein, which are responsible for its biological function, are determined by its three-dimensional structure. This structure is determined in turn by the linear sequence of the amino acids of which it is composed. The linear sequence of nucleotides in a gene must therefore somehow spell out the linear sequence of amino acids in a protein. The exact correspondence between the four-letter nucleotide alphabet of DNA and the 20-letter amino acid alphabet of proteins—the *genetic code*—is not at all obvious from the DNA structure, and it took more than a decade after the discovery of the double helix before it was worked out. In Chapter 6, we will describe this code in detail in the course of elaborating the process of *gene expression*, through which a cell converts the nucleotide sequence of a gene first into the nucleotide sequence of an RNA molecule, and then into the amino acid sequence of a protein (Figure 4–8).

The complete store of information in an organism's DNA is called its **genome**, and it specifies all the RNA molecules and proteins that the organism will ever

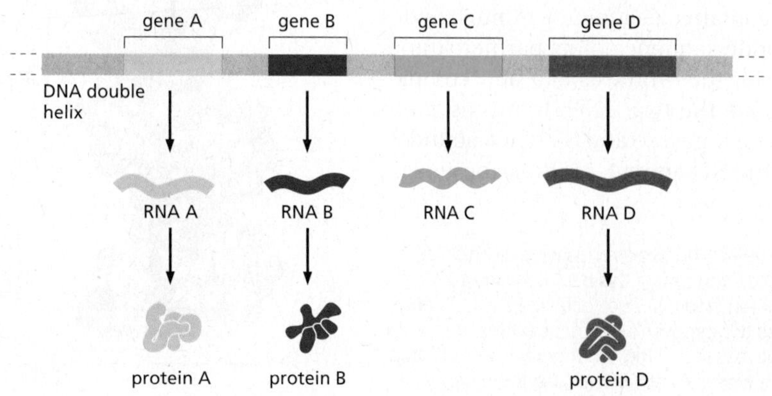

Figure 4–8 **Most genes contain information to make proteins.** As we discuss in Chapter 6, protein-coding genes each produce a set of RNA molecules (called mRNAs), which then direct the production of a specific protein molecule. Note that for a minority of genes, the final product is the RNA molecule itself, as shown here for gene C.

Figure 4–9 The nucleotide sequence of the human β-globin gene. By convention, a nucleotide sequence is written from its 5′ end to its 3′ end, and it should be read from left to right in successive lines down the page as though it were normal English text. This gene carries the information for the amino acid sequence of one of the two types of subunits of the hemoglobin molecule; a different gene, the α-globin gene, carries the information for the other. (Hemoglobin, the protein that carries oxygen in the blood, has four subunits, two of each type.) Only one of the two strands of the DNA double helix containing the β-globin gene is shown; the other strand has the exact complementary sequence.

The DNA sequences highlighted in *yellow* show the three regions of the gene that specify the amino acid sequence for the β-globin protein. We shall see in Chapter 6 how the cell splices these three sequences together at the RNA level to produce a messenger RNA molecule, and how this mRNA then guides the synthesis of a full-length β-globin protein.

```
5′ CCCTGTGGAGCCACACCCTAGGGTTGGCCA
ATCTACTCCCAGGAGCAGGGAGGGCAGGAG
CCAGGGCTGGGCATAAAAGTCAGGGCAGAG
CCATCTATTGCTTACATTTGCTTCTGACAC
AACTGTGTTCACTAGCAACTCAAACAGACA
CCATGGTGCACCTGACTCCTGAGGAGAAGT
CTGCCGTTACTGCCCTGTGGGGCAAGGTGA
ACGTGGATGAAGTTGGTGGTGAGGCCCTGG
GCAGGTTGGTATCAAGGTTACAAGACAGGT
TTAAGGAGACCAATAGAAACTGGGCATGTG
GAGACAGAGAAGACTCTTGGGTTTCTGATA
GGCACTGACTCTCTCTGCCTATTGGTCTAT
TTTCCCACCCTTAGGCTGCTGGTGGTCTAC
CCTTGGACCCAGAGGTTCTTTGAGTCCTTT
GGGGATCTGTCCACTCCTGATGCTGTTATG
GGCAACCCTAAGGTGAAGGCTCATGGCAAG
AAAGTGCTCGGTGCCTTTAGTGATGGCCTG
GCTCACCTGGACAACCTCAAGGGCACCTTT
GCCACACTGAGTGAGCTGCACTGTGACAAG
CTGCACGTGGATCCTGAGAACTTCAGGGTG
AGTCTATGGGACCCTTGATGTTTTCTTTCC
CCTTCTTTTCTATGGTTAAGTTCATGTCAT
AGGAAGGGGAGAAGTAACAGGGTACAGTTT
AGAATGGGAAACAGACGAATGATTGCATCA
GTGTGGAAGTCTCAGGATCGTTTTAGTTTC
TTTTATTTGCTGTTCATAACAATTGTTTTC
TTTTGTTTAATTCTTGCTTTCTTTTTTTTT
CTTCTCCGCAATTTTTACTATTATACTTAA
TGCCTTAACATTGTGTATAACAAAAGGAAA
TATCTCTGAGATACATTAAGTAACTTAAAA
AAAAACTTTACACAGTCTGCCTAGTACATT
ACTATTTGGAATATATGTGTGCTTATTTGC
ATATTCATAATCTCCCTACTTTATTTTCTT
TTATTTTTAATTGATACATAATCATTATAC
ATATTTATGGGTTAAAGTGTAATGTTTTAA
TATGTGTACACATATTGACCAAATCAGGGT
AATTTTGCATTTGTAATTTTAAAAAATGCT
TTCTTCTTTTAATATACTTTTTTGTTTATC
TTATTTCTAATACTTTCCCTAATCTCTTTC
TTTCAGGGCAATAATGATACAATGTATCAT
GCCTCTTTGCACCATTCTAAAGAATAACAG
TGATAATTTCTGGGTTAAGGCAATAGCAAT
ATTTCTGCATATAAATATTTCTGCATATAA
ATTGTAACTGATGTAAGAGGTTTCATATTG
CTAATAGCAGCTACAATCCAGCTACCATTC
TGCTTTTATTTTATGGTTGGGATAAGGCTG
GATTATTCTGAGTCCAAGCTAGGCCCTTTT
GCTAATCATGTTCATACCTCTTATCTTCCT
CCCACAGCTCCTGGGCAACGTGCTGGTCTG
TGTGCTGGCCCATCACTTTGGCAAAGAATT
CACCCCACCAGTGCAGGCTGCCTATCAGAA
AGTGGTGGCTGGTGTGGCTAATGCCCTGGC
CCACAAGTATCACTAAGCTCGCTTTCTTGC
TGTCCAATTTCTATTAAAGGTTCCTTTGTT
CCCTAAGTCCAACTACTAAACTGGGGGATA
TTATGAAGGGCCTTGAGCATCTGGATTCTG
CCTAATAAAAAACATTTATTTTCATTGCAA
TGATGTATTTAAATTATTTCTGAATATTTT
ACTAAAAAGGGAATGTGGGAGGTCAGTGCA
TTTAAAACATAAAGAAATGATGAGCTGTTC
AAACCTTGGGAAAATACACTATATCTTAAA
CTCCATGAAAGAAGGTGAGGCTGCAACCAG
CTAATGCACATTGGCAACAGCCCCTGATGC
CTATGCCTTATTCATCCCTCAGAAAAGGAT
TCTTGTAGAGGCTTGATTTGCAGGTTAAAG
TTTTGCTATGCTGTATTTTACATTACTTAT
TGTTTTAGCTGTCCTCATGAATGTCTTTTC 3′
```

synthesize. (The term "genome" is also used to describe the DNA that carries this information.) The amount of information contained in genomes is staggering. The nucleotide sequence of a very small human gene, written out in the four-letter nucleotide alphabet, occupies a quarter of a page of text (**Figure 4–9**), while the complete sequence of nucleotides in the human genome would fill more than a thousand books the size of this one. In addition to other critical information, the human genome contains roughly 20,000 protein-coding genes, which (through alternative splicing and other mechanisms described in Chapters 6 and 7) can give rise to a much greater number of distinct proteins.

In Eukaryotes, DNA Is Enclosed in a Cell Nucleus

As described in Chapter 1, nearly all the DNA in a eukaryotic cell is sequestered in a nucleus, which in many cells occupies about 10% of the total cell volume. This compartment is delimited by a *nuclear envelope* formed by two concentric lipid bilayer membranes (**Figure 4–10**). As illustrated, these two membranes are punctured at intervals by large nuclear pores, through which molecules move between the nucleus and the cytosol. The outer nuclear membrane is directly connected to the extensive system of intracellular membranes called the *endoplasmic reticulum*, which extend out from it into the cytoplasm. And the nuclear envelope is supported internally by a network of intermediate filaments called the *nuclear lamina*—a thin feltlike mesh just beneath the inner nuclear membrane (see Figure 4–10B).

The nuclear envelope allows the many proteins that act on DNA to be concentrated where they are needed in the cell, and, as we see in subsequent chapters, it also keeps nuclear and cytosolic proteins separate, a feature that is crucial for the proper functioning of eukaryotic cells.

Summary

Genetic information is carried in the linear sequence of nucleotides in DNA. Each molecule of DNA is a double helix formed from two complementary antiparallel strands of nucleotides held together by hydrogen bonds between G-C and A-T base pairs. Duplication of the genetic information occurs by the use of one DNA strand as a template for the formation of a complementary strand. The genetic information stored in an organism's DNA sequence contains the instructions for all the RNA molecules and proteins that the organism will ever synthesize, and it is said to be that organism's genome. In eukaryotes, DNA is contained in the cell nucleus, a large membrane-bound compartment.

CHROMOSOMAL DNA AND ITS PACKAGING IN THE CHROMATIN FIBER

The most important function of DNA is to form genes, which are the information-carrying units that specify all the RNA molecules and proteins that make up an organism—including information about when, in what types of cells, and in what quantity each RNA molecule and protein is to be made. The nuclear DNA of eukaryotes is divided up into a set of chromosomes, each of which contains one

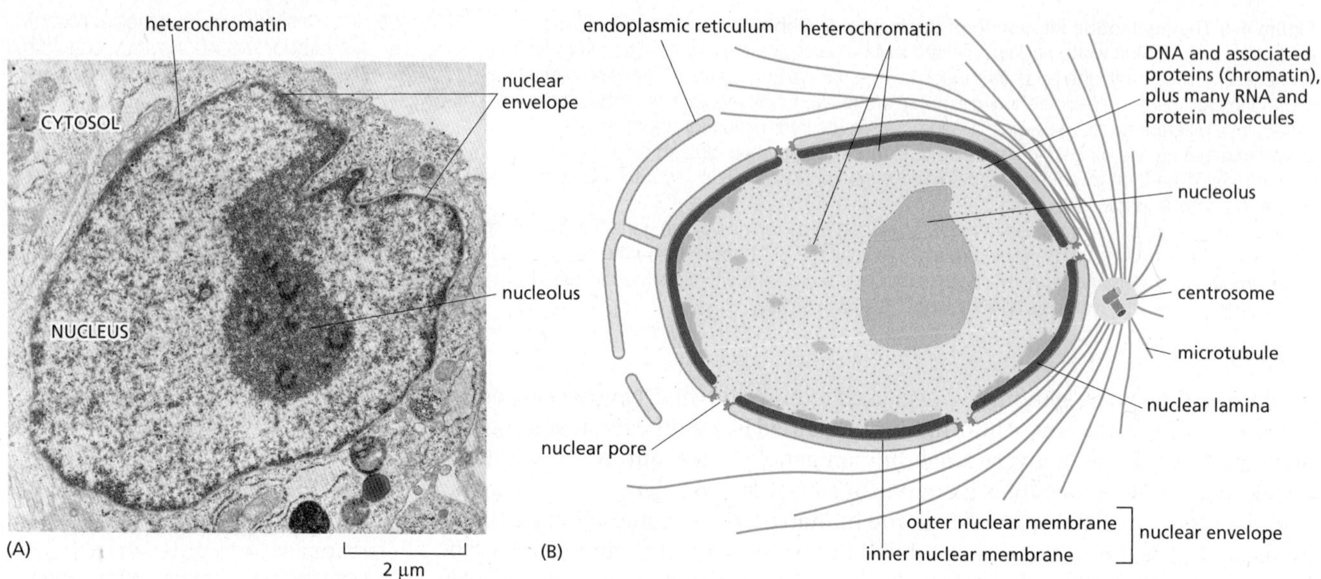

Figure 4–10 A cross-sectional view of a typical cell nucleus. (A) Electron micrograph of a thin section through the nucleus of a human fibroblast. (B) Schematic drawing, showing that the nuclear envelope consists of two membranes, the outer one being continuous with the endoplasmic reticulum (ER) membrane (see also Figure 12–54). The space inside the endoplasmic reticulum (the ER lumen) is colored *yellow*; it is continuous with the space between the two nuclear membranes. The lipid bilayers of the inner and outer nuclear membranes are connected at each nuclear pore. A sheetlike network of filaments inside the nucleus forms the nuclear lamina *(brown)*, providing mechanical support for the nuclear envelope (for details, see Chapter 12). Outside the nuclear envelope, a centrosome forms microtubules that help to organize the cytoplasm, as explained in Chapter 16. The dark-staining heterochromatin contains specially condensed regions of DNA that will be discussed later.

Prokaryotic cells (archaea and bacteria) have no nucleus. Lacking a nuclear envelope, their DNA is co-located with all other cell components in a single compartment, bounded by the cell's plasma membrane. (A, courtesy of E.G. Jordan and J. McGovern.)

enormously long DNA molecule. In this section we see how genes are arranged on chromosomes. In addition, we describe three specialized DNA sequences that are required for a chromosome to be accurately duplicated as a separate entity and passed on from one generation to the next.

We also confront the serious challenge of DNA packaging. If the double helices of all 46 chromosomes in a human cell could be laid end to end, they would reach approximately 2 meters; yet the nucleus, which contains the DNA, is only about 6 μm in diameter. This is geometrically equivalent to packing 40 km (24 miles) of extremely fine thread into a tennis ball. The complex task of packaging DNA is accomplished by specialized proteins that bind to the DNA and fold it, generating a series of organized coils and loops that prevent the DNA from becoming an unmanageable tangle. Amazingly, although the DNA is very tightly compacted, it nevertheless remains accessible to the many enzymes in the cell that replicate it, repair it, and use its genes to produce RNA molecules and proteins.

Eukaryotic DNA Is Packaged into a Set of Chromosomes

Each **chromosome** in a eukaryotic cell consists of a single, enormously long linear DNA molecule along with the proteins that fold the fine DNA thread into a more compact structure. In addition to proteins involved in packaging, chromosomes are also associated with many other proteins (as well as numerous RNA molecules). These are required for the processes of gene expression, DNA replication, and DNA repair. The complex of DNA and tightly bound protein is called *chromatin* (from the Greek *chroma*, "color," because of its staining properties).

Bacteria lack a special nuclear compartment, and they generally carry their genes on a single DNA molecule, which is often circular (see Figure 1–38). This DNA is also associated with proteins that package and condense it, but they are different from the proteins that perform these functions in eukaryotes. Although the bacterial DNA with its attendant proteins is often called the bacterial "chromosome," it does not have the same structure as the eukaryotic chromosomes that will be our main focus.

With the exception of the gametes (eggs and sperm) and a few highly specialized cell types that cannot multiply and either lack DNA altogether (for example, red blood cells) or have replicated their DNA without completing cell division (for example, megakaryocytes), each human cell nucleus contains two copies of each chromosome, one inherited from the mother and one from the father. The maternal and paternal chromosomes of a pair are called **homologous chromosomes (homologs)**. The only nonhomologous chromosome pairs are the sex chromosomes in males, where a *Y chromosome* is inherited from the father

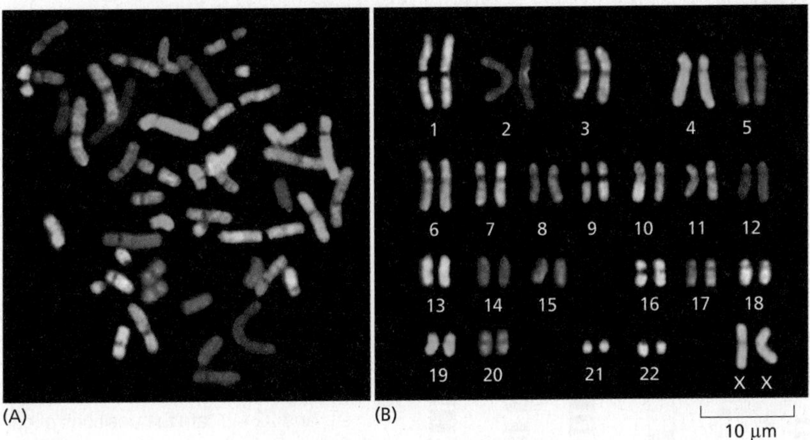

Figure 4–11 **The complete set of human chromosomes.** These chromosomes, from a female, were isolated from a cell undergoing nuclear division (mitosis) and are therefore highly compacted. Each chromosome has been "painted" a different color to permit its unambiguous identification under the fluorescence microscope, using a technique called "spectral karyotyping." Chromosome painting can be performed by exposing the chromosomes to a large collection of DNA molecules (called DNA probes) whose sequences match known DNA sequences from the human genome. A set of sequences that matches a specific chromosome is coupled to a combination of fluorescent dyes. DNA molecules derived from chromosome 1 are labeled with one specific dye combination, those from chromosome 2 with another, and so on. Because the labeled DNA can form base pairs, or hybridize, only to the chromosome from which it was derived, each chromosome becomes labeled with a different combination of dyes. For such experiments, the chromosomes are subjected to treatments that separate the two strands of double-helical DNA in a way that permits base-pairing with the single-stranded, labeled DNA but keeps the overall chromosome structure relatively intact. (A) The chromosomes visualized as they originally spilled from the lysed cell. (B) The same chromosomes artificially lined up in their numerical order. This arrangement of the full chromosome set is called a karyotype. (Adapted from N. McNeil and T. Ried, *Expert Rev. Mol. Med.* 2:1–14, 2000. With permission from Cambridge University Press.)

and an *X chromosome* from the mother. Thus, each human cell contains a total of 46 chromosomes—22 pairs common to both males and females, plus two so-called sex chromosomes (X and Y in males, two X's in females). These human chromosomes can be readily distinguished by "painting" each one a different color using a technique that is based on *DNA hybridization* (**Figure 4–11**). In this method (see p. 505), short strands of nucleic acid tagged with fluorescent dyes serve as "probes" that pick out their complementary DNA sequence, lighting up the target chromosome at any site where they bind. Chromosome painting is most frequently done at the stage in the cell cycle called mitosis, when chromosomes are especially compacted and easy to visualize (discussed shortly).

Another more traditional way to distinguish one chromosome from another is to stain them with dyes that reveal a striking and reproducible pattern of bands along each mitotic chromosome (**Figure 4–12**). These banding patterns reflect variations in chromatin structure and/or base composition, although their basis is not well understood. Nevertheless, the pattern of these bands on each type of chromosome is unique, and it provided the initial means to identify and number each human chromosome reliably.

The display of the 46 human chromosomes at mitosis shown in Figure 4–11B is called the human **karyotype**. If parts of chromosomes are lost or are switched between chromosomes, these changes can be detected either by changes in the banding patterns or—with greater sensitivity—by changes in the pattern of chromosome painting (**Figure 4–13**). Cytogeneticists use these alterations to detect inherited chromosome abnormalities and to reveal the chromosome rearrangements that occur in cancer cells as they progress to malignancy (discussed in Chapter 20).

Chromosomes Contain Long Strings of Genes

Chromosomes carry genes—the functional units of heredity. A gene is often defined as a segment of DNA that contains the instructions for making a particular protein (or a set of closely related proteins), but this definition is too narrow. Genes that code for protein are indeed the majority, and most of the genes with clear-cut mutant phenotypes fall under this heading. In addition, however, there are many "RNA genes"—segments of DNA that generate a functionally significant RNA molecule, instead of a protein, as their final product. We shall say more about these RNA genes and their products later.

As might be expected, there can be a correlation between the complexity of an organism and the number of genes in its genome (see Table 1–2, p. 29). For example, some simple bacteria have only 500 genes, compared to about 25,000 for humans. Bacteria, archaea, and some single-celled eukaryotes, such as yeast, have concise genomes, consisting of little more than strings of closely packed genes. However, the genomes of multicellular plants and animals, as well as many other eukaryotes, contain, in addition to genes, a large quantity of interspersed DNA whose function

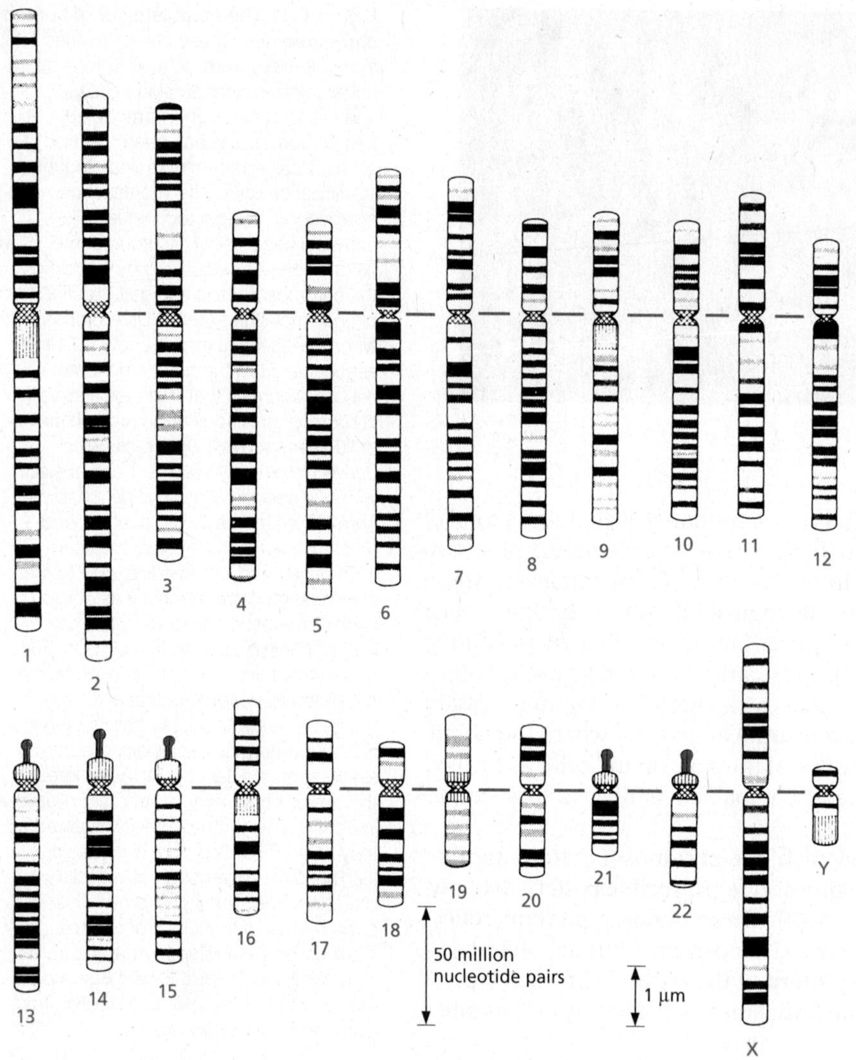

Figure 4–12 The banding patterns of human chromosomes. Chromosomes 1–22 are numbered in approximate order of size. A typical human cell contains two of each of these chromosomes, plus two sex chromosomes—two X chromosomes in a female, one X and one Y chromosome in a male. The chromosomes used to make these maps were stained at an early stage in mitosis, when the chromosomes are incompletely compacted. The *horizontal red line* represents the position of the centromere (see Figure 4–18), which appears as a constriction on mitotic chromosomes. The red knobs on chromosomes 13, 14, 15, 21, and 22 indicate the positions of genes that code for the large ribosomal RNAs and form the nucleolus (discussed in Chapter 6). These banding patterns are obtained by staining chromosomes with Giemsa stain, and they are observed under a light microscope. (Adapted from U. Francke, *Cytogenet. Cell Genet.* 31:24–32, 1981.)

is poorly understood. Some of this additional DNA is crucial for the proper control of gene expression, but this only partly explains why there is so much of it in multicellular organisms, whose genes need to be switched on and off according to complicated rules during development (discussed in Chapters 7 and 21).

Differences in the amount of noncoding DNA, most of it interspersed between genes—far more than differences in numbers of genes—account for the astonishing variations in genome size that we see when we compare one species with another (see Figure 1–30). For example, the human genome is 200 times larger than that of the yeast *Saccharomyces cerevisiae,* but 30 times smaller than that of some plants and amphibians and 200 times smaller than that of a species of amoeba. Moreover, because of differences in the amount of noncoding DNA, the genomes of closely related organisms (bony fish, for example) can vary several hundredfold in their DNA content, even though they contain roughly the same number of genes. Whatever the excess DNA may do, it seems clear that it is not a great handicap for a eukaryotic cell to carry a large amount of it.

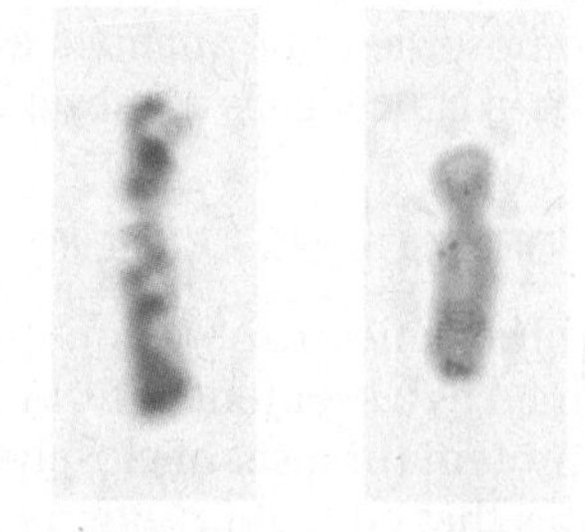

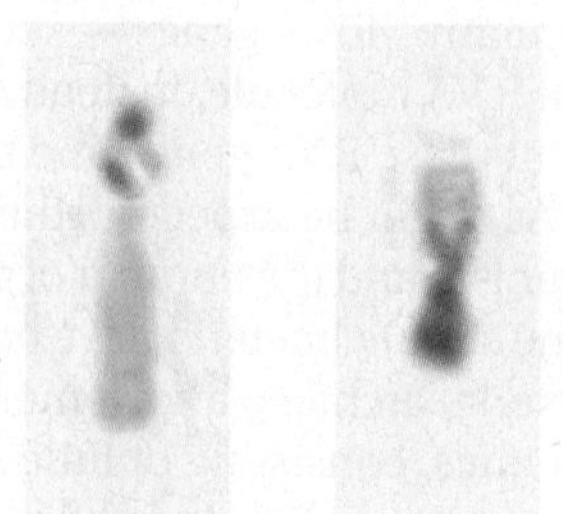

Figure 4–13 Aberrant human chromosomes. (A) Two normal human chromosomes, 4 and 6. (B) In an individual carrying a reciprocal *chromosomal translocation*, the DNA double helix in one chromosome has crossed over with the DNA double helix in the other chromosome because of an abnormal recombination event. The chromosome painting technique used on the chromosomes in each of the sets allows the identification of short pieces of chromosomes that have become translocated, a frequent event in cancer cells. (Courtesy of Zhenya Tang and the NIGMS Human Genetic Cell Repository at the Coriell Institute for Medical Research.)

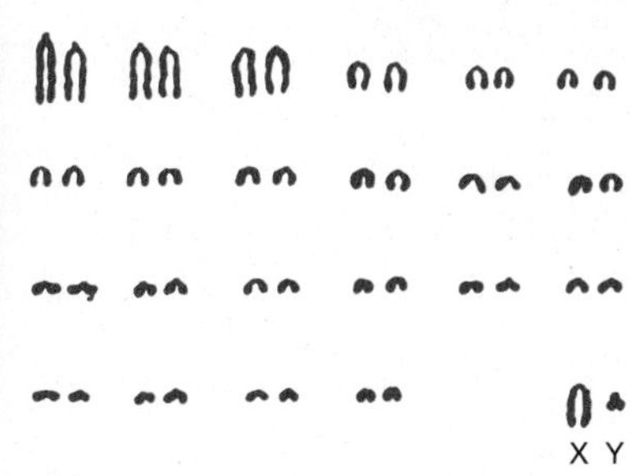

Chinese muntjac

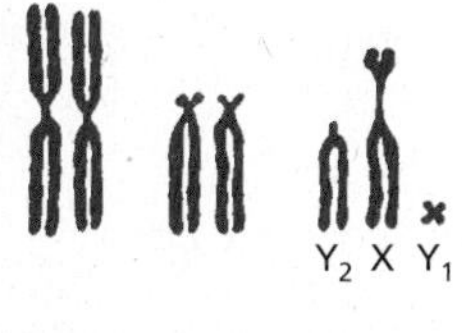

Indian muntjac

Figure 4–14 Two closely related species of deer with very different chromosome numbers. In the evolution of the Indian muntjac, initially separate chromosomes in an ancestor fused, without having a major effect on the animal. These two species contain a similar number of genes. (Chinese muntjac photo courtesy of Deborah Carreno, Natural Wonders Photography; Indian muntjac photo courtesy of Beatrice Bourgery.)

How the genome is divided into chromosomes also differs from one eukaryotic species to the next. For example, while the cells of humans have 46 chromosomes, those of some small deer have only 6, while those of the common carp contain more than 100. Even closely related species with similar genome sizes can have very different numbers and sizes of chromosomes (**Figure 4–14**). Thus, there is no simple relationship between chromosome number, complexity of the organism, and total genome size. Rather, the genomes and chromosomes of modern-day species have each been shaped by a unique history of seemingly random genetic events, acted on by poorly understood selection pressures over long evolutionary times.

The Nucleotide Sequence of the Human Genome Shows How Our Genes Are Arranged

With the determination of the full DNA sequence of the human genome, it became possible to see in detail how the genes are arranged along each of our chromosomes (**Figure 4–15**). Although many decades will pass before the information contained in the human genome sequence is fully analyzed, it has already stimulated an enormous number of new experiments that have had major effects on the content of every chapter in this book.

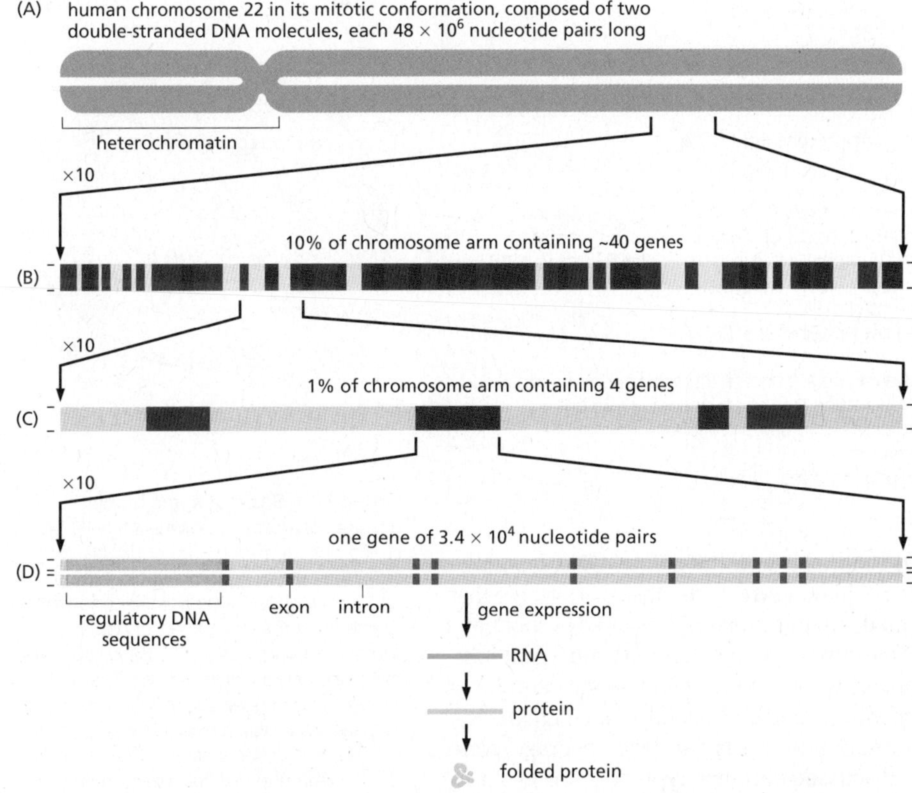

Figure 4–15 The organization of genes on a human chromosome, as determined from the initial sequencing of the human genome. (A) Chromosome 22, one of the smallest human chromosomes, contains 48×10^6 nucleotide pairs and makes up approximately 1.5% of the human genome. Most of the left arm of chromosome 22 consists of short repeated sequences of DNA that are packaged in a particularly compact form of chromatin (heterochromatin) discussed later in this chapter. (B) A tenfold expansion of a portion of chromosome 22, with about 40 genes indicated. Those in *dark brown* were known genes, and those in *red* were predicted genes. (C) An expanded portion of B showing four genes. (D) The intron–exon arrangement of a typical gene is shown after a further tenfold expansion. Each exon *(red)* codes for a portion of the protein, while the DNA sequence of the introns *(gray)* is relatively unimportant, as discussed in detail in Chapter 6. (Adapted from International Human Genome Sequencing Consortium, *Nature* 409:860–921, 2001.)

The *human genome* (3.1×10^9 nucleotide pairs) is the totality of genetic information belonging to our species. Almost all of this genome is distributed over the 22 different autosomes and 2 sex chromosomes (see Figures 4–11 and 4–12) found within the nucleus. A minute fraction of the human genome (16,569 nucleotide pairs—in multiple copies per cell) is found in the mitochondria (introduced in Chapter 1 and discussed in detail in Chapter 14). The term "human genome sequence" refers to the complete nucleotide sequence of DNA in the 24 nuclear chromosomes and the mitochondria. Being diploid, a human somatic cell nucleus contains twice the haploid amount of DNA, or 6.2×10^9 nucleotide pairs, before duplicating its chromosomes in preparation for division.

TABLE 4–1 Some Vital Statistics for the Human Genome	
Human genome DNA length	3.1×10^9 nucleotide pairs*
Number of genes coding for proteins	About 20,000 (19,116, plus hundreds of genes encoding proteins of 50 amino acids or less)**
Largest gene coding for protein	2.5×10^6 nucleotide pairs
Median size for protein-coding genes	26,000 nucleotide pairs
Median size for messenger RNAs (including 5′ and 3′ untranslated regions)	2938 nucleotides
Median size for amino-acid coding in messenger RNA sequences	1290 nucleotides
Protein-coding gene features	
Median size of protein produced	430 amino acids
Smallest number of exons per gene	1 (1068 of these unspliced genes)
Largest number of exons per gene	363
Median number of exons per gene	9.0
Median exon size	131 nucleotide pairs
Median intron size	1747 nucleotide pairs
Number of noncoding RNA genes	About 5000, with 4849 annotated***
Number of pseudogenes	More than 20,000****
Percentage of protein-coding DNA sequence (in exons)	1%
Percentage of DNA in other highly conserved sequences	3.5%
Percentage of DNA transcribed (in protein-coding plus annotated noncoding RNA genes)	45%
Percentage of DNA in high-copy-number repetitive elements	Approximately 50%

*The sequence of 2.85 billion nucleotides is known precisely (error rate of only about 1 in 100,000 nucleotides). The remaining DNA primarily consists of short sequences that are tandemly repeated many times over, with repeat numbers differing from one individual to the next. These highly repetitive blocks are difficult to sequence accurately.
**RefSeq genes, with at least one reviewed/validated mRNA (A. Piovesan et al., *BMC Res. Notes* 12:315–319, 2019). For genes encoding tiny proteins, see T.F. Martinez et al., *Nat. Chem. Biol.* 16:458–468, 2020).
***A considerable number of potential ncRNA genes are still uncharacterized and not annotated.
****A pseudogene is a DNA sequence closely resembling that of a functional gene but containing numerous mutations that prevent its proper expression or function; most are still uncharacterized and not annotated in databases.
Courtesy of Allison Piovesan, University of Bologna, Bologna, Italy.

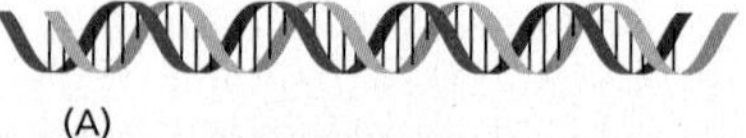

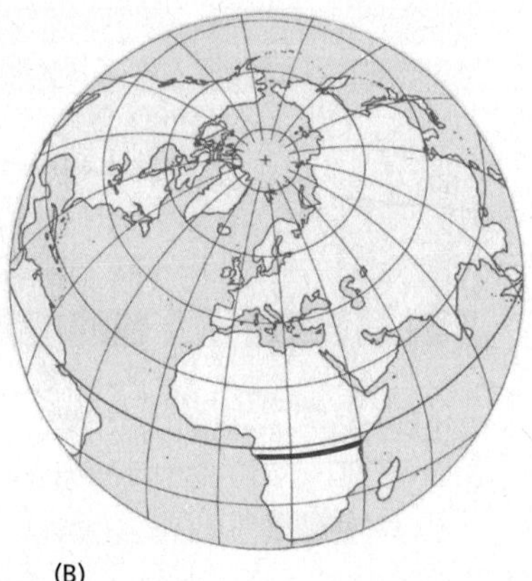

Figure 4–16 Scale of some human genome features. If drawn with a 1-mm space between each nucleotide pair (as in A), the human genome would extend 3100 km (nearly 2000 miles), far enough to stretch across the center of Africa, the site of our human origins (*red line* in B). At this scale, there would be, on average, a protein-coding gene every 160 meters. An average gene would extend for 26 meters, but the coding sequences in this gene would add up to only just over a meter.

The first striking feature of the human genome is how little of the DNA sequence (only about 1%) codes for proteins (**Table 4–1** and **Figure 4–16**). It is also notable that nearly half of the chromosomal DNA is made up of mobile pieces of DNA that have gradually inserted themselves in the chromosomes over evolutionary time, multiplying like parasites in the genome and appearing as genome-wide repeat sequences (see Figure 4–63). We discuss these *transposable elements* in detail in later chapters.

A second notable feature of the human genome is the large average gene size—about 26,000 nucleotide pairs. As discussed earlier, a typical gene carries in its linear sequence of nucleotides the information for the linear sequence of the amino acids of a protein. Only about 1300 nucleotides are required to encode a protein of average size (about 430 amino acids in humans). Most of the remaining sequence in a gene consists of long stretches of noncoding DNA that interrupt the relatively short segments of DNA that code for protein. As will be discussed in detail in Chapter 6, the coding sequences are included in the **exons**; the intervening sequences in genes are noncoding and called **introns** (see Figure 4–15 and Table 4–1). The majority of human genes thus consist of a long string of alternating exons and introns, with most of the gene consisting of introns. In contrast, the majority of genes from organisms with concise genomes either lack introns (as in prokaryotes) or contain relatively short ones. This helps to account for the much smaller size of their genes, as well as for the much higher fraction of coding DNA in their chromosomes.

In addition to introns and exons, each gene is associated with *regulatory DNA sequences*, which are responsible for ensuring that the gene is turned on or off at the proper time, expressed at the appropriate level, and only in the proper type of cell. In humans, the regulatory sequences for a typical gene are spread out over hundreds of thousands of nucleotide pairs. As would be expected, these regulatory sequences are much more compact and less numerous in organisms with concise genomes. In Chapter 7 we discuss how regulatory DNA sequences work.

Detailed studies of genome sequences have surprised biologists with the discovery that, in addition to 20,000 protein-coding genes, the human genome contains thousands of genes that encode RNA molecules that do not produce proteins, but instead have a variety of other important functions. Although the roles of some of these non-coding RNA genes have been known for decades, many more remain mysterious, as will be discussed in Chapters 6 and 7.

Last, but not least, the nucleotide sequence of the human genome has revealed that the archive of information needed to produce a human seems to be in an alarming state of chaos. As one commentator described our genome, "In some ways it may resemble your garage/bedroom/refrigerator/life: highly individualistic, but unkempt; little evidence of organization; much accumulated clutter (referred to by the uninitiated as 'junk'); virtually nothing ever discarded; and the few patently valuable items indiscriminately, apparently carelessly, scattered throughout." We shall discuss how this is thought to have come about in the final section of this chapter, which is titled "How Genomes Evolve."

Each DNA Molecule That Forms a Linear Chromosome Must Contain a Centromere, Two Telomeres, and Replication Origins

To form a functional chromosome, a DNA molecule must be able to do more than simply carry genes: it must be able to replicate, and the replicated copies must be separated and reliably partitioned into daughter cells at each cell division. This process occurs through an ordered series of stages, collectively known as the **cell cycle**, which provides for a temporal separation between the duplication of chromosomes and their segregation into two daughter cells. The eukaryotic cell cycle is briefly summarized in Figure 4–17, and it is discussed in detail in Chapter 17.

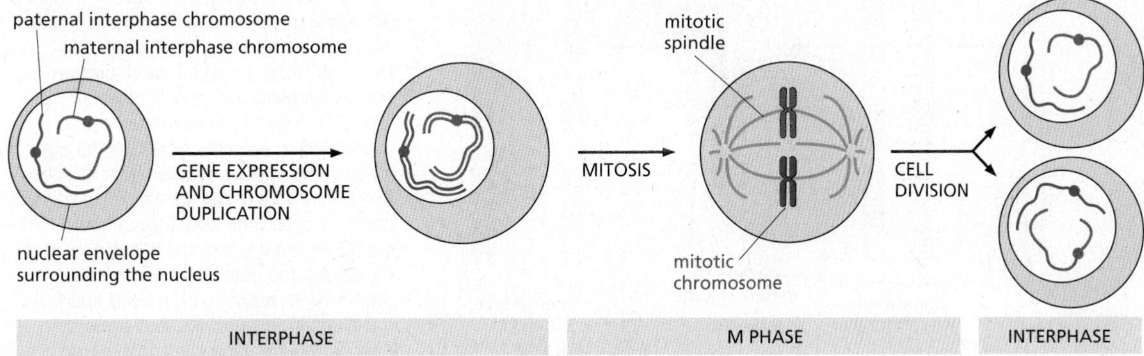

Figure 4–17 **A simplified view of the eukaryotic cell cycle.** During interphase, the cell is actively expressing its genes and is therefore synthesizing proteins. Also, during interphase and before cell division, the DNA is replicated, and each chromosome is duplicated to produce two closely paired sister DNA molecules (called sister chromatids). A cell with only one type of chromosome, present in maternal and paternal copies, is illustrated here.

Once DNA replication is complete, the cell can enter *M phase*, when mitosis occurs and the nucleus is divided into two daughter nuclei. During this stage, the chromosomes condense, the nuclear envelope breaks down, and the mitotic spindle forms from microtubules and other proteins. The condensed mitotic chromosomes are captured by the mitotic spindle, and one complete set of chromosomes is then pulled to each end of the cell by separating the members of each sister-chromatid pair. A nuclear envelope re-forms around each chromosome set, and in the final step of M phase, the cell divides to produce two daughter cells. Most of the time in the cell cycle is spent in interphase; M phase is brief in comparison, occupying only about an hour in many mammalian cells.

Briefly, during a long *interphase*, genes are expressed and chromosomes are replicated, with the two replicas remaining together as a pair of *sister chromatids*. Throughout this time the chromosomes are extended, and much of their chromatin exists as long threads in the nucleus so that individual chromosomes cannot be easily distinguished. It is only during a much briefer period of *mitosis* that each chromosome condenses so that its two sister chromatids can be separated and distributed to the two daughter nuclei. The highly condensed chromosomes in a dividing cell are known as *mitotic chromosomes* (Figure 4–18). This is the form in which chromosomes are most easily visualized; in fact, the images of chromosomes shown so far in the chapter are of chromosomes in mitosis.

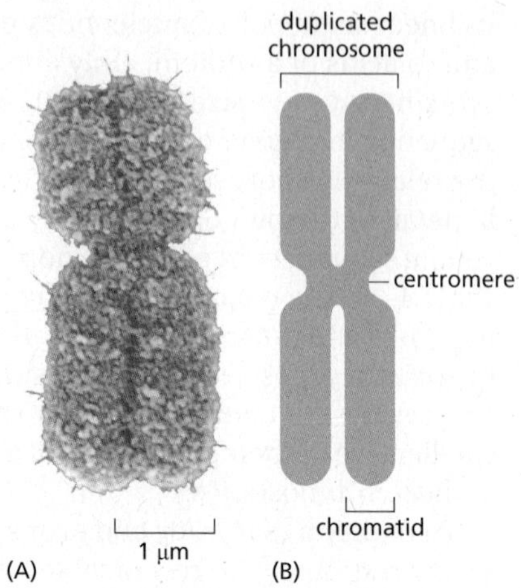

Figure 4–18 **A mitotic chromosome.** At the metaphase stage of M phase, each chromosome exists as a condensed duplicated chromosome, in which the two replicated chromosomes, called sister chromatids, are still linked together (see Figure 4–17). The constricted region indicates the position of the centromere. (A) Electron micrograph; (B) schematic drawing. (A, courtesy of Terry D. Allen.)

Each chromosome operates as a distinct structural unit. For a faithful copy of a chromosome to be passed on to each daughter cell at division, it must have been completely replicated during interphase to produce two identical DNA molecules. In addition, these newly replicated, enormously long DNA copies must then be separated and partitioned correctly into the two daughter cells. These basic functions are controlled by three types of specialized sites on the DNA, each of which binds specific proteins that guide the machinery that replicates and segregates chromosomes (Figure 4–19).

Experiments in budding yeast, whose chromosomes are relatively small and easy to manipulate, have identified the minimal DNA sequence elements responsible for each of these functions. One type of nucleotide sequence acts as a DNA **replication origin**, the location at which duplication of the DNA begins. Eukaryotic chromosomes contain many origins of replication to ensure that the entire chromosome can be replicated rapidly, as discussed in detail in Chapter 5.

After DNA replication, the two sister chromatids that form each chromosome remain attached to one another and, as the cell cycle proceeds, they are condensed further to produce mitotic chromosomes. In budding yeasts, the presence of a second specialized DNA sequence, called a **centromere**, allows one copy of each duplicated and condensed chromosome to be pulled into each daughter cell when a cell divides. A protein complex called a *kinetochore* forms at the centromere and attaches the duplicated chromosomes to the mitotic spindle, in a manner that causes the two sister chromatids to be pulled apart (discussed in Chapter 17).

The third specialized DNA sequence forms **telomeres**, the ends of a chromosome. Telomeres contain repeated nucleotide sequences that enable the ends of chromosomes to be efficiently replicated. Telomeres also perform another function: the repeated telomere DNA sequences, together with the regions adjoining them, form structures that protect the end of the chromosome from

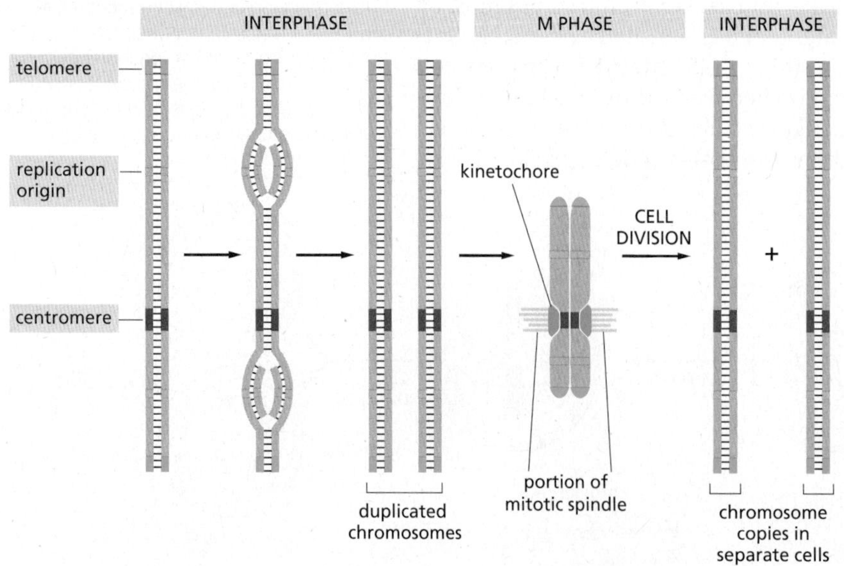

Figure 4–19 **The three specialized sites on the DNA required to produce a eukaryotic chromosome that can be replicated and then segregated accurately at mitosis.** Each chromosome has multiple origins of replication, one centromere, and two telomeres. Shown here is the sequence of events that a typical chromosome follows during the cell cycle. The DNA replicates during the portion of interphase known as S phase, beginning at the origins of replication and proceeding bidirectionally from the origins across the chromosome. In M phase, the centromere attaches the duplicated chromosomes to the mitotic spindle so that a copy of the entire genome is distributed to each daughter cell during mitosis; the special structure that attaches the centromere to the spindle is a protein complex called the kinetochore *(dark green)*. The centromere also helps to hold the duplicated chromosomes together until they are ready to be moved apart. As explained in the text, the telomeres form special caps at each chromosome end.

being mistaken by the cell for a broken DNA molecule in need of repair. We discuss both this type of repair and the structure and function of telomeres in Chapter 5.

In budding yeast cells, the three types of sequences required to propagate a chromosome are relatively short (typically fewer than 1000 base pairs each) and therefore use only a tiny fraction of the information-carrying capacity of a chromosome. Although telomere sequences are fairly simple and short in all eukaryotes, the DNA sequences that form centromeres and replication origins in more complex organisms are much longer than their yeast counterparts. For example, experiments suggest that a human centromere can contain up to a million nucleotide pairs and that it may not require a stretch of DNA with a defined nucleotide sequence. Instead, as we shall discuss later in this chapter, a human centromere is thought to consist of a large, regularly repeating protein–nucleic acid structure that can be inherited when a chromosome replicates.

DNA Molecules Are Highly Condensed in Chromosomes

All eukaryotic organisms have special ways of packaging DNA into chromosomes. Thus, if the 48 million nucleotide pairs of DNA in human chromosome 22 could be laid out as one long double helix stretched out end to end, the molecule would extend for about 1.5 cm. But chromosome 22 measures only about 2 μm in length in mitosis (see Figure 4–11), representing an end-to-end compaction ratio of more than 7000-fold. This remarkable feat of compression is performed by proteins that successively coil and fold the DNA into higher levels of organization.

Although much less condensed than mitotic chromosomes, the DNA of human interphase chromosomes is still tightly packed. But interphase chromosomes have a dynamic structure. Specific regions of interphase chromosomes decondense to allow access to specific DNA sequences for gene expression, DNA repair, and replication—and then recondense when these processes are completed. The packaging of DNA in chromosomes is therefore accomplished in a way that allows rapid localized, on-demand access to the DNA. In the next sections, we discuss the specialized proteins that make this type of packaging possible.

Nucleosomes Are a Basic Unit of Eukaryotic Chromosome Structure

The proteins that bind to the DNA to form eukaryotic chromosomes are traditionally divided into two classes: the **histones** and the *non-histone chromosomal proteins*, each contributing about the same mass to a chromosome. The complex of both classes of protein with the nuclear DNA of eukaryotic cells is known as **chromatin** (Figure 4–20; Movie 4.2).

Histones are responsible for the first and most basic level of chromosome packing, a protein–DNA complex called the **nucleosome**. When interphase nuclei are broken open very gently and their contents examined under the electron microscope, most of the chromatin appears to be in the form of a fiber

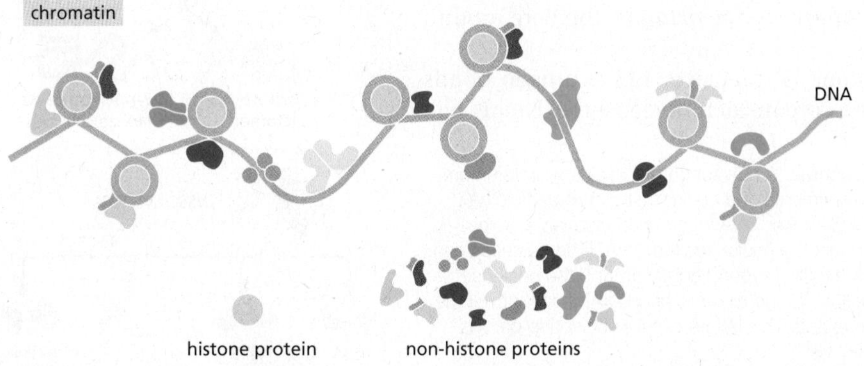

Figure 4–20 Chromatin. As illustrated, chromatin consists of DNA bound to both histone and non-histone proteins. The mass of histone protein present is about equal to the total mass of non-histone protein, but—as schematically indicated here—the latter class is composed of an enormous number of different species. In total, a chromosome is about one-third DNA and two-thirds protein by mass.

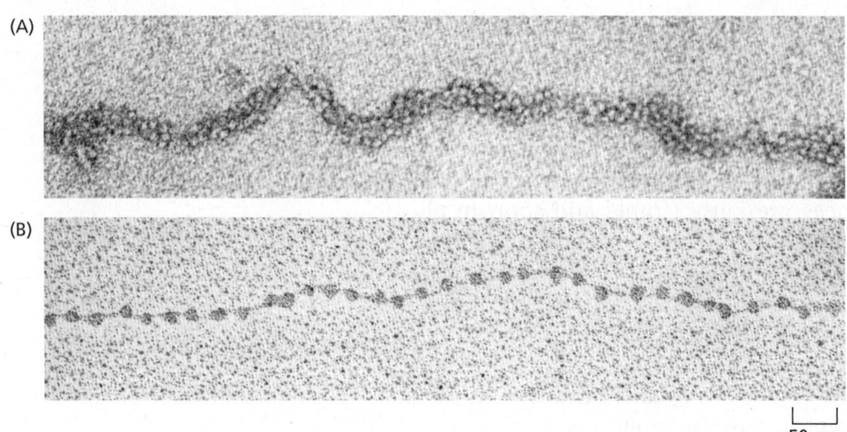

Figure 4–21 **Nucleosomes as seen in the electron microscope.** (A) Chromatin isolated directly from an interphase nucleus appears in the electron microscope as a thread of associated nucleosomes. (B) This electron micrograph shows a length of chromatin that has been experimentally decondensed after isolation to show the nucleosomes. (A, courtesy of Barbara Hamkalo; B, courtesy of Victoria Foe.)

(Figure 4–21A). If this chromatin is subjected to treatments that cause it to unfold partially, it can be seen under the electron microscope as a series of "beads on a string" (Figure 4–21B). The string is DNA, and each bead is a *nucleosome core particle* that consists of DNA wound around a histone core.

The nucleosome core particles can be isolated by digesting chromatin with particular enzymes (called nucleases) that cut DNA. After digestion for a short period, the exposed DNA between the nucleosome core particles, the *linker DNA*, is degraded. Each individual nucleosome core particle is found to consist of a complex of eight histone proteins—two molecules each of histones H2A, H2B, H3, and H4—and double-stranded DNA that is 147 nucleotide pairs long. This *histone octamer* forms a protein core around which the double-stranded DNA is wound (Figure 4–22).

The region of linker DNA that separates each nucleosome core particle from the next can vary in length from a few nucleotide pairs up to about 80. (The term "nucleosome" technically refers to a nucleosome core particle plus one of its adjacent DNA linkers, but it is often used synonymously with "nucleosome core particle.") On average, therefore, nucleosomes repeat at intervals of about 200 nucleotide pairs. For example, a diploid human cell with 6.2×10^9 nucleotide pairs contains approximately 30 million nucleosomes. The formation of nucleosomes converts a DNA molecule into a chromatin thread that is about one-third of its initial length.

The Structure of the Nucleosome Core Particle Reveals How DNA Is Packaged

The high-resolution structure of a nucleosome core particle reveals a disc-shaped histone core around which the DNA is tightly wrapped in a left-handed coil of 1.7 turns (Figure 4–23). All four of the histones that make up the core of the nucleosome are relatively small proteins (102–135 amino acids), and they share a structural motif, known as the *histone fold*, formed from three α helices connected by two loops (Figure 4–24). In assembling a nucleosome, the histone folds first bind to each other to form H3-H4 and H2A-H2B dimers, and the H3-H4 dimers combine to form tetramers. An H3-H4 tetramer then further combines with two H2A-H2B dimers to form a compact *histone octamer*, the core around which the DNA is wound.

The interface between DNA and histone is extensive: 142 hydrogen bonds are formed between DNA and the histone core in each nucleosome. Nearly half

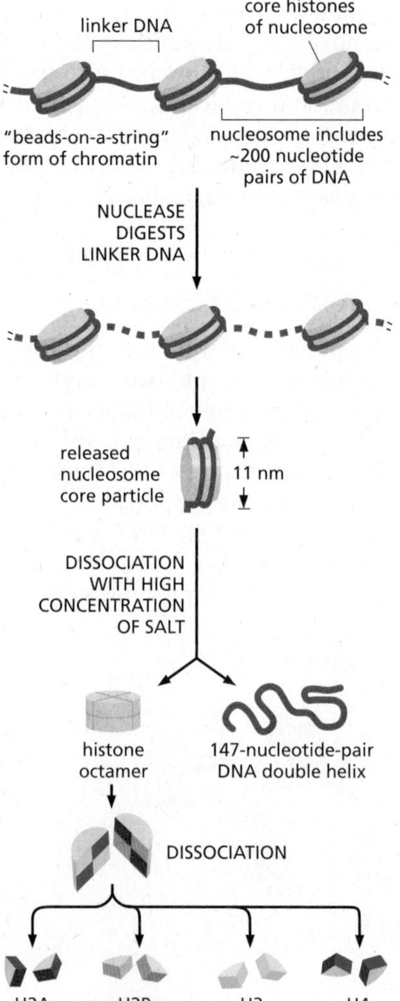

Figure 4–22 **Structural organization of the nucleosome.** A nucleosome contains a protein core made of eight histone molecules. In biochemical experiments, the nucleosome core particle can be released from isolated chromatin by digestion of the linker DNA with a nuclease, an enzyme that hydrolyzes the phosphodiester bonds that connect the nucleotides in DNA. (The nuclease can degrade the exposed linker DNA but cannot attack the DNA wound tightly around the nucleosome core.) After dissociation of the isolated nucleosome into its protein core and DNA, the length of the DNA that was wound around the core can be determined. This length of 147 nucleotide pairs is sufficient to wrap 1.7 times around the histone core.

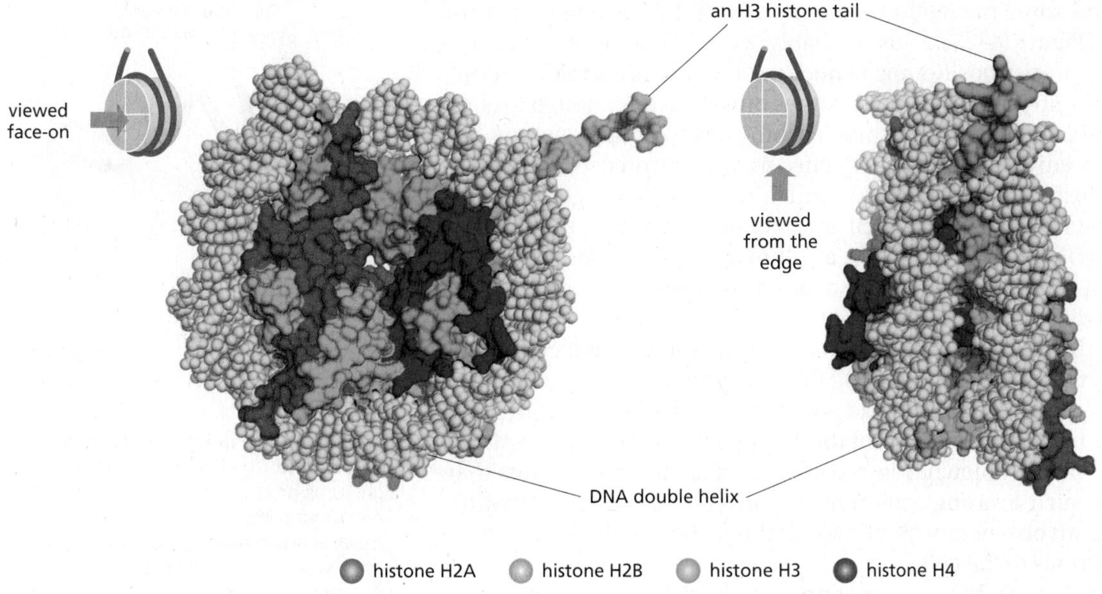

Figure 4–23 The structure of a nucleosome core particle, as determined by x-ray diffraction analyses of crystals. Each histone is colored according to the scheme in Figure 4–22, with the DNA double helix in *light gray*. One H3 N-terminal tail *(green)* can be seen extending from the nucleosome core; the positions of the tails of the other histones could not be determined, due to their disorder. (Adapted from K. Luger et al., *Nature* 389:251–260, 1997.)

of these bonds form between the amino acid backbone of the histones and the sugar–phosphate backbone of the DNA. Numerous hydrophobic interactions and salt linkages also hold DNA and protein together in the nucleosome. More than one-fifth of the amino acids in each of the core histones are either lysine or arginine (two amino acids with basic side chains), and their positive charges can effectively neutralize the negatively charged DNA backbone. These numerous interactions explain in part why DNA of virtually any sequence can be bound on a histone octamer core. The path of the DNA around the histone core is not smooth; rather, several kinks are seen in the DNA, as expected from the nonuniform surface of the core.

The bending of the DNA requires a substantial compression of the minor groove of the DNA helix. Certain dinucleotides in the minor groove are especially

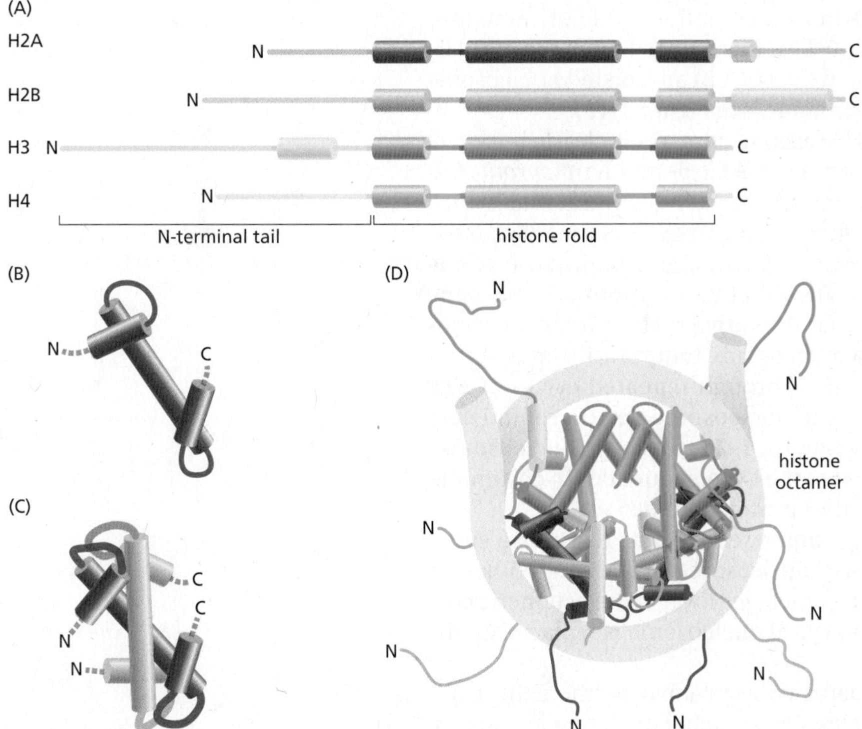

Figure 4–24 The overall structural organization of the core histones. (A) Each of the core histones contains an N-terminal tail, which is subject to several forms of covalent modification, and a histone fold region, as indicated. (B) The structure of the histone fold, which is formed by all four of the core histones. (C) Histones H2A and H2B form a dimer through an interaction known as the "handshake." Histones H3 and H4 form a dimer through the same type of interaction. (D) The final histone octamer around which the DNA is wound. Note that all eight N-terminal tails of the histones protrude from the disc-shaped core structure. Their conformations are highly flexible, and they serve as binding sites for sets of other proteins.

easy to compress, and some nucleotide sequences bind the histone core more tightly than others (**Figure 4–25**). This probably explains some striking, but unusual, cases of very precise positioning of nucleosomes along a stretch of DNA. However, the sequence preference of nucleosomes must be weak enough to allow other factors to dominate, inasmuch as nucleosomes can occupy any one of a number of positions relative to the DNA sequence in most chromosomal regions.

In addition to its histone fold, each of the core histones has a largely unstructured N-terminal amino acid "tail," which extends out from the DNA–histone core (see Figure 4–24D). These histone tails are "hot spots" for different types of covalent modifications that control critical aspects of chromatin structure and function, as we shall discuss shortly.

As a reflection of their fundamental role in DNA function through controlling chromatin structure, the histones are among the most highly conserved eukaryotic proteins. For example, the amino acid sequence of histone H4 from a pea differs from that of a human at only two of the 102 positions. Although histones H2A and H2B have been somewhat less constrained in their evolution than histones H3 and H4, such a strong evolutionary conservation suggests that the functions of histones involve nearly all of their amino acids, so that a change in any position is deleterious to the cell.

In addition to this remarkable conservation, eukaryotic organisms also produce smaller amounts of specialized variant core histones that differ in amino acid sequence from the main ones. As discussed later, these variants, combined with the surprisingly large number of covalent modifications that can be added to the histones in nucleosomes, give rise to a variety of chromatin structures in cells.

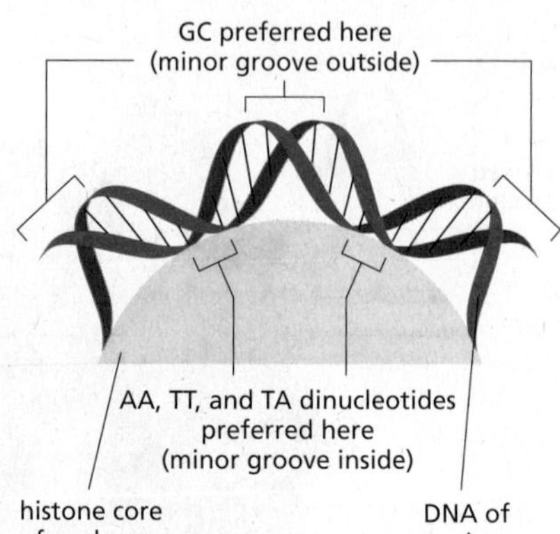

Figure 4–25 The bending of DNA in a nucleosome. The DNA helix makes 1.7 tight turns around the histone octamer. This diagram illustrates how the minor groove is compressed on the inside of the turn. Owing to structural features of the DNA molecule, the indicated dinucleotides are preferentially accommodated in such a narrow minor groove, which helps to explain why certain DNA sequences will bind more tightly than others to the nucleosome core.

Nucleosomes Have a Dynamic Structure and Are Frequently Subjected to Changes Catalyzed by ATP-dependent Chromatin-remodeling Complexes

For many years biologists thought that, once formed in a particular position on DNA, a nucleosome would remain fixed in place because of the very tight association between its core histones and DNA. If true, this would pose problems for genetic readout mechanisms, which require easy access to many specific DNA sequences. It would also hinder the rapid passage of the DNA transcription and replication machinery through chromatin. But kinetic experiments show that the DNA in an isolated nucleosome unwraps from each end at a rate of about four times per second, remaining exposed for 10–50 milliseconds before the partially unwrapped structure recloses. Thus, most of the DNA in an isolated nucleosome is in principle available for binding other proteins (see Figure 7–12).

Inside a cell, a further loosening of DNA–histone contacts is clearly required, because eukaryotic cells contain a variety of ATP-dependent *chromatin-remodeling complexes*. These abundant proteins include a subunit that hydrolyzes ATP (an ATPase evolutionarily related to the DNA helicases discussed in Chapter 5). This "motor subunit" binds both to the protein core of the nucleosome and to the double-stranded DNA that winds around it. By using the energy of ATP hydrolysis to move this DNA relative to the core, the remodeling complex changes the structure of a nucleosome temporarily, making the DNA less tightly bound to the histone core. Through repeated cycles of ATP hydrolysis that pull the DNA helix along the nucleosome core, a remodeling complex can catalyze *nucleosome sliding* (**Figure 4–26**). Because nucleosomes are frequently repositioned in this way, all of the DNA sequences in chromatin are potentially available for binding to other proteins in the cell.

In addition, some types of remodeling complexes are able to remove either all or part of the nucleosome core from a nucleosome—catalyzing either an exchange of its H2A–H2B histones or the complete removal of the octameric core from the DNA (**Figure 4–27**). As a result, a typical nucleosome is replaced on the DNA every 1 or 2 hours inside the cell.

Cells contain dozens of different ATP-dependent chromatin-remodeling complexes that are specialized for different roles, with roughly one complex present per five nucleosomes. Most are large protein complexes that can contain 10 or

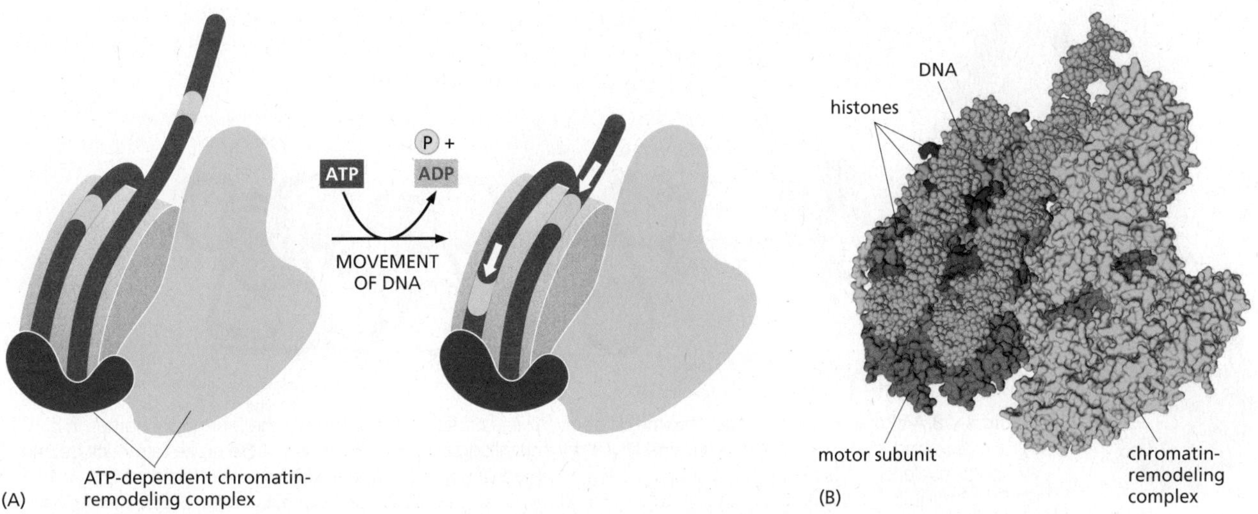

Figure 4–26 **The nucleosome movements catalyzed by ATP-dependent chromatin-remodeling complexes.** (A) Using the energy of ATP hydrolysis, a remodeling complex that catalyzes nucleosome sliding pulls on the DNA of its bound nucleosome and loosens its attachment to the histone octamer. Each cycle of ATP binding, ATP hydrolysis, and release of the ADP and phosphate products thereby moves the DNA in the direction of the arrows in this diagram. Because each cycle moves the DNA by one base pair, it requires many such cycles to produce the nucleosome sliding shown. (B) The structure of the yeast SWR1 chromatin-remodeling complex. This complex catalyzes an exchange of an H2A–H2B dimer for a dimer that contains an H2A histone variant. Its ATP-driven motor subunit is colored *purple*. (B, PDB code: 6GEJ.)

more subunits, some of which bind to specific modifications on histones. The activity of these complexes is controlled by the cell. As genes are turned on and off, chromatin-remodeling complexes are brought to specific regions of DNA where they act locally to influence chromatin structure (discussed in Chapter 7).

Although some DNA sequences bind more tightly than others to the nucleosome core (see Figure 4–25), the most important influence on nucleosome positioning appears to be the presence of other tightly bound proteins on the DNA. Some bound proteins favor the formation of a nucleosome adjacent to them. Others create obstacles that force the nucleosomes to move elsewhere. The exact positions of nucleosomes along a stretch of DNA therefore depend mainly

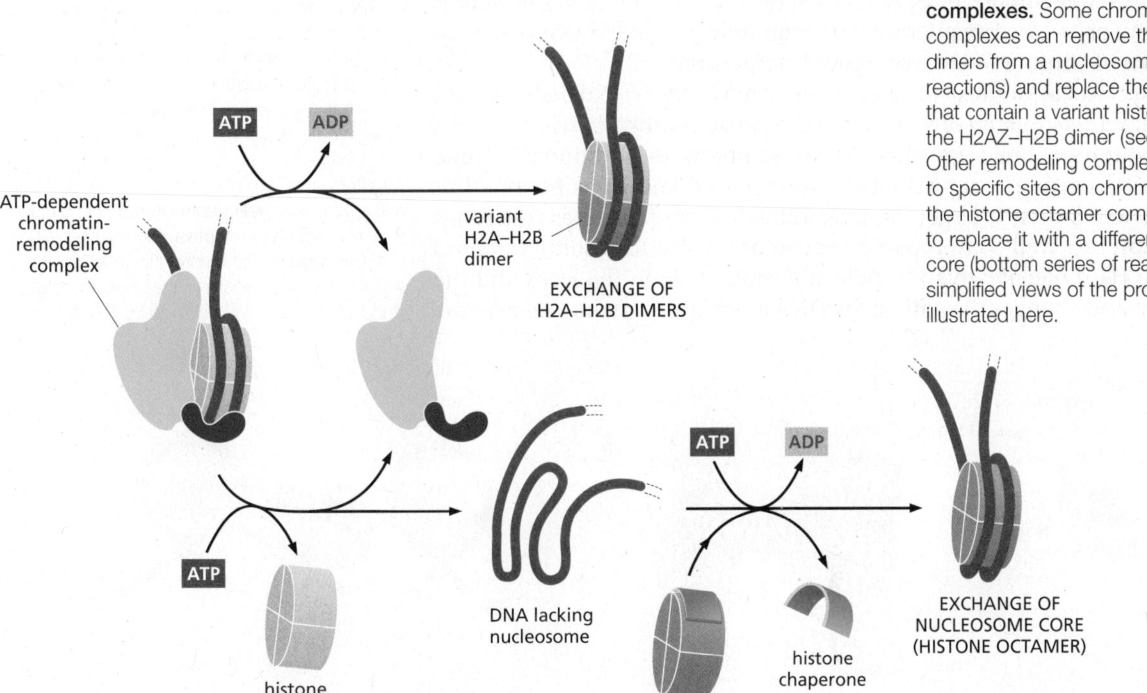

Figure 4–27 **Nucleosome removal and histone exchange catalyzed by ATP-dependent chromatin-remodeling complexes.** Some chromatin-remodeling complexes can remove the H2A–H2B dimers from a nucleosome (top series of reactions) and replace them with dimers that contain a variant histone, such as the H2AZ–H2B dimer (see Figure 4–36). Other remodeling complexes are attracted to specific sites on chromatin to remove the histone octamer completely and/or to replace it with a different nucleosome core (bottom series of reactions). Highly simplified views of the processes are illustrated here.

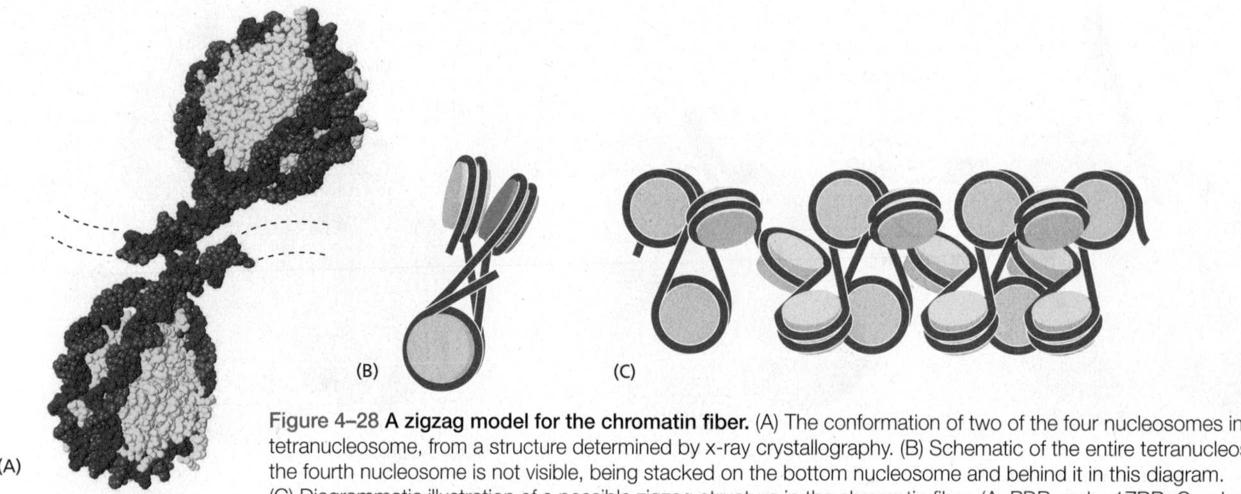

Figure 4–28 A zigzag model for the chromatin fiber. (A) The conformation of two of the four nucleosomes in a tetranucleosome, from a structure determined by x-ray crystallography. (B) Schematic of the entire tetranucleosome; the fourth nucleosome is not visible, being stacked on the bottom nucleosome and behind it in this diagram. (C) Diagrammatic illustration of a possible zigzag structure in the chromatin fiber. (A, PDB code: 1ZBB; C, adapted from C.L. Woodcock, *Nat. Struct. Mol. Biol.* 12:639–640, 2005.)

on the presence and nature of other proteins bound to the DNA. And due to the presence of ATP-dependent chromatin-remodeling complexes, the arrangement of nucleosomes on DNA can be highly dynamic, changing rapidly according to the needs of the cell.

Attractions Between Nucleosomes Compact the Chromatin Fiber

Although enormously long strings of nucleosomes form on the chromosomal DNA, chromatin in a living cell probably rarely adopts the extended "beads-on-a-string" form. Instead, the nucleosomes are packed on top of one another, generating arrays in which the DNA is more highly condensed. Thus, when nuclei are very gently lysed onto an electron microscope grid, much of the chromatin is seen to be in the form of a fiber that is considerably wider than an individual nucleosome (see Figure 4–21A).

How nucleosomes are organized into condensed arrays is unclear. The structure of a tetranucleosome (a complex of four nucleosomes) obtained by x-ray crystallography and high-resolution electron microscopy of reconstituted chromatin has been used to support a zigzag model for the stacking of nucleosomes (**Figure 4–28**). But cryo-electron microscopy of carefully prepared nuclei suggests that most regions of chromatin are less regularly structured.

What causes nucleosomes to stack on each other? Nucleosome-to-nucleosome attractions that involve histone tails, most notably the H4 tail, constitute one important factor (**Figure 4–29**). In addition, an additional histone is often present in a 1-to-1 ratio with nucleosome cores, known as **histone H1**. This so-called *linker histone* is larger than the individual core histones, and it has been considerably less well conserved during evolution. A nucleosome can bind a single histone H1 molecule; this H1 molecule contacts both the DNA and the histone octamer and changes the path of the DNA as it exits from the nucleosome

Figure 4–29 A model for the role played by histone tails in the compaction of chromatin. (A) A schematic diagram shows the approximate exit points of the eight N-terminal histone tails, one from each histone protein, that extend from each nucleosome. The actual structure is shown to its *right*. In the high-resolution structure of the nucleosome, most of the histone tails are not visible, suggesting that they are relatively unstructured and highly flexible. (B) As indicated, the histone tails are thought to be involved in interactions between nucleosomes that help to hold them together. (A, PDB code: 1KX5.)

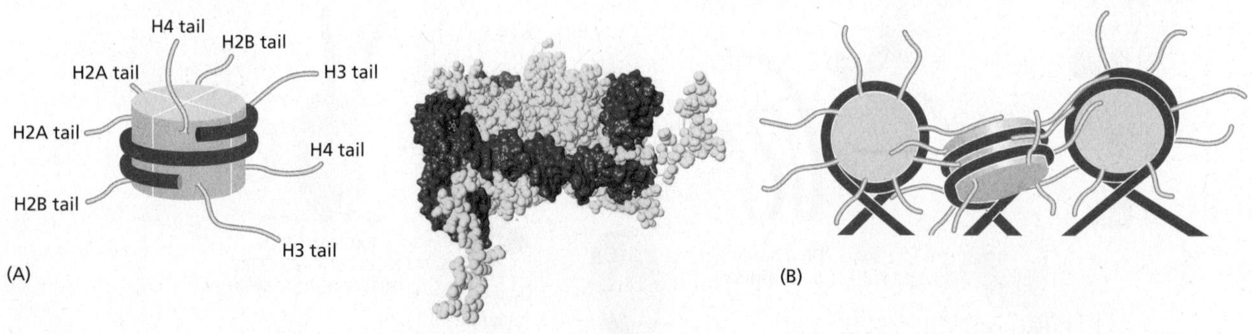

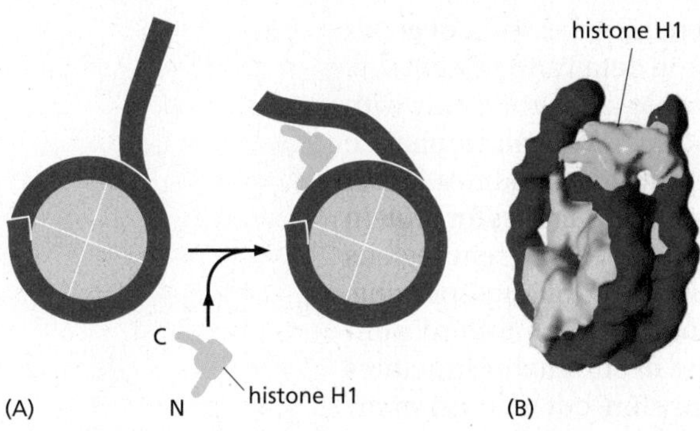

Figure 4–30 **How the linker histone binds to the nucleosome.** The position and structure of histone H1 are shown. H1 constrains the DNA where it exits from the nucleosome and thereby compacts chromatin. (A) Schematic, and (B) structure inferred for a single nucleosome from a structure determined by high-resolution electron microscopy of a reconstituted chromatin fiber. (B, adapted from F. Song et al., *Science* 344:376–380, 2014.)

(**Figure 4–30**). The change in the exit path of DNA is thought to help compact nucleosomal DNA. The presence of many other DNA-binding proteins, as well as proteins that bind directly to histones, is certain to add important additional features to any array of nucleosomes.

Summary

A gene is a nucleotide sequence in a DNA molecule that acts as a functional unit for the production of a protein, a structural RNA, or a catalytic or regulatory RNA molecule. In eukaryotes, protein-coding genes are usually composed of a string of alternating introns and exons associated with regulatory regions of DNA. A chromosome is formed from a single, enormously long DNA molecule that contains a linear array of many genes, bound to a large set of proteins. The human genome contains 3.1×10^9 DNA nucleotide pairs, divided between 22 different autosomes (present in two copies each) and 2 sex chromosomes. Only a small percentage of this DNA codes for proteins or functional RNA molecules. A chromosomal DNA molecule also contains three other types of important nucleotide sequences: replication origins and telomeres allow the DNA molecule to be efficiently replicated, while a centromere attaches the sister DNA molecules to the mitotic spindle, ensuring their accurate segregation to daughter cells during the M phase of the cell cycle.

The DNA in eukaryotes is tightly bound to an equal mass of histones, which form repeated arrays of DNA–protein particles called nucleosomes. The nucleosome is composed of an octameric core of histone proteins around which the DNA double helix is wrapped. Nucleosomes are spaced at intervals of about 200 nucleotide pairs, and they associate with neighboring nucleosomes to create a more compact chromatin fiber. Chromatin structure must be highly dynamic to allow access to the DNA. Some spontaneous DNA unwrapping and rewrapping occurs in the nucleosome itself; however, the general strategy for reversibly changing local chromatin structure in a cell features ATP-driven chromatin-remodeling complexes. Cells contain a large set of such complexes, which are targeted to specific regions of chromatin at appropriate times. The remodeling complexes allow nucleosome cores to be repositioned, reconstituted with different histones, or completely removed to expose the underlying DNA.

THE EFFECT OF CHROMATIN STRUCTURE ON DNA FUNCTION

Having described how DNA is packaged into nucleosomes to create a chromatin fiber, we next describe the mechanisms that create different chromatin structures in different regions of a cell's genome and how this affects DNA function in cells. Most strikingly, we shall see that some types of chromatin structure can be inherited; that is, the structure can be directly passed down from a cell to its descendants. Because this creates a cell memory that is not based on an inherited change in DNA sequence, it creates a form of **epigenetic inheritance**. The prefix *epi* is Greek for "on"; this is appropriate, because epigenetics represents a form of inheritance that is superimposed on the DNA-based genetic inheritance.

Chapter 7 explains the many different ways in which the expression of genes is regulated; there we will discuss epigenetic inheritance in detail and present several different mechanisms that can produce it. Here, we are concerned only with those epigenetic mechanisms based on chromatin structure. We shall emphasize some of the chemistry that makes this possible—the covalent modification of histones in nucleosomes. These modifications serve as recognition sites for protein domains that bind different non-histone protein complexes to different regions of chromatin. In total, these complexes are so abundant that, roughly speaking, chromatin consists of one-third DNA, one-third histones, and one-third non-histone proteins by mass (see Figure 4–20). The varieties of chromatin structures produced have critical effects not only on gene expression but also on many other DNA-dependent processes—playing an important role in the development, growth, and maintenance of all eukaryotic organisms, including ourselves.

Different Regions of the Human Genome Are Packaged Very Differently in Chromatin

Light-microscope studies in the 1930s distinguished two types of chromatin in the interphase nuclei of higher eukaryotic cells: a highly condensed form, called **heterochromatin**, and all the rest, called **euchromatin**, which is less condensed. Powerful new types of molecular analyses have now allowed scientists to classify these two chromatin types more precisely. Chromatin is now considered to be either "open and active" or "closed and inactive." As illustrated in Figure 4–31, roughly 80% of the human genome is in the closed form, half of which is compactly packaged in heterochromatin, with the other, less condensed half (part of the euchromatin) designated as "quiescent." Only about 20% of the genome is packaged in that portion of the euchromatin associated with the actively expressed genes that will be the focus of Chapters 6 and 7.

Heterochromatin Is Highly Condensed and Restricts Gene Expression

Heterochromatin represents an especially compact form of chromatin (see Figure 4–10), and we are finally beginning to understand its molecular properties. Some heterochromatin is concentrated in specialized chromosomal regions, most notably at the centromeres and telomeres introduced previously (see Figure 4–19). But heterochromatin is also present at many other locations along chromosomes—locations that can vary according to the developmental state of the cell.

The packaging of DNA in heterochromatin typically prevents gene expression. However, we know now that the term *heterochromatin* encompasses a number of

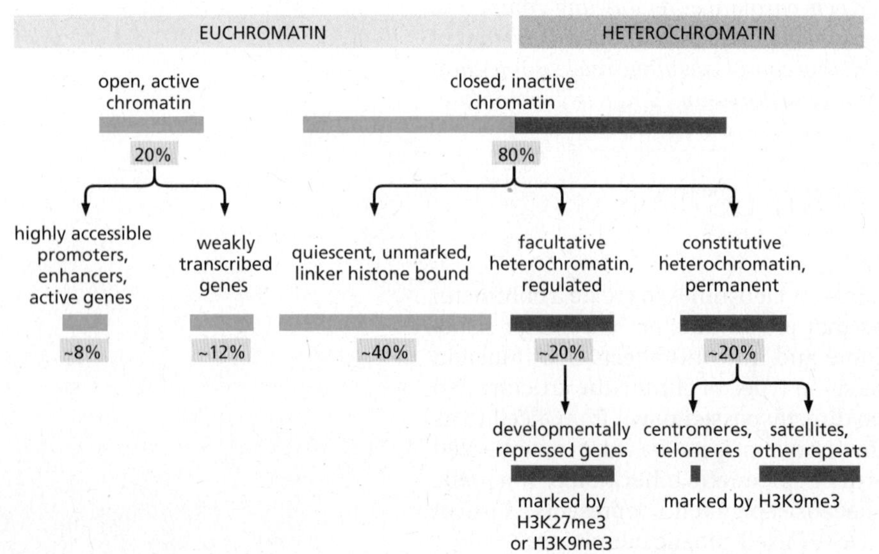

Figure 4–31 Some distinct ways that DNA is packaged in chromatin in a mammalian cell. Although all of these types of chromatin are based on long strings of nucleosomes, the nucleosomes are differently organized through association with different non-histone proteins. As we discuss shortly, these associations can depend on the particular ways that the histones in each nucleosome have been covalently "marked." For example, as indicated here, the histone H3 molecules in heterochromatin are marked either with trimethylated lysine 9 (H3K9me3) or with trimethylated lysine 27 (H3K27me3), depending on the particular structure that is formed (see Figure 4–40).

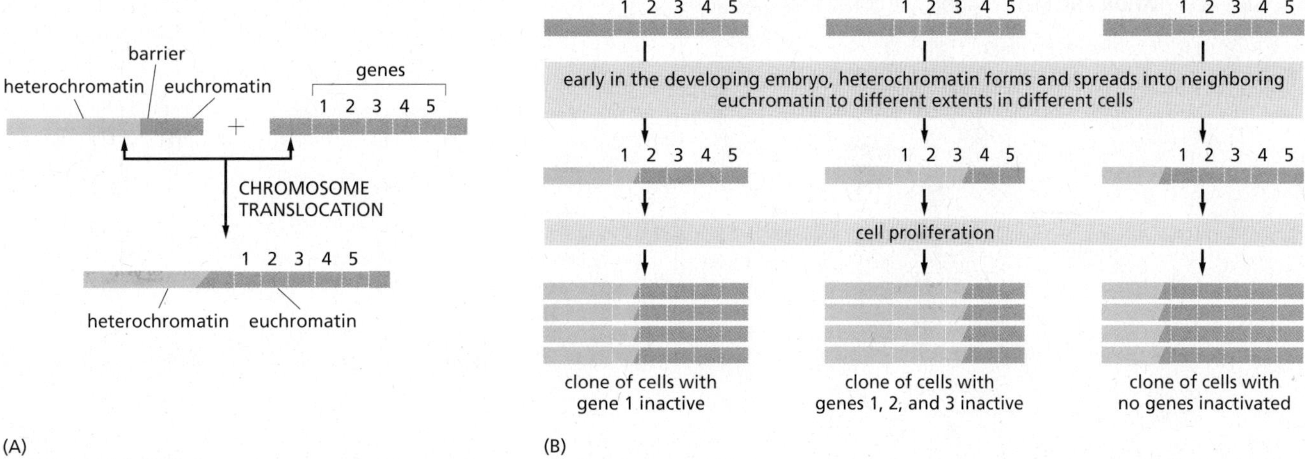

Figure 4–32 The cause of position effect variegation in *Drosophila*. (A) Heterochromatin *(green)* is normally prevented from spreading into adjacent regions of euchromatin *(red)* by *barrier* DNA sequences, which we shall discuss shortly. In flies that inherit certain chromosomal rearrangements, however, this barrier is no longer present. (B) During the early development of such flies, heterochromatin can spread into neighboring chromosomal DNA, proceeding for different distances in different cells. This spreading soon stops, but the established pattern of heterochromatin is subsequently inherited, so that large clones of progeny cells are produced that have the same neighboring genes condensed into heterochromatin and thereby inactivated (hence the "variegated" appearance of some of these flies; see Figure 4–33).

distinct modes of chromatin compaction that have different functions. There are two broad classes, distinguished as *constitutive heterochromatin* and *facultative heterochromatin* (see Figure 4–31). The constitutive class permanently condenses many regions of the genome (and hence its name), whereas the facultative class of heterochromatin can be regulated to control gene expression.

As described next, heterochromatin has the critical property of being able to self-propagate.

The Heterochromatic State Can Spread Along a Chromosome and Be Inherited from One Cell Generation to the Next

Through errors that can occur in chromosome rejoining when DNA double-strand breaks are repaired, a piece of chromosome that is normally euchromatic can be accidentally translocated into the neighborhood of heterochromatin. Remarkably, this often causes the *silencing*—inactivation—of normally active genes, a phenomenon referred to as a *position effect*. First recognized in the fruit fly *Drosophila*, such position effects have now been observed in many eukaryotes, including yeasts, plants, and humans. Position effects are caused by a spreading of the heterochromatic state into an originally euchromatic region, and detailed studies of this phenomenon have provided important clues to the mechanisms that create and maintain heterochromatin.

After the above type of chromosome breakage-and-rejoining event, the zone of silencing, where euchromatin is converted to a heterochromatic state, is found to spread for different distances in different cells of the early fly embryo. Remarkably, these differences then are perpetuated for the rest of the animal's life: in each cell, once the heterochromatic condition is established on a piece of chromatin, it tends to be stably inherited by all of that cell's progeny (**Figure 4–32**). This remarkable phenomenon, called *position effect variegation*, was first recognized as a mottled loss of red pigment in the fly eye (**Figure 4–33**).

These observations, taken together, point to a fundamental feature of heterochromatin formation: heterochromatin begets more heterochromatin. This

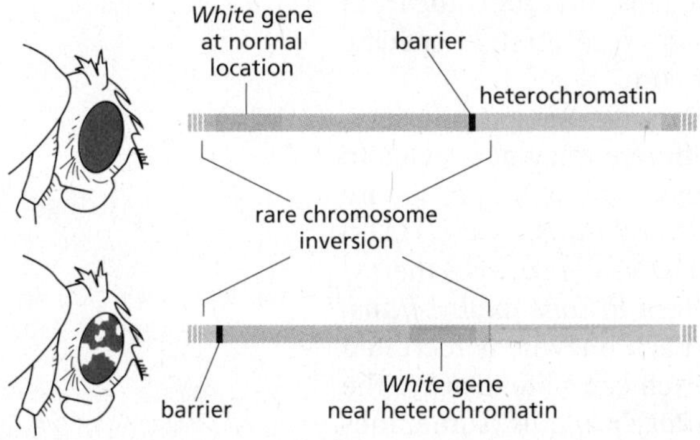

Figure 4–33 The discovery of position effects on gene expression. The *White* gene in the fruit fly *Drosophila* controls eye pigment production and is named after the mutation that first identified it. Wild-type flies with a normal *White* gene (*White*$^+$) have normal pigment production, which gives them red eyes, but if the *White* gene is mutated and inactivated, the mutant flies (*White*$^-$) make no pigment and have white eyes. In flies in which a normal *White* gene has been moved near a region of heterochromatin, the eyes are mottled, with both red and white patches. The white patches represent cell lineages in which the *White* gene has been silenced by the effects of the heterochromatin. In contrast, the red patches represent cell lineages in which the *White* gene is expressed. Early in development, when the heterochromatin is first formed, it spreads into neighboring euchromatin to different extents in different embryonic cells (see Figure 4–32). The presence of large patches of red and white cells reveals that the state of transcriptional activity, as determined by the packaging of this gene into chromatin in those ancestor cells, is inherited by all daughter cells.

(A) LYSINE ACETYLATION AND METHYLATION ARE COMPETING REACTIONS

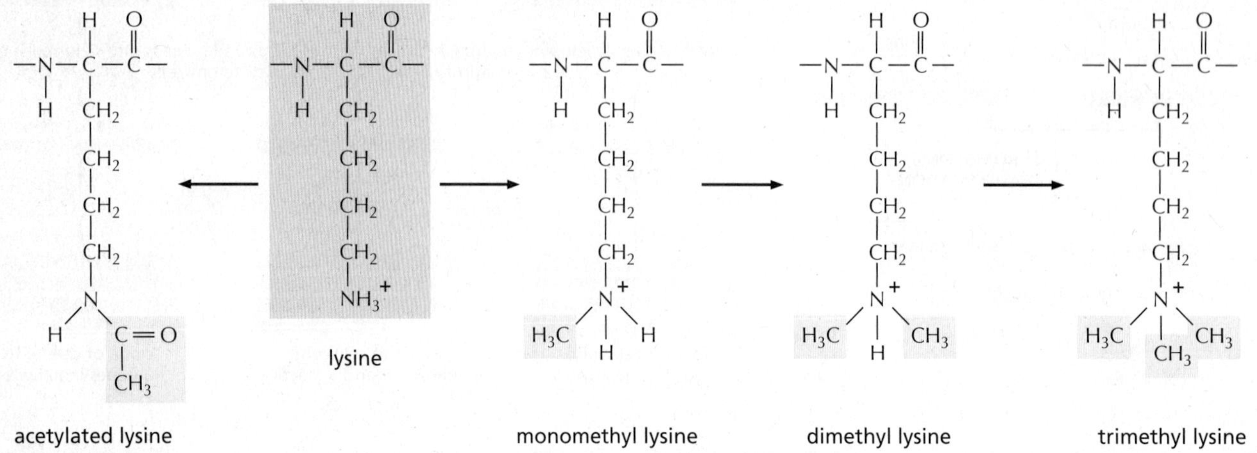

acetylated lysine ← lysine → monomethyl lysine → dimethyl lysine → trimethyl lysine

(B) SERINE PHOSPHORYLATION

serine → phosphoserine

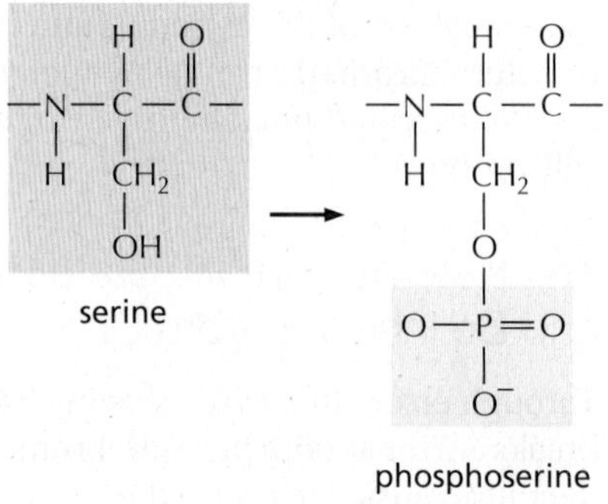

Figure 4–34 Some prominent types of covalent amino acid side-chain modifications found on nucleosomal histones. (A) Three different levels of lysine methylation are shown; each can be recognized by a different binding protein, and thus each can have a different significance for the cell. Note that acetylation removes the plus charge on lysine, and that, perhaps most important, an acetylated lysine cannot be methylated, and vice versa. (B) Serine phosphorylation adds a negative charge to a histone. Modifications of histones not shown here include the mono- or dimethylation of an arginine, the phosphorylation of a threonine, the addition of ADP-ribose to a glutamic acid, and the addition of a ubiquityl, sumoyl, or biotin group to a lysine.

positive feedback can operate both in space, causing the heterochromatic state to spread along the chromosome, and in time, across cell generations, propagating the heterochromatic state of the parent cell to its daughters. The challenge is to explain the molecular mechanisms that underlie this remarkable behavior.

As a first step, one can carry out a search for the molecules that are involved. This has been done by means of *genetic screens*, in which large numbers of mutants are generated, after which one picks out those that show an abnormality of the process in question. Extensive genetic screens in *Drosophila*, fungi, and mice have identified more than 100 genes whose products either enhance or suppress the spread of heterochromatin and its stable inheritance—in other words, genes that serve as either *enhancers* or *suppressors* of position effect variegation. Many of these genes turn out to code for non-histone chromosomal proteins that interact with histones and are involved in modifying or maintaining chromatin structure. These include genes that encode some of the enzymes that add or remove covalent modifications to histone side chains, as we discuss next.

The Core Histones Are Covalently Modified at Many Different Sites

The amino acid side chains of the four histones in the nucleosome core are subjected to a remarkable variety of covalent modifications, including the acetylation of lysines, the mono-, di-, and trimethylation of lysines, and the phosphorylation of serines (Figure 4–34). A large number of these side-chain modifications occur on the eight relatively unstructured N-terminal "histone tails" that protrude from the nucleosome (Figure 4–35). However, there are also more than 20 specific side-chain modifications on the nucleosome's globular core.

All of the above types of modifications are reversible, with one enzyme serving to create a particular type of modification and a different enzyme serving to remove it. These enzymes are highly specific. Thus, for example, acetyl groups are added to specific lysines by a set of different *histone acetyl transferases* (HATs) and removed by a set of *histone deacetylase complexes* (HDACs). Likewise, methyl groups are added to lysine side chains by a set of different *histone methyl transferases* and removed by a set of *histone demethylases*. Each enzyme is recruited to specific sites on the chromatin at defined times in each cell's life history. The initial recruitment can depend on *transcription regulatory proteins* (sometimes

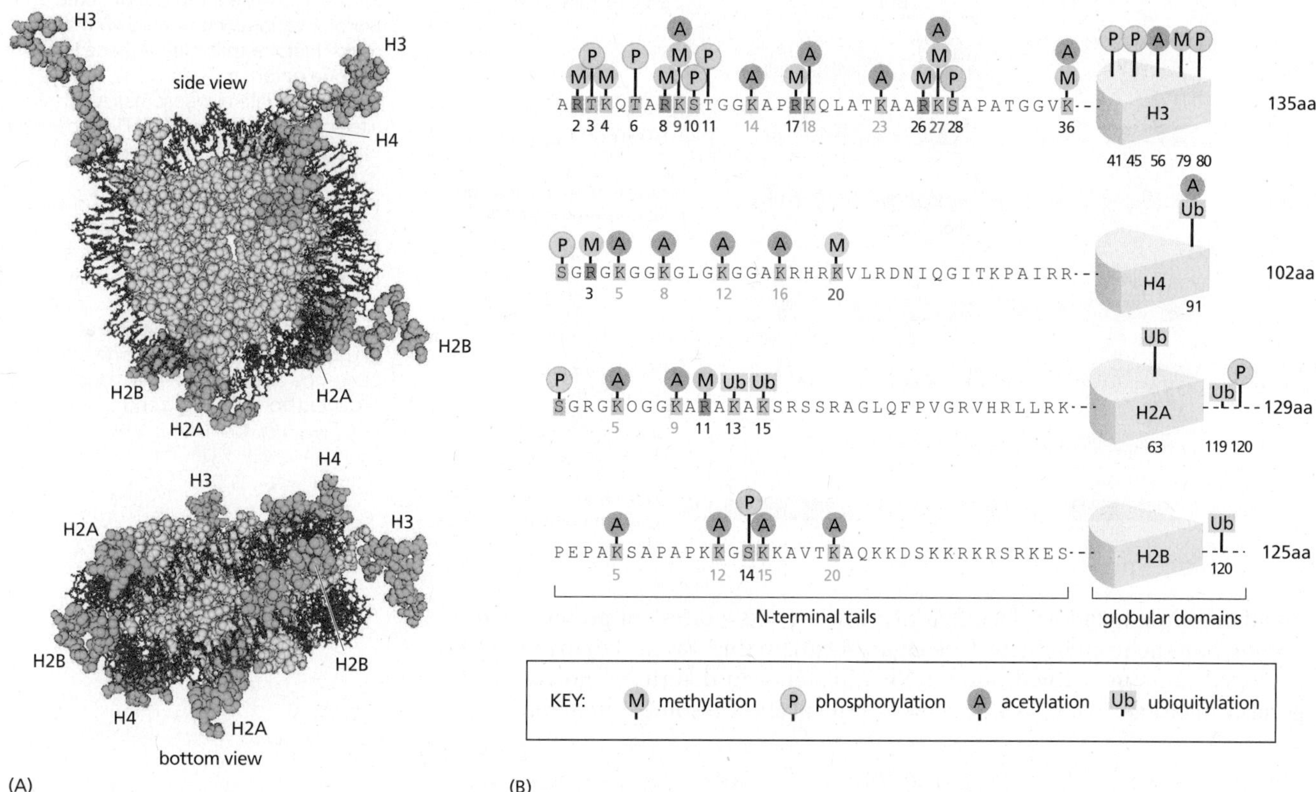

Figure 4–35 The covalent modification of core histones. (A) The structure of the nucleosome highlighting the location of the first 30 or so amino acids in each of its eight N-terminal histone tails *(green)*. These tails are unstructured and highly mobile, and thus will change their conformation depending on other bound proteins. (B) Some well-documented modifications of the four histone core proteins are indicated. Notably, although only a single symbol is used here for methylation (M), each lysine (K) or arginine (R) can be methylated in several different ways. Thus, for example, mono-, di-, or trimethyl lysine groups can have very different effects. Note also that some positions (for example, lysine 9 of H3) can be modified either by methylation or by acetylation, but not both. Most of the modifications shown add a relatively small molecule onto a histone; the exception is ubiquitin, a 76-amino-acid protein also used for other cell processes (see Figure 3–65). Note that while most of the modifications occur in unstructured regions, including on short tails at the C-terminus of histone H2A and histone H2B, there are also important modifications in structured histone folds. (A, PDB code: 1KX5; B, based on D. Allis et al., Epigenetics, 2nd ed., Overview and concepts. Cold Spring Harbor, NY: Cold Spring Harbor Laboratory Press, 2015.)

called "transcription factors"). As we shall explain in Chapter 7, these proteins recognize and bind to specific DNA sequences in the chromosomes. They are produced at different times and places in the life of an organism, thereby determining where and when the chromatin-modifying enzymes will act. In this way, the DNA sequence ultimately determines how histones are modified. But as we shall discuss shortly, the covalent modifications on nucleosomes in heterochromatin can persist long after the transcription regulatory proteins that first induced them have disappeared, thereby providing the cell with a memory of its developmental history. Most remarkably, as in the related phenomenon of position effect variegation discussed earlier, this "heterochromatin memory" can be transmitted from one cell generation to the next.

Very different patterns of covalent modification are found on different groups of nucleosomes, depending both on their exact position in the genome and on the history of the cell. The modifications of the histones are carefully controlled, and they can have important consequences. The acetylation of lysines on the N-terminal tails loosens chromatin structure, in part because adding an acetyl group to lysine removes its positive charge. However, the most profound effects of the histone modifications lie in their ability to recruit specific other proteins to the modified stretch of chromatin. For example, as we discuss shortly, the trimethylation of one specific lysine on the histone H3 tail attracts the protein HP1 and promotes the establishment and spread of one type of heterochromatin,

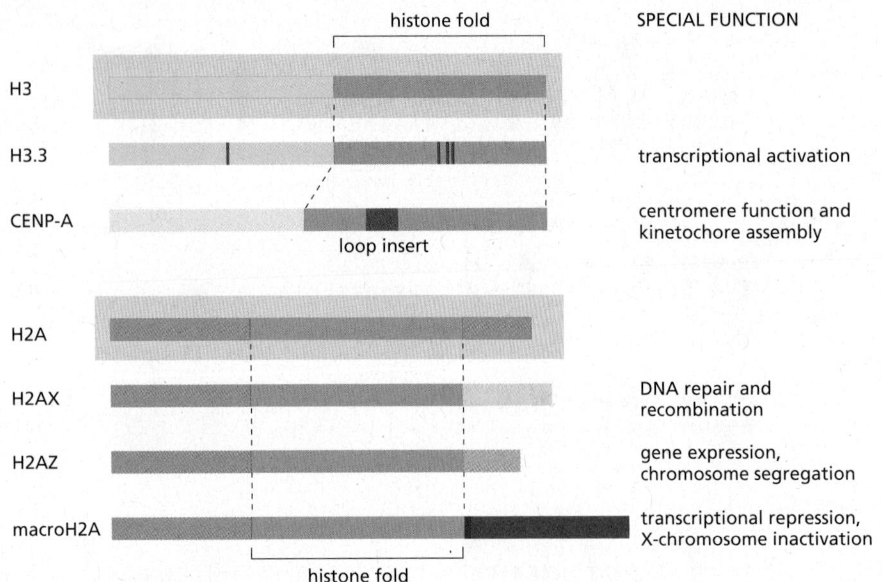

Figure 4–36 **The structure of some histone variants compared with the major histone that they replace in a histone octamer.** The histone variants are inserted into nucleosomes at specific sites on chromosomes by ATP-dependent chromatin-remodeling complexes (see Figure 4–26). The CENP-A (centromere protein-A) variant of histone H3 is discussed later in this chapter (see Figure 4–42); other variants are discussed in Chapter 7. The sequences in each variant that are colored differently (compared to the major histone above it) denote regions with an amino acid sequence different from this major histone. (Adapted from K. Sarma and D. Reinberg, *Nat. Rev. Mol. Cell Biol.* 6:139–149, 2005.)

while the trimethylation of a different lysine attracts a different protein to form a second heterochromatin type (see Figure 4–40). In this way and many others, the recruited proteins act with the modified histones to determine how and when genes will be expressed, as well as other critical chromosome functions.

Chromatin Acquires Additional Variety Through the Site-specific Insertion of a Small Set of Histone Variants

In addition to the four highly conserved standard core histones, eukaryotes also contain a few variant histones that can also assemble into nucleosomes. These histones are present in much smaller amounts than the major histones, and they have been less well conserved over long evolutionary times. Variants are known for each of the core histones with the exception of H4; some examples are shown in **Figure 4–36**.

The major histones are synthesized primarily during the S phase of the cell cycle and assembled into nucleosomes on the daughter DNA helices just behind the replication fork (see Figure 5–32). In contrast, most histone variants are synthesized throughout interphase. They are often inserted into already-formed chromatin. This requires a histone-exchange process catalyzed by a special ATP-dependent chromatin-remodeling complex that binds both the variant histone and the nucleosome containing the histone to be exchanged (see Figure 4–27). These remodeling complexes contain subunits that cause them to bind to specific sites on chromatin. As a result, each histone variant is inserted into chromatin in a highly selective manner.

Covalent Modifications and Histone Variants Can Act in Concert to Control Chromosome Functions

The number of possible distinct markings on an individual nucleosome is in principle enormous, and this potential for diversity becomes still greater when one considers the nucleosomes that contain histone variants. The histone modifications are known to occur in coordinated sets, and more than 15 such sets can be identified in mammalian cells. However, it is not yet clear how many different types of chromatin are functionally important.

Some combinations are known to have a specific meaning for the cell in the sense that they determine how and when the DNA packaged in the nucleosomes is to be accessed or manipulated—a fact that led to the idea of a "histone code." For example, one type of marking signals that a stretch of chromatin has been newly replicated, another signals that the DNA in that chromatin has been

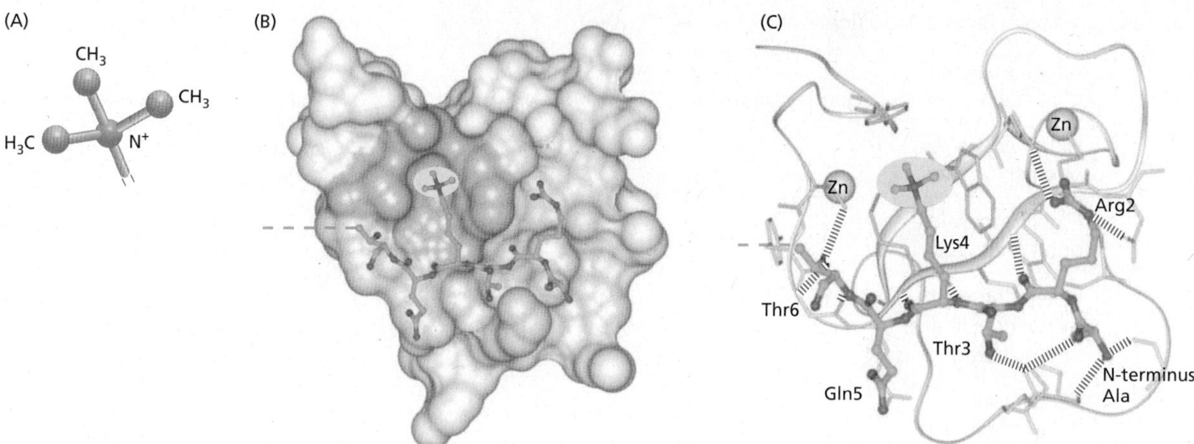

damaged and needs repair, while others signal when and how gene expression should take place. Various regulatory proteins contain small domains that bind to specific marks, recognizing, for example, a trimethylated lysine at a specific position in one of the histones (**Figure 4–37**; see also Figure 9–52). These domains are often linked together as modules in a large protein complex, such as in an ATP-dependent chromatin remodeler, allowing that complex to recognize a specific *combination* of histone modifications. These so-called *reader protein complexes* (**Figure 4–38**) allow particular combinations of markings on chromatin to attract a set of additional proteins, so as to execute an appropriate biological function at the right time.

Figure 4–37 **How a mark on a nucleosome is read.** The figure shows the structure of a protein module (called an ING PHD domain) that specifically recognizes histone H3 trimethylated on lysine 4. (A) A trimethyl group. (B) Space-filling model of an ING PHD domain bound to a histone tail (*green*, with the trimethyl group highlighted in *yellow*). (C) A ribbon model showing how the N-terminal six amino acids in the H3 tail are recognized. The *hatched red lines* represent hydrogen bonds. This is one of a family of PHD domains that recognize methylated lysines on histones; different members of the family bind tightly to lysines located at different positions, and they can discriminate between a mono-, di-, and trimethylated lysine. In a similar way, other small protein modules recognize specific histone side chains that have been marked with acetyl groups, phosphate groups, and so on. (Adapted from P.V. Peña et al., *Nature* 442:100–103, published 2006 by Nature Publishing Group. Reproduced with permission of SNCSC.)

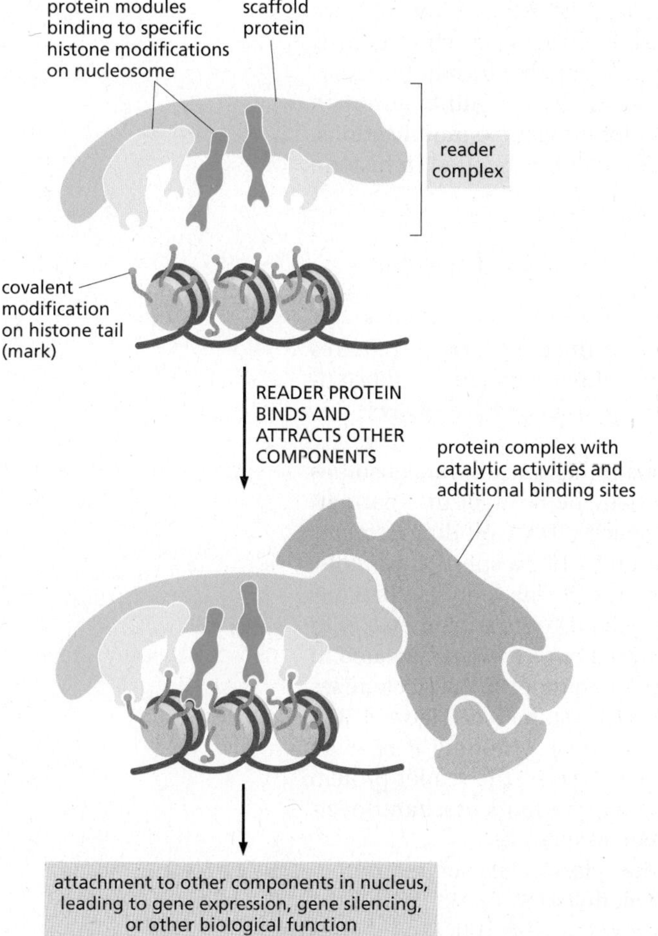

Figure 4–38 **Schematic diagram showing how a particular combination of histone modifications can be recognized by a reader protein complex.** A large protein complex that contains a series of protein modules, each of which recognizes a specific histone mark, is schematically illustrated *(green)*. This *reader complex* will bind tightly only to a region of chromatin that contains several of the different histone marks that it recognizes. Therefore, only a specific combination of marks will cause the complex to bind to chromatin and attract the additional protein complexes *(purple)* needed to catalyze a biological function.

TABLE 4–2 Some Effects of Histone Modifications

Histone modification	Association with chromatin type	Gene expression	Abundance (%)
H3K4me3	Highly accessible, open chromatin	ON	1
H3K9ac	Highly accessible, open chromatin	ON	1
H3K9me3	Heterochromatin (either constitutive or facultative)	OFF	25
H3K27me3	Facultative heterochromatin	OFF	13

In general, sets of modifications act in combination, but only a small number of their meanings are clear. For how the two different kinds of heterochromatin listed here form, see Figure 4–40.

Each of the marks on nucleosomes is denoted as follows: listed first is an abbreviation for the particular histone involved, written as H3, H4, H2A, or H2B. The marked amino acid side chain then follows, using its one-letter abbreviation followed by its distance from that histone's amino-terminus—thus, for example, H3K9 or H4K4 (see Figure 4–35). Listed last is the type of modification on that amino acid side chain, as in H3K9ac (acetylated), H3K9me2 (dimethylated), or H3K9me3 (trimethylated).

All of the covalent additions to histones are dynamic, being constantly removed and added at rates that depend both on their chromosomal locations and on specific states of the cell. Because the histone tails extend outward from the nucleosome core and are likely to be accessible even when chromatin is condensed, they would seem to provide an especially suitable format for creating marks that can be readily altered as a cell's needs change. Although much remains to be learned about the meaning of the different histone modifications, a few well-studied examples of the information that can be encoded in histone modifications are listed in **Table 4–2**.

A Complex of Reader and Writer Proteins Can Spread Specific Chromatin Modifications Along a Chromosome

The phenomenon of position effect variegation described previously requires that some modified forms of chromatin have the ability to spread for substantial distances along a chromosomal DNA molecule (see Figure 4–32). How is this possible?

The enzymes that add or remove modifications to histones in nucleosomes are part of multisubunit complexes. They can initially be brought to a particular region of chromatin by one of the sequence-specific DNA-binding proteins (transcription regulators) discussed in Chapters 6 and 7 (for a specific example, see Figure 7–23). But after a modifying enzyme "writes" its mark on one or a few neighboring nucleosomes, events that resemble a chain reaction can ensue. In such a case, the *writer enzyme* works in concert with a *reader protein* located in the same protein complex. The reader protein contains a module that recognizes the mark and binds tightly to the newly modified nucleosome (see Figure 4–37), allosterically activating an attached writer enzyme and positioning it near an adjacent nucleosome. Through many such read–write cycles, the reader protein can carry the writer enzyme along the DNA—spreading the mark in a hand-over-hand manner along the chromosome (**Figure 4–39A**; Movie 4.2).

As important examples, there are two major classes of heterochromatin in mammalian cells, one centered on the trimethylation of H3K9 and the other on the trimethylation of H3K27 (see Figure 4–31). The H3K27me3 class

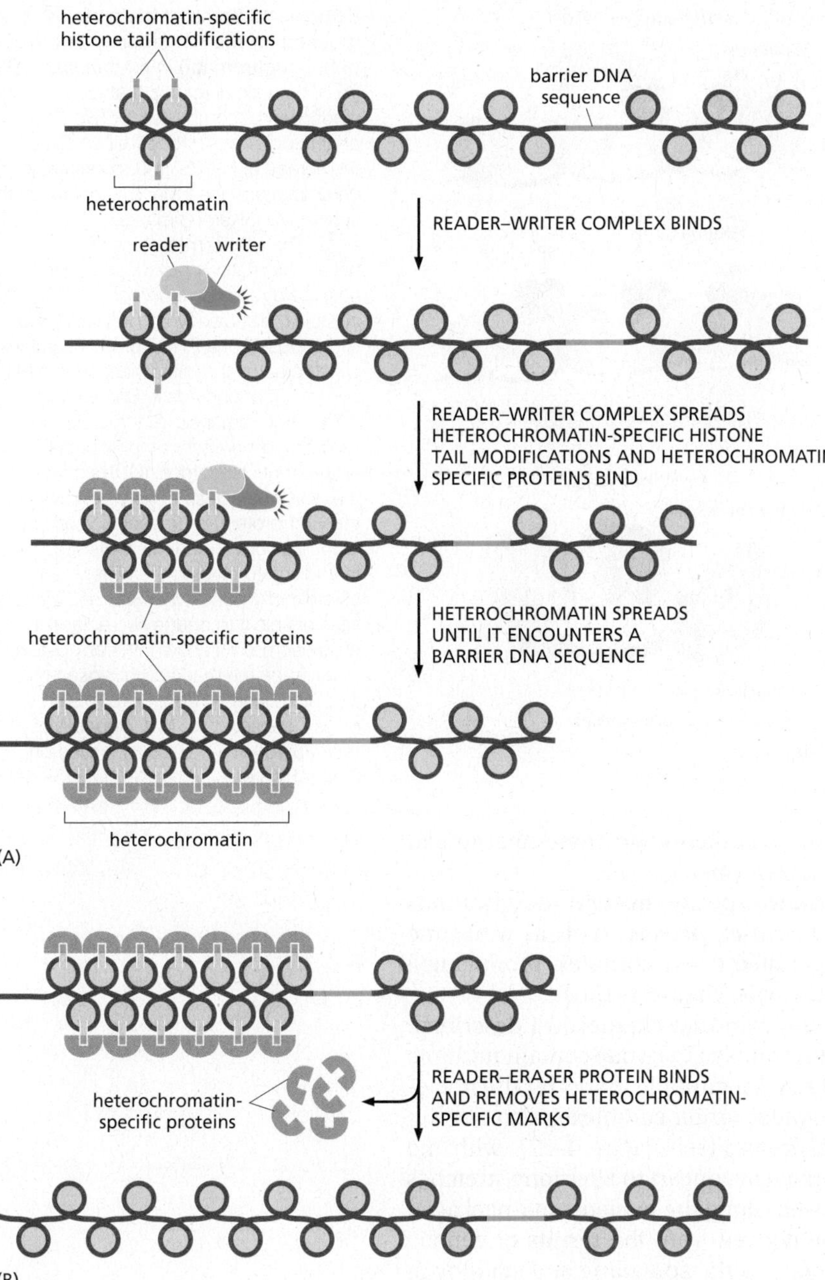

Figure 4–39 How the recruitment of reader–writer and reader–eraser complexes can spread chromatin changes along a chromosome. (A) The writer is an enzyme that creates a particular modification on one of the four nucleosomal histones. After a writer enzyme modifies a few nucleosomes following its recruitment to a specific site on a chromosome—for example, by a transcription regulatory protein—the writer collaborates with a reader protein to spread its mark from nucleosome to nucleosome by means of the indicated reader–writer complex. For this mechanism to work, the reader must recognize the same histone modification mark that the writer produces, and its binding to that mark should activate the writer. In this schematic example, a spreading wave of chromatin condensation is thereby induced, forming heterochromatin. (B) A reader–eraser complex (not shown) reverses the chromatin change illustrated in A. As described in the text, additional proteins are involved in these two types of spreading events, including ATP-dependent chromatin-remodeling complexes.

of heterochromatin is generated by the extensively studied *polycomb repressive complex* (*PRC*). Because this class is highly regulated, it is designated as facultative heterochromatin. The H3K9me3 class of heterochromatin plays an important structural role in forming the centromere (to be described shortly) and in silencing a variety of "selfish DNA" elements (see p. 465). It was once thought to remain unchanged in all cells of a multicellular organism ("constitutive heterochromatin"). However, recent data reveal that some of its forms are reversible, and that it can be used to regulate gene expression. For example, it is used to tightly repress genes that are active early in embryonic development, once they are no longer needed (see Chapter 7).

The H3K9me3 class of heterochromatin also serves to block the frequent genetic recombination that would otherwise occur between the highly repeated DNA sequences in genomes. Roughly speaking, 40% of the human genome is packaged into heterochromatin, with the ratio of the H3K9me3 class to the H3K27me3 class of heterochromatin being about 2:1 (see Figure 4–31).

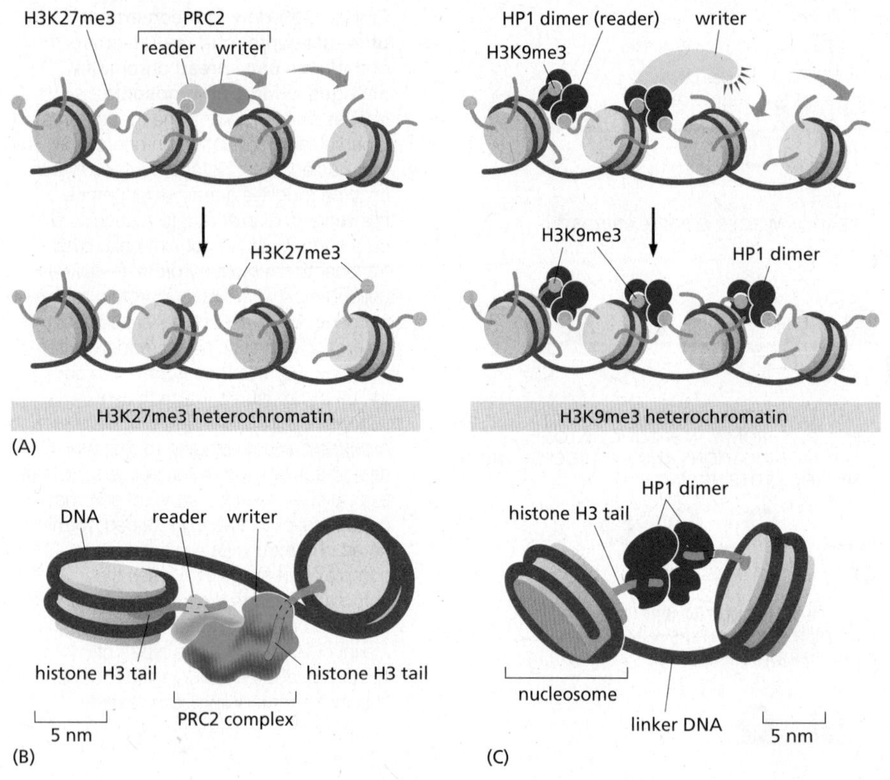

Figure 4–40 Some of the proteins required for the formation of two classes of heterochromatin in mammalian cells. (A) Schematic diagrams that compare the reader–writer complexes for two classes of heterochromatin: one that catalyzes the spread of H3K27me3 marks along chromatin and the other that catalyzes the spread of H3K9me3 marks.

(B) The H3K27me3 class of heterochromatin is produced by the *polycomb repressive complex (PRC)*. This complex, first discovered in *Drosophila*, is composed of a PRC1 protein complex, which creates the initial mark, plus a PRC2 complex that spreads it. Shown here is the three-dimensional structure of a PRC2 reader–writer complex bridging two adjacent nucleosomes. (C) The reader for the H3K9me3 class of heterochromatin, the HP1 protein, is present in very large amounts compared to its writer enzyme. In addition to binding the writer it exists as a dimer whose two H3K9me3 binding sites enable it to bridge two adjacent nucleosomes as shown, thereby helping to package the marked nucleosomes (see also panel A, above). (B, based on S. Poepsel et al., *Nat. Struct. Mol. Biol.* 25:154–162, 2018; C, based on S. Machida et al., *Mol. Cell* 69:385–397, 2018.)

Figure 4–40 outlines what is known about the two different processes that spread and maintain these abundant heterochromatin forms.

A related type of process is used to remove specific histone modifications from a region of the DNA. In this case, an *eraser enzyme*, such as a histone demethylase or a histone deacetylase, is recruited to the complex, producing a read–erase cycle that spreads along a chromosome (**Figure 4–39B**).

In reality, the process is more complicated than the schemes just described. Both readers and writers are part of a protein complex that may contain multiple readers and writers, as well as erasers, and require multiple marks on the nucleosome to spread. Moreover, many of these reader-writer complexes also contain an ATP-dependent chromatin-remodeling protein (see Figure 4–27), with the reader, writer, and remodeling proteins working in concert to alter long stretches of chromatin as the reader moves progressively along the nucleosome-packaged DNA. Some idea of the complexity can be derived from the results of genetic screens for genes that either enhance or suppress the spreading and stability of heterochromatin, as manifest for example in effects on position effect variegation in *Drosophila* (see Figure 4–33). As pointed out previously, more than 100 such genes are known, and many of them are likely to code for subunits in one or more reader-writer-remodeling protein complexes. However, nearly all of the details remain to be deciphered by future research.

Barrier DNA–Protein Complexes Block the Spread of Reader–Writer Complexes and Thereby Separate Neighboring Chromatin Domains

The above mechanisms for spreading chromatin structures raise a potential problem. Inasmuch as each chromosome contains one continuous, very long DNA molecule, what prevents a cacophony of confusing cross-talk between adjacent chromatin domains of different structure and function? Early studies of position effect variegation had suggested an answer: certain DNA sequences mark the boundaries of chromatin domains and separate one such domain from another (see Figure 4–32). Several such *barrier DNA sequences* have now

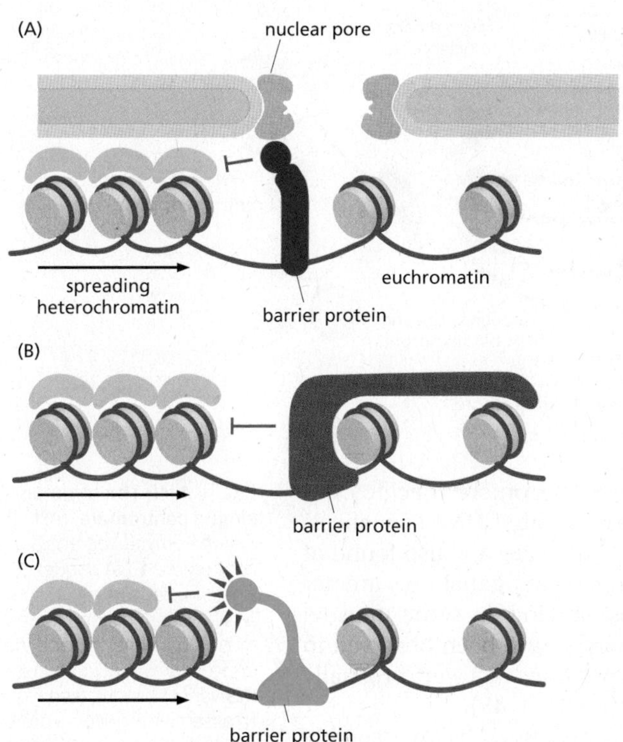

Figure 4–41 **Some mechanisms of barrier action.** These models are derived from experimental analyses of barrier action, and a combination of several of them may function at any one site. (A) The tethering of a region of chromatin to a large fixed site, such as the nuclear pore complex illustrated here, can form a barrier that stops the spread of heterochromatin. (B) The tight binding of barrier proteins to a group of nucleosomes can make this chromatin resistant to heterochromatin spreading. (C) By recruiting a group of highly active histone-modifying enzymes, barriers can erase the histone marks that are required for heterochromatin to spread. For example, a potent acetylation of lysine 9 on histone H3 will compete with lysine 9 methylation, thereby preventing the binding of the HP1 protein needed to form a major form of heterochromatin. (Based on A.G. West and P. Fraser, *Hum. Mol. Genet.* 14:R101–R111, 2005.)

been identified and characterized through the use of genetic engineering techniques that allow specific DNA segments to be deleted from or inserted into chromosomes.

Analysis of one type of barrier sequence reveals that it contains a cluster of binding sites for histone acetylase enzymes. The acetylation of a lysine side chain is incompatible with the methylation of the same side chain, and specific lysine methylations are required to spread heterochromatin (see Figure 4–40). Histone acetylases are therefore logical candidates for the formation of the barriers that stop heterochromatin spreading, as are the histone-demethylating enzymes that erase the marks on histones specific to heterochromatin. Other types of barrier mechanisms are also known (**Figure 4–41**).

Centromeres Have a Special, Inherited Chromatin Structure

There is a specialized chromatin structure at the centromere, the region of each chromosome required for its orderly attachment to the mitotic spindle. In many complex organisms, including humans, each centromere is embedded in a stretch of special *centromeric chromatin* that persists throughout interphase, even though the centromere-mediated attachment to the spindle and movement of DNA occur only during mitosis. This chromatin contains a centromere-specific variant H3 histone, known as CENP-A (centromere protein-A; see Figure 4–36), plus additional proteins that pack the nucleosomes into particularly dense arrangements and form the kinetochore, the structure required for mitotic spindle attachment (see Figure 4–19).

In the yeast *Saccharomyces cerevisiae*, a specific DNA sequence of approximately 125 nucleotide pairs is sufficient to serve as a centromere. Despite its small size, more than a dozen different proteins assemble on this DNA sequence; the proteins include the CENP-A histone H3 variant that, along with the three other core histones, forms a centromere-specific nucleosome. The additional proteins at the yeast centromere form a kinetochore, which links this nucleosome to a single microtubule from the yeast mitotic spindle (**Figure 4–42**).

The centromeres in more complex organisms are considerably larger than those in budding yeasts. For example, human centromeres extend over several million nucleotide pairs, and, while they contain multiple copies of CENP-A and bind

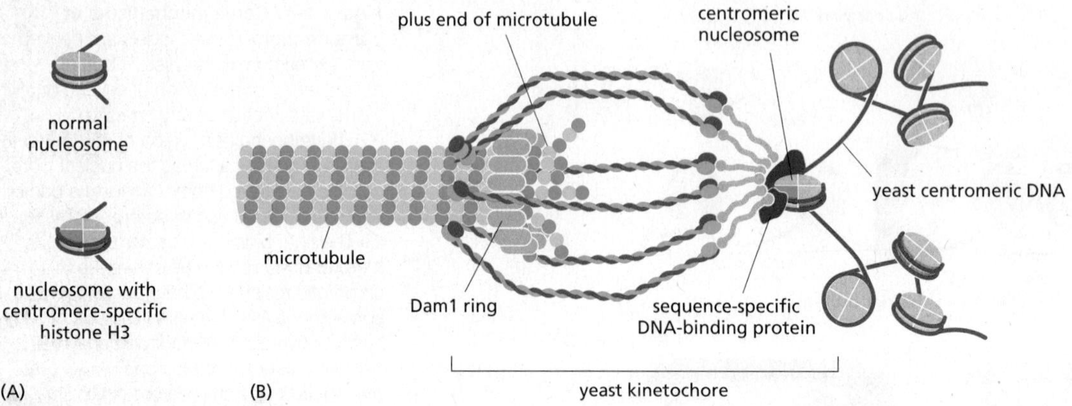

Figure 4–42 **The structure of a simple centromere.** (A) In the yeast *Saccharomyces cerevisiae*, a special centromeric DNA sequence assembles a single nucleosome in which two copies of an H3 variant histone (called CENP-A in most organisms) replace the normal H3. (B) How amino acid sequences unique to this variant histone (see Figure 4–36) help to assemble additional proteins, some of which form a kinetochore. The yeast kinetochore is unusual in capturing only a single microtubule; humans have much larger centromeres and form kinetochores that can capture 20 or more microtubules. The kinetochore is discussed in detail in Chapter 17.

about 20 microtubules, they do not seem to contain a centromere-specific DNA sequence. These centromeres largely consist of short, repeated DNA sequences, known as *alpha satellite DNA*. But the same repeat sequences are also found at other (non-centromeric) positions on chromosomes, indicating that they are not sufficient to direct centromere formation. And most striking, in some unusual cases, new human centromeres (called neocentromeres) have been observed to form spontaneously on fragmented chromosomes at positions that were originally euchromatic and lack alpha satellite DNA altogether (**Figure 4–43**).

It seems that centromeres in complex organisms are defined by an assembly of proteins rather than by a specific DNA sequence. Essential for this assembly is a set of CENP-A nucleosomes. And once a special region of centromeric chromatin forms, this assembly is thereafter faithfully inherited when a chromosome replicates, despite the fact that no special DNA sequence need be involved.

The inactivation of some centromeres and the genesis of others *de novo* occur as organisms evolve. Different species, even when quite closely related, often have different numbers of chromosomes (see Figure 4–14 for an extreme example). And, as we discuss later, detailed genome comparisons show that many changes in chromosomes have arisen through chromosome breakage-and-rejoining events; these create novel chromosomes, some of which must initially have contained abnormal numbers of centromeres—either more than one or

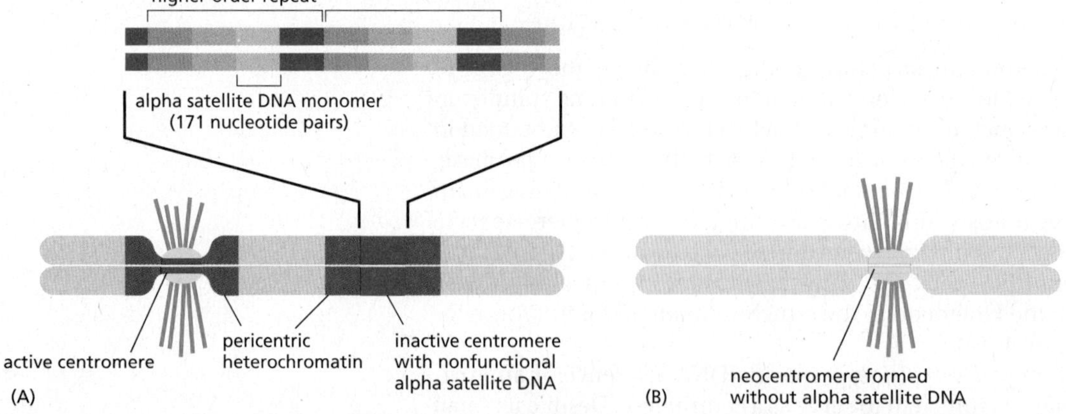

Figure 4–43 **Evidence for the plasticity of human centromere formation.** (A) A series of A-T–rich alpha satellite DNA sequences is repeated many thousands of times at each human centromere *(red)* and is surrounded by *pericentric heterochromatin (brown)*. The pericentric heterochromatin contains H3K9me3, along with HP1 protein, and it is an example of "classical" *constitutive heterochromatin* (see Figure 4–31). As indicated, some human chromosomes contain two blocks of alpha satellite DNA, each of which presumably functioned as a centromere in its original chromosome. (B) In a small fraction (1/2000) of human births, extra chromosomes are observed in cells of the offspring. Some of these extra chromosomes, which have formed from a breakage event, contain alpha satellite DNA sequences that have been co-opted to form new centromeres (neocentromeres); other neocentromeres lack alpha satellite DNA altogether and have arisen from what was originally euchromatic DNA.

none at all. Yet stable inheritance requires that each chromosome should contain one centromere, and one only. It seems that surplus centromeres must have been inactivated and/or new centromeres created, so as to allow the rearranged chromosome sets to be stably maintained.

Some Forms of Chromatin Can Be Directly Inherited

The changes in centromere activity just discussed, once established, need to be perpetuated through subsequent cell generations. What could be the mechanism of this type of epigenetic inheritance?

It has been proposed that *de novo* centromere formation requires an initial seeding event, which is followed by the formation of a specialized DNA-protein structure that contains nucleosomes formed with the CENP-A variant of histone H3. In humans, this seeding event happens more readily on arrays of alpha satellite DNA than on other DNA sequences. The entire centromere then forms as an all-or-none entity, suggesting that the creation of centromeric chromatin is a highly cooperative process, spreading out from an initial seed. And once established, this special form of chromatin is passed on to each daughter cell when a cell divides.

Both the spreading and the inheritance of centromeric chromatin mimic the phenomenon of position effect variegation that we discussed earlier (see Figure 4–32). The spreading of a particular chromatin structure can be explained by the action of reader–writer complexes (see Figure 4–39A). But how can we explain the inheritance of centromeric chromatin from one cell generation to the next?

Notably, experiments have revealed that the H3–H4 tetramers from each nucleosome on the parent DNA helix are directly inherited by both of the sister DNA helices at a replication fork, being equally partitioned between them. This is quickly followed by the addition of two H2A–H2B dimers to complete each "half-old" nucleosome, and by the deposition of the new histone octamers needed to restore the normal complement of nucleosomes (see Figure 5–32). Therefore, once a set of CENP-A-containing nucleosomes has been assembled on a stretch of DNA, it is easy to understand how a new centromere could be generated in the same place on both daughter chromosomes after each round of cell division. One need only assume that the presence of the CENP-A histone in an inherited nucleosome selectively recruits more CENP-A histone to its newly formed neighbors.

An analogous scheme is thought to explain the observation that both the H3K9me3 and H3K29me3 forms of heterochromatin, once formed, are directly inherited after each round of chromosome replication. In those cases, the H3–H4 tetramers that contain a particular modified histone will be passed to each daughter DNA helix, followed by the rapid addition of H2A–H2B to re-form specifically marked nucleosomes. The action of reader–writer complexes can then spread the identical marks to the new neighboring nucleosomes—either H3K9me3 or H3K27me3 (see Figure 4–40). Thus, the activities of reader–writer complexes can explain not only the spreading of specific forms of chromatin along a chromosome, but also the propagation of heterochromatin across cell generations—from parent cell to daughter cell (**Figure 4–44**).

In vertebrates, but not in flies, this inheritance of heterochromatin is often reinforced by a second process that silences the same genes. As described in Chapter 7, this process is based on a DNA methylation system that generates an inherited pattern of methylated C nucleotides (see Figures 7–47 and 7–48).

The Abnormal Perturbations of Heterochromatin That Arise During Tumor Progression Contribute to Many Cancers

As described in Chapter 20, cancer arises from a series of accidental changes in a cell's control systems, each of which is inherited and accumulates progressively in a clone of daughter cells. Many of these sequential changes are the result of mutations that alter the sequence of the DNA encoding an important regulatory protein, such as a protein kinase involved in cell signaling. Other

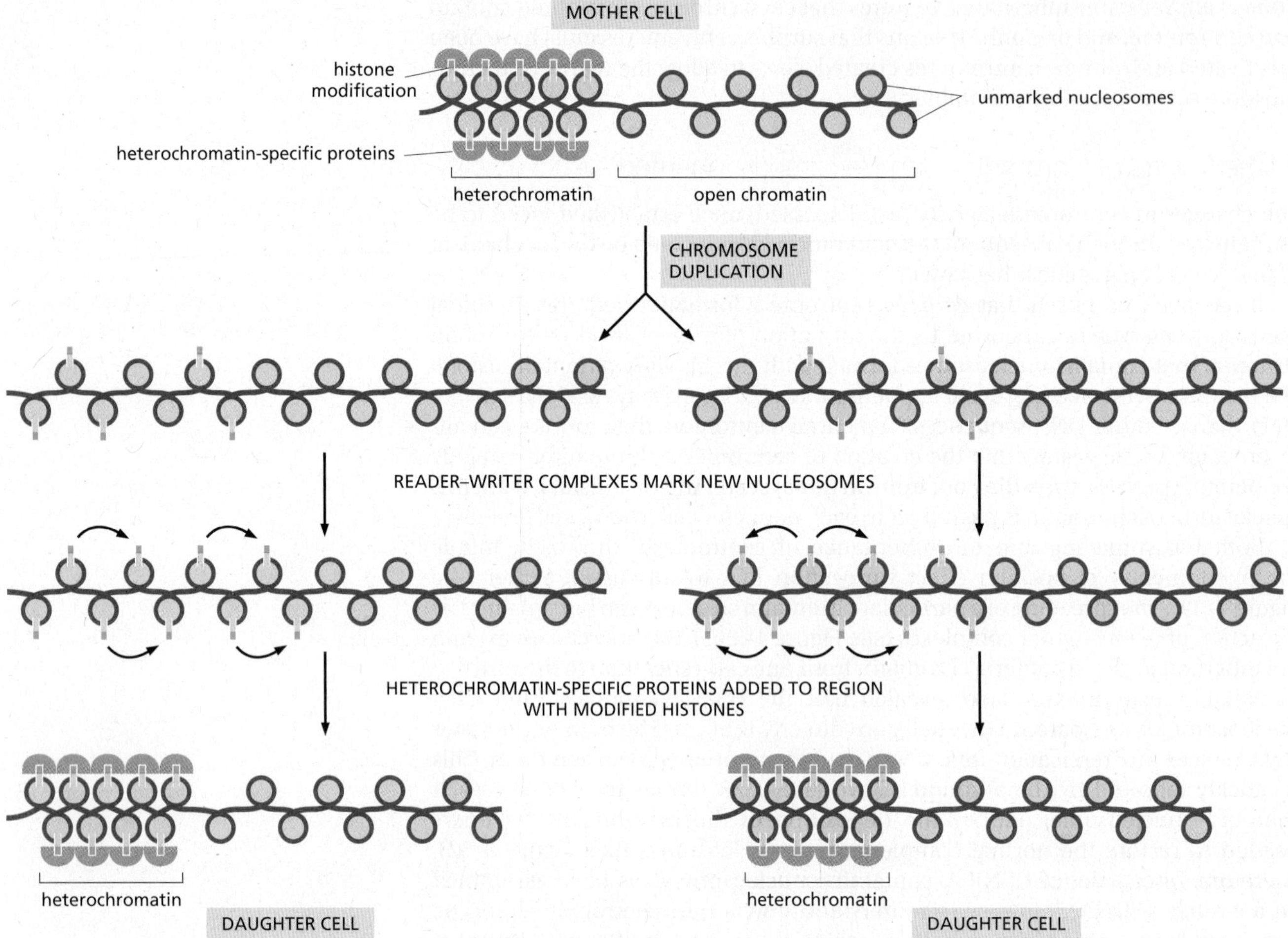

Figure 4–44 **The propagation of heterochromatin across generations.** After DNA replication, all of the old H3–H4 molecules in the parent chromatin will be passed directly to the daughter DNA helices, half to each daughter (see Figure 5–32). This causes the specially marked nucleosomes in heterochromatin to be inherited, as indicated. Reader–writer complexes can then add the same pattern of histone modification to their new, unmarked nucleosome neighbors, as previously illustrated in Figure 4–40. As the original pattern of histone marks is reestablished, binding sites are created for the heterochromatin-specific non-histone proteins that assemble to reproduce the parent chromatin structure. This type of process is thought to occur for both H3K9me3 and H3K27me3 classes of heterochromatin, as well as for the chromatin at centromeres; it is thought not to occur for the open structure of chromatin at active genes.

mutations that are known to drive tumor progression alter chromatin structure, for example by affecting the readers, writers, or erasers of histone marks—or by altering a chromatin-remodeling complex. Because changes in chromatin can alter gene expression, as we explore further in Chapter 7, this finding is not surprising.

Much more surprising is the discovery that a change in a single amino acid in a histone can cause cancer. As very abundant proteins, each histone is encoded by multiple copies of its histone gene. As a result, any mutation that changes a histone should alter less than 10% of the molecules, meaning that any effect observed must be dominant—overriding the presence of a large excess of the normal histone. Such *oncohistone* mutations appear to be present in about 4% of all tumors, and for a few special cancers they are a predominant cancer driver. In particular, a mutation that changes the lysine at position 27 of histone H3 to a methionine (denoted as H3K27M) is almost universally present in a lethal type of pediatric brain tumor (diffuse intrinsic pontine glioma; DIPG), while also being occasionally found in acute myeloid leukemias and melanomas in adults. This mutation has been shown to exert its dominant effect by binding abnormally tightly to the PCR2 protein complex, thereby altering the overall pattern and level of H3K27me3 modifications across the human genome (see Figure 4–40), a finding that illustrates the important role that heterochromatin plays in controlling genes.

Summary

In the chromosomes of eukaryotes, DNA is assembled into long strings of nucleosomes, but a variety of different chromatin structures is possible. This variety is based on a large set of reversible covalent modifications of the four histones in the nucleosome core. The modifications include the mono-, di-, and trimethylation of many different lysine side chains, an important reaction that is incompatible with the acetylation that can occur on the same lysines. Specific combinations of the modifications mark many nucleosomes, governing their interactions with other proteins. These marks are read when protein modules that are part of a larger protein complex bind to the modified nucleosomes in a region of chromatin. These reader proteins then attract additional proteins that perform various functions.

Some reader protein complexes contain a histone-modifying enzyme, such as a histone lysine methylase, that "writes" the same mark that the reader recognizes. A reader–writer–remodeling complex of this type can spread a specific form of chromatin along a chromosome. In particular, large regions of condensed heterochromatin are thought to be formed in this way. Heterochromatin is commonly found around centromeres and near telomeres, but it is also present at many other positions in chromosomes. The tight packaging of DNA into heterochromatin usually silences the genes within it.

The phenomenon of position effect variegation provides strong evidence for the inheritance of condensed states of chromatin from one cell generation to the next. A similar mechanism appears to be responsible for maintaining the specialized chromatin at centromeres. More generally, the ability to propagate specific chromatin structures across cell generations makes possible an epigenetic cell memory process that plays a role in maintaining the set of different cell states required by complex multicellular organisms.

THE GLOBAL STRUCTURE OF CHROMOSOMES

Having discussed the DNA and protein molecules from which the chromatin fiber is made, we now turn to the organization of the chromosome on a more global scale and the way in which its various domains are arranged in space. Packaged into nucleosomes, a typical human chromosome would be able to span the nucleus thousands of times. Thus, a higher level of folding is required, even in interphase chromosomes. As we shall see, this higher-order packaging involves the folding of each chromosome into a series of large loops through a process catalyzed by ring-shaped SMC (structural maintenance of chromosomes) protein complexes.

We begin this section by describing some unusual chromosomes that can be easily visualized. Exceptional though they are, these special cases reveal features that are relevant for all eukaryotic chromosomes. Next, we describe how the interphase chromosomes are arranged in the mammalian cell nucleus. Finally, we discuss the mechanisms that cause a special compaction of chromosomes during their passage from interphase to mitosis.

Chromosomes Are Folded into Large Loops of Chromatin

An early insight into the structure of the chromosomes in interphase cells came from studies of the stiff and enormously extended chromosomes in growing amphibian oocytes (immature eggs). These very unusual **lampbrush chromosomes** (the largest chromosomes known), paired in preparation for meiosis, are clearly visible in the light microscope, where they are seen to contain a

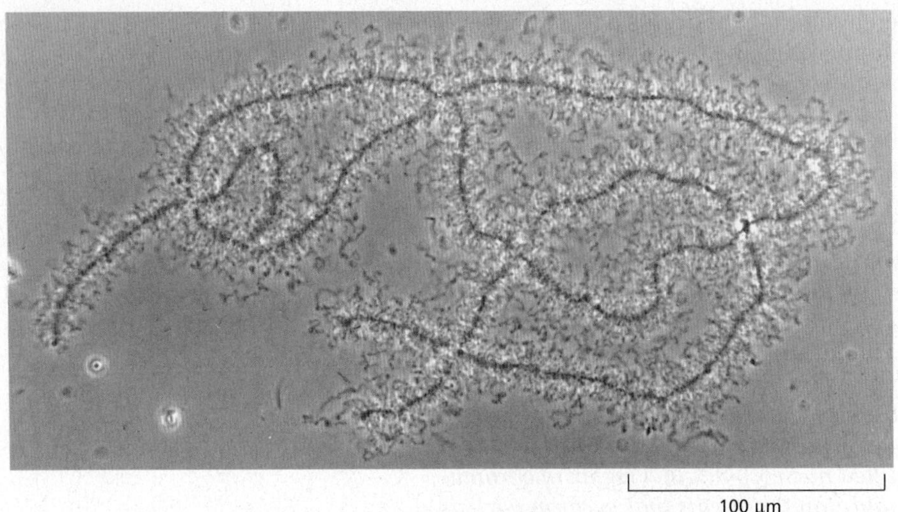

Figure 4–45 **A light micrograph of lampbrush chromosomes in an amphibian oocyte.** Early in oocyte differentiation, each chromosome replicates to begin meiosis, and the homologous replicated chromosomes pair to form this highly extended structure. Each lampbrush chromosome consists of two aligned sets of paired *sister chromatids*, with large chromatin loops as a prominent feature. The chromosome set contains thousands of loops, each containing a particular DNA sequence that remains extended in the same manner as the oocyte grows, producing huge amounts of RNA for storage in the oocyte. The lampbrush chromosome stage persists for months or years, while the oocyte builds up a supply of materials required for its ultimate development into a new individual. These chromosomes were first described in 1878. (Courtesy of Joseph G. Gall.)

series of large chromatin loops emanating from a linear chromosomal axis (Figure 4–45).

Figure 4–46 summarizes lampbrush chromosome structure. Each chromosome is formed by a very long DNA helix packed into chromatin, and it is joined to a sister DNA helix, called a *sister chromatid.* Most of the DNA is located near the junction of the two chromatids, and it is highly condensed and transcriptionally inert; this DNA serves to organize each chromatid along a linear chromosome axis. In contrast, the highly transcribed regions of the DNA extend from the axis as large loops, which range in length from tens of thousands to hundreds of thousands of nucleotide pairs.

Meiotic chromosomes are found to adopt the lampbrush chromosome state in the growing oocytes of all vertebrates, except for mammals. However, if incubated in amphibian oocyte cytoplasm, human sperm chromosomes will form lampbrush chromosomes. This reveals that chromosomes are highly dynamic structures that can restructure in different environments.

Polytene Chromosomes Are Uniquely Useful for Visualizing Chromatin Structures

Further early insight into the structure of interphase chromosomes came from a second unusual type of cell—the *polytene cells* of flies, such as the fruit fly *Drosophila.* Some special cells, in many organisms, grow abnormally large through

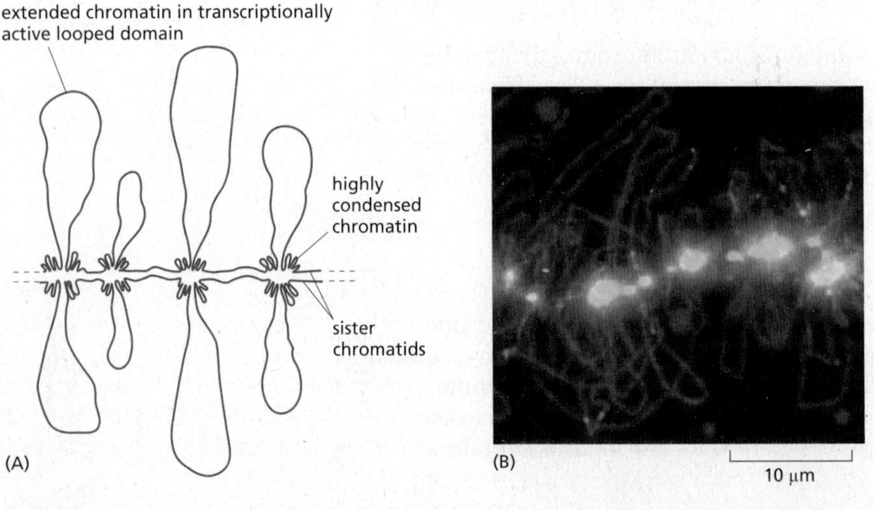

Figure 4–46 **The structure of lampbrush chromosomes.** (A) Model for a small portion of one pair of sister chromatids. Two identical DNA double helices are aligned side by side, packaged into different types of chromatin (see Figure 4–45). (B) Merged light micrographs of the lampbrush chromosomes in an axolotl (a type of salamander), stained for transcriptionally active (RNA polymerase, *red*) and inactive (DNA containing 5-methyl C, *green*) chromatin regions. The loops are stiff and extended because they are being unusually highly transcribed, with RNA polymerases spaced only about 100 nucleotide pairs apart. Most of the rest of the DNA in each chromosome (the great majority) remains condensed and is located close to the chromatid axis. (B, from G.T. Morgan et al., *Chromosome Res.* 20:925–942, 2012. Reproduced with permission of SNCSC.)

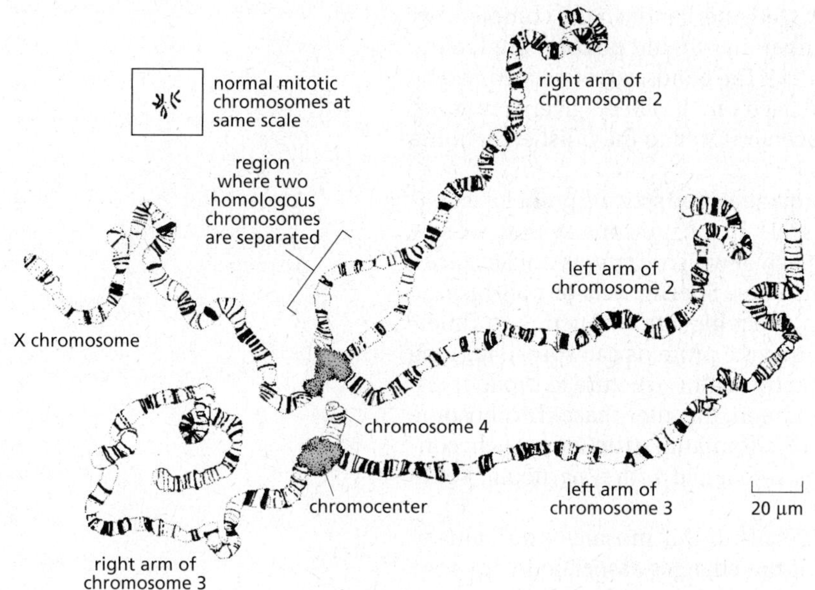

Figure 4–47 **The entire set of polytene chromosomes in one *Drosophila* salivary cell.** In this drawing of a light micrograph, the giant chromosomes have been spread out for viewing by squashing them against a microscope slide. *Drosophila* has four chromosomes, and there are four different chromosome pairs present. But each chromosome is tightly paired with its homolog (so that each pair appears as a single structure), which is not true in most nuclei (except in meiosis). Each chromosome has undergone multiple rounds of replication, and the homologs and all their duplicates have remained in exact register with each other, resulting in huge chromatin cables many DNA strands thick.

The four polytene chromosomes are normally linked together by heterochromatic regions near their centromeres that aggregate to create a single large chromocenter *(pink region)*. In this preparation, however, the chromocenter has been split into two halves by the squashing procedure used. (Adapted from T.S. Painter, *J. Hered.* 25:465–476, 1934. With permission from Oxford University Press.)

multiple cycles of DNA synthesis without cell division. Such cells, containing increased numbers of standard chromosomes, are said to be *polyploid*. In the salivary glands of fly larvae, this process is taken to an extreme degree, creating huge cells that contain hundreds or thousands of copies of the genome. Moreover, in this case, all the copies of each chromosome are aligned side by side in exact register, like drinking straws in a box, to create giant **polytene chromosomes**. These chromosomes allow features to be detected that are thought to be shared with ordinary interphase chromosomes but are normally hard to see.

When polytene chromosomes from a fly's salivary glands are viewed in the light microscope, distinct alternating dark *bands* and light *interbands* are visible (Figure 4–47), each formed from a thousand identical DNA sequences arranged side by side in register. About 95% of the DNA in polytene chromosomes is in bands, and 5% is in interbands. A very thin band can contain 3000 nucleotide pairs, while a thick band may contain 200,000 nucleotide pairs in each of its chromatin strands. The chromatin in each band appears dark because the DNA is more condensed than the DNA in interbands; it may also contain a higher concentration of proteins (Figure 4–48).

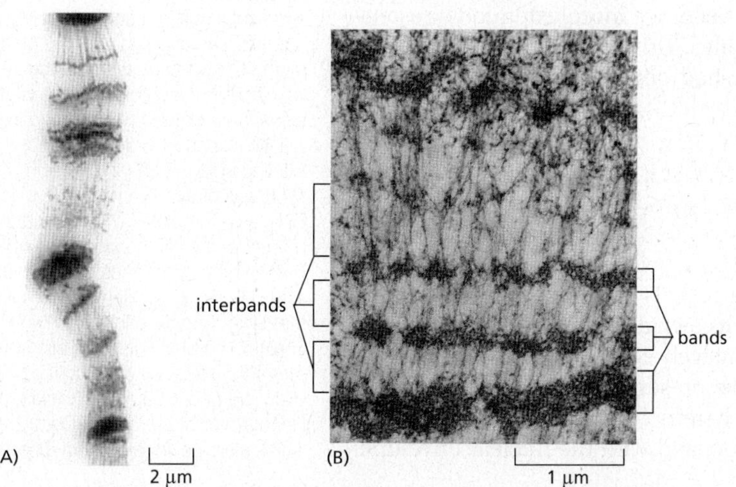

Figure 4–48 **Micrographs of polytene chromosomes from *Drosophila* salivary glands.** (A) Light micrograph of a portion of a chromosome. The DNA has been stained with a fluorescent dye, but a reverse image is presented here that renders the DNA *black* rather than *white*; the bands are clearly seen to be regions of increased DNA concentration. This chromosome has been processed by a high-pressure treatment so as to show its distinct pattern of bands and interbands more clearly. (B) An electron micrograph of a small section of a *Drosophila* polytene chromosome seen in thin section. Bands of very different thickness can be readily distinguished, separated by interbands, which contain less condensed chromatin. (A, adapted from D.V. Novikov et al., *Nat. Methods* 4:483–485, published 2007 by Nature Publishing Group. Reproduced with permission of SNCSC; B, courtesy of Veikko Sorsa.)

There are approximately 3700 bands and 3700 interbands in the complete set of *Drosophila* polytene chromosomes, a number that should be compared to the 14,000 genes in this fruit fly (see Table 1–2, p. 29). The bands can be recognized by their different thicknesses and spacings, and each one has been given a number to generate a chromosome "map" that has been indexed to the finished genome sequence of *Drosophila*.

The *Drosophila* polytene chromosomes provide a good starting point for examining how chromatin is organized on a large scale. In the previous section, we saw that there can be many forms of chromatin, each of which contains nucleosomes with a different combination of modified histones. Specific sets of non-histone proteins assemble on these nucleosomes to affect biological function in different ways. Recruitment of some of these non-histone proteins can spread for long distances along the DNA, imparting a similar chromatin structure to broad tracts of the genome (see Figure 4–39A). At low resolution, the interphase chromosome can therefore be considered as a mosaic of chromatin structures, each containing particular nucleosome modifications associated with a particular set of non-histone proteins.

Polytene chromosomes allow us to see details of this mosaic of domains in the light microscope and to observe some of the changes associated with gene expression. For example, by staining *Drosophila* polytene chromosomes with antibodies and using ChIP (chromatin immunoprecipitation) analysis (see Chapter 8), the locations of specific histone modifications and non-histone proteins in chromatin can be mapped across the entire *Drosophila* DNA sequence. These results suggest that three types of repressive chromatin predominate in this organism, along with two types of chromatin on actively transcribed genes, and that each type is associated with a different complex of non-histone proteins. In addition to these five major chromatin types, other more minor forms of chromatin appear to be present, each of which may be differently regulated and have distinct roles in the cell. Much remains to be learned about the molecular structures that underlie these findings.

Chromosome Loops Decondense When the Genes Within Them Are Expressed

When an organism containing polytene chromosomes progresses from one developmental stage to another, distinctive *chromosome puffs* arise and old puffs recede in its polytene chromosomes as new genes become expressed and old ones are turned off (**Figure 4–49**). From inspection of each puff when it is relatively small and the banding pattern is still discernible, it seems that most puffs arise from the decondensation of a single chromosome band.

The individual chromatin fibers that make up a puff can be visualized with an electron microscope. In favorable cases, loops are seen. When genes in the loop are not expressed, the loop assumes a thickened, condensed structure, but when gene expression is occurring, the loop becomes more extended—resembling the loops seen in lampbrush chromosomes. This again demonstrates that chromosome structures are dynamic, as we had concluded from lampbrush experiments.

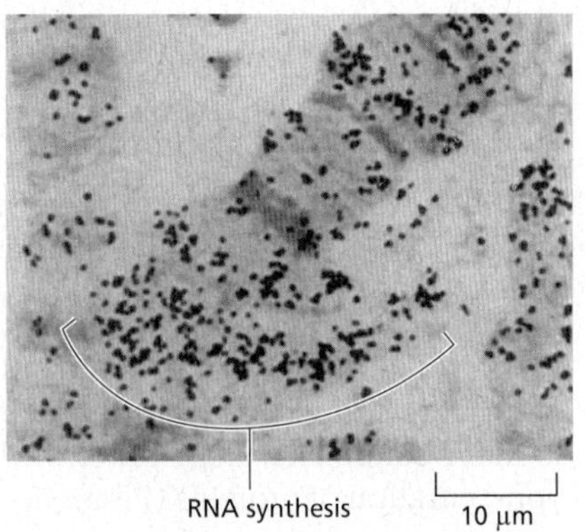

Figure 4–49 **RNA synthesis in polytene chromosome puffs.** An autoradiograph of a single puff in a polytene chromosome from the salivary glands of the freshwater midge *Chironomus tentans*, the fly in which polytene chromosomes were first discovered in 1881. As outlined in Chapter 1 and described in detail in Chapter 6, the first step in gene expression is the synthesis of an RNA molecule using the DNA as a template. The decondensed portion of the chromosome is undergoing RNA synthesis and has become labeled with ^{3}H-uridine, an RNA precursor molecule that is incorporated into growing RNA chains. The black dots are produced where the radioactivity emitted interacts with an overlying photographic emulsion. (From J.J. Bonner and M.L. Pardue, *Cell* 12:227–234, 1977. With permission from Elsevier.)

Mammalian Interphase Chromosomes Occupy Discrete Territories in the Nucleus, with Their Heterochromatin and Euchromatin Distributed Differently

Light microscopy after chromosome painting reveals that each of the 46 interphase chromosomes in a human cell occupies its own discrete territory within the nucleus; that is, the chromosomes are not extensively entangled with one another (**Figure 4–50**). However, pictures such as these present only an average view of the DNA in each chromosome. By staining a heterochromatic region of a chromosome, one finds that it is often closely associated with the nuclear envelope,

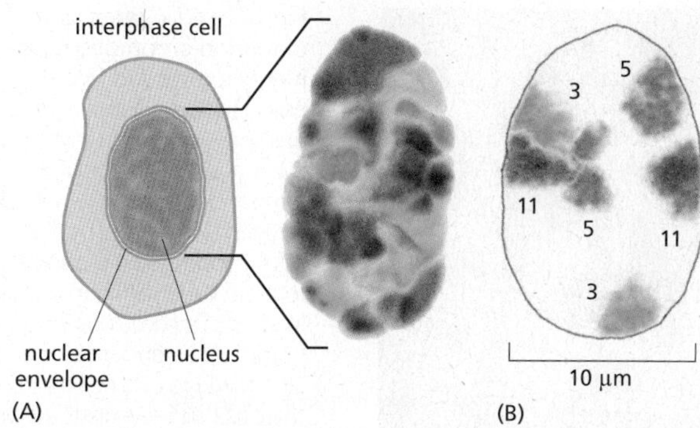

Figure 4–50 **Simultaneous visualization of the chromosome territories for all of the human chromosomes in a single interphase nucleus.** Here, the DNA probes for a specific chromosome are labeled so as to fluoresce at specific wavelengths, and a different combination of dyes is used to label the probes for each chromosome so that each chromosome fluoresces differently. This "chromosome painting" technique allows DNA–DNA hybridization to be used to detect each chromosome, as in Figure 4–11. Three-dimensional reconstructions were then produced. (A) Viewed in a fluorescence microscope, the nucleus is seen to be filled with a patchwork of discrete colors. (B) To highlight their distinct locations, three sets of chromosomes are singled out: chromosomes 3, 5, and 11. Note that pairs of homologous chromosomes, such as the two copies of chromosome 3, are not generally located in the same position. (Adapted from M.R. Hübner and D.L. Spector, *Annu. Rev. Biophys.* 39:471–489, 2010. With permission from Annual Reviews.)

regardless of the chromosome examined. And probes that preferentially stain chromosomal regions containing a high density of active genes reveal that most of these regions extend out from the territory of each chromosome into the nucleoplasm and away from the nuclear envelope (**Figure 4–51**).

Studies of mammalian cells also show that highly folded loops of chromatin expand to occupy an increased volume when a gene within them is expressed. For example, **Figure 4–52** demonstrates how a large, highly transcribed gene in a human cell, which appears as a dot in an interphase cell when inactive, is extended just like a lampbrush loop by its extensive transcription.

As we shall discuss in Chapter 6, the interior of the nucleus is very heterogeneous, with functionally different regions that are produced by *biomolecular condensates* specialized to speed different biochemical processes—such as RNA synthesis, RNA splicing, and DNA replication. The fact that a section of a chromosome can move away from its chromosome territory, as we have just seen, helps cells to use nuclear condensates to accelerate various reactions. These condensates include the nucleolus specialized for producing ribosomes and the nuclear speckles involved in RNA production (see pp. 353-357).

Although much is known about DNA transcription and how genes are turned on and off at the molecular level, we defer that discussion to Chapters 6 and 7. Here we instead concentrate on the mechanism that packages the long linear DNA molecule in each mammalian chromosome into a series of loops in the cell nucleus.

A Biochemical Technique Called Hi-C Reveals Details of Chromosome Organization

As explained in Chapter 8, new automated technologies allow scientists to determine massive amounts of DNA sequence at low cost. A powerful *chromosome conformation capture* method that exploits this ability, *Hi-C*, has made it possible

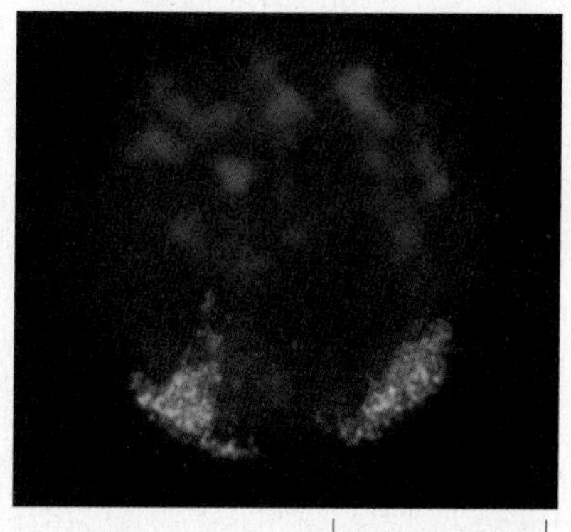

Figure 4–51 **The looping out of gene-rich regions of the genome from chromosome territories.** In this light micrograph of a cell nucleus, DNA has been fluorescently labeled with a dye *(blue)*. In addition, by using a chromosome painting technique, all of the DNA of one particular mouse chromosome has been stained *green*, and the gene-rich regions of that same chromosome have been stained *red*. (From W.A. Bickmore and B. van Steensel, *Cell* 152:1270–1284, 2013. With permission from Elsevier.)

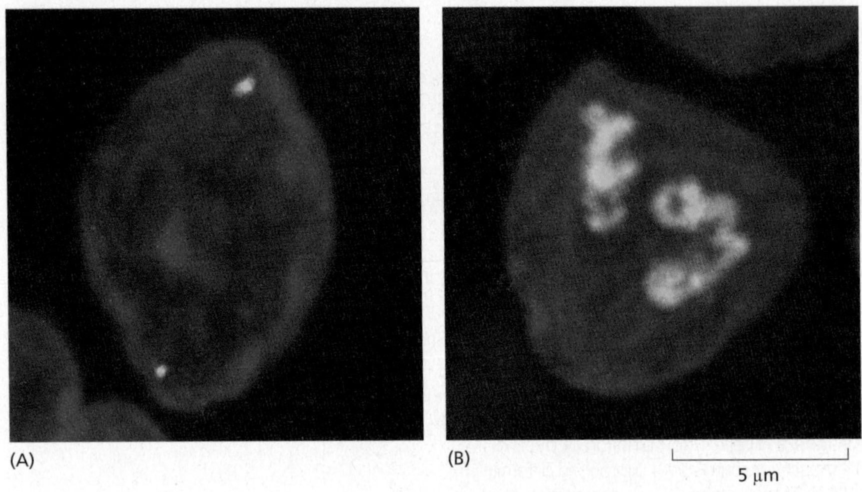

Figure 4–52 **A large gene forms a highly extended chromosome loop when highly expressed in a human interphase cell.** The thyroglobulin gene codes for the extracellular protein thyroglobulin, which is secreted by thyroid gland cells. This gene is both unusually large (270 kb) and unusually highly transcribed. DNA and RNA hybridization techniques reveal that it forms a large stiff loop that extends into the interior of the nucleus from each of the two homologous chromosomes that express it. (A) An image of a cell that does not make thyroglobulin reveals the thyroglobulin locus as a small dot. (B) An image of a thyroid gland cell expressing the gene. Experiments demonstrate that these loops retract back to a dot if RNA synthesis is blocked by drug treatment. (Courtesy of Irina Solovei, Ludwig-Maximilians University, Munich.)

to assess the frequency with which any two genomic loci—whether along a single chromosome or between chromosomes—are held together (**Figure 4–53**).

Application of the Hi-C technique has revealed some fundamental features of chromosome organization. One is that each interphase chromosome occupies its own discrete territory within the nucleus. That is, the genomic loci within a chromosome contact each other more often than they do the loci on different chromosomes, showing that the chromosomes are not extensively entangled. This was expected from the earlier findings of discrete chromosome territories by light microscopy, as just described. In addition, Hi-C and related chromosome capture methods have revealed that the interphase chromosome is folded into a long series of *topologically associated domains*, or *TADs*, in which any two

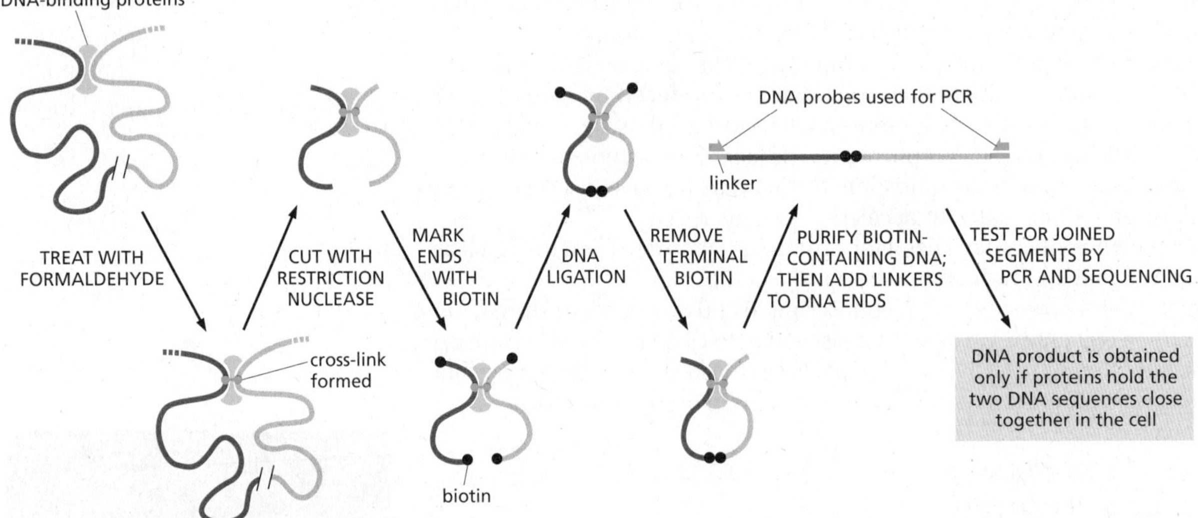

Figure 4–53 **Use of Hi-C to determine the frequency with which any two DNA sequences are adjacent to each other in chromosomes.** In the Hi-C technique, cells are treated with cross-linking agents such as formaldehyde to create covalent DNA–protein–DNA cross-links, as indicated. The DNA is then treated with an enzyme (called a restriction nuclease) that chops the DNA into many pieces, cutting at strictly defined nucleotide sequences and forming sets of identical "cohesive ends" (see Figure 8–23). These ends are then marked by the incorporation of biotinylated nucleotides, so as to enable their selective purification later. Any two DNA ends can become covalently joined if incubated with a DNA ligase enzyme. But importantly, prior to the DNA ligation step shown, the DNA is diluted so that only the fragments that have been kept in close proximity to each other (through cross-linking) are likely to join. After the ligation step, the cross-links are reversed, and all the biotin at unligated DNA ends is removed. This allows the newly ligated fragments of DNA to be selectively purified through their binding to streptavidin beads, amplified by PCR (polymerase chain reaction), and then sequenced (by methods described in Chapter 8). The results, combined with knowledge of the complete DNA sequence of each chromosome, generate detailed models for the conformation of chromosomes.

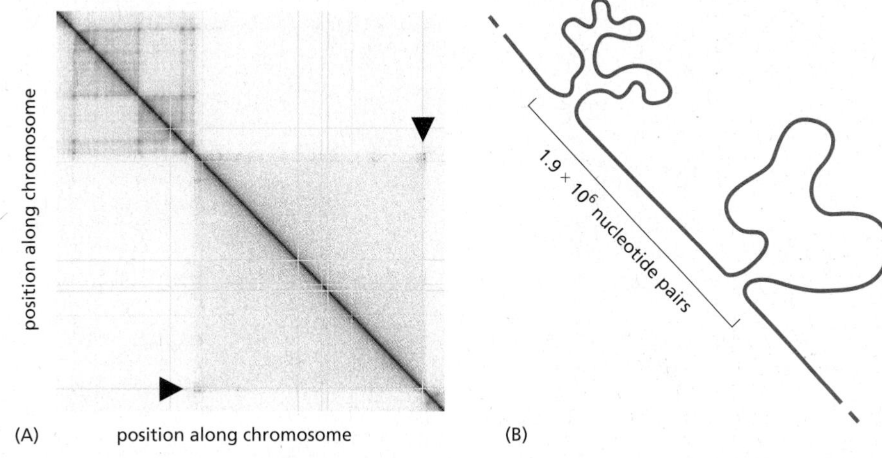

Figure 4–54 **Chromosome capture techniques reveal topologically associated, looped domains in interphase chromosomes.** (A) A plot of results for a small portion of human chromosome 8 in human fibroblasts. The intensity of the *orange* color denotes the number of times that any DNA sequence from a 2-kilobase (kb) region whose chromosome position is indicated on the X axis was found to be ligated to any DNA sequence from a 2-kb region whose chromosome position is indicated on the Y axis (see Figure 4–53). The darker the color, the more often these two regions were located in close proximity to each other inside the cell. The strong stripe along the diagonal results from the fact that DNA sequences that are very close to each other along the linear chromosome have a high probability of interacting with each other.

The largest square reveals a topologically associated domain (TAD), with the dots in the diagram (highlighted here by *black arrowheads*) showing the unusually persistent contacts between the DNA sequences located at the base of a simple looped domain that contains about a million nucleotide pairs. The box at the *upper left* shows a more complex "nested" TAD structure. Here, the multiple dots are interpreted as a "bouquet" of smaller loop subdomains gathered together to form a larger folding domain, as indicated in (B), which schematically compares the structures inferred for the two different TADs. [A, Courtesy of Oliver Rando, University of Massachusetts Medical School. The data for human chromosome 8 (125,034,652 to 126,909,657) are from N. Krietenstein et al., *Mol. Cell* 78:554–565, 2020.]

DNA sequences are much more likely to encounter each other than they are to encounter DNA sequences outside of that domain. Their locations along each chromosome can be derived from the series of squares observed in plots of results like those in Figure 4–54. These reveal that DNA segments in a contiguous set (those within a TAD) have a relatively high probability (*dark color* in the figure) of becoming ligated together. And because any one DNA segment in a TAD can become ligated to many others in that TAD, one concludes that the chromatin inside each TAD must be folded in a way that allows its DNA sequences to frequently encounter any other DNA sequence within it.

Extensive data of this type have led to the conclusion that all interphase chromosomes are organized as a long linear series of folded *looped domains* of chromatin, with the DNA in each loop being compacted, but highly mobile. Although a typical loop in a human chromosome might contain between 50,000 and 200,000 nucleotide pairs of DNA, loops of a million nucleotide pairs also exist, and there appear to be roughly 10,000 loops in the human genome. How these loops are formed is the subject that we discuss next.

Chromosomal DNA Is Organized into Loops by Large Protein Rings

The enormously long DNA molecules that form chromosomes must be organized if they are to be effective as the carriers of the genetic information that every cell requires to survive and multiply. A mechanism for creating this organization that appears to be universal involves the folding of the DNA into loops by an **SMC protein complex**, a large protein ring that both binds to and encircles the DNA double helix. These rings—which function similarly in archaea, bacteria, and eukaryotes—have the structure illustrated in Figure 4–55A. The subunits that give the complex its name are the pair of long, coiled-coil SMC (structural maintenance of chromosomes) proteins. Each long SMC protein chain folds upon itself to form a globular ATPase domain, and two of these chains join together to create a ring that is large enough for DNA that is packaged in chromatin to readily pass through it (compare ring size to nucleosomes in Figure 4–55B).

By associating with additional proteins, the two ATPase domains in an SMC protein complex allow the ring to motor rapidly along DNA. Why this type of activity is useful to the cell is perhaps most easily understood by considering the role of these rings in bacteria. The bacterial chromosome is a large circular DNA molecule that is duplicated from a single DNA replication origin to produce two identical DNA circles. Each of these two daughter chromosomes must be transferred to a different daughter cell when the parent cell divides. The required segregation process begins when a series of SMC protein complexes is loaded onto the DNA near the DNA replication origin; these protein rings then create a loop on each new DNA double helix and proceed to move continuously along the

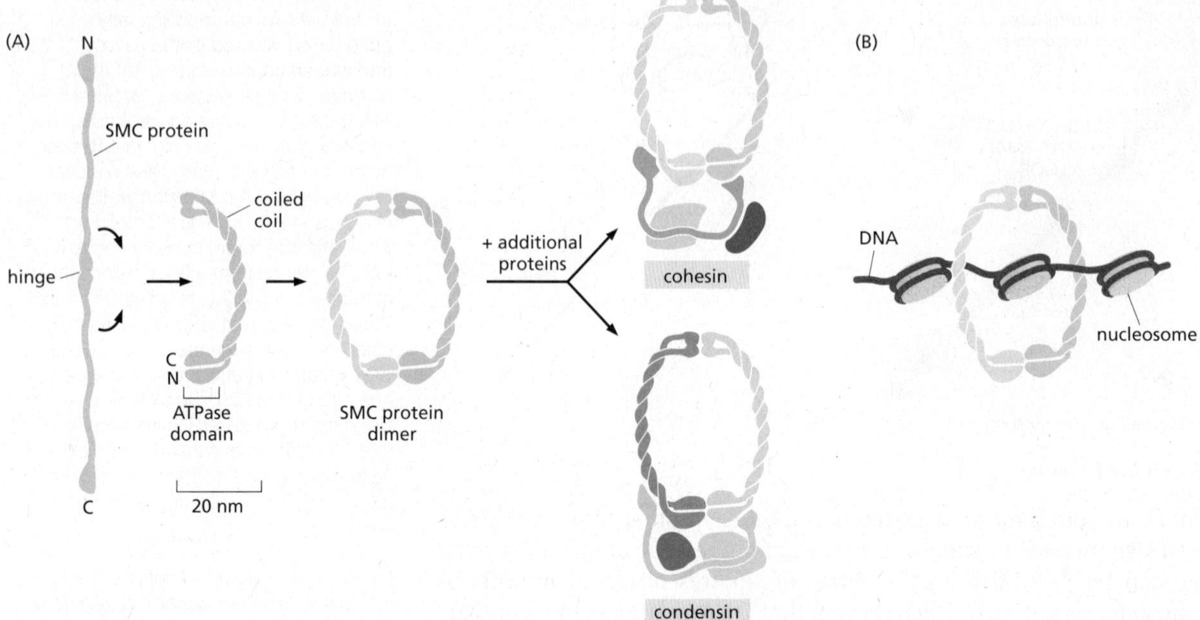

Figure 4–55 SMC protein complex formation. (A) These structures have important roles in bacteria, archaea, and eukaryotes. As indicated, their core forms from a large protein of 1000 to 1500 amino acids that folds on itself to form a antiparallel coiled-coil with a globular head; two of these molecules then pair to produce a ring structure with a flexible hinge at one end and an ATP-binding domain at the other end. Additional subunits are then added to form either cohesin or condensin, as indicated. (B) An SMC complex compared to the size of a nucleosome.

DNA, cleanly separating the two daughter chromosomes from each other in the manner illustrated in Figure 4–56A. A model proposed to explain this movement is likewise shown (Figure 4–56B).

The DNA molecules that form eukaryotic chromosomes are generally much longer than those that form bacterial chromosomes, and many different DNA replication origins are therefore needed to copy each chromosome (see Chapter 5). Moreover, the accurate separation of two daughter chromosomes involves a process that is considerably more elaborate than the process in bacteria (described in Chapter 17). The SMC protein complexes have therefore diversified during the evolution of eukaryotes to play two different roles, each of which is critical for the function of chromosomes.

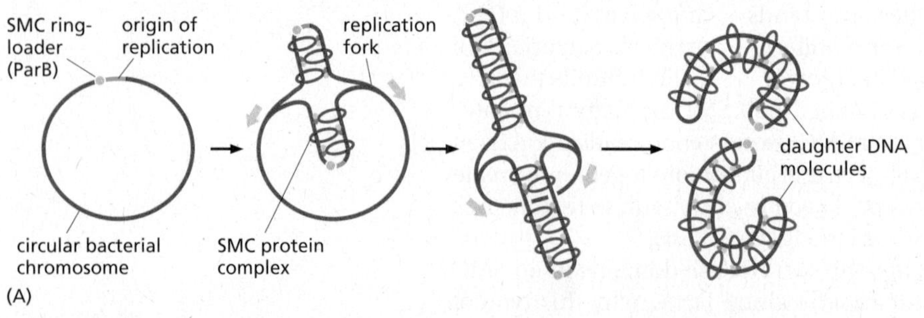

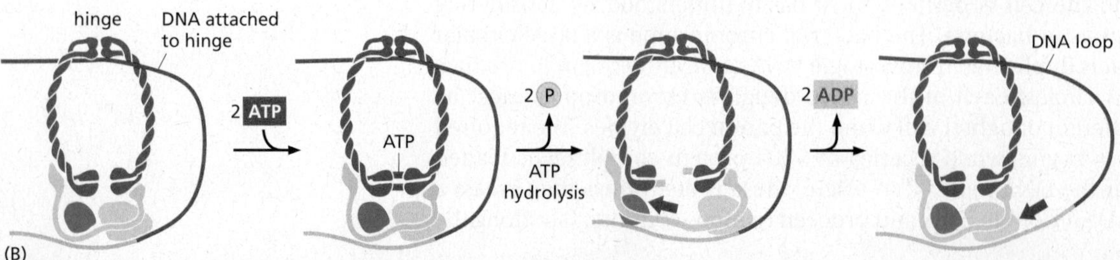

Figure 4–56 How moving SMC-containing protein rings function to separate bacterial chromosomes. (A) A DNA sequence adjacent to the origin of replication on the circular *Bacillus subtilis* chromosome binds a protein (ParB) that repeatedly loads bacterial SMC protein complexes. These complexes then use the energy released by repeated cycles of ATP hydrolysis to travel along the entire chromosome to the site where the two DNA replication forks that were formed at the origin meet, thereby partitioning the two very large daughter DNA molecules in a way that allows them to be easily segregated into two different daughter cells. (B) One model proposed to explain how SMC complexes move. These complexes function as sophisticated protein machines (see Figure 3–72), and it is still uncertain how they drive loop formation. In this "inchworm" model, the two ATPase domains separate and rejoin in sequence to motor along the DNA; for an alternative, see Figure 17–25B. (A, adapted from X. Wang et al., *Science* 355:524–527, 2017; B, adapted from M.H. Nichols and V.G. Corces, *Nat. Struct. Mol. Biol.* 25:906–910, 2018.)

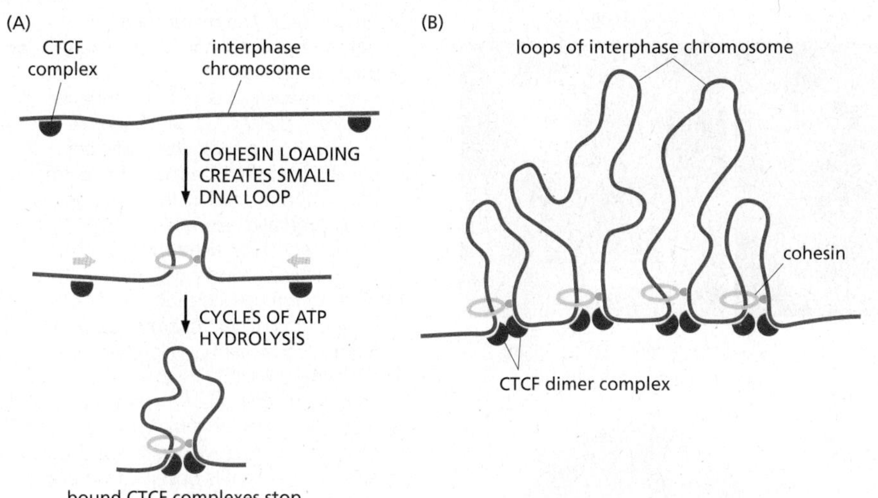

Figure 4–57 **Moving cohesin rings divide eukaryotic interphase chromosomes into a long series of looped domains.** (A) The SMC protein complex *cohesin*, aided by accessory proteins that are not shown, is loaded onto chromatin to form a small DNA loop. Then, propelled by the energy of ATP hydrolysis (see Figure 4–56B), moving cohesin rings continually enlarge the loop, stopping when they encounter a DNA-bound CTCF complex on each side. A protein complex containing the CTCF dimer then becomes the loop base. (B) Schematic illustration of a portion of an interphase chromosome, organized as a series of looped domains.

During interphase, it is the SMC protein complex named **cohesin** that is critical (we shall see that its relative, *condensin*, functions in mitosis). Cohesin is thought to function in a manner similar to that of its bacterial counterpart, except that the cohesin rings are loaded at multiple sites along a chromosome and serve to fold the linear chromosomal DNA molecule into a long series of loops. As illustrated in **Figure 4–57A**, because the traveling cohesin rings tend to stop at specific sites on the DNA, a loop is formed at a specific place on a chromosome. In vertebrates, these favored stop sites are usually marked by the sequence-specific DNA-binding protein *CTCF*, which forms a protein complex that not only stalls or stops the moving cohesin ring, but also holds the two ends of each DNA loop together.

As will be described in Chapter 7, CTCF is one of the *insulator proteins* that helps to maintain discrete domains of chromatin function (see Figure 7–28). It can also serve as part of a barrier that prevents the spreading of chromatin structures by reader–writer remodeling complexes (see Figure 4–39A). Because this protein commonly demarks the base of each looped domain in a chromosome, one can propose a general model for the structure of eukaryotic chromosomes that accounts for observations made in a diverse range of organisms. In this view, not only is the very long, linear DNA molecule that forms the core of each chromosome folded into a series of looped domains, but also the chromatin in each domain will often differ with respect to both its non-histone proteins and its covalent histone modifications—just as it does in the insect polytene chromosomes discussed earlier (**Figure 4–57B**). These differences are important because they help to control the selective expression of genes.

It is satisfying to find that a model recently derived for vertebrate interphase chromosomes is consistent with the observations made many decades ago on *Drosophila* polytene chromosomes and on the lampbrush chromosomes of amphibians. However, new methods that allow the organization of interphase chromosomes to be determined in single cells reveal that—unlike the situation in those special cases—most of the loops in a typical interphase chromosome are unstable: although those loops form in favored locations, their positions can rapidly fluctuate. In addition, loops can form within loops (see Figure 4–54). Thus, during interphase a typical eukaryotic chromosome has a structure that is highly dynamic.

Euchromatin and Heterochromatin Separate Spatially in the Nucleus

The same Hi-C experiments that reveal the boundaries of looped domains (see Figure 4–54) also detect a lower frequency of nearest neighbor contacts between DNA segments that can be either on different chromosomes or far away from

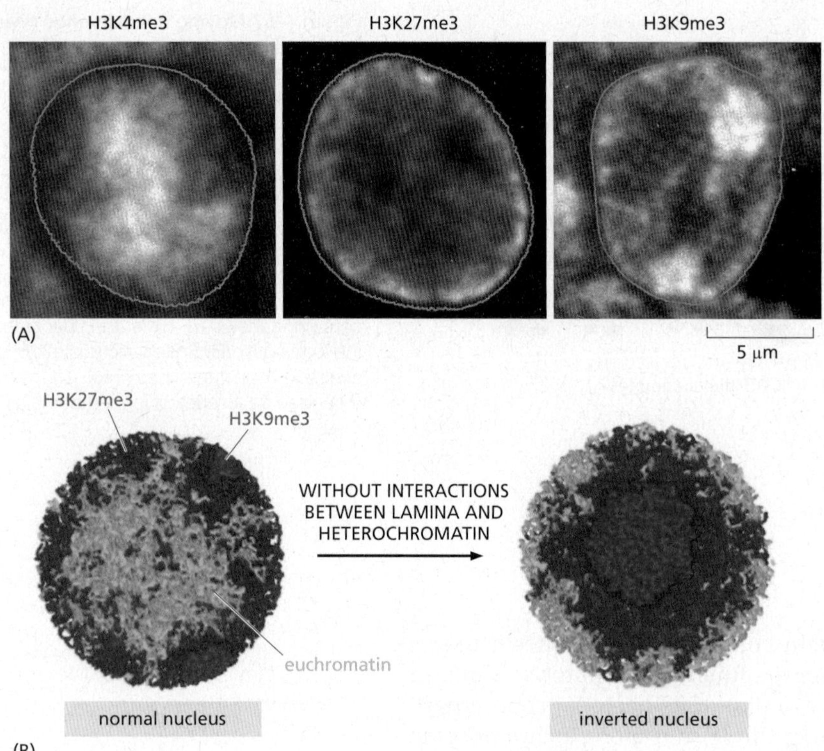

Figure 4–58 **The distribution of heterochromatin and euchromatin in an interphase nucleus.** (A) In this nucleus from a mouse fibroblast, the chromatin has been stained with antibodies to histones that contain three different covalent modifications: one recognizes the H3K4me3 mark in the active genes in euchromatin (see Table 4–2), and the others recognize either the H3K27me3 or the H3K9me3 silencing mark in heterochromatin (see Table 4–2). This segregation of heterochromatin toward the nuclear periphery is found in nearly all animal cells. (B) Polymer modeling of chromatin can explain the near-universal arrangement of the three types of chromatin in A, as well as the unusual arrangement that is observed in the special inverted nuclei of the photoreceptor cells of nocturnal animals, by assuming that the only difference in the latter nuclei is the absence of lamina–heterochromatin interactions. These models assume that chromatin–chromatin interactions are strongest between heterochromatin of the HK9me3 type, weaker between the heterochromatin of the H3K27me3 type, and nonexistent between euchromatin. (A, courtesy of Irina Solovei; B, adapted from M. Falk et al., *Nature* 570:395–399, 2019. Reproduced with permission of SNCSC.)

each other with respect to DNA sequence on the same chromosome. Analysis of these segments reveals that every chromosome can be subdivided into two compartments, within which loci preferentially interact with each other but avoid interactions with the other compartment, regardless if from the same or another chromosome. It has been shown that these two compartments closely match euchromatin and heterochromatin with regard to their gene richness, gene expression, and replication timing.

This finding is consistent with the images obtained by treating cells with antibody probes that preferentially stain these different types of chromatin; this staining produces a striking picture of the interphase nucleus, in which H3K9me3 heterochromatin is preferentially located near the nuclear periphery, H3K27me3 heterochromatin interior to that, and the open chromatin clustered in more interior regions (**Figure 4–58A**).

The tendency of heterochromatin to self-associate can be mimicked in experiments with reconstituted chromatin fragments that contain the abundant, heterochromatin-specific protein HP1 (see Figure 4-40). These experiments have led to a proposal that a phenomenon that resembles phase separation holds heterochromatin together, mediated by many fluctuating weak interactions (**Figure 4–59**). This self-associating heterochromatin is then tethered to the nuclear periphery through a set of proteins that link heterochromatic

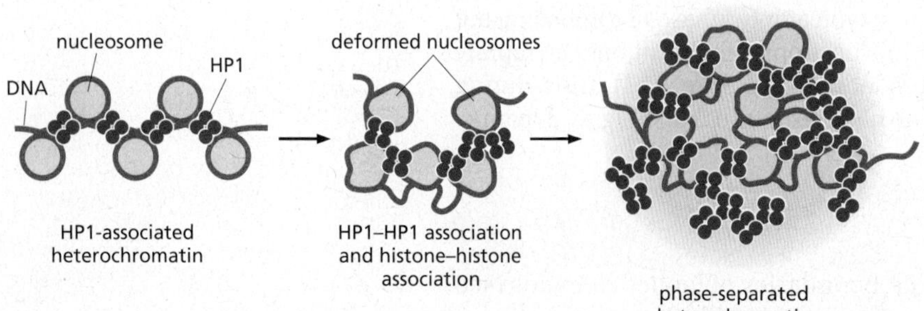

Figure 4–59 **A model for the compaction and clustering of heterochromatin through fluctuating, weak associations that drive phase separation.** According to this proposal, both associations between HP1 molecules and between transiently exposed histone amino acids in deformed nucleosomes contribute to the multiple weak interactions that condense heterochromatin. For a discussion of phase separation, see p. 173. (Adapted from S. Sanulli et al., *Cold Spring Harb. Symp. Quant. Biol.* 84:217–225, 2019.)

regions marked with specific histone modifications to the nuclear lamina. And the lamina is in turn anchored to the inner rim of the nuclear envelope by a set of transmembrane proteins embedded in the inner nuclear membrane (see Figure 16–67).

The functional relevance of heterochromatin positioning at the nuclear periphery remains unclear. However, this nuclear arrangement of chromatin prevails in eukaryotes with only one known exception—the so-called *inverted* nuclei of the rod cells of nocturnal mammals, in which euchromatin and heterochromatin have exchanged positions. The very center of the rod nucleus is occupied by highly condensed inert chromatin, whereas active euchromatin with ongoing gene expression is squeezed to the nuclear periphery (Figure 4–58B). Surprisingly, despite the inversion, the rod nuclei remain fully functional and are highly transcriptionally active.

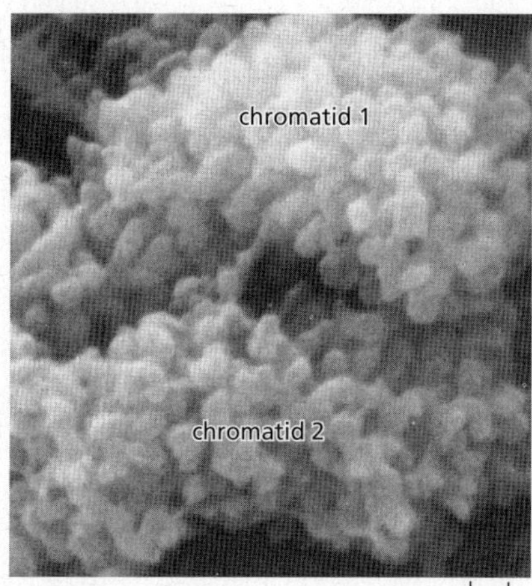

Figure 4–60 **A scanning electron micrograph of a region near one end of a typical mitotic chromosome.** Each knoblike projection is believed to represent the tip of a separate looped domain. Note that the two identical paired chromatids can be clearly distinguished. (From M.P. Marsden and U.K. Laemmli, *Cell* 17:849–858, 1979. With permission from Elsevier.)

Mitotic Chromosomes Are Highly Condensed

Having discussed the dynamic structure of interphase chromosomes, we now turn to **mitotic chromosomes**. The chromosomes from nearly all eukaryotic cells become readily visible by light microscopy during mitosis, when they coil up to form highly condensed structures. This condensation reduces the length of a typical interphase chromosome by about tenfold, and it produces a dramatic change in chromosome appearance.

A typical mitotic chromosome at the metaphase stage of mitosis was shown in Figure 4–18 (for the stages of mitosis, see Panel 17–1, pp. 1048–1049). The two DNA molecules produced by DNA replication during interphase of the cell-division cycle are separately folded to produce two *sister chromatids* held together at their centromeres. These chromosomes are normally covered with a variety of molecules, including large amounts of RNA-protein complexes. Once this covering has been stripped away, each chromatid can be seen in electron micrographs to be organized into loops of chromatin emanating from a central scaffolding (Figure 4–60).

Recent experiments have begun to decipher how interphase chromosomes condense to form mitotic chromosomes. The process is intimately connected with the progression of the cell cycle that will be the focus of Chapter 17. It begins in early M phase when gene expression shuts down, and specific modifications are made to histones that may help to reorganize the chromatin. Most of the ring-shaped *cohesin* proteins that organize the interphase chromosomes make way for their **condensin** relatives, whose ATP-driven movements—aided by the DNA cutting-and-rejoining enzyme topoisomerase II—drive the compaction and form a linear chromosome axis (Figure 4–61). This axis is highly enriched in both condensin II and topoisomerase II, and it is the combined action of these two enzymes that is believed to separate each chromosome into two sister chromatids.

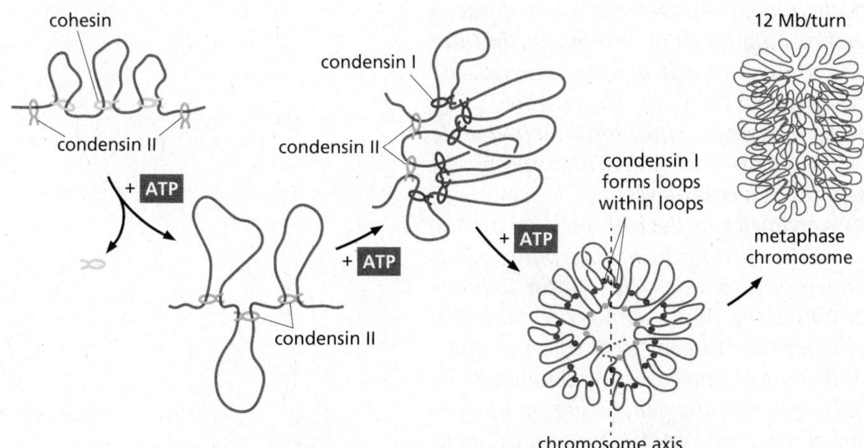

Figure 4–61 **How chromosomes fold in M phase.** There are two types of condensins in mammalian cells, designated as condensins I and II. As cells enter mitosis, the interphase organization of the chromatin loops that is created by cohesin (see Figure 4–57) is lost within minutes. A different SMC protein, condensin II, now begins to form very large new chromatin loops by a similar ATP-driven mechanism. These new loops are organized radially by a central chromosome axis, and as the condensin II loops grow ever larger in size, a second set of loops is formed by condensin I inside them. This "loops within loops" organization, when combined with an ever-tighter winding of the chromatin loops around the mitotic chromosome axis, creates the compact chromatin that is observed in the final metaphase chromosome. Not shown are the special cohesin molecules that hold the two sister chromatids together; as described in Chapter 17, these cohesins are released only at anaphase. (Adapted from J.H. Gibcus et al., *Science* 359:eaao6135, 2018.)

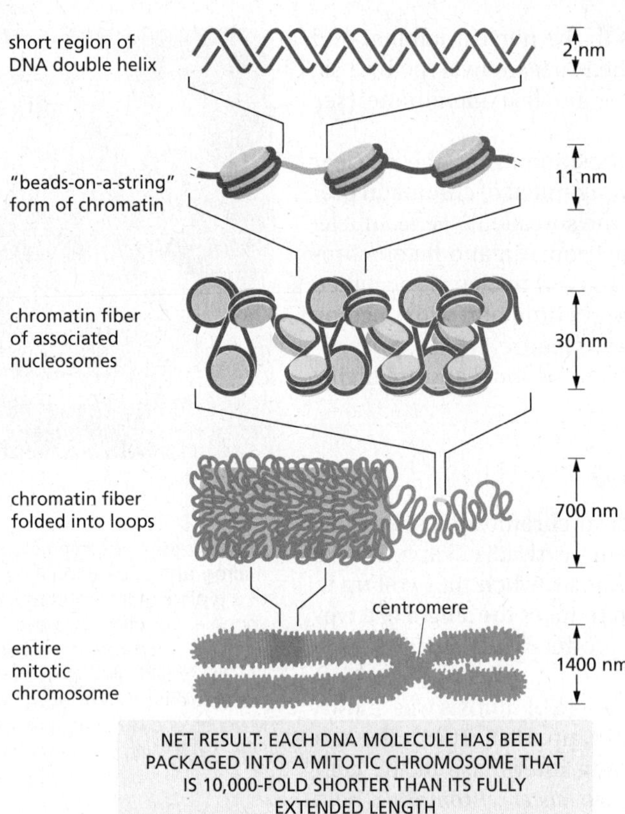

Figure 4–62 Chromosome organization. This model shows some of the levels of chromatin packaging that give rise to the highly condensed mitotic chromosome.

Mitotic chromosome condensation can be thought of as the final level in a hierarchy of chromosome packaging (Figure 4–62). This final DNA folding serves at least two important purposes. First, when condensation is complete (in metaphase), sister chromatids have been disentangled from each other and lie side by side. Thus, the sister chromatids can easily separate when the mitotic apparatus begins pulling them apart. Second, the compaction of chromosomes protects the relatively fragile DNA molecules from being broken as they are pulled to separate daughter cells by the mitotic spindle. These and other critical details of mitosis will be fully discussed in Chapter 17.

Summary

Chromosomes are generally decondensed during interphase, so that the details of their structure are difficult to visualize. Notable exceptions are the specialized lampbrush chromosomes of vertebrate oocytes and the polytene chromosomes in the giant secretory cells of insects. Studies of these two types of interphase chromosomes reveal that each long DNA molecule in a chromosome is divided into a large number of discrete domains organized as long loops of chromatin that are compacted by further folding. When genes contained in a loop are expressed, the loop unfolds and allows the cell's machinery access to the DNA. New studies using sensitive biochemical techniques show that the interphase chromosomes in a wide range of eukaryotes are also folded into a series of looped domains, and they reveal a central folding role for ring-shaped, cohesin protein complexes.

Interphase chromosomes occupy discrete territories in the cell nucleus; that is, they are not extensively intertwined. Much of the chromatin in interphase exists as loosely folded fibers of nucleosomes. However, this open chromatin is interrupted by stretches of heterochromatin, in which the nucleosomes are subjected to additional packing that renders the DNA resistant to gene expression. Heterochromatin has the important property that its type of nucleosome packing can be directly inherited during chromosome replication, thereby generating an epigenetic form of cell memory. Two major classes of heterochromatin contain different covalent histone marks propagated by distinct reader–writer enzyme complexes. Constitutive heterochromatin is formed on repeated DNA sequences; marked by H3K9me3, it is found in large blocks that remain unchanged in and around centromeres and near telomeres. This heterochromatin blocks both gene expression and genetic recombination. Facultative heterochromatin is present at many other positions on chromosomes, where its presence can be controlled so as to have a critical role in regulating genes. H3K27me3 heterochromatin is always facultative; in contrast, only some of the H3K9me3 heterochromatin is facultative, depending on its location.

The interior of the nucleus is highly dynamic, with heterochromatin often positioned near the nuclear envelope and loops of chromatin moving away from their chromosome territory when genes are very highly expressed. This reflects the existence of nuclear subcompartments, where different sets of biochemical reactions are facilitated by an increased concentration of selected proteins and RNAs.

During mitosis, gene expression shuts down and all chromosomes adopt a highly condensed, organized conformation in a process that begins early in M phase. This condensation packages the two DNA molecules of each replicated chromosome as two separately folded chromatids, and it is catalyzed by the ATP-driven movements of ring-shaped, condensin protein complexes.

HOW GENOMES EVOLVE

In this final section of the chapter, we provide an overview of some of the ways that genes and genomes have evolved over time to produce the vast diversity of modern-day life-forms on our planet. The complete sequencing of the genomes of thousands of organisms is revolutionizing our view of the process of evolution, uncovering an astonishing wealth of information about not only family relationships among organisms, but also the molecular mechanisms by which evolution has proceeded.

Much of life's chemistry is shared between all organisms, from bacteria to humans, and it is thus not surprising that genes with similar functions can be found in a diverse range of living things. But the great revelation of the past 40 years has been the extent to which the actual nucleotide sequences of many genes have been conserved. Homologous genes—that is, genes that are similar in both their nucleotide sequence and function because of a common ancestry—can often be recognized across vast phylogenetic distances. Unmistakable homologs of many human genes are present in organisms as diverse as nematode worms, fruit flies, yeasts, plants, and even bacteria. In many cases, the resemblance is so close that, for example, the protein-coding portion of a yeast gene can be substituted with its human homolog—even though humans and yeast are separated by more than a billion years of evolutionary history.

Now that human genome sequencing is being carried out at the population level, with more than a million genomes thus far available for comparison, the extent to which natural selection constrains our own DNA sequence is being analyzed in intimate detail. The results are useful for revealing exactly which bits of our genome are functional. In addition, the data are advancing the field of *precision medicine*, with health implications both for the individuals whose genomes are sequenced and for their future generations.

As emphasized in Chapter 3, the recognition of sequence similarity has become a major tool for inferring gene and protein function. Although a sequence match does not guarantee similarity in function, it has proved to be an excellent clue. Thus, it is often possible to predict the function of genes in humans for which no biochemical or genetic information is available simply by comparing their nucleotide sequences with the sequences of genes that have been characterized in other more readily studied organisms.

In general, the sequences of individual genes are much more tightly conserved than is overall genome structure. Features of genome organization such as genome size, number of chromosomes, order of genes along chromosomes, abundance and size of introns, and amount of repetitive DNA are found to differ greatly when comparing distant organisms, as does the number of genes that each organism contains.

Genome Comparisons Reveal Functional DNA Sequences by Their Conservation Throughout Evolution

A first obstacle in interpreting the sequence of the 3.1 billion nucleotide pairs in the human genome is the fact that the nucleotide sequence of most of it (about 90%) is probably functionally unimportant. The regions of the genome that code for the amino acid sequences of proteins (the exons) are typically found in short segments (median size about 130 nucleotide pairs), small islands in a sea of DNA whose exact nucleotide sequence is thought to be mostly of little consequence. This arrangement can make it difficult to identify all the exons in a stretch of DNA, and it is often hard also to determine exactly where a gene begins and ends.

One very important approach to deciphering our genome is to search for DNA sequences that are closely similar between different species, on the principle that DNA sequences that have a function are much more likely to be conserved than those without a function. For example, humans and mice are thought to have diverged from a common mammalian ancestor about 90×10^6 years ago, which is long enough for the majority of nucleotides in their genomes to have been changed by random mutational events. Consequently, the only regions that will have remained closely similar in the two genomes are those in which mutations would have impaired function and put the animals carrying them at a disadvantage, resulting in their elimination from the population by natural selection. Such closely similar pieces of DNA sequence are known as *conserved DNA regions*. In addition to revealing those DNA sequences that encode functionally important exons and RNA molecules, these conserved regions will include regulatory DNA sequences as well as DNA sequences with functions that are not yet known but which are inferred to be somehow important. In contrast, most *nonconserved DNA regions* will reflect DNA whose sequence is much less likely to be critical for function.

The power of this method can be increased by including in such comparisons the genomes of large numbers of species whose genomes have been sequenced, such as rat, chicken, fish, dog, and chimpanzee, as well as mouse and human. By revealing in this way the results of a very long natural "experiment," lasting for hundreds of millions of years, such comparative DNA sequencing studies have highlighted some of the most interesting regions in our genome. The comparisons reveal that about 4.5% of the human genome consists of *multispecies conserved sequences*. To our great surprise, only about one-fourth of these sequences code for proteins (see Table 4–1, p. 194). Most of the remaining conserved sequences consist of DNA that is thought to contain clusters of protein-binding sites involved in gene regulation, while others produce RNA molecules that are not translated into protein but are important for other reasons.

When the DNA sequences of hundreds of thousands of individual humans are compared, an additional 5% of our genome shows a reduced variation in the human population, which implies that the sequences in this 5% of the genome are also important. Taken together, these analyses suggest that only about 10% of the human genome contains nucleotide sequences that truly matter.

The important question of how much of the DNA sequence of the human genome is functionally relevant was briefly confused by a set of high-profile publications that appeared in 2012 from a large, federally funded US genome project named ENCODE. These publications, which reported the results of a massive survey using sensitive assays that can detect the presence of RNA molecules in cells at extremely low levels, reported that 76% of the total DNA sequence in human cells is transcribed to produce RNA molecules. Even though many of these transcripts were found at levels of less than a single RNA molecule per cell, the ENCODE scientists used such data to assert that most of human DNA is functional, with very little "junk." This claim received widespread publicity, along with their belief that our genome contains tens of thousands of previously undetected genes that produce RNA molecules that do not code for protein.

As previously stated, there is a strong scientific consensus that most of the human genome consists of DNA whose nucleotide sequence is not relevant to biological function—being the so-called junk. This conclusion rests on the finding that natural selection fails to preserve these sequences in the face of the inevitable random changes to genomes that occur over time, as can be seen both when different species are compared and from detailed analyses of human variation. The fact that these DNA sequences nevertheless produce an occasional RNA molecule can be explained by the occurrence of background "noise" in gene expression. Although gene expression is very accurate, it is not perfect, and biochemical errors occasionally occur. Such errors are to be expected, and so long as they are kept at a low level, they are thought to have little or no consequence for the cell.

Genome Alterations Are Caused by Failures of the Normal Mechanisms for Copying and Maintaining DNA, as Well as by Transposable DNA Elements

Evolution depends on accidents and mistakes followed by nonrandom survival. Most of the genetic changes that occur result simply from failures in the normal mechanisms by which genomes are copied or repaired when damaged, although the movement of transposable DNA elements (discussed shortly) also plays an important part. As we will explain in Chapter 5, the mechanisms that maintain DNA sequences are remarkably precise—but errors will occur. DNA sequences are inherited with such extraordinary fidelity that typically, along a given line of descent, only about one nucleotide pair in a thousand is randomly changed in the human germ line every million years. Even so, in a population of 10,000 diploid individuals, every possible nucleotide substitution will have been "tried out" on about 20 occasions in the course of a million years—a short span of time in relation to the evolution of species.

Errors in DNA replication, DNA recombination, or DNA repair can lead either to simple local changes in DNA sequence—so-called *point mutations* such as the substitution of one base pair for another—or to large-scale genome rearrangements such as deletions, duplications, inversions, and translocations of DNA from one chromosome to another. In addition to these rare failures of the genetic machinery, genomes contain mobile DNA elements that are an important source of genomic change (see Table 5–3, p. 286). These transposable DNA elements (*transposons*) are parasitic DNA sequences that can spread within the genomes they colonize. In the process, they often disrupt the function or alter the regulation of existing genes. On occasion, they have created altogether novel genes through fusions between transposon sequences and segments of existing genes. Over long periods of evolutionary time, DNA transposition events have profoundly affected genomes, so much so that nearly half of the DNA in the human genome consists of recognizable relics of past transposition events (Figure 4–63). Even more of our genome is thought to have been derived from transpositions occurring so long ago ($>10^8$ years) that, due to the accumulation of mutations, the sequences can no longer be traced to transposons.

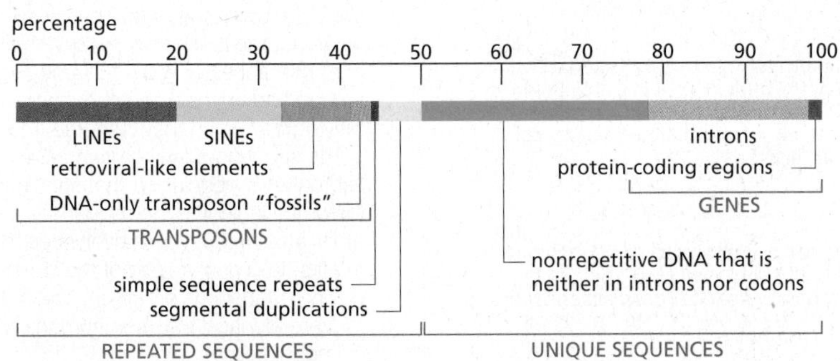

Figure 4–63 A representation of the nucleotide sequence content of the sequenced human genome. The LINEs (long interspersed nuclear elements), SINEs (short interspersed nuclear elements), retroviral-like elements, and DNA-only transposons are mobile genetic elements that have multiplied in our genome by replicating themselves and inserting the new copies in different positions. These mobile genetic elements are discussed in Chapter 5 (see Table 5–3, p. 286). Simple sequence repeats are short nucleotide sequences (fewer than 14 nucleotide pairs) that are repeated again and again for long stretches. Segmental duplications are large blocks of DNA sequence (1000–200,000 nucleotide pairs) that are present at two or more locations in the genome. Most of the highly repeated blocks of DNA in heterochromatin have not yet been completely sequenced; therefore, about 10% of human DNA sequences are not represented in this diagram. (Data courtesy of E. Margulies.)

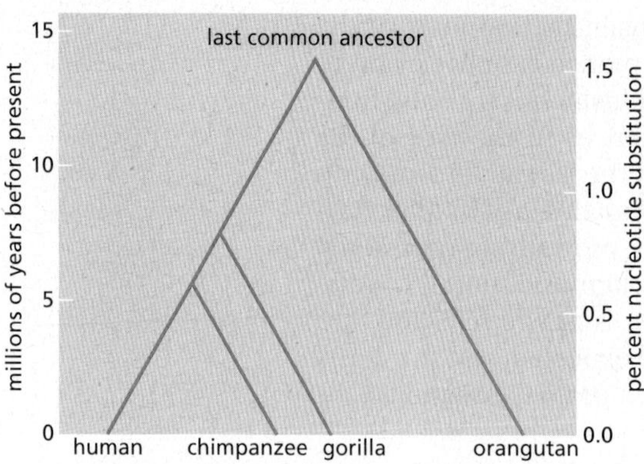

Figure 4–64 A phylogenetic tree showing the relationship between humans and the great apes based on nucleotide sequence data. As indicated, the sequences of homologous portions of the genomes of all four species are estimated to differ from the sequence of the genome of their last common ancestor by a little over 1.5%. Because changes occur independently on both diverging lineages, pairwise comparisons will reveal twice the sequence divergence from the last common ancestor. For example, human–chimpanzee comparisons show divergences of approximately 1.2%. (Modified from F.C. Chen and W.H. Li, *Am. J. Hum. Genet.* 68:444–456, 2001.)

The Genome Sequences of Two Species Differ in Proportion to the Length of Time Since They Have Separately Evolved

The differences between the genomes of all species alive today have accumulated over more than 3 billion years. Although we lack a direct record of changes over time, scientists can reconstruct the process of genome evolution from detailed comparisons of the genomes of contemporary organisms.

The basic organizing framework for comparative genomics is the phylogenetic tree. A simple example is the tree describing the divergence of humans from the great apes (**Figure 4–64**). The primary support for this tree comes from comparisons of genome sequences. For example, comparisons between the sequences of human genes or proteins and those of the great apes typically reveal the fewest differences between human and chimpanzee and the most between human and orangutan.

For closely related organisms such as humans and chimpanzees, it is relatively easy to reconstruct the gene sequences of the extinct, last common ancestor of the two species (**Figure 4–65**). The close similarity between human and chimpanzee

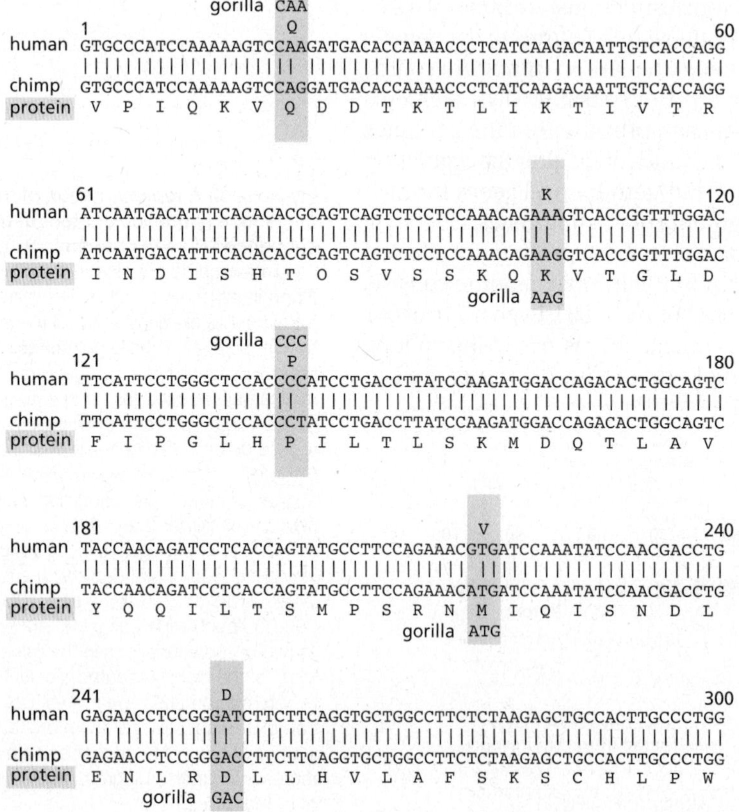

Figure 4–65 Tracing the ancestral sequence from a sequence comparison of the coding regions of human and chimpanzee leptin genes. Reading left to right and top to bottom, a continuous 300-nucleotide segment of a leptin-coding gene is illustrated. Leptin is a hormone that regulates food intake and energy utilization in response to the adequacy of fat reserves. As indicated by the codons boxed in *green*, only 5 nucleotides (of 441 total) differ between the two species. Moreover, in only one of the five positions does the difference in nucleotide lead to a difference in the encoded amino acid. For each of the five variant nucleotide positions, the corresponding sequence in the gorilla is also indicated. In two cases, the gorilla sequence agrees with the human sequence, while in three cases it agrees with the chimpanzee sequence.

What was the sequence of the leptin gene in the last common ancestor? The most economical assumption is that evolution has followed a pathway requiring the minimum number of mutations consistent with the data. Thus, it seems likely that the leptin sequence of the last common ancestor was the same as the human and chimpanzee sequences when they agree. When they disagree, the gorilla sequence can be used as a tiebreaker, a conclusion that should be tested by including the sequences from other great apes. For convenience, only the first 300 nucleotides of the leptin-coding sequences are given. The remaining 141 are identical between humans and chimpanzees.

genes is mainly due to the short time that has been available for the accumulation of mutations in the two diverging lineages, rather than to functional constraints that have kept the sequences the same. Evidence for this view comes from the observation that the human and chimpanzee genomes are nearly identical even where there is no functional constraint on the nucleotide sequence—such as in the third position of "synonymous" codons (codons specifying the same amino acid but differing in their third nucleotide).

For much less closely related organisms, such as humans and chickens (which have evolved separately for about 300 million years), the sequence conservation found in genes is almost entirely due to **purifying selection** (that is, selection that eliminates individuals carrying mutations that interfere with important genetic functions), rather than to an inadequate time for mutations to occur.

Phylogenetic Trees Constructed from a Comparison of DNA Sequences Trace the Relationships of All Organisms

Phylogenetic trees based on molecular sequence data can be compared with the fossil record, and we get our best view of evolution by integrating the two approaches. The fossil record remains essential as a source of absolute dates, which are based on radioisotope decay in the rock formations in which each fossil is found. Because the fossil record has many gaps, however, precise divergence times between species are difficult to establish, even for species that leave good fossils with a distinctive morphology.

Phylogenetic trees whose timing has been calibrated according to the fossil record suggest that changes in the sequences of particular genes or proteins tend to occur at a nearly constant rate, although rates that differ from the norm by as much as twofold are observed in particular lineages. This provides us with a *molecular clock* for evolution—or rather a set of molecular clocks corresponding to different categories of DNA sequence. As in the example in **Figure 4–66**, the clock runs most rapidly and regularly in sequences that are not subject to purifying selection. These include portions of introns that lack splicing or regulatory signals, the third position in synonymous codons, and genes that have been irreversibly inactivated by mutation (the so-called pseudogenes). The clock runs most slowly for sequences that are subject to strong functional constraints; for example, the amino acid sequences of proteins such as histones that engage in specific interactions with large numbers of other proteins and whose structure is therefore highly constrained, or the nucleotide sequences that encode the RNA subunits of the ribosome, on which all protein synthesis depends.

Occasionally, rapid change is seen in a previously highly conserved sequence. As discussed later in this chapter, such episodes are especially interesting because they are thought to reflect periods of strong positive selection for mutations that have conferred a selective advantage in the particular lineage where the rapid change occurred.

The pace at which molecular clocks run during evolution is determined not only by the degree of purifying selection but also by the mutation rate. Notably, in animals, although not in plants, clocks based on functionally unconstrained mitochondrial DNA sequences run much faster than clocks based on functionally unconstrained nuclear sequences, because the mutation rate in animal mitochondria is exceptionally high.

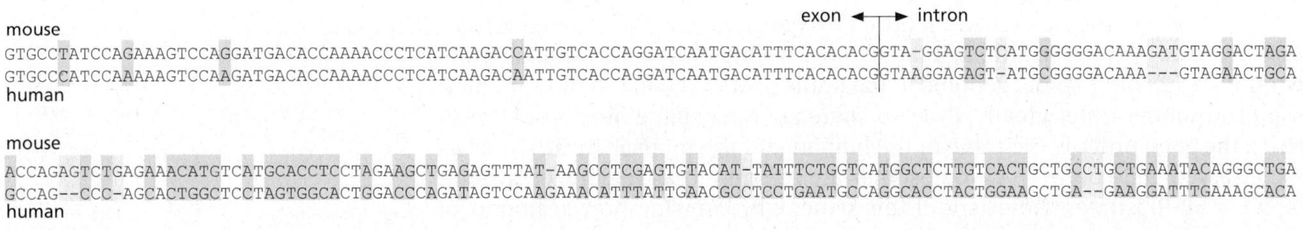

Figure 4–66 The very different rates of evolution of exons and introns, as illustrated by comparing a portion of the mouse and human leptin genes. Starting at top left and ending at bottom right, the DNA sequences of one exon and its adjacent intron are compared for human and mouse leptin genes. Positions where the sequences differ by a single nucleotide substitution are boxed in *green*, and positions that differ by the addition or deletion of nucleotides are boxed in *yellow*. Note that, thanks to purifying selection, the coding sequence of the exon is much more conserved than is the adjacent intron sequence.

exon ◄ ► intron

mouse
GTGCCTATCCAGAAAGTCCAGGATGACACCAAAACCCTCATCAAGACCATTGTCACCAGGATCAATGACATTTCACACACGGTA-GGAGTCTCATGGGGGACAAAGATGTAGGACTAGA
GTGCCCATCCAAAAAGTCCAAGATGACACCAAAACCCTCATCAAGACAATTGTCACCAGGATCAATGACATTTCACACACGGTAAGGAGAGT-ATGCGGGGACAAA---GTAGAACTGCA
human

mouse
ACCAGAGTCTGAGAAACATGTCATGCACCTCCTAGAAGCTGAGAGTTTAT-AAGCCTCGAGTGTACAT-TATTTCTGGTCATGGCTCTTGTCACTGCTGCCTGCTGAAATACAGGGCTGA
GCCAG--CCC-AGCACTGGCTCCTAGTGGCACTGGACCCAGATAGTCCAAGAAACATTTATTGAACGCCTCCTGAATGCCAGGCACCTACTGGAAGCTGA--GAAGGATTTGAAAGCACA
human

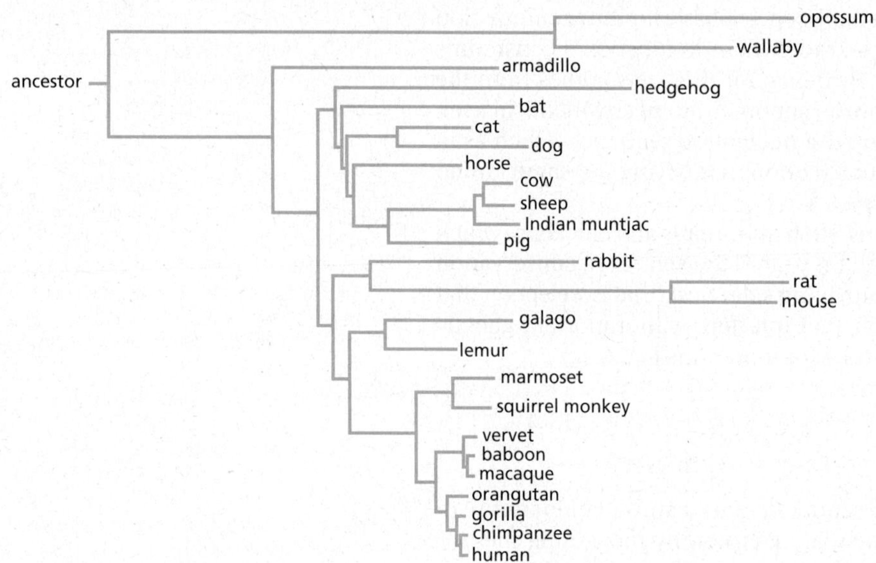

Figure 4–67 A phylogenetic tree showing the evolutionary relationships of some present-day mammals. The length of each line is proportional to the number of *neutral substitutions*; that is, nucleotide changes at sites where there is assumed to be no purifying selection. (Adapted from G.M. Cooper et al., *Genome Res.* 15:901–913, 2005. With permission from Cold Spring Harbor Laboratory Press.)

Categories of DNA for which the clock runs fast are most informative for recent evolutionary events; the mitochondrial DNA clock has been used, for example, to chronicle the divergence of the Neanderthal lineage from that of modern *Homo sapiens*. To study ancient evolutionary events, one must examine DNA for which the clock runs unusually slowly; thus, the divergence of the major branches of the tree of life—bacteria, archaea, and eukaryotes—has been deduced from study of the sequences specifying ribosomal RNA.

In general, molecular clocks, appropriately chosen, have a finer time resolution than that of the fossil record, and they are a more reliable guide to the detailed structure of phylogenetic trees than are classical methods of tree construction, which are based on family resemblances in anatomy and embryonic development. For example, the precise family tree of great apes and humans was not settled until sufficient molecular sequence data accumulated in the 1980s to produce the pedigree shown previously in Figure 4–64. And with huge amounts of DNA sequence now determined from a wide variety of mammals, much better estimates of our relationship to them are being obtained (**Figure 4–67**).

A Comparison of Human and Mouse Chromosomes Shows How the Structures of Genomes Diverge

As would be expected, the human and chimpanzee genomes are much more alike than are the human and mouse genomes, even though all three genomes are roughly the same size and contain nearly identical sets of genes. Mouse and human lineages have had approximately 90 million years to diverge through accumulated mutations, versus 6 million years for humans and chimpanzees. In addition, rodent lineages (represented by the rat and the mouse in Figure 4–67) have unusually fast molecular clocks and have diverged from the human lineage more rapidly than otherwise expected.

While the way that the genome is organized into chromosomes is almost identical between humans and chimpanzees, this organization has diverged greatly between humans and mice. According to rough estimates, a total of about 180 chromosome breakage-and-rejoining events have moved large blocks of DNA sequence in the human and mouse lineages since they last shared a common ancestor. As a result, although the number of chromosomes is similar in the two species (23 per haploid genome in the human versus 20 in the mouse), their overall structures differ greatly. Even so, there are many large blocks of DNA in which the gene order is the same in the human and the mouse. These stretches of conserved gene order in chromosomes are referred to as regions of *synteny*. **Figure 4–68** illustrates the extent of this synteny by showing how segments of

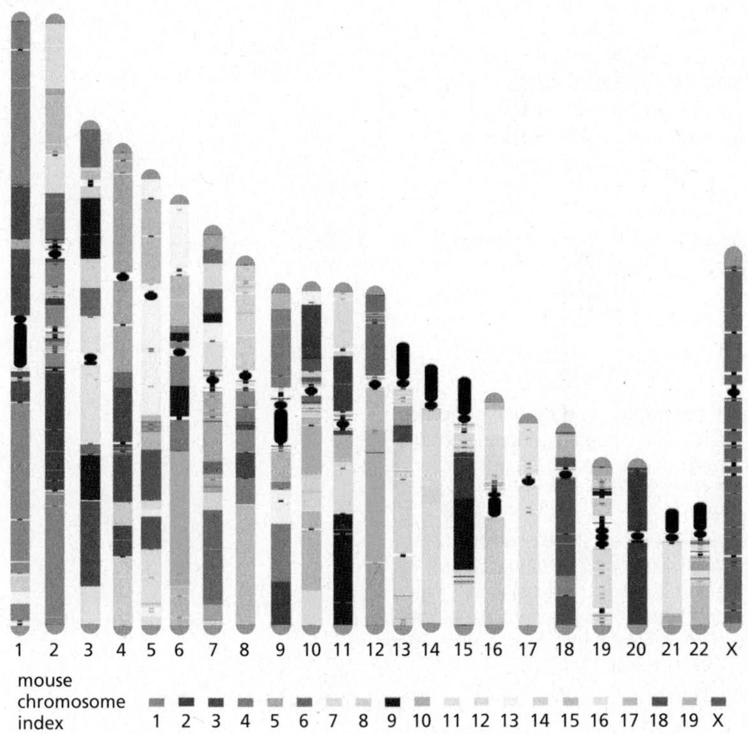

Figure 4–68 Synteny between human and mouse chromosomes. In this diagram, the human chromosome set is shown, with each part of each chromosome colored according to the mouse chromosome with which it is syntenic. The color coding used for each mouse chromosome is shown at the bottom of the figure.

Heterochromatic highly repetitive regions (such as centromeres) that are difficult to sequence cannot be mapped in this way; these are colored *black*. (Adapted from E.E. Eichler and D. Sankoff, *Science* 301:793–797, 2003. With permission from AAAS.)

mouse chromosomes map onto the human chromosome set. For much more distantly related vertebrates, such as chicken and human, the number of breakage-and-rejoining events has been much greater, and the regions of synteny are much shorter; in addition, these regions are often hard to discern because of the divergence of the DNA sequences they contain.

An unexpected conclusion from a detailed comparison of the complete mouse and human genome sequences, confirmed by subsequent comparisons between the genomes of other vertebrates, is that small blocks of DNA sequence are being deleted from and added to genomes at a surprisingly rapid rate. Thus, if we assume that our common ancestor had a genome of human size (about 3.1 billion nucleotide pairs), mice would have lost a total of about 45% of that genome from accumulated deletions during the past 90 million years, while humans would have lost about 25%. However, substantial sequence gains from many small chromosome duplications and from the multiplication of transposons have compensated for these deletions. As a result of this series of gains and losses, the size of our genome is thought to be practically unchanged from that of the last common ancestor of humans and mice, while the mouse genome is smaller by only about 0.4 billion nucleotides.

Good evidence for the loss of DNA sequences in small blocks during evolution can be obtained from a detailed comparison of regions of synteny in the human and mouse genomes. The comparative shrinkage of the mouse genome can be clearly seen from such comparisons, with the net loss of sequences scattered throughout the long stretches of DNA that are otherwise homologous (**Figure 4–69**).

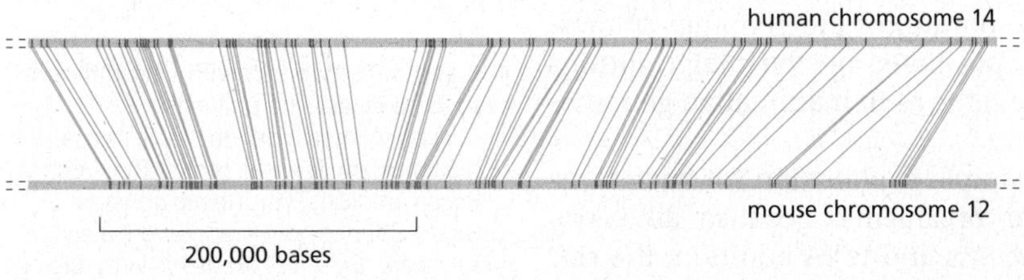

Figure 4–69 Comparison of a syntenic portion of mouse and human genomes. About 90% of the two genomes can be aligned in this way. Note that while there is an identical order of the matched index sequences *(red marks)*, there has been a net loss of DNA in the mouse lineage that is interspersed throughout the entire region. This type of net loss is typical for all such regions, and it accounts for the fact that the mouse genome contains 14% less DNA than does the human genome. (Adapted from Mouse Genome Sequencing Consortium, *Nature* 420:520–562, published 2002 by Nature Publishing Group. Reproduced with permission of SNCSC.)

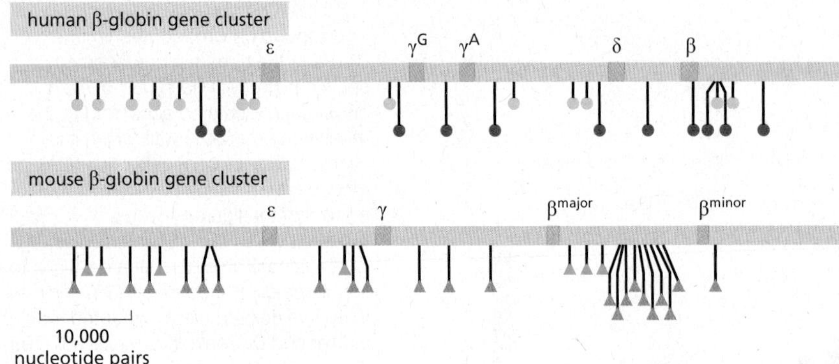

Figure 4–70 **A comparison of the β-globin gene cluster in the human and mouse genomes, showing the locations of transposable elements.** This stretch of the human genome contains five functional β-globin–like genes *(orange)*; the comparable region from the mouse genome has only four. The positions of the human *Alu* sequences are indicated by *green circles*, and the positions of the human *L1* sequences are indicated by *red circles*.

The mouse genome contains different but related transposable elements: the positions of B1 elements (which are related to the human *Alu* sequences) are indicated by *blue triangles*, and the positions of the mouse *L1* elements (which are related to the human *L1* sequences) are indicated by *orange triangles*. The absence of transposable elements from the globin structural genes can be attributed to purifying selection, which would have eliminated any insertion that compromised gene function. (Courtesy of Ross Hardison and Webb Miller.)

DNA is added to genomes both by the spontaneous duplication of chromosomal segments that are typically tens of thousands of nucleotide pairs long (as will be discussed shortly) and by insertion of new copies of active transposons. Most transposition events are duplicative, because the original copy of the transposon stays where it was when a copy inserts at the new site; see, for example, Figure 5–59. Comparison of the DNA sequences derived from transposons in the human and the mouse readily reveals some of the sequence additions (**Figure 4–70**). In contrast, the nucleotide sequences and positions of the transposons in human and chimpanzee are very similar, indicating that their movement occurred before the two species diverged.

It remains a mystery why all mammals have maintained genome sizes of roughly 3 billion nucleotide pairs that contain nearly identical sets of genes, even though 90% of this DNA appears not to be under sequence-specific functional constraints.

The Size of a Vertebrate Genome Reflects the Relative Rates of DNA Addition and DNA Loss in a Lineage

In more distantly related vertebrates, genome size can vary considerably, apparently without a drastic effect on the organism or its number of genes. Thus, the chicken genome, at 1 billion nucleotide pairs, is only about one-third the size of the mammalian genome, even though it contains nearly the same number of genes. An extreme example is the puffer fish, *Fugu rubripes* (**Figure 4–71A**), which has a tiny genome for a vertebrate (0.4 billion nucleotide pairs compared to 1 billion or more for most other fish). The small size of the *Fugu* genome is largely due to the small size of its introns and intergenic regions. Specifically, *Fugu* introns, as well as other noncoding segments of the *Fugu* genome, lack the repetitive DNA that makes up a large portion of the genomes of most well-studied vertebrates. Nevertheless, the positions of the *Fugu* introns between the exons of each gene are almost the same as in mammalian genomes (**Figure 4–72**).

While initially a mystery, we now have a simple explanation for such large differences in genome size between similar organisms: because all vertebrates experience a continual process of DNA loss and DNA addition, the size of a genome merely depends on the balance between these opposing processes

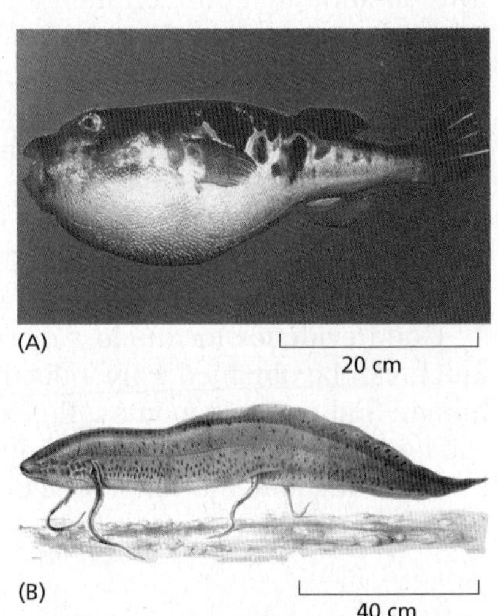

Figure 4–71 **Two fish with very different genome sizes.** The puffer fish, *Fugu rubripes* (A), has a genome size that is 300 times smaller than that of the West African lungfish, *Protopterus annectens* (B). (A, Courtesy of Byrappa Venkatesh; B, History and Art Collection/Alamy Stock Photo.)

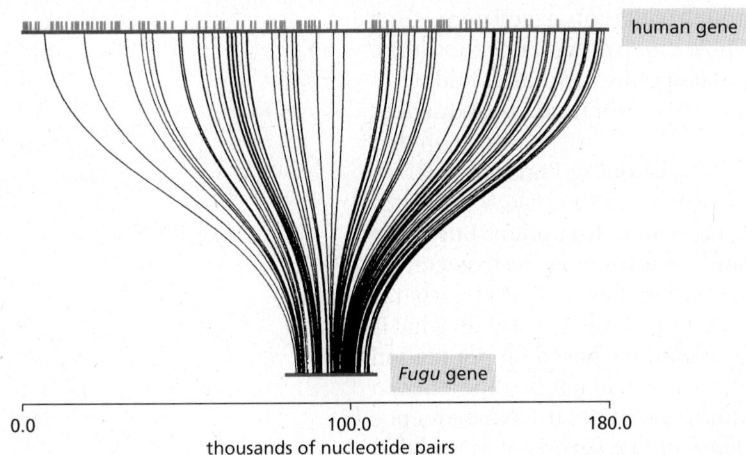

Figure 4–72 **Comparison of the genomic sequences of the human and *Fugu* genes encoding the protein huntingtin.** Both genes (indicated in *red*) contain 67 short exons that align in 1:1 correspondence to one another; these exons are connected by curved lines. The human gene is 7.5 times larger than the *Fugu* gene (180,000 versus 24,000 nucleotide pairs). The size difference is entirely due to larger introns in the human gene. The larger size of the human introns is due in part to the presence of retrotransposons (discussed in Chapter 5), whose positions are represented by *green vertical lines*; the *Fugu* introns lack retrotransposons. In humans, mutation of the huntingtin gene causes Huntington's disease, an inherited neurodegenerative disorder. (Adapted from S. Baxendale et al., *Nat. Genet.* 10:67–76, published 1995 by Nature Publishing Group. Reproduced with permission of SNCSC.)

acting over millions of years. Suppose, for example, that in the lineage leading to *Fugu*, the rate of DNA addition happened to slow greatly. Over long periods of time, this would result in a major "cleansing" from this fish genome of those DNA sequences whose loss could be tolerated. The result is an unusually compact genome, relatively free of junk and clutter, but retaining through purifying selection the vertebrate DNA sequences that are functionally important. This makes *Fugu*, with its 400 million nucleotide pairs of DNA, a valuable resource for genome research aimed at understanding humans.

At the other end of the scale, some fish—such as the primitive-looking lungfish—have enormous genomes, more than 300 times the size of that of *Fugu* and 30 times the size of that of humans (**Figure 4–71B**). Most of the extra DNA consists of transposons and other repeated DNA sequences, suggesting that genome additions have greatly exceeded losses in this lineage.

Multispecies Sequence Comparisons Identify Many Conserved DNA Sequences of Unknown Function

The mass of DNA sequence now in freely accessible databases (hundreds of billions of nucleotide pairs) provides a rich resource that scientists can mine for many purposes. This information can be used not only to unscramble the evolutionary pathways that have led to modern organisms, but also to provide insights into how cells and organisms function. Perhaps the most remarkable discovery in this realm comes from the observation that a striking amount of DNA sequence that does not code for protein has been conserved during mammalian evolution (see Table 4–1, p. 194). This is most clearly revealed when we align and compare DNA synteny blocks from many different species, thereby identifying large numbers of so-called *multispecies conserved sequences*: some of these code for protein, but most of them do not.

Many of the conserved sequences that do not code for protein are now known to produce untranslated RNA molecules, such as the thousands of *long noncoding RNAs* (*lncRNAs*) that are thought to have important functions in regulating gene transcription. As we shall also see in Chapter 7, many others are regions of DNA scattered throughout the genome that directly bind proteins involved in gene regulation. But the function of much of the conserved noncoding DNA remains a mystery. This enigma highlights how much more we need to learn about the fundamental biological mechanisms that operate in animals and other complex organisms, and its solution is certain to have profound consequences for medicine.

How can cell biologists tackle the mystery of noncoding conserved DNA? Traditionally, attempts to determine the function of a puzzling DNA sequence begin by looking at the consequences of its experimental disruption. But many DNA sequences that are crucial for an organism in the wild can be expected to have no noticeable effect on its phenotype under laboratory conditions: what is required for a mouse to survive in a laboratory cage is very much less than what is required for it to succeed in nature. Moreover, calculations based on population genetics reveal that just a tiny selective advantage—less than a 0.1% difference in survival—can be enough to strongly favor retaining a particular DNA sequence over evolutionary time spans. One should therefore not be surprised to find that many conserved DNA sequences can be deleted from the mouse genome without any noticeable effect on that mouse in a laboratory.

Changes in Previously Conserved Sequences Can Help Decipher Critical Steps in Evolution

Given genome sequence information, we can tackle another intriguing question: What alterations in our DNA have made humans so different from other animals—or for that matter, what makes any individual species so different from its relatives? For example, as soon as both the human and the chimpanzee genome sequences became available, scientists began searching for DNA sequence changes that might account for the striking differences between us and chimpanzees. With 3.1 billion nucleotide pairs to compare in the two species, this might seem an impossible task. But the job was made much easier by confining the search to 35,000 clearly defined multispecies conserved sequences (a total of about 5 million nucleotide pairs), representing parts of the genome that are most likely to be functionally important. Though these sequences are conserved strongly, they are not conserved perfectly, and when the version in one species is compared with that in another they are generally found to have drifted apart by a small amount corresponding simply to the time elapsed since the last common ancestor. In a small proportion of cases, however, one sees signs of a sudden evolutionary spurt. For example, some DNA sequences that have been highly conserved in other mammalian species are found to have accumulated nucleotide changes exceptionally rapidly during the 6 million years of human evolution since we diverged from the chimpanzees. These *human accelerated regions* (*HARs*) are thought to reflect functions that have been especially important in making us different in some useful way.

About 50 such sites were identified in one study, one-fourth of which were located near genes associated with neural development. The sequence exhibiting the most rapid change (18 changes between human and chimpanzee, compared to only two changes between chimpanzee and chicken) was examined further and found to encode a 118-nucleotide noncoding RNA molecule, HAR1F (human accelerated region 1F), that is produced in the human cerebral cortex at a critical time during brain development. The function of this HAR1F RNA is not yet known, but findings of this type are stimulating research studies that may shed light on crucial features of the human brain.

A related approach in the search for the important mutations that contributed to human evolution likewise begins with DNA sequences that have been conserved during mammalian evolution, but rather than screening for accelerated changes in individual nucleotides, it focuses instead on chromosome sites that have experienced deletions in the 6 million years since our lineage diverged from that of chimpanzees. More than 500 such sequences—conserved among other species but deleted in humans—have been discovered. Each deletion removes an average of 95 nucleotides of DNA sequence. Only one of these deletions affects a protein-coding region: the rest are thought to alter regions that affect how nearby genes are expressed, an expectation that has been experimentally confirmed in a few cases. A large proportion of the presumed regulatory regions identified in this way lie near genes that affect neural function and/or near genes involved in steroid signaling, suggesting that changes in the nervous system and in immune or reproductive functions have played an especially important role in human evolution.

Mutations in the DNA Sequences That Control Gene Expression Have Driven Many of the Evolutionary Changes in Vertebrates

The vast hoard of genomic sequence data now being accumulated can be explored in many other ways to reveal events that happened even hundreds of millions of years ago. For example, one can attempt to trace the origins of the regulatory elements in DNA that have played critical parts in vertebrate evolution. One such study began with the identification of nearly 3 million noncoding sequences, averaging 28 base pairs in length, that have been conserved in recent vertebrate evolution while being absent in more ancient ancestors. Each of these special noncoding sequences is likely to represent a functional innovation peculiar to a particular branch of the vertebrate family tree, and most of them are thought to consist of regulatory DNA that governs the expression of a neighboring gene. Given full genome sequences, one can identify the genes that appear most likely to have fallen under the sway of these novel regulatory elements. By comparing many different species, with known divergence times, one can also estimate when each such regulatory element came into existence as a conserved feature.

The findings suggest remarkable evolutionary differences between the various functional classes of genes (**Figure 4–73**). Conserved regulatory elements that originated early in vertebrate evolution—that is, more than about 300 million years ago, which is when the mammalian lineage split from the lineage leading to birds and reptiles—seem to be mostly associated with genes that code for transcription regulatory proteins and for proteins with roles in organizing embryonic development. Then came an era when the regulatory DNA innovations arose next to genes coding for receptors for extracellular signals. Finally, over the course of the past 100 million years, the regulatory innovations seem to have been concentrated in the neighborhood of genes coding for proteins (such as protein kinases) that function to modify other proteins post-translationally.

Many questions remain to be answered about these phenomena and what they mean. One possible interpretation is that the logic—the circuit diagram—of the gene regulatory network in vertebrates was established early, and that more recent evolutionary change has mainly occurred through the tuning of quantitative parameters. This could help to explain why, among the mammals, for

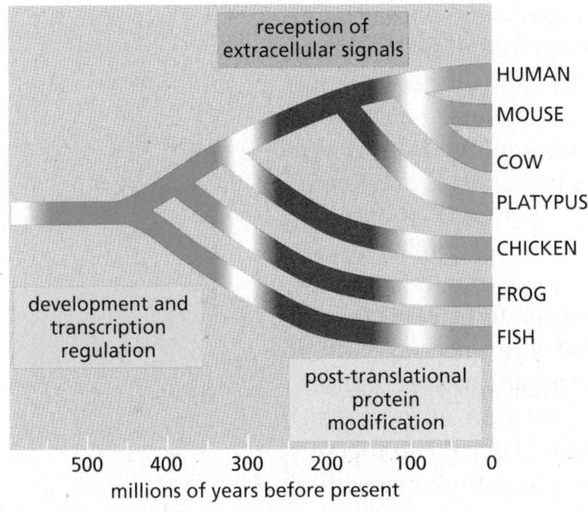

Figure 4–73 **The types of changes in gene regulation inferred to have predominated during the evolution of our vertebrate ancestors.** To produce the information summarized in this plot, wherever possible the type of gene regulated by each conserved noncoding sequence was inferred from the identity of its closest protein-coding gene. The time when each conserved sequence became fixed in the vertebrate lineage was then used to derive the conclusions shown. (Based on C.B. Lowe et al., *Science* 333:1019–1024, 2011.)

example, the basic body plan—the topology of the tissues and organs—has been largely conserved.

Gene Duplication Also Provides an Important Source of Genetic Novelty During Evolution

Evolution depends on the creation of new genes, as well as on the modification of those that already exist. How does this occur? When we compare organisms that seem very different—a primate with a rodent, for example, or a mouse with a fish—we rarely encounter genes in the one species that have no homolog in the other. Genes without homologous counterparts are relatively scarce even when we compare such divergent organisms as a mammal and a worm. On the other hand, we frequently find gene families that have different numbers of members in different species. To create such families, genes have been repeatedly duplicated, and the copies have then diverged to take on new functions that often vary from one species to another.

Gene duplication occurs at high rates in all evolutionary lineages, contributing to the vigorous process of DNA addition discussed previously. In a detailed study of spontaneous duplications in yeast, duplications of 50,000–250,000 nucleotide pairs were commonly observed, most of which were tandemly repeated. These appeared to result from DNA replication errors that led to the inexact repair of double-strand chromosome breaks. A comparison of the human and chimpanzee genomes reveals that, since the time that these two organisms diverged, such *segmental duplications* have added about 5 million nucleotide pairs to each genome every million years, with an average duplication size being about 50,000 nucleotide pairs (although there are some duplications five times larger). In fact, if one counts nucleotides, duplication events have created more differences between our two species than have single-nucleotide substitutions.

Duplicated Genes Diverge

What is the fate of newly duplicated genes? In most cases, there is presumed to be little or no selection—at least initially—to maintain the duplicated state because either copy can provide an equivalent function. Hence, many duplication events are likely to be followed by loss-of-function mutations in one or the other gene. This cycle would functionally restore the one-gene state that preceded the duplication. Indeed, there are many examples in contemporary genomes where one copy of a duplicated gene can be seen to have become irreversibly inactivated by multiple mutations. Over time, the sequence similarity between such a **pseudogene** and the functional gene whose duplication produced it would be expected to be eroded by the accumulation of many mutations in the pseudogene—the homologous relationship eventually becoming undetectable.

An alternative fate for gene duplications is for both copies to remain functional, while diverging in their sequence and pattern of expression, thus taking on different roles. This process of *duplication and divergence* almost certainly explains the presence of large families of genes with related functions in biologically complex organisms, and it is thought to play a critical role in the evolution of increased biological complexity. An examination of many different eukaryotic genomes suggests that the probability that any particular gene will undergo a duplication event that spreads to most or all individuals in a species is approximately 1% every million years.

Whole-genome duplications offer particularly dramatic examples of the duplication–divergence cycle. A whole-genome duplication can occur quite simply: all that is required is one round of genome replication in a germ-line cell lineage without a corresponding cell division. Initially, the chromosome number simply doubles. Such abrupt increases in the ploidy of an organism are common, particularly in fungi and plants. After a whole-genome duplication, all genes exist as duplicate copies. However, unless the duplication event occurred so recently that there has been little time for subsequent alterations in genome structure, the results of a series of segmental duplications—occurring at different times—are hard to distinguish from the end product of a whole-genome duplication. In mammals, for example, the role of whole-genome duplications versus a series of piecemeal duplications of DNA segments is quite uncertain. Nevertheless, it is clear that a great deal of gene duplication has occurred in the distant past.

Analysis of the genome of the zebrafish, in which at least one whole-genome duplication is thought to have occurred hundreds of millions of years ago, has cast some light on the process of gene duplication and divergence. Although many duplicates of zebrafish genes appear to have been lost by mutation, a significant fraction—perhaps as many as 30–50%—have diverged functionally while both copies have remained active. In many cases, the most obvious functional difference between the duplicated genes is that they are expressed in different tissues or at different stages of development. One attractive theory to explain such an end result imagines that different, mildly deleterious mutations occur quickly in both copies of a duplicated gene set. For example, one copy might lose expression in a particular tissue as a result of a regulatory mutation, while the other copy loses expression in a second tissue. After such an occurrence, both gene copies would be required to provide the full range of functions that were once supplied by a single gene; hence, both copies would now be protected from loss through inactivating mutations. Over a longer period, each copy could then undergo further changes through which it could acquire new, specialized features.

The Evolution of the Globin Gene Family Shows How DNA Duplications Contribute to the Evolution of Organisms

The globin gene family provides an especially good example of how DNA duplication generates new proteins, and its evolutionary history has been worked out particularly well. The unmistakable similarities in amino acid sequence and structure among the present-day globins indicate that they all must derive from a common ancestral gene, even though some are now encoded by widely separated genes in the mammalian genome.

We can reconstruct some of the past events that produced the various types of oxygen-carrying hemoglobin molecules by considering the different forms of the protein in organisms at different positions on the tree of life. A molecule like hemoglobin was necessary to allow multicellular animals to grow to a large size, because large animals cannot simply rely on the diffusion of oxygen through the body surface to oxygenate their tissues adequately. But oxygen plays a vital part in the life of nearly all living organisms, and oxygen-binding proteins homologous to hemoglobin can be recognized even in plants, fungi, and bacteria. In animals, the simplest oxygen-carrying molecule is a globin polypeptide chain of about 150 amino acids that is found in many marine worms, insects, and primitive fish. The hemoglobin molecule in more complex vertebrates, however, is composed of two kinds of globin chains. It appears that about 500 million years ago, just before fish and mammals diverged from their common ancestor, a series of gene mutations and duplications occurred. These events established two slightly different globin genes in the genome of each individual, coding for α-globin and β-globin chains that associate to form a hemoglobin molecule consisting of two α chains and two β chains (**Figure 4–74**). The four oxygen-binding sites in the $\alpha_2\beta_2$ molecule interact, allowing a cooperative allosteric change in the molecule as it binds and releases oxygen, which enables hemoglobin to take up and release oxygen more efficiently than can the single-chain version.

Still later, during the evolution of mammals, the β-chain gene apparently underwent duplication and mutation to give rise to a second β-like chain that

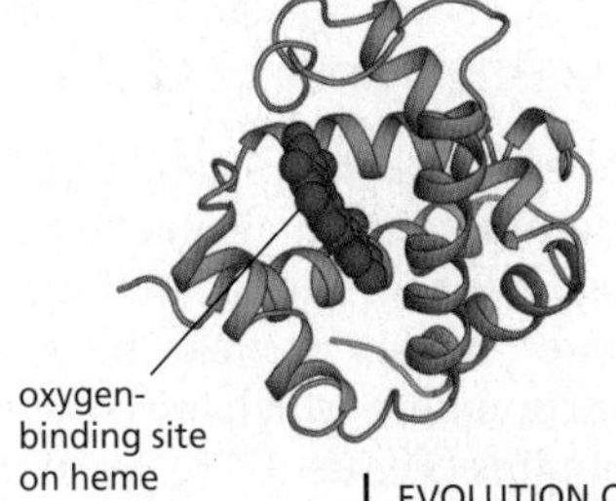
single-chain globin binds one oxygen molecule
oxygen-binding site on heme
EVOLUTION OF A SECOND GLOBIN CHAIN BY GENE DUPLICATION FOLLOWED BY MUTATION

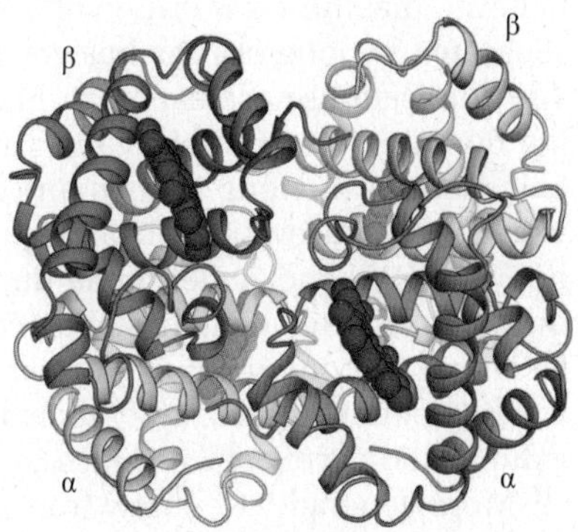
β
β
α
α
four-chain globin binds four oxygen molecules in a cooperative manner

Figure 4–74 A comparison of the structure of one-chain and four-chain globins. The four-chain globin shown is hemoglobin, which is a complex of two α-globin and two β-globin chains. The one-chain globin present in some primitive vertebrates represents an intermediate in the evolution of the four-chain globin. With oxygen bound it exists as a monomer; without oxygen it dimerizes. (PDB code: 2DHB.)

is synthesized specifically in the fetus. The resulting hemoglobin molecule has a higher affinity for oxygen than that of adult hemoglobin and thus helps in the transfer of oxygen from the mother to the fetus. The gene for the new β-like chain subsequently duplicated and mutated again to produce two new genes, ε and γ, the ε chain being produced earlier in development (to form $\alpha_2\varepsilon_2$) than the fetal γ chain, which forms $\alpha_2\gamma_2$. A duplication of the adult β-chain gene occurred still later, during primate evolution, to give rise to a δ-globin gene and thus to a minor form of hemoglobin ($\alpha_2\delta_2$) that is found only in adult primates (**Figure 4–75**).

Each of these duplicated genes has been modified by point mutations that affect the properties of the final hemoglobin molecule, as well as by changes in regulatory regions that determine the timing and level of expression of the gene. As a result, each globin is made in different amounts at different times of human development.

The history of these gene duplications is reflected in the arrangement of hemoglobin genes in the genome. In the human genome, the genes that arose from the original β gene are arranged as a series of homologous DNA sequences located within 50,000 nucleotide pairs of one another on a single chromosome. A similar cluster of human α-globin genes is located on a separate chromosome. Not only other mammals, but birds too have their α-globin and β-globin gene clusters on separate chromosomes. In the frog *Xenopus*, however, they are together, suggesting that a chromosome translocation event in the lineage of birds and mammals separated the two gene clusters about 300 million years ago, soon after our ancestors and those of amphibians diverged (see Figure 4–75).

There are several duplicated globin DNA sequences in the α-globin and β-globin gene clusters that are not functional genes but pseudogenes. These have a close sequence similarity to the functional genes but have been disabled by mutations that prevent their expression as functional proteins. The existence of such pseudogenes makes it clear that, as expected, not every DNA duplication leads to a new functional gene. Indeed, the human genome is thought to contain more pseudogenes than genes.

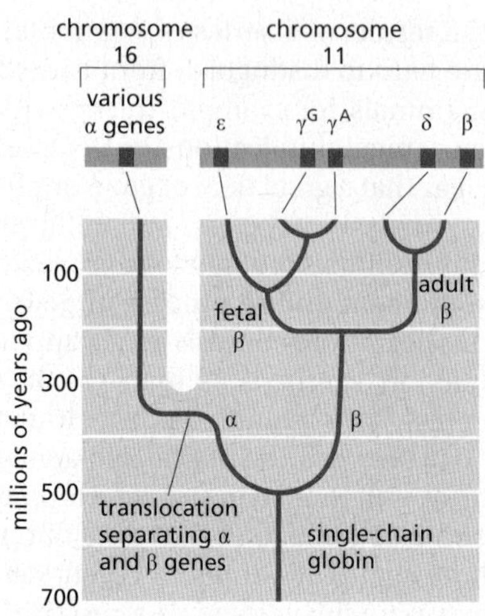

Figure 4–75 An evolutionary scheme for the globin chains that carry oxygen in the blood of animals. The scheme emphasizes the β-like globin gene family. A relatively recent gene duplication of the γ-chain gene produced γ^G and γ^A, which are fetal β-like chains of identical function. The location of the globin genes in the human genome is shown at the top of the figure.

Genes Encoding New Proteins Can Be Created by the Recombination of Exons

The role of DNA duplication in evolution is not confined to the expansion of gene families. It can also act on a smaller scale to create single genes by stringing together short duplicated segments of DNA. The proteins encoded by genes generated in this way can be recognized by the presence of repeating similar protein domains, which are covalently linked to one another in series. The immunoglobulins (**Figure 4–76**), for example, as well as most fibrous proteins (such as collagens) are encoded by genes that have evolved by repeated duplications of a primordial DNA sequence.

In genes that have evolved in this way, as well as in many other genes, each separate exon often encodes an individual protein folding unit, or domain. It is believed that the organization of DNA coding sequences as a series of such exons separated by long introns has greatly facilitated the evolution of new proteins. The duplications necessary to form a single gene coding for a protein with repeating domains, for example, can easily occur by breaking and rejoining the DNA anywhere in the long introns on either side of an exon. Without introns there would be only a few sites in the original gene at which a recombinational exchange between DNA molecules could duplicate the domain and not disrupt it. Moreover, introns often contain sequences that are repeated many times in a genome, facilitating recombination between different introns. By enabling recombination at many potential sites rather than just a few, introns increase the probability that a duplication event will produce a new protein.

More generally, we know from genome sequences that the various parts of genes—both their individual exons and their regulatory elements—have served as modular elements that have been duplicated and moved about the genome to create the great diversity of living things. Thus, for example, many present-day

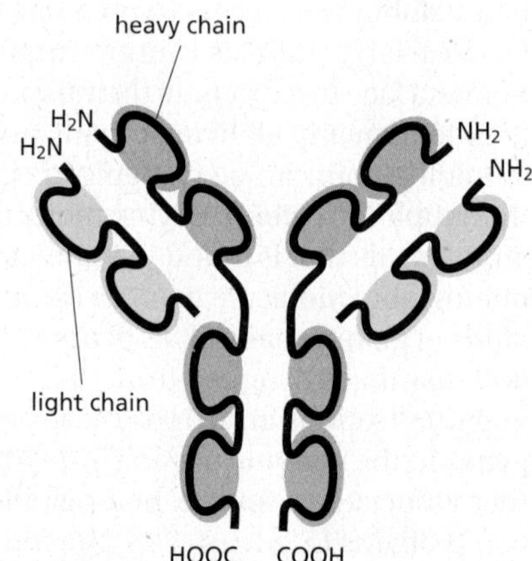

Figure 4–76 Schematic view of an antibody (immunoglobulin) molecule. This molecule is a complex of two identical heavy chains *(orange)* and two identical light chains *(blue)*. Each heavy chain contains four similar, covalently linked domains, while each light chain contains two such domains. Each of these domains is encoded by a separate exon, and all of the exons are thought to have evolved by the serial duplication of a single ancestral exon.

proteins are formed as a patchwork of domains from different origins, reflecting their complex evolutionary history (see Figure 3–15).

Neutral Mutations Often Spread to Become Fixed in a Population, with a Probability That Depends on Population Size

In comparisons between two species that have diverged from one another by millions of years, it makes little difference which individuals from each species are compared. For example, typical human and chimpanzee DNA sequences differ from one another by about 1%. In contrast, when the same region of the genome is sampled from two randomly chosen humans, the differences are typically about 0.1%. For more distantly related organisms, the interspecies differences outshine intraspecies variation even more dramatically. However, each "fixed difference" between the human and the chimpanzee (in other words, each difference that is now characteristic of all or nearly all individuals of each species) started out as a new mutation in a single individual. If the size of the interbreeding population in which the mutation occurred is N, the initial allele frequency for a new mutation would be $1/(2N)$ for a diploid organism. How does such a rare mutation become fixed in the population, and hence become a characteristic of the species rather than of a few scattered individuals?

The answer to this question depends on the functional consequences of the mutation. If the mutation has a significantly deleterious effect, it will simply be eliminated by purifying selection and will not become fixed. (In the most extreme case, the individual carrying the mutation will die without producing progeny.) Conversely, the rare mutations that confer a major reproductive advantage on individuals who inherit them can spread rapidly in the population. Because humans reproduce sexually and genetic recombination occurs each time a gamete is formed (discussed in Chapter 5), the genome of each individual who has inherited the mutation will be a unique recombinational mosaic of segments inherited from a large number of ancestors. The selected mutation along with a modest amount of neighboring sequence—ultimately inherited from the individual in which the mutation occurred—will simply be one piece of this huge mosaic.

The great majority of mutations that are not harmful are not beneficial either. These selectively *neutral* mutations can also spread and become fixed in a population. In fact, neutral mutations make a large contribution to evolutionary change in genomes. For example, as we saw earlier, they account for most of the DNA sequence differences between apes and humans. The spread of neutral mutations is not as rapid as the spread of the rare strongly advantageous mutations. It depends on a random variation in the number of mutation-bearing progeny produced by each mutation-bearing individual: through a sort of "random walk" process, the mutant allele may eventually become extinct or it may become commonplace. This can be modeled mathematically for an idealized interbreeding population, on the assumption of constant population size and random mating. While neither of these assumptions matches human population history, this idealized case reveals the general principles.

When a new neutral mutation occurs in a population of constant size N that is undergoing random mating, the probability that it will ultimately become fixed is approximately $1/(2N)$. This is because there are $2N$ copies of the gene in the diploid population, and each of them has an equal chance of becoming the predominant version in the long run. For those mutations that do become fixed, the mathematics shows that the average time to fixation is approximately $4N$ generations. Detailed analyses of data on human genetic variation have suggested an ancestral population size of approximately 10,000 at the time when the current pattern of genetic variation was largely established. With a population that has reached this size, the probability that a new, selectively neutral mutation would become fixed is small (1/20,000), while the average time to fixation would be on the order of 800,000 years (assuming a 20-year generation time). Thus, while we know that the human population has grown enormously since the development

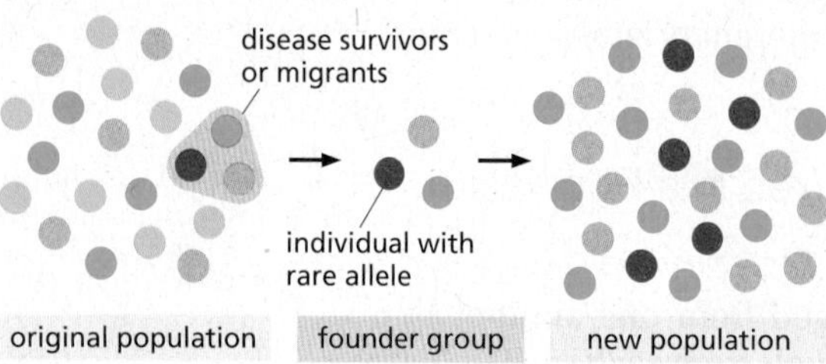

Figure 4–77 How founder effects can determine the set of genetic variants in a population of individuals belonging to the same species. In this diagram, each dot is used to represent an individual, with different colors used to denote the different variants of a particular gene. This example illustrates how, by a chance event that greatly reduces population size, an allele that is initially rare in a large population *(red)* can become established at high frequency in a new population, even though the mutation that produced it has no selective advantage—or is even mildly deleterious.

of agriculture approximately 15,000 years ago, most of the present-day set of common human genetic variants reflects a mixture of variants that was already present long before this time, when the human population was still small.

Similar arguments explain another phenomenon with important practical implications for genetic counseling. In an isolated community descended from a small group of founders, such as the people of Iceland or the Jews of Eastern Europe, a variant form of a gene that is rare in the human population as a whole can often be present at a high frequency, even if that variant is mildly deleterious (**Figure 4–77**).

We Can Trace Human History by Analyzing Genomes

The genomes of ancestral organisms can be inferred, but most can never be directly observed. DNA is very stable compared with most organic molecules, but it is not perfectly stable, and its progressive degradation, even under the best circumstances, means that it is extremely difficult to extract sequence information from fossils that are more than a million years old. Although a modern organism such as the horseshoe crab looks remarkably similar to fossil ancestors that lived 200 million years ago, there is every reason to believe that the horseshoe-crab genome has been changing during all that time in much the same way as in other evolutionary lineages, and at a similar rate. Selection must have maintained key functional properties of the horseshoe-crab genome to account for the morphological stability of the lineage. However, comparisons between different present-day organisms show that the fraction of the genome subject to purifying selection is small; hence, it is fair to assume that the sequence of the genome of the modern horseshoe crab, while preserving features critical for function, must differ greatly from that of its extinct ancestors, known to us only through the fossil record.

It is possible to get direct sequence information by examining DNA samples from ancient materials if these are not too old. In recent years, technical advances have allowed DNA sequencing from exceptionally well-preserved bone fragments that date from more than 100,000 years ago. Although any DNA this old will be imperfectly preserved, genome sequences can be reconstructed from many millions of short DNA sequences. In 2010, investigators completed their analysis of the first Neanderthal genome, obtained from DNA that was extracted from a fossilized bone fragment found in a cave in Croatia. The average difference in DNA sequence between humans and Neanderthals shows that our two lineages diverged somewhere between 270,000 and 440,000 years ago, well before the time that humans are believed to have migrated out of Africa. Neanderthals are one of our closest evolutionary relatives, and they lived side by side with the ancestors of modern humans in Europe and Western Asia. By comparing the Neanderthal genome sequence with those of people from different parts of the world, these studies revealed that many modern humans—particularly those from Europe and Asia—share about 2% of their genomes with Neanderthals. This genetic overlap indicates that our ancestors mated with Neanderthals—before outcompeting or actively exterminating them—on the way out of Africa. These and other ancient relationships still being discovered have thus left a permanent mark in the human genome (**Figure 4–78**).

But what about deciphering the genomes of much older ancestors, those for which no useful DNA samples can be isolated? For organisms that are as closely

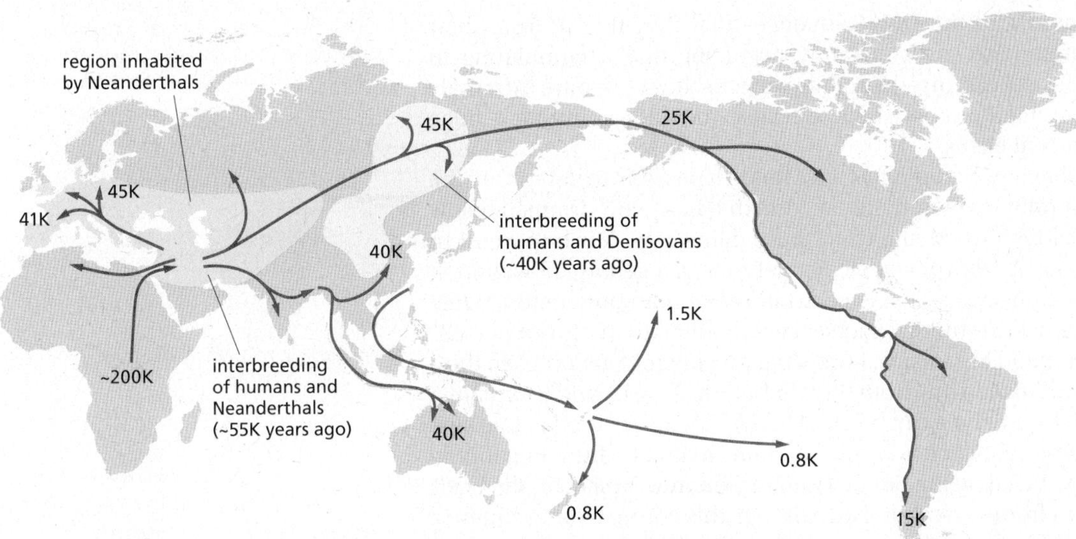

Figure 4–78 **Tracing the course of human history by analyses of genome sequences.** Modern humans arose in Africa approximately 200,000–300,000 years ago, where the earliest fossils of *Homo sapiens* are found. One small group—perhaps as few as 10,000 individuals—migrated northward, and their descendants spread across the globe. As these ancestral humans left Africa, around 60,000–80,000 years ago *(purple arrow)*, they encountered Neanderthals who inhabited the region indicated in *blue*. As a result of interbreeding, the humans who subsequently spread throughout Europe and Asia *(red arrows)* carried with them traces of Neanderthal DNA. Many of those moving eastward would later also interbreed with Denisovans *(yellow)*, a second type of ancient human. Later, ancestral humans continued their global spread to the New World, reaching North America approximately 25,000 years ago and the southern regions of South America 10,000 years later. This scenario is based on many types of data, including fossil records, anthropological studies, and the genome sequences of Neanderthals, Denisovans, and modern humans from around the world. (Adapted from M.A. Jobling et al., Human Evolutionary Genetics, 2nd ed. New York: Garland Science, 2014.)

related as human and chimpanzee, we saw that this may not be difficult: reference to the gorilla sequence can be used to sort out which of the few sequence differences between human and chimpanzee are inherited from our common ancestor some 6 million years ago (see Figure 4–65). And for an ancestor that has generated a large number of different species alive today, the DNA sequences of the existing species can be compared simultaneously to unscramble much of the ancestral sequence, allowing scientists to derive DNA sequences much farther back in time. For example, from the genome sequences obtained for dozens of modern placental mammals, it should be possible to infer much of the genome sequence of their 100-million-year-old common ancestor—the precursor of species as diverse as dog, mouse, rabbit, armadillo, and human (see Figure 4–67).

Once the amino acid sequence of an ancestral protein has been inferred, that protein can be readily produced in pure form using the recombinant DNA methods described in Chapter 8. By thus "resurrecting" the extinct protein, its biochemical properties can be compared to those of its modern counterparts. These procedures can be repeated at several branch points in the protein's evolutionary history, allowing that protein's stepwise evolution over many millions of years to be directly measured. This approach provides scientists with a powerful way to visualize evolution in action. It can also help us to understand why modern proteins look and behave the way they do.

The Sequencing of Hundreds of Thousands of Human Genomes Reveals Much Variation

Even though the common variant alleles among modern humans originate from the variants present in a comparatively tiny group of ancestors, the total number of variants that exist among modern humans is huge. This is because unavoidable errors occur in the replication and maintenance of our DNA sequences, as will be explained in Chapter 5. As a result, every person is born with 50–100 new changes, termed *de novo* mutations, in the DNA sequences that they inherit from their parents. Thus, new mutations are constantly occurring and accumulating in human populations. But the vast majority of the variants they create are extremely rare in the human population as a whole, because they arose among a huge number of relatively recent births.

Modern DNA sequencing technologies have made it possible to determine the genome sequences for millions of humans, at costs that have now dropped below $1000 per genome. Before introducing what these data show, it is important to define a few basic terms. A *variant* is typically defined as a genomic location at which two or more versions of a sequence exist in the human population, while an *allele* refers to the specific form that each such variant takes in a given person. Because humans are diploid organisms, harboring two haploid genomes in their cells (one from their mother and one from their father), if alleles A and B exist for a variant X, an individual's genotype can be either "A/A," "A/B," or "B/B" at that site.

Each human genome sequence is compared to a "reference human genome assembly" that was produced by the public Human Genome Project on the basis of the initial sequences from several individuals. For this comparison, sequence "reads" for millions or billions of DNA fragments obtained from an individual are aligned by a computer to the reference assembly (which is a haploid representation and includes only one allele at each location). The locations at which the sequenced individual displays "alternative" alleles relative to the reference are then identified.

When this is done for many individuals, single-nucleotide variants (SNVs) are found to compose the vast majority of differences among humans. *SNVs* are points in the genome sequence where some haploid human genomes have one nucleotide, while some have another. They have been extensively characterized because of their abundance, their relative ease of detection, and their value for genetic studies. Historically, the term single-nucleotide polymorphism, or SNP, was reserved for SNVs present at a frequency above 1% in the human population. Today, however, no precise threshold is used to differentiate "rare" and "common" variations, which span frequencies ranging from a singleton "private" variation (present in only one haploid genome within one person) to as high as 50% (a variant with two alleles that exist at equal frequency among human genomes).

Importantly, SNVs are only one type of variation that exists among humans. *Structural variants*, such as deletions, duplications, inversions, or rearrangements of a variety of lengths, are also prevalent features in our genomes. Additional types of variation—including mobile-element insertions and low-complexity repeat expansions or contractions—are also important components of human genetic diversity.

Two haploid human genomes sampled from the modern world population will differ from one another at roughly 1 per 1,000 nucleotides. This number, however, varies considerably between human populations, and it is affected by historical factors such as population bottlenecks, breeding patterns, and ancestral population sizes. And the more closely related two individuals are to one another, the more similar their genomes will be. The above number does not include the considerable structural diversity present among human genomes. Indeed, because they often affect tens or even hundreds of thousands of nucleotides, structural variants are a large contributor to the total number of nucleotide-level differences between any two human genomes (**Table 4–3**).

Most of the Variants Observed in the Human Population Are Common Alleles, with at Most a Weak Effect on Phenotype

A comparison of any one human genome with the reference human genome will typically find 3–5 million locations at which the sequenced individual harbors at least one allele different from that seen in the reference assembly. Some of these variants are heterozygous within that person, which is to say the individual harbors two distinct alleles, typically one that matches the reference and one "alternative" allele. More than 95% of the variant alleles observed within

TABLE 4–3 Typical Differences Between Any One Human Being's Genome Sequence and the Reference Human Genome

Type of difference	Size in nucleotide pairs	Differences per genome
Single-nucleotide variation (SNV)	1	3–4 million
Small deletion or insertion (indel)	1–49	0.4–0.5 million
Low-complexity simple sequence repeats (microsatellite and satellite DNA repeats)	1–200	100,000
Mobile-element insertion (SINE, LINE)	300–7000	2000
Structural variation (deletions, duplications, and inversions)	50 to >1,000,000	Tens of thousands; length is inversely correlated with frequency
Karyotypically visible abnormalities (e.g., aneuploidies)	Chromosome scale	Very rare; most are lethal

Courtesy of Greg Cooper and Rick Myers, HudsonAlpha Institute for Biotechnology, Huntsville, AL; based on H.J. Abel et al., *Nature* 583:83–88, 2020; gnomAD (https://www.nature.com/immersive/d42859-020-00002-x/index.html; and https://www.internationalgenome.org).

any given person are relatively common, reflecting their ancient origin. A small fraction of variant alleles will be rare; some of these, as described earlier, represent *de novo* mutations and are extremely rare, potentially existing within only that individual.

As will be described in Chapter 8, those SNVs that are frequent in the human genome have been extremely useful for genetic mapping analyses, known as genome-wide association studies (GWAS). Here, one attempts to associate specific human traits (phenotypes) with a large set of alleles that are relatively common in the human population (see p. 526). Most of these frequent SNVs have little or no effect on human fitness. This is as expected, as the deleterious SNVs will have been selected against during human evolution and should therefore be rare. Importantly, this property is generally true of all variation, including structural variation. Indeed, common structural variants also tend to not be deleterious because, if they were, they would not have arisen to appreciable frequencies. Across all forms of variation then, allele frequency is of crucial importance: common variant alleles tend to have at most weak effects on human phenotypes, and large-effect variants tend to be rare. On the other hand, it is important to note that just because a variant is rare does not mean it influences a phenotype; most rare variants, in fact, have little to no effect.

In recent years, GWAS research has identified many associations between common traits, including common diseases, and common alleles. This is a result of the fact that while common alleles tend to have at most weak effects, the combined effects of many common alleles can ultimately produce a large impact on phenotype.

Forensic Analyses Exploit Special DNA Sequences with Unusually High Mutation Rates

While mutation rates in humans are generally quite low, typically estimated to be on the order of 10^{-8} (that is, roughly one mutation every 100 million base pairs), certain sequences with exceptionally high mutation rates stand out. A dramatic example is provided by CACA repeats, which are abundant in the human genome and in the genomes of other eukaryotes. Segments of the genome with the motif $(CA)_n$ are replicated with relatively low fidelity because a slippage occurs between the template and the newly synthesized strands during DNA replication; hence, the precise value of n can vary over a considerable range from one genome to the next. These repeats make ideal DNA-based genetic markers because most humans are heterozygous at each of these loci, having inherited one repeat length (n) from their mother and a different repeat length from their father. While the value of n changes sufficiently rarely that most parent–child transmissions propagate CACA repeats faithfully, the changes are sufficiently frequent to maintain high levels of heterozygosity in the human population. Because of this, these and other simple repeats that display exceptionally high variability provide a *DNA fingerprint* that is widely used for identifying individuals by DNA analysis in crime investigations, paternity suits, and other forensic applications (see Figure 8–37).

An Understanding of Human Variation Is Critical for Improving Medicine

While most variations in the human genome are thought to have weak effects on phenotype, a subset of our genome sequence variations must be responsible for the heritable aspects of human individuality. Because it is now possible to sequence individual genomes cheaply and rapidly, we can link even rare alleles with specific phenotypes. We know that even a single nucleotide change that alters one amino acid in a protein can cause a serious disease, as for example in sickle-cell disease, which, although not rare, is caused by such a mutation in hemoglobin (**Movie 4.3**). We also know that gene dosage—a doubling or halving of the copy number of some genes—can have a profound effect on development by altering the level of gene product, as can a change in one of the many regulatory DNA sequences that are dispersed throughout the vast expanse of noncoding DNA in the human genome (see Chapter 7).

Some of the many differences between the genomes of any two human beings will have substantial effects on human health, physiology, behavior, and physique. A major challenge in modern human genetics is to discriminate those differences from the majority that are of little consequence. Significant progress is being made; for instance, we now know the nucleotide changes that give rise to thousands of rare inherited traits and diseases. These and many other results are greatly expanding our understanding of human biology. This understanding is critical for the new field of precision medicine, in which both disease prevention and disease treatments will be tailored to take account of individual genetic differences.

Summary

Comparisons of the nucleotide sequences of present-day genomes have revolutionized our understanding of gene and genome evolution. Because of the extremely high fidelity of DNA replication and DNA repair processes, random errors in maintaining the nucleotide sequences in genomes occur rarely (for example, only 50–100 new mutations for every person born). Not surprisingly, therefore, a comparison of human and chimpanzee chromosomes—which are separated by about 12 million years of evolution (double the ~6 million years since humans and chimps shared a common ancestor)—reveals fixed changes at only ~1% of base pairs. Not only are our protein-coding sequences highly similar, but also their order on each chromosome is almost identical. Although a substantial number of segmental duplications and segmental deletions have occurred, even the positions of the transposable elements that make up a major portion of our noncoding DNA are mostly unchanged.

When one compares the genomes of two more distantly related organisms—such as a human and a mouse, separated by about 90 million years—one finds many more changes. While selection leaves detectable signatures at all time scales, including for example through comparisons of human genomes to one another or through comparisons of the closely related human and chimp genomes, at greater distances the actions of selection are more pronounced. Comparison of human and mouse genomes, for example, reveal that most coding sequences (exons) and many regulatory regions have been highly conserved by the actions of selection, while huge fractions of our respective genomes have been altered to such an extent that one can detect no similarity at all among them.

Because of purifying selection, the comparison of the genome sequences of multiple related species is an especially powerful way to find DNA sequences with important functions. While it has been estimated that between 8 and 10% of the human genome has been conserved as a result of purifying selection, the function of the majority of this DNA (tens of thousands of sequence elements conserved across mammalian evolution) remains mysterious. Experiments that characterize their functions continue to teach us many new lessons about vertebrate biology.

Other sequence comparisons show that a great deal of the genetic complexity of present-day organisms is due to the expansion of ancient gene families. DNA duplication followed by sequence divergence has clearly been a major source of genetic novelty during evolution. On a more recent time scale, the diploid genomes of any two humans will differ from each other both because of nucleotide substitutions (single-nucleotide variations) and because of inherited DNA gains and DNA losses (structural variations). Deciphering the effects of these differences will improve both medicine and our understanding of human biology.

PROBLEMS

Which statements are true? Explain why or why not.

4–1 Human females have 23 different chromosomes, whereas human males have 24.

4–2 The four core histones are relatively small proteins with a very high proportion of positively charged amino acids; the positive charge helps the histones bind tightly to DNA, regardless of its nucleotide sequence.

4–3 Nucleosomes bind DNA so tightly that they cannot move from the positions where they are first assembled.

4–4 The long linear DNA molecule in an interphase chromosome is organized into loops of chromatin that appear to emanate from a central axis.

4–5 In a comparison between the DNAs of related organisms such as humans and mice, identifying the conserved DNA sequences facilitates the search for functionally important regions.

4–6 Gene duplication and divergence are thought to have played a critical role in the evolution of increased biological complexity.

Discuss the following problems.

4–7 DNA isolated from the bacterial virus M13 contains 25% A, 33% T, 22% C, and 20% G. Do these results strike you as peculiar? Why or why not? How might you explain these values?

4–8 A segment of DNA from the interior of a single strand is shown in Figure Q4–1. Should this sequence be written as ACT or TCA? Why?

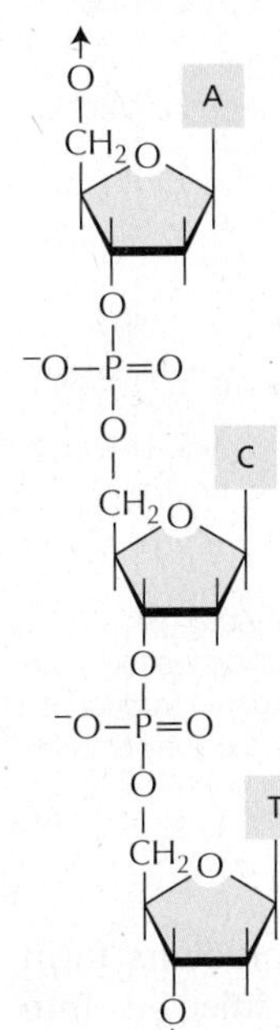

Figure Q4–1 Three nucleotides from the interior of a single strand of DNA (Problem 4–8). *Arrows* at the ends of the DNA strand indicate that the structure continues in both directions.

4–9 Human DNA contains 20% C on a molar basis. What are the mole percents of A, G, and T?

4–10 In contrast to histone acetylation, which always correlates with gene activation, histone methylation can lead to either transcriptional activation or repression. How do you suppose that the same modification—methylation—can mediate different biological outcomes?

4–11 Why is a chromosome with two centromeres (a dicentric chromosome) unstable? Would a backup centromere not be a good thing for a chromosome, giving it two chances to form a kinetochore and attach to microtubules during mitosis? Surely, a backup centromere would ensure that the chromosome did not get left behind at mitosis.

4–12 Look at the two yeast colonies in Figure Q4–2. Each of these colonies contains about 100,000 cells descended from a single yeast cell, originally somewhere in the middle of the clump. A white colony arises when the *Ade2* gene is expressed from its normal chromosomal location. When the *Ade2* gene is moved near a telomere, however, it is packed into heterochromatin and inactivated in most cells, giving rise to colonies that are mostly red. In these largely red colonies, white sectors fan out from the middle of the colony. In both the red portions and the white sectors, the *Ade2* gene is still located near telomeres. Explain why white sectors have formed near the rim of the red colony. On the basis of the patterns observed, what can you conclude about the propagation of the transcriptional state of the *Ade2* gene from mother to daughter cells in this experiment?

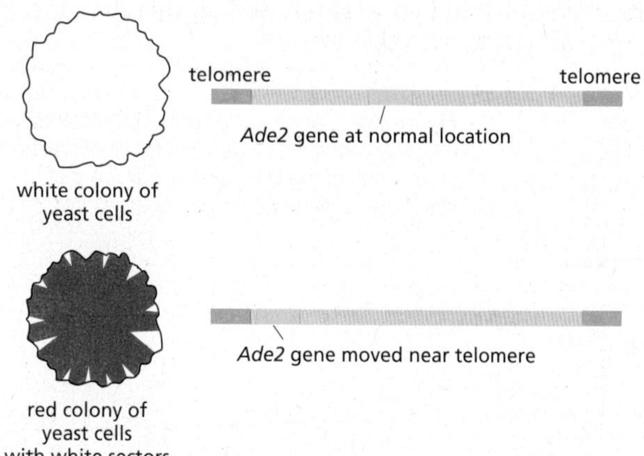

Figure Q4–2 Position effect on expression of the yeast *Ade2* gene (Problem 4–12). The *Ade2* gene codes for one of the enzymes for adenosine biosynthesis, and the absence of the *Ade2* gene product leads to the accumulation of a red pigment. Therefore, a colony of cells that express *Ade2* is *white*, and one composed of cells in which the *Ade2* gene is not expressed is *red*.

4–13 Chromosomes from different amphibians form typical lampbrush chromosomes when injected into oocytes as demembranated sperm heads. When the sperm heads from *Rana pipiens* (northern leopard frog), which forms large loops in its own oocyte chromosomes, were injected into *Xenopus laevis* oocytes, the resulting lampbrush chromosomes had the small loops typical of those found in *X. laevis* oocytes. Similarly, when the sperm heads from *X. laevis* were injected into *Notophthalmus viridescens* (red-spotted newt) oocytes, the resulting lampbrush chromosomes had the very large loop structure typical of *N. viridescens*.

Do these heterologous injection experiments support the idea that loop structure is a fixed property of a chromosome? Why or why not?

4–14 Mobile pieces of DNA—transposable elements—that insert themselves into chromosomes and accumulate during evolution make up more than 40% of the human genome. Transposable elements of four types—long interspersed nuclear elements (LINEs), short interspersed nuclear elements (SINEs), long terminal repeat (LTR) retrotransposons, and DNA transposons—are inserted more or less randomly throughout the human genome. These elements are conspicuously rare at the four homeobox gene clusters, *HoxA*, *HoxB*, *HoxC*, and *HoxD*, as illustrated for *HoxD* in Figure Q4–3, along with an equivalent region of chromosome 22, which lacks a *Hox* cluster. Each *Hox* cluster is about 100 kilobases (kb) in length and contains 9–11 genes, whose differential expression along the anteroposterior axis of the developing embryo establishes the basic body plan for humans (and for other animals). Why do you suppose that transposable elements are so rare in the *Hox* clusters?

Figure Q4–3 Transposable elements and genes in 1-megabase regions of chromosomes 2 and 22 (Problem 4–14). *Blue lines* that project *upward* indicate exons of known genes. *Red lines* that project *downward* indicate transposable elements; they are so numerous (constituting more than 40% of the human genome) that they merge into nearly a solid block outside the *Hox* clusters. (Adapted from E. Lander et al., *Nature* 409:860–921, 2001. With permission from Springer Nature.)

4–15 Chromosome 3 in orangutans differs from chromosome 3 in humans by two inversion events that occurred in the human lineage (Figure Q4–4). Draw the

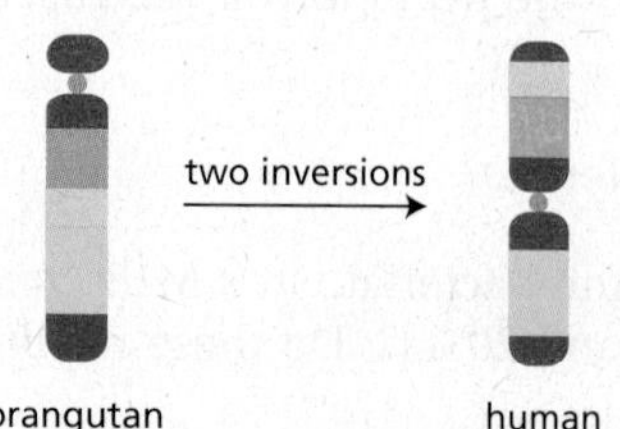

Figure Q4–4 Chromosome 3 in orangutans and humans (Problem 4–15). Differently colored blocks indicate segments of the chromosomes that are homologous in DNA sequence.

intermediate chromosome that resulted from the first inversion and explicitly indicate the segments included in each inversion.

4–16 There has been a colossal snafu in the maternity ward at your local hospital. Four sets of male twins, born within an hour of each other, were inadvertently shuffled in the excitement occasioned by that unlikely event. You have been called in to set things right. As a first step, you want to get the twins matched up. To that end, you analyze a small blood sample from each infant using a hybridization probe that detects differences in the numbers of simple sequence repeats such as $(CA)_n$ located in widely scattered regions of the genome. The results are shown in Figure Q4–5.

A. Which infants are brothers? Are they all identical twins?

B. How will you match brothers to the correct parents?

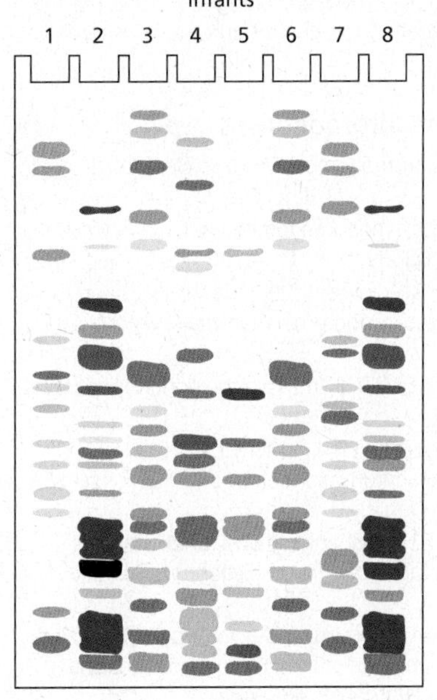

Figure Q4–5 DNA fingerprint analysis of shuffled twins (Problem 4–16).

REFERENCES

General

Hartwell L, Goldberg ML, Fischer JA & Hood L (2017) Genetics: From Genes to Genomes, 6th ed. Boston: McGraw-Hill.

Jobling M, Hollox E, Hurles M, et al. (2014) Human Evolutionary Genetics, 2nd ed. New York: Garland Science.

Strachan T & Read AP (2019) Human Molecular Genetics, 5th ed. New York: Garland Science.

The Structure and Function of DNA

Avery OT, MacLeod CM & McCarty M (1944) Studies on the chemical nature of the substance inducing transformation of pneumococcal types. *J. Exp. Med.* 79, 137–158.

Meselson M & Stahl FW (1958) The replication of DNA in *E. coli*. *Proc. Natl. Acad. Sci. USA* 44, 671–682.

Perutz MF (1993) Co-chairman's remarks: Before the double helix. *Gene* 135, 9–13.

Watson JD & Crick FHC (1953) Molecular structure of nucleic acids; a structure for deoxyribose nucleic acid. *Nature* 171, 737–738.

Chromosomal DNA and Its Packaging in the Chromatin Fiber

Bowman GD & McKnight JN (2017) Sequence-specific targeting of chromatin remodelers organizes precisely positioned nucleosomes throughout the genome. *Bioessays* 39, 1–8.

Clapier CR, Iwasa J, Cairns BR & Peterson CL (2017) Mechanisms of action and regulation of ATP-dependent chromatin-remodelling complexes. *Nat. Rev. Mol. Cell Biol.* 18, 407–422.

Luger K, Mäder AW, Richmond RK . . . Richmond TJ (1997) Crystal structure of the nucleosome core particle at 2.8 Å resolution. *Nature* 389, 251–260.

Maeshima K, Ide S & Babokhov M (2019) Dynamic chromatin organization without the 30-nm fiber. *Curr. Opin. Cell Biol.* 58, 95–104.

Miga KH, Koren S, Rhie A . . . Eichler EE & Phillippy AM (2020) Telomere-to-telomere assembly of a complete human X chromosome. *Nature* 585, 79–84.

Talbert PB & Henikoff S (2021) Histone variants at a glance. *J. Cell Sci.* 134, jcs244749.

Zhao Y & Garcia BA (2015) Comprehensive catalog of currently documented histone modifications. *Cold Spring Harb. Perspect. Biol.* 7, a025064.

Zhou CY, Johnson SL, Gamarra NI & Narlikar GJ (2016) Mechanisms of ATP-dependent chromatin remodeling motors. *Annu. Rev. Biophys.* 45, 153–181.

Zhou K, Gaullier G & Luger K (2019) Nucleosome structure and dynamics are coming of age. *Nat. Struct. Mol. Biol.* 26, 3–13.

The Effect of Chromatin Structure on DNA Function

Allis CD & Jenuwein T (2016) The molecular hallmarks of epigenetic control. *Nat. Rev. Genet.* 17, 487–500.

Black BE, Jansen LET, Foltz DR & Cleveland DW (2011) Centromere identity, function, and epigenetic propagation across cell divisions. *Cold Spring Harb. Symp. Quant. Biol.* 75, 403–418.

Elgin SCR & Reuter G (2013) Position-effect variegation, heterochromatin formation, and gene silencing in *Drosophila*. *Cold Spring Harb. Perspect. Biol.* 5, a017780.

Filion GJ, van Bemmel JG, Braunschweig U . . . van Steensel B (2010) Systematic protein location mapping reveals five principal chromatin types in *Drosophila* cells. *Cell* 143, 212–224.

Fodor BD, Shukeir N, Reuter G & Jenuwein T (2010) Mammalian *su(var)* genes in chromatin control. *Annu. Rev. Cell Dev. Biol.* 26, 471–501.

Giles KE, Gowher H, Ghirlando R . . . Felsenfeld G (2011) Chromatin boundaries, insulators, and long-range interactions in the nucleus. *Cold Spring Harb. Symp. Quant. Biol.* 75, 79–85.

Kasinath V, Poepsel S & Nogales E (2019) Recent structural insights into polycomb repressive complex 2 regulation and substrate binding. *Biochemistry* 58, 346–354.

Nicetto D & Zaret KS (2019) Role of H3K9me3 heterochromatin in cell identity establishment and maintenance. *Curr. Opin. Genet. Dev.* 55, 1–10.

Sanulli S, Gross JD & Narlikar GJ (2019) Biophysical properties of HP1-mediated heterochromatin. *Cold Spring Harb. Symp. Quant. Biol.* 84, 217–225.

Sullivan LL & Sullivan BA (2020) Genomic and functional variation of human centromeres. *Exp. Cell Res.* 389, 111896.

Xu M, Long C, Chen X . . . Zhu B (2010) Partitioning of histone H3-H4 tetramers during DNA replication-dependent chromatin assembly. *Science* 328, 94–98.

The Global Structure of Chromosomes

Callan HG (1982) Lampbrush chromosomes. *Proc. R. Soc. Lond. Biol. Sci.* 214, 417–448.

Cremer T & Cremer M (2010) Chromosome territories. *Cold Spring Harb. Perspect. Biol.* 2, a003889.

Feodorova Y, Falk M, Mirny LA & Solovei I (2020) Viewing nuclear architecture through the eyes of nocturnal mammals. *Trends Cell Biol.* 30, 276–289.

Ghirlando R & Felsenfeld G (2016) CTCF: making the right connections. *Genes Dev.* 30, 881–891.

Gibcus JH, Samejima K, Goloborodko A . . . Mirny LA & Dekker J (2018) A pathway for mitotic chromosome formation. *Science* 359, eaao6135.

Hildebrand EM & Dekker J (2020) Mechanisms and functions of chromosome compartmentalization. *Trends Biochem. Sci.* 45, 385–396.

Kharchenko PV, Alekseyenko AA, Schwartz YB . . . Karpen GH & Park PJ (2011) Comprehensive analysis of the chromatin landscape in *Drosophila melanogaster*. *Nature* 471, 480–485.

Kinoshita K & Hirano T (2017) Dynamic organization of mitotic chromosomes. *Curr. Opin. Cell Biol.* 46, 46–53.

Krietenstein N, Abraham S, Venev SV . . . Dekker J & Rando OJ (2020) Ultrastructural details of mammalian chromosome architecture. *Mol. Cell.* 78, 554–565.e7.

Kumar Y, Sengupta D & Bickmore W (2020) Recent advances in the spatial organization of the mammalian genome. *J. Biosci.* 45, 18.

Maeshima K & Laemmli UK (2003) A two-step scaffolding model for mitotic chromosome assembly. *Dev. Cell* 4, 467–480.

Terakawa T, Bisht S, Eeftens JM . . . Greene EC (2017) The condensin complex is a mechanochemical motor that translocates along DNA. *Science* 358, 672–676.

How Genomes Evolve

Doolittle WF & Brunet TDP (2017) On causal roles and selected effects: our genome is mostly junk. *BMC Biol.* 15, 116.

Friedli M & Trono D (2015) The developmental control of transposable elements and the evolution of higher species. *Annu. Rev. Cell Dev. Biol.* 31, 429–451.

Green RE, Krause J, Briggs A . . . Reich D & Pääbo S (2010) A draft sequence of the Neanderthal genome. *Science* 328, 710–722.

Harmston N, Baresic A & Lenhard B (2013) The mystery of extreme non-coding conservation. *Philos. Trans. R. Soc. Lond. B Biol. Sci.* 368, 20130021.

International Human Genome Sequencing Consortium (2001) Initial sequencing and analysis of the human genome. *Nature* 409, 860–921.

International Human Genome Sequencing Consortium (2004) Finishing the euchromatic sequence of the human genome. *Nature* 431, 931–945.

Kellis M, Wold B, Snyder MB . . . Hardison RC (2014) Defining functional DNA elements in the human genome. *Proc. Natl. Acad. Sci. USA* 111, 6131–6138.

Kostka D, Holloway AK & Pollard KS (2018) Developmental loci harbor clusters of accelerated regions that evolved independently in ape lineages. *Mol. Biol. Evol.* 35, 2034–2045.

Lander ES (2011) Initial impact of the sequencing of the human genome. *Nature* 470, 187–197.

Lappalainen T, Scott AJ, Brandt M & Hall IM (2019) Genomic analysis in the age of human genome sequencing. *Cell* 177, 70–84.

Mouse Genome Sequencing Consortium (2002) Initial sequencing and comparative analysis of the mouse genome. *Nature* 420, 520–562.

Ponting CP (2017) Biological function in the twilight zone of sequence conservation. *BMC Biol.* 15, 71.

DNA Replication, Repair, and Recombination

CHAPTER 5

The ability of cells to maintain a high degree of order in a chaotic universe depends on the accurate duplication of vast quantities of genetic information carried in chemical form as DNA. This process, called *DNA replication*, must occur before a cell can produce two genetically identical daughter cells. Maintaining order also requires the continued surveillance and repair of this genetic information, because DNA inside cells is repeatedly damaged by chemicals and radiation from the environment, as well as by thermal accidents and reactive molecules generated inside the cell. In this chapter, we describe the protein machines that replicate and repair the cell's DNA. These machines catalyze some of the most rapid and accurate processes that take place within cells, and their mechanisms provide clear illustrations of the elegance and efficiency of cell chemistry.

The short-term survival of a cell depends on preventing harmful changes in its DNA. But the long-term survival of a species requires that these same DNA sequences be changeable over many generations to permit evolutionary adaptation to changing circumstances. We shall see that, despite the great efforts that cells make to protect their DNA, occasional changes in DNA sequences are unavoidable. These changes produce the genetic variation that is required for natural selection to drive the evolution of organisms.

We begin this chapter with a brief discussion of the changes that occur in DNA as it is passed down from generation to generation. Next, we discuss the mechanisms—DNA replication and DNA repair—that are responsible for minimizing these changes. Finally, we consider some of the most intriguing pathways that alter DNA sequences—those of *DNA recombination.* These pathways include the movement within chromosomes of special DNA sequences called transposable elements.

IN THIS CHAPTER

THE MAINTENANCE OF DNA SEQUENCES

The survival of an individual organism demands a high degree of genetic stability. Only rarely do the cell's DNA-maintenance processes fail, resulting in permanent change in the DNA. Such a change is called a *mutation*, and it can destroy an organism if it occurs in a vital position in the DNA sequence.

Mutation Rates Are Extremely Low

The **mutation rate**, the rate at which changes occur in DNA sequences, can be determined directly from experiments carried out with a bacterium such as *Escherichia coli*—a resident of our intestinal tract and a commonly used laboratory organism (see Figure 1–38). Under laboratory conditions, an *E. coli* cell divides about once every 30 minutes; as a result, a single cell can generate a very large population—several billion—in less than a day. In such a population, it is possible to detect the small fraction of bacteria that have suffered a damaging mutation in a particular gene. For example, the mutation rate of a gene specifically required for cells to use the sugar lactose as an energy source can be determined by growing the cells in the presence of a different sugar, such as glucose, and testing them subsequently to see how many have lost the ability to survive on a lactose diet. The fraction of damaged genes will underestimate the actual mutation rate because many mutations are *silent* (for example, those that change a codon but not the amino acid it specifies or those that change an amino acid without affecting the activity of the protein coded for by the gene). After correcting for these silent mutations, one finds that bacteria display a mutation rate of about three nucleotide changes per 10^{10} nucleotides copied.

It is also possible to measure the germ-line mutation rate in more complex, sexually reproducing organisms such as humans. In this case, the complete genomes from a family—parents and offspring—are directly sequenced, and a careful comparison reveals that approximately 70 new single-nucleotide mutations typically arise in the germ lines of each offspring. Normalized to the size of the human genome, the mutation rate is one nucleotide change per 10^8 nucleotides per human generation. (This is a slight underestimate because some germ-line mutations will be lethal and will therefore be absent from progeny; however, because relatively little of the human genome carries critical information, this consideration has a negligible effect on the true mutation rate.) It is estimated that approximately 100 cell divisions occur in the germ line from the time of conception to the time of production of the eggs and sperm that go on to make the next generation. Thus, the human mutation rate, expressed in terms of cell divisions (instead of human generations), is approximately one nucleotide change per 10^{10} nucleotides copied.

Although *E. coli* and humans differ greatly in their modes of reproduction and in their generation times, when the mutation rates of each are normalized to a single round of DNA replication, they are both extremely low and within a factor of 3 of each other. We shall see later in the chapter that the basic mechanisms that ensure these low rates of mutation have been conserved since the very early history of cells on Earth.

Low Mutation Rates Are Necessary for Life as We Know It

Because many mutations are deleterious, no species can afford to allow them to accumulate at a high rate in its germ cells. Even though the observed mutation frequency is very low, it is thought to limit the number of essential genes that any organism can rely on to perhaps 30,000. More essential genes than this, and the probability that at least one critical component will suffer a damaging mutation becomes catastrophically high. By an extension of the same argument, a mutation frequency tenfold higher would limit an organism to about 3000 essential genes. In this case, evolution would have been limited to organisms considerably less complex than a fruit fly.

The cells of a sexually reproducing animal or plant are of two types: **germ cells** and **somatic cells**. The germ cells transmit genetic information from parent to offspring; the somatic cells form the body of the organism (**Figure 5–1**). We have seen that germ cells must be protected against high rates of mutation to maintain the species. However, the somatic cells of multicellular organisms must also be protected from genetic change to properly maintain the organized structure of the body. Nucleotide changes in somatic cells can give rise to variant cells, some of which, through "local" natural selection, proliferate rapidly at the expense of the rest of the organism. In an extreme case, the result is the uncontrolled cell proliferation that we know as cancer. This condition is due largely to an accumulation of changes in the DNA sequences of somatic cells, as discussed in Chapter 20. Any significant increase in the mutation frequency would presumably cause a disastrous increase in the incidence of cancer by accelerating the rate at which dangerous somatic-cell variants arise. Thus, both for the perpetuation of a species with a large number of genes (germ-cell stability) and for the prevention of cancer resulting from mutations in somatic cells (somatic-cell stability), multicellular organisms like ourselves absolutely depend on the remarkably high fidelity with which their DNA sequences are replicated and maintained.

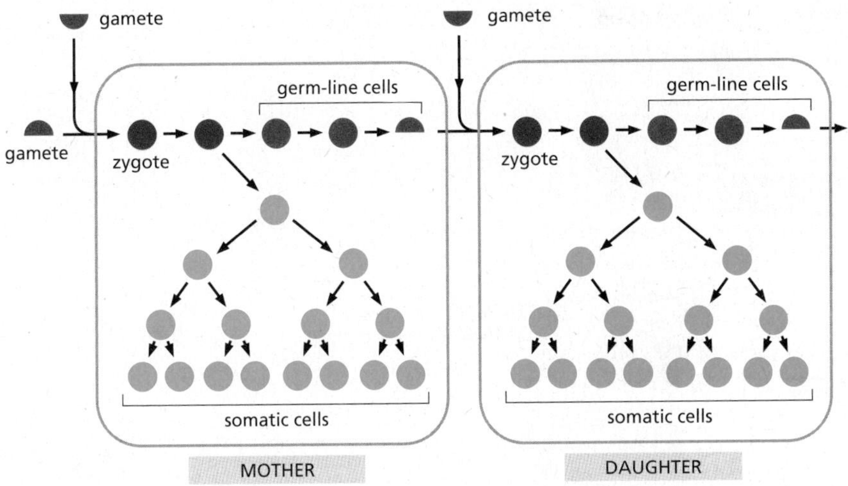

Figure 5–1 **Germ-line cells and somatic cells carry out fundamentally different functions.** In sexually reproducing organisms, genetic information is propagated into the next generation exclusively by germ-line cells *(red)*. This cell lineage includes the specialized reproductive cells—the gametes (eggs and sperm, *half circles*)—which contain only half the number of chromosomes as that contained in the other cells in the body *(full circles)*. When two gametes come together during fertilization, they form a fertilized egg, or zygote *(purple)*, which once again contains a full set of chromosomes. The zygote gives rise to both germ-line cells and somatic cells *(blue)*. Somatic cells form the body of the organism but do not contribute their DNA to the next generation.

Summary

In all cells, DNA sequences are maintained and replicated with extremely high fidelity. The mutation rate, approximately one nucleotide change per 10^{10} nucleotides each time the DNA is replicated, is very similar for organisms as different as bacteria and humans. Because of this remarkable accuracy, the sequence of the human genome (approximately 3.1×10^9 nucleotide pairs) is unchanged or changed by only a few nucleotides each time a typical human cell divides. This allows humans to pass accurate genetic instructions from one generation to the next and also—for most of us—to avoid the changes in somatic cells that lead to cancer.

DNA REPLICATION MECHANISMS

All organisms duplicate their DNA with extraordinary accuracy before each cell division. In this part of the chapter, we explore how an elaborate "replication machine" achieves this accuracy, while duplicating DNA at rates as high as 1000 nucleotides per second.

Base-pairing Underlies DNA Replication and DNA Repair

As introduced in Chapter 1, *DNA templating* is the mechanism the cell uses to copy the nucleotide sequence of one DNA strand into a complementary DNA sequence (**Figure 5–2**). This process requires the separation of the DNA helix into two template strands, and it entails the recognition of each nucleotide in the DNA *template strands* by a free (unpolymerized) complementary nucleotide. The separation of the DNA helix exposes the hydrogen-bond donor and acceptor groups on each DNA base to allow its base-pairing with the appropriate incoming free nucleotide, aligning it for its enzyme-catalyzed polymerization into a new DNA chain.

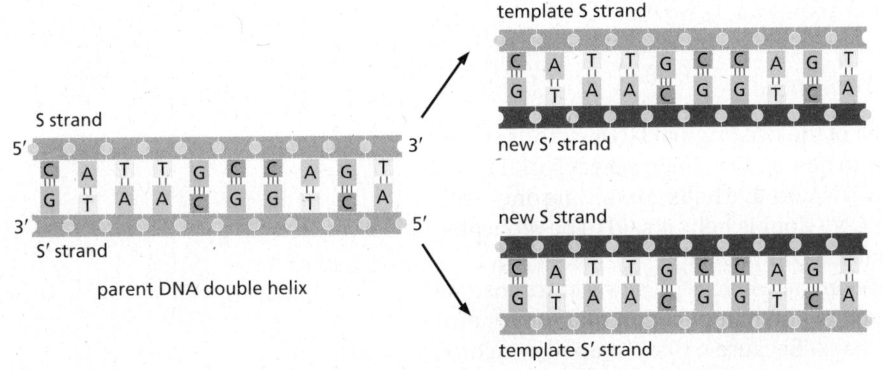

Figure 5–2 **DNA acts as a template for its own replication.** Because the nucleotide A will successfully pair only with T, and G with C, each strand of a DNA double helix—labeled here as the S strand and its complement, the S′ strand—can serve as a template to specify the sequence of nucleotides in a complementary strand. In this way, both strands of a DNA double helix can be copied with precision, producing two exact copies of the original double helix. How complementary nucleotides base-pair is shown in Figure 4–5.

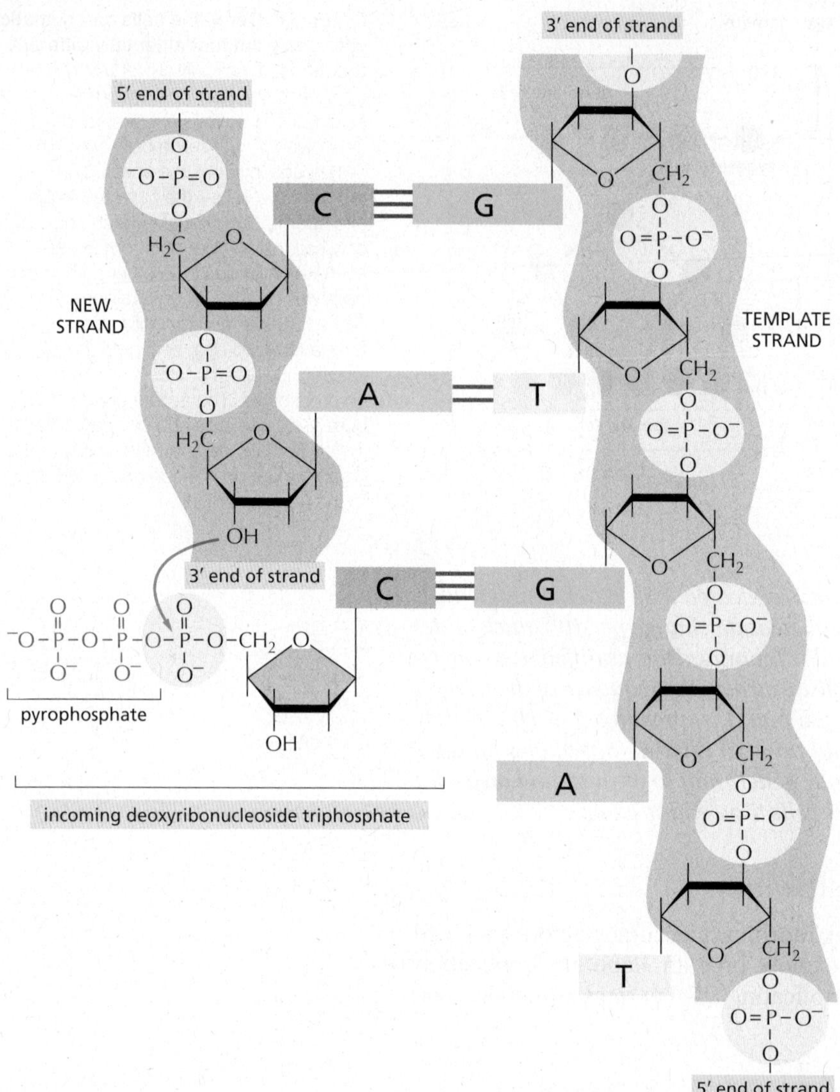

Figure 5–3 **The chemistry of DNA synthesis.** Nucleotides enter the reaction as deoxyribonucleoside triphosphates, and the addition of a deoxyribonucleotide to the 3′ end of a polynucleotide chain is the fundamental reaction by which DNA is synthesized. As shown, base-pairing between an incoming deoxyribonucleoside triphosphate and an existing strand of DNA (the *template strand*) guides the formation of the new strand of DNA and ensures that its nucleotide sequence is complementary to that of the template.

The first nucleotide-polymerizing enzyme, **DNA polymerase**, was discovered in 1957. The free nucleotides that serve as substrates for this enzyme were found to be deoxyribonucleoside triphosphates, and their polymerization into DNA required a single-strand DNA template. **Figure 5–3** and **Figure 5–4** illustrate the stepwise mechanism of this reaction.

The DNA Replication Fork Is Asymmetrical

During DNA replication inside a cell, each of the two original DNA strands serves as a template for the formation of an entire new strand. Because each of the two daughters of a dividing cell inherits a new DNA double helix containing one original and one new strand (**Figure 5–5**), the DNA double helix is said to be replicated *semiconservatively*. How is this feat actually accomplished?

Analyses carried out in the early 1960s on the whole replicating chromosome of an *E. coli* bacterium revealed a localized region of replication that moves progressively along the parent DNA double helix. Because of its Y-shaped structure,

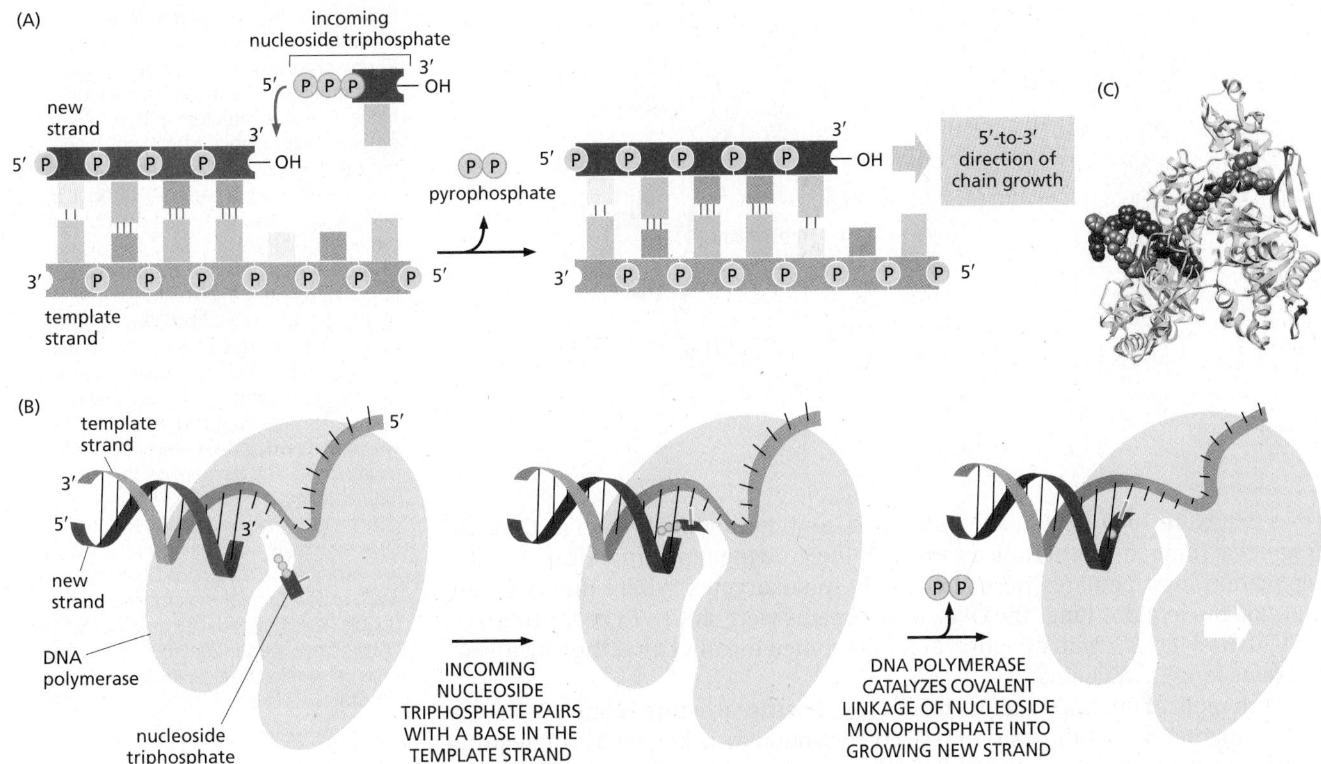

Figure 5–4 **How DNA polymerase adds a deoxyribonucleotide to the end of a growing DNA strand.** (A) An incoming deoxynucleoside triphosphate forms a base pair with its partner in the template strand. It is then covalently attached to the free 3′-hydroxyl (3′-OH) end of the growing DNA strand. The new DNA strand is therefore synthesized in the 5′-to-3′ direction. The energy for the polymerization reaction comes from the hydrolysis of a high-energy phosphate bond in the incoming nucleoside triphosphate and the release of pyrophosphate, which is subsequently hydrolyzed to yield two molecules of inorganic phosphate (not shown). (B) The reaction is catalyzed by the enzyme DNA polymerase *(light green)*. The polymerase guides the incoming nucleoside triphosphate to the template strand and positions it such that its 5′ triphosphate will be able to react with the 3′-hydroxyl group on the newly synthesized strand. The *white arrow* indicates the direction of polymerase movement. (C) Structure of DNA polymerase, as determined by x-ray crystallography, also showing the replicating DNA. The template strand is the longer, *orange* strand, and the DNA strand being synthesized is colored *red*. See **Movie 5.1**. (C, PDB code: 1KRP.)

this active region is called a **replication fork** (Figure 5–6). At the replication fork, a multienzyme complex that contains the DNA polymerase synthesizes the DNA of both new daughter strands.

Initially, the simplest mechanism of DNA replication seemed to be the continuous growth of both new strands, nucleotide by nucleotide, at the replication fork as it moves from one end of a DNA molecule to the other. But because of the antiparallel orientation of the two DNA strands in the DNA double helix (see Figure 5–2), this mechanism would require one daughter strand to polymerize in the 5′-to-3′ direction and the other in the 3′-to-5′ direction. Such a replication fork would require two distinct types of DNA polymerase enzymes. However, as attractive as this model might seem, the DNA polymerases at replication forks can synthesize only in the 5′-to-3′ direction.

How, then, can a DNA strand grow in the 3′-to-5′ direction? The answer came from an experiment performed in the late 1960s. Researchers added highly radioactive ^{3}H-thymidine to dividing bacteria for a few seconds, so that only the most recently replicated DNA—that just behind the replication fork—became radiolabeled. This experiment revealed the transient existence of pieces of

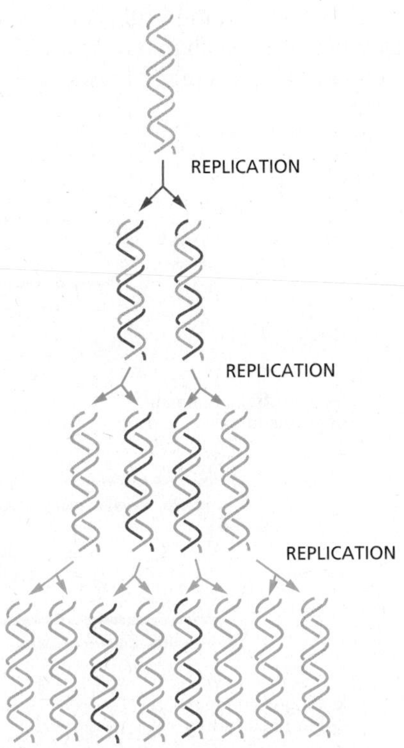

Figure 5–5 **In each round of DNA replication, each of the two strands of DNA is used as a template for the formation of a new, complementary strand.** DNA replication is *semiconservative* because each daughter DNA double helix is composed of one conserved (old) strand and one newly synthesized strand.

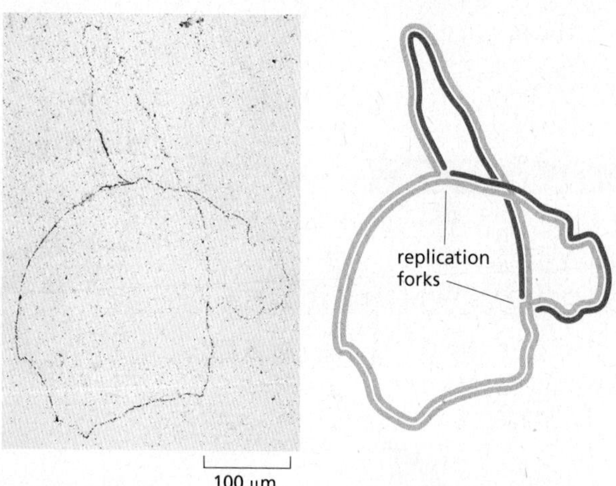

Figure 5–6 **Two replication forks moving in opposite directions on the *E. coli* chromosome, a large circular DNA molecule.** Each replication fork has a Y-shaped structure and moves progressively along the DNA, spinning out newly replicated DNA behind it. The stem of the Y is the parent DNA double helix, and the two arms of the Y contain the newly synthesized DNA. The image on the left was obtained by feeding *E. coli* radioactive thymine for several hours, gently isolating the DNA on filter paper, and placing a piece of photographic film next to the DNA. Because radioactivity exposes photographic film, an image of the DNA was captured when the film was developed and viewed under a light microscope. The diagram on the right is an interpretation of the result, with parent DNA strands in *orange* and newly synthesized DNA strands in *red*. During its isolation for this experiment, the *E. coli* DNA folded on itself, accounting for the crossing of the double helix. (From J. Cairns, *Cold Spring Harb. Symp. Quant. Biol.* 38:43–46, 1963. With permission from Cold Spring Harbor Laboratory Press.)

DNA that were 1000–2000 nucleotides long, now commonly known as *Okazaki fragments* (named for their discoverer), at the growing replication fork. Similar replication intermediates were later found in eukaryotes, where they are only 100–200 nucleotides long. The Okazaki fragments were shown to be synthesized only in the 5′-to-3′ chain direction and to be joined together after their synthesis to create long DNA chains.

Each replication fork therefore has an asymmetric structure (**Figure 5–7**). The DNA daughter strand that is synthesized continuously is known as the **leading strand**. Its synthesis slightly precedes the synthesis of the daughter strand that is synthesized discontinuously, known as the **lagging strand**. For the lagging strand, the direction of nucleotide polymerization is opposite to the overall direction of DNA chain growth. The synthesis of this strand by a discontinuous "backstitching" mechanism means that DNA replication requires only the 5′-to-3′ type of DNA polymerase.

The High Fidelity of DNA Replication Requires Several Proofreading Mechanisms

As discussed at the beginning of the chapter, the fidelity of copying DNA during replication is such that only about one mistake occurs for every 10^{10} nucleotides copied. This accuracy is much higher than one would expect solely from the

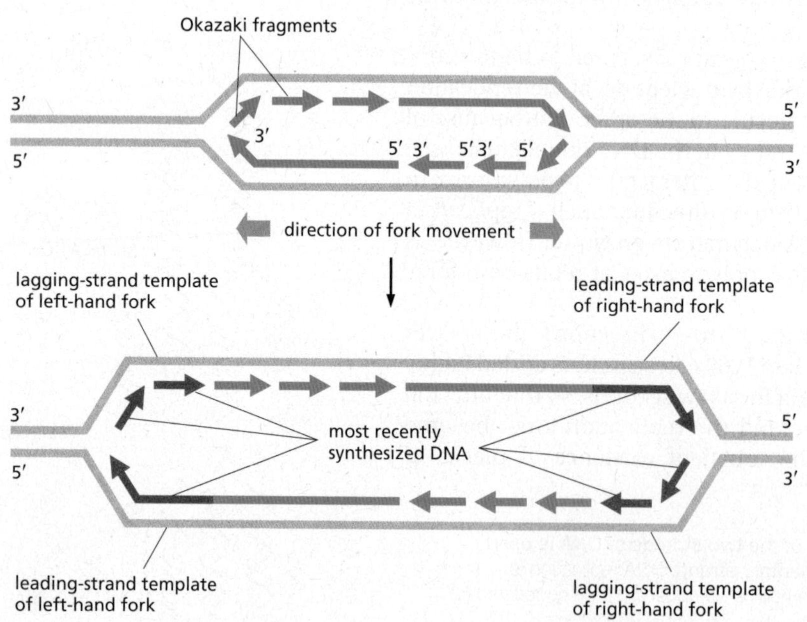

Figure 5–7 **At each replication fork, the lagging DNA strand is synthesized in pieces.** The upper diagram shows two replication forks moving in opposite directions on a double-helical DNA molecule, as in Figure 5–6; the lower diagram shows the same two forks a short time later. Because both of the new strands at a replication fork are synthesized in the 5′-to-3′ direction, the lagging strand of DNA must be made initially as a series of short DNA strands, which are later joined together. To replicate the lagging strand, the DNA polymerase molecule on that side of the fork uses a backstitching mechanism: it synthesizes a short piece of DNA in the 5′-to-3′ direction, stops, and is then moved by its protein machine back toward the fork in order to synthesize the next fragment.

properties of complementary base-pairing. The standard complementary base pairs (see Figure 4–5) are not the only ones possible. For example, with small changes in helix geometry, two hydrogen bonds can form between G and T in DNA. In addition, rare configurations of the four DNA bases (known as tautomers) occur transiently in ratios of 1 part to 10^4 or 10^5. These forms mispair without a change in helix geometry: the rare tautomeric form of C pairs with A instead of G, for example.

If the DNA polymerase did nothing special when a mispairing occurred between an incoming deoxyribonucleoside triphosphate and the DNA template, the wrong nucleotide would often be incorporated into the new DNA chain, producing frequent mutations. The high fidelity of DNA replication, however, depends not only on the initial base-pairing, but also on several "proofreading" mechanisms that act sequentially to correct any initial mispairings that might have occurred.

DNA polymerase performs the first proofreading step just before a new nucleotide is covalently added to the growing chain. After complementary nucleotide binding, but before the nucleotide is covalently added to the growing chain, the enzyme must undergo a conformational change in which its "grip" tightens around the active site. Because this change occurs more readily with correct than incorrect base-pairing, it allows the polymerase to "double-check" the exact base-pair geometry before it catalyzes the addition of the nucleotide. Incorrectly paired nucleotides are harder to add and therefore more likely to diffuse away before the polymerase can mistakenly add them.

The next error-correcting reaction, known as *exonucleolytic proofreading*, takes place immediately after those rare instances in which an incorrect nucleotide is covalently added to the growing chain. DNA polymerase enzymes are highly discriminating in the types of DNA chains they will elongate: they require a previously formed, base-paired 3′-OH end of a *primer strand* (see Figure 5–4). Those DNA molecules with a mismatched (improperly base-paired) nucleotide at the 3′-OH end of the primer strand are not effective as templates because the polymerase has difficulty extending such a strand. DNA polymerase molecules correct such a mismatched primer strand by means of a separate catalytic site (either in a separate subunit or in a separate protein domain of the polymerase molecule, depending on the polymerase). This *3′-to-5′ proofreading exonuclease* clips off any unpaired or mispaired residues at the primer terminus, continuing until enough nucleotides have been removed to regenerate a correctly base-paired 3′-OH terminus that can prime DNA synthesis. In this way, DNA polymerase functions as a "self-correcting" enzyme that removes its own polymerization errors as it moves along the DNA (**Figure 5–8** and **Figure 5–9**).

Because the self-correcting properties of DNA polymerase depend on its requirement for a perfectly base-paired primer terminus, it is apparently not possible for such an enzyme to start DNA synthesis *de novo* without an existing primer. By contrast, the RNA polymerase enzymes involved in gene transcription do not need such an efficient exonucleolytic proofreading mechanism: errors in making RNA are not passed on to the next generation, and the occasional defective RNA molecule that is produced has no long-term significance. As a result,

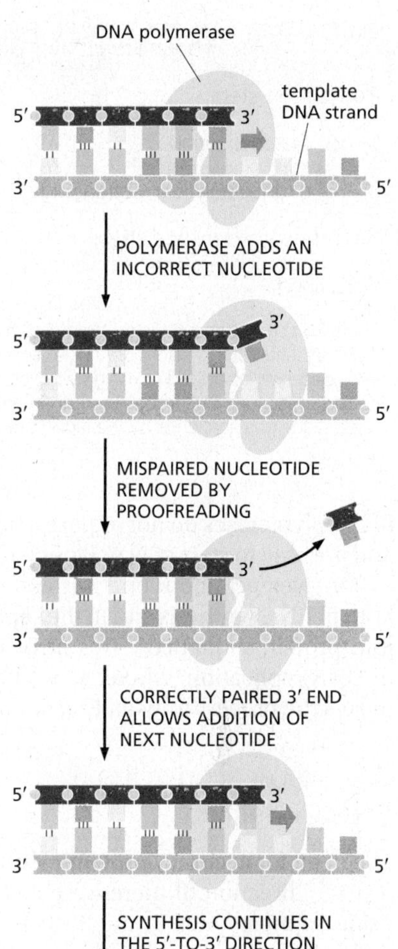

Figure 5–8 During DNA synthesis, DNA polymerase proofreads its own work. If an incorrect nucleotide is accidentally added to a growing strand, the DNA polymerase stops, cleaves it from the strand, and replaces it with the correct nucleotide before continuing.

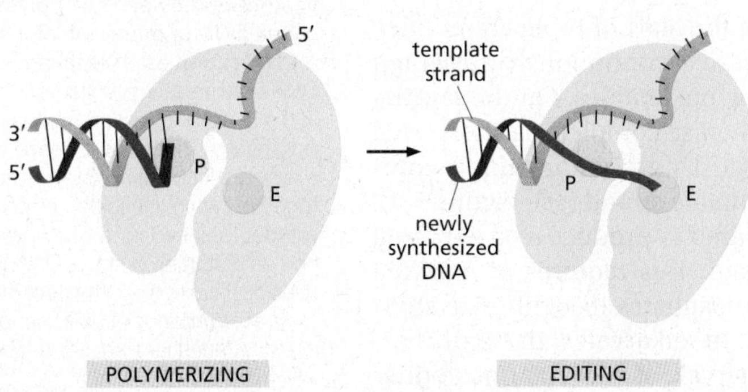

Figure 5–9 DNA polymerase contains separate sites for DNA synthesis and proofreading. The DNA polymerase, which cradles the DNA molecule being replicated, is shown in the polymerizing mode *(left)* and in the proofreading, or editing, mode *(right)*. The catalytic sites for the polymerization activity (P) and editing activity (E) are indicated. When the polymerase adds an incorrect nucleotide, the newly synthesized DNA strand *(red)* transiently unpairs from the template strand *(orange)*, and its 3′ end moves into the editing site (E) to allow the incorrect nucleotide to be removed. These diagrams are based on the structure of an *E. coli* DNA polymerase molecule, as determined by x-ray crystallography.

TABLE 5–1 The Three Steps That Give Rise to High-Fidelity DNA Synthesis

Replication step	Errors per nucleotide added
5′→3′ polymerization	1 in 10^5
3′→5′ exonucleolytic proofreading	1 in 10^2
Strand-directed mismatch repair	1 in 10^3
Combined	1 in 10^{10}

The third step, strand-directed mismatch repair, is described later in this chapter. For the polymerization step, "errors per nucleotide added" describes the probability that an incorrect nucleotide will be added to the growing chain. For the other two steps, "errors per nucleotide added" describes the probability that an error will not be corrected. Each step therefore reduces the chance of a final error by the factor shown.

RNA polymerases do not require a base-paired end 3′-OH for nucleotide addition and are able to start new polynucleotide chains without a primer.

On average, about one mistake is made for every 10^4 polymerization events both in RNA synthesis and in the separate process of translating mRNA sequences into protein sequences. This error rate is over 100,000 times greater than that in DNA replication, where, as we have seen, a series of proofreading processes makes the process unusually accurate (Table 5–1).

DNA Replication in the 5′-to-3′ Direction Allows Efficient Error Correction

The need for accuracy probably explains why DNA replication occurs only in the 5′-to-3′ direction. If there were a DNA polymerase that added deoxyribonucleoside triphosphates in the 3′-to-5′ direction, the growing 5′ end of the chain, rather than the incoming mononucleotide, would have to provide the activating triphosphate needed for the covalent linkage (see Figure 5–3). In this case, the mistakes in polymerization could not be simply hydrolyzed away, because the bare 5′ end of the chain thus created would immediately terminate DNA synthesis. It is therefore possible to correct a mismatched base only if it has been added to the 3′ end of a DNA chain. Although the backstitching mechanism for DNA replication seems complex, it preserves the 5′-to-3′ direction of polymerization that is required for exonucleolytic proofreading.

Despite these safeguards against DNA replication errors, DNA polymerases occasionally leave mistakes behind in the DNA that they produce. However, as we shall see later in this chapter, cells have yet another chance to correct these errors by a process called *strand-directed mismatch repair.* Before discussing this mechanism, however, we describe the other types of proteins that function at the replication fork as part of a large protein machine that replicates DNA.

A Special Nucleotide-polymerizing Enzyme Synthesizes Short RNA Primer Molecules

For the leading strand, a primer is needed only at the start of replication: once a replication fork is established, the DNA polymerase is continuously presented with a base-paired chain end on which to add new nucleotides. On the lagging side of the fork, however, each time the DNA polymerase completes a short DNA Okazaki fragment (which takes a few seconds), it must start synthesizing a completely new fragment at a site further along the template strand (see Figure 5–7). Each time this occurs, a special mechanism is required to produce a base-paired primer strand for the DNA polymerase to elongate. This requires an enzyme called **DNA primase** that uses ribonucleoside triphosphates to synthesize short **RNA primers** on the lagging strand (Figure 5–10). In eukaryotes, these primers are about 10 nucleotides long and are made at intervals of 100–200 nucleotides

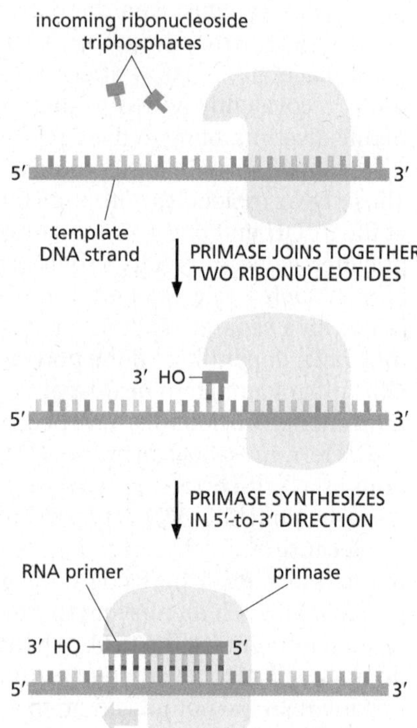

Figure 5–10 RNA primers are synthesized by an RNA polymerase called DNA primase, which uses a DNA strand as a template. Like DNA polymerase, primase synthesizes in the 5′-to-3′ direction. Unlike DNA polymerase, however, primase can start a new polynucleotide chain by joining together two nucleoside triphosphates without the need for a base-paired 3′ end as a starting point. A DNA primase uses ribonucleoside triphosphates rather than deoxyribonucleoside triphosphates, and it is much less accurate than a DNA polymerase.

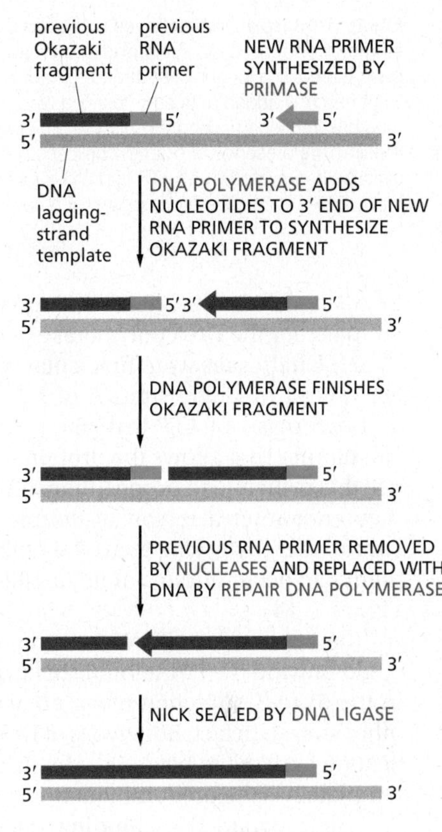

Figure 5–11 **Different enzymes act in series to synthesize DNA on the lagging strand.** In eukaryotes, RNA primers are made at intervals of about 200 nucleotides on the lagging strand, and each RNA primer is approximately 10 nucleotides long. These primers are extended by DNA polymerases at the replication fork to produce Okazaki fragments. The primers are subsequently removed by nucleases that recognize the RNA strand in an RNA–DNA hybrid helix and destroy it; this leaves gaps that are filled in by an accurate "repair" DNA polymerase that proofreads as it fills in the gaps. The completed DNA fragments are finally joined together by an enzyme called DNA ligase, which catalyzes the formation of a phosphodiester bond between the 3′-hydroxyl end of one fragment and the 5′-phosphate end of the next, thus linking up the sugar–phosphate backbones. This nick-sealing reaction requires an input of energy in the form of ATP (see Figure 5–12).

on the lagging strand. The synthesis of the leading strand also requires an RNA primer, but only at its very beginning.

RNA was introduced in Chapter 1 and is described in detail in Chapter 6. Here, we note only that RNA is very similar in structure to DNA. A strand of RNA can form base pairs with a strand of DNA, generating a DNA–RNA hybrid double helix if the two nucleotide sequences are complementary. Thus, the same templating principle used for DNA synthesis guides the synthesis of RNA primers. Because an RNA primer contains a properly base-paired nucleotide with a 3′-OH group at one end, it can be elongated by the DNA polymerase at this end to begin an Okazaki fragment.

The synthesis of each Okazaki fragment ends when this DNA polymerase runs into the RNA primer attached to the 5′ end of the previous fragment. To produce a continuous DNA chain from the many DNA fragments made on the lagging strand, a special DNA repair system acts quickly to remove the RNA primers and replace them with DNA. An enzyme called **DNA ligase** then joins the 3′ end of the new DNA fragment to the 5′ end of the previous one to complete the process (**Figure 5–11** and **Figure 5–12**).

Why might an erasable RNA primer be used instead of a DNA primer? The argument that a self-correcting polymerase cannot start chains *de novo* also implies the converse: an enzyme that starts chains anew cannot be efficient at self-correction. Thus, any enzyme that primes the synthesis of Okazaki fragments will of necessity make a relatively inaccurate copy. If these inaccurate copies were allowed to remain, the resulting increase in the overall mutation rate would be enormous. It therefore seems likely that the use of RNA rather than DNA for priming brings a powerful advantage to the cell: the ribonucleotides in the primer automatically mark these sequences as "suspect copy" to be efficiently removed and replaced by DNA produced by a highly accurate DNA polymerase.

Special Proteins Help to Open Up the DNA Double Helix in Front of the Replication Fork

For DNA synthesis to proceed, the DNA double helix must be opened up ahead of the replication fork so that the incoming deoxyribonucleoside triphosphates can form base pairs with the template strand. The DNA double helix is very stable under physiological conditions: the base pairs are locked in place so strongly that it requires temperatures approaching that of boiling water to separate the two strands in a test tube. For this reason, two additional types of replication proteins—DNA helicases and single-strand DNA-binding proteins—are needed

Figure 5–12 **DNA ligase joins together Okazaki fragments on the lagging strand during DNA synthesis.** The ligase enzyme uses a molecule of ATP to activate the 5′ phosphate of one fragment (step 1) before forming a new bond with the 3′ hydroxyl of the other fragment (step 2).

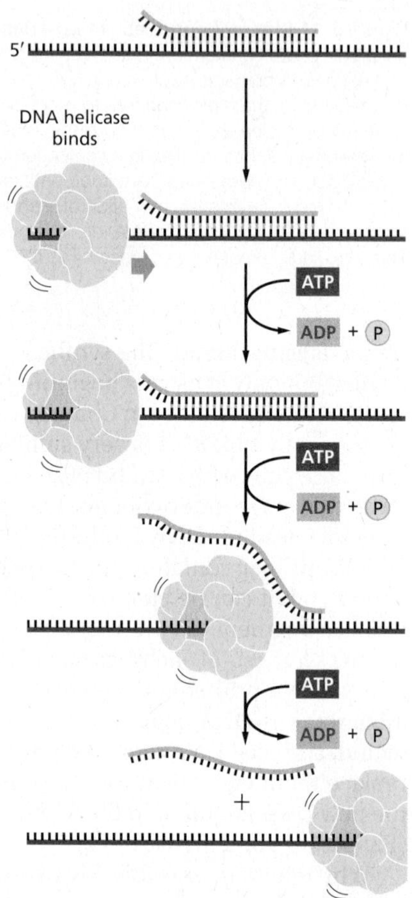

Figure 5–13 **How DNA helicase enzymes can separate strands as they move along a DNA single strand.** An experiment is diagrammed, in which a short, complementary DNA fragment is base-paired to a longer DNA strand to form a region of DNA double helix. Because the purified DNA helicase added acts as a "moving wedge," the double helix is pulled apart as the helicase runs unidirectionally along the DNA single strand, releasing the short DNA strand in a reaction that requires the presence of both the helicase protein and ATP. The rapid stepwise movement of the helicase is powered by its ATP hydrolysis (shown schematically in Figure 3–71A). As indicated, many DNA helicases are composed of a ring of six subunits.

to open the double helix and present an appropriate single-stranded DNA template for the DNA polymerase to copy.

DNA helicases were first isolated as proteins that hydrolyze ATP when they are bound to single strands of DNA. As described in Chapter 3, the binding and hydrolysis of ATP can change the shape of a protein molecule in a cyclical manner that allows the protein to perform mechanical work. DNA helicases use this principle to propel themselves rapidly along a single DNA strand. When they encounter a region of double helix, they continue to move along their strand, thereby prying apart the helix ahead of them. This unidirectional movement can occur at rates of up to 1000 nucleotides per second (**Figure 5–13** and **Figure 5–14**).

The two strands of DNA have opposite polarities, and, in principle, a helicase could unwind the DNA double helix in front of a replication fork by moving either in the 5′-to-3′ direction along one strand or in the 3′-to-5′ direction along the other strand. In fact, both types of DNA helicase exist. In the best-understood replication systems in bacteria, a helicase moving 5′-to-3′ along the lagging-strand template has the predominant role.

Single-strand DNA-binding (SSB) proteins bind tightly and cooperatively to the single-stranded DNA that is produced by helicases. Through cooperative binding, SSB proteins coat and straighten out all regions of single-stranded DNA, thereby preventing the formation of the short hairpin helices that otherwise form in these single strands (**Figure 5–15** and **Figure 5–16**). These regions occur routinely on the lagging-strand template, and if not removed, they can impede the DNA synthesis catalyzed by DNA polymerase.

A Sliding Ring Holds a Moving DNA Polymerase onto the DNA

On their own, most DNA polymerase molecules will synthesize only a short string of nucleotides before falling off the DNA template. However, an accessory protein (called PCNA in eukaryotes) forms a **sliding clamp** that keeps the polymerase firmly on the DNA when it is moving but releases the polymerase as soon as it runs into a double-strand region of DNA.

How can a sliding clamp prevent the polymerase from dissociating without impeding the polymerase's rapid movement along DNA? The three-dimensional structure of the clamp protein revealed that it forms a large ring around the DNA double helix. One face of the ring binds to the back of the DNA polymerase, and the whole ring slides freely along the DNA as the polymerase moves. The assembly of the clamp around the DNA requires a special protein complex, the **clamp loader**, that can open and close the ring in a regulated manner.

The moving DNA polymerase is tightly bound to the clamp, and, on the leading strand, the two remain associated for a very long time. The DNA polymerase on the lagging-strand template also makes use of the clamp, but each time the polymerase reaches the 5′ end of the preceding Okazaki fragment, the polymerase

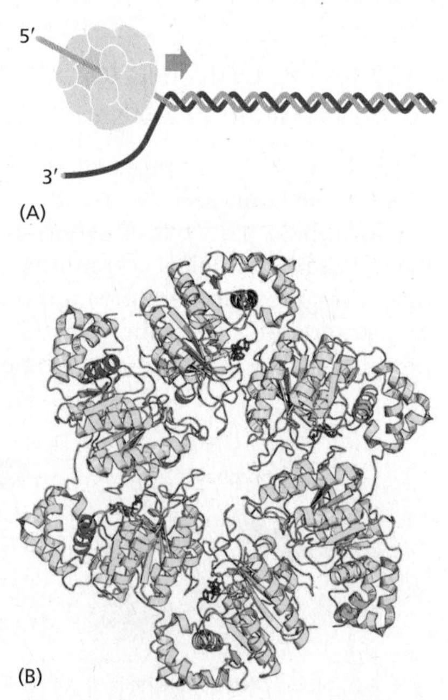

Figure 5–14 **The structure of a DNA helicase.** (A) Diagram of the protein, a hexameric ring, drawn to scale with a replication fork. (B) Detailed structure of the bacteriophage T7 replicative helicase, as determined by x-ray diffraction. Six identical subunits bind and hydrolyze ATP in an ordered fashion to propel this molecule, like a rotary engine, along a DNA single strand that passes through the central hole. Bound ATP molecules in the structure are indicated in *red* (**Movie 5.2**). (PDB code: 1E0J.)

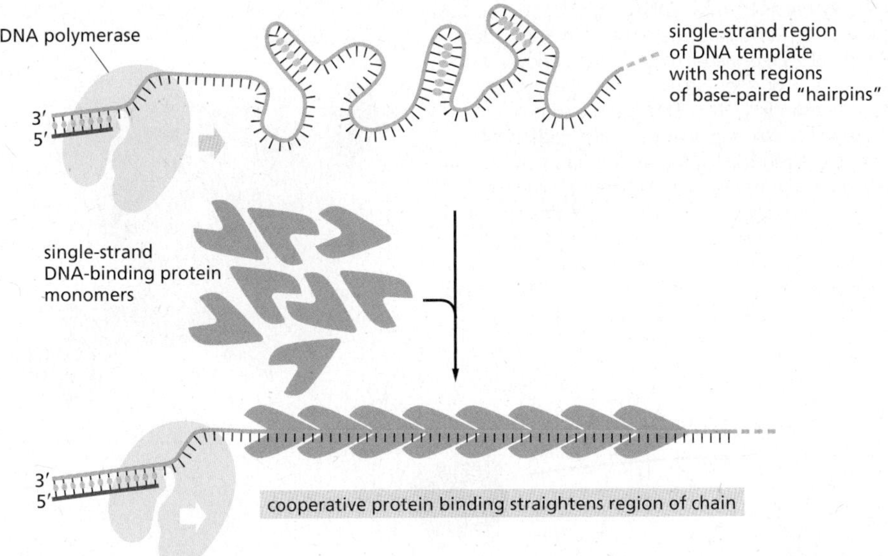

Figure 5–15 **The effect of single-strand DNA-binding proteins (SSB proteins) on the structure of single-stranded DNA.** Because each protein molecule prefers to bind next to a previously bound molecule, long rows of this protein form on a DNA single strand. This *cooperative binding* straightens out the DNA template and facilitates the DNA polymerization process. The "hairpin helices" shown in the bare, single-stranded DNA result from a chance matching of short regions of complementary nucleotide sequence.

releases itself from the clamp and dissociates from the template. With the help of the clamp loader, which hydrolyzes ATP as it loads a new clamp onto a primer–template junction (Figure 5–17), this lagging-strand polymerase molecule then associates with the new clamp that is assembled on the RNA primer of the next Okazaki fragment.

The Proteins at a Replication Fork Cooperate to Form a Replication Machine

Although we have discussed DNA replication as though it were performed by a set of proteins all acting independently, in reality most of these proteins are held together in a large and orderly multienzyme complex that rapidly synthesizes DNA. This complex can be likened to a tiny sewing machine composed of protein

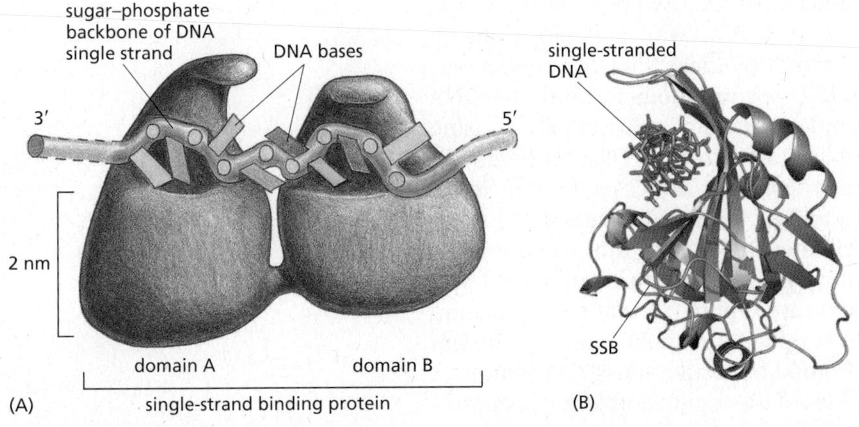

Figure 5–16 **Human single-strand binding protein bound to DNA.** (A) Front view of the two DNA-binding domains of the protein (called RPA), which cover a total of eight nucleotides. Note that the DNA bases remain exposed in this protein–DNA complex. (B) Diagram showing the three-dimensional structure, with the DNA strand *(orange)* viewed end on. (PDB code: 1JMC.)

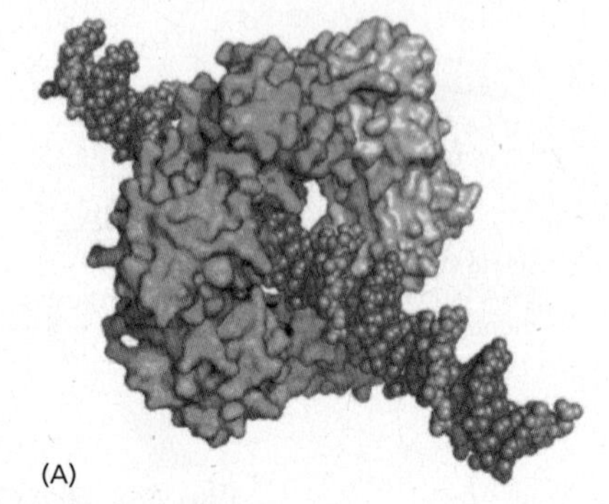

Figure 5–17 **The sliding clamp that holds DNA polymerase on the DNA.** (A) The structure of the clamp protein from *E. coli*, as determined by x-ray crystallography, with a DNA helix added to indicate how the protein fits around DNA (Movie 5.3). (B) Schematic illustration showing how the clamp is loaded onto DNA. The structure of the clamp loader *(green)* resembles a screw nut, with its threads matching the grooves of double-stranded DNA. The loader binds to a free clamp molecule, forcing a gap in its ring of subunits, which enables it to slip around DNA. The loader then "screws" the open clamp onto double-stranded DNA until it encounters the 3′ end of a primer, at which point the loader hydrolyzes ATP and releases the clamp, allowing it to close around the DNA. In the simplified reaction shown here, the clamp loader dissociates once the clamp has been assembled. At bacterial replication forks, the clamp loader remains bound to the polymerase so that, on the lagging strand, it is ready to assemble a new clamp at the start of each new Okazaki fragment. (A, from X.P. Kong et al., *Cell* 69:425–437, 1992; PDB code: 3BEP; B, adapted from B.A. Kelch et al., *Science* 334:1675–1680, 2011.)

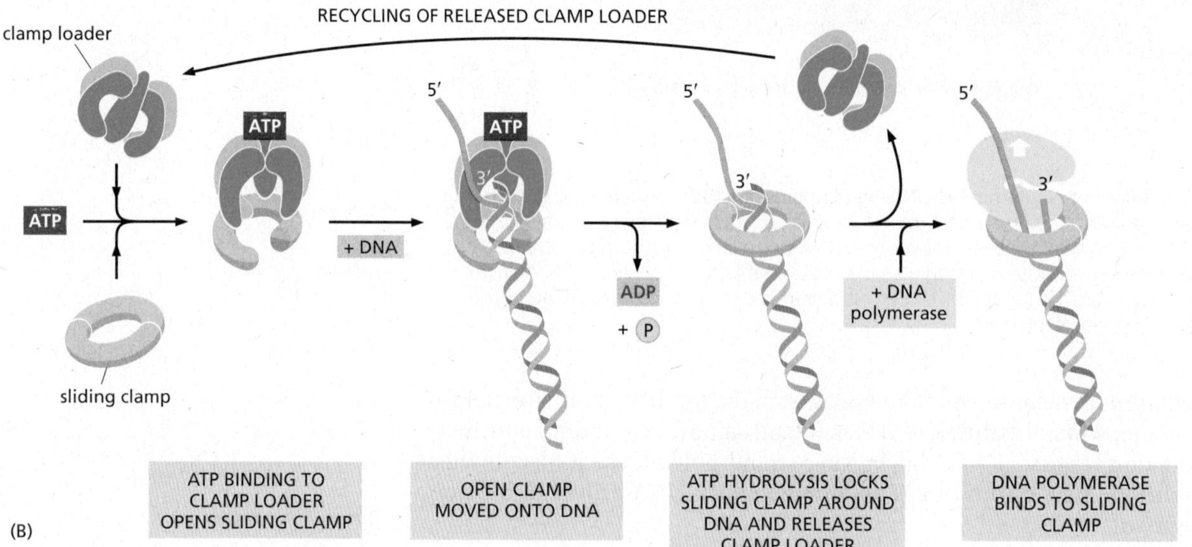

parts and powered by nucleoside triphosphate hydrolysis. Like a sewing machine, the replication complex probably remains stationary with respect to its immediate surroundings; the DNA can be thought of as a long strip of cloth being rapidly threaded through it. Although the replication complex has been most intensively studied in *E. coli* and several of its viruses, a very similar complex also operates in eukaryotes, as we shall see below.

How the different proteins at the replication fork work together in bacteria is shown in **Figure 5–18**. At the front of the replication fork, DNA helicase opens the DNA helix. Several identical DNA polymerase molecules work at the fork, one on the leading strand and two on the lagging strand. Whereas the DNA polymerase molecule on the leading strand can operate in a continuous fashion, the DNA polymerase molecules on the lagging-strand alternate at short intervals, using the short RNA primers made by DNA primase to begin each Okazaki fragment. The close association of all these protein components increases the efficiency of replication, and it is made possible by a folding back of the lagging strand as shown in the figure. This arrangement facilitates the loading of the polymerase clamp each time that an Okazaki fragment is synthesized: the clamp loader and the lagging-strand DNA polymerase molecule are kept in place at the replication fork even when they detach from their DNA template. The replication proteins are thus linked together into a single large unit (total molecular mass $>10^6$ daltons), enabling DNA to be synthesized on both sides of the replication fork in a coordinated and efficient manner.

On the lagging strand, the DNA replication machine leaves behind a series of unsealed Okazaki fragments, which still contain the RNA that primed their synthesis at their 5′ ends. As discussed earlier, this RNA is removed, and the resulting

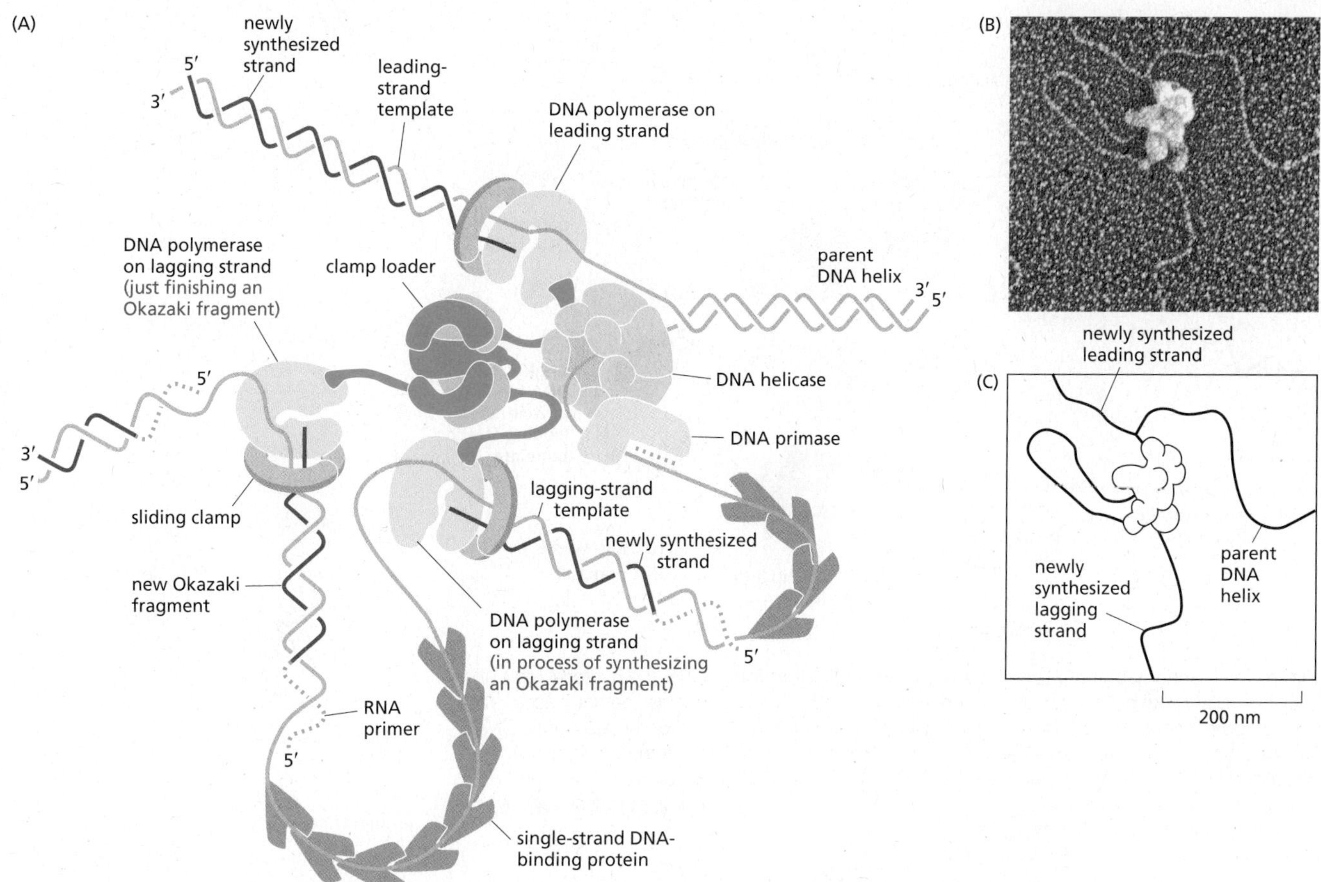

Figure 5–18 A bacterial replication fork. (A) In this case, a single DNA polymerase molecule synthesizes the leading strand while two DNA polymerases are used—in alternating fashion—for lagging-strand DNA synthesis. All of these polymerase molecules, which are identical, are held in place at the fork by flexible "arms" that extend from the clamp loader. Additional interactions (for example, between the DNA helicase and primase) ensure that all the individual components function together as a well-coordinated protein machine (**Movie 5.4**). (B) An electron micrograph showing the replication machine from the bacteriophage T4 as it moves along a template synthesizing DNA behind it. (C) An interpretation of the micrograph is given in the sketch: note especially the DNA loop on the lagging strand. Apparently, during the preparation of this sample for electron microscopy, the replication proteins became partly detached from the very front of the replication fork. (B, from P.D. Chastain et al., *J. Biol. Chem.* 278:21276–21825, 2003. With permission from American Society for Biochemistry and Molecular Biology.)

gap is filled in by DNA repair enzymes that operate behind the replication fork (see Figure 5–11).

DNA Replication Is Fundamentally Similar in Eukaryotes and Bacteria

Much of what we know about DNA replication was first derived from studies of purified bacterial and bacteriophage multienzyme systems capable of DNA replication *in vitro*. The development of these systems in the 1970s was greatly facilitated by the prior isolation of mutants in a variety of replication genes; these mutants were exploited to identify and purify the corresponding replication proteins. The first eukaryotic replication system that accurately replicated DNA *in vitro* was described in the mid-1980s, and mutations in genes encoding nearly all of the replication components have now been isolated and analyzed in the yeast *Saccharomyces cerevisiae*. As a result, much is known about the detailed enzymology of DNA replication in eukaryotes, and it is clear that the fundamental features of DNA replication—including replication-fork geometry and the use of 5′→3′ DNA polymerases, helicases, clamps, clamp loaders, and single-strand binding proteins—are similar.

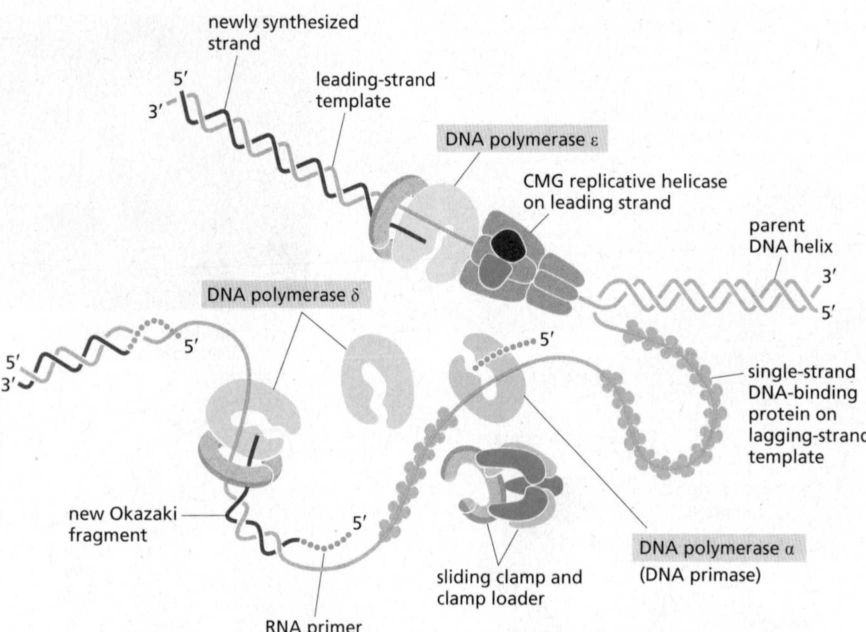

Figure 5–19 **Schematic diagram of a eukaryotic replication fork.** Unlike the bacterial replication proteins, those from eukaryotes are thought to function largely independently, perhaps accounting for the slower speed of the eukaryotic replication fork (**Movie 5.5**). Note that the eukaryotic CMG helicase moves unidirectionally along the leading-strand template, whereas the bacterial helicase discussed earlier moves in one direction along the lagging-strand template (see Figure 5–18). In both cases, the DNA duplex is rapidly pried apart at the front of the moving replication fork by harnessing the energy of ATP hydrolysis.

However, there are some important differences in how bacteria and eukaryotes replicate their DNA. Perhaps most important, eukaryotes use three different kinds of DNA polymerase at each replication fork (**Figure 5–19**). Polymerase ε (Polε) synthesizes the leading strand, whereas Polα and Polδ synthesize the lagging-strand Okazaki fragments. Each type of polymerase has special properties that make it well suited for its job. Polε binds to both the sliding clamp and the replicative helicase, allowing it to synthesize very long stretches of leading-strand DNA without dissociating. Polα includes DNA primase as one of its subunits, which begins all new chains by synthesizing a short length of RNA. This RNA is extended by a different subunit of Polα, which adds only about 20 nucleotides of DNA before dissociating. Finally, Polδ, which is loaded in conjunction with a sliding clamp, takes over and completes synthesis of each Okazaki fragment to produce a total length of about 200 nucleotides.

The use of three different kinds of DNA polymerase at the replication fork is part of a trend toward higher complexity observed for eukaryotic DNA replication compared to that of bacteria. As another example, the eukaryotic single-strand binding protein is formed from three different subunits, while only a single subunit is found in bacteria. Likewise, the eukaryotic replicative helicase (known as the CMG helicase) is composed of 11 different protein subunits, while the bacterial enzyme is a hexamer of 6 identical subunits. We do not know why the eukaryotic replication machinery is so much more complex than that of bacteria; however, there are several possibilities. In eukaryotes, DNA replication must be coordinated with the elaborate process of mitosis; it must also deal with DNA packaged into nucleosomes, topics we discuss in the next part of the chapter. It is also possible that the difference in complexity between bacteria and eukaryotes largely reflects evolutionary pressure for bacteria to make do with fewer genes.

Another important distinction between eukaryotic and bacterial replication protein complexes lies in the detailed structures of their individual protein components. With the exception of the sliding clamp, the replication proteins in bacteria have completely different structures and amino acid sequences than those of their eukaryotic counterparts. The simplest interpretation of this surprising fact is that, over hundreds of millions of years, the DNA replication machinery in eukaryotes and bacteria evolved independently, yet converged on the same basic mechanisms. This situation is in contrast to other fundamental processes in the cell, such as transcription and translation, where the fundamental components (RNA polymerase and the ribosome) are very similar between bacteria and eukaryotes—and where the structures are conserved from an ancient, common ancestor.

A Strand-directed Mismatch Repair System Removes Replication Errors That Remain in the Wake of the Replication Machine

Because bacteria such as *E. coli* are capable of dividing once every 30 minutes, it is relatively easy to screen large populations to find a rare mutant cell that is altered in a specific process. One interesting class of mutants consists of those with alterations in so-called *mutator genes*, which greatly increase the rate of spontaneous mutation. Not surprisingly, one such mutant makes a defective form of the 3′-to-5′ proofreading exonuclease that is a part of the DNA polymerase enzyme (see Figures 5–8 and 5–9). The mutant DNA polymerase no longer proofreads effectively, and many replication errors that would otherwise have been removed accumulate in the DNA.

The study of other *E. coli* mutants exhibiting abnormally high mutation rates uncovered an additional proofreading system, common to all cells on Earth, that removes those rare replication errors that were made by the polymerase and missed by its proofreading exonuclease. These errors leave mismatched base pairs behind the replication fork, which are subsequently recognized and corrected by a **strand-directed mismatch repair** system. This system picks out mismatches from normal DNA by monitoring their potential to distort the DNA double helix, which is greatly increased by the misfit between noncomplementary base pairs. However, if the repair system simply recognized a mismatch in newly replicated DNA and randomly corrected one of the two mismatched nucleotides, it would mistakenly "correct" the original template strand to match the error exactly half the time, thereby failing to lower the overall error rate. To be effective, such a proofreading system must be able to remove only the nucleotide on the newly synthesized strand, where the error occurred.

The strand-distinction mechanism used by the mismatch proofreading system in *E. coli* depends on the methylation of selected A residues in the DNA. Methyl groups are added to all A residues in the sequence GATC, but not until some time after the GATC has been synthesized. As a result, the only unmethylated GATC sequences lie in the newly synthesized strands just behind a replication fork. The recognition of these unmethylated GATCs (which are base-paired to methylated GATCs) allows the new DNA strands to be transiently distinguished from old ones, as required if their mismatches are to be selectively removed. The five-step error-correction process involves recognition of a mismatch, identification of the newly synthesized strand, excision of the portion containing the misincorporated nucleotide, resynthesis of the excised segment using the old strand as a template, and ligation to seal the DNA backbone. This strand-directed mismatch repair system reduces the number of errors made during DNA replication by an additional factor of 100–1000 (see Table 5–1, p. 260).

A similar mismatch proofreading system functions in eukaryotic cells, but it uses a different way to distinguish the newly synthesized DNA strands from the parent strands. On the lagging strand, the newly synthesized DNA will contain transient single-strand gaps before the series of Okazaki fragments are processed and ligated together. Each gap will usually carry a sliding clamp, which remains on the DNA after the DNA polymerase has dissociated from it to begin the next fragment. Together, the clamp and the single-strand break signal to the mismatch

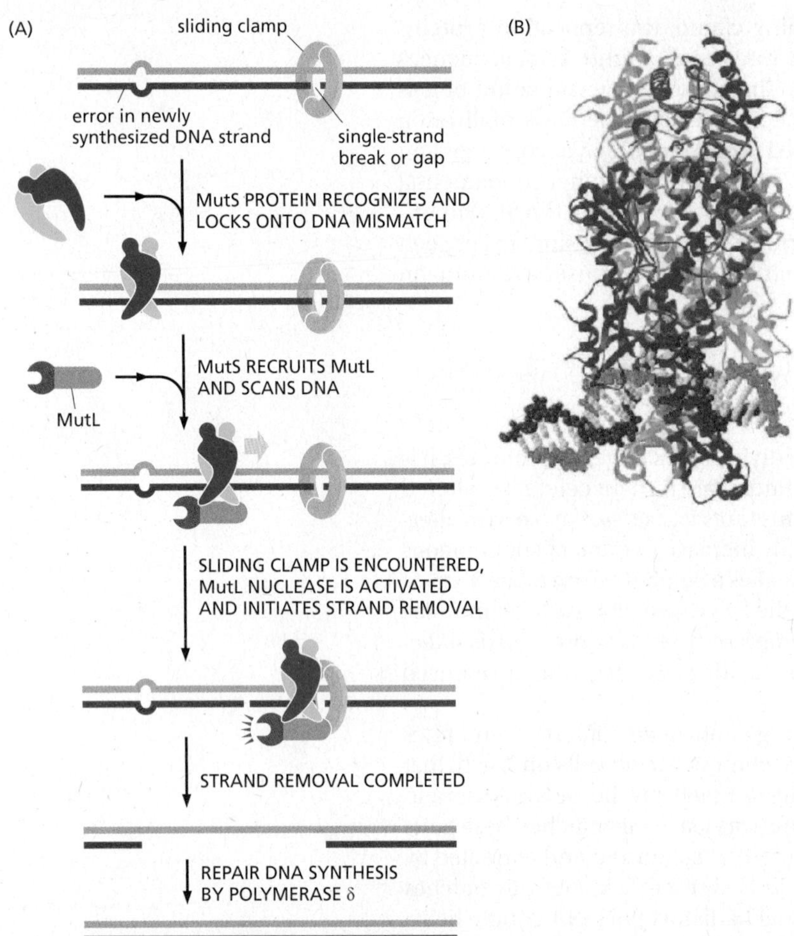
(A)
sliding clamp
error in newly synthesized DNA strand
single-strand break or gap
MutS PROTEIN RECOGNIZES AND LOCKS ONTO DNA MISMATCH
MutS RECRUITS MutL AND SCANS DNA
MutL
SLIDING CLAMP IS ENCOUNTERED, MutL NUCLEASE IS ACTIVATED AND INITIATES STRAND REMOVAL
STRAND REMOVAL COMPLETED
REPAIR DNA SYNTHESIS BY POLYMERASE δ
(B)

Figure 5–20 Strand-directed mismatch repair in eukaryotes. (A) The MutS protein binds to a mismatched base pair, recruits the MutL protein, and the complex scans the nearby DNA for a gap and a sliding clamp whose orientation determines which strand is to be cut and its nucleotides replaced. When these are encountered, MutL is activated and begins to cleave the DNA. In most organisms, MutL is joined by another nuclease and, together, they remove the newly synthesized DNA starting at the gap and extending past the mismatch. The gap is then filled in by DNA polymerase δ and sealed by DNA ligase. (B) The structure of the MutS protein bound to a DNA mismatch. This protein is a dimer, which grips the DNA double helix as shown, kinking the DNA at the mismatched base pair. It seems that the MutS protein scans the DNA for mismatches by testing for sites that can be readily kinked, which are those with an abnormal base pair. (PDB code: 1EWQ.)

repair proteins to correct the mismatch using the parent DNA strand as the template (**Figure 5–20**).

The two faces of the clamp differ, and the clamp loader always loads the clamp in the same orientation with respect to the 3′ end of the previously synthesized Okazaki fragment. Because all the clamps on the DNA "face" in the same direction relative to the replication process, the oriented clamps can be used by the mismatch repair machinery to distinguish newly synthesized DNA from parent DNA. It is not known for certain how strand discrimination occurs on the leading strand (where gaps in newly synthesized DNA should be rare), but because oriented sliding clamps are also left behind by the leading-strand polymerase, they can signal old from new DNA in the same way that they do on the lagging strand. The recent discovery of a correction system that removes misincorporated ribonucleotides suggests a further possibility for distinguishing newly synthesized DNA from parent DNA, as we discuss in the next section.

Mismatch correction is crucial for all cells; its importance for humans is seen in individuals who inherit one defective copy of a mismatch repair gene (along with a functional gene on the other copy of the chromosome). These individuals have a marked predisposition for certain types of cancers. For example, in a type of colon cancer called *hereditary nonpolyposis colorectal cancer* (*HNPCC*), a spontaneous deleterious mutation of the one functional gene will produce a clone of somatic cells that, because they are deficient in mismatch proofreading, accumulate mutations unusually rapidly. Because most cancers arise in cells that have accumulated many mutations (as discussed in Chapter 20), cells deficient in mismatch proofreading have a greatly enhanced chance of becoming cancerous. Fortunately, most of us inherit two good copies of each gene that encodes a mismatch proofreading protein; this protects us, because it is highly unlikely that both copies will become mutated in the same cell.

The Accidental Incorporation of Ribonucleotides During DNA Replication Is Corrected

We have seen that cells have several ways to correct mistakes where the wrong deoxynucleotide has been incorporated in newly replicated DNA. Occasionally, however, DNA polymerases make a different kind of mistake, one that is not caused by improper base-pairing: in this case, they accidently incorporate a ribonucleotide instead of a deoxyribonucleotide. These molecules differ by a single -OH group in the sugar portion of the nucleotide. Yet, when incorporated into DNA, they weaken the DNA chain at that point, rendering it highly susceptible to breakage. If left unrepaired, these "weak links" would cause high mutation rates and genome rearrangements. Even if it does not cause a break, an incorporated ribonucleotide distorts the DNA double helix and can stall some polymerases during the next cycle of DNA replication.

Although DNA polymerases much prefer deoxyribonucleotides over ribonucleotides (by a factor of about a million), the concentration of ribonucleotides in the cell is much higher than that of their deoxy counterparts, as much as 500-fold for ATP, which has many different uses in the cell. This concentration imbalance means that a ribonucleotide is accidentally incorporated approximately once per several thousand nucleotides of DNA synthesized. These mistakes are corrected by specific nucleases that cleave the DNA chain when they encounter a ribonucleotide, leading to the excision of the ribonucleotide and its replacement by DNA, much in the same way that RNA primers are replaced by DNA to complete lagging-strand synthesis (see Figure 5–11). Because this repair process produces gaps only in newly synthesized DNA, it has been proposed that these transient lesions help the mismatch repair system "know" which strand to repair; in particular, these cues may be especially important on the leading strand.

DNA Topoisomerases Prevent DNA Tangling During Replication

As a replication fork moves along double-stranded DNA, it creates what has been called the "winding problem." The two parent strands that are wound around each other must be unwound and separated for replication to occur. For every 10 nucleotide pairs replicated at the fork, one complete turn of the parent double helix must be unwound. In principle, this unwinding can be achieved by rapidly rotating the entire chromosome ahead of a moving fork; however, this is energetically highly unfavorable (particularly for long chromosomes). Instead, the DNA in front of a replication fork becomes overwound (**Figure 5–21**). This overwinding is continually relieved by enzymes known as *DNA topoisomerases*.

A **DNA topoisomerase** can be viewed as a reversible nuclease that adds itself covalently to a DNA backbone phosphate, thereby breaking a phosphodiester bond in a DNA strand. This reaction is reversible, and the phosphodiester bond re-forms as the protein leaves.

One type of topoisomerase, called *topoisomerase I*, produces a transient single-strand break; this break in the phosphodiester backbone allows the

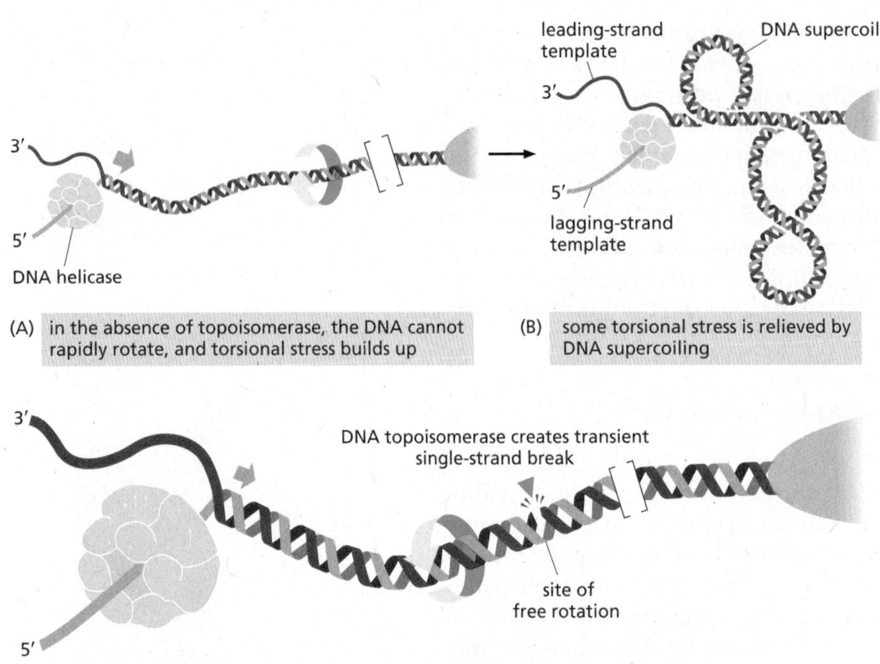
leading-strand template
DNA supercoil
3′
3′
5′
5′
lagging-strand template
DNA helicase
(A) in the absence of topoisomerase, the DNA cannot rapidly rotate, and torsional stress builds up
(B) some torsional stress is relieved by DNA supercoiling
3′
DNA topoisomerase creates transient single-strand break
site of free rotation
5′
(C) torsional stress ahead of the helicase is relieved by free rotation of DNA around the phosphodiester bond opposite the single-strand break; the same DNA topoisomerase molecule that produced the break reseals it

Figure 5–21 The "winding problem" that arises during DNA replication. (A) For a bacterial replication fork moving at 500 nucleotides per second, the parent DNA helix ahead of the fork must rotate at about 50 revolutions per second. The brackets represent about 20 turns of DNA. (B) If the ends of the DNA double helix remain fixed (or difficult to rotate), tension builds up in front of the replication fork as it becomes overwound. Some of this tension can be taken up by supercoiling, whereby the DNA double helix twists around itself. However, if the tension continues to build up, the replication fork will eventually stop because further unwinding requires more energy than the DNA helicase at the fork can provide. (C) DNA topoisomerases relieve this stress by generating temporary single-strand breaks in the DNA, which allow rapid rotation around the single strands opposite the break.

two sections of DNA helix on either side of the nick to rotate freely relative to each other, using the phosphodiester bond in the strand opposite the nick as a swivel point (**Figure 5–22**). Any tension in the DNA helix will drive this rotation in the direction that relieves the tension. As a result, DNA replication can occur with the rotation of only a short length of helix—the part just ahead of the fork. Because the covalent linkage that joins the DNA topoisomerase protein to a DNA phosphate retains the energy of the cleaved phosphodiester bond, resealing is rapid and does not require additional energy input. In this respect, the rejoining mechanism differs from that catalyzed by the enzyme DNA ligase, discussed previously (see Figure 5–12).

A second type of DNA topoisomerase, *topoisomerase II*, forms a covalent linkage to both strands of the DNA helix at the same time, making a transient *double-strand break* in the helix. These enzymes are activated by sites on chromosomes where two double helices cross over each other, such as those generated by supercoiling in front of a replication fork (see Figure 5–21B). As illustrated in **Figure 5–23**, once a topoisomerase II molecule binds to such a crossing site, the protein uses ATP hydrolysis to perform the following set of reactions: (1) it breaks one double helix reversibly to create a DNA "gate"; (2) it causes the second, nearby double helix to pass through this opening; and (3) it then reseals the break and dissociates from the DNA. At crossover points generated by supercoiling, passage of the double helix through the gate occurs in the direction that will reduce supercoiling. In this way, type II topoisomerases—like type I topoisomerases—can relieve the overwinding tension generated in front of a replication fork.

Their reaction mechanism also allows type II DNA topoisomerases to efficiently separate any intertwined DNA molecules. This ability of topoisomerase II is especially important for preventing the severe DNA tangling problems that would otherwise arise from DNA replication. This role is nicely illustrated by mutant yeast cells that produce, in place of the normal topoisomerase II, a version that is inactive above 37°C. When the mutant cells are warmed to this temperature, their daughter chromosomes remain intertwined after DNA replication and are unable to separate. The enormous usefulness of topoisomerase II for untangling

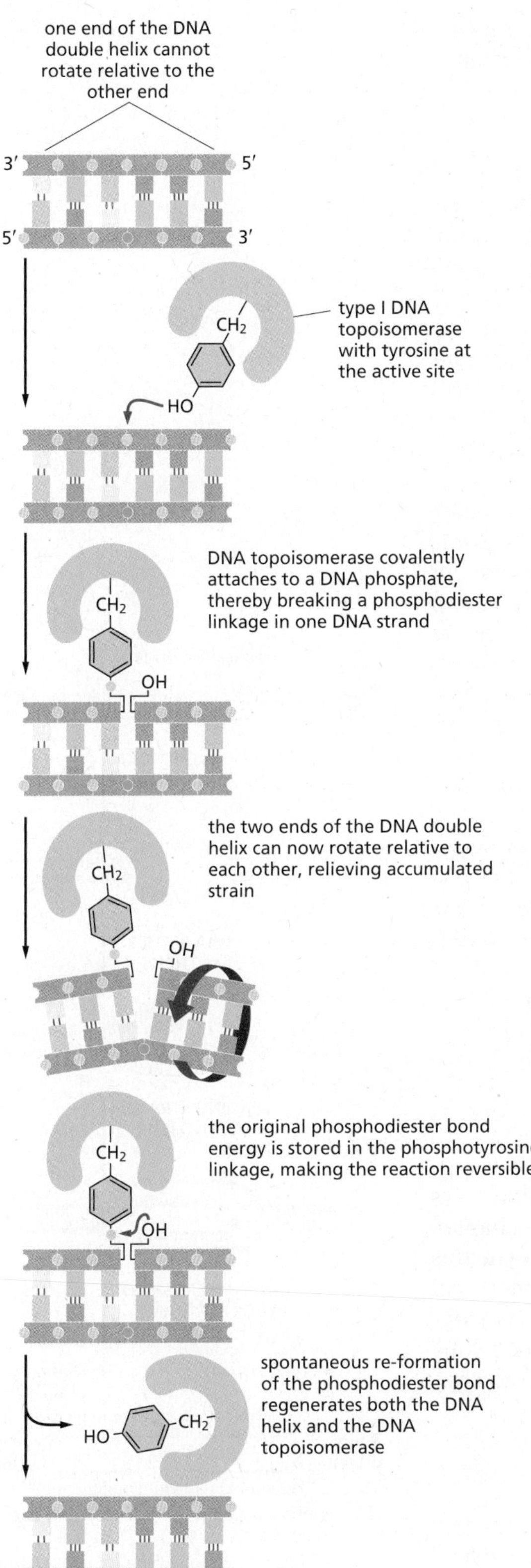
one end of the DNA double helix cannot rotate relative to the other end
3′
5′
5′
3′
type I DNA topoisomerase with tyrosine at the active site
CH2
HO
DNA topoisomerase covalently attaches to a DNA phosphate, thereby breaking a phosphodiester linkage in one DNA strand
CH2
OH
the two ends of the DNA double helix can now rotate relative to each other, relieving accumulated strain
CH2
OH
the original phosphodiester bond energy is stored in the phosphotyrosine linkage, making the reaction reversible
CH2
OH
spontaneous re-formation of the phosphodiester bond regenerates both the DNA helix and the DNA topoisomerase
HO
CH2

Figure 5–22 **The reversible DNA nicking reaction catalyzed by a DNA topoisomerase I enzyme.** As indicated, these enzymes transiently form a single covalent bond with DNA; this allows free rotation of the DNA around the covalent backbone bonds linked to the *blue* phosphate. On reversal of the reaction, the enzyme and the DNA are restored, the only difference being the relaxation of tension in the DNA.

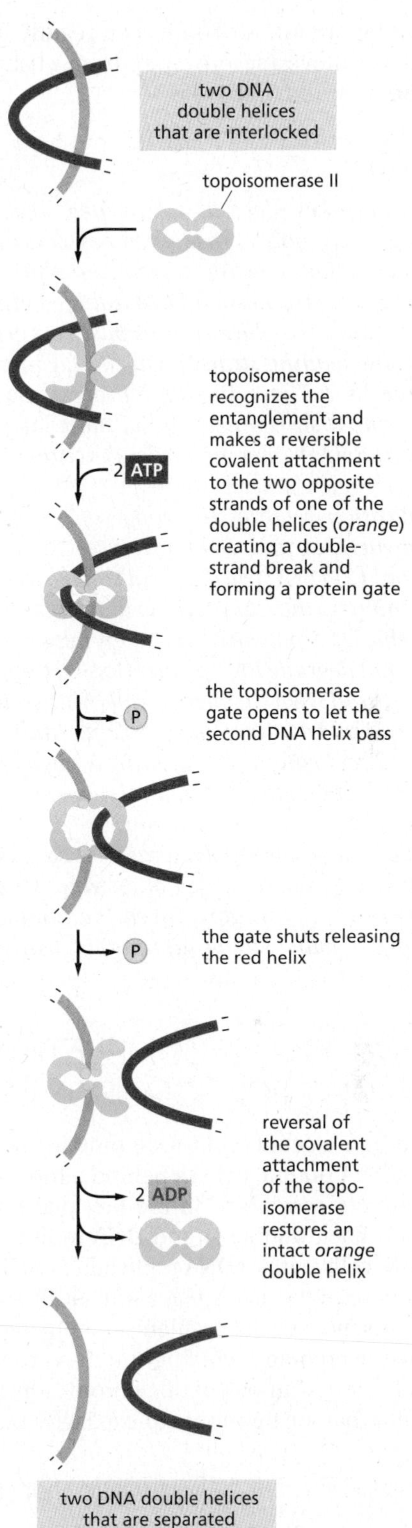
two DNA double helices that are interlocked
topoisomerase II
topoisomerase recognizes the entanglement and makes a reversible covalent attachment to the two opposite strands of one of the double helices (*orange*) creating a double-strand break and forming a protein gate
2 ATP
the topoisomerase gate opens to let the second DNA helix pass
P
the gate shuts releasing the red helix
P
reversal of the covalent attachment of the topo-isomerase restores an intact *orange* double helix
2 ADP
two DNA double helices that are separated

Figure 5–23 **The DNA-helix-passing reaction catalyzed by DNA topoisomerase II.** Unlike type I topoisomerases, type II enzymes hydrolyze ATP, which is needed to release and reset the enzyme after each cycle. The small *yellow circles* represent the 5′ phosphates in the DNA backbone that become covalently bonded to the topoisomerase. Type II topoisomerases are especially important for rapidly dividing cells; partly for that reason, they are effective targets for a large class of antibiotics, the fluoroquinolones, used to treat many different kinds of bacterial infections. These drugs inhibit bacterial topoisomerase II at the third step in the figure and thereby produce high levels of double-strand breaks that are lethal to rapidly dividing cells.

chromosomes before mitosis begins can readily be appreciated by anyone who has struggled to remove a severe tangle from a fishing line—or from a large ball of thread—without the aid of scissors.

Summary

DNA replication takes place at a Y-shaped structure called a replication fork. Self-correcting DNA polymerase enzymes catalyze nucleotide polymerization in a 5′-to-3′ direction, copying a DNA template strand with remarkable fidelity. Because the two strands of a DNA double helix are antiparallel, this 5′-to-3′ DNA synthesis can take place continuously on only one of the strands at a replication fork (the leading strand). On the lagging strand, short DNA fragments must be made by a "backstitching" process. Because the self-correcting DNA polymerases cannot start a new chain, these lagging-strand DNA fragments are primed by short RNA primer molecules that are subsequently erased and replaced with DNA.

DNA replication requires the cooperation of many proteins. These include (1) DNA polymerases and DNA primases to catalyze nucleoside triphosphate polymerization; (2) DNA helicases and single-strand DNA-binding (SSB) proteins to help in opening up the DNA helix so that it can be copied; (3) clamps and clamp loaders to enable DNA polymerases to copy longer stretches of DNA; (4) DNA ligases and enzymes that degrade RNA primers to seal together the discontinuously synthesized lagging-strand DNA fragments; and (5) DNA topoisomerases to help to relieve helical winding and DNA tangling problems. Many of these proteins associate with each other at a replication fork to form a highly efficient "replication machine," through which the activities and spatial movements of the individual components are coordinated.

The self-correcting DNA polymerases make mistakes only rarely when copying DNA; when they do, a variety of enzymes inspect the DNA shortly after it is made and correct any mishaps. Given the number of proteins dedicated to the task, copying DNA with extreme accuracy is clearly of great importance to all cells on Earth.

THE INITIATION AND COMPLETION OF DNA REPLICATION IN CHROMOSOMES

We have seen how a set of replication proteins rapidly and accurately generates two daughter DNA double helices behind a replication fork. But how is this replication machinery assembled in the first place, and how are replication forks created on an intact, double-strand DNA molecule? In this part of the chapter, we discuss how cells initiate DNA replication and how they carefully regulate this process to ensure that it takes place only at the proper time and chromosomal sites. We also discuss special problems that the replication machinery in eukaryotic cells must overcome including the need to replicate the enormously long DNA molecules found in eukaryotic chromosomes, as well as the need to copy DNA molecules that are tightly complexed with nucleosomes.

DNA Synthesis Begins at Replication Origins

As discussed previously, the DNA double helix is normally very stable: the two DNA strands are locked together firmly by the hydrogen bonds formed between the bases on each strand. To begin DNA replication, the double helix must first be opened up and the two strands separated to expose unpaired bases. As we shall see, the process of DNA replication is begun by special *initiator proteins* that bind to double-stranded DNA and pry the two strands apart, breaking the hydrogen bonds between the bases.

The positions at which the DNA helix is first opened are called **replication origins** (Figure 5–24). In simple cells like those of bacteria or budding yeast, origins are specified by DNA sequences several hundred nucleotide pairs in

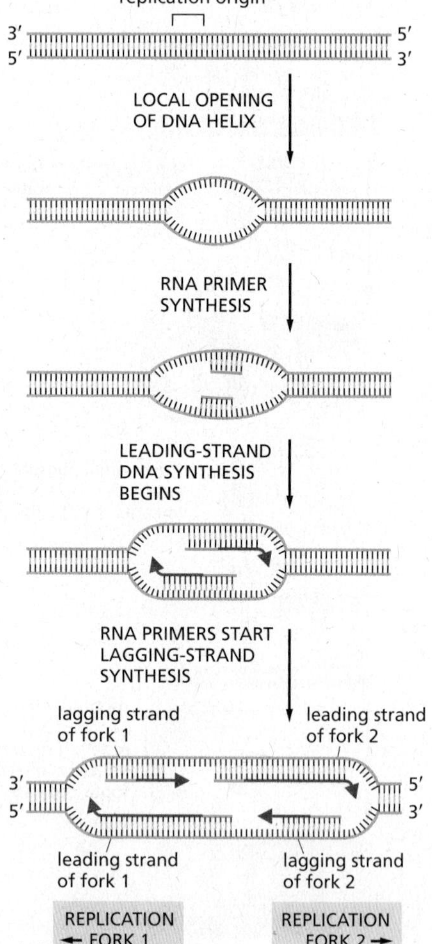

Figure 5–24 A replication bubble formed by replication-fork initiation. This diagram outlines the major steps in the initiation of replication forks at replication origins. In the last step, two replication forks move away from each other, separated by an expanding *replication bubble*.

length. This DNA contains both short sequences that attract initiator proteins and stretches of DNA that are especially easy to open. We saw in Figure 4–5A that an A-T base pair is held together by fewer hydrogen bonds than is a G-C base pair. Therefore, DNA rich in A-T base pairs is relatively easy to pull apart, and regions of DNA enriched in A-T base pairs are typically found at replication origins.

Although the basic process of replication-fork initiation depicted in Figure 5–24 is fundamentally the same for bacteria and eukaryotes, the detailed way in which this process is performed and regulated differs considerably between these two groups of organisms. We first consider the case in bacteria and then turn to the more complex situation found in yeasts, mammals, and other eukaryotes.

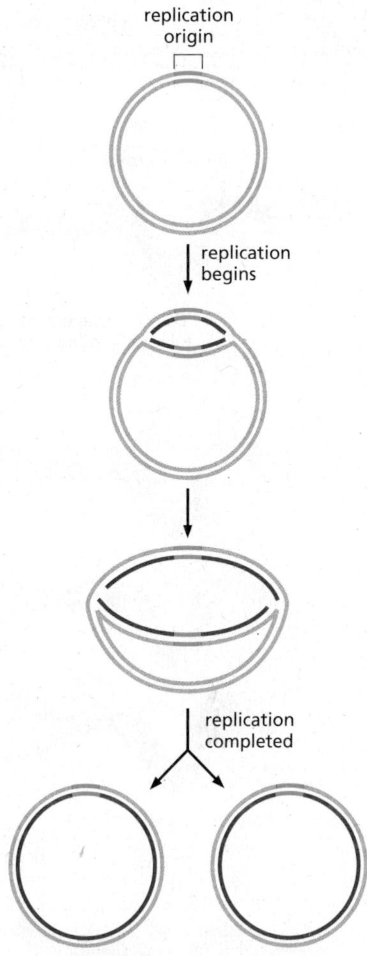

Figure 5–25 DNA replication of a bacterial genome. It takes *E. coli* about 30 minutes to duplicate its genome of 4.6×10^6 nucleotide pairs. For simplicity, Okazaki fragments are not shown on the lagging strand.

Bacterial Chromosomes Typically Have a Single Origin of DNA Replication

The genome of *E. coli* is contained in a single circular DNA molecule of 4.6×10^6 nucleotide pairs. DNA replication begins at a single origin of replication, and the two replication forks assembled there proceed (at approximately 1000 nucleotides per second) in opposite directions until they meet up roughly halfway around the chromosome (**Figure 5–25**). The only point at which *E. coli* can control DNA replication is initiation: once the forks have been assembled at the origin, they synthesize DNA at a relatively constant speed until replication is finished. Therefore, it is not surprising that the initiation step of DNA replication is tightly regulated. The process begins when specialized initiator proteins (in their ATP-bound state) bind in multiple copies to specific DNA sites located at the replication origin, wrapping the DNA around the proteins to form a large protein–DNA filament that introduces torsional stress on the DNA double helix (**Figure 5–26**). This stress is partially relieved by melting of the adjacent AT-rich sequences. The protein–DNA complex then attracts two DNA helicases, each bound to a helicase loader, and these are placed—facing in opposite directions—around adjacent DNA single strands whose bases have been exposed by the assembly of the initiator protein–DNA complex. The helicase loader is analogous to the clamp loader we encountered earlier; it has the additional job of keeping the helicase in an inactive form until it is properly loaded. Once the helicases are properly positioned on DNA, the loaders dissociate and the helicases begin to unwind DNA, exposing enough single-stranded DNA for DNA primases to synthesize the first RNA primers. This quickly leads to the assembly of the remaining replication proteins to create two replication forks that move in opposite directions away from the replication origin, each synthesizing new DNA as they travel.

In *E. coli,* the interaction of the initiator proteins with the replication origin is carefully regulated, with initiation occurring only when sufficient nutrients are available for the bacterium to complete an entire round of replication. Initiation is also controlled to ensure that only one round of DNA replication occurs for each cell division. After replication is initiated, the initiator protein is inactivated by hydrolysis of its bound ATP molecule, and the origin of replication experiences a *refractory period.* The refractory period is caused by a delay in the methylation of newly incorporated A nucleotides in the origin (**Figure 5–27**). Initiation cannot occur again until the A's are methylated and the initiator protein is restored to its ATP-bound state, conditions that are met only when the cell is capable of carrying out a new round of DNA replication.

Eukaryotic Chromosomes Contain Multiple Origins of Replication

We have seen how two replication forks begin at a single replication origin in bacteria and proceed in opposite directions, moving away from the origin until all of the DNA in the single circular chromosome is replicated. The bacterial genome is sufficiently small for these two replication forks to duplicate the genome in about

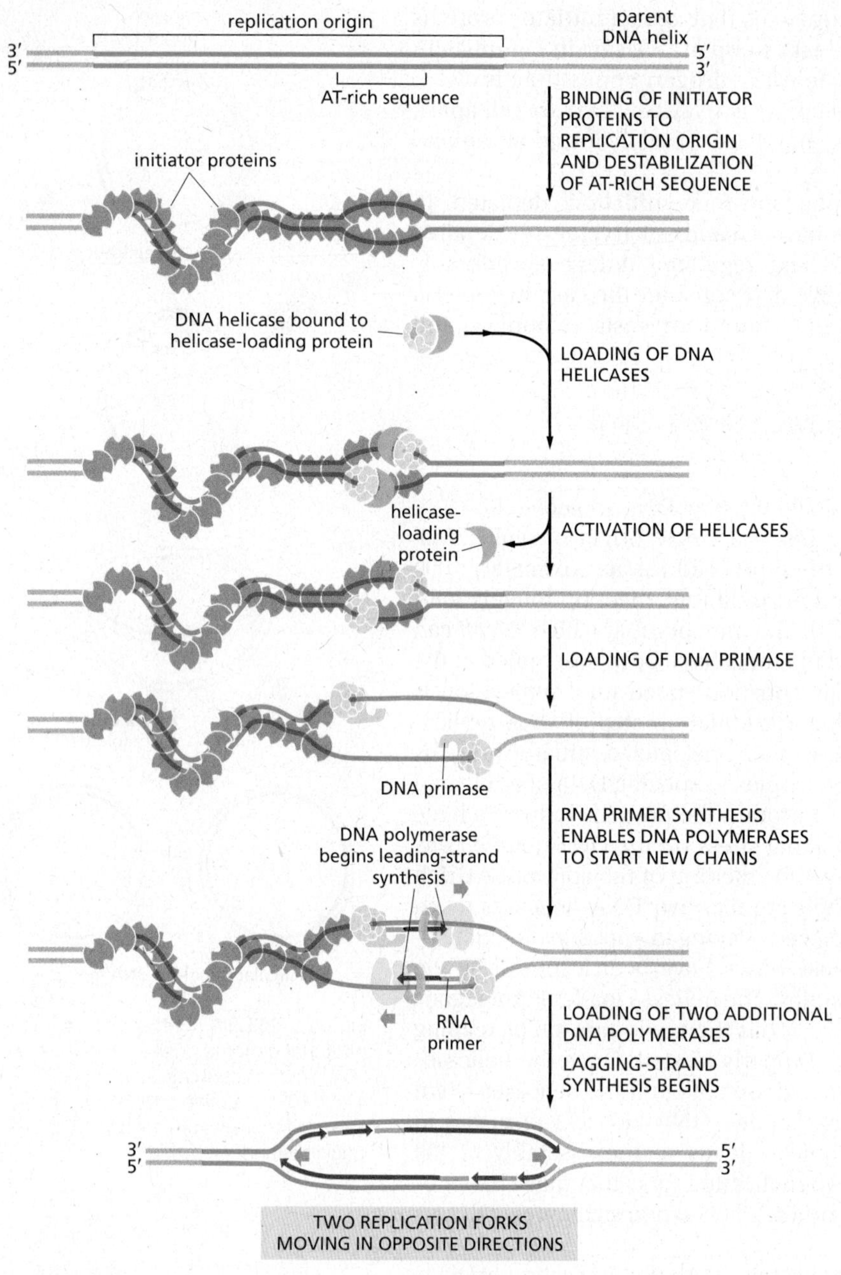

Figure 5–26 The proteins that initiate DNA replication in bacteria. The mechanism shown was established by studies *in vitro* with mixtures of highly purified proteins. For *E. coli* DNA replication, the major initiator protein *(purple)*, the helicase *(yellow)*, and the primase *(blue)* are the dnaA, dnaB, and dnaG proteins, respectively. In the first step, many molecules of the initiator protein bind to specific DNA sequences at the replication origin and destabilize the double helix by forming a filamentous structure in which the DNA is wrapped around the protein. Next, two helicases are brought in by helicase-loading proteins (the dnaC proteins; *brown*), which inhibit the helicases until they are properly loaded at the replication origin. (The helicase-loading proteins prevent the replicative DNA helices from inappropriately entering other single-strand stretches of DNA in the bacterial genome.) Aided by single-strand binding protein (not shown), the loaded helicases further separate the DNA strands, thereby enabling primases to enter and synthesize initial primers. In subsequent steps, two complete replication forks are assembled at the origin and move in opposite directions away from the replication origin. The initiator proteins are displaced as the left-hand fork moves through them.

30 minutes. Because of the much greater size of most eukaryotic chromosomes, a different strategy is required to allow their replication in a timely manner.

A method for determining the general pattern of eukaryotic chromosome replication was developed in the early 1960s that is similar to the strategy we saw earlier for visualizing bacterial replication (see Figure 5–6). Human cells growing

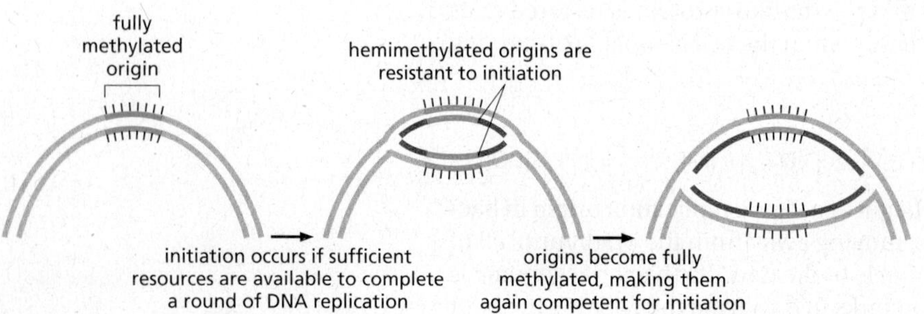

Figure 5–27 Methylation of the *E. coli* replication origin creates a refractory period for DNA initiation. DNA methylation occurs at GATC sequences, 11 of which are found in the origin of replication (spanning approximately 250 nucleotide pairs). In its hemimethylated state (that is, one strand of the DNA methylated, the other unmethylated), the origin of replication is bound by an inhibitor protein (Seq A, not shown), which blocks the ability of the initiator proteins to unwind the origin DNA. About 15 minutes after replication is initiated, the hemimethylated origins become fully methylated by a DNA methylase enzyme; Seq A then dissociates allowing the origin of replication to become active.

A single enzyme, the *Dam* methylase, is responsible for methylating all *E. coli* GATC sequences. As discussed earlier in the chapter, a lag in methylation after the replication of GATC sequences is also used by the *E. coli* mismatch proofreading system to distinguish the newly synthesized DNA strand from the parent DNA strand; in that case, the relevant GATC sequences are scattered throughout the chromosome, and they are not bound by Seq A.

in culture are labeled for a short time with ^{3}H-thymidine so that the DNA synthesized during this period becomes highly radioactive. The cells are then gently lysed, and the DNA is stretched on the surface of a glass slide coated with a photographic emulsion. Development of the emulsion in the dark reveals the pattern of labeled DNA through a technique known as *autoradiography*. The time allotted for radioactive labeling is chosen to allow each replication fork to move several micrometers along the DNA, so that the replicated DNA can be detected in the light microscope as lines of silver grains (radioactivity exposes photographic emulsion much as light does), even though the DNA molecule itself is too thin to be visible. In this way, both the rate and the direction of replication-fork movement can be determined (**Figure 5–28**). From the rate at which tracks of replicated DNA increase in length with increasing labeling time, the eukaryotic replication forks are estimated to travel at about 50 nucleotides per second. This is approximately twentyfold slower than the rate at which bacterial replication forks move, possibly reflecting the increased difficulty of replicating DNA that is packaged in chromatin.

An average-size human chromosome contains a single linear DNA molecule of about 150 million nucleotide pairs. It would take 0.02 seconds/nucleotide $\times$ 150 $\times$ 10^6 nucleotides = 3.0 $\times$ 10^6 seconds (about 35 days) to replicate such a DNA molecule from end to end with a single replication fork moving at a rate of 50 nucleotides per second. As expected, therefore, the autoradiographic experiments just described reveal that many forks, belonging to separate replication bubbles, are moving simultaneously on each eukaryotic chromosome.

Much more sophisticated methods now exist for monitoring DNA replication initiation and tracking the movement of DNA replication forks across whole genomes. If a population of cells can be synchronized so they all begin DNA replication at the same time, the amount of each segment of DNA in the genome can be determined at specific time points using one of the DNA sequencing methods described in Chapter 8. Because a segment of a genome that has been replicated will contain twice as much DNA as an unreplicated segment, replication-fork initiation and fork movement can be accurately monitored across an entire genome.

Experiments of this type have shown the following: (1) Approximately 30,000–50,000 origins of replication are used each time a human cell divides. (2) The human genome has many more (perhaps tenfold more) potential origins than this, and different cell types use different sets of origins. This excess of origins may allow a cell to coordinate its active origins with other features of its chromosomes such as which genes are being expressed. The excess origins also provide "backups" in case a primary origin fails. (3) Origins of replication do not all "fire" simultaneously; rather, they often are activated in a prescribed order in a given cell type. (4) Regardless of when a given origin fires or where on the chromosome it is located, the replication forks all move at approximately the same speed. (5) As in bacteria, replication forks are formed in pairs and create an expanding

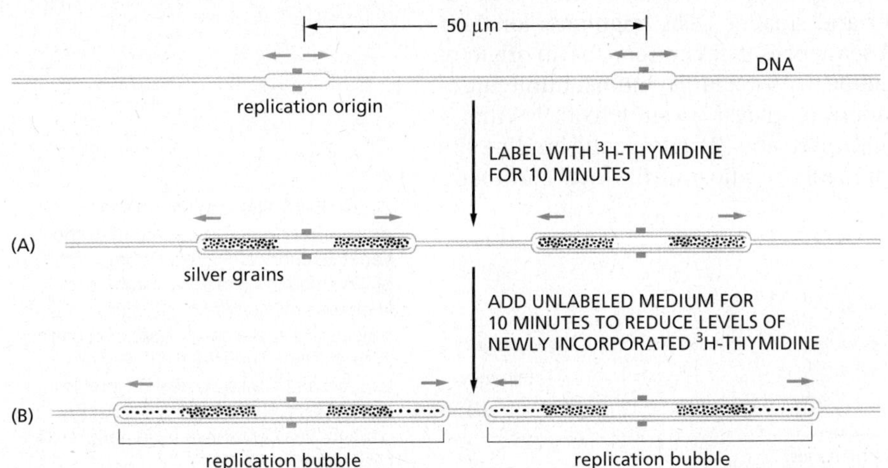

Figure 5–28 The experiments that first demonstrated the pattern in which replication forks are formed and move on eukaryotic chromosomes. The new DNA made in human cells in culture was labeled briefly with a pulse of highly radioactive thymidine (^{3}H-thymidine). (A) In this experiment, the cells were lysed, and the DNA was stretched out on a glass slide that was subsequently covered with a photographic emulsion. After several months, the emulsion was developed, revealing a line of silver grains over the radioactive DNA. The *brown* DNA in this figure is shown only to help with the interpretation of the autoradiograph; the unlabeled DNA is invisible in such experiments. (B) This experiment was the same except that a further incubation in unlabeled medium allowed additional DNA, with a lower level of radioactivity, to be replicated. The pairs of dark tracks in B were found to have silver grains tapering off in opposite directions, demonstrating bidirectional fork movement from a central replication origin where a replication bubble forms (see Figure 5–24). A replication fork is thought to stop only when it encounters a replication fork moving in the opposite direction or when it reaches the end of the chromosome; in this way, all the DNA is eventually replicated.

replication bubble as they move in opposite directions away from a common point of origin, stopping only when they meet a replication fork moving in the opposite direction or when they reach a chromosome end. In this way, many replication forks operate independently on each chromosome and yet form two complete daughter DNA helices.

In Eukaryotes, DNA Replication Takes Place During Only One Part of the Cell Cycle

When growing rapidly, bacteria replicate their DNA nearly continually. In contrast, DNA replication in most eukaryotic cells occurs only during a specific part of the cell-division cycle, called the *DNA synthesis phase*, or **S phase** (**Figure 5–29**). In a mammalian cell, the S phase typically lasts for about 8 hours; in simpler eukaryotic cells such as yeasts, the S phase can be as short as 40 minutes. By its end, each chromosome has been replicated to produce two complete copies, which remain joined together at their centromeres until the *M phase* (M for *mitosis*), which soon follows. Although different origins of replication fire at different times, all DNA replication is begun and completed during S phase. In Chapter 17, we describe the control system that runs the cell cycle, and we explain how entry into each phase of the cycle requires the cell to have successfully completed the previous phase.

In the following sections, we explore how DNA replication begins on eukaryotic chromosomes and how this event is coordinated with the cell cycle.

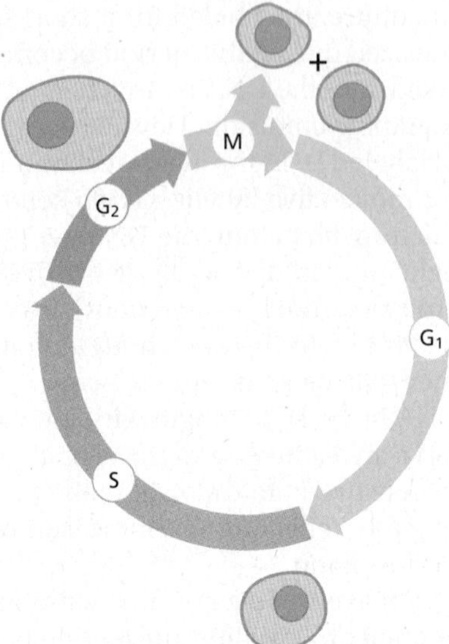

Figure 5–29 The four successive phases of a standard eukaryotic cell cycle. During the G_1, S, and G_2 phases, the cell grows continually. During M phase growth stops, the nucleus divides, and the cell divides in two. DNA replication is confined to the part of the cell cycle known as S phase. G_1 is the gap between M phase and S phase; G_2 is the gap between S phase and M phase. Many eukaryotic cells spend only a small fraction of their time in S phase.

Eukaryotic Origins of Replication Are "Licensed" for Replication by the Assembly of an Origin Recognition Complex

Having seen that a eukaryotic chromosome is replicated using many origins of replication, each of which fires at a characteristic time in S phase of the cell cycle, we turn to the nature of these origins of replication. We saw earlier in this chapter that replication origins have been precisely defined in bacteria as specific DNA sequences that attract initiator proteins, which then assemble the DNA replication machinery. We shall see that this is also the case for the single-cell budding yeast *S. cerevisiae*, but it appears not to be strictly true for many other eukaryotes.

For budding yeast, the location of every origin of replication on each chromosome has been determined. The particular chromosome shown in **Figure 5–30**—chromosome III from *S. cerevisiae*—is one of the smallest chromosomes known, with a length less than 1/100 that of a typical human chromosome. Its major origins are spaced an average of 30,000 nucleotide pairs apart, but only a subset of these origins is used by a given cell. Nonetheless, this chromosome can be replicated in about 15 minutes.

The minimal DNA sequence required for directing DNA replication initiation in *S. cerevisiae* has been determined by taking a segment of DNA that spans an origin of replication and testing smaller and smaller DNA fragments for their ability to function as origins. These DNA sequences that can serve as an origin of replication are found to contain (1) a binding site for a large, multisubunit initiator protein called ORC, for origin recognition complex; (2) a stretch of DNA that is rich in A's and T's and therefore easy to pull apart; and (3) at least one binding site for proteins that facilitate ORC binding, probably by adjusting the local chromatin structure.

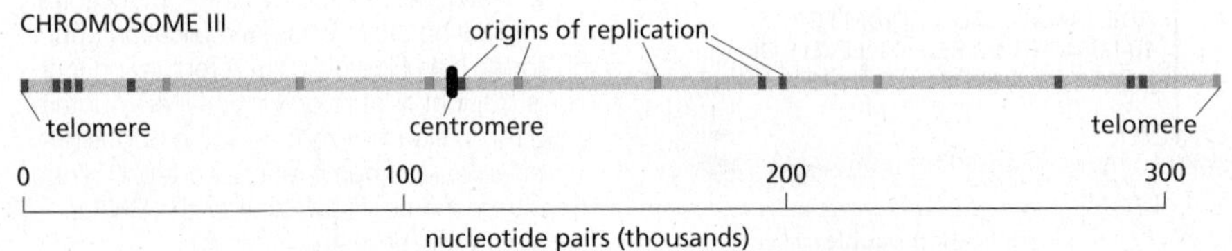

Figure 5–30 The origins of DNA replication on chromosome III of the yeast *S. cerevisiae*. This chromosome, one of the smallest eukaryotic chromosomes known, carries a total of 180 genes. As indicated, it contains 18 replication origins, although they are used with different frequencies. Those in *red* are typically used in less than 10% of cell divisions, while those in *green* are used about 90% of the time.

Features of the Human Genome That Specify Origins of Replication Remain to Be Fully Understood

Compared with the situation in budding yeast, the determinants of replication origins in humans have been difficult to discover. It has been possible to identify specific human DNA sequences, each several thousand nucleotide pairs in length, that are sufficient to serve as replication origins. These origins continue to function when moved to a different chromosomal region by recombinant DNA methods, as long as they are placed in a region where the chromatin is relatively uncondensed. However, comparisons of such DNA sequences have not revealed DNA sequences in common as in the origins of bacteria and yeasts.

Despite this, a human ORC that is very similar to the yeast ORC binds to origins of replication and initiates DNA replication in humans. Many of the other proteins that function in the initiation process in yeast likewise have central roles in humans. The yeast and human initiation mechanisms are thus similar, although some property of the genome other than a specific DNA sequence has the central role in attracting an ORC to a mammalian origin of replication. Origins of replication are often nucleosome-free, and it has been proposed that DNA that is difficult to fold onto a histone core may help define origins of replication. Nearby transcriptional activity on the genome may also play a role in activating certain origins, by altering the local chromatin structures, as we discuss in Chapter 7. This idea helps to explain why different cell types—which express different sets of genes—often use different origins. Consistent with this idea, origins that fire the earliest in S phase tend to be located near highly transcribed regions of the genome.

Finally, origins located in proximity to each other tend to fire together, and it seems likely that the three-dimensional structure of chromosomes organizes origins of replication into domains, such that all the origins in a given domain fire simultaneously. All of these influences probably work together to determine how mammalian origins of replication are selected by the cell, thereby explaining the difficulty scientists have had in precisely defining their salient features.

Properties of the ORC Ensure That Each Region of the DNA Is Replicated Once and Only Once in Each S Phase

In bacteria, once the initiator protein is properly bound to the single origin of replication, the assembly of the replication forks seems to follow more or less automatically. In eukaryotes, the situation is significantly different because of a profound problem eukaryotes have in replicating chromosomes: with so many places to begin replication, how is the process regulated to ensure that all the DNA is copied once and only once?

The answer lies in how the assembly of the replication-fork protein at the origins of replication is regulated. We discuss this process in more detail in Chapter 17, where we consider the machinery that underlies the cell-division cycle. In brief, during G_1 phase, a symmetrical complex of two incomplete helicases is loaded onto DNA by the bound ORC. Then, upon passage from G_1 phase to S phase, specialized protein kinases come into play and direct the final assembly of the two replicative helicases, positioning one on each of the two complementary DNA single strands, where they move in opposite directions to begin opening the DNA double helix. At this point, the additional replication proteins are brought to the DNA, and two complete replication forks move in opposite directions away from the origin of replication (**Figure 5–31**).

The same protein kinases that trigger the final assembly of the helicases prevent the binding of new helicases to that origin until the next M phase resets the entire cycle (for details, see pp. 1043–1045). They do this, in part, by phosphorylating ORC, rendering it unable to accept new helicases. Thus, the kinases specify a single window of opportunity for precursor helicases to be loaded at origins of replication (G_1 phase, when kinase activity is low) and a second window for

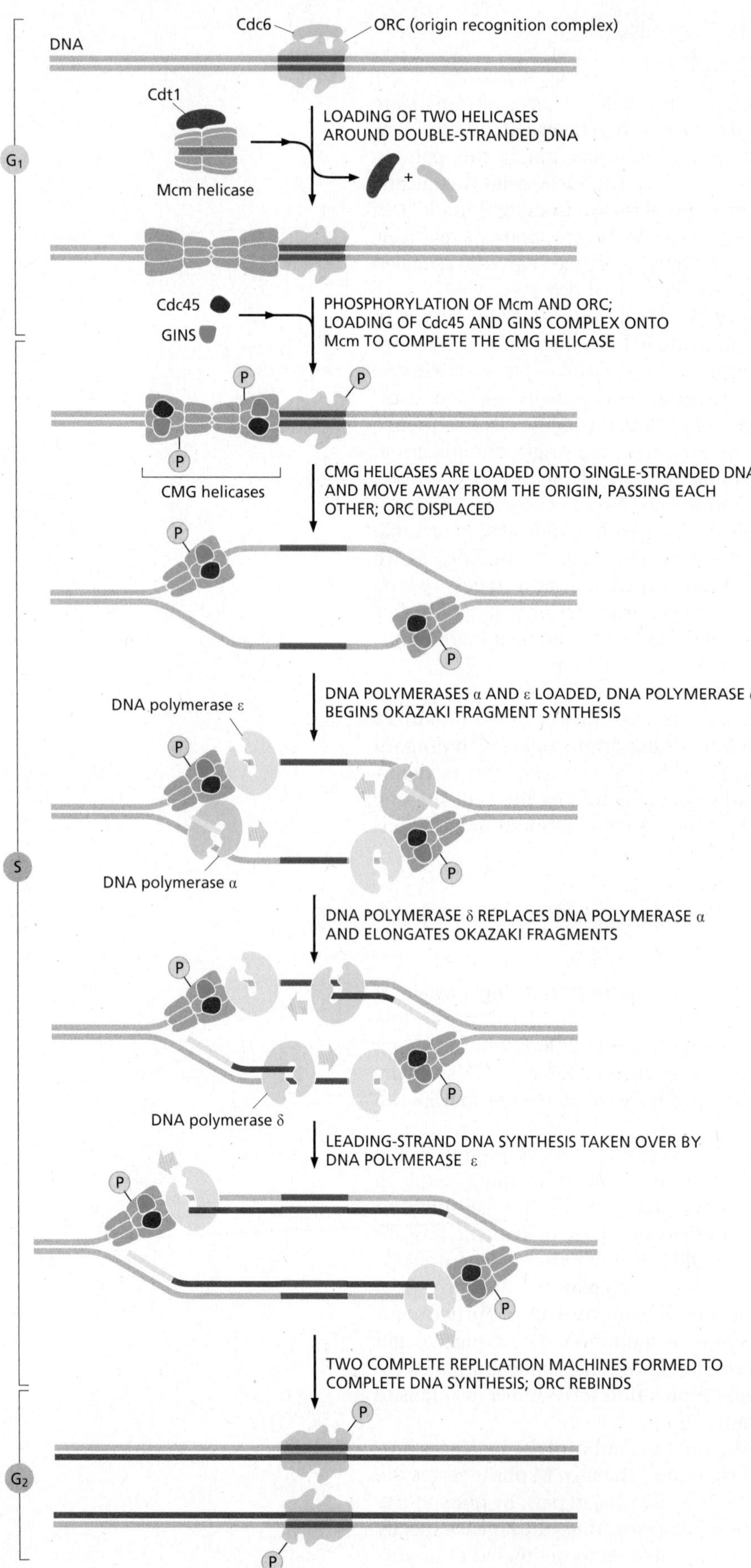

Figure 5–31 DNA replication initiation in eukaryotes. This mechanism ensures that each origin of replication is activated only once per cell cycle. An origin of replication can be used only if two Mcm helicases (which form the enzymatic cores of the replicative helicases) are loaded in G_1 phase. At the beginning of S phase, specialized kinases phosphorylate both the Mcm helicases and ORC, activating the former and inactivating the latter. These kinases also guide the assembly of additional proteins that complete the helicases to form the fully active replicative helicases, known as the CMG helicases. New Mcm helicases cannot be loaded at the origin until the cell progresses through mitosis to the next G_1 phase, when ORC is dephosphorylated. The name CMG derives from Cdc45, Mcm, and GINS, the components of the active helicase (see Figure 5–19).

them to be assembled into their active form (S phase, when kinase activity is high). Because these two phases of the cell cycle are mutually exclusive and occur in a prescribed order, each origin of replication can fire only once during each cell cycle.

Because there are many more potential replication origins on a eukaryotic chromosome than are actually used in any one cell cycle (see Figure 5-30), the DNA at many ORC-bound replication origins will be replicated by forks formed at a neighboring region of the chromosome. Thus, preventing any single origin from firing more than once during an S phase is not enough to avoid the re-replication of DNA in eukaryotes. In addition, any ORC–DNA complex that is passed by a replication fork must be inactivated, and it is the combination of the two mechanisms that guarantees that each region of the DNA is replicated once and only once in each S phase.

New Nucleosomes Are Assembled Behind the Replication Fork

Several additional aspects of DNA replication are specific to eukaryotes compared with bacteria. As discussed in Chapter 4, eukaryotic chromosomes are composed of roughly equal mixtures of DNA and protein. Chromosome duplication therefore requires not only the replication of DNA but also the synthesis of new chromosomal proteins and their assembly onto the DNA behind each replication fork. Although we are far from understanding this process in detail, we are beginning to learn how the fundamental unit of chromatin packaging, the nucleosome, is duplicated. The cell requires a large amount of new histone protein, approximately equal in mass to the newly synthesized DNA, each time it divides. For this reason, most eukaryotic organisms possess multiple copies of the gene for each histone. Vertebrate cells, for example, have about 20 repeated gene sets, most containing the genes that encode all five histones (H1, H2A, H2B, H3, and H4).

Unlike most proteins, which are made continually, histones are synthesized mainly in S phase, when the level of histone mRNA increases about fiftyfold as a result of both increased transcription and decreased mRNA degradation. The major histone mRNAs are degraded within minutes when DNA synthesis stops at the end of S phase. The mechanism depends on special properties of the 3′ ends of these mRNAs, as discussed in Chapter 7. In contrast to their mRNAs, the histone proteins themselves are remarkably stable and may survive for many generations. The tight linkage between DNA synthesis and histone synthesis appears to reflect a feedback mechanism that monitors the level of free histone to ensure that the amount of histone made exactly matches the amount of new DNA synthesized.

As a replication fork advances it must pass through the parent nucleosomes. In the cell, efficient replication requires *chromatin remodeling complexes* (discussed in Chapter 4) and *histone chaperone proteins* (discussed below) to destabilize the DNA–histone interfaces. Aided by such specialized proteins, replication forks can transit even highly condensed chromatin. As a replication fork passes through chromatin, the histones are transiently displaced leaving about 600 nucleotide pairs of "free" DNA in its wake. The reestablishment of nucleosomes behind a moving fork occurs in an intriguing way. When a nucleosome is traversed by a replication fork, the histone octamer is broken into an H3–H4 tetramer and two H2A–H2B dimers (discussed in Chapter 4), all of which are released from DNA. The H3–H4 tetramers remain in the vicinity of the fork by loosely binding to several of the proteins at the replication fork (primarily the CMG helicase) and are distributed at random to one or the other daughter duplexes as the fork moves forward. In contrast, the H2A–H2B dimers are released completely from the fork and may diffuse to entirely different chromosomes. Freshly made H3–H4 tetramers are added to the newly synthesized DNA to fill in the "spaces," and H2A–H2B dimers—half of which are old and half new—are then added at random

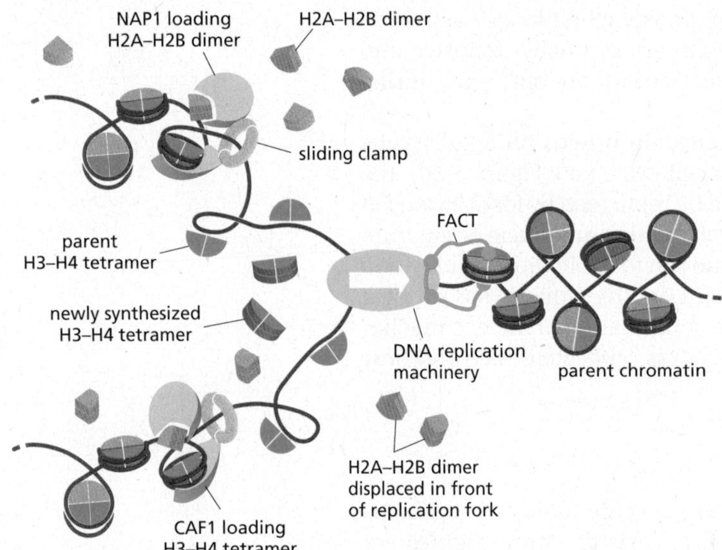

Figure 5–32 Formation of nucleosomes behind a replication fork. Parent H3–H4 tetramers remain associated with the fork and are distributed at random to the daughter DNA molecules, with roughly equal numbers inherited by each daughter. In contrast, H2A–H2B dimers are released completely from the fork as it passes. This release begins just in front of the replication fork and is facilitated by the histone chaperone FACT, which moves with the fork. FACT has several globular protein domains connected by flexible linkers and can make multiple contacts with a nucleosome to aid in its disassembly. Additional histone chaperones (NAP1 and CAF1) restore the full complement of histones to daughter molecules using both parent and newly synthesized histones. Although not shown in the figure, it has been proposed that FACT directly hands off parent H3–H4 tetramers to components of the replication machinery, which in turn hand them off to CAF1 chaperones, which deposit them evenly on the two daughter molecules. The way in which histones are distributed behind a replication fork means that some daughter nucleosomes contain only parent histones or only newly synthesized histones, but most are hybrids of old and new. For simplicity, the DNA double helix is shown as a single *red* line.

to complete the nucleosomes behind the fork (**Figure 5–32**). The formation of new nucleosomes behind a replication fork has an important consequence for the process of DNA replication itself. As DNA polymerase δ discontinuously synthesizes the lagging strand (see Figure 5–19), the length of each Okazaki fragment is determined by the point at which DNA polymerase δ is blocked by a newly formed nucleosome. This tight coupling between nucleosome duplication and DNA replication probably explains why the length of Okazaki fragments in eukaryotes (~200 nucleotides) is approximately the same as the nucleosome repeat length.

The orderly and rapid addition of new H3-H4 tetramers and H2A-H2B dimers behind a replication fork requires **histone chaperones** (also called *chromatin assembly factors*). These multisubunit complexes bind the highly basic histones and release them on DNA only in the appropriate context. For example, some of the histone chaperones, along with their histone cargoes, are directed to newly replicated DNA through a specific interaction with the sliding clamp (see Figure 5–32). As we have seen, these clamps remain on the DNA behind replication forks, and some appear to linger just long enough for the histone chaperones to complete their tasks. Because they bind so well to histones, some histone chaperones also help to disassemble nucleosomes. Of particular importance to DNA replication is the FACT chaperone, which moves at the front of the replication machinery, disassembling nucleosomes as it moves forward (see Figure 5–32).

Termination of DNA Replication Occurs Through the Ordered Disassembly of the Replication Fork

We saw earlier in this chapter that *E. coli* DNA replication begins at a single origin, and two replication forks proceed bidirectionally around the circular genome, meeting at a spot opposite to the origin of replication. Here, the two forks do not simply collide with each other running at full speed; rather, this spot on the *E. coli* genome has a special DNA sequence that slows down and stalls the movement of each fork, causing them to disassemble. The remaining gaps in the daughter DNA molecules are filled in and sealed by repair DNA polymerases and DNA ligase (see Figures 5–11 and 5–12), and the two completed bacterial genomes are separated using topoisomerases (see Figure 5–23).

As might be expected, the situation in eukaryotes is more complicated. First, each round of replication requires many termination events, roughly as many as there are initiation events at origins of replication. Thus, in mammalian cells, approximately 30,000–50,000 termination events occur in every S phase. Second, the termination of replication forks in eukaryotes is largely independent of any underlying DNA sequence in the genome. Rather, the principal termination signal is a head-on encounter with a fork moving in the opposite direction. When two forks meet, the CMG helicase at each fork is covalently modified by addition of ubiquitin (see Figure 3–65), which causes its disassembly and removal from DNA. Without the helicase, the other replication proteins rapidly dissociate from the fork. Repair DNA polymerase and DNA ligase subsequently fill in and seal any remaining gaps. Eukaryotic replication forks must also contend with the ends of chromosomes. Here, it is believed that the CMG helicase simply slides off the end of the DNA molecule, leading to the dissociation of the other fork proteins. However, replicating DNA to the very end of a chromosome presents a special challenge to the eukaryotic cell, as we describe next.

Telomerase Replicates the Ends of Chromosomes

We saw earlier in the chapter that synthesis of the lagging strand at a replication fork must occur discontinuously through a backstitching mechanism that produces short DNA fragments attached to RNA primers. The final RNA primer synthesized on the lagging-strand template cannot be replaced by DNA because there is no primer ahead of it to provide a 3′-OH end for the repair polymerase. Without a mechanism to deal with this problem, DNA would be lost from the ends of all chromosomes each time a cell divides.

Bacteria avoid this "end-replication" problem by having circular DNA molecules as chromosomes, as we have seen. Eukaryotes solve it in a different way: they have specialized nucleotide sequences at the ends of their chromosomes that are incorporated into structures called *telomeres* (discussed in Chapter 4). Telomeres contain many tandem repeats of a short sequence that is similar in organisms as diverse as protozoa, fungi, plants, and mammals. In humans, the sequence of the repeat unit is GGGTTA, and it is repeated roughly a thousand times at each telomere.

Telomere DNA sequences are recognized by sequence-specific DNA-binding proteins that attract an enzyme, called **telomerase**, that replenishes these sequences each time a cell divides. Telomerase recognizes the tip of an existing telomere DNA repeat sequence and elongates it in the 5′-to-3′ direction, using an RNA template that is a component of the enzyme itself to synthesize new DNA copies of the repeat (**Figure 5–33**). The enzymatic portion of telomerase resembles other *reverse transcriptases*, proteins that synthesize DNA using an RNA template, although, in this case, the telomerase RNA also contributes to the active site and is essential for efficient catalysis. After extension of the parent DNA strand by telomerase, replication of the lagging strand at the chromosome end can be completed by the conventional DNA polymerases, using these extensions as a template to synthesize the complementary strand (**Figure 5–34**).

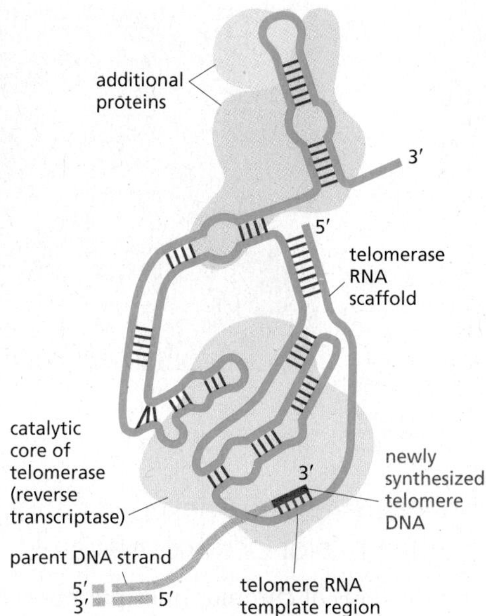

Figure 5–33 Schematic structure of human telomerase. This large enzyme is composed of 10 protein subunits and an RNA of 451 nucleotides. The RNA forms the scaffold of the complex, provides the template for synthesizing new DNA telomere repeats, and helps form the active site. The synthesis reaction itself is carried out by the reverse transcriptase domain of the protein, shown in *light green*, in conjunction with the RNA. A reverse transcriptase is a special form of polymerase enzyme that uses an RNA template to make a DNA strand; telomerase is unique in carrying its own RNA template with it. Telomerase also contains several additional protein complexes (some of which are shown in *dark green* and *blue*) that are needed to assemble the enzyme and, for many organisms but not humans, to bring it to the ends of chromosomes. (Modified from T.H.D. Nguyen et al., *Nature* 557: 190–195, 2018.)

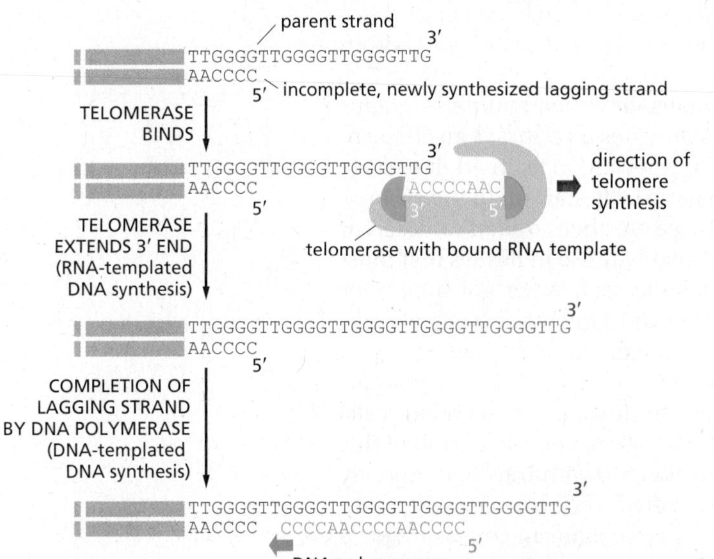

Figure 5–34 Telomere replication. Shown here is the reaction that synthesizes the repeating sequences that form the ends of the chromosomes (telomeres) of eukaryotes. The 3′ end of the parent lagging-strand template is extended by RNA-templated DNA synthesis; this allows the incomplete daughter DNA strand that is paired with it to be synthesized to the end of the chromosome. The synthesis of the final bit of lagging strand is carried out by DNA polymerase α, which carries a DNA primase as one of its subunits (**Movie 5.6**). DNA polymerase α is the same enzyme used to begin the synthesis of each Okazaki fragment on the lagging strand; it begins its synthesis with RNA (not shown) and continues with DNA *(green)*. The telomere sequence illustrated is that of the ciliate *Tetrahymena*, in which these reactions were first discovered.

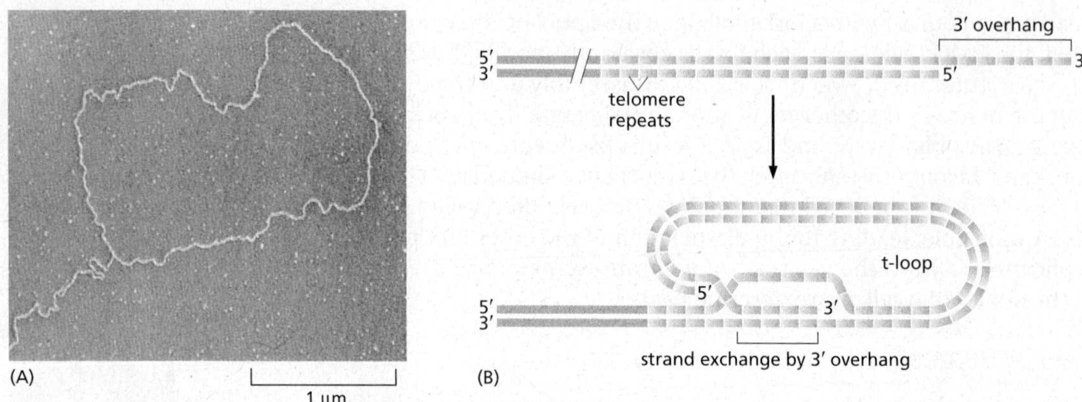

Figure 5–35 A t-loop at the end of a mammalian chromosome. (A) Electron micrograph of the DNA at the end of an interphase human chromosome. The chromosome was fixed, deproteinated, and artificially thickened before viewing. The loop seen here is approximately 15,000 nucleotide pairs in length. (B) Schematic diagram of t-loop formation. (A, from J.D. Griffith et al., *Cell* 97:503–514, 1999. With permission from Elsevier.)

Telomeres Are Packaged into Specialized Structures That Protect the Ends of Chromosomes

The ends of chromosomes present cells with an additional problem. As we will see in the next part of this chapter, when a chromosome is accidently broken into two pieces, the break is rapidly repaired. Telomeres must clearly be distinguished from these accidental breaks; otherwise, the cell will attempt to "repair" telomeres, generating chromosome fusions and other genetic abnormalities. Telomeres have several features to prevent this from happening.

A specialized nuclease chews back the 5′ end of a telomere leaving a protruding, single-strand 3′ end. This protruding end—in combination with the GGGTTA repeats in telomeres—attracts a group of proteins that form a protective chromosome cap known as *shelterin*. In particular, shelterin protects telomeres from being treated as damaged DNA. Another feature of telomeres may offer additional protection. When human telomeres are artificially cross-linked and viewed by electron microscopy, structures known as "t-loops" can be observed in which the protruding single-strand end of the telomere loops back and tucks itself into the duplex DNA of the telomere repeat sequence (**Figure 5–35**). An attractive idea is that t-loops are orchestrated by shelterin to help "hide" the very ends of chromosomes.

Telomere Length Is Regulated by Cells and Organisms

Because the processes that grow and shrink each telomere sequence are only approximately balanced, chromosome ends contain variable numbers of telomeric repeats. Not surprisingly, many cells, including stem cells and germ cells, have homeostatic mechanisms that maintain the number of these repeats within a limited range (**Figure 5–36**).

In most of the dividing somatic cells of humans, however, telomeres gradually shorten, and it has been proposed that this provides a counting mechanism that helps prevent the unlimited proliferation of wayward cells in adult tissues. In its simplest form, this idea holds that our somatic cells start off in the embryo with a full complement of telomeric repeats. These are then eroded to different extents in different cell types. Some stem cells, notably those in tissues that must be replenished at a high rate throughout life—bone marrow or gut lining, for example—retain full telomerase activity. However, in many other types of cells, the level of telomerase is reduced so that the enzyme cannot quite keep up with chromosome duplication. Such cells lose 100–200 nucleotides from each telomere every time they divide. After many cell generations, the descendant cells will inherit chromosomes that lack functioning telomeres, and, as a result of this defect, activate a DNA-damage response causing them to withdraw permanently from the cell cycle and cease dividing—a process called *replicative cell senescence* (discussed in Chapters 17 and 20). In theory, such a mechanism could provide a

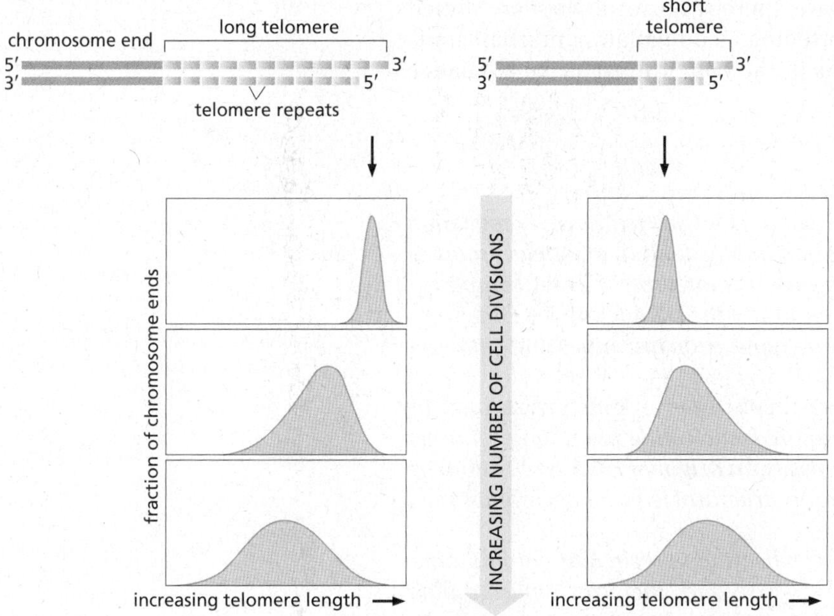

Figure 5–36 A demonstration that yeast cells control the length of their telomeres. In this experiment, the telomere at one end of a particular chromosome is artificially made either longer *(left)* or shorter *(right)* than average. After many cell divisions, the chromosome recovers, showing an average telomere length and a length distribution that is typical of the other chromosomes in the yeast cell. A similar feedback mechanism for controlling telomere length has been proposed for the germ-line cells and stem cells of mammals.

safeguard against the uncontrolled cell proliferation of abnormal cells in somatic tissues, thereby helping to protect us from cancer.

The idea that telomere length acts as a "measuring stick" to count cell divisions and thereby regulate the lifetime of the cell lineage has been tested in several ways. For certain types of human cells grown in tissue culture, the experimental results support such a theory. Human fibroblasts normally proliferate for about 60 cell divisions in culture before undergoing replicative cell senescence. Like most other somatic cells in humans, fibroblasts produce only low levels of telomerase, and their telomeres gradually shorten each time they divide. When telomerase is provided to the fibroblasts by inserting a fully active telomerase gene, telomere length is maintained and many of the cells now continue to proliferate indefinitely. Also consistent with these ideas is the observation that, in approximately 90% of cancer cells, the telomerase gene has become reactivated, thereby circumventing the normal safety mechanism (see pp. 1073–1074).

It has been proposed that this type of control on cell proliferation may contribute to the aging of animals like ourselves. These ideas have been tested by producing transgenic mice that lack telomerase entirely. The telomeres in mouse chromosomes are about five times longer than human telomeres, and the mice must therefore be bred through three or more generations before their telomeres have shrunk to the normal human length. It is therefore perhaps not surprising that the first generations of mice develop normally. However, the mice in later generations develop progressively more defects in some of their highly proliferative tissues. In addition, these mice show signs of premature aging and have a pronounced tendency to develop tumors. In these and other respects, these mice resemble humans with the genetic disease *dyskeratosis congenita*. Individuals afflicted with this disease carry one functional and one nonfunctional copy of the telomerase RNA gene; they have prematurely shortened telomeres and typically die of progressive bone marrow failure. These individuals also develop lung scarring and liver cirrhosis and show abnormalities in various epidermal structures including skin, hair follicles, and nails.

The above observations demonstrate that controlling cell proliferation by telomere shortening poses a risk to an organism, because not all of the cells that begin losing the ends of their chromosomes will stop dividing. Some apparently become genetically unstable, but continue to divide, giving rise to variant cells that can lead to cancer. As discussed above, many of these variant cells ultimately produce high levels of telomerase, thereby ensuring their continued survival. Clearly, the use of telomere shortening as a regulating mechanism is not foolproof and, like many mechanisms in the cell, it must strike a balance between benefit and risk.

Summary

The proteins that initiate DNA replication bind to DNA sequences at a replication origin to catalyze the formation of a replication bubble with two outward-moving replication forks. The process begins when an initiator protein–DNA complex is formed that subsequently loads a DNA helicase onto the DNA template. Other proteins are then added to form the multienzyme "replication machine" that catalyzes DNA synthesis at each replication fork.

In bacteria and some simple eukaryotes, replication origins are defined by specific DNA sequences that are several hundred nucleotide pairs long. In other eukaryotes, such as humans, features that specify an origin of DNA replication are less well defined, and probably depend more on structural features of chromosomes than on specific DNA sequences.

Bacteria typically have a single origin of replication in a circular chromosome. With fork speeds of up to 1000 nucleotides per second, they can replicate their genome in less than an hour. Eukaryotic DNA replication takes place in only one part of the cell cycle, the S phase. The replication fork in eukaryotes moves about 20 times more slowly than the bacterial replication fork, and the much longer eukaryotic chromosomes each require many replication origins to complete their replication in an S phase, which typically lasts for 8 hours in human cells. The different replication origins in these eukaryotic chromosomes are activated in a sequence, determined in part by which genes are currently being transcribed and the structure of chromatin across each chromosome. After the replication fork has passed, chromatin structure is re-formed by the addition of new histones to the old histones that are directly inherited by each daughter DNA molecule.

Eukaryotes solve the problem of replicating the ends of their linear chromosomes with a specialized end structure, the telomere, maintained by a special nucleotide-polymerizing enzyme called telomerase. Telomerase extends one of the DNA strands at the end of a chromosome by using an RNA template that is an integral part of the enzyme itself, producing a highly repeated DNA sequence that typically extends for thousands of nucleotide pairs at each chromosome end. Telomeres have specialized structures that distinguish them from broken ends of chromosomes, ensuring that they are not treated as damaged DNA.

DNA REPAIR

Maintaining the genetic stability that an organism needs for its survival requires not only an extremely accurate mechanism for replicating DNA but also mechanisms for repairing the many accidental lesions that DNA continually suffers. Most such spontaneous changes in DNA are temporary because they are immediately corrected by a set of processes that are collectively called **DNA repair**. Of the tens of thousands of random changes created every day in the DNA of a human cell by heat, metabolic accidents, radiation of various sorts, and exposure to substances in the environment, only a few (less than 0.02%) accumulate as permanent mutations in the DNA sequence. The rest are eliminated with remarkable efficiency by DNA repair.

The importance of DNA repair is evident from the large investment that cells make in the enzymes that carry it out: several percent of the coding capacity of most genomes is devoted solely to DNA repair functions. The importance of DNA repair is also demonstrated by the increased rate of mutation that follows the inactivation of a DNA repair gene. Many DNA repair proteins and the genes that encode them—which we now know operate in a wide range of organisms,

TABLE 5–2 Some Inherited Human Syndromes with Defects in DNA Repair

Name of syndrome or responsible genes	Phenotype	Enzyme or process affected
Msh2, *Msh3*, *Msh6*, *Mlh1*, *Pms2*	Colon cancer	Mismatch repair
Polymerase proofreading–associated polyposis	Colon cancer	Proofreading by DNA polymerase ε
Aicardi–Goutières syndrome	Encephalopathy, neurological dysfunction, genome instability	Removal of misincorporated ribonucleotides in DNA
Xeroderma pigmentosum (XP) groups A–G	Skin cancer, UV sensitivity, neurological abnormalities	Nucleotide excision repair
Cockayne syndrome	UV sensitivity, developmental abnormalities	Coupling of nucleotide excision repair to transcription
XP variant	UV sensitivity, skin cancer	Translesion synthesis by DNA polymerase η
Ataxia telangiectasia (AT)	Leukemia, lymphoma, γ-ray sensitivity, genome instability	ATM protein, a protein kinase activated by double-strand DNA breaks
Seckel syndrome	Dwarfism, microcephaly	ATR protein, a protein kinase activated by single-strand DNA breaks
Brca1	Breast and ovarian cancer	Repair by homologous recombination
Brca2	Breast, ovarian, prostate, and pancreatic cancer	Repair by homologous recombination
Ataxia-telangiectasia-like disorder (ATLD)	Leukemia, lymphoma, γ-ray sensitivity, genome instability	Mre11 protein, required for processing double-strand DNA breaks
Werner syndrome	Premature aging, cancer at several sites, genome instability	Accessory 3′-exonuclease and DNA helicase used in repair
Bloom syndrome	Cancer at several sites, stunted growth, genome instability	DNA helicase needed for recombination
Fanconi anemia groups A–W	Congenital abnormalities, leukemia, genome instability	DNA interstrand cross-link repair
46BR patient	Hypersensitivity to DNA-damaging agents, genome instability	DNA ligase I

including humans—were originally identified in bacteria by the isolation and characterization of mutants that displayed an increased mutation rate or an increased sensitivity to DNA-damaging agents.

Studies of the consequences of a diminished capacity for DNA repair in humans have linked many human diseases with decreased repair (Table 5–2). Thus, we saw previously that defects in a human gene whose product normally functions to repair the mismatched base pairs resulting from DNA replication errors can lead to an inherited predisposition to cancers of the colon and some other organs, caused by an increased mutation rate. In another human disease, *xeroderma pigmentosum* (*XP*), the afflicted individuals have an extreme sensitivity to ultraviolet radiation because they are unable to repair the damage to DNA caused by this component of sunlight. This repair defect results in an increased mutation rate that leads to serious skin lesions and a greatly increased susceptibility to skin cancers. Finally, mutations in the *Brca1* and *Brca2* genes compromise a type of DNA repair known as *homologous recombination* and are a major cause of hereditary breast and ovarian cancers.

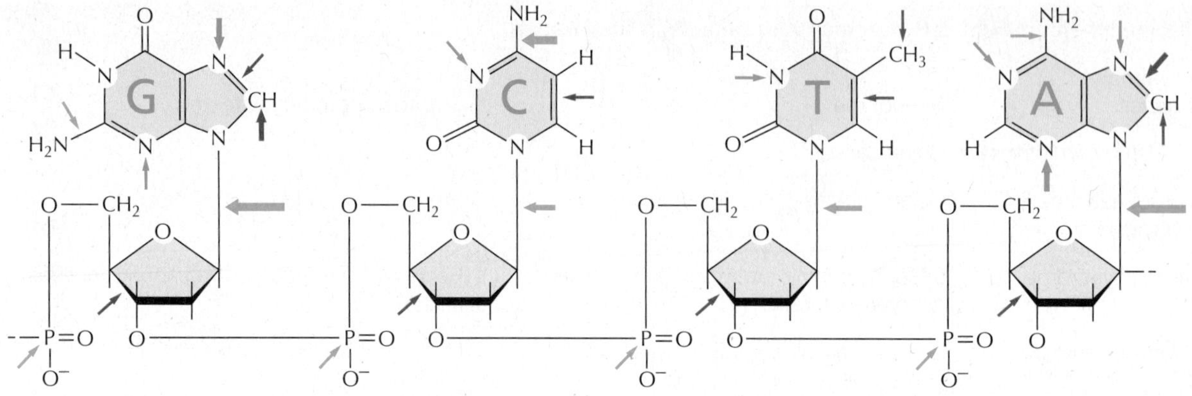

Figure 5–37 **A summary of spontaneous alterations that require DNA repair.** The sites on each nucleotide modified by spontaneous oxidative damage *(red arrows)*, hydrolytic attack *(blue arrows)*, and methylation *(green arrows)* are shown, with the width of each arrow indicating the relative frequency of each event (see Table 5–3). (After T. Lindahl, *Nature* 362:709–715, 1993.)

Without DNA Repair, Spontaneous DNA Damage Would Rapidly Change DNA Sequences

Although DNA is a highly stable material—as required for the storage of genetic information—it is a complex organic molecule that is susceptible, even under normal cell conditions, to spontaneous changes that would lead to mutations if left unrepaired (**Figure 5–37** and see **Table 5–3**). For example, the DNA of each human cell loses about 18,000 purine bases (adenine and guanine) every day because their *N*-glycosyl linkages to deoxyribose break, a spontaneous hydrolysis reaction called *depurination.* Similarly, a spontaneous *deamination* of cytosine to uracil in

TABLE 5–3 **Endogenous DNA Lesions Arising and Repaired in a Diploid Mammalian Cell in 24 Hours**

DNA lesion	Number repaired in 24 hr
Hydrolysis	
Depurination	18,000
Depyrimidination	600
Cytosine deamination	100
5-Methylcytosine deamination	10
Oxidation	
8-oxoguanine	1500
Ring-saturated pyrimidines (thymine glycol, cytosine hydrates)	2000
Lipid peroxidation products (M1G, etheno-A, etheno-C)	1000
Nonenzymatic methylation by *S*-adenosylmethionine	
7-Methylguanine	6000
3-Methyladenine	1200
Nonenzymatic methylation by nitrosated polyamines and peptides	
O^6-Methylguanine	20–100

The DNA lesions listed in the table are the result of the normal chemical reactions that take place in cells. Cells that are exposed to external chemicals and radiation suffer greater and more diverse forms of DNA damage. (From T. Lindahl and D.E. Barnes, *Cold Spring Harb. Symp. Quant. Biol.* 65:127–133, 2000.)

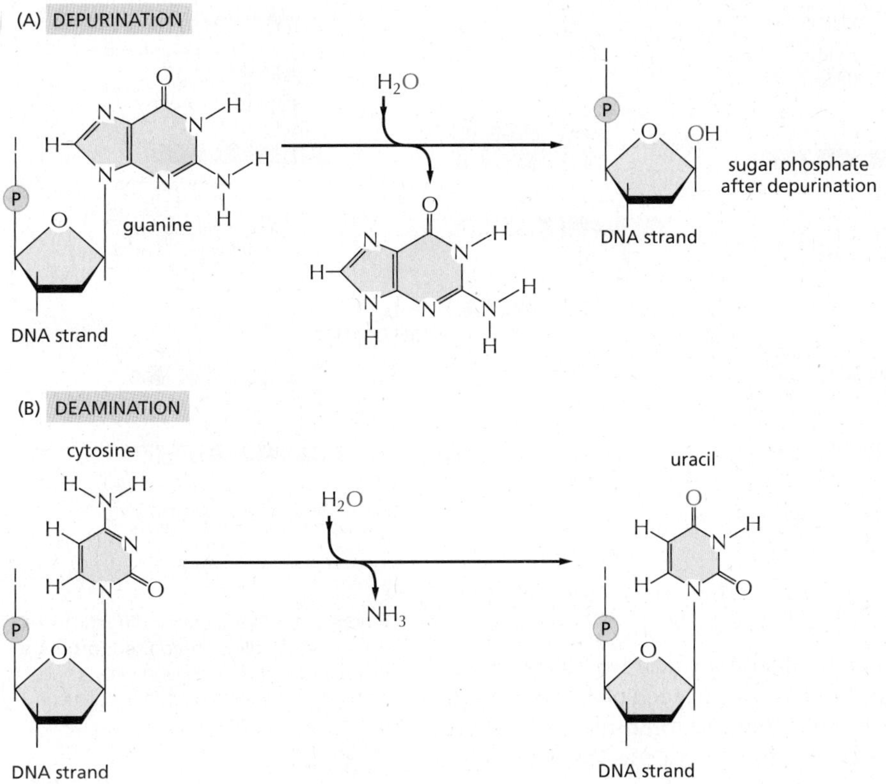

Figure 5–38 **Depurination and deamination are the most frequent spontaneous chemical reactions known to create serious DNA damage in cells.** (A) Depurination can remove guanine (or adenine) from DNA. (B) The major type of deamination reaction converts cytosine to uracil, which, as we have seen, is not normally found in DNA. However, deamination can occur on other bases as well. Both depurination and deamination take place on double-helical DNA, and neither reaction breaks the phosphodiester backbone.

DNA occurs at a rate of about 100 bases per cell per day (**Figure 5–38**). DNA bases are also occasionally damaged by encounters with reactive metabolites produced in the cell (for example, the high-energy methyl donor, *S*-adenosylmethionine) or by exposure to toxic chemicals in the environment. Likewise, ultraviolet radiation from the Sun can produce a covalent linkage between two adjacent pyrimidine bases in DNA to form, for example, thymine dimers (**Figure 5–39**). If left uncorrected, most of these changes would lead either to the deletion of one or more base pairs or to a base-pair substitution in the daughter DNA chain when the DNA is replicated (**Figure 5–40**). These mutations would then be propagated throughout all subsequent cell generations. Such a high rate of unrepaired random changes in the DNA sequence would have disastrous consequences, both in the germ line and in somatic tissues.

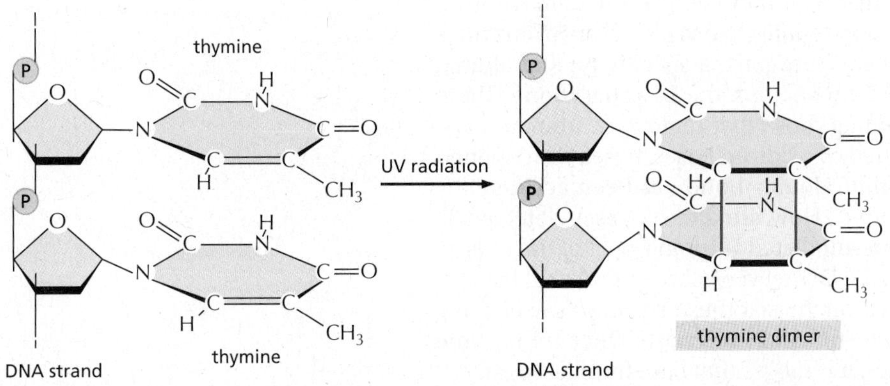

Figure 5–39 **The ultraviolet radiation in sunlight can cause the formation of thymine dimers.** Two adjacent thymine bases have become covalently attached to each other to form a thymine dimer. Skin cells that are exposed to sunlight are especially susceptible to this type of DNA damage. Dimers can also form between an adjacent thymine and cytosine.

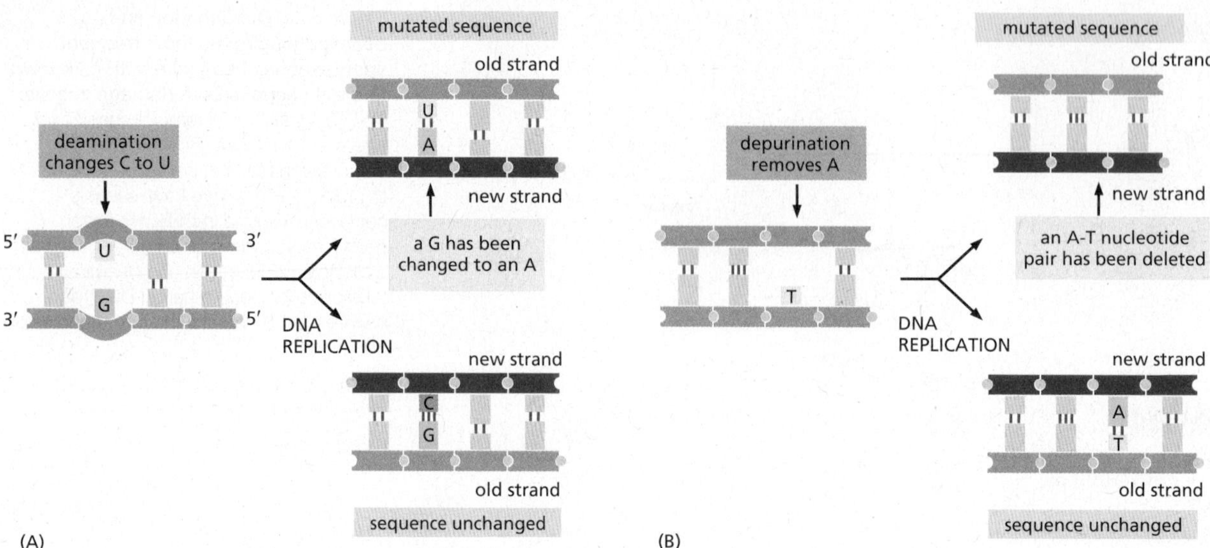

Figure 5–40 Chemical modifications of nucleotides, if left unrepaired, produce mutations. (A) Deamination of cytosine, if uncorrected, results in the substitution of one base for another when the DNA is replicated. As shown in Figure 5–43, deamination of cytosine produces uracil. Uracil differs from cytosine in its base-pairing properties and preferentially base-pairs with adenine. The DNA replication machinery therefore inserts an adenine when it encounters a uracil on the template strand. (B) Depurination, if uncorrected, can lead to the loss of a nucleotide pair. When the replication machinery encounters a missing purine on the template strand, it can skip to the next complete nucleotide, as shown, thus producing a daughter DNA molecule that is missing one nucleotide pair. In other cases, the replication machinery places an incorrect nucleotide across from the missing base, again resulting in a mutation (not shown).

The DNA Double Helix Is Readily Repaired

The double-helical structure of DNA is ideally suited for repair because it carries two separate copies of all the genetic information—one in each of its two strands. Thus, when one strand is damaged, the complementary strand retains an intact copy of the same information, and this copy is generally used as the template to restore the correct nucleotide sequences to the damaged strand.

An indication of the importance of a double-strand helix to the safe storage of genetic information is that all cells use it; only a few small viruses use single-stranded DNA or RNA as their genetic material. The types of repair processes described in this part of the chapter cannot operate on such nucleic acids, and once damaged, the chance of a permanent nucleotide change occurring in these single-strand genomes of viruses is thus very high. It seems that only tiny genomes (and therefore tiny targets for DNA damage) can have their genetic information successfully carried in any molecule other than a DNA double helix.

DNA Damage Can Be Removed by More Than One Pathway

Cells have multiple pathways to repair their DNA using different enzymes that act upon different kinds of lesions. Figure 5–41 shows two of the most common pathways. In both, the damage is excised, the original DNA sequence is restored by a high-fidelity DNA polymerase using the undamaged strand as its template, and the remaining break in the double helix is sealed by DNA ligase (see Figure 5–12).

The two pathways differ in the way in which they remove the damage from DNA. The first pathway, called **base excision repair**, involves a battery of enzymes called *DNA glycosylases*, each of which can recognize a specific type of altered base in DNA and catalyze its hydrolytic removal from the DNA backbone. There are many types of these enzymes, including those that remove deaminated C's, deaminated A's, different types of alkylated or oxidized bases, bases with opened rings, and bases in which a carbon–carbon double bond has been accidentally converted to a carbon–carbon single bond. How are altered bases detected in the double helix? A key step is an enzyme-mediated "flipping-out" of the altered nucleotide from the helix, which allows the DNA glycosylase to probe all faces of the base for damage (Figure 5–42). It is thought that these enzymes travel along DNA using base-flipping to evaluate the status of each base. Once an enzyme finds the damaged base that it recognizes, it removes that base from its sugar.

The "missing tooth" created by DNA glycosylase action is recognized by an enzyme called *AP endonuclease* (AP for *apurinic* or *apyrimidinic,* and *endo* to signify that the nuclease cleaves within the polynucleotide chain), which cuts the phosphodiester backbone, after which the resulting gap is repaired (see

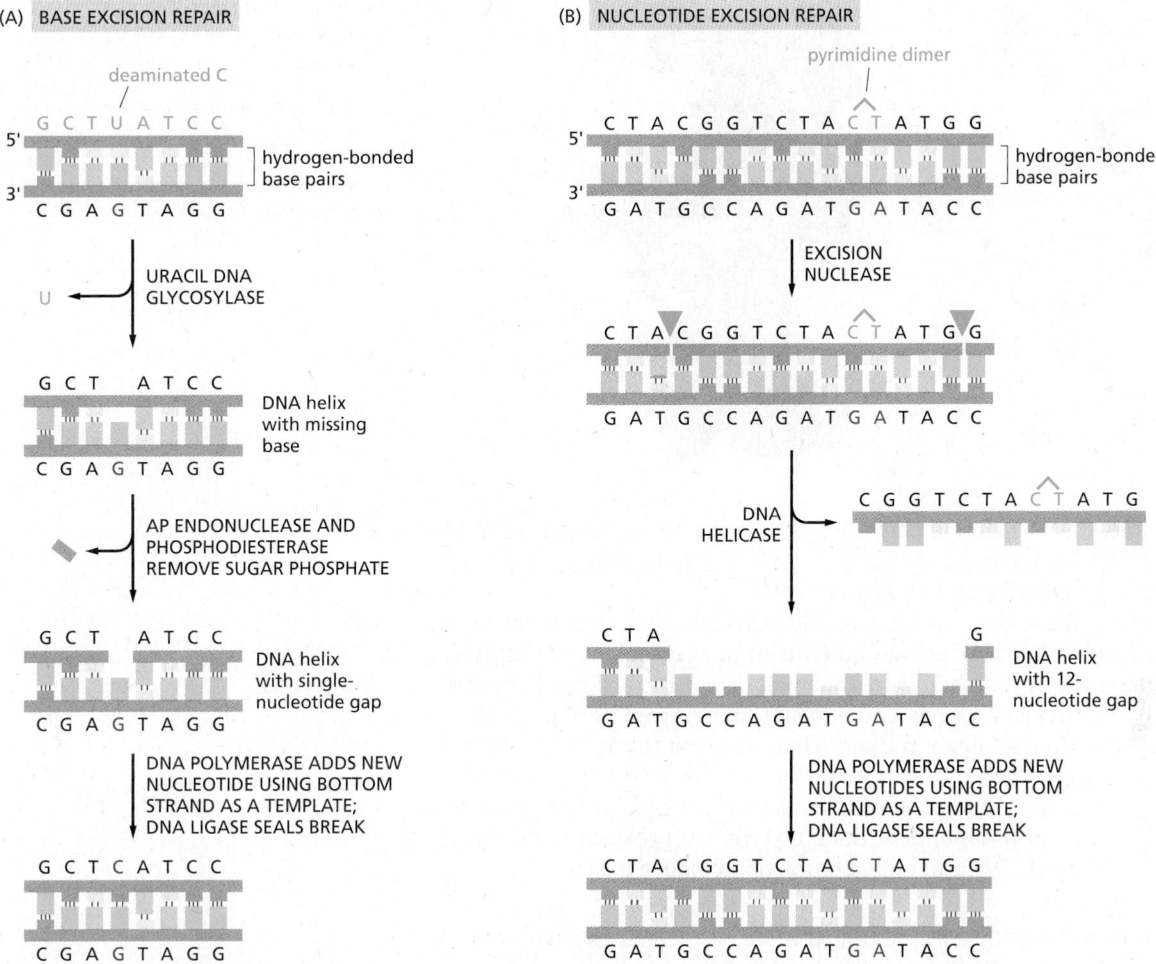

Figure 5–41 A comparison of two major DNA repair pathways. (A) *Base excision repair*. This pathway starts with a DNA glycosylase. In the example shown here, the enzyme uracil DNA glycosylase removes an accidentally deaminated cytosine in DNA. After the action of this glycosylase (or another DNA glycosylase that recognizes a different kind of damage), the sugar phosphate with the missing base is cut out by the sequential action of AP endonuclease and a phosphodiesterase. The gap of a single nucleotide is then filled by DNA polymerase and DNA ligase. The net result is that the U that was created by accidental deamination is restored to a C. The loss of a base can occur either from the actions of DNA glycosylases that recognize damaged bases or from spontaneous chemical reactions (see Figure 5–37). AP endonuclease is so named because it recognizes any site in the DNA helix that contains a deoxyribose sugar with a missing base; such sites can arise either by the loss of a purine (apurinic sites) or by the loss of a pyrimidine (apyrimidinic sites). (B) *Nucleotide excision repair*. In bacteria, after a multienzyme complex has recognized a lesion such as a pyrimidine dimer (see Figure 5–39), one cut is made on each side of the lesion, and an associated DNA helicase then removes the entire portion of the damaged strand. The excision repair machinery in bacteria operates as shown. In humans, once the damaged DNA is recognized, a helicase is recruited to locally unwind the DNA duplex. Next, the excision nuclease enters and cleaves on either side of the damage, leaving a gap of about 30 nucleotides that is subsequently filled in. The nucleotide excision repair machinery in both bacteria and humans can recognize and repair many different types of DNA damage.

Figure 5–41A). Depurination, which is by far the most frequent type of damage suffered by DNA, also leaves a deoxyribose sugar with a missing base. Depurinations are directly repaired beginning with AP endonuclease, following the bottom half of the pathway in Figure 5–41A.

The second major repair pathway is called **nucleotide excision repair**. This mechanism can repair the damage caused by almost any large change in the structure of the DNA double helix. Such "bulky lesions" include those created by the covalent reaction of DNA bases with large hydrocarbons (such as the carcinogen benzopyrene, found in tobacco smoke, coal tar, and diesel exhaust), as well as the various pyrimidine dimers (T-T, T-C, and C-C) caused by sunlight. In this pathway, a large multienzyme complex scans the DNA for a distortion in the double helix, rather than for a specific base change. Once it finds a lesion, it cleaves the phosphodiester backbone of the abnormal strand on both sides of

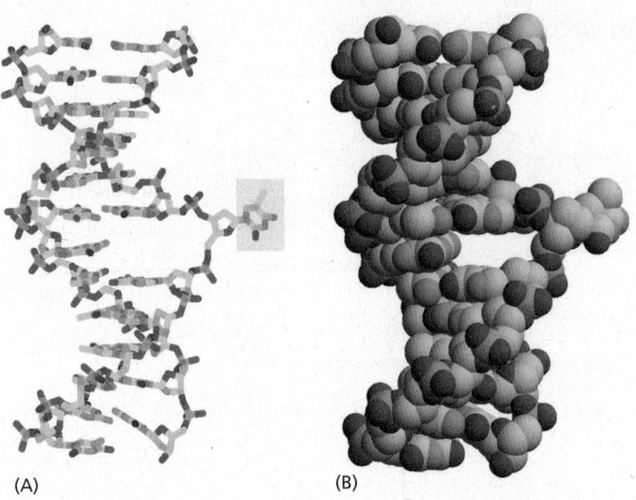

Figure 5–42 **The recognition of an unusual nucleotide in DNA by base-flipping.** The DNA glycosylase family of enzymes recognizes inappropriate bases in DNA in the conformation shown. Each of these enzymes cleaves the glycosyl bond that connects a particular recognized base *(yellow)* to the backbone sugar, removing it from the DNA. (A) Stick model of the DNA; (B) space-filling model.

the distortion, and a DNA helicase peels away the single-strand oligonucleotide containing the lesion. The large gap produced in the DNA helix is then repaired by DNA polymerase and DNA ligase (see Figure 5–41B).

An alternative to these base and nucleotide excision repair processes is the direct chemical reversal of DNA damage, and this strategy is selectively employed for the rapid removal of certain highly mutagenic or cytotoxic lesions. For example, the lesion O^6-methylguanine has its methyl group removed by direct transfer to a cysteine residue in the repair protein itself. Because the repair protein is destroyed in the process, each molecule of it can only be used once. In another example, methyl groups in the lesions 1-methyladenine and 3-methylcytosine are "burned off" by an iron-dependent demethylase, with release of formaldehyde from the methylated DNA and regeneration of the native base.

Coupling Nucleotide Excision Repair to Transcription Ensures That the Cell's Most Important DNA Is Efficiently Repaired

All of a cell's DNA is under constant surveillance for damage, and the repair mechanisms we have described act on all parts of the genome. However, cells have a way of directing DNA repair to the DNA sequences that are most needed. They do this by linking RNA polymerase, the enzyme that transcribes DNA into RNA as the first step in gene expression, to the nucleotide excision repair pathway. As discussed above, this repair system can correct many different types of DNA damage. RNA polymerase stalls at DNA lesions and, through the use of coupling proteins, directs the excision repair machinery to those sites, thereby selectively repairing genes that are in current use by the cell. In bacteria, where genes are relatively short, the stalled RNA polymerase can be dissociated from the DNA; the DNA is repaired, and the gene is transcribed again from the beginning. In eukaryotes, where genes can be enormously long, a more complex reaction is used to "back up" the RNA polymerase, repair the damage, and then restart the polymerase.

The importance of transcription-coupled excision repair is seen in people with Cockayne syndrome, which is caused by a defect in this coupling. These individuals suffer from growth retardation, skeletal abnormalities, progressive neural retardation, and severe sensitivity to sunlight. Most of these problems are thought to arise from RNA polymerase molecules that become permanently stalled at sites of DNA damage that lie in important genes.

The Chemistry of the DNA Bases Facilitates Damage Detection

The DNA double helix is well suited for repair. As noted earlier, it contains a backup copy of all genetic information. Equally importantly, the nature of the four bases in DNA makes the distinction between undamaged and damaged

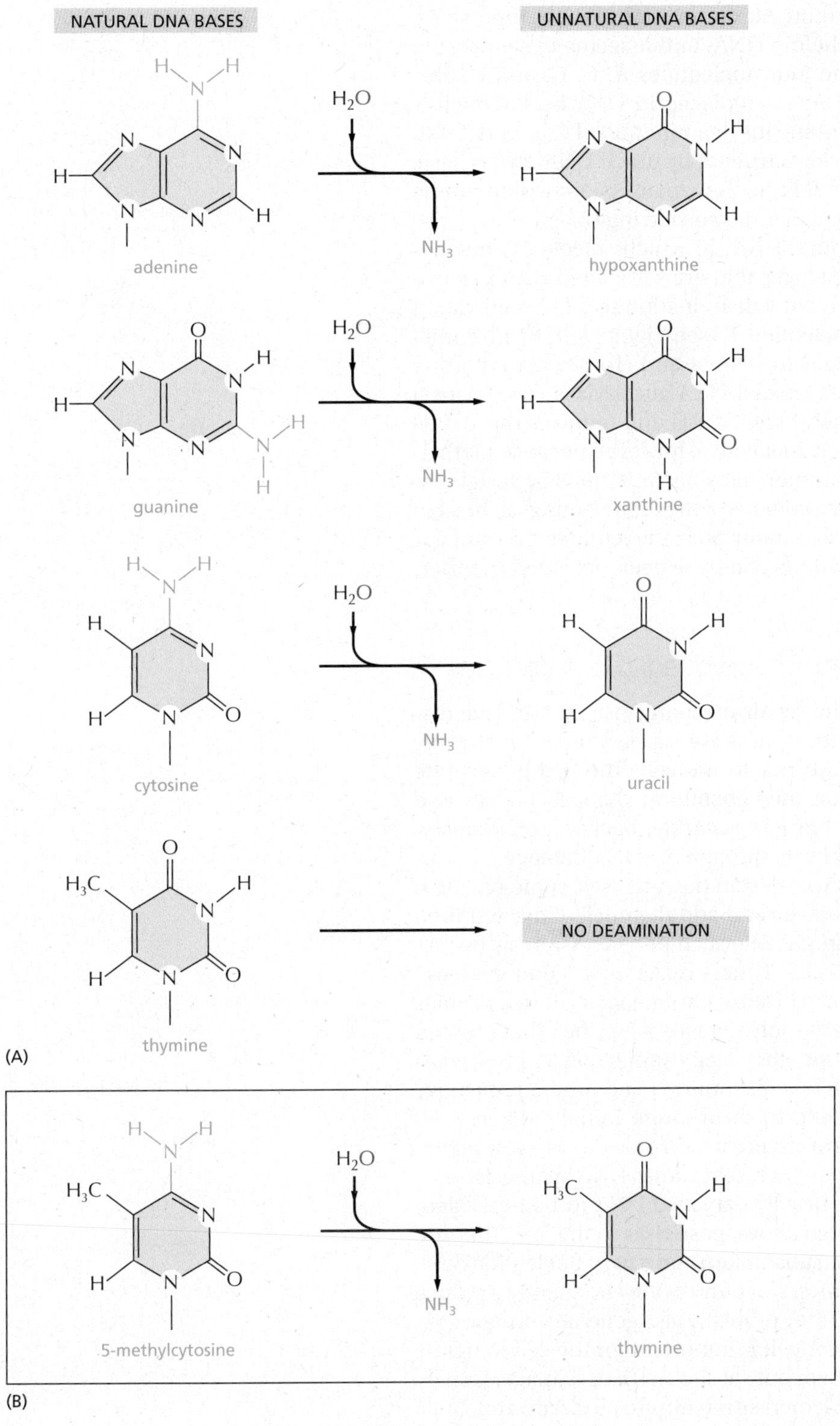

Figure 5–43 The deamination of DNA nucleotides. In each case, the oxygen atom that is added in this reaction with water is colored *red*. (A) The spontaneous deamination products of A and G are recognizable as unnatural when they occur in DNA and thus are readily found and repaired, as is the deamination of C to U; T has no amino group to remove. (B) About 3% of the C nucleotides in vertebrate DNAs are methylated to help in controlling gene expression (discussed in Chapter 7). When these 5-methyl C nucleotides are accidentally deaminated, they form the natural nucleotide T. This T will be paired with a G on the opposite strand, forming a mismatched base pair.

bases very clear. For example, every possible deamination event in DNA yields an "unnatural" base, which can be directly recognized and removed by a specific DNA glycosylase. Hypoxanthine, for example, is the simplest purine base capable of pairing specifically with C. But hypoxanthine is not used in DNA, presumably because it is the direct deamination product of A. Instead G, with a second amino group, pairs with C: G cannot form from A by spontaneous deamination, and its own deamination product (xanthine) is likewise unique (**Figure 5–43**).

As discussed in Chapter 6, RNA is thought, on an evolutionary time scale, to have served as the genetic material before DNA, and it seems likely that the genetic code was initially carried in the four nucleotides A, C, G, and U. This raises the question of why the U in RNA was replaced in DNA by T (which is 5-methyl U). We have seen that the spontaneous deamination of C converts it to U, but that this event is rendered relatively harmless by uracil DNA glycosylase. However, if DNA contained U as a natural base, the repair system would not be able to distinguish a deaminated C from a naturally occurring U.

A special situation occurs in vertebrate DNA, in which selected C nucleotides are methylated at specific CG sequences that are associated with inactive genes (discussed in Chapter 7). The accidental deamination of these methylated C nucleotides produces the natural nucleotide T (see Figure 5–43B) in a mismatched base pair with a G on the opposite DNA strand. To help in repairing deaminated methylated C nucleotides, a special DNA glycosylase recognizes a mismatched base pair involving T in the sequence T-G and removes the T. This DNA repair mechanism must be relatively ineffective, however, because methylated C nucleotides are exceptionally common sites for mutations in vertebrate DNA. It is striking that, even though only about 3% of the C nucleotides in human DNA are methylated, mutations in these methylated nucleotides account for about one-third of the single-base mutations that have been observed in inherited human diseases.

Special Translesion DNA Polymerases Are Used in Emergencies

If a cell's DNA suffers heavy damage, the repair mechanisms that we have discussed are often insufficient to cope with it. In these cases, a different strategy is called into play, one that entails some risk to the cell. The highly accurate replicative DNA polymerases stall when they encounter damaged DNA, and in emergencies cells employ versatile, but less accurate, backup polymerases, known as *translesion polymerases*, to replicate through the DNA damage.

Human cells contain seven different translesion polymerases, some of which can recognize a specific type of DNA damage and add the nucleotides required to restore the correct sequence. For example, one such polymerase adds two A's opposite a thymine dimer (see Figure 5–39). Others make only "good guesses," especially when the template base has been extensively damaged. These enzymes are not as accurate as the normal replicative polymerases even when they copy an undamaged DNA sequence. For one thing, they lack exonucleolytic proofreading activity; in addition, many are much less discriminating than the replicative polymerase in choosing which nucleotide to incorporate initially. Each such translesion polymerase is therefore given a chance to add only one or a few nucleotides before a high-fidelity replicative polymerase resumes DNA synthesis.

Despite their usefulness in allowing heavily damaged DNA to be replicated, these translesion polymerases do, as noted above, pose risks to the cell. They are probably responsible for most of the base-substitution and single-nucleotide deletion mutations that accumulate in genomes. Not only do they frequently produce mutations when copying damaged DNA, they probably also generate mutations—at a low level—on undamaged DNA. Clearly, it is important for the cell to tightly regulate these polymerases, activating them only at sites of DNA damage. Exactly how this happens for each translesion polymerase remains to be discovered, but a conceptual model is presented in **Figure 5–44**. The same principle applies to many of the DNA repair processes discussed in this chapter: because the enzymes that carry out these reactions are potentially dangerous to the genome, they must be brought into play only at the appropriate damaged sites.

Double-Strand Breaks Are Efficiently Repaired

An especially dangerous type of DNA damage occurs when both strands of the double helix are broken, leaving no intact template strand to enable accurate repair. Ionizing radiation, replication errors, oxidizing agents, and other

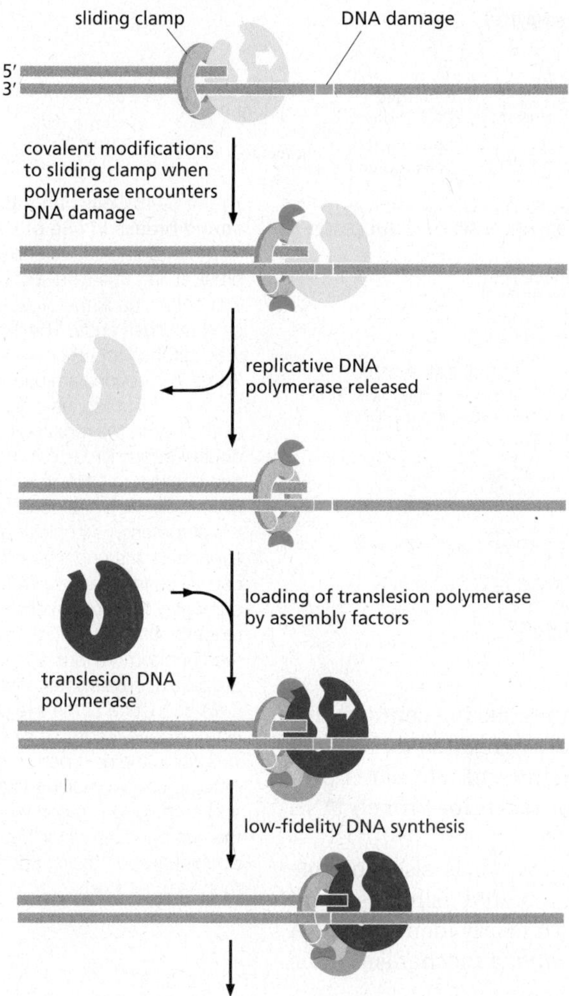

Figure 5–44 How translesion DNA polymerases are recruited to damaged templates. According to this model, a replicative polymerase stalled at a site of DNA damage is recognized by the cell as needing rescue. Specialized enzymes covalently modify the sliding clamp (typically, it is ubiquitylated—see Figure 3–65), which releases the replicative DNA polymerase and, together with the damaged DNA, attracts a translesion polymerase specific to that type of damage. Once the damaged DNA is bypassed, the covalent modification of the clamp is removed, the translesion polymerase dissociates, and the high-fidelity replicative polymerase is brought back into play.

metabolites produced in the cell cause breaks of this type. If these lesions were left unrepaired, they would quickly lead to the breakdown of chromosomes into smaller fragments and to loss of genes when the cell divides. However, two distinct mechanisms have evolved to deal with this type of damage by restoring an intact double helix: *nonhomologous end joining* and *homologous recombination* (**Figure 5–45**).

The simplest to understand is **nonhomologous end joining**, in which the broken ends are processed to remove any damaged nucleotides and simply brought together and rejoined by DNA ligation, generally with the loss of nucleotides at the site of joining (**Figure 5–46**). This end-joining mechanism, which can be seen as a "quick and dirty" solution to the repair of double-strand breaks, is the predominant way of repairing these lesions in mammalian somatic cells. Although a change in the DNA sequence (a mutation) usually results at the site of breakage, so little of the mammalian genome is essential for life that this mechanism is apparently an acceptable solution to the problem of rejoining broken chromosomes. By the time a human reaches the age of 70, the typical somatic cell contains more than 2000 such "scars," distributed throughout its genome, representing places where DNA has been inaccurately repaired by nonhomologous end joining.

But nonhomologous end joining presents another danger: nonhomologous end joining can occasionally generate rearrangements in which one broken chromosome becomes covalently attached to another. This can result

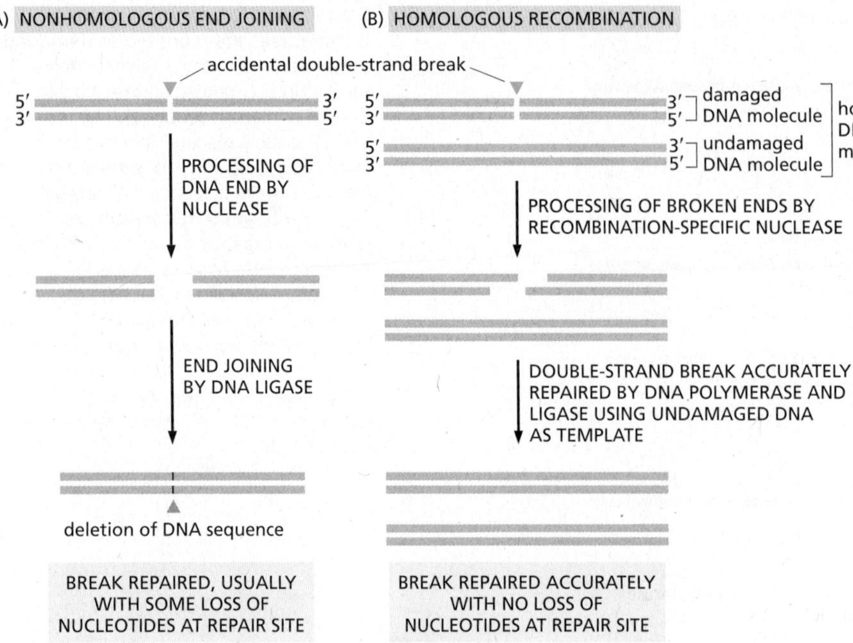

Figure 5–45 Cells can repair double-strand breaks in one of two ways. (A) In nonhomologous end joining, the break is first "cleaned" by a nuclease that chews back the broken ends to produce flush ends. The flush ends are then stitched together by a DNA ligase. Some nucleotides are usually lost in the repair process, as indicated by the *black lines* in the repaired DNA. (B) If a double-strand break occurs in one of two duplicated DNA double helices after DNA replication has occurred, but before the chromosome copies have been separated, the undamaged double helix can be used as a template to repair the damaged double helix through homologous recombination. Although more complicated than nonhomologous end joining, this process accurately restores the original DNA sequence at the site of the break. Homologous recombination is described in detail in the next part of this chapter. Although nonhomologous end joining and homologous recombination are the two principal ways that cells repair double-strand breaks, additional mechanisms exist.

in chromosomes with two centromeres and chromosomes lacking centromeres altogether; both types of aberrant chromosomes are missegregated during cell division. As previously discussed, the specialized structure of telomeres prevents the natural ends of chromosomes from being mistaken for broken DNA and "repaired" in this way.

A much more accurate type of double-strand break repair is also possible (see Figure 5–45B). Here, a damaged DNA molecule is repaired using a second DNA double helix as a template, one with an identical (or nearly identical) DNA sequence. This reaction utilizes *homologous recombination,* a mechanism to be

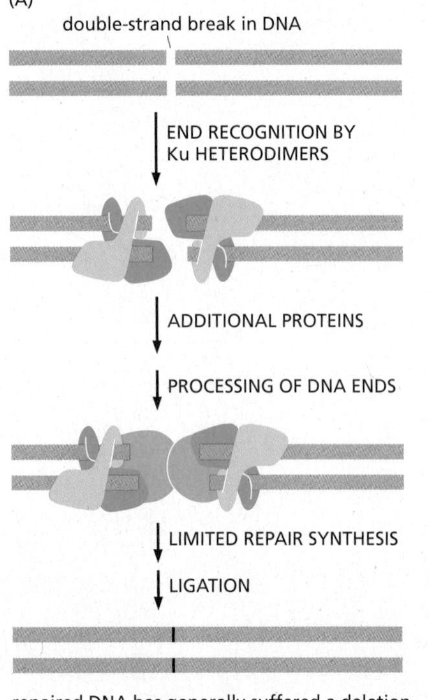

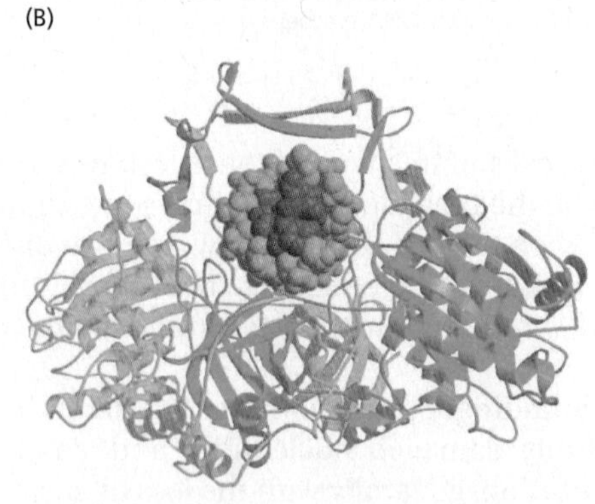

Figure 5–46 Nonhomologous end joining. (A) A central role is played by the Ku protein, a heterodimer that quickly grasps the broken chromosome ends. The additional proteins (shown in *blue*) are recruited to hold the broken ends together and remove any damaged nucleotides before the two DNA molecules are joined covalently by a specialized ligase that is dedicated to nonhomologous end joining. During this process, any single-strand gaps that arise are "filled in" by specialized repair polymerases. When DNA suffers double-strand breaks through ionizing radiation or chemical attack, the broken ends are often chemically damaged. Nonhomologous end joining is unusually versatile in being able to "clean up" just about any type of damaged end. (B) Three-dimensional structure of a Ku heterodimer bound to the end of a duplex DNA fragment. This Ku protein is also essential for V(D)J joining, a specific process through which antibody and T-cell receptor diversity is generated in developing B and T cells (discussed in Chapter 24). V(D)J joining and nonhomologous end joining share many mechanistic similarities, but the former relies on specific double-strand breaks that are produced deliberately by the cell. (From J. Walker, R. Corpina, and J. Goldberg, *Nature* 412:607–614, 2001. With permission from Springer Nature; PDB codes: 1JEQ, 1JEY.)

described later in this chapter. Most organisms employ both nonhomologous end joining and homologous recombination to repair double-strand breaks in DNA. Nonhomologous end joining predominates in humans; homologous recombination is used only in the S and G_2 cell-cycle phases, when one newly replicated daughter molecule can act as a template to repair damage to the other daughter that remains nearby.

DNA Damage Delays Progression of the Cell Cycle

We have just seen that cells contain multiple enzyme systems that can recognize and repair many types of DNA damage (**Movie 5.7**). Because of the importance of maintaining intact, undamaged DNA from generation to generation, eukaryotic cells delay the progression of their cell cycle until DNA repair is complete. As discussed in detail in Chapter 17, the orderly progression of the cell cycle is stopped when damaged DNA is detected, and it restarts only when the damage has been repaired. In mammalian cells, the presence of DNA damage can block entry from G_1 phase into S phase, it can slow S phase once it has begun, and it can block the transition from G_2 phase to M phase. These delays facilitate DNA repair by providing the time needed for the repair to reach completion.

DNA damage also results in an increased synthesis of many DNA repair enzymes. This response depends on special signaling proteins that sense DNA damage and synthesize more of the DNA repair enzymes appropriate for the damage. The importance of this mechanism is revealed by the phenotype of humans who are born with defects in the gene that encodes the *ATM protein*. These individuals have the disease *ataxia telangiectasia* (*AT*), the symptoms of which include neurodegeneration, a predisposition to cancer, and genome instability. The ATM protein is a large protein kinase that generates the intracellular signals needed to halt the cell cycle in response to many types of spontaneous DNA damage (see Figure 17–60), and individuals with defects in this protein suffer from the effects of unrepaired DNA lesions.

Summary

Genetic information can be stored stably in DNA sequences only because a large set of DNA repair enzymes continually scans the DNA double helix and replaces any damaged nucleotides. Most types of DNA repair depend on the fact that a DNA molecule carries two copies of its genetic information—one copy on each of its two complementary strands. This allows an accidental lesion on one strand to be removed by a repair enzyme and a corrected strand then resynthesized by reference to the information in the undamaged strand.

Most of the damage to DNA bases is excised by one of two major DNA repair pathways. In base excision repair, the altered base is removed by a DNA glycosylase enzyme, followed by excision of the resulting sugar phosphate. In nucleotide excision repair, a small section of the DNA strand surrounding the damage is removed from the DNA double helix. In both cases, the gap left in the DNA helix is filled in by the sequential action of DNA polymerase and DNA ligase, using the undamaged DNA strand as the template. Some types of DNA damage can be repaired by a different strategy—the direct chemical reversal of the damage—which is carried out by specialized repair proteins. Usually, all such corrections are completed prior to DNA replication. But if not, a special class of inaccurate DNA polymerases, called translesion polymerases, is used to bypass the damage, allowing the cell to survive but sometimes creating permanent mutations at the sites of damage.

Other critical repair systems—based on either nonhomologous end joining or homologous recombination—are needed to reseal the accidental double-strand breaks that occasionally occur in the DNA helix. In most cells, an elevated level of DNA damage causes a delay in the cell cycle, which helps to ensure that the damage is repaired before the cell divides.

HOMOLOGOUS RECOMBINATION

In the preceding parts of this chapter, we discussed the mechanisms that allow the DNA sequences in cells to be maintained from generation to generation with very little change. In this part, we further explore a group of repair mechanisms that depend on a process called *homologous recombination*. The key feature of **homologous recombination** (also known as *general recombination*) is an exchange of DNA strands between a pair of homologous duplex DNA sequences. Such a strand exchange between two regions of double helix that are very similar or identical in nucleotide sequence allows one stretch of duplex DNA to restore lost or damaged information on a second stretch of duplex DNA. Because the DNA sequence information that is used to correct the damage can come from a separate DNA molecule, homologous recombination can repair many types of DNA damage. It makes possible, for example, the accurate repair of double-strand breaks, as mentioned previously (see Figure 5–45B). As pointed out earlier, these double-strand breaks can result from reactive chemicals or radiation (for example, that from radon gas that accumulates in some old basements). But more frequently they arise from DNA replication accidents—when forks become stalled or broken independently of any such external cause. Homologous recombination accurately corrects these accidents, and, because they occur during nearly every round of DNA replication, this repair pathway is essential for every proliferating cell. Homologous recombination can also repair other types of DNA damage (for example, covalent cross-links between the two strands of a DNA double helix), being perhaps the most versatile DNA repair mechanism available to the cell; this probably explains why its mechanism and the proteins that carry it out have been conserved in virtually all cells on Earth.

We shall also see that homologous recombination plays an additional role in sexually reproducing organisms. During meiosis, a key step in gamete (sperm and egg) production, it catalyzes the orderly exchange of blocks of genetic information between corresponding (homologous) maternal and paternal chromosomes. This creates new combinations of DNA sequences in the chromosomes that are passed to offspring, giving the next generation unique characteristics upon which natural selection can act.

Homologous Recombination Has Common Features in All Cells

The current view of homologous recombination as a critical DNA repair mechanism in all cells developed slowly from its original discovery as a key component in the specialized process of meiosis in plants and animals. The subsequent recognition that homologous recombination also occurs in unicellular organisms made it readily amenable to molecular analyses. Thus, much of what we know about the biochemistry of genetic recombination was derived from studies of bacteria, especially of *E. coli* and its viruses, as well as from experiments with simple eukaryotes such as yeasts. For these organisms with short generation times and relatively small genomes, it was possible to isolate a large set of mutants with defects in their recombination processes. The protein altered in each mutant was then identified and, ultimately, studied biochemically. Very close relatives of these proteins were subsequently found in more complex eukaryotes including flies, mice, and humans, and it is now possible to directly analyze homologous recombination in these species as well. As a result, we now know that the fundamental processes that catalyze homologous recombination are common to all cells.

DNA Base-pairing Guides Homologous Recombination

The hallmark of homologous recombination is that it takes place only between DNA duplexes that have extensive regions of sequence similarity (homology). Not surprisingly, base-pairing underlies this requirement: before undergoing homologous recombination, two DNA helices will "sample" each other's DNA sequence by testing the potential base-pairing between a single strand from one

DNA duplex and a complementary single strand from the other. Recombination is initiated when a match is found; this match need not be perfect, but it must be very close for homologous recombination to succeed. As we shall see, the process is carefully controlled and guided by a group of specialized proteins.

Homologous Recombination Can Flawlessly Repair Double-Strand Breaks in DNA

Unlike the nonhomologous end joining discussed earlier, homologous recombination repairs double-strand breaks accurately, without any loss or alteration of nucleotides at the site of repair. For homologous recombination to do this repair job, the damaged DNA must first be brought into proximity with a homologous but undamaged DNA double helix, which can then serve as a template for repair. For this reason, homologous recombination often occurs after DNA replication, when the two daughter DNA molecules lie close together and one can serve as a template for repair of the other.

One of the simplest pathways through which homologous recombination can repair double-strand breaks is shown in Figure 5–47. In essence, the broken DNA duplex and the template duplex carry out a "strand dance" so that one of the damaged strands can use the complementary strand of the intact DNA duplex

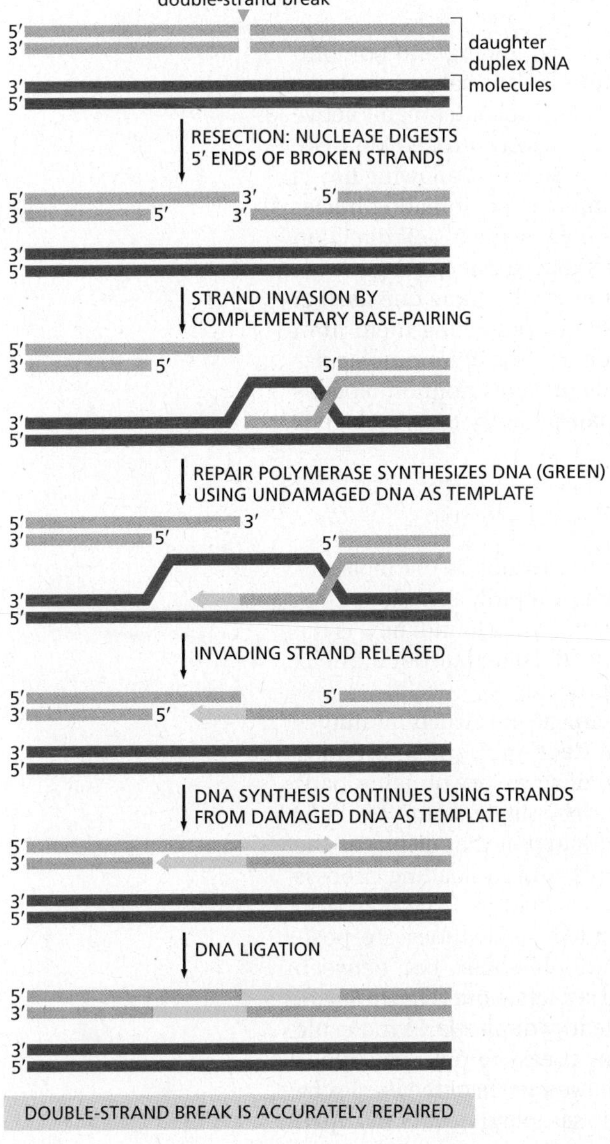

Figure 5–47 **A mechanism that repairs double-strand breaks by homologous recombination.** Homologous recombination can be regarded as a flexible series of reactions, with the exact pathway differing from one case to the next. The pathway shown here represents one of the major forms of recombinational double-strand break repair; however other, closely related pathways also exist. All share the first two steps—resection and strand invasion—but they diverge afterward. For example, recombinational repair of some double-strand breaks proceeds through a *double Holliday junction*, a structure we discuss later in this chapter.

as a template for repair. Once the damaged and template DNA double helices are in proximity (as occurs, for example, after DNA replication), the ends of the broken DNA are chewed back, or "resected," by specialized nucleases to produce overhanging, single-strand 3′ ends. The next step is **strand exchange** (also called *strand invasion*), during which one of the single-strand 3′ ends from the damaged DNA molecule searches the template duplex for homologous sequences through base-pairing. Once stable base-pairing is established (which completes the *strand-exchange* step), an accurate DNA polymerase extends the invading strand using the information provided by the undamaged template molecule, thus restoring one of the damaged DNA strands. The last steps—strand displacement, further repair synthesis, and ligation—restore the two original DNA double helices and complete the repair process, as illustrated.

Homologous recombination resembles other DNA repair reactions in that a DNA polymerase utilizes a pristine template to restore damaged DNA. However, instead of using the partner strand as a template, as occurs in most DNA repair pathways, homologous recombination makes use of a complementary strand from a separate DNA duplex. In the following sections, we discuss the steps of homologous recombination in more detail with an emphasis on the proteins that guide this remarkable process.

Specialized Processing of Double-Strand Breaks Commits Repair to Homologous Recombination

Once a double-strand break occurs, nonhomologous end joining and homologous recombination compete to repair the damage. But the specialized nuclease that resects DNA ends to begin homologous recombination becomes highly active during S and G_2 (through its phosphorylation by cell-cycle-controlled kinases), and homologous recombination usually wins out at these times, allowing use of a newly replicated daughter DNA molecule as a template. The initiating nuclease (called the Mre11 complex in eukaryotes) chews back in the 5′→3′ direction leaving protruding 3′ ends on either side of the break that can be as long as several thousand nucleotides. Single-strand binding protein (the same one used at replication forks) then coats the exposed single strands, protecting them from other nucleases in the cell and ensuring that they remain free of intramolecular base-pairing. The formation of these protruding ends prevents nonhomologous end joining from occurring, and it commits the repair pathway to homologous recombination.

Strand Exchange Is Directed by the RecA/Rad51 Protein

Of all the steps of homologous recombination, strand invasion is the most difficult to imagine. How does the invading single strand rapidly sample a DNA duplex for a complementary sequence? Once the homology is found, how is the structure stabilized? And how is the inherent stability of the template double helix overcome to allow tests for base-pairing during this process?

The answers to these questions came from biochemical and structural studies of the main protein that carries out this feat, called **RecA** in *E. coli* and **Rad51** in virtually all eukaryotic organisms. A special group of accessory proteins loads a set of RecA/Rad51 monomers onto a protruding DNA single strand (such as that in Figure 5-47), forming a cooperatively bound filament that displaces the single-strand binding protein originally present. This orderly loading process produces a protein–DNA filament in which the DNA is held by RecA/Rad51 in an unusual conformation: groups of three consecutive nucleotides are positioned as though they were in a conventional DNA double helix, but, between adjacent triplets, the DNA backbone is untwisted and stretched out (**Figure 5–48**). This unusual protein–DNA structure then grasps a nearby duplex DNA molecule in a way that stretches it, destabilizing it and making it easy to pull the strands apart. The invading single strand then can sample the sequence of the duplex by conventional base-pairing to one of its strands. This sampling occurs in triplet

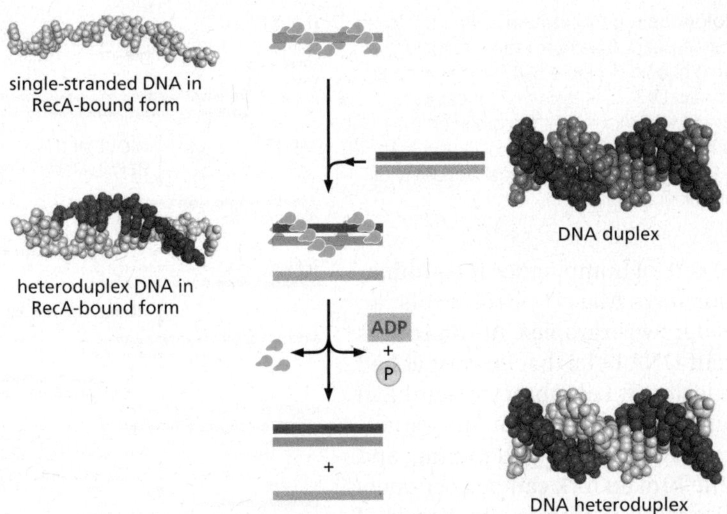

Figure 5–48 **Strand invasion catalyzed by the RecA protein.** Our understanding of this reaction is based in part on structures determined by x-ray diffraction studies of the bacterial RecA protein bound to single-stranded and double-stranded DNA. These DNA structures (illustrated with the RecA protein removed) are shown on the left side of the diagram. The reaction begins when ATP-bound RecA protein *(blue)* associates with a DNA single strand (typically a protruding 3′ end as shown in Figure 5–47), holding it in an elongated form where groups of three bases are separated from each other by a stretched and twisted backbone. The RecA-bound single strand then binds to duplex DNA, destabilizing it to allow the single strand to sample its sequence through base-pairing, three bases at a time. If an extensive match is found, the structure is disassembled through ATP hydrolysis, resulting in protein dissociation and the exchange of one single strand of DNA for another, thereby forming a new *heteroduplex* from the complementary strands of two different DNA molecules. In the vast majority of cases, no match will be found in any one binding event, in which case the RecA-bound DNA single strand rapidly dissociates to begin a new search. (PDB code: 3CMX.)

nucleotide blocks, each of which is already in a "base-pair ready" conformation in the invading strand; when a good triplet match occurs, only then is the adjacent triplet sampled, and so on. In this way, mismatches very quickly cause dissociation, so that millions of possible pairings can be tested. Only an extended stretch of base-pairing (at least 15 nucleotides) can stabilize the invading strand, leading to the next steps in homologous recombination.

RecA/Rad51 is an ATPase, and the steps described above require that each monomer along the filament be in the ATP-bound state. However, the searching itself does not require ATP hydrolysis; instead, the process occurs by simple molecular collisions, allowing an enormous number of potential sequences to be rapidly sampled. Once stable base-pairing occurs and a strand-exchange reaction is completed, ATP hydrolysis is necessary to disassemble RecA from the complex of DNA molecules. At this point, repair DNA polymerases and DNA ligase, which we encountered earlier in this chapter, complete the repair process, as shown previously in Figure 5–47.

Homologous Recombination Can Rescue Broken and Stalled DNA Replication Forks

Although accurately repairing double-strand breaks is a crucial function of homologous recombination, it can also repair other types of damage. For example, some chemicals cross-link the two strands of DNA together by covalently joining nucleotides on opposite strands. A special set of enzymes unlinks the strands and cuts out the damaged bits on both strands. At this point, the damaged DNA has been converted to a double-strand break, which can be accurately repaired by homologous recombination, as discussed earlier. Similarly, proteins can become accidently covalently linked to DNA, and these sites can also be converted by nucleases into double-strand breaks, allowing repair by homologous

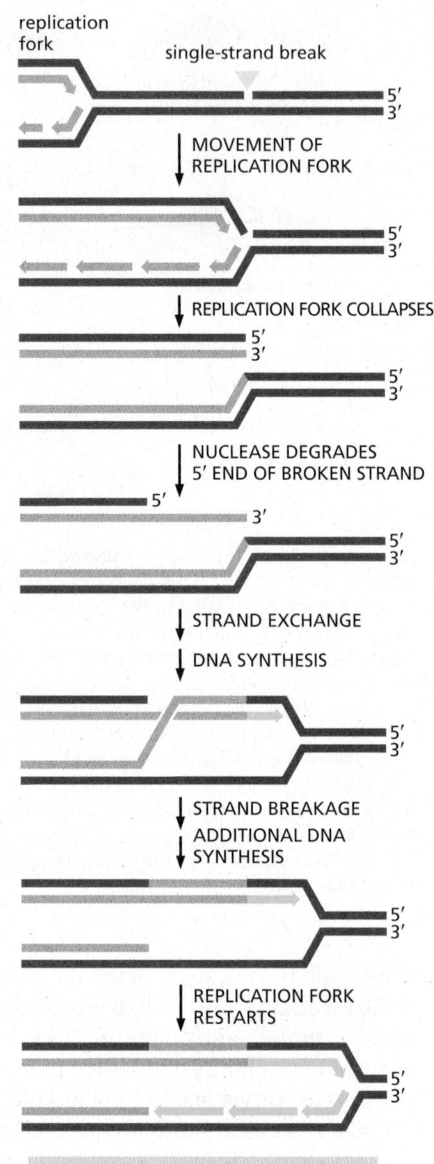

Figure 5–49 Repair of a broken replication fork by homologous recombination. When a moving replication fork encounters a single-strand break, it will collapse but can be repaired by homologous recombination. The process uses many of the same reactions shown in Figure 5–47 and proceeds through the same basic steps. *Green* strands represent the new DNA synthesis that takes place after the replication fork has broken. This pathway allows the fork to move past the break on the damaged template using the undamaged duplex as a template to synthesize DNA. (Adapted from M.M. Cox, *Proc. Natl. Acad. Sci. USA* 98:8173–8180, 2001. Copyright 2001 National Academy of Sciences, USA. With permission from National Academy of Sciences.)

recombination. But perhaps the most important role of homologous recombination is in rescuing broken or stalled DNA replication forks. Many types of events can cause a replication fork to stop, and here we consider two examples. The first arises from an accidental single-strand gap in the parent DNA helix that lies just ahead of a replication fork. When the fork reaches this lesion, it falls apart—resulting in one broken and one intact daughter chromosome. Because this is a "one-sided" double-strand break, it cannot be repaired by nonhomologous end joining, and homologous recombination becomes crucial. The broken fork can be accurately repaired using the same basic reactions we discussed earlier for the repair of double-strand breaks (**Figure 5–49**). With slight modifications, the set of reactions just depicted can accurately repair many different types of DNA damage, providing that an undamaged duplex DNA template is available.

A different type of problem arises when a replication fork attempts to move through certain types of DNA damage that clogs up the replication machinery, stalling the fork. Because such damaged DNA often ends up deeply buried in the core of the replication fork, it cannot be easily repaired. To resolve this problem, the replication machine "backs up" through a series of strand-exchange reactions similar to those we have discussed (**Figure 5–50**). This maneuver allows one newly synthesized DNA strand to act as a template for synthesis of the other new strand, thereby bypassing the damaged template and allowing replication to proceed.

DNA Repair by Homologous Recombination Entails Risks to the Cell

Although homologous recombination neatly solves the problem of accurately repairing double-strand breaks and other types of DNA damage, it sometimes "repairs" damage using the wrong bit of the genome as the template. For example, sometimes a broken human chromosome is repaired using the homolog from the other parent instead of the sister chromatid as the template. Because maternal and paternal chromosomes differ in DNA sequence at many positions along their lengths, this type of repair can convert the sequence of the repaired DNA from the maternal to the paternal sequence or vice versa. The result of this type of errant recombination is a **loss of heterozygosity**. It can have severe consequences if the homolog used for repair contains a deleterious mutation, because the recombination event destroys the "good" copy. Loss of heterozygosity, although it happens rarely, is nonetheless a critical step in the formation of many cancers (discussed in Chapter 20).

Cells go to great lengths to minimize the risk of mishaps of these types; indeed, as we have seen, nearly every step of homologous recombination is carefully regulated. Recall that the first step (resection of the broken ends) is coordinated with the cell cycle: it occurs primarily in the S and G_2 phases of the cell cycle, favoring the use of a daughter duplex (either as a partially replicated chromosome or a fully replicated sister chromatid) as a template for repair (see Figure 5–47). The close proximity of the two daughter chromosomes disfavors the use of other genome sequences in the repair process.

The loading of RecA/Rad51 onto the processed DNA ends and the subsequent strand-exchange reaction are also tightly controlled by the cell, and a host of accessory proteins is needed to regulate these steps. There are many such proteins, and exactly how all of them coordinate and control homologous recombination remains a mystery, although we do understand how a few of them work, as described below. We also know that the enzymes that catalyze

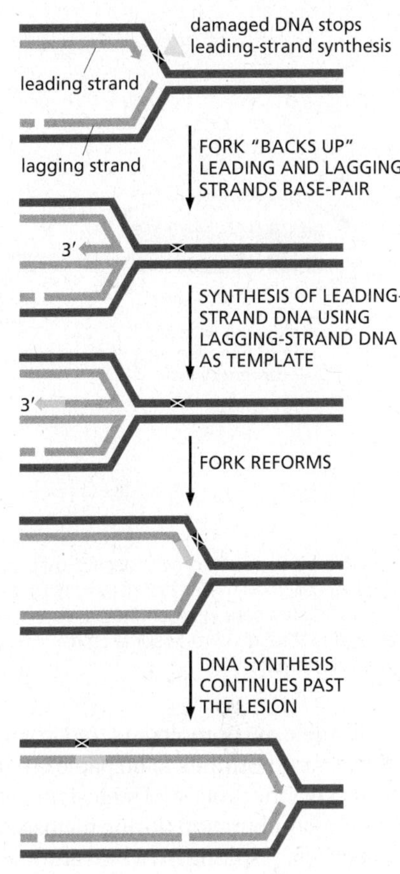

Figure 5–50 **Repair of a stalled replication fork by "fork reversal."** This mechanism is brought into play when a replication fork stalls when it encounters certain types of damaged nucleotides. A specialized helicase (not shown) peels the newly synthesized DNA strands away from their parent templates, allowing them to form complementary base pairs with each other and backing up the replication fork. At this point two outcomes are possible. In the first, because the damaged DNA has been exposed, it can be repaired by conventional repair mechanisms, and the fork can be restarted. In the second, as shown, DNA synthesis can bypass the damage using newly synthesized daughter DNA (rather than the damaged parent strand) as the template. This scheme allows the replication fork to move through the DNA damage, which can be repaired at a later time. Although the initial steps of replication fork reversal are well understood, exactly how the fork restarts afterward remains a mystery.

recombinational repair are made at relatively high levels in eukaryotes and are dispersed throughout the nucleus in an inactive form. In response to DNA damage, they rapidly converge on the sites of DNA damage, become activated, and form "repair factories" where many lesions are apparently brought together and repaired (**Figure 5–51**). Formation of these factories probably results from many weak interactions between different repair proteins and between repair proteins and damaged DNA, producing the type of biomolecular condensates discussed in Chapter 3 (see Figure 3–77). The high local concentration of the appropriate proteins and their substrates within these condensates is thought to increase the speed and efficiency of the repair process.

In Chapter 20, we shall see that both too much and too little homologous recombination can lead to cancer in humans, the former through repair using the "wrong" template (as described above) and the latter through an increased mutation rate caused by inefficient DNA repair. Clearly, a delicate balance has evolved that keeps this process in check on undamaged DNA, while still allowing it to act efficiently and rapidly on DNA lesions as soon as they arise.

Not surprisingly, mutations in the components that carry out and regulate homologous recombination are responsible for several inherited forms of cancer. Two of these, the Brca1 and Brca2 proteins, were first discovered because mutations in their genes lead to a greatly increased frequency of breast cancer. Because these mutations cause inefficient repair by homologous recombination, accumulation of DNA damage can, in a small proportion of cells, give rise to a cancer. Brca1 regulates an early step in broken-end processing; without it, such ends are not processed correctly for homologous recombination and instead damaged molecules are shunted to the error-prone nonhomologous end-joining pathway (see Figure 5–45). After resection, Brca2 is needed to correctly load the Rad51 protein onto the protruding single-strand DNA ends in preparation for strand exchange.

Homologous Recombination Is Crucial for Meiosis

We have seen that homologous recombination can use a set of reactions—including broken-end resection, strand invasion, limited DNA synthesis, and ligation—to exchange DNA sequences between two double helices with the same nucleotide sequence and thereby repair damaged DNA. We now describe how homologous recombination is used to deliberately exchange material between two different chromosomes in order to generate DNA molecules that carry novel combinations of genes. This is a frequent and necessary part of meiosis, which occurs in sexually reproducing organisms such as fungi, plants, and animals.

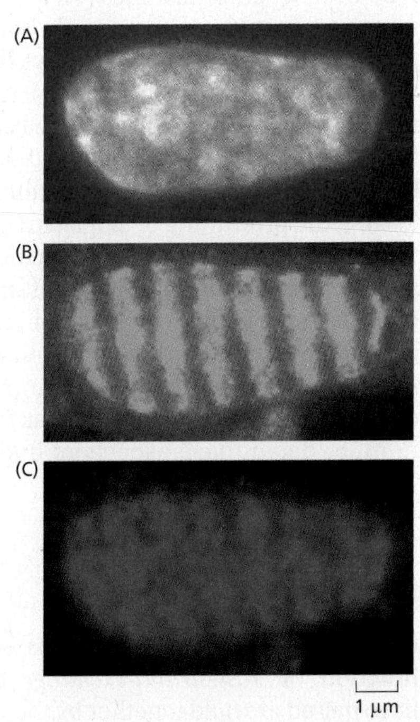

Figure 5–51 **Experiment demonstrating the rapid localization of repair proteins to DNA double-strand breaks.** Human fibroblasts were x-irradiated to produce DNA double-strand breaks. Before the x-rays struck the cells, they were passed through a microscopic grid with x-ray-absorbing "bars" spaced 1 μm apart. This produced a striped pattern of DNA damage, allowing a comparison of damaged and undamaged DNA in the same nucleus. (A) Total DNA in a fibroblast nucleus stained with the dye DAPI. (B) Sites of new DNA synthesis due to repair of DNA damage, indicated by incorporation of BrdU (a thymidine analog) and subsequent staining with fluorescently labeled antibodies to BrdU *(green)*. (C) Localization of the Mre11 complex to damaged DNA as visualized by antibodies against the Mre11 subunit *(red)*. Mre11 is the 5′→3′ nuclease that produces the protruding single-strand DNA ends needed for strand invasion (see Figure 5–47). A, B, and C were processed 30 minutes after x-irradiation. (From B.E. Nelms et al., *Science* 280:590–592, 1998. With permission from AAAS.)

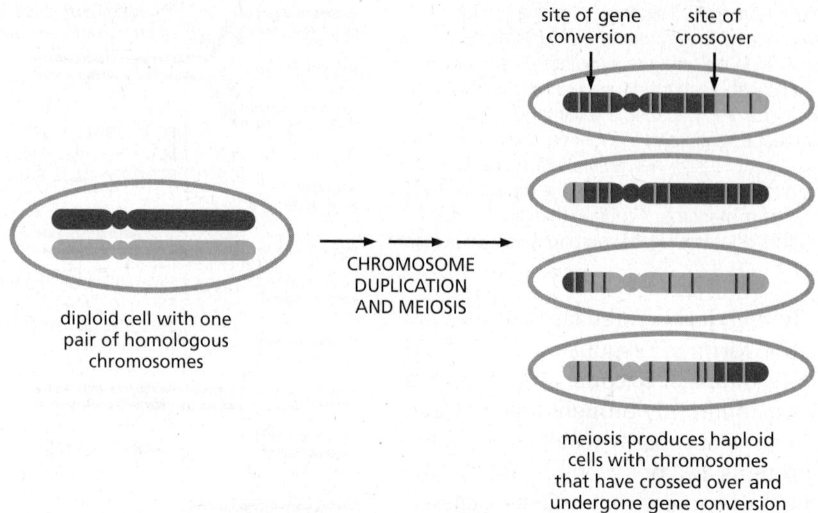

Figure 5–52 **Chromosome crossing-over occurs in meiosis.** Meiosis is the process by which a diploid cell gives rise to four haploid germ cells, as described in detail in Chapter 17. Meiosis produces germ cells in which the paternal and maternal genetic information (*red* and *blue*) has been reassorted through chromosome crossovers. In addition, many short regions of gene conversion occur, as indicated.

In meiosis, homologous recombination is an integral part of the process that allows chromosomes to be parceled out to germ cells (sperm and eggs in animals). We discuss the process of meiosis in detail in Chapter 17; here we discuss how homologous recombination during meiosis produces chromosome *crossing-over* and *gene conversion,* resulting in hybrid chromosomes that contain genetic information from both the maternal and paternal homologs (**Figure 5–52**). These mechanisms, at their core, closely resemble those used to repair double-strand breaks.

Meiotic Recombination Begins with a Programmed Double-Strand Break

Homologous recombination in meiosis starts with a bold stroke: a specialized Spo11 protein complex breaks both strands of the DNA double helix in one of the recombining chromosomes (**Figure 5–53**). After catalyzing this reaction, the protein complex remains covalently bound to the broken DNA, much like the DNA topoisomerase we encountered earlier in this chapter (see Figure 5–22). Many of the subsequent recombination reactions closely resemble those already described for the repair of double-strand breaks; indeed, some of the same proteins are used for both processes. For example, the Mre11 complex, which we encountered earlier, chews back the DNA ends, removing the proteins along with the DNA and leaving the protruding 3′ single-strand ends needed for strand invasion.

However, several meiosis-specific proteins come into play and guide the reactions somewhat differently, resulting in the distinctive outcomes observed for meiosis. A key difference is that, in meiosis, recombination occurs preferentially between maternal and paternal chromosomal homologs (which are held closely together during meiosis), rather than between newly replicated, identical DNA duplexes as in double-strand break repair. In the sections that follow, we describe in more detail those aspects of homologous recombination that are especially important for meiosis.

Holliday Junctions Are Recognized by Enzymes That Drive Branch Migration

Of special importance in meiosis is an intermediate structure known as a **Holliday junction**, or *cross-strand exchange,* in which two homologous DNA helices that have paired are held together by the reciprocal exchange of two of the four strands

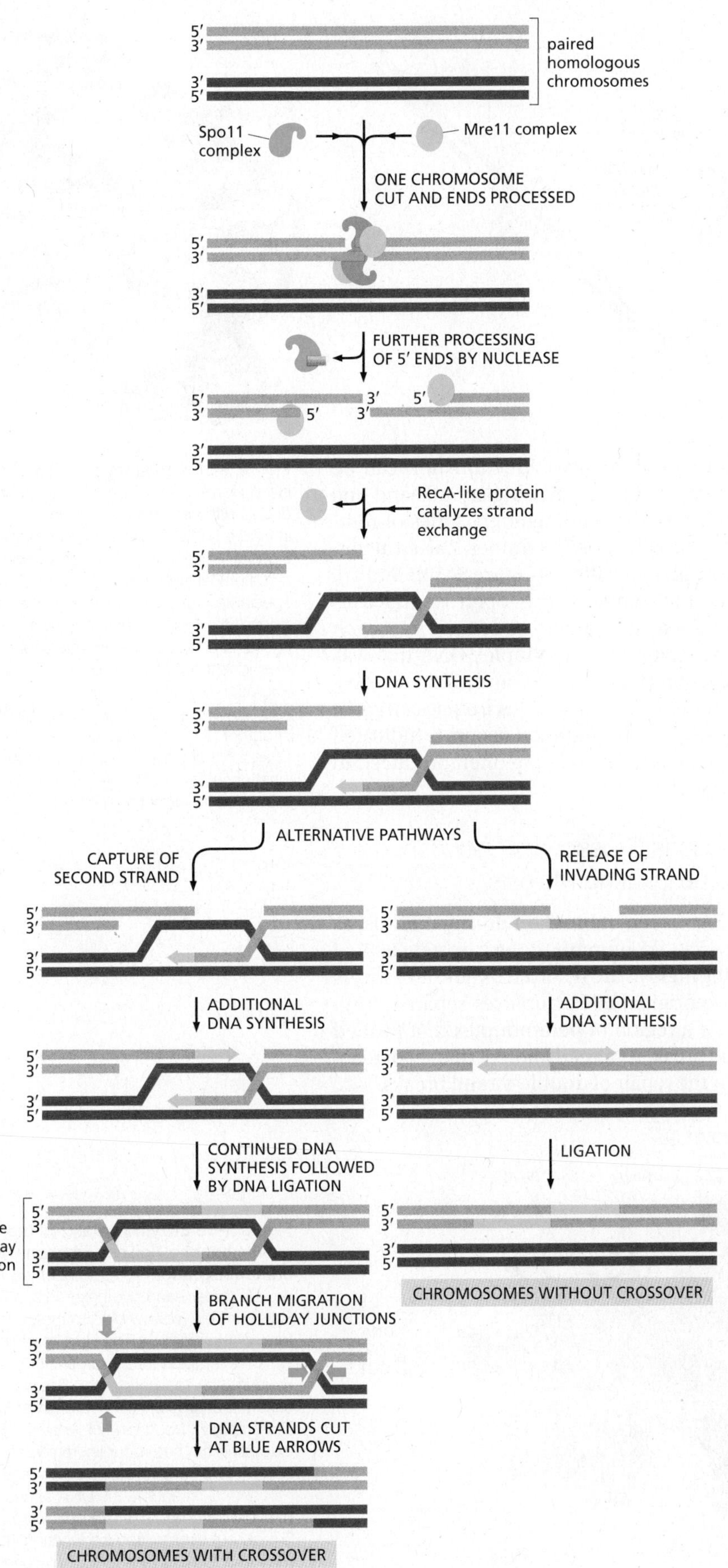

Figure 5–53 Homologous recombination during meiosis can generate chromosome crossovers. Once the Spo11 complex breaks the duplex DNA and the Mre11 complex processes the ends, homologous recombination in meiosis can proceed along alternative pathways (**Movie 5.8**). One (*right side* of figure) closely resembles the double-strand break repair reaction shown in Figure 5–47 and results in chromosomes that have been "repaired" without crossing over. The other (*left side* of figure) proceeds through a double Holliday junction and produces two chromosomes that have crossed over. During meiosis, the maternal and paternal chromosome homologs are held tightly together (see Figure 17–54), and both types of recombination occur between them.

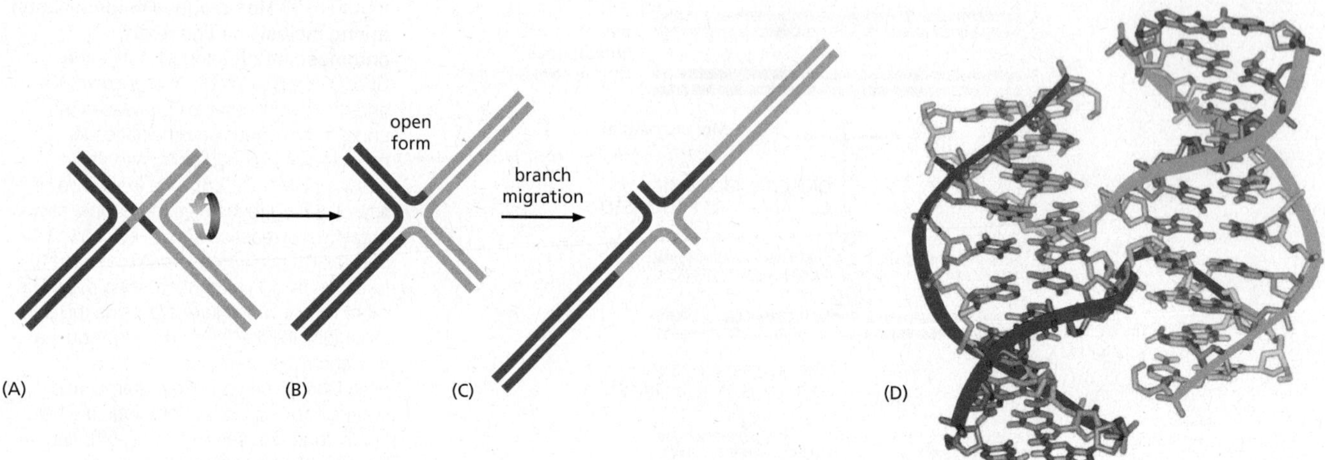

Figure 5–54 **A Holliday junction.** The initially formed structure (A) is usually drawn with two strands crossing, as in Figure 5–53. An isomerization of the Holliday junction (B) produces an open, symmetrical structure that is bound by specialized proteins. (C) These proteins "move" the Holliday junctions by a coordinated set of branch-migration reactions that involve the breaking and formation of base pairs (see Figure 5–55 and Movie 5.8). (D) Three-dimensional structure of the Holliday junction in the open form depicted in B. The Holliday junction is named for the scientist who first proposed its formation. (PDB code: 1DCW.)

present, one strand originating from each of the helices. This junction can be considered to contain two pairs of strands: one pair of crossing strands and one pair of noncrossing strands (**Figure 5–54A**). But by undergoing a series of rotational movements, it can isomerize to form an open, symmetrical structure in which both pairs of strands occupy equivalent positions (**Figure 5–54B and D**). A special set of recombination proteins that binds to this open isomer uses the energy of ATP hydrolysis to catalyze a reaction known as *branch migration* (**Figure 5–55**), which greatly expands the region of heteroduplex DNA that was initially created by a strand-exchange reaction (**Figure 5–54B and C**). In meiosis, heteroduplex regions often "migrate" thousands of nucleotides from the original site of the double-strand break. The step where this migration occurs is indicated in Figure 5–53. As shown in the figure, Holliday junctions are often produced in pairs, known as double Holliday junctions.

Homologous Recombination Produces Crossovers Between Maternal and Paternal Chromosomes During Meiosis

There are two basic outcomes of homologous recombination during meiosis, as shown previously in Figure 5–53 (Movie 5.8). In humans, approximately 90% of the double-strand breaks produced during meiosis are resolved as non-crossovers (right side of Figure 5–53). Here, the two original DNA duplexes separate from each other in a form unaltered except for a region of heteroduplex that formed near the site of the original double-strand break. As already noted, this set of reactions resembles that described earlier for the repair of double-strand breaks.

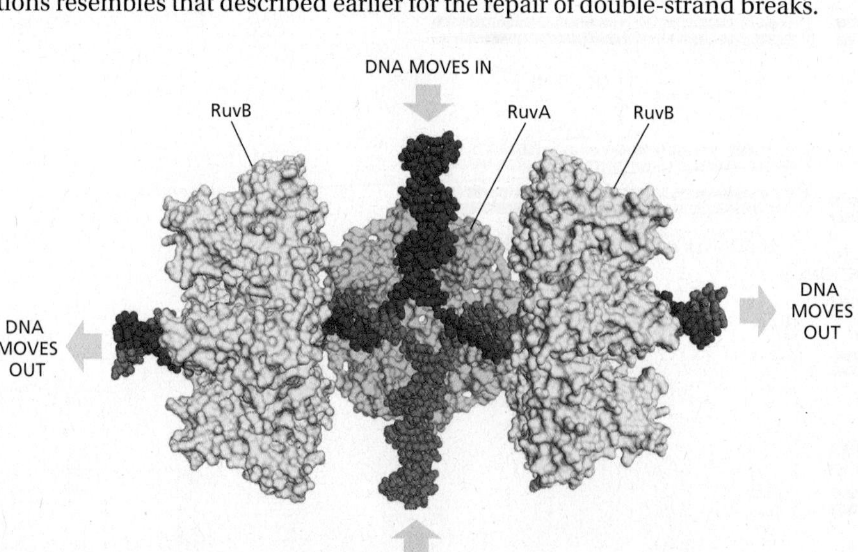

Figure 5–55 **Enzyme-catalyzed branch movement at a Holliday junction by branch migration.** A tetramer of the RuvA protein *(green)* and two hexamers of the RuvB protein *(yellow)* bind to the open form of the junction. The RuvB protein, which resembles the hexameric helicases used in DNA replication (see Figure 5–14), uses the energy of ATP hydrolysis to spool DNA rapidly through the Holliday junction, extending the heteroduplex region as shown. The RuvA protein coordinates this movement, threading the DNA strands to avoid tangling. This example is from *E. coli*, but similar proteins function in meiosis in sexually reproducing organisms. (PDB codes: 1IXR, 1C7Y.)

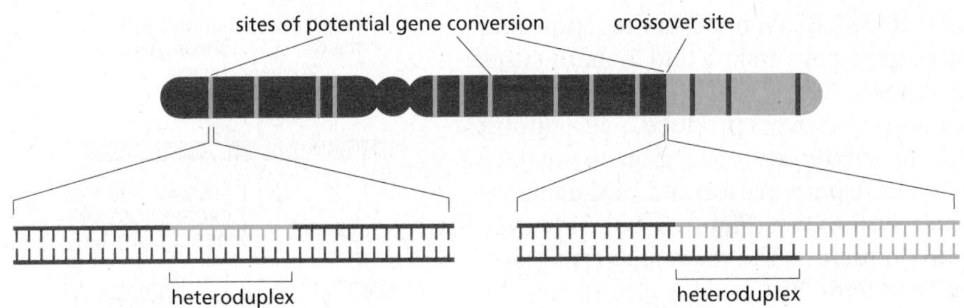

Figure 5–56 Heteroduplexes formed during meiosis. Heteroduplex DNA is present at sites of recombination that are resolved either as crossovers or non-crossovers. Because the DNA sequences of maternal and paternal chromosomes differ at many positions along their lengths, heteroduplexes often contain a small number of base-pair mismatches, and they are therefore potential sites of gene conversion (see Figure 5–57).

The other outcome is much more profound: a double Holliday junction is formed and is cleaved by specialized enzymes (*blue arrows* on the left side of Figure 5–53) to create a *crossover.* The two original portions of each chromosome upstream and downstream from the two Holliday junctions are thereby swapped, creating two chromosomes that are said to have "crossed over"—each containing a large number of both maternally inherited and paternally inherited genes.

How does the cell decide which double-strand breaks to resolve as cross-overs? The answer is not yet known, but we know the decision is not random. The relatively few crossovers that do form are distributed along chromosomes in such a way that a crossover in one position inhibits crossing-over in neighboring regions. Termed *crossover control,* this fascinating but poorly understood regulatory mechanism ensures the roughly even distribution of crossover points along chromosomes. It also ensures that each chromosome—no matter how small—undergoes at least one crossover event every meiosis. For many organisms, roughly two crossovers per chromosome occur during each meiosis, one on each arm. As discussed in detail in Chapter 17, these crossovers, in addition to producing novel DNA molecules, play an important mechanical role in the proper segregation of chromosomes during meiosis.

Whether a meiotic recombination event is resolved as a crossover or a non-crossover, the recombination machinery leaves behind a *heteroduplex region* where a strand with the DNA sequence of the paternal homolog is base-paired with a strand from the maternal homolog (Figure 5–56). These heteroduplex regions can tolerate a small percentage of mismatched base pairs, and because of branch migration, they often extend for thousands of nucleotide pairs. The many non-crossover events that occur in meiosis thereby produce scattered sites in the germ cells where short DNA sequences from one homolog have been pasted into the other homolog. Heteroduplex regions mark sites of potential *gene conversion*—where the four haploid chromosomes produced by meiosis contain three copies of a DNA sequence from one homolog and only one copy of this sequence from the other homolog, as explained next.

Homologous Recombination Often Results in Gene Conversion

In sexually reproducing organisms, it is a fundamental law of genetics that—aside from mitochondrial DNA, which is inherited only through the mother—each parent makes an equal genetic contribution to an offspring. One complete set of nuclear genes is inherited from the father and one complete set is inherited from the mother. Underlying this law is the accurate parceling out of chromosomes to the germ cells (eggs and sperm) that takes place during meiosis. Thus, when a diploid cell in a parent undergoes meiosis to produce four haploid germ cells, exactly half of the genes distributed among these four cells should be maternal (genes inherited from the mother of this parent) and the other half paternal (genes inherited from the father of this parent). In some organisms (fungi, for example), it is possible to recover and analyze all four of the haploid gametes produced from a single cell by meiosis. Studies in such organisms have revealed rare cases in which the parceling out of genes violates the standard genetic rules. Occasionally, for example, meiosis yields three copies of the maternal version of a gene and only one copy of the paternal version. Alternative versions of the same gene are called **alleles**, and it is the divergence from their expected distribution during meiosis that is known as **gene conversion** (Movie 5.8). Genetic studies show that only small sections of DNA typically undergo gene conversion, and in many cases only a part of a gene is changed. How is this possible?

We have seen that both crossovers and non-crossovers produce heteroduplex regions of DNA. If the two strands that make up a heteroduplex region do not have identical nucleotide sequences, mismatched base pairs are formed, and these are often repaired by the cell's mismatch repair system (see Figure 5–20). However, unlike what happens after DNA replication, in meiosis the mismatch repair system randomly selects the strand to be used as a template, causing one allele to be lost and the other duplicated (Figure 5–57). Thus, gene conversion (the "conversion" of one allele to the other)—originally regarded as a mysterious deviation from the rules of genetics—can be seen as a straightforward consequence of the mechanisms of homologous recombination during meiosis.

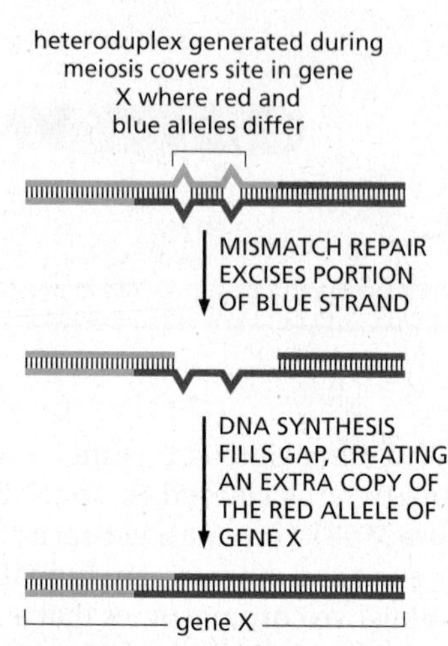
heteroduplex generated during meiosis covers site in gene X where red and blue alleles differ
MISMATCH REPAIR EXCISES PORTION OF BLUE STRAND
DNA SYNTHESIS FILLS GAP, CREATING AN EXTRA COPY OF THE RED ALLELE OF GENE X
gene X

Figure 5–57 Gene conversion caused by mismatch correction. As shown in the preceding figure, heteroduplex DNA is formed at the sites of homologous recombination between maternal and paternal chromosomes. If the maternal and paternal DNA sequences are slightly different, the heteroduplex region will include some mismatched base pairs, which may then be corrected by the DNA mismatch repair machinery (see Figure 5–20). Because neither strand of DNA is newly synthesized, such repair can "erase" nucleotide sequences on either the paternal or the maternal strand. The consequence of this mismatch repair is gene conversion, detected as a deviation from the segregation of equal copies of maternal and paternal alleles that normally occurs in meiosis.

Summary

Homologous recombination describes a flexible set of reactions resulting in the exchange of DNA sequences between a pair of identical or nearly identical duplex DNA molecules. Of special importance is a strand-exchange step whereby a single strand from one DNA duplex invades a second duplex and base-pairs with one strand while displacing the other. This reaction, catalyzed by the RecA/Rad51 family of proteins, can only occur if the invading strand can form a short stretch of consecutive nucleotide pairs with one of the strands of the duplex. This requirement ensures that homologous recombination occurs only between identical or very similar DNA sequences.

When used as a DNA repair mechanism, homologous recombination usually occurs between a damaged DNA molecule and its recently duplicated sister molecule, with the undamaged duplex acting as a template to repair the damaged copy flawlessly. In meiosis, homologous recombination is initiated by deliberate, carefully regulated double-strand breaks and occurs preferentially between the homologous chromosomes rather than the newly replicated sister chromatids. The outcome can be either two chromosomes that have crossed over (that is, chromosomes in which the DNA on either side of the site of DNA pairing originates from two different homologs) or two non-crossover chromosomes. In the latter case, the two chromosomes that result are identical to the original two homologs, except for relatively minor DNA sequence changes at the site of recombination.

TRANSPOSITION AND CONSERVATIVE SITE-SPECIFIC RECOMBINATION

We have seen that homologous recombination can result in the exchange of DNA sequences between chromosomes. However, the order of genes on the interacting chromosomes typically remains the same after homologous recombination, inasmuch as the recombining sequences must be very similar for the process to occur. In this part of the chapter, we describe two very different types of recombination—**transposition** and **conservative site-specific recombination**—that do not require substantial regions of DNA homology. These two types of recombination reactions can alter the gene order along a chromosome and introduce whole blocks of DNA sequence into the genome.

Transposition and conservative site-specific recombination are largely dedicated to moving a wide variety of specialized segments of DNA—collectively termed *mobile genetic elements*—from one position in a genome to another. We will see that mobile genetic elements can range in size from a few hundred to tens of thousands of nucleotide pairs, and each typically carries a unique set of genes. Often, one of these genes encodes a specialized enzyme that catalyzes the movement of only that element and its close relatives, thereby making this type of recombination possible.

Virtually all cells contain mobile genetic elements, known informally as "jumping genes." As explained in Chapter 4, over evolutionary time scales, they have had a profound effect on the shaping of modern genomes. For example, nearly half of the DNA in the human genome can be traced to these elements (see Figure 4–63). Over time, random mutation has altered their nucleotide sequences, and, as a result, only a few of the many copies of these elements in our DNA are still active and capable of movement. The remainder are molecular fossils whose existence provides striking clues to our evolutionary history.

Mobile genetic elements are often considered to be molecular parasites (they are also termed "selfish DNA") that persist because cells cannot get rid of them; they certainly have come close to overrunning our own genome. However, mobile DNA elements can provide benefits to the cell. For example, the genes they carry are sometimes advantageous to the host, as in the case of antibiotic resistance in bacterial cells, discussed later. The movement of mobile genetic elements also produces many of the genetic variants upon which evolution depends, because, in addition to moving themselves, mobile genetic elements occasionally rearrange neighboring sequences of the host genome. Thus, spontaneous mutations observed in bacteria, *Drosophila*, humans, and other organisms are often due to the random movement of mobile genetic elements. While many of these mutations will be deleterious to the organism, some will be advantageous and may spread throughout the population. It is almost certain that much of the variety of life we see around us originally arose from the movement of mobile genetic elements.

In this part of the chapter, we introduce mobile genetic elements and describe the mechanisms that enable them to move from place to place in a genome. As mentioned above, these elements move through a variety of different mechanisms that can be grouped into two broad categories, *transposition* and *conservative site-specific recombination.* We begin with transposition, by far the most predominant of these two processes.

Through Transposition, Mobile Genetic Elements Can Insert into Any DNA Sequence

Mobile elements that move by way of transposition are called **transposons**, or **transposable elements**. In transposition, a specific enzyme, usually encoded by the transposon itself and typically called a *transposase,* acts on specific DNA sequences at each end of the transposon, causing it to insert into a new DNA site. Most transposons are only modestly selective in choosing their target site, and they can therefore insert themselves into many different locations in a genome; in particular, there is no general requirement for sequence similarity between the ends of the element and the target sequence. Most transposons move only rarely. In bacteria, the rate is typically one transposition event once every 10^5 cell divisions, and significantly more frequent movement would probably destroy the host cell's genome. In plants and animals, the situation is different: it is common for progeny to carry tens to hundreds of new insertions relative to their parents. These high rates are tolerated, in part, because these genomes typically carry vast amounts of nonessential DNA sequences where most of the insertions are likely to occur.

On the basis of their structure and transposition mechanism, transposons can be grouped into three large classes: *DNA-only transposons, retroviral-like retrotransposons,* and *nonretroviral retrotransposons.* The differences among them are briefly outlined in **Table 5–4**, and each class will be discussed in turn.

DNA-only Transposons Can Move by a Cut-and-Paste Mechanism

DNA-only transposons, so named because they exist exclusively as DNA during their movement, predominate in bacteria, and they are largely responsible for the spread of antibiotic resistance in bacterial strains. When antibiotics such as penicillin and streptomycin first became widely available in the 1950s, most bacteria that caused human disease were susceptible to them. Now, the situation

TABLE 5–4 **Three Major Classes of Transposable Elements**

Class description and structure	Specialized enzymes required for movement	Mode of movement	Examples
		DNA-only transposons	
Short inverted repeats at each end	Transposase	Moves as DNA, either by cut-and-paste or replicative pathways	P element (*Drosophila*), Ac-Ds (maize), Tn3 (*E. coli*), Tam3 (snapdragon), Helraiser (bat)
		Retroviral-like retrotransposons	
Directly repeated long terminal repeats (LTRs) at each end	Reverse transcriptase and integrase	Moves via an RNA intermediate whose production is driven by a promoter in the LTR	Copia and Gypsy (*Drosophila*), Ty1 (yeast), HERVK (human), Bs1 (maize), EVADE (*Arabidopsis*)
		Nonretroviral retrotransposons	
AAAA TTTT Poly A at 3′ end of RNA transcript; 5′ end is often truncated	Reverse transcriptase and endonuclease	Moves via an RNA intermediate that is often synthesized from a neighboring promoter	I element (*Drosophila*), L1 (human), Cin4 (maize), Karma (rice)

These elements range in length from 1000 to about 12,000 nucleotide pairs. Each family contains many members, only a few of which are listed here. Some viruses can also move in and out of host-cell chromosomes by transpositional mechanisms. These viruses are related to the first two classes of transposons.

is different—antibiotics such as penicillin (and its modern derivatives) are no longer effective against many modern bacterial strains, including those causing gonorrhea and bacterial pneumonia. The spread of antibiotic resistance is due largely to genes that encode antibiotic-inactivating enzymes that are carried on transposons (**Figure 5–58**). Although these mobile elements can transpose only within cells that already carry them, they can be moved from one cell to another through other mechanisms known collectively as horizontal gene transfer (see Figure 1–18). Once introduced into a new cell, a transposon can insert itself into the genome and be faithfully passed on to all progeny cells through the normal processes of DNA replication and cell division.

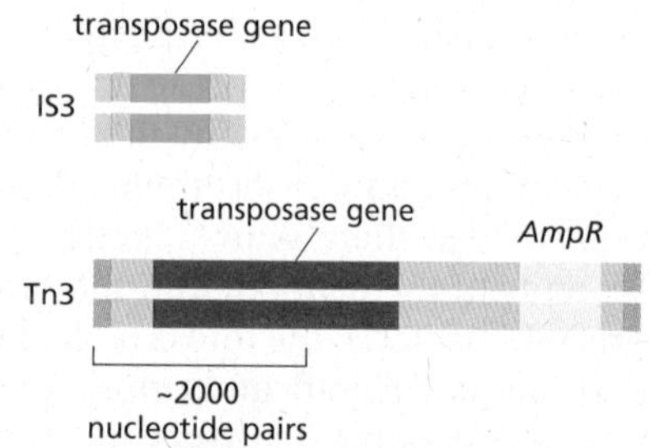

Figure 5–58 Transposons often code for the components they need for transposition. Shown here are two types of bacterial DNA-only transposons. Each carries a gene that encodes a transposase (*dark blue* and *red*)—the enzyme that catalyzes the element's movement—as well as short DNA sequences (*light blue* and *pink*) that are recognized by the matching transposase. The short sequences (two in each transposon) are usually arranged so that one is an inverted repeat of the other.

Some transposons carry additional genes (*yellow*) that encode enzymes that inactivate antibiotics such as ampicillin (*AmpR*). The spread of these transposons is a serious problem in medicine, as it has allowed many disease-causing bacteria to become resistant to the antibiotics developed in the twentieth century.

DNA-only transposons can relocate from a donor site to a target site by *cut-and-paste transposition* (**Figure 5–59**). Here, the transposon is literally excised from one spot on a genome and inserted into another. This reaction produces a short duplication of the target DNA sequence at the insertion site; these direct repeat sequences that flank the transposon serve as convenient records of a prior transposition event. Such "signatures" often provide valuable clues in identifying transposons in genome sequences.

When a cut-and-paste DNA-only transposon is excised from its original location, it leaves behind a "hole" in the chromosome. This lesion can be perfectly healed by recombinational double-strand break repair, provided that the chromosome has recently been replicated so that an identical copy of the damaged host sequence is available. Alternatively, a nonhomologous end-joining reaction can reseal the break; in this case, the DNA sequence that originally flanked the transposon is often altered, producing a mutation at the chromosomal site from which the transposon was excised (see Figure 5–45).

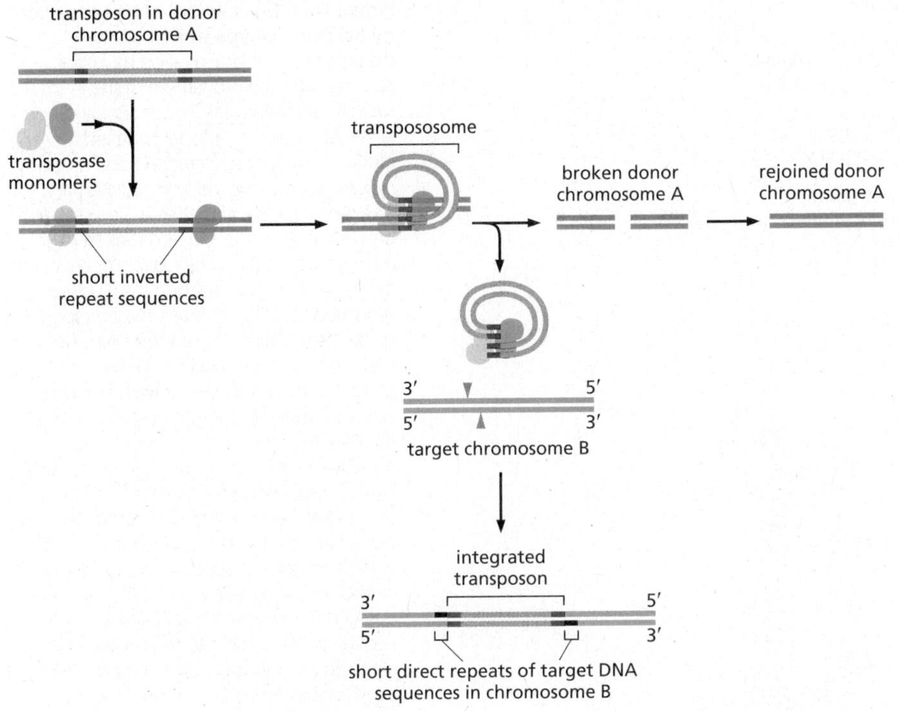

Figure 5–59 Cut-and-paste transposition. DNA-only transposons can be recognized in chromosomes by the inverted repeat DNA sequences *(red)* present at their ends. These sequences, which can be as short as 20 nucleotides, are all that is necessary for the DNA between them to be transposed by the particular transposase enzyme associated with the element. The cut-and-paste movement of a DNA-only transposable element from one chromosomal site to another begins when the transposase enzyme brings the two inverted DNA sequences together, forming a DNA loop. Insertion into the target chromosome, also catalyzed by the transposase, occurs at a random site through the creation of staggered breaks in the target chromosome *(purple arrowheads)*. After the transposition reaction, the single-strand gaps created by the staggered breaks are repaired by DNA polymerase and ligase *(black)*. As a result, the insertion site is marked by a short direct repeat of the target DNA sequence, as shown. Although the break in the donor chromosome *(green)* is repaired, this process often alters the DNA sequence, causing a mutation at the original site of the excised transposable element (not shown).

Remarkably, the same mechanism used to excise cut-and-paste transposons from DNA has been found to operate in the developing immune system of vertebrates, catalyzing the DNA rearrangements that produce antibody and T-cell receptor diversity. Known as *V(D)J recombination*, this process will be discussed in Chapter 24. Found only in vertebrates, V(D)J recombination is a relatively recent evolutionary novelty, but its mechanism was probably derived from the much more ancient cut-and-paste transposons.

Some DNA-only Transposons Move by Replicating Themselves

Although cut-and-paste transposition is common, especially in bacteria, there are other ways that DNA-only transposons can move. These involve replicating the transposon and moving the copy to a new position on the genome, leaving the original transposon intact and in its original position. There are several different ways this can occur and we discuss only one here, which is characteristic of a large class of DNA-only transposons known as *helitrons*. Found in all branches of life, these transposons are especially common in plants and animals where they can compose several percent of genomes. They carry a gene for an unusual type of transposase, one that functions as both a sequence-specific nuclease and as a helicase, thereby directing the movement of the transposon (**Figure 5–60**).

Because of the mechanism behind their movement, helitrons often transfer bits of the genome, along with themselves, to new positions. For this reason, they are thought to be especially important in reshuffling genomic information to produce variant organisms subject to natural selection over evolutionary time scales.

Some Viruses Use a Transposition Mechanism to Move Themselves into Host-Cell Chromosomes

Certain viruses are considered mobile genetic elements because they use transposition mechanisms to integrate their genomes into that of their host cell. However, unlike transposons, the nucleotide sequences that form these viruses encode proteins that package their genetic information into virus particles that can leave the original host cell to infect other cells. As discussed in Chapter 1, most viruses probably evolved from transposable elements through the capture

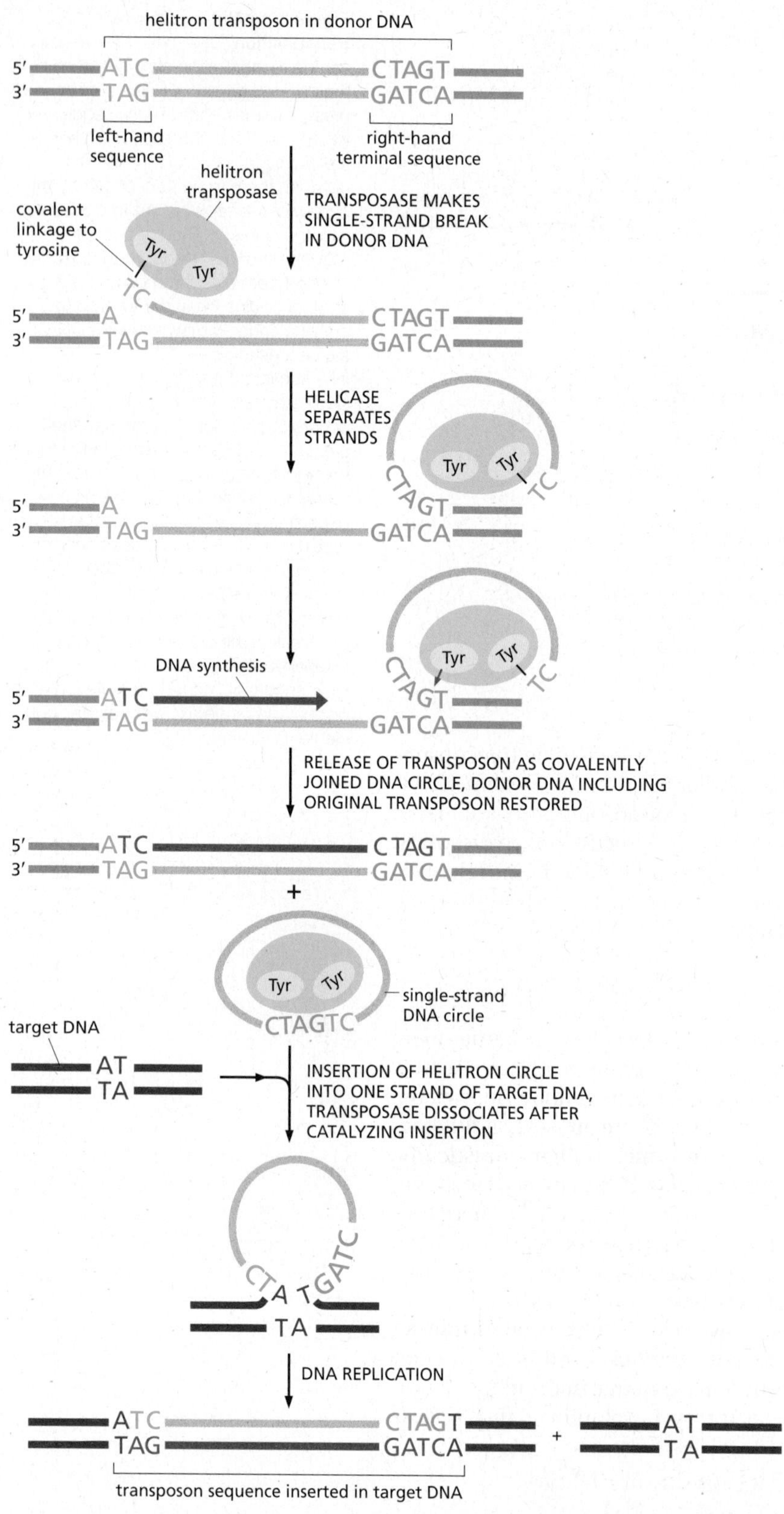

Figure 5–60 Mechanism of transposition by helitrons, a type of DNA-only transposon. Several models have been proposed for the movement of these recently discovered transposons, and one is shown here. This model is based on studies of a helitron found in bats, called *Helraiser*. The process begins when the transposase *(green)* makes a single-strand break at one end of the transposon *(blue)* and, with the aid of a helicase, "peels back" the single strand. A second transposase-mediated reaction releases the transposon in the form of single-stranded DNA, which can move to new positions in the host genome. The transposase (which travels with the single-stranded DNA) can then catalyze the covalent insertion of the transposon into a new location in the host DNA. Transposition by helitrons often moves adjacent host genome sequences along with them. This occurs when, in the third step, the transposase skips over its own CTAGT sequence and cleaves its host DNA downstream at a similar DNA sequence. According to the model, this skipping produces a single-strand DNA circle that includes both helitron and host DNA, and both are inserted into target DNA.

of genes from their host cells. Although originally serving some other purpose in the cell, such captured genes, after a long process of mutation and selection, now code for the structural proteins of viruses, allowing them to escape the cell. Viruses are among the most numerous biological entities on Earth, and we discuss them in more detail in Chapter 23. The viruses that insert themselves into host chromosomes generally do so by employing one of the first two mechanisms

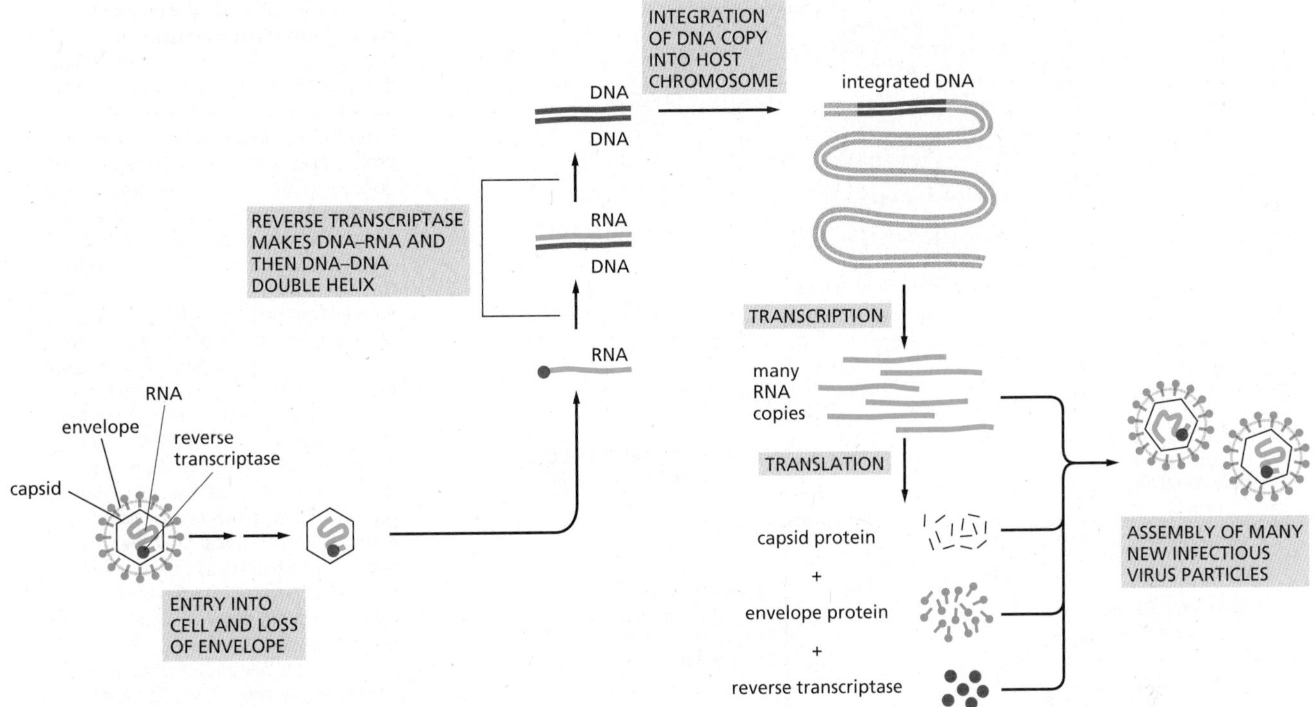

Figure 5–61 **The life cycle of a retrovirus.** The retrovirus genome consists of an RNA molecule *(blue)* that is typically between 7000 and 12,000 nucleotides in length. It is packaged inside a virus-encoded protein capsid, which is surrounded by a lipid-based envelope that contains virus-encoded envelope proteins *(green)*. Inside an infected cell, the enzyme reverse transcriptase *(red circle)* first makes a DNA copy of the viral RNA molecule and then a second DNA strand, generating a double-strand DNA copy of the RNA genome. The integration of this DNA double helix into the host chromosome is then catalyzed by a virus-encoded integrase enzyme. This integration is required for the synthesis of new viral RNA molecules by the host-cell RNA polymerase, the enzyme that transcribes DNA into RNA (discussed in Chapter 6). As indicated, this viral RNA is then used by host-cell machinery to produce the capsid, envelope, and reverse transcriptase proteins needed to form new virus particles.

listed in Table 5–4; namely, by behaving like DNA-only transposons or like retroviral-like retrotransposons. Indeed, much of our knowledge of these mechanisms has come from studies of particular viruses that employ them.

Transposition has a key role in the life cycle of many viruses. Most notable are the **retroviruses**, which include the human AIDS virus, HIV. Outside the cell, a retrovirus exists as a single-strand RNA genome packed into a protein shell, or *capsid*, along with a virus-encoded **reverse transcriptase** enzyme. During the infection process, the viral RNA enters a cell and is converted to a double-strand DNA molecule by the action of this crucial enzyme, which is able to polymerize DNA on either an RNA or a DNA template (Figure 5–61). The term *retrovirus* refers to the virus's ability to reverse the usual flow of genetic information, which normally is from DNA to RNA (see Figure 1–4).

Once the reverse transcriptase has produced a double-strand DNA molecule, specific sequences near its two ends are recognized by a virus-encoded transposase called *integrase*. Integrase then inserts the viral DNA into the chromosome by a mechanism similar to that used by the cut-and-paste DNA-only transposons (see Figure 5–59).

Some RNA Viruses Replicate and Express Their Genomes Without Using DNA as an Intermediate

Retroviruses are not the only viruses that carry their genomes in the form of RNA. Other viruses also have single-strand RNA genomes, but, unlike retroviruses, many replicate and express their genomes without ever using DNA; that is, they are *RNA-only viruses*. For example, SARS-CoV-2, the coronavirus underlying the COVID-19 pandemic, replicates its single-strand RNA genome using a special, viral-encoded *RNA-dependent RNA polymerase*. Upon entering a cell, the viral genome is directly translated by ribosomes as though it were an mRNA molecule, producing many different viral-encoded proteins, including the RNA-dependent RNA polymerase. (We discuss mRNA and the process of translation in detail in Chapter 6.) The polymerase assembles with several other viral proteins and a few host proteins to form the complete *replicase complex*. This specialized replicase, which does not require a primer to begin synthesis, starts at the 3′ end of the viral genome and makes a complementary RNA copy of the entire genome

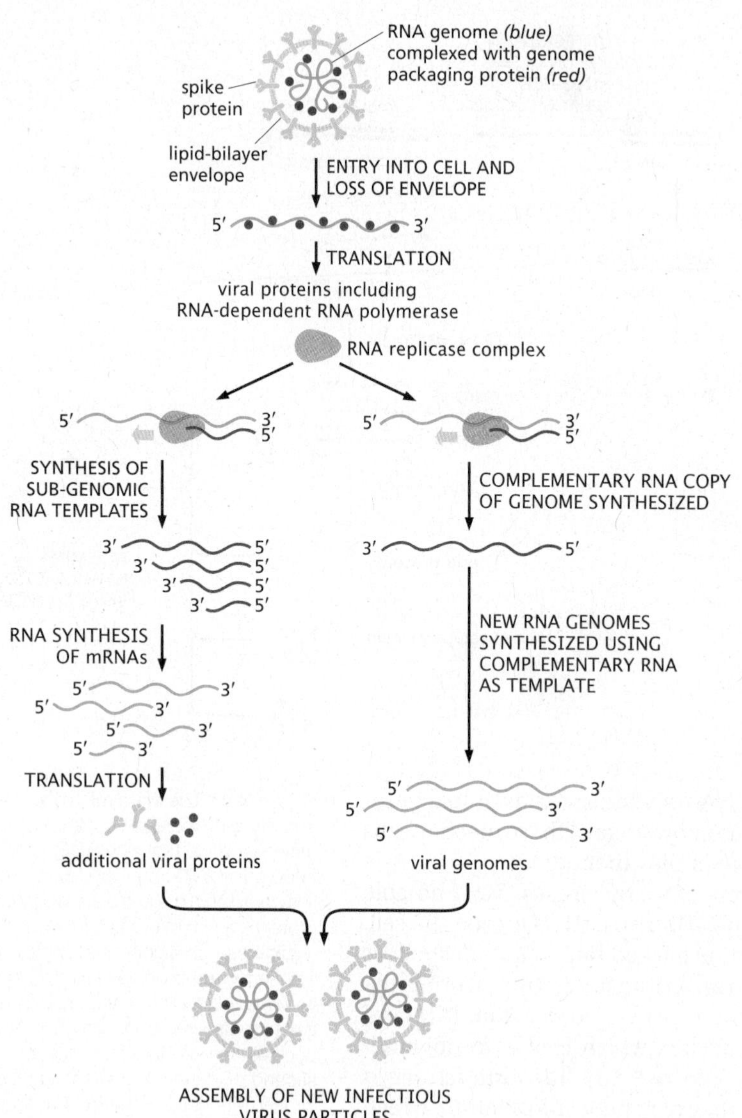

Figure 5–62 Simplified view of the coronavirus life cycle as exemplified by SARS-CoV-2. The viral genome, a single-strand RNA molecule of approximately 30,000 nucleotides, is packaged throughout its length with an RNA-binding protein *(red)* and enclosed by a lipid bilayer containing the viral spike protein. (The appearance of spike proteins, emanating from the lipid envelope, is responsible for the "corona" moniker.) As described in Chapter 23, the spike proteins bind to a receptor on the surface of susceptible cells and direct the fusion of the viral envelope with the outer cell membrane, releasing the viral genome into the cytoplasm. The genome is then directly translated into protein. Single-strand RNA viruses of this type are called *[+] strand viruses*, denoting the ability of their genomes to be immediately translated by the host-cell machinery. In contrast, the genomes of *[–] strand viruses* must first be used as templates to make complementary RNA strands that are then translated (see Table 23–1).

Among the first proteins made by coronaviruses are those that form the RNA-dependent RNA polymerase, which is responsible for producing new viral genomes by synthesizing RNA using RNA as a template. The replicase complex (which includes the polymerase and several other loosely associated proteins) first synthesizes complete noncoding copies of the viral genome. These complementary copies in turn serve as templates for the replicase complex to synthesize new genomes. The replicase also makes a series of shorter coding RNAs, which are needed to produce additional viral proteins including the spike. Once new viral genomes and proteins have been synthesized, new virus particles are assembled and exit the cell. Although only a few viral proteins are shown in the diagram, the virus codes for at least 27 different proteins; some of these organize the double-membrane structures in which the virus replicates, while others inhibit various immune system responses to the infection.

(**Figure 5–62**). Using this complementary copy as a template, the replicase then synthesizes new genomes, which are then packaged with newly synthesized viral proteins into complete virus particles. The whole process of viral replication takes about 10 hours, and a single infected cell can produce as many as 1000 new virus particles, which can spread to other cells within the same host or move in aerosols to new hosts. Because coronaviruses do not use DNA, all steps of viral replication can take place outside the nucleus. In the case of SARS-CoV-2, viral replication occurs in the cytoplasm inside double-membrane compartments that are commandeered by the virus from the endoplasmic reticulum, an organelle described in detail in Chapter 12. These virus-induced compartments are believed to protect the virus from the cell's many antiviral defenses (see pp. 1337–1338) during viral replication and assembly.

Several features of coronaviruses distinguish them from other RNA-only viruses such as those that cause influenza or polio. Perhaps the most unusual is the ability of coronavirus replicase complexes to proofread as they copy their RNA genomes. This proofreading occurs in much the same way that we saw for DNA polymerases earlier in the chapter: An incorrectly added nucleotide is excised by a 3′-to-5′ exonuclease carried in the replicase complex, giving the replicase another chance to add the correct nucleotide. This feature means that coronaviruses do not mutate as rapidly as most other RNA viruses, which lack proofreading ability. As discussed in Chapter 23, the relatively high mutation rate of influenza virus helps explain why new vaccines are needed every year, and this may not always be the case for coronaviruses.

The proofreading also has other important implications. As discussed earlier in this chapter, the lower the mutation rate, the greater the number of essential proteins that a genome can maintain. Proofreading allows coronaviruses to have a larger genome than is typical for RNA viruses; for example, compare the 30,000 nucleotides of the SARS-CoV-2 genome (coding for at least 27 proteins) to the 13,500 nucleotides of the influenza virus (coding for about 10 proteins). Proofreading also affects the development of antiviral drugs. Viral replicases are attractive targets for such drugs, in part because similar enzymes do not exist in uninfected host cells, reducing the chance of side effects. Drugs of this type (for example, remdesivir) are typically nucleoside triphosphate analogs that "fool" the RNA replicase into adding them to growing RNA chains. Once incorporated, the analogs—which have improper 3′ ends—poison further chain elongation (see Figure 8-42). Coronavirus proofreading can excise many of these analogs and thereby reduce their potency. A related strategy (exemplified by molnupiravir) employs nucleoside triphosphate analogs that are incorporated into RNA by the viral replicase, escape proofreading, but base pair incorrectly in the next round of replication, thereby introducing a lethal number of mutations.

Another striking feature of coronaviruses is the way in which they make many different proteins from a genome carried on a single RNA molecule. As described earlier, the viral genome, once it enters a cell, is treated like an mRNA molecule and translated into protein. We shall see in the next chapter, however, that most eukaryotic mRNAs can code for only a single protein. Coronavirus production of many different proteins from a single RNA genome requires a series of unusual steps, some of which appear unique to coronaviruses. We shall discuss the general topics of mRNA translation and its regulation in Chapters 6 and 7. But first, we return to our discussion of transposons, some of which closely resemble viruses in the way they move from place to place in their host genomes.

Retroviral-like Retrotransposons Resemble Retroviruses, but Cannot Move from Cell to Cell

A large family of transposons called **retroviral-like retrotransposons** (see Table 5-4) move themselves in and out of chromosomes by a mechanism similar to that used by retroviruses. These elements are present in organisms as diverse as yeasts, flies, and mammals; unlike viruses, they have no intrinsic ability to leave their resident cell but are passed along to all descendants of that cell through the normal processes of DNA replication and cell division. The first step in their transposition is the transcription of the entire transposon, producing an RNA copy of the element that is typically several thousand nucleotides long. This transcript, which is translated as a messenger RNA by the host cell, encodes a reverse transcriptase enzyme. This enzyme makes a double-strand DNA copy of the RNA molecule via an RNA–DNA hybrid intermediate, precisely mirroring the early stages of infection by a retrovirus (see Figure 5-61). Like a retrovirus, the linear, double-strand DNA molecule then integrates into a site on the chromosome using an integrase enzyme that is also encoded by the element. The structure and mechanisms of these integrases closely resemble those of the transposases of DNA-only transposons.

A Large Fraction of the Human Genome Is Composed of Nonretroviral Retrotransposons

A significant fraction of many vertebrate chromosomes is made up of repeated DNA sequences. In human chromosomes, these repeats are mostly mutated and truncated versions of **nonretroviral retrotransposons**, the third major type of transposon (see Table 5-4). Although most of these transposons in the human genome are immobile, a few retain the ability to move. Movements of the *L1 element* (sometimes referred to as a LINE, or long interspersed nuclear element) have been identified, some of which result in human disease; for example, a particular type of hemophilia results from an *L1* insertion into the gene encoding the blood-clotting protein Factor VIII (see Figure 6–25).

Nonretroviral retrotransposons are found in many organisms and move via a distinct mechanism that requires a complex of an endonuclease and a reverse transcriptase. As illustrated in **Figure 5–63**, the RNA and reverse transcriptase have a much more direct role in the recombination event than they do in the retroviral-like retrotransposons described above.

Inspection of the human genome sequence reveals that the bulk of nonretroviral retrotransposons—for example, the many copies of the *Alu* element, a member of the SINE (short interspersed nuclear element) family—do not carry their own endonuclease or reverse transcriptase genes. Nonetheless, they have successfully amplified themselves to become major constituents of our genome, presumably by pirating enzymes encoded by active *L1* elements. Together the LINEs and SINEs make up more than 30% of the human genome (see Figure 4–62); there are 500,000 copies of the former and more than a million of the latter.

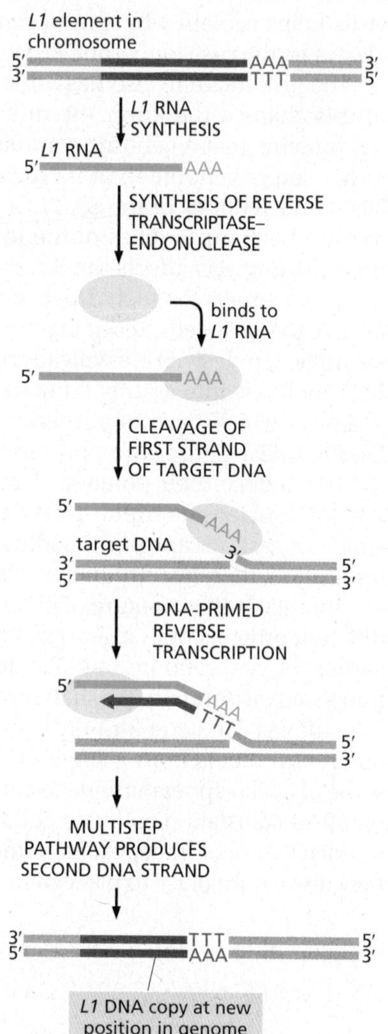

Figure 5–63 Transposition by a nonretroviral retrotransposon. Transposition of the *L1* element *(red)* begins when an endonuclease that is part of a complex with the *L1* reverse transcriptase *(green)* bound to the 3′ end of *L1* RNA *(blue)* nicks the target DNA at the point at which insertion will occur. This cleavage produces a 3′-OH DNA end in the target DNA, which is then used as a primer for the reverse transcription step shown. This generates a single-strand DNA copy of the element that is directly linked to the target DNA. In subsequent reactions, further processing of the single-strand DNA copy results in the generation of a new double-strand DNA copy of the *L1* element that is inserted at the site of the initial nick.

Different Transposable Elements Predominate in Different Organisms

We have described several types of transposable elements: (1) DNA-only transposons, the movement of which is based on DNA breaking and joining reactions; (2) retroviral-like retrotransposons, which also move via DNA breakage and joining, but where RNA has a key role as a template to generate the DNA recombination substrate; and (3) nonretroviral retrotransposons, in which an RNA copy of the element is central to the incorporation of the element into the target DNA, acting as a direct template for a DNA target-primed reverse transcription event.

Intriguingly, different types of transposons predominate in different organisms. For example, the vast majority of bacterial transposons are DNA-only types, with a few related to the nonretroviral retrotransposons also present. In yeasts, the main mobile elements are retroviral-like retrotransposons. In *Drosophila*, DNA-only, retroviral, and nonretroviral transposons are all found. Finally, the human genome contains all three types of transposon, but as discussed below, their evolutionary histories are strikingly different.

Genome Sequences Reveal the Approximate Times at Which Transposable Elements Have Moved

The nucleotide sequence of the human genome provides a rich fossil record of the activity of transposons over evolutionary time spans. By carefully comparing the nucleotide sequences of the approximately 3 million transposable element remnants in the human genome, it has been possible to broadly reconstruct the movements of transposons in our ancestors' genomes over the past several hundred million years. For example, the cut-and-paste DNA-only transposons appear to have been very active well before the divergence of humans and Old World monkeys (25–35 million years ago), but because they gradually accumulated inactivating mutations, they have been dormant in the human lineage since that time. Likewise, although our genome is littered with relics of retroviral-like retrotransposons, none appear to be active today. Only a single family of retroviral-like retrotransposons is believed to have transposed in the human genome since the divergence of human and chimpanzee approximately 6 million years ago. The nonretroviral retrotransposons are also ancient, but in contrast to other types, some are still moving in our genome, as mentioned previously. For example, it is estimated that *de novo* movement of an *Alu* element occurs once in every 100–200 human births. This movement of nonretroviral retrotransposons is responsible for a small but significant fraction of new human mutations—perhaps two mutations out of every thousand.

The situation in mice is significantly different. Although the mouse and human genomes contain roughly the same density of the three types of transposons, both types of retrotransposons are still actively transposing in the mouse genome, being responsible for approximately 10% of new mutations.

Although we are only beginning to understand how the movements of transposons have shaped the genomes of present-day mammals, it has been proposed that bursts in transposition activity could have been responsible for critical speciation events during the radiation of the mammalian lineages from a common ancestor, a process that began approximately 170 million years ago. At present, we can only wonder how many of our uniquely human qualities have been derived from the past activity of the mobile genetic elements whose remnants are found scattered throughout our chromosomes.

Conservative Site-specific Recombination Can Reversibly Rearrange DNA

A different kind of recombination mechanism, known as *conservative site-specific recombination*, rearranges other types of mobile DNA elements. In this pathway, breakage and joining occur at two special sites, one on each participating DNA molecule, with the recombination event being carried out by a specialized enzyme that breaks and rejoins the two DNA double helices at these specific sequences. The same enzyme system that joins two DNA molecules can often take them apart again, precisely restoring the sequence of the two original DNA molecules (**Figure 5–64A**). Alternatively, with a different orientation of these two sequences in a chromosome, conservative site-specific recombination produces a DNA inversion (**Figure 5–64B**).

The conservative site-specific recombination pathway illustrated in Figure 5–64A is often used by bacterial DNA viruses to move their genomes in and out of the genomes of their host cells. When integrated into its host genome, the viral DNA is replicated along with the host DNA and is faithfully passed on to all descendant cells. If the host cell suffers damage (for example, by UV irradiation), the virus can reverse the site-specific recombination reaction, excise its genome, and package it into a virus particle. In this way, many viruses can replicate themselves passively as a component of the host genome, but can also "leave the sinking ship" by excising their genomes and packaging them in a protective coat until a new, healthy host cell is encountered.

Several features distinguish conservative site-specific recombination from transposition. First, conservative site-specific recombination requires specialized

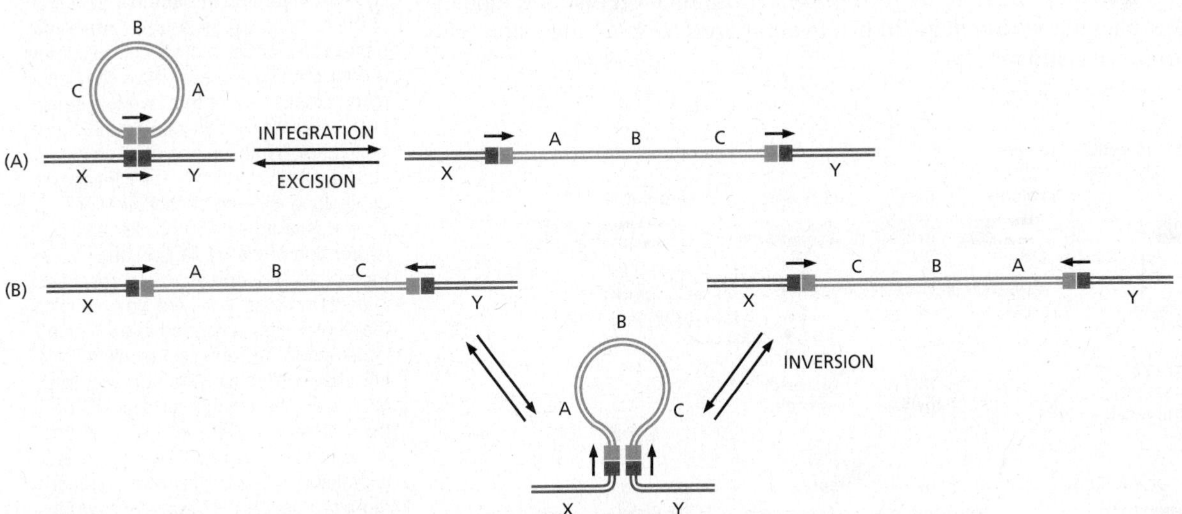

Figure 5–64 Two types of DNA rearrangement produced by conservative site-specific recombination. The only difference between the reactions in A and B is the relative orientation of the two short DNA sites (indicated by *arrows*) at which a site-specific recombination event occurs. (A) Through an integration reaction, a circular DNA molecule can become incorporated into a second DNA molecule; by the reverse reaction (excision), it can exit to re-form the original DNA circle. Many bacterial viruses move in and out of their host chromosomes in this way. (B) Conservative site-specific recombination can also invert a specific segment of DNA in a chromosome. A well-studied example of DNA inversion through site-specific recombination occurs in the bacterium *Salmonella enterica* serovar Typhimurium, as we discuss in the next section.

DNA sequences on both the donor and recipient DNA (hence the term "site-specific"). These sequences contain recognition sites for the particular recombinase that will catalyze the rearrangement. In contrast, transposition requires only that the transposon bears a specialized sequence; for most transposons, the recipient DNA can be of nearly any sequence. Second, the reaction mechanisms are fundamentally different. The recombinases that catalyze conservative site-specific recombination resemble topoisomerases in the sense that they form transient high-energy covalent bonds with the DNA and use this energy to complete all the DNA rearrangements without the need for new DNA synthesis (see Figure 5–22). Thus, all the phosphate bonds that are broken during a recombination event are restored upon its completion (hence the term "conservative"). Transposition, in contrast, typically leaves gaps in the DNA that must be repaired by DNA polymerases.

Conservative Site-specific Recombination Can Be Used to Turn Genes On or Off

Many bacteria use conservative site-specific recombination to control the expression of particular genes. A well-studied example occurs in *Salmonella* bacteria, an organism that is a major cause of food poisoning in humans. Known as **phase variation**, the switch in gene expression results from the occasional inversion of a specific 1000-nucleotide-pair piece of DNA, brought about by a conservative site-specific recombinase encoded in the *Salmonella* genome. This change alters the expression of the cell-surface protein flagellin, for which the bacterium has two different genes. The DNA inversion changes the orientation of a promoter (a DNA sequence that directs transcription of a gene) that is located within the inverted DNA segment. With the promoter in one orientation, the bacteria synthesize one type of flagellin; with the promoter in the other orientation, they synthesize the other type (**Figure 5–65**).

The recombination reaction is reversible, allowing bacterial populations to switch back and forth between the two types of flagellin. Inversions occur only rarely, and because such changes in the genome will be copied faithfully during all subsequent replication cycles, entire clones of bacteria will have one type of flagellin or the other.

Phase variation helps protect the bacterial population against the immune response of its vertebrate host. If the host makes antibodies against one type of flagellin, a few bacteria whose flagellin has been altered by gene inversion will still be able to survive and multiply.

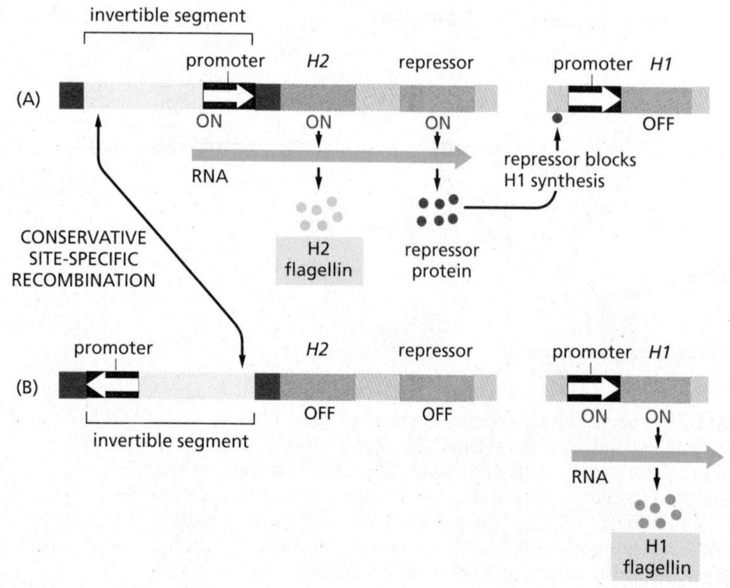

Figure 5–65 Switching gene expression by DNA inversion in bacteria. Which one of the two flagellin genes in a *Salmonella* bacterium is used to produce its flagellum is controlled by a conservative site-specific recombination event that inverts a small DNA segment containing a promoter. (A) In one orientation, the promoter activates transcription of the *H2* flagellin gene along with the transcription of a repressor protein that blocks the expression of the *H1* flagellin gene. Promoters and repressors are described in detail in Chapter 7; here we note simply that a promoter is needed to express a gene and that a repressor blocks this from happening. (B) When the promoter is inverted, it no longer turns on *H2* or the repressor, and the *H1* gene, which is thereby released from repression, is expressed instead. The inversion reaction requires specific DNA sequences *(red)* and a recombinase enzyme that is encoded in the invertible DNA segment. Because this conservative site-specific recombination mechanism is activated only rarely (about once in every 10^5 cell divisions), the production of one or the other type of flagellin tends to be faithfully inherited in each clone of cells.

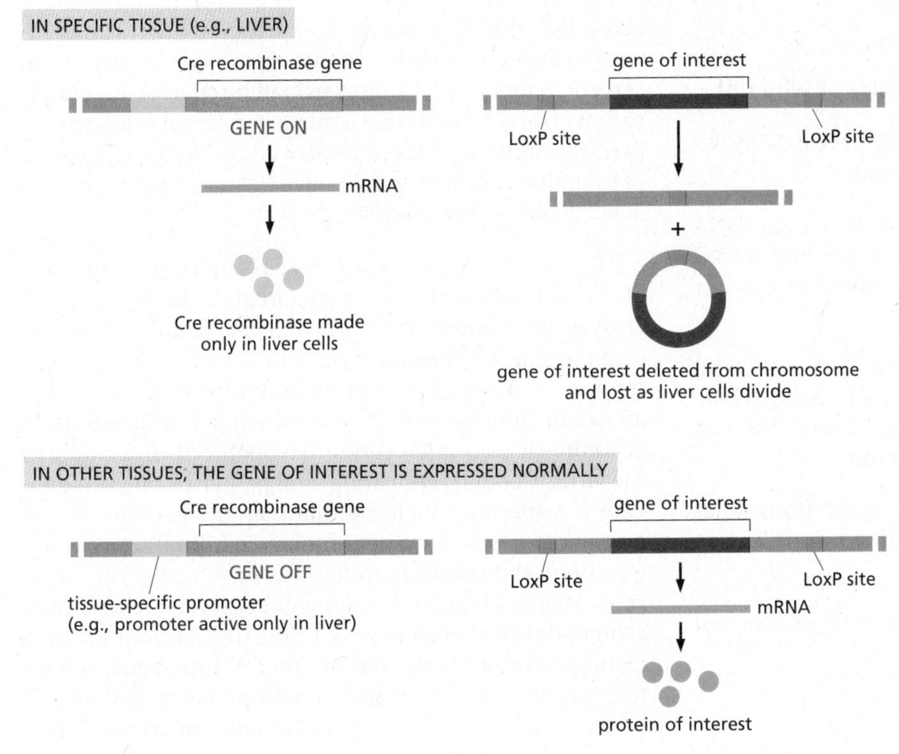

Figure 5–66 **How a conservative site-specific recombination enzyme from bacteria is used to delete a specific gene from a particular mouse tissue.** This approach requires the insertion of two specially engineered DNA molecules into the animal's germ line. The first contains the gene for a recombinase (in this case, the Cre recombinase from the bacteriophage P1) under the control of a tissue-specific promoter that ensures the recombinase is expressed only in that tissue. The second DNA molecule contains the gene of interest, flanked by the DNA sequences of the recognition sites for the recombinase (in this case, LoxP sites). The mouse has been engineered to contain only this copy of the gene of interest. Therefore, if the recombinase is expressed only in the liver, the gene of interest will be deleted there, and only there. As described in Chapter 7, many tissue-specific promoters are known; moreover, many of these promoters are active only at specific times in development. Thus, this method makes it possible to study the effects of deleting any gene of interest at specific times during the development of each tissue. For this reason, it is a powerful tool for scientists investigating the role of individual genes in animal and plant development.

Bacterial Conservative Site-specific Recombinases Have Become Powerful Tools for Cell and Developmental Biologists

Like many of the mechanisms used by cells and viruses, site-specific recombination has been put to work by scientists to aid in the study of a wide variety of problems. To decipher the roles of specific genes in complex multicellular organisms, genetic engineering techniques are used to produce worms, flies, and mice carrying both a gene encoding a site-specific recombination enzyme and a carefully designed target DNA that includes a gene of interest flanked by DNA sites recognized by the recombination enzyme. At an appropriate time, the gene encoding the enzyme can be activated to rearrange the target DNA sequence. Such a rearrangement is widely used to delete a specific gene in a particular tissue of a multicellular organism (**Figure 5–66**). This strategy is particularly useful when the gene of interest plays a key role in the early development of many tissues, and a complete deletion of the gene from the germ line would cause death very early in development. The same strategy can also be used to artificially express any specific gene in a tissue of interest; here, the triggered deletion joins a strong transcriptional promoter to the gene of interest. With this tool one can in principle determine the influence of any protein in any desired tissue of an intact animal.

Summary

The genomes of nearly all organisms contain mobile genetic elements that can move from one position in the genome to another by either transposition or conservative site-specific recombination. In most cases, this movement is random and happens at a very low frequency. There are three classes of transposons: the DNA-only transposons, the retroviral-like retrotransposons, and the nonretroviral retrotransposons. The first two classes have close relatives among the viruses, including the human retrovirus that causes AIDS, HIV. Although mobile genetic elements can be viewed as parasites, many of the new arrangements of DNA sequences that their recombination events produce have been important for creating the genetic variation required for the evolution of cells and organisms.

PROBLEMS

Which statements are true? Explain why or why not.

5–1 The majority of cells in your body have exactly the same nucleotide sequence in their genomes.

5–2 In a replication bubble, the same parent DNA strand serves as the template strand for leading-strand synthesis at one replication fork and as the template for lagging-strand synthesis at the other fork.

5–3 In *E. coli,* where the replication fork travels at 500 nucleotide pairs per second, the DNA ahead of the fork—in the absence of topoisomerase—would have to rotate at nearly 3000 revolutions per minute.

5–4 When bidirectional replication forks from adjacent origins meet, a leading strand always runs into a lagging strand.

5–5 DNA repair mechanisms all depend on the cell having two homologous chromosomes.

Discuss the following problems.

5–6 To determine the reproducibility of mutation frequency measurements, you do the following experiment. You inoculate each of 10 cultures with a single *E. coli* bacterium, allow the cultures to grow until each contains 10^6 cells, and then measure the number of cells in each culture that carry a mutation in your gene of interest. You were so surprised by the initial results that you repeated the experiment to confirm them. Both sets of results display the same extreme variability, as shown in Table Q5–1. Assuming that the rate of mutation is constant, why do you suppose there is so much variation in the frequencies of mutant cells in different cultures?

TABLE Q5–1 Frequencies of mutant cells in multiple cultures (Problem 5–6)

	Culture (mutant cells/10^6 cells)									
Experiment	1	2	3	4	5	6	7	8	9	10
1	4	0	257	1	2	32	0	0	2	1
2	128	0	1	4	0	0	66	5	0	2

5–7 Discuss the following statement: “Primase is a sloppy enzyme that makes many mistakes. Eventually, the RNA primers it makes are replaced with DNA made by a polymerase with higher fidelity. This is wasteful. It would be more energy-efficient if a DNA polymerase made an accurate copy in the first place.”

5–8 If DNA polymerase requires a perfectly paired primer in order to add the next nucleotide, how is it that any mismatched nucleotides “escape” this requirement and become substrates for mismatch repair enzymes?

5–9 DNA repair enzymes preferentially repair mismatched bases on the newly synthesized DNA strand, using the old DNA strand as a template. If mismatches were instead repaired without regard for which strand served as template, would mismatch repair reduce replication errors? Would such a mismatch repair system result in fewer mutations, more mutations, or the same number of mutations as there would have been without any repair at all? Explain your answers.

5–10 The laboratory you joined is studying the life cycle of an animal virus that uses circular, double-stranded DNA as its genome. Your project is to define the location of the origin(s) of replication and to determine whether replication proceeds in one or both directions away from an origin (unidirectional or bidirectional replication). To accomplish your goal, you broke open cells infected with the virus, isolated replicating viral genomes, cleaved them with a restriction nuclease that cuts the genome at only one site to produce a linear molecule from the circle, and examined the resulting molecules in the electron microscope. Some of the molecules you observed are illustrated schematically in Figure Q5–1. (Note that it is impossible to distinguish the orientation of one DNA molecule relative to another in the electron microscope because they land on the electron microscope grid in random orientations.)

You must present your conclusions to the rest of the lab tomorrow. How will you answer the two questions your advisor posed for you? Is there a single, unique origin of replication or several origins? Is replication unidirectional or bidirectional?

original molecule

bubbles

"H"-forms

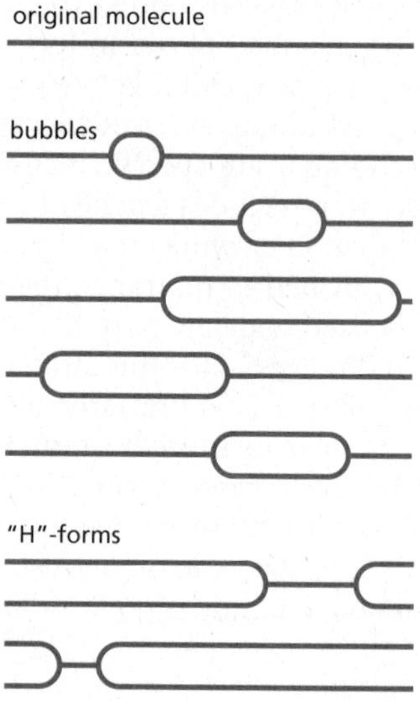

Figure Q5–1 Parent and replicating forms of an animal virus genome (Problem 5–10).

5–11 You are investigating DNA synthesis in tissue-culture cells, using ^{3}H-thymidine to radioactively label the replication forks. By breaking open the cells in a way that allows some of the DNA strands to be stretched out, very long DNA strands can be isolated intact and examined. You overlay the DNA with a photographic emulsion, and expose it for 3–6 months, a procedure known as autoradiography. Because the emulsion is sensitive to radioactive emissions, the ^{3}H-labeled DNA shows up as tracks of silver grains. Because the stretching collapses replication bubbles, the daughter duplexes lie side by side and cannot be distinguished from each other.

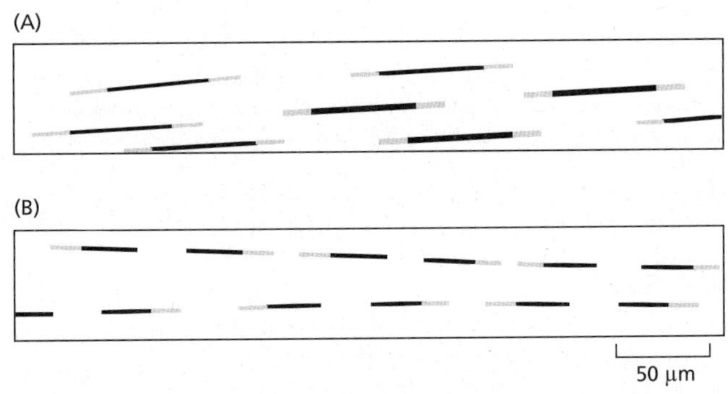

Figure Q5–2 Autoradiographic investigation of DNA replication in cultured cells (Problem 5–11). (A) Addition of ^{3}H-labeled thymidine immediately after release from the arrest. (B) Addition of ^{3}H-labeled thymidine 30 minutes after release from the arrest.

You treat the cells to arrest them at the beginning of S phase. In the first experiment, you release the arrest and add ^{3}H-thymidine immediately. After 30 minutes, you wash the cells and change the medium so that the total concentration of thymidine is the same as it was, but only one-third of it is radioactive. After an additional 15 minutes, you prepare DNA for autoradiography. The results of this experiment are shown in Figure Q5–2A. In the second experiment, you release the arrest and wait 30 minutes before adding ^{3}H-thymidine. After 30 minutes in the presence of ^{3}H-thymidine, you once again change the medium to reduce the concentration of radioactive thymidine and incubate the cells for an additional 15 minutes. The results of the second experiment are shown in Figure Q5–2B.

A. Explain why, in both experiments, some regions of the tracks are dense with silver grains (dark), whereas others are less dense (light).

B. In the first experiment, each track has a central dark section with light sections at each end. In the second experiment, the dark section of each track has a light section at only one end. Explain the reason for this difference.

C. Estimate the rate of fork movement (μm/min) in these experiments. Do the estimates from the two experiments agree? Can you use this information to gauge how long it would take to replicate the entire genome?

5–12 If you compare the frequency of the 16 possible dinucleotide sequences in the *E. coli* and human genomes, there are no striking differences except for one dinucleotide, 5′-CG-3′. The frequency of CG dinucleotides in the human genome is significantly lower than in *E. coli* and significantly lower than expected by chance. Why do you suppose that CG dinucleotides are underrepresented in the human genome?

5–13 With age, somatic cells are thought to accumulate genomic "scars" as a result of the inaccurate repair of double-strand breaks by nonhomologous end joining (NHEJ). Estimates based on the frequency of breaks in primary human fibroblasts suggest that by age 70, each human somatic cell may carry some 2000 NHEJ-induced mutations due to inaccurate repair. If these mutations were distributed randomly around the genome, how many protein-coding genes would you expect to be affected? Would you expect cell function to be compromised? Why or why not? (Assume that 2% of the genome—1.5% protein-coding and 0.5% regulatory—is crucial information.)

5–14 Draw a schematic diagram of the double Holliday junction that would result from strand invasion by both ends of the broken duplex into the intact homologous duplex shown in Figure Q5–3. Label the left end of each strand in the Holliday junction 5′ or 3′ so that the relationship to the parent and recombinant duplexes is clear. Indicate how DNA synthesis would be used to fill in any single-strand gaps in your double Holliday junction.

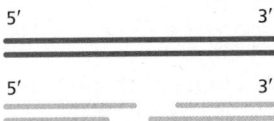

Figure Q5–3 A broken duplex with single-strand tails ready to invade an intact homologous duplex (Problem 5–14).

5–15 In addition to correcting DNA mismatches, the mismatch repair system acts to discourage homologous recombination between DNA duplexes that are only moderately similar in sequence. Why would recombination between moderately similar sequences pose a problem for human cells?

5–16 Cre recombinase is a site-specific enzyme that catalyzes recombination between two LoxP DNA sites. Cre recombinase pairs two LoxP sites in the same orientation, breaks both duplexes at the same point in each LoxP site, and joins the ends with new partners so that each LoxP site is regenerated, as shown schematically in Figure Q5–4A. On the basis of this mechanism, predict the arrangement of sequences that will be generated by Cre-mediated site-specific recombination for each of the two DNAs shown in Figure Q5–4B.

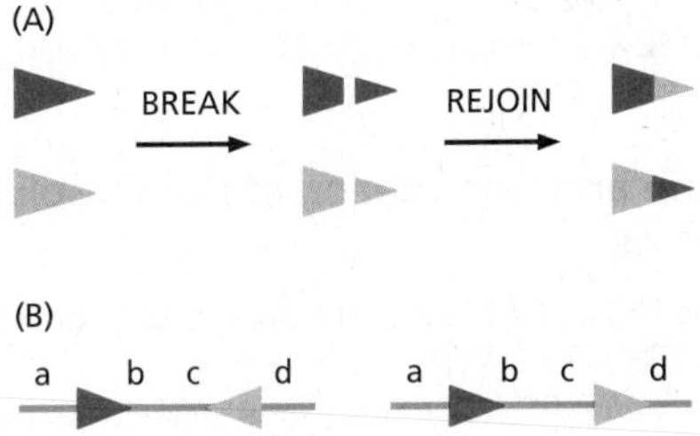

Figure Q5–4 Cre recombinase–mediated site-specific recombination (Problem 5–16). (A) Schematic representation of Cre/LoxP site–specific recombination. The LoxP sequences in the DNA are represented by *triangles* that are colored differently so that the site-specific recombination event can be followed more readily. In reality their DNA sequences are identical. (B) DNA substrates containing two arrangements of LoxP sites.

5–17 It is thought that a self-correcting polymerase cannot start chains *de novo* because the initial nucleotides will be weakly paired, thus subject to removal by an efficient proofreading exonuclease. This argument implies the converse: an enzyme that starts chains anew cannot be efficient at self-correction. Remarkably, SARS-CoV-2 encodes an RNA replicase complex that is able to do both: using RNA as a template, it starts strands *de novo* and removes errors with an efficient 3′-to-5′ proofreading exonuclease. How might this be possible?

REFERENCES

General

Brown TA (2017) Genomes 4. New York: Garland Science.

Friedberg EC, Walker GC, Siede W et al. (2005) DNA Repair and Mutagenesis. Washington, DC: ASM Press.

Haber JE (2013) Genome Stability: DNA Repair and Recombination. New York: Garland Science.

Stent GS (1971) Molecular Genetics: An Introductory Narrative. San Francisco: WH Freeman.

Watson J, Baker T, Bell S et al. (2013) Molecular Biology of the Gene, 7th ed. Menlo Park, CA: Benjamin Cummings.

The Maintenance of DNA Sequences

Catarina D & Eichler EE (2013) Properties and rates of germline mutations in humans. *Trends Genet*. 29, 575–584.

Conrad DF, Keebler J, DePristo M . . . 1000 Genomes Project (2011) Variation in genome-wide mutation rates within and between human families. *Nat. Genet.* 43, 712–714.

Hedges SB (2002) The origin and evolution of model organisms. *Nat. Rev. Genet.* 3, 838–849.

King MC & Wilson AC (1965) Evolution at two levels in humans and chimpanzees. *Science* 188, 107–116.

DNA Replication Mechanisms

Kornberg A (1960) Biological synthesis of DNA. *Science* 131, 1503–1508.

Li JJ & Kelly TJ (1984) SV40 DNA replication *in vitro*. *Proc. Natl. Acad. Sci. USA* 81, 6973–6977.

Meselson M & Stahl FW (1958) The replication of DNA in *E. coli*. *Proc. Natl. Acad. Sci. USA* 44, 671–682.

Modrich P (2016) Mechanisms in *E. coli* and human mismatch repair (Nobel Lecture). *Angew. Chem. Int. Ed. Engl*. 55(30), 8490–8501.

Okazaki R, Okazaki T, Sakabe K, Sugimoto K & Sugino A (1968) Mechanism of DNA chain growth. I. Possible discontinuity and unusual secondary structure of newly synthesized chains. *Proc. Natl. Acad. Sci. USA* 59, 598–605.

Yao NY & O'Donnell ME (2021) The DNA replication machine: structure and dynamic function. In Macromolecular Protein Complexes III: Structure and Function (Harris JR & Marles-Wright J, eds.), pp. 233–258. *Subcellular Biochemistry*, Vol. 96. Cham, Switzerland: Springer.

The Initiation and Completion of DNA Replication in Chromosomes

Bell SP & Labib K (2016) Chromosome duplication in *Saccharomyces cerevisiae*. *Genetics* 203(3), 1027–1067.

Chan SR & Blackburn EH (2004) Telomeres and telomerase. *Philos. Trans. R. Soc. Lond. B Biol. Sci.* 359, 109–121.

Dewar JM & Walter JC (2017) Mechanisms of DNA replication termination. *Nat. Rev. Mol. Cell Biol.* 18, 507–516.

On KF, Jaremko M, Stillman B & Leemor JT (2018) A structural view of the initiators for chromosome replication. *Curr. Opin. Struct. Biol.* 53, 131–139.

Roake CM & Artandi SE (2020) Regulation of human telomerase in homeostasis and disease. *Nat. Rev. Mol. Cell Biol.* 21, 384–397.

Schvartzman JB, Hernandez P, Krimer DB, Dorier J & Stasiak A (2019) Closing the DNA replication cycle: from simple circular molecules to supercoiled and knotted DNA catenanes. *Nucleic Acids Res*. 47(14), 7182–7198.

Wang Y, Sušac L & Feigon J (2019) Structural biology of telomerase. *Cold Spring Harb. Perspect. Biol*. 11(12), a032383.

DNA Repair

Her J & Bunting SF (2018) How cells ensure correct repair of DNA double-strand breaks. *J. Biol. Chem.* 293(27), 10502–10511.

Lindahl T (1993) Instability and decay of the primary structure of DNA. *Nature* 362, 709–715.

Sancar A (2016) Mechanisms of DNA repair by photolyase and excision nuclease (Nobel Lecture). *Angew. Chem. Int. Ed. Engl.* 55(30), 8502–8527.

Van den Heuvel D, Van der Weegen Y, Boer DEC, Ogi T & Luijsterburg MS (2021) Transcription-coupled DNA repair: from mechanism to human disorder. *Trends Cell Biol.* 31(5), 359–371.

Yang W & Gao Y (2018) Translesion and repair DNA polymerases: diverse structure and mechanism. *Annu. Rev. Biochem*. 87, 239–261.

Zhao B, Rothenberg E, Ramsden DA & Lieber MR (2020) The molecular basis and disease relevance of non-homologous DNA end joining. *Nat. Rev. Mol. Cell Biol.* 21, 765–781.

Homologous Recombination

Cox MM (2001) Historical overview: searching for replication help in all of the rec places. *Proc. Natl. Acad. Sci. USA* 98, 8173–8180.

Haber JE (2016) A life investigating pathways that repair broken chromosomes. *Annu. Rev. Genet.* 50, 1–28.

Holliday R (1990) The history of the DNA heteroduplex. *BioEssays* 12, 133–142.

Kreuzer KN & Brister JR (2010) Initiation of bacteriophage T4 DNA replication and replication fork dynamics: a review in the Virology Journal series on bacteriophage T4 and its relatives. *Virol. J.* 7, 358.

Szostak JW, Orr-Weaver TK, Rothstein RJ & Stahl FW (1983) The double-strand break repair model for recombination. *Cell* 33, 25–35.

Wyatt HDM & West SC (2014) Holliday junction resolvases. *Cold Spring Harb. Perspect. Biol.* 6(9), a023192.

Zickler D & Kleckner N (2015) Recombination, pairing, and synapsis of homologs during meiosis. *Cold Spring Harb. Perspect. Biol.* 7(6), a016626.

Transposition and Conservative Site-specific Recombination

Comfort NC (2001) From controlling elements to transposons: Barbara McClintock and the Nobel Prize. *Trends Biochem. Sci.* 26, 454–457.

Huang, CR, Burns KH & Boeke JD (2012) Active transposition in genomes. *Annu. Rev. Genet.* 46, 651–675.

Payer LM & Burns KH (2019) Transposable elements in human genetic disease. *Nat. Rev. Genet*. 20(12), 760–772.

Varmus H (1988) Retroviruses. *Science* 240, 1427–1435.

CHAPTER

6

How Cells Read the Genome: From DNA to Protein

IN THIS CHAPTER

Since the structure of DNA was discovered in the early 1950s, progress in cell and molecular biology has been astounding. We now know the complete genome sequences for thousands of different organisms, revealing fascinating details of their biochemistry as well as important clues as to how these organisms evolved. Complete genome sequences have also been obtained for hundreds of thousands of individual humans, as well as for a few of our now-extinct relatives, such as the Neanderthals. Knowing the maximum amount of information that is required to produce a complex organism like ourselves puts constraints on the biochemical and structural features of cells and makes it clear that biology is not infinitely complex.

As discussed in Chapter 1, most of the genetic information carried by DNA specifies the sequence of amino acids in proteins. But this DNA does not direct the synthesis of proteins directly, instead producing RNA as an intermediary. When the cell needs a particular protein, the nucleotide sequence of the appropriate portion of the DNA molecule in a chromosome is first copied into RNA (a process called *transcription*). These RNA copies of segments of the DNA sequence are then used to direct the synthesis of the protein (a process called *translation*). The genetic information in cells thereby flows from DNA to RNA to protein. All cells, from bacteria to humans, express their genetic information in this way—a principle so fundamental that it is termed the *central dogma* of molecular biology (**Figure 6–1**).

Despite this universality, there are important variations between organisms in the way information flows from DNA to protein. Most notable, the RNA transcripts in eukaryotic cells are subject to a series of processing steps in the nucleus, including *RNA splicing*, before they are permitted to exit from the nucleus and be translated into protein. As we discuss in this chapter, these processing steps can critically change the "meaning" of an RNA molecule, and they are therefore crucial for understanding how eukaryotic cells read their genome.

Although we shall focus in this chapter on the production of the proteins encoded by the genome, for some genes RNA is the final product. Like proteins, some of these RNAs fold into precise three-dimensional structures that have structural and catalytic roles in the cell. Although the functions of many noncoding RNAs are not yet known, some have been studied in great detail and are discussed in this and the following chapter.

One might have expected that the information present in genomes would be arranged in an orderly fashion, resembling a dictionary or a telephone directory. But it turns out that the genomes of most multicellular organisms are surprisingly disorderly, reflecting their chaotic evolutionary histories. The genes in these organisms largely consist of a long string of alternating short exons and long introns, as discussed in Chapter 4 (see Figure 4–15D). Moreover, small bits of DNA sequence that code for protein are interspersed with large blocks of seemingly meaningless DNA. Some sections of the genome contain many genes and others lack genes altogether. Proteins that work closely with one another in the cell usually have their genes located on different chromosomes, and adjacent genes typically encode proteins that have little to do with each other in the cell. Decoding genomes is therefore no simple matter. Even with the aid of powerful

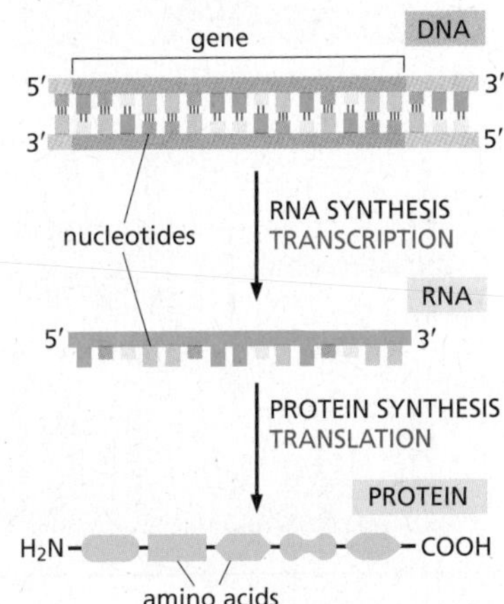

Figure 6–1 Genetic information directs the synthesis of proteins. The flow of genetic information from DNA to RNA (transcription) and from RNA to protein (translation) occurs in all living cells. As we saw in Chapter 5, DNA can also be copied—or replicated—to produce new DNA molecules.

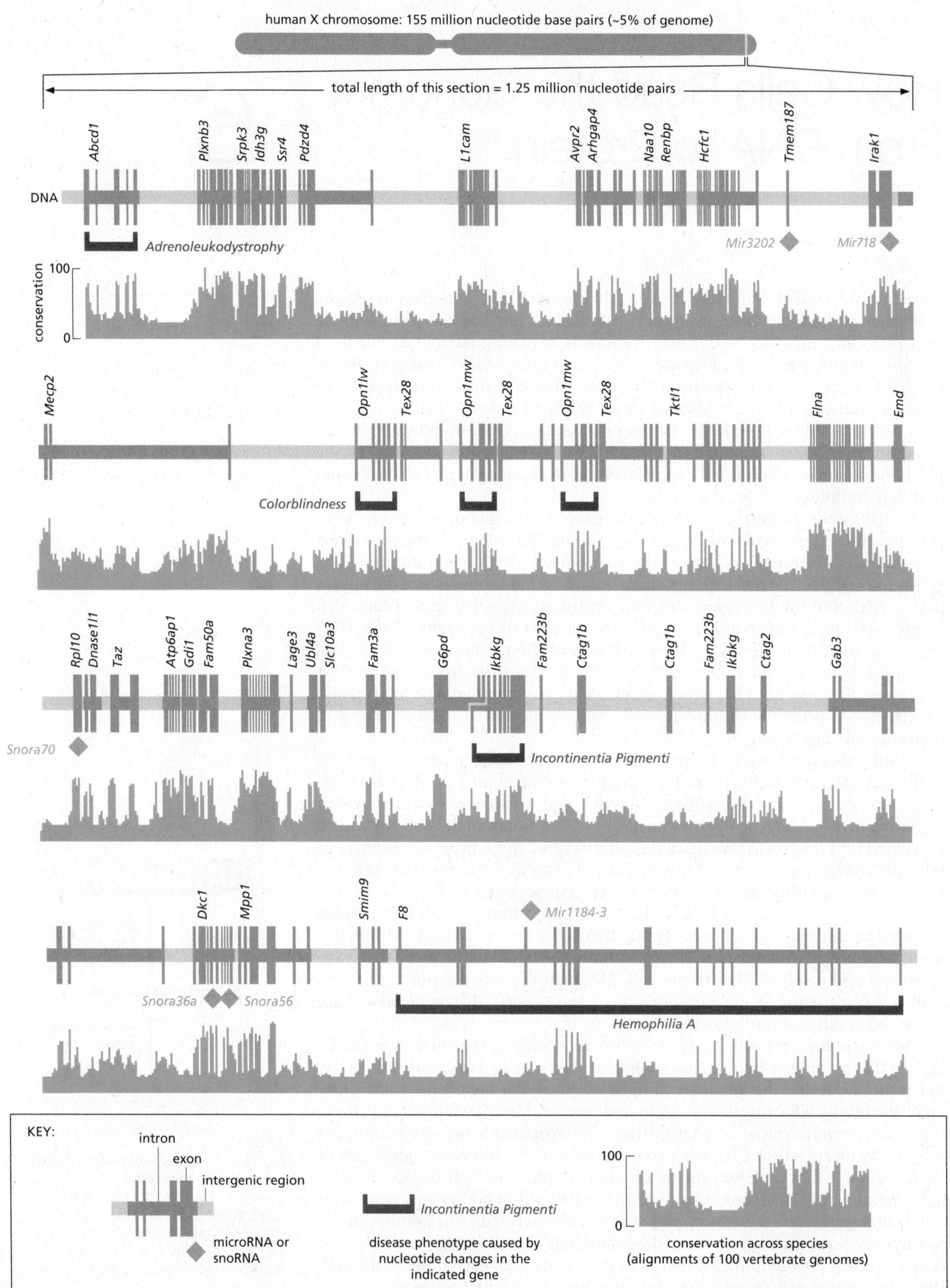
human X chromosome: 155 million nucleotide base pairs (~5% of genome)
total length of this section = 1.25 million nucleotide pairs
Abcd1
Plxnb3
Srpk3
Idh3g
Ssr4
Pdzd4
L1cam
Avpr2
Arhgap4
Naa10
Renbp
Hcfc1
Tmem187
Irak1
DNA
Adrenoleukodystrophy
Mir3202
Mir718
conservation
100
0
Mecp2
Opn1lw
Tex28
Opn1mw
Tex28
Opn1mw
Tex28
Tktl1
Flna
Emd
Colorblindness
Rpl10
Dnase1l1
Taz
Atp6ap1
Gdi1
Fam50a
Plxna3
Lage3
Ubl4a
Slc10a3
Fam3a
G6pd
Ikbkg
Fam223b
Ctag1b
Ctag1b
Fam223b
Ikbkg
Ctag2
Gab3
Snora70
Incontinentia Pigmenti
Dkc1
Mpp1
Smim9
F8
Mir1184-3
Snora36a
Snora56
Hemophilia A
KEY:
intron
exon
intergenic region
microRNA or snoRNA
Incontinentia Pigmenti
disease phenotype caused by nucleotide changes in the indicated gene
100
0
conservation across species (alignments of 100 vertebrate genomes)

Figure 6–2 **Schematic depiction of a small portion of the human X chromosome.** As summarized in the key, the known protein-coding genes (starting with *Abcd1* and ending with *F8*) are marked by a *dark gray* central line, with their coding regions (exons) indicated by *dark gray* bars that extend above and below this line. Noncoding RNAs with known functions are indicated by *purple diamonds*. The *blue* histogram indicates the extent to which portions of the human genome are conserved with other vertebrate species. It is likely that additional genes, currently unrecognized, also lie within this portion of the human genome.

Genes whose mutation causes an inherited human condition are indicated by *red brackets*. The *Abcd1* gene codes for a protein that imports fatty acids into the peroxisome; mutations in the gene cause demyelination of nerves, which can result in cognition and movement disorders. *Incontinentia pigmenti* is a disease of the skin, hair, nails, teeth, and eyes. *Hemophilia A* is a bleeding disorder caused by mutations in the Factor VIII gene, which codes for a blood-clotting protein (see Figure 6–25B). Because males have only a single copy of the X chromosome, most of the conditions shown here affect only males; females that inherit one of these defective genes are often asymptomatic because a functional protein is made from their other X chromosome. (Courtesy of Alex Williams, data obtained from the University of California, Genome Browser, http://genome.ucsc.edu.)

computers, it is difficult for researchers, in the absence of direct experimental evidence, to locate definitively the beginning and end of genes, much less to decipher when and where each gene is expressed in the life of the organism. Yet the cells in our body do this automatically, thousands of times a second.

The problems that cells face in decoding genomes can be appreciated by considering a tiny portion of the human genome (**Figure 6–2**). The region illustrated represents less than 1/2000 of our genome and includes at least 48 genes that encode proteins plus 6 genes for noncoding RNAs. When we consider the entire human genome, we can only marvel at the capacity of our cells to rapidly and accurately handle such large amounts of information.

In this chapter, we explain how cells decode and use the information in their genomes. Much has been learned about how the genetic instructions written in an alphabet of just four "letters"—the four different nucleotides in DNA—direct the formation of a bacterium, a fruit fly, or a human. Nevertheless, we still have a great deal to discover about how the information stored in an organism's genome produces even the simplest unicellular bacterium with about 500 genes, let alone how it directs the development of a human with approximately 25,000 genes. An enormous amount of ignorance remains; many fascinating challenges therefore await the next generation of cell biologists.

FROM DNA TO RNA

Transcription and translation are the means by which cells read out, or express, the genetic instructions in their genes. Because many identical RNA copies can be made from the same gene, and each RNA molecule can direct the synthesis of many identical protein molecules, cells can synthesize a large amount of protein from a single gene when necessary. Importantly, genes can be transcribed and translated with different efficiencies, allowing the cell to make vast quantities of some proteins and tiny amounts of others (**Figure 6–3**). Moreover, as we see in the next chapter, a cell can change (or regulate) the expression of each of

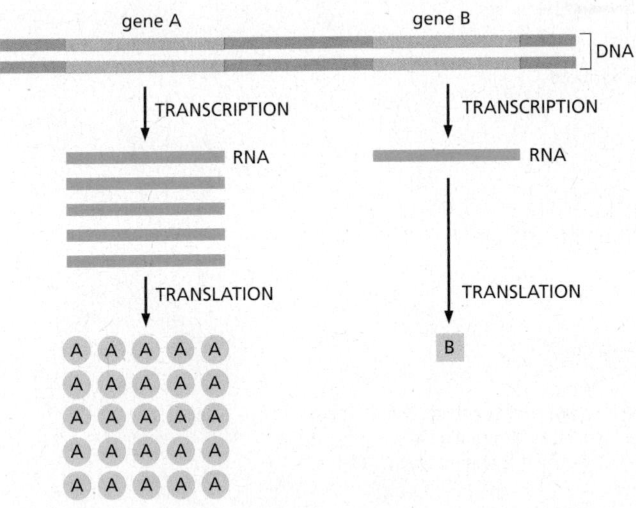

Figure 6–3 **Genes can be expressed with different efficiencies.** In this example, gene A is transcribed much more efficiently than gene B, and each RNA molecule that it produces is also translated more frequently. This causes the amount of protein A in the cell to be much greater than that of protein B. In this and later figures, the portions of DNA that are transcribed are shown in *orange*.

its genes according to its needs—commonly by controlling the production of its RNA—and many genes will not be expressed at all in some cells.

One of the central problems in producing proteins from the information carried in genomes is that most steps depend on conventional nucleic acid base-pairing, which on its own has only modest specificity. In many contexts, a correct base pair is only 10–100 times more thermodynamically stable than an incorrect base pair, so that most steps of gene expression rely on mechanisms that both improve the specificity of the base-pairing and correct the many mistakes that arise. A central theme of this chapter, therefore, is the way cells deal with the fundamentally inaccurate base-pairing process that lies at the heart of the mechanisms that they use to read their genome.

RNA Molecules Are Single-Stranded

The first step a cell takes in reading out its genetic instructions is to copy a particular portion of its DNA nucleotide sequence—a gene—into an RNA nucleotide sequence (**Figure 6–4**). The information in RNA, although copied into another chemical form, is still written in essentially the same language as it is in DNA—the language of a nucleotide sequence. Hence the name given to producing RNA molecules on DNA is *transcription*.

Like DNA, RNA is a linear polymer made of four different types of nucleotide subunits linked together by phosphodiester bonds (see Figure 6–4). It differs from DNA chemically in two respects: (1) the nucleotides in RNA are *ribonucleotides*—that is, they contain the sugar ribose (hence the name *ribo*-nucleic acid) rather than deoxyribose; (2) although, like DNA, RNA contains the bases adenine (A), guanine (G), and cytosine (C), it contains the base uracil (U) instead of the thymine (T) in DNA (**Figure 6–5**). Because U, like T, can base-pair by hydrogen-bonding with A (**Figure 6–6**), the complementary base-pairing properties described for DNA in Chapters 4 and 5 apply also to RNA (in RNA, G pairs with C, and A pairs with U).

Although these chemical differences are slight, DNA and RNA differ quite dramatically in overall structure. Whereas DNA always occurs in cells as a double-strand helix, RNA is single-stranded. An RNA chain can therefore fold up into a particular shape, just as a polypeptide chain folds up to form the final shape of a protein (**Figure 6–7**). As we see later in this chapter, the ability to fold into complex three-dimensional shapes allows some RNA molecules to have precise structural and catalytic functions.

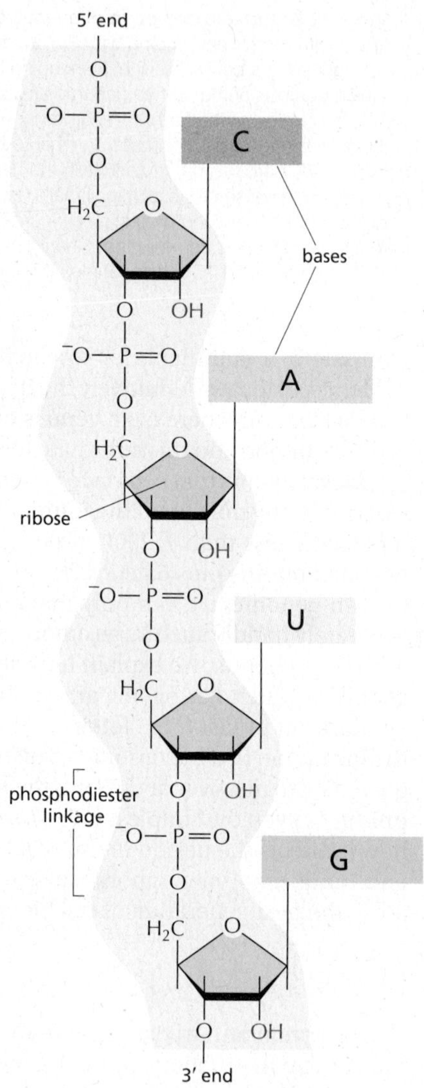

Figure 6–4 **A short length of RNA.** The phosphodiester chemical linkage between nucleotides in RNA is the same as that in DNA.

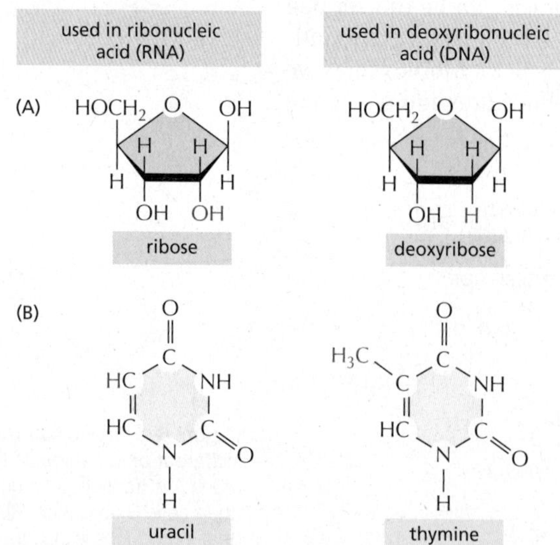

Figure 6–5 **The chemical structure of RNA.** (A) RNA contains the sugar ribose, which differs from deoxyribose, the sugar used in DNA, by the presence of an additional –OH group. (B) RNA contains the base uracil, which differs from thymine, the equivalent base in DNA, by the absence of a $-CH_3$ group.

Transcription Produces RNA Complementary to One Strand of DNA

The RNA in a cell is made by **DNA transcription**, a process that has certain similarities to the process of DNA replication discussed in Chapter 5. Transcription begins with the opening and unwinding of a small portion of the DNA double helix to expose the bases on each DNA strand. One of the two strands of the DNA double helix then acts as a template for the synthesis of an RNA molecule. As in DNA replication, the nucleotide sequence of the RNA chain is determined by the complementary base-pairing between incoming nucleotides and the DNA template. When a good match is made (A with T, U with A, G with C, and C with G), the incoming ribonucleotide is covalently linked to the growing RNA chain in an enzymatically catalyzed reaction. The RNA chain produced by transcription—the *transcript*—is therefore elongated one nucleotide at a time, and it has a nucleotide sequence that is exactly complementary to the strand of DNA used as the template (**Figure 6–8**).

Transcription, however, differs from DNA replication in several crucial ways. Unlike a newly formed DNA strand, the RNA strand does not remain hydrogen-bonded to the DNA template strand. Instead, just behind the region where the ribonucleotides are being added, the RNA chain is displaced and the DNA helix re-forms. Thus, the RNA molecules produced by transcription are released from the DNA template as single strands. In addition, because they are copied from only a limited region of the DNA, RNA molecules are much shorter than DNA molecules. A DNA molecule in a human chromosome can be up to 250 million nucleotide-pairs long; in contrast, most RNAs are no more than a few thousand nucleotides long, and many are considerably shorter.

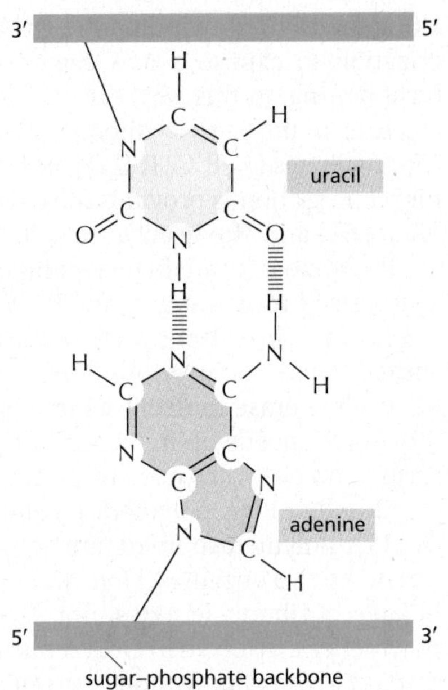

Figure 6–6 **Uracil base pairs with adenine.** The absence of a methyl group in U has no effect on base-pairing; thus, U-A base pairs closely resemble T-A base pairs (see Figure 4–5).

RNA Polymerases Carry Out DNA Transcription

The enzymes that perform transcription are called **RNA polymerases.** Like the DNA polymerase that catalyzes DNA replication (discussed in Chapter 5), RNA polymerases catalyze the formation of the phosphodiester bonds that link the nucleotides together to form a linear chain. The RNA polymerase moves stepwise

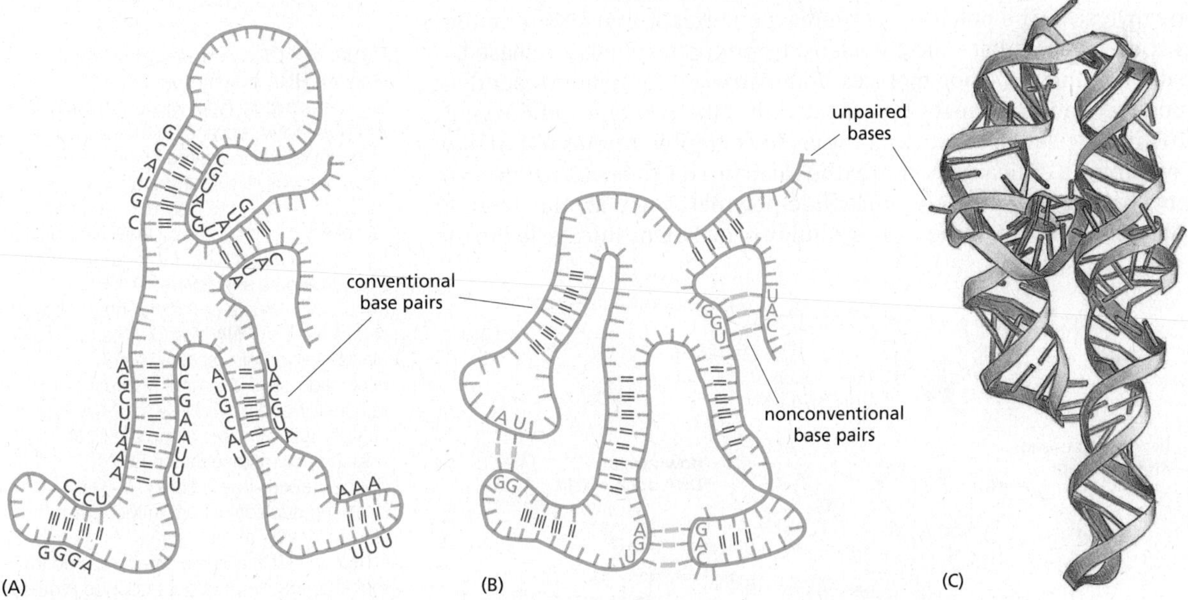

Figure 6–7 **RNA can fold into specific structures.** RNA is largely single-stranded, but it often contains short stretches of nucleotides that can form conventional base pairs with complementary sequences found elsewhere on the same molecule. These interactions, along with additional "nonconventional" base-pair interactions (for example, A-G), allow an RNA molecule to fold into a three-dimensional structure that is determined by its sequence of nucleotides (**Movie 6.1**). (A) Diagram of a folded RNA structure showing only conventional (G-C and A-U) base-pair interactions *(red)*. (B) Formation of nonconventional *(green)* base-pair interactions folds the hypothetical structure shown in A even further. (C) Structure of an actual RNA molecule, in this case one that catalyzes its own splicing (see pp. 347–348). Each conventional base-pair interaction is indicated by a "rung" in the double helix. Bases in other configurations are indicated by broken rungs.

along the DNA, unwinding the DNA helix just ahead of its active site for polymerization to expose a new region of the template strand for complementary base-pairing. In this way, the growing RNA chain is extended by one nucleotide at a time in the 5′-to-3′ direction (**Figure 6–9**). The substrates are ribonucleoside triphosphates (ATP, CTP, UTP, and GTP); as in DNA replication, the hydrolysis of high-energy bonds provides the energy needed to drive the reaction forward (see Figure 5–4 and **Movie 6.2**).

The almost immediate separation of the RNA strand from the DNA as it is synthesized means that many RNA copies can be made from the same gene in a relatively short time, with the synthesis of additional RNA molecules being started before the previous RNA molecules are completed (**Figure 6–10**). When RNA polymerase molecules follow hard on each other's heels in this way, each moving at speeds up to 50 nucleotides per second, more than a thousand transcripts can be synthesized in an hour from a single gene.

Although RNA polymerase catalyzes essentially the same chemical reaction as DNA polymerase, there are some important differences between the activities of the two enzymes. First, and most obviously, RNA polymerase catalyzes the linkage of ribonucleotides, not deoxyribonucleotides. Second, unlike the DNA polymerases involved in DNA replication (see pp. 259–260), RNA polymerases can start an RNA chain without a primer. This difference is thought possible because transcription need not be as accurate as DNA replication (see Table 5–1, p. 260). RNA polymerases make about one mistake for every 10^4 nucleotides copied into RNA (compared with an error rate for direct copying and proofreading by DNA polymerase of about one in 10^7 nucleotides), and the consequences of an error in RNA transcription are much less significant as RNA does not permanently store genetic information in cells. Finally, unlike DNA polymerases, which make their products in segments that are later stitched together, RNA polymerases are processive; that is, the same RNA polymerase that begins an RNA molecule must finish it without dissociating from the DNA template.

Although not nearly as accurate as the DNA polymerases that replicate DNA, RNA polymerases nonetheless have a modest proofreading mechanism. If an incorrect ribonucleotide is added to the growing RNA chain, the polymerase can back up, and the active site of the enzyme can perform an excision reaction that resembles the reverse of the polymerization reaction, except that a water molecule replaces the pyrophosphate and a nucleoside monophosphate is released.

Given that DNA and RNA polymerases both carry out template-dependent nucleotide polymerization, it might be expected that the two types of enzymes would be structurally related. However, x-ray crystallographic studies reveal that, other than containing a critical Mg^{2+} ion at the catalytic site, the two enzymes are quite different. Template-dependent nucleotide-polymerizing enzymes seem to have arisen at least twice during the early evolution of cells. One lineage led to the

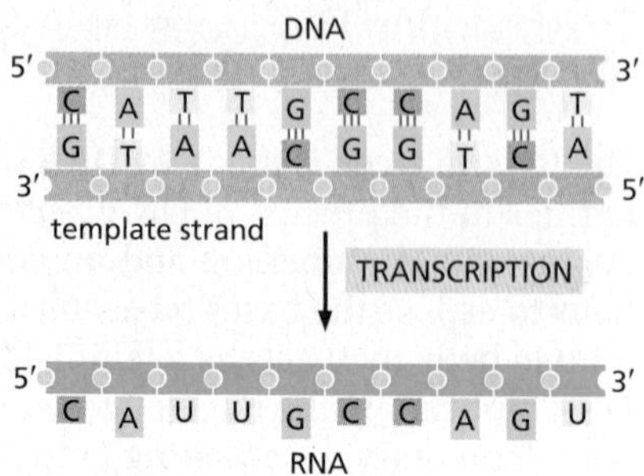

Figure 6–8 DNA transcription produces a single-strand RNA molecule that is complementary to one strand of the DNA double helix. Note that the sequence of bases in the RNA molecule produced is the same as the sequence of bases in the non-template DNA strand, except that a U replaces every T base in the DNA.

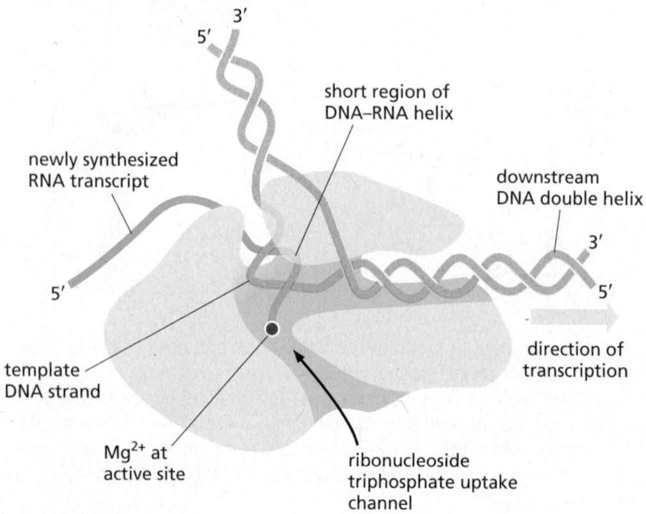

Figure 6–9 DNA is transcribed by the enzyme RNA polymerase. The RNA polymerase *(pale blue)* moves stepwise along the DNA, unwinding the DNA helix at its active site indicated by the Mg^{2+} *(red)*, which is required for catalysis. As it progresses, the polymerase adds nucleotides one by one to the RNA chain at the polymerization site, using an exposed DNA strand as a template. The RNA transcript is thus a complementary copy of one of the two DNA strands. A short region of DNA–RNA helix (approximately nine nucleotide pairs in length) is formed only transiently, and a "window" of DNA–RNA helix therefore moves along the DNA with the polymerase as the DNA double helix re-forms behind it. The incoming nucleotides are in the form of ribonucleoside triphosphates (ATP, UTP, CTP, and GTP), and the energy stored in their phosphate–phosphate bonds provides the driving force for the polymerization reaction. The figure, based on an x-ray crystallographic structure, shows a cutaway view of the polymerase: the part facing the viewer has been sliced away to reveal the interior (**Movie 6.3**). (Adapted from P. Cramer et al., *Science* 288:640–649, 2000. PDB code: 1HQM.)

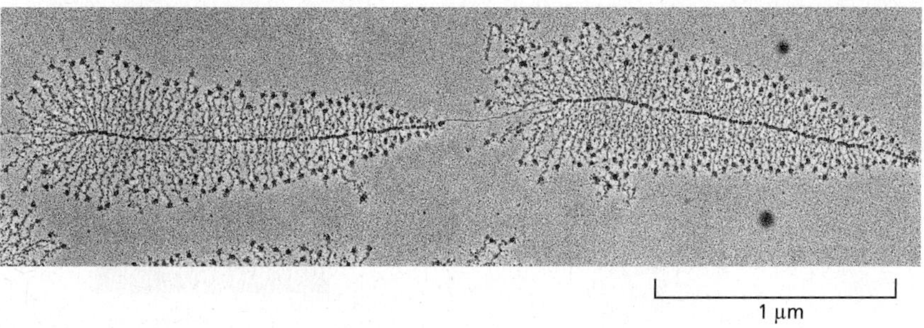

Figure 6–10 **Transcription of two genes as observed under the electron microscope.** The micrograph shows many molecules of RNA polymerase simultaneously transcribing each of two adjacent genes. Molecules of RNA polymerase are visible as a series of dots along the DNA with the newly synthesized transcripts (fine threads) attached to them. The RNA molecules (ribosomal RNAs; rRNAs) shown in this example are not translated into protein but are instead used directly as components of ribosomes, the machines on which translation takes place. The particles at the 5′ end (the free end) of each rRNA transcript are believed to reflect the beginnings of ribosome assembly. From the relative lengths of the newly synthesized transcripts, it can be deduced that the RNA polymerase molecules are transcribing from right to left. (Courtesy of Ulrich Scheer.)

modern DNA polymerases and reverse transcriptases discussed in Chapter 5, as well as to a few RNA polymerases from viruses. The other lineage formed all of the RNA polymerases that we discuss in this chapter.

Cells Produce Different Categories of RNA Molecules

The majority of genes carried in a cell's DNA specify the amino acid sequence of proteins, and the RNAs that are copied from these genes (which ultimately direct the synthesis of proteins) are called **messenger RNA (mRNA)** molecules. The final product of other genes is the RNA molecule itself. These RNAs are known as **noncoding RNAs**, because they do not code for protein. In a well-studied, single-celled eukaryote, the yeast *Saccharomyces cerevisiae*, more than 1200 genes (about 15% of the total) produce RNA as their final product. Humans produce about 5000 different noncoding RNAs. These RNAs, like proteins, serve as enzymatic, structural, and regulatory components for a wide variety of processes in the cell. In Chapter 5, we encountered one of them as the template RNA carried by the enzyme telomerase. We shall see in this chapter that *ribosomal RNA* (*rRNA*) molecules form the core of ribosomes, that *transfer RNA* (*tRNA*) molecules serve as the adaptors that select amino acids and hold them in place on a ribosome for incorporation into protein, and that *small nuclear RNA* (*snRNA*) molecules direct the splicing of pre-mRNA to form mRNA. In Chapter 7, we shall see that *microRNA* (*miRNA*) molecules and *small interfering RNA* (*siRNA*) molecules serve as key regulators of eukaryotic gene expression, and that *piwi-interacting RNAs* (*piRNAs*) protect animal germ lines from transposons; we also discuss the *long noncoding RNAs* (*lncRNAs*), a diverse set of RNAs, many of whose functions are just being discovered (**Table 6–1**).

TABLE 6–1 **Principal Types of RNAs Produced in Cells**

Type of RNA	Function
mRNAs	Messenger RNAs, code for proteins
rRNAs	Ribosomal RNAs, form the basic structure of the ribosome and catalyze protein synthesis
tRNAs	Transfer RNAs, central to protein synthesis as the adaptors between mRNA and amino acids
Telomerase RNA	Serves as the template for the telomerase enzyme that extends the ends of chromosomes
snRNAs	Small nuclear RNAs, function in a variety of nuclear processes, including the splicing of pre-mRNA
snoRNAs	Small nucleolar RNAs, help to process and chemically modify rRNAs
lncRNAs	Long noncoding RNAs, not all of which appear to have a function; some serve as scaffolds and regulate diverse cell processes, including X-chromosome inactivation
miRNAs	MicroRNAs, regulate gene expression by blocking translation of specific mRNAs and causing their degradation
siRNAs	Small interfering RNAs, turn off gene expression by directing the degradation of selective mRNAs and helping to establish repressive chromatin structures
piRNAs	Piwi-interacting RNAs, bind to piwi proteins and protect the germ line from transposable elements

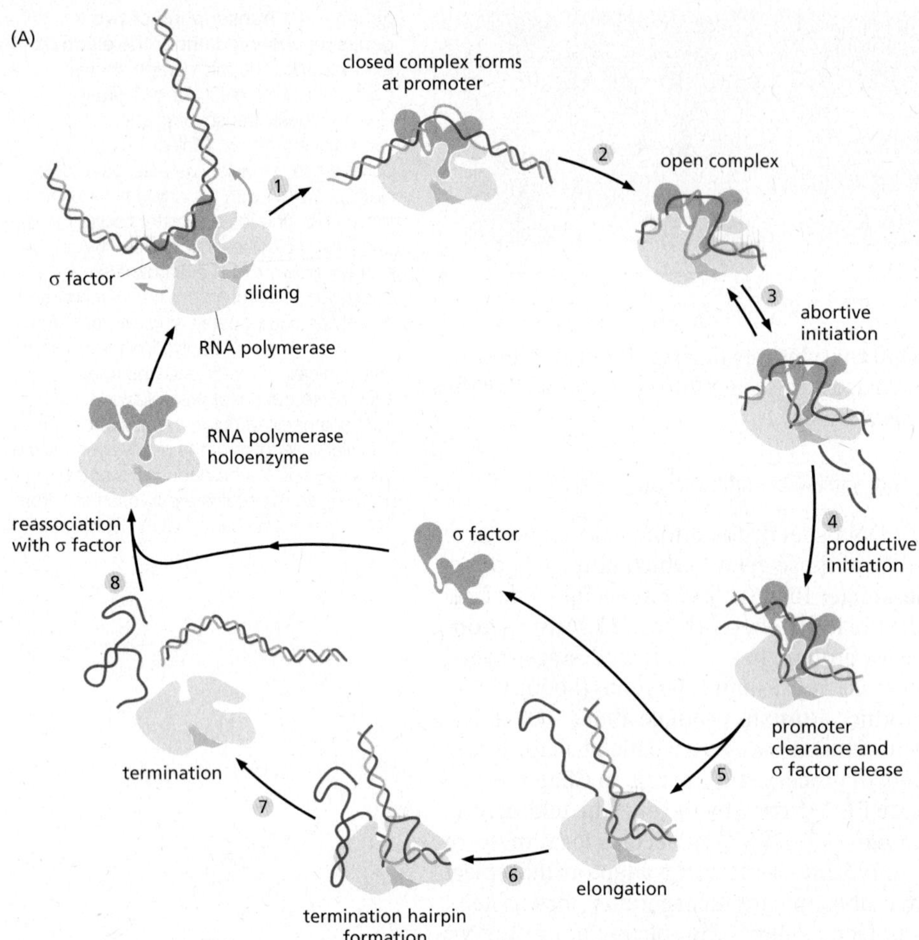

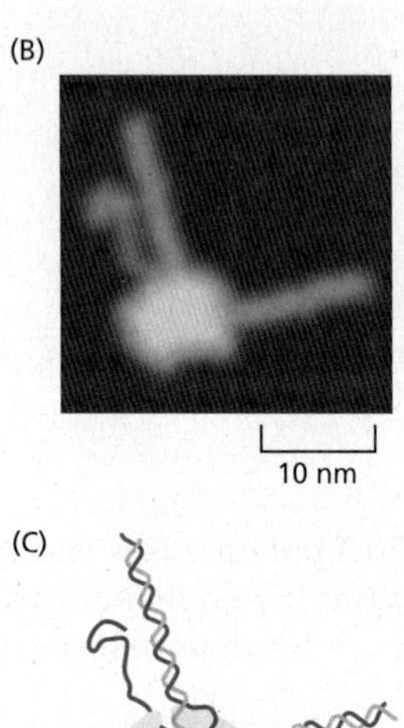

Figure 6–11 The transcription cycle of bacterial RNA polymerase. (A) In step 1, the RNA polymerase holoenzyme (polymerase core enzyme plus σ factor) assembles and then, by sliding, locates a promoter DNA sequence (see Figure 6–12). The polymerase opens (unwinds) the DNA at the position at which transcription is to begin (step 2) and begins transcribing (step 3). This initial RNA synthesis (abortive initiation) is relatively inefficient as short, unproductive transcripts are often released. However, once RNA polymerase has managed to synthesize about 10 nucleotides of RNA, it breaks its interactions with the promoter DNA (step 4) and eventually releases σ factor—as the polymerase tightens around the DNA and shifts to the elongation mode of RNA synthesis, moving along the DNA (step 5). During the elongation mode, transcription is highly processive, with the polymerase leaving the DNA template and releasing the newly transcribed RNA only when it encounters a termination signal (steps 6 and 7). Termination signals are typically encoded in DNA, and many function by forming an RNA hairpin–like structure that destabilizes the polymerase's hold on the RNA.

In bacteria, all RNA molecules are synthesized by a single type of RNA polymerase, and the cycle depicted in the figure therefore applies to the production of mRNAs as well as structural and catalytic RNAs. (B) Two-dimensional image of an elongating bacterial RNA polymerase, as determined by atomic force microscopy. (C) Interpretation of the image in B. (Adapted from K.M. Herbert et al., *Annu. Rev. Biochem.* 77:149–176, 2008. With permission from Annual Reviews.)

Each transcribed segment of DNA is called a *transcription unit*. In eukaryotes, a transcription unit typically carries the information of just one gene, and therefore codes for either a single RNA molecule or a single protein (or group of related proteins if the initial RNA transcript is spliced in more than one way to produce different mRNAs). In bacteria, a set of adjacent genes is often transcribed as a unit; the resulting mRNA molecule therefore carries the information for producing several distinct proteins.

Overall, RNA makes up a few percent of a cell's dry weight, whereas proteins compose about 50%. Most of the RNA in cells is rRNA; mRNA composes only 3–5% of the total RNA in a typical mammalian cell. The mRNA population is made up of tens of thousands of different species, and there are on average only 10–15 molecules of each species of mRNA present in each cell.

Signals Encoded in DNA Tell RNA Polymerase Where to Start and Stop

To transcribe a gene accurately, RNA polymerase must recognize where on the genome to start and where to finish. The way in which RNA polymerases perform these tasks differs somewhat between bacteria and eukaryotes. Because the processes in bacteria are simpler, we discuss them first.

The initiation of transcription is an especially important step in gene expression because it is the main point at which the cell regulates which proteins are to be produced and at what rate. The bacterial RNA polymerase core enzyme is a multisubunit complex that synthesizes RNA using the DNA template as a guide. An additional subunit called *sigma* (σ) *factor* associates with the core enzyme and assists it in reading the signals in the DNA that tell it where to begin transcribing (**Figure 6–11**). Together, σ factor and the core enzyme are known as the

RNA polymerase holoenzyme; this complex adheres only weakly to DNA when the two collide, and a holoenzyme typically slides rapidly along the long bacterial DNA molecule and then dissociates. However, when the polymerase holoenzyme slides into a special sequence of nucleotides indicating the starting point for RNA synthesis called a **promoter**, the polymerase binds tightly because its σ factor makes specific contacts with the edges of bases exposed on the outside of the DNA double helix (step 1 in Figure 6–11A).

The tightly bound RNA polymerase holoenzyme at a promoter opens up the double helix to expose a short stretch of nucleotides on each strand (step 2 in Figure 6–11A). The region of unpaired DNA (about 10 nucleotides) is called the *transcription bubble*, and it is stabilized by the binding of σ factor to the unpaired bases on one of the exposed strands. The other exposed DNA strand then acts as a template for complementary base-pairing with incoming ribonucleotides, two of which are joined together by the polymerase to begin an RNA chain (step 3 in Figure 6–11A).

The first 10 or so nucleotides of RNA are synthesized using a "scrunching" mechanism, in which RNA polymerase remains bound to the promoter and pulls the upstream DNA into its active site, thereby expanding the transcription bubble. This process creates considerable stress, and the short RNAs are often released, thereby relieving the stress and forcing the polymerase, which remains in place, to begin synthesis over again. Eventually this process of *abortive initiation* is overcome, and the stress generated by scrunching helps the core enzyme to break free of its interactions with the promoter DNA (step 4 in Figure 6–11A) and discard the σ factor (step 5 in Figure 6–11A).

At this point, the polymerase begins to move down the DNA, synthesizing RNA in a stepwise fashion: the polymerase moves forward one base pair for every nucleotide added. During this process, the transcription bubble continually expands at the front of the polymerase and contracts at its rear. Chain elongation continues (at a speed of approximately 50 nucleotides per second for bacterial RNA polymerases) until the enzyme encounters a second signal, the terminator (step 6 in Figure 6–11A), where the polymerase halts and releases both the newly made RNA molecule and the DNA template (step 7 in Figure 6–11A). The free polymerase core enzyme then reassociates with a free σ factor to form a holoenzyme that can begin the process of transcription again (step 8 in Figure 6–11A).

The process of transcription initiation is complicated and requires that the RNA polymerase holoenzyme and the DNA undergo a series of conformational changes, first opening the DNA double helix at promoters and subsequently tightening the enzyme around the DNA and RNA so that it does not dissociate before it has finished transcribing a gene. If an RNA polymerase does dissociate prematurely, it must start over again at the promoter.

How do the termination signals in the DNA stop the elongating polymerase? For most bacterial genes, a termination signal consists of a string of A-T nucleotide pairs preceded by a twofold symmetric DNA sequence, which, when transcribed into RNA, folds into a "hairpin" structure through Watson–Crick base-pairing (see Figure 6–92). As the polymerase transcribes across a terminator, the formation of the hairpin helps release the RNA transcript, which is held in place by relatively weak A-T and U-A base pairs (step 7 in Figure 6–11A). As we shall see, the folding of RNA into specific structures affects many steps in decoding the genome.

Bacterial Transcription Start and Stop Signals Are Heterogeneous in Nucleotide Sequence

As we have just seen, the processes of transcription initiation and termination involve a complicated series of structural transitions in protein, DNA, and RNA molecules. The signals encoded in DNA that specify these transitions are often difficult for researchers to recognize. Indeed, a comparison of many different bacterial promoters reveals a surprising degree of variation. Nevertheless, they all contain related sequences, reflecting aspects of the DNA that are recognized

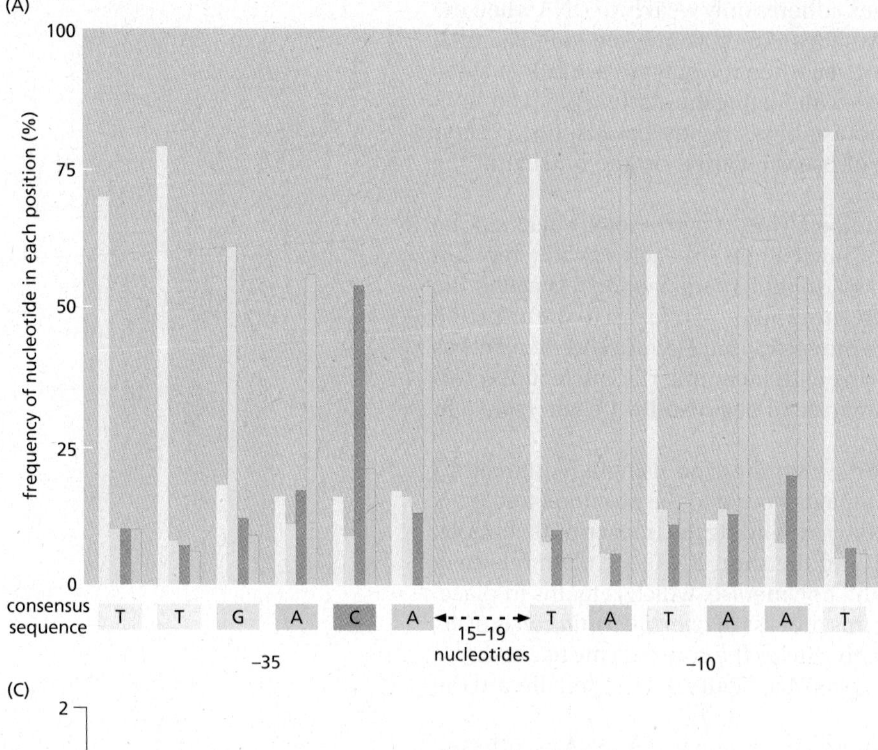

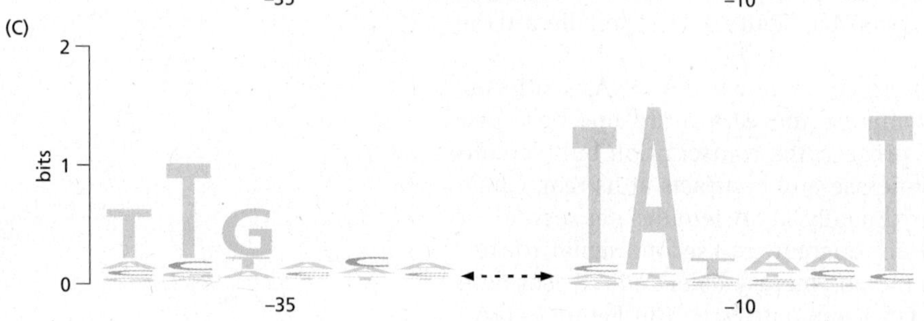

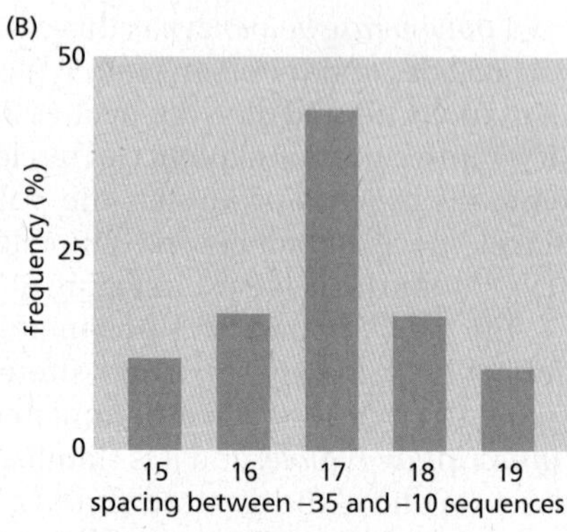

Figure 6–12 **Consensus nucleotide sequence and sequence logo for the major class of *Escherichia coli* promoters.** (A) On the basis of a comparison of 300 promoters, the frequencies of each of the four nucleotides at each position in the promoter are given. The consensus sequence, shown directly *below* the histogram, reflects the most common nucleotide found at each position in the collection of promoters. These promoters are characterized by two hexameric DNA sequences—the –35 sequence and the –10 sequence, named for their approximate location relative to the start point of transcription (designated +1). The sequence of nucleotides between the –35 and –10 hexamers shows little similarity among promoters but the spacing matters. For convenience, the nucleotide sequence of a single strand of DNA is shown; in reality, promoters are double-stranded DNA. The nucleotides shown in the figure are recognized by σ factor, a subunit of the RNA polymerase holoenzyme. (B) The distribution of spacing between the –35 and –10 hexamers found in *E. coli* promoters. (C) A *sequence logo* displaying the same information as in panel A. Here, the height of each letter is proportional to the frequency at which that base occurs at that position across a wide variety of promoter sequences. The total height of all the letters at each position is proportional to the information content (expressed in bits) at that position. For example, the total information content of a position that can tolerate several different bases is small (see the last three bases of the –35 sequences) but statistically greater than random.

directly by the σ factor. These common features are often summarized in the form of a *consensus sequence* (Figure 6–12). A **consensus nucleotide sequence** is derived by comparing many sequences with the same basic function and tallying up the most common nucleotides found at each position. It therefore serves as a summary or "average" of a large number of individual nucleotide sequences. A more accurate way of displaying the range of DNA sequences recognized by a protein is through the use of a *sequence logo,* which reveals the relative frequencies of each nucleotide at each position (Figure 6–12C).

The DNA sequences of individual bacterial promoters differ in ways that determine their strength, that is, the number of initiation events per unit time. Evolutionary processes have fine-tuned each to initiate as often as necessary and have thereby created a wide spectrum of promoter strengths. Promoters for genes that code for abundant proteins are much stronger than those associated with genes that encode rare proteins, and the nucleotide sequences of their promoters are responsible for these differences.

Like bacterial promoters, transcription terminators also have a wide range of sequences, with the potential to form a simple hairpin RNA structure being the most important common feature. Because an almost unlimited number of nucleotide sequences have this potential, terminator sequences are even more heterogeneous than promoter sequences.

We have discussed bacterial promoters and terminators in some detail to illustrate an important point regarding the analysis of genome sequences. Although we know a great deal about bacterial promoters and terminators and can construct "average" sequences that summarize their most salient features, their variation in nucleotide sequence makes it difficult to definitively locate them simply by

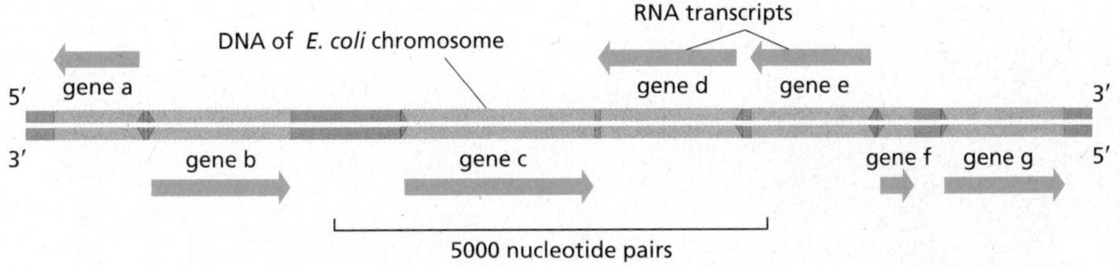

Figure 6–13 **Directions of transcription along a short portion of a bacterial chromosome.** Some genes are transcribed using one DNA strand as a template, while others are transcribed using the other DNA strand. The direction of transcription is determined by the orientation of the promoter at the beginning of each gene *(green arrowheads)*. This diagram shows approximately 0.2% (9000 base pairs) of the *E. coli* chromosome. The genes transcribed from *left* to *right* use the bottom DNA strand as the template; those transcribed from *right* to *left* use the top strand as the template. DNA that is not transcribed is indicated in *gray*.

analysis of the nucleotide sequence of a genome. It is even more difficult to locate analogous sequences in eukaryotic genomes, due in part to the excess DNA carried in these genomes. Often we need additional information, some of it from direct experimentation, to locate and accurately interpret the short DNA signals in genomes.

As shown in Figure 6–12, promoter sequences are asymmetric, ensuring that RNA polymerase can bind in only one orientation. Because the polymerase can synthesize RNA only in the 5′-to-3′ direction, the promoter orientation specifies the strand to be used as a template. Genome sequences reveal that the DNA strand that is used as the template for RNA synthesis varies from gene to gene, depending on the orientation of the promoter (**Figure 6–13**).

Having considered transcription in bacteria, we now turn to the situation in eukaryotes, where the synthesis of RNA molecules is a much more elaborate affair.

Transcription Initiation in Eukaryotes Requires Many Proteins

In contrast to bacteria, which contain a single type of RNA polymerase, eukaryotic nuclei have three: *RNA polymerase I, RNA polymerase II,* and *RNA polymerase III.* The three polymerases are structurally similar to one another and share some common subunits, but they transcribe different categories of genes (**Table 6–2**). RNA polymerases I and III transcribe the genes encoding transfer RNA, ribosomal RNA, and various small RNAs. RNA polymerase II transcribes most genes, including all those that encode proteins, and our subsequent discussion therefore focuses on this enzyme.

Eukaryotic RNA polymerase II has many structural similarities to bacterial RNA polymerase (**Figure 6–14**). But there are several important differences in the way in which the bacterial and eukaryotic enzymes function, two of which concern us immediately.

1. While bacterial RNA polymerase requires only a single transcription-initiation factor (σ) to begin transcription, eukaryotic RNA polymerases require many such factors, collectively called the *general transcription factors.*

2. Eukaryotic transcription initiation must take place on DNA that is packaged into nucleosomes and higher-order forms of chromatin structure (described in Chapter 4), features that are absent from bacterial chromosomes.

TABLE 6–2 **The Three RNA Polymerases in Eukaryotic Cells**

Type of polymerase	Genes transcribed
RNA polymerase I	5.8S, 18S, and 28S rRNA genes
RNA polymerase II	All protein-coding genes, plus snoRNA genes, miRNA genes, siRNA genes, lncRNA genes, and most snRNA genes
RNA polymerase III	tRNA genes, 5S rRNA genes, some snRNA genes, and genes for other small RNAs

The rRNAs were named according to their “S” values, which refer to their rate of sedimentation in an ultracentrifuge. The larger the S value, the larger the rRNA.

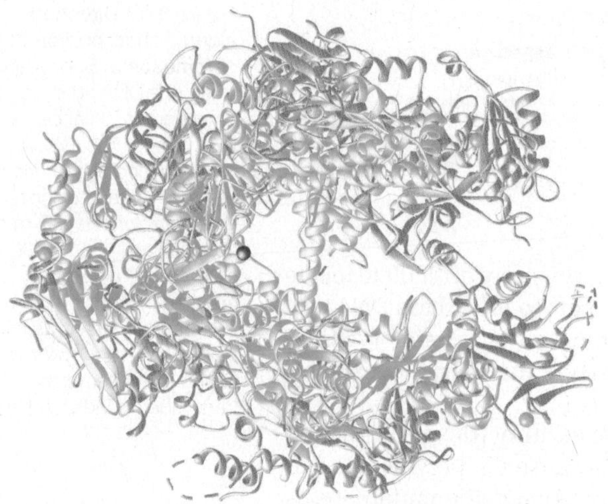

Figure 6–14 **Structural similarity between a bacterial RNA polymerase and a eukaryotic RNA polymerase II.** Regions of the two RNA polymerases that have similar structures are indicated in *green*. The eukaryotic polymerase is larger than the bacterial enzyme (12 subunits instead of 5), and some of the additional regions are shown in *gray*. The *red sphere* represents the Mg atom present at the active site, where polymerization takes place, while the *blue spheres* denote Zn atoms that serve as structural components. The RNA polymerases in all modern-day cells (bacteria, archaea, and eukaryotes) are closely related, indicating that the basic features of the enzyme were in place before the divergence of the three major branches of life. (Courtesy of P. Cramer and R. Kornberg.)

To Initiate Transcription, RNA Polymerase II Requires a Set of General Transcription Factors

The **general transcription factors** help to position eukaryotic RNA polymerase correctly at the promoter, aid in pulling apart the two strands of DNA to allow transcription to begin, and release RNA polymerase from the promoter to start its elongation mode. The proteins are "general" because they are needed at nearly all promoters used by RNA polymerase II. They consist of a set of interacting proteins denoted arbitrarily as TFIIA, TFIIB, and so on (TFII standing for transcription factor for polymerase II). In a broad sense, the eukaryotic general transcription factors carry out functions that are equivalent to those of the σ factor in bacteria.

Figure 6–15 illustrates how the general transcription factors assemble at promoters used by RNA polymerase II, and Table 6–3 summarizes their activities. The assembly process begins when TFIID binds to a short double-helical DNA sequence primarily composed of T and A nucleotides. For this reason, this sequence is known as the TATA sequence, or **TATA box**, and the subunit of TFIID that recognizes it is called TBP (for TATA-binding protein). The TATA box is typically located about 30 nucleotides upstream from the transcription start site. It is not the only DNA sequence that signals the start of transcription (Figure 6–16), but for many polymerase II promoters it is the most important. The binding of

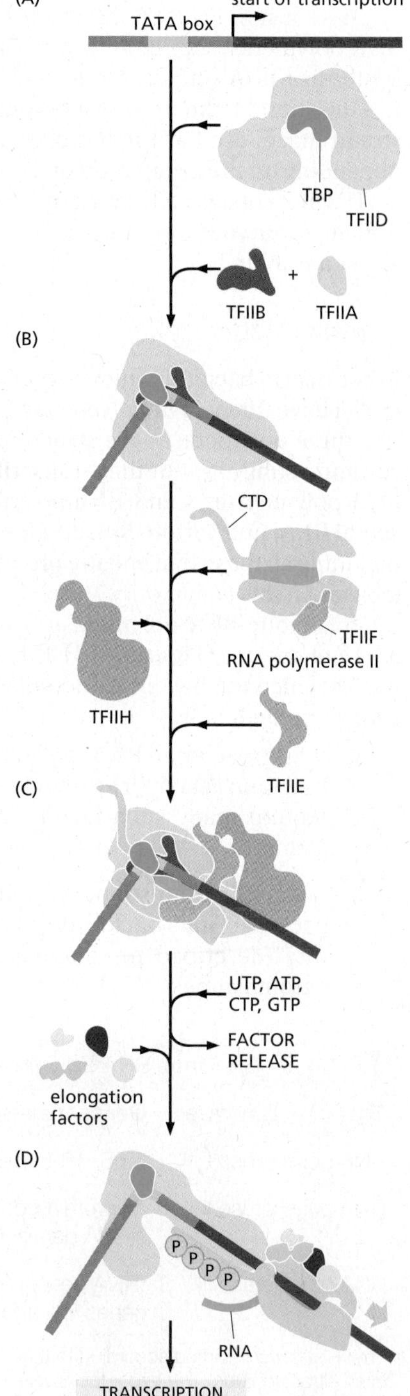

Figure 6–15 **Initiation of transcription of a eukaryotic gene by RNA polymerase II.** To begin transcription, RNA polymerase requires several general transcription factors. (A) Many promoters contain a DNA sequence called the TATA box, which, in humans, is located about 30 nucleotides away from the site at which transcription is initiated. (B) Through its subunit TBP, TFIID recognizes and binds the TATA box, which then enables the adjacent binding of TFIIB and TFIIA. (C) The RNA polymerase and the rest of the general transcription factors assemble at the promoter. (D) TFIIH then uses energy from ATP hydrolysis to pry apart the DNA double helix at the transcription start point, locally exposing the template strand. TFIIH also phosphorylates the long C-terminal polypeptide tail of RNA polymerase II, also called the C-terminal domain (CTD). This causes the polymerase to be released from the general factors and begin the elongation phase of transcription. For most genes, TFIID remains bound at the promoter whereas most of the other general transcription factors are released when the polymerase begins transcribing.

TABLE 6–3 The General Transcription Factors Needed for Transcription Initiation by Eukaryotic RNA Polymerase II

Name	Number of subunits	Roles in transition initiation
TFIID	12	Recognizes TATA box and other DNA sequences near the transcription start point
TFIIB	1	Recognizes BRE element in promoters; accurately positions RNA polymerase at the start site of transcription
TFIIA	2	Not required in all promoters; stabilizes binding of TFIID
TFIIF	3	Stabilizes RNA polymerase interaction with TFIIB; helps attract TFIIE and TFIIH
TFIIE	2	Attracts and regulates TFIIH
TFIIH	10	Unwinds DNA at the transcription start point, phosphorylates Ser5 of the RNA polymerase C-terminal domain (CTD); releases RNA polymerase from the promoter

TFIID is composed of TBP and 11 additional subunits called TAFs (TBP-associated factors).

TFIID causes a large distortion in the DNA of the TATA box (**Figure 6–17**). This distortion is thought to serve as a physical landmark for the location of an active promoter in the midst of a very large genome, and it brings DNA sequences on both sides of the distortion closer together to allow for subsequent protein assembly steps. The additional general transcription factors then assemble, along with RNA polymerase II, to form a complete *transcription initiation complex* (see Figure 6–15). The most complicated of the general transcription factors is TFIIH (shown in *pink*). Consisting of 10 subunits, it is nearly as large as RNA polymerase II itself and, as we shall see shortly, performs several enzymatic steps needed for the initiation of transcription.

After forming a transcription initiation complex on the promoter DNA, RNA polymerase II must gain access to the template strand at the transcription start point. TFIIH makes this step possible by hydrolyzing ATP and pulling apart the DNA strands at the start site, thereby exposing the template strand. Next, RNA polymerase II, like the bacterial polymerase, remains at the promoter synthesizing short lengths of RNA until it undergoes a series of conformational changes that allow it to move away from the promoter and enter the elongation phase of transcription. A key step in this transition is the addition of phosphate groups to the "tail" of the RNA polymerase (known as the CTD, or C-terminal domain). In

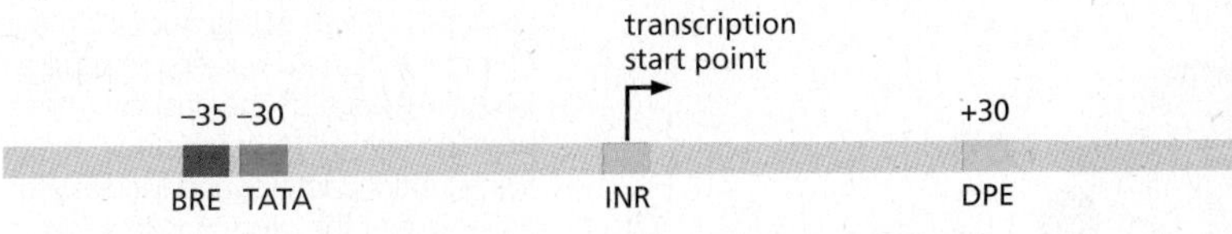

element	consensus sequence	general transcription factor
BRE	G/C G/C G/A C G C C	TFIIB
TATA	T A T A A/T A A/T	TBP subunit of TFIID
INR	C/T C/T A N T/A C/T C/T	TFIID
DPE	A/G G A/T C G T G	TFIID

Figure 6–16 **Consensus sequences found in the vicinity of eukaryotic RNA polymerase II start points.** The name given to each consensus sequence (first column) and the general transcription factor that recognizes it (last column) are indicated. N indicates any nucleotide, and two nucleotides separated by a slash indicate an equal probability of either nucleotide at the indicated position. In reality, each consensus sequence is a shorthand representation of a histogram similar to that of Figure 6–12.

For most RNA polymerase II transcription start points, only two or three of the four sequences are present. For example, many polymerase II promoters have a TATA box sequence, but those that do not typically have a "strong" INR sequence. Although most of the DNA sequences that influence transcription initiation are located upstream of the transcription start point, a few, such as the DPE shown in the figure, are located within the transcribed region.

humans, the CTD consists of 52 tandem repeats of a seven-amino-acid sequence, which extend from the RNA polymerase core structure. During transcription initiation, the serines located at the fifth position in each repeat sequence (Ser5) are phosphorylated by TFIIH, which contains a protein kinase in one of its subunits. Triggered by these phosphorylations, the polymerase disengages from the cluster of general transcription factors (see Figure 6–15D). During this process, it undergoes a series of conformational changes that tighten its interaction with DNA, and it acquires new proteins that allow it to transcribe for long distances, in some cases for many hours, without dissociating from DNA.

Once the polymerase II has begun elongating the RNA transcript, most of the general transcription factors are released from the DNA so that they are available to initiate another round of transcription with a new RNA polymerase molecule. As we see shortly, the phosphorylation of the tail of RNA polymerase II has an additional function: it causes components of the RNA-processing machinery to load onto the polymerase and thereby be positioned to modify the newly transcribed RNA as it emerges from the polymerase.

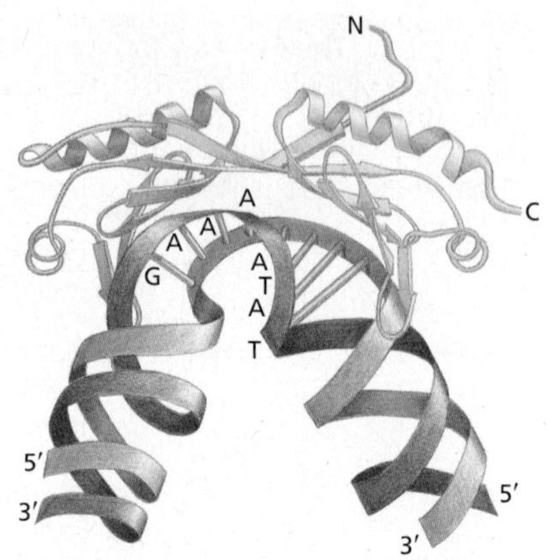

Figure 6–17 **Three-dimensional structure of TBP (TATA-binding protein) bound to DNA.** The TBP is the subunit of the general transcription factor TFIID that is responsible for recognizing and binding the TATA box sequence in the DNA. The unique DNA bending caused by TBP—kinks in the double helix separated by partly unwound DNA—is thought to serve as a landmark that helps to attract the other general transcription factors (**Movie 6.4**). TBP is a single polypeptide chain that is folded into two very similar domains (*blue* and *green*). (Adapted from J.L. Kim et al., *Nature* 365:520–527, 1993.)

In Eukaryotes, Transcription Initiation Also Requires Activator, Mediator, and Chromatin-modifying Proteins

The model for transcription initiation just described is based on experiments performed *in vitro* using purified proteins and DNA. However, as discussed in Chapter 4, DNA in eukaryotic cells is packaged into nucleosomes, which are further arranged in higher-order chromatin structures. As a result, transcription initiation in a eukaryotic cell is more complex and requires more proteins than it does on purified DNA. First, regulatory proteins known as *transcriptional activators* must bind to specific sequences in DNA (called *enhancers*) to help attract the general transcription factors and RNA polymerase II to the start point of transcription (**Figure 6–18**). We discuss the role of these activators in Chapter 7, because they are one of the main ways in which cells regulate expression of their genes. Here we simply note that their presence on DNA is required for transcription initiation in a eukaryotic cell. Second, eukaryotic transcription initiation *in vivo* requires the presence of a large protein complex known as *Mediator*, which

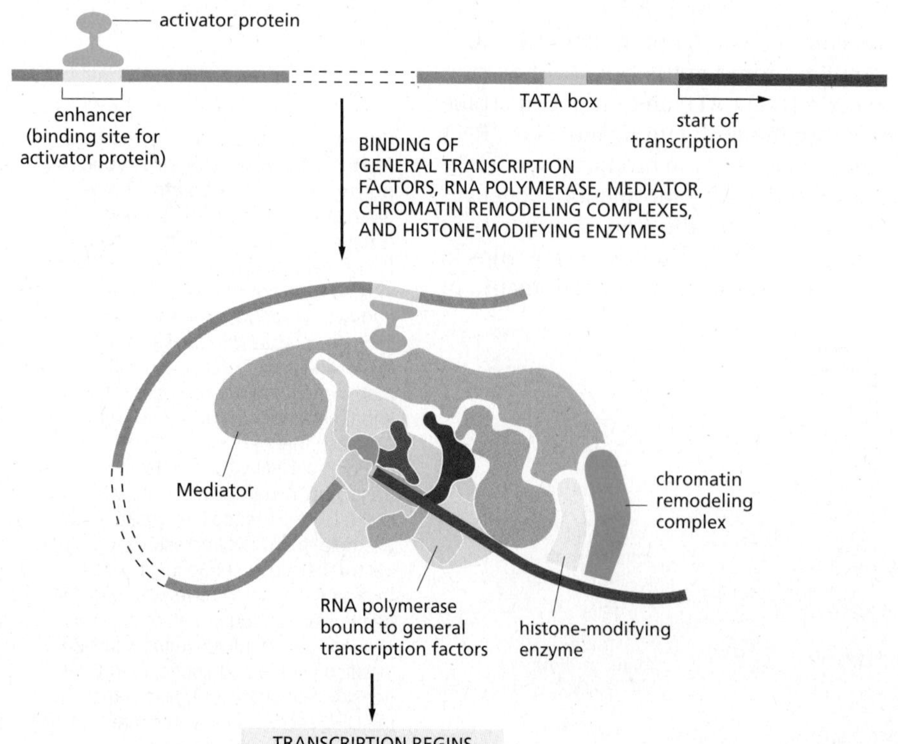

Figure 6–18 **Transcription initiation by RNA polymerase II in a eukaryotic cell.** Transcription initiation *in vivo* requires the presence of transcription activator proteins. As described in Chapter 7, these proteins bind to short, specific sequences in DNA that are located in regulatory regions called enhancers. Although only one activator is shown here (in *blue*), a typical eukaryotic gene utilizes many DNA-bound transcription activator proteins, which act together to determine that gene's rate and pattern of transcription across different cell types. Often acting from a distance of many thousand nucleotide pairs along DNA (indicated by the dashes), these proteins help RNA polymerase, the general transcription factors, and Mediator all to assemble at the promoter. In addition, ATP-dependent chromatin remodeling complexes and histone-modifying enzymes are needed at most genes. One of the main roles of Mediator is to coordinate the assembly of all these proteins at the promoter so that transcription can begin. As discussed in Chapter 4, the "default" state of eukaryotic DNA is to be packaged into nucleosomes and higher-order chromatin structures; for simplicity, these are not shown in this figure.

allows the activator proteins to communicate properly with the polymerase II and with the general transcription factors. Mediator also correctly positions TFIIH near the tail of RNA polymerase, facilitating phosphorylation of the tail and the consequent release of the polymerase from the promoter to begin synthesizing RNA. Finally, transcription initiation in a eukaryotic cell typically requires the recruitment of chromatin-modifying enzymes, including chromatin remodeling complexes and histone-modifying enzymes. As discussed in Chapter 4, both types of enzymes can increase access to the DNA in chromatin, and by doing so they facilitate the assembly of the transcription initiation machinery onto DNA.

To summarize, as illustrated schematically in Figure 6–18, many proteins (well over 100 individual subunits) must assemble at the start point of transcription to initiate transcription in a eukaryotic cell. We shall return to some of these proteins—especially transcription activator proteins, chromatin remodeling complexes, and histone-modifying enzymes—in the following chapter, where we discuss how eukaryotic cells regulate the process of transcription initiation.

Transcription Elongation in Eukaryotes Requires Accessory Proteins

Once RNA polymerase has initiated transcription, it moves jerkily, pausing at some DNA sequences and rapidly transcribing through others. Elongating RNA polymerases, both bacterial and eukaryotic, are associated with a series of *elongation factors*, proteins that decrease the likelihood that RNA polymerase will dissociate before it reaches the end of a gene. These factors typically associate with RNA polymerase shortly after initiation, and they help the polymerase move both through nucleosomes (**Figure 6–19**) and through the wide variety of different DNA sequences that are found in genes.

We will see in the next chapter that, like the process of transcription initiation, transcription elongation can be regulated by the cell; more specifically, we will see that at many human genes, RNA polymerase pauses shortly after it initiates transcription. This pause can last from several seconds to many hours, and the cell controls the duration of this pause as part of gene regulatory processes.

As RNA polymerase II moves along a gene, some of the enzymes bound to it modify the histones, leaving behind a record of where the polymerase has been. Although it is not clear exactly how the cell uses this information, it may aid in transcribing a gene over and over again once it has become active for the first time.

Transcription Creates Superhelical Tension

Nucleosomes are not the only impediment to elongating RNA polymerases, and in this section, we describe an entirely different type of barrier, one that applies to both bacterial and eukaryotic polymerases. To introduce this issue, we need first to consider a subtle property inherent in the DNA double helix called DNA supercoiling. **DNA supercoiling** is the name given to a conformation that DNA

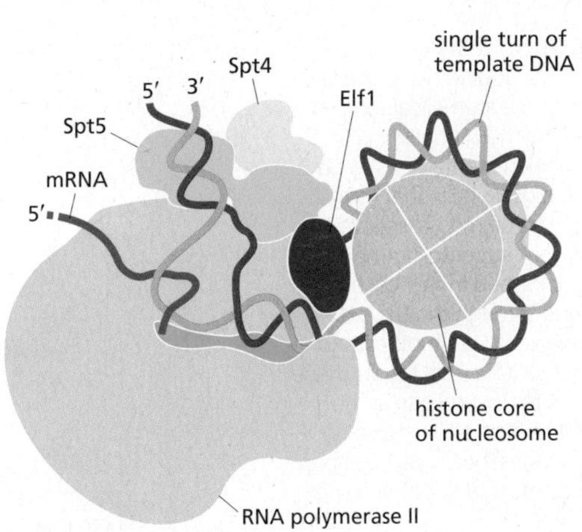

Figure 6–19 **Structure of an RNA polymerase II transcribing through a nucleosome.** In the structure diagrammed here, which was determined by cryo-electron microscopy, the polymerase has moved about halfway through the DNA of the nucleosome, leaving only one of the two loops of duplex DNA still bound to the histone core. The polymerase is shown in *blue*, associated with three elongation factors (Spt4, Spt5, and Elf1) that help the polymerase transcribe through nucleosomes. These factors act in several ways: they form a wedge to pry the DNA away from the histone core as the polymerase moves forward; they directly destabilize histone–DNA interactions by pushing a positively charged surface ahead of the RNA polymerase; and they reduce the intrinsic "stickiness" of RNA polymerase for nucleosomes. In addition to these factors, eukaryotic transcription is typically aided by ATP-dependent chromatin remodeling complexes that seek out and rescue the occasional stalled polymerase, as well as by histone chaperones that can partially disassemble nucleosomes in front of a moving RNA polymerase and reassemble them behind. (Based on PDB code 6IR9.)

can adopt in response to superhelical tension. Alternatively, the creation of loops or coils in a double-helical DNA molecule will produce such tension.

Figure 6–20 illustrates why. There are approximately 10 nucleotide pairs for every helical turn in a DNA double helix. Imagine a helix whose two ends are fixed with respect to each other (as they are in a DNA circle, such as a bacterial chromosome, or in a tightly clamped loop, as can exist in eukaryotic chromosomes). In this case, one large DNA supercoil will form to compensate for each 10 nucleotide pairs that are opened (unwound). The formation of this supercoil is energetically favorable because it restores a normal helical twist to the base-paired regions that remain, which would otherwise become overwound because of the fixed ends.

RNA polymerase creates superhelical tension as it moves along a stretch of DNA that is anchored at its ends. As illustrated in Figure 6–20C, if the polymerase is not free to rotate rapidly (and such rotation is unlikely given the size of RNA polymerases and their attached transcripts), a moving polymerase will generate positive superhelical tension in the DNA in front of it and negative helical tension

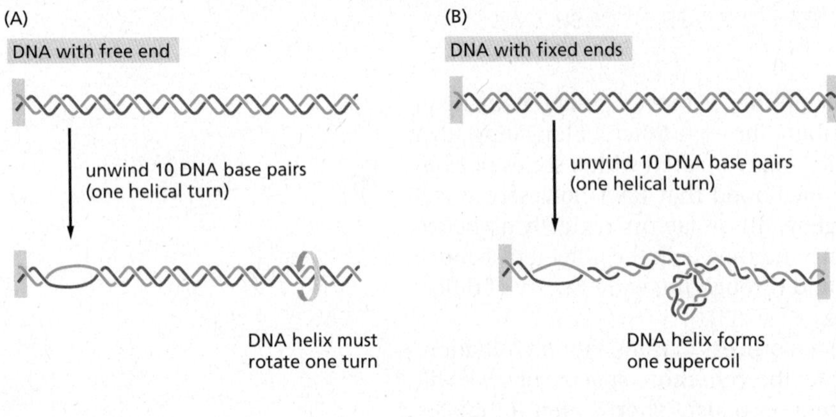

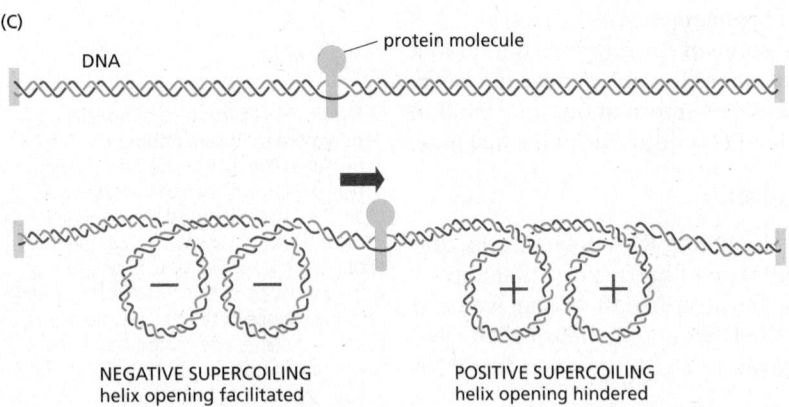

Figure 6–20 Superhelical tension in DNA causes DNA supercoiling. (A) For a DNA molecule with one free end (or a break in one strand that serves as a swivel), the DNA double helix rotates by one turn for every 10 nucleotide pairs opened. (B) If rotation is prevented, superhelical tension is introduced into the DNA by helix opening. In the example shown, the DNA helix contains 10 helical turns, one of which is opened. One way of accommodating the tension created would be to increase the helical twist from 10 to 11 nucleotide pairs per turn in the double helix that remains. The DNA helix, however, resists such a deformation in a springlike fashion, preferring to relieve the superhelical tension by bending into supercoiled loops. As a result, one DNA supercoil forms in the DNA double helix for every 10 nucleotide pairs opened. The supercoil formed in this case is a positive supercoil. (C) Supercoiling of DNA is induced by a protein tracking through the DNA double helix. The two ends of the DNA shown here are unable to rotate freely relative to each other, and the protein molecule is assumed also to be prevented from rotating freely as it moves. Under these conditions, the movement of the protein causes an excess of helical turns to accumulate in the DNA helix ahead of the protein (inducing positive supercoils) and a deficit of helical turns to arise in the DNA behind the protein (inducing negative supercoils), as shown. Because locally pulling apart the two strands of the double helix relieves the tension from negative supercoils, it is easier to do behind the moving protein than ahead of it.

(A) EUKARYOTES

cytoplasm
nucleus
introns
exons
DNA
transcription unit
TRANSCRIPTION
"primary RNA transcript"
5′ CAPPING
RNA SPLICING
3′ POLYADENYLATION
RNA PROCESSING
RNA cap
mRNA
AAAA
5′
3′
EXPORT
mRNA
AAAA
TRANSLATION
protein

(B) BACTERIA

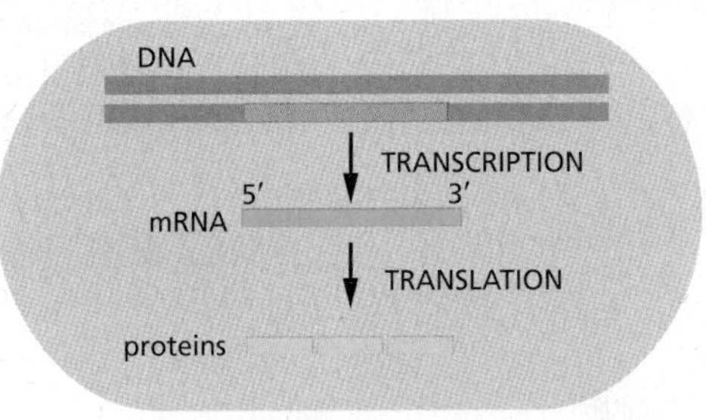

Figure 6–21 Comparison of the steps leading from gene to protein in eukaryotes and bacteria. The final amount of a protein in the cell depends on the efficiency of each step and on the rates of degradation of the RNA and protein molecules. (A) In eukaryotic cells, the mRNA molecule resulting from transcription contains both coding (exon) and noncoding (intron) sequences. Before it can be translated into protein, the two ends of the RNA are modified, the introns are removed by an enzymatically catalyzed RNA splicing reaction, and the resulting mRNA is transported from the nucleus to the cytoplasm. For convenience, the steps in this figure are depicted as occurring one at a time; in reality, many occur concurrently. For example, the RNA cap is added and splicing begins before transcription has been completed. Because of the coupling between transcription and RNA processing, intact primary transcripts—the full-length RNAs that would, in theory, be produced if no processing had occurred—are found only rarely. (B) In bacteria, the production of mRNA is much simpler. The 5′ end of an mRNA molecule is produced by the initiation of transcription, and the 3′ end is produced by the termination of transcription. Because bacteria lack a nucleus, transcription and translation take place in a common compartment, and the translation of a bacterial mRNA often begins before its synthesis has been completed. As indicated, a single bacterial mRNA typically produces several different proteins, another feature that distinguishes eukaryotes from bacteria.

behind it. If this tension is not relieved, the polymerase will grind to a halt because further unwinding requires more energy than the transcription process can provide. For eukaryotes, the mild buildup of tension is thought to provide a bonus: the positive superhelical tension ahead of the polymerase facilitates the partial unwrapping of the DNA in nucleosomes, inasmuch as the release of DNA from the histone core helps to relax this tension. The tension can also be relieved by DNA topoisomerase enzymes, as we saw in the previous chapter for the similar kind of tension generated by DNA polymerases during DNA replication (see Figure 5–21).

In bacteria (but not eukaryotes), a specialized topoisomerase called *DNA gyrase* uses the energy of ATP hydrolysis to pump supercoils continually into the DNA, thereby maintaining the DNA under constant tension. These are *negative supercoils*, having the opposite handedness from the *positive supercoils* that form when a region of DNA helix opens (see Figure 6–20B). Whenever a region of helix opens, it removes these negative supercoils from bacterial DNA, reducing the superhelical tension. DNA gyrase therefore makes the opening of the DNA helix in bacteria energetically favorable compared with helix opening in DNA that is not supercoiled. For this reason, it facilitates those genetic processes in bacteria, such as the initiation of transcription by bacterial RNA polymerase, that require helix opening (see Figure 6–11).

Transcription Elongation in Eukaryotes Is Tightly Coupled to RNA Processing

We saw earlier that bacterial mRNAs are synthesized by the RNA polymerase starting and stopping at specific spots on the genome. The situation in eukaryotes is substantially different. In particular, transcription is only the first of several steps needed to produce a mature mRNA molecule. Other critical steps are the covalent modification of the ends of the RNA and the removal of *intron sequences* that are discarded from the middle of the RNA transcript by the process of *RNA splicing* (**Figure 6–21**). As we shall see, RNA splicing not only joins together different portions of an RNA transcript to eliminate the intron sequences; it also provides eukaryotes with the ability to synthesize several related but different proteins from the same gene.

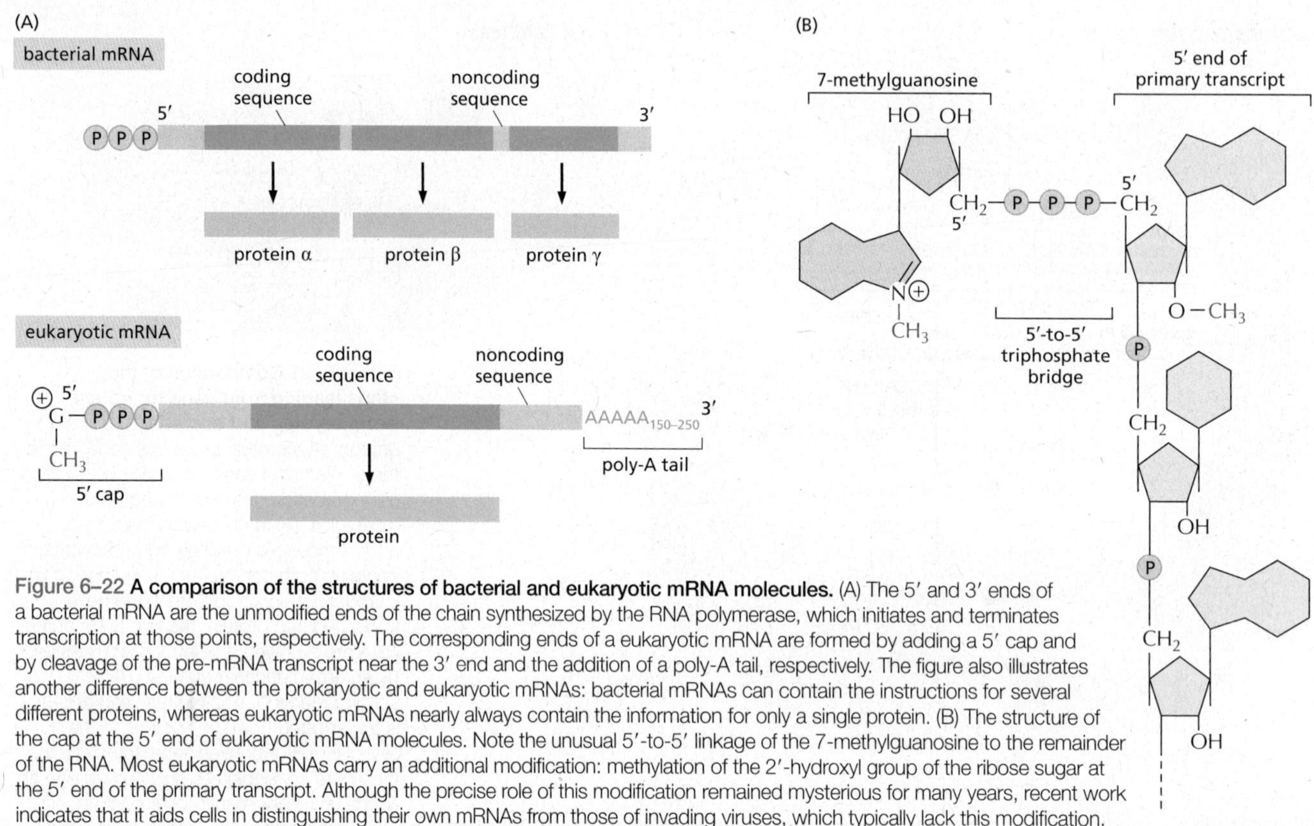

Figure 6–22 **A comparison of the structures of bacterial and eukaryotic mRNA molecules.** (A) The 5′ and 3′ ends of a bacterial mRNA are the unmodified ends of the chain synthesized by the RNA polymerase, which initiates and terminates transcription at those points, respectively. The corresponding ends of a eukaryotic mRNA are formed by adding a 5′ cap and by cleavage of the pre-mRNA transcript near the 3′ end and the addition of a poly-A tail, respectively. The figure also illustrates another difference between the prokaryotic and eukaryotic mRNAs: bacterial mRNAs can contain the instructions for several different proteins, whereas eukaryotic mRNAs nearly always contain the information for only a single protein. (B) The structure of the cap at the 5′ end of eukaryotic mRNA molecules. Note the unusual 5′-to-5′ linkage of the 7-methylguanosine to the remainder of the RNA. Most eukaryotic mRNAs carry an additional modification: methylation of the 2′-hydroxyl group of the ribose sugar at the 5′ end of the primary transcript. Although the precise role of this modification remained mysterious for many years, recent work indicates that it aids cells in distinguishing their own mRNAs from those of invading viruses, which typically lack this modification. On the basis of this difference, cells can block translation of viral RNAs, thereby defending themselves from viral attack.

Both ends of eukaryotic mRNAs are modified: by *capping* on the 5′ end and by *polyadenylation* of the 3′ end (**Figure 6–22**). These special ends allow the cell to assess whether both ends of an mRNA molecule are present (and if the message is therefore intact) before it exports the RNA from the nucleus and translates it into protein.

A simple mechanism has evolved to couple all of the above RNA-processing steps to transcription elongation. As discussed previously, a key step in transcription initiation by RNA polymerase II is the phosphorylation of the RNA polymerase II tail, also called the CTD (C-terminal domain). This phosphorylation, which proceeds gradually as the RNA polymerase initiates transcription and moves along the DNA, not only helps to dissociate the RNA polymerase II from other proteins present at the start point of transcription but also allows a new set of proteins to associate with the RNA polymerase tail, which function in transcription elongation and RNA processing, as we discuss in the following sections. Some of the processing proteins are thought to "hop" from the polymerase tail onto the nascent RNA molecule to begin their processing reactions as soon as this RNA emerges from the RNA polymerase. Thus, we can view RNA polymerase II in its elongation mode as an RNA factory that not only moves along the DNA synthesizing an RNA molecule but also processes the RNA that it produces (**Figure 6–23**). Fully extended, the CTD is nearly 10 times longer than the remainder of the RNA polymerase. As a flexible protein domain, it serves as a scaffold or tether, holding a variety of proteins close by so that they can rapidly act when needed. This scaffolding strategy, which greatly speeds up the overall rate of a series of consecutive reactions, is one that is commonly utilized in the cell (see Figure 3–76).

RNA Capping Is the First Modification of Eukaryotic Pre-mRNAs

As soon as RNA polymerase II has produced about 20 nucleotides of RNA, the 5′ end of the new RNA molecule is modified by addition of a cap that consists of a modified guanine nucleotide (see Figure 6–22B). Three enzymes, acting in

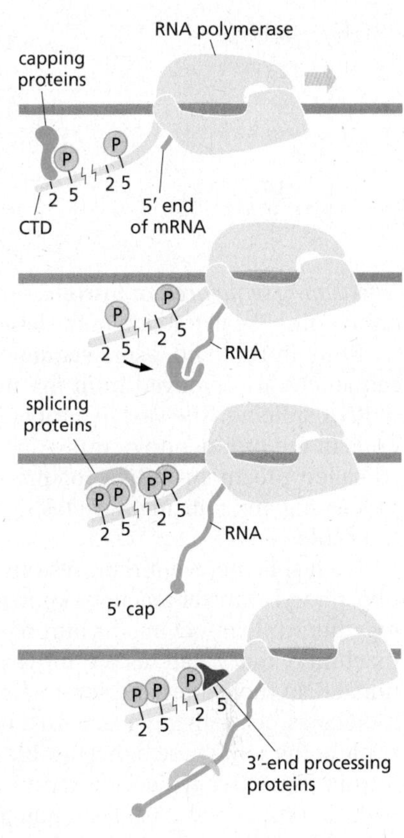

Figure 6–23 Eukaryotic RNA polymerase II as an RNA synthesis and processing machine. As the polymerase transcribes DNA into RNA, it carries RNA-processing proteins on its tail that are transferred to the nascent RNA at the appropriate time. The tail contains 52 tandem repeats of a seven-amino-acid sequence, and there are two serines in each repeat. The capping proteins first bind to the RNA polymerase tail when it is phosphorylated on Ser5 of the heptad repeats late in the process of transcription initiation (see Figure 6–15). This strategy ensures that the RNA molecule is efficiently capped as soon as its 5′ end emerges from the RNA polymerase. As the polymerase continues transcribing, its tail is extensively phosphorylated on the Ser2 positions by a kinase associated with the elongating polymerase and is eventually dephosphorylated at Ser5 positions. These further modifications attract splicing and 3′-end processing proteins to the moving polymerase, positioning them to act on the newly synthesized RNA as it emerges from the RNA polymerase. There are many RNA-processing enzymes, and not all travel with the polymerase. In RNA splicing, for example, the tail carries only a few critical components; once bound to an emerging RNA molecule, they serve as a nucleation site for the remaining components.

When RNA polymerase II finishes transcribing a gene, it is released from DNA, and protein phosphatases remove the phosphates on its tail so that it can reinitiate transcription. Only the fully dephosphorylated form of RNA polymerase II is competent to begin RNA synthesis at a promoter.

succession, perform the capping reaction: a phosphatase removes a phosphate from the triphosphate left at the 5′ end of the nascent RNA molecule, a guanyl transferase adds a GMP in a reverse linkage (5′-to-5′ instead of 5′-to-3′) to the 5′ diphosphate just produced, and a methyl transferase adds a methyl group to the guanosine (**Figure 6–24**). Because all three enzymes bind to the RNA polymerase tail phosphorylated at the Ser5 position—the modification added by TFIIH during transcription initiation—they are poised to modify the 5′ end of the nascent transcript as soon as it emerges from the polymerase.

The 7-methylguanosine cap signifies the 5′ end of eukaryotic mRNAs, and this landmark helps the cell to distinguish mRNAs from the other types of RNA molecules present in the cell. For example, RNA polymerases I and III produce uncapped RNAs during transcription, in part because these polymerases lack a CTD. In the nucleus, the cap binds a protein complex called CBC (cap-binding complex), which, as we discuss in subsequent sections, helps a future mRNA to be further processed and exported. The 5′ cap also has an important role in the translation of mRNAs in the cytosol, as we discuss later in the chapter.

Although the vast majority of eukaryotic mRNAs possess a 7-methylguanosine cap at their 5′ ends, an alternative type of cap is found on some mRNAs, specifically a nicotinamide adenine dinucleotide phosphate ($NADP^+$). We saw earlier in Chapter 2 that $NADP^+$ is an important cofactor for many biochemical reactions, and it is added to certain mRNAs by RNA polymerase itself as the first nucleotide when a new mRNA molecule chain is begun. The role of this particular type of mRNA cap is not known with certainty, but one hypothesis holds that it provides a way for the cell to link the expression of some of its genes to its overall metabolic "health."

RNA Splicing Removes Intron Sequences from Newly Transcribed Pre-mRNAs

As discussed in Chapter 4, the protein-coding sequences of eukaryotic genes are typically interrupted by noncoding intervening sequences (introns). Discovered in 1977, this feature of eukaryotic genes came as a surprise to scientists, who had been, until that time, familiar only with bacterial genes, which consist of a continuous stretch of coding DNA that is directly transcribed into mRNA. In marked contrast, eukaryotic genes were found to be broken up into small pieces of coding sequence (*expressed sequences*, or **exons**) interspersed with much longer

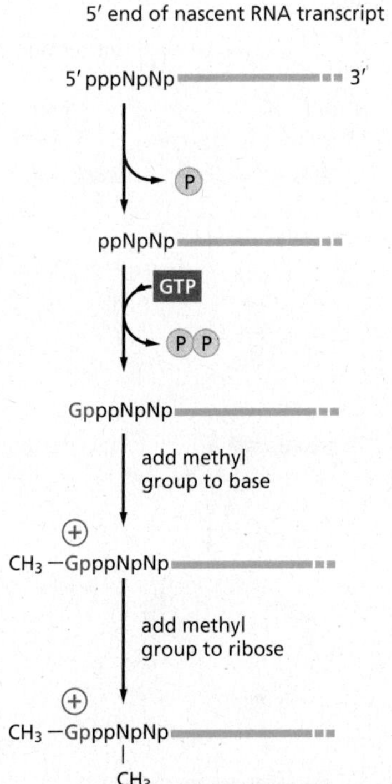

Figure 6–24 The reactions that cap the 5′ end of each RNA molecule synthesized by RNA polymerase II. The final cap contains a novel 5′-to-5′ linkage between the positively charged 7-methylguanosine residue and the 5′ end of the RNA transcript (see Figure 6–22B). The letter N represents any one of the four ribonucleotides, although the nucleotide that starts an RNA chain is usually a purine (an A or a G). (After A.J. Shatkin, *Bioessays* 7:275–277, 1987. With permission from John Wiley & Sons.)

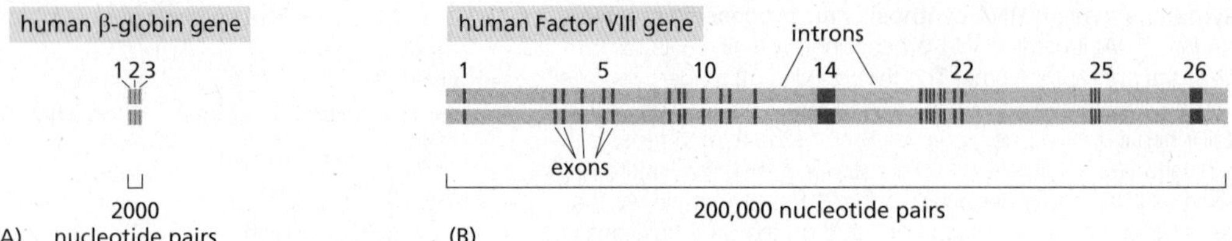

Figure 6–25 **Structure of two human genes showing the arrangement of exons and introns.** (A) The relatively small β-globin gene, which encodes a subunit of the oxygen-carrying protein hemoglobin, contains 3 exons (see also Figure 4–9). (B) The much larger Factor VIII gene contains 26 exons; it codes for a protein (Factor VIII) that functions in the blood-clotting pathway. The most prevalent form of hemophilia results from mutations in this gene.

intervening sequences, or **introns**; thus, the coding portion of a eukaryotic gene is often only a small fraction of the length of the gene (Figure 6–25).

Both intron and exon sequences are transcribed into RNA. The intron sequences are removed from the newly synthesized RNA through the process of **RNA splicing**. The vast majority of RNA splicing that takes place in cells functions in the production of mRNA, and our discussion of splicing focuses on this so-called precursor-mRNA (or pre-mRNA) splicing. Only after 5′- and 3′-end processing and splicing have taken place is an RNA transcript called an mRNA molecule.

Each splicing event removes one intron, proceeding through two sequential phosphoryl-transfer reactions known as transesterifications; these join two exons together while removing the intron between them as a "lariat" (Figure 6–26). The machinery that catalyzes pre-mRNA splicing is complex, consisting of five additional RNA molecules and several hundred proteins, and it hydrolyzes many ATP molecules per splicing event. This complexity ensures that splicing is accurate, while at the same time being flexible enough to deal with the enormous variety of introns found in a typical eukaryotic cell. On average, there are 11 introns in each of the approximately 20,000 human protein-coding genes, so the cell devotes considerable resources to this step in gene expression.

Although it seems wasteful to first produce and then remove large numbers of introns from an RNA transcript, this process provides an advantage to the cell. In many organisms, the transcript of a given gene can be spliced in more than one way, and this allows the same gene to produce a set of different but related proteins (Figure 6–27). It has been proposed that 95% of human gene transcripts are spliced in more than one way, but this is almost certainly an overestimate: many

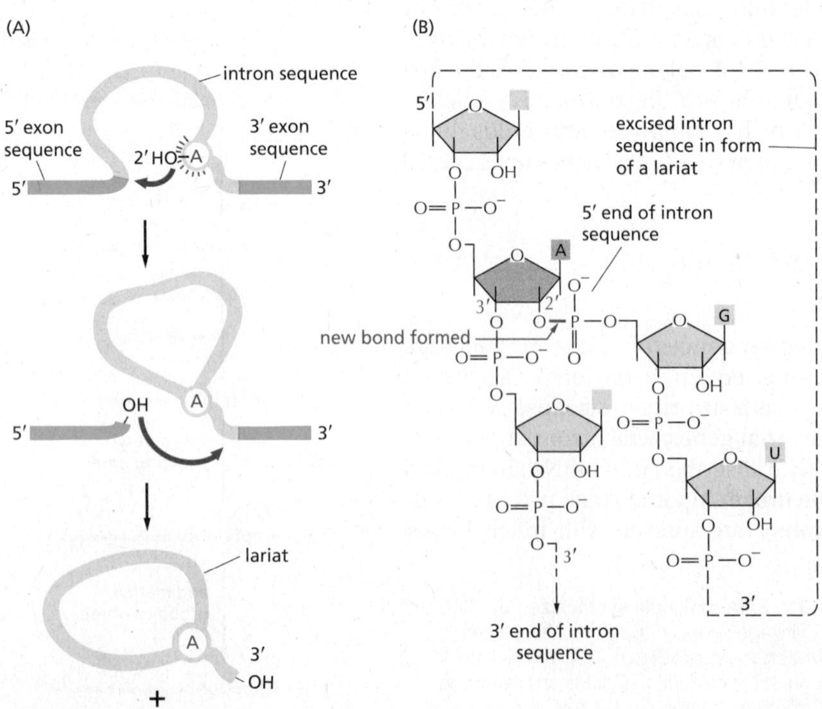

Figure 6–26 **The pre-mRNA splicing reaction.** (A) In the first step, a specific adenine nucleotide in the intron sequence (indicated in *red*) attacks the 5′ splice site and cuts the sugar–phosphate backbone of the RNA at this point. The cut 5′ end of the intron becomes covalently linked to the adenine nucleotide, as shown in detail in (B), thereby creating a loop in the RNA molecule. The released free 3′-OH end of the exon sequence then reacts with the start of the next exon sequence, joining the two exons together and releasing the intron sequence in the shape of a *lariat*. The two exon sequences thereby become joined into a continuous coding sequence. The released intron sequence is discarded, eventually being broken down into single nucleotides, which are recycled.

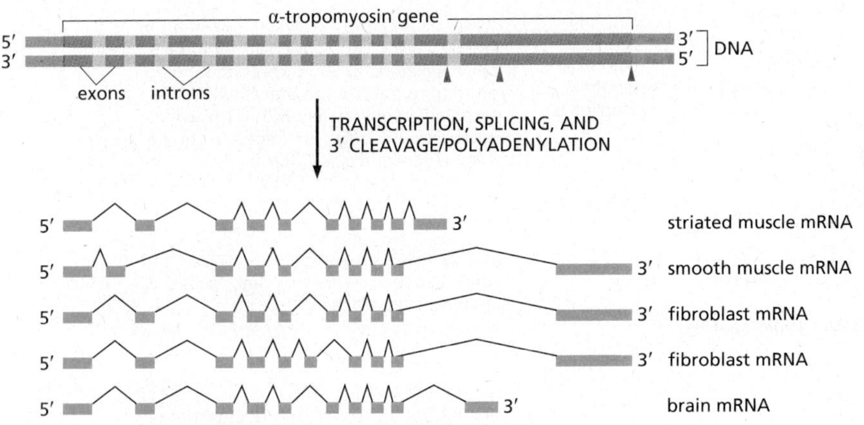

Figure 6–27 Alternative splicing of the α-tropomyosin gene from rat. α-Tropomyosin is a coiled-coil protein (see Figure 3–8) involved in the regulation of contraction in muscle cells. Its initial RNA transcript can be spliced in different ways, as indicated in the figure, to produce distinct mRNAs, which then give rise to variant proteins. Some of the splicing patterns are specific for certain types of cells. For example, the α-tropomyosin made in striated muscle is different from that made from the same gene in smooth muscle. The arrowheads in the top part of the figure mark the sites where cleavage and poly-A addition form the 3′ ends of the mature mRNAs.

of the splicing products that can be detected are the results of splicing errors and do not produce functional proteins. Nonetheless, *alternative splicing* is a key feature of gene expression in many organisms, and we shall return to this subject later, after we describe the cellular machinery that performs the basic reaction.

Nucleotide Sequences Signal Where Splicing Occurs

The mechanism of pre-mRNA splicing shown in Figure 6–26 requires that the splicing machinery recognize three portions of the precursor RNA molecule: the 5′ splice site, the 3′ splice site, and the branch point in the intron sequence that forms the base of the excised lariat. Not surprisingly, each site has a consensus nucleotide sequence that is similar from intron to intron and provides the cell with cues for where splicing is to take place (**Figure 6–28**). These consensus sequences are relatively short and can accommodate extensive sequence variability, and as we shall see shortly, the cell uses additional types of information to ultimately choose exactly where, on each RNA molecule, splicing is to take place.

RNA Splicing Is Performed by the Spliceosome

Unlike the other steps in mRNA production we have discussed, key steps in RNA splicing are performed by RNA molecules rather than proteins. Specialized RNA molecules recognize the nucleotide sequences that specify where splicing is to occur and also form the active site that catalyzes the chemistry of splicing. These RNA molecules are relatively short (less than 200 nucleotides each), and there are five of them, U1, U2, U4, U5, and U6. Known as **snRNAs (small nuclear RNAs)**, each is complexed with at least seven protein subunits to form an *snRNP* (small nuclear ribonucleoprotein). These snRNPs form the core of the **spliceosome**, the large assembly of RNA and protein molecules that performs pre-mRNA splicing in the cell. Recognition of the 5′ splice junction, the branch-point site, and the 3′ splice junction is performed through base-pairing between the snRNAs and RNA sequences in the pre-mRNA substrate.

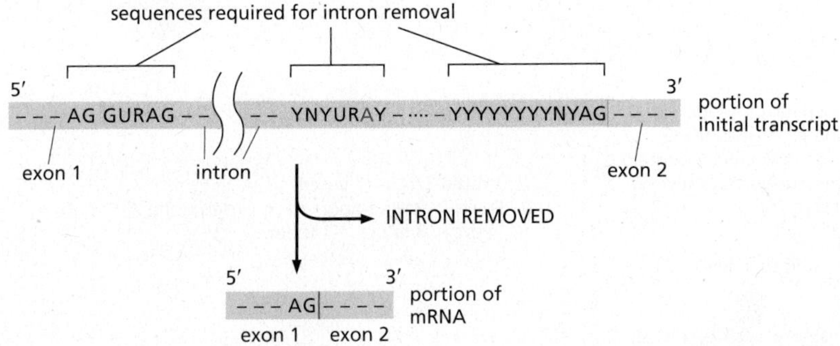

Figure 6–28 The consensus nucleotide sequences in an RNA molecule that signal the beginning and the end of most introns in humans. The three blocks of nucleotide sequences shown are required to remove an intron sequence. Here A, G, U, and C are the standard RNA nucleotides; R stands for a purine (A or G); and Y stands for a pyrimidine (C or U). The A highlighted in *red* forms the branch point of the lariat produced by splicing (see Figure 6–26). Only the GU at the start of the intron and the AG at its end are invariant nucleotides in the splicing consensus sequences. Several different nucleotides can occupy the remaining positions, although the indicated nucleotides are preferred. Although the distances along the RNA between the three splicing consensus sequences are highly variable, the distance between the branch point and 3′ splice junction is typically much shorter than the distance between the 5′ splice junction and the branch point.

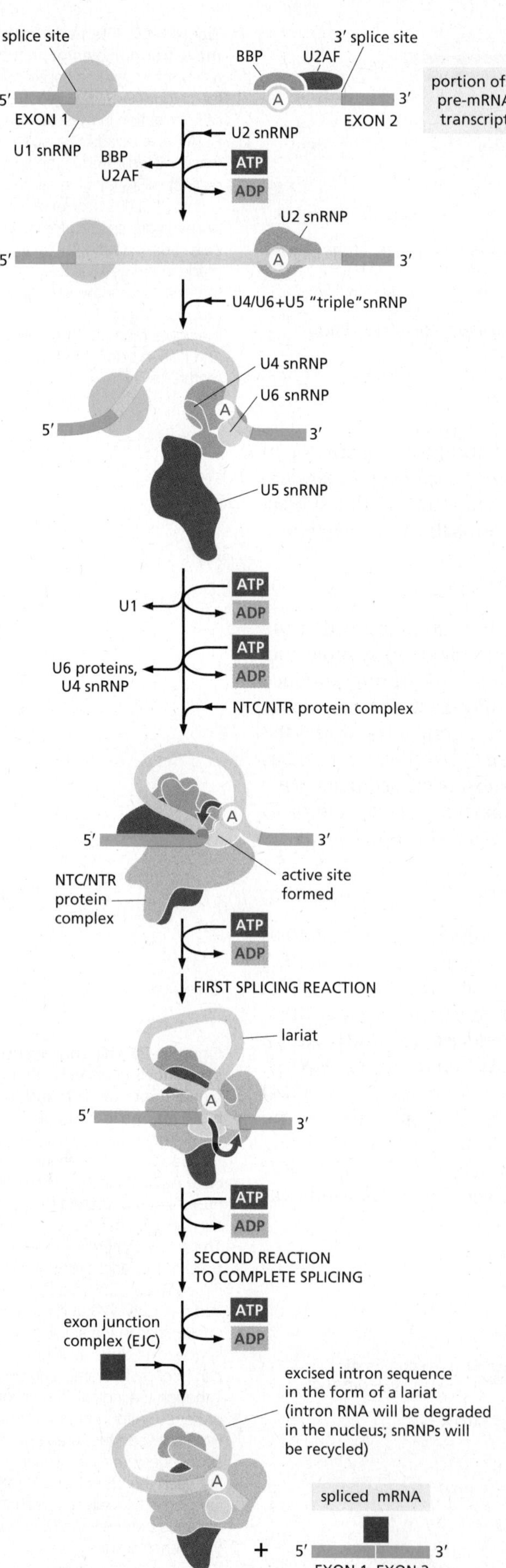

The U1 snRNP forms base pairs with the 5′ splice junction. BBP (branch-point binding protein) recognizes the branch-point site and binds cooperatively with U2AF, which recognizes the polypyrimidine tract and the 3′ splice junction (see Figure 6–28).

The U2 snRNP displaces BBP and U2AF and forms base pairs with the branch-point site consensus sequence.

The U4/U6+U5 "triple" snRNP enters the reaction. In this triple snRNP, the U4 and U6 snRNAs are held firmly together by base-pair interactions. Subsequent rearrangements break apart the U4/U6 base pairs, allowing U6 to displace U1 at the 5′ splice junction and ejecting the U4 snRNP and some of the proteins of the U6 snRNP.

Addition of the NTC/NTR protein complex positions the snRNPs to form the active site of the spliceosome and brings the branch point in proximity to the 5′ splice site.

Lariat formed. Additional rearrangements bring the two exon segments together and place them in the active site.

Spliced RNA is released.
Spliceosome components recycled using ATP hydrolysis to "reset" them.

Figure 6–29 The pre-mRNA splicing mechanism. RNA splicing is catalyzed by an assembly of snRNPs (shown as *colored shapes*) plus other proteins (most of which are not shown), which together constitute the spliceosome. The spliceosome recognizes the splicing signals on a pre-mRNA molecule, brings the two ends of the intron together, and forms the active site that catalyzes the two covalent bond-making and -breaking steps required (see Figure 6–26A and Movie 6.5). As shown, nearly every step is accompanied by hydrolysis of a molecule of ATP, which prevents the reaction from stalling or moving backwards and, as discussed below, increases the accuracy of splicing. Although only six molecules of ATP are shown in this simplified diagram, some steps require more than one, and a total of eight ATPs are consumed in each splicing reaction. As indicated in the last step, a set of proteins called the exon junction complex (EJC) is retained on the spliced mRNA molecule; its subsequent role will be discussed shortly.

The spliceosome is a complex and dynamic machine. When studied *in vitro*, a few components of the spliceosome assemble on pre-mRNA and, as the splicing reaction proceeds, new components enter and those that have already performed their tasks are jettisoned (**Figure 6–29**). However, many scientists believe that, inside the cell, the spliceosome is a preexisting, loose assembly of all the components—capturing, splicing, and releasing RNA as a coordinated unit, and undergoing extensive rearrangements each time a splice is made.

The Spliceosome Uses ATP Hydrolysis to Produce a Complex Series of RNA–RNA Rearrangements

There are many unusual features of the splicing reaction compared to the typical catalytic processes in the cell introduced in Chapter 2. First, splicing seems grossly inefficient, requiring more than a hundred proteins, five RNA molecules, and the hydrolysis of eight molecules of ATP to produce a single splice. Many genes require multiple splicing events to produce a single functional mRNA molecule (more than 20 in the example in Figure 6–25B), and the process seems inordinately complex. Second, each splicing reaction requires that the catalytic site for the reaction be assembled *de novo* on each pre-mRNA molecule through a complex, multistep process. Third, as mentioned earlier, the catalytic site of the spliceosome (which catalyzes both steps of the splicing reaction) is formed by RNA molecules (the snRNAs) rather than by proteins (**Figure 6–30**), with the proteins required to correctly position these RNAs. In the final section of this chapter, we describe the structure and chemical properties of RNA molecules that allow them to act as catalysts, as well as the proposal that the RNA-based reaction mechanisms that exist today are "leftovers" from ancient, RNA-only biological systems.

How might we rationalize the unusual biochemical complexity of splicing compared to other steps in gene expression? As we discuss shortly, pre-mRNA splicing has evolved from a much simpler, purely RNA-based process, and its complexity may in part be an example of what some evolutionary biologists term "runaway bureaucracies," accruing more and more parts over time that are now required for the process without necessarily making it any better. We do know, however, that some of the complexity of pre-mRNA splicing provides advantages. Even though ATP hydrolysis is not required for the splicing reaction per se, the numerous ATP hydrolysis steps keep the reaction from stalling or running backwards and, as we will see shortly, they also increase the accuracy of pre-mRNA splicing.

Most of the spliceosome proteins that hydrolyze ATP use the released energy to break existing RNA–RNA interactions to allow the formation of new ones. These RNA–RNA rearrangements allow the splicing signals on the pre-mRNA to be examined several times during the course of splicing. For example, the U1 snRNP initially recognizes the 5′ splice site through conventional base-pairing; as splicing proceeds, these base pairs are broken (using the energy of ATP hydrolysis) and U1 is replaced by U6 (see Figure 6–30A and B). Likewise, the branch point is "examined" twice, the first time by the branch-point binding protein and the second time by the U2 snRNP (see Figure 6-29). In this way, the spliceosome checks and rechecks the splicing signals before the active site for the two transesterification reactions forms (see Figure 6–30C), thereby increasing overall accuracy.

Note that both catalytic steps of the splicing reaction occur in the same active site (the second reaction is almost the reverse of the first). ATP-mediated

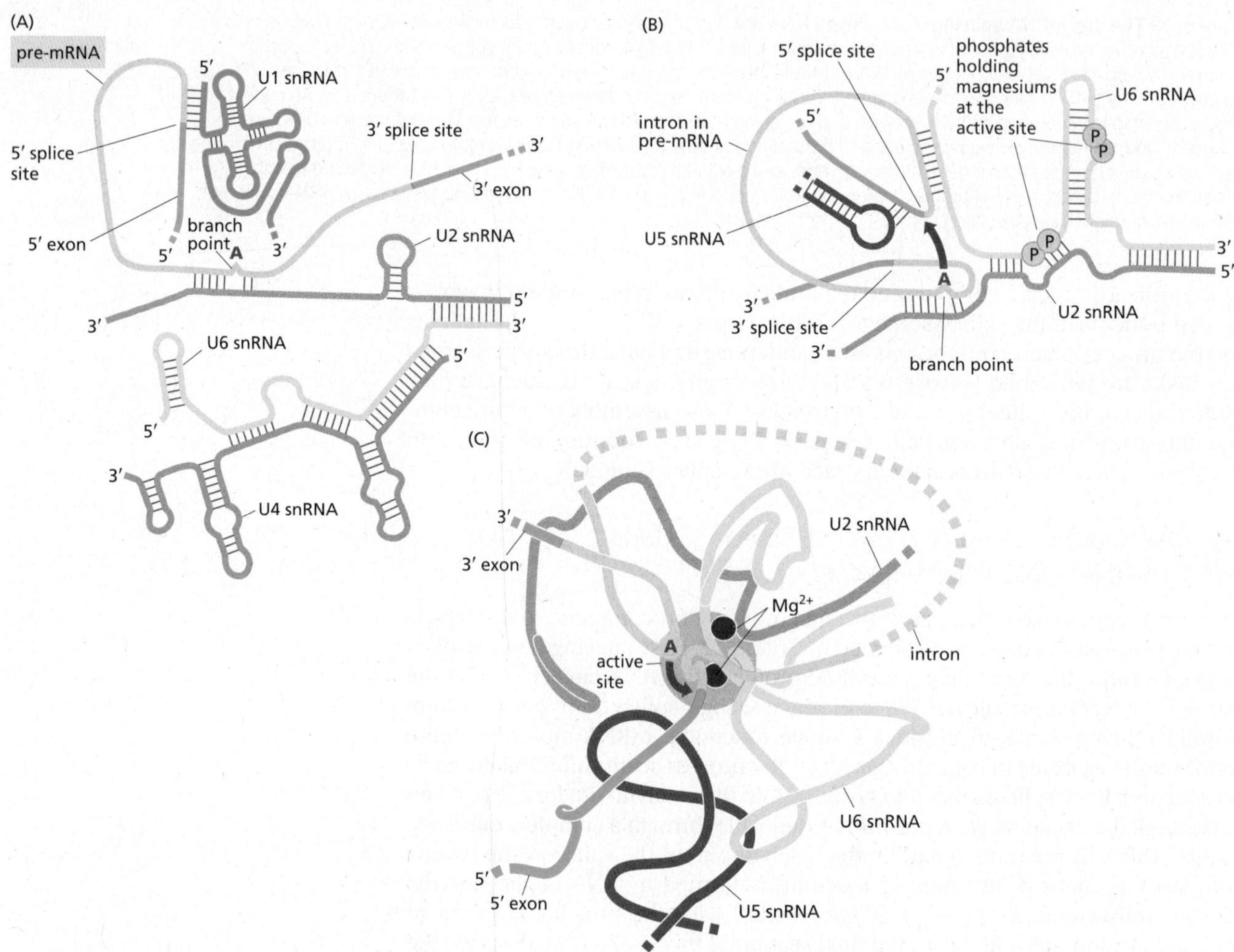

Figure 6–30 An example of an ATP hydrolysis–driven RNA–RNA rearrangement that occurs during splicing. Schematic diagram of the arrangement of RNA molecules before (A) and after (B) the active site has been formed for the two phosphoryl-transfer reactions. This rearrangement also brings the branch point and the 5′ splice site together in preparation for the first reaction. The active site (which grasps the two magnesium ions needed for the chemistry of the reactions) is formed by U2 and U6. Several sequential ATP-dependent rearrangement steps are needed to convert the configuration in A to that in B. (C) The actual structure, determined by cryo-electron microscopy, of the RNA-based catalytic core of the spliceosome that was schematically illustrated in B. For simplicity, the proteins that surround the RNAs are not shown, even though their conformations are known. High-resolution structures have also been determined for most of the other spliceosome intermediates. (B, adapted from M.E. Wilkinson, C. Charenton, and K. Nagai, *Annu. Rev. Biochem.* 89:359–398, 2020. With permission from Annual Reviews.)

rearrangements are required between the two reactions to properly reposition the pre-mRNA for the second reaction, helping to ensure that splicing accidents occur only rarely.

A general process associated with ATP hydrolysis, called *kinetic proofreading*, further increases spliceosome accuracy. Because an incorrect, "off-target" base-pairing interaction will be weaker than the correct one, incorrect interactions will dissociate more rapidly than correct ones. Each ATP-mediated rearrangement of the spliceosome takes a finite amount of time, and this time delay will favor the correct choice, because off-target interactions will often dissociate during this time window, giving the correct interaction multiple chances to form. Such kinetic proofreading is used throughout biology. For example, we saw in Chapter 5 that the initial selection of the correct nucleotides by DNA polymerase during DNA replication takes advantage of this principle. We will discuss it in more detail later in the chapter when we describe translation by the ribosome.

Once the splicing chemistry is completed, the snRNPs remain bound to the excised lariat. The disassembly of these snRNPs from the lariat (and from

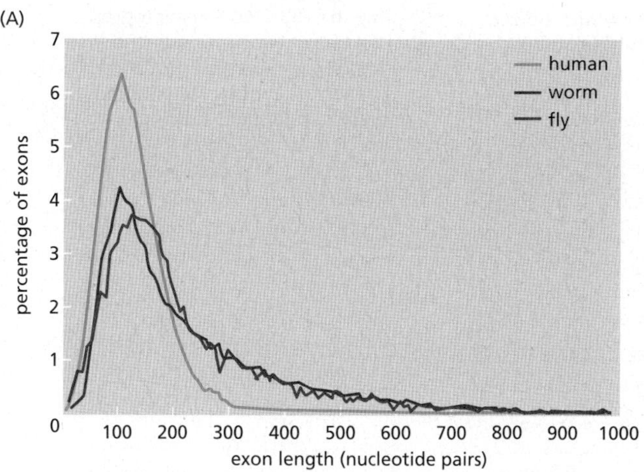

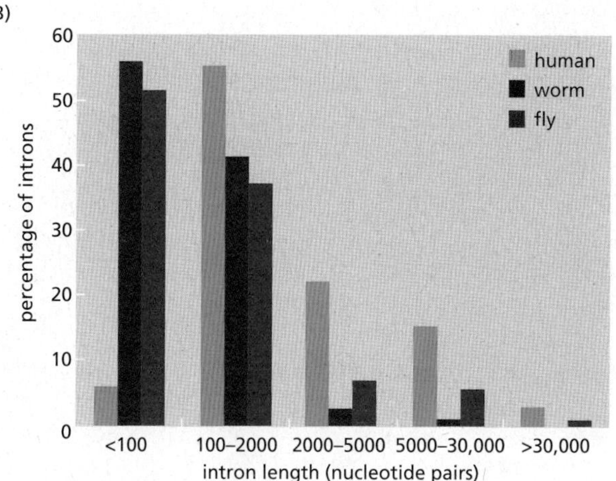

Figure 6–31 **Variation in intron and exon lengths in the human, worm, and fly genomes.** (A) Size distribution of exons. (B) Size distribution of introns. Note that exon length is much more uniform than intron length. (Adapted from International Human Genome Sequencing Consortium, *Nature* 409:860–921, published 2001 by Macmillan Magazines Ltd. Reproduced with permission of SNCSC.)

each other) requires another series of ATP-driven RNA-RNA rearrangements, thereby returning the snRNAs to their original configuration to be used again in a new reaction.

As splicing is completed, the spliceosome directs a set of proteins to bind to the mRNA near the position formerly occupied by the intron. Called the *exon junction complex* (*EJC*), these proteins mark the site of a successful splicing event, and, as we shall see later in this chapter, they influence the subsequent fate of the mRNA.

Other Properties of Pre-mRNA and Its Synthesis Help to Explain the Choice of Proper Splice Sites

As shown in **Figure 6–31**, intron sequences vary enormously in size; some are more than 100,000 nucleotides long. One might therefore expect frequent splicing mistakes—including exon skipping and the mistaken use of "cryptic" splice sites (**Figure 6–32**). To minimize these problems the fidelity mechanisms built into the spliceosome are supplemented by two additional strategies that further increase the accuracy of splicing. The first is a simple consequence of splicing being coupled to transcription (**Figure 6–33**). As transcription proceeds, the phosphorylated tail of RNA polymerase carries several components that stimulate formation of the spliceosome (see Figure 6–23), and these components are transferred directly from the polymerase to the RNA as the RNA emerges from the polymerase. This strategy helps the cell to keep track of introns and exons; for example, the snRNPs that assemble at a 5′ splice site are initially presented only with the single 3′ splice site that emerges next from the polymerase, inasmuch as the potential sites further downstream have not yet been synthesized. The coordination of transcription with splicing is thus important for preventing inappropriate exon skipping.

Another mechanism, called *exon definition*, also helps cells choose the appropriate splice sites. Exon size tends to be much more uniform than intron size,

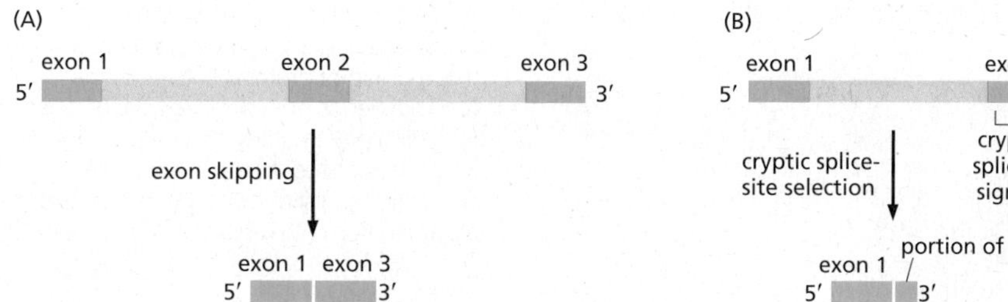

Figure 6–32 **Two types of potential splicing errors.** (A) Exon skipping. (B) Cryptic splice-site selection. Cryptic splicing signals are nucleotide sequences of RNA that closely resemble true splicing signals and are sometimes mistakenly used by the spliceosome.

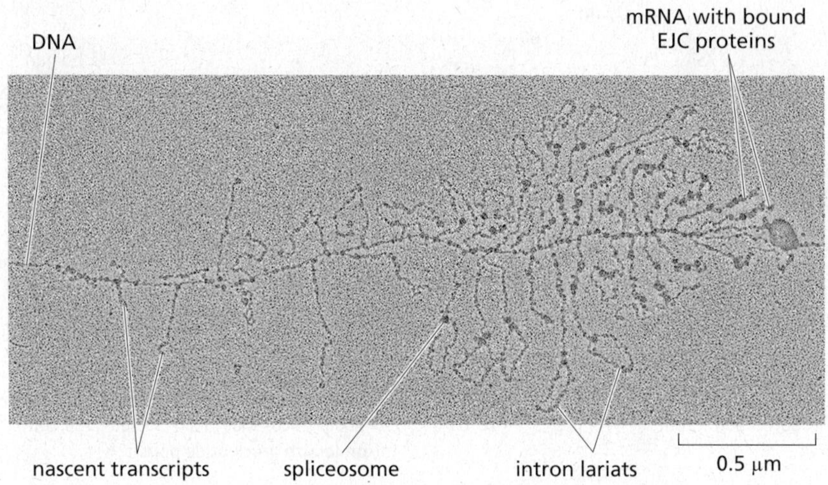

Figure 6–33 Electron micrograph of a heavily transcribed gene containing multiple introns in an early *Drosophila* embryo, illustrating the co-transcriptional splicing of its RNA transcripts. (Courtesy of Victoria Foe. Friday Harbor Laboratories, Friday Harbor WA.)

averaging about 150 nucleotide pairs across a wide variety of eukaryotic organisms (see Figure 6–31). Through exon definition, the splicing machinery seeks out the relatively homogeneously sized exon sequences rather than the intron sequences. It does this in the following way: as RNA synthesis proceeds, a group of additional components (most notably SR proteins, so-named because each contains a domain rich in serines and arginines) assemble on exon sequences and help to mark off each 3′ and 5′ splice site, starting at the 5′ end of the RNA (**Figure 6–34**). These proteins, in turn, recruit the U1 snRNP, which marks the one exon boundary, and U2AF, which, along with BBP, specifies the other. By marking out the exons in this way and thereby taking advantage of the relatively uniform size of exons, the cell increases the accuracy with which it deposits the initial splicing components on the nascent RNA and thereby avoids "near miss" splice sites. To further aid the cell in marking off exons and distinguishing them from introns, some SR proteins bind tightly to specific RNA sequences, often termed *splicing enhancers*, that are preferentially found in exons. Because several different codons specify most amino acids, it is possible for a splicing enhancer to evolve without a change in the amino acid sequence coded by the exon.

Both the marking of exon and intron boundaries and the assembly of the spliceosome begin on an RNA molecule while it is still being elongated by RNA polymerase (see Figure 6–33). However, because the actual chemistry of splicing can be delayed, intron sequences are not necessarily removed from a pre-mRNA molecule in the order in which they occur along the RNA chain.

RNA Splicing Has Remarkable Plasticity

We have seen that the choice of splice sites depends on such features of the pre-mRNA transcript as the strength of the three signals on the RNA recognized by

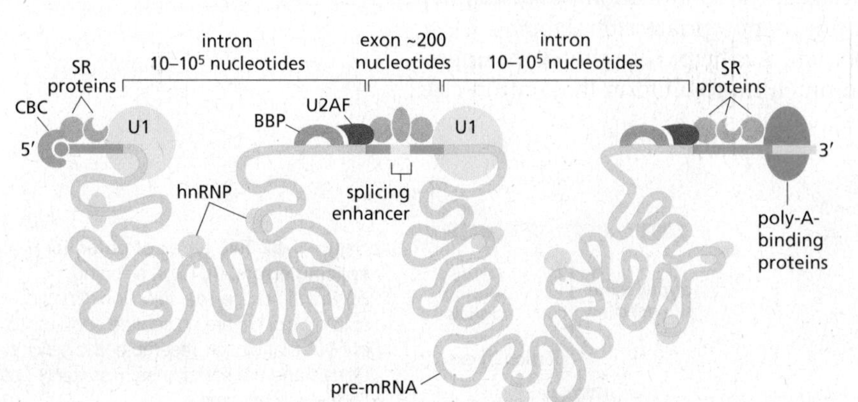

Figure 6–34 The exon definition hypothesis. SR proteins bind to each exon sequence in the pre-mRNA and thereby help to guide the snRNPs to the proper intron–exon boundaries. This demarcation of exons by the SR proteins occurs co-transcriptionally, beginning at the CBC (cap-binding complex) at the 5′ end. It has also been proposed that a group of proteins known as the heterogeneous nuclear ribonucleoproteins (hnRNPs) preferentially associates with intron sequences, further helping the spliceosome distinguish introns from exons. (Adapted from R. Reed, *Curr. Opin. Cell Biol.* 12:340–345, 2000. With permission from Elsevier.)

the splicing machinery (the 5′ and 3′ splice junctions and the branch point), the co-transcriptional assembly of the spliceosome, and the "bookkeeping" that underlies exon definition. We do not know exactly how accurate splicing normally is because, as we see later, there are several quality-control systems that rapidly destroy mRNAs whose splicing goes awry. However, we do know that, compared with other steps in gene expression, splicing is unusually flexible.

Thus, for example, a mutation in a nucleotide sequence critical for splicing of a particular intron does not necessarily prevent splicing of that intron altogether. Instead, the mutation typically creates a new pattern of splicing (**Figure 6–35**). Most common, an exon is simply skipped (Figure 6–35B). In other cases, the mutation causes a cryptic splice junction to become the default choice (Figure 6–35C). Apparently, the splicing machinery has evolved to pick out the best possible pattern of splice junctions, and if the optimal one is damaged by mutation, it will seek out the next best pattern, and so on. This inherent plasticity in the process of RNA splicing suggests that changes in splicing patterns caused by random mutations have been important in the evolution of genes and organisms. It also means that mutations that affect splicing can be severely detrimental to the organism: in addition to the β-thalassemia example presented in Figure 6–35, aberrant splicing plays important roles in the development of cystic fibrosis, frontotemporal dementia, Parkinson's disease, retinitis pigmentosa, spinal muscular atrophy, myotonic dystrophy, premature aging, and cancer. It has been estimated that of the many point mutations that cause inherited human diseases, 10% produce aberrant splicing of the gene containing the mutation.

The plasticity of RNA splicing also means that the cell can easily regulate the pattern of RNA splicing. Earlier in this section we saw that alternative splicing can give rise to different proteins from the same gene and that this is a common strategy to enhance the coding potential of genomes. Some examples of alternative splicing are constitutive; that is, the alternatively spliced mRNAs are produced continually by cells of an organism. However, in many cases, the cell regulates the splicing patterns so that different forms of the protein are produced at different times and in different tissues (see Figure 6–27). In Chapter 7, we return to this issue to discuss some specific examples of regulated RNA splicing.

(A) NORMAL ADULT β-GLOBIN RNA TRANSCRIPT

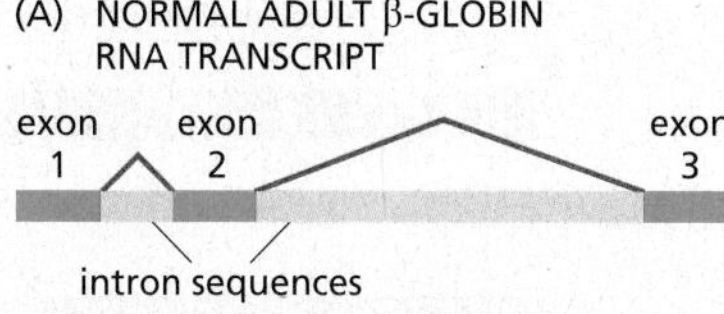

normal mRNA is formed from three exons

(B) A SINGLE-NUCLEOTIDE CHANGE THAT DESTROYS A NORMAL SPLICE SITE, THEREBY CAUSING EXON SKIPPING

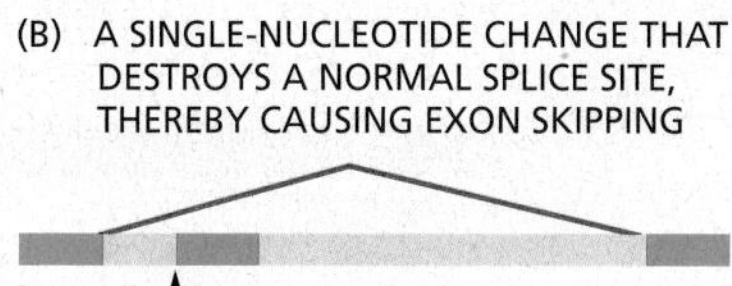

mRNA with exon 2 missing

(C) A SINGLE-NUCLEOTIDE CHANGE THAT DESTROYS A NORMAL SPLICE SITE, THEREBY ACTIVATING A CRYPTIC SPLICE SITE

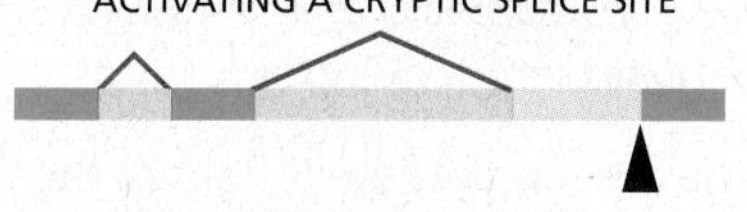

mRNA with extended exon 3

(D) A SINGLE-NUCLEOTIDE CHANGE THAT CREATES A NEW SPLICE SITE, THEREBY CAUSING A NEW EXON TO BE INCORPORATED

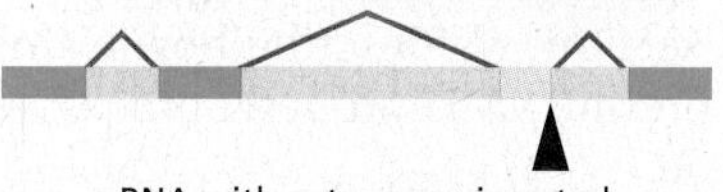

mRNA with extra exon inserted between exon 2 and exon 3

Figure 6–35 **Abnormal processing of the β-globin primary RNA transcript in humans with the disease β-thalassemia.** In the examples shown, the disease (a severe anemia due to aberrant hemoglobin synthesis) is caused by splice-site mutations found in the genomes of affected individuals. The *dark blue boxes* represent the three normal exon sequences; the *red lines* connect the 5′ and 3′ splice sites that are used. The *light blue boxes* depict new nucleotide sequences included in the final mRNA molecule as a result of the mutation denoted by the *black arrowhead*. Note that when a mutation leaves a normal splice site without a partner, an exon is skipped as in panel B or one or more abnormal cryptic splice sites nearby is used as the partner site as in panel C.

Spliceosome-catalyzed RNA Splicing Evolved from RNA Self-splicing Mechanisms

When the spliceosome was first discovered, it puzzled molecular biologists. Why do RNA molecules instead of proteins perform important roles in splice-site recognition and in the chemistry of splicing? Why is a lariat intermediate used rather than the apparently simpler alternative of bringing the 5′ and 3′ splice sites together in a single step, followed by their direct cleavage and rejoining? The answers to these questions reflect the way in which the spliceosome has evolved.

As discussed in the final section of this chapter, it is likely that early cells used RNA molecules rather than proteins as their major catalysts and that they stored their genetic information in RNA rather than in DNA sequences. RNA-catalyzed splicing reactions presumably had critical roles in these early cells. As evidence, some *self-splicing RNA* introns (that is, intron sequences in RNA whose splicing out can occur in the absence of proteins or any other RNA molecules) remain today, for example, in the nuclear rRNA genes of the ciliate *Tetrahymena*, in a few bacteriophage T4 genes, and in some mitochondrial and chloroplast genes. In these cases, the RNA molecule folds into a specific three-dimensional structure that brings the intron–exon junctions together and directly catalyzes the two transesterification reactions. A self-splicing intron sequence can be identified in a test tube by incubating a pure RNA molecule that contains the intron sequence and observing the splicing reaction. Because the basic chemistry as well as the structure of the active site of some self-splicing RNAs is so similar to those of the pre-mRNA spliceosome, the much more involved

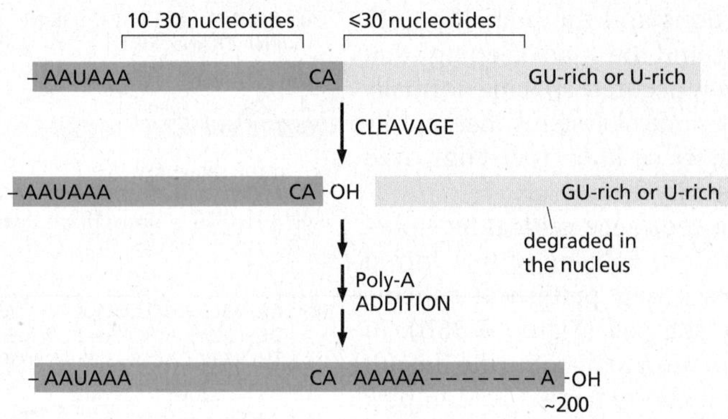

Figure 6–36 **The consensus nucleotide sequences in RNA that direct cleavage and polyadenylation to form the 3′ end of a eukaryotic mRNA.** These sequences are encoded in the genome, and specific proteins recognize them—as RNA—after they are transcribed. As shown in Figure 6–37, the hexamer AAUAAA is bound by CPSF, and the GU-rich element beyond the cleavage site is bound by CstF; the CA sequence is bound by a third protein factor required for the cleavage step. Like other consensus nucleotide sequences discussed in this chapter (see Figure 6–12), the sequences shown in the figure represent optimal sequences; in reality, a variety of related cleavage and polyadenylation signals occur in nature.

process of pre-mRNA splicing described earlier very likely evolved from a simpler, ancestral form of RNA self-splicing.

RNA-processing Enzymes Generate the 3′ End of Eukaryotic mRNAs

We have seen that the 5′ end of the pre-mRNA produced by RNA polymerase II is capped almost as soon as it emerges from the RNA polymerase. Then, as the polymerase continues its movement along a gene, spliceosomes assemble on the RNA and delineate the intron and exon boundaries. The long C-terminal tail of the RNA polymerase coordinates these processes by transferring capping and splicing components directly to the RNA as it emerges from the enzyme. In this section, we describe how a similar mechanism ensures that the 3′ end of the pre-mRNA is properly processed as RNA polymerase II reaches the end of a gene.

The position of the 3′ end of each mRNA molecule is specified by signals encoded in the DNA nucleotide sequence (**Figure 6–36**). These signals are transcribed into RNA as the RNA polymerase II moves through them, and they are then recognized (as RNA) by a series of RNA-binding proteins and RNA-processing enzymes (**Figure 6–37**). Two multisubunit proteins (called CstF, cleavage stimulation factor; and CPSF, cleavage and polyadenylation specificity factor) are of special importance. Both of these proteins travel with the RNA polymerase tail and are transferred to the 3′-end processing sequence on an RNA molecule as it emerges from the RNA polymerase.

Once the two proteins bind to their recognition sequences on the emerging RNA molecule, additional proteins assemble with them to cleave the RNA (releasing it from the RNA polymerase) and complete the 3′ end of the mRNA. Once the RNA is cleaved, an enzyme called *poly-A polymerase (PAP)* adds approximately 200 A nucleotides, one at a time, to the 3′ end produced by the cleavage (see Figure 6-37). The precursor for these additions is ATP, and the same type of 5′-to-3′ bonds are formed as in conventional RNA synthesis. But unlike other RNA polymerases, poly-A polymerase does not require a template; hence, the poly-A tail of eukaryotic mRNAs is not directly encoded in the genome. As the poly-A tail is synthesized, proteins called poly-A-binding proteins assemble onto it and, by a poorly understood mechanism, they help determine the final length of the tail.

The RNA polymerase II continues to transcribe after the 3′ end of a eukaryotic pre-mRNA molecule has been cleaved, in some cases for hundreds of nucleotides. However, two factors increase the likelihood that an RNA polymerase will terminate transcription shortly after it has synthesized the RNA signal for 3′-end

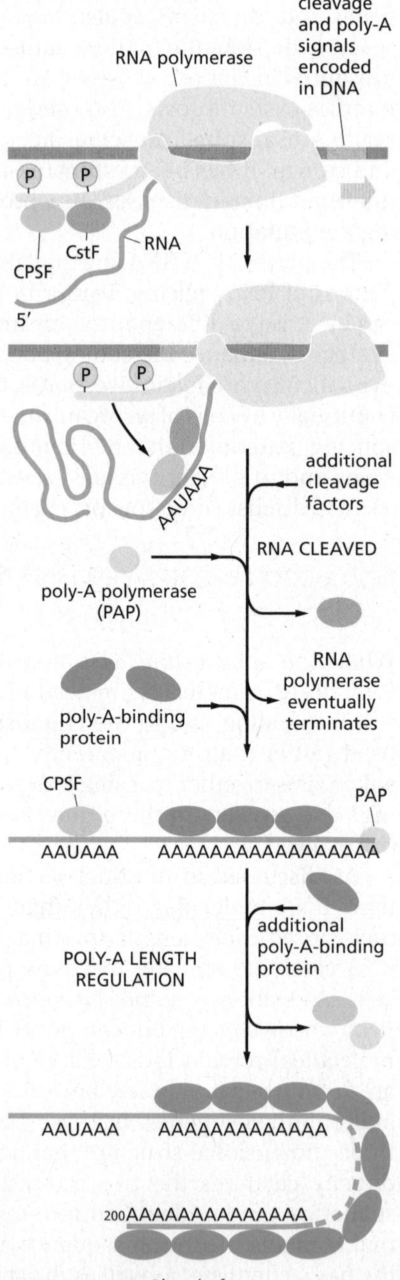

Figure 6–37 **Some of the major steps in generating the 3′ end of a eukaryotic mRNA.** This process is much more complicated than the analogous process in bacteria, where the RNA polymerase simply stops at a termination signal and releases both the 3′ end of its transcript and the DNA template (see Figure 6–11).

cleavage and polyadenylation. First, the recruitment of the many proteins needed for 3′-end processing (which occurs while some of them are still bound to the RNA polymerase tail) causes a conformational change in the polymerase, slowing it down and decreasing its processivity. Second, once 3′-end cleavage has occurred, the newly synthesized RNA emerging from the polymerase lacks a 5′ cap; this unprotected RNA is rapidly degraded by a 5′→3′ exonuclease, and, when it catches up to the polymerase, it causes the RNA polymerase to release its grip on the template and terminate transcription.

In the simplest case, a gene carries a single site for 3′ RNA cleavage and polyadenylation. However, many genes have several such sites and can therefore produce a variety of mRNAs that differ in their 3′ ends. As will be discussed in the next chapter, cells can regulate 3′-end processing to produce different proteins from the same gene in a manner analogous to alternative splicing.

Mature Eukaryotic mRNAs Are Selectively Exported from the Nucleus

Eukaryotic pre-mRNA synthesis and processing take place in an orderly fashion within the cell nucleus. But of the pre-mRNA that is synthesized, only a small fraction—the mature mRNA—is of further use to the cell. Most of the rest—excised introns, broken RNAs, aberrantly processed pre-mRNAs, and accidently transcribed portions of the genome—is not only useless but potentially dangerous. How does the cell distinguish between the relatively rare mature mRNA molecules it wishes to keep and the overwhelming amount of useless debris?

The answer is that the RNAs are distinguished by the proteins bound to them. For example, we have seen that acquisition of cap-binding complexes, exon junction complexes, and poly-A-binding proteins marks the completion of capping, splicing, and poly-A addition, respectively. A properly completed mRNA molecule is also distinguished by the proteins it lacks. For example, the long-term presence of an snRNP protein would signify incomplete or aberrant splicing. Only when the proteins present on an mRNA molecule collectively signify that processing was successfully completed is the mRNA exported from the nucleus into the cytosol, where it can be translated into protein. Improperly processed mRNAs and other RNA debris are retained in the nucleus, where they are eventually degraded by the nuclear **RNA exosome**, a large protein complex whose interior is rich in 3′-to-5′ RNA exonucleases (**Figure 6–38**). Indeed, the default fate of RNAs in the nucleus is degradation; only those bearing the proper constellation of proteins are spared.

Of all the proteins that assemble on pre-mRNA molecules as they emerge from transcribing RNA polymerases, the most abundant are the hnRNPs (heterogeneous nuclear ribonucleoproteins). Some of these proteins (there are approximately 30 different ones in humans) unwind the hairpin helices in the RNA so that splicing and other signals on the RNA can be read more easily. Others

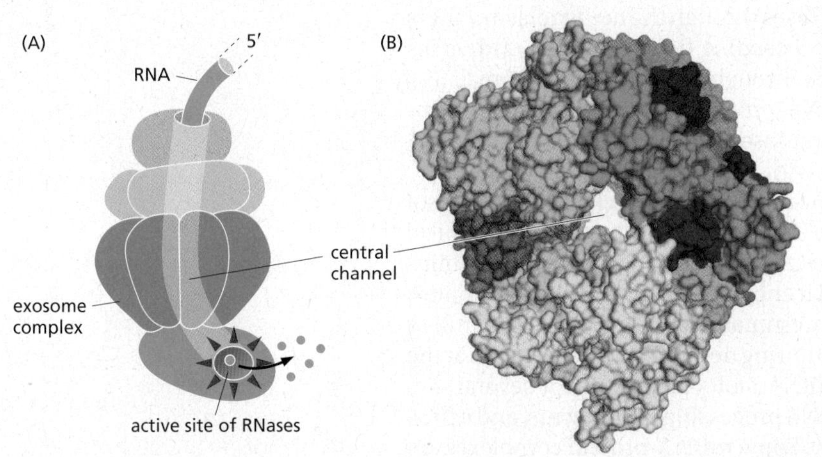

Figure 6–38 **Structure of the nuclear RNA exosome.** (A) RNA is fed into one end, passes through the central channel, and is degraded by RNases at the other end. (B) Structure of the central channel of the human RNA exosome viewed end-on. Nine different protein subunits (each represented by a different color) make up this large ring structure. Eukaryotic cells have both a nuclear exosome and a cytoplasmic exosome; both forms include the central channel but differ in their additional subunits. The nuclear RNA exosome degrades aberrant RNAs (including excised intron sequences and incorrectly spliced RNAs) before they are exported to the cytosol. It also processes certain types of RNA (for example, the ribosomal RNAs) to produce their final form. The cytoplasmic form of the RNA exosome is responsible for degrading mRNAs in the cytosol and is thus crucial in determining the lifetime of each mRNA molecule. (A, adapted from C. Kilchert et al., *Nat. Rev. Mol. Cell Biol.* 17:227, 2016. With permission from Springer Nature; B, PDB code: 2NN6.)

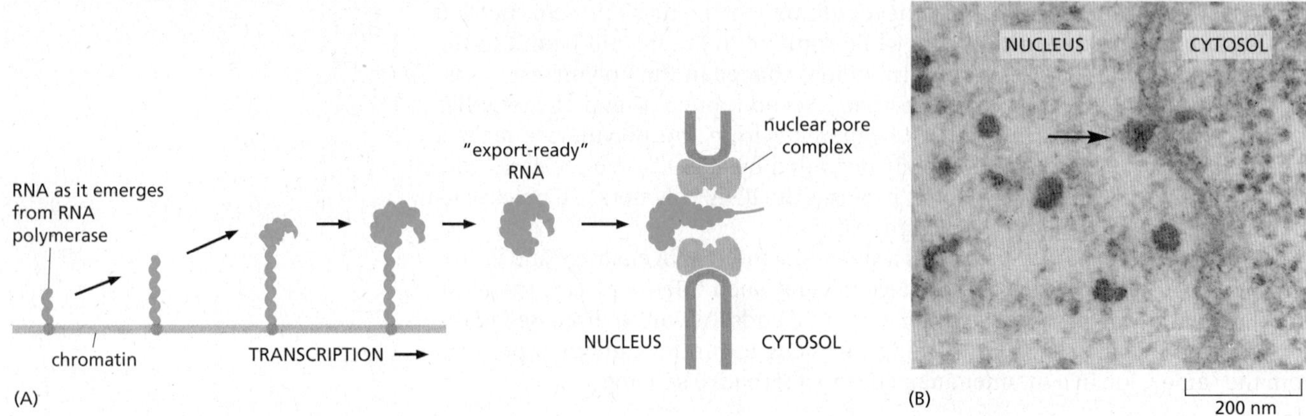

Figure 6–39 Transport of a large mRNA molecule through the nuclear pore complex. (A) The maturation of an mRNA molecule as it is synthesized by RNA polymerase and packaged by a variety of nuclear proteins. This drawing of an unusually large and abundant insect RNA, called the Balbiani Ring mRNA, is based on electron microscope micrographs such as that shown in (B). (A, adapted from B. Daneholt, *Cell* 88:585–588, 1997; B, © 1966 B.J. Stevens and H. Swift. Originally published in *J. Cell Biol.* https:doi.org/10.1083/jcb.31.1.55. With permission from Rockefeller University Press.)

preferentially package the RNA contained in the very long intron sequences found in complex organisms (see Figure 6–31); these may also help to distinguish the debris left over from RNA processing from fully mature mRNAs.

Successfully processed mRNAs are guided through the **nuclear pore complexes (NPCs)**—aqueous channels in the nuclear membrane that directly connect the nucleoplasm and cytosol (**Figure 6–39**). Small molecules (less than 40,000 daltons or about 5 nm in diameter) can diffuse freely through these channels. However, most of the macromolecules in cells, including mRNAs complexed with proteins, are far too large to pass through the channels without a special process. The cell uses energy to actively transport such macromolecules in both directions through the nuclear pore complexes.

As explained in detail in Chapter 12, macromolecules are moved through nuclear pore complexes by *nuclear transport receptors*, which, depending on the identity of the macromolecule, escort it from the nucleus to the cytoplasm or vice versa. For mRNA export to occur, a specific nuclear transport receptor must be loaded onto the mRNA, a step that, in many organisms, takes place in concert with 3′ cleavage and polyadenylation.

The export of mRNA-protein complexes from the nucleus can be readily observed with the electron microscope for the unusually abundant mRNA of the insect *Balbiani Ring genes*. As these genes are transcribed, the newly formed pre-mRNA is seen to be packaged by proteins, including hnRNPs, SR proteins, and components of the spliceosome. This protein-pre-mRNA complex undergoes a series of structural transitions, probably reflecting RNA-processing events, culminating in a curved fiber. This curved fiber diffuses through the nucleoplasm, enters the nuclear pore complex (with its 5′ cap proceeding first), and then undergoes additional structural transitions as it moves through the pore (see Figure 6–39). Such observations reveal that the pre-mRNA-protein and mRNA-protein complexes are dynamic structures that gain and lose specific proteins during RNA synthesis, processing, and export (**Figure 6–40**).

The journey of an individual mRNA molecule from the nucleus to the cytosol can also be tracked by fluorescently labeling it and observing it over time. A typical mRNA molecule that is released from its site of transcription spends several minutes randomly diffusing in the nucleus until it encounters a nuclear pore complex. During this time, RNA-processing events presumably continue, with the mRNA shedding previously bound proteins and acquiring new ones. Once it arrives at the entrance to the pore, the "export-ready" mRNA molecule hovers for several seconds, during which time the completion of RNA processing likely occurs, and it then is transported through the pore very rapidly. Some mRNA–protein complexes are

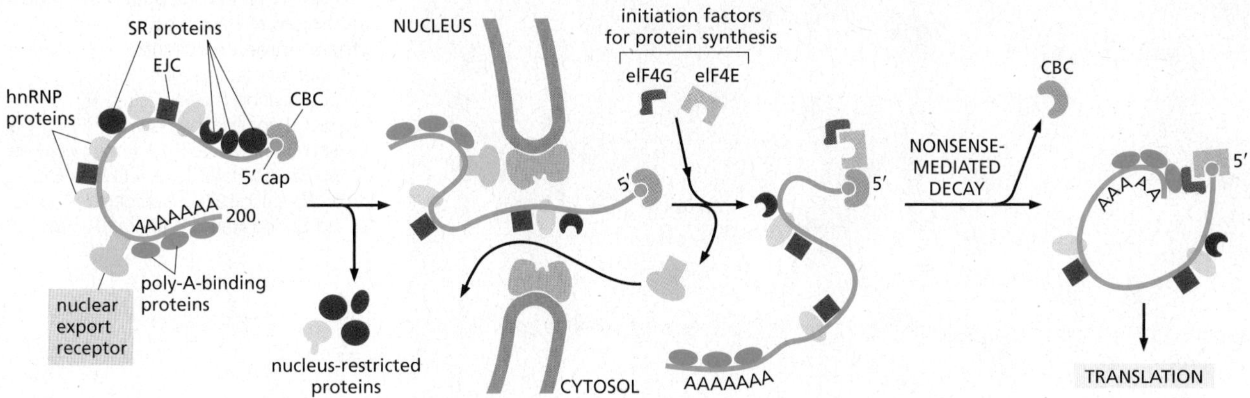

Figure 6–40 **Schematic illustration of an export-ready mRNA molecule and its transport through the nuclear pore.** As indicated, some proteins travel with the mRNA as it moves through the pore, whereas others remain in the nucleus. Some of the nuclear proteins that are lost are eventually replaced by cytosolic versions, such as those that bind the 5′ cap. In some species (humans, for example), there are different poly-A-binding proteins in the nucleus and the cytosol; other species have only a single poly-A-binding protein. The nuclear export receptor for mRNAs is a complex of proteins that binds to an mRNA molecule once it has been correctly spliced and polyadenylated. After the mRNA has been exported to the cytosol, this export receptor dissociates from the mRNA and is re-imported into the nucleus, where it can be used again.

very large, and how they are moved through the nuclear pore complexes so rapidly (in about 10 milliseconds) remains a mystery.

Some of the proteins deposited on the mRNA while it is still in the nucleus can affect the fate of the mRNA after it is transported to the cytosol. Thus, the stability of an mRNA in the cytosol, the efficiency with which it is translated into protein, and its ultimate destination in the cell can all be determined by proteins acquired in the nucleus that remain bound to the mRNA in the cytosol.

Before discussing what next happens to the exported mRNAs, we briefly consider how the synthesis and processing of noncoding RNA molecules occur. There are many types of noncoding RNAs produced by cells (see Table 6–1, p. 327), but here we focus on the rRNAs, which are critically important for the translation of mRNAs into protein.

Noncoding RNAs Are Also Synthesized and Processed in the Nucleus

Of all the RNAs in a typical cell, only a few percent are mRNA. The bulk of RNA performs structural and catalytic functions (see Table 6–1). The most abundant RNAs in cells are the ribosomal RNAs (rRNAs), constituting approximately 80% of the RNA in rapidly dividing cells. As discussed later in this chapter, these RNAs form the core of the ribosome. Unlike bacteria—in which a single RNA polymerase synthesizes all RNAs in the cell—eukaryotes have a separate, specialized polymerase, RNA polymerase I, that is dedicated to producing rRNAs. RNA polymerase I is similar structurally to the RNA polymerase II discussed previously; however, the absence of a C-terminal tail in polymerase I helps to explain why its transcripts are neither capped nor polyadenylated.

Because multiple rounds of translation of each mRNA molecule can provide an enormous amplification in the production of protein molecules, many of the proteins that are very abundant in a cell can be synthesized from genes that are present in a single copy per haploid genome (see Figure 6–3). In contrast, the RNA components of the ribosome are final gene products, and a growing mammalian cell must synthesize approximately 10 million copies of each type of ribosomal RNA in each cell generation to construct its 10 million ribosomes. The cell can produce adequate quantities of ribosomal RNAs only because it contains multiple copies of the **rRNA genes** that code for **ribosomal RNAs (rRNAs)**. Even *E. coli* needs seven copies of its rRNA genes to meet the cell's need for ribosomes. Human cells contain about 200 rRNA gene copies per haploid genome, spread

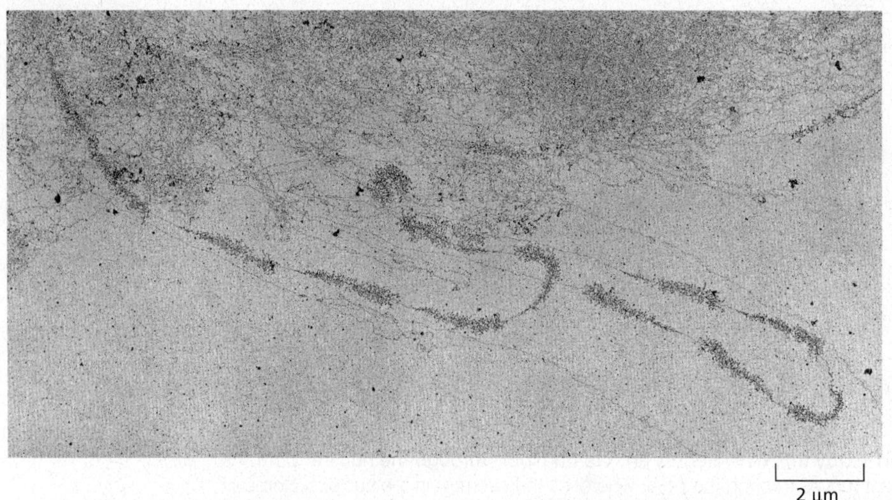

Figure 6–41 **Transcription from tandemly arranged rRNA genes, as seen in the electron microscope.** The pattern of alternating transcribed gene and nontranscribed spacer is readily seen. A higher-magnification view of rRNA genes is shown in Figure 6–10. (From V.E. Foe, *Cold Spring Harb. Symp. Quant. Biol.* 42:723–740, 1978. With permission from Cold Spring Harbor Laboratory Press.)

out in small clusters on five different chromosomes (see Figure 4–12), while cells of the frog *Xenopus* contain about 600 rRNA gene copies per haploid genome in a single cluster on one chromosome (**Figure 6–41**).

There are four types of eukaryotic rRNAs, each present in one copy per ribosome. Three of the four rRNAs (18S, 5.8S, and 28S) are made by chemically modifying and cleaving a single large precursor rRNA (**Figure 6–42**); the fourth (5S RNA) is synthesized from a separate cluster of genes by a different polymerase, RNA polymerase III, and it does not require chemical modification.

Extensive chemical modifications occur in the 13,000-nucleotide-long precursor rRNA before the mature 18S, 5.8S, and 28S rRNAs are cleaved out of it. These include about 100 methylations of the 2′-OH positions on nucleotide sugars and 100 isomerizations of uridine nucleotides to pseudouridine (**Figure 6–43A**). The functions of these modifications are not understood in detail, but they probably aid in ribosome assembly, and they may also subtly affect the operation of completed ribosomes. Each modification is made at a specific position in the precursor rRNA, specified by "guide RNAs," which position themselves on the precursor rRNA through base-pairing and thereby bring an RNA-modifying enzyme to the appropriate position (**Figure 6–43B**). Other guide RNAs promote cleavage of the precursor rRNAs into the mature rRNAs, probably by causing conformational changes in the precursor rRNA that expose these sites to nucleases. All of these guide RNAs are members of a large class of RNAs called **small nucleolar RNAs (snoRNAs)**, so named because these RNAs perform their functions in a subcompartment of the nucleus called the nucleolus. Many snoRNAs are encoded in the introns of other genes, especially those encoding ribosomal proteins. They are synthesized by RNA polymerase II and processed from excised intron sequences.

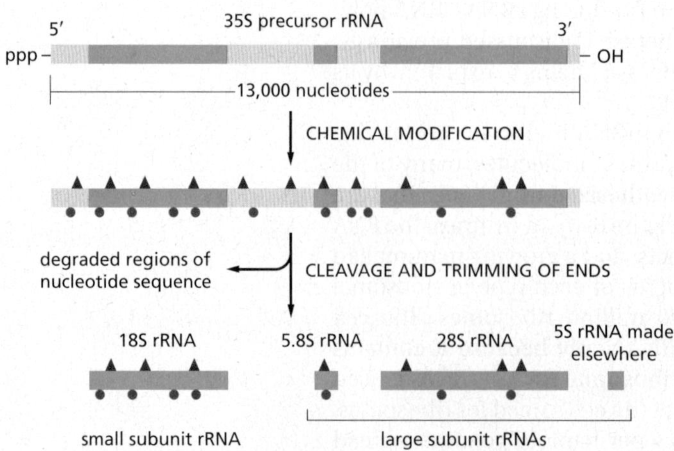

Figure 6–42 **The chemical modification and nucleolytic processing of a eukaryotic precursor rRNA molecule into three separate ribosomal RNAs.** Two types of chemical modifications (see Figure 6–43) are made to the precursor rRNA before it is cleaved. Nearly half of the nucleotide sequences in this precursor rRNA are discarded and degraded in the nucleus by the RNA exosome. The processing of the ribosomal RNAs begins while they are still being transcribed; the nascent transcripts also begin to be assembled with ribosomal proteins (see Figure 6–10). The rRNAs are named according to their "S" values, which refer to their rate of sedimentation in an ultracentrifuge. The larger the S value, the larger the rRNA.

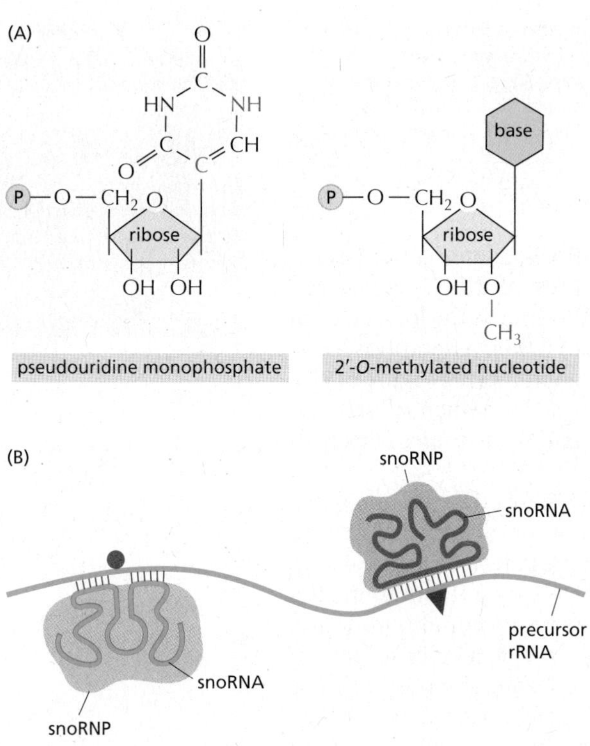

Figure 6–43 Modifications of the precursor rRNA by guide RNAs. (A) Two prominent covalent modifications made to rRNA; the differences from the initially incorporated nucleotide are indicated by *red* atoms. Pseudouridine is an isomer of uridine; the base has been "rotated" and is attached to the *red* C rather than to the *red* N of the sugar (compare to Figure 6–5B). (B) As indicated, snoRNAs determine the sites of modification by base-pairing to complementary sequences on the precursor rRNA. The snoRNAs are bound to proteins, and the complexes are called snoRNPs (small nucleolar ribonucleoproteins). The snoRNPs contain both the guide sequences and the enzymes that modify the rRNA.

The Nucleolus Is a Ribosome-producing Factory

The nucleolus is the most obvious structure seen in the nucleus of a eukaryotic cell when viewed in the light microscope. It was so closely scrutinized by early cytologists that an 1898 review could list some 700 references. We now know that the **nucleolus** is the site for the synthesis and processing of rRNAs and the assembly of ribosomes. Unlike many of the major organelles in the cell, the nucleolus is not bound by a membrane (Figure 6–44); instead, it is a huge *biomolecular condensate* of macromolecules, including the rRNA genes themselves, precursor

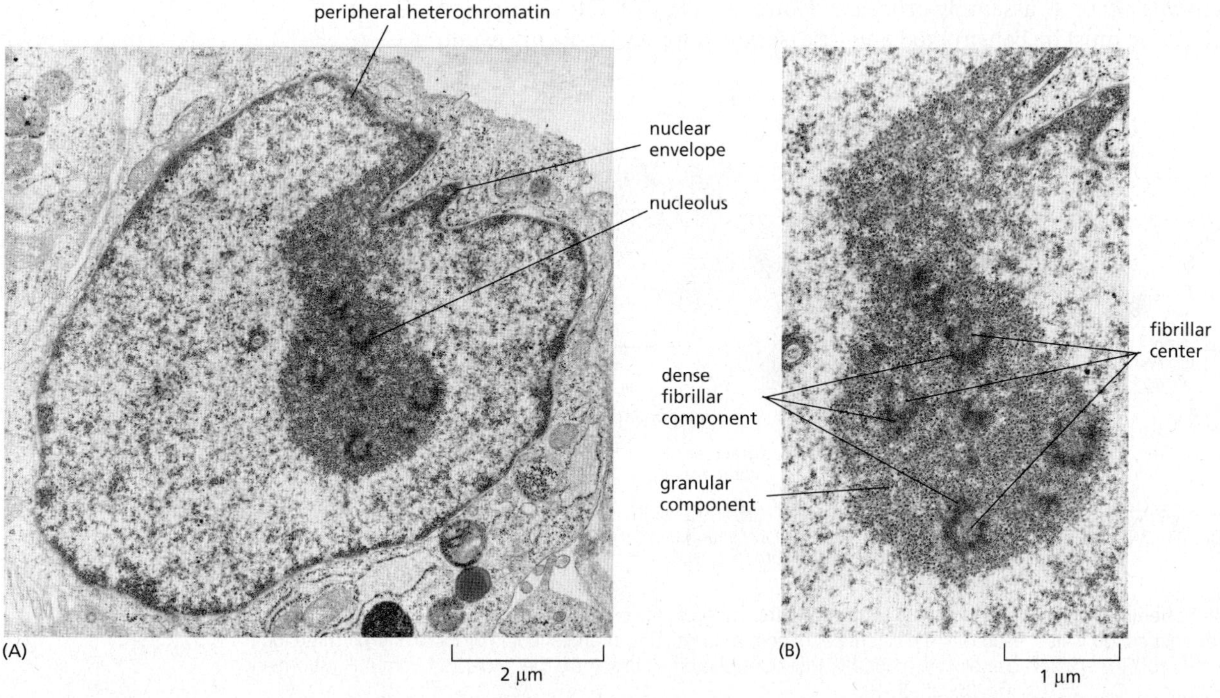

Figure 6–44 Electron micrograph of a thin section of a nucleolus in a human fibroblast, showing its three distinct zones. (A) View of entire nucleus. (B) Higher-power view of the nucleolus. It is believed that processing of the rRNAs and their assembly into the two subunits of the ribosome proceeds outward from the dense fibrillar component to the surrounding granular components (see Figure 6–46). (Courtesy of E.G. Jordan and J. McGovern.)

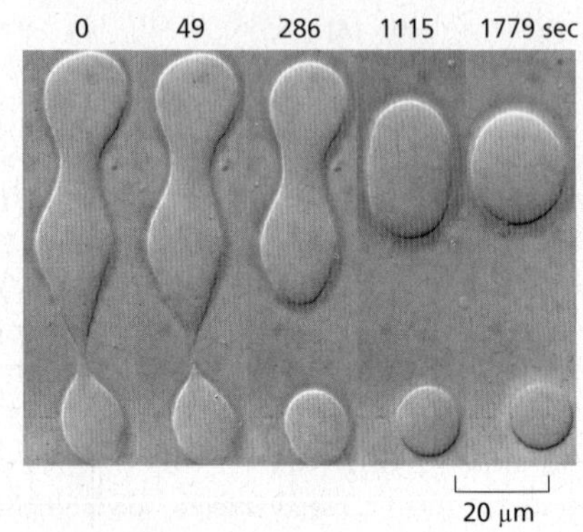

Figure 6–45 **Nucleoli exhibit fluidlike behavior when observed *in vitro*.** A time course showing the fate of three nucleoli from frog oocytes that have begun to fuse with one another. The lower fusion joint eventually breaks while the other enlarges, completing the fusion event. This experiment was carried out *in vitro* under mineral oil, and the nucleoli were observed using differential-interference-contrast microscopy (see Chapter 9). (Courtesy of C.P. Brangwynne et al., *Proc. Natl. Acad. Sci. USA* 108:4334–4339, 2011).

rRNAs, mature rRNAs, rRNA-processing enzymes, snoRNPs, a large set of assembly factors (including ATPases, GTPases, protein kinases, and RNA helicases), ribosomal proteins, and partly assembled ribosomes. We discuss the formation of membraneless organelles in Chapter 12; here, we note that their assembly is likely driven by the type of phase transitions discussed in Chapter 3 (see Figure 3–77). The close, but loose, association of all these components, which allows the assembly of ribosomes to occur rapidly and smoothly, endows the nucleolus with liquid-like properties (**Figure 6–45**).

The rRNA genes themselves have an important role in forming the nucleolus (**Figure 6–46**). In a diploid human cell, the rRNA genes are distributed into 10 clusters, located near the tips of five different chromosome pairs (see Figure 4–12). During interphase, these 10 chromosomes contribute DNA loops (containing the rRNA genes) to the nucleolus; in M phase, when the chromosomes condense, the nucleolus fragments and then disappears. Then, in the telophase part of mitosis, as chromosomes return to their semi-dispersed state, the tips of the 10 chromosomes re-form small nucleoli, which progressively coalesce into a single nucleolus (**Figure 6–47** and **Figure 6–48**). As might be expected, the size of the nucleolus reflects the number of ribosomes that the cell is producing. Its size therefore varies greatly in different cells and can change in a single cell, occupying nearly 25% of the total nuclear volume in cells that are making unusually large amounts of protein.

Ribosome assembly is a complex process, requiring, in addition to the proteins and RNA molecules that compose the finished ribosome, more than 200 proteins that aid the assembly of the finished ribosome. These include *chaperones* (discussed later in this chapter), ATP-dependent RNA helicases, nucleases, and a wide variety of RNA-binding proteins. In addition, assembly requires a number of small RNA molecules, such as the two snoRNAs of Figure 6–43. In many respects, building a ribosome resembles the process by which a spliceosome is formed, as we discussed earlier in the chapter. In particular, many ATP-driven RNA structural rearrangements occur as assembly proceeds. However, a key difference is that a new spliceosome must be constructed and disassembled for each splicing event,

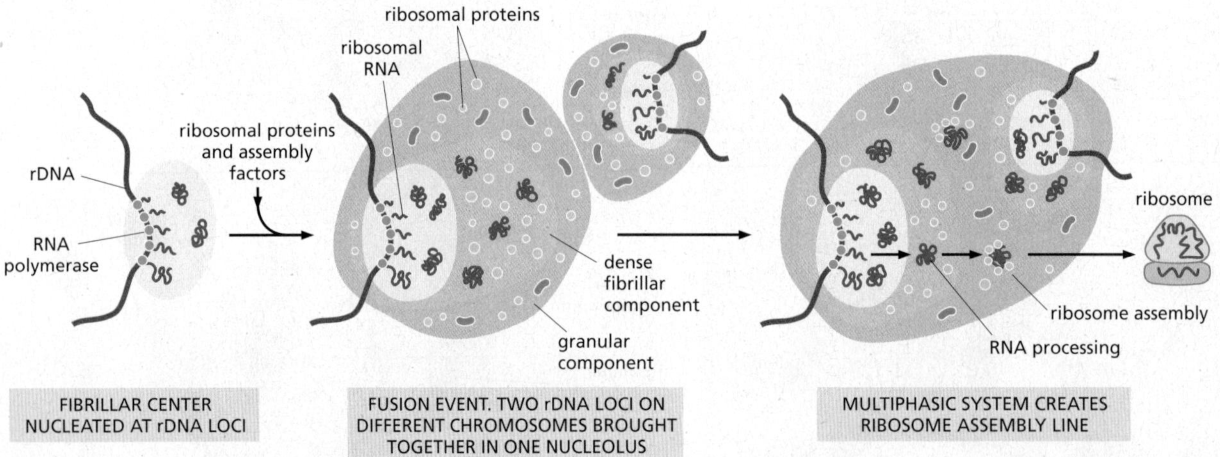

Figure 6–46 **Schematic diagram of nucleolus formation after mitosis.** According to this model, the nucleolus is formed from three distinct condensates, each with a different set of components. This arrangement is proposed to promote the orderly assembly of RNA–protein complexes, much like that observed on an assembly line. (Adapted from A.R. Strom and C.P. Brangwynne, *J. Cell Sci.* 132:jcs235093, 2019.)

while a ribosome, once formed, is stable and is used repeatedly to translate many mRNAs into protein. In human cells, it is estimated that each ribosome makes, on average, 3000 individual proteins in its lifetime. Ribosome assembly is understood in great detail, and only a few of the key features are summarized in Figure 6–49.

In addition to its central role in ribosome biogenesis, the nucleolus is the site where other noncoding RNAs are produced and other RNA–protein complexes are assembled. For example, the U6 snRNP, a key component in pre-mRNA splicing (see Figure 6–29), is composed of one RNA molecule and seven proteins. The U6 snRNA is chemically modified by snoRNAs and assembled with its proteins in the nucleolus. Other important RNA–protein complexes, including telomerase (encountered in Chapter 5) and the signal-recognition particle (which we discuss in Chapter 12), are also assembled in the nucleolus. Finally, the tRNAs (transfer RNAs) that carry the amino acids for protein synthesis are processed there as well; like the rRNA genes, the genes encoding tRNAs are clustered in the nucleolus. Thus, the nucleolus can be thought of as a large factory at which different noncoding RNAs are transcribed, processed, and assembled with proteins to form a large variety of ribonucleoprotein complexes.

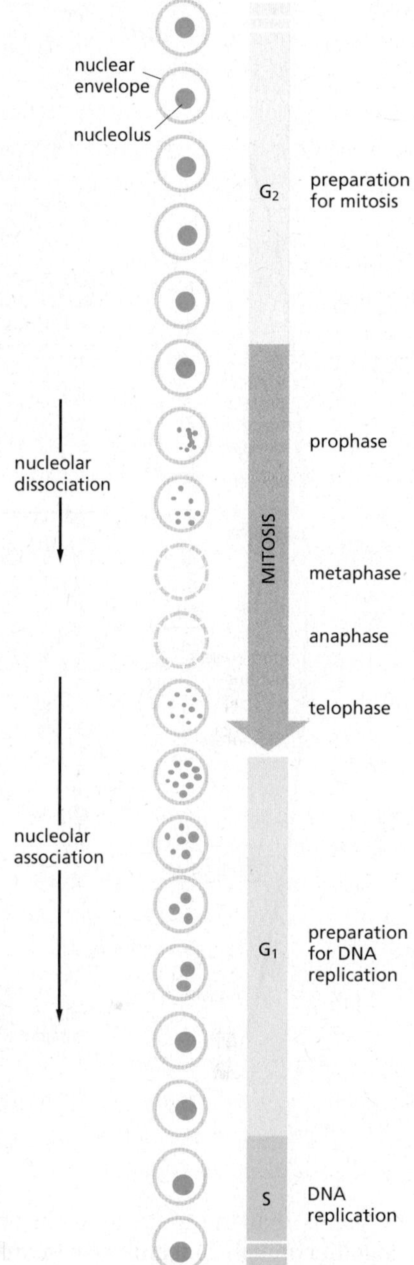

Figure 6–47 **Changes in the appearance of the nucleolus in a human cell during the cell cycle.** Only the cell nucleus is represented in this diagram. In most eukaryotic cells, the nuclear envelope breaks down during mitosis, as indicated by the *dashed circles*.

The Nucleus Contains a Variety of Subnuclear Biomolecular Condensates

Although the nucleolus is the most prominent structure in the nucleus, several other membraneless compartments have been observed and studied (Figure 6–50). These include Cajal bodies (named for the scientist who first described them in 1906) and interchromatin granule clusters (also called "speckles"). Like the nucleolus, these other compartments are highly dynamic depending on the needs of the cell, and their assembly is likely the result of the association of protein and RNA components involved in the synthesis, assembly, and storage of macromolecules involved in gene expression. Cajal bodies are sites where the snRNPs and snoRNPs undergo their final maturation steps, and where the snRNPs are recycled and their RNAs are "reset" after the rearrangements that occur during splicing (see pp. 344–345). In contrast, the interchromatin granule clusters are stockpiles of fully mature snRNPs and other RNA-processing components that are ready to be used in the production of mRNA.

Scientists have had difficulties in working out the exact function of these small compartments, in part because their appearances can change dramatically as cells traverse the cell cycle or respond to changes in their environment. Moreover, disrupting a particular type of nuclear body often has little effect on cell viability.

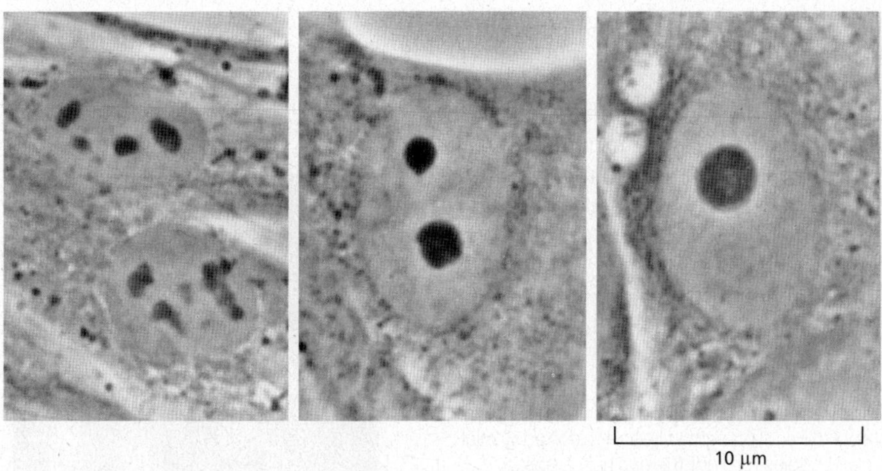

Figure 6–48 **Nucleolar fusion *in vivo*.** These light micrographs of human fibroblasts grown in culture show various stages of nucleolar fusion. After mitosis, each of the 10 human chromosomes that carry a cluster of rRNA genes begins to form a tiny nucleolus, but these rapidly coalesce as they grow to form the single large nucleolus typical of many interphase cells. (Courtesy of E.G. Jordan and J. McGovern.)

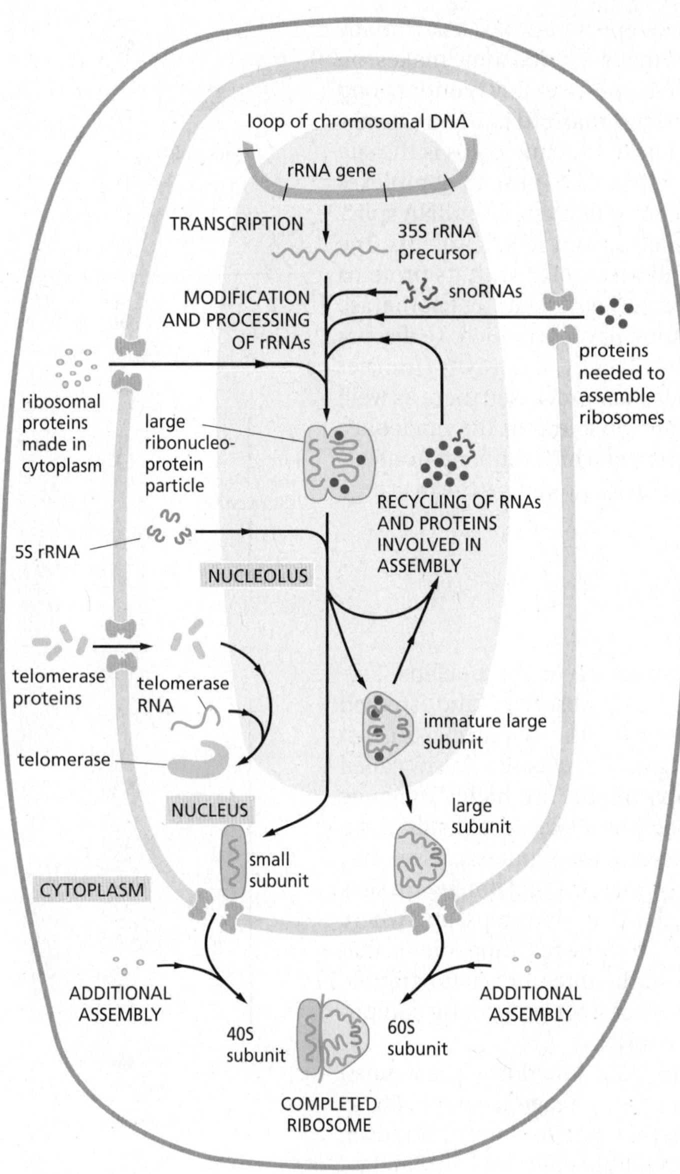

Figure 6–49 The function of the nucleolus in ribosome and other ribonucleoprotein synthesis. The 35S precursor rRNA is packaged in a large ribonucleoprotein particle containing many ribosomal proteins imported from the cytoplasm. While this particle remains at the nucleolus, selected components are added and others discarded as it is processed into immature large and small ribosomal subunits. The two ribosomal subunits attain their final functional form only after each is individually transported through the nuclear pores into the cytoplasm. Other ribonucleoprotein complexes, including telomerase shown here, are also assembled in the nucleolus. (Adapted from A.R. Strom and C.P. Brangwynne, *J. Cell Sci.* 132:jcs235093, 2019. With permission from the Company of Biologists.)

It seems that the main function of these aggregates is to bring components together at high concentration in order to speed up their assembly. For example, it is estimated that assembly of the U4/U6 snRNP (see Figure 6–29) occurs 10 times more rapidly in Cajal bodies than would be the case if the same number of components were dispersed throughout the nucleus. Consequently, Cajal bodies appear dispensable in many types of cells but are absolutely required in situations where cells must proliferate rapidly, such as in early vertebrate development. Here, protein synthesis (which depends on RNA splicing) must occur especially rapidly, and delays can be lethal.

Given the prominence of nuclear compartments in RNA processing, it might be expected that pre-mRNA splicing would occur in a particular location in the

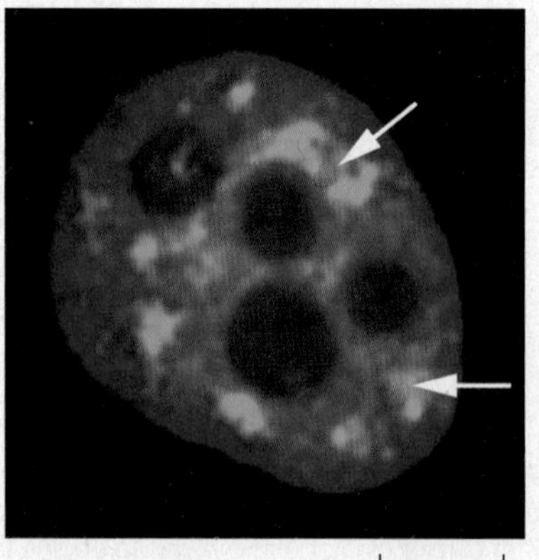

Figure 6–50 Visualization of some prominent membraneless compartments in the nucleus. The protein fibrillarin *(red)*, a component of several snoRNPs, is present in both nucleoli and Cajal bodies; the latter are indicated by the arrows. The Cajal bodies (but not the nucleoli) are also highlighted by staining one of their main components, the protein coilin; the superposition of the snoRNP and coilin stains appears *pink*. Interchromatin granule clusters *(green)* have been revealed by using antibodies against a protein involved in pre-mRNA splicing. DNA is stained *blue* by the dye DAPI. (From J.R. Swedlow and A.I. Lamond, *Genome Biol.* 2:1–7, 2001. Micrograph courtesy of Judith Sleeman.)

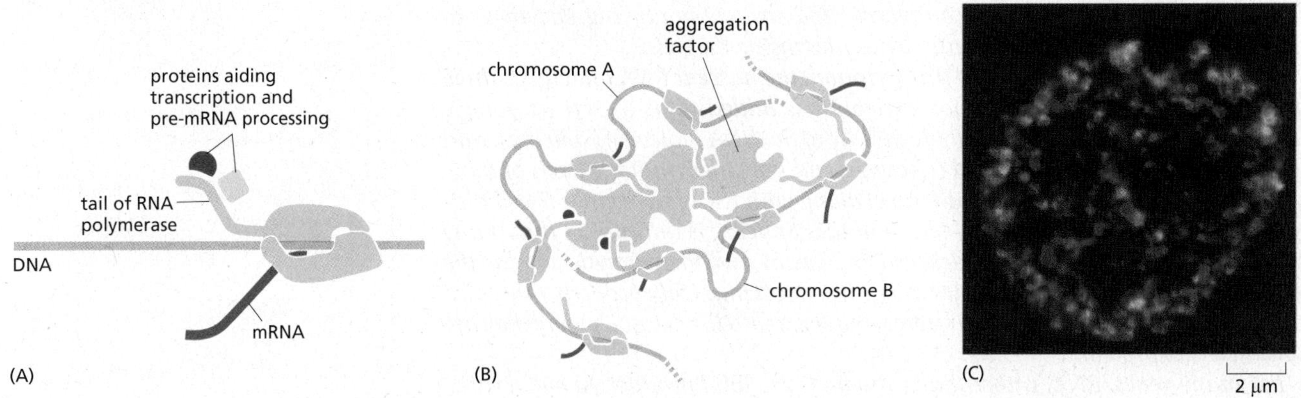

Figure 6–51 A model for an mRNA production factory. mRNA production is made more efficient in the nucleus by an aggregation of the many components needed for transcription and pre-mRNA processing, thereby producing a specialized biochemical factory. In (A), various components in the proximity of a transcribing RNA polymerase are carried on the tail (see Figure 6–23). In (B), a large number of RNA polymerase tails have been brought together to form a condensate that is highly enriched in the many components needed for the synthesis and processing of pre-mRNAs. Such a model can account for the several thousand sites of active RNA transcription and processing typically observed in the nucleus of a mammalian cell, each of which has a diameter of roughly 100 nm and is estimated to contain, on average, about 10 RNA polymerase II molecules in addition to many other proteins. (C) Here, mRNA production factories and DNA replication factories have been visualized in a mammalian cell by briefly incorporating differently modified nucleotides into each nucleic acid and detecting the RNA and DNA produced using antibodies, one *(green)* detecting the newly synthesized DNA and the other *(red)* detecting the newly synthesized RNA. (C, from D.G. Wansink et al., *J. Cell Sci.* 107:1449–1456, 1994. With permission from the Company of Biologists.)

nucleus, as it requires numerous RNA and protein components. However, as we have seen, the assembly of splicing components on pre-mRNA is co-transcriptional; thus, splicing must occur at many locations along chromosomes. Although a typical mammalian cell may be expressing on the order of 15,000 genes, transcription and RNA splicing takes place in only several thousand sites in the nucleus. These sites are highly dynamic and probably result from the association of transcription and splicing components to create small *factories*, the name given to specific condensates containing a high local concentration of selected components that create biochemical assembly lines (**Figure 6–51**). Indeed, it is thought that initial rounds of transcription and RNA processing are very slow and perhaps error-prone due to limiting concentrations of key components; only when a factory becomes fully assembled does mRNA production become rapid and accurate. Interchromatin granule clusters—which contain stockpiles of RNA-processing components—are often observed next to these sites of transcription, as though poised to replenish supplies. We can thus view the nucleus as organized into dynamic condensates of different sizes, with snRNPs, snoRNPs, and other nuclear components diffusing rapidly among them, so as to maintain high concentrations of the many components needed for each step of RNA production.

Summary

Before the synthesis of a particular protein can begin, the corresponding mRNA molecule must be produced by transcription. Bacteria contain a single type of RNA polymerase (the enzyme that carries out the transcription of DNA into RNA). An mRNA molecule is produced after this enzyme initiates transcription at a promoter, synthesizes the RNA by chain elongation, stops transcription at a terminator, and releases both the DNA template and the completed mRNA molecule. In eukaryotic cells, the process of transcription is much more complex, and there are three RNA polymerases—polymerase I, II, and III—that are related evolutionarily to one another and to the bacterial polymerase.

RNA polymerase II synthesizes eukaryotic mRNA. This enzyme requires a set of additional proteins, the general transcription factors, to initiate transcription on a DNA template. It requires still more proteins (including transcription activator proteins, chromatin remodeling complexes, and histone-modifying enzymes) to initiate transcription on its chromatin templates inside the cell.

During the elongation phase of transcription, the nascent RNA undergoes three types of processing events: a special nucleotide is added to its 5′ end (capping), intron sequences are removed from the middle of the RNA molecule (splicing), and the 3′ end of the RNA is generated (cleavage and polyadenylation). Each of these processes is initiated by proteins that travel along with RNA polymerase II by binding to sites on its long, extended C-terminal tail. Splicing is unusual in that many assembly steps are required for each splicing event, and the catalytic site for the reaction is formed by RNA molecules rather than proteins. Only properly processed mRNAs are passed through nuclear pore complexes into the cytosol, where they are translated into protein.

For many genes, RNA, rather than protein, is the final product. In eukaryotes, the most abundant of these non-coding RNAs are transcribed by either RNA polymerase I or RNA polymerase III. RNA polymerase I makes the ribosomal RNAs, which are by far the most abundant RNAs in a cell. The rRNAs are chemically modified, cleaved, and assembled into the two ribosomal subunits in the nucleolus—a distinct membraneless organelle that also helps to process some smaller RNA–protein complexes in the cell. Additional biomolecular condensates in the nucleus (including Cajal bodies and interchromatin granule clusters) are sites where components involved in RNA processing are assembled, stored, and recycled. The high concentration of components in these and other biomolecular condensates ensures that the processes being catalyzed are rapid and efficient.

FROM RNA TO PROTEIN

In the preceding section, we saw that the final product of some genes is an RNA molecule itself, such as the RNAs present in the snRNPs and in ribosomes. However, most genes in a cell produce mRNA molecules that serve as intermediaries on the pathway to proteins. In this section, we examine how the cell converts the information carried in an mRNA molecule into a protein molecule. This feat of translation was a strong focus of attention for biologists in the late 1950s, when it was posed as the "coding problem": How is the information in a linear sequence of nucleotides in RNA translated into the linear sequence of a chemically quite different set of units—the amino acids in proteins? This fascinating question stimulated great excitement. Here was a cryptogram set up by nature that, after more than 3 billion years of evolution, could finally be solved by one of the products of evolution—human beings. And indeed, not only was the code cracked step by step, but in the year 2000 the structure of the elaborate machinery by which cells read this code—the ribosome—was finally revealed in atomic detail.

An mRNA Sequence Is Decoded in Sets of Three Nucleotides

Once an mRNA has been produced by transcription and processing, the information present in its nucleotide sequence is used to synthesize a protein. Transcription is simple to understand as a means of information transfer: because DNA and RNA are chemically and structurally similar, the DNA can act as a direct template for the synthesis of RNA by complementary base-pairing. As the term "transcription" signifies, it is as if a message written out by hand is being converted, say, into a typewritten text. The language itself and the form of the message do not change, and the symbols used are closely related.

In contrast, the conversion of the information in RNA into protein represents a **translation** of the information into another language that uses quite different symbols. Moreover, because there are only 4 different nucleotides in mRNA and 20 different types of amino acids in a protein, this translation cannot be accounted for by a direct one-to-one correspondence between a nucleotide in RNA and an amino acid in protein. The nucleotide sequence of a gene, through the intermediary of mRNA, is instead translated into the amino acid sequence of a protein by rules that are known as the **genetic code**. This code was deciphered in the early 1960s.

	AGA									UUA					AGC					
	AGG									UUG					AGU					
GCA	CGA						GGA			CUA				CCA	UCA	ACA			GUA	
GCC	CGC						GGC		AUA	CUC				CCC	UCC	ACC			GUC	UAA
GCG	CGG	GAC	AAC	UGC	GAA	CAA	GGG	CAC	AUC	CUG	AAA		UUC	CCG	UCG	ACG		UAC	GUG	UAG
GCU	CGU	GAU	AAU	UGU	GAG	CAG	GGU	CAU	AUU	CUU	AAG	AUG	UUU	CCU	UCU	ACU	UGG	UAU	GUU	UGA
Ala	Arg	Asp	Asn	Cys	Glu	Gln	Gly	His	Ile	Leu	Lys	Met	Phe	Pro	Ser	Thr	Trp	Tyr	Val	stop
A	R	D	N	C	E	Q	G	H	I	L	K	M	F	P	S	T	W	Y	V	

Figure 6–52 **The genetic code.** The standard one-letter abbreviation for each amino acid is presented below its three-letter abbreviation (see Panel 3–1, pp. 118–119, for the full name of each amino acid and its structure). By convention, codons are always written with the 5′-terminal nucleotide to the left. Note that most amino acids are represented by more than one codon, and that there are some regularities in the set of codons that specifies each amino acid: codons for the same amino acid tend to contain the same nucleotides at the first and second positions, and vary at the third position. Three codons do not specify any amino acid but act as termination sites (stop codons), signaling the end of the protein-coding sequence. One codon—AUG—acts both as an initiation codon, signaling the start of a protein-coding message, and also as the codon that specifies methionine.

The sequence of nucleotides in the mRNA molecule is read in consecutive groups of three. RNA is a linear polymer of four different nucleotides, so there are 4 × 4 × 4 = 64 possible combinations of three nucleotides: the triplets AAA, AUA, AUG, and so on. However, only 20 different amino acids are commonly found in proteins. Either some nucleotide triplets are never used or the code is redundant and some amino acids are specified by more than one triplet. The second possibility is, in fact, the correct one, as shown by the completely deciphered genetic code in Figure 6–52. Each group of three consecutive nucleotides in mRNA is called a **codon**, and each codon specifies either one amino acid or a stop to the translation process.

In principle, an RNA sequence can be translated in any one of three different **reading frames**, depending on where the decoding process begins (Figure 6–53). However, only one of the three possible reading frames in an mRNA encodes the required protein. We see later how a special punctuation signal at the beginning of each RNA message sets the correct reading frame at the start of protein synthesis.

tRNA Molecules Match Amino Acids to Codons in mRNA

The codons in an mRNA molecule do not directly recognize the amino acids they specify: the group of three nucleotides does not, for example, bind directly to the amino acid. Rather, the translation of mRNA into protein depends on *adaptor* molecules that can recognize and bind both to the codon and, at another site on their surface, to the amino acid. These adaptors consist of a set of small RNA molecules known as **transfer RNAs (tRNAs)**, each about 80 nucleotides in length.

We saw earlier in this chapter that RNA molecules can fold into precise three-dimensional structures, and the tRNA molecules provide striking examples. Four short segments of each folded tRNA are double-helical, producing a molecule that looks like a cloverleaf when drawn schematically (Figure 6–54). For example, a 5′-GCUC-3′ sequence in one part of a polynucleotide chain can form a relatively strong base-pairing association with a 5′-GAGC-3′ sequence in another region of the same molecule. The cloverleaf undergoes further folding to form a compact L-shaped structure that is held together by additional hydrogen bonds between different regions of the molecule (see Figure 6–54B and C).

Two regions of unpaired nucleotides situated at either end of the L-shaped molecule are crucial to the function of tRNA in protein synthesis. One of these regions forms the **anticodon**, a set of three consecutive nucleotides that pairs with the complementary codon in an mRNA molecule. The other is a short single-strand region at the 3′ end of the molecule; this is the site where the amino acid that matches the codon is attached to the tRNA.

We saw above that the genetic code is redundant; that is, several different codons can specify a single amino acid. This redundancy implies either that there is more than one tRNA for many of the amino acids or that some tRNA molecules can base-pair with more than one codon. In fact, both situations occur. Some

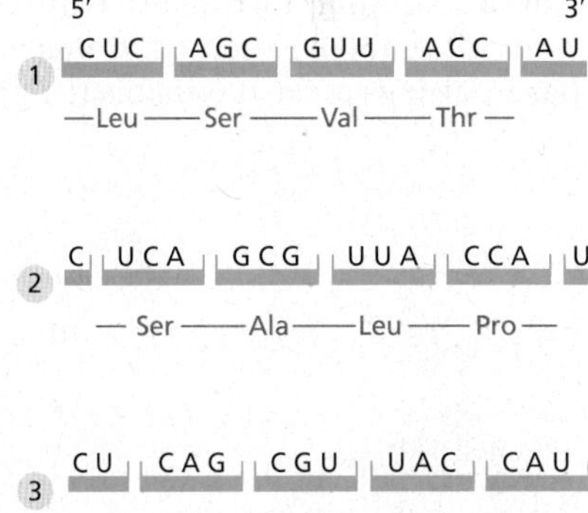

Figure 6–53 **The three possible reading frames in protein synthesis.** In the process of translating a nucleotide sequence *(blue)* into an amino acid sequence *(red)*, the sequence of nucleotides in an mRNA molecule is read from the 5′ end to the 3′ end in consecutive sets of three nucleotides. In principle, therefore, the same RNA sequence can specify three completely different amino acid sequences, depending on the reading frame. In reality, however, only one of these reading frames contains the actual message.

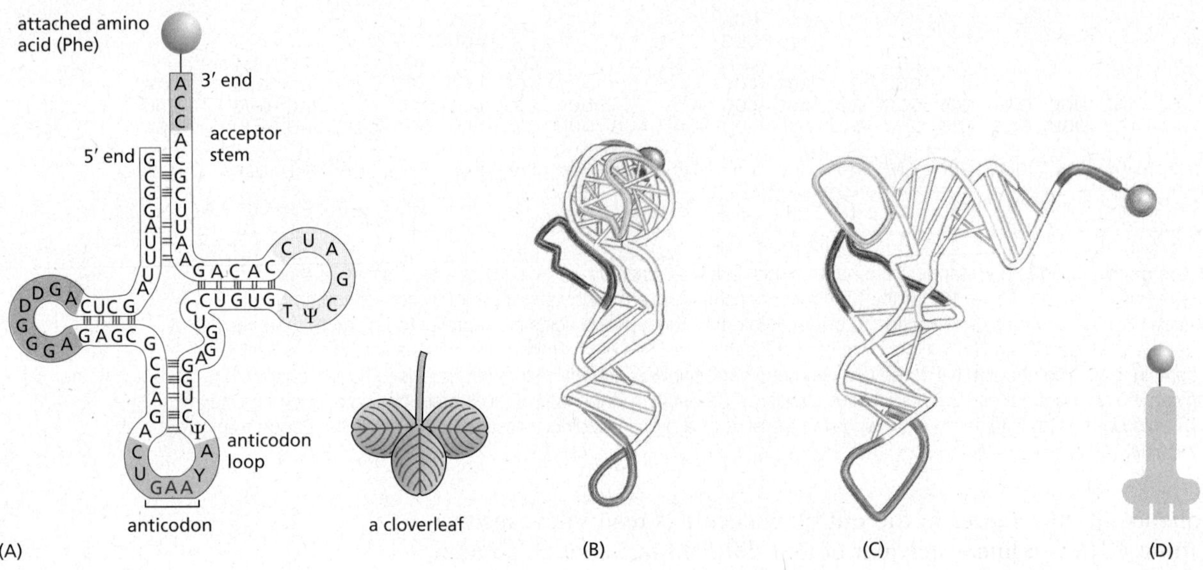

5′ GCGGAUUUAGCUCAGDDGGGAGAGCGCCAGACUGAAYAΨCUGGAGGUCCUGUGTΨCGAUCCACAGAAUUCGCACCA 3′

(E) anticodon

Figure 6–54 **A tRNA molecule.** A tRNA specific for the amino acid phenylalanine (Phe) is depicted in various ways. (A) The cloverleaf structure showing the complementary base-pairing *(red lines)* that creates the double-helical regions of the molecule. The anticodon is the sequence of three nucleotides that base-pairs with a codon in mRNA. The amino acid matching the codon–anticodon pair is attached at the 3′ end of the tRNA. tRNAs contain some unusual bases, which are produced by chemical modification after the tRNA has been synthesized. For example, the bases denoted Ψ (pseudouridine—see Figure 6–43) and D (dihydrouridine—see Figure 6–57) are derived from uracil. (B and C) Views of the L-shaped molecule that are based on x-ray diffraction analysis. Although this diagram shows the tRNA for the amino acid phenylalanine, all other tRNAs have similar structures. (D) The tRNA icon we use in this book. (E) The linear nucleotide sequence of the molecule, color-coded to match the illustrations in A, B, and C.

amino acids have more than one tRNA, and some tRNAs are constructed so that they require accurate base-pairing only at the first two positions of the codon and can tolerate a mismatch (or *wobble*) at the third position (**Figure 6–55**). This wobble base-pairing explains why so many of the alternative codons for an amino acid differ only in their third nucleotide (see Figure 6–52). In bacteria, wobble base-pairings make it possible to fit the 20 amino acids to their 61 codons with as

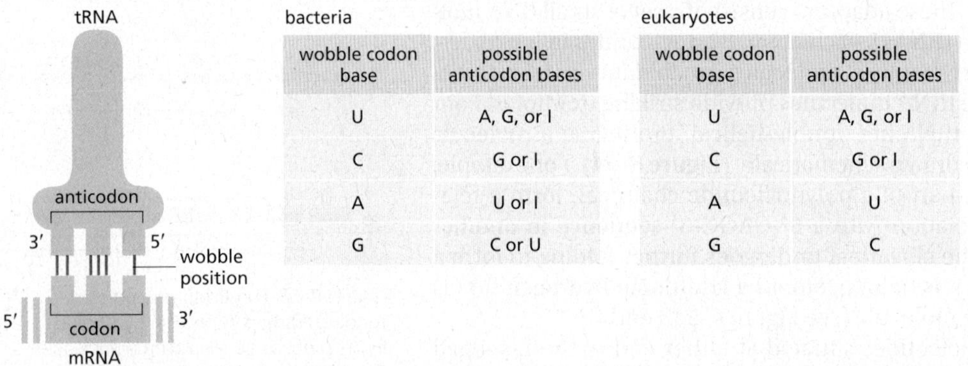

bacteria		eukaryotes	
wobble codon base	possible anticodon bases	wobble codon base	possible anticodon bases
U	A, G, or I	U	A, G, or I
C	G or I	C	G or I
A	U or I	A	U
G	C or U	G	C

Figure 6–55 **Wobble base-pairing between codons and anticodons.** If the nucleotide listed in the first column is present at the third, or wobble, position of the codon, it can base-pair with any of the nucleotides listed in the second column. Thus, for example, when inosine (I) is present in the wobble position of the tRNA anticodon, the tRNA can recognize any one of three different codons in bacteria and either of two codons in eukaryotes. The inosine in tRNAs is formed from the deamination of adenosine (see Figure 6–57), a chemical modification that takes place after the tRNA has been synthesized. The nonstandard base pairs, including those made with inosine, are generally weaker than conventional base pairs. Codon–anticodon base-pairing is more stringent at positions 1 and 2 of the codon, where only conventional base pairs are permitted. The differences in wobble base-pairing interactions between bacteria and eukaryotes presumably result from subtle structural differences between bacterial and eukaryotic ribosomes, the molecular machines that perform protein synthesis. (Adapted from C. Guthrie and J. Abelson, in The Molecular Biology of the Yeast *Saccharomyces*: Metabolism and Gene Expression, pp. 487–528. Cold Spring Harbor, NY: Cold Spring Harbor Laboratory Press, 1982.)

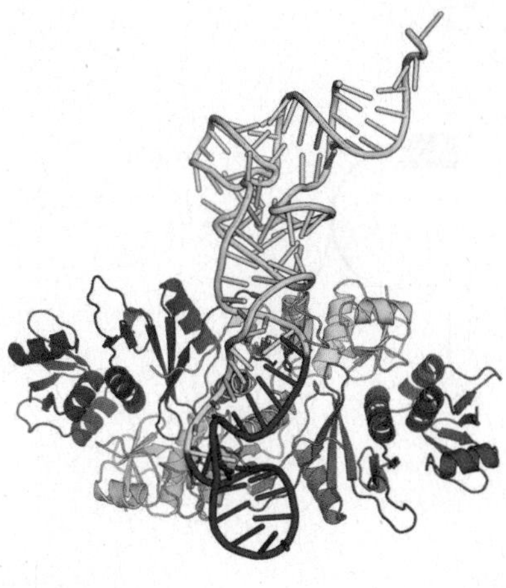

Figure 6–56 Structure of a tRNA-splicing endonuclease docked to a precursor tRNA. The endonuclease (a four-subunit enzyme) removes the tRNA intron *(dark blue, bottom)*. A second enzyme, a multifunctional tRNA ligase (not shown), then joins the two tRNA halves together. (Courtesy of H. Li, C. Trotta, and J. Abelson. PDB code: 2A9L.)

few as 31 kinds of tRNA molecules. The exact number of different kinds of tRNAs, however, differs from one species to the next. For example, humans have nearly 500 tRNA genes that encode tRNAs with 48 different anticodons.

tRNAs Are Covalently Modified Before They Exit from the Nucleus

Like most other eukaryotic RNAs, tRNAs are covalently modified before they are allowed to exit from the nucleus. Eukaryotic tRNAs are synthesized by RNA polymerase III. Both bacterial and eukaryotic tRNAs are typically synthesized as larger precursor tRNAs, which are then trimmed to produce the mature tRNA. In addition, some tRNA precursors (from both bacteria and eukaryotes) contain introns that must be spliced out. This splicing reaction differs chemically from that of pre-mRNA splicing discussed earlier in the chapter; rather than generating a lariat intermediate, tRNA splicing uses a cut-and-paste mechanism that is catalyzed by proteins (**Figure 6–56**). Trimming and splicing both require the precursor tRNA to be correctly folded in its cloverleaf configuration. Because misfolded tRNA precursors will not be processed properly, the trimming and splicing reactions serve as quality-control steps in the generation of tRNAs. Those that do not pass the tests are degraded by the nuclear exosome (see Figure 6–38).

All tRNAs are modified chemically—nearly 1 in 10 nucleotides in each mature tRNA molecule is an altered version of a standard G, U, C, or A ribonucleotide. More than 50 different types of tRNA modifications are known; a few are shown in **Figure 6–57**. Some of the modified nucleotides lie within the anticodon—most notably inosine, produced by the deamination of adenosine—and affect the base-pairing of the anticodon, thereby facilitating the recognition of the appropriate mRNA codon by the tRNA molecule (see Figure 6–55). Other modifications affect the accuracy with which the tRNA is attached to the correct amino acid.

Specific Enzymes Couple Each Amino Acid to Its Appropriate tRNA Molecule

We have seen that, to read the genetic code in DNA, cells make a series of different tRNAs. We now consider how each tRNA molecule becomes linked to the one amino acid in 20 that is its appropriate partner. Recognition and attachment of

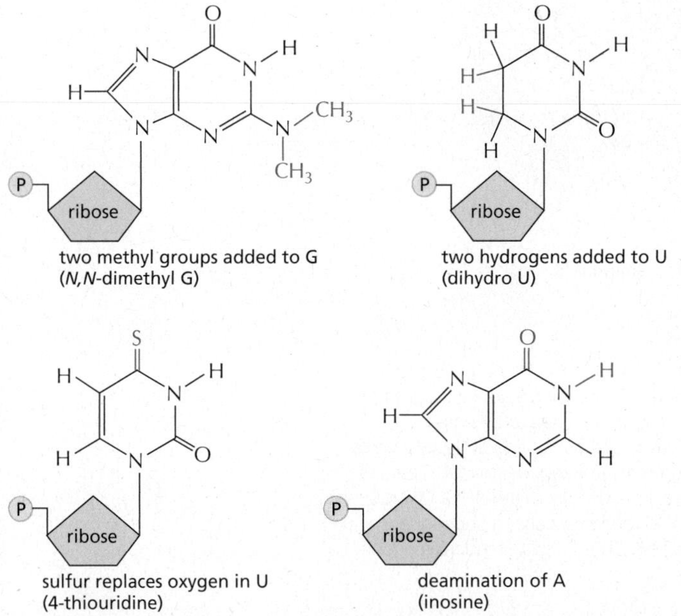

two methyl groups added to G
(*N,N*-dimethyl G)

two hydrogens added to U
(dihydro U)

sulfur replaces oxygen in U
(4-thiouridine)

deamination of A
(inosine)

Figure 6–57 A few of the unusual nucleotides found in tRNA molecules. These nucleotides are produced by covalent modification of a normal nucleotide after it has been incorporated into an RNA chain. Two other types of modified nucleotides are shown in Figure 6–43. In most tRNA molecules, about 10% of the nucleotides are modified (see Figure 6–54). As shown in Figure 6–55, inosine is sometimes present at the wobble position in the tRNA anticodon.

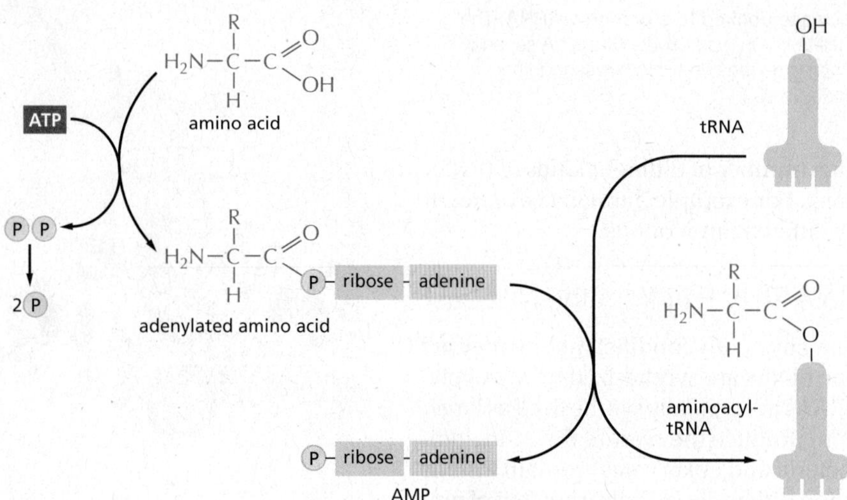

Figure 6–58 **Amino acid activation by synthetase enzymes.** An amino acid is attached to its corresponding tRNA in two steps by an aminoacyl-tRNA synthetase enzyme. As indicated, the energy of ATP hydrolysis is used in the reaction to produce a high-energy linkage. The amino acid is first activated through attachment of its carboxyl group directly to AMP, forming an *adenylated amino acid*; the linkage of the AMP, normally an unfavorable reaction, is driven by the hydrolysis of the ATP molecule that donates the AMP. Without leaving the synthetase enzyme, the AMP-linked carboxyl group on the amino acid is then transferred to a hydroxyl group on the sugar at the 3′ end of the tRNA molecule. This transfer joins the amino acid by an activated ester linkage to the tRNA and forms the final aminoacyl-tRNA molecule. The synthetase enzyme is not shown in this diagram.

the correct amino acid depends on enzymes called **aminoacyl-tRNA synthetases**, which covalently couple each amino acid to its appropriate set of tRNA molecules (**Figure 6–58** and **Figure 6–59**). Most cells have a different synthetase enzyme for each amino acid (that is, 20 synthetases in all); one attaches glycine to all tRNAs that recognize codons for glycine, another attaches alanine to all tRNAs that recognize codons for alanine, and so on. Many bacteria, however, have fewer than 20 synthetases, and the same synthetase enzyme is responsible for coupling more

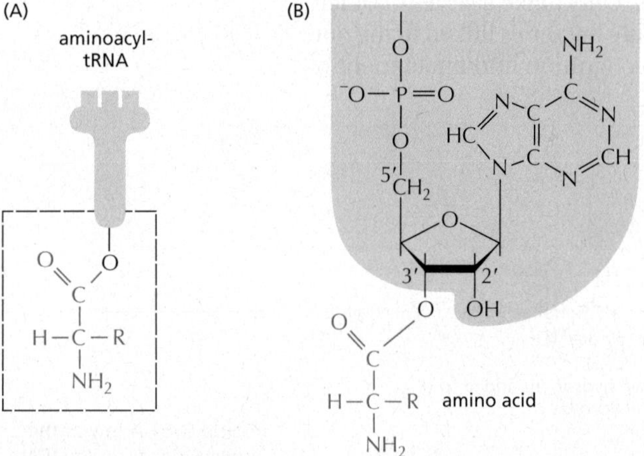

Figure 6–59 **The structure of the aminoacyl-tRNA linkage.** The carboxyl end of the amino acid forms an ester bond to ribose. Because the hydrolysis of this ester bond is associated with a large favorable change in free energy, an amino acid held in this way is said to be activated. (A) Schematic drawing of the structure. The amino acid is linked to the nucleotide at the 3′ end of the tRNA (see Figure 6–54). (B) Actual structure corresponding to the boxed region in A. There are two major classes of synthetase enzymes: one links the amino acid directly to the 3′-OH group of the ribose, and the other links it initially to the 2′-OH group. In the latter case, a subsequent transesterification reaction shifts the amino acid to the 3′ position. The "R" is a standard symbol used to represent the side chain of an amino acid.

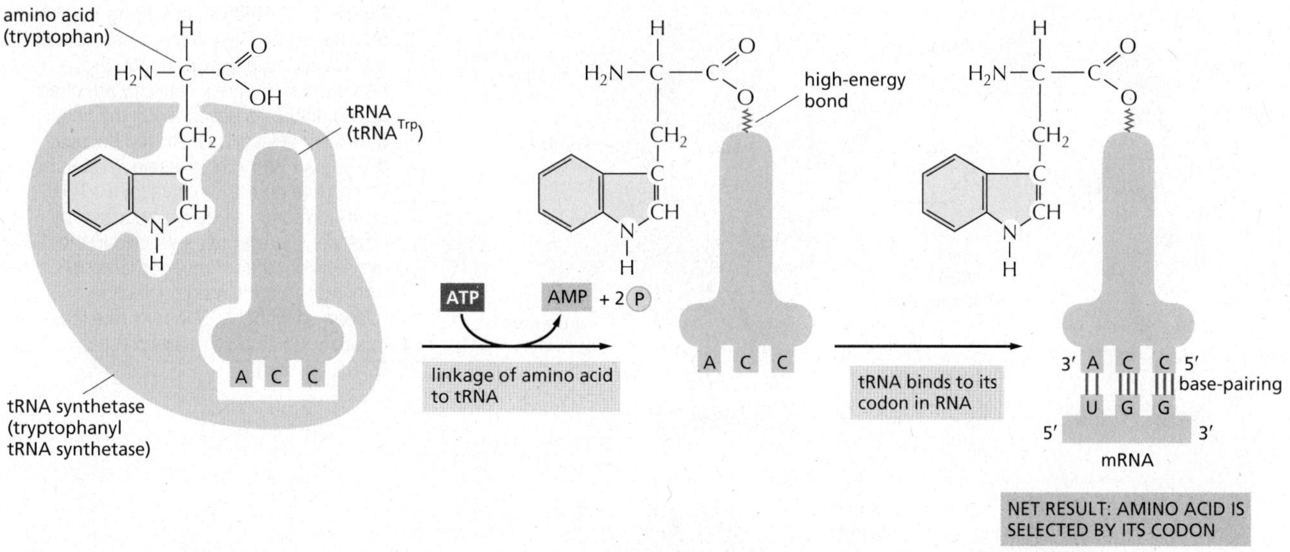

Figure 6–60 The genetic code is translated by means of two adaptors that act one after another. The first adaptor is the aminoacyl-tRNA synthetase, which couples a particular amino acid to its corresponding tRNA; the second adaptor is the tRNA molecule itself, whose *anticodon* forms base pairs with the appropriate *codon* on the mRNA. An error in either step would cause the wrong amino acid to be incorporated into a protein chain (**Movie 6.6**). In the sequence of events shown, the amino acid tryptophan (Trp) is selected by the codon UGG on the mRNA.

than one amino acid to the appropriate tRNAs. In these cases, a single synthetase places the identical amino acid on two different types of tRNAs, only one of which has an anticodon that matches the amino acid. A second enzyme then chemically modifies each "incorrectly" attached amino acid so that it now corresponds to the anticodon displayed by its covalently linked tRNA.

The synthetase-catalyzed reaction that attaches the amino acid to the 3′ end of the tRNA is one of many reactions coupled to the energy-releasing hydrolysis of ATP (see pp. 70–72), and it produces a high-energy bond between the tRNA and the amino acid. The energy of this bond is used at a later stage in protein synthesis to link the amino acid covalently to the growing polypeptide chain.

The aminoacyl-tRNA synthetase enzymes and the tRNAs are equally important in the decoding process (**Figure 6–60**). This was established by an experiment in which one amino acid (cysteine) was chemically converted into a different amino acid (alanine) after it already had been attached to its specific tRNA. When such "hybrid" aminoacyl-tRNA molecules were used for protein synthesis in a cell-free system, the wrong amino acid was inserted at every point in the protein chain where that tRNA was used. Although, as we shall see, cells have several quality-control mechanisms to avoid this type of mishap, the experiment did establish that the genetic code is translated by two sets of adaptors that act sequentially. Each matches one molecular surface to another with great specificity, and it is their combined action that associates each sequence of three nucleotides in the mRNA molecule—that is, each codon—with its particular amino acid.

Editing by tRNA Synthetases Ensures Accuracy

Several mechanisms working together ensure that an aminoacyl-tRNA synthetase links the correct amino acid to each tRNA. Most synthetase enzymes select the correct amino acid by a two-step mechanism. The correct amino acid has the highest affinity for the active-site pocket of its synthetase and is therefore favored over the other 19; in particular, amino acids larger than the correct one are excluded from the active site. However, accurate discrimination between two similar amino acids, such as isoleucine and valine (which differ by only a methyl group), is very difficult to achieve in a single step. A second

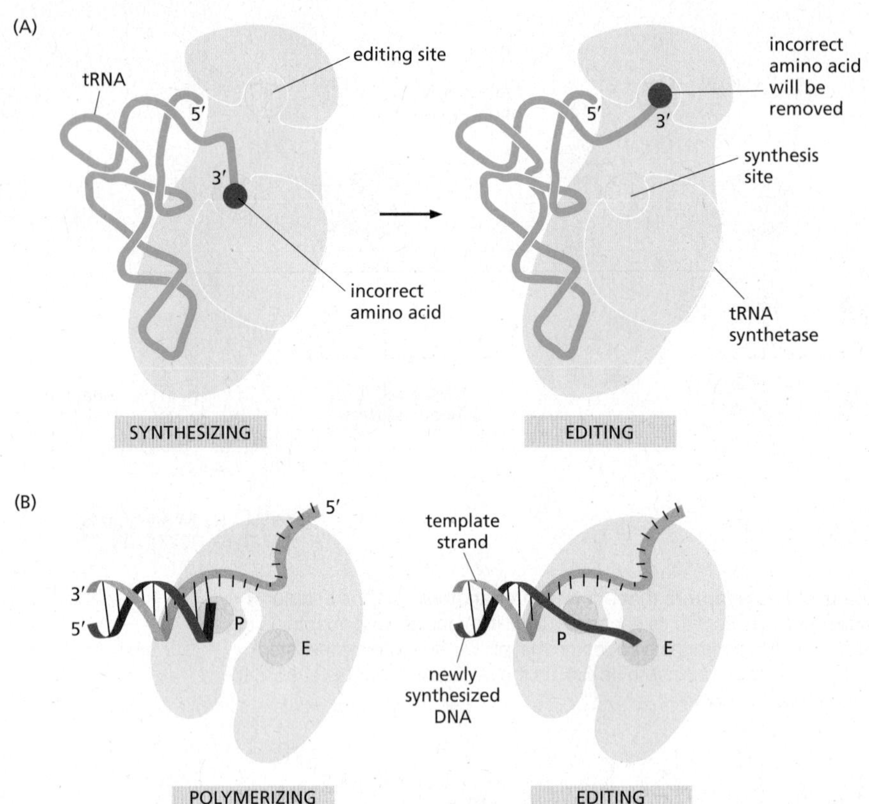

Figure 6–61 Hydrolytic editing in biology. (A) Aminoacyl-tRNA synthetases correct their own coupling errors through the hydrolytic editing of incorrectly attached amino acids. As described in the text, errors are selectively removed because the correct amino acid is rejected by the editing site on the synthetase. (B) The error-correction process performed by DNA polymerase, as described in the previous chapter. Here error removal depends on the wrong nucleotide mispairing with the DNA template (see Figure 5–9). (P, polymerization site; E, editing site.)

discrimination step occurs after the amino acid has been covalently linked to AMP (see Figure 6–58): when tRNA binds, the synthetase tries to force the adenylated amino acid into a second editing pocket in the enzyme. The precise dimensions of this pocket exclude the correct amino acid, while allowing access by closely related amino acids. In the editing pocket, an amino acid is removed from the AMP (or from the tRNA itself if the aminoacyl-tRNA bond has already formed) by hydrolysis. This hydrolytic editing, which is analogous to the exonucleolytic proofreading by DNA polymerases, increases the overall accuracy of tRNA charging so that only about one mistake is made in 40,000 couplings (**Figure 6–61**).

The tRNA synthetase must also recognize the correct set of tRNAs, and extensive structural and chemical complementarity between the synthetase and the tRNA allows the synthetase to probe various features of the tRNA (**Figure 6–62**). Most tRNA synthetases directly recognize the matching tRNA anticodon; these synthetases contain three adjacent nucleotide-binding pockets, each of which is complementary in shape and charge to a nucleotide in the anticodon. For other synthetases, the nucleotide sequence of the amino acid–accepting arm (acceptor stem) is the key recognition determinant. In most cases, however, the synthetase "reads" the nucleotides at several different positions on the tRNA, thereby increasing the accurate linking of amino acids to their appropriate tRNAs.

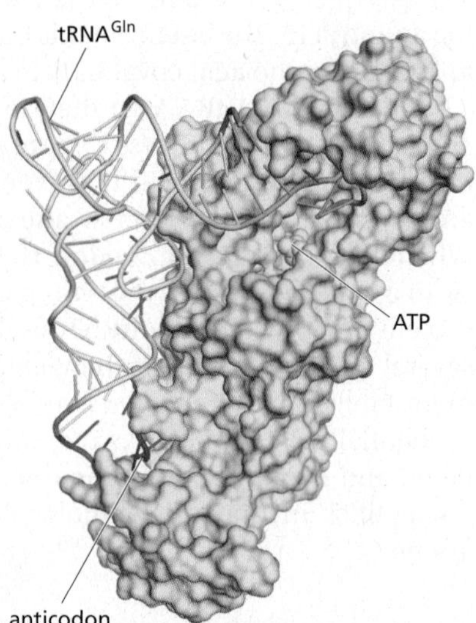

Figure 6–62 The recognition of a tRNA molecule by its aminoacyl-tRNA synthetase. For this tRNA ($tRNA^{Gln}$), specific nucleotides in both the anticodon *(dark blue)* and the amino acid–accepting arm *(green)* allow the correct tRNA to be recognized by the synthetase enzyme *(yellow-green).* The ATP molecule used in the coupling reaction is *yellow.* (Courtesy of Tom Steitz. PDB code: 1QRS.)

Amino Acids Are Added to the C-terminal End of a Growing Polypeptide Chain

Having seen that amino acids are first coupled to specific tRNA molecules that serve as adaptors, we now turn to the mechanism that joins these amino acids together to form proteins. The fundamental reaction of protein synthesis is the formation of a peptide bond between the carboxyl group at the end of a growing polypeptide chain and a free amino group on an incoming amino acid.

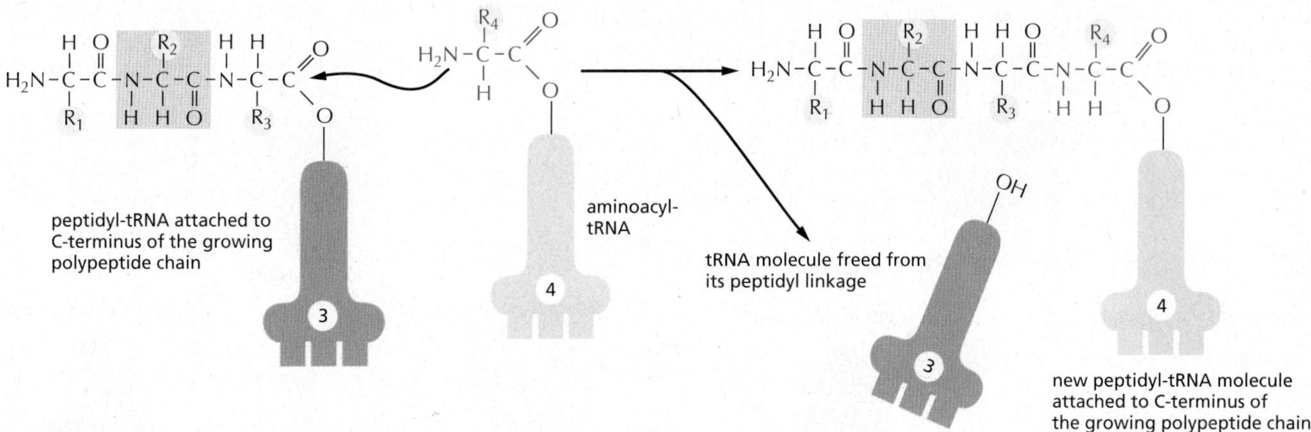

Figure 6–63 The incorporation of an amino acid into a protein. A polypeptide chain grows by the stepwise addition of amino acids to its C-terminal end. The formation of each peptide bond is energetically favorable because the growing C-terminus has been activated by the covalent attachment of a tRNA molecule. Note that the peptidyl-tRNA linkage that activates the growing end is regenerated during each addition. The amino acid side chains have been abbreviated as R_1, R_2, R_3, and R_4; as a reference point, all of the atoms in the second amino acid in the polypeptide chain are shaded *gray*. The figure shows the addition of the fourth amino acid *(red)* to the growing chain.

Consequently, a protein is synthesized from its N-terminal end to its C-terminal end, one amino acid at a time. Throughout the entire process, the growing carboxyl end of the polypeptide chain remains activated by its covalent attachment to a tRNA molecule (forming a *peptidyl-tRNA*). Each addition disrupts this high-energy covalent linkage but immediately replaces it with an identical linkage on the most recently added amino acid (**Figure 6–63**). In this way, each amino acid added carries with it the activation energy for the addition of the next amino acid rather than the energy for its own addition—an example of the *polymer-end activation* mechanism for polymer synthesis described in Figure 2–44.

The RNA Message Is Decoded in Ribosomes

The synthesis of proteins is guided by information carried by mRNA molecules. To maintain the correct reading frame and to ensure accuracy (about 1 mistake every 10,000 amino acids), protein synthesis is performed by the **ribosome**, a complex catalytic machine made from more than 50 different proteins (the *ribosomal proteins*) and several RNA molecules, the ribosomal RNAs (rRNAs). A typical eukaryotic cell contains millions of ribosomes in its cytoplasm (**Figure 6–64**), and it takes approximately 1 minute to synthesize an average-sized

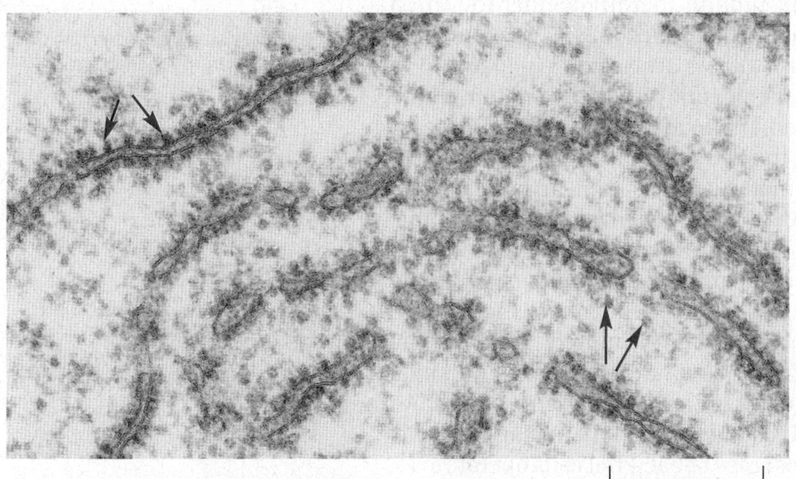

Figure 6–64 Ribosomes in the cytoplasm of a eukaryotic cell. This electron micrograph shows a thin section of a small region of cytoplasm. The ribosomes appear as black dots *(red arrows)*. Some are free in the cytosol; others are attached to membranes of the endoplasmic reticulum. (Courtesy of Daniel S. Friend, by permission of E.L. Bearer.)

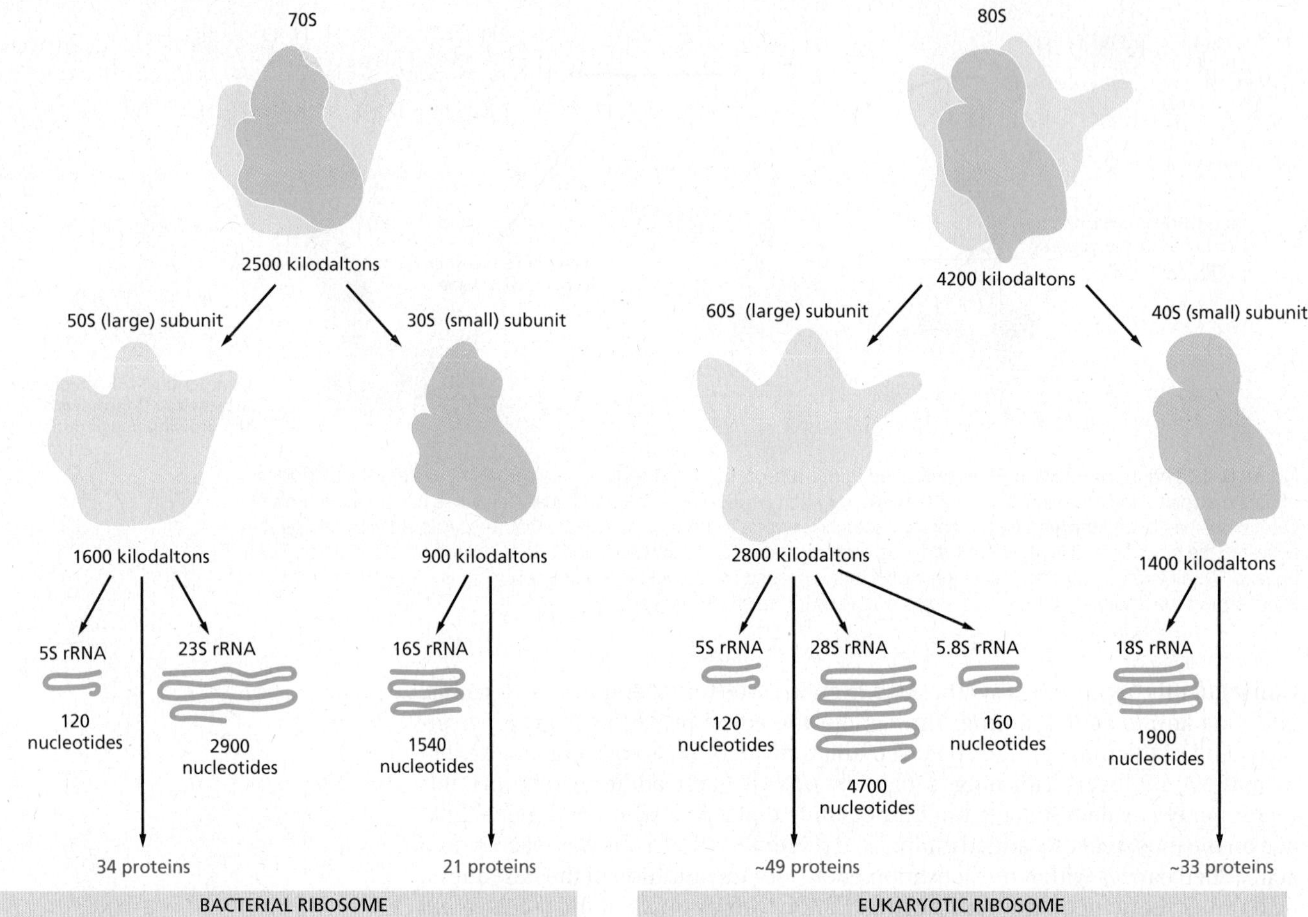

Figure 6–65 A comparison of bacterial and eukaryotic ribosomes. Despite differences in the number and size of their rRNA and protein components, both bacterial and eukaryotic ribosomes have nearly the same structure and they function similarly. Although the 18S and 28S rRNAs of the eukaryotic ribosome contain many nucleotides not present in their bacterial counterparts, these extra nucleotides are present as multiple insertions that form extra domains and leave the basic structure of the rRNA largely unchanged.

protein. As we saw earlier in the chapter, the large and small ribosome subunits of eukaryotes are assembled in the nucleolus, where newly transcribed and modified rRNAs are brought into association with the ribosomal proteins that have been transported into the nucleus after their synthesis in the cytoplasm. These two ribosomal subunits are then exported to the cytoplasm, where they each undergo further assembly and join together as fully mature ribosomes to synthesize proteins (see Figure 6-49).

Eukaryotic and bacterial ribosomes have similar structures and functions, being composed of one large and one small subunit that fit together to form a complete ribosome with a mass of several million daltons (**Figure 6–65**). The small subunit provides the framework on which the tRNAs are accurately matched to the codons of the mRNA, while the large subunit catalyzes the formation of the peptide bonds that link the amino acids together into a polypeptide chain.

When not actively synthesizing proteins, the two subunits of the ribosome are separate. They join together on an mRNA molecule, usually near its 5′ end, to initiate the synthesis of a protein. The mRNA is then pulled through the ribosome, three nucleotides at a time. As its codons enter the core of the ribosome, the mRNA nucleotide sequence is translated into an amino acid sequence using the tRNAs as adaptors to add each amino acid in the correct sequence to the growing end of the polypeptide chain. When a stop codon is encountered, the ribosome releases the finished protein, and its two subunits separate again. These subunits can then be used to start the synthesis of another protein on the same or another mRNA molecule. On average, a eukaryotic ribosome adds about four amino acids to a polypeptide chain every second; the ribosomes of bacterial cells operate even faster, at a rate of about 20 amino acids per second.

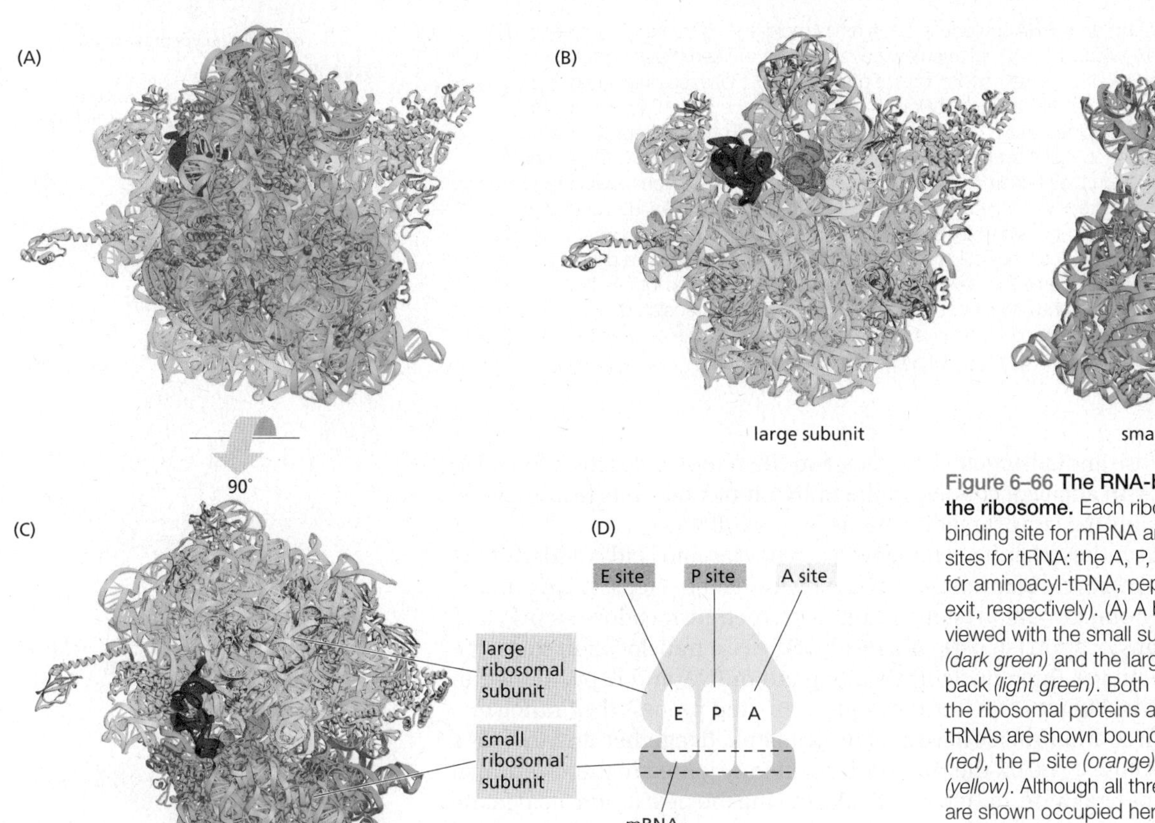

Figure 6–66 **The RNA-binding sites in the ribosome.** Each ribosome has one binding site for mRNA and three binding sites for tRNA: the A, P, and E sites (short for aminoacyl-tRNA, peptidyl-tRNA, and exit, respectively). (A) A bacterial ribosome viewed with the small subunit in the front *(dark green)* and the large subunit in the back *(light green)*. Both the rRNAs and the ribosomal proteins are illustrated. tRNAs are shown bound in the E site *(red)*, the P site *(orange)*, and the A site *(yellow)*. Although all three tRNA sites are shown occupied here, during the process of protein synthesis not more than two of these sites are thought to contain tRNA molecules at any one time (see Figure 6–68). (B) Large and small ribosomal subunits shown separately, and arranged as though the ribosome in panel A were opened like a book. (C) The entire ribosome in panel A rotated through 90° and viewed with the large subunit on top and small subunit on the bottom. (D) Schematic representation of the ribosome in the same orientation as that in panel C, which is how the ribosome will be depicted in subsequent figures. (A, B, and C, adapted from M.M. Yusupov et al., *Science* 292:883–896, 2001. With permission from AAAS.)

To choreograph the many coordinated movements required for efficient translation, a ribosome contains four binding sites for RNA molecules: one is for the mRNA and three (called the A site, the P site, and the E site) are for tRNAs (**Figure 6–66**). A tRNA molecule is held tightly at the A and P sites only if its anticodon forms base pairs with a complementary codon (allowing for wobble) on the mRNA molecule that is threaded through the ribosome (**Figure 6–67**). The A

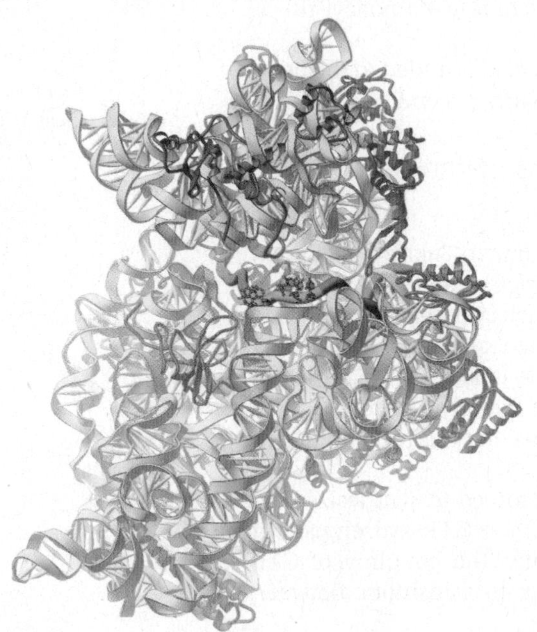

Figure 6–67 **The path of mRNA *(blue)* through the small ribosomal subunit.** The orientation is the same as that for the small subunit in the right-hand panel of Figure 6–66B, allowing the position of the three tRNA-binding sites to be compared to that of the mRNA shown here. (Courtesy of Harry F. Noller, based on data in G.Z. Yusupova et al., *Cell* 106:233–241, 2001.)

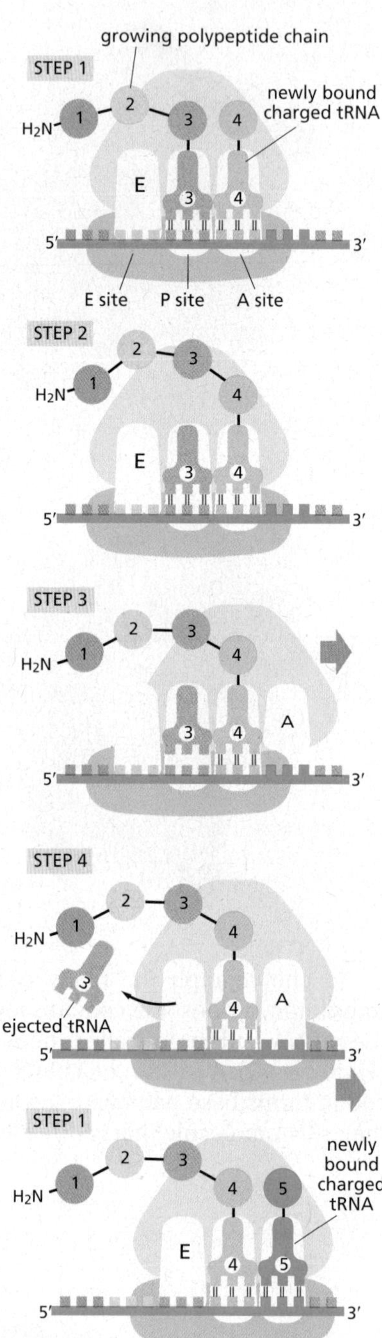

Figure 6–68 **Translating an mRNA molecule.** Each amino acid added to the growing end of a polypeptide chain is selected by complementary base-pairing between the anticodon on its attached tRNA molecule and the next codon on the mRNA chain. Because only one of the many types of tRNA molecules in a cell can base-pair with each codon, the codon determines the specific amino acid to be added to the growing polypeptide chain. The four-step cycle shown is repeated over and over during the synthesis of a protein. In step 1, an aminoacyl-tRNA molecule binds to a vacant A site on the ribosome. In step 2, a new peptide bond is formed. In step 3, the large subunit translocates relative to the small subunit, leaving the two tRNAs in hybrid sites: P on the large subunit and A on the small, for one tRNA; E on the large subunit and P on the small, for the other tRNA. In step 4, the small subunit translocates carrying its mRNA a distance of three nucleotides through the ribosome. This "resets" the ribosome with a fully empty A site, ready for the next aminoacyl-tRNA molecule to bind. As indicated, the mRNA is translated in the 5′-to-3′ direction, and the N-terminal end of a protein is made first, with each cycle adding one amino acid to the C-terminus of the polypeptide chain (Movie 6.7 and Movie 6.8).

and P sites are close enough together for their two tRNA molecules to be forced to form base pairs with adjacent codons on the mRNA molecule. This feature of the ribosome maintains the correct reading frame on the mRNA.

Once protein synthesis has been initiated, each new amino acid is added to the elongating chain in a cycle of reactions containing four major steps: tRNA binding (step 1), peptide bond formation (step 2), large subunit translocation (step 3), and small subunit translocation (step 4). As a result of the two translocation steps, the entire ribosome moves three nucleotides along the mRNA and is positioned to start the next cycle. Figure 6–68 illustrates this four-step process, beginning at a point at which three amino acids have already been linked together, and there is a tRNA molecule in the P site on the ribosome, covalently joined to the C-terminal end of the short polypeptide. In step 1, a tRNA carrying the next amino acid in the chain binds to the ribosomal A site by forming base pairs with the mRNA codon positioned there, so that the P site and the A site contain adjacent bound tRNAs. In step 2, the carboxyl end of the polypeptide chain is released from the tRNA at the P site (by breakage of the high-energy bond between the tRNA and its amino acid) and joined to the free amino group of the amino acid linked to the tRNA at the A site, forming a new peptide bond. This central reaction of protein synthesis is catalyzed by a *peptidyl transferase* contained in the large ribosomal subunit. In step 3, the large subunit moves relative to the mRNA held by the small subunit, thereby shifting the acceptor stems of the two tRNAs to the E and P sites of the large subunit. In step 4, another series of conformational changes moves the small subunit and its bound mRNA exactly three nucleotides, ejecting the spent tRNA from the E site and resetting the ribosome so it is ready to receive the next aminoacyl-tRNA. Step 1 is then repeated with a new incoming aminoacyl-tRNA, and so on.

This four-step cycle is repeated each time an amino acid is added to the polypeptide chain, as the chain grows from its amino to its carboxyl end.

Elongation Factors Drive Translation Forward and Improve Its Accuracy

The basic cycle of polypeptide elongation shown in outline in Figure 6–68 has an additional feature that makes translation especially efficient and accurate. Two *elongation factors* enter and leave the ribosome during each cycle, each hydrolyzing GTP to GDP and undergoing conformational changes in the process. These factors are called EF-Tu and EF-G in bacteria, and EF1 and EF2 in eukaryotes. Figure 6–69 illustrates how their cycles of ribosome association, GTP hydrolysis, and ribosome dissociation contribute to the process of protein synthesis that was just outlined in Figure 6–68.

Under some conditions *in vitro,* ribosomes can be forced to synthesize proteins without the aid of these elongation factors and their GTP hydrolysis; but this synthesis is very slow, inefficient, and inaccurate. The coupling of GTP hydrolysis-driven changes in these elongation factors to transitions between

Figure 6–69 **Detailed view of the translation cycle.** The outline of translation presented in Figure 6–68 has been expanded to show the roles of the two elongation factors EF-Tu and EF-G, which drive translation in the forward direction. As explained in the text, EF-Tu provides two opportunities for proofreading of the codon–anticodon match. In this way, incorrectly paired tRNAs are preferentially rejected, and the accuracy of translation is improved. The binding of a molecule of EF-G to the ribosome and that elongation factor's subsequent hydrolysis of GTP lead to a rearrangement of the ribosome structure, moving the mRNA being decoded exactly three nucleotides through it (Movie 6.9).

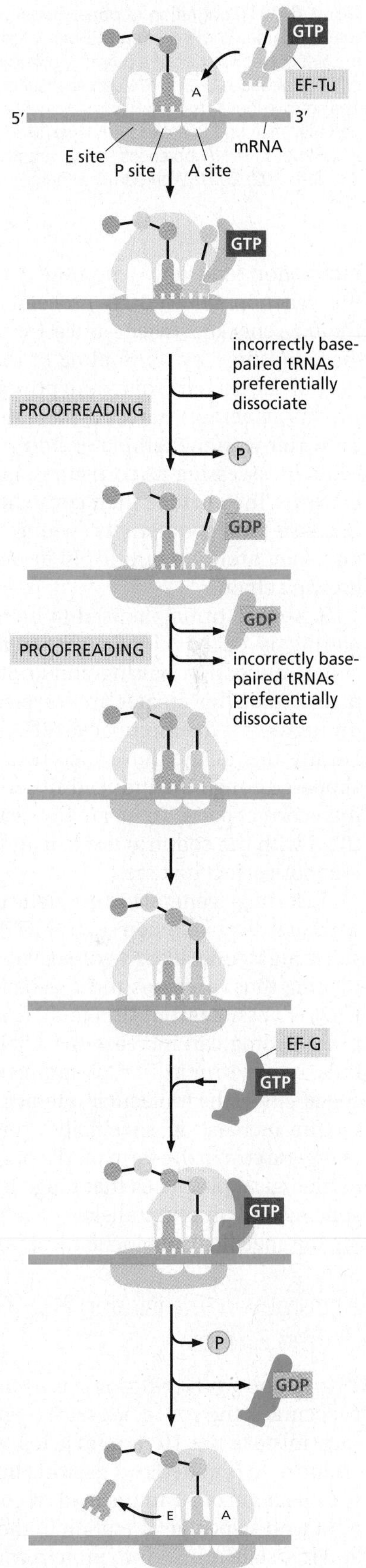

different states of the ribosome speeds up protein synthesis enormously, ensuring that the many changes required occur only in the "forward" direction.

In addition to driving translation forward, EF-Tu increases its accuracy. As we discussed in Chapter 3, EF-Tu can simultaneously bind GTP and aminoacyl-tRNAs (see Figures 3–68 and 3–69), and it is in this form that the initial codon–anticodon interaction occurs in the A site of the ribosome. Because of the free-energy change associated with base-pair formation, a correct codon–anticodon match will bind more tightly than an incorrect interaction. However, this difference in affinity is relatively modest, and it cannot by itself account for the high accuracy of translation.

To increase the accuracy of aminoacyl-tRNA binding in the first step of protein synthesis, the ribosome and EF-Tu work together in the following ways. First, the 16S rRNA in the small subunit of the ribosome assesses the "correctness" of the codon–anticodon match by folding around it and probing its molecular details (**Figure 6–70**). When a correct match is found, the rRNA closes tightly around the codon–anticodon pair, causing a conformational change in the ribosome that triggers GTP hydrolysis by EF-Tu. Only when GTP is hydrolyzed does EF-Tu release its grip on the aminoacyl-tRNA and allow it to be used in protein synthesis. Because incorrect codon–anticodon matches do not readily trigger this conformational change, most of these errant tRNAs fall off the ribosome before they can be used in protein synthesis (see the first proofreading step in Figure 6–69).

After GTP is hydrolyzed and EF-Tu dissociates from the ribosome, there is a second opportunity for the ribosome to prevent an incorrect amino acid from being added to the growing chain. This arises due to a time delay before the amino acid carried by the tRNA moves into its correct position on the ribosome. Not only is this time delay shorter for correct than incorrect codon–anticodon pairs, but incorrectly matched tRNAs dissociate more rapidly than those that are correctly bound. Thus, most of the incorrectly bound tRNA molecules that remain (as well as a significant number of correctly bound molecules) will leave the ribosome without being used for protein synthesis (see the second proofreading step in Figure 6–69). The two proofreading steps, acting in series, are largely responsible for the 99.99% accuracy of the ribosome in translating RNA into protein.

Induced Fit and Kinetic Proofreading Help Biological Processes Overcome the Inherent Limitations of Complementary Base-Pairing

We have seen in this and the previous chapter that DNA replication, repair, transcription, RNA splicing, and translation all rely on complementary base-pairing—G with C, and A with T (or U). However, if only the difference in hydrogen bonding is considered, a correct versus incorrect match should differ in affinity only by a factor of 10- to 100-fold. These processes have an accuracy much higher than can be accounted for by this difference. Although the mechanisms used to "squeeze out" additional specificity from complementary base-pairing differ from one process to the next, two principles exemplified by the ribosome appear to be general.

The first is **induced fit**. We have seen that, before an amino acid is added to a growing polypeptide chain, the ribosome folds around the codon–anticodon

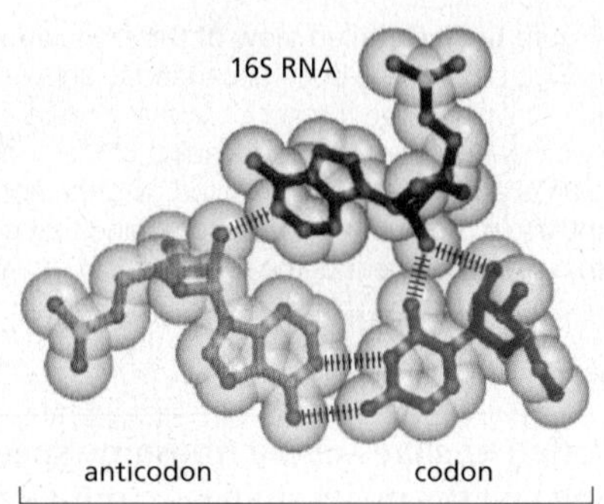

Figure 6–70 **Recognition of correct codon–anticodon matches by the small-subunit rRNA of the ribosome.** Shown here is the interaction between a nucleotide of the small-subunit rRNA and the first nucleotide pair of a correctly paired codon–anticodon. Similar interactions form between other nucleotides of the rRNA and the second and third positions of the codon–anticodon pair. The small-subunit rRNA can form this network of hydrogen bonds only when an anticodon is correctly matched to a codon. As explained in the text, this codon–anticodon monitoring by the small-subunit rRNA increases the accuracy of protein synthesis. (From J.M. Ogle et al., *Science* 292:897–902, 2001. With permission from AAAS.)

interaction, and only when the match is correct is this folding completed and the reaction allowed to proceed. Thus, the codon-anticodon interaction is thereby checked twice—once by the initial complementary base-pairing and a second time by the folding of the ribosome, which depends on the correctness of the match. This same principle of induced fit is seen in transcription by RNA polymerase; here, an incoming nucleoside triphosphate initially forms a base pair with the template; at this point the enzyme folds around the base pair (thereby assessing its correctness) and, in doing so, creates the active site of the enzyme. The enzyme then covalently adds the nucleotide to the growing chain. Because their geometry is "wrong," incorrect base pairs impair this induced fit, and they are therefore likely to dissociate before being incorporated into the growing chain.

A second principle used to increase the specificity of complementary base-pairing is called **kinetic proofreading**. We have seen that after the initial codon-anticodon pairing and conformational change of the ribosome, GTP is hydrolyzed. This creates an irreversible step and starts the clock on a time delay during which the aminoacyl-tRNA moves into the proper position for catalysis. During this delay, those incorrect codon-anticodon pairs that have somehow slipped through the induced-fit scrutiny have a higher likelihood of dissociating than correct pairs. There are two reasons for this: (1) the interaction of the wrong tRNA with the codon is weaker, and (2) the delay is longer for incorrect matches than for correct matches.

In its most general form, kinetic proofreading refers to a time delay that begins with an irreversible step such as ATP or GTP hydrolysis, during which an incorrect substrate is more likely to dissociate than a correct one. In this case, kinetic proofreading thus increases the specificity of complementary base-pairing far above what is possible from simple thermodynamic associations alone. In fact, kinetic proofreading can increase the fidelity of a reaction from one error in 10^x to as little as one error in 10^{2x}, assuming that dissociation rate differences underlie the specificity of the molecular interactions involved.

The increase in specificity produced by kinetic proofreading comes at an energetic cost in the form of ATP or GTP hydrolysis. Kinetic proofreading operates in biological processes that range from DNA replication and DNA repair to RNA splicing and protein translation, helping to make life possible by greatly increasing the specificity of biochemical reactions.

Accuracy in Translation Requires a Large Expenditure of Free Energy

Translation by the ribosome is a compromise between the opposing constraints of accuracy and speed. We have seen, for example, that the accuracy of translation (one mistake per 10^4 amino acids joined) requires time delays each time a new amino acid is added to a growing polypeptide chain, producing an overall speed of translation of 20 amino acids incorporated per second in bacteria. Mutant bacteria with a specific alteration in the small ribosomal subunit have longer delays and translate mRNA into protein with an accuracy considerably higher than this; however, protein synthesis is so slow in these mutants that the bacteria are barely able to survive.

We have also seen that attaining the observed accuracy of protein synthesis requires the expenditure of a great deal of free energy; this is expected, because, as discussed in Chapter 2, there is a price to be paid for any increase in order in the cell. In most cells, protein synthesis consumes more energy than any other biosynthetic process. At least four high-energy phosphate bonds are split to make each new peptide bond: two are consumed in charging a tRNA molecule with an amino acid (see Figure 6–58), and two more drive steps in the cycle of reactions occurring on the ribosome during protein synthesis itself (see Figure 6–69). In addition, extra energy is consumed each time that an incorrect amino acid linkage is hydrolyzed by a tRNA synthetase (see Figure 6–61) and whenever an incorrect tRNA enters the ribosome, triggers GTP hydrolysis, and is rejected (see Figure 6–69). To be effective, any proofreading mechanism must also allow an appreciable fraction of correct interactions to be removed, and this adds an even greater energy cost to proofreading.

The Ribosome Is a Ribozyme

The ribosome is a large complex composed, by mass, of two-thirds RNA and one-third protein. The determination, in 2000, of the entire three-dimensional conformation of its large and small subunits was a major triumph of modern structural biology. The ribosomal RNAs are folded into highly compact, precise three-dimensional structures that form the compact core of the ribosome and determine its overall shape (**Figure 6–71**). These observations confirmed earlier evidence that rRNAs—and not proteins—are responsible for the ribosome's overall structure, its ability to position tRNAs on the mRNA, and its catalytic activity in forming covalent peptide bonds.

In marked contrast to the central positions of the rRNAs, the ribosomal proteins are generally located on the surface and fill in the gaps and crevices of the

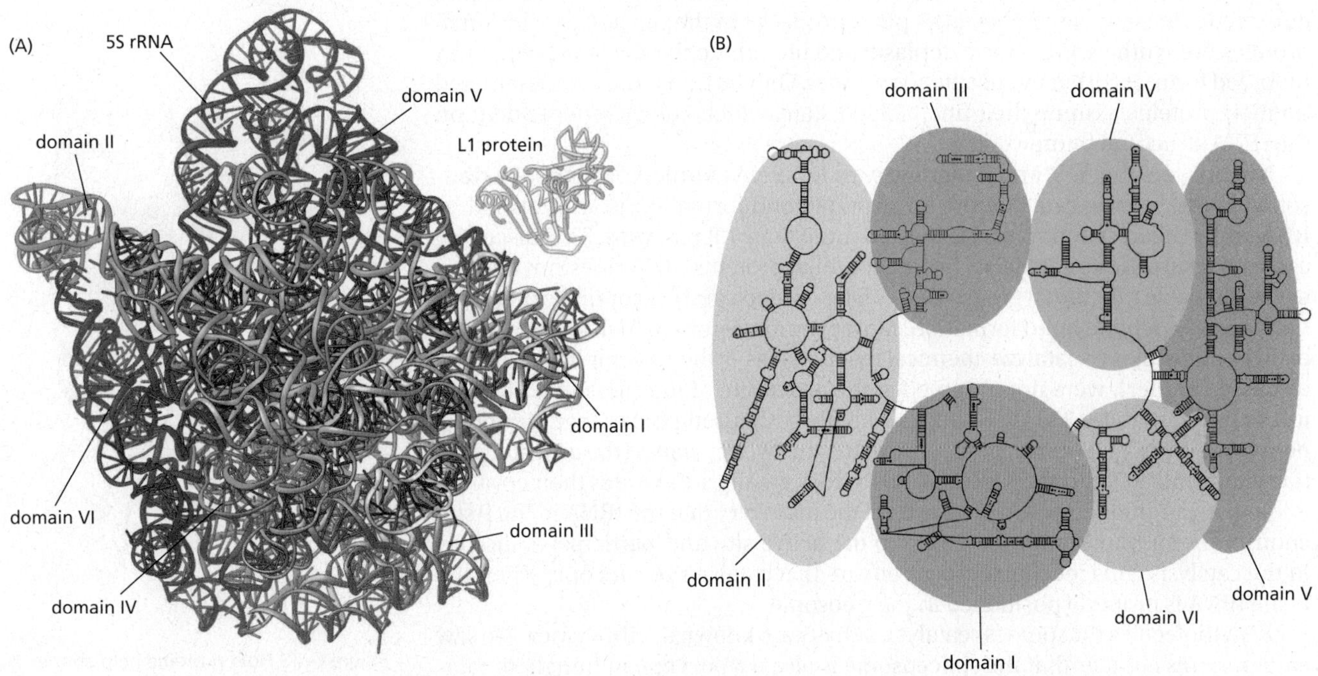

Figure 6–71 **Structure of the rRNAs in the large subunit of a bacterial ribosome, as determined by x-ray crystallography.** (A) Three-dimensional conformations of the large-subunit rRNAs (5S and 23S) as they appear in the ribosome. One of the protein subunits of the ribosome (L1) is also shown as a reference point, because it forms a characteristic protrusion on the ribosome. (B) Schematic diagram of the secondary structure of the 23S rRNA, showing the extensive network of base-pairing. The structure has been divided into six "domains" whose colors correspond to those in A. The secondary-structure diagram is highly schematized to represent as much of the structure as possible in two dimensions. To do this, several discontinuities in the RNA chain have been introduced, although in reality the 23S rRNA is a single RNA molecule. For example, the base of domain III is continuous with the base of domain IV even though a gap appears in the diagram. (Adapted from N. Ban et al., *Science* 289:905–920, 2000. With permission from AAAS.)

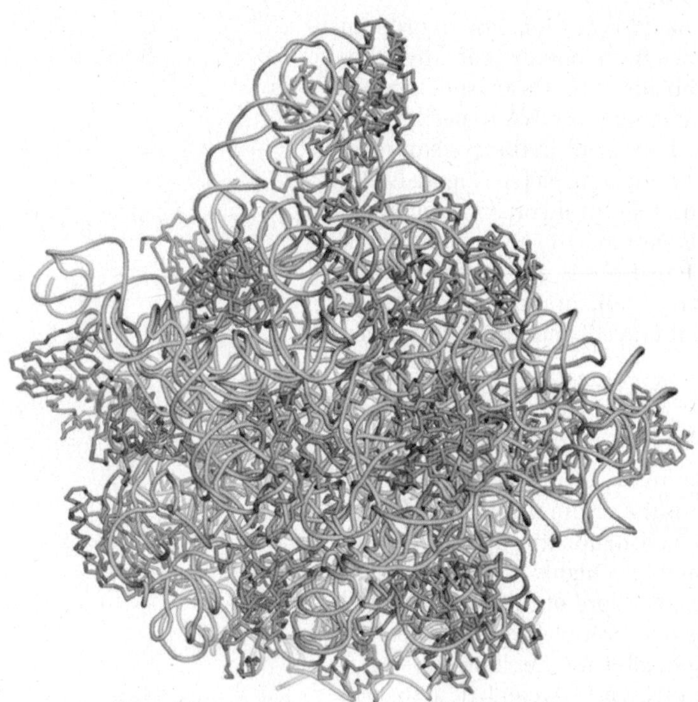

Figure 6–72 **Location of the protein components of the bacterial large ribosomal subunit.** The rRNAs (5S and 23S) are shown in *blue* and the proteins of the large subunit in *green*. This view is toward the backside of the ribosome relative to Figure 6–66A; the interface with the small subunit is facing into the page. (PDB code: 1FFK.)

folded RNA (Figure 6–72). Some of these proteins send out extended regions of polypeptide chain that penetrate short distances into holes in the RNA core (Figure 6–73). The main role of the ribosomal proteins seems to be to stabilize the RNA core, while permitting the changes in rRNA conformation that are necessary for this RNA to catalyze efficient protein synthesis. As discussed earlier, eukaryotic ribosome assembly takes place primarily in the nucleolus; ribosomal proteins are synthesized in the cytoplasm and brought to the nucleolus largely in unfolded form, escorted by specific *chaperones*. Only as ribosomes are assembled do their proteins assume their final, folded state, which is highly dependent on the rRNA structural framework.

Not only are the A, P, and E binding sites for tRNAs formed primarily by ribosomal RNAs, but the catalytic site for peptide bond formation is also formed by RNA, as the nearest amino acid is located more than 1.8 nm away. This discovery came as a surprise to biologists because, unlike proteins, RNA does not contain easily ionizable functional groups that can be used to catalyze sophisticated reactions such as peptide bond formation. Moreover, metal ions, which are often used by RNA molecules to catalyze chemical reactions (as is the case for RNA splicing, discussed earlier), were not observed at the active site of the ribosome. Instead, it is believed that the 23S rRNA forms a highly structured pocket that, through a network of hydrogen bonds, precisely orients the two reactants (the growing peptide chain and an aminoacyl-tRNA) and thereby greatly accelerates their covalent joining. An additional surprise came from the discovery that the tRNA in the P site contributes an important -OH group to the active site and participates directly in the catalysis. This mechanism may ensure that catalysis occurs only when the P-site tRNA is properly positioned in the ribosome.

RNA molecules that possess catalytic activity are known as **ribozymes**. We saw earlier in this chapter that the spliceosome is also a ribozyme, although its catalytic site is formed from several different RNA molecules rather than a single RNA, as is the case in the ribosome. In the final section of this chapter, we consider what the ability of RNA molecules to function as catalysts might mean for the early evolution of living cells. For now, we merely note that there is good reason to suspect that RNA rather than protein molecules served as the first catalysts for living cells. Thus the ribosome, with its RNA core, is suspected to be a relic of an earlier time in life's history—when protein synthesis evolved in cells that were run almost entirely by ribozymes.

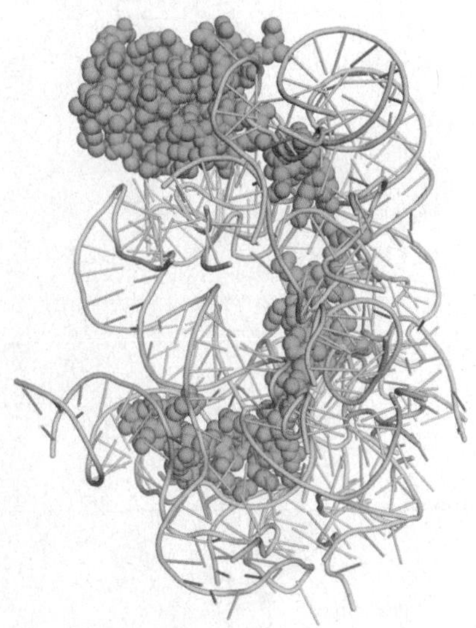

Figure 6–73 **How proteins help shape ribosomal RNA.** Shown here is the L15 protein in the large subunit of the bacterial ribosome. The globular domain of the protein lies on the surface of the ribosome, and an extended region penetrates deeply into the RNA core of the ribosome. The protein is shown in *green* and a portion of the ribosomal RNA core is shown in *blue*. (From D. Klein et al., *J. Mol. Biol.* 340:141–177, 2004. PDB code: 1S72.)

Nucleotide Sequences in mRNA Signal Where to Start Protein Synthesis

The initiation and termination of translation share properties of the translation elongation cycle described earlier but have additional features. The site at which protein synthesis begins on the mRNA is especially crucial, because it sets the reading frame for the whole length of the message. An error of one nucleotide either way at this stage would cause every subsequent codon in the message to be misread, resulting in a nonfunctional protein with a garbled sequence of amino acids. The initiation step is also important because, for most genes, it is the last point at which the cell can decide whether the mRNA is to be translated to produce a protein. The efficiency of this step is thus one determinant of the rate at which any particular protein will be synthesized. We shall see in Chapter 7 how regulation of this step occurs.

The translation of an mRNA usually begins with the codon AUG, and a special tRNA is required to start translation. This **initiator tRNA** always carries the amino acid methionine (in bacteria, a modified form of methionine—formylmethionine—is used), with the result that all newly made proteins have methionine as the first amino acid at their N-terminus, the end of a protein that is synthesized first. (This methionine is usually removed later by a specific protease.) The initiator tRNA is specially recognized by initiation factors because it has a nucleotide sequence distinct from that of the tRNA that normally carries methionine.

In eukaryotes, the initiator tRNA–methionine complex (Met–tRNAi) is first loaded into the small ribosomal subunit along with additional proteins called **eukaryotic initiation factors**, or **eIFs**. Of all the aminoacyl-tRNAs in the cell, only the methionine-charged initiator tRNA is capable of tightly binding the small ribosomal subunit without the complete ribosome being present, and unlike other tRNAs, it binds directly to the P site (**Figure 6–74**). Next, the small ribosomal subunit binds to the 5′ end of an mRNA molecule, which is recognized by virtue of its 5′ cap that has previously bound two initiation factors, eIF4E and eIF4G (see Figure 6–40). The small ribosomal subunit then moves forward (5′ to 3′) along the mRNA, searching for the first AUG; additional initiation factors that act as ATP-powered helicases facilitate this movement. In about 90% of mRNAs, translation begins at the first AUG encountered by the small subunit. At this point, the initiation factors dissociate, allowing the large ribosomal subunit to assemble with the complex and complete the ribosome. The initiator tRNA remains at the P site, leaving the A site vacant. Protein synthesis is therefore ready to begin (see Figure 6–69).

The nucleotides immediately surrounding the start site in eukaryotic mRNAs influence the efficiency of AUG recognition during the above scanning process. If this recognition site differs substantially from the consensus recognition sequence (5′-ACCAUGG-3′, known as the Kozak sequence after its discoverer), scanning ribosomal subunits will sometimes ignore the first AUG codon in the mRNA and move to the second or third AUG codon instead. Cells frequently use this phenomenon, known as "leaky scanning," to produce two or more proteins, differing in their N-termini, from the same mRNA molecule. This mechanism allows some genes to produce the same protein with and without a signal sequence attached at its N-terminus, for example, so that the protein is directed to two different compartments in the cell.

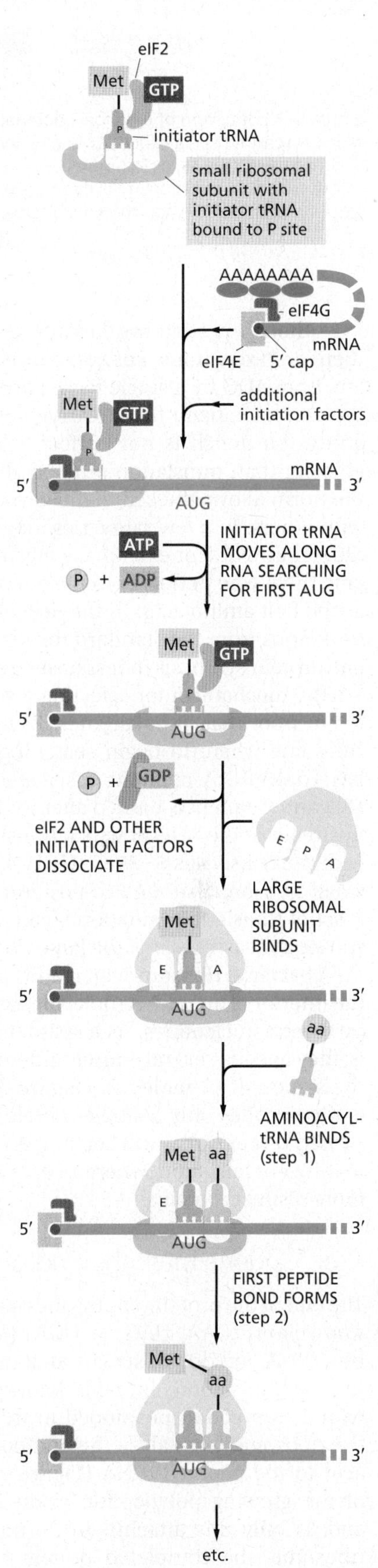

Figure 6–74 The initiation of protein synthesis in eukaryotes. Only three of the many translation initiation factors required for this process are shown. In addition to an initiating AUG codon, efficient translation initiation requires the poly-A tail of the mRNA bound by poly-A-binding proteins. In this way, the translation apparatus ascertains that both ends of the mRNA are intact before initiating protein synthesis. The communication between the 5′ and 3′ ends of the mRNA is mediated, at least in part, by interactions between the poly-A-binding proteins and eIF4G, as shown. This interaction appears to be transient; once translation begins, the 5′ and 3′ ends of the mRNA dissociate. Although only one GTP-hydrolysis event is shown in the figure, a second is known to occur just before the large and small ribosomal subunits join. In the last two steps shown in the figure, the ribosome has begun the standard elongation cycle, depicted in Figure 6–68.

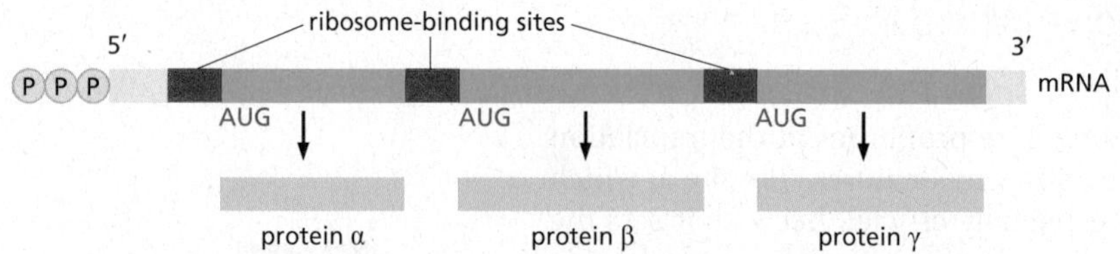

Figure 6–75 Structure of a typical bacterial mRNA molecule. Unlike eukaryotic ribosomes, which typically require a capped 5′ end on the mRNA, prokaryotic ribosomes initiate translation at ribosome-binding sites (Shine–Dalgarno sequences), which can be located anywhere along an mRNA molecule. This property of their ribosomes permits bacteria to synthesize more than one type of protein from a single mRNA molecule.

Although protein synthesis in eukaryotes usually begins at an AUG codon, there are exceptions. For some proteins, translation begins at a codon that differs from AUG by a single base, particularly in the third position. In these cases, the normal initiator tRNA (carrying methionine) is used, but because the codon–anticodon match is not perfect, translation of these proteins is typically less efficient than translation of those that begin with an AUG. This deviation from the norm allows the cell to make very low amounts of some proteins compared with others. In a few rare cases, however, translation can begin with an entirely different tRNA. For example, a few proteins begin with a CUG codon, and a leucine tRNA (the perfect codon–anticodon match) begins translation with leucine as the first amino acid. In the next chapter, we will discuss how these and other deviations from the standard translation initiation process can be used to regulate protein synthesis in response to signals from the environment.

The mechanism for selecting a start codon in bacteria is fundamentally different from that in eukaryotes. Bacterial mRNAs have no 5′ caps to signal the ribosome where to begin searching for the start of translation. Instead, each bacterial mRNA contains a specific ribosome-binding site (called the Shine–Dalgarno sequence, named after its discoverers) that is located a few nucleotides upstream of the AUG at which translation is to begin. This nucleotide sequence, with the consensus 5′-AGGAGGU-3′, forms base pairs with the 16S rRNA of the small ribosomal subunit to position the initiating AUG codon in the ribosome. A set of translation initiation factors orchestrates this interaction, as well as the subsequent assembly of the large ribosomal subunit to complete the ribosome.

A bacterial ribosome can readily assemble directly on a start codon that lies in the interior of an mRNA molecule, so long as a ribosome-binding site precedes it by several nucleotides. As a result, bacterial mRNAs are often *polycistronic*; that is, they encode several entirely different proteins, each of which is translated from the same mRNA molecule (**Figure 6–75**). In contrast, a eukaryotic mRNA generally encodes only a single protein, or more accurately, a single set of closely related proteins. We will see in the next chapter that there are some exceptions to this generalization, where a eukaryotic mRNA can carry information for two or more distinct proteins.

Stop Codons Mark the End of Translation

The end of the protein-coding message is signaled by the presence of one of three *stop codons* (UAA, UAG, or UGA) (see Figure 6–52). These are not recognized by a tRNA and do not specify an amino acid, but instead signal to the ribosome to stop translation. Proteins known as *release factors* bind to any ribosome with a stop codon positioned in the A site, forcing the peptidyl transferase in the ribosome to catalyze the addition of a water molecule instead of an amino acid to the peptidyl-tRNA (**Figure 6–76**). This reaction frees the carboxyl end of the growing polypeptide chain from its attachment to a tRNA molecule, and as only this attachment normally holds the growing polypeptide to the ribosome, the completed protein chain is released into the cytoplasm. The ribosome then releases its bound mRNA molecule and separates into the large

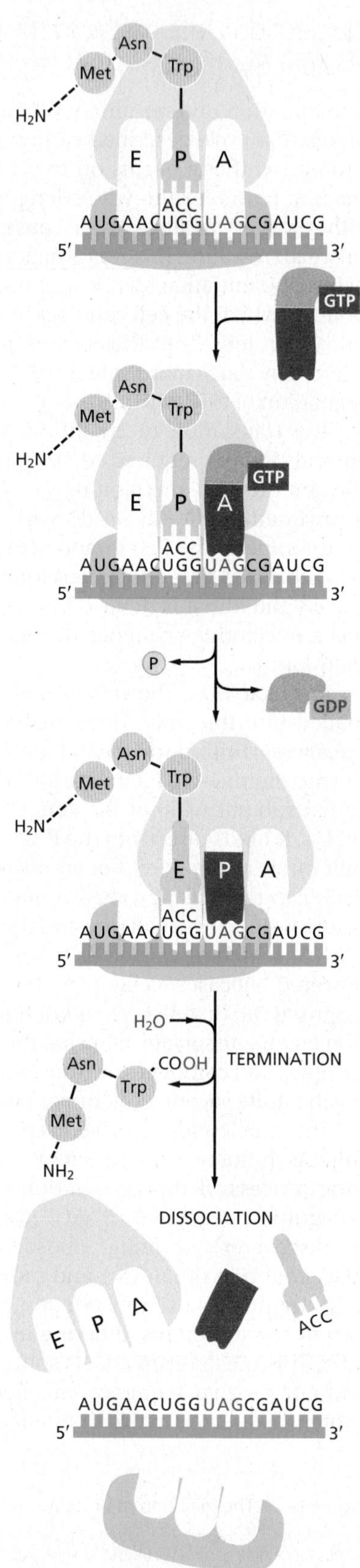

Figure 6–76 The final phase of eukaryotic protein synthesis. The binding of a two-subunit release factor to an A site bearing a stop codon terminates translation. Release factors are proteins that resemble tRNAs in their overall shape and charge distribution.

and small subunits. These subunits can then assemble on this or another mRNA molecule to begin a new round of protein synthesis.

During translation, the nascent polypeptide moves through a large, water-filled exit tunnel (approximately 10 nm × 1.5 nm) in the large subunit of the ribosome. The walls of this tunnel, made primarily of 23S rRNA, are a patchwork of tiny hydrophobic surfaces embedded in a more extensive hydrophilic surface. This structure is not complementary to any peptide and thus provides a "Teflon" coating through which a polypeptide chain can easily slide. We will see later in the chapter that, although some protein folding can occur in the exit tunnel, most of it takes place as the newly synthesized protein emerges from the ribosome.

Proteins Are Made on Polyribosomes

The synthesis of most protein molecules takes between 20 seconds and several minutes. During this short period, multiple initiations generally take place on each mRNA molecule being translated. As soon as the preceding ribosome has translated enough of the nucleotide sequence to move out of the way, the 5′ end of the mRNA is threaded into a new ribosome. For this reason, most mRNA molecules, and particularly those being translated at high rates, are found in the form of *polyribosomes* (or *polysomes*): large cytoplasmic assemblies made up of several ribosomes spaced as close as 80 nucleotides apart along a single mRNA molecule (**Figure 6–77**). The multiple initiations allow the cell to make many more protein molecules in a given time than would be possible if each protein had to be completed before the next could start. It is estimated that, in a typical human cell, about one-third of the mRNAs lack ribosomes altogether, and the remainder have 10–20 ribosomes per mRNA.

There Are Minor Variations in the Standard Genetic Code

As discussed in Chapter 1, the genetic code (shown in Figure 6–52) applies to all three major branches of life, providing important evidence for the common ancestry of all life on Earth. Although rare, there are exceptions to this code. For example, *Candida albicans*, the most prevalent fungal pathogen of humans, translates the codon CUG as serine, whereas nearly all other organisms translate it as leucine. And in some ciliates (unicellular eukaryotes that propel themselves using cilia), the three conventional stop codons specify particular amino

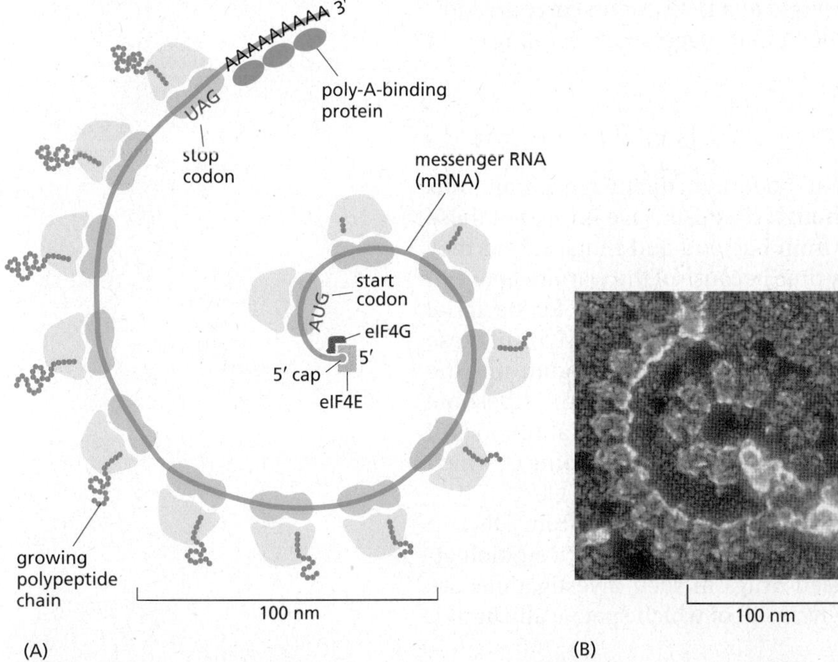

Figure 6–77 A polyribosome. (A) Schematic drawing showing how a series of ribosomes can simultaneously translate the same eukaryotic mRNA molecule. (B) Electron micrograph of a polyribosome from a eukaryotic cell (**Movie 6.10**). (B, courtesy of John Heuser.)

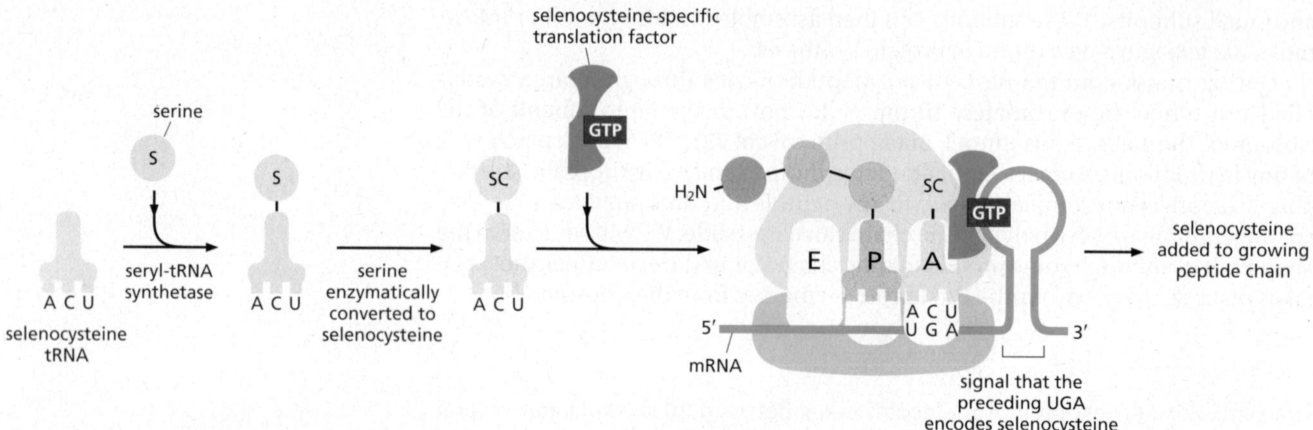

Figure 6–78 **Incorporation of selenocysteine into a growing polypeptide chain.** A specialized tRNA is charged with serine by the normal seryl-tRNA synthetase, and the serine is subsequently converted enzymatically to selenocysteine. A specific RNA structure in the mRNA (a stem-and-loop structure with a particular nucleotide sequence) signals that selenocysteine is to be inserted at the neighboring UGA codon. As indicated, this event requires the participation of a selenocysteine-specific translation factor. After the addition of selenocysteine, translation continues until a conventional stop codon is encountered.

acids, and the end of translation is instead signaled by the 3′ end of the mRNA. Mitochondria of many species (which have their own genomes and encode much of their translational apparatus) routinely deviate from the standard code. For example, in mammalian mitochondria AUA is translated as methionine, whereas in the cytosol of the cell it is translated as isoleucine (see Table 14–4, p. 864).

A common type of variation, sometimes called *translation recoding,* is used to incorporate selenocysteine into proteins. In this case, neighboring nucleotide sequence information present in an mRNA changes the meaning of the genetic code at a particular site in the mRNA molecule. The standard code allows cells to manufacture proteins using only 20 amino acids. However, bacteria, archaea, and eukaryotes have selenocysteine available as a twenty-first amino acid that can be incorporated directly into a growing polypeptide chain. Selenocysteine, which contains a selenium atom in place of the sulfur atom of cysteine, is essential for the efficient function of a variety of enzymes. It is enzymatically produced from a serine attached to a special tRNA molecule that base-pairs with the UGA codon, a codon normally used to signal a translation stop. The mRNAs for those proteins in which selenocysteine is to be inserted at a UGA codon carry an additional nearby nucleotide sequence in the mRNA that triggers this recoding event (**Figure 6–78**).

Inhibitors of Prokaryotic Protein Synthesis Are Useful as Antibiotics

Many of the most effective antibiotics used in modern medicine are compounds that inhibit bacterial protein synthesis. Although chemists have improved these compounds, most were originally isolated from bacteria and fungi, where they are thought to have arisen over evolutionary time because of the warfare between competing microbes. Because many of these drugs are selective for bacterial ribosomes, humans can take high dosages without undue toxicity. Many of these antibiotics lodge in pockets in the ribosomal RNAs and simply "gum up" the smooth operation of the ribosome; others block specific parts of the ribosome such as the exit tunnel (**Figure 6–79**). **Table 6–4** lists some common antibiotics of this kind, along with several other inhibitors of protein synthesis, some of which act on eukaryotic cells and therefore cannot be used as antibiotics.

Because they block specific steps in the processes that lead from DNA to protein, many of the compounds listed in Table 6–4 are useful for cell biological studies. Among the most commonly used drugs in such investigations are *chloramphenicol, cycloheximide,* and *puromycin,* all of which specifically inhibit

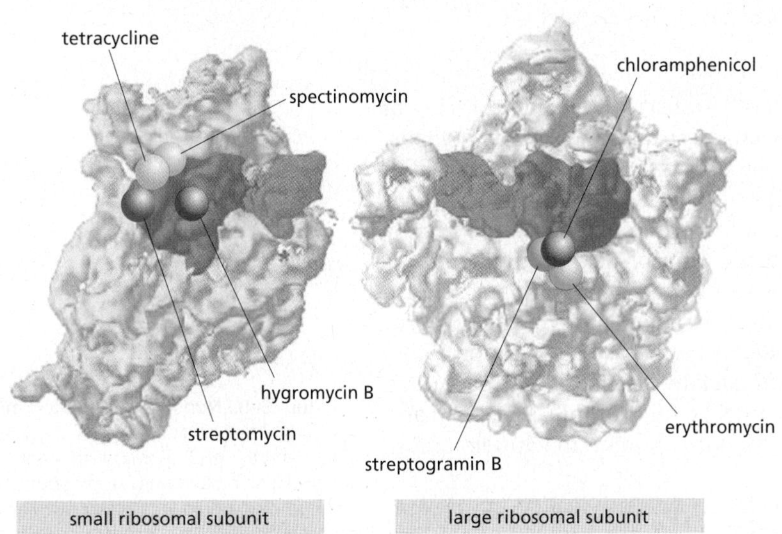

Figure 6–79 **Binding sites for antibiotics on the bacterial ribosome.** The small *(left)* and large *(right)* subunits of the ribosome are arranged as though the ribosome has been opened like a book. Antibiotic-binding sites are marked with colored spheres, and the bound tRNA molecules are shown in *purple* (see Figure 6–66). Most of the antibiotics shown bind directly to pockets formed by the ribosomal RNA molecules. Hygromycin B induces errors in translation, spectinomycin blocks the translocation of the peptidyl-tRNA from the A site to the P site, and streptogramin B prevents elongation of nascent peptides. Table 6–4 lists the inhibitory mechanisms of additional commonly used antibiotics. (Adapted from J. Poehlsgaard and S. Douthwaite, *Nat. Rev. Microbiol.* 3:870–881, 2005.)

protein synthesis. In a eukaryotic cell, for example, chloramphenicol inhibits protein synthesis on ribosomes only in mitochondria (and in chloroplasts in plants), presumably reflecting the bacterial origins of these organelles (discussed in Chapter 14). Cycloheximide, in contrast, affects only ribosomes in the cytoplasm. Puromycin is especially interesting because it is a structural analog of a tRNA molecule linked to an amino acid and is therefore a fascinating example of molecular mimicry. The ribosome mistakes it for an authentic charged tRNA and covalently incorporates it at the C-terminus of the growing peptide chain, causing premature termination and release of the polypeptide. As might be expected, puromycin inhibits protein synthesis in both bacteria and eukaryotes.

TABLE 6–4 **Inhibitors of Protein or RNA Synthesis**

Inhibitor	Specific effect
	Acting only on bacteria
Tetracycline	Blocks binding of aminoacyl-tRNA to the A site of the ribosome
Streptomycin	Prevents the transition from translation initiation to chain elongation and also causes miscoding
Chloramphenicol	Blocks the peptidyl transferase reaction on ribosomes (step 2 in Figure 6–68)
Erythromycin	Binds in the exit tunnel of the ribosome and thereby inhibits elongation of the peptide chain
Rifamycin	Blocks initiation of RNA chains by binding to RNA polymerase (prevents RNA synthesis)
	Acting on bacteria and eukaryotes
Puromycin	Causes the premature release of nascent polypeptide chains by its addition to the growing chain end
Actinomycin D	Binds to DNA and blocks the movement of RNA polymerase (prevents RNA synthesis)
	Acting on eukaryotes but not bacteria
Harringtonine	Blocks the A site of the 60S ribosome subunit after translation initiation, but before elongation (see Figure 6–74)
Anisomycin	Blocks the peptidyl transferase reaction on ribosomes (step 2 in Figure 6–68)
α-Amanitin	Blocks mRNA synthesis by binding preferentially to RNA polymerase II

The ribosomes of eukaryotic mitochondria (and chloroplasts) often resemble those of bacteria in their sensitivity to inhibitors. Therefore, some of these antibiotics can have a deleterious effect on human mitochondria.

Quality-Control Mechanisms Act to Prevent Translation of Damaged mRNAs

In eukaryotes, mRNA production involves both transcription and a series of elaborate RNA-processing steps; as we have seen, these take place in the nucleus, segregated from ribosomes, and only when the processing is complete and the mRNAs deemed "export-ready" are they transported to the cytosol to be translated (see Figure 6–40). However, this scheme is not foolproof, and some incorrectly processed mRNAs are inadvertently sent to the cytosol. In addition, mRNAs that were flawless when they left the nucleus can become broken or otherwise damaged in the cytosol. The danger of translating damaged or incompletely processed mRNAs (which would produce truncated or otherwise aberrant proteins) is apparently so great that the cell has several backup measures to prevent this from happening. To avoid translating broken mRNAs, for example, the 5′ cap and the poly-A tail are both recognized by the translation initiation machinery before translation begins (see Figure 6–74).

One of the most powerful mRNA surveillance systems, called **nonsense-mediated mRNA decay**, prevents defective mRNAs from escaping from the nuclear envelope. This mechanism is brought into play when the cell determines that an mRNA molecule has a nonsense (stop) codon (UAA, UAG, or UGA) in the "wrong" place. This situation is likely to arise in an mRNA molecule that has been improperly spliced, because aberrant splicing will usually result in the random introduction of a nonsense codon into the reading frame of the mRNA—especially in organisms, such as humans, that have large introns.

The nonsense-mediated mRNA decay mechanism begins as an mRNA molecule is being transported from the nucleus to the cytosol. As soon as its 5′ end emerges from a nuclear pore, the mRNA is met by a ribosome, which begins to translate it. As translation proceeds, the exon junction complexes (EJCs) that are bound to the mRNA at each completed splice site (see Figure 6–29) are displaced by the moving ribosome. The normal stop codon should lie within the last exon, so by the time the ribosome reaches it, the mRNA should be free of EJCs. In this case, the mRNA "passes inspection" and is released to the cytosol where it can be translated in earnest (Figure 6–80). However, if the ribosome reaches a stop codon earlier, when EJCs remain bound, the mRNA molecule is rapidly degraded. In this way, the first round of translation allows the cell to test the fitness of each mRNA molecule as it exits the nucleus.

Figure 6–80 **Nonsense-mediated mRNA decay.** As shown on the *right*, the failure to correctly splice a pre-mRNA often introduces a premature stop codon into the reading frame for the protein. These abnormal mRNAs are destroyed by the nonsense-mediated decay mechanism. To activate this mechanism, an mRNA molecule, bearing the exon junction complexes (EJCs) that mark successfully completed splices, is quickly met by a ribosome that performs a "test" round of translation. As the mRNA passes through the ribosome, the EJCs are stripped off, and successful mRNAs are released to undergo multiple rounds of translation *(left side)*. However, if an in-frame stop codon is encountered before the final EJC is reached *(right side)*, the mRNA undergoes nonsense-mediated decay, which is triggered by the Upf proteins *(green* that bind to each EJC. This mechanism ensures that nonsense-mediated decay is triggered only when the premature stop codon is in the same reading frame as that of the normal protein. (Adapted from J. Lykke-Andersen et al., *Cell* 103:1121–1131, 2000.)

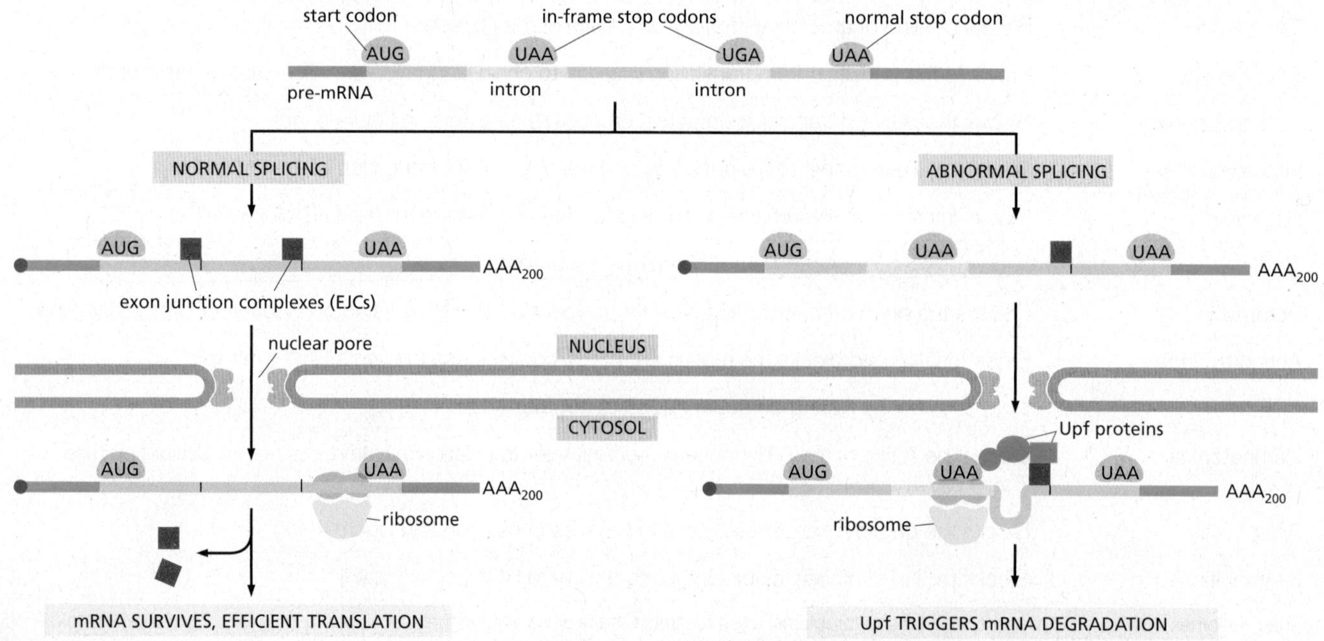

Nonsense-mediated decay may have been especially important in evolution, allowing eukaryotic cells to more easily explore new genes formed by DNA rearrangements, mutations, or alternative patterns of splicing—by selecting only those mRNAs for translation that can produce a full-length protein. Nonsense-mediated decay is also important in cells of the developing immune system, where the extensive DNA rearrangements that occur (see Figure 24–28) often generate premature termination codons. The surveillance system degrades the mRNAs produced from these faulty rearranged genes, thereby avoiding the potential toxic effects of truncated proteins.

The nonsense-mediated surveillance pathway also plays an important role in mitigating the symptoms of many inherited human diseases. As we have seen, inherited diseases are usually caused by mutations that spoil the function of a key protein, such as hemoglobin or one of the blood-clotting factors. Approximately one-third of all genetic disorders in humans result from mutations that change a normal codon into a stop (nonsense) codon or mutations (such as frameshift mutations or splice-site mutations) that place nonsense mutations into the gene's reading frame. In individuals that carry one mutant and one functional gene, nonsense-mediated decay eliminates the aberrant mRNA, and thereby prevents a potentially toxic protein from being made. Without this safeguard, individuals with one functional and one mutant "disease gene" would likely suffer much more severe symptoms.

Stalled Ribosomes Can Be Rescued

Even when an mRNA molecule passes all the initial inspections, additional problems can arise as it is being translated in the cytosol, and the cell has several ways to deal with them. In a sense, translating ribosomes constantly act as sensors for mRNA "health," and when one stalls, those upstream of it collide with it (and each other), generating a string of nonfunctioning ribosomes. For example, if an mRNA becomes broken and thereby lacks an in-frame stop codon, the ribosome will translate to the 3′ end of the RNA but will not be released. Damaged RNA bases (which cannot form stable codon–anticodon interactions), stable mRNA secondary structures, and stretches of rare codons (that is, codons that have very low concentrations of their matching tRNAs) can also stall ribosomes. Although some of these problems can be overcome simply by allowing the ribosome sufficient time, many are dealt with more aggressively by a set of mechanisms known collectively as *ribosome-associated quality control* (**Figure 6–81**). Although the detailed pathways differ for the different types of barriers that stall ribosomes, three steps are common: the mRNA is degraded, the nascent protein is degraded, and the stalled ribosome is disengaged so it can be used again.

All of these steps make conceptual sense: the first removes a damaged mRNA so it will cause no further problems, the second prevents an aberrant protein from

Figure 6–81 **Eukaryotic ribosome-associated quality control.** A ribosome can stall at damaged bases, rare codons, or other barriers *(top left)*; here, the signal to correct the problem is generated by a collision between the stalled ribosome and the one immediately upstream of it. Ribosomes also stall at the ends of broken mRNAs *(bottom left)*; in this case the empty A site is the cue for rescue. Although the mechanisms differ somewhat depending on the nature of the stall, in general a stalled ribosome is split (immediately releasing the 40S subunit for reuse), the mRNA is degraded, and the 60S subunit, which contains a tRNA lodged in its P site attached to a partially synthesized protein, is rescued by the ribosome quality control (RQC) complex (indicated in *blue*). The nascent protein is ubiquitylated and the tRNA cleaved from it, freeing the protein from the ribosome. As will be described shortly, this ubiquitylation serves as a signal for the protein to be destroyed. (Adapted from C.A.P. Joazeiro, *Nat. Rev. Mol. Cell Biol.* 20:368–383, 2019. With permission from Springer Nature.)

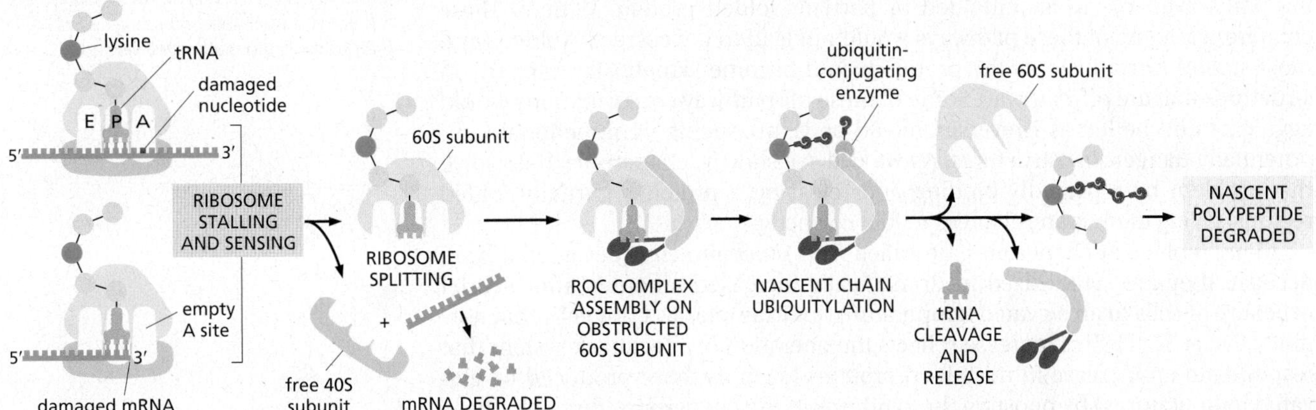

being released into the cell, and the third rescues a valuable RNA–protein machine complex that requires many resources to assemble. It is estimated that each ribosome in a human cell synthesizes approximately 3000 protein molecules, and the ability to rescue stalled ribosomes (as opposed to degrading them) makes a major contribution to their longevity.

The Ribosome Coordinates the Folding, Enzymatic Modification, and Assembly of Newly Synthesized Proteins

The process of gene expression is not over when the genetic code has been used to create the sequence of amino acids that constitutes a protein. To be useful to the cell, this new polypeptide chain must fold up into its unique three-dimensional conformation, bind any small-molecule cofactors required for its activity, be appropriately modified by protein kinases and other protein-modifying enzymes, and assemble correctly with the other protein subunits with which it functions (**Figure 6–82**).

The information needed for all of the steps listed above is ultimately contained in the sequence of amino acids that the ribosome produces when it translates an mRNA molecule into a polypeptide chain. As discussed in Chapter 3, when a protein folds into a compact structure, it buries most of its hydrophobic residues in an interior core. In addition, large numbers of noncovalent interactions form between various parts of the molecule. It is the sum of all of these energetically favorable arrangements that determines the final folding pattern of the polypeptide chain—as the conformation of lowest free energy (see p. 121).

For some proteins, folding begins in the exit tunnel of the ribosome. Although too narrow to accommodate the folding of complete proteins, the exit tunnel widens toward its end, and simple structures such as short α helices can form as the growing peptide is pushed through by the translation process. But the major portion of protein folding begins as the protein emerges from the ribosome exit tunnel, where the new protein is also met by enzymes that modify its N-terminus. We saw earlier in the chapter that translation of a typical mRNA always begins with a methionine (see Figure 6–74); however, in most eukaryotic proteins (70% of cytosolic proteins), this amino acid is cleaved off and the resulting N-terminus modified by acetylation. The enzymes that carry out these and other modifications have a weak affinity for the ribosome and "hover" (that is, rapidly associate and dissociate with the ribosome) near the exit tunnel—in position to act on the growing protein as it first emerges.

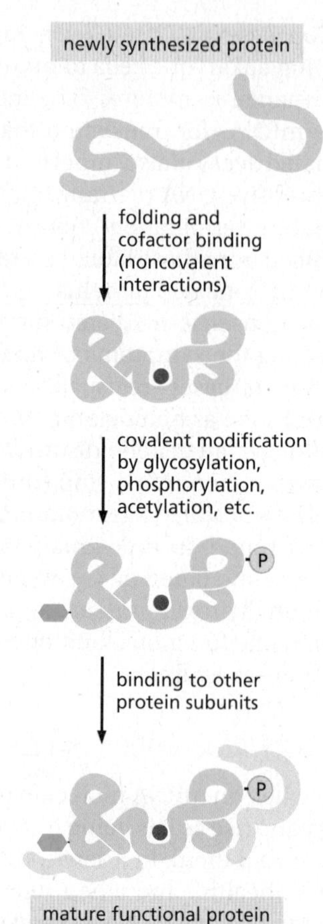

Figure 6–82 **Steps in the creation of a functional protein.** As indicated, translation of an mRNA sequence into an amino acid sequence on the ribosome is not the end of the process of forming a protein. To function, the completed polypeptide chain must fold correctly into its three-dimensional conformation, bind any cofactors required, and assemble with its partner protein chains, if any. Noncovalent bond formation drives these changes. As indicated, many proteins also require covalent modifications of selected amino acids. Although the most frequent modifications are protein glycosylation and protein phosphorylation, more than 200 different types of covalent modifications are known (see pp. 175–176). As described in the text, many of these steps begin while the protein is still being synthesized.

Molecular Chaperones Help Guide the Folding of Most Proteins

Most proteins on their own do not fold correctly during their synthesis; instead, they require a special class of proteins called **molecular chaperones** to do so. Molecular chaperones are useful for cells because there are many different folding paths available to an unfolded or partially folded protein. Without these chaperones, some of these pathways would not lead to the correctly folded (and most stable) form; instead, the protein would become "kinetically trapped" in structures that are off-pathway. Some of these off-pathway conformations would aggregate and be left as irreversible dead ends, producing nonfunctional (and potentially dangerous) structures. As we will see shortly, chaperones help solve this problem by repeatedly binding and releasing a protein's partially folded regions, giving them many chances to fold correctly.

Many molecular chaperones are called *heat-shock proteins* (designated *hsp*), because they are synthesized in dramatically increased amounts after a brief exposure of cells to an elevated temperature (for example, 42°C for cells that normally live at 37°C). This increase reflects the operation of a feedback system that responds to an increase in misfolded proteins (such as those produced by elevated temperatures) by boosting the synthesis of the chaperones that help these proteins refold. However, even under normal conditions, the heat-shock proteins are so abundant that all proteins should be considered to be embedded in a rich soup of these molecules.

There are several major families of molecular chaperones, including the hsp60 and hsp70 proteins, with different family members functioning in different organelles. Thus, as discussed in Chapter 12, mitochondria contain their own hsp60 and hsp70 molecules that are distinct from those that function in the cytosol, and a special hsp70 (called *BiP*) helps to fold proteins in the endoplasmic reticulum.

In all, humans have thirteen hsp70 proteins, including a group that associates with the ribosome and helps nearly all emerging proteins fold correctly. They do this by rapidly binding and releasing short sequences (approximately five amino acids each) as a new protein is pushed through the ribosome exit tunnel. When one of these chaperones encounters a sequence rich in hydrophobic amino acids, which typically form the core of folded proteins, it clamps down on it, thereby delaying the folding of the emerging protein until enough of it has been made to begin to fold correctly. Hydrophobic stretches that emerge early from the ribosome are thereby prevented from aggregating with other hydrophobic surfaces until they can be properly folded into the core of the nascent protein. The hsp70 clamping and release reactions required are regulated by ATP hydrolysis catalyzed by the nucleotide-binding domain of hsp70 (**Figure 6–83**).

Most proteins are at least partially folded when they are released from the ribosome, but the final folding for many occurs as additional hsp70 molecules associate with them away from the ribosome. By undergoing multiple cycles of ATP hydrolysis, these hsp70 molecules allow proteins to complete their folding.

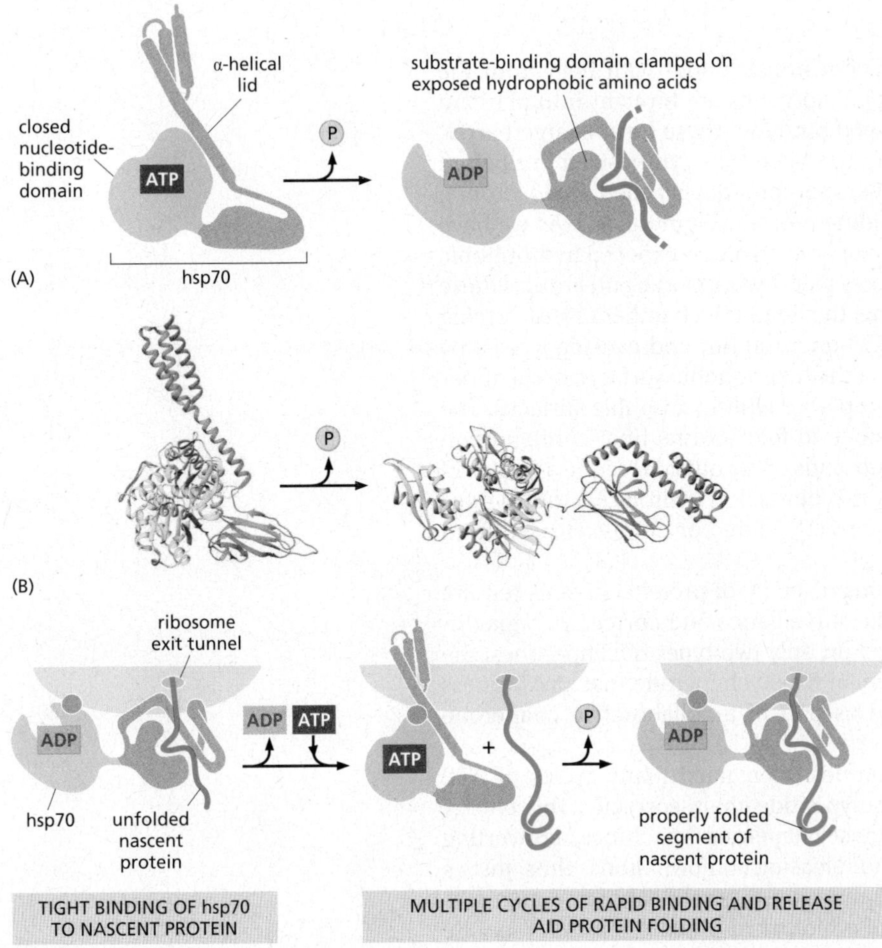

Figure 6–83 **The hsp70 family of molecular chaperones.** These are among the most abundant of all proteins, constituting several percent of total cell protein, and they thus play prominent roles inside the cell. (A) Short stretches of hydrophobic amino acids that are abnormally exposed in misfolded proteins *(green)* trigger the hydrolysis of ATP to ADP, causing hsp70 to close down on its substrate, trapping it in an extended conformation. The rebinding of another molecule of ATP will open the hsp70 "clamp," releasing the substrate in its extended form. Repeated cycles of hsp70 clamping and release, often involving multiple hsp70 molecules on the same target, help fold the target protein by keeping its hydrophobic regions from aggregating. This can continue until these regions are assimilated into the core of the properly folded protein. (B) The three-dimensional structures of hsp70 that produce the mechanism schematically illustrated in A. (C) Some hsp70 family members bind to the ribosome and act early in the life of a newly synthesized protein. Aided by other proteins (not shown), ATP-bound hsp70 molecules rapidly bind and release the nascent protein as it emerges from the ribosome. (B, PDB codes: 2KHO and 4JNE.)

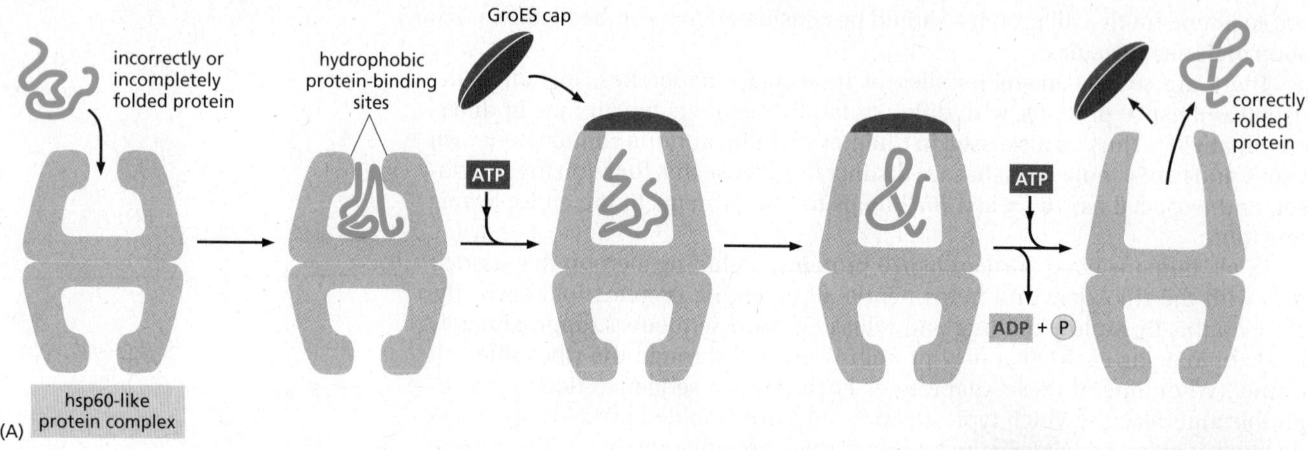

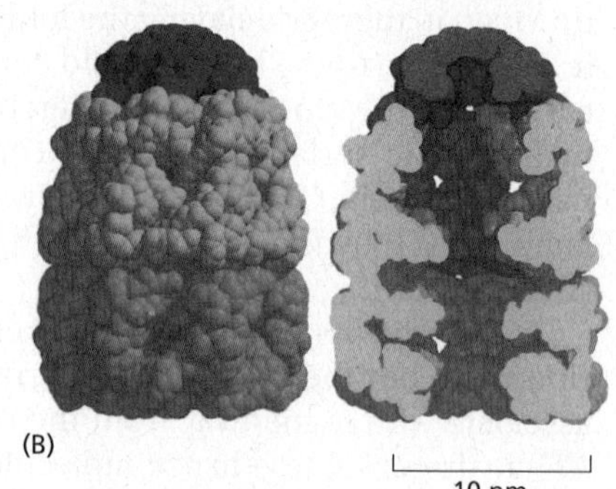

Figure 6–84 **The structure and function of the hsp60 family of molecular chaperones.** (A) A misfolded protein is initially captured by hydrophobic interactions with the exposed surface of the opening. This initial binding often helps to unfold a misfolded protein. The subsequent binding of ATP and a cap releases the substrate protein into an enclosed space, where it has a new opportunity to fold. After about 10 seconds, ATP hydrolysis occurs, weakening the binding of the cap. Subsequent binding of additional ATP molecules ejects the cap, and the protein is released. As indicated, only half of the symmetric barrel operates on a client protein at any one time. This hsp type of chaperone, known also as a *chaperonin*, is designated as hsp60 in mitochondria, TRiC in the cytosol of vertebrate cells, and GroEL in bacteria. (B) The structure of GroEL bound to its GroES cap, as determined by x-ray crystallography. On the left is shown the outside of the barrel-like structure, and on the right is a cross section through its center. (B, adapted from B. Bukau and A.L. Horwich, *Cell* 92:351–366, 1998. With permission from Elsevier.)

Some proteins produced by the cell cannot properly fold with help from the hsp70 proteins alone, and other types of chaperones are brought into play. An important class is exemplified by the hsp60 proteins. These form a large barrel-shaped structure that acts after a protein has been fully synthesized but before it has folded correctly. This type of chaperone, sometimes called a *chaperonin*, forms an "isolation chamber" for the folding process (**Figure 6–84**). As we have seen, incorrectly folded proteins are often characterized by exposed hydrophobic patches. The entrance to the hsp chamber is itself hydrophobic and attracts these patches. In conjunction with ATP binding, the lid to the chamber closes, forcing the incorrectly folded protein into the chamber interior and causing a twisting of the chamber subunits whereby some of the hydrophobic surfaces of the inner chamber are moved out of the way and replaced with hydrophilic surfaces. The substrate protein is thereby given a chance to fold into its final configuration (which generally favors hydrophilic amino acids on its outside) in the absence of any other proteins with which to aggregate. When ATP is hydrolyzed, the lid pops off, and the substrate protein, whether correctly folded or not, is released from the chamber.

Folding and maintaining the enormous diversity of proteins in cells require a wide range of chaperones with versatile surveillance and correction capabilities. Although our discussion has focused on only two types of chaperones, the cell has a variety of others. These include an hsp90 chaperone that can harness mechanical forces to help proteins fold as part of a collaborative chaperone network (**Figure 6–85**).

The hsp70, hsp60, and hsp90 chaperones often need many cycles of ATP binding and hydrolysis to fold a single polypeptide chain correctly. This energy is used to create movements in each of these chaperone "machines," converting them back and forth between binding and releasing conformations. Thus, just as we saw for transcription, splicing, and translation, a great deal of free energy is used by cells to improve the accuracy of a biological process—in this case the correct folding of proteins (**Movie 6.11**).

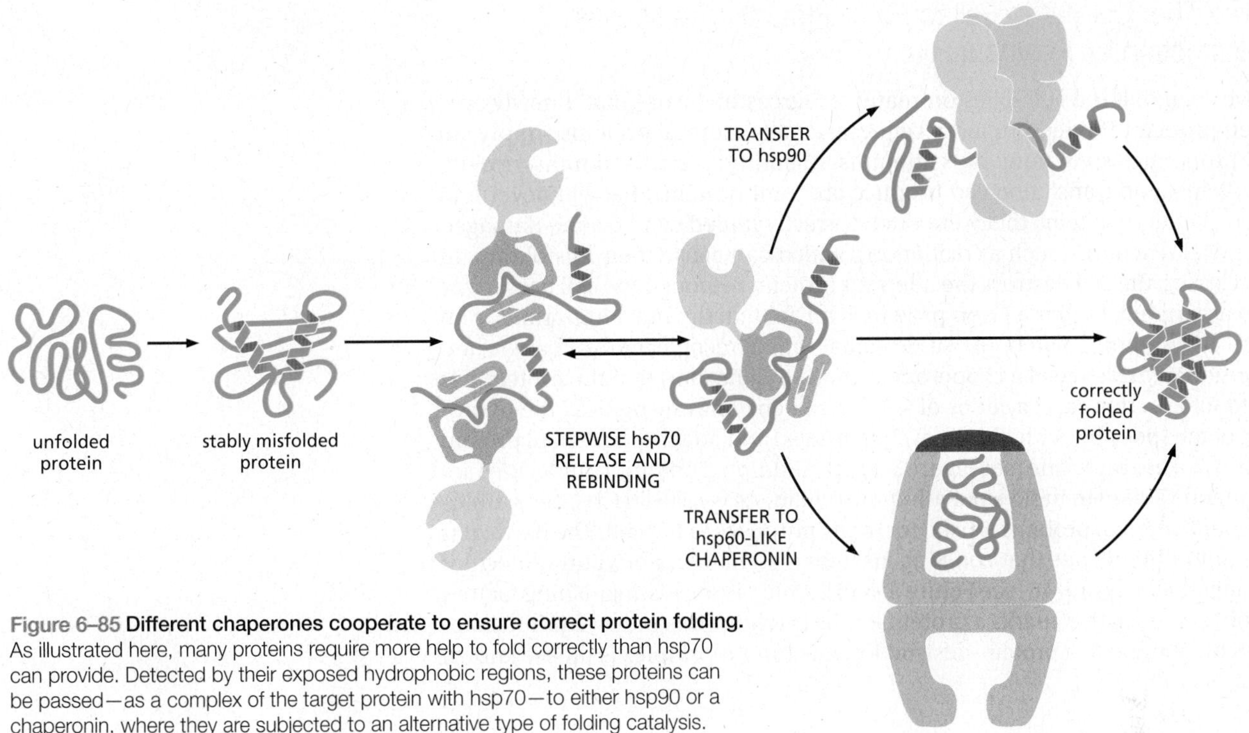

Figure 6–85 Different chaperones cooperate to ensure correct protein folding. As illustrated here, many proteins require more help to fold correctly than hsp70 can provide. Detected by their exposed hydrophobic regions, these proteins can be passed—as a complex of the target protein with hsp70—to either hsp90 or a chaperonin, where they are subjected to an alternative type of folding catalysis.

Proper Folding of Newly Synthesized Proteins Is Also Aided by Translation Speed and Subunit Assembly

In addition to the highly abundant chaperones, two additional mechanisms help to correctly fold newly synthesized proteins. First, translation is not a smooth process; the ribosome moves in fits and starts, translating some RNA sequences rapidly and pausing at others. The pauses are typically caused by a series of rare codons (which require extra time for the corresponding low-abundance charged tRNAs to enter the ribosome by diffusion) or by mRNA secondary structures that form ahead of the ribosome, slowing down or temporarily blocking its smooth passage. It is thought that many of these pauses are not accidental; rather, they are spaced to allow extra time for "problem" portions of newly synthesized proteins to fold as they exit the ribosome. If true, it means that an mRNA molecule, in addition to coding for a protein, specifies changes in the speed of translation matched to its protein's folding characteristics.

A second mechanism reflects the fact that a protein molecule does not typically work on its own; instead, it usually assembles with other subunits to form multisubunit structures. As a protein emerges from the ribosome, it often begins assembling with one or more of its fully folded partner subunits, which, by acting as complementary surfaces, help the newly synthesized protein adopt its correct three-dimensional structure. We have seen in both bacteria and eukaryotes that a given mRNA molecule can be simultaneously translated by many ribosomes. For protein complexes made up of identical subunits, fully synthesized protein molecules will therefore always be in proximity to aid in the folding of those being synthesized. In bacteria, different proteins can be translated from the same mRNA, providing an opportunity for different subunits of a protein complex to efficiently help each other fold. In eukaryotes, where only a single protein is typically produced from each mRNA, it has been proposed that mRNAs that code for different subunits of a protein complex are held in proximity, perhaps by information in the untranslated portion of the message, providing a high local concentration of mature subunits to aid in the assembly and folding of newly synthesized subunits.

Proteins That Ultimately Fail to Fold Correctly Are Marked for Destruction by Polyubiquitin

We have seen that the cell relies on many resources to ensure that a newly synthesized protein is folded properly. Yet occasionally, normal proteins simply fail to fold properly despite numerous attempts. In addition, errors in transcription, RNA splicing, and translation can produce aberrant proteins that will never fold properly. Finally, proteins that were once correctly folded can become damaged by chemical reactions (such as oxidation), which can cause their misfolding. In all these cases, the cell destroys the aberrant protein. Although the cell uses many cues to ascertain whether a given protein is misfolded, the most important is an exposed hydrophobic patch, the same signal that is recognized by chaperones. The continued presence of a chaperone is thus an indication that the protein has failed to fold. In this case, a series of specialized enzymes recognizes the hydrophobic patch (possibly with the aid of chaperones) and attaches the small protein ubiquitin to a nearby lysine (see Figure 3–65A). Additional ubiquitin molecules are subsequently added to build a polyubiquitin chain. As we saw in Chapter 3, ubiquitin modification of proteins is used for many purposes in the cell. The particular type of ubiquitin linkage that concerns us here is a chain of ubiquitin molecules linked together at lysine 48 (see Figure 3–65B), which is the distinguishing feature of the ubiquitin tag that marks a protein for destruction. This polyubiquitin chain delivers the protein to a protein-destruction machine, a complex protease known as the *proteasome*.

The Proteasome Is a Compartmentalized Protease with Sequestered Active Sites

Proteasomes are very abundant, constituting approximately 1% of the total proteins in cells. They are dispersed throughout the cytosol and the nucleus, but the proteasome also destroys aberrant proteins that have entered the endoplasmic reticulum (ER). In this case, an ER-based surveillance system detects proteins that have failed either to fold or to be assembled properly after they enter the ER and *retrotranslocates* them back to the cytosol for degradation by the proteasome (discussed in Chapter 12).

Each proteasome consists of a central hollow cylinder (the 20S core proteasome) formed from multiple protein subunits that assemble as a stack of four heptameric rings. Some of the subunits are proteases whose active sites face the cylinder's inner chamber, thus preventing them from running rampant through the cell. Each end of the cylinder is normally associated with a large protein complex (the 19S cap) that contains a six-subunit protein ring through which target proteins are threaded into the proteasome core, where they are degraded (**Figure 6–86**). The threading reaction, driven by ATP hydrolysis, unfolds the target

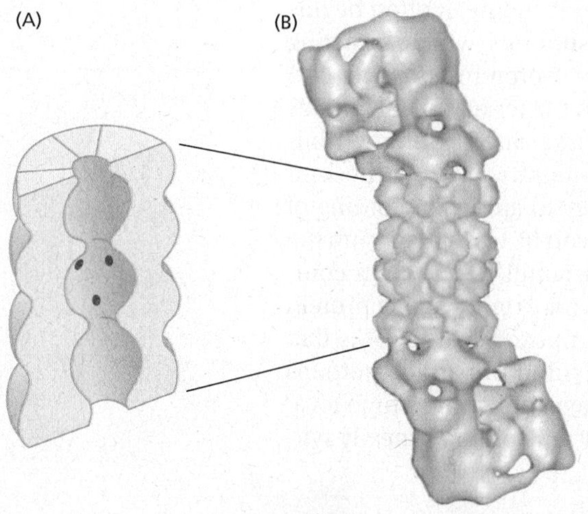

Figure 6–86 The proteasome. (A) A cutaway view of the structure of the central 20S cylinder, as determined by x-ray crystallography, with the active sites of the proteases indicated by *red dots*. (B) The entire proteasome, in which the central cylinder *(yellow)* is supplemented by a 19S cap *(blue)* at each end. The complex cap selectively binds proteins that are marked by ubiquitin for destruction; if the protein also contains a loosely structured region, the cap uses ATP hydrolysis to further unfold its polypeptide chain and feed it through the narrow channel (see Figure 6–87) into the inner chamber of the 20S cylinder for digestion to short peptides. (B, from W. Baumeister et al., *Cell* 92:367–380, 1998. With permission from Elsevier.)

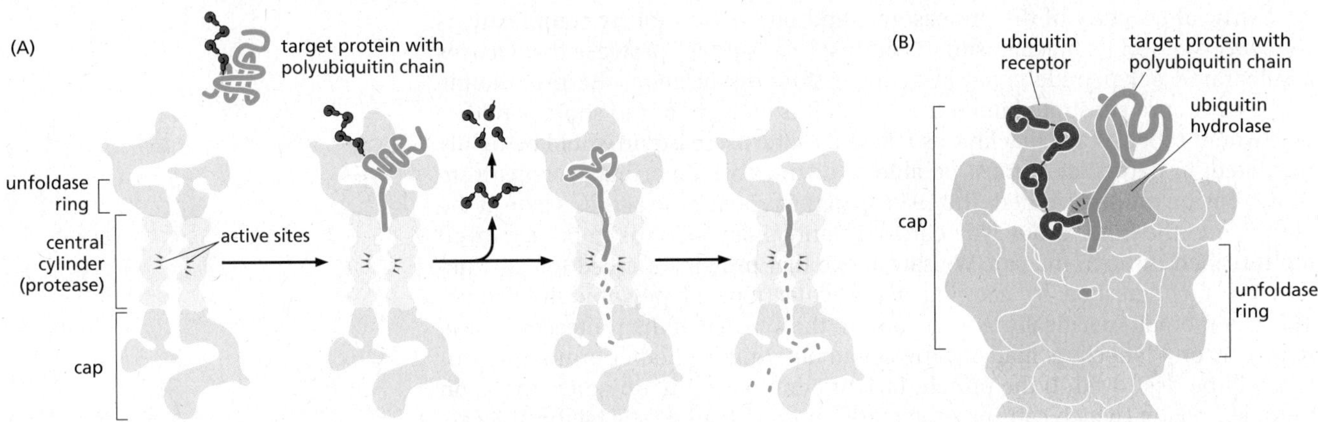

Figure 6–87 Processive protein digestion by the proteasome. (A) The proteasome cap recognizes proteins marked by a polyubiquitin chain (see Figure 3–65B). Most of these proteins also contain an unfolded region, and the cap translocates these proteins into the proteasome core, where they are digested. Before moving through the proteasome cap, the ubiquitin is cleaved from the substrate protein and is recycled. Translocation into the core of the proteasome is mediated by a ring of ATPases that unfold the substrate protein as it is threaded through the ring and into the proteasome core. (B) Detailed structure of the proteasome cap. The cap includes a ubiquitin receptor, which holds a ubiquitylated protein in place while attempts are made to pull it into the proteasome core, and a ubiquitin hydrolase, which cleaves ubiquitin from the protein. (A, from S. Prakash and A. Matouschek, *Trends Biochem. Sci.* 29:593–600, 2004. With permission from Elsevier; B, adapted from G.C. Lander et al., *Nature* 482:186–191, 2012.)

proteins as they move through the cap, exposing them to the proteases lining the proteasome core (Figure 6–87). The proteins that make up the ring structure in the proteasome cap belong to a large class of protein "unfoldases" known as *AAA proteins.* Many of them function as hexamers, and they share mechanistic features with the ATP-dependent DNA helicases that unwind DNA (Figure 6–88; see also Figure 5–14).

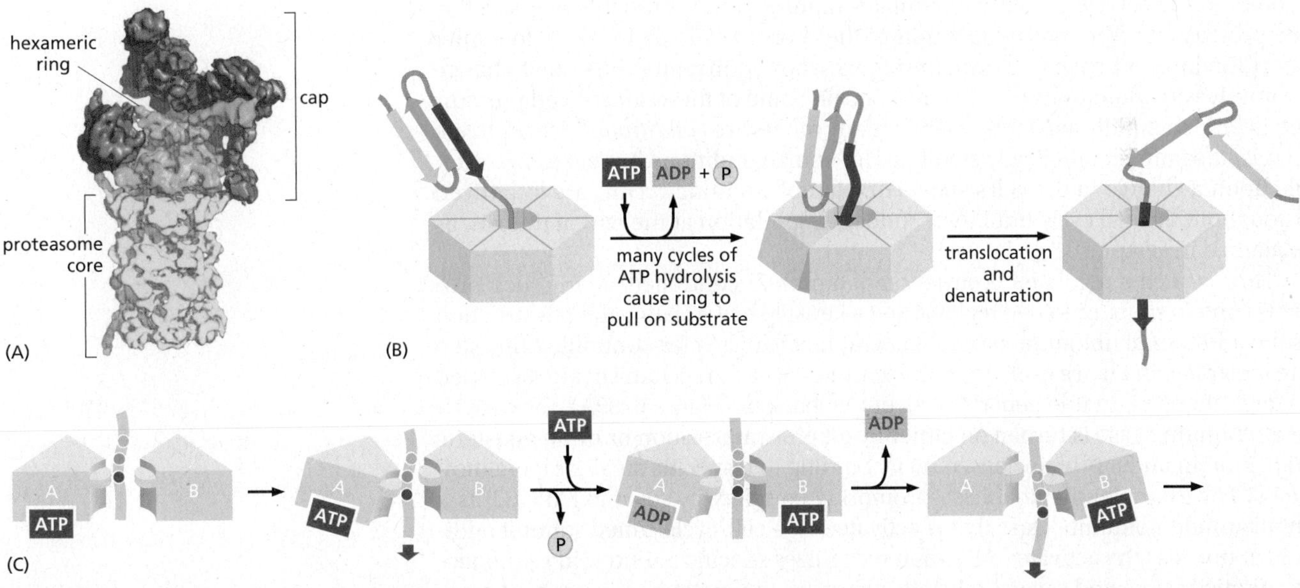

Figure 6–88 A hexameric protein unfoldase. (A) The proteasome cap includes a hexameric ring *(blue)* through which proteins to be destroyed are threaded. The hexameric ring (also called the unfoldase ring) is formed from six subunits, each belonging to the AAA family of proteins. (B) Model for the ATP-dependent unfoldase activity of AAA proteins. The ATP-bound form of a hexameric ring of AAA proteins grasps a substrate protein, and a conformational change, driven by ATP hydrolysis, further pulls the substrate and strains the ring structure. At this point, the substrate protein, which is being tugged upon, can partially unfold and enter further into the pore or it can maintain its structure and partially withdraw. Some protein substrates may require hundreds of cycles of ATP hydrolysis and dissociation before they are successfully pulled through the AAA protein ring; some proteins continue to resist these efforts and are ultimately de-ubiquitylated and released. (C) How the successive tilting of adjacent subunits (only two of which are shown), driven by ATP hydrolysis, is thought to pull the unfolded polypeptide chain through the hexameric ring into the proteasome. The motions of the ring subunits resemble a hand-over-hand pulling motion on the substrate protein, with the hands often slipping until the substrate is unfolded. Once unfolding occurs, the substrate protein moves relatively quickly through the ring by successive rounds of ATP hydrolysis. (A, adapted from G.C. Lander et al., *Nature* 482:186–191, published 2012 by Macmillan Publishers Ltd. Reproduced with permission of SNCSC; B, adapted from R.T. Sauer et al., *Cell* 119:9–18, 2004.)

A crucial property of the proteasome, and one reason for its complexity, is the *processivity* of its mechanism: in contrast to a "simple" protease that cleaves a substrate's polypeptide chain just once before dissociating, the proteasome keeps the entire substrate bound until all of it is converted into short peptides. One would expect that a machine as efficient as the proteasome would be tightly regulated; in particular, it must be able to distinguish abnormal proteins from those that are properly folded. The 19S cap of the proteasome acts as a gate at the entrance to the inner proteolytic core, and only those proteins to be destroyed are threaded through the cap. We saw above that proteins marked for destruction are distinguished by carrying a particular type of polyubiquitin chain. This chain binds specifically to a receptor in the 19S cap of the proteasome, and here a second "check" is made by the cell to determine whether or not the protein will be destroyed. If the protein has, in addition to the ubiquitin mark, an unfolded region (which can span the mark), it is grasped tightly by the 19S cap, de-ubiquitylated, and pulled through the cap into the proteasome core. Ubiquitylated proteins that lack such a region are typically de-ubiquitylated and released back into solution.

As might be expected, there is competition for misfolded proteins between chaperones and the protein degradation machinery. Proteins that are folded quickly escape destruction (at least early in their life before they accumulate damage), whereas those that undergo many rounds of chaperone-assisted folding are more likely to be degraded. Some chaperones can directly hand off those proteins that remain improperly folded, after many attempts, to the protein destruction machinery. It is estimated that between 1 and 5% of all newly synthesized proteins fail to fold properly and are degraded by the proteasome.

Many Proteins Are Controlled by Regulated Destruction

One function of intracellular proteolytic mechanisms is to recognize and eliminate misfolded or otherwise abnormal proteins, as just described. Indeed, every protein in the cell eventually accumulates damage and is probably degraded by the proteasome. Yet another function of these proteolytic pathways is to confer short lifetimes on specific normal proteins whose concentrations must change promptly with alterations in the state of a cell. Some of these short-lived proteins are degraded rapidly at all times, while many others are *conditionally* short-lived; that is, they are metabolically stable under some conditions but become unstable upon a change in the cell's state. For example, mitotic cyclins are long-lived throughout the cell cycle until their sudden degradation at the end of mitosis, as explained in Chapter 17.

How is such a regulated destruction of a protein controlled? Many such proteins contain short, unfolded regions, and a key step in regulating their destruction is the addition of ubiquitin. Several general mechanisms for controlling this step are illustrated in **Figure 6–89**. Specific examples of each mechanism are discussed in later chapters. In one general class of mechanism (Figure 6–89A), the activity of a ubiquitin ligase is turned on either by E3 phosphorylation or by an allosteric transition in an E3 protein caused by its binding to a specific small or large molecule. For example, the anaphase-promoting complex/cyclosome (APC/C) is a multisubunit ubiquitin ligase that is activated by a cell-cycle-timed subunit addition at mitosis. The activated APC then recognizes specific amino acid sequences in mitotic cyclins and several other regulators of the metaphase–anaphase transition and ubiquitylates them, thereby sending them to the proteasome (see Figure 17–18). We saw another example of regulated destruction earlier in this chapter, where a nascent polypeptide extending from a stalled ribosome is ubiquitylated, targeting it for destruction. Here, the signal for bringing the ubiquitin ligase into proximity to its target is a "gummed up" large ribosome subunit (see Figure 6–81).

Alternatively, in response either to intracellular signals or to signals from the environment, a ubiquitylation site can be created in a protein (Figure 6–89B). One common way to create such a signal is by phosphorylation of a specific amino

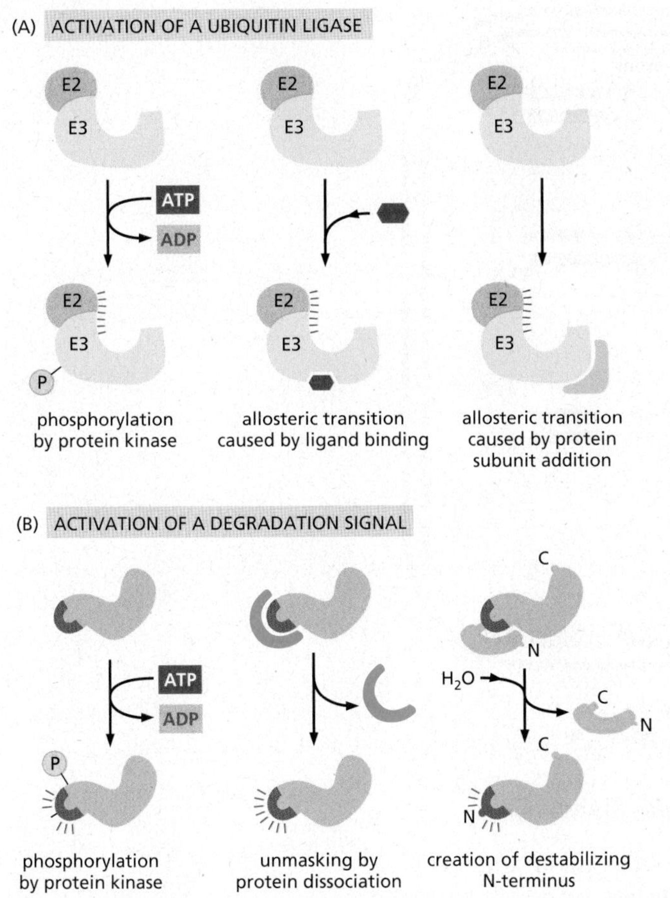

Figure 6–89 Two general ways of inducing the destruction of a specific protein. (A) Activation of a specific E3 molecule creates a new ubiquitin ligase. Eukaryotic cells have many different E3 molecules, each activated by a different signal. (B) Creation of an exposed destruction signal in the protein to be degraded. This signal binds a ubiquitin ligase, causing the addition of a polyubiquitin chain to a nearby lysine on the target protein. All six pathways shown are known to be used by cells to induce the movement of selected proteins into the proteasome for destruction.

acid sequence, which completes a ubiquitin ligase recognition site on the protein. Another is preexisting degradation signals that are unmasked by the regulated dissociation of a protein subunit. Finally, powerful degradation signals can be created by cleaving a single peptide bond, provided that this cleavage creates a new N-terminus that is recognized by a specific E3 protein as a "destabilizing" N-terminal residue. This E3 protein recognizes only certain amino acids at the N-terminus of a protein; thus, not all protein-cleavage events will lead to degradation of the C-terminal fragment produced.

In humans, nearly 70% of cytosolic proteins are acetylated on their N-terminal residue, and we now know that this modification is recognized by a specific E3 enzyme, which directs the ubiquitylation of the protein and sends it to the proteasome for degradation. Thus, the majority of human proteins carry their own signals for destruction. It has been proposed that when a protein is properly folded (and, before that, when it is in contact with a chaperone), this acetylated N-terminus is buried and therefore inaccessible to the E3. According to this idea, as a protein ages and becomes damaged (or if it fails to fold correctly from the start), this destruction signal becomes exposed, and the protein is destroyed.

There Are Many Steps from DNA to Protein

We have seen in this chapter that many different types of chemical reactions are required to produce a properly folded protein from the information contained in a gene (**Figure 6–90**). The final level of a useful protein in a cell therefore depends on the efficiency with which each of the many steps is performed. We also now know that the cell devotes enormous resources to selectively degrading proteins, particularly those that fail to fold properly or accumulate damage as they age. It is

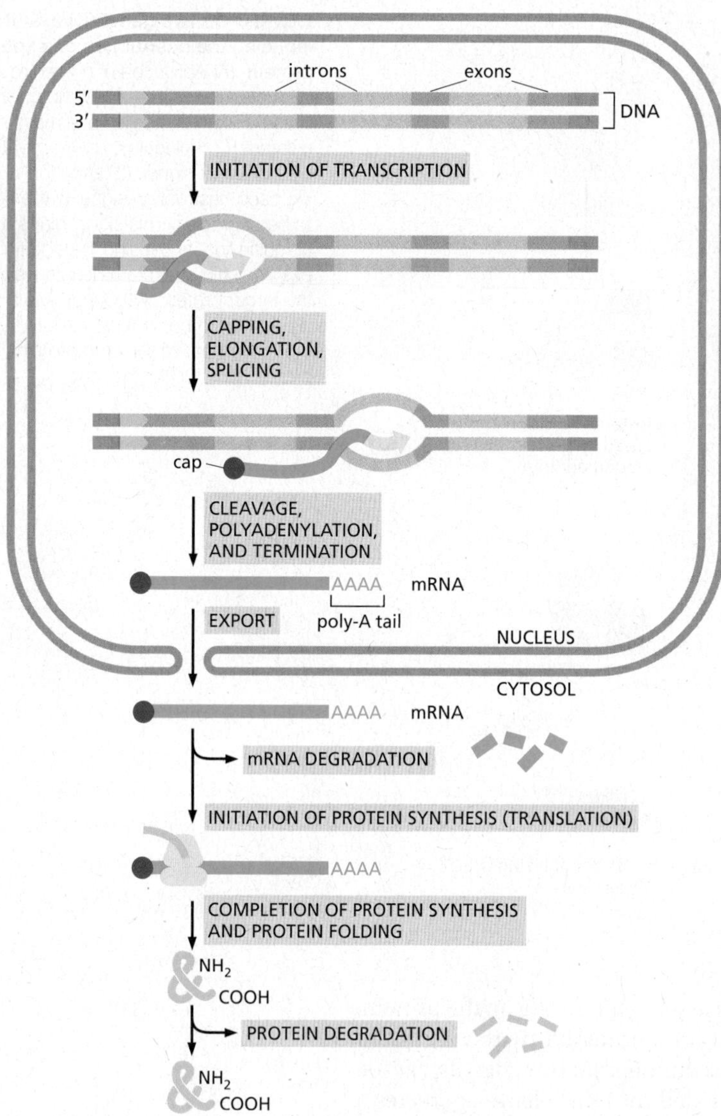

Figure 6–90 **The production of a protein by a eukaryotic cell: a summary.** The final level of each protein in a eukaryotic cell depends on the efficiency of each step depicted.

the balance between the rates of synthesis and degradation that determines the final amount of every protein in the cell.

In the following chapter, we shall see that cells have the ability to change the levels of their proteins according to their needs. In principle, any or all of the steps in Figure 6–90 could be regulated for each individual protein. As we shall see, there are examples of regulation at each step from gene to protein.

Summary

The translation of the nucleotide sequence of an mRNA molecule into protein takes place in the cytosol on a large ribonucleoprotein assembly called a ribosome. Each amino acid used for protein synthesis is first attached to a tRNA molecule that recognizes, by complementary base-pair interactions, a particular set of three nucleotides (codons) in the mRNA. As an mRNA is threaded through a ribosome, its sequence of nucleotides is then read from one end to the other in sets of three according to the genetic code.

To initiate translation, a small ribosomal subunit binds to the mRNA molecule at a start codon (AUG) that is recognized by a unique initiator tRNA molecule.

A large ribosomal subunit then binds to complete the ribosome and begin protein synthesis. During this phase, aminoacyl-tRNAs—each bearing a specific amino acid—bind sequentially to the appropriate codons in mRNA through complementary base-pairing between tRNA anticodons and mRNA codons. Each amino acid is added to the C-terminal end of the growing polypeptide in four sequential steps: aminoacyl-tRNA binding, followed by peptide bond formation, followed by two ribosome translocation steps. Elongation factors use GTP hydrolysis both to drive these reactions forward and to improve the accuracy of amino acid selection. The mRNA molecule progresses codon by codon through the ribosome in the 5′-to-3′ direction until it reaches one of three stop codons. A release factor then binds to the ribosome, terminating translation and releasing the completed polypeptide.

Eukaryotic and bacterial ribosomes are closely related, despite differences in the number and size of their rRNA and protein components. The rRNA has the dominant role in translation, determining the overall structure of the ribosome, forming the binding sites for the tRNAs, matching the tRNAs to codons in the mRNA, and creating the active site of the peptidyl transferase enzyme that links amino acids together during translation.

Several distinct types of molecular chaperones, including hsp60 and hsp70, illustrate how the energy of ATP hydrolysis is used to help newly synthesized proteins assume their correct three-dimensional conformations. This protein-folding process competes with a control mechanism that destroys proteins that are abnormally folded by recognizing exposed hydrophobic patches and other unstructured regions. In this case, ubiquitin is covalently added to a misfolded protein by a ubiquitin ligase, and the resulting polyubiquitin chain is recognized by the cap on a proteasome that unfolds the protein as it is threaded into the interior of the proteasome for proteolytic degradation. A closely related proteolytic mechanism, based on special degradation signals recognized by ubiquitin ligases, is used to determine the lifetimes of many normally folded proteins, as well as to remove selected proteins from the cell in response to specific signals.

THE RNA WORLD AND THE ORIGINS OF LIFE

We have seen that the expression of hereditary information requires extraordinarily complex machinery and proceeds from DNA to protein through an RNA intermediate. This machinery presents a central paradox: if nucleic acids are required to synthesize proteins and proteins are required, in turn, to synthesize nucleic acids, how did such a system of interdependent components ever arise? A widely held view is that an **RNA world** existed on Earth before modern cells arose (Figure 6–91). According to this hypothesis, RNA not only stored genetic information but also directly catalyzed the chemical reactions in primitive cells—thus serving as enzymes ("ribozymes"). Only later in evolutionary time did DNA take over as the genetic material and proteins become the major catalysts and structural components of cells. But the transition out of the RNA world was never complete; as we have seen in this chapter, RNA still catalyzes several fundamental reactions in modern-day cells, which can be viewed as molecular holdovers from an earlier world.

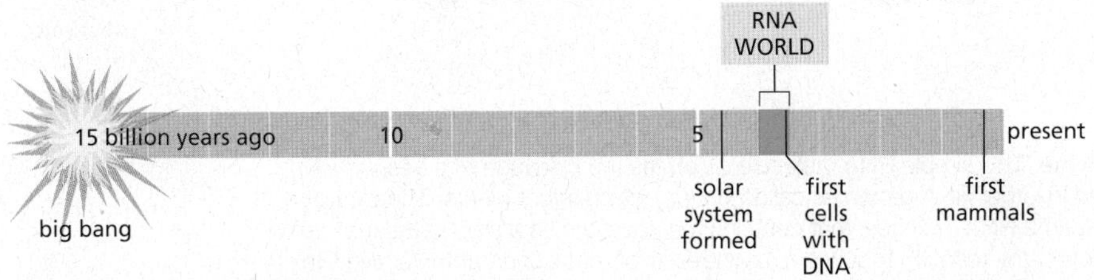

Figure 6–91 Timeline for the universe, highlighting the possible early existence of an "RNA world" in the evolution of Earth's living systems.

The RNA world hypothesis relies on the fact that, among present-day biological molecules, RNA is unique in being able to act as both a carrier of genetic information and as a catalyst for chemical reactions. In this section, we discuss these properties of RNA and how they may have been especially important in early cells.

Single-Strand RNA Molecules Can Fold into Highly Elaborate Structures

We have seen in this chapter that RNA can carry genetic information in mRNAs, and we saw in Chapter 5 that the genomes of some viruses are composed solely of RNA. We have also seen that complementary base-pairing and other types of hydrogen bonds can occur between nucleotides in the same chain of RNA, causing an RNA molecule to fold up in a unique way that is determined by its nucleotide sequence (see, for example, Figures 6–54 and 6–71). Comparisons of many RNA structures have revealed conserved structural motifs, short elements that often appear as parts of larger structures (**Figure 6–92**).

Protein catalysts require a surface with unique contours and chemical properties on which a given set of substrates can react (discussed in Chapter 3). In exactly the same way, an RNA molecule with an appropriately folded shape can serve as a catalyst (**Figure 6–93**). Like some proteins, many of these ribozymes work by positioning metal ions at their active sites. This feature gives them a wider range of catalytic activities than can be provided by the limited chemical groups of a polynucleotide chain.

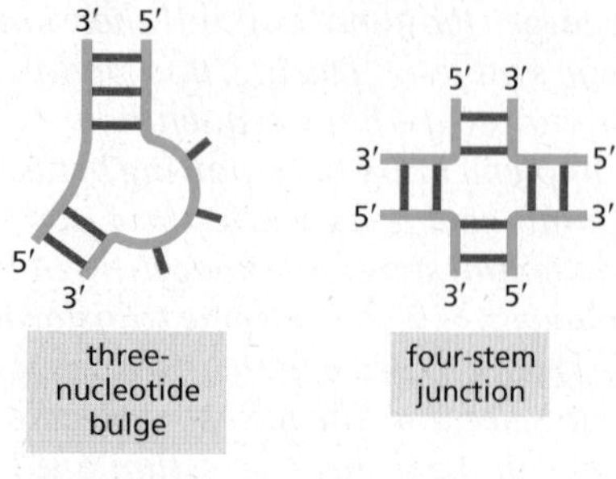

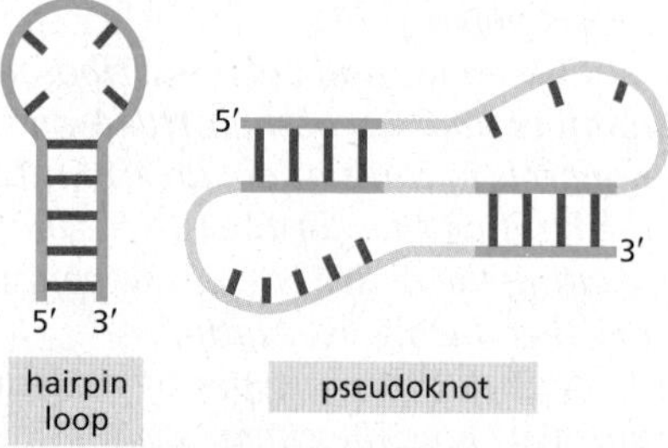

Figure 6–92 Some common elements of RNA structure. Conventional, complementary base-pairing interactions are indicated by *red* "rungs" in double-helical portions of the RNA.

Ribozymes Can Be Produced in the Laboratory

Much of our inference about the RNA world has come from experiments in which large pools of RNA molecules of random nucleotide sequences are generated in the laboratory. Those rare RNA molecules with a property specified by the experimenter are then selected out and studied (**Figure 6–94**). Experiments of this type have created RNAs that can catalyze a wide variety of biochemical reactions (**Table 6–5**), with reaction rate enhancements only a few orders of magnitude lower than those of the "fastest" protein enzymes. Given these findings, it is not clear why protein catalysts greatly outnumber ribozymes in modern cells. Experiments have shown, however, that RNA molecules may have more difficulty than proteins in binding to flexible, hydrophobic substrates and in forming pockets specific for different small molecules. The availability of 20 types of amino acids

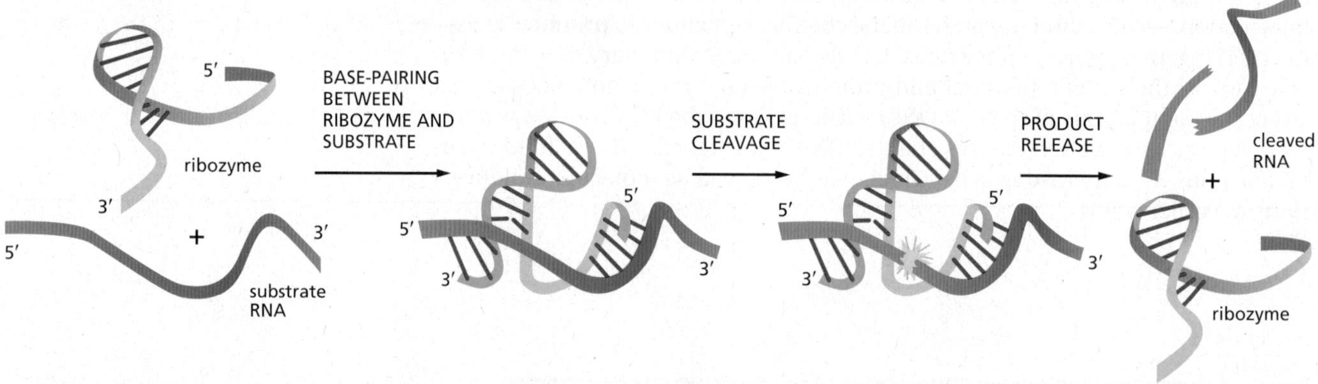

Figure 6–93 The activity of a well-studied ribozyme. This simple RNA molecule catalyzes the cleavage of a second RNA at a specific site. This ribozyme is found embedded in larger RNA genomes called viroids, which infect plants. The cleavage, which occurs in nature at a distant location on the same RNA molecule that contains the ribozyme, is a step in the replication of the viroid genome. Although not shown in the figure, the reaction requires a magnesium ion positioned at the active site. (Adapted from T.R. Cech and O.C. Uhlenbeck, *Nature* 372:39–40, 1994.)

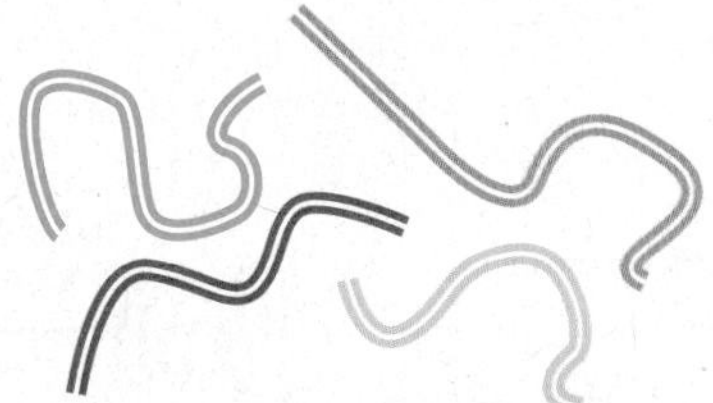

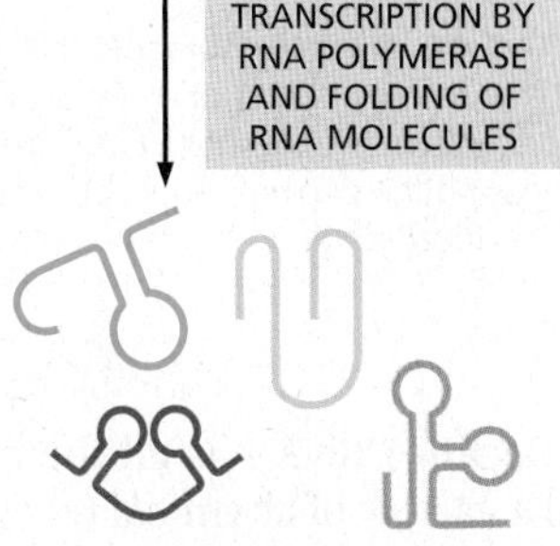

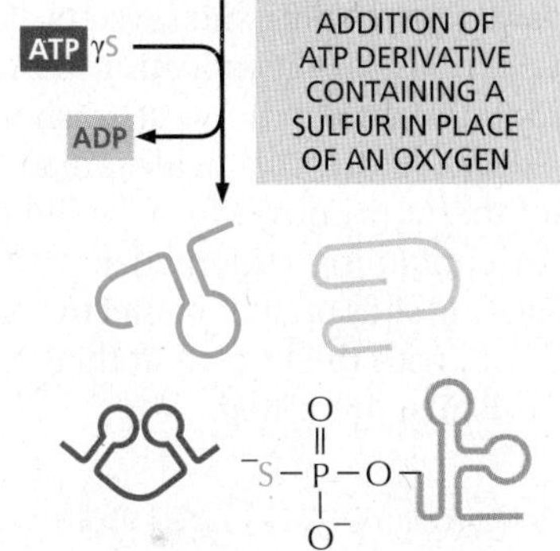

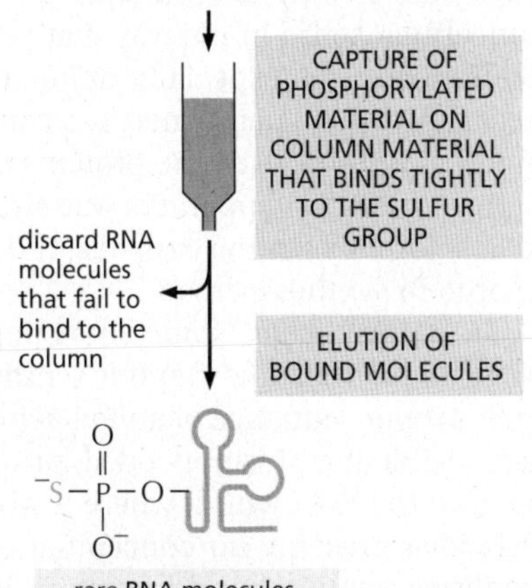

Figure 6–94 *In vitro* selection of a synthetic ribozyme. Beginning with a large pool of different nucleic acid molecules synthesized in the laboratory, those rare RNA molecules that possess a specified catalytic activity can be isolated and studied. Although only one specific example (that of an autophosphorylating ribozyme) is shown, variations of this procedure have been used to generate many of the ribozymes listed in Table 6–5. For the strategy shown here, the RNA molecules are kept sufficiently dilute during the phosphorylation step to prevent the "cross-phosphorylation" of other RNA molecules. In reality, several repetitions of this procedure are necessary to select the very rare RNA molecules with this catalytic activity. Thus, the material initially eluted from the column is converted back into DNA, amplified many-fold (using reverse transcriptase and PCR, as explained in Chapter 8), transcribed back into RNA, and subjected to repeated rounds of selection. (Adapted from J.R. Lorsch and J.W. Szostak, *Nature* 371:31–36, 1994.)

presumably provides proteins with a much greater number of binding and catalytic strategies.

RNA Can Both Store Information and Catalyze Chemical Reactions

RNA molecules have one property that contrasts with those of proteins: they can directly guide the formation of copies of their own sequence. This capacity depends on complementary base-pairing of their nucleotide subunits, which enables one RNA to act as a template for the formation of another. As we have seen in this and the preceding chapter, these complementary templating mechanisms lie at the heart of DNA replication and transcription in modern-day cells. But the efficient synthesis of RNA by such complementary templating mechanisms requires catalysts to promote the polymerization reaction: without catalysts, polymer formation would be slow, error-prone, and inefficient.

TABLE 6–5 Some Biochemical Reactions That Can Be Catalyzed by Ribozymes

Activity	Ribozymes
Peptide bond formation in protein synthesis	Ribosomal RNA
RNA cleavage, RNA ligation	Self-splicing RNAs; RNase P; also *in vitro* selected RNA
DNA cleavage	Self-splicing RNAs
RNA splicing	Self-splicing RNAs; RNAs of the spliceosome
RNA polymerization	*In vitro* selected RNA
RNA and DNA phosphorylation	*In vitro* selected RNA
RNA aminoacylation	*In vitro* selected RNA
RNA alkylation	*In vitro* selected RNA
Amide bond formation	*In vitro* selected RNA
Glycosidic bond formation	*In vitro* selected RNA
Oxidation–reduction reactions	*In vitro* selected RNA
Carbon–carbon bond formation	*In vitro* selected RNA
Phosphoamide bond formation	*In vitro* selected RNA
Disulfide exchange	*In vitro* selected RNA

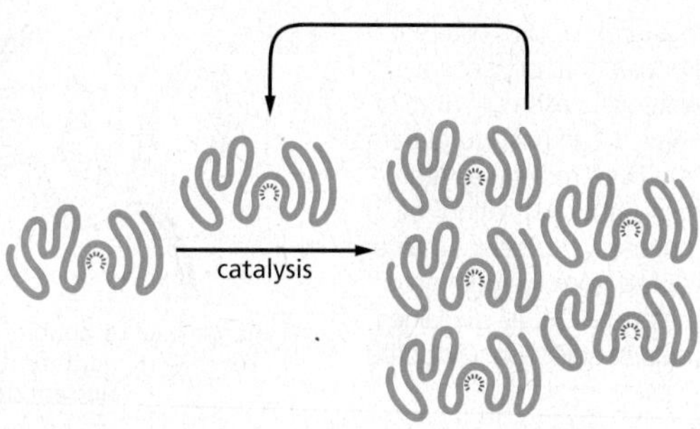

Figure 6–95 **An RNA molecule that can catalyze its own synthesis.** This hypothetical process would require synthesis of a second RNA strand complementary to the original strand (not shown) and the use of this second RNA molecule as a template to synthesize many molecules of RNA possessing the original sequence. The *red* rays represent the active site of this hypothetical RNA enzyme (ribozyme).

Because RNA has all the properties required of a molecule that could catalyze a variety of chemical reactions, including those that lead to its own synthesis (**Figure 6–95**), it has been proposed that RNAs served long ago as the catalysts for their own template-dependent RNA synthesis. Although self-replicating systems of RNA molecules have not been found in nature, scientists have made progress toward constructing such systems in the laboratory. Experiments of this type cannot prove that self-replicating RNA molecules were central to the origin of life on Earth, but they can help establish whether such a scenario is plausible.

Today, there is a widespread interest in investigating the exciting possibility that a primitive form of life once existed, or may even still exist, in some water-containing regions below the surface of Mars. Vehicles are being sent to promising sites on that planet to collect subterranean samples for eventual return to Earth, with the hope that their analysis will allow scientists to refine scenarios like that just described.

How Did Protein Synthesis Evolve?

The molecular processes underlying protein synthesis in present-day cells seem inextricably complex. Although we understand most of them, they do not make conceptual sense in the way that DNA transcription, DNA repair, and DNA replication do. It is especially difficult to imagine how protein synthesis evolved because it is now performed by a complex interlocking system of protein and RNA molecules; obviously, the proteins could not have existed until an early version of the translation apparatus was already in place. As attractive as the RNA world idea is for envisioning early life, it does not explain how the modern-day system of protein synthesis arose.

In modern cells, some short peptides (such as antibiotics) are synthesized without the ribosome; peptide synthetase enzymes assemble these peptides, with their proper sequence of amino acids, without mRNAs to guide their synthesis. It is plausible that this noncoded, primitive version of protein synthesis first developed in the RNA world, where it would have been catalyzed by RNA molecules. This idea presents no conceptual difficulties because, as we have seen, rRNA catalyzes peptide bond formation in present-day cells. Moreover, short simple peptides (for example, polylysine) have been shown to enhance the function of ribozymes created in the laboratory, raising the possibility that the first peptides were "selected" for their ability to help RNA molecules fold, assemble with each other, and catalyze reactions. These ideas, however, leave unexplained how the genetic code—which lies at the core of protein synthesis in today's cells—might have arisen. We know that ribozymes created in the laboratory can perform specific aminoacylation reactions; that is, they can match specific amino acids to specific tRNAs. It is therefore possible that tRNA-like adaptors, each matched to a specific amino acid, could have arisen in the RNA world, marking the beginnings

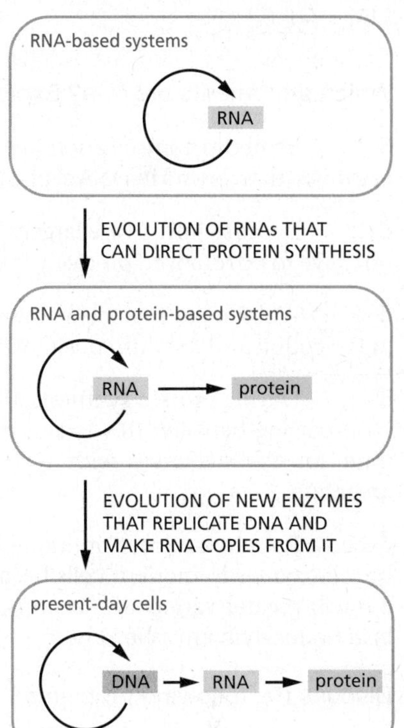

Figure 6–96 The hypothesis that RNA preceded DNA and proteins in evolution. In the earliest cells, RNA molecules (or their close analogs) would have had combined genetic, structural, and catalytic functions. In present-day cells, DNA is the repository of genetic information, and proteins perform the vast majority of catalytic functions in cells. RNA primarily functions today as a go-between in protein synthesis, although it remains a catalyst for a small number of crucial reactions.

of a genetic code. And, although we can only speculate on the code's evolution, scientists have provided plausible scenarios, in which the first codons specified only 10 or so of today's 20 amino acids.

Once coded protein synthesis evolved, the transition to a protein-dominated world could proceed, with proteins eventually taking over the majority of catalytic and structural tasks because of their greater versatility, with 20 rather than 4 different subunits. Although these ideas are highly speculative, they are consistent with the known properties of RNA and protein molecules.

All Present-Day Cells Use DNA as Their Hereditary Material

If the evolutionary speculations embodied in the RNA world hypothesis are correct, early cells would have differed fundamentally from the cells we know today in having their hereditary information stored in RNA rather than in DNA (**Figure 6–96**).

Evidence that RNA arose before DNA in evolution can be found in the chemical differences between them. Ribose, like glucose and other simple carbohydrates, can be formed from formaldehyde (HCHO), a simple chemical that is readily produced in laboratory experiments that attempt to simulate conditions on the primitive Earth. The sugar deoxyribose is harder to make, and in present-day cells it is produced from ribose in a reaction catalyzed by a protein enzyme, suggesting that ribose predates deoxyribose in cells. Presumably, DNA appeared on the scene later, but then proved more suitable than RNA as a permanent repository of genetic information. In particular, the deoxyribose in its sugar-phosphate backbone makes chains of DNA chemically more stable than chains of RNA, so that much greater lengths of DNA can be maintained without breakage. Because DNA and RNA use similar base-pairing rules for their template-dependent synthesis, the transition between the two is not difficult to envision. Consistent with this idea, a ribozyme has been created in the laboratory that can synthesize both RNA using a DNA template and DNA using an RNA template.

The other differences between RNA and DNA—the double-helical structure of DNA and the use of thymine rather than uracil—further enhance DNA stability by making the many unavoidable accidents that occur to the molecule much easier to repair, as discussed in detail in Chapter 5 (p. 288).

Summary

From our knowledge of present-day organisms and the molecules they contain, it seems likely that the development of the distinctive autocatalytic mechanisms fundamental to living systems began with the evolution of families of RNA molecules that could catalyze their own replication. DNA is thought to have been a later addition: as the accumulation of protein catalysts allowed more efficient and complex cells to evolve, the DNA double helix replaced RNA as a much more stable molecule for storing the increased amounts of genetic information required by such cells.

PROBLEMS

Which statements are true? Explain why or why not.

6–1 Errors in transcription are less dangerous to an organism than errors in DNA replication.

6–2 Because introns are largely genetic "junk," they do not have to be removed precisely during RNA splicing.

6–3 Wobble pairing occurs between the first position in the codon and the third position in the anticodon.

6–4 During protein synthesis, the thermodynamics of base-pairing between tRNAs and mRNAs sets the upper limit for the accuracy with which protein molecules are made.

6–5 Protein enzymes are thought to greatly outnumber ribozymes in modern cells because they can catalyze a much greater variety of reactions, and all of them have faster rates than any ribozyme.

Discuss the following problems.

6–6 You have attached an RNA polymerase molecule to a glass slide and have allowed it to initiate transcription on a template DNA that has both strands tethered to a magnetic bead as shown in Figure Q6–1. If the DNA with its attached magnetic bead moves relative to the RNA polymerase as indicated by the arrows in the figure, in which direction will the bead rotate?

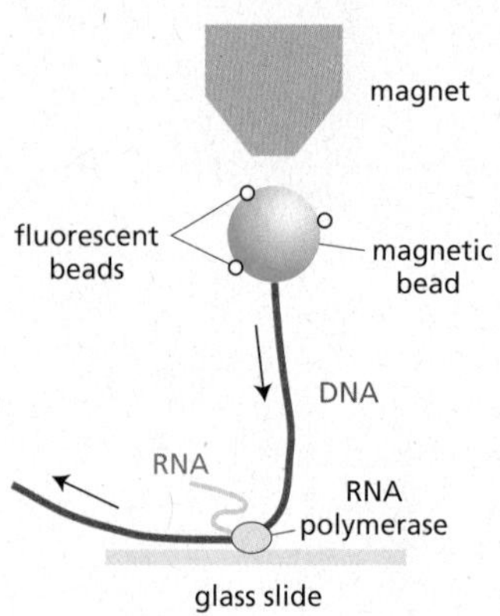

Figure Q6–1 System for measuring the rotation of DNA caused by RNA polymerase (Problem 6–6). The magnet holds the bead upright (but doesn't interfere with its rotation), and the attached tiny fluorescent beads allow the direction of motion to be visualized under the microscope. RNA polymerase is held in place by attachment to the glass slide. (Adapted from Y. Harada et al., *Nature* 409: 113–115, 2001. With permission from Springer Nature.)

6–7 In which direction along the template must the RNA polymerase in Figure Q6–2 be moving to have generated the supercoiled structures that are shown? Would you expect supercoils to be generated if the RNA polymerase were free to rotate about the axis of the DNA as it progressed along the template?

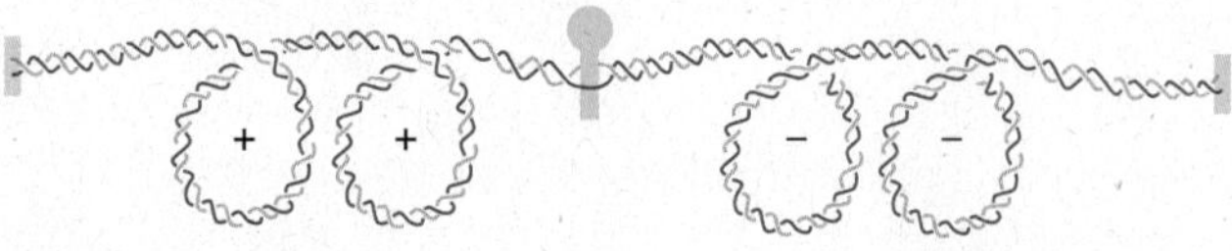

Figure Q6–2 Supercoils produced by a moving RNA polymerase (Problem 6–7).

6–8 The human α-tropomyosin gene is alternatively spliced to produce different forms of α-tropomyosin mRNA in different cell types (Figure Q6–3). All forms of the mRNA contain the protein sequences encoded by exon 1 and exon 10. Exons 2, 3, 7, and 8 are alternative exons used in different mRNAs. Which one of the following statements about exons 2 and 3 is the most accurate? Is that statement also the most accurate one for exons 7 and 8? Explain your answers.

A. Exons 2 and 3 must have the same number of nucleotides.

B. Exons 2 and 3 must each contain an integral number of codons (that is, the number of nucleotides divided by 3 must be an integer).

C. Exons 2 and 3 must each contain a number of nucleotides that when divided by 3 leaves the same remainder (that is, 0, 1, or 2).

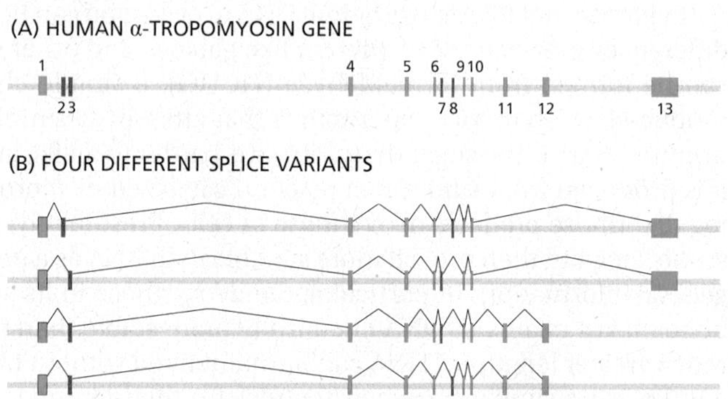

Figure Q6–3 Alternatively spliced mRNAs from the human α-tropomyosin gene (Problem 6–8). (A) Exons in the human α-tropomyosin gene. The locations and relative sizes of exons are shown by the *blue* and *red rectangles*, with alternative exons in *red.* (B) Splicing patterns for four α-tropomyosin mRNAs. Splicing is indicated by *lines* connecting the exons present in each mRNA.

6–9 After treating cells with a chemical mutagen, you isolate two mutants. One carries alanine and the other carries methionine at a site in a protein that normally contains valine (Figure Q6–4). After treating these two mutants again with the mutagen, you isolate mutants from each that now carry threonine at the site of the original valine (Figure Q6–4). Assuming that all mutations involve single-nucleotide changes, which codons are used for valine, methionine, threonine, and alanine at this site in the protein? Would you expect to be able to isolate valine-to-threonine mutants in one step?

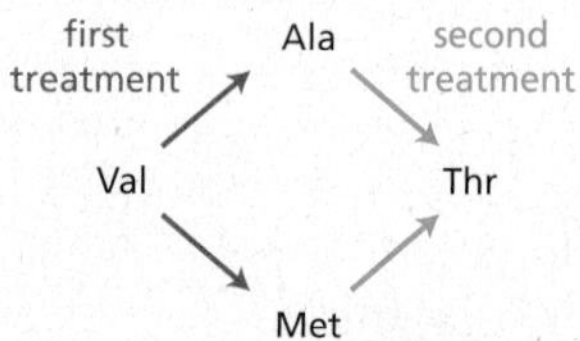

Figure Q6–4 Two rounds of mutagenesis and the altered amino acids at a single position in a protein (Problem 6–9).

6–10 Which one of the following mutational changes would you predict to be the most deleterious to gene function? Explain your answers.

1. Insertion of a single nucleotide near the end of the coding sequence.
2. Removal of a single nucleotide near the beginning of the coding sequence.
3. Deletion of three consecutive nucleotides in the middle of the coding sequence.
4. Substitution of one nucleotide for another in the middle of the coding sequence.

6–11 Prokaryotes and eukaryotes both protect against the dangers of translating broken mRNAs. What dangers do partial mRNAs pose for the cell?

6–12 Both hsp60-like and hsp70 molecular chaperones share an affinity for exposed hydrophobic patches on proteins, using them as indicators of incomplete folding. Why do you suppose hydrophobic patches serve as critical signals for the folding status of a protein?

6–13 Most proteins require molecular chaperones to assist in their correct folding. How do you suppose the chaperones themselves manage to fold correctly?

6–14 If an RNA molecule could form a hairpin with a symmetric internal loop, as shown in **Figure Q6–5**, could the complement of this RNA form a similar structure? If so, would there be any regions of the two structures that are identical? Which ones?

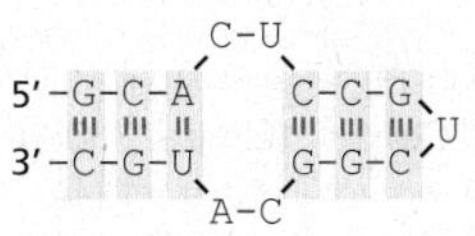

Figure Q6–5 An RNA hairpin with a symmetric internal loop (Problem 6–14).

6–15 What is so special about RNA that it is hypothesized to be an evolutionary precursor to DNA and protein? What is it about DNA that makes it a better material than RNA for storage of genetic information?

6–16 Imagine a warm pond on primordial Earth. Chance processes have just assembled a single copy of an RNA molecule with a catalytic site that can carry out RNA replication. This RNA molecule folds into a structure that is capable of linking nucleotides according to instructions in an RNA template. Given an adequate supply of nucleotides, will this single RNA molecule be able to use itself as a template to catalyze its own replication? Why or why not?

REFERENCES

General

Berg JM, Tymoczko JL, Gatto G & Stryer L (2019) Biochemistry, 9th ed. New York: Macmillan Learning.

Brown TA (2018) Genomes 4. New York: Garland Science.

Cold Spring Harbor Symposia on Quantitative Biology, Vol 31: The Genetic Code (1966). Cold Spring Harbor, NY: Cold Spring Harbor Laboratory Press.

Cold Spring Harbor Symposia on Quantitative Biology, Vol 66: The Ribosome (2001). Cold Spring Harbor, NY: Cold Spring Harbor Laboratory Press.

Cold Spring Harbor Symposia on Quantitative Biology, Vol 84: RNA Control and Regulation (2019). Cold Spring Harbor, NY: Cold Spring Harbor Laboratory Press.

Craig N, Green R, Greider C, Storz G, Wolberger C & Cohen-Fix O (2021) Molecular Biology: Principles of Genome Function, 2nd ed. Oxford: Oxford University Press.

Darnell J (2011) RNA: Life's Indispensable Molecule. Cold Spring Harbor, NY: Cold Spring Harbor Laboratory Press.

Goldberg ML, Fischer J, Hood L & Hartwell L (2021) Genetics: From Genes to Genomes, 7th ed. Boston: McGraw-Hill.

Judson HF (1996) The Eighth Day of Creation, 25th anniversary ed. Cold Spring Harbor, NY: Cold Spring Harbor Laboratory Press.

Lodish H, Berk A, Kaiser C et al. (2021) Molecular Cell Biology, 9th ed. New York: Macmillan Learning.

Stent GS (1971) Molecular Genetics: An Introductory Narrative. San Francisco: WH Freeman.

From DNA to RNA

Berget SM, Moore C & Sharp PA (1977) Spliced segments at the 5′ terminus of adenovirus 2 late mRNA. *Proc. Natl. Acad. Sci. USA* 74, 3171–3175.

Brenner S, Jacob F & Meselson M (1961) An unstable intermediate carrying information from genes to ribosomes for protein synthesis. *Nature* 190, 576–581.

Bresson S & Tollervey D (2018) Surveillance-ready transcription: nuclear RNA decay as a default fate. *Open Biol.* 8(3), 170270.

Chow LT, Gelinas RE, Broker TR & Roberts RJ (1977) An amazing sequence arrangement at the 5′ ends of adenovirus 2 messenger RNA. *Cell* 12, 1–8.

Cramer P (2019) Organization and regulation of gene transcription. *Nature* 573, 45–54.

Fica SM, Tuttle N, Novak T . . . Piccirilli JA (2013) RNA catalyses nuclear pre-mRNA splicing. *Nature* 503, 229–234.

Gamarra N & Narlikar GJ (2021) Collaboration through chromatin: motors of transcription and chromatin structure. *J. Mol. Biol.* 433, 166876.

Kumar A, Clerici M, Muckenfuss LM . . . Jinek M (2019) Mechanistic insights into mRNA 3′-end processing. *Curr. Opin. Struct. Biol.* 59, 143–150.

Neugebauer KM (2019) Nascent RNA and the coordination of splicing with transcription. *Cold Spring Harb Perspect Biol.* 11(8), a032227.

Proudfoot NJ (2016) Transcriptional termination in mammals: stopping the RNA polymerase II juggernaut. *Science* 352, aad9926.

Roberts JW (2019) Mechanisms of bacterial transmission termination. *J. Mol Biol.* 431(20), 4030–4039.

Roeder RG (2019) 50+ years of eukaryotic transcription: an expanding universe of factors and mechanisms. *Nat. Struct. Mol. Biol.* 26, 783–791.

Ruskin B, Krainer AR, Maniatis T & Green MR (1984) Excision of an intact intron as a novel lariat structure during pre-mRNA splicing *in vitro*. *Cell* 38, 317–331.

Sato H, Das S, Singer RH & Vera M (2020) Imaging of DNA and RNA in living eukaryotic cells to reveal spatiotemporal dynamics of gene expression. *Annu. Rev. Biochem.* 89, 159–187.

Staley JP & Woolford JL (2009) Assembly of ribosomes and spliceosomes: complex ribonucleoprotein machines. *Curr. Opin. Cell Biol.* 21, 109–118.

Stewart M (2019) Polyadenylation and nuclear export of mRNAs. *J. Biol. Chem.* 294, 2977–2987.

Strom AR & Brangwynne CP (2019) The liquid nucleome—phase transitions in the nucleus at a glance. *J. Cell Sci.* 132, jcs235093.

Wan R, Bai R & Shi Y (2019) Molecular choreography of pre-mRNA splicing by the spliceosome. *Curr. Opin. Struct. Biol.* 59, 124–133.

Wilkinson ME, Charenton C & Nagai K (2020) RNA splicing by the spliceosome. *Annu. Rev. Biochem.* 89, 1.1–1.30.

Zinder JC & Lima CD (2017) Targeting RNA for processing or destruction by the eukaryotic RNA exosome and its cofactors. *Genes Dev.* 31, 88–100.

From RNA to Protein

Anfinsen CB (1973) Principles that govern the folding of protein chains. *Science* 181, 223–230.

Bard JAM, Goodall EA, Greene ER . . . Martin A (2018) Structure and function of the 26S proteasome. *Annu. Rev. Biochem.* 87, 697–724.

Cohue P, Hurt E & Panse VG (2017) Eukaryotic ribosome assembly, transport and quality control. *Nat. Struct. Mol. Biol.* 24, 689–699.

Crick FHC (1966) The genetic code: III. *Sci. Am.* 215, 55–62.

Finley D, Chen X & Walters KJ (2016) Gates, channels, and switches: elements of the proteasome machine. *Trends Biochem. Sci.* 41, 77–93.

Gomez MAR & Ibba M (2020) Aminoacyl-tRNA synthetases. *RNA* 26, 910–936.

Hershko A, Ciechanover A & Varshavsky A (2000) The ubiquitin system. *Nat. Med.* 6, 1073–1081.

Klinge S & Woolford JL (2019) Ribosome assembly coming into focus. *Nat. Rev. Mol. Cell Biol.* 20, 116–131.

Kramer G, Shiber A & Bukau B (2019) Mechanisms of cotranslational maturation of newly synthesized proteins. *Annu. Rev. Biochem.* 88, 337–364.

Kurosaki T, Popp MW & Maquat LE (2019) Quality and quantity control of gene expression by nonsense-mediated mRNA decay. *Nat. Rev. Mol. Cell Biol.* 20, 406–420.

Lin J, Zhou D, Steitz TA . . . Gagnon MG (2018) Ribosome-targeting antibiotics: modes of action, mechanisms of resistance, and implications for drug design. *Annu. Rev. Biochem*. 87, 451–478.

Noller HF (2017) The parable of the caveman and the Ferrari: protein synthesis and the RNA world. *Philos. Trans. R. Soc. Lond. B Biol. Sci.* 372(1716), 20160187.

Schuller AP & Green R (2018) Roadblocks and resolutions in eukaryotic translation. *Nat. Rev. Mol. Cell Biol*. 19, 526–541.

Simonovic M & Steitz TA (2009) A structural view on the mechanism of the ribosome-catalyzed peptide bond formation. *Biochim. Biophys. Acta* 1789(9–10), 612–623.

Voorhees RM & Ramakrishnan V (2013) Structural basis of the translational elongation cycle. *Annu. Rev. Biochem.* 82, 203–236.

The RNA World and the Origins of Life

Cech TR (2009) Crawling out of the RNA world. *Cell* 136, 599–602.

Joyce GF & Szostak JW (2018) Protocells and RNA self-replication. *Cold Spring Harb. Perspect. Biol*. 10, a034801.

Kruger K, Grabowski P, Zaug P . . . Cech TR (1982) Self-splicing RNA: autoexcision and autocyclization of the ribosomal RNA intervening sequence of *Tetrahymena*. *Cell* 31, 147–157.

Orgel L (2000) Origin of life. A simpler nucleic acid. *Science* 290, 1306–1307.

Tjhung KF, Shokhirev MN, Horning DP & Joyce GF (2020) An RNA polymerase ribozyme that synthesizes its own ancestor. *Proc. Natl. Acad. Sci. USA* 117, 2906–2913.

CHAPTER 7

Control of Gene Expression

IN THIS CHAPTER

An organism's DNA encodes all of the RNA and protein molecules required to construct its cells. Yet a complete description of the DNA sequence of an organism—be it the few million nucleotides of a bacterium or the few billion nucleotides of a human—no more enables us to reconstruct the organism than a list of English words enables us to reconstruct a play by Shakespeare. In both cases, the problem is to know how the elements in the DNA sequence or the words on the list are used. Under what conditions is each gene product made, and, once made, what does it do?

In this chapter, we focus on the first half of this problem—the rules and mechanisms that enable a subset of genes to be selectively expressed in each cell and also determine the amount of each gene product. These mechanisms operate at many levels, and we shall discuss each level in turn. But first we present some of the basic principles involved.

AN OVERVIEW OF GENE CONTROL

The different cell types in a multicellular organism differ dramatically in both structure and function. If we compare a mammalian neuron with a liver cell, for example, the differences are so extreme that it is difficult to imagine that the two cells contain the same genome (**Figure 7–1**). For this reason, and because cell differentiation often seemed irreversible, biologists originally suspected that genes might be selectively lost when a cell differentiates. We now know, however, that cell differentiation generally occurs without changes in the nucleotide sequence of a cell's genome.

The Different Cell Types of a Multicellular Organism Contain the Same DNA

The cell types in a multicellular organism become different from one another because they synthesize and accumulate different sets of RNA and protein molecules. The initial evidence that they do this without altering the sequence of their DNA came from a classic set of experiments in frogs. When the nucleus of a fully differentiated frog cell is injected into a frog egg whose nucleus has been

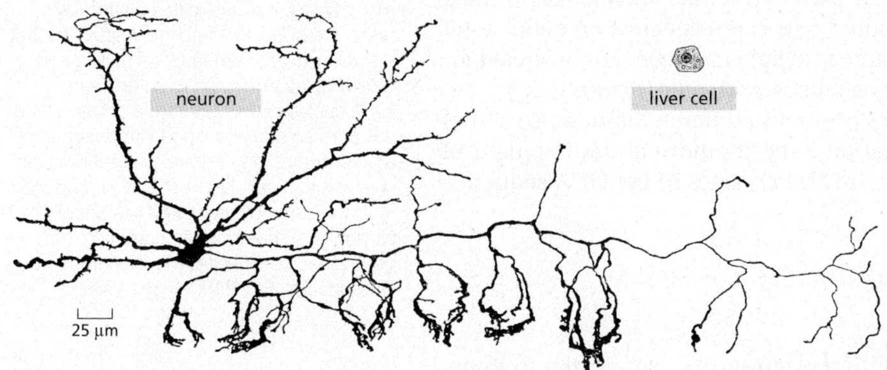

Figure 7–1 A neuron and a liver cell share the same genome. The long branches of this neuron from the retina enable it to receive electrical signals from many other neurons and convey them to neighboring neurons. The liver cell, which is drawn to the same scale, is involved in many metabolic processes, including digestion and the detoxification of alcohol and other drugs. Both of these mammalian cells contain the same genome, but they express different sets of RNAs and proteins. (Neuron adapted from S. Ramón y Cajal, Histologie du Système Nerveux de l'Homme et de Vertébrés, 1909–1911. Paris: A. Maloine Éditeur; reprinted, Madrid: C.S.I.C., 1972.)

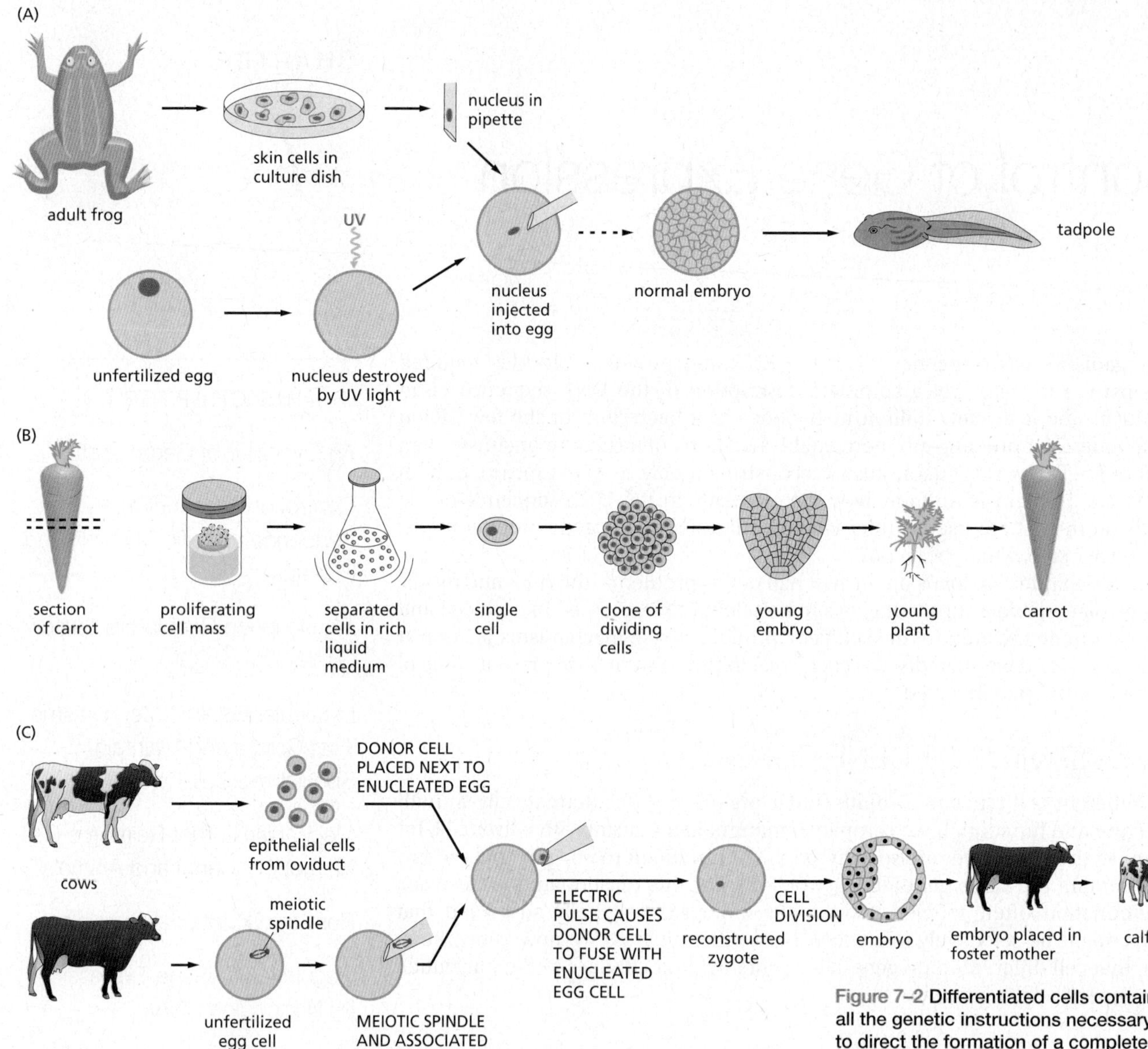

Figure 7–2 Differentiated cells contain all the genetic instructions necessary to direct the formation of a complete organism. (A) The nucleus of a skin cell from an adult frog transplanted into an enucleated egg can give rise to an entire tadpole. The *broken arrow* indicates that, to give the transplanted genome time to adjust to an embryonic environment, a further transfer step is required in which one of the nuclei is taken from an early embryo that begins to develop and is put back into a second enucleated egg. (B) In many types of plants, differentiated cells retain the ability to "de-differentiate," so that a single cell can form a clone of progeny cells that later give rise to an entire plant. (C) A nucleus removed from a differentiated cell from an adult cow and introduced into an enucleated egg from a different cow can give rise to a calf. Different calves produced from the same differentiated cell donor are all clones of the donor and are therefore genetically identical. (A, modified from J.B. Gurdon, *Sci. Am.* 219:24–35, 1968.)

removed, the injected donor nucleus is capable of directing the recipient egg to produce a normal tadpole (**Figure 7–2A**). Because this tadpole contains a full range of differentiated cells, each of which derived their DNA sequences from the nucleus of the original donor skin cell, that differentiated cell cannot have lost any important DNA sequences. Experiments performed with plants produced a similar conclusion. When differentiated pieces of plant tissue are placed in culture and then dissociated into single cells, often one of these individual cells can regenerate an entire adult plant (**Figure 7–2B**). More recently, the same principle has been demonstrated for mammals that include sheep, mice, pigs, goats, dogs, and cattle (**Figure 7–2C**).

Detailed DNA sequencing of genomes present in different tissues also shows that the changes in gene expression that underlie the normal development of multicellular organisms do not generally involve changes in the DNA sequence of the genome.

Different Cell Types Synthesize Different Sets of RNAs and Proteins

As a first step in understanding cell differentiation, we would like to know how many differences there are between any one cell type and another.

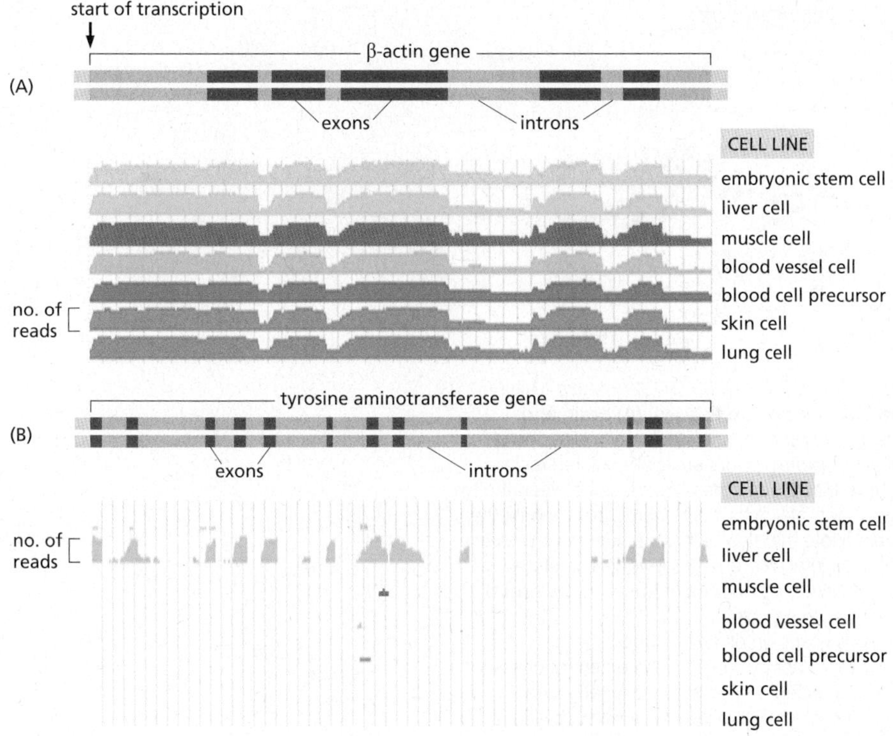

Figure 7–3 **Differences in RNA levels for two human genes in seven different tissues.** To obtain RNA data by the technique known as RNA-seq (see pp. 514–516), RNA was collected from seven different human cell lines grown in culture, each derived from a different tissue. Millions of "sequence reads" were obtained for each RNA sample and mapped by matching RNA sequences to the DNA sequence of the human genome. At each position along the genome, the height of the colored trace is proportional to the number of sequence reads that match the genome sequence at that point. As seen in the figure, the exon sequences in transcribed genes are present at high levels, reflecting their presence in mature mRNAs. Intron sequences are present at much lower levels and reflect pre-mRNA molecules that have not yet been spliced, plus intron sequences that have been spliced out but not yet degraded. (A) The data for one of the genes coding for actin, a major component of the cytoskeleton in all cells. Note that the left-hand end of the mature β-actin mRNA is not translated into protein. As explained later in this chapter, many mRNAs have 5′ untranslated regions that regulate their translation into protein. (B) The same type of data displayed for the enzyme tyrosine aminotransferase, which is highly expressed in liver cells but not in the other cell types tested. [Information for both panels from the University of California, Santa Cruz, Genome Browser (https://genome.ucsc.edu), which provides this type of information for every human gene. See also S. Djebali et al., *Nature* 489:101–108, 2012.]

Although we still do not have an exact answer for each cell type, we can make several general statements.

1. Many processes are common to all cells, and any two cells in a single organism therefore have many gene products in common. These include the structural proteins of chromosomes, RNA and DNA polymerases, DNA repair enzymes, ribosomal proteins and RNAs, the enzymes that catalyze the central reactions of metabolism, and many of the proteins that form the cytoskeleton such as actin (**Figure 7–3A**).

2. Some RNAs and proteins are abundant in the specialized cells in which they function and cannot be detected elsewhere, even by sensitive tests. Hemoglobin, for example, is expressed specifically in red blood cells, where it carries oxygen, whereas the enzyme tyrosine aminotransferase (which breaks down tyrosine in food) is expressed in liver but not in most other tissues (**Figure 7–3B**).

3. Analyses of RNAs reveal that, at any one time, a typical human cell expresses 30–60% of its approximately 25,000 genes at some meaningful level. There are about 20,000 protein-coding genes and an estimated 5000 noncoding RNA genes in humans. When the patterns of RNA expression in different human cell lines are compared, the level of expression of almost every gene is found to vary from one cell type to another. A few of these differences are striking, like those of hemoglobin and tyrosine aminotransferase noted above, but most are much more subtle. But even those genes that are expressed in all cell types usually vary in their *level* of expression from one cell type to the next.

4. Although there are striking differences in the protein-coding RNAs (mRNAs) in specialized cell types, they underestimate the full range of differences in the final pattern of protein production. As we shall discuss later in this chapter, there are many steps after RNA production at which gene expression can be regulated. And, as we saw in Chapter 3, proteins are often covalently modified after they are synthesized. The differences in gene expression between cell types are therefore most fully revealed through methods that directly display the levels of proteins, along with their post-translational modifications (**Figure 7–4**).

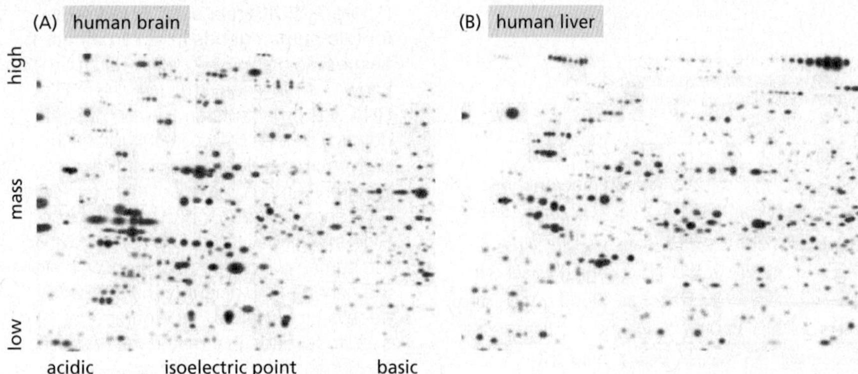

Figure 7–4 **Differences in the proteins expressed by two human tissues, (A) brain and (B) liver.** The proteins have been separated by size (*top* to *bottom*) and isoelectric point, the pH at which the protein has no net charge (*right* to *left*). The protein spots artificially colored *red* are common to both samples; those in *blue* are specific to that tissue. The differences between the two tissue samples vastly outweigh their similarities: even for proteins that are shared between the two tissues, their relative abundances are usually different. Note that this technique separates proteins by both size and charge; therefore, a protein that has several different phosphorylation states will appear as a series of *horizontal spots* (see *upper right-hand* portion of *right* panel). Only a small portion of the complete protein spectrum is shown for each sample.

The method used to display proteins in these panels is known as *two-dimensional gel electrophoresis* (see Figure 8–16). Although it is useful for easily visualizing the extent of protein differences between the two cell types, newer methods based on mass spectrometry (see pp. 491–492) provide much more detailed information, including the identity of each protein, the position of each modification, and the nature of the modification. (Courtesy of Tim Myers and Leigh Anderson, Large Scale Biology Corporation.)

The Spectrum of mRNAs Present in a Cell Can Be Used to Accurately Identify the Cell Type

We have seen that each cell type produces a characteristic set of mRNAs. Therefore, if all the mRNAs present in a cell are known, the cell type can be unambiguously identified, using prior knowledge from cell lines or analyses of tissues. This approach is made possible by the ability to determine the nucleotide sequence of all the mRNAs produced by a single cell (see pp. 537–538). Thus, for example, because human cells have approximately 20,000 mRNA-producing genes, this strategy provides very fine resolution of the differences among our different individual cells.

In general, the mRNA approach agrees well with the traditional categorization of cell types that is based on staining and microscopy, but the mRNA strategy has also revealed that many cells that "look" the same can differ significantly in their mRNA content and therefore in their function. This strategy has thereby identified many new cell types, most of which are subdivisions of cell types that had been classically defined (**Figure 7–5**). The ability to determine the mRNA content of individual cells also provides a new appreciation for how cells present in a tissue (liver, for example) differ according to their positions in the tissue.

External Signals Can Cause a Cell to Change the Expression of Its Genes

Although the specialized cells in a multicellular organism have characteristic patterns of gene expression, each cell is capable of altering its pattern of gene expression in response to extracellular cues. If a liver cell is exposed to a glucocorticoid hormone, for example, the production of a set of proteins is dramatically increased, and once the hormone is no longer present, the production of these proteins drops back to its normal, unstimulated level. Glucocorticoids are released in the body during periods of starvation or intense exercise, and they signal the liver to increase the production of energy from amino acids and other

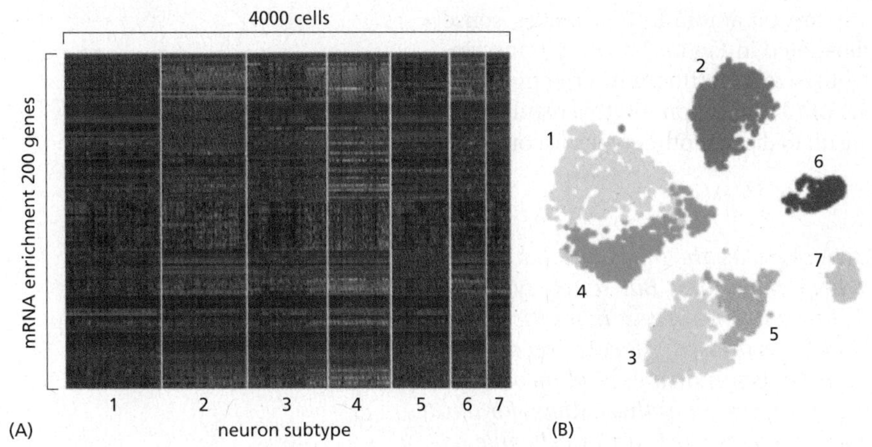

Figure 7–5 **Classification of a group of neurons in the mouse brain into seven different subtypes by single-cell mRNA sequencing.** For this experiment, approximately 4000 individual neurons (which were activated in response to a particular stimulus) were dissected from the brain and separated from each other. The mRNAs produced by each cell were isolated and their sequences determined by the methods described in Chapter 8. On the basis of the spectrum of the mRNAs produced by each cell, the 4000 different cells could be grouped into seven distinctive subtypes. Within each subtype, the mRNAs were similar from cell to cell, but between subtypes they differed significantly. (A) Here, the level of the mRNAs detected for approximately 200 different genes is plotted for each cell as a tiny rectangle, whose color intensity is proportional to the amount of that mRNA in that cell, with *red* indicating increased expression and *green* decreased expression, relative to all the samples. These data are plotted for each of the 4000 cells along the X axis. The cells have been arranged so that similar cells are located next to each other. In this way it is possible, using mRNA sequence data alone, to recognize seven distinctive types of neurons, as indicated. To highlight similarities, the data for the 25 mRNAs specifically enriched in each of the seven subtypes is indicated by *red blocks*. (B) By analyzing the mRNA data using a mathematical method known as *unsupervised clustering* (see Figure 8–66), the seven different subtypes can readily be distinguished on a two-dimensional "cluster diagram," with each dot representing a single cell. In addition, information regarding the extent of differences among the subgroups can also be ascertained. For example, subtypes 1 and 4 are more closely related to each other in the mRNAs they make than are subtypes 1 and 7. This type of analysis helps us to understand how the brain processes sensory information and indicates that, even though neurons may look the same under the microscope, they can differ significantly in gene expression patterns and therefore in their functions. (A and B, from M.B. Chen et al., *Nature* 587:437–442, 2020, doi 10.1038/s41586-020-2905-5. With permission from Springer Nature.)

small molecules. The set of induced proteins includes the enzyme tyrosine aminotransferase, mentioned earlier.

Other cell types respond to glucocorticoids differently. Fat cells, for example, reduce the production of tyrosine aminotransferase, while some other cell types do not respond to glucocorticoids at all. These examples illustrate a general feature of cell specialization: different cell types can respond very differently to the same extracellular signal. Other features of the gene expression pattern do not change and give each cell type its permanently distinctive character.

Gene Expression Can Be Regulated at Many of the Steps in the Pathway from DNA to RNA to Protein

We have seen that differences among the various cell types of an organism depend on the particular genes that the cells express. But at what level does this control of gene expression occur? As we saw in the previous chapter, there are many steps in the pathway leading from DNA to protein, and all of them can in principle be regulated. Thus, as illustrated in Figure 7–6, a cell can control the proteins it makes by (1) controlling when and how often a given gene is transcribed (**transcriptional control**), (2) controlling the splicing and processing of RNA transcripts (**RNA-processing control**), (3) selecting which completed mRNAs are exported from the nucleus to the cytosol and determining where in the cytosol they are localized (**RNA transport and localization control**), (4) selecting which mRNAs in the cytoplasm are translated by ribosomes (**translational control**), (5) selectively destabilizing certain mRNA molecules in the cytoplasm (**mRNA degradation control**), (6) selectively degrading specific protein molecules (**protein degradation control**), and (7) activating, inactivating, or localizing specific protein molecules (**protein activity control**).

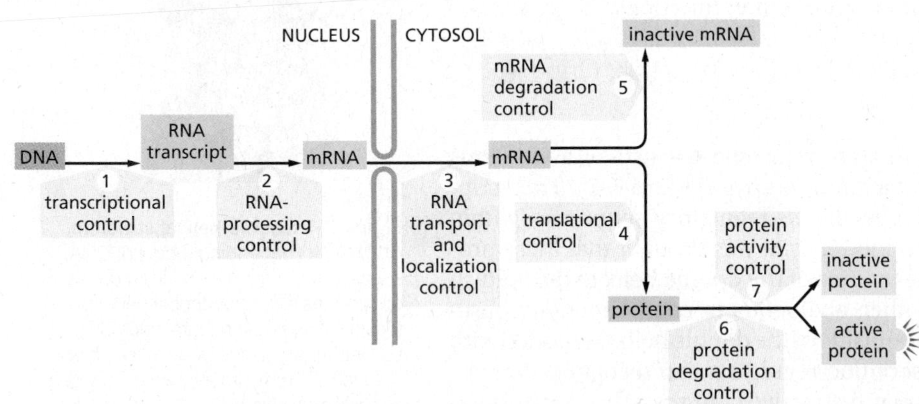

Figure 7–6 **Seven steps at which eukaryotic gene expression can be controlled.** Controls that operate at steps 1 through 5 are discussed in this chapter. Step 7, the regulation of protein activity, occurs largely through covalent post-translational modifications including phosphorylation, acetylation, and ubiquitylation (see Table 3–4, p. 175). Steps 6 and 7 were introduced in Chapters 3 and 6 and will be subsequently discussed in other chapters throughout the book.

For many genes, transcriptional controls are paramount. This makes sense because, of all the possible control points illustrated in Figure 7–6, only transcriptional control ensures that the cell will not synthesize superfluous intermediates. In the following sections, we discuss the DNA and protein components that regulate the initiation of gene transcription, before moving on to discuss other types of controls.

Summary

The genome of a cell contains in its entire DNA sequence the information to make many thousands of different protein and RNA molecules. But a cell typically expresses only a fraction of its genes, and the different types of cells in multicellular organisms arise because different sets of genes are expressed. All cells can change the pattern of genes they express in response to changes in their environment, such as signals from other cells. The regulation of gene expression is thus crucial for life. Although all of the many steps involved in expressing a gene can in principle be regulated, for most genes it is the initiation of RNA transcription that provides the most important point of control.

CONTROL OF TRANSCRIPTION BY SEQUENCE-SPECIFIC DNA-BINDING PROTEINS

How does a cell determine which of its thousands of genes to transcribe? Perhaps the most important concept, one that applies to all species on Earth, is based on a group of proteins known as **transcription regulators**. These proteins recognize specific sequences of DNA (typically 5–12 nucleotide pairs in length) that are often called ***cis*-regulatory sequences**, because they must be on the same chromosome (that is, *in cis*) to the genes they control. Transcription regulators bind to these sequences, which are dispersed throughout genomes, and this binding puts into motion a series of reactions that ultimately specify which genes are to be transcribed and at what rate. Approximately 10% of the protein-coding genes of most organisms are devoted to transcription regulators, making them one of the largest classes of proteins in the cell. A given transcription regulator typically recognizes a specific *cis*-regulatory sequence that is different from those recognized by the other transcription regulators in the cell.

The transcription of each gene is, in turn, controlled by its unique collection of *cis*-regulatory DNA sequences, which thus constitute a crucial part of the information coded in genomes. These sequences typically lie near the gene, often in the intergenic region directly upstream from the transcription start point of the gene. Although a few genes are controlled by a single *cis*-regulatory sequence that is recognized by a single transcription regulator, the majority have complex arrangements of *cis*-regulatory sequences, each of which is recognized by a different transcription regulator. It is therefore the positions, identity, and arrangement of *cis*-regulatory sequences that ultimately determine the time and place that each gene is transcribed. We begin our discussion by describing how transcription regulators "read" the information present in *cis*-regulatory sequences; later in the chapter, we shall discuss how they carry out their functions.

The Sequence of Nucleotides in the DNA Double Helix Can Be Read by Proteins

As discussed in Chapter 4, the DNA in a chromosome consists of a very long double helix that has both a major and a minor groove (Figure 7–7). Transcription regulators must recognize short, specific *cis*-regulatory sequences within this structure. When first discovered in the 1960s, it was thought that these proteins might require direct access to the interior of the double helix to distinguish between one DNA sequence and another, analogous to complementary base-pairing. It is now clear, however, that the outside of the double helix is studded with DNA sequence information that transcription regulators can recognize directly: the outside edges of each base pair display distinctive patterns of hydrogen-bond

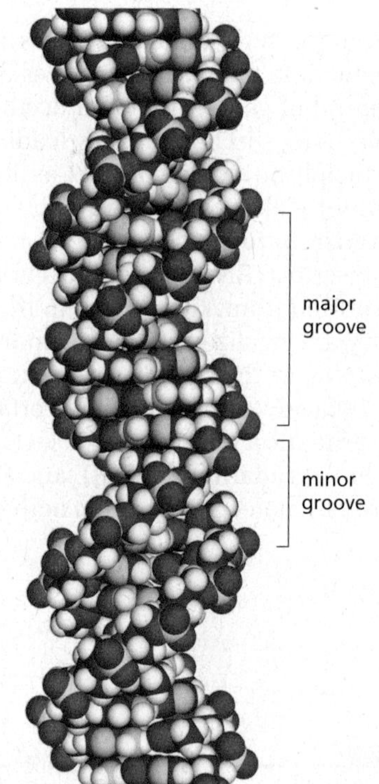

Figure 7–7 **Double-helical structure of DNA.** A space-filling model of DNA showing the major and minor grooves on the outside of the double helix (see Movie 4.1). The atoms are colored conventionally as follows: carbon, *black*; nitrogen, *blue*; hydrogen, *white*; oxygen, *red*; phosphorus, *yellow*.

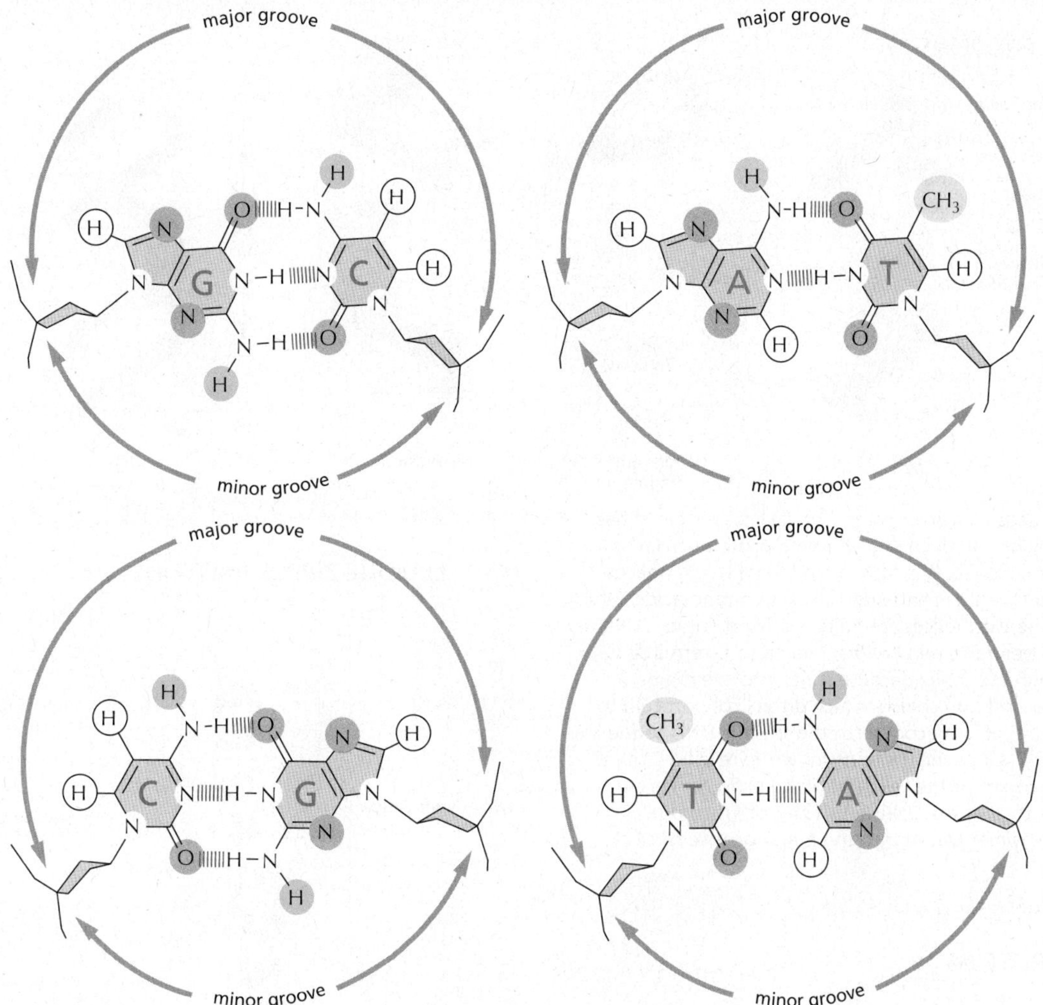

Figure 7–8 **How the different base pairs in DNA can be recognized from their edges without the need to open the double helix.** The four possible configurations of base pairs are shown, with potential hydrogen-bond donors indicated in *blue*, potential hydrogen-bond acceptors in *red*, and hydrogen bonds of the base pairs themselves as a series of short, parallel *red* lines. Methyl groups, which form hydrophobic protuberances, are shown in *yellow*, and hydrogen atoms that are attached to carbons, and are therefore unavailable for hydrogen-bonding, are *white*. From the major groove, each of the four base-pair configurations projects a unique pattern of features. (From C. Branden and J. Tooze, Introduction to Protein Structure, 2nd ed. New York: Garland Publishing, 1999. With permission from Taylor & Francis.)

donors, hydrogen-bond acceptors, and hydrophobic patches in both the major and minor grooves, allowing each base to be distinguished from the other three (**Figure 7–8**). Because the major groove is wider and displays more molecular features than does the minor groove, nearly all transcription regulators make the majority of their contacts with the major groove—as we shall see.

Transcription Regulators Contain Structural Motifs That Can Read DNA Sequences

Molecular recognition in biology generally relies on an exact fit between the surfaces of two molecules, and the study of transcription regulators provides some of the clearest examples of this principle. Thus, a transcription regulator recognizes its specific *cis*-regulatory sequence because the surface of the protein is complementary to surface features of the double helix that displays that sequence. Each transcription regulator makes a series of contacts with the DNA, involving hydrogen bonds, ionic bonds, and hydrophobic interactions. Although each individual contact is weak, the 20 or so contacts that are typically formed at the protein–DNA interface add together to ensure that the interaction is both highly specific and very strong (**Figure 7–9**). In fact, DNA–protein interactions include some of the tightest and most specific molecular interactions known in biology.

Although each example of protein–DNA recognition is unique in detail, x-ray crystallographic and nuclear magnetic resonance (NMR) spectroscopic studies of hundreds of transcription regulators reveal that many contain one or another of a small set of DNA-binding structural motifs (**Panel 7–1**). These motifs generally

HELIX–TURN–HELIX PROTEINS

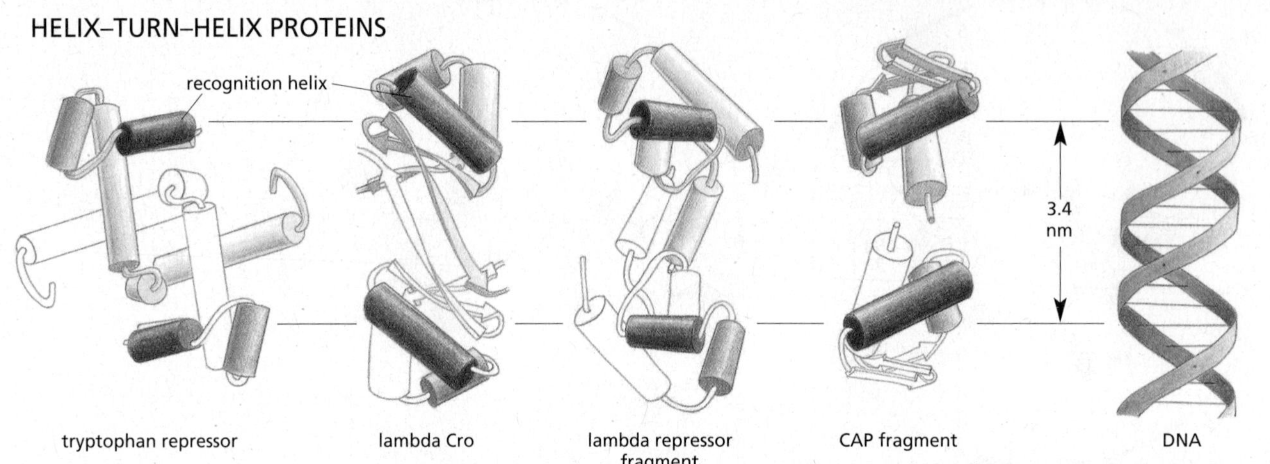

Originally identified in bacterial transcription regulators, this motif has since been found in many hundreds of DNA-binding proteins from eukaryotes, bacteria, and archaea. It is constructed from two α helices (*blue* and *red*) connected by a short extended chain of amino acids, which constitutes the "turn." The two helices are held at a fixed angle, primarily through interactions between the two helices. The more C-terminal helix (in *red*) is called the *recognition helix* because it fits into the major groove of DNA; its amino acid side chains, which differ from protein to protein, play an important part in recognizing the specific DNA sequence to which the protein binds. All of the proteins shown here bind DNA as dimers in which the two copies of the recognition helix (in *red*) are separated by exactly one turn of the DNA helix (3.4 nm); thus both recognition helices of the dimer can fit into the major groove of DNA.

HOMEODOMAIN PROTEINS

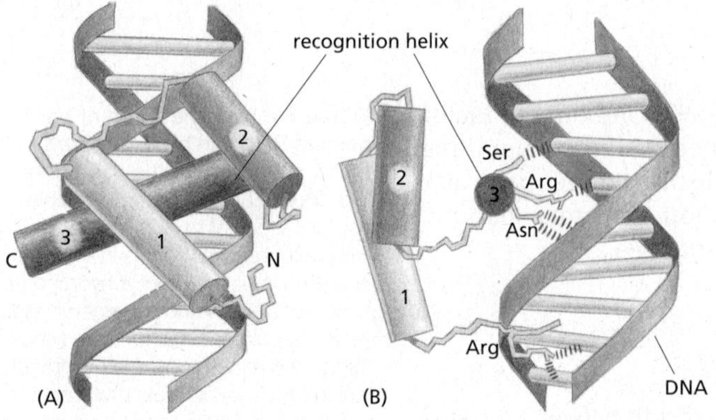

Not long after the first transcription regulators were discovered in bacteria, genetic analyses of the fruit fly *Drosophila* led to the characterization of an important class of genes, the *homeotic selector genes*, that play a critical part in orchestrating fly development (discussed in Chapter 21). It was later shown that these genes coded for transcription regulators that bound DNA through a structural motif named the homeodomain. Two different views of the same structure are shown. (A) The homeodomain is folded into three α helices, which are packed tightly together by hydrophobic interactions. The part containing helices 2 and 3 closely resembles the bacterial helix–turn–helix motif. (B) The recognition helix (helix 3, *red*) forms important contacts with the major groove of DNA. The asparagine (Asn) of helix 3, for example, contacts an adenine, as shown in Figure 7–9. A flexible arm attached to helix 1 forms contacts with nucleotide pairs in the minor groove (**Movie 7.1**).

LEUCINE ZIPPER PROTEINS

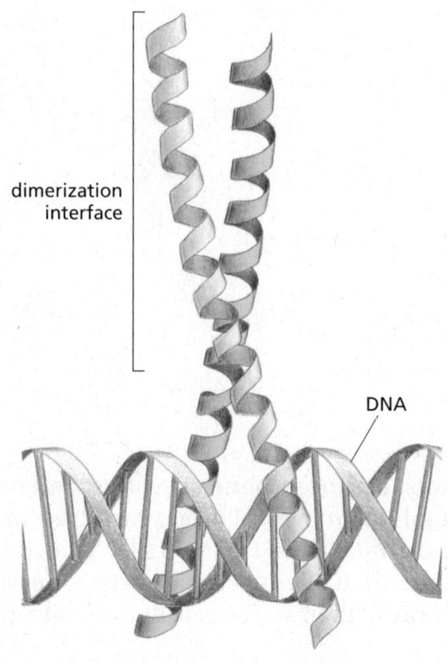

The *leucine zipper* motif is named because of the way the two α helices, one from each monomer, are joined together to form a short coiled-coil. These proteins bind DNA as dimers where the two long α helices are held together by interactions between hydrophobic amino acid side chains (often on leucines) that extend from one side of each helix. Just beyond the dimerization interface, the two α helices separate from each other to form a Y-shaped structure, which allows their side chains to contact the major groove of DNA. The dimer thus grips the double helix like a clothespin on a clothesline (**Movie 7.2**).

β-SHEET DNA RECOGNITION PROTEINS

In the other DNA-binding motifs displayed in this panel, α helices are the primary mechanism used to recognize specific DNA sequences. In one group of transcription regulators, however, a two-stranded β sheet, with amino acid side chains extending from the sheet toward the DNA, reads the information on the surface of the major groove. As in the case of a recognition α helix, this β-sheet motif can be used to recognize many different DNA sequences; the exact DNA sequence recognized depends on the sequence of amino acids that make up the β sheet. Shown is a transcription regulator that binds two molecules of *S*-adenosyl methionine *(red)*. On the left is a dimer of the protein; on the right is a simplified diagram showing just the two-stranded β sheet bound to the major groove of DNA. *S*-adenosyl methionine is needed for this protein to bind DNA. Thus, the small molecule regulates the activity of the DNA-binding protein.

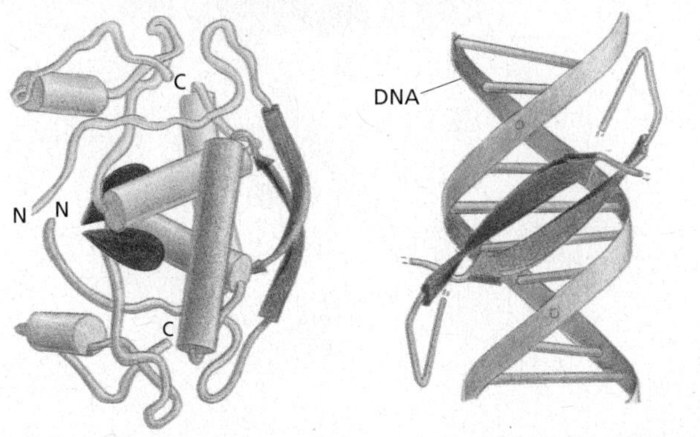

ZINC FINGER PROTEINS

This group of DNA-binding motifs includes one or more zinc atoms as structural components. All such zinc-coordinated DNA-binding motifs are called zinc fingers, referring to their appearance in early schematic drawings *(left)*. They fall into several distinct structural groups, only one of which we consider here. It has a simple structure, in which the zinc atom holds an α helix and a β sheet together *(middle)*. This type of zinc finger is often found in clusters with the α helix of each finger contacting the major groove of the DNA, forming a nearly continuous stretch of α helices along that groove (**Movie 7.3**). In this way, a strong and specific DNA–protein interaction is built up through a repeating basic structural unit. Three such fingers are shown on the *right*.

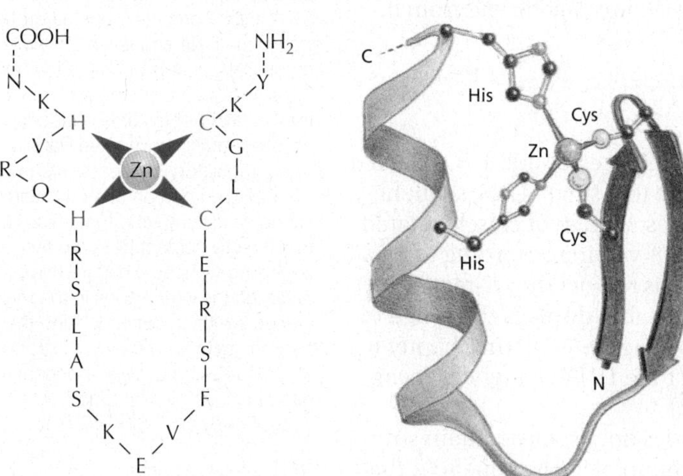

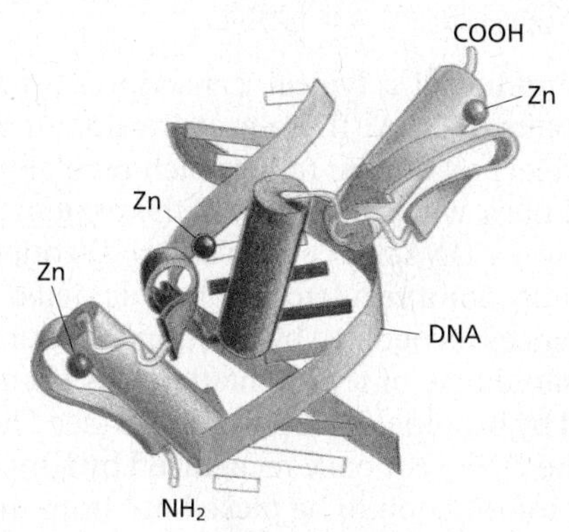

HELIX–LOOP–HELIX PROTEINS

Related to the leucine zipper, the helix–loop–helix motif consists of a short α helix connected by a loop *(red)* to a second, longer α helix. The flexibility of the loop allows one helix to fold back and park against the other thereby forming the dimerization surface. As shown, this two-helix structure binds both to DNA and to the two-helix structure of a second protein to create either a homodimer or a heterodimer. Two α helices that extend from the dimerization interface make specific contacts with the major groove of DNA.

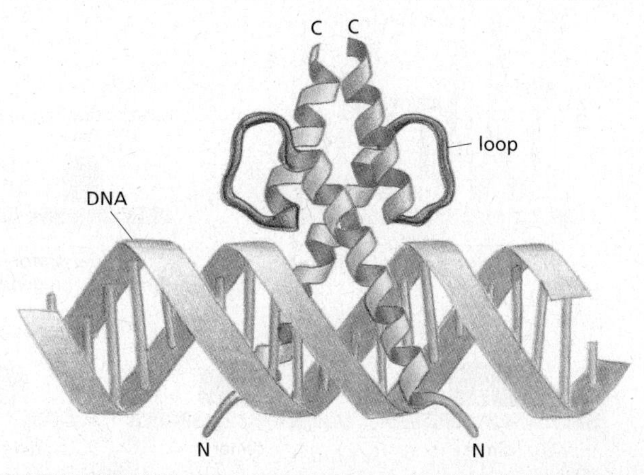

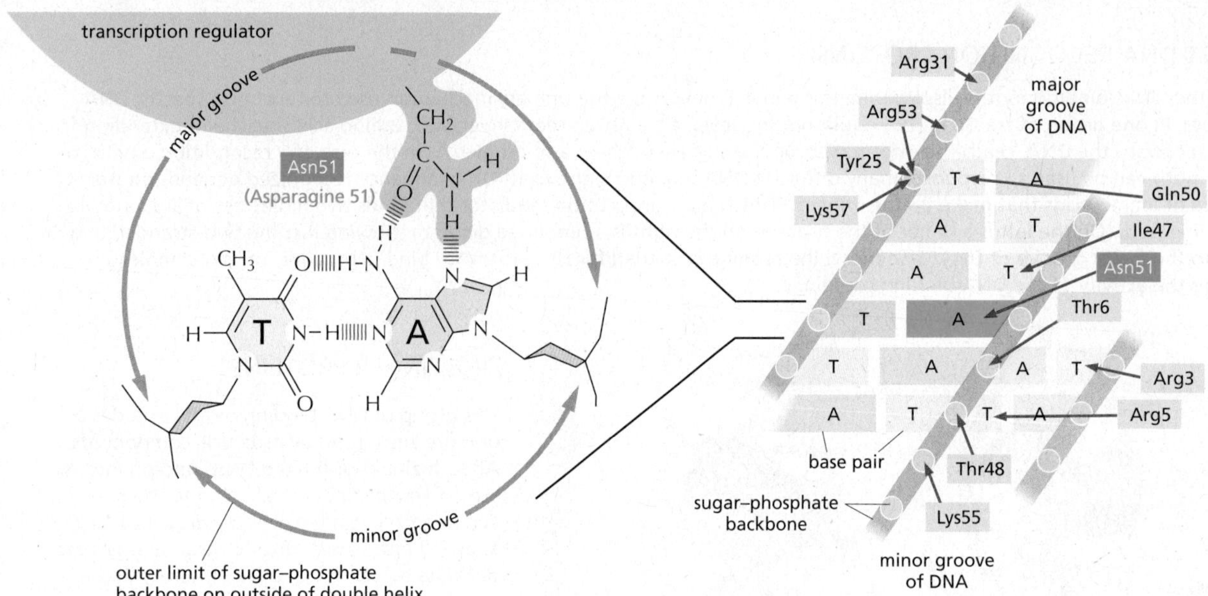

Figure 7–9 The binding of a transcription regulator to a specific DNA sequence. On the *left*, a single contact is shown between a transcription regulator and DNA; such contacts allow the protein to "read" the DNA sequence from the outside of the DNA double helix. On the *right*, the complete set of contacts between a transcription regulator (a member of the homeodomain family—see Panel 7–1) and its *cis*-regulatory sequence is shown. The DNA-binding portion of the protein is 60 amino acids long, and the amino acids that directly contact DNA are numbered beginning with the amino terminus. Although the interactions in the major groove are the most important, the protein also contacts both the minor groove and phosphates in the sugar–phosphate DNA backbone, as shown. (See C. Wolberger et al., *Cell* 67:517–528, 1991.)

use either α helices or β sheets to bind to the major groove of DNA, with amino acid side chains that extend from these motifs making their specific DNA contacts. Thus, a given structural motif can be used to recognize many different *cis*-regulatory sequences depending on the specific side chains that extend from it.

Dimerization of Transcription Regulators Increases Their Affinity and Specificity for DNA

A monomer of a typical transcription regulator recognizes about 4–8 nucleotide pairs of DNA. These proteins do not bind tightly to a single DNA sequence and reject all others; rather, each regulator recognizes a range of closely related sequences, with the affinity of the protein for the DNA varying according to how closely the DNA matches its optimal sequence. For this reason, the *cis*-regulatory sequence for a regulator is often depicted by a "logo" that displays the range of sequences recognized by that transcription regulator (**Figure 7–10**). In Chapter 6, this same type of representation was used to depict the DNA sequences recognized by bacterial RNA polymerase (see Figure 6–12).

The DNA sequence recognized by a monomer does not usually contain sufficient information to be picked out from the background of such sequences that would occur at random across the genome. For example, an exact six-nucleotide DNA sequence would be expected to occur by chance approximately once every

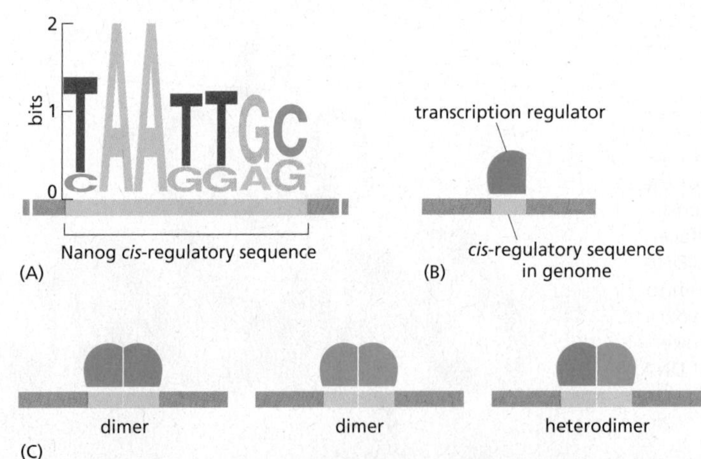

Figure 7–10 Transcription regulators and *cis*-regulatory sequences. (A) Depiction of the *cis*-regulatory sequence for Nanog, a homeodomain family member that is a key transcription regulator in embryonic stem cells. This "logo" form (see Figure 6–12) shows that the protein can recognize a collection of closely related DNA sequences and gives the preferred nucleotide pair at each position. *Cis*-regulatory sequences are almost always "read" as double-stranded DNA, but only one strand typically is shown in a logo. (B) Representation of the *cis*-regulatory sequence as a *green box* embedded in a longer DNA molecule *(gray)*. (C) Many transcription regulators form dimers (homodimers and heterodimers). In the example shown, three different DNA-binding specificities are formed from two transcription regulators.

4096 nucleotides (4^6), and the range of six-nucleotide sequences described by a typical logo would be expected to occur by chance much more often, perhaps every 1000 nucleotides. Clearly, for a bacterial genome of 4.6×10^6 nucleotide pairs, not to mention a mammalian genome of 3×10^9 nucleotide pairs, this is insufficient information to accurately control the transcription of individual genes. Additional contributions to DNA-binding specificity must therefore be present.

Many transcription regulators form dimers, with both monomers making nearly identical contacts with DNA (see Figure 7–10C). This arrangement doubles the length of the *cis*-regulatory sequence recognized and greatly increases both the affinity and the specificity of transcription regulator binding. Because the DNA sequence recognized by the protein has increased from approximately 6 nucleotide pairs to 12 nucleotide pairs, there are many fewer random occurrences of matching sequences. In many cases, heterodimers can form between two different transcription regulators, and this configuration also increases both affinity and specificity by expanding the DNA sequence recognized. Some transcription regulators can form heterodimers with more than one partner protein; in this way, the same transcription regulator can be "reused" to create several distinct DNA-binding specificities (see Figure 7–10C).

Many Transcription Regulators Bind Cooperatively to DNA

In the simplest case, the collection of noncovalent bonds that holds dimers or heterodimers together is so extensive that these structures form obligatorily and virtually never fall apart. In this case, the unit of binding is the dimer or heterodimer, and the binding curve for the transcription regulator (the fraction of DNA bound as a function of protein concentration) has a standard exponential shape (**Figure 7–11A**).

In many cases, however, the dimers and heterodimers are held together very weakly; they exist predominantly as monomers in solution, and yet dimers are observed on the appropriate DNA sequence. In this case, the proteins are said to bind to DNA cooperatively, and the curve describing their binding is S-shaped (**Figure 7–11B**). *Cooperative binding* means that, over a range of concentrations of the transcription regulator, binding is more of an all-or-none phenomenon than for noncooperative binding; that is, at most protein concentrations, the *cis*-regulatory sequence is either nearly empty or nearly fully occupied and is rarely somewhere in between. A discussion of the mathematics behind cooperative binding is given in Chapter 8 (see Figure 8–81).

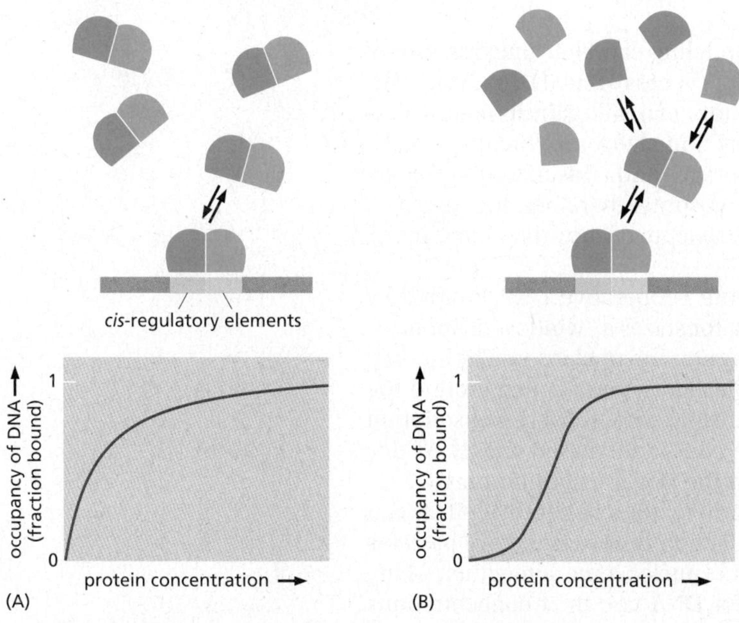

Figure 7–11 **Occupancy of a *cis*-regulatory sequence by a transcription regulator.** (A) Noncooperative binding by a stable heterodimer. (B) Cooperative binding by components of a heterodimer that are predominantly monomers in solution. The shape of the curve differs from that of panel A because the fraction of protein in a form competent to bind DNA (the heterodimer) increases with increasing protein concentration.

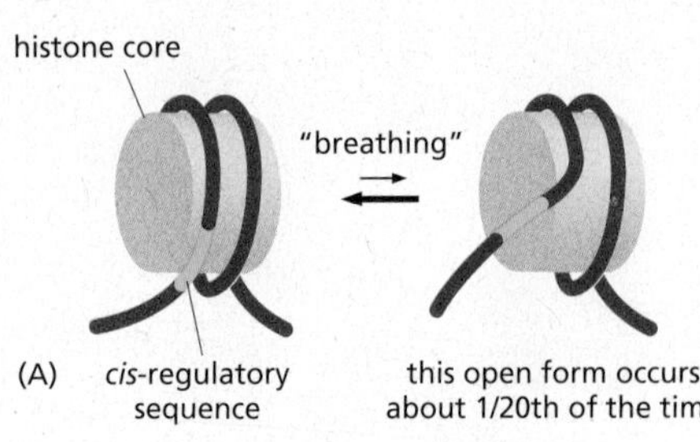

this open form occurs about 1/20th of the time

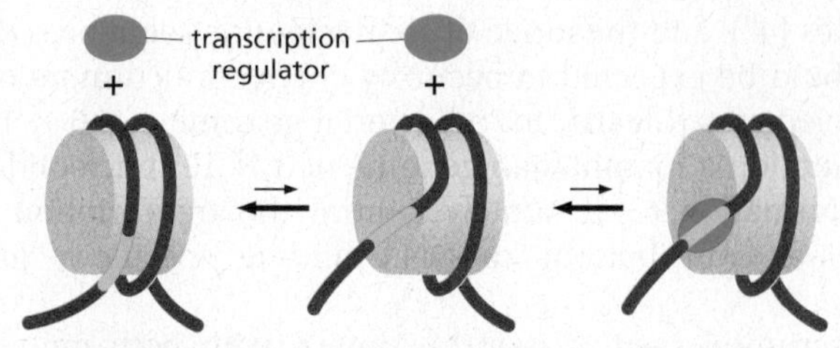

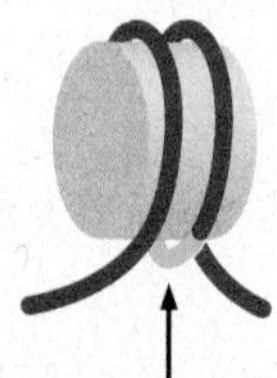

(C) a typical transcription regulator will bind with roughly 200-fold less affinity if its cis-regulatory sequence is located in the middle of a nucleosome

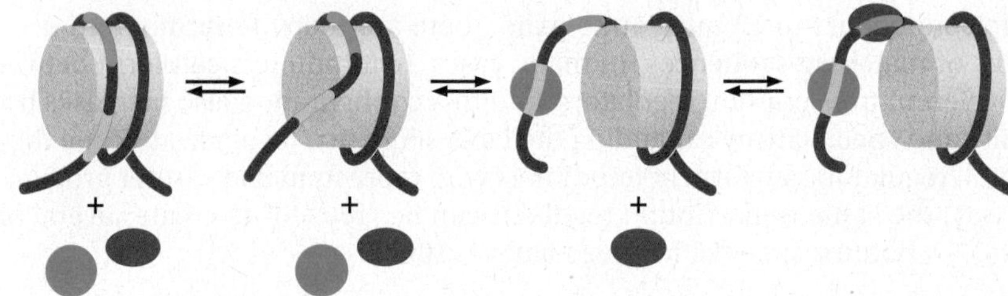

Figure 7–12 How nucleosomes affect the binding of transcription regulators.

Nucleosome Structure Promotes Cooperative Binding of Transcription Regulators

As we have just seen, cooperative binding of transcription regulators to DNA often occurs because the proteins involved have only a weak affinity for each other. However, there is a second, indirect mechanism for cooperative binding in eukaryotes, one that arises from the nucleosome structure of their chromosomes.

In general, transcription regulators bind to DNA in nucleosomes with lower affinity than they do to naked DNA. There are two reasons for this difference. First, the surface of the *cis*-regulatory sequence recognized by the transcription regulator may be facing inward on the nucleosome, toward the histone core, and therefore not be readily available to the regulatory protein. Second, even if the face of the *cis*-regulatory sequence is exposed on the outside of the nucleosome, many transcription regulators subtly alter the conformation of the DNA when they bind, and these changes are generally opposed by the tight wrapping of the DNA around the histone core. For example, many transcription regulators induce a bend or kink in the DNA when they bind.

We saw in Chapter 4 that nucleosome remodeling can alter the structure of the nucleosome, allowing transcription regulators access to the DNA. Even without remodeling, however, transcription regulators can still gain limited access to DNA in a nucleosome. The DNA at the end of a nucleosome "breathes," transiently exposing the DNA and allowing regulators to bind. This breathing occurs at a much lower rate in the middle of the nucleosome; therefore, the positions where the DNA exits the nucleosome are much easier to occupy than those in the middle of the nucleosome (**Figure 7–12**).

These properties of the nucleosome promote cooperative DNA binding by transcription regulators. If a transcription regulator seizes a "window of opportunity" provided by nucleosome breathing, it can enter the nucleosome by binding to the exposed DNA and prevent the DNA from tightly rewrapping around the nucleosome core. When this happens, the affinity of a second transcription regulator for a nearby *cis*-regulatory sequence can be increased simply by this loosening of the DNA from the histone core. If the two transcription regulators also interact with each other (as described earlier), the cooperative effect can be even greater. In some cases, the combined action of the regulatory proteins can eventually displace the histone core of the nucleosome altogether. Many transcription regulators, when their affinities for DNA and their concentrations are sufficiently high, can take advantage of nucleosome breathing and thereby "invade" nucleosomes. Moreover, as we saw in Chapter 5, passing replication forks, which transiently displace histones, offer additional windows of opportunity for transcription regulators to bind to DNA.

Although nucleosomes generally inhibit the DNA binding of transcription regulators, some regulators—if their *cis*-regulatory sequences are exposed on the nucleosome surface—can bind with nearly the same affinity as they do on naked DNA, occupying their binding sites while the DNA is still tightly wrapped around the histone core (Figure 7–13). Transcription regulators with this property are sometimes called *pioneer factors*, because they are often the first proteins to bind DNA when a previously silent gene becomes transcriptionally active. Although their binding typically destabilizes the nucleosome, pioneer factors probably exert their major effects by attracting additional proteins that alter chromatin structure, such as nucleosome remodeling complexes. If one transcription regulator binds its *cis*-regulatory sequence on a nucleosome and attracts a chromatin remodeling complex, the localized action of the remodeling complex can allow a second transcription regulator to efficiently bind nearby.

Our discussion has emphasized how transcription regulators can work together in pairs. But in reality, larger numbers often cooperate by repeated use of the same principles. It is the cooperative formation of clusters of transcription regulators on DNA that probably explains why many key regulatory sequences in eukaryotic genomes are found to be "nucleosome free."

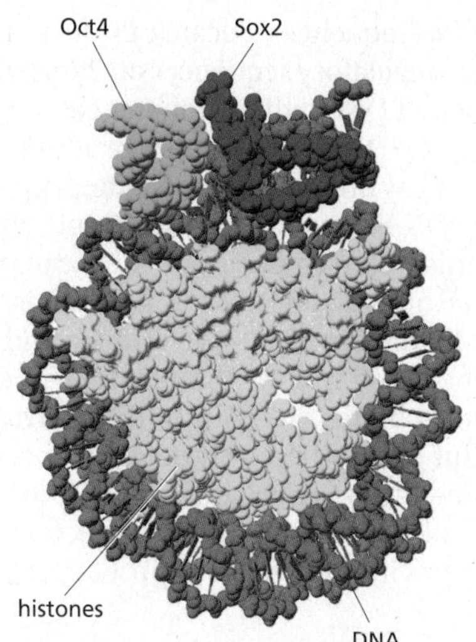

Figure 7–13 **Two cooperating transcription regulators, Oct4 *(green)* and Sox2 *(blue)*, bound to a nucleosome.** These two transcription regulators work together and play key roles in maintaining embryonic stem cells (see Figures 7–36 and 7–37). Only the DNA-binding portion of each regulator protein is shown. (Courtesy of Nicolas H. Thomä and Alicia K. Michael. PDB code: 6T90.)

DNA-Binding by Transcription Regulators Is Dynamic

Thus far, we have treated transcription regulators as static—we have considered them as either bound to DNA or free in solution. But in reality, the situation is highly dynamic, with transcription regulator molecules in constant motion, rapidly binding and dissociating from DNA. In most cases a given transcription regulator molecule stays on its *cis*-regulatory sequence for only a short time, but it is rapidly replaced by other molecules of the same regulator. Thus, when we consider a *cis*-regulatory sequence being fully bound by its matching transcription regulator, this state is an average, over time, of many individual association and dissociation events.

By attaching a transcription regulator to a bright fluorescent tag, it is possible to follow single regulator molecules in live cells, as they diffuse randomly within the nucleus, bind to their *cis*-regulatory sequences, and then dissociate from them. In these *single-molecule tracking experiments*, different states for the regulator can be distinguished on the basis of the tagged protein's mobility over short time periods. A high-mobility regulator state is observed for the free protein diffusing in the nucleoplasm. At the other extreme, a very low-mobility state is attributed to the regulator bound to DNA, inasmuch as its restrained motions are similar to that of a histone molecule that has been labeled in the same way (Figure 7–14).

Whereas a histone remains stably bound in a nucleosome, transcription regulators remain in a low-mobility, DNA-bound state only transiently. Individual regulator molecules are observed to leave their DNA-bound state at a wide variety of rates—some molecules persist for only a fraction of a second, while others remain for minutes. How can we explain these differences? We saw earlier in the chapter (see Figure 7-10) that each transcription regulator has a preferred *cis*-regulatory sequence, but that it can also bind—albeit with lower affinity—to related

Figure 7–14 **Tracking single molecules of a transcription regulator in the nucleus of a living cell.** By conjugating a fluorescent tag to the glucocorticoid receptor (see pp. 573–575), the behavior of this transcription regulator can be followed in living cells, using a microscope that follows its fluorescence. Computational methods then allow the observed behavior of such molecules to be classified into sets of distinct mobility groups, two of which are shown here. (A) Sample tracks observed for individual molecules of the glucocorticoid receptor in the freely diffusing mobility group. The positions illustrated were determined for a total of 10 seconds. (B) Tracks of individual molecules bound to DNA, with positions determined over a 120-second interval. (A and B, courtesy of D.A. Garcia and G.L. Hager.)

DNA sequences. Because the forward rates at which regulatory proteins "find" their *cis*-regulatory sequences are largely independent of the exact nucleotide sequence of that DNA, affinity differences are reflected in how long a protein remains bound on the DNA—the higher the affinity, the longer the protein stays bound.

Any protein, such as a transcription regulator, that binds tightly to a specific set of DNA sequences will also bind, albeit much more weakly, to any DNA sequence. This weak binding is useful because it allows a regulator to search for its target by "scanning" the DNA in the vicinity of the initial chromosomal site that it binds. Most such regulators will fail to find a matching *cis*-regulatory DNA sequence, and it is these that are thought to dissociate within seconds. The minority that persist for minutes are likely to have engaged with a matching *cis*-regulatory sequence. But because even these regulators do not remain on DNA for long periods, they need to be constantly replaced by another such molecule. Thus, as always, the static pictures in this textbook fail to do justice to the frantic state of motion that exists inside a cell (see pp. 65–66).

Summary

Transcription regulators recognize short stretches of double-helical DNA of defined sequence called cis-regulatory sequences, and they thereby determine which of the thousands of genes in a cell will be transcribed. Transcription regulators determine many cell properties, and their importance is reflected by the fact that approximately 10% of the protein-coding genes in most organisms produce them. Although each transcription regulator has unique features, most bind to DNA as homodimers or heterodimers and recognize DNA through one of a small number of structural motifs. Transcription regulators typically work in groups and bind to DNA cooperatively, a feature that is explained by several underlying mechanisms, some of which exploit the packaging of DNA in nucleosomes.

TRANSCRIPTION REGULATORS SWITCH GENES ON AND OFF

Having seen how transcription regulators bind to *cis*-regulatory sequences embedded in the genome, we can now discuss how, once bound, these proteins influence the transcription of genes. The situation in bacteria is simpler than in eukaryotes (for one thing, chromatin structure is not an issue), and we therefore discuss bacterial mechanisms before proceeding to the more complex situation in eukaryotes.

The Tryptophan Repressor Switches Genes Off

The genome of the bacterium *Escherichia coli* consists of a single, circular DNA molecule of about 4.6×10^6 nucleotide pairs that encodes approximately 4300 proteins. Only a fraction of these proteins are made at any one time. For example, all bacteria regulate the expression of many of their genes according to the food sources that are available in the environment. Thus in *E. coli*, five genes code for enzymes that manufacture the amino acid tryptophan. These genes are arranged in a cluster on the chromosome and are transcribed from a single promoter as one long mRNA molecule; such coordinately transcribed clusters are called *operons* (**Figure 7–15**). Such operons are common in bacteria but rare in

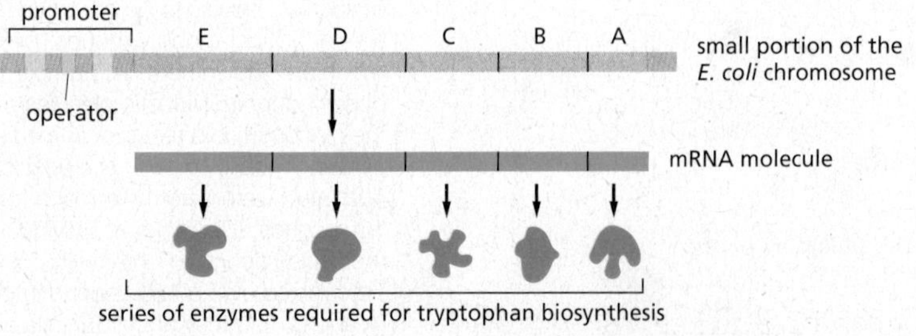

Figure 7–15 A cluster of bacterial genes can be transcribed from a single promoter. Each of these five genes encodes a different enzyme, and all of these enzymes are needed to synthesize the amino acid tryptophan from simpler molecules. The genes are transcribed as a single mRNA molecule, a feature that allows their expression to be coordinated. Clusters of genes transcribed as a single mRNA molecule are common in bacteria. Each of these clusters is called an *operon* because its expression is controlled by a *cis*-regulatory sequence called the *operator (green)*, situated within the promoter. (In this and subsequent figures, the *yellow blocks* in the promoter represent DNA sequences that bind RNA polymerase; see Figure 6–12).

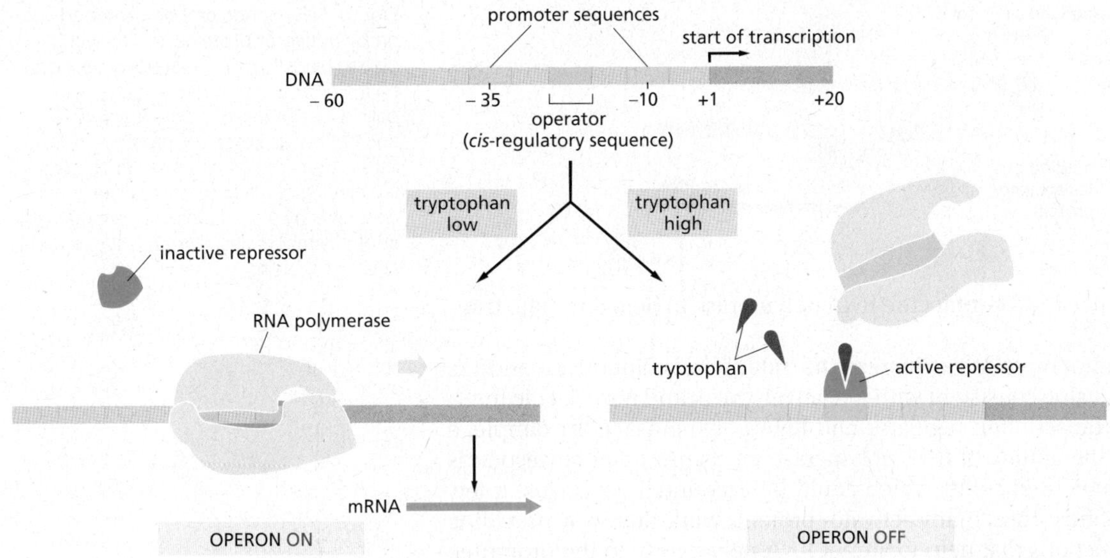

Figure 7–16 **Genes can be switched off by repressor proteins.** If the concentration of tryptophan inside a bacterium is low *(left)*, RNA polymerase *(blue)* binds to the promoter and transcribes the five genes of the tryptophan operon. However, if the concentration of tryptophan is high *(right)*, the tryptophan repressor protein *(dark green)* becomes active and binds to the operator *(light green)*, where it blocks the binding of RNA polymerase to the promoter. Whenever the concentration of intracellular tryptophan drops, this transcription regulator falls off the DNA, allowing the polymerase to again transcribe the operon. Although not shown in the figure, the tryptophan repressor exists as a stable protein dimer.

eukaryotes, where genes are typically transcribed and regulated individually (see Figures 6–75 and 6–90).

When tryptophan concentrations are low, the operon is transcribed; the resulting mRNA is translated to produce a full set of biosynthetic enzymes, which work in tandem to synthesize tryptophan from much simpler molecules. When tryptophan is abundant, however—for example, when the bacterium is in the gut of a mammal that has just eaten a protein-rich meal—the amino acid is imported into the cell and shuts down production of the enzymes, which are no longer needed.

We now understand exactly how this repression of the tryptophan operon comes about. Within the operon's promoter is a *cis*-regulatory sequence that is recognized by a transcription regulator. When this regulator binds to this sequence, it blocks access of RNA polymerase to the promoter, thereby preventing transcription of the operon (and thus production of the tryptophan-producing enzymes). The transcription regulator is known as the *tryptophan repressor*, and its *cis*-regulatory sequence is called the *tryptophan operator*. These components are controlled in a simple way: the repressor can bind to DNA only if it has also bound several molecules of tryptophan (**Figure 7–16**).

The tryptophan repressor is an allosteric protein, and the binding of tryptophan causes a subtle change in its three-dimensional structure so that the protein can bind tightly to the operator sequence. Whenever the concentration of free tryptophan in the bacterium drops, tryptophan dissociates from the repressor, the repressor no longer binds to DNA, and the tryptophan operon is transcribed. The repressor is thus a simple device that switches production of a set of biosynthetic enzymes on and off according to the availability of the end product of the pathway that the enzymes catalyze.

The tryptophan repressor protein itself is always present in the cell. The gene that encodes it is continually transcribed at a low level, so that a small amount of the repressor protein is always being made. Thus the bacterium can respond very rapidly to a rise or fall in tryptophan concentration.

Repressors Turn Genes Off and Activators Turn Them On

The tryptophan repressor, as its name suggests, is a *transcription repressor* protein: in its active form, it switches genes off, or *represses* them. Some bacterial transcription regulators do the opposite: they switch genes on, or *activate* them. These *transcription activator* proteins work on promoters that—in contrast to the promoter for the tryptophan operon—are only marginally able to bind and position RNA polymerase on their own. However, these poorly functioning promoters can be made fully functional by activator proteins that bind to nearby

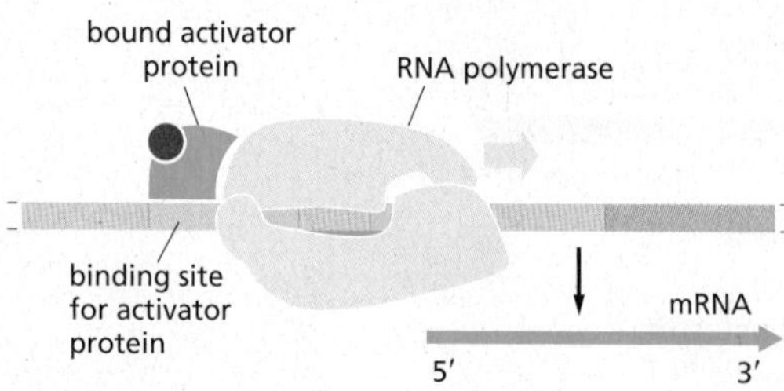

Figure 7–17 **Genes can be switched on by activator proteins.** An activator protein binds to its *cis*-regulatory sequence on the DNA and interacts with the RNA polymerase to help it initiate transcription. Without the activator, the promoter fails to initiate transcription efficiently. In bacteria, the binding of the activator to DNA is often controlled by the interaction of a metabolite or other small molecule *(red circle)* with the activator protein.

cis-regulatory sequences and contact the RNA polymerase to help it initiate transcription (**Figure 7–17**).

DNA-bound activator proteins can increase the rate of transcription initiation as much as 1000-fold, a value consistent with a relatively weak and nonspecific interaction between the transcription regulator and RNA polymerase. For example, a 1000-fold change in the affinity of RNA polymerase for its promoter corresponds to a change in ΔG of ~18 kJ/mole, which could be accounted for by just a few weak, noncovalent bonds. Thus, many activator proteins work simply by providing a few favorable interactions that help to attract RNA polymerase to the promoter. To provide this assistance, however, the activator protein must be bound to its *cis*-regulatory sequence, and this sequence must be positioned precisely so that these favorable interactions can occur with an RNA polymerase molecule at its promoter.

Like the tryptophan repressor, activator proteins often have to interact with a second molecule to be able to bind DNA. For example, the bacterial activator protein CAP has to bind cyclic AMP (cAMP) before it can bind to DNA. Genes activated by CAP are switched on in response to an increase in intracellular cAMP concentration, which rises when glucose, the bacterium's preferred carbon source, is no longer available. CAP then drives the production of enzymes that allow the bacterium to digest other sugars.

Both an Activator and a Repressor Control the *Lac* Operon

The activity of a single bacterial promoter is often controlled by several different transcription regulators. The *Lac* operon in *E. coli*, for example, is controlled by both the Lac repressor and the CAP activator just discussed. The *Lac* operon encodes proteins required to import and digest the disaccharide lactose, a key nutrient in milk. In the absence of glucose (the cell's favorite energy source), the bacterium makes cAMP, which activates CAP to switch-on genes that allow the cell to utilize alternative sources of carbon—including lactose. It would be wasteful, however, for CAP to induce expression of the *Lac* operon if lactose itself were not present. Thus the Lac repressor shuts off the operon in the absence of lactose. This arrangement enables the control region of the *Lac* operon to integrate two different signals, so that the operon is highly expressed only when two conditions are met: glucose must be absent and lactose must be present (**Figure 7–18**). This genetic circuit thus behaves much like a switch that carries out a logic operation in a computer. When lactose is present AND glucose is absent, the cell executes the appropriate program—in this case, transcription of the genes that permit the uptake and utilization of lactose.

All transcription regulators, whether they are repressors or activators, must be bound to DNA to exert their effects. In this way, each regulatory protein acts selectively, controlling only those genes that bear a *cis*-regulatory sequence recognized by it. The logic of the *Lac* operon first attracted the attention of biologists more than 60 years ago. The way it works was uncovered by a combination of genetics and biochemistry, providing some of the first insights into how transcription is controlled in any organism.

DNA Looping Can Occur During Bacterial Gene Regulation

We have seen that transcription activators help RNA polymerase to initiate transcription and repressors hinder it. Otherwise the two types of transcription regulators are similar: both the tryptophan repressor and the CAP activator

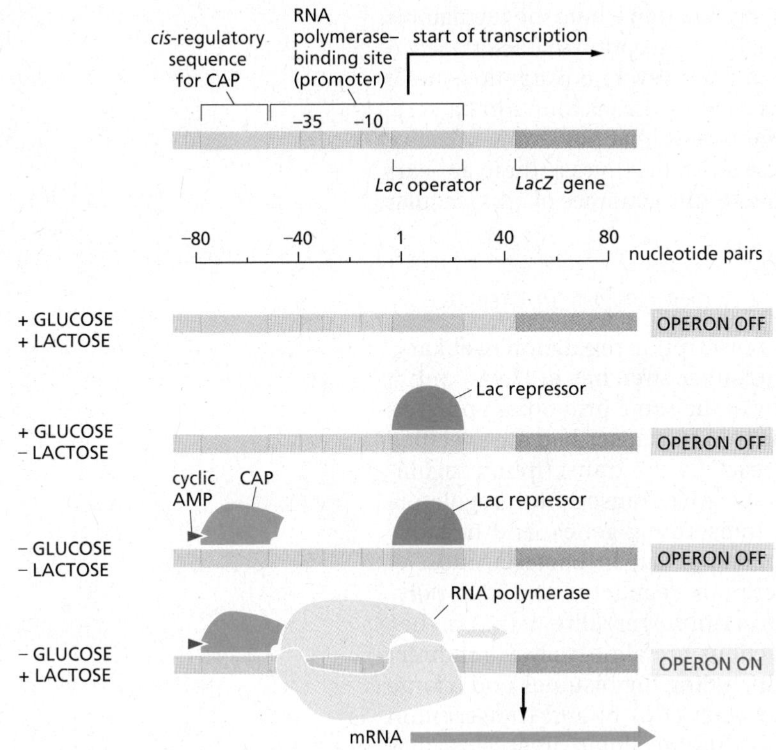

Figure 7–18 **How the *Lac* operon is controlled by two transcription regulators, causing it to be expressed only when needed.** *LacZ*, the first gene of the operon, encodes the enzyme β-galactosidase, which breaks down lactose to galactose and glucose. When lactose is absent, the Lac repressor binds to a *cis*-regulatory sequence, called the *Lac* operator, and shuts off expression of the operon (**Movie 7.4**). Addition of lactose increases the intracellular concentration of a related compound, allolactose; allolactose binds to the Lac repressor, causing it to undergo a conformational change that releases its grip on the operator DNA (not shown). This removes a block to expression of the *Lac* operon, but the operon can turn on only if the sugar glucose, a preferred carbon source, is absent. This is because cyclic AMP *(red triangle)* is produced by the cell in the absence of glucose, and this small molecule is required for CAP to bind to DNA and activate transcription.

protein must bind a small molecule to occupy their *cis*-regulatory sequences, and both recognize these DNA sequences using the same structural motif (the helix-turn-helix shown in Panel 7–1). Some proteins (for example, the CAP protein) can act either as a repressor or an activator, depending on the exact placement of a binding site relative to the promoter: if this site overlaps the promoter, CAP binding can prevent the assembly of RNA polymerase at the promoter, thus serving as a repressor.

Most bacteria have small, compact genomes, and the *cis*-regulatory sequences that control the transcription of a gene are typically located very near to the start point of transcription. But there are some exceptions to this generalization—*cis*-regulatory sequences can be located hundreds and even thousands of nucleotide pairs from the bacterial genes they control. In these cases, the intervening DNA loops out, allowing a transcription regulator bound at a distant site along the DNA to contact RNA polymerase (**Figure 7–19**). Here, the DNA is serving as a tether, enormously increasing the probability that the regulator will collide with a

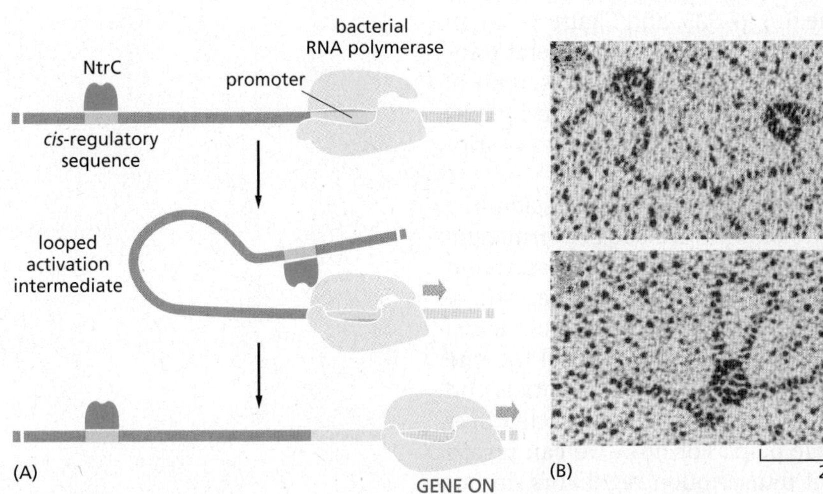

Figure 7–19 **Transcriptional activation by DNA looping in bacteria.** (A) The NtrC protein is a bacterial transcription regulator that activates transcription by directly contacting RNA polymerase. (B) The interaction of NtrC and RNA polymerase, with the intervening DNA looped out, can be seen in the electron microscope. (B, courtesy of Harrison Echols and Sydney Kustu.)

promoter-bound polymerase, compared with the situation where the regulator is free in solution. We will see shortly that, although the exception in bacteria, DNA looping is thought to occur in the regulation of nearly every eukaryotic gene. It has been proposed that the compact, simple genetic switches found in bacteria evolved in response to a severe competition for growth that put strong selective pressure on bacteria to maintain small genome sizes. In contrast, there appears to have been little selective pressure to "streamline" the genomes of multicellular organisms.

Complex Switches Control Gene Transcription in Eukaryotes

When compared to the situation in bacteria, transcription regulation in eukaryotes involves many more proteins and much longer stretches of DNA—and it often seems bewilderingly complex. Yet many of the same principles apply. As in bacteria, the time and place that each gene is to be transcribed are specified by its *cis*-regulatory sequences, which are "read" by the transcription regulators that bind to them. Once bound to DNA, positive transcription regulators (activators) help RNA polymerase to begin transcribing genes, and negative regulators (repressors) block this from happening. But in bacteria, most of the interactions between DNA-bound transcription regulators and RNA polymerases (whether they activate or repress transcription) are direct; that is, they contact each other. In contrast, these interactions are almost always indirect in eukaryotes: many intermediate proteins, including the histones and a large protein complex known as *Mediator*, act between DNA-bound transcription regulators and RNA polymerase. Moreover, in multicellular organisms, it is common for dozens of transcription regulators to control a single gene and for *cis*-regulatory sequences to be spread over tens of thousands of nucleotide pairs. DNA looping allows the DNA-bound regulatory proteins to interact with each other and ultimately to control RNA polymerase at the promoter. Many of the protein–protein interactions involved are of low affinity and are thought to trigger the formation of biomolecular condensates, which can facilitate reactions requiring such a large number of different components (see pp. 171–173). Finally, because nearly all of the DNA in eukaryotic organisms is organized in nucleosomes and higher-order chromatin structures, transcription initiation in eukaryotes must overcome this inherent block. In the next sections, we discuss each of these features of transcription initiation in eukaryotes, emphasizing how they provide extra levels of control not found in bacteria.

A Eukaryotic Gene Control Region Includes Many *cis*-Regulatory Sequences

In eukaryotes, RNA polymerase II transcribes all the protein-coding genes and many noncoding RNA genes. This polymerase requires five general transcription factors (with 27 subunits *in toto*; see Table 6–3, p. 333, and Figure 6–15), in contrast to bacterial RNA polymerase, which needs only a single general transcription factor (the σ subunit). As we saw in Chapter 6, the stepwise assembly of the general transcription factors at a eukaryotic promoter provides, in principle, multiple steps at which the cell can speed up or slow down the rate of transcription initiation in response to transcription regulators.

Because the many *cis*-regulatory sequences that control the expression of a typical gene are often spread over long stretches of DNA, we use the term **gene control region** to describe the whole expanse of DNA involved in regulating and initiating transcription of a eukaryotic gene. This includes the **promoter**, where the general transcription factors and the polymerase assemble, plus all of the *cis*-regulatory sequences to which transcription regulators bind to control the rate of the gene activation processes at the promoter (**Figure 7–20**). In animals and plants, it is not unusual to find the regulatory sequences of a gene dotted over stretches of DNA as large as 100,000 nucleotide pairs. For now, we can regard much of this DNA as "spacer" sequences that transcription regulators do not

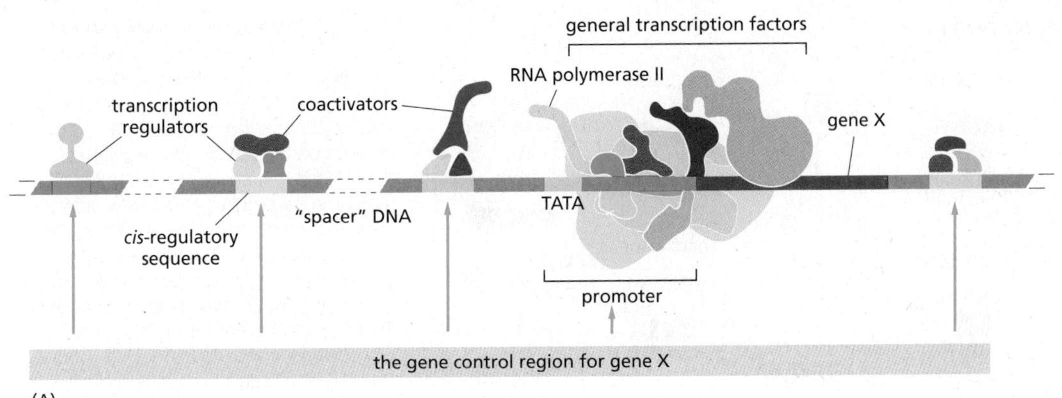

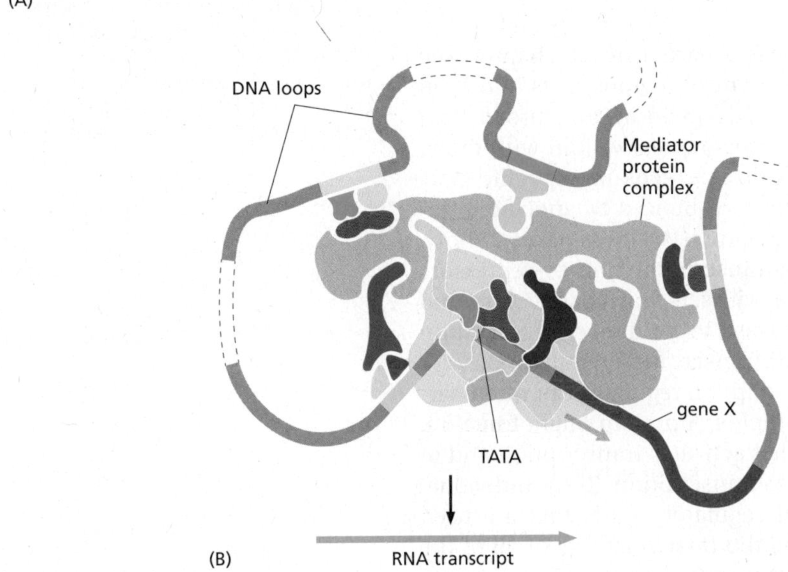

Figure 7–20 Transcription is controlled by gene control regions. (A) The gene control region of a typical eukaryotic gene depicted with the DNA arranged in a straight line. The *promoter* is the DNA sequence where the general transcription factors and the polymerase assemble (see Figure 6–15). The *cis-regulatory sequences* are binding sites for transcription regulators, whose presence on the DNA ultimately affects the rate of transcription initiation. These sequences can be located adjacent to the promoter, far upstream of it, or even within introns or entirely downstream of the gene. The broken stretches of DNA signify that the length of DNA between the *cis*-regulatory sequences and the start of transcription varies, sometimes reaching tens of thousands of nucleotide pairs in length. The TATA box is a DNA recognition sequence for the general transcription factor TFIID (see Figures 6–15 and 6–17). (B) DNA looping allows transcription regulators bound at many positions to "communicate" with the proteins that assemble at the promoter. As shown in this schematic diagram, many transcription regulators act through Mediator (described in Chapter 6), while some interact with the general transcription factors and RNA polymerase directly. Transcription regulators also act by recruiting proteins that alter the chromatin structure of the promoter (not shown here but discussed later in the chapter). Whereas Mediator and the general transcription factors are the same for all RNA polymerase II–transcribed genes, the transcription regulators and the locations of their binding sites relative to the promoter differ for each gene. At especially complex gene control regions, the many proteins that assemble can, by virtue of large numbers of low-specificity interactions, undergo phase transitions that further coalesce the protein and DNA components needed to initiate transcription—presumably accelerating the process.

directly recognize. We will see later in this chapter that some of this DNA is transcribed (but not translated) into long noncoding RNAs (lncRNAs), which have diverse functions in the cell.

In this chapter, we shall loosely use the term **gene** to refer to a segment of DNA that is transcribed into a functional RNA molecule, one that either codes for a protein or has a different role in the cell (see Table 6–1, p. 327). However, the classical view of a gene includes the gene control region as well, because mutations in it can produce an altered phenotype. Alternative RNA splicing further complicates the definition of a gene—a point we shall return to later.

In contrast to the small number of general transcription factors, which are abundant proteins that assemble on the promoters of all genes transcribed by RNA polymerase II, there are thousands of different transcription regulators devoted to turning individual genes on and off. As we have seen, each eukaryotic gene is usually transcribed individually. Not surprisingly, the regulation of each eukaryotic gene is different in detail from that of every other gene, and it is difficult to formulate simple rules for gene regulation that apply in every case. We can, however, make some generalizations about how transcription regulators, once bound to gene control regions on DNA, set in motion the series of events that lead to gene activation or repression.

Eukaryotic Transcription Regulators Work in Groups

In bacteria, we saw that proteins such as the tryptophan repressor, the Lac repressor, and the CAP protein bind to DNA on their own and directly affect RNA polymerase at the promoter. Eukaryotic transcription regulators, in contrast, usually assemble together in groups at their *cis*-regulatory sequences. Often two

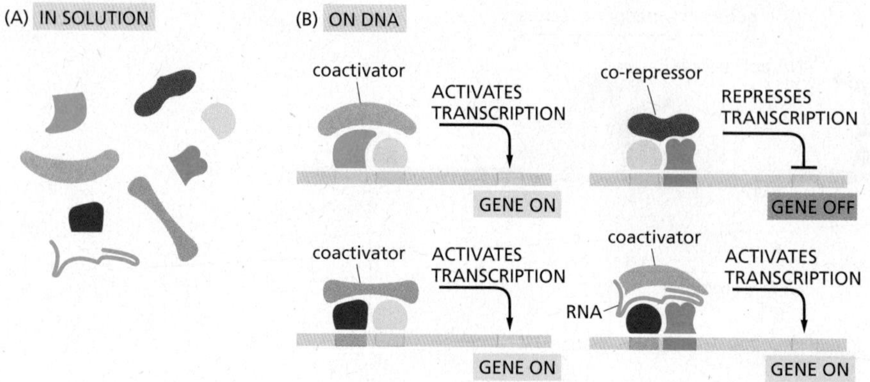

Figure 7–21 Eukaryotic transcription regulators assemble into complexes on DNA. (A) Seven different proteins and an RNA molecule are shown. The nature and function of the complex they form depend on the specific *cis*-regulatory sequences that seed their assembly. (B) Some assembled complexes activate gene transcription, while another represses transcription. Note that the *light green* and *dark green* proteins are shared by both activating and repressing complexes. Proteins that do not themselves bind DNA but assemble on other DNA-bound transcription regulators are termed coactivators or co-repressors. In some cases *(lower right)*, long, noncoding RNA molecules are also found in these assemblies. As described later in this chapter, these RNAs often act as scaffolds to hold groups of proteins together.

or more regulators bind cooperatively, as discussed earlier in the chapter (see Figure 7–10). In some especially complex gene control regions, tens and even hundreds of such proteins may coassemble on DNA. In addition, a broad class of multisubunit proteins termed *coactivators* and *co-repressors* join with them. Typically, these coactivators and co-repressors do not recognize specific DNA sequences themselves; they are brought to those sequences by specific interactions with the DNA-bound transcription regulators. As their names imply, coactivators are typically involved in activating transcription and co-repressors in repressing it. In the following sections, we will see that coactivators and co-repressors can act in a variety of different ways to influence transcription once they have been localized on the genome by transcription regulators.

As shown in **Figure 7–21**, an individual transcription regulator can often participate in more than one type of regulatory complex. A protein might function, for example, in one case as part of a complex that activates transcription and in another case as part of a complex that represses transcription. Thus, individual eukaryotic transcription regulators function as regulatory parts that are used to build complexes whose function depends on the final assembly of all of the individual components. Each eukaryotic gene is therefore regulated by a "committee" of proteins, all of which must be present to express the gene at its proper level. Often the protein–protein interactions between transcription regulators and between regulators and coactivators are too weak for them to assemble in solution; however, the appropriate combination of *cis*-regulatory sequences can "crystallize" the assembly of these complexes on DNA. In very large and complex gene control regions, this assembly may be accompanied by a phase transition to form a biomolecular condensate, whereby all the components are held together even more efficiently by keeping them in rough proximity even when individual proteins disassociate from DNA.

Activator Proteins Promote the Assembly of RNA Polymerase at the Start Point of Transcription

The *cis*-regulatory sequences to which eukaryotic transcription activator proteins bind were originally called *enhancers* because their presence "enhanced" the rate of transcription initiation. It initially came as a surprise when it was discovered that these sequences could be found tens of thousands of nucleotide pairs away from the promoter; as we have seen, DNA looping, which was not widely appreciated at the time, can now explain this initially puzzling observation.

Once bound to DNA, how do assemblies of activator proteins increase the rate of transcription initiation? At most genes, several mechanisms work in concert. Their ultimate function is to attract and position RNA polymerase II at the promoter and to release it so that transcription can begin.

Some activator proteins bind directly to one or more of the general transcription factors, accelerating their assembly on a promoter that has been brought in proximity—through DNA looping—to that activator. Most transcription activators, however, attract coactivators that then perform the biochemical tasks needed to initiate transcription. As we have seen, one of the most prevalent coactivators is the large *Mediator* protein complex, composed of more than 30 subunits. About the same size as RNA polymerase itself, Mediator serves as a bridge between DNA-bound transcription activators, RNA polymerase, and the general transcription factors, facilitating their assembly at the promoter (see Figure 7–20).

Eukaryotic Transcription Activators Direct the Modification of Local Chromatin Structure

The eukaryotic general transcription factors and RNA polymerase are unable, on their own, to assemble on a promoter that is packaged in nucleosomes. Thus, in addition to directing the assembly of the transcription machinery at the promoter, eukaryotic transcription activators—once bound to their *cis*-regulatory sequences—promote transcription by triggering changes to the chromatin structure of the promoters, rendering the underlying DNA more accessible. The enzymes that alter chromatin structure are usually carried as subunits of coactivators, which are typically multiprotein complexes, with different subunits carrying out different functions. For example, such a coactivator might carry one subunit that associates with specific DNA-bound transcription regulators, another that associates with one of the general transcription factors, and several more that alter chromatin structure in different ways.

The most important ways of locally altering chromatin are through covalent histone modifications, nucleosome remodeling, nucleosome removal, and histone replacement (all discussed in Chapter 4). Eukaryotic transcription activators use all four of these mechanisms: thus they attract coactivators that include histone modification enzymes, ATP-dependent chromatin remodeling complexes, and histone chaperones. These proteins often act cooperatively to alter the chromatin structure of promoters, providing greater access to the DNA (**Figure 7–22**).

Often a series of individual events, ultimately directed by transcription regulators, must occur before RNA polymerase can be assembled onto a promoter,

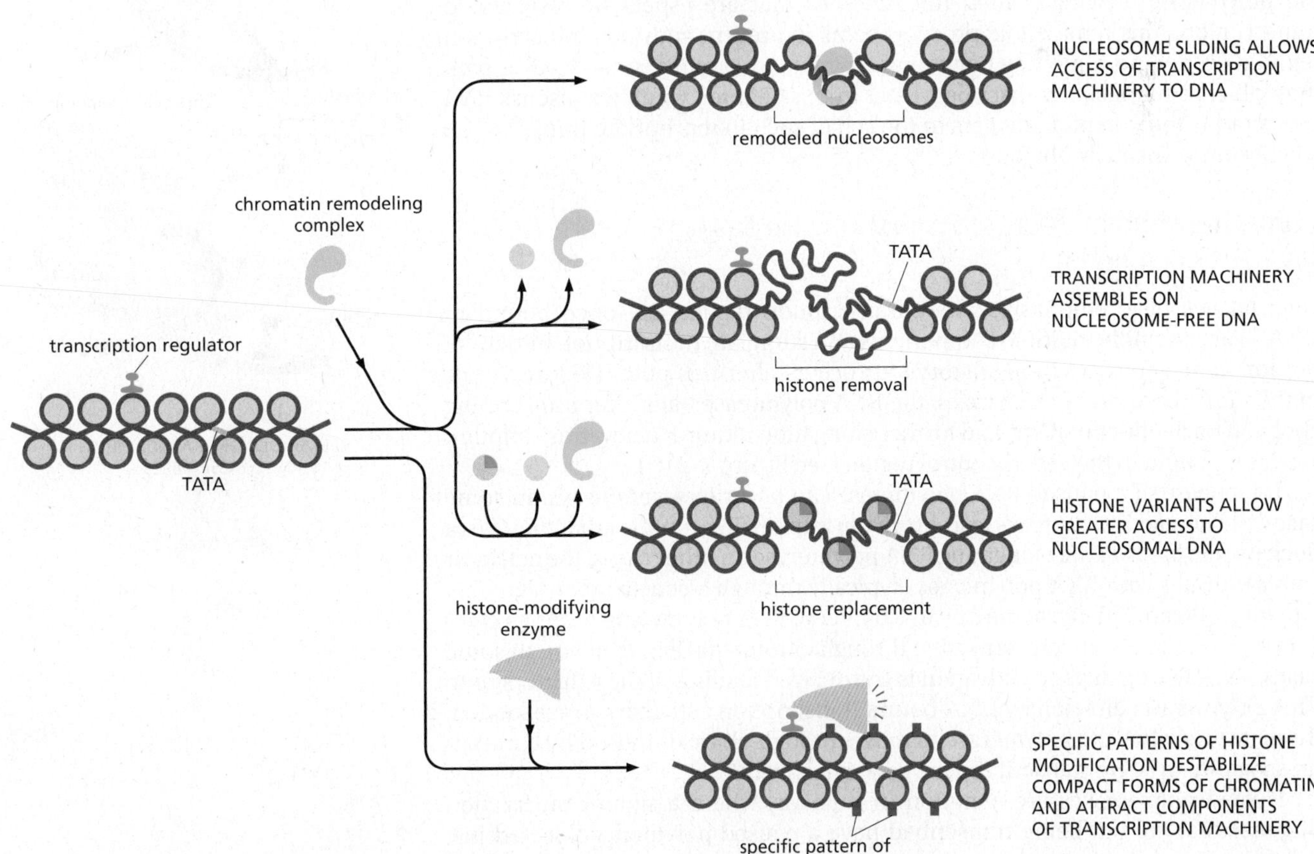

Figure 7–22 **Eukaryotic transcription activator proteins direct local alterations in chromatin structure.** Nucleosome remodeling, nucleosome removal, histone replacement, and certain types of histone modifications favor transcription initiation (see Table 4–2, p. 210). As illustrated, some of these changes are driven by different types of ATP-dependent chromatin remodeling complexes (see Figures 4–26 and 4–27); most also involve histone chaperones (not shown). Such alterations increase the accessibility of DNA and facilitate the binding of RNA polymerase and the general transcription factors.

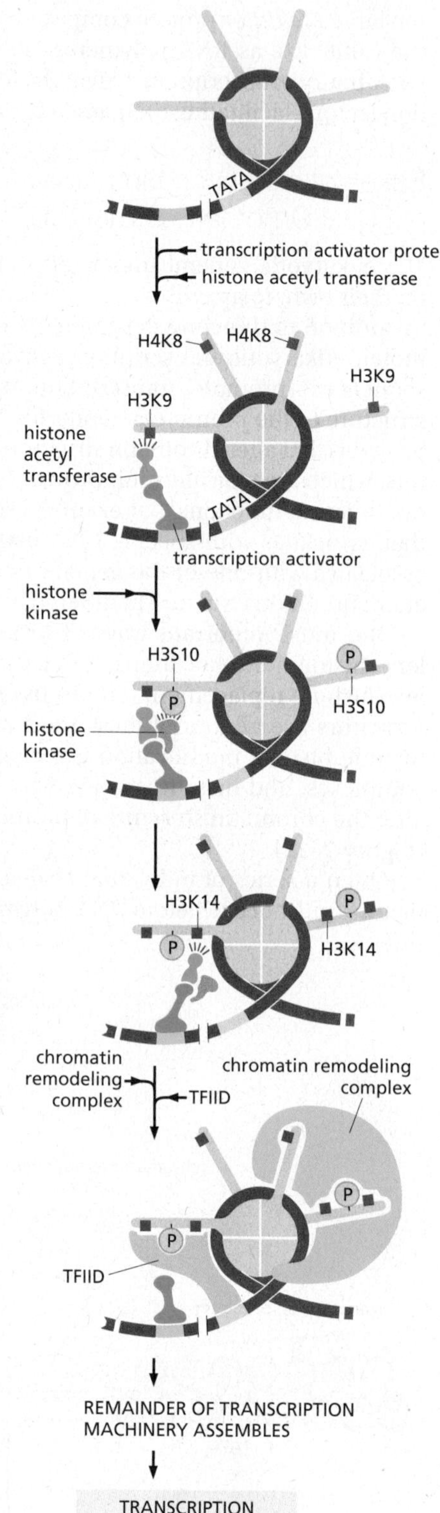

Figure 7–23 Successive histone modifications during transcription initiation. In this example, taken from the human interferon-β gene promoter, a transcription activator binds to DNA packaged into chromatin and attracts a histone acetyl transferase that acetylates lysine 9 of histone H3 and lysine 8 of histone H4 (see Figure 4–35). Next, a histone kinase, part of a different coactivator attracted by the same transcription activator, phosphorylates serine 10 of histone H3, but it can only do so after lysine 9 has been acetylated. This serine modification signals the original histone acetyl transferase to acetylate position K14 of histone H3. Next, the general transcription factor TFIID and a chromatin remodeling complex come into play to promote the subsequent steps of transcription initiation. TFIID and the remodeling complex both recognize acetylated histone tails through a *bromodomain*, a protein domain specialized to read this particular mark on histones; a bromodomain is carried in a subunit of each protein complex. Binding of TFIID causes a sharp bend in the DNA (not shown but see Figure 6–17), which facilitates sliding of the nucleosome to a new position, thereby freeing the start site of transcription for binding by RNA polymerase II.

The histone acetyl transferase, the histone kinase, and the chromatin remodeling complex are all subunits of coactivators. The order of events shown applies to a specific promoter; at other genes, the steps may occur in a different order or individual steps may be omitted altogether. (Adapted from T. Agalioti et al., *Cell* 111:381–392, 2002.)

with details that depend on the gene being regulated. In the example illustrated in **Figure 7–23**, a series of specific histone tail modifications is triggered by a transcription activator; these modifications then attract additional proteins to the promoter, including both a chromatin remodeling complex and a general transcription factor. Those proteins can in turn recruit additional proteins to the promoter, while also destabilizing adjacent nucleosomes.

Because the local chromatin changes directed by one transcription regulator often allow the binding of additional proteins—both directly (see Figure 7–12) and indirectly as just described—a cascade of events typically takes place on the control regions of eukaryotic genes to regulate their transcription.

As RNA polymerase II transcribes through a gene a different type of chromatin modification occurs. The histones just ahead of the polymerase are acetylated by enzymes carried by the polymerase, removed by histone chaperones, and deposited behind the moving polymerase. These histones are then rapidly deacetylated and methylated, also by complexes that are carried by the polymerase, leaving behind nucleosomes that are especially resistant to transcription. This remarkable process seems to prevent spurious transcription reinitiation behind a moving polymerase, which, in essence, must clear a path through chromatin as it transcribes. Later in this chapter, when we discuss *RNA interference*, the potential dangers to the cell of such inappropriate transcription will become especially obvious.

Some Transcription Activators Work by Releasing Paused RNA Polymerase

Thus far, we have emphasized how transcription regulators—once bound to DNA—can assemble multiple components and stimulate transcription initiation. But for some genes, a key regulatory step occurs after this point (**Figure 7–24**). In the most common of these cases, the RNA polymerase halts after transcribing about 50 nucleotides of RNA, and further elongation requires a new transcription activator to bind to the gene's control region (see Figure 7–24C).

The release of a paused RNA polymerase can occur in several ways. In some cases, the new activator brings in a chromatin remodeling complex that removes a nucleosome block to the elongating RNA polymerase. In other cases, the activator communicates with RNA polymerase (typically through a coactivator), signaling it to forge ahead. Finally, as we saw in Chapter 6, RNA polymerase requires *elongation factors* to effectively transcribe through chromatin (Figure 6–19). In some cases, the key step in gene activation is the delayed loading of these factors onto RNA polymerase, directed by DNA-bound transcription activators. Once loaded, these factors allow the polymerase to move through blocks imposed by chromatin structure to begin transcribing the gene effectively.

Paused polymerases are common in humans, where a significant fraction of genes that are not being transcribed have a paused polymerase located just

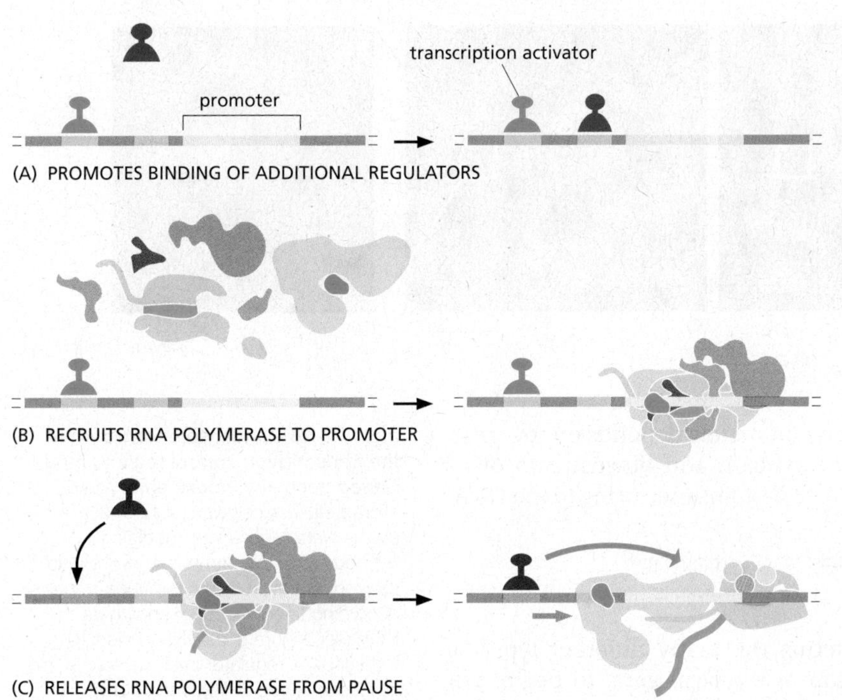

Figure 7–24 Different transcription regulators can act at different steps. (A) As described earlier in this chapter (see Figure 7–12), a DNA-bound transcription activator can promote DNA binding by additional transcription regulators. (B) As shown in more detail in Figures 7–20 and 7–22, most transcription activators direct assembly of RNA polymerase at promoters; this can occur by a variety of mechanisms. (C) Some other transcription activators, once bound to DNA, release RNA polymerase molecules that are paused after transcribing about 50 nucleotides of RNA. For simplicity, many of the additional proteins required for transcription initiation are not shown.

downstream from the promoter. Having RNA polymerase already poised on a promoter in the beginning stages of transcription bypasses the step of assembling many components at the promoter, which is often slow. This mechanism is therefore thought to allow cells to begin transcribing a gene in rapid response to an extracellular signal.

Transcription Activators Work Synergistically

We have seen that complexes of transcription activators and coactivators assemble cooperatively on DNA. We have also seen that these assemblies can promote different steps in transcription initiation. In general, where several factors work together to enhance a reaction rate, the joint effect is not merely the sum of the enhancements that each factor alone contributes, but the product of them. If, for example, factor A lowers the free-energy barrier for a reaction by a certain amount and thereby speeds up the reaction 100-fold, and factor B, by acting on that reaction, does likewise, then A and B acting in parallel can lower the energy barrier by a double amount and speed up the reaction 10,000-fold. Even if A and B work simply by attracting the same protein, the affinity of that protein for the reaction site increases multiplicatively. Thus, transcription activators often exhibit *transcriptional synergy*, where several DNA-bound activator proteins working together produce a transcription rate that is much higher than the sum of their transcription rates working alone (Figure 7–25).

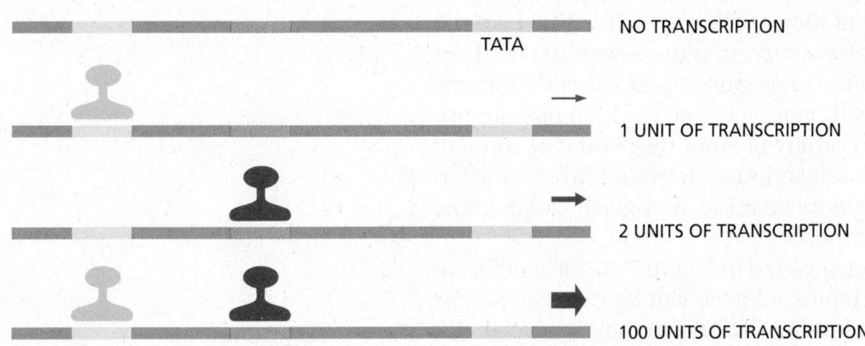

Figure 7–25 Transcriptional synergy. This experiment compares the rate of transcription produced by three experimentally constructed regulatory regions in a eukaryotic cell and reveals transcriptional synergy, a greater than additive effect of multiple activators working together. Such transcriptional synergy is not only observed between different transcription activators from the same organism; it is also seen between activator proteins from different eukaryotic species when they are experimentally introduced into the same cell. This last observation reflects the high degree of conservation of the machinery responsible for eukaryotic transcription initiation.

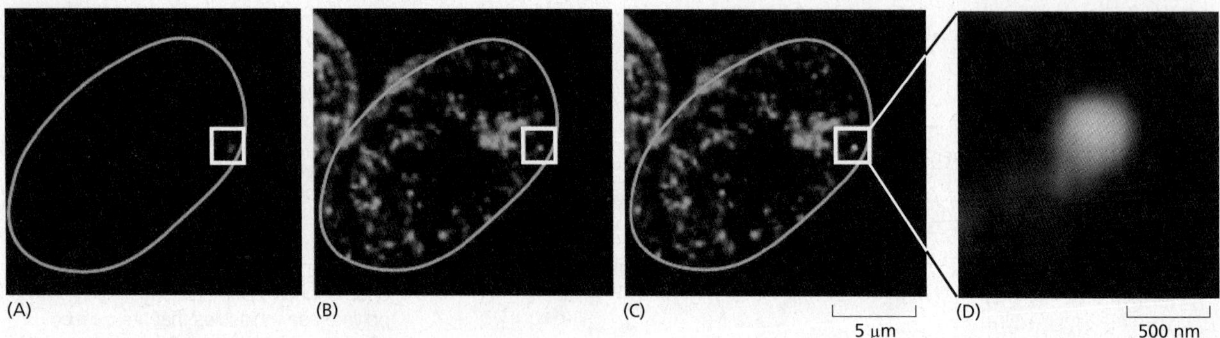

Figure 7–26 **Condensate formation at the transcription control region of the *Nanog* gene in a mouse embryonic stem cell.** The cell was fixed, and in (A) the *Nanog* gene was identified by hybridizing a complementary nucleotide sequence attached to a *red* fluorophore, according to a procedure known as FISH (see Figure 8–32). Nanog is a key transcription regulator in embryonic stem cells (see Figure 7–10), and its own regulatory region is one of the most complex in the mouse genome. The nucleus is indicated by the *blue oval*. (B) A subunit of Mediator fused to a green fluorescent protein (see Figure 9–16) was visualized. (C) The two preceding images have been merged, and in (D) the portion of the image in the *white square* is magnified tenfold. The size and diffuse nature of the "blob" suggest a condensate containing a large number of proteins. Note that additional condensates of Mediator are visible throughout the nucleus and may represent condensates at other enhancers. These condensates are much smaller than those of the nuclear "organelles," such as the nucleolus, discussed in Chapter 6. (From B.R. Sabari et al., *Science* 361:eaar3958, 2018. With permission from AAAS.)

As a result, the rate of transcription of a gene ultimately depends on the spectrum of regulatory proteins that are bound upstream and downstream of its transcription start site, along with the coactivator proteins they bring to the DNA.

Condensate Formation Likely Increases the Efficiency of Transcription Initiation

We have discussed in broad, conceptual terms the many different types of proteins that must assemble for transcription of a typical gene to begin. For especially complex gene control regions, such as those of key human genes that orchestrate development, several hundred individual subunits are involved and, as they begin to assemble on DNA, they become involved in networks that create phase transitions, forming small biomolecular condensates. As described in Chapter 3, such condensates hold their proteins in loose proximity, such that, when one disassociates from the assembly, it can be retained nearby by a network of fluctuating weak interactions (see pp. 171–173). Consistent with this idea, many transcription regulators, coactivators, and co-repressors contain the type of low-complexity, unstructured regions that help to drive condensate formation.

How might this aid transcription? At least some of these transcription condensates contain additional copies of key proteins, including the Mediator complex (**Figure 7–26**). The presence of these extra copies in the same condensate is proposed to make transcription initiation an efficient but highly dynamic process, with proteins within the condensate rapidly exchanging on and off DNA. According to this view, Figure 7–20B represents only a frozen moment in transcription initiation. Whether such condensates form on most eukaryotic genes that are being transcribed—or on just those whose regulation is especially complex—remains to be determined.

Eukaryotic Transcription Repressors Can Inhibit Transcription in Several Ways

Although the "default" state of eukaryotic DNA packaged into nucleosomes is resistant to transcription, eukaryotes nonetheless use transcription regulators to repress the transcription of individual genes. These transcription repressors can rapidly turn off a gene that is being actively transcribed, and they can depress the rate of transcription even below that of the very low default value. Like the transcription activators discussed earlier, transcription repressors often work on a gene-by-gene basis. But unlike the bacterial repressors discussed earlier in this chapter, eukaryotic repressors do not directly compete with the RNA polymerase for access to the DNA. Instead, they use a variety of other mechanisms, some of which are illustrated in **Figure 7–27**. Like transcription activation, transcription repression can act through more than one mechanism at a given target gene, thereby ensuring especially efficient repression.

The different mechanisms of repression depicted in Figure 7–27 have different consequences for the ease with which a repressed gene can be reactivated. For most of the strategies, the repressed state is relatively easy to rapidly reverse, for

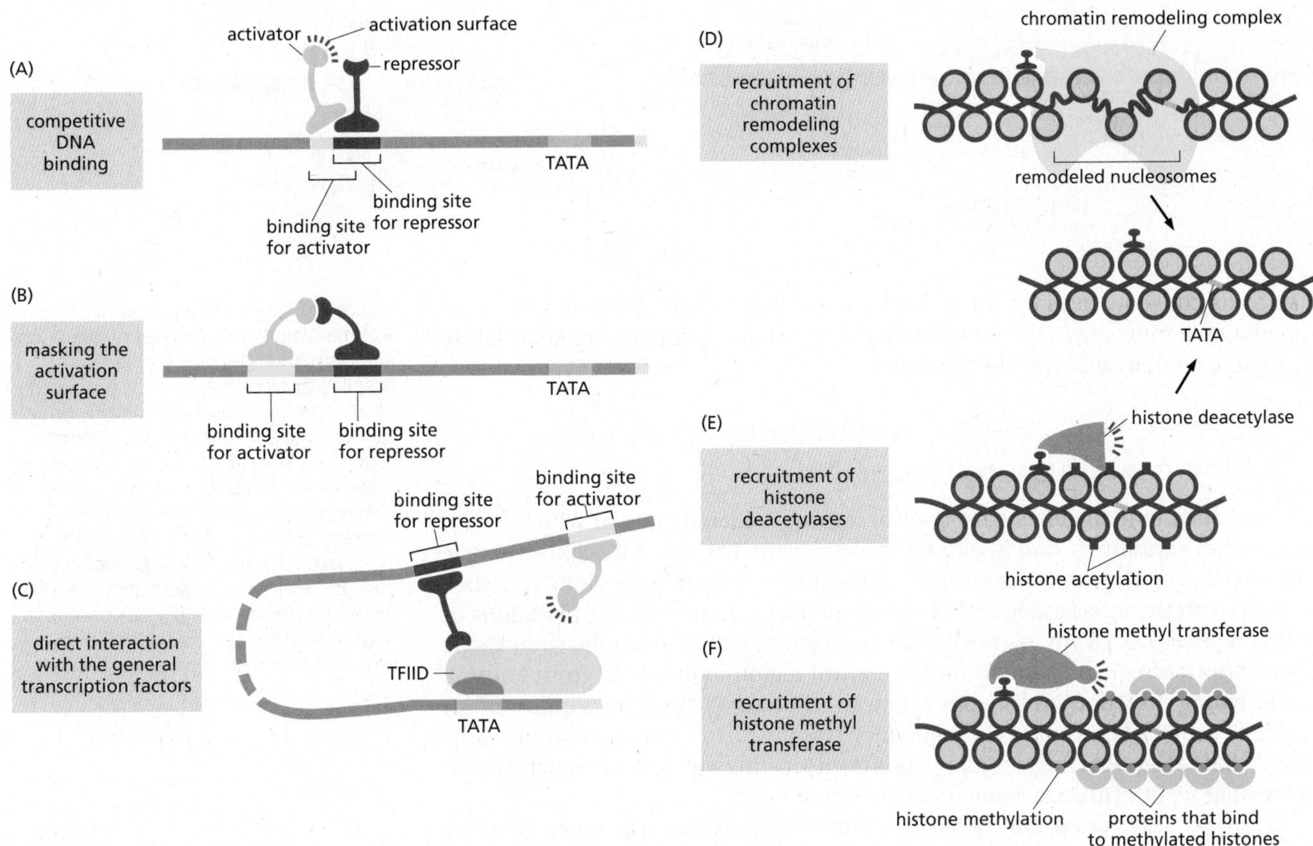

Figure 7–27 Six of the ways in which eukaryotic repressor proteins can operate. (A) A repressor protein outcompetes activator proteins for binding to the same regulatory DNA sequence. (B) Both activator and repressor proteins bind close to each other on DNA, and the repressor "quenches" the activator, preventing it from functioning (for example, by blocking the recruitment of its coactivators). (C) The repressor "poisons" assembly of the general transcription factors by binding to and stabilizing an intermediate. (D) The repressor recruits a chromatin remodeling complex that restores the nucleosomal state of the promoter region to its pre-transcriptional, default form. (E) The repressor attracts a histone deacetylase to the promoter, removing the histone acetylation needed for transcription initiation (see Figure 7–23). (F) Heterochromatin formation is triggered when a repressor attracts a specific histone methyl transferase that trimethylates either lysine 9 or lysine 27 on histone H3, thereby creating either H3K9me3- or H3K27me3-marked nucleosomes. "Read–write" mechanisms then spread each type of methylated nucleosome for thousands of nucleotide pairs along the DNA; they also help the methylation pattern to be inherited across cell divisions (see Figures 4–40 and 4–44). The final step in heterochromatin formation occurs when each type of modified nucleosome attracts additional proteins that condense the DNA and maintain it in a transcriptionally silent form.

example, by simply inactivating the repressor. But, the last mechanism—a directed methylation of specific histone amino acids that creates an unusually highly condensed form of chromatin, known as heterochromatin—is self-reinforcing and can propagate even when the initiating signal is no longer present (see Figure 4–44). As discussed in Chapter 4, chromatin that is marked by H3K9me3 (trimethylation of the lysine at position 9 of histone H3) appears to be the most difficult to transcribe. Typically located around centromeres and repeated DNA sequences such as inactive transposons, this type of heterochromatin strongly suppresses both genetic recombination and transcription. A different histone H3 modification (H3K27me3) is associated with a second form of heterochromatin that is also resistant to transcription. Although apparently easier to activate than the H3K9me3 form, this form of chromatin is also self-propagating and can persist across cell divisions, after the initiating signal has disappeared.

These two types of heterochromatin are used to tightly repress genes active in early development, presumably to make sure that these genes are not expressed in the mature organism. Tight, heritable gene repression is especially important to animals and plants whose growth depends on elaborate and complex developmental programs. Misexpression of a single gene at a critical time can have

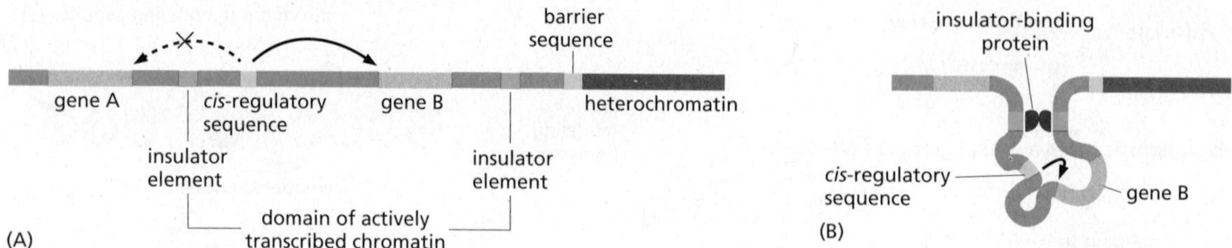

Figure 7–28 Schematic diagram summarizing the properties of insulators and barrier sequences. (A) Insulators directionally block the action of enhancers, whereas barrier sequences prevent the spread of heterochromatin. How barrier sequences likely function is depicted in Figure 4–41. (B) Insulator-binding proteins *(purple)* hold chromatin in loops that favor "correct" enhancer–promoter associations. Thus, gene B is properly regulated, and gene B's *cis*-regulatory sequences can be prevented from influencing the transcription of gene A. The major insulator-binding protein in mammals is denoted CTCF.

disastrous consequences for the individual. For this reason, many of the genes encoding the most important developmental regulatory proteins are kept tightly repressed, often by multiple mechanisms.

Insulator DNA Sequences Prevent Eukaryotic Transcription Regulators from Influencing Distant Genes

We have seen that all genes have control regions, which dictate at which times, under what conditions, and in what tissues the gene will be expressed. We have also seen that eukaryotic transcription regulators can act across very long stretches of DNA, with the intervening DNA looped out. How, then, are control regions of different genes kept from interfering with one another? For example, what keeps a transcription regulator bound on the control region of one gene from looping in the wrong direction and inappropriately influencing the transcription of an adjacent gene? And, if complex regulatory regions form biomolecular condensates, what keeps all of the control regions from forming a giant condensate where the regulatory information would become scrambled?

To avoid such cross-talk between control regions, several types of DNA elements compartmentalize the genome into discrete regulatory domains. In Chapter 4, we discussed *barrier sequences* that prevent the spread of heterochromatin into genes that need to be expressed (see Figure 4–41). A second type of DNA element, called an *insulator*, prevents *cis*-regulatory sequences from running amok and activating inappropriate genes (**Figure 7–28**). As we saw in Chapter 4, insulator sequences function by forming loops of chromatin, an effect mediated by specialized proteins that recognize them (see Figures 4–57 and 7–28B). The loops are thought to keep a gene and its control region in rough proximity and help to prevent the control region from "spilling over" to adjacent genes. More generally, the distribution of insulators and barrier sequences in a genome helps to divide it into independent domains of gene regulation and chromatin structure (see pp. 223–225).

The distribution of the more than 10,000 loops on the collection of mammalian chromosomes can change as cells differentiate or as they respond to changes in their environment. In addition, these loops formed by insulators are not static; rather, they undergo a continual process of loop extrusion and release that is driven by cohesion protein rings (see Figure 4–57). It has been proposed that the extrusion process itself helps to juxtapose enhancers with their matching promoters by sliding them past one another, while helping to break up inappropriate enhancer–promoter connections by physically separating them.

Although chromosomes are dynamically organized into domains that discourage control regions from acting indiscriminately, there are special circumstances where a control region located on one chromosome has been found to deliberately activate a gene located on a different chromosome. Although there is much we do not understand about this mechanism, it reflects the extreme versatility of transcription regulation strategies.

Summary

Transcription regulators switch the transcription of individual genes on and off in cells. In prokaryotes, these proteins typically bind to specific DNA sequences close to the RNA polymerase start site and, depending on the nature of the regulatory protein and the precise location of its binding site relative to the start site, either activate or repress transcription of the gene. The flexibility of the DNA helix, however, also allows transcription regulators bound at distant sites to affect the RNA polymerase at the promoter by the looping out of the intervening DNA. The regulation of higher eukaryotic genes is much more complex, commensurate with a larger genome size and the large variety of cell types that are formed. A single eukaryotic gene is typically controlled by many transcription regulators bound to sequences that can be tens or even hundreds of thousands of nucleotide pairs from the promoter that directs transcription of the gene. Eukaryotic activators and repressors act by a wide variety of mechanisms—generally both altering chromatin structure and controlling the assembly of the general transcription factors and RNA polymerase at the promoter. They do this by attracting coactivators and co-repressors, protein complexes that perform the necessary biochemical reactions. The time and place that each gene is transcribed, as well as its rates of transcription under different conditions, are determined by the particular spectrum of transcription regulators present in the cell that bind to the control region of the gene.

MOLECULAR GENETIC MECHANISMS THAT CREATE AND MAINTAIN SPECIALIZED CELL TYPES

Although all cells must be able to switch genes on and off in response to changes in their environments, the cells of multicellular organisms have evolved this capacity to an extreme degree. In particular, once a cell in a multicellular organism becomes committed to differentiate into a specific cell type, the cell maintains this choice through many subsequent cell generations, which means that it remembers the changes in gene expression involved in the choice. This phenomenon of *cell memory* is a prerequisite for the creation of organized tissues and for the maintenance of stably differentiated cell types. In contrast, other changes in gene expression in eukaryotes, as well as most such changes in bacteria, are only transient. The tryptophan repressor, for example, switches off the tryptophan genes in bacteria only in the presence of tryptophan; as soon as tryptophan is removed from the medium, the genes are switched back on, and the descendants of the cell will have no memory that their ancestors had been exposed to tryptophan.

In this section, we shall examine some specific examples that illustrate how cell types are specified and maintained and how simple gene regulatory devices can be combined to create the "logic circuits" through which cells integrate signals and remember events in their past. We begin by considering one such complex gene control region that has been studied in great detail.

Complex Genetic Switches That Regulate *Drosophila* Development Are Built Up from Smaller Modules

We have seen that transcription regulators can be positioned at multiple sites along long stretches of DNA and that these proteins can bring into play coactivators and co-repressors that ultimately position and activate RNA polymerase to begin transcription. Here, we discuss how the numerous transcription regulators that bind to the control region of a gene can integrate external information, so as to cause the gene to be transcribed at the proper place and time.

The expression of the *Drosophila Even-skipped* (*Eve*) gene plays an important part in the development of the *Drosophila* embryo. If this gene is inactivated by mutation, many parts of the embryo fail to form, and the embryo dies early in development. At the stage of development when *Eve* begins to be expressed, the embryo is a single giant cell containing multiple nuclei in a common cytoplasm. This cytoplasm contains a mixture of transcription regulators that are distributed unevenly along the length of the embryo, thus providing *positional information* that distinguishes one part of the embryo from another

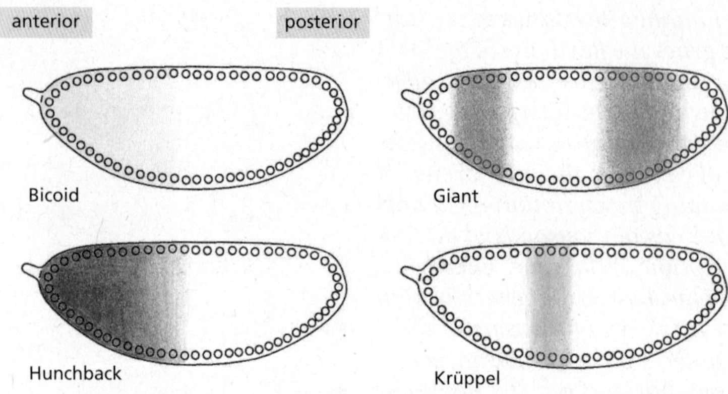

Figure 7–29 **The nonuniform distribution of transcription regulators in an early *Drosophila* embryo.** At this stage, the embryo is a syncytium; that is, multiple nuclei are contained in a common cytoplasm. Although the nuclei are shown in only a slice of the embryo, in reality, they are arranged in three dimensions around the inner surface of the giant cell.

(**Figure 7–29**). Although the nuclei are initially identical, they rapidly begin to express different genes because they are exposed to different transcription regulators: the nuclei near the anterior end of the developing embryo are exposed to a set of transcription regulators that is different from the set present at the middle and that present at the posterior end of the embryo.

The regulatory DNA sequences that control the *Eve* gene have evolved to "read" the concentrations of transcription regulators at each position along the length of the embryo, so as to cause the *Eve* gene to be expressed in seven precisely positioned stripes, each initially five to six nuclei wide. How is this remarkable feat of information processing carried out? Although there is still much to learn, several general principles have emerged from studies of *Eve* and other genes that are similarly regulated.

The control region of the *Eve* gene is very large (approximately 20,000 nucleotide pairs). It is formed from a series of relatively simple regulatory modules, each of which contains multiple *cis*-regulatory sequences and is responsible for specifying a particular stripe of *Eve* expression along the embryo. This modular organization of the *Eve* gene control region was revealed by experiments in which a particular regulatory module (say, that specifying stripe 2) is removed from its normal setting upstream of the *Eve* gene, placed in front of a reporter gene, and reintroduced into the *Drosophila* genome. When developing embryos derived from flies carrying this genetic construct are examined, the reporter gene is found to be expressed in precisely the position of stripe 2 but not in the other normal stripe positions (**Figure 7–30**). Similar experiments reveal the existence of other regulatory modules, which specify other stripes.

The *Drosophila Eve* Gene Is Regulated by Combinatorial Controls

A detailed study of the stripe 2 regulatory module has provided insights into how it reads and interprets positional information. The module contains recognition sequences for two transcription regulators that activate *Eve* transcription (Bicoid and Hunchback) and for two that repress it (Krüppel and Giant) (**Figure 7–31**).

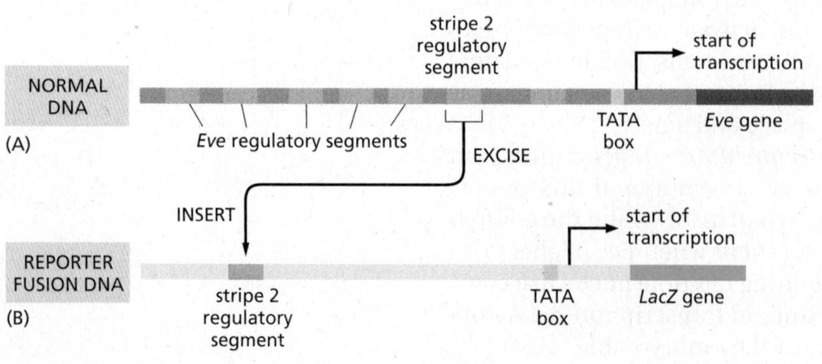

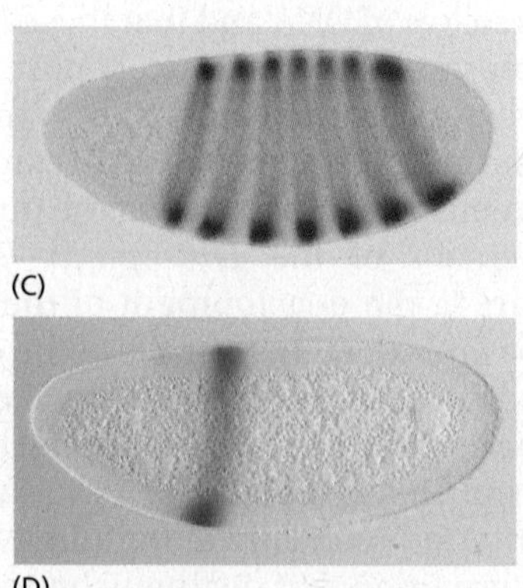

Figure 7–30 **Experiment demonstrating the modular construction of the *Eve* gene regulatory region.** (A) A 480-nucleotide-pair section of the *Eve* regulatory region was removed and (B) inserted upstream of a test promoter that directs the synthesis of the enzyme β-galactosidase (the product of the *E. coli LacZ* gene—see Figure 7–18). (C, D) When this artificial construct was reintroduced into the genome of *Drosophila* embryos, the embryos (D) expressed β-galactosidase (detectable by histochemical staining) precisely in the position of the second of the seven *Eve* stripes. (C) The complete set of *Eve* stripes was detected using antibodies directed against the Eve protein. β-Galactosidase is simple to detect and thus provides a convenient way to monitor the expression specified by a gene control region. As used here, β-galactosidase is said to serve as a reporter, because it "reports" the activity of a gene control region. (C and D, courtesy of Stephen Small and Michael Levine.)

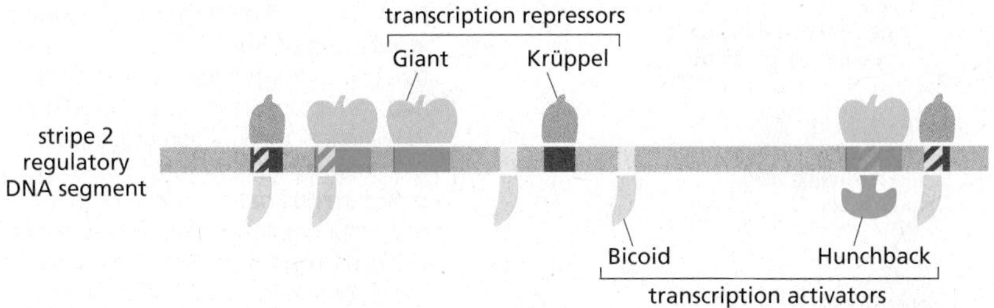

Figure 7–31 **The *Eve* stripe 2 unit.** The segment of the *Eve* gene control region identified in Figure 7–30 contains *cis*-regulatory sequences for four transcription regulators. It is known from genetic experiments that these four regulatory proteins are responsible for the proper expression of *Eve* in stripe 2. Flies that are deficient in the two gene activators Bicoid and Hunchback, for example, fail to efficiently express *Eve* in stripe 2. In flies deficient in either of the two gene repressors, Giant and Krüppel, stripe 2 expands and covers an abnormally broad region of the embryo. As indicated, in some cases the binding sites for the transcription regulators overlap, and the proteins can compete for binding to the DNA. For example, binding of Krüppel and binding of Bicoid to the site at the far right is mutually exclusive.

The relative concentrations of these four proteins determine whether the protein complexes that form at the stripe 2 module activate transcription of the *Eve* gene. Figure 7–32 shows the distributions of the four transcription regulators across the region of a *Drosophila* embryo where stripe 2 forms. It is thought that either of the two repressor proteins, when bound to the DNA, will turn off the stripe 2 module, whereas both Bicoid and Hunchback must bind for this module's maximal activation. This simple regulatory scheme suffices to turn on the stripe 2 module (and therefore the expression of the *Eve* gene) only in those nuclei located where the levels of both Bicoid and Hunchback are high and both Krüppel and Giant are absent—a combination that occurs in only one region of the early embryo. It is not known exactly how these four transcription regulators interact with coactivators and co-repressors to specify the final level of transcription across the stripe, but the outcome very likely relies on competition between activators and repressors that act by the mechanisms outlined in Figures 7–21, 7–22, and 7–27.

The stripe 2 element is autonomous, inasmuch as it specifies stripe 2 when isolated from its normal context (see Figure 7–30). The other stripe regulatory modules are thought to be constructed similarly, reading positional information provided by other combinations of transcription regulators. The entire *Eve* gene control region binds more than 20 different transcription regulators. Seven combinations of regulators—one combination for each stripe—specify *Eve* expression, while many other combinations (all those found in the interstripe regions of the embryo) keep all the stripe elements silent. A large and complex control region is thereby built from a series of smaller modules, each of which consists of a unique arrangement of short *cis*-regulatory sequences recognized by specific transcription regulators.

The *Eve* gene itself encodes a transcription regulator, which, after its pattern of expression is set up in seven stripes, controls the expression of other *Drosophila* genes. As development proceeds, the embryo is thus subdivided into finer and finer regions that eventually give rise to the different body parts of the adult fly, as discussed in Chapter 21.

Eve exemplifies the complexity of transcription control regions in plants and animals. As this example shows, control regions can respond to many different inputs, integrate this information, and produce a complex spatial and temporal output as

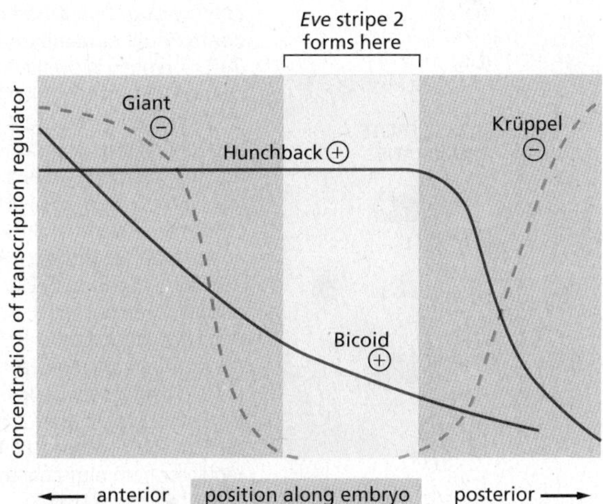

Figure 7–32 **Distribution of the transcription regulators responsible for ensuring that *Eve* is expressed in stripe 2.** The distributions of these proteins were visualized by staining a developing *Drosophila* embryo with antibodies directed against each of the four proteins, and a graph of the staining intensities is shown. The expression of *Eve* in stripe 2 occurs only at the position where the two activators (Bicoid and Hunchback) are present and the two repressors (Giant and Krüppel) are absent. In fly embryos that lack Krüppel, for example, stripe 2 expands posteriorly. Likewise, stripe 2 expands posteriorly if the DNA-binding sites for Krüppel in the stripe 2 module are inactivated by mutation (see also Figure 7–31).

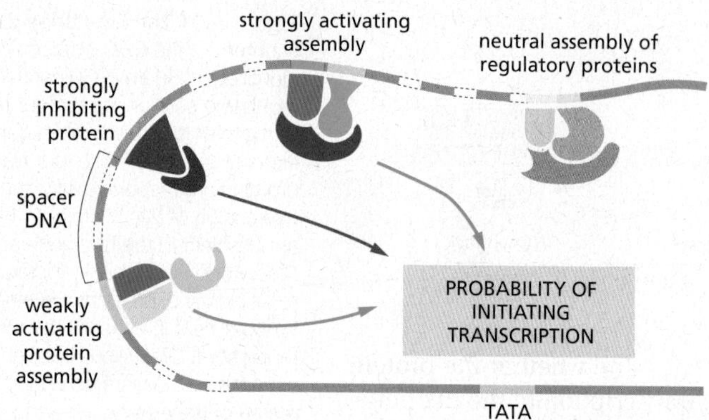

Figure 7–33 The integration of multiple inputs at a promoter. Multiple sets of transcription regulators, coactivators, and co-repressors can work together to influence transcription initiation at a promoter, as they do in the *Eve* stripe 2 module illustrated in Figure 7–31. It is not yet understood in detail how the cell achieves integration of multiple inputs, but it is likely that the final transcriptional activity of the gene results from competitions between activators and repressors that act by the mechanisms summarized in Figures 7–21, 7–22, and 7–27. As we saw earlier, for especially complex gene control regions, it has been proposed that these competitions take place and are "summed up" in localized biomolecular condensates formed by networks of weak interactions.

development proceeds. However, exactly how all these mechanisms work together to produce the final output is understood only in broad outline (**Figure 7–33**).

Transcription Regulators Are Brought into Play by Extracellular Signals

The above example from *Drosophila* clearly illustrates the power of combinatorial control, but this case is unusual in that the nuclei are exposed directly to positional cues in the form of concentrations of transcription regulators. In embryos of most other organisms and in all adults, individual nuclei are in separate cells, and extracellular information (including positional cues) must be passed across the plasma membrane so as to generate signals in the cytosol that cause different transcription regulators to become active in different cell types. Some of the different mechanisms that are known to be used to activate transcription regulators are diagrammed in **Figure 7–34**; in Chapter 15, we discuss how extracellular signals trigger these changes.

Like the fly example discussed earlier, mammalian enhancers are also modular. An example is the control region responsible for regulating the α-globin gene, which codes for one of the subunits of hemoglobin (see Figure 3–20). Here, five

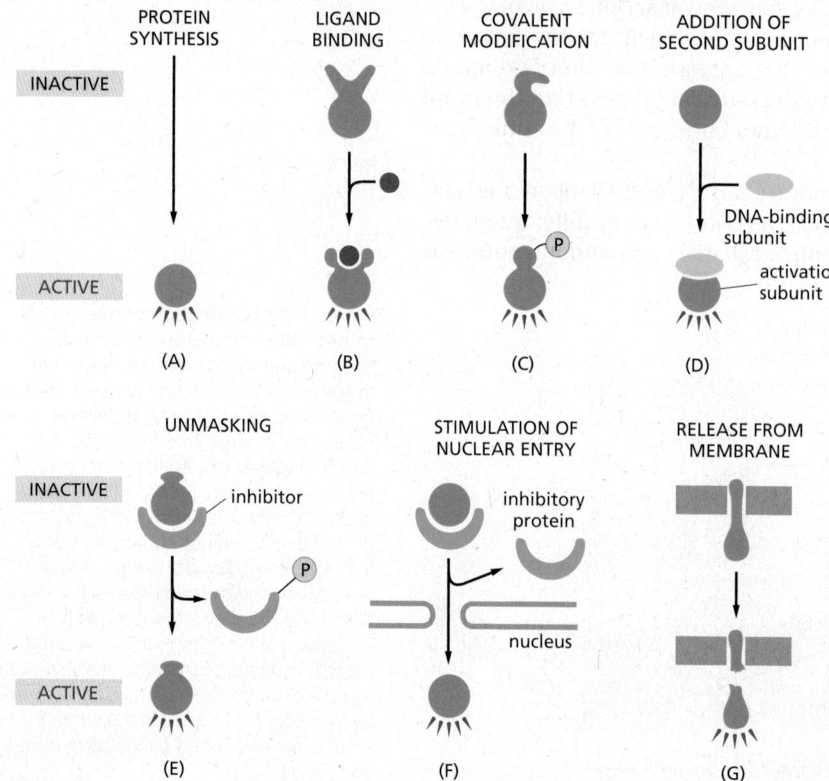

Figure 7–34 Some ways in which the activity of transcription regulators is controlled inside eukaryotic cells. (A) The protein is synthesized only when needed. (B) Activation by ligand binding. (C) Activation by covalent modification; phosphorylation is shown here, but many other modifications are possible (see Table 3–4, p. 175). (D) Formation of a complex between a DNA-binding protein and a separate protein with a transcription-activating domain. (E) Unmasking of an activation domain by the phosphorylation of an inhibitor protein. (F) Stimulation of nuclear entry by removal of an inhibitory protein that otherwise keeps the regulatory protein from entering the nucleus. (G) Release of a transcription regulator from a membrane bilayer by regulated proteolysis.

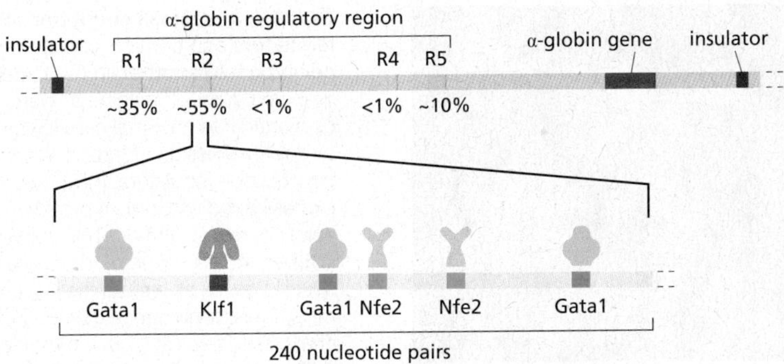

Figure 7–35 Modular structure of the control region for the mouse α-globin gene. Each of the five modules (R1–R5) can independently act as an enhancer, that is, they can each activate transcription of a reporter construct (see Figure 7–30B). However, the patterns of expression in a developing embryo are somewhat different for different modules. As indicated by the percentage designations, each module differs in the quantitative contributions it makes to the overall transcription rate in erythroid cells, with the total amount of mRNA being roughly equal to that of the sum of that produced by the individual modules. The additive properties of this control region suggest that the modules all affect the same step in transcription.

The combination of transcription regulators that recognize the R2 module, the most active of the five, is shown in the expanded view. These three transcription regulators are made in erythroid cells and are absent in most other cell types, explaining why expression of the globin gene occurs only in erythroid cells. Most of these same proteins also bind to the other α-globin regulatory modules, consistent with the modules working additively. As shown, insulator sequences flank the gene (including its control region), allowing the α-globin gene to be regulated independently of other genes on the same chromosome (see Figure 7–28). It is thought that modules R3 and R4 make no significant contribution to the overall transcription of the α-globin gene, but are once-functional modules that are in the slow evolutionary process of disappearing due to a gradual accumulation of mutations. (Courtesy of Helena Francis and Douglas Higgs.)

different modules are spread out over about 25,000 nucleotide pairs (**Figure 7–35**). Each of the five modules, when experimentally separated from the other four, can act as an independent enhancer to specify production of α-globin; but they do so only in erythroid cells, the precursors to red blood cells, because only erythroid cells express the appropriate transcription regulators. Red blood cells, which contain high concentrations of hemoglobin, are unusual in that they lack DNA and rely on their precursor cells to synthesize this protein.

Combinatorial Gene Control Creates Many Different Cell Types

We have seen that transcription regulators usually act in combination to control the expression of an individual gene. It is also generally true that each transcription regulator in an organism contributes to the control of many genes. This point is illustrated schematically in **Figure 7–36**, which shows how combinatorial gene

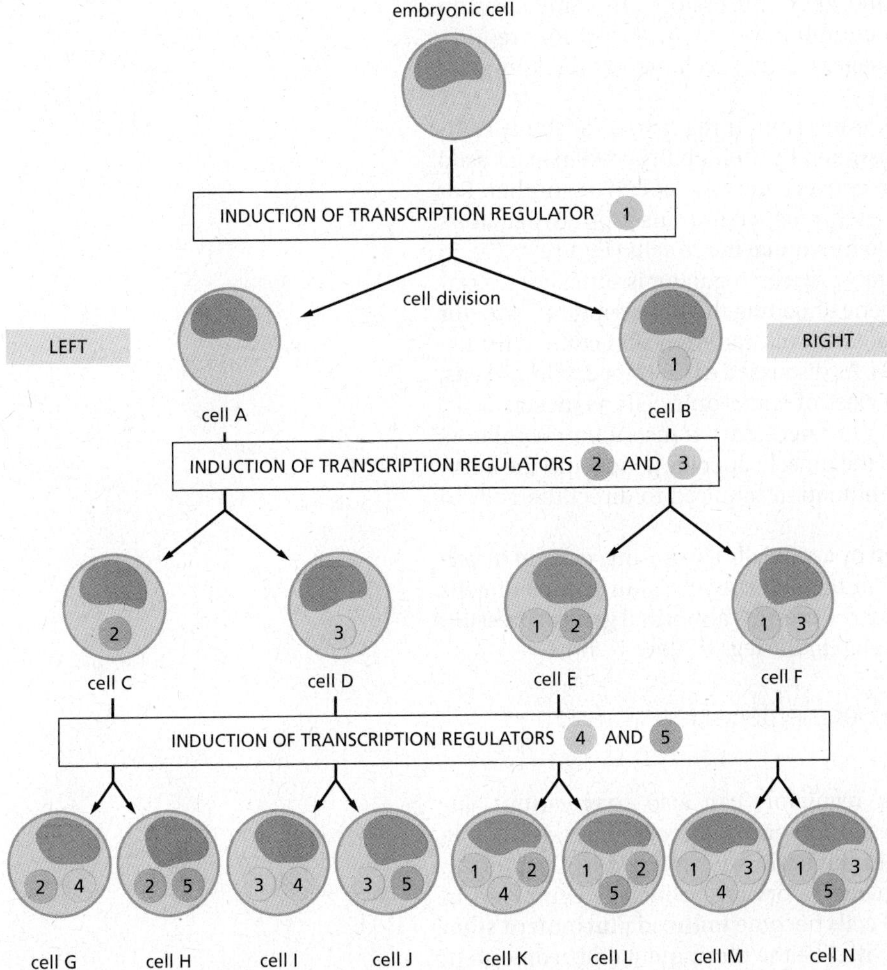

Figure 7–36 The importance of combinatorial gene control for development. Combinations of a few transcription regulators can generate many cell types during development. In this simple, idealized scheme, a "decision" to make one of a pair of different transcription regulators (shown as numbered circles) is made after each cell division. Sensing its relative position in the embryo, the daughter cell toward the *left side* of the embryo is always induced to synthesize the even-numbered protein of each pair, while the daughter cell toward the *right side* of the embryo is induced to synthesize the odd-numbered protein. The production of each transcription regulator is assumed to be self-perpetuating once it has become initiated (see Figure 7–42). In this way, through cell memory, the final combinatorial specification is built up step by step. In this purely hypothetical example, five different transcription regulators have created eight final cell types (G–N).

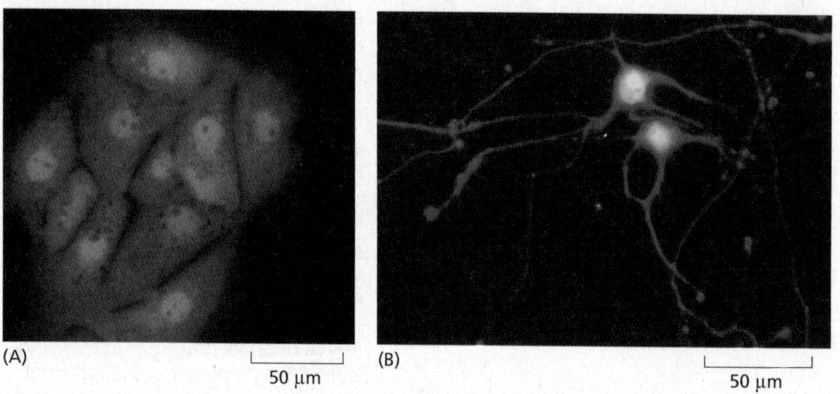

Figure 7–37 A small set of transcription regulators can convert one differentiated cell type into another. In this experiment, liver cells grown in culture (A) were converted into neuronal cells (B) by the artificial expression of three neuron-specific transcription regulators. (Both types of cells express a red fluorescent protein, which helps to visualize them.) This conversion involves the activation of many neuron-specific genes as well as the repression of many liver-specific genes. (From S. Marro et al., *Cell Stem Cell* 9:374–382, 2011. With permission from Elsevier.)

control makes it possible to generate a great deal of biological complexity even with relatively few transcription regulators.

Because of such combinatorial control, a given transcription regulator need not have a single, simply definable function as commander of a particular battery of genes or specifier of a particular cell type. Rather, transcription regulators can be likened to the words of a language: they are used with different meanings in a variety of contexts and rarely alone; it is the well-chosen combination that conveys the information that specifies a gene regulatory event.

Because of combinatorial gene control, the effect of adding a new transcription regulator to a cell will depend on that cell's past history, inasmuch as this history determines the transcription regulators already present. Thus, during embryonic development, a cell can accumulate a series of transcription regulators that may not initially alter gene expression. Only the addition of the final members of a requisite combination of transcription regulators will complete the regulatory message, leading to large changes in gene expression.

The importance of a combination of transcription regulators for the specification of cell types is most easily demonstrated by their ability—when expressed artificially in a specific combination—to convert one type of cell to another. For example, the artificial expression of three neuron-specific transcription regulators in liver cells can convert the liver cells into functional nerve cells (**Figure 7–37**). In some cases, expression of even a single transcription regulator is sufficient to convert one cell type to another: when the gene encoding the transcription regulator MyoD is artificially introduced into fibroblasts cultured from skin connective tissue, the fibroblasts form muscle-like cells. As discussed in Chapter 22, fibroblasts, which are derived from the same broad class of embryonic cells as muscle cells, have already accumulated many of the other necessary transcription regulators required for the combinatorial control of the muscle-specific genes, and the addition of MyoD completes the unique combination required to direct the cells to become muscle.

An even more striking example is seen by artificially expressing, early in development, a single *Drosophila* transcription regulator (Eyeless) in groups of cells that would normally go on to form leg parts. Here, this abnormal gene expression change causes eye-like structures to develop in the legs (**Figure 7–38**).

Specialized Cell Types Can Be Experimentally Reprogrammed to Become Pluripotent Stem Cells

Artificial manipulation of transcription regulators can also coax various differentiated cells to *de-differentiate* into pluripotent stem cells that are capable of giving rise to the different cell types in the body, as discussed in Chapter 22. Thus, when three specific transcription regulators are artificially expressed in cultured mouse fibroblasts, a number of cells become **induced pluripotent stem cells (iPS cells)**—cells that look and behave like the pluripotent embryonic stem

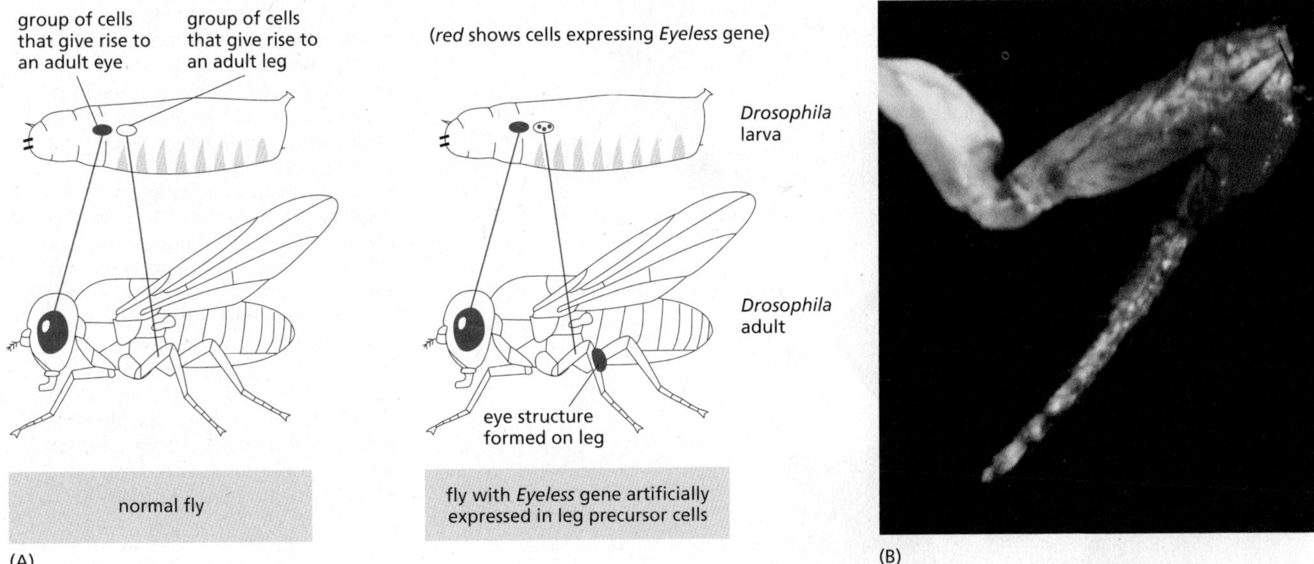

Figure 7–38 Expression of the *Drosophila Eyeless* gene in precursor cells of the fly leg triggers the development of an eye on the leg. (A) Simplified diagrams showing the result when a fruit fly larva contains either the normally expressed *Eyeless* gene *(left)* or an *Eyeless* gene that is additionally expressed artificially in cells that normally give rise to leg tissue *(right)*. (B) Photograph of an abnormal leg that contains a misplaced eye (see also Figure 21–2). The transcription regulator was named Eyeless because its inactivation in otherwise normal flies causes the loss of eyes (see Figure 21–32). (B, courtesy of Walter Gehring.)

(ES) cells that are derived from embryos (**Figure 7–39**). This approach has been adapted to produce iPS cells from a variety of specialized cell types, including cells taken from humans. Such human iPS cells can then be directed to generate a population of differentiated cells for use in the study or treatment of disease, a topic discussed in detail in Chapter 22.

Although it was once thought that cell differentiation was irreversible, it is now clear that by manipulating combinations of transcription regulators, cell types and differentiation pathways can be readily reversed and otherwise altered.

Combinations of Master Transcription Regulators Specify Cell Types by Controlling the Expression of Many Genes

As we saw in the introduction to this chapter, different cell types of multicellular organisms differ enormously in the proteins and RNAs they express. For example, only muscle cells express special types of actin and myosin that form the contractile apparatus, while nerve cells must make and assemble all the proteins needed to form dendrites and synapses. We have seen that these patterns of cell-type-specific expression are orchestrated by a combination of so-called **master transcription regulators**. In many cases, these proteins bind directly to *cis*-regulatory sequences of the genes particular to that cell type. Thus, MyoD binds directly to *cis*-regulatory sequences located in the control regions of the muscle-specific genes. In other cases, the master regulators control the

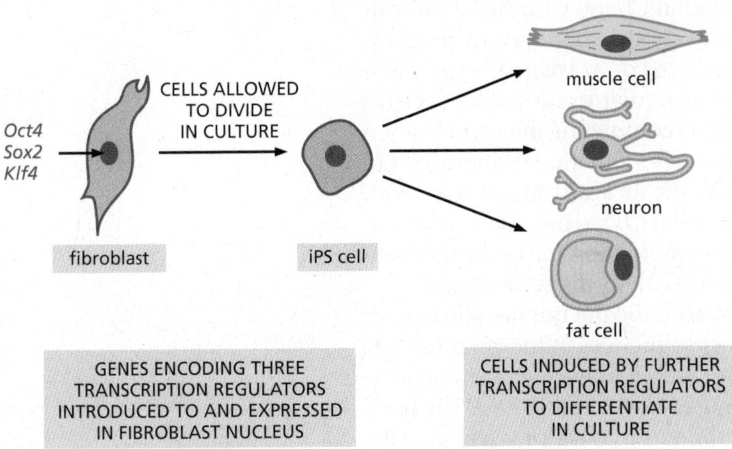

Figure 7–39 A combination of transcription regulators can induce a differentiated cell to de-differentiate into a pluripotent cell. The artificially induced expression of a set of three genes, each of which encodes a transcription regulator, can reprogram a fibroblast into a pluripotent cell with embryonic stem (ES) cell–like properties. Like ES cells, such induced pluripotent stem (iPS) cells can proliferate indefinitely in culture and can be stimulated by appropriate extracellular signal molecules to differentiate into almost any cell type found in the body. Transcription regulators such as Oct4, Sox2, and Klf4 are often called *master transcription regulators* because their expression is sufficient to trigger a change in cell identity. How two of these transcription regulators interact with DNA in a nucleosome is shown in Figure 7–13.

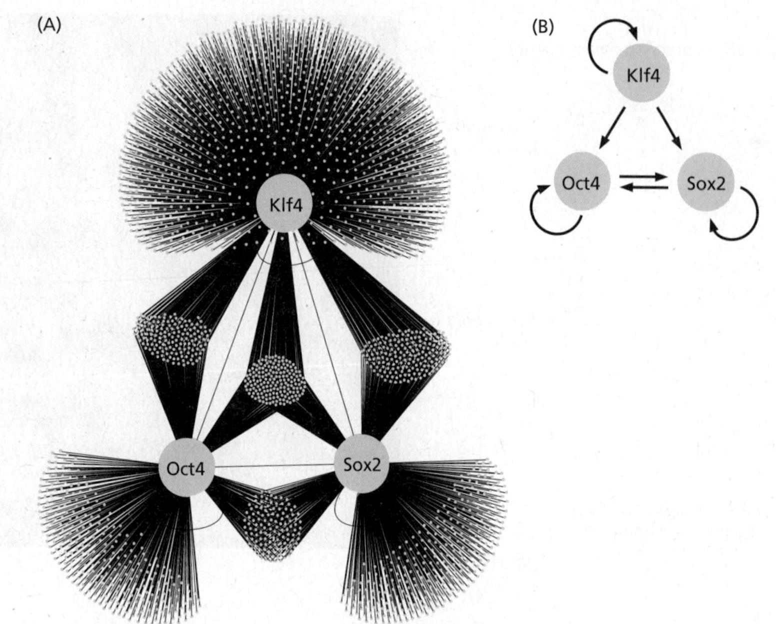

Figure 7–40 A portion of the transcription network specifying embryonic stem cells. (A) The three master transcription regulators in Figure 7–39 are shown as *large circles*. Genes whose *cis*-regulatory sequences are bound by each regulator in embryonic stem cells are indicated by a small *green dot* (representing the gene) connected by a thin line (representing the binding interaction). Note that many of the target genes are bound by more than one of the regulators. (B) The master regulators control their own expression. As shown here, the three transcription regulators bind to their own control regions (indicated by feedback loops), as well as those of the other master regulators (indicated by *straight arrows*). (Courtesy of Trevor Sorrells, based on data from J. Kim et al., *Cell* 132:1049–1061, 2008.)

expression of "downstream" transcription regulators that, in turn, bind to the control regions of other cell-type-specific genes and control their synthesis.

The specification of a particular cell type typically involves changes in the expression of several thousand genes. Genes whose protein products are required in the cell type are expressed at high levels, while those not needed are typically down-regulated. As might be imagined, the pattern of binding between the master regulators and all of the regulated genes can be extremely elaborate (**Figure 7–40**). When we consider that many of these regulated genes have control regions that span tens of thousands of nucleotide pairs, commensurate with the *Eve* example discussed earlier, we can begin to appreciate the enormous complexity of cell-type specification.

An outstanding question in biology is how the information in a genome is used to specify a multicellular organism. Although we have the general outline of the answer, we are far from understanding how a single cell type is completely specified, let alone a whole organism.

Specialized Cells Must Rapidly Turn Some Genes On and Off

Although they generally maintain their identities, specialized cells must constantly respond to changes in their environment. Among the most important changes are signals from other cells that coordinate the behavior of the whole organism. Many of these signals induce transient changes in gene transcription, and we discuss the nature of these signals in detail in Chapter 15. Here, we consider how specialized cell types rapidly and decisively switch groups of genes on and off in response to their environment. Even though control of gene expression is combinatorial, the effect of a single transcription regulator can still be decisive in switching any particular gene on or off, simply by completing the combination needed to maximally activate or repress that gene. This situation is analogous to dialing in the final number of a combination lock: the lock will spring open with only this simple addition if all of the other numbers have been previously entered. And just as the same number can complete the combination for many different locks, the addition of a particular protein can turn on many different genes.

An example is the rapid control of gene expression by the human glucocorticoid receptor protein. To bind to its *cis*-regulatory sequences in the genome, this transcription regulator must first form a complex with a molecule of a glucocorticoid steroid hormone, such as cortisol (see Figures 15–65 and 15–66). The body releases this hormone during times of starvation and intense physical activity,

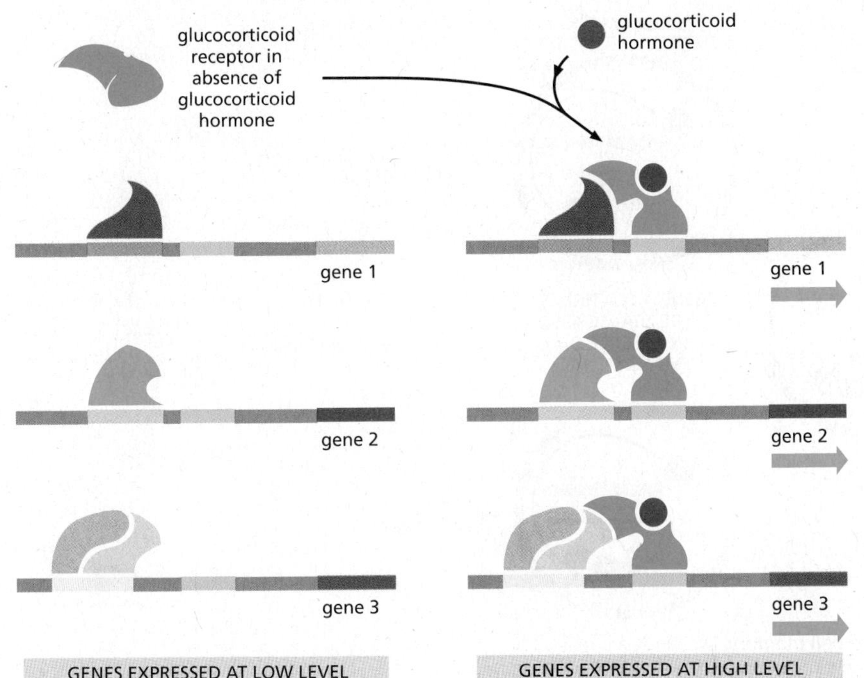

Figure 7–41 **A single transcription regulator can coordinate the expression of many different genes.** The action of the glucocorticoid receptor is illustrated schematically. On the *left* is a series of genes, each of which has various transcription regulators bound to its regulatory region. However, these bound proteins are not sufficient on their own to fully activate transcription. On the *right* is shown the effect of adding an additional transcription regulator—the glucocorticoid receptor in a complex with glucocorticoid hormone—that has a *cis*-regulatory sequence in the control region of each gene. The glucocorticoid receptor completes the combination of transcription regulators required for maximal initiation of transcription, and the genes are now maximally switched on as a set. When the hormone is no longer present, the glucocorticoid receptor dissociates from DNA, and the genes return to their prestimulated levels.

and among its other activities, it stimulates liver cells to increase the production of glucose from amino acids and other small molecules. To respond in this way, liver cells increase the expression of many different genes that code for metabolic enzymes, such as tyrosine aminotransferase, as we discussed earlier in this chapter (see Figure 7-3). Although these genes all have different and complex control regions, their maximal expression depends on the binding of the hormone-glucocorticoid receptor complex to its *cis*-regulatory sequence, which is present in the control region of each gene. When the body has recovered and the hormone is no longer present, the expression of each of these genes drops to its normal level in the liver. In this way, a single transcription regulator can rapidly control the expression of many different genes (**Figure 7–41**).

The effects of the glucocorticoid receptor are not confined to cells of the liver. In other cell types, activation of this transcription regulator by hormone also causes changes in the expression levels of many genes; the genes affected, however, are usually different from those affected in liver cells. As we have seen, each cell type has an individualized set of transcription regulators, and because of combinatorial control, these critically influence the action of the glucocorticoid receptor. Because the receptor is able to assemble with different sets of cell-type-specific transcription regulators, switching it on with hormone produces a different spectrum of effects in each cell type.

Differentiated Cells Maintain Their Identity

Once a cell has become differentiated into a particular cell type, it will generally remain differentiated, and all its progeny cells will remain that same cell type. Some highly specialized cells, including skeletal muscle cells and neurons, never divide again once they have differentiated; that is, they are *terminally differentiated* (as discussed in Chapter 17). But many other differentiated cells—such as fibroblasts, smooth muscle cells, and liver cells—will divide many times in the life of an individual. When they do, these specialized cell types give rise only to cells like themselves: smooth muscle cells do not give rise to liver cells, nor liver cells to fibroblasts.

For a proliferating cell to maintain its identity—a property called **cell memory**—the patterns of gene expression responsible for that identity must be remembered and passed on to its daughter cells through subsequent cell divisions. Thus, in the model we discussed in Figure 7-36, the production of each

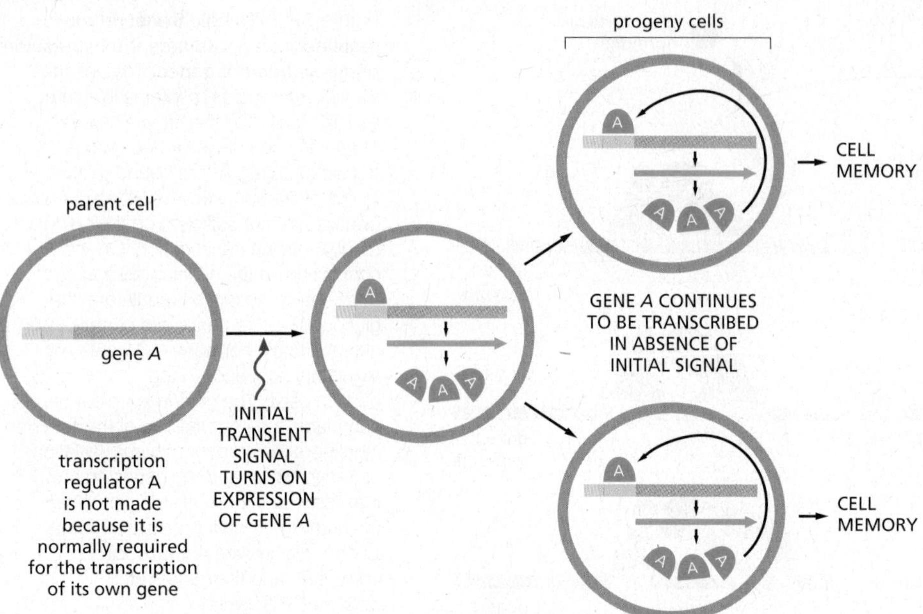

Figure 7–42 A positive feedback loop can create cell memory. Protein A is a master transcription regulator that activates the transcription of its own gene—as well as other cell-type-specific genes (not shown). All of the descendants of the original cell will therefore "remember" that the progenitor cell had experienced a transient signal that initiated the production of protein A.

transcription regulator, once begun, has to be continued in the daughter cells of each cell division. How is such perpetuation accomplished?

Cells have several ways of ensuring that their daughters "remember" what kind of cells they are. One of the simplest and most important is through a positive feedback loop, where a master cell-type transcription regulator activates transcription of its own gene, in addition to that of the other cell-type-specific genes needed to maintain the cell type. Each time a cell divides, the regulator is distributed to both daughter cells, where it continues to stimulate the positive feedback loop, making more of itself and the cell-type proteins it controls each division. Positive feedback is crucial for establishing "self-sustaining" circuits of gene expression that allow a cell to commit to a particular fate—and then to transmit that information to its progeny (**Figure 7–42**).

As was previously indicated in Figure 7–40B, the master regulators needed to maintain the pluripotency of iPS cells bind to *cis*-regulatory sequences in their own control regions, providing examples of this type of positive feedback loop. In addition, most of these pluripotent stem cell regulators also activate transcription of other master regulators, resulting in a complex series of indirect feedback loops. For example, if A activates B, and B activates A, this forms a positive feedback loop where A activates its own expression, albeit indirectly. The series of direct and indirect feedback loops observed in the iPS circuit is typical of other specialized cell circuits. Such a network structure strengthens cell memory, increasing the probability that a particular pattern of gene expression is transmitted through successive generations. For example, if the level of A drops below the critical threshold to stimulate its own synthesis, regulator B can rescue it. By successive application of this mechanism, a complex series of positive feedback loops among multiple transcription regulators can stably maintain a differentiated state through many cell divisions.

Positive feedback loops formed by transcription regulators are probably the most prevalent way of ensuring that daughter cells remember what kind of cells they are meant to be, and they are found in all species on Earth. For example, many bacteria and single-cell eukaryotes form different types of cells, and positive feedback loops lie at the heart of mechanisms that maintain their cell types through many rounds of cell division. Plants and animals also make extensive use of transcription feedback loops; but as we saw in Chapter 4 and shall discuss again later in the chapter, they have additional, more specialized mechanisms for making cell memory even stronger (see, for example, Figure 4–44). We will return

mechanisms to ensure this cell memory. Direct or indirect positive feedback loops, which enable transcription regulators to perpetuate their own synthesis, provide one of the simplest mechanisms for producing a cell memory. Transcription circuits also provide the cell with the means to carry out many other types of logic operations. Simple transcription circuits combined into large regulatory networks drive highly sophisticated programs of embryonic development that will require new approaches to fully decipher.

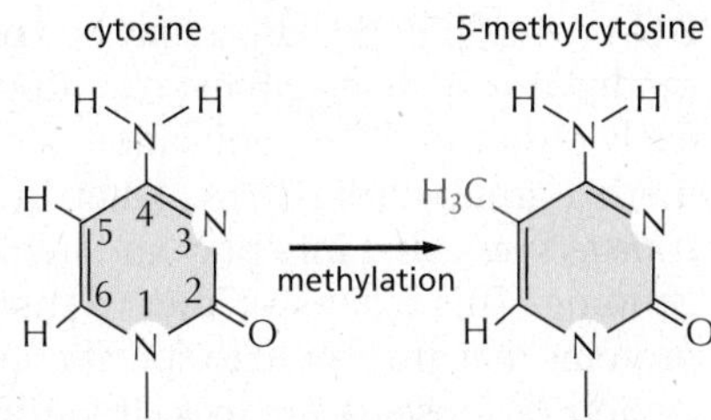

Figure 7–46 Formation of 5-methylcytosine occurs by methylation of a cytosine base in the DNA double helix. In vertebrates, this event is largely confined to selected cytosine (C) nucleotides located in the sequence CG. CG sequences are sometimes denoted as CpG sequences, where the p indicates a phosphate linkage to distinguish it from a CG base pair. In this chapter, we will continue to use the simpler nomenclature CG to indicate this dinucleotide.

MECHANISMS THAT REINFORCE CELL MEMORY IN PLANTS AND ANIMALS

Thus far in this chapter, we have emphasized the regulation of gene transcription by proteins that associate either directly or indirectly with DNA. However, DNA itself can be covalently modified, and, as we saw in Chapter 4, certain types of chromatin states can be inherited. In this section, we shall see how these phenomena provide additional opportunities for the regulation of gene expression, particularly in mammals. Near the end of this section, we discuss how a whole chromosome can be transcriptionally shut down using such mechanisms, and how this state can be maintained through many cell divisions.

Patterns of DNA Methylation Can Be Inherited When Vertebrate Cells Divide

In vertebrate cells, the methylation of cytosine provides one mechanism through which gene expression patterns can be passed on to progeny cells. The methylated form of cytosine, 5-methylcytosine (5-methyl C), has the same relation to cytosine that thymine has to uracil, and the modification likewise has no effect on base-pairing (**Figure 7–46**). **DNA methylation** in vertebrate DNA occurs on cytosine (C) nucleotides largely in the sequence CG, which is base-paired to exactly the same sequence (in opposite orientation) on the other strand of the DNA helix. Consequently, a simple mechanism permits the existing pattern of DNA methylation to be inherited directly by the daughter DNA strands. An enzyme called *maintenance methyl transferase* acts preferentially on those CG sequences that are base-paired with a CG sequence that is already methylated. As a result, the pattern of DNA methylation on the parent DNA strand serves as a template for the methylation of the daughter DNA strand, causing this pattern to be inherited directly after DNA replication (**Figure 7–47**).

Although DNA methylation patterns can be maintained in differentiated cells by the mechanism shown in Figure 7–47, methylation patterns are dynamic during mammalian development. Shortly after fertilization, there is a genome-wide wave of demethylation, when the vast majority of methyl groups are lost from the

Figure 7–47 How DNA methylation patterns are faithfully inherited. In vertebrate DNA, a large fraction of the cytosine nucleotides in the sequence CG is methylated (see Figure 7–46). Because of the existence of a methyl-directed methylating enzyme (the maintenance methyl transferase), once a pattern of DNA methylation is established, that pattern of methylation is inherited in the progeny DNA, as shown.

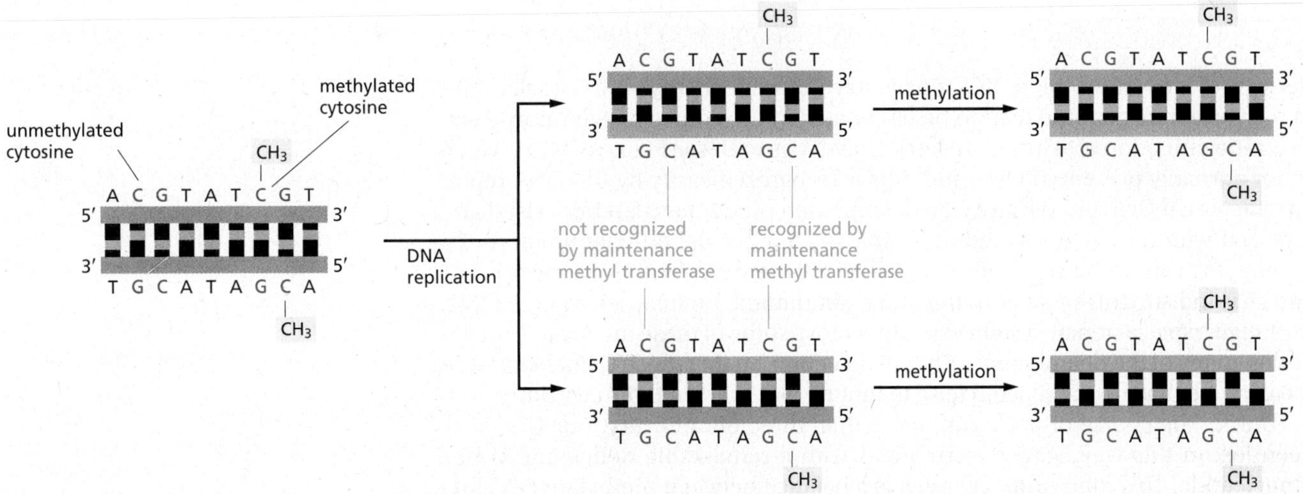

DNA. This demethylation may occur either by suppression of maintenance DNA methyl transferase activity, resulting in the passive loss of methyl groups during each round of DNA replication, or by *DNA demethylases* that actively remove methyl groups from DNA. Later in development, several *de novo DNA methyl transferases* come into play and methylate about 70% of the CG sequences in the genome. This extensive methylation occurs largely indiscriminately, although proteins that are bound to specific sequences on the genome can block the methylation of those sequences. In addition, some sequence-specific DNA-binding proteins direct DNA methylases to specific locations in genomes, resulting in very high local densities of methylation in the neighborhoods of those DNA-bound proteins. Conversely, DNA demethylases can also be directed to certain regions of the genome, resulting in loss of methyl groups in those regions. Despite these selective mechanisms, the patterns of overall methylation across differentiated cell types are broadly similar, and many methylated positions—on their own—appear to have little or no impact on gene expression.

DNA methylation has several uses in the vertebrate cell. A very important role of dense methylation is to work in conjunction with other gene expression control mechanisms to establish a particularly efficient form of gene repression. This combination of mechanisms enables unneeded eukaryotic genes to be repressed to a very high degree. The rate at which a vertebrate gene is transcribed can vary 10^6-fold between one tissue and another, and unexpressed vertebrate genes are much less "leaky" in terms of transcription than bacterial genes, in which the largest known differences in transcription rates between expressed and unexpressed gene states are only about 1000-fold.

Dense DNA methylation helps to repress transcription in several ways. The methyl groups on methylated cytosines lie in the major groove of DNA and interfere directly with the binding of some proteins (transcription regulators as well as the general transcription factors) required for transcription initiation. In addition, the cell contains a repertoire of proteins that bind specifically to methylated DNA. The best characterized of these also associate with histone-modifying enzymes, leading to a repressive, heterochromatin state where chromatin structure and DNA methylation act synergistically (**Figure 7–48**).

Many genes that are needed only in differentiated cells are tightly repressed in this way in embryonic cells. As differentiation proceeds, they become activated, although this process typically requires many steps, often involving "pioneer factors" (see Figure 7–13), histone demethylases, and DNA demethylases. The latter enzymes convert 5-methyl C to 5-hydroxymethyl C, which is later replaced by C either through DNA repair (see Figure 5–41A) or, passively, through multiple rounds of DNA replication. In addition, many genes active in embryonic tissues become repressed during differentiation by the mechanisms shown in Figure 7-48. The reactivation of these genes is one of the key steps in converting differentiated cells back into stem cells, as explained in Chapter 22.

CG-Rich Islands Are Associated with Many Genes in Mammals

Because of the way in which DNA repair enzymes work, methylated C nucleotides in the vertebrate genome tend to be eliminated in the course of evolution. Accidental deamination of an unmethylated C gives rise to U (see Figure 5–38B), which is not normally present in DNA and thus is recognized easily by the DNA repair enzyme uracil DNA glycosylase. The deamination product is thereby excised and replaced with a C, as discussed in Chapter 5. But accidental deamination of a 5-methyl C cannot be repaired in this way, for the deamination product is a T and so is indistinguishable from the other, nonmutant T nucleotides in the DNA. Although a special repair system exists to remove some of these incorrect T nucleotides, many of the deaminations escape detection, so that those C nucleotides in the genome that are methylated tend to mutate to T over evolutionary time.

During the course of evolution, more than three out of every four CGs have been lost in this way, leaving vertebrates with a remarkable deficiency of this dinucleotide. This ratio probably reflects a balance between methylated CG loss

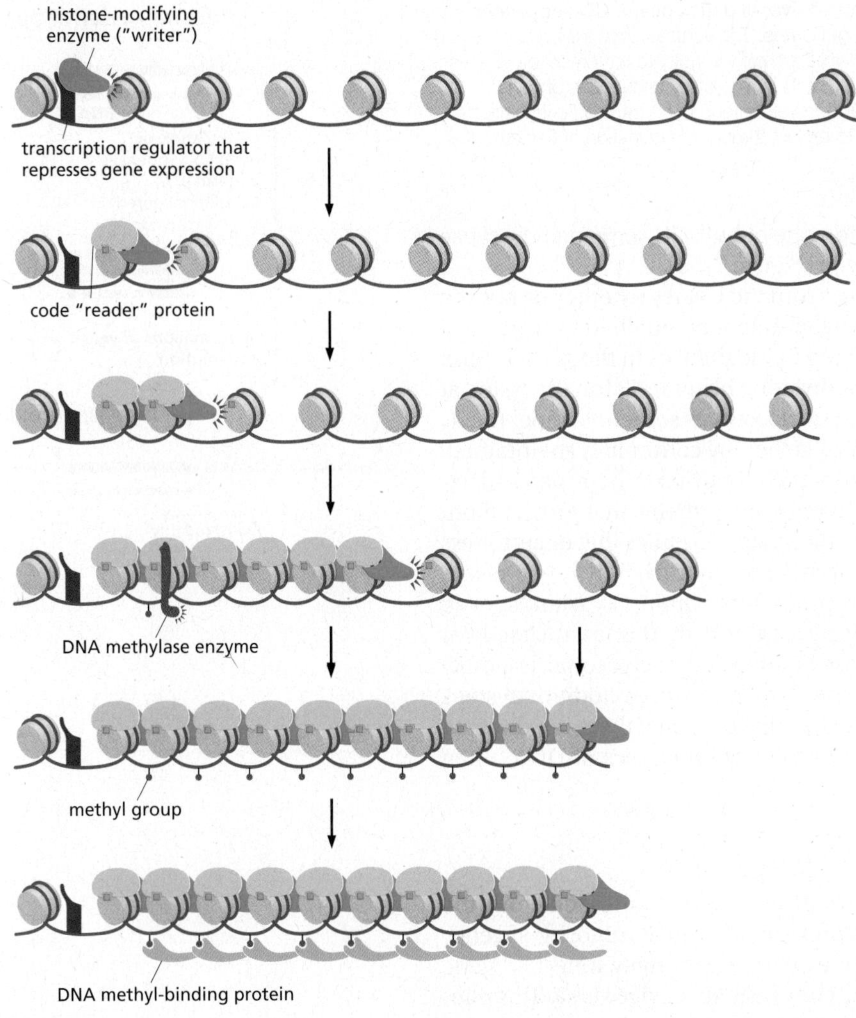

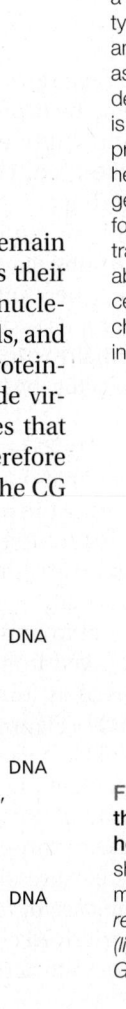

Figure 7–48 Multiple mechanisms can produce especially stable gene repression. In this schematic example, histone reader and writer proteins (discussed in Chapter 4), triggered by transcription regulators, establish a repressive form of chromatin whose nucleosomes are marked by the trimethylation of specific lysine amino acids in histones (see Figure 4–35, as well as Figure 7–27F). An additional layer of repression can occur when a *de novo* DNA methylase is attracted by the modified histones and methylates nearby cytosines in DNA; and these are, in turn, bound by DNA methyl-binding proteins. During DNA replication, some of the modified *(blue dot)* histones will be inherited by one daughter chromosome, some by the other, and in each daughter they can induce reconstruction of the same pattern of chromatin modifications (see Figure 4–44). At the same time, the mechanism shown in Figure 7–47 will cause both daughter chromosomes to inherit the same methylation pattern. This makes the two mechanisms for inheriting a repressed gene mutually reinforcing, accounting for the inheritance by daughter cells of both the histone and the DNA modifications. It can also explain the tendency of some chromatin modifications to spread along a chromosome (see Figure 4–39). This type of heterochromatin is assembled and disassembled on different genes as mammalian development proceeds, depending on whether the gene product is needed. For example, when endoderm precursor cells differentiate into the hepatocytes of the liver, an estimated 6000 genes are unpackaged from this repressive form of chromatin and become actively transcribed. At roughly the same time, about 1600 genes active in endoderm cells become packaged into this type of chromatin and are thereby tightly repressed in hepatocytes.

by DNA repair and CG gain by random mutation. The CG sequences that remain are very unevenly distributed in the genome; they are present at 10 times their average density in selected regions, called **CG islands**, which average 1000 nucleotide pairs in length. The human genome contains roughly 20,000 CG islands, and they usually include promoters of genes. For example, 60% of human protein-coding genes have promoters embedded in CG islands, and these include virtually all the promoters of the so-called *housekeeping genes*—those genes that code for the many proteins that are essential for cell viability and are therefore expressed in nearly all cells (Figure 7–49). Over evolutionary time scales, the CG

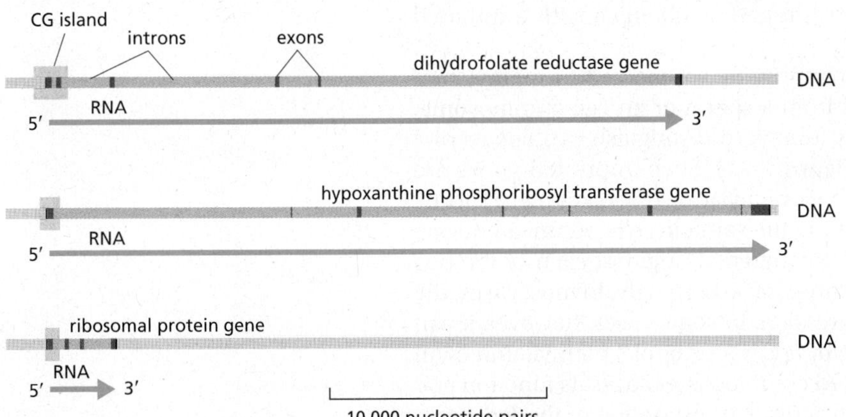

Figure 7–49 The CG islands surrounding the promoter in three mammalian housekeeping genes. The *yellow boxes* show the extent of each island. As for most genes in mammals, the exons *(dark red)* are very short relative to the introns *(light red)*. (Adapted from A.P. Bird, *Trends Genet.* 3:342–347, 1987.)

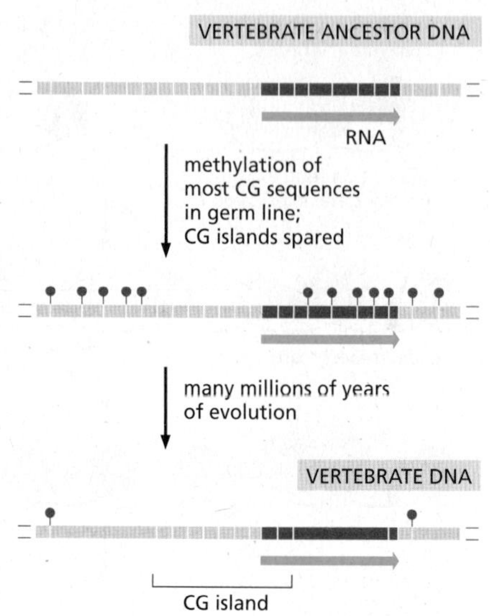

Figure 7–50 **A mechanism to explain both the marked overall deficiency of CG sequences and their clustering into CG islands in vertebrate genomes.** The vertical *white lines* mark the location of CG dinucleotides in the DNA sequences, while *red circles* indicate the presence of a methyl group on the CG dinucleotide. CG sequences that lie in regulatory sequences of genes that are transcribed in germ cells are unmethylated and therefore tend to be retained in evolution. Methylated CG sequences, on the other hand, tend to be lost through deamination of 5-methyl C to T, unless the CG sequence is critical for survival.

islands were spared the accelerated mutation rate of bulk CG sequences because they remained unmethylated in the germ line (Figure 7–50).

CG islands remain unmethylated in most somatic tissues whether or not the associated gene is expressed. The unmethylated state is maintained by a group of proteins that bind specifically to unmethylated CG sequences in the genome and modify the neighboring nucleosomes by methylating histone H3 (on the lysine at position 4; see Figure 4–35). These modified nucleosomes somehow repel the *de novo* methylases, and the unmethylated state is thereby continually maintained. Unmethylated CG islands have several properties that make them particularly suitable for promoters. For example, some of the same proteins that protect them from methylation recruit additional histone-modifying enzymes that decompress the chromatin, making the islands particularly "promoter friendly." As a result, RNA polymerase is often found bound to promoters within CG islands, even when the associated gene is not being actively transcribed. At unmethylated CG islands, the competition between polymerase binding and nucleosome assembly at promoters is thus always tipped toward the former. However, additional steps are needed for the final "push" to transcribe the adjacent gene; these are directed by transcription regulators that bind to *cis*-regulatory sequences of DNA, often well upstream from the CG islands.

Genomic Imprinting Is Based on DNA Methylation

Mammalian cells are diploid, containing one set of genes inherited from the father and one set from the mother. The expression of a small minority of genes depends on which parent they came from: when the paternally inherited gene copy is active, the maternally inherited gene copy is silent, or vice versa. This phenomenon is called **genomic imprinting**.

Roughly 300 genes are imprinted in humans. Because only one copy of an imprinted gene is expressed, imprinting can "unmask" harmful mutations that would normally be covered by the other, functional copy. For example, Angelman syndrome, a disorder of the nervous system in humans that causes reduced mental ability and severe speech impairment, results from a gene deletion on one chromosomal homolog and the silencing, by imprinting, of the intact gene on the other homolog.

The *insulin-like growth factor-2* (*Igf2*) gene in the mouse provides a well-studied example of imprinting. Mice that do not express *Igf2* at all are born half the size of normal mice. However, only the paternal copy of *Igf2* is transcribed, and only this gene copy matters for the phenotype. As a result, mice with a mutated paternally derived *Igf2* gene are stunted, while mice with a mutated maternally derived *Igf2* gene are normal.

In the early embryo, genes subject to imprinting are marked by methylation according to whether they were derived from a sperm or an egg chromosome. In this way, DNA methylation is used as a mark to distinguish two copies of a gene that can be otherwise identical (Figure 7–51). Such imprinted genes are somehow protected from the wave of DNA demethylation that takes place shortly after fertilization (see pp. 435–436), enabling the somatic cells produced during embryonic development to "remember" the parental origin of each of the two copies of the gene and to regulate their expression accordingly. In most cases, the methyl imprint silences nearby gene expression. In some cases, however, it can activate expression of a gene. In the case of *Igf2*, for example, methylation of an insulator element on the paternally derived chromosome blocks its function and allows distant *cis*-regulatory sequences to activate transcription of the *Igf2* gene.

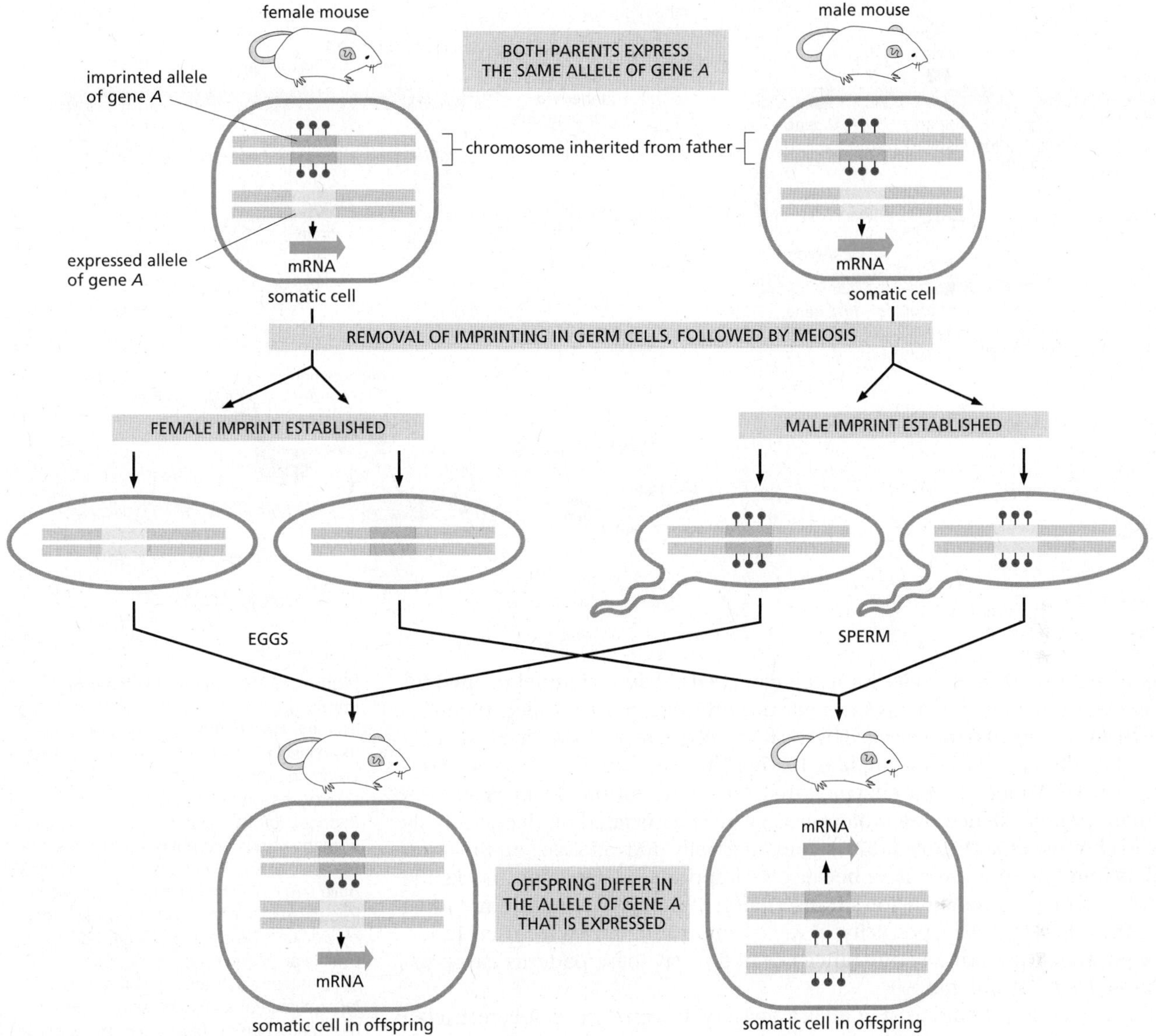

Figure 7–51 **How imprinting can cause a non-Mendelian pattern of inheritance.** The *top* portion of the figure shows a pair of homologous chromosomes in the somatic cells of two adult mice, one male and one female. In this example, both mice have inherited the top homolog from their father and the bottom homolog from their mother, and the paternal copy of a gene subject to imprinting (indicated in *orange*) is methylated, preventing its expression. The maternally derived copy of the same gene *(yellow)* is expressed. The remainder of the figure shows the outcome of a cross between these two mice. During germ-cell formation, but before meiosis, the imprints are erased and then, much later in germ-cell development, they are reimposed in a sex-specific pattern (*middle* portion of figure). In eggs produced from the female, neither allele of the *A* gene is methylated. In sperm from the male, both alleles of gene *A* are methylated. Shown at the *bottom* of the figure are two of the possible imprinting patterns inherited by the progeny mice; the mouse on the *left* has the same imprinting pattern as each of the parents, whereas the mouse on the *right* has the opposite pattern. If the two alleles of gene *A* are distinct (for example, if one codes for a mutant protein), the different imprinting patterns can cause phenotypic differences in the progeny mice, even though they carry exactly the same DNA sequences of the two *A* gene alleles.

Imprinting provides an important exception to classical "Mendelian" genetic behavior, and several hundred mouse genes are thought to be affected in this way. However, the majority of mouse genes are not imprinted, and therefore the rules of Mendelian inheritance apply to most of the mouse genome.

On the maternally derived chromosome, the insulator is not methylated, and the *Igf2* gene is therefore not transcribed (**Figure 7–52A**).

Other cases of imprinting are also based on DNA methylation, but they employ different "downstream" mechanisms. Some involve *long noncoding RNAs* (*lncRNAs*), which are defined as RNA molecules more than 200 nucleotides in length that do not code for proteins. We discuss lncRNAs broadly at the end of this chapter; here, we focus on the role of a specific lncRNA in imprinting. In the case

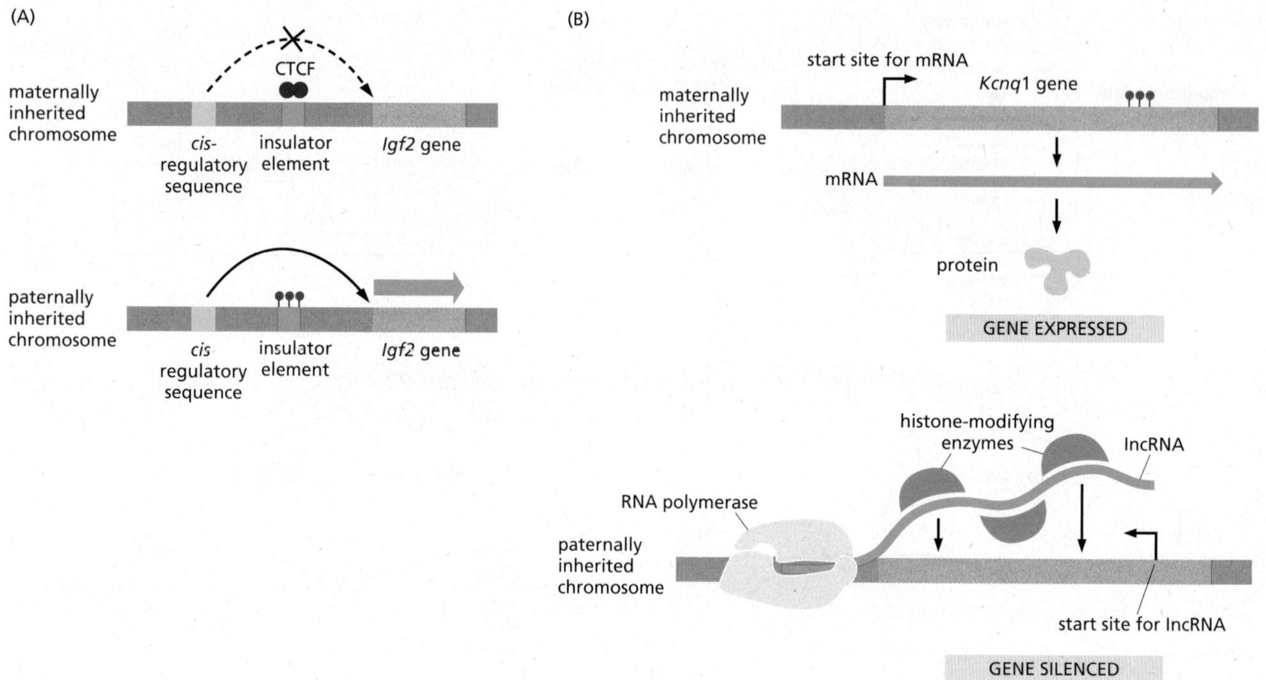

Figure 7–52 Some mechanisms of imprinting. (A) On chromosomes inherited from the female, the CTCF protein binds to an insulator (see Figure 7–28), blocking communication between *cis*-regulatory sequences *(green)* and the *Igf2* gene *(orange)*. *Igf2* is therefore not expressed from the maternally inherited chromosome. Because of imprinting, the insulator on the male-derived chromosome is methylated *(red circles)*; this inactivates the insulator by blocking the binding of the CTCF protein and allows the *cis*-regulatory sequences to activate transcription of the *Igf2* gene. In other examples of imprinting, methylation simply blocks gene expression by interfering with the binding of proteins required for a gene's transcription. (B) Imprinting of the mouse *Kcnq1* gene. On the maternally derived chromosome, synthesis of the lncRNA is blocked by methylation of the DNA *(red circles)*, and the *Kcnq1* gene is expressed. On the paternally derived chromosome, the lncRNA is synthesized, remains in place, and by directing alterations in chromatin structure blocks expression of the *Kcnq1* gene. Although shown as directly binding to lncRNA, the histone-modifying enzymes are likely to be recruited indirectly, through additional proteins.

of the *Kcnq1* gene, which codes for a voltage-gated calcium channel needed for proper heart function, the lncRNA is made only from the paternal allele (which is unmethylated), and it is not released by the RNA polymerase, remaining instead at its site of synthesis on the DNA template. This RNA in turn recruits the histone-modifying and DNA-methylating enzymes that direct the formation of repressive chromatin, which silences the protein-coding gene associated on the paternally derived chromosome (**Figure 7–52B**). The maternally derived gene, on the other hand, is immune to these effects because its imprinted methylation blocks the synthesis of the lncRNA but allows transcription of the adjacent protein-coding gene. Thus, like *Igf2*, the specificity of *Kcnq1* imprinting arises from an inherited methylation pattern; the difference lies in the way these patterns cause the differential gene expression.

Why imprinting should exist at all is a mystery. In vertebrates, it is restricted to mammals that develop within the mother, and many of the imprinted genes are involved in fetal development. One idea is that imprinting reflects a middle ground in the evolutionary struggle between males to produce larger offspring and females to limit offspring size by "halving" the dosage of certain gene products that might accelerate growth. Whatever its purpose might be, imprinting provides startling evidence that features of DNA other than its sequence of nucleotides can be inherited.

A Chromosome-wide Alteration in Chromatin Structure Can Be Inherited

We have seen that DNA methylation and certain types of chromatin structure can be heritable, preserving patterns of gene expression across cell generations. Perhaps the most striking example of this effect occurs in mammals, in which an alteration in the chromatin structure of an entire chromosome can modulate the levels of expression of most genes on that chromosome.

Males and females differ in their *sex chromosomes*. Females have two X chromosomes, whereas males have one X and one Y chromosome. In humans, the X and Y sex chromosomes differ radically in gene content: the X chromosome is three times larger and contains about 900 protein-coding genes compared to the Y chromosome's 55 protein-coding genes. Mammals have evolved a *dosage compensation* mechanism to ensure that the same amount of most of the

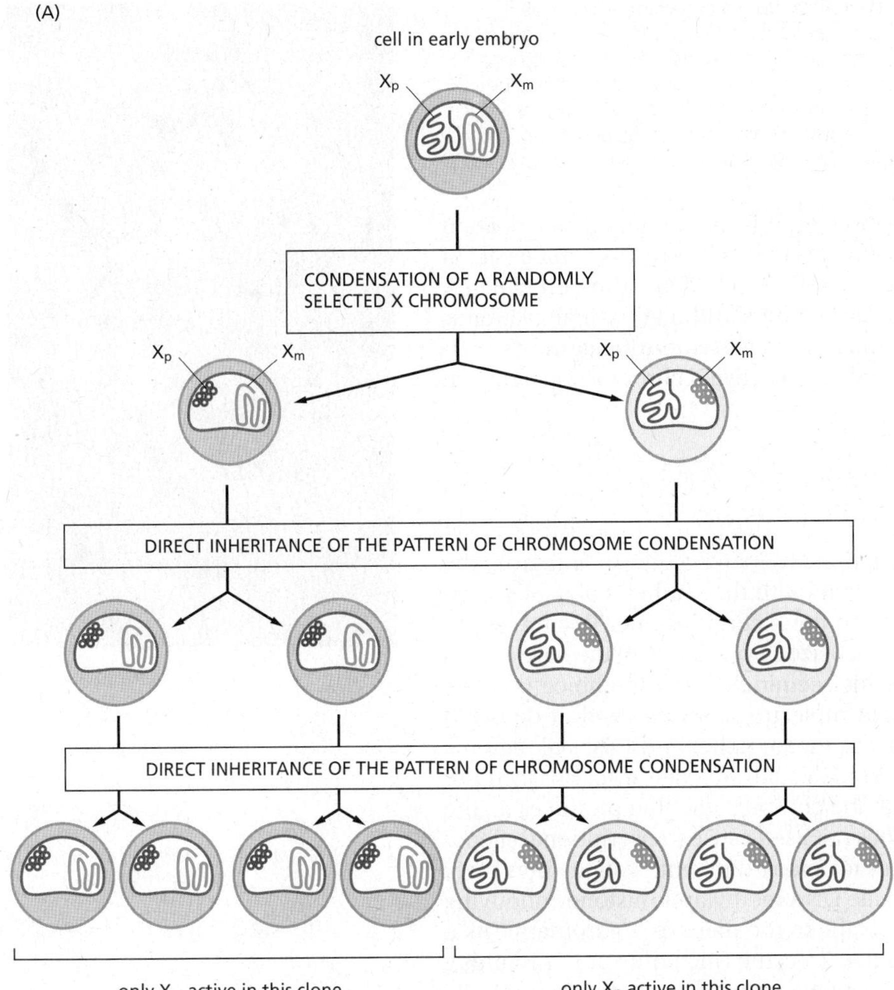

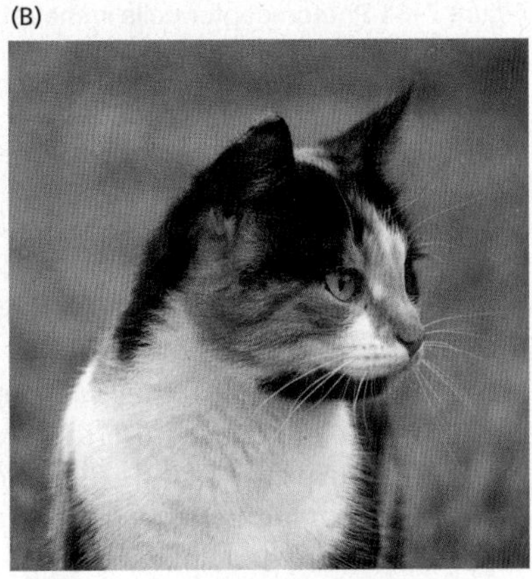

Figure 7–53 **X-inactivation.** (A) The clonal inheritance in female mammals of a condensed, inactive X chromosome. (B) A calico cat, whose patches of color reflect the random nature of the X-inactivation process. (B, bluecaterpillar/Deposit photos.)

X-chromosome gene products is made in both male and female cells, despite the fact that females contain twice as many X-chromosome genes. Mutations that interfere with this dosage compensation are generally lethal.

Mammals achieve dosage compensation by the transcriptional inactivation of one of the two X chromosomes in female somatic cells, a process known as **X-inactivation**. As a result of X-inactivation, two X chromosomes can coexist within the same nucleus, be exposed to the same diffusible transcription regulators, and yet differ entirely in their expression.

Early in the development of a female embryo, when it consists of a few hundred cells, one of the two X chromosomes in each cell becomes highly condensed into a type of heterochromatin. In placental mammals, the initial choice of which X chromosome to inactivate—the maternally inherited one (X_m) or the paternally inherited one (X_p)—appears to be random. And once either X_p or X_m has been inactivated, it remains silent throughout all subsequent cell divisions of that cell and its progeny, indicating that the inactive state is faithfully maintained through many cycles of DNA replication and mitosis. Because X-inactivation is random and takes place after several hundred cells have already formed in the embryo, every female is a mosaic of clonal groups of cells in which either X_p or X_m is silenced (**Figure 7–53**), distributed in small clusters in the adult animal because sister cells tend to remain close together during later stages of development (**Figure 7–54**).

X-inactivation creates the orange and black coat coloration of some female cats (see Figure 7–53B). In these "calico" cats, one X chromosome carries a gene that produces orange hair color, and the other X chromosome carries an allele

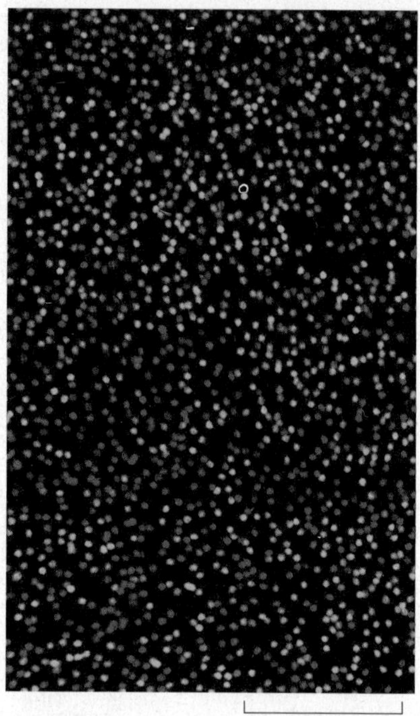

Figure 7–54 **Photoreceptor cells in the retina of a female mouse showing patterns of X-inactivation.** Using genetic engineering techniques (described in Chapter 8), the germ line of a mouse was modified so that one copy of the X chromosome (if active) makes a green fluorescent protein and the other (if active) a red fluorescent protein. Both proteins concentrate in the nucleus, and, in the field of cells shown here, it is clear that only one of the two X chromosomes is active in each cell. If both chromosomes were active, the nuclei would fluoresce both red and green, and therefore appear yellow. (From H. Wu et al., *Neuron* 81:103–119, 2014. With permission from Elsevier.)

of the same gene that results in black hair color; it is the random X-inactivation that produces patches of cells of two distinctive colors. In contrast, male cats of this genetic stock are either solid orange or solid black, depending on which X chromosome they have inherited from their mothers. Although X-inactivation is maintained over thousands of cell divisions, it is reversed during germ-cell formation, so that all the haploid oocytes contain an active X chromosome and can express X-linked gene products.

The Mammalian X-Inactivation in Females Is Triggered by the Synthesis of a Long Noncoding RNA

How is an entire chromosome transcriptionally inactivated? In humans, the chromosome-wide inactivation process begins with the synthesis of a long noncoding RNA, called **Xist**, whose gene lies on the X chromosome. This transcript (about 20,000 nucleotides in length) is synthesized by only one of the two X chromosomes in females, and exactly how this seemingly random choice is made remains to discovered. Once an Xist RNA molecule is synthesized, it does not leave the X chromosome from which it was made; rather, it diffuses along only that chromosome. Ultimately, about 2000 molecules of Xist are synthesized per X chromosome, and they eventually coat the chromosome that produces it. The spread of Xist across the chromosome does not itself cause transcriptional silencing; this long RNA contains binding sites for many different proteins that carry out the actual gene silencing. These include DNA methylases, histone-modifying enzymes, and structural components specific to the inactive X chromatin. As a result, extensive methylation of the inactive X occurs (including at CG islands), and the chromosome is folded into compact structures that are generally resistant to transcription (**Figure 7–55**). These multiple layers, each of which can be self-propagating (see Figure 7–48), ensure that the randomly chosen X chromosome remains inactive through multiple cell divisions.

Not every gene on the inactive X chromosome is transcriptionally silenced. Of the approximately 900 protein-coding genes on the human X chromosome, 15–20% remain actively expressed after the chromosome-wide inactivation process has been completed. And for many of these genes, both copies—one from the active X and one from the inactive X—must be expressed to obtain sufficient levels of their gene products for proper development to occur.

How do select genes escape silencing after the majority of the X chromosome is rendered transcriptionally inactive? As we saw in Chapter 4 and earlier in this chapter, transcriptionally active genes generally occur in DNA loops that are held in place by insulator proteins such as CTCF (see Figure 7–28), and this is the case for the "escapees" of the inactive X chromosome. These loops are believed to extend from the bulk of the tightly packaged chromosome. In contrast, most of the inactive genes lie in the interior of the inactive X chromosome, which is depleted for CTCF. It has been proposed that X-inactivation is accompanied by the formation of a specialized biomolecular condensate, where the proteins and RNAs needed for gene repression are kept at high local concentrations; according to this model, the loops of active genes would extend outward, beyond the boundary of the condensate.

We have described the way that placental mammals deal with dosage compensation on the X chromosome, but the details of this process differ from those in most other animals in important ways. For example, in marsupials, the choice of which X chromosome to inactivate is not random; instead, the X chromosome inherited from the father is automatically silenced. And in flies, dosage

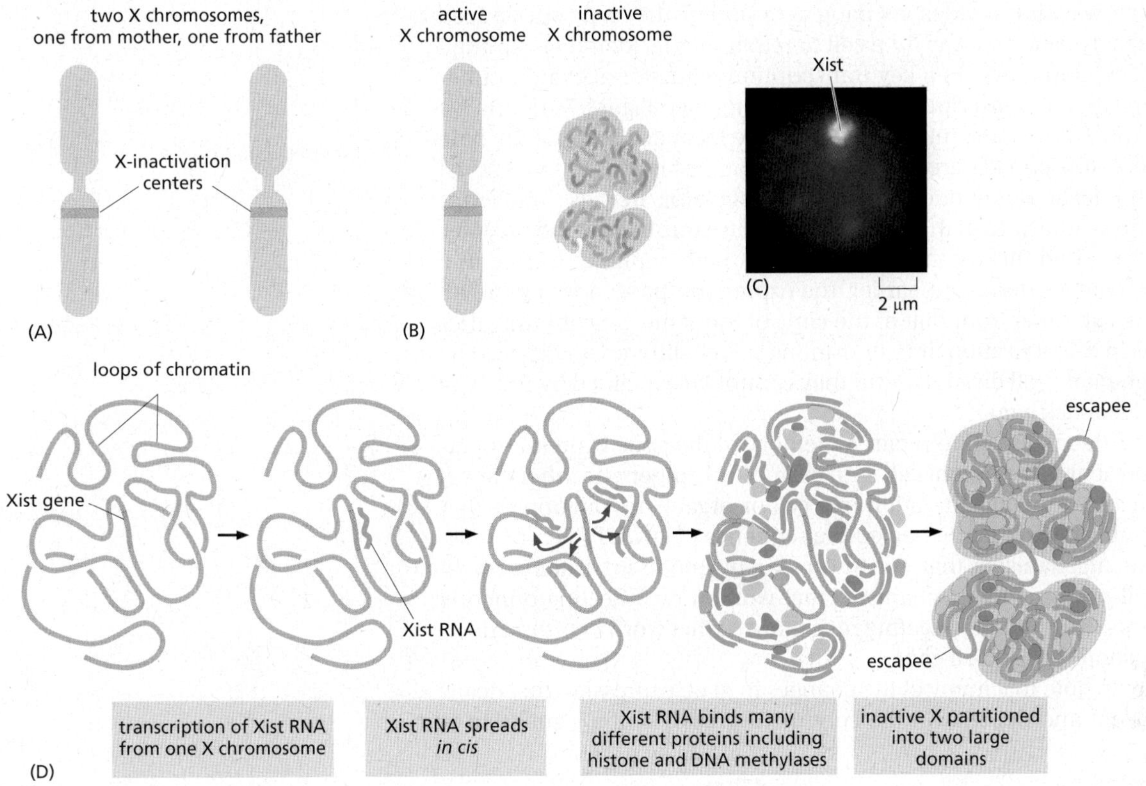

Figure 7–55 Mammalian X-inactivation. The two X chromosomes in a female mammal (A) before and (B) after X-inactivation. (C) At an early stage of X-chromosome inactivation, mouse chromosomes have been hybridized with a fluorescent probe that is complementary to the Xist RNA, which coats only the inactive X chromosome; the remaining DNA has been stained blue with a dye. (D) A schematic illustration of how the continuing synthesis of Xist RNA at the Xist locus moves Xist molecules outward, across the chromosome. As Xist molecules coat the chromosome, they begin to associate with a variety of structural proteins and enzymes that modify histones and DNA. [Although some of these proteins are bound to the chromosome prior to Xist spreading (not shown), most are brought in by direct association with Xist.] The two major chromosome domains that are created at the completion of the inactivation process have been proposed to be biomolecular condensates. Genes that escape the inactivation process are shown as loops, extending from the compact domains. (B and D, based on a figure supplied by Agnese Loda and Edith Heard; C, from L. Giorgetti et al., *Nature* 535:575–579, 2016. Reproduced with permission from SNCSC.)

compensation takes place in the male, where the single X chromosome is upregulated approximately twofold to match the female dose. Finally, in nematode worms, the hermaphrodites reduce gene expression by roughly half on both X chromosomes to match the single X-chromosome dosage in males.

The fundamentally different mechanisms of dosage compensation among animals suggest that it has been a relatively recent evolutionary innovation. We have some clues for its origin in humans. Some of the key components in X-inactivation also function to repress the many transposons in the human genome, a process we discuss later in the chapter. It has been proposed that Xist evolved from multiple transposons that inserted into our X chromosome, eventually "tricking" the cell into inactivating the whole chromosome.

Stable Patterns of Gene Expression Can Be Transmitted to Daughter Cells

Imprinting and X-inactivation are examples of **monoallelic gene expression**, where only one of the two copies of a gene is expressed in a diploid genome. In addition to the silenced genes on the X chromosome and the 300 or so genes that are imprinted, there are another 1000–2000 human genes that exhibit monoallelic expression. Like X-inactivation (but unlike imprinting), the choice of which copy of the gene is expressed and which is silenced appears random. Yet once the choice is made, it can persist for many cell divisions. Because the choice is often made relatively late in development, cells of the same tissue in the same individual can express different copies of a given gene. In other words, somatic tissues are often mosaics, where different clones of cells have subtly different patterns of gene expression. The mechanisms responsible for this type of monoallelic expression and its memory through cell divisions are not known in detail, and its general purpose—if any—is poorly understood. However, several different mechanisms are known that may contribute to such inheritance, as we now discuss.

In considering the general question of cell memory, it is useful to return to our discussion of the different cell types in an organism. As we have seen, once a cell in an organism differentiates into a particular cell type, it generally remains specialized in that way; if it divides, its daughters inherit the same specialized character. Perhaps the simplest way for a cell to remember its identity is through a positive feedback loop in which a key transcription regulator activates, either directly or indirectly, the transcription of its own gene (see Figure 7–42). As we discussed earlier in this chapter, interlocking positive feedback loops of the type shown in Figure 7–40B provide greater stability by buffering the circuit against fluctuations in the level of any one transcription regulator. Because transcription regulators are synthesized in the cytosol and diffuse throughout the nucleus, feedback loops based on this mechanism will affect both copies of a gene in a diploid cell. However, as discussed earlier, the expression pattern of a gene on one chromosome can differ from that of the copy of the same gene on the other chromosome (as in X-inactivation or in imprinting). Such differences can also be inherited through many cell divisions, and they cannot be explained by this type of transcription feedback loop.

The ability of a daughter cell to retain a memory of the gene expression patterns that were present in the parent cell is an example of **epigenetic inheritance**, which we define as a heritable alteration in a cell or organism's phenotype that does not result from changes in the nucleotide sequence of DNA. In **Figure 7–56**, we illustrate four mechanisms that can produce epigenetic inheritance, contrasting those self-propagating mechanisms that work *in cis*, affecting only one chromosomal copy, with self-propagating mechanisms that work *in trans*, affecting both chromosomal copies of a gene.

It is important to note that many of the changes in gene expression that occur in cells are transient and depend on the continued presence of a signal that is

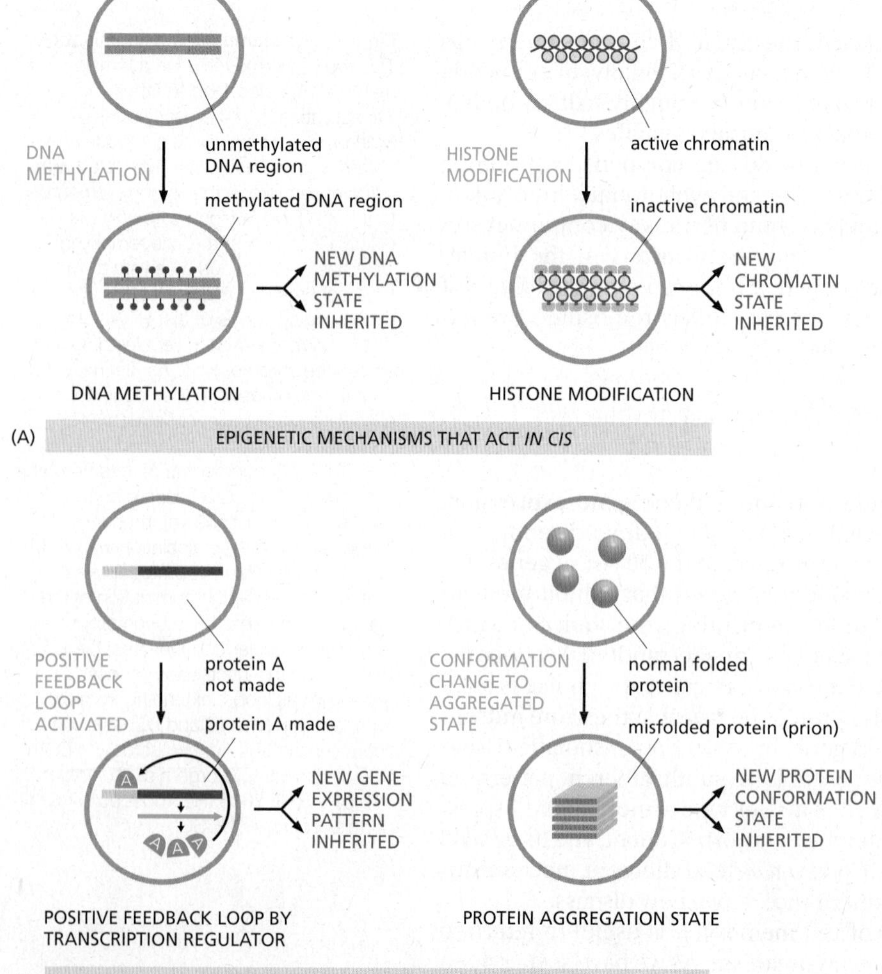

Figure 7–56 **Four distinct mechanisms that can produce an epigenetic form of inheritance in an organism.** (A) Two epigenetic mechanisms that act *in cis*. As discussed in this chapter, a maintenance methylase can propagate specific patterns of cytosine methylation (see Figure 7–47). Alternatively, as discussed in Chapter 4, a histone-modifying enzyme that replicates the same covalent modification that attracts it to chromatin can result in a chromatin structure being self-propagating (see Figure 4–44). Note that the term epigenetic is sometimes misused to refer to all covalent modifications of histones, whether or not they are self-propagating. But many histone modifications are erased each time a cell divides, and they therefore do not fit our definition. (B) Two epigenetic mechanisms that act *in trans*. Positive feedback loops formed by transcription regulators are found in all species and are probably the most common form of cell memory. As discussed in Chapter 3, some proteins can form self-propagating prions (see Figure 3–33). When these proteins are involved in gene expression, prions can transmit a particular pattern of gene expression to daughter cells.

external to the cell. When the signal disappears, so does the new gene expression pattern; in other words, the pattern is not directly heritable (see Chapter 15). Gene expression changes of both types—both heritable and non-heritable—are crucial for the function of all cells on earth. And the discovery more than 60 years ago that gene expression can be regulated by cells ranks as one of the fundamental principles of biology.

Summary

In addition to the positive feedback loops created by transcription regulators, eukaryotic cells can use both inherited forms of DNA methylation and inherited states of chromatin condensation as mechanisms for generating a cell memory of gene expression patterns. An especially dramatic case that involves chromatin condensation is the inactivation of an entire X chromosome in female mammals. DNA methylation underlies the phenomenon in mammals of genomic imprinting, in which the expression of a gene depends on whether that gene was inherited from the mother or the father. All of these mechanisms allow cells to pass on gene expression patterns to their progeny cells, contributing to the epigenetic inheritance that makes complex multicellular life possible.

POST-TRANSCRIPTIONAL CONTROLS

In principle, every step required for the process of gene expression can be controlled. Indeed, one can find examples of each type of regulation, and many genes are known to be regulated by multiple mechanisms. As we have seen, controls on the initiation of gene transcription are one critical form of regulation for all genes. But other, equally important controls often act later in the pathway from DNA to protein to change the amount of gene product that is made—and in some cases, even to alter the amino acid sequence of a protein product. These **post-transcriptional controls**, which operate after RNA polymerase has bound to the gene's promoter and has begun its RNA synthesis, are crucial for the regulation of many genes.

In the following sections, we consider the varieties of post-transcriptional regulation in a temporal order, following the sequence of events that an RNA molecule might experience after its transcription has begun (**Figure 7–57**).

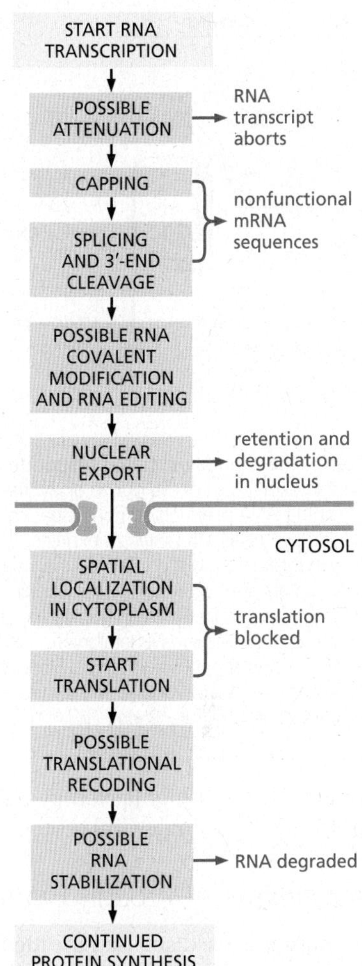

Figure 7–57 Post-transcriptional controls of gene expression. The final synthesis rate of a protein can, in principle, be controlled at any of the steps listed in capital letters, although only a few of the steps depicted here are likely to be critical for the regulation of any one particular protein. As we shall discuss, the 3′ end cleavage, splicing, editing, and translation recoding steps also make it possible for the cell to produce more than one protein variant from the same gene.

Transcription Attenuation Causes the Premature Termination of Some RNA Molecules

It has long been known that the expression of some genes is inhibited by premature termination of transcription, a phenomenon called *transcription attenuation*. In some of these cases, the nascent RNA chain adopts a structure that causes it to interact with the RNA polymerase in such a way as to abort its transcription. When the gene product is required, regulatory proteins bind to the nascent RNA chain to remove the attenuation, allowing the transcription of a complete RNA molecule.

A well-studied example of transcription attenuation occurs during the life cycle of HIV, the human immunodeficiency virus that is the causative agent of acquired immune deficiency syndrome, or AIDS. Once the HIV genome has been integrated into the host genome, the viral DNA is transcribed by the cell's RNA polymerase II (see Figure 5–61). However, this polymerase usually terminates transcription after synthesizing transcripts of several hundred nucleotides and thus fails to efficiently transcribe the entire viral genome. But when conditions for viral growth are optimal, a virus-encoded protein called Tat, which binds to a specific stem–loop structure in the nascent RNA that contains a "bulged base," prevents this premature termination (see Figure 6–92). Once bound to this specific RNA structure (called TAR), Tat assembles several host-cell proteins that allow the RNA polymerase to continue transcribing. The normal role of at least some of these proteins is to prevent pausing and premature termination by RNA polymerase when it transcribes normal cell genes. A normal cell mechanism has

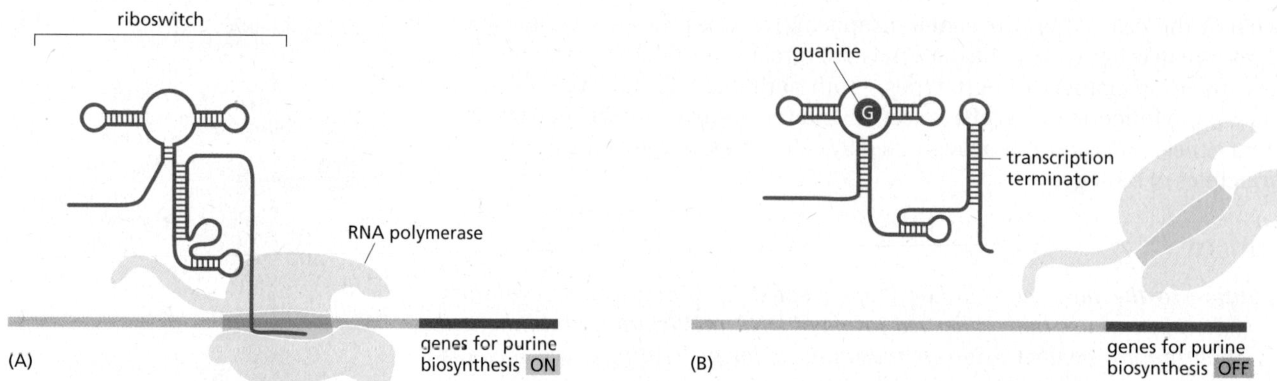

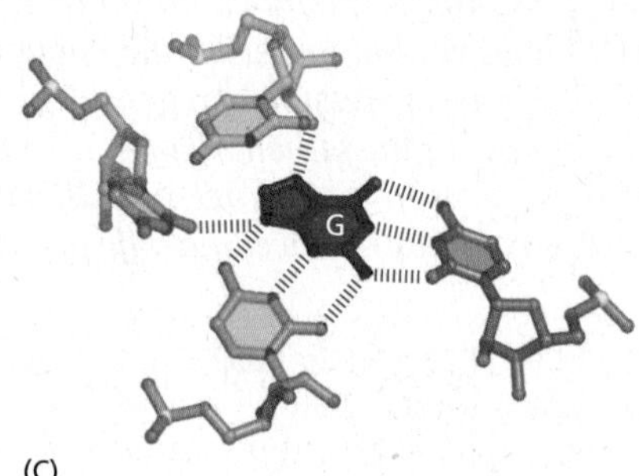

Figure 7–58 A riboswitch that responds to guanine. In this example from bacteria, the riboswitch controls expression of the purine biosynthetic genes. (A) When guanine levels in cells are low, an elongating RNA polymerase transcribes the purine biosynthetic genes, and the enzymes needed for guanine synthesis are therefore expressed. (B) When guanine is abundant, it binds the riboswitch, causing it to undergo a conformational change that forces the RNA polymerase to terminate transcription (see Figure 6–11). (C) Guanine *(red)* bound to the riboswitch. Only those nucleotides that form the guanine-binding pocket are shown. Many other riboswitches exist, including those that recognize *S*-adenosylmethionine, coenzyme B_{12}, flavin mononucleotide, adenine, lysine, and glycine. (A and B, adapted from M. Mandal and R.R. Breaker, *Nat. Rev. Mol. Cell Biol.* 5:451–463, 2004; and C.K. Vanderpool and S. Gottesman, *Mol. Microbiol.* 54:1076–1089, 2004; C, adapted from A. Serganov et al., *Chem. Biol.* 11:1729–1741, 2004. PDB code: 1Y27.)

apparently been highjacked by HIV to permit transcription of its genome to be controlled by a single viral protein.

Riboswitches Probably Represent Ancient Forms of Gene Control

In Chapter 6, we discussed the idea that, before modern cells arose on Earth, RNA played the role of both DNA and proteins, storing hereditary information and catalyzing chemical reactions (see pp. 389–393). The discovery of *riboswitches* shows that RNA can also form control devices. Riboswitches are short sequences of RNA that change their conformation when they bind a specific small molecule, such as a metabolite. Riboswitches are often located near the 5′ end of mRNAs, and they fold while the mRNA is being synthesized, blocking or permitting progress of the RNA polymerase according to whether the regulatory small molecule is bound (**Figure 7–58**).

Riboswitches are particularly common in bacteria, where they sense key small metabolites in the cell and adjust gene expression accordingly. Each recognizes only the appropriate small molecule with high specificity. In many cases, every chemical feature of the small molecule is read by the RNA, and the binding affinities observed are as tight as those typically observed between small molecules and proteins (see Figure 7–58C).

Riboswitches are perhaps the most economical examples of gene control devices, inasmuch as they completely bypass the need for regulatory proteins. In the example illustrated (see Figure 7–58), the riboswitch controls transcription elongation, but riboswitches can also regulate other steps in gene expression, as we shall see later in this chapter. The fact that highly sophisticated gene control devices can be made from short sequences of RNA provides important support for the early "RNA world" hypothesis.

Alternative RNA Splicing Can Produce Different Forms of a Protein from the Same Gene

As discussed in Chapter 6, RNA splicing shortens the transcripts of many eukaryotic genes by removing the intron sequences from mRNA precursors. A cell can splice an RNA transcript differently and thereby make different

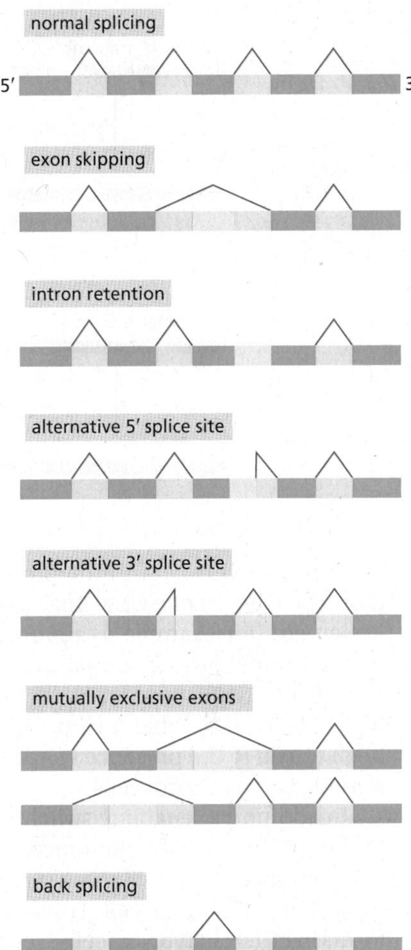

Figure 7–59 **Patterns of alternative RNA splicing.** *Dark blue boxes* mark exon sequences that are retained in spliced mRNAs. *Light blue boxes* mark possible exon sequences that are included only in the indicated mRNAs. The boxes are joined by *red lines* to indicate where intron sequences *(yellow)* are removed. In back splicing (discussed later in the chapter), a single exon is removed as a circular RNA molecule. (Adapted from H. Keren et al., *Nat. Rev. Genet.* 11:345–355, 2010.)

polypeptide chains from the same gene—a process called **alternative RNA splicing** (Figure 7–59; see also Figure 6–27). Many animal genes produce multiple proteins in this way.

When different splicing possibilities exist at several positions in the transcript, a single gene can produce dozens of different proteins. In one extreme case, a single *Drosophila* gene can, in principle, produce as many as 38,000 different proteins through alternative splicing (Figure 7–60), although only a fraction of these forms have thus far been experimentally observed. Considering that the *Drosophila* genome has approximately 14,000 protein-coding genes, it is clear that the protein complexity of an organism can greatly exceed the number of its genes. This example also illustrates the perils in equating gene number with an organism's complexity. For example, alternative splicing is rare in single-celled budding yeasts but very common in flies. Budding yeast has about 6200 genes, only about 300 of which are subject to splicing, and nearly all of these have only a single intron. The fact that flies have only 2–3 times as many genes as yeasts greatly underestimates the difference in complexity of these two genomes.

In some cases, alternative RNA splicing occurs because there is an *intron sequence ambiguity*: the standard spliceosome mechanism for removing intron sequences (discussed in Chapter 6) is unable to distinguish clearly between two or more alternative pairings of 5′ and 3′ splice sites, so that different choices are made by chance on different individual transcripts. Where such *constitutive* alternative splicing occurs, several versions of the protein encoded by the gene are made in all cells in which the gene is expressed.

In many cases, however, alternative RNA splicing is regulated. In the simplest examples, regulated splicing is used to switch from the production of a

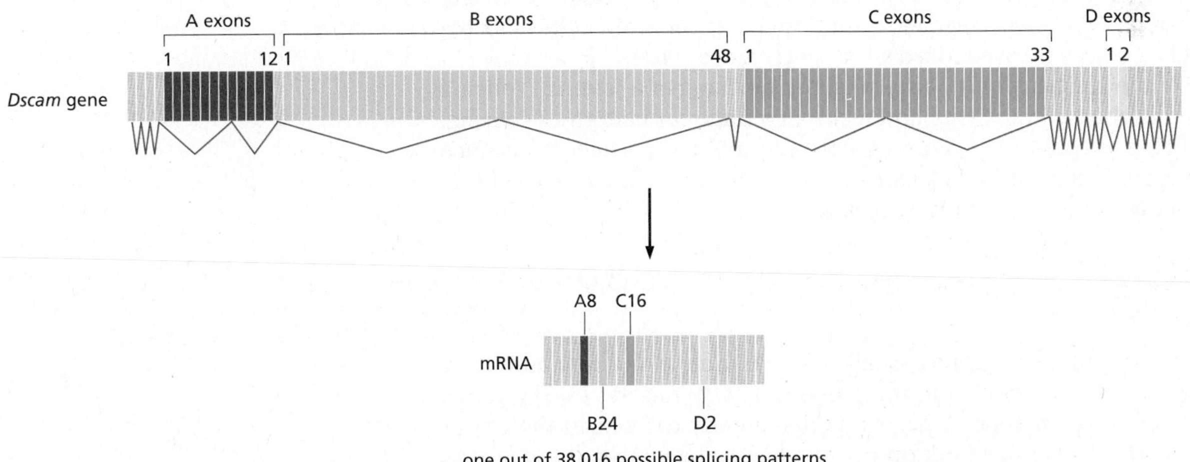

Figure 7–60 **Alternative splicing of RNA transcripts of the *Drosophila Dscam* gene.** Dscam proteins have several different functions. In cells of the fly immune system, they mediate the phagocytosis of bacterial pathogens. In cells of the nervous system, they are needed for proper wiring of neurons. Each mature mRNA contains 24 exons, four of which (denoted A, B, C, and D) are present in the *Dscam* gene as arrays of alternative exons. Each RNA contains 1 of 12 alternatives for exon A *(red)*, 1 of 48 alternatives for exon B *(green)*, 1 of 33 alternatives for exon C *(blue)*, and 1 of 2 alternatives for exon D *(yellow)*. This figure shows only one of the many possible splicing patterns (indicated by the *red line* and by the mature mRNA below it). Each variant Dscam protein folds into roughly the same structure (predominantly a series of extracellular immunoglobulin-like domains linked to a membrane-spanning region; see Figure 24–48), but the amino acid sequences of the domains vary according to the splicing pattern. The diversity of Dscam variants contributes to the plasticity of the immune system, as well as to the formation of complex neural circuits. (Adapted from D.L. Black, *Cell* 103:367–370, 2000. With permission from Elsevier.)

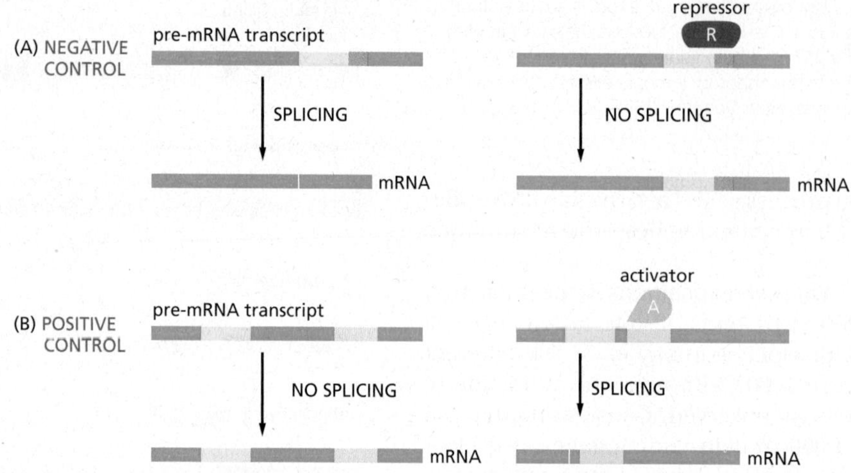

Figure 7–61 Negative and positive control of alternative RNA splicing. (A) In negative control, a repressor protein binds to a specific sequence in the pre-mRNA transcript and blocks access of the splicing machinery to a splice junction. This often results in the use of a secondary splice site, thereby producing an altered pattern of splicing (see Figure 7–59). (B) In positive control, the splicing machinery is unable to remove a particular intron sequence efficiently without assistance from an activator protein. Because an RNA molecule is flexible, the nucleotide sequences that bind these activators can be located many nucleotide pairs from the splice junctions they control, and they are often called *splicing enhancers*, by analogy with the transcription enhancers mentioned earlier in this chapter.

nonfunctional protein to the production of a functional one (or the other way around). The transposase that catalyzes the transposition of the *Drosophila* P element, for example, is produced in a functional form in germ cells and a nonfunctional form in somatic cells of the fly, allowing the P element to spread throughout the genome of the fly without causing damage in somatic cells (see Figure 5–59 and Table 5–4, p. 308). This difference in transposon activity has been traced to the presence of an intron sequence in the transposase RNA that is removed only in germ cells.

In addition to enabling switching from the production of a functional protein to the production of a nonfunctional one (or vice versa), the regulation of RNA splicing can generate different versions of a protein in different cell types, according to the needs of the cell. Tropomyosin, for example, is produced in specialized forms in different types of cells (see Figure 6–27). Cell-type-specific forms of many other proteins are produced in the same way.

RNA splicing can be regulated either negatively, by a regulatory molecule that prevents the splicing machinery from gaining access to a particular splice site on the RNA, or positively, by a regulatory molecule that helps direct the splicing machinery to an otherwise overlooked splice site (**Figure 7–61**).

Because of the plasticity of RNA splicing, the blocking of a "strong" splicing site will often expose a "weak" site and result in a different pattern of splicing. Thus, the splicing of a pre-mRNA molecule can be thought of as a delicate balance between competing splice sites—a balance that can easily be tipped by the effects on splicing of RNA-bound regulatory proteins.

The Definition of a Gene Has Been Modified Since the Discovery of Alternative RNA Splicing

The discovery that eukaryotic genes usually contain introns and that their coding sequences can be assembled in more than one way raised new questions about the definition of a gene. A gene was first clearly defined in molecular terms in the early 1940s from work on the biochemical genetics of the fungus *Neurospora*. Until then, a gene had been defined as a region of the genome that segregates as a single unit during meiosis and gives rise to a definable phenotypic trait—such as a red or a white eye in *Drosophila* or a round or wrinkled seed in peas. The *Neurospora* findings revealed that most genes correspond to a region of the genome that directs the synthesis of a single enzyme, leading to the view that each gene encodes one polypeptide chain. As more was learned about the mechanism of gene expression in the 1960s, a gene became identified as that stretch of DNA that was transcribed into the RNA coding for either a single polypeptide chain or a single structural RNA such as a tRNA or an rRNA molecule. The discovery of introns in the late 1970s could be readily accommodated by the original definition of a gene, provided that a single polypeptide chain was specified by the RNA transcribed from any one DNA sequence. But now that it is clear that many DNA sequences in eukaryotic cells can produce a set of distinct (but related) proteins by means of alternative RNA splicing, how should a gene be defined?

It is relatively rare that a single transcription unit produces two very different eukaryotic proteins, and in those cases, the two proteins are considered to be produced by distinct genes that overlap on the chromosome. It seems unnecessarily complex, however, to consider most of the protein variants produced by alternative RNA splicing as being derived from overlapping genes. A more sensible alternative is to modify the original definition to consider any DNA sequence that is transcribed as a single unit and encodes a set of closely related polypeptide chains (protein isoforms) as a single protein-coding gene. This definition of a **gene** also accommodates those DNA sequences that encode protein variants produced by post-transcriptional processes other than RNA splicing, such as the transcript cleavage and RNA editing discussed shortly.

Back Splicing Can Produce Circular RNA Molecules

We have seen that pre-mRNA splicing is remarkably plastic, and recent discoveries have revealed a new surprise. Some pre-mRNAs undergo what is termed "back splicing" where a 3′ splice site is joined to a downstream 5′ splice site, thereby reversing the normal joining order (see Figure 7–59). This process typically releases a single exon sequence as a covalently closed, circular RNA molecule. These unusual RNAs are exported from the nucleus but are rarely translated into protein. Instead, they have been proposed to "soak up" complementary RNAs as well as RNA-binding proteins and to provide scaffolds for multisubunit RNA–protein complexes. Because they lack free ends, which are the normal substrates for RNA-degrading enzymes, these circular RNAs are much more stable than typical mRNAs. Although usually made in small amounts, their stability can allow them to accumulate to high concentrations in cells, and several specific circular RNAs are especially prominent in cells of the mammalian brain and immune systems. Although we still have much to learn about these peculiar RNAs, they attest to the many surprises that RNA biology has in store for us. We shall revisit this general issue at the end of the chapter, when we discuss the diversity of noncoding RNAs.

A Change in the Site of RNA Transcript Cleavage and Poly-A Addition Can Change the C-terminus of a Protein

We saw in Chapter 6 that the 3′ end of a eukaryotic mRNA molecule is not formed by the termination of RNA synthesis by the RNA polymerase, as it is in bacteria. Instead, it results from an RNA cleavage reaction that is catalyzed by additional proteins while the transcript is elongating (see Figure 6–36). A cell can control the site of this cleavage so as to change the C-terminus of the resultant protein. In the simplest cases *of alternate cleavage and polyadenylation*, one protein variant is simply a truncated version of the other; in many other cases, however, the alternative cleavage and polyadenylation sites lie within intron sequences, and the pattern of splicing is thereby altered. This process can produce two closely related proteins that differ only in the amino acid sequences at their C-terminal ends. An analysis of RNAs produced from the human genome in a variety of cell types indicates that as many as half of all human protein-coding genes produce mRNA species with more than one site of polyadenylation.

A well-studied example of regulated polyadenylation is the switch from the synthesis of membrane-bound to secreted antibody molecules that occurs during the development of B lymphocytes (see Figure 24–22). Early in the life history of a B lymphocyte, the antibody it produces is anchored in the plasma membrane, where it serves as a receptor for antigen. Antigen stimulation causes B lymphocytes to multiply and to begin secreting their antibody. The secreted form of the

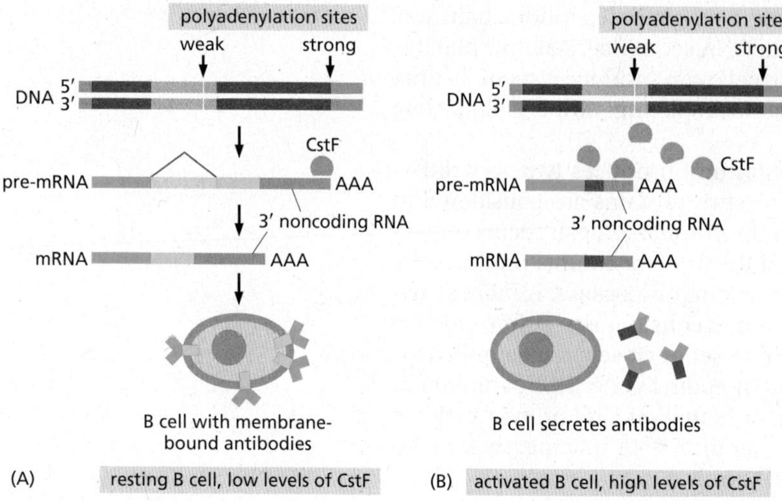

Figure 7–62 Regulation of the site of RNA cleavage and poly-A addition determines whether an antibody molecule is secreted or remains membrane-bound. (A) In unstimulated B lymphocytes, a long RNA transcript is produced, and the intron sequence *(yellow)* near its 3′ end is removed by RNA splicing to provide an mRNA molecule that codes for a membrane-bound antibody molecule. Only a portion of the antibody gene is shown in the figure; the actual gene and its mRNA would extend further to the left of the diagram. (B) After antigen stimulation, the RNA transcript is cleaved and polyadenylated upstream from the intron's 3′ splice site. As a result, some of the intron sequence remains as a coding sequence in the short transcript, specifying the hydrophilic C-terminal portion of the secreted antibody molecule *(brown)*. (Adapted from D. Di Giammartino et al., *Mol. Cell* 43:853–866, 2011.)

antibody is identical to the membrane-bound form except at the extreme C-terminus. In this part of the protein, the membrane-bound form has a long string of hydrophobic amino acids that traverses the lipid bilayer of the membrane, whereas the secreted form has a much shorter string of hydrophilic amino acids. The switch from membrane-bound to secreted antibody is generated through a change in the site of RNA cleavage and polyadenylation, as shown in **Figure 7–62**.

The change is caused by an increased concentration of a subunit of a protein (CstF) that promotes RNA cleavage (see Figure 6–36). The first cleavage/poly-A addition site that a transcribing RNA polymerase encounters is suboptimal and is usually skipped in unstimulated B lymphocytes, leading to production of the longer RNA transcript. But when activated to produce antibodies, the B lymphocyte produces more CstF; as a result, cleavage now occurs at the suboptimal site, and the shorter transcript is produced. In this way, a change in concentration of a general RNA-processing factor can have a dramatic effect on the expression of a specific gene.

Nucleotides in mRNA Can Be Covalently Modified

In the previous chapter, we saw how specialized proteins modify the 5′ and 3′ ends of eukaryotic mRNAs and how a complex assembly of proteins and RNA molecules removes intron sequences. However, mRNA molecules are subject to more than 100 additional kinds of covalent changes, predominantly chemical modifications of individual bases, a few of which are shown in **Figure 7–63**. The reasons for most of these modifications of individual mRNAs remain a mystery. We do not know what they might do or even if they are biologically meaningful, inasmuch as many of them may simply represent "spillover" from the processes that modify the highly abundant tRNA and rRNA molecules (see Figures 6–43 and 6–57).

One of the most prominent and best understood mRNA modifications is the methylation of the amino group on adenine to produce N^6-methyladenosine

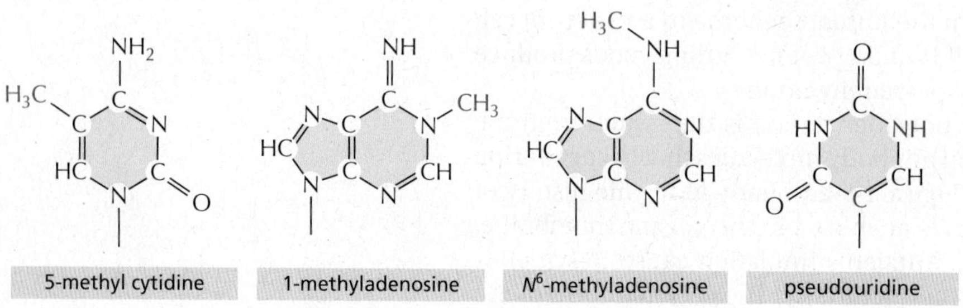

Figure 7–63 Four of the most prominent of the many types of covalent base modifications found in mRNA. Differences from the normal nucleosides are indicated in *red*. Each base is joined to a ribose sugar (not shown) by the indicated bond to form the nucleoside.

(see Figure 7–63). This addition is constantly being removed by protein complexes that contain demethylases, or "erasers," making the modification temporary. The methylases responsible for this modification typically act as the RNA is being transcribed; they recognize short sequences in the emerging RNA (often with the help of other proteins) and methylate the adenosines adjacent to these sequences.

In humans an average of 1–3 N^6-methyladenosine modifications occur on each mRNA molecule. What are their consequences? One effect is the destabilization of the hairpin helices that are formed by intramolecular base-pairing. This modification can thereby change the secondary structure of mRNA, which in some cases alters the splicing pattern of transcripts. In other cases, the modification promotes destruction of mRNAs through "reader" proteins that attract the RNA degradation machinery. A rapid destruction of certain mRNAs is especially important during cell differentiation, when the mRNAs produced earlier need to be cleared out. Finally, other specific N^6-methyladenosine modifications are known to promote translation of the modified mRNA. In this case, reader proteins that attract the translation machinery come into play. The many other mRNA modifications, some of which are shown in Figure 7–63, are more poorly understood, but some may likewise help to determine exactly how each mRNA is to be handled by the cell.

RNA Editing Can Change the Meaning of the RNA Message

The molecular mechanisms used by cells are providing scientists with a continual source of surprises. An example is a covalent modification of mRNA that alters its nucleotide sequence and thereby changes the coded message it carries—a process known as **RNA editing.**

In animals, two principal types of such RNA editing occur: the deamination of adenine to produce inosine (A-to-I editing) and, less frequently, the deamination of cytosine to produce uracil (C-to-U editing; see Figure 5–43). Because these chemical modifications alter the pairing properties of the bases (I pairs with C, and U pairs with A), they can have profound effects on the meaning of the RNA. If the edit occurs in a coding region, it can either change the amino acid sequence of the protein or produce a truncated protein by creating a premature stop codon. Edits that occur outside coding sequences can affect the pattern of pre-mRNA splicing, the transport of mRNA from the nucleus to the cytosol, the efficiency with which the RNA is translated, or the base-pairing between microRNAs (miRNAs) and their mRNA targets, a form of gene regulation that will be discussed later in the chapter.

The process of A-to-I editing is particularly prevalent in humans, where it occurs for approximately 1000 genes. Enzymes called *ADARs* (*adenosine deaminases acting on RNA*) perform this type of editing; these enzymes recognize a double-strand RNA structure that is formed through base-pairing between the site to be edited and a complementary sequence located elsewhere on the same RNA molecule, typically in an intron (**Figure 7–64**). The structure of

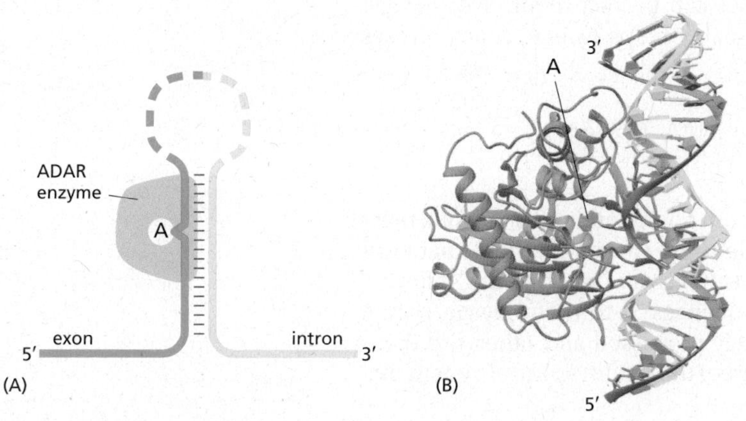

Figure 7–64 **Mechanism of A-to-I RNA editing in mammals.** (A) Typically, a sequence complementary to the position of the edit is present in an intron, and the resulting double-strand RNA structure attracts an A-to-I editing enzyme (ADAR). In the case illustrated, the edit is made in an exon; in most cases, however, it occurs in noncoding portions of the mRNA. Editing by ADAR takes place in the nucleus, before the pre-mRNA has been fully processed. Mice and humans have two ADAR genes: *ADR1* is expressed in many tissues and is required in the liver for proper red blood cell development; *ADR2* is expressed only in the brain, where it is required for proper brain development. (B) The human ADR2 enzyme bound to double-stranded RNA. The adenine to be edited is seen to be flipped out of the RNA double helix and buried deep in the catalytic pocket of the enzyme. Base flipping, which allows the enzyme access to the entire base, is also observed in enzymes that repair DNA (see Figure 5–42). (PDB code: 5ED1.)

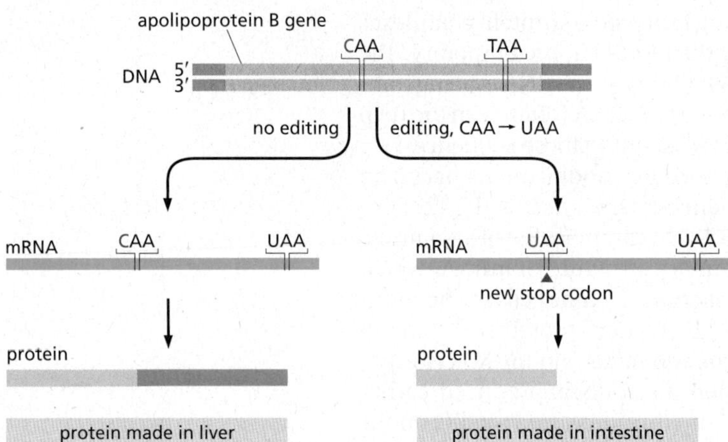

Figure 7–65 **C-to-U RNA editing produces a truncated form of apolipoprotein B.** As indicated, a tissue-specific edit in the middle of a coding sequence creates a truncated version of this protein in the intestine.

the double-stranded RNA specifies whether the mRNA is to be edited, and if so, where the edit should be made. An especially important example of A-to-I editing takes place in the mRNA that codes for a transmitter-gated ion channel in the brain. A single edit changes a glutamine to an arginine; the affected amino acid lies on the inner wall of the channel, and the editing change alters the Ca^{2+} permeability of the channel. Mutant mice that cannot make this edit are prone to epileptic seizures and die during or shortly after weaning, showing that editing of the ion channel RNA is normally crucial for proper brain development.

C-to-U editing, which is carried out by a different set of enzymes, is also important in mammals. For example, in certain cells of the gut, the mRNA for apolipoprotein B undergoes a C-to-U edit that creates a premature stop codon and therefore produces a shorter form of the protein. In cells of the liver, the editing enzyme is not expressed, and the full-length apolipoprotein B is produced. The two protein isoforms have different properties, and each plays a role in lipid metabolism that is specific to the organ that produces it (**Figure 7–65**).

Why RNA editing exists at all is a mystery. One idea is that it arose in evolution to correct "mistakes" in the genome. Another is that it arose as a somewhat slapdash way for the cell to produce subtly different proteins from the same gene. A third possibility is that RNA editing originally evolved as a defense mechanism against retroviruses and retrotransposons and was only later adapted by the cell to change the meanings of certain mRNAs. The last explanation receives support from the fact that RNA editing plays important roles in cell defense. The RNA genomes of some retroviruses, including HIV, are extensively edited after they infect cells. This hyperediting creates many harmful mutations in the viral RNA genome and also causes viral mRNAs to be retained in the nucleus, where they are eventually degraded. Although some modern retroviruses can protect themselves against this defense mechanism, RNA editing presumably helps to hold many viruses in check.

The Human AIDS Virus Illustrates How RNA Transport from the Nucleus Can Be Regulated

It has been estimated that in mammals only about one-twentieth of the total mass of RNA that is synthesized ever leaves the nucleus. We saw in Chapter 6 that most mammalian RNA molecules undergo extensive processing and that the "leftover" RNA fragments (excised introns and RNA sequences 3′ to the cleavage/poly-A site) are degraded in the nucleus. Incompletely processed and otherwise damaged RNAs are also eventually degraded as part of the quality-control system that acts on RNA production.

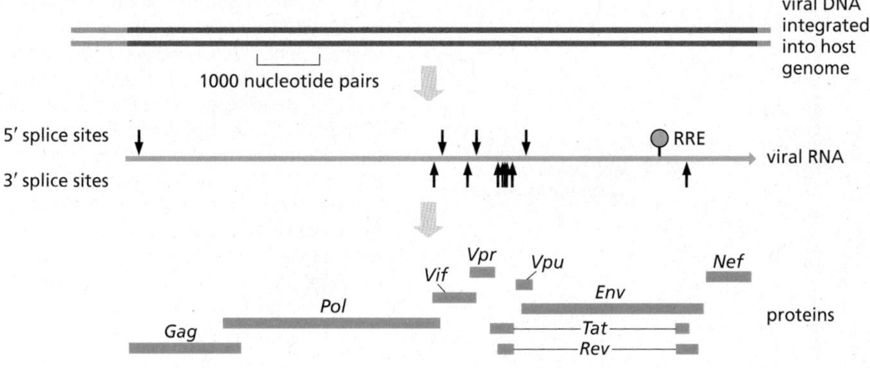

Figure 7–66 The compact genome of HIV, the human AIDS virus. The positions of the nine HIV genes are shown in *green.* The *red double line* indicates a DNA copy of the viral genome that has become integrated into the host DNA *(gray).* Note that the coding regions of many HIV genes overlap, and that those for *Tat* and *Rev* are split by introns. The *blue line* in the middle of the figure represents the pre-mRNA transcript of the viral DNA and shows the locations of all the possible splice sites *(arrows).* There are many alternative ways of splicing the viral transcript; for example, the *Env* mRNAs retain the intron that has been spliced out of the *Tat* and *Rev* mRNAs. The Rev response element (RRE) is indicated by a *blue* ball and stick. It is a 234-nucleotide-long stretch of RNA that folds into a defined structure; Rev recognizes a particular hairpin within this larger structure.

The *Gag* gene codes for a protein that is cleaved into several smaller proteins that form the viral capsid. The *Pol* gene codes for a protein that is cleaved to produce reverse transcriptase (which transcribes RNA into DNA), as well as the integrase involved in integrating the viral genome (as double-stranded DNA) into the host genome. The *Env* gene codes for the envelope proteins (see Figure 5–61). Tat, Rev, Vif, Vpr, Vpu, and Nef are small proteins with a variety of functions. As discussed in the text, Rev regulates nuclear export (see Figure 7–67), and Tat regulates the elongation of transcription across the integrated viral genome (see pp. 445–446).

As described in Chapter 6, the export of RNA molecules from the nucleus is delayed until processing has been completed. However, mechanisms that deliberately override this control point can be used to regulate gene expression. This strategy forms the basis for one of the best-understood examples of regulated nuclear transport of mRNA, which occurs in the human AIDS virus, HIV.

As we saw in Chapter 5, HIV, once inside the cell, directs the formation of a double-strand DNA copy of its single-strand RNA genome, which is then inserted into the genome of the host (see Figure 5–61). Once inserted, the viral DNA can be transcribed as one long RNA molecule by the host cell's RNA polymerase II. This transcript is then spliced in many different ways to produce more than 30 different species of mRNA, which in turn are translated into a variety of different proteins (**Figure 7–66**). In order to make progeny virus, entire unspliced viral transcripts must be exported from the nucleus to the cytosol, where they are packaged into viral capsids and serve as the viral genome. This large transcript, as well as certain alternatively spliced HIV mRNAs that are needed to produce viral proteins, still carry complete introns. The host cell's normal block to the nuclear export of unspliced RNAs therefore presents a special problem for HIV.

The block is overcome by a viral-coded protein (called Rev) that binds to a specific RNA sequence (called the Rev response element; RRE) located within a viral intron. The Rev protein interacts with a nuclear export receptor (Crm1), which directs the movement of viral RNAs through nuclear pores into the cytosol despite the presence of intron sequences. (How export receptors function is discussed in detail in Chapter 12.) The regulation of nuclear export by Rev has several important consequences for HIV growth and pathogenesis. In addition to ensuring the nuclear export of specific unspliced RNAs, it divides the viral infection into an early phase (in which Rev is translated from a fully spliced RNA, and all of the intron-containing viral RNAs are retained in the nucleus and degraded) and a late phase (in which unspliced RNAs are exported because of Rev function). This timing helps the virus replicate by providing the gene products in roughly the order in which they are needed (**Figure 7–67**).

Regulation by Rev and by Tat, the HIV protein that counteracts premature transcription termination (see pp. 445–446), allows the virus to achieve latency, a condition in which the HIV genome has become integrated into the host-cell genome, but the production of viral proteins has temporarily ceased. If, after the virus's initial entry into a host cell, conditions are unfavorable for viral replication, Rev and Tat are made at levels too low to promote transcription and export of unspliced RNA. This stalls the viral growth cycle until conditions improve, whereupon Rev and Tat levels increase and the virus enters the replication cycle.

mRNAs Can Be Localized to Specific Regions of the Cytosol

Once a newly made eukaryotic mRNA molecule has passed through a nuclear pore and entered the cytosol, it is typically met by ribosomes, which translate it into a polypeptide chain. Once the first round of translation "passes" the

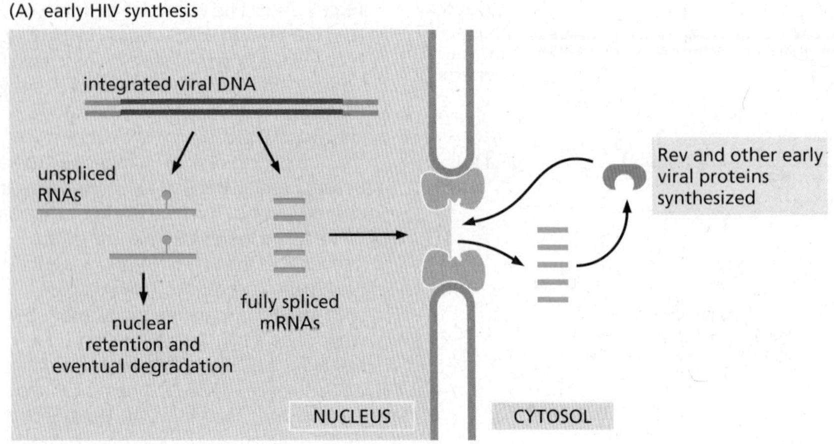

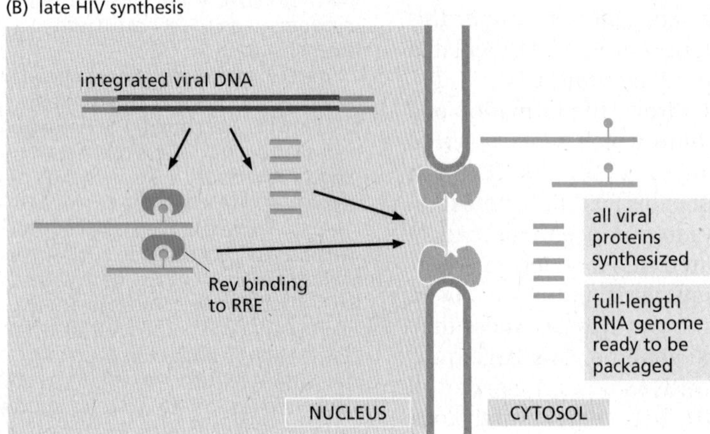

Figure 7–67 Regulation of nuclear export by the HIV Rev protein. (A) Early in HIV infection, only the fully spliced RNAs (which contain the coding sequences for Rev, Tat, and Nef) are exported from the nucleus and translated. (B) Once sufficient Rev protein has accumulated and been transported into the nucleus, unspliced viral RNAs can be exported from the nucleus. Many of these RNAs are translated into protein, and the full-length transcripts are packaged into new viral particles.

nonsense-mediated decay test (see Figure 6–80), the mRNA is usually translated in earnest. If the mRNA encodes a protein that is destined to be secreted or expressed on the cell surface, a signal sequence at the protein's N-terminus will direct it to the endoplasmic reticulum (ER). In this case, as discussed in Chapter 12, components of the cell's protein-sorting apparatus recognize the signal sequence as soon as it emerges from the ribosome and direct the entire complex of ribosome, mRNA, and nascent protein to the membrane of the ER, where the remainder of the polypeptide chain is synthesized. In other cases, free ribosomes in the cytosol synthesize the entire protein, and signals in the completed polypeptide chain may then direct the protein to other sites in the cell.

Many mRNAs are themselves directed to specific intracellular locations before their efficient translation begins, allowing the cell to position its mRNAs close to the sites where the encoded protein is needed. RNA localization has been observed in many organisms, including unicellular fungi, plants, and animals, and it appears to be a common mechanism that cells use to concentrate high-level production of proteins at specific sites. This strategy also provides the cell with other advantages. For example, it allows the establishment of asymmetries in the cytosol of the cell, a key step in many stages of development.

The localization of mRNA, coupled with translational control, also allows the cell to regulate gene expression independently in different regions. This feature is particularly important in large, highly polarized cells such as neurons; in those cells, specific mRNAs are transported for long distances along axons and dendrites to synapses, and the translation of the mRNAs that become localized there is often controlled by synaptic activity.

The mechanisms for mRNA localization that have been discovered all require specific signals in the mRNA itself (**Figure 7–68**). These signals are usually

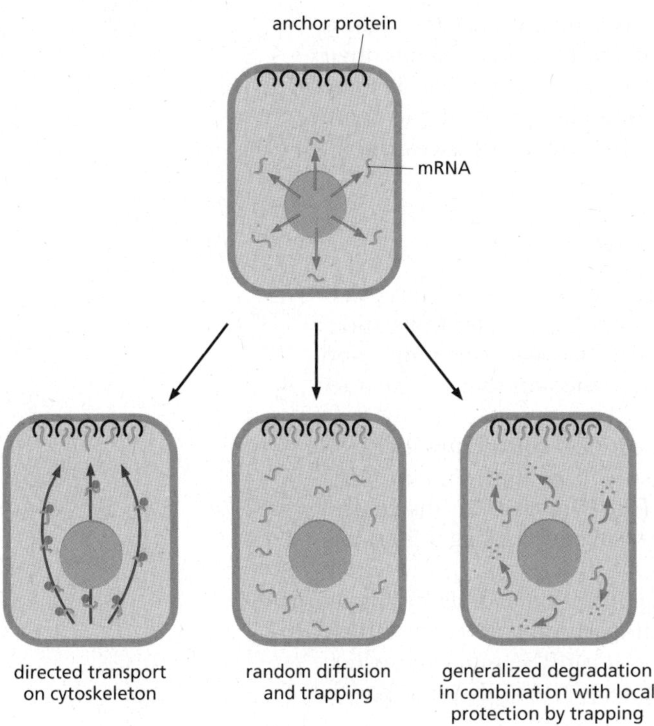

Figure 7–68 **Mechanisms for the localization of mRNAs.** The mRNA to be localized leaves the nucleus through nuclear pores *(top)*. Some localized mRNAs *(left diagram)* travel to their destination by associating with cytoskeletal motors, which use the energy of ATP hydrolysis to move the mRNAs unidirectionally along filaments in the cytoskeleton *(red)* (see Chapter 16). At their destination, the mRNAs are held in place by anchor proteins *(black)*. Other mRNAs randomly diffuse through the cytosol and are simply trapped by anchor proteins at their sites of localization *(center diagram)*. As an additional feature, many mRNAs *(right diagram)* are degraded in the cytosol unless they have bound, through random diffusion, a localized anchor protein complex that protects the mRNA from degradation *(black)*. These mechanisms require specific signals on the mRNA, which are typically located in the 3′ UTR. In all cases, other RNA-bound components block the translation of the mRNA until it is properly localized. Even then, additional signals are often needed to begin translation. (Adapted from H.D. Lipshitz and C.A. Smibert, *Curr. Opin. Genet. Dev.* 10:476–488, 2000.)

concentrated in the 3′ *untranslated region* (*UTR*), the region of RNA that extends from the stop codon that terminates protein synthesis to the start of the poly-A tail (**Figure 7–69**). As in neurons, mRNA localization is usually coupled with translational controls to ensure that the localized mRNA remains quiescent until it is needed.

The *Drosophila* egg provides an especially striking example of mRNA localization. The mRNA encoding the Bicoid transcription regulator is localized by attachment to the cytoskeleton at the anterior tip of the developing egg. When fertilization triggers the translation of this mRNA, it generates a gradient of the Bicoid protein that plays a crucial part in directing the development of the anterior part of the embryo (see Figures 7–29 and 21–19). Many mRNAs in somatic cells are also localized in a similar way. The mRNA that encodes actin, for example, is localized to the actin-filament-rich cell cortex in mammalian fibroblasts by means of a 3′ UTR signal.

We saw in Chapter 6 that mRNA molecules exit from the nucleus bearing numerous markings in the form of RNA modifications (the 5′ cap and the 3′ poly-A tail) and bound proteins (exon junction complexes, for example) that signify the successful completion of the different pre-mRNA processing steps. As just described, the 3′ UTR of an mRNA can be thought of as a "ZIP code" that directs mRNAs to different places in the cell. Shortly, we will see that mRNAs also carry information specifying their average lifetime in the cytosol and the efficiency with

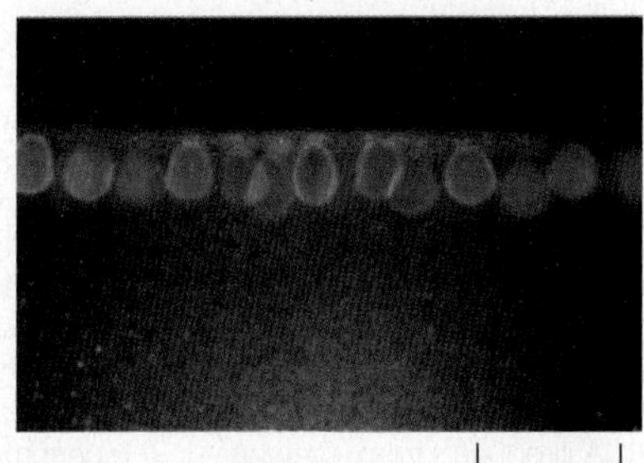

Figure 7–69 **An experiment demonstrating the importance of the 3′ UTR in localizing mRNAs to specific regions of the cytoplasm.** For this experiment, two different fluorescently labeled RNAs were prepared by transcribing DNA *in vitro* in the presence of fluorescently labeled derivatives of ribonucleoside triphosphates. One RNA (labeled with a *red* fluorochrome) contains the coding region for the *Drosophila* Hairy protein and includes the adjacent 3′ UTR. The other RNA (labeled *green*) contains the Hairy coding region with the 3′ UTR deleted. The two RNAs were mixed and injected into a *Drosophila* embryo at a stage of development when multiple nuclei reside in a common cytoplasm (see Figure 7–29). When the fluorescent RNAs were visualized 10 minutes later, the full-length *hairy* RNA *(red)* was localized to the apical side of nuclei *(blue)*, whereas the transcript missing the 3′ UTR *(green)* failed to localize and is seen as a diffuse cloud. Hairy is one of many transcription regulators that specify positional information in the developing *Drosophila* embryo, and the localization of its mRNA (shown in this experiment to depend on its 3′ UTR) is critical for proper fly development. (Courtesy of Simon Bullock and David Ish-Horowicz.)

which they are translated into protein. In a broad sense, the untranslated regions of eukaryotic mRNAs resemble the transcriptional control regions of genes: their nucleotide sequences contain information specifying the way the RNA is to be used, and proteins interpret this information by binding specifically to these sequences. Thus, in addition to the specification of amino acid sequences, mRNA molecules are rich with other types of information.

Untranslated Regions of mRNAs Control Their Translation

Once an mRNA has been synthesized, one of the most common ways of regulating the levels of its protein product is to control the step that initiates translation. Even though the details of translation initiation differ between eukaryotes and bacteria (as we saw in Chapter 6), they each use some of the same basic regulatory strategies.

In bacterial mRNAs, a conserved stretch of nucleotides—the *Shine-Dalgarno sequence*—is always found a few nucleotides upstream of the initiating AUG codon, and it is required to start protein synthesis (see Figure 6–75). The control of bacterial translation generally involves either exposing or blocking this critical sequence (**Figure 7–70**).

Eukaryotic mRNAs do not contain such a sequence. Instead, as discussed in Chapter 6, the selection of an AUG codon as a translation start site is largely determined by its proximity to the cap at the 5′ end of the mRNA molecule, which is the site at which the small ribosomal subunit binds to the mRNA and begins

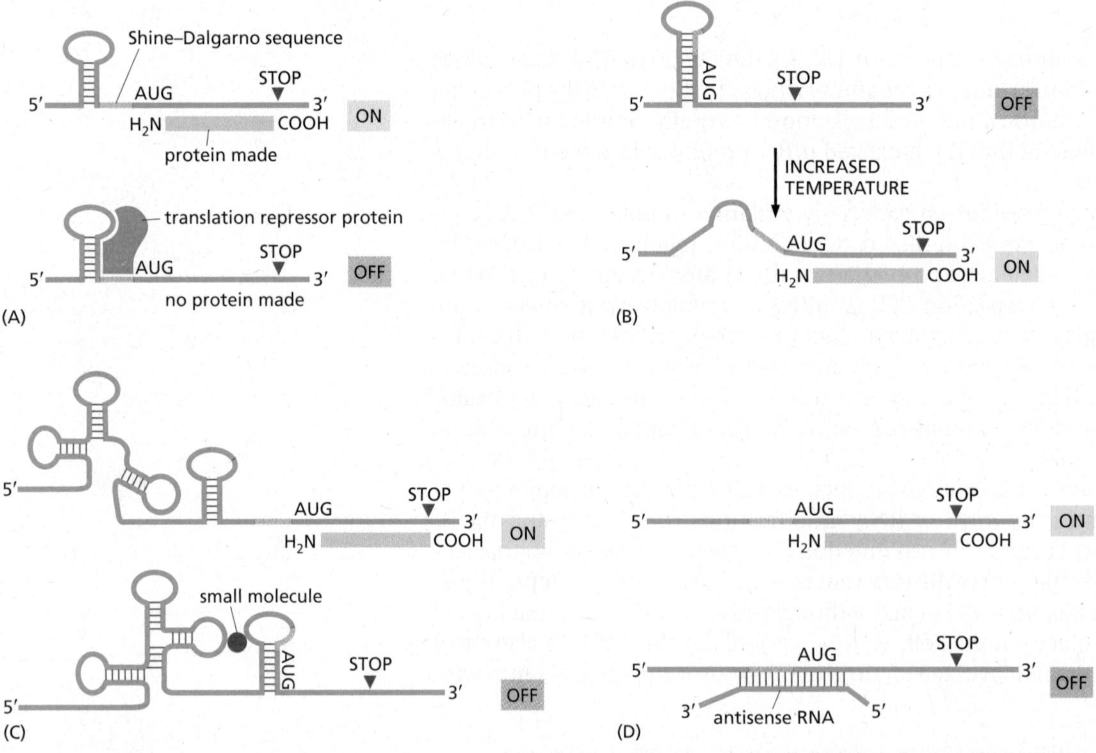

Figure 7–70 Mechanisms of translational control. Although these examples are from bacteria, many of the same principles operate in eukaryotes. (A) Sequence-specific RNA-binding proteins repress translation of specific mRNAs by blocking access of the ribosome to the Shine–Dalgarno sequence *(orange)*. For example, some ribosomal proteins repress translation of their own mRNA. This negative feedback mechanism allows the cell to maintain balanced quantities of the various components needed to form ribosomes. (B) An RNA "thermosensor" permits efficient translation initiation only at elevated temperatures at which the stem–loop structure has been melted. An example occurs in the human pathogen *Listeria monocytogenes*, in which the translation of its virulence genes increases at 37°C, the temperature of the host. (C) Binding of a small molecule to a riboswitch causes a major rearrangement of RNA structure, creating a different set of stem–loop structures. In the bound structure, the Shine–Dalgarno sequence *(orange)* is sequestered, and translation initiation is thereby blocked. In many bacteria, *S*-adenosylmethionine acts in this manner to block production of the enzymes that synthesize it. (D) An "antisense" RNA produced from elsewhere in the genome base-pairs with a specific mRNA and blocks its translation. Many bacteria regulate expression of iron-storage proteins in this way.

scanning for an initiating AUG codon. In eukaryotes, translational repressors can bind to the 5′ end of the mRNA and thereby inhibit translation initiation (see Figure 7–74). A particularly important type of translational control in eukaryotes relies on small RNAs (termed *microRNAs*, or *miRNAs*) that bind to mRNAs and reduce protein output, as described later in this chapter.

The Phosphorylation of an Initiation Factor Regulates Protein Synthesis Globally

Eukaryotic cells decrease their overall rate of protein synthesis in response to a variety of situations, including deprivation of growth factors or nutrients, infection by viruses, and sudden increases in temperature. (This response is coordinated by the TOR signaling pathway, which is described in Chapter 17; see Figure 17–61.) Much of the decrease in translation is caused by the phosphorylation of the translation initiation factor eIF2 by specific protein kinases that respond to the changes in conditions.

The normal function of eIF2 was outlined in Chapter 6 (see Figure 6–74). It forms a complex with GTP and mediates the binding of the methionyl initiator tRNA to the small ribosomal subunit, which then binds to the 5′ end of the mRNA and begins scanning along the mRNA. When an AUG codon is recognized, the eIF2 protein hydrolyzes the bound GTP to GDP, causing a conformational change in the protein and releasing it from the small ribosomal subunit. The large ribosomal subunit then joins the small one to form a complete ribosome that begins protein synthesis.

Because eIF2 binds very tightly to GDP, a guanine nucleotide exchange factor (see p. 880) designated eIF2B is required to release the GDP from eIF2 so that a new GTP molecule can bind—as required for eIF2 reuse (**Figure 7–71A**). When eIF2 is phosphorylated, it binds to eIF2B unusually tightly, inactivating this exchange factor. Because there is more eIF2 than eIF2B in cells, even a fraction of phosphorylated eIF2 can trap nearly all of the eIF2B. Without this exchange factor, GDP remains bound to nearly all of the nonphosphorylated eIF2, greatly slowing protein synthesis (**Figure 7–71B**).

Regulation of the level of active eIF2 is especially important in mammalian cells. As described in Chapter 17, eIF2 down-regulation is part of the mechanism that allows these cells to enter a nonproliferating, resting state (called G_0) in which the rate of total protein synthesis is reduced to about one-fifth the rate in proliferating cells.

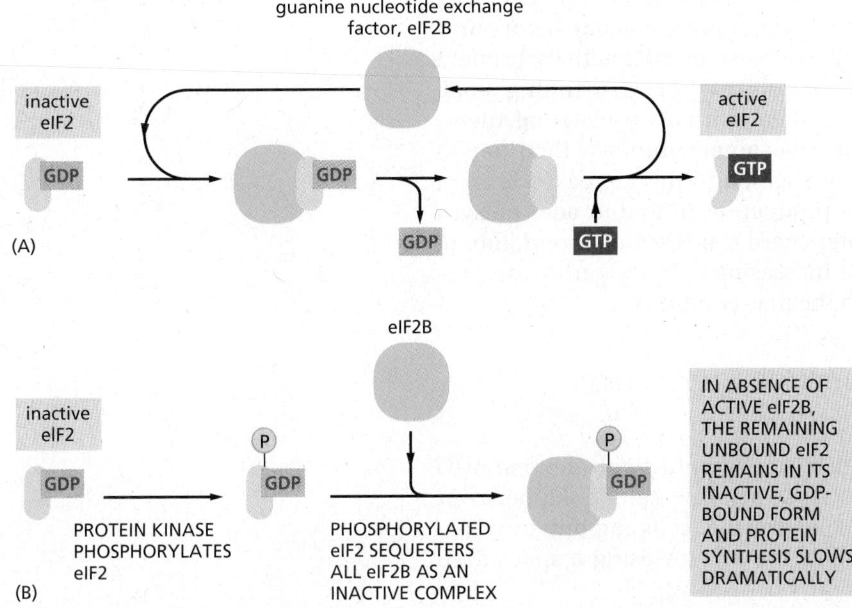

Figure 7–71 The eIF2 cycle. (A) The recycling of used eIF2 by a guanine nucleotide exchange factor (eIF2B). (B) How eIF2 phosphorylation controls protein synthesis rates by sequestering eIF2B.

Initiation at AUG Codons Upstream of the Translation Start Can Regulate Eukaryotic Translation Initiation

We saw in Chapter 6 that eukaryotic translation typically begins at the first AUG downstream of the 5′ end of the mRNA, which is the first AUG encountered by a scanning small ribosomal subunit. But, as we have seen, the nucleotides immediately surrounding the AUG also influence the efficiency of translation initiation. If the recognition site is poor enough, scanning ribosomal subunits will sometimes ignore the first AUG codon in the mRNA and skip to the second or third AUG codon instead. This phenomenon, known as "leaky scanning," is a strategy frequently used to produce two or more closely related proteins, differing only in their N-termini, from the same mRNA. A particularly important use of this mechanism is the production of the same protein with and without a signal sequence attached at its N-terminus. This allows the protein to be directed to two different locations in the cell (for example, to both mitochondria and the cytosol). Cells can regulate the relative abundance of the protein isoforms produced by leaky scanning; for example, a cell-type-specific increase in the abundance of the initiation factor eIF4F favors the use of the AUG closest to the 5′ end of the mRNA, even if it is surrounded by nonoptimal nucleotides.

Another type of control found in eukaryotes uses one or more short *open reading frames*—short stretches of DNA that begin with a start codon (ATG) and end with a stop codon, with no stop codons in between—that lie between the 5′ end of the mRNA and the beginning of a gene. Often, the amino acid sequences coded by these upstream open reading frames (uORFs) are not important; instead, the uORFs serve a purely regulatory function. A uORF present on an mRNA molecule will generally decrease translation of the downstream gene by trapping a scanning ribosome initiation complex and causing the ribosome to translate the uORF and dissociate from the mRNA before it reaches the bona fide protein-coding sequence.

When the activity of a general translation factor (such as eIF2 discussed earlier) is reduced, one might expect that the translation of all mRNAs would be reduced equally. Contrary to this expectation, however, the phosphorylation of eIF2 can have selective effects, even enhancing the translation of specific mRNAs that contain uORFs. This can enable cells, for example, to adapt to starvation for specific nutrients by shutting down the synthesis of all proteins except those that are required for synthesis of the missing nutrients. The details of this mechanism have been worked out for the yeast mRNA that encodes a protein called Gcn4, a transcription regulator that activates many genes that encode proteins that are important for amino acid synthesis.

The *Gcn4* mRNA encodes several short uORFs, and when amino acids are abundant, ribosomes translate the uORFs and generally dissociate before they reach the *Gcn4* coding region. But a global decrease in eIF2 activity brought about by amino acid starvation makes it more likely that a scanning small ribosomal subunit will move across the uORFs (without translating them) before it acquires a molecule of eIF2. This ribosomal subunit is then free to initiate translation on the actual *Gcn4* sequences, and the increased level of this transcription regulator increases the production of amino acid biosynthetic enzymes. Thus, when cells encounter "hard times," phosphorylation of eIF2 globally decreases translation while increasing synthesis of those proteins most needed by the cell to cope with the new conditions.

Internal Ribosome Entry Sites Also Provide Opportunities for Translational Control

Although most eukaryotic mRNAs are translated beginning with the first AUG downstream from the 5′ cap, certain AUGs, as we just saw, can be skipped over during the scanning process. There is a second way that cells can initiate translation at positions distant from the 5′ end of the mRNA, using a specialized

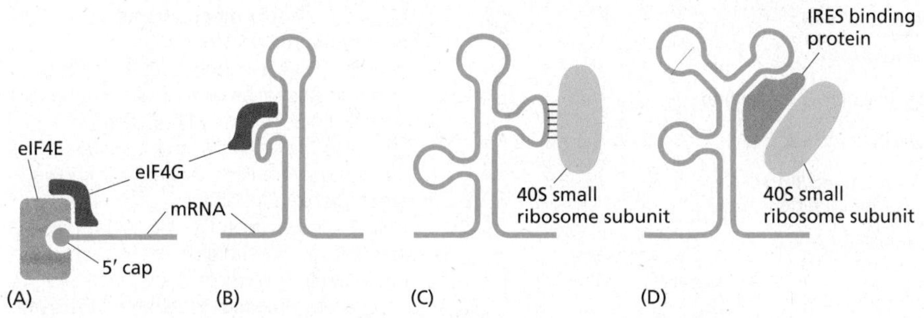

Figure 7–72 Internal ribosome entry sites (IRESs) can promote translation initiation by a variety of mechanisms. (A) The normal cap-dependent mechanism requires eIF4G binding to the cap to begin assembly of the other translation components (see Figure 6–74). (B) The cap and eIF4E are bypassed by direct binding of eIF4G to a specific RNA structure formed by the IRES. (C) The small ribosome subunit binds directly to the IRES through base-pairing between sequences in the IRES and the 18S rRNA, positioning it to begin translation. (D) Specialized proteins bind to an IRES and then attract the small ribosome subunit.

type of RNA sequence called an **internal ribosome entry site (IRES)**. In some cases, two distinct protein-coding sequences are carried in tandem on the same eukaryotic mRNA; translation of the first occurs by the usual scanning mechanism utilizing the first AUG encountered, and translation of the second occurs by means of an IRES located much further into the mRNA. IRESs are typically several hundred nucleotides in length, and they fold into specific structures that bypass the need for a 5′ cap and the translation factor that recognizes it, eIF4E (**Figure 7–72**).

It is estimated that 10% of all mammalian mRNAs contain an IRES. Some of these protein synthesis start sites are specifically activated by external signals such as stress. But the best-understood examples occur with viruses, which use IRESs as part of a strategy to get their own mRNA molecules translated while blocking the normal 5′ cap-dependent translation of host mRNAs. On infection, these viruses produce a protease (encoded in the viral genome) that cleaves the host-cell translation factor eIF4G, rendering it unable to bind to eIF4E, the cap-binding complex (see Figure 6–74). This shuts down most of the host cell's translation and effectively diverts the translation machinery to the IRES sequences present on the viral mRNAs. (The truncated eIF4G remains competent to initiate translation at these internal sites.)

The many ways in which viruses manipulate their host's protein-synthesis machinery for their own advantage continue to surprise cell biologists. Studying the many results of this evolutionary "arms race" between humans and pathogens has led to many fundamental insights into the workings of the cell, and we revisit this topic in Chapter 23.

Changes in mRNA Stability Can Control Gene Expression

Most mRNAs in a bacterial cell are very unstable, having half-lives of less than a couple of minutes. Exonucleases, which degrade in the 3′-to-5′ direction, are usually responsible for the rapid destruction of these mRNAs. Because its mRNAs are both rapidly synthesized and rapidly degraded, a bacterium can adapt quickly to environmental changes.

As a general rule, the mRNAs in eukaryotic cells are more stable. Some, such as that encoding β-globin, have half-lives of more than 10 hours. But most are considerably less stable, with half-lives of less than 30 minutes. The mRNAs that code for proteins such as growth factors and transcription regulators, whose production rates need to change rapidly in cells, are especially short-lived.

We saw in Chapter 6 that the cell has several mechanisms that rapidly destroy incorrectly processed RNAs. But now we focus on the ultimate fate of a typical "normal" eukaryotic mRNA molecule. Two general mechanisms exist for eventually destroying it, both of which begin with a gradual shortening of the poly-A tail by an exonuclease, a process that starts as soon as the mRNA reaches the cytosol. In a broad sense, this poly-A shortening acts as a timer that counts down the lifetime of each mRNA. Once the poly-A tail is reduced to a critical length (about 25 nucleotides in humans), the two destruction pathways converge. In one, the 5′ cap is removed (a process called decapping), and the "exposed" mRNA is rapidly degraded from its 5′ end. In the other, the mRNA continues to

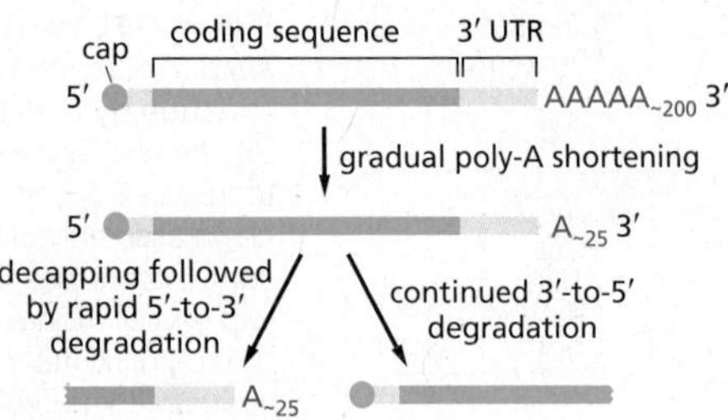

Figure 7–73 **Two mechanisms of eukaryotic mRNA decay.** Once a mature mRNA is exported to the cytosol, enzymes known as deadenylases gradually shorten its poly-A tail. When a critical threshold of poly-A tail length occurs, the two degradation mechanisms shown are triggered, probably by loss of the poly-A–binding proteins. Although 5′-to-3′ and 3′-to-5′ degradation are shown here on separate RNA molecules, these two processes can occur together on the same molecule. (Adapted from C.A. Beelman and R. Parker, *Cell* 81:179–183, 1995.)

be degraded from the 3′ end, through the poly-A tail into the coding sequences (Figure 7–73).

Nearly all mRNAs are subject to both types of decay, which can occur simultaneously on the same mRNA molecule. Specific nucleotide sequences determine how fast each step occurs and therefore how long each mRNA will persist in the cell and be able to produce protein. The 3′ UTR sequences are especially important in controlling mRNA lifetimes, and they often carry binding sites for specific proteins that increase or decrease the rates of poly-A shortening, decapping, or 3′-to-5′ degradation. The half-life of an mRNA is also affected by how efficiently it is translated. Poly-A shortening and decapping compete directly with the machinery that translates the mRNA; therefore, any factors that increase the translation efficiency for an mRNA will tend to reduce its degradation.

Although poly-A shortening controls the half-life of most eukaryotic mRNAs, some mRNAs can be degraded by a specialized mechanism that bypasses this step altogether. In these cases, specific endonucleases cleave the mRNA internally, effectively decapping one end and removing the poly-A tail from the other, so that both halves are rapidly degraded. The mRNAs that are destroyed in this way carry specific nucleotide sequences—often in their 3′ UTRs—that serve as recognition sequences for these endonucleases. This strategy makes it simple to tightly regulate the stability of these mRNAs by blocking or exposing the endonuclease site in response to extracellular signals. For example, the addition of iron to cells decreases the stability of the mRNA that encodes the receptor protein that binds the iron-transporting protein transferrin, causing less of this receptor to be made. This effect is mediated by the iron-sensitive RNA-binding protein aconitase. During iron starvation, aconitase binds the 3′ UTR of the transferrin receptor mRNA and increases receptor production by blocking endonucleolytic cleavage of the mRNA (Figure 7–74A). On the addition of iron, aconitase is released from

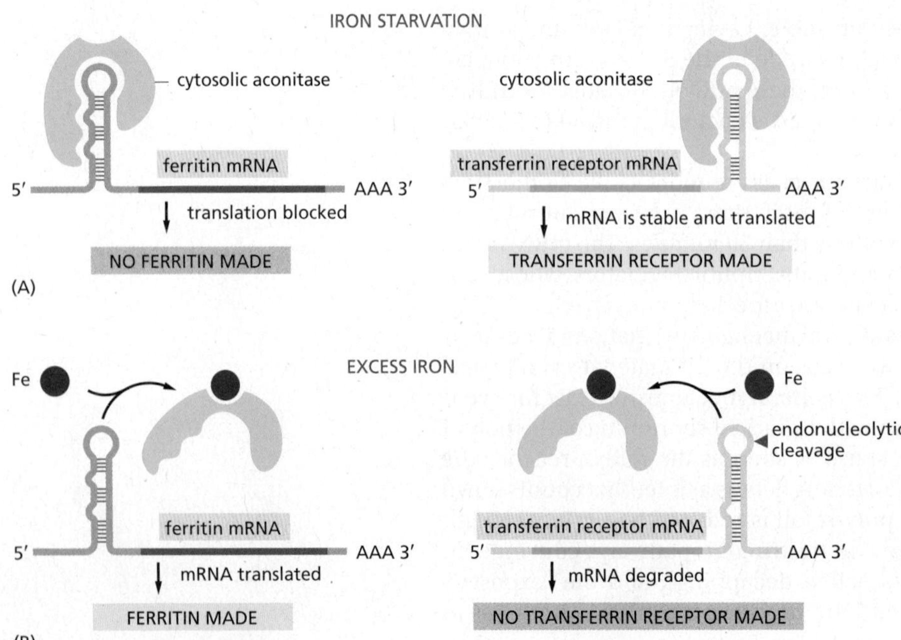

Figure 7–74 **Two post-translational controls mediated by iron.** (A) During iron starvation, the binding of aconitase to the 5′ UTR of the ferritin mRNA blocks translation initiation; its binding to the 3′ UTR of the transferrin receptor mRNA blocks an endonuclease cleavage site and thereby stabilizes the mRNA. (B) In response to an increase in iron concentration in the cytosol, a cell increases its synthesis of ferritin in order to bind the extra iron and decreases its synthesis of transferrin receptors in order to import less iron across the plasma membrane. Both responses are mediated by the same iron-responsive regulatory protein, aconitase, which recognizes common features in a stem–loop structure in the mRNAs encoding ferritin and the transferrin receptor. Aconitase dissociates from the mRNA when it binds iron. But because the transferrin receptor and ferritin are regulated by different types of mechanisms, their levels respond oppositely to iron concentrations even though they are regulated by the same iron-responsive regulatory protein. (Adapted from M.W. Hentze et al., *Science* 238:1570–1573, 1987; and J.L. Casey et al., *Science* 240:924–928, 1988.)

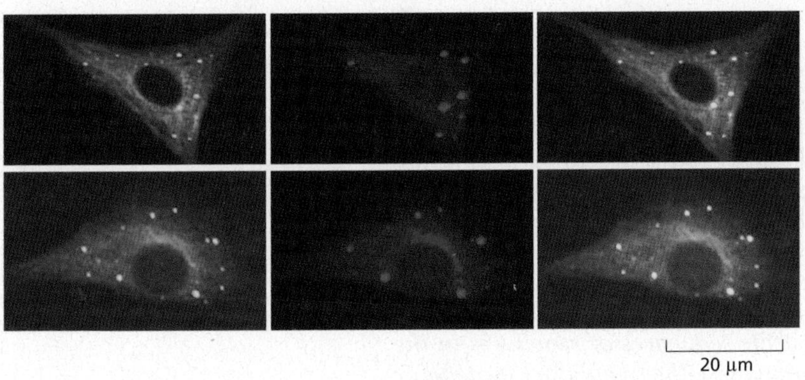

Figure 7–75 **Visualization of P-bodies.** Human cells were stained with antibodies to a component of the mRNA decapping enzyme Dcp1a *(left panels)* and to the Argonaute protein *(middle panels)*. As described later in this chapter, Argonaute is a key component of RNA interference pathways and both it and the decapping enzyme destabilize mRNAs. The merged images *(right panels)* show that the two proteins co-localize to P-bodies in the cytoplasm. (Adapted from J. Liu et al., *Nat. Cell Biol.* 7:719–723, 2005. Reproduced with permission from SNCSC.)

the mRNA, exposing the cleavage site and thereby decreasing mRNA stability (**Figure 7–74B**).

Regulation of mRNA Stability Involves P-bodies and Stress Granules

We saw in Chapter 6 that large aggregates of RNA and protein can form membraneless compartments in the nucleus, such as nucleoli and Cajal bodies. The cytosol also contains such biomolecular condensates, and here we discuss two of them, *Processing* or *P-bodies* and *stress granules*, each of which has a role in handling mRNAs (**Figure 7–75**). When an mRNA in the cytosol is no longer actively translated, it often moves to P-bodies where several fates are possible. P-bodies are rich in mRNA-degrading enzymes, and mRNAs that have already undergone significant poly-A shortening can continue to be degraded within P-bodies. Alternatively, some intact mRNAs can be stored in P-bodies in a translationally repressed form. According to the needs of the cell, these mRNAs can then be moved back to the cytosol and "reactivated" to begin translation again (**Figure 7–76**). mRNAs stored in this way often code for proteins that the cell needs quickly, and this strategy bypasses the time-consuming steps of *de novo* mRNA production.

Stress granules are dynamic membraneless organelles that form when the cell undergoes a sudden block to translation, whether by starvation, small-molecule inhibitors, or genetic manipulation. These treatments allow ongoing translation to be completed but block new translation initiation. The resulting ribosome-free mRNAs accumulate in stress granules that grow in size as more and more mRNAs enter them. As the stressful conditions are relieved, the stress granules shrink along with the release of the stored mRNAs to the cytosol where they resume being translated. Clearly, once a cell has made the large investment in producing a properly processed mRNA molecule, it carefully controls its subsequent fate.

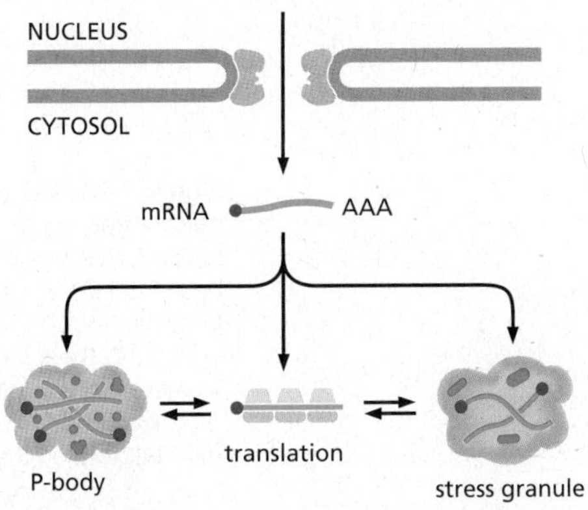

Figure 7–76 **Possible fates of an intact mRNA molecule.** An mRNA molecule released from the nucleus can be actively translated *(center)*, stored in P-bodies *(left)*, or, if the cell is stressed, moved into stress granules *(right)*. As the needs of the cell change, stored mRNAs can be reactivated and returned to the cytosol to be translated into protein. Although not shown, all mRNA molecules are eventually degraded, and some of the final steps take place in P-bodies.

Summary

Many steps in the pathway from RNA to protein are regulated by cells in order to control gene expression. Most genes are regulated at multiple levels, in addition to being controlled at the initiation stage of transcription. The regulatory mechanisms include (1) attenuation of the RNA transcript by its premature termination, (2) alternative RNA splice-site selection, (3) control of 3′-end formation by cleavage and poly-A addition, (4) RNA covalent modifications including editing, (5) control of transport from the nucleus to the cytosol, (6) localization of mRNAs to particular parts of the cytoplasm, (7) control of translation initiation, and (8) regulated mRNA degradation. Most of these control processes require the recognition of specific sequences or structures in the RNA molecule being regulated, a task performed by either regulatory proteins or regulatory RNA molecules.

REGULATION OF GENE EXPRESSION BY NONCODING RNAs

In the previous chapter, we introduced the *central dogma*, according to which the flow of genetic information proceeds from DNA through RNA to protein (see Figure 6–1). But we have seen throughout this book that RNA molecules perform many critical tasks in the cell besides serving as intermediate carriers of genetic information. Among these noncoding RNAs are the rRNA and tRNA molecules, which are responsible for reading the genetic code and synthesizing proteins. The RNA molecule in telomerase serves as a template for the replication of chromosome ends, snoRNAs modify ribosomal RNA, and snRNAs direct RNA splicing. And earlier in this chapter we saw that Xist RNA has an important role in inactivating one copy of the X chromosome in female mammals. In this section, we introduce several additional classes of noncoding RNAs that have important roles in regulating gene expression and in protecting the genome from viruses and transposable elements. These RNAs also make possible powerful new experimental techniques in genome editing.

Small Noncoding RNA Transcripts Regulate Many Animal and Plant Genes Through RNA Interference

We begin our discussion with a group of short RNAs that carry out **RNA interference**, or **RNAi**. Here, short single-stranded RNAs (20–30 nucleotides) serve as guide RNAs that selectively bind—through complementary base-pairing—other RNAs in the cell. When the target is a mature mRNA, the small noncoding RNAs can inhibit its translation or catalyze its rapid destruction. If the target RNA molecule is in the process of being transcribed, the small noncoding RNA can bind to it and direct the formation of repressive chromatin on its attached DNA template to block further transcription (Figure 7–77).

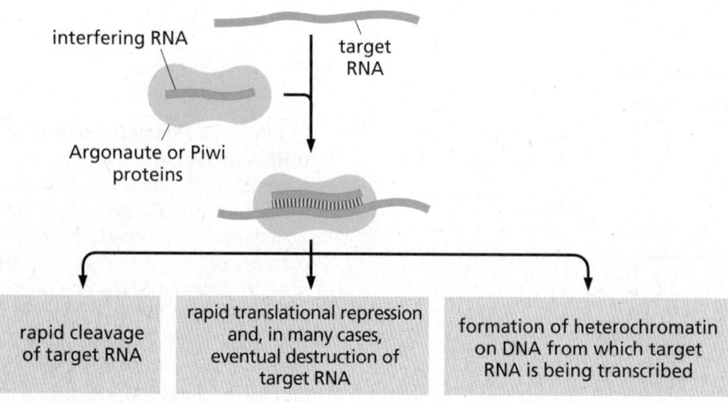

Figure 7–77 **RNA interference in eukaryotes.** Single-strand interfering RNAs locate target RNAs through complementary base-pairing, and, at this point, several fates are possible, as shown. As described in the text, there are several types of RNA interference; the way that interfering RNA is produced and the ultimate fate of the target RNA depend on the particular system.

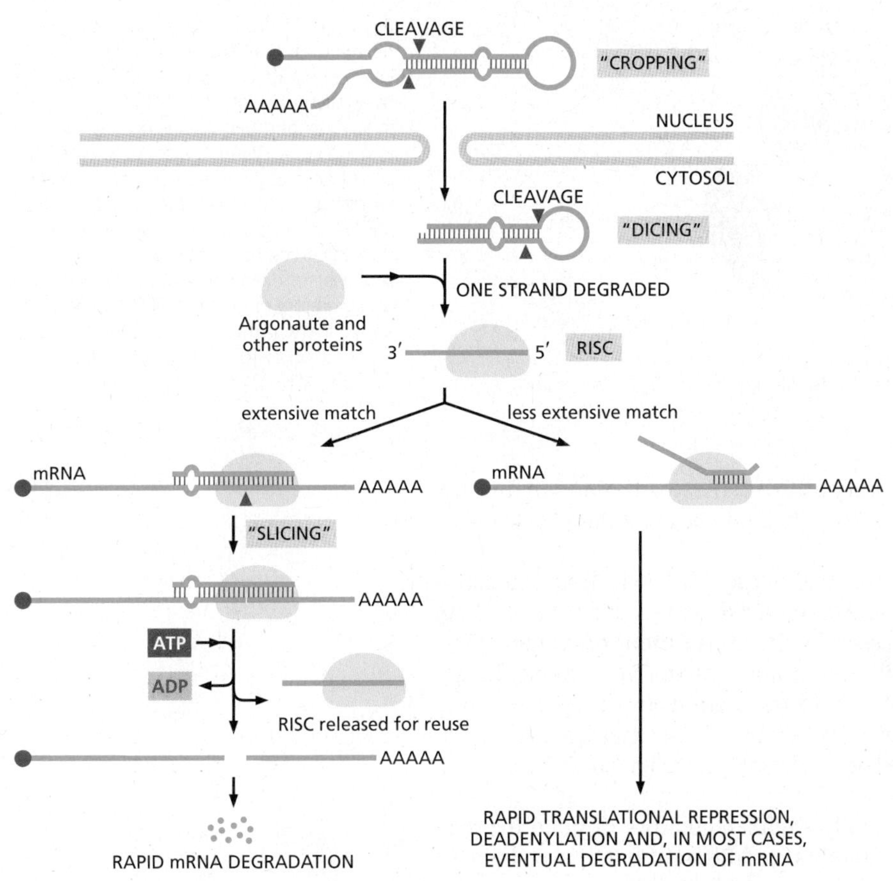

Figure 7–78 miRNA processing and mechanism of action. The precursor miRNA, through complementary base-pairing between one part of its sequence and another, forms a double-strand structure. This RNA is "cropped" while still in the nucleus and then exported to the cytosol, where it is further cleaved ("diced") by the Dicer enzyme to form the miRNA proper. Argonaute, in conjunction with other components of RISC, initially associates with both strands of the miRNA and then cleaves and discards one of them. The other strand guides RISC to specific mRNAs through base-pairing. If the RNA–RNA match is extensive, as is commonly seen in plants, Argonaute cleaves the target mRNA ("slicing"), causing its rapid degradation. In mammals, the miRNA–mRNA match often does not extend beyond a short seven-nucleotide "seed" region near the 5′ end of the miRNA. This less extensive base-pairing leads to a rapid inhibition of translation and, in most cases, eventual destruction of the mRNA.

Three classes of small noncoding RNAs work in this way—*microRNAs* (*miRNAs*), *small interfering RNAs* (*siRNAs*), and *piwi-interacting RNAs* (*piRNAs*)—and we discuss them in turn in the next sections. Although they differ in both the way the short pieces of single-stranded RNA are generated and in their ultimate functions, all three types of RNAs locate their targets through RNA–RNA base-pairing, and they generally cause reductions in gene expression.

miRNAs Regulate mRNA Translation and Stability

More than 1000 different **microRNAs (miRNAs)** are produced from the human genome, and these appear to regulate at least one-half of all human protein-coding genes. Once made, miRNAs base-pair with specific mRNAs and fine-tune their translation and stability. The miRNA precursors are synthesized by RNA polymerase II and are capped and polyadenylated. They then undergo a special type of processing, after which the miRNA (typically 23 nucleotides in length) is assembled with a set of proteins to form an *RNA-induced silencing complex,* or *RISC.* Once formed, the RISC seeks out its target mRNAs by searching for complementary nucleotide sequences (Figure 7–78). This search is greatly facilitated by the Argonaute protein, a component of RISC, which holds the 5′ region of the miRNA so that it is optimally positioned for base-pairing to another RNA molecule (Figure 7–79). In animals, the extent of base-pairing is typically at least seven nucleotide pairs, and this pairing most often occurs in the 3′ UTR of the target mRNA.

Once an mRNA has been bound by an miRNA, several outcomes are possible. If the base-pairing is extensive (which is unusual in humans but common in many plants), the mRNA is cleaved (*sliced*) by the Argonaute protein, effectively removing the mRNA's poly-A tail and exposing it to exonucleases (see Figure 7–73). After cleavage of the mRNA, the RISC with its associated miRNA is released, and it can seek out additional mRNAs (see Figure 7–78). Thus, a single miRNA can act

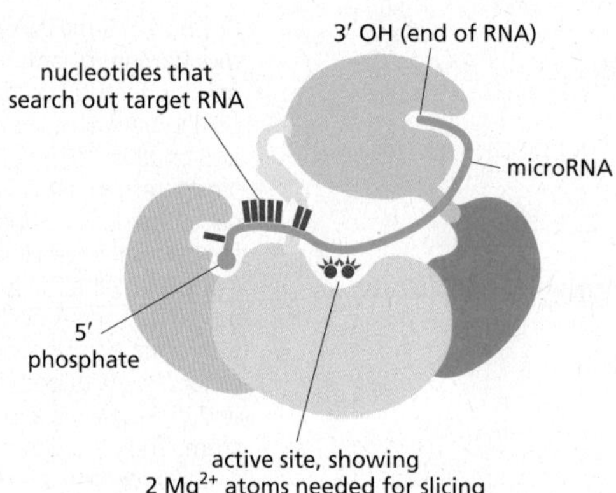

Figure 7–79 Human Argonaute protein carrying an miRNA. The protein is folded into four structural domains, each indicated by a different color. The miRNA is held in an extended form that is optimal for forming RNA–RNA base pairs. The active site of Argonaute that slices a target RNA, when it is extensively base-paired with the miRNA, is indicated in *red*. Many Argonaute proteins (three out of the four human proteins, for example) lack the catalytic site and therefore bind target RNAs without slicing them. (Adapted from C.D. Kuhn and L. Joshua-Tor, *Trends Biochem. Sci.* 38:263–271, 2013.)

catalytically to destroy many complementary mRNAs. These miRNAs can thus be thought of as guide sequences that repeatedly bring destructive nucleases into contact with specific mRNAs.

If the base-pairing between the miRNA and the mRNA is less extensive (as observed for most human miRNAs), Argonaute does not slice the mRNA; rather, translation of the mRNA is repressed by the recruitment of deadenylase enzymes—which shorten the poly-A tail—and other proteins that directly block access of the mRNA to the proteins needed to translate it. In many cases, the "blocked" mRNAs are shuttled to P-bodies (see Figure 7–76) where, sequestered from ribosomes, they are either degraded or, at a later time, released to the cytosol to be translated again.

Several features make miRNAs especially useful regulators of gene expression. First, a single miRNA can regulate a whole set of different mRNAs, so long as the mRNAs carry a short complementary sequence in their UTRs. This situation is common in humans, where a single miRNA can control hundreds of different mRNAs. Second, regulation by miRNAs can be combinatorial. As discussed earlier for transcription regulators, combinatorial control greatly expands the possibilities available to the cell by linking gene expression to a combination of different regulators rather than a single regulator. Like many transcription regulators, different miRNAs can bind cooperatively to their target mRNAs if their recognition sites are spaced appropriately. The basis for the cooperative binding is a scaffold protein that weakly holds two different RISCs together at a fixed spacing, thereby coupling their individual miRNA–mRNA binding energies. Third, an miRNA occupies relatively little space in the genome when compared with a protein. Indeed, their small size is one reason that miRNAs were discovered only recently. Although we are only beginning to appreciate the full impact of miRNAs, it is clear that they represent an important part of the cell's repertoire for regulating the expression of genes. We shall discuss specific examples of miRNAs with key roles in development in Chapter 21.

RNA Interference Also Serves as a Cell Defense Mechanism

Many of the proteins that participate in the miRNA regulatory mechanisms just described also serve a second function as a defense mechanism: they orchestrate the degradation of foreign RNA molecules, specifically those that occur in double-strand form. Many transposable elements and viruses produce double-stranded RNA at least transiently in their life cycles, and RNA interference helps to keep these potentially dangerous invaders in check. As we shall see, this form of RNAi also provides scientists with a powerful experimental technique to turn off the expression of individual genes.

The presence of double-stranded RNA in the cell triggers RNAi by attracting a protein complex containing *Dicer*, the same nuclease that processes miRNAs (see Figure 7–78). This protein cleaves the double-stranded RNA into small fragments (of approximately 23 nucleotide pairs) called **small interfering RNAs (siRNAs)**.

These double-stranded siRNAs are then bound by Argonaute and other components of RISC. As we saw earlier for miRNAs, one strand of the duplex RNA is then cleaved by Argonaute and discarded. The single-strand siRNA molecule that remains directs RISC back to complementary RNA molecules produced by the virus or transposable element. Because the match is usually exact, Argonaute also cleaves these molecules, leading to their rapid destruction.

Each time RISC cleaves a new RNA molecule, the RISC is released; thus, as we saw for miRNAs, a single RNA molecule can act catalytically to destroy many complementary RNAs. Some organisms employ an additional mechanism that amplifies the RNAi response even further. In these organisms, RNA-dependent RNA polymerases use siRNAs as primers to produce additional copies of double-stranded RNAs that are then cleaved into siRNAs. This amplification ensures that, once initiated, RNA interference can continue even after all the initiating double-stranded RNA has been degraded or diluted out. For example, it permits progeny cells to continue carrying out the specific RNA interference that was provoked in the parent cells.

In some organisms, the RNA interference activity can be spread by the transfer of RNA fragments from cell to cell. This is particularly important in plants (whose cells are linked by fine connecting channels, as discussed in Chapter 19), because it allows an entire plant to become resistant to an RNA virus after only a few of its cells have been infected. In a broad sense, the RNAi response resembles certain aspects of the animal immune system; in both, an invading organism elicits a customized response, and—through amplification of the "attack" molecules—the host becomes systemically protected.

We have seen that although miRNAs and siRNAs are generated in different ways, they rely on some of the same proteins and seek out their targets in a fundamentally similar manner. Because siRNAs are found in widespread species, they are believed to be the most ancient form of RNA interference, with miRNAs being a later evolutionary refinement. The siRNA-mediated defense mechanisms are especially crucial for plants, worms, and insects. In mammals, a protein-based immune system (described in Chapter 24) has largely taken over the task of fighting off viruses.

RNA Interference Can Direct Heterochromatin Formation

The siRNA interference pathway just described does not necessarily stop with the inactivation of target RNA molecules. In some cases, the RNA interference machinery can also selectively shut off the *synthesis* of the target RNAs. For this to occur, the short siRNAs produced by the Dicer protein are assembled with a group of proteins (including Argonaute) to form an RITS (RNA-induced transcriptional silencing) complex. Using single-stranded siRNA as a guide sequence, this complex binds complementary RNA transcripts as they emerge from a transcribing RNA polymerase II (**Figure 7–80**). Positioned on the genome in this way, the RITS complex then attracts enzymes that covalently modify nearby histones and DNA causing the formation of a "constitutive" form of heterochromatin. As described in Chapter 4, this form of heterochromatin is distinguished by the H3K9me3 mark, and, in many cases, it also includes DNA methylation (see Figure 7–48). Although low levels of transcription probably persist (and may be important to continually signal where the heterochromatin should be formed), this form of heterochromatin, as we have seen, is generally resistant to transcription and effectively shuts off the genes that lie within it. In some cases, an RNA-dependent RNA polymerase and a Dicer enzyme are also recruited by the RITS complex to continually generate additional siRNAs *in situ*. This positive feedback loop ensures continued repression of the target gene even after the original, initiating siRNA molecules have disappeared.

RNAi-directed heterochromatin formation is an especially important cell defense mechanism; it limits the spread of transposable elements in genomes by maintaining their DNA sequences in a transcriptionally silent form. However, this same mechanism is also used in some normal processes in the cell. For example,

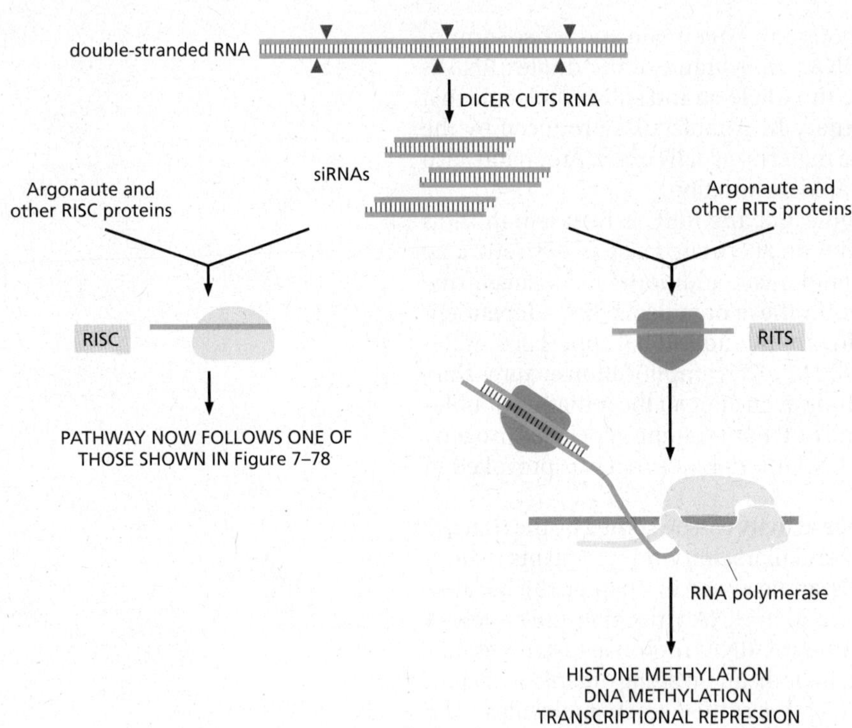

Figure 7–80 RNA interference directed by siRNAs. In many organisms, double-stranded RNA can trigger both the destruction of complementary mRNAs *(left)* and transcriptional silencing *(right)*. The change in chromatin structure induced by the bound RITS (RNA-induced transcriptional silencing) complex resembles that of Figure 7–48.

in many organisms, the RNA interference machinery maintains the heterochromatin formed around centromeres. Centromeric DNA sequences are transcribed in both directions, producing complementary RNA transcripts that can base-pair to form double-stranded RNA. This double-stranded RNA triggers the RNA interference pathway and stimulates formation of the heterochromatin that surrounds centromeres, which is necessary for the centromeres to segregate chromosomes accurately during mitosis.

piRNAs Protect the Germ Line from Transposable Elements

A third system of RNA interference relies on **piRNAs (piwi-interacting RNAs,** named for Piwi, a class of proteins related to Argonaute). piRNAs are found in many organisms, and they carry out a diverse set of functions. Here, we describe one of their most important roles, which is to hold transposable elements (transposons) in check in the germ line of animals. The germ line is especially susceptible to transposon movement because many of the histone modifications and methylated DNA sites are "erased" during gametogenesis, temporarily releasing transposons from their normal constraints. piRNAs cover this vulnerability. Unlike miRNAs and siRNAs, they are synthesized from specialized piRNA "clusters" in the genome as long, single-strand RNA molecules that are then broken up and trimmed by specialized processing enzymes (different from the Dicer enzymes discussed earlier) into fragments that are slightly longer than miRNAs and siRNAs. These RNAs are covalently modified at their 3′ ends by a 2′-*O*-methyl group (see Figure 6–43A) and assembled with Piwi proteins. Once complexed with their proteins, piRNAs seek out RNA targets by complementary base-pairing and, much like siRNAs, they both cleave the complementary RNAs and package the DNA on which they are being transcribed into repressive forms of chromatin. The piRNA clusters in the genome are rich with sequence fragments from transposons, and the piRNAs attack any transposon whose sequence is represented in the piRNA cluster. In this way, the genome contains a "hit list" of transposons that need to be inactivated during the vulnerable period of gametogenesis. It has been proposed that piRNA clusters are unusually attractive landing sites for transposons and, for this reason, they carry a record of all past bursts of transposon activity.

Although the genome carries a linear record of transposons to be inactivated, piRNAs have an additional way to attack those transposons that are most active during gametogenesis. In brief, once a piRNA and its associated proteins cleave a complementary, transposon-coded mRNA, additional piRNAs can be created from nearby sequences in the transposon mRNA. This mechanism not only amplifies the original response but extends its breadth by incorporating additional sequence information from active transposons, information that might not be carried in the piRNA clusters themselves.

Many mysteries surround piRNAs. More than a million piRNA species are coded in the genomes of many mammals and expressed in the testes, yet only a fraction seem to be directed against the transposons present in those genomes. Are the other piRNAs remnants of past invaders? Do they cover so much "sequence space" that they are broadly protective for any foreign DNA? Another curious feature of piRNAs is that many of them (particularly if base-pairing does not have to be perfect) should, in principle, attack the normal mRNAs made by the organism, yet they do not. It has been proposed that these large numbers of piRNAs may form a system to distinguish "self" RNAs from "foreign" RNAs and attack only the latter. If this is the case, there must be a special way for the cell to spare its own RNAs. One idea is that RNAs produced in the previous generation of an organism are somehow registered and set aside from piRNA attack in subsequent generations. Another idea holds that all legitimate mRNAs carry specialized sequences that spare them from attack. Whether or not this mechanism truly exists, and, if so, how it might work, are questions that demonstrate our incomplete understanding of the full range of RNA interference.

RNA Interference Has Become a Powerful Experimental Tool

Although it likely first arose in evolution as a defense mechanism against viruses and transposable elements, RNA interference, as we have seen, has become thoroughly integrated into many aspects of normal cell biology, ranging from the control of gene expression to a fine tuning of chromosome structure. RNA interference has also been developed by scientists into a powerful experimental tool that allows almost any gene to be inactivated by evoking an RNAi response to it. This technique, which can be readily carried out in cultured cells and, in many cases, whole animals and plants, has made possible new genetic approaches in cell and molecular biology. We shall discuss it in detail in the following chapter when we cover the modern genetic methods used to study cells (see pp. 533–534). RNAi also has potential in treating human disease. Because many human disorders result from the misexpression of genes, the ability to turn these genes off by experimentally introducing complementary siRNA molecules into cells holds great medical promise. Although delivery of RNA molecules to the appropriate tissue has been a persistent problem in using RNAi as a human therapy, the strategy is currently used to treat a rare disease called transthyretin amyloidosis. This inherited disease, which affects heart and nerve function, is caused by the accumulation of a mutated protein, and siRNAs directed by complementary base-pairing to the mutated mRNA relieve its symptoms. In this case, the siRNAs are delivered to the liver (the key site of synthesis of the mutated protein) by a special combination of lipids that forms tiny vesicles to encase the siRNA.

Cells Have Additional Mechanisms to Hold Transposons and Integrated Viral Genomes in Check

From the preceding sections, it should be clear that cells are locked in an eternal "arms race" with parasitic DNA elements, such as transposons and viruses. Indeed, it seems that our own genome came close to being overrun with such elements; even with our many defense mechanisms, they still make up nearly half our DNA (see Figure 4–63). Most of these elements have accumulated mutations that prevent them from being active, but this process likely occurred after host-cell mechanisms came into play to hold them in check.

We have seen how siRNAs and piRNAs constitute surveillance systems to monitor transcription from transposable elements, to destroy their transcripts, and to package their DNA into repressive forms of chromatin. Although these overlapping defense mechanisms may seem highly effective, cells have at least one additional strategy for recognizing transposons and integrated viruses and silencing them. In contrast to the RNA-based strategies, which utilize complementary base-pairing to recognize these genome invaders, this additional system employs a special set of sequence-specific DNA-binding proteins to monitor the genome. When these proteins recognize a DNA sequence present in a transposon or integrated viral genome, they bind directly to that sequence and recruit both histone "writers" that place H3K9me3 marks on nearby histones and DNA methylases that heavily methylate the surrounding DNA. As discussed earlier in the chapter, this repressive form of chromatin can then spread and render the underlying DNA resistant to transcription and recombination (see Figure 7–48). Our genome codes for hundreds of different sequence-specific DNA proteins that carry out this surveillance (called KRAB-ZPF proteins), and they cover a wide variety of transposable element DNA and viral sequences. Most recognize a DNA sequence that is crucial for that element to transpose (or in the case of integrated viral genomes, for the virus to multiply), making it difficult for the element to escape through a mutation. However, such escape does apparently occur, because KRAB-ZPF proteins are evolving rapidly (compared with other human genes), and they appear to be "keeping up" with mutated versions of resident transposable elements. Their rapid evolution also suggests that the KRAB-ZPF proteins can easily adapt through mutation to attack new parasitic elements that might enter the genome.

Transposable elements, if left unchecked, present many challenges to the cell: their sequences can serve as recombination sites leading to crossovers between nonhomologous chromosomes, double-strand DNA breaks are produced in the host genome when they move, and they can disrupt coding or regulatory sequences when they insert into a new position. On the other hand, their movement has provided a source of variation that is necessary for natural selection to occur. But the many different strategies host cells have evolved to neutralize these invaders suggest that the short-term dangers must far outweigh any long-term advantages.

Bacteria Use Small Noncoding RNAs to Protect Themselves from Viruses

In the previous sections, we emphasized the defense systems of animals and plants, but it is important to keep in mind that bacteria and archaea make up the vast majority of Earth's diversity. Not surprisingly, the viruses that infect these single-cell organisms greatly outnumber plant and animal viruses. Many species of bacteria (and almost all species of archaea) use a repository of small noncoding RNA molecules to seek out and destroy invading viruses. Many features of this defense mechanism, known as **CRISPR**, resemble those of miRNAs, siRNAs, and piRNAs that we saw earlier. When bacteria and archaea are first infected by a virus, short fragments of that viral DNA become integrated into their genomes by a process that is only beginning to be understood. These serve as "vaccinations," in the sense that they become the templates for producing small noncoding RNAs known as **crRNAs** (CRISPR RNAs) that will thereafter destroy the virus should it reinfect the descendants of the original cell. This aspect of the CRISPR system resembles both human adaptive immunity and piRNA-based surveillance, insofar as the cell carries a record of past exposures that is used to protect against future exposures.

In most cases, crRNAs associate with special proteins that allow them to seek out and destroy invading viral genomes, which are typically composed of double-stranded DNA. Many distinct CRISPR systems exist across different species of

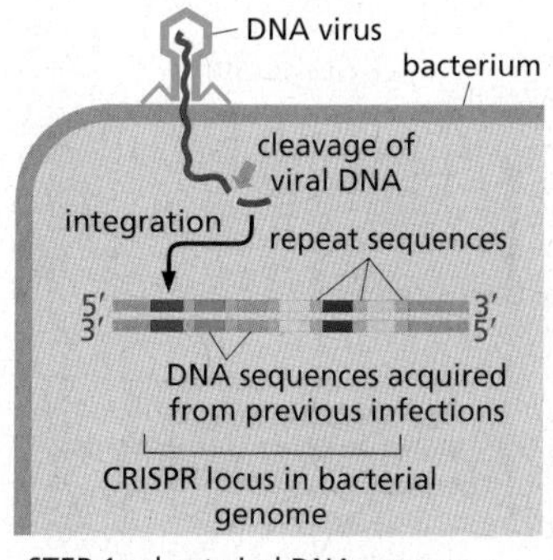

STEP 1: short viral DNA sequence is integrated into CRISPR locus

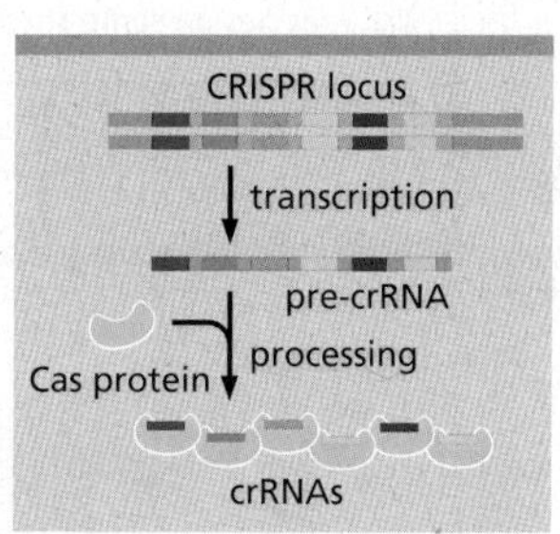

STEP 2: RNA is transcribed from CRISPR locus, processed, and bound to Cas protein

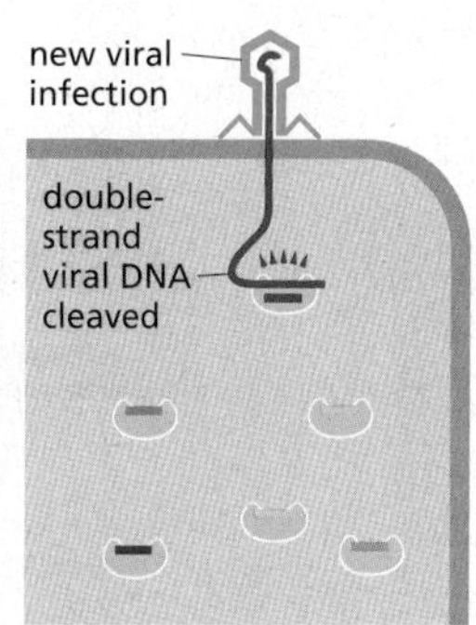

STEP 3: small crRNA in complex with Cas seeks out and destroys viral sequences

Figure 7–81 CRISPR-mediated immunity in bacteria and archaea. After infection by a virus *(left panel)*, a small bit of DNA from the viral genome is inserted into the CRISPR locus. For this to happen, a small fraction of infected cells must survive the initial viral infection. The surviving cells, or more generally their descendants, transcribe the CRISPR locus and process the transcript into crRNAs *(middle panel)*. Upon reinfection with a virus that the population has already been "vaccinated" against, the incoming viral DNA is destroyed by a complementary crRNA *(right panel)*.

For a CRISPR system to be effective, the crRNAs must not destroy the CRISPR locus itself, even though the crRNAs are complementary in sequence to it. How is this possible? In many species, there must be additional short nucleotide sequences carried by the target molecule in order for crRNAs to attack it. Because these sequences, known as PAMs (protospacer adjacent motifs), lie outside the crRNA sequences, the host CRISPR locus is spared (see Figure 8–57).

bacteria and archaea. Here we merely outline one of the most common and best understood, describing its three steps (**Figure 7–81**). In the first step, viral DNA sequences are integrated into special regions of the bacterial genome known as CRISPR (clustered regularly interspersed short palindromic repeat) loci, named for the peculiar DNA sequences that first drew the attention of scientists. In its simplest form, a CRISPR locus consists of several hundred repeats of a host DNA sequence interspersed with a large collection of DNA sequences (typically 25–70 nucleotide pairs each) derived from prior exposures to viruses and other foreign DNA. The newest viral sequence is always integrated at the 5′ end of the CRISPR locus, the end that is transcribed first. Each locus, therefore, carries a temporal, ordered record of prior infections. Many bacterial and archaeal species carry several large CRISPR loci in their genomes and are thus immune to a wide range of viruses.

In the second step, the CRISPR locus is transcribed to produce a long RNA molecule, which is then processed into the much shorter (approximately 30 nucleotides) crRNAs. These crRNAs become complexed with *Cas* (*CRISPR-associated*) proteins, and, in the final step, they seek out complementary viral DNA sequences and direct their destruction by nucleases. Although structurally dissimilar, Cas proteins are analogous to the Argonaute and Piwi proteins discussed earlier: they hold small single-stranded RNAs in an extended configuration that is optimized, in this case, for seeking and forming complementary base pairs with double-stranded DNA.

We still have much to learn about CRISPR-based immunity in bacteria and archaea. For example, the mechanism through which viral sequences are first identified and integrated into the host genome is poorly understood. Moreover, in different species of bacteria and archaea, crRNAs are processed in different ways, and in some cases, the crRNAs can attack viral RNAs as well as DNAs. As might be predicted, many viruses have evolved anti-CRISPR systems to counteract the defense systems of their hosts. These anti-CRISPRs range from viral proteins that bind to and inactivate the Cas proteins to special coats that form around the viral DNA and protect it from CRISPR attack during replication, gene expression, and virus assembly.

In Chapter 8, we describe how bacterial CRISPR systems have been artificially "moved" into plants and animals, where they have revolutionized our ability to manipulate genomes.

Long Noncoding RNAs Have Diverse Functions in the Cell

In this and the preceding chapters, we have seen that noncoding RNA molecules have many functions in the cell. Yet there remain many noncoding RNAs whose functions are still unknown. Many of these RNAs belong to a group known as **long noncoding RNA (lncRNA)**, arbitrarily defined as RNAs longer than 200 nucleotides that do not code for protein. The sheer number of lncRNAs (an estimated 5000 for the human genome, for example) came as a surprise to scientists. Most of these lncRNAs are transcribed by RNA polymerase II and have 5′ caps and

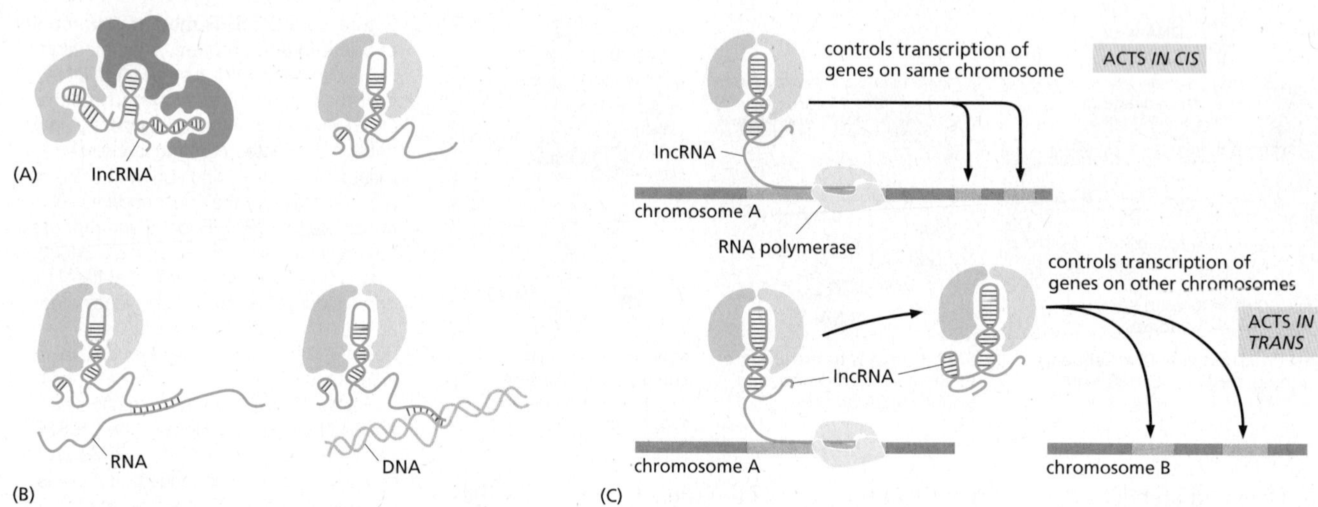

Figure 7–82 **Roles of long noncoding RNA (lncRNA).** (A) As described in Chapter 6, RNAs can fold into short, specific three-dimensional structures whose specific features can be recognized by proteins. Thus, lncRNAs can serve as scaffolds, bringing together proteins that function together in the same process and thereby facilitating their interactions and speeding the reactions that they catalyze. (B) lncRNAs can also, through formation of complementary base pairs, localize the proteins that they bind near specific nucleotide sequences on RNA or DNA molecules. (C) In some cases, lncRNAs act only *in cis* at their sites of synthesis—as, for example, when the RNA is held in place by the RNA polymerase that produced them *(top)*. But as shown, other lncRNAs diffuse from their sites of synthesis and are said to act *in trans*.

poly-A tails, and, in many cases, they are spliced. It has been difficult to accurately annotate lncRNAs, in part because low levels of RNA are now known to be made from about 75% of the human genome. Most of these RNAs are thought to result from a background "noise" of leaky transcription, and they are rapidly degraded. According to this idea, such nonfunctional RNAs provide no fitness advantage or disadvantage to the organism and are a tolerated by-product of the complex patterns of gene expression that need to be produced in multicellular organisms. For these reasons, it is difficult to estimate the number of lncRNAs that are likely to have a function in the cell and to distinguish them from the background of transcription noise.

In terms of biological function, lncRNA should be considered a catch-all phrase encompassing a great diversity of functions. We have already encountered a few notable lncRNAs, including the RNA in telomerase (see Figure 5–33), Xist RNA (see Figure 7–55), and an RNA involved in imprinting (see Figure 7–52). Other lncRNAs have been implicated in controlling the enzymatic activity of proteins, inactivating transcription regulators, affecting splicing patterns, and blocking translation of certain mRNAs through complementary base-pairing. However, there are three unifying features of lncRNAs that can account for their many roles in the cell. The first is that they can function as *scaffold RNA molecules*, holding together groups of proteins to coordinate their functions (**Figure 7–82A**; see also Figure 7–21). We have already seen examples in telomerase, the ribosome, and X-inactivation, where an RNA molecule holds together and organizes protein components. These RNA-based scaffolds are analogous to protein scaffolds we discussed in Chapter 3 (see Figure 3–76). RNA molecules are well suited to act as scaffolds: small bits of RNA sequence, often those portions that form stem–loop structures, can serve as binding sites for proteins, and these can be strung together with random sequences of RNA in between. This property may be one reason that many lncRNAs show relatively little primary-sequence conservation across species.

A second key feature of lncRNAs is their ability to serve as guide sequences, binding to specific RNA or DNA target molecules through base-pairing. By doing so, they bring proteins that are bound to them into close proximity with the DNA and RNA sequences (**Figure 7–82B**). This behavior is similar to that of snoRNAs (see Figure 6–43), miRNAs (see Figure 7–78), siRNAs (see Figure 7–80), and crRNAs (see Figure 7–81), all of which act in this way to guide protein enzymes to specific nucleic acid sequences. A third characteristic of RNA in general is its ability to organize biomolecular condensates, the non-membrane-bound assemblies of proteins and nucleic acids discussed in this and previous chapters. For example, rRNA is crucial for formation of the nucleolus, and untranslated mRNA provides the framework for P-bodies and stress granules. The propensity of RNA to form condensates derives in part from its ability to bind multiple proteins, as discussed above, but also because many RNAs can form multiple weak intramolecular interactions that, on their own, can lead to condensation. Some lncRNAs are thought to function solely by organizing and driving formation of such condensates.

In some of the simplest cases, lncRNAs work simply by base-pairing, without bringing in enzymes or other proteins. For example, a number of lncRNA genes are embedded in protein-coding genes, but they are transcribed in the "wrong direction." These *antisense RNAs* can form complementary base pairs with the mRNA (transcribed in the "correct" direction) and block its translation into protein (see Figure 7–70D). Other antisense lncRNAs base-pair with pre-mRNAs as they are synthesized and change the pattern of RNA splicing by masking the preferred splice-site sequences. Still others act as "sponges," base-pairing with miRNAs and thereby reducing their effects.

Finally, we note that some lncRNAs act only *in cis*; that is, they affect only the chromosome from which they are transcribed. This readily occurs when the transcribed RNA has not yet been released from RNA polymerases (**Figure 7–82C**) or when the completed RNA molecule does not diffuse away from the chromosome as for the case of Xist (see Figure 7–55). Many lncRNAs, however, leave their site of synthesis and act *in trans*. Although the best-understood lncRNAs work in the nucleus, many are found in the cytosol. The functions—if any—of the great majority of these cytosolic lncRNAs remain undiscovered.

Summary

RNA molecules have many uses in the cell besides carrying the information needed to specify the order of amino acids during protein synthesis. Although we have encountered noncoding RNAs in other chapters (tRNAs, rRNAs, snoRNAs, for example), the sheer number of noncoding RNAs produced by cells has surprised scientists. One well-understood use of noncoding RNAs occurs in RNA interference, where guide RNAs (miRNAs, siRNAs, piRNAs) base-pair with mRNAs. RNA interference can cause mRNAs to be either destroyed or translationally repressed. It can also cause specific genes to become packaged into heterochromatin suppressing their transcription. In bacteria and archaea, RNA interference is used as an adaptive immune response to destroy viruses that infect them. A large family of large noncoding RNAs (lncRNAs) has recently been discovered through detailed genomic analyses. Although the function (if any) of most of these RNAs is unknown, some serve as RNA scaffolds to bring specific proteins and RNA molecules together to speed up needed reactions.

PROBLEMS

Which statements are true? Explain why or why not.

7–1 When the nucleus of a fully differentiated carrot cell is injected into a frog egg whose nucleus has been removed, the injected donor nucleus is capable of programming the recipient egg to produce a normal carrot.

7–2 In terms of the way it interacts with DNA, the helix-loop-helix motif is more closely related to the leucine zipper motif than it is to the helix-turn-helix motif.

7–3 Many transcription regulators in eukaryotes can act even when they are bound to DNA thousands of nucleotide pairs away from the promoter they influence.

7–4 Once cells have differentiated to their final specialized forms, they never again alter expression of their genes.

7–5 CG islands are thought to have arisen during evolution because they were associated with portions of the genome that remained unmethylated in the germ line.

7–6 In one extreme case, a single gene in *Drosophila*—the *Dscam* gene—has the potential to produce more than 38,000 different proteins by alternative splicing; thus, the complexity of this one gene rivals the complexity of the whole human genome.

7–7 crRNAs in bacteria and piRNAs in animals serve analogous functions; they defend against foreign invaders.

Discuss the following problems.

7–8 Comparisons of the patterns of mRNA abundance across different human cell types show that the level of expression of almost every active gene is different. The patterns of mRNA abundance are so characteristic of cell type that they can be used to determine the tissue of origin of cancer cells, even though the cells may have metastasized to different parts of the body. By definition, however, cancer cells are different from their noncancerous precursor cells. How do you suppose then that patterns of mRNA expression might be used to determine the tissue source of a human cancer?

7–9 What are the two fundamental components that allow a gene to be switched on and off?

7–10 The nucleus of a eukaryotic cell is much larger than a bacterium, and it contains much more DNA. As a consequence, a transcription regulator in a eukaryotic cell must be able to select its specific binding site from among many more unrelated sequences than does a transcription regulator in a bacterium. Does this present any special problems for eukaryotic gene regulation?

Consider the following situation. Assume that the eukaryotic nucleus and the bacterial cell each have a single copy of the same DNA binding site. In addition, assume that the nucleus is 500 times the volume of the bacterium and has 500 times as much DNA. If the concentration of the transcription regulator that binds the site were the same in the nucleus and in the bacterium, would the regulator occupy its binding site equally as well in the eukaryotic nucleus as it does in the bacterium? Explain your answer.

7–11 The genes encoding the enzymes for arginine biosynthesis are located at several positions around the genome of *E. coli*. The ArgR transcription regulator coordinates their expression. The activity of ArgR is modulated by arginine. Upon binding arginine, ArgR dramatically changes its affinity for the *cis*-regulatory sequences in the promoters of the genes for the arginine biosynthetic enzymes. Given that ArgR is a transcription repressor, would you expect that ArgR would bind more tightly or less tightly to the regulatory sequences when arginine is abundant? If ArgR functioned instead as a transcription activator, would you expect the binding of arginine to increase or to decrease its affinity for its regulatory sequences? Explain your answers.

7–12 Some transcription regulators bind to DNA and cause the double helix to bend at a sharp angle. Such "bending proteins" can affect the initiation of transcription without directly contacting any other protein. Can you devise a plausible explanation for how such proteins might work to modulate transcription? Draw a diagram that illustrates your explanation.

7–13 How is it that protein-protein interactions that are too weak to cause proteins to assemble in solution can nevertheless allow the same proteins to assemble into complexes on DNA?

7–14 Imagine the two situations shown in **Figure Q7–1**. In cell 1, a transient signal induces the synthesis of protein A, which is a transcription activator that turns on many genes including its own. In cell 2,

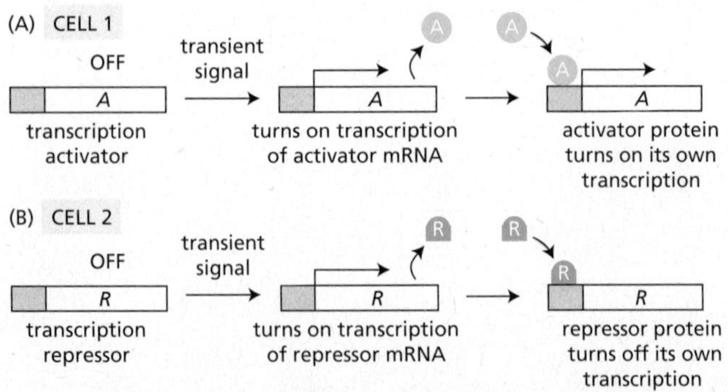

Figure Q7–1 Gene regulatory circuits and cell memory (Problem 7–14). (A) Induction of synthesis of transcription activator A by a transient signal. (B) Induction of synthesis of transcription repressor R by a transient signal.

a transient signal induces the synthesis of protein R, which is a transcription repressor that turns off many genes including its own. In which, if either, of these situations will the descendants of the original cell "remember" that the progenitor cell had experienced the transient signal? Explain your reasoning.

7–15 Examine the two pedigrees shown in Figure Q7–2. One results from deletion of a maternally imprinted autosomal gene. The other pedigree results from deletion of a paternally imprinted autosomal gene. In both pedigrees, affected individuals (*red symbols*) are heterozygous for the deletion. These individuals are affected because one copy of the chromosome carries an imprinted, inactive gene, while the other carries a deletion of the gene. *Dotted yellow symbols* indicate individuals that carry the deleted locus but do not display the mutant phenotype. Which pedigree is based on paternal imprinting and which on maternal imprinting? Explain your answer.

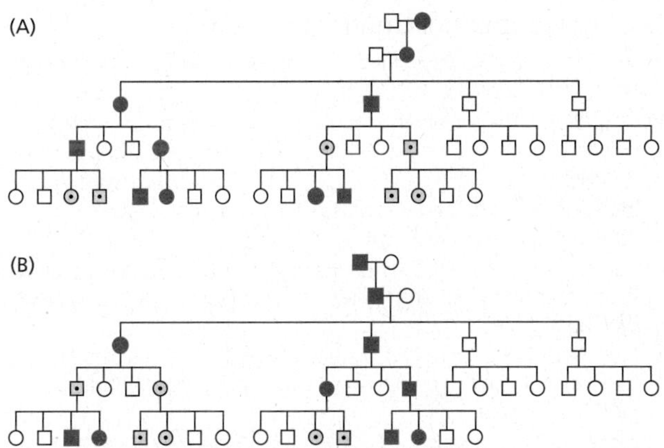

Figure Q7–2 Pedigrees reflecting maternal and paternal imprinting (Problem 7–15). In one pedigree, the gene is paternally imprinted; in the other, it is maternally imprinted. In generations 3 and 4, only one of the two parents in the indicated matings is shown; the other parent is a normal individual from outside this pedigree. Affected individuals are represented by *red circles* for females and *red squares* for males. *Dotted yellow symbols* indicate individuals that carry the deletion but do not display the phenotype.

7–16 To determine the role of the *Xist* gene in X-inactivation, scientists generated embryonic stem cells that carried one normal X chromosome and one mutant X chromosome with a nonfunctional *Xist* gene. Sequence differences allowed them to distinguish the two X chromosomes. What pattern of X-inactivation do you predict was observed in mice derived from these embryonic stem cells? Explain your reasoning.

A. Only the normal X chromosomes were inactivated.
B. Only the mutant X chromosomes were inactivated.
C. None of the X chromosomes were inactivated.
D. The X chromosomes were randomly inactivated.

7–17 The level of β-tubulin gene expression in cells is controlled by an unusual regulatory pathway, in which the intracellular concentration of free tubulin dimers (composed of one α-tubulin and one β-tubulin subunit) regulates the rate of new β-tubulin synthesis at the level of β-tubulin mRNA stability. The first 12 nucleotides of the coding portion of the mRNA were found to contain the site responsible for this autoregulatory control. Because the critical segment of the mRNA involves a coding region, it was not clear whether the regulation of mRNA stability resulted from the interaction of tubulin dimers with the RNA or with the nascent protein. Either interaction might plausibly trigger a nuclease that would destroy the mRNA.

These two possibilities were tested by mutagenizing the regulatory region on a cloned version of the gene. The mutant genes were then expressed in cells, and the stability of their mRNAs was assayed in the presence of excess free tubulin dimers. The results from a dozen mutants that affect the regulatory region of the mRNA are shown in Figure Q7–3. Does the regulation of β-tubulin mRNA stability result from an interaction with the RNA or from an interaction with the encoded protein? Explain your reasoning. (The genetic code is inside the back cover of this book.)

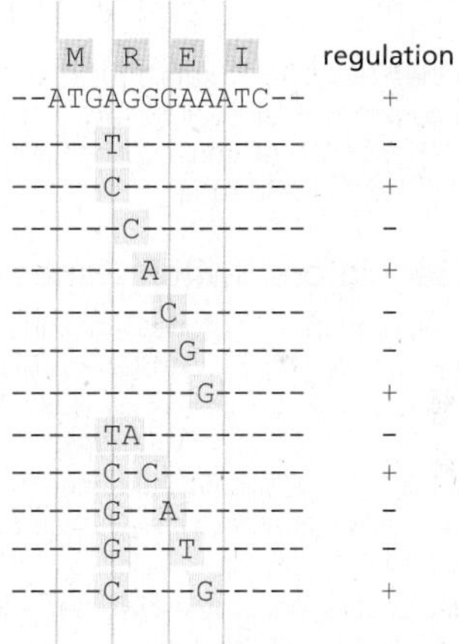

Figure Q7–3 Effects of mutations on the regulation of β-tubulin mRNA stability (Problem 7–17). The wild-type sequence for the first 12 nucleotides of the coding portion of the gene is shown at the top, and the first four amino acids beginning with methionine (M) are indicated above the codons. The nucleotide changes in the 12 mutants are shown below; only the altered nucleotides are indicated. Regulation of mRNA stability is shown on the right: + indicates wild-type response to changes in intracellular tubulin concentration, and – indicates no response to changes. Vertical lines mark the position of the first nucleotide in each codon.

7–18 If you insert a *β-galactosidase* gene lacking its own transcription control region into a cluster of piRNA genes in *Drosophila*, you find that β-galactosidase expression from a normal copy elsewhere in the genome is strongly inhibited in the fly's germ cells. If the inactive *β-galactosidase* gene is inserted outside the piRNA gene cluster, the normal gene is properly expressed. What do you suppose is the basis for this observation? How would you test your hypothesis?

REFERENCES

General

Barressi MJF & Gilbert SF (2020) Developmental Biology, 12th ed. Sunderland, MA: Sinauer Associates.

Brown TA (2017) Genomes 4. New York: Garland Science.

Craig N, Green R, Greider C, Storz G, Wolberger C & Cohen-Fix O (2021) Molecular Biology: Principles of Genome Function, 2nd ed. Oxford: Oxford University Press.

Goldberg ML, Fischer J, Hood L & Hartwell L (2021) Genetics: From Genes to Genomes, 7th ed. Boston: McGraw-Hill.

Watson J, Baker T, Bell S et al. (2013) Molecular Biology of the Gene, 7th ed. Menlo Park, CA: Benjamin Cummings.

An Overview of Gene Control

Davidson EH (2006) The Regulatory Genome: Gene Regulatory Networks in Development and Evolution. Burlington, MA: Elsevier.

Gurdon JB (1992) The generation of diversity and pattern in animal development. *Cell* 68, 185–199.

Ptashne M & Gann A (2002) Genes and Signals. Cold Spring Harbor, NY: Cold Spring Harbor Laboratory Press.

Control of Transcription by Sequence-specific DNA-binding Proteins

Gilbert W & Müller-Hill B (1967) The *lac* operator is DNA. *Proc. Natl. Acad. Sci. USA* 58, 2415.

McKnight SL (1991) Molecular zippers in gene regulation. *Sci. Am.* 264, 54–64.

Weirauch MT & Hughes TR (2011) A catalogue of eukaryotic transcription factor types, their evolutionary origin, and species distribution. In A Handbook of Transcription Factors (Hughes TR, ed.), pp. 25–74. New York: Springer.

Transcription Regulators Switch Genes On and Off

Beckwith J (1987) The operon: an historical account. In *Escherichia coli* and *Salmonella typhimurium*: Cellular and Molecular Biology (Neidhart FC, Ingraham JL, Low KB et al., eds.), Vol. 2, pp. 1439–1443. Washington, DC: ASM Press.

Goldstein I & Hager GL (2018) Dynamic enhancer function in the chromatin context. *Wiley Interdiscip. Rev. Syst. Biol. Med.* 10(1), 10.1002/wsbm.1390.

Hahn S (2018) Phase separation, protein disorder, and enhancer function. *Cell* 175(7), 1723–1725.

Hnisz D, Shrinivas K, Young RA . . . Sharp PA (2017) A phase separation model for transcriptional control. *Cell* 169(1), 13–23.

Jacob F & Monod J (1961) Genetic regulatory mechanisms in the synthesis of proteins. *J. Mol. Biol.* 3, 318–356.

McSwiggen DT, Mir M, Darzacq X & Tjian R (2019) Evaluating phase separation in live cells: diagnosis, caveats, and functional consequences. *Genes Dev*. 33(23-24), 1619–1634.

Patel AB, Greber BJ & Nogales E (2020) Recent insights into the structure of TFIID, its assembly, and its binding to core promoter. *Curr. Opin. Struct. Biol*. 61, 17–24.

Ptashne M (2004) A Genetic Switch: Phage Lambda Revisited, 3rd ed. Cold Spring Harbor, NY: Cold Spring Harbor Laboratory Press.

Soutourina J (2018) Transcription regulation by the Mediator complex. *Nat. Rev. Mol. Cell Biol*. 19(4), 262–274.

Molecular Genetic Mechanisms That Create and Maintain Specialized Cell Types

Alon U (2007) Network motifs: theory and experimental approaches. *Nature* 8, 450–461.

Hobert O (2011) Regulation of terminal differentiation programs in the nervous system. *Annu. Rev. Cell Dev. Biol.* 27, 681–696.

Lawrence PA (1992) The Making of a Fly: The Genetics of Animal Design. New York: Blackwell Scientific.

Rickels R & Shilatifard A (2018) Enhancer logic and mechanics in development and disease. *Trends Cell Biol*. 28(8), 608–630.

Mechanisms That Reinforce Cell Memory in Plants and Animals

Brahma S & Henikoff S (2020) Epigenome regulation by dynamic nucleosome unwrapping. *Trends Biochem. Sci.* 45(1), 13–26.

Deaton AM & Bird A (2011) CpG islands and the regulation of transcription. *Genes Dev*. 25(10), 1010–1022.

Galupa R & Heard E (2018) X-chromosome inactivation: a crossroads between chromosome architecture and gene regulation. *Annu. Rev. Genet.* 52, 535–566.

Klemm SL, Shipony Z & Greenleaf WJ (2019) Chromatin accessibility and the regulatory epigenome. *Nat. Rev. Genet.* 20(4), 207–220.

Ondičová M, Oakey RJ & Walsh CP (2020) Is imprinting the result of "friendly fire" by the host defense system? *PLoS Genet.* 16(4), e1008599.

Schmitz RJ, Lewis ZA & Goll MG (2019) DNA methylation: shared and divergent features across eukaryotes. *Trends Genet.* 35(11), 818–827.

Post-transcriptional Controls

Breaker RR (2018) Riboswitches and translation control. *Cold Spring Harb. Perspect. Biol*. 10(11), a032797.

Chen LL (2020) The expanding regulatory mechanisms and cellular functions of circular RNAs. *Nat. Rev. Mol. Cell Biol.* 21(8), 475–490.

Das S, Vera M, Gandin V . . . Tutucci E (2021) Intracellular mRNA transport and localized translation. *Nat. Rev. Mol. Cell Biol.* 22, 483–504.

Gottesman S & Storz G (2011) Bacterial small RNA regulators: versatile roles and rapidly evolving variations. *Cold Spring Harb. Perspect. Biol.* 3, a003798.

Hershey JWB, Sonenberg N & Mathews MB (2012) Principles of translational control: an overview. *Cold Spring Harb. Perspect. Biol.* 4, a011528.

Kortmann J & Narberhaus F (2012) Bacterial RNA thermometers: molecular zippers and switches. *Nat. Rev. Microbiol.* 10, 255–265.

Schwartz S (2016) Cracking the epitranscriptome. *RNA* 22(2), 169–174.

Tauber D, Tauber G & Parker R (2020) Mechanisms and regulation of RNA condensation in RNP granule formation. *Trends Biochem. Sci*. 45(9), 764–778.

Thompson SR (2012) Tricks an IRES uses to enslave ribosomes. *Trends Microbiol.* 20, 558–566.

Tian B & Manley JL (2017) Alternative polyadenylation of mRNA precursors. *Nat. Rev. Mol. Cell Biol*. 18(1), 18–30.

Yablonovitch AL, Deng P, Jacobson D & Li JB (2017) The evolution and adaptation of A-to-I RNA editing. *PLoS Genet.* 13(11), e1007064.

Regulation of Gene Expression by Noncoding RNAs

Bartel DP (2018) Metazoan microRNAs. *Cell* 173(1), 20–51.

Czech B, Munafò M, Ciabrelli F . . . Hannon GJ (2018) piRNA-guided genome defense: from biogenesis to silencing. *Annu. Rev. Genet.* 52, 131–157.

Fire A, Xu S, Montgomery MK . . . Mello CC (1998) Potent and specific genetic interference by double-stranded RNA in *Caenorhabditis elegans*. *Nature* 391, 806–811.

Geis FK & Goff SP (2020) Silencing and transcriptional regulation of endogenous retroviruses: an overview. *Viruses* 12(8), 884.

Ozata DM, Gainetdinov I, Zoch A . . . Zamore PD (2019) PIWI-interacting RNAs: small RNAs with big functions. *Nat. Rev. Genet*. 20(2), 89–108.

Statello L, Guo CJ, Chen LL & Huarte M (2021) Gene regulation by long non-coding RNAs and its biological functions. *Nat. Rev. Mol. Cell Biol.* 22(2), 96–118.

Wiedenheft B, Sternberg SH & Doudna JA (2012) RNA-guided genetic silencing systems in bacteria and archaea. *Nature* 482, 331–338.

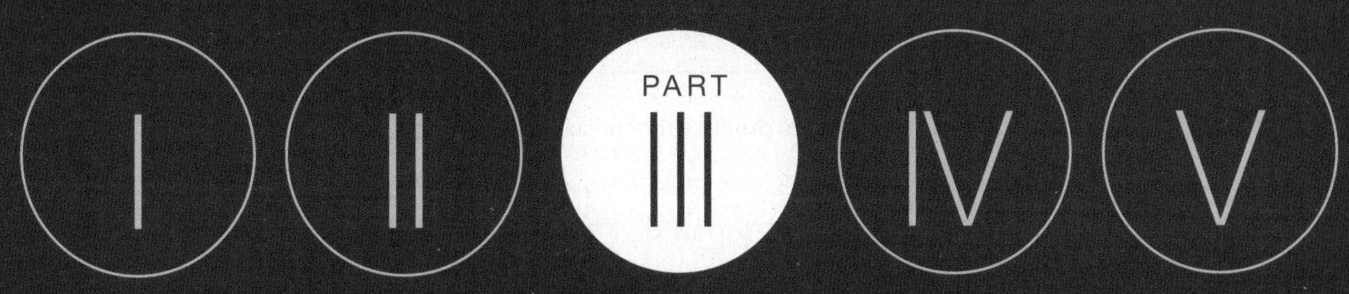

WAYS OF WORKING WITH CELLS

CHAPTER 8

Analyzing Cells, Molecules, and Systems

Progress in science is often driven by advances in technology. The entire field of cell biology, for example, came into being when optical craftsmen learned to grind small lenses of sufficiently high quality to observe cells and their substructures. Innovations in lens grinding, rather than any conceptual or philosophical advance, allowed Hooke and van Leeuwenhoek to discover a previously unseen cellular world, where tiny creatures tumble and twirl in a droplet of water (**Figure 8–1**).

The twenty-first century is a particularly exciting time for biology. New methods for analyzing cells, proteins, DNA, and RNA are fueling an information explosion and allowing scientists to study cells and their macromolecules in previously unimagined ways. We now have access to the sequences of many billions of nucleotides, providing the complete molecular blueprints for hundreds of organisms—from microbes and mustard weeds to worms, flies, mice, dogs, chimpanzees, and humans. And powerful new techniques are helping us to decipher that information, allowing us not only to compile huge, detailed catalogs of genes and proteins but also to begin to unravel how these components work together to form functional cells and organisms. The long-range goal is nothing short of obtaining a complete understanding of what takes place inside a cell as it responds to its environment and interacts with its neighbors.

In this and the following chapter, we present some of the principal methods used to study cells and their molecular components. Chapter 9 describes the remarkable advances in microscopy that have helped fuel our understanding of the structure and function of cells. In this chapter, we first discuss the rapidly developing methods for analysis of the molecules and genes that drive cell behavior. We present the techniques used to determine protein structure, function, and interactions, and we discuss the breakthroughs in DNA technology that continue to revolutionize our understanding of cell function. We end the chapter with an overview of some of the mathematical approaches that are helping us understand the enormous complexity of cells. By considering cells as dynamic systems with many moving parts, mathematical approaches can reveal hidden insights into

IN THIS CHAPTER

how the many components of cells work together to produce the special qualities of life.

ISOLATING CELLS AND GROWING THEM IN CULTURE

As described in Chapter 1, numerous model organisms are used by researchers to unravel the molecular basis of cell biology. In the first section of this chapter, we concern ourselves with the organisms and cells best suited for biochemical studies of proteins in the cell. Typically, we gain access to these proteins by obtaining large numbers of cells and physically breaking them open. Unicellular organisms, such as bacteria and yeast, are easy to produce in large amounts in the laboratory and are rich sources of the proteins involved in fundamental cell processes. But a deep understanding of human proteins in specific cell types requires human cells, or at the very least cells from a mammal. Specific animal tissues can be a useful solution, but these tend to be composed of a heterogeneous mixture of cell types. To obtain as much information as possible about specific cell types in a tissue, biologists have developed ways of dissociating cells from tissues and separating them according to type. These manipulations result in a relatively homogeneous population of cells that can then be analyzed—either directly or after their number has been greatly increased by allowing the cells to proliferate in culture.

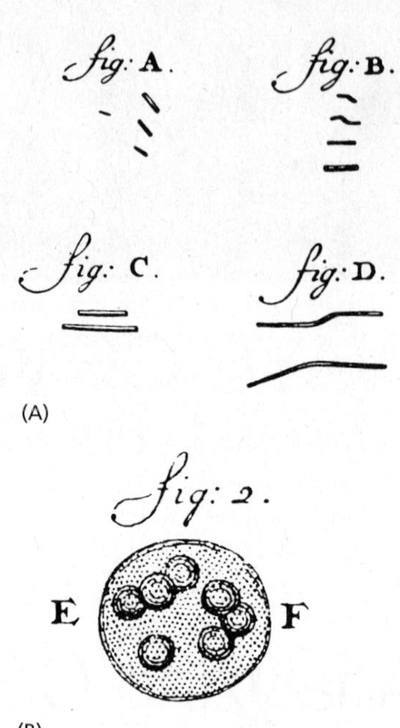

Figure 8–1 Microscopic life. A sample of "diverse animalcules" seen by van Leeuwenhoek using his simple microscope. (A) Bacteria seen in material he excavated from between his teeth. Those in fig. B he described as "swimming first forward and then backwards" (1692). (B) The eukaryotic green alga *Volvox* (1700). (Courtesy of the John Innes Foundation.)

Cells Can Be Isolated from Tissues and Grown in Culture

Intact tissues provide the most realistic source of material, as they represent the actual cells found within the body. For some biochemical preparations, the protein of interest can be obtained in sufficient quantity without having to separate the tissue or organ into cell types. Examples include the preparation of histones from calf thymus, actin from rabbit muscle, or tubulin from cow brain. For the majority of proteins, however, tissues are not an ideal source and it is preferable to use specific cell types grown in culture. Cultured cells provide a more homogeneous population of cells from which to extract material, and they are also much more convenient to work with in the laboratory. Given appropriate surroundings, most animal cells can live, multiply, and even express differentiated properties in a culture dish. The cells can be watched continually under the microscope or analyzed biochemically, and the effects of adding or removing specific molecules, such as hormones or growth factors, can be systematically explored.

Experiments performed on cultured cells are sometimes said to be carried out *in vitro* (literally, "in glass") to contrast them with experiments using intact organisms, which are said to be carried out *in vivo* (literally, "in the living organism"). These terms can be confusing, however, because they are often used in a very different sense by biochemists. In the biochemistry lab, *in vitro* refers to reactions carried out in a test tube in the absence of living cells, whereas *in vivo* refers to any reaction taking place inside a living cell, even if that cell is growing in culture.

Tissue culture began in 1907 with an experiment designed to settle a controversy in neurobiology. The hypothesis under examination was known as the neuronal doctrine, which proposed that each nerve fiber is the outgrowth of a single nerve cell and not the product of the fusion of many cells. To test this contention, small pieces of spinal cord were placed in lymphatic fluid in a warm, moist chamber and observed at regular intervals under the microscope. After a day or so, individual nerve cells could be seen extending long, thin filaments (axons) into the clot. Thus, the neuronal doctrine received strong support, and the foundation was laid for the cell-culture revolution.

These original experiments on nerve fibers used cultures of small tissue fragments called explants. Today, cultures are more commonly made from suspensions of cells dissociated from tissues. The first step in isolating individual cells is to disrupt the extracellular matrix and cell–cell junctions that hold the cells together. For this purpose, a tissue sample is typically treated with proteolytic enzymes (such as trypsin and collagenase) to digest proteins in the extracellular

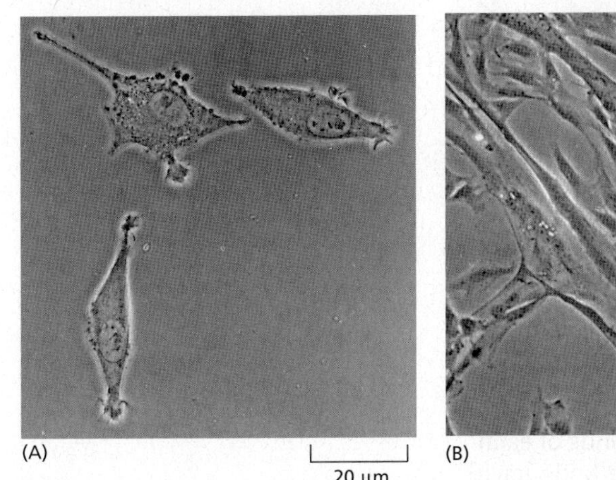

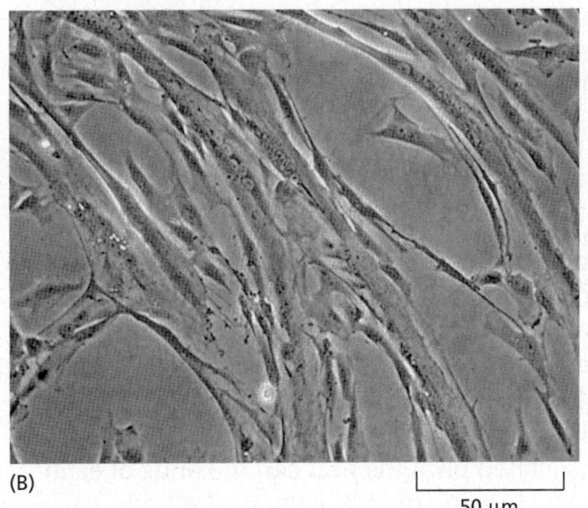

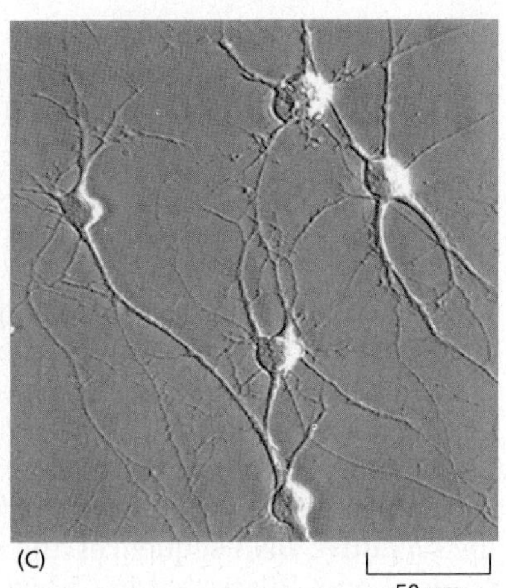

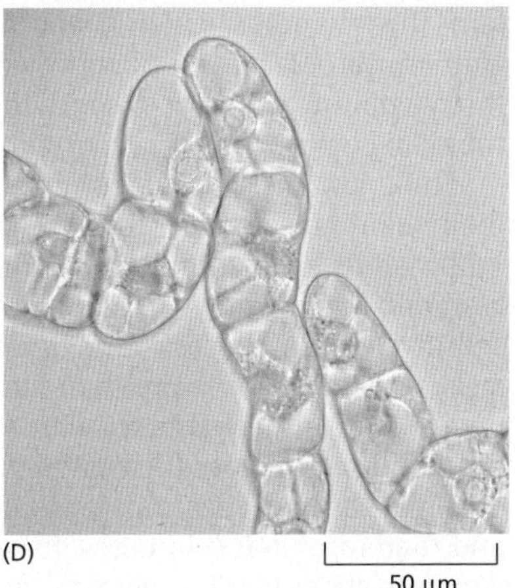

Figure 8-2 Light micrographs of cells in culture. (A) Mouse fibroblasts. (B) Chick myoblasts fusing to form multinucleate muscle cells. (C) Purified rat retinal ganglion nerve cells. (D) Tobacco cells in liquid culture. (A, courtesy of Daniel Zicha; B, courtesy of Rosalind Zalin; C, from A. Meyer-Franke et al., *Neuron* 15:805–819, 1995. With permission from Elsevier; D, courtesy of Gethin Roberts.)

matrix and with agents (such as ethylenediaminetetraacetic acid, or EDTA) that bind, or chelate, the Ca^{2+} on which cell-cell adhesion depends. The tissue can then be teased apart into single cells by gentle agitation.

Unlike yeast, most tissue cells are not adapted to living suspended in fluid and require a solid surface on which to grow and divide. For cell cultures, this support is usually provided by the surface of a plastic culture dish. Cells vary in their requirements, however, and many do not proliferate or differentiate unless the culture dish is coated with materials that cells adhere to, such as polylysine or extracellular matrix components.

Cultures prepared directly from the tissues of an organism are called *primary cultures*. These can be made with or without an initial fractionation step to separate different cell types. In most cases, cells in primary cultures can be removed from the culture dish and recultured repeatedly in so-called secondary cultures; in this way, they can be repeatedly subcultured (*passaged*) for weeks or months. Such cells often display many of the differentiated properties appropriate to their origin (**Figure 8–2**): fibroblasts continue to secrete collagen; cells derived from embryonic skeletal muscle fuse to form muscle fibers that contract spontaneously in the culture dish; nerve cells extend axons that are electrically excitable and make synapses with other nerve cells; and epithelial cells form extensive sheets with many of the properties of an intact epithelium. Because these properties are maintained in culture, they are accessible to study in ways that are often not possible in intact tissues.

Embryonic stem cells are an important cell type isolated from the early mammalian embryo. As described in Chapter 22, these cells are *pluripotent*; that is, they have the potential to differentiate into any cell type in the body. When cultured in the presence of the appropriate extracellular signaling factors and nutrients, stem cells can be directed to differentiate into a wide range of specific cell types. Under some conditions, it is even possible to stimulate these cells to assemble into three-dimensional multicellular structures that are miniature versions of certain organs, such as the gut. These *organoids* provide a powerful tool for the analysis of tissue function (see Chapter 22).

Cell culture is not limited to animal cells. When a piece of plant tissue is cultured in a sterile medium containing nutrients and appropriate growth regulators, many of the cells are stimulated to proliferate indefinitely in a disorganized manner, producing a mass of relatively undifferentiated cells called a *callus*. If the nutrients and growth regulators are carefully manipulated, one can induce the formation of a shoot and then root apical meristems within the callus, and, in many species, regenerate a whole new plant. Similar to animal cells, callus cultures can be dissociated into single cells, which will grow and divide as a suspension culture (see Figure 8–2D).

Eukaryotic Cell Lines Are a Widely Used Source of Homogeneous Cells

The cell cultures obtained by disrupting tissues tend to suffer from a problem—eventually the cells die. Most vertebrate cells stop dividing after a finite number of cell divisions in culture, a process called *replicative cell senescence* (discussed in Chapter 17). Normal human fibroblasts, for example, typically divide only 25–40 times in culture before they stop. In these cells, the limited proliferation capacity reflects a progressive shortening and uncapping of the cell's telomeres, the repetitive DNA sequences and associated proteins that cap the ends of each chromosome (discussed in Chapter 5). Human somatic cells in the body have turned off production of the enzyme, called *telomerase*, that normally maintains the telomeres, which is why their telomeres shorten with each cell division. Human fibroblasts can often be coaxed to proliferate indefinitely by providing them with the gene that encodes the catalytic subunit of telomerase; in this case, they can be propagated as an *immortalized* cell line.

Some human cells, however, cannot be immortalized by this trick. Although their telomeres remain long, they still stop dividing after a limited number of divisions because culture conditions cause excessive stimulation of cell proliferation, which activates a poorly understood protective mechanism that stops cell division—a process sometimes called *culture shock*. To immortalize these cells, one has to do more than introduce telomerase. One must also inactivate the protective mechanisms, which can be done by introducing certain cancer-promoting oncogenes (discussed in Chapter 20).

Unlike human cells, most rodent cells do not turn off production of telomerase, and therefore their telomeres do not shorten with each cell division. Therefore, if culture shock can be avoided, some rodent cell types will divide indefinitely in culture. In addition, rodent cells often undergo spontaneous genetic changes in culture that inactivate their protective mechanisms, thereby producing immortalized cell lines.

Cell lines can often be most easily generated from cancer cells, but these cultures—referred to as *transformed cell lines*—differ from those prepared from normal cells in several ways. Transformed cell lines often grow without attaching to a surface, for example, and they can proliferate to a much higher density in a culture dish. Similar properties can be induced experimentally in normal cells by transforming them with a tumor-inducing virus or chemical. The resulting transformed cell lines can usually cause tumors if injected into a susceptible animal.

Transformed and nontransformed cell lines are extremely useful in cell research as sources of very large numbers of cells of a uniform type, especially because they can be stored in liquid nitrogen at −196°C for an indefinite period and retain their viability when thawed. It is important to keep in mind, however, that cell lines nearly always differ in important ways from their normal progenitors in the tissues from which they were derived.

Some widely used cell lines are listed in **Table 8–1**. Different lines have different advantages; for example, the PtK epithelial cell lines derived from the rat kangaroo remain flat during mitosis (unlike many other cell types), allowing the mitotic apparatus to be readily observed in action.

TABLE 8–1 Some Commonly Used Cell Lines

Cell line*	Cell type and origin
NIH-3T3	Fibroblast (mouse)
MDCK	Kidney epithelial cell (dog)
HeLa	Cervical epithelial cell (human)
PtK	Kidney epithelial cell (rat kangaroo)
L6	Myoblast (rat)
PC12	Chromaffin cell (rat)
COS	Kidney fibroblast (monkey)
HEK293	Kidney epithelial cell (human)
CHO	Ovary epithelial cell (Chinese hamster)
RPE	Retinal pigment epithelial cell (human)
Vero	Kidney epithelial cell (African green monkey)
Jurkat	White blood cell (human)

*Many of these cell lines were derived from tumors. All of them are capable of indefinite replication in culture and express at least some of the special characteristics of their cells of origin.

Hybridoma Cell Lines Are Factories That Produce Monoclonal Antibodies

As we see in this chapter and throughout this book, antibodies are particularly useful tools for cell biology. Their great specificity allows precise detection of selected proteins among the many thousands that each cell typically produces.

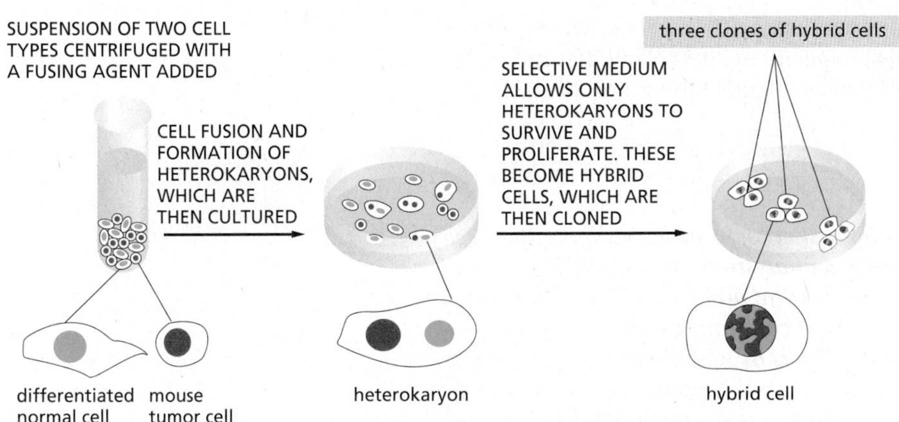
SUSPENSION OF TWO CELL TYPES CENTRIFUGED WITH A FUSING AGENT ADDED
CELL FUSION AND FORMATION OF HETEROKARYONS, WHICH ARE THEN CULTURED
SELECTIVE MEDIUM ALLOWS ONLY HETEROKARYONS TO SURVIVE AND PROLIFERATE. THESE BECOME HYBRID CELLS, WHICH ARE THEN CLONED
three clones of hybrid cells
differentiated normal cell
mouse tumor cell
heterokaryon
hybrid cell

Figure 8–3 The production of hybrid cells. It is possible to fuse one cell with another to form a *heterokaryon*, a combined cell with two separate nuclei. Typically, a suspension of cells is treated with certain inactivated viruses or with polyethylene glycol, each of which alters the plasma membranes of cells in a way that induces them to fuse. Eventually, a heterokaryon proceeds to mitosis and produces a hybrid cell in which the two separate nuclear envelopes have been disassembled, allowing all the chromosomes to be brought together in a single large nucleus. Such hybrid cells can give rise to immortal hybrid cell lines. If one of the parent cells was from a tumor cell line, the hybrid cell is called a hybridoma.

Antibodies are often produced by inoculating animals with the protein of interest and subsequently isolating the antibodies specific to that protein from the serum of the animal. However, only limited quantities of antibodies can be obtained from a single inoculated animal, and the *polyclonal* antibodies produced will be a heterogeneous mixture of antibodies that recognize a variety of different antigenic sites on the protein. Moreover, antibodies specific for the antigen will constitute only a fraction of the antibodies found in the serum. An alternative technology, which allows the production of an unlimited quantity of identical antibodies and greatly increases the specificity and convenience of antibody-based methods, is the production of monoclonal antibodies by hybridoma cell lines.

This procedure involves propagating a clone of cells from a single antibody-secreting B lymphocyte to obtain a homogeneous preparation of antibodies in large quantities. B lymphocytes normally have a limited life span in culture, but individual antibody-producing B lymphocytes from an immunized mouse, when fused with cells derived from a transformed B lymphocyte cell line, can give rise to hybrids that have both the ability to make a particular antibody and the ability to multiply indefinitely in culture (Figure 8–3). These **hybridomas** are propagated as individual clones, each of which provides a permanent and stable source of a single type of **monoclonal antibody**. Each type of monoclonal antibody recognizes a single type of antigenic site; for example, a particular cluster of five or six amino acid side chains on the surface of a protein. Their uniform specificity makes monoclonal antibodies much more useful than conventional antisera for many purposes.

An important advantage of the hybridoma technique is that monoclonal antibodies can be made against molecules that constitute only a minor component of a complex mixture. In an ordinary antiserum made against such a mixture, the proportion of antibody molecules that recognize the minor component would be too small to be useful. But if the B lymphocytes that produce the various components of this antiserum are made into hybridomas, it becomes possible to screen individual hybridoma clones from the large mixture to select one that produces the desired type of monoclonal antibody and to propagate the selected hybridoma indefinitely so as to produce that antibody in unlimited quantities. In principle, therefore, a monoclonal antibody can be made against any protein in a biological sample. Once an antibody has been made, it can be used to localize the protein in cells and tissues, to follow its movement, and to purify the protein to study its structure and function.

Monoclonal antibodies are not just useful research tools but are valuable as treatments for a number of human diseases. Certain cancers, for example, can be treated by intravenous infusion of monoclonal antibodies that bind and inhibit signaling receptors on the cancer cell surface, thereby reducing proliferation of the tumor cells. In other cases, monoclonal antibodies that bind specific cell-surface immune regulators can promote immunological attack of cancer cells. Monoclonal antibodies that bind and inhibit specific immune-stimulatory molecules provide useful therapies in autoimmune diseases such as rheumatoid arthritis. Monoclonal antibodies against specific viral proteins can reduce infection. Given their exquisite specificity, it is likely that monoclonal antibodies will continue to be developed as effective therapies for these and other diseases.

Summary

Tissues can be dissociated into their component cells, from which individual cell types can be purified and used for biochemical analysis or for the establishment of cell cultures. Many animal and plant cells survive and proliferate in a culture dish if they are provided with a suitable culture medium containing nutrients and appropriate signal molecules. Although many animal cells stop dividing after a finite number of cell divisions, cells that have been immortalized through spontaneous mutations or genetic manipulation can be maintained indefinitely as cell lines. Hybridoma cells are widely employed to produce unlimited quantities of uniform monoclonal antibodies, which are used to detect and purify cell proteins, as well as to diagnose and treat diseases.

PURIFYING PROTEINS

The challenge of isolating a single type of protein from the thousands of other proteins in a cell is formidable but must be overcome to study protein function *in vitro*. As we shall see later in this chapter, *recombinant DNA technology* can enormously simplify this task by engineering cells to produce large quantities of a given protein, thereby making its purification much easier. Whether the source of the protein is an engineered cell or a natural tissue, a purification procedure usually starts with subcellular fractionation to reduce the complexity of the material and is then followed by purification steps of increasing specificity.

Cells Can Be Separated into Their Component Fractions

To purify a protein, it must first be extracted from inside the cell. Cells can be broken up in various ways: they can be subjected to osmotic shock or ultrasonic vibration, forced through a small orifice, or ground up in a blender. These procedures break many of the membranes of the cell (including the plasma membrane and endoplasmic reticulum) into fragments that immediately reseal to form small closed vesicles. If carefully carried out, however, the disruption procedures leave organelles such as nuclei, mitochondria, the Golgi apparatus, lysosomes, and peroxisomes largely intact. The suspension of cells is thereby reduced to a thick slurry (called a *homogenate* or *extract*) that contains a variety of membrane-enclosed organelles, each with a distinctive size and density. Provided that the homogenization medium has been carefully chosen (by trial and error for each organelle), the various components—including the vesicles derived from the endoplasmic reticulum, called microsomes—retain most of their original biochemical properties.

The different components of the homogenate must then be separated. Such cell fractionations became possible only after the commercial development in the early 1940s of an instrument known as the *preparative ultracentrifuge*, which rotates extracts of broken cells at high speeds (**Figure 8–4**). This treatment separates cell components by size and density: in general, the largest objects experience the largest centrifugal force and move the most rapidly. At relatively low speed, large components such as nuclei sediment to form a pellet at the bottom of the centrifuge tube; at slightly higher speed, a pellet of mitochondria is deposited; and at even higher speeds and with longer periods of centrifugation, first vesicles and then ribosomes can be collected (**Figure 8–5**). All of these fractions are impure, but many of the contaminants can be removed by resuspending the pellet and repeating the centrifugation procedure several times.

Centrifugation is the first step in most fractionations, but it separates only components that differ greatly in size. A finer degree of separation can be achieved by

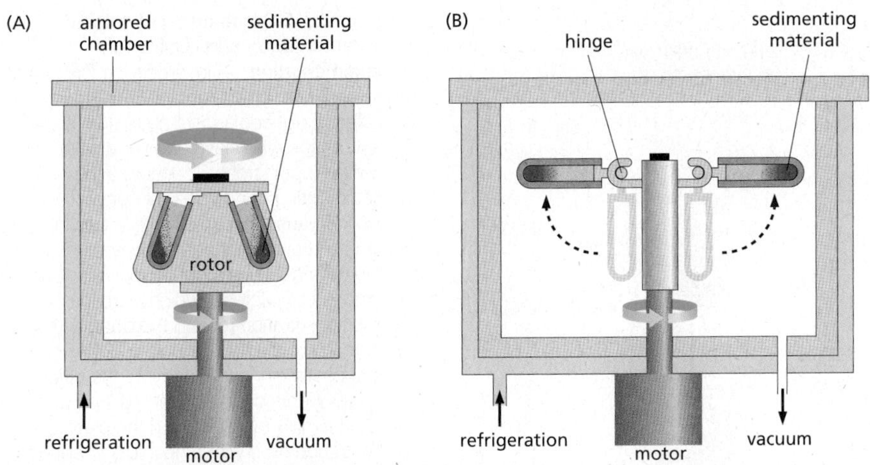

Figure 8–4 **The preparative ultracentrifuge.** (A) The sample is contained in tubes that are inserted into a ring of angled cylindrical holes in a metal *rotor*. Rapid rotation of the rotor generates enormous centrifugal forces, which cause particles in the sample to sediment against the bottom sides of the sample tubes, as shown here. The vacuum reduces friction, preventing heating of the rotor and allowing the refrigeration system to maintain the sample at 4°C. (B) Some fractionation methods require a different type of rotor called a *swinging-bucket rotor*. In this case, the sample tubes are placed in metal tubes on hinges that allow the tubes to swing outward when the rotor spins. Sample tubes are therefore horizontal during spinning, and samples are sedimented toward the bottom, not the sides, of the tube, providing better separation of differently sized components (see Figures 8–5 and 8–6).

layering the homogenate in a thin band on top of a salt solution that fills a centrifuge tube. When centrifuged, the various components in the mixture move as a series of distinct bands through the solution, each at a different rate, in a process called *velocity sedimentation* (Figure 8-6A). For the procedure to work effectively, the bands must be protected from convective mixing, which would normally occur whenever a denser solution (for example, one containing organelles) finds itself on top of a lighter one (the salt solution). This is achieved by augmenting the solution in the tube with a shallow gradient of sucrose prepared by a special mixing device. The resulting density gradient—with the dense end at the bottom of the tube—keeps each region of the solution denser than any solution above it, and it thereby prevents convective mixing from distorting the separation.

When sedimented through sucrose gradients, different cell components separate into distinct bands that can be collected individually. The relative rate at which each component sediments depends primarily on its size and shape—normally being described in terms of its sedimentation coefficient, or S value. Present-day ultracentrifuges rotate at speeds of up to 80,000 rpm and produce forces as high as 500,000 times gravity. These enormous forces drive even small macromolecules, such as tRNA molecules and simple enzymes, to sediment at an appreciable rate and allow them to be separated from one another by size.

The ultracentrifuge is also used to separate cell components on the basis of their buoyant density, independently of their size and shape. In this case, the sample is sedimented through a steep density gradient that contains a very high concentration of sucrose or cesium chloride. Each cell component begins to move down the gradient as in Figure 8-6A, but it eventually reaches a position where the density of the solution is equal to its own density. At this point, the component

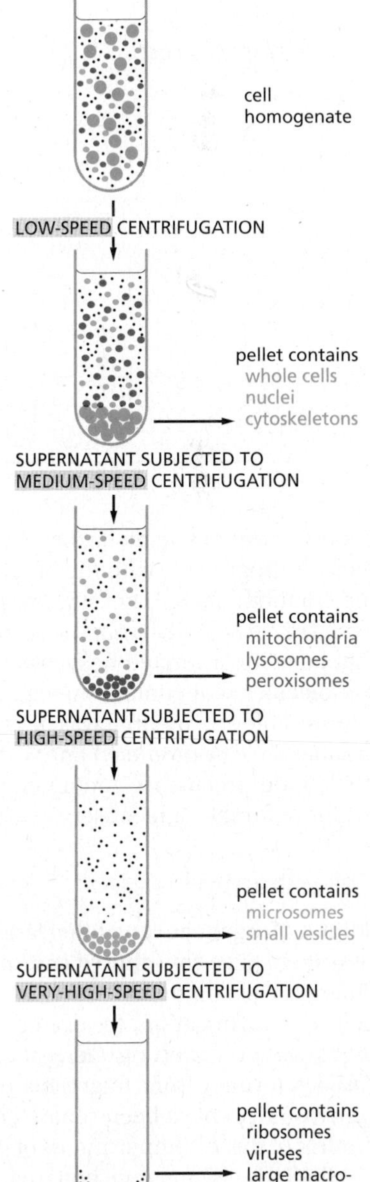

Figure 8–5 **Cell fractionation by centrifugation.** Repeated centrifugation at progressively higher speeds will fractionate homogenates of cells into their components. In general, the smaller the subcellular component, the greater the centrifugal force required to sediment it. Typical values for the various centrifugation steps referred to in the figure are:
low speed: 1000 times gravity for 10 minutes
medium speed: 20,000 times gravity for 20 minutes
high speed: 80,000 times gravity for 1 hour
very high speed: 150,000 times gravity for 3 hours

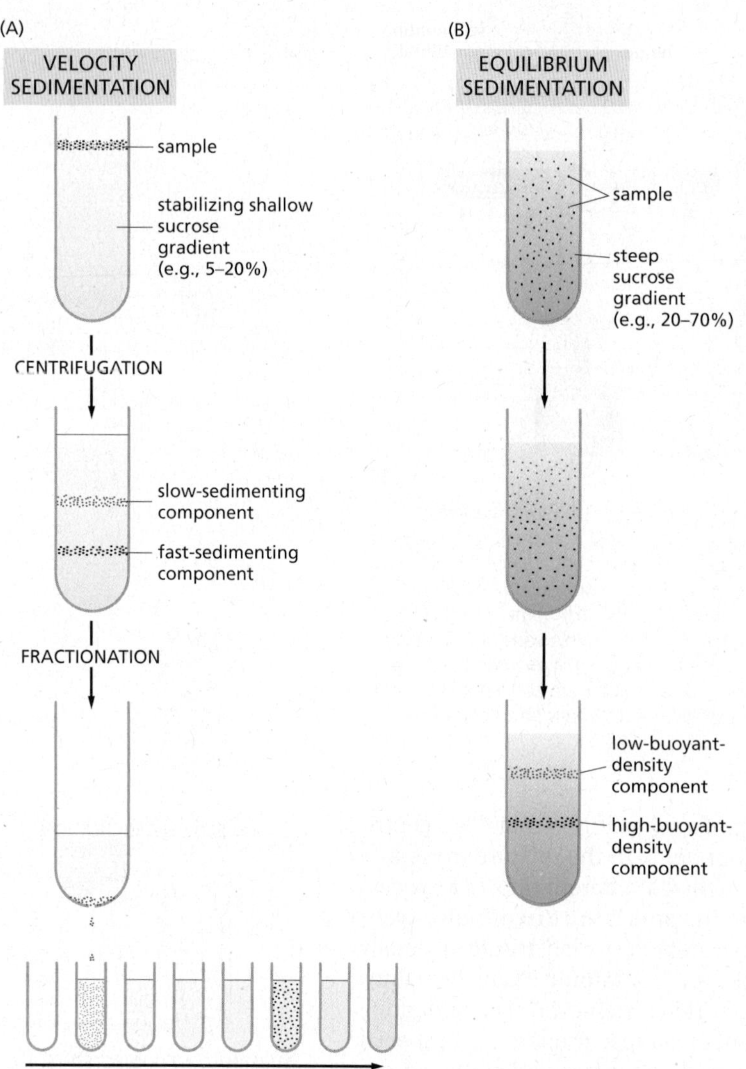

Figure 8–6 Comparison of velocity sedimentation and equilibrium sedimentation. (A) In velocity sedimentation, subcellular components sediment at different speeds according to their size and shape when layered over a solution containing sucrose. To stabilize the sedimenting bands against convective mixing caused by small differences in temperature or solute concentration, the tube contains a continuous shallow gradient of sucrose, which increases in concentration toward the bottom of the tube (typically from 5 to 20% sucrose). After centrifugation, the different components can be collected individually, most simply by puncturing the plastic centrifuge tube with a needle and collecting drops from the bottom, as illustrated here. (B) In equilibrium sedimentation, subcellular components move up or down when centrifuged in a gradient until they reach a position where their density matches that of their surroundings. Although a sucrose gradient is shown here, denser gradients, which are especially useful for protein and nucleic acid separation, can be formed from cesium chloride. The final bands, at equilibrium, can be collected as in A.

floats and can move no farther. A series of distinct bands is thereby produced in the centrifuge tube, with the bands closest to the bottom of the tube containing the components of highest buoyant density (**Figure 8–6B**). This method, called *equilibrium sedimentation,* is so sensitive that it can separate macromolecules that have incorporated heavy isotopes, such as ^{13}C or ^{15}N, from the same macromolecules that contain the lighter, common isotopes (^{12}C or ^{14}N). In fact, the cesium-chloride method was developed in 1957 to separate the labeled from the unlabeled DNA produced after exposure of a growing population of bacteria to nucleotide precursors containing ^{15}N; this classic experiment provided direct evidence for the semiconservative replication of DNA (see Figure 5–5).

Cell Extracts Provide Accessible Systems to Study Cell Functions

Studies of organelles and other large subcellular components isolated in the ultracentrifuge have contributed enormously to our understanding of the functions of different cell components. Experiments on mitochondria and chloroplasts purified by centrifugation, for example, demonstrated the central function of these organelles in converting energy into forms that the cell can use. Similarly, resealed vesicles formed from fragments of rough and smooth endoplasmic reticulum (microsomes) have been separated from each other and analyzed as functional models of these compartments of the intact cell.

Similarly, highly concentrated cell extracts, especially undiluted cytoplasm from *Xenopus laevis* (African clawed frog) oocytes and fertilized eggs, have played

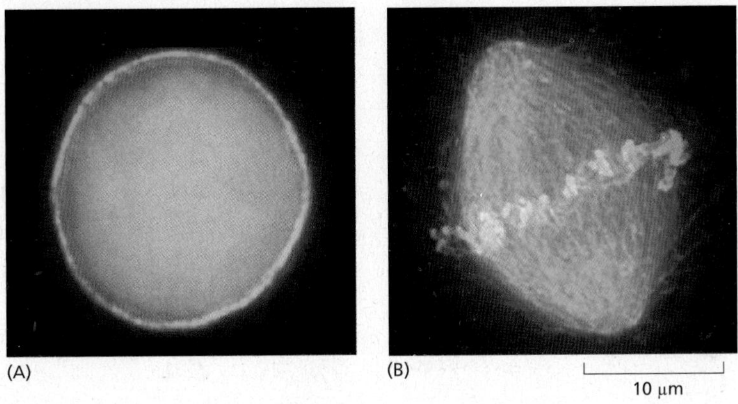

Figure 8–7 **The formation of subcellular structures in a cytoplasmic extract.** Concentrated cytoplasm was prepared from unfertilized eggs of the frog *Xenopus laevis*. After addition of *Xenopus* sperm chromosomes, these extracts were stimulated to progress through the cell cycle. (A) In an interphase extract, a nuclear envelope *(green)* forms around the sperm chromosomes *(pink)*. (B) In a metaphase extract, the nuclear envelope breaks down and the microtubules *(green)* form a mitotic spindle that is attached to the condensed chromosomes *(pink)*. [B, T. Maresca et al., *J Cell Biol.* 169(6):859–69, 2005, doi 10.1083/jcb.200503031. With permission from Rockefeller University Press.]

a critical role in the study of such complex and highly organized processes as the cell-division cycle, the assembly and function of the mitotic spindle (**Figure 8–7**), and the vesicular-transport steps involved in the movement of proteins through the secretory pathway.

Cell extracts also provide, in principle, the starting material for the complete separation of all of the individual macromolecular components of the cell. We now consider how this separation is achieved, focusing on proteins.

Proteins Can Be Separated by Chromatography

Proteins are most often fractionated by **column chromatography**, in which a mixture of proteins in solution is passed through a column containing a porous gel matrix. Different proteins are retarded to different extents by their interaction with the matrix, and they can be collected separately as they flow out of the bottom of the column (**Figure 8–8**). Depending on the choice of matrix, proteins can be separated according to their charge (*ion-exchange chromatography*), their hydrophobicity (*hydrophobic chromatography*), their size (*gel-filtration*

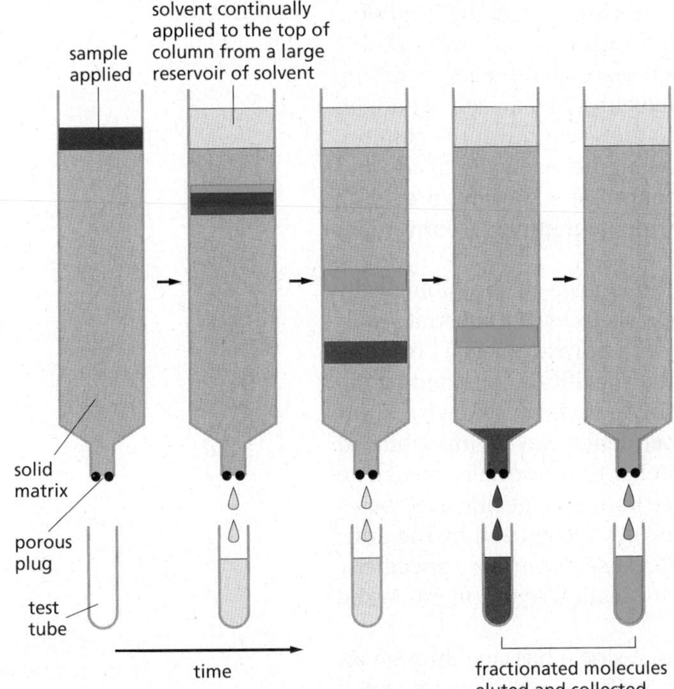

Figure 8–8 **The separation of molecules by column chromatography.** The sample, a solution containing a mixture of different molecules, is applied to the top of a cylindrical glass or plastic column filled with a permeable gel matrix, such as cellulose. A large amount of solvent is then passed slowly through the column and collected in separate tubes as it emerges from the bottom. Because various components of the sample travel at different rates through the column, they are fractionated into different tubes.

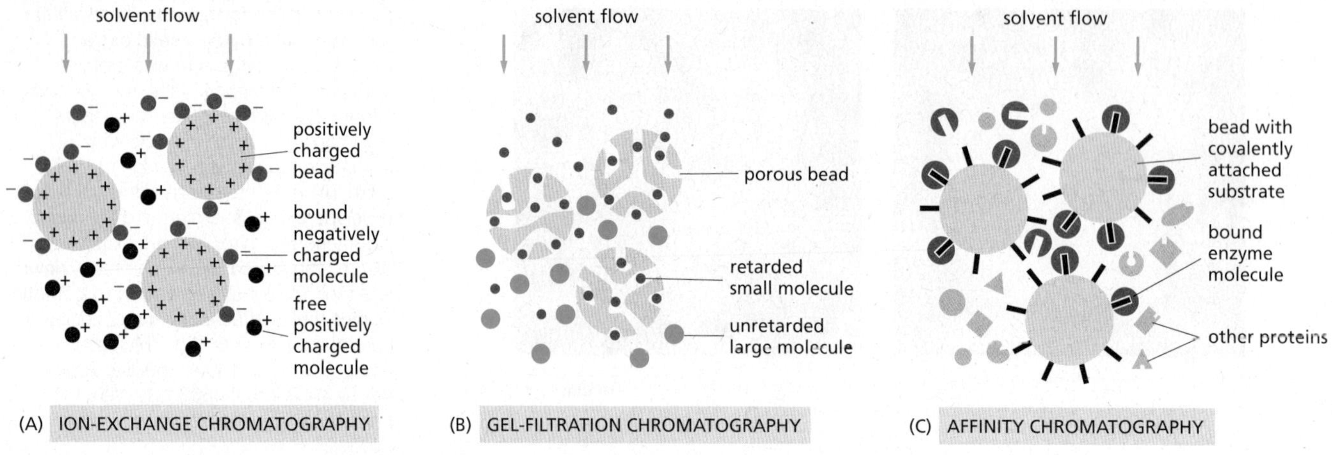

Figure 8–9 Three types of matrices used for chromatography. (A) In ion-exchange chromatography, the insoluble matrix carries ionic charges that retard the movement of molecules of opposite charge. Matrices used for separating proteins include diethylaminoethylcellulose (DEAE-cellulose), which is positively charged, and carboxymethylcellulose (CM-cellulose) and phosphocellulose, which are negatively charged. Analogous matrices based on agarose or other polymers are also frequently used. The strength of the association between the dissolved molecules and the ion-exchange matrix depends on both the ionic strength and the pH of the solution that is passing down the column, which may therefore be varied systematically (as in Figure 8–10) to achieve an effective separation. (B) In gel-filtration chromatography, the small beads that form the matrix are inert but porous. Molecules that are small enough to penetrate into the matrix beads are thereby delayed and travel more slowly through the column than do larger molecules that cannot penetrate. Beads of cross-linked polysaccharide (dextran, agarose, or acrylamide) are available commercially in a wide range of pore sizes, making them suitable for the fractionation of molecules of various masses, from less than 500 daltons to more than 5×10^6 daltons. (C) Affinity chromatography uses an insoluble matrix that is covalently linked to a specific ligand, such as an antibody molecule or an enzyme substrate, that will bind a specific protein. Enzyme molecules that bind to immobilized substrates on such columns can be eluted with a concentrated solution of the free form of the substrate molecule, while molecules that bind to immobilized antibodies can be eluted by dissociating the antibody–antigen complex with concentrated salt solutions or solutions of high or low pH. High degrees of purification can be achieved in a single pass through an affinity column.

chromatography), or their ability to bind to particular small molecules or to other macromolecules (*affinity chromatography*).

Many types of matrices are available. Ion-exchange columns (**Figure 8–9A**) are packed with small beads that carry either a positive or a negative charge, so that proteins are fractionated according to the arrangement of charges on their surface. Hydrophobic columns are packed with beads from which hydrophobic side chains protrude, selectively retarding proteins with exposed hydrophobic regions. Gel-filtration columns (**Figure 8–9B**), which separate proteins according to their size, are packed with tiny porous beads: molecules that are small enough to enter the pores linger inside successive beads as they pass down the column, while larger molecules remain in the solution flowing between the beads and therefore move more rapidly, emerging from the column first. Besides providing a means of separating molecules, gel-filtration chromatography is a convenient way to estimate their size.

Affinity chromatography (**Figure 8–9C**) takes advantage of the biologically important binding interactions that occur on protein surfaces. If a substrate molecule is covalently coupled to an inert matrix such as a polysaccharide bead, the enzyme that operates on that substrate will often be specifically retained by the matrix and can then be eluted (washed out) in nearly pure form. Likewise, short DNA oligonucleotides of a specifically designed sequence can be immobilized in this way and used to purify DNA-binding proteins that normally recognize this sequence of nucleotides in chromosomes. Alternatively, specific antibodies can be coupled to a matrix to purify protein molecules recognized by the antibodies (called *immunoaffinity chromatography*). Because of the great specificity of all such affinity columns, 1000- to 10,000-fold purifications can sometimes be achieved in a single pass.

If one starts with a complex mixture of proteins, a single passage through an ion-exchange or a gel-filtration column does not produce very highly purified

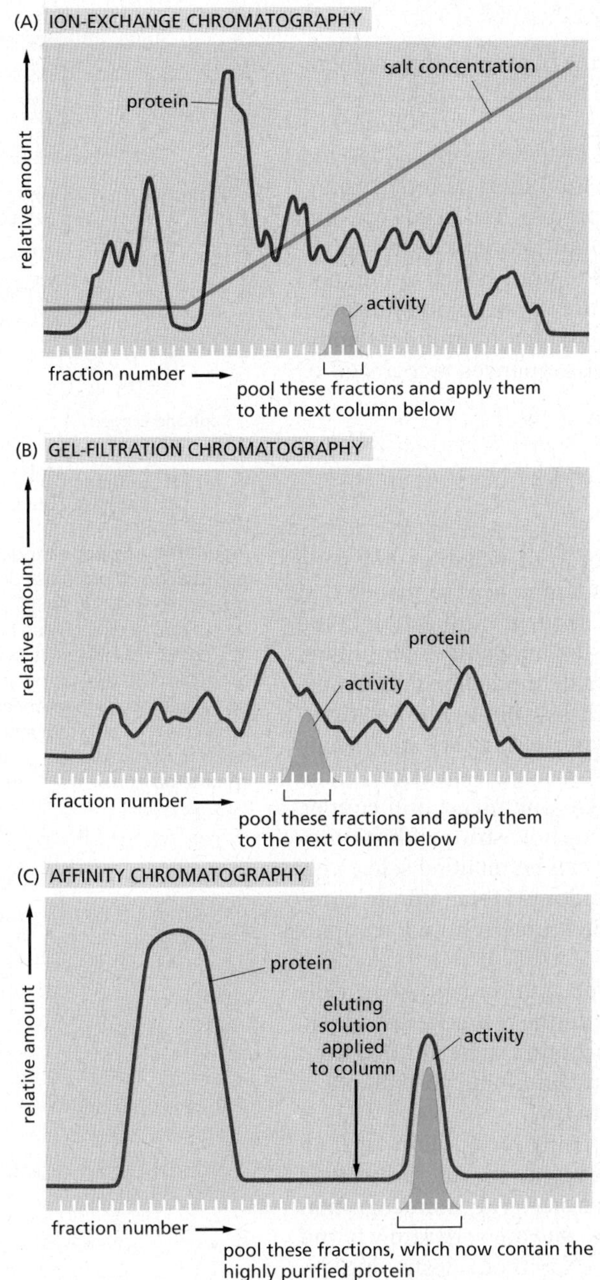

Figure 8–10 **Protein purification by chromatography.** Typical results obtained when three different chromatographic steps are used in succession to purify a protein. In this example, a homogenate of cells was first fractionated by allowing it to percolate through an ion-exchange resin packed into a column (A). The column was washed to remove all unbound contaminants, and the bound proteins were then eluted by pouring a solution containing a gradually increasing concentration of salt onto the top of the column. Proteins with the lowest affinity for the ion-exchange resin passed directly through the column and were collected in the earliest fractions eluted from the bottom of the column. The remaining proteins were eluted in sequence according to their affinity for the resin—those proteins binding most tightly to the resin requiring the highest concentration of salt to remove them. The protein of interest was eluted in several fractions and was detected by its enzymatic activity. The fractions with activity were pooled and then applied to a gel-filtration column (B). The elution position of the still-impure protein was again determined by its enzymatic activity, and the active fractions were pooled and purified to homogeneity on an affinity column (C) that contained an immobilized substrate of the enzyme.

fractions, because these methods individually increase the proportion of a given protein in the mixture no more than twentyfold. Because most individual proteins represent less than 1/1000 of the total cell protein, it is usually necessary to use several different types of columns in succession to attain sufficient purity, with affinity chromatography being the most efficient (Figure 8–10).

Imperfections in the matrices (such as cellulose), which cause an uneven flow of solvent through the column, limit the resolution of conventional column chromatography. Special chromatography resins (usually silica-based) composed of tiny spheres (3–10 μm in diameter) can be packed with a special apparatus to form a uniform column bed. Such **high-performance liquid chromatography (HPLC)** columns attain a high degree of resolution. In HPLC, the solutes are passed through the column at high pressure and equilibrate very rapidly with the interior of the tiny spheres, and so solutes with different affinities for the matrix are efficiently separated from one another even at very fast flow rates. HPLC is therefore the method of choice for separating many proteins and small molecules.

Immunoprecipitation Is a Rapid Affinity Purification Method

Immunoprecipitation is a useful variation on the theme of affinity chromatography. Specific antibodies that recognize the protein to be purified are attached to small agarose beads. Rather than being packed into a column, as in affinity chromatography, a small quantity of the antibody-coated beads is simply added to a protein extract in a test tube and mixed in suspension for a short period of time—thereby allowing the antibodies to bind the desired protein. The beads are then collected by low-speed centrifugation, and the unbound proteins in the supernatant are discarded. This method is commonly used to purify small amounts of enzymes from cell extracts for analysis of enzymatic activity or for identification of associated proteins. As we describe later in this chapter, immunoprecipitation also provides a method to identify DNA or RNA sequences recognized by specific proteins.

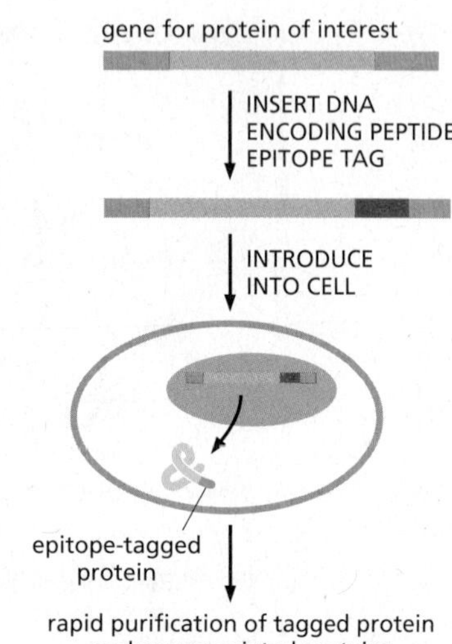

Figure 8–11 **Epitope tagging for the purification of proteins.** Using standard genetic-engineering techniques, a short amino acid sequence can be added to a protein of interest. If the tag is an antigenic determinant, or *epitope*, it can be targeted by an appropriate antibody, which can be used to purify the protein by immunoprecipitation or affinity chromatography.

Genetically Engineered Tags Provide an Easy Way to Purify Proteins

Using the recombinant DNA methods discussed later in this chapter, any gene can be modified to produce its protein with an extra amino acid sequence that provides a specific recognition tag, so as to make subsequent purification of the protein simple and rapid. Often the recognition tag is an antigenic determinant, or *epitope*, which can be recognized by a highly specific monoclonal antibody. The antibody can then be used to purify the protein by affinity chromatography or immunoprecipitation (Figure 8–11). Other types of tags are specifically designed for protein purification. For example, a repeated sequence of the amino acid histidine binds to certain metal ions, including nickel and copper. If genetic-engineering techniques are used to attach a short string of histidines to one end of a protein, the slightly modified protein can be retained selectively on an affinity column containing immobilized nickel ions. Metal affinity chromatography can thereby be used to purify the modified protein from a complex molecular mixture.

In other cases, an entire protein is used as the recognition tag. When cells are engineered to synthesize the small enzyme glutathione *S*-transferase (GST) attached to a protein of interest, the resulting **fusion protein** can be purified from the other contents of the cell with an affinity column containing glutathione, a substrate molecule that binds specifically and tightly to GST.

As a further refinement of purification methods using recognition tags, an amino acid sequence that forms a cleavage site for a highly specific proteolytic enzyme can be engineered between the protein of choice and the recognition tag. Because the amino acid sequences at the cleavage site are very rarely found by chance in proteins, the tag can later be cleaved off without destroying the purified protein.

This type of specific cleavage is used in an especially powerful purification methodology known as *tandem affinity purification tagging* (*TAP-tagging*). Here, one end of a protein is engineered to contain two recognition tags that are separated by a protease cleavage site. The tag on the very end of the construct is chosen to bind irreversibly to an affinity column, allowing the column to be washed extensively to remove all contaminating proteins. Protease cleavage then releases the protein, which is then further purified using the second tag. Because this two-step strategy provides an especially high degree of protein purification with relatively little effort, it is used extensively in cell biology.

Purified Cell-free Systems Are Required for the Precise Dissection of Molecular Functions

Purified cell-free systems provide a means of studying biological processes free from all of the complex side reactions that occur in a living cell. To make this possible, cell homogenates are fractionated with the aim of purifying each of the individual macromolecules that are needed to catalyze a biological process of interest. For example, the experiments to decipher the mechanisms of protein synthesis began with a cell homogenate that could translate RNA molecules to produce proteins. Fractionation of this homogenate, step by step, produced in turn the ribosomes, tRNAs, and various enzymes that together constitute the protein-synthetic machinery. Once individual pure components were available, each could be added or withheld separately to define its exact role in the overall process.

A major goal for cell biologists is the reconstitution of every biological process in a purified cell-free system. Only in this way can we define all of the components needed for the process and control their concentrations, which is required to work out their precise mechanism of action. Although much remains to be done, a great deal of what we know today about the molecular biology of the cell has been discovered by studies in such cell-free systems. They have been used, for example, to decipher the molecular details of DNA replication and DNA transcription, RNA splicing, protein translation, muscle contraction, particle transport along microtubules, and many other processes that occur in cells.

Summary

Populations of cells can be analyzed biochemically by disrupting them and fractionating their contents, allowing functional cell-free systems to be developed. Highly purified cell-free systems are needed for determining the molecular details of complex cell processes, and the development of such systems requires extensive purification of all the proteins and other components involved. The proteins in soluble cell extracts can be purified by column chromatography; depending on the type of column matrix, biologically active proteins can be separated on the basis of their molecular weight, hydrophobicity, charge characteristics, or affinity for other molecules. In a typical purification, the sample is passed through several different columns in turn, with the enriched fractions obtained from one column being applied to the next. Recombinant DNA techniques (described later) allow special recognition tags to be attached to proteins, thereby greatly simplifying their purification.

ANALYZING PROTEINS

Proteins perform most cellular processes: they catalyze metabolic reactions, use nucleotide hydrolysis to do mechanical work, and serve as the major structural elements of the cell. The great variety of protein structures and functions has stimulated the development of a multitude of techniques to study them.

Proteins Can Be Separated by SDS Polyacrylamide-Gel Electrophoresis

Proteins usually possess a net positive or negative charge, depending on the mixture of charged amino acids they contain. An electric field applied to a solution containing a protein molecule causes the protein to migrate at a rate that depends on its net charge and on its size and shape. The most useful application of this property is **sodium dodecyl sulfate polyacrylamide-gel electrophoresis (SDS-PAGE)**. It uses a highly cross-linked gel of polyacrylamide as the inert matrix through which the proteins migrate. The gel is prepared by polymerization of monomers; the pore size of the gel can be adjusted so that it is small enough to retard the migration of the protein molecules of interest. The proteins are dissolved in a solution that includes a powerful negatively charged detergent, sodium dodecyl sulfate, or SDS (Figure 8–12). Because this detergent binds to hydrophobic regions of the protein molecules, causing them to unfold into extended polypeptide chains, the individual protein molecules are released from their associations with other proteins or lipid molecules and rendered freely soluble in the detergent solution. In addition, a reducing agent such as β-mercaptoethanol

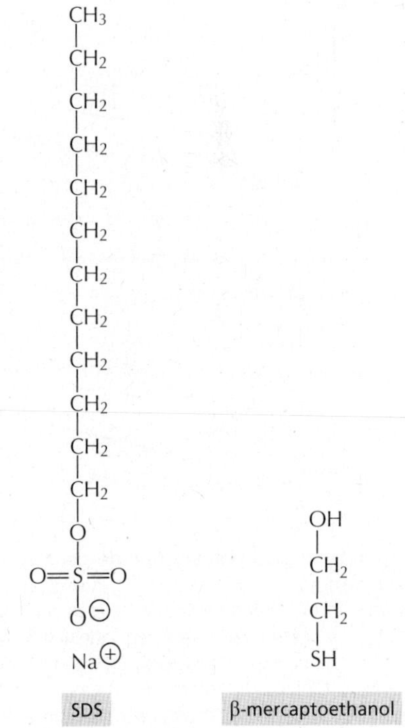

Figure 8–12 **The detergent sodium dodecyl sulfate (SDS) and the reducing agent β-mercaptoethanol.** These two chemicals are used to solubilize proteins for SDS polyacrylamide-gel electrophoresis. The SDS is shown here in its ionized form.

(see Figure 8–12) is usually added to break any disulfide linkages in the proteins, so that all of the constituent polypeptides in multisubunit proteins can be analyzed separately.

What happens when a mixture of SDS-solubilized proteins is run through a slab of polyacrylamide gel? Each protein molecule binds large numbers of the negatively charged detergent molecules, which mask the protein's intrinsic charge and cause it to migrate toward the positive electrode when a voltage is applied. Proteins of the same size tend to move through the gel with similar speeds because (1) their native structure is completely unfolded by the SDS, so that their shapes are the same, and (2) they bind the same amount of SDS and therefore have the same amount of negative charge. Larger proteins, with more charge, are subjected to larger electrical forces but also to a larger drag. In free solution, the two effects would cancel out, but, in the mesh of the polyacrylamide gel, which acts as a molecular sieve, large proteins are retarded much more than small ones. As a result, a complex mixture of proteins is fractionated into a series of discrete protein bands mostly according to their mass (**Figure 8–13**). The major proteins are readily detected by staining the proteins in the gel with a dye such as Coomassie blue. When small amounts of protein are present and more sensitive methods are required, gels can be treated with a silver stain, which will detect as little as 10 ng of protein in a band. For some purposes, specific proteins can also be labeled with a radioactive isotope tag; exposure of the gel to film or a radiation detector results in an *autoradiograph* on which the labeled proteins are visible (see Figure 8–16).

SDS-PAGE is widely used because it can separate all types of proteins, including those that are normally insoluble in water—such as the many proteins in

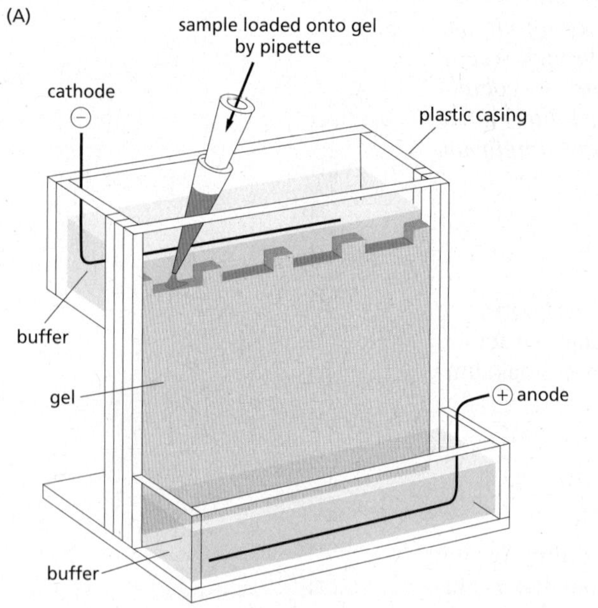

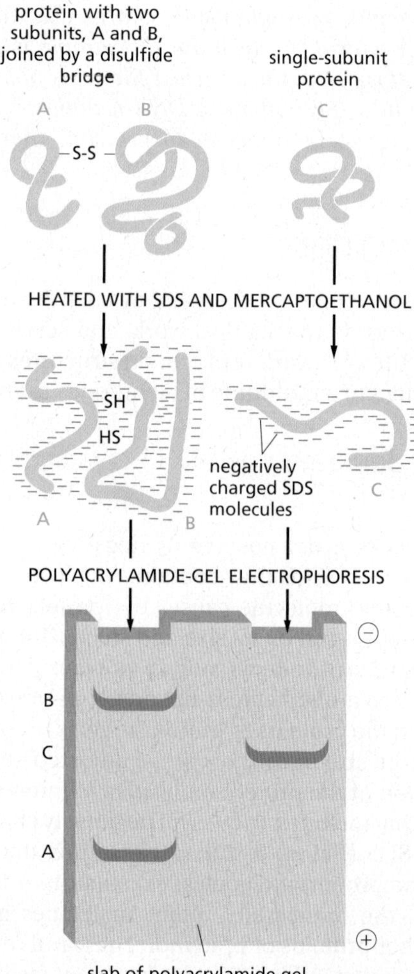

Figure 8–13 SDS polyacrylamide-gel electrophoresis (SDS-PAGE). (A) An electrophoresis apparatus, in which a polyacrylamide gel is sandwiched between two glass plates, with each end of the gel immersed in a buffer connected to an electrode. (B) Individual polypeptide chains form a complex with negatively charged molecules of sodium dodecyl sulfate (SDS) and therefore migrate as a negatively charged SDS–protein complex through a porous gel of polyacrylamide. Because smaller polypeptides move more quickly through the gel, this technique can be used to determine the approximate mass of a polypeptide chain as well as the subunit composition of a protein complex. If the protein contains a large amount of carbohydrate, however, it will move anomalously on the gel, and its apparent mass estimated by SDS-PAGE will be misleading. Other modifications, such as phosphorylation, can also cause small changes in a protein's migration in the gel.

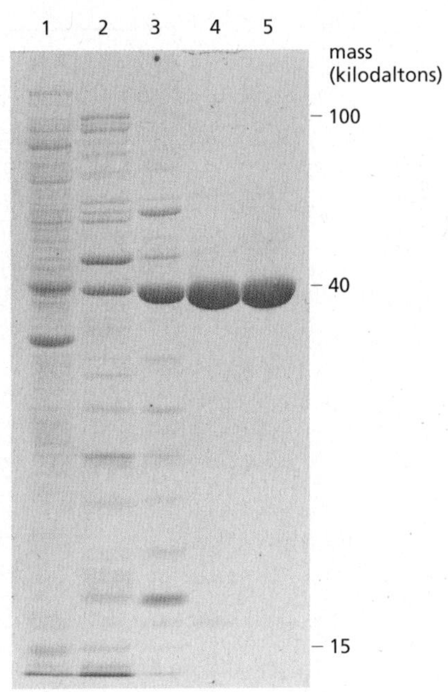

Figure 8–14 **Analysis of protein samples by SDS polyacrylamide-gel electrophoresis.** The photograph shows a Coomassie blue–stained gel that has been used to detect the proteins present at successive stages in the purification of an enzyme. The leftmost lane (lane 1) contains the complex mixture of proteins in the starting cell extract, and each succeeding lane analyzes the proteins obtained after a chromatographic fractionation of the protein sample analyzed in the previous lane (see Figure 8–10). The same total amount of protein (10 μg) was loaded onto the gel at the top of each lane. Individual proteins normally appear as sharp, dye-stained bands; a band broadens, however, when it contains a large amount of protein. (From T. Formosa and B.M. Alberts, *J. Biol. Chem.* 261:6107–6118, 1986.)

membranes. And because the method separates polypeptides by size, it provides information about the mass and subunit composition of proteins. Figure 8–14 presents a photograph of a gel that has been used to analyze each of the successive stages in the purification of a protein.

Two-dimensional Gel Electrophoresis Provides Greater Protein Separation

Because different proteins can have similar sizes, shapes, masses, and overall charges, most separation techniques such as SDS polyacrylamide-gel electrophoresis or ion-exchange chromatography cannot typically separate all the proteins in a cell or even in an organelle. In contrast, **two-dimensional gel electrophoresis**, which combines two different separation procedures, can resolve up to 2000 proteins in the form of a two-dimensional protein map.

In the first step, the proteins are separated by their intrinsic charges. The sample is dissolved in a small volume of a solution containing a nonionic (uncharged) detergent, together with β-mercaptoethanol and the denaturing reagent urea. This solution solubilizes, denatures, and dissociates all the polypeptide chains but leaves their intrinsic charge unchanged. The polypeptide chains are then separated in a pH gradient by a procedure called *isoelectric focusing*, which takes advantage of the variation in the net charge on a protein molecule with the pH of its surrounding solution. Every protein has a characteristic isoelectric point, the pH at which the protein has no net charge and therefore does not migrate in an electric field. In isoelectric focusing, proteins are separated electrophoretically in a narrow tube of polyacrylamide gel in which a gradient of pH is established by a mixture of special buffers. Each protein moves to a position in the gradient that corresponds to its isoelectric point and remains there (Figure 8–15). This is the first dimension of two-dimensional polyacrylamide-gel electrophoresis.

In the second step, the narrow tube of gel containing the separated proteins is soaked in SDS and placed along the top edge of an SDS polyacrylamide-gel slab. Electrophoresis is then carried out as in one-dimensional SDS-PAGE, and each polypeptide chain migrates into the gel to form a discrete spot. The only proteins left unresolved are those that have both identical sizes and identical isoelectric points, a relatively rare situation. Even trace amounts of each polypeptide chain

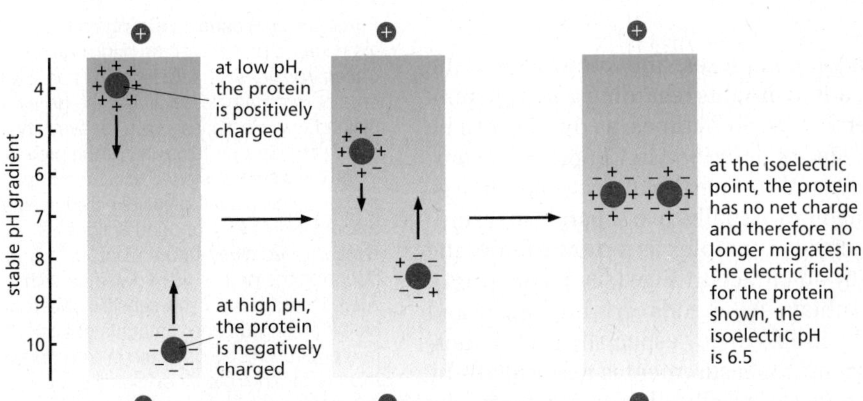

Figure 8–15 **Separation of protein molecules by isoelectric focusing.** At low pH (high H^+ concentration), the carboxylic acid groups of proteins tend to be uncharged (–COOH) and their nitrogen-containing basic groups fully charged (for example, $-NH_3^+$), giving most proteins a net positive charge. At high pH, the carboxylic acid groups are negatively charged ($-COO^-$) and the basic groups tend to be uncharged (for example, $-NH_2$), giving most proteins a net negative charge. At its *isoelectric point*, a protein has no net charge because the positive and negative charges balance. Thus, when a tube containing a fixed pH gradient is subjected to a strong electric field in the appropriate direction, each protein species migrates until it forms a sharp band at its isoelectric point, as shown.

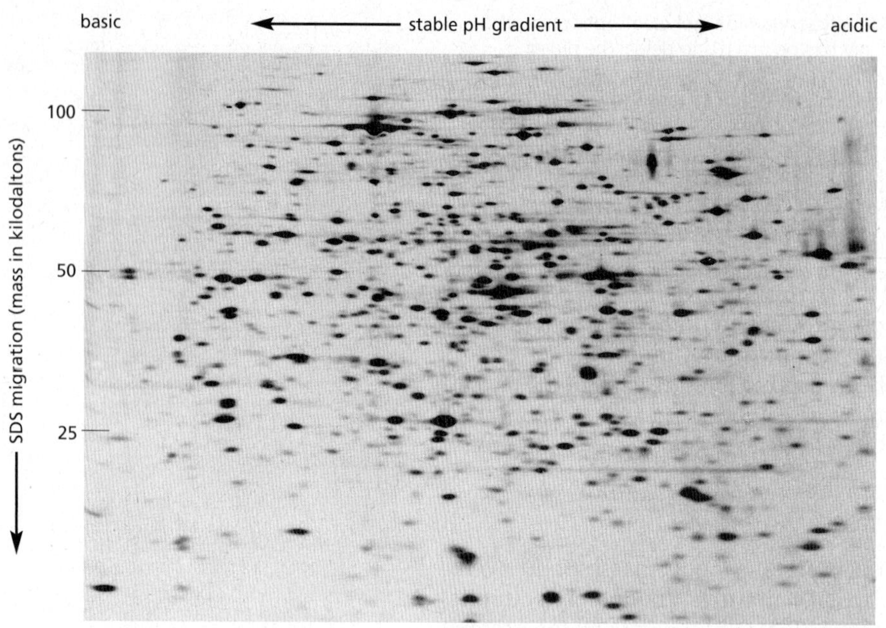

Figure 8–16 Two-dimensional polyacrylamide-gel electrophoresis. All the proteins in an *Escherichia coli* bacterial cell are separated in this gel, in which each spot corresponds to a different polypeptide chain. The proteins were first separated on the basis of their isoelectric points by isoelectric focusing in the horizontal dimension. They were then further fractionated according to their mass by electrophoresis from top to bottom in the presence of SDS. Note that different proteins are present in very different amounts. The bacteria were fed with a mixture of radioisotope-labeled amino acids so that all of their proteins were radioactive and could be detected by autoradiography. (Courtesy of Patrick O'Farrell.)

can be detected on the gel by various staining procedures—or by autoradiography if the protein sample was initially labeled with a radioisotope (**Figure 8–16**). The technique has such great resolving power that it can distinguish between two proteins that differ in only a single charged amino acid or a single negatively charged phosphorylation site.

Specific Proteins Can Be Detected by Blotting with Antibodies

A specific protein can be identified after its fractionation on a polyacrylamide gel by exposing all the proteins on the gel to a specific antibody that has been labeled with a radioactive isotope or a fluorescent dye. This procedure is normally carried out after transferring all of the separated proteins in the gel onto a sheet of nitrocellulose paper or nylon membrane. Placing the membrane against the gel and driving the proteins out of the gel with a strong electric current transfers the protein onto the membrane. The membrane is then soaked in a solution of labeled antibody to reveal the protein of interest. This method of detecting proteins is called **Western blotting**, or **immunoblotting** (**Figure 8–17**). Sensitive Western-blotting methods can detect very small amounts of a specific protein (1 nanogram or less) in a total cell extract or some other heterogeneous protein mixture. The method can be very useful when assessing the amounts of a specific protein in the cell or when measuring changes in those amounts under various conditions.

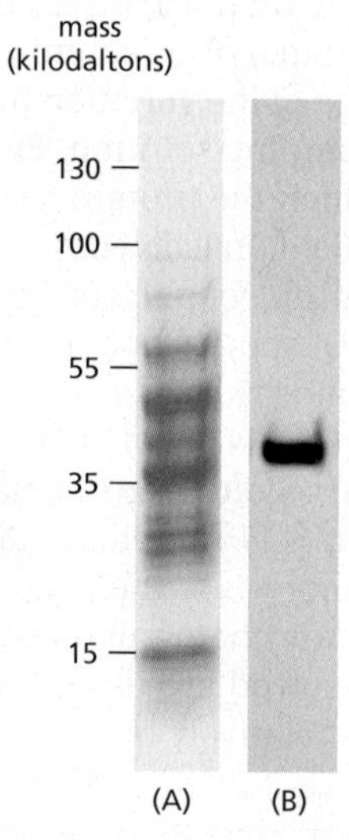

Figure 8–17 Western blotting. In this experiment, proteins of the budding yeast *Saccharomyces cerevisiae* were separated on a polyacrylamide gel. In (A), the gel was stained with Coomassie blue to reveal the most abundant proteins. In (B), the proteins in the gel were transferred to a membrane and exposed to antibodies directed against a specific protein. Unbound antibodies were washed away, and antibodies bound to the protein were detected with a fluorescent label. By this sensitive Western blotting method, small amounts of a single rare protein can be detected in a complex mixture of other proteins. (Courtesy of Jonathan Asfaha.)

Hydrodynamic Measurements Reveal the Size and Shape of a Protein Complex

Most proteins in a cell are subunits of larger complexes, and knowledge of the size and shape of these complexes often leads to insights regarding their function. This information can be obtained in several ways. Sometimes, a complex can be directly visualized using electron microscopy, as described in Chapter 9. A complementary approach relies on the hydrodynamic properties of a complex; that is, its behavior as it moves through a liquid medium. Usually, two separate measurements are made. One measure is the velocity of a complex as it moves under the influence of a centrifugal field produced by an ultracentrifuge (see Figure 8–6A). The sedimentation coefficient (or S value) obtained depends on both the size and the shape of the complex and does not, by itself, convey especially useful information. However, once a second hydrodynamic measurement is performed—by charting the migration of a complex through a gel-filtration chromatography column (see Figure 8–9B)—both the approximate shape of a complex and its mass can be calculated.

The mass of a protein complex can also be determined more directly by using an analytical ultracentrifuge, a complex device that allows protein absorbance measurements to be made on a sample while it is subjected to centrifugal forces. In this approach, the sample is centrifuged until it reaches equilibrium, where the centrifugal force on a protein complex exactly balances its tendency to diffuse away. Because this balancing point is dependent on a complex's mass but not on its particular shape, the mass can be directly calculated.

Mass Spectrometry Provides a Highly Sensitive Method for Identifying Unknown Proteins

A frequent problem in cell biology and biochemistry is the identification of a protein or collection of proteins that has been obtained by one of the purification procedures discussed in the preceding pages. Because the genome sequences of most experimental organisms are known, catalogs of all the proteins produced in those organisms are available. The task of identifying an unknown protein (or collection of unknown proteins) thus reduces to matching some of the amino acid sequences present in the unknown sample with known cataloged genes. This task is now performed almost exclusively by mass spectrometry in conjunction with computer searches of databases.

Charged particles have very precise dynamics when subjected to electric and magnetic fields in a vacuum. *Mass spectrometry* exploits this principle to separate ions according to their mass-to-charge (m/z) ratio. It is an enormously sensitive technique. It requires very little material and is capable of determining the precise mass of intact proteins and of peptides derived from them by enzymatic or chemical cleavage. Masses can be obtained with great accuracy, often with an error of less than one part in a million.

Mass spectrometry is performed using complex instruments with three major components (**Figure 8–18A**). The first is the *ion source*, which transforms tiny amounts of a peptide sample into a gas containing individual charged peptide molecules. These ions are accelerated by an electric field into the second component, the *mass analyzer*, where electric or magnetic fields are used to separate the ions on the basis of their mass-to-charge ratios. Finally, the separated ions collide with a *detector*, which generates a mass spectrum containing a series of peaks representing the masses of the molecules in the sample.

There are many different types of mass spectrometer, varying mainly in the nature of their ion sources and mass analyzers. One of the most common ion sources depends on a technique called *matrix-assisted laser desorption ionization* (*MALDI*). In this approach, the proteins in the sample are first cleaved into short peptides by a protease such as trypsin. These peptides are mixed with an organic acid and then dried onto a metal or ceramic slide. A brief laser burst is directed toward the sample, producing a gaseous puff of ionized peptides, each carrying one or more positive charges. In many cases, the MALDI ion source is coupled to a mass analyzer called a *time-of-flight* (*TOF*) analyzer, which is a long chamber through which the ionized peptides are accelerated by an electric field toward a detector. Their mass and charge determine the time it takes them to reach the detector: large peptides move more slowly, and more highly charged molecules move more quickly. By analyzing those ionized peptides that bear a single charge, the precise masses of peptides present in the original sample can be determined. This information is then used to search genomic databases, in which the masses of all proteins and of all their predicted peptide fragments have been tabulated from the genomic sequences of the organism. An unambiguous match to a particular open reading frame can often be made by knowing the mass of only a few peptides derived from a given protein.

By employing two mass analyzers in tandem (an arrangement known as tandem mass spectrometry, or MS/MS; **Figure 8–18B**), it is possible to directly determine the amino acid sequences of individual peptides in a complex mixture.

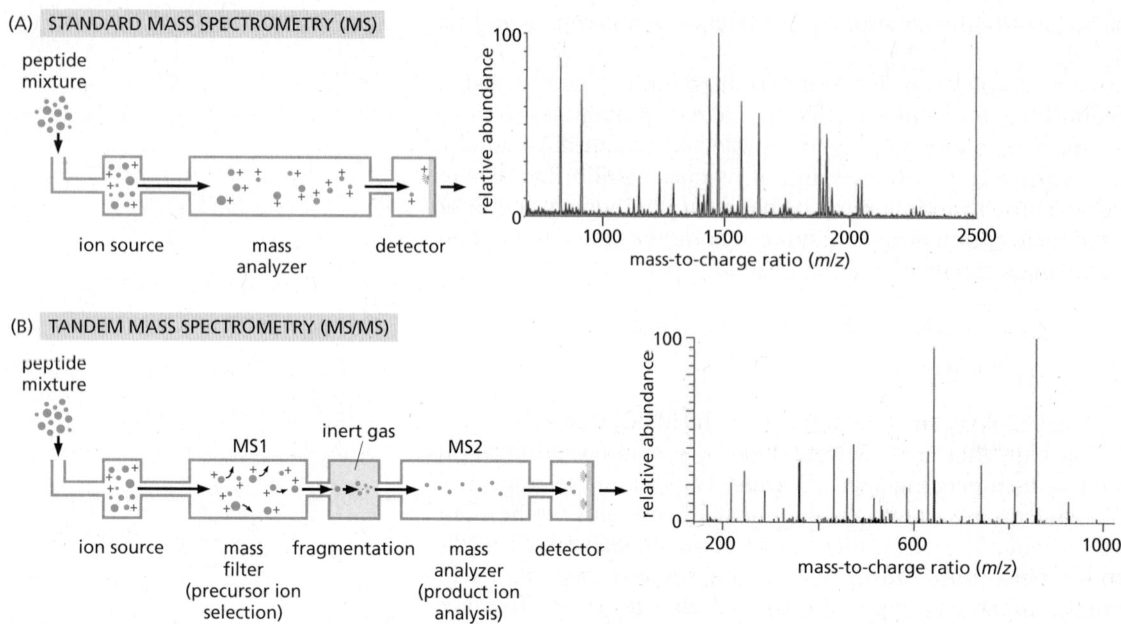

Figure 8–18 **The mass spectrometer.** (A) Mass spectrometers used in biology contain an ion source that generates gaseous peptides or other molecules under conditions that render most molecules positively charged. The two major types of ion source are MALDI and electrospray, as described in the text. Ions are accelerated into a mass analyzer, which separates the ions on the basis of their mass and charge by one of three major methods: (1) *Time-of-flight* (*TOF*) analyzers determine the mass-to-charge ratio of each ion in the mixture from the rate at which it travels from the ion source to the detector. (2) *Quadrupole* mass filters contain a long chamber lined by four electrodes that produce oscillating electric fields that govern the trajectory of ions; by varying the properties of the electric field over a wide range, a spectrum of ions with specific mass-to-charge ratios is allowed to pass through the chamber to the detector, while other ions are discarded. (3) *Ion traps* contain doughnut-shaped electrodes producing a three-dimensional electric field that traps all ions in a circular chamber; the properties of the electric field can be varied over a wide range to eject a spectrum of specific ions to a detector. (B) Tandem mass spectrometry typically involves two mass analyzers separated by a collision chamber containing an inert, high-energy gas. The electric field of the first mass analyzer is adjusted to select a specific peptide ion, called a *precursor ion*, which is then directed to the collision chamber. Collision of the peptide with gas molecules causes random peptide fragmentation, primarily at peptide bonds, resulting in a highly complex mixture of fragments containing one or more amino acids from throughout the original peptide. The second mass analyzer is then used to measure the masses of the fragments (called *product* or *daughter ions*). With computer assistance, the pattern of fragments can be used to deduce the amino acid sequence of the original peptide.

The MALDI-TOF instrument described above is not ideal for this method. Instead, MS/MS typically involves an *electrospray* ion source, which produces a continuous thin stream of peptides that are ionized and accelerated into the first mass analyzer. The mass analyzer is typically either a *quadrupole* or *ion trap*, which employs large electrodes to produce oscillating electric fields inside the chamber containing the ions. These instruments act as *mass filters*: the electric field is adjusted over a broad range to select a single peptide ion and discard all the others in the peptide mixture. In tandem mass spectrometry, this single ion is then exposed to an inert, high-energy gas, which collides with the peptide, resulting in fragmentation, primarily at peptide bonds. The second mass analyzer then determines the masses of the peptide fragments, which can be used by computational methods to determine the amino acid sequence of the original peptide and thereby identify the protein from which it came.

Tandem mass spectrometry is also useful for detecting and precisely mapping post-translational modifications of proteins, such as phosphorylation or acetylation. Because these modifications impart a characteristic mass increase to an amino acid, they are easily detected during the analysis of peptide fragments in the second mass analyzer, and the precise site of the modification can often be deduced from the spectrum of peptide fragments.

A powerful, "two-dimensional" mass spectrometry technique can be used to determine all of the proteins present in an organelle or another complex mixture of proteins. First, the mixture of proteins present is digested with trypsin to produce short peptides. Next, these peptides are separated by an automated high-performance form of liquid chromatography (LC). Every peptide fraction from the chromatographic column is injected directly into an electrospray ion source on a tandem mass spectrometer, providing the amino acid sequence and post-translational modifications for every peptide in the mixture. This method, often called LC-MS/MS, is used to identify hundreds or thousands of proteins in complex protein mixtures from specific organelles or from whole cells. It can also be used to map all of the phosphorylation sites in the cell, or all of the proteins targeted by other post-translational modifications such as acetylation or ubiquitylation.

Sets of Interacting Proteins Can Be Identified by Biochemical Methods

Because most proteins in the cell function as part of complexes with other proteins, an important way to begin to characterize the biological role of an unknown protein is to identify all of the other proteins to which it specifically binds.

A key method for identifying proteins that bind to one another tightly is *co-immunoprecipitation.* A target protein is immunoprecipitated from a cell lysate using specific antibodies coupled to beads, as described earlier. If the target protein is associated tightly enough with another protein when it is captured by the antibody, the partner precipitates as well and can be identified by mass spectrometry. This method is useful for identifying proteins that are part of a complex inside cells, including those that interact only transiently; for example, when extracellular signal molecules stimulate cells (discussed in Chapter 15).

Optical Methods Can Monitor Protein Interactions

Once two proteins—or a protein and a small molecule—are known to associate, it becomes important to characterize their interaction in more detail. Proteins can associate with each other more or less permanently (like the subunits of RNA polymerase or the proteasome), or engage in transient encounters that may last only a few milliseconds (like a protein kinase and its substrate). To understand how a protein functions inside a cell, we need to determine how tightly it binds to other proteins and how covalent modifications, small molecules, or other proteins influence these interactions.

As we discussed in Chapter 3 (see Figure 3–42), the extent to which two proteins interact is determined by the rates at which they associate and dissociate. These rates depend, respectively, on the association rate constant (k_{on}) and dissociation rate constant (k_{off}). The kinetic rate constant k_{off} is a particularly useful number because it provides valuable information about how long two proteins remain bound to one another. The ratio of the two kinetic constants (k_{on}/k_{off}) yields another very useful number called the equilibrium constant (K, also known as K_{eq} or K_a), the inverse of which is the more commonly used dissociation constant, K_d. The equilibrium constant is useful as a general indicator of the affinity of the interaction, and it can be used to estimate the amount of bound complex at different concentrations of the two protein partners—thereby providing insights into the importance of the interaction at the protein concentrations found inside the cell.

A wide range of methods can be used to determine binding constants for a two-protein complex. In a simple *equilibrium binding experiment,* two proteins are mixed at a range of concentrations, allowed to reach equilibrium, and the amount of bound complex is measured; half of the protein complex will be bound at a concentration that is equal to K_d. Equilibrium experiments often involve the use of radioactive or fluorescent tags on one of the protein partners, coupled with biochemical or optical methods for measuring the amount of bound protein. In a more complex *kinetic binding experiment,* the kinetic rate constants are determined using rapid methods that allow real-time measurement of the formation of a bound complex over time (to determine k_{on}) or the dissociation of a bound complex over time (to determine k_{off}).

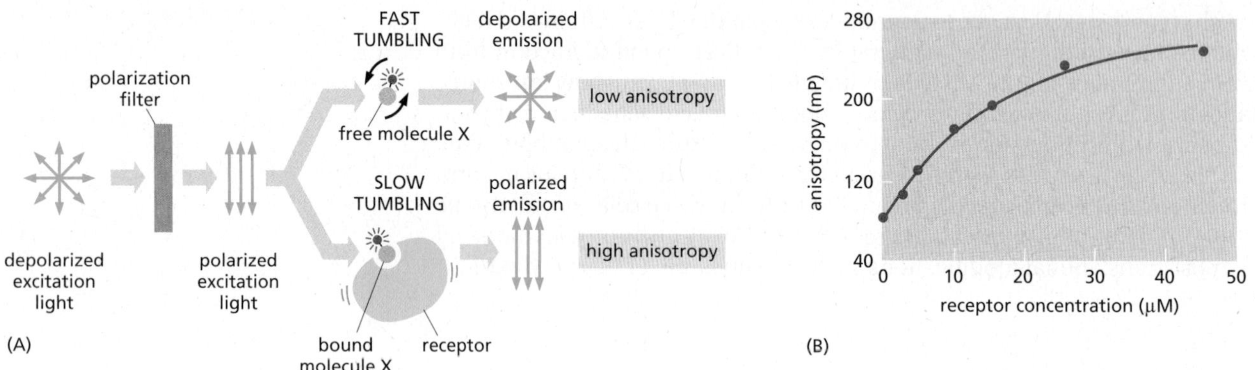

Figure 8–19 **Measurement of binding with fluorescence anisotropy.** This method depends on a fluorescently tagged protein that is illuminated with polarized light at the appropriate wavelength for excitation; a fluorimeter is used to measure the intensity and polarization of the emitted light. If the fluorescent protein is fixed in position and therefore does not rotate during the brief period between excitation and emission, then the emitted light will be polarized at the same angle as the excitation light. This directional effect is called *fluorescence anisotropy*. Protein molecules in solution rotate or tumble rapidly, however, so that there is a decrease in the amount of anisotropic fluorescence. Larger molecules tumble at a slower rate and therefore have higher fluorescence anisotropy. (A) To measure the binding between a small molecule and a large receptor protein, the smaller molecule is labeled with a fluorophore. In the absence of its binding partner, the molecule tumbles rapidly, resulting in low fluorescence anisotropy *(top)*. When the small molecule binds to its larger partner, however, it tumbles less rapidly, resulting in an increase in fluorescence anisotropy *(bottom)*. (B) In the equilibrium binding experiment shown here, a small, fluorescent peptide ligand was present at a low concentration, and the amount of fluorescence anisotropy (in millipolarization units; mP) was measured after incubation with various concentrations of a larger protein receptor for the ligand. From the hyperbolic curve that fits the data, it can be seen that 50% binding occurred at about 10 μM, which is equal to the dissociation constant K_d for the binding interaction.

Optical techniques provide particularly rapid, convenient, and accurate binding measurements, and in some cases the proteins do not even need to be labeled. Certain amino acids (tryptophan, for example) exhibit weak fluorescence that can be detected with sensitive fluorimeters. In many cases, the fluorescence intensity, or the emission spectrum of fluorescent amino acids located in a protein–protein interface, will change when two proteins associate. When this change can be detected by fluorimetry, it provides a simple and sensitive measure of protein binding that is useful in both equilibrium and kinetic binding experiments. A related but more widely useful optical binding technique is based on *fluorescence anisotropy*, a change in the polarized light that is emitted by a fluorescently tagged protein in the bound and free states (**Figure 8–19**).

Another optical method for probing protein interactions uses *green fluorescent protein* (discussed in detail later) and its derivatives of different colors. In this application, two proteins of interest are each labeled with a different fluorescent protein, such that the emission spectrum of one fluorescent protein overlaps the absorption spectrum of the second. If the two proteins come very close to each other (within about 1–5 nm), the energy of the absorbed light is transferred from one fluorescent protein to the other. The energy transfer, called *fluorescence resonance energy transfer* (*FRET*), is determined by illuminating the first fluorescent protein and measuring emission from the second (see Figure 9–19). When combined with fluorescence microscopy, this method can be used to characterize protein–protein interactions at specific locations inside living cells (discussed in Chapter 9).

Protein Structure Can Be Determined Using X-ray Diffraction

A deep understanding of protein function in the cell requires knowledge of its three-dimensional structure at atomic resolution, revealing the precise position of every amino acid in the protein. Structural analysis provides powerful insights into the mechanisms underlying a protein's enzymatic activity and its interactions with other proteins or small molecules. Numerous methods are used to unravel protein structure. In this chapter, we discuss well-established methods that depend on x-ray crystallography and nuclear magnetic resonance spectroscopy. Chapter 9 describes recently developed methods by which electron microscopy is being used to determine the high-resolution structures of larger proteins and protein complexes.

For many decades, the primary technique for protein structural analysis has been **x-ray crystallography**. X-rays, like light, are a form of electromagnetic radiation, but they have a much shorter wavelength, typically around 0.1 nm (the diameter of a hydrogen atom). If a narrow beam of parallel x-rays is directed at a sample of a pure protein, most of the x-rays pass straight through it. A small fraction, however, is scattered by the atoms in the sample. If the sample is a well-ordered crystal, the scattered waves reinforce one another at certain points and appear as diffraction spots when recorded by a suitable detector (**Figure 8–20**).

The slowest step in this technique is likely to be the generation of suitable protein crystals. This step requires large amounts of very pure protein and often involves

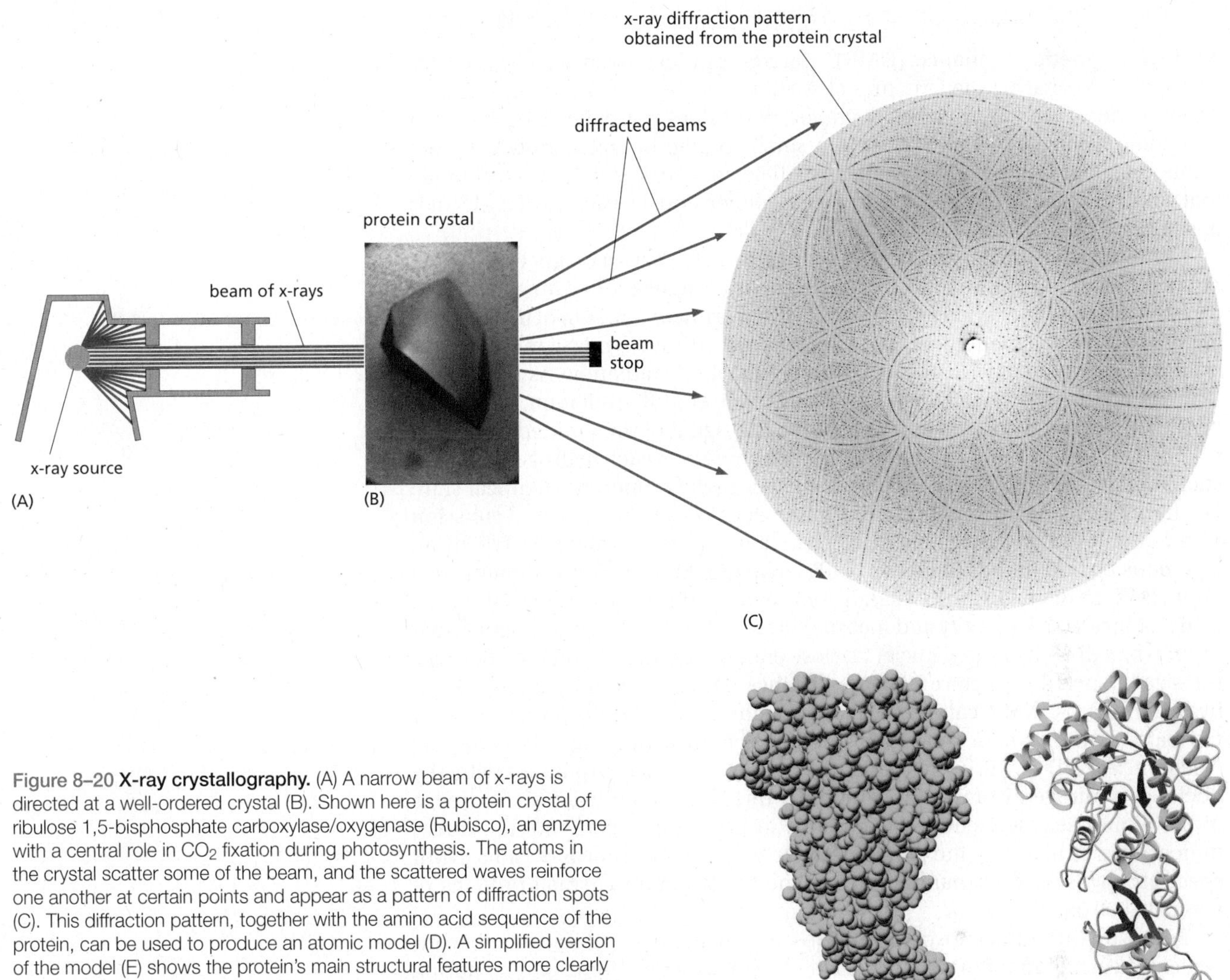

Figure 8–20 **X-ray crystallography.** (A) A narrow beam of x-rays is directed at a well-ordered crystal (B). Shown here is a protein crystal of ribulose 1,5-bisphosphate carboxylase/oxygenase (Rubisco), an enzyme with a central role in CO_2 fixation during photosynthesis. The atoms in the crystal scatter some of the beam, and the scattered waves reinforce one another at certain points and appear as a pattern of diffraction spots (C). This diffraction pattern, together with the amino acid sequence of the protein, can be used to produce an atomic model (D). A simplified version of the model (E) shows the protein's main structural features more clearly (α helices, *green*; β strands, *red*). The components pictured in A–E are not shown to scale. (B, courtesy of C. Branden; C, courtesy of J. Hajdu and I. Andersson; D and E, PDB code: 1RBL.)

years of trial and error to discover the proper crystallization conditions; the pace has greatly accelerated with the use of recombinant DNA techniques to produce pure proteins and robotic techniques to test large numbers of crystallization conditions.

The position and intensity of each spot in the x-ray diffraction pattern contain information about the locations of the atoms in the crystal that gave rise to it. Computer-assisted computational methods process the diffraction pattern to generate a three-dimensional electron-density map. Interpreting this map—translating its contours into a three-dimensional structure—can be a laborious procedure. Largely by trial and error, the sequence and the electron-density map are correlated by computer to give the best possible fit. The reliability of the final atomic model depends on the resolution of the original crystallographic data: 0.5-nm resolution might produce a low-resolution map of the polypeptide backbone, whereas a resolution of 0.15 nm allows all of the non-hydrogen atoms in the molecule to be reliably positioned.

A complete atomic model is often too complex to appreciate directly, but simplified versions that show a protein's essential structural features can be readily derived from it. The three-dimensional structures of tens of thousands of different proteins have been determined—enough to allow the grouping of common structures into families (**Movie 8.1**). These structures or protein folds often seem to be more conserved in evolution than are the amino acid sequences that form them (see Figure 3–13).

NMR Can Be Used to Determine Protein Structure in Solution

Nuclear magnetic resonance (NMR) spectroscopy has been widely used for many years to analyze the structure of small molecules, small proteins, or protein domains. Unlike x-ray crystallography, NMR does not depend on having a crystalline sample. It simply requires a small volume of concentrated protein solution that is placed in a strong magnetic field; indeed, it is the main technique that yields detailed evidence about the three-dimensional structure of molecules in solution.

Certain atomic nuclei, particularly hydrogen nuclei, have a magnetic moment or spin; that is, they have an intrinsic magnetization, like a bar magnet. When exposed to a strong magnetic field in an NMR experiment, the spin of these nuclei aligns with the magnetic field, but it can be forced into a misaligned, excited state by radiofrequency (RF) pulses of electromagnetic radiation. As the excited hydrogen nuclei return to their aligned state, they emit RF radiation, which can be measured and displayed as a spectrum. The signal of the emitted radiation, called the *chemical shift*, depends on the environment of each hydrogen nucleus, such that every nucleus in a protein displays a slightly different chemical shift. Furthermore, if one nucleus is excited, it influences the absorption and emission of radiation by other nuclei that lie close to it. It is consequently possible, by an ingenious elaboration of the basic NMR technique known as two-dimensional (2D) NMR, to distinguish the signals from hydrogen nuclei in different amino acid residues and to identify and measure the small changes in these signals that occur when these hydrogen nuclei lie close enough together to interact. Because the size of such a change reveals the distance between the interacting pair of hydrogen atoms, NMR can provide information about the distances between the parts of the protein molecule. Additional information can be gained by *heteronuclear* 2D NMR, which analyzes hydrogen nuclei in parallel with the nuclei of a nitrogen isotope. By combining this information with a knowledge of the amino acid sequence, it is possible in principle to compute the three-dimensional structure of the protein (**Figure 8–21**). For technical reasons, NMR spectroscopy is used primarily to determine the structure of small proteins of about 30,000 daltons or less.

Because NMR studies are performed in solution, this method also offers a convenient means of monitoring changes in protein structure; for example, during protein folding or when the protein binds to another molecule. NMR is also used widely to investigate molecules other than proteins and is valuable, for example, as a method to determine the three-dimensional structures of RNA molecules and the complex carbohydrate side chains of glycoproteins.

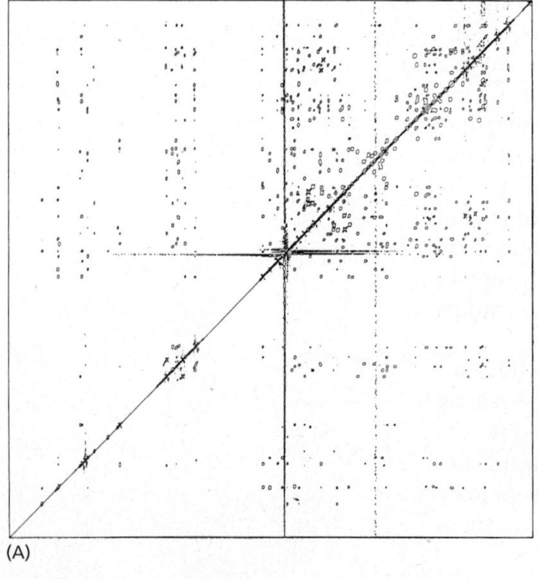

(A)

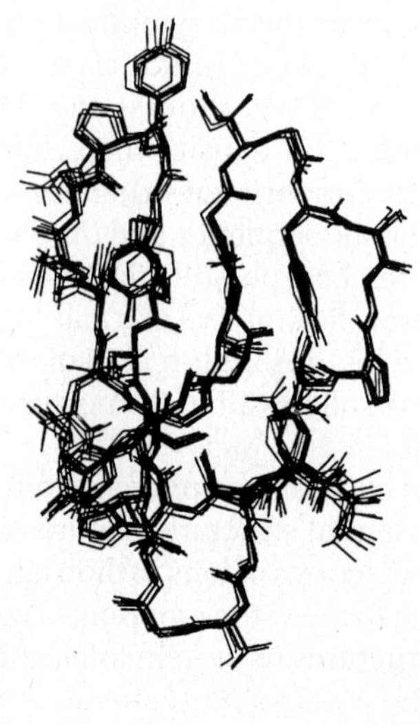

(B)

Figure 8–21 NMR spectroscopy. (A) An example of the data from an NMR machine. This NMR spectrum, displaying chemical shifts of hydrogen nuclei in two dimensions, is derived from the C-terminal domain of the enzyme cellulase. The spots away from the diagonal represent interactions between hydrogen atoms that are near neighbors in the protein; their positions reflect the distance that separates them. Complex computing methods, in conjunction with the known amino acid sequence, enable possible compatible structures to be derived. (B) Ten structures of the enzyme, all of which satisfy the distance constraints equally well, are shown superimposed on one another, giving a good indication of the probable three-dimensional structure. (Courtesy of P. Kraulis.)

A major problem in structural biology is the analysis of large protein complexes, which are difficult to crystallize and too large for NMR analysis. Single-particle analysis by cryo-electron microscopy provides a relatively straightforward approach to the analysis of large macromolecular assemblies, as we describe in Chapter 9.

Protein Sequence and Structure Provide Clues About Protein Function

Having discussed methods for purifying and analyzing proteins, we now turn to a common situation in cell and molecular biology: an investigator has identified a gene important for a biological process but has no direct knowledge of the biochemical properties of its protein product.

Thanks to the proliferation of protein and nucleic acid sequences that are cataloged in genome databases, the function of a gene—and its encoded protein—can often be predicted by simply comparing its sequence with those of previously characterized genes. Because amino acid sequence determines protein structure, and structure dictates biochemical function, proteins that share a similar amino acid sequence usually have the same structure and usually perform similar biochemical functions, even when they are found in distantly related organisms. Thus, the study of a newly discovered protein usually begins with a search for previously characterized proteins that are similar in their amino acid sequences.

Searching a collection of known sequences for similar genes or proteins simply involves selecting a database and entering the desired sequence. A sequence-alignment program—the most popular is *BLAST (Basic Local Alignment Search Tool)*—scans the database for similar sequences by sliding the submitted sequence along the archived sequences until a cluster of residues falls into full or partial alignment (**Figure 8–22**).

As was explained in Chapter 3, many proteins that adopt the same conformation and have related functions are too distantly related to be identified from a comparison of their amino acid sequences alone (see Figure 3–13). Thus, an ability to reliably predict the three-dimensional structure of a protein from its amino acid sequence would improve our ability to infer protein function from the sequence information in genomic databases. In recent years, major progress has been made in predicting the precise structure of a protein. These predictions are based, in part, on our knowledge of the thousands of protein structures that have already been determined by x-ray crystallography and NMR spectroscopy and, in part, on computations using our knowledge of the physical forces acting on the atoms. However, it remains a substantial and important challenge to predict

```
 Score =  399 bits (1025), Expect = e-111
 Identities = 198/290 (68%), Positives = 241/290 (82%), Gaps = 1/290

Query: 57   MENFQKVEKIGEGTYGVVYKARNKLTGEVVALKKIRLDTETEGVPSTAIREISLLKELNH 116
            ME ++KVEKIGEGTYGVVYKA +K T E +ALKKIRL+ E EGVPSTAIREISLLKE+NH
Sbjct: 1    MEQYEKVEKIGEGTYGVVYKALDKATNETIALKKIRLEQEDEGVPSTAIREISLLKEMNH 60

Query: 117  PNIVKLLDVIHTENKLYLVFEFLHQDLKKFMDASALTGIPLPLIKSYLFQLLQGLAFCHS 176
             NIV+L DV+H+E ++YLVFE+L  DLKKFMD+          LIKSYL+Q+L G+A+CHS
Sbjct: 61   GNIVRLHDVVHSEKRIYLVFEYLDLDLKKFMDSCPEFAKNPTLIKSYLYQILHGVAYCHS 120
                              ▼
Query: 177  HRVLHRDLKPQNLLINTE-GAIKLADFGLARAFGVPVRTYTHEVVTLWYRAPEILLGCKY 235
            HRVLHRDLKPQNLLI+    A+KLADFGLARAFG+PVRT+THEVVTLWYRAPEILLG +
Sbjct: 121  HRVLHRDLKPQNLLIDRRTNALKLADFGLARAFGIPVRTFTHEVVTLWYRAPEILLGARQ 180

Query: 236  YSTAVDIWSLGCIFAEMVTRRALFPGDSEIDQLFRIFRTLGTPDEVVWPGVTSMPDYKPS 295
            YST VD+WS+GCIFAEMV ++ LFPGDSEID+LF+IFR LGTP+E  WPGV+ +PD+K +
Sbjct: 181  YSTPVDVWSVGCIFAEMVNQKPLFPGDSEIDELFKIFRILGTPNEQSWPGVSCLPDFKTA 240

Query: 296  FPKWARQDFSKVVPPLDEDGRSLLSQMLHYDPNKRISAKAALAHPFFQDV 345
            FP+W  QD + VVP LD  G  LLS+ML Y+P+KRI+A+ AL H +F+D+
Sbjct: 241  FPRWQAQDLATVVPNLDPAGLDLLSKMLRYEPSKRITARQALEHEYFKDL 290
```

Figure 8–22 Results of a BLAST search. Sequence databases can be searched to find similar amino acid or nucleic acid sequences. Here, a search for proteins similar to the human cell-cycle regulatory protein Cdk1 *(Query)* locates maize Cdk1 *(Sbjct)*, which is 68% identical to human Cdk1 in its amino acid sequence. The alignment begins at residue 57 of the Query protein, suggesting that the human protein has an N-terminal region that is absent from the maize protein. The *green blocks* indicate differences in sequence, and the *yellow bars* summarize the similarities: when the two amino acid sequences are identical, the residue is shown; similar amino acid substitutions are indicated by a plus sign (+). Only one small gap has been introduced—indicated by the *red arrowhead* at position 194 in the Query sequence—to align the two sequences maximally. The alignment score *(Score)*, which is expressed in two different types of units, takes into account penalties for substitutions and gaps; the higher the alignment score, the better the match. The significance of the alignment is reflected in the *Expectation* (E) value, which specifies how often a match this good would be expected to occur by chance. The lower the E value, the more significant the match; the extremely low value here (e^{-111}) indicates certain significance. E values much higher than 0.1 are unlikely to reflect true relatedness. For example, an E value of 0.1 means there is a 1 in 10 likelihood that such a match would arise solely by chance.

the structures of proteins that are large or have multiple domains or to predict structures at the very high levels of resolution needed to assist in computer-based drug discovery.

While finding related sequences and structures for a new protein will provide many clues about its function, it is usually necessary to test these insights through direct experimentation. However, the clues generated from sequence comparisons typically point the investigator in the correct experimental direction, and their use has therefore become one of the most important strategies in modern cell biology.

Summary

Many methods exist for identifying proteins and analyzing their biochemical properties, structures, and interactions with other proteins. The most powerful and commonly used methods include protein separation by polyacrylamide gel electrophoresis, protein analysis by mass spectrometry, and high-resolution structural determination. Because proteins with similar structures often have similar functions, the biochemical activity of a protein can often be predicted by searching databases for previously characterized proteins that are similar in their amino acid sequences.

ANALYZING AND MANIPULATING DNA

Until the early 1970s, DNA was the most difficult biological molecule for the biochemist to analyze. Enormously long and chemically monotonous, the string of nucleotides that forms the genetic material of an organism could be determined only indirectly, from protein sequence or by genetic analysis. Today, the situation has changed entirely. From being the most difficult macromolecule of the cell to analyze, DNA has become the easiest. It is now possible to determine the entire nucleotide sequence of a bacterial or fungal genome in a matter of hours and the sequence of an individual human genome in less than a day. Once the nucleotide sequence of a genome is known, any individual gene can be easily isolated, and large quantities of the gene product (be it RNA or protein) can be made either by introducing the gene into bacteria or animal cells and coaxing these cells to overexpress the foreign gene or by synthesizing the gene product *in vitro*. In this way, proteins and RNA molecules that might be present in only tiny amounts in living cells can be produced in large quantities for biochemical and structural analyses. And this approach can also be used to produce large quantities of human proteins (such as insulin, or interferon, or blood-clotting proteins) for use as human pharmaceuticals. As we will see later in this chapter, it is also possible for scientists to alter an isolated gene and transfer it back into the germ line of an animal or plant, so as to become a functional and heritable part of the organism's genome. In this way, the biological roles of any gene can be assessed by observing—in the whole organism—the results of modifying it.

The ability to manipulate DNA with precision in a test tube or an organism, known as **recombinant DNA technology**, has had a dramatic impact on all aspects of cell and molecular biology. We now describe the key features of these techniques.

Restriction Nucleases Cut Large DNA Molecules into Specific Fragments

Unlike a protein, a gene does not exist as a discrete entity in cells, but rather as a small region of a much longer DNA molecule. Although the DNA molecules in a cell can be randomly broken into small pieces by mechanical force, a fragment containing a single gene in a mammalian genome would still be only one among a hundred thousand or more DNA fragments, indistinguishable in their average size. How could such a gene be separated from all the others? Because all DNA molecules consist of an approximately equal mixture of the same four nucleotides, they cannot be readily separated, as proteins can, on the basis of their different charges and biochemical properties. The solution to this problem began to emerge with the discovery of **restriction nucleases**. These enzymes, which are purified from bacteria, cut the DNA double helix at specific sites defined by the local nucleotide sequence, thereby cleaving a long, double-strand DNA molecule into fragments of strictly defined sizes.

Like many of the tools of recombinant DNA technology, restriction nucleases were discovered by researchers trying to understand an intriguing biological phenomenon. It had been observed that certain bacteria always degraded "foreign" DNA that was introduced into them experimentally. A search for the mechanism responsible revealed a then unanticipated class of bacterial nucleases that cleave DNA at specific nucleotide sequences. The bacterium's own DNA is protected from cleavage by methylation of these same sequences, thereby protecting the bacterium from being overrun by foreign DNA. Because these enzymes restrict the transfer of DNA into bacteria, they were called restriction nucleases. The pursuit of this seemingly arcane biological puzzle set off the development of technologies that have forever changed the way cell and molecular biologists study living things.

Different bacterial species produce different restriction nucleases, each cutting at a different, specific nucleotide sequence (**Figure 8–23**). Because these target sequences are short—generally four to eight nucleotide pairs—many sites of cleavage will occur, purely by chance, in any long DNA molecule. The reason restriction nucleases are so useful in the laboratory is that each enzyme will always cut a particular DNA molecule at the same sites. Thus for a given sample of DNA (which contains many identical molecules), a particular restriction nuclease will reliably generate the same set of DNA fragments.

The size of the resulting fragments depends on the length of the target sequences of the restriction nucleases. As shown in Figure 8–23, the enzyme HaeIII cuts at a sequence of four nucleotide pairs; a sequence this long would be expected to occur purely by chance approximately once every 256 nucleotide pairs (1 in 4^4). In comparison, a restriction nuclease with a target sequence that is eight nucleotides long would be expected to cleave DNA on average once every 65,536 nucleotide pairs (1 in 4^8). This difference in sequence selectivity makes it possible to cleave a long DNA molecule into the fragment sizes that are most suitable for a given application.

Gel Electrophoresis Separates DNA Molecules of Different Sizes

The same types of gel-electrophoresis methods that have proved so useful in the analysis of proteins (see Figure 8–13) can be applied to DNA molecules. The procedure is actually simpler than for proteins: because each nucleotide in a nucleic acid molecule carries a single negative charge (on the phosphate group), there is no need to add the negatively charged detergent SDS that is required to make protein molecules move uniformly toward the positive electrode. Larger DNA

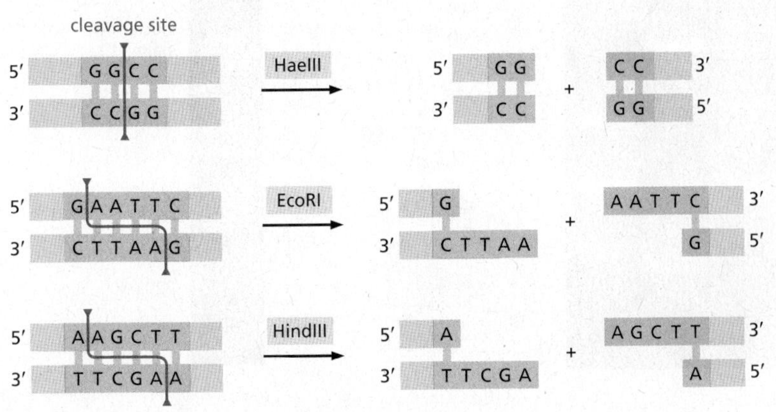

Figure 8–23 Restriction nucleases cleave DNA at specific nucleotide sequences. Like the sequence-specific DNA-binding proteins we encountered in Chapter 7 (see Figure 7–9), restriction enzymes often work as dimers, and the DNA sequence that each restriction enzyme recognizes and cleaves is often symmetrical around a central point. Here, both strands of the DNA double helix are cut at specific points within the target sequence *(orange)*. Some enzymes, such as HaeIII, cut straight across the double helix and leave two blunt-ended DNA molecules; with others, such as EcoRI and HindIII, the cuts on each strand are staggered. These staggered cuts generate "sticky ends"—short, single-strand overhangs that help the cut DNA molecules join back together through complementary base-pairing. This rejoining of DNA molecules becomes important for DNA cloning, as we discuss later. Restriction nucleases are usually obtained from bacteria, and their names reflect their origins; for example, the enzyme EcoRI comes from *Escherichia coli*. Hundreds of different restriction enzymes are commercially available.

fragments will migrate more slowly because their progress is impeded to a greater extent by the gel matrix. In less than an hour, the DNA fragments become spread out across the gel according to size, forming a ladder of discrete bands, each composed of a collection of DNA molecules of identical length (**Figure 8–24A and B**). To separate DNA molecules longer than 500 nucleotide pairs, the gel is made of a diluted solution of agarose (a polysaccharide isolated from seaweed). For DNA fragments less than 500 nucleotides long, specially designed polyacrylamide gels allow the separation of molecules that differ in length by as little as a single nucleotide (**Figure 8–24C**).

A variation of agarose-gel electrophoresis, called *pulsed-field gel electrophoresis*, makes it possible to separate extremely long DNA molecules, even those found in whole chromosomes. Ordinary gel electrophoresis fails to separate very large DNA molecules because the steady electric field stretches them out so that they travel end-first through the gel in snake-like configurations at a rate that is independent of their length. In pulsed-field gel electrophoresis, by contrast, the direction of the electric field changes periodically, which forces the molecules to reorient before continuing to move snake-like through the gel. This reorientation takes much more time for larger molecules, so that longer molecules move more slowly than shorter ones. As a consequence, entire bacterial or yeast chromosomes separate into discrete bands in pulsed-field gels and so can be sorted and identified on the basis of their size (**Figure 8–24D**). Although a typical mammalian chromosome of 10^8 nucleotide pairs is still too long to be sorted even in this way, large segments of these chromosomes are readily separated and identified if the chromosomal

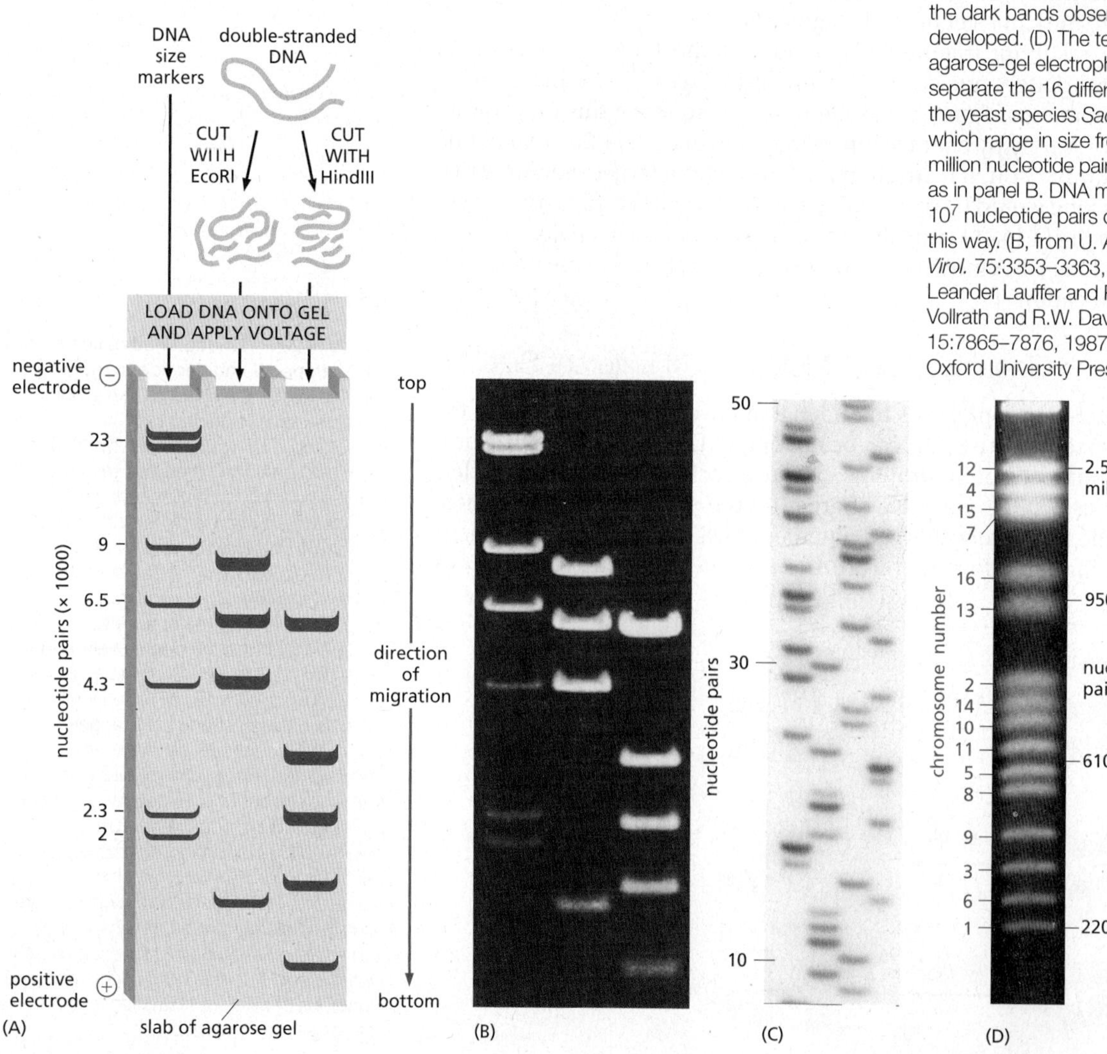

DNA size markers; double-stranded DNA; CUT WITH EcoRI; CUT WITH HindIII; LOAD DNA ONTO GEL AND APPLY VOLTAGE; negative electrode; positive electrode; nucleotide pairs (× 1000); 23; 9; 6.5; 4.3; 2.3; 2; slab of agarose gel; top; direction of migration; bottom; (A); (B); nucleotide pairs; 50; 30; 10; (C); chromosome number; 12; 4; 15; 7; 16; 13; 2; 14; 10; 11; 5; 8; 9; 3; 6; 1; 2.5 million; 950,000; nucleotide pairs; 610,000; 220,000; (D)

Figure 8–24 **DNA molecules can be separated by size using gel electrophoresis.** (A) Schematic illustration comparing the results of cutting the same DNA molecule (in this case, the genome of a virus that infects wasps) with two different restriction nucleases, EcoRI *(middle)* and HindIII *(right)*. The fragments are then separated by gel electrophoresis using a gel matrix of agarose. Because larger fragments migrate more slowly than smaller ones, the lowermost bands on the gel contain the smallest DNA fragments. The sizes of the fragments can be estimated by comparing them to a set of DNA fragments of known sizes *(left)*. (B) Photograph of an actual agarose gel showing DNA bands that have been stained with ethidium bromide. (C) A polyacrylamide gel with small pores was used to separate short DNA molecules that differ by only a single nucleotide. Shown here are the results of a dideoxy sequencing reaction, explained later in this chapter. From left to right, the bands in the four lanes were produced by adding G, A, T, and C chain-terminating nucleotides (see Figure 8-42). The DNA molecules were labeled with ^{32}P, and the image shown was produced by laying a piece of photographic film over the gel and allowing the ^{32}P to expose the film, producing the dark bands observed when the film was developed. (D) The technique of pulsed-field agarose-gel electrophoresis was used to separate the 16 different chromosomes of the yeast species *Saccharomyces cerevisiae*, which range in size from 220,000 to 2.5 million nucleotide pairs. The DNA was stained as in panel B. DNA molecules as large as 10^7 nucleotide pairs can be separated in this way. (B, from U. Albrecht et al., *J. Gen. Virol.* 75:3353–3363, 1994; C, courtesy of Leander Lauffer and Peter Walter; D, from D. Vollrath and R.W. Davis, *Nucleic Acids Res.* 15:7865–7876, 1987. With permission from Oxford University Press.)

DNA is first cut with a restriction nuclease selected to recognize sequences that occur only rarely.

The DNA bands on agarose or polyacrylamide gels are invisible unless the DNA is labeled or stained in some way. A particularly sensitive method of staining DNA is to soak the gel in the dye *ethidium bromide*, which fluoresces under ultraviolet light when it is bound to DNA (see Figure 8–24B and D). Even more sensitive detection methods incorporate a radioisotope or chemical marker into the DNA molecules before electrophoresis, as we next describe.

Purified DNA Molecules Can Be Specifically Labeled with Radioisotopes or Chemical Markers *in Vitro*

The DNA polymerases that synthesize and repair DNA (discussed in Chapter 5) have become important tools in experimentally manipulating DNA. They are often used in the test tube to create exact copies of existing DNA molecules. In some experiments, the copies can include specially modified nucleotides (**Figure 8–25**). To synthesize DNA in this way, the DNA polymerase is presented with a template and a pool of nucleotide precursors that contain the modification. As long as the polymerase can use these precursors, it automatically makes new, modified molecules that match the sequence of the template. Modified DNA molecules have many uses. DNA labeled with the radioisotope ^{32}P can be detected after gel electrophoresis by exposing the gel to photographic film or a radiation detector (see Figure 8–24C). Other types of modified DNA, such as that labeled by digoxigenin (see Figure 8–25B), are useful for visualizing DNA molecules in whole cells, a topic we discuss later in this chapter.

Genes Can Be Cloned Using Bacteria

Any DNA fragment can be cloned. In molecular biology, the term **DNA cloning** refers to the act of making many identical copies (typically billions) of a DNA molecule; that is, the amplification of a specific DNA sequence (often a particular gene) from the rest of the cell's genome. We note that elsewhere in the book, cloning can also refer to the generation of many genetically identical cells starting from a single cell or even to the generation of genetically identical organisms (see,

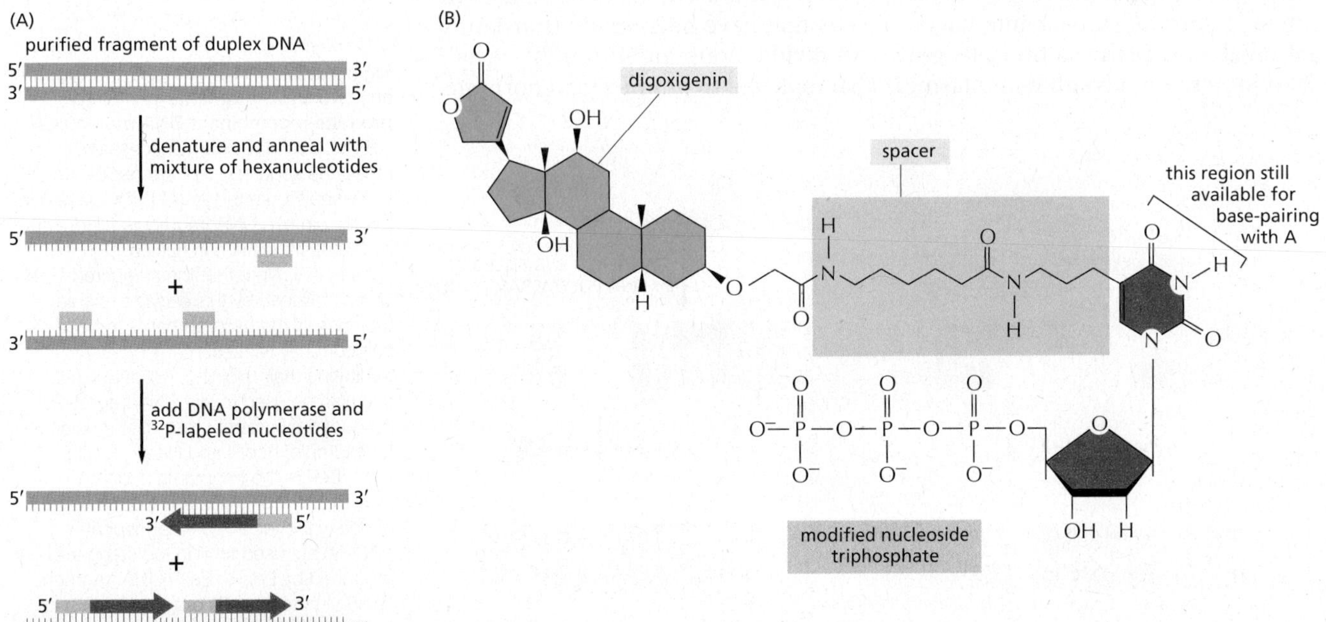

Figure 8–25 Methods for labeling DNA molecules *in vitro*. (A) A purified DNA polymerase enzyme can incorporate radiolabeled nucleotides as it synthesizes new DNA molecules. In this way, radiolabeled versions of any DNA sequence can be prepared in the laboratory. (B) The method in panel A is also used to produce nonradioactive DNA molecules that carry a specific chemical marker that can be detected with an appropriate antibody. The base on the nucleoside triphosphate shown is an analog of thymine, in which the methyl group on T has been replaced by a spacer arm linked to the plant steroid digoxigenin. An anti-digoxigenin antibody coupled to a visible marker such as a fluorescent dye is then used to visualize the DNA. Other chemical labels, such as biotin, can be attached to nucleotides and used in the same way. The only requirements are that the modified nucleotides properly base-pair and appear "normal" to the DNA polymerase.

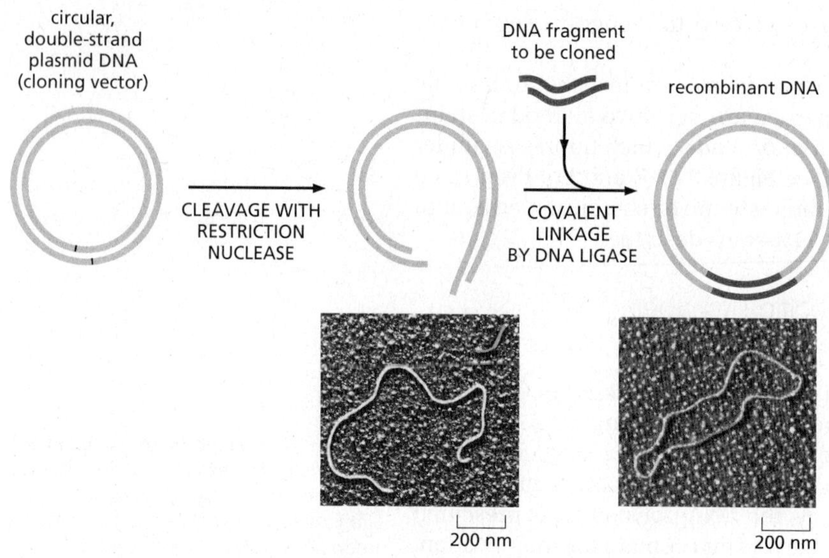

Figure 8–26 The insertion of a DNA fragment into a bacterial plasmid with the enzyme DNA ligase. The plasmid is cut open with a restriction nuclease (in this case, one that produces staggered ends) and is mixed with the DNA fragment to be cloned (which has been prepared with the same restriction nuclease). DNA ligase and ATP are added. The staggered ends base-pair, and DNA ligase seals the nicks in the DNA backbone, producing a complete recombinant DNA molecule. In the accompanying micrographs, the inserted DNA is colored *red*. (Micrographs courtesy of Huntington Potter and David Dressler.)

for example, Figure 7–2). In all cases, cloning refers to the act of making many identical copies, and here we use the term to refer to methods designed to generate many identical copies of a defined segment of nucleic acid.

DNA cloning can be accomplished in several ways. One of the simplest involves inserting a particular fragment of DNA into the purified DNA of a self-replicating genetic element—usually a plasmid. The **plasmid vectors** most widely used for gene cloning are small, circular molecules of double-stranded DNA derived from plasmids that occur naturally in bacterial cells. They generally account for only a minor fraction of the total host bacterial cell DNA, but owing to their small size, they can easily be separated from the much larger chromosomal DNA molecules. For use as cloning vectors, the purified plasmid DNA circles are first cut with a restriction nuclease to create linear DNA molecules. The DNA to be cloned is added to the cut plasmid and then covalently joined using the enzyme DNA ligase (Figure 8–26 and Figure 8–27). As discussed in Chapter 5, this enzyme is used by the cell to stitch together the Okazaki fragments produced during DNA replication. The recombinant DNA circle is introduced back into bacterial cells that have been made transiently permeable to DNA. As the cells grow and divide, doubling in number every 30 minutes, the recombinant plasmids also replicate to produce an enormous

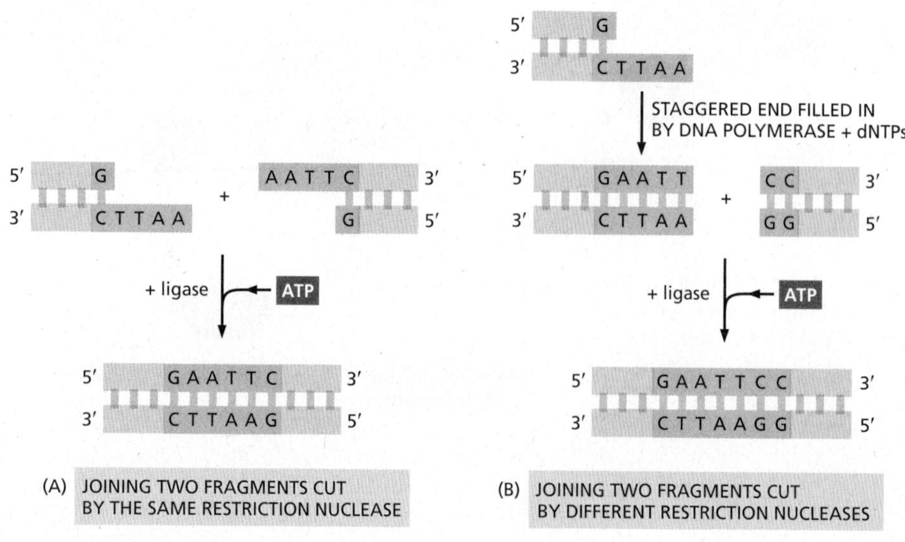

Figure 8–27 DNA ligase can join together any two DNA fragments *in vitro* to produce recombinant DNA molecules. ATP provides the energy necessary to reseal the sugar–phosphate backbone of DNA (see Figure 5–12). (A) DNA ligase can readily join two DNA fragments produced by the same restriction nuclease, in this case EcoRI. Note that the staggered ends produced by this enzyme enable the ends of the two fragments to base-pair correctly with each other, greatly facilitating their rejoining. (B) DNA ligase can also be used to join DNA fragments produced by different restriction nucleases; for example, EcoRI and HaeIII. In this case, before the fragments undergo ligation, DNA polymerase plus a mixture of deoxyribonucleoside triphosphates (dNTPs) are used to fill in the staggered cut produced by EcoRI. Each DNA fragment shown in the figure is oriented so that its 5′ ends are at the left end of the upper strand and the right end of the lower strand, as indicated.

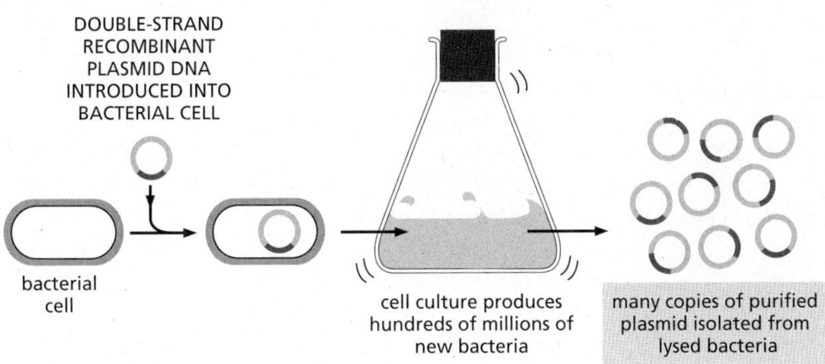

Figure 8–28 **A DNA fragment can be replicated inside a bacterial cell.** To clone a particular fragment of DNA, it is first inserted into a plasmid vector, as shown in Figure 8–26. The resulting recombinant plasmid DNA is then introduced into a bacterium, where it is replicated many millions of times as the bacterium multiplies. For simplicity, the genome of the bacterial cell is not shown.

number of copies of DNA circles containing the foreign DNA (**Figure 8–28**). Once the cells are lysed and the plasmid DNA isolated, the cloned DNA fragment can be readily recovered by cutting it out of the plasmid DNA with the same restriction nuclease that was used to insert it, and then separating it from the plasmid DNA by gel electrophoresis. Together, these steps allow the amplification and purification of any segment of DNA from the genome of any organism.

A particularly useful plasmid vector is based on the naturally occurring F plasmid of *E. coli*. Unlike smaller bacterial plasmids, the F plasmid—and its engineered derivative, the **bacterial artificial chromosome (BAC)**—is present in only one or two copies per *E. coli* cell. The fact that BACs are kept in such low numbers means that they can stably maintain very long DNA sequences, up to 1 million nucleotide pairs in length. With only a few BACs present per bacterium, it is less likely that the cloned DNA fragments will become scrambled by recombination with sequences carried on other copies of the plasmid. Because of their stability, ability to accept large DNA inserts, and ease of handling, BACs are the preferred vector for handling large fragments of foreign DNA.

An Entire Genome Can Be Represented in a DNA Library

Often it is useful to break up a genome into much smaller fragments and clone every fragment, separately, using a plasmid vector. This approach is useful because it allows scientists to work with easily managed, discrete pieces of a genome instead of whole, unwieldy chromosomes.

This strategy involves cleaving genomic DNA into small pieces using a restriction nuclease (or, in some cases, by mechanically shearing the DNA) and ligating the entire collection of DNA fragments into plasmid vectors, using conditions that favor the insertion of a single DNA fragment into each plasmid molecule. These recombinant plasmids are then introduced into *E. coli* at a concentration that ensures that no more than one plasmid molecule is taken up by each bacterium. The collection of cloned plasmid molecules is known as a **DNA library**. Because the DNA fragments were derived directly from the chromosomal DNA of the organism of interest, the resulting collection—called a **genomic library**—will represent the entire genome of that organism (**Figure 8–29**), spread out over tens of thousands of individual bacterial colonies.

An alternative strategy, one that enriches for protein-coding genes, is to begin the cloning process by selecting only those DNA sequences that are transcribed into mRNA and thus correspond to protein-encoding genes. This is done by extracting the mRNA from cells and then making a DNA copy of each mRNA

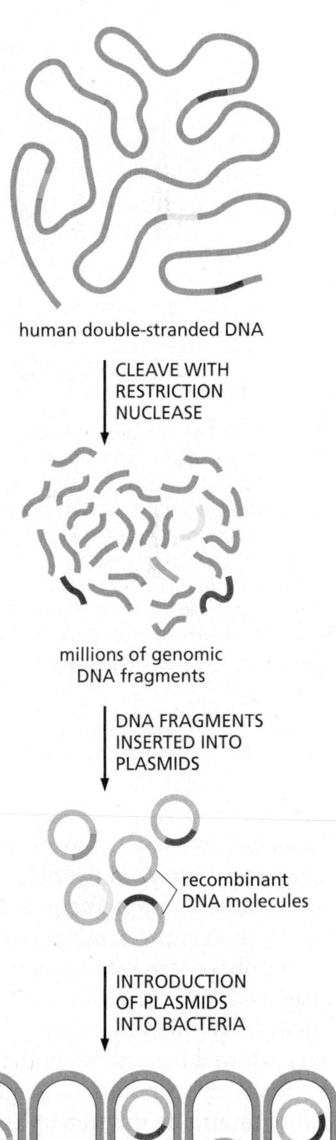

Figure 8–29 **Human genomic libraries containing DNA fragments that represent the whole human genome can be constructed using restriction nucleases and DNA ligase.** Such a genomic library consists of a set of bacteria, each carrying a different fragment of human DNA. For simplicity, only the *colored* DNA fragments are shown in the library; in reality, all of the different *gray* fragments will also be represented.

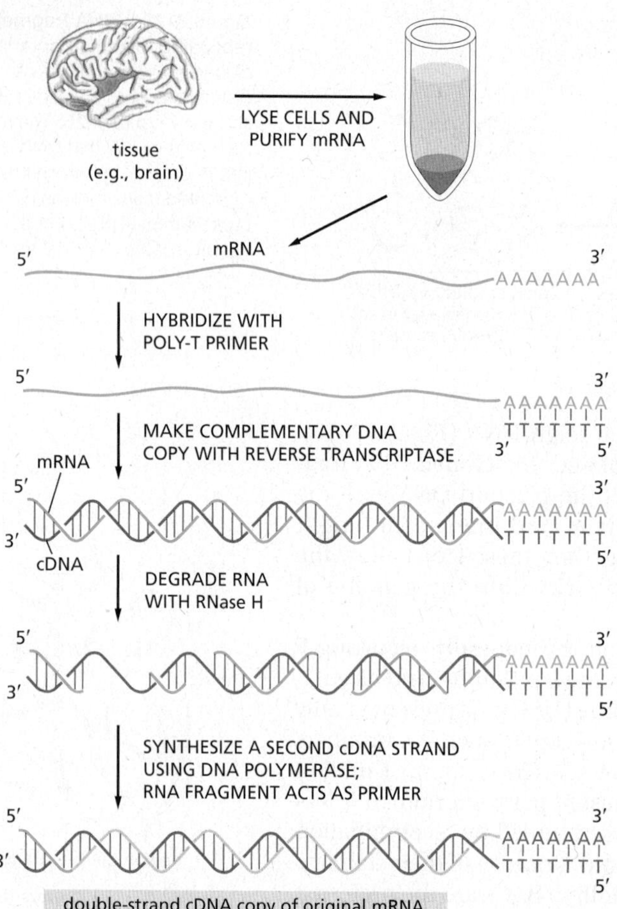

Figure 8–30 **The synthesis of cDNA.** Total mRNA is extracted from a particular tissue, and the enzyme reverse transcriptase (see Figure 5–61) is used to produce DNA copies (cDNA) of the mRNA molecules. For simplicity, the copying of just one of these mRNAs into cDNA is illustrated. A short oligonucleotide complementary to the poly-A tail at the 3′ end of the mRNA is first hybridized to the RNA to act as a primer for the reverse transcriptase, which then copies the RNA into a complementary DNA chain, thereby forming a DNA–RNA hybrid helix. Treating the DNA–RNA hybrid with a specialized ribonuclease (RNase H) that attacks only the RNA produces nicks and gaps in the RNA strand. DNA polymerase then copies the remaining single-stranded cDNA into double-stranded cDNA. Because DNA polymerase can synthesize through the bound RNA molecules, the RNA fragment that is base-paired to the 3′ end of the first DNA strand usually acts as the primer for the second strand synthesis, as shown. Any remaining RNA is eventually degraded during subsequent cloning steps. As a result, the nucleotide sequences at the extreme 5′ ends of the original mRNA molecules are often absent from cDNA libraries.

molecule present—a so-called *complementary DNA*, or *cDNA*. The copying reaction is catalyzed by the reverse transcriptase enzyme of retroviruses, which synthesizes a complementary DNA chain on an RNA template. The single-strand cDNA molecules synthesized by the reverse transcriptase are converted by DNA polymerase into double-strand cDNA molecules, and these molecules are inserted into a plasmid or virus vector and cloned (**Figure 8–30**). Each clone obtained in this way is called a **cDNA clone**, and the entire collection of clones derived from one mRNA preparation constitutes a **cDNA library**.

The most important advantage of cDNA clones, over genomic clones, is that they contain the uninterrupted coding sequence of a gene. When the aim of the cloning is to produce the protein in large quantities by expressing the cloned gene in a bacterial or other cell type, it is best to start with cDNA.

Cloning DNA in bacteria revolutionized the study of genomes and is still in wide use today. However, there is an even simpler way to clone DNA and produce

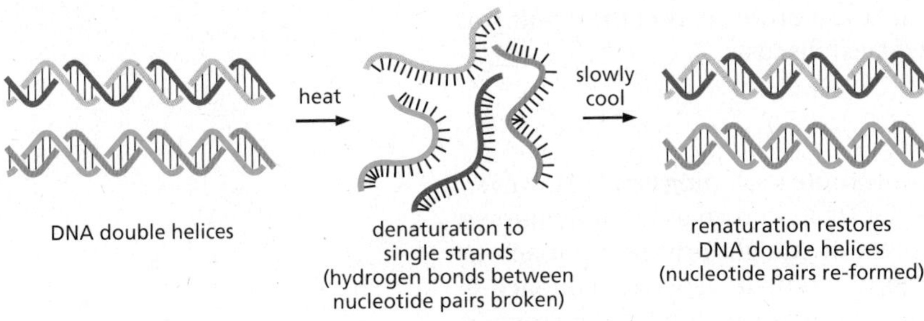

Figure 8–31 **A molecule of DNA can undergo denaturation and renaturation (hybridization).** For two single-strand molecules to hybridize, they must have complementary nucleotide sequences that allow base-pairing. In this example, the *red* and *orange* strands are complementary to each other, and the *blue* and *green* strands are complementary to each other. Although denaturation by heating is shown, DNA can also be renatured after being denatured by alkali treatment.

genomic libraries, entirely *in vitro*. We discuss this approach, called the *polymerase chain reaction*, shortly. However, first we need to review a fundamental, far-reaching property of DNA and RNA called *hybridization*.

Hybridization Provides a Powerful but Simple Way to Detect Specific Nucleotide Sequences

Under normal conditions, the two strands of a DNA double helix are held together by hydrogen bonds between the complementary base pairs (see Figure 4–3). These relatively weak, noncovalent bonds are easy to break. Such *DNA denaturation* releases the two strands from each other but does not break the covalent bonds that link together the nucleotides within each strand. The simplest way to achieve this separation is to heat the DNA to around 90°C. When the conditions are reversed—by slowly lowering the temperature—the complementary strands will readily come back together to re-form a double helix. This **hybridization**, or *DNA renaturation*, is driven by the re-formation of the hydrogen bonds between complementary base pairs (**Figure 8-31**).

This fundamental capacity of a single-strand nucleic acid molecule, either DNA or RNA, to form a double helix with a single-strand molecule of a complementary sequence provides a powerful and sensitive technique for detecting specific nucleotide sequences. Today, one simply designs a short, single-strand DNA molecule (often called a *DNA probe*) that is complementary to the nucleotide sequence of interest. Because the nucleotide sequences of so many genomes are known—and are stored in publicly accessible databases—designing a probe to hybridize anywhere in a genome is straightforward. Probes are single-stranded, typically 30 nucleotides in length, and are usually synthesized chemically by a commercial service for pennies per nucleotide. A DNA sequence of 30 nucleotides will occur by chance only once every 1×10^{18} nucleotides (4^{30}); so, even in the human genome of 3×10^9 nucleotide pairs, a DNA probe designed to match a unique 30-nucleotide sequence will be highly unlikely to hybridize—by chance—anywhere else on the genome. The hybridization conditions can be set so that even a single mismatch will prevent hybridization to "near-miss" sequences. The exquisite specificity of nucleic acid hybridization can be easily appreciated by the *in situ* (Latin for "in place") *hybridization* experiment shown in **Figure 8-32**. As we will see throughout this chapter, nucleic acid hybridization has many uses in

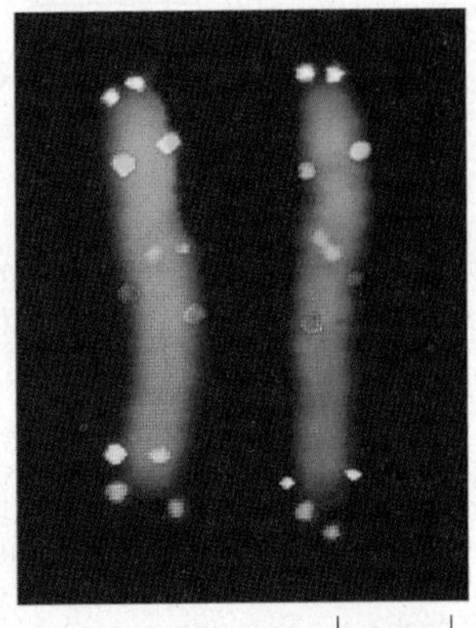

Figure 8–32 ***In situ* hybridization can be used to locate genes on isolated chromosomes.** Here, six different DNA probes have been used to mark the locations of their complementary nucleotide sequences on human chromosome 5, isolated from a mitotic cell in metaphase. The DNA probes have been labeled with different chemical groups (see Figure 8–25B) and are detected using fluorescent antibodies specific for those groups. The chromosomal DNA has been partially denatured to allow the probes to base-pair with their complementary sequences. Both the maternal and paternal copies of chromosome 5 are shown, aligned side by side. Each probe produces two dots on each chromosome because chromosomes undergoing mitosis have already replicated their DNA; therefore, each chromosome contains two identical DNA helices. The technique employed here is nicknamed FISH, for *fluorescence in situ hybridization*. (Courtesy of David C. Ward.)

modern cell and molecular biology; one of the most powerful is in the cloning of DNA by the polymerase chain reaction, as we next discuss.

Genes Can Be Cloned *in Vitro* Using PCR

Genomic and cDNA libraries were once the only route to cloning genes. However, a powerful and versatile method for amplifying DNA, known as the **polymerase chain reaction (PCR)**, provides a more rapid and straightforward approach to DNA cloning, particularly in organisms whose complete genome sequence is known. Today, because genome sequences are abundant, cloning is often carried out by PCR.

Invented in the 1980s, PCR revolutionized the way that DNA and RNA are analyzed. The technique can amplify any nucleotide sequence selectively and is performed entirely in a test tube. Eliminating the need for bacteria makes PCR convenient and rapid—billions of copies of a nucleic acid sequence can be generated in a matter of hours. Starting with an entire genome, PCR allows DNA from a specified region—selected by the experimenter—to be greatly amplified, effectively purifying this DNA away from the remainder of the genome, which remains unamplified. Because of its power to greatly amplify nucleic acids, PCR is remarkably sensitive: the method can be used to detect the trace amounts of DNA in a drop of blood left at a crime scene or in a few copies of a viral genome in a sample of someone's saliva.

The success of PCR depends both on the selectivity of DNA hybridization and on the ability of DNA polymerase to copy a DNA template faithfully through repeated rounds of replication *in vitro*. As discussed in Chapter 5, this enzyme adds nucleotides to the 3′ end of a growing strand of DNA (see Figure 5–4). To copy DNA, the polymerase requires a *DNA primer*—a short nucleotide sequence that provides a 3′ end from which synthesis can begin. For PCR, the primers are designed by the experimenter, synthesized chemically, and, by hybridizing to genomic DNA, guide the polymerase to the part of the genome to copy. DNA primers can be designed to uniquely locate any position on a genome.

PCR is an iterative process in which the cycle of amplification is repeated many times. At the start of each cycle, the two strands of the double-strand DNA template are separated, and a different primer is annealed to each. These primers mark the right and left boundaries of the DNA to be amplified. DNA polymerase is then allowed to replicate each strand independently (**Figure 8–33**). In subsequent cycles, all the newly synthesized DNA molecules produced by the

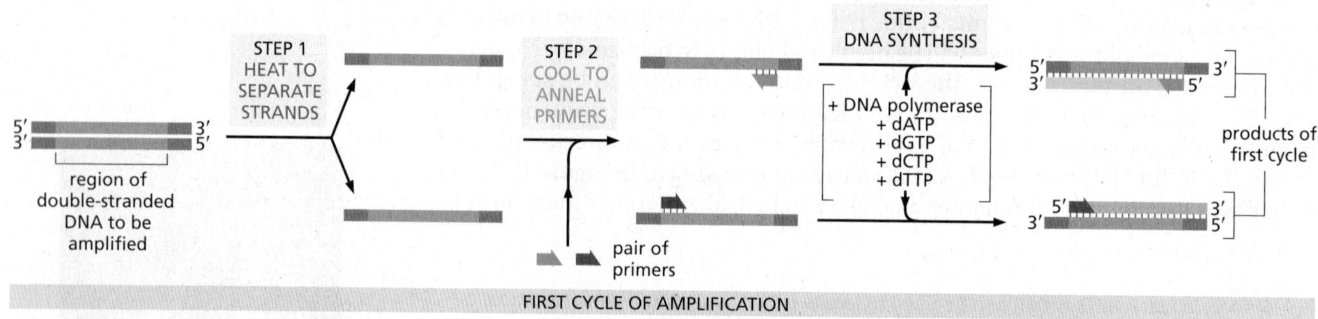

Figure 8–33 **A pair of primers directs the synthesis of a desired segment of DNA in a test tube.** Each cycle of PCR includes three steps: (1) The double-stranded DNA is heated briefly to separate the two strands. (2) The DNA is exposed to a large excess of a pair of specific primers—designed to bracket the region of DNA to be amplified—and the sample is cooled to allow the primers to hybridize to complementary sequences in the two DNA strands. (3) This mixture is incubated with DNA polymerase and the four deoxyribonucleoside triphosphates so that DNA can be synthesized, starting from the two primers. To amplify the DNA, the cycle is repeated many times by reheating the sample to separate the newly synthesized DNA strands (see Figure 8–34).

The technique depends on the use of a special DNA polymerase isolated from a thermophilic bacterium; this polymerase is stable at much higher temperatures than eukaryotic DNA polymerases, so it is not denatured by the heat treatment shown in step 1. The enzyme therefore does not have to be added again after each cycle.

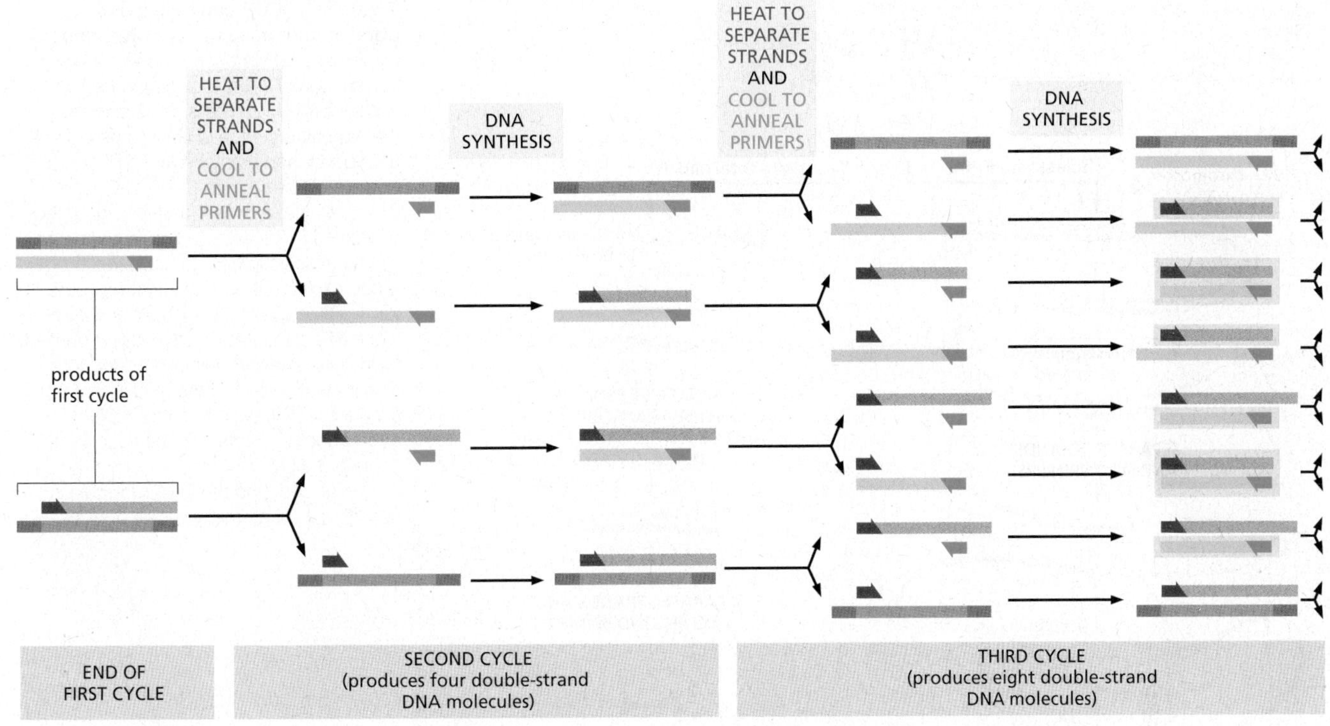

Figure 8–34 **PCR uses repeated rounds of strand separation, hybridization, and synthesis to amplify DNA.** As the procedure outlined in Figure 8–33 is repeated, all the newly synthesized fragments serve as templates in their turn. Because the polymerase and the primers remain in the sample after the first cycle, PCR involves simply heating and then cooling the same sample, in the same test tube, again and again. Each cycle doubles the amount of DNA synthesized in the previous cycle, so that within a few cycles, the predominant DNA is identical to the sequence bracketed by and including the two primers in the original template. In the example illustrated here, three cycles of reaction produce 16 DNA chains, 8 of which (boxed in *yellow*) correspond exactly to one or the other strand of the original bracketed sequence. After four more cycles, 240 of the 256 DNA chains will correspond exactly to the original sequence, and after several more cycles, essentially all of the DNA strands will be this length. Typically, 20–30 cycles are carried out to effectively clone a region of DNA starting from genomic DNA; the rest of the genome remains unamplified, and its concentration is therefore negligible compared with that of the amplified region (**Movie 8.2**).

polymerase serve as templates for the next round of replication (**Figure 8–34**). Through this iterative amplification process, many copies of the original sequence can be made—billions after about 20–30 cycles.

PCR is now the method of choice for cloning relatively short DNA fragments (say, under 10,000 nucleotide pairs). Each cycle takes only about 5 minutes, and automation of the whole procedure enables cell-free cloning of a DNA fragment in a few hours. The original template for PCR can be either DNA or RNA, so this method can be used to obtain either a genomic clone (complete with introns and exons) or a cDNA copy of an mRNA (**Figure 8–35**).

PCR Is Also Used for Diagnostic and Forensic Applications

The PCR method is extraordinarily sensitive; it can detect a single DNA molecule in a sample if at least part of the sequence of that molecule is known. Trace amounts of RNA can be analyzed in the same way by first transcribing them into DNA with reverse transcriptase. For these reasons, PCR is frequently employed for uses that go beyond simple cloning. For example, it can be used to detect invading pathogens at very early stages of infection. In this case, short sequences complementary to a segment of the infectious agent's genome are used as primers, and, after many cycles of amplification, even a few copies of an invading bacterial or viral genome in a human sample can be detected (**Figure 8–36**). For many infections, PCR has replaced the use of antibodies against microbial molecules to detect the presence of the invader. It is also used to verify the

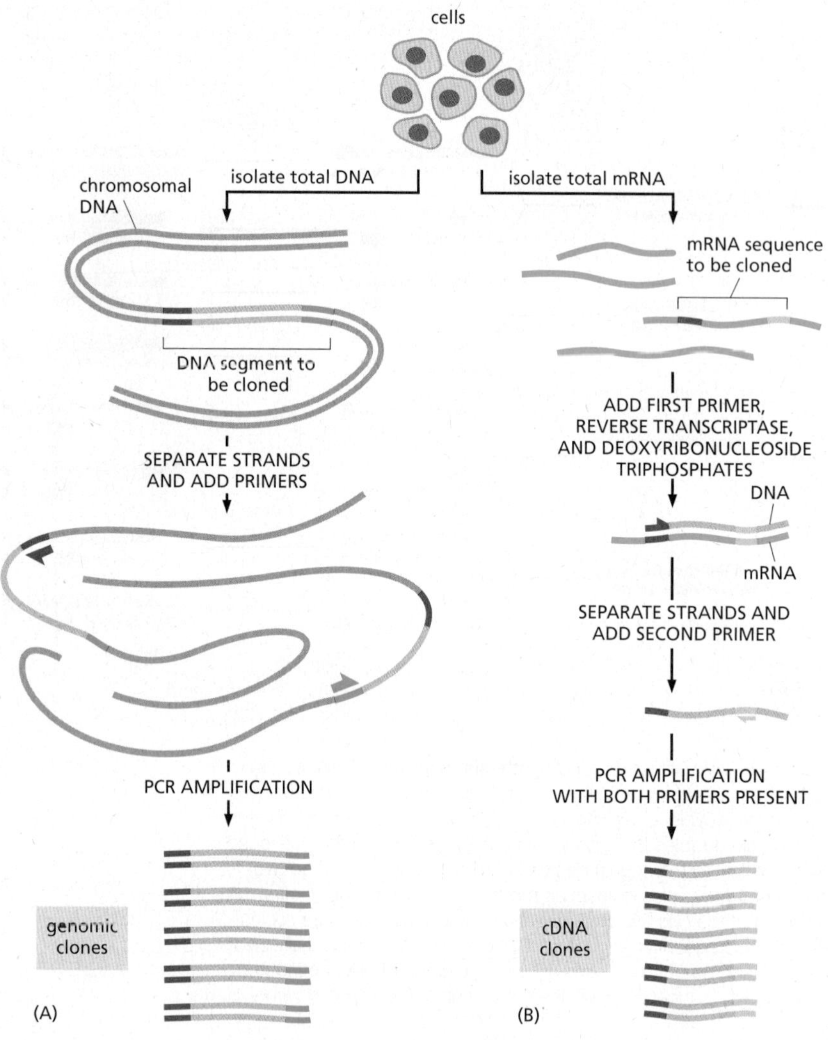

Figure 8–35 PCR can be used to obtain either genomic or cDNA clones. (A) To use PCR to clone a segment of chromosomal DNA, total genomic DNA is first purified from cells. PCR primers that flank the stretch of DNA to be cloned are added, and many cycles of PCR are completed (see Figure 8–34). Because only the DNA between (and including) the primers is amplified, PCR provides a way to obtain selectively any short stretch of chromosomal DNA in an effectively pure form. (B) To use PCR to obtain a cDNA clone of a gene, total mRNA is first purified from cells. The first primer is added to the population of mRNAs, and reverse transcriptase is used to make a DNA strand complementary to the specific RNA sequence of interest. The second primer is then added, and the DNA molecule is amplified through many cycles of PCR.

authenticity of a food source; for example, whether a sample of beef actually came from a cow.

Finally, PCR is now widely used in forensics. The method's extreme sensitivity allows forensic investigators to isolate DNA from minute traces of human blood or other tissue to obtain a *DNA fingerprint* of the person who left the sample behind. With the possible exception of identical twins, the genome of each human differs in DNA sequence from that of every other person on Earth. Using primer pairs targeted at genome sequences that are known to be highly variable in the human population, PCR makes it possible to generate a distinctive DNA fingerprint for any individual (Figure 8–37). Such forensic analyses can be used not only to help

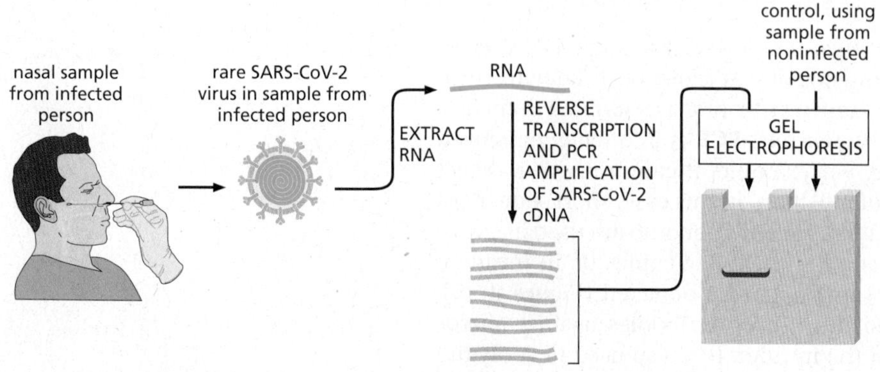

Figure 8–36 PCR can be used to detect the presence of a viral genome in a nasal sample. Because of its ability to amplify enormously the signal from a single molecule of nucleic acid, PCR is an extraordinarily sensitive method for detecting trace amounts of virus in a sample of saliva, blood, or other tissue, without the need to purify the virus. For the coronavirus SARS-CoV-2, the virus that causes COVID-19, the genome is a single-strand molecule of RNA, as illustrated here. Typically, only a short segment of the viral genome (100–200 nucleotides) is amplified. Although it is possible to visualize the amplified DNA by gel electrophoresis as shown here, it is usually detected by rapid optical methods described later in the chapter (see Figure 8–64). Many other viruses that infect humans—such as HIV—are detected using this strategy.

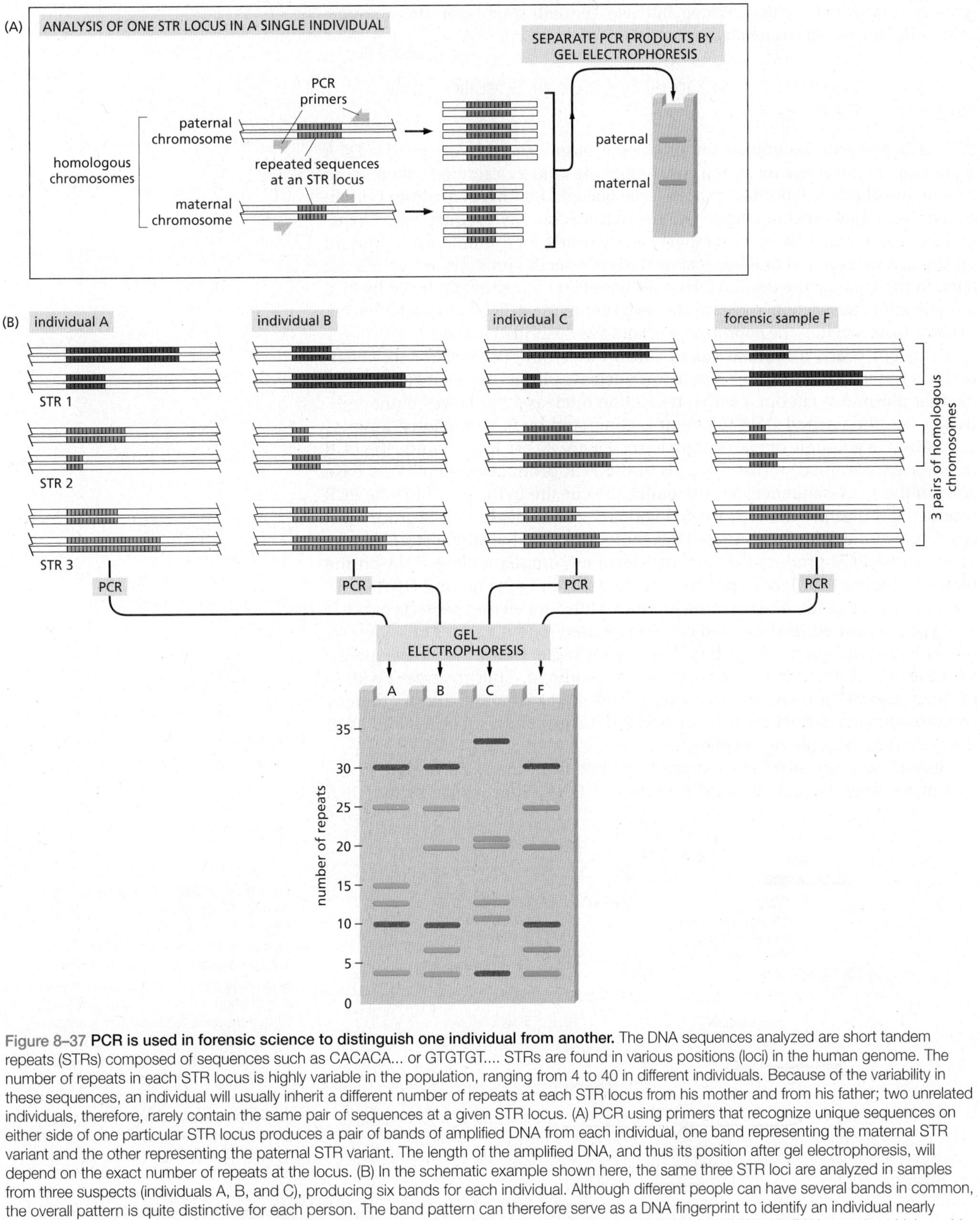

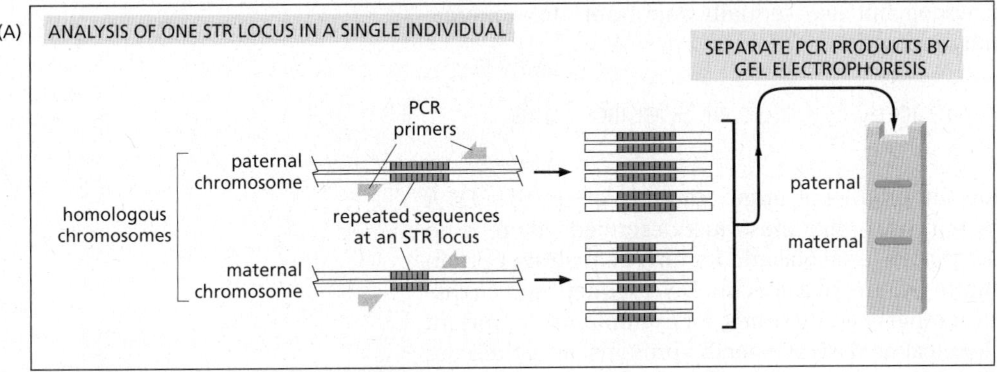

Figure 8–37 PCR is used in forensic science to distinguish one individual from another. The DNA sequences analyzed are short tandem repeats (STRs) composed of sequences such as CACACA... or GTGTGT.... STRs are found in various positions (loci) in the human genome. The number of repeats in each STR locus is highly variable in the population, ranging from 4 to 40 in different individuals. Because of the variability in these sequences, an individual will usually inherit a different number of repeats at each STR locus from his mother and from his father; two unrelated individuals, therefore, rarely contain the same pair of sequences at a given STR locus. (A) PCR using primers that recognize unique sequences on either side of one particular STR locus produces a pair of bands of amplified DNA from each individual, one band representing the maternal STR variant and the other representing the paternal STR variant. The length of the amplified DNA, and thus its position after gel electrophoresis, will depend on the exact number of repeats at the locus. (B) In the schematic example shown here, the same three STR loci are analyzed in samples from three suspects (individuals A, B, and C), producing six bands for each individual. Although different people can have several bands in common, the overall pattern is quite distinctive for each person. The band pattern can therefore serve as a DNA fingerprint to identify an individual nearly uniquely. The fourth lane (F) contains the products of the same PCR amplifications carried out on a hypothetical forensic DNA sample, which could have been obtained from a single hair or a tiny spot of blood left at a crime scene.

The more loci that are examined, the more confident one can be about the results. When examining the variability at 5–10 different STR loci, the odds that two random individuals would share the same fingerprint by chance are approximately 1 in 10 billion. In the case shown here, individuals A and C can be eliminated from inquiries, while B is a clear suspect. A similar approach is used routinely in paternity testing.

identify those who have done wrong, but also—equally important—to exonerate those who have been wrongfully accused.

PCR and Synthetic DNA Are Ideal Sources of Specific Gene Sequences for Cloning

PCR is a powerful technique for producing large amounts of specific DNA sequences *in vitro*. For many purposes—like those just described—there is no need to insert the PCR product in a bacterial plasmid for further analysis. Certain experiments, however, do require cloning of a specific DNA sequence in a plasmid for production in bacteria or other cells. Plasmids, for example, are important for the production and biochemical analysis of specific proteins, as we discuss later. In these cases, the desired DNA sequence can be readily produced by PCR amplification. Restriction nuclease cleavage sites in the PCR product are then cut to allow insertion into the appropriate plasmid as described earlier.

In recent years, a method called Gibson assembly (named for its inventor) has reduced the need for restriction nucleases in plasmid construction. A circular plasmid is cut open with a restriction nuclease that leaves blunt ends. In parallel, the desired DNA fragment is generated by PCR-mediated amplification from a genomic or cDNA source (see Figure 8–35). Importantly, the PCR primers are designed so that the ends of the PCR product contain 15–40 base pairs of the same sequence that surrounds the cut site in the plasmid. The PCR product and plasmid are then treated with a 5′ exonuclease that degrades just one DNA strand at the ends of all DNA molecules, leaving single-strand sticky ends on the PCR product that will hybridize to complementary DNA on the plasmid (**Figure 8–38**). DNA polymerase then fills in the gaps, and DNA ligase joins the DNA strands. The resulting product is introduced into bacteria or other cell types as desired. This method can also be used to join a series of DNA fragments in a single plasmid. Each end of a DNA fragment must share sequence with the end of the fragment to which it will be joined. After treatment with 5′ exonuclease, all the complementary sticky ends find each other and are thereby assembled in the correct order in the plasmid. Large genes can be assembled in this way from multiple subfragments.

Another key advance in recent years resulted from the rapid decline in the cost of methods for the chemical synthesis of DNA. Short single-strand DNA

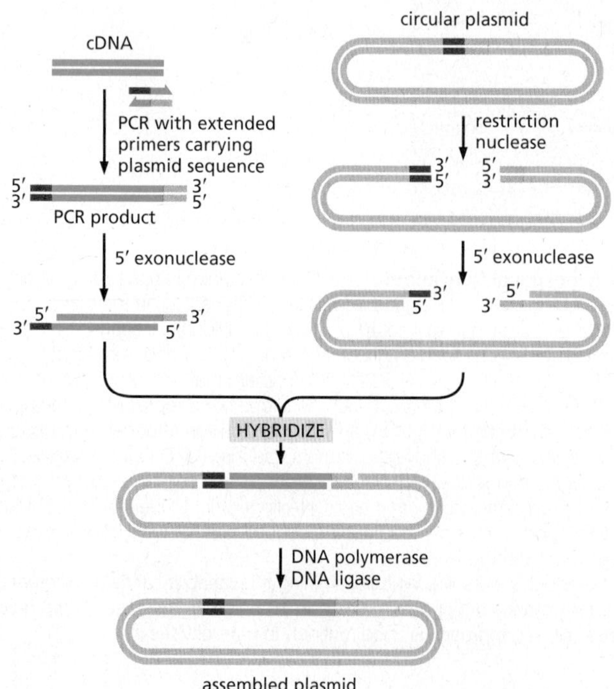

Figure 8–38 DNA cloning by Gibson assembly. It is often useful to insert a DNA fragment in a circular bacterial plasmid, as described earlier (see Figure 8–26). In Gibson assembly, as shown here, a plasmid is cleaved at a specific site with a restriction nuclease. In parallel, the DNA sequence to be cloned is amplified by PCR of a cDNA or other source. In addition to sequences complementary to the ends of the DNA to be amplified, the PCR primers include 15–40 nucleotides of sequence that matches the sequence on either side of the cut site in the linearized plasmid. The PCR product and plasmid are treated with a 5′ exonuclease, which partially digests the DNA from the 5′ end, resulting in single-strand 3′ overhangs. These single-strand overhangs hybridize with their complementary sequence, neatly inserting the PCR product in the plasmid as shown. DNA polymerase is then added to fill in any gaps in the sequence, and DNA ligase seals the nicks to provide the fully assembled plasmid.

molecules, called *oligonucleotides*, are cheap to make and have been used for decades as probes and primers for PCR and other methods. Recently, it has become possible to cheaply synthesize much larger double-strand DNA molecules up to a few thousand base pairs in length. Thus, for many purposes, it is easier to order the desired DNA fragment from a DNA synthesis company than it is to produce it by PCR. If the ends of the synthetic DNA fragment are identical to those of a cut plasmid, then it is straightforward to insert the synthetic DNA in a plasmid by Gibson assembly (see Figure 8–38). Using multiple synthetic DNA fragments with interlocking ends, it is possible to assemble very large stretches of DNA (approaching the size of small genomes) from purely synthetic DNA.

We will see later in this chapter that the study of gene and protein function often requires methods to produce a mutant gene or protein that carries specific point mutations that alter its function. In molecular biology, the production of a mutant DNA sequence, called *site-directed mutagenesis*, is readily achieved through the clever application of PCR and synthetic DNA. If the goal is a single point mutation, then one of the two PCR primers can be designed to include the mutation while still having sufficient flanking sequence to hybridize to the nonmutant source DNA. PCR then generates a DNA product with the mutation near one end. This mutant DNA fragment can be assembled in a plasmid with other portions of the gene. If multiple mutations in a DNA sequence are needed, then the simplest approach is to purchase a synthetic DNA containing the mutations and insert that into the desired plasmid.

DNA Cloning Allows Any Protein to Be Produced in Large Amounts

Using the genetic code (and assuming all intron and exon boundaries are known), the amino acid sequence of any protein coded in a genome can be deduced. As was discussed earlier, this sequence can often provide an important clue to the protein's function if found to be similar to the amino acid sequence of a protein that has already been studied (see Figure 8–22). Although this strategy is often successful, it typically provides only the likely biochemical function of the protein; for example, whether the protein resembles a kinase or a protease. It usually remains for the experimenter to verify (or refute) this assignment and, most important, to discover the protein's biological function in the whole organism.

An important approach in determining gene function is to alter the gene (or in some cases, its expression pattern), place the altered copy back into the organism, and deduce the function of the normal gene by the changes caused by its alteration. Various techniques to implement this strategy are discussed in the next section of this chapter. But it is equally important to study the biochemical and structural properties of a gene product, as outlined earlier in this chapter. One of the most important contributions of DNA cloning to cell and molecular biology is the ability to produce any protein, even the rare ones, in nearly unlimited amounts. Such high-level production is usually carried out in living cells using expression vectors (**Figure 8–39**). These are generally plasmids that have been designed to produce a large amount of stable mRNA that can be efficiently translated into protein when the plasmid is introduced into bacterial, yeast, insect, or

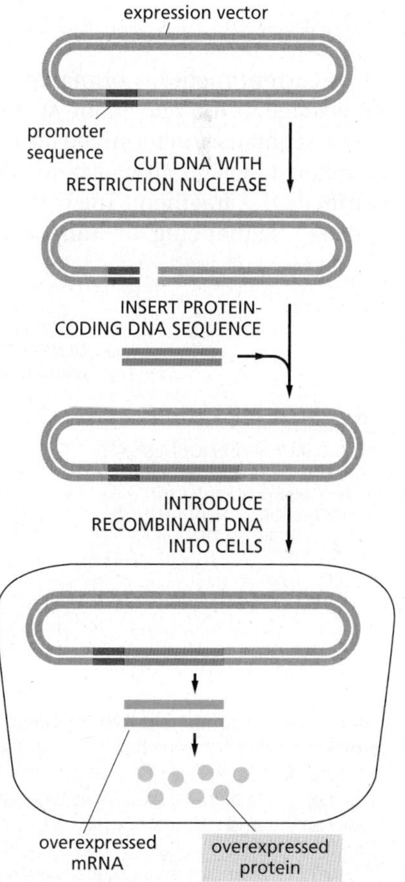

Figure 8–39 Production of large amounts of a protein from a protein-coding DNA sequence cloned into an expression vector and introduced into cells. A plasmid vector has been engineered to contain a highly active promoter, which causes unusually large amounts of mRNA to be produced from an adjacent protein-coding gene inserted into the plasmid vector. Depending on the characteristics of the cloning vector, the plasmid is introduced into bacterial, yeast, insect, or mammalian cells, where the inserted gene is efficiently transcribed and translated into protein. If the gene to be overexpressed has no introns (typical for genes from bacteria, archaea, and simple eukaryotes), it can simply be cloned from genomic DNA by PCR. For cloned animal and plant genes, it is often more convenient to obtain the gene as cDNA, either from a cDNA library (see Figure 8–30) or cloned directly by PCR from RNA isolated from the organism (see Figure 8–35). Alternatively, the DNA coding for the protein can be made by chemical synthesis.

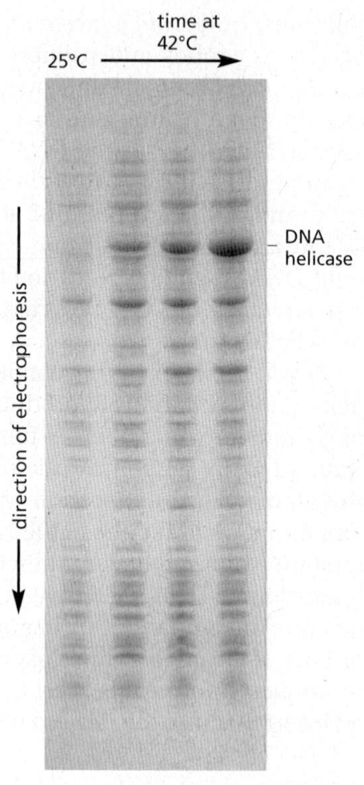

Figure 8–40 **Production of large amounts of a protein by using a plasmid expression vector.** In this example, an expression vector that overproduces a DNA helicase has been introduced into bacteria. In this expression vector, transcription from this coding sequence is under the control of a viral promoter that becomes active only at a temperature of 37°C or higher. The total cell protein, either from bacteria grown at 25°C (no helicase protein made) or after a shift of the same bacteria to 42°C for up to 2 hours (helicase protein has become the most abundant protein species in the lysate), has been analyzed by SDS polyacrylamide-gel electrophoresis. (Courtesy of Kevin Hacker.)

mammalian cells. To prevent the high level of the foreign protein from interfering with the cell's growth, the expression vector is often designed to delay the synthesis of the foreign mRNA and protein until shortly before the cells are harvested and lysed (**Figure 8–40**).

Because the desired protein made from an expression vector is produced inside a cell, it must be purified away from the host-cell proteins by chromatography after cell lysis. However, because the protein is such a plentiful species in the cell (often 1–10% of the total cell protein), the purification is usually easy to accomplish in only a few steps. As we saw earlier, it is also possible to fuse a molecular tag—a cluster of histidine residues or a small marker protein—to the expressed protein to facilitate easy purification by affinity chromatography (see Figure 8–11). A variety of expression vectors is available, each engineered to function in the type of cell in which the protein is to be made.

This technology is also used to make large amounts of many medically useful proteins, including hormones (such as insulin and growth factors) used as pharmaceuticals, and viral proteins for use in vaccines. Expression vectors also allow scientists to produce many proteins of biological interest in large enough amounts for detailed structural studies. Nearly all three-dimensional protein structures depicted in this book are of proteins produced in this way. Recombinant DNA techniques thus allow scientists to move with ease from protein to gene, and vice versa, so that the functions of both can be explored on multiple fronts (**Figure 8–41**).

DNA Can Be Sequenced Rapidly by Dideoxy Sequencing

Most current methods of manipulating DNA, RNA, and proteins rely on prior knowledge of the nucleotide sequence of the genome of interest. But how are these sequences determined in the first place? In the late 1970s, researchers developed several strategies for determining the nucleotide sequence of any purified DNA fragment. The method that became the most widely used is called **dideoxy sequencing** or **Sanger sequencing** (named after the scientist who

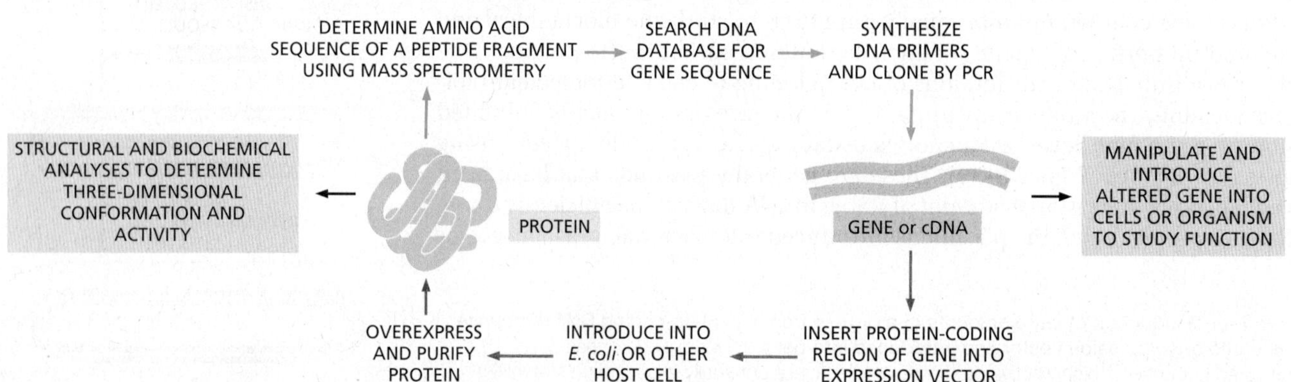

Figure 8–41 **Recombinant DNA techniques make it possible to move experimentally from gene to protein and from protein to gene.** If a gene has been identified *(right)*, its protein-coding sequence can be inserted into an expression vector to produce large quantities of the protein (see Figure 8–39), which can then be studied biochemically or structurally. If a protein has been purified on the basis of its biochemical properties, mass spectrometry (see Figure 8–18) can be used to obtain a partial amino acid sequence, which is used to search a genome sequence for the corresponding nucleotide sequence. The complete gene can then be cloned by PCR from a sequenced genome (see Figure 8–35). The gene can also be manipulated and introduced into cells or organisms to study its function, a topic covered in the next section of this chapter.

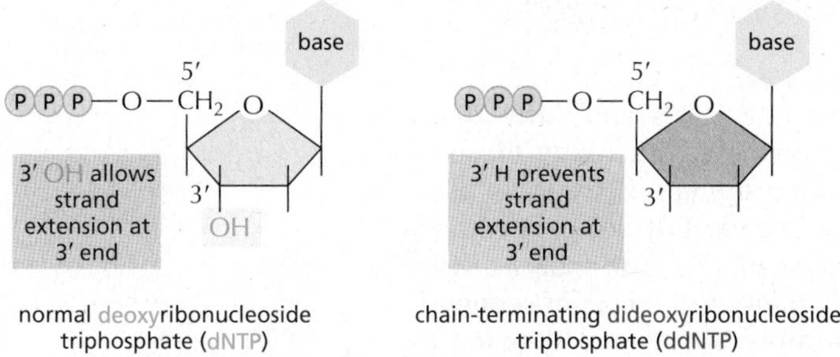

Figure 8–42 **The dideoxy method of sequencing DNA relies on chain-terminating dideoxyribonucleoside triphosphates (ddNTPs).** These ddNTPs are derivatives of the normal deoxyribonucleoside triphosphates (dNTPs) that lack the 3′-hydroxyl group. When incorporated into a growing DNA strand, they block further elongation of that strand.

invented it). This method uses DNA polymerase, along with special chain-terminating nucleotides called dideoxyribonucleoside triphosphates (**Figure 8–42**). Dideoxy sequencing reactions produce a collection of different DNA copies that terminate at every position in the original DNA sequence. In the original form of the method, four separate sequencing reactions were performed, each with a different dideoxyribonucleotide; the DNA copies were labeled with radioactivity and separated on polyacrylamide gels, which were then exposed to film to produce four ladders of bands that were read manually to reveal the sequence (see Figure 8–24C). This laborious method was replaced, beginning in the late 1980s, with technologies that are simpler, safer, and fully automated: robotic devices mix the reagents—including the four different chain-terminating dideoxyribonucleotides, each tagged with a different-colored fluorescent dye—and load the reaction samples onto long, thin capillary gels, which separate the reaction products into a series of distinct bands. A detector then records the color of each band, and a computer translates the information into a nucleotide sequence (**Figure 8–43**).

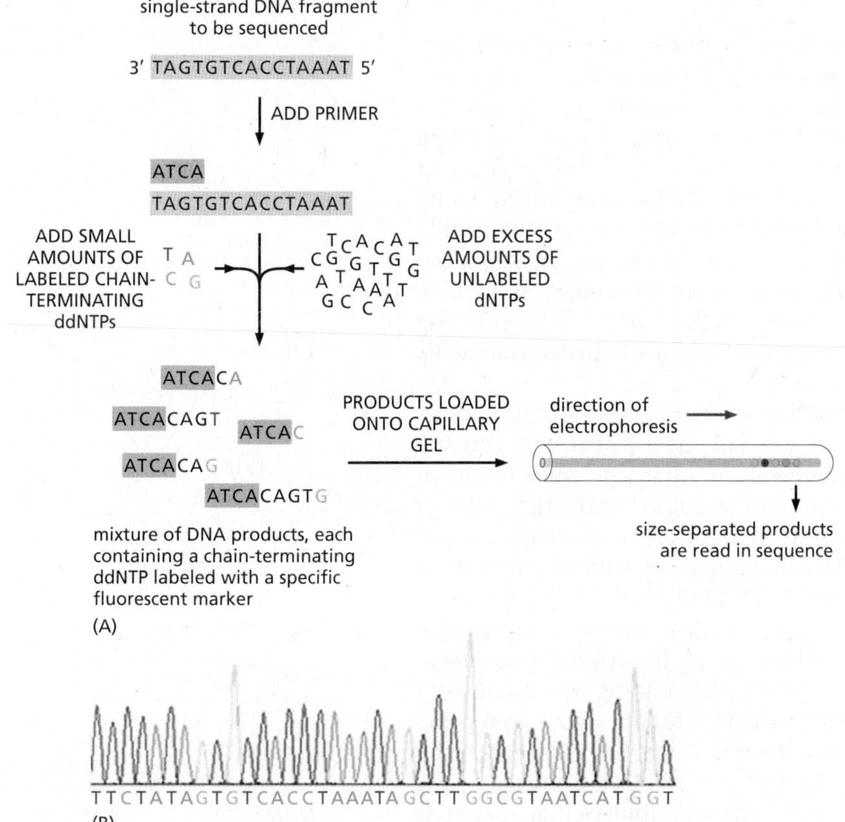

Figure 8–43 **Automated dideoxy sequencing relies on a set of four ddNTPs, each bearing a uniquely colored fluorescent tag.** (A) To determine the complete sequence of a single-strand fragment of DNA *(gray)*, the DNA is first hybridized with a short DNA primer *(orange)*. The DNA is then mixed with DNA polymerase (not shown), an excess amount of normal dNTPs, and a mixture containing small amounts of all four chain-terminating ddNTPs, each of which is labeled with a fluorescent tag of a different color. Because the chain-terminating ddNTPs will be incorporated only occasionally, each reaction produces a diverse set of DNA copies that terminate at different points in the sequence. The reaction products are loaded onto a long, thin capillary gel and separated by electrophoresis. A camera reads the color of each band on the gel and feeds the data to a computer that assembles the sequence (not shown). The sequence read from the gel will be complementary to the sequence of the original DNA molecule. (B) A tiny part of the data from such an automated sequencing run. Each colored peak represents a nucleotide in the DNA sequence.

Next-Generation Sequencing Methods Have Revolutionized DNA and RNA Analysis

Automated dideoxy sequencing was used in the late 1990s and early 2000s to determine the nucleotide sequences of many genomes, including those of *E. coli*, yeast, fruit flies, nematode worms, and humans. It continues to be used today as a low-cost approach to small-scale sequencing. But newer methods, developed since 2005, are now used for most large-scale genomic analysis. With these so-called *second-generation sequencing technologies*, the cost of sequencing DNA has decreased dramatically, and the number of sequenced genomes has increased enormously. These rapid methods allow multiple genomes to be sequenced in a matter of weeks, enabling investigators to examine thousands of individual human genomes, catalog the variation in nucleotide sequences from people around the world, and uncover the mutations that increase the risk of various diseases, from cancer to autism. These methods have also made it possible to determine the genome sequence of extinct species, including Neanderthal man and the woolly mammoth (**Movie 8.3**). By sequencing genomes from many closely related species, they have also helped us understand the molecular basis of key evolutionary events in the tree of life. The ability to rapidly sequence DNA has had major effects on all branches of biology, agriculture, and medicine; it is almost impossible to imagine where we would be without it.

Several second-generation sequencing methods are now in wide use. The most common is *Illumina sequencing*, named for the company that manufactures the equipment and reagents. This approach begins with the construction of libraries of small DNA fragments that represent the DNA of the entire genome. Instead of using bacterial cells to generate these libraries, they are made using PCR amplification of billions of DNA fragments, each attached to the glass surface of a flow cell. The amplification is carried out so that the PCR-generated copies of an original DNA fragment, instead of floating away in solution, remain bound in proximity to that original DNA fragment—resulting in a cluster of about 1000 identical copies of that small bit of the genome. These clusters—a billion of which can fit in a single flow cell—are then sequenced at the same time; that is, in parallel.

Sequencing is achieved using chain-terminating nucleotides with uniquely colored fluorescent tags. Unlike conventional dideoxy sequencing, however, the fluorescent tag and the chemical group that blocks elongation are both removable. Once DNA polymerase has added the fluorescent, chain-terminating nucleotide, a photo of the reaction records the color to reveal the identity of the nucleotide that was added. The colored label and the chain-terminating group are then removed, allowing DNA polymerase to add the next nucleotide (**Figure 8–44**). This cycle is repeated hundreds of times to provide the sequence of the DNA in each cluster. Billions of these clusters are sequenced in parallel. The full genome sequence is then reconstructed in the computer by stitching together the sequences of all fragments, using the overlaps between fragments as a guide.

Illumina sequencing provides short DNA sequences of a few hundred nucleotides, which can sometimes be difficult to assemble into a complete genome sequence because of the many repeated sequences that are often found in genomes. Recently developed *third-generation sequencing methods* are capable of sequencing much longer DNA molecules. Two methods are particularly promising. The first is *single-molecule real-time* (*SMRT*) sequencing, which is carried out in an array of tiny wells, each containing a single DNA polymerase anchored to its bottom surface. The key to SMRT sequencing is that it uses deoxyribonucleoside triphosphates in which the fluorescent dye is attached to the terminal phosphate. As the DNA polymerase copies the template DNA, the binding of a fluorescent nucleotide generates a color signal that reveals its identity. The signal disappears when the fluorescent terminal phosphate is released during incorporation of the nucleotide into the growing DNA chain (see Figure 5–4). The sequence of the DNA is thus revealed by the colors of the brief fluorescent pulses that appear as

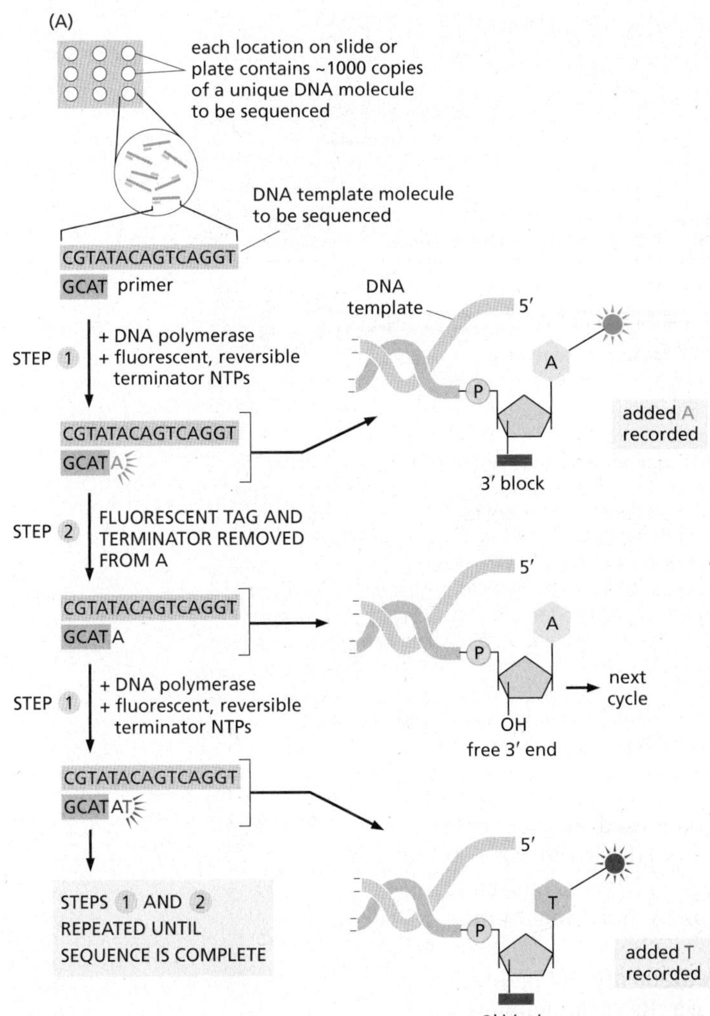

Figure 8–44 Principles of Illumina sequencing. (A) A genome or other large DNA sample is broken into millions of short fragments. These fragments are attached to the surface of a flow cell and amplified by PCR to generate DNA clusters, each containing about a thousand copies of a single DNA fragment. The large number of clusters provides complete coverage of the genome. In the first step, the anchored DNA clusters are incubated with DNA polymerase and a special set of all four nucleoside triphosphates (NTPs), each with two reversible chemical modifications: a uniquely colored fluorescent marker and a 3′ chemical group that terminates DNA synthesis. Normal dNTPs are not present. After a nucleotide is added by DNA polymerase, a high-resolution digital camera records the color of the fluorescence at each DNA cluster. In the second step, the DNA is chemically treated to remove the fluorescent markers and chemical blockers. A new batch of fluorescent, reversible terminator NTPs is then added to initiate another round of DNA synthesis. These steps are repeated until the sequence is complete. The snapshots of each round of synthesis are compiled by computer to yield the sequence of each DNA fragment. The sequence of the millions of overlapping DNA fragments can then be used to reconstruct the complete genome sequence. (B) An image of the surface of the Illumina flow cell, showing individual DNA clusters after a round of DNA synthesis with colored NTPs. (B, courtesy of Illumina, Inc.)

each nucleotide binds (Figure 8–45). Because very long DNA fragments (tens of kilobases) can be read by this method, and tens of thousands of reactions can be analyzed in parallel, complete genome sequences are well within reach in a short period of time. It is also possible to use circular DNA templates that are sequenced repeatedly on both strands, greatly improving the accuracy of the resulting sequence (see Figure 8–45C).

Another third-generation sequencing method, called *nanopore sequencing*, does not require DNA synthesis at all, but instead involves the transport of a single-strand DNA molecule through a tiny protein pore in a membrane. Voltage is applied across the membrane, resulting in current through the pore. The passage of nucleotides through the pore generates tiny shifts in electric current across the membrane, and the unique shape of each nucleotide base results in a slightly different disruption of the current. Measurement of these tiny current changes reveals the identity of each nucleotide as it passes through the pore. As in SMRT sequencing, extremely long DNAs can be sequenced in this manner. Another advantage is that modified nucleotides (such as 5-methylcytosine, depicted in Figure 7–46) can be identified because their effect on the current differs slightly from that of the unmodified nucleotide. Efforts are under way to allow direct sequencing of RNA by this approach as well. A major advantage of this method is that it can be performed with portable, handheld instruments that can be taken into the field, opening up exciting possibilities for DNA and RNA sequence analysis in global health and biology.

The development of cheaper and faster DNA sequencing methods has led to vast improvements in our ability to obtain and analyze genomic information. The

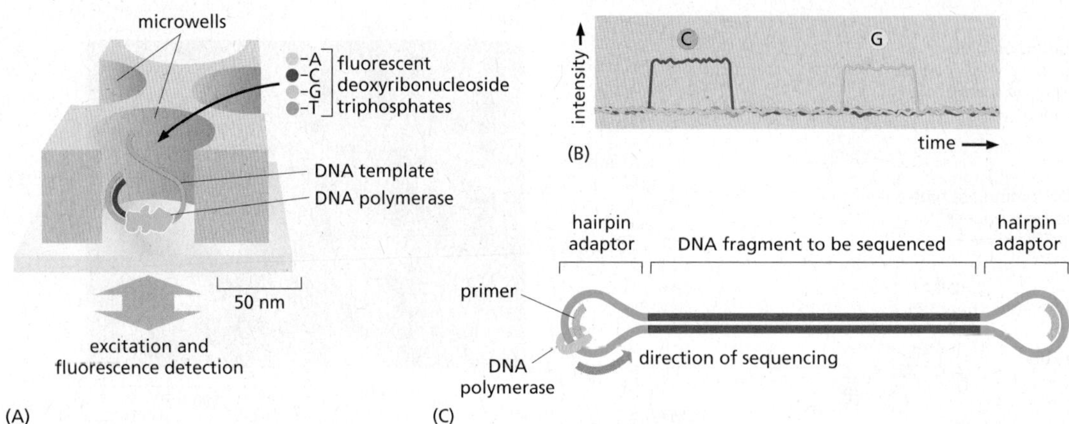

Figure 8–45 Single-molecule real-time (SMRT) sequencing. (A) SMRT sequencing uses a flow cell with thousands of tiny wells, each containing a single DNA polymerase, a single DNA template, and four fluorescently tagged deoxyribonucleoside triphosphates. Initial binding of a nucleotide to the template generates a local fluorescent signal that disappears when the terminal phosphates are removed during incorporation of the nucleotide into the DNA. To reduce background fluorescence from unbound nucleotides, only the bottom 30 nm of the well is illuminated, so that fluorescence is detected in a tiny volume (20 zeptoliters, or 20×10^{-21} liters). (B) Detection of fluorescent signals in the well reveals transient pulses of a single color, indicating the nucleotide that has been incorporated. (C) SMRT sequencing is often performed with a circular DNA template that is constructed by attaching hairpin adaptor DNAs to each end of the DNA to be sequenced. Using a primer that matches the adaptor, DNA polymerase can then replicate the template as shown in panel A. The enzyme used in this method is a *strand-displacing* polymerase that separates double-stranded DNA as it moves along the template, allowing it to continue around the entire circular molecule many times. Thus, both strands of the DNA are repeatedly sequenced, allowing the experimenter to eliminate sequence errors that arise from random mistakes made by the polymerase.

original "reference" sequence of the human genome, completed in 2003, cost more than $1 billion and required many scientists from around the world working together for 13 years. The enormous progress made in the past 15 years has made it possible for a single person to complete the sequence of an individual human genome in less than a day, at a cost of less than $1000.

As mentioned above, next-generation sequencing methods are being developed for the direct sequencing of RNA. Currently, however, RNA sequencing is typically carried out by converting the RNA to cDNA (using reverse transcriptase) and using one of the methods described above for DNA sequencing. It is important to keep in mind that although genomes remain the same from cell to cell and from tissue to tissue, the RNA produced from the genome can vary enormously. We will see later in this chapter that sequencing the entire repertoire of RNA from a cell or tissue (known as **deep RNA sequencing**, or **RNA-seq**) is a powerful way to understand how the information present in the genome is used by different cells under different circumstances. RNA-seq is also a valuable tool for annotating genomes, as we discuss next.

To Be Useful, Genome Sequences Must Be Annotated

Long strings of nucleotides, at first glance, reveal nothing about how this genetic information directs the development of a living organism—or even what types of DNA, protein, and RNA molecules are produced by a genome. The process of **genome annotation** attempts to mark out all the genes (both protein-coding and noncoding) in a genome and ascribe a role to each. It also seeks to understand more subtle types of genome information, such as the *cis*-regulatory sequences that specify the time and place that a given gene is expressed and whether its mRNA undergoes alternative splicing to produce different protein isotypes. Clearly, this is a daunting task, and we are far short of completing it for any form of life, even the simplest bacterium. For many organisms, we know the approximate number of genes, and, for very simple organisms, we understand the functions of about half their genes.

How does one begin to make sense of a genome sequence? The first step is usually to translate *in silico* the entire genome into protein. There are six different

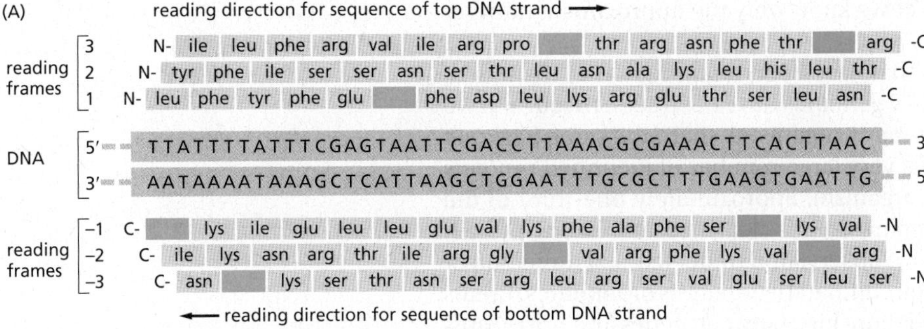

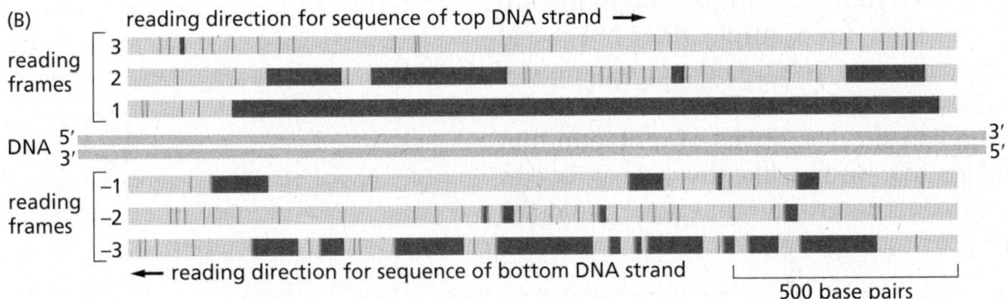

Figure 8–46 **Finding the regions in a DNA sequence that encode a protein.** (A) Any region of the DNA sequence can, in principle, code for six different amino acid sequences, because any one of three different reading frames can be used to interpret the nucleotide sequence on each strand. Note that a nucleotide sequence is always read in the 5′-to-3′ direction and encodes a polypeptide from the N-terminus to the C-terminus. For a random nucleotide sequence read in a particular frame, a stop signal for protein synthesis is encountered, on average, about once every 20 amino acids. In this sample sequence of 48 base pairs, each such signal (*stop codon*) is colored *blue*, and only reading frame 2 lacks a stop signal. (B) Search of a 1700-base-pair DNA sequence for a possible protein-encoding sequence. The information is displayed as in panel A, with each stop signal for protein synthesis denoted by a *blue line*. In addition, all of the regions between possible start and stop signals for protein synthesis are displayed as *red bars*. Only reading frame 1 actually encodes a protein, which is 475 amino acid residues long.

reading frames for any piece of double-stranded DNA (three on each strand). We saw in Chapter 6 that a random sequence of nucleotides, read in frame, will contain a stop codon about every 20 amino acids. In contrast, protein-coding regions will usually contain much longer stretches without stop codons (**Figure 8–46**). Known as **open reading frames (ORFs)**, these usually signify bona fide protein-coding genes. This assignment is often "double-checked" by comparing the ORF amino acid sequence to the many databases of documented proteins from other species. If a match is found, even an imperfect one, it is very likely that the ORF will code for a functional protein (see Figure 8–22).

This strategy works very well for compact genomes, where intron sequences are rare and ORFs often extend for many hundreds of amino acids. However, in many animals and plants, the average exon size is 150–200 nucleotide pairs, and additional information is usually required to unambiguously locate all the exons of a gene. Although it is possible to search genomes for splicing signals and other features that help to identify exons (codon bias, for example), one of the most powerful methods is simply to sequence the total RNA produced from the genome in living cells. As can be seen in Figure 7–3, this RNA-seq information, when mapped onto the genome sequence, can be used to accurately locate all the introns and exons of even complex genes. By sequencing total RNA from different cell types, it is also possible to identify cases of alternative splicing.

RNA-seq also identifies noncoding RNAs produced by a genome. Although the function of some of them can be readily recognized (tRNAs or snoRNAs, for example), many have unknown functions and still others probably have no function at all. The existence of the many noncoding RNAs and our relative ignorance of their function is the main reason that we know only the approximate number of genes in the human genome.

But even for protein-coding genes that have been unambiguously identified, we still have much to learn. Thousands of genomes have been sequenced, and we know from *comparative genomics* that many organisms share the same basic set of proteins. However, the functions of a very large number of identified proteins remain unknown. Depending on the organism, approximately one-third of the proteins encoded by a sequenced genome do not clearly resemble any protein that has been studied biochemically. This observation underscores a limitation of the emerging field of genomics: although comparative analysis of genomes reveals a great deal of information about the relationships between genes and organisms, it often does not provide immediate information about how these genes function or what roles they have in the physiology of an organism. Comparison of the full gene complement of several thermophilic bacteria, for example, does not reveal why these bacteria thrive at temperatures exceeding 70°C. And examination of the genome of the incredibly radiation-resistant bacterium *Deinococcus radiodurans* does not explain how this organism can survive a blast of radiation that can shatter glass. Further biochemical and genetic studies, like those described in the other sections of this chapter, are required to determine how genes, and the proteins they produce, function in the context of living organisms.

Summary

DNA cloning allows a copy of any specific part of a DNA or RNA sequence to be selected from the millions of other sequences in a cell and produced in unlimited amounts in pure form. DNA sequences can be amplified by inserting the desired DNA fragment into a self-replicating genetic element such as a bacterial plasmid. Bypassing cloning vectors and bacterial cells altogether, the polymerase chain reaction (PCR) allows DNA cloning to be performed directly with a DNA polymerase and DNA primers—provided that the DNA sequence of interest is already known.

The procedures used to obtain DNA clones that correspond in sequence to mRNA molecules are the same, except that a DNA copy of the mRNA sequence, called cDNA, is first made. Unlike genomic DNA clones, cDNA clones lack intron sequences, making them the clones of choice for analyzing the protein product of a gene.

Nucleic acid hybridization reactions provide a sensitive means of detecting any nucleotide sequence of interest. The enormous specificity of this hybridization reaction allows any single-strand sequence of nucleotides to be labeled with a radioisotope or chemical and used as a probe to find a complementary partner strand, even in a cell or cell extract that contains millions of different DNA and RNA sequences. DNA hybridization also makes it possible to use PCR to amplify any section of any genome once its sequence is known.

The nucleotide sequence of any genome can be determined rapidly and simply by using highly automated techniques that are based on several different strategies. Comparison of the genome sequences of different organisms allows us to trace the evolutionary relationships among genes and organisms, and it has proved valuable for discovering new genes and predicting their functions.

STUDYING GENE FUNCTION AND EXPRESSION

Ultimately, our goal is to understand how genes—and the proteins they encode—function in the intact organism. Although it may seem counterintuitive, one of the most direct ways to find out what a gene does is to see what happens to the organism when that gene is missing. Studying mutant organisms that have acquired changes or deletions in their nucleotide sequences is a time-honored practice in biology and forms the basis of the important field of **genetics**. Because mutations can disrupt cell processes, mutants often hold the key to understanding gene function. In the classical genetic approach, one begins by isolating mutants that have an interesting or unusual appearance: fruit flies with white eyes or curly wings, for example. Working backward from the **phenotype**—the appearance or behavior of the individual—one then determines the organism's **genotype**, the form of the gene responsible for that characteristic (Panel 8–1).

Today, with numerous genome sequences available, the exploration of gene function often begins with a DNA sequence. Here, the challenge is to translate sequence into function. One approach, discussed earlier in the chapter, is to search databases for well-characterized proteins that have similar amino acid sequences to the protein encoded by a new gene. From there, the protein can be overexpressed and purified, and the methods described earlier in this chapter can be employed to study its biochemical properties and three-dimensional structure. But to determine directly a gene's function in a cell or organism, the most effective approach involves studying mutants that either lack the gene or express an altered version of it. Determining which cell processes have been disrupted or compromised in such mutants will usually shed light on a gene's biological role.

In this section, we describe several approaches to determining a gene's function, starting either from an individual with an interesting phenotype or from a DNA sequence. We begin with the classical genetic approach, which starts with a *genetic screen* for isolating mutants of interest and then proceeds toward identification of the gene or genes responsible for the observed phenotype. We then describe the set of techniques that are sometimes called reverse genetics, in which one begins with a gene or gene sequence and attempts to determine its function. This approach often involves some intelligent guesswork—searching for similar sequences in other organisms or determining when and where a gene is expressed—as well as generating mutant organisms and characterizing their phenotype.

Classical Genetic Screens Identify Random Mutants with Specific Abnormalities

Before the advent of gene cloning technology, most genes were identified by the abnormalities produced when the gene was mutated. Indeed, the very concept of the gene was deduced from the heritability of such abnormalities. This classical genetic approach—identifying the genes responsible for mutant phenotypes—is most easily performed in organisms that reproduce rapidly and are amenable to genetic manipulation, such as bacteria, yeasts, nematode worms, and fruit flies. Although spontaneous mutants can sometimes be found by examining extremely large populations—thousands or tens of thousands of individual organisms—isolating mutant individuals is much more efficient if one generates mutations with chemicals or radiation that damage DNA. By treating organisms with such mutagens, very large numbers of mutant individuals can be created quickly and then screened for a particular defect of interest.

An alternative approach to chemical or radiation mutagenesis is called *insertional mutagenesis.* This method relies on the fact that exogenous DNA inserted randomly into the genome can produce mutations if the inserted fragment interrupts a gene or its regulatory sequences. The inserted DNA, whose sequence is known, then serves as a molecular tag that aids in the subsequent identification and cloning of the disrupted gene (Figure 8–47). In *Drosophila,* the use of the transposable P element to inactivate genes has revolutionized the study of gene function in the fly. Transposable elements (see Table 5–4, p. 308) have also been used to generate mutations in bacteria, yeast, mice, and the flowering plant *Arabidopsis.*

Once a collection of mutants in a model organism has been produced, one generally must examine thousands of individuals to find the altered phenotype of interest. Such a search is called a **genetic screen**, and the larger the genome, the less likely it is that any particular gene will be mutated. Therefore, the larger the genome of an organism, the bigger the screening task becomes. The phenotype being screened for can be simple or complex. Simple phenotypes are easiest to detect: one can screen many organisms rapidly, for example, for mutations

Figure 8–47 **Insertional mutant of the snapdragon, *Antirrhinum*.** A mutation in a single gene coding for a regulatory protein causes leafy shoots *(left)* to develop in place of flowers, which occur in the normal plant *(right)*. The mutation causes cells to adopt a character that would be appropriate to a different part of the normal plant, so instead of a flower, the cells produce a leafy shoot. (Courtesy of Enrico Coen and Rosemary Carpenter.)

GENES AND PHENOTYPES

Gene: a functional unit of inheritance, usually corresponding to the segment of DNA coding for a single protein.

Genome: all of an organism's DNA sequences.

locus: the site of the gene in the genome

alleles: alternative forms of a gene

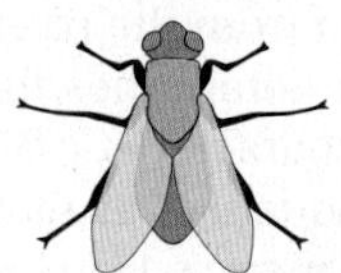

Wild type: the normal, naturally occurring type

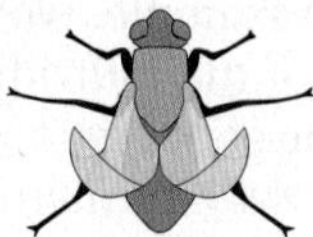

Mutant: differing from the wild type because of a genetic change (a mutation)

GENOTYPE: the specific set of alleles forming the genome of an individual

PHENOTYPE: the visible character of the individual

homozygous *A/A*

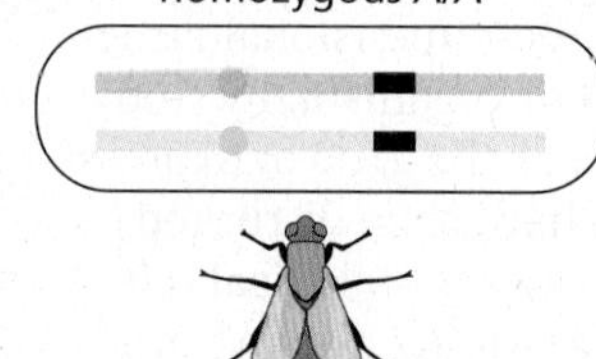

heterozygous *a/A*

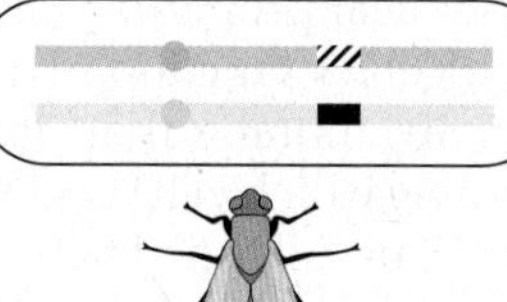

homozygous *a/a*

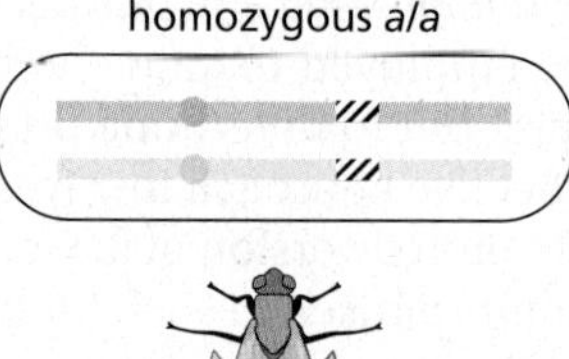

allele *A* is dominant (relative to *a*); allele *a* is recessive (relative to *A*)

In the example above, the phenotype of the heterozygote is the same as that of one of the homozygotes; in cases where it is different from both, the two alleles are said to be codominant.

CHROMOSOMES

centromere

a chromosome at the beginning of the cell cycle, in G_1 phase; the single long bar represents one long double helix of DNA

short "p" arm

long "q" arm

a chromosome near the end of the cell cycle, in metaphase; it is duplicated and condensed, consisting of two identical sister chromatids (each containing one DNA double helix) joined at the centromere.

short "p" arm

long "q" arm

pair of autosomes

maternal 1

paternal 1

maternal 3

paternal 3

paternal 2

maternal 2

X

Y

sex chromosomes

A normal diploid chromosome set, as seen in a metaphase spread, prepared by bursting open a cell at metaphase and staining the scattered chromosomes. In the example shown schematically here, there are three pairs of autosomes (chromosomes inherited symmetrically from both parents, regardless of sex) and two sex chromosomes—an X from the mother and a Y from the father. The numbers and types of sex chromosomes and their role in sex determination are variable from one class of organisms to another, as is the number of pairs of autosomes.

THE HAPLOID–DIPLOID CYCLE OF SEXUAL REPRODUCTION

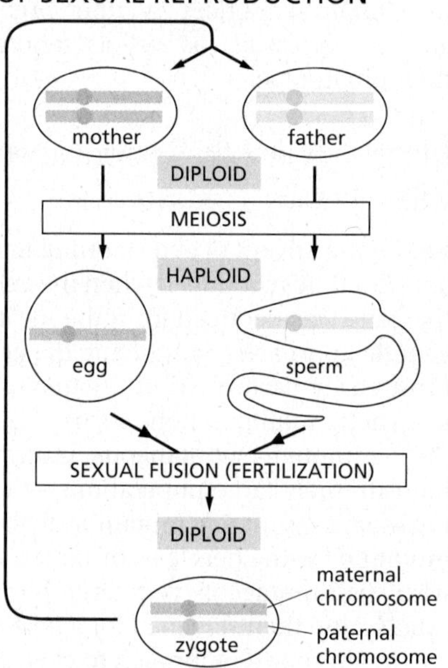

For simplicity, the cycle is shown for only one chromosome/chromosome pair.

MEIOSIS AND GENETIC RECOMBINATION

maternal chromosome

A *B*

paternal chromosome

a *b*

diploid germ cell

genotype $\frac{AB}{ab}$

MEIOSIS AND RECOMBINATION

genotype *Ab*

A *b*

site of crossing-over

genotype *aB*

a *B*

haploid gametes (eggs or sperm)

The greater the distance between two loci on a single chromosome, the greater is the chance that they will be separated by crossing-over occurring at a site between them. If two genes are thus reassorted in *x*% of gametes, they are said to be separated on a chromosome by a genetic map distance of *x* map units (or *x* centimorgans).

TYPES OF MUTATIONS

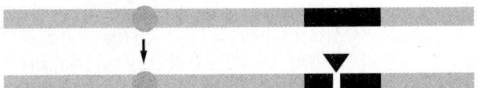

POINT MUTATION: maps to a single site in the genome, corresponding to a single nucleotide pair or a very small part of a single gene

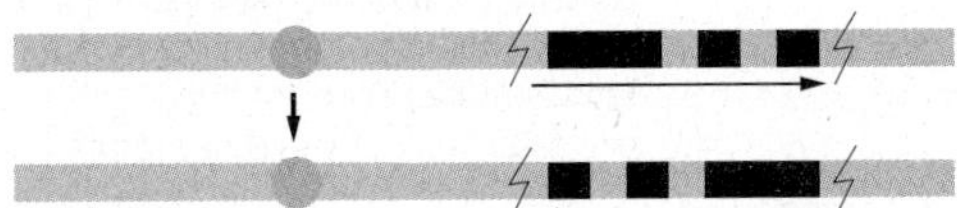

INVERSION: inverts a segment of a chromosome

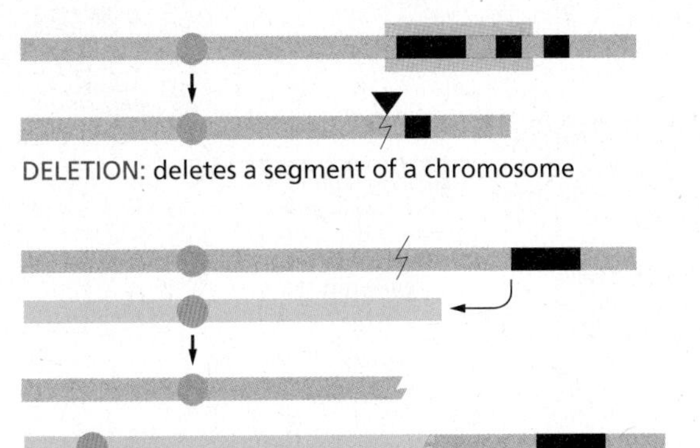

DELETION: deletes a segment of a chromosome

TRANSLOCATION: breaks off a segment from one chromosome and attaches it to another

lethal mutation: causes the developing organism to die prematurely.

conditional mutation: produces its phenotypic effect only under certain conditions, called the *restrictive* conditions. Under other conditions—the *permissive* conditions—the effect is not seen. For a *temperature-sensitive* mutation, the restrictive condition typically is high temperature, while the permissive condition is low temperature.

loss-of-function mutation: either reduces or abolishes the activity of the gene. These are the most common class of mutations. Loss-of-function mutations are usually *recessive*—the organism can usually function normally as long as it retains at least one normal copy of the affected gene.

null mutation: a loss-of-function mutation that completely abolishes the activity of the gene.

gain-of-function mutation: increases the activity of the gene or makes it active in inappropriate circumstances; these mutations are usually *dominant.*

dominant-negative mutation: dominant-acting mutation that blocks gene activity, causing a loss-of-function phenotype even in the presence of a normal copy of the gene. This phenomenon occurs when the mutant gene product interferes with the function of the normal gene product.

suppressor mutation: suppresses the phenotypic effect of another mutation, so that the double mutant seems normal. An *intragenic* suppressor mutation lies within the gene affected by the first mutation; an *extragenic* suppressor mutation lies in a second gene—often one whose product interacts directly with the product of the first.

TWO GENES OR ONE?

Given two mutations that produce the same phenotype, how can we tell whether they are mutations in the same gene? If the mutations are recessive (as they most often are), the answer can be found by a complementation test.

In the simplest type of complementation test, an individual who is homozygous for one mutation is mated with an individual who is homozygous for the other. The phenotype of the offspring gives the answer to the question.

COMPLEMENTATION: MUTATIONS IN TWO DIFFERENT GENES

homozygous mutant mother

homozygous mutant father

a a b b a b

hybrid offspring shows normal phenotype: one normal copy of each gene is present

NONCOMPLEMENTATION: TWO INDEPENDENT MUTATIONS IN THE SAME GENE

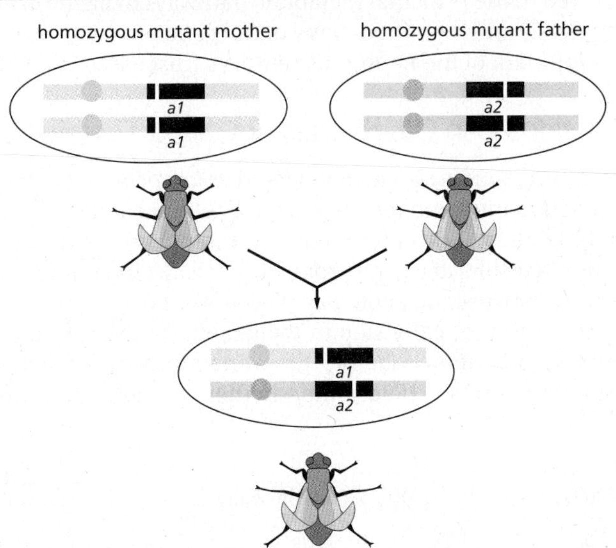

hybrid offspring shows mutant phenotype: no normal copies of the mutated gene are present

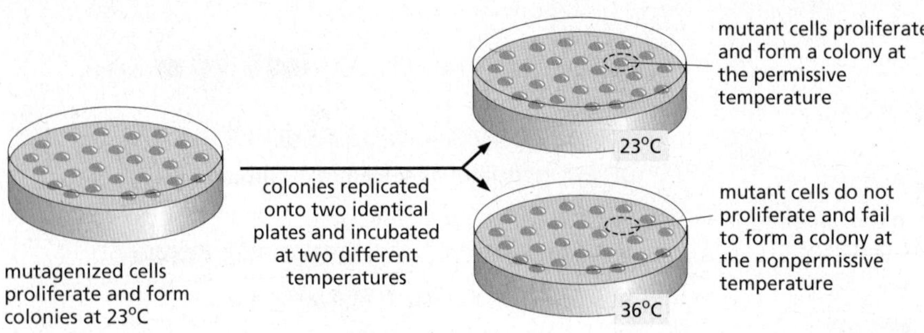

Figure 8–48 Screening for temperature-sensitive bacterial or yeast mutants. Mutagenized cells are plated out at the permissive temperature. They divide and form colonies, which are transferred to two identical Petri dishes by replica plating. One of these plates is incubated at the permissive temperature, the other at the restrictive temperature. Cells containing a temperature-sensitive mutation in a gene essential for proliferation can divide at the normal (permissive) temperature but fail to divide at the elevated (restrictive) temperature.

that make it impossible for the organism to survive in the absence of a particular amino acid or nutrient.

Because defects in genes that are required for fundamental cell processes—RNA synthesis and processing or cell-cycle control, for example—are usually lethal, the functions of these genes are often studied in individuals with **conditional mutations**. The mutant individuals function normally as long as *permissive* conditions prevail but demonstrate abnormal gene function when subjected to *restrictive* (*nonpermissive*) conditions. In organisms with *temperature-sensitive mutations*, for example, the abnormality can be switched on and off experimentally simply by changing the ambient temperature; thus, a cell containing a temperature-sensitive mutation in a gene essential for survival will die at a restrictive temperature but proliferate normally at a permissive temperature (**Figure 8–48**). The temperature-sensitive gene in such a mutant usually contains a point mutation that causes a subtle change in its protein product; for example, the mutant protein may function normally at low temperatures but unfold at higher temperatures.

Temperature-sensitive mutations were crucial to find the bacterial genes that encode the proteins required for DNA replication. The mutants were identified by screening populations of mutagen-treated bacteria for cells that stop making DNA when they are warmed from 30°C to 42°C. These mutants were later used to identify and characterize the corresponding DNA replication proteins (discussed in Chapter 5). Similarly, screens for temperature-sensitive mutations in yeast led to the identification of many proteins involved in regulating the cell cycle, as well as many proteins involved in moving proteins through the secretory pathway. Related screening approaches demonstrated the function of enzymes involved in the principal metabolic pathways of bacteria and yeast (discussed in Chapter 2) and identified many of the gene products responsible for the orderly development of the *Drosophila* embryo (discussed in Chapter 21).

Mutations Can Cause Loss or Gain of Protein Function

Gene mutations are generally classed as *loss of function* or *gain of function*. A loss-of-function mutation results in a gene product that either does not work or works too little; thus, it can reveal the normal function of the gene. A gain-of-function mutation results in a gene product that works too much, works at the wrong time or place, or works in a new way (**Figure 8–49**).

An important early step in the genetic analysis of any mutant cell or organism is to determine whether the mutation causes a loss or a gain of function. A standard test is to determine whether the mutation is *dominant* or *recessive*. A

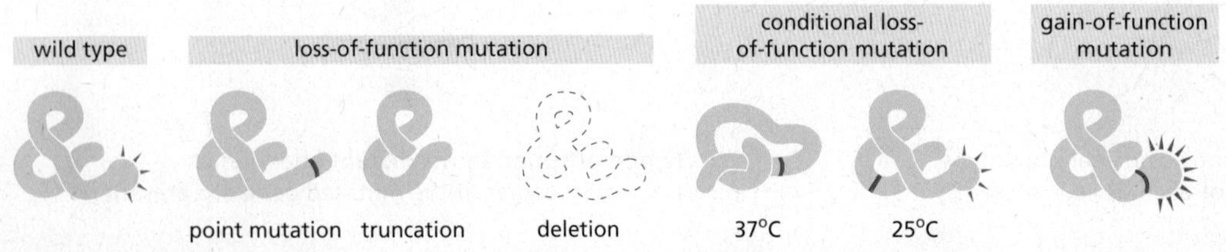

Figure 8–49 Gene mutations that affect their protein product in different ways. In this example, the wild-type protein has a specific cell function denoted by the *red rays*. Mutations that eliminate this function or inactivate it at higher temperatures are shown. The conditional mutant protein carries an amino acid substitution *(red)* that prevents its proper folding at 37°C but allows the protein to fold and function normally at 25°C. Such temperature-sensitive conditional mutations are especially useful for studying essential genes (see Figure 8–48). In some cases, a mutation increases the activity of the mutant protein.

dominant mutation is one that still causes the mutant phenotype in the presence of a single copy of the wild-type gene. A recessive mutation is one that is no longer able to cause the mutant phenotype in the presence of a single wild-type copy of the gene. In the majority of cases, recessive mutations are loss of function and dominant mutations are gain of function—although cases have been described in which a loss-of-function mutation is dominant or a gain-of-function mutation is recessive. It is easy to determine if a mutation is dominant or recessive. One simply mates a mutant with a wild type to obtain diploid cells or organisms. The progeny from the mating will be heterozygous for the mutation. If the mutant phenotype is no longer observed, one can conclude that the mutation is recessive and is very likely to be a loss-of-function mutation (see Panel 8–1).

Complementation Tests Reveal Whether Two Mutations Are in the Same Gene or Different Genes

A large-scale genetic screen can turn up many different mutations that show the same phenotype. These defects might lie in different genes that function in the same process or they might represent different mutations in the same gene. Alternative forms of the same gene are known as **alleles**. The most common difference between alleles is a substitution of a single nucleotide pair, but different alleles can also bear deletions, substitutions, and duplications. How can we tell, then, whether two mutations that produce the same phenotype occur in the same gene or in different genes? If the mutations are recessive—if, for example, they represent a loss of function of a particular gene—a **complementation test** can be used to ascertain whether the mutations fall in the same gene or in different genes. To test complementation in a diploid organism, an individual that is homozygous for one mutation—that is, it possesses two identical alleles of the mutant gene in question—is mated with an individual that is homozygous for the other mutation. If the two mutations are in the same gene, the offspring show the mutant phenotype, because they still will have no normal copies of the gene in question. If, in contrast, the mutations fall in different genes, the resulting offspring show a normal phenotype, because they retain one normal copy (and one mutant copy) of each gene; the mutations thereby complement one another and restore a normal phenotype (Figure 8–50). Complementation testing of mutants identified during genetic screens has revealed, for example, that 5 different genes are required for yeast to digest the sugar galactose, 20 genes are needed for *E. coli* to build a functional flagellum, 48 genes are involved in assembling bacteriophage T4 viral particles, and hundreds of genes are involved in the development of an adult nematode worm from a fertilized egg.

Figure 8–50 **A complementation test can reveal that mutations in two different genes are responsible for the same abnormal phenotype.** When an albino (white) bird from one strain is bred with an albino from a different strain, the resulting offspring (bottom) have normal coloration. This restoration of the wild-type plumage indicates that the two white breeds lack color because of recessive mutations in different genes. (From W. Bateson, Mendel's Principles of Heredity, 1st ed. Cambridge, UK: Cambridge University Press, 1913.)

Gene Products Can Be Ordered in Pathways by Epistasis Analysis

Once a set of genes involved in a particular biological process has been identified, it is helpful to determine the order in which the genes function. Gene order is perhaps easiest to explain for metabolic pathways, where, for example, enzyme A is necessary to produce the substrate for enzyme B. In this case, we would say that the gene encoding enzyme A acts before (upstream of) the gene encoding enzyme B in the pathway. Similarly, where one protein regulates the activity of another protein, we would say that the former gene acts before the latter. Gene order can, in many cases, be determined purely by genetic analysis without any knowledge of the mechanism of action of the gene products involved.

Suppose we have a biosynthetic process consisting of a sequence of steps, such that performance of step B requires completion of the preceding step A; and suppose gene *A* is required for step A, and gene *B* is required for step B. Then a null mutation (a mutation that abolishes function) in gene *A* will arrest the process at step A, regardless of whether gene *B* is functional or not, whereas a null mutation in gene *B* will cause arrest at step B only if gene *A* is still active. In such a case, gene *A* is said to be *epistatic* to gene *B*. By comparing the phenotypes of different combinations of mutations, we can therefore discover the order in which

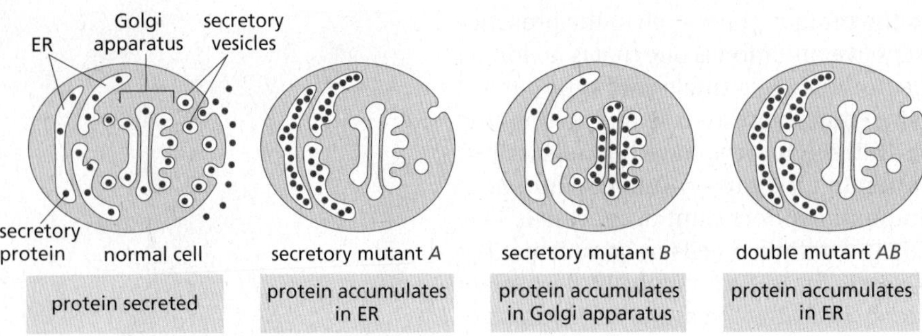

Figure 8–51 Using genetics to determine the order of function of genes. In normal cells, secretory proteins are loaded into vesicles, which fuse with the plasma membrane to secrete their contents into the extracellular medium. Two mutants, *A* and *B*, fail to secrete proteins. In mutant *A*, secretory proteins accumulate in the ER. In mutant *B*, secretory proteins accumulate in the Golgi. In the double mutant *AB*, proteins accumulate in the ER; this indicates that the gene defective in mutant *A* acts before the gene defective in mutant *B* in the secretory pathway.

the genes act. This type of analysis is called **epistasis analysis**. As an example, the pathway of protein secretion in yeast has been analyzed in this way. Different mutations in this pathway cause proteins to accumulate aberrantly in the endoplasmic reticulum (ER) or in the Golgi apparatus. When a yeast cell is engineered to carry both a mutation that blocks protein processing in the ER *and* a mutation that blocks processing in the Golgi apparatus, proteins accumulate in the ER. This indicates that proteins must pass through the ER before being sent to the Golgi before secretion (**Figure 8–51**). Strictly speaking, an epistasis analysis can only provide information about gene order in a pathway when both mutations are null alleles. When the mutations retain partial function, their epistasis interactions can be difficult to interpret.

Sometimes, a double mutant will show a new or more severe phenotype than either single mutant alone. This type of genetic interaction is called a *synthetic* phenotype, and if the phenotype is death of the organism, it is called *synthetic lethality*. In most cases, a synthetic phenotype indicates that the two genes act in two different parallel pathways, either of which is capable of mediating the same cell process. Thus, when both pathways are disrupted in the double mutant, the process fails altogether, and the synthetic phenotype is observed.

Mutations Responsible for a Phenotype Can Be Identified Through DNA Analysis

Once a collection of mutant organisms with interesting phenotypes has been obtained, the next task is to identify the gene or genes responsible for the altered phenotype. If the phenotype has been produced by insertional mutagenesis, locating the disrupted gene is fairly simple. DNA fragments containing the insertion (a transposon or a retrovirus, for example) are amplified by PCR, and the nucleotide sequence of the flanking DNA is determined. The gene affected by the insertion can then be identified by a computer-aided search of the complete genome sequence of the organism.

If a DNA-damaging chemical was used to generate the mutations, identifying the inactivated gene is often more laborious, but there are several powerful strategies available. With recent advances in DNA sequencing technology, it is possible to simply determine the genome sequence of the mutant organism and identify the affected gene by comparison with the wild-type sequence. Because of the continual accumulation of neutral mutations, there will probably be differences between the two genome sequences in addition to the mutation responsible for the phenotype. One way of proving that a mutation is causative is to introduce the putative mutation back into a normal organism and determine whether or not it causes the mutant phenotype. We will discuss how this is accomplished later in the chapter.

Rapid and Cheap DNA Sequencing Has Revolutionized Human Genetic Studies

Genetic screens in model experimental organisms have been spectacularly successful in identifying genes and relating them to various phenotypes, including many that are conserved between these organisms and humans. But how can we study humans directly? They do not reproduce rapidly, cannot be treated with mutagens, and, if they have a defect in an essential process such as DNA replication, would die long before birth.

Despite their limitations compared to model organisms, humans are becoming increasingly attractive subjects for genetic studies. Because the human population is so large, spontaneous nonlethal mutations have arisen many times in all human genes. A substantial proportion of these mutations remains in the genomes of present-day humans. Deleterious mutations are discovered when the mutant individuals call attention to themselves by seeking medical help.

With the recent advances that have enabled the sequencing of entire human genomes cheaply and quickly, we can now identify such mutations and study their evolution and inheritance in ways that were impossible even a few years ago. By comparing the sequences of thousands of human genomes from all around the world, we can begin to identify directly the DNA differences that distinguish one individual from another. These differences hold clues to our evolutionary origins and can be used to explore the roots of disease.

Linked Blocks of Polymorphisms Have Been Passed Down from Our Ancestors

When we compare the sequences of multiple human genomes, we find that any two individuals will differ in roughly 1 nucleotide pair in 1000. As described in Chapter 4, most sequence variation results from substitution of a single nucleotide, called a *single-nucleotide variant* (*SNV*), while other variation is due to structural chromosome changes such as deletions and rearrangements. Human genetic studies have benefited greatly from a particularly common type of sequence variants, present in more than 1% of the population, that are called **polymorphisms**—most of which are **single-nucleotide polymorphisms**, or **SNPs** (Figure 8–52). Although these common variants can be found throughout the genome, they are not scattered randomly—or even independently. Instead, they tend to travel in groups called **haplotype blocks**—combinations of polymorphisms that are inherited as a unit.

To understand why such haplotype blocks exist, we need to consider our evolutionary history. It is thought that modern humans expanded from a relatively small population—perhaps around 10,000 individuals—that existed in Africa about 200,000 years ago. Among that small group of our ancestors, some individuals will have carried one set of genetic variants, others a different set. The chromosomes of a present-day human represent a shuffled combination of chromosome segments from different members of this small ancestral group of people. Because only about 2000 generations separate us from them, large segments of these ancestral chromosomes have passed from parent to child, unbroken by the crossover events that occur during meiosis. As described in Chapter 5, only a few crossovers occur between each set of homologous chromosomes during each meiosis (see Figure 5–52).

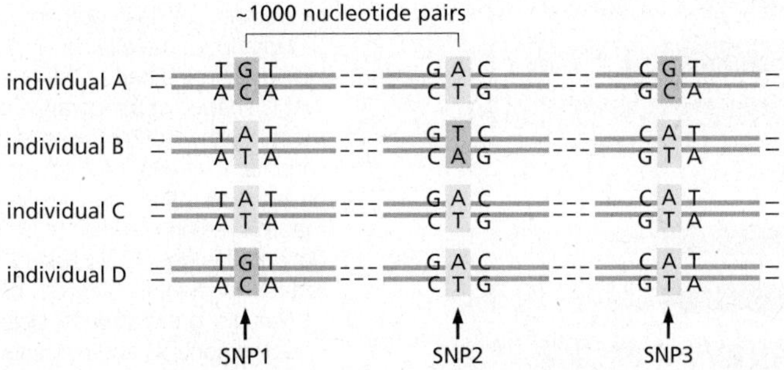

Figure 8–52 **Single-nucleotide polymorphisms (SNPs) are sites in the genome where two or more alternative choices of a nucleotide are common in the population.** Most such variations in the human genome occur at locations where they do not significantly affect a gene's function.

As a result, certain sets of DNA sequences—and their associated polymorphisms—have been inherited in linked groups, with little genetic rearrangement across the generations. These are the haplotype blocks. Like genes that exist in different allelic forms, haplotype blocks also come in a limited number of variants that are common in the human population, each representing a combination of DNA polymorphisms passed down from a particular ancestor long ago.

Sequence Variants Can Aid the Search for Mutations Associated with Disease

Mutations that give rise, in a reproducible way, to rare but clearly defined differences, such as albinism, hemophilia, or congenital deafness, can often be identified by studies of affected families. Such single-gene, or monogenic, disorders are referred to as *Mendelian* because their pattern of inheritance is easy to track. Moreover, individuals who inherit the causative mutation will exhibit the disorder irrespective of environmental factors such as diet or exercise. But for many common disorders, the genetic roots are more complex. Instead of a single allele of a single gene, such disorders stem from a combination of contributions from multiple genes. And often, environmental factors have strong influences on the severity of the disorder. For these *multigenic* conditions, such as diabetes or arthritis, population studies are often helpful in tracking down the genes that increase the risk of getting the disease.

In population studies, investigators collect DNA samples from a large number of people who have the disease and compare them to samples from a group of people who do not have the disease. They look for variants—SNPs, for example—that are more common among the people who have the disease. Because DNA sequences that are close together on a chromosome tend to be inherited together, the presence of such SNPs could indicate that an allele that increases the risk of the disease might lie nearby (**Figure 8–53**). Although, in principle, the disease could be caused by the SNP itself, the culprit is much more likely to be a change that is merely linked to the SNP as part of a haplotype block.

Such *genome-wide association studies* have been used to search for genes that predispose individuals to common diseases, including diabetes, coronary artery disease, rheumatoid arthritis, and even depression. For many of these conditions, the DNA polymorphisms identified increase the risk of disease only slightly. Moreover, environmental factors (diet and exercise, for example) play an important role in the onset and severity of the disease. Nonetheless, the identification of potential disease genes linked to polymorphisms is leading to a mechanistic understanding of some of our most common disorders.

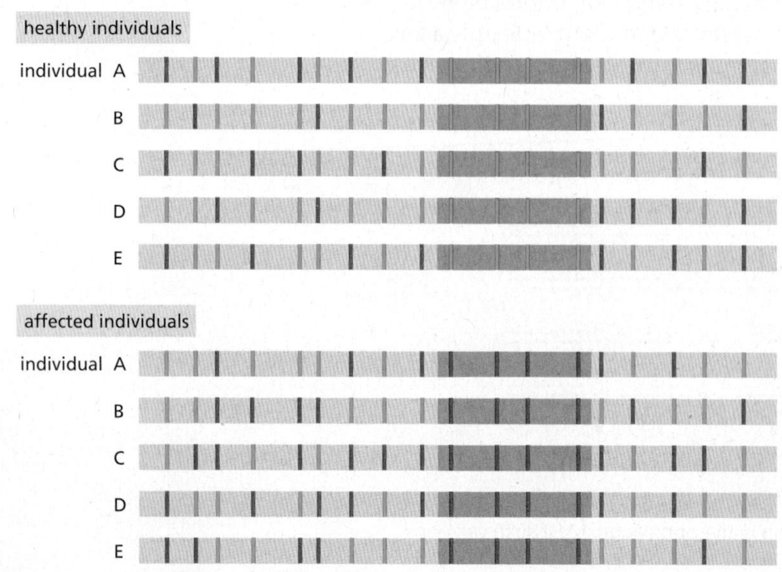

Figure 8–53 **Genes that affect the risk of developing a common disease can often be tracked down through linkage to SNPs.** Here, the patterns of SNPs are compared between two sets of individuals—a set of healthy controls and a set affected by a particular common disease. A segment of a typical chromosome is shown. For most polymorphic sites in this segment, it is a random matter whether an individual has one SNP variant *(red vertical bars)* or another *(blue vertical bars)*; this same randomness is seen both for the control group and for the affected individuals. However, in the part of the chromosome that is shaded in *dark gray*, a bias is seen: most healthy individuals have the blue SNP variants, whereas most affected individuals have the red SNP variants. This suggests that this region contains or is close to a gene that is genetically linked to these red SNP variants and which predisposes individuals to the disease. Using carefully selected controls and thousands of affected individuals, this approach can help track down disease-related genes, even when they confer only a slight increase in the risk of developing the disease.

Genomics Is Accelerating the Discovery of Rare Mutations That Predispose Us to Serious Disease

The polymorphisms that have allowed us to identify some of the genes that increase our risk of disease are common. They arose long ago in our evolutionary past and are now present, in one form or another, in a substantial portion of the population. Such polymorphisms are thought to account for about 90% of the differences between one person's genome and another. But when we try to tie these common variants to differences in disease susceptibility or other heritable traits, such as height, we find that they do not have as much predictive power as we had anticipated: thus, for example, most confer relatively small increases—less than twofold—in the risk of developing a common disease.

Part of the problem is that many of the mutations that are directly responsible for complex human diseases appeared more recently in our evolutionary history—during a period when the human population expanded from the few million individuals who existed 10,000 years ago to the more than 7 billion who exist today. Because recent mutations occur more rarely than the ancient polymorphisms that are common in the human population, they could slip through the genome-wide association studies just described.

In contrast to polymorphisms, rare DNA variants—those much less frequent in humans than SNPs—can have large effects on the risk of developing some common diseases. For example, numerous loss-of-function mutations, each individually rare, have been found to increase greatly the predisposition to autism and schizophrenia. Many of these are *de novo* mutations, which arose spontaneously in the germ-line cells of one or the other parent. The fact that these mutations arise spontaneously with some frequency could help explain why these common disorders—each observed in about 1% of the population—remain with us, even though the affected individuals might leave few or no descendants. These rare mutations can arise in any one of hundreds of different genes, which could explain much of the clinical variability of autism and schizophrenia.

Now that DNA sequencing has become fast and inexpensive, the most efficient way to identify these rare, large-effect mutations is by comparing the genomes of large numbers of affected individuals with those of unaffected controls. When the key variants are identified, the major challenge is then to determine how they affect the individuals who carry them and how small variations in multiple genes produce the disease phenotype.

The Cellular Functions of a Known Gene Can Be Studied with Genome Engineering

As we have seen, classical genetics starts with a mutant phenotype and identifies the mutations, and consequently the genes, responsible for it. Recombinant DNA technology has made possible a different type of genetic approach that is used widely in a variety of species. Instead of beginning with a mutant organism and using it to identify a gene and its protein, an investigator can start with a particular gene and proceed to make mutations in it, creating mutant cells or organisms so as to analyze the gene's function. Because this approach reverses the traditional direction of genetic discovery—proceeding from genes to mutations, rather than vice versa—it is sometimes referred to as *reverse genetics*. And because the genome of the organism is deliberately altered in a particular way, this approach is also called *genome engineering* or *genome editing*. We shall see in this chapter that this approach can be scaled up so that whole collections of organisms can be created, each of which has a different gene altered.

There are several ways a gene of interest can be altered. In the simplest, the gene can simply be deleted from the genome, although in a diploid organism, this requires that both copies—one on each chromosome homolog—be deleted. Such *gene knockouts* are especially useful if the gene is not essential. The gene in question (even if it is essential) can also be replaced by one that is expressed in the wrong tissue or at the wrong time in development; this type of manipulation

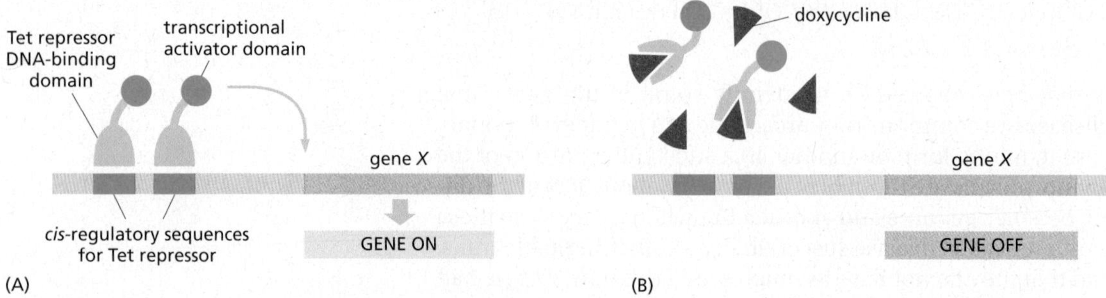

Figure 8–54 Engineered genes can be turned on and off with small molecules. Here, the DNA-binding portion of a bacterial protein (the tetracycline, Tet, repressor) has been fused to a portion of a mammalian transcriptional activator and expressed in cultured mammalian cells. The engineered gene *X*, present in place of the normal gene, has its usual gene control region replaced by *cis*-regulatory sequences recognized by the tetracycline repressor. In the absence of doxycycline (a particularly stable version of tetracycline), the engineered gene is expressed; in the presence of doxycycline, the gene is turned off because the drug causes the tetracycline repressor to dissociate from the DNA. This strategy can also be used in mice by incorporating the engineered genes into the germ line. In many tissues, the gene can be turned on and off simply by adding doxycycline to or removing it from the animal's water. If the tetracycline repressor construct is placed under the control of a tissue-specific gene control region, the engineered gene will be turned on and off only in that tissue.

often provides important clues to the gene's normal function. In a particularly powerful approach, a gene of interest can be modified to be expressed at will by the experimenter (**Figure 8–54**). Finally, genes can also be engineered so that they are expressed normally in most cell types and tissues but deleted in certain cell types or tissues selected by the experimenter (see Figure 5–66). This approach is especially useful when a gene has different roles in different tissues.

It is also possible to make subtler changes to a gene. It is sometimes useful to make slight changes in a protein's structure so that one can begin to dissect which portions of a protein are important for its function. The activity of an enzyme, for example, can be studied by changing a single amino acid in its active site. It is also possible, through genome engineering, to create new types of proteins in an animal. For example, a gene can be fused to the gene for a fluorescent protein. When this altered gene is introduced into the genome, the protein can be tracked in the living organism by monitoring its fluorescence.

Altered genes can be created in several ways. Perhaps the simplest is to chemically synthesize the DNA that makes up the gene. In this way, the investigator can specify any type of variant of the normal gene. It is also possible to construct altered genes using recombinant DNA technology, as described earlier in this chapter. Once obtained, altered genes can be introduced into cells in a variety of ways. DNA can be microinjected into mammalian cells with a glass micropipette or introduced by a virus that has been engineered to carry foreign genes. In plant cells, genes can be introduced by a technique called particle bombardment: DNA samples are painted onto tiny gold beads and then literally shot through the cell wall with a specially modified gun. *Electroporation* is sometimes used for introducing DNA into bacteria and some other cells. In this technique, a brief electric shock renders the cell membrane temporarily permeable, allowing foreign DNA to enter the cytoplasm.

To be most useful to experimenters, the altered gene, once it is introduced into a cell, must recombine with the cell's genome so that the normal gene is replaced. In simple organisms such as bacteria and yeasts, this process occurs with high frequency using the cell's own homologous recombination machinery, as described in Chapter 5. In more complex organisms that have elaborate developmental programs, the procedure is more complicated because the altered gene must be introduced into the germ line, as we next describe.

Animals and Plants Can Be Genetically Altered

Animals and plants that have been genetically engineered by gene deletion or gene replacement are called **transgenic organisms**, and any foreign or modified genes that are added are called **transgenes**. We discuss transgenic plants later

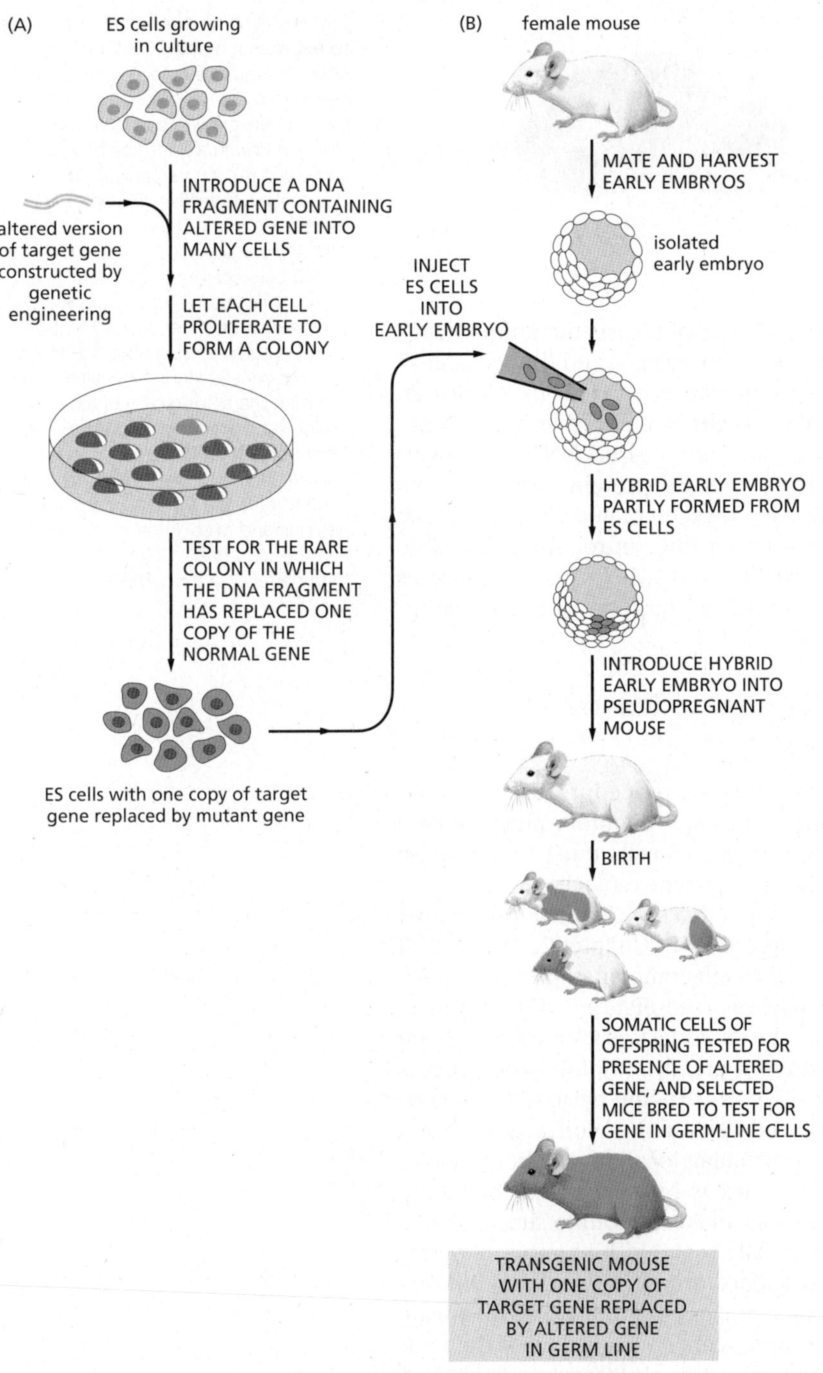

Figure 8–55 **Summary of the procedures used for making gene replacements in mice.** (A) In the first step, an altered version of the gene is introduced into cultured embryonic stem (ES) cells (described in Chapter 22). Only a few ES cells will have their corresponding normal genes replaced by the altered gene through a homologous recombination event. These cells can be identified by PCR and cultured to produce many descendants, each of which carries an altered gene in place of one of its two normal corresponding genes. (B) In the next step of the procedure, these altered ES cells are injected into a very early mouse embryo; the cells are incorporated into the growing embryo, and a mouse produced by such an embryo will contain some somatic cells (indicated by *orange*) that carry the altered gene. Some of these mice will also contain germ-line cells that contain the altered gene; when bred with a normal mouse, some of the progeny of these mice will contain one copy of the altered gene in all of their cells.

The mice with the transgene in their germ line are then bred to produce both a male and a female animal, each heterozygous for the gene replacement (that is, they have one normal and one mutant copy of the gene). When these two mice are mated (not shown), one-fourth of their progeny will be homozygous for the altered gene.

in this chapter and, for now, concentrate our discussion on transgenic mice. If a DNA molecule carrying a mutated mouse gene is transferred into a mouse cell, it is possible to direct the mutant gene to replace the normal gene by homologous recombination. By exploiting these gene-targeting events, any specific gene can be altered or inactivated in a mouse cell by a direct gene replacement. In the case in which both copies of the gene of interest are completely inactivated or deleted, the resulting animal is called a *knockout* mouse. The technique is summarized in **Figure 8–55**.

The ability to prepare transgenic mice lacking a known normal gene was a major advance, and the technique has been used to determine the functions of many mouse genes (**Figure 8–56**). If the gene functions in early development, a knockout mouse will usually die before it reaches adulthood. These lethal defects can be carefully analyzed to help determine the function of the missing gene.

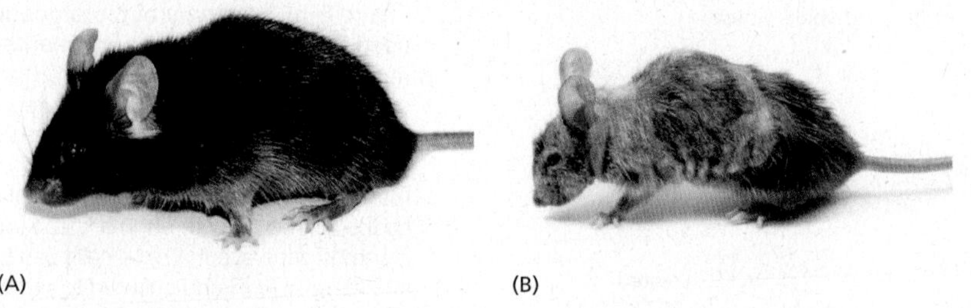

Figure 8–56 **Transgenic mice engineered to express a mutant DNA helicase show premature aging.** The helicase, encoded by the *Xpd* gene, is involved in both transcription and DNA repair. Compared with a wild-type mouse of the same age (A), a transgenic mouse that expresses a defective version of *Xpd* (B) exhibits many of the symptoms of premature aging, including osteoporosis, emaciation, early graying, infertility, and reduced life span. The mutation in *Xpd* used here impairs the activity of the helicase and mimics a mutation that in humans causes trichothiodystrophy, a disorder characterized by brittle hair, skeletal abnormalities, and a very reduced life expectancy. These results indicate that an accumulation of DNA damage can contribute to the aging process in both humans and mice. (From J. de Boer et al., *Science* 296:1276–1279, 2002. With permission from AAAS.)

As described in Chapter 5, an especially useful type of transgenic animal takes advantage of a site-specific recombination system to excise—and thus disable—the target gene in a particular place or at a particular time (see Figure 5–66). In this case, the target gene in embryonic stem (ES) cells is replaced by a fully functional version of the gene that is flanked by a pair of the short DNA sequences, called *lox sites*, that are recognized by the *Cre recombinase* protein. The transgenic mice that result are phenotypically normal. They are then mated with transgenic mice that express the Cre recombinase gene under the control of an inducible promoter. In the specific cells or tissues in which Cre is switched on, it catalyzes recombination between the lox sequences—excising a target gene and eliminating its activity (see Figure 22–7).

The Bacterial CRISPR System Has Been Adapted to Edit Genomes in a Wide Variety of Species

One of the difficulties in making transgenic mice by the procedure just described is that the introduced DNA molecule (bearing the experimentally altered gene) often inserts at random in the genome, and many ES cells must therefore be screened individually to find one that has the correct gene replacement.

Creative use of the CRISPR system, discovered in bacteria as a defense against viruses, has largely solved this problem. As described in Chapter 7, the CRISPR system uses a guide RNA sequence to target (through complementary base-pairing) double-stranded DNA, which it then cleaves (see Figure 7–81). The gene coding for the key component of this system, the bacterial Cas9 protein, has been transferred into a variety of organisms, where it greatly simplifies the process of making transgenic organisms (**Figure 8–57A and B**). The basic strategy is as follows: Cas9 protein is expressed in cultured cells along with a guide RNA designed by the experimenter to target a particular location on the genome. The Cas9 and guide RNA associate, the complex is brought to the matching sequence on the genome, and the Cas9 protein makes a double-strand break. As we saw in Chapter 5, these breaks are usually repaired by nonhomologous end joining, which often results in small sequence errors or deletions that disrupt gene function. In many cases, this repair process is sufficient to inactivate the gene, particularly if it produces a frameshift near the beginning of the coding sequence. If the goal is a precise gene knockout or replacement, then Cas9 and guide RNA can be co-expressed in ES cells with an altered homologous gene sequence, which the cell uses to repair the double-strand break by homologous recombination. In this way, the normal gene can be selectively damaged by the CRISPR system and replaced at high efficiency by an experimentally altered gene.

The CRISPR system has a variety of other uses. Its particular power lies with its ability to target Cas9 to thousands of different positions across a genome through the simple rules of complementary base-pairing. Thus, if a catalytically inactive Cas9 protein is fused to a transcription activator or repressor, it is possible, in principle, to turn any gene on or off by providing a guide RNA that matches a unique sequence in the gene promoter (**Figure 8–57C and D**; **Movie 8.4**).

The CRISPR system has several advantages over other strategies for experimentally manipulating gene expression. First, it is relatively easy for the experimenter to design the guide RNA: it simply follows standard base-pairing

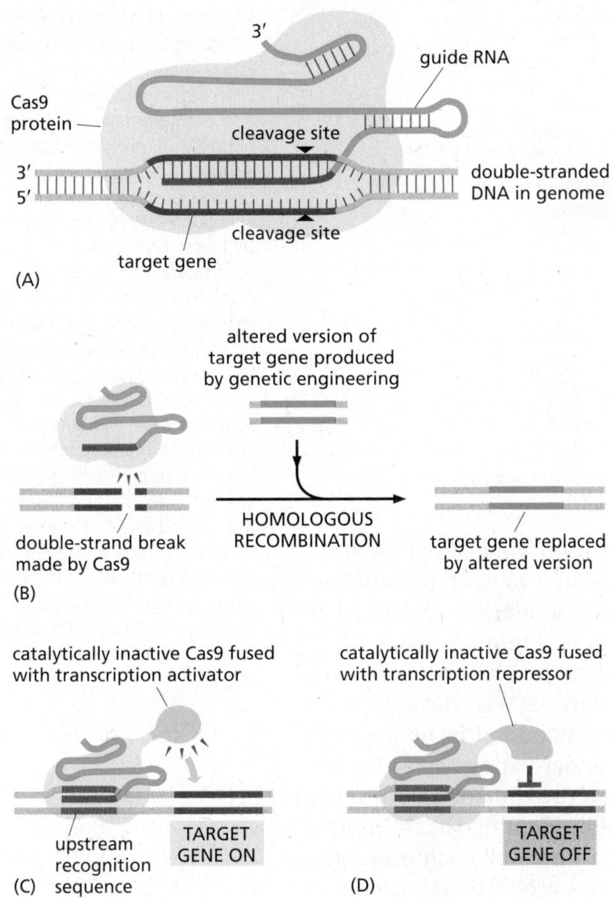

Figure 8–57 Use of CRISPR to study gene function in a wide variety of species. (A) The Cas9 protein (artificially expressed in the species of interest) binds to a guide RNA, designed by the experimenter and also expressed. The portion of RNA in *light blue* is needed for associations with Cas9; that in *dark blue* is specified by the experimenter to match a position on the genome. The only other requirement is that the adjacent genome sequence includes a short PAM (protospacer adjacent motif, not shown) that is needed for Cas9 to cleave. As described in Chapter 7, this sequence allows the CRISPR system in a bacterium to distinguish its own genome from that of invading viruses. (B) When Cas9 is directed to make a double-strand break in a gene, the break is generally repaired by nonhomologous end joining (not shown), which introduces local sequence errors that can disrupt gene function. A more precise mutation can be made as shown here, where the double-strand break is repaired by homologous recombination with an altered gene provided by the experimenter. (C, D) By using a mutant form of Cas9 that can no longer cleave DNA, Cas9 can be used to activate a normally dormant gene (C) or turn off an actively expressed gene (D). (Adapted from P. Mali et al., *Nat. Methods* 10:957–963, 2013.)

convention. Second, the gene to be controlled does not have to be modified; the CRISPR strategy exploits DNA sequences already present in the genome. Third, numerous genes can be controlled simultaneously. Cas9 has to be expressed only once, but many guide RNAs can be expressed in the same cell; this strategy allows the experimenter to turn on or off a whole set of genes at once.

The export of the CRISPR system from bacteria to virtually all other experimental organisms (including mice, zebrafish, worms, flies, rice, and wheat) has revolutionized the study of gene function. Like the earlier discovery of restriction nucleases, this breakthrough came from scientists studying a fascinating phenomenon in bacteria without—at first—realizing the enormous impact these discoveries would have on all aspects of biology.

Large Collections of Engineered Mutations Provide a Tool for Examining the Function of Every Gene in an Organism

Extensive collaborative efforts have produced comprehensive libraries of mutations in a variety of model organisms, including *S. cerevisiae, Caenorhabditis elegans, Drosophila, Arabidopsis,* and even the mouse. The ultimate aim in each case is to produce a collection of mutant strains in which every gene in the organism—one at a time—has been systematically deleted or altered in such a way that it can be conditionally disrupted. Collections of this type provide an invaluable resource for investigating gene function on a genomic scale. For example, a large collection of mutant organisms can be screened for a particular phenotype. Like the classic genetic approaches described earlier, this is one of the most powerful ways to identify the genes responsible for a particular phenotype. Unlike the classical genetic approach, however, the set of mutants is "pre-engineered," so that there is no need to rely on chance events such as

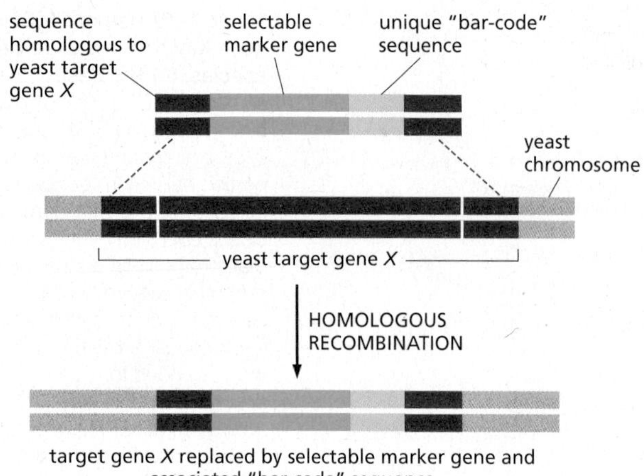

Figure 8–58 Making bar-coded collections of mutant organisms. A deletion construct for use in yeast contains DNA sequences *(red)* homologous to each end of a target gene *X*, a selectable marker gene *(blue)*, and a unique "bar-code" sequence approximately 20 nucleotide pairs in length *(green)*. This DNA is introduced into yeast cells, where it readily replaces the target gene by homologous recombination. Cells that carry a successful gene replacement are identified by expression of the selectable marker gene, typically a gene that provides resistance to a drug. By using a collection of such constructs, each specific for one gene, a library of yeast mutants was constructed containing a mutant for every gene. Essential genes cannot be studied this way, as their deletion from the genome causes the cells to die. In this case, the target gene is replaced by a version of the gene that can be regulated by the experimenter (see Figure 8–54). The gene can then be turned off, and the effect of this can be monitored before the cells die.

spontaneous mutations or transposon insertions. In addition, each of the individual mutations within the collection is often engineered to contain a distinct molecular "bar code"—in the form of a unique DNA sequence—designed to make identification of the altered gene rapid and routine (**Figure 8–58**).

In *S. cerevisiae*, the task of generating a complete set of 6000 mutants, each missing only one gene, was accomplished in the early 2000s. Because each mutant strain has an individual bar-code sequence embedded in its genome, a large mixture of engineered strains can be grown under various selective test conditions—such as nutritional deprivation, a temperature shift, or the presence of various drugs—and the cells that survive can be rapidly identified by the unique sequence tags present in their genomes. By assessing how well each mutant in the mixture fares, one can begin to discern which genes are essential, useful, or irrelevant for growth under the various conditions (**Figure 8–59**).

Similar methods can be applied to human cells using the CRISPR system described earlier. Using viral expression vectors, a large library of different guide RNAs can be expressed in a cell population, such that only one guide RNA is expressed in each cell, along with Cas9. After growth of the cells under various conditions, surviving cells are subjected to genomic sequencing to measure the abundance of guide RNAs in the population. Guide RNAs that target genes essential for survival will disappear from the population, whereas those that enhance survival will be enriched—providing important clues about the function of those genes under the conditions tested.

The insights generated by examining mutant libraries can be considerable. For example, studies of an extensive collection of mutants in *Mycoplasma genitalium*—the organism with the smallest known genome—have identified the minimum complement of genes essential for cellular life. Growth under laboratory conditions requires about three-quarters of the 480 protein-coding genes in *M. genitalium*. Approximately 100 of these essential genes are of unknown function, which suggests that a surprising number of the basic molecular mechanisms that underlie life have yet to be discovered.

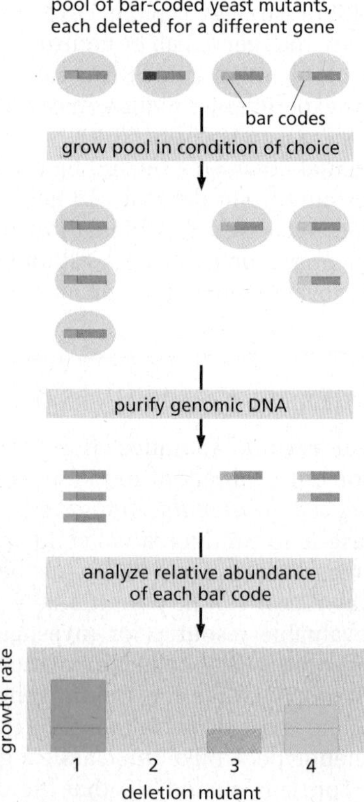

Figure 8–59 Genome-wide screens for fitness using a large pool of bar-coded yeast deletion mutants. A large pool of yeast mutants, each with a different gene deleted and present in equal amounts, is grown under conditions selected by the experimenter. Some mutants *(blue)* grow normally, but others show reduced growth (*orange* and *green*) or no growth at all *(red)*. The fitness of each mutant is experimentally determined in the following way. After the growth phase is completed, genomic DNA (isolated from the mixture of strains) is purified, and the relative abundance of each mutant is determined by quantifying the level of the DNA bar code matched to each deletion. This can be done by sequencing the pooled genomic DNA. In this way, the contribution of every gene to growth under the specified condition can be rapidly ascertained. This type of study has revealed that of the approximately 6000 coding genes in yeast, only about 1000 are essential under standard growth conditions.

Collections of mutant organisms are also available for many animal and plant species. For example, it is possible to "order," by phone or e-mail from a consortium of investigators, a deletion or insertion mutant for almost all coding genes in *Drosophila*. Likewise, a nearly complete set of mutants exists for the model plant *Arabidopsis*. And the adaptation of the CRISPR system for use in mice means that, in the near future, we can expect to be able to turn on or off—at will—each gene in the mouse genome at different stages of development. Although we are still ignorant about the function of most genes in most organisms, these technologies allow an exploration of gene function on a scale that was unimaginable a decade ago.

RNA Interference Is a Simple and Rapid Way to Test Gene Function

Although knocking out (or conditionally expressing) a gene in an organism and studying the consequences is the most powerful approach for understanding the functions of the gene, *RNA interference* (*RNAi*, for short) is an alternative, particularly convenient approach. As discussed in Chapter 7, this method exploits a natural mechanism used in many plants, animals, and fungi to protect themselves against viruses and transposable elements. The technique introduces into a cell or organism a double-strand RNA molecule whose nucleotide sequence matches that of part of the gene to be inactivated. After the RNA is processed, it hybridizes with the target-gene RNA (either mRNA or noncoding RNA) and reduces its expression by the mechanisms shown in Figure 7–78.

RNAi is frequently used to inactivate genes in *Drosophila* and mammalian cell culture lines. Indeed, a set of 15,000 *Drosophila* RNAi molecules (one for every coding gene) allows researchers, in several months, to test the role of every fly gene in any process that can be monitored using cultured cells. RNAi has also been widely used to study gene function in whole organisms, including the nematode *C. elegans*. When working with worms, introducing the double-stranded RNA is quite simple: either the RNA can be injected directly into the intestine of the worm or the worm can be fed with *E. coli* engineered to produce the RNA (**Figure 8–60**). The RNA is amplified and distributed throughout the body of the worm, where it inhibits expression of the target gene in different tissue types. RNAi is being used to help in assigning functions to the entire complement of worm genes (**Figure 8–61**).

A related technique has also been applied to mice. In this case, the RNAi molecules are not injected or fed to the mouse; rather, recombinant DNA techniques are used to make transgenic animals that express the RNAi under the control of an inducible promoter. Often this is a specially designed RNA that can fold back on itself and, through base-pairing, produce a double-strand region that is recognized by the RNAi machinery. In the simplest cases, the process inactivates only the genes that exactly match the RNAi sequence. Depending on the

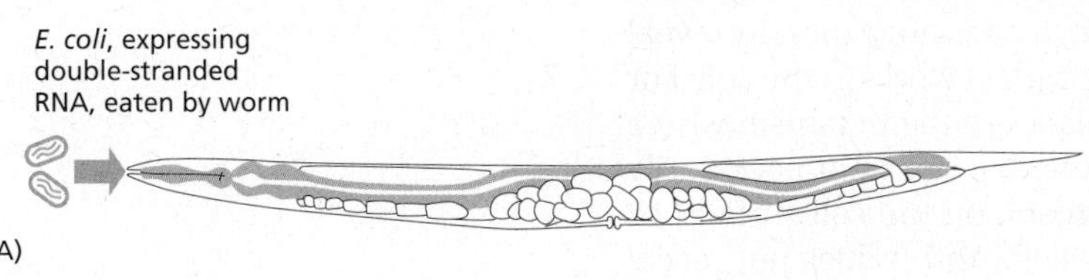

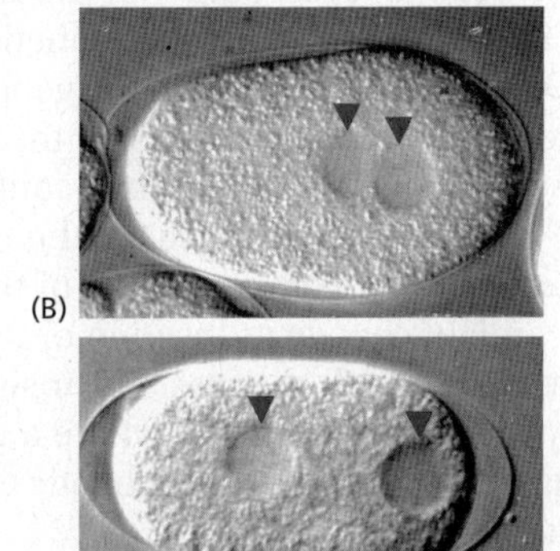

Figure 8–60 Gene function can be tested by RNA interference. (A) Double-stranded RNA (dsRNA) can be introduced into *C. elegans* by feeding the worms *E. coli* that express the dsRNA. (B) In a wild-type worm embryo, the egg and sperm pronuclei *(red arrowheads)* come together in the posterior half of the embryo shortly after fertilization. (C) In an embryo in which a particular gene has been inactivated by RNAi, the pronuclei fail to migrate. This experiment revealed an important but previously unknown function of this gene in embryonic development. (B and C, from P. Gönczy et al., *Nature* 408:331–336, 2000. Reproduced with permission of SNCSC.)

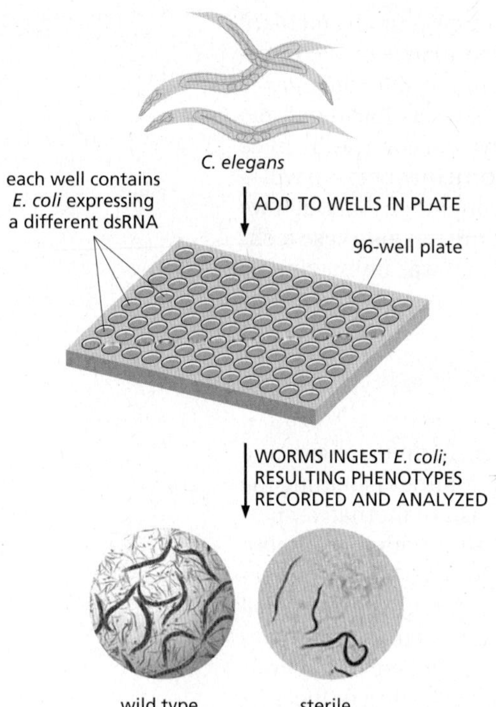

Figure 8–61 **RNA interference provides a convenient method for conducting genome-wide genetic screens.** In this experiment, each well in this 96-well plate is filled with *E. coli* that produce a different double-stranded RNA than that produced by *E. coli* in other wells. Each interfering RNA matches the nucleotide sequence of a single *C. elegans* gene, thereby inactivating it. About 10 worms are added to each well, where they ingest the genetically modified bacteria. The plate is incubated for several days, which gives the RNAs time to inactivate their target genes—and the worms time to grow, mate, and produce offspring. The plate is then examined in a microscope, which can be controlled robotically, to screen for genes that affect the worms' ability to survive, reproduce, develop, and behave. Shown here are normal worms alongside worms that show an impaired ability to reproduce because of inactivation of a particular "fertility" gene. (From B. Lehner et al., *Nat. Genet.* 38:896–903, 2006. With permission from Nature.)

inducible promoter used, the RNAi can be produced only in a specified tissue or only at a particular time in development, allowing the functions of the target genes to be analyzed in elaborate detail.

RNAi is a simple and efficient tool for analysis of gene function in many organisms, but it has several potential limitations compared with true genetic knockouts. For unknown reasons, RNAi does not efficiently inactivate all genes. Moreover, within whole organisms, certain tissues may be resistant to the action of RNAi (for example, neurons in nematodes). Another problem arises because many organisms contain large gene families, the members of which exhibit sequence similarity. RNAi therefore sometimes produces "off-target" effects, inactivating related genes in addition to the targeted gene. One strategy to avoid such problems is to use multiple small RNA molecules matched to different regions of the same gene. Ultimately, the results of any RNAi experiment must be viewed as a strong clue to, but not necessarily a proof of, normal gene function.

Reporter Genes Reveal When and Where a Gene Is Expressed

We have just discussed some of the approaches that can be used to assess a gene's function in cultured cells or, even better, in the intact organism. Although this information is crucial to understanding gene function, it does not generally reveal the molecular mechanisms through which the gene product works in the cell. For example, genetics on its own rarely tells us all the places in the organism where the gene is expressed or how its expression is controlled. It does not necessarily reveal whether the gene acts in the nucleus, the cytosol, on the cell surface, or in one of the numerous other compartments of the cell. And it does not reveal how a gene product might change its location or its expression pattern when the external environment of the cell changes. Key insights into gene function can be obtained by simply observing when and where a gene is expressed. A variety of approaches, most involving some form of genetic engineering, can easily provide this critical information.

As discussed in detail in Chapter 7, *cis*-regulatory DNA sequences, located upstream or downstream of the coding region, control gene transcription. These regulatory sequences, which determine precisely when and where the gene is

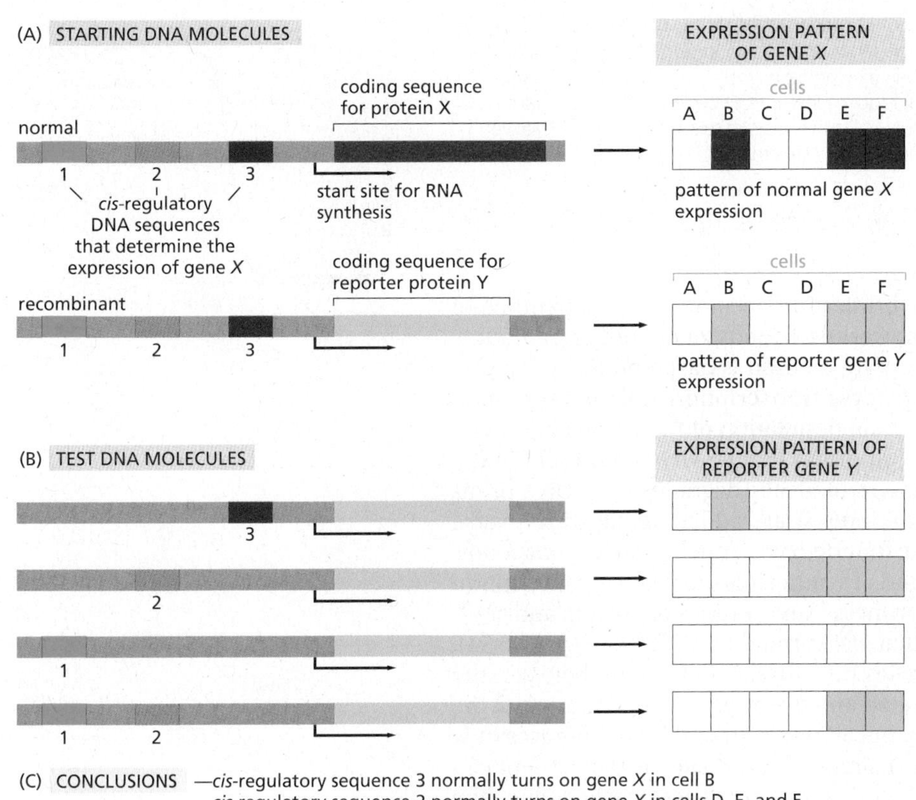

Figure 8–62 Using a reporter protein to determine the pattern of a gene's expression. (A) In this example, the coding sequence for protein X is replaced by the coding sequence for reporter protein Y. The expression patterns for X and Y are the same. (B) Various fragments of DNA containing candidate *cis*-regulatory sequences are added in combinations to produce test DNA molecules encoding reporter gene *Y*. These recombinant DNA molecules are then tested for expression after introducing them into a variety of different types of mammalian cells. The results are summarized in (C). For experiments in eukaryotic cells, two commonly used reporter proteins are the enzyme β-galactosidase (β-gal) (see Figure 7–30) and green fluorescent protein (GFP) (see Figure 9–16).

expressed, can be easily studied by placing a reporter gene under their control and introducing these recombinant DNA molecules into cells (**Figure 8–62**). In this way, the normal expression pattern of a gene can be determined, as well as the contribution of individual *cis*-regulatory sequences in establishing this pattern.

Reporter genes also allow any protein to be tracked over time in living cells. Here, the reporter gene typically encodes a fluorescent protein, often **green fluorescent protein (GFP)**, the molecule that gives luminescent jellyfish their greenish glow. The GFP is simply attached—in the coding frame—to the protein-coding gene of interest. The resulting *GFP fusion protein* often behaves in the same way the normal protein does, and its location can be monitored by fluorescence microscopy, a topic that is discussed in the next chapter (see Figure 9–10). GFP fusion has become a standard strategy for tracking not only the location but also the movement of specific proteins in living cells.

In Situ Hybridization Can Reveal the Location of mRNAs and Noncoding RNAs

It is also possible to directly observe the time and place that an RNA product of a gene is expressed using *in situ hybridization*. For protein-coding genes, this strategy often provides the same general information as the reporter gene approaches described above; however, it is crucial for genes whose final product is RNA rather than protein. We encountered *in situ* hybridization earlier in the chapter (see Figure 8–32); it relies on the basic principles of nucleic acid hybridization. Typically, tissues are gently fixed so that their RNA is retained in an exposed form that can hybridize with a labeled complementary DNA or RNA probe. In this way, the patterns of differential gene expression can be observed in tissues, and the location of specific RNAs can be determined (**Figure 8–63**). An advantage of *in situ* hybridization over other approaches is that genetic engineering is not required. Thus, it is often simpler and faster and can be used for genetically intractable species.

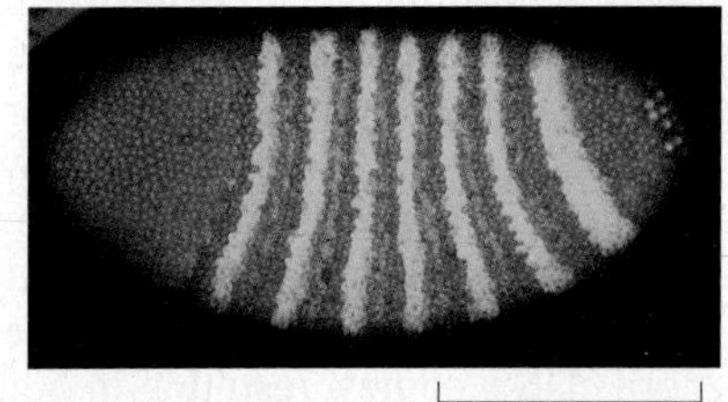

Figure 8–63 *In situ* hybridization to mRNAs reveals patterns of gene expression during development. Expression of specific regulatory genes is localized in the early *Drosophila* embryo, forming a series of stripes (see Figures 7–29 and 7–30). In this image, expression of the genes *eve (magenta)* and *ftz (green)* has been revealed by *in situ* hybridization with differently colored fluorescent probes. DNA is lightly stained with a white fluorescent dye to label all nuclei in the embryo. (Courtesy of Erik Clark and Angela DePace.)

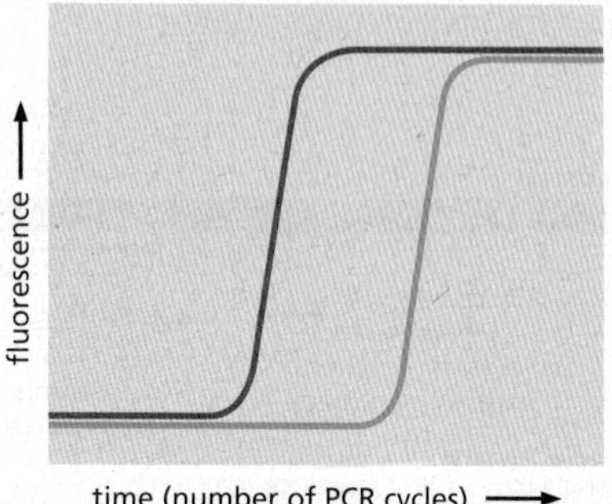

Figure 8–64 RNA levels can be measured by quantitative RT-PCR. The fluorescence measured is generated by a dye that fluoresces only when bound to the double-strand DNA products of the RT-PCR (see Figure 8–34). The *red sample* has a higher concentration of the mRNA being measured than does the *blue sample*, because it requires fewer PCR cycles to reach the same half-maximal concentration of double-stranded DNA. On the basis of this difference, the relative amounts of the mRNA in the two samples can be precisely determined.

Expression of Individual Genes Can Be Measured Using Quantitative RT-PCR

Although reporter genes and *in situ* hybridization accurately reveal patterns of gene expression, they are not the most powerful methods for quantifying amounts of individual RNAs in cells. A more accurate method is based on the principles of PCR. Called **quantitative RT-PCR (reverse transcription–polymerase chain reaction)**, this method begins with the total population of RNA molecules purified from a tissue or a cell culture. It is important that no DNA be present in the preparation; it must be purified away or enzymatically degraded. Two DNA primers that specifically match the mRNA of interest are added, along with reverse transcriptase, DNA polymerase, and the four deoxyribonucleoside triphosphates needed for DNA synthesis. The first round of synthesis is the reverse transcription of the RNA into DNA using one of the primers. Next, a series of heating and cooling cycles allows the amplification of that DNA strand by PCR (see Figure 8–34). The quantitative part of this method relies on a direct relationship between the rate at which the PCR product is generated and the original concentration of the mRNA species of interest. By adding chemical dyes to the PCR that fluoresce only when bound to double-stranded DNA, a simple fluorescence measurement can be used to track the progress of the reaction and thereby accurately deduce the starting concentration of the mRNA that is amplified (**Figure 8–64**). This technique is relatively fast and simple to perform in the laboratory and is the preferred method for accurate measurement of mRNA levels from a specific gene or small group of genes. It is also used routinely to detect rare viral RNAs to determine if a person is infected.

Global Analysis of mRNAs by RNA-seq Provides a Snapshot of Gene Expression

As discussed in Chapter 7, a cell expresses only a subset of the many thousands of genes available in its genome; moreover, this subset differs from one cell type to another or, in the same cell, from one extracellular environment to the next. One way to determine which genes are being expressed by a population of cells or a tissue is to measure all of the RNAs that are being produced.

Global sequencing of all RNAs, or *RNA-seq*, provides the most direct approach for cataloging the RNAs produced by a cell or tissue. As mentioned earlier in this chapter, this approach uses reverse transcriptase to copy all RNAs into cDNAs, which are then fragmented and sequenced by next-generation sequencing methods such as Illumina sequencing (see Figure 8–44). More abundant RNAs will have more cDNA copies, resulting in higher numbers of "sequence reads" for those RNAs. Thus, as we have seen in Chapter 7 (see Figure 7–3), RNA-seq does not simply identify the RNAs in a sample but also provides information about their relative abundance. RNA-seq has other important benefits as well: it can detect alternative RNA splicing, RNA editing, and the many noncoding RNAs produced from a complex genome.

Comprehensive studies of gene expression often provide information that is useful for predicting gene function. Earlier in this chapter, we discussed how identifying a protein's interaction partners can yield clues about that protein's function. A similar principle holds true for genes: information about a gene's function can be deduced by identifying genes that share its expression pattern. Using a computational approach called *cluster analysis*, one can identify sets of genes that are coordinately regulated. Genes that are turned on or turned off together under different circumstances are likely to work in concert in the

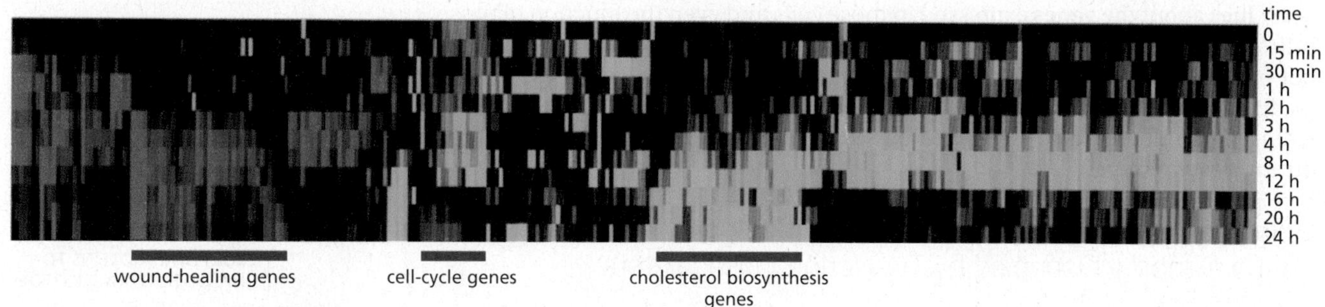

Figure 8–65 **Using cluster analysis to identify sets of genes that are coordinately regulated.** Genes that have the same expression pattern are likely to be involved in common pathways or processes. To perform a cluster analysis, gene expression data are obtained from cell samples exposed to a variety of different conditions, and genes that show coordinate changes in their expression pattern are grouped together. In this experiment, human fibroblasts were deprived of serum for 48 hours; serum was then added back to the cultures at time 0, and the cells were harvested for mRNA measurements at different time points. Of the 8600 genes depicted here (each represented by a thin vertical line), just over 300 showed threefold or greater variation in their expression patterns in response to serum reintroduction. Here, *red* indicates an increase in expression; *green* is a decrease in expression. On the basis of the results of many other experiments, the 8600 genes have been grouped in clusters according to similar patterns of expression. The results of this analysis show that genes involved in wound healing are turned on in response to serum, while genes involved in regulating cell-cycle progression and cholesterol biosynthesis are shut down. (From M.B. Eisen et al., *Proc. Natl. Acad. Sci. USA* 95:14863–14868, 1998. With permission from National Academy of Sciences.)

cell: they may encode proteins that are part of the same multiprotein machine or proteins that are involved in a complex coordinated activity, such as DNA replication or RNA splicing. Characterizing a gene of unknown function by grouping it with known genes that share its transcriptional behavior is sometimes called "guilt by association." Cluster analyses have been used to analyze the gene expression profiles that underlie many interesting biological processes (**Figure 8–65**).

Major new insights into gene expression patterns have come from the development of sensitive methods for sequencing RNA from single cells. In this method, a complex tissue, tumor, embryo, or even organism is dissociated into single cells, after which microfluidics systems are used to separate the cell population into individual cells in single droplets that are each processed for RNA-seq. Sophisticated computational methods are then used to process the vast quantities of sequence data. These methods include cluster analysis algorithms that categorize cells into groups with similar gene expression patterns (**Figure 8–66**). The exciting outcome of these methods is a visualization of how individual cells fall into groups with similar expression patterns. In many cases, the cell groups identified in this fashion can be identified as previously known cell types, thereby revealing

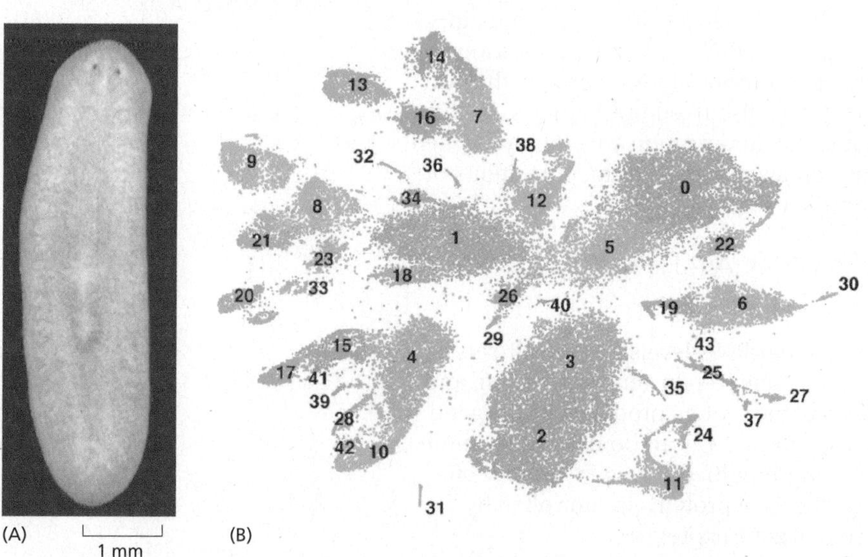

Figure 8–66 **Using single-cell RNA-seq to identify the cell types in a whole animal.** (A) The planarian flatworm is a small (3–5 mm) model organism used in studies of regeneration (see Figure 22–20). (B) In this experiment, flatworms were cut in small pieces and dissociated into individual cells. About 67,000 cells were analyzed by single-cell RNA sequencing, providing a complete list of all major mRNAs expressed in every cell. Sophisticated computational methods were then used to cluster cells into groups with similar gene expression patterns. These clusters are typically displayed in two dimensions on a plot as shown here, with each dot in a cluster representing a single cell. About 44 distinct cell types, highlighted with distinct colors, were identified in this cluster analysis. The pattern of gene expression in each cluster provides clues about the function of the cells in that cluster. Further insights can be obtained using *in situ* hybridization of intact flatworms, which reveals the precise location of cells and tissues expressing specific mRNAs from each cluster. Such analyses reveal that cell clusters 2 and 3, for example, represent cells from the epidermis. The end result is a comprehensive catalog of the genes expressed in every major cell type of the animal's body. (From C.T. Fincher et al., *Science* 360:eaaq1736, 2018. With permission from AAAS.)

new clues about the genes expressed in those cells and even the function of those cells. In other cases, this approach can reveal the existence of cell types that were not previously known. Similar methods can be used to analyze gene expression in distinct cell types during multiple stages of embryonic development, providing comprehensive insights into cell lineage and gene expression patterns during development.

Genome-wide Chromatin Immunoprecipitation Identifies Sites on the Genome Occupied by Transcription Regulators

We have discussed several strategies to measure the levels of individual RNAs in a cell and to monitor changes in their levels in response to external signals. But this information does not tell us how such changes are brought about. We saw in Chapter 7 that transcription regulators, by binding to *cis*-regulatory sequences in DNA, are responsible for establishing and changing patterns of transcription. Typically, these proteins do not occupy all of their potential *cis*-regulatory sequences in the genome under all conditions. For example, in some cell types, the regulatory protein may not be expressed, or it may be present but lack an obligatory partner protein, or it may be excluded from the nucleus until an appropriate signal is received from the cell's environment. Even if the protein is present in the nucleus and is competent to bind DNA, other transcription regulators or components of chromatin can occupy overlapping DNA sequences and thereby occlude some of its *cis*-regulatory sequences in the genome.

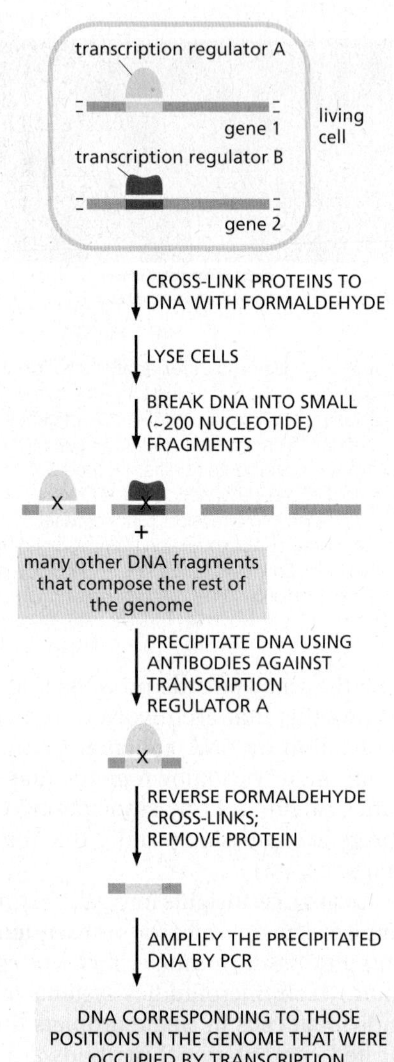

Figure 8–67 Chromatin immunoprecipitation. This method allows the identification of all the sites in a genome that a transcription regulator occupies *in vivo*. The identities of the precipitated, amplified DNA fragments are determined by DNA sequencing.

Chromatin immunoprecipitation provides a way to experimentally determine all the *cis*-regulatory sequences in a genome that are occupied by a given transcription regulator under a particular set of conditions (**Figure 8–67**). In this approach, proteins are covalently cross-linked to DNA in living cells, the cells are broken open, and the DNA is mechanically sheared into small fragments. Antibodies directed against a given transcription regulator are then used to purify the DNA that became covalently cross-linked to that protein in the cell. This DNA is then sequenced using the rapid methods discussed earlier; the precise location of each precipitated DNA fragment along the genome is determined by comparing its DNA sequence to that of the whole genome sequence (**Figure 8–68**). In this way, all of the sites occupied by the transcription regulator in the cell sample can be mapped across the cell's genome. In combination with RNA-seq information, chromatin immunoprecipitation can identify the key transcriptional regulator responsible for specifying a particular pattern of gene expression.

Chromatin immunoprecipitation can also be used to deduce the *cis*-regulatory sequences recognized by a given transcription regulator. Here, all the DNA sequences precipitated by the regulator are lined up (by computer), and features in common are tabulated to produce the spectrum of *cis*-regulatory sequences recognized by the protein (see Figure 7–10A). Chromatin immunoprecipitation is also used routinely to identify the positions along a genome that are bound by the various types of modified histones discussed in Chapter 4. In this case, antibodies specific to the particular histone modification are employed (see Figure 8–68). A variation of the technique can also be used to map positions of chromosomes that are in physical proximity (see Figure 4–53).

Ribosome Profiling Reveals Which mRNAs Are Being Translated in the Cell

We have learned that there are several ways that RNA levels in the cell can be monitored. But for mRNAs, this represents only one step in gene expression, and we are often more interested in the final level of the protein produced by the gene. As described earlier in this chapter, mass spectrometry can be used to monitor the levels of all proteins in the cell, including modified forms of the proteins. However, if we want to understand *how* synthesis of proteins is controlled by the cell, we need to consider the translation step of gene expression.

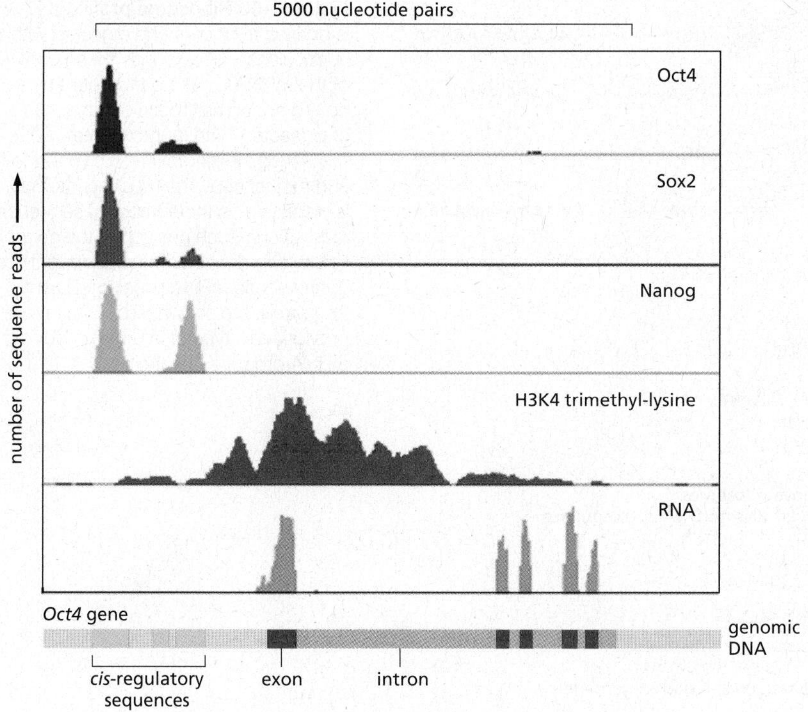

Figure 8–68 Chromatin immunoprecipitations showing proteins bound to the genomic region that controls expression of the *Oct4* gene. In this series of chromatin immunoprecipitation experiments, antibodies directed against a transcription regulator (first three panels) or a particular histone modification (fourth panel) were used to precipitate bound, cross-linked DNA. Precipitated DNA was sequenced, and the positions across the genome were mapped. (Only the small part of the mouse genome containing the *Oct4* gene is shown.) The results show that, in the embryonic stem cells analyzed in these experiments, Oct4 binds upstream of its own gene, and Sox2 and Nanog are bound in close proximity. Oct4, Sox2, and Nanog are key regulators in embryonic stem cells (discussed in Chapter 22), and this experiment reveals the position on the genome through which they exert their effects on *Oct4* expression. In the fourth panel, the positions of a histone modification associated with actively transcribed genes are shown (see Figure 4–35). Finally, the bottom panel shows the RNA produced from the *Oct4* gene under the same conditions used for the chromatin immunoprecipitations. Note that the introns and exons are relatively easy to identify from these RNA-seq data.

An approach called *ribosome profiling* provides an instantaneous map of the positions of ribosomes on each mRNA in the cell and thereby identifies those mRNAs that are being actively translated. To accomplish this, total RNA from a cell line or tissue is exposed to ribonucleases under conditions where only those RNA sequences covered by ribosomes are spared. The protected RNAs are released from ribosomes, converted to DNA, and the nucleotide sequence of each is determined (**Figure 8–69**). When these sequences are mapped on the genome, the positions of ribosomes across each mRNA species can be ascertained.

Ribosome profiling has revealed many cases where mRNAs are abundant but are not translated until the cell receives an external signal. It has also shown that many open reading frames (ORFs) that were too short to be annotated as genes are actively translated and probably encode functional, albeit very small, proteins (**Figure 8–70**). Finally, ribosome profiling has revealed the ways that cells rapidly and globally change their translation patterns in response to sudden changes in temperature, nutrient availability, or chemical stress.

Recombinant DNA Methods Have Revolutionized Human Health

We have seen that nucleic acid methodologies developed in the past 40 years have completely changed the study of cell biology. But they have also had a profound effect on our day-to-day lives. Many pharmaceuticals in routine use (insulin, human growth hormone, blood-clotting factors, and interferon, for example) are based on cloning human genes and expressing the encoded proteins in large amounts. As DNA sequencing continues to drop in cost, more and more individuals will elect to have their genome sequenced; this information can be used to predict susceptibility to diseases (often with the option of minimizing this possibility by appropriate behavior) or to predict the way an individual will respond to a given drug. The genomes of tumor cells from an individual can be sequenced to determine the best type of anticancer treatment. And mutations that cause or greatly increase the risk of disease continue to be identified at an unprecedented pace. Using the recombinant DNA technologies discussed in this chapter, these mutations can then be introduced into animals, such as mice, that can be studied in the laboratory. The resulting transgenic animals, which often mimic some of

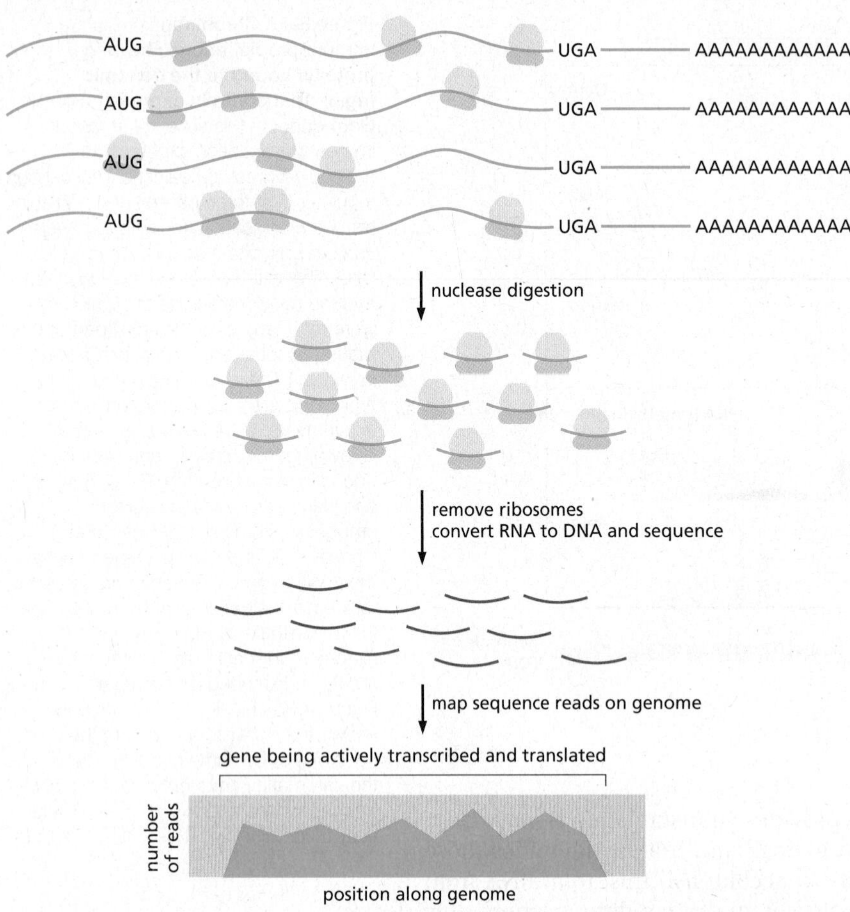

Figure 8–69 Ribosome profiling. RNA is purified from cells and digested with a ribonuclease to leave only those portions of the mRNAs that are protected by a bound ribosome. These short pieces of protected RNA (approximately 20 nucleotides in length) are converted to DNA and sequenced. The resulting information is displayed as the number of sequence reads along each position of the genome. In the diagram here, the data for only one gene, whose mRNA is being efficiently translated, are shown. Ribosome profiling provides this type of information for every mRNA produced by the cell.

the phenotypic abnormalities associated with the condition in humans, can be used to explore the cellular and molecular basis of the disease and to screen for drugs that could potentially be used for human therapy.

Transgenic Plants Are Important for Agriculture

Recombinant DNA technology has had a profound impact on the study of plants. In fact, certain features of plants make them especially amenable to recombinant DNA methods.

When a piece of plant tissue is cultured in a sterile medium containing nutrients and appropriate growth regulators, some of the cells are stimulated to proliferate indefinitely in a disorganized manner, producing a mass of relatively undifferentiated cells called a *callus*. If the nutrients and growth regulators are carefully manipulated, one can induce the formation of a shoot within the callus, and in many species a whole new plant can be regenerated from such shoots. In a

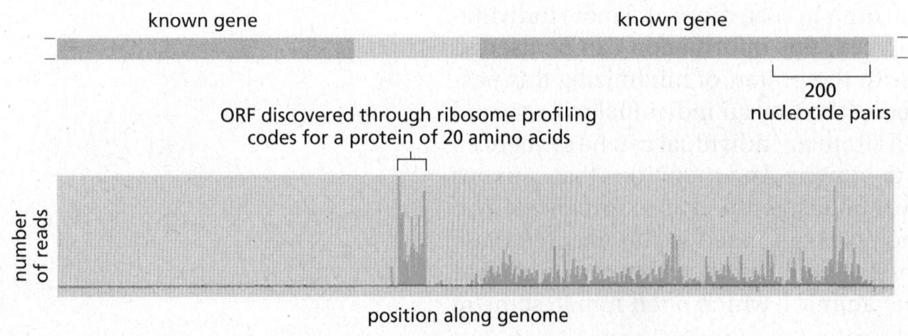

Figure 8–70 Ribosome profiling can identify new genes. This experiment shows the discovery of a previously unrecognized gene—one that encodes a protein of only 20 amino acids. At the top is shown a portion of a viral genome with two previously annotated genes. Below are the results of a ribosome profiling experiment, displayed across the same section of the genome, after human cells were infected by the virus. The results show that the left-hand gene is not expressed under these conditions, the right-hand gene is expressed at low levels, and a previously unrecognized gene that lies between them is expressed at high levels.

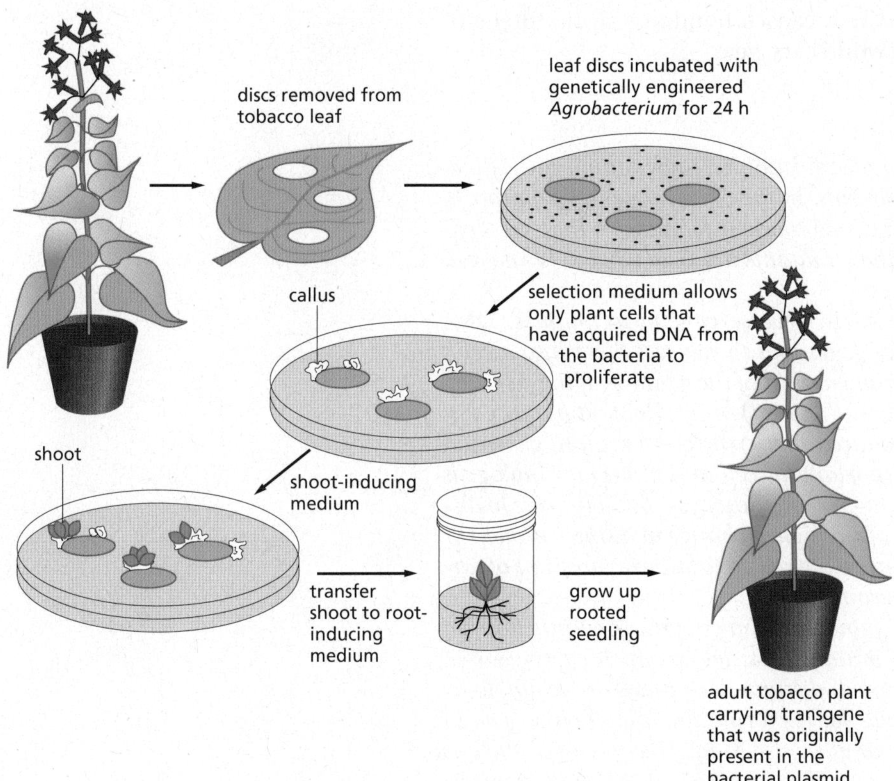

Figure 8–71 Transgenic plants can be made using recombinant DNA techniques optimized for plants. A disc is cut out of a leaf and incubated in a culture of *Agrobacterium* that carries a recombinant plasmid with both a selectable marker and a desired genetically engineered gene. The wounded plant cells at the edge of the disc release substances that attract the bacteria, which inject their DNA into the plant cells. Only those plant cells that take up the appropriate DNA and express the selectable marker gene survive and proliferate and form a callus. The manipulation of growth factors supplied to the callus induces it to form shoots, which subsequently root and grow into adult plants carrying the engineered gene.

number of plants—including tobacco, petunia, carrot, potato, and *Arabidopsis*—a single cell from such a callus (known as a *totipotent cell*) can be grown into a small clump of cells from which a whole plant can be regenerated (see Figure 7–2B). Just as mutant mice can be derived by the genetic manipulation of embryonic stem cells in culture, so transgenic plants can be created from plant cells transfected with DNA in culture (**Figure 8–71**).

The ability to produce transgenic plants has greatly accelerated progress in many areas of plant cell biology. It has played an important part, for example, in isolating receptors for growth regulators and in analyzing the mechanisms of morphogenesis and of gene expression. These techniques have also opened up many new possibilities in agriculture that could benefit both the farmer and the consumer. They have made it possible, for example, to modify the ratio of lipid, starch, and protein in seeds, to impart pest and virus resistance to plants, and to create modified plants that tolerate extreme habitats such as salt marshes or water-stressed soil. One variety of rice has been genetically engineered to produce β-carotene, the precursor of vitamin A (**Figure 8–72**). If it replaced conventional rice, this "golden rice"—so called because of its yellow color—could help to

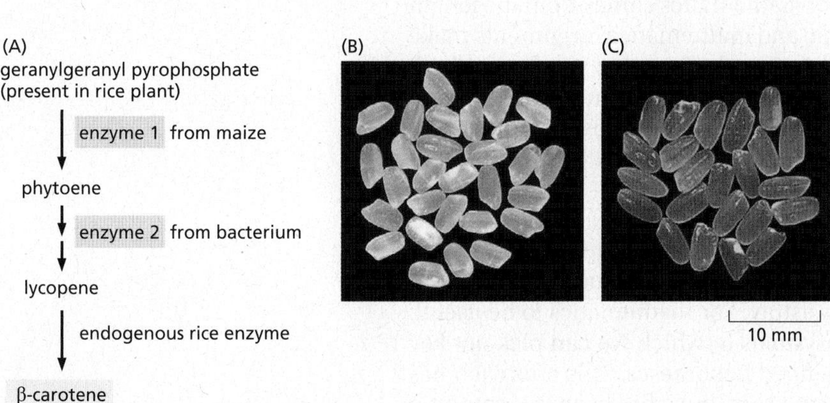

Figure 8–72 DNA technology allows the production of rice grains with high levels of β-carotene. To help reduce vitamin A deficiency in the developing world, a strain of rice, called "golden rice," was developed in which the edible part of the grain (called the endosperm) contains large amounts of β-carotene, which is converted in the human gut to vitamin A. (A) Rice plants, like most other plants, can synthesize β-carotene in their leaves from an abundant precursor (geranylgeranyl pyrophosphate) found in all plant tissues. However, the genes that code for two of the enzymes that act early in this biosynthetic pathway are turned off in the endosperm, preventing the production of β-carotene in rice grains. To produce golden rice, the genes for these two enzymes were obtained from organisms that produce large amounts of β-carotene: one from maize and the other from a bacterium. Using DNA technology, these genes were connected to a promoter that drives gene expression in rice endosperm. Using the method outlined in Figure 8–71, this engineered DNA was then used to generate a transgenic rice plant that expresses these enzymes in endosperm, resulting in rice grains that contain high levels of β-carotene. Compared to the milled grains of wild-type rice (B), the grains of the transgenic rice are a deep yellow/orange due to the presence of β-carotene (C). (B and C, from J.A. Paine et al., *Nat. Biotechnol.* 23:482–487, 2005. Reproduced with permission of SNCSC.)

alleviate severe vitamin A deficiency, which causes blindness in hundreds of thousands of children in the developing world each year.

Summary

Genetics and genetic engineering provide powerful tools for understanding the function of individual genes in cells and organisms. In the classical genetic approach, random mutagenesis is coupled with screening to identify mutants that are deficient in a particular biological process. These mutants are then used to locate and study the genes responsible for that process.

Gene function can also be ascertained by reverse genetic techniques. DNA engineering methods can be used to alter genes and to reinsert them into a cell's chromosomes so that they become a permanent part of the genome. If the cell used for this gene transfer is a fertilized egg (for an animal) or a totipotent plant cell in culture, transgenic organisms can be produced that express the mutant gene and pass it on to their progeny. Especially important for cell and molecular biology is the ability to alter cells and organisms in highly specific ways—allowing one to discern the effect on the cell or the organism of a designed change in a single protein or RNA molecule. For example, genomes can be altered so that the expression of any gene can be switched on or off by the experimenter.

Many of these methods are being expanded to investigate gene function on a genome-wide scale. The generation of mutant libraries in which every gene in an organism has been systematically deleted, disrupted, or made controllable by the experimenter provides invaluable tools for exploring the role of each gene in the elaborate molecular collaboration that gives rise to life. Technologies such as RNA-seq can monitor the expression of tens of thousands of genes simultaneously, providing detailed, comprehensive snapshots of the dynamic patterns of gene expression that underlie complex cell processes.

MATHEMATICAL ANALYSIS OF CELL FUNCTION

Quantitative experiments combined with mathematical theory mark the beginning of modern science. Galileo, Kepler, Newton, and their contemporaries did more than set out some rules of mechanics and offer an explanation of the movements of the planets around the Sun: they showed how a quantitative mathematical approach could provide a depth and precision of understanding, at least for physical systems, that had never before been dreamed to be possible.

What is it that gives mathematics this almost magical power to explain the natural world, and why has mathematics played so much more important a part in physical sciences than in biology? What do biologists need to know about mathematics?

Mathematics can be viewed as a tool for deriving logical consequences from propositions. It differs from ordinary intuitive reasoning in its insistence on rigorous, accurate logic and the precise treatment of quantitative information. If the initial propositions are correct, then the deductions drawn from them by mathematics will be true. The surprising power of mathematics comes from the length of the chains of reasoning that rigorous logic and mathematical arguments make possible, and from the unexpectedness of the conclusions that can be reached, often revealing connections that one would not otherwise have guessed at. Reversing the argument, mathematics provides a way to test experimental hypotheses: if mathematical reasoning from a given hypothesis leads to a prediction that is not true, then the hypothesis is not true.

Clearly, mathematics is not much use unless we can frame our ideas—our initial hypotheses—about the given system in a precise, quantitative form. A mathematical edifice raised on a rickety or—even worse—a vague or overcomplicated set of propositions is likely to lead us astray. For mathematics to be useful, we must focus our analysis on simple subsystems in which we can pick out key quantitative parameters and frame well-defined hypotheses. This approach has been used with great success in physics for centuries, but it has been less common in biology. But times are changing, and more and more it is becoming possible for biologists to exploit the power of quantitative mathematical analysis.

In this final section of our methods chapter, we do not attempt to teach readers every way in which mathematics can be fruitfully applied to biological problems. Rather, we simply aim to give a sense of what mathematics and quantitative approaches can do for us in modern biology. We focus primarily on the important principles that mathematics teaches us about the dynamics of molecular interactions and how mathematics can unveil surprising and useful features of complex systems containing feedback. We will illustrate these principles using the regulation of gene expression by transcription regulators like those discussed in Chapter 7. The same principles apply to the post-transcriptional regulatory systems that govern cell signaling (see Chapter 15), cell-cycle control (see Chapter 17), and essentially all cell processes.

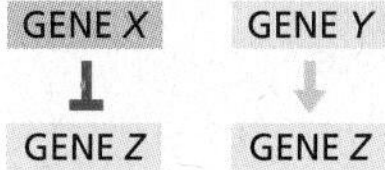

Figure 8–73 Diagrams that summarize biochemical relationships. Here, a simple cartoon indicates that gene *X* represses gene *Z* *(left)* whereas gene *Y* activates gene *Z* *(right)*.

Regulatory Networks Depend on Molecular Interactions

Cell function and regulation depend on transient interactions among thousands of different macromolecules in the cell. We often summarize these interactions in this book with schematic cartoons. These diagrams are useful, but a complete picture requires a deeper, more *quantitative* level of understanding. To meaningfully assess the biological impact of any interaction in the cell, we need to know in precise terms how the molecules interact, how they catalyze reactions, and, most important, how the behaviors of the molecules change over time. If a cartoon shows that protein A activates protein B, for example, we cannot judge the importance of this relationship without quantitative details about the concentrations, affinities, and kinetic behaviors of proteins A and B.

Let us begin by defining two different types of regulatory interaction in our cartoons: one designating inhibition and the other designating activation. If the protein product of gene *X* is a transcription repressor that inhibits the expression of gene *Z*, we depict the relationship as a *red bar-headed line* (⊣) drawn between genes *X* and *Z* (**Figure 8–73**). If the protein product of gene *Y* is a transcription activator that induces the expression of gene *Z*, then a *green arrow* (→) is drawn between genes *Y* and *Z*.

The regulation of one gene's expression by another is more complicated than a single arrow connecting them, and a complete understanding of this regulation requires that we tease apart the underlying biochemical processes. **Figure 8–74A** sketches some of the biochemical steps in the activation of gene expression by a transcription activator. A gene encoding the activator, designated as gene *A*, will produce its product, protein *A*, via an RNA intermediate. This protein *A* will then bind to p_X, the regulatory *promoter* of gene *X*, to form the complex $A{:}p_X$. Once the $A{:}p_X$ complex forms, it stimulates the production of an RNA transcript that is subsequently translated to produce protein *X*.

We will focus here on the binding interaction that lies at the heart of this regulatory system: the interaction between protein *A* and the promoter p_X. This interaction is reversible: any molecule of protein *A* that is bound to p_X can also dissociate from it. The steps represented by the green activation arrow in Figure 8–74A include both the binding of *A* to p_X and the dissociation of the complex $A{:}p_X$ to re-form *A* and p_X, as illustrated by the notation in **Figure 8–74B**. This reaction notation is more informative than the diagrams in our figures but has its own limitations. Suppose that the concentration of *A* increases by a factor of 10 as a response to an environmental input. If *A* increases, we intuitively know that $A{:}p_X$ should increase too, but we cannot determine the amount of the increase without additional information. We need to know the affinity of the binding interaction and the concentrations of the two binding partners. With this information in hand, we can rigorously derive the answer.

As discussed earlier and in Chapter 3 (see Figure 3–42), we know that the formation of a complex between two binding partners, such as *A* and p_X, depends on a rate constant k_{on}, which describes how many productive collisions occur per unit time per protein at a given concentration of p_X. The rate of complex

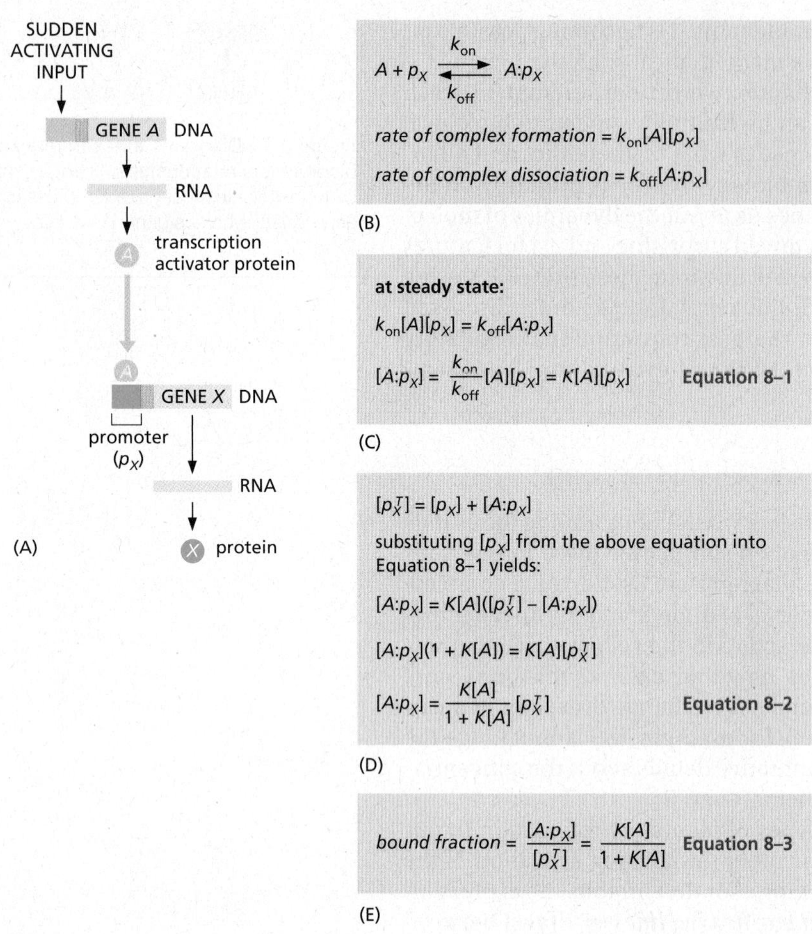

SUDDEN ACTIVATING INPUT
GENE A DNA
RNA
transcription activator protein
GENE X DNA
promoter (p_X)
RNA
protein
(A)
$A + p_X \underset{k_{off}}{\overset{k_{on}}{\rightleftharpoons}} A{:}p_X$
rate of complex formation = $k_{on}[A][p_X]$
rate of complex dissociation = $k_{off}[A{:}p_X]$
(B)
at steady state:
$k_{on}[A][p_X] = k_{off}[A{:}p_X]$
$[A{:}p_X] = \frac{k_{on}}{k_{off}}[A][p_X] = K[A][p_X]$ Equation 8–1
(C)
$[p_X^T] = [p_X] + [A{:}p_X]$
substituting $[p_X]$ from the above equation into Equation 8–1 yields:
$[A{:}p_X] = K[A]([p_X^T] - [A{:}p_X])$
$[A{:}p_X](1 + K[A]) = K[A][p_X^T]$
$[A{:}p_X] = \frac{K[A]}{1 + K[A]}[p_X^T]$ Equation 8–2
(D)
bound fraction = $\frac{[A{:}p_X]}{[p_X^T]} = \frac{K[A]}{1 + K[A]}$ Equation 8–3
(E)

Figure 8–74 A simple transcriptional interaction. (A) Genes *A* and *X* each produce a protein, with the product of gene *A* serving as a transcription activator to stimulate expression of gene *X*. As indicated by the *green arrow*, stimulation depends in part on the binding of protein *A* to the promoter region of gene *X*, designated as p_X. (B) The binding of protein *A* to the gene promoter is determined by the concentrations of the two binding partners (denoted as $[A]$ and $[p_X]$, in units of mol/liter, or M), the association rate constant k_{on} (in units of M^{-1} sec^{-1}), and the dissociation rate constant k_{off} (in units of sec^{-1}). (C) At steady state, the rates of association and dissociation are equal, and the concentration of the bound complex is determined by Equation 8–1, in which the two rate constants are combined in the equilibrium constant *K*. (D) Equation 8–2 can be derived to calculate the steady-state concentration of bound complex at a known total concentration of the promoter $[p_X^T]$. (E) Rearrangement of Equation 8–2 yields Equation 8–3, which allows calculation of the fraction of promoter p_X that is occupied by protein *A*.

formation equals the product of this rate constant k_{on} and the concentrations of *A* and p_X (see Figure 8–74B). Complex dissociation occurs at a rate k_{off} multiplied by the concentration of the complex. The rate constant k_{off} can differ by orders of magnitude for different DNA sequences because it depends on the strength of the noncovalent bonds formed between *A* and p_X.

We are primarily interested in understanding the amount of bound promoter complex at equilibrium or *steady state*, where the rate of complex formation equals the rate of complex dissociation. Under these conditions, the concentration of the promoter complex is specified by a simple equation that combines the two rate constants into a single equilibrium constant $K = k_{on}/k_{off}$ (Equation 8–1; **Figure 8–74C**). *K* is sometimes called the association constant, K_a. The larger this constant *K*, the stronger the interaction between *A* and p_X. The reciprocal of *K* is the dissociation constant, K_d.

To calculate the steady-state concentration of the promoter complex using Equation 8–1, we need to account for another complication: both *A* and p_X exist in two forms—free in solution and bound to each other. In most cases, we know the total concentration of p_X and not the free or bound concentrations, so we must find a way to use the total concentration in our calculations. To do this, we first specify that the total concentration of p_X ($[p_X^T]$) is the sum of the concentrations of free ($[p_X]$) and bound ($[A{:}p_X]$) forms (**Figure 8–74D**). This leads to a new equation that allows us to use $[p_X^T]$ to calculate the steady-state concentration of the promoter complex ($[A{:}p_X]$) (Equation 8–2; Figure 8–74D).

Protein *A* also exists in two forms: free ($[A]$) and bound to p_X ($[A{:}p_X]$). In a cell, there are typically one or two copies of p_X (assuming there is only one gene *X* per haploid genome) and multiple copies of *A*. As a result, we can safely assume that from the viewpoint of *A*, $[A{:}p_X]$ is negligible relative to the total $[A^T]$. This means $[A] \approx [A^T]$, and we can just plug in the values of total $[A^T]$ in Equation 8–2 without incurring appreciable error in the calculation of $[A{:}p_X]$.

Now, we are ready to determine the effects of increasing the concentration of A. Suppose that $K = 10^8\ M^{-1}$, which is a typical value for many such interactions. The starting concentration of A is $[A^T] = 10^{-9}$ M, and $[p_X^T] = 10^{-10}$ M (assuming there is one copy of gene X in a haploid yeast cell, for example, with a volume of around 2×10^{-14} L). Using Equation 8–2, we find that a tenfold increase in the concentration of A causes the amount of promoter complex $[A{:}p_X]$ to increase 5.5-fold, from 0.09×10^{-10} M to 0.5×10^{-10} M at steady state. The effects of a tenfold increase in the concentration of A will vary dramatically depending on its starting concentration relative to the equilibrium constant. Only through this mathematical approach can we achieve a thorough understanding of what these effects will be and what impact they will have on the biological response.

To assess the biological impact of a change in transcription activator levels, it is also important in many cases to determine the fraction of the target gene promoter that is bound by the activator, because this number will be directly proportional to the activity of the gene's promoter. In our case, we can calculate the fraction of the gene X promoter, p_X, that has protein A bound to it by rearranging Equation 8–2 (Equation 8–3; **Figure 8–74E**). This fraction can be viewed as the probability that promoter p_X is occupied, averaged over time. It is also equal to the average occupancy across a large population of cells at any instant in time. When there is no protein A present, p_X is always free, the bound fraction is zero, and transcription is off. When $[A] = 1/K$, the promoter p_X has a 50% chance of being occupied. When $[A]$ greatly exceeds $1/K$, the bound fraction is almost equal to one, meaning that p_X is fully occupied and transcription is maximal.

Differential Equations Help Us Predict Transient Behavior

The most important and basic insights for which we, as biologists, depend on mathematics concern the behavior of regulatory systems over time. This is the central theme of dynamics, and it was for the solution of problems in dynamics that the techniques of calculus were developed, by Newton and Leibniz, in the seventeenth century. Briefly, the general problem is this: if we are given the rates of change of a set of variables that characterize the system at any instant, how can we compute its future state? The problem becomes especially interesting, and the predictions often remarkable, when the rates of change themselves depend on the values of the state variables, as in systems with feedback.

Let us return to Equation 8–2 (Figure 8–74D), which tells us that when $[A]$ changes, $[A{:}p_X]$ at steady state will also change to a new concentration that we can calculate with precision. However, $[A{:}p_X]$ does not change instantaneously to this value. If we hope to understand the behavior of this system in detail, we must also ask how long it takes $[A{:}p_X]$ to get to its new steady-state value inside the cell. Equation 8–2 cannot answer this question. We need calculus.

The most common strategy for solving this problem is to use ordinary differential equations. The equations that describe biochemical reactions have a simple premise: the rate of change in the concentration of any molecular species X (that is, $d[X]/dt$) is given by the balance of the rate of its appearance with that of its disappearance. For our example, the rate of change in the concentration of the bound promoter complex, $[A{:}p_X]$, is determined by the rates of complex assembly and disassembly. We can incorporate these rates into the differential equation shown in **Figure 8–75A** (Equation 8–4). When $[A]$ changes, Equation 8–4 can be solved to generate the concentration of $[A{:}p_X]$ as a function of time. Notice that when $k_{on}[A][p_X] = k_{off}[A{:}p_X]$, then $d[A{:}p_X]/dt = 0$ and $[A{:}p_X]$ stops changing. At this point, the system has reached the steady state.

Calculation of all $[A{:}p_X]$ values as a function of time, using Equation 8–4, allows us to determine the rate at which $[A{:}p_X]$ reaches its steady-state value. Because this value is attained asymptotically, it is often most useful to compare the times needed to get to 50%, 90%, or 99% of this new steady state. The simplest way to determine these values is to solve Equation 8–4 with a method called numerical integration, which involves plugging in values for all of the parameters (k_{on}, k_{off},

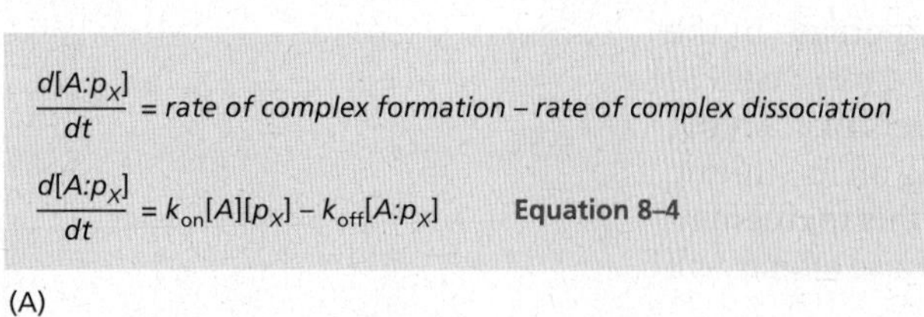

(A)

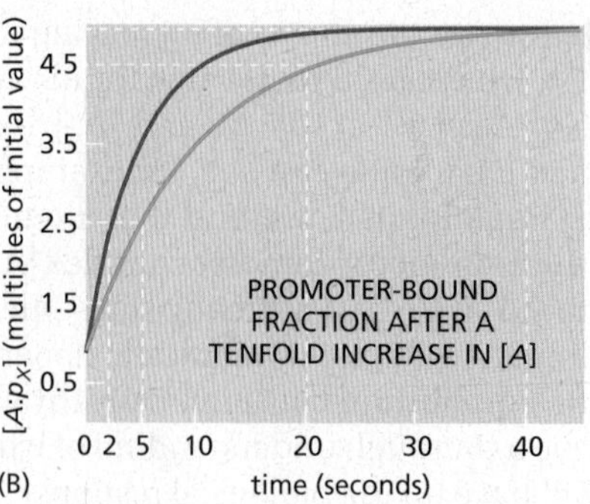

(B)

Figure 8–75 Using differential equations to study the dynamics and steady-state behavior of a biological system. (A) Equation 8–4 is an ordinary differential equation for calculating the rate of change in the formation of bound promoter complex in response to a change in other components. (B) Formation of $[A{:}p_X]$ after a tenfold increase in $[A]$, as determined by solving Equation 8–4. In *blue* is the solution corresponding to $k_{on} = 0.5 \times 10^7\ M^{-1}\ sec^{-1}$ and $k_{off} = 0.5 \times 10^{-1}\ sec^{-1}$. In this case, it takes $[A{:}p_X]$ about 5, 20, and 40 seconds to reach 50%, 90%, and 99% of the new steady-state value. For the *red curve*, the k_{on} and k_{off} values are doubled, and the system reaches the same steady state more rapidly.

etc.) and then using a computer to determine the values of $[A{:}p_X]$ over time, starting from given initial concentrations of $[A]$ and $[p_X]$. For $k_{on} = 0.5 \times 10^7\ M^{-1}\ sec^{-1}$, $k_{off} = 0.5 \times 10^{-1}\ sec^{-1}$ ($K = 10^8\ M^{-1}$ as above), and $[p_X^T] = 10^{-10}$ M, it takes $[A{:}p_X]$ about 5, 20, and 40 seconds to reach 50%, 90%, and 99% of the new steady-state value after a sudden tenfold change in $[A]$ (**Figure 8–75B**). Thus, a sudden jump in $[A]$ does not have instantaneous effects, as we might have assumed from looking at the cartoon in Figure 8–74A.

Differential equations therefore allow us to understand the transient dynamics of biochemical reactions. This tool is critical for achieving a deep understanding of cell behavior, in part because it allows us to determine the dependence of the dynamics inside cells on parameters that are specific to the particular molecules involved. For example, if we double the values of both k_{on} and k_{off}, then Equation 8–1 (Figure 8–74C) indicates that the steady-state value of $[A{:}p_X]$ does not change. However, the time it takes to reach 50% of this steady state after a tenfold change in $[A]$ in our example changes from about 5 seconds to 2 seconds (see Figure 8–75B). These insights are not accessible from either cartoons or equilibrium equations. This is an unusually simple example; mathematical descriptions such as differential equations become more indispensable for understanding biological interactions as the number of interactions increases.

Promoter Activity and Protein Degradation Affect the Rate of Change of Protein Concentration

To understand our gene regulatory system further, we also need to describe the dynamics of protein *X* production in response to changes in the amount of transcription activator protein *A*. Here again, we use an ordinary differential equation for the rate of change of protein *X* concentration—determined by the balance of the rate of production of protein *X* through expression of gene *X* and the protein's rate of degradation.

Let us begin with the rate of protein *X* production, which is determined primarily by the occupancy of the promoter of gene *X* by protein *A*. The binding and dissociation of a transcription regulator at a promoter generally occur on a much faster time scale than transcription initiation, causing many binding and unbinding events to occur before transcription proceeds. As a result, we can assume that the binding reaction is at equilibrium on the time scale of transcription, and we can calculate promoter occupancy by protein *A* using the equilibrium equation discussed earlier (Equation 8–3; Figure 8–74E). To determine the transcription rate, we simply multiply the occupied promoter fraction by a *transcription rate constant*, β, that represents the binding of RNA polymerase and the subsequent

$$\text{transcription rate} = \beta \frac{K[A]}{1 + K[A]}$$

$$\text{protein production rate} = \beta \cdot m \frac{K[A]}{1 + K[A]}$$

$$\text{protein degradation rate} = \frac{[X]}{\tau_X}$$

(A)

$$\frac{d[X]}{dt} = \text{protein production rate} - \text{protein degradation rate}$$

$$\frac{d[X]}{dt} = \beta \cdot m \frac{K[A]}{1 + K[A]} - \frac{[X]}{\tau_X} \qquad \textbf{Equation 8–5}$$

(B)

at steady state:

$$[X_{st}] = \beta \cdot m \frac{K[A]}{1 + K[A]} \cdot \tau_X \qquad \textbf{Equation 8–6}$$

(C)

$$[X](t) = [X_{st}](1 - \exp(-t/\tau_X))$$

(D)

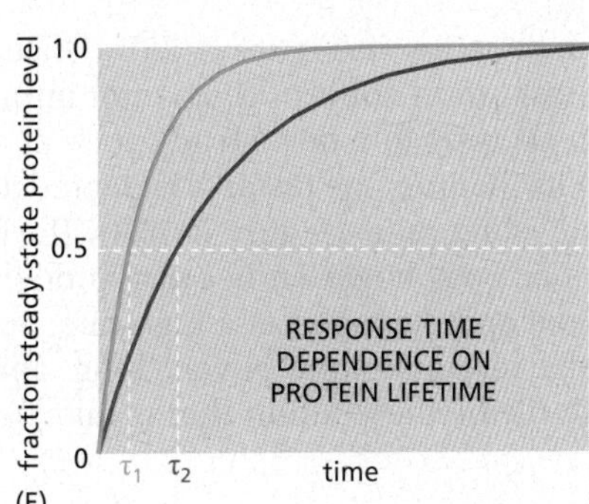

Figure 8–76 Effect of protein lifetime on the timing of the response. (A) Equations for calculation of the rates of gene *X* transcription, protein *X* production, and protein *X* degradation, as explained in the text. (B) Equation 8–5 is an ordinary differential equation for calculating the rate of change in protein *X* in response to changes in other components. (C) When the rate of change in protein *X* is zero (steady state), its concentration can be calculated with Equation 8–6, revealing a direct relationship with protein lifetime (τ). (D) The solution of Equation 8–5 specifies the concentration of protein *X* over time as it approaches its steady-state concentration. (E) Response time depends on protein lifetime. As described in the text, the time that it takes a protein to reach a new steady state is greater when the protein is more stable. Here, the *blue line* corresponds to a protein with a lifetime that is 2.5-fold shorter than the lifetime of the protein in *red*.

steps that lead to production of mRNA and protein (**Figure 8–76A**). If each mRNA molecule produces, on average, *m* molecules of protein product, then we can determine the protein production rate by multiplying the transcription rate by *m* (Figure 8–76A).

Now let us consider the factors that influence protein *X* degradation and its dilution due to cell growth. Degradation generally results in an exponential decline in protein levels, and the average time required for a specific protein to be degraded is defined as its mean lifetime, τ. In our current example, the rate of degradation of protein *X* depends on its mean lifetime τ_X, which takes into account active degradation as well as its dilution as the cell grows. The degradation rate depends on the concentration of protein *X* and is calculated by dividing this concentration by the lifetime (see Figure 8–76A).

With equations for rates of production and degradation in hand, we can now generate a differential equation to determine the rate of change of protein *X* as a function of time (Equation 8–5; **Figure 8–76B**). This equation can be solved by the numerical methods mentioned earlier. According to the solution of this equation, when transcription begins, the concentration of protein *X* rises to a steady-state level at which the concentration of *X* is not changing anymore; that is, its rate of change is zero. When this occurs, rearrangement of Equation 8–5 yields an equation that can be used to determine the steady-state value of *X*, $[X_{st}]$ (Equation 8–6; **Figure 8–76C**). An important concept emerges from the mathematics: the steady-state concentration of a gene product is directly proportional to its lifetime. If lifetime doubles, protein concentration doubles as well.

The Time Required to Reach Steady State Depends on Protein Lifetime

We can see from Equation 8–6 (see Figure 8–76C) that when the concentration of protein *A* rises, protein *X* increases to a new steady-state value, $[X_{st}]$. But this cannot happen instantaneously. Instead, *X* changes dynamically according to the solution of its differential rate equation (Equation 8–5). The solution of this equation reveals that the concentration of *X* over time is related to its steady-state concentration according to the equation in Figure 8–76D. Once again, mathematics uncovers a simple but important concept that is not intuitively obvious: after a sudden increase in $[A]$, $[X]$ rises to a new steady state at an exponential rate that is inversely related to its lifetime; the faster X is degraded, the less time it takes it to reach its new steady-state value (Figure 8–76E). The faster response time comes at a higher metabolic cost, however, because proteins with a rapid response time must be produced and degraded at a high rate. For proteins that are not rapidly turned over, the response time is very long, and protein concentration is determined primarily by the dilution that results from cell growth and division.

Quantitative Methods Are Similar for Transcription Repressors and Activators

Positive control is not the only mechanism that cells use to regulate the expression of their genes. As we discussed in Chapter 7, cells also actively shut off genes, often by employing transcription repressor proteins that bind to specific sites on target genes, thereby blocking access to RNA polymerase. We can analyze the function of these repressors by the same quantitative methods described above for transcription activators. If a repressor protein R binds to the regulatory region of gene X and represses its transcription, then the fraction of gene-binding sites occupied by the repressor is specified by the same equation we used earlier for the transcription activator (Figure 8–77A). In this case, however, it is only when the DNA is free that RNA polymerase can bind to the promoter and transcribe the gene. Thus, the quantity of interest is the unbound fraction, which can be viewed as the probability that the site is free, averaged over multiple binding and unbinding events. When the repressor concentration is zero, the unbound fraction is 1 and the promoter is fully active; when the repressor concentration greatly exceeds $1/K$, the unbound fraction approaches zero. Figure 8–77B and Figure 8–77C compare these relationships for a transcription activator and a transcription repressor.

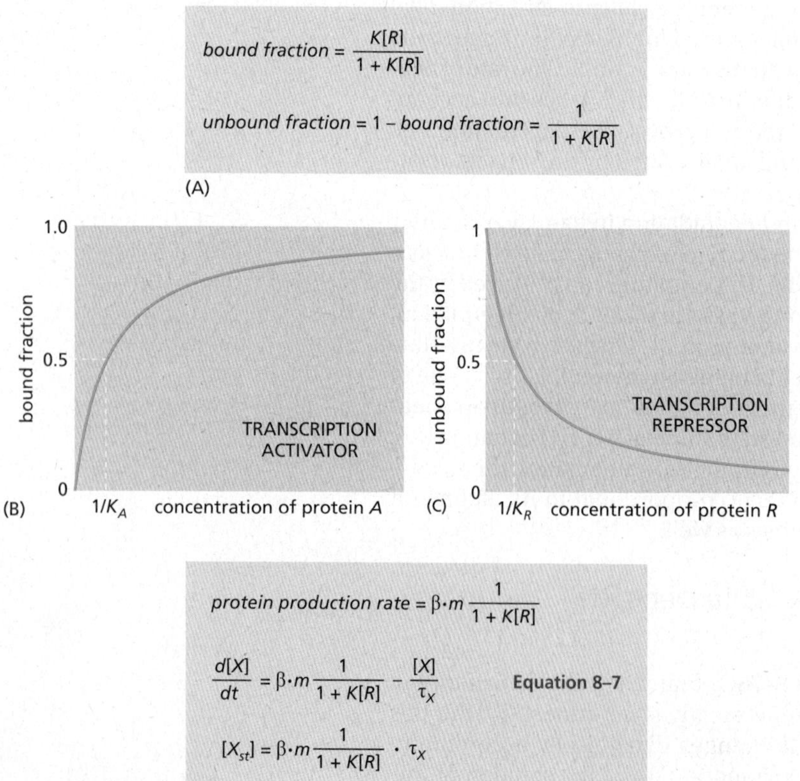

$$\text{bound fraction} = \frac{K[R]}{1 + K[R]}$$
$$\text{unbound fraction} = 1 - \text{bound fraction} = \frac{1}{1 + K[R]}$$
(A)
1.0
0.5
0
bound fraction
TRANSCRIPTION ACTIVATOR
(B)
$1/K_A$
concentration of protein A
1
0.5
0
unbound fraction
TRANSCRIPTION REPRESSOR
(C)
$1/K_R$
concentration of protein R
$$\text{protein production rate} = \beta \cdot m \frac{1}{1 + K[R]}$$
$$\frac{d[X]}{dt} = \beta \cdot m \frac{1}{1 + K[R]} - \frac{[X]}{\tau_X}$$
Equation 8–7
$$[X_{st}] = \beta \cdot m \frac{1}{1 + K[R]} \cdot \tau_X$$
(D)

Figure 8–77 **How promoter occupancy depends on the binding affinity of a transcription regulator protein.** (A) The fraction of a binding site that is occupied by a transcription repressor R is determined by an equation that is similar to the one we used for a transcription activator (see Figure 8–74E), except that in the case of a repressor we are interested primarily in the unbound fraction. (B) For a transcription activator A, half of the promoters are occupied when $[A] = 1/K_A$. Gene activity is proportional to this bound fraction. (C) For a transcription repressor R, gene activity is proportional to the unbound fraction of promoters. As indicated, this fraction is reduced to half of its maximal value when $[R] = 1/K_R$. (D) As in the case of the transcription activator A (see Figure 8–76), we can derive equations to assess the timing of protein X production as a function of repressor concentrations.

We can create a differential equation that provides the rate of change in protein X when repressor concentrations change (Equation 8–7; **Figure 8–77D**). As in the case of the transcription activator, the steady-state concentration of protein X increases as its lifetime increases, but it decreases as the concentration of the transcription repressor increases.

Negative Feedback Is a Powerful Strategy in Cell Regulation

Thus far, we have considered simple regulatory systems of just a few components. In most of the complex regulatory systems that govern cell behaviors, multiple modules are linked to produce larger circuits that we call *network motifs*, which can produce surprisingly complex and biologically useful responses whose properties become apparent only through mathematical analysis. A particularly common and important network motif is the negative feedback loop, which can have dramatically different functions depending on how it is structured.

We take as a first example a network motif consisting of two linked modules (**Figure 8–78A**). Here, an input signal initiates the transcription of gene A, which produces a transcription activator protein A. This activates gene R, which synthesizes a transcription repressor protein R. Protein R in turn binds to the promoter of gene A to inhibit its expression. This cyclical organization creates a negative feedback loop that one can intuitively understand as a mechanism to prevent proteins from accumulating to high levels. But what can we learn about negative feedback loops, and their value in biology, by using mathematics to model them?

The negative feedback loop in Figure 8–78A can be modeled using Equation 8–7 (see Figure 8–77D) for the repression of gene A and Equation 8–5 (see Figure 8–76B) for the activation of gene R. Thus, for proteins A and R, we use the set of differential equations (Equation set 8–8) shown in **Figure 8–78B**. The two equations in this set are coupled, which means that they must be solved together to describe the behavior of A and R over time for any value of the input. As before, we plug in values for the parameters (β_R, τ_R, etc.) and then use a computer to determine the values of $[A]$ and $[R]$ as a function of time after a sudden input activates gene A.

The results reveal several important properties of negative feedback. First, rather surprisingly, negative feedback increases the speed of the response to the activating input. As shown in **Figure 8–78C**, the system with negative feedback reaches its new steady state faster than the system with no feedback.

Second, negative feedback is useful for protecting cells from perturbations that continually arise in the cell's internal environment—due either to random variations in the birth and death of molecules or to fluctuations in environmental variables such as temperature and nutritional supplies. Let us imagine, for example, that β_A, the transcription rate constant for gene A, fluctuates by 25% of its value and ask whether and how much the levels of protein R are affected. The results, shown in **Figure 8–79**, reveal that a change in β_A causes a smaller change in the steady-state value of R when the network has negative feedback.

Delayed Negative Feedback Can Induce Oscillations

A beautiful thing happens when a negative feedback loop contains some delay mechanism that slows the feedback signal through the loop: rather than generating a new stable state as in a rapid negative feedback loop, a delayed loop

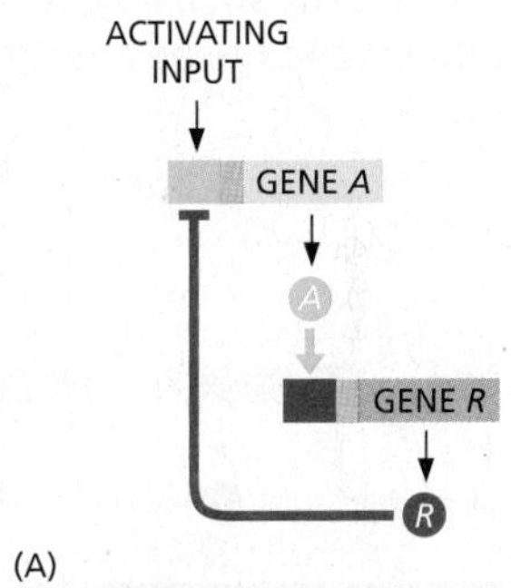

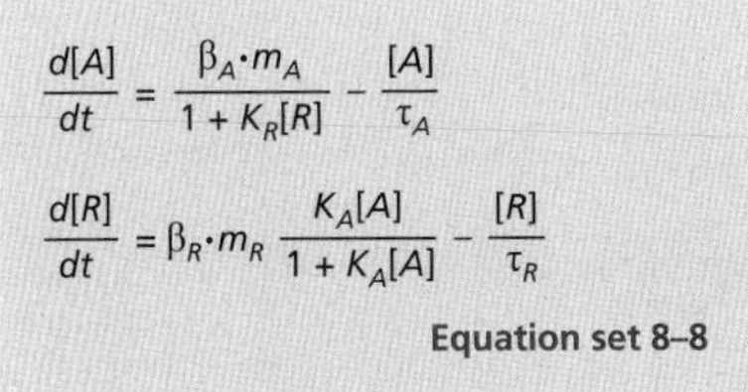

$$\frac{d[A]}{dt} = \frac{\beta_A \cdot m_A}{1 + K_R[R]} - \frac{[A]}{\tau_A}$$

$$\frac{d[R]}{dt} = \beta_R \cdot m_R \frac{K_A[A]}{1 + K_A[A]} - \frac{[R]}{\tau_R}$$

Equation set 8–8

(B)

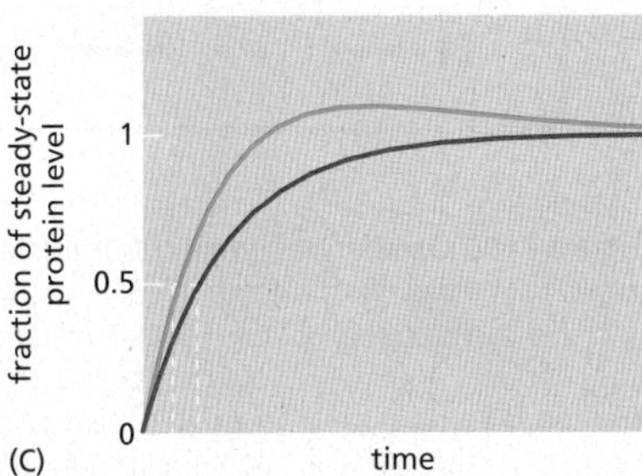

Figure 8–78 A simple negative feedback motif. (A) Gene A negatively regulates its own expression by activating gene R. The product of gene R is a transcription repressor that inhibits gene A. (B) Equation set 8–8 can be solved to determine the dynamics of system components over time. (C) A system with negative feedback *(blue)* reaches its steady state faster than a system with no feedback *(red)*. The plots indicate the levels of protein A, expressed as a fraction of the steady-state level. The *blue line* reflects the solution of Equation set 8–8, which includes negative feedback of gene A by the repressor R. The *red line* represents the solution when the rate of synthesis of A was set to a constant value that is unaffected by the repressor R.

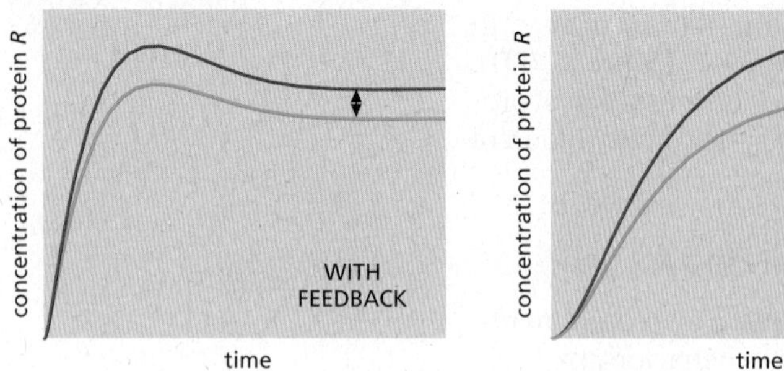

Figure 8–79 The effect of fluctuations in kinetic rate constants on a system with negative feedback compared to one without feedback. The plot at *left* represents the levels of protein *R* after a sudden activating stimulus, according to the regulatory scheme in Figure 8–78A and determined by the solution of Equation set 8–8 (see Figure 8–78B). A perturbation was induced by changing β_A from 4 M/min *(red line)* to 3 M/min *(blue line)*. The plot at *right* shows the results when negative feedback was removed. The system with negative feedback deviates less from its normal operation as β changes than does the system with no feedback. Notice that, as in Figure 8–78C, the system with negative feedback also reaches its steady state more rapidly.

generates pulses, or *oscillations*, in the levels of its components. This can be seen, for example, if the number of components in a negative feedback loop increases, which leads to delays in the amount of time required for the cycle of signals to be completed. Figure 8–80 compares the behavior of two network motifs—one with a three-stage and one with a five-stage negative feedback loop. Using the same kinetic parameters at each stage in the two loops, one finds that stable oscillations arise in the longer loop, while in the shorter loop the same parameters lead to relatively rapid convergence to a stable steady state.

Changes in the parameters of a delayed negative feedback loop—binding affinities, transcription rates, or protein stabilities, for example—can change the amplitude and period of the oscillations, providing a remarkably versatile mechanism for generating all sorts of oscillators that can be used for various purposes in the cell. Indeed, many naturally occurring oscillators, including the calcium oscillators described in Chapter 15 and the cell-cycle network described in Chapter 17, use delayed negative feedback as the basis for biologically important oscillations. Not all of the oscillations observed in cells are thought to have a function, however. Oscillations become inevitable in a highly complex, multicomponent biochemical pathway such as glycolysis, due simply to the large number of feedback loops that appear to be required for its regulation.

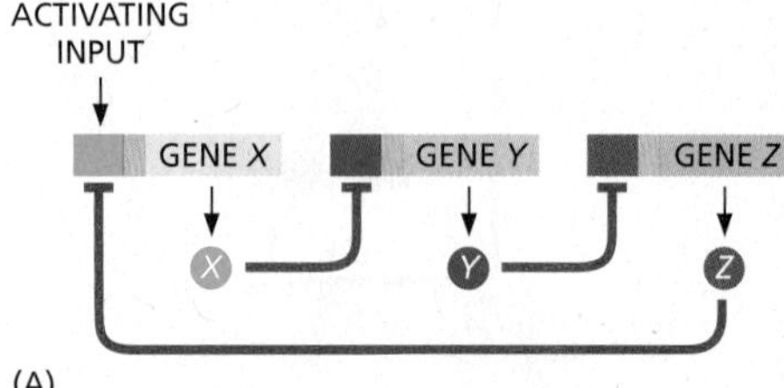

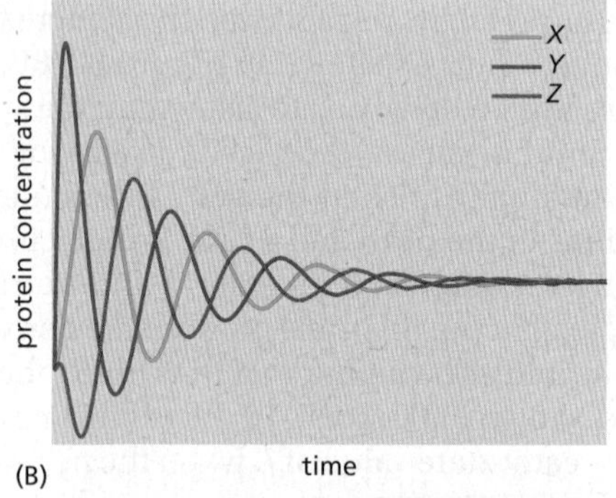

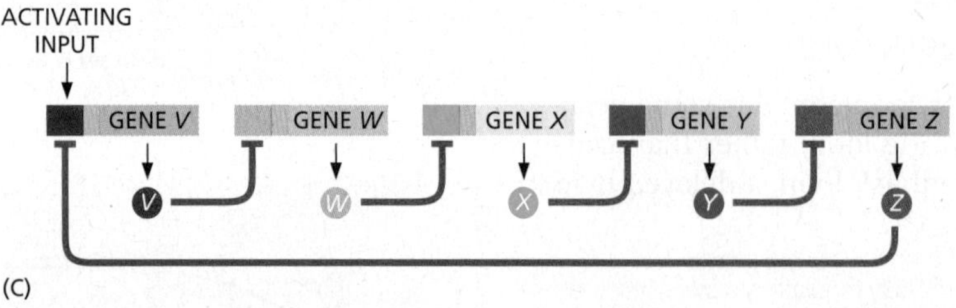

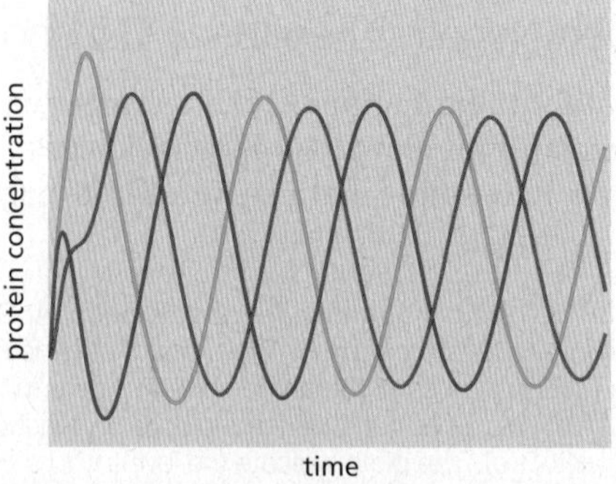

Figure 8–80 Oscillations arising from delayed negative feedback. A transcriptional circuit with three components (A, B) is less likely to oscillate than a transcriptional circuit with five components (C, D). The *X (light blue)*, *Y (dark blue)*, and *Z (brown)* here represent transcription regulatory proteins. For the simulations in panels B and D, the system was initiated from random initial conditions for *X*, *Y*, and *Z*. Oscillations are produced by a delay induced as the signal propagates through the loop.

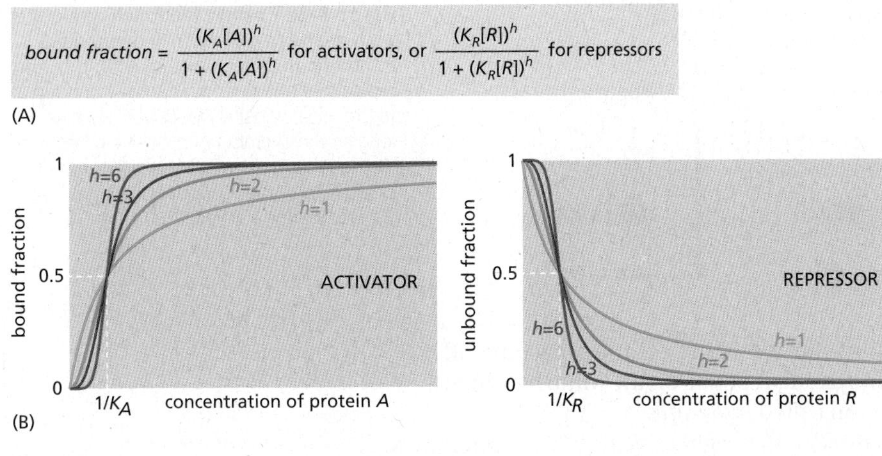

Figure 8–81 **How the cooperative binding of transcription regulatory proteins affects the fraction of promoters bound.** (A) Cooperativity is incorporated into our mathematical models by including a Hill coefficient (*h*) in the equations used previously to determine the fraction of bound promoter (see Figures 8–74E and 8–77A). When *h* is 1, the equations shown here become identical to the equations used previously, and there is no cooperativity. (B) The *left panel* depicts a cooperatively bound transcription activator, and the *right panel* depicts a cooperatively bound transcription repressor. Recall from Figure 8–77B that gene activity is proportional to bound activator *(left panel)* or unbound repressor *(right panel)*. Note that the plots get steeper as the Hill coefficient increases.

DNA Binding by a Repressor or an Activator Can Be Cooperative

We have focused thus far on the binding of a single transcription regulator to a single site in a gene promoter. Many promoters, however, contain multiple adjacent binding sites for the same transcription regulator, and it is not uncommon for these regulators to interact with each other on the DNA to form dimers or larger oligomers. These interactions can result in a *cooperative* form of DNA binding, such that DNA-binding affinity increases at higher concentrations of the transcription regulator. Cooperativity produces a steeper transcriptional response to increasing regulator concentration than the response that can be generated by the binding of a monomeric protein to a single site. A steep transcriptional response of this sort, when present in conjunction with positive feedback, is an important ingredient for producing systems with the ability to switch between different discrete phenotypic states. To begin to understand how this occurs, we need to modify our equations to include cooperativity.

Cooperative binding events can produce steep S-shaped (or *sigmoidal*) relationships between the concentration of regulatory protein and the amount bound on the DNA (see Figure 7–11 and Figure 15–17). In this case, a number called the *Hill coefficient* (*h*) describes the degree of cooperativity, and we can include this coefficient in our equations for calculating the bound fraction of promoter (**Figure 8–81A**). As the Hill coefficient increases, the dependence of binding on protein concentration becomes steeper (**Figure 8–81B**). In principle, the Hill coefficient is similar to the number of molecules that must come together to generate a reaction. In practice, however, cooperativity is rarely complete, and the Hill coefficient does not reach this number.

Positive Feedback Is Important for Switchlike Responses and Bistability

We turn now to positive feedback and its very important consequences. First and foremost, positive feedback can make a system *bistable*, enabling it to persist in either of two (or more) alternative steady states. The idea is simple and can be conveyed by drawing an analogy with a candle, which can exist either in a burning state or in an unlit state. The burning state is maintained by positive feedback: the heat generated by burning keeps the flame alight. The unlit state is maintained by the absence of this feedback signal: so long as sufficient heat has never been applied, the candle will stay unlit.

For the biological system, as for the candle, bistability has an important corollary: it means that the system has a memory, such that its present state depends on its history. If we start with the system in an Off state and gradually rack up the concentration of the activator protein, there will come a point where autostimulation becomes self-sustaining (the candle lights), and the system moves rapidly to an On state. If we now intervene to decrease the level of activator, there will come a point where the same thing happens in reverse, and the system moves rapidly

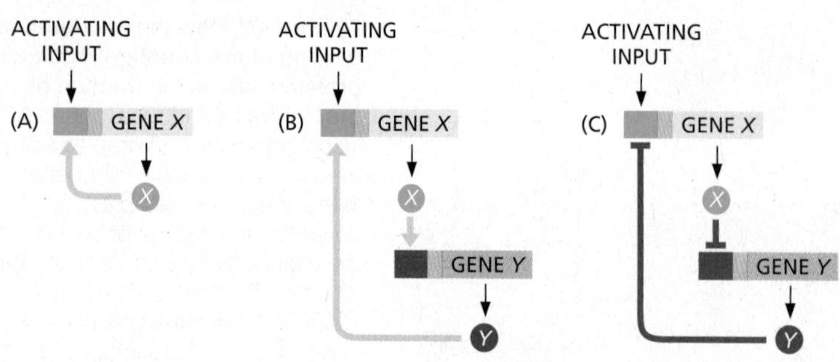

Figure 8–82 Positive feedback of a gene onto itself through serially connected interactions. A sequence of activators and repressors of any length can be connected to produce a positive feedback loop, as long as the overall sign is positive. Because the negative of a negative is positive, not only circuit (A) and (B) but also circuit (C) create positive feedback.

back to an Off state. But the transition points for switching on and switching off are different, and so the current state of the system depends on the route by which it has been taken in the past—a phenomenon called *hysteresis*.

A simple case of positive feedback can be seen in a regulatory system in which a transcription regulator activates (directly or indirectly) its own expression, as in **Figure 8–82A**. Positive feedback can also arise in a circuit with many intervening repressors or activators, so long as the net overall effect of the interactions is activation (**Figure 8–82B and C**).

To illustrate how positive feedback can generate stable states, let us focus on a simple positive feedback loop containing two repressors, *X* and *Y*, each of which inhibits expression of the other (**Figure 8–83A**). As we saw with Equation set 8–8 (Figure 8–78B) earlier, we can create differential equations describing the rate of change of [*X*] and [*Y*] (Equation set 8–9; **Figure 8–83B**). We can further modify these equations to include cooperativity by adding Hill coefficients. As we did earlier, we can then create equations for calculating the concentrations of [*X*] and [*Y*] when the system reaches a steady state—that is, when $(d[X]/dt) = 0$ and $(d[Y]/dt) = 0$ (Equations 8–10 and 8–11, **Figure 8–83C**).

Equations 8–10 and 8–11 can be used to carry out an intriguing mathematical procedure called a *nullcline* analysis. These equations define the relationships between the concentration of *X* at steady state, $[X_{st}]$, and the concentration of *Y* at steady state, $[Y_{st}]$, which must be simultaneously satisfied. We can plug in different values for $[Y_{st}]$ in Equation 8–10 and calculate the corresponding $[X_{st}]$ for each of these values. We can then graph $[X_{st}]$ as a function of $[Y_{st}]$. Next, we repeat the process by varying $[X_{st}]$ in Equation 8–11 to graph the resulting $[Y_{st}]$. The intersections of these two graphs determine the theoretically possible steady states of the system. For systems in which the Hill coefficients h_X and h_Y are much larger than 1, the lines in the two graphs intersect at three locations (**Figure 8–83D**). In other systems that have the same arrangement of regulators but different parameters, there might only be one intersection, indicating the presence of only a single

GENE X
GENE Y
(A)

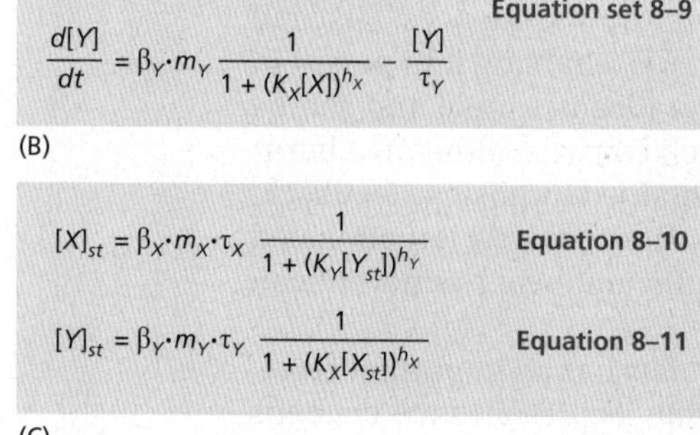

$$\frac{d[X]}{dt} = \beta_X \cdot m_X \frac{1}{1 + (K_Y[Y])^{h_Y}} - \frac{[X]}{\tau_X}$$
$$\frac{d[Y]}{dt} = \beta_Y \cdot m_Y \frac{1}{1 + (K_X[X])^{h_X}} - \frac{[Y]}{\tau_Y}$$
Equation set 8–9
(B)
$$[X]_{st} = \beta_X \cdot m_X \cdot \tau_X \frac{1}{1 + (K_Y[Y_{st}])^{h_Y}}$$
Equation 8–10
$$[Y]_{st} = \beta_Y \cdot m_Y \cdot \tau_Y \frac{1}{1 + (K_X[X_{st}])^{h_X}}$$
Equation 8–11
(C)

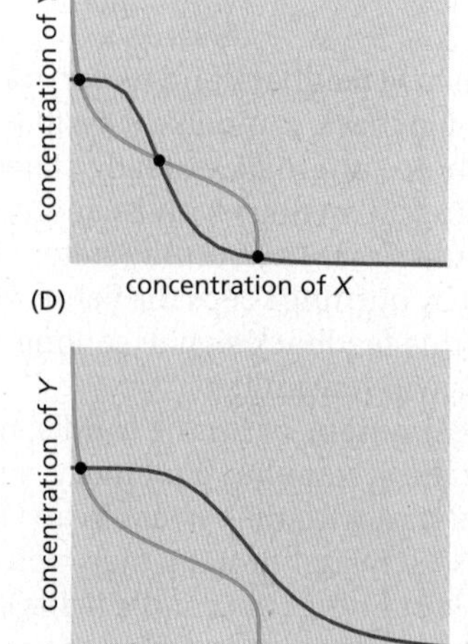

Figure 8–83 A graphical nullcline analysis. (A) *X* inhibits *Y* and *Y* inhibits *X*, resulting in a positive feedback loop. (B) Equation set 8–9 can be used to determine the rate of change in the concentrations of proteins *X* and *Y*. (C) Equations 8–10 and 8–11 provide the concentrations of proteins *X* and *Y*, respectively, when these concentrations reach a steady state. (D, E) *Blue curves* (called nullclines) are plots of $[X_{st}]$ calculated from Equation 8–10 over a range of concentrations of $[Y_{st}]$. *Red curves* are nullclines that indicate values of $[Y_{st}]$ calculated from Equation 8–11 over a range of concentrations of $[X_{st}]$. At an intersection of the two lines, both [*X*] and [*Y*] are at steady state. For plot D, the binding of both proteins to their target gene promoters was cooperative (h_X and h_Y much larger than 1), resulting in the presence of multiple intersections of the nullclines—suggesting that the system can assume multiple discrete steady states. In plot E, the binding of protein *X* to the promoter of gene *Y* was not cooperative (h_X close to 1), resulting in only one nullcline intersection and thus just one likely steady state.

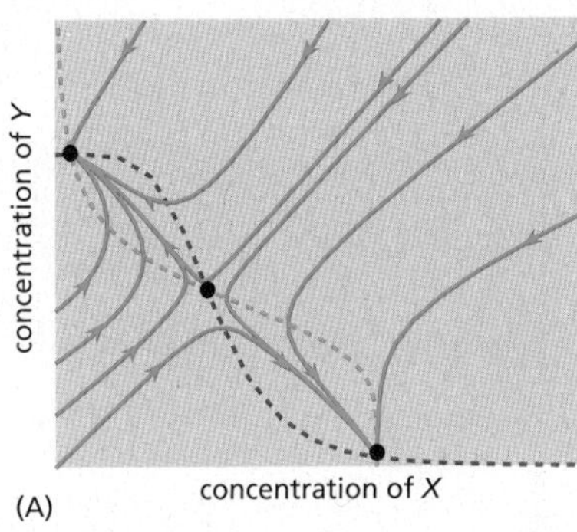

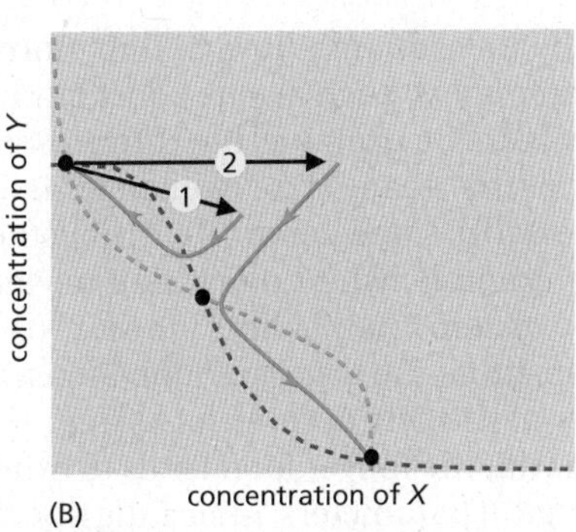

Figure 8–84 Analysis of the stability of a system's steady states. (A) The dotted lines are the nullclines for the system shown in Figure 8–83. Also shown are dynamic trajectories *(green)* that show the changes over time in [X] and [Y], starting at a variety of different initial concentrations (determined by solution of Equation set 8–9; see Figure 8–83B). By plotting [X] versus [Y] at each time point, we find that, although there are three possible steady states in this system, the dynamic trajectories converge on only two of them. The middle steady state is avoided: it is unstable, being unable to attract any trajectories. (B) Imagine that the system is at the upper-left steady state and experiences a perturbation *(black arrows)*, such as a random fluctuation in the production rates of *X* and/or *Y*. If the perturbation is small *(arrow 1)*, the system will return to the same steady state. On the other hand, a perturbation that drives the system beyond the unstable (middle) steady state *(arrow 2)* causes it to switch to the lower-right steady state. The set of perturbations that a system can withstand without switching from one steady state to the other is known as the region of attraction of that steady state.

steady state. For example, when there is a low cooperativity of protein *X* binding to the promoter of gene *Y* (that is, a small Hill coefficient, h_X, in Equation 8–11), the plot of [*Y*] is less curved (**Figure 8–83E**), and it is less likely that there will be multiple intersections of the two curves.

We emphasized earlier that positive feedback typically generates a bistable system with two stable steady states. Why does the system modeled in Figure 8–83D have three? This conundrum can be explained by solving the reaction rate equations (Equation set 8–9; Figure 8–83B) for various different starting conditions of [*X*] and [*Y*], determining all values of [*X*] and [*Y*] as a function of time. Starting with each set of initial concentrations of [*X*] and [*Y*], these calculations produce a so-called *trajectory* of points, each indicated by a curved green line on **Figure 8–84A**. A fascinating pattern emerges: each trajectory moves across the plot and settles in one of two steady states, but never in the third (middle steady state). We conclude that the middle steady state is *unstable* because it cannot "attract" any trajectories. The system therefore has only two *stable* steady states. Thus, the number of stable steady states in a system need not be equal to the total number of its theoretically possible steady states. In fact, stable steady states are usually separated by unstable ones, as in our example.

Once this system adopts a fate by settling in one of the two steady states, does it have the ability to switch to the other state? The numerical solution of Equation set 8–9 can again provide an answer. In **Figure 8–84B**, we show the solution of this equation set for two perturbations from the upper-left steady state. For a small perturbation, the system returns to its original steady state. But the larger perturbation causes the system to switch to the alternate steady state. Thus, this system can be switched from one stable steady state to the other by subjecting it to an input (or a perturbation) that is large enough to make the other steady state more attractive. More generally, every stable steady state has a corresponding *region of attraction*, which can be intuitively thought of as the range of perturbations (of [*X*] or [*Y*] in this example) for which the dynamic trajectories converge back to that particular steady state, rather than switch to the other one.

The concept of a region of attraction has interesting implications for the heritability of transcriptional states and the transition rates between them. If the region of attraction around one steady state is large, for example, then most cells in the population will assume this particular state. Furthermore, this state is likely to be inherited by daughter cells, because minor perturbations, like those ensuing from an asymmetric distribution of molecules during cell division, will rarely be sufficient to induce switching to the other steady state. We should expect that the use of positive feedback, coupled to cooperativity, will quite often be associated with systems requiring stable cell memory.

Robustness Is an Important Characteristic of Biological Networks

Biological regulatory systems are exposed to frequent and sometimes extreme variations in external conditions or the concentrations or activities of key components. The ability of these systems to function normally in the face of such perturbations is called **robustness.** If we understand a complex system to the extent that we can reproduce its behavior with a computational model, then the robustness of the system can be assessed by determining how well its normal function persists after changes in various parameters, such as rate constants and component concentrations. We have already seen, for example, how the presence of negative feedback reduces the sensitivity of the steady state to changes in the values of the system's parameters (see Figure 8–79). Considerations of robustness also apply to dynamic behaviors. Thus, for example, when discussing negative feedback, we described how the behavior of a system tends to become more oscillatory as the number of components that constitute the feedback loop increases. If we use different values of the parameters in models derived for systems like those in Figure 8–80, we find that the system with the longer loop tends to exhibit stable oscillations within a much broader range of parameters, indicating that this system provides a more robust oscillator. We can perform similar calculations to determine the ability of different systems to achieve robust bistability arising from positive feedback. Thus, one benefit of computational models is that they allow us to probe the robustness of biological networks in a systematic and rigorous way.

Two Transcription Regulators That Bind to the Same Gene Promoter Can Exert Combinatorial Control

Thus far, we have discussed how one transcription regulator can modulate the expression level of a gene. Most genes, however, are controlled by more than one type of transcription regulator, providing *combinatorial control* that allows two or more inputs to influence the expression of one gene. We can use computational methods to unveil some of the important regulatory features of combinatorial control systems.

Consider a gene whose promoter contains binding sites for two regulatory proteins, *A* and *R*, which bind to their individual sites independently. There are four possible binding configurations (**Figure 8–85A**). Suppose that *A* is a transcription activator, *R* is a transcription repressor, and the gene is only active when *A* is bound and *R* is not bound. We learned earlier that the probability that *A* is bound and the probability that *R* is not bound can be determined by the equations in **Figure 8–86A**. The product of these two probabilities gives us the probability of gene activation.

This example illustrates an AND NOT logic function (*A* and not *R*) (see Figure 8–85A). Maximal activation of this gene is accomplished when [*A*] is high and [*R*] is zero. However, intermediate levels of gene activation are also possible depending on the levels of *A* and *R* and also on the binding affinities of *A* and *R* for their respective sites (that is, K_A and K_R). When $K_A > K_R$, a small concentration of *A* is capable of overcoming repression by *R*. Conversely, if $K_A < K_R$, then much more *A* is needed to activate the gene (**Figure 8–86B and C**).

Many other logic functions can govern combinatorial gene regulation. For example, an AND logic gate results when two activators, *A1* and *A2*, are both required for a gene to be transcribed (**Figure 8–85B** and **Figure 8–86D**). In *E. coli* cells, the *AraJ* gene controls some aspects of arabinose sugar metabolism: its

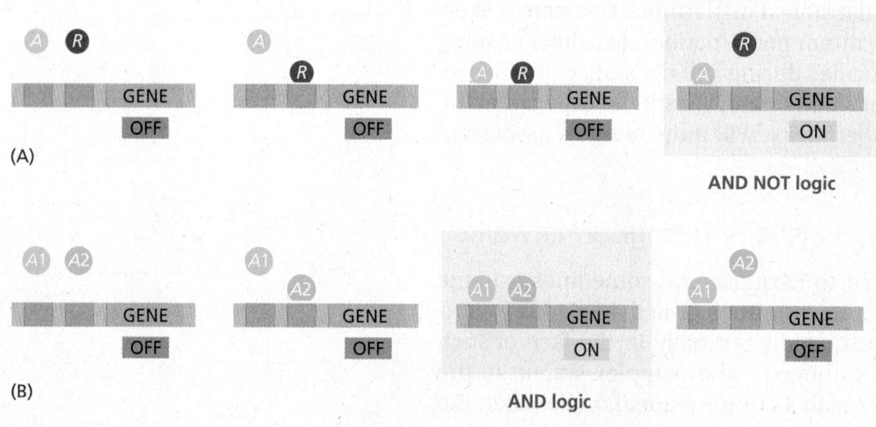

Figure 8–85 **Combinatorial control of gene expression.** There are many ways in which gene expression can be controlled by two transcription regulators. To define precisely the relationship between the two inputs and the gene expression output, a regulatory circuit is often described as a specific type of *logic gate*, a term borrowed from electronic circuit design. A simple example is the OR logic gate (not shown here), in which a gene is controlled by two transcription activators, and one or the other can activate gene expression. (A) In a system with an activator *A* and repressor *R*, if transcription is turned on only when *A* is bound and *R* is not, then the result is an AND NOT logic gate. We saw an example of this logic in Chapter 7 (Figure 7–18). (B) An AND gate results when two transcription activators, *A1* and *A2*, are both required to turn on a gene.

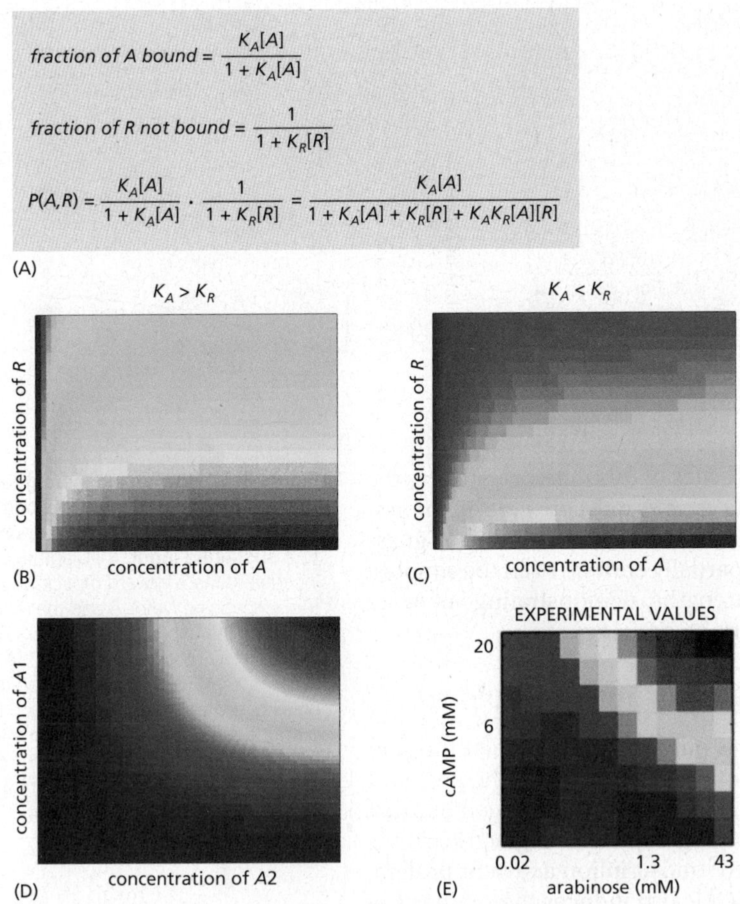

Figure 8–86 How the quantitative output of a gene depends on both its combinatorial logic and the affinities of transcription regulators. (A) In a combinatorial gene regulatory system like that illustrated in Figure 8–85A, the fraction of promoters bound by activator *A* and the fraction not bound by repressor *R* are each determined as shown here. The product of these probabilities provides the probability, *P*(*A*, *R*), that a gene promoter is active. (B–E) In these four panels, *red* indicates high gene expression and *blue* indicates low gene expression. (B, C) Depictions of gene expression from the system described in panel A. The two panels demonstrate how the system behaves when the relative affinities of the two transcription regulators change as indicated above each panel. (D) Gene expression in a case where the gene turns on only at high levels of both activating inputs (*A*1 and *A*2), as shown in Figure 8–85B. (E) Experimental data showing measured expression of a gene in *E. coli* that is combinatorially regulated by two inputs: arabinose and cAMP. Note the close resemblance to panel D. (E, adapted from S. Kaplan et al., *Mol. Cell* 29:786–792, 2008. With permission from Elsevier.)

expression requires two transcription regulators, one activated by arabinose and the other activated by the small molecule cAMP (**Figure 8–86E**).

An Incoherent Feed-forward Interaction Generates Pulses

Imagine that a sudden input signal immediately activates a transcription activator *A* and that the same input signal induces the much slower synthesis of a transcription repressor protein *R* that acts on the same gene *X*. If *A* and *R* control gene expression by an AND NOT logic function like that described above, our intuition tells us that this system should be able to generate a pulse of transcription: when *A* is activated (and *R* is absent), the transcription of gene *X* will begin and cause an increase in the concentration of protein *X*, but then transcription will shut off when the concentration of *R* increases to a sufficiently high value.

Arrangements of this type are common in the cell. In *E. coli*, for example, galactose metabolic genes are positively regulated by the catabolite activator protein (CAP), which is activated at high levels of cAMP. The same genes are repressed by the GalS repressor protein, which is encoded by a gene whose transcription is likewise activated by CAP. Thus, an increase in input (cAMP) activates *A* (CAP), and transcription of the galactose genes begins. But activation of *A* also causes a subsequent buildup of *R* (GalS), which causes the same genes to be repressed after a delay. This results in an *incoherent feed-forward motif* (**Figure 8–87A**).

The response of the incoherent feed-forward motif will vary, depending on the parameters of the system. Suppose, for example, that the transcription activator protein *A* binds more weakly to the gene regulatory region than does the transcription repressor protein *R* ($K_A << K_R$). In this case, there will be a transient burst of protein synthesized by the affected gene (gene *X*) in response to a sudden activating input (**Figure 8–87B**). In contrast, the output will be more sustained if K_A is much larger than K_R, because the repression will be too weak to overcome

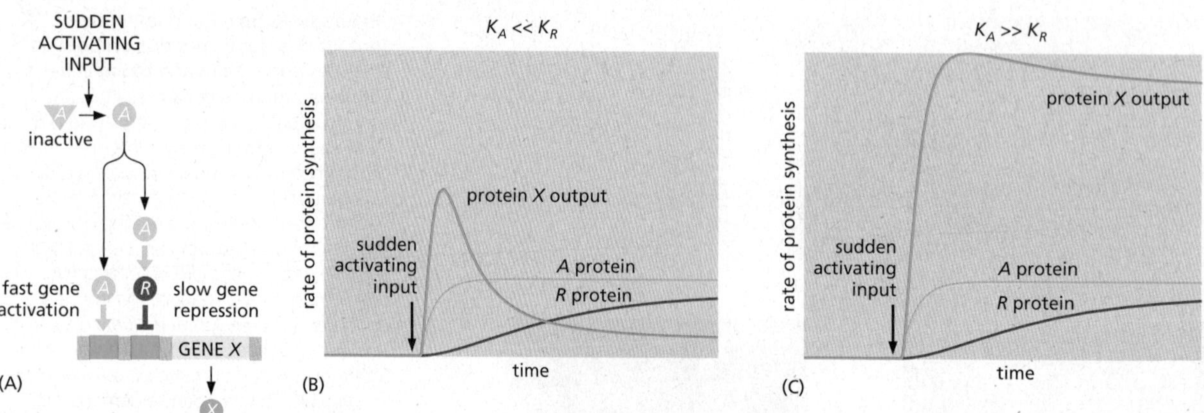

Figure 8–87 **How an incoherent feed-forward motif can generate a brief pulse of gene activation in response to a sustained input.** (A) Diagram of an incoherent feed-forward motif in which the transcription activator *A* and the repressor *R* control the expression of gene *X* using the AND NOT logic of Figure 8–85A. (B) When $K_A << K_R$, this motif generates a pulse of protein *X* expression, such that the output goes back down even if the input remains high. (C) When $K_A >> K_R$, the same motif responds to a sustained input by generating a sustained output.

the gene activation (Figure 8–87C). Other properties of this network, such as the dependence of the amplitude of the pulse on the various rate constants in the system, can be explored with the same computational tools. Thus, our intuitive guess about how this system would behave was only partially correct; even the simplest of networks depends on precise interaction strengths, demonstrating yet again why mathematics is needed to complement cartoon drawings.

A Coherent Feed-forward Interaction Detects Persistent Inputs

In the bacterium *E. coli*, the sugar arabinose is only consumed when the preferred sugar, glucose, is scarce. The strategy that cells use to assess the presence of arabinose and absence of glucose involves a feed-forward arrangement that is different from the one just described. In this case, depletion of glucose causes an increase of cAMP, which is sensed by the CAP transcription activator protein, as described previously. In this case, however, CAP also induces the synthesis of a second transcription activator, AraC. Both activator proteins are necessary to activate arabinose metabolic genes (the AND logic function in Figure 8–85B).

This arrangement, known as a *coherent feed-forward motif*, has the interesting characteristics illustrated in Figure 8–88. Imagine that two activators, *A1* and *A2*, are both required to initiate transcription of a gene. The input to the network activates *A1* directly, but only activates *A2* through this *A1* activation. Thus, for a protein to be synthesized from this gene, long-term inputs are required that allow both *A1* and *A2* to be produced in active form. Brief input pulses are either ignored or produce small outputs. The requirement for a long input is important if assurances about a signal are needed before a costly cellular program is triggered. For example, glucose is the sugar on which *E. coli* cells grow best. Before cells trigger arabinose metabolism in the example above, it might be beneficial to be sure that glucose has been depleted (a sustained CAP pulse), rather than inducing the arabinose program during a transient glucose fluctuation.

Figure 8–88 **How a coherent feed-forward motif responds to various inputs.** (A) Diagram of a coherent feed-forward motif in which the transcription activators *A1* and *A2* together activate expression of gene *X* using the AND logic of Figure 8–85B. (B) The response to a brief input can be either weak (as shown) or nonexistent. This allows the motif to ignore random fluctuations in the concentration of signaling molecules. (C) A prolonged input produces a strong response that can turn off rapidly.

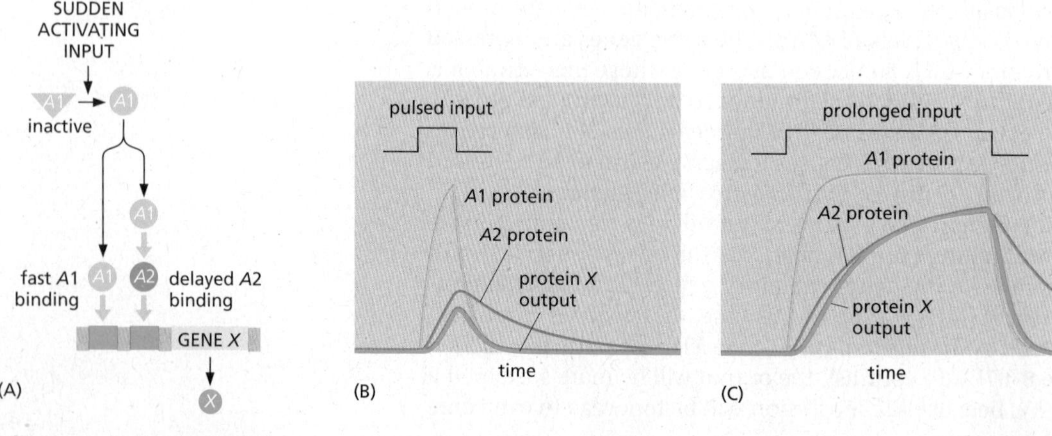

The Same Network Can Behave Differently in Different Cells Because of Stochastic Effects

Up to this point, we have assumed that all cells in a population produce identical behaviors if they contain the same network. It is important, however, to account for the fact that cells often show considerable individuality in their responses. Consider a situation in which a single mother cell divides into two daughter cells of equal volume. If the mother cell has only one molecule of a given protein, then only one daughter will inherit it. The daughters, though genetically identical, are already different. This variability is most pronounced for molecules that are present in small numbers. Nevertheless, even when there are many copies of a particular protein (or RNA), it is very unlikely that both daughter cells will end up with exactly the same number of molecules.

Figure 8–89 **Different levels of gene expression in individual cells within a population of *E. coli* bacteria.** For this experiment, two different reporter proteins (one fluorescing *green*, the other *red*), controlled by a copy of the same promoter, have been introduced into all of the bacteria. Some cells express only one gene copy, and so appear either *red* or *green*, while others express both gene copies, and so appear *yellow*. This experiment reveals variable levels of fluorescence, indicating variable levels of gene expression within an apparently uniform population of cells. (From M.B. Elowitz et al., *Science* 297:1183–1186, 2002. With permission from AAAS.)

This is just one illustration of a universal feature of cells: their behaviors are often **stochastic**, meaning that they display variability in their protein content and therefore exhibit variations in phenotypes. In addition to the asymmetric partitioning of molecules after cell division, variability can originate from many chemical reactions. Imagine, for example, that our mother cell contains a simple gene regulatory circuit with a positive feedback loop like that shown in Figure 8–82B. Even if both daughter cells receive a copy of this circuit, including one copy of the initial transcription activator protein, there will be variability in the time required for promoter binding—and it will be statistically nearly impossible for the genes in the two daughter cells to become activated at precisely the same time. If the system is bistable and poised near a switching point, then variability in the response might flip the switch in only one daughter cell. Two daughter cells that were born identical can thereby acquire, by chance, a dramatic difference in phenotype.

More generally, isogenic populations of cells grown in the same environment display diversity in size, shape, cell-cycle position, and gene expression. These differences arise because biochemical reactions require probabilistic collisions between randomly moving molecules, with each event resulting in changes in the number of molecular species by integer amounts. The amplified effect of fluctuations in a molecular reactant, or the compounded effects of fluctuations across many molecular reactants, often accumulates as an observable phenotype. This can endow a cell with individuality and generate nongenetic cell-to-cell variability in a population.

Nongenetic variability can be studied in the laboratory by single-cell measurements of fluorescent proteins expressed from genes under the control of a specific promoter. Live cells can be mounted on a slide and viewed through a fluorescence microscope, revealing the striking variability in protein expression levels (**Figure 8–89**). Another approach is to use flow cytometry, which works by streaming a dilute suspension of cells past an illuminator and measuring the fluorescence of individual cells as they flow past the detector. Fluorescence values can be used to build histograms that reveal the variability in a process across a population of cells, with a broad histogram indicating higher variability.

Several Computational Approaches Can Be Used to Model the Reactions in Cells

We have focused primarily on the use of ordinary differential equations to model the dynamics of simple regulatory circuits. These models are called *deterministic*, because they do not incorporate stochastic variability and will always produce the same result from a specific set of parameters. As we have seen, such models can provide useful insights, particularly in the detailed mechanistic analysis of small regulatory circuits. However, other types of computational approaches are also needed to comprehend the great complexity of cell behavior. *Stochastic models*, for example, attempt to account for the very important problem of random variability in molecular networks. These models do not provide deterministic predictions about the behavior of molecules; instead, they incorporate random variation into molecule numbers and interactions, and the purpose of these models is to obtain a better understanding of the probability that a system will exist in a certain state over time.

Numerous other modeling strategies have been or are being developed. *Boolean networks* are used for the qualitative analysis of complex gene regulatory networks containing large numbers of interacting components. In these models, each molecule is a node that can exist in either the active or inactive state, thereby affecting the state of the nodes it is linked to. Models of this sort provide insights into the flow of information through a network, and they were useful in helping us understand the complex gene regulatory network that controls the early development of the sea urchin (see Figure 7–45). Boolean networks therefore reduce complex networks to a highly simplified (and potentially inaccurate) form. At the other extreme are *agent-based simulations*, in which thousands of molecules (or "agents") in a system are modeled individually, and their probable behaviors and interactions with each other over time are calculated on the basis of predicted physical and chemical behaviors, often while taking stochastic variation into account. Agent-based approaches are computationally demanding but have the potential to generate highly life-like simulations of real biological systems.

Statistical Methods Are Critical for the Analysis of Biological Data

Dynamics, differential equations, and theoretical modeling are not the be-all and end-all of mathematics. Other branches of the subject are no less important for biologists. Statistics—the mathematics of probabilistic processes and noisy data sets—is an inescapable part of every biologist's life.

This is true in two main ways. First, imperfect measurement devices and other errors generate experimental noise in our data. Second, all cell-biological processes depend on the stochastic behavior of individual molecules, as we just discussed, and this results in biological noise in our results. How, in the face of all this noise, do we come to conclusions about the truth of hypotheses? The answer is statistical analysis, which shows how to move from one level of description to another: from a set of erratic individual data points to a simpler description of the key features of the data.

Statistics teaches us that the more times we repeat our measurements, the better and more refined the conclusions we can draw from them. Given many repetitions, it becomes possible to describe our data in terms of variables that summarize the features that matter: the mean value of the measured variable, taken over the set of data points; the magnitude of the noise (the standard deviation of the set of data points); the likely error in our estimate of the mean value (the standard error of the mean); and, for specialists, the details of the probability distribution describing the likelihood that an individual measurement will yield a given value. For all these things, statistics provides recipes and quantitative formulas that biologists must understand if they are to make rigorous conclusions on the basis of variable results.

Summary

Quantitative mathematical analysis can provide a powerful extra dimension in our understanding of cell regulation and function. Cell regulatory systems often depend on macromolecular interactions, and mathematical analysis of the dynamics of these interactions can unveil important insights into the importance of binding affinities and protein stability in the generation of transcriptional or other signals. Regulatory systems often employ network motifs that generate useful behaviors: a rapid negative feedback loop dampens the response to input signals; a delayed negative feedback loop creates a biochemical oscillator; positive feedback yields a system that alternates between two stable states; and feed-forward motifs provide systems that generate transient signal pulses or respond only to sustained inputs. The dynamic behavior of these network motifs can be dissected in detail with deterministic and stochastic mathematical modeling.

PROBLEMS

Which statements are true? Explain why or why not.

8–1 Because a monoclonal antibody recognizes a specific antigenic site (epitope), it binds only to the specific protein against which it was made.

8–2 Given the inexorable march of technology, it seems inevitable that the sensitivity of detection of molecules will ultimately be pushed beyond the yoctomole level (10^{-24} mole).

8–3 If each cycle of PCR doubles the amount of DNA synthesized in the previous cycle, then 10 cycles will give a 10^3-fold amplification, 20 cycles will give a 10^6-fold amplification, and 30 cycles will give a 10^9-fold amplification.

8–4 To judge the biological importance of an interaction between protein *A* and protein *B*, we need to know quantitative details about their concentrations, affinities, and kinetic behaviors.

8–5 The rate of change in the concentration of any molecular species *X* is given by the balance between its rate of appearance and its rate of disappearance.

8–6 After a sudden increase in its rate of synthesis, a protein with a slow rate of degradation will reach a new steady-state level more quickly than a protein with a rapid rate of degradation.

Discuss the following problems.

8–7 A common step in the isolation of cells from a sample of animal tissue is to treat the tissue with trypsin, collagenase, and EDTA. Why is such a treatment necessary, and what does each component accomplish? And why does this treatment not kill the cells?

8–8 Tropomyosin, at 93 kilodaltons, sediments at 2.6S, whereas the 65-kilodalton protein, hemoglobin, sediments at 4.3S. (The sedimentation coefficient S is a linear measure of the rate of sedimentation.) These two proteins are drawn to scale in Figure Q8–1. Why does the bigger protein sediment more slowly than the smaller one? Can you think of an analogy from everyday experience that might help you with this problem?

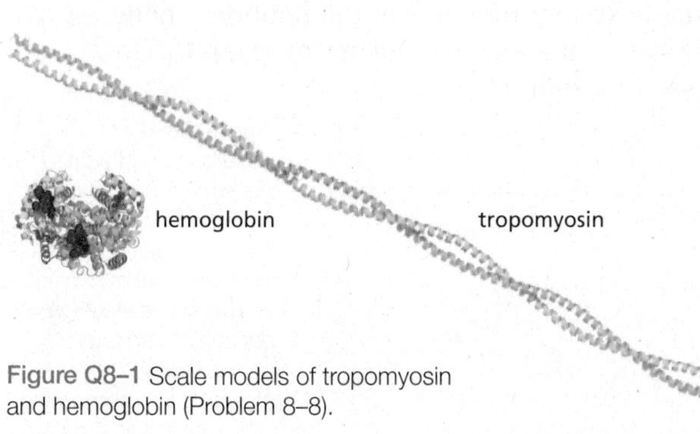

Figure Q8–1 Scale models of tropomyosin and hemoglobin (Problem 8–8).

8–9 Hybridoma technology allows one to generate monoclonal antibodies to virtually any protein. Why is it, then, that genetically tagging proteins with epitopes is such a commonly used technique, especially as an epitope tag has the potential to interfere with the function of the protein?

8–10 How many copies of a protein need to be present in a cell in order for it to be visible as a band on an SDS gel? Assume that you can load 100 μg of cell extract onto a gel and that you can detect 10 ng in a single band by silver staining the gel. The concentration of protein in cells is about 200 mg/mL, and a typical mammalian cell has a volume of about 1000 μm^3 and a typical bacterium a volume of about 1 μm^3. Given these parameters, calculate the number of copies of a 120-kilodalton protein that would need to be present in a mammalian cell and in a bacterium in order to give a detectable band on a gel.

8–11 You have isolated the proteins from two adjacent spots after two-dimensional polyacrylamide-gel electrophoresis and digested them with trypsin. When the masses of the peptides were measured by MALDI-TOF mass spectrometry, the peptides from the two proteins were found to be identical except for one (Figure Q8–2). For this peptide, the mass-to-charge (m/z) values differed by 80, a value that does not correspond to a difference in amino acid sequence. (For example, glutamic acid instead of valine at one position would give an m/z difference of around 30.) Can you suggest a possible difference between the two peptides that might account for the observed m/z difference?

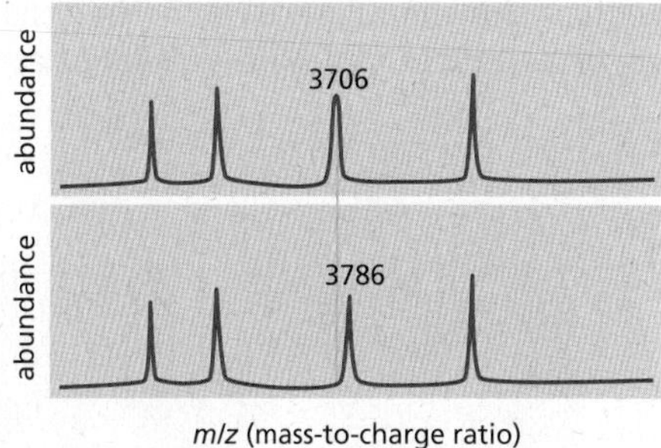

Figure Q8–2 Masses of peptides measured by MALDI-TOF mass spectrometry (Problem 8–11). Only the numbered peaks differ between the two protein samples.

8–12 You want to amplify the DNA between the two stretches of sequence shown in **Figure Q8–3**. Of the listed primers, choose the pair that will allow you to amplify the DNA by PCR.

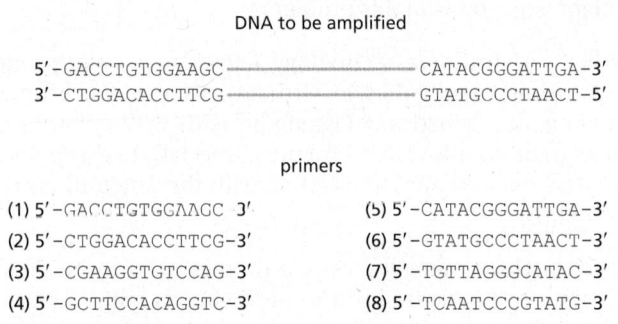

DNA to be amplified

```
5′-GACCTGTGGAAGC————————CATACGGGATTGA-3′
3′-CTGGACACCTTCG————————GTATGCCCTAACT-5′
```

primers

(1) 5′-GACCTGTGGAAGC-3′	(5) 5′-CATACGGGATTGA-3′
(2) 5′-CTGGACACCTTCG-3′	(6) 5′-GTATGCCCTAACT-3′
(3) 5′-CGAAGGTGTCCAG-3′	(7) 5′-TGTTAGGGCATAC-3′
(4) 5′-GCTTCCACAGGTC-3′	(8) 5′-TCAATCCCGTATG-3′

Figure Q8–3 DNA to be amplified and potential PCR primers (Problem 8–12).

8–13 In the very first round of PCR using genomic DNA, the DNA primers prime synthesis that terminates only when the cycle ends (or when a random end of DNA is encountered). Yet, by the end of 20–30 cycles—a typical amplification—the only visible product is defined precisely by the ends of the DNA primers. In what cycle is a double-strand fragment of the correct size first generated?

8–14 Explain the difference between a gain-of-function mutation and a dominant-negative mutation. Why are both these types of mutation usually dominant?

8–15 Discuss the following statement: "We would have no idea today of the importance of insulin as a regulatory hormone if its absence were not associated with the human disease diabetes. It is the dramatic consequences of its absence that focused early efforts on the identification of insulin and the study of its normal role in physiology."

8–16 You have received the results from an RNA-seq analysis of mRNAs from liver. You had anticipated counting the number of reads of each mRNA to determine the relative abundance of different mRNAs. But you are puzzled because many of the mRNAs have given you results like those shown in **Figure Q8–4**. How is it that different parts of an mRNA can be represented at different levels?

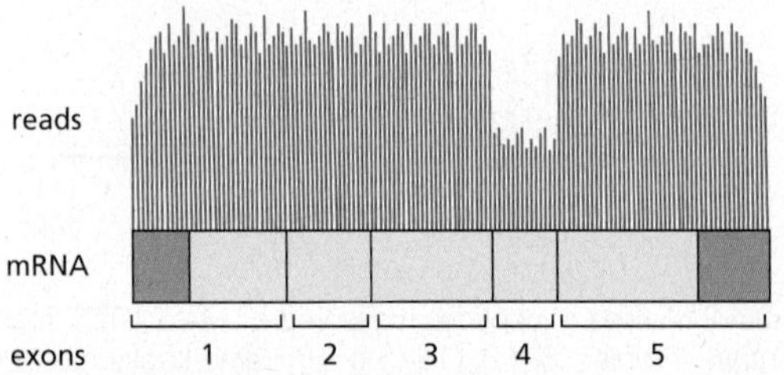

Figure Q8–4 RNA-seq reads for a liver mRNA (Problem 8–16). The exon structure of the mRNA is indicated, with protein-coding segments indicated in *light blue* and untranslated regions in *dark blue*. The numbers of sequencing reads are indicated by the heights of the vertical lines above the mRNA.

8–17 Examine the network motifs in **Figure Q8–5**. Decide which ones are negative feedback loops and which are positive. Explain your reasoning.

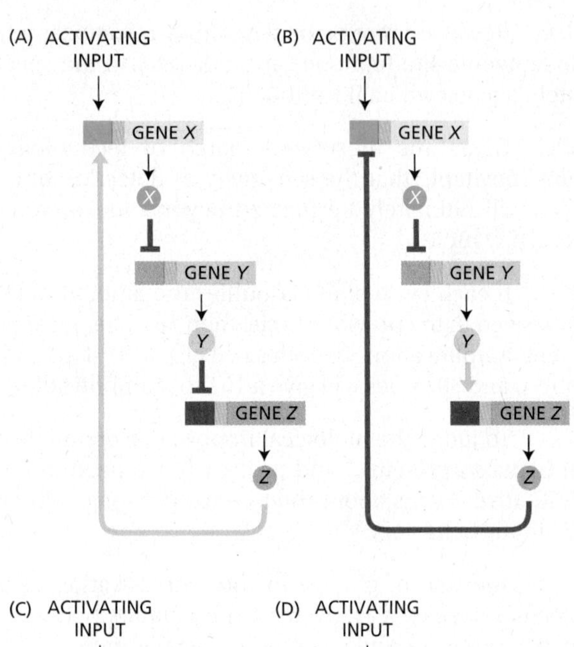

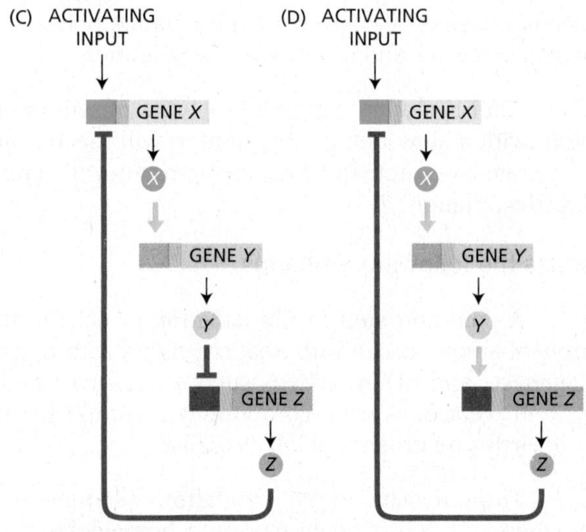

Figure Q8–5 Network motifs composed of transcription activators and repressors (Problem 8–17).

8–18 Imagine that a random perturbation positions a bistable system precisely at the boundary between two stable states (at the *orange dot* in **Figure Q8–6**). How would the system respond?

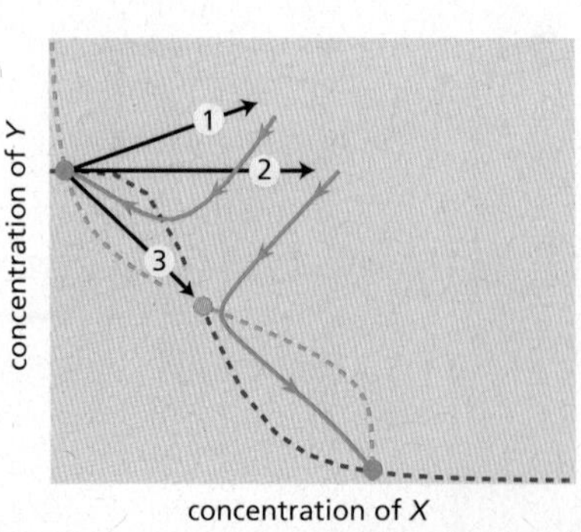

Figure Q8–6 Perturbations of a bistable system (Problem 8–18). As shown by the *green lines*, after perturbation 1 the system returns to its original stable state (*green dot* at left), and after perturbation 2, the system moves to the other stable state (*green dot* at right). Perturbation 3 moves the system to the precise boundary between the two stable states *(orange dot)*.

8–19 Detailed analysis of the regulatory region of the *Lac* operon has revealed surprising complexity. Instead of a single binding site for the Lac repressor, as might be expected, there are three sites termed operators (O_1, O_2, and O_3) arrayed along the DNA as shown in Figure Q8–7.

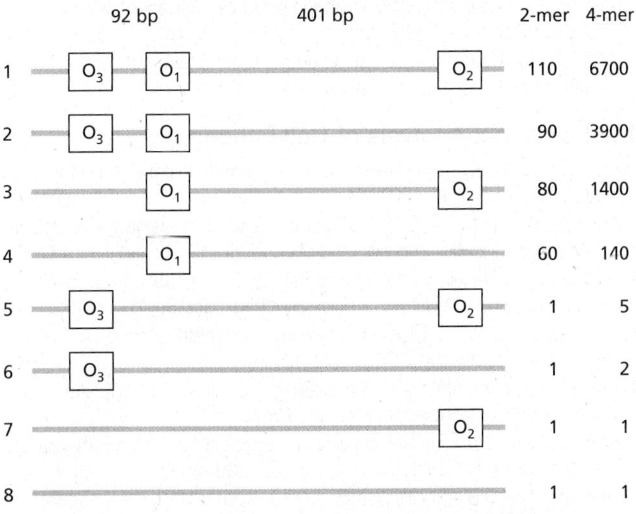

Figure Q8–7 Repression of β-galactosidase by promoter regions that contain different combinations of Lac repressor binding sites (Problem 8–19). The base-pair (bp) separation of the three operator sites is shown. Numbers at *right* refer to the level of repression, with higher numbers indicating more effective repression by dimeric (2-mer) or tetrameric (4-mer) repressors. (From S. Oehler et al., *EMBO J.* 9:973–979, 1990. With permission from John Wiley & Sons.)

To probe the functions of these three sites, you make a series of constructs in which various combinations of operator sites are present. You examine their ability to repress expression of β-galactosidase, using either tetrameric (wild type) or dimeric (mutant) forms of the Lac repressor. The dimeric form of the repressor can bind to a single operator (with the same affinity as the tetramer) with each monomer binding to half the site. The tetramer, the form normally expressed in cells, can bind to two sites simultaneously. When you measure repression of β-galactosidase expression, you find the results shown in Figure Q8–7, with higher numbers indicating more effective repression.

A. Which single operator site is the most important for repression? How can you tell?

B. Do combinations of operator sites (Figure Q8–7, constructs 1, 2, 3, and 5) substantially increase repression by the dimeric repressor? Do combinations of operator sites substantially increase repression by the tetrameric repressor? If the two repressors behave differently, offer an explanation for the difference.

C. The wild-type repressor binds O_3 very weakly when it is by itself on a segment of DNA. However, if O_1 is included on the same segment of DNA, the repressor binds O_3 quite well. How can that be?

REFERENCES

Isolating Cells and Growing Them in Culture

Greenfield E (2014) Antibodies: A Laboratory Manual, 2nd ed. Cold Spring Harbor, NY: Cold Spring Harbor Laboratory Press.

Sherr CJ & DePinho RA (2000) Cellular senescence: mitotic clock or culture shock? *Cell* 102, 407–410.

Spector DL, Goldman RD & Leinwand LA (eds.) (1998) Cells: A Laboratory Manual. Cold Spring Harbor, NY: Cold Spring Harbor Laboratory Press.

Todaro GJ & Green H (1963) Quantitative studies of the growth of mouse embryo cells in culture and their development into established lines. *J. Cell Biol.* 17, 299–313.

Purifying Proteins

Burgess R & Deutcher MP (eds.) (2009) Guide to Protein Purification, 2nd ed. *Methods in Enzymology*, Vol. 463. Cambridge, MA: Academic Press.

De Duve C & Beaufay H (1981) A short history of tissue fractionation. *J. Cell Biol.* 91, 293s–299s.

Scopes RK (1994) Protein Purification: Principles and Practice, 3rd ed. New York: Springer-Verlag.

Simpson RJ, Adams PD & Golemis EA (2008) Basic Methods in Protein Purification and Analysis: A Laboratory Manual. Cold Spring Harbor, NY: Cold Spring Harbor Laboratory Press.

Wood DW (2014) New trends and affinity tag designs for recombinant protein purification. *Curr. Opin. Struct. Biol.* 26, 54–61.

Analyzing Proteins

Aebersold R & Mann M (2016) Mass-spectrometric exploration of proteome structure and function. *Nature* 537, 347–355.

Baxevanis AD, Bader GD & Wishart DS (eds.) (2020) Bioinformatics: A Practical Guide to the Analysis of Genes and Proteins, 4th ed. Hoboken, NJ: Wiley.

Goodrich JA & Kugel JF (2007) Binding and Kinetics for Molecular Biologists. Cold Spring Harbor, NY: Cold Spring Harbor Laboratory Press.

Mehmood S, Allison TM & Robinson CV (2015) Mass spectrometry of protein complexes: from origins to applications. *Annu. Rev. Phys. Chem.* 66, 453–474.

O'Connell MR, Gamsjaeger R & Mackay JP (2009) The structural analysis of protein-protein interactions by NMR spectroscopy. *Proteomics* 9, 5224–5232.

Pollard TD (2010) A guide to simple and informative binding assays. *Mol. Biol. Cell* 21, 4061–4067.

Wuthrich K (2003) NMR studies of structure and function of biological macromolecules (Nobel Lecture). *J. Biomol. NMR* 27, 13–39.

Analyzing and Manipulating DNA

Gibson DG, Young L, Chuang RY . . . Smith HO (2009) Enzymatic assembly of DNA molecules up to several hundred kilobases. *Nat. Methods* 6, 343–345.

Green MR & Sambrook J (2012) Molecular Cloning: A Laboratory Manual, 4th ed. Cold Spring Harbor, NY: Cold Spring Harbor Laboratory Press.

Kosuri S & Church GM (2014) Large-scale de novo DNA synthesis: technologies and applications. *Nat. Methods* 11, 499–507.

Metzker ML (2010) Sequencing technologies—the next generation. *Nat. Rev. Genet.* 11, 31–46.

Mullis K, Faloona F, Scharf S . . . Erlich H (1986) Specific enzymatic amplification of DNA in vitro: the polymerase chain reaction. *Cold Spring Harb. Symp. Quant. Biol.* 51 Pt 1, 263–273.

Mullis KB (1990) The unusual origin of the polymerase chain reaction. *Sci. Am.* 262, 56–61, 64–55.

Nathans D & Smith HO (1975) Restriction endonucleases in the analysis and restructuring of DNA molecules. *Annu. Rev. Biochem.* 44, 273–293.

Sanger F, Nicklen S & Coulson AR (1977) DNA sequencing with chain-terminating inhibitors. *Proc. Natl. Acad. Sci. USA* 74, 5463–5467.

Shendure J & Lieberman Aiden E (2012) The expanding scope of DNA sequencing. *Nat. Biotechnol.* 30, 1084–1094.

Watson JD, Myers RM & Caudy AA (2007) Recombinant DNA: Genes and Genomes—A Short Course, 3rd ed. Cold Spring Harbor, NY: Cold Spring Harbor Laboratory Press.

Studying Gene Function and Expression

Agrotis A & Ketteler R (2015) A new age in functional genomics using CRISPR/Cas9 in arrayed library screening. *Front. Genet.* 6, 300.

D'Haeseleer P (2005) How does gene expression clustering work? *Nat. Biotechnol.* 23, 1499–1501.

Doudna JA (2020) The promise and challenge of therapeutic genome editing. *Nature* 578, 229–236.

Fellmann C & Lowe SW (2014) Stable RNA interference rules for silencing. *Nat. Cell Biol.* 16, 10–18.

Ingolia NT, Hussmann JA & Weissman JS (2019) Ribosome profiling: global views of translation. *Cold Spring Harb. Perspect. Biol.* 11.

Luecken MD & Theis FJ (2019) Current best practices in single-cell RNA-seq analysis: a tutorial. *Mol. Syst. Biol.* 15, e8746.

Mali P, Esvelt KM & Church GM (2013) Cas9 as a versatile tool for engineering biology. *Nat. Methods* 10, 957–963.

Rio DC, Ares M, Hannon GJ & Wilson TW (2011) RNA: A Laboratory Manual. Cold Spring Harbor, NY: Cold Spring Harbor Laboratory Press.

Stark R, Grzelak M & Hadfield J (2019) RNA sequencing: the teenage years. *Nat. Rev. Genet.* 20, 631–656.

Strachan T & Read AP (2018) Human Molecular Genetics, 5th ed. Boca Raton, FL: CRC Press.

Wilson RC & Doudna JA (2013) Molecular mechanisms of RNA interference. *Annu. Rev. Biophys.* 42, 217–239.

Mathematical Analysis of Cell Function

Alon U (2007) Network motifs: theory and experimental approaches. *Nat. Rev. Genet.* 8, 450–461.

Alon U (2019) An Introduction to Systems Biology: Design Principles of Biological Circuits, 2nd ed. Boca Raton, FL: Chapman & Hall/CRC.

Ferrell JE, Jr. & Ha SH (2014) Ultrasensitivity part III: cascades, bistable switches, and oscillators. *Trends Biochem. Sci.* 39, 612–618.

Gunawardena J (2014) Models in biology: "accurate descriptions of our pathetic thinking." *BMC Biol.* 12, 29.

Ingalis BP (2013) Mathematical Modeling in Systems Biology: An Introduction. Cambridge, MA: MIT Press.

Lewis J (2008) From signals to patterns: space, time, and mathematics in developmental biology. *Science* 322, 399–403.

Novak B & Tyson JJ (2008) Design principles of biochemical oscillators. *Nat. Rev. Mol. Cell Biol.* 9, 981–991.

Xia PF, Ling H, Foo JL & Chang MW (2019) Synthetic genetic circuits for programmable biological functionalities. *Biotechnol. Adv.* 37, 107393.

CHAPTER 9

Visualizing Cells and Their Molecules

IN THIS CHAPTER

Understanding the structural organization of cells, and the macromolecules that build and animate them, is essential for learning how they function. In this chapter, we briefly describe some of the principal light and electron microscopy methods used to study cells and molecules. In the past decade or so, there have been major technical developments in both methods that allow us to see biological structures with increasing resolution and clarity. Optical microscopy will be our starting point because cell biology began with the light microscope, and it is still an indispensable tool. The development of methods for the specific labeling and imaging of individual cellular constituents and the reconstruction of their three-dimensional architecture has meant that, far from falling into disuse, optical microscopy continues to increase in importance. One advantage of optical microscopy is that light is relatively nondestructive. By tagging specific cell components with fluorescent probes, such as intrinsically fluorescent proteins, we can watch their movement, dynamics, and interactions in living cells.

Although conventional optical microscopy is limited in resolution by the wavelength of visible light, new methods cleverly bypass this limitation and allow the exact position of even single molecules to be mapped. By using a beam of electrons instead of visible light, electron microscopy can image the interior of cells, and their macromolecular components, at almost atomic resolution and in three dimensions. But all imaging methods involve trade-offs; in this case, the higher resolution means only small objects are imaged and only in fixed, dead cells. There is now a bewildering variety of imaging technologies for the cell biologist to choose from, and when some of these are described later in the chapter, it is worth considering why you might use one rather than another. Trade-offs will always have to be made between thin and thick specimens, living and fixed cells, high and low resolution, fast and slow imaging, signal and noise, or cells and molecules.

This chapter is intended as a companion, rather than an introduction, to the chapters that follow; readers may wish to refer back to it as applications of microscopy to basic biological problems are encountered in other chapters of the book.

LOOKING AT CELLS AND MOLECULES IN THE LIGHT MICROSCOPE

A typical animal cell is 10–20 μm in diameter, which is just less than a tenth the size of the smallest object that we can normally see with the naked eye. Only after good light microscopes became available in the early part of the nineteenth century did Matthias Schleiden and Theodor Schwann propose that all plant and animal tissues were aggregates of individual cells. Their proposal in 1838, known as the **cell doctrine**, marks the formal birth of cell biology.

Animal cells are not only tiny, but they are also colorless and translucent. The discovery of their main internal features, therefore, depended on the development, in the late nineteenth century, of a variety of stains that provided sufficient color and contrast to make those features visible. Similarly, the far

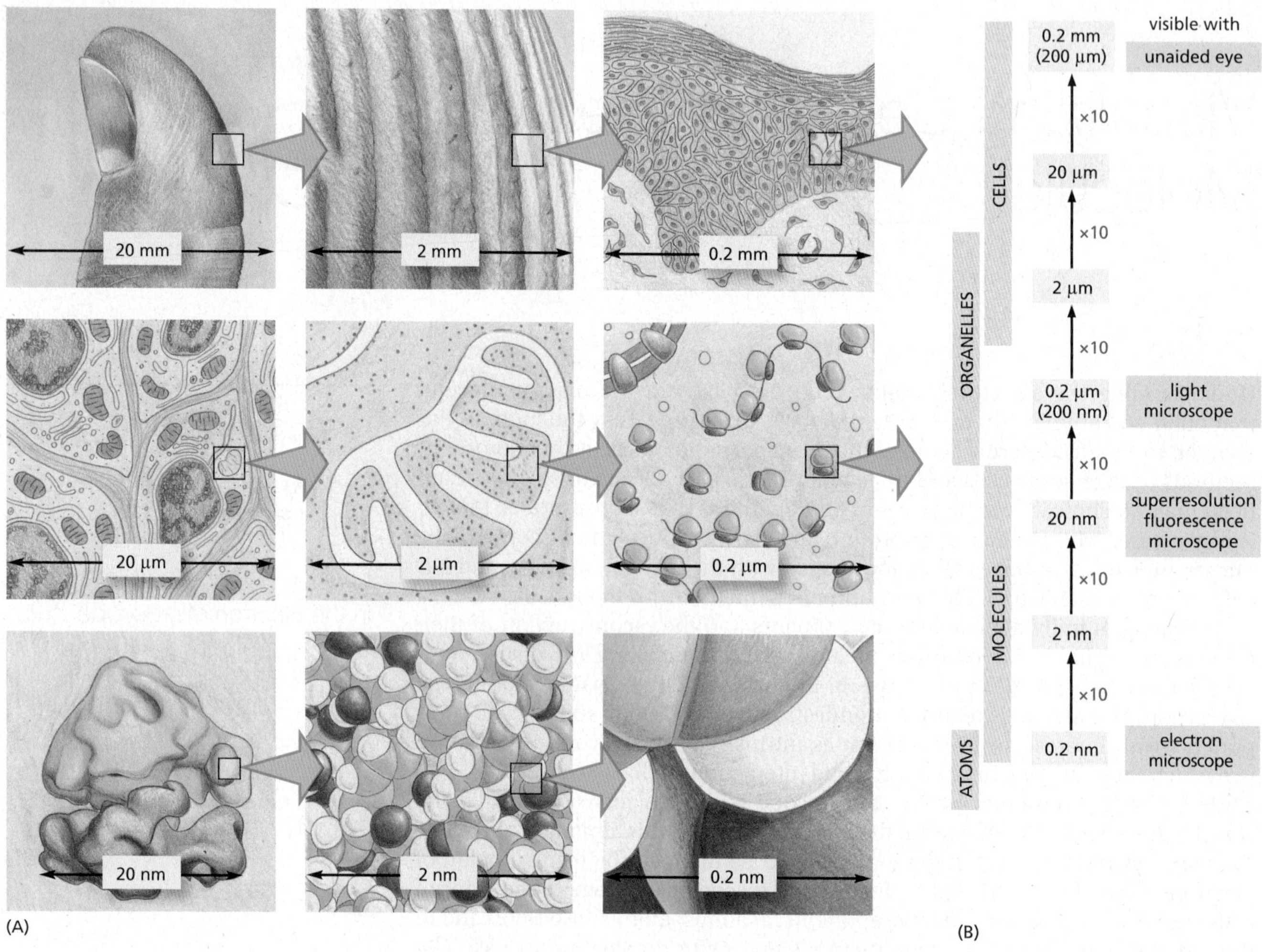

Figure 9–1 A sense of scale between living cells and atoms. (A) Each diagram shows an image magnified by a factor of 10 in an imaginary progression from a thumb, through skin cells, to a ribosome, to a cluster of atoms forming part of one of the many protein molecules in the ribosome. Atomic details of biological macromolecules, as shown in the last two panels, are just within the power of the electron microscope. While color has been used here in all the panels, it is not a feature of objects much smaller than the wavelength of light, so the last five panels should really be in black and white. (B) Sizes of cells and their components are shown on a logarithmic scale, indicating the range of objects that can readily be resolved by the naked eye and in the light and electron microscopes. Note that new superresolution microscopy techniques, discussed in detail later, allow an improvement in resolution by an order of magnitude compared with conventional light microscopy.

more powerful electron microscope introduced in the early 1940s required the development of new techniques for preserving and staining cells before the full complexities of their internal fine structure could begin to emerge. To this day, microscopy often relies as much on techniques for preparing the specimen as on the performance of the microscope itself. In the following discussions, we therefore consider both instruments and specimen preparation, beginning with the light microscope.

The images in Figure 9–1A illustrate a stepwise progression from a thumb to a cluster of atoms. Each successive image represents a tenfold increase in magnification. The naked eye can see features in the first two panels, the light microscope allows us to see details corresponding to about the fifth panel, and the electron microscope takes us to about the eighth or ninth panel. Figure 9–1B shows the sizes of various cellular and subcellular structures and the ranges of size that different types of microscopes can visualize.

The Conventional Light Microscope Can Resolve Details 0.2 μm Apart

For well over 100 years, all microscopes were constrained by a fundamental limitation: that a given type of radiation cannot be used to probe structural details much smaller than its own wavelength. A limit to the resolution of a light microscope was therefore set by the wavelength of visible light, which ranges from about 0.4 μm (for violet) to 0.7 μm (for deep red). In practical terms, bacteria and mitochondria, which are about 500 nm (0.5 μm) wide, are generally the smallest objects whose shape we can clearly discern in a standard **light microscope**;

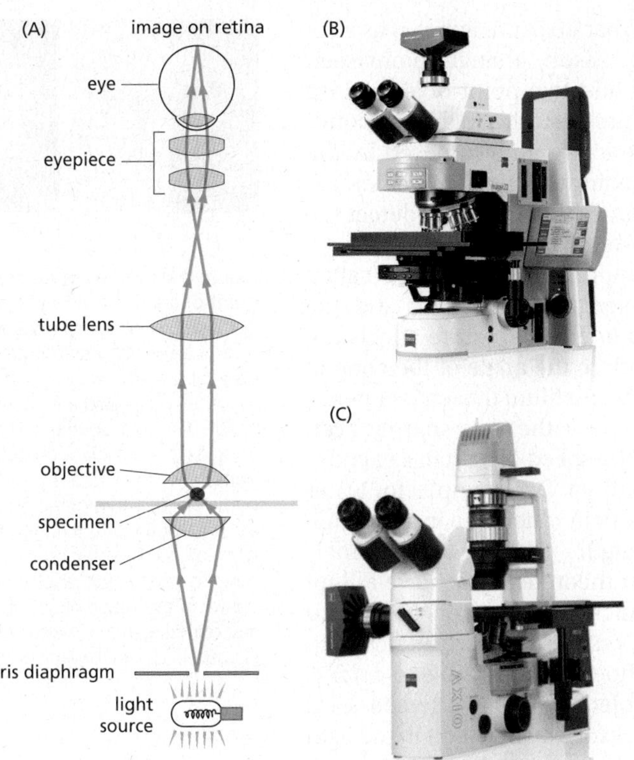

Figure 9–2 **A light microscope.** (A) Diagram showing the light path in an upright compound microscope. Light is focused on the specimen by lenses in the condenser. A combination of objective lenses, tube lenses, and eyepiece lenses is arranged to focus an image of the illuminated specimen in the eye. (B) A modern upright research light microscope. (C) A modern inverted microscope, particularly useful for looking at cells in culture. Both microscopes are equipped for fluorescence imaging (B and C, courtesy of Carl Zeiss Microscopy, GmbH.)

details smaller than this are obscured by effects resulting from the wave-like nature of light. Let us follow the behavior of a beam of light as it passes through the lenses of a microscope (**Figure 9–2**).

The following units of length are commonly employed in microscopy:

μm (micrometer) = 10^{-6} m

nm (nanometer) = 10^{-9} m

Å (angstrom) = 10^{-10} m

Because of its wave nature, light does not follow the idealized straight ray paths that geometrical optics predicts. Instead, light waves travel through an optical system by many slightly different routes, like ripples in water, so that they interfere with one another and cause *optical diffraction* effects. If two trains of waves reaching the same point by different paths are precisely *in phase*, with crest matching crest and trough matching trough, they will reinforce each other so as to increase brightness. In contrast, if the trains of waves are *out of phase*, they will interfere with each other in such a way as to cancel each other partly or entirely (**Figure 9–3**). The interaction of light with an object changes the phase relationships of the light waves in a way that produces complex interference effects. At high magnification, for example, the shadow of an edge that is evenly illuminated with light of uniform wavelength appears as a set of parallel lines (**Figure 9–4A**),

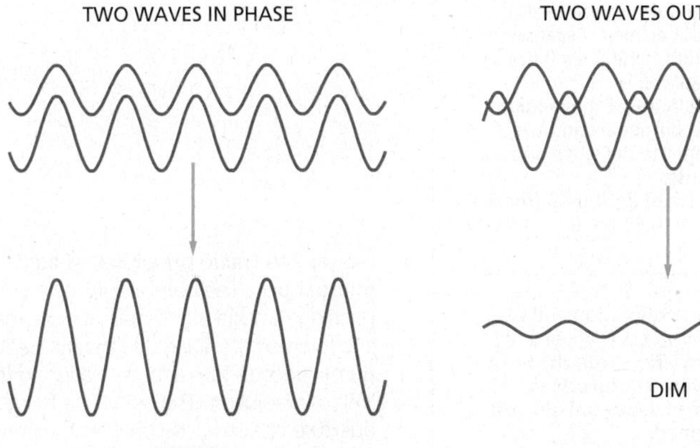

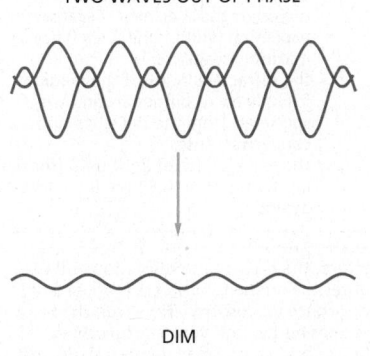

Figure 9–3 **Interference between light waves.** When two light waves combine in phase, the amplitude of the resultant wave is larger, and the brightness is increased. Two light waves that are out of phase cancel each other partly and produce a wave whose amplitude, and therefore brightness, is decreased.

whereas the smallest focused image of a bright circular aperture appears as a set of concentric rings (Figure 9–4B). For the same reason, a single point seen through a microscope appears as a blurred disc, and two point objects close together give overlapping images and may merge into one. Although no amount of refinement of the lenses can overcome the diffraction limit imposed by the wave-like nature of light, other ways of cleverly bypassing this limit have emerged, creating so-called superresolution imaging techniques that can even detect the position of single molecules. These are discussed later in the chapter.

The limiting separation at which two objects appear distinct—the so-called **limit of resolution**—depends on both the wavelength of the light and the *numerical aperture* of the lens system used. The numerical aperture affects the light-gathering ability of the lens and is related both to the angle of the cone of light that can enter it and to the refractive index of the medium the lens is operating in; the wider the microscope opens its eye, so to speak, the more sharply it can see (Figure 9–5). The *refractive index* is the ratio of the speed of light in a vacuum to the speed of light in a particular transparent medium. For example, for water this is 1.33, meaning that light travels 1.33 times slower in water than in a vacuum. Under the best conditions, with violet light (wavelength = 0.4 μm) and a numerical aperture of 1.4, the basic light microscope can theoretically achieve a limit of resolution of about 0.2 μm, or 200 nm. Some microscope makers at the end of the nineteenth century achieved this resolution, but it is routinely matched in contemporary, factory-produced microscopes. Although it is possible to *enlarge* an image as much as we want—for example, by projecting it onto a screen—it is not possible, in a conventional light microscope, to resolve two objects in the light microscope that are separated by less than about 0.2 μm; they will always appear as a single object. It is important, however, to distinguish between *resolution* and *detection*. If a small object, below the resolution limit, itself emits light, then we may still be able to see or detect it. Thus, we can see a single fluorescently labeled microtubule even though it is about 10 times thinner than the resolution limit of the light microscope. Diffraction effects, however, will cause it to appear blurred and at least 0.2 μm thick (see Figure 9–14). In a similar way, we can see the stars in the night sky, even though their diameters are far below the angular resolution of our unaided eyes: they all appear as similar, slightly blurred points of light, differing only in their color and brightness.

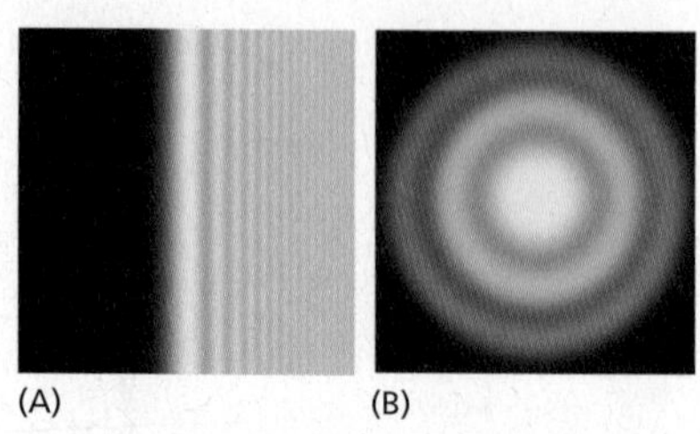

Figure 9–4 Images of an edge and of a point of light. (A) The interference effects, or fringes, seen at high magnification when light of a specific wavelength passes the edge of a solid object placed between the light source and the observer. (B) The image of a point source of light. Diffraction spreads this out into a complex, circular pattern, whose width depends on the numerical aperture of the optical system: the smaller the aperture, the bigger (more blurred) the diffracted image. Two point sources can be just resolved when the center of the image of one lies within the first dark ring in the image of the other: this is used to define the limit of resolution.

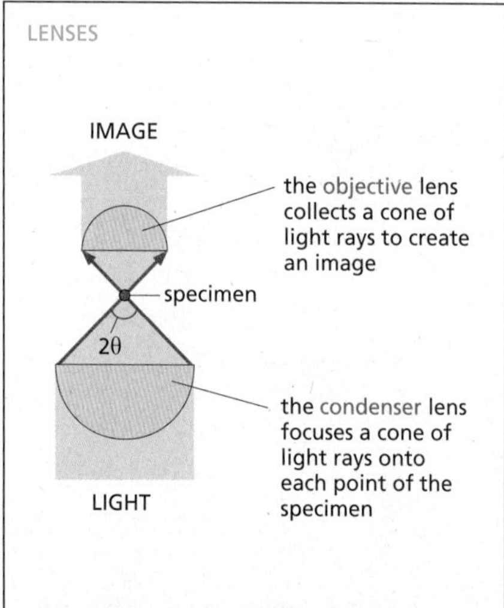

RESOLUTION: the resolving power of the microscope depends on the width of the cone of illumination and therefore on both the condenser and the objective lens. It is calculated using the formula

$$\text{resolution} = \frac{0.61\,\lambda}{n \sin \theta}$$

where:

θ = half the angular width of the cone of rays collected by the objective lens from a typical point in the central region of the specimen (because the maximum width is 180°, sin θ has a maximum value of 1)

n = the refractive index of the medium (usually air or oil) separating the specimen from the objective and condenser lenses

λ = the wavelength of light used (for white light a figure of 0.53 μm is commonly assumed)

NUMERICAL APERTURE: $n \sin \theta$ in the equation above is called the numerical aperture of the lens and is a function of its light-collecting ability. For dry lenses this cannot be more than 1, but for oil-immersion lenses it can be as high as 1.4. The higher the numerical aperture, the greater the resolution and the brighter the image (brightness is important in fluorescence microscopy). However, this advantage does necessitate very short working distances and a very small depth of field, just as in a conventional camera.

Figure 9–5 Basic principles of light microscopy. The path of light rays passing through a transparent specimen in a microscope illustrates the concept of numerical aperture and its relation to the limit of resolution. The higher the numerical aperture of a lens, the brighter the image it forms and the higher its resolution.

Photon Noise Creates Additional Limits to Resolution When Light Levels Are Low

Any image, whether produced by an electron microscope or by an optical microscope, is made by particles—electrons or photons—striking a detector of some sort. But these particles are governed by quantum mechanics, so the numbers reaching the detector are predictable only in a statistical sense. Finite samples, collected by imaging for a limited period of time (that is, by taking a snapshot), will show random variation: successive snapshots of the same scene will not be exactly identical. Moreover, every detection method has some level of background signal or noise, adding to the statistical uncertainty. With bright illumination, corresponding to very large numbers of photons or electrons, the features of the imaged specimen are accurately determined on the basis of the distribution of these particles at the detector. However, with smaller numbers of particles, the structural details of the specimen are obscured by the statistical fluctuations in the numbers of particles detected in each region, which give the image a speckled appearance and limit its precision. The term *noise* describes this random variability. Because noise in the image is proportional to the square root of the number of photons that are detected (or electrons in electron microscopy), then as the number of photons or electrons recorded increases, the absolute noise also increases, but because of the square root relationship, the percentage of noise decreases, in other words the signal-to-noise ratio improves. A poor signal-to-noise ratio is an important consideration when weak fluorescent light signals are recorded or low, but less damaging, electron doses are required.

Living Cells Are Seen Clearly in a Phase-Contrast or a Differential-Interference-Contrast Microscope

There are many ways in which contrast in a specimen can be generated (**Figure 9–6**). While fixing and staining a specimen can generate contrast through color (Figure 9–6A), microscopists have always been challenged by the possibility that some components of the cell may be lost or distorted during specimen preparation. The only certain way to avoid the problem is to examine cells while they are alive, without fixing or freezing. For this purpose, light microscopes with special optical systems are especially useful.

In the normal **bright-field microscope**, light passing through a cell in culture forms the image directly. Another system, **dark-field microscopy**, exploits the fact that light rays can be scattered in all directions by small objects in their path.

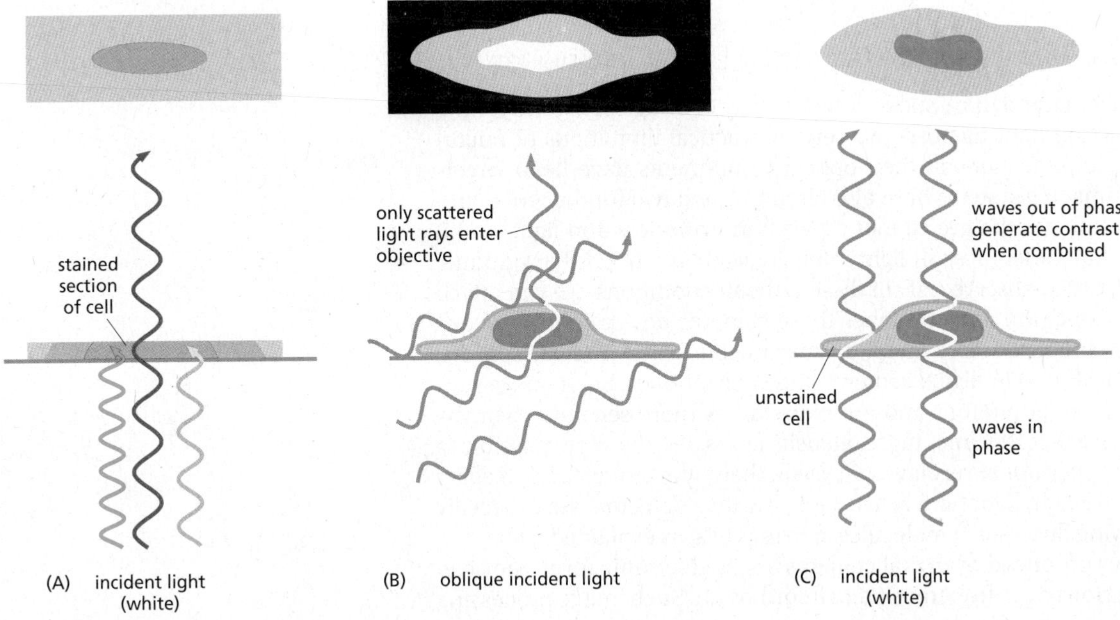

Figure 9–6 **Contrast in light microscopy.** (A) The stained portion of the cell will absorb light of some wavelengths, which depends on the stain, but will allow other wavelengths to pass through it. A colored image of the cell is thereby obtained that is visible in the normal bright-field light microscope. (B) In the dark-field microscope, oblique rays of light focused on the specimen do not enter the objective lens, but light that is scattered by components in the living cell can be collected to produce a bright image on a dark background. (C) Light passing through the unstained living cell experiences very little change in amplitude, and the structural details cannot be seen even if the image is highly magnified. The phase of the light, however, is altered by its passage through either thicker or denser parts of the cell, and small phase differences can be made visible by exploiting interference effects using a phase-contrast or a differential-interference-contrast microscope.

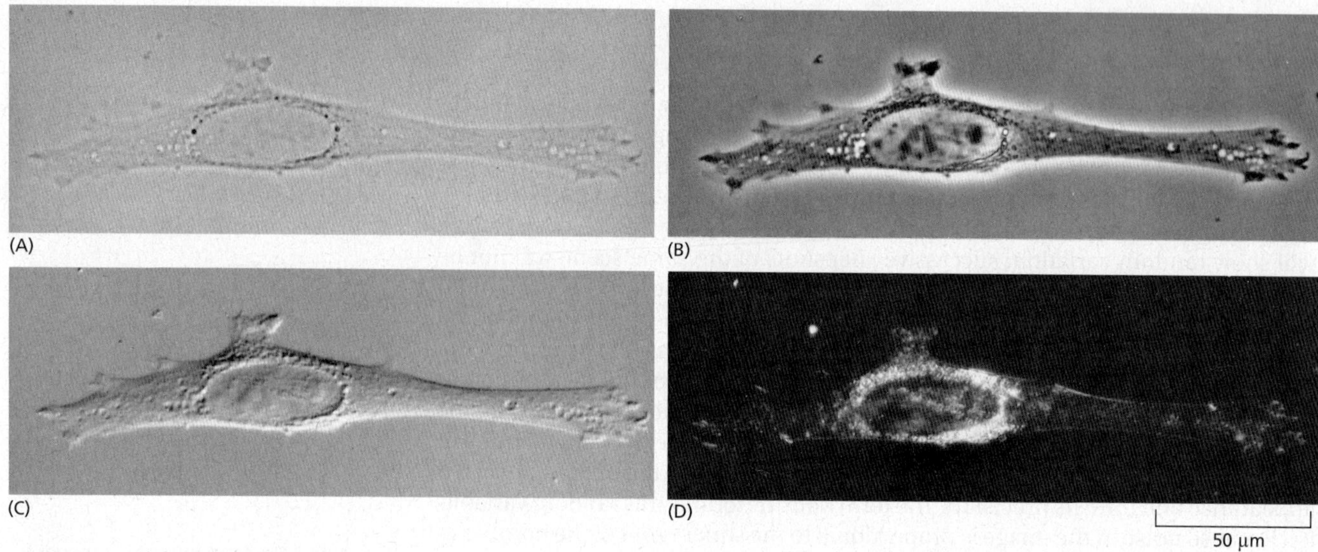

Figure 9–7 Four types of light microscopy. Four images are shown of the same fibroblast cell in culture. All images can be obtained with most modern microscopes by interchanging optical components. (A) Bright-field microscopy, in which light is transmitted straight through the specimen. (B) Phase-contrast microscopy, in which phase alterations of light transmitted through the specimen are translated into brightness changes. (C) Differential-interference-contrast microscopy, which highlights edges where there is a steep change of refractive index. (D) Dark-field microscopy, in which the specimen is lit from the side and only the scattered light is seen.

If oblique lighting from the condenser is used, which does not directly enter the objective, unstained objects in a living cell can scatter the rays, some of which then enter the objective to create a bright image against a black background (Figure 9–6B).

When light passes through a living cell, the phase of the light wave is changed according to the cell's refractive index: a relatively thick or dense part of the cell, such as a nucleus, slows the light passing through it. The phase of the light, consequently, is shifted relative to light that has passed through an adjacent thinner region of the cytoplasm (Figure 9–6C). The **phase-contrast microscope** and, in a more complex way, the **differential-interference-contrast microscope** increase these phase differences to produce amplitude differences, or contrast, when the sets of waves recombine, thereby creating an image of the cell's structure. Both types of light microscopy are widely used to look at living cells (see Movie 17.2). **Figure 9–7** compares images of the same cell obtained by four kinds of light microscopy.

Phase-contrast, differential-interference-contrast, and dark-field microscopy make it possible to watch the movements involved in such processes as mitosis and cell migration. Because many cellular motions are too slow to be seen in real time, it is often helpful to make time-lapse videos in which the camera records successive frames separated by a short time delay, so that when the resulting picture series is played at normal speed, events appear greatly speeded up.

Images Can Be Enhanced and Analyzed by Digital Techniques

Digital imaging systems, and the associated technology of **image processing**, have had a major impact on light microscopy. Certain practical limitations of microscopes relating to imperfections in their optical components have been largely overcome. Digital imaging systems have also circumvented two fundamental limitations of the human eye: the eye cannot see well in extremely dim light, and it cannot perceive small differences in light intensity against a bright background. To increase our ability to observe cells in these difficult conditions, we can attach a sensitive digital camera to a microscope. These cameras detect light by means of high-sensitivity complementary metal-oxide semiconductor (CMOS) sensors, similar to those now found in digital cameras and smartphones. Such image sensors can count individual photons and are many times more sensitive than the human eye and can detect 100 times more intensity levels. It is therefore possible to observe cells for long periods at very low light levels, thereby avoiding the damaging effects of prolonged bright light (and heat). Such sensitive detectors are especially important for viewing fluorescent molecules in living cells, as explained later.

Because images produced by digital cameras are in electronic form, they can be processed in various ways to extract latent information. Such image processing makes it possible to compensate for several aberrations in the lenses of microscopes. Moreover, by digital image processing, contrast can be greatly enhanced to overcome the eye's limitations in detecting small differences in light intensity, and background irregularities in the optical system can be digitally subtracted. This procedure reveals small transparent objects that were previously impossible to distinguish from the background.

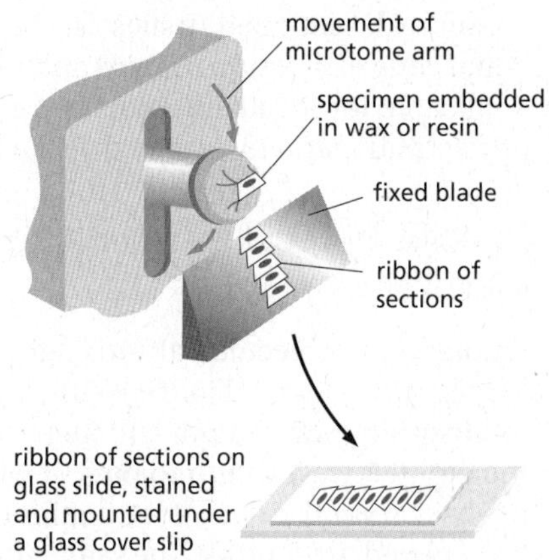

Figure 9–8 **Making tissue sections.** This illustration shows how an embedded tissue is sectioned with a microtome in preparation for examination in the light microscope. Very rapidly frozen samples can also be sectioned, and these better preserve the structure of cells in their native state.

Intact Tissues Are Usually Fixed and Sectioned Before Microscopy

Looking at individual living cells in culture is relatively easy, but most cells are found in complex tissues and organs, and this forces another trade-off when we want to look at them. Because most tissue samples are too thick for their individual cells to be examined directly at high resolution, they are often cut into very thin transparent slices, or *sections*. To preserve the cells within the tissue they must first be treated with a *fixative*. A common fixative is glutaraldehyde, which forms covalent bonds with the free amino groups of proteins, cross-linking them so they are stabilized and locked into position.

Because tissues are generally soft and fragile, even after fixation, they need to be either frozen or embedded in a supporting medium before they can be sectioned. The usual embedding media are waxes or resins. In liquid form, these media both permeate and surround the fixed tissue before being hardened (by cooling or by polymerization) to form a solid block, which is readily sectioned with a microtome. This is a machine with a sharp blade, usually of steel or glass, which operates like a meat slicer (**Figure 9–8**). The sections (typically 0.5–10 μm thick) are then laid flat on the surface of a glass microscope slide.

There is little in the contents of most cells (which are 70% water by weight) to impede the passage of light rays. Thus, most cells in their natural state, particularly if fixed and sectioned, are almost invisible in an ordinary light microscope. We have seen that cellular components can be made visible by techniques such as phase-contrast and differential-interference-contrast microscopy, but these methods tell us almost nothing about the underlying chemistry. There are three main approaches to working with thin tissue sections that reveal differences in the types of molecules that are present.

First, and traditionally, sections can be stained with organic dyes that have some specific affinity for particular subcellular components. The dye hematoxylin, for example, has an affinity for negatively charged molecules and therefore reveals the general distribution of DNA, RNA, and acidic proteins in a cell (**Figure 9–9**). The chemical basis for the specificity of many dyes, however, is not known, although they are used widely in hospital laboratories.

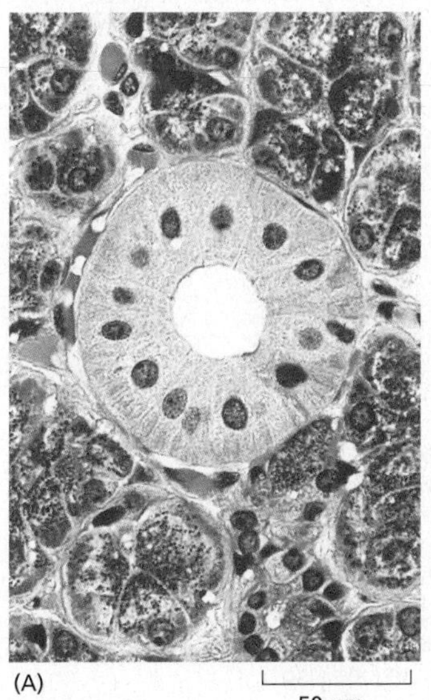

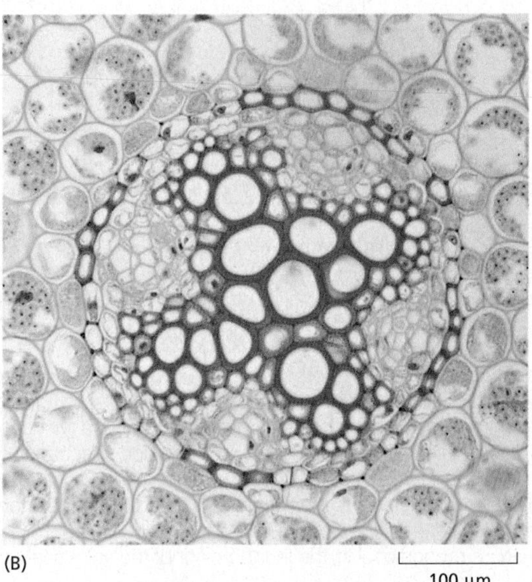

Figure 9–9 **Staining of cell components.** (A) This section of cells in a salivary gland was stained with hematoxylin and eosin, two dyes commonly used in histology. The central duct is made of closely packed cells with nuclei stained *purple* and cytoplasm stained *red*. The duct is surrounded by groups of saliva-secreting cells. (B) This section of a young plant root is stained with two dyes, safranin and fast green. Fast green stains the cellulosic cell walls, while the safranin stains the lignified xylem cell walls *red*. (A, from R.L. Sorenson and T.C. Brelje, Atlas of Human Histology: A Guide to Microscopic Structure of Cells, Tissues and Organs, 3rd ed., 2014. With permission from the authors; B, courtesy of University of Wisconsin Plant Teaching Collection.)

Second, sectioned tissues can be used to visualize specific patterns of differential gene expression. A third and very sensitive approach, generally and widely applicable for localizing proteins of interest, depends on the use of fluorescent probes and markers, as we explain next.

Specific Molecules Can Be Located in Cells by Fluorescence Microscopy

Fluorescent molecules absorb light at one wavelength and emit it at another, longer wavelength (**Figure 9–10A and B**). If we illuminate such a molecule at its absorbing wavelength and then view it through a filter that allows only light of the emitted wavelength to pass, it will glow against a dark background. Because the background is dark, even a minute amount of the glowing fluorescent dye can be detected. In contrast, the same number of molecules of a nonfluorescent stain, viewed conventionally, would be practically indiscernible because the absorption of light by molecules in the stain would result in only the faintest tinge of color in the light transmitted through that part of the specimen.

The fluorescent dyes used for staining cells are visualized with a **fluorescence microscope**. This microscope is similar to an ordinary upright or inverted light microscope except that the illuminating light, from a very powerful source, is passed through two sets of filters—one to filter the light before it reaches the specimen, and one to filter the light obtained from the specimen. The first filter passes only the wavelengths that excite the particular fluorescent dye, while the second filter blocks out this light and passes only those wavelengths emitted when the dye fluoresces (**Figure 9–10C**).

Fluorescence microscopy is most often used to detect specific proteins or other molecules in cells and tissues. For example, when using fluorescent nucleotide probes, *in situ* hybridization, discussed earlier (see Figure 8–63), can reveal

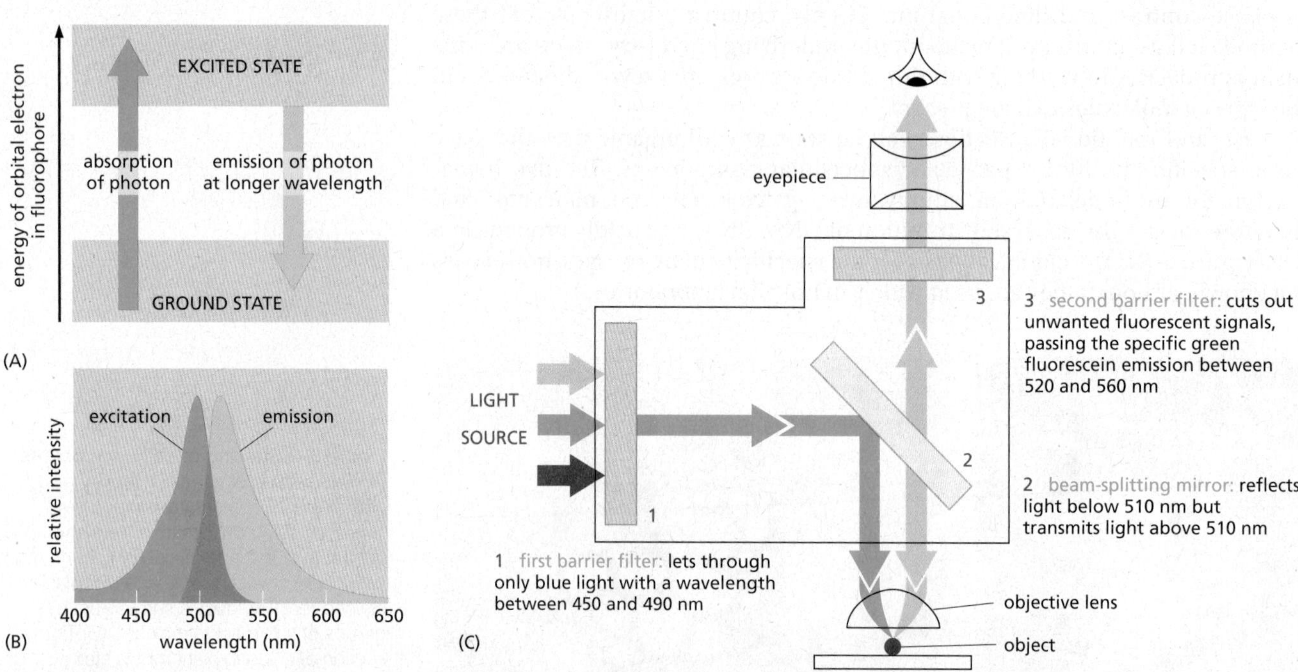

Figure 9–10 Fluorescence and the fluorescence microscope. (A) An orbital electron of a fluorochrome molecule can be raised to an excited state after the absorption of a photon. Fluorescence occurs when the electron returns to its ground state and emits a photon of light at a longer wavelength. Too much exposure to light or too bright a light can destroy the fluorochrome molecule in a process called *photobleaching*. (B) The excitation and emission spectra for the common fluorescent dye fluorescein isothiocyanate (FITC). (C) In the fluorescence microscope, a filter set consists of two barrier filters (1 and 3) and a dichroic (beam-splitting) mirror (2). This example shows the filter set for detection of the fluorescent molecule fluorescein. High-numerical-aperture objective lenses are especially important in this type of microscopy because, for a given magnification, the brightness of the fluorescent image is proportional to the fourth power of the numerical aperture (see also Figure 9–5).

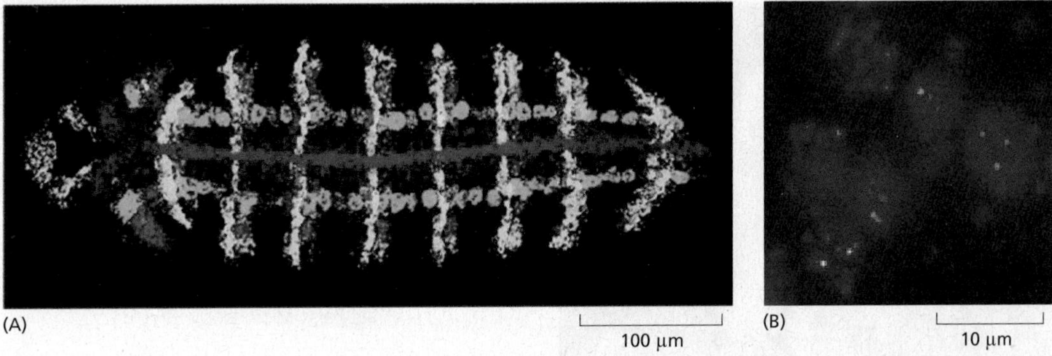

Figure 9–11 **RNA *in situ* hybridization.** (A) As described in Chapter 8 (see Figure 8–63), it is possible to visualize the distribution of different RNAs in tissues using *in situ* hybridization. Here, the transcription pattern of five different genes involved in patterning the early fruit fly embryo is revealed in a single embryo. Each RNA probe has been fluorescently labeled, and the resulting images are displayed each in a different color ("false-colored") and then combined to give an image where different color combinations represent different sets of genes expressed. The genes whose expression pattern is revealed here are *wingless (yellow)*, *engrailed (blue)*, *short gastrulation (red)*, *intermediate neuroblasts defective (green)*, and *muscle specific homeobox (purple)*. (B) Individual RNA transcripts can be detected in a single cell. Each of these six yeast cells is expressing less than 20 transcripts of a particular gene. Using multiple DNA oligonucleotide probes to that particular gene, each labeled with many fluorescent Cy5 molecules, each individual RNA transcript can be detected as a *red spot*. [A, from D. Kosman et al., *Science* 305:846, 2004. With permission from AAAS; B, from G.M. Wadsworth, R.Y. Parikh, and H.D. Kim, Single-probe RNA FISH in yeast. *Bio Protoc.* 8(11):e2868, 2018, doi 10.21769/BioProtoc.2868.]

the cellular distribution and abundance of specific expressed RNA molecules in sectioned material or in whole mounts of small organisms, organs, or cells (**Figure 9–11**).

A versatile and widely used technique is to couple fluorescent dyes to antibody molecules, which then serve as highly specific and versatile staining reagents that bind selectively to the particular macromolecules they recognize in cells or in the extracellular matrix. Two fluorescent dyes that have been commonly used for this purpose are *fluorescein*, which emits an intense green fluorescence when excited with blue light, and *rhodamine*, which emits a deep red fluorescence when excited with green-yellow light (**Figure 9–12**). By coupling one antibody to fluorescein and another to rhodamine, the distributions of different molecules can be compared in the same cell; the two molecules are visualized separately in the microscope by switching back and forth between two sets of filters, each specific for one dye. As shown in **Figure 9–13**, multiple fluorescent dyes can be used in the same way to clearly distinguish several different types of molecules in the same cell. Many fluorescent dyes, such as Cy3, Cy5, and the Alexa dyes, have been specifically developed for fluorescence microscopy, but, like many organic fluorochromes, they fade fairly rapidly when continually illuminated. Later in the chapter, additional fluorescence microscopy methods will be discussed that can be used to monitor changes in the concentration and location of specific molecules inside

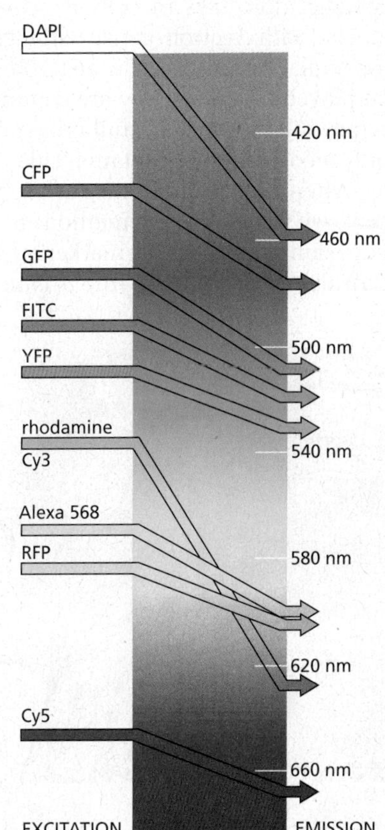

Figure 9–12 **Fluorescent probes.** The maximum excitation and emission wavelengths of several commonly used fluorescent probes are shown in relation to the corresponding colors of the spectrum. The photon emitted by a fluorescent molecule is necessarily of lower energy (longer wavelength) than the absorbed photon, and this accounts for the difference between the excitation and emission peaks. CFP, GFP, YFP, and RFP are cyan, green, yellow, and red fluorescent proteins, respectively. DAPI is widely used as a general fluorescent DNA probe, which absorbs ultraviolet light and fluoresces bright blue. FITC is an abbreviation for fluorescein isothiocyanate—a widely used derivative of fluorescein—which fluoresces bright green. The other probes are all commonly used to fluorescently label antibodies and other proteins. Note that although the true fluorescence emission colors are shown here, the actual color seen in the microscope will depend on the second barrier filter used (see Figure 9–10), and these are usually optimized so as to allow as many different non-overlapping colored probes to be seen in the same specimen. Thus although YFP emits in the green spectrum, it actually appears as a yellow-green in the microscope because of the filter used. The use of fluorescent proteins will be discussed later in the chapter.

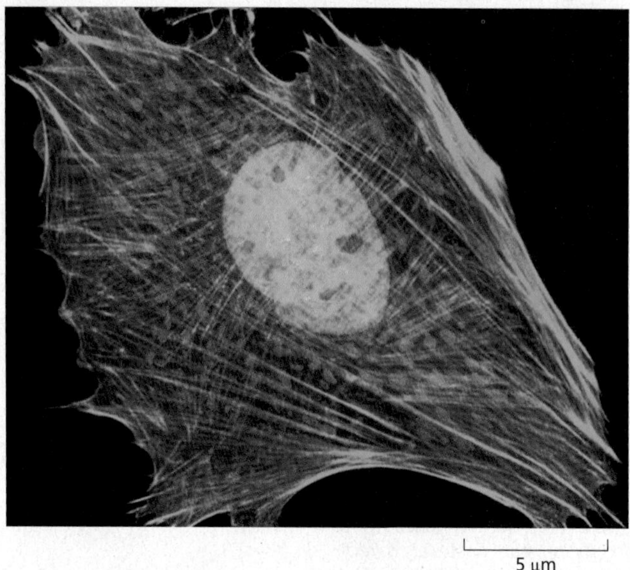

Figure 9–13 **Different fluorescent probes can be visualized in the same cell.** In this composite micrograph of an epithelial cell in culture, three different fluorescent probes have been used to label three different cellular components. The actin filaments of the cytoskeleton are revealed with a *green* fluorescent probe, the numerous mitochondria with a *red* fluorescent dye that accumulates inside the organelles, and the nucleus with a *blue* fluorescent dye that binds to DNA. (Courtesy of Carl Zeiss Microscopy, GmbH.)

living cells. As with all microscopy methods there are trade-offs to consider. In all fluorescence microscopes, the only molecules that can be imaged are those that are fluorescently labeled; all the other molecules in the cell remain hidden.

Antibodies Can Be Used to Detect Specific Proteins

Antibodies are proteins produced by the vertebrate immune system as a defense against infection (discussed in Chapter 24). They are unique among proteins in that they are made in billions of different forms, each with a different binding site that recognizes a specific target molecule (or *antigen*). The precise antigen specificity of antibodies makes them powerful tools for the cell biologist. When chemically coupled to fluorescent dyes, antibodies are invaluable for locating specific molecules in cells by fluorescence microscopy (**Figure 9–14**). When labeled with electron-dense particles such as colloidal gold spheres, they are used for similar purposes in the electron microscope (discussed later). The antibodies employed in microscopy are commonly either purified from antiserum so as to remove all nonspecific antibodies or they are specific monoclonal antibodies that only recognize the target molecule.

When we use antibodies as probes to detect and assay specific molecules in cells, we frequently use methods to amplify the fluorescent signal they produce. For example, although a marker molecule such as a fluorescent dye can be linked directly to an antibody—the *primary antibody*—a stronger signal is achieved by

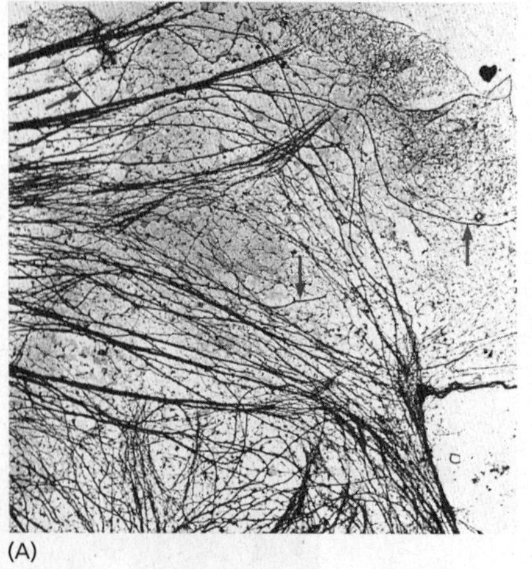

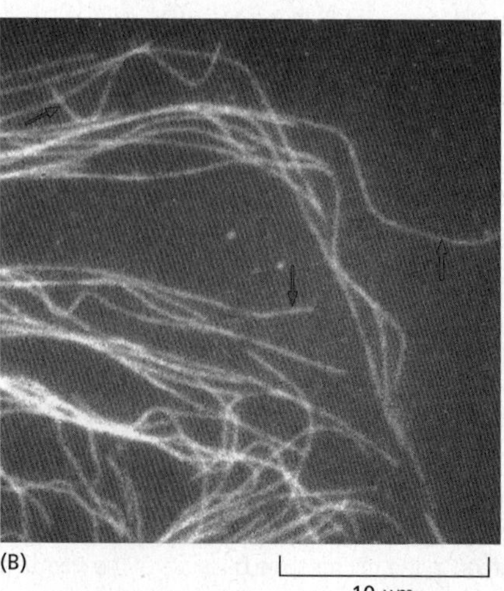

Figure 9–14 **Immunofluorescence.** (A) A transmission electron micrograph of the periphery of a cultured epithelial cell showing the distribution of microtubules and other filaments. (B) The same area stained with fluorescent antibodies against tubulin, the protein that assembles to form microtubules, using the technique of indirect immunocytochemistry (see Figure 9–15). *Red arrows* indicate individual microtubules that are readily recognizable in both images. Note that, because of diffraction effects, the microtubules in the light microscope appear 0.2 μm wide rather than their true width of 0.025 μm. (© 1978 M. Osborn et al. Originally published in *J. Cell Biol.* doi 10.1083/jcb.77.3.R27. With permission from Rockefeller University Press.)

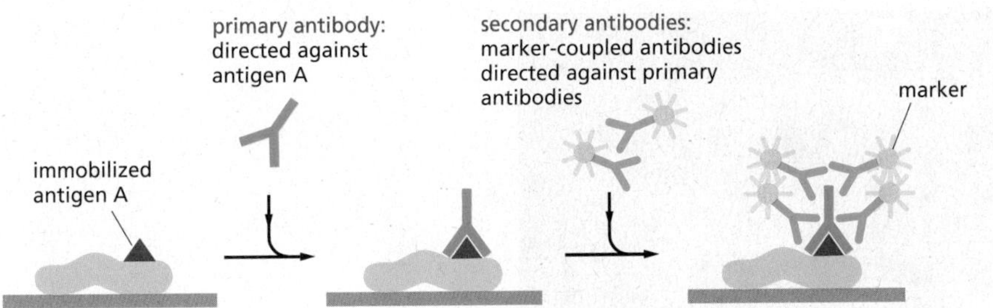

Figure 9–15 Indirect immunocytochemistry. This detection method is very sensitive because many molecules of the secondary antibody recognize each primary antibody. The secondary antibody is covalently coupled to a marker molecule that makes it readily detectable. Commonly used marker molecules include fluorescent dyes (for fluorescence microscopy) and colloidal gold spheres (for electron microscopy).

using an unlabeled primary antibody and then detecting it with a group of labeled *secondary antibodies* that bind to it (**Figure 9–15**). This process is called *indirect immunocytochemistry.*

Individual Proteins Can Be Fluorescently Tagged in Living Cells and Organisms

Even the most stable cell structures must be assembled, disassembled, and reorganized during the cell's life cycle. Other structures, often enormous on the molecular scale, rapidly change, move, and reorganize themselves as the cell conducts its internal affairs and responds to its environment. Complex, highly organized pieces of molecular machinery move components around the cell, controlling traffic into and out of the nucleus, from one organelle to another, and into and out of the cell itself.

Various techniques have been developed to visualize the specific components involved in such dynamic phenomena, and many of these methods use fluorescent proteins. All of the fluorescent molecules discussed so far are made outside the cell and then artificially introduced into it. But the use of genes encoding protein molecules that are themselves inherently fluorescent also enables the creation of organisms and cell lines that make their own visible tags and labels, without the introduction of foreign molecules. These cellular exhibitionists display their inner workings in glowing fluorescent color.

Foremost among the fluorescent proteins used for these purposes by cell biologists is the **green fluorescent protein (GFP)**, isolated from the jellyfish *Aequorea victoria*. This protein is encoded by a single gene, which can be cloned and introduced into cells of other species. The freshly translated protein is not fluorescent, but within an hour or so (less for some alleles of the gene, more for others) some of the amino acids undergo a self-catalyzed post-translational modification to generate an efficient fluorochrome, shielded within the interior of a barrel-like protein, which will now fluoresce green when illuminated appropriately with blue light (**Figure 9–16**). Extensive site-directed mutagenesis performed on the

Figure 9–16 Green fluorescent protein (GFP). (A) The structure of GFP, shown here schematically, highlights the eleven β strands that form the staves of a barrel, buried within which is the active chromophore *(dark green).* (B) The chromophore is formed post-translationally from the protruding side chains of two amino acid residues in a series of autocatalytic steps. (A, PDB code: 1EMA, from M. Ormö et al., *Science* 273:1392–1395, 1996. With permission from AAAS.)

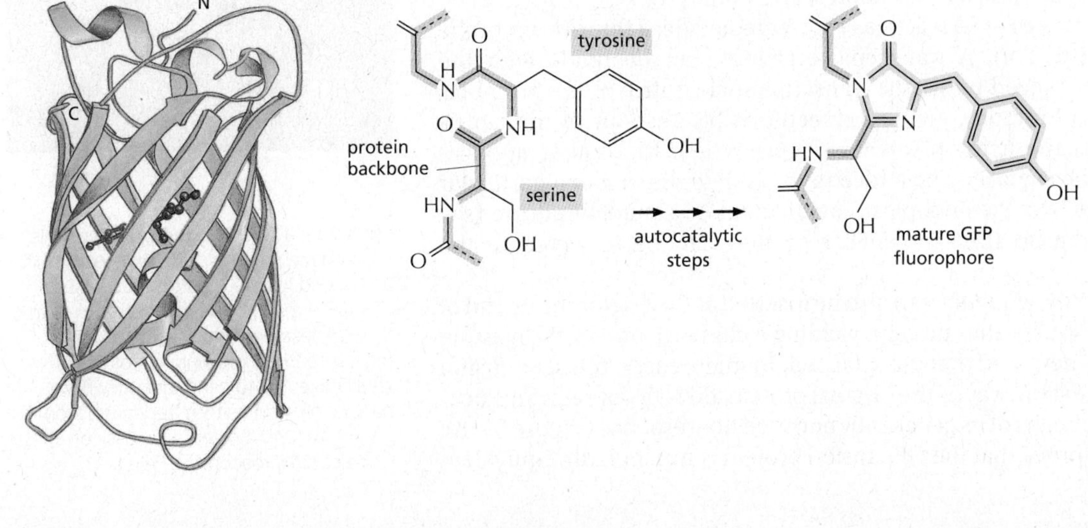

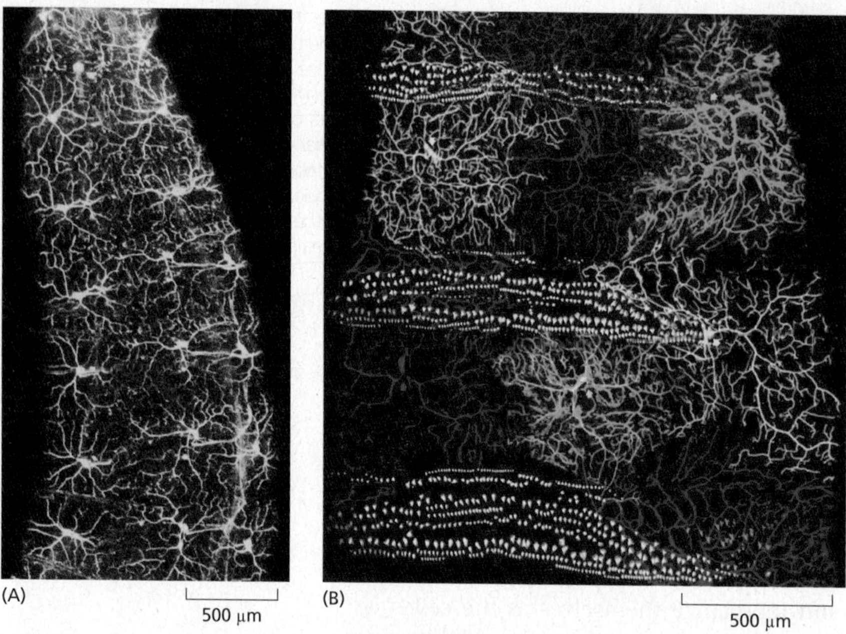

Figure 9–17 **Fluorescent proteins as reporter molecules.** (A) For this experiment, carried out in the fruit fly, the GFP gene was joined (using recombinant DNA techniques) to a fly promoter that is active only in a specialized set of neurons. This image of a live fly embryo was captured by a fluorescence microscope and shows approximately 20 neurons, each with long projections (axons and dendrites) that communicate with other (nonfluorescent) cells. These neurons are located just under the surface of the animal and allow it to sense its immediate environment. (B) In a variation of this method, three different fluorescent proteins, red, yellow, and cyan, can be expressed at random in neurons of the live fly embryo. The genetic constructs have been arranged such that a strong pulse of blue light will activate the expression of one or other of the three fluorescent proteins at random in neuronal cells, where they are then targeted to the plasma membrane. This noninvasive control of the timing of cell labeling allows the behavior of individual cells to be followed subsequently over time. The fine detail of all the dendrites of individual sensory neurons can be clearly seen. The lines of *pale dots* arise from the autofluorescence of the bands of denticles in the cuticle that define the segments of the embryo (see Figure 21–24). (A, from W.B. Grueber et al., *Curr. Biol.* 13:618–626, 2003. With permission from Elsevier; B, from M. Boulina et al., *Development* 140:1605–1613, 2013, doi 10.1242/dev.088930. © 2013. Published by the Company of Biologists Ltd.)

original gene sequence has resulted in multiple variants that can be used effectively in organisms ranging from animals and plants to fungi and microbes. The fluorescence efficiency has also been improved, and variants have been generated with altered absorption and emission spectra from the blue-green, like blue fluorescent protein (BFP), to the far visible red. Other, related fluorescent proteins have since been discovered (for example, in corals) that also extend the range into the red region of the spectrum, like red fluorescent protein (RFP).

One of the simplest uses of GFP is as a reporter molecule, a fluorescent probe to monitor gene expression. A transgenic organism can be made with the GFP-coding sequence placed under the transcriptional control of the promoter belonging to a gene of interest, giving a directly visible readout of the gene's expression pattern in the living organism (**Figure 9–17**). In another application, a peptide location signal can be added to the GFP to direct it to a particular cell compartment, such as the endoplasmic reticulum or a mitochondrion (see Figure 9–25B), lighting up these organelles so they can be observed in the living state.

The GFP DNA-coding sequence can also be inserted at the beginning or end of the coding sequence for another protein, yielding a chimeric product consisting of that protein with a new GFP domain attached. In many cases, this GFP fusion protein behaves in the same way as the original protein, directly revealing its location and activities by means of its genetically encoded fluorescence (**Figure 9–18**). It is often possible to prove that the GFP fusion protein is functionally equivalent

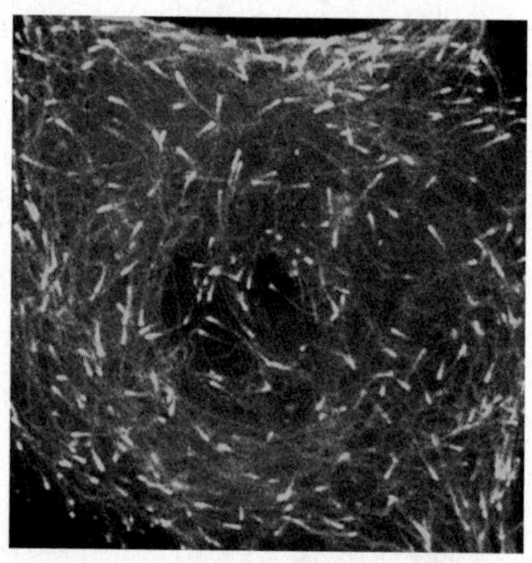

Figure 9–18 **GFP-tagged proteins.** This cultured mammalian cell is expressing EB3, a plus-end tracking protein that is fused to a GFP-derived blue fluorescent protein (BFP). These proteins associate with the plus ends of growing microtubules (see Figure 16–49), and their dynamics can be followed as they appear to zoom brightly around the cell. (Courtesy of Carl Zeiss Microscopy, GmbH.)

to the untagged protein, for example by using it to rescue a mutant lacking that protein. GFP tagging is the clearest and most unequivocal way of showing the distribution and dynamics of a protein in a living organism (see Movie 16.8).

Protein Dynamics Can Be Followed in Living Cells

Fluorescent proteins are now exploited not just to see where in a cell a particular protein is located, but also to uncover its kinetic properties and to find out whether it might interact with other molecules. We now describe techniques in which fluorescent proteins are used in this way.

First, interactions between one protein and another can be monitored by **Förster resonance energy transfer**, also called **fluorescence resonance energy transfer** but both abbreviated to **FRET**. In this technique, two molecules of interest are each labeled with a different fluorochrome, chosen so that the emission spectrum of one fluorochrome, the donor, overlaps with the absorption spectrum of the other, the acceptor. If the two proteins interact in such a way as to bring their fluorochromes into very close proximity (closer than about 5 nm), one fluorochrome, when excited, can transfer energy from the absorbed light directly (by resonance, nonradiatively) to the other. Thus, when the complex is illuminated at the excitation wavelength of the first fluorochrome, fluorescent light is produced at the emission wavelength of the second (**Figure 9–19**). This method can be used with two different spectral variants of GFP as fluorochromes to monitor processes such as the interaction of signaling molecules with their receptors (see Figure 15–49) or proteins in macromolecular complexes at specific locations inside living cells. The FRET can be measured by quantifying the reduction of the donor fluorescence in the presence of the acceptor. The efficiency of FRET is inversely proportional to the sixth power of the distance between the donor and acceptor molecules and so is extremely sensitive to small changes in distance.

The genes encoding GFP and related fluorescent proteins can be engineered to produce protein variants, usually with one or more amino acid changes, that fluoresce only weakly under normal excitation conditions but can be induced to fluoresce either more strongly or with a color shift (for example, from green to red) by activating them with a strong pulse of light at a different wavelength in a process called **photoactivation**. In principle, the biologist can then follow the local *in vivo* behavior of any protein that can be expressed as a fusion with one of

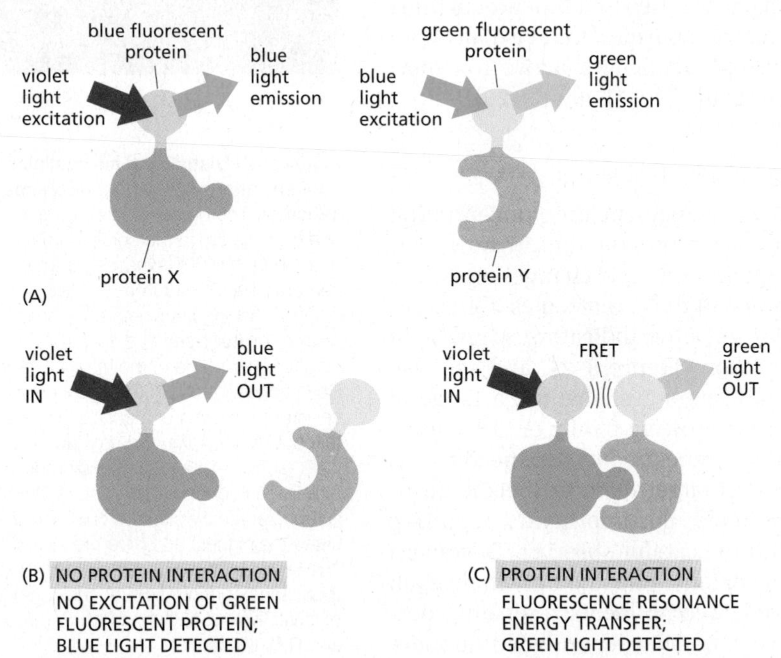

Figure 9–19 Fluorescence resonance energy transfer (FRET). To determine whether (and when) two proteins interact inside a cell, the proteins are first produced as fusion proteins attached to different color variants of green fluorescent protein (GFP). (A) In this example, protein X is coupled to a blue fluorescent protein, which is excited by violet light and emits blue light; protein Y is coupled to a green fluorescent protein, which is excited by blue light and emits green light. (B) If protein X and Y do not interact, illuminating the sample with violet light yields fluorescence only from the blue fluorescent protein. (C) When protein X and protein Y interact, the resonance transfer of energy, FRET, can now occur. Illuminating the sample with violet light excites the blue fluorescent protein, which transfers its energy to the green fluorescent protein, resulting in an emission of green light. The fluorochromes must be quite close together—within about 1–5 nm of one another—for FRET to occur. Because not every molecule of protein X and protein Y is bound at all times, some blue light may still be detected. But as the two proteins begin to interact, emission from the donor blue fluorescent protein falls as the emission from the acceptor GFP rises.

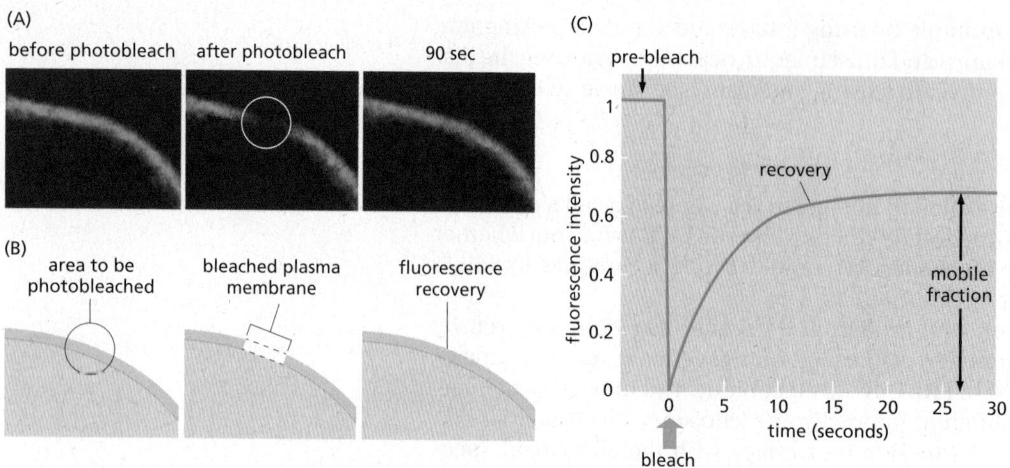

Figure 9–20 **Fluorescence recovery after photobleaching (FRAP).** A strong focused pulse of laser light will extinguish, or bleach, fluorescent proteins. By selectively photobleaching a set of fluorescently tagged protein molecules within a defined region of a cell, the microscopist can monitor recovery over time, as the remaining fluorescent molecules move into the bleached region (see Movie 10.6). (A) This cultured mammalian cell is expressing an integral membrane protein called CD86, which is fused with a fluorescent protein. CD86 is a co-stimulatory protein present in the plasma membrane of antigen-presenting cells and is required for the activation of T cells (see Figure 24–34). After a small region of the plasma membrane is selectively photobleached, the remaining fluorescent molecules diffuse rapidly within the plane of the membrane and populate the bleached region. This recovery can be followed as a function of time. (B) Schematic diagram of the experiment shown in A. (C) Measurements of the fluorescence intensity in the bleached area as a function of time can be plotted as a fluorescence recovery curve. From such graphs quantitative data can be derived about the rate of recovery and the fraction of fluorescent protein molecules that are either mobile or immobile. (A, from S. Dorsch et al., *Nat. Methods* 6:225–230, 2009.)

these GFP variants. These genetically encoded, photoactivatable, fluorescent proteins allow the lifetime and behavior of any protein to be studied independently of other newly synthesized proteins.

Another way to exploit GFP fused to a protein of interest is known as **fluorescence recovery after photobleaching (FRAP)**. Here, a strong focused beam of light from a laser is used to extinguish the GFP fluorescence in a specified region of the cell, after which one can analyze the way in which remaining unbleached fluorescent protein molecules move into the bleached area as a function of time. This technique, like photoactivation, can deliver valuable quantitative data about a protein's kinetic parameters, such as diffusion coefficients, active transport rates, or binding and dissociation rates from other proteins (**Figure 9–20**).

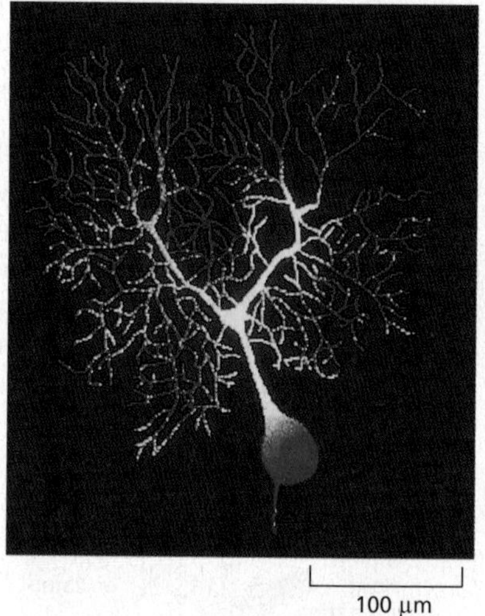

Figure 9–21 **Visualizing intracellular Ca^{2+} concentrations by using a fluorescent indicator.** The branching tree of dendrites of a Purkinje cell in the cerebellum receives more than 100,000 synapses from other neurons. The output from the cell is conveyed along the single axon seen leaving the cell body at the bottom of the picture. This image of the intracellular Ca^{2+} concentration in a single Purkinje cell (from the brain of a guinea pig) was taken with a low-light camera and the Ca^{2+}-sensitive fluorescent indicator fura-2. The concentration of free Ca^{2+} is represented by different colors, *red* being the highest and *blue* the lowest. The highest Ca^{2+} levels are present in the thousands of dendritic branches. (Courtesy of D.W. Tank, J.A. Connor, M. Sugimori, and R.R. Llinas.)

Fluorescent Biosensors Can Monitor Cell Signaling

Extracellular signals cause rapid and transient changes in the concentration of intracellular signaling molecules that play an important role in how cells respond. But how to see and measure such dynamic and rapid changes remains a challenge. Changes in the concentration of some of these molecules, for example Ca^{2+} ions, can be analyzed using simple ion-sensitive indicators, whose light emission reflects the local Ca^{2+} ion concentration (**Figure 9–21**, and see also Figure 15–31). However, the most sensitive indicators available are a range of genetically encoded biosensors, all based on the growing family of fluorescent proteins described earlier. These sensors can be synthesized by the specific cells of interest and easily fused with protein tags that target them to specific destinations within the cell. Here they can act as molecular informants, reporting back like spies on transient signaling events to the careful observer. To convert information about changes in the level of a signaling molecule into changes in observable fluorescence intensity requires two key components: a sensing component that responds to the target signaling event, and a reporting component

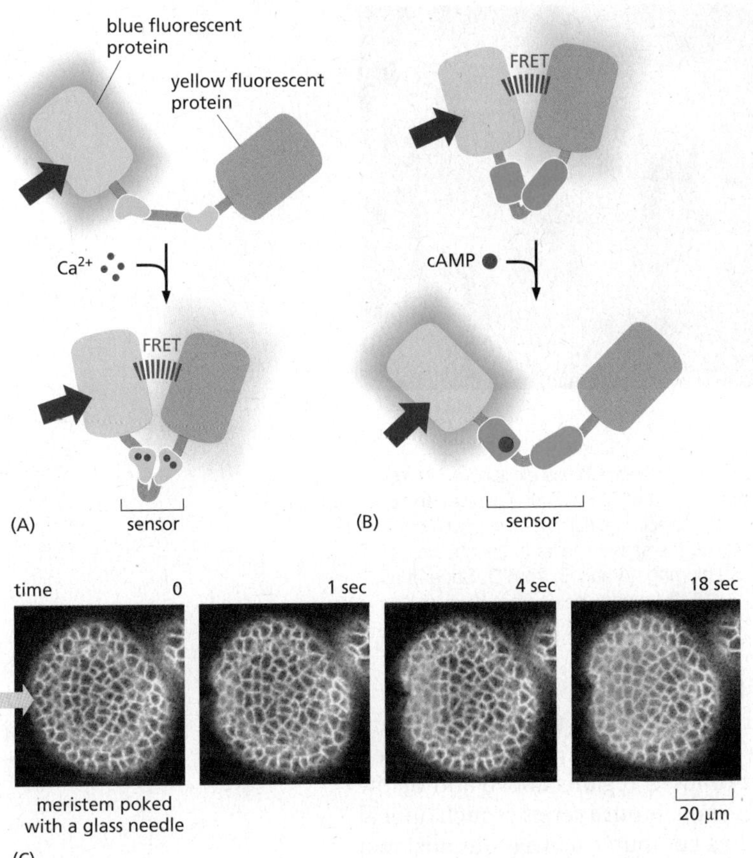

Figure 9–22 Genetically encoded fluorescent biosensors. (A) Here we show one strategy for constructing a fluorescent biosensor for calcium ions. A sensor, in this case calmodulin (see Figure 15–34), undergoes a large conformational change on binding Ca^{2+}. This change brings together the blue and yellow fluorescent proteins to which each end of the sensor is attached, close enough to undergo Förster resonance energy transfer (FRET) and to change the wavelength of the fluorescence emission to yellow in response to a violet excitation light. By measuring the ratio of fluorescence intensity at two wavelengths, blue and yellow, we can determine the concentration ratio of the Ca^{2+}-bound indicator to the Ca^{2+}-free indicator, thereby providing an accurate measurement of the free Ca^{2+} concentration. (B) This panel illustrates a similar strategy used to construct a biosensor for cAMP. In this case the sensor is a cAMP-regulated guanine nucleotide exchange factor, which again undergoes a large conformational change, enough to move the two attached fluorescent proteins farther apart, thus abolishing their FRET. Hence the emitted light is switched from yellow to blue. (C) A calcium biosensor, similar to that shown in A, is genetically encoded and expressed in an *Arabidopsis* seedling. When a cell in the epidermis, on the side of the shoot apical meristem, is pricked with a small glass needle, calcium enters the cell from the extracellular environment, and this response is rapidly propagated as a wave of calcium entering cells across the entire surface of the meristem. Mechanical signals help pattern plant morphogenesis, and transient calcium responses affect cell polarity. (C, from T. Li et al., *Nat. Commun.* 10:726–735, 2019. Reproduced with permission of SNCSC.)

that translates that response into a visible and quantifiable output. Many biosensors use two connected fluorescent proteins that can be brought close enough together to undergo Förster resonance energy transfer (see Figure 9–19). Bringing them together, or indeed moving them apart, is a connecting sensor module. The sensor is usually a protein or protein domain that undergoes a large conformational change on binding to the target molecule. The general principle used to construct a genetically encoded biosensor is shown in **Figure 9–22**. Measuring the ratio of the intensities of light emitted by the two fluorescent proteins in the biosensor provides a quantitative measure of the concentration of the target molecule of interest. Many hundreds of such biosensors have been created. Some can monitor and measure small molecules in living cells, such as Ca^{2+}, cAMP, IP_3, NADPH (and hence redox state), H^+ ions (and hence pH), and neurotransmitters such as acetylcholine and glutamate. Others can measure the activity of kinases, phosphatases, active caspases, and even temperature.

Imaging of Complex Three-dimensional Objects Is Possible with the Optical Microscope

For ordinary light microscopy, as we have seen, a tissue has to be sliced into thin sections to be examined; the thinner the section, the crisper the image. Because information about the third dimension is lost upon sectioning, how, then, can we get a picture of the three-dimensional architecture of a cell or tissue, and how can we view the microscopic structure of a specimen that, for one reason or another, cannot first be sliced into sections? Although an optical microscope is focused on a particular focal plane within a three-dimensional specimen, all the other parts of the specimen, above and below the plane of focus, are also illuminated, and the light originating from these regions contributes to the image as out-of-focus blur. This can make it very hard to interpret the image in detail and can lead to fine image structure being obscured by the out-of-focus light.

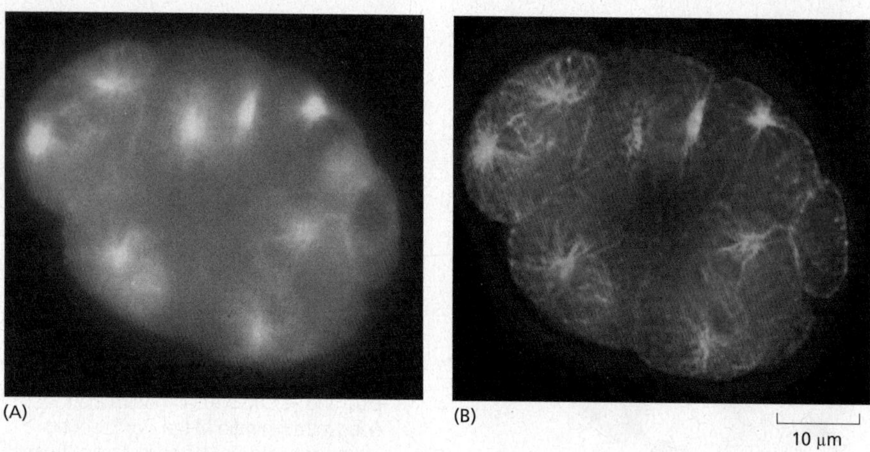

Figure 9–23 Image deconvolution. (A) A light micrograph of a *Caenorhabditis elegans* embryo, fluorescently labeled for microtubules *(green)*, mitochondria *(red)*, and DNA *(blue)*. Detail at any one level of focus is blurred by light from out-of-focus levels of the specimen. (B) After deconvolution of the three-dimensional stack of images, an optical section at the same level of focus shows a much crisper image with more contrast and much reduced blurring. (A and B, from D. Sage et al., *Methods* 115:28–41, 2017, doi 10.1016/j.ymeth.2016.12.015. With permission from Elsevier.)

Two distinct but complementary approaches help to solve this problem: one is computational, the other optical. These three-dimensional microscopic imaging methods make it possible to focus on a chosen plane in a thick specimen while rejecting the light that comes from out-of-focus regions above and below that plane. Thus, one sees a crisp, thin *optical section*. From a series of such optical sections taken at different depths and stored in a computer, a three-dimensional image can be reconstructed (**Movie 9.1**). The methods do for the microscopist what the computed tomography (CT) scanner does (by different means) for the radiologist investigating a human body: both machines give detailed sectional views of the interior of an intact structure.

The computational approach is often called *image deconvolution*. To understand how it works, remember that the wave-like nature of light means that the microscope lens system produces a small blurred disc as the image of a point source of light (see Figure 9–4), with increased blurring if the point source lies above or below the focal plane. This blurred image of a point source is called the *point spread function* (see Figure 9–29). An image of a complex object can then be thought of as being built up by replacing each point of the three-dimensional specimen by a corresponding blurred disc, resulting in an image that is blurred overall. For deconvolution, a computer program uses the measured point spread function of a point source of light from that particular microscope to determine what the effect of the blurring would have been on the image, and then applies an equivalent "deblurring" (deconvolution), turning the blurred three-dimensional image into a series of clean optical sections, albeit still constrained by the diffraction limit (**Figure 9–23**).

The Confocal Microscope Produces Optical Sections by Excluding Out-of-Focus Light

The confocal microscope achieves a result similar to that of deconvolution, but does so by manipulating the light before it is measured; it is an analog technique rather than a digital one. The optical details of the **confocal microscope** are complex, but the basic idea is simple, as illustrated in **Figure 9–24**, and the results are far superior to those obtained by conventional light microscopy.

The confocal microscope is generally used with fluorescence optics (see Figure 9–10C), but instead of illuminating the whole specimen at once, in the usual way, the optical system at any instant focuses a spot of light onto a single point at a specific depth in the specimen. This requires a source of pinpoint illumination

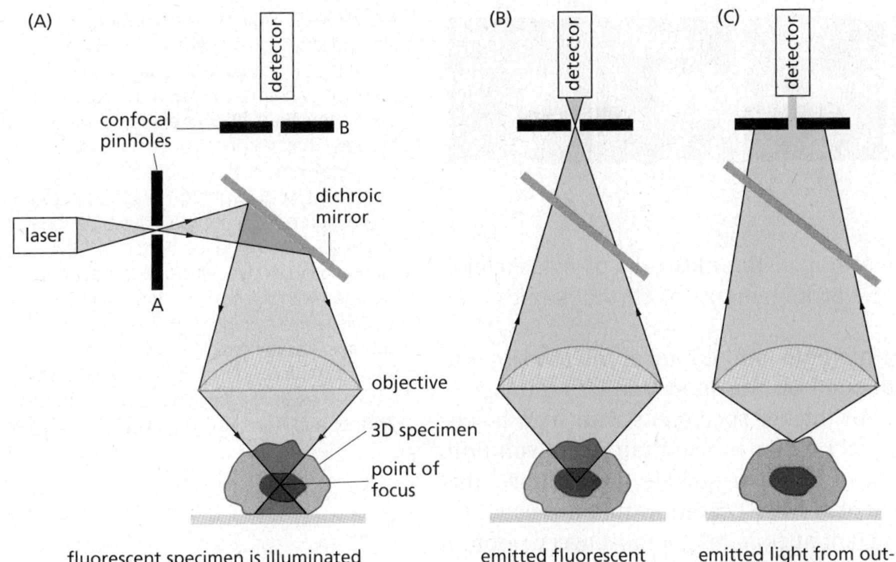

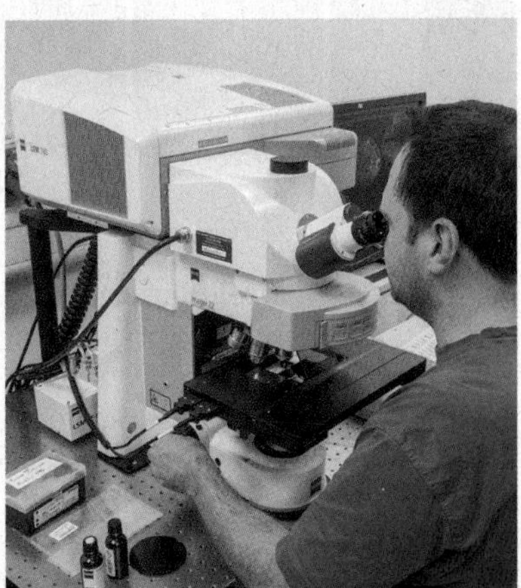

Figure 9–24 The confocal fluorescence microscope. (A) This simplified diagram shows that the basic arrangement of optical components is similar to that of the standard fluorescence microscope shown in Figure 9–10C, except that a laser is used to illuminate a small pinhole whose image is focused at a single point in the three-dimensional (3D) specimen. (B) Emitted fluorescence from this focal point in the specimen is focused at a second (confocal) pinhole. (C) Emitted light from elsewhere in the specimen is not focused at the pinhole and therefore does not contribute to the final image. By scanning the beam of light across the specimen, a very sharp two-dimensional image of the exact plane of focus is built up that is not significantly degraded by light from other regions of the specimen. (D) Commercial versions of laser scanning confocal microscopes can be configured for both upright and inverted microscopes. Shown here is a standard upright confocal microscope. (D, courtesy of Andrew Davis.)

that is usually supplied by a laser whose light has been passed through a pinhole. The fluorescence emitted from the illuminated material is collected at a suitable light detector and used to generate an image. A pinhole aperture is placed in front of the detector, at a position that is *confocal* with the illuminating pinhole; that is, precisely where the rays emitted from the illuminated point in the specimen come to a focus. Thus, the light from this point in the specimen converges on this aperture and enters the detector.

By contrast, the light emitted from regions of the specimen that are out of focus is also out of focus at the pinhole aperture and is therefore largely excluded from the detector. To build up a two-dimensional image, data from each point in the plane of focus are collected sequentially by scanning across the field from one side to the other in a regular pattern of pixels and are displayed on a computer screen. Although not shown in Figure 9–24, the scanning is usually done by deflecting the beam with an oscillating mirror placed between the dichroic (beam-splitting) mirror and the objective lens in such a way that the illuminating spot of light and the confocal pinhole at the detector remain strictly in register. Variations in design now allow the rapid collection of data at video rates.

The confocal microscope has been used to resolve the structures of numerous complex three-dimensional objects (**Figure 9–25**), from large multicellular

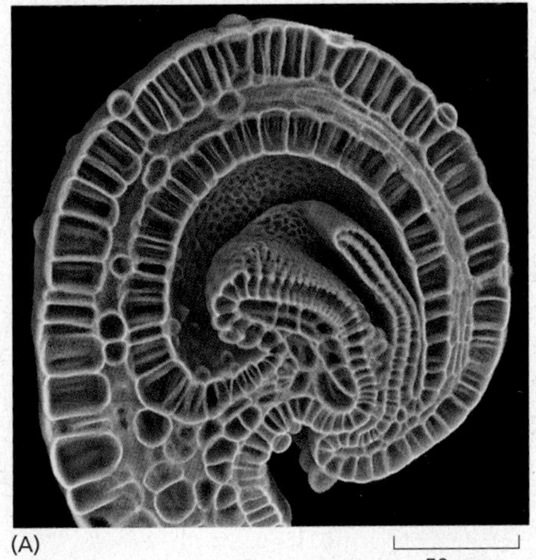

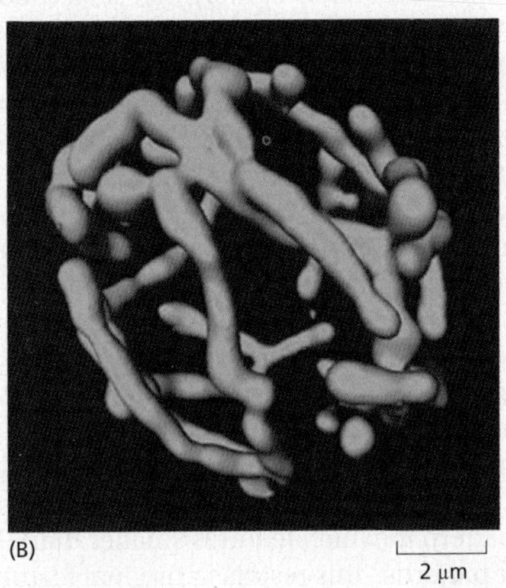

Figure 9–25 Confocal fluorescence microscopy produces clear optical sections and three-dimensional data sets. (A) The elaborate cup-shaped trap of the carnivorous water plant, *Utricularia gibba*. A stack of 452 separate confocal images using a fluorescent label for the cell walls was assembled to produce the image. (B) A reconstruction of an object can be assembled from a stack of optical sections. In this case, and at a vastly different scale, the complex branching structure of the mitochondrial compartment in a single live yeast cell is shown. (A, courtesy of Karen Lee, Claire Bushell, and Enrico Coen; B, courtesy of Stefan Hell.)

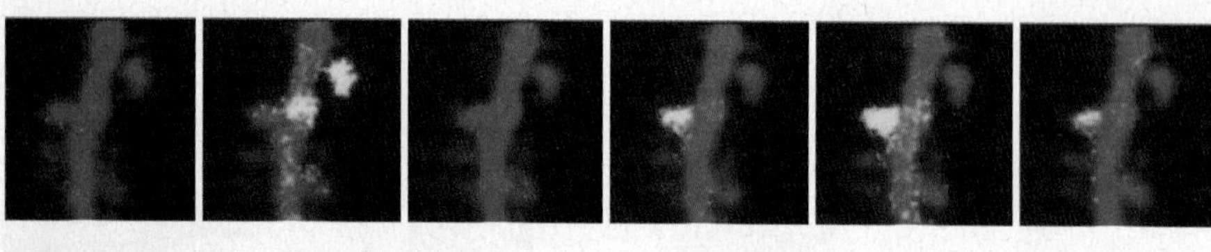

Figure 9–26 **Multiphoton imaging.** Infrared laser light causes less damage to living cells than does visible light and can also penetrate farther, allowing microscopists to peer deeper into living tissues. The two-photon effect, in which a fluorochrome can be excited by two coincident infrared photons instead of a single high-energy photon, allows us to see nearly 0.5 mm inside the cortex of a live mouse brain. A dye, whose fluorescence changes with the calcium concentration, reveals active synapses *(yellow)* on the dendritic spines *(red)* that change as a function of time; in this case, there is a day between each image. (Courtesy of Thomas Oertner and Karel Svoboda.)

structures to subcellular structures; for example, the networks of cytoskeletal fibers, the dynamics of organelles, and the arrangements of chromosomes and genes in the nucleus.

The relative merits of deconvolution methods and confocal microscopy for three-dimensional optical microscopy depend on the specimen being imaged. Confocal microscopes tend to be better for thicker specimens with high levels of out-of-focus light. They are also generally easier to use than deconvolution systems, and the final optical sections can be seen quickly. In contrast, the complementary metal-oxide semiconductor (CMOS) cameras that are used for deconvolution systems are extremely efficient at collecting almost every photon emitted, and they can be used to make detailed three-dimensional images from specimens that are too weakly stained or too easily damaged by the bright light used for confocal microscopy.

Both methods, however, have another drawback; neither is good at coping with very thick specimens. Deconvolution methods quickly become ineffective any deeper than about 40 μm into a specimen, while confocal microscopes can only obtain images up to a depth of about 150 μm. Special microscopes can now take advantage of the way in which fluorescent molecules are excited, to probe even deeper into a specimen. Fluorescent molecules are usually excited by a single high-energy photon, of shorter wavelength than the emitted light, but they can in addition be excited by the absorption of two (or more) photons of lower energy, as long as they both arrive within a femtosecond or so of each other. The use of this longer-wavelength excitation has some important advantages. In addition to reducing background noise, red or near-infrared light can penetrate deeper into a specimen. Multiphoton microscopes, constructed to take advantage of this *two-photon* effect, can obtain sharp images, sometimes even at a depth of half a millimeter within a specimen. This is particularly valuable for studies of living tissues, notably in imaging the dynamic activity of synapses and neurons just below the surface of living brains (**Figure 9–26**).

Superresolution Fluorescence Techniques Can Overcome Diffraction-limited Resolution

The variations on light microscopy we have described so far are all constrained by the classic diffraction limit to resolution described earlier; that is, to about 0.2 μm, or 200 nm (see Figure 9–5). Yet many cellular structures—from nuclear pores and ribosomes to nucleosomes and clathrin-coated pits—are much smaller than this and so are unresolvable by conventional light microscopy. However, several approaches are now available that bypass the limit imposed by the diffraction of light, and some can now successfully resolve objects as small as 10 nm, a remarkable, twentyfold improvement.

The first of these so-called **superresolution** approaches, *structured illumination microscopy* (*SIM*), is a fluorescence imaging method with a resolution of about 100 nm, or twice the resolution of conventional bright-field microscopy. SIM overcomes the diffraction limit by using a grated or structured pattern of light to illuminate the sample. The microscope's physical setup and operation are quite complex, but the general principle can be thought of as similar to creating a moiré pattern, an interference pattern created by overlaying two grids with different angles or mesh sizes (**Figure 9–27**). The illuminating grid and the sample features combine into an interference pattern in which features smaller than the grid spacing are transformed into larger patterns. This results in original features

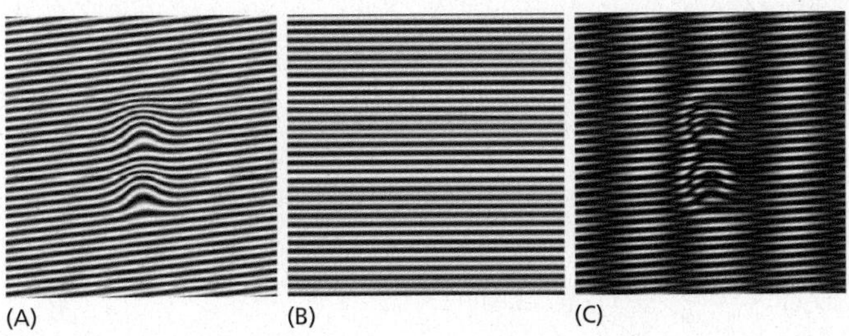

Figure 9–27 Structured illumination microscopy. The principle, illustrated here, is to illuminate a sample with patterned light and measure the moiré pattern. Shown are (A) the pattern from an unknown structure and (B) a defined grid pattern. (C) When these are combined, the resulting moiré pattern contains more information than is easily seen in A, the original pattern. If the known pattern (B) has higher spatial frequencies, then better resolution will result. However, because the spatial patterns that can be created optically are also diffraction-limited, SIM can only improve the resolution by about a factor of 2. (From B.O. Leung and K.C. Chou, *Appl. Spectrosc.* 65:967–980, 2011. With permission from SAGE.)

beyond the classical limit being transformed so that they can now be imaged by the optical system. Computer image processing can then be used to restore them into an image that has a resolution up to twice the classical limit. Illumination by a grid means that the parts of the sample in the dark stripes of the grid are not illuminated and therefore not imaged, so the imaging is repeated several times (usually three) after translating the grid through a fraction of the grid spacing between each image. As the interference effect is strongest for image components close to the direction of the grid bars, the whole process is repeated with the grid pattern rotated through a series of angles to obtain an equivalent enhancement in all directions. Finally, mathematically combining all these separate images by computer creates an enhanced superresolution image. SIM is versatile because it can be used with any fluorescent dye or protein, and combining SIM images captured at consecutive focal planes can create three-dimensional data sets (**Figure 9–28**).

To get around the diffraction limit, two other superresolution techniques exploit aspects of the point spread function, a property of the optical system mentioned earlier. The **point spread function** is the distribution of light intensity within the three-dimensional, blurred image that is formed when a single point source of light is brought to a focus with a lens. Instead of being identical to the point source, the image has an intensity distribution that is approximately described by a Gaussian distribution, which in turn determines the resolution of the lens system. Two points that are closer than the width at half-maximum

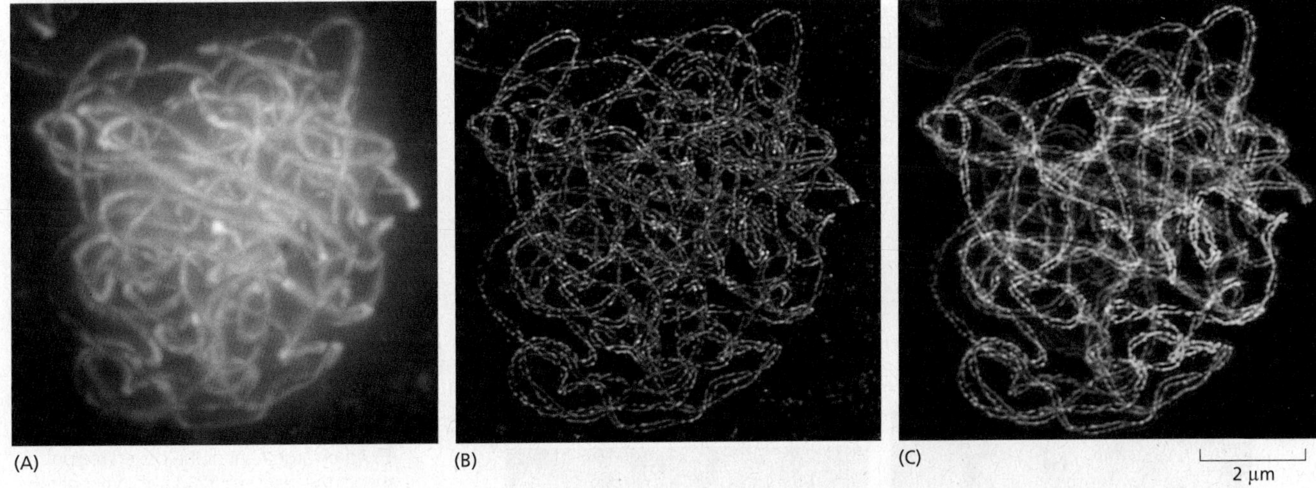

Figure 9–28 Structured illumination microscopy can be used to create three-dimensional data. These three-dimensional projections of the meiotic chromosomes at pachytene in a maize cell show the paired lateral elements of the synaptonemal complexes. (A) The chromosome set has been stained with a fluorescent antibody to cohesin and is viewed here by conventional fluorescence microscopy. Because the distance between the two lateral elements is about 200 nm, the diffraction limit, the two lateral elements that make up each complex are not resolved. (B) In the three-dimensional SIM image, the improved resolution enables each lateral element, about 100 nm across, to be clearly resolved, and the two chromosomes of each separate pair can be seen to coil around each other. (C) Because the complete three-dimensional data set for the whole nucleus is available, the path of each separate pair of chromosomes can be traced and artificially assigned a different color. (Courtesy of C.J. Rachel Wang, Peter Carlton, and Zacheus Cande.)

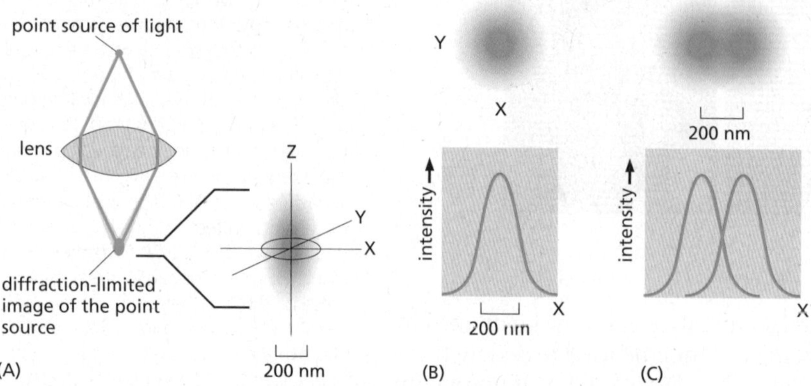

Figure 9–29 **The point spread function of a lens determines resolution.** (A) When a point source of light is brought to a focus by a lens system, diffraction effects mean that, instead of being imaged as a point, it is blurred in all dimensions. As shown, the point spread function is elongated, meaning that the resolution is better in the XY axes than along the Z axis. (B) In the plane of the image, the distribution of light approximates a Gaussian distribution, whose width at half-maximum height under ideal conditions is about 200 nm. (C) Two separate point sources that are about 200 nm apart can still just be distinguished as separate objects in the image, but if they are any nearer than that, their images will overlap and not be resolvable.

height of this distribution will become hard to resolve because their images overlap too much (**Figure 9–29**).

In fluorescence microscopy, the excitation light is focused to a spot on the specimen by the objective lens, which then captures the photons emitted by any fluorescent molecule that the beam has raised from a ground state to an excited state. Because the excitation spot is blurred according to the point spread function, fluorescent molecules that are closer than about 200 nm will be imaged as a single blurred spot. One approach to increasing the resolution is to switch all the fluorescent molecules at the periphery of the blurry excitation spot back to their ground state or to a state where they no longer fluoresce in the normal way, leaving only those at the very center to be recorded. This can be done in practice by adding a second, very bright laser beam that wraps around the excitation beam like a torus. The wavelength and intensity of this second beam are adjusted so as to switch the fluorescent molecules off everywhere except at the very center of the point spread function, a region that can be as small as 20 nm across (**Figure 9–30**).

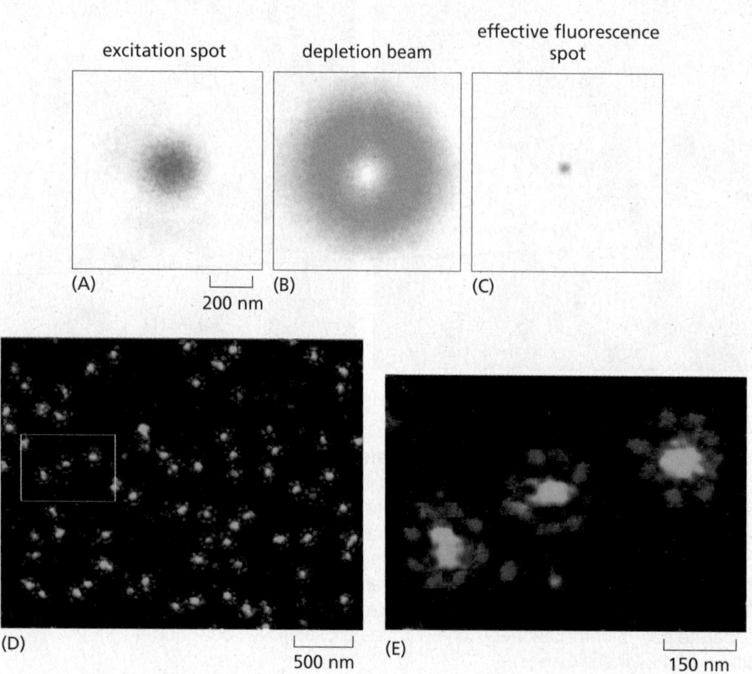

Figure 9–30 **Superresolution microscopy can be achieved by reducing the size of the point spread function.** (A) The size of a normal focused beam of excitatory light. (B) An extremely strong superimposed laser beam, at a different wavelength and in the shape of a torus, or doughnut, depletes emitted fluorescence everywhere in the specimen except right in the center of the beam, reducing the effective width of the point spread function (C). As the specimen is scanned, this small point spread function can then build up a crisp image in a process called STED (stimulated emission depletion) microscopy. (D) Here, STED microscopy is used to examine the structure of the nuclear pore. Fixed samples of the nuclear envelope have been stained by indirect immunofluorescence, using antibodies to different nuclear pore components. Membrane ring proteins (see Figure 12–55) have been stained *red* while the FC repeat proteins that form fibrils in the center of the pore are stained *green*. (E) An enlargement of the boxed region shows the clear eightfold symmetry of the membrane ring proteins and the central fibrillar region with a resolution of about 20 nm. [A, B, and C, from G. Donnert et al., *Proc. Natl. Acad. Sci. USA* 103: 11440–11445, 2006. Copyright 2006 National Academy of Sciences. With permission from National Academy of Sciences; D and E, from F. Gottfert et al., *Biophysical Journal* 105(1):PL01–L03, 2013. With permission from Elsevier.]

The fluorescent probes used must be in a special class that is photoswitchable: their emission can be reversibly switched on and off with lights of different wavelengths. As the specimen is scanned with this arrangement of lasers, in much the same way as in a confocal microscope, fluorescent molecules are switched on and off, and the small point spread function at each location is recorded. The diffraction limit is breached because the technique ensures that similar but very closely spaced molecules are in one of two different states, either fluorescing or dark. This approach is called *STED (stimulated emission depletion) microscopy,* and various microscopes using versions of the general method are now in wide use. Resolutions of 20 nm have been achieved in biological specimens (see Figure 9–30).

Single-Molecule Localization Microscopy Also Delivers Superresolution

If a single fluorescent molecule is imaged, it appears as a circular blurry disc about 200 nm across, but if sufficient photons have contributed to this image, then the precise mathematical center of the disc-like image, and therefore the position of that fluorescent molecule, can be determined very accurately, often to within a few nanometers (**Figure 9–31**). But the problem with a specimen that contains a large number of adjacent fluorescent molecules, as we saw earlier, is that they each contribute blurry, overlapping point spread functions to the image, making the exact position of any one molecule impossible to resolve. Another way around this limitation is to arrange for only a very few, clearly separated molecules to actively fluoresce at any one moment. The exact position of each of these can then be computed, before subsequent sets of molecules are examined.

In practice, this can be achieved by using lasers to sequentially switch on a sparse subset of fluorescent molecules in a specimen containing switchable fluorescent labels. There are now hundreds of such labels, and they fall into three classes: *photoactivated* labels, which switch for example from dark to green; *photoconvertible* labels, which switch for example from green to red; and *photoswitchable* labels, which switch back and forth. Labels are activated, for example, by illumination with near-ultraviolet light, which modifies a small subset of molecules so that they fluoresce when exposed to an excitation beam at another wavelength. These are then imaged before bleaching quenches their fluorescence, and a new subset is activated. Each molecule emits a few thousand photons in response to the excitation before switching off, and the switching process can be repeated tens or even hundreds of thousands of times, allowing the exact coordinates of a very large set of single molecules to be determined. The full set can be combined and digitally displayed as an image in which the computed

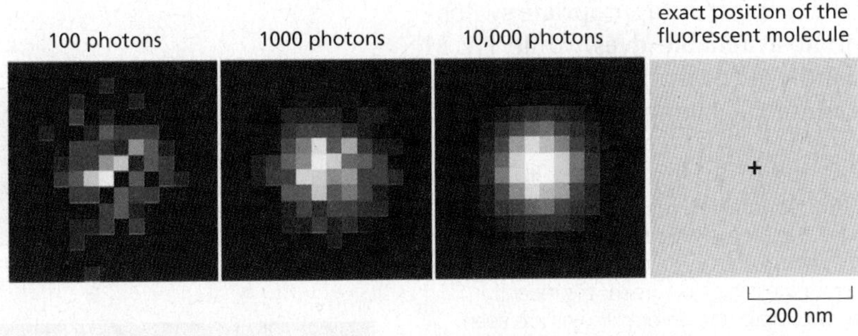

Figure 9–31 Single fluorescent molecules can be located with great accuracy. Determination of the exact mathematical center of the blurred image of a single fluorescent molecule becomes more accurate as more photons contribute to the final image. The point spread function described in the text dictates that the size of the molecular image is about 200 nm across, but in very bright specimens, the position of its center can be pinpointed to within a nanometer or so. (From A.L. McEvoy et al., *BMC Biol.* 8:106, 2010. With permission of the authors.)

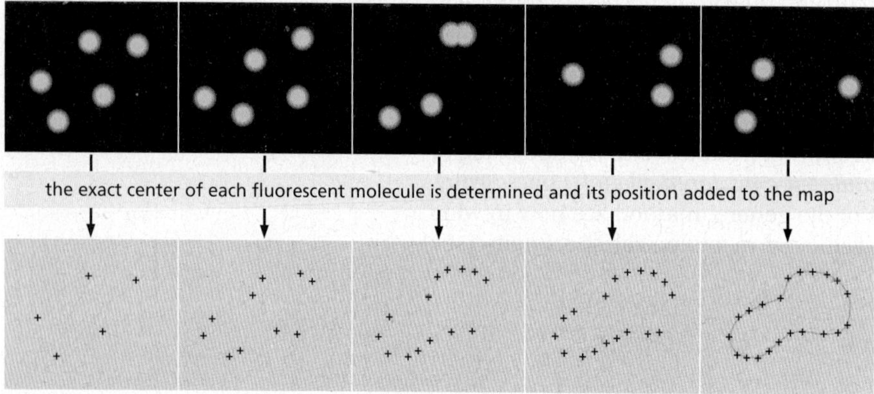

successive cycles of activation and bleaching allow well-separated single fluorescent molecules to be detected
the exact center of each fluorescent molecule is determined and its position added to the map
a superresolution image of the fluorescent structure is built up as the positions of tens of thousands of successive small groups of molecules are added to the map
(A)

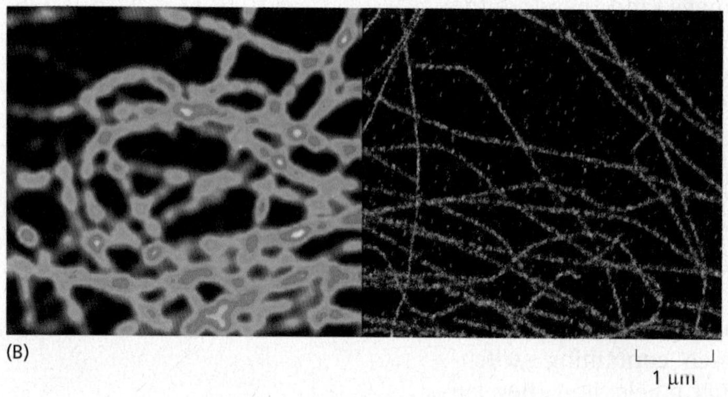

Figure 9–32 **Single-molecule localization microscopy (SMLM).** (A) In this imaginary specimen, sparse subsets of fluorescent molecules are individually switched on briefly and then bleached. The exact positions of all these well-spaced molecules can be gradually added together and built up into an image at superresolution. (B) In this portion of a cell, the microtubules have been fluorescently labeled and imaged *(left)* in a TIRF microscope (see Figure 9–38) and *(right)* at superresolution in a PALM microscope. The diameter of each microtubule on the *right* now resembles its true size, about 25 nm, rather than the 250 nm for each microtubule in the blurred diffraction-limited image on the *left*. (B, courtesy of Shinsuke Niwa.)

location of each individual molecule is exactly marked (**Figure 9–32**). The two main methods of **single-molecule localization microscopy (SMLM)** have been variously termed *photoactivated localization microscopy* (*PALM*) or *stochastic optical reconstruction microscopy* (*STORM*).

By switching the fluorophores off and on sequentially in different regions of the specimen as a function of time, all the superresolution imaging methods described above allow the resolution of molecules that are much closer together than the 200-nm diffraction limit. In STED, the locations of the molecules are determined by using optical methods to define exactly where their fluorescence will be on or off. In PALM and STORM, individual fluorescent molecules are switched on and off at random over a period of time, allowing their positions to be accurately determined. PALM and STORM techniques have depended on the development of novel fluorescent probes that exhibit the appropriate switching behavior. STORM originally relied on photoswitchable dyes, while PALM used photoswitchable fluorescent proteins, but the general principle is the same for both. All these methods can incorporate multicolor imaging (**Figure 9–33**),

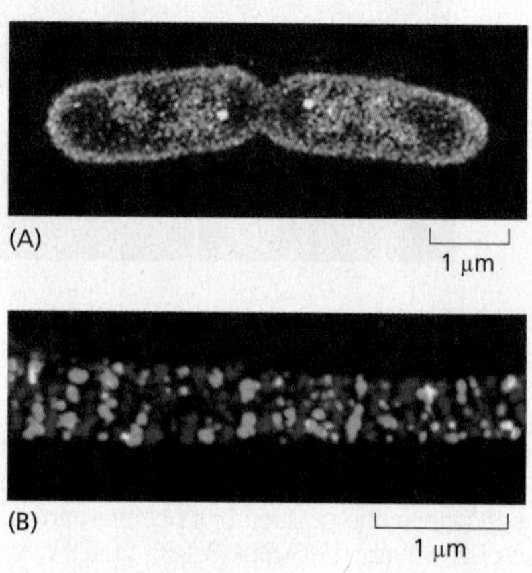

Figure 9–33 **Multiple structures that are below the diffraction-limited resolution can be imaged by single-molecule localization microscopy.** (A) Two recently divided *Escherichia coli* cells imaged in a STORM microscope with a resolution of about 20 nm. The cells are stained with three separate switchable fluorescent labels: the membrane is labeled green, the recently segregated DNA molecules are blue, and the ends of the two replicated chromosomes are seen as two bright white dots. (B) In this nerve cell, evenly spaced ring-like structures of actin *(red)* are wrapped around the circumference of the axon with a periodicity of about 190 nm, just smaller than the diffraction limit to resolution. In between are similarly spaced structures of spectrin *(blue)*. This periodic actin–spectrin cytoskeletal framework helps support the long thin axons of nerve cells. Such images depend heavily on the development of new, very fast-switching, and extremely bright fluorescent probes. (A, from C.K. Spahn et al., *Sci. Rep.* 8:14768, 2018, doi.org/10.1038/s41598-018-33052-3; B, from K. Xu et al., *Science* 339:452–456, 2013, doi 10.1126/science.1232251.)

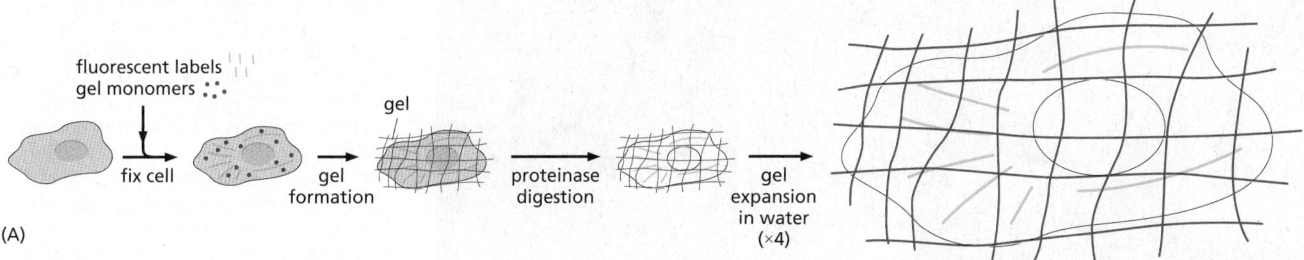

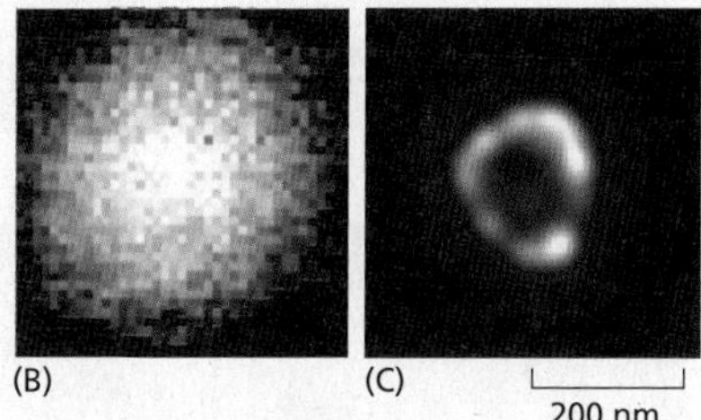

Figure 9–34 **Expansion microscopy.** (A) Although the technique has numerous variations, the essential features are that the fluorescently labeled sample is embedded in a polymer gel to which the fluorescent labels are covalently attached. After a proteinase digestion step, the gel is immersed in water and everything in the sample expands equally in every direction, usually by between 4 and 10 times, thus allowing details to be seen far more easily. (B) A peroxisome, whose membrane has been labeled with a fluorescent probe, appears in a confocal microscope as a blurred, diffraction-limited disc. (C) After expansion by a factor of 10, the image is captured with a standard epifluorescence microscope and, after deconvolution, shows the peroxisomal membrane well resolved and with a resolution of 25 nm. (From S. Truckenbrodt et al., *EMBO Rep.* 19:e45836, 2018, doi 10.15252/embr.201845836.)

and to some extent live-cell imaging in real time. Ending the long reign of the diffraction limit has reinvigorated light microscopy and its place in cell biology research.

Expanding the Specimen Can Offer Higher Resolution, but with a Conventional Microscope

All the approaches to improvement of resolution that we have discussed so far have centered on increasingly sophisticated and expensive developments of the microscope itself. Looking at the problem from the other end, the specimen end, swelling the sample to physically make it larger would in theory allow higher-resolution imaging, while still using a conventional fluorescence microscope. A new specimen preparation technique, called **expansion microscopy (ExM)**, does exactly that. The process starts by staining the fixed sample with fluorescent labels such as antibodies that target the molecules of interest. The labeled specimen is then treated with a chemical cross-linker and incubated with acrylate and acrylamide monomers. These monomers then polymerize to form a polyelectrolyte gel that simultaneously incorporates the cross-linked labels. With the labels covalently cross-linked to the polymer gel, and locked in their original relative positions, cellular material in the sample, predominantly proteins that might hinder subsequent expansion, is then carefully digested away. The gel containing the labeled specimen is now gently swollen by removing the buffer salts with water, so that it expands equally in all directions by between 4 and 10 times (**Figure 9–34A**). Two fluorochromes that were initially 100 nm apart, and consequently below the diffraction-limited resolution of a standard microscope, will now be 0.4–1.0 μm apart and are therefore easily resolved (**Figure 9–34B and C**). "Blowing-up" the sample allows effective superresolution to be enjoyed at up to 25 nm and without costly hardware (**Figure 9–35A and B**).

Expansion microscopy is proving valuable for detecting and quantitating which RNA transcripts are expressed in which individual cells in the brain. If all RNA molecules present are anchored firmly to the polymer gel before the expansion step, then the sample can be washed and re-probed sequentially with multiple fluorescent RNA probes using *in situ* hybridization (**Figure 9–35C and D**). Expansion takes place in all directions, and so depth information is also retrievable at higher resolution. Expansion microscopy samples can still be imaged by either confocal or light-sheet microscopy (discussed next), and deconvolution methods can still be used on the images—both help to improve three-dimensional imaging of large specimens.

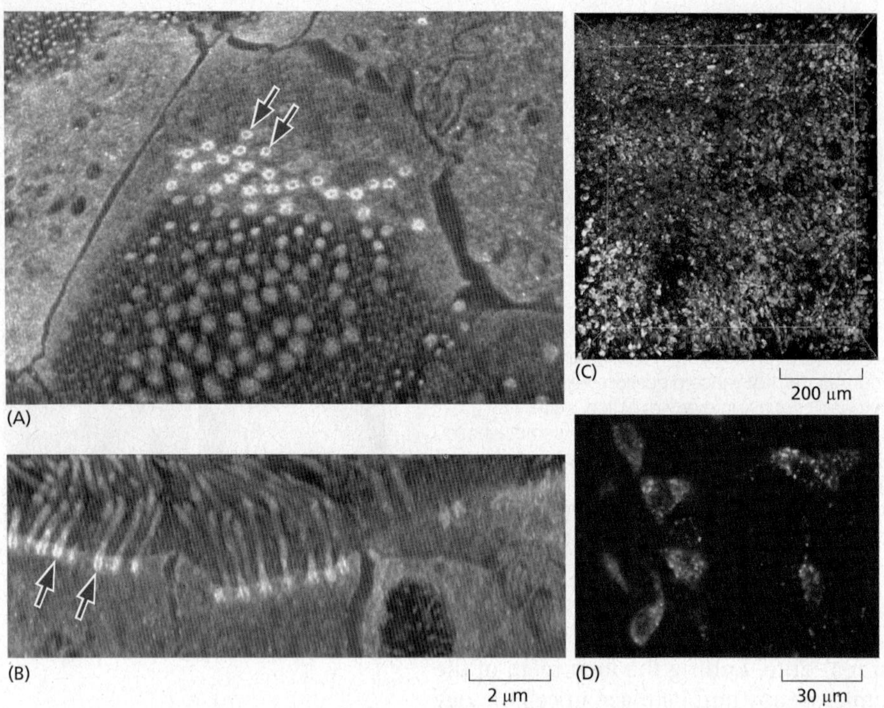

Figure 9–35 Expansion microscopy. (A and B) Two orthogonal views of the same cultured human nasal epithelial cells that have been stained with a fluorescent dye, expanded by ten times, and then imaged by conventional confocal microscopy. The hollow centers of ciliary basal bodies, which are not resolvable by conventional microscopy, are clearly visible in both top view (A) and side view (B) *(red arrows)* (see Movie 9.1). (C and D) A segment of mouse brain lateral hypothalamus, 800 × 800 × 300 μm, that has been expanded by a factor of 2, probed by sequential rounds of *in situ* RNA hybridization, and imaged by light-sheet microscopy. The cellular expression patterns of six different genes are shown: *Gad1 (red)*, *Slc17a6 (green)*, *Hcrt (blue)*, *Trh (yellow)*, *Calb2 (magenta)*, and *Meis2 (cyan)*. (A and B, courtesy of Hugo Damstra, Lukas Kapitein, and Paul Tillberg; C and D, courtesy of Yuhan Wang, Mark Eddison, Scott Sternson, and Paul Tillberg.)

Large Multicellular Structures Can Be Imaged Over Time

Many problems in cell biology involve being able to follow the movement and behavior of cells in multicellular living organisms, in early embryo development for example. Other problems require the ability to disentangle the complexity of cellular interactions in large and dense tissues, for example the millions of connections between the neurons of the brain. The side effects of prolonged exposure to high levels of light in the first case, and depth and out-of-focus fluorescence in the second, mean that most of the techniques we have discussed so far cannot help. One way of eliminating a lot of the out-of-focus fluorescence is to arrange for the beam of light from the excitation laser to illuminate the specimen from a direction perpendicular to the axis from which the emitted fluorescence is viewed. In this arrangement, called *light-sheet microscopy*, a thin sheet of laser light, less than a micrometer thick, is scanned through the specimen, exciting only the labeled molecules at that depth in the sample to emit their fluorescence (**Figure 9–36**). There are many advantages to this method: it results in high-contrast images with very low photobleaching or photodamage, and three-dimensional information is readily obtained. It is also quick. Variants of the method allow ultrathin light sheets to scan through successive planes of a sample at a rate of hundreds of planes a second. The long-term, three-dimensional observation of living cells is a major application, for example in following early embryonic development in flies or zebrafish over a period of days. In fixed brain samples, the complex architecture of all the cells and their interconnections can be disentangled

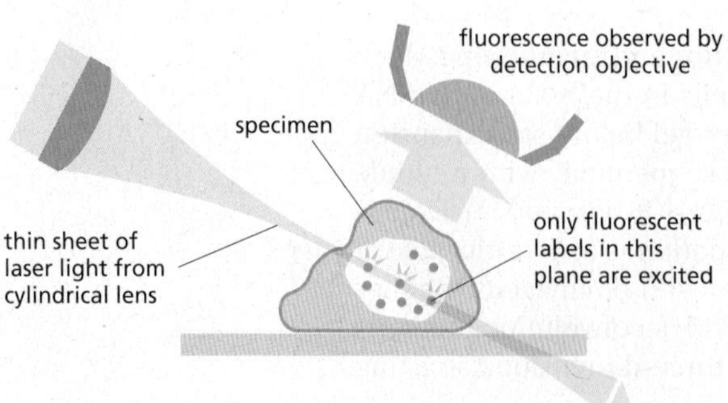

Figure 9–36 Light-sheet microscopy. A simple diagram showing how a very thin sheet of light that is projected (usually from a special cylindrical microscope objective lens) through a large specimen excites only those fluorescent labels in the thin plane that is illuminated. The resulting fluorescence is observed by an objective lens that is placed perpendicular to the light sheet. This means that, by progressively moving the specimen stage, multiple, sequential, and very sharp optical sections can be obtained rapidly and then digitally combined into a three-dimensional image.

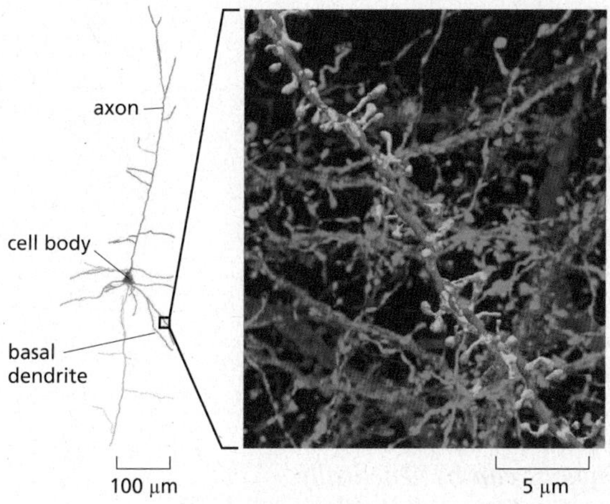

Figure 9–37 Light-sheet microscopy in the brain. A 1-mm-thick portion of a mouse brain has been prepared for expansion microscopy and then imaged with a light-sheet microscope. Reconstructions of thousands of optical sections allow the tracing of individual neurons and all their connections, such as this pyramidal neuron *(left)* from the visual cortex. On the *right* is shown the complex cellular context *(green)* for a short region of one of the neuron's basal dendrites (*orange* dendrite with its spines shown in *yellow*). (From R. Gao et al., *Science* 363:245–261, 2019, doi 10.1126/science.aau8302. With permission from AAAS.)

(**Figure 9–37** and Movie 9.1). Light-sheet microscopy can also be combined with other techniques. Coupled with STED imaging, for example, superresolution is attainable, and higher-resolution images can also be obtained by preparing the sample for expansion microscopy.

Single Molecules Can Be Visualized by Total Internal Reflection Fluorescence Microscopy

As we have seen, the strong background fluorescence due to light emitted or scattered by out-of-focus molecules tends to blot out the fluorescence from any one particular molecule of interest. This problem can be solved by the use of a special optical technique called *total internal reflection fluorescence* (*TIRF*) microscopy. In a TIRF microscope, laser light shines onto the cover-slip surface at the precise critical angle at which total internal reflection occurs (**Figure 9–38A**). Because of total internal reflection, the light does not enter the sample, and the majority of fluorescent molecules are not, therefore, illuminated. However, electromagnetic energy does extend, as an evanescent field, for a very short distance beyond the surface of the cover slip and into the specimen, allowing just those molecules in the layer closest to the surface to become excited. When these molecules fluoresce, their emitted light is no longer competing with out-of-focus light from the overlying molecules and can now be detected. TIRF has allowed several dramatic experiments, for instance imaging of single motor proteins moving along microtubules or actin filaments. At present, the technique is restricted to a thin layer about 200 nm below the cell surface. Although not strictly TIRF, decreasing the angle of the incident light so that it is almost parallel to the cover slip can increase the depth into the cell that can be examined, albeit not so uniformly, a feature useful in cells with an outer wall, such as those of plants and fungi (**Figure 9–38B and C**).

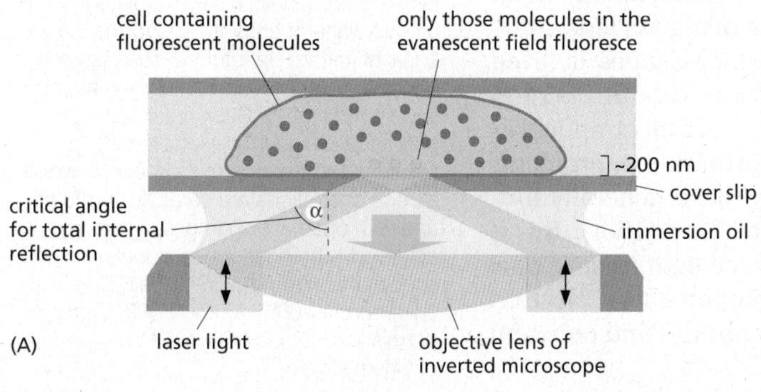

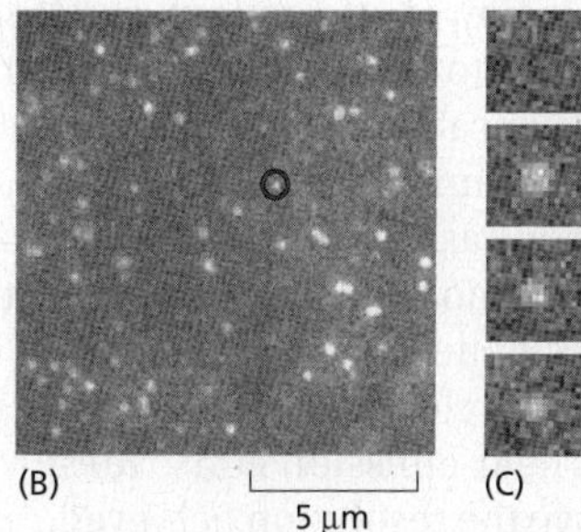

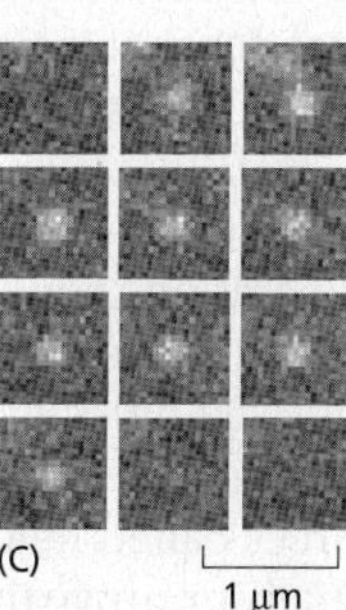

Figure 9–38 TIRF microscopy allows the detection of single fluorescent molecules near the cell surface. (A) TIRF microscopy uses excitatory laser light to illuminate the cover-slip surface at the critical angle at which all the light is reflected by the glass–water interface. Some electromagnetic energy extends a short distance across the interface as an evanescent wave that excites just those molecules that are attached to the cover slip or are very close to its surface. (B) TIRF microscopy is used to follow the formation of an individual clathrin-coated pit and its subsequent endocytosis. In this image of the surface of the plasma membrane of an *Arabidopsis* root cell, a clathrin adaptor protein is tagged with GFP. Individual pits can be followed over time. (C) The pit ringed in B is shown at 1-second intervals, demonstrating that its appearance and its removal at the plasma membrane by endocytosis takes place in about 10 seconds. (B and C, from A. Johnson and G. Vert, *Front. Plant Sci.* 8:612, 2017, doi 10.3389/fpls.2017.00612.)

Summary

Many light-microscope techniques are available for observing cells. Cells that have been fixed and stained can be studied in a conventional light microscope, whereas antibodies coupled to fluorescent dyes can be used to locate specific molecules in cells in a fluorescence microscope. Living cells can be seen with phase-contrast, differential-interference-contrast, dark-field, or bright-field microscopes. All forms of light microscopy are facilitated by digital image-processing techniques, which enhance sensitivity and refine the image. Confocal microscopy and image deconvolution both provide thin optical sections and can be used to reconstruct three-dimensional images.

Techniques are now available for detecting, measuring, and following almost any desired molecule in a living cell. Fluorescent labels can be introduced to measure the concentrations of specific ions or signaling molecules in individual cells or in different parts of a cell. Virtually any protein of interest can be genetically engineered as a fluorescent fusion protein and then imaged in living cells by fluorescence microscopy. The dynamic behavior and interactions of many molecules can be followed in living cells by variations on the use of fluorescent protein tags, in some cases at the level of single molecules. Various superresolution techniques can circumvent the diffraction limit in different ways and resolve molecules separated by distances as small as 20 nm.

LOOKING AT CELLS AND MOLECULES IN THE ELECTRON MICROSCOPE

Light microscopy is limited in the fineness of detail that it can reveal. Microscopes using other types of radiation—in particular, electron microscopes—can resolve much smaller structures than is possible with visible light. This higher resolution comes at a cost: specimen preparation for electron microscopy is complex and it is harder to be sure that what we see in the image corresponds precisely to the original living structure. It is possible, however, to use very rapid freezing to preserve structures faithfully for electron microscopy. Digital image analysis can be used to reconstruct three-dimensional objects by combining information either from many individual particles or from multiple tilted views of a single object. Together, these approaches extend the resolution and scope of electron microscopy to the point at which we can faithfully image the detailed structures of individual macromolecules and the complexes they form, even inside cells.

The Electron Microscope Resolves the Fine Structure of the Cell

The formal relationship between the diffraction limit to resolution and the wavelength of the illuminating radiation (see Figure 9–5) holds true for any form of radiation, whether it is a beam of light or a beam of electrons. With electrons, however, the limit of resolution is very small. The wavelength of an electron decreases as its velocity increases. In an **electron microscope** with an accelerating voltage of 100,000 V, the wavelength of an electron is 0.004 nm. In theory, the resolution of such a microscope should be about 0.002 nm, which is 100,000 times that of the conventional light microscope. Because the aberrations of an electron lens are considerably harder to correct than those of a glass lens, however, the practical resolving power of modern electron microscopes is, even with careful image processing to correct for lens aberrations, about 0.05 nm (0.5 Å) (**Figure 9–39**). This is because only the very center of the electron lenses can be used, and the effective numerical aperture is tiny. Furthermore, problems of specimen preparation, contrast, and radiation damage have generally limited the normal effective resolution for biological objects to 1 nm (10 Å). This is nonetheless about 200 times better than the resolution of the light microscope. Moreover, the performance of electron microscopes is improved by electron illumination sources called field-emission guns. These very bright and coherent sources substantially improve the resolution achieved.

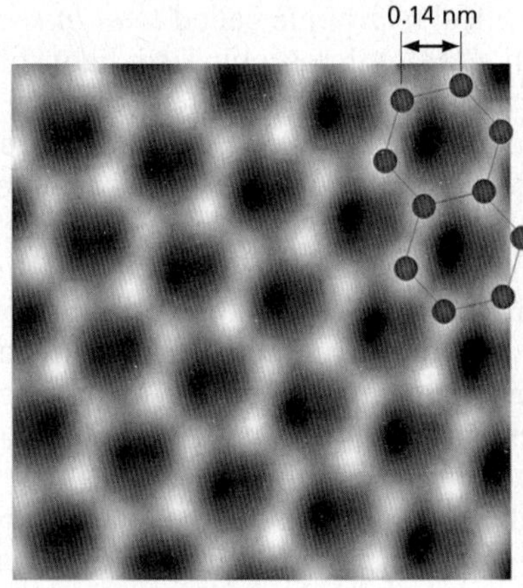

Figure 9–39 **The resolution of the electron microscope.** This transmission electron micrograph of a monolayer of graphene resolves the individual carbon atoms as bright spots in a hexagonal lattice. Graphene is a single isolated atomic plane of graphite and forms the basis of carbon nanotubes. The distance between adjacent bonded carbon atoms is 0.14 nm (1.4 Å). Such resolution can only be obtained in a specially built transmission electron microscope in which all lens aberrations are carefully corrected, and with optimal specimens; it is rarely achieved with most conventional biological specimens. (From A. Dato et al., *Chem. Commun.* 40:6095–6097, 2009. With permission from the Royal Society of Chemistry.)

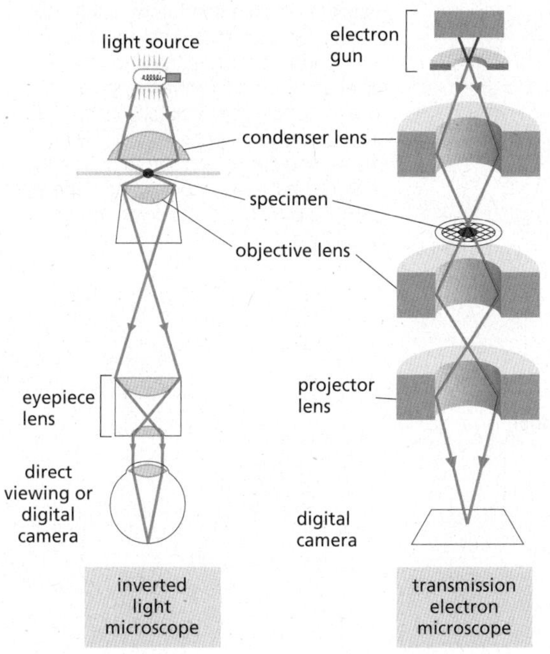

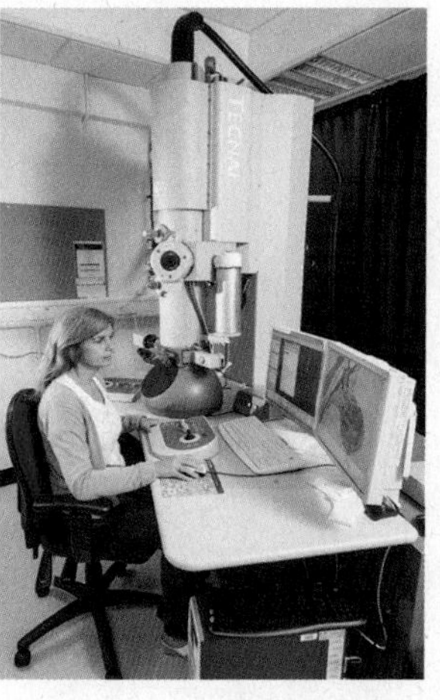

Figure 9–40 **The principal features of an inverted light microscope and a transmission electron microscope.** These drawings emphasize the similarities of overall design. Whereas the lenses in the light microscope are made of glass, those in the electron microscope are magnetic coils. The electron microscope requires that the specimen be placed in a vacuum. The inset shows a routine transmission electron microscope in use. (Photograph courtesy of Andrew Davis.)

In overall design, the transmission electron microscope (TEM) is similar to an inverted light microscope, albeit much larger (**Figure 9–40**). The source of illumination is a filament or cathode that emits electrons at the top of a cylindrical column about 2 m high. Because electrons are scattered by collisions with air molecules, air must first be pumped out of the column to create a vacuum. The electrons are then accelerated from the filament by a nearby anode and allowed to pass through a tiny hole to form an electron beam that travels down the column. Magnetic coils placed at intervals along the column focus the electron beam, just as glass lenses focus the light in a light microscope. The specimen is put into the vacuum, through an airlock, into the path of the electron beam. As in light microscopy, the specimen can be stained—in this case, with *electron-dense* material. Some of the electrons passing through the specimen are scattered by structures stained with the electron-dense material; the remainder are focused to form an image. The image can be observed on a monitor or is typically recorded with a sensitive CMOS electron detector. Because the scattered electrons are lost from the beam, the dense regions of the specimen show up in the image as areas of reduced electron flux, which look dark.

Biological Specimens Require Special Preparation for Electron Microscopy

In the early days of its application to biological materials, the electron microscope revealed many previously unimagined structures in cells. But before these discoveries could be made, electron microscopists had to develop new procedures for embedding, cutting, and staining tissues.

Because the specimen is exposed to a very high vacuum in the electron microscope, living tissue is usually killed and preserved by chemical fixation. As electrons have very limited penetrating power, the fixed tissues normally have to be cut into extremely thin sections (25–100 nm thick, about 1/200 the thickness of a single cell) before they are viewed. This is achieved by dehydrating the specimen, permeating it with a monomeric resin that polymerizes to form a solid block of plastic, then cutting the block with a fine glass or diamond knife on a special microtome. The resulting *ultrathin sections*, free of water and other volatile solvents, are supported on a small metal grid for viewing in the microscope (**Figure 9–41**).

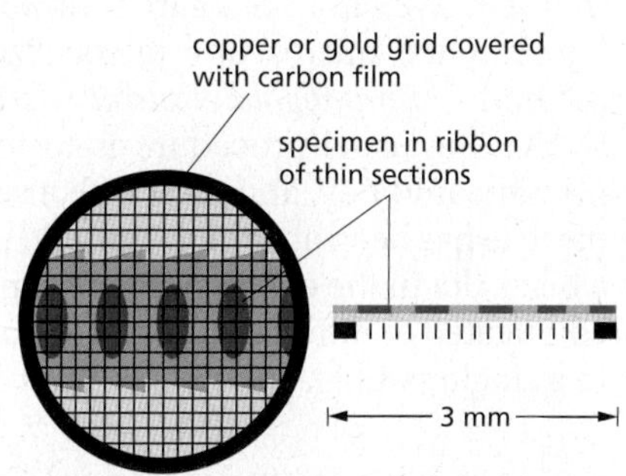

Figure 9–41 **Specimen support.** The metal grid that supports the thin sections of a specimen in a transmission electron microscope.

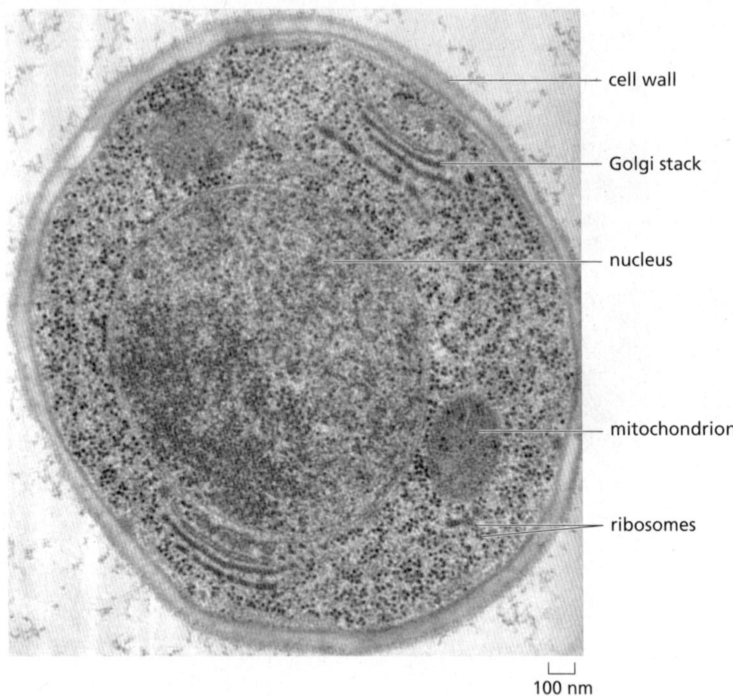

Figure 9–42 **Thin section of a cell.** This thin section is of a yeast cell that has been very rapidly frozen and the vitreous ice replaced by organic solvents and then by plastic resin (freeze substitution). The nucleus, mitochondria, cell wall, Golgi stacks, and ribosomes can all be readily seen in a state that is presumed to be as lifelike as possible. (Courtesy of Andrew Staehelin.)

The steps required to prepare biological material for electron microscopy are challenging. How can we be sure that the image of the fixed, dehydrated, resin-embedded specimen bears any relation to the delicate, aqueous biological system present in the living cell? The best current approaches to this problem depend on rapid freezing. If an aqueous system is cooled fast enough and to a low enough temperature, the water and other components in it do not have time to rearrange themselves or crystallize into ice. Instead, the water is supercooled into a rigid but noncrystalline state—a "glass"—called vitreous ice. This rapid freezing is usually performed by plunging the sample into a coolant such as liquid ethane or by cooling it at very high pressure.

Some rapidly frozen specimens can be examined directly in the electron microscope using a special cooled specimen holder. In other cases, the frozen block can be fractured to reveal interior cell surfaces or the surrounding ice can be sublimed away to expose external surfaces. However, we often want to examine thin sections, and the frozen tissue can be sectioned directly in a cooled microtome. A compromise is to rapidly freeze the tissue, replace the water with organic solvents, embed the tissue in plastic resin, and finally cut sections. This approach, called *freeze substitution*, stabilizes and preserves the tissue in a condition very close to its original living state (**Figure 9–42**).

Molecules in all kinds of thin sections can be labeled to identify and localize them. We have seen earlier how antibodies can be used in conjunction with fluorescence microscopy to localize specific macromolecules. An analogous method—*immunogold electron microscopy*—can be used in the electron microscope. The usual procedure is to incubate a thin section first with a specific primary antibody, and then with a secondary antibody to which a colloidal gold particle has been attached. The gold particle is electron-dense and can be seen as a black dot in the electron microscope (**Figure 9–43**). Different antibodies can be conjugated to different-sized gold particles so multiple proteins can be localized in a single sample.

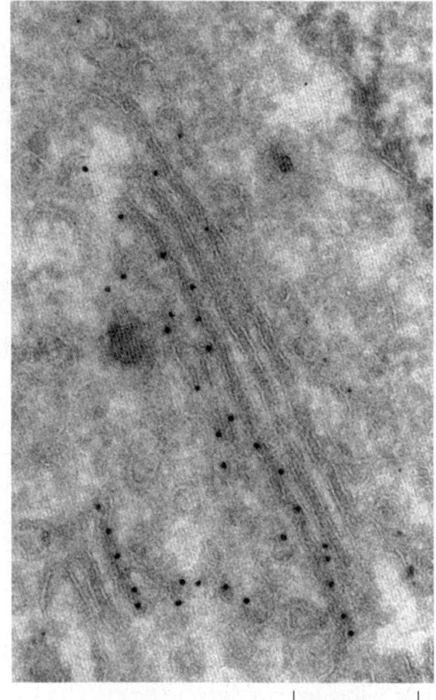

Figure 9–43 **Localizing proteins in electron microscopy.** Immunogold electron microscopy is used here to find the specific location of a protein that is targeted to the Golgi apparatus. The protein has been tagged with a genetically encoded fluorescent protein and is localized to the trans-Golgi network. The protein is seen in this thin section using an antibody to the fluorescent protein coupled to 10-nm colloidal gold particles, seen in the electron microscope as black dots. The cell has been frozen under high pressure and freeze substituted before embedding and sectioning. (Courtesy of Charlotta Funaya and M. Teresa Alonso.)

Heavy Metals Can Provide Additional Contrast

Although phase contrast can make unstained specimens more visible, image clarity in an electron micrograph usually depends on having a range of electron densities to provide amplitude contrast within the specimen. Electron density in

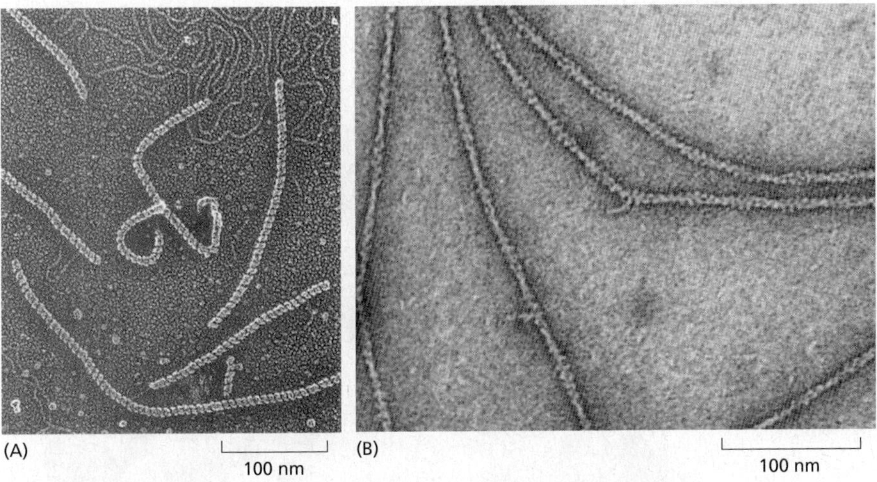

Figure 9–44 **Heavy metals provide contrast in the electron microscope.** (A) This transmission electron micrograph shows RecA protein together with *E. coli* DNA adsorbed to flakes of mica, frozen, carefully dried, and then shadowed with evaporated platinum atoms. The RecA protein clearly forms tight, right-handed helices around the bacterial DNA molecules, some of which can be seen free at the top of the image (see also Figure 5–48). (B) In this transmission electron micrograph of actin filaments, negatively stained with uranyl acetate, each filament is about 8 nm in diameter and is seen, on close inspection, to be composed of a helical chain of globular actin molecules (see also Figure 16–8). (A, from J. Heuser, *J. Electron Microsc. Tech.* 13:244–263, 1989; B, courtesy of Roger Craig.)

turn depends on the atomic number of the atoms that are present: the higher the atomic number, the more electrons are scattered and the darker that part of the image. Biological tissues are composed mainly of atoms of very low atomic number (primarily carbon, oxygen, nitrogen, and hydrogen). To make them more readily visible, tissues are often impregnated (before or after sectioning) with the salts of heavy metals such as uranium, lead, and osmium. The degree of impregnation, or "staining," with these salts will vary for different cell constituents. Lipids, for example, tend to stain darkly after osmium fixation, revealing the location of cell membranes (see, for example, Figure 12–2 or Figure 12–15).

Alternatively, if isolated molecules are "shadowed" by platinum or other heavy metals evaporated from a heated filament, macromolecules such as DNA or large proteins can be visualized with high contrast in the electron microscope (**Figure 9–44A**). **Negative staining** is a similar approach that also allows fine detail to be seen in isolated molecules or macromolecular machines. In this technique, the molecules are supported on the thin film of carbon on a grid and mixed with a solution of a heavy-metal salt such as uranyl formate or acetate. After the sample has dried, a very thin film of metal salt covers the carbon film everywhere except where it has been excluded by the presence of an adsorbed macromolecule. Because the macromolecule allows electrons to pass through it much more readily than does the surrounding heavy-metal stain, a reverse or negative image of the molecule is created. Negative staining is especially useful for quickly and cheaply viewing large macromolecular aggregates such as viruses or ribosomes and for seeing the subunit structure of protein filaments (**Figure 9–44B**). Shadowing and negative staining can provide high-contrast surface views of small macromolecular assemblies, but the size of the smallest metal particles in the shadow or stain limits the resolution of both techniques to about 2 nm.

Images of Surfaces Can Be Obtained by Scanning Electron Microscopy

A **scanning electron microscope (SEM)** directly produces an image of the three-dimensional structure of the surface of a specimen. The SEM is usually smaller, simpler, and cheaper than a transmission electron microscope. Whereas

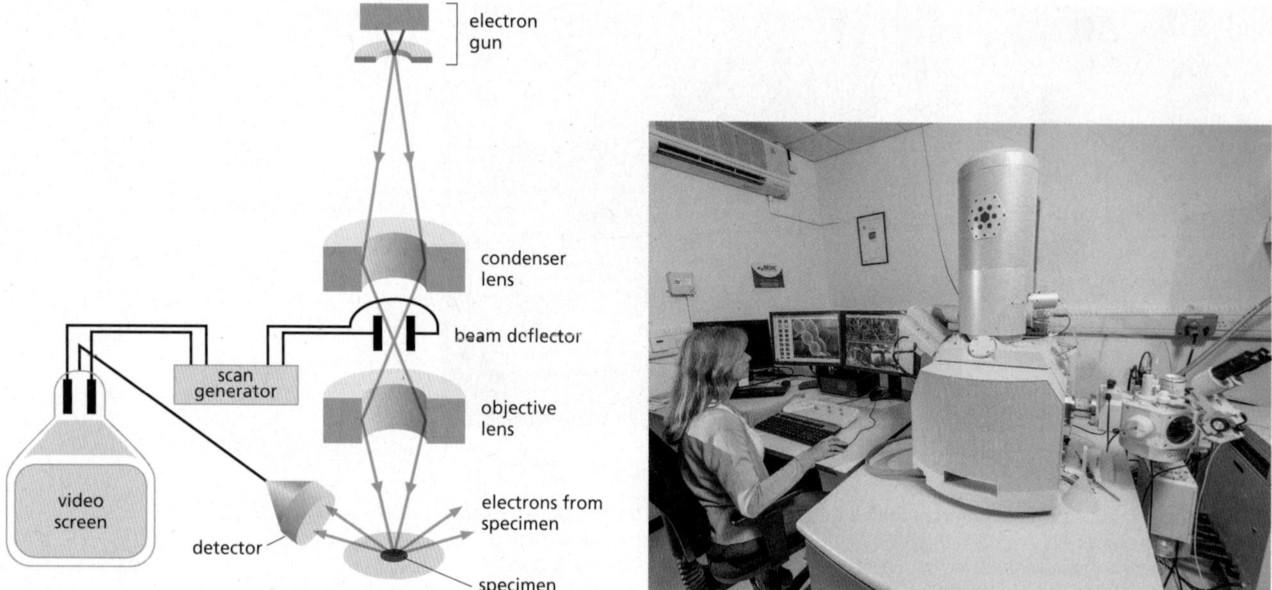

Figure 9–45 **The scanning electron microscope.** In an SEM, the specimen is scanned by a beam of electrons brought to a focus on the specimen by the electromagnetic coils that act as lenses. The detector measures the quantity of electrons scattered or emitted as the beam bombards each successive point on the surface of the specimen and records the intensity of successive points in an image built up on a screen. The SEM creates striking images of three-dimensional objects with great depth of focus and a resolution between 0.5 nm and 10 nm depending on the kind of instrument. (Photograph courtesy of Andy Davis.)

the TEM uses the electrons that have passed through the specimen to form an image, the SEM uses electrons that are scattered or emitted from the specimen's surface. The specimen to be examined is usually either fixed, dried, and coated with a thin layer of heavy metal or alternatively rapidly frozen and then transferred to a cooled specimen stage for coating and direct examination in the microscope (**Figure 9–45**). The specimen is scanned with a very narrow beam of electrons. The quantity of electrons scattered or emitted as this primary beam bombards each successive point of the metallic surface is measured and builds up an image on a computer screen. Often an entire plant part or small animal can be put into the microscope with very little preparation (**Figure 9–46**).

The SEM technique provides great depth of field, thus objects both near and far in the field of view are imaged sharply. Moreover, because the amount of electron scattering depends on the angle of the surface relative to the beam, the image has highlights and shadows that give it a three-dimensional appearance

Figure 9–46 **The scanning electron microscope produces surface images with great depth of field.** SEM micrographs taken at a wide range of magnifications. (A) A developing wheat flower, or spike. This delicate flower spike was rapidly frozen, coated with a thin metal film, and examined in the frozen state with an SEM. This low-magnification micrograph demonstrates the large depth of focus of an SEM, even with a large specimen like this. (B) These pollen grains from a hellebore flower reveal their sculpted cell walls in the SEM. The shapes and patterns are specific for each species of pollen grain. (C) Chains of bacteria living in the blue veins of a Stilton cheese. (A, B, and C, courtesy of Kim Findlay.)

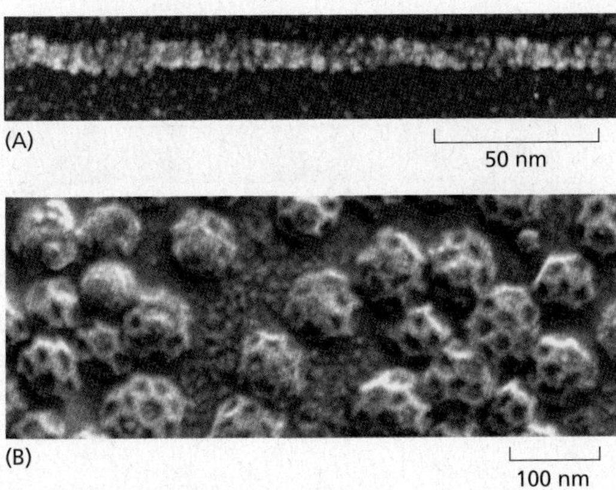

Figure 9–47 **Higher-resolution SEM.** Macromolecular assemblies, shadowed with a very thin coating of tungsten and imaged in an SEM equipped with a field-emission electron gun. (A) An actin filament showing the helical arrangement of actin monomers. (B) Clathrin-coated vesicles. [A and B, from R. Wepf et al., in Biological Field Emission Scanning Electron Microscopy (R. Fleck and B. Humbel, eds.), pp. 269–298. Hoboken, NJ: Wiley, 2019.]

(see Figure 9–46). Only surface features can be examined, however, and in most forms of SEM, the resolution attainable is not very high (about 10 nm). As a result, the technique is usually used to study whole cells and tissues rather than subcellular organelles (see Movie 21.3). However, very-high-resolution SEMs have been developed with a bright, coherent, field-emission gun as the electron source. As resolution in the SEM depends not on the wavelength of the electron beam but on the size of the electron spot that is scanned across the specimen, this type of SEM can produce images that rival the resolution possible with a negatively stained specimen in a TEM (**Figure 9–47**).

Electron Microscope Tomography Allows the Molecular Architecture of Cells to Be Seen in Three Dimensions

The SEM can only provide a surface view of an object, which tells us little about the important three-dimensional relationships between macromolecules and organelles within a living cell. Moreover, thin sections viewed in a TEM also often fail to convey the three-dimensional arrangement of cellular components, and the images can be misleading. It is possible to reconstruct the third dimension from serial sections, but this is a lengthy and tedious process. But even thin sections have a significant depth compared with the resolution of the electron microscope, so the TEM image can also be misleading in an opposite way, through the superimposition of objects that lie at different depths.

Because of the large depth of field of electron microscopes, all the parts of the three-dimensional specimen are in focus, and the resulting image is a projection (a superimposition of layers) of the structure along the viewing direction. The lost information in the third dimension can be recovered if we have views of the same specimen but from many different directions. The computational methods for this technique are widely used in medical CT scans. In a CT scan, the imaging equipment is moved around the patient to generate the different views. In **electron microscope (EM) tomography**, the specimen holder is tilted in the microscope, which achieves the same result. The specimen is usually tilted to a maximum of 60° in every direction, and in this way we can arrive at a three-dimensional reconstruction, in a chosen standard orientation, by combining different views of a single object. Each individual view will be very noisy, but by combining them in three dimensions and taking an average, the noise can be significantly reduced. Thick plastic sections of embedded material have been used to create three-dimensional reconstructions, or *tomograms* (**Movie 9.2**), of cells, but increasingly microscopists are applying EM tomography to unstained, frozen, hydrated sections, and even to rapidly frozen whole cells or organelles. Individual macromolecular assemblies that

SINGLE-PARTICLE RECONSTRUCTION BY CRYOEM

X-ray crystallography is one way to determine a protein structure. However, large macromolecular machines are often hard to crystallize, as are many integral membrane proteins, and for dynamic proteins and assemblies it is hard to access different conformations through crystallography alone. To get around these problems, investigators are increasingly turning to cryo-electron microscopy (cryoEM) to solve macromolecular structures.

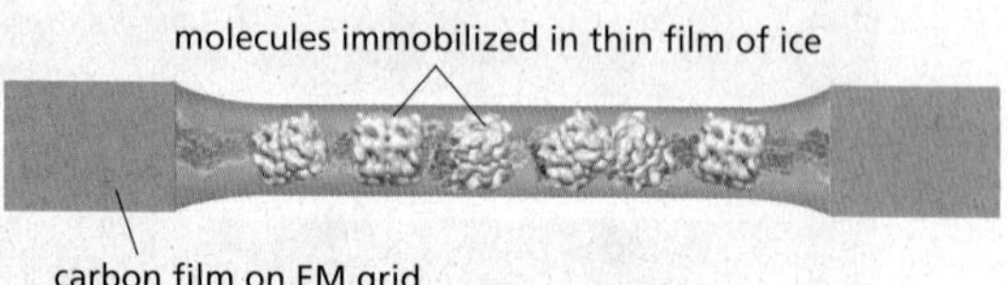

In this technique, a droplet of the pure protein in water is placed on a small EM grid that is plunged into a vat of liquid ethane at −180°C. This freezes the proteins in a thin film of ice and the rapid freezing ensures that the surrounding water molecules have no time to form ice crystals, which would damage the protein's shape.

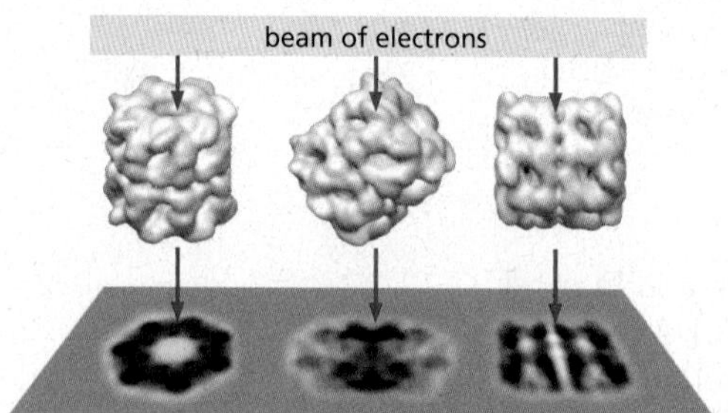

The sample is examined, still frozen, by high-voltage transmission electron microscopy. To avoid damage, it is important that only a few electrons pass through each part of the specimen. Sensitive detectors are therefore deployed to capture every electron that passes through the specimen. Much EM specimen preparation and data collection is now fully automated and many thousands of micrographs are typically captured, each of which will contain hundreds or thousands of individual molecules all arranged in random orientations within the ice.

Algorithms then sort the molecules into sets where each set contains molecules that are all oriented in the same direction. The thousands of images in each set are all then superimposed and averaged to improve the signal-to-noise ratio.

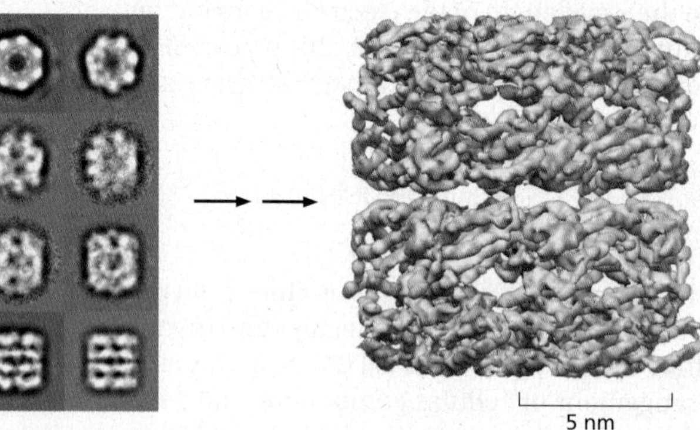

This crisper two-dimensional image set, which represents different views of the particle, are then combined and converted via a series of complex iterative steps into a high-resolution three-dimensional structure.

Model of GroEL (Courtesy of Gabriel Lander.)

CRYOEM STRUCTURE OF THE RIBOSOME

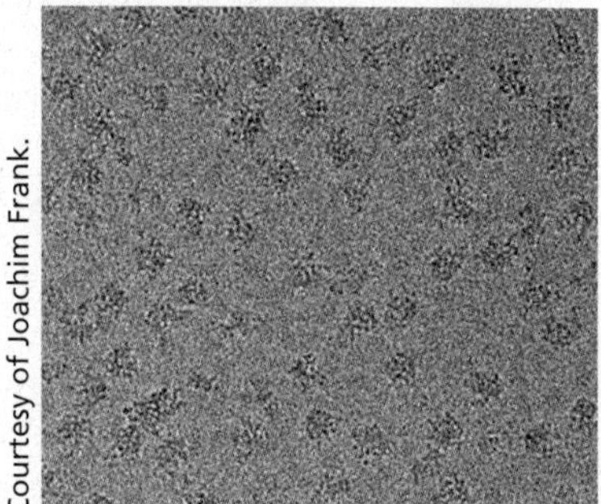

Courtesy of Joachim Frank.

60S ribosomal subunits randomly oriented in a thin film of ice 100 nm

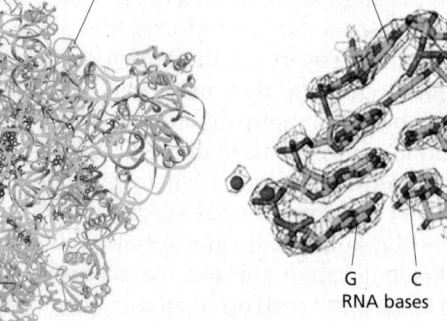

Although by no means routine, big improvements in image-processing algorithms, modeling tools and sheer computing power all mean that structures of macromolecular complexes are now becoming attainable with resolutions in the 0.2- to 0.3-nm range.

This resolving power now approaches that of x-ray crystallography, and the two techniques thrive together, each bootstrapping the other to obtain ever more useful and dynamic structural information. A good example is the structure of the ribosome shown here at a resolution of 0.25 nm.

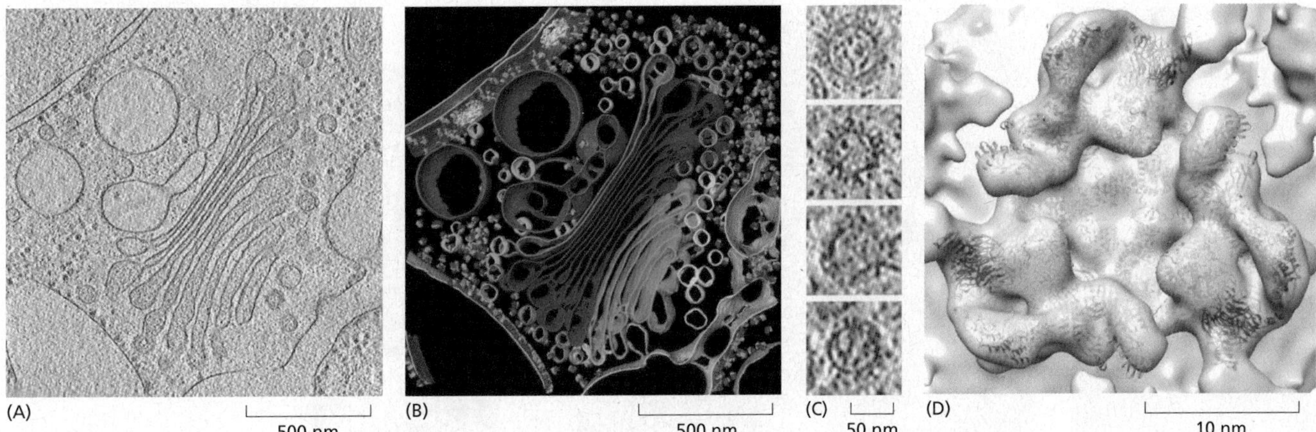

appear as multiple copies in the tomogram can be identified, and with a computational process called *subtomogram averaging* to reduce noise and gain structural information, molecular structures inside cells can now be obtained at a resolution of better than 2 nm (**Figure 9–48**). Electron microscopy now provides a robust bridge between the scale of the single molecule and that of its cellular environment.

Figure 9–48 EM tomography. The COP1 coat mediates vesicle traffic within the Golgi apparatus and retrograde traffic to the endoplasmic reticulum (ER) (see Figures 13–4 and 13–5). EM tomography has helped visualize the details of COP1 coats *in situ* on buds and vesicles in rapidly frozen *Chlamydomonas* cells. (A) One slice through a three-dimensional tomogram of a complete Golgi apparatus. (The tomogram can be seen in Movie 9.2.) (B) Using the information from several such tomograms, a portion of the Golgi is shown here, color coded to show ER *dark yellow*, the *cis* vesicles *yellow*, the four *cis* cisternae *green*, the four medial cisternae *red*, the *trans* cisterna *blue*, medial vesicles *pink*, *trans* vesicles *light blue*, and the *trans* Golgi network *purple*. Ribosomes can also be seen as small gray blobs. (C) Individual slices through COP1 vesicles in the tomogram; the bottom one is partially uncoated. (D) By identifying and averaging more than 10,000 COP1 subunits on vesicles in the tomograms, a molecular structure was obtained by subtomogram averaging at a resolution of 2 nm. Structures of the various proteins in the COP1 coat have been solved, and they can be fitted neatly into the electron-density envelope of the EM structure. A surface view of a triad of COP1 subunits on the surface of a vesicle is shown here together with the molecular structures (in color) of the individual components that have been fitted into the EM structure. (Adapted from Y.S. Bykov et al., *eLife* 6:e32493, 2017, doi 10.7554/eLife.32493.)

Cryo-electron Microscopy Can Determine Molecular Structures at Atomic Resolution

As we saw earlier (p. 567), noise is important in light microscopy at low light levels, but it is a particularly severe problem for electron microscopy of unstained macromolecules. A protein molecule can tolerate a dose of only a few hundreds of electrons per square nanometer without damage, and this dose is orders of magnitude below what is needed to define an image at atomic resolution.

The solution is to obtain images of many identical molecules—perhaps hundreds of thousands of images of individual particles—and combine them to produce an averaged image, revealing structural details that are hidden by the noise in the original images. This procedure is called **single-particle reconstruction** (**Panel 9–1**). Before combining all the individual images, however, they must be aligned with each other. With the help of a computer, the digital images of randomly distributed and unaligned molecules can be processed and combined to yield high-resolution reconstructions (see Movie 13.1). Although structures that have some intrinsic symmetry, such as dimers or helical repeats, are somewhat easier to solve (**Figure 9–49**), this technique has also been used for huge macromolecular machines, such as ribosomes, that have no symmetry (see Panel 9–1).

Cryo-electron microscopy (cryoEM) depends crucially on very rapidly freezing the aqueous specimen to form vitreous ice, which does not allow ice crystals to form and therefore does not damage the specimen. A very thin (about 100 nm) film of an aqueous suspension of purified macromolecular complex is prepared on a microscope grid and is then rapidly frozen by being plunged into a coolant. A special sample holder keeps this hydrated specimen at -160°C in the vacuum of the microscope, where it can be viewed directly without fixation, staining, or drying. Unlike negative staining, in which what we see is the envelope of stain exclusion around the particle, cryoEM produces an image from the macromolecular structure itself. The specialized transmission electron microscopes required operate with much higher electron accelerating voltages than that of a routine TEM and typically run at 300,000 V. However, as very low electron doses are used to obtain cryoEM images, the intrinsic contrast in the images produced is very low, and to extract the maximum amount of structural information, special image-processing techniques must be used. Huge advances in direct electron detectors and faster, more efficient image-processing techniques that involve image alignment routines, motion correction, and contrast transfer function corrections mean that the structures of molecules as small as 100 kilodaltons can now be solved. The smaller the molecule, the noisier the image, and the main

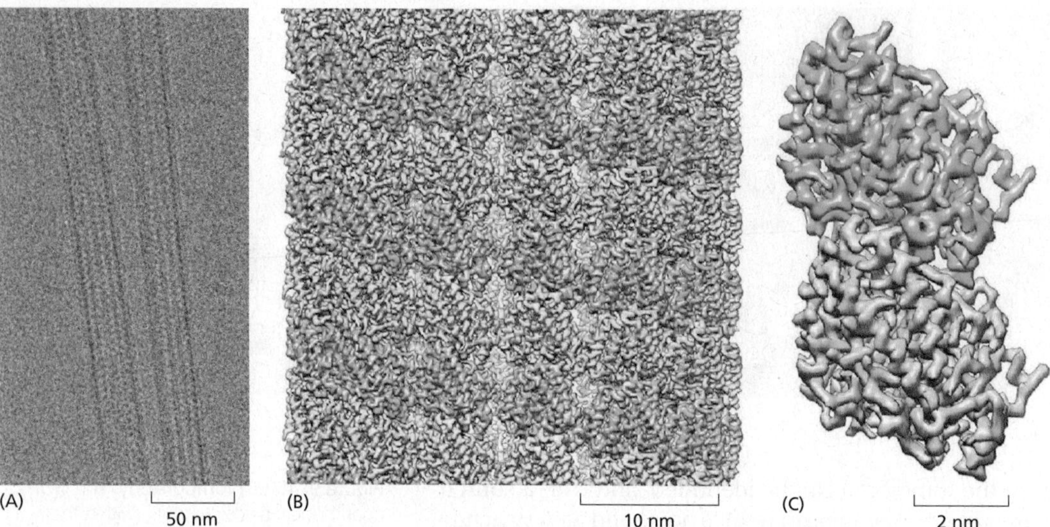

Figure 9–49 CryoEM structure of microtubules. This cryoEM reconstruction of the structure of a microtubule was helped by the intrinsic symmetry of the microtubule itself (see Figure 16–37). This detailed model of the whole microtubule has allowed an examination of the way in which the protofilaments interact and the way in which the whole lattice and associated proteins are assembled. (A) CryoEM image of two intact microtubules embedded in vitreous ice. (B) A reconstruction of the surface lattice of a single microtubule at a resolution of 0.35 nm (3.5 Å). (C) The detailed electron-density map of the tubulin dimer extracted from the structure of the intact microtubule. α-Tubulin is *darker green*, and β-tubulin is *lighter green*. (From E. Nogales, *Mol. Biol. Cell* 27:3202–3204, 2016, doi 10.1091/mbc.E16-06-0372. With permission from Elsevier.)

advantages of the method are best seen with large and sometimes flexible macromolecular complexes such as viruses, ribosomes, and large integral membrane proteins that are hard to crystallize (**Figure 9–50**).

A remarkable resolution of 0.12 nm (1.2 Å) has been achieved in a particularly stable protein by cryoEM, enough to see clearly the detailed atomic structure and to rival x-ray crystallography in resolution (**Figure 9–51**). Electron microscopy, however, also has some very clear additional advantages over x-ray crystallography (discussed in Chapter 8) as a method for macromolecular structure determination. First, it does not require crystalline specimens. Second, it can deal with extremely large complexes—structures that may be too large or too variable to crystallize satisfactorily; for example, membrane proteins. Third, it allows the rapid analysis of different conformations of protein machines; for example, the different states of the F_1 ATPase proton pump shown in Figure 14–31. Fourth, the glycosylation patterns and mobile loops on the surface of proteins, which are often impossible to see in x-ray structures, are more readily resolved in cryoEM structures. And fifth, only a minute amount of sample is required compared with that needed to make crystals.

The analysis of large and complex macromolecular structures is helped considerably if the atomic structure of one or more of the subunits is known, for

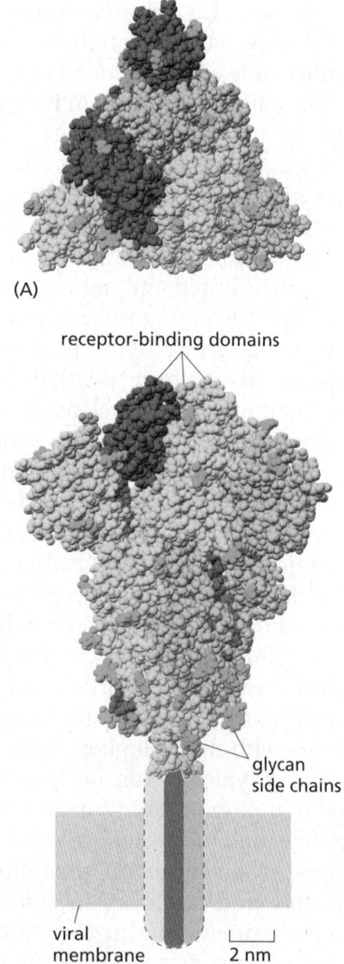

Figure 9–50 The spike protein on the SARS-CoV-2 virus. The SARS-CoV-2 virus was responsible for the COVID-19 pandemic. Protruding from the viral membrane are many trimeric spike proteins that mediate binding of the virus to a receptor on cells in our respiratory tract and its subsequent entry into the cell. The trimeric spike protein is a target both of our immune system and of vaccine developers. The closed conformation of the trimeric spike protein shown here, both from the top (A) and from the side (B), was obtained from rapidly frozen intact virus particles. Spike proteins were identified by computer from multiple tilted images of the viruses and subtomogram averaging applied to them. The final electron-density map was determined to a resolution of 0.35 nm, good enough for the molecular model (shown here) to be accurately fitted within its envelope, although the details of the membrane-spanning portion of the trimeric spike protein are not revealed. The proteins are heavily *N*-glycosylated, and these surface glycans are shown in *green*, while the three spike proteins are shown in *dark green*, *light blue*, and *light brown*. (PDB code: 6ZWV.)

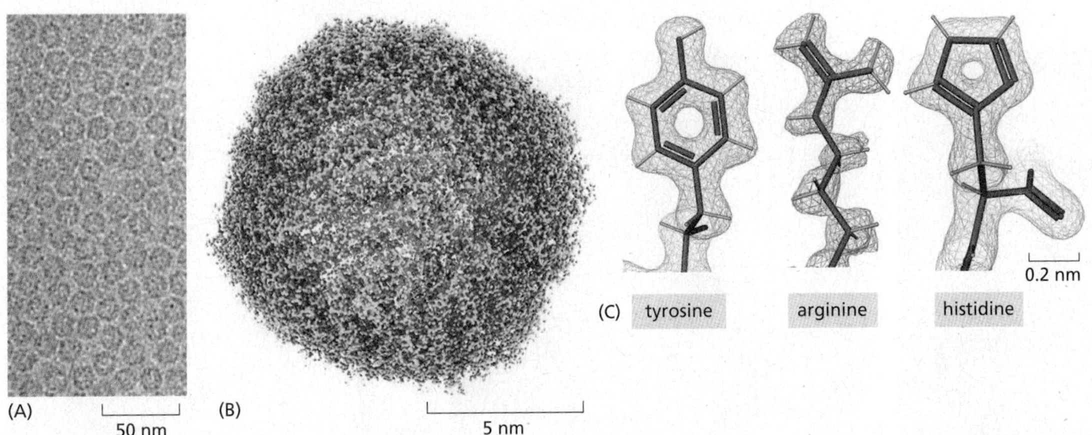

Figure 9–51 Atomic resolution by cryoEM. Apoferritin is a cytosolic protein, present in almost all living organisms, that reversibly stores iron in a nontoxic form. It is a large (474 kilodaltons) and particularly stable molecule. Its hollow globular cage has 24 symmetrical subunits, which means that a structure can be determined with relatively few particles. (A) Cryo-electron micrograph of cage-like apoferritin particles. (B) By use of every possible new technical advance in single-particle reconstruction, the complete cryoEM structure shown here is at the remarkable resolution of 0.12 nm (1.2 Å). (C) When the known amino acid sequence is modeled into the electron-density map, clear electron densities can be seen associated with hydrogen atoms in the three amino acid side chains. The molecular model is fitted into the final electron-density envelope that is shown as a *gray cage*. (A, from T. Nakane et al., *Nature* 587:152–156, 2020, doi 10.1038/s41586-020-2829-0; B, EMD-11668; C, adapted from K.M. Yip et al., *Nature* 587:157–161, 2020. With permission from Nature.)

example from x-ray crystallography (**Figure 9–52**). Molecular models can then be mathematically "fitted" or docked into the envelope of the structure determined at lower resolution using the electron microscope. X-ray and cryoEM approaches often combine profitably together to determine molecular structures.

Light Microscopy and Electron Microscopy Are Mutually Beneficial

The interior of the cell is a confusing place, with molecules crowded together in the cytosol and intricate and complex membrane-bounded compartments. To discover which molecules are located exactly where and in which tiny vesicles or subcompartments of the cell is not straightforward, even with the genetically encoded labels that can target almost any protein. We have seen that superresolution light microscopy can be used to very accurately locate specific molecules within a cell. A major disadvantage, however, of all fluorescence imaging techniques is that it is only the tagged molecules that are imaged—their cellular context remains invisible. When fluorescence imaging is combined, however, with looking at the same specimen in the electron microscope, this correlative light microscopy and electron microscopy technique, or *CLEM*, can allow specific target molecules to be examined in their full cellular context. Although this can be achieved using fixed and sectioned material, most such approaches now use rapidly frozen material to co-localize target molecules both in the light and in the

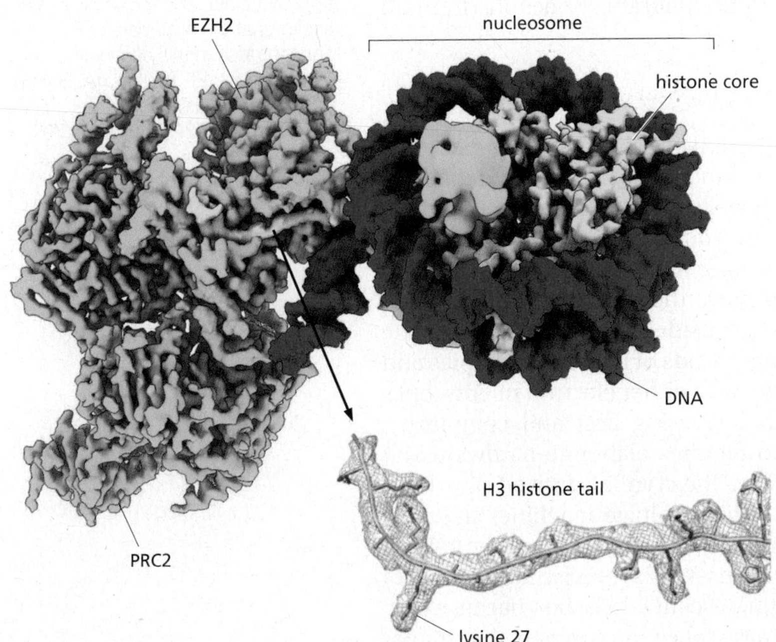

Figure 9–52 PRC2, a large macromolecular machine. Polycomb repressive complex 2 (PRC2) is a large protein complex involved in establishing heterochromatin and the epigenetic regulation of gene expression (see Figure 4–40). PRC2 interacts with a nucleosome through the binding of the nucleosomal DNA by one of its subunits, EZH2, which also engages the extended tail of histone H3 to direct its lysine 27 (K27) to the active site for methylation. The density map of PRC2 and two essential cofactors bound to a single nucleosome was produced by single-particle cryo-electron microscopy reconstruction at a resolution of 0.35 nm. The long arm of histone H3 is shown in more detail with the protein backbone modeled into the density map. (Courtesy of Vignesh Kasinath and Eva Nogales and based on EMDB-21707. From V. Kasinath et al., *Science* 371:eabc3393, 2021. With permission from AAAS.)

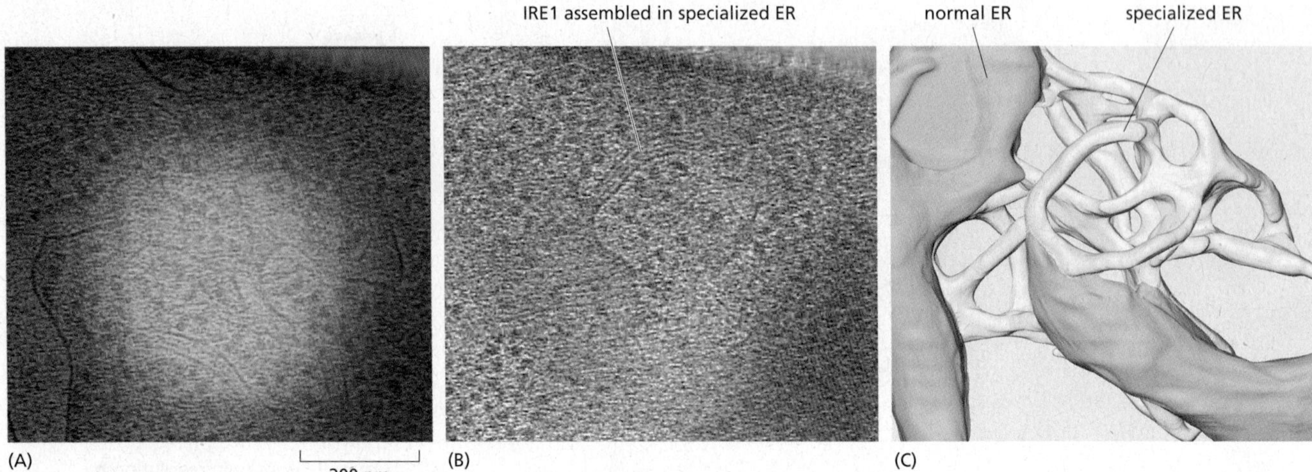

Figure 9–53 **Correlated light and electron microscopy (CLEM).** The correct folding of proteins in the endoplasmic reticulum (ER) is sensed by a major transmembrane protein called IRE1 (see Figures 12–36 and 12–37). If IRE1 is activated, it forms oligomers that are visible in fluorescence microscopy as bright foci. Here, stressed cells, expressing fluorescent IRE1 and growing on an EM grid, are rapidly frozen and subsequently imaged by EM tomography. The resulting tomograms can be directly correlated with the light micrographs. (A) A fluorescent spot of labeled IRE1 is shown here precisely superimposed on a slice through its corresponding EM tomogram that contains a network of ER. (B) Another slice through the tomogram at a different level shows IRE1 is localized as aggregates in a complex network of specialized, narrow ER tubules. (C) The outlines of the ER membranes in each slice of the tomogram are manually defined (in a process called segmentation), and the drawing here shows that the oligomers of IRE1 are concentrated in this convoluted network of specialized ER tubules. (A, B, and C, adapted from S.D. Carter et al., 2021, doi 10.1101/2021.02.24.432779.)

electron microscope, and of these two general approaches are common. The first is to freeze the cell or tissue, locate the positions of the target molecule with fluorescence light microscopy, and then, after transferring the frozen specimen to an electron microscope, tilting it, using EM tomography to find the exact point in the tomogram that corresponds to the fluorescent signal (**Figure 9–53**).

A second approach, and a demanding one too, is again to rapidly freeze the cell and locate fluorescent molecules at high resolution by single-molecule localization microscopy. The frozen cell is then transferred to a modified SEM that incorporates a separate focused ion beam, usually of gallium ions, that can be scanned across the frozen block face like a miniature milling machine, removing about 10 nm of the sample at a time. The SEM records a two-dimensional image of the scattered electrons from the surface of the block face at each step, and a three-dimensional image of the cell is gradually built up that can be correlated with the original localization data, all with a final resolution of about 5 nm (**Figure 9–54**). The technique is called *focused ion beam-scanning electron microscopy*, or FIB-SEM for short. The same technique, but without the fluorescent labels, can be used on much larger specimens that have been conventionally fixed, stained with heavy-metal salts, and embedded in plastic. Although the structural preservation may not be so good as with frozen specimens, the approach, although very time consuming, is proving useful in analyzing complex cellular interactions; for example, in mapping the neural connections in brain tissue (see Movie 9.1).

Using Microscopy to Study Cells Always Involves Trade-Offs

The history of cell biology has been tightly interlinked with that of microscopy. What we now know about the structure and function of cells has depended crucially on being able to image cells, organelles, and the molecules they contain—seeing is indeed believing. But for the young biologist today, there is, as we have seen, a bewildering variety of imaging technologies from which to choose, and knowing which is best suited to solve the problem at hand is not easy. All imaging approaches have trade-offs to consider. At an obvious level, the dynamics of cells are only accessible with certain kinds of light microscopies and with living cells. If higher resolution is required, with either electron microscopes or light microscopes, then that comes with increasing cost and complexity. Single-molecule localization microscopy also requires elaborate hardware and also takes many minutes to acquire each image. The cryoEM-derived structures of large protein complexes require the use of high-voltage machines that cost many millions of dollars. Such resources are usually confined to large centralized microscopy facilities that can be shared by many users. The precise localization of molecules within the cell requires the use of fluorescent labels, but, because only the labels themselves can be detected in a fluorescence microscope, the cellular

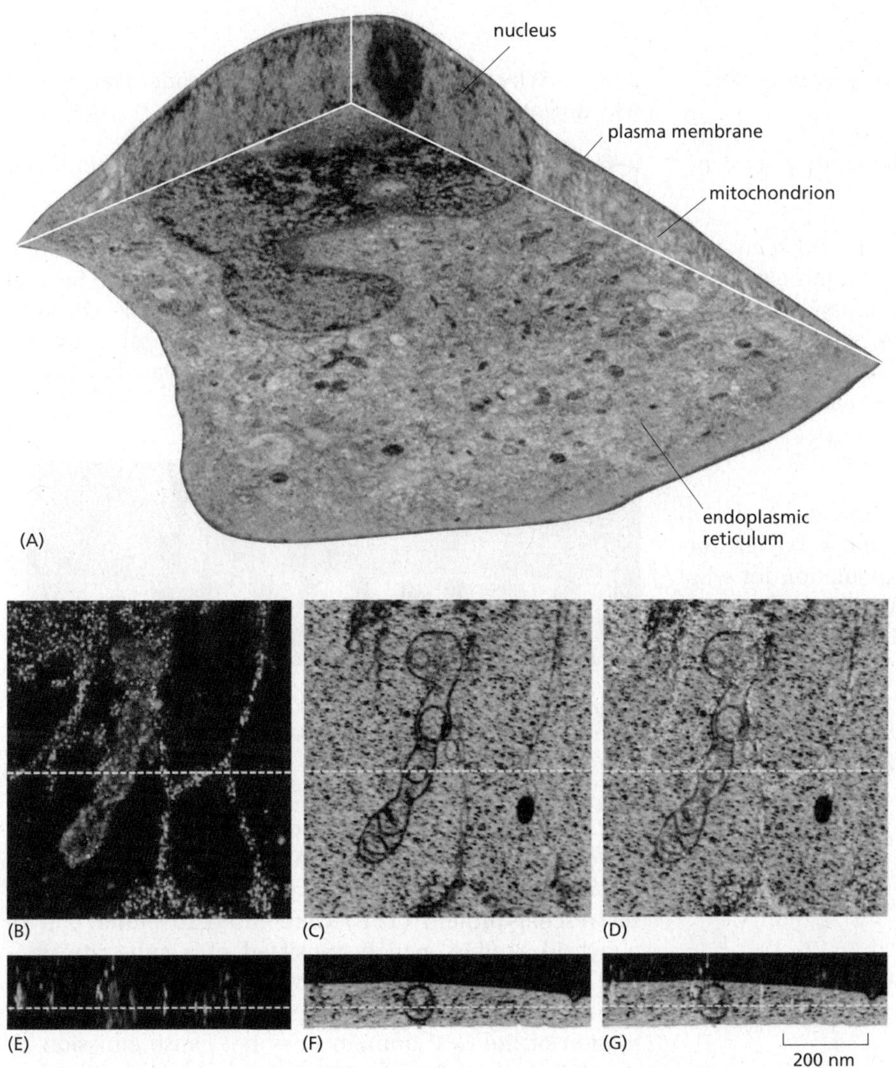

Figure 9–54 **Focused ion beam–scanning electron microscopy (FIB–SEM).** Superresolution light microscopy is combined here with three-dimensional electron microscopy of rapidly frozen cells to enable the high-resolution localization of target molecules throughout the entire volume of a cell. Sequential slices through the frozen cell are obtained by steadily milling the surface of the frozen block face with a focused ion beam, while images of the surface are collected at each step in an SEM. This particular cell has been labeled with fluorescent markers for the lumen of the endoplasmic reticulum *(green)* and for the outer membrane of mitochondria *(magenta)*. (A) Three orthogonal slices through the cell show the combined electron microscope and fluorescence light microscope images. (B) A small region of the same cell imaged with a structured illumination microscope (SIM) is used to define mitochondrion and ER. (C) The corresponding block face image in the SEM. (D) The correlated electron microscope and light microscope images identify the position of the fluorescent labels in the electron micrograph. (E, F, and G) Because the three-dimensional SEM data set is of the entire cell, different views of the same area can be readily obtained. Here, the three corresponding vertical sections along the *yellow dotted lines* on the images above are shown. (From D.P. Hoffman et al., *Science* 367:265–277, 2020. With permission from AAAS.)

context is sacrificed. Imaging itself involves several trade-offs to be considered. An improvement in any one parameter—image contrast, resolution, signal-to-noise ratio, specimen damage by photons or electrons, the depth of specimen that can be imaged, or the speed of image recording—will inevitably require a sacrifice in one or more of the others, and understanding these trade-offs will help determine which approach is best for the cell biology problem being tackled.

Summary

Discovering the detailed structure of cells and their molecules requires the higher resolution attainable in a transmission electron microscope. Three-dimensional views of the surfaces of cells and tissues are obtained by scanning electron microscopy. Specific macromolecules can be localized by combining electron microscopy with fluorescence light microscopy. EM tomography enables three-dimensional information about cellular architecture to be obtained. The shapes of isolated molecules can be roughly determined by electron microscopy techniques involving negative staining or heavy-metal shadowing, but detailed molecular structures require cryoEM and single-particle reconstruction using computational manipulations of data obtained from multiple images and multiple viewing angles to produce detailed reconstructions of macromolecules and molecular complexes. The resolution obtained with these methods means that atomic structures of individual macromolecules can be "fitted" to the images derived by electron microscopy. CryoEM can often determine the structures of molecules that are inaccessible to x-ray crystallography.

PROBLEMS

Which statements are true? Explain why or why not.

9–1 A fluorescent molecule, having absorbed a single photon of light at one wavelength, always emits it at a longer wavelength.

9–2 Transmission electron microscopy and scanning electron microscopy can both be used to examine a structure in the interior of a thin section: transmission electron microscopy provides a projection view, while scanning electron microscopy captures electrons scattered from the structure and gives a more three-dimensional view.

Discuss the following problems.

9–3 The diagrams in **Figure Q9–1** show the paths of light rays passing through a specimen into a dry lens or into an oil-immersion lens. Offer an explanation for why oil-immersion lenses should give better resolution. Air, glass, and oil have refractive indices of 1.00, 1.51, and 1.51, respectively.

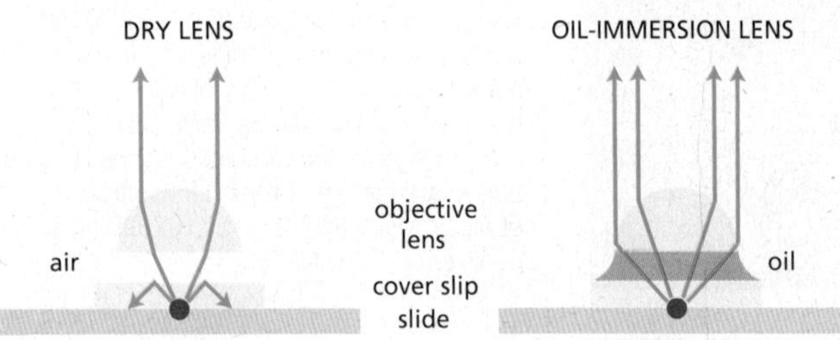

Figure Q9–1 Paths of light rays through dry and oil-immersion lenses (Problem 9–3). The *red circle* at the origin of the light rays is the specimen.

9–4 **Figure Q9–2** shows a diagram of the human eye. The refractive indices of the components in the light path are air, 1.00; cornea, 1.38; aqueous humor, 1.33; crystalline lens, 1.41; and vitreous humor, 1.38. Where does the main refraction—the main focusing—occur? What role do you suppose the lens plays?

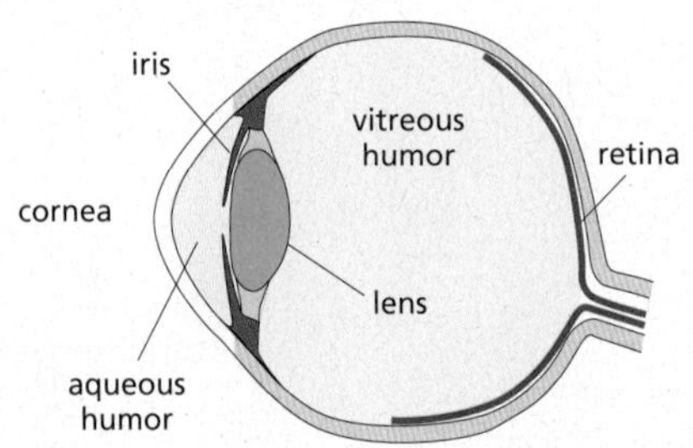

Figure Q9–2 Diagram of the human eye (Problem 9–4).

9–5 Why do humans see so poorly under water? And why do goggles help?

9–6 Explain the difference between resolution and magnification.

9–7 **Figure Q9–3** shows a series of modified fluorescent proteins that emit light in a range of colors. Several of these fluorescent proteins contain the same chromophore, yet they fluoresce at different wavelengths. How do you suppose the exact same chromophore can fluoresce at several different wavelengths?

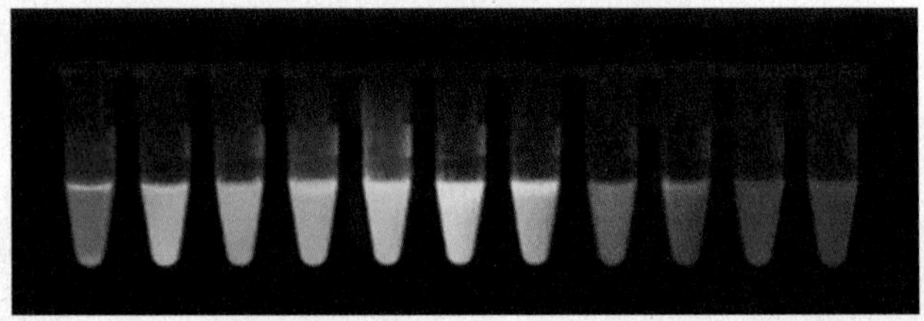

Figure Q9–3 A rainbow of colors produced by modified fluorescent proteins (Problem 9–7). (Courtesy of Nathan Shaner, Paul Steinbach, and Roger Tsien.)

9–8 A fluorescent biosensor was designed to report the cellular location of active Abl protein tyrosine kinase. A blue (cyan) fluorescent protein (CFP) and a yellow fluorescent protein (YFP) were fused to either end of a hybrid protein, which consisted of a substrate peptide recognized by the Abl protein tyrosine kinase and a phosphotyrosine-binding domain (**Figure Q9–4A**). Stimulation of the CFP domain does not cause emission by the YFP domain when the domains are separated. When the CFP and YFP domains are brought close together, however, fluorescence resonance energy transfer (FRET)

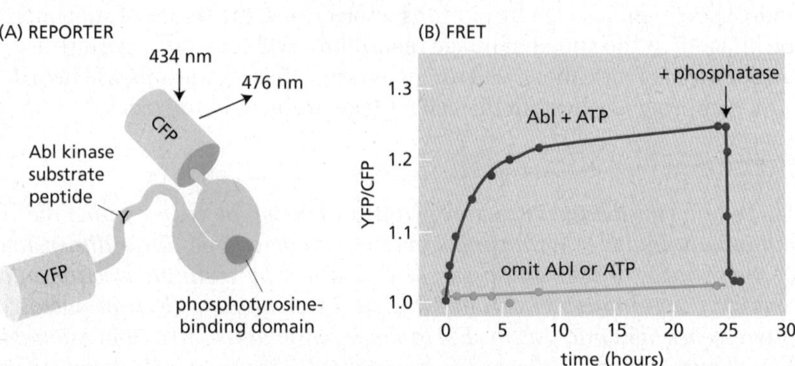

Figure Q9–4 Fluorescent biosensor designed to detect tyrosine phosphorylation (Problem 9–8). (A) Domain structure of the biosensor. Four domains are indicated: CFP, YFP, tyrosine kinase substrate peptide, and a phosphotyrosine-binding domain. (B) FRET assay. YFP/CFP is normalized to 1.0 at time zero. The biosensor was incubated in the presence (or absence) of Abl and ATP for the indicated times. *Arrow* indicates time of addition of a tyrosine phosphatase. (From A.Y. Ting et al., *Proc. Natl. Acad. Sci. USA* 98:15003–15008, 2001. With permission from National Academy of Sciences.)

allows excitation of CFP to stimulate emission by YFP. FRET shows up experimentally as an increase in the ratio of emission at 526 nm (from YFP) versus 476 nm (from CFP) when CFP is excited by 434-nm light.

Incubation of the biosensor protein with Abl protein tyrosine kinase in the presence of ATP gave an increase in the ratio of YFP/CFP emission (Figure Q9–4B). In the absence of ATP or the Abl protein, no FRET occurred. FRET was also eliminated by addition of a tyrosine phosphatase (Figure Q9–4B). Describe as best you can how this biosensor detects active Abl protein tyrosine kinase.

9–9 Under ideal conditions, with the simplest of specimens (a monolayer of carbon atoms, for example) and careful image processing, the practical resolving power of modern electron microscopes is about 0.05 nm, some 25-fold above the theoretical limit of 0.002 nm. This is because only the very center of the electron lens can be used, and the effective numerical aperture ($n \sin \theta$) is limited by θ (half the angular width of rays collected at the objective lens). Assuming that the wavelength (λ) of the electrons is 0.004 nm and that the refractive index (n) is 1.0, calculate the value for θ, where resolution (0.05 nm) = $0.61\ \lambda/n \sin \theta$. How does this value of θ compare with that for a conventional light microscope (60°)?

9–10 Aquaporin water channels in the plasma membrane play a major role in water metabolism and osmoregulation in many cells. To determine their structural organization in the membrane, you use immunogold electron microscopy. You prepare a membrane sample, incubate it with primary antibodies against aquaporin then with gold-tagged secondary antibodies that bind to the primary antibodies. You then examine it by electron microscopy (Figure Q9–5). Are the gold particles (black dots) consistently associated with any particular structure?

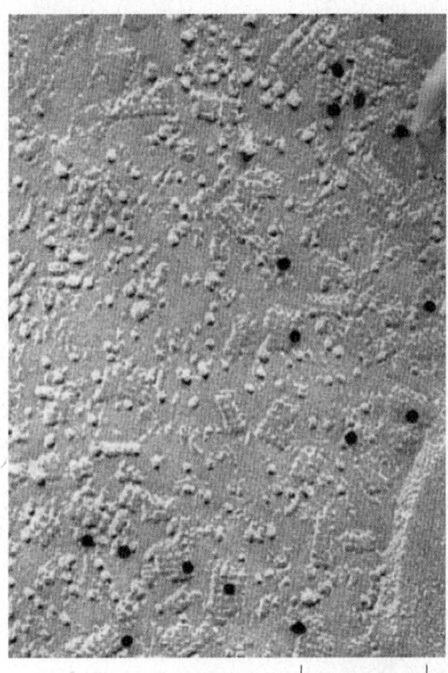

Figure Q9–5 An astrocyte membrane labeled with primary antibodies against aquaporin and then with secondary antibodies to which colloidal gold particles have been attached (Problem 9-10). (From J.E. Rash et al., *Proc. Natl. Acad. Sci. USA* 95:11981–11986, 1998. With permission from National Academy of Sciences.)

9–11 The technique of negative staining uses heavy metals such as uranium to provide contrast. If these heavy metals do not actually bind to defined biological structures (which they do not), how is it that they can help to make such structures visible?

REFERENCES

General

Vale R, Stuurman N & Thorn K (2006–2021) Microscopy Series (https://www.ibiology.org/online-biology-courses/microscopy-series/) A major online video series starts with the basics of optical microscopy and concludes with some of the latest techniques, such as superresolution microscopy.

Looking at Cells and Molecules in the Light Microscope

Agard DA, Hiraoka Y, Shaw P & Sedat JW (1989) Fluorescence microscopy in three dimensions. In Fluorescence Microscopy of Living Cells in Culture, part B (Taylor DL & Wang Y-L, eds.). *Methods in Cell Biology*, Vol. 30. San Diego: Academic Press.

Boulina M, Samarajeewa H, Baker JD . . . Chiba A (2013) Live imaging of multicolor-labeled cells in *Drosophila*. *Development* 140, 1605–1613.

Burnette DT, Sengupta P, Dai Y . . . Kachar B (2011) Bleaching/blinking assisted localization microscopy for superresolution imaging using standard fluorescent molecules. *Proc. Natl. Acad. Sci. USA* 108, 21081–21086.

Chalfie M, Tu Y, Euskirchen G . . . Prasher DC (1994) Green fluorescent protein as a marker for gene expression. *Science* 263, 802–805.

Chen F, Tillberg PW & Boyden ES (2015) Expansion microscopy. *Science* 347, 543–548.

Giepmans BN, Adams SR, Ellisman MH & Tsien RY (2006) The fluorescent toolbox for assessing protein location and function. *Science* 312, 217–224.

Greenfield EA (ed.) (2014) Antibodies: A Laboratory Manual, 2nd ed. Cold Spring Harbor, NY: Cold Spring Harbor Laboratory Press.

Greenwald EC, Mehta S & Zhang J (2018) Genetically encoded fluorescent biosensors illuminate the spatiotemporal regulation of signaling networks. *Chem. Rev*. 118, 11707–11794.

Hell S (2009) Microscopy and its focal switch. *Nat. Methods* 6, 24–32.

Huang B, Babcock H & Zhuang X (2010) Breaking the diffraction barrier: super-resolution imaging of cells. *Cell* 143, 1047–1058.

Jacquemet G, Carisey AF, Hamidi H . . . Leterrier C (2020) The cell biologist's guide to super-resolution microscopy. *J. Cell Sci.* 133, jcs240713.

Jonkman J, Brown CM, Wright GD . . . North AJ (2020) Tutorial: guidance for quantitative confocal microscopy. *Nat. Protoc*. 15, 1585–1611.

Klar TA, Jakobs S, Dyba M . . . Hell SW (2000) Fluorescence microscopy with diffraction resolution barrier broken by stimulated emission. *Proc. Natl. Acad. Sci. USA* 97, 8206–8210.

Lippincott-Schwartz J & Patterson GH (2003) Development and use of fluorescent protein markers in living cells. *Science* 300, 87–91.

Lippincott-Schwartz J, Altan-Bonnet N & Patterson G (2003) Photobleaching and photoactivation: following protein dynamics in living cells. *Nat. Cell Biol.* 5(suppl.), S7–S14.

Liu T-L, Upadhyayula S, Milkie DE . . . Betzig E (2018) Observing the cell in its native state: imaging subcellular dynamics in multicellular organisms. *Science* 360, eaaq1392.

Lu C-H, Tang W-C, Liu Y-T . . . Chen B-C (2019) Lightsheet localization microscopy enables fast, large-scale, and three-dimensional super-resolution imaging. *Commun. Biol.* 2, 177.

Minsky M (1988) Memoir on inventing the confocal scanning microscope. *Scanning* 10, 128–138.

Parton RM & Read ND (1999) Calcium and pH imaging in living cells. In Light Microscopy in Biology: A Practical Approach, 2nd ed. (Lacey AJ, ed.). Oxford: Oxford University Press.

Patterson G, Davidson M, Manley S & Lippincott-Schwartz J (2010) Superresolution imaging using single-molecule localization. *Annu. Rev. Phys. Chem.* 61, 345–367.

Pawley BP (ed.) (2006) Handbook of Biological Confocal Microscopy, 3rd ed. New York: Springer Science.

Rust MJ, Bates M & Zhuang X (2006) Sub-diffraction-limit imaging by stochastic optical reconstruction microscopy (STORM). *Nat. Methods* 3, 793–795.

Sako Y & Yanagida T (2003) Single-molecule visualization in cell biology. *Nat. Rev. Mol. Cell Biol.* 4(suppl.), SS1–SS5.

Schermelleh L, Heintzmann R & Leonhardt H (2010) A guide to super-resolution fluorescence microscopy. *J. Cell Biol.* 190, 165–175.

Shaner NC, Steinbach PA & Tsien RY (2005) A guide to choosing fluorescent proteins. *Nat. Methods* 2, 905–909.

Sigal YM, Zhou R & Zhuang X (2018) Visualizing and discovering cellular structures with super-resolution microscopy. *Science* 361, 880–887.

Sluder G & Wolf DE (2007) Digital Microscopy, 3rd ed. *Methods in Cell Biology*, Vol. 81. San Diego, CA: Academic Press.

Tsien RY (2008) Constructing and exploiting the fluorescent protein paintbox (Nobel Lecture). *Angew. Chem. Int. Ed. Engl.* 48, 5612–5626.

Wayne R (2014) Light and Video Microscopy. San Diego, CA: Academic Press.

White JG, Amos WB & Fordham M (1987) An evaluation of confocal versus conventional imaging of biological structures by fluorescence light microscopy. *J. Cell Biol.* 105, 41–48.

Zernike F (1955) How I discovered phase contrast. *Science* 121, 345–349.

Looking at Cells and Molecules in the Electron Microscope

Allen TD & Goldberg MW (1993) High-resolution SEM in cell biology. *Trends Cell Biol.* 3, 205–208.

Baumeister W (2002) Electron tomography: towards visualizing the molecular organization of the cytoplasm. *Curr. Opin. Struct. Biol.* 12, 679–684.

Beck M & Baumeister W (2016) Cryo-electron tomography: can it reveal the molecular sociology of cells in atomic detail? *Trends Cell Biol.* 26, 825–837.

Böttcher B, Wynne SA & Crowther RA (1997) Determination of the fold of the core protein of hepatitis B virus by electron cryomicroscopy. *Nature* 386, 88–91.

Cheng Y, Grigorieff N, Penczek PA & Walz T (2015) A primer to single-particle cryo-electron microscopy. *Cell* 161, 438–449.

Dubochet J, Adrian M, Chang J-J . . . Schultz P (1988) Cryo-electron microscopy of vitrified specimens. *Q. Rev. Biophys.* 21, 129–228.

Frank J (2003) Electron microscopy of functional ribosome complexes. *Biopolymers* 68, 223–233.

Frank J (2018) Single-particle reconstruction of biological molecules—story in a sample (Nobel Lecture). *Angew. Chem. Int. Ed. Engl.* 57, 10826–10841.

Hayat MA (2000) Principles and Techniques of Electron Microscopy, 4th ed. Cambridge: Cambridge University Press.

Henderson R (2015) Overview and future of single particle electron cryomicroscopy. *Arch. Biochem. Biophys.* 581, 19–24.

Henderson R (2018) From electron crystallography to single particle cryo-EM (Nobel Lecture). *Angew. Chem. Int. Ed. Engl.* 57, 10804–10825.

Heuser J (1981) Quick-freeze, deep-etch preparation of samples for 3-D electron microscopy. *Trends Biochem. Sci.* 6, 64–68.

Hoffman DP, Shtengel G, Shan Xu C . . . Hess HF (2020) Correlative three-dimensional super-resolution and block face electron microscopy of whole vitreously frozen cells. *Science* 367, eaaz5357.

Kukulski W, Schorb M, Kaksonen M & Briggs JAG (2012) Plasma membrane reshaping during endocytosis is revealed by time-resolved electron tomography. *Cell* 150, 508–520.

Lin DH, Stuwe T, Stillbach S, . . . Hoelz A (2016) Architecture of the symmetric core of the nuclear pore. *Science* 352, aaf1015.

McDonald KL & Auer M (2006) High-pressure freezing, cellular tomography, and structural cell biology. *Biotechniques* 41, 137–139.

McIntosh JR (2007) Cellular Electron Microscopy, 3rd ed. *Methods in Cell Biology*, Vol. 79. San Diego, CA: Academic Press.

McIntosh R, Nicastro D & Mastronarde D (2005) New views of cells in 3D: an introduction to electron tomography. *Trends Cell Biol.* 15, 43–51.

Nakane T, Kotecha A, Sente A . . . Scheres SHW (2020) Single-particle cryo-EM at atomic resolution. *Nature* 587, 152–156.

Orlova EWV & Saibil HR (2011) Structural analysis of macromolecular assemblies by electron microscopy. *Chem. Rev.* 111, 7710–7748.

Pease DC & Porter KR (1981) Electron microscopy and ultramicrotomy. *J. Cell Biol.* 91, 287s–292s.

Unwin PNT & Henderson R (1975) Molecular structure determination by electron microscopy of unstained crystalline specimens. *J. Mol. Biol.* 94, 425–440.

Yip KM, Fischer N, Paknia E . . . Stark H (2020) Atomic-resolution protein structure determination by cryo-EM. *Nature* 587, 157–164.

Zhou ZH (2008) Towards atomic resolution structural determination by single particle cryo-electron microscopy. *Curr. Opin. Struct. Biol.* 18, 218–228.

INTERNAL ORGANIZATION OF THE CELL

CHAPTER 10

Membrane Structure

IN THIS CHAPTER

Cell membranes are crucial to the life of the cell. The **plasma membrane** encloses the cell, defines its boundaries, and maintains the essential differences between the cytosol and the extracellular environment. Without plasma membranes, cells could not have evolved as individual self-replicating units. Inside eukaryotic cells, the membranes of the nucleus, endoplasmic reticulum, Golgi apparatus, mitochondria, and other membrane-enclosed organelles maintain the characteristic differences between the contents of each organelle and the cytosol. Ion gradients across membranes, established by the activities of specialized membrane proteins, can be used to synthesize ATP, to drive the transport of selected solutes across the membrane, or, as in nerve and muscle cells, to produce and transmit electrical signals. In all cells, the plasma membrane also contains proteins that act as sensors of external signals, allowing the cell to change its behavior in response to environmental cues, including signals from other cells; these protein sensors, or *receptors*, transfer information—rather than molecules—across the membrane.

Despite their differing functions, all biological membranes have a common general structure: each is a very thin film of lipid and protein molecules, held together mainly by noncovalent interactions (Figure 10–1). Cell membranes are

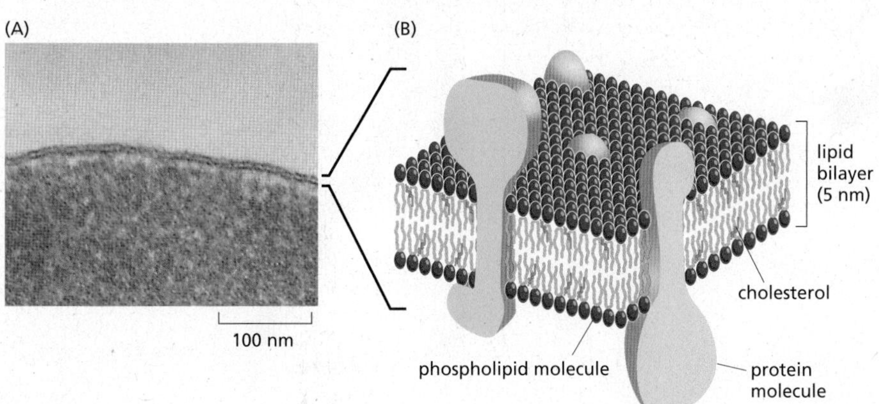

Figure 10–1 **Two views of a cell membrane.** (A) An electron micrograph of a segment of the plasma membrane of a human red blood cell seen in cross section, showing its bilayer structure. (B) A three-dimensional schematic view of a cell membrane and the general disposition of its lipid and protein constituents. (A, courtesy of Daniel S. Friend, reused by permission of E.L. Bearer.)

dynamic, fluid structures, and most of their molecules move about in the plane of the membrane. The lipid molecules are arranged as a continuous double layer about 5 nm thick. This *lipid bilayer* provides the basic fluid structure of the membrane and serves as an essentially impermeable barrier to the passage of most water-soluble molecules. Most *membrane proteins* span the lipid bilayer and mediate nearly all of the other functions of the membrane, including the transport of specific molecules across it, and the catalysis of membrane-associated reactions such as ATP synthesis. In the plasma membrane, some transmembrane proteins serve as structural links that connect the cytoskeleton through the lipid bilayer to either the extracellular matrix or an adjacent cell, while others serve as receptors to detect and transduce chemical signals in the cell's environment. It takes many kinds of membrane proteins to enable a cell to function and interact with its environment, and it is estimated that about 30% of the proteins encoded in an animal's genome are membrane proteins.

In this chapter, we consider the structure and organization of the two main constituents of biological membranes—the lipids and the proteins. Although we focus mainly on the plasma membrane, most concepts discussed apply to the various internal membranes of eukaryotic cells as well. The functions of cell membranes are considered in later chapters: their role in energy conversion and ATP synthesis, for example, is discussed in Chapter 14; their role in the transmembrane transport of small molecules in Chapter 11; and their roles in cell signaling and cell adhesion in Chapters 15 and 19, respectively. In Chapters 12 and 13, we discuss the internal membranes of the cell and the protein traffic through and between them.

THE LIPID BILAYER

The **lipid bilayer** provides the basic structure for all cell membranes. It is easily seen by electron microscopy, and its bilayer structure is attributable exclusively to the special properties of the lipid molecules, which assemble spontaneously into bilayers even under simple artificial conditions. In this section, we discuss

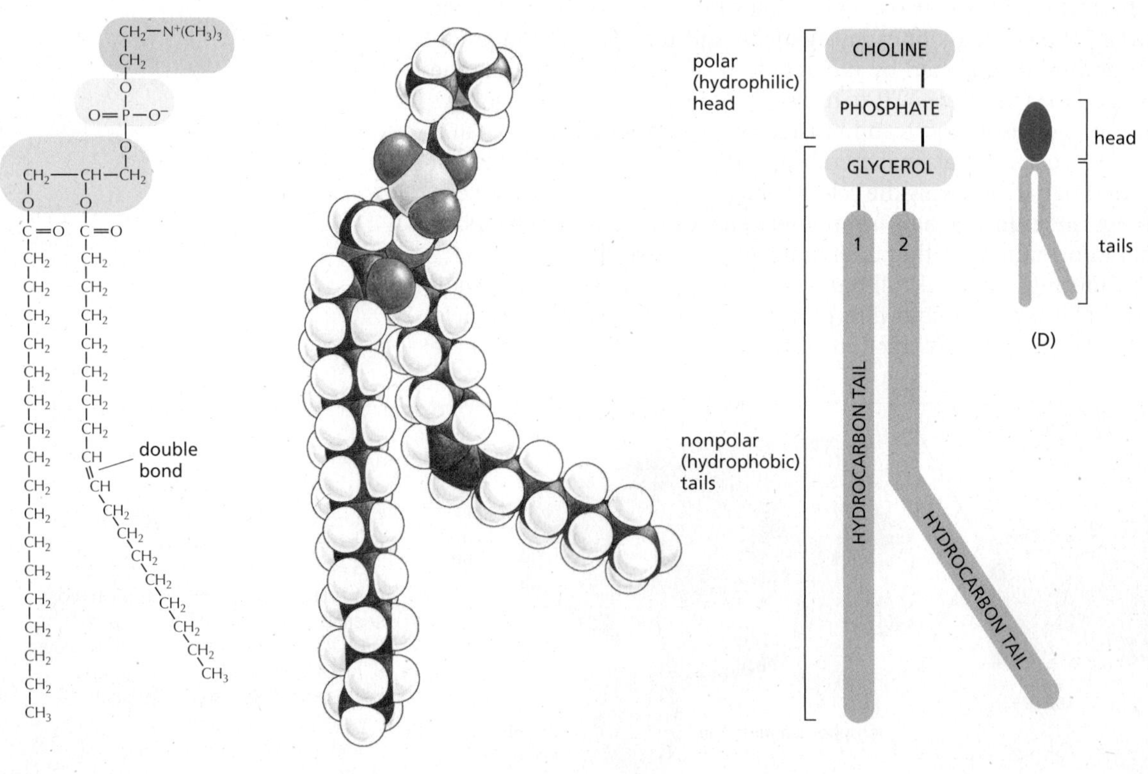

Figure 10–2 **The parts of a typical phospholipid molecule.** This example is phosphatidylcholine, represented (A) by a formula, (B) as a space-filling model (Movie 10.1), (C) schematically, and (D) as a symbol.

the different types of lipid molecules found in cell membranes and the general properties of lipid bilayers.

Glycerophospholipids, Sphingolipids, and Sterols Are the Major Lipids in Cell Membranes

Lipid molecules constitute about 50% of the mass of most animal cell membranes, nearly all of the remainder being protein. There are approximately 5×10^6 lipid molecules in a 1 μm × 1 μm area of lipid bilayer, or about 7×10^8 lipid molecules in the plasma membrane of a red blood cell. All of the lipid molecules in cell membranes are **amphiphilic**; that is, they have a **hydrophilic** ("water-loving") or *polar* end and a **hydrophobic** ("water-fearing") or *nonpolar* end.

The most abundant membrane lipids are the **phospholipids**. These have a polar head group, which includes a phosphate group, and two hydrophobic *hydrocarbon tails*. In animal, plant, and bacterial cells, the tails are usually fatty acids, and they can differ in length (they normally contain between 14 and 24 carbon atoms). One tail typically has one or more *cis*-double bonds (that is, it is *unsaturated*), while the other tail does not (that is, it is *saturated*). As shown in Figure 10–2, each *cis*-double bond creates a kink in the tail. Differences in the length and saturation of the fatty acid tails influence how phospholipid molecules pack against one another, thereby affecting the fluidity of the membrane, as we discuss later.

The main phospholipids in most animal cell membranes are the **glycerophospholipids**, which have a three-carbon *glycerol* backbone (see Figure 10–2). Two long-chain fatty acids are linked through ester bonds to adjacent carbon atoms of the glycerol, and the third carbon atom of the glycerol is attached to a phosphate group, which in turn is linked to one of several types of head group. By combining several different fatty acids and head groups, cells make many different glycerophospholipids. *Phosphatidylethanolamine, phosphatidylserine,* and *phosphatidylcholine* are the most abundant ones in mammalian cell membranes (**Figure 10–3A, B, and C**).

Another important class of phospholipids is the *sphingolipids*, which are built from *sphingosine* rather than glycerol (**Figure 10–3D and E**). Sphingosine is a long fatty acid tail with an amino group (NH_2) and two hydroxyl groups (OH) at one end. In sphingomyelin, the most common sphingolipid, a fatty acid tail

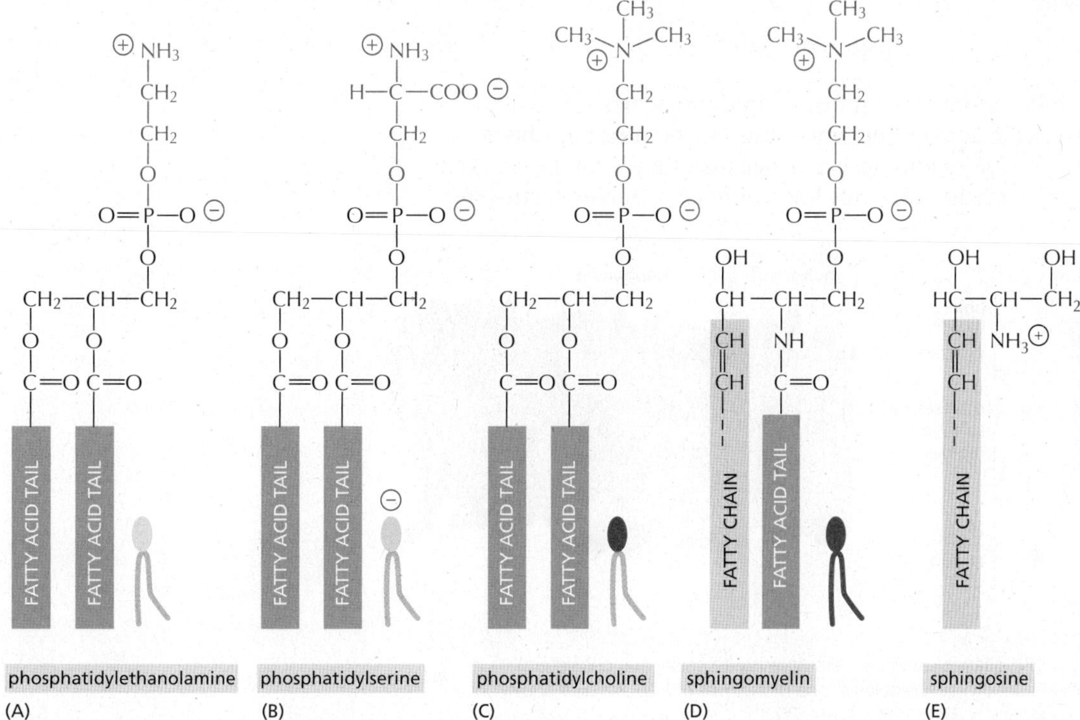

Figure 10–3 Four major phospholipids in mammalian plasma membranes. Different head groups are represented by different colors in the symbols. The lipid molecules shown in (A–C) are glycerophospholipids, which are derived from glycerol. The molecule in (D) is sphingomyelin, which is derived from (E) sphingosine and is therefore a sphingolipid. Note that only phosphatidylserine carries a net negative charge, the importance of which we discuss later; the other three are electrically neutral at physiological pH, carrying one positive and one negative charge.

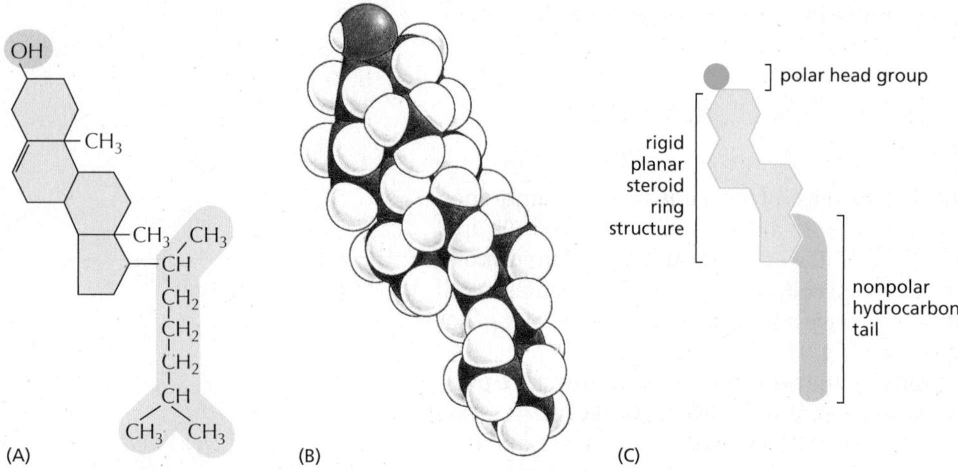

Figure 10–4 **The structure of cholesterol.** Cholesterol, a sterol, is represented (A) by a formula, (B) as a space-filling model, and (C) by a schematic drawing.

is attached to the amino group, and a phosphocholine group is attached to the terminal hydroxyl group. Together, the phospholipids phosphatidylcholine, phosphatidylethanolamine, phosphatidylserine, and sphingomyelin constitute more than half the mass of lipid in most mammalian cell membranes (see Table 10–1, p. 610).

In addition to phospholipids, the lipid bilayers in many cell membranes contain *glycolipids* and *sterols*. Glycolipids resemble sphingolipids, but, instead of a phosphate-linked head group, they have sugars attached. We discuss glycolipids later. Sterols are rigid ring structures related to steroids but containing a single polar hydroxyl group and a short nonpolar hydrocarbon chain (**Figure 10–4**). Different types of sterols, distinguished primarily by the side chain attached to the ringed scaffold, are found in fungi, plants, and animal cells. Eukaryotic plasma membranes contain especially large amounts of sterols—up to one molecule for every phospholipid molecule. **Cholesterol** is the major sterol found in animal cells. The cholesterol molecules orient themselves in the bilayer with their hydroxyl group close to the polar head groups of adjacent phospholipid molecules (**Figure 10–5**).

Phospholipids Spontaneously Form Bilayers

The shape and amphiphilic nature of the phospholipid molecules cause them to form bilayers spontaneously in aqueous environments. As discussed in Chapter 2, hydrophilic molecules dissolve readily in water because they contain charged groups or uncharged polar groups that can form either favorable electrostatic

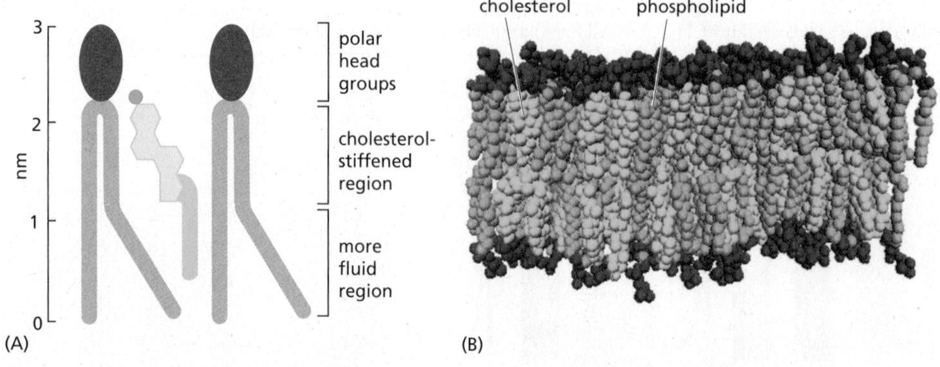

Figure 10–5 **Cholesterol in a lipid bilayer.** (A) Schematic drawing (to scale) of a cholesterol molecule interacting with two phospholipid molecules in one monolayer of a lipid bilayer shown in (B) (B from H.L. Scott, *Curr. Opin. Struct. Biol, 12:495-502, 2022.*).

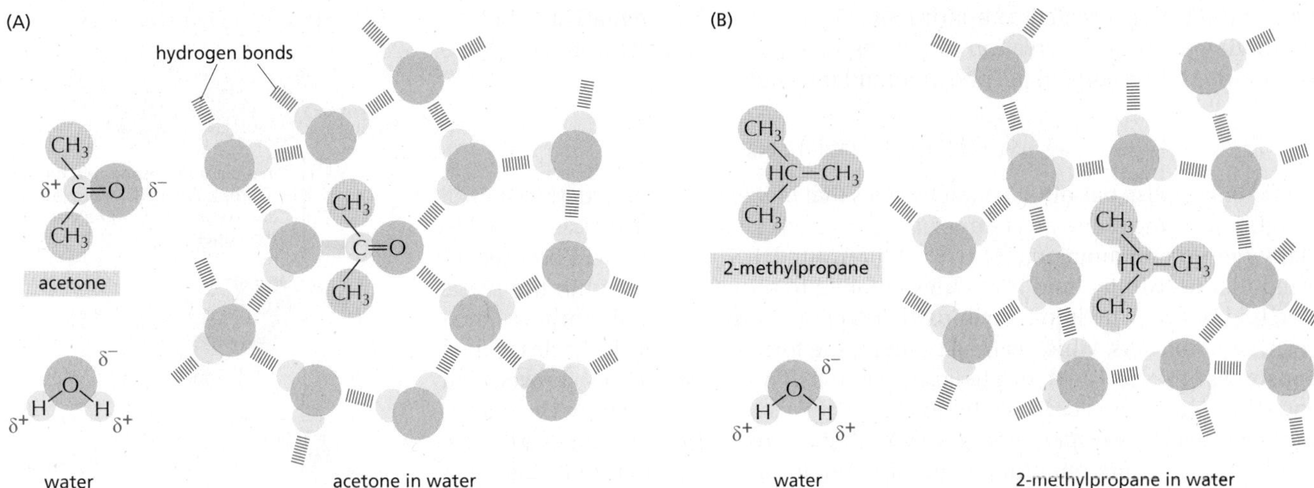

Figure 10–6 **How hydrophilic and hydrophobic molecules interact differently with water.** (A) Because acetone is polar, it can form hydrogen bonds *(red)* and favorable electrostatic interactions *(yellow)* with water molecules, which are also polar. Thus, acetone readily dissolves in water. (B) By contrast, 2-methylpropane is entirely hydrophobic. Because it cannot form favorable interactions with water, it forces adjacent water molecules to reorganize into icelike cage structures, which increases the free energy. This compound is therefore virtually insoluble in water. The symbol δ^- indicates a partial negative charge, and δ^+ indicates a partial positive charge. Polar atoms are shown in color, and nonpolar groups are shown in *gray*.

interactions or hydrogen bonds with water molecules (Figure 10–6A). Hydrophobic molecules, by contrast, are insoluble in water because all, or almost all, of their atoms are uncharged and nonpolar and therefore cannot form energetically favorable interactions with water molecules. If dispersed in water, this forces the adjacent water molecules to reorganize into icelike cages that surround the hydrophobic molecule (Figure 10–6B). Because these cage structures are more ordered than the surrounding water, their formation increases the free energy. This entropic free-energy cost is minimized, however, if the hydrophobic molecules (or the hydrophobic portions of amphiphilic molecules) cluster together so that the smallest number of water molecules is affected.

When phospholipid molecules are exposed to an aqueous environment, they behave as you would expect from the above discussion. They spontaneously pack together to minimize exposure of their hydrophobic tails to water and maximize exposure of their hydrophilic heads to water. Depending on their shape, the optimal packing arrangement is achieved in either of two ways: they can form spherical *micelles*, with the tails inward, or they can form double-layered sheets, or *bilayers*, with the hydrophobic tails sandwiched between the hydrophilic head groups (Figure 10–7).

The same forces that drive phospholipids to form bilayers also provide a self-sealing property. A small tear in the bilayer creates a free edge exposed to water; because this is energetically unfavorable, the lipids will rearrange spontaneously to eliminate the free edge. The prohibition of free edges has a profound consequence: the only way for a bilayer to avoid having edges is by closing in on

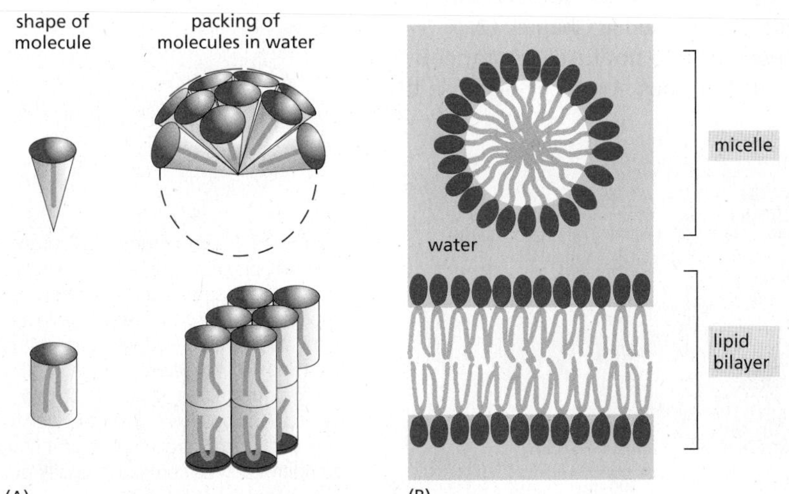

Figure 10–7 **Packing arrangements of amphiphilic molecules in an aqueous environment.** (A) These molecules spontaneously form micelles or bilayers in water, depending on their shape. Cone-shaped amphiphilic molecules *(above)* form micelles, whereas cylinder-shaped amphiphilic molecules such as phospholipids *(below)* form bilayers. (B) A micelle and a lipid bilayer seen in cross section. Note that micelles of amphiphilic molecules are thought to be much more irregular than drawn here (see Figure 10–26C).

itself and forming a sealed compartment (Figure 10–8). This remarkable behavior, fundamental to the creation of a living cell, follows directly from the shape and amphiphilic nature of the phospholipid molecule.

ENERGETICALLY UNFAVORABLE

in a planar phospholipid bilayer, hydrophobic tails (white layer) are exposed to water along the edges

formation of a sealed compartment shields hydrophobic tails from water

ENERGETICALLY FAVORABLE

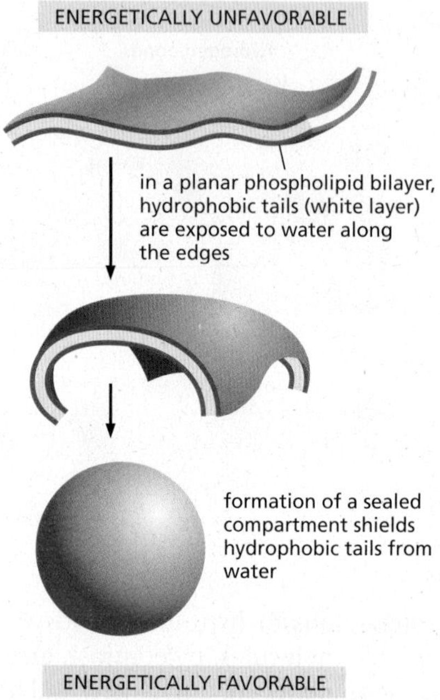

Figure 10–8 **The spontaneous closure of a phospholipid bilayer to form a sealed compartment.** The closed structure is stable because it avoids the exposure of the hydrophobic hydrocarbon tails to water, which would be energetically unfavorable.

The Lipid Bilayer Is a Two-dimensional Fluid

A lipid bilayer also has other characteristics that make it an ideal structure for cell membranes. One of the most important of these is its fluidity, which is crucial to many membrane functions (Movie 10.2). Around 1970, researchers first recognized that individual lipid molecules are able to diffuse freely within the plane of a lipid bilayer. The initial demonstration came from studies of synthetic (artificial) lipid bilayers, which can be made in the form of spherical vesicles, called **liposomes** (Figure 10–9), or planar lipid films. Biophysical studies showed that phospholipid molecules in synthetic bilayers very rarely migrate from the monolayer (also called a *leaflet*) on one side to that on the other. This process, known as "flip-flop," occurs on a time scale of hours for any individual molecule. It is slow because during flip-flop, the hydrophilic head groups must transiently enter and pass through the hydrophobic core of the bilayer, which is energetically disfavored. Cholesterol is an exception to this rule and can flip-flop rapidly, having only a single hydroxyl group to accommodate transiently in the hydrophobic core. By contrast to flip-flop, all lipid molecules rapidly exchange places with their neighbors *within* a monolayer (~10^7 times per second). This gives rise to a rapid lateral diffusion, with a diffusion coefficient (D) of about 10^{-8} cm^2/sec, which means that an average lipid molecule diffuses the length of a large bacterial cell (~2 μm) in about 1 second. These studies have also shown that individual lipid molecules rotate very rapidly about their long axis and have flexible hydrocarbon chains. Computer simulations show that lipid molecules in synthetic bilayers are very disordered, presenting an irregular, ragged surface of variously spaced and oriented head groups to the water phase on either side of the bilayer (Figure 10–10). The liquidity of membranes allows lipids to rapidly patch transient holes that may appear in the bilayer through mechanical or other stresses.

Similar mobility studies on labeled lipid molecules in isolated biological membranes and in living cells give results similar to those in synthetic bilayers. They demonstrate that the lipid component of a biological membrane is a two-dimensional liquid in which the constituent molecules are free to move laterally. As in synthetic bilayers, individual phospholipid molecules are normally confined to their own monolayer. This confinement creates a problem for the growth of biological membranes. Phospholipid molecules are manufactured in only one monolayer of a membrane, mainly in the cytosolic monolayer of the endoplasmic reticulum membrane. If none of these newly made molecules could migrate reasonably promptly to the noncytosolic monolayer, the membrane would expand asymmetrically. The problem is solved by a special class of membrane proteins called *phospholipid translocators*, or *flippases* and *scramblases*, which catalyze the rapid flip-flop of phospholipids from one monolayer to the other, as discussed in Chapter 12.

Despite the fluidity of the lipid bilayer, liposomes do not fuse spontaneously with one another when suspended in water. Fusion does not occur because the

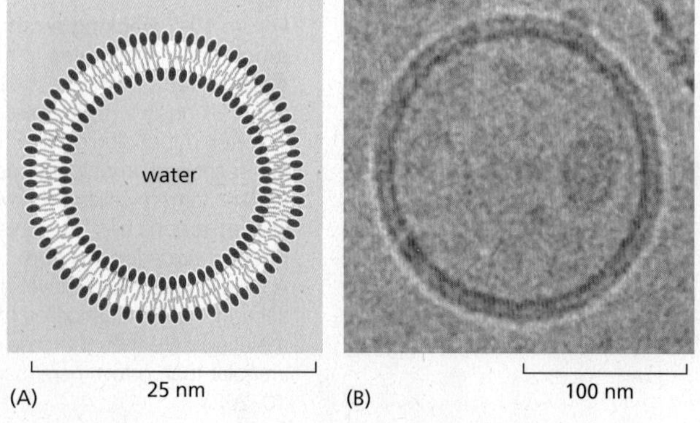

Figure 10–9 **Liposomes.** (A) A drawing of a small spherical liposome seen in cross section. Liposomes are commonly used as model membranes in experimental studies, especially to study incorporated membrane proteins. (B) An electron micrograph of an unfixed, unstained, synthetic phospholipid vesicle—a liposome—in water, which has been rapidly frozen at liquid-nitrogen temperature. (B, from J. Kotouček et al., *Sci. Rep.* 10:5595, 2020.)

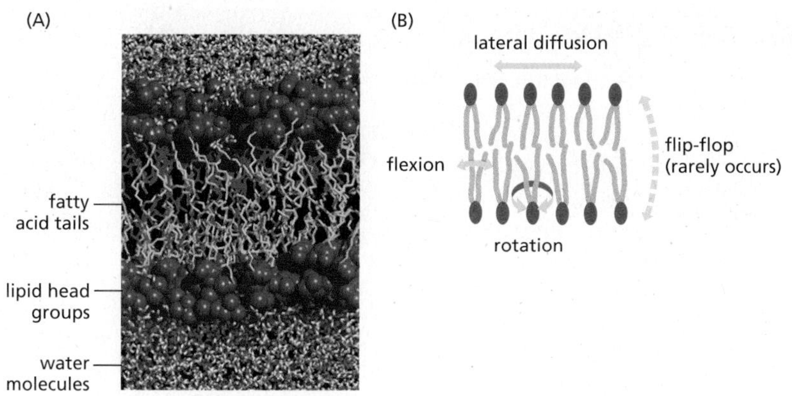

Figure 10–10 **The mobility of phospholipid molecules in an artificial lipid bilayer.** (A) Starting with a model of 100 phosphatidylcholine molecules arranged in a regular bilayer, a computer calculated the position of every atom after 300 picoseconds of simulated time. From these theoretical calculations, a model of the lipid bilayer emerges that accounts for almost all of the measurable properties of a synthetic lipid bilayer, including its thickness, number of lipid molecules per membrane area, depth of water penetration, and unevenness of the two surfaces. Note that the tails in one monolayer can interact with those in the other monolayer, if the tails are long enough. (B) The different motions of a lipid molecule in a bilayer. (A, based on S.W. Chiu et al., *Biophys. J.* 69:1230–1245, 1995.)

polar lipid head groups bind water molecules and ions that need to be displaced for the bilayers of two different liposomes to come into sufficiently close contact to fuse. Biological membranes have an even larger hydration shell due to the proteins embedded or associated with them. This hydration shell insulates the many internal membranes in a eukaryotic cell and prevents their uncontrolled fusion, thereby maintaining the compartmental integrity of membrane-enclosed organelles. All cell membrane fusion events are catalyzed by tightly regulated tethers that bring appropriate membranes close together and fusion proteins that force out the water layer that keeps the bilayers apart, as we discuss in Chapter 13.

The Fluidity of a Lipid Bilayer Depends on Its Composition

The fluidity of cell membranes has to be precisely regulated. It allows membrane proteins to interact rapidly and transiently, and certain membrane transport processes and enzyme activities, for example, cease when the bilayer viscosity is experimentally increased beyond a threshold level.

The fluidity of a lipid bilayer depends on both its composition and its temperature, as is readily demonstrated in studies of synthetic lipid bilayers. A synthetic bilayer made from a single type of phospholipid changes from a liquid state to a two-dimensional rigid crystalline (or gel) state at a characteristic temperature. This change of state is called a *phase transition*, and the temperature at which it occurs is lower (that is, the membrane becomes more difficult to freeze) if the hydrocarbon chains are short or have double bonds. A shorter chain length reduces the tendency of the hydrocarbon tails to interact with one another in both the same and opposite monolayer so that the membrane remains fluid at lower temperatures. Fluidity is also favored by *cis*-double bonds because they produce kinks in the chains that make them more difficult to pack together (Figure 10–11). The makeup of membranes as a complex mix of many different lipid species further adjusts most membranes so that they remain liquids just above the phase-transition point. Bacteria, yeasts, and other organisms whose temperature fluctuates with that of their environment adjust the fatty acid composition of their membrane lipids to maintain a relatively constant fluidity. As the temperature falls, for instance, the cells of those organisms synthesize fatty acids with more *cis*-double bonds, thereby avoiding the decrease in bilayer fluidity that would otherwise result from the temperature drop.

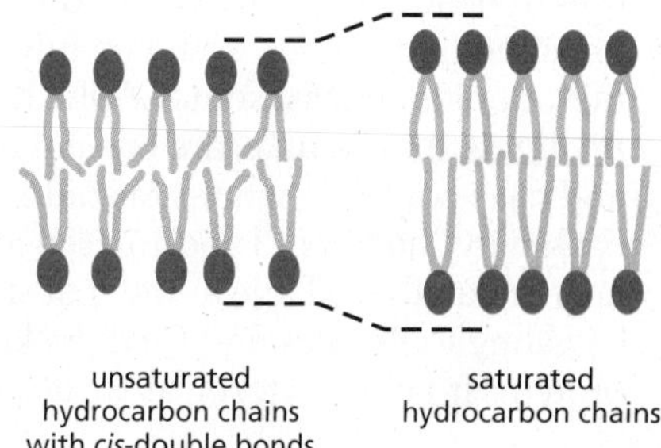

Figure 10–11 **The influence of *cis*-double bonds in hydrocarbon chains.** The double bonds make it more difficult to pack the chains together, thereby making the lipid bilayer more difficult to freeze. In addition, because the hydrocarbon chains of unsaturated lipids are more spread apart, lipid bilayers containing them are thinner than bilayers formed exclusively from saturated lipids.

Sterols, such as cholesterol, modulate the properties of lipid bilayers. When mixed with phospholipids, they enhance the permeability-barrier properties of the lipid bilayer. Cholesterol inserts into the bilayer with its hydroxyl group close to the polar head groups of the phospholipids, so that its rigid, platelike steroid rings interact with—and stiffen—those regions of the hydrocarbon chains closest to the polar head groups (see Figure 10–5 and Movie 10.3). By decreasing the mobility of the first few CH_2 groups of the hydrocarbon chains of the phospholipid molecules, cholesterol makes the lipid bilayer less deformable in this region and thereby decreases the permeability of the bilayer to small water-soluble molecules. Although cholesterol tightens the packing of the lipids in a bilayer, it does

TABLE 10–1 **Approximate Lipid Compositions of Different Cell Membranes**

Lipid	Percentage of total lipid by weight					
	Liver cell plasma membrane	Red blood cell plasma membrane	Myelin	Mitochondrion (inner and outer membranes)	Endoplasmic reticulum	*Escherichia coli* bacterium
Cholesterol	17	23	22	3	6	0
Phosphatidylethanolamine	7	18	15	28	17	70
Phosphatidylserine	4	7	9	2	5	Trace
Phosphatidylcholine	24	17	10	44	40	0
Sphingomyelin	19	18	8	0	5	0
Glycolipids	7	3	28	Trace	Trace	0
Others	22	14	8	23	27	30

not make membranes any less fluid because it also prevents the hydrocarbon chains from coming together and crystallizing.

Table 10–1 compares the lipid compositions of several biological membranes. Note that bacterial plasma membranes are often composed of one main type of phospholipid and contain no cholesterol. In archaea, lipids usually contain 20- to 25-carbon-long prenyl chains instead of fatty acids; prenyl and fatty acid chains are similarly hydrophobic and flexible (see Figure 10–18F). In thermophilic archaea, the longest lipid chains span both leaflets, making the membrane particularly stable to heat. Thus, lipid bilayers can be built from molecules with similar features but different molecular designs. The plasma membranes of most eukaryotic cells are more varied than those of prokaryotes and archaea, not only in containing large amounts of sterols but also in containing a mixture of different phospholipids.

Analysis of membrane lipids by mass spectrometry has revealed that the lipid composition of a typical eukaryotic cell membrane is much more complex than originally thought. These membranes contain a bewildering variety of perhaps 500–2000 different lipid species with even the simple plasma membrane of a red blood cell containing well over 150. Lipid heterogeneity antagonizes phase transitions and may help membrane-spanning proteins to fit better in the bilayer, avoiding leaks. While some of this complexity reflects the combinatorial variation in head groups, hydrocarbon chain lengths, and desaturation of the major phospholipid classes, some membranes also contain many structurally distinct minor lipids, at least some of which have important functions. The *inositol phospholipids*, for example, are present in small quantities in animal cell membranes and have crucial functions in guiding membrane traffic and in cell signaling (discussed in Chapters 13 and 15, respectively). Their local synthesis and destruction are regulated by a large number of enzymes, which create both small intracellular signaling molecules and lipid docking sites on membranes that recruit specific proteins from the cytosol, as we discuss later.

Despite Their Fluidity, Lipid Bilayers Can Form Domains of Different Compositions

Because a lipid bilayer is a two-dimensional fluid, we might expect most types of lipid molecules in it to be well mixed and randomly distributed in their own monolayer. The van der Waals attractive forces between neighboring hydrocarbon tails are not selective enough to hold groups of phospholipid molecules together. With certain lipid mixtures in artificial bilayers, however, one can observe phase transitions that lead to the lateral segregation of lipids with specific lipids coming

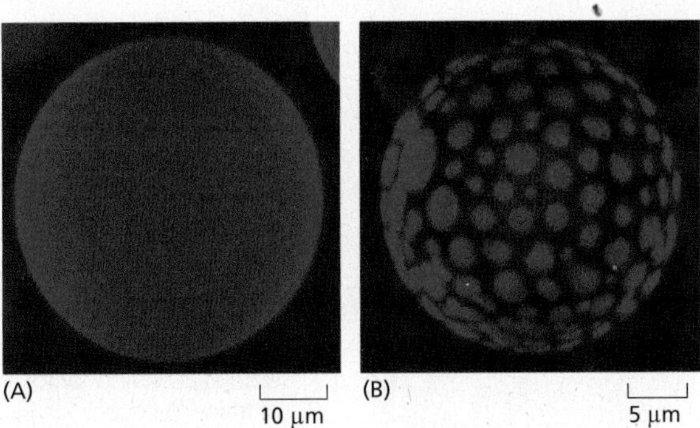

Figure 10–12 **Lateral phase separation in artificial lipid bilayers.** (A) Giant liposomes produced from a 1:1 mixture of phosphatidylcholine and sphingomyelin form uniform bilayers. (B) By contrast, liposomes produced from a 1:1:1 mixture of phosphatidylcholine, sphingomyelin, and cholesterol form bilayers with two separate phases. The liposomes are stained with trace concentrations of a fluorescent dye that preferentially partitions into one of the two phases. The average size of the domains formed in these giant artificial liposomes is much larger than that expected in cell membranes, where *lipid rafts* (see text) may be as small as a few nanometers in diameter. (A, from N. Kahya et al., *J. Struct. Biol.* 147:77-89, 2004. With permission from Elsevier; B, courtesy of Schwille Lab, MPG.)

together in separate domains (**Figure 10–12**). In these cases, attractive forces between lipid molecules must outweigh the entropic cost associated with concentrating them. Phase transitions thus break the homogeneity of the bilayer into a patchwork of domains with different properties.

There has been a long debate among cell biologists about whether the lipid molecules in the plasma membrane of living cells similarly segregate into specialized domains, called **lipid rafts**. Although many lipids and membrane proteins are not distributed uniformly, large-scale lipid phase segregations are seen rarely in living cell membranes. Instead, specific membrane proteins and lipids are seen to concentrate in a more temporary, dynamic fashion facilitated by protein–protein interactions that allow the transient formation of specialized membrane regions (**Figure 10–13**). Such clusters can be tiny nanoclusters on a scale of a few molecules or larger assemblies that can be seen with electron microscopy, such as the *caveolae* (discussed in Chapter 13). The tendency of mixtures of lipids to undergo phase transitions, as seen in artificial bilayers (see Figure 10–12), may help create rafts in living cell membranes—organizing and concentrating membrane proteins either for transport in membrane vesicles (discussed in Chapter 13) or for working together in protein assemblies, such as when they convert extracellular signals into intracellular ones (discussed in Chapter 15).

Lipid Droplets Are Surrounded by a Phospholipid Monolayer

Most eukaryotic cells store an excess of lipids in **lipid droplets**, from where they can be retrieved as building blocks for membrane synthesis or as a food source fueling metabolic energy generation. Fat cells, or *adipocytes*, are specialized for lipid storage. They contain a giant lipid droplet that fills up most of their cytoplasm. Most other cells have many smaller lipid droplets, the number and size varying with the cell's metabolic state. Fatty acids can be liberated from lipid droplets on demand and exported to other cells through the bloodstream. Lipid droplets store neutral lipids, such as triacylglycerols and cholesterol esters, which are synthesized from fatty acids and cholesterol by enzymes in the endoplasmic reticulum membrane. Because these lipids do not contain hydrophilic head groups, they are exclusively hydrophobic molecules, and therefore aggregate into three-dimensional droplets rather than into bilayers.

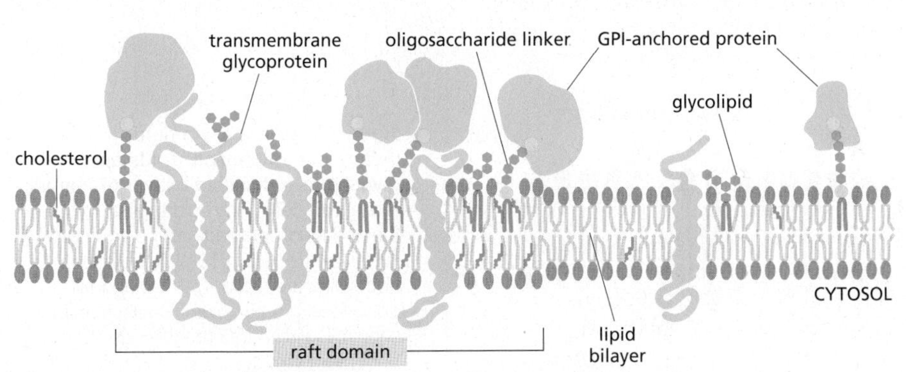

Figure 10–13 **A model of a raft domain.** Weak protein–protein, protein–lipid, and lipid–lipid interactions reinforce one another to partition the interacting components into raft domains. Cholesterol, sphingolipids, glycolipids, glycosylphosphatidylinositol (GPI)-anchored proteins, and some transmembrane proteins are enriched in these domains. Note that because of their composition, raft domains are thought to have an increased membrane thickness. We discuss glycolipids, GPI-anchored proteins, and oligosaccharide linkers later. (Adapted from D. Lingwood and K. Simons, *Science* 327:46–50, 2010.)

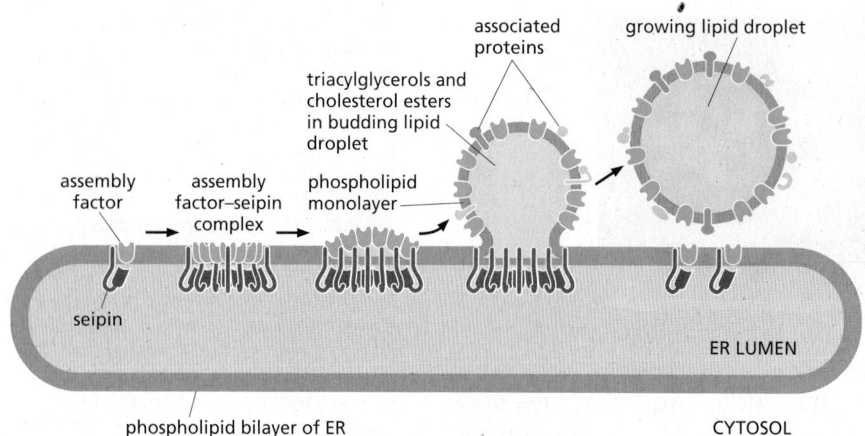

Figure 10–14 A model for the formation of lipid droplets. Neutral lipids are deposited between the two monolayers of the endoplasmic reticulum (ER) membrane forming a lenslike structure between the two monolayers. The lens forms as triacylglycerides and cholesterol esters are made and accumulate in the ER membrane and self-aggregate. Multiple copies of the transmembrane protein seipin assemble into a ring together with a lipid droplet assembly factor. In the presence of triacylglycerides, seipin dissociates from the assembly factor, which migrates to the cytosolic monolayer of the ER membrane where it facilitates a process in which the droplet buds, fills up with nonpolar lipids, and pinches off as a unique organelle that is surrounded by a single monolayer of phospholipids and associated proteins. In some cells, such as adipocytes, droplets fuse and can reach a gigantic size. (Adapted from J. Chung et al., *Dev. Cell* 51:551–563, 2019.)

In order for these hydrophobic droplets to reside in the aqueous cytosol of the cell, their surface is covered by phospholipids oriented with their hydrophobic acyl chains facing the lipid droplet and hydrophilic head groups facing the cytosol. This is why lipid droplets are surrounded by a monolayer of phospholipids rather than the bilayer that defines all other membrane-bounded compartments of the cell. The surface of lipid droplets contains a large variety of proteins, some of which are enzymes involved in lipid metabolism. Lipid droplets form rapidly when cells are exposed to high concentrations of fatty acids. They form from the endoplasmic reticulum membrane where many enzymes of lipid metabolism are localized. Figure 10–14 shows one model of how lipid droplets form and acquire their surrounding monolayer of phospholipids and proteins. In some specialized cells, such as liver cells and enterocytes (the absorptive cells of the gut), droplets bud into the lumen of the endoplasmic reticulum from where they are secreted as lipoprotein particles that move metabolic energy in the form of triglycerides through the body.

The Asymmetry of the Lipid Bilayer Is Functionally Important

The lipid compositions of the two monolayers of the lipid bilayer in many membranes are strikingly different. In the human red blood cell (erythrocyte) membrane, for example, almost all of the phospholipid molecules that have choline—$(CH_3)_3N^+CH_2CH_2OH$—in their head group (phosphatidylcholine and sphingomyelin) are in the outer monolayer, whereas almost all that contain a terminal primary amino group (phosphatidylethanolamine and phosphatidylserine) are in the inner monolayer (Figure 10–15). Because the negatively charged phosphatidylserine is located in the inner monolayer, there is a significant difference in charge between the two halves of the bilayer. We discuss in Chapter 12 how membrane-bound phospholipid translocators generate and maintain lipid asymmetry.

Lipid asymmetry is functionally important, especially in converting extracellular signals into intracellular ones (discussed in Chapter 15). Many cytosolic proteins bind to specific lipid head groups found in the cytosolic monolayer of the lipid bilayer. The enzyme *protein kinase C* (*PKC*), for example, which is

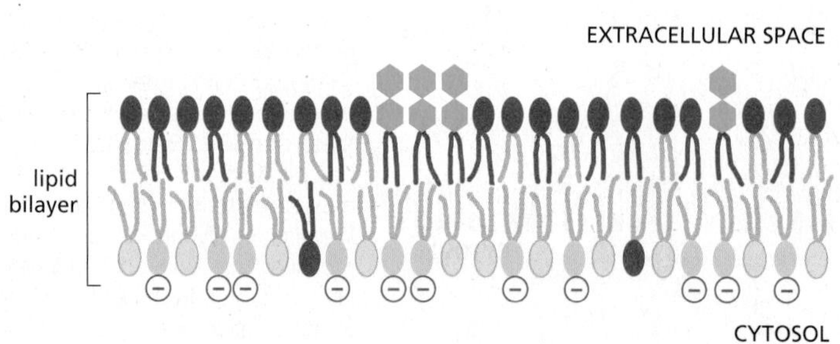

Figure 10–15 The asymmetric distribution of phospholipids and glycolipids in the lipid bilayer of human red blood cells. The colors used for the phospholipid head groups are those introduced in Figure 10–3. In addition, glycolipids are drawn with hexagonal polar head groups *(blue)*. Cholesterol (not shown) is distributed roughly equally in both monolayers.

activated in response to various extracellular signals, binds to the cytosolic face of the plasma membrane, where phosphatidylserine is concentrated, and requires this negatively charged phospholipid for its activity.

In other cases, specific lipid head groups must first be modified to create protein-binding sites at a particular time and place. One example is *phosphatidylinositol* (*PI*), one of the minor phospholipids that are concentrated in the cytosolic monolayer of cell membranes (see Figure 13–10A, B, and C). Various lipid kinases can add phosphate groups at distinct positions on the inositol ring, creating binding sites that recruit specific proteins from the cytosol to the membrane. An important example of such a lipid kinase is *phosphoinositide 3-kinase* (*PI 3-kinase*), which is activated in response to extracellular signals and helps to recruit specific intracellular signaling proteins to the cytosolic face of the plasma membrane (see Figure 15–53). Similar lipid kinases phosphorylate inositol phospholipids in intracellular membranes and thereby help to recruit proteins that guide membrane transport.

Phospholipids in the plasma membrane are used in yet another way to convert extracellular signals into intracellular ones. The plasma membrane contains various *phospholipases* that are activated by extracellular signals to cleave specific phospholipid molecules, generating fragments of these molecules that act as short-lived intracellular messengers. *Phospholipase C*, for example, cleaves an inositol phospholipid in the cytosolic monolayer of the plasma membrane to generate two fragments, one of which remains in the membrane and helps activate protein kinase C, while the other is released into the cytosol and stimulates the release of Ca^{2+} from the endoplasmic reticulum (see Figure 15–29).

Animals exploit the phospholipid asymmetry of their plasma membranes to distinguish between live and dead cells. When animal cells undergo apoptosis (discussed in Chapter 18), phosphatidylserine, which is normally confined to the cytosolic (or inner) monolayer of the plasma membrane lipid bilayer, rapidly translocates to the extracellular (or outer) monolayer. The phosphatidylserine exposed on the cell surface signals neighboring cells, such as macrophages, to phagocytose the dead cell and digest it. The translocation of the phosphatidylserine in apoptotic cells occurs because the active mechanisms that generate and maintain lipid bilayer asymmetry are impaired.

Glycolipids Are Found on the Surface of All Eukaryotic Plasma Membranes

Sugar-containing lipid molecules called **glycolipids** have the most extreme asymmetry in their membrane distribution: these molecules, whether in the plasma membrane or in intracellular membranes, are found exclusively in the monolayer facing away from the cytosol. In animal cells, they are made from sphingosine, just like sphingomyelin (see Figure 10–3). These intriguing molecules tend to self-associate, partly through hydrogen bonds between their sugars and partly through van der Waals forces between their long and straight hydrocarbon chains, which causes them to partition preferentially into lipid raft phases (see Figure 10–13). The asymmetric distribution of glycolipids in the bilayer results from the addition of sugar groups to the lipid molecules in the lumen of the Golgi apparatus. Thus, the compartment in which they are manufactured is topologically equivalent to the exterior of the cell (discussed in Chapter 12). As they are delivered to the plasma membrane, the sugar groups are exposed at the cell surface (see Figure 10–15), where they have important roles in interactions of the cell with its surroundings.

Glycolipids probably occur in all eukaryotic cell plasma membranes, where they generally constitute about 5% of the lipid molecules in the outer monolayer. They are also found in some intracellular membranes. The most complex of the glycolipids, the **gangliosides**, contain oligosaccharides with one or more sialic acid moieties, which give gangliosides a net negative charge (**Figure 10–16**). The most abundant of the more than 40 different gangliosides that have been identified are in the plasma membrane of nerve cells, where gangliosides constitute 5–10% of the total lipid mass; they are also found in much smaller quantities in other cell types.

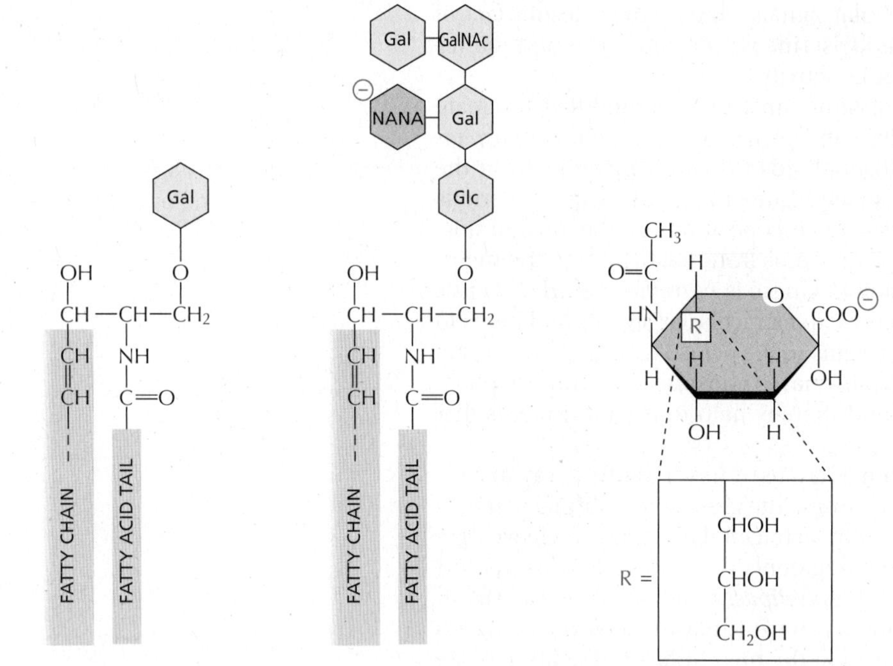

(A) galactocerebroside (B) G_{M1} ganglioside (C) a sialic acid (NANA)

Figure 10–16 Glycolipid molecules. (A) Galactocerebroside is called a *neutral glycolipid* because the sugar that forms its head group is uncharged. (B) A ganglioside always contains one or more negatively charged sialic acid moieties. There are various types of sialic acid; in human cells, it is mostly *N*-acetylneuraminic acid, or NANA, whose structure is shown in (C). Whereas in bacteria and plants almost all glycolipids are derived from glycerol, as are most phospholipids, in animal cells almost all glycolipids are based on sphingosine, as is the case for sphingomyelin (see Figure 10–3). Gal = galactose, Glc = glucose, GalNAc = *N*-acetylgalactosamine; these three sugars are uncharged.

Hints as to the functions of glycolipids come from their localization. In the plasma membrane of epithelial cells, for example, glycolipids are confined to the exposed apical surface, where they may help to protect the membrane against the harsh conditions frequently found there (such as low pH and high concentrations of degradative enzymes). Charged glycolipids, such as gangliosides, may be important because of their electrical effects: their presence alters the electrical field across the membrane and the concentrations of ions—especially Ca^{2+}—at the membrane surface. Glycolipids also function in cell-recognition processes, in which membrane-bound carbohydrate-binding proteins (*lectins*) bind to the sugar groups on both glycolipids and glycoproteins in the process of cell–cell adhesion (discussed in Chapter 19). Mutant mice that are deficient in all of their complex gangliosides show abnormalities in the nervous system, including axonal degeneration and reduced myelination.

The ubiquitous presence of glycolipids on the cell surface has been exploited by a number of bacterial toxins and viruses as a means to enter cells. For example, influenza virus interacts with sialic acid sugars on gangliosides during its entry into cells (see Figure 10–16). Polyomaviruses also enter the cell after binding initially to gangliosides. Similarly, the ganglioside G_{M1} acts as a cell-surface receptor for the bacterial toxin that causes the debilitating diarrhea of cholera. Cholera toxin binds to and enters only those cells that have G_{M1} on their surface, including intestinal epithelial cells. Its entry into a cell leads to a prolonged increase in the concentration of intracellular cyclic AMP (discussed in Chapter 15), which in turn causes a large efflux of Cl^-, leading to the secretion of Na^+, K^+, HCO_3^-, and water into the intestine.

Summary

Biological membranes consist of a continuous double layer of lipid molecules in which membrane proteins are embedded. This lipid bilayer is fluid, with individual lipid molecules able to diffuse rapidly within their own monolayer. The membrane lipid molecules are amphiphilic. When placed in water, they assemble spontaneously into bilayers, which form sealed compartments.

Although cell membranes can contain hundreds of different lipid species, the plasma membrane in animal cells contains three major classes—phospholipids, cholesterol, and glycolipids. Because of their different backbone structure, phospholipids fall into two subclasses—glycerophospholipids and sphingolipids. The lipid compositions of the inner and outer monolayers are different, reflecting the different functions of the two faces of a cell membrane. Different mixtures of lipids are found in the membranes of cells of different types, as well as in the various membranes of a single eukaryotic cell. Inositol phospholipids are a minor class of phospholipids, which in the cytosolic leaflet of the plasma membrane lipid bilayer play an important part in cell signaling: in response to extracellular signals, specific lipid kinases phosphorylate the head groups of these lipids to form docking sites for cytosolic signaling proteins, whereas specific phospholipases cleave certain inositol phospholipids to generate small intracellular signaling molecules.

MEMBRANE PROTEINS

Although the lipid bilayer provides the basic structure of biological membranes, the membrane proteins perform most of the membrane's specific tasks and therefore give each type of cell membrane its characteristic functional properties. Accordingly, the amounts and types of proteins in a membrane are highly variable. In the myelin membrane, which serves mainly as electrical insulation for nerve-cell axons, less than 25% of the membrane mass is protein. By contrast, in the membranes involved in ATP production (such as the internal membranes of mitochondria and chloroplasts), approximately 75% is protein. A typical plasma membrane is somewhere in between, with protein accounting for about half of its mass. Because lipid molecules are small compared with protein molecules, however, there are always many more lipid molecules than protein molecules in cell membranes—about 50 lipid molecules for each protein molecule in cell membranes that are 50% protein by mass. Membrane proteins vary widely in structure and in the way they associate with the lipid bilayer, which reflects their diverse functions.

Membrane Proteins Can Be Associated with the Lipid Bilayer in Various Ways

Figure 10–17 shows the different ways in which proteins can associate with the membrane. Like their lipid neighbors, **membrane proteins** are amphiphilic, having hydrophobic and hydrophilic regions. Many membrane proteins extend through the lipid bilayer, and hence are called **transmembrane proteins**, with part of their mass extruding from the membrane on both sides (Figure 10–17, examples 1, 2, and 5). Other transmembrane proteins are inserted with the bulk of their mass exposed almost exclusively on one or the other side of the membrane (Figure 10–17, examples 3 and 4). In all cases their hydrophobic regions pass through the membrane and interact with the hydrophobic tails of the lipid molecules in the interior of the bilayer, where they are sequestered away from water. Their hydrophilic regions are exposed to water on either side of the membrane.

Figure 10–17 Various ways in which proteins associate with the lipid bilayer. Most membrane proteins are thought to extend across the bilayer as (1) a single α helix, (2) as multiple α helices, or (5) as a rolled-up β sheet (a β barrel). Some of these *single-pass* and *multipass* proteins have a covalently attached fatty acid chain inserted in the cytosolic lipid monolayer (1). Other membrane proteins are exposed at only one side of the membrane (3, 4). These classes include glycosyltransferases that carry out glycosylation reactions in the Golgi apparatus (3) and SNARE proteins that catalyze membrane fusion (4), both discussed in Chapter 13. (6) Some of these are anchored to the cytosolic surface by an amphiphilic α helix that partitions into the cytosolic monolayer of the lipid bilayer through the hydrophobic face of the helix. (7) Others are attached to the bilayer solely by a covalently bound lipid chain—either a fatty acid chain or a prenyl group (see Figure 10–18)—in the cytosolic monolayer or, (8) via an oligosaccharide linker, to phosphatidylinositol in the noncytosolic monolayer—called a GPI anchor. (9, 10) Finally, membrane-associated proteins are attached to the membrane only by noncovalent interactions with other membrane proteins. The way in which the structure in (7) is formed is illustrated in Figure 10–18, while the way in which the GPI anchor shown in (8) is formed is illustrated in Figure 12–30. The details of how membrane proteins become associated with the lipid bilayer are discussed in Chapter 12.

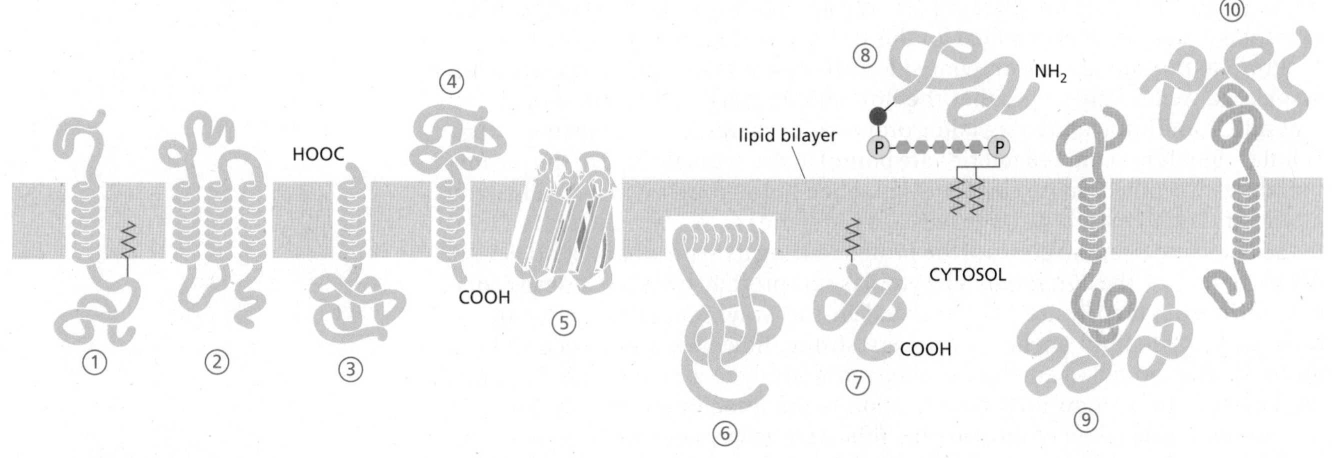

Other membrane proteins are located entirely in the cytosol and are attached to the cytosolic monolayer of the lipid bilayer, either by an amphiphilic α helix exposed on the surface of the protein (Figure 10–17, example 6) or by one or more covalently attached lipid chains (Figure 10–17, example 7). The lipid-linked proteins in example 7 in Figure 10–17 are made as soluble proteins in the cytosol and are subsequently anchored to the membrane by the covalent attachment of the lipid group. Lipids can also be attached to the cytosolic facing domains of transmembrane proteins as an additional means of anchoring them to the membrane (see Figure 10–17, example 1).

Yet other membrane proteins are entirely exposed at the external cell surface, being attached to the lipid bilayer only by a covalent linkage (via a specific oligosaccharide) to a lipid anchor in the outer monolayer of the plasma membrane (Figure 10–17, example 8). These proteins are initially made and inserted into the endoplasmic reticulum (ER) by a single transmembrane segment at the C-terminus (similar to example 4 in Figure 10-17). While still in the ER, the transmembrane segment of the protein is cleaved off and a *glycosylphosphatidylinositol (GPI) anchor* is added, leaving the protein bound to the noncytosolic surface of the ER membrane solely by this anchor (discussed in Chapter 12); transport vesicles eventually deliver the protein to the plasma membrane (discussed in Chapter 13).

By contrast to these examples, **membrane-associated proteins** do not extend into the hydrophobic interior of the lipid bilayer at all; they are instead bound to either face of the membrane by noncovalent interactions with other membrane proteins (Figure 10–17, examples 9 and 10). Many of the proteins of this type can be released from the membrane by relatively gentle extraction procedures, such as exposure to solutions of very high or low ionic strength or of extreme pH, which interfere with protein–protein interactions but leave the lipid bilayer intact; these proteins are often referred to as *peripheral membrane proteins*, and their association with membranes is often regulated by the cell as we discuss next. Transmembrane proteins and many proteins held in the bilayer by lipid groups or hydrophobic polypeptide regions that insert into the hydrophobic core of the lipid bilayer cannot be released in these ways.

Lipid Anchors Control the Membrane Localization of Some Signaling Proteins

How a membrane protein is associated with the lipid bilayer reflects the function of the protein. Only transmembrane proteins can function on both sides of the bilayer or transport molecules across it. Cell-surface receptors, for example, are usually transmembrane proteins that bind signal molecules in the extracellular space and generate different intracellular signals on the opposite side of the plasma membrane, as we discuss in Chapter 15. To transfer small hydrophilic molecules across a membrane, a membrane transport protein must provide a path for the molecules to cross the hydrophobic permeability barrier of the lipid bilayer; the molecular architecture of multipass transmembrane proteins (Figure 10–17, examples 2 and 5) is ideally suited for this task, as we discuss in Chapter 11.

Proteins that function on only one side of the lipid bilayer, by contrast, are often associated exclusively with either the lipid monolayer or a protein domain on that side. Some intracellular signaling proteins, for example, that help relay extracellular signals into the cell interior are bound to the cytosolic half of the plasma membrane by one or more covalently attached lipid groups, which can be fatty acid chains or *prenyl groups* (**Figure 10–18**). In some cases, myristic acid is added to the N-terminal amino group of the protein during its synthesis on a ribosome. All members of the *Src family* of cytoplasmic protein tyrosine kinases (discussed in Chapter 15) are myristoylated in this way. Membrane attachment through a single lipid anchor is not very strong, however, and a second lipid group is often added to anchor proteins more firmly to a membrane. For most Src kinases, the second lipid modification is the attachment of palmitic acid to a cysteine side chain of the protein. This modification occurs in response to

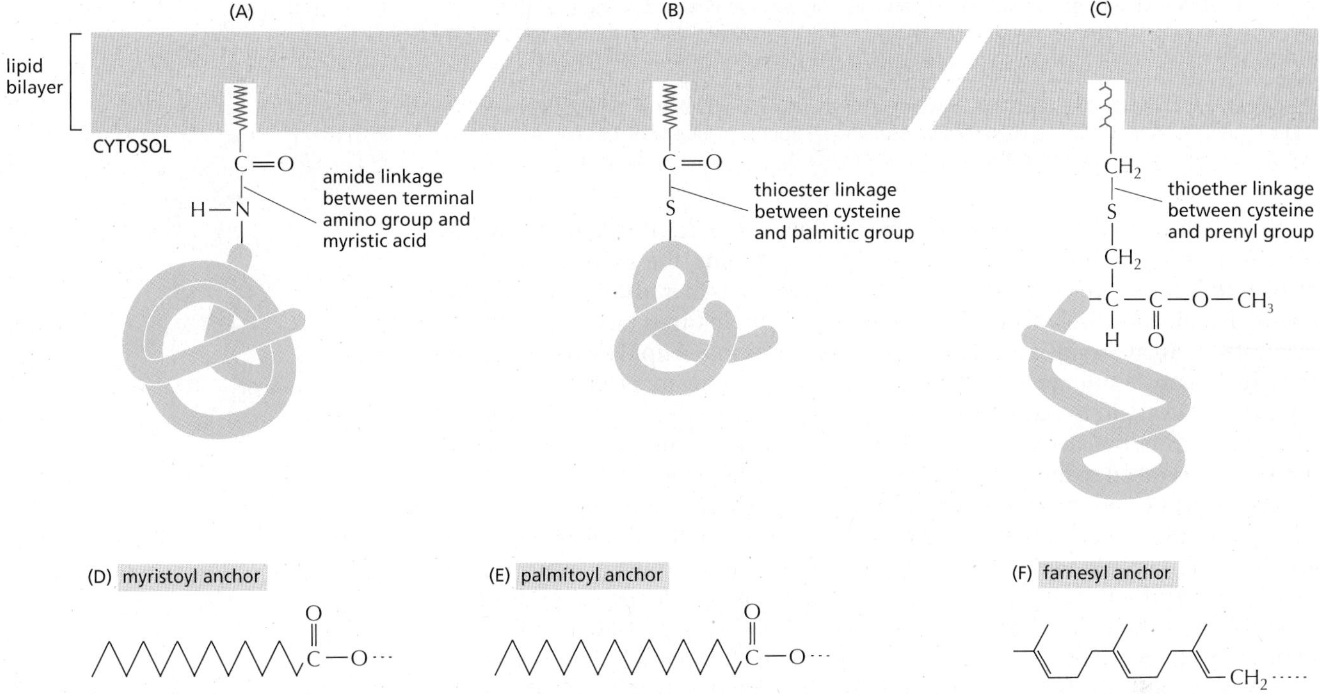

Figure 10–18 Membrane protein attachment by a fatty acid chain or a prenyl group. The covalent attachment of either type of lipid can help localize a water-soluble protein to a membrane after its synthesis in the cytosol. (A) A fatty acid chain (myristic acid) is attached via an amide linkage to an N-terminal glycine. (B) A fatty acid chain (palmitic acid) is attached via a thioester linkage to a cysteine. (C) A prenyl chain (either farnesyl or a longer geranylgeranyl chain) is attached via a thioether linkage to a cysteine residue that is initially located four residues from the protein's C-terminus. After prenylation, the terminal three amino acids are cleaved off, and the new C-terminus is methylated before insertion of the anchor into the membrane (not shown). The structures of the lipid anchors are shown below: (D) a myristoyl anchor (derived from a 14-carbon saturated fatty acid chain), (E) a palmitoyl anchor (a 16-carbon saturated fatty acid chain), and (F) a farnesyl anchor (a 15-carbon unsaturated hydrocarbon chain composed of three 5-carbon isoprenoid repeats).

an extracellular signal and helps recruit the kinases to the plasma membrane. When the signaling pathway is turned off, the palmitic acid is removed, allowing the kinase to return to the cytosol. Other intracellular signaling proteins, such as the Ras family small GTPases (discussed in Chapter 15), use a combination of prenyl group and palmitic acid attachment to recruit the proteins to the plasma membrane.

Many proteins attach to membranes transiently. Some are classical peripheral membrane proteins that associate with membranes by regulated protein–protein interactions. Others undergo a transition from soluble to membrane protein by a conformational change that exposes a hydrophobic peptide or covalently attached lipid anchor. Many of the small GTPases of the Rab protein family that regulate intracellular membrane traffic (discussed in Chapter 13), for example, switch depending on the nucleotide that is bound to the protein. In their GDP-bound state they are soluble in the cytosol, often stabilized by binding to a GDP dissociation inhibitor, or GDI, whereas in their GTP-bound state their lipid anchor is exposed and tethers them to membranes. They are membrane proteins at one moment and soluble proteins at the next. Such highly dynamic interactions greatly expand the repertoire of membrane functions.

In Most Transmembrane Proteins, the Polypeptide Chain Crosses the Lipid Bilayer in an α-Helical Conformation

A transmembrane protein always has a unique orientation in the membrane. This reflects both the asymmetric manner in which it is inserted into the lipid bilayer in the ER during its biosynthesis (discussed in Chapter 12) and the different functions of its cytosolic and noncytosolic domains. These domains are separated by the membrane-spanning segments of the polypeptide chain, which contact the hydrophobic environment of the lipid bilayer and are composed largely of amino acids with nonpolar side chains. Because the peptide bonds themselves are polar and because water is absent in the bilayer, all peptide bonds in the membrane-spanning segments of a polypeptide are driven to form hydrogen bonds with one another (discussed in Chapter 3).

There are two ways that hydrogen-bonding between peptide bonds can be maximized. The most common way, found in the majority of transmembrane

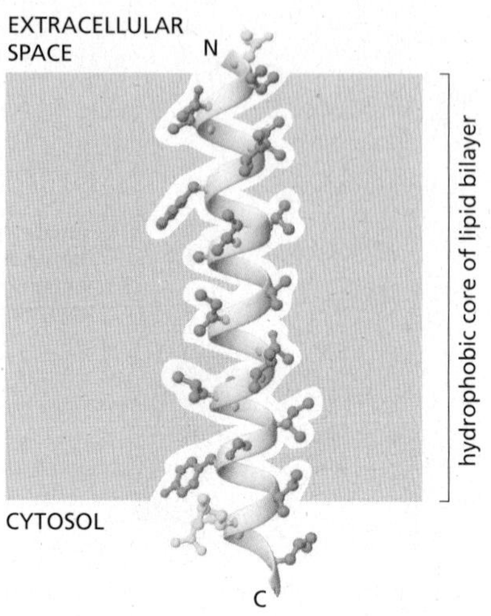

Figure 10–19 **A segment of a membrane-spanning polypeptide chain crossing the lipid bilayer as an α helix.** Only the α-carbon backbone of the polypeptide chain is shown, with the hydrophobic amino acids in *green* and the hydrophilic amino acids in *yellow*. The polypeptide segment shown is part of the single membrane-spanning protein glycophorin, which is found abundantly in the red blood cell plasma membrane. Glycophorin normally is a dimer of two identical subunits whose transmembrane segments cross in the hydrophobic interior of the bilayer. Only a single transmembrane segment is shown. (PDB code: 5EH4.)

proteins, is if the polypeptide chain forms a regular α helix as it crosses the bilayer (Figure 10–19). In **single-pass transmembrane proteins**, the polypeptide chain crosses only once (see Figure 10–17, example 1), whereas in **multipass transmembrane proteins**, the polypeptide chain crosses multiple times (see Figure 10–17, example 2). An alternative way for the peptide bonds in the lipid bilayer to satisfy their hydrogen-bonding requirements is for multiple transmembrane strands of a polypeptide chain to be arranged as a β sheet that is rolled up into a cylinder (a so-called *β barrel*; see Figure 10–17, example 5, and Figure 3–7). This protein architecture is seen in the *porin proteins* that we discuss later.

Progress in x-ray crystallography and single-particle cryo-electron microscopy of membrane proteins has enabled the determination of the three-dimensional structure of many of them. The structures confirm that it is often possible to predict from the protein's amino acid sequence which parts of the polypeptide chain extend across the lipid bilayer. Segments containing about 20–30 amino acids, with a high degree of hydrophobicity, are long enough to span a lipid bilayer as an α helix, and they can often be identified in *hydropathy plots* (Figure 10–20). From such plots, it is estimated that about 30% of an organism's proteins are transmembrane proteins, emphasizing their importance. Hydropathy plots cannot identify the membrane-spanning segments of a β barrel, as 10 amino acids or fewer are sufficient to traverse a lipid bilayer as an extended β strand, and only every other amino acid side chain is hydrophobic.

The strong drive to maximize hydrogen-bonding in the absence of water means that most transmembrane helices span the membrane completely. But multipass transmembrane proteins can also contain regions that fold into the membrane from either side, squeezing into spaces between transmembrane α helices without contacting the hydrophobic core of the lipid bilayer. Because such regions interact only with other polypeptide regions, they do not need to maximize hydrogen-bonding; they can therefore have a variety of secondary structures, including helices that extend only partway across the lipid bilayer (Figure 10–21). Such regions are important for the function of some membrane proteins, including water channel and ion channel proteins, in which the regions contribute to the walls of the pores traversing the membrane and confer substrate specificity on the channels, as we discuss in Chapter 11. These regions cannot be identified

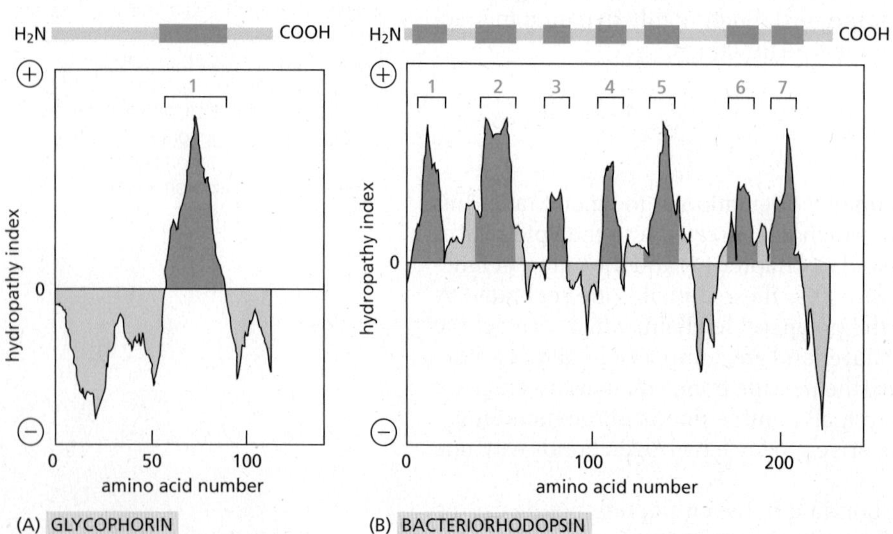

Figure 10–20 **Using hydropathy plots to localize potential α-helical membrane-spanning segments in a polypeptide chain.** The free energy needed to transfer successive segments of a polypeptide chain from a nonpolar solvent to water is calculated from the amino acid composition of each segment using data obtained from model compounds. This calculation is made for segments of a fixed size (usually around 10–20 amino acids), beginning with each successive amino acid in the chain. The *hydropathy index* of the segment is plotted on the Y axis as a function of its location in the chain. A positive value indicates that free energy is required for transfer to water (that is, the segment is hydrophobic), and the value assigned is an index of the amount of energy needed. Peaks in the hydropathy index appear at the positions of hydrophobic segments in the amino acid sequence. (A and B) Hydropathy plots for two membrane proteins that are discussed later in this chapter. Glycophorin (A) has a single membrane-spanning α helix and one corresponding peak in the hydropathy plot. Bacteriorhodopsin (B) has seven membrane-spanning α helices and seven corresponding peaks in the hydropathy plot. (A, adapted from D. Eisenberg, *Annu. Rev. Biochem*. 53:595–624, 1984.)

in hydropathy plots and are only revealed by determining the protein's three-dimensional structure or by sequence alignment with homologous proteins whose structures are known.

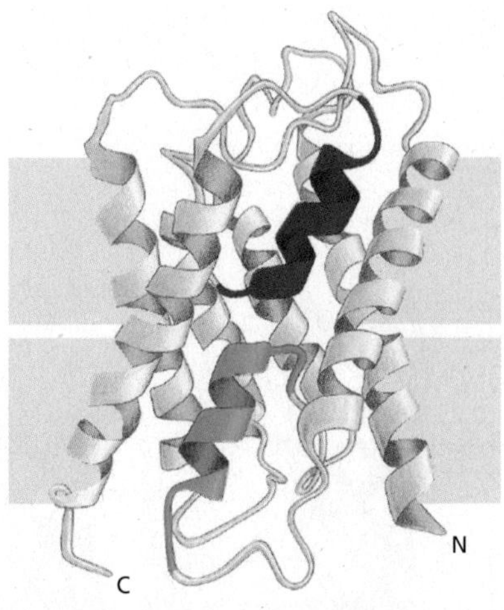

Figure 10–21 **Two short α helices in the aquaporin water channel, each of which spans only halfway through the lipid bilayer.** In the plasma membrane, four monomers, one of which is shown here, form a tetramer. Each monomer has a hydrophilic pore at its center, which allows water molecules to cross the membrane in single file (see Figure 11–20 and Movie 11.6). The two short, colored helices are buried at an interface formed by protein–protein interactions. The mechanism by which the channel allows the passage of water molecules is discussed in more detail in Chapter 11. (PDB code: 1H6I.)

Transmembrane α Helices Often Interact with One Another

The transmembrane α helices of many single-pass membrane proteins do not contribute to the folding of the protein domains on either side of the membrane. As a consequence, it is often possible to engineer cells to produce just the cytosolic or extracellular domains of these proteins as water-soluble molecules. This approach has been invaluable for studying the structure and function of these domains, especially the domains of transmembrane receptor proteins (discussed in Chapter 15). A transmembrane α helix, even in a single-pass membrane protein, however, often does more than just anchor the protein to the lipid bilayer. Many single-pass membrane proteins form homodimers or heterodimers that are held together by noncovalent, but strong and highly specific, interactions between the two transmembrane α helices; the sequence of the amino acids of these helices contains the information that directs the protein–protein interaction.

Similarly, the transmembrane α helices in multipass membrane proteins occupy specific positions in the folded protein structure that are determined by interactions between the neighboring helices (Figure 10–22). These interactions are crucial for the structure and function of the many receptors, channels, and transporters that communicate or move molecules across cell membranes. In these proteins, each transmembrane helix shields regions of neighboring transmembrane helices from membrane lipids. When all the helices are packed together into the final folded structure, the outer surface of the helical bundle that is exposed to lipids is composed primarily of hydrophobic amino acids. By contrast, the interior of the helical bundle, which is not exposed directly to lipids, can contain polar and even charged amino acids that would ordinarily be disfavored in the membrane. The ability to accommodate hydrophilic amino acids within a bundle of transmembrane helices means that multipass membrane proteins can contain binding sites and channels across the membrane for hydrophilic molecules. This property affords multipass membrane proteins considerable functional diversity and probably explains why they represent the majority of membrane proteins.

Some β Barrels Form Large Channels

Unlike a bundle of α helices that can be arranged in numerous ways, β-barrel membrane proteins are always arranged as a cylinder. This is because all hydrogen bonds must be satisfied, so a sheet that exposes an edge is energetically

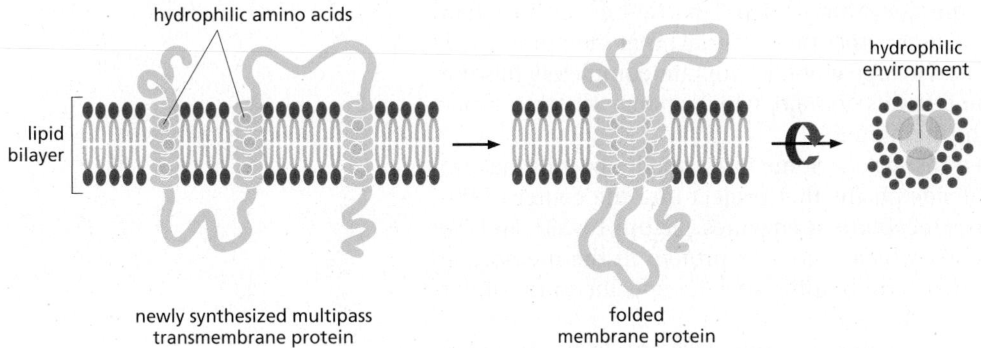

Figure 10–22 **Steps in the folding of a multipass transmembrane protein.** Polar and charged amino acids contained in transmembrane helices are energetically disfavored in the hydrophobic environment of the lipid bilayer. They become buried in the interface between spatially adjacent helices in folded membrane proteins. In membrane protein complexes, these contacts can occur between helices from different protein subunits, as is the case for many ion channels as discussed in Chapter 11. In this way, multipass membrane proteins can provide a hydrophilic path across the hydrophobic barrier of the bilayer.

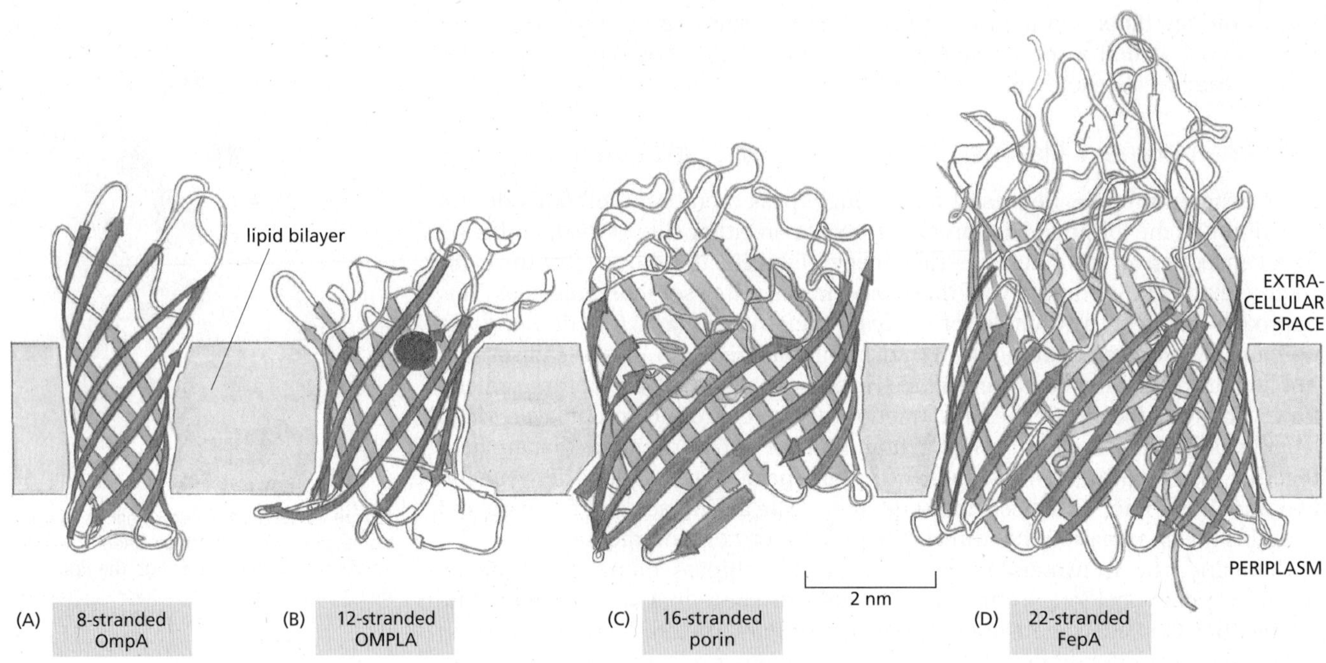

Figure 10–23 β barrels formed from different numbers of β strands. (A) The *Escherichia coli* OmpA protein serves as a receptor for a bacterial virus. (B) The *E. coli* OMPLA protein is an enzyme (a lipase) that hydrolyzes lipid molecules. The amino acids that catalyze the enzymatic reaction (indicated in *red*) protrude from the outside surface of the barrel. (C) A porin from the bacterium *Rhodobacter capsulatus* forms a water-filled pore across the outer membrane. The diameter of the channel is restricted by loops (shown in *yellow*) that protrude into the channel. (D) The *E. coli* FepA protein transports iron ions. The inside of the barrel is completely filled by a globular protein domain (shown in *yellow*) that contains an iron-binding site (not shown).

disfavored in the membrane. This means that β barrels all share the same overall architecture in which the amino acids facing the outside of the cylinder are hydrophobic. By contrast, the size of the barrel, and the content inside the barrel, are highly variable and suited to the function of each β-barrel membrane protein. For example, the number of β strands in a β barrel varies widely, from as few as 8 strands to as many as 22 (**Figure 10–23**).

β-Barrel proteins are abundant in the outer membranes of bacteria, mitochondria, and chloroplasts. Some are pore-forming proteins, which create water-filled channels that allow selected small hydrophilic molecules to cross the membrane. The porins are well-studied examples (Figure 10–23C). Many porin barrels are formed from a 16-strand, antiparallel β sheet rolled up into a cylindrical structure. Polar amino acid side chains line the aqueous channel on the inside, while nonpolar side chains project from the outside of the barrel to interact with the hydrophobic core of the lipid bilayer. Loops of the polypeptide chain often protrude into the **lumen** of the channel, narrowing it so that only certain solutes can pass. Some porins are therefore highly selective: *maltoporin*, for example, preferentially allows maltose and maltose oligomers to cross the outer membrane of *E. coli*.

The *FepA protein* is a more complex example of a β-barrel transport protein (Figure 10–23D). It transports iron ions across the bacterial outer membrane. It is constructed from 22 β strands, and a large globular domain completely fills the inside of the barrel. Iron ions bind to this domain, which changes its conformation to transfer the iron across the membrane.

Not all β-barrel proteins are transporters. Some form smaller barrels that are completely filled by amino acid side chains that project into the center of the barrel. These proteins function as receptors or enzymes (Figure 10–23A and B); the barrel serves as a rigid anchor, which holds the protein in the membrane and orients the cytosolic loops that form binding sites for specific intracellular molecules.

Most multipass membrane proteins in eukaryotic cells and in the bacterial plasma membrane are constructed from transmembrane α helices. The helices can slide against each other, allowing conformational changes in the protein that can open and shut ion channels, transport solutes, or transduce extracellular signals into intracellular ones. In β-barrel proteins, by contrast, hydrogen bonds bind each β strand rigidly to its neighbors, making conformational changes within the wall of the barrel unlikely. This rigidity makes β barrels remarkably stable, allowing them to be more easily purified and crystallized than α-helical membrane proteins. This is one reason why β-barrel proteins were among the first multipass proteins whose structures were determined.

Many Membrane Proteins Are Glycosylated

The plasma membrane of a cell is exposed to the often harsh and constantly changing extracellular environment. How are the membrane and its embedded proteins protected from damage? One answer is that many cells have a layer of oligosaccharide and polysaccharide chains that are attached to the lipids and protein domains facing the outside. Most transmembrane proteins in animal cells are glycosylated. As in glycolipids, the sugar residues are added in the lumen of the endoplasmic reticulum and the Golgi apparatus (discussed in Chapters 12 and 13). For this reason, the oligosaccharide chains are always present on the noncytosolic side of the membrane. Another important difference between proteins (or parts of proteins) on the two sides of the membrane results from the reducing environment of the cytosol. This environment decreases the likelihood that intrachain or interchain disulfide (S–S) bonds will form between cysteines on the cytosolic side of membranes. These bonds form on the noncytosolic side, where they can help stabilize either the folded structure of the polypeptide chain or its association with other polypeptide chains (Figure 10–24).

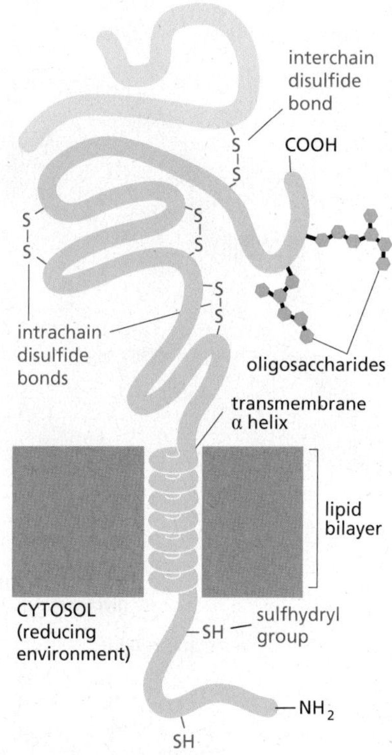

Figure 10–24 **A single-pass transmembrane protein.** Note that the polypeptide chain traverses the lipid bilayer as a right-handed α helix and that the oligosaccharide chains and disulfide bonds are all on the noncytosolic surface of the membrane. As shown, sulfhydryl groups within a transmembrane protein can form either intrachain disulfide bonds with each other or interchain disulfide bonds with sulfhydryl groups in other proteins. The sulfhydryl groups in the cytosolic domain of the protein do not normally form disulfide bonds because the reducing environment in the cytosol maintains these groups in their reduced (–SH) form.

Because the extracellular parts of most plasma membrane proteins are glycosylated, carbohydrates extensively coat the surface of all eukaryotic cells. These carbohydrates occur as oligosaccharide chains covalently bound to membrane proteins (glycoproteins) and lipids (glycolipids). They also occur as the polysaccharide chains of integral membrane *proteoglycan* molecules. Proteoglycans, which consist of long polysaccharide chains linked covalently to a protein core, are found mainly outside the cell, as part of the extracellular matrix (discussed in Chapter 19). But, for some proteoglycans, the protein core either extends across the lipid bilayer or is attached to the bilayer by a glycosylphosphatidylinositol (GPI) anchor.

The terms *cell coat* or *glycocalyx* are sometimes used to describe the carbohydrate-rich zone on the cell surface. This **carbohydrate layer** can be visualized by various stains, such as ruthenium red (Figure 10–25A), as well as by its affinity for carbohydrate-binding proteins called **lectins**, which can be labeled with a fluorescent dye or some other visible marker. Although most of the sugar groups are attached to intrinsic plasma membrane molecules, the carbohydrate layer also contains both glycoproteins and proteoglycans that have been secreted into the extracellular space and then adsorbed onto the cell surface (Figure 10–25B). Many of these adsorbed macromolecules are components of the extracellular matrix, so that the boundary between the plasma membrane and the extracellular matrix is often not sharply defined. One of the many functions of a slippery carbohydrate layer is to protect cells against mechanical and chemical damage; it also keeps various other cells at a distance, preventing unwanted cell-cell interactions.

The oligosaccharide side chains of glycoproteins and glycolipids are enormously diverse in their arrangement of sugars. Although they usually contain fewer than 15 sugars, the chains are often branched, and the sugars can be bonded together by various kinds of covalent linkages—unlike the amino acids in a polypeptide chain, which are all linked by identical peptide bonds. Even three sugars can be put together to form hundreds of different trisaccharides. How sugars can form such a vast variety of different structures is discussed in Chapter 2. Both the diversity and the exposed position of the oligosaccharides on the cell surface make them especially well suited to function in specific cell-recognition processes. Plasma-membrane-bound lectins that recognize specific oligosaccharides on cell-surface glycolipids and glycoproteins mediate a variety of transient cell-cell adhesion processes, including those occurring in lymphocyte recirculation and inflammatory responses (see Figure 19–28).

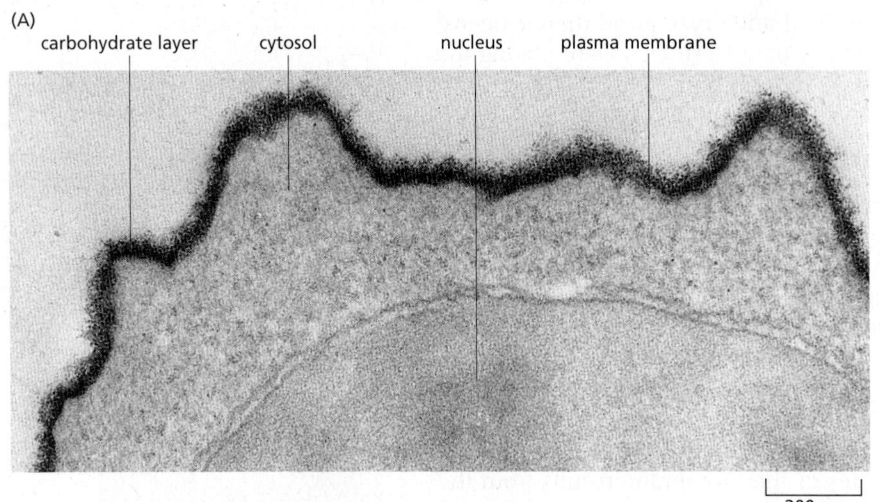

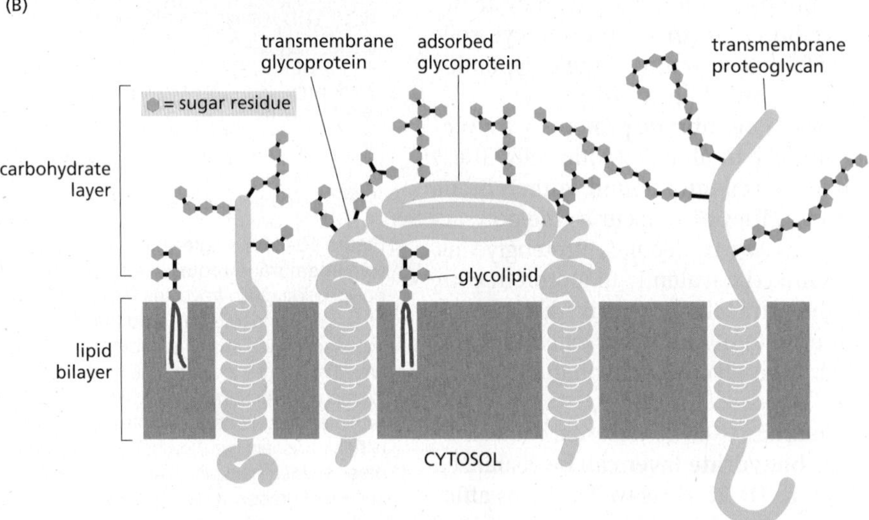

Figure 10–25 The carbohydrate layer on the cell surface. (A) This electron micrograph of the surface of a lymphocyte stained with ruthenium red emphasizes the thick carbohydrate-rich layer surrounding the cell. (B) The carbohydrate layer is made up of the oligosaccharide side chains of membrane glycolipids and membrane glycoproteins and the polysaccharide chains on membrane proteoglycans. In addition, adsorbed glycoproteins, and adsorbed proteoglycans (not shown), contribute to the carbohydrate layer in many cells. Note that all of the carbohydrate is on the extracellular surface of the membrane. (A, courtesy of Audrey M. Gluaert and G.M.W. Cook.)

Membrane Proteins Can Be Solubilized and Purified in Detergents

In general, only agents that disrupt hydrophobic associations and disassemble the lipid bilayer can liberate membrane proteins in a soluble form. The most useful of these for the membrane biochemist are **detergents**, which are small amphiphilic molecules of variable structure (**Movie 10.4**). Detergents are much more soluble in water than in lipids. Their polar (hydrophilic) ends can be either charged (ionic), as in *sodium dodecyl sulfate* (*SDS*), or uncharged (nonionic), as in *β-octylglucoside* and Triton X-100 (**Figure 10–26A**). At low concentration, detergents are monomeric in solution, but when their concentration is increased above a threshold, called the *critical micelle concentration* (*CMC*), they aggregate to form micelles (**Figure 10–26B, C, and D**). Above the CMC, detergent molecules rapidly diffuse in and out of micelles, keeping the concentration of monomer in the solution constant, no matter how many micelles are present. Because of this dynamic behavior, the structure of a micelle changes constantly: at any moment most, but not all, of the hydrophilic ends of the detergent molecules will be external facing the water phase and most, but not all, of the hydrophobic ends will be internal to the micelle. Both the CMC and the average number of detergent molecules in a micelle are characteristic properties of each detergent, but they also depend on the temperature, pH, and salt concentration. Detergent solutions are therefore complex systems and are difficult to study.

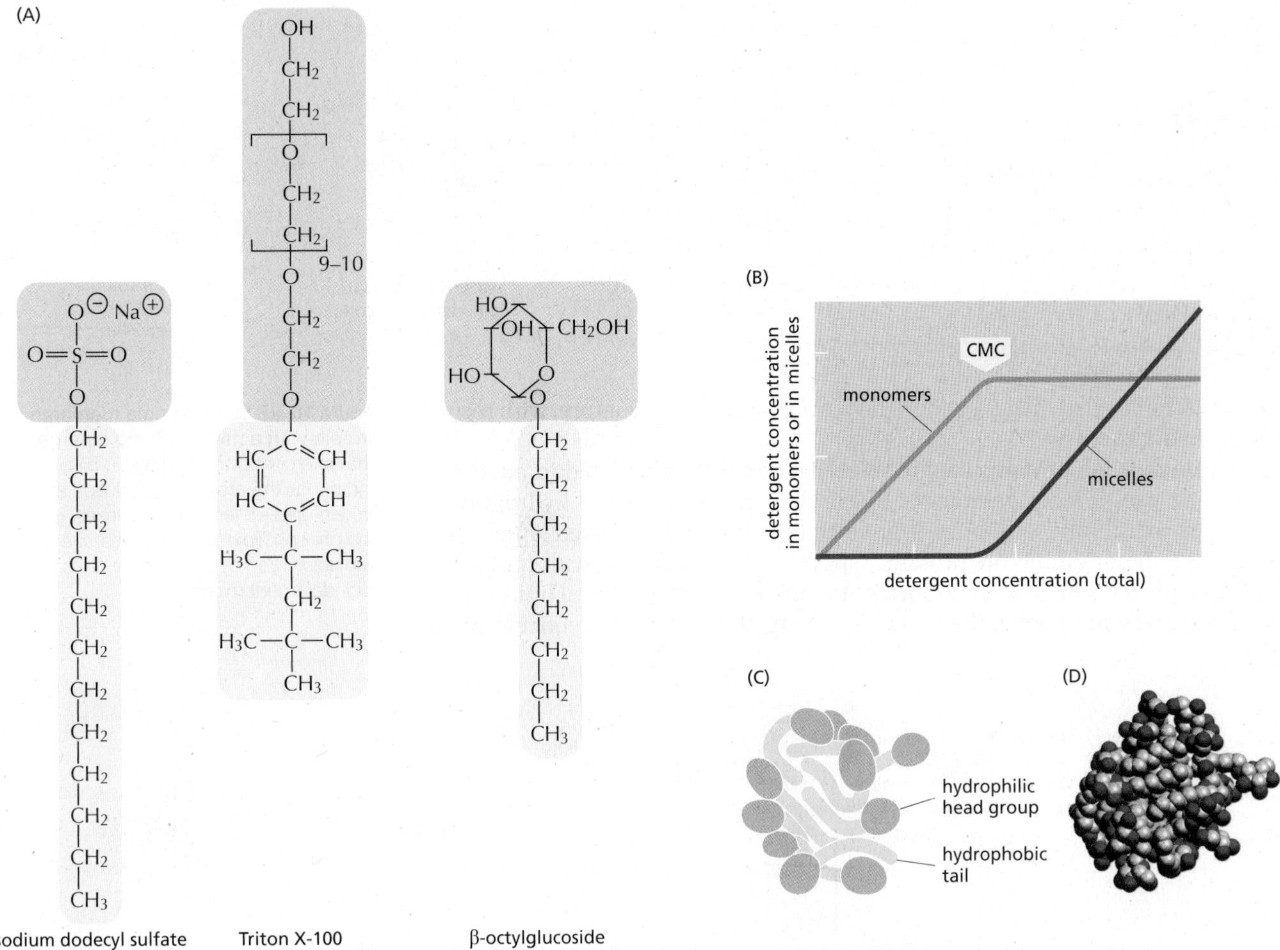

Figure 10–26 **The structure and function of detergents.** (A) Three commonly used detergents are sodium dodecyl sulfate (SDS), an anionic detergent, and Triton X-100 and β-octylglucoside, two nonionic detergents. Triton X-100 is a mixture of compounds in which the region in brackets is repeated between 9 and 10 times. The hydrophobic portion of each detergent is shown in *yellow*, and the hydrophilic portion is shown in *orange*. (B) At low concentration, detergent molecules are monomeric in solution. As their concentration is increased beyond the critical micelle concentration (CMC), some of the detergent molecules form micelles. Note that the concentration of detergent monomer stays constant above the CMC. (C) Because they have both polar and nonpolar ends, detergent molecules are amphiphilic; and because they are cone-shaped, they form micelles rather than bilayers (see Figure 10–7). Detergent micelles are thought to have constantly changing, irregular shapes. Because of packing constraints, the hydrophobic tails are partially exposed to water. (D) The space-filling model shows a snapshot in time of a micelle composed of 20 β-octylglucoside molecules, predicted by *molecular dynamics calculations*. The head groups are shown in *red* and the hydrophobic tails in *gray*. Note that the hydrophobic regions are transiently exposed. (B, adapted with permission from G. Gunnarsson, B. Jönsson, and H. Wennerström, *J. Phys. Chem. A* 84:3114–3121, 1980. Copyright 1980 American Chemical Society; C, from S. Bogusz, R.M. Venable, and R.W. Pastor, *J. Phys. Chem. B* 104:5462–5470, 2000.)

When mixed with membranes, the hydrophobic ends of detergents bind to the hydrophobic regions of the membrane proteins, where they displace lipid molecules with a collar of detergent molecules. Because the other end of the detergent molecule is polar, this binding tends to bring the membrane proteins into solution as detergent-protein complexes (**Figure 10–27**). Usually, some lipid molecules also remain attached to the protein.

Strong ionic detergents, such as SDS, can solubilize even the most hydrophobic membrane proteins. This allows the proteins to be analyzed by *SDS polyacrylamide-gel electrophoresis* (discussed in Chapter 8). Such strong detergents, however, unfold (denature) proteins by disrupting their internal hydrophobic cores, thereby rendering the proteins inactive and unusable for functional studies. Nonetheless, proteins can be readily separated and purified in their SDS-denatured form. In

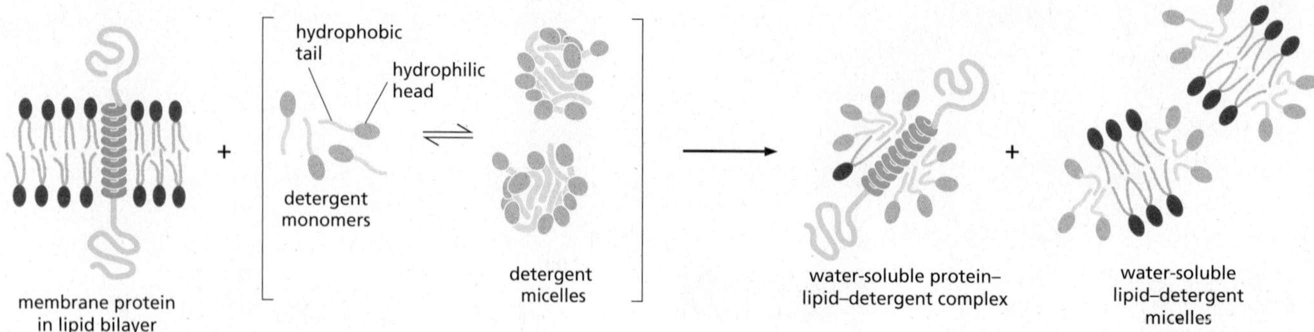

Figure 10–27 **Solubilizing a membrane protein with a mild nonionic detergent.** The detergent disrupts the lipid bilayer and brings the protein into solution as protein–lipid–detergent complexes. The phospholipids in the membrane are also solubilized by the detergent, as lipid–detergent micelles.

some cases, removal of the SDS allows the purified protein to renature, with recovery of functional activity.

Many membrane proteins can be solubilized and then purified in an active form by the use of mild detergents. These detergents cover the hydrophobic regions on membrane-spanning segments that become exposed after lipid removal but do not unfold the protein. Especially when working with multipass membrane proteins, it is often important to maintain a thin layer of lipids upon detergent extraction to retain the protein's activity. If the detergent concentration

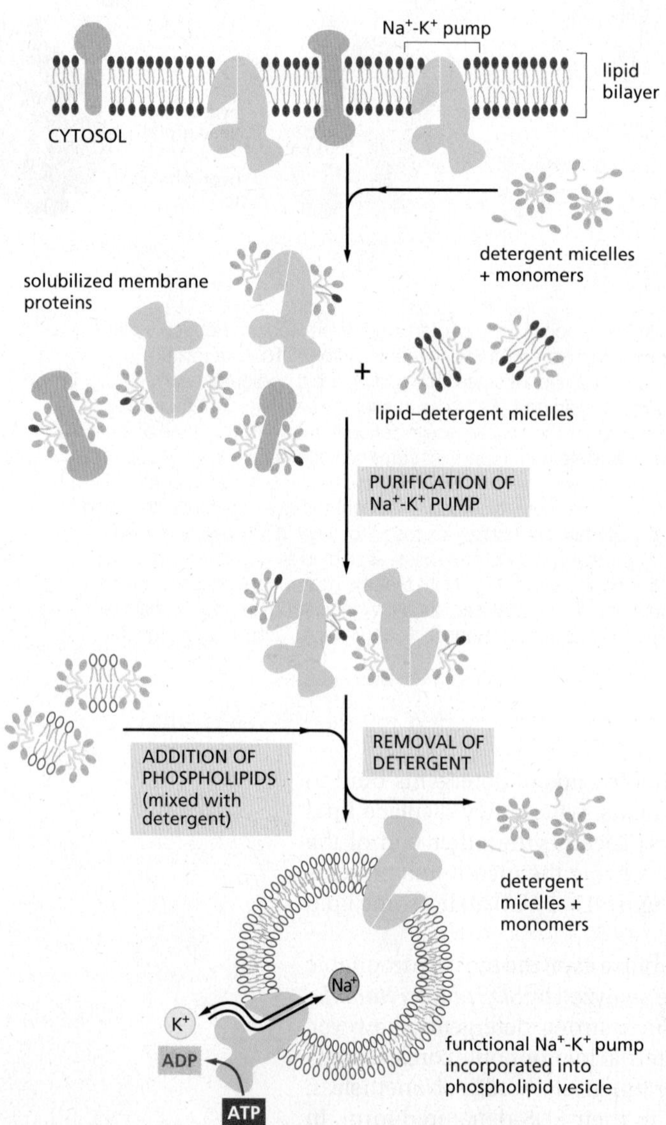

Figure 10–28 **The use of mild nonionic detergents for solubilizing, purifying, and reconstituting functional membrane protein systems.** In this example, functional Na^+-K^+ pump molecules are purified and incorporated into phospholipid vesicles. This pump is present in the plasma membrane of most animal cells, where it uses the energy of ATP hydrolysis to pump Na^+ out of the cell and K^+ in, as discussed in Chapter 11. The phospholipids that are newly added in the reconstitution experiments are shown with white polar head groups.

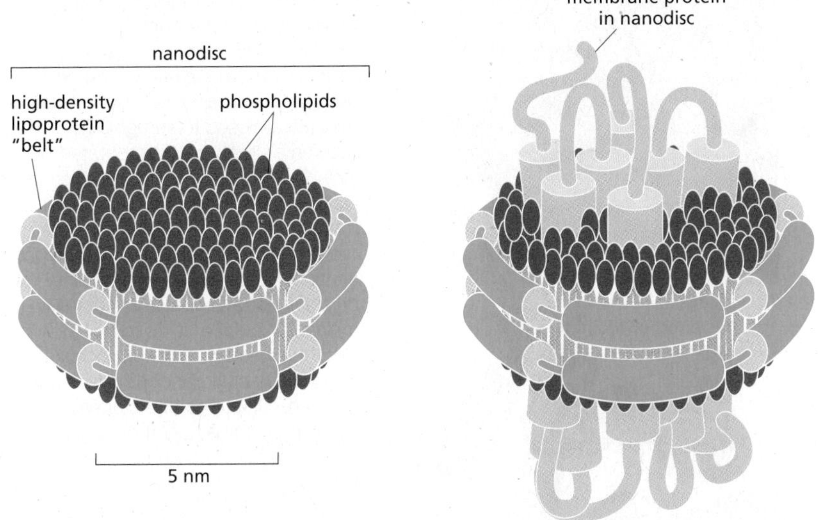

Figure 10–29 Model of a membrane protein reconstituted into a nanodisc. When detergent is removed from a solution containing a multipass membrane protein, lipids, and a protein subunit of the high-density lipoprotein (HDL), the membrane protein becomes embedded in a small patch of lipid bilayer, which is surrounded by a belt of the HDL protein. In such nanodiscs, the hydrophobic edges of the bilayer patch are shielded by the protein belt, which renders the assembly water-soluble.

of a solution of solubilized membrane proteins is reduced (by dilution, for example), membrane proteins do not remain soluble. In the presence of an excess of phospholipid molecules in such a solution, however, membrane proteins incorporate into small liposomes that form spontaneously. In this way, functionally active membrane protein systems can be reconstituted from purified components, providing a powerful means of analyzing the activities of membrane transporters, ion channels, signaling receptors, and so on (**Figure 10–28**). Such functional reconstitution can determine which proteins are both necessary and sufficient for a particular cell function. For example, the approach provided proof for the hypothesis that the enzymes that make ATP (ATP synthases) use H^+ gradients in mitochondrial, chloroplast, and bacterial membranes to produce ATP.

Membrane proteins can also be reconstituted from detergent solution into nanodiscs, which are small, uniformly sized patches of membrane that are surrounded by a belt of a specially designed protein, which covers the exposed edge of the bilayer to keep the patch in solution (**Figure 10–29**). The belt protein is derived from high-density lipoproteins (HDLs), whose normal function is to keep lipids soluble for transport in the blood. In nanodiscs the membrane protein of interest can be studied in its native lipid environment and is experimentally accessible from both sides of the bilayer, which is useful, for example, for ligand-binding experiments. Proteins contained in nanodiscs can also be analyzed by single-particle electron microscopy techniques to determine their structure. By this technique (discussed in Chapter 9), the structure of a membrane protein can be determined to high resolution without a requirement of the protein of interest to crystallize into a regular lattice, which is often hard to achieve for membrane proteins. These developments have led to a rapid increase in the number of three-dimensional structures of membrane proteins and protein complexes that are known, although they are still few compared to the known structures of water-soluble proteins and protein complexes.

Bacteriorhodopsin Is a Light-driven Proton (H^+) Pump That Traverses the Lipid Bilayer as Seven α Helices

In Chapter 11, we consider how multipass transmembrane proteins mediate the selective transport of small hydrophilic molecules across cell membranes. But a detailed understanding of how such a membrane transport protein works requires precise information about its three-dimensional structure in the bilayer. *Bacteriorhodopsin* was the first membrane transport protein whose structure was determined, and it emerged as the prototype of many multipass membrane proteins that have a similar structure.

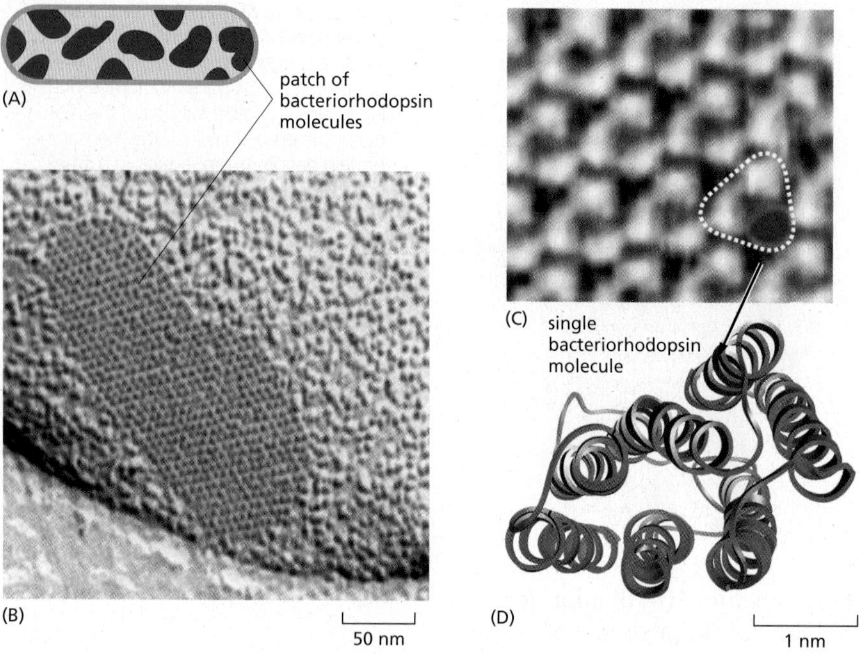

Figure 10–30 Patches of purple membrane, which contain bacteriorhodopsin in the archaeon *Halobacterium salinarum*. (A) These archaea live in saltwater pools, where they are exposed to sunlight. They have evolved a variety of light-activated proteins, including bacteriorhodopsin, which is a light-activated H^+ pump in the plasma membrane. (B) The bacteriorhodopsin molecules in the purple membrane patches are tightly packed into two-dimensional crystalline arrays. (C) Details of the molecular surface visualized by atomic force microscopy. With this technique, individual bacteriorhodopsin molecules can be seen. (D) Outline of the approximate location of the bacteriorhodopsin monomer and the individual α helices in the image shown in C. (B–C, courtesy of Dieter Oesterhelt; D, PDB code: 2BRD.)

The "purple membrane" of the archaeon *Halobacterium salinarum* is a specialized patch in the plasma membrane that contains a single species of protein molecule, **bacteriorhodopsin** (**Figure 10–30A**). The protein functions as a light-activated H^+ pump that transfers H^+ out of the archaeal cell. The ability of bacteriorhodopsin molecules to tightly pack with each other into a planar two-dimensional crystal (**Figure 10–30B, C, and D**) facilitated the determination of its three-dimensional structure.

Each bacteriorhodopsin molecule is folded into seven closely packed transmembrane α helices and contains a single light-absorbing group, or chromophore (in this case, *retinal*), which gives the protein its purple color. Retinal is vitamin A in its aldehyde form and is identical to the chromophore found in *rhodopsin* of the photoreceptor cells of the vertebrate eye (discussed in Chapter 15). Retinal is covalently linked to a lysine side chain of the bacteriorhodopsin protein. When activated by a single photon of light, the excited chromophore changes its shape and causes a series of small conformational changes in the protein, resulting in the transfer of one H^+ from the inside to the outside of the cell (**Figure 10–31A**). In bright light, each bacteriorhodopsin molecule can pump several hundred protons per second. The light-driven proton transfer establishes an H^+ gradient across the plasma membrane, which in turn drives the production of ATP by a second protein in the cell's plasma membrane. The energy stored in the

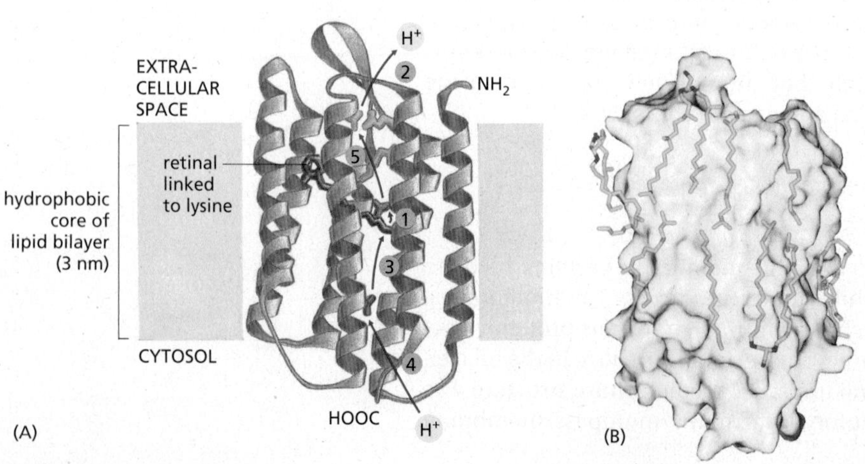

Figure 10–31 The three-dimensional structure of a bacteriorhodopsin molecule. (Movie 10.5) (A) The polypeptide chain crosses the lipid bilayer seven times as α helices. The location of the retinal chromophore *(purple)* and the probable pathway taken by H^+ during the light-activated pumping cycle *(red arrows)* are shown. The first and key step is the passing of an H^+ from the chromophore to the side chain of aspartic acid 85 *(red*, located next to the chromophore) that occurs upon absorption of a photon by the chromophore. Subsequently, other H^+ transfer steps—in the numerical order indicated and utilizing the hydrophilic amino acid side chains that line a path through the membrane—complete the pumping cycle and return the enzyme to its starting state. (Movie 10.5 explains how the individual transfer steps are linked mechanistically.) Color code: glutamic acid *(orange)*, aspartic acid *(red)*, arginine *(blue)*. (B) The high-resolution crystal structure of bacteriorhodopsin shows many lipid molecules *(yellow* with *red* head groups) that are tightly bound to specific places on the surface of the protein. (A, adapted from H. Luecke et al., *Science* 286:255–261, 1999. B, from H. Luecke et al., *J. Mol. Biol.* 291:899–911, 1999. With permission from Elsevier.)

H^+ gradient also drives other energy-requiring processes in the cell. Thus, bacteriorhodopsin converts solar energy into an H^+ gradient, which provides energy to the archaeal cell.

The high-resolution crystal structure of bacteriorhodopsin reveals many lipid molecules bound in specific places on the protein surface (**Figure 10–31B**). Interactions with specific lipids are thought to help stabilize many membrane proteins, which work best and sometimes crystallize more readily if some of the lipids remain bound during detergent extraction or if specific lipids are added back to the proteins in detergent solutions. The specificity of these lipid–protein interactions helps explain why eukaryotic membranes contain such a variety of lipids, with head groups that differ in size, shape, and charge. This layer of lipids also helps to fill gaps between the jagged hydrophobic surface of the membrane protein and the hydrophobic core of the lipid bilayer and to maintain the permeability barrier of the membrane for ions and other solutes. These lipids are in dynamic association with the protein surface, binding and dissociating at a millisecond time scale. We can think of the membrane lipids as constituting a two-dimensional solvent for the proteins in the membrane, just as water constitutes a three-dimensional solvent for proteins in an aqueous solution: some membrane proteins can function only in the presence of specific lipid head groups, just as many enzymes in aqueous solution require a particular ion for activity.

Bacteriorhodopsin is a member of a large superfamily of membrane proteins with similar structures but different functions. For example, rhodopsin in rod cells of the vertebrate retina and many cell-surface receptor proteins that bind extracellular signal molecules are also built from seven transmembrane α helices. These proteins function as signal transducers rather than as transporters: each responds to an extracellular signal by activating a GTP-binding protein (G protein) inside the cell, and they are therefore called *G-protein-coupled receptors* (*GPCRs*), as we discuss in Chapter 15 (see Figure 15–6B). Although the structures of bacteriorhodopsins and GPCRs are strikingly similar, they show no sequence similarity and thus probably belong to two evolutionarily distant branches of an ancient protein family. A related class of membrane proteins, the *channelrhodopsins* that green algae use to detect light, form ion channels when they absorb a photon. When engineered so that they are expressed in animal brains, these proteins have become invaluable tools in neurobiology because they allow specific neurons to be stimulated experimentally by shining light on them, as we discuss in Chapter 11 (Figure 11–47).

Membrane Proteins Often Function as Large Complexes

Many membrane proteins function as part of multicomponent complexes. One is a bacterial photosynthetic reaction center, which was the first membrane protein complex to be crystallized and analyzed by x-ray diffraction. In Chapter 14, we discuss how such photosynthetic complexes function to capture light energy and use it to pump H^+ across the membrane. Many of the membrane protein complexes involved in photosynthesis, proton pumping, and electron transport are even larger than the photosynthetic reaction center. The enormous photosystem II complex from cyanobacteria, for example, contains 19 protein subunits and well over 60 transmembrane helices (see Figure 14–49). Membrane proteins are often arranged in large complexes, not only for harvesting various forms of energy but also for transducing extracellular signals into intracellular ones (discussed in Chapter 15).

Many Membrane Proteins Diffuse in the Plane of the Membrane

Like most membrane lipids, membrane proteins do not tumble (flip-flop) across the lipid bilayer. Tumbling would require large hydrophilic domains to pass through the membrane's hydrophobic core, which is energetically prohibitive. But just like membrane lipids, proteins can rotate rapidly about an axis perpendicular to the plane of the bilayer (*rotational diffusion*) and move laterally within the membrane (*lateral diffusion*). An experiment in which mouse cells were

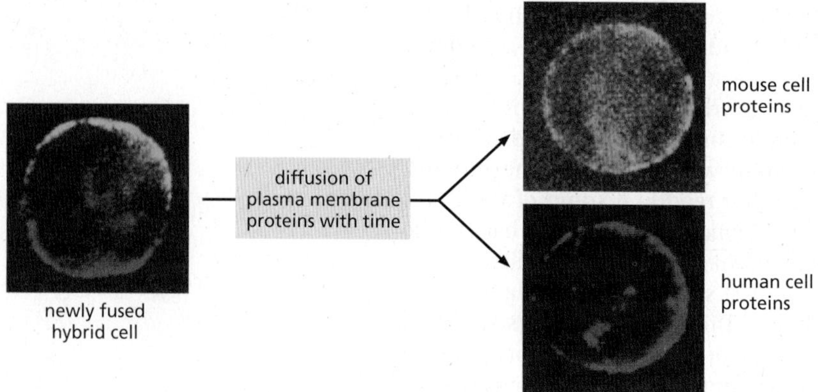

Figure 10–32 **An experiment demonstrating the diffusion of proteins in the plasma membrane of mouse–human hybrid cells.** In this experiment, a mouse and a human cell were fused to create a hybrid cell, which was then stained with two fluorescently labeled antibodies. One antibody (labeled with a *green* dye) detects mouse plasma membrane proteins, the other antibody (labeled with a *red* dye) detects human plasma membrane proteins. When cells are stained immediately after fusion, mouse and human plasma membrane proteins are still found in the membrane domains originating from the mouse and human cell, respectively. After a short time, however, the plasma membrane proteins diffuse over the entire cell surface and completely intermix. (From L.D. Frye and M. Edidin, *J. Cell Sci.* 7:319–335, 1970. With permission from the Company of Biologists.)

artificially fused with human cells to produce hybrid cells (*heterokaryons*) provided the first direct evidence that some plasma membrane proteins are mobile in the plane of the membrane. Two differently labeled antibodies were used to distinguish selected mouse and human plasma membrane proteins. Although at first the mouse and human proteins were confined to their own halves of the newly formed heterokaryon, the two sets of proteins diffused and mixed over the entire cell surface in about half an hour (Figure 10–32). Lateral membrane protein mobility is important as it allows many cell signaling proteins to assemble and disassemble into protein complexes in response to extracellular ligands, turning their signaling functions on and off.

The lateral diffusion rates of membrane proteins and lipids can be measured by using the technique of *fluorescence recovery after photobleaching* (*FRAP*). The method usually involves marking the membrane protein of interest with a specific fluorescent group. This can be done either with a fluorescent ligand such as a fluorophore-labeled antibody that binds to the protein or with recombinant DNA technology to express the protein fused to a fluorescent protein such as green fluorescent protein (GFP; discussed in Chapter 9). The fluorescent group is then bleached in a small area of membrane by a laser beam, and the time taken for adjacent membrane proteins carrying unbleached ligand or GFP to diffuse into the bleached area is measured (Figure 10–33). From FRAP measurements, we can estimate the diffusion coefficient for the marked cell-surface protein. The values of the diffusion coefficients for different membrane proteins in different cells are highly variable, because interactions with other proteins impede the diffusion of the proteins to varying degrees. Measurements of proteins that are minimally impeded in this way indicate that cell membranes have a viscosity comparable to that of olive oil.

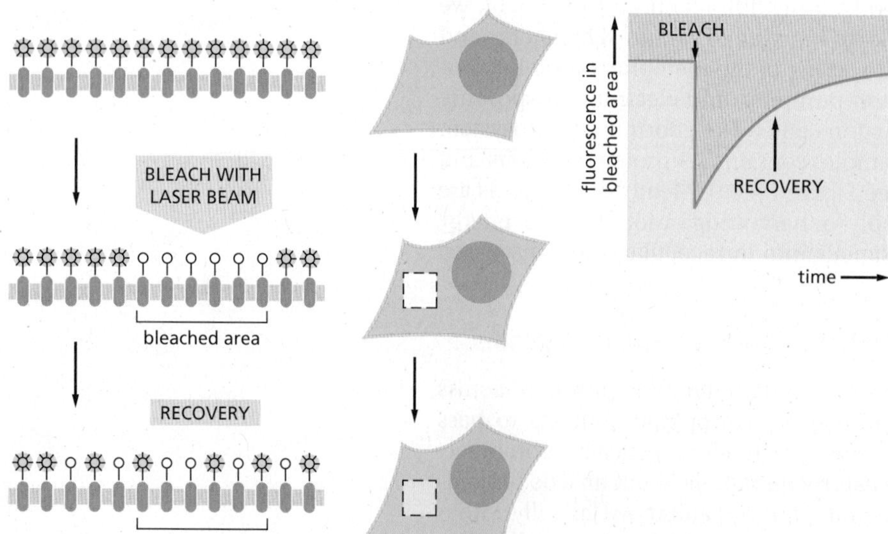

Figure 10–33 **Measuring the rate of lateral diffusion of a membrane protein by fluorescence recovery after photobleaching.** A specific protein of interest can be expressed as a fusion protein with green fluorescent protein (GFP), which is intrinsically fluorescent. The fluorescent molecules are bleached in a small area using a laser beam. The fluorescence intensity recovers as the bleached molecules diffuse away and unbleached molecules diffuse into the irradiated area (shown here in side and top views). The diffusion coefficient is calculated from a graph of the rate of recovery: the greater the diffusion coefficient of the membrane protein, the faster the recovery (see Figure 9–20 and Movie 10.6).

One drawback to the FRAP technique is that it monitors the movement of large populations of molecules in a relatively large area of membrane; one cannot follow individual protein molecules. If a protein fails to migrate into a bleached area, for example, one cannot tell whether the molecule is truly immobile or just restricted in its movement to a very small region of membrane—perhaps by cytoskeletal proteins. *Single-particle tracking* techniques overcome this problem by labeling individual membrane molecules with antibodies coupled to fluorescent dyes or tiny gold particles and tracking their movement by video microscopy. Using single-particle tracking, one can record the diffusion path of a single membrane protein molecule over time. Results from all of these techniques indicate that plasma membrane proteins differ widely in their diffusion characteristics, as we now discuss.

Cells Can Confine Proteins and Lipids to Specific Domains Within a Membrane

The recognition that biological membranes are two-dimensional fluids was a major advance in understanding membrane structure and function. It has become clear, however, that the picture of a membrane as a lipid sea in which all proteins float freely is greatly oversimplified. Most cells confine membrane proteins to specific regions in a continuous lipid bilayer. We have already discussed how bacteriorhodopsin molecules in the purple membrane of *Halobacterium* assemble into large two-dimensional crystals, in which individual protein molecules are relatively fixed in relationship to one another (see Figure 10–30). ATP synthase complexes in the inner mitochondrial membrane associate into long double rows, as we discuss in Chapter 14 (see Figure 14–33). Large aggregates of this kind diffuse very slowly.

In epithelial cells, such as those that line the gut or the tubules of the kidney, certain plasma membrane enzymes and transport proteins are confined to the apical surface of the cells, whereas others are confined to the basal and lateral surfaces (**Figure 10–34**). This asymmetric distribution of membrane proteins is often essential for the function of the epithelium, as we discuss in Chapter 11 (see Figure 11–11). The lipid compositions of these two membrane domains are also different, demonstrating that epithelial cells can prevent the diffusion of lipid as well as protein molecules between the domains. The barriers set up by a specific type of intercellular junction (called a *tight junction*, discussed in Chapter 19; see Figure 19–18) maintain the separation of both protein and lipid molecules.

A cell can also create membrane domains without using intercellular junctions. As we already discussed, regulated protein–protein interactions in membranes can create nanometer-scale raft domains that are thought to function in signaling and membrane trafficking. A more extreme example is seen in the mammalian spermatozoon, a single cell that consists of several structurally and functionally distinct parts covered by a continuous plasma membrane. When a sperm cell is examined

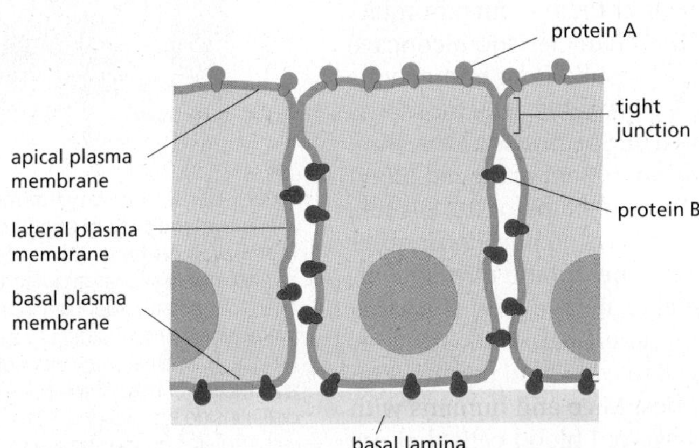

Figure 10–34 **How membrane molecules can be restricted to a particular membrane domain.** In this drawing of an epithelial cell, protein A (in the apical domain of the plasma membrane) and protein B (in the basal and lateral domains) can diffuse laterally in their own domains but are prevented from entering the other domain, at least partly by the specialized cell–cell junction called a tight junction. Lipid molecules in the outer (extracellular) monolayer of the plasma membrane are likewise unable to diffuse between the two domains; lipids in the inner (cytosolic) monolayer, however, are able to do so (not shown). The basal lamina is a thin mat of extracellular matrix that separates epithelial sheets from other tissues (discussed in Chapter 19).

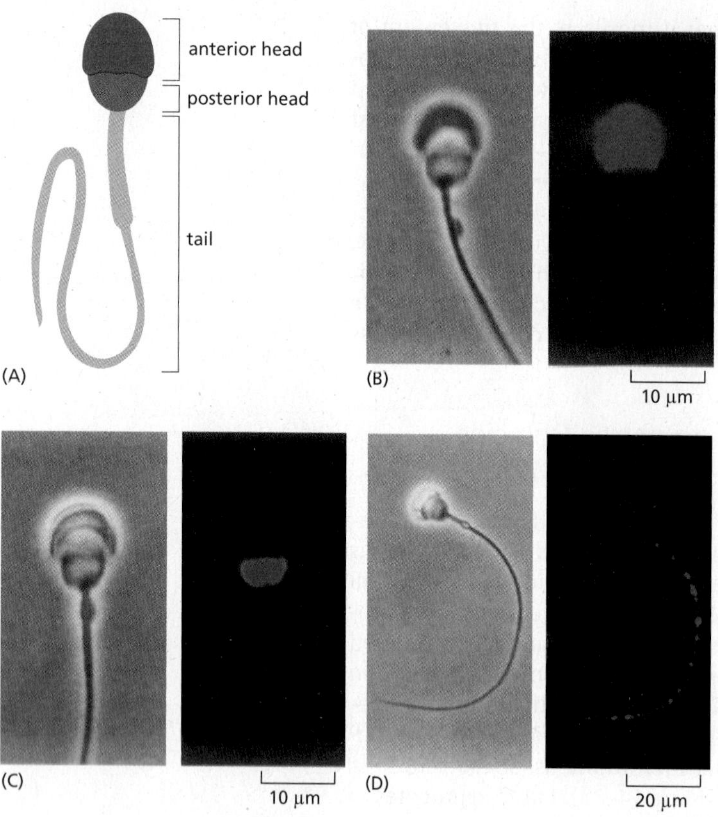

Figure 10–35 **Three domains in the plasma membrane of a guinea pig sperm.** (A) A drawing of a guinea pig sperm. (B–D) In the three pairs of micrographs, phase-contrast micrographs are on the *left*, and the same cell is shown with cell-surface immunofluorescence staining on the *right*. Different monoclonal antibodies selectively label cell-surface molecules on (B) the anterior head, (C) the posterior head, and (D) the tail. (From D.G. Myles, P. Primakoff, and A.R. Belvé, *Cell* 23:433–439, 1981. With permission from Elsevier.)

by immunofluorescence microscopy with a variety of antibodies, each of which reacts with a specific cell-surface molecule, the plasma membrane is found to consist of at least three distinct domains (Figure 10–35). Some of the membrane molecules can diffuse freely within the confines of their own domain. While the molecular nature of most "fences" that prevent the molecules from leaving their domain is not known, some molecular mechanisms for restricting membrane protein movements are understood. The plasma membrane of nerve cells, for example, contains a domain enclosing the cell body and dendrites, and another enclosing the axon; in this case a belt of actin filaments tightly associates with the plasma membrane at the cell-body–axon junction and forms part of the barrier.

Figure 10–36 shows four common ways of immobilizing specific membrane proteins through protein–protein interactions.

The Cortical Cytoskeleton Gives Membranes Mechanical Strength and Restricts Membrane Protein Diffusion

As shown in Figure 10–36B and C, a common way in which a cell restricts the lateral mobility of specific membrane proteins is to tether them to macromolecular assemblies on either side of the membrane. The characteristic biconcave shape of a red blood cell (Figure 10–37), for example, results from interactions of its plasma membrane proteins with an underlying *cytoskeleton*, which consists mainly of a meshwork of the filamentous protein **spectrin**. Spectrin is a long, thin, flexible rod about 100 nm in length. As the principal component of the red blood cell cytoskeleton, it maintains the structural integrity and shape of the plasma membrane, which is the cell's only membrane, as the cell has no nucleus or other organelles. The spectrin cytoskeleton is attached to the membrane through various membrane proteins. The final result is a deformable, netlike meshwork that covers the entire cytosolic surface of the cell membrane (Figure 10–38). This spectrin-based cytoskeleton enables the red blood cell to withstand the stress on its membrane as it is forced through narrow capillaries. Mice and humans with genetic abnormalities in spectrin are anemic and have red blood cells that are

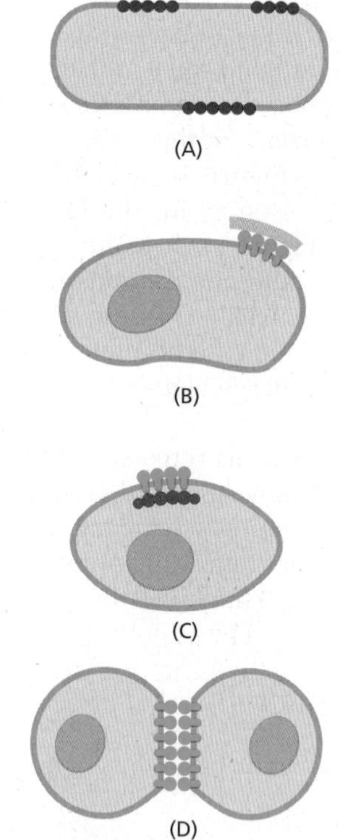

Figure 10–36 **Four ways of restricting the lateral mobility of specific plasma membrane proteins.** (A) The proteins can self-assemble into large aggregates (as seen for bacteriorhodopsin in the purple membrane of *Halobacterium salinarum*); they can be tethered by interactions with assemblies of macromolecules (B) outside or (C) inside the cell; or (D) they can interact with proteins on the surface of another cell.

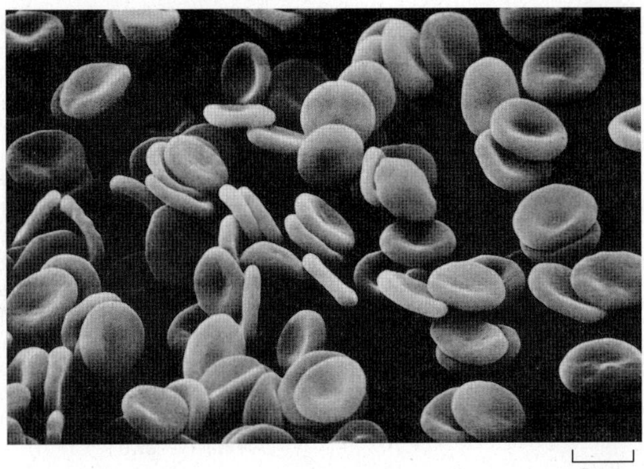

Figure 10–37 **A scanning electron micrograph of human red blood cells.** The cells have a biconcave shape and lack a nucleus and other organelles (Movie 10.7). (Courtesy of Bernadette Chailley.)

spherical (instead of concave) and fragile; the severity of the anemia increases with the degree of spectrin deficiency.

An analogous but much more elaborate and highly dynamic cytoskeletal network exists beneath the plasma membrane of most other cells in our body. This network, which constitutes the **cortex** of the cell, is rich in actin filaments, which are attached to the plasma membrane in numerous ways. The dynamic

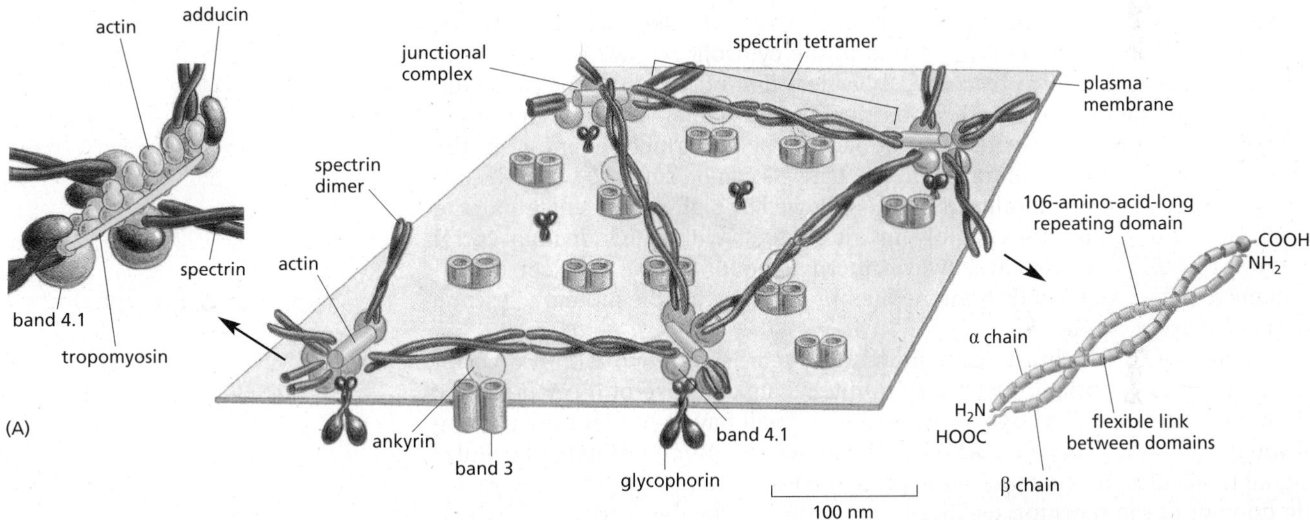

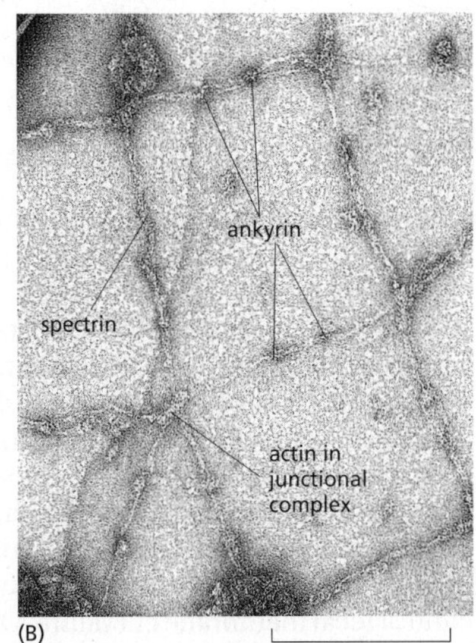

Figure 10–38 **The spectrin-based cytoskeleton on the cytosolic side of the human red blood cell plasma membrane.** (A) The arrangement shown in the drawing has been deduced mainly from studies on the interactions of purified proteins *in vitro.* Spectrin heterodimers (enlarged in the drawing on the *right*) are linked together into a netlike meshwork by *junctional complexes* (enlarged in the drawing on the *left*). Each spectrin heterodimer consists of two antiparallel, loosely intertwined, flexible polypeptide chains called α and β. The two spectrin chains are attached noncovalently to each other at multiple points, including at both ends. Both the α and β chains are composed largely of repeating domains. Two spectrin heterodimers join end-to-end to form tetramers.

The junctional complexes are composed of short actin filaments (containing 13 actin monomers) and the proteins *band 4.1* and *adducin*, as well as a *tropomyosin* molecule that probably determines the length of the actin filaments. The cytoskeleton is linked to the membrane through two transmembrane proteins: a multipass protein called band 3 and a single-pass protein called glycophorin. The spectrin tetramers bind to some band 3 proteins via *ankyrin* molecules, and to glycophorin and band 3 (not shown) via band 4.1 proteins.

(B) The electron micrograph shows the cytoskeleton on the cytosolic side of a red blood cell membrane after fixation and negative staining. The spectrin meshwork has been purposely stretched out to allow the details of its structure to be seen. In a normal cell, the meshwork shown would be much more crowded and occupy only about one-tenth of this area. (B, courtesy of T. Byers and D. Branton, *Proc. Natl. Acad. Sci. USA* 82:6153–6157, 1985.)

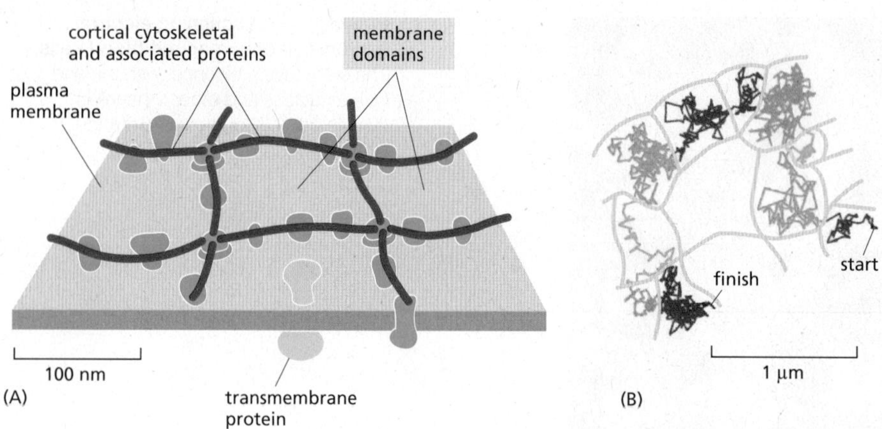

Figure 10–39 **Corralling plasma membrane proteins by cortical cytoskeletal filaments.** (A) The filaments are thought to provide diffusion barriers that divide the membrane into small domains, or corrals. (B) High-speed, single-particle tracking was used to follow the path of a single fluorescently labeled membrane protein of one type over time. The trace shows that an individual protein molecule diffuses within tightly delimited membrane domains and only infrequently escapes into a neighboring domain (highlighted by a switch of color). (Adapted from A. Kusumi et al., *Annu. Rev. Biophys. Biomol. Struct.* 34:351–378, 2005. With permission from Annual Reviews.)

remodeling of the cortical actin network provides a driving force for many essential cell functions, including cell movement, endocytosis, and the formation of transient, mobile plasma membrane structures such as filopodia and lamellipodia discussed in Chapter 16. The cortex of nucleated cells also contains proteins that are structurally homologous to spectrin and the other components of the red cell cytoskeleton. We discuss the cortical cytoskeleton in nucleated cells and its interactions with the plasma membrane in Chapter 16.

The cortical cytoskeletal network restricts diffusion of plasma membrane proteins beyond those that are directly anchored to it. Because the cytoskeletal filaments are often closely apposed to the cytosolic surface of the plasma membrane, they can form mechanical barriers that obstruct the free diffusion of proteins in the membrane. These barriers partition the membrane into small domains, or *corrals* (**Figure 10–39A**), which can be either permanent, as in the sperm (see Figure 10–35), or transient. The barriers can be detected when the diffusion of individual membrane proteins is followed by high-speed, single-particle tracking. The proteins diffuse rapidly but are confined within an individual corral (**Figure 10–39B**); occasionally, however, thermal motions cause a few cortical filaments to detach transiently from the membrane, allowing the protein to escape into an adjacent corral.

The extent to which a transmembrane protein is confined within a corral depends on its association with other proteins and the size of its cytoplasmic domain; proteins with a large cytosolic domain will have a harder time passing through cytoskeletal barriers. When a cell-surface receptor binds its extracellular signal molecules, for example, large protein complexes build up on the cytosolic domain of the receptor, making it more difficult for the receptor to escape from its corral. It is thought that corralling helps concentrate such signaling complexes, increasing the speed and efficiency of the signaling process (discussed in Chapter 15).

Membrane-bending Proteins Deform Bilayers

Cell membranes assume many different shapes, as illustrated by the elaborate and varied structures of cell-surface protrusions and membrane-enclosed organelles in eukaryotic cells. Flat sheets, narrow tubules, round vesicles, fenestrated sheets, and pita-bread-shaped cisternae are all part of the repertoire. Often, a variety of shapes will be present in different regions of the same continuous bilayer. Membrane shape is controlled dynamically, as many essential cell processes—including vesicle budding, cell movement, and cell division—require elaborate transient membrane deformations. In many cases, membrane shape is influenced by dynamic pushing and pulling forces exerted by cytoskeletal or extracellular structures, as we discuss in Chapters 13 and 16. A crucial part in producing these deformations is played by **membrane-bending proteins**, which control local membrane curvature. Often, cytoskeletal dynamics and membrane-bending-protein forces work together. Membrane-bending proteins attach to

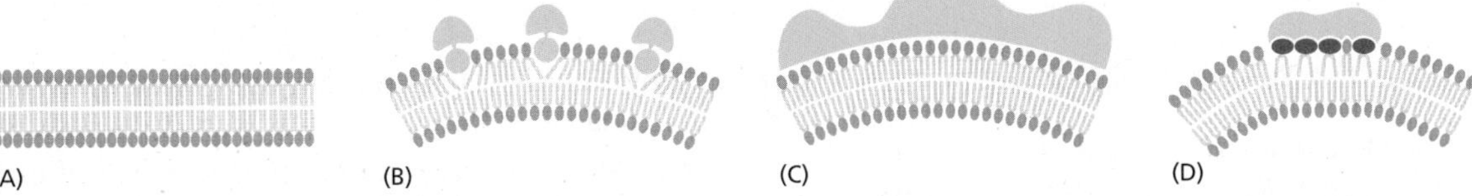

Figure 10–40 **Three ways in which membrane-bending proteins shape membranes.** Lipid bilayers are *gray* and proteins are *green*. (A) Bilayer without protein bound. (B) A hydrophobic region of the protein can insert as a wedge into one monolayer to pry lipid head groups apart. Such regions can either be amphiphilic helices as shown or hydrophobic hairpins. (C) The curved surface of the protein can bind to lipid head groups and deform the membrane or stabilize its curvature. (D) A protein can bind to and cluster lipids that have large head groups and thereby bend the membrane. (Adapted from W.A. Prinz and J.E. Hinshaw, *Crit. Rev. Biochem. Mol. Biol.* 44:278–291, 2009.)

specific membrane regions as needed and act by one or more of three principal mechanisms (Figure 10–40):

1. Some insert hydrophobic protein domains or attached lipid anchors into one of the leaflets of a lipid bilayer. Increasing the area of only one leaflet causes the membrane to bend (Figure 10–40B). The proteins that shape the convoluted network of narrow endoplasmic reticulum tubules work in this way (discussed in Chapter 12).
2. Some membrane-bending proteins form rigid scaffolds that deform the membrane or stabilize an already bent membrane (Figure 10–40C). The coat proteins that shape the budding vesicles in intracellular transport fall into this class (discussed in Chapter 13).
3. Asymmetric distribution of cone-shaped and inverted cone-shaped lipids in the inner or outer leaflets can cause membrane bending. Some membrane-bending proteins cause particular membrane lipids to cluster together, thereby inducing membrane curvature. The ability of a lipid to induce positive or negative membrane curvature is determined by the relative cross-sectional areas of its head group and its hydrocarbon tails. For example, the large head group of phosphoinositides make these lipid molecules wedge-shaped, and their accumulation in a domain of one leaflet of a bilayer therefore induces positive curvature (Figure 10–40D). By contrast, phospholipases that remove lipid head groups produce inversely shaped lipid molecules that induce negative curvature.

Often, different membrane-bending proteins collaborate to achieve a particular curvature, as in shaping a budding transport vesicle, as we discuss in Chapter 13.

Summary

Whereas the lipid bilayer determines the basic structure of biological membranes, proteins are responsible for most membrane functions, serving as specific receptors, enzymes, transporters, and so on. Transmembrane proteins extend across the lipid bilayer. Some of these membrane proteins are single-pass proteins, in which the polypeptide chain crosses the bilayer as a single α helix. Others are multipass proteins, in which the polypeptide chain crosses the bilayer multiple times—either as a series of α helices or as a β sheet rolled up into the shape of a barrel. All proteins responsible for the transport of ions and other small water-soluble molecules through the membrane are multipass proteins. Some membrane proteins do not span the bilayer but instead are attached to either side of the membrane: some are attached to the cytosolic side by an amphipathic α helix on the protein surface or by the covalent attachment of one or more lipid chains, others are attached to the noncytosolic side by a GPI anchor. Some membrane-associated proteins are bound by noncovalent interactions with transmembrane proteins. In the plasma membrane of all eukaryotic cells, most of the proteins exposed on the cell surface and some of the lipid molecules in the outer lipid monolayer have oligosaccharide chains covalently attached to them. Like the lipid molecules in the bilayer, many membrane proteins are able to diffuse rapidly in the plane of the membrane. However, cells have ways of immobilizing specific membrane proteins, as well as ways of confining both membrane protein and lipid molecules to particular domains in a continuous lipid bilayer. The dynamic association of membrane-bending proteins confers on membranes their characteristic three-dimensional shapes.

PROBLEMS

Which statements are true? Explain why or why not.

10–1 It is estimated that about 30% of the proteins encoded in an animal's genome are membrane proteins that are required for a cell to function and interact with its environment.

10–2 All of the common phospholipids—phosphatidylcholine, phosphatidylethanolamine, phosphatidylserine, and sphingomyelin—carry a positively charged moiety on their head group, but none carry a net positive charge.

10–3 Although phospholipid molecules are free to diffuse in the plane of the bilayer, they cannot flip-flop across the bilayer unless enzyme catalysts called phospholipid translocators are present in the membrane.

10–4 Whereas all the carbohydrate in the plasma membrane faces outward on the external surface of the cell, all the carbohydrate on internal membranes faces toward the cytosol.

10–5 Although membrane domains with different protein compositions are well known, there are at present no examples of membrane domains that differ in lipid composition.

Discuss the following problems.

10–6 When a lipid bilayer is torn, why does it not seal itself by forming a "hemi-micelle" cap at the edges, as shown in Figure Q10–1?

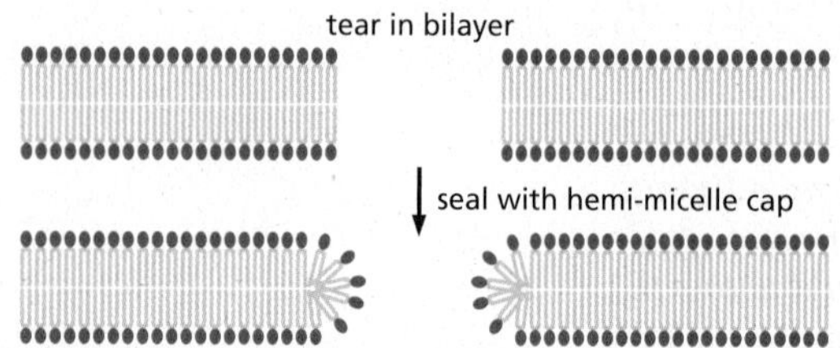

Figure Q10–1 A torn lipid bilayer sealed with a hypothetical "hemi-micelle" cap (Problem 10–6).

10–7 Which one of the following changes is energetically favorable and occurs spontaneously in an aqueous solution?

A. Conversion of a membrane vesicle to a flat bilayer
B. Dispersion of one oil droplet into many small ones
C. Formation of a bilayer from phospholipid molecules
D. Formation of a long tear in a phospholipid bilayer

10–8 Hydrophobic solutes are said to "force the adjacent water molecules to reorganize into icelike cages" (Figure Q10–2). It seems paradoxical that water molecules do not interact with hydrophobic solutes, yet they seem to "know" about the presence of a hydrophobic solute and change their behavior to interact differently with one another. Why would such an icelike cage be energetically unfavorable relative to pure water?

A. An icelike cage gives water a higher temperature.
B. An icelike cage is less organized than pure water.
C. An icelike cage is unstable and easily breaks down.
D. An icelike cage reduces the entropy of the system.

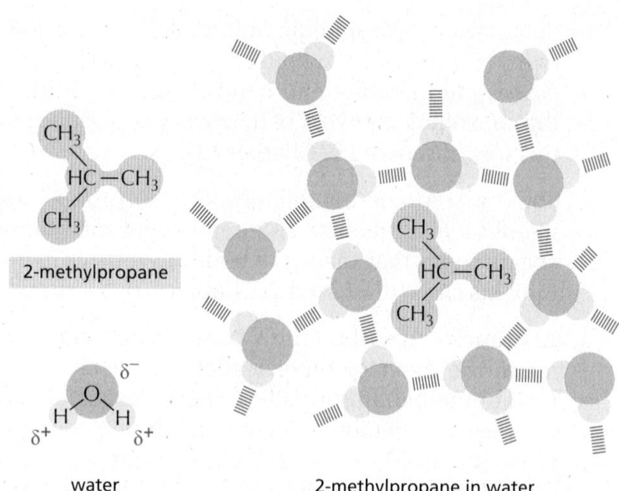

Figure Q10–2 An icelike cage of water molecules around a hydrophobic solute (Problem 10–8).

10–9 Margarine is made from vegetable oil by a chemical process. Do you suppose this process converts saturated fatty acids to unsaturated ones or vice versa? Explain your answer.

10–10 If a lipid raft is typically 70 nm in diameter and each lipid molecule has a diameter of 0.5 nm, about how many lipid molecules would there be in a lipid raft composed entirely of lipid?

10-11 Each of the following lipid anchors is used to attach intracellular proteins to membranes *except*:

A. A farnesyl anchor
B. A GPI anchor
C. A myristoyl anchor
D. A palmitoyl anchor

10–12 Monomeric single-pass transmembrane proteins span a membrane with a single α helix that has characteristic chemical properties in the region of the bilayer. Which of the three 19-amino-acid sequences listed below is the most likely candidate for such a transmembrane segment? Explain the reasons for your choice. (See back of book for one-letter amino acid code; FAMILY VW is a convenient mnemonic for hydrophobic amino acids.)

A. I T E I Y F G R M A G V I G T D L I S
B. I T L I Y F G V M A G V I G T I L I S
C. I T P I Y F G P M A G V I G T P L I S

10–13 You are studying the binding of proteins to the cytoplasmic face of cultured neuroblastoma cells and have found a method that gives a good yield of inside-out vesicles from the plasma membrane. Unfortunately, your preparations are contaminated with variable amounts of right-side-out vesicles. Nothing you have tried avoids this problem. A friend suggests that you pass your vesicles over an affinity column made of lectin coupled to solid beads. What is the point of your friend's suggestion?

10–14 Glycophorin, a protein in the plasma membrane of red blood cells, normally exists as a homodimer that is held together entirely by interactions between its transmembrane domains. As transmembrane domains are hydrophobic, how is it that they can associate with one another so specifically?

10–15 Three mechanisms by which membrane-binding proteins bend a membrane are illustrated in **Figure Q10-3A, B, and C**. As shown, each of these cytosolic membrane-bending proteins would induce an invagination of the plasma membrane. Could similar kinds of cytosolic proteins induce a protrusion of the plasma membrane (**Figure Q10–3D**)? Which ones? Explain how they might work.

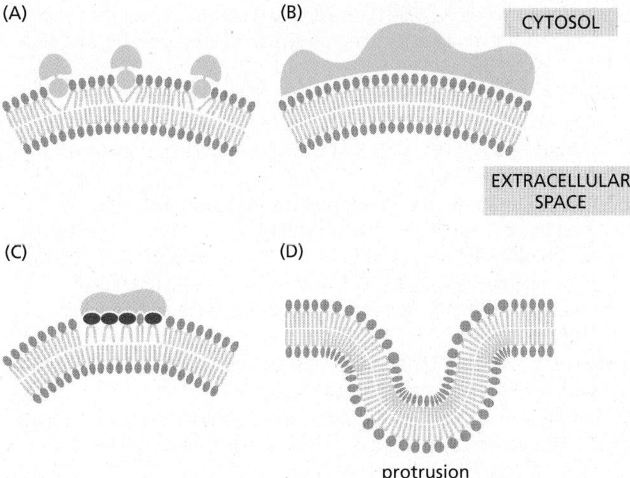

Figure Q10–3 Bending of the plasma membrane by cytosolic proteins (Problem 10–15). (A) Insertion of a protein "finger" into the cytosolic leaflet of the membrane. (B) Binding of lipids to the curved surface of a membrane-binding protein. (C) Binding of membrane proteins to membrane lipids with large head groups. (D) A segment of the plasma membrane showing a protrusion.

REFERENCES

General

Bretscher MS (1973) Membrane structure: some general principles. *Science* 181, 622–629.

Edidin M (2003) Lipids on the frontier: a century of cell-membrane bilayers. *Nat. Rev. Mol. Cell Biol.* 4, 414–418.

Goñi FM (2014) The basic structure and dynamics of cell membranes: an update of the Singer-Nicolson model. *Biochim. Biophys. Acta* 1838, 1467–1476.

Langmuir I (1917) The constitution and fundamental properties of solids and liquids II: liquids. *J. Am. Chem. Soc.* 39, 1848–1906.

Pockets A (1891) Surface tension. *Nature* 43, 437–439.

Singer SJ & Nicolson GL (1972) The fluid mosaic model of the structure of cell membranes. *Science* 175, 720–731.

Tanford C (1980) The Hydrophobic Effect: Formation of Micelles and Biological Membranes. New York: Wiley.

Tanford C (2004) Ben Franklin Stilled the Waves. New York: Oxford University Press.

The Lipid Bilayer

Antonny B, Vanni S, Shindou H & Ferreira T (2016) From zero to six double bonds: phospholipid unsaturation and organelle function. *Trends Cell Biol.* 25, 427–436.

Brügger B (2014) Lipidomics: analysis of the lipid composition of cells and subcellular organelles by electrospray ionization mass spectrometry. *Annu. Rev. Biochem.* 83, 79–98.

Chung J, Wu X, Lambert TJ . . . Farese RV, Jr (2019) LDAF1 and seipin form a lipid droplet assembly complex. *Dev. Cell* 51, 551–563.

Clarke RJ, Hossain KR & Cao K (2020) Physiological roles of transverse lipid asymmetry of animal membranes. *Biochim. Biophys. Acta Biomembr.* 1862, 183382.

Cornell CE, Mileant A, Thakkar N . . . Keller SL (2020) Direct imaging of liquid domains in membranes by cryo-electron tomography. *Proc. Natl. Acad. Sci. USA* 117, 19713–19719.

Hakomori SI (2002) The glycosynapse. *Proc. Natl. Acad. Sci. USA* 99, 225–232.

Harayama T & Riezman H (2018) Understanding the diversity of membrane lipid composition. *Nat. Rev. Mol. Cell Biol.* 19, 281–296.

Heberle FA, Doktorova M, Scott HL . . . Levental I (2020) Direct label-free imaging of nanodomains in biomimetic and biological membranes by cryogenic electron microscopy. *Proc. Natl. Acad. Sci. USA* 117, 19943–19952.

Ichikawa S & Hirabayashi Y (1998) Glucosylceramide synthase and glycosphingolipid synthesis. *Trends Cell Biol.* 8, 198–202.

Kornberg RD & McConnell HM (1971) Lateral diffusion of phospholipids in a vesicle membrane. *Proc. Natl. Acad. Sci. USA* 68, 2564–2568.

Lingwood D & Simons K (2010) Lipid rafts as a membrane-organizing principle. *Science* 327, 46–50.

Mansilla MC, Cybulski LE, Albanesi D & de Mendoza D (2004) Control of membrane lipid fluidity by molecular thermosensors. *J. Bacteriol.* 186, 6681–6688.

Maxfield FR & van Meer G (2010) Cholesterol, the central lipid of mammalian cells. *Curr. Opin. Cell Biol.* 22, 422–429.

Munro S (2003) Lipid rafts: elusive or illusive? *Cell* 115, 377–388.

Pomorski TG & Menon AK (2016) Lipid somersaults: uncovering the mechanisms of protein-mediated lipid flipping. *Prog. Lipid Res.* 64, 69–84.

Rothman JE & Lenard J (1977) Membrane asymmetry. *Science* 195, 743–753.

Walther TC & Farese RV, Jr (2017) Lipid droplet biogenesis. *Annu. Rev. Cell Dev. Biol.* 33, 491–510.

Membrane Proteins

Agasid MT & Robinson CV (2021) Probing membrane protein-lipid interactions. *Curr. Opin. Struct. Biol.* 69, 78–85.

Bennett V & Baines AJ (2001) Spectrin and ankyrin-based pathways: metazoan inventions for integrating cells into tissues. *Physiol. Rev.* 81, 1353–1392.

Bijlmakers MJ & Marsh M (2003) The on-off story of protein palmitoylation. *Trends Cell Biol.* 13, 32–42.

Bretscher MS & Raff MC (1975) Mammalian plasma membranes. *Nature* 258, 43–49.

Buchanan SK (1999) Beta-barrel proteins from bacterial outer membranes: structure, function and refolding. *Curr. Opin. Struct. Biol.* 9, 455–461.

Chen Y, Lagerholm BC, Yang B & Jacobson K (2006) Methods to measure the lateral diffusion of membrane lipids and proteins. *Methods* 39, 147–153.

Corin K & Bowie JU (2020) How bilayer properties influence membrane protein folding. *Protein Sci.* 29(12), 2348–2362.

Curran AR & Engelman DM (2003) Sequence motifs, polar interactions and conformational changes in helical membrane proteins. *Curr. Opin. Struct. Biol.* 13, 412–417.

Deisenhofer J & Michel H (1991) Structures of bacterial photosynthetic reaction centers. *Annu. Rev. Cell Biol.* 7, 1–23.

Drickamer K & Taylor ME (1998) Evolving views of protein glycosylation. *Trends Biochem. Sci.* 23, 321–324.

Giménez-Andrés M, Čopič A & Antonny B (2018) The many faces of amphipathic helices. *Biomolecules* 8, 45.

Helenius A & Simons K (1975) Solubilization of membranes by detergents. *Biochim. Biophys. Acta* 415, 29–79.

Henderson R & Unwin PN (1975) Three-dimensional model of purple membrane obtained by electron microscopy. *Nature* 257, 28–32.

Hite RK, Gonen T, Harrison SC & Walz T (2008) Interactions of lipids with aquaporin-0 and other membrane proteins. *Pflugers Arch.* 456, 651–661.

Kyte J & Doolittle RF (1982) A simple method for displaying the hydropathic character of a protein. *J. Mol. Biol.* 157, 105–132.

Lee AG (2003) Lipid-protein interactions in biological membranes: a structural perspective. *Biochim. Biophys. Acta* 1612, 1–40.

Marchesi VT, Furthmayr H & Tomita M (1976) The red cell membrane. *Annu. Rev. Biochem.* 45, 667–698.

Nakada C, Ritchie K, Oba Y . . . Kusumi A (2003) Accumulation of anchored proteins forms membrane diffusion barriers during neuronal polarization. *Nat. Cell Biol.* 5, 626–632.

Oesterhelt D (1998) The structure and mechanism of the family of retinal proteins from halophilic archaea. *Curr. Opin. Struct. Biol.* 8, 489–500.

Popot J-L (2010) Amphipols, nanodiscs, and fluorinated surfactants: three nonconventional approaches to studying membrane proteins in aqueous solution. *Annu. Rev. Biochem.* 79, 737–775.

Prinz WA & Hinshaw JE (2009) Membrane-bending proteins. *Crit. Rev. Biochem. Mol. Biol.* 44, 278–291.

Sezgin E, Levental I, Mayor S & Eggeling C (2017) The mystery of membrane organization: composition, regulation and roles of lipid rafts. *Nat. Rev. Mol. Cell Biol.* 18, 361–374.

Sharon N & Lis H (2004) History of lectins: from hemagglutinins to biological recognition molecules. *Glycobiology* 14, 53R–62R.

Sheetz MP (2001) Cell control by membrane-cytoskeleton adhesion. *Nat. Rev. Mol. Cell Biol.* 2, 392–396.

Shibata Y, Hu J, Kozlov MM & Rapoport TA (2009) Mechanisms shaping the membranes of cellular organelles. *Annu. Rev. Cell Dev. Biol.* 25, 329–354.

Steck TL (1974) The organization of proteins in the human red blood cell membrane. A review. *J. Cell Biol.* 62, 1–19.

Subramaniam S (1999) The structure of bacteriorhodopsin: an emerging consensus. *Curr. Opin. Struct. Biol.* 9, 462–468.

Vinothkumar KR & Henderson R (2010) Structures of membrane proteins. *Q. Rev. Biophys.* 43, 65–158.

von Heijne G (2011) Membrane proteins: from bench to bits. *Biochem. Soc. Trans.* 39, 747–750.

White SH & Wimley WC (1999) Membrane protein folding and stability: physical principles. *Annu. Rev. Biophys. Biomol. Struct.* 28, 319–365.

Zhang Y, Daday C, Gu RX . . . Walz T (2021) Visualization of the mechanosensitive ion channel MscS under membrane tension. *Nature* 590, 509–514.

Small-Molecule Transport and Electrical Properties of Membranes

CHAPTER

11

IN THIS CHAPTER

Because of its hydrophobic interior, the lipid bilayer of cell membranes restricts the passage of most polar molecules. This barrier function allows the cell to maintain concentrations of solutes in its cytosol that differ from those in the extracellular fluid and in each of the intracellular membrane-enclosed compartments. To benefit from this barrier, however, cells have had to evolve ways of transferring specific water-soluble molecules and ions across their membranes in order to ingest essential nutrients, excrete metabolic waste products, and regulate intracellular ion concentrations. Cells use specialized *membrane transport proteins* to accomplish this goal. The importance of such small-molecule transport is reflected in the large number of genes in all organisms that code for the transmembrane transport proteins involved, which make up 15–30% of the membrane proteins in all cells. Some mammalian cells, such as nerve and kidney cells, devote up to two-thirds of their total metabolic energy consumption to such transport processes.

Cells can also transfer macromolecules and even large particles, such as cell debris, viruses, and bacteria, across their membranes, but the mechanisms involved in most of these cases differ from those used for transferring small molecules, and they are discussed in Chapters 12 and 13.

We begin this chapter by describing some general principles of how small water-soluble molecules traverse cell membranes. We then consider, in turn, the two main classes of membrane proteins that mediate this transmembrane traffic: *transporters,* which undergo sequential conformational changes to transport specific small molecules across membranes, and *channels,* which form narrow pores, allowing passive transmembrane movement, primarily of water and small inorganic ions. Transporters can be coupled to a source of energy to catalyze active transport, which, together with selective passive permeability, creates large differences in the composition of the cytosol compared with that of either the extracellular fluid (**Table 11–1**) or the fluid within membrane-enclosed organelles. By generating inorganic ion-concentration differences across the lipid bilayer, cell membranes can store potential energy in the form of electrochemical gradients, which drive various transport processes, convey electrical signals in electrically excitable cells, and (in mitochondria, chloroplasts, and bacteria) are harnessed to make most of the cell's ATP. We focus our discussion mainly on transport across the plasma membrane, but similar mechanisms operate across the other membranes of the eukaryotic cell, as discussed in later chapters.

In the last part of the chapter, we concentrate mainly on the functions of ion channels in neurons (nerve cells). In these cells, channel proteins perform at their highest level of sophistication, enabling networks of neurons to carry out all the astonishing feats your brain is capable of.

PRINCIPLES OF MEMBRANE TRANSPORT

We begin this section by describing the permeability properties of an artificial membrane—a synthetic lipid bilayer made solely from lipids without proteins present. We then introduce some of the terms used to describe the various forms of membrane transport and some strategies for characterizing the proteins and processes involved.

TABLE 11–1 A Comparison of Inorganic Ion Concentrations Inside and Outside a Typical Mammalian Cell*

Component	Cytoplasmic concentration (mM)	Extracellular concentration (mM)
	Cations	
Na^+	5–15	145
K^+	140	5
Mg^{2+}	0.5	1–2
Ca^{2+}	10^{-4}	1–2
H^+	7×10^{-5} ($10^{-7.2}$ M or pH 7.2)	4×10^{-5} ($10^{-7.4}$ M or pH 7.4)
	Anions	
Cl^-	5–15	110

*The cell must contain equal quantities of positive and negative charges (that is, it must be electrically neutral). Thus, in addition to Cl^-, the cell contains many other anions not listed in this table; in fact, most cell constituents are negatively charged (HCO_3^-, PO_4^{3-}, nucleic acids, metabolites carrying phosphate and carboxyl groups, etc.). The concentrations of Ca^{2+} and Mg^{2+} given are for the free ions: although there is a total of about 20 mM Mg^{2+} and 1–2 mM Ca^{2+} in cells, both ions are mostly bound to other substances (such as proteins, free nucleotides, RNA, etc.) and, for Ca^{2+}, stored within various organelles (such as the endoplasmic reticulum and mitochondria).

Protein-free Lipid Bilayers Are Impermeable to Ions

Given enough time, virtually any molecule will diffuse down its concentration gradient across a protein-free lipid bilayer. The rate of diffusion, however, varies enormously, depending partly on the size of the molecule but mostly on its relative hydrophobicity (solubility in oil). In general, the smaller the molecule and the more hydrophobic, or nonpolar, it is, the more easily it will diffuse across a lipid bilayer. Small nonpolar molecules, such as O_2 and CO_2, readily dissolve in lipid bilayers and therefore diffuse rapidly across them. Small uncharged polar molecules, such as water or urea, also diffuse across a bilayer, albeit much more slowly (**Figure 11–1** and see Movie 10.3). By contrast, lipid bilayers are essentially impermeable to charged molecules (ions), no matter how small: the charge and high degree of hydration of such molecules prevent them from entering the hydrocarbon phase of the bilayer (**Figure 11–2**).

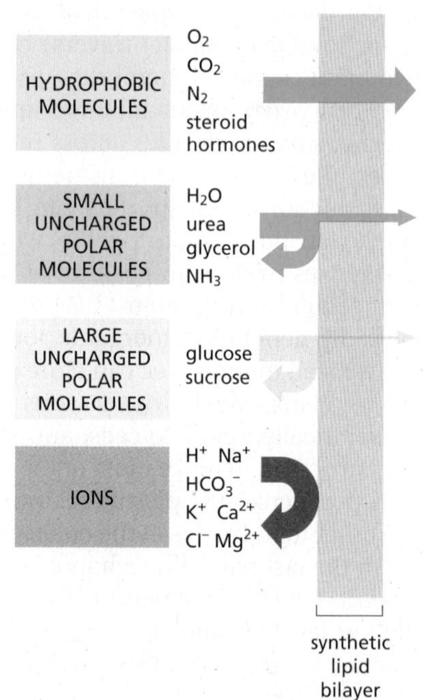

Figure 11–1 The relative permeability of a synthetic lipid bilayer to different classes of molecules. The smaller the molecule and, more important, the less strongly it associates with water, the more rapidly the molecule diffuses across the bilayer.

There Are Two Main Classes of Membrane Transport Proteins: Transporters and Channels

Like synthetic lipid bilayers, cell membranes allow small nonpolar molecules to permeate by diffusion. Cell membranes, however, also have to allow the passage of various polar molecules, such as ions, sugars, amino acids, nucleotides, water, and many cell metabolites that cross synthetic lipid bilayers only very slowly. Special **membrane transport proteins** transfer such solutes across cell membranes. These proteins occur in many forms and in all types of biological membranes. Each protein often transports only a specific molecular species or sometimes a class of molecules (such as ions, sugars, or amino acids). Early studies found that bacteria with a single-gene mutation were unable to transport a particular class of sugars across their plasma membrane, thereby demonstrating the specificity of membrane transport proteins. We now know that humans with similar mutations suffer from various inherited diseases that hinder the transport of a specific solute or solute class in the kidney, intestine, or other cell type. Individuals with the inherited disease *cystinuria*, for example, cannot transport certain amino acids

Figure 11–2 **Permeability coefficients for the passage of various molecules through synthetic, protein-free lipid bilayers.** The rate of flow of a solute across the bilayer is directly proportional to the difference in its concentration on the two sides of the membrane. Multiplying this concentration difference (in mol/cm^3) by the permeability coefficient (in cm/sec), which is an experimentally determined constant characteristic of each solute, gives the flow of solute in moles per second per square centimeter of bilayer. A concentration difference of tryptophan of 10^{-4} mol/cm^3 (10^{-4} mol/10^{-3} L = 0.1 M), for example, would cause a flow of 10^{-4} mol/cm^3 $\times$ 10^{-7} cm/sec = 10^{-11} mol/sec through 1 cm^2 of bilayer, or 6 $\times$ 10^4 molecules/sec through 1 μm^2 of bilayer.

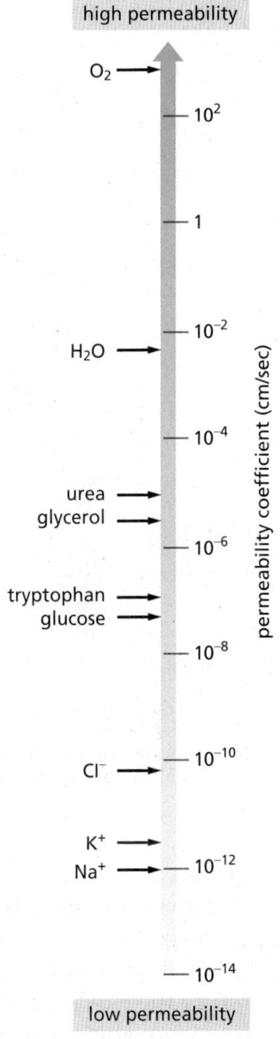

(including cystine, the disulfide-linked dimer of cysteine) from either the urine or the intestine into the blood; the resulting accumulation of cystine in the urine leads to the formation of cystine stones in the kidneys.

All membrane transport proteins that have been studied in detail are multipass transmembrane proteins; that is, their polypeptide chains traverse the lipid bilayer two or more times. By forming a protein-lined pathway across the membrane, these proteins enable specific hydrophilic solutes to cross the membrane without coming into direct contact with the hydrophobic interior of the lipid bilayer.

Transporters and channels are the two major classes of membrane transport proteins (Figure 11–3). **Transporters** (sometimes called *carriers* or *permeases*) bind the specific solute to be transported and undergo a series of conformational changes that alternately expose solute-binding sites on one side of the membrane and then on the other side to transfer the solute across it. **Channels**, by contrast, interact much more transiently with the solute to be transported. When opened by conformational changes, channels form continuous pores that extend across the lipid bilayer. The pores allow specific solutes (such as inorganic ions of appropriate size and charge, and in some cases small molecules, including water, glycerol, and ammonia) to pass through them and thereby cross the membrane. Because no stepwise conformational changes are required once a channel is opened, it is not surprising that transport through channels occurs at a much faster rate than transport mediated by transporters. Although water can slowly diffuse across synthetic lipid bilayers, cells use dedicated channel proteins (called *water channels*, or *aquaporins*) that greatly increase the permeability of their membranes to water, as we discuss later.

Active Transport Is Mediated by Transporters Coupled to an Energy Source

All channels and some transporters allow solutes to cross the membrane only passively ("downhill"), a process called **passive transport**. In the case of transport of a single uncharged molecule, the difference in the concentration on the two sides of the membrane—its *concentration gradient*—drives passive transport and determines its direction (Figure 11–4A). If the solute carries a net charge, however, both its concentration gradient and the electrical potential difference across the membrane, the *membrane potential*, influence its transport. The concentration gradient and the electrical gradient combine to form a net driving force, the **electrochemical gradient**, for each charged solute (Figure 11–4B). We discuss electrochemical gradients in more detail later and in Chapter 14. In fact, almost all plasma membranes have an electrical potential difference (that is, a voltage) across them, with the inside usually negative with respect to the outside. This

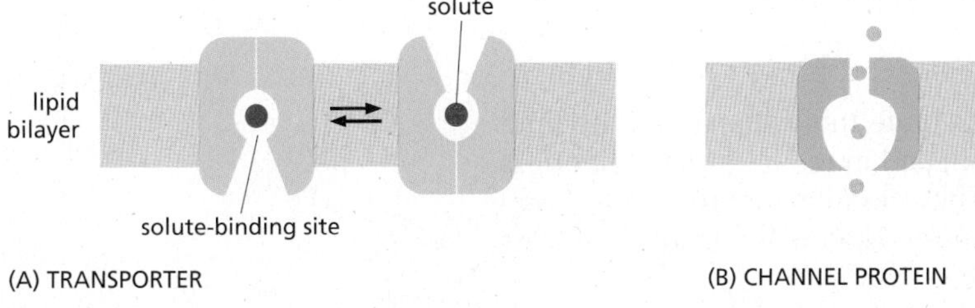

Figure 11–3 **Transporters and channel proteins.** (A) A transporter alternates between two conformations, so that the solute-binding site of the transporter is sequentially accessible on one side of the bilayer and then on the other. (B) In contrast, a channel protein forms a pore across the bilayer through which specific solutes can passively diffuse.

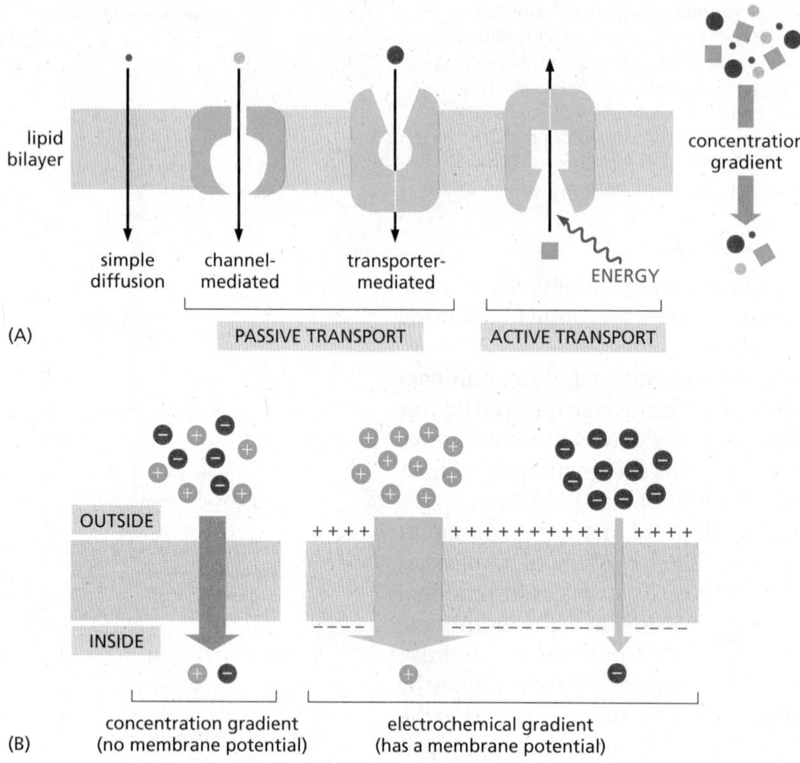

Figure 11–4 Different forms of membrane transport and the influence of the membrane. (A) Passive transport down a concentration gradient (or an electrochemical gradient—see panel B) occurs spontaneously, by diffusion, either through the lipid bilayer directly or through channels or passive transporters. By contrast, active transport involves movement of the solute against its concentration or electrochemical gradient and hence requires an input of metabolic energy. (B) The electrochemical gradient of a charged solute (an ion) affects its transport. This gradient *(green)* combines the membrane potential and the concentration gradient of the solute. The electrical and chemical gradients can work additively to increase the driving force on an ion across the membrane *(middle)* or they can work against each other *(right)*.

potential favors the entry of positively charged ions into the cell but opposes the entry of negatively charged ions (see Figure 11–4B); it also opposes the efflux of positively charged ions.

In addition to passive transport, cells need to be able to actively pump certain solutes across the membrane "uphill," against their electrochemical gradients. Such **active transport** is mediated by transporters whose pumping activity is directional because it is tightly coupled to a source of metabolic energy, such as an ion gradient or ATP hydrolysis, as discussed later. Transmembrane movement of small molecules mediated by transporters can be either active or passive, whereas that mediated by channels is always passive (see Figure 11–4A).

Summary

Lipid bilayers are virtually impermeable to most polar molecules. To transport small water-soluble molecules into or out of cells or intracellular membrane-enclosed compartments, cell membranes contain various membrane transport proteins, each of which is responsible for transferring a particular solute or class of solutes across the membrane. There are two types of membrane transport proteins—transporters and channels. Both form protein pathways across the lipid bilayer. Whereas transmembrane movement mediated by transporters can be either active or passive, solute flow through channel proteins is always passive. Ion transport across the membrane is influenced by the ion's concentration gradient and the membrane potential; that is, its electrochemical gradient.

TRANSPORTERS AND ACTIVE MEMBRANE TRANSPORT

The process by which a transporter transfers a solute molecule across the lipid bilayer resembles an enzyme-substrate reaction, and in many ways transporters behave like enzymes. In contrast to ordinary enzyme-substrate reactions, however, the transporter does not modify the transported solute but instead delivers it unchanged to the other side of the membrane.

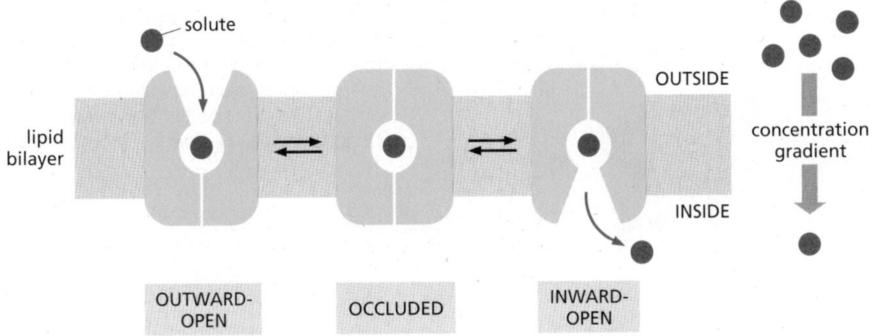

Figure 11–5 **A model of how a conformational change in a transporter mediates the passive movement of a solute.** The transporter is shown in three conformational states: in the outward-open state, the binding sites for solute are exposed on the outside; in the occluded state, the same sites are not accessible from either side; and in the inward-open state, the sites are exposed on the inside. The transitions between the states occur randomly. They are completely reversible and do not depend on whether the solute-binding site is occupied. Therefore, if the solute concentration is higher on the outside of the bilayer, more solute binds to the transporter in the outward-open conformation than in the inward-open conformation, and there is a net transport of solute down its concentration gradient (or, if the solute is an ion, down its electrochemical gradient).

Each type of transporter has one or more specific binding sites for its solute (substrate). It transfers the solute across the lipid bilayer by undergoing reversible conformational changes that alternately expose the solute-binding site first on one side of the membrane and then on the other side—but never on both sides at the same time. The transition occurs through an intermediate state in which the solute is inaccessible, or occluded, from either side of the membrane (**Figure 11–5**). When the transporter is saturated (that is, when all solute-binding sites are occupied), the rate of transport is maximal. This rate, referred to as V_{max} (V for velocity), is characteristic of the specific carrier. V_{max} is the maximal rate at which the carrier can flip between its conformational states. In addition, each transporter has a characteristic affinity for its solute, reflected in the K_m of the reaction, which is equal to the concentration of solute when the transport rate is half its maximum value (**Figure 11–6**). As with enzymes, the binding of solute can be blocked by either competitive inhibitors (which compete for the same binding site and may or may not be transported) or noncompetitive inhibitors (which bind elsewhere and alter the structure of the transporter).

As we discuss shortly, conceptually it requires only a relatively minor modification of the mechanism shown in Figure 11–5 to link a transporter to a source of energy in order to pump a solute uphill against its electrochemical gradient. Cells carry out such active transport in three main ways (**Figure 11–7**):

1. *Coupled transporters* harness the energy stored in concentration gradients to couple the uphill transport of one solute across the membrane to the downhill transport of another.
2. *ATP-driven pumps* couple uphill transport to the hydrolysis of ATP.
3. *Light- or redox-driven pumps*, which are known in bacteria, archaea, mitochondria, and chloroplasts, couple uphill transport to an input of energy from light, as with bacteriorhodopsin and photosystem II (discussed in Chapters 10 and 14, respectively), or from a redox reaction, as with cytochrome *c* oxidase (discussed in Chapter 14).

Comparisons of amino acid sequences and three-dimensional structures suggest that, in many cases, there are strong similarities in structure between transporters that mediate active transport and those that mediate passive transport. Some bacterial transporters, for example, that use the energy stored in the H^+ gradient across the plasma membrane to drive the active uptake of various sugars are structurally similar to the transporters that mediate passive glucose transport into most animal cells. There is thus a clear evolutionary relationship between various transporters. Given the cell's essential need to transport small metabolites across membranes, it comes as no surprise that the superfamily of transporters is a large and ancient one.

We begin our discussion of active membrane transport by considering a class of coupled transporters that are driven by ion-concentration gradients. These proteins have a crucial role in the transport of small metabolites across membranes in all cells. We then discuss ATP-driven pumps, including the Na^+-K^+ pump that is found in the plasma membrane of most animal cells. Examples of the third class of active transport—light- or redox-driven pumps—are discussed in Chapter 14.

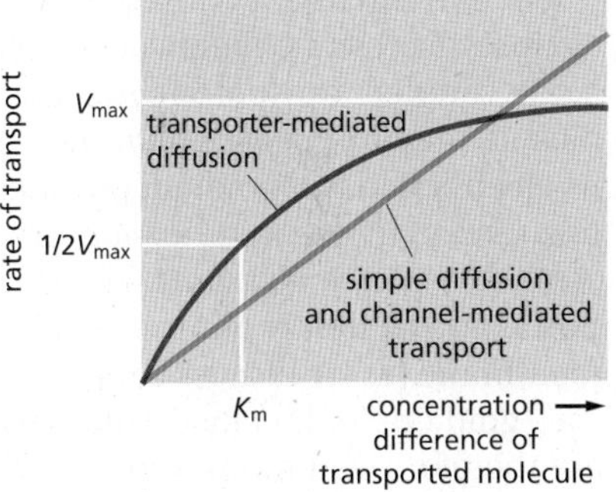

Figure 11–6 **The kinetics of simple diffusion compared with transporter-mediated diffusion.** Whereas the rate of simple diffusion and of channel-mediated transport is directly proportional to the solute concentration (within the physical limits imposed by total surface area or total channels available), the rate of transporter-mediated diffusion approaches a maximum (V_{max}) as the transporter reaches saturation. The solute concentration when the transport rate is at half its maximal value is termed its K_m and is analogous to the K_m of an enzyme for its substrate. Note that, while the graph illustrates the shape of the curves, their absolute scale on the Y axis is very different. A typical channel conducts water or ions at rates of up to 10^8 per second, whereas a typical transporter moves solutes at rates between 10^2 and 10^4 molecules per second.

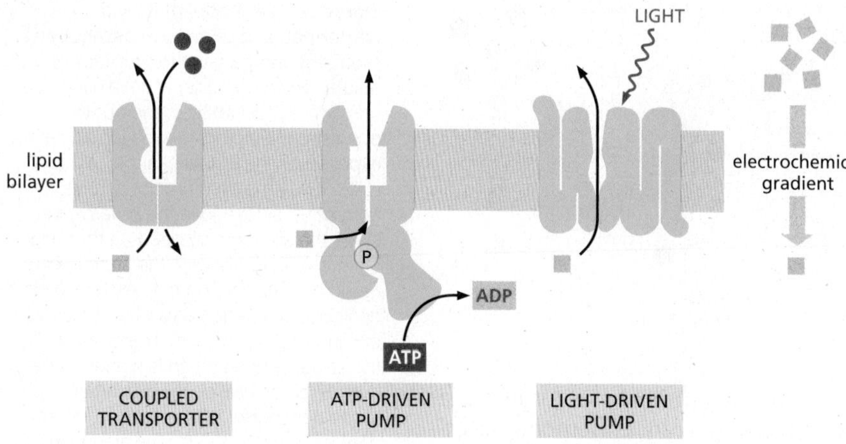

Figure 11–7 Three ways of driving active transport. The actively transported molecule is shown in *orange*, and the energy source is shown in *red*. Redox-driven active transport is discussed in Chapter 14 (see Figures 14–18 and 14–19).

Active Transport Can Be Driven by Ion-Concentration Gradients

Some transporters simply facilitate the passive movement of a single solute from one side of the membrane to the other at a rate determined by their V_{max} and K_m; they are called **uniporters.** Others function as *coupled transporters,* in which the transfer of one solute strictly depends on the transport of a second. Coupled transport involves either the intimately coupled transfer of a second solute in the same direction, performed by **symporters** (also called *co-transporters*), or the transfer of a second solute in the opposite direction, performed by **antiporters** (also called *exchangers*) (**Figure 11–8**).

The tight coupling between the transfer of two solutes allows the coupled transporters to harvest the energy stored in the electrochemical gradient of one solute, typically an inorganic ion, to transport the other. In this way, the free energy released during the movement of an inorganic ion or H^+ down an electrochemical gradient is used as the driving force to pump other solutes uphill, against their electrochemical gradient. This strategy can work in either direction; some coupled transporters function as symporters, others as antiporters. In the plasma membrane of animal cells, Na^+ is the usual co-transported ion because its electrochemical gradient provides a large driving force for the active transport of a second molecule. Such ion-driven coupled transporters are said to mediate *secondary active transport.* The Na^+ that enters the cell during coupled transport is subsequently pumped out by an ATP-driven Na^+-K^+ pump in the plasma membrane (as we discuss later), which, by exchanging K^+ for Na^+, maintains the Na^+ gradient, indirectly driving the coupled transport. Such ATP-driven pumps are therefore said to mediate *primary active transport* because in these the free energy of ATP hydrolysis is used to directly drive the transport of a solute against its electrochemical gradient. The energy stored in the gradient is then used to fuel the secondary active transport processes.

Intestinal and kidney epithelial cells contain a variety of symporters that are driven by the Na^+ gradient across the plasma membrane. Each Na^+-driven

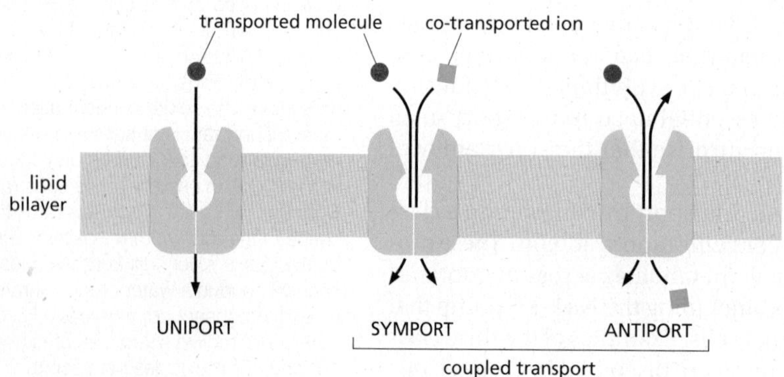

Figure 11–8 This schematic diagram shows transporters functioning as uniporters, symporters, and antiporters (Movie 11.1).

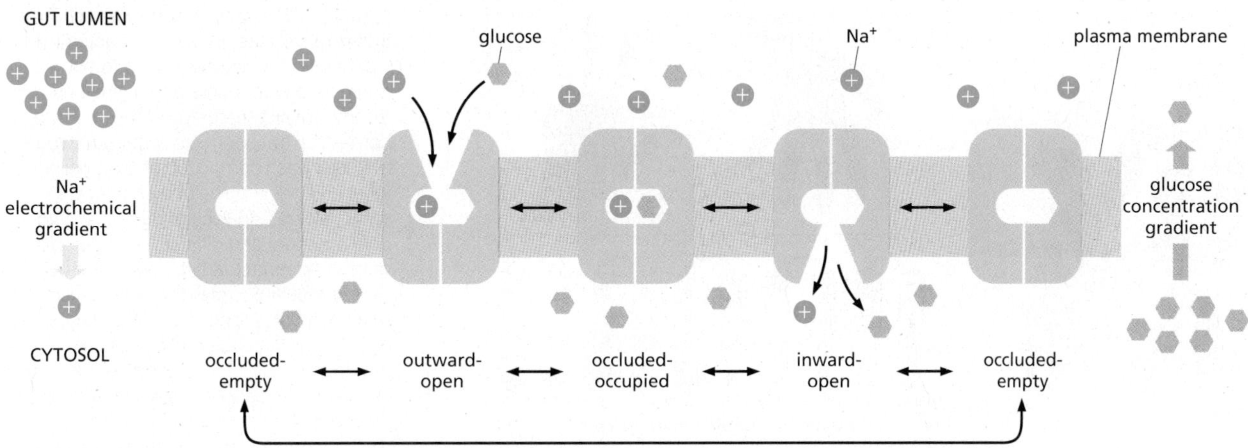

Figure 11–9 **Mechanism of glucose transport fueled by an Na^+ gradient.** As in the model shown in Figure 11–5, the transporter alternates between inward-open and outward-open states via occluded intermediate states. Binding of Na^+ and glucose is cooperative; that is, the binding of either solute increases the protein's affinity for the other. Because the Na^+ concentration is much higher in the extracellular space than in the cytosol, glucose is more likely to bind to the transporter in the outward-facing state. The transition to the occluded state occurs only when both Na^+ and glucose are bound; their precise interactions in the solute-binding sites slightly stabilize the occluded state and thereby make this transition energetically favorable. Stochastic fluctuations caused by thermal energy drive the transporter randomly into the inward-open or outward-open conformation. If it opens outwardly, nothing is achieved, and the process starts all over. However, whenever it opens inwardly, Na^+ dissociates quickly in the low-Na^+-concentration environment of the cytosol. Glucose dissociation is likewise enhanced when Na^+ is lost, because of cooperativity in binding of the two solutes. The overall result is the net transport of both Na^+ and glucose into the cell. Because the occluded state is not formed when only one of the solutes is bound, the transporter switches conformation only when it is fully occupied or fully empty, thereby ensuring strict coupling of the transport of Na^+ and glucose.

symporter is specific for importing a small group of related sugars or amino acids into the cell. Because the Na^+ tends to move into the cell down its electrochemical gradient, the sugar or amino acid is, in a sense, "dragged" into the cell with it. The greater the electrochemical gradient for Na^+, the more solute is transported into the cell (**Figure 11–9**). Neurotransmitters (released by nerve cells to signal at synapses—as we discuss later) are taken up again by Na^+ symporters after their release. This both terminates their signaling to postsynaptic cells and recycles them for reuse. These neurotransmitter transporters are important drug targets: stimulants, such as cocaine and antidepressants, inhibit them and thereby prolong signaling by the neurotransmitters because they are not cleared efficiently.

Despite their great variety, transporters share structural features that can explain how they function and how they evolved. Transporters are typically built from bundles of 10 or more α helices that span the membrane. Solute- and ion-binding sites are located midway through the membrane, where some helices are broken or distorted and amino acid side chains and polypeptide backbone atoms form ion- and solute-binding sites. In the inward-open and outward-open conformations, these binding sites are accessible by passageways from one side of the membrane but not the other. In switching between the two conformations, the transporter protein transiently adopts an occluded conformation, in which both passageways are closed; this prevents the driving ion and the transported solute from crossing the membrane unaccompanied, which would deplete the cell's energy store to no purpose. Because only transporters with both types of binding sites appropriately filled change their conformation, tight coupling between ion and solute transport is ensured.

Like enzymes, transporters can work in the reverse direction if ion and solute gradients are appropriately adjusted experimentally. This chemical symmetry is mirrored in their physical structure. Protein structural analyses have revealed that many transporters are built from *inverted repeats*: the packing of the transmembrane α helices in one half of the helix bundle is structurally similar to the packing in the other half, but the two halves are inverted in the membrane relative to each other. Transporters are therefore said to be pseudosymmetric, and the

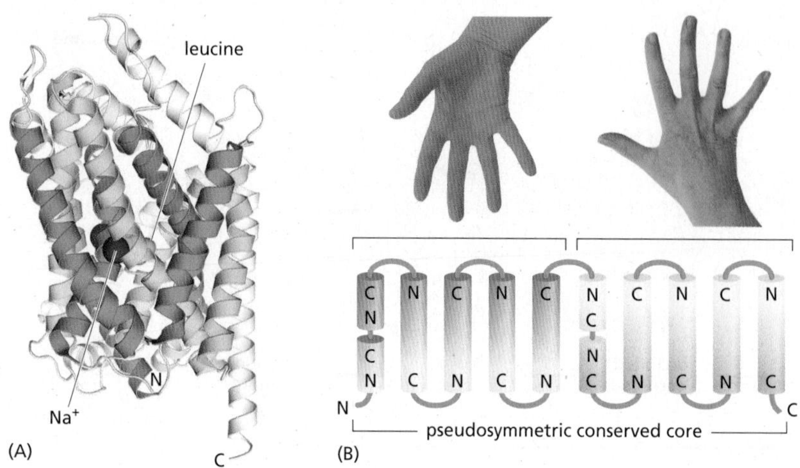

Figure 11–10 Transporters are built from inverted repeats. (A) LeuT, a bacterial Na^+/leucine symporter related to human neurotransmitter transporters, such as the serotonin transporter, is shown. The core of the transporter is built from two bundles, each composed of six α helices (*blue* and *yellow*). The helices shown in *light gray* are additions to the conserved core structure and differ among members of this transporter family. They are thought to play regulatory roles that are specific to a particular transporter. (B) Both core helix bundles are packed in a similar arrangement, but the second bundle is inverted with respect to the first (shown as two right hands, with the broken helices as the thumbs). The transporter's structural pseudosymmetry reflects its functional symmetry: the transporter can work in either direction, depending on the direction of the ion gradient. (A, PDB code: 3F3E.)

passageways that open and close on either side of the membrane have closely similar geometries, allowing alternating access to the ion- and solute-binding sites in the center (Figure 11–10). It is thought that the two halves evolved by gene duplication of a smaller ancestor protein.

Some other types of important membrane transport proteins are also built from inverted repeats. Examples even include channel proteins such as the aquaporin water channel (discussed later) and the Sec61 channel through which nascent polypeptides move into the endoplasmic reticulum (discussed in Chapter 12). It is thought that these channels evolved from coupled transporters in which the gating functions were lost, allowing them to open toward both sides of the membrane simultaneously to provide a continuous path across the membrane.

In bacteria, yeasts, and plants, as well as in many membrane-enclosed organelles of animal cells, most ion-driven active transport systems depend on H^+ rather than Na^+ gradients, reflecting the predominance of H^+ pumps in these membranes. An electrochemical H^+ gradient across the bacterial plasma membrane, for example, drives the inward active transport of many sugars and amino acids.

Transporters in the Plasma Membrane Regulate Cytosolic pH

Most proteins operate optimally at a particular pH. Lysosomal enzymes, for example, function best at the low pH (~5) found in lysosomes, whereas cytosolic enzymes function best at the close-to-neutral pH (~7.2) found in the cytosol. It is therefore crucial that cells control the pH of their intracellular compartments.

Most cells have one or more types of Na^+-driven antiporters in their plasma membrane that help to maintain the cytosolic pH at about 7.2. These transporters use the energy stored in the Na^+ gradient to pump out excess H^+, which either leaks in or is produced in the cell by acid-forming reactions. Two mechanisms are used: either H^+ is directly transported out of the cell or HCO_3^- is brought into the cell to neutralize H^+ in the cytosol (according to the reaction $HCO_3^- + H^+ \rightarrow H_2O + CO_2$). One of the antiporters that uses the first mechanism is an *Na^+-H^+ exchanger*, which couples an influx of Na^+ to an efflux of H^+. Another, which uses a combination of the two mechanisms, is an *Na^+-driven Cl^--HCO_3^- exchanger* that couples an influx of Na^+ and HCO_3^- to an efflux of Cl^- and H^+ (so that $NaHCO_3$ comes in and HCl goes out). The Na^+-driven Cl^-–HCO_3^- exchanger is twice as effective as the Na^+–H^+ exchanger: it pumps out one H^+ and neutralizes another for each Na^+ that enters the cell. If HCO_3^- is available, as is usually the case, this antiporter is the most important transporter regulating the cytosolic pH. The pH inside the cell regulates both exchangers; when the pH in the cytosol falls, both exchangers sense the change and increase their activity.

An *Na^+-independent Cl^--HCO_3^- exchanger* adjusts the cytosolic pH in the reverse direction. Like the Na^+-dependent transporters, pH regulates the Na^+-independent Cl^-–HCO_3^- exchanger, but the exchanger's activity increases as the cytosol becomes too alkaline. The movement of HCO_3^- in this case is normally out of the cell, down its electrochemical gradient, which decreases the pH of the cytosol. An Na^+-independent Cl^-–HCO_3^- exchanger in the membrane of red blood cells (called band 3 protein—see Figure 10–38) facilitates the quick discharge of CO_2 (as HCO_3^-) as the cells pass through capillaries in the lung.

The intracellular pH is not entirely regulated by transporters in the plasma membrane: ATP-driven H^+ pumps are used to control the pH of many intracellular compartments. As discussed in Chapter 13, H^+ pumps maintain the low pH in lysosomes, as well as in endosomes and secretory vesicles. These H^+ pumps use the energy of ATP hydrolysis to pump H^+ into these acidic organelles from the cytosol. An advantage of using electrochemical H^+ gradients to power intracellular transport events is that they can dissipate and regenerate quickly, thus affording more opportunity to switch transport reactions on and off. To create an electrochemical H^+ gradient of similar energy to that of the Na^+ gradient at the plasma membrane requires the movement of far fewer H^+. This is because the H^+ concentration is many orders of magnitude smaller (0.1 μM at pH 7) than that of Na^+ and K^+ (~100 mM).

An Asymmetric Distribution of Transporters in Epithelial Cells Underlies the Transcellular Transport of Solutes

In epithelial cells, such as those that absorb nutrients from the gut, transporters are distributed nonuniformly between the apical and basolateral plasma membranes and thereby contribute to the **transcellular transport** of absorbed solutes. By the actions of the transporters in these cells, solutes are moved across the epithelial-cell layer into the extracellular fluid from where they pass into the blood. As shown in Figure 11–11, Na^+-linked symporters located in the apical (absorptive) domain of the plasma membrane actively transport nutrients into the cell, building up substantial concentration gradients for these solutes across the plasma membrane. Uniporters in the basal and lateral (basolateral) domain allow the nutrients to leave the cell passively down these concentration gradients to enter the bloodstream for use in the rest of the body.

Figure 11–11 Transcellular transport. The transcellular transport of glucose across an intestinal epithelial cell depends on the nonuniform distribution of transporters in the cell's plasma membrane. The process shown here results in the transport of glucose from the intestinal lumen to the extracellular fluid (from where it passes into the blood). Glucose is pumped into the cell through the apical domain of the membrane by an Na^+-powered glucose symporter (see Figure 11–9). Glucose passes out of the cell (down its concentration gradient) by passive movement through a glucose uniporter in the basal and lateral membrane domains. The Na^+ gradient driving the glucose symport is maintained by the Na^+-K^+ pump in the basal and lateral plasma membrane domains, which keeps the internal concentration of Na^+ low (Movie 11.2). Adjacent cells are connected by impermeable tight junctions, which have a dual function in the transport process illustrated: they prevent solutes from crossing the epithelium between cells, allowing a concentration gradient of glucose to be maintained across the cell sheet (see Figure 19–19). They also serve as diffusion barriers (fences) within the plasma membrane, which help confine the various transporters to their respective membrane domains (see Figure 10–34). (Micrograph from Dennis Kunkel/Science Source.)

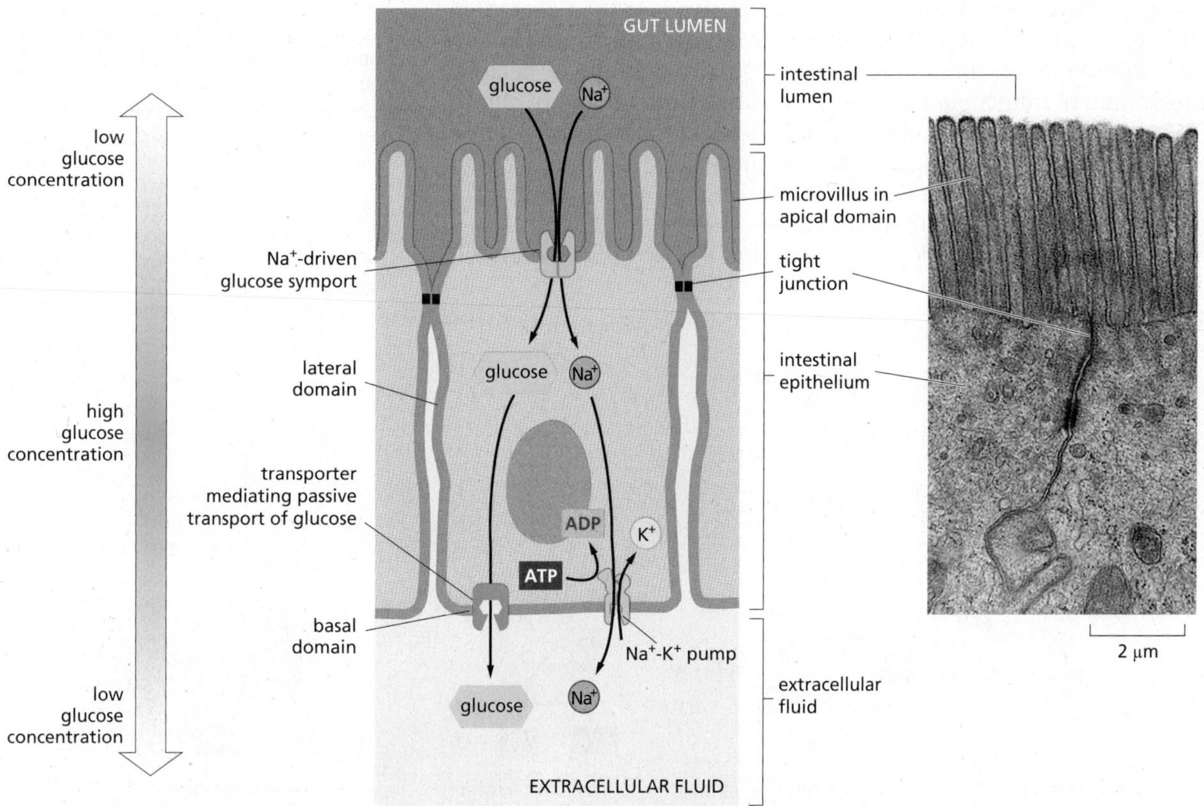

In many of these epithelial cells, the plasma membrane area is greatly increased by the formation of thousands of microvilli, which extend as thin, fingerlike projections from the apical surface of each cell. Such microvilli can increase the total absorptive area of a cell as much as 25-fold, thereby enhancing its transport capabilities.

As we have seen, ion gradients have a crucial role in driving many essential transport processes in cells. Ion pumps that use the energy of ATP hydrolysis establish and maintain these gradients, as we discuss next.

There Are Three Classes of ATP-driven Pumps

ATP-driven pumps are often called *transport ATPases* because they hydrolyze ATP to ADP and phosphate and use the energy released to pump ions or other solutes across a membrane. There are three principal classes of ATP-driven pumps (**Figure 11–12**), and representatives of each are found in all prokaryotic and eukaryotic cells.

1. **P-type pumps** are structurally and functionally related multipass transmembrane proteins. They are called "P-type" because they phosphorylate themselves during the pumping cycle. This class includes many of the ion pumps that are responsible for setting up and maintaining gradients of Na^+, K^+, H^+, and Ca^{2+} across cell membranes.

2. **ABC transporters** (ATP-binding cassette transporters) differ structurally from P-type ATPases and primarily pump small organic molecules across cell membranes.

3. **V-type pumps** are turbine-like protein machines constructed from multiple different subunits. The V-type proton pump transfers H^+ into organelles, such as lysosomes, synaptic vesicles, and plant or yeast vacuoles (V = vacuolar), to acidify the interior of these organelles (see Figures 13–46 and 13–47).

Structurally related to the V-type pumps is a distinct subclass of *F-type ATPases*, more commonly called *ATP synthases* because they normally work in reverse: instead of using ATP hydrolysis to drive H^+ transport, they use the H^+ gradient across the membrane to drive the synthesis of ATP from ADP and phosphate (see Figure 14–31). ATP synthases are found in the plasma membrane of bacteria, the inner membrane of mitochondria, and the thylakoid membrane of chloroplasts. The H^+ gradient is generated either during the electron-transport steps of oxidative phosphorylation (in aerobic bacteria and mitochondria), during photosynthesis (in chloroplasts), or by the light-driven H^+ pump (bacteriorhodopsin) in *Halobacterium*. We discuss some of these proteins in detail in Chapter 14.

For the remainder of this section, we focus on P-type pumps and ABC transporters.

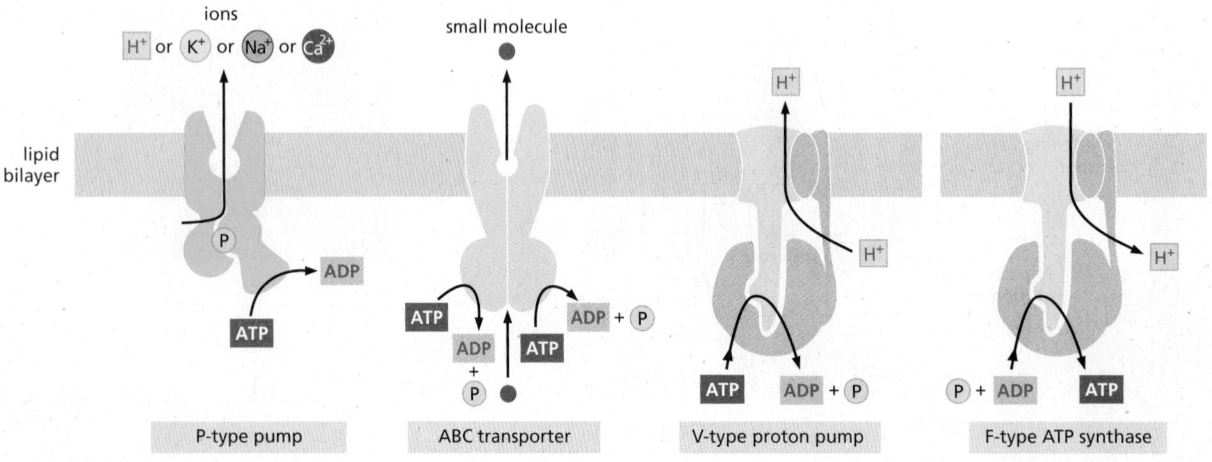

Figure 11–12 Three types of ATP-driven pumps. Like any enzyme, all ATP-driven pumps can work in either direction, depending on the electrochemical gradients of their solutes and the ATP/ADP ratio. When the ATP/ADP ratio is high, they hydrolyze ATP; when the ATP/ADP ratio is low, they can synthesize ATP. The F-type ATPases in mitochondria and chloroplasts normally work in this "reverse" mode to make most of the cell's ATP.

A P-type ATPase Pumps Ca^{2+} into the Sarcoplasmic Reticulum in Muscle Cells

Eukaryotic cells maintain very low concentrations of free Ca^{2+} in their cytosol (~10^{-7} M) in the face of a very much higher extracellular Ca^{2+} concentration (~10^{-3} M). Therefore, even a small influx of Ca^{2+} significantly increases the concentration of free Ca^{2+} in the cytosol, and the flow of Ca^{2+} down its steep concentration gradient in response to extracellular signals is one means of transmitting these signals rapidly across the plasma membrane (discussed in Chapter 15). It is thus important that the cell maintains a steep Ca^{2+} gradient across its plasma membrane. Ca^{2+} transporters that actively pump Ca^{2+} out of the cytosol help maintain the gradient. One of these is a P-type Ca^{2+} ATPase; the other is an antiporter (called an *Na^+–Ca^{2+} exchanger*) that is driven by the Na^+ electrochemical gradient (discussed in Chapter 15).

The **Ca^{2+} pump**, or **Ca^{2+} ATPase**, in the *sarcoplasmic reticulum* (SR) membrane of skeletal muscle cells is a well-understood P-type transport ATPase. The SR is a specialized type of endoplasmic reticulum that forms a network of tubular sacs in the muscle-cell cytoplasm, and it serves as an intracellular store of Ca^{2+}. When an action potential depolarizes the muscle-cell plasma membrane, Ca^{2+} is released into the cytosol from the SR through *Ca^{2+}-release channels*, stimulating the muscle to contract (discussed in Chapters 15 and 16). The Ca^{2+} pump, which accounts for about 90% of the membrane protein of the SR, moves Ca^{2+} from the cytosol back into the SR. The endoplasmic reticulum of non-muscle cells also stores Ca^{2+} using a closely homologous Ca^{2+} pump and Ca^{2+}-release channels.

Enzymatic studies and analyses of the three-dimensional structures of transport intermediates of the SR Ca^{2+} pump and related pumps have revealed the molecular mechanism of P-type transport ATPases in great detail. They all have similar structures, containing 10 transmembrane α helices connected to three cytosolic domains (Figure 11–13). In the Ca^{2+} pump, amino acid side chains protruding from the transmembrane helices form two centrally positioned binding sites for Ca^{2+}. As shown in Figure 11–14, in the pump's ATP-bound nonphosphorylated state, these binding sites are accessible only from the cytosolic side of the SR membrane. Ca^{2+} binding triggers a series of conformational changes that close the passageway to the cytosol and activate a phosphotransfer reaction in which the terminal phosphate of the ATP is transferred to an aspartate that is conserved among all P-type ATPases. The ADP then dissociates and is replaced with a fresh ATP, causing another conformational change that opens a passageway to the SR lumen through which the two Ca^{2+} ions exit. They are replaced by two H^+ ions and water molecules that stabilize the empty Ca^{2+}-binding sites and close the passageway to the SR lumen, switching the pump to the occluded conformation. Hydrolysis of the labile phosphoryl–aspartate bond opens the passageway to the cytosol. H^+ is released as the pump opens the passageway to the cytosol,

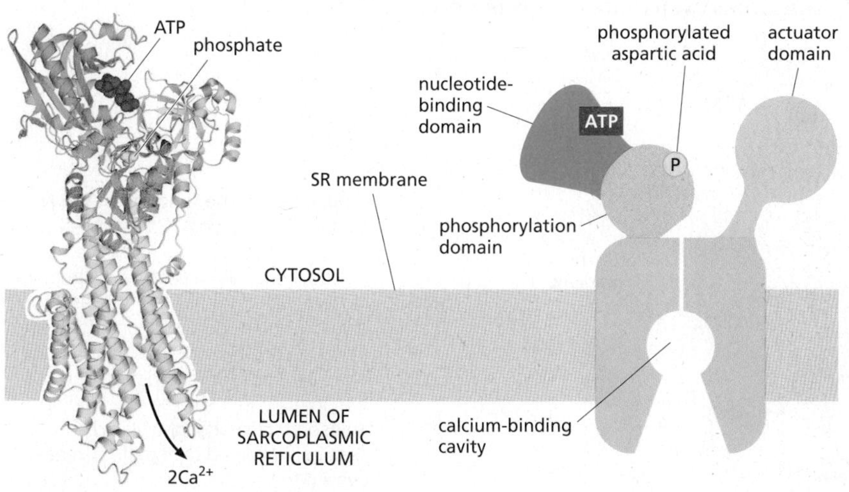

Figure 11–13 The structure of the sarcoplasmic reticulum Ca^{2+} pump. The ribbon model *(left)*, derived from x-ray crystallographic analyses, shows the pump in its phosphorylated, ATP-bound state. The three globular cytosolic domains of the pump—the nucleotide-binding domain *(dark green)*, the actuator domain *(blue)*, and the phosphorylation domain *(pink)*, also shown schematically on the *right*—change conformation dramatically during the pumping cycle. These changes in turn alter the arrangement of the transmembrane helices, which allows the Ca^{2+} to be released from its binding cavity into the SR lumen (Movie 11.3). (PDB code: 3B9B.)

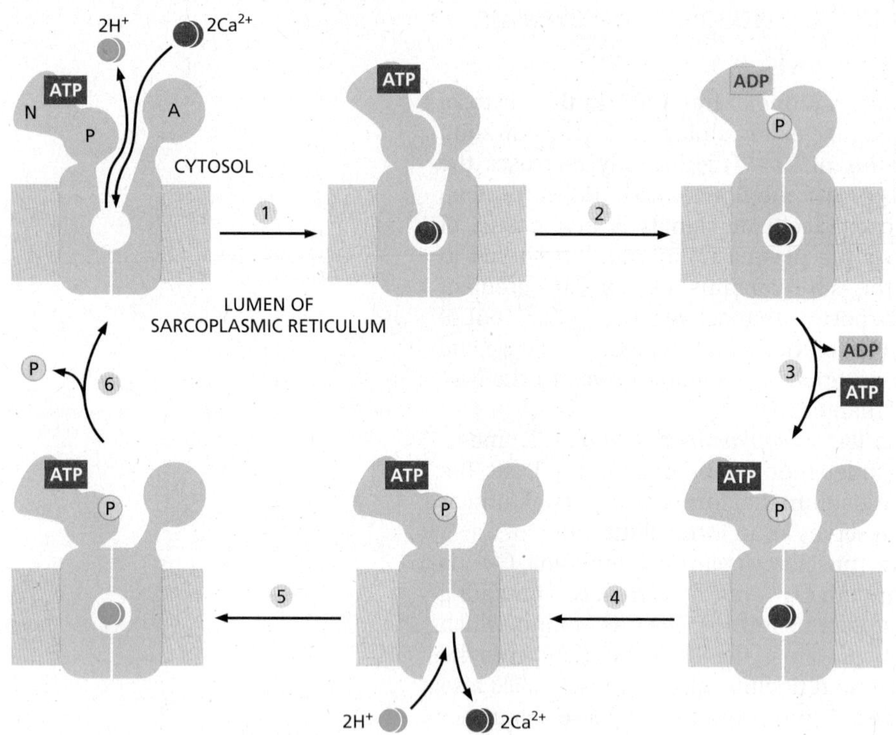

Figure 11–14 The pumping cycle of the sarcoplasmic reticulum Ca^{2+} pump. Ion pumping proceeds by a series of stepwise conformational changes in which movements of the pump's three cytosolic domains [the nucleotide-binding domain (N), the phosphorylation domain (P), and the actuator domain (A)] are mechanically coupled to movements of the transmembrane α helices. Helix movement opens and closes passageways through which Ca^{2+} enters from the cytosol and binds to the two centrally located Ca^{2+}-binding sites. The two Ca^{2+} then exit into the SR lumen and are replaced by two H^+, which are transported in the opposite direction. The ion-dependent (Ca^{2+} and H^+ in the case of the SR Ca^{2+} pump) phosphorylation and dephosphorylation of an aspartate are universally conserved steps in the reaction cycle of all P-type pumps: they cause the conformational transitions to occur in an orderly manner, enabling the proteins to do useful work. (Adapted from C. Toyoshima et al., *Nature* 432:361–368, 2004; and J.V. Møller et al., *Q. Rev. Biophys.* 43:501–566, 2010.)

returning it to the initial conformation, and the cycle starts again. The transient self-phosphorylation of the pump during its cycle is an essential characteristic of all P-type pumps.

The Plasma Membrane Na^+-K^+ Pump Establishes Na^+ and K^+ Gradients Across the Plasma Membrane

The concentration of K^+ is typically 10–30 times higher inside cells than outside, whereas the reverse is true of Na^+ (see Table 11–1, p. 638). An **Na^+-K^+ pump**, or **Na^+-K^+ ATPase**, found in the plasma membrane of virtually all animal cells, maintains these concentration differences. Like the Ca^{2+} pump, the Na^+-K^+ pump belongs to the family of P-type ATPases and operates as an ATP-driven antiporter, actively pumping Na^+ out of the cell against its steep electrochemical gradient and pumping K^+ in (**Figure 11–15**).

We mentioned earlier that the Na^+ gradient produced by the Na^+-K^+ pump drives the transport of most nutrients into animal cells and also has a crucial role in regulating cytosolic pH. A typical animal cell devotes almost one-third of its energy to fueling this pump, and the pump consumes even more energy in nerve

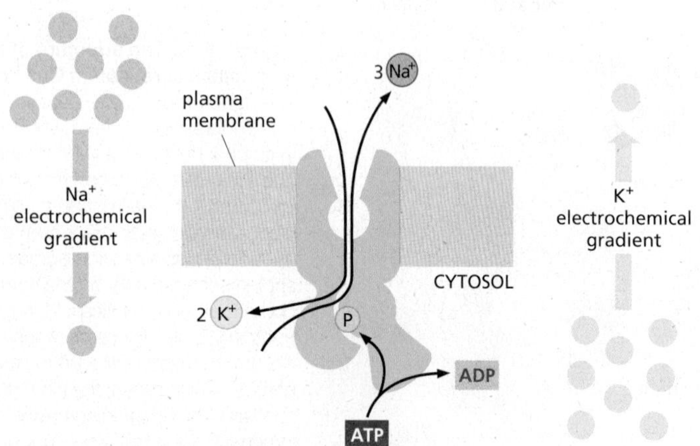

Figure 11–15 The function of the Na^+-K^+ pump. This P-type ATPase actively pumps Na^+ out of and K^+ into a cell against their electrochemical gradients. It is structurally closely related to the Ca^{2+} ATPase but differs in its selectivity for ions: for every molecule of ATP hydrolyzed by the pump, three Na^+ are pumped out and two K^+ are pumped in. As in the Ca^{2+} pump, an aspartate is phosphorylated and dephosphorylated during the pumping cycle (**Movie 11.4**).

cells and in cells that are dedicated to transport processes, such as those forming kidney tubules.

Because the Na^+-K^+ pump drives three positively charged ions out of the cell for every two it pumps in, it is *electrogenic*: it drives a net electric current across the membrane, tending to create an electrical potential, with the cell's inside being negative relative to the outside. This electrogenic effect of the pump, however, seldom directly contributes more than 10% to the membrane potential. The remaining 90%, as we discuss later, depends only indirectly on the Na^+-K^+ pump.

ABC Transporters Constitute the Largest Family of Membrane Transport Proteins

The last type of transport ATPase that we discuss here is the family of ABC transporters, so named because each member contains two highly conserved ATPase domains, or ATP-binding "cassettes," on the cytosolic side of the membrane. ATP binding brings together the two ATPase domains, and ATP hydrolysis leads to their dissociation (**Figure 11–16**). These movements of the cytosolic domains are transmitted to the transmembrane segments, driving cycles of conformational changes that alternately expose solute-binding sites on one side of the membrane and then on the other side, as we have seen for other transporters. In this way, ABC transporters harvest the energy released upon ATP binding and hydrolysis to drive transport of solutes across the bilayer. The transport is directional toward the inside or toward the outside, depending on the particular conformational change in the solute-binding site that is linked to ATP hydrolysis (see Figure 11–16).

ABC transporters constitute the largest family of membrane transport proteins and are of great clinical importance. The first of these proteins to be characterized was found in bacteria. We have already mentioned that the plasma membranes of all bacteria contain transporters that use the H^+ gradient across the membrane to actively transport a variety of nutrients into the cell. In addition, bacteria use ABC transporters to import certain small molecules. In Gram-negative bacteria, such

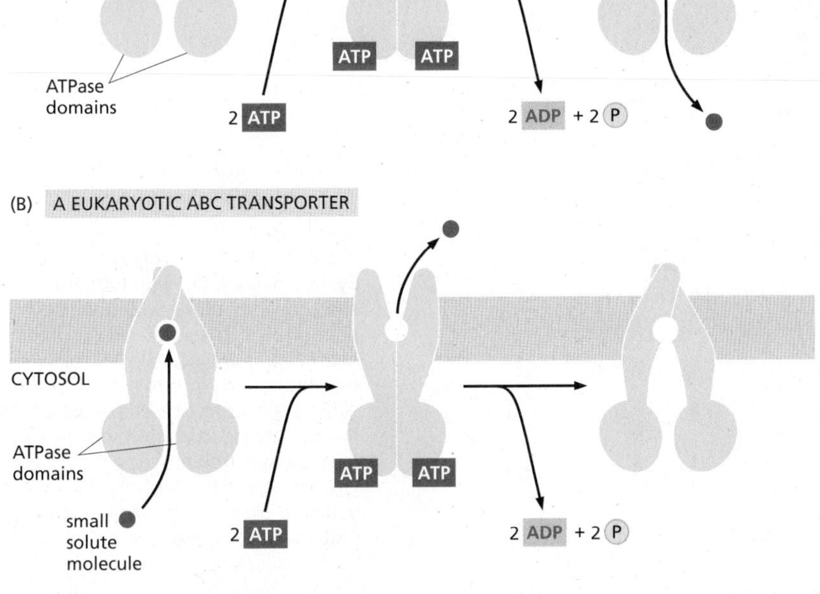

Figure 11–16 **Small-molecule transport by typical ABC transporters.** ABC transporters consist of multiple domains. Typically, two hydrophobic domains, each built of six membrane-spanning α helices, together form the translocation pathway and provide substrate specificity. Two ATPase domains protrude into the cytosol. In some cases, the two halves of the transporter are formed by a single polypeptide, whereas in other cases they are formed by two or more separate polypeptides that assemble into a similar structure. Without ATP bound, the transporter exposes a substrate-binding site on one side of the membrane. ATP binding induces a conformational change that exposes the substrate-binding site on the opposite side; ATP hydrolysis followed by ADP dissociation returns the transporter to its original conformation. Most individual ABC transporters are unidirectional. (A) Both importing and exporting ABC transporters are found in bacteria; an ABC importer is shown in this diagram. (B) In eukaryotes, most ABC transporters export substances—either from the cytosol to the extracellular space or from the cytosol to a membrane-bound intracellular compartment such as the endoplasmic reticulum—or from the mitochondrial matrix to the intermembrane space.

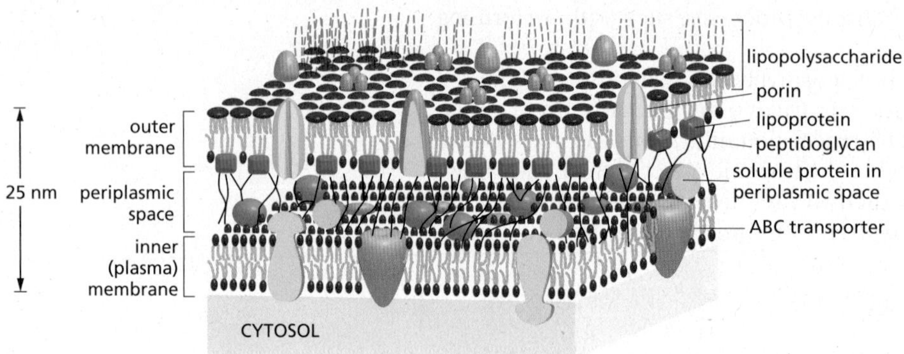

Figure 11–17 A small section of the double membrane of an *E. coli* bacterium. The inner membrane is the cell's plasma membrane. Between the inner and outer membranes is a highly porous, rigid peptidoglycan layer, composed of protein and polysaccharide that constitute the bacterial cell wall. It is attached to lipoprotein molecules in the outer membrane and fills the *periplasmic space* (only a little of the peptidoglycan layer is shown). This space also contains a variety of soluble protein molecules. The dashed threads (shown in *green*) at the top represent the polysaccharide chains of the special lipopolysaccharide molecules that form the external monolayer of the outer membrane; for clarity, only a few of these chains are shown. Bacteria with double membranes are called *Gram negative* because they do not retain the dark blue dye used in Gram staining. Bacteria with single membranes (but thicker peptidoglycan cell walls), such as staphylococci and streptococci, retain the dark blue dye and are therefore called *Gram positive*; their single membrane is analogous to the inner (plasma) membrane of Gram-negative bacteria.

as *Escherichia coli,* that have double membranes (Figure 11–17), the ABC transporters are located in the inner membrane, and auxiliary proteins in the periplasm typically capture the nutrients and deliver them to the transporters (Figure 11–18).

In *E. coli,* 78 genes (an amazing 5% of the bacterium's genes) encode ABC transporters, and animal genomes encode an even larger number. Although each transporter is thought to be specific for a particular molecule or class of molecules, the variety of substrates transported by this superfamily is great and includes inorganic ions, amino acids, monosaccharides and polysaccharides, peptides, lipids, drugs, and, in some cases, even proteins that can be larger than the transporter itself.

The first eukaryotic ABC transporters identified were discovered because they pump hydrophobic drugs out of the cytosol. One of these transporters is the **multidrug resistance (MDR) protein**, also called P-glycoprotein. It is present at elevated levels in many human cancer cells and makes the cells simultaneously resistant to a variety of chemically unrelated cytotoxic drugs that are widely used in cancer chemotherapy. Treatment with any one of these drugs can result in the selective survival and overgrowth of those cancer cells that express an especially large amount of the MDR transporter. These cells pump drugs out of the cell very efficiently and are therefore relatively resistant to the drugs' toxic effects (Movie 11.5). Selection for cancer cells with resistance to one drug can thereby lead to resistance to a wide variety of anticancer drugs. Some studies indicate that up to 40% of human cancers develop multidrug resistance, making it a major hurdle in the battle against cancer.

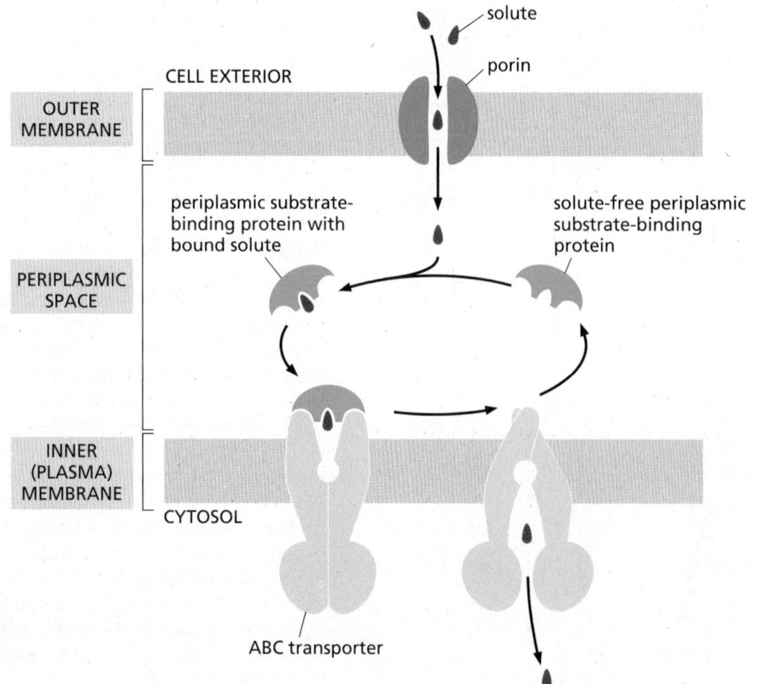

Figure 11–18 The auxiliary transport system associated with transport ATPases in bacteria with double membranes. The solute diffuses through channel proteins (porins) in the outer membrane and binds to a *periplasmic substrate-binding protein* that delivers it to the ABC transporter, which pumps it across the plasma membrane. The peptidoglycan layer is omitted for simplicity; its porous structure allows the substrate-binding proteins and water-soluble solutes to move freely through it by diffusion.

A related and equally sinister phenomenon occurs in the protist *Plasmodium falciparum*, which causes malaria. More than 200 million people are infected worldwide with this parasite, which remains a major cause of human death, killing almost a million people every year. The development of resistance to the antimalarial drug *chloroquine* has hampered the control of malaria. The resistant *P. falciparum* have amplified a gene encoding an ABC transporter that pumps out the chloroquine.

In most vertebrate cells, an ABC transporter in the endoplasmic reticulum (ER) membrane (named *transporter associated with antigen processing*, or *TAP transporter*) actively pumps a wide variety of peptides from the cytosol into the ER lumen. These peptides are produced by protein degradation in proteasomes (discussed in Chapter 6). They are carried from the ER to the cell surface, where they are displayed for scrutiny by cytotoxic T lymphocytes, which kill the cell if the peptides are derived from a virus or other microorganism lurking in the cytosol of an infected cell (discussed in Chapter 24).

Yet another member of the ABC transporter family is the *cystic fibrosis transmembrane conductance regulator* (CFTR) protein, which was discovered through studies of the common genetic disease *cystic fibrosis*. This disease is caused by a mutation in the gene encoding CFTR, a Cl^- transport protein in the plasma membrane of epithelial cells. CFTR regulates ion concentrations in the extracellular fluid, especially in the lung. One in 27 Caucasians carries a gene encoding a mutant form of this protein; in 1 in 2900, both copies of the gene are mutated, causing the disease. In contrast to other ABC transporters, ATP binding and hydrolysis in the CFTR protein do not drive the transport process. Instead, they control the opening and closing of a continuous channel, which provides a passive conduit for Cl^- to move down its electrochemical gradient. Thus, some ABC proteins can function as transporters and others as gated channels. Even though we treat transporters and channels as distinct in this chapter, the border is not absolute as this example illustrates. Indeed, the structure of some other Cl^- channels reveals that they also resemble transporters more than they resemble most other ion channels.

Summary

Transporters bind specific solutes and transfer them across the lipid bilayer by undergoing conformational changes that alternately expose the solute-binding site on one side of the membrane and then on the other side. Some transporters move a single solute "downhill," whereas others can act as pumps to move a solute "uphill" against its electrochemical gradient, by using energy provided by ATP hydrolysis, by a downhill flow of another solute (such as Na^+ or H^+), or by light to drive the requisite series of conformational changes in an orderly manner. Transporters belong to a small number of protein families. Each family evolved from a common ancestral protein, and its members all operate by a similar mechanism. The family of P-type transport ATPases, which includes Ca^{2+} and Na^+-K^+ pumps, is an important example; each of these ATPases sequentially phosphorylates and dephosphorylates itself during the pumping cycle. The superfamily of ABC transporters is the largest family of membrane transport proteins and is especially important clinically. It includes proteins that are responsible for drug resistance in both cancer cells and cells infected with malaria-causing parasites and for pumping pathogen-derived peptides into the ER for cytotoxic lymphocytes to reorganize on the surface of infected cells, and mutations in an ABC transporter cause cystic fibrosis.

CHANNELS AND THE ELECTRICAL PROPERTIES OF MEMBRANES

Unlike transporters, channels form pores across membranes. One class of channel proteins found in virtually all animals forms *gap junctions* between adjacent cells; each plasma membrane contributes equally to the formation of the channel, which connects the cytoplasm of the two cells. In plants, *plasmodesmata* fulfill many of the same functions. These channels are discussed in Chapter 19 and will not be considered further here. Both gap junctions and *porins*, the channels in the outer membranes of bacteria, mitochondria, and chloroplasts (discussed in Chapter 10), have relatively large and permissive pores, and it would be disastrous if they directly connected the inside of a cell to an extracellular space. Indeed, many bacterial toxins do exactly that to kill other cells (discussed in Chapter 24).

In contrast, most channels in the plasma membrane of animal and plant cells that connect the cytosol to the cell exterior necessarily have narrow, highly selective pores that can open and close rapidly. Because these proteins are concerned specifically with inorganic ion transport, they are referred to as **ion channels**. For transport efficiency, ion channels have an advantage over transporters, in that they can pass up to 100 million ions through one open channel each second—a rate 10^5 times greater than even the fastest transporter. As discussed earlier, however, channels cannot be coupled to an energy source to perform active transport, so the conductance they mediate is always passive (downhill). Thus, the function of ion channels is to allow specific inorganic ions—primarily Na^+, K^+, Ca^{2+}, or Cl^-—to diffuse rapidly down their electrochemical gradients across the lipid bilayer. In this section, we will see that the ability to control ion fluxes through these channels is essential for many cell functions. Nerve cells (neurons), in particular, have made a specialty of using ion channels, and we will consider how they use many different ion channels to receive, conduct, and transmit signals. Before we discuss ion channels, however, we briefly consider the aquaporin water channels that we mentioned earlier.

Aquaporins Are Permeable to Water but Impermeable to Ions

Because cells are mostly water (typically ~70% by weight), water movement across cell membranes is fundamentally important for life. Cells also contain a high concentration of solutes, including numerous negatively charged organic molecules that are confined inside the cell (the so-called *fixed anions*) and their accompanying cations that are required for charge balance. This creates an osmotic gradient, which mostly is balanced by an opposite osmotic gradient due to a high concentration of inorganic ions—chiefly Na^+ and Cl^-—in the extracellular fluid. The small remaining osmotic force tends to "pull" water into the cell, causing it to swell until the forces are balanced. Because all biological membranes are moderately permeable to water (see Figure 11–2), cell volume equilibrates in minutes or less in response to an osmotic gradient. For most animal cells, however, osmosis has only a minor role in regulating cell volume. This is because most of the cytoplasm is in a gel-like state and resists large changes in its volume in response to changes in osmolarity.

In addition to the direct diffusion of water across the lipid bilayer, some prokaryotic and eukaryotic cells have **water channels**, or **aquaporins**, embedded in their plasma membranes to allow water to move more rapidly. Aquaporins are particularly abundant in animal cells that must transport water at high rates, such as the epithelial cells of the kidney or exocrine cells that must transport or secrete large volumes of fluids (**Figure 11–19**). Water flow is highly regulated in these tissues. In the kidney, hormones such as antidiuretic hormone (vasopressin) regulate the concentration of aquaporin in the plasma membrane.

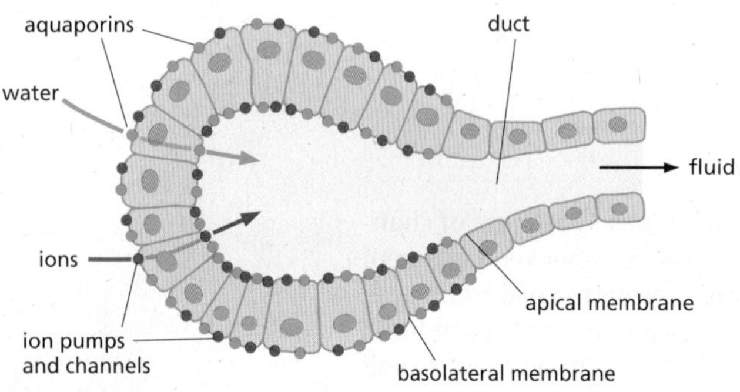

Figure 11–19 **The role of aquaporins in fluid secretion.** Cells lining the ducts of exocrine glands (as found, for example, in the pancreas and liver, and in mammary, sweat, and salivary glands) secrete large volumes of body fluids. These cells are organized into epithelial sheets in which their apical plasma membrane faces the lumen of the duct. Ion pumps and channels situated in the basolateral and apical plasma membranes move ions (mostly Na^+ and Cl^-) into the ductal lumen, creating an osmotic gradient between the surrounding tissue and the duct. Water molecules rapidly follow the osmotic gradient through aquaporins that are present in high concentrations in both the apical and basolateral plasma membranes.

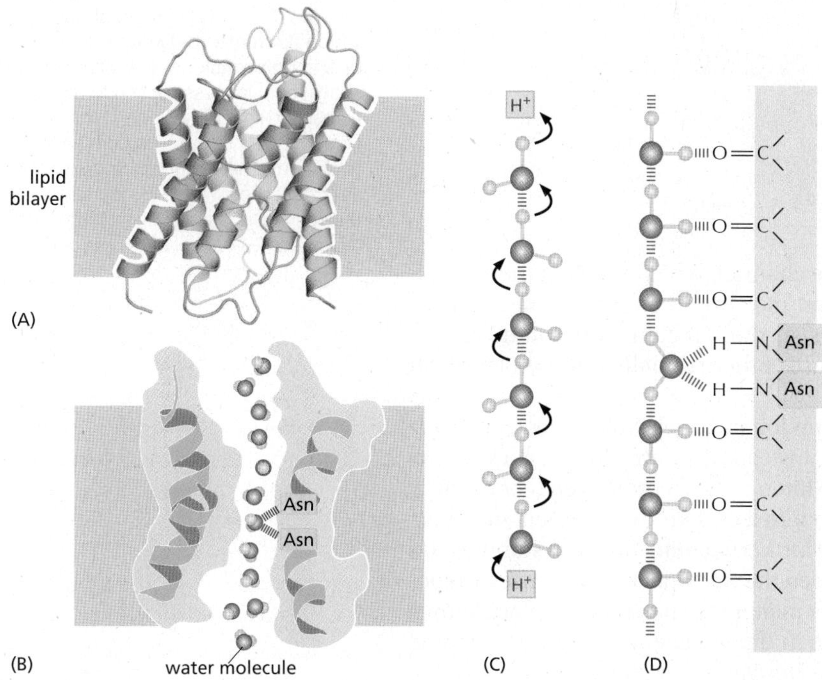

Figure 11–20 **The structure of aquaporins.** (A) A ribbon diagram of an aquaporin monomer. In the membrane, aquaporins form tetramers, with each monomer containing an aqueous pore in its center (not shown). Each individual aquaporin channel passes about 10^9 water molecules per second. (B) A longitudinal cross section through one aquaporin monomer, in the plane of the central pore. One face of the pore is lined with hydrophilic amino acids, which provide transient hydrogen bonds to water molecules; these bonds help line up the transiting water molecules in a single row and orient them as they traverse the pore. (C and D) A model explaining why aquaporins are impermeable to H^+. (C) In water, H^+ diffuses extremely rapidly by being relayed from one water molecule to the next. (D) Carbonyl groups (C=O) lining the hydrophilic face of the pore align water molecules, and two strategically placed asparagines in the center are thought to tether a central water molecule such that both valences on its oxygen are occupied. This arrangement bipolarizes the entire column of water molecules, with each water molecule acting as a hydrogen-bond acceptor from its inner neighbor (Movie 11.6). (PDB code: 3GD8.)

Aquaporins must solve a problem that is opposite to that facing ion channels. To avoid disrupting ion gradients across membranes, they have to allow the rapid passage of water molecules while completely blocking the passage of ions. The three-dimensional structure of an aquaporin reveals how it achieves this remarkable selectivity. The channels have a narrow pore that allows water molecules to traverse the membrane in single file, following the path of carbonyl oxygens that line one side of the pore (**Figure 11–20A and B**). Hydrophobic amino acids line the other side of the pore. The pore is too narrow for any hydrated ion to enter, and the energy cost of dehydrating an ion would be enormous because the hydrophobic wall of the pore cannot interact with a dehydrated ion to compensate for the loss of water. This design readily explains why the aquaporins cannot conduct K^+, Na^+, Ca^{2+}, or Cl^- ions. These channels are also impermeable to H^+, which is mainly present in cells as H_3O^+. These hydronium ions diffuse through water extremely rapidly, using a molecular relay mechanism that requires the making and breaking of hydrogen bonds between adjacent water molecules (**Figure 11–20C**). Aquaporins contain two strategically placed asparagines, which bind to the oxygen atom of the central water molecule in the line of water molecules traversing the pore, imposing a bipolarity on the entire single-file column of water molecules (**Figure 11–20C and D**). Because both valences of this central oxygen are unavailable for hydrogen-bonding, the central water molecule cannot participate in an H^+ relay. This makes it impossible for the "making and breaking" sequence of hydrogen bonds (shown in Figure 11–20C) to get past the central asparagine-bonded water molecule, and the pore is therefore impermeable to H^+.

We now turn to ion channels, the subject of the rest of the chapter.

Ion Channels Are Ion-selective and Fluctuate Between Open and Closed States

Two important properties distinguish ion channels from aqueous pores. First, they show *ion selectivity*, permitting some inorganic ions to pass, but not others. This suggests that their pores must be narrow enough in places to force permeating ions into intimate contact with the walls of the channel so that only ions of appropriate size and charge can pass. In some cases permeating ions have to shed most or all of their associated water molecules to pass, whereas in other cases hydrated or partially hydrated ions pass through the channel. Passage occurs in

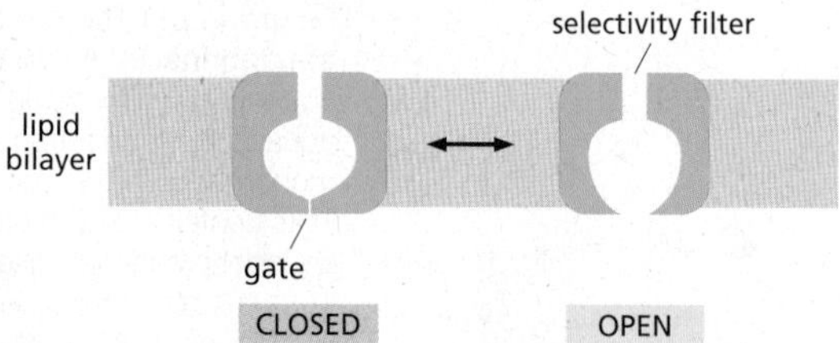

Figure 11–21 **A typical ion channel, which fluctuates between closed and open conformations.** The ion channel shown here in cross section forms a pore across the lipid bilayer only in the "open" conformational state. The pore narrows to atomic dimensions in one region (the selectivity filter), where the ion selectivity of the channel is largely determined. Another region of the channel forms the gate.

single file, through the narrowest part of the channel, which is called the *selectivity filter*, and limits the channel's rate of ion movement and determines which ions can pass through (Figure 11–21). Thus, as the ion concentration increases, the flux of the ion through a channel increases proportionally but then levels off (saturates) at a maximum rate.

The second important distinction between ion channels and aqueous pores is that ion channels are not continually open. Instead, they are *gated*, which allows them to open briefly and then close again. Moreover, with prolonged (chemical or electrical) stimulation, most ion channels go into a closed "desensitized," or "inactivated," state, in which they are refractory to further opening until the stimulus has been removed, as we discuss later. In most cases, the gate opens in response to a specific stimulus. As shown in Figure 11–22, the main types of stimuli that are known to cause ion channels to open are a change in the voltage across the membrane (*voltage-gated channels*), a mechanical stress (*mechanically gated channels*), or the binding of a ligand (*ligand-gated channels*). The ligand can be either an extracellular mediator—specifically, a neurotransmitter (*transmitter-gated channels*)—or an intracellular mediator such as an ion (*ion-gated channels*) or a nucleotide (*nucleotide-gated channels*). In addition, protein phosphorylation and dephosphorylation regulate the activity of many ion channels; this type of channel regulation is discussed, together with nucleotide-gated ion channels, in Chapter 15.

More than 200 types of ion channels have been identified thus far, and new ones are still being discovered, each characterized by the ions it conducts, the mechanism by which it is gated, and its abundance and localization in the cell and in specific cells. Ion channels are responsible for the electrical excitability of muscle cells, and they mediate most forms of electrical signaling in the nervous system. A single neuron typically contains 10 or more kinds of ion channels, located in different domains of its plasma membrane. But ion channels are not restricted to electrically excitable cells. They are present in all animal cells and are found in plant cells and microorganisms: they propagate the leaf-closing response of the mimosa plant, for example (Movie 11.7), and allow the single-celled motile *Paramecium* to reverse direction after a collision.

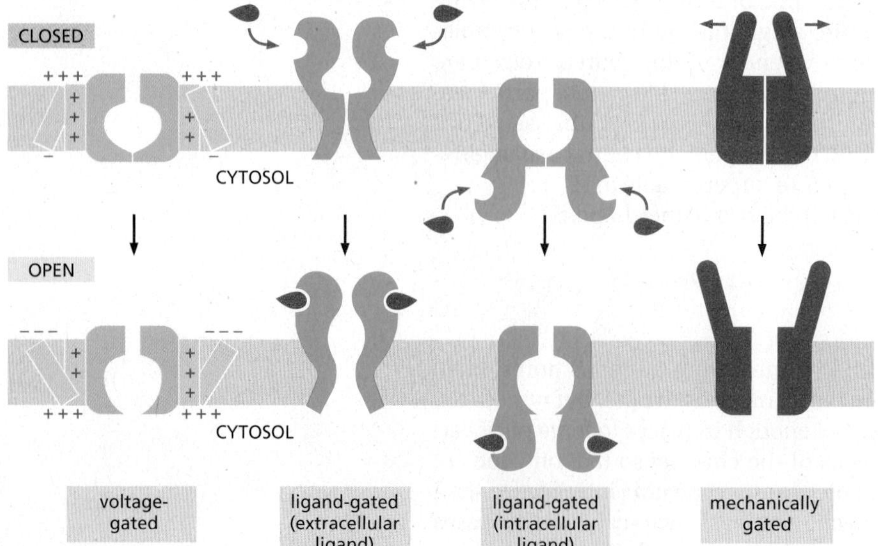

Figure 11–22 **The gating of ion channels.** This schematic drawing shows several kinds of stimuli that open ion channels. Mechanically gated channels often have cytoplasmic extensions (not shown) that link the channel to the cytoskeleton.

Ion channels that are permeable to K^+ are found in the plasma membrane of almost all cells. An important subset of K^+ channels opens even in an unstimulated or "resting" cell, and hence these are called **K^+ leak channels**. Although this term applies to many different K^+ channels, depending on the cell type, they serve a common purpose: by making the plasma membrane much more permeable to K^+ than to other ions, they have a crucial role in maintaining the membrane potential across all plasma membranes, as we discuss next.

The Membrane Potential in Animal Cells Depends Mainly on K^+ Leak Channels and the K^+ Gradient Across the Plasma Membrane

A **membrane potential** arises when there is a difference in the electrical charge on the two sides of a membrane because of a minute excess of positive ions over negative ones on one side and a minute deficit on the other side. Such charge differences can result both from active electrogenic pumping (see p. 648) and from passive ion diffusion in channels. As we discuss in Chapter 14, electrogenic H^+ pumps in the mitochondrial inner membrane generate most of the membrane potential across this membrane. Electrogenic pumps also generate most of the electrical potential across the plasma membrane of plants and fungi. In typical animal cells, however, passive ion movements make the largest contribution to the electrical potential across the plasma membrane.

As explained earlier, because of the action of the Na^+-K^+ pump, there is little Na^+ inside the cell, and other intracellular inorganic cations have to be plentiful enough to balance the charge carried by the cell's fixed anions—the negatively charged organic molecules that are confined inside the cell. This balancing role is performed largely by K^+, which is actively pumped into the cell by the Na^+-K^+ pump and can also move in or out through the *K^+ leak channels* in the plasma membrane. Because of the presence of these channels, K^+ comes almost to equilibrium, where an electrical force exerted by an excess of negative charges attracting K^+ into the cell balances the tendency of K^+ to leak out down its concentration gradient. The membrane potential (of the plasma membrane) is the manifestation of this electrical force, and we can calculate its equilibrium value from the steepness of the K^+ concentration gradient. The following argument may help to make this clear.

Suppose that initially there is no voltage gradient across the plasma membrane (the membrane potential is zero), but the concentration of K^+ is high inside the cell and low outside. K^+ will tend to leave the cell through the K^+ leak channels, driven solely by its concentration gradient. As K^+ begins to move out, each ion leaves behind an unbalanced negative charge, thereby creating an electrical field, or membrane potential, which will tend to oppose the further efflux of K^+. The net efflux of K^+ halts when the membrane potential reaches a value at which this electrical driving force on K^+ exactly balances the effect of its concentration gradient; that is, when the electrochemical gradient for K^+ is zero.

The equilibrium condition, in which there is no net flow of ions across the plasma membrane, defines the **resting membrane potential** for this idealized cell. A simple but very important formula, the **Nernst equation**, quantifies the equilibrium condition and, as explained in **Panel 11–1** (p. 656), makes it possible to calculate the theoretical resting membrane potential if we know the ratio of internal and external ion concentrations. As the plasma membrane of a real cell is not exclusively permeable to K^+, however, the actual resting membrane potential is usually not exactly equal to that predicted by the Nernst equation for K^+.

The Resting Potential Decays Only Slowly When the Na^+-K^+ Pump Is Stopped

Movement of only a minute number of inorganic ions across the plasma membrane through ion channels suffices to set up the membrane potential. Thus, we can think of the membrane potential as arising from movements of charge that

THE NERNST EQUATION AND ION FLOW

The flow of any inorganic ion through a membrane channel is driven by the electrochemical gradient for that ion. This gradient represents the combination of two influences: the voltage gradient and the concentration gradient of the ion across the membrane. When these two influences just balance each other, the electrochemical gradient for the ion is zero, and there is no *net* flow of the ion through the channel. The voltage gradient (membrane potential) at which this equilibrium is reached is called the equilibrium potential for the ion. It can be calculated from an equation that will be derived below, called the Nernst equation.

The Nernst equation is $V = \frac{RT}{zF} \ln \frac{C_o}{C_i}$

where

- V = the equilibrium potential in volts (internal potential minus external potential)
- C_o and C_i = outside and inside concentrations of the ion, respectively
- R = the gas constant (8.3 J mol^{-1} K^{-1})
- T = the absolute temperature (K)
- F = Faraday's constant (9.6×10^4 J V^{-1} mol^{-1})
- z = the valence (charge) of the ion
- ln = logarithm to the base e

The Nernst equation is derived as follows:

A molecule in solution (a solute) tends to move from a region of high concentration to a region of low concentration simply due to the random movement of molecules, which results in their equilibrium. Consequently, movement down a concentration gradient is accompanied by a favorable free-energy change ($\Delta G < 0$), whereas movement up a concentration gradient is accompanied by an unfavorable free-energy change ($\Delta G > 0$). (Free energy is introduced in Chapter 2 and discussed in the context of redox reactions in Panel 14–1, p. 825.)

The free-energy change per mole of solute moved across the plasma membrane (ΔG_{conc}) is equal to $-RT\ln(C_o/C_i)$.

If the solute is an ion, moving it into a cell across a membrane whose inside is at a voltage V relative to the outside will cause an additional free-energy change (per mole of solute moved) of $\Delta G_{volt} = zFV$.

At the point where the concentration and voltage gradients just balance,

$$\Delta G_{conc} + \Delta G_{volt} = 0$$

and the ion distribution is at equilibrium across the membrane.

Thus,

$$zFV - RT \ln \frac{C_o}{C_i} = 0$$

and, therefore,

$$V = \frac{RT}{zF} \ln \frac{C_o}{C_i}$$

or, using the constant that converts natural logarithms to base 10,

$$V = 2.3 \frac{RT}{zF} \log_{10} \frac{C_o}{C_i}$$

For a univalent cation,

$$2.3 \frac{RT}{F} = 58 \text{ mV at } 20°\text{C} \quad \text{and} \quad 61.5 \text{ mV at } 37°\text{C}$$

Thus, for such an ion at 37°C,

$$V = +61.5 \text{ mV when } C_o/C_i = 10$$

whereas

$$V = 0 \text{ when } C_o/C_i = 1$$

The K^+ equilibrium potential (V_K), for example, is

$$61.5 \log_{10}([K^+]_o/[K^+]_i) \text{ millivolts}$$

(–89 mV for a typical cell, where $[K^+]_o$ = 5 mM and $[K^+]_i$ = 140 mM).

At V_K, there is no net flow of K^+ across the membrane.

Similarly, when the membrane potential has a value of

$$61.5 \log_{10}([Na^+]_o/[Na^+]_i)$$

which is the Na^+ equilibrium potential (V_{Na}), there is no net flow of Na^+.

For any particular membrane potential, V_M, the net force tending to drive a particular type of ion out of the cell is proportional to the difference between V_M and the equilibrium potential for the ion; hence,

for K^+ it is $V_M - V_K$

and for Na^+ it is $V_M - V_{Na}$

When there is a voltage gradient across the membrane, the ions responsible for it—the excess positive ions on one side and the excess negative ions on the other—are concentrated in thin layers on either side of the membrane because of the attraction between positive and negative electric charges. The number of ions that go to form the layer of charge adjacent to the membrane is minute compared with the total number inside the cell. For example, the movement of 6000 Na^+ ions across 1 μm^2 of membrane will carry sufficient charge to shift the membrane potential by about 100 mV.

Because there are about 6×10^6 Na^+ ions in 1 μm^3 of bulk cytoplasm, such a movement of charge will generally have a negligible effect on the ion concentration gradients across the membrane.

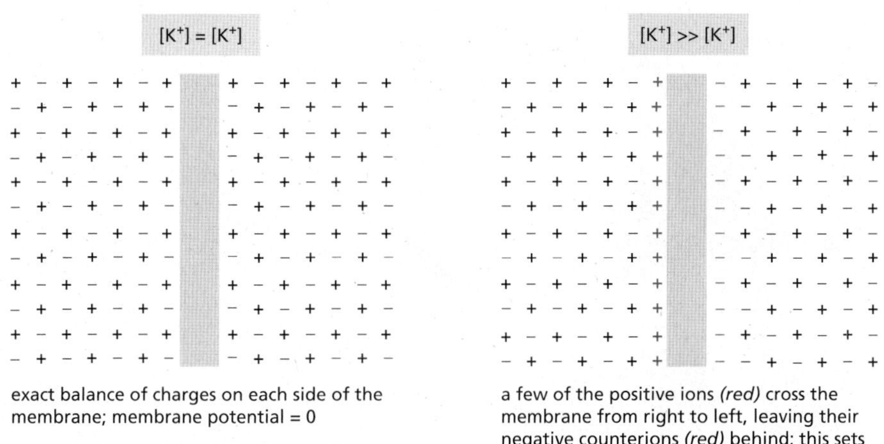
[K+] = [K+]
[K+] >> [K+]
exact balance of charges on each side of the membrane; membrane potential = 0
a few of the positive ions (red) cross the membrane from right to left, leaving their negative counterions (red) behind; this sets up a nonzero membrane potential

Figure 11–23 The ionic basis of a membrane potential. A small flow of inorganic ions through an ion channel carries sufficient charge to cause a large change in the membrane potential. The ions that give rise to the membrane potential lie in a thin (<1 nm) surface layer close to the membrane, held there by their electrical attraction to their oppositely charged counterparts (counterions) on the other side of the membrane. For a typical cell, 1 microcoulomb of charge (6×10^{12} monovalent ions) per square centimeter of membrane, transferred from one side of the membrane to the other side, changes the membrane potential by roughly 1 V. This means, for example, that in a spherical cell of diameter 10 μm, the number of K^+ ions that have to flow out to alter the membrane potential by 100 mV is only about 1/100,000 of the total number of K^+ ions in the cytosol. This amount is so minute that the intracellular K^+ concentration remains virtually unchanged.

leave ion *concentrations* practically unaffected and result in only a very slight discrepancy in the number of positive and negative ions on the two sides of the membrane (Figure 11–23). Moreover, these movements of charge are generally rapid, taking only a few milliseconds or less.

Consider the change in the membrane potential in a real cell after the sudden inactivation of the Na^+-K^+ pump. A slight drop in the membrane potential may occur immediately. This is because the pump is electrogenic and, when active, makes a small direct contribution to the membrane potential by pumping out three Na^+ for every two K^+ that it pumps in (see Figure 11–15). However, switching off the pump does not abolish the major component of the resting potential, which is generated by the K^+ equilibrium mechanism just described. This component of the membrane potential persists as long as the Na^+ concentration inside the cell stays low and the K^+ ion concentration high—typically for many minutes. But the plasma membrane is somewhat permeable to all small ions, including Na^+. Therefore, without the Na^+-K^+ pump, the ion gradients set up by the pump will eventually run down, and the membrane potential established by diffusion through the K^+ leak channels will fall as well. As Na^+ enters, the cell eventually comes to a new resting state where Na^+, K^+, and Cl^- are all at equilibrium across the membrane. The membrane potential in this state is much less than it was in the normal cell with an active Na^+-K^+ pump.

The resting potential of an animal cell varies between –20 mV and –120 mV, depending on the organism and cell type. Although the K^+ gradient always has a major influence on this potential, the gradients of other ions (and the disequilibrating effects of ion pumps) also have a significant effect: the more permeable the membrane for a given ion, the more strongly the membrane potential tends to be driven toward the equilibrium value for that ion. Consequently, changes in a membrane's relative permeability to different ions can cause significant changes in the membrane potential. This is one of the key principles relating the electrical excitability of cells to the activities of ion channels.

To understand how ion channels select their ions and how they open and close, we need to know their three-dimensional molecular structure. The first ion channel to be crystallized and studied by x-ray diffraction was a bacterial K^+ channel. The details of its structure revolutionized our understanding of ion channels.

The Three-dimensional Structure of a Bacterial K^+ Channel Shows How an Ion Channel Can Work

Scientists were puzzled by the remarkable ability of ion channels to combine exquisite ion selectivity with a high conductance. K^+ leak channels, for example, conduct K^+ 10,000-fold faster than Na^+, yet the two ions are both featureless spheres and have similar radii (0.133 nm and 0.095 nm, respectively). A single amino acid substitution in the pore of an animal cell K^+ channel can result in a loss

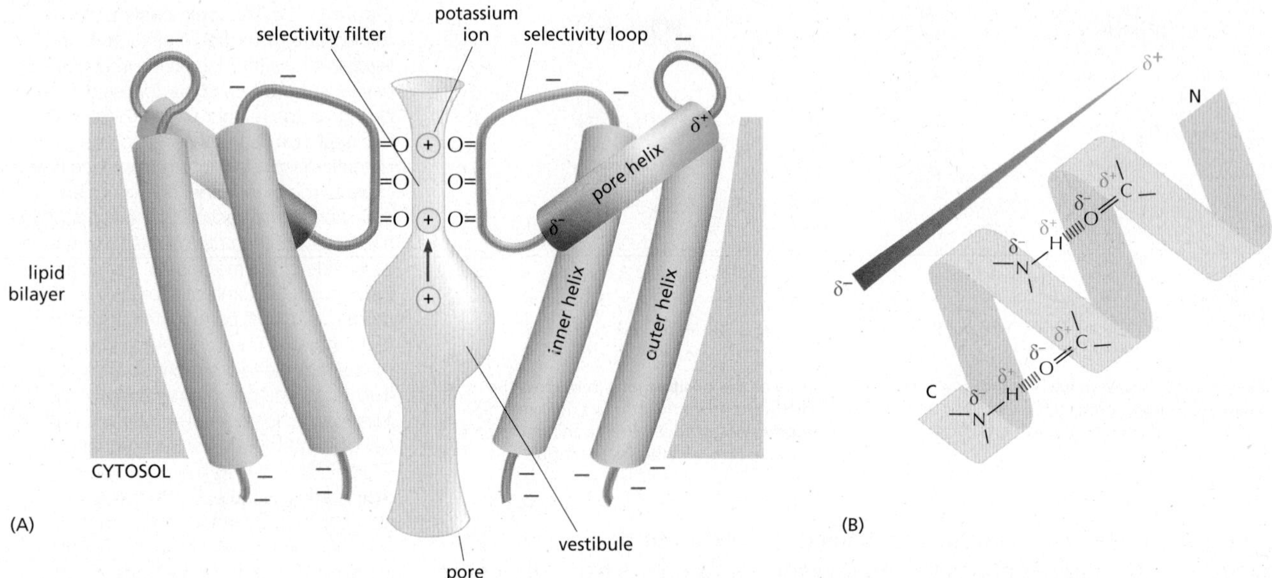

Figure 11–24 The structure of a bacterial K⁺ channel. (A) The transmembrane α helices from only two of the four identical subunits are shown. From the cytosolic side, the pore (schematically shaded in *blue*) opens up into a vestibule in the middle of the membrane. The pore vestibule facilitates transport by allowing the K^+ ions to remain hydrated even though they are more than halfway across the membrane. The narrow selectivity filter of the pore links the vestibule to the outside of the cell. Carbonyl oxygens line the walls of the selectivity filter and form transient binding sites for partially dehydrated K^+ ions. Two K^+ ions occupy different sites in the selectivity filter, while a third K^+ ion is located in the center of the vestibule, where it is stabilized by electrical interactions including those contributed by the more negatively charged ends of the pore helices. The ends of the four short "pore helices" (only two of which are shown) point precisely toward the center of the vestibule, thereby guiding K^+ ions into the selectivity filter (**Movie 11.8**). (B) Peptide bonds have an electric dipole, with more negative charge accumulated at the oxygen of the C=O bond and at the nitrogen of the N—H bond. In an α helix, hydrogen bonds *(red)* align the dipoles. As a consequence, every α helix has an electric dipole along its axis, resulting from summation of the dipoles of the individual peptide bonds, with a more negatively charged C-terminal end (δ^-) and a more positively charged N-terminal end (δ^+). (A, adapted from D.A. Doyle et al., *Science* 280:69–77, 1998.)

of ion selectivity and cell death. We cannot explain the normal K^+ selectivity by pore size alone, because Na^+ is smaller than K^+. Moreover, its high conductance rate is incompatible with the channel having selective, high-affinity K^+-binding sites, as the binding of K^+ ions to such sites would greatly slow their passage.

The puzzle was solved when the structure of a *bacterial K^+ channel* was determined by x-ray crystallography. The channel is made from four identical transmembrane subunits, which together form a central pore through the membrane. Each subunit contributes two transmembrane α helices, which are tilted outward in the membrane and together form a cone, with its wide end facing the outside of the cell where K^+ ions exit from the channel (**Figure 11–24**). The polypeptide chain that connects the two transmembrane helices forms a short α helix (the *pore helix*) and a crucial loop that protrudes into the wide section of the cone to form the **selectivity filter**. The selectivity loops from the four subunits form a short, rigid, narrow pore, which is lined by the carbonyl oxygen atoms of their polypeptide backbones. Because the selectivity loops of all known K^+ channels have similar amino acid sequences, it is likely that they form a closely similar structure.

The structure of the selectivity filter explains the ion selectivity of the channel. A K^+ ion must lose almost all of its bound water molecules to enter the filter, where it interacts instead with the carbonyl oxygens lining the filter; the oxygens are rigidly spaced at the exact distance to accommodate a K^+ ion. An Na^+ ion, in contrast, cannot enter the filter because the carbonyl oxygens are too far away from the smaller Na^+ ion to compensate for the high energy expense associated with the loss of water molecules required for entry (**Figure 11–25**).

Structural studies of K^+ channels and other ion channels have also indicated some general principles of how these channels open and close. The gating involves movement of the helices in the membrane so that they either obstruct or open the path for ion movement. Depending on the particular type of channel,

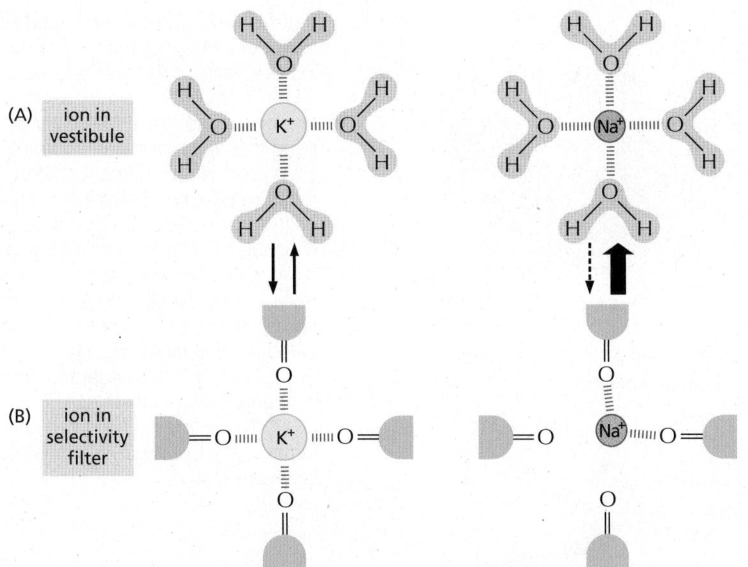

Figure 11–25 **K^+ specificity of the selectivity filter in a K^+ channel.** The drawings show K^+ and Na^+ ions (A) in the vestibule and (B) in the selectivity filter of the pore, viewed in cross section. In the vestibule, the ions are hydrated. In the selectivity filter, they have lost their water, and the carbonyl oxygens are placed to accommodate a dehydrated K^+ ion. The dehydration of the K^+ ion requires energy, which is precisely balanced by the energy regained by the interaction of the ion with all of the carbonyl oxygens that serve as surrogate water molecules. Because the Na^+ ion is too small to interact with all the oxygens, it can enter the selectivity filter only at a great energetic expense. The filter therefore selects K^+ ions with high specificity. (A, adapted from Y. Zhou et al., *Nature* 414:43–48, 2001.)

helices tilt, rotate, or bend during gating. The structure of a closed K^+ channel shows that by tilting the inner helices, the pore constricts like a diaphragm at its cytosolic end (Figure 11–26). Bulky hydrophobic amino acid side chains block the small opening that remains, preventing the entry of ions.

Many other ion channels operate on similar principles: the channel's pore helices are allosterically coupled to sensor domains that in response, say, to ligand binding or altered membrane potential bring about conformational change in the ion-conducting pathway, either opening it or blocking it off.

Mechanosensitive Channels Allow Cells to Sense Their Physical Environment

All organisms, from single-cell bacteria to multicellular animals and plants, must sense and respond to mechanical forces both in their external environment (such as sound, touch, pressure, shear forces, and gravity) and in their internal environment (such as osmotic pressure and membrane bending). Numerous proteins are known to be capable of responding to such mechanical forces, and a large subset of those proteins has been identified as possible **mechanosensitive channels**, but very few of the candidate proteins have been shown directly to be mechanically activated ion channels. One reason for this dearth in our knowledge is that most such channels are extremely rare. Auditory hair cells in the human cochlea, for example, contain extraordinarily sensitive mechanically gated ion channels, but each of the approximately 15,000 individual hair cells is thought to have a total of only 50–100 of them (Movie 11.9 and Movie 11.10). Additional difficulties arise because the gating mechanisms of many mechanosensitive channel types require the channels to

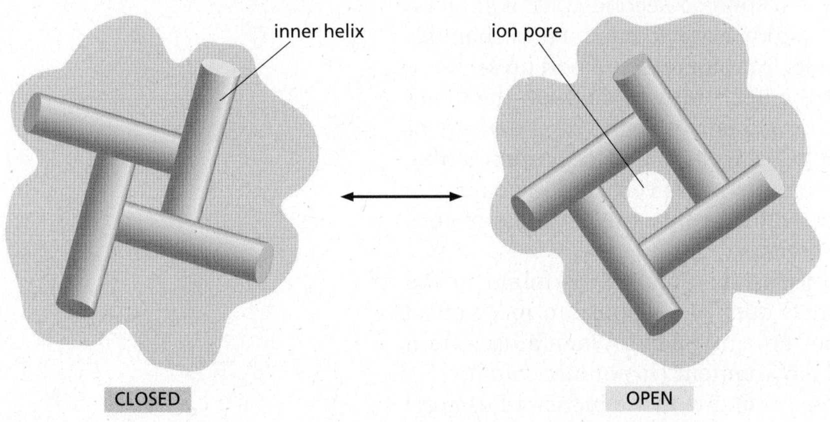

Figure 11–26 **A model for the gating of a bacterial K^+ channel.** The channel is viewed in cross section. To adopt the closed conformation, the four inner transmembrane helices that line the pore on the cytosolic side of the selectivity filter (see Figure 11–24) rearrange to close the cytosolic entrance to the channel. (Adapted from E. Perozo et al., *Science* 285:73–78, 1999.)

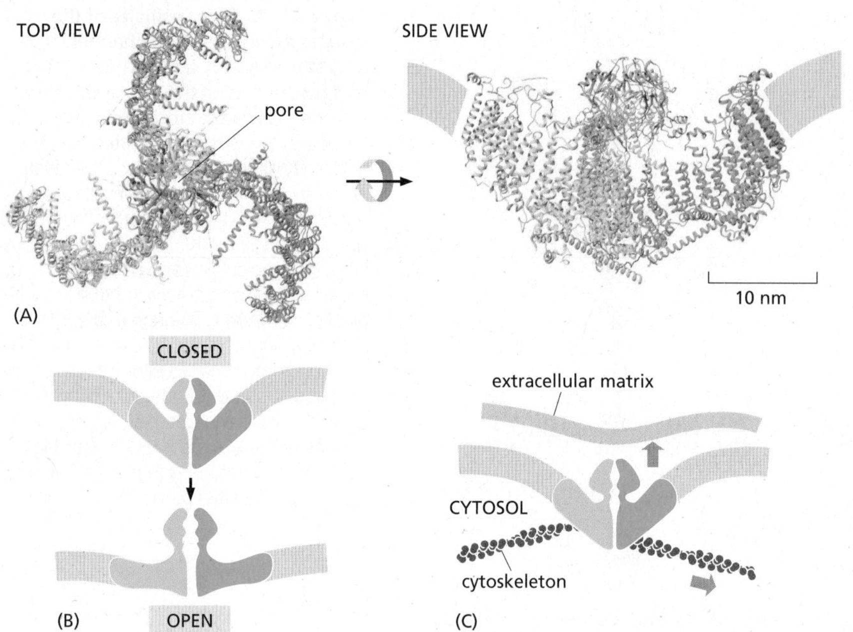

Figure 11–27 Stretch-activated Piezo channels. (A) The structure of Piezo1. Three identical subunits surround the central pore *(left)*. The three large arm domains extend from the center and bend the membrane into a dome *(right)*. (B) When the membrane is stretched, the dome-like protrusion flattens and the central pore opens. (C) By coupling Piezo channels to tethers on either side of the membrane (for example, to the extracellular matrix or the cytoskeleton), Piezo channels can be opened in response to extracellular or intracellular mechanical forces *(arrows)*. (A, PDB code: 6B3R. Adapted from Y.R. Guo and R. MacKinnon, *eLife* 6:e33660, 2017. This article is distributed under the terms of the Creative Commons Attribution License.)

be embedded in complex architectures that require attachment to the extracellular matrix or to the cytoskeleton, and this is difficult to reconstitute in the test tube.

To fill this gap in our knowledge, scientists used patch clamping to identify cell lines that would open an ion channel when mechanical pressure was applied to them. They then reasoned that a mechanosensitive channel would need to contain at least two membrane-spanning helices and proceeded to systematically disrupt expression of genes encoding such proteins. One such disruption caused the cells to lose their response to mechanical stimuli and identified two genes that code for related proteins. Each of these proteins contains more than 30 membrane-spanning helices, but neither shows any sequence similarity to other known ion channels. Three identical copies of either of these proteins can assemble into a mechanosensitive *Piezo* ion channel that lies in the plasma membranes of numerous types of animal cells (**Figure 11–27**). Each of the three subunits of a Piezo channel contains a large arm built from 36 transmembrane helices. The arms extend outward from a central hub that contains the channel's ion-conducting pore. The particular packing of the wing helices deforms the membrane into a dome. In this resting state, the Piezo channel is closed. When the membrane becomes stretched by mechanically pushing on the plasma membrane elsewhere, the resulting membrane tension pulls the deformation flat, which opens the channel's central pore.

Piezo channels in skin cells are required for touch sensation, and in bladder cells they detect when the bladder is full. But their importance is far greater than that. Animals that lack Piezo channels die in development because many developmental processes, such as formation of the vasculature, rely on mechanical stretching cues. Piezo channels also provide second-to-second control of blood pressure. Nerve cells in the aortic arch and carotid artery contain Piezo channels and are highly sensitive to stretch that results from increased blood pressure. In response, they signal to immediately slow the heart rate. In this way, the blood pressure falls, preventing life-threatening consequences from vessel damage or rupture. Conversely, acutely reduced blood pressure, as might occur upon getting up or bending over, stops the firing of these same nerve cells and increases the heart rate and peripheral vasoconstriction, which in turn prevents loss of consciousness by increasing blood flow to the brain.

Another well-studied class of mechanosensitive channels is found in the bacterial plasma membrane. These channels open in response to mechanical stretching of the lipid bilayer in which they are embedded. When a bacterium experiences a low-ionic-strength external environment (hypotonic conditions), such as rainwater, the cell swells as water seeps in due to the increased osmotic

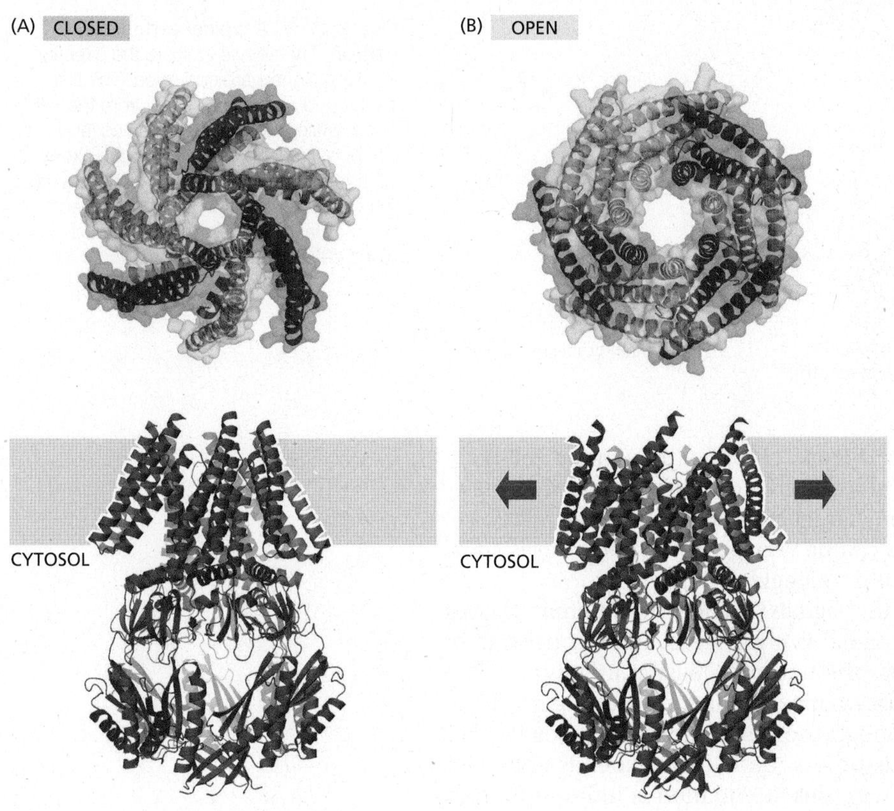

Figure 11–28 The structure of bacterial mechanosensitive channels. The crystal structures of MscS in its (A) closed and (B) open conformation are shown. The side views (lower panels) show the entire protein, including the large intracellular domain. The face views (upper panels) show the transmembrane domains only. The open structure occupies more area in the lipid bilayer and is energetically favored when a membrane is stretched. This may explain why the MscS channel opens as pressure builds up inside the cell. (PDB codes: 2OAU, 2VV5.)

pressure. If the pressure rises to dangerous levels, the cell opens mechanosensitive channels that allow small molecules to leak out. Bacteria that are experimentally placed in fresh water can rapidly lose more than 95% of their small molecules in this manner, including amino acids, sugars, and potassium ions. However, they keep their macromolecules safely inside and thus can recover quickly after environmental conditions return to normal.

Mechanical gating has been demonstrated using biophysical techniques in which force is exerted on pure lipid bilayers containing the bacterial mechanosensitive channels; for example, by applying suction with a micropipette. Such measurements demonstrate that the cell has several different channels that open at different levels of pressure. The mechanosensitive channel of small conductance, called the MscS channel, opens at low and moderate pressures (**Figure 11–28**). It is composed of seven identical subunits, which in the open state form a pore about 1.3 nm in diameter—just big enough to pass ions and small molecules. Large cytoplasmic domains limit the size of molecules that can reach the pore. The mechanosensitive channel of large conductance, called the MscL channel, opens to more than 3 nm in diameter when the pressure gets so high that the cell might burst.

The Function of a Neuron Depends on Its Elongated Structure

The cells that make the most sophisticated use of channels are neurons. Before discussing how they do so, we digress briefly to describe how a typical neuron is organized.

The fundamental task of a **neuron**, or **nerve cell**, is to receive, integrate, conduct, and transmit signals. To perform these functions, neurons are often extremely elongated. In humans, for example, a single neuron extending from the spinal cord to a muscle in the foot may be as long as 1 meter. Every neuron consists of a cell body (containing the nucleus) with a number of thin processes radiating outward from it. Usually one long **axon** conducts signals away from the cell body toward distant targets, and several shorter, branching **dendrites** extend from the cell body like antennae, providing an enlarged surface area to receive

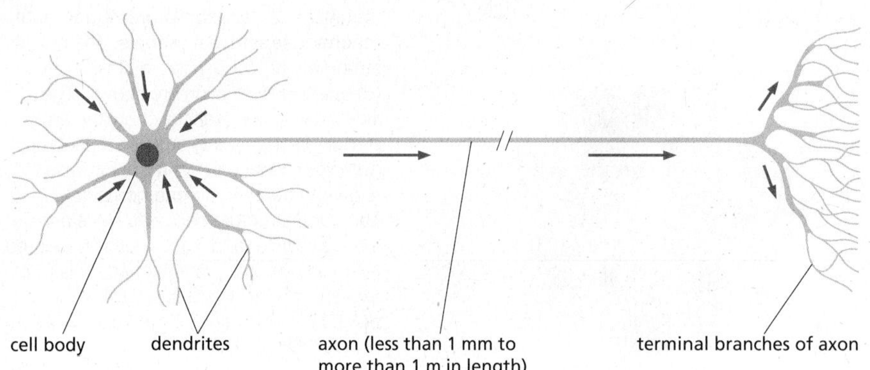

Figure 11–29 **A typical vertebrate neuron.** The arrows indicate the direction in which signals are conveyed. The single axon conducts signals away from the cell body, while the multiple dendrites (and the cell body) receive signals from the axons of other neurons. The axon terminals end on the dendrites or cell body of other neurons or on other cell types, such as muscle or gland cells.

signals from the axons of other neurons (**Figure 11–29**), although the cell body itself also receives such signals. A typical axon divides at its far end into many branches, passing on its message to many target cells simultaneously. Likewise, the extent of branching of the dendrites can be very great—in some cases sufficient to receive as many as 100,000 inputs on a single neuron.

Despite the varied significance of the signals carried by different classes of neurons, the form of the conducted signal is always the same, consisting of changes in the electrical potential across the neuron's plasma membrane. The signal spreads because an electrical disturbance produced in one part of the membrane spreads to other parts, although the disturbance becomes weaker with increasing distance from its source, unless the neuron expends energy to amplify it as it travels. Over short distances, this attenuation is unimportant; in fact, many small neurons conduct their signals passively, without amplification. For long-distance communication, however, such passive spread is inadequate. Thus, larger neurons employ an active signaling mechanism, which is one of their most striking features. An electrical stimulus that exceeds a certain threshold strength triggers an explosion of electrical activity that propagates rapidly along the neuron's plasma membrane and is sustained by automatic amplification all along the way. This traveling wave of electrical excitation, known as an **action potential**, or *nerve impulse*, can carry a message without attenuation from one end of a neuron to the other at speeds of 100 meters per second or more. Action potentials are the direct consequence of the properties of voltage-gated cation channels, as we now discuss.

Voltage-gated Cation Channels Generate Action Potentials in Electrically Excitable Cells

The plasma membrane of all electrically excitable cells—not only neurons, but also muscle, endocrine, and egg cells—contains **voltage-gated cation channels**, which are responsible for generating the action potentials. An action potential is triggered by a **depolarization** of the plasma membrane; that is, by a shift in the membrane potential to a less negative value inside. (We shall see later how the action of some neurotransmitters causes depolarization.) In nerve and skeletal muscle cells, a stimulus that causes sufficient depolarization promptly opens the **voltage-gated Na^+ channels**, allowing a small amount of Na^+ to enter the cell down its electrochemical gradient. The influx of positive charge depolarizes the membrane further, thereby opening more Na^+ channels, which admit more Na^+ ions, causing still further depolarization. This self-amplification process (an example of *positive feedback*, discussed in Chapters 8 and 15) continues until, within a fraction of a millisecond, the electrical potential in the local region of the membrane has shifted from its resting value of about –70 mV to almost as far as the Na^+ equilibrium potential of about +50 mV (see Panel 11–1, p. 656). At this point, when the net electrochemical driving force for the flow of Na^+ is almost zero, the cell would come to a new resting state, with all of its Na^+ channels permanently open, if the open conformation of the channel were stable. Two mechanisms act in concert to save the cell from such a permanent electrical spasm: the Na^+

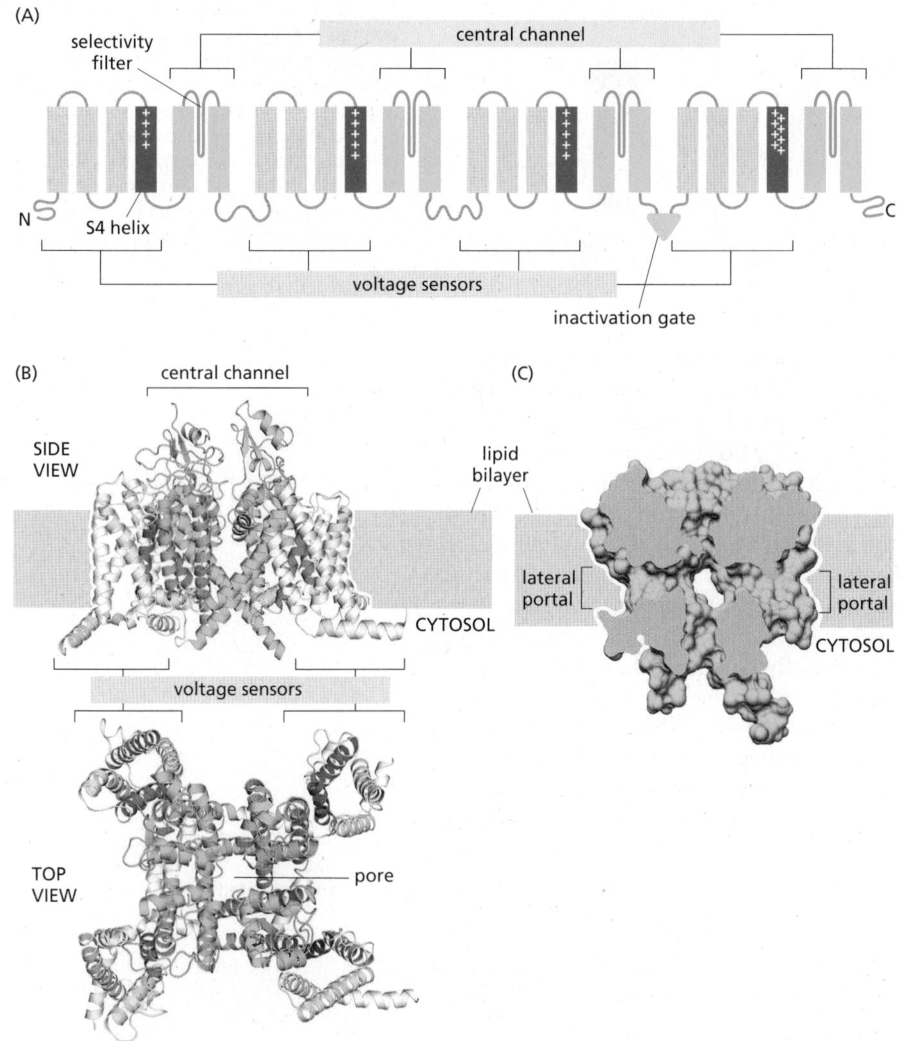

Figure 11–30 Structural models of voltage-gated Na^+ channels. (A) The channel in animal cells is built from a single polypeptide chain that contains four homologous domains. Each domain contains two transmembrane α helices *(green)* that surround the central ion-conducting pore. They are separated by sequences *(dark green)* that form the selectivity filter. Four additional α helices *(blue and gray)* in each domain constitute the voltage sensor. The S4 helices *(blue)* are unique in that they contain an abundance of positively charged arginines. An inactivation gate that is part of a flexible loop connecting the third and fourth domains acts as a plug that obstructs the pore in the channel's inactivated state, as shown in Figure 11–32C. (B) Side and top views of a channel protein showing its arrangement within the membrane. (C) A cross section of the pore domain of the channel shown in B shows lateral portals, through which the central cavity is accessible from the hydrophobic core of the lipid bilayer. In the crystals, lipid acyl chains were found to intrude into the pore. These lateral portals are large enough to allow entry of small, hydrophobic, pore-blocking drugs that are commonly used as local anesthetics and block ion conductance. (PDB code: 6AGF.)

channels automatically inactivate, and **voltage-gated K^+ channels** open to restore the membrane potential to its initial negative value.

The Na^+ channel is built from a single polypeptide chain that contains four structurally very similar domains. It is thought that these domains evolved by gene duplication followed by fusion into a single large gene (**Figure 11–30A**). In bacteria, in fact, the Na^+ channel is a tetramer of four identical polypeptide chains, supporting this evolutionary idea. The structure of the voltage-gated Na^+ channel provides insights into how the structural elements are arranged in the membrane (**Figure 11–30B and C**).

Each domain contributes to the central channel, which is very similar to the K^+ channel. Each domain also contains a *voltage sensor* that is characterized by an unusual transmembrane helix, S4, that contains many positively charged amino acids. When the membrane is polarized, these charges are exposed on the negatively charged cytosolic side of the membrane. Because of the charge compensation, this conformation is thermodynamically stabilized in a polarized membrane. As the membrane becomes depolarized, the S4 helices experience the electrostatic force that now stabilizes an alternative conformation in which the helices have twisted and expose the charges to the opposite, extracellular side of the membrane. The resulting conformational change opens the channel (**Figure 11–31**). It is estimated from energetic calculations that the exposure to the alternate side of the membrane can change the probability of the channel being in an open or closed state by many orders of magnitude.

The Na^+ channels also have an automatic inactivating mechanism, which causes the channels to reclose rapidly even though the membrane is still depolarized (see Figure 11–32). The Na^+ channels remain in this *inactivated* state, unable

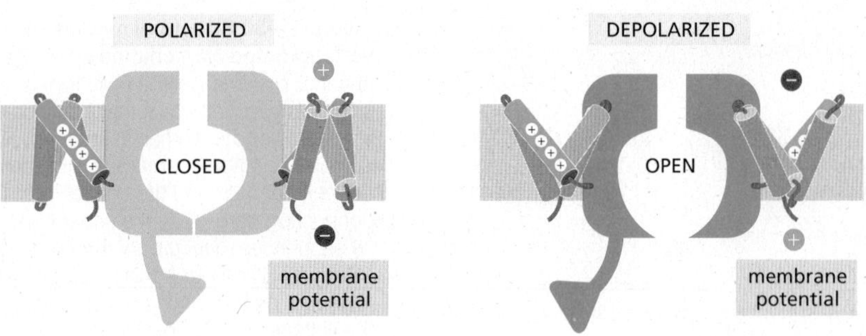

Figure 11–31 **Model for the mechanism of voltage-gating.** The voltage-sensing domain oscillates between two conformational states that expose positively charged arginines on one helix (corresponding to the S4 helix in Figure 11–30) to alternate sides of the membrane. Exposure to the more negatively charged side of the membrane is thermodynamically favored and hence stabilizes that state. Conformational coupling closes or opens the gate in the central pore domain of the channel. (Adapted from A.F. Kintzer et al., *Proc. Natl. Acad. Sci. USA* 115:E9095–E9104, 2018.)

to reopen, until after the membrane potential has returned to its initial negative value. Repolarization of the membrane requires opening of additional channels, the **delayed K^+ channels**. These channels also open in response to membrane depolarization (that is, they are also voltage-gated), but because of their slower kinetics they open only during the falling phase of the action potential, when the Na^+ channels are inactive, but with a delay of a few hundred microseconds. When these channels open, K^+ rushes out of the cell until the cell's resting potential is restored within a millisecond or two. The time necessary for a sufficient number of Na^+ channels to recover from inactivation to support a new action potential, termed the *refractory period*, limits the repetitive firing rate of a neuron and ensures that an action potential propagates unidirectionally. The Na^+ channel can therefore exist in three distinct states—closed, open, and inactivated—which contribute to the rise and fall of the action potential (**Figure 11–32**). The cycle from initial stimulus to the return to the original resting state takes a few milliseconds or less.

This description of an action potential applies only to a small patch of plasma membrane. The self-amplifying depolarization of the patch, however, is sufficient

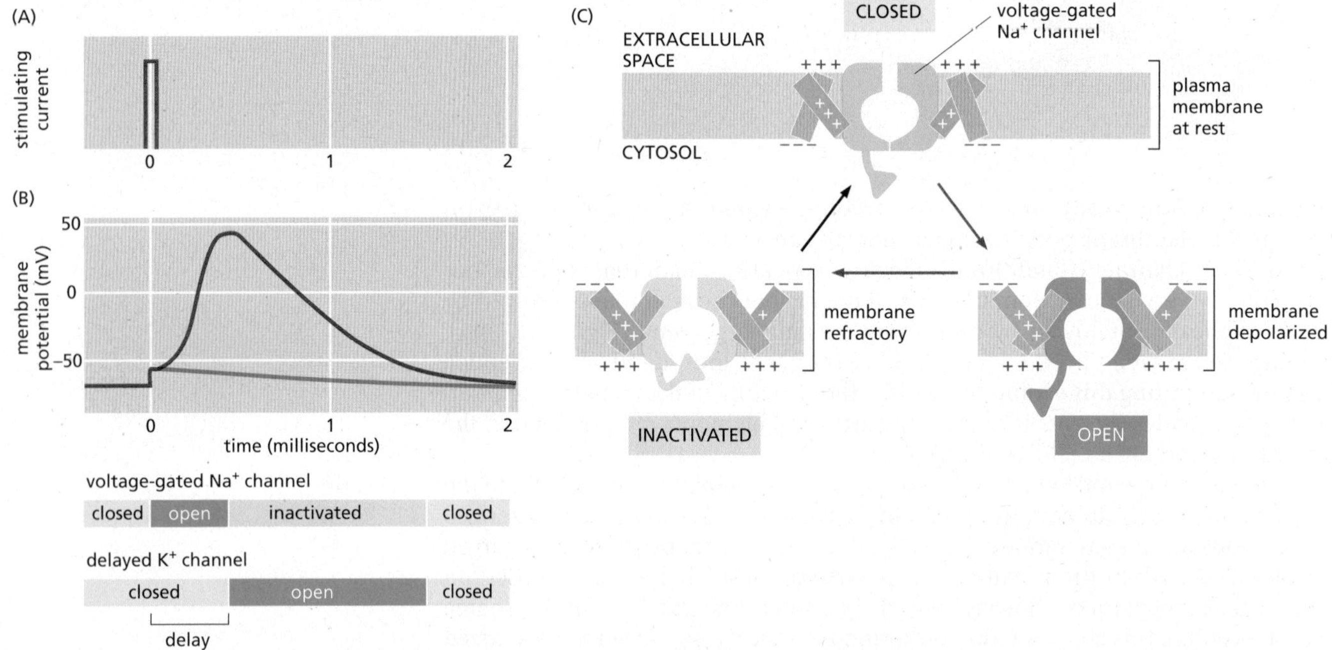

Figure 11–32 **Na^+ channels and an action potential.** (A) An action potential is triggered by a brief pulse of current, which (B) partially depolarizes the membrane, as shown in the plot of membrane potential versus time. The *green curve* shows how the membrane potential would have simply relaxed back to the resting value after the initial depolarizing stimulus if there had been no voltage-gated Na^+ channels in the membrane. The ascending part of the *red curve* shows the course of the action potential that is caused by the opening of voltage-gated Na^+ channels. Delayed voltage-gated K^+ channels open and Na^+ channels become inactivated, allowing membrane repolarization (the descending part of the *red curve*). The states of the Na^+ and K^+ channels are indicated in the schematic in the bottom part of panel B. The membrane cannot fire a second action potential until the Na^+ channels have returned from the inactivated to the closed conformation; until then, the membrane is refractory to stimulation. (C) The three states of the Na^+ channel. When the membrane is at rest (highly polarized), the closed conformation of the channel has the lowest free energy and is therefore most stable; when the membrane is depolarized, the energy of the *open* conformation is lower, so the channel has a high probability of opening. But the free energy of the *inactivated* conformation is lower still; therefore, after a randomly variable period spent in the open state, the channel becomes inactivated. Thus, the open conformation corresponds to a metastable state that can exist only transiently when the membrane depolarizes (**Movie 11.11**).

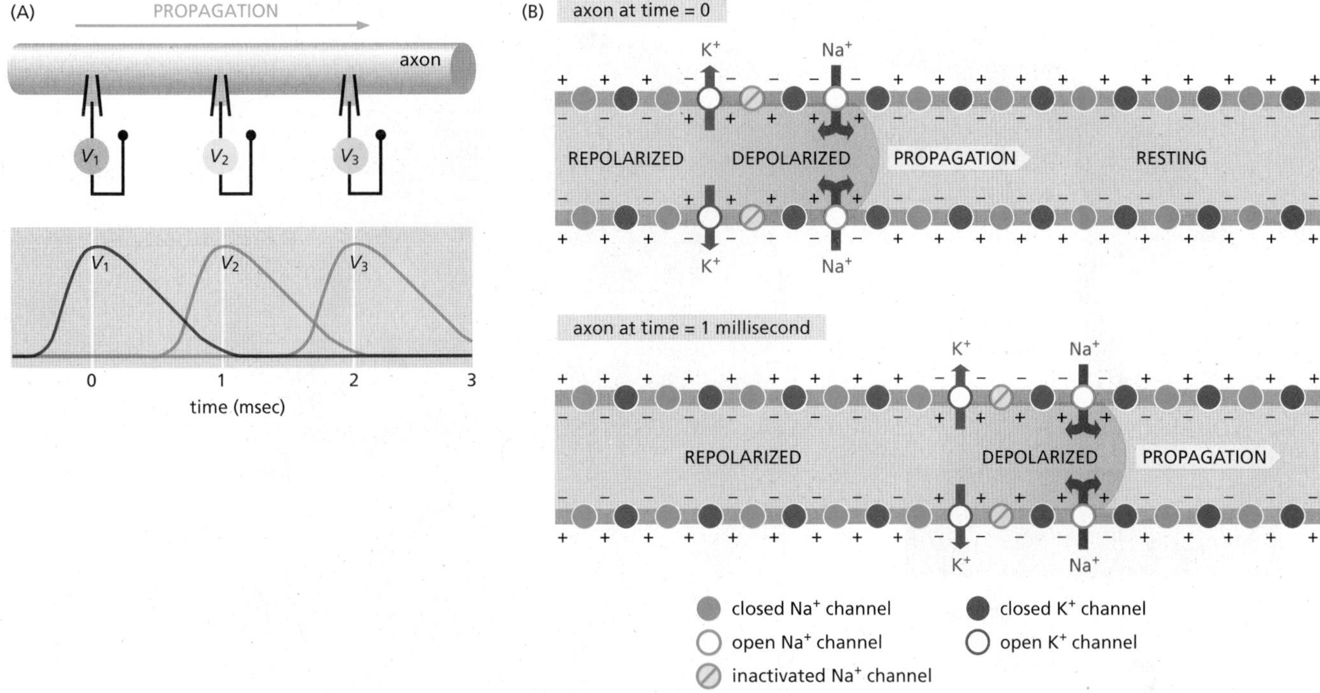

Figure 11–33 **The propagation of an action potential along an axon.** (A) The voltages that would be recorded from a set of intracellular electrodes placed at intervals along the axon. (B) The changes in the Na^+ channels, voltage-gated K^+ channels, and current flows (*red* and *blue arrows*) that give rise to a traveling action potential. The region of the axon with a depolarized membrane is shaded in *orange.* Note that once an action potential has started to progress, it has to continue in the same direction, traveling only away from the site of depolarization, because Na^+-channel inactivation prevents the depolarization from spreading backward.

to depolarize neighboring regions of membrane, which then go through the same cycle. In this way, the action potential sweeps like a wave from the initial site of depolarization over the entire plasma membrane, as shown in Figure 11–33.

Albeit small, the ion flow during an action potential is not negligible. Especially during rapid firing and in small axon branches with a high surface-to-volume ratio, the Na^+-K^+ pump speeds up locally to restore the electrochemical gradients. Brain positron emission tomography (for example, in a PET scan) visualizes regions of the brain whose metabolism is raised when performing a particular task. In this procedure, neurons that have a high energy consumption due to their accelerated ion pumping activity preferentially take up glucose, supplied as a radioactive tracer, to replenish the ATP expended by the Na^+-K^+ pump. The enrichment of the radioactive glucose is then imaged in a three-dimensional brain scan. In this way, the Na^+-K^+ pump serves as a sentinel in the noninvasive mapping of brain regions associated with particular brain activities (Figure 11–34).

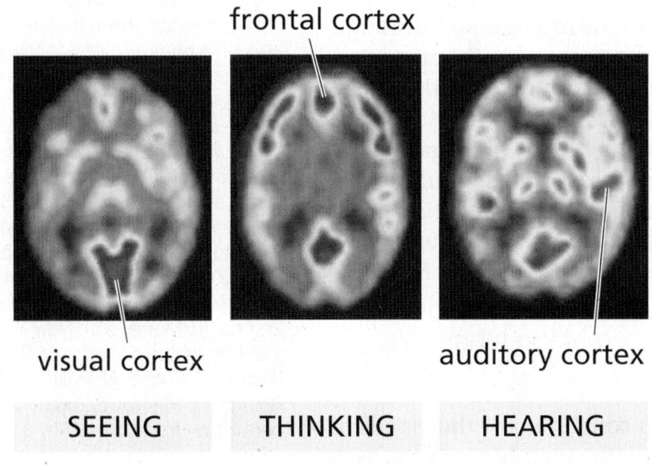

Figure 11–34 **These PET scans show the increases in glucose metabolism due to the activation of specific areas of the human brain due to specific tasks.** (Courtesy of Drs. John Mazziotta and Michael Phelps, David Geffen School of Medicine at UCLA.)

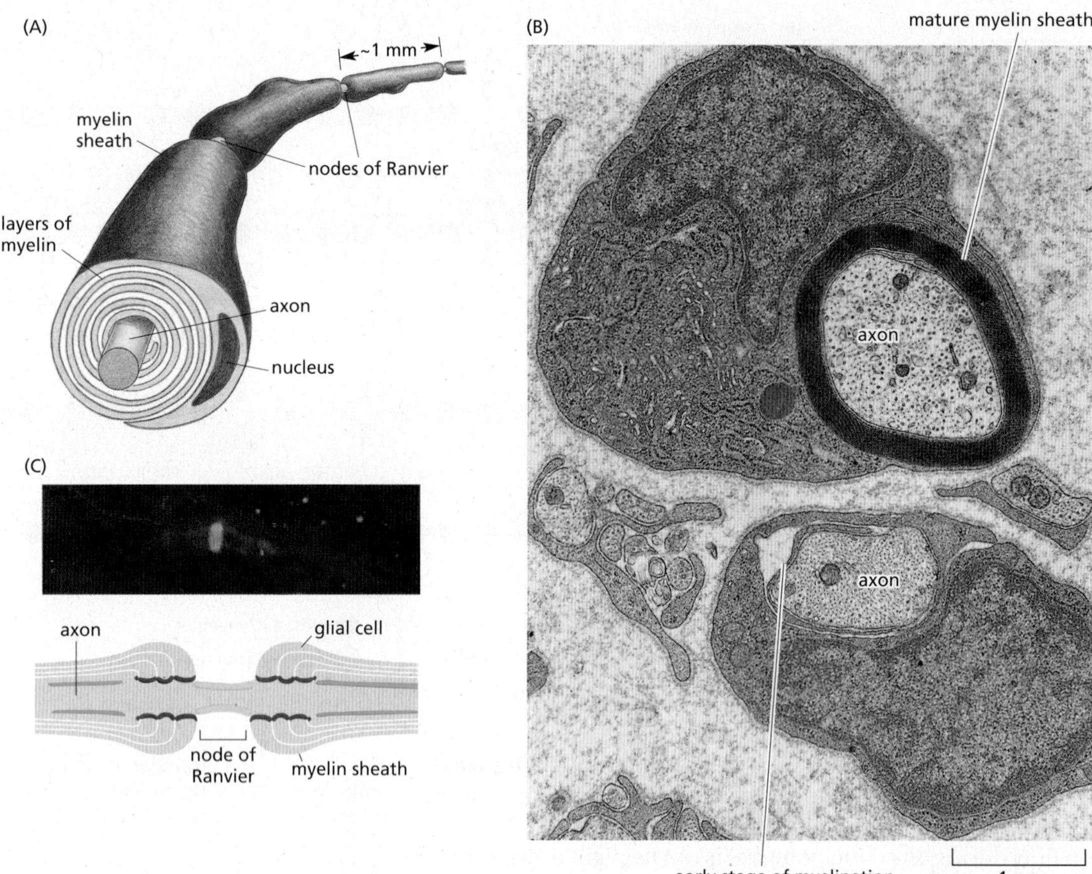

Figure 11–35 Myelination. (A) A myelinated axon from a peripheral nerve. Each Schwann cell wraps its plasma membrane spirally around the axon to form a segment of myelin sheath about 1 mm long. For clarity, the membrane layers of the myelin are shown less compacted than they are in reality (see panel B). (B) An electron micrograph of a nerve in the leg of a young rat. Two Schwann cells can be seen: one near the bottom is just beginning to myelinate its axon; the one above it has formed an almost mature myelin sheath. (C) Fluorescence micrograph and diagram of an individual myelinated axon teased apart in a rat optic nerve, showing the confinement of the voltage-gated Na^+ channels *(green)* to the axonal membrane at the node of Ranvier. A protein called Caspr *(red)* marks the junctions where the myelinating glial-cell plasma membrane tightly abuts the axon on either side of the node. Voltage-gated K^+ channels *(blue)* localize to regions in the axonal plasma membrane well away from the node. [B, from C.S. Raine, in Myelin, 2nd ed. (P. Morell, ed.). New York: Plenum, 1984. With permission from Springer Nature; C, from M.N. Rasband and P. Shrager, *J. Physiol.* 525:63–73, 2000. With permission from John Wiley & Sons.]

Myelination Increases the Speed and Efficiency of Action Potential Propagation in Nerve Cells

The axons of many vertebrate neurons are insulated by a **myelin sheath**, which greatly increases the speed at which an axon can conduct an action potential. The importance of myelination is dramatically demonstrated by the demyelinating disease *multiple sclerosis*, in which the immune system destroys myelin sheaths in some regions of the central nervous system; in the affected regions, nerve impulse propagation greatly slows or fails, often with devastating neurological consequences.

Myelin is formed by specialized non-neuronal supporting cells called **glial cells. Schwann cells** are the glial cells that myelinate axons in peripheral nerves, and **oligodendrocytes** do so in the central nervous system. These myelinating glial cells wrap layer upon layer of their own plasma membrane in a tight spiral around the axon (**Figure 11–35A and B**), thereby insulating the axonal membrane so that little current can leak across it. The myelin sheath is interrupted at regularly spaced *nodes of Ranvier*, where almost all the Na^+ channels in the axon are concentrated (**Figure 11–35C**). This arrangement allows an action potential to propagate along a myelinated axon by jumping from node to node, a process called *saltatory conduction*. This type of conduction has two main advantages: action potentials travel very much faster, and metabolic energy is conserved because the active excitation is confined to the small regions of axonal plasma membrane at nodes of Ranvier.

Patch-Clamp Recording Indicates That Individual Ion Channels Open in an All-or-Nothing Fashion

Neuron and skeletal muscle cell plasma membranes contain many thousands of voltage-gated Na^+ channels, and the current crossing the membrane is the sum of the currents flowing through all of these. An intracellular microelectrode can record this aggregate current. Remarkably, however, it is also possible to record

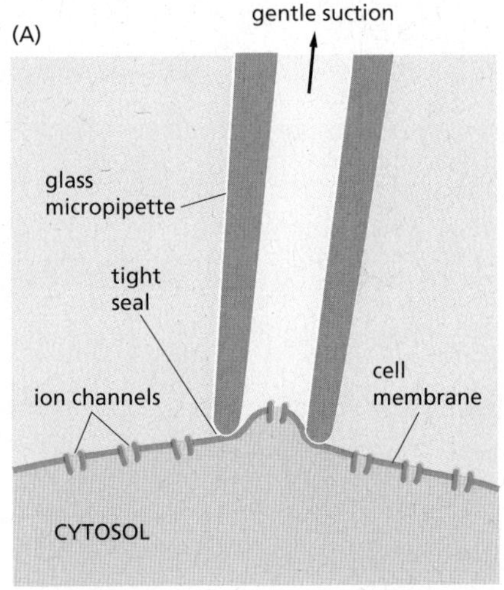

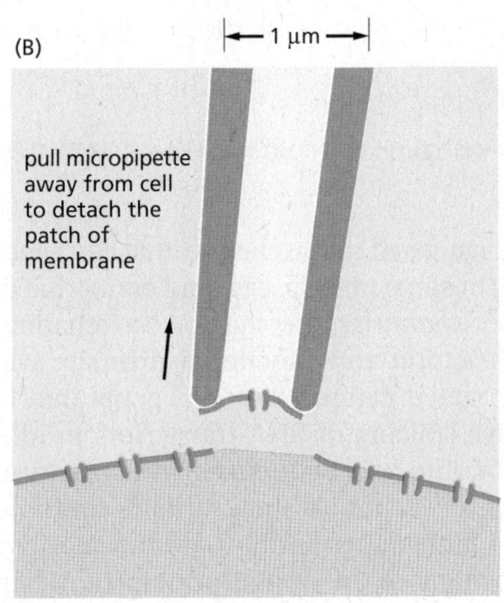

Figure 11–36 The technique of patch-clamp recording. Because of the extremely tight seal between the micropipette and the membrane, current can enter or leave the micropipette only by passing through the ion channels in the *patch* of membrane covering its tip. The term *clamp* is used because an electronic device is employed to maintain, or "clamp," the membrane potential at a set value while recording the ionic current through individual channels. The current through these channels can be recorded with the patch still attached to the rest of the cell, as in (A), or detached, as in (B). The advantage of the detached patch is that it is easy to alter the composition of the solution on either side of the membrane to test the effect of various solutes on channel behavior. A detached patch can also be produced with the opposite orientation, so that the cytoplasmic surface of the membrane faces the inside of the pipette.

current flowing through individual channels. **Patch-clamp recording**, developed in the 1970s and 1980s, revolutionized the study of ion channels and made it possible to examine transport through a single channel in a small patch of membrane covering the mouth of a micropipette (**Figure 11–36**). With this simple but powerful technique, one can study the detailed properties of ion channels in all sorts of cell types. This work led to the discovery that even cells that are not electrically excitable usually have a variety of ion channels in their plasma membrane. Many of these cells, such as yeasts, are too small to be investigated by the traditional electrophysiologist's method of impalement with an intracellular microelectrode.

Patch-clamp recording indicates that individual ion channels open in an all-or-nothing fashion. For example, a voltage-gated Na^+ channel opens and closes stochastically with a voltage-dependent probability, but when open, the channel always has the same large conductance, allowing more than 1000 ions to pass per millisecond (**Figure 11–37**). Therefore, the aggregate current crossing the membrane of an entire cell does not indicate the *degree* to which a typical individual channel is open but rather the *total number* of channels in its membrane that are open at any one time.

Some simple physical principles allow us to refine our understanding of voltage-gating from the perspective of a single Na^+ channel. The interior of the resting neuron or muscle cell is at an electrical potential about 40–100 mV more negative than the external medium. Although this potential difference seems small, it exists across a plasma membrane only about 5 nm thick, so that the resulting voltage gradient is about 100,000 V/cm. Charged proteins in the membrane such as Na^+ channels are thus subjected to a very large electrical field that can profoundly affect their conformation. Each conformation can "flip" to another conformation if given a sufficient jolt by the random thermal movements of the surroundings, and it is the relative stability of the closed, open, and inactivated

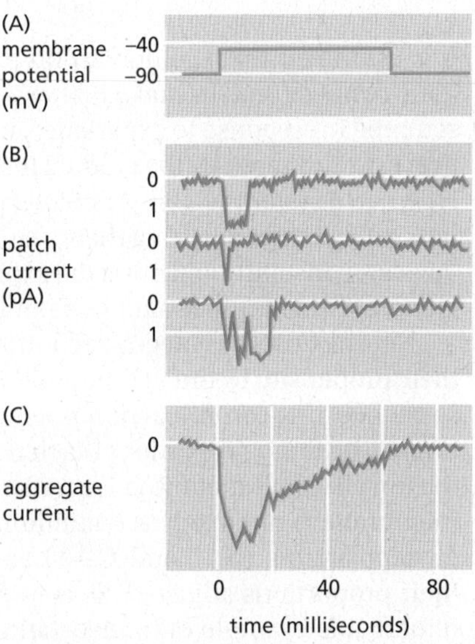

Figure 11–37 Patch-clamp measurements for a single voltage-gated Na^+ channel. A tiny patch of plasma membrane was detached from an embryonic rat muscle cell, as in Figure 11–36. (A) The membrane was depolarized by an abrupt shift of potential from –90 to about –40 mV. (B) Three current records from three experiments performed on the same patch of membrane. Each major current step in panel B represents the opening and closing of a single channel. A comparison of the three records shows that, whereas the durations of channel opening and closing vary greatly, the rate at which current flows through an open channel (its conductance) is practically constant. The minor fluctuations in the current records arise largely from electrical noise in the recording apparatus. Current flowing into the cell, measured in picoamperes (pA), is shown as a downward deflection of the curve. By convention, the electrical potential on the outside of the cell is defined as zero. (C) The sum of the currents measured in 144 repetitions of the same experiment. This aggregate current is equivalent to the usual Na^+ current that would be observed flowing through a relatively large region of membrane containing 144 channels. A comparison of panels B and C reveals that the time course of the aggregate current reflects the probability that any individual channel will be in the open state; this probability decreases with time as the channels in the depolarized membrane adopt their inactivated conformation. (Data from J. Patlak and R. Horn, *J. Gen. Physiol.* 79:333–351, 1982.)

conformations against flipping that is altered by changes in the membrane potential (see Figure 11–32C).

Voltage-gated Cation Channels Are Evolutionarily and Structurally Related

Na^+ channels are not the only kind of voltage-gated cation channel that can generate an action potential. The action potentials in some muscle, egg, and endocrine cells, for example, depend on *voltage-gated Ca^{2+} channels* rather than on Na^+ channels.

There is a surprising amount of structural and functional diversity within each of the different classes of voltage-gated cation channels, generated both by multiple genes and by the alternative splicing of RNA transcripts produced from the same gene. Nonetheless, the amino acid sequences and structures of the known voltage-gated Na^+, K^+, and Ca^{2+} channels show striking similarities, demonstrating that they all belong to a large superfamily of evolutionarily and structurally related proteins and share many of the design principles. Whereas the single-celled yeast *Saccharomyces cerevisiae* contains a single gene that codes for a voltage-gated K^+ channel, the genome of the worm *Caenorhabditis elegans* contains 68 genes that encode different but related K^+ channels. This complexity indicates that even a simple nervous system made up of only 302 neurons uses a large number of different ion channels to compute its responses.

Humans carrying mutations in genes encoding ion channels can suffer from a variety of nerve, muscle, brain, kidney, or heart diseases, depending on the cells in which the mutant channel is expressed. Mutations in genes that encode voltage-gated Na^+ channels in skeletal muscle cells, for example, can cause *myotonia,* a condition in which there is a delay in muscle relaxation after voluntary contraction, causing painful muscle spasms. In some cases, this occurs because the abnormal channels fail to inactivate normally; as a result, Na^+ entry persists after an action potential finishes and repeatedly reinitiates membrane depolarization and muscle contraction. Similarly, mutations that affect Na^+ or K^+ channels in the brain can cause *epilepsy,* in which excessive synchronized firing of large groups of neurons causes epileptic seizures (convulsions, or fits).

The particular combination of ion channels conducting Na^+, K^+, and Ca^{2+} that are expressed in a neuron largely determines how the cell fires repetitive sequences of action potentials. Some neurons can repeat action potentials up to 300 times per second; other neurons fire short bursts of action potentials separated by periods of silence; while others rarely fire more than one action potential at a time. There is a remarkable diversity of neurons in the brain.

Different Neuron Types Display Characteristic Stable Firing Properties

It is estimated that the human brain contains about 10^{11} neurons and 10^{14} synaptic connections. To make matters more complex, neural circuitry is continually sculpted in response to experience, modified as we learn and store memories, and irreversibly altered by the gradual loss of neurons and their connections as we age. How can a system so complex be subject to such change and yet continue to function stably? One emerging theory suggests that individual neurons are self-tuning devices, constantly adjusting the expression of ion channels and neurotransmitter receptors in order to maintain a stable function. How might this work?

Neurons can be categorized into functionally different types, based in part on their propensity to fire action potentials and their pattern of firing. For example, some neurons can fire action potentials at high frequencies, whereas others fire rarely. The firing properties of each neuron type are determined to a large extent by the ion channels that the cell expresses. The number of ion channels in a neuron's membrane is not fixed: as conditions change, a neuron can modify the numbers of depolarizing (Na^+ and Ca^{2+}) and hyperpolarizing (K^+) channels and keep their proportions adjusted so as to maintain its characteristic firing behavior—a remarkable example of homeostatic control. Deciphering the molecular mechanisms involved remains an important challenge.

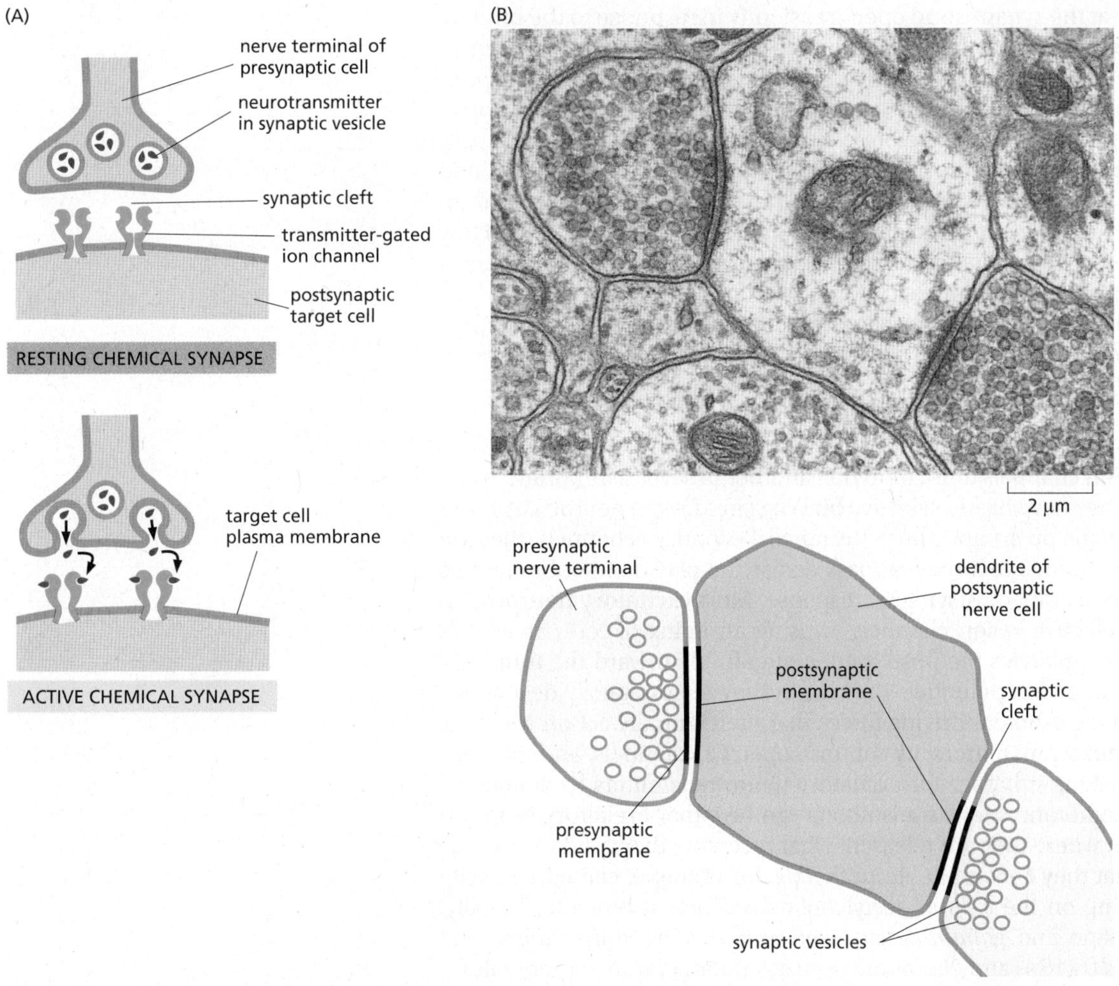

Figure 11–38 A chemical synapse. (A) When an action potential reaches the nerve terminal in a presynaptic cell, it stimulates the terminal to release its neurotransmitter. The neurotransmitter molecules are contained in synaptic vesicles and are released to the cell exterior when the vesicles fuse with the plasma membrane of the nerve terminal. The released neurotransmitter binds to and opens the transmitter-gated ion channels concentrated in the plasma membrane of the postsynaptic target cell at the synapse. The resulting ion flows alter the membrane potential of the postsynaptic membrane, thereby transmitting a signal from the excited nerve (**Movie 11.12**). (B) A thin-section electron micrograph and an interpretive drawing of two nerve terminal synapses on a dendrite of a postsynaptic cell. (B, micrograph courtesy of Cedric Raine.)

Transmitter-gated Ion Channels Convert Chemical Signals into Electrical Ones at Chemical Synapses

Neuronal signals are transmitted from cell to cell at specialized sites of contact known as **synapses**. The usual mechanism of transmission is indirect. The cells are anatomically and electrically isolated from one another, the *presynaptic cell* being separated from the *postsynaptic cell* by a narrow *synaptic cleft*. When an action potential arrives at the presynaptic site, the depolarization of the membrane opens voltage-gated Ca^{2+} channels that are clustered in the presynaptic membrane. Ca^{2+} influx triggers the release into the cleft of small signal molecules known as **neurotransmitters** that are stored in membrane-enclosed *synaptic vesicles* and released by exocytosis (discussed in Chapter 13). The neurotransmitter diffuses rapidly across the synaptic cleft and provokes an electrical change in the postsynaptic cell by binding to and opening *transmitter-gated ion channels* (**Figure 11–38**). After the neurotransmitter has been secreted, it is rapidly removed: it is either destroyed by specific enzymes in the synaptic cleft or, more commonly, taken up by the presynaptic nerve terminal or by surrounding glial cells. Reuptake is mediated by a variety of Na^+-dependent neurotransmitter symporters (see Figure 11–8); in this way, neurotransmitters are recycled, allowing cells to keep up with high rates of release. Rapid removal ensures both spatial and temporal precision of signaling at a synapse. It decreases the chances that the neurotransmitter will influence neighboring cells, and it clears the synaptic cleft before the next pulse of neurotransmitter is released, so that the timing of repeated, rapid signaling events can be accurately communicated to the postsynaptic cell.

Transmitter-gated ion channels, also called **ionotropic receptors**, rapidly convert extracellular chemical signals into electrical signals at chemical synapses. The channels are concentrated in a specialized region of the postsynaptic plasma membrane at the synapse and open transiently in response to the binding of neurotransmitter molecules, thereby producing a brief permeability change in the membrane (see Figure 11–38A). Unlike the voltage-gated channels responsible for action potentials, transmitter-gated channels are relatively insensitive to the membrane potential and therefore cannot by themselves produce a self-amplifying excitation. Instead, they produce local permeability increases, and hence changes of membrane potential, that are graded according to the amount of neurotransmitter released at the synapse and how long it persists there. Only if the summation of small depolarizations at this site opens sufficient numbers of nearby voltage-gated cation channels can an action potential be triggered. This may require the opening of transmitter-gated ion channels at numerous synapses in close proximity on the target nerve cell.

Chemical Synapses Can Be Excitatory or Inhibitory

Transmitter-gated ion channels differ from one another in several important ways. First, as receptors, they have highly selective binding sites for the neurotransmitter that is released from the presynaptic nerve terminal. Second, as channels, they are selective in the type of ions that they let pass across the plasma membrane; this determines the nature of the postsynaptic response. Most **excitatory neurotransmitters** open nonselective cation channels, causing an influx of Na^+, as well as Ca^{2+} and K^+, that depolarizes the postsynaptic membrane toward the threshold potential for firing an action potential. The reason they are excitatory despite K^+ efflux is because of the different driving forces that preferentially act on Na^+ and Ca^{2+}. **Inhibitory neurotransmitters**, by contrast, open Cl^- channels, and this suppresses firing by making it harder for excitatory neurotransmitters to depolarize the postsynaptic membrane. Many transmitters can be either excitatory or inhibitory, depending on where they are released, what receptors they bind to, and the ionic conditions that they encounter. *Acetylcholine*, for example, can either excite or inhibit, depending on the type of acetylcholine receptor it binds to. Usually, however, acetylcholine and *glutamate* are used as excitatory transmitters, and *γ-aminobutyric acid* (*GABA*) and *glycine* are used as inhibitory transmitters. Glutamate, for instance, mediates most of the excitatory signaling in the vertebrate brain.

We have already discussed how the opening of Na^+ or Ca^{2+} channels depolarizes a membrane. The opening of K^+ channels has the opposite effect because the K^+ concentration gradient is in the opposite direction—high concentration inside the cell, low outside. The movement of the cell's membrane potential to a more negative value is called **hyperpolarization**. Opening K^+ channels tends to keep the cell close to the equilibrium potential for K^+, which, as we discussed earlier, is normally close to the resting membrane potential because at rest K^+ channels are the main type of channel that is open. When additional K^+ channels open, it becomes harder to drive the cell away from the resting potential. We can understand the effect of opening Cl^- channels similarly. The concentration of Cl^- is much higher outside the cell than inside (see Table 11–1, p. 638), but the membrane potential opposes its influx. In fact, for many neurons, the equilibrium potential for Cl^- is close to the resting potential—or even more negative. For this reason, opening of Cl^- channels tends to buffer the membrane potential; as the membrane starts to depolarize, more negatively charged Cl^- ions enter the cell and counteract the depolarization. Thus, the opening of Cl^- channels makes it more difficult to depolarize the membrane and hence to excite the cell. Some powerful toxins act by blocking the action of inhibitory neurotransmitters: strychnine, for example, binds to glycine receptors and prevents their inhibitory action, causing muscle spasms, convulsions, and death.

However, not all chemical signaling in the nervous system operates through these ionotropic ligand-gated ion channels. In fact, most neurotransmitter molecules that are secreted by nerve terminals, including a large variety of neuropeptides, bind to **metabotropic receptors**, which regulate ion channels only indirectly through the action of small intracellular signal molecules (discussed in Chapter 15). All neurotransmitter receptors fall into one or other of these two major classes—ionotropic or metabotropic—on the basis of their signaling mechanisms:

1. Ionotropic receptors are ion channels that appear at fast chemical synapses. Acetylcholine, glycine, glutamate, and GABA all act on transmitter-gated ion channels, mediating excitatory or inhibitory signaling that is generally immediate, simple, and brief.

2. Metabotropic receptors are *G-protein-coupled receptors* (discussed in Chapter 15) that bind to all other neurotransmitters (and, confusingly, also to acetylcholine, glutamate, and GABA). Signaling mediated by ligand-binding to metabotropic receptors tends to be far slower and more complex than that at ionotropic receptors, and longer-lasting in its consequences.

The Acetylcholine Receptors at the Neuromuscular Junction Are Excitatory Transmitter–gated Cation Channels

A well-studied example of a transmitter-gated ion channel is the **acetylcholine receptor** of vertebrate skeletal muscle cells. This channel is opened transiently by acetylcholine released from the nerve terminal at a **neuromuscular junction**—the specialized chemical synapse between a motor neuron and a skeletal muscle cell (**Figure 11–39**). This synapse has been intensively investigated because it is readily accessible to electrophysiological study, unlike most of the synapses in the central nervous system (that is, the brain and spinal cord in vertebrates). Moreover, the acetylcholine receptors are densely packed in the muscle-cell plasma membrane at a neuromuscular junction (about 20,000 such receptors per square micrometer), with relatively few receptors elsewhere in the same membrane.

The muscle acetylcholine receptors are composed of five transmembrane polypeptides, two of one kind and three others, encoded by four separate genes (**Figure 11–40A**). The four genes are strikingly similar in sequence, implying that they evolved from a single ancestral gene. The two identical polypeptides in the pentamer each contribute one acetylcholine-binding site. When two acetylcholine molecules bind to the pentameric complex, they induce a conformational change that opens the channel. With ligand bound, the channel still flickers between open and closed states, but now it has a high probability of being open. This state continues—with acetylcholine binding and unbinding—until hydrolysis of the free acetylcholine by the enzyme *acetylcholinesterase* lowers its concentration at the neuromuscular junction sufficiently. Once freed of its bound neurotransmitter, the acetylcholine receptor reverts to its initial resting state. If the presence of acetylcholine persists for a prolonged time as a result of excessive nerve stimulation, the channel inactivates. Normally, the acetylcholine is rapidly hydrolyzed and the channel closes within about 1 millisecond, well before significant desensitization occurs.

The five subunits of the acetylcholine receptor are arranged in a ring, forming a water-filled transmembrane channel that consists of a narrow pore through the

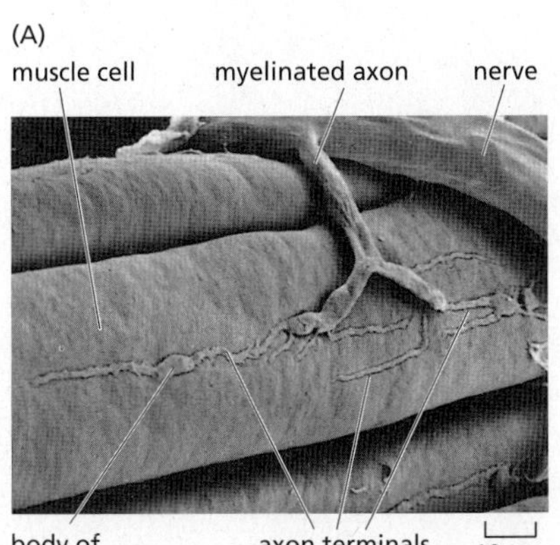

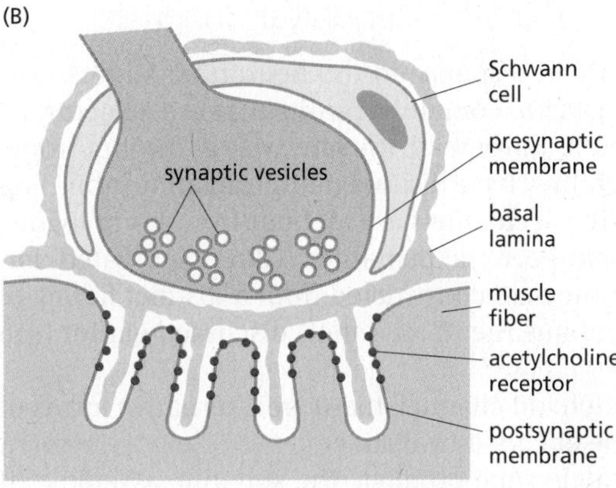

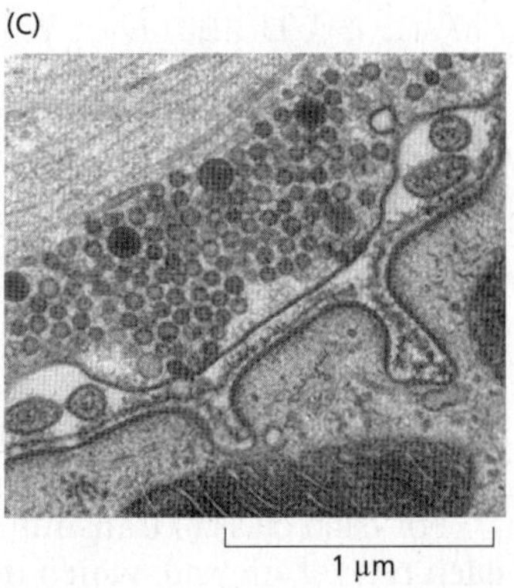

Figure 11–39 The neuromuscular junction. (A) A low-magnification scanning electron micrograph of a neuromuscular junction in a frog. The termination of a single axon on a skeletal muscle cell is shown. (B) A schematic of the specialized anatomy of the neuron–muscle synaptic cleft. Schwann (glial) cells wrap around the neuron outside the region of synaptic contact to provide myelination. (C) Transmission electron micrograph of the region of synaptic contact. [A, from J. Desaki and Y. Uehara, *J. Neurocytol.* 10:101–110, 1981. With permission from Springer Nature; C, from J. Heuser, *J. Electron Microsc.* (*Tokyo*) 60 (Suppl. 1): S3–S29, 2011.]

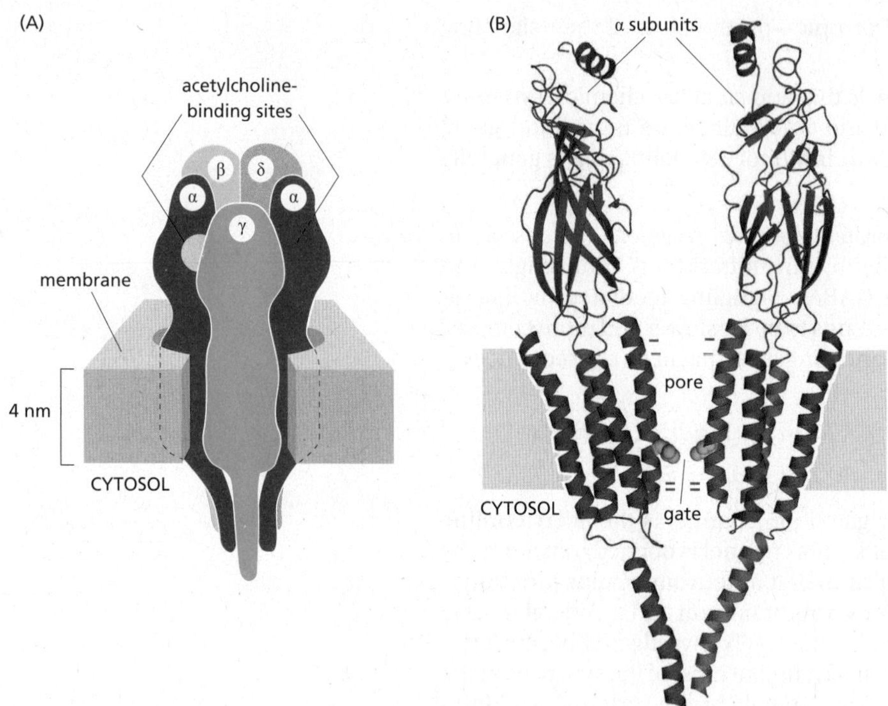

Figure 11–40 A model for the structure of the skeletal muscle acetylcholine receptor. (A) Five homologous subunits (α, α, β, γ, δ) combine to form a transmembrane pore. Both of the α subunits contribute an acetylcholine-binding site nestled between adjoining subunits. (B) The pore is lined by a ring of five transmembrane α helices, one contributed by each subunit (just the two α subunits are shown). In its closed conformation, the pore is occluded by the hydrophobic side chains of five leucines *(green)*, one from each α helix, which form a gate near the middle of the lipid bilayer. When both α subunits bind acetylcholine, the channel undergoes a conformational change that opens the gate by an outward rotation of the helices containing the occluding leucines. Negatively charged side chains (indicated by the "–" signs) at either end of the pore ensure that only positively charged ions pass through the channel. (PDB code: 2BG9.)

lipid bilayer, which widens into vestibules at both ends. Acetylcholine binding opens the channel by causing the helices that line the pore to rotate outward, thus disrupting a ring of hydrophobic amino acids that blocks ion flow in the closed state. Clusters of negatively charged amino acids at either end of the pore help to exclude negative ions and encourage any positive ion of diameter less than 0.65 nm to pass through (**Figure 11–40B**). The normal through-traffic consists chiefly of Na^+ and K^+, together with some Ca^{2+}. Thus, unlike voltage-gated cation channels, such as the K^+ channel discussed earlier, there is little selectivity among cations, and the relative contributions of the different cations to the current through the channel depend chiefly on their concentrations and on the electrochemical driving forces. When the muscle-cell membrane is at its resting potential, the net driving force for K^+ is near zero, because the voltage gradient nearly balances the K^+ concentration gradient across the membrane (see Panel 11–1, p. 656). For Na^+, in contrast, the voltage gradient and the concentration gradient both act in the same direction to drive the ion into the cell. (The same is true for Ca^{2+}, but the extracellular concentration of Ca^{2+} is so much lower than that of Na^+ that Ca^{2+} makes only a small contribution to the total inward current.) Therefore, the opening of the acetylcholine-receptor channels leads to a large net influx of Na^+ (a peak rate of about 30,000 ions per channel each millisecond). This influx causes a membrane depolarization that signals the muscle to contract, as discussed later.

Neurons Contain Many Types of Transmitter-gated Channels

The ion channels that open directly in response to the neurotransmitters acetylcholine, serotonin, GABA, and glycine contain subunits that are structurally similar and probably form transmembrane pores in the same way as the ionotropic acetylcholine receptor, even though they have distinct neurotransmitter-binding specificities and ion selectivities. These channels are all built from homologous polypeptide subunits, which assemble as a pentamer. Glutamate-gated ion channels are an exception, in that they are constructed from a distinct family of subunits and form tetramers resembling the K^+ channels discussed earlier (see Figure 11–24A).

For each class of transmitter-gated ion channel, there are alternative forms of each type of subunit, which may be encoded by distinct genes or else generated by alternative RNA splicing of a single gene product. The subunits assemble in different combinations to form an extremely diverse set of distinct channel subtypes, with different ligand affinities, different channel conductances, different rates of opening and closing, and different sensitivities to drugs and toxins. Some vertebrate neurons, for example, have acetylcholine-gated ion channels that differ from those of muscle cells in that they are formed from two subunits of one type and three of another; but there are at least nine genes coding for different versions of the first type of subunit and at least three coding for different versions of the second. Subsets of such neurons performing different functions in the brain express different combinations of the genes for these subunits. In principle, and already to some extent in practice, it is possible to design drugs targeted against these narrowly defined subsets, thereby specifically influencing particular brain functions.

Many Psychoactive Drugs Act at Synapses

Transmitter-gated ion channels have for a long time been important drug targets. A surgeon, for example, can relax muscles for the duration of an operation by blocking the acetylcholine receptors on skeletal muscle cells with *curare*, a plant-derived drug that was originally used by South American Indians to make poison arrows. Most drugs used to treat insomnia, anxiety, depression, and schizophrenia exert their effects at chemical synapses, and many of these act by binding to transmitter-gated channels. Barbiturates, tranquilizers such as Valium, and sleeping pills such as Ambien, for example, bind to GABA receptors, potentiating the inhibitory action of GABA by allowing lower concentrations of this neurotransmitter to open Cl^- channels. Our increasing understanding of the molecular biology of ion channels should allow us to design a new generation of psychoactive drugs that will act still more selectively to alleviate the miseries of mental illness.

In addition to ion channels, many other components of the synaptic signaling machinery are potential targets for psychoactive drugs. As mentioned earlier, after release into the synaptic cleft, many neurotransmitters are cleared by reuptake mechanisms mediated by Na^+-driven symports. Inhibiting such transporters prolongs the effect of the neurotransmitter, thereby strengthening synaptic transmission. Many antidepressant drugs, including Prozac, inhibit the reuptake of serotonin; others inhibit the reuptake of both serotonin and norepinephrine.

Ion channels are the basic molecular units from which neuronal devices for signaling and computation are built. To provide a glimpse of how sophisticated these devices can be, we consider several examples that demonstrate how the coordinated activities of groups of ion channels allow you to move, feel, and remember.

Neuromuscular Transmission Involves the Sequential Activation of Five Different Sets of Ion Channels

The following process, in which a nerve impulse stimulates a muscle cell to contract, illustrates the importance of ion channels to electrically excitable cells. This apparently simple response requires the sequential activation of at least five different sets of ion channels, all within a few milliseconds (**Figure 11–41**).

1. Neuromuscular transmission is initiated when a nerve impulse reaches the nerve terminal and depolarizes the plasma membrane of the terminal. The depolarization transiently opens voltage-gated Ca^{2+} channels in this presynaptic membrane. As the Ca^{2+} concentration outside cells is more than 1000 times greater than the free Ca^{2+} concentration inside, Ca^{2+} flows into the nerve terminal. The increase in Ca^{2+} concentration in the cytosol of the nerve terminal triggers the local release of acetylcholine by exocytosis into the synaptic cleft.

2. The released acetylcholine binds to acetylcholine receptors in the muscle-cell plasma membrane, transiently opening the cation channels associated with them. The resulting influx of Na^+ causes a local membrane depolarization in the muscle.

3. The local depolarization opens voltage-gated Na^+ channels in this membrane, allowing more Na^+ to enter, which further depolarizes the

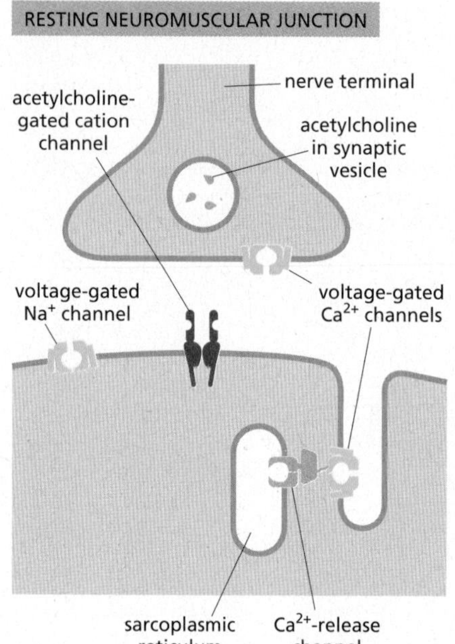

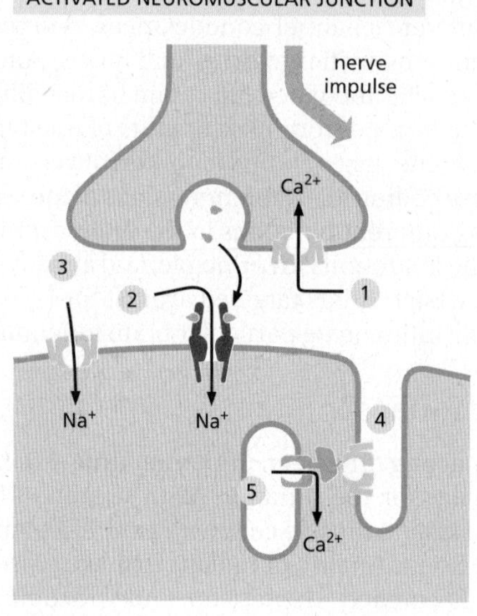

Figure 11–41 The system of ion channels at a neuromuscular junction. These gated ion channels are essential for the stimulation of muscle contraction by a nerve impulse. The various channels are numbered in the sequence in which they are activated, as described in the text.

membrane. This, in turn, opens neighboring voltage-gated Na^+ channels and results in a self-propagating depolarization (an action potential) that spreads to involve the entire muscle plasma membrane (see Figure 11–33).

4. The generalized depolarization of the muscle-cell plasma membrane activates voltage-gated Ca^{2+} channels in the transverse tubules (T tubules—discussed in Chapter 16) of this membrane.
5. This in turn causes *Ca^{2+}-release channels* in an adjacent region of the sarcoplasmic reticulum (SR) membrane to open transiently and release Ca^{2+} stored in the SR into the cytosol. The T-tubule and SR membranes are closely apposed, and the two types of channel are joined together in a specialized structure, in which activation of the voltage-sensitive Ca^{2+} channel in the T-tubule plasma membrane causes a channel conformational change that is mechanically transmitted to the Ca^{2+}-release channel in the SR membrane, opening it and allowing Ca^{2+} to flow from the SR lumen into the cytoplasm (see Figure 16–30). The sudden increase in the cytosolic Ca^{2+} concentration causes the myofibrils in the muscle cell to contract.

Whereas the initiation of muscle contraction by a motor neuron is complex, an even more sophisticated interplay of ion channels is required for a neuron to integrate a large number of input signals at its synapses and compute an appropriate output, as we now discuss.

Single Neurons Are Complex Computation Devices

In the central nervous system, a single neuron can receive inputs from thousands of other neurons, and it can in turn form synapses with many thousands of other cells. Several thousand nerve terminals, for example, make synapses on an average motor neuron in the spinal cord, almost completely covering its cell body and dendrites (**Figure 11–42**). Some of these synapses transmit signals from the brain or spinal cord; others bring sensory information from muscles or from the skin. The motor neuron must combine the information received from all these sources and react, either by firing action potentials along its axon or by remaining quiet.

Of the many synapses on a neuron, some tend to excite it, while others inhibit it. Neurotransmitter released at an excitatory synapse causes a small depolarization in the postsynaptic membrane called an *excitatory postsynaptic potential* (*excitatory PSP*), whereas neurotransmitter released at an inhibitory synapse generally causes a small hyperpolarization called an *inhibitory PSP*. The plasma membrane of the dendrites and cell body of most neurons contains a relatively

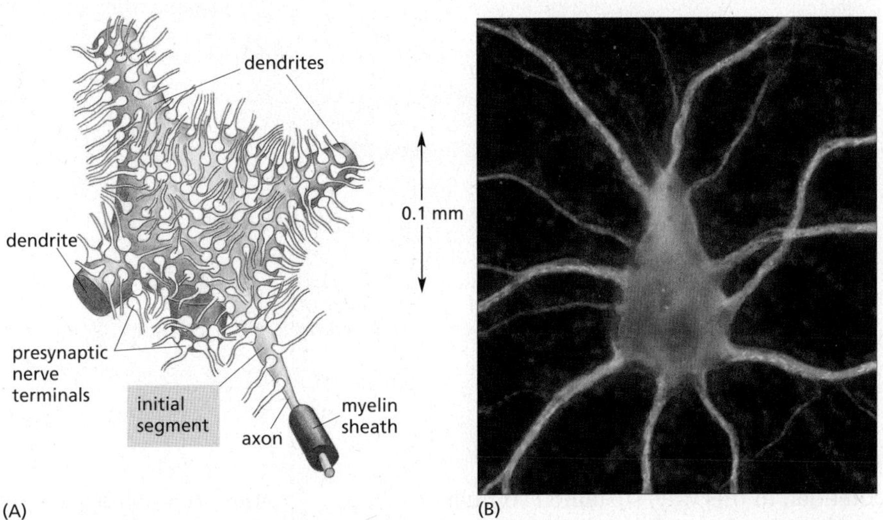

Figure 11–42 A motor neuron in the spinal cord. (A) Many thousands of nerve terminals synapse on the cell body and dendrites. These deliver signals from other parts of the organism to control the firing of action potentials along the single axon of this large cell. (B) Fluorescence micrograph showing a nerve-cell body in cell culture. Its dendrites are stained with a fluorescent antibody that recognizes a cytoskeletal protein *(green)* that is not present in axons. Thousands of axon terminals *(red)* from other nerve cells (not visible) make synapses on the cell body and dendrites; the terminals are stained with a fluorescent antibody that recognizes a protein in synaptic vesicles. (B, courtesy of Olaf Mundigl and Pietro de Camilli.)

low density of voltage-gated Na^+ channels, and so an individual excitatory PSP is generally too small to trigger an action potential. Instead, each incoming signal initiates a local PSP, which decreases with distance from the site of the synapse. If signals arrive simultaneously at several synapses in the same region of the dendritic tree, the total PSP in that neighborhood will be roughly the sum of the individual PSPs, with inhibitory PSPs making a negative contribution to the total. The PSPs from each neighborhood spread passively and converge on the cell body. For long-distance transmission, the combined magnitude of the PSP is then translated, or *encoded*, into the *frequency* of firing of action potentials: the greater the stimulation (depolarization), the higher the frequency of action potentials.

Neuronal Computation Requires a Combination of at Least Three Kinds of K^+ Channels

The intensity of stimulation that a neuron receives is encoded by that neuron into action potential frequency for long-distance transmission. The encoding takes place at a specialized region of the axonal membrane known as the **initial segment**, or *axon hillock*, at the junction of the axon and the cell body (see Figure 11–42). This membrane is rich in voltage-gated Na^+ channels, but it also contains at least four other classes of ion channels—three selective for K^+ and one selective for Ca^{2+}—all of which contribute to the axon hillock's encoding function. The three varieties of K^+ channels have different properties; we shall refer to them as *delayed, rapidly inactivating,* and *Ca^{2+}-activated K^+ channels.*

To understand the need for multiple types of channels, consider first what would happen if the only voltage-gated ion channels present in the nerve cell were the Na^+ channels. Below a certain threshold level of synaptic stimulation, the depolarization of the initial-segment membrane would be insufficient to trigger an action potential. With gradually increasing stimulation, the threshold would be crossed, the Na^+ channels would open, and an action potential would fire. The action potential would be terminated by inactivation of the Na^+ channels. Before another action potential could fire, these channels would have to recover from their inactivation. But that would require a return of the membrane voltage to a very negative value, which would not occur as long as the strong depolarizing stimulus (from PSPs) was maintained. An additional channel type is needed, therefore, to repolarize the membrane after each action potential to prepare the cell to fire again.

The delayed K^+ channels perform this task, as discussed previously in relation to the propagation of the action potential (see Figure 11–32). Their opening permits an efflux of K^+ that drives the membrane back toward the K^+ equilibrium potential, which is so negative that the Na^+ channels rapidly recover from their inactivated state. Repolarization of the membrane also closes the delayed K^+ channels. The initial segment is now reset so that the depolarizing stimulus from

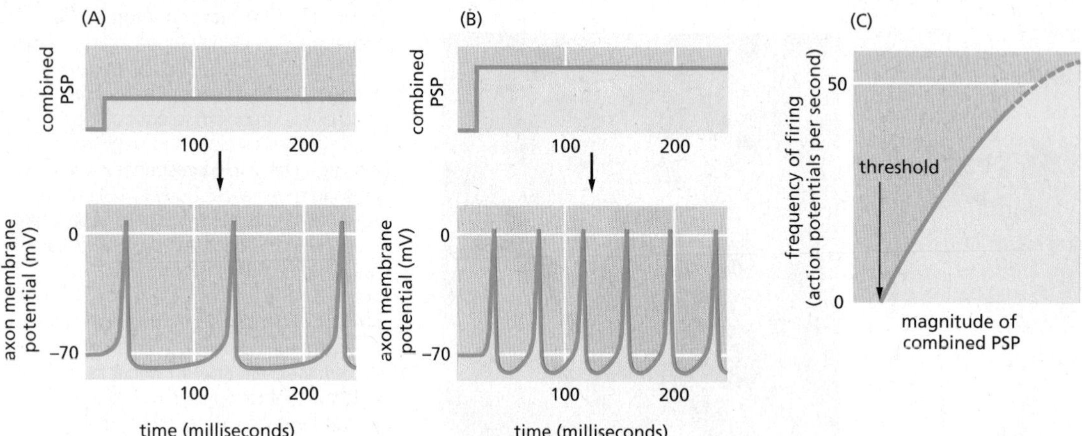

Figure 11–43 **The magnitude of the combined postsynaptic potential (PSP) is reflected in the frequency of firing of action potentials.** The mix of excitatory and inhibitory PSPs produces a *summed PSP* at the initial segment. A comparison of (A) and (B) shows how the firing frequency of an axon increases with an increase in the combined PSP, while (C) summarizes the general relationship.

synaptic inputs can fire another action potential. In this way, sustained stimulation of the dendrites and cell body leads to repetitive firing of the axon.

Repetitive firing in itself, however, is not enough. The frequency of firing has to reflect the intensity of stimulation, and a simple system of Na^+ channels and delayed K^+ channels is inadequate for this purpose. Below a certain threshold level of steady stimulation, the cell will not fire at all; above that threshold level, it will abruptly begin to fire at a relatively rapid rate. The **rapidly inactivating K^+ channels** solve the problem. These, too, are voltage-gated and open when the membrane is depolarized, but their specific voltage sensitivity and kinetics of inactivation are such that they act to reduce the rate of firing at levels of stimulation that are only just above the threshold required for firing. Thus, they remove the discontinuity in the relationship between the firing rate and the intensity of stimulation. The result is a firing rate that is proportional to the strength of the depolarizing stimulus over a very broad range (Figure 11–43).

The process of encoding is usually further modulated by the two other types of ion channels in the initial segment that were mentioned earlier—voltage-gated Ca^{2+} channels and Ca^{2+}-activated K^+ channels. They act together to decrease the response of the cell to an unchanging, prolonged stimulation—a process called **adaptation**. These Ca^{2+} channels are similar to the Ca^{2+} channels that mediate the release of neurotransmitter from presynaptic axon terminals; they open when an action potential fires, transiently allowing Ca^{2+} into the axon cytosol at the initial segment.

The **Ca^{2+}-activated K^+ channel** opens in response to a raised concentration of Ca^{2+} at the channel's cytoplasmic face (Figure 11–44). Prolonged, strong depolarizing stimuli will trigger a long train of action potentials, each of which permits a brief influx of Ca^{2+} through the voltage-gated Ca^{2+} channels, so that local cytosolic Ca^{2+} concentration gradually builds up to a level high enough to open the Ca^{2+}-activated K^+ channels. Because the resulting increased permeability of the membrane to K^+ makes the membrane harder to depolarize, the delay between one action potential and the next is increased. In this way, a neuron that is stimulated continually for a prolonged period becomes gradually less responsive to the constant stimulus.

Such adaptation, which can also occur by other mechanisms, allows a neuron—indeed, the nervous system generally—to react sensitively to *change*, even against a high background level of steady stimulation. It is one of the computational strategies that help us, for example, to feel a light touch on the shoulder and yet ignore the constant pressure of our clothing. We discuss adaptation as a general feature in cell signaling processes in more detail in Chapter 15.

Other neurons do different computations, reacting to their synaptic inputs in myriad ways, reflecting the different assortments of ion channels in their membrane. There are several hundred genes that code for ion channels in the human genome, with more than 70 encoding voltage-gated channels alone. Further complexity is introduced by alternative splicing of RNA transcripts and assembly of channel subunits in different combinations. Moreover, ion channels are selectively

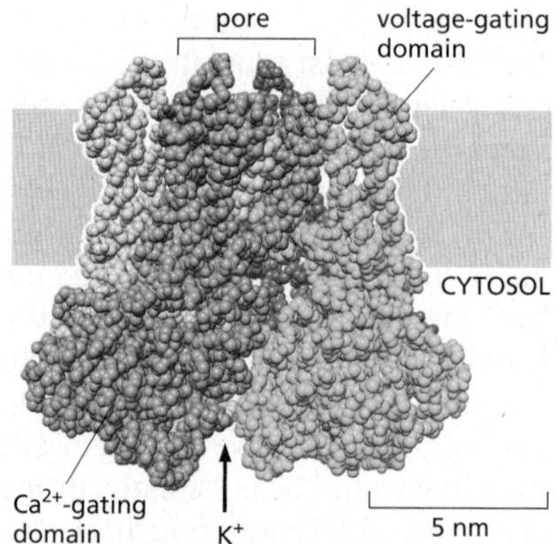

Figure 11–44 **Structure of a Ca^{2+}-activated K^+ channel.** The channel contains four identical subunits (which are shown in different colors for clarity). It is both voltage- and Ca^{2+}-gated. The structure shown is a composite of the cytosolic and membrane portions of the channel that were separately crystallized. Note the large ligand-binding domains in the cytosol. (PDB code: 1LNQ.)

localized to different sites in the plasma membrane of a neuron. Some K^+ and Ca^{2+} channels are concentrated in the dendrites and participate in processing the input that a neuron receives. As we have seen, other ion channels are located at the axon's initial segment, where they control action potential firing; and some ligand-gated channels are distributed over the cell body and, depending on their ligand occupancy, modulate the cell's general sensitivity to synaptic inputs. The multiplicity of ion channels and their locations evidently allows each of the many types of neurons to tune their electrical behavior to the particular tasks they perform.

Voltage-gated Ca^{2+} channels are ubiquitous in neurons, and the Ca^{2+} that flows in through them is a good reflection of each nerve cell's activity. Today it is therefore quite common to express optical Ca^{2+} reporters in specific subtypes of neurons to follow their activities in transgenic animals. It is possible to image and track Ca^{2+} signals of 10,000 neurons in the brain simultaneously with a sensitive camera looking at the surface of the cortex or looking deeper into the brain with optical fibers. In this way, we begin to see neuronal circuits in living animals.

One of the crucial properties of the nervous system is its ability to learn and remember. This property depends in part on the ability of individual synapses to strengthen or weaken depending on their use—a process called **synaptic plasticity**. We will next consider a remarkable type of ion channel that has a special role in some forms of synaptic plasticity. It is located at many excitatory synapses in the central nervous system, where it is doubly gated by voltage and the excitatory neurotransmitter glutamate. It is also the site of action of the psychoactive drug phencyclidine, or angel dust.

Long-term Potentiation in the Mammalian Hippocampus Depends on Ca^{2+} Entry Through NMDA-Receptor Channels

Practically all animals can learn, but mammals seem to learn exceptionally well (or so we like to think). In a mammal's brain, the region called the *hippocampus* has a special role in learning. When it is destroyed on both sides of the brain, the ability to form new memories is largely lost, although previous long-established memories remain. Some synapses in the hippocampus show a striking form of synaptic plasticity with repeated use: whereas occasional single action potentials in the presynaptic cells leave no lasting trace, a short burst of repetitive firing causes **long-term potentiation (LTP)**, such that subsequent single action potentials in the presynaptic cells evoke a greatly enhanced response in the postsynaptic cells. The effect lasts hours, days, or weeks, according to the number and intensity of the bursts of repetitive firing. Only the synapses that were activated exhibit LTP; synapses that have remained quiet on the same postsynaptic cell are not affected. However, while the cell is receiving a burst of repetitive stimulation via one set of synapses, if a single action potential is delivered at *another* synapse on its surface, that latter synapse also will undergo LTP, even though a single action potential delivered there at another time would leave no such lasting trace.

The underlying rule in such events seems to be that *LTP occurs on any occasion when a presynaptic cell fires (once or more) at a time when the postsynaptic membrane is strongly depolarized* (either through recent repetitive firing of the same presynaptic cell or by other means). This rule reflects the behavior of a particular class of ion channels in the postsynaptic membrane. Glutamate is the main excitatory neurotransmitter in the mammalian central nervous system, and glutamate-gated ion channels are the most common of all transmitter-gated channels in the brain. In the hippocampus, as elsewhere, most of the depolarizing current responsible for excitatory PSPs is carried by glutamate-gated ion channels called **AMPA receptors**, which operate in the standard way (**Figure 11–45**, **Figure 11–46**). But the current has, in addition, a second and more intriguing component, which is mediated by a separate subclass of glutamate-gated ion channels known as **NMDA receptors**, so named because they can be selectively activated by the artificial glutamate analog *N*-methyl-D-aspartate. The NMDA-receptor channels are doubly gated, opening only when two conditions are satisfied simultaneously: glutamate must be bound to the receptor, and the membrane must

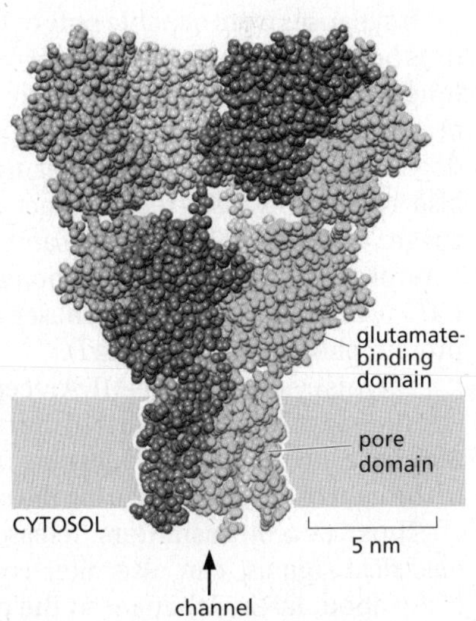

Figure 11–45 **The structure of the AMPA receptor.** This ionotropic glutamate receptor (named after the glutamate analog α-amino-3-hydroxy-5-methyl-4-isoxazolepropionic acid) is the most common mediator of fast, excitatory synaptic transmission in the central nervous system (CNS). Note the large ligand-binding domains in the extracellular space. (PDB code: 3KG2.)

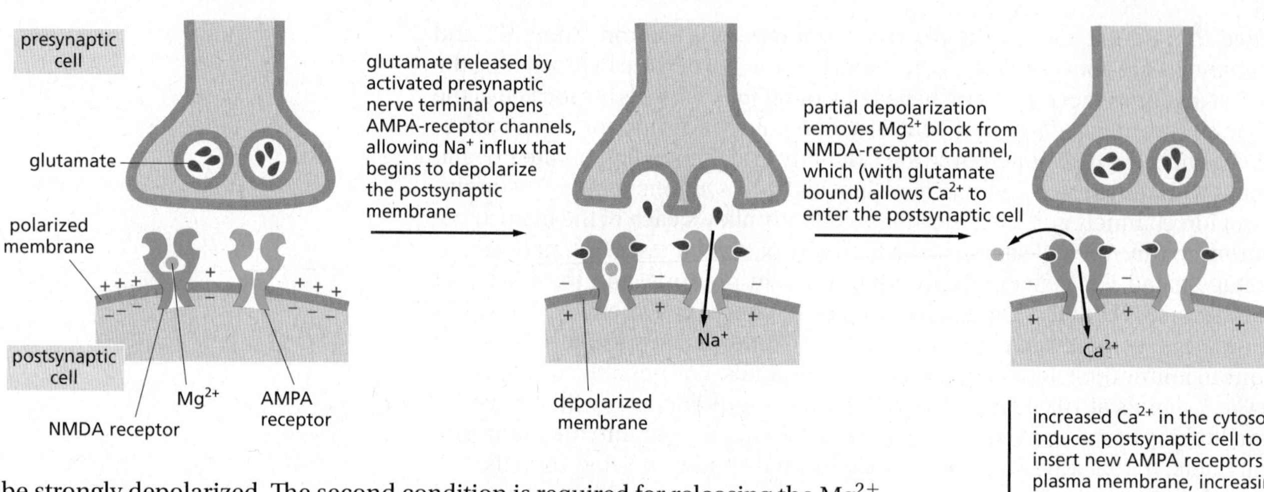

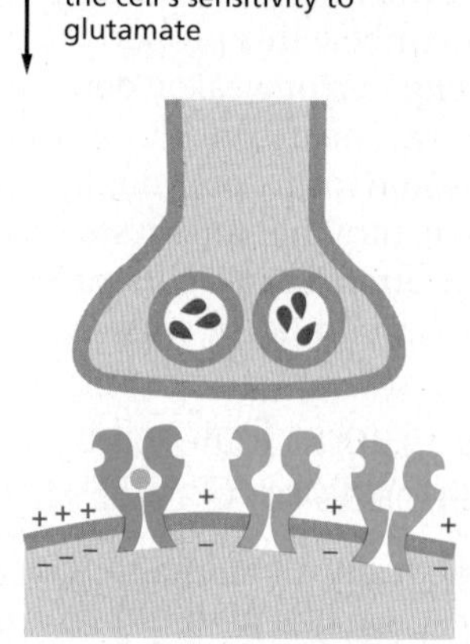

Figure 11–46 The signaling events in long-term potentiation. Although not shown, transmission-enhancing changes can also occur in the presynaptic nerve terminals in LTP, which may be induced by retrograde signals from the postsynaptic cell.

be strongly depolarized. The second condition is required for releasing the Mg^{2+} that normally blocks the resting channel. This means NMDA receptors are normally activated only when AMPA receptors are activated as well and depolarize the membrane (see Figure 11–43). The NMDA receptors are critical for LTP. When they are selectively blocked with a specific inhibitor or inactivated genetically, LTP does not occur, even though ordinary synaptic transmission continues, indicating the importance of NMDA receptors for LTP induction. Such animals exhibit specific deficits in their learning abilities but behave almost normally otherwise.

How do NMDA receptors mediate LTP? The answer is that these channels, when open, are highly permeable to Ca^{2+}, which acts as an intracellular signal in the postsynaptic cell, triggering a cascade of changes that are responsible for LTP. Thus, LTP is prevented when Ca^{2+} levels are held artificially low in the postsynaptic cell by injecting the Ca^{2+} chelator EGTA into it, and LTP can be induced by artificially raising intracellular Ca^{2+} levels in the cell. Among the long-term changes that increase the sensitivity of the postsynaptic cell to glutamate is the insertion of new AMPA receptors into the plasma membrane (see Figure 11–46). In some forms of LTP, changes occur in the presynaptic cell as well, so that it releases more glutamate than normal when it is activated subsequently.

If synapses were capable only of LTP, they would quickly become saturated and thus be of limited value as an information-storage device. In fact, they also exhibit **long-term depression (LTD)**, with the long-term effect of reducing the number of AMPA receptors in the postsynaptic membrane. This feat is accomplished by degrading AMPA receptors after their selective endocytosis. Surprisingly, LTD also requires NMDA-receptor activation and a rise in Ca^{2+}. How does Ca^{2+} trigger opposite effects at the same synapse? It turns out that this bidirectional control of synaptic strength depends on the magnitude of the rise in Ca^{2+}: high Ca^{2+} levels activate protein kinases and LTP, whereas modest Ca^{2+} levels activate protein phosphatases and LTD.

There is evidence that NMDA receptors have an important role in synaptic plasticity and learning in other parts of the brain, as well as in the hippocampus. Moreover, they have a crucial role in adjusting the anatomical pattern of synaptic connections in the light of experience during the development of the nervous system.

Thus, neurotransmitters released at synapses, besides relaying transient electrical signals, can also alter concentrations of intracellular mediators that bring about lasting changes in the efficacy of synaptic transmission. However, it is still uncertain how these changes endure for weeks, months, or a lifetime in the face of the normal turnover of cell constituents.

The Use of Channelrhodopsins Has Revolutionized the Study of Neural Circuits

Channelrhodopsins are photosensitive ion channels that open in response to light. They evolved as sensory receptors in photosynthetic green algae to allow the algae to swim toward light. The structure of channelrhodopsin closely resembles that of bacteriorhodopsin (see Figure 10–31). It contains a covalently bound

Figure 11–47 Optogenetic control of aggression neurons in a living mouse. A gene encoding channelrhodopsin was introduced into a subpopulation of neurons in the hypothalamus of a mouse. When the neurons were exposed to flashing blue light using a tiny, implanted fiber-optic cable, the channelrhodopsin channels opened, depolarizing and activating the cells. When the light was switched on, the mouse immediately became aggressive and attacked the inflated rubber glove; when the light was switched off, its behavior immediately returned to normal (Movie 11.13). (Adapted from D. Lin et al., *Nature* 470:221–226, 2011. With permission from the authors.)

retinal group that absorbs light and undergoes an isomerization reaction, which triggers a conformational change in the protein, opening an ion channel in the plasma membrane. In contrast to bacteriorhodopsin, which is a light-driven proton pump, channelrhodopsin is a light-activated cation channel.

Using genetic engineering techniques, channelrhodopsin can be expressed in virtually any cell type in vertebrates and invertebrates. Researchers first introduced the gene into cultured neurons and showed that flashing light could now activate the channelrhodopsin and induce the neurons to fire action potentials. Because the frequency of the light flashes determined the frequency of the action potentials, one can control the frequency of neuronal firing with millisecond precision.

Next, neurobiologists used the approach to activate specific neurons in the brain of experimental animals. Using a tiny fiber-optic cable implanted near the relevant brain region, they could flash light to specifically activate the channelrhodopsin-containing neurons to fire action potentials. One group of researchers expressed channelrhodopsin in a subset of mouse neurons thought to be involved in aggression: when these cells were activated by light, the mouse immediately attacked anything in its environment—including other mice or even an inflated rubber glove (Figure 11–47); when the light was switched off, the neurons fell silent and the mouse's behavior returned to normal.

Since these pioneering studies, researchers have engineered additional light-responsive ion channels and transporters, including some that can rapidly inactivate specific neurons. It is therefore now possible to transiently activate or inhibit specific neurons in the brains of awake animals with remarkable spatial and temporal precision. In this way, the rapidly expanding new field of **optogenetics** is revolutionizing neurobiology, allowing neuroscientists to analyze the neurons and circuits underlying even the most complex behaviors in experimental animals, including nonhuman primates.

Summary

Ion channels form aqueous pores across the lipid bilayer and allow inorganic ions of appropriate size and charge to cross the membrane down their electrochemical gradients at rates about 1000 times greater than those achieved by any known transporter. The channels are "gated" and usually open transiently in response to a specific perturbation in the membrane, such as a change in membrane potential (voltage-gated channels), or the binding of a neurotransmitter to the channel (transmitter-gated channels).

K^+-selective leak channels have an important role in determining the resting membrane potential across the plasma membrane in most animal cells. Voltage-gated cation channels are responsible for the amplification and propagation of action potentials in electrically excitable cells, such as neurons and heart and skeletal muscle cells. Transmitter-gated ion channels convert chemical signals to electrical signals at chemical synapses. Excitatory neurotransmitters, such as acetylcholine and glutamate, open transmitter-gated cation channels and thereby depolarize the postsynaptic membrane toward the threshold level for firing an action potential. Inhibitory neurotransmitters, such as GABA and glycine, open transmitter-gated Cl^- or K^+ channels and thereby suppress firing by keeping the postsynaptic membrane polarized. A subclass of glutamate-gated ion channels, called NMDA-receptor channels, is highly permeable to Ca^{2+}, which can trigger the long-term changes in synapse efficacy (synaptic plasticity) such as LTP and LTD that are thought to be involved in some forms of learning and memory.

PROBLEMS

Which statements are true? Explain why or why not.

11–1 Transport by transporters can be either active or passive, whereas transport by channels is always passive.

11–2 A symporter would function as an antiporter if its orientation in the membrane were reversed; that is, if the portion of the protein normally exposed to the cytosol faced the outside of the cell instead.

11–3 Excitatory synapses generally cause a small hyperpolarization of the postsynaptic membrane, whereas inhibitory synapses generally cause a small depolarization of the postsynaptic membrane.

11–4 Transporters approach saturation at high concentrations of the transported molecule when all their binding sites are occupied; channels, on the other hand, do not bind the ions they transport and thus the flux of ions through a channel does not saturate.

11–5 The membrane potential arises from movements of charge that leave ion concentrations practically unaffected, causing only a very slight discrepancy in the number of positive and negative ions on the two sides of the membrane.

Discuss the following problems.

11–6 Order Ca^{2+}, CO_2, glucose, RNA, and H_2O according to their ability to diffuse through a lipid bilayer, beginning with the one that crosses the bilayer most readily. Explain the basis for your ranking.

11–7 How is it possible for some molecules to be at equilibrium across a biological membrane and yet not be at the same concentration on both sides?

11–8 Suppose a membrane contains a single passive transporter with a K_m of 0.1 mM for its solute. How effective would the transporter be at equalizing the concentrations of solute across the membrane if the starting concentrations were 0.01 mM inside and 0.05 mM outside? What if the concentrations were 100 mM inside and 500 mM outside?

A. Effective at both low and high solute concentrations

B. Effective at low solute levels but ineffective at high levels

C. Ineffective at both low and high solute concentrations

D. Ineffective at low solute levels but effective at high levels

11–9 Microvilli increase the surface area of intestinal cells, providing more efficient absorption of nutrients. Microvilli are shown in profile and cross section in **Figure Q11–1**. From the dimensions given in the figure, estimate the increase in surface area that microvilli provide (for the portion of the plasma membrane in contact with the lumen of the gut) relative to the corresponding surface of a cell with a "flat" plasma membrane.

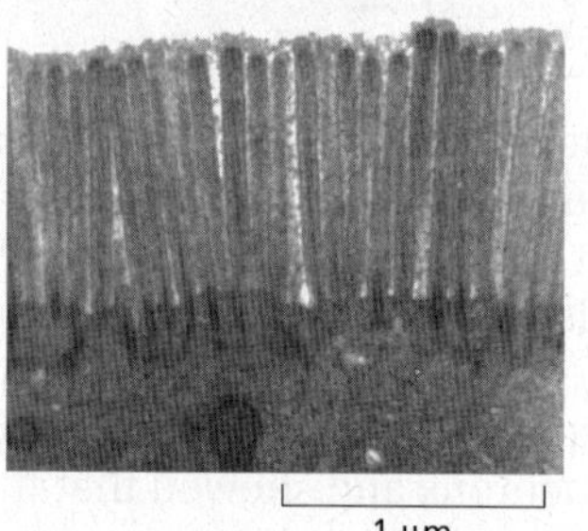

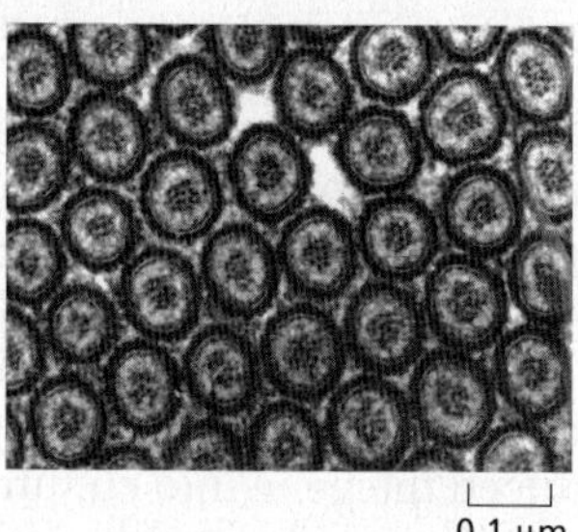

Figure Q11–1 Microvilli of intestinal epithelial cells in profile and cross section (Problem 11–9). (Left panel, from Rippel Electron Microscope Facility, Dartmouth College; right panel, from David Burgess.)

11–10 Ion transporters are "linked" together—not physically, but as a consequence of their actions. For example, cells can raise their intracellular pH, when it becomes too acidic, by exchanging external Na^+ for internal H^+, using an Na^+-H^+ antiporter. The change in internal Na^+ is then redressed using the Na^+-K^+ pump.

A. Can these two transporters, operating together, return both the H^+ and the Na^+ concentrations to their normal levels inside the cell?

B. Does the linked action of these two pumps cause imbalances in either the K^+ concentration or the membrane potential? Why or why not?

11–11 According to Newton's laws of motion, an ion exposed to an electric field in a vacuum would experience a constant acceleration from the electric driving force, just as a falling body in a vacuum constantly accelerates due to gravity. In water, however, an ion moves at constant velocity in an electric field. Why do you suppose that is?

11–12 In a subset of voltage-gated K^+ channels, the N-terminus of each subunit acts like a tethered ball that occludes the cytoplasmic end of the pore soon after it opens, thereby inactivating the channel. This "ball-and-chain" model for the rapid inactivation of voltage-gated K^+ channels has been elegantly supported for the *shaker* K^+ channel from *Drosophila melanogaster*. (The *shaker* K^+ channel in *Drosophila* is named after a mutant form that causes excitable behavior—even anesthetized flies keep twitching.) Deletion of the N-terminal amino acids from the normal *shaker* channel gives rise to a channel that opens in response to membrane depolarization but stays open instead of rapidly closing as the normal channel does. A peptide (MAAVAGLYGLGEDRQHRKKQ) that corresponds to the deleted N-terminus can inactivate the open channel at 100 μM.

Is the concentration of free peptide (100 μM) that is required to inactivate the defective K^+ channel anywhere near the local concentration of the tethered ball on a normal channel? Assume that the tethered ball can explore a hemisphere [volume = $(2/3)\pi r^3$] with a radius of 21.4 nm, which is the length of the polypeptide "chain" (Figure Q11–2). Calculate the concentration for one ball in this hemisphere. How does that value compare with the concentration of free peptide needed to inactivate the channel?

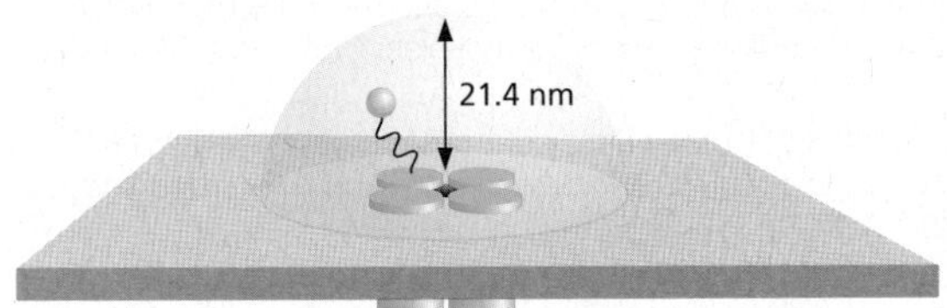

Figure Q11–2 A "ball" tethered by a "chain" to a voltage-gated K^+ channel (Problem 11–12).

11–13 The giant axon of the squid (Figure Q11–3) occupies a unique position in the history of our understanding of cell membrane potentials and nerve action. When an electrode is stuck into an intact giant axon, the membrane potential registers –70 mV. When the axon, suspended in a bath of seawater, is stimulated to conduct a nerve impulse, the membrane potential changes transiently from –70 mV to +40 mV.

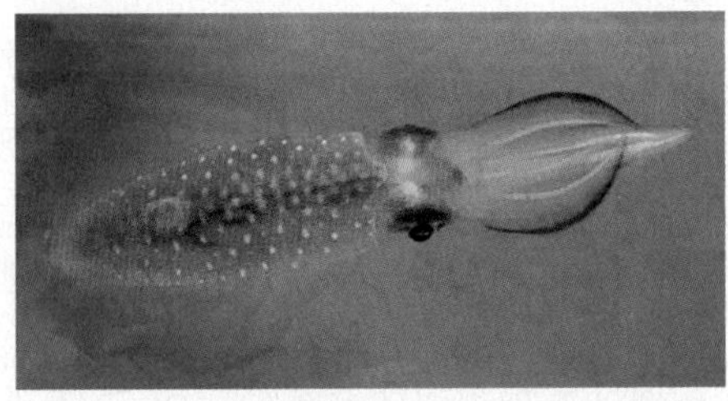

Figure Q11–3 The squid *Sepioteuthis lessoniana* (Problem 11–13). This squid can grow up to 30 cm in length.

For univalent ions and at 20°C (293 K), the Nernst equation reduces to

$$V = 58 \text{ mV} \times \log(C_o/C_i)$$

where C_o and C_i are the concentrations outside and inside, respectively.

Using this equation, calculate the potential across the resting membrane (1) assuming that it is due solely to K^+ and (2) assuming that it is due solely to Na^+. (The Na^+ and K^+ concentrations in the axon cytosol and in seawater are given in Table Q11–1.) Which calculation is closer to the measured resting potential? Which calculation is closer to the measured action potential? Explain why these assumptions approximate the measured resting and action potentials.

TABLE Q11–1 Ionic Composition of Seawater and of the Cytosol in the Squid Giant Axon (Problem 11–13)

Ion	Cytosol	Seawater
Na^+	65 mM	430 mM
K^+	344 mM	9 mM

11–14 If the resting membrane potential of a cell is –70 mV and the thickness of the lipid bilayer is 5 nm, what is the strength of the electric field across the membrane in V/cm? What do you suppose would happen if you applied this voltage to two metal electrodes separated by a 1-cm air gap?

11–15 Acetylcholine-gated cation channels at the neuromuscular junction open in response to acetylcholine released by the nerve terminal and allow Na^+ ions to enter the muscle cell, which causes membrane depolarization and ultimately leads to muscle contraction.

A. Patch-clamp measurements show that young rat muscles have cation channels that respond to acetylcholine (Figure Q11–4). How many kinds of channel are there? How can you tell?

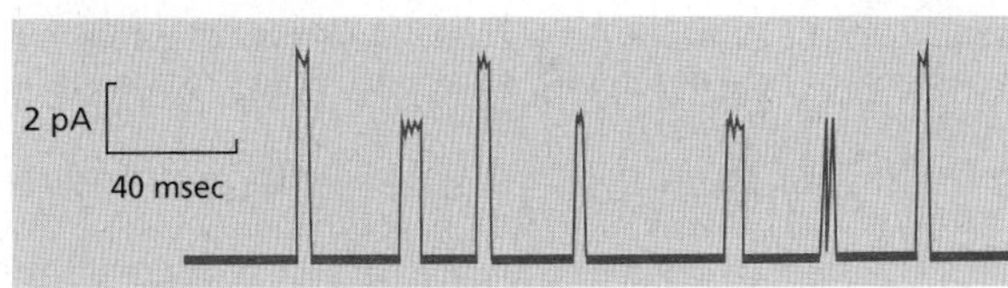

Figure Q11–4 Patch-clamp measurements of acetylcholine-gated cation channels in young rat muscle (Problem 11–15).

B. For each kind of channel, calculate the number of ions that enter in 1 millisecond. (One ampere is a current of 1 coulomb per second; 1 pA equals 10^{-12} ampere. An ion with a single charge such as Na^+ carries a charge of 1.6×10^{-19} coulomb.)

REFERENCES

Principles of Membrane Transport

Engel A & Gaub HE (2008) Structure and mechanics of membrane proteins. *Annu. Rev. Biochem.* 77, 127–148.

Hille B (2001) Ion Channels of Excitable Membranes, 3rd ed. Sunderland, MA: Sinauer.

Li F, Egea PF, Vecchio AJ . . . Stroud RM (2021) Highlighting membrane protein structure and function: a celebration of the Protein Data Bank. *J. Biol. Chem.* 296, 100557.

Luo L (2020) Principles of Neurobiology, 2nd ed. Boca Raton, FL: CRC Press.

Mitchell P (1977) Vectorial chemiosmotic processes. *Annu. Rev. Biochem.* 46, 996–1005.

Stein WD (2014) Channels, Carriers, and Pumps: An Introduction to Membrane Transport, 2nd ed. San Diego: Academic Press.

Vinothkumar KR & Henderson R (2010) Structures of membrane proteins. *Q. Rev. Biophys.* 43, 65–158.

Transporters and Active Membrane Transport

Baldwin SA & Henderson PJ (1989) Homologies between sugar transporters from eukaryotes and prokaryotes. *Annu. Rev. Physiol.* 51, 459–471.

Drew D & Boudker O (2016) Shared molecular mechanisms of membrane transporters. *Annu. Rev. Biochem.* 85, 543–572.

Dyla M, Kjærgaard M, Poulsen H & Nissen P (2020) Structure and mechanism of P-type ATPase ion pumps. *Annu. Rev. Biochem.* 89, 583–603.

Forrest LR & Rudnick G (2009) The rocking bundle: a mechanism for ion-coupled solute flux by symmetrical transporters. *Physiology (Bethesda)* 24, 377–386.

Gadsby DC (2009) Ion channels versus ion pumps: the principal difference, in principle. *Nat. Rev. Mol. Cell Biol.* 10, 344–352.

Gouaux E & MacKinnon R (2005) Principles of selective ion transport in channels and pumps. *Science* 310, 1461–1465.

Higgins CF (2007) Multiple molecular mechanisms for multidrug resistance transporters. *Nature* 446, 749–757.

Kaback HR & Guan L (2019) It takes two to tango: the dance of the permease. *J. Gen. Physiol.* 151, 878–886.

Kanai R, Ogawa O, Vilsen B . . . Toyoshima C (2013) Crystal structure of the Na^+-bound Na^+,K^+-ATPase preceding the E1P state. *Nature* 502, 201–206.

Kefauver JM, Ward AB & Patapoutian A (2020) Discoveries in structure and physiology of mechanically activated ion channels. *Nature* 587, 567–576.

Khademi S, O'Connell J, 3rd, Remis J . . . Stroud RM (2004) Mechanism of ammonia transport by Amt/MEP/Rh: structure of AmtB at 1.35Å. *Science* 305, 1587–1594.

Kim Y & Chen J (2018) Molecular structure of the human P-glycoprotein in the ATP-bound, outward facing conformation. *Science* 359, 915–919.

Kühlbrandt W (2004) Biology, structure and mechanism of P-type ATPases. *Nat. Rev. Mol. Cell Biol.* 5, 282–295.

Kühlbrandt W (2019) Structure and mechanisms of F-type ATP synthases. *Annu. Rev. Biochem.* 88, 515–549.

Kumar H, Finer-Moore JS, Kaback HR & Stroud RM (2015) Structure of LacY with an alpha-substituted galactoside: connecting the binding site and the protonation site. *Proc. Natl. Acad. Sci. USA* 112, 9004–9009.

Perez C, Koshy C, Yildiz O & Ziegler C (2012) Alternating-access mechanism in conformationally asymmetric trimers of the betaine transporter BetP. *Nature* 490, 126–130.

Rees D, Johnson E & Lewinson O (2009) ABC transporters: the power to change. *Nat. Rev. Mol. Cell Biol.* 10, 218–227.

Rudnick G (2011) Cytoplasmic permeation pathway of neurotransmitter transporters. *Biochemistry* 50, 7462–7475.

Ruprecht JJ & Kunji ERS (2021) Structural mechanism of transport of mitochondrial carriers. *Annu. Rev. Biochem.* 90, 535–558.

Saier MH, Jr (2000) Vectorial metabolism and the evolution of transport systems. *J. Bacteriol.* 182, 5029–5035.

Stein WD (2002) Cell volume homeostasis: ionic and nonionic mechanisms. The sodium pump in the emergence of animal cells. *Int. Rev. Cytol.* 215, 231–258.

Thomas C & Tampé R (2020) Structural and mechanistic principles of ABC transporters. *Annu. Rev. Biochem.* 89, 605–636.

Toyoshima C (2009) How Ca^{2+}-ATPase pumps ions across the sarcoplasmic reticulum membrane. *Biochim. Biophys. Acta* 1793, 941–946.

Yamashita A, Singh SK, Kawate T . . . Gouaux E (2005) Crystal structure of a bacterial homologue of Na^+/Cl^--dependent neurotransmitter transporters. *Nature* 437, 215–223.

Channels and the Electrical Properties of Membranes

Armstrong C (1998) The vision of the pore. *Science* 280, 56–57.

Arnadóttir J & Chalfie M (2010) Eukaryotic mechanosensitive channels. *Annu. Rev. Biophys.* 39, 111–137.

Atlas D (2013) The voltage-gated calcium channel functions as the molecular switch of synaptic transmission. *Annu. Rev. Biochem.* 82, 607–635.

Bezanilla F (2008) How membrane proteins sense voltage. *Nat. Rev. Mol. Cell Biol.* 9, 323–332.

Carbrey JM & Agre P (2009) Discovery of the aquaporins and development of the field. *Handb. Exp. Pharmacol.* 190, 3–28.

Catterall WA (2017) Forty years of sodium channels: structure, function, pharmacology, and epilepsy. *Neurochem. Res.* 42, 2495–2504.

Chen S & Gouaux E (2019) Structure and mechanism of AMPA receptor—auxiliary protein complexes. *Curr. Opin. Struct. Biol.* 54, 104–111.

Davis GW (2006) Homeostatic control of neural activity: from phenomenology to molecular design. *Annu. Rev. Neurosci.* 29, 307–323.

Greengard P (2001) The neurobiology of slow synaptic transmission. *Science* 294, 1024–1030.

Hodgkin AL & Huxley AF (1952a) Currents carried by sodium and potassium ions through the membrane of the giant axon of *Loligo*. *J. Physiol.* 116, 449–472.

Hodgkin AL & Huxley AF (1952b) A quantitative description of membrane current and its application to conduction and excitation in nerve. *J. Physiol.* 117, 500–544.

Jessell TM & Kandel ER (1993) Synaptic transmission: a bidirectional and self-modifiable form of cell–cell communication. *Cell* 72(Suppl), 1–30.

Jin P, Jan LY & Jan J-N (2020) Mechanosensitive ion channels: structural features relevant to mechanotransduction mechanisms. *Annu. Rev. Neurosci*. 43, 207–229.

Julius D (2013) TRP channels and pain. *Annu. Rev. Cell Dev. Biol.* 29, 355–384.

Katz B (1966) Nerve, Muscle and Synapse. New York: McGraw-Hill.

King LS, Kozono D & Agre P (2004) From structure to disease: the evolving tale of aquaporin biology. *Nat. Rev. Mol. Cell Biol.* 5, 687–698.

Kintzer AF, Green EM, Dominik PK . . . Stroud RM (2018) Structural basis for activation of voltage sensor domains in an ion channel TPC1. *Proc. Natl. Acad. Sci. USA* 115, E9095–E9104.

Lee CH & MacKinnon R (2019) Voltage sensor movements during hyperpolarization in the HCN channel. *Cell* 179, 1582–1589e7.

Liao M, Cao E, Julius D & Cheng Y (2014) Single particle electron cryo-microscopy of a mammalian ion channel. *Curr. Opin. Struct. Biol.* 27, 1–7.

Long SB, Tao X, Campbell EB & MacKinnon R (2007) Atomic structure of a voltage-dependent K+ channel in a lipid membrane-like environment. *Nature* 450, 376–382.

MacKinnon R (2003) Potassium channels. *FEBS Lett.* 555, 62–65.

Miesenböck G (2011) Optogenetic control of cells and circuits. *Annu. Rev. Cell Dev. Biol.* 27, 731–758.

Moss SJ & Smart TG (2001) Constructing inhibitory synapses. *Nat. Rev. Neurosci.* 2, 240–250.

Neher E & Sakmann B (1992) The patch clamp technique. *Sci. Am.* 266, 44–51.

Nicholls JG, Fuchs PA, Martin AR & Wallace BG (2000) From Neuron to Brain, 4th ed. Sunderland, MA: Sinauer.

Payandeh J, Scheuer T, Zheng N & Catterall WA (2011) The crystal structure of a voltage-gated sodium channel. *Nature* 475, 353–358.

Savage DF, O'Connell JD, 3rd, Miercke LJW . . . Stroud RM (2010) Structural context shapes the aquaporin selectivity filter. *Proc. Natl. Acad. Sci. USA* 107, 17164–17169.

Scannevin RH & Huganir RL (2000) Postsynaptic organization and regulation of excitatory synapses. *Nat. Rev. Neurosci.* 1, 133–141.

Stevens CF (2004) Presynaptic function. *Curr. Opin. Neurobiol.* 14, 341–345.

Szczot M, Nickolls AR, Lam RM & Chesler AT (2021) The form and function of PIEZO2. *Annu. Rev. Biochem.* 90, 507–534.

Intracellular Organization and Protein Sorting

CHAPTER 12

IN THIS CHAPTER

The many thousands of macromolecules and their associated biochemical activities inside a cell are spatially segregated to different regions. Intracellular organization is a particularly prominent feature of eukaryotic cells, which unlike bacteria are elaborately subdivided into functionally distinct, membrane-enclosed compartments. Many of these compartments define the cell's major **organelles** such as the endoplasmic reticulum, Golgi apparatus, lysosome, plastid, and mitochondrion, the last two of which are still further subcompartmentalized by internal membranes. Other subsets of the cell's macromolecules can organize into dynamic and reversible assemblies, called **biomolecular condensates**, that can serve as specialized biochemical factories or temporary storage depots. To understand the eukaryotic cell, it is essential to know how the cell creates and maintains its complex intracellular organization.

An animal cell contains about 10 billion (10^{10}) protein molecules of perhaps 10,000 kinds, and the synthesis of almost all of them begins in the **cytosol**, the space of the cytoplasm outside the membrane-enclosed organelles. Each newly synthesized protein is then delivered specifically to the organelle that requires it. The unique protein and lipid composition on the surface of each organelle is used as a cue to direct new deliveries of proteins and lipids to sustain that organelle's identity. The characteristic set of proteins and other specialized molecules define each organelle's structural and functional properties. They catalyze the reactions that occur there and selectively transport molecules into and out of the organelle. The intracellular transport of proteins is the central theme of both this chapter and the next. By tracing the protein traffic from one part of the cell to another, one can begin to make sense of the otherwise bewildering maze of intracellular membranes and other subcellular structures.

THE COMPARTMENTALIZATION OF CELLS

In this brief overview of the compartments of the cell and the relationships between them, we organize the cell's organelles conceptually into a small number of discrete families, discuss how proteins are directed to specific organelles, and explain how proteins cross organelle membranes.

All Eukaryotic Cells Have the Same Basic Set of Membrane-enclosed Organelles

Many vital biochemical processes take place in membranes or on their surfaces. Membrane-bound enzymes, for example, catalyze lipid metabolism, and oxidative phosphorylation and photosynthesis both require a membrane to couple the transport of H^+ to the synthesis of ATP. In addition to providing increased membrane area to host biochemical reactions, intracellular membrane systems form enclosed compartments that are separate from the cytosolic compartment, thus creating functionally specialized aqueous spaces within the cell. In these spaces, subsets of molecules (proteins, reactants, ions) are concentrated to optimize the biochemical reactions in which they participate. By having multiple types of compartments inside the same cell, biochemical reactions that require very different conditions, or would compete with each other, can nevertheless occur simultaneously. Because the lipid bilayer of cell membranes is impermeable to most

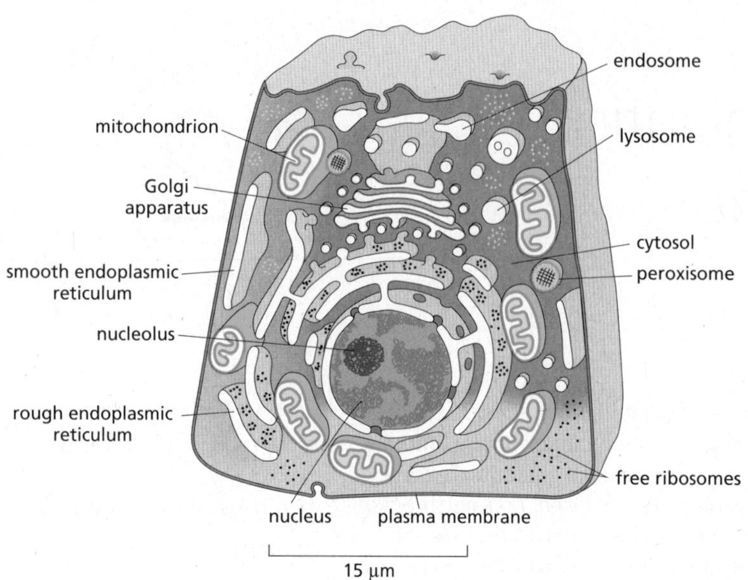

Figure 12–1 The major intracellular compartments of an animal cell. The cytosol *(gray)*, rough endoplasmic reticulum, smooth endoplasmic reticulum, Golgi apparatus, nucleus, mitochondrion, endosome, lysosome, and peroxisome are distinct compartments isolated from the rest of the cell by at least one selectively permeable membrane (see Movie 9.2). By contrast, the nucleolus is not enclosed by a membrane and represents one example of a biomolecular condensate.

hydrophilic molecules, the membrane of an organelle must contain membrane transport proteins to import and export specific metabolites. Each organellar membrane must also have a mechanism for importing, and incorporating into the organelle, the specific proteins that make the organelle unique.

Figure 12–1 illustrates the major intracellular compartments common to eukaryotic cells. The *nucleus* contains the genome (aside from mitochondrial and chloroplast DNA), and it is the principal site of DNA and RNA synthesis. The surrounding **cytoplasm** consists of the cytosol and the cytoplasmic organelles suspended in it. The cytosol constitutes a little more than half the total volume of the cell, and it is the main site of protein synthesis and degradation. It also performs most of the cell's intermediary metabolism; that is, the many reactions that degrade some small molecules and synthesize others to provide the building blocks for macromolecules (discussed in Chapter 2).

About half the total area of membrane in a eukaryotic cell encloses the labyrinthine spaces of the *endoplasmic reticulum (ER)*. Most soluble and integral membrane proteins destined for the cell exterior or for other organelles are initially assembled at the ER. These proteins are transported into the ER as they are synthesized by ribosomes. The distinctive appearance in electron micrographs of ribosomes studding the surface of these regions of the ER is the reason they are termed the *rough ER*. The ER also produces most of the lipid and sterols for the rest of the cell and functions as a store for Ca^{2+} ions. These regions of the ER typically lack bound ribosomes and are called *smooth ER*.

The ER sends many of its proteins and lipids to the *Golgi apparatus*, which often consists of organized stacks of disc-like compartments called *Golgi cisternae*. The Golgi apparatus receives lipids and proteins from the ER and dispatches them to various destinations, usually covalently modifying them *en route*. *Lysosomes* contain digestive enzymes that degrade defunct intracellular organelles, as well as macromolecules and particles taken in from outside the cell by endocytosis. On the way to lysosomes, endocytosed material must first pass through a series of organelles called *endosomes*. As we will see, the ER, Golgi apparatus, lysosomes, endosomes, and plasma membrane are linked by the cell's major pathways of membrane traffic.

Mitochondria and *chloroplasts* generate most of the ATP that cells use to drive reactions requiring an input of free energy; chloroplasts are a specialized version of *plastids* (present in plants, algae, and some protozoa), which can also have other functions, such as the storage of food or pigment molecules. Finally, *peroxisomes* are small vesicular compartments that contain enzymes used in various oxidative reactions.

On average, the membrane-enclosed compartments together occupy nearly half the volume of a cell (Table 12–1), and a large amount of intracellular

TABLE 12–1 Relative Volumes Occupied by the Major Intracellular Compartments in a Liver Cell (Hepatocyte)

Intracellular compartment	Percentage of total cell volume
Cytosol	54
Mitochondria	22
Rough ER cisternae	9
Smooth ER cisternae	5
Golgi cisternae	1
Nucleus	6
Peroxisomes	1
Lysosomes	1
Endosomes	1

TABLE 12–2 Relative Amounts of Membrane Types in Two Kinds of Eukaryotic Cells

	Percentage of total cell membrane	
Membrane type	Liver hepatocyte*	Pancreatic exocrine cell*
Plasma membrane	2	5
Rough ER membrane	35	60
Smooth ER membrane	16	<1
Golgi apparatus membrane	7	10
Mitochondria		
Outer membrane	7	4
Inner membrane	32	17
Nucleus		
Inner membrane**	0.2	0.7
Secretory vesicle membrane	Not determined	3
Lysosome membrane	0.4	Not determined
Peroxisome membrane	0.4	Not determined
Endosome membrane	0.4	Not determined

*These two cells are of very different sizes: the average hepatocyte has a volume of about 5000 μm^3 compared with 1000 μm^3 for the pancreatic exocrine cell. Total cell membrane areas are estimated at about 110,000 μm^2 and 13,000 μm^2, respectively.
**The outer nuclear membrane is included in the measurement of the rough ER and is roughly equal to the inner membrane.

membrane is required to make them. In liver and pancreatic cells, for example, the endoplasmic reticulum has a total membrane surface area that is, respectively, 25 times and 12 times that of the plasma membrane (**Table 12–2**). The membrane-enclosed organelles are packed tightly in the cytoplasm, and, in terms of area and mass, the plasma membrane is only a minor membrane in most eukaryotic cells (**Figure 12–2**).

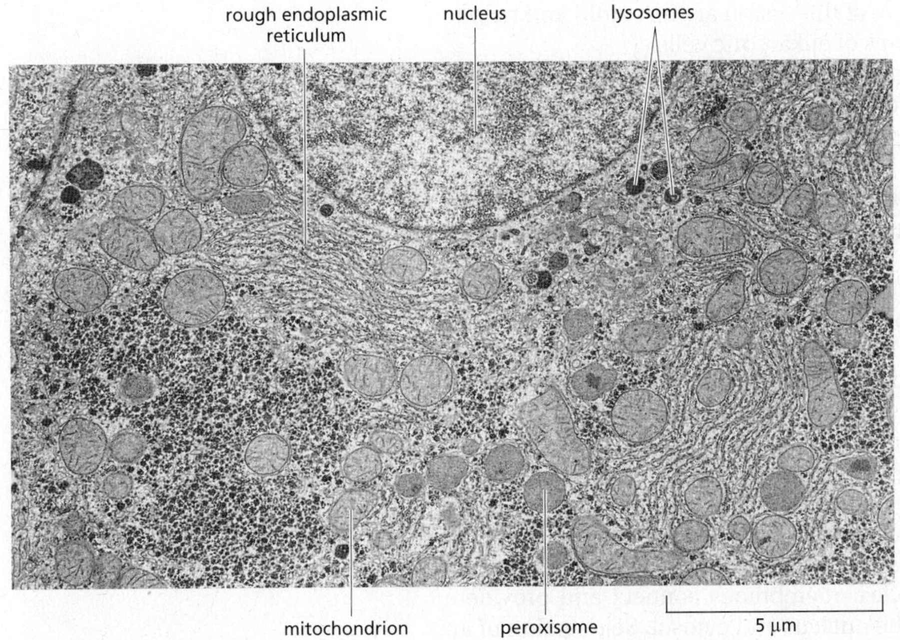

Figure 12–2 An electron micrograph of part of a liver cell seen in cross section. Examples of most of the major intracellular organelles are indicated. (Reused by permission of E.L. Bearer and Daniel S. Friend.)

In general, each membrane-enclosed organelle performs the same set of basic functions in all cell types. But to serve the specialized functions of cells, these organelles vary in abundance and can have additional properties that differ from cell type to cell type. This is particularly apparent in cells that are highly specialized and therefore disproportionately rely on specific organelles. Plasma cells, for example, which daily secrete their own weight in antibody molecules into the bloodstream, contain vastly amplified amounts of rough ER, which is found in large, flat sheets. Cardiac muscle cells instead expand and specialize their smooth ER for Ca^{2+} storage and proliferate their mitochondria for energy production. Moreover, membrane-enclosed organelles are often found in characteristic positions in the cytoplasm. In most cells, for example, the Golgi apparatus is located close to the nucleus, whereas the network of ER tubules extends from the nucleus throughout the entire cytosol. These characteristic distributions depend on interactions of the organelles with the cytoskeleton (discussed in Chapter 16).

Evolutionary Origins Explain the Topological Relationships of Organelles

To understand the relationships between the compartments of the cell, it is helpful to consider how they might have originated. The precursors of the first eukaryotic cells are thought to have been relatively simple cells that—like most bacterial and archaeal cells—had a plasma membrane but no internal membranes. The plasma membrane in such cells provided all membrane-dependent functions, including the pumping of ions, ATP synthesis, protein secretion, and lipid synthesis. These ancestral precursors, like their modern-day prokaryotic counterparts,probably had a 1000- to 10,000-fold smaller volume than present-day eukaryotic cells. To increase in volume, the ancestral cells would have needed to maintain their surface area to volume ratio to sustain the many vital functions that membranes perform.

On the basis of the appearance of modern-day archaeal cells (see Figure 1–26), the membrane surface area might have initially increased by plasma membrane protrusions. The increased capacity to exchange metabolites with the surrounding environment via these protrusions would have facilitated symbiotic relationships with other organisms. Increased resource availability due to a combination of symbioses and membrane expansion may have allowed the evolution of progressively larger cells (**Figure 12–3**). Ultimately, the network of spaces between the numerous expanded protrusions would have become sealed off from the surrounding environment because of membrane fusion between protrusions. The consequences of this fusion are threefold and help to explain the major distinguishing features of eukaryotic cells.

First, the cell now has a set of internal membranes that are derived from an ancestral prokaryotic plasma membrane. These internal membranes enclose interior spaces that are said to be *topologically equivalent* to each other and to the exterior of the cell (**Figure 12–4**), because they can communicate with one another, in the sense that molecules can get from one to the other without having to cross a membrane. We shall see that this topological relationship holds for all of the organelles involved in the secretory and endocytic pathways, including the ER, Golgi apparatus, endosomes, lysosomes, and peroxisomes. As we discuss in detail in the next chapter, their interiors communicate extensively with one another and with the outside of the cell via *transport vesicles,* which bud off from one organelle and fuse with another. In this way, proteins that enter the **lumen** of the ER can be secreted outside the cell.

Second, the ancestral plasma membrane that surrounded the genome is now an internal membrane that becomes the *inner nuclear membrane*. Because of how it originated, the inner nuclear membrane is continuous with other plasma membrane–derived internal membranes, including the *outer nuclear membrane*. Specialized structures, the *nuclear pore complexes*, are located at points where the inner and outer nuclear membranes connect and provide a conduit for communication between the nucleus and cytosol. Segregation of an

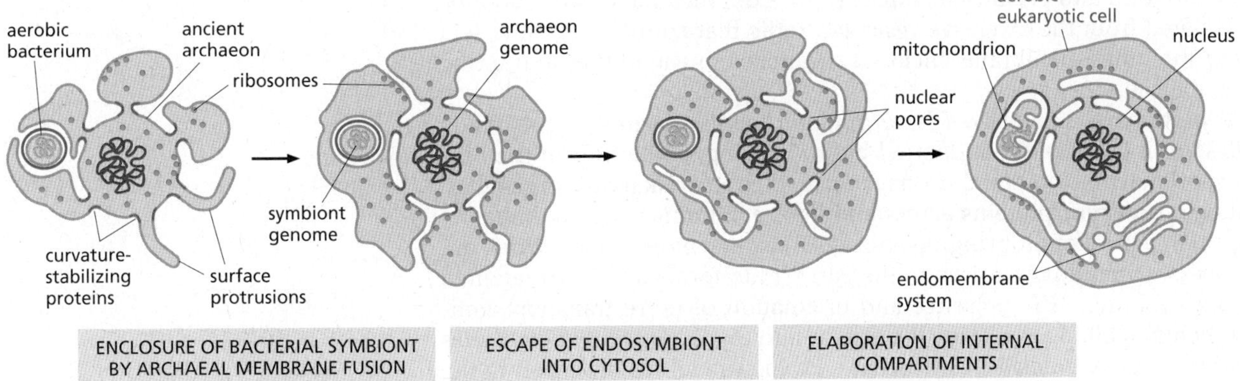

Figure 12–3 **Evolutionary origins of the major internal membrane systems of a eukaryotic cell.** As discussed in Chapter 1, there is evidence that the first eukaryotic cells arose when an ancient anaerobic archaeon joined forces with an aerobic bacterium roughly 1.6 billion years ago. An early step in this process was expansion of the archaeon's plasma membrane, probably through protrusions and blebs. The highly curved membrane at the necks of these protrusions might have been stabilized by proteins that eventually became part of the nuclear pore. The added surface area of these protrusions facilitated metabolite exchange with the environment and with neighboring cells. A fruitful symbiotic relationship with an aerobic bacterium might have allowed the archaeon to increase in volume. These protrusions eventually fused with each other to pinch off internal membrane-enclosed compartments, some of which contained the symbiotic bacteria. This intermediate now begins to resemble modern-day eukaryotes, with a primordial nucleus and nuclear pores, internal compartments, and an endosymbiont destined to become the mitochondrion. The lumen of the internal compartments is topologically equivalent to the extracellular space (see Figure 12–4). The membrane-enclosed endosymbiont subsequently escaped the enclosing membrane into the cytosol where it evolved into modern-day mitochondria. The internal compartments expanded and became progressively specialized to form the major intracellular compartments of a eukaryotic cell. Their common origin from a primordial intracellular compartment explains why all of these compartments can exchange material with each other through vesicular transport. The nucleus was formerly the cytosol in the ancient archaeon, explaining why the cytosol and nucleus are topologically equivalent compartments that can intermix during mitosis. (Adapted from J. Martijn and T.J.G. Ettema, *Biochem. Soc. Trans.* 41:451–457, 2013; D. Baum and B. Baum, *BMC Biol.* 12:76, 2014.)

organism's genetic material into a nucleus separate from the plasma membrane probably afforded greater protection from the environment. Furthermore, an expanded cytosol segregated from the nucleus would have facilitated the spatial separation of transcription from translation, thereby allowing greater regulation of gene expression by several mechanisms distinctive to eukaryotic cells.

Third, symbionts that were originally outside the cell were trapped inside the cell and became endosymbionts. At some point, endosymbionts escaped from their membrane enclosure into the cytosol where they eventually became mitochondria and plastids that contain their own genomes. The nature of these genomes and the close resemblance of the proteins in these organelles to those in some present-day bacteria provide strong evidence for their endosymbiont origins (see Figure 14–55). Like the bacteria from which they were derived,

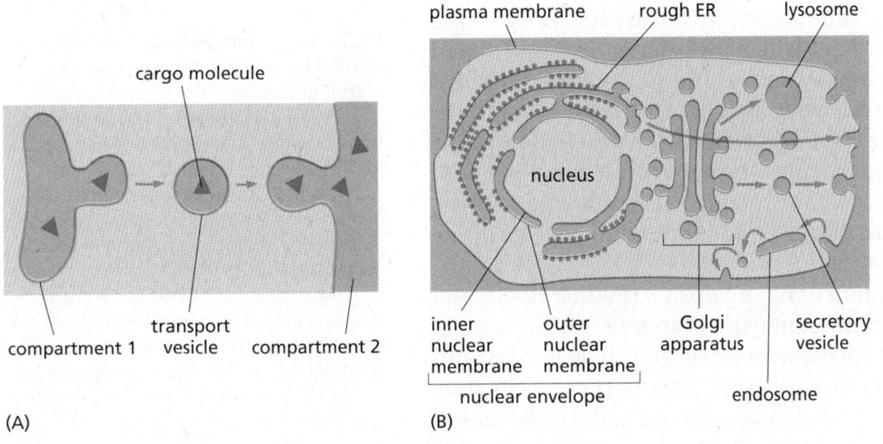

Figure 12–4 **Topologically equivalent compartments in the secretory and endocytic pathways in a eukaryotic cell.** Topologically equivalent spaces are shown in *red*. (A) Molecules can be carried from one compartment to another topologically equivalent compartment by transport vesicles that bud from one and fuse with the other. (B) In principle, cycles of membrane budding and fusion permit the lumen of any of the organelles shown to communicate with any other and with the cell exterior by means of transport vesicles. *Blue arrows* indicate the extensive outbound and inbound vesicular traffic (discussed in Chapter 13). Some organelles, most notably mitochondria and (in plant cells) plastids, do not take part in this communication and are isolated from the vesicular traffic between organelles shown here.

both mitochondria and plastids are enclosed by a double membrane, and they remain isolated from the extensive vesicular traffic that connects the interiors of most of the other membrane-enclosed organelles to each other and to the outside of the cell.

The evolutionary schemes we have outlined for the origins of eukaryotic organelles are most strongly supported by the striking similarities of the protein transport machinery of modern-day prokaryotes and eukaryotic organelles. The ability to transport proteins across and into membranes is a fundamental and essential feature of all living organisms. Thus, machinery that carries out these processes would have arisen in the earliest life-forms and been retained throughout evolution. The presence and orientation of these transport components therefore allow us to trace the origins and topology of the membranes within which they now reside. Consistent with the model for evolution of the endomembrane system of eukaryotic cells, the components that mediate protein import into the ER are homologous to the proteins that mediate export across the archaeal plasma membrane. Similarly, membrane protein insertion machinery in the outer and inner membranes of mitochondria and plastids contains homologous components found in the outer and inner membranes of various modern-day bacteria.

The major intracellular compartments in eukaryotic cells can therefore be categorized into three distinct families: (1) the nucleus and the cytosol, which are topologically equivalent (although functionally distinct) and connected by nuclear pore complexes; (2) all organelles that function in the secretory and endocytic pathways—including the ER, Golgi apparatus, endosomes, lysosomes, and the transport vesicles that move between them—and peroxisomes; (3) the endosymbiont-derived organelles: mitochondria and the plastids (in plants only).

Macromolecules Can Be Segregated Without a Surrounding Membrane

A membrane barrier is not the only way subsets of macromolecules can selectively segregate within cells. As we discussed in Chapter 3, one or more interacting proteins or nucleic acids can serve as *scaffolds* in *biomolecular condensates* (see Figure 3–77). These scaffold macromolecules create the condensate through multiple weak, fluctuating binding interactions among themselves; in addition, they recruit specific proteins and nucleic acids into the condensate as *client macromolecules* (**Figure 12–5**). Once recruited, the clients typically remain within the condensate because the local concentration of its binding sites on the interacting scaffold molecule is very high. Thus, when the client dissociates from a scaffold molecule, it is more likely to re-bind to another site on the scaffold molecule or to a neighboring one within the condensate than to diffuse away altogether. In this way, a specific set of proteins and nucleic acids can be concentrated into a cellular structure that excludes other surrounding macromolecules.

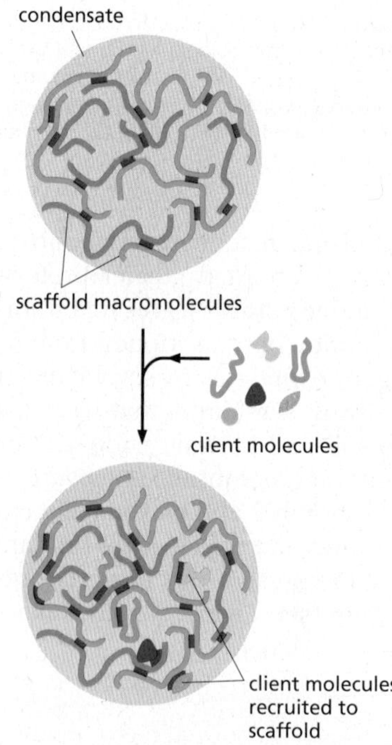

Figure 12–5 Biomolecular condensates formed by scaffold macromolecules recruit clients. As described in Chapter 3 (see Figure 3–77), a set of macromolecules that participate in weak, dynamic, and multivalent interactions (shown in *red*) with each other can form a biomolecular condensate (Movie 12.1). The macromolecules that directly participate in formation of the condensate are termed "scaffolds." The scaffold proteins and RNA can recruit other macromolecules, termed "clients," via specific interactions. Condensate formation does not depend on these clients, but they are part of the condensate because of their specific scaffold interactions.

The largest and most conspicuous condensate in eukaryotic cells is the *nucleolus*, the structure within the nucleus where ribosomes are assembled (**Movie 12.1**). The central scaffolding component of the nucleolus is nascent pre-rRNA that is actively transcribed from arrays of rRNA genes. Nascent pre-rRNA recruits numerous proteins and small nucleolar RNAs (snoRNAs) required for pre-rRNA processing. These macromolecules further recruit other scaffold proteins—plus clients that include ribosomal proteins, assembly chaperones, and modification enzymes. In total, more than 400 proteins and RNAs contribute to the formation of this enormous condensate.

Biomolecular condensates can be found in all organisms (**Figure 12–6**), and eukaryotic cells contain a dozen or more different types (**Table 12–3**). The sizes of the known condensates range from ~50 nm in diameter (slightly bigger than ribosomes) to a micrometer or more in the case of nucleoli. These structures include many different types of ribonucleoprotein condensates (some in the nucleus and

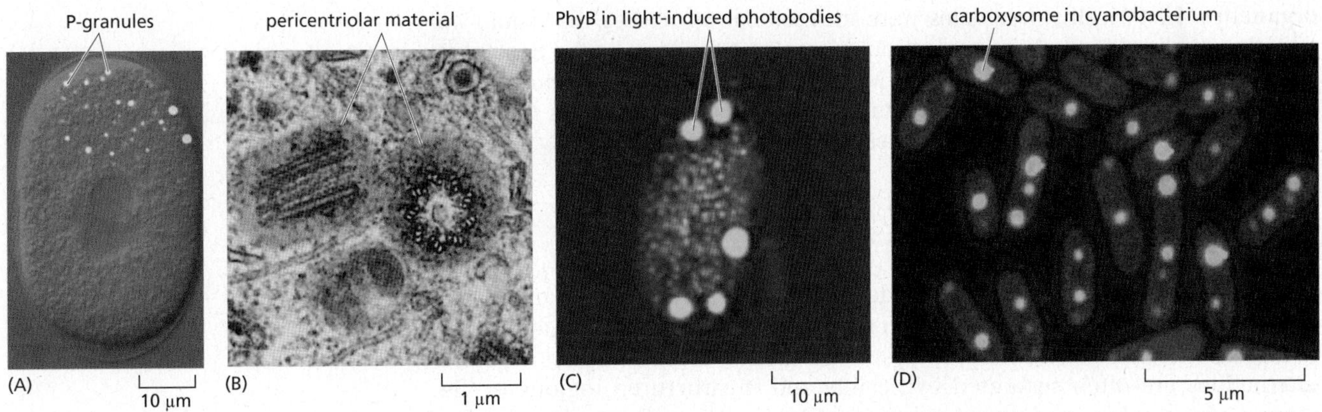

Figure 12–6 **Examples of biomolecular condensates in different organisms.** (A) Fluorescent image of the nematode *Caenorhabditis elegans* at the single-cell stage showing P-granules that are asymmetrically distributed and inherited at cell division. P-granules contain certain mRNAs and other macromolecules. (B) The pericentriolar material in the centrosome that nucleates the assembly of microtubules in an animal cell as seen by electron microscopy. (C) Green fluorescent protein (GFP)-labeled PhyB concentrates in condensates known as photobodies in the plant nucleus when the cells are exposed to bright light. Photobodies might be sites of light-mediated signaling and gene regulation. (D) Rubisco and carbonic anhydrase are concentrated in carboxysomes *(green)* in photosynthetic bacteria (cyanobacteria). Chlorophyll is shown in *red*. Carboxysomes are analogous to the pyrenoids of algae and facilitate carbon fixation. (A, from P. Brangwynne et al., *Science* 324:1729–1732, 2009. B, from M. McGill et al., *J. Ultrastruct. Res.* 57:43–53, 1976. C, from E.K. Van Buskirk et al., *Plant Physiol*. 158:52–60, 2012. D, from Y. Fang et al., *Front. Plant Sci*. 9:article 739, 2018. Courtesy of Liu Luning.)

others in the cytoplasm), a variety of different signaling protein clusters that can form tethered to the plasma membrane, and the "biochemical factories" that form as needed to catalyze DNA repair, DNA replication, and DNA transcription in the nucleus.

As illustrated by the nucleolus, each type of condensate is enriched for a characteristic complement of proteins (and in many cases, nucleic acids) that interact with each other to maintain the condensate's identity and integrity. The specificity of at least a subset of macromolecular interactions within the condensate ensures that it remains distinct in its composition and function. Thus, biomolecular condensates and membrane-enclosed compartments represent two different mechanisms that are used by eukaryotic cells to segregate subsets of macromolecules that execute specialized biochemistry (see Table 3-3). Because of this conceptual similarity, condensates are sometimes referred to as **membraneless**

TABLE 12–3 **Examples of Eukaryotic Biomolecular Condensates**

Biomolecular condensate	Location	Proposed associated function(s)
Nucleolus	Nucleus	rRNA transcription and ribosome assembly
Pyrenoid	Chloroplast	Carbon fixation from CO_2 in algae
Stress granules	Cytosol	Temporary storage, particularly of translation-related components
P-granules	Cytosol	RNA metabolism and inheritance
Balbiani body	Cytosol	Localization and inheritance of mRNAs and organelles
Cajal body	Nucleus	mRNA processing
Paraspeckles	Nucleus	Regulation of gene expression
RNA transport granule	Neuron	RNA localization to subcellular locations in development and in neurons
PML body	Nucleus	Storage of nuclear factors; regulation of gene expression
Postsynaptic density	Dendrite	Organization of macromolecules needed for neuronal transmission

organelles. Historically, organelles were intracellular structures that could be directly visualized in the light or electron microscope. This is why the nucleolus and centrosome are called organelles. Most condensates are not organelles by this historical definition; nevertheless, they are cellular structures that segregate and concentrate specific macromolecules.

Multivalent Interactions Mediate Formation of Biomolecular Condensates

The formation of a biomolecular condensate requires that at least one of its constituent macromolecules engage in a set of weak, multivalent interactions with either itself or other constituents (see Figure 3–77A). The sites of these interactions are often separated by flexible and unstructured regions of the macromolecule. For example, nascent pre-rRNA in the nucleolus is a flexible molecule that binds a variety of clients and other scaffold molecules at numerous points along its length. Similarly, the scaffolding proteins that generate condensates of signaling proteins under the plasma membrane typically contain multiple protein–protein interaction domains separated by flexible *intrinsically disordered regions* that lack secondary structure. Experiments with artificial multivalent proteins have shown that this is the minimal element needed to drive condensate formation in a test tube and in cells.

Each individual interaction within a condensate is often very weak, allowing the macromolecules to rapidly exchange their relative positions. These dynamic and frequent rearrangements, together with the structural flexibility of many condensate constituents, means that the molecules within the condensate are highly mobile and do not have fixed positions relative to each other. This property causes the condensate to behave as a liquid. As discussed in Chapter 3 (pp. 171–173), despite this liquidlike property, the condensate does not dissolve into its surroundings because the interaction energy within the condensate offsets the entropy that would be gained if the molecules were dispersed. This is how the condensate can remain a liquid that stably resides within another liquid (the cytosol), a phenomenon termed "liquid–liquid phase separation."

Different noncovalent chemical bonds can form the weak interactions between macromolecules that drive condensate formation. Cation–pi interactions, pi–pi interactions, charge–charge interactions, short regions of crossed β sheets, and short stretches of nucleic acid base-pairing can all contribute to condensate formation. The key requirement is that the interactions have sufficient binding energy to offset the loss of entropy caused by association, while being sufficiently dynamic to give the condensate a liquid character. When these requirements are met, condensates have a spherical shape, deform and flow under shear force, and can undergo fusion and fission.

If the interactions in a condensate become less dynamic as incrementally more stable interactions form over time, its properties can change to that of a gel and eventually a solid where the binding interactions remain fixed. There is a continuum across the spectrum of physical properties that characterize different condensates. Cells can exploit such differences by forming a condensate within a condensate. This can occur if a subset of macromolecules within a condensate has slightly higher affinity among its macromolecules than for other macromolecules of the condensate. This subset then forms a new condensate with distinct physical properties (**Figure 12–7**). This is how the nucleolus is thought to be segregated into morphologically different concentric shells, each of which is enriched for subsets of nucleolar proteins dedicated to different aspects of ribosome assembly.

Biomolecular Condensates Create Biochemical Factories

The fluctuating network of interactions among the macromolecules inside of a condensate excludes other macromolecules from the surrounding environment. By contrast, nucleotides, metabolites, cofactors, and other small

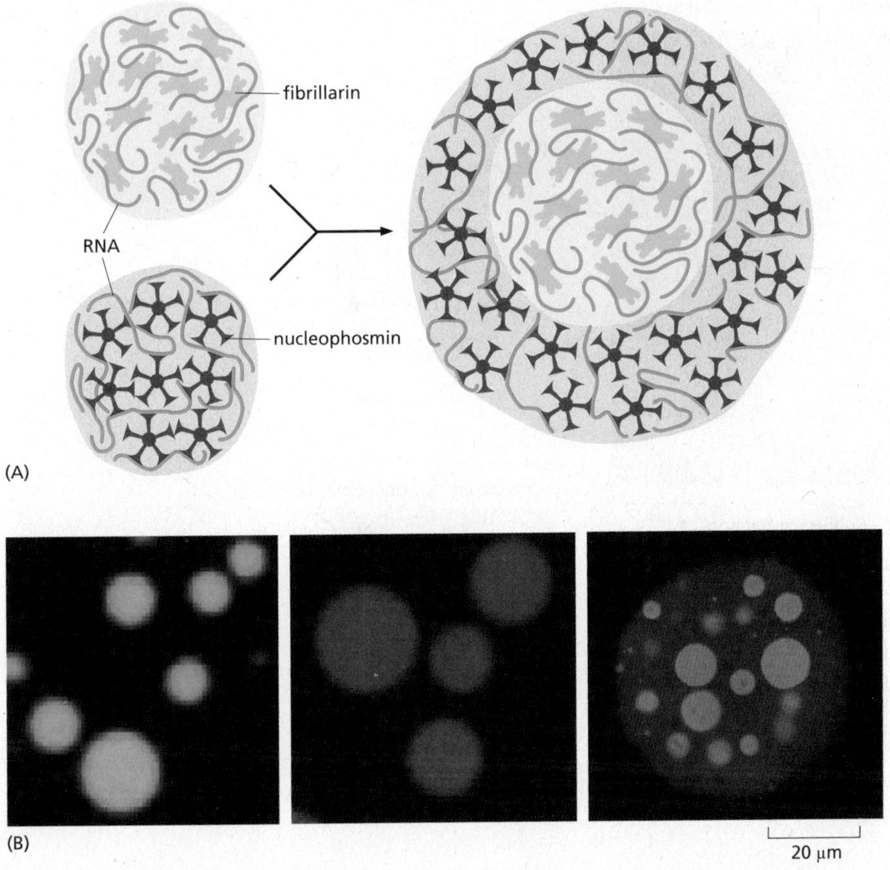

Figure 12–7 **Condensates with different properties can coexist as part of a larger structure.** The nucleolus is a condensate that is composed of three morphologically and functionally distinct regions, one inside the other, each formed by a different set of scaffold macromolecules. (A) In the experiment shown, the scaffolds from two nucleolar substructures have been purified—fibrillarin from the nucleolus's fibrillar component and nucleophosmin from its granular component. Both of these scaffold proteins contain binding sites for RNA, and when mixed with RNA in a test tube, each purified scaffold will assemble into an RNA–protein condensate, as illustrated at *left*. However, as illustrated at *right*, when mixed together they instead form a multilayered structure that has one type of condensate encased inside the other type of condensate. (B) The condensates, formed either separately or together, viewed by fluorescence microscopy; fibrillarin is *green* and nucleophosmin is *red*. (Adapted from M. Feric et al., *Cell* 165:1686–1697, 2016.)

molecules can rapidly diffuse into the condensate where they can engage with the enzymes that reside there. The product of these enzymes within the condensate can be used by other enzymes that are coresident in the condensate before the product diffuses away. In this way, multistep reactions can be accelerated to rates beyond that possible without co-segregation of the enzymes inside a condensate.

Consider for example, the *pyrenoid*, a complex structure found in many algae that contains a condensate enriched in the enzyme ribulose 1,5-bisphosphate carboxylase/oxygenase (Rubisco) and a pyrenoid-specific scaffolding protein. The Rubisco-containing condensate is interwoven with membrane tubules. Carbonic anhydrase inside the membrane tubule converts bicarbonate (HCO_3^-) into CO_2, which Rubisco uses to carboxylate ribulose 1,5-bisphosphate (**Figure 12–8**). This carboxylation reaction is a critical early step in carbon fixation during photosynthesis (discussed in Chapter 14). If Rubisco were not in a condensate in close proximity to carbonic anhydrase, the low free CO_2 concentration combined with a competition by oxygen for Rubisco's active site would favor reaction with oxygen (termed "photorespiration"; see Chapter 14, p. 847) over carbon fixation. Land plants do not need a pyrenoid to fix carbon because

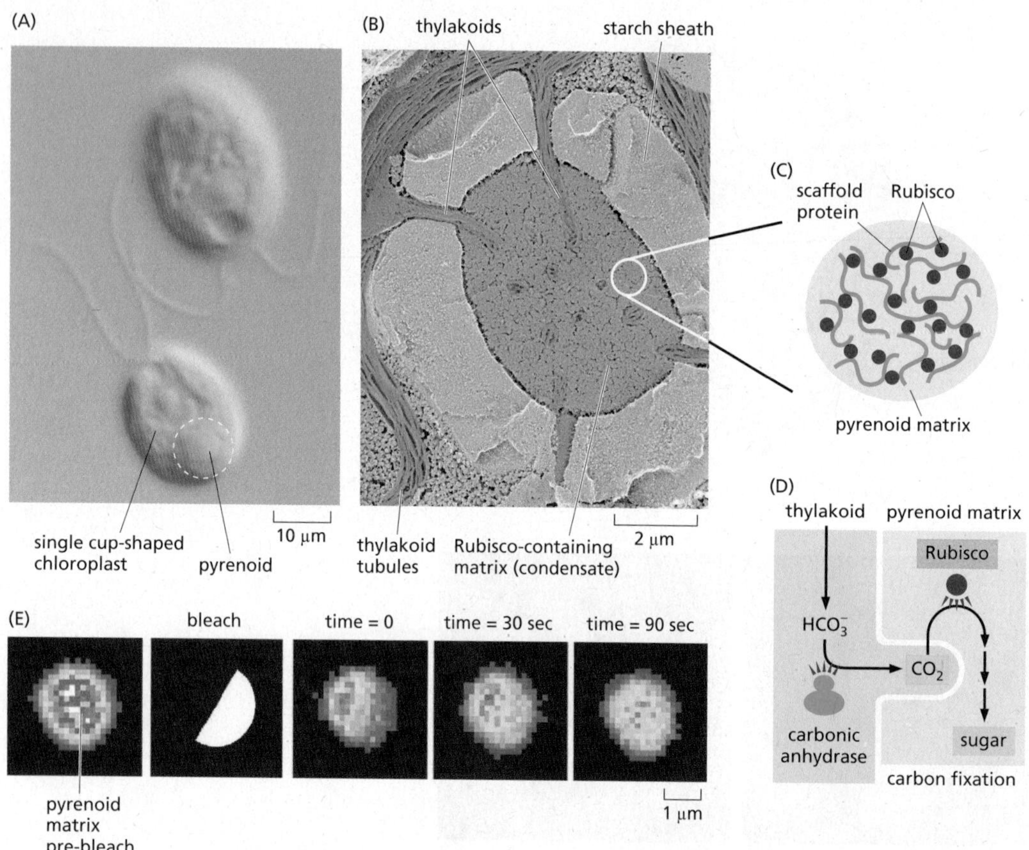

Figure 12–8 **The pyrenoid contains a condensate of Rubisco that exploits locally generated CO_2.** (A) The pyrenoid as seen by light microscopy of the single-celled alga *Chlamydomonas reinhardtii*. (B) Scanning electron micrograph of the *C. reinhardtii* pyrenoid showing how the Rubisco-containing condensate is penetrated by tubules from the surrounding thylakoid. (C) Simplified depiction of the condensate containing the enzyme Rubisco and the multivalent scaffold protein EPYC1. (D) The tubules of thylakoid membranes in the pyrenoid contain the enzyme carbonic anhydrase. HCO_3^- is used by carbonic anhydrase to generate CO_2, which is used by the surrounding Rubisco in the pyrenoid matrix. The use of locally generated CO_2 is thought to favor the carboxylation reaction performed by Rubisco rather than the competing oxygenation reaction that would be favored when O_2 is used by Rubisco instead of CO_2. (E) Experiment demonstrating that the contents inside the pyrenoid matrix are highly dynamic. Shown is a pseudocolored fluorescence image of a pyrenoid containing a Rubisco subunit tagged with a fluorescent protein. *Red* and *white* indicate the areas of brightest fluorescence, and *blue* indicates areas of dim fluorescence. The right half of the pyrenoid was photobleached, after which fluorescence was monitored over time. Fluorescent molecules from the non-bleached half intermix with the bleached molecules within 90 seconds. (A, from Carnegie Institution for Science. B, courtesy of Ursula Goodenough. E, adapted from E.S. Freeman Rosenzweig et al., *Cell* 171:148–162, 2017.)

CO_2 in the air is more plentiful than the very low concentration of dissolved CO_2 available as HCO_3^- to algae living in aquatic environments. The pyrenoid illustrates how the segregation of sequential biochemical reactions into a condensate can both speed reactions and minimize alternate off-pathway outcomes. In the same way, carrying out the highly complex and ordered process of ribosome assembly within the nucleolus prevents unwanted side reactions while promoting the desired ones.

Scientists can produce artificial condensates that contain a desired set of macromolecules by engineering them with multivalent interaction domains. This approach can be used to experimentally enhance the efficiency of an otherwise unfavorable reaction. In one experiment, all of the factors required to misread a UAG stop codon as a sense codon were engineered with artificial multivalent interaction modules to generate a condensate inside the cell. The UAG codon of the mRNA within this condensate was efficiently interpreted as a sense codon, while other mRNAs in the surrounding cytosol terminated at UAG codons. This experiment illustrates the minimal features needed to produce a condensate-based biochemical factory within a cell and that such features can be rationally designed and engineered.

Biomolecular Condensates Form and Disassemble in Response to Need

As we have seen, the formation and stability of a biomolecular condensate rely on weak interactions among its constituents overcoming the entropy of a well-mixed system. This means that even small changes in the strength of interactions can influence the formation and physical properties of the condensate. The formation and dissolution of a condensate can therefore be readily controlled by changing the strength of the multivalent interactions that mediate its assembly. This is often accomplished by post-translational modifications, such as phosphorylation, and this mechanism is commonly used to rapidly form and disassemble large signaling clusters at the plasma membrane (**Figure 12–9**). Condensate formation and disassembly can also be induced by a change in a cellular condition such as temperature, pH, or osmolarity. The reversibility of condensate formation is used by cells to regulate condensates in response to need, and it affords the cell flexibility and speed in adapting to changing needs.

For example, the condensates called *stress granules* only form during certain types of cellular stress, and they dissolve when the stress is alleviated. These condensates are enriched in translationally inactive mRNAs, various translation factors, ribosomal subunits, and various RNA-binding proteins. They

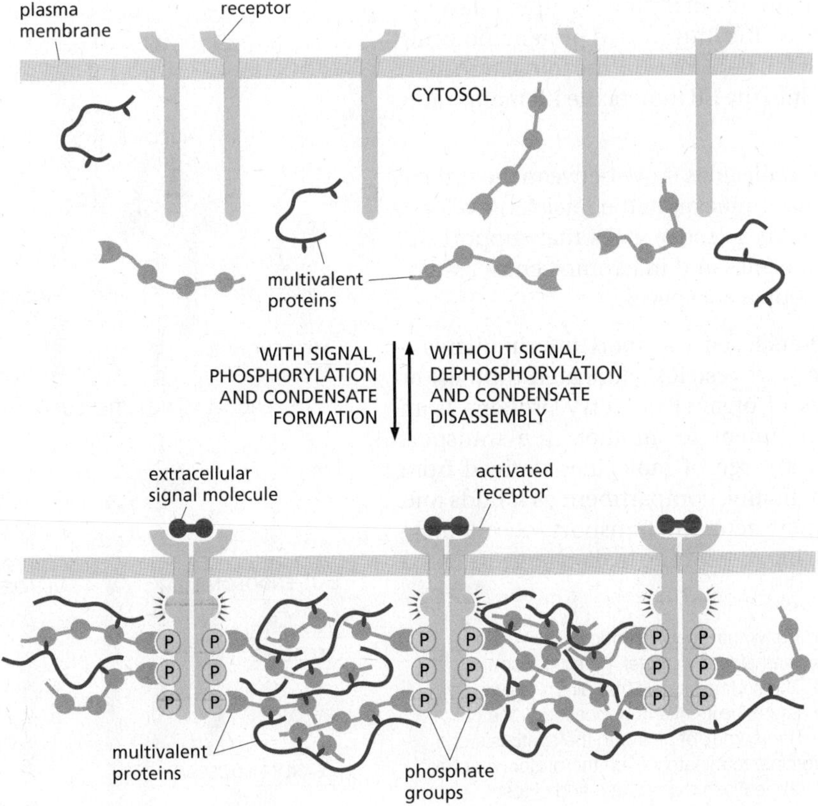

Figure 12–9 Phosphorylation regulates the formation and dissolution of condensates during signaling. When a receptor at the plasma membrane is engaged by its ligand, its cytosolic tail and associated proteins become phosphorylated. This modification, along with surrounding amino acids, forms a specific binding site for various cytosolic and membrane proteins, many of which are multivalent. The multivalent proteins interact with each other to drive the formation of a condensate that has distinctive signaling properties. When the key sites become dephosphorylated, the condensate disassembles and signaling stops. Examples of signaling clusters that form and disassemble in response to ligand are discussed in Chapter 15.

form when a block in translation initiation exposes mRNA regions that would normally be covered by translating ribosomes. When these mRNAs become exposed, they can interact with each other and with RNA-binding proteins to nucleate a condensate. It is thought that the resulting condensates serve as a storage depot for these mRNAs and factors when they are not being actively used. By temporarily sequestering these macromolecules during stress rather than degrading them, the cell can avoid the need to produce them *de novo* once the stress has been resolved.

Proteins Can Move Between Compartments in Different Ways

Nearly all proteins, except a few inside mitochondria and plastids, begin their synthesis on ribosomes in the cytosol. The final location of each protein depends on its amino acid sequence, which can contain one or more **sorting signals** that direct its delivery to different parts of the cell. The sorting signals in the transported protein are recognized by complementary **sorting receptors** that mediate movement between compartments. By contrast, proteins that do not have any sorting signals remain in the cytosol as permanent residents. There are four fundamentally different ways a protein is moved from one compartment to another. These four mechanisms are described below, and the transport steps at which they operate are outlined in Figure 12–10. We discuss protein translocation and gated transport in this chapter, vesicular transport in Chapter 13, and engulfment in both this chapter and the next.

1. In **protein translocation**, transmembrane *protein translocators* directly transport specific proteins from the cytosol into a space that is topologically distinct: either the other side of a membrane or within the lipid bilayer in the case of integral membrane proteins. The transported protein molecule usually must unfold to snake through the translocator. The initial transport of selected proteins from the cytosol into the ER lumen, the ER membrane, or mitochondria occurs in this way.

2. In **gated transport**, proteins and RNA molecules move between the cytosol and the nucleus through nuclear pore complexes in the nuclear envelope. The nuclear pore complexes function as selective gates that support the active transport of specific macromolecules and macromolecular assemblies between the two topologically equivalent spaces.

3. In **vesicular transport**, membrane-enclosed transport intermediates—which may be small, spherical transport vesicles, elongated tubules, or larger, irregularly shaped fragments of organelles—ferry proteins from one topologically equivalent compartment to another. The transport intermediate becomes loaded with a cargo of molecules derived from the lumen and membrane of the originating compartment as it buds and pinches off. At the destination compartment, the transport intermediate fuses with the compartment's enclosing membrane to discharge its cargo

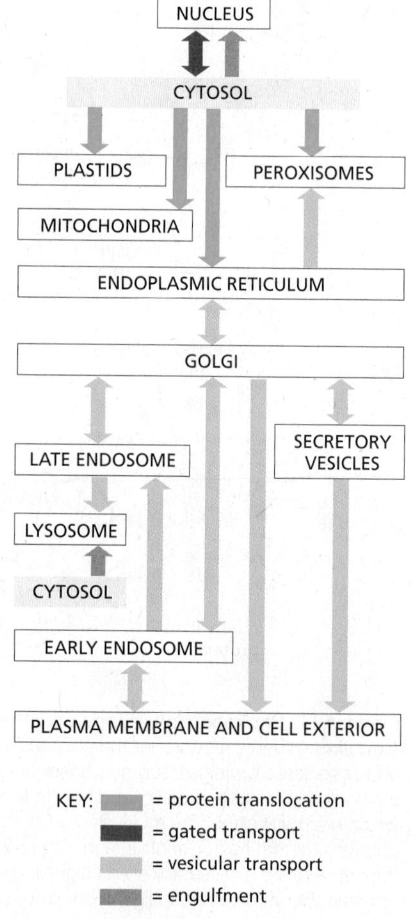

Figure 12–10 **A simplified "road map" of protein traffic within a eukaryotic cell.** Proteins can move from one compartment to another by protein translocation *(blue)*, gated transport *(red)*, vesicular transport *(green)*, or engulfment *(gray)*. The sorting signals that direct a given protein's movement through the system, and thereby determine its eventual location in the cell, are contained in each protein's amino acid sequence. The journey begins with the synthesis of a protein on a ribosome in the cytosol and, for many proteins, terminates when the protein reaches its final destination. Other proteins shuttle back and forth between the nucleus and cytosol. At each intermediate station *(boxes)*, a decision is made as to whether the protein is to be retained in that compartment or transported further. A sorting signal may direct either retention in or exit from a compartment. A special transport process termed "engulfment" is used to move proteins from the cytosol into the lysosome in autophagy or used to enclose chromosomes inside the nucleus during nuclear envelope re-formation after mitosis. The movement of macromolecules into and out of a condensate is not shown here. This process does not involve crossing a membrane barrier and is mediated by direct physical interactions among the macromolecules that form the condensate. We shall refer to this figure often as a guide in this chapter and the next, highlighting in color the particular pathway being discussed.

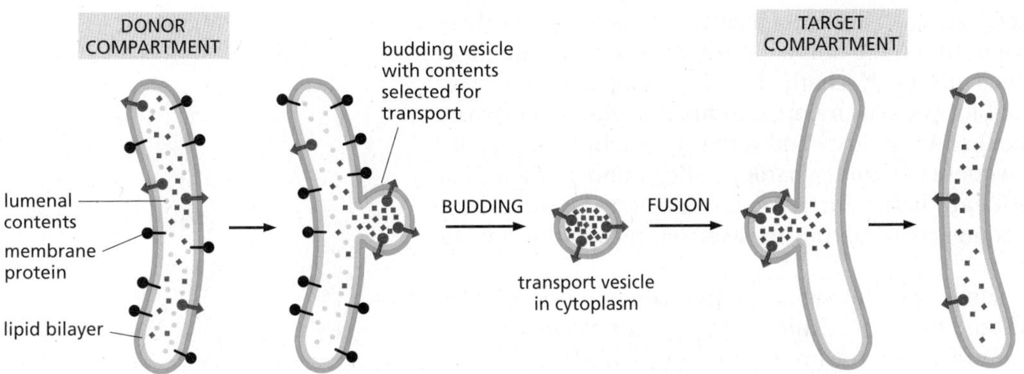

Figure 12–11 **Vesicle budding and fusion during vesicular transport.** Transport vesicles bud from one compartment (donor) and fuse with another topologically equivalent (target) compartment. In the process, a subset of soluble components *(red dots)* are transferred from lumen to lumen. Note that membrane is also transferred and that the original orientation of both proteins and lipids in the donor compartment membrane is preserved in the target compartment membrane. Thus, membrane proteins retain their asymmetrical orientation, with the same domains always facing the cytosol.

(Figure 12–11). The transfer of soluble and membrane-embedded proteins from the ER to the Golgi apparatus, for example, occurs in this way. The proteins transported by vesicular transport never cross a membrane during the process, and therefore retain their topological relationships within the cell.

4. In **engulfment**, such as autophagy (discussed in Chapter 13), double-membrane sheets wrap around portions of the cytoplasm often including fragments of organelles or even entire organelles (Figure 12–12). This membrane structure then seals by membrane fusion to enclose a separate compartment, the autophagosome. The re-formation of the nuclear envelope after mitosis (discussed later in this chapter) follows a conceptually similar process. ER tubes and sheets wrap around decondensing chromosomes and then fuse laterally with one another to form a sealed double-membrane envelope only traversed by the nuclear pores.

In addition to these mechanisms for protein movement into and between membrane-enclosed compartments, a simpler mechanism based on direct physical binding is used by macromolecules to enter biomolecular condensates. In this mechanism, the macromolecule specifically binds to another protein or RNA that is already part of the condensate to which it is specifically recruited. Once recruited, the macromolecule remains within the condensate because of persistent and repeated interactions with its partner. The interaction between a macromolecule and its binding partner in the condensate is analogous to the interaction between a sorting signal and its cognate sorting receptor; in both cases, the interaction specifies the macromolecule's destination.

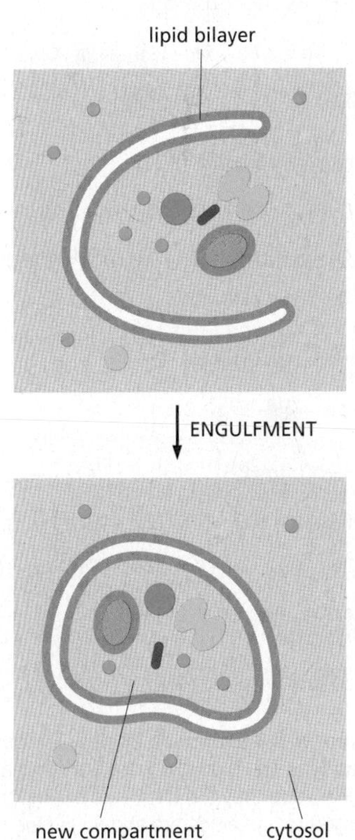

Figure 12–12 **Formation of a new compartment by engulfment of contents inside of a membrane.**

Sorting Signals and Sorting Receptors Direct Proteins to the Correct Cell Address

Sorting signals are usually composed of amino acid side chains in a protein and come in two general varieties: a linear sequence of amino acids (called a signal sequence) or a specific three-dimensional arrangement of amino acids (called a signal patch). Sorting signals for protein translocation into organelles are linear **signal sequences**, while examples of linear signals and **signal patches** are known for nuclear and vesicular transport. The linear signal sequences for protein translocation are often found at the N-terminus of the polypeptide chain. These N-terminal signal sequences are usually removed from the finished protein by specialized **signal peptidases** once the sorting process is complete. Other types of signal sequences are not removed and remain part of the final mature protein.

Each signal sequence specifies a particular destination in the cell. The signal sequence for initial transfer to the ER usually includes a linear sequence of about 5–10 predominantly hydrophobic amino acids. Many of these proteins will in turn pass from the ER to the Golgi apparatus, but those with a specific signal sequence of four amino acids at their C-terminus are recognized as ER residents and are returned to the ER. Proteins destined for mitochondria have signal sequences of yet another type, in which positively charged amino acids alternate with hydrophobic ones. The signal for protein import into the nucleus is composed primarily of positively charged amino acids. Finally, many proteins destined for peroxisomes have a signal sequence of three characteristic amino acids at their C-terminus. A sorting signal for any particular destination needs to be sufficiently distinctive from all other sequences to permit its selective recognition by the appropriate sorting receptor.

Figure 12–13 presents some specific signal sequences. Experiments in which the peptide is transferred from one protein to another by genetic engineering techniques have demonstrated the importance of each of these signal sequences for protein targeting. Placing the N-terminal ER signal sequence at the beginning of a cytosolic protein, for example, redirects the protein to the ER; removing or mutating the signal sequence of an ER protein causes its retention in the cytosol. Signal sequences are therefore both necessary and sufficient for protein targeting. Even though their amino acid sequences can vary greatly, the signal sequences of proteins having the same destination are often functionally interchangeable; in these instances, physical properties, such as hydrophobicity, are more important in the signal-recognition process than the exact amino acid sequence.

import into nucleus

– Pro – Pro – Lys – Lys – Lys – Arg – Lys – Val –

export from nucleus

– Met – Glu – Glu – Leu – Ser – Gln – Ala – Leu – Ala – Ser – Ser – Phe –

import into mitochondria

N – Met – Leu – Ser – Leu – Arg – Gln – Ser – Ile – Arg – Phe – Phe – Lys – Pro – Ala – Thr – Arg – Thr –

Leu – Cys – Ser – Ser – Arg – Tyr – Leu – Leu –

import into plastids

N – Met – Val – Ala – Met – Ala – Met – Ala – Ser – Leu – Gln – Ser – Ser – Met – Ser – Ser – Leu – Ser –

Leu – Ser – Ser – Asn – Ser – Phe – Leu – Gly – Gln – Pro – Leu – Ser – Pro – Ile – Thr – Leu – Ser – Pro –

Phe – Leu – Gln – Gly –

import into peroxisomes

– Ser – Lys – Leu – C

import into ER

N – Met – Met – Ser – Phe – Val – Ser – Leu – Leu – Leu – Val – Gly – Ile – Leu – Phe – Trp – Ala – Thr –

Glu – Ala – Glu – Gln – Leu – Thr – Lys – Cys – Glu – Val – Phe – Gln –

return to ER

– Lys – Asp – Glu – Leu – C

Figure 12–13 Examples of signal sequences that direct proteins to different intracellular locations. The primary characteristic features of each type of signal sequence are highlighted in color. Where they are known to be important for the function of the signal sequence, negatively charged amino acids are shown in *blue* and positively charged amino acids are shown in *red*. Similarly, important hydrophobic amino acids are shown in *green* and important uncharged polar amino acids are shown in *yellow*. N– indicates the N-terminus of a protein; –C indicates the C-terminus.

Sorting signals are recognized by complementary sorting receptors that guide proteins to their appropriate destination, where the receptors unload their cargo. The receptors function catalytically: after completing one round of targeting, they return to their point of origin to be reused. Most sorting receptors recognize classes of proteins rather than an individual protein species. They can therefore be viewed as public transportation systems, dedicated to delivering numerous different components to their correct locations in the cell.

Construction of Most Organelles Requires Information in the Organelle Itself

When a cell reproduces by division, it has to duplicate its chromosomes, its enclosing plasma membrane, and its organelles. In general, cells do this by expanding the plasma membrane and organelles with new proteins and lipids before division and segregation to the two daughter cells. The delivery of new proteins for growth of the ER, mitochondria, and plastids requires preexisting organelle-specific protein translocators. Because the incorporation of new protein translocators requires a preexisting protein translocator, a cell must already have at least some functional ER to make more ER; the same applies to mitochondria and plastids. Thus, two types of information are required to construct these organelles: the DNA that specifies an organelle's proteins and preexisting protein translocator(s) in the organellar membrane for incorporating new deliveries of protein. Both types of information are passed from parent cell to daughter cells to maintain the cell's compartmental organization.

Some organelles, such as lysosomes, acquire all of their proteins and membrane by vesicular transport from other organelles (see Chapter 13). Because it is possible, in principle, to construct such organelles *de novo*, they do not necessarily have to be inherited at cell division. Similarly, biomolecular condensates can be constructed *de novo* by self-assembly of the constituent proteins and nucleic acids. Thus, during cell division, a condensate can be disassembled, its constituents distributed stochastically among the two daughter cells, then reassembled into a condensate. This is how the nucleolus is acquired by daughter cells.

Summary

Eukaryotic cells contain intracellular membrane-enclosed organelles that make up nearly half the cell's total volume. The main ones present in all eukaryotic cells are the endoplasmic reticulum, Golgi apparatus, nucleus, mitochondria, lysosomes, endosomes, and peroxisomes; plant cells also contain plastids such as chloroplasts. All organelles contain distinct sets of proteins, which mediate each organelle's unique function.

Cells can also segregate subsets of their macromolecules into biomolecular condensates such as the nucleolus. The components inside these condensates can work together to carry out specialized biochemical reactions. The cell contains a dozen or more condensates that vary widely in size and can assemble and disassemble in response to need.

Each newly synthesized organellar protein must find its way from a ribosome in the cytosol, where the protein is made, to the organelle where it functions. It does so by using sorting signals in its amino acid sequence that are recognized by complementary sorting receptors, which deliver the protein to the appropriate target organelle. Proteins that function in the cytosol do not contain sorting signals and therefore remain there after they are synthesized.

During cell division, organelles such as the ER and mitochondria are distributed to each daughter cell. These organelles contain information that is required for their construction, and so they cannot be made de novo. Biomolecular condensates can be constructed de novo because they self-assemble from components that are encoded genetically.

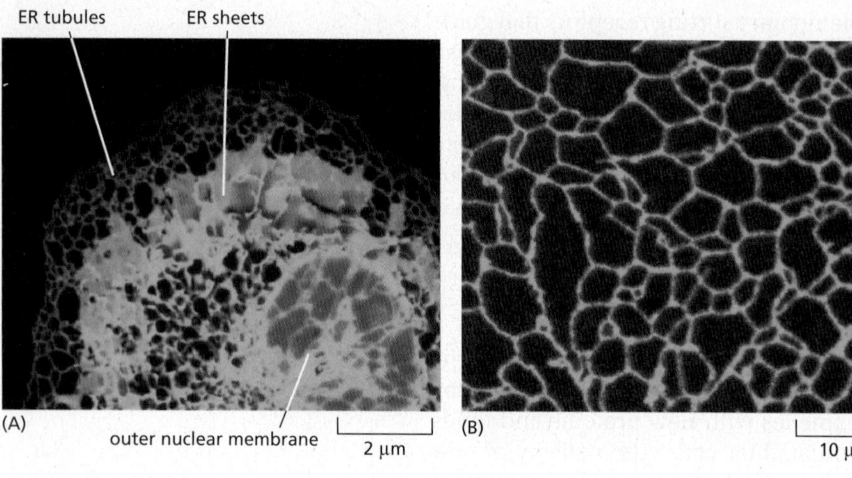

Figure 12–14 Fluorescence micrographs of the endoplasmic reticulum. (A) An animal cell in tissue culture that was genetically engineered to express an ER membrane protein fused to a fluorescent protein. The ER extends as a network of tubules and sheets throughout the entire cytosol, so that all regions of the cytosol are close to some portion of the ER membrane. The outer nuclear membrane, which is continuous with the ER, is also stained. (B) Part of an ER network in a living plant cell that was genetically engineered to express a fluorescent protein in the ER. (A, courtesy of Patrick Chitwood and Gia Voeltz. B, courtesy of Petra Boevink and Chris Hawes.)

THE ENDOPLASMIC RETICULUM

The membrane of the **endoplasmic reticulum (ER)** typically constitutes more than half of the total membrane of an average animal cell (see Table 12–2). The ER is organized into a netlike labyrinth of branching tubules and flattened sacs that extends throughout the cytosol (**Figure 12–14** and **Movie 12.2**). The tubules and sacs interconnect, and their membrane is continuous with the outer nuclear membrane. This membrane system encloses a single internal space, called the **ER lumen**, which is continuous with the space between the inner and outer nuclear membranes. The ER often occupies more than 10% of the total cell volume (see Table 12–1).

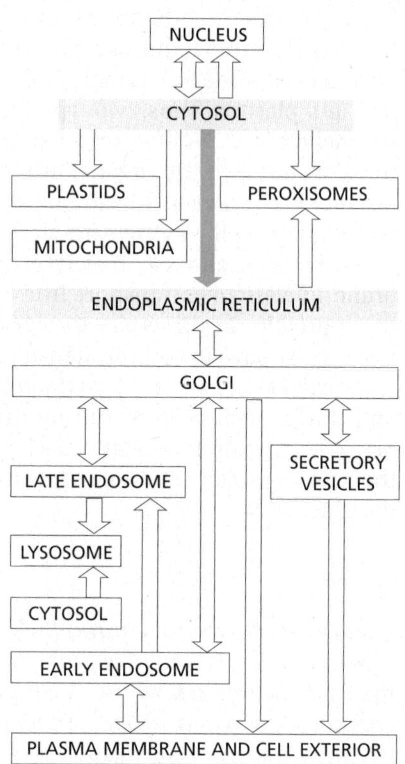

The ER has a central role in the biosynthesis of both lipids and proteins, and the ER lumen stores intracellular Ca^{2+} that is mobilized in many cell signaling responses (discussed in Chapter 15). The ER membrane is the site of production of many of the transmembrane proteins and lipids of the cell's organelles, including the ER itself, the Golgi apparatus, lysosomes, endosomes, secretory vesicles, peroxisomes, and the plasma membrane. The ER membrane is also the site at which most of the lipids for mitochondrial and plastid membranes are made. In addition, almost all of the proteins that will be secreted to the cell exterior—plus those destined for the lumen of the ER, Golgi apparatus, or lysosomes—are initially delivered to the ER lumen.

The ER Is Structurally and Functionally Diverse

While the various functions of the ER are essential to every cell, their relative importance varies greatly between individual cell types. To meet different functional demands, distinct regions of the ER become highly specialized. Functional specialization entails dramatic changes in the proportional abundance of different parts of the ER. These changes are observed as characteristically different types of ER membrane in different types of cells. The most visually remarkable specializations are the **rough ER** and **smooth ER** (**Figure 12–15**). The rough appearance is due to the abundance of ribosomes engaged in protein synthesis bound to the surface of this part of the ER. By contrast, regions of smooth ER lack ribosomes and are dedicated to other ER functions such as the biosynthesis and metabolism of lipids. All cells have both rough and smooth ER, but their relative abundance can vary enormously in specialized cells.

Most secreted proteins are synthesized by the ribosomes that stud the surface of the rough ER. Thus, cells specialized to secrete vast amounts of protein are packed with an abundance of rough ER. For example, exocrine cells of the pancreas secrete their own weight in digestive enzymes every day, explaining why the rough ER makes up 60% of these cells' membranes (see Table 12–2). Similarly, antibody-secreting plasma cells and insulin-secreting β cells also contain a markedly expanded rough ER. This correlation between highly secretory cells and an

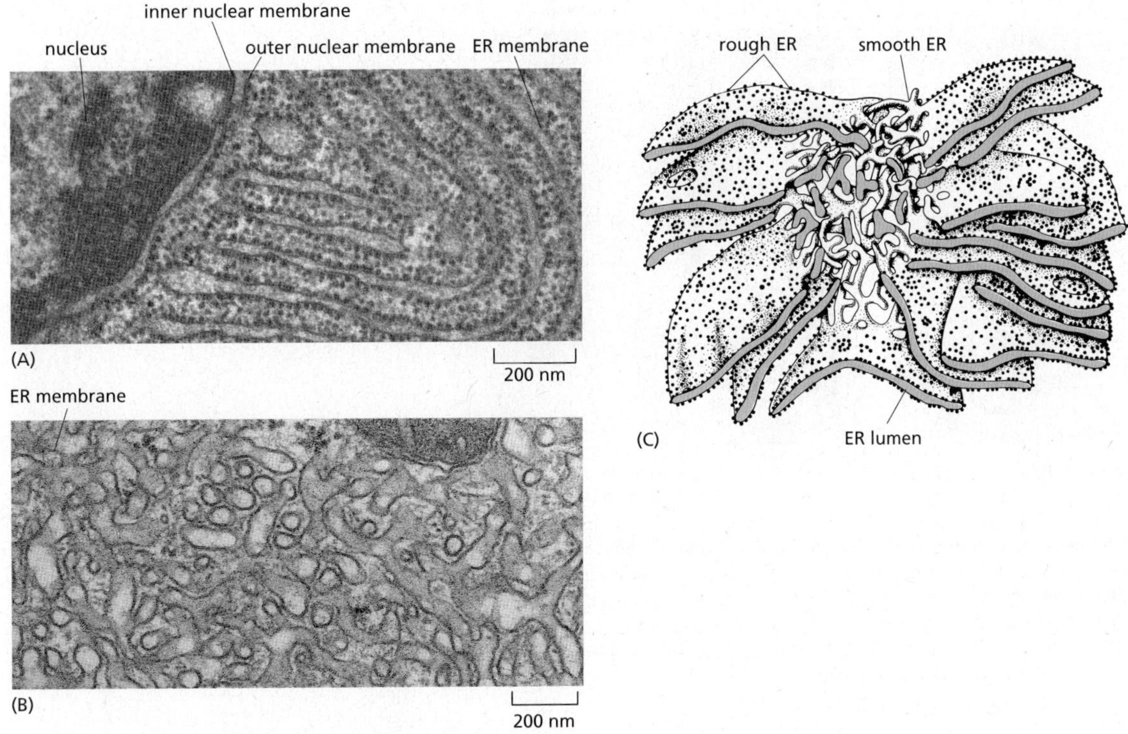

Figure 12–15 The rough and smooth ER. (A) An electron micrograph of the rough ER in a pancreatic exocrine cell that makes and secretes large amounts of digestive enzymes every day. The cytosol is filled with closely packed sheets of ER membrane that are studded with ribosomes. At the *top left* is a portion of the nucleus and its nuclear envelope; note that the outer nuclear membrane, which is continuous with the ER, is also studded with ribosomes. (B) Abundant smooth ER in a cell that secretes steroid hormone. This electron micrograph is of a testosterone-secreting Leydig cell in the human testis. (C) A three-dimensional reconstruction of a region of smooth ER and rough ER in a liver cell. The rough ER forms oriented stacks of flattened cisternae, each having a lumenal space 20–30 nm wide. The smooth ER membrane is connected to these cisternae and forms a fine network of tubules 30–60 nm in diameter. The ER lumen is colored *green*. (A, courtesy of Lelio Orci. B, courtesy of Daniel S. Friend, by permission of E.L. Bearer. C, after R.V. Kristić, Ultrastructure of the Mammalian Cell. New York: Springer-Verlag, 1979. With permission from Springer Nature.)

abundance of rough ER provided biologists the first clue that the ER is responsible for the synthesis and assembly of secreted proteins.

In contrast to rough ER, functions for the smooth ER are more diverse and can become highly specialized. A type of smooth ER found in all cells is called *transitional ER*, from which transport vesicles carrying newly synthesized proteins and lipids bud off for transport to the Golgi apparatus. In certain specialized cells, the smooth ER has additional functions that warrant its expansion. For example, cells that synthesize steroid hormones contain prominent smooth ER to accommodate the enzymes that make cholesterol and modify it to form a variety of steroid hormones (see Figure 12–15B).

The main cell type in the liver, the *hepatocyte*, also has expanded amounts of smooth ER (see Table 12–2) serving two separate purposes. The hepatocyte is the principal site of production of *lipoprotein particles*, which carry lipids via the bloodstream to other parts of the body. The enzymes that synthesize the lipid components of the particles are enriched in the membrane of the smooth ER. In addition, these membranes contain enzymes that catalyze a series of reactions to detoxify drugs and various harmful compounds produced by metabolism. The most extensively studied of these *detoxification reactions* are carried out by the *cytochrome P450* family of enzymes. They catalyze a series of reactions in which water-insoluble drugs or metabolites that would otherwise accumulate to toxic levels in cell membranes are rendered sufficiently water soluble to leave the cell and be excreted in the urine or bile.

Another crucially important function of the ER in most eukaryotic cells is to sequester Ca^{2+} from the cytosol. The release of Ca^{2+} into the cytosol from the

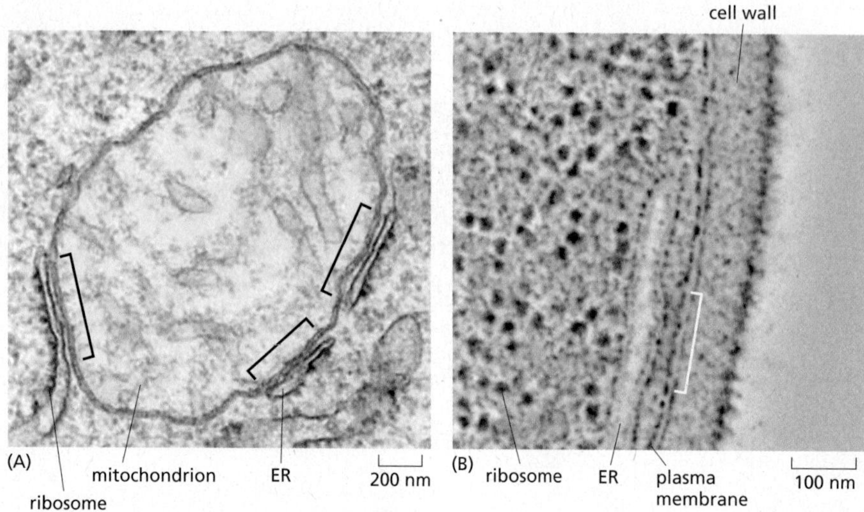

Figure 12–16 **The ER makes close contacts with the mitochondria and plasma membrane.** Electron micrographs of organelle contact sites between the ER and other membranes. (A) Region of a mouse embryonic fibroblast showing a mitochondrion that is closely apposed by sections of ER *(black brackets)*. (B) Yeast cell showing a section of the ER that is closely juxtaposed with the plasma membrane *(white bracket)*. [A, from P. Cosson et al., *PLOS ONE* 7(9):e46293, 2012. Courtesy of Pierre Cosson. B, courtesy of Wanda Kukulski.]

ER, and its subsequent reuptake, occur in many rapid responses to extracellular signals, as discussed in Chapter 15. A Ca^{2+} pump transports Ca^{2+} from the cytosol into the ER lumen. A high concentration of Ca^{2+}-binding proteins in the ER facilitates Ca^{2+} storage. In some cell types, specific regions of the ER are specialized for Ca^{2+} storage. Muscle cells have an abundant, modified smooth ER called the *sarcoplasmic reticulum*. The release and reuptake of Ca^{2+} by the sarcoplasmic reticulum trigger myofibril contraction and relaxation, respectively, during each round of muscle contraction (discussed in Chapter 16).

Finally, the smooth ER can be specialized in regions that make intimate contacts with other organelles, most notably the mitochondria, plastids, endosomes, and the plasma membrane (Figure 12–16). These **organelle contact sites** are enriched for proteins involved in communication or transport of key metabolites between the juxtaposed membranes. For example, the transport of lipids from their site of synthesis in the ER to the mitochondrion is thought to occur at ER-mitochondria contact sites. Contact of ER with the plasma membrane modulates levels of plasma membrane phosphoinositides, which are lipids that participate in numerous signaling pathways (discussed in Chapters 13 and 15). Contacts between other combinations of organelles have also been observed, and it is likely that these are also involved in the selective transfer of lipids and other metabolites.

To study the functions and biochemistry of the ER, it is necessary to isolate it. This may seem to be a hopeless task because the ER is intricately interleaved with other components of the cytoplasm. Fortunately, when tissues or cells are disrupted by homogenization, the ER breaks into fragments, which reseal to form small (~100–200 nm in diameter) closed vesicles called **microsomes** (Figure 12–17). To the biochemist, microsomes represent small authentic versions of the ER, still capable of protein translocation, protein glycosylation (discussed later), Ca^{2+} uptake and release, and lipid synthesis. *Rough microsomes*, derived from rough ER, contain ribosomes on their outside surface and enclose a small part of the ER lumen. *Smooth microsomes*, which lack ribosomes, are derived from vesiculated fragments of the smooth ER, plasma membrane, Golgi apparatus, endosomes, and mitochondria. The ribosomes attached to rough microsomes make them denser than smooth microsomes. As a result, scientists use equilibrium density centrifugation to separate the rough and smooth microsomes (Figure 12–17). Smooth microsomes derived from different organelles can

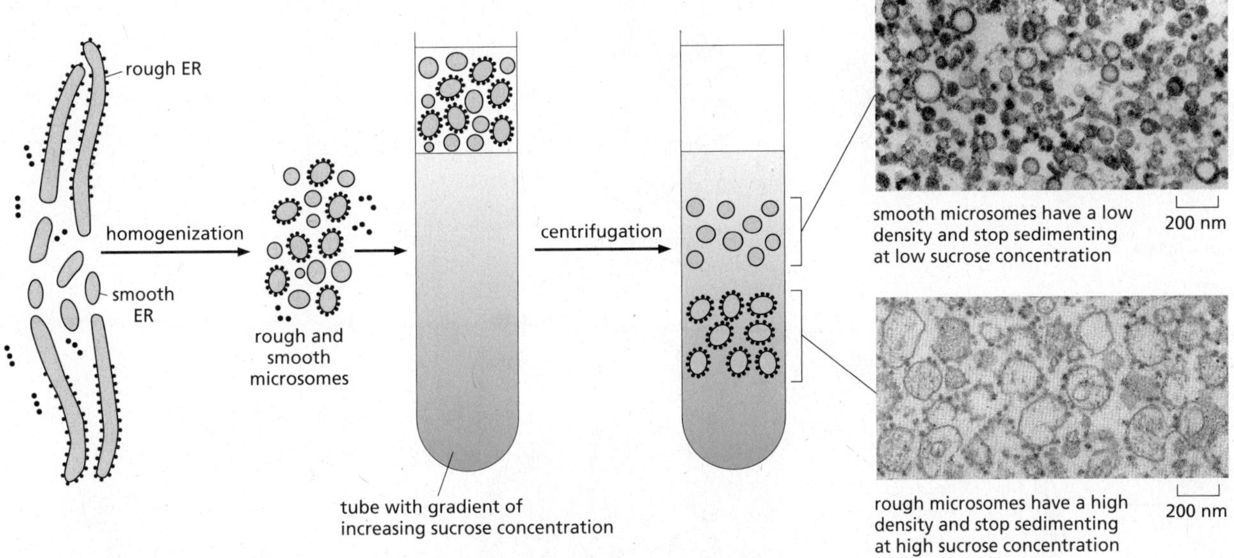

Figure 12–17 **The isolation of purified rough and smooth microsomes from the ER.** When an intact cell or tissue is homogenized, many of its membrane-enclosed compartments form small sealed vesicles called microsomes. When this mixture of vesicles is sedimented to equilibrium through a gradient of sucrose, the two types of microsomes separate from each other on the basis of their different densities. Note that the smooth fraction contains non-ER–derived material. Thin section electron micrographs of the purified smooth and rough microsome fractions show an abundance of ribosome-studded vesicles in the rough microsome fraction that originated from the rough ER. These are not seen in the smooth microsome fraction, which primarily contains ribosome-free vesicles originating from the smooth ER, Golgi cisternae, and other organelles. (Electron micrographs courtesy of George Palade.)

be further separated on the basis of differences in their protein content. Microsomes have been invaluable in elucidating the molecular aspects of ER function, as we discuss next.

Signal Sequences Were First Discovered in Proteins Imported into the Rough ER

The ER captures selected proteins from the cytosol as they are being synthesized. These proteins are of two types: *transmembrane proteins*, which become embedded in the ER membrane, and *water-soluble proteins*, which are fully translocated across the ER membrane into the ER lumen. Some of these proteins function in the ER, but many are destined to reside in another organelle, to reside in the plasma membrane, or to be secreted outside the cell. All of these proteins, regardless of their subsequent fate, are initially directed to the ER membrane by an **ER signal sequence**.

Signal sequences (and the signal sequence strategy of protein sorting) were discovered in secreted water-soluble proteins that are first translocated across the ER membrane. In the key experiment, the mRNA encoding a secreted protein was added to cytosol extracted from cells. In this cell-free reaction, ribosomes in the cytosol translated the mRNA into a protein that was slightly larger than the normal secreted protein (**Figure 12–18**). When the reaction was repeated in the presence of microsomes derived from the rough ER, a protein of the correct size was produced and located inside the microsomes (Figure 12–18). By contrast, mRNA encoding a cytosolic protein produced the correctly sized product regardless of the presence or absence of rough microsomes. The *signal hypothesis* was formulated to explain these observations. According to this model, the mRNA for the secretory protein codes for a protein that is bigger than the protein that is eventually secreted. It was proposed that the extra polypeptide is a signal sequence that directs the secreted protein to the ER membrane. After the signal sequence has served its function, it is cleaved off by a *signal peptidase* in the ER membrane before the polypeptide chain has been completed.

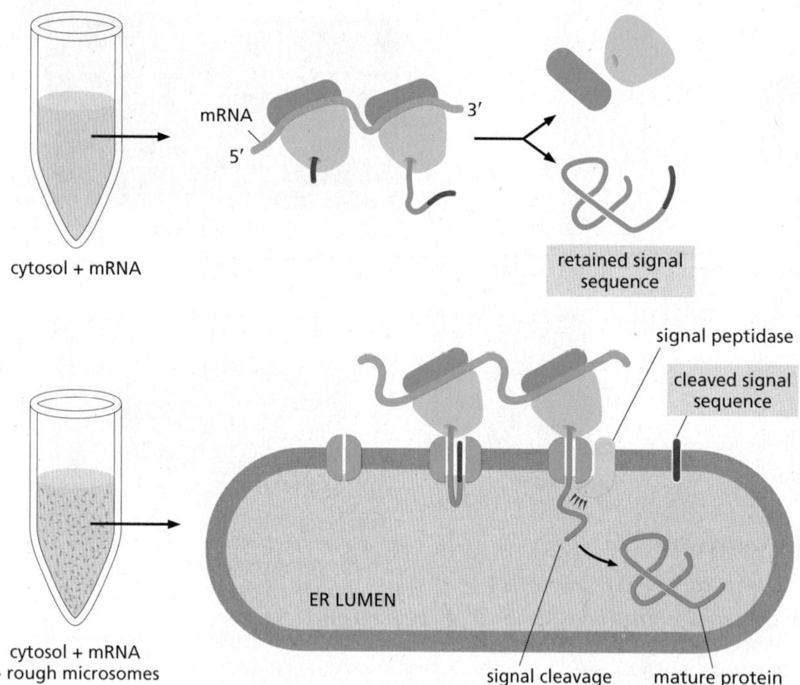

Figure 12–18 Experimental basis for the signal hypothesis. In a test tube, cytosol is mixed with mRNA that codes for a secreted protein. Two versions of this reaction are performed: one lacking and the other containing rough microsomes derived from the ER (see Figure 12–17). In both reactions, ribosomes in the cytosol translate the mRNA to produce a protein. The protein produced in the reaction lacking microsomes was observed to be slightly larger than the protein produced in the reaction containing microsomes. This difference in size was shown to be due to a small segment of protein at the N-terminus that was selectively removed only in the reaction containing microsomes. Additional analysis showed that the protein produced in the presence of microsomes was located in the microsome lumen. This collection of results was used to formulate the signal hypothesis. This sequence at the N-terminus was postulated to represent a signal sequence. When the ER signal sequence emerges from the ribosome, it directs the ribosome to a translocator on the ER membrane that forms a pore in the membrane through which the polypeptide is translocated. A signal peptidase is closely associated with the translocator and clips off the signal sequence during translation, and the mature protein is released into the lumen of the ER immediately after its synthesis is completed. The translocator is closed until the ribosome has bound, so that the permeability barrier of the ER membrane is maintained at all times.

These experiments highlight how a complex cellular process such as ER import can be reconstituted in a cell-free system by mixing together requisite cellular components such as mRNA, cytosol, and microsomes. By combining the constituent parts in different ways, the existence of signal sequences on secreted proteins was deduced long before it became possible to directly sequence their mRNAs. The ease with which this cell-free system could be manipulated proved indispensable for identifying, purifying, and studying the various components of the molecular machinery responsible for ER import. Similar systems were later established to dissect protein transport into and out of the nucleus, protein import into mitochondria and chloroplasts, and vesicular transport.

A Signal-Recognition Particle (SRP) Directs the ER Signal Sequence to a Specific Receptor at the ER

The ER signal sequence is guided to the ER membrane by at least two components: a **signal-recognition particle (SRP)**, which binds to the signal sequence, and an **SRP receptor** in the ER membrane. SRP is a large complex; in animal cells, it consists of six different polypeptide chains bound to a single RNA molecule (Figure 12–19A). While SRP and SRP receptor have fewer subunits in bacteria, homologs of both components are present in all living organisms.

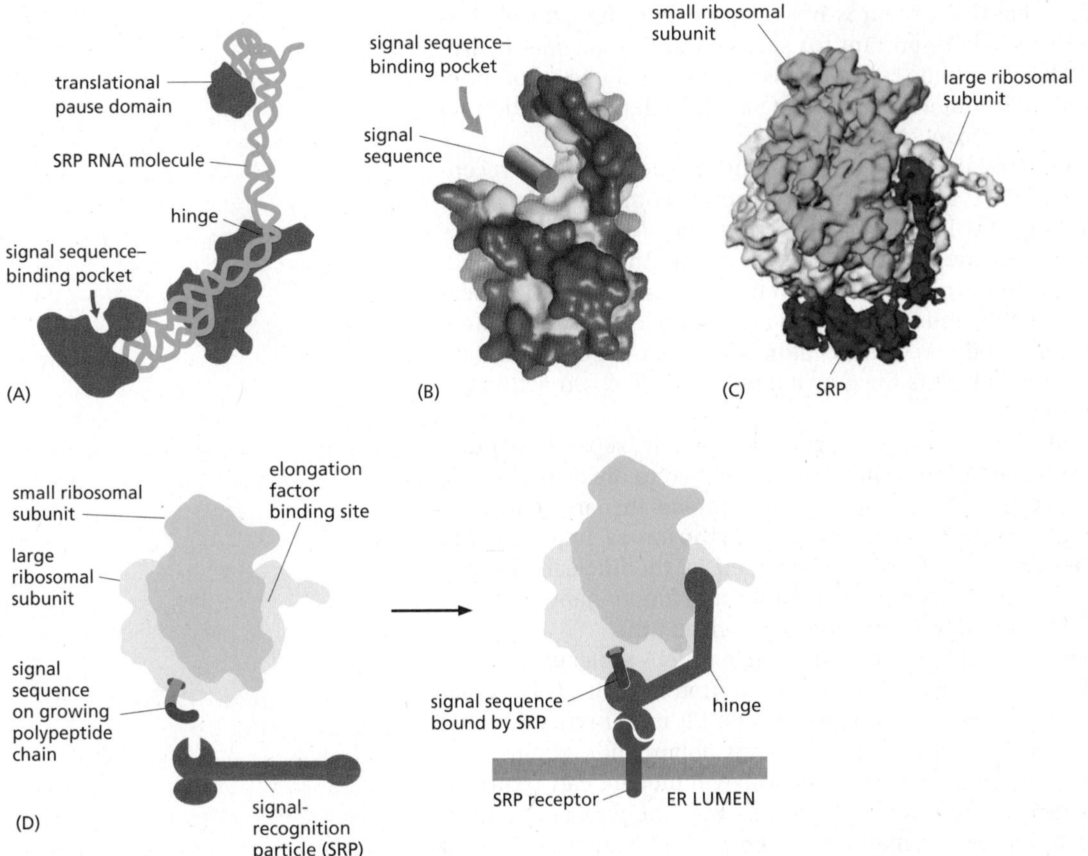

Figure 12–19 **The signal-recognition particle (SRP).** (A) A mammalian SRP is a rodlike ribonucleoprotein complex containing six protein subunits *(brown)* and one RNA molecule *(blue)*. The SRP RNA forms a backbone that links the protein domain containing the signal sequence–binding pocket to the domain responsible for slowing translation. (B) Structure of the signal sequence–binding domain of SRP bound to the hydrophobic region of a signal sequence *(red cylinder)*. The surfaces of SRP that are hydrophobic are depicted in *yellow*. (C) The three-dimensional outline of the SRP bound to a ribosome was determined by cryo-electron microscopy. SRP binds to the large ribosomal subunit so that its signal sequence–binding pocket is positioned near the growing polypeptide chain exit site. Its translational pause domain is positioned at the interface between the ribosomal subunits, where it interferes with elongation factor binding. (D) As a signal sequence emerges from the ribosome and binds to the SRP, a conformational change in the SRP exposes a binding site for the SRP receptor. (C, adapted from M. Halic et al., *Nature* 427:808–814, 2004.)

This protein-targeting mechanism therefore arose early in evolution and has been conserved.

ER signal sequences vary greatly in amino acid sequence, but each has eight or more nonpolar amino acids at its center (see Figure 12–13). How can SRP bind specifically to so many different sequences? The answer has come from structures of one of the SRP proteins, which shows that the signal sequence-binding site is a large hydrophobic pocket enriched in methionines (**Figure 12–19B**). Because methionines have unbranched, flexible side chains, the pocket is sufficiently plastic to accommodate different hydrophobic signal sequences of various sizes and shapes.

In eukaryotic cells, SRP is a hinged rodlike structure that can wrap along the large ribosomal subunit (**Figure 12–19C**). The end of SRP that contains the signal sequence-binding pocket is positioned near the ribosomal tunnel through which newly made polypeptides emerge. This allows SRP to engage a signal sequence as it emerges from the ribosome. Once SRP engages a signal sequence, the other end of SRP can bind at the interface between the large and small ribosomal subunits (**Figure 12–19D**). This is the same site where translation elongation factors bind, so a ribosome engaged by SRP will translate proteins more slowly than normal. Slower translation presumably gives the ribosome enough time to bind to the ER membrane before completion of the polypeptide chain, thereby ensuring that the protein is not released into the cytosol. This safety device may be especially important for secreted and lysosomal hydrolases, which could wreak havoc in the cytosol; cells that secrete large amounts of hydrolases, however, take the added precaution of having high concentrations of hydrolase inhibitors in their cytosol.

When a signal sequence binds, SRP exposes a binding site for an SRP receptor (see Figure 12–19D), which is a transmembrane protein complex in the rough ER membrane. The binding of SRP to its receptor brings the SRP–ribosome complex to an unoccupied **protein translocator** in the ER membrane. The part of SRP bound near the ribosomal tunnel moves to a different site, allowing the translocator to occupy this position. SRP and SRP receptor are then released, and protein synthesis resumes at full speed. The translocator, which is now tightly bound to the translating ribosome, transfers the growing polypeptide chain across the membrane (**Figure 12–20**).

This **co-translational** transfer process creates two spatially separate populations of ribosomes. **Membrane-bound ribosomes**, attached to the cytosolic side of the ER membrane, are engaged in the synthesis of proteins that are being concurrently translocated across the ER membrane. **Free ribosomes**, unattached to any membrane, synthesize all other proteins encoded by the nuclear genome. Membrane-bound and free ribosomes are structurally and functionally identical. They differ only in the proteins they are making at any given time.

Because many ribosomes can engage with a single mRNA molecule, a **polyribosome** is usually formed. If the mRNA encodes a protein with an ER signal sequence, the polyribosome becomes attached to the ER membrane, directed there by the signal sequences on multiple growing polypeptide chains. The individual ribosomes associated with such an mRNA molecule can return to the cytosol when they finish translation and intermix with the pool of free ribosomes. The mRNA itself, however, remains attached to the ER membrane by a

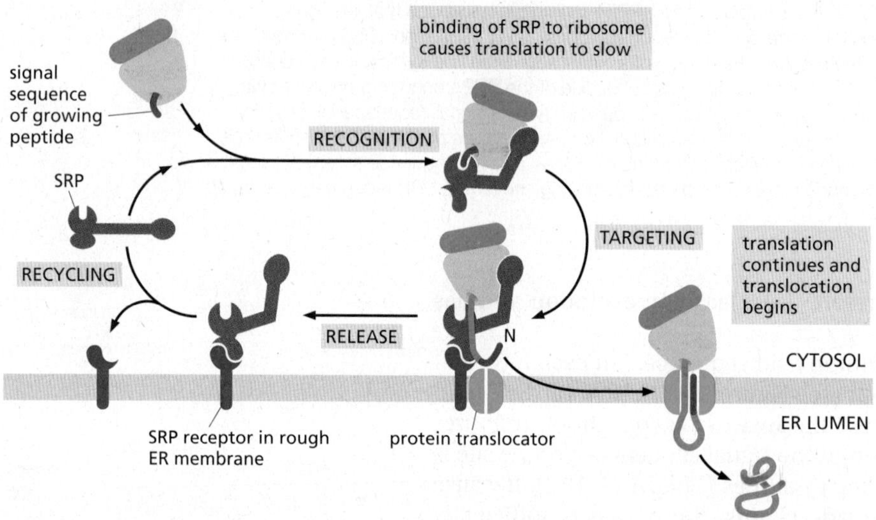

Figure 12–20 How ER signal sequences and SRP direct ribosomes to the ER membrane. The SRP and its receptor act in concert. The SRP binds to both the exposed ER signal sequence and the ribosome, thereby causing translation to slow. The SRP receptor in the ER membrane, which in animal cells is composed of two different polypeptide chains, binds the SRP–ribosome complex and directs it to the translocator. The SRP (in complex with SRP receptor) then moves away from its binding site on the ribosome, which is then occupied by the translocator in the ER membrane. SRP then releases the signal sequence, which inserts into the translocator to initiate polypeptide chain transfer across the lipid bilayer. The SRP and SRP receptor dissociate from each other and are recycled for the next round of protein targeting. Although not shown in the figure, one of the SRP proteins and both chains of the SRP receptor contain GTP-binding domains. Conformational changes that occur during cycles of GTP binding and hydrolysis (discussed in Chapter 15) ensure that SRP preferentially binds a signal sequence in the cytosol and releases it only after SRP successfully engages the SRP receptor at the ER membrane. The energy of GTP hydrolysis is therefore used to impart directionality to the cycle of SRP-mediated protein targeting.

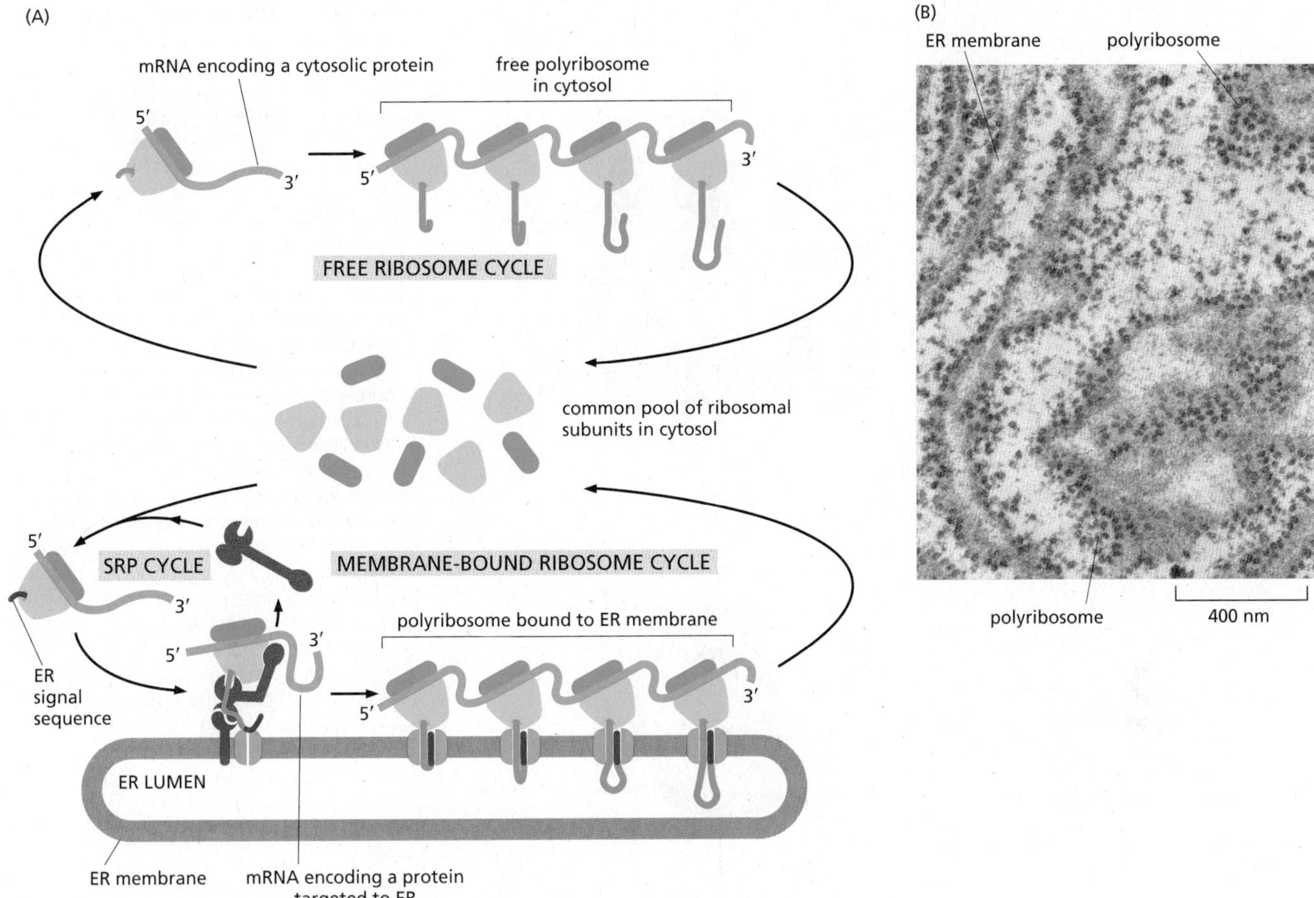

Figure 12–21 **Free and membrane-bound polyribosomes.** (A) A common pool of ribosomes synthesizes the proteins that stay in the cytosol and those that are transported into the ER. The ER signal sequence on a newly formed polypeptide chain binds to SRP, which directs the translating ribosome to the ER membrane. The mRNA molecule remains bound to the ER as part of a polyribosome, while the ribosomes that move along it are recycled; at the end of each round of protein synthesis, the ribosomal subunits are released and rejoin the common pool in the cytosol. (B) A thin section electron micrograph of polyribosomes attached to the ER membrane. The plane of the section in some places cuts through the ER roughly parallel to the membrane, giving a face-on view of the circular or spiral pattern of the polyribosomes. (B, courtesy of George Palade.)

changing population of ribosomes, each transiently held at the membrane by the translocator (**Figure 12–21**).

The Polypeptide Chain Passes Through a Signal Sequence–gated Aqueous Channel in the Translocator

It had long been debated whether polypeptide chains are transferred across the ER membrane in direct contact with the lipid bilayer or through a channel in a protein translocator. The debate ended with the identification of the translocator, which was shown to form a water-filled channel across the membrane through which the polypeptide chain passes. The core of the translocator, called the **Sec61 complex**, is built from three subunits that are highly conserved from bacteria to eukaryotic cells. The structure of the Sec61 translocator revealed that 10 α helices surround a central channel (**Figure 12–22**). The channel is plugged by a short α helix that keeps the translocator closed when it is idle. It is important to keep the channel closed to prevent ions, such as Ca^{2+}, from leaking out of the ER. During translocation, the plug moves out of the way so the polypeptide can pass through the channel.

The Sec61 translocator only opens for proteins containing a signal sequence. The ability of the Sec61 translocator to recognize signal sequences provides a proofreading step to ensure that only proteins truly intended for the ER are

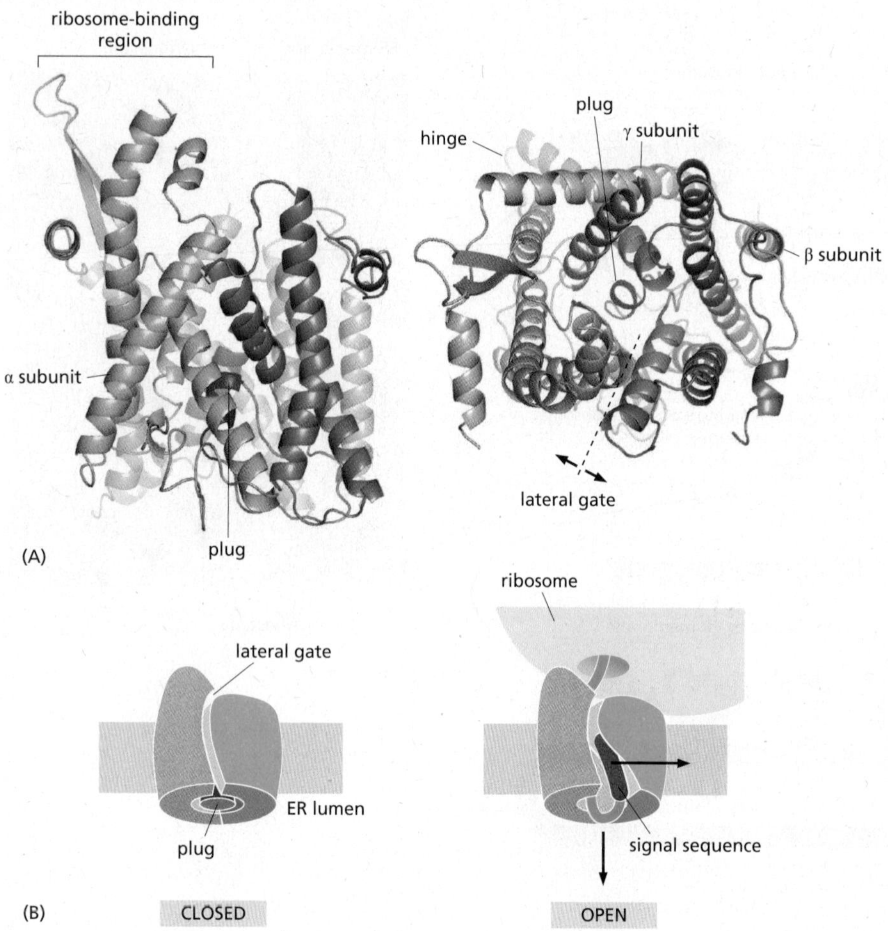

Figure 12–22 Structure of the Sec61 translocator. (A) A side view (*left*, seen from the membrane) and a top view (*right*, seen from the cytosol) of the structure of the Sec61 translocator of the archaeon *Methanococcus jannaschii* (where it is called the SecY translocator). The Sec61 α subunit has an inverted repeat structure (see Figure 11–10) and is shown in *blue* and *orange* to indicate this pseudosymmetry; the two smaller β and γ subunits are shown in *gray*. Some regions of the Sec61 α subunit that protrude into the cytosol bind to the ribosome during protein translocation. The *purple* short helix forms a plug that seals the pore when the translocator is closed. When the translocator is open, the plug helix moves out of the way. The Sec61/SecY translocator can also open sideways toward the membrane at a lateral gate. (B) Models of the closed and functionally active states of the Sec61/SecY translocator. In the active state, a protein chain can either translocate across the membrane through the central channel in the translocator or move sideways into the lipid bilayer through the lateral gate. (A, PDB code: 1RH5.)

allowed to enter. **Cryo-electron microscopy** structures of the Sec61 translocator before and after signal sequence recognition show that the signal sequence wedges into a lateral gate, or seam, in Sec61 with its N-terminus facing the cytosol (**Figure 12–23A**). Insertion of the signal sequence at this lateral gate widens the central channel and releases the plug. The open translocator then readily accommodates the segment of polypeptide following the signal sequence inside the channel. The signal sequence, which is hydrophobic, laterally exits the gate into the membrane where it is cleaved off by signal peptidase and then rapidly degraded to amino acids by other proteases in the ER membrane and cytosol. As this mechanism illustrates, the lateral gate in the Sec61 translocator provides the access route from Sec61's central channel into the hydrophobic core of the membrane. In addition to its role in recognition of signal sequences, the lateral gate guides the integration of transmembrane proteins into the ER, as we discuss later.

Once the signal sequence has opened the Sec61 translocator and threaded the ensuing polypeptide into the channel, translocation occurs concurrently with continued translation. During translocation, the polypeptide tunnel inside the ribosomal large subunit is aligned with the channel within the Sec61 translocator (**Figure 12–23B**). This configuration provides a continuous path for the polypeptide from the peptidyl-transferase center in the ribosome, where new amino acids are added to the growing protein chain, to the ER lumen 15 nm away. In this way, the energy used for polypeptide elongation is indirectly harnessed to also drive translocation across the ER membrane.

When translation terminates, the C-terminus of the polypeptide is released from the ribosome and slips through the Sec61 translocator, whose plug returns to close the channel. Thus, the entire process of ER import, from signal sequence recognition by SRP to translocation through the Sec61 translocator, occurs co-translationally before the polypeptide has a chance to fold. This pathway provides one solution

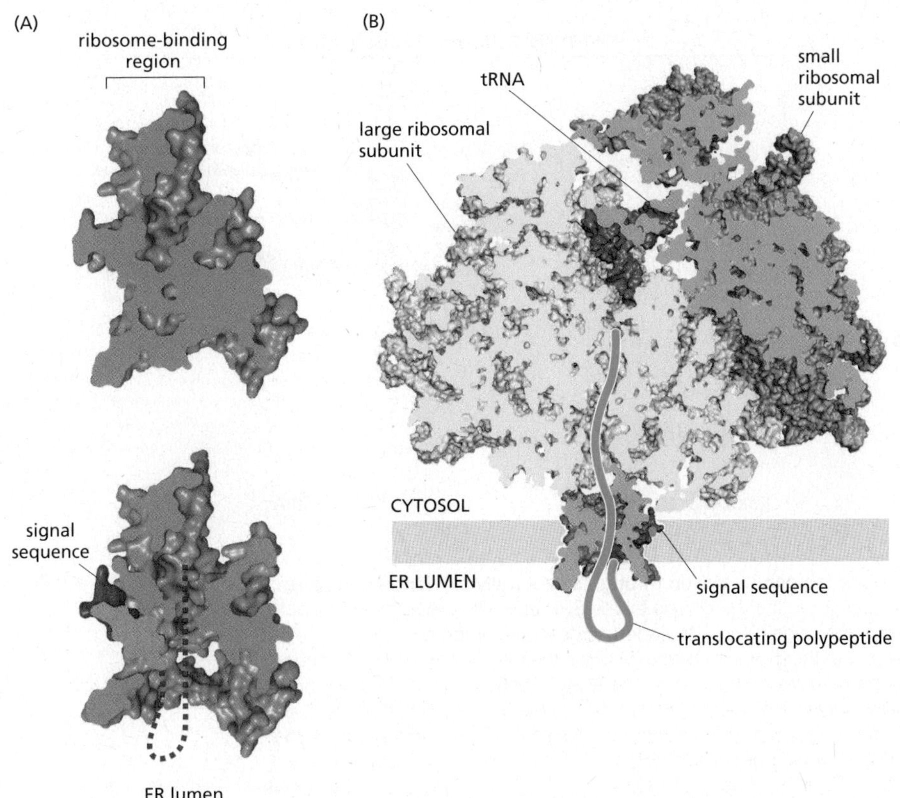

Figure 12–23 A signal sequence opens the Sec61 translocator. (A) Cross section through the Sec61 translocator before and after a signal sequence has inserted into the lateral gate. Insertion of the signal sequence causes the central channel in the translocator to widen and the plug to move out of this channel; hence, a continuous path across the membrane is now apparent *(dashed line)*. (B) Cross section through the structure of a translating ribosome *(green)* bound to a Sec61 translocator *(blue)* that has been opened by a signal sequence *(red)*. A translocating polypeptide is shown passing through the tunnel within the large ribosomal subunit and the Sec61 translocator. (A, PDB codes: 3J7Q and 3JC2; B, PDB code: 3JC2.)

to the problem of how to move a large protein across a membrane barrier without leakage of much smaller ions and metabolites during the process.

Translocation Across the ER Membrane Does Not Always Require Ongoing Polypeptide Chain Elongation

Some proteins are completely synthesized in the cytosol as precursors before they are imported into the ER, demonstrating that translocation does not always require ongoing translation (**Figure 12–24**). This is termed **post-translational** translocation. Post-translational protein translocation is more common across the yeast ER membrane and the evolutionarily related bacterial plasma membrane. In both cases, the Sec61 translocator (called SecY in bacteria) is used as the

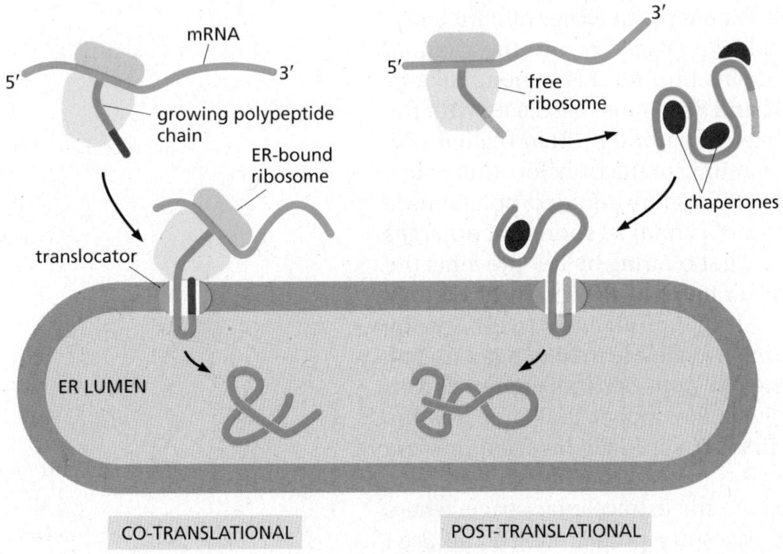

Figure 12–24 Co-translational and post-translational protein translocation. Ribosomes bind to the ER membrane during co-translational translocation. By contrast, cytosolic ribosomes complete the synthesis of a protein and release it prior to post-translational translocation. The released protein is kept unfolded in the cytosol by chaperones that dissociate before the protein is translocated across the membrane. In both cases, the protein is directed to the ER by an ER signal sequence (*red* and *orange*). See **Movie 12.3**.

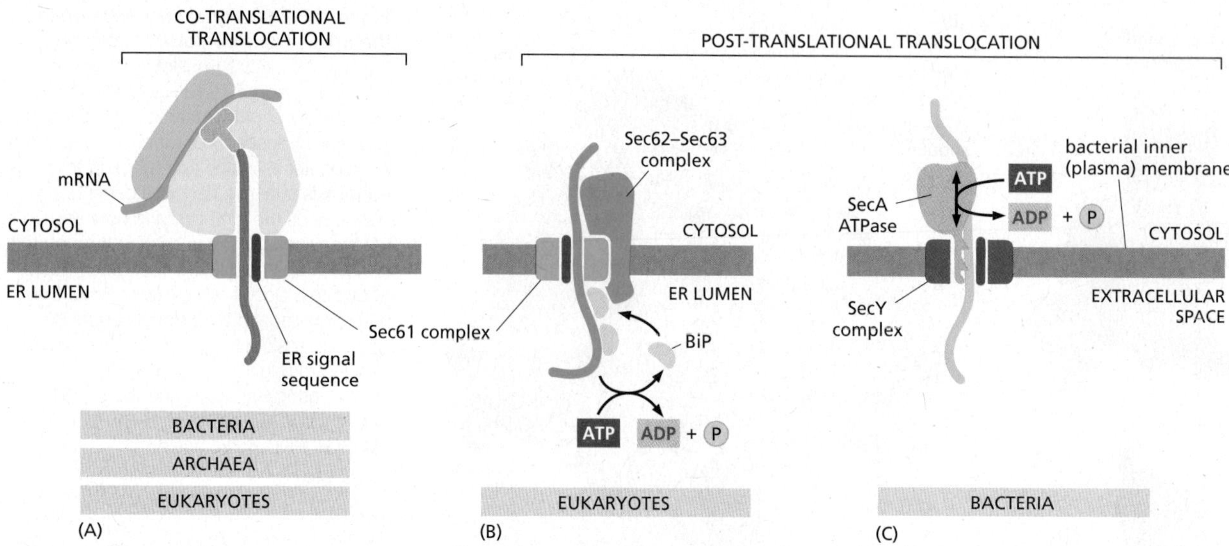

Figure 12–25 **Three ways in which protein translocation can be driven through structurally similar translocators.** (A) Co-translational translocation. The ribosome is brought to the membrane by the SRP and SRP receptor and then engages with the Sec61 translocator. The growing polypeptide chain is threaded across the membrane as it is made. No additional energy is needed, as the only path available to the growing chain is to cross the membrane. (B) Post-translational translocation in eukaryotic cells requires an additional complex composed of Sec62 and Sec63 proteins. This complex is attached to the Sec61 translocator and positions BiP molecules where they can bind to the translocating chain as it emerges from the translocator in the lumen of the ER. ATP-driven cycles of BiP binding and release pull the protein into the lumen. (C) Post-translational translocation in bacteria. The completed polypeptide chain is fed from the cytosolic side into the bacterial homolog of the Sec61 translocator (called SecY) in the plasma membrane by the SecA ATPase. ATP hydrolysis–driven conformational changes drive a pistonlike motion in SecA. The piston not only pushes several amino acids of the protein chain through the pore of the translocator but also prevents backsliding of the chain into the cytosol. Whereas the Sec61 translocator, SRP, and SRP receptor are found in all organisms, SecA is found exclusively in bacteria, and the Sec62–Sec63 complex is found exclusively in eukaryotic cells. (Adapted from P. Walter and A.E. Johnson, *Annu. Rev. Cell Biol.* 10:87–119, 1994.)

translocator; its narrow channel means that precursors can only be translocated as unfolded polypeptides. Thus, precursor proteins do not fold after their initial synthesis in the cytosol. Instead, they interact with other cytosolic proteins that prevent precursor folding or aggregation before they engage the Sec61 translocator. These interacting proteins typically are general chaperone proteins, such as those of the hsp70 family (discussed in Chapter 6), and must dissociate as the unfolded polypeptide is threaded through the translocator.

Just as in co-translational translocation discussed earlier, the signal peptide of a precursor directly engages the Sec61 translocator to open the channel. However, the next step of translocation across the membrane occurs differently and relies on accessory proteins that use cellular energy to either pull the polypeptide across the channel from the lumenal side or feed it into the channel from the cytosol (**Figure 12–25**). To pull proteins into the ER lumen, eukaryotic cells use accessory proteins called Sec62 and Sec63 that associate with the Sec61 translocator and position an hsp70-like chaperone protein (called *BiP*, for binding protein) adjacent to the lumenal opening of the translocation channel. Like its cytosolic cousin, BiP has a high affinity for unfolded polypeptide chains, and it binds tightly to an imported protein chain as soon as it emerges from the Sec61 translocator in the ER lumen. Tight binding by BiP prevents the protein chain from sliding backwards, favoring more of the chain to emerge into the lumen where it can bind another molecule of BiP. ATP hydrolysis by BiP causes it to release the polypeptide, making it available to bind again to any newly emerged segments of the translocating polypeptide. This energy-driven cycle of binding and release serves as a molecular ratchet that provides the driving force for protein import after a precursor has initially inserted into the Sec61 translocator.

Because bacteria transport proteins directly to the extracellular space, where energy is not available, they use a cytosolic accessory protein called the *SecA*

ATPase. SecA binds to the precursor polypeptide and attaches to the cytosolic side of the translocator, where it undergoes cyclic conformational changes fueled by ATP hydrolysis. Each time an ATP is hydrolyzed, a portion of the SecA protein inserts into the pore of the translocator, pushing a short segment of the precursor protein with it. As a result of this pistonlike ratchet mechanism, the SecA ATPase progressively pushes the polypeptide chain of the transported protein across the membrane.

Transmembrane Proteins Contain Hydrophobic Segments That Are Recognized Like Signal Sequences

All of the transmembrane proteins that populate the ER, Golgi apparatus, lysosomes, endosomes, secretory vesicles, and plasma membrane are inserted into the ER membrane before moving to their final destination. Transmembrane proteins made at the ER span the lipid bilayer via one or more α-helical hydrophobic **transmembrane segments** (see Figure 10–17). Thus, the biosynthesis of membrane proteins requires some parts of the polypeptide chain to be translocated across the lipid bilayer, other parts to remain in the cytosol, and the transmembrane segments to be integrated into the membrane. Despite this additional complexity, the same factors (SRP, SRP receptor, and the Sec61 translocator) just described for transferring a soluble protein into the ER lumen also mediate transmembrane protein integration into the ER membrane. The same factors can be used because the transmembrane segments that define a transmembrane protein resemble the hydrophobic ER signal sequences that direct soluble protein translocation.

In the simplest case, a transmembrane protein contains a single transmembrane segment that will ultimately be embedded in the lipid bilayer as a membrane-spanning α helix. When this transmembrane segment emerges from the ribosome during synthesis, SRP recognizes its hydrophobic α-helical features as a signal sequence and brings this ribosome to the Sec61 translocator at the ER membrane. The transmembrane segment then inserts into the lateral gate of the Sec61 translocator, which is the same site where signal sequences bind. The orientation in which the transmembrane segment inserts into the lateral gate determines whether the protein segment preceding or the one following the transmembrane segment is moved across the membrane into the ER lumen (**Figure 12–26**). If the N-terminus is short and unfolded, orientation of the transmembrane segment depends on features of the polypeptide chain such as the distribution of nearby charged amino acids and the length of the transmembrane segment. If the preceding N-terminal segment is long and stably folded, it does not cross the membrane through the narrow Sec61 channel. In this case, the C-terminal segment that is still being synthesized, and therefore unfolded, is translocated across the membrane.

Figure 12–26 **A transmembrane segment directs membrane protein insertion into the ER membrane.** Many single-pass membrane proteins use their transmembrane segment to direct insertion into the ER membrane (Movie 12.3). The transmembrane segment is recognized by SRP (not shown) and delivered via the SRP receptor (not shown) to the Sec61 translocator at the ER membrane. The transmembrane segment then inserts into the lateral gate of the Sec61 translocator in one of two orientations. (A) Some transmembrane segments insert into the lateral gate such that the N-terminal domain is retained on the cytosolic side of Sec61. This orientation is favored for proteins whose N-terminal domains are very long or folded, and for transmembrane segments whose flanking amino acids have a net positive charge on the N-terminal side. (B) Some transmembrane segments insert into the lateral gate such that the C-terminal flanking region is retained on the cytosolic side of Sec61. In this case, the N-terminal flanking region is thought to translocate across the membrane through the Sec61 channel. This orientation is favored for transmembrane segments whose flanking amino acids have a net positive charge on the C-terminal side.

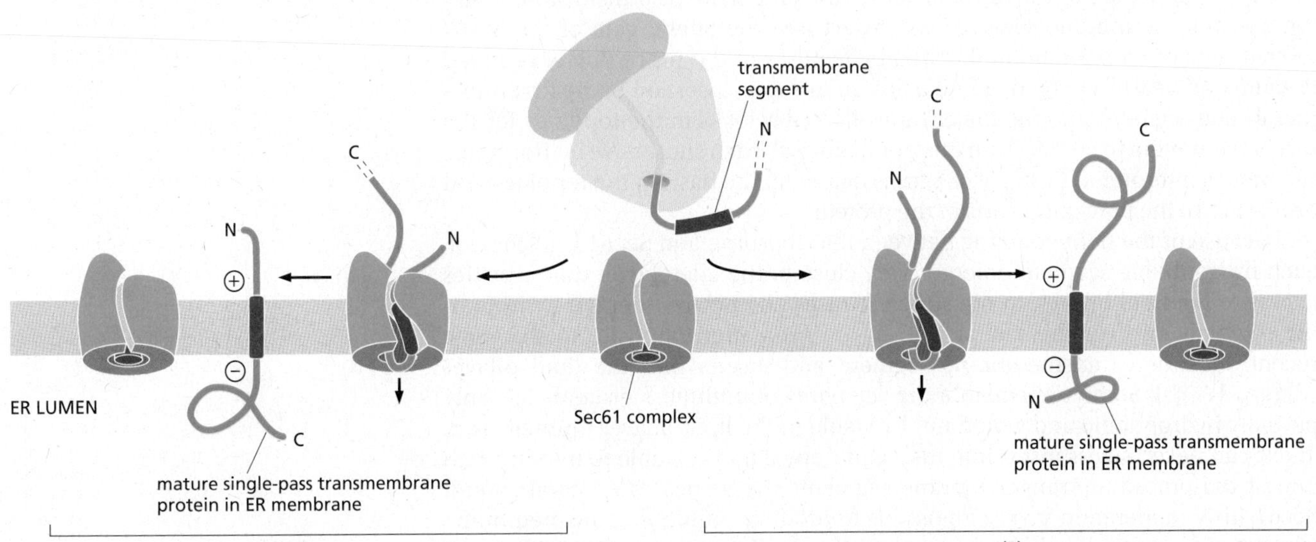

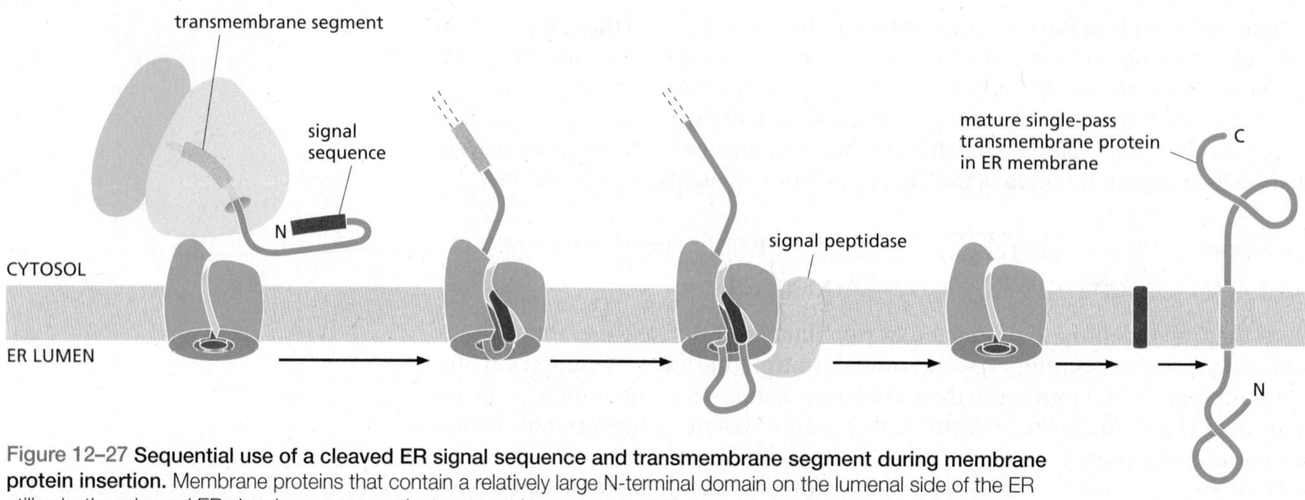

Figure 12–27 **Sequential use of a cleaved ER signal sequence and transmembrane segment during membrane protein insertion.** Membrane proteins that contain a relatively large N-terminal domain on the lumenal side of the ER utilize both a cleaved ER signal sequence and a transmembrane segment. Targeting to the ER membrane, initiation of translocation through Sec61, and cleavage of the signal sequence all occur exactly as for a secretory protein (see Figure 12–20). However, when the transmembrane segment enters the Sec61 translocator, translocation stops and the transmembrane segment moves through the lateral gate into the lipid bilayer. The remainder of the protein continues to be synthesized on the cytosolic side of the membrane until translation terminates.

Many transmembrane proteins contain large N-terminal lumenal domains. In this case, an N-terminal signal sequence is used to initiate translocation, just as for a soluble protein. In this way, the N-terminus of the mature polypeptide is committed to the ER lumen by the signal sequence, and the remainder of the polypeptide begins translocation through the Sec61 translocator. When a hydrophobic segment in the polypeptide emerges from the ribosome, it inserts into the lateral gate to gain access to the lipid bilayer. Because the hydrophobic segment is more stable in the membrane than in the aqueous channel, it exits the channel laterally, translocation stops, and the rest of the protein is synthesized on the cytosolic side of the ER membrane (**Figure 12–27**).

Hydrophobic Segments of Multipass Transmembrane Proteins Are Interpreted Contextually to Determine Their Orientation

In multipass transmembrane proteins, the polypeptide chain passes back and forth repeatedly across the lipid bilayer as hydrophobic α helices (see Figure 10–17). Synthesis of multipass transmembrane proteins up to the first transmembrane segment occurs as we have just described for single-pass transmembrane proteins. Hence, SRP will deliver the protein to the translocator, where the first transmembrane segment will insert into the lateral gate of the Sec61 translocator in an orientation dictated by features of the preceding N-terminal domain and nearby charged amino acids. In this way, insertion of the first transmembrane segment into the membrane effectively locks in the topology for the rest of the protein to come. From this point onward, each successive hydrophobic segment is interpreted by the Sec61 translocator on the basis of the topology and properties of the preceding parts of the protein.

Because of the tight coupling between the ribosome and Sec61 translocator, each hydrophobic segment emerges very close to the lateral gate that provides access to the lipid bilayer. In the simplest cases, the newly emerged hydrophobic segment engages the lateral gate in an orientation opposite to the most recently inserted transmembrane segment and inserts into the lipid bilayer (**Figure 12–28**). Some transmembrane segments of multipass proteins are only partially hydrophobic and would not be stable in the lipid bilayer on their own. These can nevertheless insert into the membrane if they are able to interact with one of the preceding transmembrane segments that is near the lateral gate of Sec61. This cooperation makes it possible to produce multipass transmembrane proteins that contain hydrophilic parts within the lipid bilayer, which is crucial

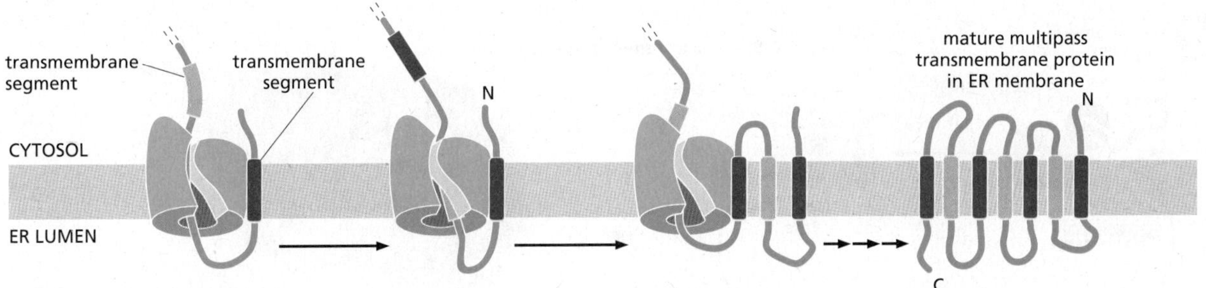

Figure 12–28 **The insertion of a multipass transmembrane protein into the ER membrane.** The events up to the insertion of the first transmembrane segment follow the steps for single-pass membrane proteins (see Figures 12–26 and 12–27). The orientation of this first transmembrane segment depends on the characteristics of the transmembrane segment and flanking regions just as for single-pass membrane proteins. When the next transmembrane segment emerges from the ribosome, it inserts into the lateral gate of Sec61 in an orientation opposite to that of the first transmembrane segment, then moves into the lipid bilayer. Each successive transmembrane segment is similarly inserted into the membrane via the lateral gate in an orientation opposite to that of the transmembrane segment that immediately preceded it. This proceeds until all transmembrane segments have been inserted into the membrane.

for many important proteins such as transporters and channels (discussed in Chapter 11). The hydrophilic sequences between the transmembrane segments are either synthesized into the cytosol or threaded through the Sec61 translocator, depending on the orientation of the preceding transmembrane segment. In this way, a multipass protein is woven into the membrane with successive hydrophobic segments achieving opposite orientations until all of them have been inserted into the membrane as transmembrane α helices.

Because membrane proteins are always inserted from the cytosolic side of the ER in this programmed manner, all copies of the same polypeptide chain will have the same orientation in the lipid bilayer. This generates an asymmetrical ER membrane in which the protein domains exposed on one side are different from those exposed on the other side. This asymmetry is maintained during the many membrane budding and fusion events that transport the proteins made in the ER to other cell membranes (discussed in Chapter 13). Thus, the way in which a newly synthesized protein is inserted into the ER membrane determines the orientation of the protein in all of the other membranes as well.

Some Proteins Are Integrated into the ER Membrane by a Post-translational Mechanism

Many important cytosol-facing membrane proteins are anchored in the membrane by a single transmembrane α helix very close to the C-terminus. These **tail-anchored proteins** include a large number of SNARE protein subunits that guide vesicular traffic (discussed in Chapter 13). When a tail-anchored protein is translated, the ribosome reaches the termination codon while the polypeptide sequence destined to become a transmembrane α helix is still inside the ribosome exit tunnel. Recognition by SRP is therefore not possible, and the protein is released from the ribosome into the cytosol. The hydrophobic segment is recognized by a specialized chaperone complex that transfers it to a targeting factor called Get3 (**Figure 12–29**). Although unrelated to SRP, Get3 also contains a hydrophobic pocket lined by many methionine side chains to help it recognize diverse hydrophobic segments independent of their exact sequence. Two proteins at the ER membrane called Get1 and Get2 serve not only as the receptor for Get3 but also as the translocator that inserts the hydrophobic segment of the tail-anchored protein into the lipid bilayer. This post-translational targeting mechanism is therefore conceptually similar to SRP-dependent protein targeting (see Figure 12–20). Some tail-anchored proteins are targeted to mitochondria or peroxisomes instead of the ER, but the mechanism of their targeting is not known.

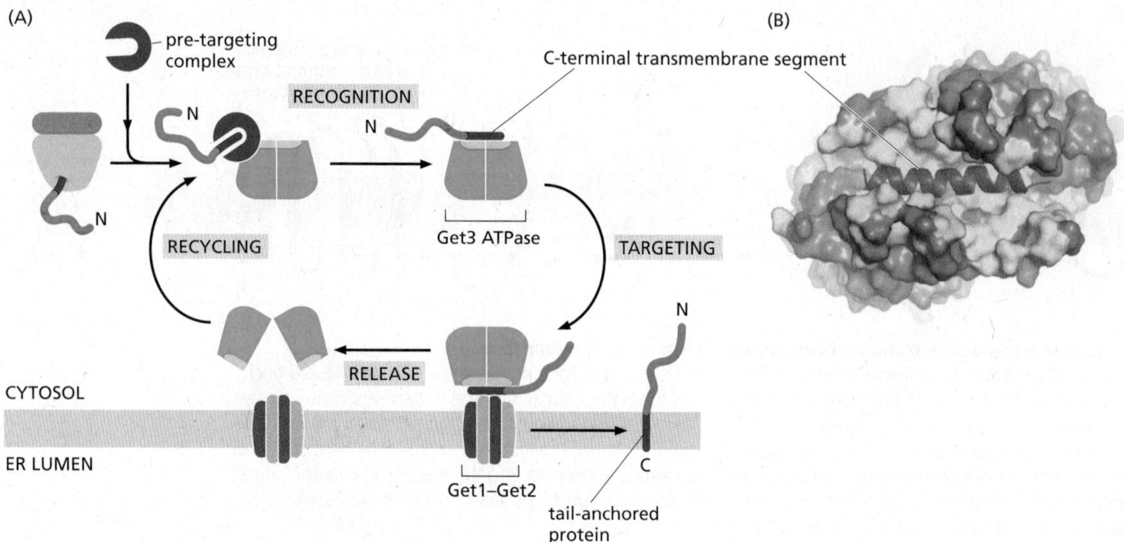

Figure 12–29 The insertion mechanism for tail-anchored proteins. (A) In this post-translational pathway for the insertion of tail-anchored membrane proteins into the ER, a soluble pre-targeting complex captures the hydrophobic C-terminal transmembrane segment *(red)* after it emerges from the ribosomal exit tunnel and loads it onto the Get3 targeting factor. The resulting complex is targeted to the ER membrane by interaction with the Get1–Get2 receptor complex, which functions as a membrane protein insertion machine. After the tail-anchored protein is released from Get3 and inserted into the ER membrane, Get3 is recycled back to the cytosol. This targeting cycle is conceptually similar to protein targeting by SRP (see Figure 12–20). Although not shown in the figures, both Get3 and SRP bind and hydrolyze nucleoside triphosphates to provide directionality to the targeting cycle. ATP is used by Get3, and GTP is used by SRP. (B) Crystal structure of the Get3 targeting factor bound to a transmembrane segment *(red helix)*. The hydrophobic transmembrane segment binds to a deep groove in Get3 lined by hydrophobic amino acids *(yellow)*, including many flexible methionines. (PDB code: 4XTR.)

Some Membrane Proteins Acquire a Covalently Attached Glycosylphosphatidylinositol (GPI) Anchor

Another way that proteins are attached to the membrane is by a **glycosylphosphatidylinositol (GPI) anchor** that is covalently linked to the C-terminus of some proteins destined for the plasma membrane. GPI-anchored proteins are initially made with an N-terminal signal sequence to direct them to the ER and a hydrophobic segment very close to the C-terminus. This hydrophobic segment is selectively recognized by a transamidase enzyme in the ER membrane that simultaneously cleaves off the hydrophobic segment and attaches a preformed GPI anchor to the rest of the protein (**Figure 12–30**). Many plasma membrane proteins are modified in this way. Because they are attached to the exterior of the plasma membrane only by their GPI anchors, they can be released from cells in soluble form in response to signals that activate a specific phospholipase in the plasma membrane. Trypanosome parasites, for example, use this mechanism to shed their coat of GPI-anchored surface proteins when attacked by the immune system. GPI anchors also participate in directing some plasma membrane proteins into specialized domains, such as lipid rafts, thus laterally segregating them from other membrane proteins (see Figure 10–13).

Translocated Polypeptide Chains Fold and Assemble in the Lumen of the Rough ER

Proteins enter the ER lumen as unfolded polypeptides. They must therefore fold and assemble into their correct three-dimensional structures just as newly made proteins in the cytosol must fold (discussed in Chapter 3). To meet this demand, the lumen of the ER contains a high concentration of resident chaperones and other protein-folding catalysts. These **ER resident proteins** contain an **ER retention signal** of four amino acids at their C-terminus that is responsible for retaining the protein in the ER (see Figure 12–13; discussed in Chapter 13, p. 768).

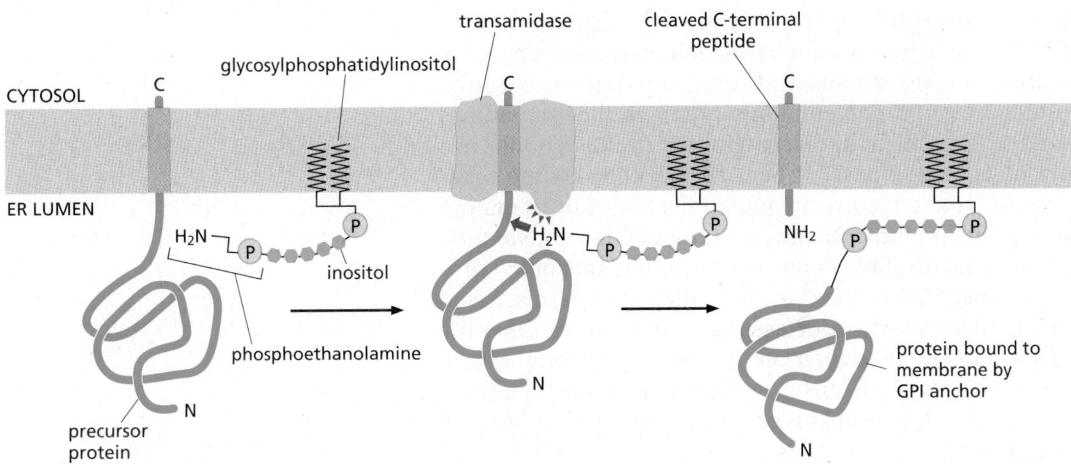

Figure 12–30 **The attachment of a GPI anchor to a protein in the ER.** GPI-anchored proteins are targeted to the ER membrane by an N-terminal signal sequence (not shown), integrated into the membrane, and processed by signal peptidase similarly to a single-pass transmembrane protein (see Figure 12–27). Immediately after the completion of protein synthesis, the precursor protein remains anchored in the ER membrane by a hydrophobic C-terminal sequence of 15–20 amino acids; the rest of the protein is in the ER lumen. Within less than a minute, a transamidase enzyme in the ER cleaves the protein from its membrane-bound C-terminus and simultaneously attaches the new C-terminus to an amino group on a preassembled GPI intermediate. The sugar chain contains an inositol attached to the lipid from which the GPI anchor derives its name. It is followed by a glucosamine and three mannoses. The terminal mannose links to a phosphoethanolamine that provides the amino group to attach the protein through an amide bond. The signal that specifies this modification is contained within the hydrophobic C-terminal sequence and a few amino acids adjacent to it on the lumenal side of the ER membrane; if this signal is added to other proteins, they too become modified in this way. Because of the covalently linked lipid anchor, the protein remains membrane-bound, with all of its amino acids exposed initially on the lumenal side of the ER and eventually on the exterior of the plasma membrane.

The protein **BiP**, a member of the hsp70 family of chaperone proteins, is a major component of the ER folding machinery. We have already discussed how BiP pulls proteins post-translationally into the ER through the Sec61 ER translocator. Like other chaperones (discussed in Chapter 6), BiP recognizes incorrectly folded proteins, as well as protein subunits that have not yet assembled into their final oligomeric complexes. It does so by binding to exposed hydrophobic amino acid sequences that would normally be buried in the interior of correctly folded or assembled polypeptide chains. The bound BiP both prevents the protein from aggregating and helps keep it in the ER (and thus out of the Golgi apparatus and later parts of the secretory pathway). BiP hydrolyzes ATP to shuttle between high- and low-affinity polypeptide-binding states. In this way, BiP periodically lets go of its substrate proteins to allow them an opportunity to fold, and then re-binds them if folding is not yet achieved.

The ER resident protein *protein disulfide isomerase* (*PDI*) catalyzes the oxidation of free sulfhydryl (SH) groups on cysteines to form disulfide (S—S) bonds (**Figure 12–31**). Almost all cysteines in protein domains exposed to either the

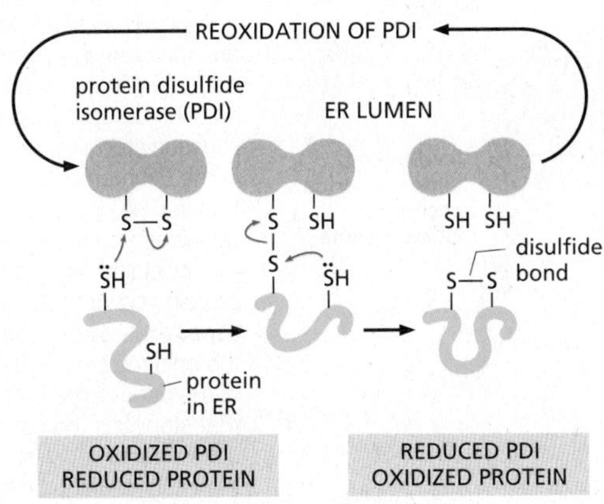

Figure 12–31 **The formation of disulfide bonds in the ER.** Proteins that contain free sulfhydryl (SH) groups on cysteines are oxidized during protein folding to incorporate disulfide (S—S) bonds. Protein disulfide isomerase (PDI) contains an intramolecular disulfide bond that accepts electrons from a free sulfhydryl group in the substrate protein to be oxidized. This leads to the formation of an intermolecular mixed disulfide bond between PDI and its substrate. A second free sulfhydryl group in the substrate then donates its electrons to the mixed disulfide bond, resulting in an oxidized substrate and reduced PDI. Reoxidation of PDI is carried out by other ER enzymes (not shown).

extracellular space or the lumen of organelles in the secretory and endocytic pathways are disulfide bonded. Disulfide bonds stabilize the folded state of a protein, enabling it to better withstand a harsh, variable, and chaperone-free extracellular environment. Because proteins often contain multiple cysteines, they sometimes pair incorrectly. PDI resolves this problem by rearranging the disulfide bonds in a protein until it is correctly folded. This is possible because PDI enzymes are capable of operating in reverse to reduce incorrectly paired disulfides of immature proteins. The ER lumen contains multiple members of the PDI family, some of which are specialized for reducing disulfide bonds to fully unfold misfolded proteins that need to be translocated back to the cytosol for degradation (discussed later). All PDI enzymes are therefore oxidoreductases that can catalyze either the formation or breakage of disulfide bonds in their client proteins. The formation of disulfide bonds relies on maintaining an oxidizing environment in the ER lumen. Disulfide bonds form only very rarely in domains exposed to the cytosol because of the reducing environment there.

Most Proteins Synthesized in the Rough ER Are Glycosylated by the Addition of a Common *N*-Linked Oligosaccharide

The covalent addition of oligosaccharides to proteins is one of the major biosynthetic functions of the ER. About half of the soluble and membrane-bound proteins that are processed in the ER—including those destined for transport to the Golgi apparatus, lysosomes, plasma membrane, or extracellular space—are **glycoproteins** that are modified in this way. Some proteins in the cytosol and nucleus are also glycosylated, but not with large oligosaccharides: they instead carry a much simpler sugar modification, in which a single *N*-acetylglucosamine group is added to a serine or threonine of the protein.

During the most common form of **protein glycosylation** in the ER, a preformed *precursor oligosaccharide* (containing 14 sugars composed of 2 *N*-acetylglucosamines, 9 mannoses, and 3 glucoses) is transferred as a complete unit to proteins. Because this oligosaccharide is transferred to the side-chain NH_2 group of an asparagine in the protein, it is said to be *N-linked*, or *asparagine-linked* (**Figure 12–32A**). A special lipid molecule called **dolichol** (see Panel 2–5, pp. 102–103) anchors the precursor oligosaccharide in the ER membrane. The precursor oligosaccharide is transferred to the target asparagine in a single enzymatic step by an **oligosaccharyl transferase.** This membrane-bound enzyme associates with the Sec61 translocator and has its active site exposed on the lumenal side

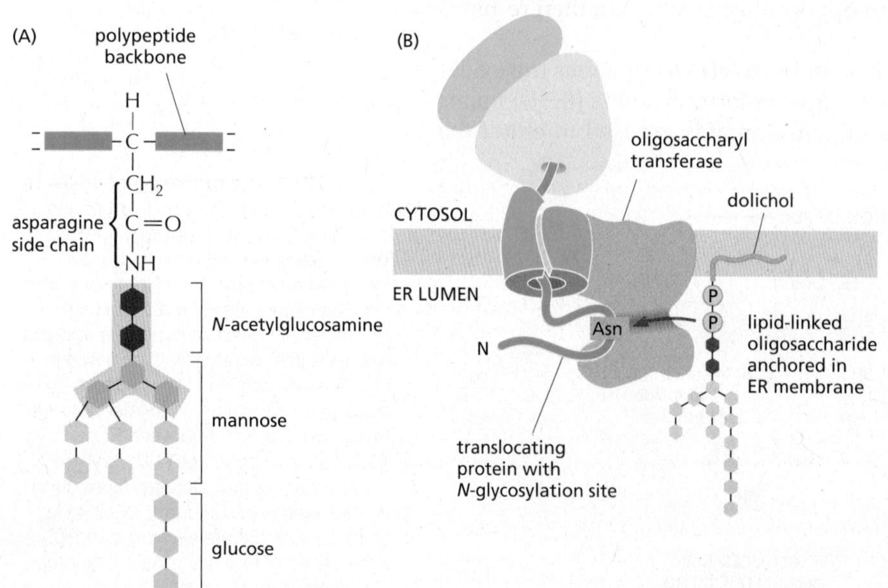

Figure 12–32 ***N*-linked protein glycosylation in the rough ER.** (A) Almost as soon as a polypeptide chain enters the ER lumen, it is glycosylated on target asparagine amino acids. The precursor oligosaccharide (shown in color) is attached only to asparagine side chains in the sequences Asn-X-Ser and Asn-X-Thr (where X is any amino acid except proline). These sequences occur much less frequently in glycoproteins than in nonglycosylated cytosolic proteins. Evidently there has been selective pressure against these sequences during protein evolution, presumably because glycosylation at inappropriate sites would interfere with protein folding. The five sugars in the *gray box* form the core region of this oligosaccharide. For many glycoproteins, only the core sugars survive the extensive oligosaccharide trimming that takes place in the Golgi apparatus (Movie 13.4). (B) The precursor oligosaccharide is transferred from a dolichol lipid anchor to the asparagine as an intact unit in a reaction catalyzed by a transmembrane *oligosaccharyl transferase* enzyme complex. One copy of this enzyme is associated with each protein translocator in the ER membrane. Oligosaccharyl transferase contains 13 transmembrane α helices and a large ER lumenal domain that contains binding sites for the nascent protein and dolichol–oligosaccharide. The asparagine binds a tunnel that penetrates the enzyme interior. There, the amino group of the asparagine is twisted out of the plane that stabilizes the otherwise poorly reactive amide bond, activating it for reaction with the dolichol–oligosaccharide.

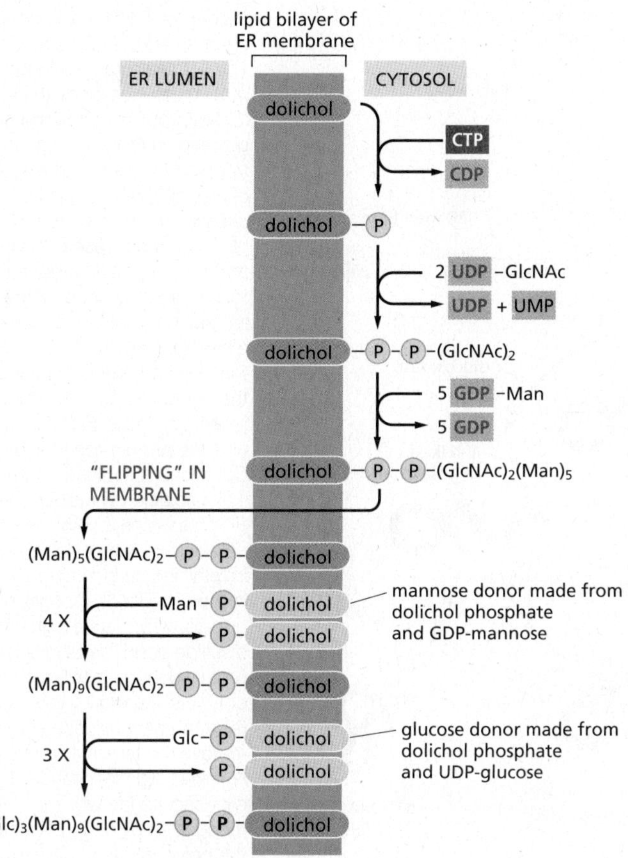

Figure 12–33 **Synthesis of the lipid-linked precursor oligosaccharide in the rough ER membrane.** The oligosaccharide is assembled sugar by sugar onto the carrier lipid dolichol (a polyisoprenoid; see Panel 2–5, pp. 102–103). Dolichol is long and very hydrophobic: its 22 five-carbon units can span the thickness of a lipid bilayer more than three times, so that the attached oligosaccharide is firmly anchored in the membrane. The first sugar is linked to dolichol by a pyrophosphate bridge. This high-energy bond activates the oligosaccharide for its eventual transfer from the lipid to an asparagine side chain of a growing polypeptide on the lumenal side of the rough ER. As indicated, the synthesis of the oligosaccharide starts on the cytosolic side of the ER membrane and continues on the lumenal face after the $(Man)_5(GlcNAc)_2$ lipid intermediate is flipped across the bilayer by a transporter (which is not shown). All the subsequent glycosyl transfer reactions on the lumenal side of the ER involve transfers from dolichol-P-glucose and dolichol-P-mannose; these activated, lipid-linked monosaccharides are synthesized from dolichol phosphate and UDP-glucose or GDP-mannose (as appropriate) on the cytosolic side of the ER and are then flipped across the ER membrane. GlcNAc = *N*-acetylglucosamine; Man = mannose; Glc = glucose.

of the ER membrane. This allows the oligosaccharyl transferase to modify newly made proteins immediately after the target asparagine enters the ER lumen during protein translocation (**Figure 12–32B**).

The precursor oligosaccharide is built up sugar by sugar on the membrane-bound dolichol lipid. The sugars are first activated in the cytosol by the formation of *nucleotide (UDP or GDP)-sugar intermediates,* which then donate their sugar first to the dolichol lipid and then to the partially assembled oligosaccharide tree in an orderly sequence. Partway through this process, the lipid-linked oligosaccharide is flipped, with the help of a transporter, from the cytosolic to the lumenal side of the ER membrane (**Figure 12–33**).

The *N*-linked oligosaccharides are by far the most common oligosaccharides, being found on 90% of all glycoproteins. Less frequently, oligosaccharides are linked to the hydroxyl group on the side chain of a serine, threonine, hydroxylysine, or hydroxyproline amino acid. The first sugar of these *O-linked oligosaccharides* is added in the ER. *N*-linked and *O*-linked oligosaccharides undergo extensive processing, modification, and extension in the Golgi apparatus (Chapter 13), producing the diversity of oligosaccharide structures observed on mature glycoproteins.

Oligosaccharides Are Used as Tags to Mark the State of Protein Folding

It has long been debated why glycosylation is such a common modification of proteins that enter the ER. One particularly puzzling observation has been that some proteins require *N*-linked glycosylation for proper folding in the ER, yet the precise location of the oligosaccharides attached to the protein's surface does not seem to matter. A clue to the role of glycosylation in protein folding came from studies of two ER chaperone proteins, which are called **calnexin** and **calreticulin** because they require Ca^{2+} for their activities. These chaperones are carbohydrate-binding proteins, or *lectins,* which bind to oligosaccharides on incompletely folded proteins

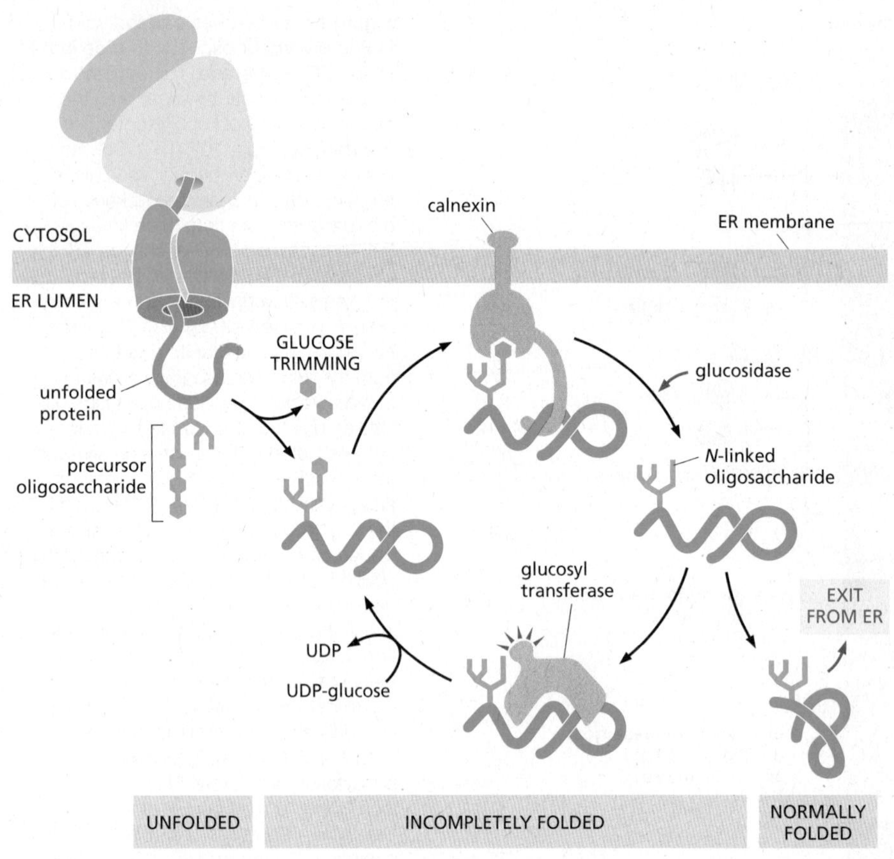

Figure 12–34 **The role of *N*-linked glycosylation in ER protein folding.** The ER membrane–bound chaperone protein calnexin binds to incompletely folded proteins containing one terminal glucose on *N*-linked oligosaccharides, trapping the protein in the ER. Removal of the terminal glucose by a glucosidase releases the protein from calnexin. A glucosyl transferase is the crucial enzyme that determines whether the protein is folded properly or not: if the protein is still incompletely folded, the enzyme transfers a new glucose from UDP-glucose to the *N*-linked oligosaccharide, renewing the protein's affinity for calnexin and retaining it in the ER. The cycle repeats until the protein has folded completely. Calreticulin functions similarly, except that it is a soluble ER resident protein. Another ER chaperone, ERp57 (not shown), collaborates with calnexin and calreticulin in retaining an incompletely folded protein in the ER. ERp57 recognizes free sulfhydryl groups, which are a sign of incomplete disulfide bond formation. The longer a protein spends in this cycle without folding correctly, the more likely it is that ER-resident mannosidase enzymes (not shown) remove the terminal mannoses from the *N*-linked oligosaccharide. The trimmed oligosaccharide with reduced mannoses is recognized by other ER lectins that route the polypeptide for degradation. Thus, only proteins that fold promptly and exit the ER avoid trimming by mannosidases and escape degradation.

and retain them in the ER. Like other chaperones, they prevent incompletely folded proteins from irreversibly aggregating. Both calnexin and calreticulin also promote the association of incompletely folded proteins with another ER chaperone, which binds to cysteines that have not yet formed disulfide bonds.

How do calnexin and calreticulin distinguish properly folded from incompletely folded proteins? The answer lies in the structure of the oligosaccharide attached to the protein. Shortly after a newly made protein acquires an *N*-linked precursor oligosaccharide, ER glucosidases rapidly remove two glucoses, leaving behind a single terminal glucose. This singly glucosylated oligosaccharide is recognized by calnexin and calreticulin, ensuring that all newly made (and hence, likely to be not yet folded) glycoproteins bind to one of these chaperones. This last glucose is removed over time, leaving a de-glucosylated glycoprotein that no longer binds to calnexin or calreticulin. If the glycoprotein is folded, it can leave the ER. However, yet another ER enzyme, a glucosyl transferase, re-adds the terminal glucose selectively to glycoproteins that have not yet folded completely. The terminal glucose then causes re-association of the unfolded protein with calnexin or calreticulin. Thus, glucose trimming (by glucosidases) and glucose addition (by the glucosyl transferase) drive cycles of dissociation and re-association with calnexin and calreticulin until a newly made unfolded protein has achieved its fully folded state (**Figure 12–34**).

Improperly Folded Proteins Are Exported from the ER and Degraded in the Cytosol

Despite all the help from chaperones, many protein molecules translocated into the ER fail to achieve their properly folded or oligomeric state. Such proteins are exported from the ER back into the cytosol, where they are degraded in proteasomes (discussed in Chapter 6). In many ways, the mechanism of such retrotranslocation is similar to other post-translational modes of translocation.

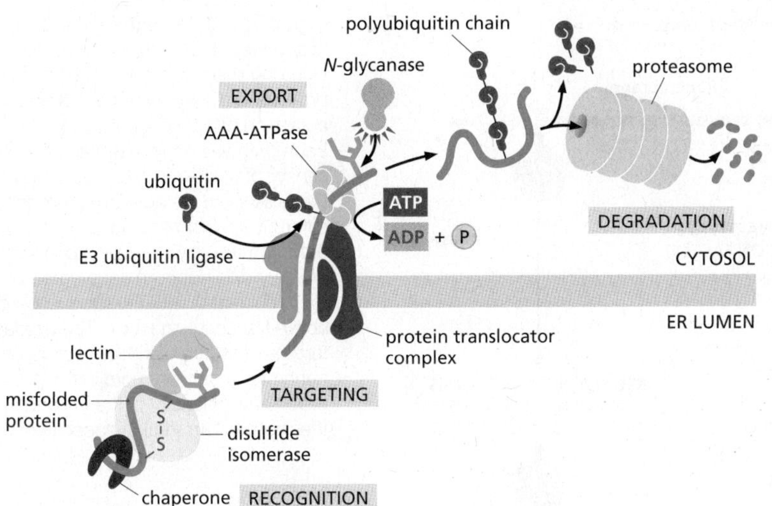

Figure 12–35 The export and degradation of misfolded ER proteins. Misfolded soluble proteins in the ER lumen are recognized and targeted to a translocator complex in the ER membrane. They first interact in the ER lumen with chaperones, disulfide isomerases, and lectins. The chaperones maintain the misfolded protein in an unfolded conformation and prevent their aggregation. The disulfide isomerases reduce disulfide bonds to fully unfold the protein. The lectins selectively recognize trimmed *N*-linked oligosaccharides that are generated when a protein spends too long in the ER. The lectins have binding sites on a membrane-embedded protein translocator built around an E3 ubiquitin ligase. The unfolded protein is then exported into the cytosol through the translocator. The E3 ubiquitin ligase ubiquitylates the unfolded protein as it emerges on the cytosolic side of the translocator. The ubiquitin prevents backsliding of the protein into the ER and provides a molecular handle for an AAA-ATPase that completes the extraction reaction. The unfolded protein is then de-glycosylated and degraded in proteasomes. Misfolded membrane proteins follow a similar pathway but are thought to engage the translocator sideways within the lipid bilayer. Multiple translocator complexes containing different E3 ubiquitin ligases reside in the ER. They are thought to handle different subsets of proteins that are misfolded in different ways.

For example, like post-translational import into the ER, chaperone proteins are necessary to keep the polypeptide chain in an unfolded state prior to and during translocation. Similarly, a source of energy is required to provide directionality to the transport and to pull the protein into the cytosol. Finally, a translocator is necessary.

Selecting proteins from the ER for degradation is a challenging process: misfolded proteins or unassembled protein subunits should be degraded, but folding intermediates of newly made proteins should not. Help in making this distinction comes from the *N*-linked oligosaccharides, which serve as timers that measure how long a protein has spent in the ER. The slow trimming of a particular mannose on the core oligosaccharide tree by an enzyme (a mannosidase) in the ER creates a new oligosaccharide structure that ER-lumenal lectins of the retrotranslocation apparatus recognize. A protein that folds and exits from the ER faster than the mannosidase can remove its target mannose therefore escapes degradation.

In addition to the lectins in the ER that recognize the oligosaccharides, chaperones and protein disulfide isomerases associate with the proteins that must be degraded. The chaperones prevent the unfolded proteins from aggregating, and the disulfide isomerases break disulfide bonds that may have formed incorrectly, so that a linear polypeptide chain can be translocated back into the cytosol.

Multiple translocator complexes move different proteins from the ER membrane or lumen into the cytosol. Translocator complexes always contain an E3 ubiquitin ligase enzyme (Chapter 6), which attaches polyubiquitin tags to the unfolded proteins as they emerge into the cytosol, marking them for destruction. Fueled by the energy derived from ATP hydrolysis, a hexameric ATPase of the family of AAA-ATPases (see Figure 6–88) pulls the unfolded protein through the translocator into the cytosol. An *N*-glycanase removes *en bloc* any oligosaccharide chains attached to the retrotranslocated protein. Guided by its ubiquitin tag, the de-glycosylated polypeptide is rapidly fed into proteasomes, where it is degraded (**Figure 12–35**).

Misfolded Proteins in the ER Activate an Unfolded Protein Response

Cells carefully monitor the amount of misfolded protein in various compartments. An accumulation of misfolded proteins in the cytosol, for example, triggers a *heat-shock response* (discussed in Chapter 6), which stimulates the transcription of genes encoding cytosolic chaperones that help to refold the proteins. Similarly, an accumulation of misfolded proteins in the ER triggers an **unfolded**

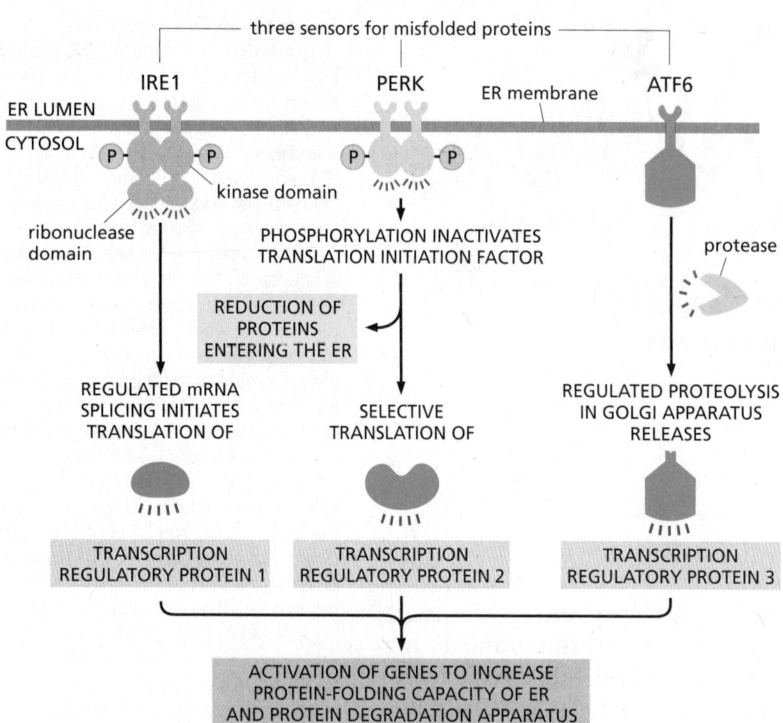

Figure 12–36 **The unfolded protein response.** Three parallel intracellular signaling pathways sense misfolded proteins in the ER lumen and lead to the activation of transcription in the nucleus. Each pathway begins with an ER-resident sensor of misfolded proteins. When these sensors are activated, they initiate different downstream signaling pathways. Although the downstream mechanisms are very different from each other, all of them culminate with the production of an active transcription factor. The overlapping targets of the transcription factors produce gene products that improve the protein-processing capacity of the ER and increase the protein degradation capacity of the cell.

protein response, which stimulates transcription of genes that collectively improve the protein-folding capacity of the ER. The stimulated genes code for ER chaperones, the machinery for protein retrotranslocation and degradation, factors for protein transport out of the ER, and factors for expansion of the ER. This multipronged response operates by coupling the detection of misfolded proteins in the ER lumen to the production of transcription regulatory proteins that enter the nucleus to tune the transcription of hundreds of genes.

How do misfolded proteins in the ER signal to the nucleus? There are three parallel pathways that execute the unfolded protein response (Figure 12–36). The first pathway, which was initially discovered in yeast cells, is conserved in all eukaryotic cells and is particularly remarkable. Misfolded proteins in the ER cause IRE1, a transmembrane protein kinase in the ER, to oligomerize and phosphorylate itself. This mechanism of activation is similar to how some cell-surface receptor kinases in the plasma membrane are activated (discussed in Chapter 15). Oligomeric and phosphorylated IRE1 enables its cytosolic endoribonuclease domain to remove an intron from a specific cytosolic mRNA molecule. IRE1 accomplishes this task by cleaving the mRNA at two positions that are then joined together by an RNA ligase. The mRNA produced by this splicing reaction is translated to produce an active transcription regulatory protein that increases expression of a subset of the genes of the unfolded protein response (Figure 12–37). The regulated splicing of a cytosolic mRNA by IRE1 is a unique exception to the rule that all mRNA splicing occurs in the nucleus and is catalyzed by the spliceosome.

Misfolded proteins also activate a second transmembrane kinase in the ER, PERK. The target of activated PERK is a translation initiation protein whose phosphorylation has two consequences. First, translation of new proteins is reduced throughout the cell, thereby reducing the load of proteins that need to be folded in the ER. Second, some proteins are preferentially translated when translation initiation factors are scarce, and one of these is a transcription regulator that helps activate the transcription of the genes that execute the unfolded protein response.

Finally, a third transcription regulator, ATF6, is initially synthesized as a transmembrane ER protein. Because it is embedded in the ER membrane, it cannot

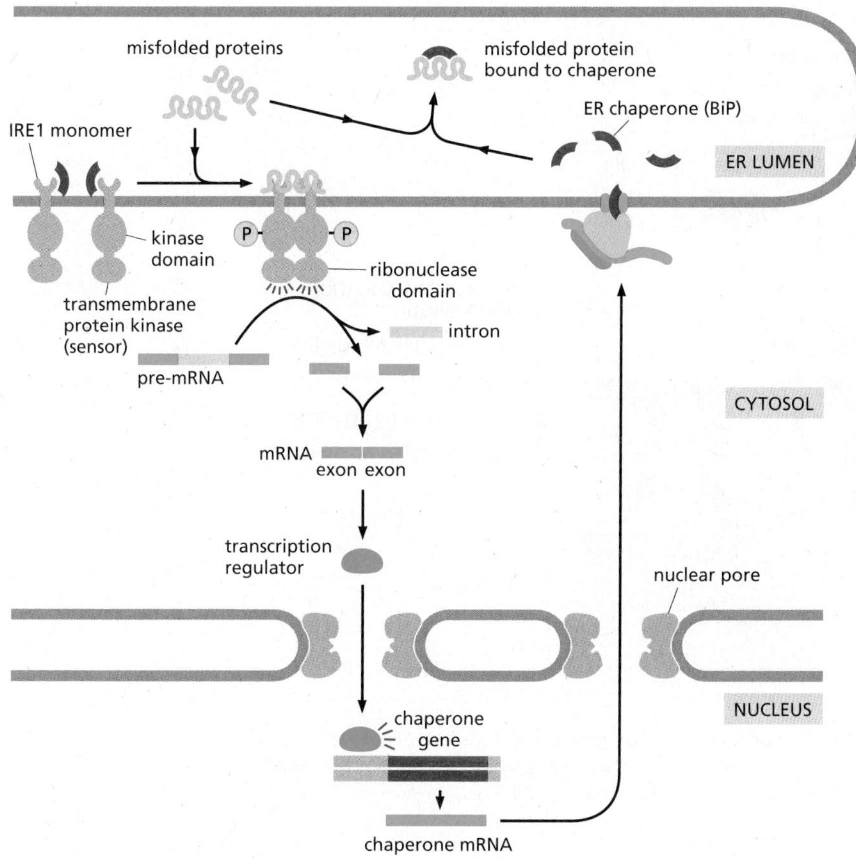

Figure 12–37 **The IRE1 limb of the unfolded protein response.** Regulated RNA splicing is a key regulatory switch in the unfolded protein response pathway initiated by IRE1 (Movie 12.4). During normal conditions, IRE1 is maintained in an inactive state by its association with the ER-lumenal chaperone BiP. Elevated levels of misfolded proteins activate IRE1 by a combination of two mechanisms. First, BiP dissociates from IRE1 to bind and protect misfolded proteins from aggregation. Second, misfolded proteins bind to the lumenal domain of IRE1 facilitating the formation of IRE1 oligomers. The oligomerized IRE1 phosphorylates itself on the cytosolic side, activating its ribonuclease domain. The activated ribonuclease catalyzes the splicing of a pre-mRNA that codes for a transcription factor that ultimately activates numerous genes in the nucleus including those coding for chaperones. Elevated chaperones help reduce the level of misfolded proteins in the ER lumen, eventually turning off IRE1 signaling.

activate the transcription of genes in the nucleus. When misfolded proteins accumulate in the ER, the ATF6 protein is transported to the Golgi apparatus. Resident proteases in the Golgi apparatus membrane cleave off the cytosolic domain of ATF6, which can now migrate to the nucleus and help activate the transcription of genes encoding proteins involved in the unfolded protein response. This mechanism of activation of a latent membrane-embedded transcription factor is similar to how the transcription regulator that controls cholesterol biosynthesis is activated (discussed later in this chapter). The relative importance of each of these three pathways in the unfolded protein response differs in different cell types, enabling each cell type to tailor the unfolded protein response to its particular needs.

The signaling pathways that execute the unfolded protein response are used during normal physiological conditions to adjust ER capacity to closely match demand for the ER. For example, insulin production increases substantially in pancreatic β cells in response to eating a meal. The elevated demand for the processing capacity of the ER, where insulin is initially assembled, partially activates PERK so cells can adjust insulin synthesis rates to avoid overburdening the ER. In another example, IRE1 is activated when B cells begin differentiating into antibody-secreting plasma cells. IRE1 activation dramatically expands the ER content of the cell in preparation for the very high levels of immunoglobulins that will soon be assembled there.

The unfolded protein response ultimately increases the production of proteins that improve protein processing in the ER and reduce the burden of misfolded proteins. As homeostasis is restored, the activities of IRE1, PERK, and ATF6 abate. If homeostasis cannot be restored, persistently active signaling from the ER, particularly via PERK, activates genes that initiate apoptosis. In multicellular organisms, it is often less detrimental to eliminate a persistently dysfunctional cell than risk its aberrant interactions with neighboring cells.

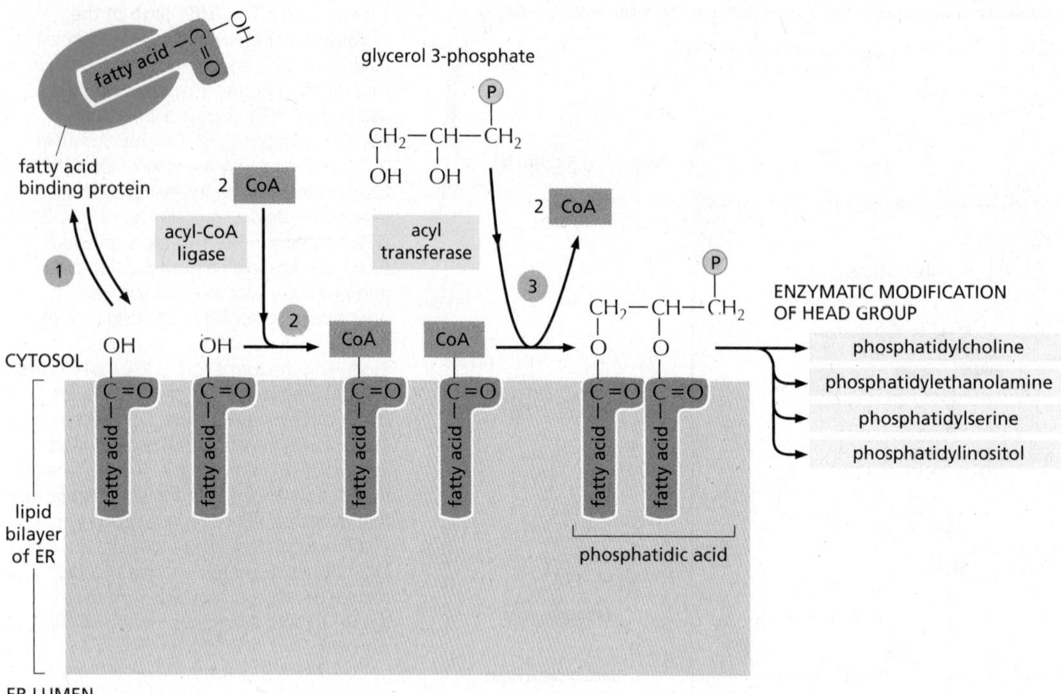

Figure 12–38 **The synthesis of phospholipids at the ER membrane.** As illustrated, fatty acids delivered to the ER by a cytosolic fatty acid binding protein are linked to glycerol 3-phosphate to produce phosphatidic acid, which serves as a precursor to make other phospholipids that differ in the structures of their polar head groups.

The ER Assembles Most Lipid Bilayers

The ER membrane is the site of synthesis of nearly all of the cell's major classes of lipids, including both phospholipids and cholesterol, required for the production of new cell membranes. The major phospholipid made is *phosphatidylcholine*, which can be formed in three steps from choline, two fatty acids, and glycerol phosphate (**Figure 12–38**). Each step is catalyzed by enzymes in the ER membrane, which have their active sites facing the cytosol, where all of the required metabolites are found. Thus, phospholipid synthesis occurs exclusively in the cytosolic leaflet of the ER membrane. Because fatty acids are not soluble in water, they are shepherded from their sites of synthesis in the cytosol to the ER by a fatty acid binding protein. After arrival in the ER membrane and activation with CoA, acyl transferases successively add two fatty acids to glycerol phosphate to produce phosphatidic acid. Phosphatidic acid is sufficiently water-insoluble to remain in the lipid bilayer; it cannot be extracted from the bilayer by the fatty acid binding proteins. It is therefore this first step that enlarges the ER lipid bilayer. The later steps determine the head group of a newly formed lipid molecule and therefore the chemical nature of the bilayer, but they do not result in net membrane growth. The two other major membrane phospholipids—phosphatidylethanolamine and phosphatidylserine (see Figure 10–3)—as well as the minor phospholipid phosphatidylinositol (PI), are all synthesized in this way.

Because phospholipid synthesis takes place in the cytosolic leaflet of the ER lipid bilayer, there needs to be a mechanism that transfers some of the newly formed phospholipid molecules to the lumenal leaflet of the bilayer. In synthetic lipid bilayers, lipids do not "flip-flop" in this way (see Figure 10–10). In the ER, however, phospholipids equilibrate across the membrane within minutes, which is almost 100,000 times faster than can be accounted for by spontaneous "flip-flop." This rapid trans-bilayer movement is mediated by a poorly characterized phospholipid translocator called a *scramblase*, which nonselectively equilibrates

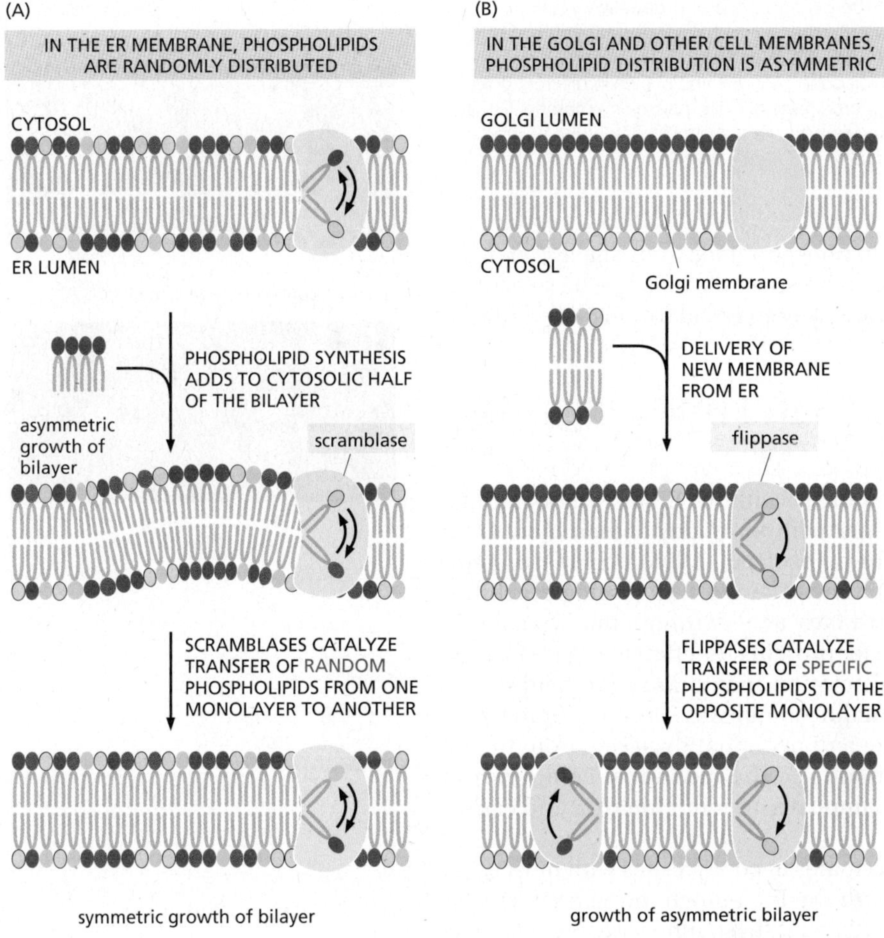

Figure 12–39 The role of phospholipid translocators in lipid bilayer synthesis. (A) Because new lipid molecules are added only to the cytosolic half of the ER membrane bilayer and lipid molecules do not flip spontaneously from one monolayer to the other, a transmembrane phospholipid translocator (called a scramblase) is required to transfer lipid molecules from the cytosolic half to the lumenal half so that the membrane grows as a bilayer. The scramblase is not specific for particular phospholipid head groups and therefore equilibrates the different phospholipids between the two monolayers. Scramblases do not need energy to catalyze phospholipid flipping and probably function by providing a hydrophilic path for passive movement of the phospholipid head group through the hydrophobic interior of the membrane. (B) The membranes of the Golgi apparatus, cell surface, and other compartments of the secretory and endocytic pathways are asymmetric. When new membrane is delivered via transport vesicles, the incoming lipids must be segregated to the appropriate side of the lipid bilayer to maintain its asymmetry. This is accomplished by enzymes called flippases, which move selective phospholipids unidirectionally from one side of the bilayer to the other. Flippases typically couple the transport of their substrate (the phospholipid head group) to ATP hydrolysis, and are therefore considered active transporters (see Chapter 11).

phospholipids between the two leaflets of the lipid bilayer (Figure 12–39). Thus, the different types of phospholipids are thought to be equally distributed between the two leaflets of the ER membrane.

The ER also produces cholesterol and ceramide (Figure 12–40). *Ceramide* is made by condensing the amino acid serine with a fatty acid to form the amino alcohol *sphingosine* (see Figure 10–3); a second fatty acid is then covalently added to form ceramide. The ceramide is exported to the Golgi apparatus, where it serves as a precursor for the synthesis of two types of lipids. *Glycosphingolipids* (glycolipids; see Figure 10–16) are formed when oligosaccharides are added to ceramide, while *sphingomyelin* (discussed in Chapter 10) results from the addition of phosphocholine. Because glycolipids and sphingomyelin are both produced by enzymes that have their active sites exposed to the lumen of the Golgi apparatus, they are restricted to the noncytosolic leaflet of the lipid bilayers that contain them.

As discussed in Chapter 13, the plasma membrane and the membranes of the Golgi apparatus, lysosomes, and endosomes all form part of a membrane system that communicates with the ER by means of transport vesicles. A large proportion of the lipids that compose the membranes of these organelles is acquired via the membranes delivered by transport vesicles. Despite exchange of membrane lipids through vesicular transport, the lipid composition of each organellar membrane is distinct and contributes to its unique identity and functional properties. This specialization is achieved by a combination of three mechanisms. First, a transport vesicle can have a different lipid composition than the organelle it is departing, thereby delivering only a subset of lipids to its destination. Second, proteins in an organelle's membrane can modify the head groups of certain lipids to change their identity (such as production of sphingomyelin from

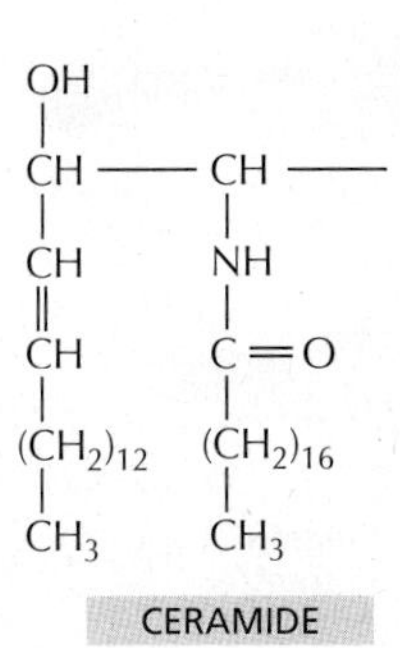

OH — CH — CH(=CH—$(CH_2)_{12}$—CH_3) ; CH—NH—C(=O)—$(CH_2)_{16}$—CH_3 ; CH_2—OH

CERAMIDE

Figure 12–40 The structure of ceramide.

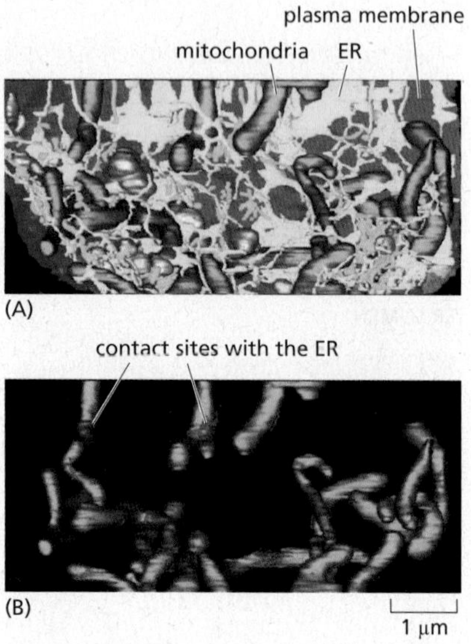

Figure 12–41 The spatial relationships between the ER and several organelles within a mouse neuron. A section of the cell body of a neuron in the mouse brain was analyzed by focused ion beam-scanning electron microscopy. (A) The three-dimensional positions of the major organelles reconstructed from the serial electron microscopy images and shown in different colors. The ER *(yellow)* makes close contacts with all major organelles and the plasma membrane. (B) The mitochondria *(green)* from the reconstruction are shown with the areas that contact the ER *(red)*. (A and B, from Y. Wu et al., *Proc. Natl. Acad. Sci. USA* 114:E4859–E4867, 2017.)

ceramide) or use *flippases* to move certain phospholipids from one leaflet of the membrane to the other (Figure 12–39B). Third, specific lipids can be selectively transferred from one membrane to another by nonvesicular transport routes as discussed next.

Membrane Contact Sites Between the ER and Other Organelles Facilitate Selective Lipid Transfer

Mitochondria and plastids do not communicate with the ER by vesicular transport, so they require different mechanisms to import many of their lipids from the ER for growth. Carrier proteins in the cytosol called *lipid transfer proteins* ferry individual lipid molecules between membranes, functioning much like fatty acid binding proteins that shepherd fatty acids through the cytosol (see Figure 12–38). In many cases, lipid transfer proteins function at organelle contact sites where the originating and destination membranes are held within ~10–30 nm of each other by specific junction complexes. Different lipid transfer proteins shuttle phosphatidylcholine and phosphatidylserine from the ER to mitochondria at contact sites. Disruption of the junctional complexes or the lipid transfer proteins impairs lipid import into mitochondria and causes their dysfunction.

The extensive network of the ER participates in contact sites with most other cellular organelles (**Figure 12–41**). As with the ER-mitochondria contact sites (see Figure 12–16), one of the main functions of these other organelle contact sites is to exchange lipids (**Figure 12–42**). Cells contain several families of lipid transfer proteins. Each of these can typically bind one molecule of a specific lipid (or in some cases multiple related lipids) and has additional domains that can interact with specific cellular membranes. In this manner, they serve as shuttling proteins that have distinctive specificities for the donor and acceptor membranes and the lipid they transport. Contact sites between two organellar membranes favor recruitment of the lipid transfer protein that binds these membranes, thereby enhancing the efficiency of lipid exchange. Cholesterol uses a specialized transport system from lysosomes, where it is delivered as cholesterol esters in lipoproteins, to the plasma membrane and other locations in the cell (as we discuss in Chapter 13).

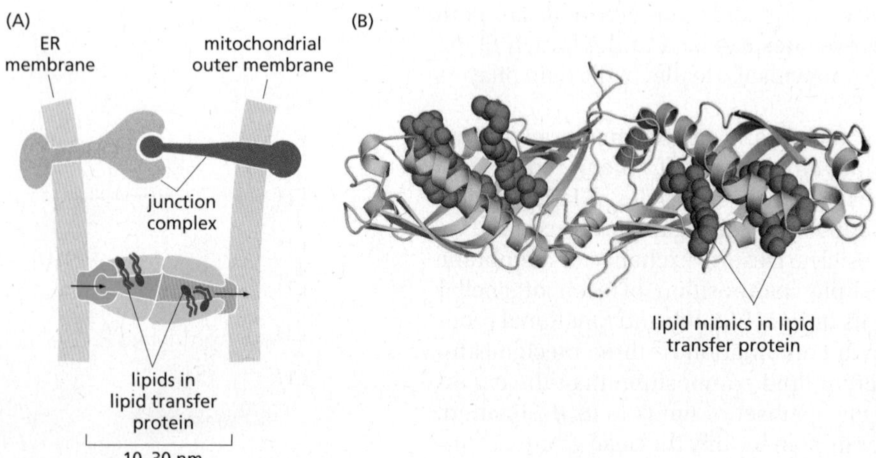

Figure 12–42 The transfer of lipids at organelle contact sites. (A) Proteins anchored to two different membranes (the ER and mitochondrion in the depicted example) interact with each other to hold the membranes 10–30 nm apart. Specialized lipid transfer proteins are recruited to these contact sites or in some cases are part of the junction complex. These transfer proteins have cavities that can bind lipids and facilitate their movement from one membrane to the other. (B) The structure of one such transfer protein is shown with lipid-like molecules bound inside its cavity. (PDB code: 4P42.)

Summary

The extensive ER network serves as a factory for the production of almost all of the cell's lipids. In addition, a major portion of the cell's protein synthesis occurs on the cytosolic surface of the rough ER: virtually all proteins destined for secretion or for the ER itself, the Golgi apparatus, the lysosomes, the endosomes, and the plasma membrane are first imported into the ER from the cytosol. In the ER lumen, the proteins fold and oligomerize, disulfide bonds are formed, and N-linked oligosaccharides are added. The pattern of N-linked glycosylation is used to indicate the extent of protein folding, so that proteins leave the ER only when they are properly folded. Proteins that do not fold or oligomerize correctly are translocated back into the cytosol, where they are de-glycosylated, polyubiquitylated, and degraded in proteasomes. If misfolded proteins accumulate in excess in the ER, they trigger an unfolded protein response, which activates appropriate genes in the nucleus to help the ER cope.

Only proteins that carry a special ER signal sequence are imported into the ER. The signal sequence is recognized by a signal-recognition particle (SRP), which binds both the growing polypeptide chain and the ribosome and directs them to a receptor protein on the cytosolic surface of the rough ER membrane. This binding to the ER membrane initiates the translocation process that threads a loop of polypeptide chain across the ER membrane through the hydrophilic pore of a protein translocator.

Soluble proteins—destined for the ER lumen, for secretion, or for transfer to the lumen of other organelles—pass completely into the ER lumen. Transmembrane proteins destined for the ER or for other cell membranes become anchored in the ER membrane by one or more membrane-spanning α-helical segments in their polypeptide chains. As these hydrophobic portions of the protein emerge from the ribosome, they are recognized by the protein translocator, which provides a passageway into the membrane. When a polypeptide contains multiple hydrophobic segments, it will pass back and forth across the bilayer multiple times as a multipass transmembrane protein.

The asymmetry of protein insertion and glycosylation in the ER establishes the sidedness of the membranes of all the other organelles that the ER supplies with membrane proteins. Lipids are synthesized at the cytosolic face of the ER, equilibrate between both leaflets of the lipid bilayer, and are transported to other organelles often at interorganelle junctions by lipid transfer proteins localized there. Specific flippases establish and maintain lipid asymmetry in the plasma membrane, further contributing to its sidedness.

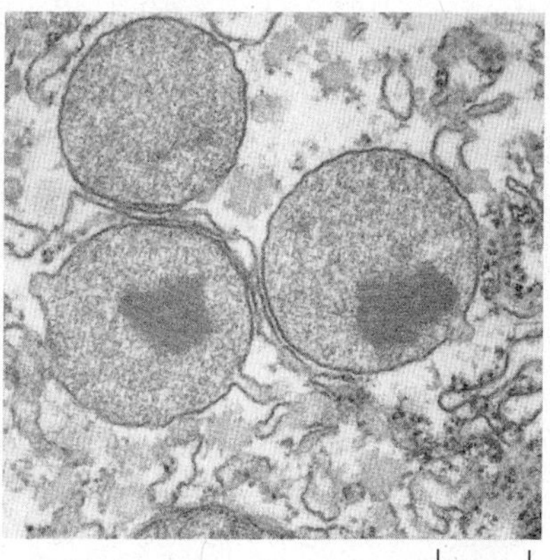

Figure 12–43 An electron micrograph of three peroxisomes in a rat liver cell. The paracrystalline, electron-dense inclusions are composed primarily of the enzyme urate oxidase. (Courtesy of Daniel S. Friend, by permission of E.L. Bearer.)

PEROXISOMES

Peroxisomes are major sites of oxygen utilization and are found in virtually all eukaryotic cells. They contain oxidative enzymes, such as *catalase* and *urate oxidase*, at such high concentrations that, in some cells, the peroxisomes stand out in electron micrographs because of the presence of a crystalloid protein core (**Figure 12–43**). The evolutionary origin of peroxisomes is not firmly established, but they are generally thought to represent a specialized offshoot of the membrane system that composes the secretory and endocytic pathways. One hypothesis is that peroxisomes are a vestige of an ancient organelle that performed all the oxygen metabolism in the primitive ancestors of eukaryotic cells. When the oxygen produced by photosynthetic bacteria first accumulated in the atmosphere, it would have been highly toxic to most cells. Peroxisomes might have lowered the intracellular concentration of oxygen, while also exploiting its chemical reactivity to perform useful oxidation reactions. According to this view, the later development of mitochondria rendered peroxisomes less critical for cellular metabolism because many of the same biochemical reactions—which had formerly been carried out in peroxisomes without producing energy—were now coupled to ATP formation by means of oxidative phosphorylation. The oxidation reactions performed by peroxisomes in present-day cells could therefore partly be those whose functions were not taken over by mitochondria.

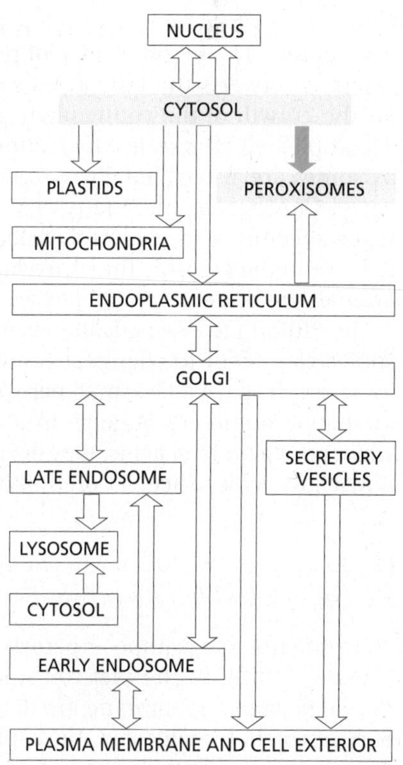

Peroxisomes Use Molecular Oxygen and Hydrogen Peroxide to Perform Oxidation Reactions

Peroxisomes are so named because they usually contain one or more enzymes that use molecular oxygen to remove hydrogen atoms from specific organic substrates (designated here as R) in an oxidation reaction that produces *hydrogen peroxide* (H_2O_2):

$$RH_2 + O_2 \rightarrow R + H_2O_2$$

Catalase uses the H_2O_2 generated by other enzymes in the organelle to oxidize a variety of substrates—including formic acid, formaldehyde, and alcohol—by the "peroxidation" reaction: $H_2O_2 + R'H_2 \rightarrow R' + 2H_2O$. This type of oxidation reaction is particularly important in liver and kidney cells, where the peroxisomes detoxify various harmful molecules that enter the bloodstream. About 25% of the ethanol we drink is oxidized to acetaldehyde in this way. In addition, when excess H_2O_2 accumulates in the cell, catalase converts it to H_2O through the reaction

$$2H_2O_2 \rightarrow 2H_2O + O_2$$

A major function of the oxidation reactions performed in peroxisomes is the breakdown of fatty acid molecules. The process, called *β oxidation*, shortens the alkyl chains of fatty acids sequentially in blocks of two carbon atoms at a time, thereby converting the fatty acids to acetyl CoA. The peroxisomes then export the acetyl CoA to the cytosol for use in biosynthetic reactions. In mammalian cells, β oxidation occurs in both mitochondria and peroxisomes; in fungi and plant cells, however, this essential reaction occurs exclusively in peroxisomes.

An essential biosynthetic function of animal peroxisomes is to catalyze the first reactions in the formation of *plasmalogens*. This abundant class of phospholipids is found in all human cells but is particularly enriched in brain, where it is a major constituent of myelin (**Figure 12–44**). Plasmalogen deficiencies cause profound abnormalities in the myelination of nerve-cell axons, which is one reason why many peroxisomal disorders lead to neurological disease.

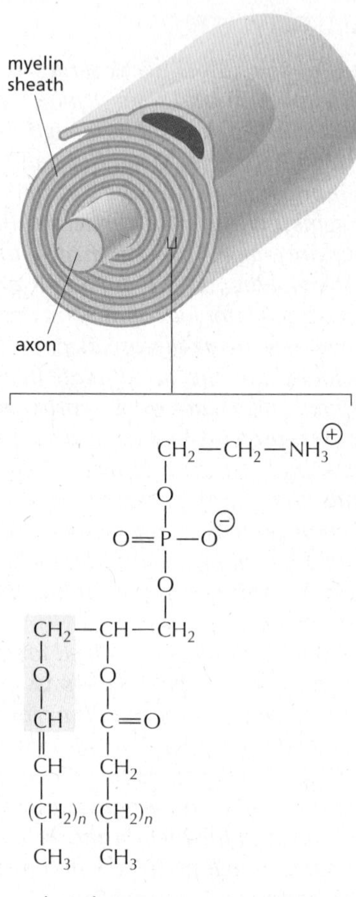

Figure 12–44 The structure of a plasmalogen. Plasmalogens are very abundant in the myelin sheaths that insulate the axons of nerve cells. They make up some 80–90% of the myelin membrane phospholipids. In addition to an ethanolamine head group and a long-chain fatty acid attached to the same glycerol phosphate backbone used for phospholipids, plasmalogens contain an unusual fatty alcohol that is attached through an ether linkage highlighted in *yellow (bottom left)*.

Peroxisomes are unusually diverse organelles, and even in the various cell types of a single organism they may contain different sets of enzymes. For example, most plants have two major types of peroxisomes (**Figure 12–45**). One is present in leaves, where it participates in *photorespiration* (discussed in Chapter 14). The other type of peroxisome is present in germinating seeds, where it converts the fatty acids stored in seed lipids into the sugars needed for the growth of the young plant. Because this conversion of fats to sugars is accomplished by a series of reactions known as the *glyoxylate cycle*, these peroxisomes are also called *glyoxysomes*. In the glyoxylate cycle, two molecules of acetyl CoA produced by fatty acid breakdown in the peroxisome are used to make succinic acid, which then leaves the peroxisome and is converted into glucose in the cytosol. The glyoxylate cycle does not occur in animal cells, and animals are therefore unable to convert fats into carbohydrates.

In addition to diversification across different cell types or organisms, peroxisomes can adapt to changing conditions within a cell. Yeasts grown on sugar, for example, have a few small peroxisomes. But when some yeasts are grown on methanol, numerous large peroxisomes are formed that oxidize methanol; and when grown on fatty acids, they develop numerous large peroxisomes that break down fatty acids to acetyl CoA by β oxidation.

Short Signal Sequences Direct the Import of Proteins into Peroxisomes

The proteins that compose peroxisomes are delivered by two different routes (**Figure 12–46**). In the first route, some of the integral membrane proteins of the peroxisomal membrane are first inserted into the ER using the ER-resident Sec61 protein translocator. These peroxisome-destined proteins are then packaged into specialized peroxisomal precursor vesicles. New precursor vesicles

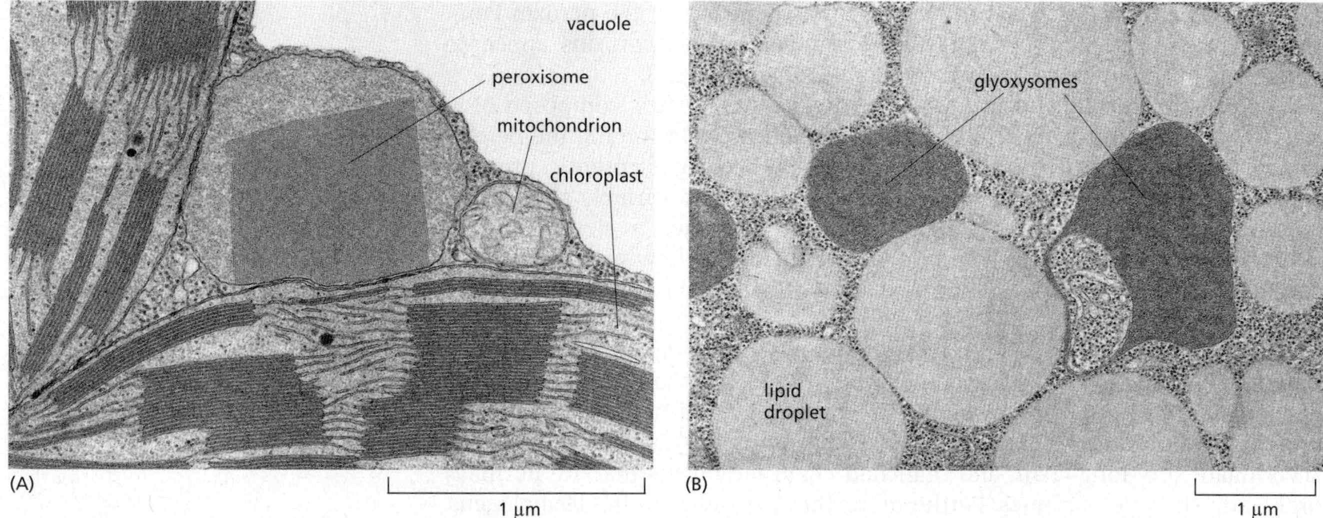

Figure 12–45 Electron micrographs of two types of peroxisomes found in plant cells. (A) A peroxisome with a paracrystalline core in a tobacco leaf mesophyll cell. Its close association with chloroplasts is thought to facilitate the exchange of materials between these organelles during photorespiration. The vacuole in plant cells is equivalent to the lysosome in animal cells. (B) Peroxisomes in a fat-storing cotyledon cell of a tomato seed 4 days after germination. Here the peroxisomes (glyoxysomes) are associated with the lipid droplets that store fat, reflecting their central role in fat mobilization and gluconeogenesis during seed germination. (A, © 1969 S.E. Frederick and E.H. Newcomb. Originally published in *J. Cell Biol.* https://doi.org/10.1083/jcb.43.2.343. With permission from Rockefeller University Press. B, from W.P. Wergin et al., *J. Ultrastruct. Res.* 30:533–557, 1970. With permission from Elsevier.)

then fuse with one another to form a new peroxisome or fuse with an existing peroxisome to facilitate its growth. In the second route, peroxisomal proteins can be imported into preexisting peroxisomes directly from the cytosol. A specific sequence of three amino acids (Ser-Lys-Leu) located at the C-terminus of many peroxisomal proteins functions as an import signal (see Figure 12–13). Other peroxisomal proteins contain a slightly longer and partially hydrophobic signal sequence near the N-terminus. If either sequence is attached to a cytosolic protein, the protein is imported into peroxisomes.

Peroxisomal protein import is driven by ATP hydrolysis and utilizes a collection of proteins, called **peroxins**, that catalyze the import cycle. C-terminal peroxisomal sorting signals are recognized by the peroxin Pex5 in the cytosol. This import receptor accompanies its cargo all the way into a protein translocator in the peroxisomal membrane. After cargo release inside the peroxisome, Pex5 is recycled back to the cytosol. This recycling step requires modification of Pex5 with ubiquitin, which is used as a handle by an ATPase complex composed of Pex1 and Pex6. The Pex1–Pex6 complex harnesses the energy of ATP hydrolysis to release Pex5 from peroxisomes so it can pick up the next cargo molecule.

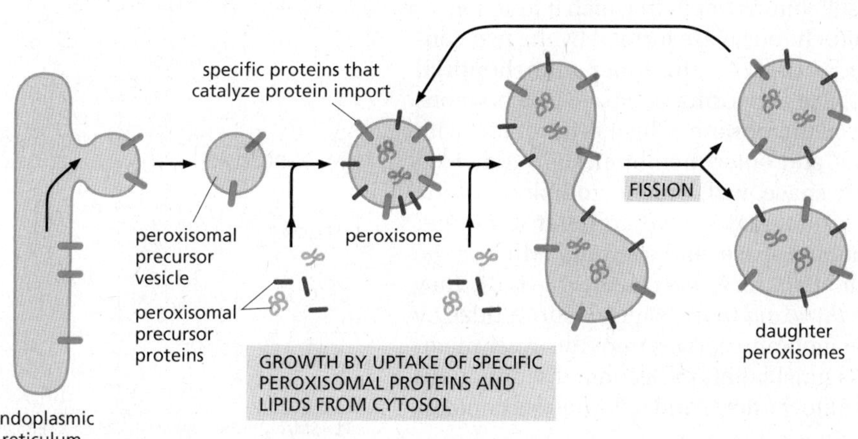

Figure 12–46 A model that explains how peroxisomes proliferate and how new peroxisomes arise. Peroxisomal precursor vesicles bud from the ER. At least two peroxisomal membrane proteins, Pex3 and Pex15, follow this route. The machinery that drives the budding reaction and that selects only peroxisomal proteins for packaging into these vesicles depends on Pex19 and other cytosolic proteins that are still unknown. Peroxisomal precursor vesicles may then fuse with one another or with preexisting peroxisomes. The peroxisomal membrane contains import receptors and protein translocators that are required for the import of peroxisomal proteins made on cytosolic ribosomes, including new copies of the import receptors and translocator components. Presumably, the lipids required for growth are also imported, although some may derive directly from the ER in the membrane of peroxisomal precursor vesicles.

N-terminal peroxisomal signal sequences are recognized by the peroxin Pex7. The Pex7–cargo complex, together with additional accessory peroxins, appear to participate in an import cycle similar to that mediated by Pex5.

The protein translocator in the peroxisomal membrane is composed of at least six different peroxins. Unlike protein translocators in the ER, the peroxisomal translocator can transport fully folded and even oligomeric proteins across the membrane. To allow the passage of large and variably sized cargo molecules, the transporter is thought to dynamically adapt in size to the particular cargo molecules to be transported. It is not known how such a large pore can be utilized for transport without leakage of contents between the cytosol and peroxisome.

The importance of protein import into peroxisomes is demonstrated by the inherited human disease *Zellweger syndrome*. Mutations in any of a dozen different peroxins, the most common being Pex1, cause an impairment in peroxisomal protein import. These individuals, whose cells contain "empty" peroxisomes, accumulate very-long-chain and branched-chain fatty acids that are normally broken down in peroxisomes. Furthermore, they are deficient in plasmalogens. These metabolic impairments cause severe abnormalities in the brain, liver, and kidneys of individuals, and they die soon after birth.

Summary

Peroxisomes are specialized for carrying out oxidation reactions using molecular oxygen. They generate hydrogen peroxide, which they employ for oxidative purposes—and contain catalase to destroy the excess. All peroxisomal proteins are encoded in the cell nucleus. Some of these proteins are conveyed to peroxisomes via peroxisomal precursor vesicles that bud from the ER, but most are synthesized in the cytosol and directly imported. A specific sequence of three amino acids near the C-terminus of many of the latter proteins functions as a peroxisomal import signal that is recognized by a complementary import receptor in the cytosol. Import proceeds through a protein translocator in the peroxisomal membrane, which differs from the protein translocators in the ER in that large and fully folded proteins are imported from the cytosol without unfolding.

THE TRANSPORT OF PROTEINS INTO MITOCHONDRIA AND CHLOROPLASTS

Mitochondria and chloroplasts (a specialized form of plastids in green algae and plant cells) are double membrane–enclosed organelles. They specialize in ATP synthesis, using energy derived from electron transport and oxidative phosphorylation in mitochondria and from photosynthesis in chloroplasts (discussed in Chapter 14). Although both organelles contain their own DNA, ribosomes, and other components required for protein synthesis, almost all of their proteins are encoded in the cell nucleus and imported from the cytosol. Each imported protein must reach the particular organelle subcompartment in which it functions.

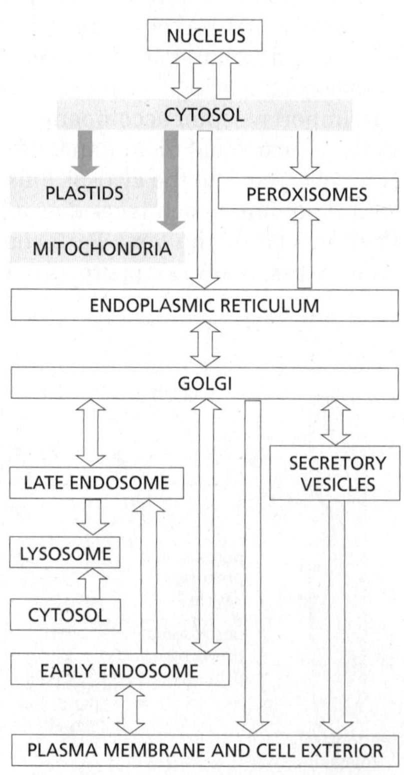

The different subcompartments in **mitochondria** are formed by the two concentric mitochondrial membranes (Figure 12–47A): the **inner mitochondrial membrane**, which encloses the **matrix space** and forms extensive invaginations called *cristae*, and the **outer mitochondrial membrane**, which is in contact with the cytosol. The space between the inner and outer membranes is subdivided into the crista space and **intermembrane space**, with protein complexes at the junctions where the cristae invaginate. Chloroplasts have an outer and inner membrane, which enclose an intermembrane space, and a stroma, which is the chloroplast equivalent of the mitochondrial matrix space (Figure 12–47B). They have an additional subcompartment, the *thylakoid space*, which is surrounded by the *thylakoid membrane*. The thylakoid membrane derives from the inner membrane during plastid development and is pinched off to become discontinuous with it. Each of the subcompartments in mitochondria and chloroplasts contains a distinct set of proteins.

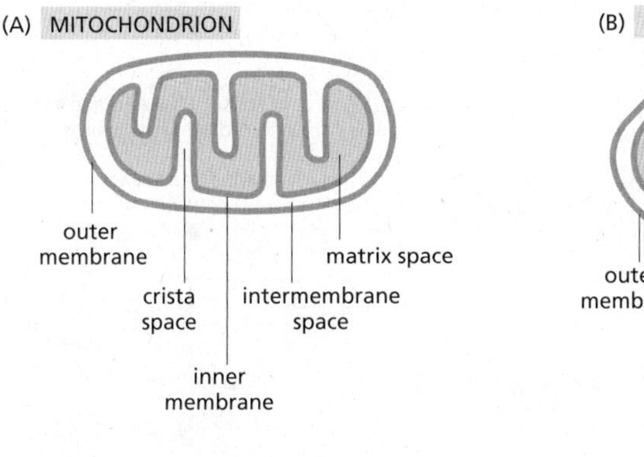

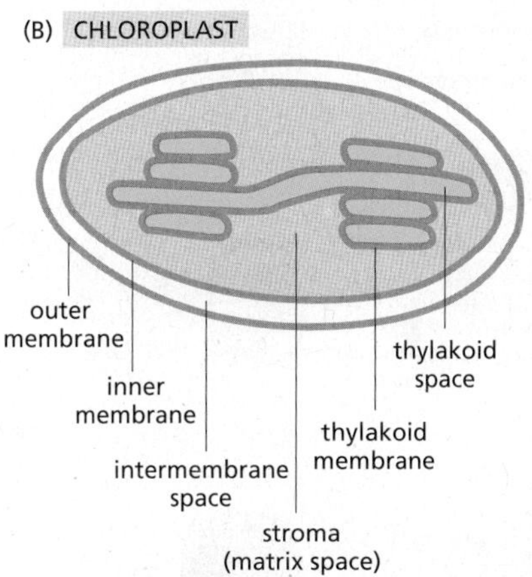

Figure 12–47 The subcompartments of mitochondria and chloroplasts. In contrast to the cristae of mitochondria (A), the thylakoids of chloroplasts (B) are not connected to the inner membrane and therefore form a sealed continuous compartment with a separate internal space.

New mitochondria and chloroplasts are produced by the growth of preexisting organelles, followed by fission (discussed in Chapter 14). The growth depends mainly on the import of proteins from the cytosol. Many of the core principles of protein import into mitochondria and chloroplasts are similar to the analogous process of protein import into the ER we discussed earlier. However, the presence of multiple membranes and subcompartments adds to the complexity of delivering newly imported proteins to the correct location. This section explains how it occurs.

Translocation into Mitochondria Depends on Signal Sequences and Protein Translocators

One or more signal sequences direct all mitochondrial precursor proteins to their appropriate mitochondrial subcompartment. Many proteins entering the matrix space contain a signal sequence at their N-terminus that a signal peptidase rapidly removes after import. Other imported proteins, including all outer membrane and many inner membrane and intermembrane-space proteins, have internal signal sequences that are not removed. The signal sequences are both necessary and sufficient for the import and correct localization of the proteins: when genetic engineering techniques are used to link these signals to a cytosolic protein, the signals direct the protein to the correct mitochondrial subcompartment. Thus, the principles of the signal hypothesis, conceived to explain how proteins are segregated to the ER, also apply to mitochondria.

Multisubunit protein complexes that function as protein translocators mediate protein movement across or into mitochondrial membranes (**Figure 12–48A**). To provide access to each mitochondrial subcompartment, protein translocator complexes are located in both the inner and outer mitochondrial membranes. In general, each translocator has the capacity to recognize particular types of signals and serves as a conduit across or into the membrane within which it resides. Together, these translocators direct ~1500 different precursor proteins from the cytosol to the appropriate subcompartment of mitochondria: the outer membrane, the intermembrane space and crista space, the inner membrane, and the matrix space.

The organization of signals in a precursor protein ultimately controls which translocator(s) the precursor protein engages and the order in which the signals are used to reach the protein's final destination inside mitochondria. This combinatorial system means that there is sometimes more than one route to reach a particular destination, the same way that different subway lines can take you from Brooklyn to Times Square in New York City. For example, membrane proteins residing in the inner mitochondrial membrane use at least three routes to arrive there. **Figure 12–48B** shows the possible routes to each mitochondrial subcompartment and the translocator complexes that direct proteins there.

The **TOM complex** is required for the import of nearly all nucleus-encoded mitochondrial proteins. It initially recognizes their signal sequences and transports

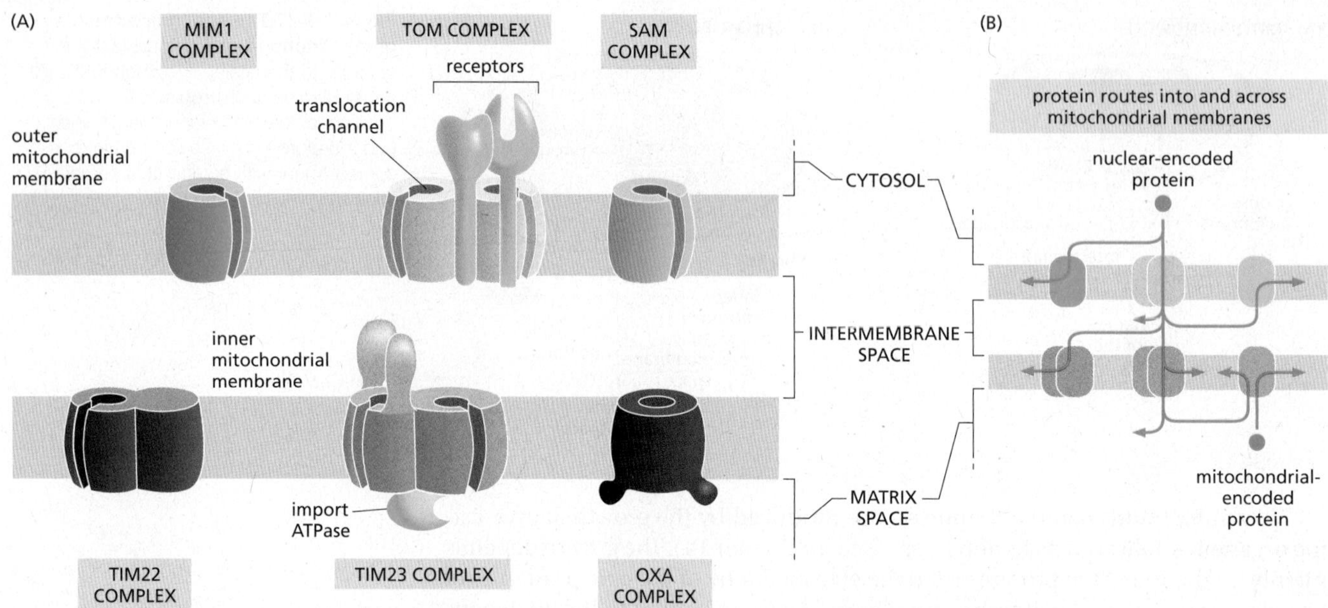

Figure 12–48 The protein translocators in the mitochondrial membranes. (A) The TOM, TIM, SAM, MIM, and OXA complexes are multimeric membrane protein assemblies that catalyze protein translocation across mitochondrial membranes. The protein components of the TIM22 and TIM23 complexes that line the import channel are structurally related, suggesting a common evolutionary origin of both TIM complexes. On the matrix side, the TIM23 complex is bound to a multimeric protein complex containing mitochondrial hsp70, which acts as an import ATPase, using ATP hydrolysis to pull proteins through the pore. In animal cells, subtle variations exist in the subunit composition of the translocator complexes to adapt the mitochondrial import machinery to the particular needs of specialized cell types. (B) Newly made proteins synthesized in the cytosol can use multiple routes to arrive at their final destination. The known routes via the different protein complexes are shown as *green lines*. There are multiple routes for a protein to become embedded into the inner and outer mitochondrial membranes, including one route for mitochondrially encoded proteins synthesized in the matrix space. SAM = sorting and assembly machinery; OXA = cytochrome oxidase activity; TIM = translocator of the inner mitochondrial membrane; TOM = translocator of the outer membrane; MIM = mitochondrial import machinery.

them from the cytosol into the intermembrane space. From here, different mitochondrial proteins follow different itineraries depending on sequence features encoded in the protein. β-Barrel proteins, which are particularly abundant in the outer membrane, are passed to the **SAM complex** for insertion and folding in the outer membrane. Two different **TIM complexes** mediate protein transport at the inner membrane. Matrix proteins use the TIM23 complex for transport, while inner membrane proteins use the TIM22 complex, the TIM23 complex, or the **OXA complex** for insertion. The remainder of proteins stay in the intermembrane space where they function.

In addition to the ~99% of mitochondrial proteins that must be imported from the cytosol, a handful of membrane proteins are encoded by the mitochondrial genome in all eukaryotes. These proteins are synthesized by mitochondrial ribosomes and inserted into the inner membrane by the OXA complex. Mitochondrially encoded membrane proteins are assembled with nuclear-encoded membrane proteins imported from the cytosol to form functional protein complexes such as the respiratory-chain complexes used for energy production (see Chapter 14). How cells communicate between the mitochondria and nucleus to ensure equal expression of the proteins that build inner membrane complexes is not understood.

Mitochondrial Proteins Are Imported Post-translationally as Unfolded Polypeptide Chains

As we learned in an earlier section, protein translocation into the ER usually occurs as the protein is being synthesized by ribosomes that are tightly coupled to the ER protein translocator. The binding of ribosomes to the translocator during protein import is what gives the rough ER its characteristic appearance.

In contrast, the protein translocators in the mitochondrial outer membrane do not bind to ribosomes, and most mitochondrial proteins are imported by a post-translational mechanism. This is why very few ribosomes are observed on the surface of mitochondria.

As with ER translocation, mitochondrial protein import can be reconstituted in a cell-free reaction in the test tube. In such experiments, a radioactively labeled mitochondrial precursor protein is mixed with purified mitochondria to permit import into the organelle. By changing the conditions in the test tube, it is possible to establish the biochemical requirements for import, to trap intermediates in the process, and to identify which translocators are used. Most of our knowledge about the molecular mechanism of mitochondrial import comes from analysis in cell-free reactions.

Mitochondrial precursor proteins do not immediately fold into their native structures after they are synthesized; instead, they remain unfolded in the cytosol through interactions with other proteins. Some of these interacting proteins are general *chaperones* of the *hsp70 family* (discussed in Chapter 6), whereas others are dedicated to mitochondrial precursor proteins and bind directly to their signal sequences. All the interacting proteins help to prevent the precursor proteins from aggregating or folding up spontaneously before they engage with the TOM complex in the outer mitochondrial membrane. As a first step in the import process, the import receptors of the TOM complex bind the signal sequence of the mitochondrial precursor protein. The unfolded polypeptide chain is then fed—signal sequence first—into the translocation channel within the TOM complex as the cytosolic interacting proteins are stripped off.

Once the translocating protein protrudes into the intermembrane space, sequences within the polypeptide chain determine what happens next. For example, proteins destined for the matrix or inner membrane engage one of the TIM complexes and are either translocated across or inserted into the inner membrane. It is possible to rapidly cool a cell-free mitochondrial import reaction to arrest the proteins at an intermediate step during translocation. Experiments examining an arrested protein destined for the matrix show that it spans both the inner and outer mitochondrial membranes: its N-terminal signal sequence has been removed by the signal peptidase located in the matrix, while the C-terminal part of the protein is still exposed outside the mitochondria. We can therefore conclude that precursor proteins can pass through both mitochondrial membranes at once to enter the matrix space (**Figure 12–49**).

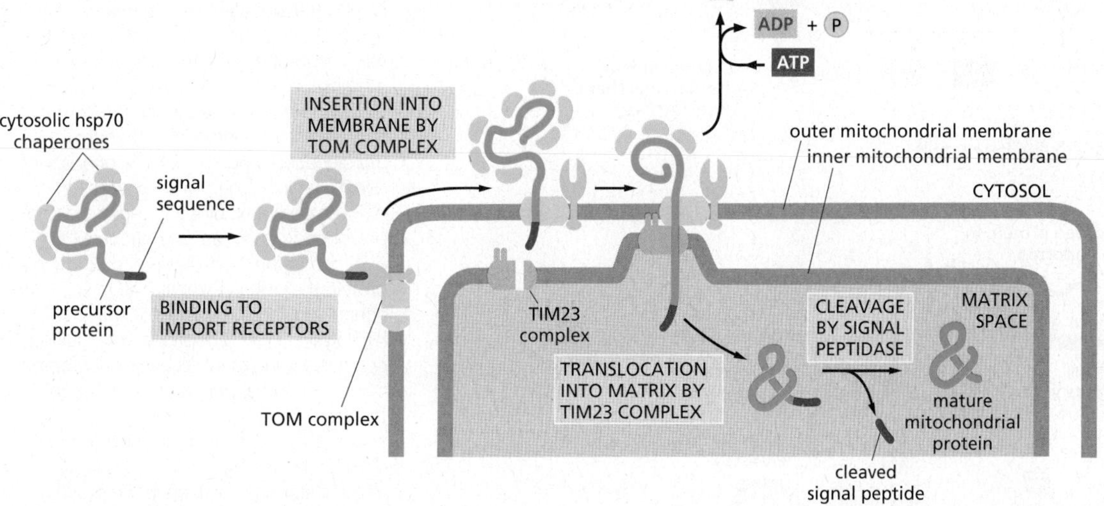

Figure 12–49 **Protein import by mitochondria.** The N-terminal signal sequence of the mitochondrial precursor protein is recognized by receptors of the TOM complex while the remainder of the protein is kept unfolded by cytosolic hsp70 chaperones. The protein is then translocated through the TIM23 complex so that it transiently spans both mitochondrial membranes (**Movie 12.5**). The signal sequence is cleaved off by a signal peptidase in the matrix space to form the mature protein. The free signal sequence is then rapidly degraded (not shown).

Although the TOM and TIM complexes usually work together to translocate precursor proteins across both membranes at the same time, they are capable of operating independently. In isolated outer membranes, for example, the TOM complex can translocate the signal sequence of precursor proteins across the membrane. Similarly, if the outer membrane is experimentally removed from isolated mitochondria, the exposed TIM23 complex can efficiently import precursor proteins into the matrix space. The experimental uncoupling of ordinarily linked processes allows each step and translocator system to be studied and understood in greater detail.

Protein Import Is Powered by ATP Hydrolysis, a Membrane Potential, and Redox Potential

Directional transport of proteins requires energy (**Figure 12–50**). Mitochondrial protein import utilizes three different sources of energy at four discrete sites. ATP, a common fuel in most biological systems, is used at two of these sites: outside the mitochondria and inside the matrix. The other two energy sources are contributed by the membrane potential across the inner mitochondrial membrane and the redox potential of the electron-transport chain. Not all mitochondrial precursor proteins need each of these energy sources to arrive at their final destination.

The initial use of energy, needed by most mitochondrial precursor proteins at the initial stage of the translocation process, serves to maintain the polypeptide in an unfolded state prior to import (see Figure 12–49). As discussed in Chapter 6,

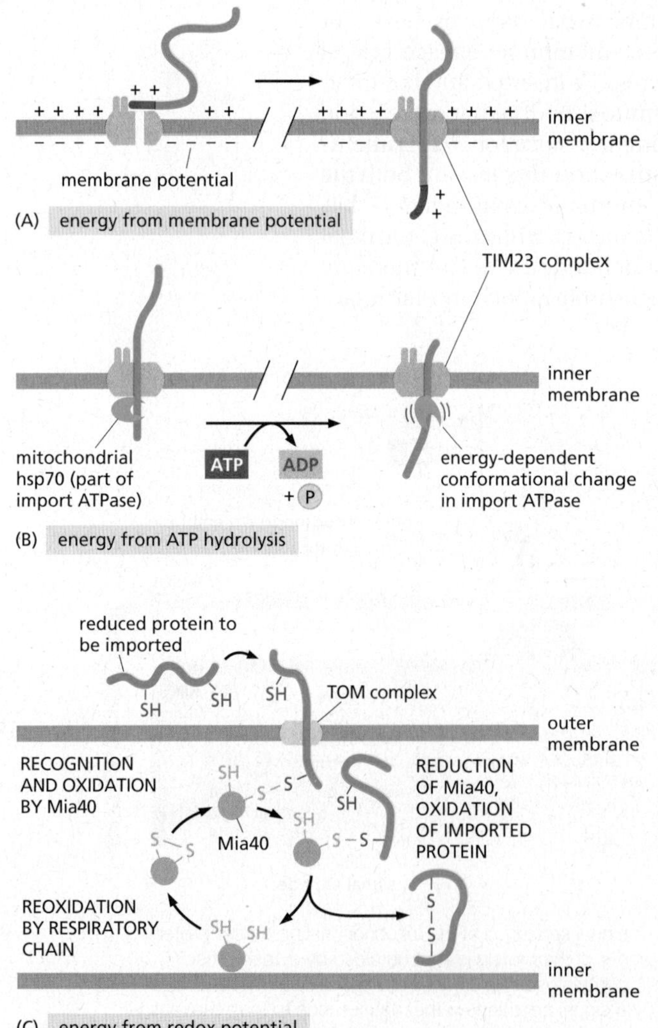

Figure 12–50 The role of energy in protein import into mitochondria. Three different sources of energy are used to import protein into the mitochondria. (A) After initial insertion of the signal sequence and of adjacent portions of the polypeptide chain into the TOM complex translocation channel (not shown), the signal sequence interacts with a TIM23 complex *(orange)*. The signal sequence is then translocated into the matrix space in a process that requires the energy in the membrane potential across the inner membrane. Positively charged amino acids in the signal sequence facilitate this membrane potential–dependent translocation reaction. (B) *Mitochondrial hsp70*, which is part of an import ATPase complex, binds to regions of the polypeptide chain as they become exposed in the matrix space, pulling the protein through the translocation channel, using the energy of ATP hydrolysis. (C) Polypeptides with multiple cysteines can sample the intermembrane space via partial translocation through the TOM complex. In the intermembrane space, these cysteines are oxidized to disulfide bonds by the enzyme Mia40, which becomes reduced in the process. The oxidized polypeptide is now partially folded, preventing it from sliding back into the cytosol. Reduced Mia40 is reoxidized by electrons provided by the respiratory chain so it can function again in the import reaction.

the chaperones that carry out this task use cycles of ATP binding and hydrolysis to control their interactions with newly synthesized polypeptides. Chaperone interaction is required to prevent premature folding in the cytosol, while chaperone dissociation is needed to permit transport through the TOM complex.

Once the signal sequence has passed through the TOM complex and is bound to a TIM complex, further translocation through the TIM translocation channel requires the membrane potential (Figure 12–50A), which is the electrical component of the electrochemical H^+ gradient across the inner membrane (see Figure 11–4). Pumping of H^+ from the matrix space to the intermembrane space, driven by electron-transport processes in the inner membrane (discussed in Chapter 14), maintains the electrochemical gradient. The energy in the electrochemical H^+ gradient across the inner membrane drives the translocation of the positively charged signal sequences through the TIM complexes by electrophoresis. The same H^+ gradient also powers most of the cell's ATP synthesis by ATP synthase complexes in the inner mitochondrial membrane.

Once the initial segment of a precursor protein reaches the matrix, **mitochondrial hsp70** is crucial for completing the import process similar to how BiP is needed for post-translation protein import into the ER. The mitochondrial hsp70 is bound to the matrix side of the TIM23 complex and acts as a motor to pull the precursor protein into the matrix space. Like its cytosolic cousin, mitochondrial hsp70 has a high affinity for unfolded polypeptide chains, and it binds tightly to an imported protein chain as soon as the chain emerges from the TIM translocator in the matrix space. The hsp70 then undergoes an ATP-dependent conformational change that exerts a pulling force on the protein being imported before releasing it. This energy-driven cycle of binding, pulling, and release continues until the protein has completed import through the TIM23 complex (Figure 12–50B). Many imported matrix proteins are passed on to another chaperone protein, *mitochondrial hsp60,* to assist their folding through cycles of ATP hydrolysis (see Chapter 6).

Certain intermembrane-space proteins that contain cysteine motifs use the difference in redox potential between the cytosol and mitochondria as a source of energy. When a portion of these proteins initially emerges into the intermembrane space, they form a transient covalent disulfide bond to the Mia40 protein (Figure 12–50C). This interaction prevents backsliding of the protein through the TOM complex into the cytosol. The imported proteins are eventually released from Mia40 in an oxidized form containing intrachain disulfide bonds, resulting in a folded protein that is now trapped in the intermembrane space. Mia40 becomes reduced in the process and is then reoxidized by passing electrons to the electron-transport chain in the inner mitochondrial membrane. In this way, the energy stored in the redox potential in the mitochondrial electron-transport chain is tapped to drive protein import.

Transport into the Inner Mitochondrial Membrane Occurs Via Several Routes

The three different translocators in the inner mitochondrial membrane (see Figure 12–48) are all capable of membrane protein insertion. Different subsets of inner mitochondrial membrane proteins take different routes to reach one of these translocators for insertion into the membrane.

In the most common translocation route, a precursor that begins in the cytosol uses the TOM and TIM23 complexes to begin import into the matrix. However, only the N-terminal signal sequence of the transported protein actually enters the matrix space (**Figure 12–51A**). A hydrophobic amino acid sequence, strategically located after the N-terminal signal sequence, is recognized as a transmembrane domain by the TIM23 complex. This allows insertion of the transmembrane domain into the inner membrane and prevents further translocation into the matrix, perhaps through a lateral gate analogous to that found in the ER-resident Sec61 translocator. The remainder of the protein enters the intermembrane space through the TOM complex, and the signal sequence is cleaved off in the matrix.

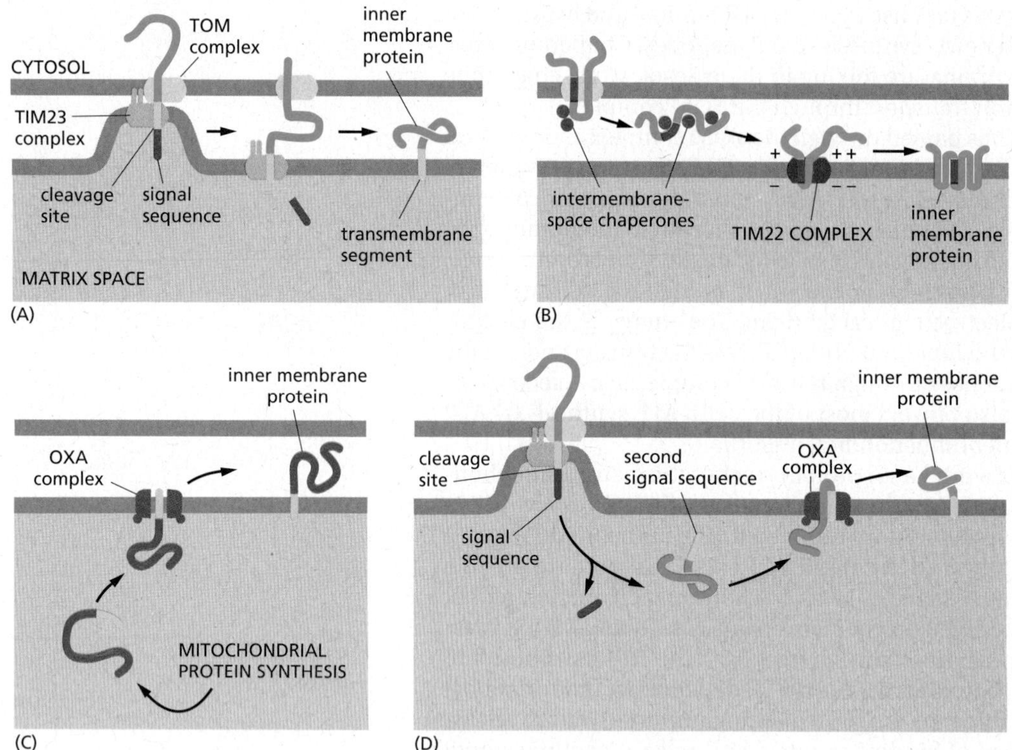

Figure 12–51 **Routes for the production of inner mitochondrial membrane proteins.** (A) The N-terminal signal sequence *(red)* initiates import into the matrix space (see Figure 12–49). A hydrophobic transmembrane segment *(blue)* that follows the matrix-targeting signal sequence binds to the TIM23 translocator *(orange)* in the inner membrane and stops translocation. The remainder of the protein is then pulled into the intermembrane space through the TOM translocator in the outer membrane, and the transmembrane segment is released into the inner membrane, anchoring the protein there. (B) Multipass inner membrane proteins that function as metabolite transporters contain internal signal sequences and snake through the TOM complex as a loop. They then bind to the chaperones in the intermembrane space, which guide the proteins to the TIM22 complex. The TIM22 complex is specialized for the insertion of multipass inner membrane proteins. (C) The OXA complex mediates membrane protein insertion into the inner membrane for proteins that are encoded by the mitochondrial genome and translated in the matrix space. (D) The OXA complex in the inner membrane can mediate protein insertion from the matrix space. To access this route, nuclear-encoded proteins must first translocate completely into the matrix space via the TOM and TIM23 complexes. Cleavage of the signal sequence *(red)* used for the initial translocation unmasks an adjacent hydrophobic signal sequence *(blue)* at the new N-terminus. This signal then directs the protein into the inner membrane.

The second transport route to the inner membrane is specialized for a family of metabolite-specific transporters that transfer a vast number of small molecules across the inner membrane. These transporters supply substrates for metabolic enzymes in the mitochondrial matrix, such as those of the citric acid cycle, and export their products back to the cytosol. These multipass transmembrane proteins use internal signal sequences to enter the intermembrane space through the TOM complex. They engage intermembrane-space chaperones that guide them to the TIM22 complex, where hydrophobic transmembrane regions partition into the inner membrane. This insertion process requires the membrane potential to ensure that appropriate regions of the protein are transported to the matrix side so that the transporter acquires the correct topology (**Figure 12–51B**).

The final insertion route into the inner membrane uses the OXA complex. As mentioned earlier, the OXA complex also inserts the few membrane proteins that are encoded and translated in the mitochondrial matrix (**Figure 12–51C**). Thus, the OXA complex can only be accessed from the matrix side of the membrane. For this reason, nuclear-encoded membrane proteins that rely on the OXA complex for insertion must first use TIM23 to translocate into the matrix (**Figure 12–51D**). Here, the N-terminal signal sequence is removed to expose a hydrophobic signal sequence that is then used by the OXA complex for insertion into the inner membrane.

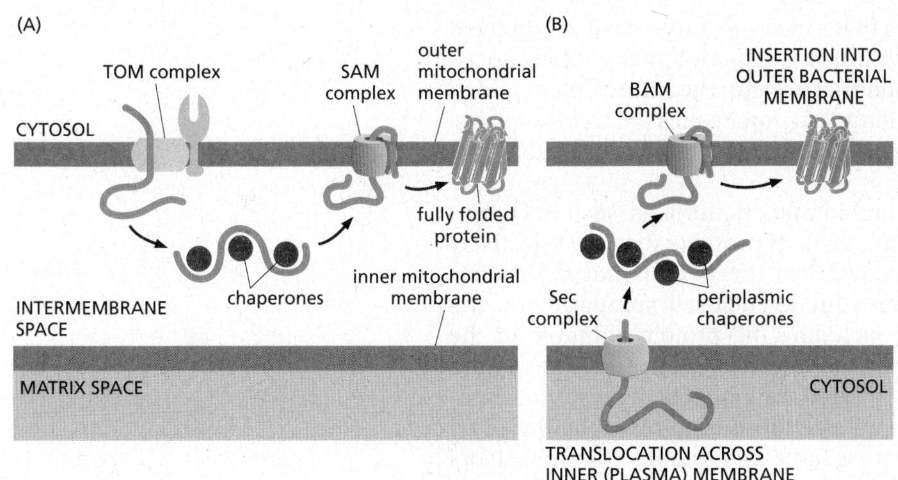

Figure 12–52 Integration of porins into the outer mitochondrial and bacterial membranes. (A) After translocation through the TOM complex in the outer mitochondrial membrane, β-barrel proteins bind to chaperones in the intermembrane space. The SAM complex then inserts the unfolded polypeptide chain into the outer membrane and helps the chain fold. (B) A structurally related BAM complex in the outer membrane of Gram-negative bacteria catalyzes β-barrel protein insertion and folding (see Figure 11–17).

Bacteria and Mitochondria Use Similar Mechanisms to Insert β Barrels into Their Outer Membrane

As discussed earlier in this chapter, mitochondria evolved from an ancestral endosymbiont bacterium inside the primordial eukaryotic cell. The outer mitochondrial membrane is therefore evolutionarily related to the outer membrane of Gram-negative bacteria (see Figure 11–17). Both membranes contain **porins**, abundant pore-forming β-barrel proteins that are permeable to inorganic ions and metabolites (but not to most proteins). The TOM complex only allows proteins containing hydrophobic α helices to exit laterally and thus cannot integrate porins or other β-barrel proteins into the lipid bilayer. Instead, they are first transported through the TOM complex as unfolded proteins into the intermembrane space. Specialized chaperone proteins in the intermembrane space keep the β-barrel proteins from aggregating (**Figure 12–52A**) until they are inserted and folded by the SAM complex in the outer membrane.

One of the central subunits of the SAM complex is homologous to a bacterial outer membrane protein that helps insert β-barrel proteins into the bacterial outer membrane. In bacteria, β-barrel proteins are inserted from the periplasmic space, which is the topological equivalent of the intermembrane space in mitochondria (**Figure 12–52B**). This conserved pathway for inserting β-barrel proteins further underscores the endosymbiotic origin of mitochondria. Notably, the central subunits of the TOM and SAM complexes are themselves β-barrel proteins. Thus, preexisting TOM and SAM complexes are required to make more copies of these essential protein translocators.

Two Signal Sequences Direct Proteins to the Thylakoid Membrane in Chloroplasts

Protein transport into **chloroplasts** resembles transport into mitochondria. Both processes occur post-translationally, use separate translocation complexes in each membrane, require energy, and use multiple types of signal sequences to direct a precursor to the appropriate organelle subcompartment. However, many of the protein components that form the translocation complexes differ. Moreover, whereas mitochondria harness the electrochemical H^+ gradient across their inner membrane to drive transport, chloroplasts, which have an electrochemical H^+ gradient across their thylakoid membrane but not their inner membrane, use GTP and ATP hydrolysis to power import across their double-membrane envelope. The functional similarities thus result from convergent evolution, reflecting the common requirements for translocation across a double membrane.

Although the signal sequences for import into chloroplasts superficially resemble those for import into mitochondria, a plant cell can have both mitochondria and chloroplasts, so proteins must partition appropriately between the two organelles. Experiments have shown that a cytosolic protein can be directed specifically to a plant cell's mitochondria if it is experimentally joined to an N-terminal signal sequence of a mitochondrial protein; the same protein joined to an N-terminal signal sequence of a chloroplast protein ends up in chloroplasts. Thus, the import receptors on each organelle distinguish between the different signal sequences.

The same compartments that are found in mitochondria are also in chloroplasts, and each has its distinctive complement of proteins that are selectively delivered there using mechanisms analogous to the mitochondrial systems. However, chloroplasts have an extra membrane-enclosed compartment, the **thylakoid**. Many chloroplast proteins, including the protein subunits of the photosynthetic system and of the ATP synthase (discussed in Chapter 14), are located in the thylakoid membrane. Many of the components of these vital complexes are encoded in the nuclear genome, and those residing in the thylakoid lumen therefore have to be imported across three membranes. The precursors of these proteins are translocated from the cytosol to their final destination in two steps using bipartite signal sequences. First, they pass across the outer and inner membranes into the **stroma** guided by an N-terminal chloroplast signal sequence. There, a stromal signal peptidase removes the N-terminal chloroplast signal sequence, unmasking a thylakoid signal sequence that follows it in the sequence of the precursor protein. The thylakoid signal sequence initiates integration into the thylakoid membrane or translocation into the thylakoid space (**Figure 12–53A**).

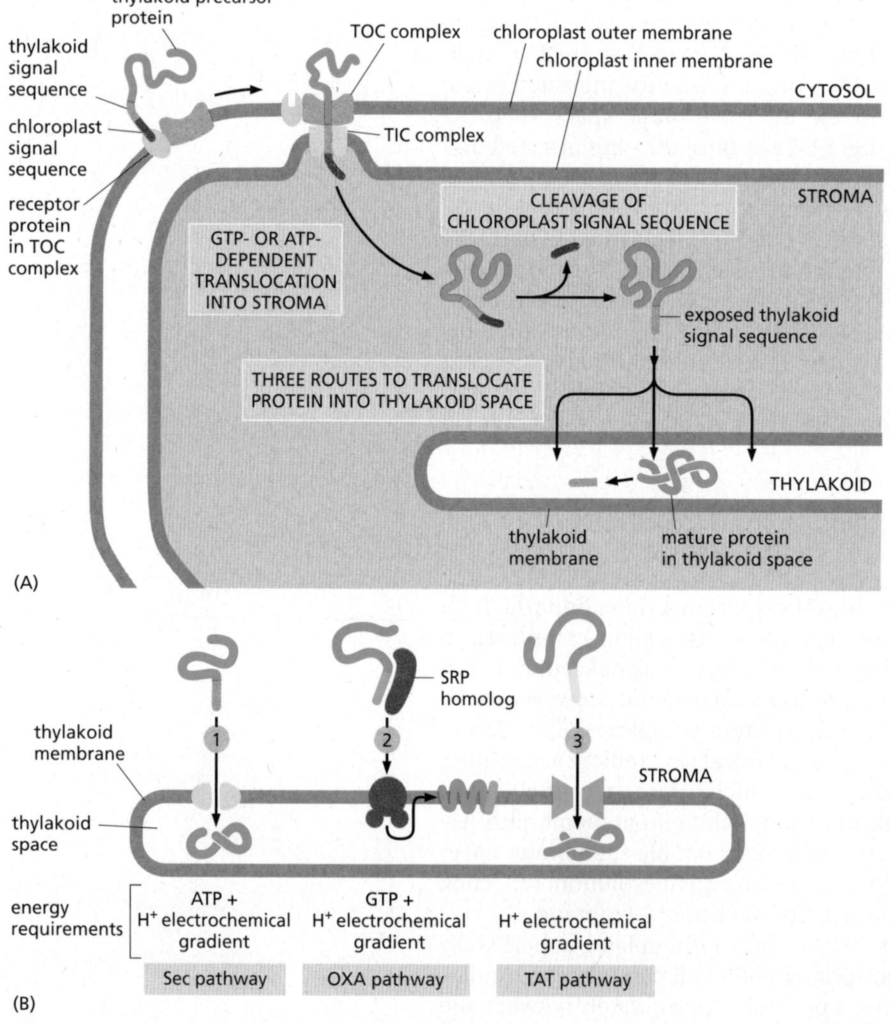

Figure 12–53 Translocation of chloroplast precursor proteins into the thylakoid space. (A) The precursor protein contains an N-terminal chloroplast signal sequence *(red)*, followed immediately by a thylakoid signal sequence *(brown)*. The chloroplast signal sequence initiates translocation into the stroma by a mechanism similar to that used for the translocation of mitochondrial precursor proteins into the matrix space, although the translocator complexes, named TOC and TIC (for translocator in the outer and inner chloroplast membrane, respectively), are different. The signal sequence is then cleaved off, unmasking the thylakoid signal sequence, which initiates translocation across the thylakoid membrane. (B) Translocation into the thylakoid space or thylakoid membrane can occur by any one of at least three routes: (1) a *Sec pathway*, so called because it uses components that are homologs of Sec proteins, which mediate protein translocation across the ER and bacterial plasma membrane; (2) an *OXA-like pathway*, so called because it uses a chloroplast homolog of the OXA translocase; (3) a *TAT* (twin arginine translocation) *pathway*, so called because two arginines are critical in the signal sequences that direct proteins into this pathway, which depends on the H^+ gradient across the thylakoid membrane. The OXA-like pathway makes use of a chloroplast SRP that lacks an RNA subunit. This specialized SRP located in the stroma recognizes a thylakoid-directed signal sequence and functions exclusively post-translationally because it is found in a separate compartment from the ribosome that made the thylakoid precursor protein.

There are three different protein translocators in the thylakoid membrane, each of which recognizes a different type of signal sequence, handles a different subset of thylakoid precursors, and uses energy in different ways (**Figure 12–53B**). As we saw earlier, the thylakoid membrane is developmentally derived from the inner chloroplast membrane, which is evolutionarily related to the bacterial inner membrane. It is therefore not surprising that each of the three translocators in the thylakoid membrane has homologs that are used for translocation or membrane insertion in bacteria.

Summary

Although mitochondria and chloroplasts have their own genetic systems, they produce less than 1% of their own proteins. Instead, the two organelles import most of their proteins from the cytosol, using similar mechanisms. In both cases, multiple protein translocator complexes in the outer and inner membranes recognize different types of signal sequences to direct a precursor to the correct organelle subcompartment. Proteins are transported in an unfolded state by a post-translational mechanism. Chaperone proteins of the cytosolic hsp70 family maintain the precursor proteins in an unfolded state prior to translocation, and a second set of hsp70 proteins in the matrix space or stroma pulls the polypeptide chain across the inner membrane. Translocation into mitochondria is powered by ATP hydrolysis, a membrane potential across the inner membrane, and the redox potential of the electron-transport chain. Translocation into chloroplasts is powered by GTP and ATP hydrolysis and a membrane potential across the thylakoid membrane. In chloroplasts, import from the stroma into the thylakoid can occur by several routes, distinguished by the protein translocator complex and energy source used.

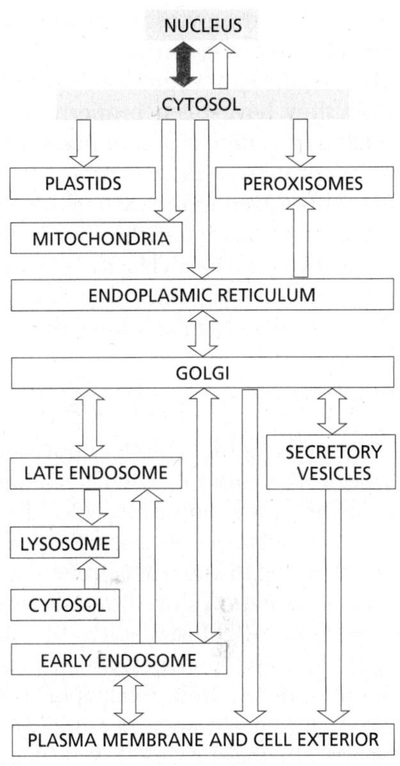

THE TRANSPORT OF MOLECULES BETWEEN THE NUCLEUS AND THE CYTOSOL

The **nuclear envelope** encloses the DNA and defines the *nuclear compartment*. This envelope consists of two concentric membranes, which are perforated by nuclear pore complexes (**Figure 12–54**). Although the inner and outer nuclear membranes are continuous, they maintain distinct protein compositions. The **inner nuclear membrane** contains proteins that act as binding sites for the **nuclear lamina**, a meshwork of polymerized protein subunits called **nuclear lamins**. The lamin proteins are members of the intermediate filament family of cytoskeletal proteins (see Chapter 16). The lamina provides structural support for

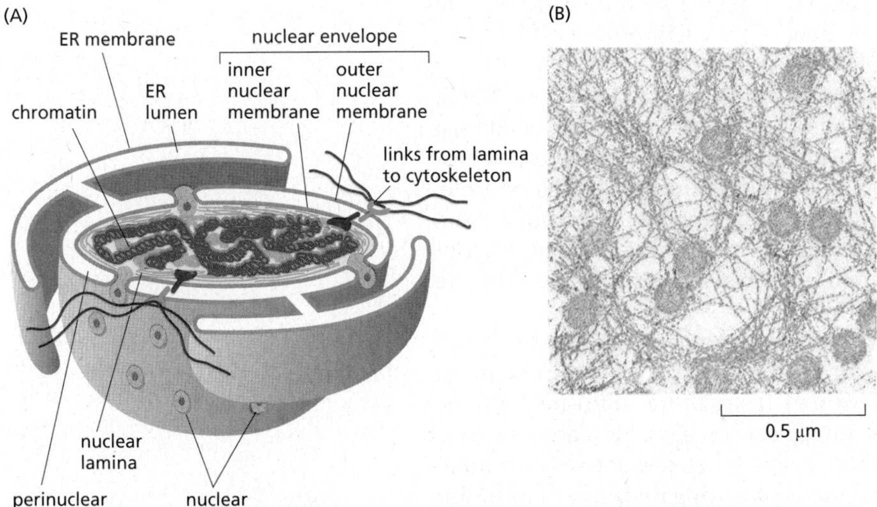

Figure 12–54 The nuclear envelope. (A) The double membrane of the nuclear envelope is penetrated by nuclear pore complexes. Transmembrane proteins in the inner and outer nuclear membranes link the nuclear lamina to the cytosolic cytoskeleton. The outer nuclear membrane is continuous with the endoplasmic reticulum (ER). The ribosomes that are normally bound to the cytosolic surface of the ER membrane and outer nuclear membrane are not shown. (B) The nuclear lamina is a fibrous protein meshwork underlying the inner membrane. Nuclear pores are seen in *light brown*. (B, from Y. Turgay et al., *Nature* 543:261–264, 2017.)

the nuclear envelope and acts as an anchoring site for chromosomes and nuclear pore complexes. The lamina is also connected to the cytoplasmic cytoskeleton via protein complexes that span the nuclear envelope, thereby providing structural links between the DNA, nuclear envelope, and cytoskeleton. The **outer nuclear membrane** is continuous with the membrane of the ER and is studded with ribosomes engaged in protein synthesis (see Figure 12–15). The proteins made on these ribosomes are transported into the space between the inner and outer nuclear membranes (the *perinuclear space*), which is continuous with the ER lumen.

Nuclear pores conduct extensive bidirectional traffic between the cytosol and the nucleus. The many proteins that function in the nucleus—including histones, DNA polymerases, RNA polymerases, transcriptional regulators, and RNA-processing proteins—are selectively imported into the nuclear compartment from the cytosol, where they are made. At the same time, all RNAs that function in the cytosol—including mRNAs, rRNAs, tRNAs, and miRNAs—are exported after they are synthesized and processed in the nucleus. Like the import process, the export process is selective; mRNAs, for example, are exported only after they have been properly modified by RNA-processing reactions in the nucleus. In some cases, multiple selective transport steps are needed to assemble a complex structure. Ribosomes, for instance, are made from proteins that are synthesized in the cytosol, imported into the nucleus, and exported back to the cytosol only after their assembly with newly made ribosomal RNA. These pre-ribosomal particles then complete their assembly into functional ribosomes in the cytosol, with certain assembly and transport factors returning to the nucleus to help assemble the next ribosome.

Nuclear Pore Complexes Perforate the Nuclear Envelope

Large and elaborate **nuclear pore complexes (NPCs)** perforate the nuclear envelope in all eukaryotes. Each NPC is composed of a set of approximately 30 different proteins, or **nucleoporins**. NPCs display eightfold rotational symmetry, with axial symmetry of the central core. Hence, each nucleoporin is present in multiple copies, resulting in 500–1000 protein molecules in the fully assembled NPC, with an estimated mass of 66 million daltons in yeast and 125 million daltons in vertebrates (**Figure 12–55**). Most nucleoporins are composed of repetitive protein domains of only a few different types, which have evolved through extensive gene duplication. Some of the scaffold nucleoporins that abut the membrane (see Figure 12–55) are evolutionarily and structurally related to vesicle coat protein complexes, such as clathrin and the COPII coat (discussed in Chapter 13), which shape transport vesicles. One protein is even used as a common building block in both NPCs and vesicle coats. It appears that an ancestral membrane-bending protein that helped shape the elaborate membrane systems of eukaryotic cells evolved into a family of proteins that stabilize the sharp membrane bends at nuclear pores and budding transport vesicles.

The nuclear envelope of a typical mammalian cell contains 3000–4000 NPCs, although that number varies widely, from a few hundred in glial cells to almost 20,000 in Purkinje neurons. Each NPC can transport a staggering 1000 macromolecules per second and can transport in both directions at the same time. The internal diameter of each NPC is ~40 nm, large enough to accommodate ribosomal subunits and even viral particles. However, this enormous pore is not empty; instead, it is filled with unstructured protein regions contributed by the channel nucleoporins.

These unstructured domains contain numerous repeats of phenylalanine-glycine (FG) motifs whose weak affinity for each other creates a gel-like mesh inside the NPC. This mesh acts as a sieve that restricts the diffusion of large macromolecules while allowing smaller molecules to pass. Researchers have determined the effective size of the sieve by injecting labeled water-soluble molecules of different sizes into the cytosol and then measuring their rate of diffusion into the nucleus. Small molecules (5000 daltons or less) diffuse in so fast that we

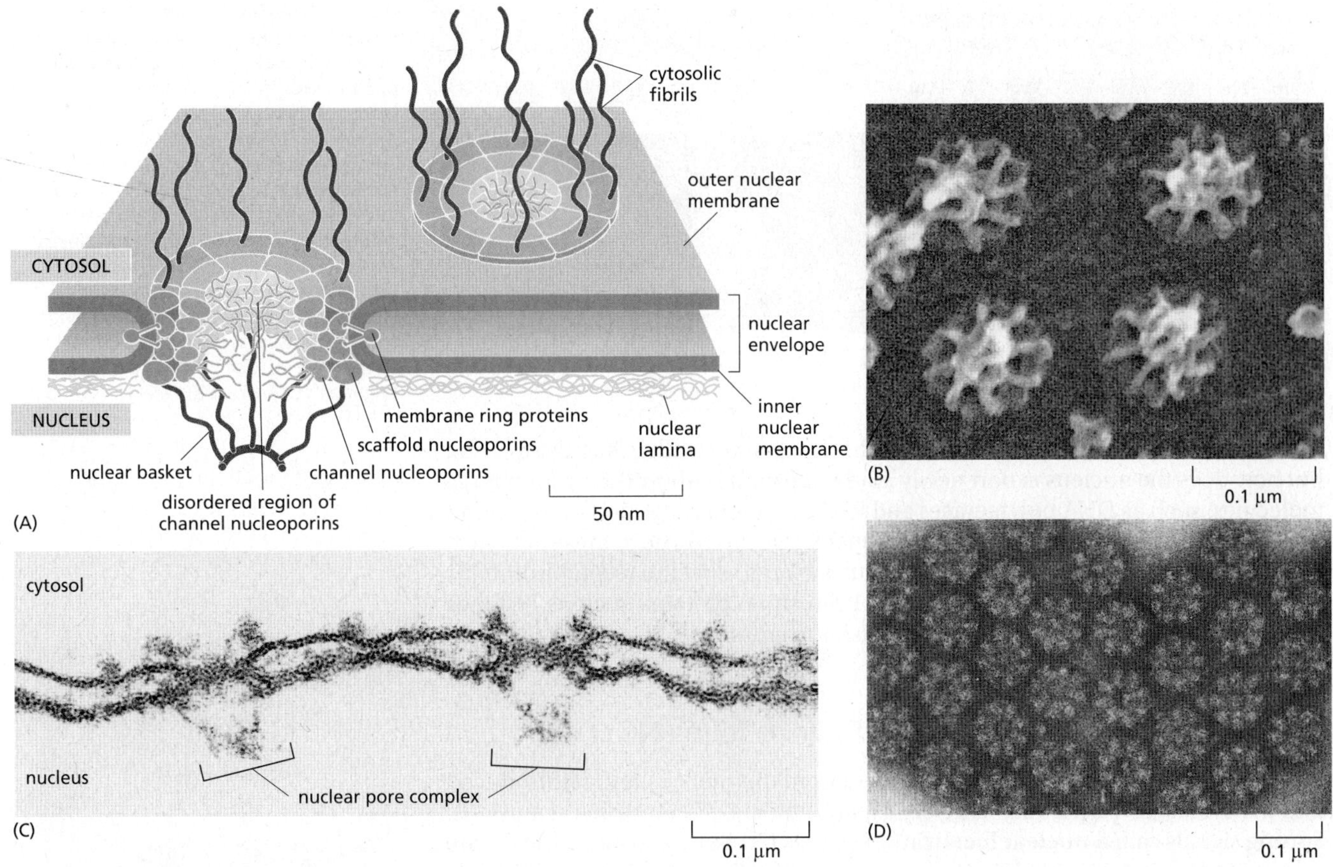

Figure 12–55 **The arrangement of NPCs in the nuclear envelope.** (A) In a vertebrate NPC, nucleoporins are arranged with striking eightfold rotational symmetry. In addition, immunoelectron microscope studies show that the proteins that make up the central portion of the NPC are oriented symmetrically across the nuclear envelope, so that the nuclear and cytosolic sides look identical. The eightfold rotational and twofold transverse symmetry explains how such a huge structure can be formed from only about 30 different proteins: many of the nucleoporins are present in 8, 16, or 32 copies. On the basis of their approximate localization in the central portion of the NPC, nucleoporins can be classified into (1) transmembrane ring proteins that span the nuclear envelope and anchor the NPC to the envelope; (2) scaffold nucleoporins that form layered ring structures (some scaffold nucleoporins are membrane-bending proteins that stabilize the sharp membrane curvature where the nuclear envelope is penetrated); and (3) channel nucleoporins that line a central pore. In addition to folded domains that anchor the proteins in specific places, many channel nucleoporins contain extensive unstructured regions, where the polypeptide chains are intrinsically disordered. The central pore is filled with a high concentration of these disordered domains whose weak interactions with each other form a gel that blocks the passive diffusion of large macromolecules. The disordered regions contain a large number of phenylalanine–glycine (FG) repeats. Fibrils protrude from both the cytosolic and the nuclear sides of the NPC. By contrast to the twofold transverse symmetry of the NPC core, the fibrils facing the cytosol and nucleus are different: on the nuclear side, the fibrils converge at their distal end to form a basketlike structure. The precise arrangement of individual nucleoporins in the assembled NPC is still a matter of intense debate, because atomic resolution analyses have been hindered by the sheer size and flexible nature of the NPC and by difficulties in purifying sufficient amounts of homogeneous material. A combination of electron microscopy, computational analyses, and crystal structures of nucleoporin subcomplexes has been used to develop the current models of the NPC architecture. (B) A scanning electron micrograph of the nuclear side of the nuclear envelope of an oocyte, showing NPCs with their basketlike fibrils. (C) An electron micrograph showing a side view of two NPCs *(brackets)*; note that the inner and outer nuclear membranes are continuous at the edges of the pore. (D) An electron micrograph showing face-on views of negatively stained NPCs. The membrane has been removed by detergent extraction. Note that some of the NPCs contain material in their center, which is thought to be trapped macromolecules in transit through these NPCs. (A, adapted from A. Hoelz et al., *Annu. Rev. Biochem.* 80:613–643, 2011. B, © 1992 M.W. Goldberg and T.D. Allen. Originally published in *J. Cell Biol.* https://doi.org/10.1083/jcb.119.6.1429. With permission from Rockefeller University Press. C, courtesy of Werner Franke and Ulrich Scheer. D, courtesy of Ron Milligan.)

can consider the nuclear envelope freely permeable to them. The barrier is progressively restrictive to larger molecules such that proteins greater than ~40,000 daltons or ~5 nm in diameter cannot enter by passive diffusion.

Because many cell proteins are too large to diffuse passively through the NPCs, the nuclear compartment and the cytosol can maintain different protein compositions. Mature cytosolic ribosomes, for example, are about 30 nm in diameter and

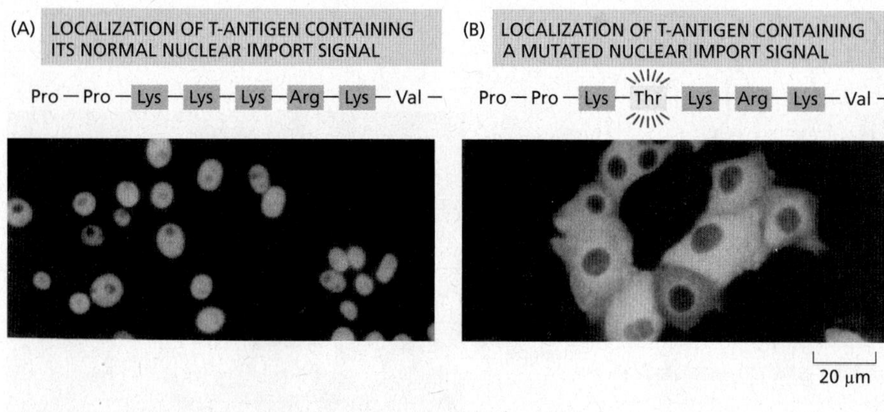

Figure 12–56 The function of a nuclear localization signal. Immunofluorescence micrographs showing the cell location of SV40 virus T-antigen containing or lacking a short sequence that serves as a nuclear localization signal. (A) The normal T-antigen protein contains the lysine-rich sequence indicated and is imported to its site of action in the nucleus, as indicated by immunofluorescence staining with antibodies against the T-antigen. (B) T-antigen with an altered nuclear localization signal (a threonine replacing a lysine) remains in the cytosol. (From D. Kalderon et al., *Cell* 39:499–509, 1984. With permission from Elsevier.)

thus cannot diffuse through the NPC, confining protein synthesis to the cytosol. But how does the nucleus export newly made ribosomal subunits or import large molecules, such as DNA polymerases and RNA polymerases, which have subunit molecular masses of 100,000–200,000 daltons? As we discuss next, these and most other transported protein and RNA molecules bind to specific receptor proteins that ferry large molecules through NPCs. Even small proteins such as histones frequently use receptor-mediated mechanisms to cross the NPC, thereby increasing transport efficiency.

Nuclear Localization Signals Direct Proteins to the Nucleus

When proteins are experimentally extracted from the nucleus and reintroduced into the cytosol, even the very large ones reaccumulate efficiently in the nucleus. Sorting signals called **nuclear localization signals (NLSs)** are responsible for the selectivity of this active nuclear import process. The signals have been precisely defined by using recombinant DNA technology for numerous proteins that are imported into the nucleus (**Figure 12–56**). The most commonly used signal consists of one or two short sequences that are rich in the positively charged amino acids lysine and arginine (see Figure 12–13), with the precise sequence varying for different proteins. Some nuclear proteins contain different types of signals, some of which are not yet characterized.

NLSs can be located almost anywhere in the amino acid sequence and are thought to form loops or patches on the protein surface. Many NLSs function even when linked as short peptides to the surface of a cytosolic protein, suggesting that the precise location of the signal within the amino acid sequence of a nuclear protein is not important. Moreover, as long as one of the protein subunits of a multicomponent complex displays a nuclear localization signal, the entire complex will be imported into the nucleus.

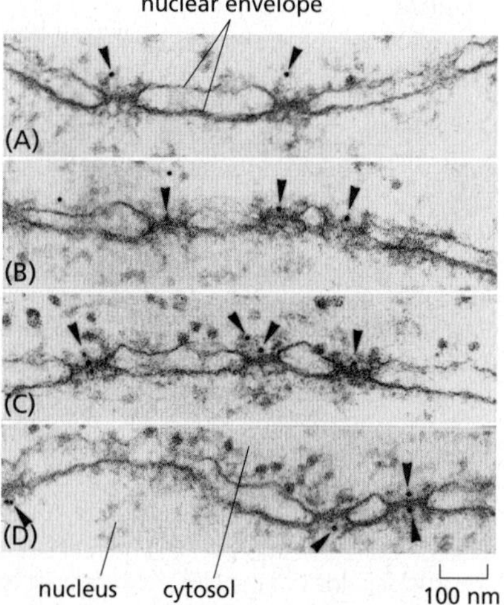

Figure 12–57 Visualizing active import through NPCs. This series of electron micrographs shows 5- to 10-nm-diameter colloidal gold spheres *(arrowheads)* coated with peptides containing nuclear localization signals entering the nucleus through NPCs. The gold particles were injected into the cytosol of living cells, which then were fixed and prepared for electron microscopy at various times after injection. (A) Gold particles are first seen in proximity to the cytosolic fibrils of the NPCs. (B, C) They are then seen at the center of the NPCs, exclusively on the cytosolic face. (D) They then appear on the nuclear face. These gold particles have much larger diameters than those of the diffusion channels in the NPC and are imported by active transport. (From N. Panté and U. Aebi, *Science* 273:1729–1732, 1996. With permission from AAAS.)

Macromolecular transport across NPCs differs fundamentally from the transport of proteins across the membranes of other organelles: NPC transport occurs through a large, constitutively open, mesh-filled pore, rather than through a much smaller protein translocator whose aqueous pore is typically gated by the protein being transported. For this reason, fully folded proteins and large multiprotein complexes can be transported in either direction through the nuclear pore. By contrast, transport through organellar protein translocators of the ER, mitochondria, and chloroplasts is unidirectional and usually requires the protein to be extensively unfolded.

One can visualize the transport of nuclear proteins through NPCs by coating tiny colloidal gold particles with a nuclear localization signal, injecting the particles into the cytosol, and then following their fate by electron microscopy (**Figure 12–57**). The particles first arrive at the tentacle-like fibrils that extend from the scaffold nucleoporins at the rim of the NPC into the cytosol, and then proceed through the center of the NPC. This observation illustrates that NLSs impart the ability of large particles to navigate through the otherwise impermeable diffusion barrier posed by the disordered mesh inside the nuclear pore.

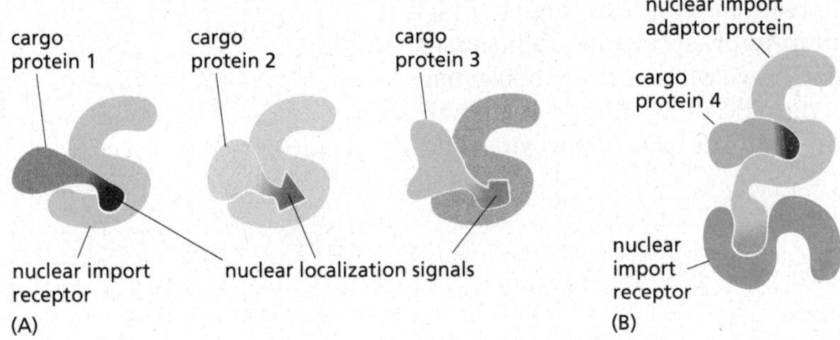

Figure 12–58 **Nuclear import receptors.** (A) Different nuclear import receptors bind different nuclear localization signals and thereby different cargo proteins. (B) Cargo protein 4 requires an adaptor protein to bind to its nuclear import receptor. The adaptors are structurally related to nuclear import receptors and recognize nuclear localization signals on cargo proteins. They also contain a nuclear localization signal that binds them to an import receptor, but this signal only becomes exposed when they are loaded with a cargo protein.

Nuclear Import Receptors Bind to Both Nuclear Localization Signals and NPC Proteins

To initiate nuclear import, nuclear localization signals must be recognized by **nuclear transport receptors**. Most of these receptors are part of a large family of proteins called *karyopherins*. In yeast, there are 14 genes encoding karyopherins; in animal cells, the number is significantly larger. Karyopherin family members that mediate nuclear import are called **nuclear import receptors**, while those for nuclear export (discussed later) are called **nuclear export receptors**. Each import receptor can bind and transport the subset of cargo proteins containing the appropriate nuclear localization signal (**Figure 12–58A**). Nuclear import receptors sometimes use adaptor proteins that form a bridge between the import receptors and the nuclear localization signals on the proteins to be transported (**Figure 12–58B**). Some adaptor proteins are structurally related to nuclear import receptors, suggesting a common evolutionary origin. By using a variety of import receptors and adaptors, cells are able to recognize the broad repertoire of nuclear localization signals that are displayed on nuclear proteins.

The import receptors are soluble cytosolic proteins that contain multiple low-affinity binding sites for the FG repeats found in the unstructured domains of several nucleoporins. The FG repeats in the fibrils of cytosol-facing nucleoporins serve to initially recruit import receptors and their bound cargo proteins to NPCs. The import receptors can then bind the FG repeats that form the mesh inside the nuclear pore to disrupt interactions between the repeats. In this way, the receptor–cargo complex locally dissolves the gel-like mesh and can diffuse into and within the NPC pore (**Figure 12–59**).

It is possible to re-create in a test tube a gel consisting of unstructured polypeptides containing FG repeats. This gel displays restricted diffusion of inert cargoes in a size-dependent manner similar to diffusion through NPCs. Diffusion into this artificial gel is more than 1000-fold faster for cargoes bound to an import receptor. At this rate, a cargo in complex with an import receptor could traverse the distance across an NPC in a few milliseconds, consistent with the rate

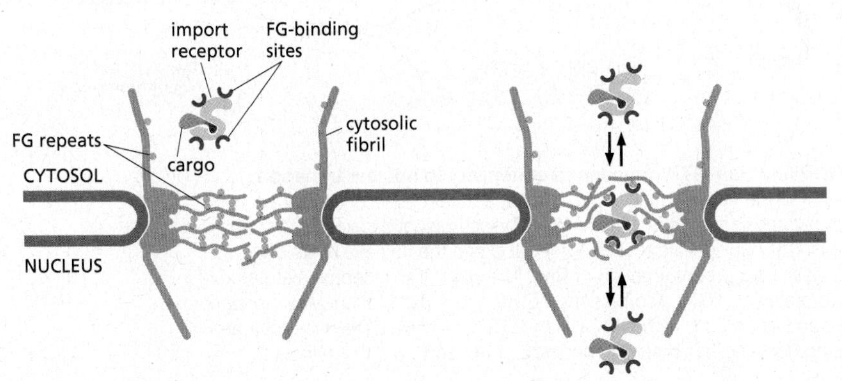

Figure 12–59 **Interaction of nuclear import receptors with FG repeats.** *Left:* Nuclear import receptors contain various low-affinity FG repeat–binding sites on their surface. This facilitates their initial recruitment to NPCs because of interactions with FG repeats found on the cytosolic fibrils of the NPCs. The interior of the NPC is filled with a mesh of FG repeat–containing proteins whose weak interactions with each other restrict nonspecific diffusion of proteins and other macromolecules through the pore. *Right:* Cargo receptors can rapidly partition into the FG repeat mesh by interacting with the FG repeats and locally melting the mesh. This partitioning into and out of the mesh substantially accelerates diffusion of the cargo receptor (and its bound cargo) through the NPC. Proteins without surface FG repeat–binding sites cannot melt the mesh, and their diffusion through the NPC is comparatively slow.

of transport observed in cells. It is important to realize that in this model, diffusion is not directional; instead, the import receptor simply accelerates diffusion to provide cargo access to the nuclear compartment. As we will see, it is the selective dissociation of cargo only on the nuclear side of the NPC that confers directionality to the import process. The import receptor then returns back to the cytosol for transport of the next cargo.

The Ran GTPase Imposes Directionality on Nuclear Import Through NPCs

The import of nuclear proteins through NPCs concentrates specific proteins in the nucleus and thereby increases order in the cell. The cell fuels this ordering process by harnessing the energy of GTP hydrolysis by the GTPase **Ran**, which is required for both nuclear import and export.

Like other GTPases, Ran is a molecular switch that can exist in two conformational states, depending on whether GDP or GTP is bound (Figure 3–63). Two Ran-specific regulatory proteins trigger the conversion between the two states: a cytosolic *GTPase-activating protein* (*GAP*) triggers GTP hydrolysis and thus converts Ran-GTP to Ran-GDP, and a nuclear *guanine nucleotide exchange factor* (*GEF*) promotes the exchange of GDP for GTP and thus converts Ran-GDP to Ran-GTP. Because *Ran GAP* is located in the cytosol and *Ran GEF* is located in the nucleus, the cytosol contains mainly Ran-GDP, and the nucleus contains mainly Ran-GTP (**Figure 12–60A**). The partitioning of the GAP and GEF between the cytosol and nucleus in a cell is due to their preferential association with the cytosolic cytoskeleton and nuclear chromatin, respectively.

The gradient of the two conformational forms of Ran drives nuclear transport in the appropriate direction. Import receptors, facilitated by FG-repeat binding, accelerate diffusion through the mesh inside the NPC channel. When an import receptor reaches the nuclear side of the pore complex, Ran-GTP binds to it and causes the receptor to release its cargo (**Figure 12–60B**). Because this occurs only

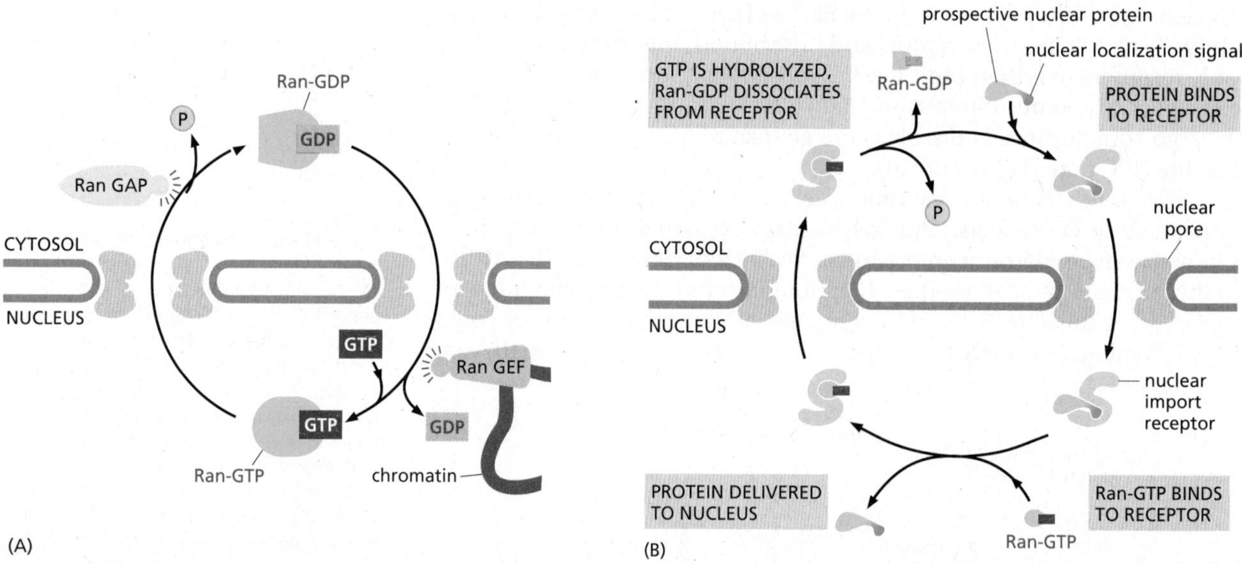

Figure 12–60 The compartmentalization of Ran-GDP and Ran-GTP provides directionality to nuclear transport. (A) Localization of Ran-GDP in the cytosol and Ran-GTP in the nucleus results from the localization of two Ran regulatory proteins: Ran GTPase-activating protein (Ran GAP) is located in the cytosol, and Ran guanine nucleotide exchange factor (Ran GEF) binds to chromatin and is therefore located in the nucleus. Ran-GDP is imported into the nucleus by its own import receptor (not shown), which is specific for the GDP-bound conformation of Ran. The Ran-GDP receptor is structurally unrelated to the main family of nuclear transport receptors. However, it also binds to FG repeats in NPC channel nucleoporins. (B) The interaction between a nuclear import receptor and its cargo is reversed by Ran-GTP. This means the receptor–cargo interaction is favored in the cytosol but disfavored in the nucleus. This results in net cargo transport from the cytosol to the nucleus.

on the nuclear side of the pore where the Ran-GTP concentration is high, the import process becomes rectified (that is, unidirectional), even though diffusion of the cargo–import receptor complex through the pore is governed by random back-and-forth diffusion.

Having discharged its cargo in the nucleus, the empty import receptor with Ran-GTP bound is transported back through the pore complex by the same mechanism of facilitated diffusion. When the complex of Ran-GTP and the import receptor reaches the cytosol, Ran GAP triggers Ran-GTP to hydrolyze its bound GTP. The resulting Ran-GDP lacks affinity for the import receptor, releasing it for another cycle of nuclear import. Thus, Ran-GDP permits cargo binding in the cytosol, while Ran-GTP stimulates cargo discharge in the nucleus, thereby imparting directionality to the import process.

Nuclear Export Works Like Nuclear Import, but in Reverse

The nuclear export of large molecules, such as new ribosomal subunits and RNA molecules, occurs through NPCs and also depends on a selective transport system. The transport system relies on **nuclear export signals** on the macromolecules to be exported. Export receptors bind to both the export signal, either directly or via an adaptor, and to NPC proteins to guide their cargo to the cytosol. As might be expected from the structural and evolutionary similarity of import receptors and export receptors, the import and export transport systems work in similar ways but in opposite directions: the import receptors bind their cargo molecules in the cytosol, release them in the nucleus, and are then exported to the cytosol for reuse, while the export receptors function in the opposite fashion (**Figure 12–61**).

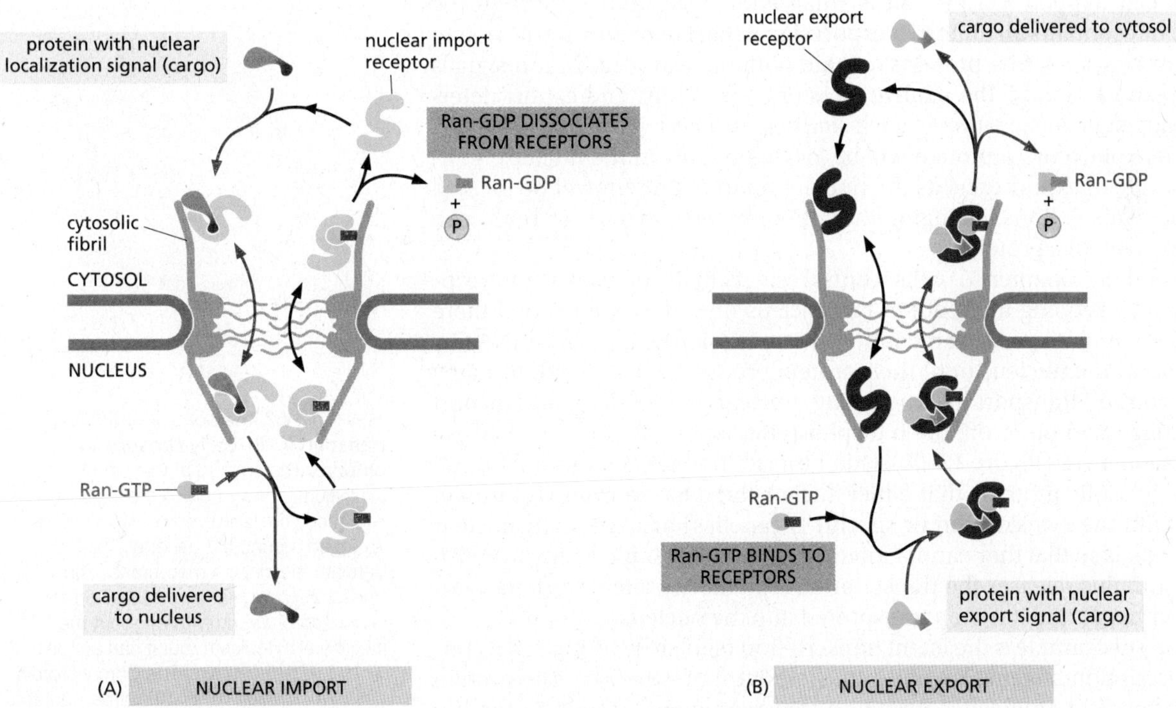

Figure 12–61 Nuclear import and nuclear export both use the Ran GTPase cycle. Movement through the NPC of loaded nuclear transport receptors occurs along the FG repeats displayed by certain NPC proteins. The differential localization of Ran-GTP in the nucleus and Ran-GDP in the cytosol provides directionality *(red arrows)* to both nuclear import (A) and nuclear export (B). Ran GAP stimulates the hydrolysis of GTP to produce Ran-GDP on the cytosolic side of the NPC (see Figure 12–60A). The critical difference between Ran-mediated nuclear import and nuclear export is the nature of cargo binding by the cargo receptor. In nuclear import, cargo binding is mutually exclusive of Ran-GTP; in nuclear export, cargo binding requires Ran-GTP. Thus, the locations where cargo is picked up and released are exactly reversed in nuclear export compared to nuclear import.

The ability of export receptors to work in reverse derives from the way they interact with the Ran GTPase. Ran-GTP in the nucleus promotes cargo binding to the export receptor, rather than promoting cargo dissociation as in the case of import receptors. Once the export receptor moves through the pore to the cytosol, it encounters Ran GAP, which induces the receptor to hydrolyze its GTP to GDP. As a result, the export receptor flips its conformation and releases both its cargo and Ran-GDP in the cytosol. Free export receptors and free Ran-GDP use the nuclear import pathway to enter the nucleus and complete the cycle.

As we discuss in detail in Chapter 6, cells control the export of RNAs from the nucleus. snRNAs, miRNAs, and tRNAs bind to nuclear export receptors, and they use the Ran-GTP gradient to fuel the transport process. By contrast, the export of mRNAs out of the nucleus uses a different mechanism that does not use export receptors or the Ran GTPase system. Instead, the spliced and processed mRNA is assembled with several nuclear RNA-binding proteins, some of which can bind the nuclear side of NPCs and others that bind FG repeats (see Figure 6–40). This export-competent mRNA ribonucleoprotein (mRNP) complex can then navigate through the FG repeat mesh within the NPC. A helicase complex that resides on the cytosolic side of NPCs uses the energy of ATP hydrolysis to strip several proteins from the mRNP, including the FG repeat–binding protein. This prevents the exported mRNA from reentering the NPC, making the export process unidirectional. The stripped RNA-binding proteins are rapidly imported back to the nucleus (using the import receptor and Ran GTPase system) for another round of transport.

Transport Through NPCs Can Be Regulated by Controlling Access to the Transport Machinery

Some proteins continually shuttle back and forth between the nucleus and the cytosol. This can happen if a protein is small enough to diffuse through the nuclear pore but contains an import or export signal that constantly retrieves it to the nucleus or cytosol. Other proteins contain both nuclear localization signals and nuclear export signals. The relative rates of their import and export determine the steady-state localization of such *shuttling proteins*: if the rate of import exceeds the rate of export, a protein will be located mainly in the nucleus; conversely, if the rate of export exceeds the rate of import, a protein will be located mainly in the cytosol. Thus, changing the rate of import, export, or both, can change the location of a protein.

As discussed in Chapter 7, cells control the activity of some transcription regulators by keeping them out of the nucleus until they are needed there (**Figure 12–62**); similarly, cells can control the translation of certain mRNAs by retaining them in the nucleus until their protein products are needed. In many cases, cells control transport by regulating nuclear localization and export signals—turning them on or off, often by phosphorylation of amino acids close to the signal sequences (**Figure 12–63**). Other transcription regulators are bound to inhibitory cytosolic proteins that either anchor them in the cytosol (through interactions with the cytoskeleton or specific organelles) or mask their nuclear localization signals so that they cannot interact with nuclear import receptors. An appropriate stimulus releases the transcription regulatory protein from its cytosolic anchor or mask, and it is then transported into the nucleus.

One important example is the latent transcription regulatory protein that controls the transcription of genes involved in cholesterol metabolism. The protein is made and stored in an inactive form as a transmembrane protein in the ER. When a cell is deprived of cholesterol, the protein is transported from the ER to the Golgi apparatus where it encounters specific proteases that cleave off the cytosolic domain, releasing it into the cytosol. This domain is then imported into the nucleus, where it activates the transcription of genes required for both cholesterol uptake and synthesis (**Figure 12–64**). Earlier in this chapter, we discussed a similar mechanism that controls the activation of the ATF6 arm of the unfolded protein response (see Figure 12–36).

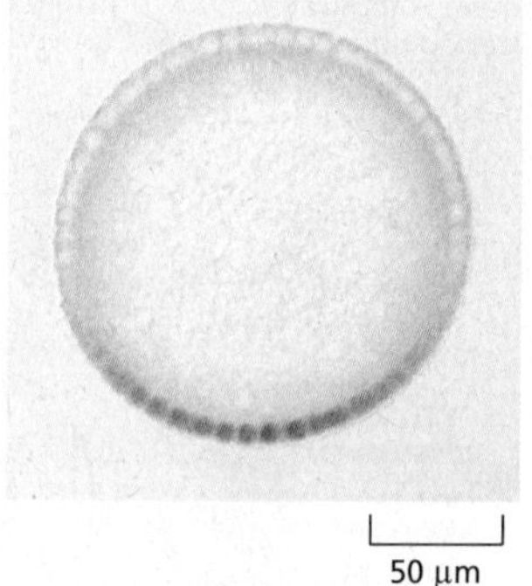

Figure 12–62 **The control of nuclear transport in the early *Drosophila* embryo.** The embryo at this stage is a syncytium, shown here in cross section, with many nuclei in a common cytoplasm, arranged around the periphery, just beneath the plasma membrane. The transcription regulatory protein Dorsal is produced uniformly throughout the peripheral cytoplasm, but it can act only when inside the nuclei. The Dorsal protein has been stained with an enzyme-coupled antibody that yields a brown product, revealing that Dorsal is excluded from the nuclei at the dorsal side *(top)* of the embryo but is concentrated in the nuclei toward the ventral side *(bottom)* of the embryo. The regulated traffic of Dorsal into the nuclei controls the differential development between the back and belly of the animal. (Courtesy of Siegfried Roth.)

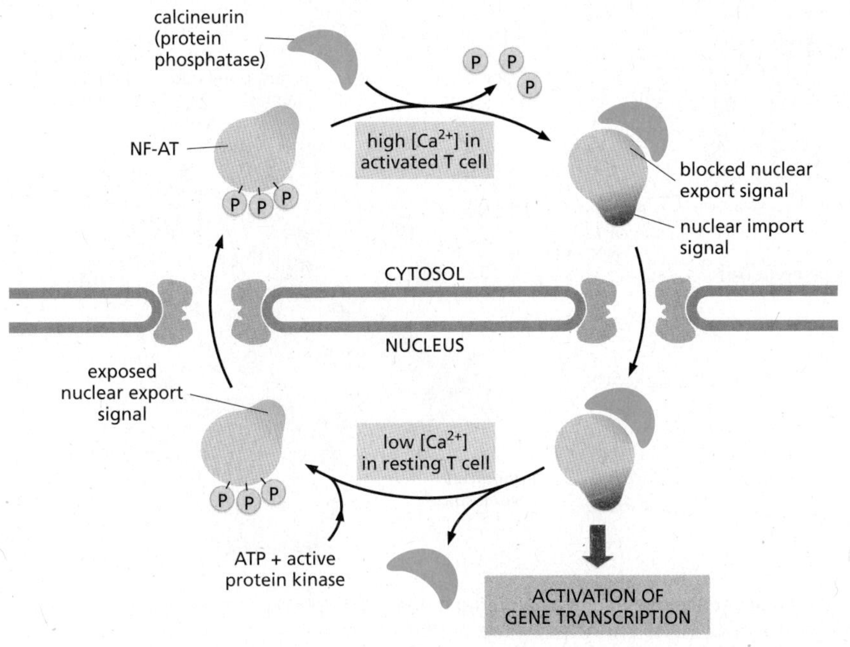

Figure 12–63 **The control of nuclear import during T cell activation.** The nuclear factor of activated T cells (NF-AT) is a transcription regulatory protein that, in the resting T cell, is found in the cytosol in a phosphorylated state. When T cells are activated by foreign antigen (discussed in Chapter 24), the intracellular Ca^{2+} concentration increases. At high concentrations of Ca^{2+}, the protein phosphatase calcineurin binds to NF-AT and dephosphorylates it. The dephosphorylation exposes nuclear import signals and blocks a nuclear export signal. The complex of NF-AT and calcineurin is therefore imported into the nucleus, where NF-AT activates the transcription of numerous genes required for T cell activation. The response shuts off when Ca^{2+} levels decrease, releasing NF-AT from calcineurin. Rephosphorylation of NF-AT inactivates the nuclear import signals and reexposes the nuclear export signal, causing NF-AT to relocate to the cytosol. Some of the most potent immunosuppressive drugs, including cyclosporin A and FK506, inhibit the ability of calcineurin to dephosphorylate NF-AT and thereby block the nuclear accumulation of NF-AT and T cell activation (**Movie 12.6**).

The Nuclear Envelope Disassembles and Reassembles During Mitosis

In animal cells, the nuclear envelope is dismantled during mitosis so that microtubules can access the replicated chromosomes for segregation between the two daughter cells (discussed in Chapter 17). At the end of mitosis, the nuclear envelope reassembles, and the asymmetrical distribution of cellular contents between the cytosol and nucleus is reestablished. The major structures that must be reversibly disassembled are the nuclear lamina, the NPCs, and the membranes of the nuclear envelope.

The dismantling process is initiated by the cyclin-dependent kinase (Cdk) that is activated at the onset of mitosis (discussed in Chapter 17). Cdk

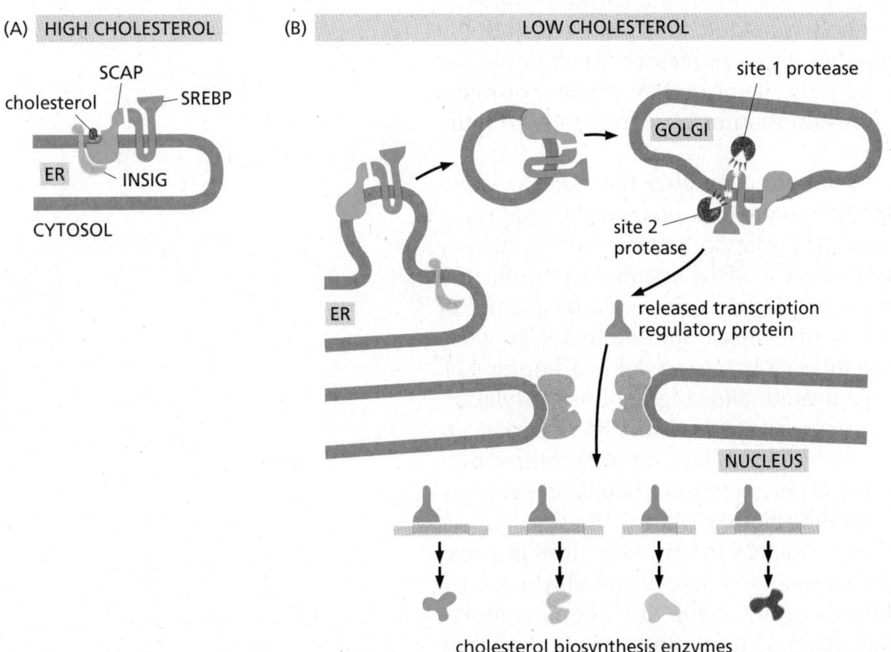

Figure 12–64 **Feedback regulation of cholesterol biosynthesis.** (A) SREBP (sterol response element binding protein), a latent transcription regulator that controls expression of cholesterol biosynthetic enzymes, is initially synthesized as an ER membrane protein. It is anchored in the ER if there is sufficient cholesterol in the membrane by interaction with a membrane protein complex composed of the proteins INSIG and SCAP (SREBP cleavage activation protein), which binds cholesterol. (B) If the cholesterol-binding site on SCAP is empty (at low cholesterol concentrations), SCAP changes conformation and dissociates from INSIG. Dissociation from INSIG frees the SCAP–SREBP complex so it can be packaged together into transport vesicles that are delivered to the Golgi apparatus. In the Golgi apparatus, two Golgi-resident proteases cleave SREBP to free its cytosolic domain from the membrane. The cytosolic domain, which is a transcription regulatory protein, then moves into the nucleus, where it binds to the promoters of genes that encode proteins involved in cholesterol biosynthesis and activates their transcription. In this way, more cholesterol is made when its concentration falls below a threshold.

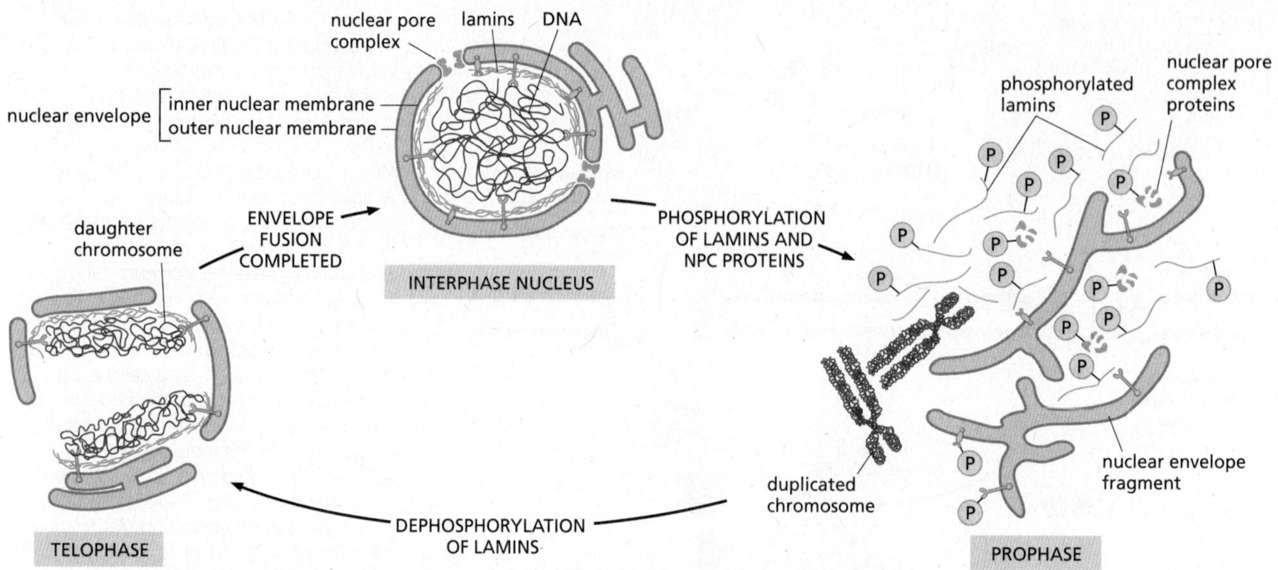

Figure 12–65 **The breakdown and re-formation of the nuclear envelope and lamina during mitosis.** Phosphorylation of the lamins triggers the disassembly of the nuclear lamina, which initiates the breakup of the nuclear envelope. Dephosphorylation of the lamins reverses the process. An analogous phosphorylation and dephosphorylation cycle occurs for some nucleoporins and proteins of the inner nuclear membrane, and some of these dephosphorylation events are also involved in the reassembly process. The lamin network beings to re-form around regions of individual decondensing daughter chromosomes. The lamins recruit membranes that contain interacting lamin receptors that were in the inner nuclear membrane. Eventually, as decondensation progresses, these membrane structures fuse to form a single complete nucleus. Mitotic breakdown of the nuclear envelope occurs in all metazoan cells. However, in many other species, such as yeasts, the nuclear envelope remains intact during mitosis, and the nucleus divides by fission.

phosphorylates nucleoporins, lamins, and inner nuclear membrane proteins to disrupt their interactions with each other and with chromatin. During this process, some NPC proteins become bound to nuclear import receptors, which play an important part in the reassembly of NPCs at the end of mitosis. Nuclear envelope membrane proteins—no longer tethered to the pore complexes, lamina, or chromatin—disperse throughout the ER membrane. The dynein motor protein, which moves along microtubules (discussed in Chapter 16), is recruited to the outer nuclear membrane early in mitosis and exerts a pulling force. Transmembrane proteins that tether the outer nuclear membrane to the inner nuclear membrane and lamina help transduce this force and pull the nuclear envelope off the chromatin. Together, these processes break down the barriers that normally separate the nucleus and cytosol, and the nuclear proteins that are not bound to membranes or chromosomes intermix completely with the proteins of the cytosol (**Figure 12–65**).

One protein that remains bound to chromatin even after the nuclear envelope breaks down is Ran GEF. This means Ran molecules close to chromatin are mainly in their GTP-bound conformation. By contrast, Ran molecules further away are in their GDP-bound conformation because of the action of cytosolic Ran GAP. As a result, the chromosomes in mitotic cells are surrounded by a cloud of Ran-GTP, which is important for assembling the mitotic spindle that segregates chromosome into the newly forming daughter cells (discussed in Chapter 17). After chromosome segregation, Cdk is inactivated, allowing dephosphorylation of nucleoporins, lamins, and nuclear membrane proteins. This triggers reassembly of the nuclear envelope on the surface of the complete set of chromosomes in each daughter cell. The positional marker for recruitment of nuclear envelope components to chromosomes is the surrounding cloud of Ran-GTP.

Ran-GTP releases the NPC proteins from nuclear import receptors in proximity to the chromosomes. The free NPC proteins attach to the chromosome surface, where they assemble into new NPCs. At the same time, dephosphorylated lamins bind again to chromatin and recruit ER membranes via the inner nuclear membrane proteins that reside within them. The ER progressively wraps around the entire group of chromosomes until the ER forms a sealed nuclear envelope, engulfing the chromosomes and proteins bound to them (**Movie 12.7**). The newly formed inner nuclear envelope is closely applied to the surface of the chromosomes, is enriched for inner nuclear membrane proteins, and excludes all proteins except those initially bound to the mitotic chromosomes, thus conferring a high level of selectivity to the engulfment process. Because Ran-GTP is inside the nucleus and Ran-GDP remains outside, unidirectional import of proteins that contain nuclear localization signals can occur through NPCs. In this way, the nuclear protein content is replenished, while all other large proteins, including ribosomes, are kept out of the newly assembled nucleus.

Summary

The nuclear envelope consists of an inner and an outer nuclear membrane that are connected with each other at perforations formed by nuclear pore complexes (NPCs). The outer nuclear membrane is continuous with the ER membrane, and the space between the inner and outer nuclear membranes is continuous with the ER lumen. RNA molecules, which are made in the nucleus, and ribosomal subunits, which are assembled there, are exported to the cytosol; in contrast, all the proteins that function in the nucleus are synthesized in the cytosol and are then imported. The extensive traffic of materials between the nucleus and cytosol occurs through NPCs, which provide a direct passageway across the nuclear envelope. The interior of NPCs contains a mesh of unstructured proteins that allows passage of small molecules but imposes a diffusion barrier that requires large macromolecules to be actively transported.

Nuclear localization signals and nuclear export signals on proteins to be transported through NPCs are recognized by corresponding nuclear transport receptors. These receptors function by binding their cargoes selectively on one side of the nuclear envelope, increasing the diffusion rate through NPCs, and releasing cargoes selectively on the other side. The free energy of GTP hydrolysis by the monomeric GTPase Ran is harnessed to provide the directionality for nuclear transport. Messenger RNAs are exported from the nucleus through NPCs as parts of large ribonucleoprotein complexes; they use a different transport route that uses ATP hydrolysis to remodel the complexes at the cytosolic side of NPCs. Cells regulate the transport of nuclear proteins and RNA molecules through the NPCs by controlling the access of these molecules to the transport machinery. Because nuclear localization signals are not removed, nuclear proteins can be imported repeatedly, as is required each time that the nucleus reassembles after mitosis.

PROBLEMS

Which statements are true? Explain why or why not.

12–1 Like the lumen of the ER, the interior of the nucleus is topologically equivalent to the outside of the cell.

12–2 ER-bound and free ribosomes, which are structurally and functionally identical, differ only in the proteins they happen to be making at a particular time.

12–3 The signal sequence binds to a hydrophobic site on the ribosome causing a slowdown in protein synthesis, which resumes when SRP binds to the signal sequence.

12–4 Peroxisomes are found in only a few specialized types of eukaryotic cell.

12–5 The two signal sequences required for insertion of nucleus-encoded proteins into the mitochondrial inner membrane via the TIM23 complex are cleaved off the protein in different mitochondrial compartments.

12–6 To avoid the collisions that would occur if two-way traffic through a single pore were allowed, nuclear pore complexes are specialized so that some mediate import while others mediate export.

Discuss the following problems.

12–7 Biomolecular condensates form just under the membrane during T cell receptor signal transduction in immune responses. Three components are critical: the transmembrane protein LAT (linker for activation of T cells), which is phosphorylated at three cytoplasmic tyrosine residues (pY); Grb2, which contains one SH2 domain and two SH3 domains; and Sos1, which contains four binding sites for SH3 domains (**Figure Q12–1A**).

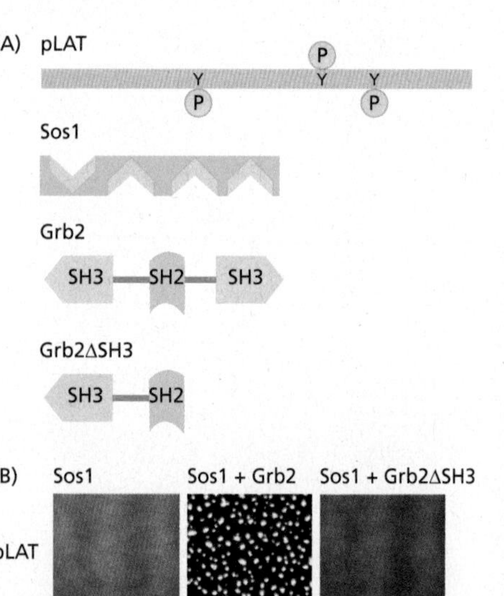

Figure Q12–1 Condensate formation during T cell receptor signal transduction (Problem 12–7). (A) Key molecular components. (B) Appearance of small condensates (bright spots) when the critical components are present. (Modified from Figure 1A and D in X. Su et al., *Science* 352:595–599, 2016. With permission from AAAS.)

When you insert phosphorylated LAT (pLAT) in an artificial lipid bilayer and add Grb2 and Sos1, micrometer-sized condensates form just below the bilayer (**Figure Q12–1B**). However, if you use a form of Grb2 that contains just one SH3 domain (Grb2ΔSH3), instead of two, the condensates do not form (Figure Q12–1B). Why do you suppose that Grb2 supports condensate formation, while Grb2ΔSH3, which is still multivalent, does not?

12–8 What is the fate of a protein with no sorting signal?

12–9 Are proteins bound for the plasma membrane common or rare among all ER membrane proteins? A few simple considerations allow one to answer this question. In a typical growing cell that is dividing once every 24 hours, the equivalent of one new plasma membrane must transit the ER every day. If the ER membrane is 20 times the area of a plasma membrane, what is the ratio of plasma membrane proteins to other membrane proteins in the ER? (Assume that all proteins on their way to the plasma membrane remain in the ER for 30 minutes on average before exiting, and that the ratio of proteins to lipids in the ER and plasma membranes is the same.)

12–10 A multipass transmembrane protein with several membrane-spanning segments is shown schematically in **Figure Q12–2**. The boxes represent membrane-spanning segments, and the arrow represents the site for cleavage of the signal sequence. In which compartments—cytosol or ER lumen—will the N- and C-termini of the mature protein be located?

Figure Q12–2 A multipass transmembrane protein with a cleavable signal sequence (Problem 12–10).

12–11 All new phospholipids are added to the cytosolic leaflet of the ER membrane, yet the ER membrane has a symmetrical distribution of different phospholipids in its two leaflets. By contrast, the plasma membrane, which receives all its membrane components ultimately from the ER, has a very asymmetrical distribution of phospholipids in the two leaflets of its lipid bilayer. How is the symmetry generated in the ER membrane, and how is the asymmetry generated and maintained in the plasma membrane?

12–12 Cells with functional peroxisomes incorporate 9-(1′-pyrene)nonanol (P9OH) into membrane lipids. Exposure of such cells to ultraviolet (UV) light causes cell death by generating reactive oxygen species, which are toxic. Cells that do not make peroxisomes lack a critical enzyme responsible for incorporating P9OH into membrane lipids. How might you make use of P9OH to select for cells that are missing peroxisomes?

12–13 Components of the TIM complexes were initially identified using a genetic trick. The yeast *Ura3* gene, which encodes an enzyme that is normally located in the cytosol where it is essential for synthesis of uracil, was modified so that the protein carried an import signal for the mitochondrial matrix. A population of cells carrying the modified *Ura3* gene in place of the normal gene was then grown in the absence of uracil. Most cells died, but the rare cells that grew were shown to be defective for mitochondrial import. Explain how this selection system works. Why do most of the cells die, and why do the import-defective cells grow?

12–14 If the enzyme dihydrofolate reductase (DHFR), which is normally located in the cytosol, is engineered to carry a mitochondrial targeting sequence at its N-terminus, it is efficiently imported into mitochondria. If the modified DHFR is first incubated with methotrexate, which binds tightly to the active site, the enzyme remains in the cytosol. How do you suppose that the binding of methotrexate interferes with mitochondrial import?

12–15 Why do mitochondria need a special translocator to import proteins across the outer membrane, when the membrane already has large pores formed by porins?

12–16 Assuming that 32 million histone octamers are required to package the human genome, how many histone molecules must be transported per second per nuclear pore complex in cells whose nuclei contain 3000 nuclear pores and are dividing once per day?

12–17 Selective permeability of the nuclear pore complex (NPC) is controlled by protein components with unstructured tails that extend into the central pore. These tails are characterized by periodic repeats of the hydrophobic amino acids phenylalanine (F) and glycine (G). In a test tube at a concentration of 50 mM, the FG repeat domains of these proteins form a gel, which is held together by weak interactions between the hydrophobic FG repeats. These gels allow passive diffusion of small molecules, which fit through the holes in the mesh, but they prevent entry of larger proteins such as the fluorescent protein mCherry fused to maltose-binding protein (MBP) (**Figure Q12–3A**). However, if the nuclear import receptor, importin, is fused to a similar protein, MBP-GFP, the importin-MBP-GFP fusion readily enters the gel (**Figure Q12–3B**).

Is diffusion of importin-MBP-GFP through the FG repeat gel fast enough to account for the efficient flow of materials between the nucleus and cytosol? From experiments of the type shown in Figure Q12–3B, the diffusion coefficient (D) of importin-MBP-GFP through the FG repeat gel was determined to be about 0.1 μm^2/sec. The equation for diffusion is $t = x^2/2D$, where t is time and x is distance. About how long would it take importin-MBP-GFP to diffuse through a yeast nuclear pore (a distance of 30 nm) if the pore consisted of a gel of FG repeats?

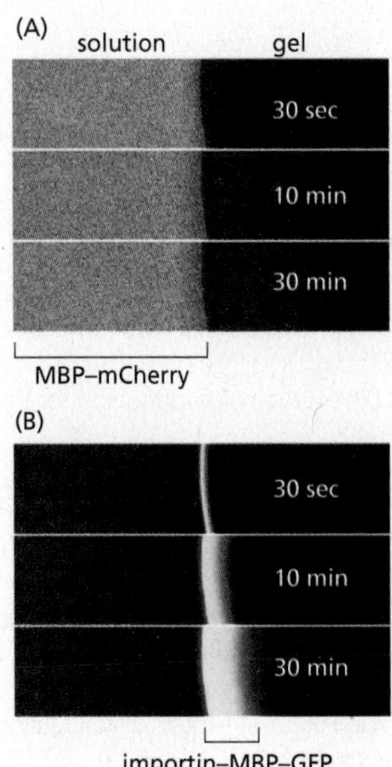

Figure Q12–3 Diffusion of proteins through FG repeat gels (Problem 12–17). Diffusion of MBP–mCherry (A) and importin–MBP–GFP (B) into a gel of FG repeats (on the *right*). The bright areas indicate regions that contain the fluorescent proteins. (Modified from Figure 2 of S. Frey & D. Görlich, *EMBO J.* 28:2554–2567, 2009.)

12–18 A classic experiment that addressed whether nuclear proteins diffused passively into the nucleus or were actively imported used several forms of radioactive nucleoplasmin, which is a large pentameric protein involved in chromatin assembly. In this experiment, either the intact protein or the nucleoplasmin heads, tails, or heads with a single tail were injected into the cytosol or the nucleus of a frog oocyte (**Figure Q12–4**). All forms of nucleoplasmin, except heads, accumulated in the nucleus when injected into the cytoplasm, and all forms were retained in the nucleus when injected there.

How do these experiments distinguish between active transport, in which a nuclear localization signal triggers transport by the nuclear pore complex, and passive diffusion, in which a binding site for a nuclear component allows accumulation in the nucleus?

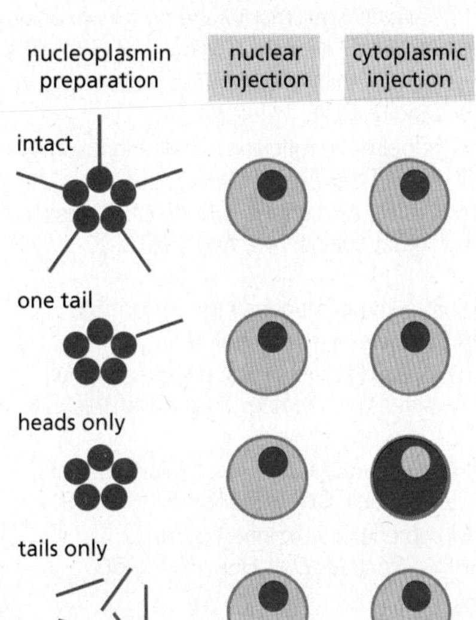

Figure Q12–4 Cellular location of injected nucleoplasmin components (Problem 12–18). The location of nucleoplasmin is indicated by the *red* areas.

REFERENCES

General

Palade G (1975) Intracellular aspects of the process of protein synthesis. *Science* 189, 347–358.

The Compartmentalization of Cells

Banani SF, Lee HO, Hyman AA & Rosen MK (2017) Biomolecular condensates: organizers of cellular biochemistry. *Nat. Rev. Mol. Cell Biol.* 18, 285–298.

Baum DA & Baum B (2014) An inside-out origin for the eukaryotic cell. *BMC Biol.* 12, 76.

Blobel G (1980) Intracellular protein topogenesis. *Proc. Natl. Acad. Sci. USA* 77, 1496–1500.

Devos DP, Gräf R & Field MC (2014) Evolution of the nucleus. *Curr. Opin. Cell Biol.* 28, 8–15.

Knoblach B & Rachubinski RA (2015) Motors, anchors, and connectors: orchestrators of organelle inheritance. *Annu. Rev. Cell Dev. Biol.* 31, 55–81.

Wickner W & Schekman R (2005) Protein translocation across biological membranes. *Science* 310, 1452–1456.

The Endoplasmic Reticulum

Akopian D, Shen K, Zhang X & Shan SO (2013) Signal recognition particle: an essential protein-targeting machine. *Annu. Rev. Biochem.* 82, 693–721.

Blobel G & Dobberstein B (1975) Transfer of proteins across membranes. I. Presence of proteolytically processed and unprocessed nascent immunoglobulin light chains on membrane-bound ribosomes of murine myeloma. *J. Cell Biol.* 67, 835–851.

Braakman I & Bulleid NJ (2011) Protein folding and modification in the mammalian endoplasmic reticulum. *Annu. Rev. Biochem.* 80, 71–99.

Chen S, Novick P & Ferro-Novick S (2013) ER structure and function. *Curr. Opin. Cell Biol.* 25, 428–433.

Christianson JC & Ye Y (2014) Cleaning up in the endoplasmic reticulum: ubiquitin in charge. *Nat. Struct. Mol. Biol.* 21, 325–335.

Cymer F, von Heijne G & White SH (2015) Mechanisms of integral membrane protein insertion and folding. *J. Mol. Biol.* 427, 999–1022.

Deshaies RJ, Sanders SL, Feldheim DA & Schekman R (1991) Assembly of yeast Sec proteins involved in translocation into the endoplasmic reticulum into a membrane-bound multisubunit complex. *Nature* 349, 806–808.

Görlich D, Prehn S, Hartmann E, . . . , Rapoport TA (1992) A mammalian homolog of SEC61p and SECYp is associated with ribosomes and nascent polypeptides during translocation. *Cell* 71, 489–503.

Hegde RS & Keenan RJ (2011) Tail-anchored membrane protein insertion into the endoplasmic reticulum. *Nat. Rev. Mol. Cell Biol.* 12, 787–798.

Mamathambika BS & Bardwell JC (2008) Disulfide-linked protein folding pathways. *Annu. Rev. Cell Dev. Biol.* 24, 211–235.

Marciniak SJ & Ron D (2006) Endoplasmic reticulum stress signaling in disease. *Physiol. Rev.* 86, 1133–1149.

Milstein C, Brownlee GG, Harrison TM & Mathews MB (1972) A possible precursor of immunoglobulin light chains. *Nat. New Biol.* 239, 117–120.

Pelham HR (1986) Speculations on the functions of the major heat shock and glucose-related proteins. *Cell* 46, 959–961.

Pomorski TG & Menon AK (2016) Lipid somersaults: uncovering the mechanisms of protein-mediated lipid flipping. *Prog. Lipid Res.* 64, 69–84.

Rapoport TA, Li L & Park E (2017) Structural and mechanistic insights into protein translocation. *Annu. Rev. Cell Dev. Biol.* 33, 369–390.

Shao S & Hegde RS (2011) Membrane protein insertion at the endoplasmic reticulum. *Annu. Rev. Cell Dev. Biol.* 27, 25–56.

Van den Berg B, Clemons Jr WM, Collinson I, . . . , Rapoport TA (2004) X-ray structure of a protein-conducting channel. *Nature* 427, 36–44.

Walter P & Ron D (2011) The unfolded protein response: from stress pathway to homeostatic regulation. *Science* 334, 1081–1086.

Wu H, Carvalho P & Voeltz GK (2018) Here, there, and everywhere: the importance of ER membrane contact sites. *Science* 361, eaan5835.

Peroxisomes

Agrawal G & Subramani S (2016) De novo peroxisome biogenesis: evolving concepts and conundrums. *Biochim. Biophys. Acta* 1863, 892–901.

Fujiki Y, Yagita Y & Matsuzaki T (2012) Peroxisome biogenesis disorders. *Biochim. Biophys. Acta* 1822, 1337–1342.

Walter T & Erdmann R (2019) Current advances in protein import into peroxisomes. *Protein J.* 38, 351–362.

The Transport of Proteins into Mitochondria and Chloroplasts

Araiso Y, Imai K & Endo T (2020) Structural snapshot of the mitochondrial protein import gate. *FEBS J.*, doi: 10.1111/febs.15661.

Jarvis P & Robinson C (2004) Mechanisms of protein import and routing in chloroplasts. *Curr. Biol.* 14, R1064–R1077.

Kessler F & Schnell DJ (2009) Chloroplast biogenesis: diversity and regulation of the protein import apparatus. *Curr. Opin. Cell Biol.* 21, 494–500.

Schleiff E & Becker T (2011) Common ground for protein translocation: access control for mitochondria and chloroplasts. *Nat. Rev. Mol. Cell Biol.* 12, 48–59.

Wiedemann N & Pfanner N (2017) Mitochondrial machineries for protein import and assembly. *Annu. Rev. Biochem.* 86, 685–714.

The Transport of Molecules Between the Nucleus and the Cytosol

Adam SA & Gerace L (1991) Cytosolic proteins that specifically bind nuclear location signals are receptors for nuclear import. *Cell* 66, 837–847.

Burke B & Stewart CL (2013) The nuclear lamins: flexibility in function. *Nat. Rev. Mol. Cell Biol.* 14, 13–24.

Cole CN & Scarcelli JJ (2006) Transport of messenger RNA from the nucleus to the cytoplasm. *Curr. Opin. Cell Biol.* 18, 299–306.

Güttinger S, Laurell E & Kutay U (2009) Orchestrating nuclear envelope disassembly and reassembly during mitosis. *Nat. Rev. Mol. Cell Biol.* 10, 178–191.

Hetzer MW & Wente SR (2009) Border control at the nucleus: biogenesis and organization of the nuclear membrane and pore complexes. *Dev. Cell* 17, 606–616.

Hülsmann BB, Labokha AA & Görlich D (2012) The permeability of reconstituted nuclear pores provides direct evidence for the selective phase model. *Cell* 150, 738–751.

Köhler A & Hurt E (2007) Exporting RNA from the nucleus to the cytoplasm. *Nat. Rev. Mol. Cell Biol.* 8, 761–773.

Lin DH & Hoelz A (2019) The structure of the nuclear pore complex (an update). *Annu. Rev. Biochem.* 88, 725–783.

Moore MS & Blobel G (1993) The GTP-binding protein Ran/TC4 is required for protein import into the nucleus. *Nature* 365, 661–663.

Rothballer A & Kutay U (2013) Poring over pores: nuclear pore complex insertion into the nuclear envelope. *Trends Biochem. Sci.* 38, 292–301.

Strambio-De-Castilla C, Niepel M & Rout MP (2010) The nuclear pore complex: bridging nuclear transport and gene regulation. *Nat. Rev. Mol. Cell Biol.* 11, 490–501.

Tran EJ & Wente SR (2006) Dynamic nuclear pore complexes: life on the edge. *Cell* 125, 1041–1053.

CHAPTER 13

Intracellular Membrane Traffic

IN THIS CHAPTER

Every cell must eat, communicate with the world around it, and quickly respond to changes in its environment. To help accomplish these tasks, cells continually adjust the composition of their plasma membrane and internal compartments in response to need. Eukaryotic cells use an elaborate internal membrane system to add and remove cell-surface proteins, such as receptors, ion channels, and transporters (**Figure 13–1**). Through the process of *exocytosis*, the secretory pathway delivers newly synthesized proteins, carbohydrates, and lipids either to the plasma membrane or to the extracellular space. By the converse process of *endocytosis*, cells take in components of the plasma membrane and extracellular space and deliver them to internal compartments called *endosomes*.

The proteins, nutrients, lipids, and receptors delivered by endocytosis to endosomes are sorted and either recycled to the plasma membrane or delivered to *lysosomes* where they are broken down into building blocks and transported to the cytosol for use in various biosynthetic processes. Lysosomes also break down and recycle intracellular macromolecules through a process called *autophagy*. This pathway engulfs parts of the cytosol or whole organelles into a newly assembled compartment, which then fuses with lysosomes to deliver its contents for degradation. In development, cells often use autophagy to remodel their cytoplasm as they differentiate and adapt to new physiological tasks.

The interior space, or lumen, of each membrane-enclosed compartment along the secretory and endocytic pathways is topologically equivalent to the cell exterior. This means proteins can travel from the lumen of one compartment to another by means of numerous membrane-enclosed transport containers without ever having to cross a membrane (**Figure 13–2**). These containers are formed by membrane budding from a compartment and are either small spherical *vesicles*, larger irregular vesicles, or tubules. We shall use the term **transport vesicle** to apply to all forms of these containers.

Transport vesicles continually bud off from one membrane compartment and fuse with another, carrying membrane components and soluble lumenal molecules, which are referred to as **cargo**. This vesicular traffic flows along highly

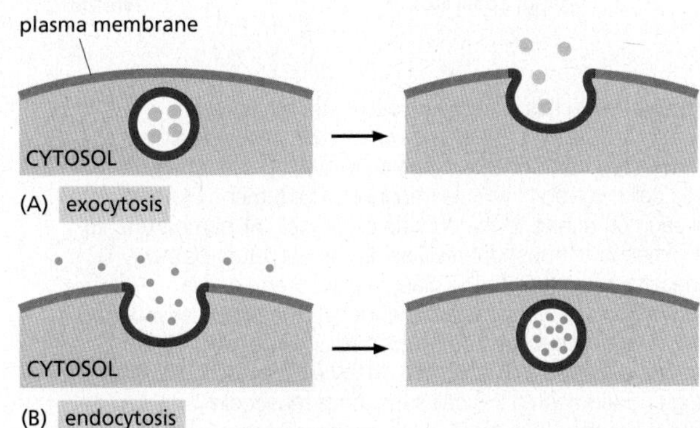

Figure 13–1 Exocytosis and endocytosis. (A) In exocytosis, a transport vesicle fuses with the plasma membrane. Its content is released into the extracellular space, while the vesicle membrane *(red)* becomes continuous with the plasma membrane. (B) In endocytosis, a plasma membrane patch *(red)* is internalized, forming a transport vesicle. Its content derives from the extracellular space. The interior of the transport vesicles in panels A and B is topologically equivalent to the extracellular space.

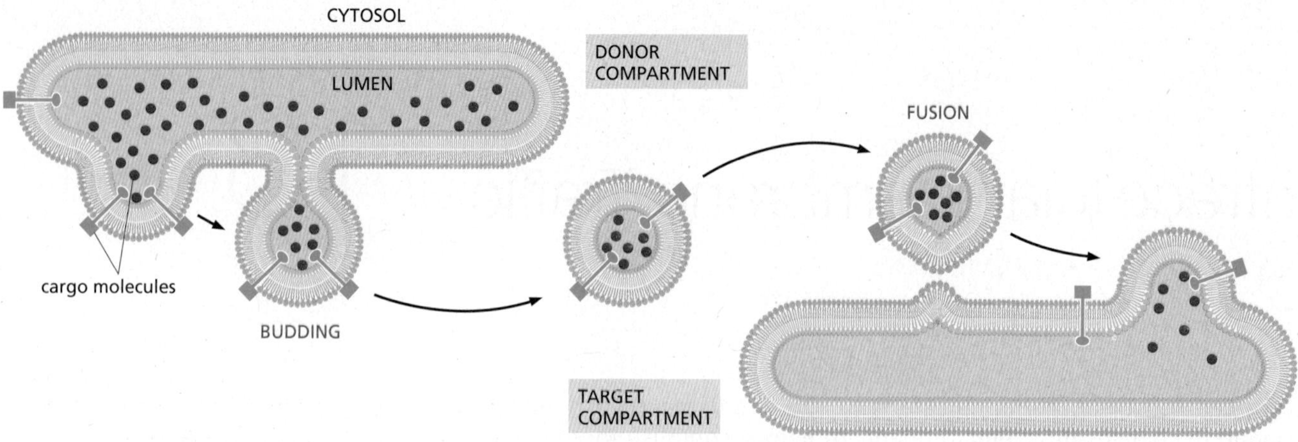

Figure 13–2 **Vesicle transport.** Transport vesicles bud off from one compartment and fuse with another. As they do so, they carry material as cargo from the *lumen* (the space within a membrane-enclosed compartment) and membrane of the donor compartment to the lumen and membrane of the target compartment, as shown.

organized, directional routes, which allow the cell to secrete, eat, and remodel its plasma membrane and organelles (Figure 13–3). The *secretory pathway* leads outward from the endoplasmic reticulum (ER) toward the Golgi apparatus and cell surface, with a side route leading to endosomes, while the *endocytic pathway* leads inward from the plasma membrane. In each case, *retrieval pathways* bring membrane and selected proteins back to the compartment of origin to balance the flow of materials between compartments.

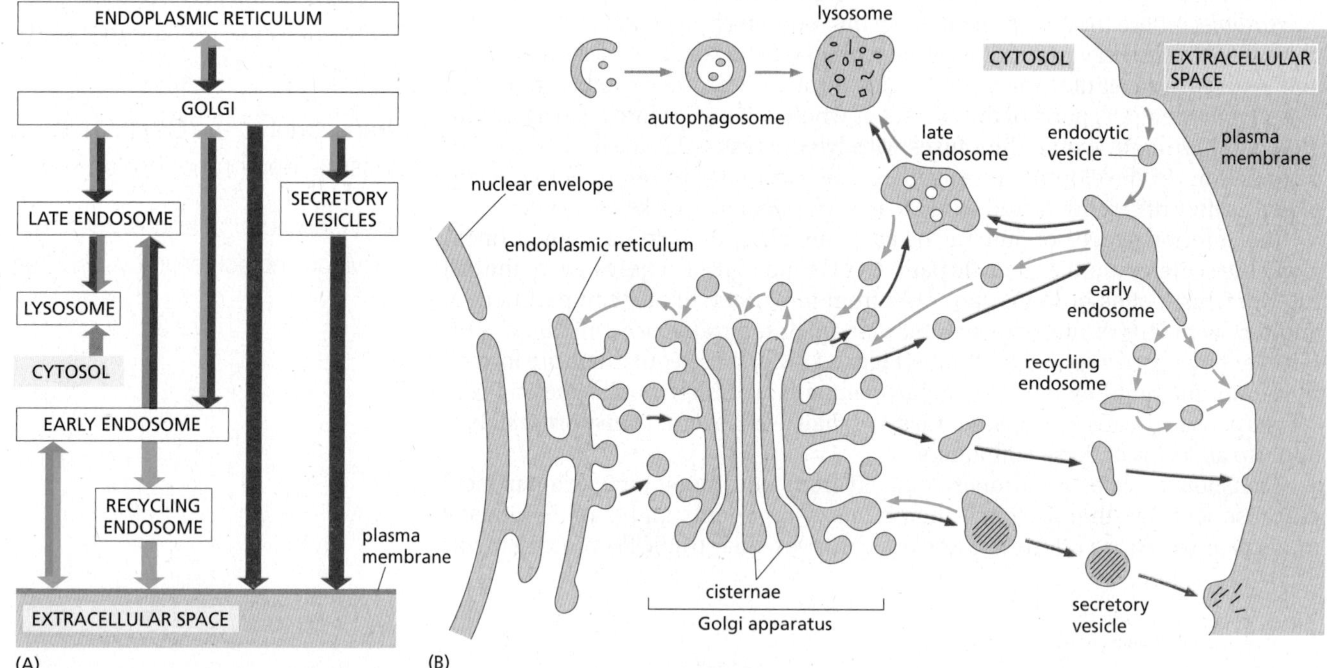

Figure 13–3 **A "road map" of the secretory and endocytic pathways.** (A) In this schematic road map, which was introduced in Chapter 12, the endocytic and secretory pathways are illustrated with *green* and *red arrows*, respectively. In addition, *blue arrows* denote retrieval pathways for the backflow of selected components. Engulfment during autophagy is illustrated with a *gray arrow*. (B) The compartments of the eukaryotic cell involved in vesicle transport. The lumens of most membrane-enclosed compartments are topologically equivalent to each other and to the outside of the cell. All compartments shown communicate with one another and the outside of the cell by means of transport vesicles. In the secretory pathway *(red arrows)*, protein molecules are transported from the endoplasmic reticulum (ER) to the plasma membrane or (via endosomes) to lysosomes. In the endocytic pathway *(green arrows)*, molecules are ingested in endocytic vesicles derived from the plasma membrane and delivered to early endosomes and then (via late endosomes) to lysosomes. In autophagy *(gray arrows)*, cytoplasmic components engulfed into an autophagosome are delivered to lysosomes. Many endocytosed molecules are retrieved from early endosomes and returned (some via recycling endosomes) to the cell surface for reuse; similarly, some molecules are retrieved from the early and late endosomes and returned to the Golgi apparatus, and some are retrieved from the Golgi apparatus and returned to the ER. All of these retrieval pathways are shown with *blue arrows*, as in panel A.

To perform its function, each transport vesicle that buds from a compartment must be selective. It must take up only the appropriate molecules and must fuse only with the appropriate target membrane. A vesicle carrying cargo from the ER to the Golgi apparatus, for example, must exclude most other proteins that are to stay in the ER, and it must fuse only with the Golgi apparatus and not with any other organelle.

We begin this chapter by considering the molecular mechanisms of budding and fusion that underlie all vesicle transport. We then discuss the fundamental problem of how, in the face of this transport, the cell maintains the molecular and functional differences between its compartments. Finally, we consider the function of the Golgi apparatus, secretory vesicles, endosomes, and lysosomes as we trace the pathways that connect these organelles.

MECHANISMS OF MEMBRANE TRANSPORT AND COMPARTMENT IDENTITY

Vesicle transport mediates a continual exchange of components between the 10 or more chemically distinct, membrane-enclosed compartments that collectively compose the secretory and endocytic pathways. In this section, we discuss how transport vesicles form, how they concentrate cargo within them, and how they deliver their contents selectively to another compartment. Transport begins when a special coat of proteins is assembled on a region of the cytosolic face of a membrane compartment. The coat is used to collect specific cargo components from the membrane and compartment lumen for delivery to another compartment. The coat, with the help of additional proteins, shapes the membrane into a transport vesicle that buds from the originating compartment. These vesicles selectively dock at the appropriate destination membrane and then fuse with it to deliver their cargo.

Despite the constant exchange of components between membrane-enclosed compartments, each compartment maintains its special identity of molecular markers, such as proteins or specific lipids, that are displayed on the cytosolic surface of the membrane. Cells achieve this by tightly controlling the membrane components that are packaged into departing transport vesicles. The identity markers of a compartment serve as guidance cues for outgoing traffic by recruiting the appropriate coat and for incoming traffic to ensure that transport vesicles fuse only with the correct compartment. Many of these membrane markers, however, are found on more than one compartment, and it is the specific combination of marker molecules that gives each compartment its molecular address.

There Are Various Types of Coated Vesicles

Most transport vesicles form from specialized, coated regions of membranes. They bud off as **coated vesicles**, which have a distinctive cage of proteins covering their cytosolic surface. Before the vesicles fuse with a target membrane, they shed their coat so that the membrane surfaces of the vesicle and destination compartment can interact directly and fuse.

The coat performs two main functions that are reflected in a common two-layered structure. First, an inner coat layer concentrates specific membrane proteins in a specialized patch, which then gives rise to the vesicle membrane. In this way, the inner layer selects the appropriate membrane molecules for transport. Second, an outer coat layer assembles into a curved, basketlike lattice that deforms the membrane patch and thereby shapes the vesicle.

There are four well-characterized types of coated vesicles, distinguished by their major coat proteins: *clathrin-coated*, *COPI-coated*, *COPII-coated*, and *retromer-coated* (**Figure 13–4**). Each type is used for different transport steps (**Figure 13–5**). Clathrin-coated vesicles mediate transport originating from the Golgi apparatus, endosome, and the plasma membrane. COPI-coated and COPII-coated vesicles mediate transport originating from the Golgi cisternae and the ER, respectively. Retromer forms coats on transport vesicles for a retrieval

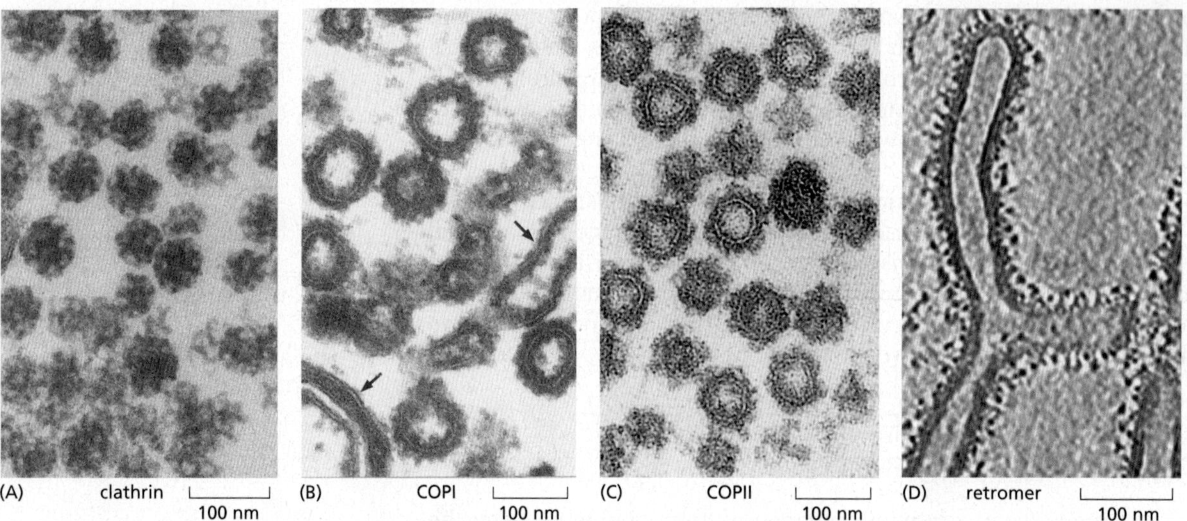

Figure 13–4 Electron micrographs of clathrin-coated, COPI-coated, COPII-coated, and retromer-coated vesicles. All coated vesicles are shown in electron micrographs at the same scale. (A) Clathrin-coated vesicles. (B) COPI-coated vesicles and Golgi cisternae *(red arrows)* from a cell-free system in which COPI-coated vesicles bud in the test tube. (C) COPII-coated vesicles. (D) Retromer-coated tubules formed in a cell-free system containing large membrane vesicles and purified retromer. (A and B, from L. Orci et al., *Cell* 46:171–184, 1986. With permission from Elsevier; C, courtesy of Charles Barlowe and Lelio Orci; D, courtesy of John Briggs.)

pathway from endosomes to the Golgi apparatus. There is, however, much more variety in coated vesicles and their functions than this short list suggests. As we discuss shortly, there are several types of clathrin-coated vesicles, each specialized for a different transport step, and the COPI-coated and COPII-coated vesicles may be similarly diverse. We discuss clathrin-coated vesicles first, as they provide a good example of how vesicles form.

The Assembly of a Clathrin Coat Drives Vesicle Formation

The major protein component of **clathrin-coated vesicles** is clathrin, which forms the outer layer of the coat. **Clathrin** is composed of a large subunit (the heavy chain) and a small subunit (the light chain). Three heavy chains and three light chains assemble into a three-legged structure called a *triskelion* (**Figure 13–6A and B**). Clathrin triskelions assemble into a basketlike framework of hexagons and pentagons on the cytosolic surface of membranes. Clathrin assembly induces the formation of coated buds (called coated pits when on the plasma membrane), which eventually pinch off to become clathrin-coated vesicles (**Figure 13–7**). Under appropriate conditions, isolated triskelions spontaneously self-assemble into typical polyhedral cages in a test tube, even

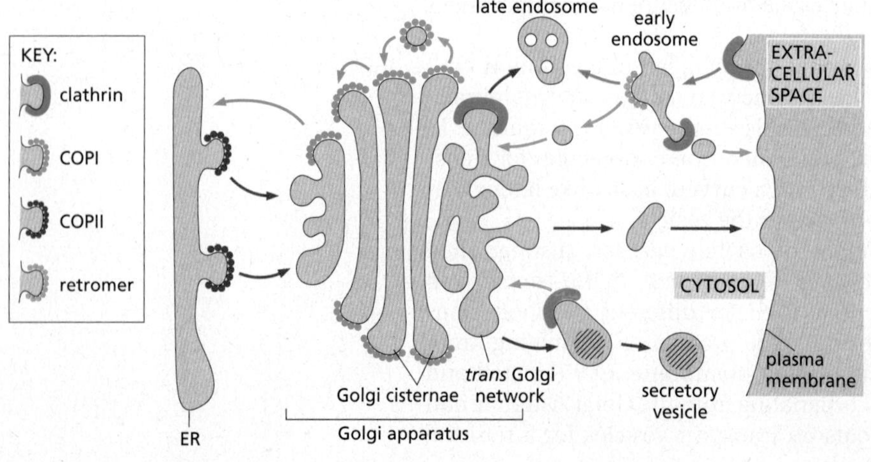

Figure 13–5 Use of different coats for different steps in vesicle traffic. Different coat proteins select different cargo and shape the transport vesicles that mediate the various steps in the secretory and endocytic pathways. When the same coats function in different places in the cell, they usually incorporate different coat protein subunits that modify their properties (not shown). Many differentiated cells have additional pathways besides those shown here, including a sorting pathway from the *trans* Golgi network to the apical surface of epithelial cells and a specialized recycling pathway for proteins of synaptic vesicles in the nerve terminals of neurons (see Figure 11–38). The arrows are colored as in Figure 13–3.

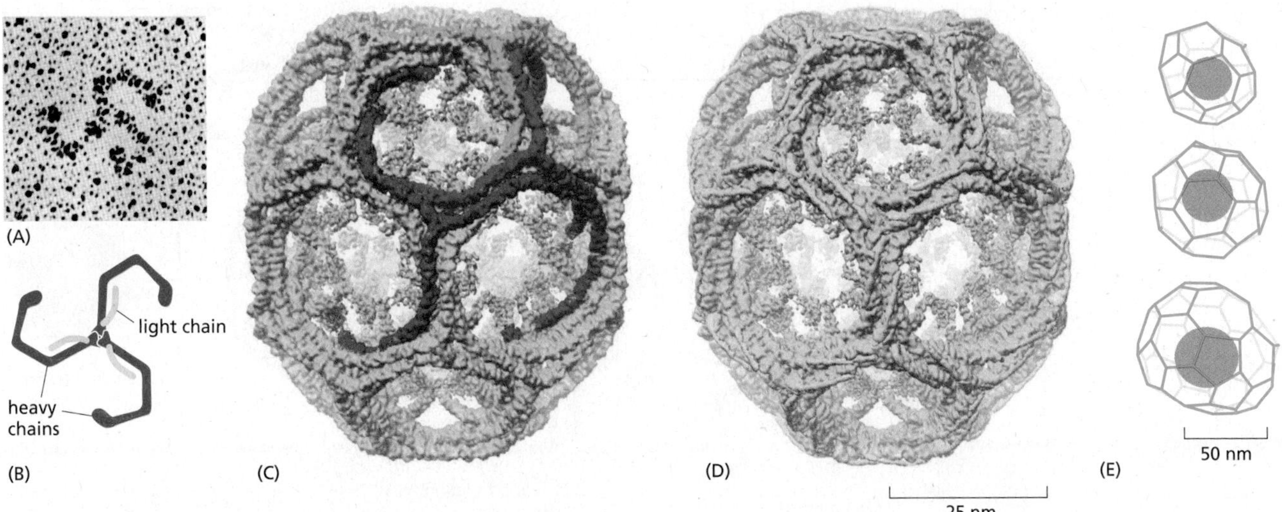

Figure 13–6 **The structure of a clathrin coat.** (A) Electron micrograph of a clathrin triskelion shadowed with platinum. (B) Each triskelion is composed of three clathrin heavy chains and three clathrin light chains, as shown in the diagram. (C and D) A cryo-electron micrograph taken of a clathrin coat composed of 36 triskelions organized in a network of 12 pentagons and 6 hexagons, with some heavy chains (C) and light chains (D) highlighted (**Movie 13.1**). The light chains link to the actin cytoskeleton, which helps generate force for membrane budding and vesicle movement, and their phosphorylation regulates clathrin coat assembly. The interwoven legs of the clathrin triskelions form an outer shell from which the N-terminal domains of the triskelions protrude inward. These domains bind to the adaptor proteins shown in Figure 13–8. The coat shown was assembled biochemically from pure clathrin triskelions and is too small to enclose a membrane vesicle. (E) Structures of clathrin-coated vesicles isolated from bovine brain. The clathrin coats are constructed similarly to but in a less regular way than the coat of panels C and D, utilizing pentagons, a larger number of hexagons, and sometimes heptagons, resembling the architecture of deformed soccer balls. The structures were determined by cryo-electron microscopy and tomographic reconstruction. (A, from E. Ungewickell and D. Branton, *Nature* 289:420–422, published 1981 by Nature Publishing Group; reproduced with permission of SNCSC; C and D, from A. Fotin et al., *Nature* 432:573–579, published 2004 by Nature Publishing Group; all reproduced with permission of SNCSC; E, from Y. Cheng et al., *J. Mol. Biol.* 365:892–899, 2007. With permission from Elsevier.)

in the absence of the membrane vesicles that these baskets normally enclose (**Figure 13–6C and D**). Thus, the clathrin triskelions determine the geometry of the clathrin cage (**Figure 13–6E**).

Adaptor Proteins Select Cargo into Clathrin-coated Vesicles

Adaptor proteins, another major coat component in clathrin-coated vesicles, form a discrete inner layer of the coat, positioned between the clathrin cage and the cytosolic face of the membrane. They bind to various transmembrane protein cargoes and transmembrane receptors that capture soluble cargo molecules inside the vesicle—so-called *cargo receptors*. Adaptor proteins also bind to clathrin and recruit it to the membrane surface where it assembles and bends the membrane. In this way, the specific set of transmembrane and soluble cargoes selected by adaptor proteins is packaged into a newly formed clathrin-coated transport vesicle (**Figure 13–8**).

The assembly of adaptor proteins on the membrane is tightly controlled, in part by the cooperative interaction of the adaptor proteins with the membrane, transmembrane cargoes, and other components of the coat. The adaptor protein AP2 serves as a well-understood example. When it binds to a specific phosphorylated phosphatidylinositol lipid (a *phosphoinositide*), AP2 acquires a different conformation that exposes binding sites for cargo receptors in the membrane. The simultaneous binding to the cargo receptors and lipid head groups greatly enhances the binding of AP2 to the membrane (**Figure 13–9**). Upon binding, AP2 induces membrane curvature, which makes the binding of additional AP2 proteins in its proximity more likely. The cooperative assembly of the AP2 coat layer then is further amplified by clathrin binding, which leads to the formation and budding of a transport vesicle.

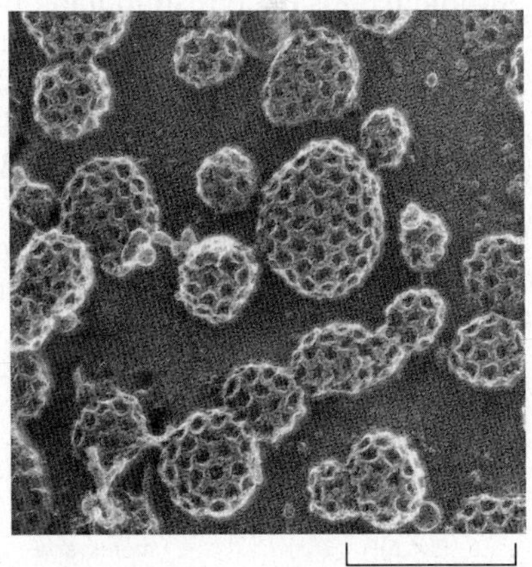

Figure 13–7 **Clathrin-coated pits and vesicles.** This rapid-freeze, deep-etch electron micrograph shows numerous clathrin-coated pits and vesicles on the inner surface of the plasma membrane of cultured fibroblasts. The cells were rapidly frozen in liquid helium, fractured, and deep-etched to expose the cytoplasmic surface of the plasma membrane. (Courtesy of John Heuser.)

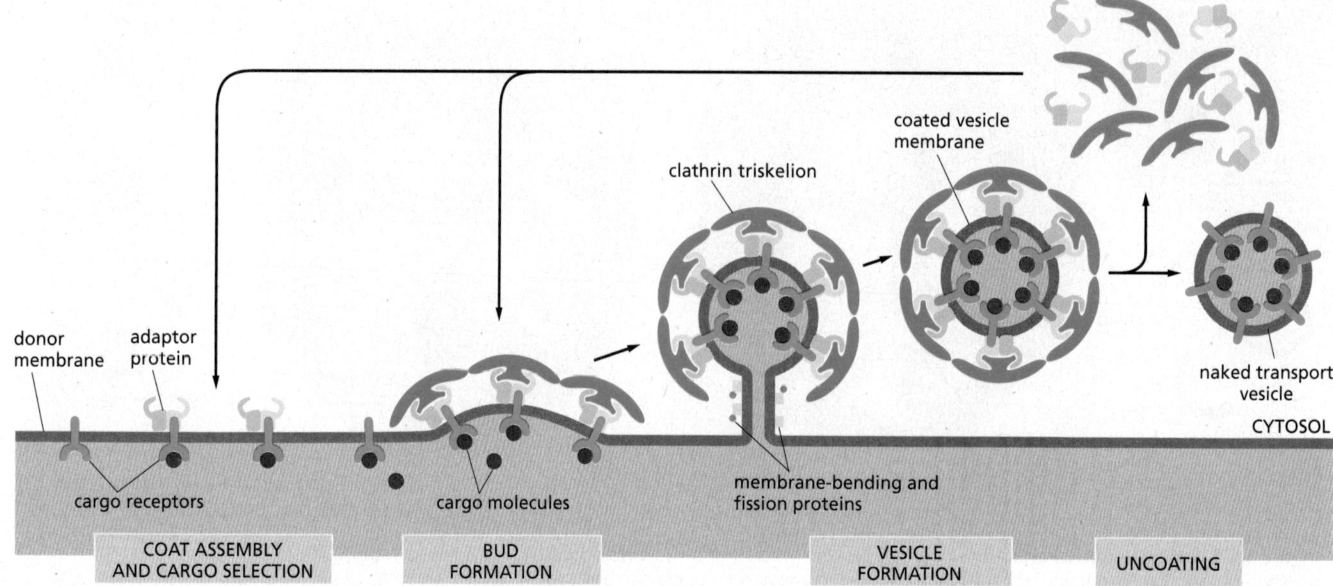

Figure 13–8 The assembly and disassembly of a clathrin coat. The assembly of the coat introduces curvature into the membrane, which leads in turn to the formation of a coated bud (called a coated pit if it is in the plasma membrane). The adaptor proteins bind both clathrin triskelions and membrane-bound cargo receptors, thereby mediating the selective recruitment of both membrane and soluble cargo molecules into the vesicle. Other membrane-bending and fission proteins are recruited to the neck of the budding vesicle, where sharp membrane curvature is introduced. The coat is rapidly lost shortly after the vesicle buds off.

There are several types of adaptor proteins. The best characterized, like AP2, have four different protein subunits; others are single-chain proteins. Many of the adaptor proteins bind to phosphoinositides. As we will see next, different types of phosphoinositides are located in different membrane compartments, serving as one of the molecular markers of that compartment's identity. Each type of adaptor protein is specific for transmembrane cargoes and cargo receptors that share a particular amino acid sequence motif displayed on the cytosolic side of the membrane. Because different adaptor proteins have different specificities for both the type of phosphoinositide and the sequence motif they recognize, each type of adaptor protein directs assembly of a clathrin-coated vesicle only at particular membranes.

Phosphoinositides Mark Organelles and Membrane Domains

Although inositol phospholipids typically compose less than 10% of the total phospholipids in a membrane, they have important regulatory functions. They can undergo rapid cycles of phosphorylation and dephosphorylation at the 3′, 4′, and 5′ positions of their inositol sugar head groups to produce various types of **phosphoinositides (phosphatidylinositol phosphates, or PIPs)**. The interconversion of phosphatidylinositol (PI) and PIPs is highly compartmentalized: different organelles in the endocytic and secretory pathways have distinct sets of PI and PIP kinases and PIP phosphatases (**Figure 13–10**). The distribution, regulation, and local balance of these enzymes determine the steady-state distribution

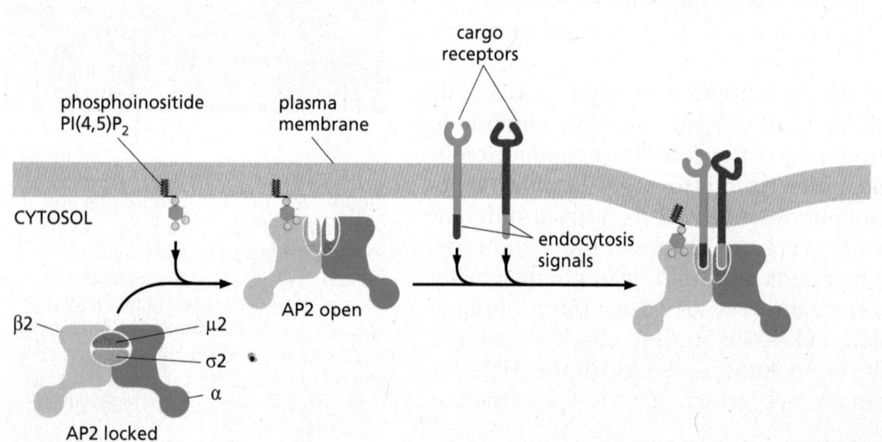

Figure 13–9 Lipid-induced conformation switching of AP2. The AP2 adaptor protein complex has four subunits (α, β2, μ2, and σ2). Upon interaction with the phosphoinositide PI(4,5)P_2 (see Figure 13–10) in the cytosolic leaflet of the plasma membrane, AP2 rearranges so that binding sites for cargo receptors become exposed. Each AP2 complex binds four PI(4,5)P_2 molecules (for clarity, only one is shown). In the open AP2 complex, the μ2 and σ2 subunits bind the cytosolic tails of cargo receptors that display the appropriate endocytosis signals. These signals consist of short amino acid sequence motifs. When AP2 binds tightly to the membrane, it induces curvature, which favors the binding of additional AP2 complexes in the vicinity.

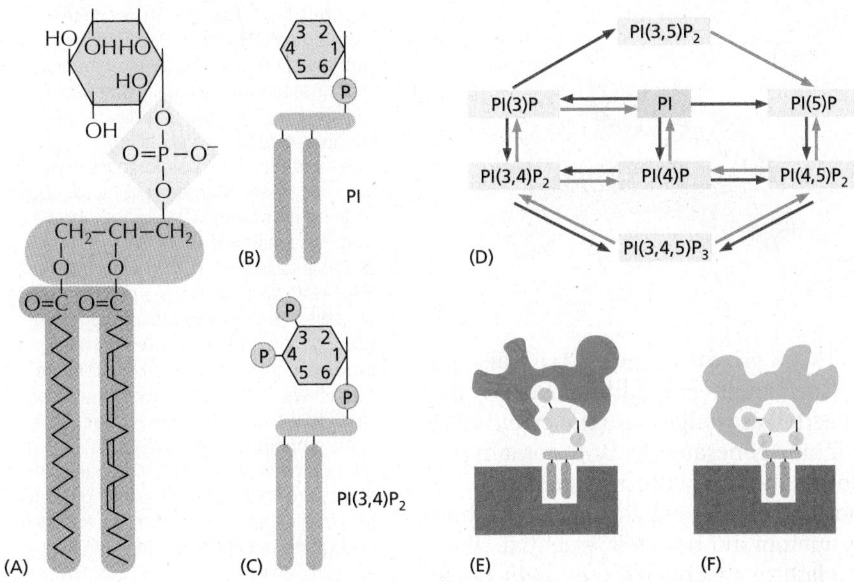

Figure 13–10 **Phosphatidylinositol (PI) and phosphoinositides (phosphatidylinositol phosphates, or PIPs).** (A, B) The structure of PI shows the free hydroxyl groups in the inositol sugar that can be modified. (C) Phosphorylation of one, two, or three of the hydroxyl groups on PI by PI and PIP kinases produces a variety of PIP species. They are named according to the ring position (in parentheses) and the number of phosphate groups (subscript) added to PI. Phosphatidylinositol 3,4-bisphosphate [$PI(3,4)P_2$] is shown. (D) Animal cells have several PI and PIP kinases and a similar number of PIP phosphatases, which are localized to different organelles, where they are regulated to catalyze the production of particular PIPs. The *red* and *green arrows* show the kinase and phosphatase reactions, respectively. (E, F) Phosphoinositide head groups are recognized by protein domains that discriminate between the different forms. In this way, select groups of proteins containing such domains are recruited to regions of membrane in which these phosphoinositides are present. Phosphatidylinositol 3-phosphate [PI(3)P] in the endosome membrane, and phosphatidylinositol 4,5-bisphosphate [$PI(4,5)P_2$] in the plasma membrane, are shown. (D, modified from M.A. De Matteis and A. Godi, *Nat. Cell Biol.* 6:487–492, 2004.)

of each PIP species. As a consequence, the distribution of PIPs varies from organelle to organelle, and often within a continuous membrane from one region to another, thereby defining specialized membrane domains.

Many proteins involved at different steps in vesicle transport contain domains that bind with high specificity to the head groups of particular PIPs, distinguishing one phosphorylated form from another (see Figure 13–10). Local control of the PI and PIP kinases and PIP phosphatases can therefore be used to rapidly control the binding of proteins to a membrane or membrane domain. The production of a particular type of PIP recruits proteins containing matching PIP-binding domains. The PIP-binding proteins then help regulate vesicle formation and other steps in the control of vesicle traffic (**Figure 13–11**). The same strategy is widely used to recruit specific intracellular signaling proteins to the plasma membrane in response to extracellular signals (discussed in Chapter 15).

Membrane-bending Proteins Help Deform the Membrane During Vesicle Formation

Although vesicle-budding is similar at various locations in the cell, each cell membrane poses its own special challenges. The plasma membrane, for example, is comparatively flat and stiff, owing to its cholesterol-rich lipid composition and underlying actin-rich cortex. Thus, the forces generated by clathrin coat assembly alone are not sufficient to shape and pinch off a vesicle from the plasma membrane. Other membrane-bending and force-generating proteins participate at every stage of the process.

Membrane-bending proteins that contain crescent-shaped domains, called *BAR domains*, bind to and impose their shape on the underlying membrane

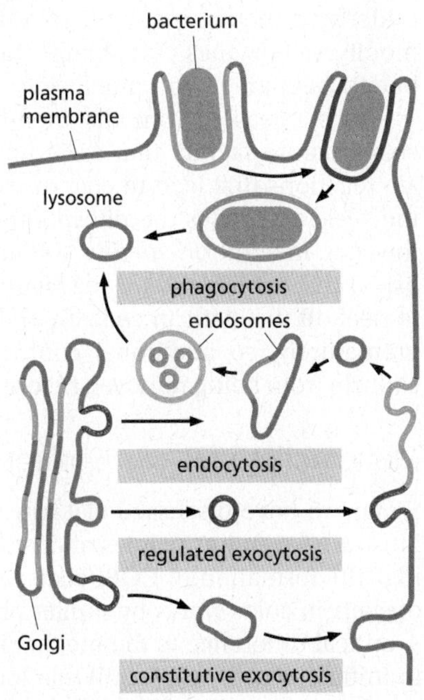

Figure 13–11 **The intracellular location of phosphoinositides.** Different types of PIPs are located in different membranes and membrane domains, where they are often associated with specific vesicle transport events. The membrane of secretory vesicles, for example, contains PI(4)P. When the vesicles fuse with the plasma membrane, a phosphoinositide 5-kinase (PI 5-kinase) that is localized there converts the PI(4)P into $PI(4,5)P_2$. The $PI(4,5)P_2$, in turn, helps recruit adaptor proteins, which initiate the formation of a clathrin-coated pit, as the first step in clathrin-mediated endocytosis. Once the clathrin-coated vesicle buds off from the plasma membrane, a PI(5)P phosphatase hydrolyzes $PI(4,5)P_2$, which weakens the binding of the adaptor proteins, promoting vesicle uncoating. We discuss phagocytosis and the distinction between regulated and constitutive exocytosis later in the chapter. (Modified from M.A. De Matteis and A. Godi, *Nat. Cell Biol.* 6: 487–492, 2004.)

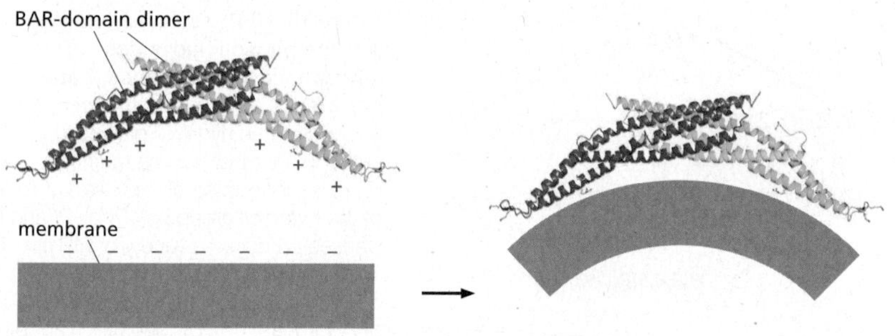

Figure 13–12 **The structure of BAR domains.** BAR-domain proteins are diverse and enable many membrane-bending processes in the cell. BAR domains are built from coiled coils that dimerize into modules that have a positively charged inner surface, which preferentially interacts with negatively charged lipid head groups to bend membranes. Local membrane deformations caused by BAR-domain proteins facilitate the binding of additional BAR-domain proteins, thereby generating a positive feedback cycle for curvature propagation. Individual BAR-domain proteins contain a distinctive curvature and often have additional features that adapt them to their specific tasks: some have short amphiphilic helices that cause further membrane deformation by wedge insertion; others are flanked by PIP-binding domains that direct them to membranes enriched in cognate phosphoinositides. (PDB code: 1ZWW.)

via electrostatic interactions with the lipid head groups (Figure 13–12; also see Figure 10–40). Some of these proteins also contain amphiphilic helices that induce membrane curvature after being inserted as wedges into the cytoplasmic leaflet of the membrane. The curved membrane generated by BAR-domain proteins is thought to help AP2 nucleate the formation of a clathrin-coated bud.

Other BAR-domain proteins are important in shaping the neck of a budding vesicle, where stabilization of sharp membrane bends is essential. These BAR-domain proteins, together with the clathrin machinery they help nucleate, stimulate the local assembly of actin filaments (Figure 13–13). The growing filaments push on the membrane surrounding the budding vesicle and further help propel it away from the membrane.

Cytoplasmic Proteins Regulate the Pinching off and Uncoating of Coated Vesicles

As a clathrin-coated bud grows, soluble cytoplasmic proteins, including **dynamin**, assemble at the neck of the bud and ultimately pinch off the membrane to release the fully formed clathrin-coated vesicle (Figure 13–14). Dynamin contains a phosphoinositide-binding domain, which tethers the protein to the membrane, and a GTPase domain, which regulates the rate at which vesicles pinch off from the membrane. The pinching-off process brings the two noncytosolic leaflets of the membrane at the bud neck into close proximity and seals off the forming vesicle (see Figure 13–2). To perform this task, dynamin assembles in a ring around the neck, then undergoes a conformational change when it hydrolyzes its bound GTP. This constricts the dynamin ring together with the underlying membrane at the bud neck. In addition, dynamin may recruit lipid-modifying enzymes that change the lipid composition locally at the neck of the bud to facilitate membrane fusion.

Once released from the membrane, the vesicle rapidly loses its clathrin coat because factors that are co-packaged into a clathrin-coated vesicle initiate reactions that lead to coat disassembly. A phosphoinositide phosphatase in the vesicle depletes the phosphoinositide that binds to the adaptor proteins of the coat. In addition, *auxilin*, another vesicle protein, activates the ATPase of an hsp70 chaperone protein (see Figure 6–80) that uses the energy of ATP hydrolysis to peel off the clathrin coat. The release of the coat, however, must not happen prematurely, so additional control mechanisms must somehow prevent the clathrin from being removed before it has formed a complete vesicle.

Monomeric GTPases Control Coat Assembly

So far we have discussed clathrin-coated vesicles at the plasma membrane to illustrate several principles of coat formation, membrane budding, and uncoating. The formation of COPI coats, COPII coats, retromer coats, and other types of clathrin coats works by similar principles but differs in many important ways. A critical difference is the mechanism cells use to determine when and where to initiate coat formation. While local production of PIPs plays a major part in regulating the assembly of clathrin coats on the plasma membrane and Golgi apparatus, *coat-recruitment GTPases* control the assembly of COPI coats on Golgi

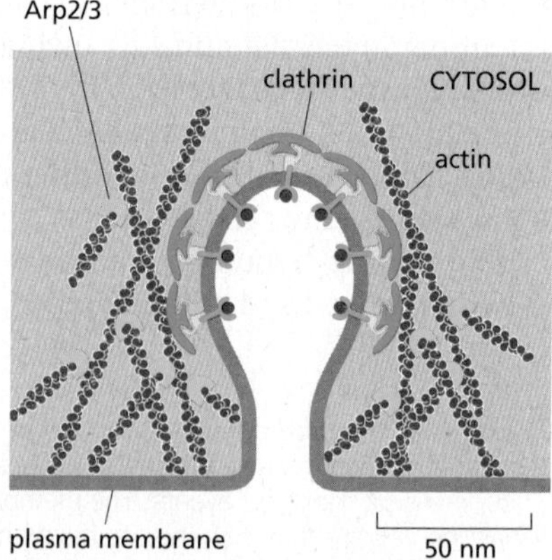

Figure 13–13 **Local actin polymerization helps drive budding of membrane vesicles.** Polymerization of actin filaments occurs near the vesicle neck, helping propel the budding vesicle away from the plasma membrane.

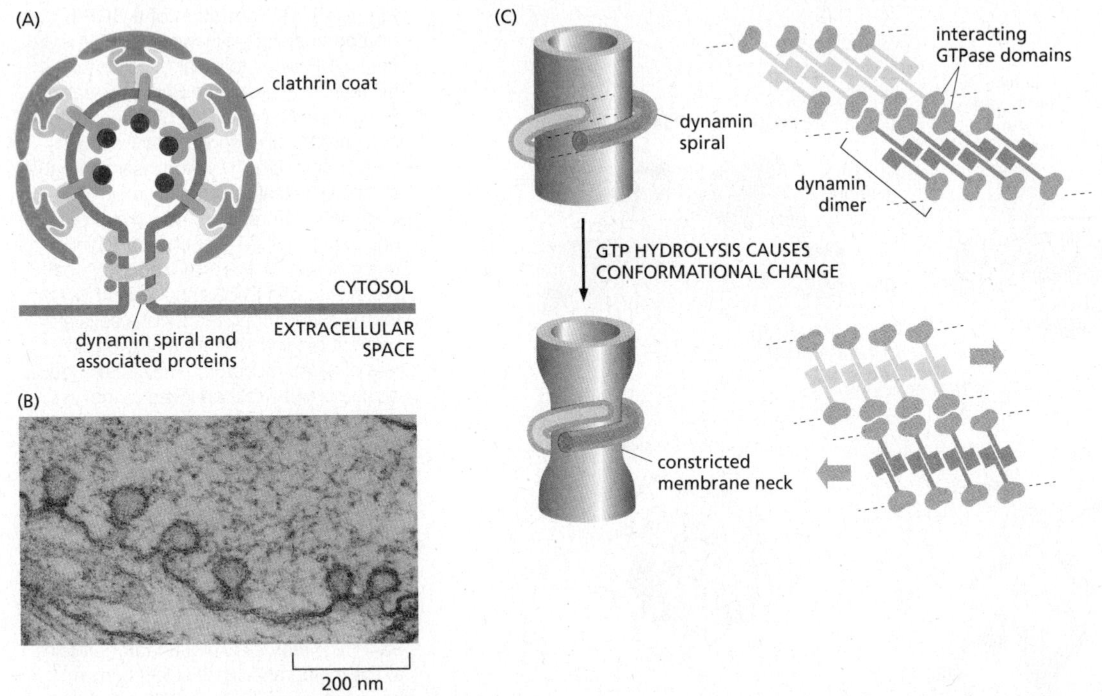

Figure 13–14 The role of dynamin in pinching off clathrin-coated vesicles. (A) Multiple dynamin molecules assemble into a spiral around the neck of the forming bud. The dynamin spiral is thought to recruit other proteins to the bud neck, which, together with dynamin, destabilize the interacting lipid bilayers so that the noncytoplasmic leaflets flow together. The newly formed vesicle then pinches off from the membrane. Specific mutations in dynamin can either enhance or block the pinching-off process. (B) Dynamin was discovered as the protein defective in the *shibire* mutant of *Drosophila*. These mutant flies become paralyzed because clathrin-mediated endocytosis stops, and the synaptic vesicle membrane fails to recycle, blocking neurotransmitter release. Deeply invaginated clathrin-coated pits form in the nerve endings of the fly's nerve cells, with a belt of mutant dynamin assembled around the neck, as shown in this thin-section electron micrograph. The pinching-off process fails because the required membrane fusion does not take place. (C) A model of how conformational changes in the GTPase domains of membrane-assembled dynamin can power a conformational change that constricts the neck of the bud. In this model, dynamin dimers polymerize to form the dynamin spiral. Where two rungs of the spiral meet, the GTPase domains of one rung's dynamins interact with the GTPase domains of the other rung's dynamins. GTP hydrolysis by the GTPase domains induces conformational changes that cause compaction of individual dynamin molecules, leading to partial constriction of the dynamin spiral. Loss of GTP weakens the interactions of the GTPase domains with each other, and rebinding of GTP reverses the compacted state to allow the GTPase domains to bind at the adjacent position on the opposite rung (this step is not shown). In this way, the dynamin molecules can undergo a stepping motion in the direction indicated by the *pink arrows*. Cycles of dynamin compaction and stepping, powered by cycles of GTP binding and hydrolysis, is one model for how dynamin constricts the neck of a budding vesicle. (B, from J.H. Koenig and K. Ikeda, *J. Neurosci.* 9:3844–3860, 1989. Copyright 1989 Society for Neuroscience. With permission from the Society for Neuroscience.)

membranes, COPII coats on ER membranes, and retromer and clathrin coats on endosomes.

Coat-recruitment GTPases are members of a family of monomeric GTPases. They include the **ARF proteins**, which are responsible for the assembly of both COPI and clathrin coats at Golgi membranes, the **Sar1 protein**, which is responsible for the assembly of COPII coats at the ER membrane, and Rab7, which initiates the assembly of retromer coats at the endosome membrane. As discussed in Chapter 3, GTP-binding proteins regulate many processes in eukaryotic cells. They act as molecular switches, which toggle between an active state with GTP bound and an inactive state with GDP bound. Two classes of proteins regulate the toggling: *guanine nucleotide exchange factors* (*GEFs*) activate the proteins by catalyzing the exchange of GDP for GTP, and *GTPase-activating proteins* (*GAPs*) inactivate the proteins by triggering the hydrolysis of the bound GTP to GDP (see Figures 3–68 and 15–7).

Coat-recruitment GTPases are usually found in high concentration in the cytosol in an inactive, GDP-bound state. When a **COPII-coated vesicle** is to bud from the ER membrane, a specific **Sar1 GEF** embedded in the ER membrane binds to cytosolic Sar1, causing the Sar1 to release its GDP. Because GTP is present in

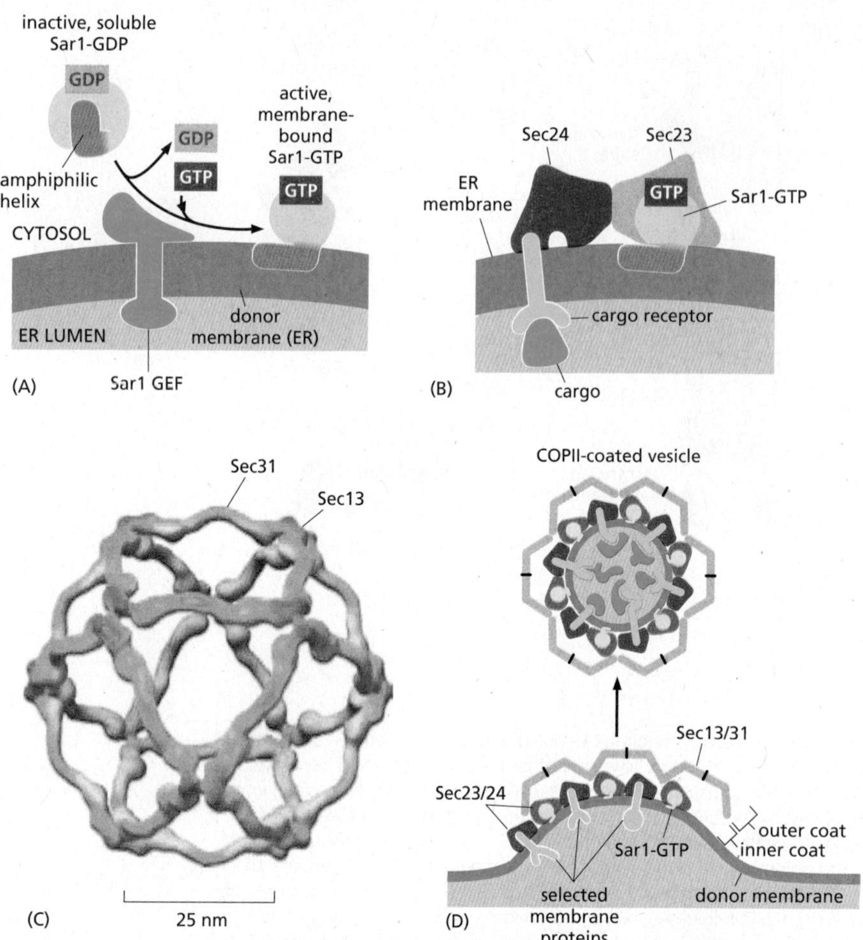

Figure 13–15 Formation of a COPII-coated vesicle. (A) Inactive, soluble Sar1-GDP binds to a Sar1 GEF in the ER membrane, causing Sar1 to release its GDP and bind GTP. A GTP-triggered conformational change in Sar1 exposes an amphiphilic helix, which inserts into the cytoplasmic leaflet of the ER membrane, initiating membrane bending (which is not shown). (B) GTP-bound Sar1 binds to a complex of two COPII adaptor coat proteins, called Sec23 and Sec24, which form the inner coat. Sec24 has several different binding sites for the cytosolic tails of cargo receptors. The entire surface of the complex that attaches to the membrane is gently curved, matching the diameter of COPII-coated vesicles. (C) A complex of two additional COPII coat proteins, called Sec13 and Sec 31, forms the outer shell of the coat. Like clathrin, they can assemble on their own into symmetrical cages with appropriate dimensions to enclose a COPII-coated vesicle. (D) Membrane-bound, active Sar1-GTP recruits COPII adaptor proteins to the membrane. They select certain transmembrane proteins and cause the membrane to deform. The adaptor proteins then recruit the outer coat proteins, which help form a bud. A subsequent sealing event pinches off the coated vesicle. Other coated vesicles are thought to form in a similar way. (C, modified from S.M. Stagg et al., *Nature* 439:234–238, published 2006 by Nature Publishing Group. Reproduced with permission of SNCSC.)

much higher concentration in the cytosol than GDP, Sar1 binds GTP as soon as GDP is released. In its GTP-bound state, the Sar1 protein exposes an amphiphilic helix, which inserts into the cytoplasmic leaflet of the lipid bilayer of the ER membrane. The tightly bound Sar1 now recruits adaptor coat protein subunits to the ER membrane to initiate budding (**Figure 13–15**). Other GEFs and coat-recruitment GTPases operate in a similar way on other membranes (**Movie 13.2**). Some of the small monomeric GTPases use an amphiphilic helix, whereas others use an attached lipid to anchor them to membranes. Thus, GEFs located in different compartments serve as important spatial cues that control where different coat-recruitment GTPases are activated to initiate the formation of different types of transport vesicles.

Coat-recruitment GTPases Participate in Coat Disassembly

As with clathrin-coated vesicles at the plasma membrane, other types of coats must also disassemble once the transport vesicle has budded off the originating compartment. Without coat disassembly, the vesicle membrane could not fuse with that of its target compartment, and coated vesicles would permanently accumulate in the cell with no place to go. Budding vesicles therefore incorporate proteins that initiate coat disassembly only after the vesicle has fully formed. This critical switch from coat formation to coat disassembly is triggered by coat-recruitment GTPases. The hydrolysis of bound GTP to GDP causes the GTPase to change its conformation so that its hydrophobic tail pops out of the membrane, causing the vesicle's coat to disassemble. Thus, the rate at which coat-recruitment GTPases hydrolyze GTP determines the length of time their associated coats stay assembled.

COPII coats accelerate GTP hydrolysis by Sar1, and a fully formed vesicle will be produced only when bud formation occurs faster than the timed disassembly process; otherwise, disassembly will be triggered before a vesicle pinches off, and the process will have to start again, perhaps at a more appropriate time and place. Once a vesicle pinches off, GTP hydrolysis releases Sar1, but the sealed coat is sufficiently stabilized through many cooperative interactions, including binding to the cargo receptors in the membrane, that it may stay on the vesicle until the vesicle arrives at a target membrane. There, a kinase phosphorylates the coat proteins, which completes coat disassembly and readies the vesicle for fusion.

Clathrin-coated and COPI-coated vesicles, by contrast, shed their coat soon after they pinch off. For **COPI-coated vesicles**, the curvature of the vesicle membrane serves as a trigger to begin uncoating. An ARF GAP that is recruited to the COPI coat as it assembles senses the lipid packing density. When the curvature of the membrane approaches that of a transport vesicle, the ARF GAP is activated. It then stimulates ARF to hydrolyze its GTP, causing the coat to disassemble.

The Shape and Size of Transport Vesicles Are Diverse

The types of cargoes that need to be transported through the cell are diverse in size, shape, and topology. Transport vesicles can similarly be diverse in their morphology to accommodate the cargoes they carry. Collagen, for example, is assembled in the ER as 300-nm-long, stiff procollagen rods that then are secreted from the cell where they are eventually embedded into the extracellular matrix (discussed in Chapter 19). Procollagen rods do not fit into the 60- to 80-nm-diameter COPII vesicles that normally carry smaller cargoes. To circumvent this problem, the procollagen cargo molecules bind to transmembrane *packaging proteins* in the ER, which control the assembly of the COPII coat components (**Figure 13–16**). These events are thought to drive the local assembly of much larger COPII vesicles that accommodate the oversized cargo. Human mutations in genes encoding such packaging proteins result in collagen defects with severe consequences, such as skeletal abnormalities and other developmental defects. Similar mechanisms must regulate the sizes of vesicles required to secrete other large macromolecular complexes, including the lipoprotein particles that transport lipids out of cells.

Another variation on small spherical transport vesicles is thin membrane tubules. Tubules have a higher surface-to-volume ratio than vesicles or the larger organelles from which they form. They are therefore relatively enriched in membrane proteins compared with soluble cargo proteins. As we discuss later, this property of tubules is an important feature for sorting proteins in endosomes. The retromer coat, which is specialized for transporting membrane proteins from endosomes to the Golgi apparatus, preferentially drives the formation of tubular transport vesicles (Figure 13–4).

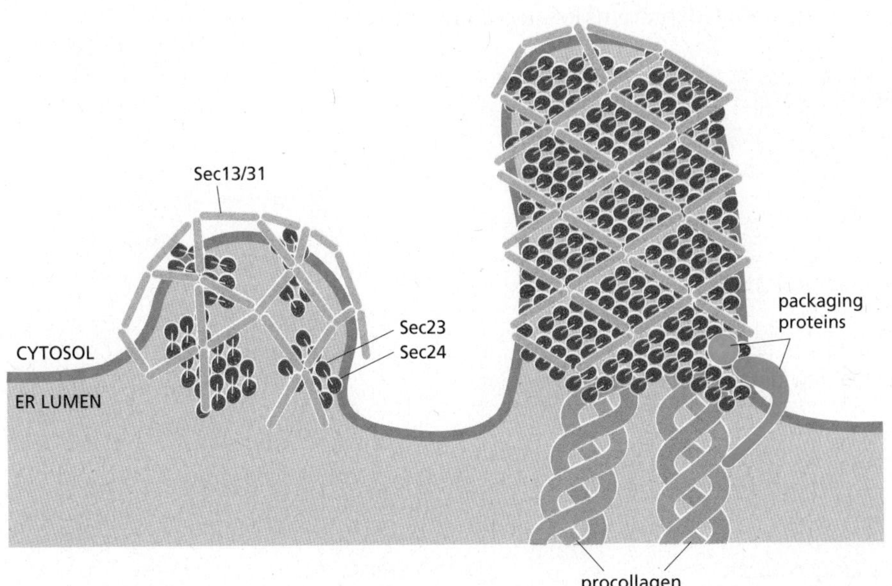

Figure 13–16 **Packaging of procollagen into large tubular COPII-coated vesicles.** The diagrams show models for two COPII coat assembly modes. The models are based on cryo-electron tomography images of reconstituted COPII vesicles. On a spherical membrane *(left)*, the Sec23/24 inner COPII coat proteins assemble in patches that anchor the Sec13/31 outer coat COPII protein cage. The Sec13/31 rods assemble a cage of triangles, squares, and pentagons. When procollagen needs to be packaged *(right)*, special packaging proteins sense the cargo and modify the coat assembly process. This interaction recruits the COPII inner coat protein Sec24 and locally enhances the rate with which Sar1 cycles on and off the membrane (not shown). In addition, a monoubiquitin (not shown) is added to the Sec31 protein, changing the assembly properties of the outer cage. Sec23/24 proteins arrange in larger arrays, and Sec13/31 proteins arrange in a regular lattice of diamond shapes. As a result, a large tubular vesicle is formed that can accommodate the large cargo molecules. The packaging proteins are not part of the budding vesicle but remain in the ER. (Modified from G. Zanetti et al., *eLife* 2:e00951, 2013. With permission from the authors.)

Rab Proteins Guide Transport Vesicles to Their Target Membrane

To ensure an orderly flow of vesicle traffic, transport vesicles must be highly accurate in recognizing the correct target membrane with which to fuse. Because of the diversity and crowding of membrane systems in the cytoplasm, a vesicle is likely to encounter many potential target membranes before it finds the correct one. Specificity in targeting is ensured because all transport vesicles display surface markers that identify them according to their origin and type of cargo, and target membranes display complementary receptors that recognize the appropriate markers. Two types of markers act sequentially to ensure the specificity of vesicle targeting. First, *Rab proteins* direct the vesicle to specific spots on the correct target membrane. Second, *SNARE proteins* enable the fusion of the lipid bilayers.

Like the coat-recruitment GTPases discussed earlier (see Figure 13–15), **Rab proteins** are also monomeric GTPases. With more than 60 known members in mammalian cells, the Rab subfamily is the largest of the monomeric GTPase subfamilies. Each Rab protein is associated with one or more membrane-enclosed organelles of the secretory or endocytic pathway, and each of these organelles has at least one Rab protein on its cytosolic surface (**Table 13–1**). Different Rab proteins are also found on the different types of transport vesicles that ferry cargoes between organelles. Their selective distribution on these membrane systems makes Rab proteins ideal molecular markers for identifying each type of transport vesicle and target membrane in order to guide vesicle traffic.

Rab proteins cycle between a membrane and the cytosol and regulate the reversible assembly of protein complexes on the membrane. In their GDP-bound state, they are inactive and bound to another protein (*GDP dissociation inhibitor*, or *GDI*) that keeps them soluble in the cytosol. Membrane-bound Rab GEFs activate Rab proteins by catalyzing the exchange of GDP for GTP. Once in the GTP-bound state, the Rab protein's lipid anchor inserts into the membrane where the Rab binds to a diverse set of proteins called **Rab effectors** (**Figure 13–17**). The rate of GTP hydrolysis sets the concentration of active Rab and, consequently, the concentration of its effectors on the membrane.

A Rab protein is activated on a transport vesicle when a specific component of the vesicle, often a coat component, recruits a Rab GEF. Rab protein activation has several consequences. First, the Rab protein itself serves as a specific molecular cue that can be recognized by *tethering proteins* localized at the target membrane. Tethering proteins are typically protein complexes that often contain threadlike domains that serve as "fishing lines" capable of capturing a vesicle up to 200 nm away. Second, the Rab protein can interact with *motor proteins*, common Rab effectors that propel vesicles along actin filaments or microtubules to their target membrane. Third, the Rab protein can recruit a Rab effector that selectively binds proteins on the target membrane, such as the SNARE proteins located there.

TABLE 13–1 Subcellular Locations of Some Rab Proteins

Protein	Organelle
Rab1	ER and Golgi complex
Rab2	*cis* Golgi network
Rab3A	Synaptic vesicles, secretory vesicles
Rab4/Rab11	Recycling endosomes
Rab5	Early endosomes, plasma membrane, clathrin-coated vesicles
Rab6	Medial and *trans* Golgi cisternae
Rab7	Late endosomes
Rab8	Cilia
Rab9	Late endosomes, *trans* Golgi network

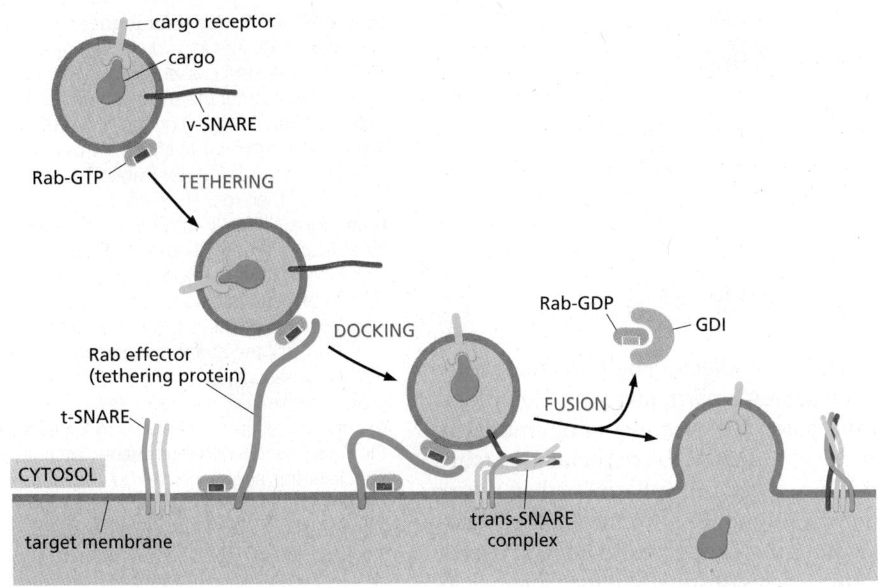

Figure 13–17 Tethering of a transport vesicle to a target membrane. Rab effector proteins interact with active Rab proteins (Rab-GTPs, *brown*) located on the target membrane, vesicle membrane, or both, to establish the first connection between the two membranes that are going to fuse. In the example shown here, the Rab effector is a filamentous tethering protein *(dark green)*. Next, SNARE proteins on the two membranes (*red* and *blue*) pair, docking the vesicle to the target membrane and catalyzing the fusion of the two apposed lipid bilayers. During docking and fusion, a Rab GAP (not shown) induces the Rab protein to hydrolyze its bound GTP to GDP, causing the Rab to dissociate from the membrane and return to the cytosol as Rab-GDP, where it is bound by a GDP dissociation inhibitor (GDI) protein that keeps the Rab soluble and inactive.

Through one or more of these mechanisms, a Rab protein selectively activated on a transport vesicle guides and docks it at the correct target membrane. Some Rab proteins, such as Rab7 discussed earlier, also function as coat-recruitment GTPases that initiate new budding events as organelles mature, as we discuss next.

Rab Proteins Create and Change the Identity of an Organelle

In addition to acting on vesicles, Rab proteins also function on organelle membranes. As on vesicles, a specific Rab GEF at the organelle catalyzes Rab protein activation and insertion at the membrane surface. Many of the effector proteins recruited by an activated Rab protein help give the organelle its identity by directly controlling incoming and outgoing transport vesicles. These effectors include tethering proteins mentioned previously, SNAREs that mediate membrane fusion of incoming vesicles, and enzymes that generate or modify specific phosphoinositides.

The assembly of Rab proteins and their effectors on an organelle membrane can be cooperative and results in the formation of large, specialized membrane patches that define the identity of that organelle. Active Rab5 on the endosome membrane, for example, recruits more copies of the same Rab5 GEF that initially activated Rab5. This stimulates the recruitment of more Rab5 to the same site. At the same time, active Rab5 activates a PI 3-kinase, which locally converts PI to PI(3)P, which in turn binds some of the Rab effectors including tethering proteins and stabilizes their local membrane attachment (**Figure 13–18**). This type of positive feedback greatly amplifies the assembly process and helps to establish functionally distinct membrane domains within a continuous organelle membrane.

Figure 13–18 The formation of a Rab5-associated patch on the endosome membrane. A Rab5 GEF on the endosome membrane binds a Rab5 protein and induces it to exchange GDP for GTP. GDI is lost, and GTP binding alters the conformation of the Rab protein to expose a covalently attached lipid group, which anchors the Rab5-GTP to the membrane. Active Rab5 activates PI 3-kinase, which converts PI into PI(3)P. PI(3)P and active Rab5 together bind a variety of Rab effector proteins that contain PI(3)P-binding sites, including filamentous tethering proteins that catch incoming clathrin-coated endocytic vesicles from the plasma membrane. With the help of another effector, active Rab5 also recruits more Rab5 GEF, further enhancing the assembly of the Rab5-associated patch on the membrane. Controlled cycles of GTP hydrolysis and GDP–GTP exchange dynamically regulate the size and activity of such Rab-associated membrane patches. Unlike SNAREs, which are integral membrane proteins, the GDP–GTP cycle, coupled to the membrane–cytosol translocation cycle, endows the Rab machinery with the ability to undergo assembly and disassembly on the membrane. (Adapted from M. Zerial and H. McBride, *Nat. Rev. Mol. Cell Biol.* 2:107–117, 2001.)

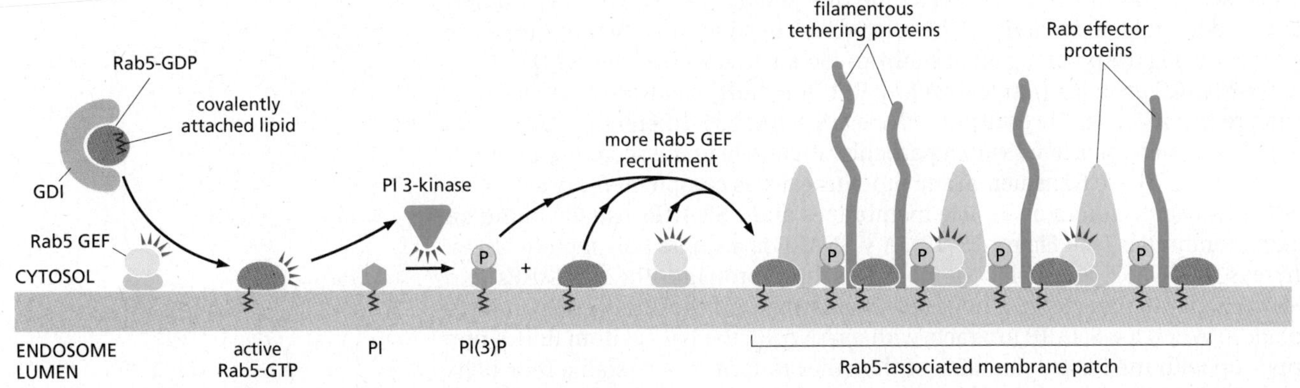

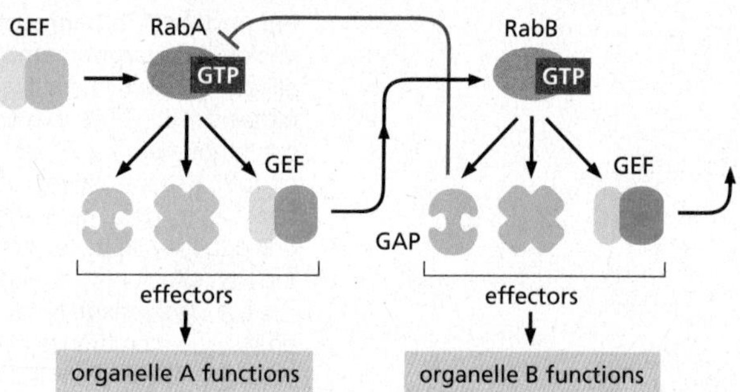

Figure 13–19 A model for a generic Rab cascade. The local activation of a RabA GEF leads to assembly of a RabA-associated membrane patch (sometimes called a "Rab domain") on the membrane. Active RabA recruits its effector proteins, one of which is a GEF for RabB. The RabB GEF then recruits RabB to the membrane, which in turn begins to recruit its effectors, among them a GAP for RabA. The RabA GAP activates RabA-GTP hydrolysis leading to the inactivation of the RabA and the disassembly of the RabA-associated membrane patch as the RabB-associated membrane patch grows. In this way, the RabA-associated membrane patch is irreversibly replaced by the RabB-associated membrane patch. In principle, this sequence can be continued by the recruitment of a next GEF by RabB. (Adapted from A.H. Hutagalung and P.J. Novick, *Physiol. Rev.* 91:119–149, 2011.)

It is thought that different Rab proteins and their effectors help to create multiple specialized membrane domains, each fulfilling a particular set of functions. Thus, while the Rab5-associated membrane patch receives incoming endocytic vesicles from the plasma membrane, distinct Rab11- and Rab4-associated patches in the same endosome organize the budding of recycling vesicles that return proteins from the endosome to the plasma membrane. As we have already discussed, Rab7 on the endosome membrane serves as a coat-recruitment GTPase for retromer, initiating the formation of transport vesicles destined for the Golgi.

One Rab protein can be replaced by a different Rab protein, and this can change the identity of its associated organelle. This is accomplished by one Rab protein selectively recruiting and activating a different Rab protein whose complement of effectors includes proteins that inactivate the first Rab protein and thereby disassemble its associated membrane patch. Such ordered recruitment of sequentially acting Rab proteins is called a **Rab cascade** (Figure 13–19). Over time, for example, Rab5-associated membrane patches are replaced by Rab7-associated membrane patches on endosomal membranes. This converts an early endosome, marked by Rab5, into a late endosome, marked by Rab7. Because the set of Rab effectors recruited by Rab7 is different from that recruited by Rab5, this change reprograms the compartment including the incoming and outgoing traffic and repositions the organelle away from the plasma membrane toward the cell interior. All of the cargo contained in the early endosome that has not been recycled to the plasma membrane is now part of a late endosome. This process is also referred to as *endosome maturation.* The self-amplifying nature of the Rab-associated membrane patches renders the process of endosome maturation unidirectional and irreversible.

SNAREs Mediate Membrane Fusion

Once a transport vesicle has budded from its originating compartment and shed its coat, membrane fusion allows it to unload its cargo at its destination compartment. Membrane fusion requires bringing the lipid bilayers of two membranes to within 1.5 nm of each other so that they can merge. When the membranes are in such close proximity, lipids can flow from one bilayer to the other. For this close approach, water must be displaced from the hydrophilic surface of the membrane—a process that is highly energetically unfavorable and requires specialized *fusion proteins* that overcome this energy barrier. We have already discussed the role of dynamin in the related task of squeezing membranes close together during the pinching off of clathrin-coated vesicles (see Figure 13–14).

The **SNARE proteins** (also called **SNAREs**, for short) catalyze the membrane fusion reactions in vesicle transport. There are at least 35 different SNAREs in an animal cell, each associated with a particular organelle in the secretory or endocytic pathway. These transmembrane proteins exist as complementary sets, with **v-SNAREs** usually found on vesicle membranes and **t-SNAREs** usually found on target membranes (see Figure 13–17). A v-SNARE is a single polypeptide chain, whereas a t-SNARE is usually composed of three proteins. The v-SNAREs and t-SNAREs have characteristic helical domains that are mostly unstructured in isolation. When a v-SNARE interacts with a t-SNARE, the helical domains of one zipper up with the helical domains of the other to form a very stable four-helix

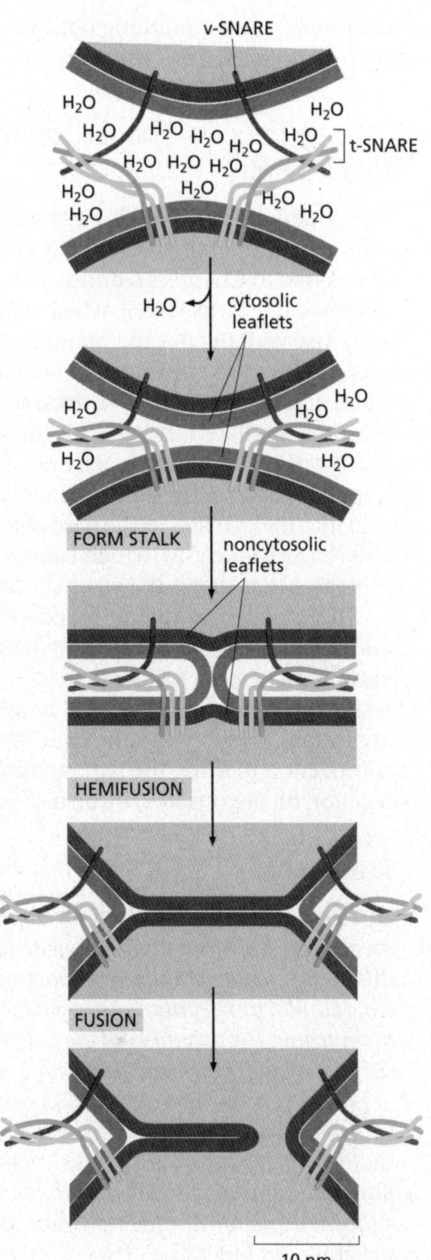

Figure 13–20 A model for how SNARE proteins catalyze membrane fusion. Bilayer fusion occurs in multiple steps. A tight pairing between v- and t-SNAREs forces lipid bilayers into close apposition and expels water molecules from the interface. Lipid molecules in the two interacting (cytosolic) leaflets of the bilayers then flow between the membranes to form a connecting stalk. Lipids of the two noncytosolic leaflets then contact each other, forming a new bilayer, which widens the fusion zone (*hemifusion*, or half-fusion). Rupture of the new bilayer completes the fusion reaction.

bundle. The resulting *trans-SNARE complex* locks the two membranes together. Biochemical membrane fusion assays with all different SNARE combinations show that v- and t-SNARE pairing is highly specific. The SNAREs thus provide an additional layer of specificity in the transport process by helping to ensure that vesicles fuse only with their correct target membrane.

The extremely high stability of the trans-SNARE complex means that its assembly from initially unstructured v- and t-SNAREs is energetically favorable. This energy is exploited to pull the membrane faces together, simultaneously squeezing out water molecules from the interface to initiate lipid bilayer fusion (**Figure 13–20**). When liposomes containing purified v-SNAREs are mixed with liposomes containing complementary t-SNAREs, their membranes fuse, albeit slowly. In the cell, fusion is greatly accelerated by factors that interact with v-SNARE and t-SNARE pairs to align them precisely so they can initiate zippering. Fusion does not always follow immediately after v-SNAREs and t-SNAREs pair. As we discuss later, in the process of regulated exocytosis, zippering of the last part of the trans-SNARE complex is delayed until secretion is triggered by a specific extracellular signal.

Interacting SNAREs Need to Be Pried Apart Before They Can Function Again

After SNARE proteins have participated in membrane fusion, the highly stable trans-SNARE complexes have to disassemble before the SNAREs can mediate new rounds of transport. A crucial protein called **NSF** cycles between membranes and the cytosol and catalyzes the disassembly process. NSF is a hexameric ATPase of the family of AAA-proteins (see Figure 6–88) that uses the energy of ATP hydrolysis to unravel the intimate interactions between the helical domains of paired SNARE proteins (**Figure 13–21**). After disassembly, the SNARE proteins can again exploit the energy gained by their assembly to drive another fusion reaction. Thus, the energy for SNARE-mediated fusion reactions ultimately comes from the ATP consumed by NSF to pry them apart. After trans-SNARE complex disassembly at the destination compartment, v-SNAREs are selectively retrieved and returned to their compartment of origin so that they can be reused in newly formed transport vesicles. Such selective retrieval pathways (discussed later) are critical for maintaining the identity of each compartment in the face of constant outgoing traffic.

Membrane fusion is important in other processes besides vesicle transport. The plasma membranes of a sperm and an egg fuse during fertilization, myoblasts fuse with one another during the development of multinucleate muscle fibers (discussed in Chapter 22), and the epithelial cells in the human placenta fuse into a giant syncytium that separates the mother from the fetus. Likewise, the ER network and mitochondria fuse and fragment in a dynamic way (discussed in Chapters 12 and 14). All cell membrane fusions require special proteins and are tightly regulated to ensure that only appropriate membranes fuse. The controls

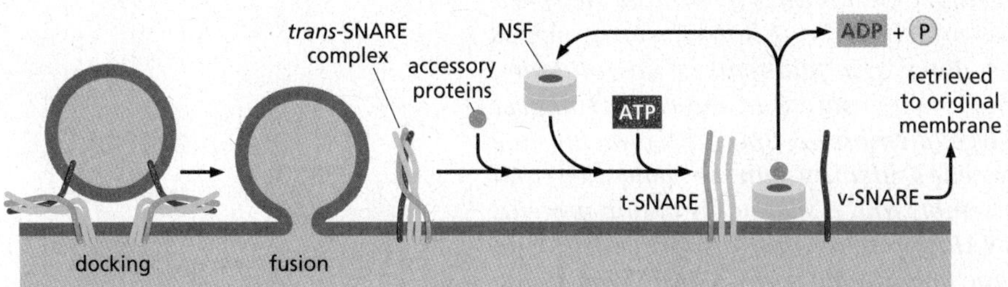

Figure 13–21 Dissociation of SNARE pairs by NSF after a membrane fusion cycle. After a v-SNARE and t-SNARE have effected the fusion of a transport vesicle with a target membrane, NSF binds to the SNARE complex and, with the help of accessory proteins, hydrolyzes ATP to pry the SNAREs apart.

are crucial for maintaining both the identity of cells and the individuality of each type of intracellular compartment.

Viruses Encode Specialized Membrane Fusion Proteins Needed for Cell Entry

Enveloped viruses, which have a lipid bilayer-based membrane coat, enter the cells that they infect when the viral membrane fuses with a cell's membrane (discussed in Chapters 5 and 23). For example, viruses such as the human immunodeficiency virus (HIV), which causes AIDS, bind to cell-surface receptors and then fuse with the plasma membrane of the target cell (**Figure 13–22**). This fusion event allows the viral nucleic acid inside the nucleocapsid to enter the cytosol, where it replicates. Other viruses, such as the influenza virus, first enter the cell by receptor-mediated endocytosis (discussed later) and are delivered to endosomes; the low pH in endosomes activates a fusion protein in the viral envelope that catalyzes the fusion of the viral and endosomal membranes, releasing the viral nucleic acid into the cytosol. In the case of the severe acute respiratory syndrome coronavirus 2 (SARS-CoV-2), which causes COVID-19, the fusion reaction requires host proteases that cleave the virus surface protein to activate its fusion activity.

The membrane fusion reactions catalyzed by viral fusion proteins are well understood. Unlike SNARE-mediated fusion, which involves proteins in both membranes, viral fusion typically requires only the viral protein. These viral fusion proteins unfurl in the appropriate environment and insert a partially hydrophobic patch into the host membrane. The fusion protein then undergoes compaction to bring the two membranes close together to drive their fusion in a reaction analogous to SNARE-mediated fusion.

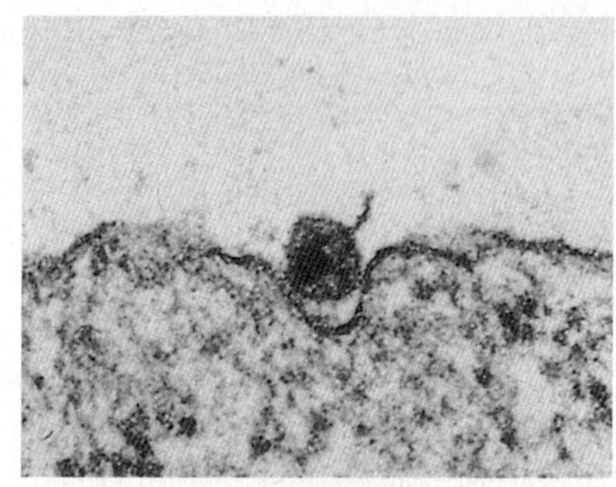

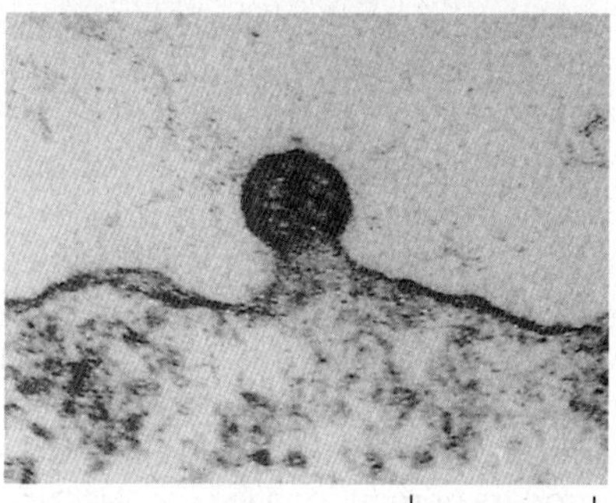

Figure 13–22 The entry of enveloped viruses into cells. Electron micrographs showing how HIV enters a cell by fusing its membrane with the plasma membrane of the cell. (From B.S. Stein et al., *Cell* 49:659–668, 1987. With permission from Elsevier.)

Summary

Directed and selective transport of particular membrane components from one membrane-enclosed compartment to another in a eukaryotic cell maintains the differences between those compartments. Transport vesicles, which can be spherical, tubular, or irregularly shaped, bud from specialized coated regions of the donor membrane. The assembly of the coat helps to collect specific membrane and soluble cargo molecules for transport and to drive the formation of the vesicle.

There are various types of coated vesicles. Clathrin-coated vesicles mediate transport from the plasma membrane, endosomes, and the trans Golgi network. COPI-coated and COPII-coated vesicles mediate transport between Golgi cisternae and between the ER and the Golgi apparatus. Retromer forms a coat at the endosome membrane for transport to the Golgi. Coats have a common two-layered structure: an inner layer formed of adaptor proteins traps specific cargo molecules for packaging into the vesicle and an outer layer that forms a cage and helps deform the membrane into a vesicle. The coat is shed before the vesicle fuses with its appropriate target membrane.

The specificity of membrane transport is mediated by several types of molecular markers that determine where transport vesicles originate and where they deliver their cargo. Local synthesis of specific phosphoinositides creates binding sites that trigger clathrin coat assembly and vesicle budding. In addition, the coat-recruitment GTPases, including Sar1 and the ARF proteins, regulate coat assembly and disassembly. Rab proteins are a large family of GTPases that function on both transport vesicles and target membranes to control the specificity of membrane transport. Active Rab proteins recruit Rab effectors, such as motor proteins, which transport vesicles along actin filaments or microtubules, and filamentous tethering proteins, which help ensure that the vesicles deliver their contents only to the appropriate target membrane. Specialized membrane domains that help determine an organelle's identity can be generated and changed in a dynamic manner by the assembly and disassembly of Rab proteins and their effectors. Complementary v-SNARE proteins on transport vesicles and t-SNARE proteins on the target membrane form stable trans-SNARE complexes, which force the two membranes into close apposition so that their lipid bilayers can fuse.

TRANSPORT FROM THE ENDOPLASMIC RETICULUM THROUGH THE GOLGI APPARATUS

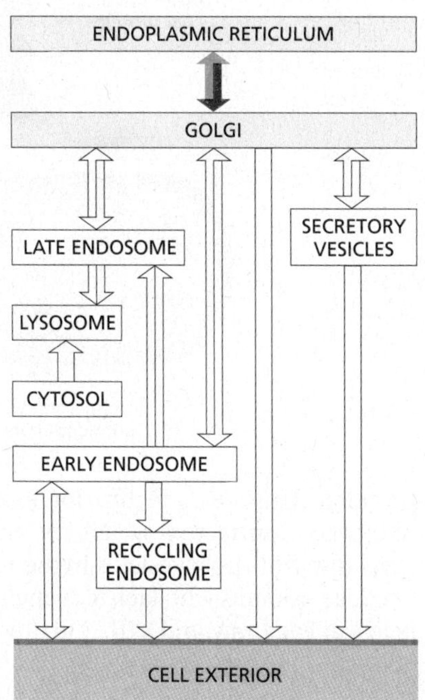

As discussed in Chapter 12, newly synthesized proteins cross the endoplasmic reticulum (ER) membrane from the cytosol to enter the secretory pathway. These proteins are successively modified as they pass through a series of compartments from the ER to the Golgi apparatus and from the Golgi apparatus to the cell surface and elsewhere. Transfer from one compartment to the next involves a delicate balance between forward and backward (retrieval) transport pathways. Some transport vesicles select cargo molecules and move them to the next compartment in the pathway, while others retrieve escaped proteins and return them to a previous compartment where they normally function. Thus, the pathway from the ER to the cell surface consists of many sorting steps, which continually select membrane and soluble lumenal proteins for packaging and transport.

In this section, we focus mainly on the **Golgi apparatus** (also called the Golgi complex). It is a major site of carbohydrate synthesis, as well as a sorting and dispatching station for products delivered to it from the ER. The cell makes many polysaccharides in the Golgi apparatus, including the pectin and hemicellulose of the cell wall in plants and most of the glycosaminoglycans of the extracellular matrix in animals (discussed in Chapter 19). The Golgi apparatus also builds and attaches oligosaccharide chains to the many proteins and lipids that the ER sends to it. Some of these oligosaccharides serve as tags to direct specific proteins carrying them into vesicles that are then transported to endosomes for eventual delivery to lysosomes. But most proteins and lipids, once they have acquired their appropriate oligosaccharides in the Golgi apparatus, are recognized in other ways for targeting into the transport vesicles going to the cell surface and other destinations.

Proteins Leave the ER in COPII-coated Transport Vesicles

To initiate their journey along the secretory pathway, proteins that have entered the ER and are destined for the Golgi apparatus or beyond are first packaged into COPII-coated transport vesicles. These vesicles bud from specialized regions of the ER called *ER exit sites*, whose membrane lacks bound ribosomes. Most animal cells have ER exit sites dispersed throughout the ER network.

Entry into vesicles that leave the ER can be a selective process or can happen by default. Many transmembrane proteins are actively recruited into such vesicles, where they become concentrated. These transmembrane proteins display exit (transport) signals on their cytosolic surface that adaptor proteins of the inner COPII coat recognize (**Figure 13–23**). Soluble cargo proteins in the ER lumen have exit signals that are recognized by some of these transmembrane proteins, which serve as cargo receptors. These receptors are recycled back to the ER after they have delivered their cargo to the Golgi apparatus. Proteins without exit signals can also enter transport vesicles, including protein molecules that normally function in the ER (so-called *ER resident proteins*). These resident proteins slowly leak out of the ER and need retrieval pathways to bring them back from the Golgi apparatus. Different cargo proteins enter the transport vesicles with substantially different rates and efficiencies. These differences can be due to their folding and oligomerization efficiencies and kinetics, as well as their different capacities to engage cargo receptors and the COPII coat. The exit step from the ER is a major checkpoint at which quality control is exerted on the proteins that a cell secretes or displays on its surface, as we discussed in Chapter 12.

The exit signals that direct soluble proteins out of the ER for transport to the Golgi apparatus and beyond are not well understood. Some transmembrane proteins that serve as cargo receptors for packaging some secretory proteins into COPII-coated vesicles are lectins that bind to oligosaccharides on the secreted

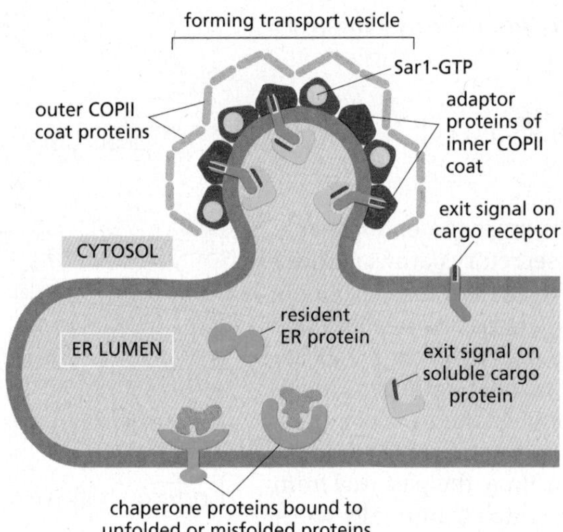

Figure 13–23 The recruitment of membrane and soluble cargo molecules into ER transport vesicles. Transmembrane proteins are packaged into budding transport vesicles through interactions of exit signals on their cytosolic tails with adaptor proteins of the inner COPII coat. Some of these transmembrane proteins function as cargo receptors, binding soluble proteins in the ER lumen and helping to package them into vesicles. Other proteins may enter the vesicle by bulk flow. A typical 50-nm transport vesicle contains about 200 transmembrane proteins, which can be of many different types. As indicated, unfolded or incompletely assembled proteins are bound to chaperones and transiently retained in the ER compartment.

proteins. One such lectin, for example, binds to mannose on two secreted blood-clotting factors (Factor V and Factor VIII), thereby packaging the proteins into transport vesicles in the ER; its role in protein transport was identified because humans who lack it owing to an inherited mutation have lowered serum levels of Factors V and VIII, and they therefore bleed excessively.

Only Proteins That Are Properly Folded and Assembled Can Leave the ER

To exit from the ER, proteins must be properly folded, and, if they are subunits of multiprotein complexes, they need to be completely assembled. Those that are misfolded or incompletely assembled transiently remain in the ER, where they are bound to chaperone proteins (discussed in Chapter 6) such as *BiP* or *calnexin*. The chaperones may cover up the exit signals or somehow anchor the proteins in the ER. Such failed proteins are eventually transported back into the cytosol, where they are degraded by proteasomes (discussed in Chapters 6 and 12). This quality-control step prevents the onward transport of misfolded or misassembled proteins that could potentially interfere with the functions of normal proteins. Such failures are surprisingly common. Most of the newly synthesized subunits of the T cell receptor (discussed in Chapter 24) and of the acetylcholine receptor (discussed in Chapter 11), for example, are normally degraded without ever reaching the cell surface where they function. Thus, cells must make a large excess of some protein molecules to produce a select few that fold, assemble, and function properly.

Sometimes, however, there are drawbacks to the stringent quality-control mechanism. The predominant mutations that cause cystic fibrosis, a common inherited disease, result in the production of a slightly misfolded form of a plasma membrane protein important for Cl^- transport. Although the mutant protein would function almost normally if it reached the plasma membrane, it is retained in the ER and then is degraded by cytosolic proteasomes. This devastating disease thus results not because the mutation inactivates the protein but because potentially active protein is discarded before it reaches the plasma membrane.

Vesicular Tubular Clusters Mediate Transport from the ER to the Golgi Apparatus

After transport vesicles have budded from ER exit sites and have shed their coat, they begin to fuse with one another. The fusion of membranes from the same compartment is called *homotypic fusion*, to distinguish it from *heterotypic fusion*, in which a membrane from one compartment fuses with the membrane of a different compartment. As with heterotypic fusion, homotypic fusion requires a set

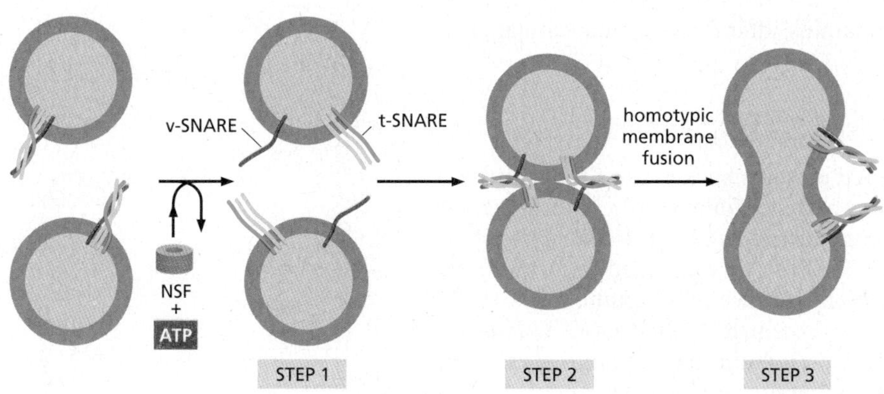

Figure 13–24 Homotypic membrane fusion. In step 1, NSF pries apart identical pairs of v-SNAREs and t-SNAREs in both membranes (see Figure 13–21). In steps 2 and 3, the separated matching SNAREs on adjacent identical membranes interact, which leads to membrane fusion and the formation of one continuous compartment. Subsequently, the compartment grows by further homotypic fusion with vesicles from the same kind of membrane, displaying matching SNAREs. Homotypic fusion occurs when ER-derived transport vesicles fuse with one another, but also when endosomes fuse to generate larger endosomes. Rab proteins help regulate the extent of homotypic fusion and hence the size of a cell's compartments (not shown).

of matching SNAREs. In this case, however, the interaction is symmetrical, with both membranes contributing v-SNAREs and t-SNAREs (**Figure 13–24**).

The structures formed when ER-derived vesicles fuse with one another are called *vesicular tubular clusters,* because they have a convoluted appearance in the electron microscope (**Figure 13–25A**). These clusters constitute a compartment that is separate from the ER and lacks many of the proteins that function in the ER. They are generated continually and function as transport containers that bring material from the ER to the Golgi apparatus. The clusters move quickly along microtubules to the Golgi apparatus with which they fuse (**Figure 13–25B** and **Movie 13.3**).

As soon as vesicular tubular clusters form, they begin to bud off transport vesicles of their own. Unlike the COPII-coated vesicles that bud from the ER, these vesicles are COPI-coated (see Figure 13–25B). COPI-coated vesicles are unique in that the components that make up the inner and outer coat layers are recruited as a preassembled complex, called *coatomer*. They function as a *retrieval pathway,* carrying back ER resident proteins that have escaped, as well as proteins such as cargo receptors and SNAREs that participated in the ER budding and vesicle fusion reactions. This retrieval process demonstrates the exquisite control mechanisms that regulate coat assembly reactions. The COPI coat assembly begins only seconds after the COPII coats have been shed; it remains a mystery how this switch in coat assembly is controlled.

The retrieval (or retrograde) transport continues as the vesicular tubular clusters move toward the Golgi apparatus. Thus, the clusters continually mature, gradually changing their composition as selected proteins are returned to the ER.

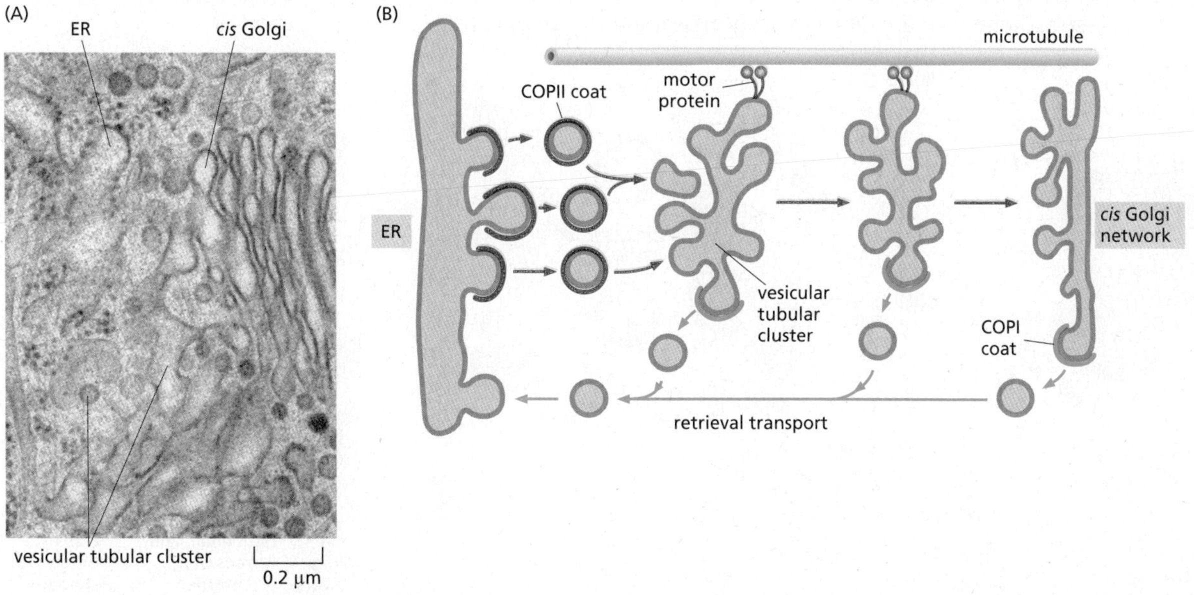

Figure 13–25 Vesicular tubular clusters. (A) An electron micrograph of vesicular tubular clusters forming around an exit site. Many of the vesicle-like structures seen in the micrograph are cross sections of tubules that extend above and below the plane of this thin section and are interconnected. (B) Vesicular tubular clusters move along microtubules to carry proteins from the ER to the Golgi apparatus. COPI coats mediate the budding of vesicles that return to the ER from these clusters (and from the Golgi apparatus). (A, courtesy of Judith Klumperman, from J.A. Martínez-Menárguez et al., *Cell* 98:81–90, 1999.)

The retrieval continues from the Golgi apparatus, after the vesicular tubular clusters have delivered their cargo.

The Retrieval Pathway to the ER Uses Sorting Signals

The retrieval pathway for returning escaped proteins back to the ER depends on *ER retrieval signals*. Resident ER membrane proteins, for example, contain signals that bind directly to COPI coats and are thus packaged into COPI-coated transport vesicles for retrograde delivery to the ER. The best-characterized retrieval signal of this type consists of two lysines, followed by any two other amino acids, at the extreme C-terminal end of the ER membrane protein. It is called a *KKXX sequence*, based on the single-letter amino acid code. Most membrane proteins that function at the interface between the ER and Golgi apparatus, including v- and t-SNAREs and some cargo receptors, use this retrieval pathway to come back to the ER.

Soluble ER resident proteins, such as BiP, also contain a short ER retrieval signal at their C-terminal end, but it is different: it consists of a Lys-Asp-Glu-Leu or a similar sequence. If this signal (called the *KDEL sequence*) is removed from BiP by genetic engineering, the protein is slowly secreted from the cell. If the signal is transferred to a protein that is normally secreted, the protein is now efficiently returned to the ER, where it accumulates. Unlike the retrieval signals on ER membrane proteins, which can interact directly with the COPI coat, soluble ER resident proteins must bind to specialized receptor proteins such as the *KDEL receptor*—a multipass transmembrane protein that binds to the KDEL sequence and packages any protein displaying it into COPI-coated retrograde transport vesicles (**Figure 13–26**).

The KDEL receptor accomplishes this task by cycling between the ER and the Golgi apparatus, selectively binding proteins with the KDEL sequence in the Golgi apparatus and releasing them in the ER. The markedly different affinity between the receptor and the KDEL sequence in these two compartments is due to the lower pH in the Golgi apparatus, which is regulated by H^+ pumps. A critical histidine in the KDEL receptor is protonated in the lower-pH environment of the Golgi apparatus, strongly favoring its interaction with the KDEL sequence. As we discuss later, pH-sensitive protein–protein interactions form the basis for many of the protein-sorting steps in the cell.

Many Proteins Are Selectively Retained in the Compartments in Which They Function

The KDEL retrieval pathway only partly explains how ER resident proteins are maintained in the ER. As mentioned, cells that express genetically modified ER resident proteins, from which the KDEL sequence has been experimentally removed, secrete these proteins. But the rate of secretion is much slower than that

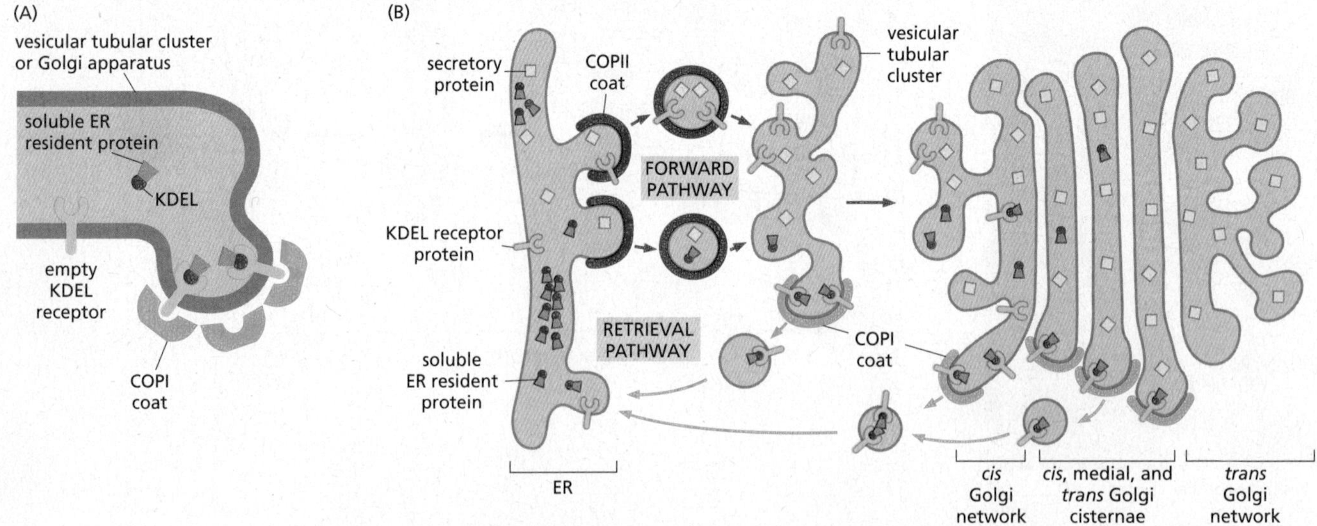

Figure 13–26 Retrieval of soluble ER resident proteins. ER resident proteins that escape from the ER are returned by vesicle transport. (A) The KDEL receptor present in both vesicular tubular clusters and the Golgi apparatus captures the soluble ER resident proteins and carries them in COPI-coated transport vesicles back to the ER. (Recall that the COPI-coated vesicles shed their coats as soon as they are formed.) Upon binding its ligands in the tubular cluster or Golgi apparatus, the KDEL receptor may change conformation, so as to facilitate its recruitment into budding COPI-coated vesicles. (B) The retrieval of ER proteins begins in vesicular tubular clusters and continues in later parts of the Golgi apparatus. In the environment of the ER, the ER resident proteins dissociate from the KDEL receptor, which is then returned to the Golgi apparatus for reuse. We discuss the different compartments of the Golgi apparatus shortly.

for a normal secretory protein. It seems that a mechanism that is independent of their KDEL signal normally retains ER resident proteins and that only those proteins that escape this retention mechanism are captured and returned via the KDEL receptor.

A suggested retention mechanism is that ER resident proteins bind to one another, thus forming complexes that are too big to enter transport vesicles efficiently. Because ER resident proteins are present in the ER at very high concentrations (estimated to be millimolar), relatively low-affinity interactions would suffice to retain most of the proteins in such complexes. Aggregation of proteins that function in the same compartment is a general mechanism that compartments use to organize and retain their resident proteins. Golgi enzymes that function together, for example, also bind to each other and are thereby restrained from entering transport vesicles leaving the Golgi apparatus.

The Golgi Apparatus Consists of an Ordered Series of Compartments

Because it could be selectively visualized by silver stains, the Golgi apparatus was one of the first organelles described by early light microscopists. It consists of a collection of flattened, membrane-enclosed compartments called *cisternae*, that somewhat resemble a stack of pita breads. Each Golgi stack typically consists of four to six cisternae (**Figure 13–27**), although some unicellular flagellates can have more than 20. In animal cells, tubular connections between corresponding cisternae link many stacks, thus forming a single complex, which is usually located near the cell nucleus and close to the centrosome (**Figure 13–28A**). This localization depends on microtubules. If microtubules are experimentally depolymerized, the Golgi apparatus reorganizes into individual stacks that are found throughout the cytoplasm, adjacent to ER exit sites. Some cells, including most plant cells, have hundreds of individual Golgi stacks

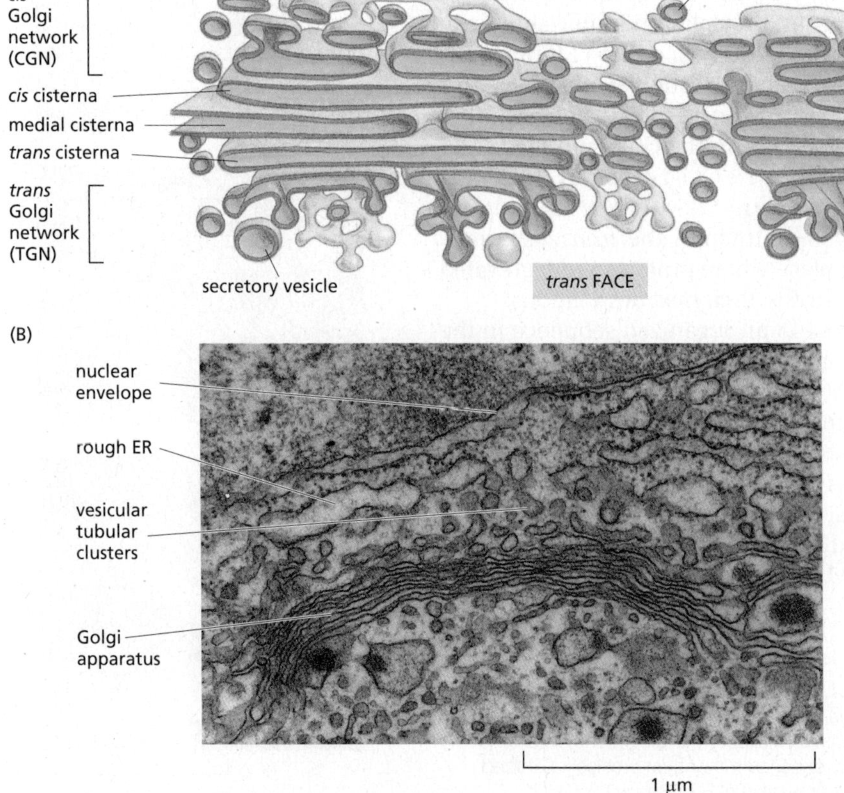

Figure 13–27 The Golgi apparatus. (A) Three-dimensional reconstruction from electron micrographs of the Golgi apparatus in a secretory animal cell. The *cis* face of the Golgi stack is that closest to the ER. (B) A thin-section electron micrograph of an animal cell. In plant cells, the Golgi apparatus is generally more distinct and more clearly separated from other intracellular membranes than in animal cells. (A, redrawn from A. Rambourg and Y. Clermont, *Eur. J. Cell Biol.* 51:189–200, 1990; B, courtesy of Brij J. Gupta.)

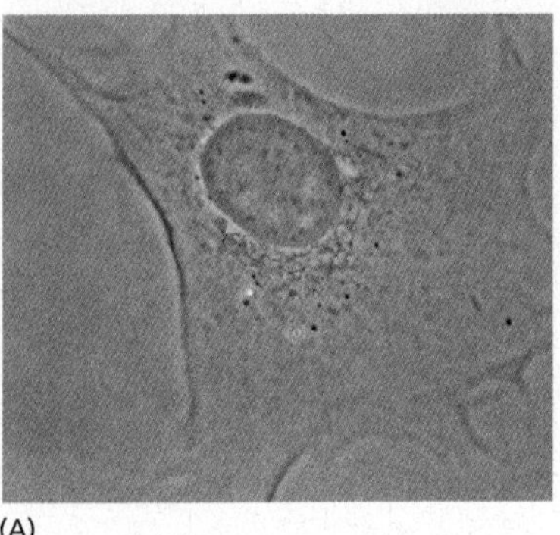
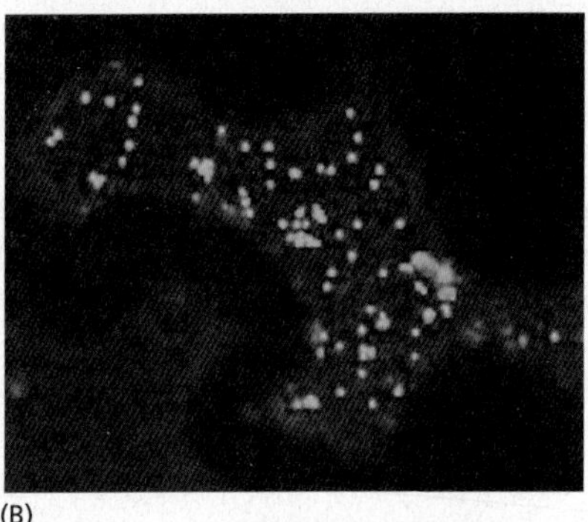

Figure 13–28 **Localization of the Golgi apparatus in animal and plant cells.** (A) The Golgi apparatus in a cultured fibroblast stained with a fluorescent antibody that recognizes a Golgi resident protein *(bright orange)*. The Golgi apparatus is polarized, facing the direction in which the cell was crawling before fixation. (B) The Golgi apparatus in a plant cell that is expressing a fusion protein consisting of a resident Golgi enzyme fused to green fluorescent protein. (A, courtesy of John Henley and Mark McNiven; B, courtesy of Chris Hawes.)

dispersed throughout the cytoplasm where they are typically found adjacent to ER exit sites (**Figure 13–28B**).

During their passage through the Golgi apparatus, many transported molecules undergo an ordered series of covalent modifications. Each Golgi stack has two distinct faces: a ***cis* face** (or entry face) and a ***trans* face** (or exit face). Both *cis* and *trans* faces are closely associated with special compartments, each composed of a network of interconnected tubular and cisternal structures: the ***cis* Golgi network (CGN)** and the ***trans* Golgi network (TGN)**, respectively. The CGN is a collection of fused vesicular tubular clusters arriving from the ER. Proteins and lipids enter the *cis* Golgi network and exit from the *trans* Golgi network, bound for the cell surface or another compartment. Both networks are important for protein sorting: proteins entering the CGN can either move onward in the Golgi apparatus or be returned to the ER. Similarly, proteins exiting from the TGN move onward and are sorted according to their next destination: endosomes, secretory vesicles, or the cell surface. They also can be returned to an earlier compartment. Some membrane proteins are retained in the part of the Golgi apparatus where they function.

As described in Chapter 12, a single species of *N-linked oligosaccharide* is attached *en bloc* to many proteins in the ER and then trimmed while the protein is still in the ER. The oligosaccharide intermediates created by the trimming reactions serve to help proteins fold and to help transport misfolded proteins to the cytosol for degradation in proteasomes. Thus, they play an important role in controlling the quality of proteins exiting from the ER. Once these ER functions have been fulfilled, the cell reutilizes the oligosaccharides for new functions. This begins in the Golgi apparatus, which generates the heterogeneous oligosaccharide structures seen in mature proteins. After arrival in the CGN, proteins enter the first of the Golgi processing compartments (the *cis* Golgi cisterna). They then move to the next compartment (the medial cisterna) and finally to the *trans* cisterna, where glycosylation is completed. The lumen of the *trans* cisterna is thought to be continuous with the TGN, the place where proteins are segregated into different transport packages and dispatched to their next destinations.

The oligosaccharide-processing steps occur in an organized sequence in the Golgi apparatus, with each cisterna containing a characteristic mixture of processing enzymes. Proteins are modified in successive stages as they move from cisterna to cisterna across the stack, so that the stack forms a multistage processing unit.

Investigators discovered the functional differences between the *cis*, medial, and *trans* subdivisions of the Golgi apparatus by localizing the enzymes involved in processing *N*-linked oligosaccharides in distinct regions of the organelle, both by physical fractionation of the organelle and by labeling the enzymes in electron microscope sections with antibodies (**Figure 13–29**). The removal of mannose and

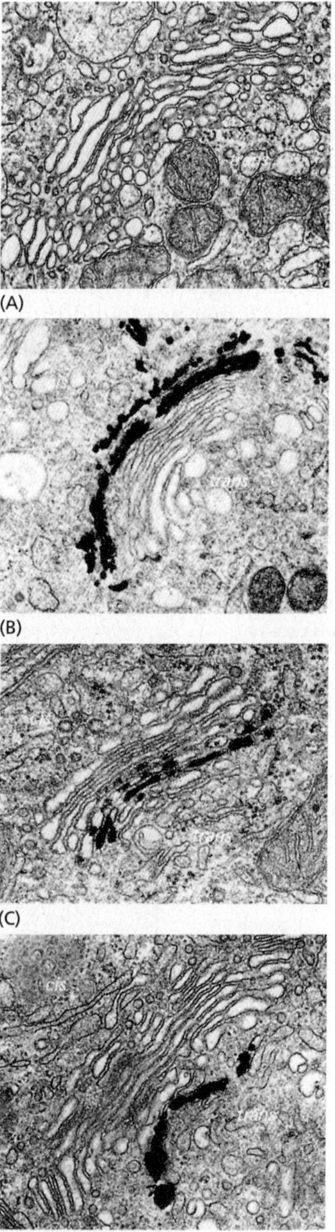

Figure 13–29 **Molecular compartmentalization of the Golgi apparatus.** A series of electron micrographs shows the Golgi apparatus (A) unstained, (B) stained with osmium, which preferentially labels the cisternae of the *cis* compartment, and (C and D) stained to reveal the location of specific enzymes. Nucleoside diphosphatase is found in the *trans* Golgi cisternae (C), while acid phosphatase is found in the *trans* Golgi network (D). Note that usually more than one cisterna is stained. The enzymes are therefore thought to be highly enriched rather than precisely localized to a specific cisterna. (Courtesy of Daniel S. Friend, by permission of E.L. Bearer.)

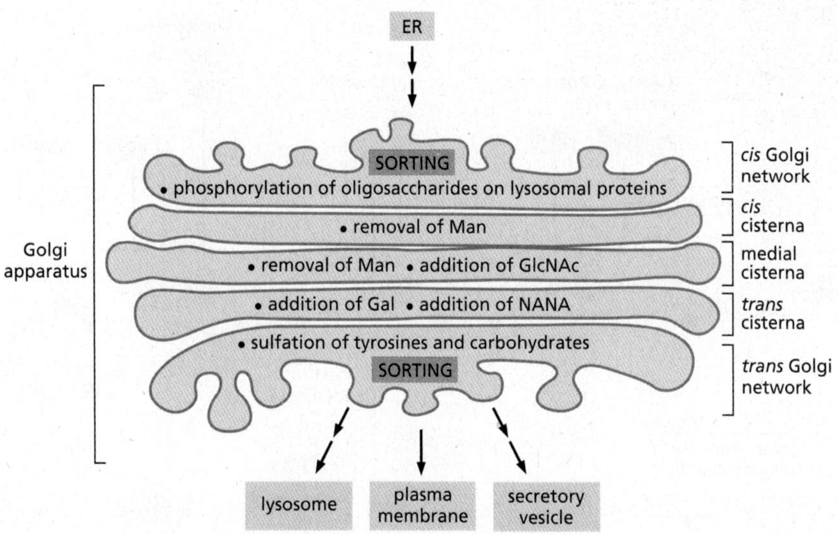

Figure 13–30 Oligosaccharide processing in Golgi compartments. The localization of each processing step shown was determined by a combination of techniques, including biochemical subfractionation of the Golgi apparatus membranes and electron microscopy after staining with antibodies specific for some of the processing enzymes. Processing enzymes are not restricted to a particular cisterna; instead, their distribution is graded across the stack, such that early-acting enzymes are present mostly in the *cis* Golgi cisternae and later-acting enzymes are mostly in the *trans* Golgi cisternae. Man, mannose; GlcNAc, *N*-acetylglucosamine; Gal, galactose; NANA, *N*-acetylneuraminic acid (sialic acid).

the addition of *N*-acetylglucosamine, for example, occur in the *cis* and medial cisternae, while the addition of galactose and sialic acid occurs in the *trans* cisterna and *trans* Golgi network. **Figure 13–30** summarizes the functional compartmentalization of the Golgi apparatus.

Oligosaccharide Chains Are Processed in the Golgi Apparatus

Whereas the ER lumen is full of soluble resident proteins and enzymes, the resident proteins in the Golgi apparatus are all membrane bound. All of the Golgi glycosidases and glycosyl transferases, for example, are single-pass transmembrane proteins, many of which are organized in multienzyme complexes.

Two broad classes of *N*-linked oligosaccharides, the **complex oligosaccharides** and the **high-mannose oligosaccharides**, are attached to mammalian glycoproteins. Sometimes, both types are attached (in different places) to the same polypeptide chain. Complex oligosaccharides are generated when the original *N*-linked oligosaccharide added in the ER is trimmed and further sugars are added; by contrast, high-mannose oligosaccharides are trimmed but have no new sugars added to them in the Golgi apparatus (**Figure 13–31**). The sialic acids

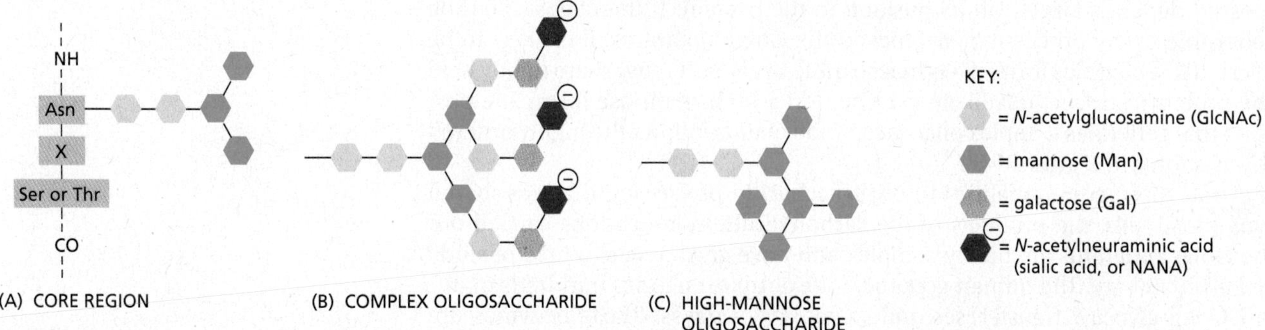

Figure 13–31 The two main classes of asparagine-linked (*N*-linked) oligosaccharides found in mature mammalian glycoproteins. (A) Both complex oligosaccharides and high-mannose oligosaccharides share a common *core region* derived by trimming the original *N*-linked oligosaccharide added in the ER (see Figure 12–32) and typically containing two *N*-acetylglucosamines (GlcNAc) and three mannoses (Man). The three amino acids constitute the sequence recognized by the oligosaccharyl transferase enzyme that adds the initial oligosaccharide to the protein. Asn, asparagine; Ser, serine; Thr, threonine; X, any amino acid, except proline. (B) Each complex oligosaccharide consists of a *core region*, together with a *terminal region* that contains a variable number of copies of a special trisaccharide unit (*N-acetylglucosamine–galactose–sialic acid*) linked to the three core mannoses. Frequently, the terminal region is truncated and contains only GlcNAc and galactose (Gal) or just GlcNAc. In addition, a fucose may be added, usually to the core GlcNAc attached to the asparagine (Asn). Thus, although the steps of processing and subsequent sugar addition are rigidly ordered, complex oligosaccharides can be heterogeneous. Moreover, although the complex oligosaccharide shown has three terminal branches, two and four branches are also common, depending on the glycoprotein and the cell in which it is made (**Movie 13.4**). (C) High-mannose oligosaccharides are not trimmed back all the way to the core region and contain additional mannoses. Hybrid oligosaccharides with one Man branch and one GlcNAc and Gal branch are also found (not shown).

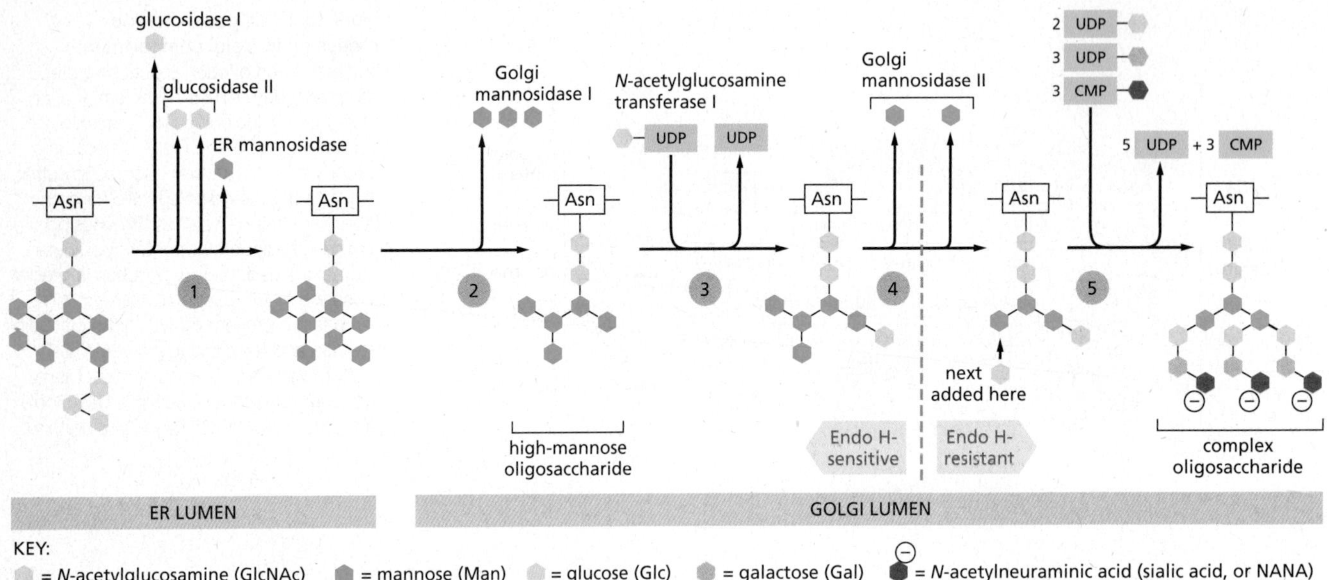

Figure 13–32 Oligosaccharide processing in the ER and the Golgi apparatus. The processing pathway is highly ordered, so that each step shown depends on the previous one. Step 1: Processing begins in the ER with the removal of the glucoses from the oligosaccharide initially transferred to the protein. Then a mannosidase in the ER membrane removes a specific mannose. The remaining steps occur in the Golgi stack. Step 2: Golgi mannosidase I removes three more mannoses. Step 3: *N*-acetylglucosamine transferase I then adds an *N*-acetylglucosamine. Step 4: Golgi mannosidase II then removes two additional mannoses. This yields the final core of three mannoses that is present in a complex oligosaccharide. At this stage, the bond between the two *N*-acetylglucosamines in the core becomes resistant to attack by a highly specific endoglycosidase (*Endo H*). Because all later structures in the pathway are also Endo H–resistant, treatment with this enzyme is widely used to distinguish complex from high-mannose oligosaccharides. Step 5: Finally, as shown in Figure 13–31, additional *N*-acetylglucosamines, galactoses, and sialic acids are added. These final steps in the synthesis of a complex oligosaccharide occur in the cisternal compartments of the Golgi apparatus: three types of glycosyl transferase enzymes act sequentially, using sugar substrates that have been activated by linkage to the indicated nucleotide; the membranes of the Golgi cisternae contain specific carrier proteins that allow each sugar nucleotide to enter in exchange for the nucleoside phosphates that are released after the sugar is attached to the protein on the lumenal face.

Note that, as a biosynthetic organelle, the Golgi apparatus differs from the ER: all sugars in the Golgi are assembled inside the lumen from sugar nucleotides, whereas in the ER, the *N*-linked precursor oligosaccharide is assembled partly in the cytosol and partly in the lumen, and most lumenal reactions use dolichol-linked sugars as their substrates (see Figure 12–33).

in the complex oligosaccharides are of special importance because they bear a negative charge. Whether a given oligosaccharide remains high-mannose or is processed depends largely on its position in the protein. If the oligosaccharide is accessible to the processing enzymes in the Golgi apparatus, it is likely to be converted to a complex form; if it is inaccessible because its sugars are tightly held to the protein's surface, it is likely to remain in a high-mannose form. The processing that generates complex oligosaccharide chains follows the highly ordered pathway shown in **Figure 13–32**.

Beyond these commonalities in oligosaccharide processing that are shared among most cells, the products of the carbohydrate modifications carried out in the Golgi apparatus are highly complex and have given rise to a field of study called glycobiology. The human genome, for example, encodes hundreds of different Golgi glycosyl transferases and many glycosidases. These enzymes are expressed differently from one cell type to another and at different times during development, resulting in a variety of glycosylated forms of a given protein or lipid in different cell types and at varying stages of differentiation. The complexity of modifications is not limited to *N*-linked oligosaccharides but also occurs on *O-linked sugars*, as we discuss next.

Proteoglycans Are Assembled in the Golgi Apparatus

In addition to the *N*-linked oligosaccharide alterations, many proteins are modified in the Golgi apparatus in other ways as they pass through the Golgi cisternae *en route* from the ER to their final destinations. Some proteins have sugars added

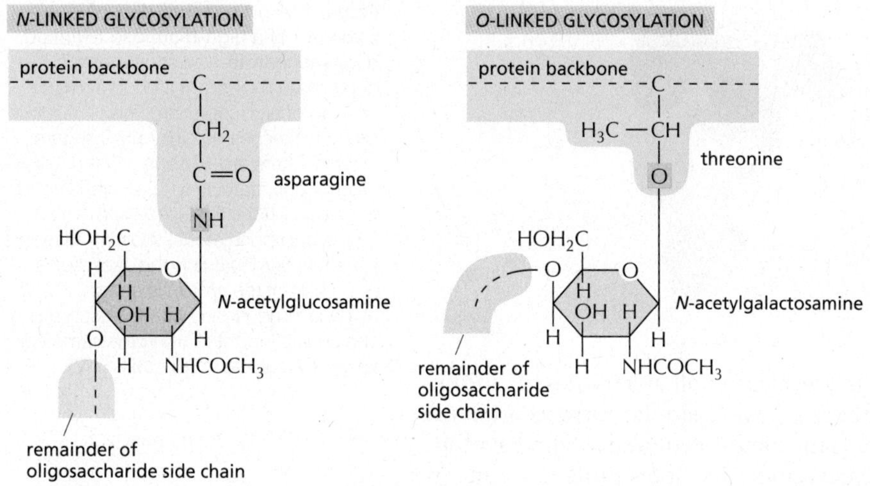

Figure 13–33 *N*- and *O*-linked glycosylation. In each case, only the single sugar group that is directly attached to the protein chain is shown.

to the hydroxyl groups of selected serines or threonines or, in some cases (such as collagens), to hydroxylated proline and lysine side chains. This ***O*-linked glycosylation** (Figure 13–33), like the extension of *N*-linked oligosaccharide chains, is catalyzed by a series of glycosyl transferase enzymes that use the sugar nucleotides in the lumen of the Golgi apparatus to add sugars to a protein one at a time. Usually, *N*-acetylgalactosamine is added first, followed by a variable number of additional sugars, ranging from just a few to 10 or more.

The Golgi apparatus confers the heaviest *O*-linked glycosylation of all on *mucins*, the glycoproteins in mucus secretions, and on *proteoglycan core proteins*, which it modifies to produce **proteoglycans**. As discussed in Chapter 19, this process involves the polymerization of one or more *glycosaminoglycan chains* (long, unbranched polymers composed of repeating disaccharide units; see Figure 19–35) onto serines on a core protein. Many proteoglycans are secreted and become components of the extracellular matrix, while others remain anchored to the extracellular face of the plasma membrane. Still others form a major component of slimy materials, such as the mucus that is secreted to form a protective coating on the surface of many epithelia.

The sugars incorporated into glycosaminoglycans are heavily sulfated in the Golgi apparatus immediately after these polymers are made, thus adding a significant portion of their characteristically large negative charge. Some tyrosines in proteins also become sulfated shortly before they exit from the Golgi apparatus. In both cases, the sulfation depends on the sulfate donor 3′-phosphoadenosine-5′-phosphosulfate (PAPS) (Figure 13–34), which is transported from the cytosol into the lumen of the *trans* Golgi network.

What Is the Purpose of Glycosylation?

There is an important difference between the construction of an oligosaccharide and the synthesis of other macromolecules such as DNA, RNA, and protein. Whereas nucleic acids and proteins are copied from a template in a repeated series of identical steps using the same enzyme or set of enzymes, complex carbohydrates require a different enzyme at each step. The product of each enzyme is recognized as the exclusive substrate for the next enzyme in the series. The vast abundance of glycoproteins and the complicated pathways that have evolved to synthesize them emphasize that the oligosaccharides on glycoproteins and glycosphingolipids have very important functions. A large family of genetic human diseases known as congenital disorders of glycosylation is caused by inherited mutations in individual enzymes involved in glycan modification of proteins and lipids.

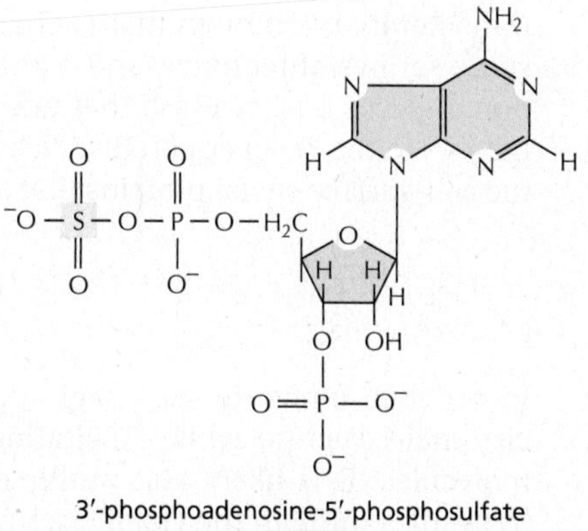

3′-phosphoadenosine-5′-phosphosulfate (PAPS)

Figure 13–34 The structure of PAPS.

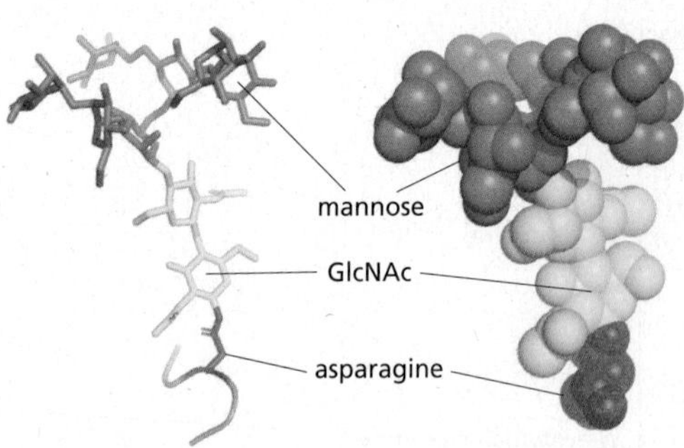

Figure 13–35 The three-dimensional structure of a high-mannose *N*-linked oligosaccharide. The structure was determined by x-ray crystallographic analysis of a glycoprotein. This oligosaccharide contains only 9 sugars, whereas there are 14 sugars in the *N*-linked oligosaccharide that is initially transferred to proteins in the ER (see Figure 12–32). *Left:* a backbone model showing all atoms except hydrogens; only the asparagine side chain of the protein is shown. *Right:* a space-filling model, with the asparagine and sugars indicated using the same color scheme as at left. (PDB code: 5KZC.)

N-linked glycosylation, for example, is prevalent in all eukaryotes, including yeasts. *N*-linked oligosaccharides also occur in a very similar form in archaeal cell-wall proteins, suggesting that the whole machinery required for their synthesis is evolutionarily ancient. *N*-linked glycosylation promotes protein folding in two ways. First, it has a direct role in making folding intermediates more soluble, thereby preventing their aggregation. Second, the sequential modifications of the *N*-linked oligosaccharide establish a "glyco-code" that marks the progression of protein folding. This glyco-code is used by chaperones and lectins in the ER to guide protein folding and degradation (discussed in Chapter 12) and by other lectins that guide ER-to-Golgi transport. As we discuss later, oligosaccharides also participate in protein sorting in the *trans* Golgi network.

Because chains of sugars have limited flexibility, even a small *N*-linked oligosaccharide protruding from the surface of a glycoprotein (**Figure 13–35**) can limit the approach of other macromolecules to the protein surface. In this way, for example, the presence of oligosaccharides tends to make a glycoprotein more resistant to digestion by proteolytic enzymes. It may be that the oligosaccharides on cell-surface proteins originally provided an ancestral cell with a protective coat; compared to the rigid bacterial cell wall, such a sugar coat has the advantage that it leaves the cell with the freedom to change shape and move.

The sugar chains have since been adapted to serve other purposes as well. The mucus coat of lung and intestinal cells, for example, protects against many pathogens. The recognition of sugar chains by *lectins* in the extracellular space is important in many developmental processes and in cell–cell recognition: *selectins*, for example, are transmembrane lectins that function in cell–cell adhesion during blood-cell migration, as discussed in Chapter 19. The presence of oligosaccharides may modify a protein's antigenic and functional properties, making glycosylation an important factor in the production of proteins for pharmaceutical purposes.

Glycosylation can also have important regulatory roles. Signaling through the cell-surface signaling receptor Notch, for example, is an important factor in determining the cell's fate in development (discussed in Chapter 21). Notch is a transmembrane protein that is *O*-glycosylated by addition of a single fucose to some serines, threonines, and hydroxylysines. Some cell types express an additional glycosyl transferase that adds an *N*-acetylglucosamine to each of these fucoses in the Golgi apparatus. This addition changes the specificity of Notch for the cell-surface signal proteins that activate it.

Transport Through the Golgi Apparatus Occurs by Multiple Mechanisms

In order to function, the Golgi apparatus must maintain its polarized multicisternal structure while facilitating the transit of a large number of diverse molecules. It is likely that multiple mechanisms are used to transport cargo molecules through the Golgi cisternae while efficiently retaining Golgi resident proteins. One mechanism involves the movement of cargo in transport vesicles from one compartment to the next while retrieving any escaped resident

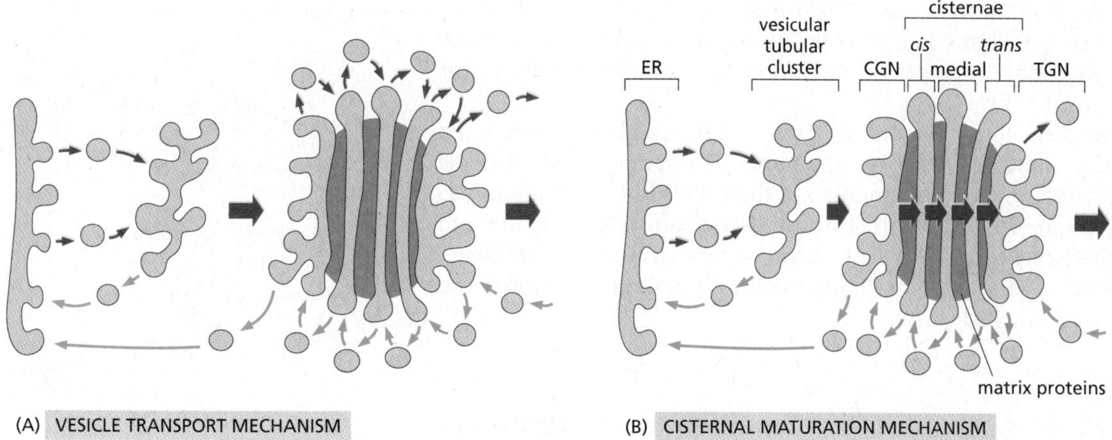

Figure 13–36 Two mechanisms explaining the organization of the Golgi apparatus and how proteins move through it. It is likely that transport of cargo molecules through the Golgi apparatus in the forward direction *(red arrows)* involves elements of both mechanisms. (A) In the vesicle transport mechanism, Golgi cisternae are static compartments, which contain a characteristic complement of resident enzymes. The passing of molecules from *cis* to *trans* through the Golgi is accomplished by forward-moving transport vesicles, which bud from one cisterna and fuse with the next in a *cis*-to-*trans* direction. (B) In the cisternal maturation mechanism, each Golgi cisterna matures as it migrates outward through the stack. At each stage, the Golgi resident proteins that are carried forward in a maturing cisterna are moved backwards *(blue arrows)* to an earlier compartment in COPI-coated vesicles. When a newly formed cisterna moves to a medial position, for example, "leftover" *cis* Golgi enzymes would be extracted and transported retrogradely to a new *cis* cisterna behind. Likewise, the medial enzymes would be received by retrograde transport from the cisternae just ahead. In this way, a *cis* cisterna would mature to a medial and then *trans* cisterna as it moves outward.

proteins using different transport vesicles (Figure 13–36A). This *vesicle transport mechanism* is conceptually similar to how proteins and lipids are transported from the ER to the Golgi, except that only COPI-coated vesicles are used. Although both forward- and backward-moving vesicles would likely be COPI-coated, the coats may contain different adaptor proteins that confer selectivity on the packaging of cargo molecules.

A different way for cargo to move through the Golgi apparatus involves the **cisternal maturation mechanism**. According to this view, new *cis* cisternae continually form as vesicular tubular clusters arrive from the ER and fuse with transport vesicles containing Golgi resident proteins and enzymes. As the cargo within a *cis* cisterna is modified, the enzymes leave in transport vesicles that will fuse with newly arriving vesicular tubular clusters. At the same time, the cisterna accepts transport vesicles containing enzymes from later Golgi cisternae, converting it into a medial cisterna. In this way, a cisterna full of cargo moves through the Golgi stack while different subsets of Golgi resident proteins transit backwards in COPI-coated vesicles from later to earlier cisternae (Figure 13–36B). When a cisterna finally moves forward to become part of the *trans* Golgi network, various types of coated vesicles bud off it until this network disappears, to be replaced by a maturing cisterna just behind. At the same time, other transport vesicles are continually retrieving membrane from post-Golgi compartments and returning it to the *trans* Golgi network.

It is likely that aspects of both mechanisms are used to varying degrees depending on the type of cell and the nature of cargo molecules that need to be transported. A stable core of long-lasting cisternae might exist in the center of each Golgi cisterna, while regions at the rim may undergo continual maturation, perhaps utilizing Rab cascades that change their identity. As matured pieces of the cisternae are formed, they might break off and fuse with downstream cisternae by homotypic fusion mechanisms, taking large cargo molecules such as procollagen rods and lipoprotein particles with them. In addition, COPI-coated vesicles might transport small cargo in the forward direction and retrieve escaped Golgi enzymes to their appropriate upstream cisternae.

Golgi Matrix Proteins Help Organize the Stack

The unique architecture of the Golgi apparatus depends on both the microtubule cytoskeleton, as already mentioned, and cytoplasmic Golgi matrix proteins. The Golgi reassembly and stacking proteins (called GRASPs) form a scaffold between adjacent cisternae and give the Golgi stack its structural integrity. Other matrix proteins, called *golgins*, form long tethers composed of stiff coiled-coil domains with interspersed hinge regions. Golgins form a forest of tentacles that can extend 100–400 nm from the surface of the Golgi stack. Different members of the golgin family are found in different regions of the Golgi stack and contain binding sites for different Rab proteins. Because transport vesicles arriving from different locations have their characteristic Rab proteins on them, golgins are thought to function as tethers that initially select which part of the Golgi stack a transport vesicle engages (Figure 13–37).

When the cell prepares to divide, mitotic protein kinases phosphorylate the Golgi matrix proteins, causing the Golgi apparatus to fragment and disperse throughout the cytosol. The Golgi fragments are then distributed evenly to the two daughter cells, where the matrix proteins are dephosphorylated, leading to the reassembly of the Golgi stack. Similarly, during apoptosis, proteolytic cleavage of golgins by caspases (discussed in Chapter 18) leads to fragmentation of the Golgi apparatus as the cell self-destructs.

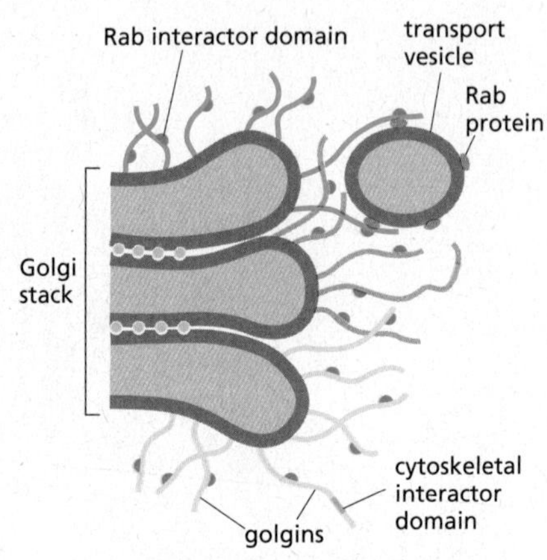

Figure 13–37 A model of golgin function. Filamentous golgins anchored to Golgi membranes capture transport vesicles by binding to Rab proteins on the vesicle surface. Different members of the golgin family of proteins are localized to different regions of the Golgi apparatus. GRASPs are shown tethering adjacent cisternae to each other.

Summary

Correctly folded and assembled proteins in the ER are packaged into COPII-coated transport vesicles that pinch off from the ER membrane. Shortly thereafter, the vesicles shed their coat and fuse with one another to form vesicular tubular clusters. In animal cells, the clusters then move on microtubule tracks to the Golgi apparatus, where they fuse with one another to form the cis Golgi network. Any resident ER proteins that escape from the ER are returned there from the vesicular tubular clusters and Golgi apparatus by retrograde transport in COPI-coated vesicles.

The Golgi apparatus, unlike the ER, contains many sugar nucleotides, which glycosyl transferase enzymes use to glycosylate lipid and protein molecules as they pass through the Golgi apparatus. The mannoses on the N-linked oligosaccharides that are added to proteins in the ER are often initially removed, and further sugars are added. Moreover, the Golgi apparatus is the site where O-linked glycosylation occurs and where glycosaminoglycan chains are added to core proteins to form proteoglycans. Sulfation of the sugars in proteoglycans and of selected tyrosines on proteins also occurs in a late Golgi compartment.

The Golgi apparatus modifies the many proteins and lipids that it receives from the ER and then distributes them to the plasma membrane, endosomes, and secretory vesicles. The Golgi apparatus is a polarized organelle, consisting of one or more stacks of disc-shaped cisternae. Each stack is organized as a series of at least three functionally distinct compartments, termed cis, medial, and trans cisternae. The cis and trans cisternae are each connected to special sorting stations, called the cis Golgi network and the trans Golgi network, respectively. Proteins and lipids move through the Golgi stack in the cis-to-trans direction. This movement may occur by vesicle transport, by progressive maturation of the cis cisternae as they migrate continuously through the stack, or by a combination of these two mechanisms. Continual retrograde vesicle transport from later to earlier cisternae keeps the enzymes concentrated in the cisternae where they are needed. The finished new proteins end up in the trans Golgi network, which packages them in transport vesicles and dispatches them to their specific destinations in the cell.

TRANSPORT FROM THE *TRANS* GOLGI NETWORK TO THE CELL EXTERIOR AND ENDOSOMES

After transiting the Golgi cisternae, cargo molecules that arrive at the *trans* Golgi network (TGN) are sorted and packaged into transport vesicles that depart for different destinations. Transport vesicles destined for the *cell surface* normally leave the TGN in a steady stream as irregularly shaped tubules. The membrane proteins and the lipids in these vesicles provide new components for the cell's plasma membrane, while the soluble proteins inside the vesicles are secreted to the extracellular space. The fusion of the vesicles with the plasma membrane is called **exocytosis**. This is the route, for example, by which cells secrete most of the proteoglycans and glycoproteins of the extracellular matrix, as discussed in Chapter 19.

All cells require this **constitutive secretory pathway**, which operates continually (Movie 13.5). Specialized secretory cells, however, have a second secretory pathway in which soluble proteins and other substances are destined to be

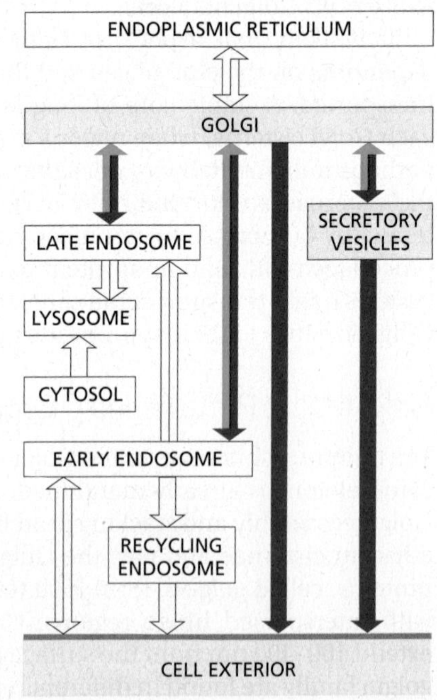

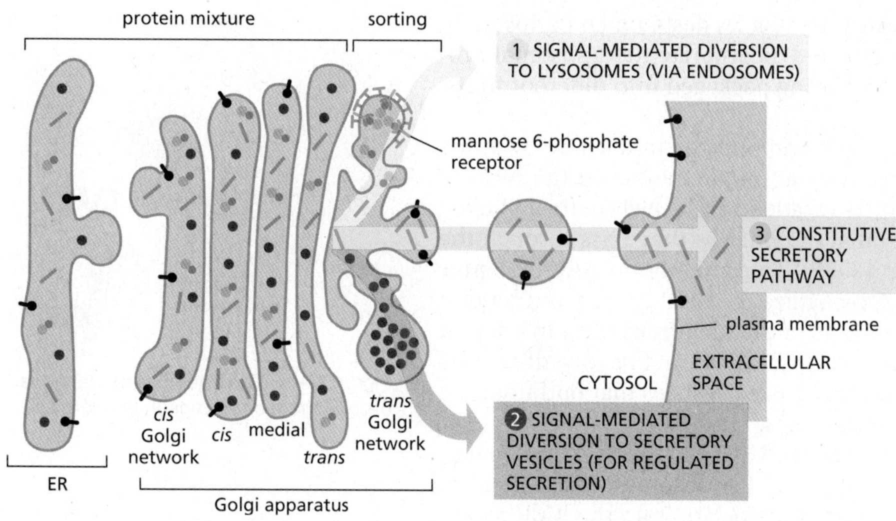

Figure 13–38 The three best-understood pathways of protein sorting in the *trans* Golgi network. (1) Proteins with the mannose 6-phosphate (M6P) marker (see Figure 13–40) are diverted to lysosomes (via endosomes) in clathrin-coated transport vesicles. (2) Proteins with signals directing them to secretory vesicles are concentrated in such vesicles as part of a regulated secretory pathway that is present only in specialized secretory cells. (3) In unpolarized cells, a constitutive secretory pathway delivers proteins with no special features to the cell surface. In polarized cells, such as epithelial cells, however, secreted and plasma membrane proteins are selectively directed to either the apical or the basolateral plasma membrane domain, so a specific signal must mediate at least one of these two pathways, as we discuss later.

initially stored in *secretory vesicles* for later release by exocytosis. This is the **regulated secretory pathway**, found mainly in cells specialized for secreting products rapidly on demand—such as hormones, neurotransmitters, or digestive enzymes.

The third major destination from the TGN is *endosomes*. Hydrolases that function in the lumen of lysosomes use this pathway to first arrive at endosomes, which progressively mature into lysosomes (discussed later). The sorting mechanism at the TGN for lysosomal hydrolase proteins is especially well understood and provides an example of how cargo molecules in the TGN are segregated among different types of transport vesicles. In this section, we consider the role of the Golgi apparatus in sorting proteins between these three pathways and compare the mechanisms of constitutive and regulated secretion.

Many Proteins and Lipids Are Carried Automatically from the *Trans* Golgi Network to the Cell Surface

A cell capable of regulated secretion must separate at least three classes of proteins before they leave the TGN—those destined for lysosomes (via endosomes), those destined for secretory vesicles, and those destined for immediate delivery to the cell surface (**Figure 13–38**). Specific signals are needed to direct *secretory proteins* into secretory vesicles and lysosomal proteins into different specialized transport vesicles. The nonselective constitutive secretory pathway transports most other proteins directly to the cell surface. Because entry into this pathway does not require a particular signal, it is also called the **default pathway**. Thus, in an unpolarized cell such as a white blood cell or a fibroblast, it seems that any protein in the lumen of the Golgi apparatus is automatically carried by the constitutive pathway to the cell surface unless it is specifically returned to the ER, retained as a resident protein in the Golgi apparatus itself, or selected for the pathways that lead to regulated secretion or to endosomes. In polarized cells, where different products have to be delivered to different domains of the cell surface, we shall see that the options are more complex.

A Mannose 6-Phosphate Receptor Sorts Lysosomal Hydrolases in the *Trans* Golgi Network

The best-understood mechanism for sorting of cargo molecules at the TGN operates on lysosomal hydrolases. Lysosomes are membrane-enclosed organelles filled with about 40 hydrolytic enzymes responsible for digesting all the macromolecules delivered there. Lysosomes are therefore a major site for degradation and recycling of proteins, nucleic acids, lipids, and even whole organelles. The function of lysosomes and the various transport routes leading to this organelle are considered later. For now, we address the pathway that selectively packages lysosomal hydrolases at the TGN into transport vesicles destined for endosomes. The vesicles that leave the TGN for endosomes incorporate the lysosomal proteins and exclude the many other proteins being packaged into different transport vesicles for delivery elsewhere.

How are lysosomal hydrolases recognized and selected in the TGN with the required accuracy? In animal cells they carry a unique marker in the form of *mannose 6-phosphate* (*M6P*) groups, which are added exclusively to the *N*-linked oligosaccharides of these soluble lysosomal enzymes as they pass through the lumen of the *cis* Golgi network (**Figure 13–39**). Transmembrane **M6P receptor proteins**, which are present in the TGN, recognize the M6P groups and bind to the lysosomal hydrolases on the lumenal side of the membrane and to adaptor proteins in assembling clathrin coats on the cytosolic side. In this way, the receptors help package the hydrolases into clathrin-coated vesicles that bud from the TGN and deliver their contents to early endosomes.

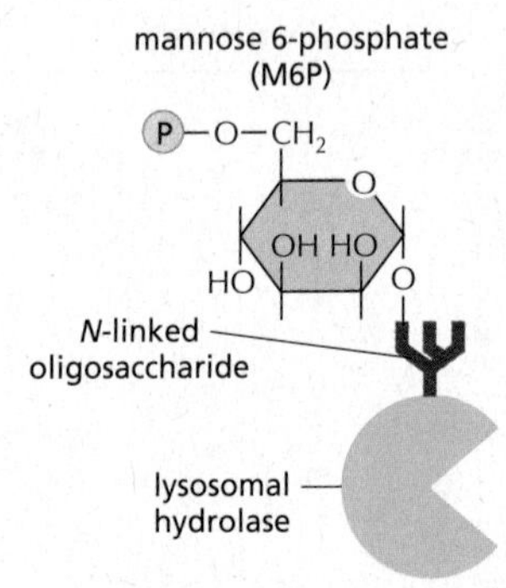

Figure 13–39 The structure of mannose 6-phosphate on a lysosomal hydrolase.

The M6P receptor protein binds to M6P at pH 6.5–6.7 in the TGN lumen and releases it at pH 6, which is the pH in the lumen of endosomes. Thus, after the receptor is delivered, the lysosomal hydrolases dissociate from the M6P receptors, which are retrieved into transport vesicles that bud from endosomes. These vesicles are coated with retromer, a coat protein complex specialized for endosome-to-TGN transport, which returns the receptors to the TGN for reuse (**Figure 13–40**).

Transport in either direction requires signals in the cytoplasmic tail of the M6P receptor that direct this protein to the endosome or back to the TGN. An adaptor protein of the clathrin coat recognizes the tail at the TGN, while retromer recognizes it at the endosome. The assembly of different coats at different membranes for the same receptor is ensured by organelle-specific markers, such as Rab7 and PI(3)P at the endosome. The recycling of the M6P receptor resembles the recycling of the KDEL receptor discussed earlier, although it differs in the type of coated vesicles that mediate the transport.

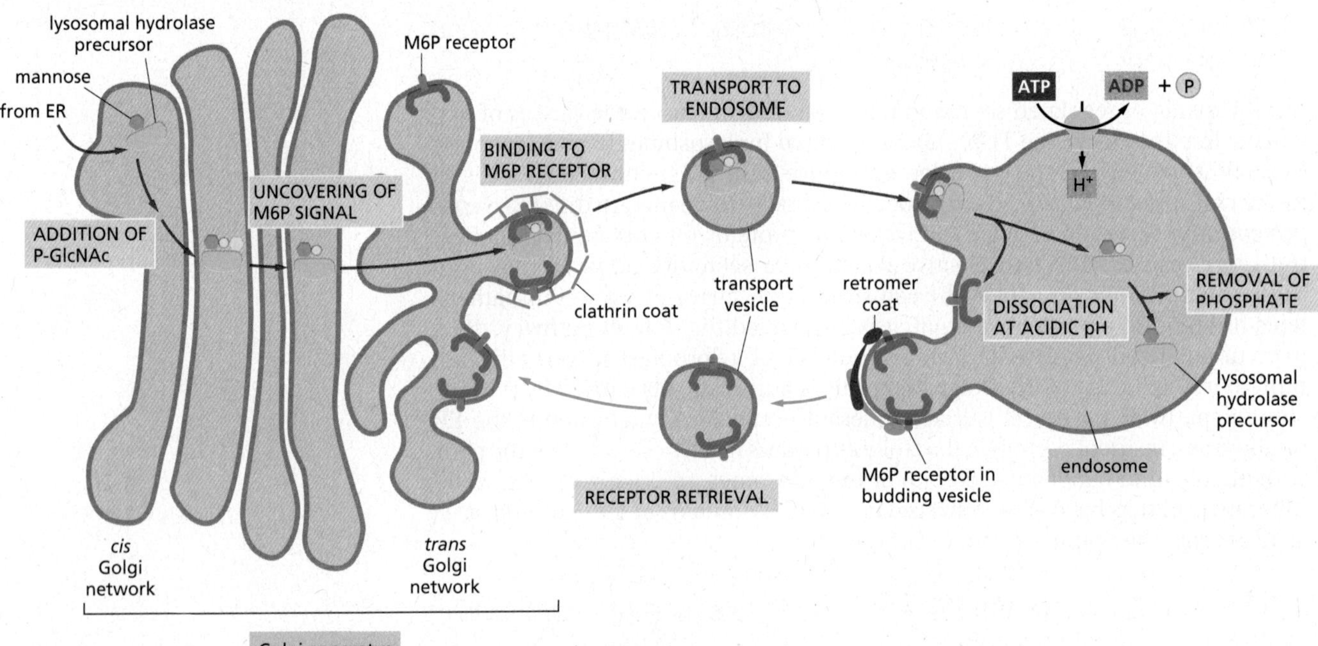

Figure 13–40 The transport of newly synthesized lysosomal hydrolases to endosomes. The sequential action of two enzymes in the *cis* and *trans* Golgi network adds mannose 6-phosphate (M6P) groups to the precursors of lysosomal enzymes (see Figure 13–41). The M6P-tagged hydrolases then segregate from all other types of proteins in the TGN because adaptor proteins (not shown) in the clathrin coat bind the M6P receptors, which, in turn, bind the M6P-modified lysosomal hydrolases. The clathrin-coated vesicles bud off from the TGN, shed their coat, and fuse with early endosomes. At the lower pH of the endosome, the hydrolases dissociate from the M6P receptors, and the empty receptors are retrieved in retromer-coated vesicles to the TGN for further rounds of transport. In the endosomes, the phosphate is removed from the M6P attached to the hydrolases, which may further ensure that the hydrolases do not return to the TGN with the receptor.

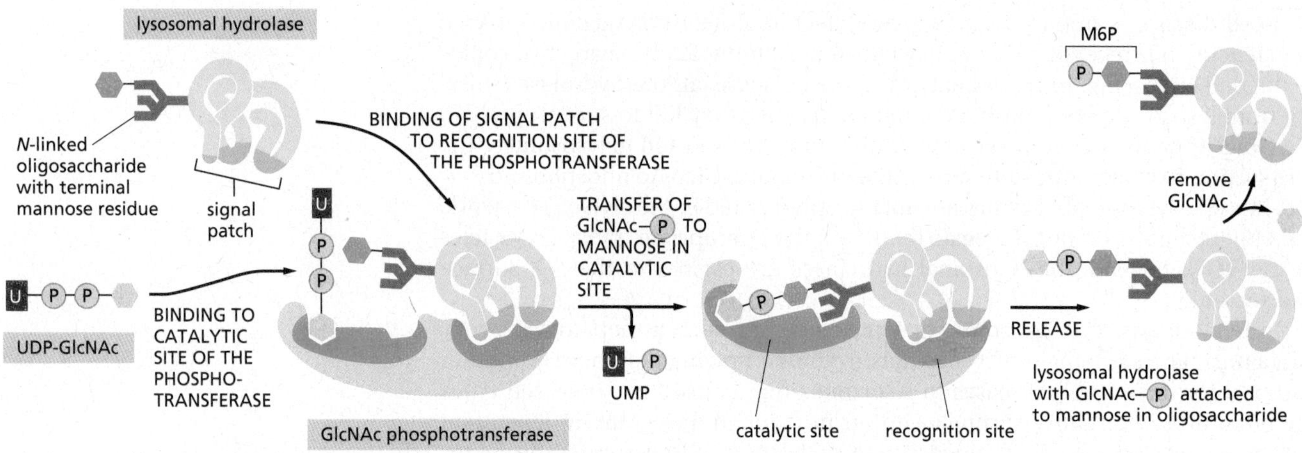

Figure 13–41 **The recognition of a lysosomal hydrolase.** A GlcNAc phosphotransferase recognizes lysosomal hydrolases in the Golgi apparatus. The enzyme has separate catalytic and recognition sites. The catalytic site binds both high-mannose *N*-linked oligosaccharides and UDP-GlcNAc. The recognition site binds to a signal patch that is present only on the surface of lysosomal hydrolases. A second enzyme cleaves off the GlcNAc, leaving the mannose 6-phosphate exposed.

Not all the hydrolase molecules that are tagged with M6P get to lysosomes. Some escape the normal packaging process in the *trans* Golgi network and are transported by the constitutive secretory pathway to the cell surface, where they are secreted into the extracellular fluid. Some M6P receptors, however, also take a detour to the plasma membrane, where they recapture the escaped lysosomal hydrolases and return them by *receptor-mediated endocytosis* (discussed later) to lysosomes via early and late endosomes. As lysosomal hydrolases require an acidic milieu to work, they can do little harm in the extracellular fluid, which usually has a neutral pH of 7.4.

For the sorting system that segregates lysosomal hydrolases and dispatches them to endosomes to work, the M6P groups must be added only to the appropriate glycoproteins in the Golgi apparatus. This requires specific recognition of the hydrolases by the Golgi enzymes responsible for adding M6P. Because all glycoproteins leave the ER with identical *N*-linked oligosaccharide chains, the signal for adding the M6P units to oligosaccharides must reside somewhere in the polypeptide chain of each hydrolase. Genetic engineering experiments have revealed that the recognition signal is a cluster of neighboring amino acids on each protein's surface, known as a *signal patch* (**Figure 13–41**). Because most lysosomal hydrolases contain multiple oligosaccharides, they acquire many M6P groups, providing a high-affinity signal for the M6P receptor.

Defects in the GlcNAc Phosphotransferase Cause a Lysosomal Storage Disease in Humans

Genetic defects that affect one or more of the lysosomal hydrolases cause a number of human **lysosomal storage diseases**. The defects result in an accumulation of undigested substrates in lysosomes, with severe pathological consequences, most often in the nervous system. In most cases, there is a mutation in a structural gene that codes for an individual lysosomal hydrolase. This occurs in *Hurler's disease*, for example, in which the enzyme required for the breakdown of certain types of glycosaminoglycan chains is defective or missing. The most severe form of lysosomal storage disease, however, is a very rare inherited metabolic disorder called *inclusion-cell disease* (*I-cell disease*). In this condition, almost all of the hydrolytic enzymes are missing from the lysosomes of many cell types, and their undigested substrates accumulate in these lysosomes, which consequently form large *inclusions* in the cells. The consequent pathology is complex, affecting all organ systems, skeletal integrity, and mental development; individuals rarely live beyond 6 or 7 years.

I-cell disease is due to a single gene defect and, like most genetic enzyme deficiencies, it is recessive; that is, it occurs only in individuals having two copies of the defective gene. In individuals with I-cell disease, all the hydrolases missing from lysosomes are found in the blood: because they fail to sort properly in the Golgi apparatus, they are secreted rather than transported to lysosomes. The mis-sorting has been traced to a defective or missing GlcNAc phosphotransferase. Because lysosomal enzymes are not phosphorylated in the *cis* Golgi network, the M6P receptors do not segregate them into the appropriate transport vesicles in the TGN. Instead, the lysosomal hydrolases are carried to the cell surface and secreted.

In I-cell disease, the lysosomes in some cell types, such as hepatocytes, contain a normal complement of lysosomal enzymes, implying that there is another pathway for directing hydrolases to lysosomes that is used by some cell types but not others. Alternative sorting receptors function in these M6P-independent pathways. Similarly, an M6P-independent pathway in all cells sorts the membrane proteins of lysosomes from the TGN for transport to late endosomes, and those proteins are therefore normal in I-cell disease.

Secretory Vesicles Bud from the *Trans* Golgi Network

Cells that are specialized for secreting some of their products rapidly on demand concentrate and store these products in **secretory vesicles** (often called *dense-core secretory granules* because they have dense cores when viewed in the electron microscope). As we discussed (see Figure 13–38), secretory vesicles form from the TGN, and they release their contents to the cell exterior by exocytosis in response to specific signals. The secreted product can be either a small molecule (such as histamine or a neuropeptide) or a protein (such as a hormone or digestive enzyme).

Proteins destined for secretory vesicles (called *secretory proteins*) are packaged into appropriate vesicles in the TGN by a mechanism that involves the selective aggregation of the secretory proteins. Clumps of aggregated, electron-dense material can be detected by electron microscopy in the lumen of the TGN. The signals that direct secretory proteins into such aggregates are not well defined and may be quite diverse. When a gene encoding a secretory protein is artificially expressed in a secretory cell that normally does not make the protein, the foreign protein is appropriately packaged into secretory vesicles. This observation shows that, although the proteins that an individual cell expresses and packages in secretory vesicles differ, they contain common sorting signals, which function properly even when the proteins are expressed in cells that do not normally make them.

It is unclear how the aggregates of secretory proteins are segregated into secretory vesicles. Secretory vesicles have unique proteins in their membrane, some of which might serve as receptors for aggregated protein in the TGN. The aggregates are much too big, however, for each molecule of the secreted protein to be bound by its own cargo receptor, as occurs for transport of the lysosomal enzymes. Instead, the aggregate might cause the membrane region containing the cargo receptor to zipper up around the aggregate, thereby enclosing it within the budding vesicle.

Initially, the membrane of the secretory vesicles that leave the TGN is only loosely wrapped around the clusters of aggregated secretory proteins. Morphologically, these *immature secretory vesicles* resemble dilated *trans* Golgi cisternae that have pinched off from the Golgi stack. As immature secretory vesicles mature, clathrin-coated transport vesicles bud from them and go back to the TGN (**Figure 13–42**). This recycling process not only returns Golgi components to the Golgi apparatus, but also serves to concentrate the contents of secretory vesicles. The sum of all the retrieval pathways during the transit of a secretory protein from the ER through the Golgi cisternae to a mature secretory vesicle results in a 200- to 400-fold increase in net concentration.

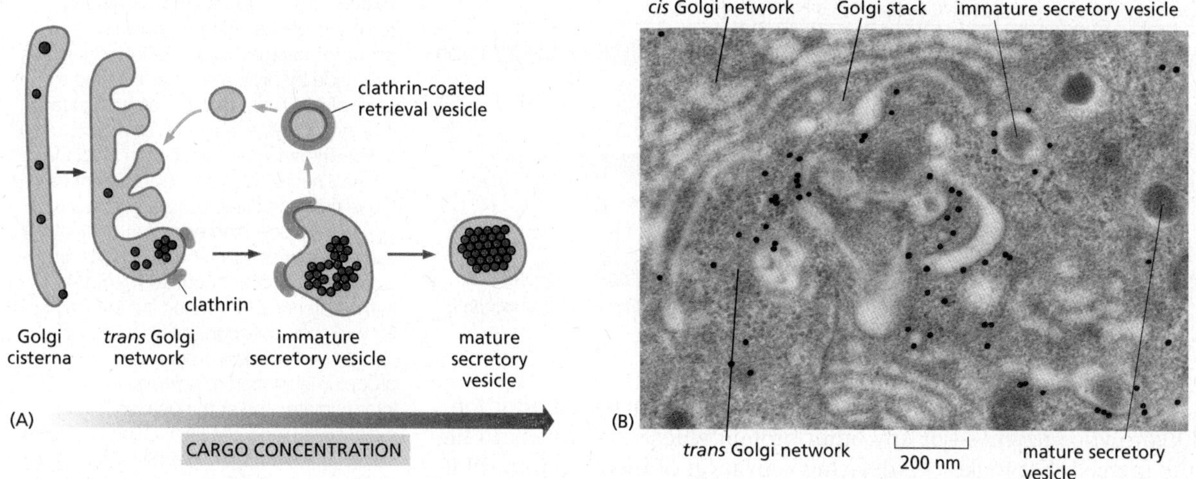

Figure 13–42 The formation of secretory vesicles. (A) Secretory proteins become segregated and highly concentrated in secretory vesicles by two mechanisms. First, they aggregate in the ionic environment of the TGN; often, the aggregates become more condensed as a secretory vesicle matures and its lumen becomes more acidic. Second, clathrin-coated vesicles retrieve excess membrane and lumenal content present in immature secretory vesicles as the secretory vesicles mature. (B) This electron micrograph shows secretory vesicles forming from the TGN in an insulin-secreting β cell of the pancreas. Anti-clathrin antibodies conjugated to gold spheres *(black dots)* have been used to locate clathrin molecules. The immature secretory vesicles, which contain insulin precursor protein (proinsulin), contain clathrin patches, which are no longer seen on the mature secretory vesicle. (B, courtesy of Lelio Orci.)

Immature secretory vesicles also fuse with one another, and the lumens become progressively more acidic from the increasing concentration of V-type ATPases in the vesicle membrane. Acidification of the lumen further condenses the secretory protein aggregate within a vesicle whose excess membrane has now been retrieved back to the TGN. Because the final mature secretory vesicles are so densely filled with contents, the secretory cell can disgorge large amounts of material promptly by exocytosis when triggered to do so (**Figure 13–43**).

Precursors of Secretory Proteins Are Proteolytically Processed During the Formation of Secretory Vesicles

Concentration is not the only process to which secretory proteins are subjected as the secretory vesicles mature. Many protein hormones and small neuropeptides, as well as many secreted hydrolytic enzymes, are synthesized as inactive precursors. Proteolysis is necessary to liberate the active molecules from these precursor proteins. The cleavages occur in the secretory vesicles and sometimes in the extracellular fluid after secretion. Additionally, many of the precursor proteins exit the ER with an N-terminal *propeptide* that is cleaved off only later in the secretory pathway to yield the mature protein. These proteins are initially synthesized as *pre-pro-proteins*, with the ER signal peptide (sometimes referred to as a *pre-peptide*) cleaved off earlier as the protein enters the rough ER (see Figure 12–18). In other cases, peptide signaling molecules are made as *polyproteins* that contain multiple copies of the same amino acid sequence. In still more complex cases, a variety of peptide signaling molecules are synthesized as parts of a single polyprotein that acts as a precursor for multiple end products, which are individually cleaved from the initial polypeptide chain. The same polyprotein may be processed in various ways to produce different peptides in different cell types (**Figure 13–44**).

Why is proteolytic processing so common in the secretory pathway? Some of the peptides produced in this way, such as the *enkephalins* (five-amino-acid neuropeptides with morphine-like activity), are undoubtedly too short in their mature forms to be co-translationally transported into the ER lumen or to

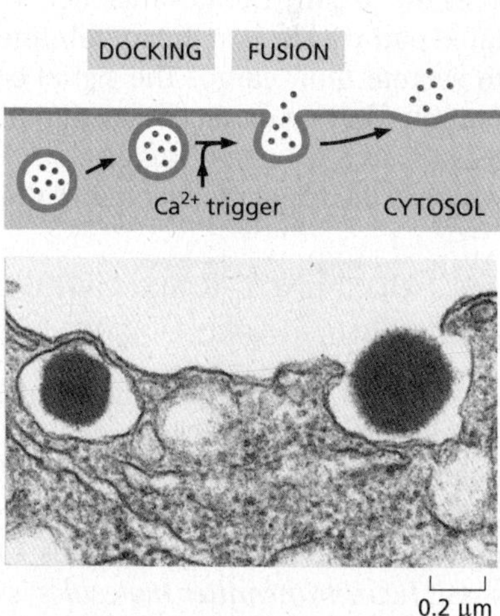

Figure 13–43 Exocytosis of secretory vesicles. The process is illustrated schematically *(top)* and in an electron micrograph that shows the release of insulin from a secretory vesicle of a pancreatic β cell. (Courtesy of Lelio Orci, from L. Orci et al., *Sci. Am.* 259:85–94, 1988.)

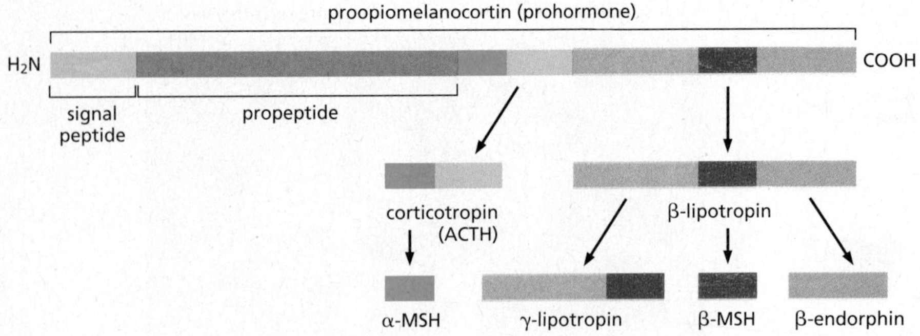

Figure 13–44 Processing pathways for the prohormone polyprotein proopiomelanocortin. The initial cleavages are made by proteases that cut next to pairs of positively charged amino acids (Lys-Arg, Lys-Lys, Arg-Lys, or Arg-Arg pairs). Trimming reactions then produce the final secreted products. Different cell types produce different concentrations of individual processing enzymes, so that the same prohormone precursor is cleaved to produce different peptide hormones. In the anterior lobe of the pituitary gland, for example, only corticotropin (ACTH) and β-lipotropin are produced from proopiomelanocortin, whereas in the intermediate lobe of the pituitary gland, mainly α-melanocyte stimulating hormone (α-MSH), γ-lipotropin, β-MSH, and β-endorphin are produced—α-MSH from ACTH and the other three from β-lipotropin, as shown.

include the necessary signal for packaging into secretory vesicles. In addition, for secreted hydrolytic enzymes—or any other protein whose activity could be harmful inside the cell that makes it—delaying activation of the protein until it reaches a secretory vesicle or until after it has been secreted has a clear advantage: the delay prevents the protein from acting prematurely inside the cell in which it is synthesized.

Secretory Vesicles Wait Near the Plasma Membrane Until Signaled to Release Their Contents

Once loaded, a secretory vesicle has to reach the site of secretion, which in some cells is far away from the TGN. Nerve cells are the most extreme example. Secretory proteins, such as peptide neurotransmitters (neuropeptides), which will be released from nerve terminals at the end of the axon, are made and packaged into secretory vesicles in the cell body. They then travel along the axon to the nerve terminals, which can be a meter or more away. As discussed in Chapter 16, motor proteins propel the vesicles along axonal microtubules, whose uniform orientation guides the vesicles in the proper direction. Microtubules also guide transport vesicles to the cell surface for constitutive exocytosis.

Whereas transport vesicles containing materials for constitutive release fuse with the plasma membrane once they arrive there, secretory vesicles in the regulated pathway wait at the membrane until the cell receives a signal for the vesicles to secrete their cargo. The signal can be an electrical nerve impulse (an action potential) or an extracellular signal molecule, such as a hormone. In either case, it leads to a transient increase in the concentration of free Ca^{2+} in the cytosol, which is the trigger for secretory vesicle fusion.

For Rapid Exocytosis, Synaptic Vesicles Are Primed at the Presynaptic Plasma Membrane

Nerve cells (and some endocrine cells) contain two types of secretory vesicles. As for all secretory cells, these cells package proteins and neuropeptides in dense-cored secretory vesicles in the standard way for release by the regulated secretory pathway. In addition, however, they use another specialized class of tiny (~50 nm diameter) secretory vesicles called **synaptic vesicles**. These vesicles store small *neurotransmitter molecules*, such as acetylcholine, glutamate, glycine, and γ-aminobutyric acid (GABA), which mediate rapid signaling from a nerve cell to its target cell at chemical synapses as we discussed in Chapter 11. When an action potential arrives at a nerve terminal, it causes an influx of Ca^{2+} through voltage-gated Ca^{2+} channels, which triggers the synaptic vesicles to fuse with the plasma membrane and release their contents to the extracellular space (see Figure 11–38). Some neurons fire more than 1000 times per second, releasing neurotransmitters each time.

The speed of transmitter release (taking only milliseconds) indicates that the proteins mediating the fusion reaction do not undergo complex, multistep

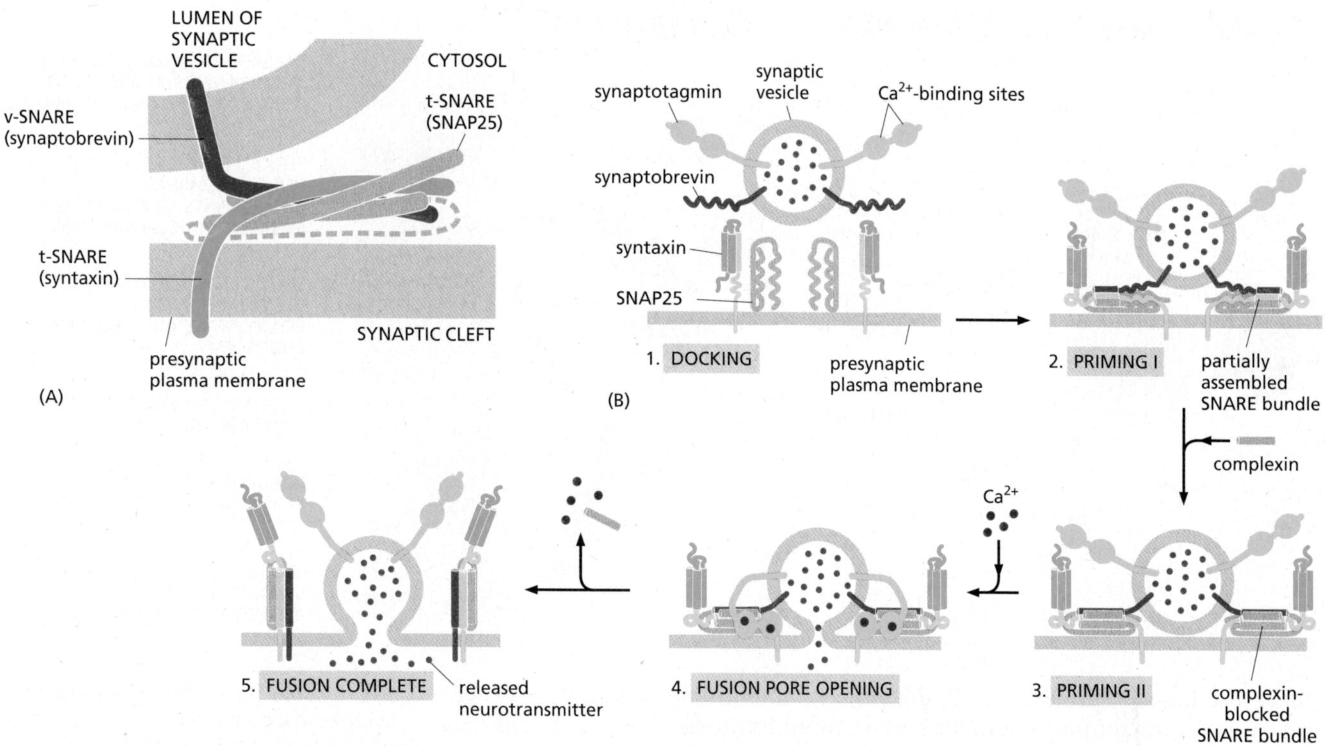

Figure 13–45 Exocytosis of synaptic vesicles. For orientation at a synapse, see Figure 11–38. (A) The trans-SNARE complex responsible for docking synaptic vesicles at the plasma membrane of nerve terminals consists of three proteins. The v-SNARE *synaptobrevin* and the t-SNARE *syntaxin* are both transmembrane proteins, and each contributes one α helix to the complex. By contrast to other SNAREs discussed earlier, the t-SNARE *SNAP25* is a peripheral membrane protein that contributes two α helices to the four-helix bundle; the two helices are connected by a loop *(dashed line)* that lies parallel to the membrane and has fatty acyl chains (not shown) attached to anchor it there. The four α helices are shown as rods for simplicity. (B) At the synapse, the basic SNARE machinery is modulated by the Ca^{2+} sensor *synaptotagmin* and an additional protein called *complexin*. Synaptic vesicles first dock at the membrane (step 1), and the SNARE bundle partially assembles (step 2), resulting in a "primed vesicle" that is already drawn close to the membrane. The SNARE bundle assembles further, but the additional binding of complexin prevents fusion (step 3). Upon arrival of an action potential, Ca^{2+} enters the cell and binds to synaptotagmin, which releases the block and opens a fusion pore (step 4). Further rearrangements complete the fusion reaction (step 5) and release the fusion machinery, which now can be reused. This elaborate arrangement allows the fusion machinery to respond on the millisecond time scale essential for rapid and repetitive synaptic signaling. (A, adapted from R.B. Sutton et al., *Nature* 395:347–353, 1998; B, adapted from J. Tang et al., *Cell* 126:1175–1187, 2006. With permission from Elsevier.)

rearrangements. Rather, after vesicles have been docked at the presynaptic plasma membrane, they undergo a priming step, which prepares them for rapid fusion. In the primed state, the SNAREs are partly paired but their helices are not fully wound into the final four-helix bundle required for fusion (**Figure 13–45**). Proteins called *complexins* freeze the SNARE complexes in this metastable state. The brake imposed by the complexins is released by another synaptic vesicle protein, synaptotagmin, which contains Ca^{2+}-binding domains. A rise in cytosolic Ca^{2+} triggers binding of synaptotagmin to the SNAREs, displacing the complexins. As the SNARE bundle zippers up completely, a fusion pore opens and the neurotransmitters are released. At a typical synapse, only a small number of the docked vesicles are primed and ready for exocytosis. The use of only a small fraction of primed vesicles at a time allows each synapse to fire over and over again in quick succession. With each firing, new synaptic vesicles dock and become primed to replace those that have fused and released their contents.

Synaptic Vesicles Can Be Recycled Locally After Exocytosis

For the nerve terminal to respond rapidly and repeatedly, synaptic vesicles need to be replenished very quickly after they discharge. This is achieved by local recycling of synaptic vesicles from the presynaptic plasma membrane in the

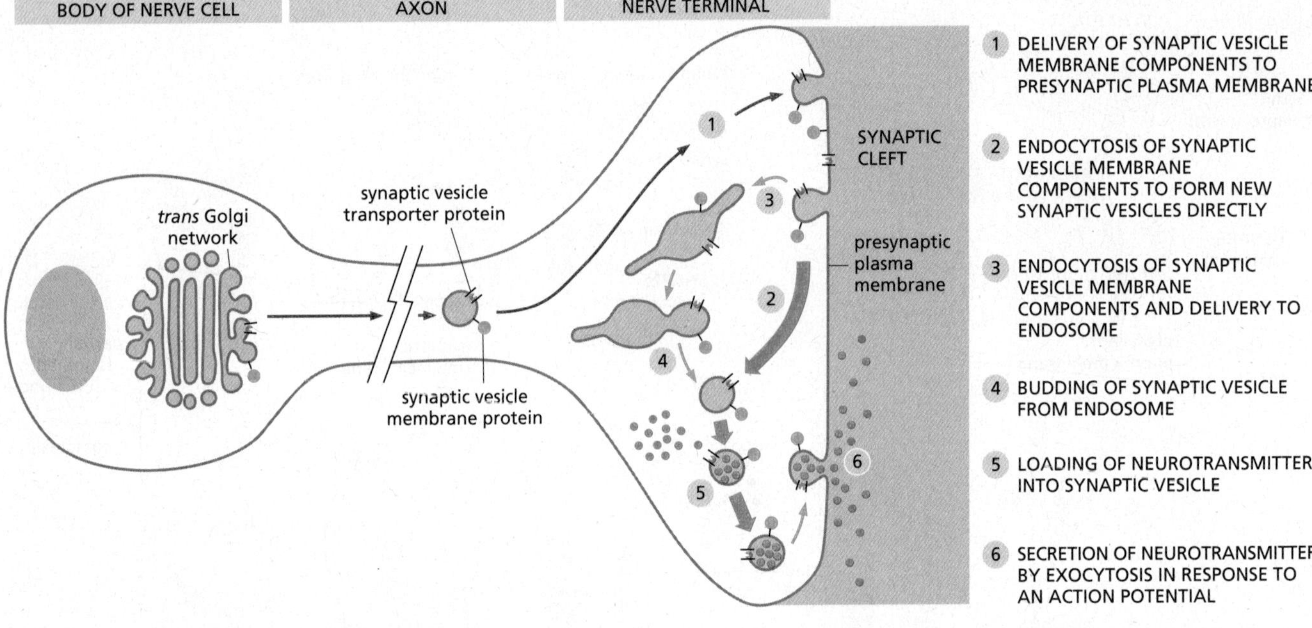

BODY OF NERVE CELL
AXON
NERVE TERMINAL
trans Golgi network
synaptic vesicle transporter protein
synaptic vesicle membrane protein
SYNAPTIC CLEFT
presynaptic plasma membrane
1 DELIVERY OF SYNAPTIC VESICLE MEMBRANE COMPONENTS TO PRESYNAPTIC PLASMA MEMBRANE
2 ENDOCYTOSIS OF SYNAPTIC VESICLE MEMBRANE COMPONENTS TO FORM NEW SYNAPTIC VESICLES DIRECTLY
3 ENDOCYTOSIS OF SYNAPTIC VESICLE MEMBRANE COMPONENTS AND DELIVERY TO ENDOSOME
4 BUDDING OF SYNAPTIC VESICLE FROM ENDOSOME
5 LOADING OF NEUROTRANSMITTER INTO SYNAPTIC VESICLE
6 SECRETION OF NEUROTRANSMITTER BY EXOCYTOSIS IN RESPONSE TO AN ACTION POTENTIAL

Figure 13–46 **The formation of synaptic vesicles in a nerve cell.** These tiny uniform vesicles are found only in nerve cells and in some endocrine cells, where they store and secrete small-molecule neurotransmitters. The import of neurotransmitter directly into the small endocytic vesicles that form from the plasma membrane is mediated by membrane transporters that function as antiports and are driven by an H^+ gradient maintained by V-type ATPase H^+ pumps in the vesicle membrane (discussed in Chapter 11).

nerve terminals (**Figure 13–46**). In this process, membrane components of synaptic vesicles are removed from the surface by endocytosis almost as fast as they are added by exocytosis. Similarly, newly made membrane components of the synaptic vesicles are initially delivered to the plasma membrane by the constitutive secretory pathway and then retrieved by endocytosis. The membrane components of a synaptic vesicle include transporters specialized for the uptake of neurotransmitter from the cytosol, where the small-molecule neurotransmitters that mediate fast synaptic signaling are synthesized (**Figure 13–47**). Most of the endocytic vesicles immediately fill with neurotransmitter to become synaptic vesicles. Once filled with neurotransmitter, the synaptic vesicles can be used again (see Figure 13–46).

Secretory Vesicle Membrane Components Are Quickly Removed from the Plasma Membrane

When a secretory vesicle fuses with the plasma membrane, its contents are discharged from the cell by exocytosis and its membrane becomes part of the plasma membrane. Although this should increase the surface area of the plasma membrane, it does so only transiently, because an equivalent amount of membrane is removed from the surface by endocytosis almost as fast as it is added by exocytosis, a process reminiscent of the endocytic–exocytic cycle discussed later. The proteins of the secretory vesicle membrane that are endocytosed from the plasma membrane are either recycled or shuttled to lysosomes for degradation through mechanisms discussed later. The amount of secretory vesicle membrane that is temporarily added to the plasma membrane can be enormous: in a pancreatic acinar cell discharging digestive enzymes for delivery to the gut lumen, about 900 μm^2 of vesicle membrane is inserted into the apical plasma membrane (whose area is only 30 μm^2) when the cell is stimulated to secrete.

Control of membrane traffic thus has a major role in maintaining the composition of the various membranes of the cell. To maintain each membrane-enclosed compartment in the secretory and endocytic pathways at a constant size, the balance between the outward and inward flows of membrane needs to be precisely regulated. For cells to grow, however, the forward flow needs to be greater than the retrograde flow, so that the membrane can increase in area. For cells to maintain a constant size, the forward and retrograde flows must be equal. We still know very little about the mechanisms that coordinate these flows.

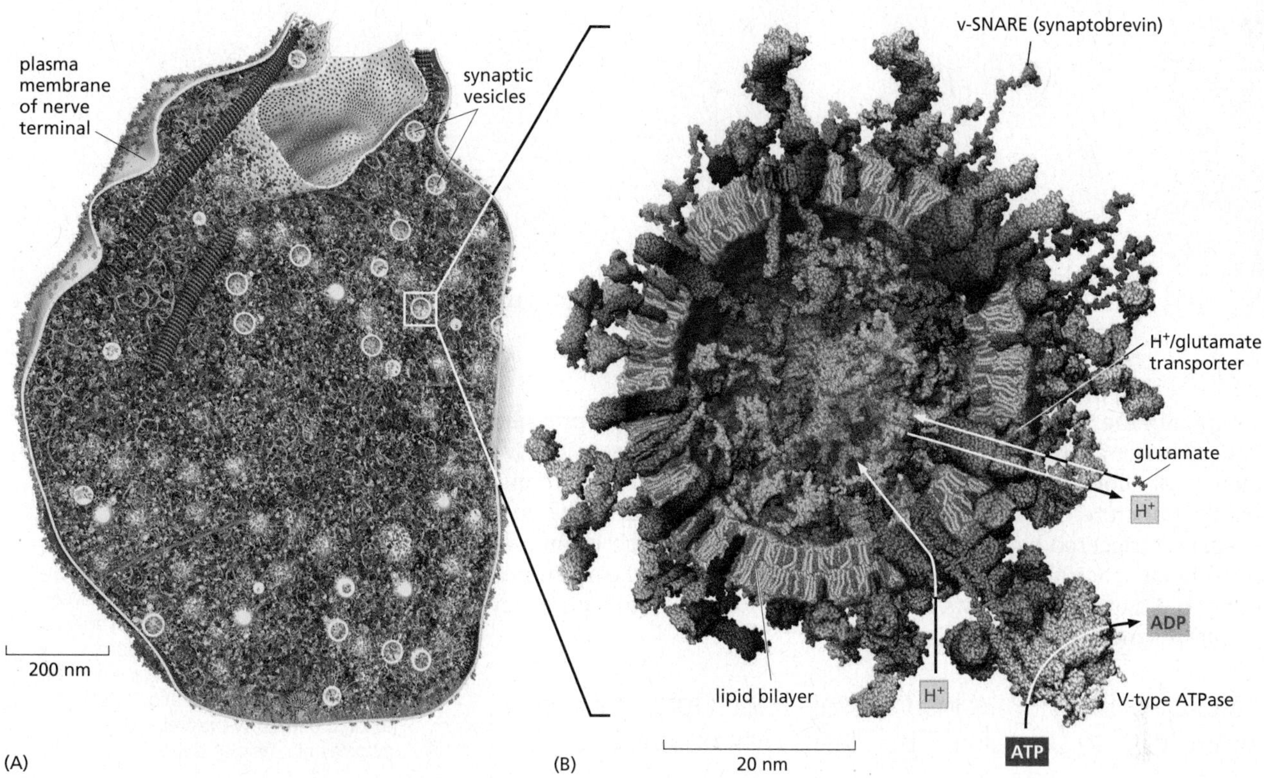

Figure 13–47 **Scale models of a brain presynaptic terminal and a synaptic vesicle.** The illustrations show sections through a presynaptic terminal (A) and a synaptic vesicle (B) in which proteins and lipids are drawn to scale on the basis of their known stoichiometry and either known or approximated structures. The relative localization of protein molecules in different regions of the presynaptic terminal was inferred from superresolution imaging and electron microscopy. The model in A contains 300,000 proteins of 60 different kinds that vary in abundance from 150 copies to 20,000 copies. In the model in B, only 70% of the membrane proteins present in the membrane are shown; a complete model would therefore show a membrane that is even more crowded than this picture suggests (**Movie 13.6**). Each synaptic vesicle membrane contains 7000 phospholipid molecules and 5700 cholesterol molecules. Each also contains close to 50 different integral membrane protein molecules, which vary widely in their relative abundance and together contribute about 600 transmembrane α helices. The transmembrane v-SNARE synaptobrevin is the most abundant protein in the vesicle (~70 copies per vesicle). By contrast, the V-type ATPase, which uses ATP hydrolysis to pump H^+ into the vesicle lumen, is present in 1–2 copies per vesicle. The H^+ gradient provides the energy for neurotransmitter import by an H^+/neurotransmitter antiport, which loads each vesicle with 1800 neurotransmitter molecules, such as glutamate, one of which is shown to scale. (A, from B.G. Wilhelm et al., *Science* 344:1023–1028, 2014. With permission from AAAS; B, adapted from S. Takamori et al., *Cell* 127:831–846, 2006. With permission from Elsevier.)

Some Regulated Exocytosis Events Serve to Enlarge the Plasma Membrane

An important task of regulated exocytosis is to deliver more membrane to enlarge the surface area of a cell's plasma membrane when such a need arises. A spectacular example is the plasma membrane expansion that occurs during the cellularization process in a fly embryo, which initially is a syncytium—a single cell containing about 6000 nuclei surrounded by a single plasma membrane (see Figure 21–14). Within tens of minutes, the embryo is converted into the same number of cells. This process of *cellularization* requires a vast amount of new plasma membrane, which is added by a carefully orchestrated fusion of cytoplasmic vesicles, eventually forming the plasma membranes that enclose the separate cells. Similar vesicle fusion events are required to enlarge the plasma membrane when other animal cells or plant cells divide during *cytokinesis* (discussed in Chapter 17).

Many animal cells, especially those subjected to mechanical stresses, frequently experience small ruptures in their plasma membrane. In a remarkable process thought to involve both homotypic vesicle–vesicle fusion and exocytosis, a temporary cell-surface patch is quickly fashioned from locally available

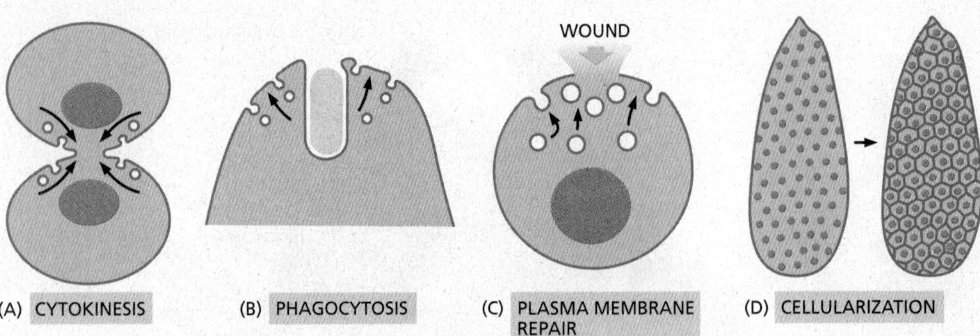

Figure 13–48 Four examples of regulated exocytosis leading to plasma membrane enlargement. The vesicles fusing with the plasma membrane during cytokinesis (A) (discussed in Chapter 17) and phagocytosis (B) (discussed later in this chapter) are thought to be derived from endosomes, whereas those involved in wound repair (C) may be derived from plasma membranes and lysosomes. The vast amount of new plasma membrane inserted during cellularization in a fly embryo occurs by the fusion of cytoplasmic vesicles (D).

internal-membrane sources, such as lysosomes. In addition to providing an emergency barrier against leaks, the patch reduces membrane tension over the wounded area, allowing the bilayer to flow back together to restore continuity and seal the puncture. The fusion and exocytosis of vesicles that mediated membrane repair is triggered by the sudden increase of Ca^{2+}, which is abundant in the extracellular space and rushes into the cell as soon as the plasma membrane is punctured. **Figure 13–48** shows four examples in which regulated exocytosis leads to plasma membrane expansion.

Polarized Cells Direct Proteins from the *Trans* Golgi Network to the Appropriate Domain of the Plasma Membrane

Most cells in tissues are *polarized*, with two or more molecularly and functionally distinct plasma membrane domains. This raises the general problem of how the delivery of membrane from the Golgi apparatus is organized so as to maintain the differences between one cell-surface domain and another. A typical epithelial cell, for example, has an *apical domain*, which faces either an internal cavity or the outside world and often has specialized features such as cilia or a brush border of microvilli. It also has a *basolateral domain*, which covers the rest of the cell. The two domains are separated by a ring of *tight junctions* (see Figure 19–20), which prevents proteins and lipids from diffusing between the two domains, so that the differences between the two domains are maintained.

Different subsets of proteins are secreted from the apical and basolateral surfaces of the cell. Epithelial cells lining the gut, for example, secrete digestive enzymes and mucus at their apical surface and components of the basal lamina at their basolateral surface. Such cells must have ways of directing vesicles carrying different cargoes to different plasma membrane domains. Proteins destined for different domains travel together from the ER until they reach the TGN, where they are separated and dispatched in secretory or transport vesicles to the appropriate plasma membrane domain (**Figure 13–49**). These routes are known as the *direct pathways* for polarized secretion because cargo destined for the apical and basolateral domains is delivered there directly.

The apical plasma membrane of most epithelial cells is greatly enriched in glycosphingolipids, which help protect this exposed surface from damage; for example, from the digestive enzymes and low pH in sites such as the gut or stomach, respectively. Similarly, plasma membrane proteins that are linked to the lipid bilayer by a glycosylphosphatidylinositol (GPI) anchor (see Figure 12–30) are found predominantly in the apical plasma membrane. If recombinant DNA techniques are used to attach a GPI anchor to a protein that would normally be delivered to the basolateral surface, the protein is delivered to the apical surface instead. GPI-anchored proteins are thought to be directed to the apical membrane because they associate with glycosphingolipids in lipid rafts that form in the membrane of the TGN. As discussed in Chapter 10, lipid rafts form in the TGN and plasma membrane when glycosphingolipids and cholesterol molecules self-associate (see Figure 10–13). Having selected a unique set of cargo molecules, the rafts then bud from the TGN into transport vesicles destined for the apical plasma membrane.

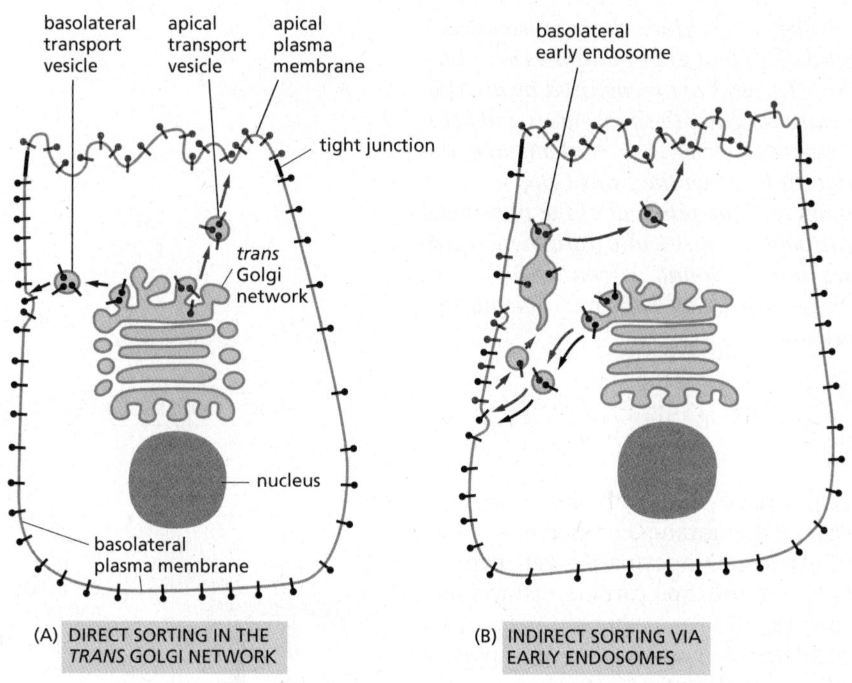

Figure 13–49 Two ways of sorting plasma membrane proteins in a polarized epithelial cell. (A) In the direct pathway, proteins destined for different plasma membrane domains are sorted and packaged into different transport vesicles. The lipid raft–dependent delivery system to the apical domain described in the text is an example of the direct pathway. (B) In the indirect pathway, a protein is retrieved from the inappropriate plasma membrane domain by endocytosis and then transported to the correct domain via early endosomes; that is, by transcytosis. The indirect pathway, for example, is used in liver hepatocytes to deliver proteins to the apical domain that lines bile ducts.

While secretory and GPI-anchored proteins rely on the direct pathways, membrane proteins can sometimes use an indirect route to arrive at the appropriate membrane surface (see Figure 13–49B). In this route, both apical and basolateral cargo travel together in transport vesicles from the TGN to the basolateral membrane. Membrane proteins that do not belong in that region of the plasma membrane are retrieved by endocytosis and are transported via early endosomes to the correct region. Membrane proteins destined for delivery to the basolateral membrane contain sorting signals in their cytosolic tail. When present in an appropriate structural context, these signals are recognized by coat proteins that package them into appropriate transport vesicles in the TGN. The same basolateral signals that are recognized in the TGN also function in early endosomes to redirect the proteins back to the basolateral plasma membrane after they have been endocytosed. A combination of direct and indirect deliveries ensures that the apical and basolateral membranes retain their distinct identities.

Summary

Transport vesicles departing the TGN carry their contents to one of two major destinations: the plasma membrane for exocytosis or endosomes for eventual delivery to lysosomes. Vesicle transport from the TGN to the plasma membrane is further divided into a constitutive pathway or regulated pathways. Proteins follow the constitutive pathway unless they are diverted into other pathways or retained in the Golgi apparatus. In polarized cells, the transport pathways from the TGN to the plasma membrane operate selectively to ensure that different sets of membrane proteins, secreted proteins, and lipids are delivered to the different domains of the plasma membrane.

The regulated pathways operate only in specialized secretory cells and neurons. The molecules for regulated secretion are stored either in secretory vesicles or in synaptic vesicles, which do not fuse with the plasma membrane to release their contents until they receive an appropriate signal. Secretory vesicles containing proteins for secretion bud from the TGN. The secretory proteins become concentrated during the formation and maturation of the secretory vesicles. Synaptic vesicles, which are confined to nerve cells and some endocrine cells, form from both endocytic vesicles and from endosomes, and they mediate the regulated secretion of small-molecule neurotransmitters at the axon terminals of nerve cells.

Newly synthesized lysosomal proteins are carried from the TGN to endosomes by means of clathrin-coated transport vesicles before moving on to lysosomes.

The lysosomal hydrolases contain N-linked oligosaccharides that are covalently modified in a unique way in the cis Golgi network so that their mannoses are phosphorylated. These mannose 6-phosphate (M6P) groups are recognized by an M6P receptor protein in the trans Golgi network that segregates the hydrolases and helps package them into budding transport vesicles that deliver their contents to endosomes. The M6P receptors shuttle back and forth between the trans Golgi network and the endosomes. The low pH in endosomes and the removal of the phosphate from the M6P group cause the lysosomal hydrolases to dissociate from these receptors, making the transport of the hydrolases unidirectional. A separate transport system uses clathrin-coated vesicles to deliver resident lysosomal membrane proteins from the trans Golgi network to endosomes.

TRANSPORT INTO THE CELL FROM THE PLASMA MEMBRANE: ENDOCYTOSIS

The routes that lead inward from the cell surface start with the process of **endocytosis**, by which cells take up plasma membrane components, fluid, solutes, macromolecules, and particulate substances. Endocytosed cargo includes receptor–ligand complexes, a spectrum of nutrients and their carriers, extracellular matrix components, cell debris, bacteria, viruses, and, in specialized cases, even other cells. Through endocytosis, the cell regulates the composition of its plasma membrane in response to changing extracellular conditions.

In endocytosis, the material to be ingested is progressively enclosed by a small portion of the plasma membrane, which first invaginates and then pinches off to form an **endocytic vesicle** containing the ingested substance or particle. Most eukaryotic cells constantly form endocytic vesicles, a process called *pinocytosis* ("cell drinking"); in addition, some specialized cells contain dedicated pathways that take up large particles on demand, a process called *phagocytosis* ("cell eating"). Endocytic vesicles form at the plasma membrane by multiple mechanisms that differ in both the molecular machinery used and how that machinery is regulated.

Once generated at the plasma membrane, most endocytic vesicles fuse with a common receiving compartment, the *early endosome*, where internalized cargo is sorted: some cargo molecules are returned to the plasma membrane, either directly or via a *recycling endosome*, and others remain as the early endosome changes into a *late endosome* by a process termed *endosome maturation* (**Figure 13–50**). This conversion process changes the protein composition of the endosome membrane, patches of which invaginate and become incorporated within the organelles as *intralumenal vesicles*, while the endosome itself moves from the cell periphery to a location close to the nucleus. As an endosome matures, it ceases to recycle material to the plasma membrane. Instead, late

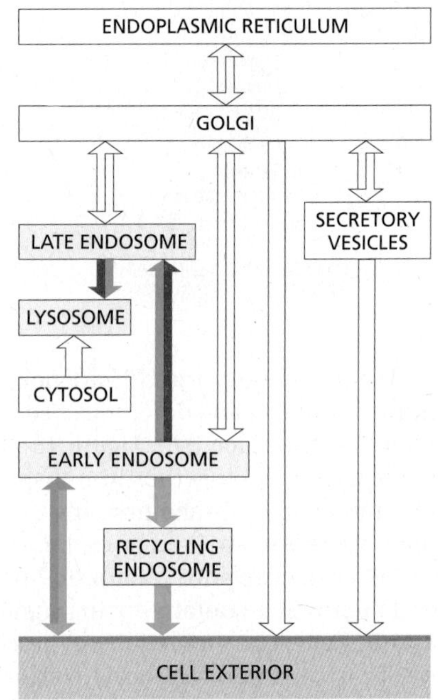

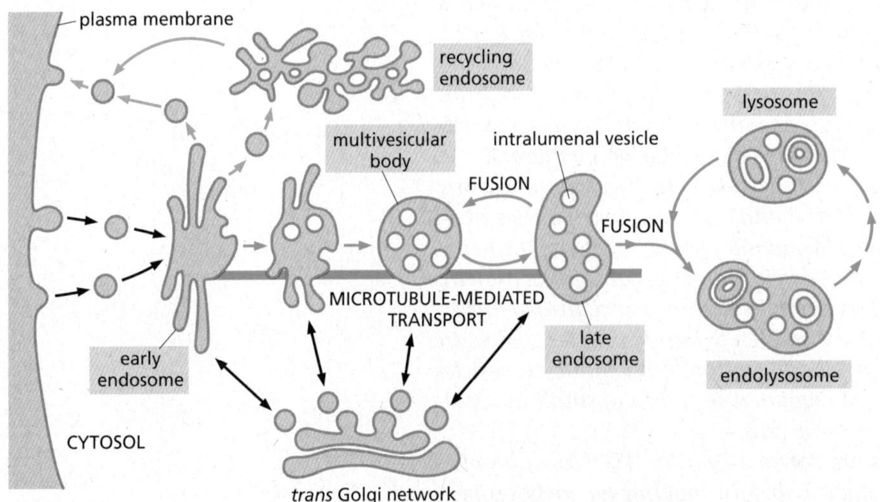

Figure 13–50 **Endosome maturation: the endocytic pathway from the plasma membrane to lysosomes.** Near the cell periphery, endocytic vesicles fuse with an early endosome, which is the primary sorting station. Tubular portions of the early endosome bud off vesicles that recycle endocytosed cargo back to the plasma membrane—either directly or indirectly via recycling endosomes. Recycling endosomes can store proteins until they are needed. Conversion of early endosomes to late endosomes is accompanied by loss of the tubular projections. Membrane proteins destined for degradation are internalized in intralumenal vesicles. The developing late endosome, or multivesicular body, moves on microtubules to the cell interior. Fully matured late endosomes no longer send vesicles to the plasma membrane, and they fuse with one another and with endolysosomes and lysosomes to degrade their contents. Each stage of endosome maturation is connected to the TGN via transport vesicles, providing a continual supply of newly synthesized lysosomal proteins.

endosomes fuse with one another and with lysosomes to form endolysosomes, which degrade their contents.

Each of the stages of endosome maturation—from the early endosome to the endolysosome—is connected to the TGN through bidirectional vesicle transport pathways. These pathways allow insertion of newly synthesized materials, such as lysosomal enzymes arriving from the ER, and the retrieval of components, such as the M6P receptor, back into the early parts of the secretory pathway. We later discuss how the cell uses and controls the various features of endocytic trafficking.

Pinocytic Vesicles Form from Coated Pits in the Plasma Membrane

Virtually all eukaryotic cells continually ingest portions of their plasma membrane in the form of small pinocytic (endocytic) vesicles. The rate at which plasma membrane is internalized in this process of **pinocytosis** varies between cell types, but it is usually surprisingly high. A macrophage, for example, ingests 25% of its own volume of fluid each hour. This means it must ingest 3% of its plasma membrane each minute, or 100% in about half an hour. Fibroblasts endocytose at a somewhat lower rate (1% of their plasma membrane per minute), whereas some amoebae ingest their plasma membrane even more rapidly. Because a cell's surface area and volume remain unchanged during this process, it is clear that the same amount of membrane being removed by endocytosis is being added to the cell surface by the converse process of *exocytosis*. In this sense, endocytosis and exocytosis are linked processes that can be considered to constitute an *endocytic–exocytic cycle*. The coupling between exocytosis and endocytosis is particularly strict in specialized structures characterized by high membrane turnover, such as a nerve terminal.

The endocytic part of the cycle often begins at **clathrin-coated pits**. These specialized regions typically occupy about 2% of the total plasma membrane area. The lifetime of a clathrin-coated pit is short: within a minute or so of being formed, it invaginates into the cell and pinches off to form a clathrin-coated vesicle (**Figure 13–51**). About 2500 clathrin-coated vesicles pinch off from the plasma membrane of a cultured fibroblast every minute. The coated vesicles are even more transient than the coated pits: within seconds of being formed, they shed their coat and fuse with early endosomes.

Not All Membrane Invaginations and Pinocytic Vesicles Are Clathrin Coated

In addition to clathrin-coated pits and vesicles, cells can form other types of pinocytic vesicles and membrane invaginations. Most of these clathrin-independent membrane invaginations are poorly understood, and the molecules that mediate

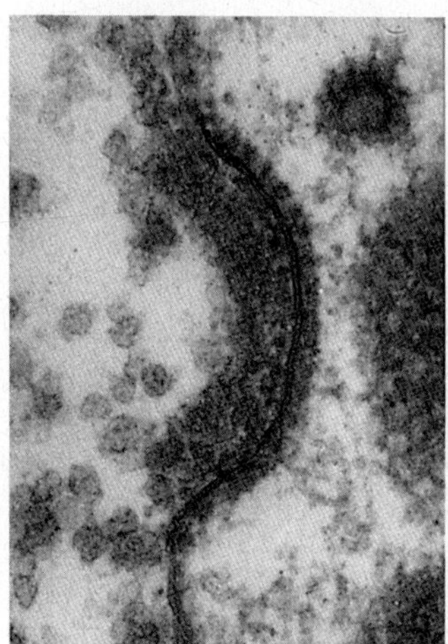
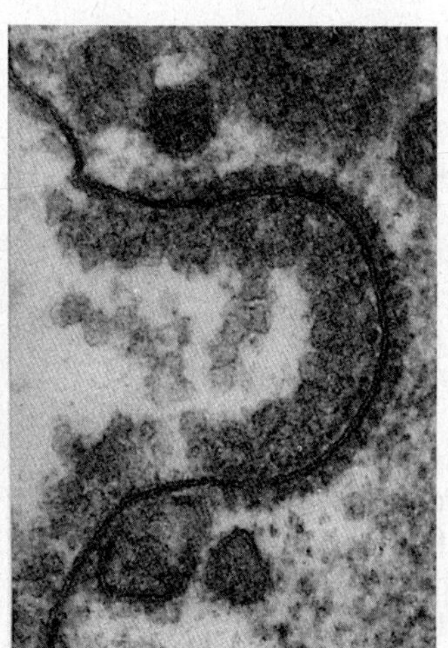
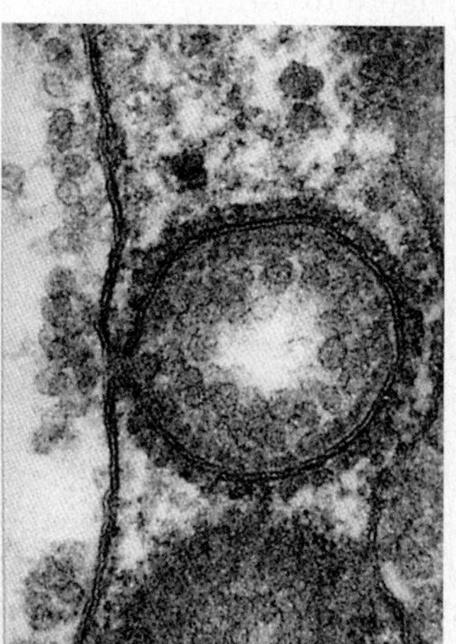
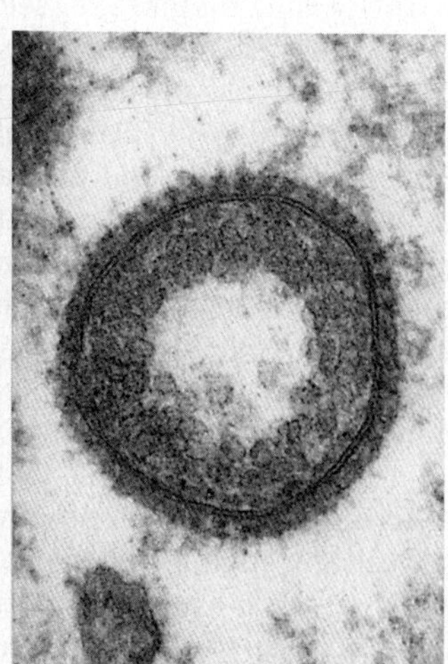

Figure 13–51 **The formation of clathrin-coated vesicles from the plasma membrane.** These electron micrographs illustrate the probable sequence of events in the formation of a clathrin-coated vesicle from a clathrin-coated pit. The clathrin-coated pits and vesicles shown are larger than those seen in normal-sized cells; they are from a very large hen oocyte, and they take up lipoprotein particles to form yolk. The lipoprotein particles bound to their membrane-bound receptors appear as a dense, fuzzy layer on the extracellular surface of the plasma membrane—which is the inside surface of the coated pit and vesicle. (Courtesy of M.M. Perry and A.B. Gilbert, *J. Cell Sci.* 39:257–272, 1979. With permission from the Company of Biologists.)

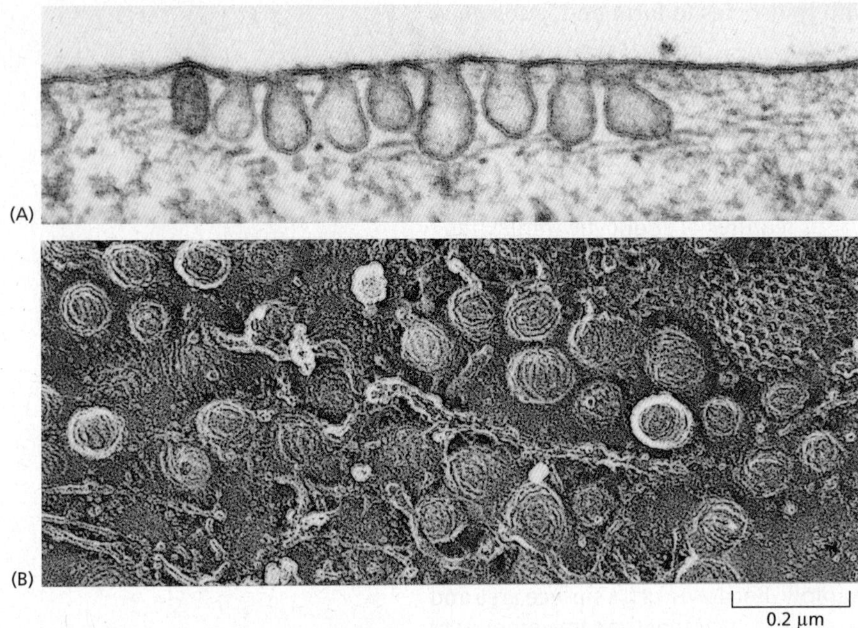

Figure 13–52 **Caveolae in the plasma membrane of a fibroblast.** (A) This electron micrograph shows a plasma membrane with a very high density of caveolae. (B) This rapid-freeze deep-etch image demonstrates the characteristic "cauliflower" texture of the cytosolic face of the caveolae membrane. The characteristic texture is thought to result from aggregates of caveolins and cavins. A clathrin-coated pit is also seen at the upper right. (From K.G. Rothberg et al., *Cell* 68:673–682, 1992. With permission from Elsevier.)

membrane bending and vesicle formation are often not defined fully. The best-understood clathrin-independent invaginations are called **caveolae** (from the Latin for "little cavities"), originally observed as a prominent feature of the endothelial cells that form the inner lining of blood vessels.

Caveolae, sometimes seen in the electron microscope as deeply invaginated flasks, are present in the plasma membrane of most vertebrate cell types (**Figure 13–52**). The major structural proteins in caveolae are **caveolins**, a family of unusual integral membrane proteins that each insert a hydrophobic loop into the membrane from the cytosolic side but do not extend across the membrane. On their cytosolic side, caveolins are bound to large protein complexes of cavin proteins, which are thought to stabilize the membrane curvature. Caveolae are especially rich in cholesterol, glycosphingolipids, and glycosylphosphatidylinositol (GPI)-anchored membrane proteins and might represent a type of *lipid raft* in the plasma membrane (see Figure 10–13).

In contrast to clathrin-coated and COPI-coated or COPII-coated vesicles, caveolae are usually static structures that can serve as a reservoir of additional plasma membrane. It is thought that cells subjected to dynamic changes in shear forces, such as the endothelial cells of arteries, exploit this reservoir to provide their plasma membranes greater resilience to stretch. This is accomplished by the rapid disassembly of the cavin protein scaffold in response to mechanical force, thereby allowing the underlying membrane to temporarily increase the surface area of the cell. The ability to rapidly change membrane surface area is thought to be important for accommodating dynamic changes in blood flow to different parts of the brain.

Two other endocytosis pathways are known, neither of which uses clathrin. **Macropinocytosis** is a process whereby the plasma membrane protrudes from the cell and engulfs a portion of the surrounding extracellular fluid into a *macropinosome*. This is a nonselective process for bringing fluid into the cell under certain conditions. In *phagocytosis*, the plasma membrane is directed to wrap around the particle to be engulfed until it fuses with itself, resulting in an enclosed *phagosome* inside the cell. Both processes utilize actin polymerization underneath the plasma membrane to mediate the large-scale membrane deformations required to engulf a large particle or a high volume of fluid. Both macropinosomes and phagosomes are destined to fuse with lysosomes so their internal contents can be degraded. These routes of degradation will be discussed later when we consider the function of lysosomes.

Cells Use Receptor-mediated Endocytosis to Import Selected Extracellular Macromolecules

In most animal cells, clathrin-coated pits and vesicles provide an efficient pathway for taking up specific macromolecules from the extracellular fluid. In this process, called **receptor-mediated endocytosis**, the macromolecules bind to complementary transmembrane receptor proteins, which accumulate in coated pits, and then enter the cell as receptor–macromolecule complexes in clathrin-coated vesicles (see Figure 13–51). Because ligands are selectively captured by receptors, receptor-mediated endocytosis provides a selective concentrating mechanism that increases the efficiency of internalization of particular ligands more than a hundredfold. In this way, even minor components of the extracellular fluid can be efficiently taken up in large amounts. A particularly well-understood and physiologically important example is the process that mammalian cells use to import cholesterol.

Many animal cells take up cholesterol through receptor-mediated endocytosis and, in this way, acquire most of the cholesterol they require to make new membrane. If the uptake is blocked, cholesterol accumulates in the blood and can contribute to the formation in blood vessel (artery) walls of *atherosclerotic plaques*, deposits of lipid and fibrous tissue that can cause strokes and heart attacks by blocking arterial blood flow. In fact, it was a study of humans with a strong genetic predisposition for *atherosclerosis* that first revealed the mechanism of receptor-mediated endocytosis.

Most cholesterol is transported in the blood as cholesterol esters in the form of lipid–protein particles known as **low-density lipoproteins (LDLs)** that, architecturally, resemble lipid droplets bearing a core of triacylglycerol, free cholesterol, and cholesterol esters. The droplet is stabilized by a single molecule of apolipoprotein B, a very large protein that wraps around the LDL particle (**Figure 13–53**). When a cell needs cholesterol for membrane synthesis, it makes transmembrane receptor proteins for LDL at the ER and transports them to the plasma membrane. Once in the plasma membrane, the *LDL receptor* diffuses until an endocytosis signal in its cytoplasmic tail binds the adaptor protein AP2 after AP2's conformation has been locally unlocked by binding to $PI(4,5)P_2$ on the plasma membrane. This

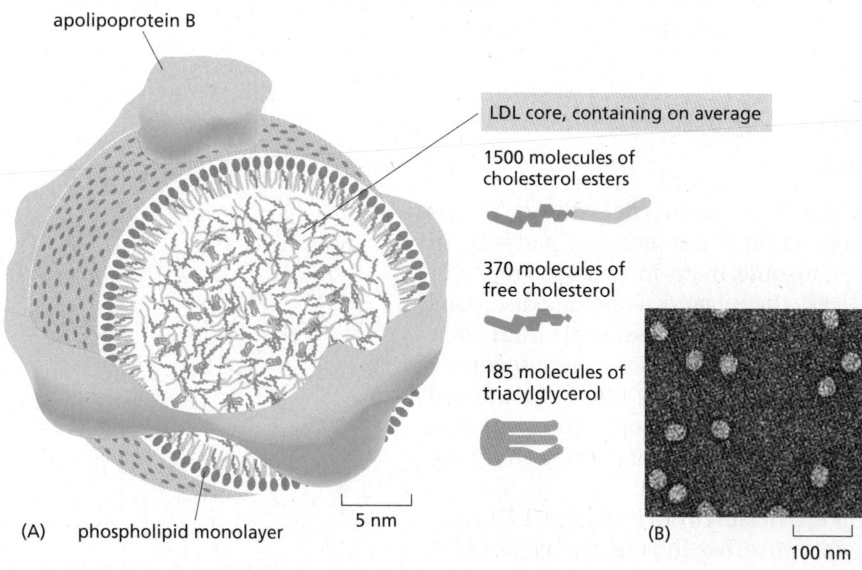

Figure 13–53 A low-density lipoprotein (LDL) particle. (A) Each roughly spherical particle has a mass of 3×10^6 daltons. It contains a core of about 1500 cholesterol molecules esterified to long-chain fatty acids and smaller amounts of free cholesterol and triacylglycerol molecules. A lipid monolayer composed of about 800 phospholipid and 500 unesterified cholesterol molecules surrounds the core of cholesterol esters. A single molecule of apolipoprotein B, a 500,000-dalton beltlike protein, organizes the particle and mediates the specific binding of LDL to cell-surface LDL receptors. (B) Purified LDL particles seen by negative staining in the electron microscope. (B, from L. Zhang et al., *J. Lipid Res.* 52:175–184, 2011. With permission from Elsevier.)

two-step mechanism of AP2 binding to an endocytosis signal (see Figure 13–9) imparts both efficiency and selectivity to the process. AP2 then recruits clathrin to initiate endocytosis.

Because coated pits constantly pinch off to form coated vesicles, any LDL particles bound to LDL receptors in the coated pits are rapidly internalized in coated vesicles. After shedding their clathrin coats, the vesicles deliver their contents to early endosomes. Once the LDL and LDL receptors encounter the low pH in early endosomes, LDL is released from its receptor and is delivered via late endosomes to lysosomes. There, the cholesterol esters in the LDL particles are hydrolyzed to free cholesterol, which is now available to the cell for new membrane synthesis (**Movie 13.7**). If too much free cholesterol accumulates in a cell, the cell simultaneously shuts off endogenous cholesterol synthesis (Figure 12–64) and reduces exogenous cholesterol intake by shutting off the synthesis of LDL receptors.

This regulated pathway for cholesterol uptake is disrupted in individuals who inherit defective genes encoding LDL receptors. The resulting high levels of blood cholesterol predispose these individuals to develop atherosclerosis prematurely, and many would die at an early age of heart attacks resulting from coronary artery disease if they were not treated with drugs such as statins that lower the level of blood cholesterol. In some cases, the receptor is lacking altogether. In others, the receptors are defective—in either the extracellular binding site for LDL or the intracellular binding site for the AP2 adaptor protein in clathrin-coated pits. In the latter case, normal numbers of LDL receptors are present, but they fail to become localized in clathrin-coated pits. Although LDL binds to the surface of these mutant cells, it is not internalized, directly demonstrating the importance of clathrin-coated pits for the receptor-mediated endocytosis of cholesterol.

More than 25 distinct receptors are known to participate in receptor-mediated endocytosis of different types of molecules. They all apparently use clathrin-dependent internalization routes and are guided into clathrin-coated pits by signals in their cytoplasmic tails that bind to adaptor proteins in the clathrin coat. Many of these receptors, like the LDL receptor, enter coated pits irrespective of whether they have bound their specific ligands. Others enter preferentially when bound to a specific ligand, suggesting that a ligand-induced conformational change is required for them to activate the signal sequence that guides them into the pits. Because most plasma membrane proteins fail to become concentrated in clathrin-coated pits, the pits serve as molecular filters, preferentially collecting certain plasma membrane proteins (receptors) over others.

Electron microscopy of cultured cells exposed simultaneously to different labeled ligands demonstrates that many kinds of receptors can cluster in the same clathrin-coated pit, whereas some other receptors cluster in different clathrin-coated pits. The plasma membrane of one clathrin-coated pit can accommodate more than 100 receptors of assorted varieties.

Specific Proteins Are Retrieved from Early Endosomes and Returned to the Plasma Membrane

Early endosomes are the main sorting stations in the endocytic pathway, just as the *cis* and *trans* Golgi networks serve this function in the secretory pathway. In the mildly acidic environment of the early endosome, many internalized receptor proteins change their conformation and release their ligand, as already discussed for the M6P receptors. Those endocytosed ligands that dissociate from their receptors in the early endosome are usually destined for delivery to lysosomes, where they are either degraded and recycled into building blocks or utilized directly by the cell (such as the cholesterol just discussed). Some other endocytosed ligands, however, remain bound to their receptors and thereby share the fate of the receptors.

In the early endosome, the LDL receptor dissociates from its ligand, LDL, and is recycled back to the plasma membrane for reuse, leaving the discharged LDL

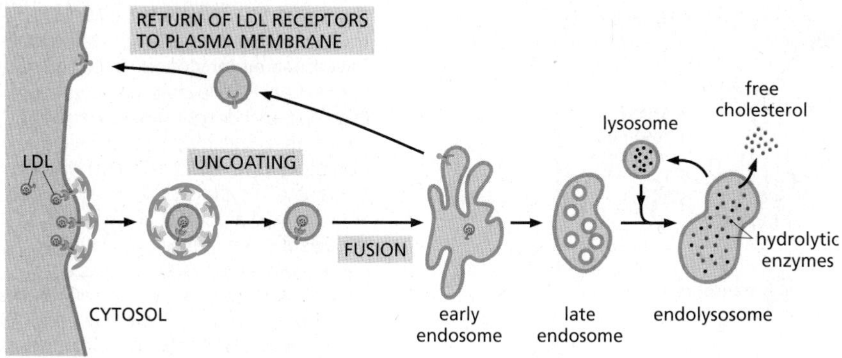

Figure 13–54 The receptor-mediated endocytosis of LDL. Note that the LDL dissociates from its receptors in the acidic environment of the early endosome. After a number of steps, the LDL ends up in endolysosomes and lysosomes, where it is degraded to release free cholesterol. In contrast, the LDL receptors are returned to the plasma membrane via transport vesicles that bud off from the tubular region of the early endosome, as shown. For simplicity, only one LDL receptor is shown entering the cell and returning to the plasma membrane. Whether it is occupied or not, an LDL receptor typically makes one round trip into the cell and back to the plasma membrane every 10 minutes, making up to several hundred trips in its 20-hour life span.

to be carried to lysosomes (Figure 13–54). The recycling transport vesicles bud from long, narrow tubules that extend from the early endosomes (Figure 13–55). It is likely that the geometry of these tubules helps the sorting process: because tubules have a large membrane area enclosing a small volume, membrane proteins become enriched over soluble proteins. The transport vesicles return the LDL receptor directly to the plasma membrane.

The **transferrin receptor** follows a similar recycling pathway as the LDL receptor, but unlike the LDL receptor it also recycles its ligand. Transferrin is a soluble protein that carries iron in the blood. Cell-surface transferrin receptors deliver transferrin with its bound iron to early endosomes by receptor-mediated endocytosis. The low pH in the endosome induces transferrin to release its bound iron, but the iron-free transferrin itself (called apotransferrin) remains bound to its receptor. The receptor–apotransferrin complex enters the tubular extensions of the early endosome and from there is recycled back to the plasma membrane. When the apotransferrin returns to the neutral pH of the extracellular fluid, it dissociates from the receptor and is thereby freed to pick up more iron and begin the cycle again. Thus, transferrin shuttles back and forth between the extracellular fluid and early endosomes, avoiding lysosomes and delivering iron to the cell interior, as needed for cells to grow and proliferate.

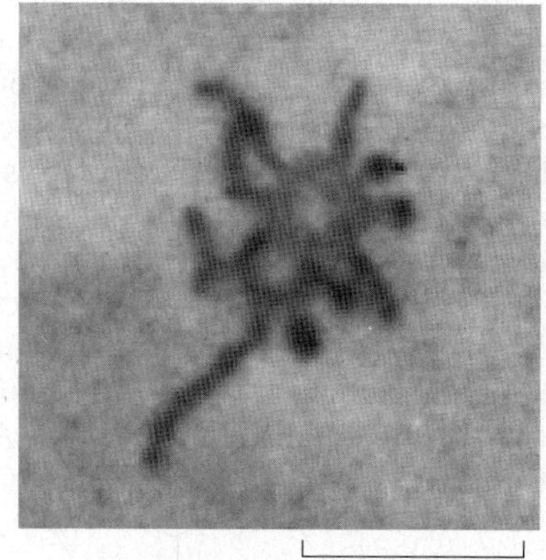

Figure 13–55 Electron micrograph of an early endosome. The endosomal compartments can be made visible in the electron microscope by adding a readily detectable tracer molecule, such as the enzyme peroxidase, to the extracellular medium and allowing the cell to endocytose the tracer. Within a minute or so after adding the tracer, it starts to appear in early endosomes, just beneath the plasma membrane. The image shows an electron-dense reaction product of peroxidase in the early endosome that has been visualized in the electron microscope. Many tubular extensions protrude from the central vacuolar space of the early endosome, which will later mature to give rise to a late endosome. (© 1992 J. Tooze and M. Hollinshead. Originally published in *J. Cell Biol.* 118:813–830, 1992, doi.org/10.1083/jcb.118.4.813. With permission from Rockefeller University Press.)

Recycling Endosomes Regulate Plasma Membrane Composition

The fates of endocytosed receptors—and of any ligands remaining bound to them—vary according to the specific type of receptor. As we discussed, most receptors are recycled and returned to the same plasma membrane domain from which they came; some proceed to a different domain of the plasma membrane, thereby mediating **transcytosis**; and some remain in the endosomal system and progress to lysosomes, where they are degraded, as we discuss next.

Receptors on the surface of polarized epithelial cells can transfer specific macromolecules from one extracellular space to another by transcytosis. A newborn, for example, obtains antibodies from its mother's milk (which help protect it against infection) by transporting them across the epithelium of its gut. The lumen of the gut is acidic, and, at this low pH, the antibodies in the milk bind to specific receptors on the apical (absorptive) surface of the gut epithelial cells. The receptor–antibody complexes are internalized via clathrin-coated pits and vesicles and are delivered to early endosomes. The complexes remain intact and are retrieved in transport vesicles that bud from the early endosome and subsequently fuse with the basolateral domain of the plasma membrane. On exposure to the neutral pH of the extracellular fluid that bathes the basolateral surface of the cells, the antibodies dissociate from their receptors and eventually enter the baby's bloodstream.

The transcytotic pathway from the early endosome back to the plasma membrane is not direct. The receptors first move from the early endosome to the **recycling endosome**. The variety of pathways that different receptors follow from early endosomes implies that, in addition to binding sites for their ligands and

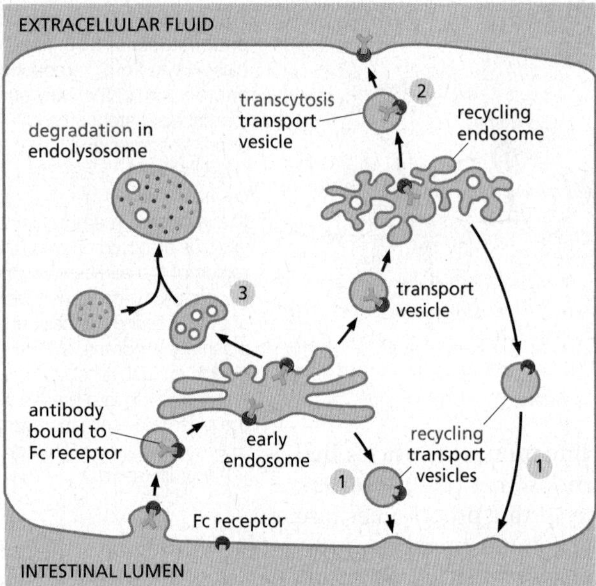

Figure 13–56 **Possible fates for transmembrane receptor proteins that have been endocytosed.** Three pathways from the early endosomal compartment in an epithelial cell are shown. Retrieved receptors are returned (1) to the same plasma membrane domain from which they came (*recycling*) or (2) via a recycling endosome to a different domain of the plasma membrane (*transcytosis*). (3) Receptors that are not specifically retrieved from early or recycling endosomes follow the pathway from the endosomal compartment to lysosomes, where they are degraded (*degradation*). If the ligand that is endocytosed with its receptor stays bound to the receptor in the acidic environment of the endosome, it shares the same fate as the receptor; otherwise, it is delivered to lysosomes. Recycling endosomes are a way station on the transcytotic pathway. In the transcytosis example shown here, an antibody Fc receptor on a gut epithelial cell binds antibody and is endocytosed, eventually carrying the antibody to the basolateral plasma membrane. The receptor is called an Fc receptor because it binds the Fc part of the antibody (discussed in Chapter 24).

binding sites for coated pits, many receptors also possess sorting signals that guide them into the appropriate transport pathway (**Figure 13–56**).

Cells can regulate the release of membrane proteins from recycling endosomes, thus adjusting the flux of proteins through the transcytotic pathway according to need. This regulation, the mechanism of which is uncertain, allows recycling endosomes to play an important part in adjusting the concentration of specific plasma membrane proteins. Fat cells and muscle cells, for example, contain large intracellular pools of the glucose transporters that are responsible for the uptake of glucose across the plasma membrane. These membrane transport proteins are stored in specialized recycling endosomes until the hormone *insulin* stimulates the cell to increase its rate of glucose uptake. In response to the insulin signal, transport vesicles rapidly bud from the recycling endosome and deliver large numbers of glucose transporters to the plasma membrane, thereby greatly increasing the rate of glucose uptake into the cell (**Figure 13–57**). Similarly, kidney cells regulate the insertion of aquaporins and V-type ATPase into the plasma membrane to increase water reabsorption and acid excretion, respectively, both in response to hormones.

Plasma Membrane Signaling Receptors Are Down-regulated by Degradation in Lysosomes

The final potential fate for endocytosed receptors is to remain in the endosome as it matures and eventually fuses with lysosomes, where the receptor is degraded. Many signaling receptors, including opioid receptors and the receptor that binds

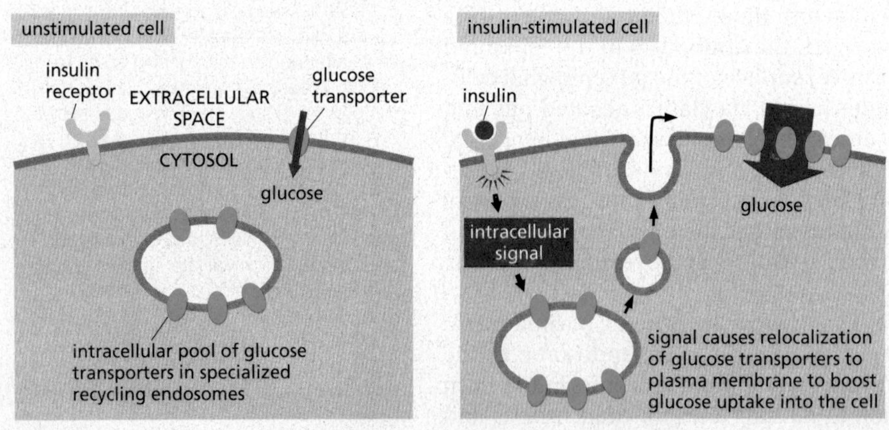

Figure 13–57 **Storage of plasma membrane proteins in recycling endosomes.** Recycling endosomes can serve as an intracellular storage site for specialized plasma membrane proteins that can be mobilized when needed. In the example shown, insulin binding to the insulin receptor triggers an intracellular signaling pathway that causes the rapid insertion of glucose transporters into the plasma membrane of a fat or muscle cell, greatly increasing its glucose intake.

epidermal growth factor (*EGF*), follow this route. EGF is a small, extracellular signal protein that stimulates epidermal and various other cells to divide. Unlike LDL receptors, EGF receptors accumulate in clathrin-coated pits only after binding their ligand, and most do not recycle but are degraded in lysosomes, along with the ingested EGF. EGF binding therefore first activates intracellular signaling pathways and then leads to a decrease in the concentration of EGF receptors on the cell surface, a process called *receptor down-regulation*, that reduces the cell's subsequent sensitivity to EGF (see Figure 15–21).

Receptor down-regulation is highly regulated. The activated receptors are first covalently modified on the cytosolic face with the small protein ubiquitin. Unlike *polyubiquitylation*, which adds a chain of ubiquitins that typically targets a protein for degradation in proteasomes (discussed in Chapter 6), ubiquitin tagging for sorting into the clathrin-dependent endocytic pathway adds just one or a few single ubiquitin molecules to the protein—a process called *monoubiquitylation* or *multiubiquitylation*, respectively. Ubiquitin-binding proteins recognize the attached ubiquitin and help direct the modified receptors into clathrin-coated pits. The ubiquitylated receptor does not get recycled back to the plasma membrane from the early endosome. Instead, it remains there as the endosome matures. During the maturation process, which we discuss next, the ubiquitin tag is used to selectively sort the receptor and its bound ligand into intralumenal vesicles. Receptor signaling is terminated when the receptor is sequestered into intralumenal vesicles, which ultimately are degraded in lysosomes. In this way, addition of ubiquitin blocks receptor recycling to the plasma membrane and directs the receptors into the degradation pathway.

Early Endosomes Mature into Late Endosomes

Early endosomes are mainly derived from incoming endocytic vesicles that fuse with one another (**Movie 13.8**). Typically, an **early endosome** receives incoming vesicles for about 10 minutes before beginning its maturation into a **late endosome**. Early endosomes have tubular and vacuolar domains (see Figure 13–55). Most of the membrane surface is in the tubules, and most of the volume is in the vacuolar domain.

Many changes occur during the maturation process. (1) The endosome changes shape and location as the tubular domains are mostly recycled back to the plasma membrane, the vacuolar domains are thoroughly modified, and the endosome is moved by motors along microtubules toward the nucleus. (2) Rab proteins drive changes in phosphoinositide lipids and fusion machinery (SNAREs and tethers) on the cytosolic face of the endosome membrane to change the functional characteristics of the organelle. (3) Lysosome proteins, including lumenal hydrolases and a membrane-embedded V-type ATPase, are delivered from the TGN to the maturing endosome. (4) The V-type ATPase pumps H^+ from the cytosol into the endosome lumen and further acidifies the organelle. Crucially, the increasing acidity that accompanies maturation renders lysosomal hydrolases increasingly more active, influencing many receptor–ligand interactions, thereby controlling receptor loading and unloading. (5) Intralumenal vesicles sequester endocytosed signaling receptors inside the endosome, thus halting the receptor signaling activity. Most of these events occur gradually but eventually lead to a complete transformation of the endosome into an early endolysosome.

In addition to committing selected cargo for degradation, the maturation process is important for lysosome maintenance. The continual delivery of lysosome components from the TGN to maturing endosomes ensures a steady supply of new lysosome proteins. The endocytosed materials mix in early endosomes with newly arrived acid hydrolases. Although mild digestion may start here, many hydrolases are synthesized and delivered as proenzymes, called *zymogens*, which contain extra inhibitory domains that keep the hydrolases inactive until these domains are proteolytically removed at later stages of **endosome maturation**. Moreover, the pH in early endosomes is not low enough to activate lysosomal

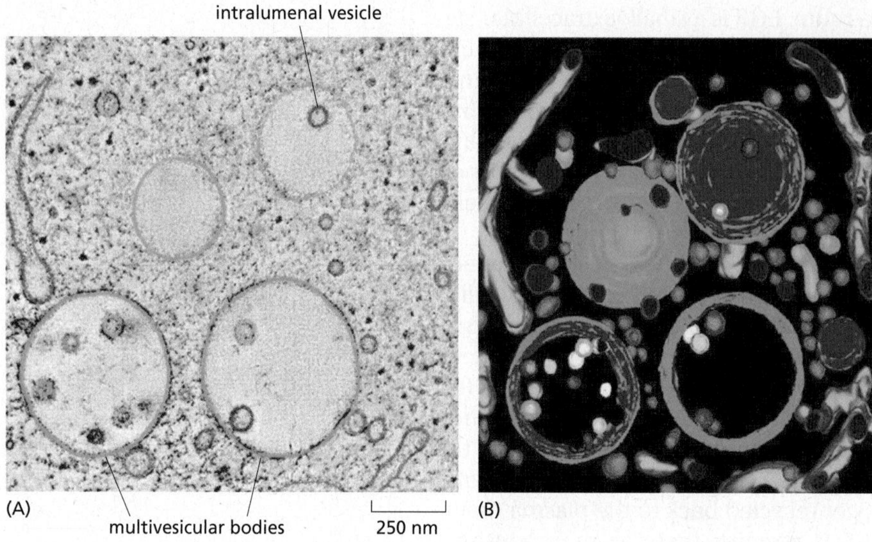

Figure 13–58 **Cryo-electron microscopy (cryo-EM) tomogram of multivesicular bodies in cultured human lymphocytes.** The tracing of membranes in single sections (A) allowed reconstruction of the three-dimensional arrangement of the organelles (B). The enclosing membranes of the multivesicular bodies are traced in *green*. (From J.L.A.N. Murk et al., *Proc. Natl. Acad. Sci. USA* 100:13332–13337, 2003.)

hydrolases optimally. By these means, cells can retrieve membrane proteins intact from early endosomes and recycle them back to the plasma membrane.

ESCRT Protein Complexes Mediate the Formation of Intralumenal Vesicles in Multivesicular Bodies

As endosomes mature, patches of their membrane invaginate into the endosome lumen and pinch off to form intralumenal vesicles. Because of their appearance in the electron microscope, such maturing endosomes are also called **multivesicular bodies** (Figure 13–58).

The multivesicular bodies carry endocytosed membrane proteins that are to be degraded. As part of the protein-sorting process, receptors destined for degradation, such as the occupied EGF receptors described previously, selectively partition into the invaginating membrane of the multivesicular bodies. In this way, both the receptors and any signaling proteins strongly bound to them are sequestered away from the cytosol where they might otherwise continue signaling. They also are made fully accessible to the digestive enzymes that eventually will degrade them (Figure 13–59). In addition to endocytosed membrane proteins, multivesicular bodies include the soluble content of early endosomes destined for late endosomes and digestion in lysosomes.

As discussed earlier, sorting into intralumenal vesicles requires one or multiple ubiquitin tags, which are added to the cytosolic domains of membrane proteins. These tags initially help guide the proteins into clathrin-coated vesicles in the plasma membrane. Once delivered to the endosomal membrane, the ubiquitin tags are recognized again, this time by a series of cytosolic **ESCRT protein complexes** (*ESCRT-0, -I, -II,* and *-III*), which bind sequentially and ultimately mediate

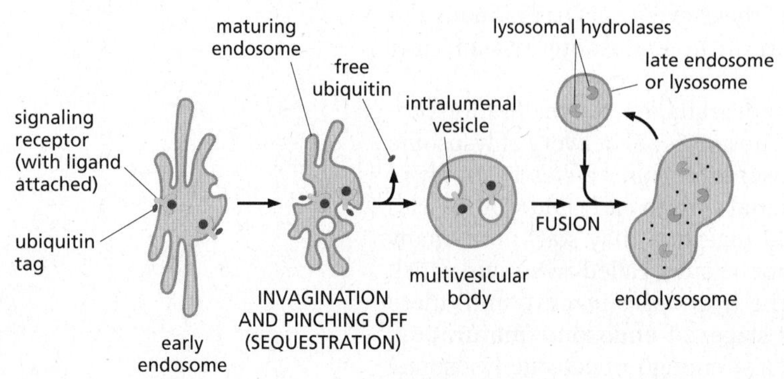

Figure 13–59 **The sequestration of endocytosed proteins into intralumenal vesicles of multivesicular bodies.** Ubiquitylated membrane proteins are sorted into domains on the endosome membrane, which invaginate and pinch off to form intralumenal vesicles. The ubiquitin marker is removed and returned to the cytosol for reuse before the intralumenal vesicle closes. Eventually, lysosomal hydrolases (such as proteases and lipases) in lysosomes digest all of the internal membranes. The invagination processes are essential for complete digestion of endocytosed membrane proteins. Because the outer membrane of the multivesicular body becomes continuous with the lysosomal membrane, the hydrolases only digest the cytosolic domains of endocytosed transmembrane proteins when the protein becomes localized in intralumenal vesicles.

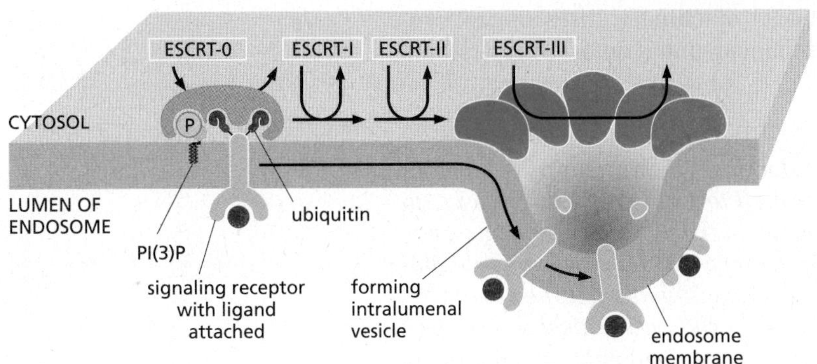

Figure 13–60 Sorting of endocytosed membrane proteins into the intralumenal vesicles of a multivesicular body. A series of complex binding events passes ubiquitylated membrane proteins, such as the signaling receptor with its ligand shown here, sequentially from one ESCRT complex (ESCRT-0) to the next, eventually concentrating them in membrane areas that bud into the lumen of the endosome to form intralumenal vesicles. ESCRT-III assembles into expansive multimeric structures and mediates invagination. The mechanisms of how cargo molecules are shepherded into the vesicles and how the vesicles are formed without including the ESCRT complexes themselves remain unknown. ESCRT complexes are soluble in the cytosol, are recruited to the membrane sequentially as needed, and are then released back into the cytosol as the vesicle pinches off.

the sorting process into the intralumenal vesicles. Membrane invagination into multivesicular bodies also depends on a lipid kinase that phosphorylates phosphatidylinositol to produce PI(3)P, which serves as an additional docking site for the ESCRT complexes. For docking and vesicle invagination, ESCRT complexes require both PI(3)P and the presence of ubiquitylated cargo proteins to bind to the endosomal membrane. ESCRT-III forms large multimeric assemblies on the membrane that bend the membrane (**Figure 13–60**).

Mutant cells compromised in ESCRT function display signaling defects. In such cells, activated receptors cannot be down-regulated by endocytosis and packaging into multivesicular bodies. The still-active receptors therefore mediate prolonged signaling, which can lead to uncontrolled cell proliferation and cancer.

The ESCRT machinery that drives the internal budding from the endosome membrane to form intralumenal vesicles is also used in animal-cell cytokinesis and virus budding, which are topologically equivalent. In all three processes, budding occurs in a direction away from the cytosolic surface of the membrane (**Figure 13–61A**). ESCRT complexes are thought to have originated from similar components that mediate cell-membrane deformation during cytokinesis in archaea.

Although some viruses such as HIV hijack the host ESCRT machinery to bud directly out of the cell, other viruses escape using different mechanisms. For example, SARS-CoV-2, the virus that causes COVID-19, buds into the vesicular tubular clusters between the ER and Golgi apparatus, then uses the secretory pathway to exit the cell (**Figure 13–61B**). This budding reaction deforms membranes away

Figure 13–61 Conserved mechanism in multivesicular body formation, virus budding, and cytokinesis. (A) In the three topologically equivalent processes indicated by the arrows, ESCRT complexes *(green)* shape membranes into buds that bulge away from the cytosol. (B) Electron micrographs of a cultured cell infected with SARS-CoV-2. The top panel shows virus particles budding away from the cytosol into vesicular tubular clusters between the ER and the Golgi apparatus. The virus particles are carried through the secretory pathway until they are released to the outside of the cell by exocytosis (bottom panel). (B, from N.S. Ogando et al., *J. Gen. Virol.* 101:925–940, 2020.)

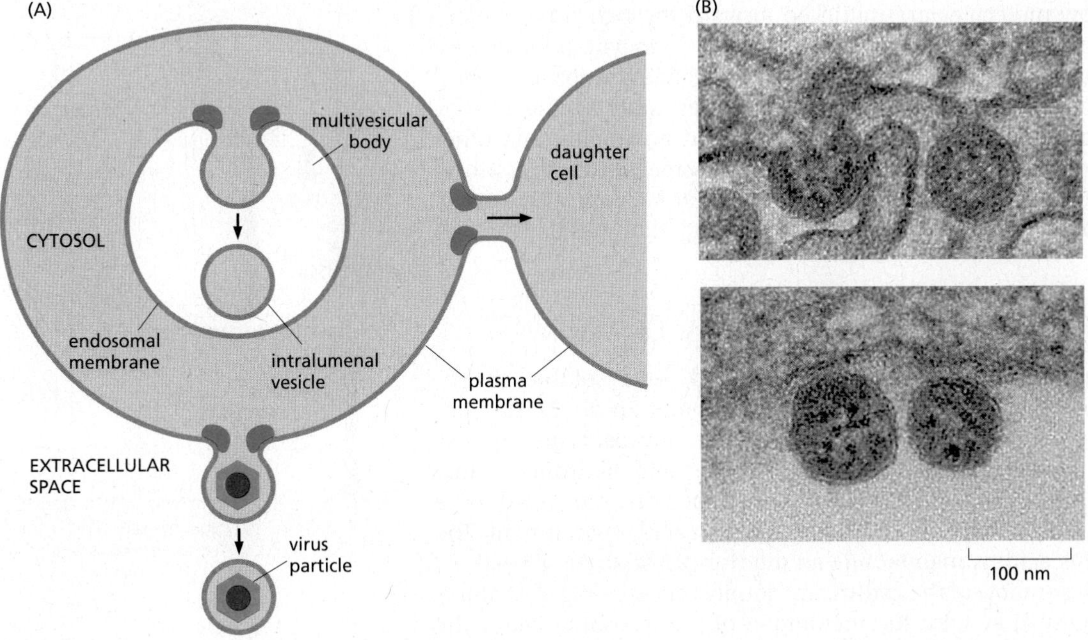

from the cytosol but does not seem to use ESCRT machinery. Instead, viral proteins are thought to facilitate budding by a mechanism that is not well understood.

Summary

Cells ingest fluid, molecules, and particles by endocytosis, in which localized regions of the plasma membrane invaginate and pinch off to form endocytic vesicles. In most cells, endocytosis internalizes a large fraction of the plasma membrane every hour. The cells remain the same size because most of the plasma membrane components (proteins and lipids) that are endocytosed are continually returned to the cell surface by exocytosis. This large-scale endocytic–exocytic cycle is mediated largely by clathrin-coated pits and vesicles, but clathrin-independent endocytic pathways also contribute.

While many of the endocytosed molecules are quickly recycled to the plasma membrane, others eventually end up in lysosomes, where they are degraded. Most of the ligands that are endocytosed with their receptors dissociate from their receptors in the acidic environment of the endosome and eventually end up in lysosomes, while most of the receptors are recycled via transport vesicles back to the cell surface for reuse. Many cell-surface signaling receptors become tagged with ubiquitin when activated by binding their extracellular ligands. Ubiquitylation guides activated receptors into clathrin-coated pits, and they and their ligands are efficiently internalized and delivered to early endosomes.

Early endosomes rapidly mature into late endosomes. During maturation, patches of the endosomal membrane containing ubiquitylated receptors invaginate and pinch off to form intralumenal vesicles. This process is mediated by ESCRT complexes and sequesters the receptors away from the cytosol, which terminates their signaling activity. Late endosomes migrate along microtubules toward the interior of the cell where they fuse with one another and with lysosomes to form endolysosomes, where degradation occurs.

In some cases, both receptor and ligand are transferred to a different plasma membrane domain, causing the ligand to be released at a different surface from where it originated, a process called transcytosis. In some cells, endocytosed plasma membrane proteins and lipids can be stored in recycling endosomes for as long as necessary until they are needed.

THE DEGRADATION AND RECYCLING OF MACROMOLECULES IN LYSOSOMES

Having discussed how molecules are trafficked forward through the secretory pathway and how material enters the cell through the endocytic pathway, we now consider lysosomes. The lysosome is a terminal destination for the degradation of proteins, microorganisms, dead cells, and other materials ingested by endocytosis and phagocytosis. In addition, proteins, old organelles, and other components in the cytosol can also be degraded in lysosomes through a process termed autophagy. In this section, we begin with a brief account of lysosome structure and function, then discuss how material for degradation is delivered to lysosomes.

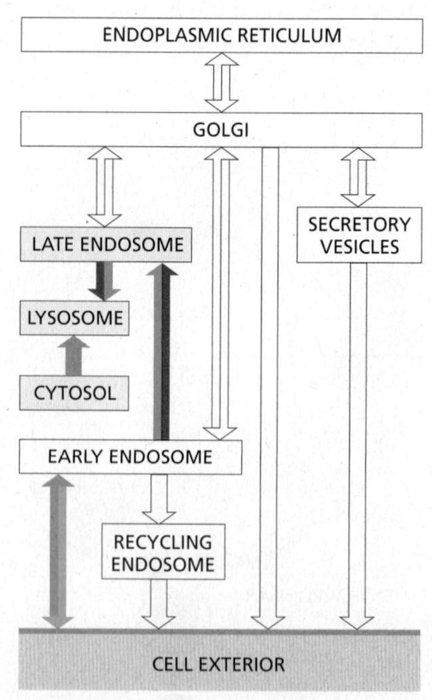

Lysosomes Are the Principal Sites of Intracellular Digestion

Lysosomes are membrane-enclosed organelles filled with soluble hydrolytic enzymes that digest macromolecules. Lysosomes contain about 40 types of hydrolytic enzymes, including proteases, nucleases, glycosidases, lipases, phospholipases, phosphatases, and sulfatases. All are **acid hydrolases**; that is, hydrolases that work best at acidic pH. For optimal activity, they need to be activated by proteolytic cleavage, which also requires an acid environment. The lysosome provides this acidity, maintaining an interior pH of about 4.5–5.0. By this arrangement, the contents of the cytosol are doubly protected against attack by the cell's own digestive system: the membrane of the lysosome keeps the digestive enzymes out of the cytosol, but, even if they leak out, they can do little damage at the cytosolic pH of about 7.2.

Like all other membrane-enclosed organelles, the lysosome not only contains a unique collection of enzymes but also has a unique surrounding membrane. Most of the lysosome membrane proteins, for example, are highly glycosylated, which helps to protect them from the lysosome proteases in the lumen. Transport proteins in the lysosome membrane carry the final products of the digestion of macromolecules—such as amino acids, sugars, and nucleotides—to the cytosol, where the cell can either reuse or excrete them.

A *vacuolar H^+ ATPase* in the lysosome membrane uses the energy of ATP hydrolysis to pump H^+ into the lysosome, thereby maintaining the lumen at its acidic pH (**Figure 13–62**). The lysosome H^+ pump belongs to the family of *V-type ATPases* and has a similar architecture to the mitochondrial and chloroplast ATP synthases (F-type ATPases), which convert the energy stored in H^+ gradients into ATP (see Figure 11–12). By contrast to these enzymes, however, the vacuolar H^+ ATPase exclusively works in reverse, pumping H^+ into the organelle. Similar or identical V-type ATPases acidify all endocytic and exocytic organelles, including lysosomes, endosomes, some compartments of the Golgi apparatus, and many transport and secretory vesicles. In addition to providing a low-pH environment that is suitable for reactions occurring in the organelle lumen, the H^+ gradient provides a source of energy that drives the transport of small metabolites across the organelle membrane.

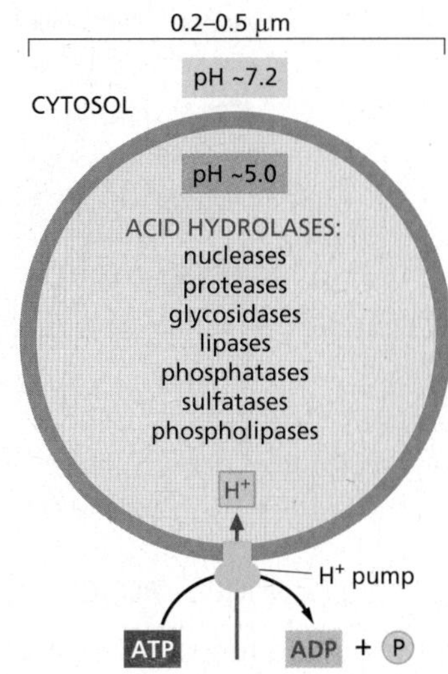

Figure 13–62 **Lysosomes.** The acid hydrolases are hydrolytic enzymes that are active under acidic conditions. An H^+ ATPase in the membrane pumps H^+ into the lysosome, maintaining its lumen at an acidic pH.

Lysosomes Are Heterogeneous

Lysosomes are found in all eukaryotic cells. They were initially discovered by the biochemical fractionation of cell extracts; only later were they seen clearly in the electron microscope. Although extraordinarily diverse in shape and size, staining them with specific antibodies shows they are members of a single family of organelles. They can also be identified by histochemical techniques that reveal which organelles contain acid hydrolase (**Figure 13–63**).

The heterogeneous morphology of lysosomes contrasts with the relatively uniform structures of many other cell organelles. The diversity reflects the wide variety of digestive functions that acid hydrolases mediate, including the breakdown of intracellular and extracellular debris, the destruction of phagocytosed microorganisms, and the production of nutrients for the cell. Their morphological diversity, however, also reflects the way lysosomes form. Late endosomes containing material received from the plasma membrane by endocytosis and containing newly synthesized lysosomal hydrolases fuse with preexisting lysosomes to form structures that are referred to as *endolysosomes*, which then fuse with one another (**Figure 13–64**). When the majority of the endocytosed material within an endolysosome has been digested so that only resistant or slowly digestible residues remain, these organelles become "classical" lysosomes. These are relatively dense, round, and small, but they can enter the cycle again by fusing with late endosomes or endolysosomes. Thus, there is no real distinction between endolysosomes and lysosomes: they are the same except that they are in different stages of a maturation cycle. For this reason, lysosomes are sometimes viewed as a heterogeneous collection of distinct organelles, the common feature of which is a high content of hydrolytic enzymes. It is especially hard to apply a narrower definition than this in plant cells, as we discuss next.

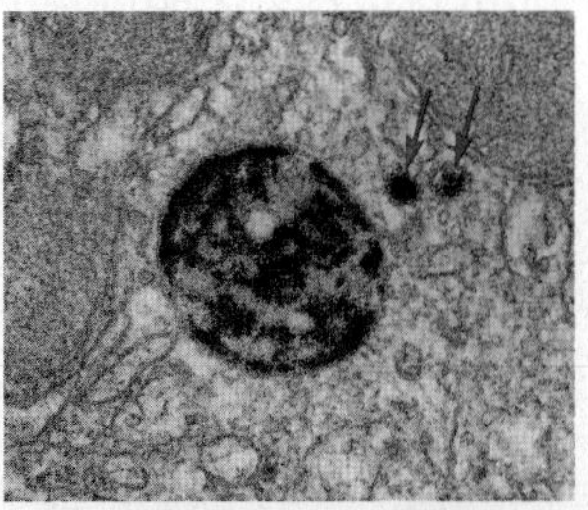

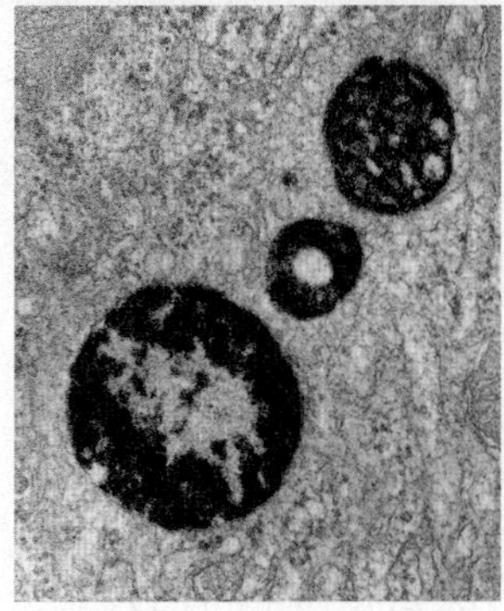

Figure 13–63 **Histochemical visualization of lysosomes.** These electron micrographs show two sections of a cell stained to reveal the location of acid phosphatase, a marker enzyme for lysosomes. The larger membrane-enclosed organelles, containing dense precipitates of lead phosphate, are lysosomes. Their diverse morphology reflects variations in the amount and nature of the material they are digesting. The precipitates are produced when tissue fixed with glutaraldehyde (to fix the enzyme in place) is incubated with a phosphatase substrate in the presence of lead ions. *Red arrows* in the top panel indicate two small vesicles thought to be carrying acid hydrolases from the Golgi apparatus. (Courtesy of Daniel S. Friend, and by permission of E.L. Bearer.)

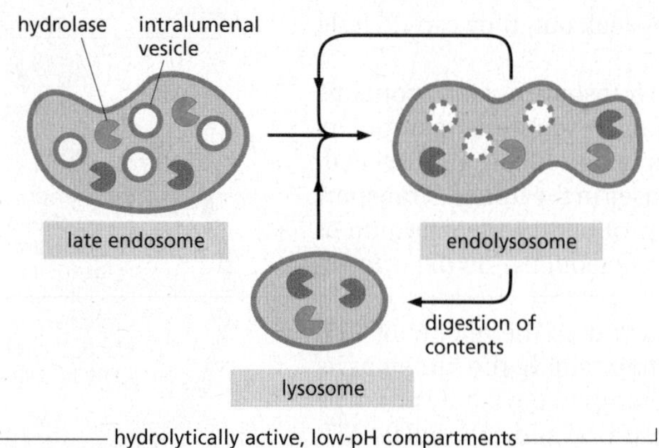

Figure 13–64 A model for lysosome maturation. Late endosomes fuse with preexisting lysosomes *(bottom)* or preexisting endolysosomes *(top right)*. Endolysosomes eventually mature into lysosomes as hydrolases complete the digestion of their contents, which can include intralumenal vesicles. Lysosomes also fuse with phagosomes, as we discuss later.

Plant and Fungal Vacuoles Are Remarkably Versatile Lysosomes

Most plant and fungal cells (including yeasts) contain one or several very large, fluid-filled vesicles called **vacuoles**. They typically occupy more than 30% of the cell volume and as much as 90% in some cell types (**Figure 13–65**). Vacuoles are related to animal-cell lysosomes and contain a variety of hydrolytic enzymes, but their functions are remarkably diverse. The plant vacuole can act as a storage organelle for both nutrients and waste products, as a degradative compartment, as an economical way of increasing cell size, and as a controller of *turgor pressure* (the osmotic pressure that pushes outward on the cell wall and keeps the plant from wilting) (**Figure 13–66**). The same cell may have different vacuoles with distinct functions, such as digestion and storage.

The vacuole is important as a homeostatic device, enabling plant cells to withstand wide variations in their environment. When the pH in the environment drops, for example, the flux of H^+ into the cytosol is balanced, at least in part, by an increased transport of H^+ into the vacuole, which tends to keep the pH in the cytosol constant. Similarly, many plant cells maintain an almost constant turgor pressure despite large changes in the tonicity of the fluid in their immediate environment. They do so by changing the osmotic pressure of the cytosol and vacuole—in part by the controlled breakdown and resynthesis of polymers such as polyphosphate in the vacuole, and in part by altering the transport rates of sugars, amino acids, and other metabolites across the plasma membrane and the vacuolar membrane. The turgor pressure regulates the activities of distinct transporters in each membrane to control these fluxes.

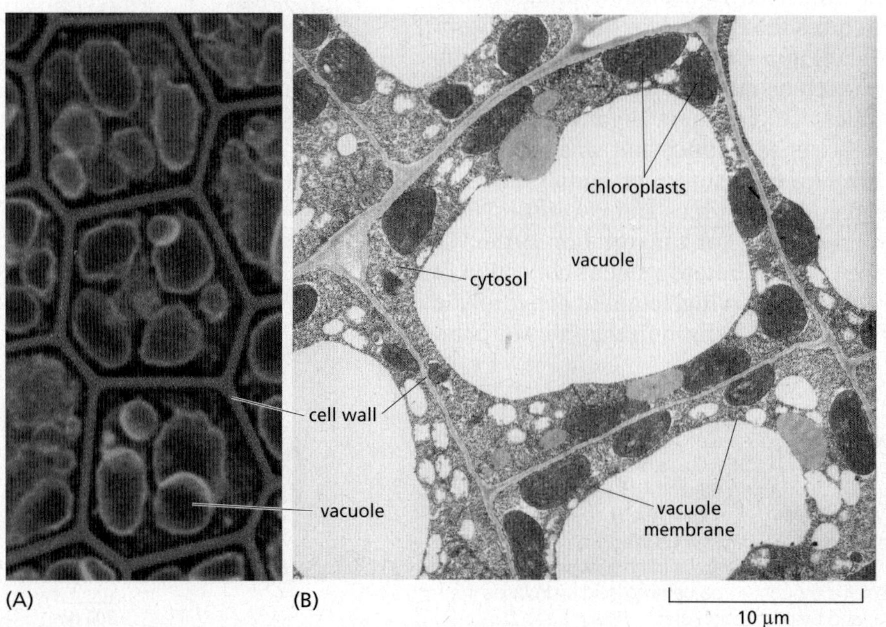

Figure 13–65 The plant-cell vacuole. (A) A confocal image of cells from an *Arabidopsis* embryo that is expressing an aquaporin—YFP (yellow fluorescent protein) fusion protein in its vacuole membrane. YFP fluorescence and the cell walls have been false colored *green* and *orange*, respectively. Each cell contains several large vacuoles. (B) This electron micrograph of cells in a young tobacco leaf shows the cytosol as a thin layer, containing chloroplasts, pressed against the cell wall by the enormous vacuole. (A, courtesy of C. Carroll and L. Frigerio, based on S. Gattolin et al., *Mol. Plant* 4:180–189, 2011; B, courtesy of J. Burgess.)

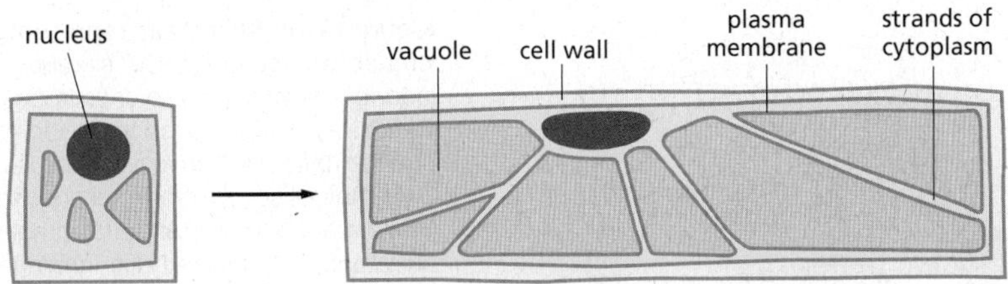

Figure 13–66 **The role of the vacuole in controlling the size of plant cells.** A plant cell can achieve a large increase in volume without increasing the volume of the cytosol. Localized weakening of the cell wall orients a turgor-driven cell enlargement that accompanies the uptake of water into an expanding vacuole. The cytosol is eventually confined to a thin peripheral layer, which is connected to the nuclear region by strands of cytosol stabilized by bundles of actin filaments (not shown).

Humans often harvest substances stored in plant vacuoles—from rubber to opium to the flavoring of garlic. Many stored products have a metabolic function. Proteins, for example, can be preserved for years in the vacuoles of the storage cells of many seeds, such as those of peas and beans. When the seeds germinate, these proteins are hydrolyzed, and the resulting amino acids provide a food supply for the developing embryo. Anthocyanin pigments stored in vacuoles color the petals of many flowers so as to attract pollinating insects, while noxious molecules released from vacuoles when a plant is eaten or damaged provide a defense against predators.

Multiple Pathways Deliver Materials to Lysosomes

Lysosomes are meeting places where several streams of intracellular traffic converge. A route that leads outward from the ER via the Golgi apparatus delivers most of the lysosome's digestive enzymes, as we discussed earlier. In addition, at least three paths from the cell surface and extracellular space feed substances into lysosomes for digestion, while a fourth route called *autophagy* originates in the cytoplasm and is used to digest intracellular macromolecules and organelles.

We have already discussed how macromolecules taken up from plasma membrane and extracellular fluid by endocytosis can reside in endosomes until they mature and fuse with lysosomes. A second pathway called *macropinocytosis* specializes in the nonspecific uptake of fluids, membrane, and particles attached to the plasma membrane. A third pathway found in phagocytic cells, such as macrophages and neutrophils in vertebrates, is dedicated to the engulfment, or *phagocytosis,* of large particles and microorganisms to form *phagosomes.* In contrast to these routes originating from the plasma membrane, autophagy is used to digest cytosol, worn-out organelles, and microbes that invade the cytosol. The four paths to degradation in lysosomes are illustrated in **Figure 13–67**.

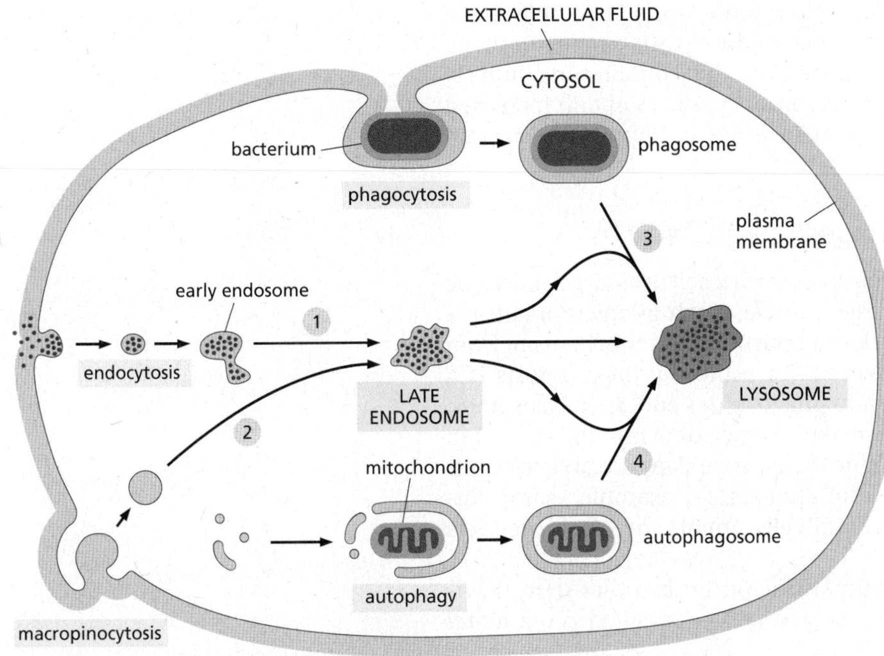

Figure 13–67 **Four pathways to degradation in lysosomes.** Materials in each pathway are derived from a different source. Note that the autophagosome has a double membrane, as we explain later in this chapter. In all cases, the final step is the fusion with lysosomes.

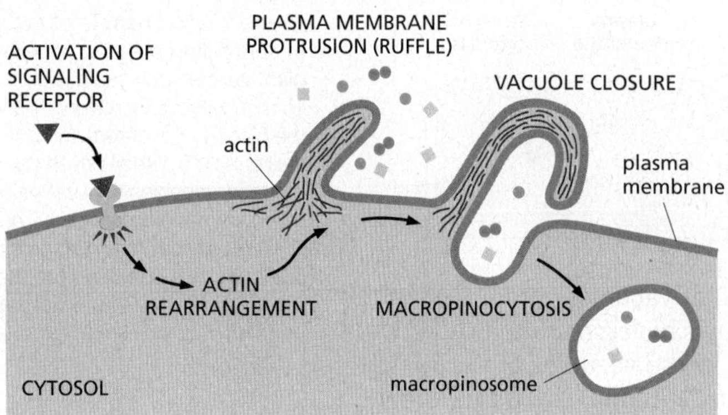

Figure 13–68 Schematic representation of macropinocytosis. Cell signaling events lead to a reprogramming of actin dynamics, which in turn triggers the formation of cell-surface ruffles. As the ruffles collapse back onto the cell surface, they nonspecifically trap extracellular fluid and macromolecules and particles contained in it, forming large vacuoles, or macropinosomes, as shown.

As we have seen, the endocytic pathway not only routes macromolecules for degradation but is also used to regulate the location and trafficking of macromolecules. By contrast, phagocytosis, macropinocytosis, and autophagy are pathways dedicated to degradation. We consider each of the latter three pathways in turn, highlighting the materials each one is responsible for degrading and the mechanisms used to deliver these materials to lysosomes.

Cells Can Acquire Nutrients from the Extracellular Fluid by Macropinocytosis

Macropinocytosis was among the first types of endocytosis to be described because it is visible by light microscopy, where cells can be seen taking up the surrounding fluid into large vesicles called *macropinosomes* (**Figure 13–68**). In most cell types, macropinocytosis does not operate continually but rather is induced for a limited time in response to cell-surface receptor activation by specific cargoes, including growth factors, integrin ligands, apoptotic-cell remnants, and some viruses. These ligands activate a complex signaling pathway, resulting in a change in actin dynamics and the formation of cell-surface protrusions, called *ruffles* (discussed in Chapter 16). Macropinosomes form when the protruding ends of ruffles fuse with each other or the cell membrane, thereby trapping a portion of extracellular content.

Macropinocytosis is a dedicated degradative pathway: macropinosomes acidify and then fuse with late endosomes or endolysosomes, without recycling their cargo back to the plasma membrane. Micropinocytosis is stimulated by activation of the oncogene *Ras*. Induction of macropinocytosis can increase the bulk fluid uptake of a cell by up to tenfold. Cancer cells that contain constitutively active Ras (see Chapter 15) use enhanced micropinocytosis to obtain increased nutrients from the surrounding environment in order to support their rapid growth and division.

Specialized Phagocytic Cells Can Ingest Large Particles

Phagocytosis is a special form of endocytosis in which a cell uses large endocytic vesicles called **phagosomes** to ingest large particles such as microorganisms and dead cells. Phagocytosis is distinct, both in purpose and mechanism, from macropinocytosis, which we discussed earlier. In protozoa, phagocytosis is a form of feeding: large particles taken up into phagosomes end up in lysosomes, and the products of the subsequent digestive processes pass into the cytosol to be used as food. However, few cells in multicellular organisms are able to ingest such large particles efficiently. In the gut of animals, for example, extracellular processes break down food particles, and cells import the small products of hydrolysis.

Phagocytosis is important in most animals for purposes other than nutrition, and it is carried out mainly by specialized cells—so-called *professional*

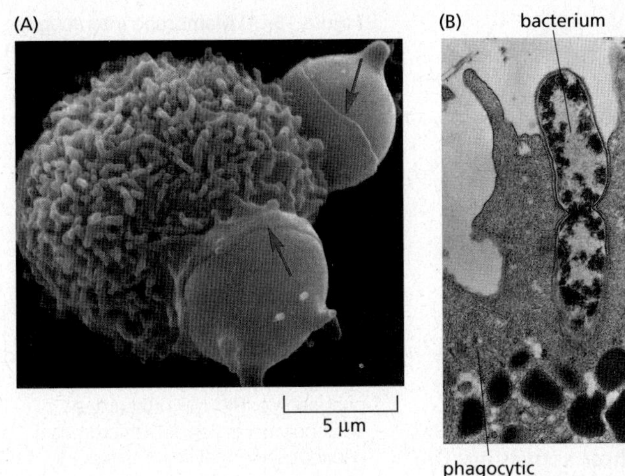

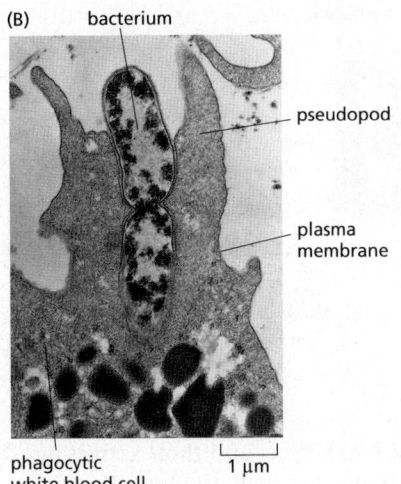

Figure 13–69 **Phagocytosis by a macrophage.** (A) Scanning electron micrograph of a mouse macrophage phagocytosing two chemically altered red blood cells. The *red arrows* point to edges of thin processes (pseudopods) of the macrophage that are extending as collars to engulf the red cells. (B) An electron micrograph of a neutrophil phagocytosing a bacterium, which is in the process of dividing. (A, courtesy of Jean Paul Revel; B, courtesy of Dorothy F. Bainton, Phagocytic Mechanisms in Health and Disease. New York: Intercontinental Medical Book Corporation, 1971.)

phagocytes. In mammals, two important classes of white blood cells that act as professional phagocytes are **macrophages** and **neutrophils** (Movie 13.9). These cells develop from hemopoietic stem cells (discussed in Chapter 22), and they ingest invading microorganisms to defend us against infection. Macrophages also have an important role in scavenging senescent cells and cells that have died by apoptosis (discussed in Chapter 18). In quantitative terms, the clearance of senescent and dead cells is by far the most important: our macrophages, for example, phagocytose more than 10^{11} senescent red blood cells in each of us every day.

The diameter of a phagosome is determined by the size of its ingested particles, and those particles can be almost as large as the phagocytic cell itself (Figure 13–69). Phagosomes fuse with lysosomes, and the ingested material is then degraded. Indigestible substances remain in the lysosomes, forming *residual bodies* that can be excreted from cells by exocytosis Some of the internalized plasma membrane components never reach the lysosome, because they are retrieved from the phagosome in transport vesicles and returned to the plasma membrane.

Some pathogenic bacteria have evolved elaborate mechanisms to prevent phagosome–lysosome fusion. The bacterium *Legionella pneumophila,* for example, which causes Legionnaires' disease (discussed in Chapter 23), injects into its unfortunate host a Rab-modifying enzyme that causes certain Rab proteins to misdirect membrane traffic, thereby preventing phagosome–lysosome fusion. The bacterium, thus spared from lysosomal degradation, remains in the modified phagosome, growing and dividing as an intracellular pathogen, protected from the host's adaptive immune system.

Cargo Recognition by Cell-surface Receptors Initiates Phagocytosis

Phagocytosis is a cargo-triggered process. That is, it requires the activation of cell-surface receptors that transmit signals to the cell interior. Thus, to be phagocytosed, particles must first bind to the surface of the phagocyte (although not all particles that bind are ingested). Phagocytes have a variety of cell-surface receptors that are functionally linked to the phagocytic machinery of the cell. The best-characterized triggers of phagocytosis are antibodies, which protect us by binding to the surface of infectious microorganisms (pathogens) and initiating a series of events that culminate in the invader being phagocytosed. When antibodies initially attack a pathogen, they coat it with antibody molecules that bind to *Fc receptors* on the surface of macrophages and neutrophils, activating the receptors to induce the phagocytic cell to extend pseudopods, which engulf

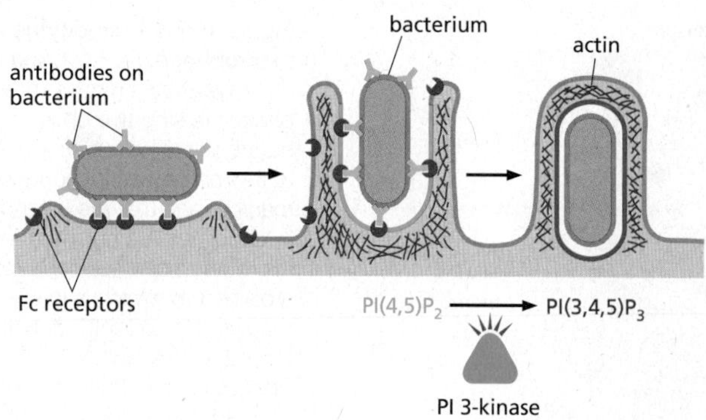

Figure 13–70 Membrane interactions and dynamics during phagocytosis. A bacterium in the body is recognized by antibodies that coat its surface. The Fc receptor on the surface of phagocytic cells recognizes the antibody, recruiting the bacterium to the plasma membrane. This initiates phagocytosis by triggering the formation of pseudopods that begin to surround the bacterium. Pseudopod extension and phagosome formation are driven by actin polymerization and reorganization, which respond to the accumulation of specific phosphoinositides in the membrane of the forming phagosome: $PI(4,5)P_2$ stimulates actin polymerization, which promotes pseudopod formation, and then $PI(3,4,5)P_3$ depolymerizes actin filaments at the base.

the particle and fuse at their tips to form a phagosome (Figure 13–70). Localized actin polymerization, initiated by Rho family GTPases and their activating Rho GEFs (discussed in Chapters 15 and 16), shapes the pseudopods. The activated Rho GTPases switch on the kinase activity of local PI kinases to produce $PI(4,5)P_2$ in the membrane (see Figure 13–11), which stimulates actin polymerization. To seal off the phagosome and complete the engulfment, actin is depolymerized by a PI 3-kinase that converts the $PI(4,5)P_2$ to $PI(3,4,5)P_3$, which is required for closure of the phagosome and may also contribute to reshaping the actin network to help drive the invagination of the forming phagosome (Figure 13–70). In this way, the ordered generation and consumption of specific phosphoinositides guides sequential steps in phagosome formation.

Several other classes of receptors that promote phagocytosis have been characterized. Some recognize *complement* components, which collaborate with antibodies in targeting microbes for destruction (discussed in Chapter 24). Others directly recognize oligosaccharides on the surface of certain pathogens. Still others recognize cells that have died by apoptosis. Apoptotic cells lose the asymmetric distribution of phospholipids in their plasma membrane. As a consequence, negatively charged phosphatidylserine, which is normally confined to the cytosolic leaflet of the lipid bilayer, is now exposed on the outside of the cell, where it helps to trigger the phagocytosis of the dead cell.

Remarkably, macrophages will also phagocytose a variety of inanimate particles—such as glass or latex beads and asbestos fibers—yet they do not phagocytose live cells in their own body. The living cells display "don't-eat-me" signals in the form of cell-surface proteins that bind to inhibitory receptors on the surface of macrophages. The inhibitory receptors recruit tyrosine phosphatases that antagonize the intracellular signaling events required to initiate phagocytosis, thereby locally inhibiting the phagocytic process. Thus phagocytosis, like many other cell processes, depends on a balance between positive signals that activate the process and negative signals that inhibit it. Apoptotic cells are thought both to gain "eat-me" signals (such as extracellularly exposed phosphatidylserine) and to lose their "don't-eat-me" signals, causing them to be very rapidly phagocytosed by macrophages.

Autophagy Degrades Unwanted Proteins and Organelles

All eukaryotic cells can carry out a process called **autophagy**, or "self-eating." During autophagy, a portion of the cytoplasm is engulfed into a membrane structure called the **autophagosome** that subsequently fuses with the lysosome where the autophagosome's contents are degraded (Figure 13–71). Autophagy can be either nonselective or selective. In *nonselective autophagy,* a bulk portion of cytoplasm is sequestered in autophagosomes. In *selective autophagy,* autophagosomes tightly enclose specific cargo and mostly exclude the surrounding cytosol.

Autophagy serves several important roles in the cell. During normal cell growth and in development, autophagy helps restructure differentiating cells by removing unwanted organelles or other cellular contents. When cells experience stress or starvation, nonselective autophagy is used to recycle existing

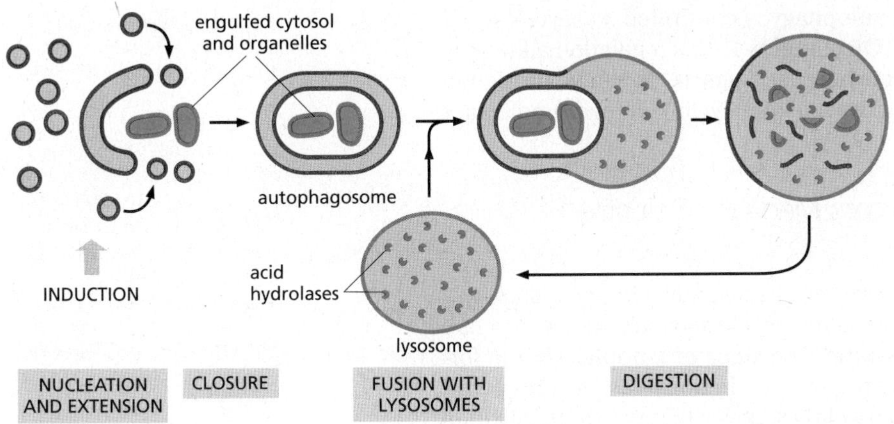

Figure 13–71 A model of autophagy. Activation of a signaling pathway initiates a nucleation event in the cytoplasm. A crescent of autophagosomal membrane grows by fusion of vesicles of unknown origin and eventually fuses to form a double-membrane-enclosed autophagosome, which sequesters a portion of the cytoplasm. The autophagosome then fuses with lysosomes containing acid hydrolases that digest its content.

proteins and macromolecules into building blocks that are used for other priorities. Selective autophagy can be used to degrade bacteria and viruses that invade the cytosol, as well as damaged proteins, protein aggregates, and damaged whole organelles. The range of cargoes degraded by autophagy explains why dysregulated autophagy contributes to diseases ranging from infectious disorders to neurodegeneration and cancer.

The autophagosome assembles by the fusion of small vesicles of unknown origin. The process begins when a phosphoinositide lipid kinase complex (termed the ATG1 complex) locally produces PI(3)P (see Figure 13–10) to mark a potential membrane site for the recruitment of several autophagy-related factors. These factors catalyze the covalent attachment of the membrane lipid phosphatidylethanolamine to a ubiquitin-like protein called ATG8. The ATG8-marked vesicles undergo homotypic fusion with each other and heterotypic fusion with vesicles containing ATG9. This results in expansion of a membrane whose identity is provided by ATG8 and other autophagy-related proteins.

For reasons that are not known, the growing membrane structure formed by vesicle fusion is not spherical. Instead, it is a flattened disc (similar to a Golgi apparatus cisterna) that curls into a cup-shaped structure. Fusion of the lips of this cup encloses the contents inside a compartment that is now surrounded by two membranes. The outer membrane of this newly formed autophagosome then fuses with the lysosome in a SNARE-mediated process. The inner membrane and its enclosed cargo are released into the lysosome where they are degraded by the acid hydrolases.

The Rate of Nonselective Autophagy Is Regulated by Nutrient Availability

The activity of the ATG1 kinase complex that initiates autophagosome formation is tightly regulated. Most of the time, it is kept inactive because of phosphorylation by another protein kinase called the mTOR complex 1 (discussed in Chapter 15). The activity of mTOR complex 1 is dependent on the availability of certain amino acids generated by the recycling of proteins in the lysosome. When these amino acids become limiting, mTOR complex 1 activity is reduced, relieving its inhibition of the ATG1 complex. Activation of the ATG1 complex initiates nonselective autophagy to degrade bulk cytosol in lysosomes, thereby generating amino acids that activate the mTOR complex 1. Through this feedback loop, the rate of nonselective autophagy is dynamically regulated by the nutrient status of the cell.

Starvation-induced nonselective autophagy is particularly important in mammals in the hours immediately after birth. During this transition period, the constant nutrient supply from the womb is abruptly lost, and feeding by mouth has not yet started. Rapid acquisition of amino acids via autophagy is used to sustain critical cellular functions until a steady source of food from the mother is established. In mammals, the rate of autophagy is controlled by several other signaling pathways in addition to the mTOR complex 1. This regulation allows the cell to integrate multiple external and internal cues, such as growth factor signaling and ATP levels, into a decision about the rate of recycling of macromolecules in the cytosol.

A Family of Cargo-specific Receptors Mediates Selective Autophagy

Selective autophagy mediates the degradation of invading microbes, damaged or otherwise unwanted organelles, and protein complexes or aggregates that are too large for degradation by the proteasome. The steps of autophagosome formation and its fusion with lysosomes are the same as in nonselective autophagy. Selective autophagy differs only at the very early stages when the cargo destined for degradation is recruited to the concave surface of the forming autophagosome membrane. This recruitment is mediated by specialized receptor proteins that recognize the cargo and have a binding site for the autophagosome-specific protein ATG8.

To accommodate the large range of potential cargoes, cells have evolved numerous cargo-specific autophagy receptors. In most cases, autophagy receptors recognize their cargo via a mark that is acquired only when the cargo is destined for degradation. The most commonly used mark is ubiquitin. For example, bacteria that escape from phagosomes and invade the cytosol are recognized by cytosolic proteins that ubiquitylate proteins on the bacterial surface. Several cargo receptors recognize the ubiquitin and other bacteria-specific proteins and recruit ATG8-containing vesicles. These vesicles fuse together, and the forming autophagosome effectively zippers around the cargo to engulf it without trapping bulk cytosol in the process (**Figure 13–72A**). This is why the shape of an autophagosome during selective autophagy typically reflects the shape of its cargo (**Figure 13–72B**).

The selective autophagy of worn out or damaged mitochondria is called *mitophagy*. As discussed in Chapters 12 and 14, when mitochondria function normally, the inner mitochondrial membrane is energized by an electrochemical H^+ gradient that drives ATP synthesis and the import of mitochondrial precursor proteins and metabolites. Damaged mitochondria cannot maintain the gradient, so protein import is blocked. As a consequence, a protein kinase called Pink1, which is normally imported into mitochondria, is instead retained on the mitochondrial surface where it recruits the ubiquitin ligase Parkin from the cytosol. Parkin ubiquitylates mitochondrial outer membrane proteins, which serves as a mark for a ubiquitin-dependent cargo receptor for autophagy. Mutations in Pink1 or Parkin cause a form of early-onset Parkinson's disease, a degenerative disorder of the central nervous system. It is not known why the neurons that die prematurely in this disease are particularly reliant on mitophagy.

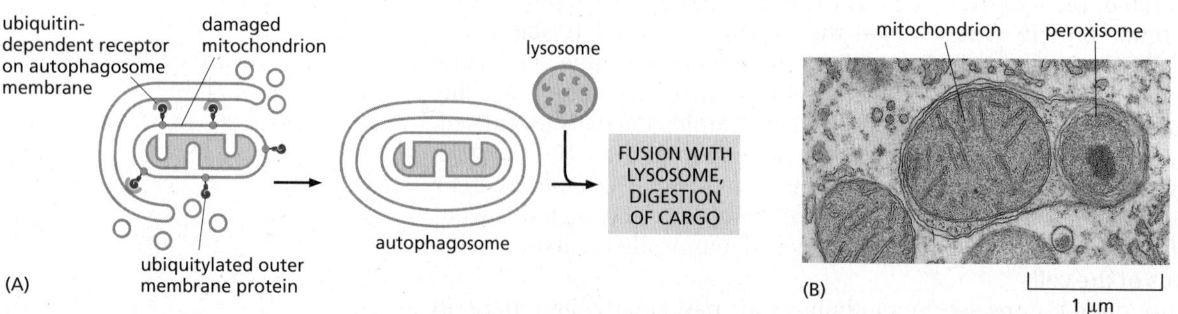

Figure 13–72 **Selective autophagy is mediated by receptors that recruit cargo to the autophagosome membrane.** (A) Diagram illustrating the concept of a cargo receptor *(blue)* in the autophagosome membrane that directs it to a specific cargo, in this case a damaged mitochondrion. (B) An electron micrograph of an autophagosome containing a mitochondrion and a peroxisome. (B, courtesy of Daniel S. Friend and by permission of E.L. Bearer.)

Some Lysosomes and Multivesicular Bodies Undergo Exocytosis

Targeting of material to lysosomes is not necessarily the end of the pathway. *Lysosomal secretion* of undigested content enables cells to eliminate indigestible debris. For most cells, this seems to be a minor pathway, used only when the cells are stressed. Some cell types, however, contain specialized lysosomes that have acquired the necessary machinery for fusion with the plasma membrane. *Melanocytes* in the skin, for example, produce and store pigments in their lysosomes. These pigment-containing *melanosomes* release their pigment into the extracellular space of the epidermis by exocytosis. The melanosomes are then taken up by keratinocytes, leading to normal skin pigmentation. In some genetic disorders, defects in melanosome exocytosis block this transfer process, leading to forms of hypopigmentation (albinism). Under certain conditions, multivesicular bodies can also fuse with the plasma membrane. If that occurs, their intralumenal vesicles are released from cells. Circulating small vesicles, also called *exosomes*, have been observed in the blood and may be used to transport components between cells, although the importance of such a mechanism of potential communication between distant cells is unknown. Some exosomes may derive from direct vesicle budding events at the plasma membrane similar to how some viruses bud and are released from the cell (see Figure 13–61).

Summary

Lysosomes are specialized for the intracellular digestion of macromolecules. They contain unique membrane proteins and a wide variety of soluble hydrolytic enzymes that operate best at pH 5, which is the internal pH of lysosomes. An ATP-driven H^+ pump in the lysosomal membrane maintains this low pH. In plants and fungi, lysosome-related compartments called vacuoles are adapted for other functions including storage, regulation of cell volume, and maintenance of turgor pressure. Lysosomes are the end product of endosome maturation. Cell-surface proteins that are endocytosed into endosomes are degraded in lysosomes unless they are retrieved back to the plasma membrane. Extracellular contents internalized by micropinocytosis are also delivered via endosomes to lysosomes and provide a source of nutrients under some conditions.

Lysosomes can receive content from two other routes. In phagocytosis, cells can engulf large particles including bacteria and even other cells into phagosomes that fuse with lysosomes. Phagocytosis is usually a receptor-mediated process and is especially prominent in some specialized cell types such as macrophages. In autophagy, cells engulf parts of their own cytoplasm into a double-membrane structure called an autophagosome. The autophagosome can engulf random parts of cytosol nonselectively or specific cargoes identified by selective receptors. Nonselective autophagy is used to recycle macromolecules into their building blocks during nutrient starvation. Selective autophagy is used to destroy invading microbes, protein aggregates, and damaged or unwanted organelles.

PROBLEMS

Which statements are true? Explain why or why not.

13–1 In all events involving fusion of a vesicle to a target membrane, the cytosolic leaflets of the vesicle and target bilayers always fuse together, as do the leaflets that are not in contact with the cytosol.

13–2 In order for a protein to exit from the ER, it must be correctly folded and, if part of a multiprotein complex, properly assembled.

13–3 When a foreign gene encoding a secretory protein is introduced into a secretory cell that normally does not make the protein, the alien secretory protein is not packaged into secretory vesicles.

13–4 More than 25 different receptors, including the low-density lipoprotein (LDL) receptor, participate in receptor-mediated endocytosis. In all cases, they enter coated pits only after they have bound their specific ligands.

13–5 Lysosomal membranes contain a proton pump that utilizes the energy of ATP hydrolysis to pump protons out of the lysosome, thereby maintaining the lumen at a low pH.

Discuss the following problems.

13–6 In a nondividing cell such as a liver cell, why must the flow of membrane between compartments be balanced, so that the retrieval pathways match the outward flow? Would you expect the same balanced flow in a gut epithelial cell, which is actively dividing?

13–7 For fusion of a vesicle with its target membrane to occur, the membranes have to be brought to within 1.5 nm so that the two bilayers can join (**Figure Q13–1**). Assuming that the relevant portions of the two membranes at the fusion site are circular regions 1.5 nm in diameter, calculate the number of water molecules that would remain between the membranes. (Water is 55.5 M and the volume of a cylinder is $\pi r^2 h$.) Given that an average phospholipid occupies a membrane surface area of 0.2 nm^2, how many phospholipids would be present in each of the opposing monolayers at the fusion site? Are there sufficient water molecules to bind to the hydrophilic head groups of this number of phospholipids? (It is estimated that 10–12 water molecules are normally associated with each phospholipid head group at the exposed surface of a membrane.)

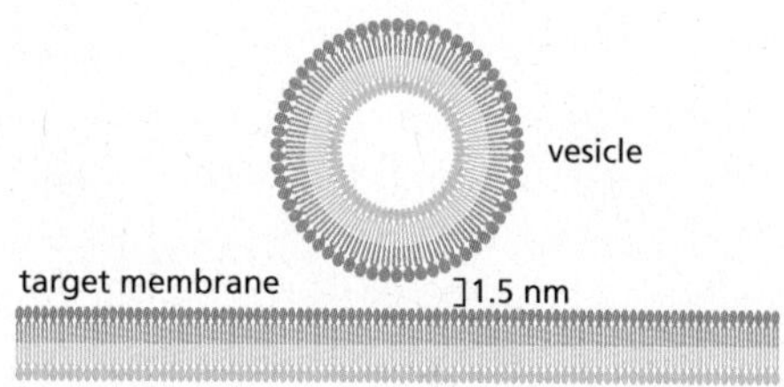

Figure Q13–1 Close approach of a vesicle and its target membrane in preparation for fusion (Problem 13–7).

13–8 SNAREs exist as complementary partners that carry out membrane fusions between appropriate vesicles and their target membranes. In this way, a vesicle with a particular variety of v-SNARE will fuse only with a membrane that carries the complementary t-SNARE. In some instances, however, fusions of identical membranes (homotypic fusions) are known to occur. For example, when a yeast cell forms a bud, vesicles derived from the mother cell's vacuole move into the bud where they fuse with one another to form a new vacuole. These vesicles carry both v-SNAREs and t-SNAREs. Are both types of SNAREs essential for this homotypic fusion event?

To test this point, you have developed an ingenious assay for fusion of vacuolar vesicles. You prepare vesicles from two different mutant strains of yeast: strain B has a defective gene for vacuolar alkaline phosphatase (Pase); strain A is defective for the protease that converts the precursor of alkaline phosphatase (pro-Pase) into its active form (Pase) (**Figure Q13–2A**). Neither strain has

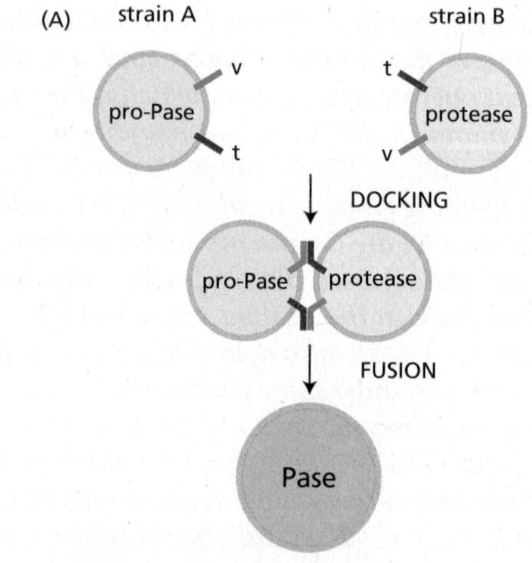

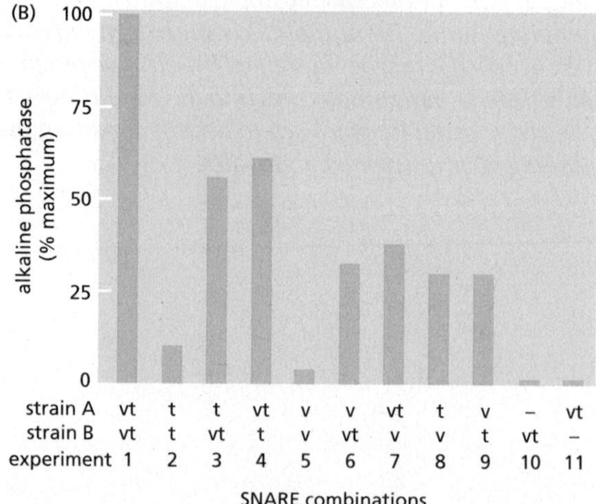

Figure Q13–2 SNARE requirements for vesicle fusion (Problem 13–8). (A) Scheme for measuring the fusion of vacuolar vesicles. (B) Results of fusions of vesicles with different combinations of v-SNAREs and t-SNAREs. The SNAREs present on the vesicles of the two strains are indicated as v (v-SNARE) and t (t-SNARE). (Adapted from B.J. Nichols et al., *Nature* 387:199-202, 1997.)

active alkaline phosphatase, but when extracts of the strains are mixed, vesicle fusion generates active alkaline phosphatase, which can be easily measured.

Now you delete the genes for the vacuolar v-SNARE, t-SNARE, or both, in each of the two yeast strains. You prepare vacuolar vesicles from each and test them for their ability to fuse, as measured by the alkaline phosphatase assay (**Figure Q13–2B**).

What do these data say about the requirements for v-SNAREs and t-SNAREs in the fusion of vacuolar vesicles? Does it matter which kind of SNARE is on which vesicle?

13–9 Enveloped viruses, which have a membrane coat, gain access to the cytosol by fusing with a cell membrane. Why do you suppose that these viruses carry their own special fusion protein, rather than making use of a cell's SNAREs?

13–10 If you were to remove the ER retrieval signal from protein disulfide isomerase (PDI), which is normally a soluble resident of the ER lumen, where would you expect the modified PDI to be located?

13–11 The KDEL receptor must shuttle back and forth between the ER and the Golgi apparatus to accomplish its task of ensuring that soluble ER proteins are retained in the ER lumen. In which compartment does the KDEL receptor bind its ligands more tightly? In which compartment does it bind its ligands more weakly? What is thought to be the basis for its different binding affinities in the two compartments? If you were designing the system, in which compartment would you have the highest concentration of KDEL receptor? Would you predict that the KDEL receptor, which is a transmembrane protein, would itself possess an ER retrieval signal?

13–12 *Drosophila shibire* mutants, which carry a temperature-sensitive mutation in the dynamin gene, are rapidly paralyzed when the temperature is elevated. They recover quickly once the temperature is lowered. The complete paralysis at the elevated temperature suggests that synaptic transmission between nerve and muscle cells is blocked. Electron micrographs of nerve terminals from the paralyzed flies showed a loss of synaptic vesicles and a tremendously increased number of coated pits relative to normal synapses (**Figure Q13–3**). What step in synaptic transmission is defective in *shibire* mutants at high temperature?

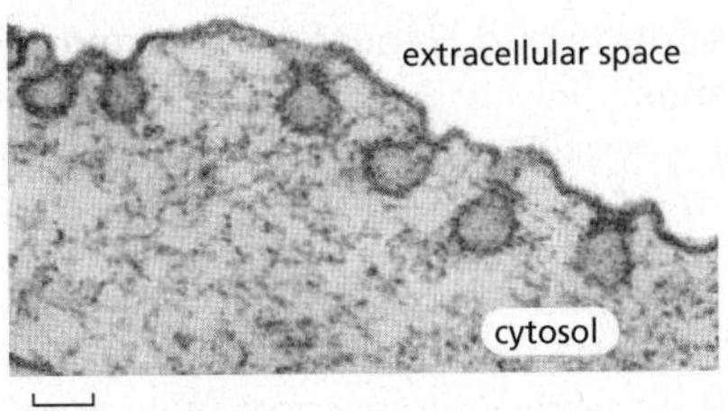

Figure Q13–3 Electron micrograph of a nerve terminal from a *shibire* mutant fly at elevated temperature (Problem 13–12). (From J.H. Koenig and K. Ikeda, *J. Neurosci.* 9:3844–3860, 1989. With permission from the Society of Neuroscience.)

13–13 A macrophage ingests the equivalent of 100% of its plasma membrane each half hour by endocytosis. What is the rate at which membrane is returned by exocytosis?

13–14 The recycling of transferrin receptors has been studied by covalently attaching radioactive iodine (^{125}I) to the receptors on the cell surface at 0°C and then following their fate at 0°C and 37°C. If the labeled cells were kept at 0°C and treated with trypsin to digest the receptors on the cell surface, all the labeled transferrin receptors were completely degraded. If the cells were first warmed to 37°C for 1 hour, then cooled back to 0°C and treated with trypsin, about 70% of the labeled receptors were resistant to trypsin. Why did the transferrin receptors respond differently to trypsin digestion depending on whether they were kept at 0°C or first warmed to 37°C before they were returned to 0°C?

13–15 How does the low pH of lysosomes protect the rest of the cell from lysosomal enzymes in case the lysosome breaks?

13–16 Melanosomes are specialized lysosomes that store pigments for eventual release by exocytosis. Various cells such as skin and hair cells then take up the pigment, which accounts for their characteristic pigmentation. Mouse mutants that have defective melanosomes often have pale or unusual coat colors. One such light-colored mouse, the *Mocha* mouse (**Figure Q13–4**), has a defect in the gene for one of the subunits of the adaptor protein complex AP3, which is associated with coated vesicles budding from the *trans* Golgi network. How might the loss of AP3 cause a defect in melanosomes?

Figure Q13–4 A normal mouse and the *Mocha* mouse (Problem 13–16). In addition to its light coat color, the *Mocha* mouse has a poor sense of balance. (Courtesy of Margit Burmeister.)

REFERENCES

General

Harrison SC & Kirchhausen T (2010) Structural biology: conservation in vesicle coats. *Nature* 466, 1048–1049.

Hernandez-Gonzalez M, Larocque G & Way M (2021) Viral use and subversion of membrane organization and trafficking. *J. Cell Sci.* 134(5), jcs252676.

Pfeffer SR (2013) A prize for membrane magic. *Cell* 155, 1203–1206.

Thor F, Gautschi M, Geiger R & Helenius A (2009) Bulk flow revisited: transport of a soluble protein in the secretory pathway. *Traffic* 10, 1819–1830.

Mechanisms of Membrane Transport and Compartment Identity

Antonny B (2011) Mechanisms of membrane curvature sensing. *Annu. Rev. Biochem.* 80, 101–123.

Burd C & Cullen PJ (2014) Retromer: a master conductor of endosome sorting. *Cold Spring Harb. Perspect. Biol.* 6(2), a016774.

Faini M, Beck R, Weiland FT & Briggs JAG (2013) Vesicle coats: structure, function, and general principles of assembly. *Trends Cell Biol.* 23(6), 279-288.

Ferguson SM & De Camilli P (2012) Dynamin, a membrane-remodelling GTPase. *Nat. Rev. Mol. Cell Biol.* 13, 75–88.

Frost A, Unger VM & De Camilli P (2009) The BAR domain superfamily: membrane-molding macromolecules. *Cell* 137, 191–196.

Grosshans BL, Ortiz D & Novick P (2006) Rabs and their effectors: achieving specificity in membrane traffic. *Proc. Natl. Acad. Sci. USA* 103, 11821–11827.

Jackson LP, Kelly BT, McCoy AJ . . . Owen DJ (2010) A large-scale conformational change couples membrane recruitment to cargo binding in the AP2 clathrin adaptor complex. *Cell* 141, 1220–1229.

Jahn R & Scheller RH (2006) SNAREs—engines for membrane fusion. *Nat. Rev. Mol. Cell Biol.* 7, 631–643.

Jean S & Kiger AA (2012) Coordination between RAB GTPase and phosphoinositide regulation and functions. *Nat. Rev. Mol. Cell Biol.* 13, 463–470.

Jin L, Pahuja KB, Wickliffe KE . . . Rape M (2012) Ubiquitin-dependent regulation of COPII coat size and function. *Nature* 482, 495–500.

Martens S & McMahon HT (2008) Mechanisms of membrane fusion: disparate players and common principles. *Nat. Rev. Mol. Cell Biol.* 9, 543–556.

McNew JA, Parlati F, Fukuda R . . . Rothman JE (2000) Compartmental specificity of cellular membrane fusion encoded in SNARE proteins. *Nature* 407, 153–159.

Miller EA & Schekman R (2013) COPII—a flexible vesicle formation system. *Curr. Opin. Cell Biol.* 25, 420–427.

Müller MP & Goody RS (2018) Molecular control of Rab activity by GEFs, GAPs and GDI. *Small GTPases* 9(1–2), 5–21.

Pfeffer SR (2013) Rab GTPase regulation of membrane identity. *Curr. Opin. Cell Biol.* 25, 414–419.

Saito K, Chen M, Bard F . . . Malhotra V (2009) TANGO1 facilitates cargo loading at endoplasmic reticulum exit sites. *Cell* 136, 891–902.

Transport from the Endoplasmic Reticulum Through the Golgi Apparatus

Ellgaard L & Helenius A (2003) Quality control in the endoplasmic reticulum. *Nat. Rev. Mol. Cell Biol.* 4, 181–191.

Emr S, Glick BS, Linstedt AD . . . Wieland FT (2009) Journeys through the Golgi—taking stock in a new era. *J. Cell Biol.* 187, 449–453.

Farquhar MG & Palade GE (1998) The Golgi apparatus: 100 years of progress and controversy. *Trends Cell Biol.* 8, 2–10.

Gillingham AK & Munro S (2016) Finding the Golgi: golgin coiled-coil proteins show the way. *Trends Cell Biol.* 26(6), 399–408.

Ladinsky MS, Mastronarde DN, McIntosh JR . . . Staehelin LA (1999) Golgi structure in three dimensions: functional insights from the normal rat kidney cell. *J. Cell Biol.* 144, 1135–1149.

Pfeffer S (2010) How the Golgi works: a cisternal progenitor model. *Proc. Natl. Acad. Sci. USA* 107, 19614–19618.

Varki A (2011) Evolutionary forces shaping the Golgi glycosylation machinery: why cell surface glycans are universal to living cells. *Cold Spring Harb. Perspect. Biol.* 3, a005462.

Transport From the *Trans* Golgi Network to the Cell Exterior and Endosomes

Burgess TL & Kelly RB (1987) Constitutive and regulated secretion of proteins. *Annu. Rev. Cell Biol.* 3, 243–293.

Martin TF (1997) Stages of regulated exocytosis. *Trends Cell Biol.* 7, 271–276.

Mellman I & Nelson WJ (2008) Coordinated protein sorting, targeting and distribution in polarized cells. *Nat. Rev. Mol. Cell Biol.* 9, 833–845.

Mostov K, Su T & ter Beest M (2003) Polarized epithelial membrane traffic: conservation and plasticity. *Nat. Cell Biol.* 5, 287–293.

Pang ZP & Südhof TC (2010) Cell biology of Ca^{2+}-triggered exocytosis. *Curr. Opin. Cell Biol.* 22, 496–505.

Rizo J & Xu J (2015) The synaptic vesicle release machinery. *Annu. Rev. Biophys.* 44, 339–367.

Schuck S & Simons K (2004) Polarized sorting in epithelial cells: raft clustering and the biogenesis of the apical membrane. *J. Cell Sci.* 117, 5955–5964.

Transport Into the Cell from the Plasma Membrane: Endocytosis

Bonifacino JS & Traub LM (2003) Signals for sorting of transmembrane proteins to endosomes and lysosomes. *Annu. Rev. Biochem.* 72, 395–447.

Brown MS & Goldstein JL (1986) A receptor-mediated pathway for cholesterol homeostasis. *Science* 232, 34–47.

Conner SD & Schmid SL (2003) Regulated portals of entry into the cell. *Nature* 422, 37–44.

Henne WM, Stenmark H & Emr SD (2013) Molecular mechanisms of the membrane sculpting ESCRT pathway. *Cold Spring Harb. Perspect. Biol.* 5(9), a016766.

Howes MT, Mayor S & Parton RG (2010) Molecules, mechanisms, and cellular roles of clathrin-independent endocytosis. *Curr. Opin. Cell Biol.* 22, 519–527.

Huotari J & Helenius A (2011) Endosome maturation. *EMBO J.* 30, 3481–3500.

Kelly BT & Owen DJ (2011) Endocytic sorting of transmembrane protein cargo. *Curr. Opin. Cell Biol.* 23, 404–412.

Maxfield FR & McGraw TE (2004) Endocytic recycling. *Nat. Rev. Mol. Cell Biol.* 5, 121–132.

McMahon HT & Boucrot E (2011) Molecular mechanism and physiological functions of clathrin-mediated endocytosis. *Nat. Rev. Mol. Cell Biol.* 12, 517–533.

Schöneberg J, Lee I-H, Iwasa JH & Hurley JH (2017) Reverse-topology membrane scission by the ESCRT proteins. *Nat. Rev. Mol. Cell Biol.* 18(1), 5–17.

Sorkin A & von Zastrow M (2009) Endocytosis and signalling: intertwining molecular networks. *Nat. Rev. Mol. Cell Biol.* 10, 609–622.

The Degradation and Recycling of Macromolecules in Lysosomes

Andrews NW (2000) Regulated secretion of conventional lysosomes. *Trends Cell Biol.* 10, 316–321.

Bloomfield A & Kay RR (2016) Uses and abuses of micropinocytosis. *J. Cell Sci.* 129(14), 2697–2705.

de Duve C (2005) The lysosome turns fifty. *Nat. Cell Biol.* 7, 847–849.

Flannagan RS, Jaumouillé V & Grinstein S (2012) The cell biology of phagocytosis *Annu. Rev. Pathol.* 7, 61–98.

Futerman AH & van Meer G (2004) The cell biology of lysosomal storage disorders. *Nat. Rev. Mol. Cell Biol.* 5, 554–565.

Levine B & Kroemer G (2019) Biological functions of autophagy genes: a disease perspective. *Cell* 176(1–2), 11–42.

Mizushima N, Yoshimori T & Ohsumi Y (2011) The role of Atg proteins in autophagosome formation. *Annu. Rev. Cell Dev. Biol.* 27, 107–132.

Energy Conversion and Metabolic Compartmentation: Mitochondria and Chloroplasts

CHAPTER 14

IN THIS CHAPTER

To maintain their high degree of organization in a universe that is constantly drifting toward chaos, cells have a constant need for a plentiful supply of ATP, as explained in Chapter 2. In eukaryotic cells, most of the ATP that powers life processes is produced by specialized, membrane-enclosed, *energy-converting organelles*. These are of two types. **Mitochondria** burn food molecules to produce ATP by *oxidative phosphorylation*, and they are present in virtually all cells of animals, plants, and fungi. In contrast, **chloroplasts** harness solar energy to produce ATP by *photosynthesis*, and they occur only in plants and green algae. In electron micrographs, the most striking features of both mitochondria and chloroplasts are their extensive internal membrane systems. These internal membranes contain sets of membrane protein complexes that work together to harvest energy and then use that energy to catalyze the production of most of the cell's ATP.

Comparisons of DNA sequences suggest that the energy-converting organelles in present-day eukaryotes originated from prokaryotic cells that entered symbiotic relationships during the evolution of eukaryotes (discussed in Chapter 1). This would explain why mitochondria and chloroplasts contain their own DNA, which still encodes a subset of their proteins. Over time, the majority of the genes originally encoded in the prokaryotic genome appear to have been transferred to the nuclear genome. As a result, these organelles have become heavily dependent on those nuclear-encoded proteins being synthesized in the cytosol and then imported into the organelle. And eukaryotic cells rely on these organelles not only for the ATP they need for biosynthesis, solute transport, and movement, but also for many important biosynthetic reactions that occur inside each organelle.

The common evolutionary origin of the energy-converting machinery in mitochondria, chloroplasts, and prokaryotes (archaea and bacteria) is reflected in the fundamental mechanism that they share for harnessing energy. This mechanism is known as **chemiosmotic coupling**, signifying a link between the chemical bond-forming reactions that generate ATP ("chemi") and membrane transport processes ("osmotic"). The chemiosmotic process, which occurs in two linked stages, is performed by protein complexes embedded in the membrane that is colored in the bacterium, mitochondrion, and chloroplast shown in **Figure 14–1**.

Stage 1: High-energy electrons (derived from the oxidation of food molecules, from pigments excited by sunlight, or from other sources described later) are transferred along a series of electron-transport protein complexes that form an *electron-transport chain* embedded in a membrane. Each electron transfer releases a small amount of energy that is used to pump

Figure 14–1 The membranes of bacteria, mitochondria, and chloroplasts that carry out chemiosmotic energy-conversion processes. Mitochondria and chloroplasts are cell organelles that have originated from bacteria and have retained the bacterial energy-conversion mechanisms as well as other features of their ancestry. Like their bacterial ancestors, mitochondria and chloroplasts have an outer and an inner membrane. For each type of cell or organelle, it is the membrane that is colored in the figure that contains its energy-harvesting electron-transport chains. The deep invaginations of the mitochondrial inner membrane and the internal membrane system of the chloroplast harbor the machinery for cellular respiration and photosynthesis, respectively.

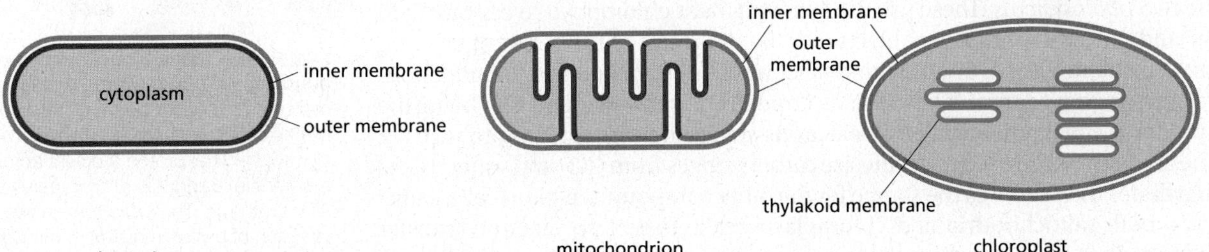

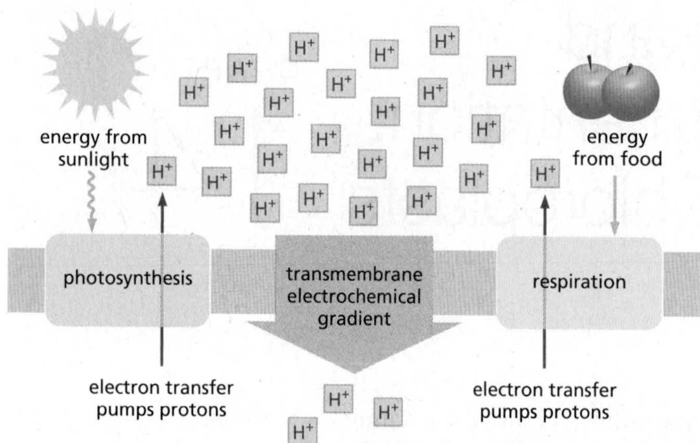

Figure 14–2 **Stage 1 of chemiosmotic coupling.** Energy from either sunlight or the oxidation of food compounds is captured by special, membrane-embedded protein complexes to generate an electrochemical proton gradient across a membrane. The electrochemical gradient serves as a versatile energy store that drives energy-requiring reactions in mitochondria, chloroplasts, and prokaryotes (bacteria and archaea).

protons (H^+) and thereby generate a large *electrochemical gradient* across the membrane (Figure 14–2). As discussed in Chapter 11, such an electrochemical gradient provides a way of temporarily storing energy, and it can be harnessed to do useful work when ions flow back across the membrane.

Stage 2: The protons flow back down their electrochemical gradient through an elaborate membrane protein machine called *ATP synthase*, which uses this energy to catalyze the production of ATP from ADP and inorganic phosphate. This ubiquitous enzyme works like a turbine in the membrane, driven by a flow of protons, to synthesize ATP (Figure 14–3). In this way, the energy derived from either food or sunlight in stage 1 is converted into the chemical energy of a phosphate bond in ATP.

Electrons can move through protein complexes in biological systems via tightly bound metal ions or other carriers that take up and release electrons easily or by special small molecules that pick electrons up at one location and deliver them to another. For mitochondria, one critical electron carrier is NAD^+, a water-soluble small molecule that takes up two electrons and one H^+ derived from food molecules (fats and carbohydrates) to become NADH (see Figure 2–36). NADH transfers these electrons from food-derived molecules to the inner mitochondrial membrane. There, the electrons from the energy-rich NADH are passed from one membrane protein complex to the next, transitioning to a lower-energy state at each step, until they reach a final complex in which they combine with molecular oxygen (O_2) plus protons to produce water. The energy released at each step as the electrons flow down this path from the energy-rich NADH to the low-energy water molecule drives H^+ pumps that are embedded in three different protein complexes in the inner mitochondrial membrane. Together, these three complexes generate the proton gradient (or proton-motive force) that is harnessed by ATP synthase to produce ATP—the molecule that serves as the universal energy currency throughout the cell.

Figure 14–4 compares the electron-transport processes in mitochondria, which harness energy from food molecules, with those in chloroplasts, which harness energy from sunlight. The energy-conversion systems of mitochondria and chloroplasts can be described in similar terms, and we shall see later in the chapter that two of their key components are closely related. One of these is the ATP synthase, and the other is a proton pump.

Among the crucial constituents that are unique to photosynthetic organisms are the two *photosystems*. These use the green pigment chlorophyll to capture light energy and power the transfer of electrons. The net result of the series of electron transfers in chloroplasts is opposite to the net result in mitochondria. In mitochondria, electrons derived from food are transferred to O_2, with water and CO_2 being the final products. But in photosynthesis, electrons are taken from water to produce O_2, and these electrons are used to synthesize carbohydrates from CO_2 and water. These carbohydrates then serve as the source for the other compounds a plant cell needs.

Thus, both mitochondria and chloroplasts make use of an electron-transfer chain to produce an H^+ gradient that powers reactions that are critical for the cell.

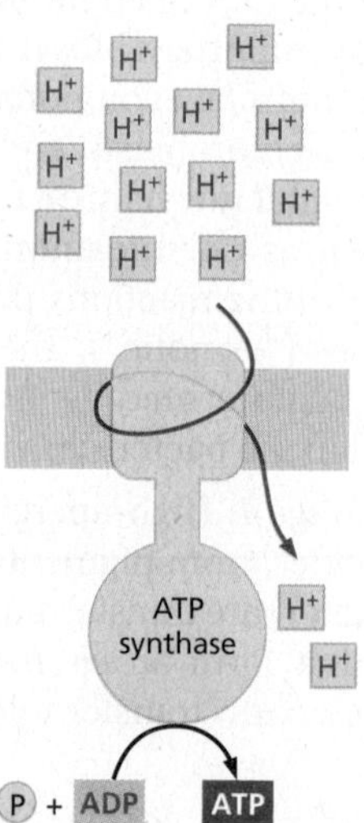

Figure 14–3 **Stage 2 of chemiosmotic coupling.** An ATP synthase protein machine *(yellow)* embedded in the lipid bilayer of a membrane harnesses the electrochemical proton gradient across the membrane, using this energy store to drive ATP synthesis. The *red arrow* shows the direction of proton movement through the ATP synthase.

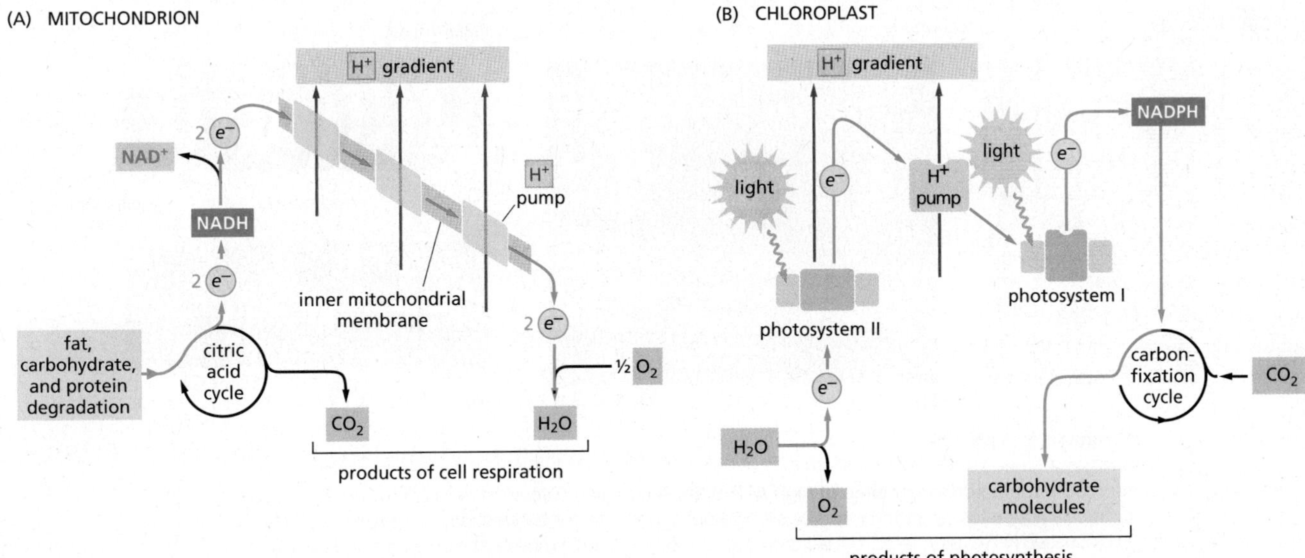

Figure 14–4 A comparison of two electron-conversion processes. The mitochondrion converts energy from chemical fuels, whereas the chloroplast converts energy from sunlight. In both cases, electron flow is indicated by *blue arrows*. As indicated, the electrons pass through a series of protein complexes embedded in a membrane. (A) In the mitochondrion, energy-rich molecules derived from fat, carbohydrate, or protein degradation are fed into the citric acid cycle. This cycle provides electrons to generate the energy-rich compound NADH from NAD^+. The NADH electrons then flow down an energy gradient as they pass from one protein complex to the next in the mitochondrial electron-transport chain, until they combine with molecular O_2 and H^+ in the final complex to produce water. The energy released is harnessed by the three protein complexes to pump H^+ across the membrane. (B) In the chloroplast, by contrast, light-derived energy is used to extract electrons from water in the photosystem II complex, and molecular O_2 is released. The high-energy electrons produced by photosystem II are passed to the next protein complex in the chain, which uses some of the energy derived from electron transfer to pump protons across the membrane, before passing the electrons to photosystem I, where sunlight generates high-energy electrons that combine with $NADP^+$ to produce NADPH (NADPH is a small molecule closely related to the NADH used in mitochondria). This NADPH then enters the *carbon-fixation cycle* along with CO_2 to generate the carbohydrates that provide both the carbon and the energy required for cell processes. Note that the products in A are the inputs for B, while the products in B are inputs for A.

However, chloroplasts generate O_2 and take up CO_2, whereas mitochondria consume O_2 and release CO_2 (see Figure 14–4). Plants provide the food for animals, and the complementary chemistry that is performed by mitochondria and chloroplasts has been fundamental for producing a sustainable ecosystem on Earth—balancing O_2 with CO_2 and carbohydrate production with carbohydrate consumption.

THE MITOCHONDRION

Mitochondria can occupy up to 20% of the cytoplasmic volume of a eukaryotic cell. Although they are often depicted as short, bacterium-like bodies with a diameter of 0.5–1 μm, they are in fact remarkably dynamic and plastic, moving about the cell, constantly changing shape, dividing, and fusing (**Movie 14.1**). Mitochondria are often associated with the cytoskeleton, which determines their distribution in different cell types. Thus, in highly polarized cells such as neurons, mitochondria can move long distances (up to a meter or more in the extended axons of neurons), being propelled along microtubule tracks from where they are formed in the cell body to sites of energy demand like synapses (**Movie 14.2**). In other cells, mitochondria remain fixed at points of high energy demand; for example, in skeletal or cardiac muscle cells, they pack between myofibrils, and in sperm cells they wrap tightly around the flagellum (**Figure 14–5**).

In addition to their directional movement along cytoskeletal filaments, mitochondria undergo highly dynamic interactions with other membrane systems in the cell, most notably the endoplasmic reticulum (ER). Contacts between mitochondria and ER define specialized domains thought to facilitate the exchange of lipids, calcium, and potentially other molecules between the two

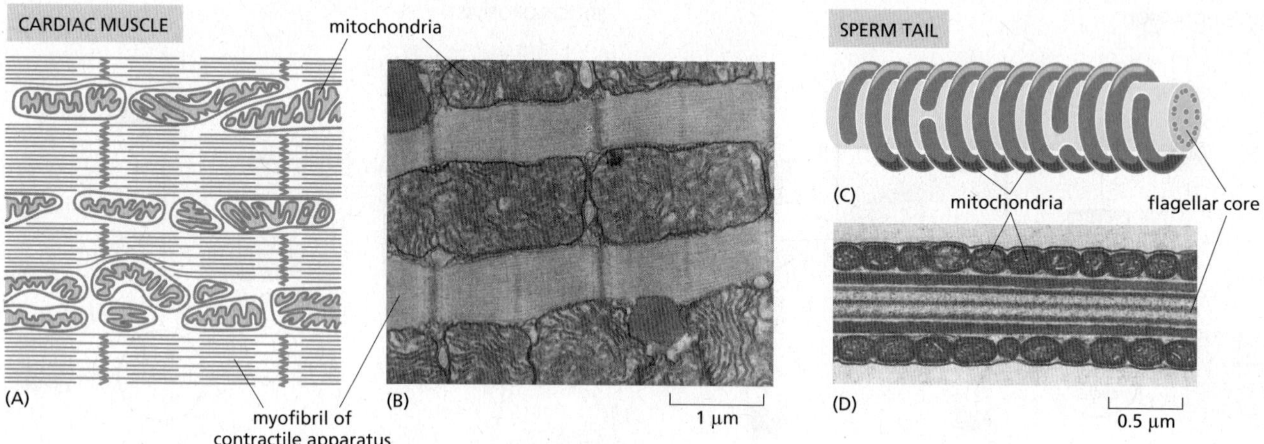

Figure 14–5 Localization of mitochondria near sites of high ATP demand. (A) In a cardiac muscle cell, mitochondria are located close to the contractile apparatus, where ATP hydrolysis provides the energy for contraction. The structure of the contractile apparatus is discussed in Chapter 16. (B) An electron micrograph of cardiac muscle shows a preponderance of mitochondria. (C) In a sperm, mitochondria are located in the tail, wrapped around a portion of the motile flagellum that requires ATP for its movement. The internal structure of the flagellar core is discussed in Chapter 16. (D) Micrograph showing a flagellum that has been thinly sliced to reveal the internal core structure as well as the surrounding mitochondria. (B, Keith Porter papers, Center for Biological Sciences Archives, University of Maryland, Baltimore County; D, from W. Bloom and D.W. Fawcett, A Textbook of Histology, 10th ed. Philadelphia, PA: W.B. Saunders Company, 1975. Reprinted with permission from the Estate of D.W. Fawcett.)

membrane systems. These regions of ER contact appear to be specialized domains that are also the sites of mitochondrial fission, which aids the distribution and partitioning of the mitochondria within cells, as we discuss later.

The acquisition of mitochondria was a prerequisite for the evolution of complex animals. Without mitochondria, present-day animal cells would have had to generate all of their ATP through glycolysis. When glycolysis converts glucose to pyruvate, it releases only a small fraction of the total free energy that is potentially available from glucose oxidation (see Chapter 2). In mitochondria, the metabolism of sugars is complete: pyruvate is imported into the mitochondrion and ultimately oxidized to CO_2 and H_2O, which allows 15 times more ATP to be made from a sugar than by glycolysis alone. As explained later, this energy source became available only after enough molecular oxygen had accumulated in Earth's atmosphere to allow organisms to take advantage, via respiration, of the energy potentially available from the complete oxidation of organic compounds.

The Mitochondrion Has an Outer Membrane and an Inner Membrane

Like the bacteria from which they likely originated, mitochondria have an outer and an inner membrane. The two membranes have distinct functions and properties, and they delineate separate compartments within the organelle. The inner membrane, which surrounds the internal **mitochondrial matrix** compartment (**Figure 14–6**), is highly folded to form invaginations known as **cristae** (the singular is crista), which contain in their membranes the proteins of the electron-transport chain. Where the inner membrane runs parallel to the outer membrane, between the cristae, it is known as the *inner boundary membrane*. The narrow (20–30 nm) gap between the inner boundary membrane and the outer membrane is known as the **intermembrane space**. The cristae are about 20-nm-wide membrane discs or tubules that protrude deeply into the matrix and enclose the *crista space*. The *crista membrane* is continuous with the inner boundary membrane, and where their membranes join, the membrane forms narrow membrane tubes or slits, known as *crista junctions*.

Like the bacterial outer membrane, the **outer mitochondrial membrane** is freely permeable to ions and to small molecules as large as 5000 daltons. This is because it contains many porin molecules, a special class of β-barrel-type

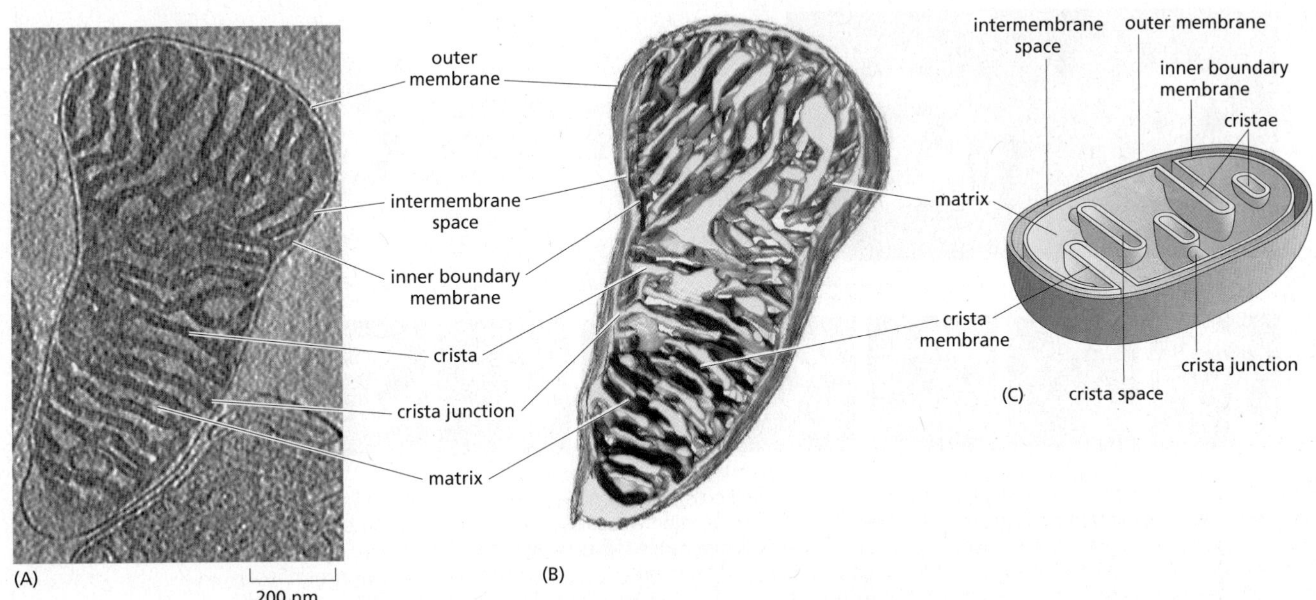

Figure 14–6 **Structure of a mitochondrion.** (A) Tomographic slice through a three-dimensional map of a mouse heart mitochondrion determined by electron microscope tomography. The outer membrane envelops the inner boundary membrane. The inner membrane is highly folded into tubular or lamellar cristae, which crisscross the matrix. The dense matrix, which contains most of the mitochondrial protein, appears dark in the electron microscope, whereas the intermembrane space and the crista space appear light because of their lower protein content. The inner boundary membrane follows the outer membrane closely at a distance of ~20 nm. The inner membrane turns sharply at the crista junctions, where the cristae join the inner boundary membrane. (B) Tomographic surface-rendered portion of a mouse heart mitochondrion, showing how flattened cristae project into the matrix from the inner membrane (**Movie 14.3**). (C) Schematic drawing of a mitochondrion showing the outer membrane *(gray)* and the inner membrane *(yellow)*. Note that the inner membrane is compartmentalized into the inner boundary membrane and the crista membrane. There are three distinct spaces: the intermembrane space, the crista space, and the matrix. (A and B, courtesy of Tobias Brandt.)

membrane protein that creates aqueous pores across the membrane (see Figure 10–21). As a consequence, the intermembrane space between the outer and inner membrane has a pH and ionic composition very similar to that of the cytoplasm, and there is no electrochemical gradient across the outer membrane.

Fission, Fusion, Distribution, and Degradation of Mitochondria

In mammalian cells, mitochondrial DNA makes up less than 1% of the total cellular DNA. In other cells, however, a larger fraction of the cellular DNA may be present in mitochondria or chloroplasts (**Table 14–1**), and a large fraction of the total RNA and protein synthesis takes place in the organelles.

TABLE 14–1 Quantity of Organelles and Organelle DNA in Some Cells and Tissues

Organism	Tissue or cell type	DNA molecules per organelle	Organelles per cell	Organelle DNA as percentage of total cellular DNA (%)
		Mitochondrial DNA		
Rat	Liver	5–10	1000	1
Yeast*	Vegetative	2–50	1–50	15
		Chloroplast DNA		
Chlamydomonas	Vegetative	80	1	7
Maize	Leaves	0–300**	20–40	0–15**

*The large variation in the number and size of mitochondria per cell in yeasts is due to mitochondrial fusion and fission.
**In maize, the amount of chloroplast DNA drops precipitously in mature leaves, after cell division ceases: the chloroplast DNA is degraded, and stable mRNAs persist to provide for protein synthesis.

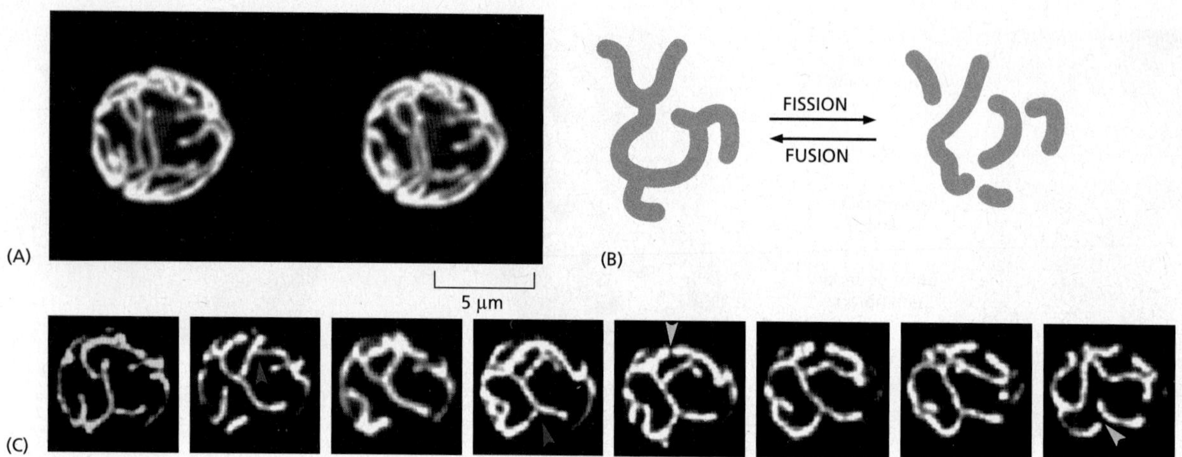

Figure 14–7 The mitochondrial reticulum is dynamic. (A) In yeast cells, mitochondria form a continuous reticulum on the cytoplasmic side of the plasma membrane (stereo pair). (B) A balance between fission and fusion determines the arrangement of the mitochondria in different cells. (C) Time-lapse fluorescence microscopy shows the dynamic behavior of the mitochondrial network in a yeast cell. In addition to shape changes, fission *(green arrowheads)* and fusion *(red arrowheads)* constantly remodel the network. These pictures were taken at 3-minute intervals. (A and C, from J. Nunnari et al., *Mol. Biol. Cell* 8:1233–1242, 1997. With permission from the American Society for Cell Biology.)

Mitochondria and chloroplasts are large enough to be visible by light microscopy in living cells. For example, mitochondria can be visualized by expressing in cells a genetically engineered fusion of a mitochondrial protein linked to green fluorescent protein (GFP), or cells can be incubated with a fluorescent dye that is specifically taken up by mitochondria because of their membrane potential. Such images demonstrate that the mitochondria in living cells are dynamic—frequently dividing by fission, fusing, and changing shape (**Figure 14–7** and **Movie 14.4**). The fission of mitochondria may be necessary so that small parts of the network can pinch off and reach remote regions of the cell; for example, in the thin, extended axon and dendrites of a neuron, for them to be distributed to each of two daughters, or for them to be degraded by mitochondrial autophagy (see Chapter 13).

The fission and fusion of mitochondria are topologically complex processes that must ensure the integrity of the separate mitochondrial compartments defined by the inner and outer membranes. These processes control the number and shape of mitochondria, which can vary dramatically in different cell types, ranging from multiple spherical or wormlike organelles to a highly branched, net-shaped single organelle called a *reticulum*. Each process depends on its own special set of proteins. The mitochondrial fission machine works by assembling dynamin-related GTPases (discussed in Chapter 13) into helical oligomers that cause local constrictions in tubular mitochondria. GTP hydrolysis then generates the mechanical force that severs the inner and outer mitochondrial membranes in one step (**Figure 14–8**). Mitochondrial fusion requires two separate machineries, one each for the outer and the inner membrane (**Figure 14–9**).

Mitochondria that have become nonfunctional as assessed by loss of the proton electrochemical gradient typically become overly fragmented because of ongoing fission but loss of fusion. This state is typically coupled to the highly specific and organized degradation of such mitochondria through a

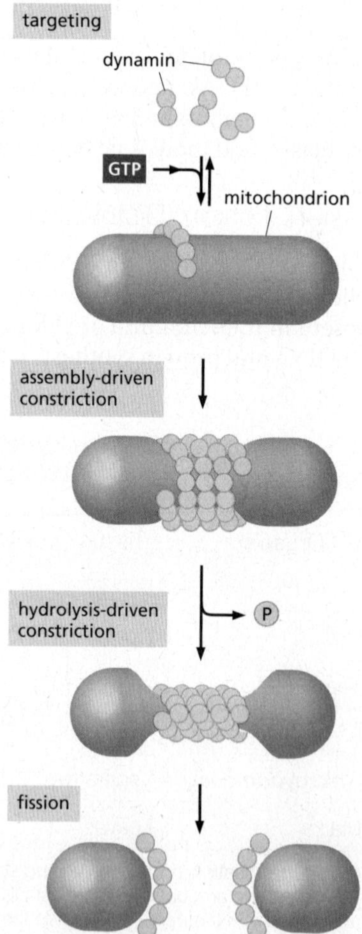

Figure 14–8 A model for mitochondrial division. Dynamin *(yellow)* exists as dimers in the cytosol, which form larger oligomeric structures in a process that requires GTP hydrolysis. At special sites of ER contact (not shown), dynamin assemblies interact with the outer mitochondrial membrane through special adaptor proteins, forming a spiral of GTP–dynamin around the mitochondrion that causes a constriction. A concerted GTP-hydrolysis event in the dynamin subunits is then thought to produce the conformational changes that result in fission. (Adapted from S. Hoppins et al., *Annu. Rev. Biochem*. 76:751–780, 2007. With permission from Annual Reviews.)

specialized organelle-specific form of autophagy known as mitophagy (see p. 806 in Chapter 13). One mechanism that induces autophagy is based on the fact that mitochondrial protein import becomes inefficient upon loss of the electrochemical gradient, causing the Pink1 protein kinase, which is normally imported into mitochondria, to be retained on the mitochondrial outer membrane. Pink1 then phosphorylates both mitochondria-associated ubiquitin and the mitophagy regulator Parkin to initiate the formation of an autophagosome that engulfs the damaged mitochondrion. The Parkin protein is so-named because mutations in the gene encoding it (as well as the gene encoding Pink1) are associated with Parkinson's disease, which can be characterized by an accumulation of nonfunctional mitochondria.

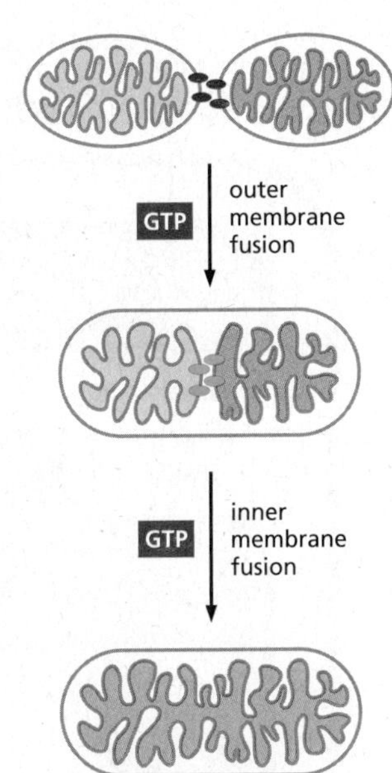

Figure 14–9 **A model for mitochondrial fusion.** The fusions of the outer and inner mitochondrial membranes are coordinated sequential events, each of which requires a separate set of protein factors. Outer-membrane fusion is brought about by an outer-membrane GTPase *(purple)*, which forms an oligomeric complex that includes subunits anchored in each of the two membranes to be fused. Fusion of outer membranes requires GTP and an H^+ gradient across the inner membrane. For fusion of the inner membrane, a dynamin-related protein forms an oligomeric tethering complex *(blue)* that includes subunits anchored in the two inner membranes to be fused. Fusion of the inner membranes requires GTP and an electrical potential across the inner membrane. (Adapted from S. Hoppins et al., *Annu. Rev. Biochem*. 76:751–780, 2007. With permission from Annual Reviews.)

The Inner Membrane Cristae Contain the Machinery for Electron Transport and ATP Synthesis

Unlike the outer mitochondrial membrane, the **inner mitochondrial membrane** is a diffusion barrier to ions and small molecules, just like the inner membrane of a bacterium. However, selected ions, most notably protons and phosphate, as well as essential metabolites such as ATP, ADP, and pyruvate, can pass through it by means of specific transport proteins.

The inner mitochondrial membrane is highly differentiated into functionally distinct regions with different protein compositions. As discussed in Chapter 10, the lateral segregation of membrane regions with different protein and lipid compositions is a key feature of cells. The boundary membrane region of the inner mitochondrial membrane contains the machinery for protein import, for new membrane protein insertion, and for the assembly of the respiratory-chain complexes. The membranes of the cristae, despite being continuous with the boundary membrane, have a distinct composition. The crista membrane contains the ATP synthase enzyme that produces most of the cell's ATP: this large protein machine is also thought to structurally support the highly curved, crista membrane structure. The crista membrane also contains the large protein complexes of the **respiratory chain**—the name given to the mitochondrion's electron-transport chain. Cristae membranes have one of the highest protein densities of all biological membranes, with a lipid content of 25% and a protein content of 75% by weight.

At the cristae junctions, where the membranes of the cristae join the boundary membrane, specialized protein complexes provide a diffusion barrier that segregates the membrane proteins in the two regions of the inner membrane; these complexes are also thought to anchor the cristae to the outer membrane, thus maintaining the highly folded topology of the inner membrane. The folding of the inner membrane into cristae greatly increases the membrane area available for oxidative phosphorylation. In highly active cardiac muscle cells, for example, the total area of cristae membranes can be up to 20 times larger than the area of the cell's plasma membrane. In total, the surface area of cristae membranes in each human body adds up to roughly the size of a football field.

The Citric Acid Cycle in the Matrix Produces NADH

Together with the cristae that project into it, the matrix is the principal location for the majority of mitochondrial metabolic proteins. Mitochondria can use pyruvate, fatty acids, as well as other substrates as fuel. Pyruvate is derived from glucose and other sugars, whereas fatty acids are derived from fats. Unlike pyruvate and fatty acids, amino acids derived from protein degradation and the ketone bodies that can be produced from fatty acids are typically minor mitochondrial fuels, but they become very important substrates under specific situations such as prolonged fasting. All of these fuel molecules are transported across the inner mitochondrial membrane by specialized transport proteins, and they are then converted to the crucial metabolic intermediate *acetyl CoA* by enzymes located in the mitochondrial matrix (see Chapter 2).

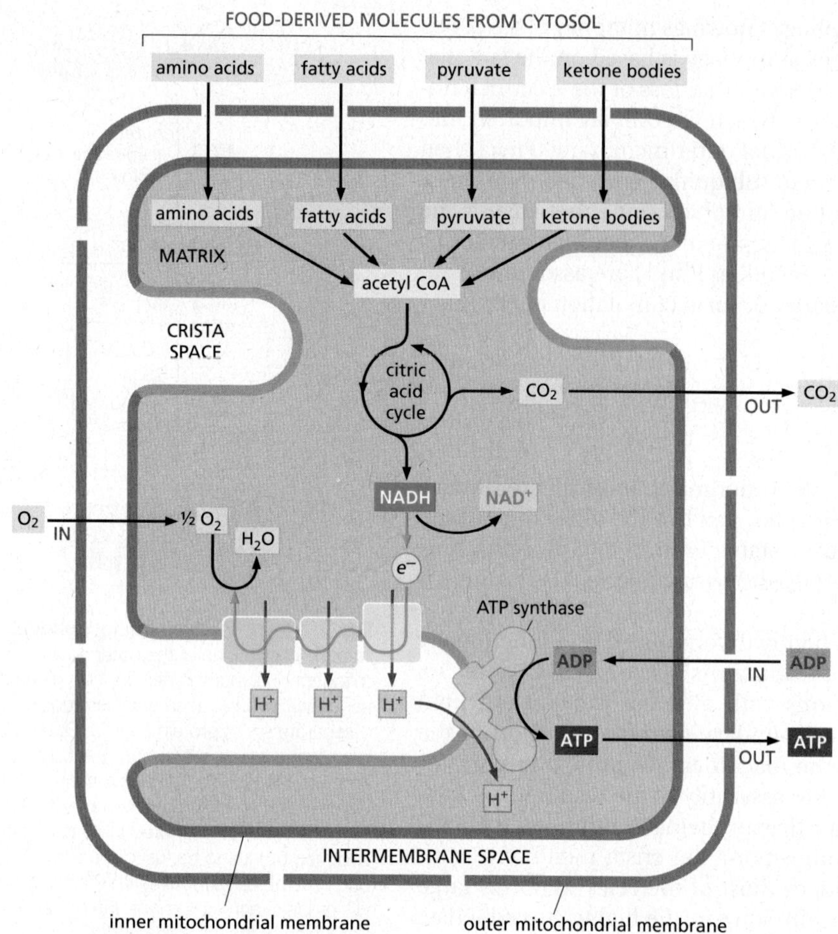

Figure 14–10 A summary of the energy-converting metabolism in mitochondria. Pyruvate, fatty acids, amino acids, and ketone bodies enter the mitochondrion *(top of the figure)* and are broken down to acetyl CoA. The acetyl CoA is oxidized through a series of steps by the citric acid cycle, passing electrons to NAD^+ thereby generating NADH, which then passes its high-energy electrons to the first of three large protein complexes in the electron-transport chain. In the process of oxidative phosphorylation, these electrons pass along the electron-transport chain in the inner membrane cristae to oxygen (O_2). This electron transport generates a proton gradient, which drives the production of ATP by the ATP synthase (see Figure 14–3).

The acetyl groups in acetyl CoA are oxidized in the matrix via the *citric acid cycle,* also called the Krebs cycle (see Figure 2–58 and Movie 2.6). The oxidation of these carbon atoms in acetyl CoA produces CO_2, which diffuses out of the mitochondrion to be released to the environment as a waste product. More important, the citric acid cycle saves a great deal of the bond energy released by this oxidation in the form of electrons carried by NADH. These electrons from NADH are transferred from the matrix to the electron-transport chain in the inner mitochondrial membrane, where—through the *chemiosmotic coupling* process described previously (see Figures 14–2 and 14–3)—the energy that was carried by NADH electrons is converted into phosphate-bond energy in ATP. **Figure 14–10** outlines this sequence of reactions schematically.

The matrix contains the genetic system of the mitochondrion, including the mitochondrial DNA and the ribosomes. The large number of enzymes required for the maintenance of the mitochondrial genetic system, as well as for many other essential reactions to be outlined next, accounts for the very high protein concentration in the matrix; at more than 500 mg/mL, this concentration is close to that in a protein crystal.

Mitochondria Have Many Essential Roles in Cellular Metabolism

Mitochondria not only generate most of the cell's ATP; they also provide many other essential resources for biosynthesis and cell growth (**Table 14–2**). Before describing in detail the remarkable machinery of the respiratory chain, we diverge briefly to touch on some of these important roles.

Mitochondria are critical for buffering the redox potential in the cytosol. Cells need a constant supply of the electron acceptor NAD^+ for the central reaction in glycolysis that converts glyceraldehyde 3-phosphate to 1,3-bisphosphoglycerate

TABLE 14–2 Mitochondrial Functions

Function	Description
Production of ATP	Oxidative phosphorylation in mitochondria produces most of the ATP used by eukaryotic cells
Regeneration of NAD^+	NAD^+ is required for glycolysis and other reactions; under aerobic conditions, this NAD^+ is regenerated when NADH donates electrons to oxygen via the respiratory chain (see Chapter 2)
Provision of precursors for biosynthesis of amino acids, nucleotides, fatty acids	Intermediates produced by the citric acid cycle, which takes place in the mitochondrial matrix, serve as precursors for the synthesis of many macromolecules (see Figure 2–60)
Participation in synthesis of heme and iron–sulfur clusters	These metal-containing components are synthesized in mitochondria and play a central role in respiration and other cellular processes
Cell signaling	Mitochondria buffer the concentration of Ca^{2+}, an ion that plays a role in many signaling processes, including muscle contraction (see Chapters 15 and 16)
Generation of reactive oxygen species	Although reactive oxygen species can damage macromolecules, they are also involved in signaling
Regulation of apoptosis	Molecules released from mitochondria trigger a proteolytic cascade that leads to cell death (see Chapter 18)

(see Figure 2–47). This NAD^+ is converted to NADH in the process, and the NAD^+ needs to be regenerated by transferring the high-energy NADH electrons somewhere. The NADH electrons will eventually be used to help drive oxidative phosphorylation inside the mitochondrion. But the inner mitochondrial membrane is impermeable to NADH. The electrons are therefore passed from the NADH to smaller molecules in the cytosol that can move through the inner mitochondrial membrane. Once in the matrix, these smaller molecules transfer their electrons to the NAD^+ located there to form mitochondrial NADH, after which they are returned to the cytosol for recharging—creating a so-called *shuttle system* for the NADH electrons.

In cells with adequate access to oxygen, this shuttle system and the electron-transport chain provide the cell with a nearly boundless sink for electrons. But under conditions of low oxygen, such as in strenuously exercised muscle or in the center of a poorly vascularized tumor, a cell must change its metabolic program in major ways in order to generate its ATP (see p. 1175 in Chapter 20).

The biosynthesis needed in the cytosol for cell division, cell growth, and normal cell maintenance requires, in addition to ATP, both a constant supply of reducing power and small carbon-rich molecules to serve as the building blocks for the synthesis of nucleotides, amino acids, lipids, and other specialized molecules (discussed in Chapter 2). The reducing power comes from NADPH, a close relative of NADH (see Figure 2–36). Most of this NADPH is produced in the cytosol by a side pathway for the breakdown of sugars (the *pentose phosphate pathway*, an alternative to glycolysis). The needed carbon-rich molecules are almost completely derived from either intermediates of glycolysis or intermediates of the mitochondrial citric acid cycle (the "carbon skeletons" in Panel 2–1, pp. 94–95). For example, citrate produced in the mitochondrial matrix by the citric acid cycle is transported down its electrochemical gradient to the cytosol, where it is metabolized to produce the acetyl CoA that is required in the cytosol to support the production of the fatty acids and sterols that build new membranes (described in Chapter 10). Rapidly dividing normal cells and cancer cells frequently adapt their metabolism in ways that enhance this and other biosynthetic pathways, as part of their program of cell growth (see Figure 20–30).

While the majority of the reactions involved in nucleotide biosynthesis occur in the cytosol, important steps also occur in the mitochondria. For example, the molecules synthesized there provide the single-carbon units required for both purine and pyrimidine biosynthesis.

Mitochondria also play a particularly important role in the biosynthesis of protein building blocks. Mitochondrial acetyl-CoA, α-ketoglutarate, and oxaloacetate are all required for the *de novo* synthesis of amino acids, and, therefore, mitochondrial perturbations can affect amino acid metabolism. Conversely, perturbations in amino acid metabolism can have profound effects on mitochondrial energetics and function through indirect effects on the production and consumption of citric acid cycle intermediates and on the $NADH^+$/NAD ratio.

The urea cycle is a central metabolic pathway in mammals that converts the ammonia (NH_4^+) produced by the breakdown of nitrogen-containing compounds (such as amino acids) to the urea excreted in urine. Two critical steps of the urea cycle are carried out in the mitochondria of liver cells, while the remaining steps occur in the cytosol.

The biosynthesis of *heme groups*—which, as we shall see in the next part of this chapter, play a central role in electron transfer—is another critical process that is shared between the mitochondrion and the cytoplasm. Iron–sulfur clusters, which are essential not only for electron transfer in the respiratory chain (see p. 826) but also for the maintenance and stability of the nuclear genome, are produced in mitochondria (and chloroplasts). Nuclear genome instability, a hallmark of cancer, can sometimes be linked to a decreased function of cellular proteins that contain iron–sulfur clusters. In fact, for those cells that can make sufficient ATP from glycolysis, it has been proposed that iron–sulfur cluster synthesis is the mitochondrial function most essential for viability.

Mitochondria also have a central role in the biosynthesis of membrane lipids. Cardiolipin is a two-headed phospholipid (**Figure 14–11**) that is confined to the mitochondrial membranes, where it is also produced. But mitochondria are also a major source of phospholipids for the biogenesis of other cell membranes. Phosphatidylethanolamine, phosphatidylglycerol, and phosphatidic acid are synthesized in the mitochondrion, while phosphatidylinositol, phosphatidylcholine, and phosphatidylserine are primarily synthesized in the endoplasmic reticulum (ER). As described in Chapter 12, most of the cell's membranes are assembled in the ER, and a critical exchange of lipids between ER and mitochondria is thought to occur at special sites of close contact.

Finally, mitochondria are important for specialized cellular signaling mechanisms. Mitochondria play a critical regulatory role in the major controlled process of eukaryotic cell death, known as apoptosis, as detailed in Chapter 18. In addition, mitochondria buffer calcium concentrations by taking up calcium from the ER (and the sarcoplasmic reticulum in muscle cells) at special membrane junctions.

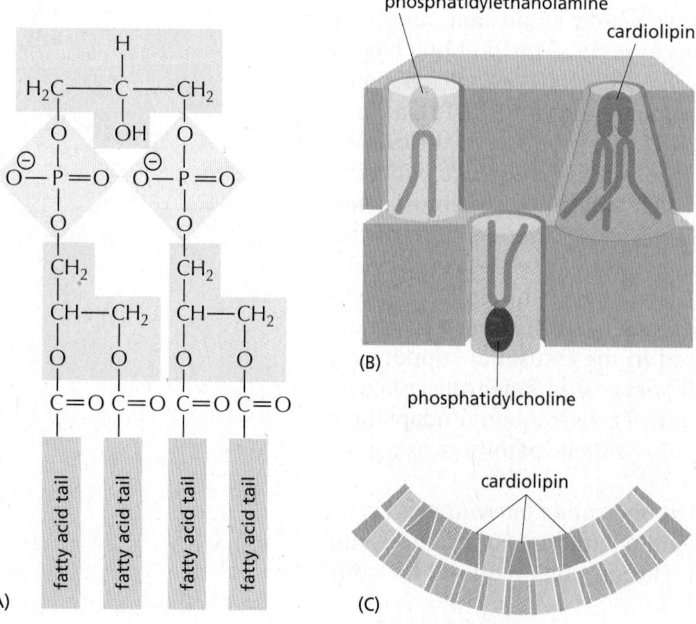

Figure 14–11 The structure of cardiolipin. (A) Cardiolipin consists of two covalently linked phospholipid units, with a total of four rather than the usual two fatty acid chains. (B) Phosphatidylcholine and phosphatidylethanolamine, which are conventional phospholipids and the most abundant lipid species in mitochondria, are shown for comparison. (C) Cardiolipin is produced in the mitochondrial membranes, where it interacts closely with membrane proteins involved in oxidative phosphorylation and ATP transport. In cristae, its unusual structure is thought to support the high curvature of this membrane, as indicated.

Intracellular calcium levels not only control muscle contraction (see Chapter 16); alterations of calcium levels are also implicated in neurodegeneration and apoptosis. Mitochondria are also the principal site of production of reactive oxygen species, which—while they can cause damage—also play important signaling roles. And the concentrations of a number of metabolites with regulatory roles are controlled by mitochondrial metabolism and transport. Clearly, cells and organisms depend on mitochondria in many different ways, and the co-housing of diverse functions in this organelle has important implications for cell function and survival.

We now return to the central function of the mitochondrion in respiratory ATP generation.

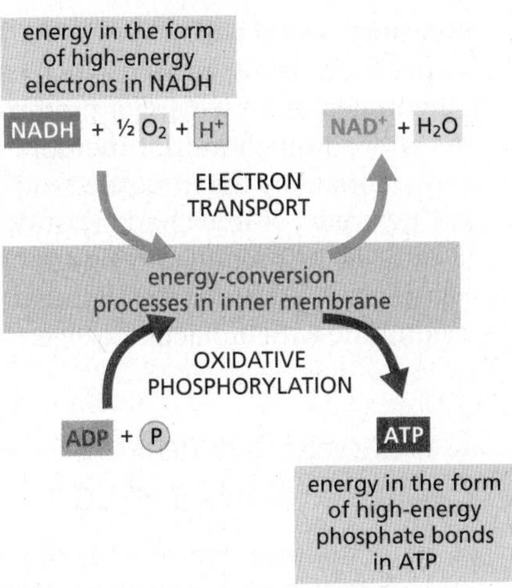

Figure 14–12 **The major net energy conversion catalyzed by the mitochondrion.** In the process of oxidative phosphorylation, the mitochondrial inner membrane serves as a device that changes one form of chemical-bond energy to another, converting a major part of the energy of NADH oxidation into phosphate-bond energy in ATP.

A Chemiosmotic Process Couples Oxidation Energy to ATP Production

Although the citric acid cycle that takes place in the mitochondrial matrix is considered to be part of aerobic metabolism, it does not itself use oxygen. Only the final step of oxidative metabolism consumes molecular oxygen (O_2) directly.

Nearly all the energy available from metabolizing carbohydrates, fats, and other foodstuffs in earlier stages is saved in the form of energy-rich compounds that feed electrons into the respiratory chain in the inner mitochondrial membrane. These electrons, most of which are carried by NADH, finally combine with O_2 at the end of the respiratory chain to form water. The energy released during the complex series of electron transfers from NADH to O_2 is harnessed in the inner membrane to generate an electrochemical gradient that drives the conversion of ADP + phosphate to ATP. For this reason, the term **oxidative phosphorylation** is used to describe this final series of reactions (**Figure 14–12**).

The total amount of energy released by biological oxidation in the respiratory chain is equivalent to that released by the explosive combustion of hydrogen when it combines with oxygen in a single step to form water. But the combustion of hydrogen in a single-step chemical reaction, which has a strongly negative ΔG, releases this large amount of energy unproductively as heat. In the respiratory chain, the same energetically favorable reaction $H_2 + \frac{1}{2}O_2 \rightarrow H_2O$ is divided into small steps (**Figure 14–13**). This stepwise process allows the cell to capture

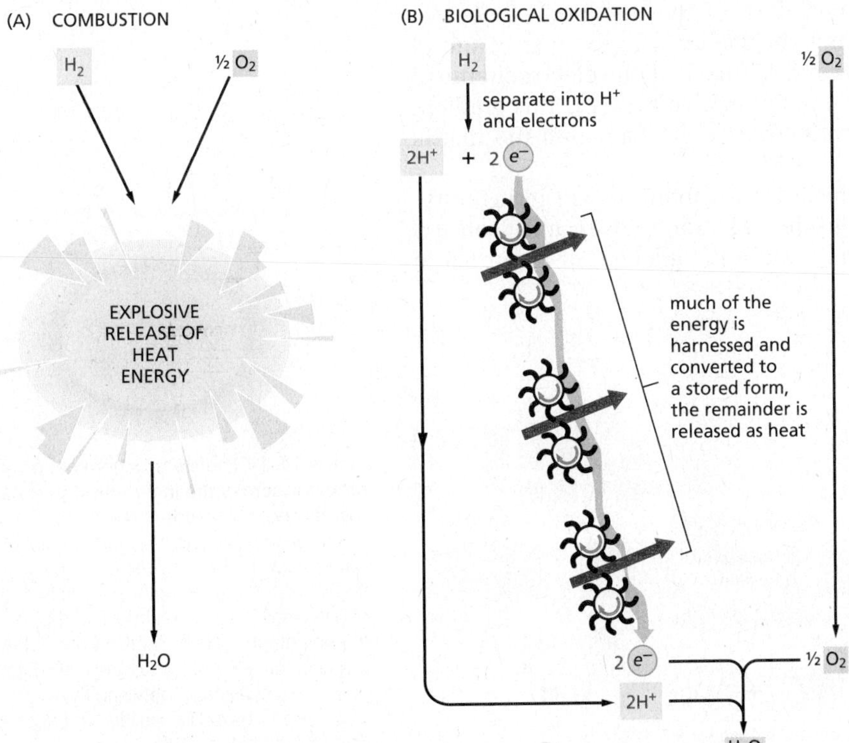

Figure 14–13 **A comparison between the biological oxidation catalyzed by the respiratory chain and combustion.** (A) If hydrogen were simply burned, nearly all of the energy would be released in the form of heat. (B) In biological oxidation, about half of the released energy is stored in a form useful to the cell by means of the electron-transport chain (the respiratory chain) in the crista membrane of the mitochondrion. The rest of the energy is released as heat. In the respiratory chain, the protons and electrons, shown here as being derived from H_2, are removed from hydrogen atoms that are covalently linked to NADH molecules.

and store nearly half of the total energy that is released in a useful form. At each step, the electrons, each of which can be thought of as having been removed from a hydrogen atom leaving a proton, pass through a series of electron carriers in the inner mitochondrial membrane. At each of three distinct steps along the way (marked by the three electron-transport complexes of the respiratory chain; see next part of this chapter), much of the energy released by electron transfer is utilized to produce an electrochemical gradient across the membrane (see Figure 14–2). At the end of the electron-transport chain, the electrons and protons recombine with molecular oxygen to produce water, a very low-energy molecule.

The Energy Derived from Oxidation Is Stored as an Electrochemical Gradient

In mitochondria, the process of electron transport begins when two electrons and a proton are removed from NADH (to regenerate NAD^+). These electrons are passed to the first of about 20 different electron carriers in the respiratory chain. The electrons start at a large *negative* redox potential (see **Panel 14–1**, p. 825)—that is, at a high energy level—which gradually drops as they pass along the chain. Most of the electron carriers are housed in the three large multiprotein *respiratory enzyme complexes*, each composed of protein subunits that sit in the inner mitochondrial membrane. Each complex in the chain has a higher affinity for electrons than its predecessor, and electrons pass sequentially from one complex to the next until they are finally transferred to molecular oxygen, which has the highest electron affinity of all.

The net result is the pumping of H^+ out of the matrix across the inner membrane, driven by the energetically favorable flow of electrons. This transmembrane movement of H^+ has two major consequences:

1. It generates a pH gradient across the inner mitochondrial membrane, with a high pH in the matrix (close to 8) and a lower pH in the crista space.
2. It generates a voltage gradient across the inner mitochondrial membrane, creating a *membrane potential* with the matrix side negative and the crista space positive.

The pH gradient (ΔpH) reinforces the effect of the membrane potential (ΔV), because the latter acts to attract any positive ion into the matrix and to push any negative ion out. Together, ΔpH and ΔV make up the **electrochemical proton gradient**, which is measured in units of millivolts (mV). This gradient exerts a **proton-motive force**, which tends to drive H^+ back into the matrix (**Figure 14–14**).

The electrochemical gradient across the inner membrane of a respiring mitochondrion is typically about 180 mV (the inside is electronegative), and it consists of a membrane potential of about 150 mV and a pH gradient of about 0.5 to

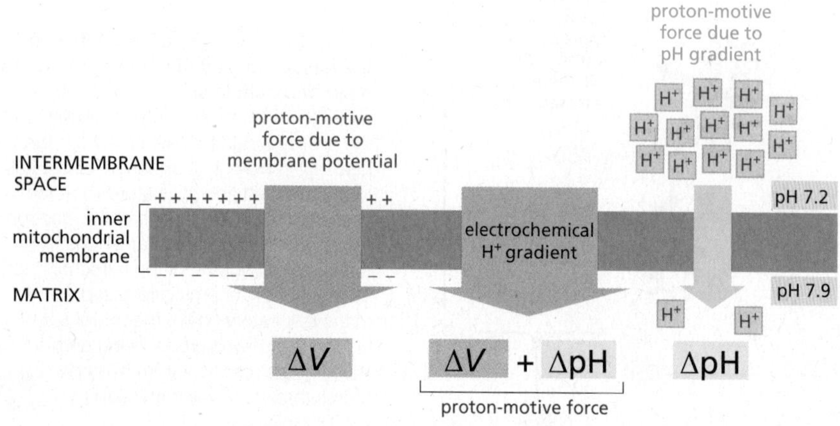

Figure 14–14 The electrochemical proton gradient across the inner mitochondrial membrane. This gradient is composed of a large force due to the membrane potential (ΔV) and a smaller force due to the H^+ concentration gradient; that is, the pH gradient (ΔpH). Both forces combine to generate the proton-motive force, which pulls H^+ back into the mitochondrial matrix. The exact relationship between these forces is expressed by the Nernst equation (see Panel 11–1, p. 656).

0.6 pH units (each ΔpH of 1 pH unit is equivalent to a membrane potential of about 60 mV). The electrochemical gradient drives not only ATP synthesis but also the energetically unfavorable transport of selected molecules across the inner mitochondrial membrane, including the import of nuclear-encoded proteins from the cytosol (discussed in Chapter 12).

Summary

The mitochondrion performs most cellular oxidations and produces the bulk of an animal cell's ATP. A mitochondrion has two separate membranes: the outer membrane and the inner membrane. The inner membrane surrounds the innermost space (the matrix) of the mitochondrion, and it forms the cristae that project into the matrix and contain the electron-transport chain (the respiratory chain). The mitochondrial matrix and the inner membrane crista are the major sites of mitochondrial metabolism.

The mitochondrial matrix contains a large variety of enzymes, including those that convert pyruvate and fatty acids to acetyl CoA and those that oxidize this acetyl CoA to CO_2 through the citric acid cycle. These oxidation reactions produce large amounts of NADH, whose high-energy electrons are passed to the respiratory chain. The respiratory chain then uses the energy derived from transporting electrons from NADH to molecular oxygen to pump H^+ out of the matrix. This produces a large electrochemical proton gradient across the inner mitochondrial membrane, which is composed of contributions from both a membrane potential and a pH difference. This electrochemical gradient exerts a force to drive H^+ back into the matrix. This proton-motive force is harnessed both to produce ATP and to drive the selective transport of metabolites across the inner mitochondrial membrane.

THE PROTON PUMPS OF THE ELECTRON-TRANSPORT CHAIN

Having considered in general terms how a mitochondrion uses electron transport to generate a proton-motive force, we now turn to the molecular mechanisms that underlie this membrane-based energy-conversion process. In describing the respiratory chain of mitochondria, we accomplish the larger purpose of explaining how an electron-transport process can pump protons across a membrane. As stated at the beginning of this chapter, mitochondria, chloroplasts, archaea, and bacteria use very similar chemiosmotic mechanisms. In fact, these mechanisms underlie the function of all living organisms—including anaerobes that derive all their energy from electron transfers between two inorganic molecules, as we shall see later.

We start with some of the basic principles on which all of these processes depend.

The Redox Potential Is a Measure of Electron Affinities

In chemical reactions, any electrons removed from one molecule are always passed to another, so that whenever one molecule is oxidized, another is reduced. As with any other chemical reaction, the tendency of such **redox reactions** to proceed spontaneously depends on the free-energy change (ΔG) for the electron transfer, which in turn depends on the relative affinities of the two molecules for electrons.

Because electron transfers provide most of the energy for life, it is worth taking the time to understand them. As discussed in Chapter 2, acids donate protons and bases accept them (see Panel 2–2, pp. 96–97). Acids and bases exist in conjugate acid–base pairs, in which the acid is readily converted into the base by the loss of a proton. For example, acetic acid (CH_3COOH) is converted into its conjugate base, the acetate ion (CH_3COO^-), in the reaction:

$$CH_3COOH \rightleftharpoons CH_3COO^- + H^+$$

In an exactly analogous way, pairs of compounds such as NADH and NAD^+ are called redox pairs, as NADH is converted to NAD^+ by the loss of electrons in the reaction:

$$NADH \rightleftharpoons NAD^+ + H^+ + 2e^-$$

NADH is a strong electron donor: because two of its electrons are engaged in a covalent bond, which releases energy when broken, the free-energy change for passing these electrons to many other molecules is favorable. Energy is required to form this bond from NAD^+, two electrons, and a proton (the same amount of energy that is released when the bond is broken). Therefore NAD^+, the redox partner of NADH, is of necessity a weak electron acceptor.

We can measure the tendency to transfer electrons from any redox pair experimentally. All that is required is the formation of an electrical circuit linking a 1:1 (equimolar) mixture of the redox pair to a second redox pair that has been arbitrarily selected as a reference standard, so that we can measure the voltage difference between them (see Panel 14–1). This voltage difference is defined as the redox potential; electrons move spontaneously from a redox pair like NADH/NAD^+ with a lower redox potential (a lower affinity for electrons) to a redox pair like O_2/H_2O with a higher redox potential (a higher affinity for electrons). Thus, NADH is a good molecule for donating electrons to the respiratory chain, while O_2 is well suited to act as the "sink" for electrons at the end of the chain. As explained in Panel 14–1, the difference in redox potential, $\Delta E'_0$, is a direct measure of the standard free-energy change ($\Delta G°$) for the transfer of an electron from one molecule to another.

While the standard redox potential is a useful tool to understand redox reactions, the true redox potential is also influenced by environmental factors in a cell such as pH, temperature, and electrostatic influences, which are often different from standardized conditions in cells. This issue is of particular relevance when the redox pair is embedded within a protein, where the environment is heavily influenced by nearby amino acid residues.

Electron Transfers Release Large Amounts of Energy

As just discussed, those pairs of compounds that have the most negative redox potentials have the weakest affinity for electrons and therefore are useful as carriers with a strong tendency to donate electrons. Conversely, those pairs that have the most positive redox potentials have the greatest affinity for electrons and therefore are useful as carriers for accepting electrons. A 1:1 mixture of NADH and NAD^+ has a redox potential of –320 mV, indicating that NADH has a strong tendency to donate electrons; a 1:1 mixture of H_2O and $\frac{1}{2}O_2$ has a redox potential of +820 mV, indicating that O_2 has a strong tendency to accept electrons. The difference in redox potential is 1140 mV, which means that the transfer of each electron from NADH to O_2 under these standard conditions is enormously favorable, with $\Delta G° = -109$ kJ/mole. Twice this amount of energy is gained for the two electrons transferred per NADH molecule (see Panel 14–1). If we compare this free-energy change with that for the formation of the phosphoanhydride bonds in ATP, where $\Delta G° = 30.6$ kJ/mole (see Figure 2–49), we see that, under standard conditions, the oxidation of one NADH molecule releases more than enough energy to synthesize seven molecules of ATP from ADP and phosphate. (In the cell, the number of ATP molecules generated will be lower because the standard conditions are far from the physiological ones; in addition, energy is inevitably dissipated as heat due to imperfect efficiency in energy transfers.)

Transition Metal Ions and Quinones Accept and Release Electrons Readily

The electron-transport properties of the membrane protein complexes in the respiratory chain depend on electron-carrying *cofactors*, most of which utilize *transition metals* such as Fe, Cu, Ni, and Mn bound to proteins in the complexes.

HOW REDOX POTENTIALS ARE MEASURED

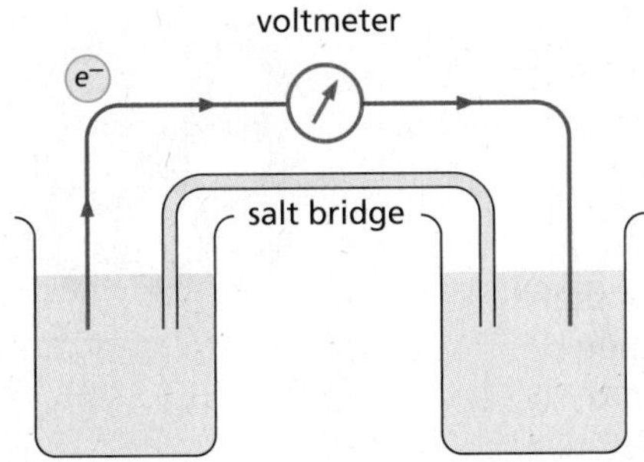

One beaker (*left*) contains substance A with an equimolar mixture of the reduced ($A_{reduced}$) and oxidized ($A_{oxidized}$) members of its redox pair. The other beaker contains the hydrogen reference standard ($2H^+ + 2e^- \rightleftharpoons H_2$), whose redox potential is arbitrarily assigned as zero by international agreement. (A salt bridge formed from a concentrated KCl solution allows K^+ and Cl^- to move between the beakers, as required to neutralize the charges when electrons flow between the beakers.) The metal wire (*dark blue*) provides a resistance-free path for electrons, and a voltmeter then measures the redox potential of substance A. If electrons flow from $A_{reduced}$ to H^+, as indicated here, the redox pair formed by substance A is said to have a negative redox potential. If they instead flow from H_2 to $A_{oxidized}$, the redox pair is said to have a positive redox potential.

THE STANDARD REDOX POTENTIAL, E'_0

The standard redox potential for a redox pair, defined as E_0, is measured for a standard state where all of the reactants are at a concentration of 1 M, including H^+. Since biological reactions occur at pH 7, biologists instead define the standard state as $A_{reduced} = A_{oxidized}$ and $H^+ = 10^{-7}$ M. This standard redox potential is designated by the symbol E'_0, in place of E_0.

examples of redox reactions	standard redox potential E'_0
$NADH \rightleftharpoons NAD^+ + H^+ + 2e^-$	–320 mV
reduced ubiquinone ⇌ oxidized ubiquinone $+ 2H^+ + 2e^-$	+30 mV
reduced cytochrome *c* ⇌ oxidized cytochrome *c* $+ e^-$	+230 mV
$H_2O \rightleftharpoons \frac{1}{2}O_2 + 2H^+ + 2e^-$	+820 mV

CALCULATION OF $\Delta G°$ FROM REDOX POTENTIALS

To determine the energy change for an electron transfer, the $\Delta G°$ of the reaction (kJ/mole) is calculated as follows:

$\Delta G° = -n(0.096)\Delta E'_0$, where *n* is the number of electrons transferred across a redox potential change of $\Delta E'_0$ millivolts (mV), and

$\Delta E'_0 = E'_0\text{ (acceptor)} - E'_0\text{ (donor)}$

EXAMPLE:

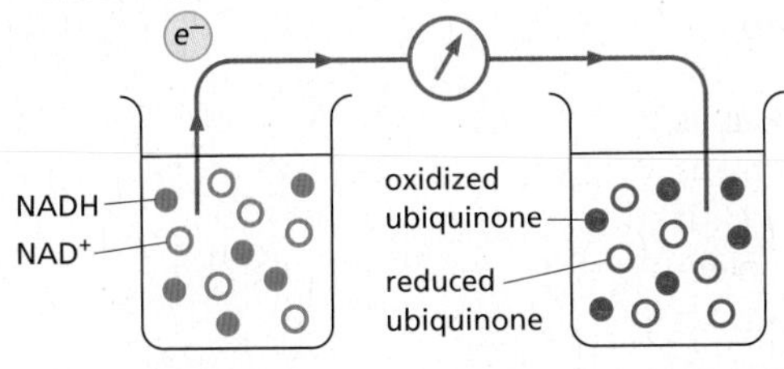

For the transfer of one electron from NADH to ubiquinone:

$$\Delta E'_0 = +30 - (-320) = +350 \text{ mV}$$

$$\Delta G° = -n(0.096)\Delta E'_0 = -1(0.096)(350) = -34 \text{ kJ/mole}$$

The same calculation reveals that the transfer of one electron from ubiquinone to oxygen has an even more favorable $\Delta G°$ of –76 kJ/mole. The $\Delta G°$ value for the transfer of one electron from NADH to oxygen is the sum of these two values, –110 kJ/mole.

EFFECT OF CONCENTRATION CHANGES

As explained in Chapter 2, the actual free-energy change for a reaction, ΔG, depends on the concentration of the reactants and generally will be different from the standard free-energy change, $\Delta G°$. The standard redox potentials are for a 1:1 mixture of the redox pair. For example, the standard redox potential of –320 mV is for a 1:1 mixture of NADH and NAD^+. But when there is an excess of NADH over NAD^+, electron transfer from NADH to an electron acceptor becomes more favorable. This is reflected by a more negative redox potential and a more negative ΔG for electron transfer.

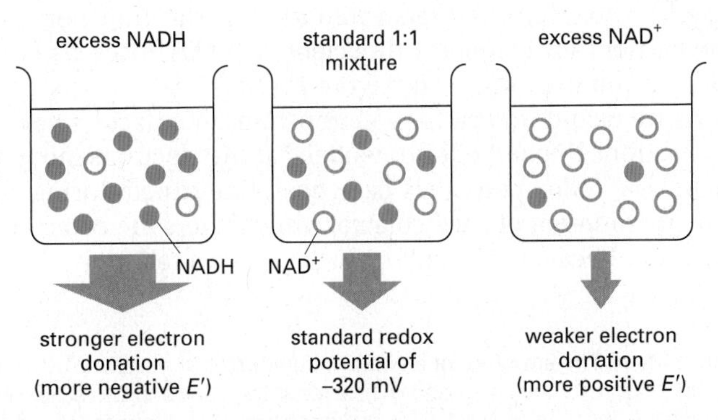

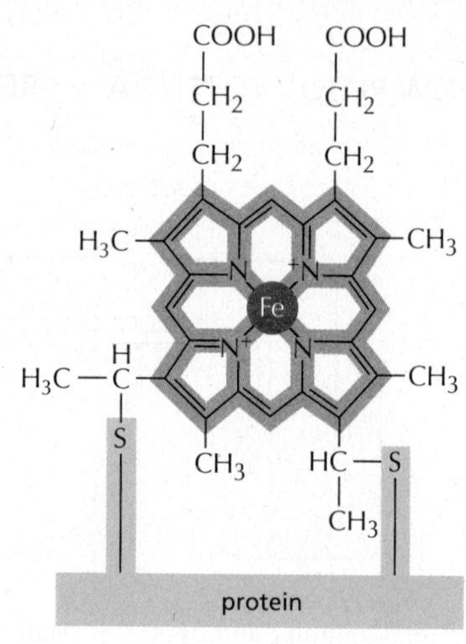

Figure 14–15 The structure of the heme group attached covalently to cytochrome *c*. The porphyrin ring of the heme is shown in *light red*. There are six different cytochromes in the respiratory chain. Because the hemes in different cytochromes have slightly different structures and are kept in different local environments by their respective proteins, each has a different affinity for an electron and a slightly different spectroscopic signature. Note that many heme-containing proteins, including others in the electron-transport chain, bind to heme noncovalently unlike the heme in cytochrome *c* shown here. They do so through noncovalent interactions of the protein with the porphyrin ring and the heme iron.

These metals have special properties that allow them to promote both enzyme catalysis and electron-transfer reactions. Most relevant here is the fact that their ions exist in several different oxidation states with closely spaced redox potentials, which enables them to accept or give up electrons readily; this property is exploited by the membrane protein complexes in the respiratory chain to move electrons both within and between complexes.

Unlike the colorless atoms H, C, N, and O that constitute the bulk of biological molecules, transition metal ions are often brightly colored, which makes the proteins that contain them easy to study by spectroscopic methods using visible light. One family of such colored proteins, the **cytochromes**, contains a bound *heme group*, in which an iron atom is tightly held by four nitrogen atoms at the corners of a square in a *porphyrin ring* (**Figure 14–15**). Similar porphyrin rings are responsible both for the red color of blood and for the green color of leaves, binding an iron in hemoglobin or a magnesium in chlorophyll, respectively.

Iron–sulfur proteins contain a second major family of electron-transfer cofactors. In this case, either two or four iron atoms are bound to an equal number of sulfur atoms and to cysteine side chains, forming iron–sulfur clusters in the protein (**Figure 14–16**). Like the cytochrome hemes, these clusters carry one electron at a time through redox reactions with the iron atoms.

The simplest of the electron-transfer cofactors in the respiratory chain—and the only one that is not always bound to a protein—is a **quinone** (called *ubiquinone*, or *coenzyme Q*). A quinone (Q) is a small hydrophobic molecule that is freely mobile in the lipid bilayer. This *electron carrier* can accept or donate either one or two electrons. Upon reduction (note that reduced quinones are called quinols), it picks up a proton from water along with each electron (**Figure 14–17**).

In the mitochondrial electron-transport chain, six different cytochrome hemes, eight iron–sulfur clusters, three copper atoms, a flavin mononucleotide (another electron-transfer cofactor), and ubiquinone work in a defined sequence to carry electrons from NADH to O_2. In total, this pathway involves more than 60 different polypeptides; these are arranged in three large membrane protein complexes, each of which binds several of the above electron-carrying cofactors (**Figure 14–18**).

As we will discuss later, there is an unusual additional respiratory chain complex that is an important component of the citric acid cycle, where it is known as *succinate dehydrogenase*. This membrane-embedded enzyme captures electrons during the conversion of succinate to fumarate (see Panel 2–9, pp. 110–111). It passes these electrons directly into the electron-transport chain via a flavin electron carrier (flavin adenine dinucleotide; FAD), instead of utilizing NAD^+, and it does not pump protons (see Figure 14–18).

As we would expect, the electron-transfer cofactors have increasing affinities for electrons (higher redox potentials) as the electrons move along the respiratory chain. The redox potentials have been fine-tuned during evolution by the protein environment of each cofactor, which alters the cofactor's normal affinity for electrons. Because iron–sulfur clusters have a relatively low affinity for electrons,

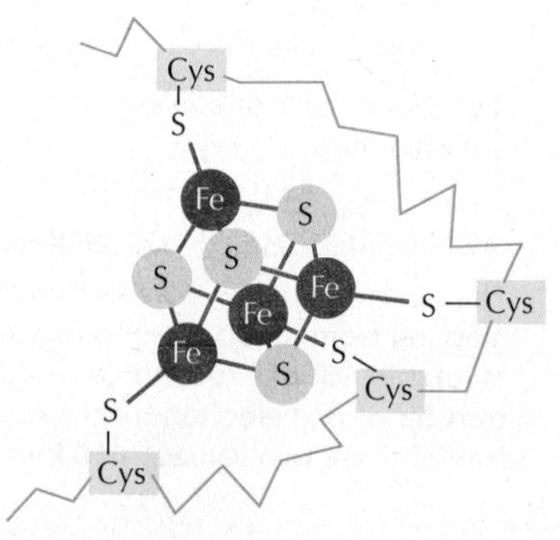

Figure 14–16 The structure of an iron–sulfur cluster. Iron–sulfur clusters consist either of four iron and four sulfur atoms, as shown here, or of two irons and two sulfurs linked to cysteines in the polypeptide chain via covalent sulfur bridges; alternatively they may be linked to histidines. Although they contain several iron atoms, each iron–sulfur cluster can carry only one electron at a time. Nine different iron–sulfur clusters participate in electron transport in the respiratory chain.

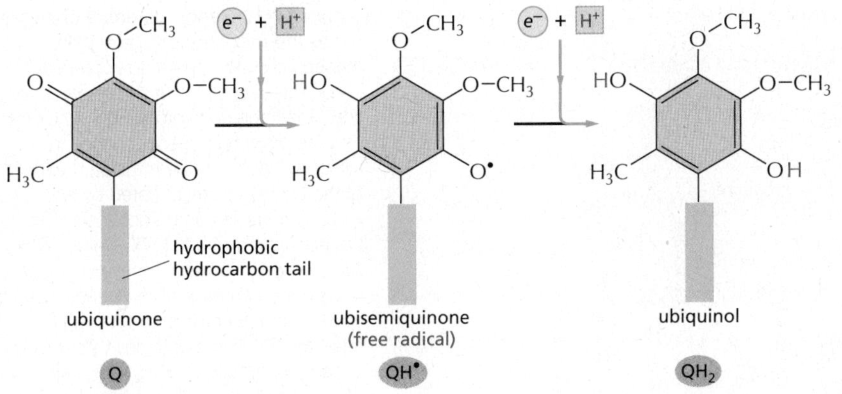

Figure 14–17 Quinone electron carriers. Ubiquinone in the lipid bilayer picks up one H^+ *(red)* from the aqueous environment for each electron *(blue)* it accepts from respiratory-chain complexes. The first step in this process involves the acquisition of a proton and an electron and converts the ubiquinone into an unstable ubisemiquinone radical. With the transfer of the second electron, it becomes a fully reduced ubiquinone (called ubiquinol), which is freely mobile as an electron carrier in the lipid bilayer of the membrane. When the ubiquinol donates its electrons to the next complex in the chain, the two protons are released. The long hydrophobic tail *(green)* that confines ubiquinone to the membrane consists of 6 to 10 five-carbon isoprene units, depending on the organism. The corresponding electron carrier in the photosynthetic membranes of chloroplasts is plastoquinone, which has almost the same structure and works in the same way. For simplicity, we refer to both ubiquinone and plastoquinone in this chapter as quinone (abbreviated as Q).

they predominate in the first half of the respiratory chain; in contrast, the heme cytochromes predominate further down the chain, where a higher electron affinity is required.

NADH Transfers Its Electrons to Oxygen Through Three Large Enzyme Complexes Embedded in the Inner Membrane

Membrane proteins are difficult to purify because they are insoluble in aqueous solutions, and they are easily disrupted by the detergents that are required to solubilize them. But by using mild nonionic detergents, such as digitonin or dodecyl maltoside, they can be solubilized and purified in their native form. The three large, detergent-solubilized respiratory complexes can be reinserted into artificial lipid bilayer vesicles, and each protein complex can be shown to pump protons across the membrane as electrons pass through it.

In the mitochondrion, the three complexes are linked in series, serving as electron transport–driven H^+ pumps that pump protons out of the matrix to acidify the crista space (see Figure 14–18):

1. The **NADH dehydrogenase complex** (typically called *Complex I*) is the largest of these respiratory enzyme complexes. It accepts electrons from NADH and passes them through a flavin mononucleotide and eight iron-sulfur clusters to ubiquinone. The reduced ubiquinol then transfers its electrons to cytochrome *c* reductase.

2. The **cytochrome *c* reductase** (also called the *cytochrome b-c$_1$ complex* and typically called *Complex III*) is a large membrane protein assembly that functions as a dimer. Each monomer contains three cytochrome hemes and an iron-sulfur cluster. The complex accepts electrons from ubiquinol

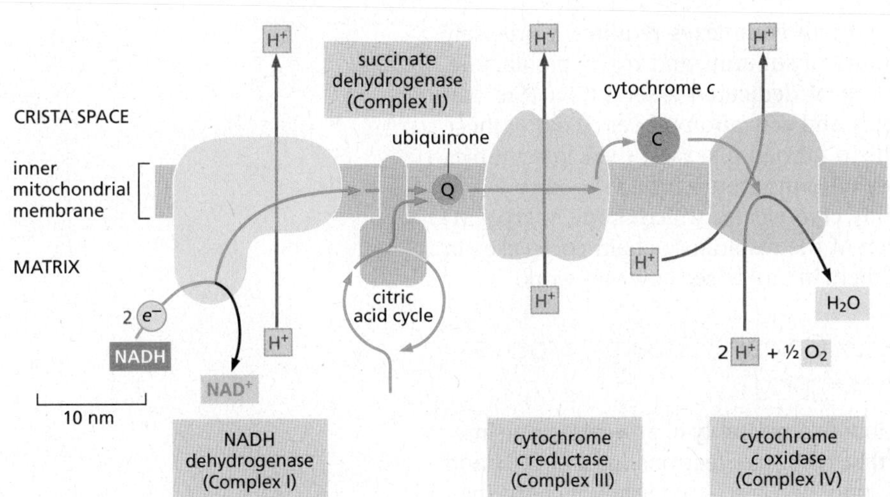

Figure 14–18 The path of electrons through the three respiratory-chain proton pumps (Movie 14.5). The approximate size and shape of each protein complex are shown. During the transfer of electrons from NADH to oxygen *(blue arrows)*, ubiquinone and cytochrome *c* serve as mobile carriers that ferry electrons from one complex to the next. During the electron-transfer reactions, protons are pumped across the membrane by each of the respiratory enzyme complexes, as indicated *(red arrows)*.

The three proton pumps in the respiratory chain are typically denoted as Complex I, Complex III, and Complex IV, according to the order in which electrons pass through them from NADH. Electrons from the oxidation of succinate by succinate dehydrogenase (designated as Complex II) are fed into the electron-transport chain in the form of reduced ubiquinone. Although embedded in the crista membrane, succinate dehydrogenase does not pump protons and thus does not contribute to the proton-motive force.

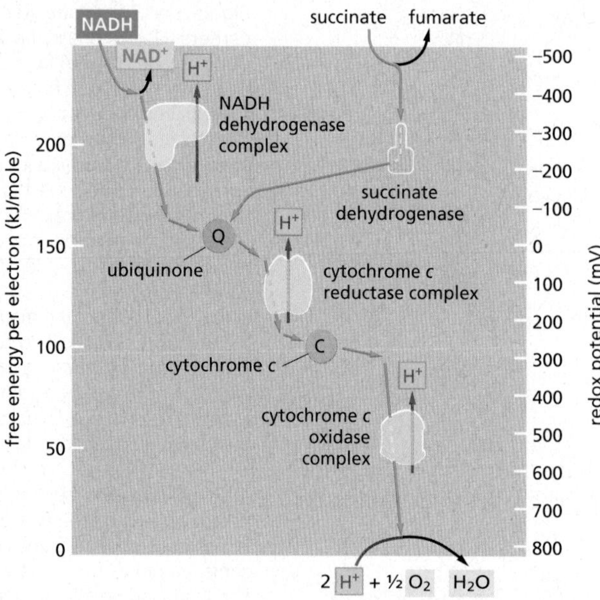

Figure 14–19 **Redox potential changes along the mitochondrial electron-transport chain.** The redox potential (designated $\Delta E'_0$) increases as electrons flow down the respiratory chain to oxygen. The standard free-energy change in kilojoules, ΔG°, for the transfer of each of the two electrons donated by an NADH molecule can be obtained from the left-hand ordinate [$\Delta G^\circ = -n(0.096)\,\Delta E'_0$, where n is the number of electrons transferred across a redox potential change of $\Delta E'_0$ mV]. Electrons flow through a respiratory enzyme complex by passing in sequence through the multiple electron carriers in each complex *(dotted portion of blue arrows)*. As indicated, part of the favorable free-energy change is harnessed by each enzyme complex to pump H^+ across the inner mitochondrial membrane *(red arrows)*. The NADH dehydrogenase pumps up to four H^+ per electron, the cytochrome *c* reductase complex pumps two per electron, whereas the cytochrome *c* oxidase complex pumps one per electron.

Succinate dehydrogenase (Complex II) also passes electrons into the ubiquinone pool by oxidizing succinate and passing those electrons through a stably bound $FAD^+/FADH_2$ redox pair. Fatty acid oxidation (see Figure 2–57) also leads to the generation of $FADH_2$, which also passes its two electrons directly to ubiquinone, bypassing NADH dehydrogenase. No proton pumping accompanies these electron transfers.

and passes them on to the small, soluble protein cytochrome *c*, which is located in the crista space and carries electrons one at a time to cytochrome *c* oxidase.

3. The **cytochrome *c* oxidase complex** (typically called *Complex IV*) contains two cytochrome hemes and three copper atoms. The complex accepts electrons one at a time from cytochrome *c* and passes them to molecular oxygen. In total, four electrons and four protons are needed to convert one molecule of oxygen to water.

We have previously discussed how the redox potential reflects electron affinities. Figure 14–19 presents an outline of the redox potentials measured along the respiratory chain. These potentials change in three large steps, one across each proton-translocating respiratory complex. The change in redox potential between any two electron carriers is directly proportional to the free energy released when an electron transfers between them. Each complex acts as an energy-conversion device by harnessing some of this free-energy change to pump H^+ across the inner membrane, thereby creating an electrochemical proton gradient as electrons pass along the chain. In addition to these proton-pumping complexes, the *succinate dehydrogenase* complex, which catalyzes the oxidation of succinate to fumarate in the citric acid cycle, is often also considered to be a component of the electron-transport chain. It is typically called *Complex II* because, like Complex I, it passes electrons to ubiquinone.

The assembly of each of these four protein complexes requires a precisely managed program because of their number of subunits and reactive cofactors. This is accomplished through the function of dedicated assembly factors that chaperone particular steps in the assembly and activation process. Loss of these assembly factors through mutation leads to human diseases; their loss mimics mutations that directly affect the complex subunits themselves.

X-ray crystallography (and, more recently, cryo-electron microscopy, or cryoEM) has elucidated the final structures of each of the respiratory-chain complexes in great detail, and we next examine each of them in turn to see how they work.

The NADH Dehydrogenase Complex Contains Separate Modules for Electron Transport and Proton Pumping

The NADH dehydrogenase complex is a massive assembly of proteins, some integral to the membrane and others not, that receives electrons from NADH and passes them to ubiquinone. In animal mitochondria, it consists of more than

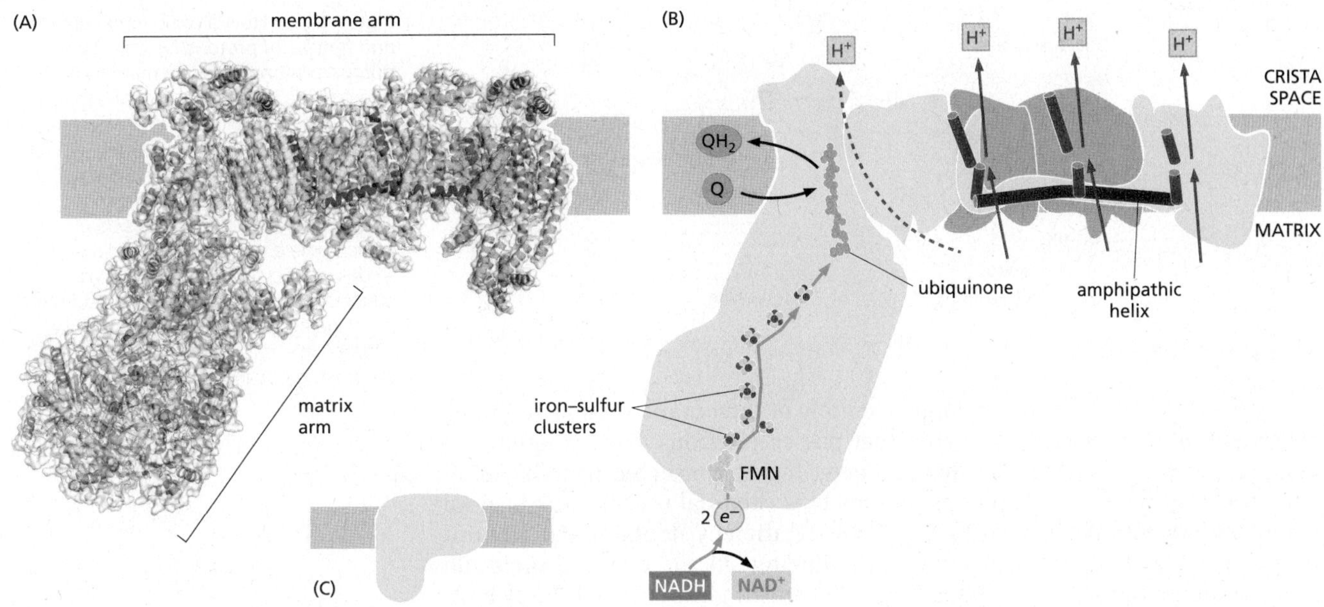

Figure 14–20 The structure of NADH dehydrogenase (also known as Complex I). (A) The structure of Complex I is shown. The matrix arm of NADH dehydrogenase contains one flavin mononucleotide (FMN) and eight iron–sulfur (FeS) clusters that appear to participate in electron transport. The membrane contains more than 70 transmembrane helices, forming three distinct proton-pumping modules. (B) Schematic of NADH dehydrogenase electron transport–coupled proton pumping. NADH donates two electrons, via a bound FMN *(yellow)*, to a chain of seven iron–sulfur clusters *(red and yellow spheres)*. From the terminal iron–sulfur cluster, the electrons pass to ubiquinone *(orange)*. Electron transfer results in conformational changes that are thought to be transmitted to a long amphipathic α helix *(purple)* on the matrix side of the membrane arm, which pulls on discontinuous transmembrane helices *(red)* in three membrane subunits, each of which resembles an antiporter (see Chapter 11). This movement is thought to change the conformation of charged residues in the three proton channels, resulting in the translocation of three protons out of the matrix. A fourth proton may be translocated at the interface of the two arms *(dotted line)*. (C) This shows the symbol for NADH dehydrogenase used throughout this chapter. (A, adapted from R.G. Efremov et al., *Nature* 465:441–445, 2010. PDB code: 3M9S.)

40 different protein subunits, with a molecular mass of nearly a million daltons. The x-ray crystallography and cryoEM structures of the NADH dehydrogenase complex show that it is L-shaped, with both a hydrophobic membrane arm and a hydrophilic arm that projects into the mitochondrial matrix (**Figure 14–20**).

Electron transfer and proton pumping are physically separated in the NADH dehydrogenase complex, with electron transfer occurring in the matrix arm and proton pumping in the membrane arm. The NADH docks near the tip of the matrix arm, where it transfers its electrons via a bound flavin mononucleotide to a string of iron–sulfur clusters that runs down the arm, acting like a wire to carry electrons to a protein-bound molecule of ubiquinone. Electron transfer to the quinone is thought to trigger proton translocation in a set of proton pumps in the membrane arm, and for this to happen the two processes must be energetically and mechanically linked. One hypothesized mechanism is for this link to be provided by a 6-nm-long, amphipathic α helix that runs parallel to the membrane surface on the matrix side of the membrane arm. This helix may act like the connecting rod in a steam engine to generate a mechanical, energy-transducing power stroke that links the quinone-binding site to the proton-translocating modules in the membrane (see Figure 14–20).

The reduction of each quinone by the transfer of its two electrons, followed by its release from the complex into the membrane, can cause four protons to be pumped out of the matrix into the crista space. In this way, NADH dehydrogenase generates roughly half of the total proton-motive force in mitochondria.

Cytochrome *c* Reductase Takes Up and Releases Protons on Opposite Sides of the Crista Membrane, Thereby Pumping Protons

As described previously, when a quinone molecule (Q) accepts its two electrons, it also takes up two protons to form a quinol (QH_2; see Figure 14–17). In the respiratory chain, ubiquinol transfers electrons to cytochrome *c* reductase, after picking them up from either NADH dehydrogenase or succinate dehydrogenase. Because the protons in this QH_2 molecule are obtained from the matrix and released on the opposite side of the crista membrane, two protons are transferred from the matrix into the crista space per pair of electrons transferred (**Figure 14–21**). This simple vectorial transfer of protons supplements the electrochemical proton gradient that is created by the NADH dehydrogenase proton pumping just discussed.

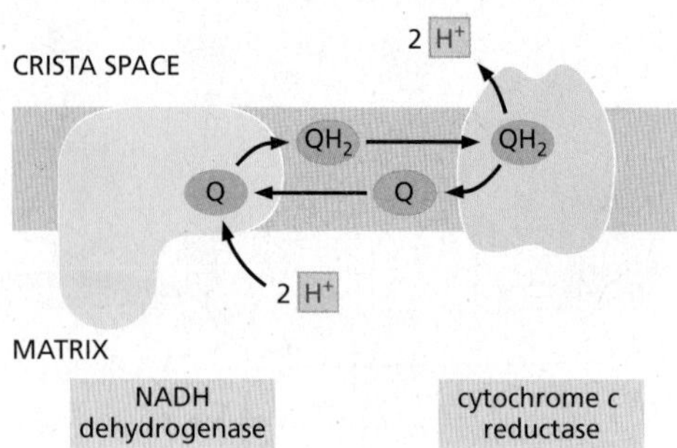

Figure 14–21 How a directional release and uptake of protons by a quinone pumps protons across a membrane. Two protons are picked up on the matrix side of the inner mitochondrial membrane when the reaction $Q + 2e^- + 2H^+ \rightarrow QH_2$ is catalyzed by the NADH dehydrogenase complex. This molecule of ubiquinol (QH_2) then binds to the crista side of cytochrome *c* reductase. When its oxidation by cytochrome *c* reductase generates two protons and two electrons (see Figure 14–17), the two protons are released into the crista space. The flow of electrons is not shown in this diagram.

Cytochrome *c* reductase is a large assembly of membrane protein subunits. Three subunits form a catalytic core that passes electrons from ubiquinol to cytochrome *c*, with a structure that has been highly conserved from bacterial ancestors (**Figure 14–22**). It pumps protons by a vectorial transfer that is more complex than the mechanism in Figure 14–21, thereby doubling the amount of useful energy harvested. This involves a binding site for a second molecule of ubiquinone; the elaborate redox loop mechanism used is called the *Q cycle* because one of the electrons received from each QH_2 molecule is transferred from

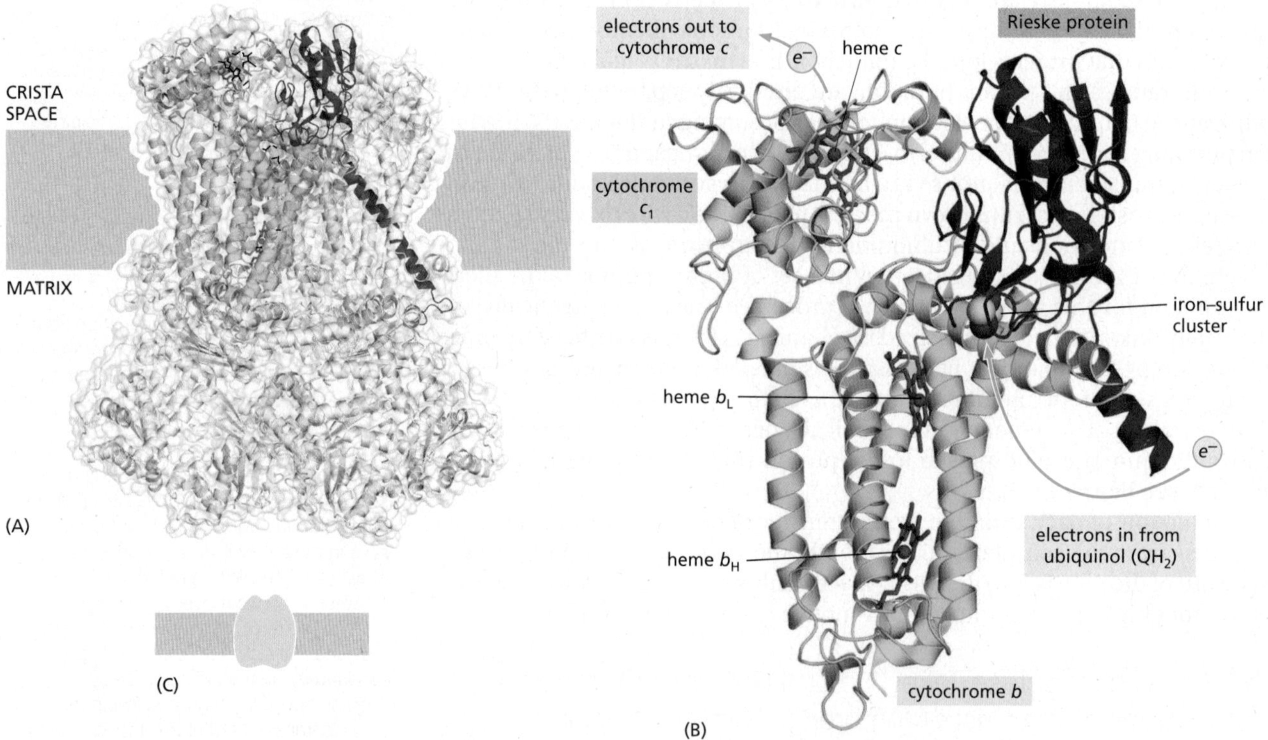

Figure 14–22 The structure of cytochrome c reductase. Cytochrome *c* reductase (also known as the cytochrome *b*-c_1 complex) is a dimer of two identical 240,000-dalton halves, each composed of 11 different protein molecules in mammals. (A) A structural model of the entire dimer based on x-ray crystallography, showing in color the three proteins that form the functional core of the enzyme complex: cytochrome *b* *(green)* and cytochrome c_1 *(blue)* are colored in one half, and the Rieske protein *(purple)* containing an Fe_2S_2 iron–sulfur cluster *(red and yellow)* is colored in the other. These three protein subunits interact across the two halves. (B) Transfer of electrons through cytochrome *c* reductase to the small, soluble carrier protein cytochrome *c*. Electrons entering from ubiquinol near the matrix side of the membrane are captured by the iron–sulfur cluster of the Rieske protein, which moves its iron–sulfur group back and forth to transfer these electrons to heme *c* *(red)*. Heme *c* then transfers them to the carrier molecule cytochrome *c*.

As detailed in Figure 14–23, only one of the two electrons from each ubiquinol is transferred through this path. To increase proton pumping, the second ubiquinol electron is passed to a molecule of ubiquinone bound to cytochrome *c* reductase on the opposite side of the membrane—near the matrix. (C) This shows the symbol for cytochrome *c* reductase used throughout this chapter. (A and B, PDB code: 1EZV.)

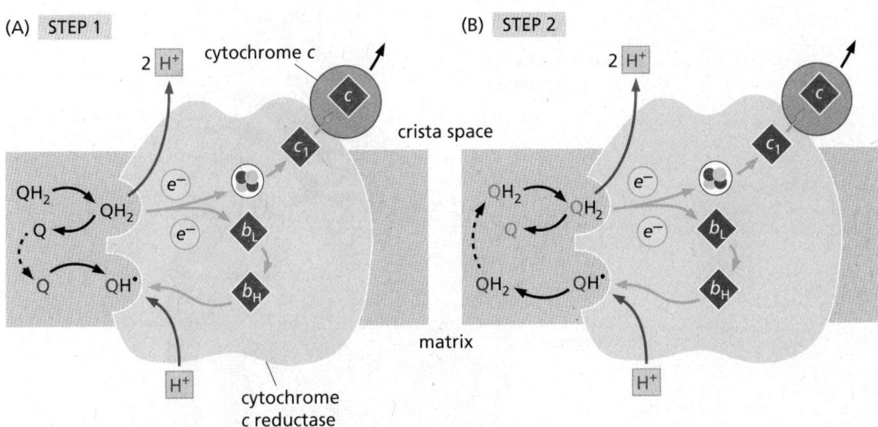

Figure 14–23 **The two-step mechanism of the cytochrome *c* reductase Q cycle.** A clear grasp of the processes illustrated here can be obtained from the step-by-step video provided in Movie 14.6. (A) In step 1, ubiquinol reduced by NADH dehydrogenase docks to the cytochrome *c* reductase complex. Oxidation of the quinol produces two protons and two electrons. The protons are released into the crista space. One electron passes via an iron–sulfur cluster to heme c_1, and then to the soluble electron carrier protein cytochrome *c* on the membrane surface. The second electron passes via hemes b_L and b_H to a ubiquinone (*red* Q) bound at a separate site near the matrix side of the protein. Uptake of a proton from the matrix produces a ubisemiquinone radical (see Figure 14–17), which remains bound to this site (*red* QH•).

(B) In step 2, a second ubiquinol (*blue* QH_2) docks and releases two protons and two electrons, as described for step 1. One electron is passed to a second molecule of cytochrome *c*, whereas the other electron is accepted by the ubisemiquinone. The ubisemiquinone then takes up a proton from the matrix and is released into the lipid bilayer as fully reduced ubiquinol (*red* QH_2), from whence it can subsequently rebind to the complex and donate electrons and protons as in step 1.

On balance, the oxidation of one ubiquinol in the Q cycle pumps two protons through the membrane by a directional release and uptake of protons (see Figure 14–18), while releasing another two into the crista space. In addition, in each of the two steps (A and B), one electron is transferred to a cytochrome *c* carrier protein.

ubiquinone through the complex to the carrier protein cytochrome *c* while the other electron is recycled back into the quinone pool. Through the mechanism illustrated in **Figure 14–23** and **Movie 14.6**, the **Q cycle** increases the total amount of redox energy that can be stored in the electrochemical proton gradient: for every electron that is transferred from NADH dehydrogenase to cytochrome *c*, two protons are pumped across the crista membrane into the crista space.

The Cytochrome *c* Oxidase Complex *Pumps* Protons and Reduces O_2 Using a Catalytic Iron–Copper Center

The final link in the mitochondrial electron-transport chain is cytochrome *c* oxidase, or Complex IV. The cytochrome *c* oxidase complex accepts electrons from the soluble electron carrier cytochrome *c*, and it uses yet a different, third mechanism to pump protons across the inner mitochondrial membrane. The structure of the mammalian complex is illustrated in **Figure 14–24**.

Because oxygen has a high affinity for electrons, it can release a large amount of free energy when it is reduced to form water. Thus, the evolution of cellular respiration, in which O_2 is converted to water, enabled organisms to harness much more energy than can be derived from anaerobic metabolism. As we discuss later, the availability of the large amount of energy released by the reduction of molecular oxygen to form water is thought to have been essential to the emergence of multicellular life, thereby explaining why all large organisms respire. The ability of biological systems to use O_2 in this way, however, requires sophisticated chemistry. Once a molecule of O_2 has picked up one electron, it forms a superoxide radical anion ($O_2^{\bullet -}$) that is dangerously reactive and rapidly takes up an additional three electrons wherever it can get them, with destructive effects on its immediate environment. We can tolerate oxygen in the air we

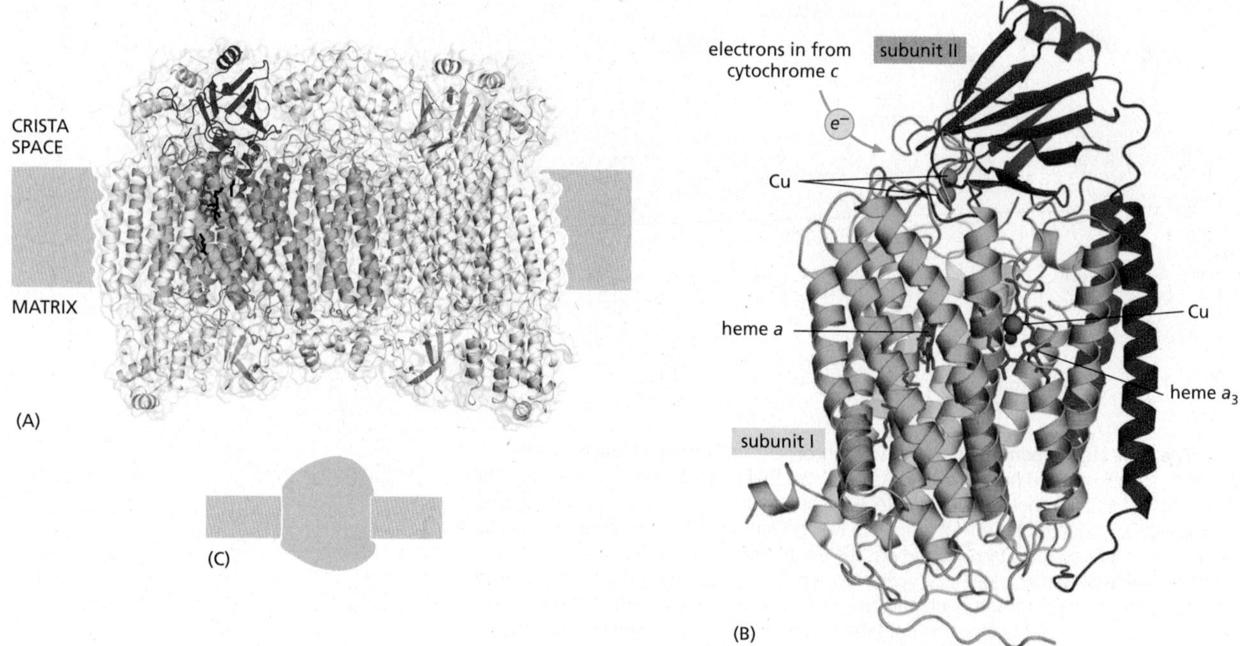

Figure 14–24 The structure of cytochrome *c* oxidase. The final complex in the human mitochondrial electron-transfer chain consists of ~13 different protein subunits, depending on the cell type, with a total mass of approximately 204,000 daltons. (A) The entire dimeric complex is shown, positioned in the crista membrane. The highly conserved subunits I *(green)*, II *(purple)*, and III *(blue)* are encoded by the mitochondrial genome, and they form the functional core of the enzyme. (B) The functional core of the complex. Electrons pass through this structure from cytochrome *c* via bound copper ions *(blue spheres)* and hemes *(red)* to an O_2 molecule bound between heme a_3 and a copper ion. The four protons needed to reduce O_2 to water are taken up from the matrix; see also Figure 14–25. (C) This shows the symbol for cytochrome *c* oxidase used throughout this chapter. (A and B, PDB code: 2OCC.)

breathe only because the uptake of the first electron by a free O_2 molecule is slow and inefficient, allowing cells to use enzymes to control electron uptake by oxygen. Thus, cytochrome *c* oxidase holds on to oxygen at a special bimetallic center, where it remains clamped between a heme-linked iron atom and a copper ion until it has picked up a total of four electrons. Only then are the two oxygen atoms of the oxygen molecule safely released as two molecules of water (**Figure 14–25**).

The cytochrome *c* oxidase reaction accounts for about 90% of the total oxygen uptake in most cells. This protein complex is therefore crucial for all aerobic life. Oxygen limitation is life-threatening to obligate aerobic organisms because of an impaired activity of cytochrome *c* oxidase and these organisms respond rapidly to low oxygen levels to decrease their dependence on mitochondrial respiration. Cyanide is extremely toxic because it binds to the heme iron atoms in cytochrome *c* oxidase much more tightly than does oxygen, thereby greatly reducing mitochondrial ATP production.

Succinate Dehydrogenase Acts in Both the Electron-Transport Chain and the Citric Acid Cycle

In addition to the three proton pumps just discussed, one of the enzymes in the citric acid cycle, *succinate dehydrogenase* (**Figure 14–26**), is embedded in the mitochondrial crista membrane. In the course of oxidizing succinate to fumarate in the matrix, this enzyme complex captures electrons in the form of a tightly bound $FADH_2$ molecule (see Panel 2–9, pp. 110–111) and passes them through three iron–sulfur clusters to a molecule of ubiquinone. The reduced ubiquinol then enters the pool of ubiquinol generated by Complex I and passes its two electrons to cytochrome *c* reductase in the respiratory chain (see Figure 14–18). Succinate dehydrogenase is not a proton pump, and

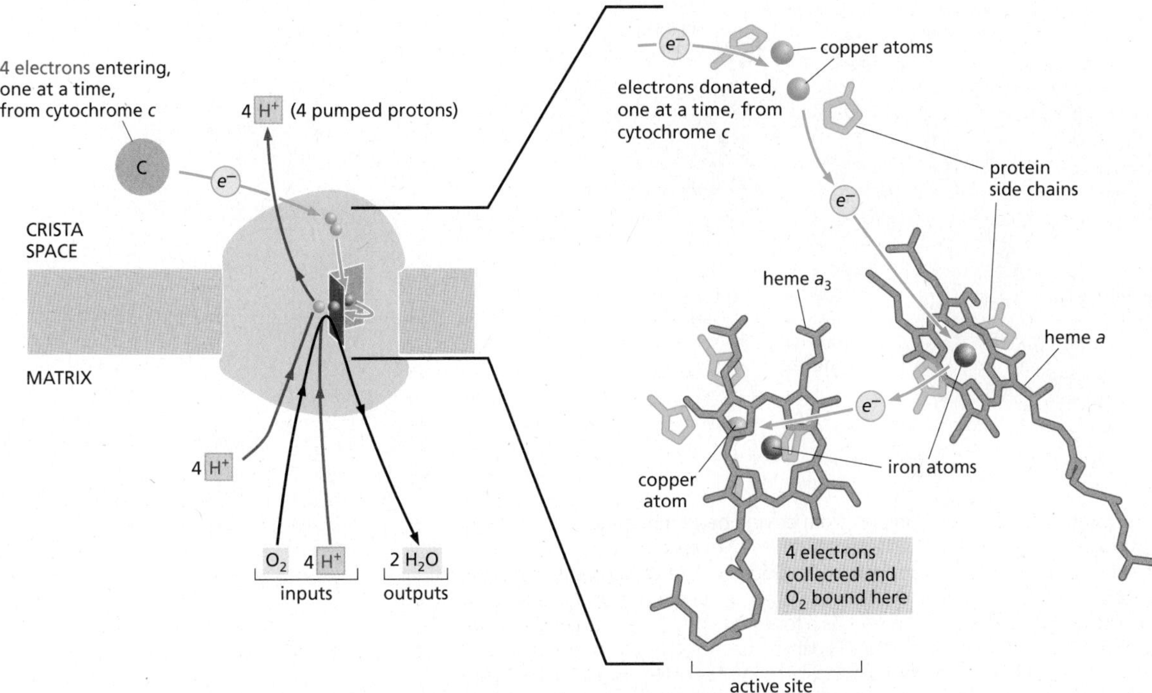

Figure 14–25 **The reaction of O_2 with electrons in cytochrome *c* oxidase.** Electrons from cytochrome *c* pass through the complex via bound copper ions *(blue spheres)* and hemes *(red)* to an O_2 molecule bound between heme a_3 and a copper ion. Iron ions are shown as *red spheres*. The iron atom in heme *a* serves as an electron queuing point where electrons are held so that they can be released to an O_2 molecule (not shown) that is held at the bimetallic center active site, which is formed by the central iron of the other heme (heme a_3) and a closely apposed copper atom. The four protons needed to reduce O_2 to water are removed from the matrix. For each O_2 molecule that undergoes the reaction $4e^- + 4H^+ + O_2 \rightarrow 2H_2O$, an additional four protons are pumped out of the matrix by mechanisms driven by allosteric changes in protein conformation (see Figure 14–29).

it does not contribute directly to the electrochemical potential utilized for ATP production in mitochondria. Its dual action in the citric acid cycle and in electron transport has implications for human disease that are different from those of the other three respiratory complexes. For example, mutations that cause a decreased activity of Complex II contribute to specific forms of cancer.

The Respiratory Chain Forms a Supercomplex in the Crista Membrane

When the three large mitochondrial respiratory complexes that pump protons in the electron-transport chain are gently isolated, they are found in even larger *supercomplexes*. Such observations support the hypothesis that this massive structure in the crista membrane helps the mobile electron carriers ubiquinone (in the crista membrane) and cytochrome *c* (in the crista space) transfer electrons more efficiently than could be accomplished with dissociated, freely diffusing electron carriers (**Figure 14–27**). The supercomplex allows a cell to capture as much of the free energy of electron transfer from NADH to O_2 as possible and to avoid potentially damaging, redox side reactions. The formation of this structure depends on the presence of the mitochondrial lipid cardiolipin (see Figure 14–11) and on special proteins that are thought to hold the components together.

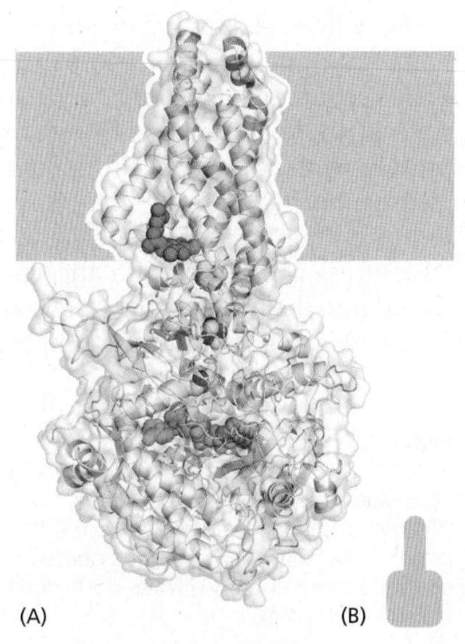

Figure 14–26 **The structure of succinate dehydrogenase.** (A) This membrane-embedded enzyme is composed of four subunits. One subunit, Sdh1, contains a covalently bound FAD cofactor *(green)* that receives an electron directly from succinate oxidation and passes it successively through the three iron–sulfur clusters *(red and yellow)* contained within Shd2 to ubiquinone. The membrane component of this complex, formed by the other two subunits, has a bound heme of unknown function *(orange)*. (B) This shows the symbol for succinate dehydrogenase used throughout this chapter. (A, PDB code: 1NEK.)

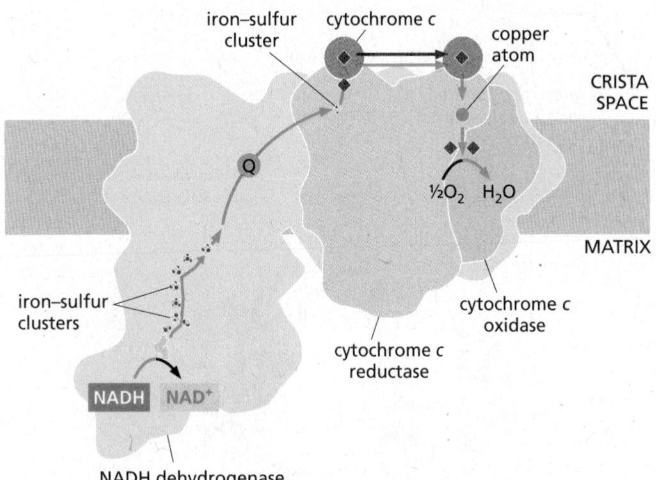

Figure 14–27 **The respiratory-chain supercomplex from bovine heart mitochondria.** The three proton-pumping complexes of the mitochondrial respiratory chain of mammalian mitochondria assemble into large supercomplexes in the crista membrane. Supercomplexes can be isolated by mild detergent treatment of mitochondria, and their structure has been deciphered by single-particle cryo-electron microscopy. The bovine heart supercomplex has a total mass of 1.7 megadaltons. Shown is a schematic of such a complex that consists of NADH dehydrogenase, cytochrome *c* reductase, and cytochrome *c* oxidase, as indicated. The facing quinol-binding sites of NADH dehydrogenase and cytochrome *c* reductase, plus the short distance between the cytochrome *c*–binding sites in cytochrome *c* reductase and cytochrome *c* oxidase, facilitate fast, efficient electron transfer. Cofactors active in electron transport are marked as a *yellow icon* (flavin mononucleotide), *red and yellow spheres* (iron–sulfur clusters), Q (quinone), *red diamonds* (hemes), and a *blue sphere* (copper atom). Only cofactors participating in the linear flow of electrons from NADH to water are shown. *Blue arrows* indicate the path of the electrons through the supercomplex. (Adapted from T. Athoff et al., *EMBO J.* 30:4652–4664, 2011.)

Protons Can Move Rapidly Through Proteins Along Predefined Pathways

The protons in water are highly mobile: by rapidly dissociating from one water molecule and associating with its neighbor, they can rapidly flit through a hydrogen-bonded network of water molecules (see Figure 2–5). But how can a proton move through the hydrophobic interior of a protein embedded in the lipid bilayer? Proton-translocating proteins contain so-called *proton wires*, which are rows of polar or ionic side chains or water molecules spaced at short distances, so that the protons can jump from one to the next (**Figure 14–28**). Along such predefined pathways, protons move up to 40 times faster than through bulk water.

How does electron transport cause allosteric changes in protein conformations that pump protons? From the most basic point of view, if electron transport drives sequential allosteric changes in protein conformation that alter the redox state of the components, these conformational changes can be connected to protein wires that allow the protein to pump H^+ across the crista membrane. This type of H^+ pumping requires at least three distinct conformations for the pump protein, as schematically illustrated in **Figure 14–29**. Atomic-resolution structures, combined with the effects of specific amino acid changes introduced into the proteins by genetic engineering, are helping to reveal the detailed mechanisms of the proton pumping driven by electron transfer, but how these

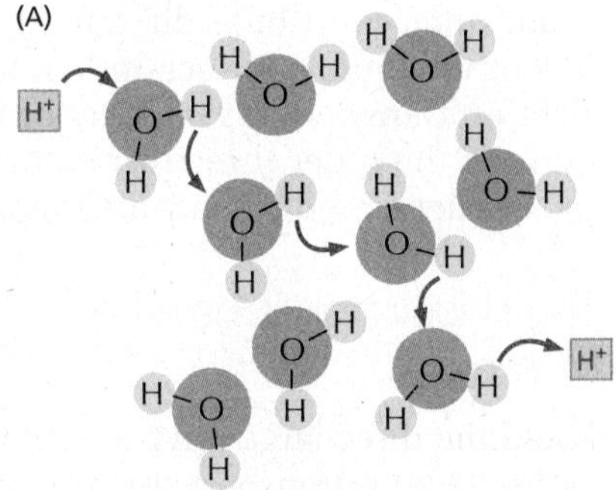

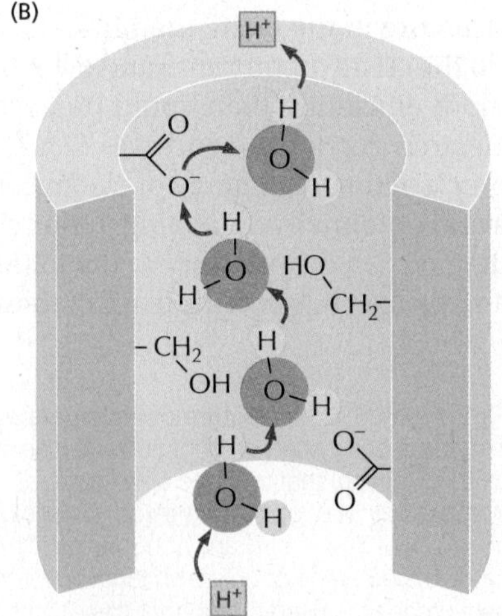

Figure 14–28 **Proton movement through water and proteins.** (A) Protons move rapidly through water, hopping from one H_2O molecule to the next by the continuous formation and dissociation of hydronium ions, H_3O^+ (see Chapter 2). In this diagram, proton jumps are indicated by *red arrows*. (B) Protons can move even more rapidly through a protein along *proton wires*. These are predefined proton paths consisting of suitably spaced amino acid side chains that accept and release protons easily (Asp, Glu) or carry a water-like hydroxyl group (Ser, Thr), along with water molecules trapped in the protein interior.

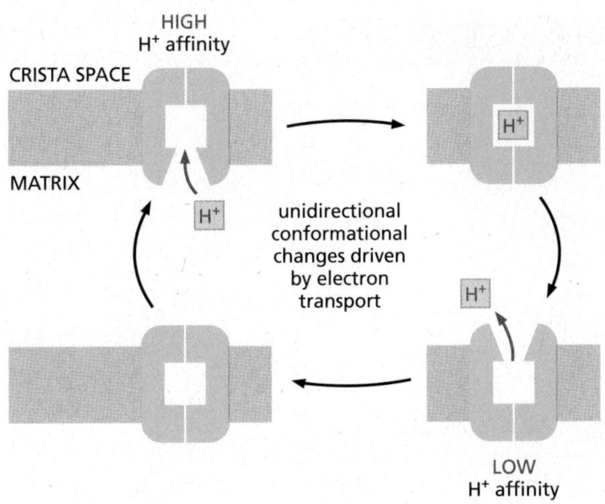

Figure 14–29 **A general model for H^+ pumping coupled to electron transport.** This type of mechanism for H^+ pumping by a transmembrane protein is thought to be used by NADH dehydrogenase and cytochrome *c* oxidase, and by many other proton pumps. The protein is driven through a cycle of three conformations. In one of these conformations, the protein has a high affinity for H^+, causing it to pick up an H^+ on the inside of the membrane. In another conformation, the protein has a low affinity for H^+, causing it to release an H^+ on the outside of the membrane. As indicated, the transitions from one conformation to another occur only in one direction, because they are driven by being allosterically coupled to an energetically favorable process—in this case by the free energy released by electron transport (see also Chapter 11).

conformations are formed and regulated in the respiratory complexes remain incompletely understood.

Summary

The respiratory chain is embedded in the crista membrane portion of the inner mitochondrial membrane. It contains three respiratory enzyme complexes through which electrons pass on their way from NADH to O_2. Within these large protein complexes, electrons are transferred along a series of protein-bound electron carriers that include hemes and iron-sulfur clusters. The energy released as the electrons move to lower and lower energy levels is used by each protein complex to pump protons out of the mitochondrial matrix, coupling lateral electron transport to vectorial proton transport across the membrane. Electrons are shuttled between the enzyme complexes by the mobile electron carriers ubiquinone and cytochrome c to complete the electron-transport chain. The path of electron flow from NADH is as follows: NADH → NADH dehydrogenase complex → ubiquinone → cytochrome c reductase → cytochrome c → cytochrome c oxidase complex → molecular oxygen (O_2). In addition, electrons derived from a citric acid cycle intermediate, succinate, enter this pathway by flowing through a fourth membrane-embedded, respiratory enzyme complex, succinate dehydrogenase, to ubiquinone.

ATP PRODUCTION IN MITOCHONDRIA

As we have just discussed, the three proton pumps of the respiratory chain each contribute to the formation of an electrochemical proton gradient across the inner mitochondrial membrane. This gradient drives ATP synthesis by ATP synthase, a large membrane-embedded protein complex that performs the extraordinary feat of converting the energy contained in this electrochemical gradient into biologically useful, chemical-bond energy in the form of ATP (see Figure 14–10). Protons flow down their electrochemical gradient through the membrane part of this proton turbine, and this drives the synthesis of ATP from ADP and phosphate in the part of the complex that protrudes into the mitochondrial matrix. As discussed in Chapter 2, the formation of ATP from ADP and phosphate is highly unfavorable energetically. As we shall see, ATP synthase can produce ATP only because of allosteric shape changes in this protein complex that directly couple ATP synthesis to the energetically favorable flow of protons across the membrane.

The Large Negative Value of ΔG for ATP Hydrolysis Makes ATP Useful to the Cell

An average person turns over roughly 50 kg of ATP per day. In athletes running a marathon, this figure can go up to several hundred kilograms. The ATP produced in mitochondria is derived from the energy available in the intermediates NADH,

TABLE 14–3 Product Yields from the Oxidation of Sugars and Fats
A. Net products from oxidation of one molecule of glucose
In cytosol (glycolysis)
1 glucose → 2 pyruvate + 2 NADH + 2 ATP
In mitochondrion (pyruvate dehydrogenase and citric acid cycle)
2 pyruvate → 2 acetyl CoA + 2 NADH
2 acetyl CoA → 6 NADH + 2 $FADH_2$ + 2 GTP
Net result in mitochondrion
2 pyruvate → 8 NADH + 2 $FADH_2$ + 2 GTP
B. Net products from oxidation of one molecule of palmitoyl CoA (activated form of palmitate, a fatty acid)
In mitochondrion (fatty acid oxidation and citric acid cycle)
1 palmitoyl CoA → 8 acetyl CoA + 7 NADH + 7 $FADH_2$
8 acetyl CoA → 24 NADH + 8 $FADH_2$ + 8 GTP
Net result in mitochondrion
1 palmitoyl CoA → 31 NADH + 15 $FADH_2$ + 8 GTP

$FADH_2$, and GTP. These three energy-rich compounds are produced by the oxidation of glucose (Table 14–3, part A), fats (Table 14–3, part B), and other fuels.

Glycolysis alone can produce only two molecules of ATP for every molecule of glucose that is metabolized, and this is the total energy yield for the fermentation processes that occur in the absence of O_2 (discussed in Chapter 2). In oxidative phosphorylation, each pair of electrons donated by the NADH produced in mitochondria can provide energy for the formation of about 2.5 molecules of ATP. Oxidative phosphorylation also produces about 1.5 ATP molecules per electron pair from the $FADH_2$ produced by succinate dehydrogenase in the mitochondrial matrix and from the NADH molecules produced by glycolysis in the cytosol. Combining this information with the product yields of glycolysis and the citric acid cycle, we can calculate that the complete oxidation of one molecule of glucose—starting with glycolysis and ending with oxidative phosphorylation—gives a net yield of about 30 molecules of ATP. Nearly all of this ATP is produced by the mitochondrial ATP synthase.

In Chapter 2, we introduced the concept of free energy (G). The free-energy change for a reaction, ΔG, determines whether that reaction will occur in a cell. We showed on pages 67–69 that the ΔG for a given reaction can be written as the sum of two parts: the first, called the standard free-energy change, ΔG°, depends only on the intrinsic characters of the reacting molecules; the second depends only on their concentrations. For the simple reaction A → B,

$$\Delta G = \Delta G^\circ + RT \ln \frac{[B]}{[A]}$$

where [A] and [B] denote the concentrations of A and B, and ln is the natural logarithm. ΔG° is the standard reference value, which can be seen to be equal to the value of ΔG when the molar concentrations of A and B are equal (as ln 1 = 0).

In Chapter 2, we discussed how the large, favorable free-energy change (large negative ΔG) for ATP hydrolysis is used, through coupled reactions, to drive many other chemical reactions in the cell that would otherwise not occur (see pp. 71–73). The ATP hydrolysis reaction produces two products, ADP and phosphate; it is therefore of the type A → B + C, where, as demonstrated in Figure 14–30,

$$\Delta G = \Delta G^\circ + RT \ln \frac{[B][C]}{[A]}$$

When ATP is hydrolyzed to ADP and phosphate under the conditions that normally exist in a cell, the free-energy change is roughly –46 to –54 kJ/mole

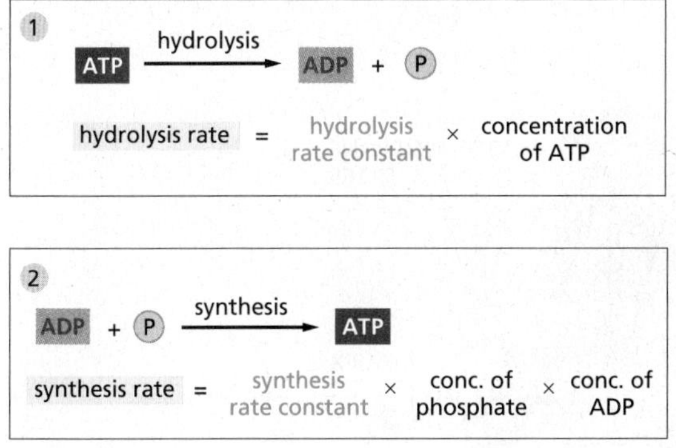

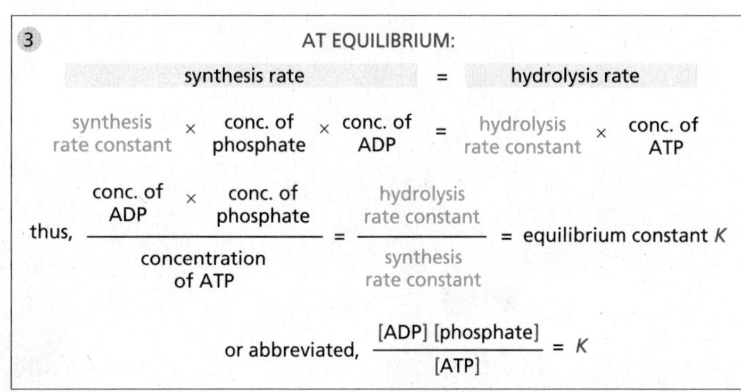

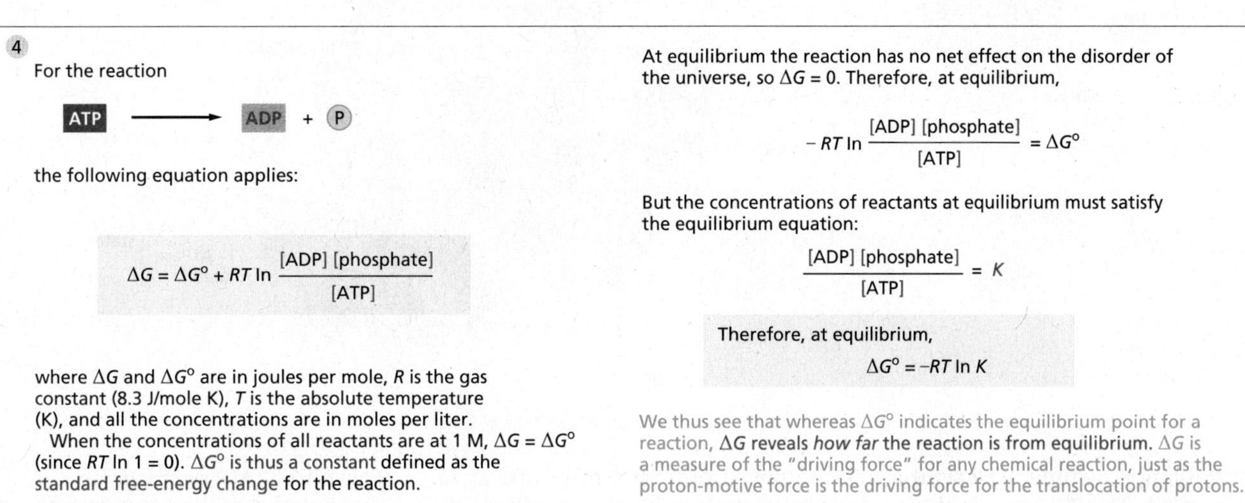

Figure 14–30 The basic relationship between free-energy changes and equilibrium in the ATP hydrolysis reaction. The rate constants in boxes 1 and 2 are determined from experiments in which product accumulation is measured as a function of time (conc., concentration). The equilibrium constant shown here, K, is in units of moles per liter. (See Panel 2–7, pp. 106–107, for a discussion of free energy, and see Figure 3–42 for a discussion of the equilibrium constant.)

(–11 to –13 kcal/mole). This extremely favorable ΔG depends on maintaining a high concentration of ATP compared with the concentrations of ADP and phosphate. When ATP, ADP, and phosphate are all present at the same concentration of 1 mole/liter (so-called standard conditions), the ΔG for ATP hydrolysis drops to the standard free-energy change (ΔG°), which is only –30.5 kJ/mole (–7.3 kcal/mole). At much lower concentrations of ATP relative to ADP and phosphate, ΔG becomes zero. At this point, the rate at which ADP and phosphate will join to form ATP will be equal to the rate at which ATP hydrolyzes to form ADP and phosphate. In other words, when $\Delta G = 0$, the reaction is at *equilibrium* (see Figure 14–30).

It is ΔG, not ΔG°, that indicates how far a reaction is from equilibrium and determines whether it can drive other reactions. Because the efficient conversion of ADP to ATP in mitochondria maintains such a high concentration of ATP relative to ADP and phosphate, the ATP hydrolysis reaction in cells is kept very far from equilibrium, and ΔG is correspondingly very negative. Without this large disequilibrium, ATP hydrolysis could not be used to drive the reactions of the cell. At low ATP concentrations, many essential reactions would become energetically unfavorable, run backward, and the cell would die.

The ATP Synthase Is a Nanomachine That Produces ATP by Rotary Catalysis

The **ATP synthase** is a finely tuned nanomachine composed of 23 or more separate protein subunits, with a total mass of about 600,000 daltons. The ATP synthase can work both in the forward direction, producing ATP from ADP and phosphate by consuming an electrochemical gradient, or in reverse, generating an electrochemical gradient by ATP hydrolysis. To distinguish it from other enzymes that hydrolyze ATP, it is also called an F_1F_0 ATP synthase or F-type ATPase.

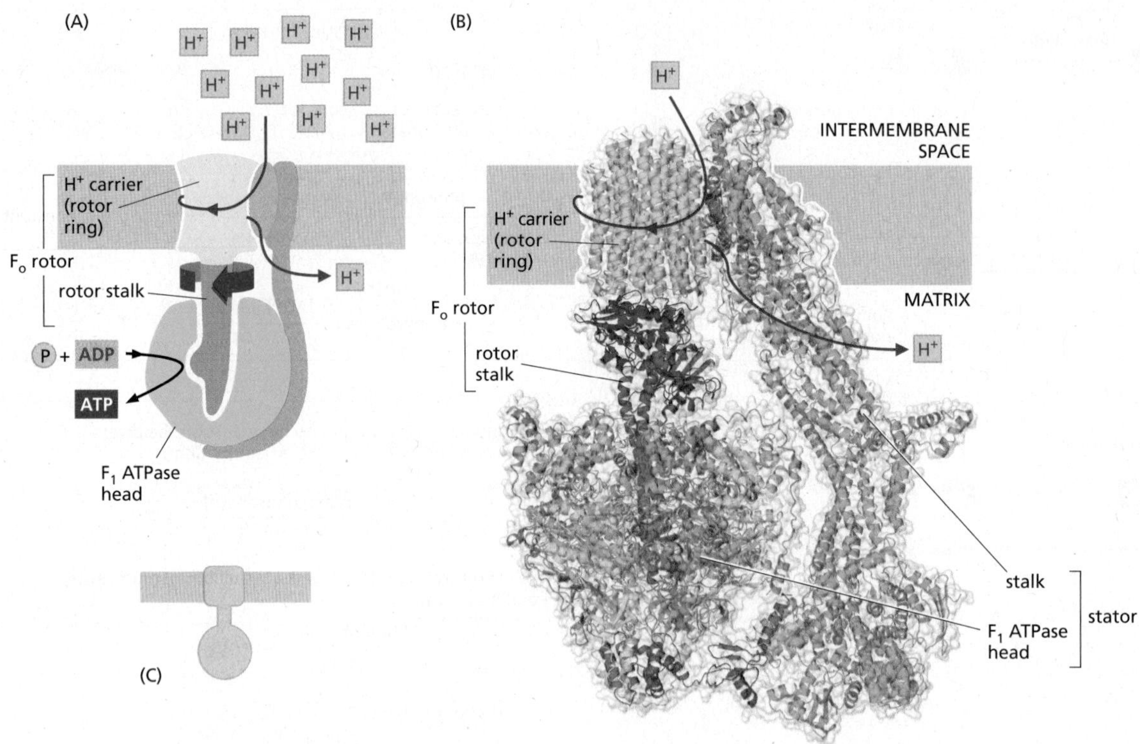

Figure 14–31 ATP synthase. The three-dimensional structure of the F_1F_o ATP synthase determined by x-ray crystallography. Also known as an F-type ATPase, it consists of an F_o part (from "oligomycin-sensitive factor") in the membrane and the large, catalytic F_1 head extending into the matrix. (A) Diagram of the enzyme complex showing how its globular head portion *(green)* is kept stationary as proton flow across the membrane drives a rotor *(light blue)* that turns the rotor stalk *(dark blue)* inside the head. (B) In bovine heart mitochondria, the F_o rotor ring in the membrane *(light blue)* has eight *c* subunits. It is attached to the γ subunit of the rotor stalk *(dark blue)* by the ε subunit *(purple)*. The catalytic F_1 head consists of a ring of three α and three β subunits *(light and dark green)*, and it directly converts mechanical energy into chemical-bond energy in ATP, as described in the text. The elongated peripheral stalk of the stator *(orange)* is connected to the F_1 head by the small δ subunit *(red)* at one end and to the *a* subunit in the membrane at the other. Together with the *c* subunits of the ring rotating past it, the *a* subunit creates a path for protons through the membrane. (C) The symbol for ATP synthase used throughout this book. (B, PDB codes: 2WPD, 2CLY, 2WSS, 2BO5.)

Resembling a turbine, ATP synthase is composed of both a rotor and a stator (**Figure 14–31**). To prevent the catalytic head from rotating, a stalk at the periphery of the complex (the stator stalk) connects the head to stator subunits embedded in the membrane. A second stalk in the center of the assembly (the rotor stalk) is connected to the rotor ring in the membrane, which turns as protons flow through it, driven by the electrochemical gradient across the membrane. As a result, proton flow makes the rotor stalk rotate inside the stationary head, where the catalytic sites that assemble ATP from ADP and phosphate are located. Three α and three β subunits of similar structure alternate to form the head. Each of the three β subunits has a catalytic nucleotide-binding site at the α–β interface. These catalytic sites are all in different conformations, depending on their interaction with the rotor stalk. This stalk acts like a camshaft, the device that opens and closes the valves in a combustion engine. As it rotates within the head, the stalk changes the conformations of the β subunits sequentially. One of the possible conformations of the catalytic sites has high affinity for ADP and phosphate, and as the rotor stalk pushes the binding site into a different conformation, these two substrates are driven to form ATP. As the rotor stalk continues to rotate, the binding site releases ATP, allowing the ATP-producing cycle to restart. In this way, the mechanical force exerted by the central rotor stalk is directly converted into the chemical energy of the ATP phosphate bond.

Serving as a proton-driven turbine, the ATP synthase is driven by H^+ flow into the matrix to spin at about 8000 revolutions per minute, generating three molecules of ATP per turn. In this way, each ATP synthase can produce roughly 400 molecules of ATP per second.

The closely related ATP synthases of mitochondria, chloroplasts, and bacteria synthesize ATP by harnessing the proton-motive force across a membrane. This powers the rotation of the rotor against the stator in a counterclockwise direction, as seen from the F_1 head. The same enzyme complex can also pump protons against their electrochemical gradient by hydrolyzing ATP, which then drives the clockwise rotation of the rotor. The direction of operation depends on the net free-energy change (ΔG) for the coupled processes of H^+ translocation across the membrane and the synthesis of ATP from ADP and phosphate (**Movie 14.7** and

Movie 14.8). Measurement of the torque that the ATP synthase can produce by ATP hydrolysis reveals that the ATP synthase is 60 times more powerful than a diesel engine of equal size.

Proton-driven Turbines Are Ancient and Critical for Energy Conversion

The membrane-embedded rotors of ATP synthases consist of a ring of identical *c* subunits (Figure 14–32). Each *c* subunit is a hairpin of two membrane-spanning α helices that contain a proton-binding site defined by a glutamate or aspartate in the middle of the lipid bilayer. The *a* subunit, which is part of the stator (see Figure 14–31), makes two narrow channels at the interface between the rotor and stator, each spanning half of the membrane and converging on the proton-binding site at the middle of the rotor subunit. Protons flow through the two half-channels down their electrochemical gradient from the crista space back into the matrix. The negatively charged glutamate or aspartate side chain in the binding site accepts a proton arriving from the crista space through the first half-channel, as it rotates past the *a* subunit. The bound proton then rides around in the ring for a full cycle, whereupon it is thought to be displaced by a positively charged arginine in the *a* subunit to escape through the second half-channel into the matrix. This directional proton flow causes the rotor ring to spin against the stator like a proton-driven turbine.

The mitochondrial ATP synthase is of ancient origin: essentially the same enzyme exists in plant chloroplasts and in the plasma membrane of bacteria or archaea. The main difference between them is the number of *c* subunits in the rotor ring. In mammalian mitochondria, the ring has 8 subunits. In yeast mitochondria, the number is 10; in bacteria and archaea, it ranges from 11 to 13; in plant chloroplasts, there are 14; and the rings of some cyanobacteria contain 15 *c* subunits.

The number of *c* subunits in the ring determines how many protons need to pass through this marvelous device to make each molecule of ATP. These subunits can be thought of as cogs in the gear wheels of a bicycle. A high gear, with a small number of cogs in the wheel, is advantageous when the supply of protons is limited, as in mitochondria, but a low gear, with a large number of cogs in the wheel, is preferable when the proton gradient is high. This is the case in chloroplasts and cyanobacteria, where protons produced through the action of sunlight are plentiful. Because each rotation produces three molecules of ATP in the head, the synthesis of one ATP requires around three protons in mitochondria but up to five in photosynthetic organisms—allowing the latter organisms to create a higher ratio of ATP to ADP and thus to maintain a greater ΔG for ATP hydrolysis.

Even under scenarios where mitochondrial ATP production is not possible or necessary, mitochondria must maintain a proton gradient to power the proper import of essential proteins and metabolites. In such a situation, ATP synthase can also run in reverse as an ATP-powered proton pump that converts the energy of ATP back into a proton gradient across the membrane. Moreover, in many bacteria, the rotor of the ATP synthase in the plasma membrane changes direction routinely, from ATP synthesis mode in aerobic respiration to ATP hydrolysis mode in anaerobic metabolism. In this latter case, ATP hydrolysis serves to maintain the proton gradient across the plasma membrane, which is used to power many other essential cell functions including nutrient transport and the rotation of bacterial flagella. The V-type ATPases that acidify certain cellular organelles are

(A)

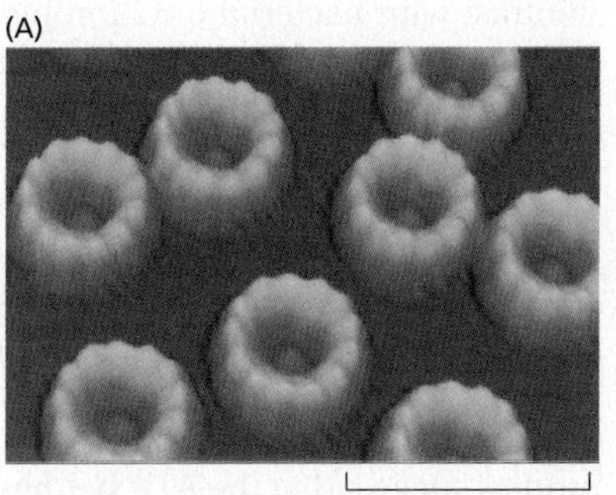

(B)

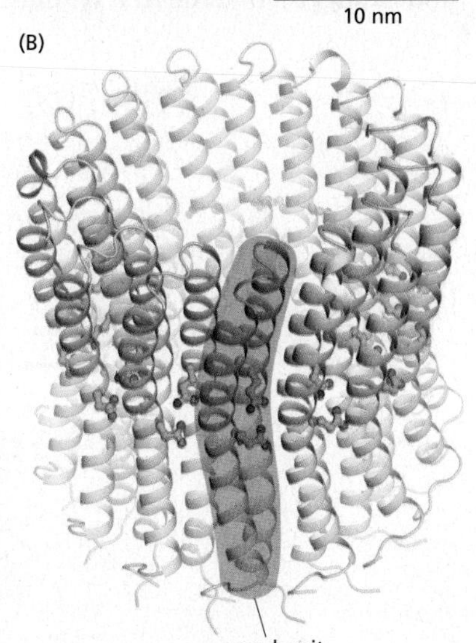

Figure 14–32 **F_o ATP synthase rotor rings.** (A) Atomic force microscopy image of ATP synthase rotors from the cyanobacterium *Synechococcus elongatus* in a lipid bilayer. Whereas 8 *c* subunits form the rotor in Figure 14–31, there are 13 *c* subunits in this ring. (B) The x-ray structure of the F_o ring of the ATP synthase from *Spirulina platensis*, another cyanobacterium, shows that this rotor has 15 *c* subunits. In all ATP synthases, the *c* subunits are hairpins of two membrane-spanning α helices (one subunit is highlighted in *gray*). The helices are highly hydrophobic, except for glutamine and glutamate side chains *(yellow)* that create proton-binding sites in the membrane. (A, courtesy of Thomas Meier and Denys Pogoryelov; B, PDB code: 2WIE.)

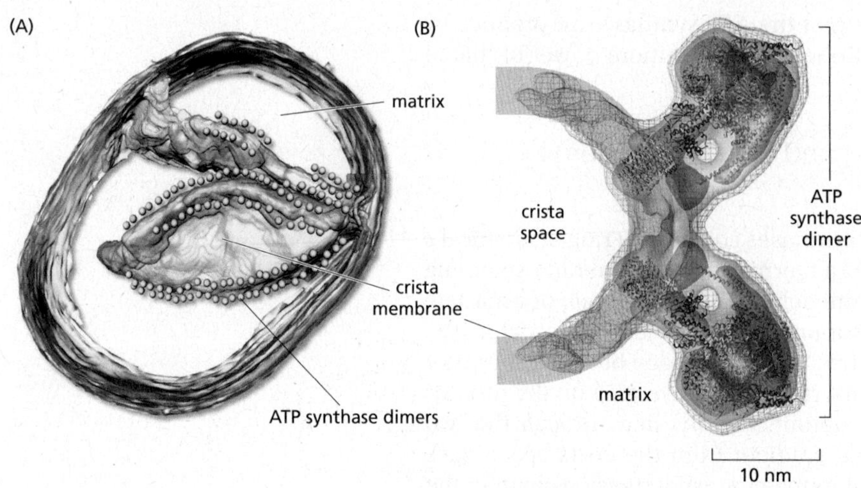

Figure 14–33 Dimers of mitochondrial ATP synthase in cristae membranes. (A) A three-dimensional map of a small mitochondrion obtained by electron microscope tomography shows that ATP synthases form long paired rows along cristae ridges. The outer membrane is *gray*, and the inner membrane and cristae membranes have been colored *light blue*. Each head of an ATP synthase is indicated by a *yellow sphere*. (B) A three-dimensional map of a mitochondrial ATP synthase dimer in the crista membrane obtained by sub-tomogram averaging, with fitted x-ray structures (**Movie 14.9**). (A, from K. Davies et al., *Proc. Natl. Acad. Sci. USA* 108:14121–14126, 2011. With permission from the National Academy of Sciences. B, from K. Davies et al., *Proc. Natl. Acad. Sci. USA* 109:13602–13607, 2012. With permission from the National Academy of Sciences.)

architecturally similar to the F-type ATP synthases, but they normally function in this ATP-dependent proton-pumping mode (see Figure 13–62).

Mitochondrial Cristae Help to Make ATP Synthesis Efficient

In the electron microscope, the mitochondrial ATP synthase complexes can be seen to project like lollipops on the matrix side of cristae membranes. Recent studies by cryo-electron microscopy and tomography have shown that this large complex is not distributed randomly in the membrane, but forms long rows of dimers along the cristae ridges (**Figure 14–33**). The dimer rows induce or stabilize these regions of high membrane curvature, which are otherwise energetically unfavorable. Indeed, the formation of ATP synthase dimers and their assembly into rows are required for cristae formation and have far-reaching consequences for cellular fitness. By contrast with bacterial or chloroplast ATP synthases, which do not form dimers, the mitochondrial complex contains additional subunits, located mostly near the membrane end of the stator stalk. If these dimer-specific subunits are mutated in yeast, the ATP synthase in the membrane remains monomeric, the mitochondria have no cristae, cellular respiration drops by half, and the cells grow more slowly.

Electron tomography suggests that the respiratory enzyme complexes that pump protons are located in the crista membrane at either side of the dimer rows. Protons pumped into the crista space by these respiratory-chain complexes are thought to diffuse very rapidly along the membrane surface, with the ATP synthase rows creating a proton "sink" at the cristae tips (**Figure 14–34**). *In vitro* studies suggest that the ATP synthase needs a proton gradient of about 2 pH units

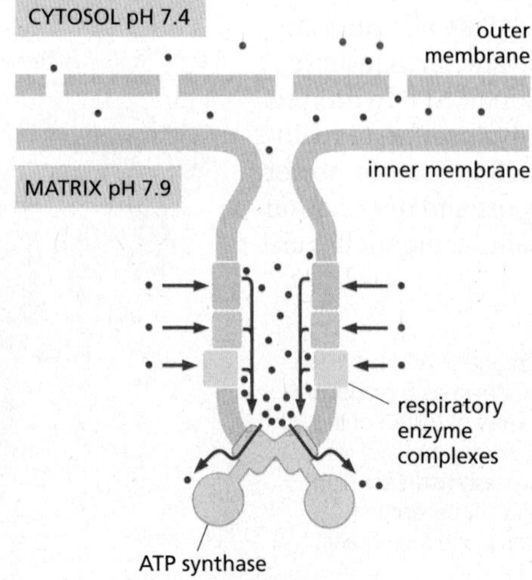

Figure 14–34 ATP synthase dimers at cristae ridges and ATP production. At the cristae ridges, the ATP synthases *(yellow)* form a sink for protons *(red)*. The proton pumps of the electron-transport chain *(blue and green)* are located in the membrane regions on either side of the crista. As illustrated, protons tend to be enriched in the crista space and diffuse along the membrane from their source to the proton sink created by the ATP synthase. This allows efficient ATP production despite the small overall H^+ gradient between the cytosol and matrix. *Red arrows* show the direction of the proton flow.

to produce ATP at the rate required by the cell, irrespective of the membrane potential. The H^+ gradient across the inner mitochondrial membrane is only 0.5–0.6 pH units. The cristae thus seem to work as proton traps that enable the ATP synthase to make efficient use of the protons pumped out of the mitochondrial matrix. As we shall see in the next part of this chapter, this elaborate arrangement of membrane protein complexes is absent in chloroplasts, where the H^+ gradient is much larger.

Special Transport Proteins Move Solutes Through the Inner Membrane

Like all biological membranes, the inner mitochondrial membrane contains numerous specific *transport proteins* that allow particular substances to pass through. More than 50 of these transporters are members of a single protein family, known as mitochondrial carriers. One of the most abundant of these family members is the *ADP/ATP carrier protein* (**Figure 14–35**). This carrier shuttles the ATP produced in the matrix through the inner membrane to the intermembrane space, from where it diffuses through the outer mitochondrial membrane to the cytosol. In exchange, ADP passes from the cytosol into the matrix for recycling into ATP. ATP^{4-} has one more negative charge than ADP^{3-}, and the exchange of ATP and ADP is driven by the electrochemical gradient across the inner membrane, so that the more negatively charged ATP is pushed out of the matrix, and the less negatively charged ADP is pulled in. Indeed, most of these carriers utilize the electrochemical proton gradient to drive directional transport of solutes across the inner membrane (see Chapter 11). In addition to members of the mitochondrial carrier family, all of which share significant sequence and structural similarity, several unrelated proteins transport pyruvate, calcium, serine, and other critical solutes.

In some specialized fat cells, mitochondrial respiration is uncoupled from ATP synthesis by the *uncoupling protein,* another member of the mitochondrial carrier family. In these cells, known as brown or beige fat cells, most of the energy of oxidation is dissipated as heat rather than being converted into ATP. In the inner

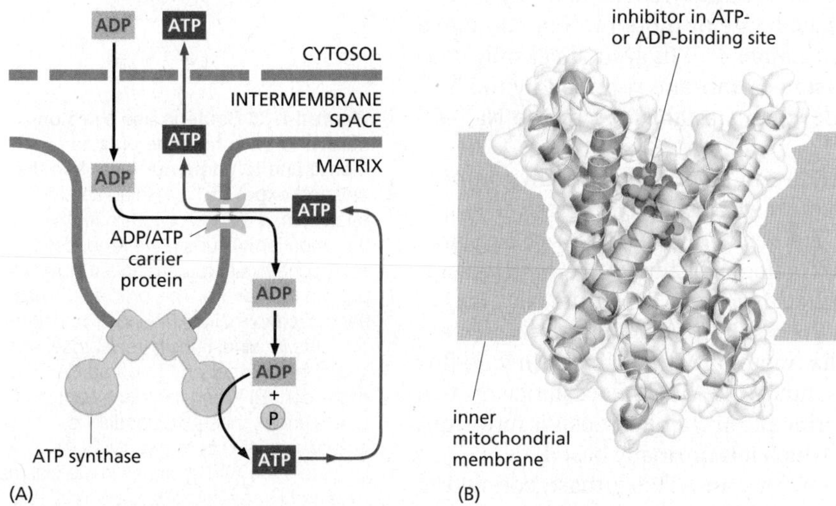

Figure 14–35 The ADP/ATP carrier protein. (A) The ADP/ATP carrier protein is a small membrane protein that carries the ATP produced on the matrix side of the inner membrane to the intermembrane space and carries the ADP that is needed for ATP synthesis into the matrix. (B) In the ADP/ATP carrier, six transmembrane α helices define a cavity that binds either ADP or ATP. In this x-ray structure, the substrate is replaced by a tightly bound inhibitor instead *(colored)*. When ADP binds from outside the inner membrane, it triggers a conformational change and is released into the matrix. In exchange, a molecule of ATP quickly binds to the matrix side of the carrier and is transported to the intermembrane space. From there the ATP diffuses through the outer mitochondrial membrane to the cytoplasm, where it powers the energy-requiring processes in the cell. (PDB code: 1OKC.)

membranes of the large mitochondria in these cells, the uncoupling protein allows protons to move down their electrochemical gradient without passing through ATP synthase. This process is switched on when heat generation is required, causing the cells to oxidize their fat stores at a rapid rate and produce heat rather than ATP. Activating this uncoupling process protects newborn human babies against the cold and has been shown to protect against obesity and diabetes in genetically engineered mice.

Chemiosmotic Mechanisms First Arose in Bacteria

Bacteria use enormously diverse energy sources. Some, like animal cells, are aerobic; they synthesize ATP from sugars they oxidize to CO_2 and H_2O by glycolysis, the citric acid cycle, and a respiratory chain in their plasma membrane that is similar to the one in the inner mitochondrial membrane. Others are strict anaerobes, deriving their energy either from glycolysis alone (by fermentation, see Figure 2–50) or from an electron-transport chain that employs a molecule other than oxygen as the final electron acceptor. The alternative electron acceptor can be a nitrogen compound (nitrate or nitrite), a sulfur compound (sulfate or sulfite), or a carbon compound (fumarate or carbonate), as examples. A series of electron carriers in the plasma membrane that are comparable to those in mitochondrial respiratory chains transfers the electrons to these acceptors.

Despite this diversity, the plasma membrane of the vast majority of bacteria contains an ATP synthase that is very similar to the one in mitochondria. In bacteria that use an electron-transport chain to harvest energy, the electron-transport chain pumps H^+ out of the cell and thereby establishes a proton-motive force across the plasma membrane that drives the ATP synthase to make ATP. In other bacteria, the ATP synthase works in reverse, using the ATP produced by glycolysis to pump H^+ and establish a proton gradient across the plasma membrane.

Bacteria, including the strict anaerobes, maintain a proton gradient across their plasma membrane that is harnessed to drive many other processes. It can be used to drive a flagellar motor, for example. The gradient is also harnessed to pump Na^+ out of the bacterium via an Na^+–H^+ antiporter that takes the place of the Na^+-K^+ pump of eukaryotic cells. In addition, the gradient is used for the active inward transport of nutrients, such as most amino acids and many sugars: each nutrient is dragged into the cell along with one or more protons through a specific symporter (**Figure 14–36**; see also Chapter 11). In animal cells, by contrast, most inward transport across the plasma membrane is driven by the Na^+ gradient (high Na^+ outside, low Na^+ inside) that is established by the Na^+-K^+ pump (see Figure 11–15).

Some unusual bacteria have adapted to live in a very alkaline environment and yet must maintain their cytoplasm at a physiological pH. For these cells, any attempt to generate an electrochemical H^+ gradient would be opposed by a large H^+ concentration gradient in the wrong direction (H^+ higher inside than outside). Presumably for this reason, some of these bacteria substitute Na^+ for H^+ in all of their chemiosmotic mechanisms. The respiratory chain pumps Na^+ out of the cell, the transport systems and flagellar motor are driven by an inward flux of Na^+, and an Na^+-driven ATP synthase synthesizes ATP. The existence of such bacteria demonstrates a critical point: the principle of chemiosmosis is more fundamental than the proton-motive force on which it is normally based.

As we discuss in the next part of this chapter, an ATP synthase coupled to chemiosmotic processes is also a central feature of plants, where it plays critical roles in both mitochondria and chloroplasts.

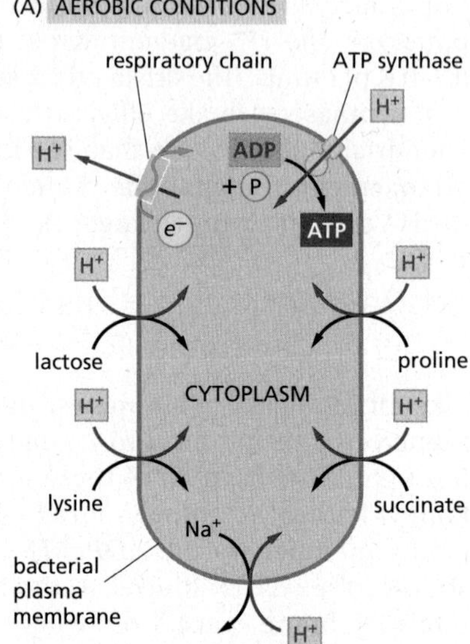

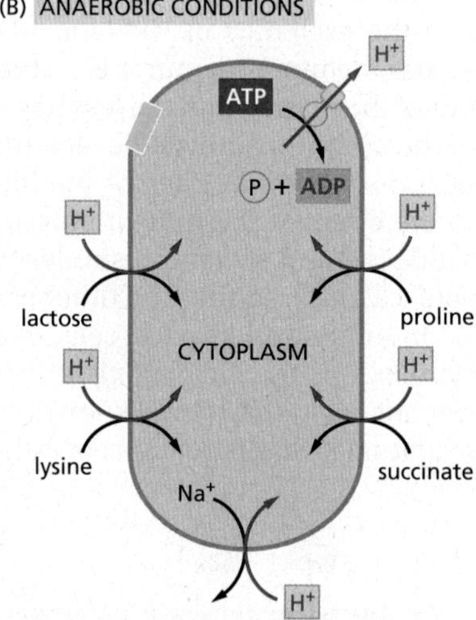

Figure 14–36 Bacteria use a proton-motive force across the plasma membrane to pump nutrients into the cell and expel Na^+. (A) In an aerobic bacterium, a respiratory chain fed by the oxidation of substrates produces an electrochemical proton gradient across the plasma membrane. This gradient is then harnessed to make ATP and to transport nutrients (proline, succinate, lactose, and lysine) into the cell and to pump Na^+ out of the cell. (B) When the same bacterium grows under anaerobic conditions, it derives its ATP from glycolysis. As indicated, the ATP synthase in the plasma membrane then hydrolyzes some of this ATP to establish an electrochemical proton gradient that drives the same transport processes that depend on respiratory chain proton-pumping in A.

Summary

The large amount of free energy released when H^+ flows back into the matrix from the cristae provides the basis for ATP production on the matrix side of mitochondrial cristae membranes by a remarkable protein machine—the ATP synthase. The ATP synthase functions like a miniature turbine, and it is a reversible device that *can couple proton flow to either ATP synthesis or ATP hydrolysis. The transmembrane electrochemical gradient that drives ATP production in mitochondria also drives the active transport of selected metabolites across the inner mitochondrial membrane, including an efficient ADP/ATP exchange between the mitochondrion and the cytosol that keeps the cell's ATP pool highly charged. The resulting high concentration of ATP inside the cell makes the free-energy change for ATP hydrolysis extremely favorable, allowing this hydrolysis reaction to drive a large number of energy-requiring processes throughout the cell. The universal presence of an ATP synthase in bacteria, mitochondria, and chloroplasts testifies to the central importance of chemiosmotic mechanisms for life.*

CHLOROPLASTS AND PHOTOSYNTHESIS

All animals and most microorganisms rely on the continual uptake of large amounts of organic compounds from their environment. These compounds provide both the carbon-rich building blocks for biosynthesis and the metabolic energy for life. It has been proposed that the first organisms on the primitive Earth had access to an abundance of organic compounds produced by geochemical processes, but these would have been used up early in life's evolution. Since then, the vast majority of the organic materials required by living cells has been produced by *photosynthetic organisms,* including plants and photosynthetic bacteria. The core machinery that drives all photosynthesis appears to have evolved more than 3 billion years ago in the ancestors of present-day bacteria; today it provides the major solar energy storage mechanism on Earth.

The most advanced photosynthetic bacteria are the cyanobacteria, which have minimal nutrient requirements. They use electrons from water and the energy of sunlight to convert atmospheric CO_2 into organic compounds—a process called *carbon fixation.* In the course of the overall reaction

$$n\text{H}_2\text{O} + n\text{CO}_2 \xrightarrow{\text{Light}} (\text{CH}_2\text{O})_n + n\text{O}_2$$

they also liberate into the atmosphere the molecular oxygen needed for oxidative phosphorylation. In this way, it is thought that the evolution of an organism similar to modern cyanobacteria from more primitive photosynthetic bacteria eventually made possible the development of the many different aerobic life-forms that populate Earth today.

Chloroplasts Resemble Mitochondria but Have a Separate Thylakoid Compartment

Plants (including algae) developed much later than cyanobacteria, and their photosynthesis occurs in a specialized intracellular organelle—the **chloroplast** (**Figure 14–37**). There is good evidence that the chloroplast evolved after an endosymbiotic event between a eukaryotic cell and a photosynthetic

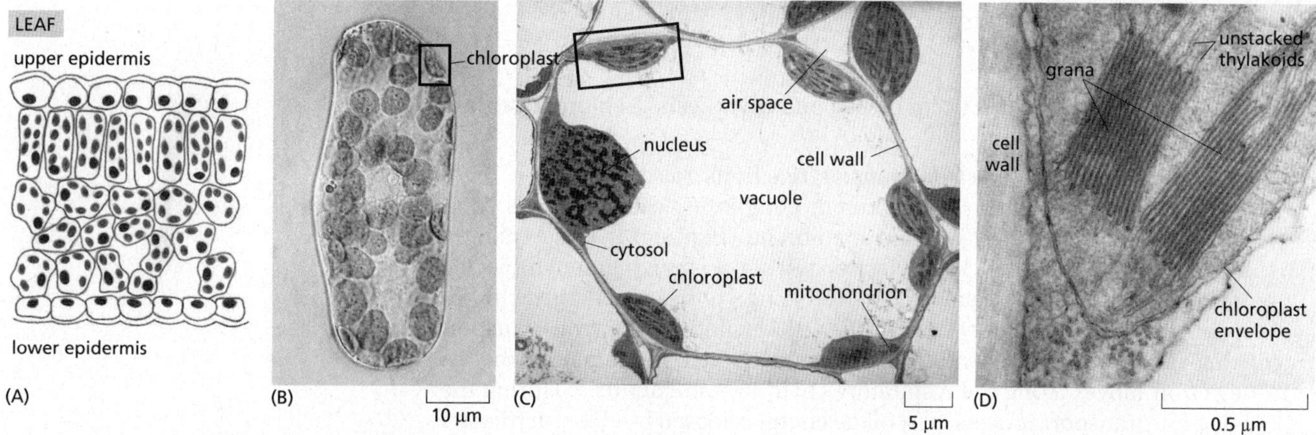

Figure 14–37 Chloroplasts in a plant cell. (A) Schematic cross section through the leaf of a green plant. (B) Light microscopy of a plant leaf cell—here, a mesophyll cell from *Zinnia elegans*—shows chloroplasts as bright green bodies, measuring several micrometers across, in the transparent cell interior. (C) An electron micrograph of a thin, stained section through a wheat leaf cell shows a thin rim of cytoplasm—containing chloroplasts, the nucleus, and mitochondria—surrounding a large, water-filled vacuole. (D) At higher magnification, electron microscopy reveals the chloroplast envelope membrane and the thylakoid membrane within the chloroplast that is highly folded into grana stacks (**Movie 14.10** and **Movie 14.11**). (B, courtesy of Preeti Dahiya; C and D, courtesy of K. Plaskitt.)

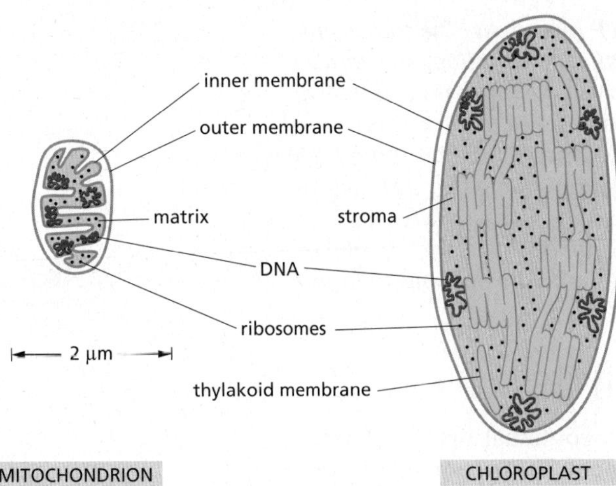

Figure 14–38 **A mitochondrion and chloroplast compared.** Chloroplasts are generally larger than mitochondria. In addition to an outer and inner envelope membrane, they contain the thylakoid membrane with its internal thylakoid lumen. The chloroplast thylakoid membrane, which is the site of solar energy conversion in plants and algae, corresponds to the mitochondrial cristae, which are the sites of energy conversion by cellular respiration. Unlike the crista membrane, which is continuous with the inner mitochondrial membrane at cristae junctions, the thylakoid membrane is not connected to the inner chloroplast membrane at any point.

cyanobacteria-like organism. Chloroplasts use chemiosmotic mechanisms to carry out their energy interconversions in much the same way that mitochondria do. Although much larger than mitochondria, they are organized on the same principles. They have a highly permeable outer membrane; a much less permeable inner membrane, in which membrane transport proteins are embedded; and a narrow intermembrane space in between. Together, these two membranes form the chloroplast envelope (Figure 14–37D). The inner chloroplast membrane surrounds a large space called the **stroma**, which is analogous to the mitochondrial matrix and to the bacterial cytoplasm. The stroma contains many metabolic enzymes, and, as for the mitochondrial matrix, it is the place where ATP is made by the heads of ATP synthase protein machines. Like the mitochondrion, the chloroplast has its own genome and genetic system. The stroma therefore also contains a special set of ribosomes, RNAs, and the chloroplast DNA.

An important difference between the organization of mitochondria and chloroplasts is highlighted in Figure 14–38. The inner membrane of the chloroplast is not folded into cristae and does not contain electron-transport chains. Instead, the electron-transport chains, photosynthetic light-capturing systems, and ATP synthase are all contained in the **thylakoid membrane**, a separate, distinct membrane that forms a set of flattened, disc-like sacs, the *thylakoids*. The thylakoid membrane is highly folded into numerous local stacks of flattened vesicles called *grana*, interconnected by nonstacked thylakoids. The lumen of each thylakoid is connected with the lumen of other thylakoids, thereby defining a third internal compartment called the *thylakoid lumen*. This space represents a separate compartment in each chloroplast that is not connected to either the intermembrane space or the stroma. In addition to serving an analogous function, the thylakoid membrane has other similarities to the mitochondrial crista membrane, including a very high protein density and a unique lipid composition.

Chloroplasts Capture Energy from Sunlight and Use It to Fix Carbon

We can group the reactions that occur during photosynthesis in chloroplasts into two broad categories:

1. The **photosynthetic electron-transfer reactions** (also called the "light reactions") occur in two large protein complexes, called *reaction centers*, embedded in the thylakoid membrane (Figure 14–39). A photon (a quantum of light) knocks an electron out of the green pigment molecule *chlorophyll* in the first reaction center, creating a positively charged chlorophyll ion. This high-energy electron then moves along an electron-transport chain and through a second reaction center in much the same way that an electron moves along the respiratory chain in mitochondria. During the electron-transport process, part of the energy released by electron transfer

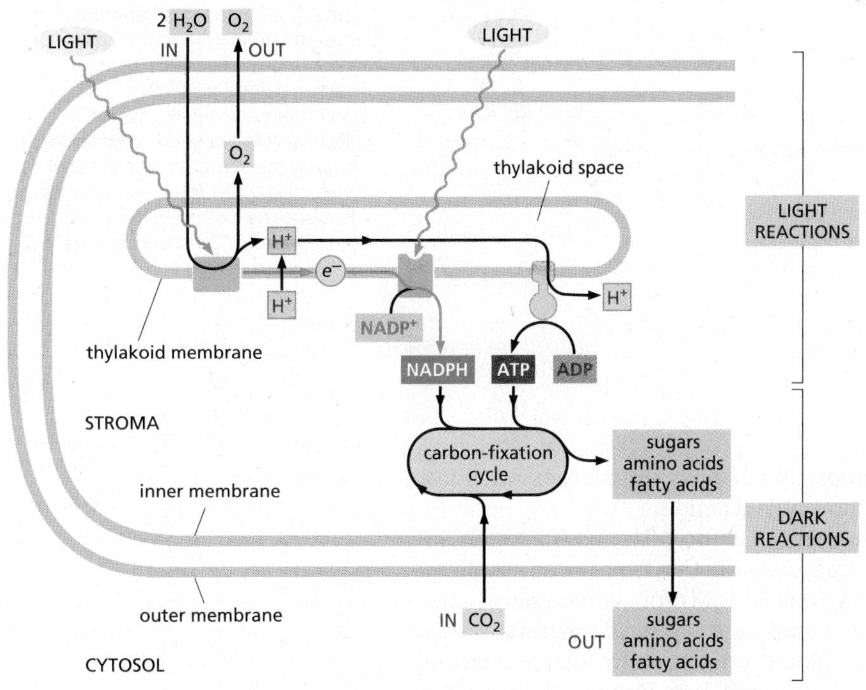

Figure 14–39 A summary of the energy-converting metabolism in chloroplasts. Chloroplasts require only water and carbon dioxide as inputs for their light-driven photosynthesis reactions, and they produce the nutrients for most other organisms on the planet. Each oxidation of two water molecules by a photochemical reaction center in the thylakoid membrane produces one molecule of oxygen, which is released into the atmosphere. At the same time, protons are concentrated in the thylakoid space. These protons create a large electrochemical gradient across the thylakoid membrane, which is utilized by the chloroplast ATP synthase to produce ATP from ADP and phosphate. The electrons withdrawn from water are transferred to a second type of photochemical reaction center to produce NADPH from $NADP^+$. As indicated, the NADPH and ATP are fed into the *carbon-fixation cycle* to reduce carbon dioxide, thereby producing the precursors for sugars, amino acids, and fatty acids. The CO_2 that is taken up from the atmosphere here is the source of the carbon atoms for most organic molecules on Earth.

In a plant cell, a variety of metabolites produced in the chloroplast are exported to the cytoplasm for biosyntheses. Some of the sugar produced is stored in the form of starch granules in the chloroplast, but the rest is transported throughout the plant as sucrose or converted to starch in special storage tissues. These storage tissues serve as a major food source for animals.

is harnessed to pump H^+ ions (protons) across the thylakoid membrane. The resulting electrochemical proton gradient is then used by ATP synthase to drive the synthesis of ATP in the stroma. As the final step in this series of reactions, electrons are loaded (together with H^+) onto $NADP^+$, converting it to the energy-rich NADPH molecule. Because the positively charged chlorophyll in the first reaction center quickly regains its electrons from water (H_2O), H^+ and O_2 gas are produced as by-products. All of these reactions are confined to the chloroplast.

2. The **carbon-fixation reactions** (also called the "dark reactions") do not require sunlight. Here the ATP and NADPH generated by the above light reactions serve as the source of energy and reducing power, respectively, to drive the conversion of CO_2 to carbohydrate. These carbon-fixation reactions occur in the chloroplast stroma, where they generate the three-carbon sugar *glyceraldehyde 3-phosphate.* This simple sugar is an intermediate in the glycolysis and gluconeogenesis pathways. Upon export to the cytosol, it is used to produce sucrose and many other organic metabolites in the leaves of the plant. The sucrose is then exported to meet the metabolic needs of the nonphotosynthetic plant tissues, serving as a source of both carbon skeletons and energy for growth.

Thus, the formation of ATP, NADPH, and O_2 (which requires light energy directly) and the conversion of CO_2 to carbohydrate (which requires light energy only indirectly) are separate processes (Figure 14–39). But, they are linked by elaborate feedback mechanisms that allow a plant to manufacture sugars only when it is appropriate to do so. Several of the chloroplast enzymes required for carbon fixation, for example, are inactive in the dark and reactivated by light-stimulated electron-transport processes.

Carbon Fixation Uses ATP and NADPH to Convert CO_2 into Sugars

We have seen earlier in this chapter how mitochondria in eukaryotic cells produce ATP by using the large amount of free energy released when carbohydrates are oxidized to CO_2 and H_2O. The reverse reaction, in which plants make carbohydrate from CO_2 and H_2O, takes place in the chloroplast stroma. The large amounts

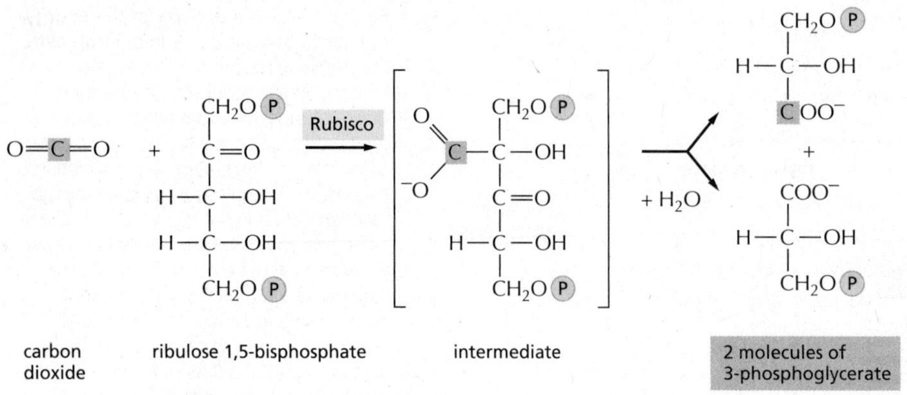

Figure 14–40 The initial reaction in carbon fixation. This carboxylation reaction allows one molecule each of carbon dioxide and water to be incorporated into organic carbon molecules. It is catalyzed in the chloroplast stroma by the abundant enzyme ribulose 1,5-bisphosphate carboxylase/oxygenase, or Rubisco. As indicated, the product is two molecules of 3-phosphoglycerate.

of ATP and NADPH produced by the photosynthetic electron-transfer reactions are required to drive this energetically unfavorable reaction.

Figure 14–40 illustrates the central reaction of carbon fixation, in which an atom of inorganic carbon is converted to organic carbon: CO_2 from the atmosphere combines with the five-carbon compound ribulose 1,5-bisphosphate plus water to yield two molecules of the three-carbon compound 3-phosphoglycerate. This carboxylation reaction is catalyzed in the chloroplast stroma by a large enzyme called *ribulose 1,5-bisphosphate carboxylase/oxygenase*, or *Rubisco.* Because the reaction is so slow (each Rubisco molecule turns over only about 3 molecules of substrate per second, compared to 1000 molecules per second for a typical enzyme), a very large number of enzyme molecules are needed. Rubisco constitutes up to 50% of the chloroplast protein mass, and it is one of the most abundant proteins on Earth. Of great importance to the global environment, Rubisco also helps to reduce the amount of the greenhouse gas CO_2 in the atmosphere.

Although the production of carbohydrates from CO_2 and H_2O is energetically unfavorable, the fixation of CO_2 catalyzed by Rubisco is an energetically favorable reaction. Carbon fixation is energetically favorable because a continuous supply of the energy-rich ribulose 1,5-bisphosphate is fed into the process via the carbon-fixation cycle described below. This compound is consumed by the addition of CO_2, and it must be replenished. As we describe next, the energy and reducing power needed to regenerate ribulose 1,5-bisphosphate come from the ATP and NADPH produced by the photosynthetic light reactions.

The elaborate series of reactions in which CO_2 combines with ribulose 1,5-bisphosphate to produce a simple sugar—a portion of which is used to regenerate ribulose 1,5-bisphosphate—forms a cycle, called the *carbon-fixation cycle*, or the Calvin–Benson–Bassham cycle (Figure 14–41). This cycle was one of the first metabolic pathways to be defined by applying radioisotopes as tracers in biochemistry. As indicated, each turn of the cycle converts six molecules of 3-phosphoglycerate to three molecules of ribulose 1,5-bisphosphate plus one molecule of glyceraldehyde 3-phosphate. The net result of each turn of the cycle is that 9 ATP and 6 NADPH molecules are used to power the conversion of 3 CO_2 molecules into one molecule of glyceraldehyde 3-phosphate.

Glyceraldehyde 3-phosphate, the three-carbon sugar produced by the cycle, occupies a pivotal point in carbohydrate biochemistry as an intermediate in glycolysis and gluconeogenesis, enabling it to serve as the starting material for the synthesis of many other sugars and all of the other organic molecules that form the plant.

Carbon Fixation in Some Plants Is Compartmentalized to Facilitate Growth at Low CO_2 Concentrations

Although ribulose 1,5-bisphosphate carboxylase/oxygenase preferentially adds CO_2 to ribulose 1,5-bisphosphate, it can also use O_2 as a substrate in place of CO_2, and if the concentration of CO_2 is low, it will add O_2 to ribulose 1,5-bisphosphate

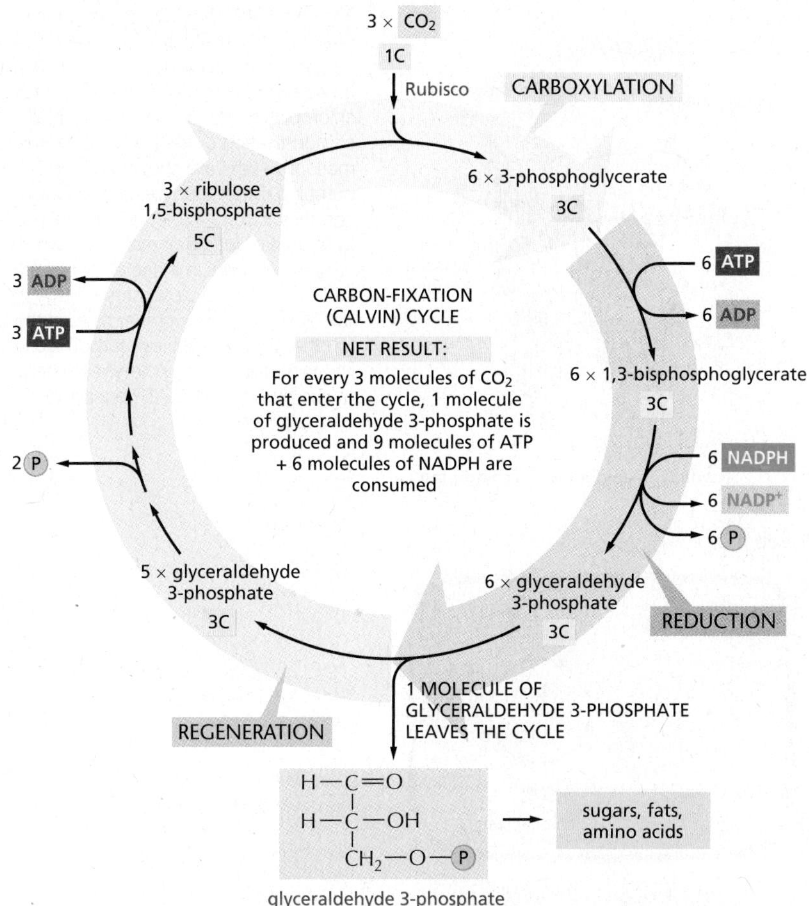

Figure 14–41 **The Calvin–Benson–Bassham cycle (Calvin cycle) for carbon fixation.** This central metabolic pathway allows organic molecules to be produced from CO_2 and H_2O. In the first stage of the cycle (carboxylation), CO_2 is added to ribulose 1,5-bisphosphate, as shown in Figure 14–40. In the second stage (reduction), ATP and NADPH are consumed to produce glyceraldehyde 3-phosphate molecules. In the final stage (regeneration), some of the glyceraldehyde 3-phosphate produced is used to regenerate ribulose 1,5-bisphosphate. Other glyceraldehyde 3-phosphate molecules are either converted to starch and fat in the chloroplast stroma or transported out of the chloroplast into the cytosol. The number of carbon atoms in each type of molecule is indicated in *yellow*. There are many intermediates between glyceraldehyde 3-phosphate and ribulose 1,5-bisphosphate, but they have been omitted here for clarity. The entry of water into the cycle is also not shown (but see Figure 14–40).

instead (see Figure 14–40). This is the first step in a pathway called **photorespiration**, whose ultimate effect is to use up O_2 and liberate CO_2 without the production of useful energy stores. In many plants, about one-third of the CO_2 fixed is lost again as CO_2 because of photorespiration.

Photorespiration can be a serious liability for plants in hot, dry conditions, which cause them to close their stomata (the gas-exchange pores in their leaves, each of which is called a stoma) to avoid excessive water loss. This in turn causes the CO_2 levels in the leaf to fall precipitously, thereby favoring photorespiration. A special adaptation, however, occurs in the leaves of many plants, such as corn and sugarcane, that grow in hot, dry environments. In these plants, the carbon-fixation cycle occurs only in the chloroplasts of specialized *bundle-sheath cells,* which contain all of the plant's ribulose 1,5-bisphosphate carboxylase/oxygenase. These cells are protected from the air and are surrounded by a specialized layer of *mesophyll cells* that use the energy harvested by their chloroplasts to "pump" CO_2 into the bundle-sheath cells. This supplies the ribulose 1,5-bisphosphate carboxylase/oxygenase with a high concentration of CO_2, thereby greatly reducing photorespiration.

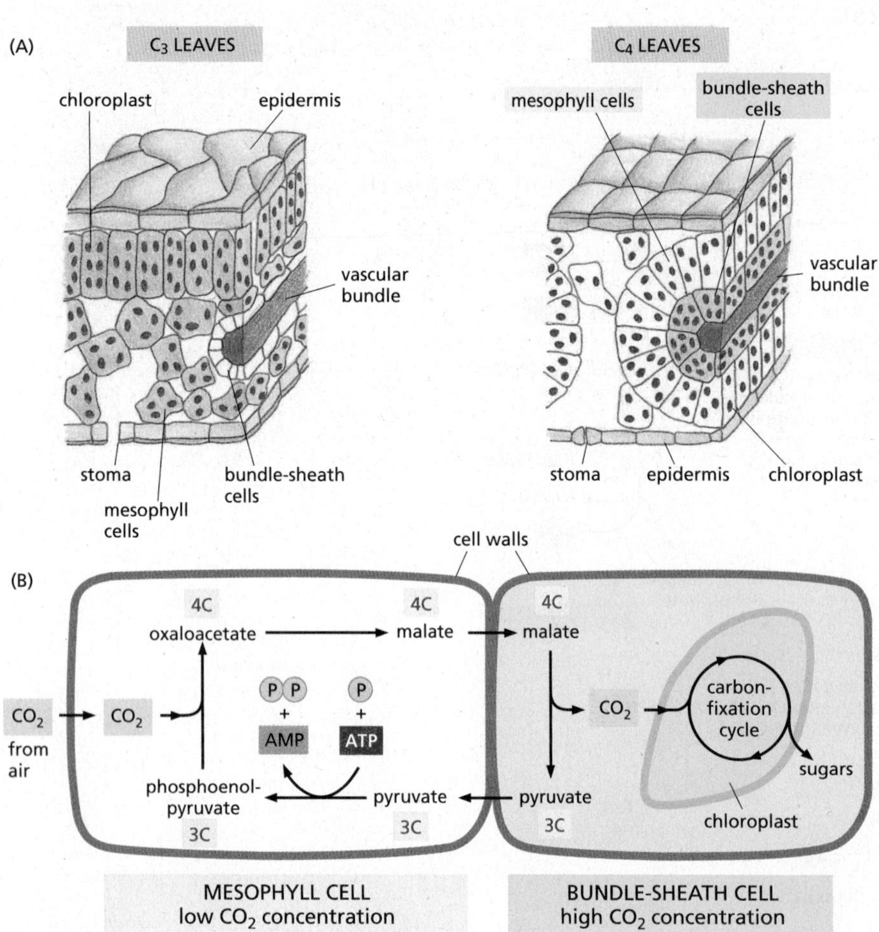

Figure 14–42 The CO_2 pumping in C_4 plants. (A) Comparative leaf anatomy in a C_3 plant and a C_4 plant. The cells with *green* cytosol in the leaf interior contain chloroplasts that perform the normal carbon-fixation cycle. In C_4 plants, the mesophyll cells are specialized for CO_2 pumping rather than for carbon fixation, and they thereby create a high ratio of CO_2 to O_2 in the bundle-sheath cells, which are the only cells in these plants where the carbon-fixation cycle occurs. The vascular bundles carry the sucrose made in the leaf to other tissues. (B) How carbon dioxide is concentrated in bundle-sheath cells by the harnessing of ATP energy in mesophyll cells.

The CO_2 pump mechanism depends on a reaction cycle that begins in the cytosol of the mesophyll cells. An initial CO_2-fixation step is catalyzed by an enzyme that binds CO_2 (as bicarbonate) and combines it with an activated three-carbon molecule (phosphoenolpyruvate) to produce a four-carbon molecule. The four-carbon molecule diffuses into the bundle-sheath cells, where it is broken down to release the CO_2 and generate a molecule with three carbons. The pumping cycle is completed when this three-carbon molecule (pyruvate) is returned to the mesophyll cells and converted back to its original activated form. Because the CO_2 is initially captured by converting it into a compound containing four carbons, the CO_2-pumping plants are called *C_4 plants*. All other plants are called *C_3 plants* because they capture CO_2 into the three-carbon compound 3-phosphoglycerate (**Figure 14–42**).

As with any vectorial transport process, pumping CO_2 into the bundle-sheath cells in C_4 plants costs energy (ATP is hydrolyzed; see Figure 14–42B). In hot, dry environments, however, this cost can be much less than the energy lost by photorespiration in C_3 plants, so C_4 plants have a potential advantage. Moreover, because C_4 plants can perform photosynthesis at a lower concentration of CO_2 inside the leaf, they need to open their stomata less often and therefore can fix about twice as much net carbon as C_3 plants per unit of water lost. This type of carbon fixation has evolved independently in several different plant lineages. Although the vast majority of plant species are C_3 plants, C_4 plants such as corn and sugarcane are much more effective at converting sunlight energy into biomass than C_3 plants such as cereal grains. They are therefore of special importance in world agriculture.

Many algae have a different adaptation to manage the challenge of limited CO_2 concentration, which is often a problem in the aquatic environment in which algae grow. The pyrenoid is a complex structure that contains a biomolecular

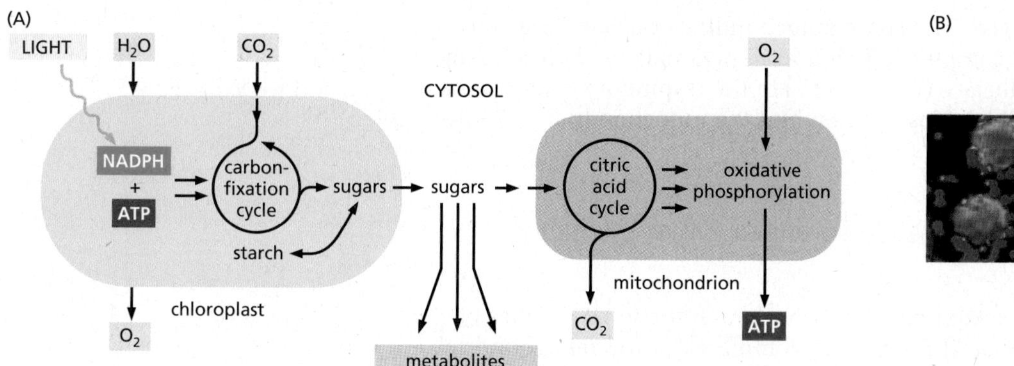

(B)

10 μm

Figure 14–43 How chloroplasts and mitochondria collaborate to supply cells with both metabolites and ATP. (A) The inner chloroplast membrane is impermeable to the ATP and NADPH that are produced in the stroma during the light reactions of photosynthesis. These molecules are instead funneled into the carbon-fixation cycle, where they are used to make sugars. The resulting sugars and their metabolites are either stored within the chloroplast—in the form of starch or fat—or exported to the rest of the plant cell. There, they can enter the energy-generating pathway that ends in ATP synthesis linked to oxidative phosphorylation inside the mitochondrion. Unlike the chloroplast, mitochondrial membranes contain a specific transporter that makes them permeable to ATP (see Figure 14–35). Note that the O_2 released to the atmosphere by photosynthesis in chloroplasts is used for oxidative phosphorylation in mitochondria; similarly, the CO_2 released by the citric acid cycle in mitochondria is used for carbon fixation in chloroplasts. (B) In a leaf, mitochondria *(red)* tend to cluster close to the chloroplasts *(green)*, as seen in this light micrograph. (B, courtesy of Olivier Grandjean.)

condensate highly enriched in Rubisco along with a scaffold protein (see Figure 12–8). This condensate is interwoven with membrane tubules that contain carbonic anhydrase, an enzyme that converts the bicarbonate ions (HCO_3^-) dissolved in water into CO_2. The pyrenoid therefore provides CO_2 for Rubisco, thereby favoring carbon fixation over deleterious photorespiration.

The Sugars Generated by Carbon Fixation Can Be Stored as Starch or Consumed to Produce ATP

The glyceraldehyde 3-phosphate generated by carbon fixation in the chloroplast stroma can be used in a number of ways, depending on the needs of the plant. During periods of excess photosynthetic activity, much of it is converted to glucose and stored as *starch* in the chloroplast stroma. Like glycogen in animal cells, starch is a storage polymer of glucose that is found as large granules. Starch forms an important part of the diet of all animals that eat plants. Other glyceraldehyde 3-phosphate molecules are converted to different important biosynthetic or storage molecules. For example, fat, which accumulates as triglycerides and other storage esters in fat droplets, serves as an additional energy reserve.

At night, stored starch and fat can be broken down to glucose and fatty acids, which are exported to the cytosol to help support the metabolic needs of the plant. Some of the exported sugar enters the glycolytic pathway (see Figure 2–46), where it is converted to pyruvate. Both that pyruvate and the fatty acids can enter the plant cell mitochondria and be catabolized to acetyl CoA that is fed into the citric acid cycle, ultimately leading to the efficient production of ATP by oxidative phosphorylation (**Figure 14–43**). In this way, the metabolic transitions of the light/dark cycle of the plant are reminiscent of the fasting/feeding cycles typical of many animals.

In the cytosol, glucose exported from chloroplasts after starch hydrolysis or generated from gluconeogenesis from glyceraldehyde 3-phosphate can also be converted into the disaccharide *sucrose*. Sucrose is the major form in which sugar is transported between the cells of a plant. Just as glucose is transported in the blood of animals, so sucrose is exported from the leaves to provide carbohydrate to the rest of the plant.

The Thylakoid Membranes of Chloroplasts Contain the Protein Complexes Required for Photosynthesis and ATP Generation

We next need to explain how the large amounts of ATP and NADPH required for carbon fixation are generated in the chloroplast. Chloroplasts make use of chemiosmotic energy conversion in much the same way as mitochondria. As we saw in Figure 14–38, chloroplasts and mitochondria are organized on the same principles, although the chloroplast contains a separate thylakoid membrane system in which its chemiosmotic mechanisms occur. The thylakoid membranes contain two large membrane protein complexes, called *photosystems*, which endow photosynthetic organisms with the ability to capture and convert solar energy into usable forms of energy. Two other protein complexes in the thylakoid membrane work together with the photosystems to catalyze photophosphorylation—the generation of ATP with sunlight. These have mitochondrial equivalents: a heme-containing cytochrome b_6-f complex, which both functionally and structurally resembles cytochrome c reductase (Complex III) in the respiratory chain; and a chloroplast ATP synthase, which closely resembles the mitochondrial ATP synthase and works in the same way.

Chlorophyll–Protein Complexes Can Transfer Either Excitation Energy or Electrons

The photosystems in the thylakoid membrane are multiprotein assemblies of a complexity comparable to that of the protein complexes in the mitochondrial electron-transport chain. They contain large numbers of specifically bound chlorophyll molecules, in addition to cofactors that will be familiar from our discussion of mitochondria (heme, iron–sulfur clusters, and quinones). **Chlorophyll**, the green pigment of photosynthetic organisms, has a long hydrophobic tail that makes it behave like a lipid, plus a porphyrin ring that has an extensive system of delocalized electrons in conjugated double bonds and a central Mg atom (structurally analogous to heme and its central Fe atom; **Figure 14–44**).

When a chlorophyll molecule absorbs a quantum of sunlight (a photon), the energy of the photon causes one of these electrons to move from a low-energy orbital to another orbital of higher energy. A photon of red light (~600–700 nm wavelength) moves an electron to a low-level excited state. A higher-energy photon of blue light (~400–500 nm) kicks an electron to a higher-energy excited state, which rapidly decays to the low-level excited state by converting the extra energy into heat (molecular motion). The excited electrons in the low-energy state tend to return to their ground state in one of three ways:

1. By converting the extra energy into heat (molecular motion) or to some combination of heat and light of a longer wavelength (fluorescence); this is what usually happens when light is absorbed by an isolated chlorophyll molecule in solution.
2. By transferring the energy—but not the electron—directly to a neighboring chlorophyll molecule by a process called *resonance energy transfer*.
3. By transferring the excited electron with its negative charge to another nearby molecule, an *electron acceptor*, after which the positively charged chlorophyll returns to its original state by taking up an electron from some other molecule, an *electron donor*.

Only the latter two mechanisms are useful for the capturing of energy from sunlight, and they are enabled by the chlorophylls being precisely positioned within a *chlorophyll–protein complex*. The protein coordinates the central Mg atom in the chlorophyll porphyrin, most often through a histidine side chain located in the hydrophobic interior of a membrane, causing each of the chlorophylls in the protein complex to be held at exactly defined distances and orientations. The flow of excitation energy or electrons then depends on both the precise spatial arrangement and the local protein environment of the protein-bound chlorophylls.

When excited by a photon, most protein-bound chlorophylls simply transmit the absorbed energy to another nearby chlorophyll by the process of resonance energy transfer. However, in a few specially positioned chlorophylls, the energy difference between the ground state and the excited state is just right for the photon to trigger a light-induced chemical reaction. The special state of such chlorophyll molecules derives from their close interaction with a second chlorophyll molecule in the same chlorophyll–protein complex. Together, these two chlorophylls form a *special pair*.

The photosynthetic electron-transfer process starts when a photon of suitable energy ionizes a chlorophyll molecule in such a special pair, dissociating it into an electron and a positively charged chlorophyll ion. The energized electron is then passed rapidly to a quinone in the same protein complex, preventing its unproductive reassociation with the chlorophyll ion. This light-induced transfer of

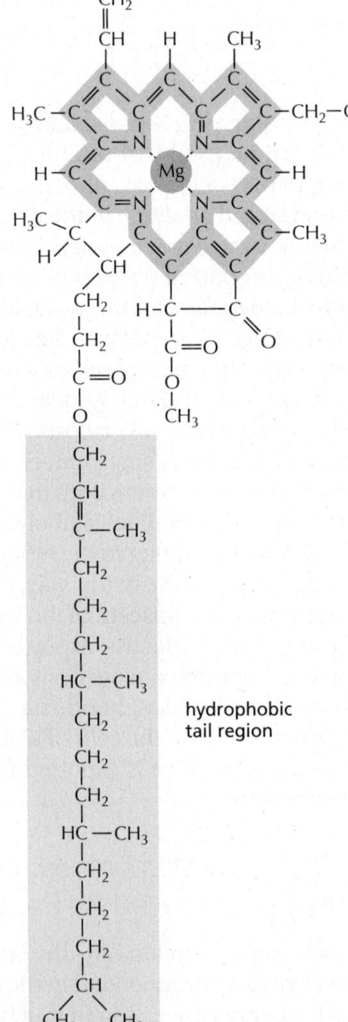

Figure 14–44 The structure of chlorophyll *a*. A magnesium atom is held in a porphyrin ring, which is related to the porphyrin ring that binds iron in heme (see Figure 14–15). Electrons are delocalized over the bonds shaded in *blue*, and the hydrophobic tail is shaded in *green*.

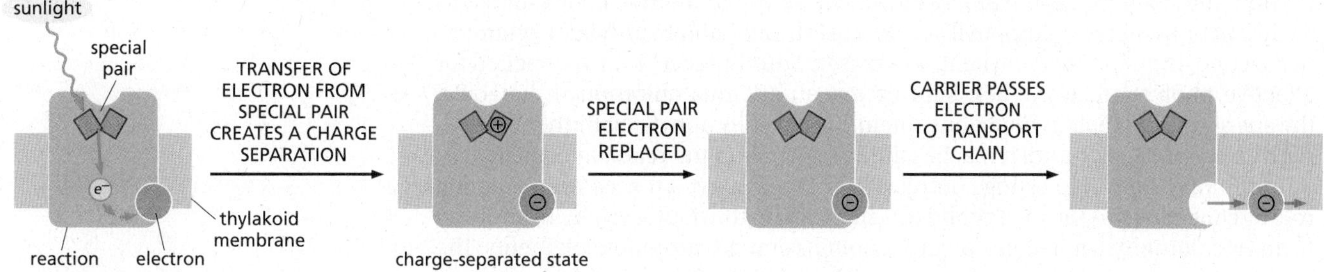

Figure 14–45 A general scheme for the charge-separation step in a photosynthetic reaction center. In a reaction center, light energy is harnessed to transfer electrons from chlorophyll to mobile electron carriers in which the electrons are held in a high-energy state. Light energy is thereby converted to chemical energy. The process starts when a photon absorbed by the special pair of chlorophylls in the reaction center knocks an electron out of one of the chlorophylls. This electron is rapidly taken up by a mobile electron carrier *(orange)* bound at the opposite membrane surface, because a set of intermediary carriers embedded in the reaction center (not shown) provides the path from the special pair to this carrier. The physical distance between the positively charged chlorophyll ion and the negatively charged electron carrier stabilizes the charge-separated state for a short time, during which the chlorophyll ion, a strong oxidant, withdraws an electron from a suitable compound (for example, from water, an event we will discuss in detail shortly). The mobile electron carrier then diffuses away from the reaction center as a strong electron donor that will transfer its electron to an electron-transport chain.

an electron from a chlorophyll to a mobile electron carrier is the central *charge-separation* step in photosynthesis, in which a chlorophyll becomes positively charged and an electron carrier becomes negatively charged (**Figure 14–45**). A positively charged chlorophyll ion is a very strong oxidant that is able to withdraw an electron from a low-energy substrate; in the first step of oxygenic photosynthesis, this low-energy substrate is water.

Upon transfer to a mobile carrier in the electron-transport chain, the electron has been stabilized as part of a strong electron donor and made available for subsequent reactions. These subsequent reactions require more time to complete, and we shall see how they cause the production of light-generated energy-rich compounds.

A Photosystem Contains Chlorophylls in Antennae and a Reaction Center

There are two distinct types of chlorophyll–protein complexes in the photosynthetic membrane. A *photochemical reaction center* contains the special pair of chlorophylls just described. The other type engages exclusively in light absorption and resonance energy transfer and contains *antenna chlorophylls*. Together, the two types of chlorophyll–protein complexes help to create a **photosystem** (**Figure 14–46**).

Most of the chlorophylls in a photosystem are present in *light-harvesting complexes*, whose role is to efficiently collect the energy of photons for photosynthesis.

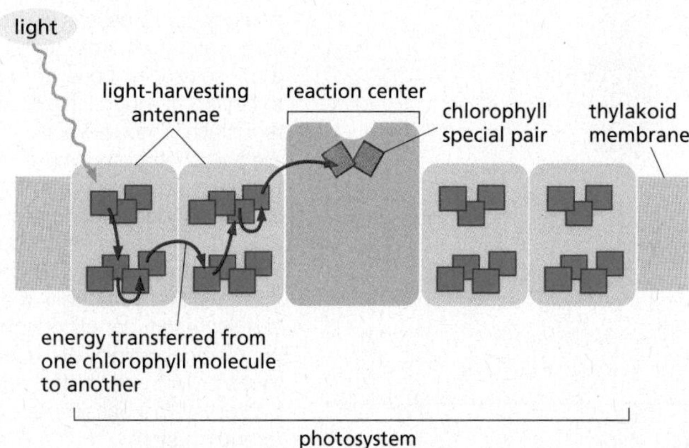

Figure 14–46 A photosystem. Each photosystem consists of a reaction center plus a large number of light-harvesting protein-bound chlorophylls (more than depicted in this figure). The solar energy for photosynthesis is collected by the antenna chlorophylls, which account for most of the chlorophyll in a plant cell. The energy hops randomly by resonance energy transfer *(red arrows)* from one chlorophyll molecule to another, until it reaches the reaction center complex, where it ionizes a chlorophyll in the special pair. The chlorophyll special pair holds its electrons at a lower energy than that of the electrons of the antenna chlorophylls, causing the energy transferred to it to become trapped there. Note that it is only energy, not electrons, that moves from one antenna chlorophyll molecule to another (**Movie 14.12**).

Without these antennae, the process would be slow and inefficient, inasmuch as each reaction-center chlorophyll would absorb only about one light quantum per second, even in broad daylight, whereas hundreds per second are needed for effective photosynthesis. When light excites an antenna chlorophyll molecule, the energy passes rapidly from one protein-bound chlorophyll to another by resonance energy transfer until it reaches the special pair in the reaction center.

Additional light-harvesting complexes, or LHCs, serve an accessory antenna role by collecting photon energy and transferring it to the photosystem. In addition to many chlorophyll molecules, an LHC contains orange carotenoid pigments. The carotenoids collect light of a different wavelength than that absorbed by chlorophylls, helping to make the antennae more efficient. They also have an important protective role in preventing the formation of harmful oxygen radicals in the photosynthetic membrane.

The Thylakoid Membrane Contains Two Different Photosystems Working in Series

As just described, the excitation energy collected by the antenna chlorophylls is delivered to the special pair in the **photochemical reaction center**. The reaction center is a transmembrane chlorophyll–protein complex that lies at the heart of photosynthesis. It harbors the special pair of chlorophyll molecules, which acts as an irreversible trap for excitation energy (see Figure 14–46).

Chloroplasts contain two functionally different although structurally related photosystems, each of which feeds electrons generated by the action of sunlight into an electron-transfer chain. In the chloroplast thylakoid membrane, *photosystem I* is confined to the unstacked stroma thylakoids, while the stacked grana thylakoids contain *photosystem II*. The two photosystems were named in order of their discovery, not of their actions in the photosynthetic pathway, and electrons are first activated in photosystem II before being transferred to photosystem I (**Figure 14–47**). The path of the electron through the two photosystems can be

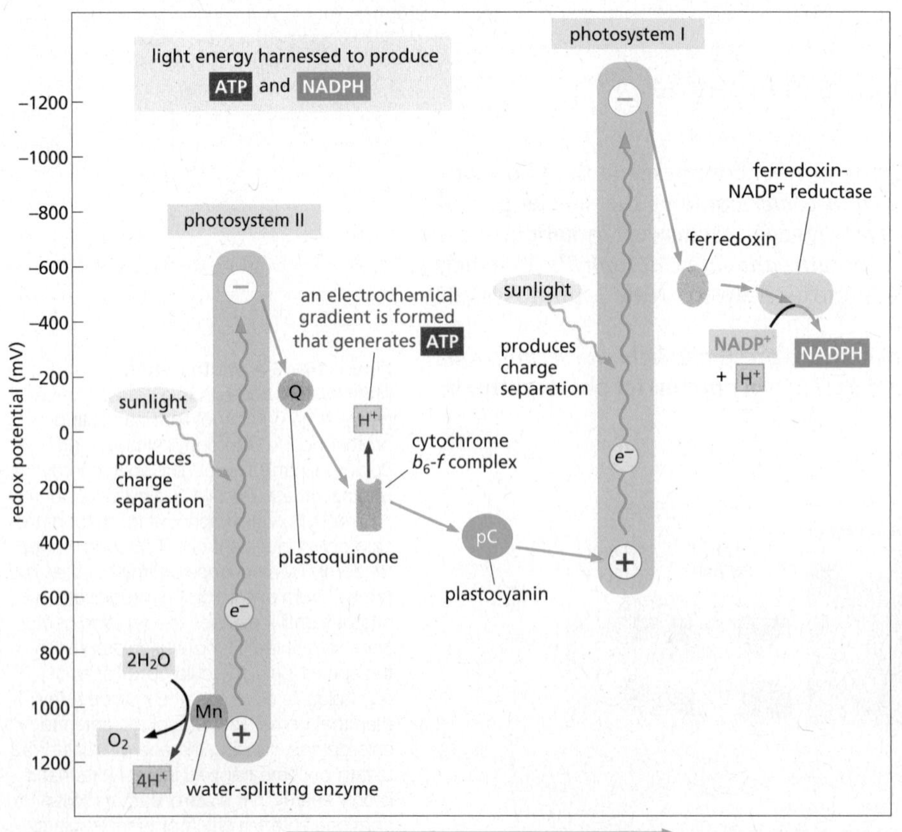

Figure 14–47 Changes in redox potential during photosynthesis. The redox potential for each molecule is indicated by its position along the vertical axis. Using excitation energy from sunlight, photosystem II passes electrons derived from water to photosystem I, which also uses sunlight energy to pass them to $NADP^+$ through ferredoxin-$NADP^+$ reductase. The net electron flow through the two photosystems is from water to $NADP^+$, and it produces NADPH as well as an electrochemical proton gradient. This proton gradient is used by the ATP synthase to produce ATP. Details in this figure will be explained in the subsequent text. Note that, for historical reasons, the two photosystems were named opposite to the order in which they act, with photosystem II passing its electrons to photosystem I.

described as a Z-like trajectory and is known as the *Z scheme*. In the Z scheme, the reaction center of *photosystem II* first withdraws an electron from water. The electron passes via an electron-transport chain (composed of the electron carrier plastoquinone, the cytochrome b_6-f complex, and the protein plastocyanin) to *photosystem I*, which propels the electron across the membrane in a second light-driven charge-separation reaction that leads to NADPH production. In the process, an electrochemical proton gradient is established across the thylakoid membrane that is used to fuel ATP production.

The Z scheme is necessary to bridge the very large energy gap between water and NADPH (Figure 14–47). A single quantum of visible light does not contain enough energy both to withdraw electrons from water, which holds on to its electrons very tightly (redox potential +820 mV) and therefore is a very poor electron donor, and to force them on to $NADP^+$, which is a very poor electron acceptor (redox potential -320 mV). The Z scheme first evolved in cyanobacteria to enable them to use water as a universally available electron source. Other, simpler photosynthetic bacteria have only one photosystem. As we shall see, they cannot use water as an electron source and must rely on other, more energy-rich substrates instead, from which electrons are more readily withdrawn. The ability to extract electrons from water (and thereby to produce molecular oxygen) was acquired by plants when their ancestors took up the endosymbiotic cyanobacteria that later evolved into chloroplasts (see Figure 1–29).

Photosystem II Uses a Manganese Cluster to Withdraw Electrons from Water

In biology, only photosystem II is able to withdraw electrons from water and to generate molecular oxygen as a by-product. This remarkable specialization of photosystem II is conferred by the unique properties of one of the two chlorophyll molecules of its special pair and by a *manganese cluster* linked to the protein. The special pair of chlorophyll molecules plus the manganese cluster form the catalytic core of the photosystem II reaction center, whose mechanism is outlined in **Figure 14–48**.

Water is an inexhaustible source of electrons, but it is also extremely stable; therefore, a large amount of energy is required to make it part with its electrons. The only compound in living organisms that is able to achieve this feat, after its ionization by light, is the chlorophyll special pair called P_{680} ($P_{680}/P_{680}{}^+$ redox potential = +1270 mV). The reaction $2H_2O + 4\text{ photons} \rightarrow 4H^+ + 4e^- + O_2$ is catalyzed by its adjacent manganese cluster. Except for the electrons, the intermediates remain firmly attached to the manganese cluster until two water molecules have been fully oxidized to O_2, thus ensuring that no dangerous oxygen radicals are released as the reaction proceeds. The four protons released by the two water molecules are discharged to the thylakoid space, contributing to the proton gradient across the thylakoid membrane (pH lower in the thylakoid

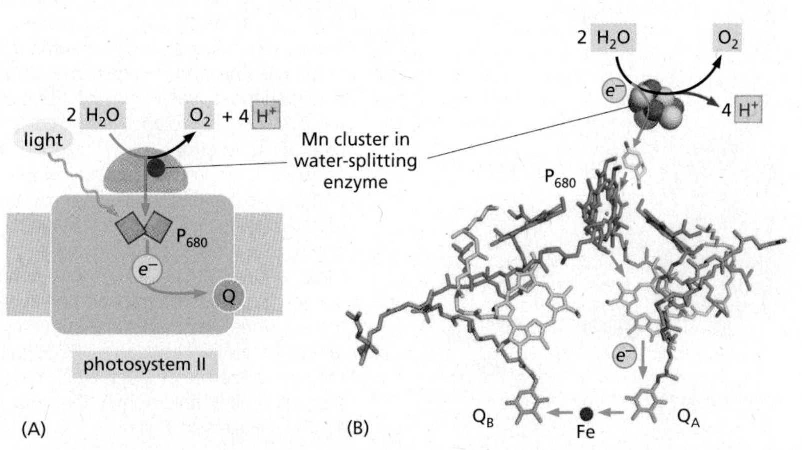

Figure 14–48 **The conversion of light energy to chemical energy in the photosystem II complex.** (A) Schematic diagram of the photosystem II reaction center, whose special pair of chlorophyll molecules is designated as P_{680} on the basis of the wavelength of its absorbance maximum (680 nm). (B) Cofactors and pigments at the core of the reaction center. Shown are the manganese (Mn) cluster, the tyrosine side chain that connects it to the P_{680} special pair, four chlorophylls *(green)*, two pheophytins *(light blue)*, two plastoquinones *(orange)*, and an iron atom *(red)*. The path of electrons is shown by *blue arrows*. In the manganese cluster, four manganese atoms *(light blue)*, one calcium atom *(purple)*, and five oxygen atoms *(red)* work together to catalyze the oxidation of water. The water-splitting reaction occurs in four successive steps, each requiring the energy of one photon. Each photon turns a P_{680} reaction-center chlorophyll into a positively charged chlorophyll ion. Through an ionized tyrosine side chain *(yellow)*, this chlorophyll ion pulls an electron away from a water molecule bound at the manganese cluster. In this way, a total of four electrons are withdrawn from two water molecules to generate molecular oxygen, which is released into the atmosphere (see also Figure 4–49).

Each electron that is energized by light passes from the special pair to an electron-transfer chain inside the complex, along the indicated path to the permanently bound plastoquinone Q_A and then to plastoquinone Q_B. Once Q_B has picked up two electrons (plus two protons; see Figure 14–17), it dissociates from its binding site in the complex and enters the lipid bilayer as a mobile electron carrier, being immediately replaced by a new, nonreduced molecule of plastoquinone. Note that the chlorophylls and pheophytins form two symmetrical branches of a potential electron-transport chain. Only one branch is active, thus ensuring that the plastoquinones become fully reduced in minimum time.

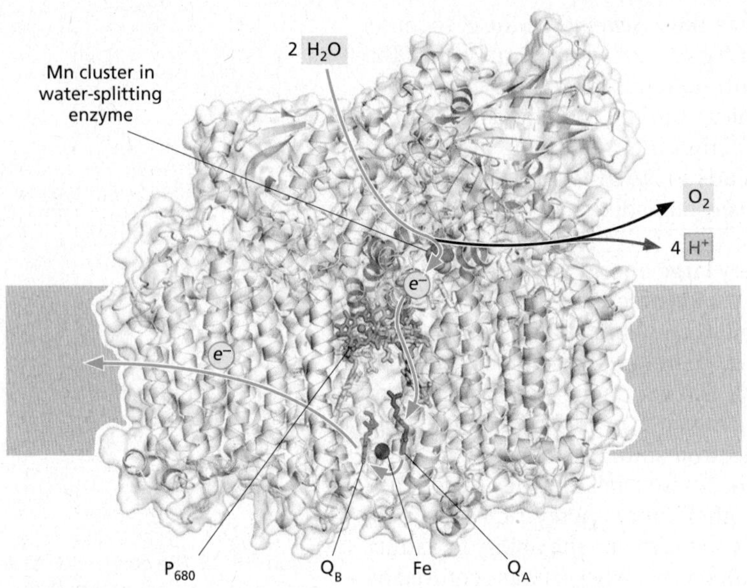

Figure 14–49 **The structure of the complete photosystem II complex.** This photosystem contains at least 16 protein subunits, along with 36 chlorophylls, two pheophytins, two hemes, and a number of protective carotenoids *(colored)*. Most of these pigments and cofactors are deeply buried, tightly complexed to protein *(gray)*. The path of electrons is indicated by the *blue arrows* and is explained in Figure 14–48B. The photosystem II complex presented here is the cyanobacterial complex, which is simpler and more stable than the plant complex, which works in the same way. (PDB code: 3WU2.)

space than in the stroma). The unique protein environment that endows life with this all-important ability to oxidize water has remained essentially unchanged throughout billions of years of evolution (**Figure 14–49**). All of the oxygen in Earth's atmosphere is believed to have been generated in this way.

The Cytochrome b_6-f Complex Connects Photosystem II to Photosystem I

Following the path shown previously in Figure 14–49, the electrons extracted from water by photosystem II are transferred to plastoquinol, a strong electron donor similar to ubiquinol in mitochondria. This quinol, which can diffuse rapidly in the lipid bilayer of the thylakoid membrane, transfers its electrons to the *cytochrome b_6-f complex,* whose structure is similar to the cytochrome *c* reductase in mitochondria. The cytochrome b_6-f complex pumps H^+ into the thylakoid space using the same Q cycle that is utilized in mitochondria (see Figure 14–21), thereby adding to the proton gradient across the thylakoid membrane.

The cytochrome b_6-f complex forms the connecting link between photosystems II and I in the chloroplast electron-transport chain. It passes its electrons one at a time to the mobile electron carrier plastocyanin (a small copper-containing protein that takes the place of the cytochrome *c* in mitochondria), which transfers those electrons to photosystem I (**Figure 14–50**). As we discuss next, photosystem I then uses a second photon of light to boost the electrons that it receives to an even higher energy level.

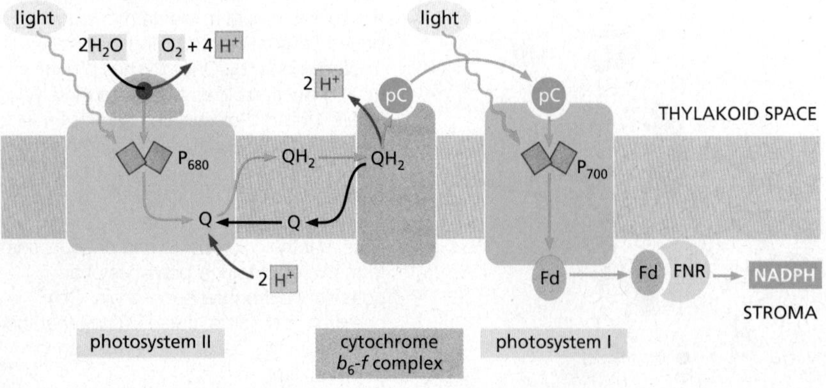

Figure 14–50 **Electron flow through the cytochrome b_6-f complex to NADPH.** The cytochrome b_6-f complex is the functional equivalent of cytochrome *c* reductase (the cytochrome *b*-c_1 complex) in mitochondria (see Figure 14–22). Like its mitochondrial homolog, the b_6-f complex receives its electrons from a quinone and engages in a complicated Q cycle that pumps two protons across the membrane (details not shown). It passes its electrons, one at a time, to plastocyanin (pC). Plastocyanin diffuses along the membrane surface to photosystem I and transfers the electrons via ferredoxin (Fd) to the ferredoxin-NADP$^+$ reductase (FNR), where they are utilized to produce NADPH. P_{700} is a special pair of chlorophylls that absorbs light of wavelength 700 nm.

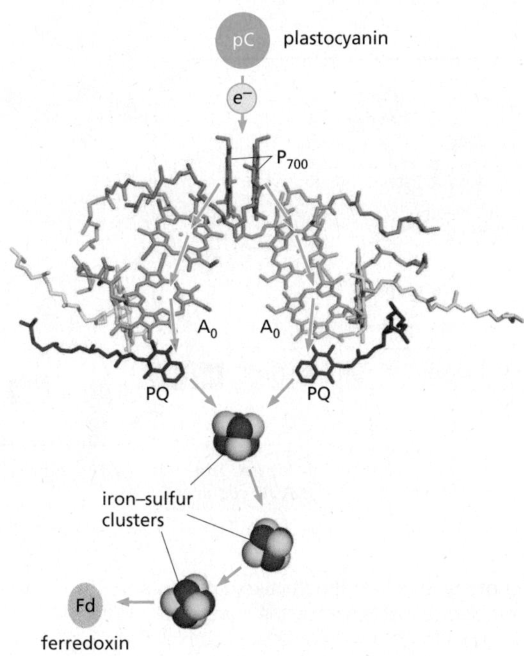

Figure 14–51 Structure and function of photosystem I. At the heart of the photosystem I complex assembly is the electron-transfer chain shown. At one end is a special pair of chlorophylls called P_{700} (because it absorbs light of 700-nm wavelength), receiving electrons from plastocyanin (pC). At the other end are the A_0 chlorophylls, which hand the electrons on to ferredoxin via two plastoquinones (PQ; *purple*) and three iron–sulfur clusters. Even though the roles of photosystems I and II in photosynthesis are very different, their central electron-transfer chains are structurally similar, suggesting a common evolutionary origin (see Figure 14–53). Note that in photosystem I both branches of the electron-transfer chain are active, unlike in photosystem II (see Figure 14–48). (PDB code: 3LW5.)

Photosystem I Carries Out the Second Charge-Separation Step in the Z Scheme

Photosystem I receives electrons from plastocyanin in the thylakoid space and transfers them, via a second charge-separation reaction, to the small protein ferredoxin on the opposite membrane surface (**Figure 14–51**). Then, in a final step, ferredoxin feeds its electrons to a membrane-associated enzyme complex, the *ferredoxin-NADP$^+$ reductase*, which uses the electrons to reduce NADP$^+$ to NADPH (see Figure 14–50). The reduced NADPH is released into the chloroplast stroma, where it is used for biosynthesis of glyceraldehyde 3-phosphate, amino acid precursors, and fatty acids.

The redox potential of the NADP$^+$/NADPH pair (–320 mV) is already very low, and the reduction of NADP$^+$ therefore requires a compound with an even lower redox potential. This turns out to be a chlorophyll molecule near the stromal membrane surface of photosystem I that has a redox potential of –1000 mV (chlorophyll A_0), making it the strongest known electron donor in biology.

The Chloroplast ATP Synthase Uses the Proton Gradient Generated by the Photosynthetic Light Reactions to Produce ATP

The sequence of events that results in light-driven production of ATP and NADPH in chloroplasts and cyanobacteria is summarized in **Figure 14–52**. Starting with the withdrawal of electrons from water, the light-driven charge-separation steps in photosystems II and I enable the energetically unfavorable flow of electrons from water to NADPH (see Figure 14–47). Three small mobile electron carriers—plastoquinone, plastocyanin, and ferredoxin—participate in this process. Together with the electron-driven proton pump of the cytochrome b_6-f complex, the photosystems generate a large proton gradient across the thylakoid membrane. The ATP synthase molecules embedded in the thylakoid membranes then harness this proton gradient to produce ATP in the chloroplast stroma, analogous to the synthesis of ATP in the mitochondrial matrix.

The linear Z scheme for photosynthesis thus far discussed can switch to a circular mode of electron flow through photosystem I and the b_6-f complex. Here, the electrons held in the reduced ferredoxin are transferred back to the b_6-f complex to reduce plastoquinone, instead of being passed to the ferredoxin-NADP$^+$ reductase enzyme complex. This, in effect, turns photosystem I into a light-driven

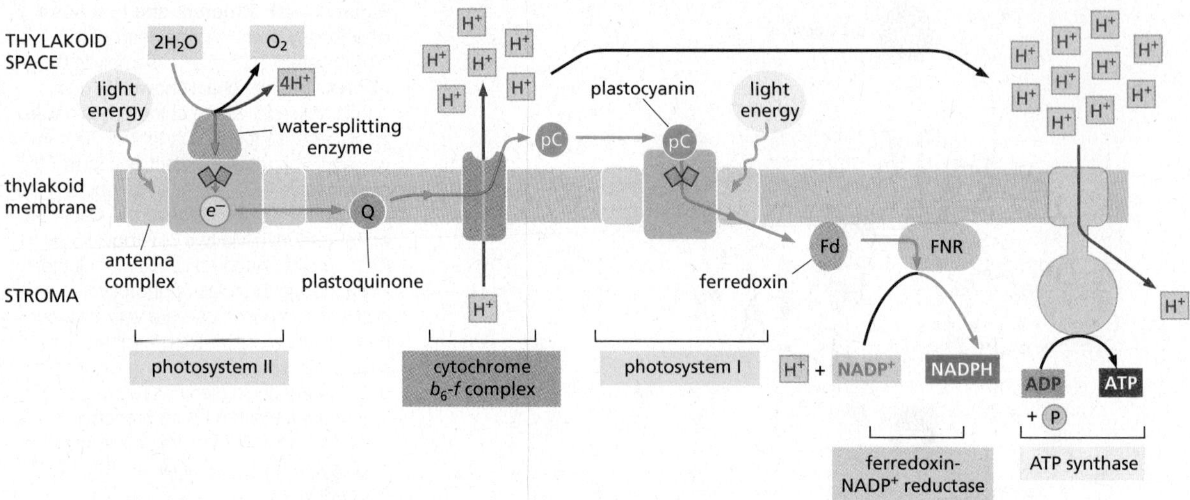

Figure 14–52 Summary of electron and proton movements during photosynthesis in the thylakoid membrane. Electrons are withdrawn, through the action of light energy, from a water molecule that is held by the manganese cluster in photosystem II. The electrons pass on to plastoquinone, which delivers them to the cytochrome b_6-f complex that resembles the cytochrome *c* reductase of mitochondria and the cytochrome *b*-*c* complex of bacteria. They are then carried to photosystem I by the soluble electron carrier plastocyanin, the functional equivalent of cytochrome *c* in mitochondria. From photosystem I they are transferred to ferredoxin-NADP$^+$ reductase (FNR) by the soluble carrier ferredoxin (Fd; a small protein containing an iron–sulfur center). Protons are pumped into the thylakoid space by the cytochrome b_6-f complex, in the same way that protons are pumped into mitochondrial cristae by cytochrome *c* reductase (see Figure 14–23). In addition, the H^+ released into the thylakoid space by water oxidation, and the H^+ consumed during NADPH formation in the stroma, contribute to the generation of the electrochemical H^+ gradient across the thylakoid membrane. As illustrated, this gradient drives ATP synthesis by an ATP synthase that sits in the same membrane (see Figure 14–47).

proton pump, thereby increasing the proton gradient and thus the amount of ATP made by the ATP synthase. An elaborate set of regulatory mechanisms controls this switch, which enables the chloroplast to generate either more NADPH (linear mode) or more ATP (circular mode), depending on the metabolic needs of the cell.

The Proton-Motive Force for ATP Production in Mitochondria and Chloroplasts Is Essentially the Same

The proton gradient across the thylakoid membrane depends both on the proton-pumping activity of the cytochrome b_6-f complex and on the photosynthetic activity of the two photosystems. In chloroplasts exposed to light, H^+ is pumped out of the stroma (pH around 8, similar to the mitochondrial matrix) into the thylakoid space (pH 5–6), creating a gradient of 2–3 pH units across the thylakoid membrane, representing a proton-motive force of about 180 mV. This is very similar to the proton-motive force in respiring mitochondria. However, a membrane potential across the inner mitochondrial membrane makes the largest contribution to the proton-motive force that drives the mitochondrial ATP synthase to make ATP, whereas an H^+ gradient predominates for chloroplasts.

Chemiosmotic Mechanisms Evolved in Stages

The first living cells on Earth may have consumed geochemically produced organic molecules and generated their ATP by fermentation. Because oxygen was not yet present in the atmosphere, such anaerobic fermentation reactions would have dumped organic acids—such as lactic or formic acids, for example—into the environment (see Figure 2–50). Perhaps such acids lowered the pH of the environment, favoring the survival of cells that evolved transmembrane proteins that could pump H^+ out of the cytosol, thereby preventing the cell from becoming too acidic (stage 1 in Figure 14–53). One of these pumps may have used the energy available from ATP hydrolysis to eject H^+ from the cell; such a proton pump could have been the ancestor of present-day ATP synthases.

As Earth's supply of geochemically produced nutrients began to dwindle, organisms that could find a way to pump H^+ without consuming ATP would have been at an advantage: they could save the small amounts of ATP they derived from the fermentation of increasingly scarce foodstuffs to fuel other important activities. This need to conserve resources might have led to the evolution of electron-transport proteins that allowed cells to use the movement of electrons between molecules of different redox potentials as a source of energy for pumping H^+ across the plasma membrane (stage 2 in Figure 14–53). Some of these cells might have used the nonfermentable organic acids that neighboring cells had excreted as waste to provide the electrons needed to feed this electron-transport system. Some present-day bacteria grow on formic acid, for example, using the small amount of redox energy derived from the transfer of electrons from formic acid to fumarate to pump H^+.

Eventually, some bacteria would have developed H^+-pumping electron-transport systems that were so efficient that the bacteria could harvest more redox energy than they needed to maintain their internal pH. Such cells would probably have generated large electrochemical proton gradients, which they could then use to produce ATP. Protons could leak back into the cell through the ATP-driven H^+ pumps, essentially running them in reverse so that they synthesized ATP (stage 3 in Figure 14–53). Because such cells would require much less of the dwindling supply of fermentable nutrients, they would have proliferated at the expense of their neighbors.

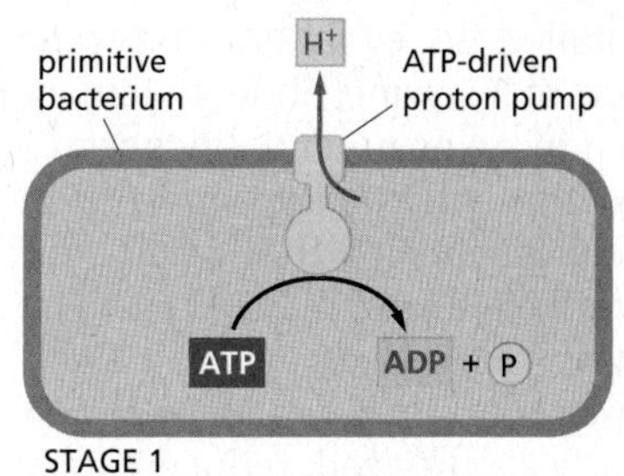

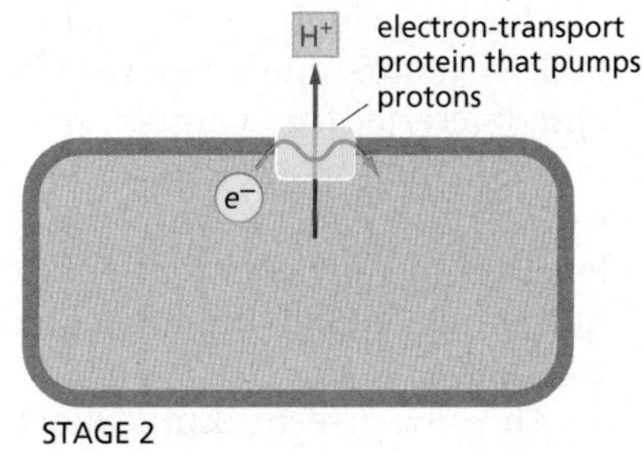

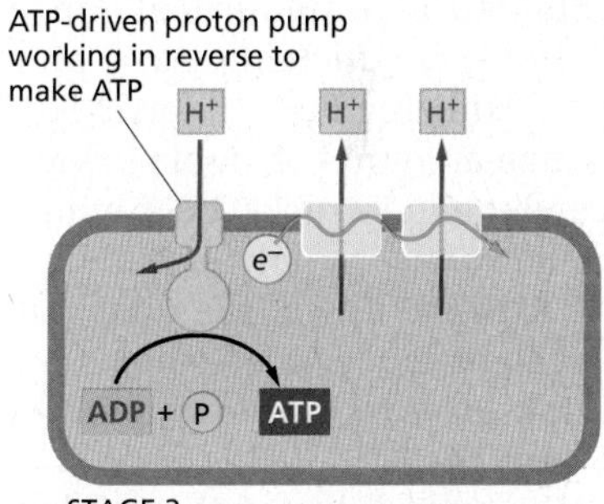

Figure 14–53 How ATP synthesis by chemiosmosis might have evolved in stages. The first stage could have involved the evolution of an ATPase that pumped protons out of the cell using the energy of ATP hydrolysis. Stage 2 could have involved the evolution of a different proton pump, driven by an electron-transport chain. Stage 3 would then have linked these two systems together to generate a primitive ATP synthase that used the protons pumped by the electron-transport chain to synthesize ATP. An early bacterium with this final system would have had a selective advantage over bacteria with neither of the systems or only one.

By Providing an Inexhaustible Source of Reducing Power, Photosynthetic Bacteria Overcame a Major Evolutionary Obstacle

The gradual depletion of nutrients from the environment on the early Earth meant that organisms had to find some alternative source of carbon to make the sugars that serve as the precursors for so many other cell components. Although the CO_2 in the atmosphere provides an abundant potential carbon source, to convert it into an organic molecule such as a carbohydrate requires reducing the fixed CO_2 with a strong electron donor, such as NADPH, which can generate (CH_2O) units from CO_2 (see Figure 14–41). Early in cellular evolution, strong reducing agents (electron donors) are thought to have been plentiful. But once an ancestor of ATP synthase began to generate most of the ATP, it paved the way for cells to evolve a new way of generating strong reducing agents.

A major evolutionary breakthrough in energy metabolism came with the development of photochemical reaction centers that could use the energy of sunlight to produce molecules such as NADPH. It is thought that this occurred early in the process of cellular evolution in the ancestors of the green sulfur bacteria. Present-day green sulfur bacteria use light energy to transfer hydrogen atoms (as an electron plus a proton) from H_2S to NADPH, thereby producing the strong reducing power required for carbon fixation. Because the redox potential of H_2S is much lower than that of H_2O (–230 mV for H_2S compared with +820 mV for H_2O), one quantum of light absorbed by the single photosystem in these bacteria is sufficient to generate NADPH via a relatively simple photosynthetic electron-transport chain.

The Photosynthetic Electron-Transport Chains of Cyanobacteria Produced Atmospheric Oxygen and Permitted New Life-Forms

The next evolutionary step, which is thought to have occurred with the development of the cyanobacteria perhaps 3 billion years ago, was the evolution of organisms capable of using water as the electron source for CO_2 reduction. This entailed the evolution of a water-splitting enzyme and also required the addition of a second photosystem, acting in series with the first, to bridge the large gap in redox potential between H_2O and NADPH. The biological consequences of this evolutionary step were far-reaching. For the first time, there would have been organisms that could survive on water, CO_2, and sunlight (plus several other elements; see Figure 2–1). These cells would have been able to spread and evolve in ways denied to the earlier photosynthetic bacteria, which needed H_2S or organic acids as a source of electrons. Consequently, large amounts of biologically synthesized, reduced organic materials accumulated, and oxygen entered the atmosphere for the first time.

Oxygen is highly toxic because the oxidation of biological molecules alters their structure and properties indiscriminately and irreversibly. Most anaerobic bacteria, for example, occupy environments devoid of oxygen and are rapidly killed when exposed to air. Thus, organisms on the primitive Earth would have had to evolve mechanisms to protect them from the rising O_2 levels in the environment. Late evolutionary arrivals, such as ourselves, have numerous detoxifying mechanisms that protect our cells from the ill effects of oxygen.

The increase in atmospheric O_2 was very slow at first and would have allowed a gradual evolution of protective devices. For example, the early seas contained large amounts of iron in its reduced, ferrous state (Fe^{2+}), and nearly all the O_2 produced by early photosynthetic bacteria would have been used up in oxidizing Fe^{2+} to ferric Fe^{3+}. This conversion caused the precipitation of huge amounts of stable oxides, and the extensive banded iron formations in sedimentary rocks, beginning about 2.7 billion years ago, help to date the spread of the cyanobacteria. By about 2 billion years ago, the supply of Fe^{2+} was exhausted, and the deposition of further iron precipitates ceased. Geological evidence reveals how O_2 levels in the atmosphere have changed over billions of years, approximating current levels only about 0.5 billion years ago (**Figure 14–54**).

The availability of O_2 enabled the rise of bacteria that developed an aerobic metabolism to make their ATP. These organisms could harness the

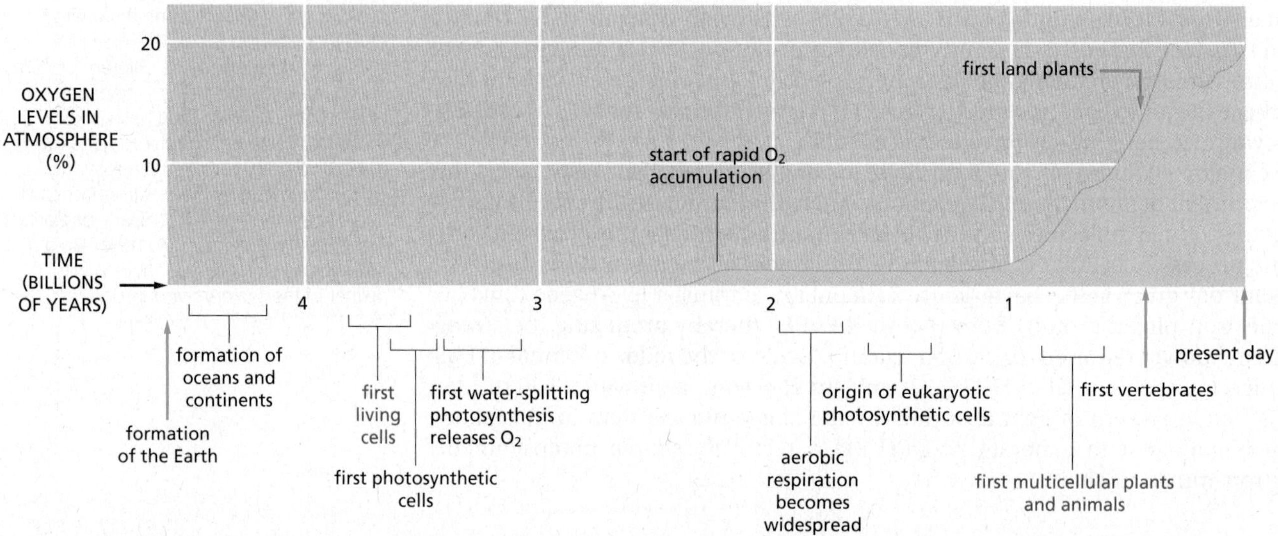

Figure 14–54 **Major events during the evolution of living organisms on Earth.** With the evolution of the membrane-based process of photosynthesis, organisms were able to make their own organic molecules from CO_2 gas. The delay of more than a billion years between the appearance of bacteria that split water and released O_2 during photosynthesis and the accumulation of high levels of O_2 in the atmosphere is thought to be due to the initial reaction of the oxygen with the abundant ferrous iron (Fe^{2+}) that was dissolved in the early oceans. Only when the ferrous iron was used up would oxygen have started to accumulate in the atmosphere. In response to the rising oxygen levels, nonphotosynthetic oxygen-consuming organisms evolved, and the concentration of oxygen in the atmosphere equilibrated at its present-day level.

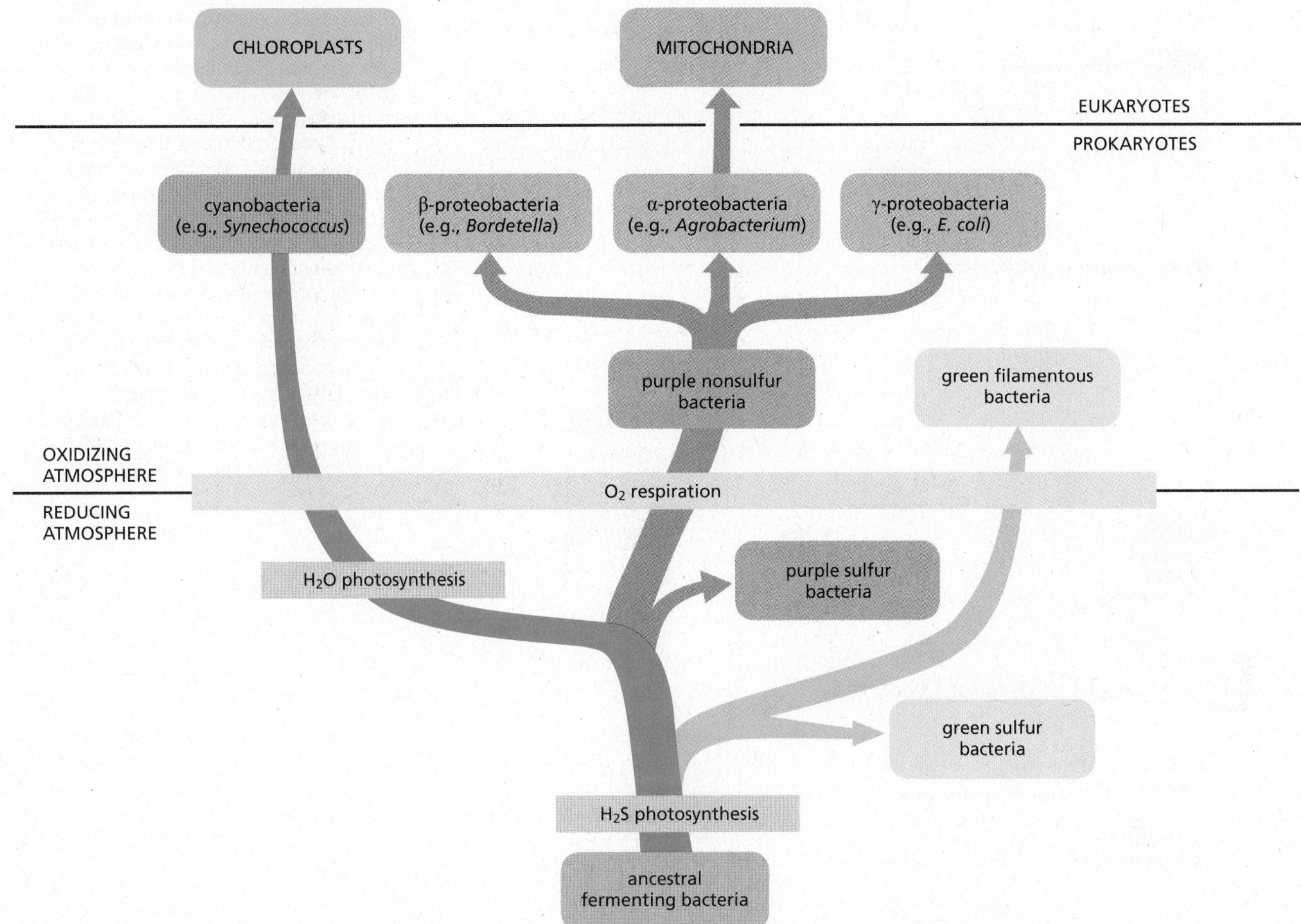

Figure 14–55 **Evolutionary scheme showing the postulated origins of mitochondria and chloroplasts and their bacterial ancestors.** The consumption of oxygen by respiration is thought to have first developed about 2 billion years ago. Nucleotide-sequence analyses suggest that an endosymbiotic oxygen-evolving cyanobacterium gave rise to chloroplasts *(green)*, while mitochondria *(pink)* arose from an α-proteobacterium. The nearest relatives of mitochondria are members of three closely related groups of α-proteobacteria—the rhizobacteria, agrobacteria, and rickettsias—known to form intimate associations with present-day eukaryotic cells. Proteobacteria are *pink*, purple photosynthetic bacteria are *purple*, and other photosynthetic bacteria are *light green*.

large amount of energy released by breaking down carbohydrates and other reduced organic molecules all the way to CO_2 and H_2O, as explained when we discussed mitochondria. Components of preexisting electron-transport complexes were modified to produce a cytochrome oxidase, so that the electrons obtained from organic or inorganic substrates could be transported to O_2 as the terminal electron acceptor. Some present-day purple photosynthetic bacteria can switch between photosynthesis and respiration depending on the availability of light and O_2, with only relatively minor reorganizations of their electron-transport chains.

In **Figure 14–55**, we relate these postulated evolutionary pathways to different types of bacteria. By necessity, evolution is always conservative, taking parts of the old and building on them to create something new. Thus, parts of the electron-transport chains that were derived to service anaerobic bacteria 3–4 billion years ago survive today, in altered form, in the mitochondria and chloroplasts of higher eukaryotes. A good example is the overall similarity in structure and function between the cytochrome *c* reductase that pumps H^+ in the central segment of the mitochondrial respiratory chain and the analogous cytochrome b_6-f complex in the electron-transport chains of both bacteria and chloroplasts, revealing their common evolutionary origin (**Figure 14–56**).

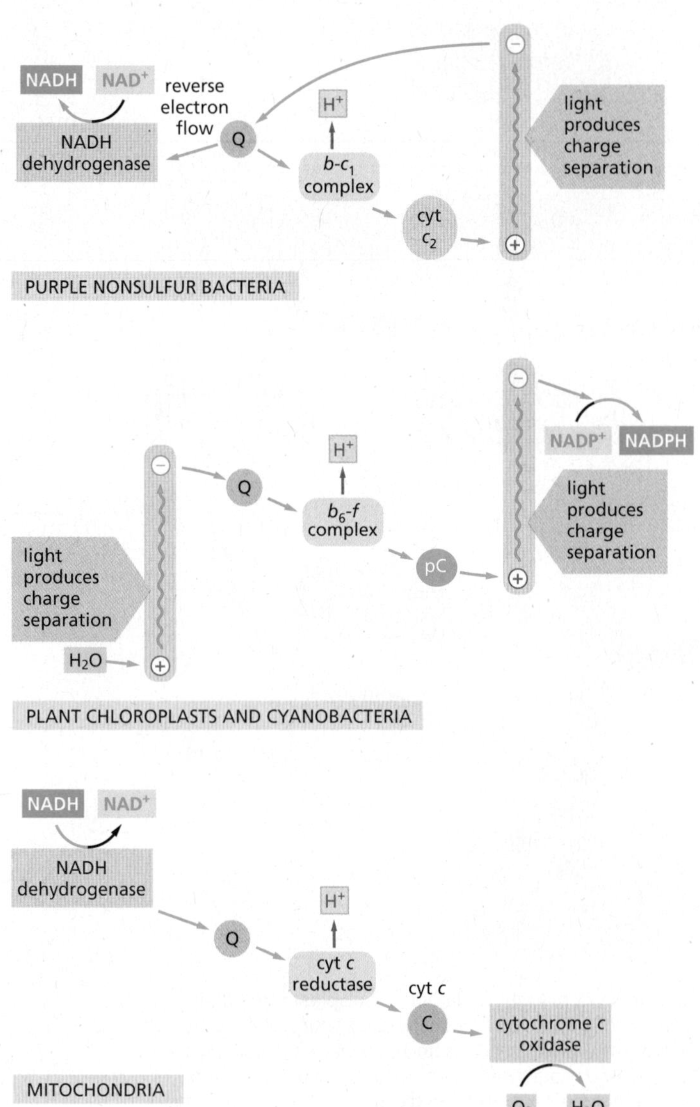

Figure 14–56 A comparison of three electron-transport chains discussed in this chapter. Bacteria, chloroplasts, and mitochondria all contain a membrane-bound enzyme complex that resembles the cytochrome *c* reductase of mitochondria. These complexes all accept electrons from a quinone carrier (Q) and pump H^+ across their respective membranes. Moreover, in reconstituted *in vitro* systems, the different complexes can substitute for one another, and the structures of their protein components reveal that they are evolutionarily related. Note that the purple nonsulfur bacteria use a cyclic flow of electrons to produce a large electrochemical proton gradient that drives a *reverse electron flow* through NADH dehydrogenase to produce NADH from $NAD^+ + H^+ + e^-$.

Summary

Chloroplasts and photosynthetic bacteria have the unique ability to harness the energy of sunlight to produce energy-rich compounds. This is achieved by the photosystems, in which chlorophyll molecules attached to proteins are excited when hit by a photon. Photosystems are composed of light-harvesting antennae that collect solar energy and a photochemical reaction center, in which the collected energy is funneled to a chlorophyll molecule held in a special position, enabling it to withdraw electrons from an electron donor. Chloroplasts and cyanobacteria contain two distinct photosystems. The two photosystems are normally linked in series in the Z scheme, and they transfer electrons from water to $NADP^+$ *to form NADPH, while also generating a transmembrane electrochemical potential. One of the two photosystems—photosystem II—can split water by removing electrons from this ubiquitous, low-energy compound. It is believed that essentially all of the molecular oxygen (*O_2*) in our atmosphere is a by-product of the water-splitting reaction in this photosystem. The three-dimensional structures of photosystems I and II are strikingly similar to those of the photosystems of purple photosynthetic bacteria, demonstrating a remarkable degree of conservation over billions of years of evolution.*

The two photosystems and the cytochrome b_6*-f complex reside in the thylakoid membrane, a separate membrane system in the central stroma compartment of the chloroplast that is differentiated into stacked grana and unstacked stroma thylakoids. Electron-transport processes in the thylakoid membrane cause protons to be released* *into the thylakoid space. The backflow of protons through the chloroplast ATP synthase then generates ATP. This ATP is used in conjunction with the NADPH produced by photosynthesis to drive many biosynthetic reactions in the chloroplast stroma, including the carbon-fixation cycle, which generates large amounts of carbohydrates from* CO_2.

In the early evolution of life, cyanobacteria overcame a major obstacle by devising a way to use solar energy to split water and fix carbon dioxide. By proliferating widely on Earth, the cyanobacteria eventually produced both abundant organic nutrients and the molecular oxygen that enabled the rise of a multitude of aerobic life-forms. The chloroplasts in plants have evolved from a cyanobacterium that was endocytosed long ago by an aerobic eukaryotic host organism.

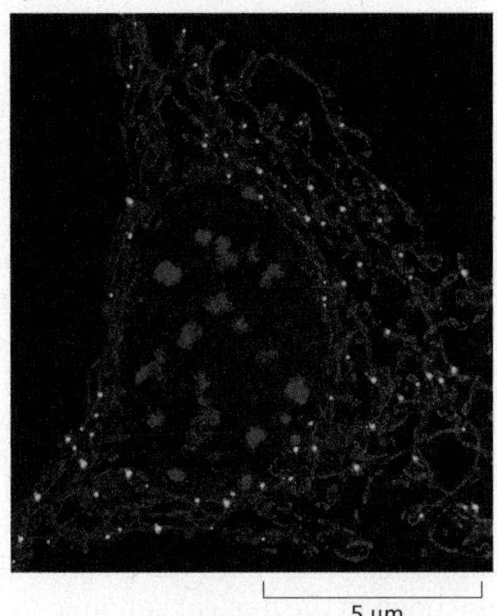

Figure 14–57 **Staining of nuclear and mitochondrial DNA.** In this confocal micrograph of a single fibroblast cell, the nuclear DNA is stained with a fluorescent dye *(blue)* while the mitochondrial DNA is visualized indirectly using a tagged mitochondrial transcription factor *(green)*. The mitochondrial network is stained with a fluorescent mitochondrial matrix marker *(red)*. The image was acquired using a structured illumination microscope (SIM), which yields approximately double the resolution of a confocal microscope. Numerous copies of the mitochondrial genome can be seen distributed in distinct nucleoids throughout the mitochondria that snake through the cytoplasm. [From C. Kukat et al., *Proc. Natl. Acad. Sci. USA* 108(33):13534–13539, 2011. Copyright 2011 National Academy of Sciences, USA. With permission from National Academy of Sciences.]

THE GENETIC SYSTEMS OF MITOCHONDRIA AND CHLOROPLASTS

As we discussed in Chapter 1, mitochondria and chloroplasts are thought to have evolved from endosymbiotic bacteria (see Figures 1–27 and 1–29). Both types of organelles still contain their own genomes (**Figure 14–57**). As we will discuss shortly, they also retain their own machinery for transcribing that DNA into RNA and for translating mRNAs into organelle proteins.

Like bacteria, mitochondria and chloroplasts proliferate by growth and division of an existing organelle. In actively dividing cells, each type of organelle must double in mass in each cell generation and then be distributed into each daughter cell. In addition, nondividing cells must replenish organelles that are degraded as part of the continual process of organelle turnover or produce additional organelles as the need arises. Organelle growth and proliferation are therefore carefully controlled. The process is complicated because mitochondrial and chloroplast proteins are encoded in two places: the nuclear genome and the separate genomes harbored in the organelles themselves. The biogenesis of mitochondria and chloroplasts thus requires contributions from two separate genetic systems, which must be closely coordinated (see pp. 867–868).

Most organellar proteins are encoded by the nuclear DNA. The organelle imports these proteins from the cytosol, after they have been synthesized on cytosolic ribosomes. In Chapter 12, we discuss how this process is catalyzed by mitochondrial protein translocase complexes of the outer and inner mitochondrial membrane. Here, we describe the organellar genomes and genetic systems and consider the consequences of separate organelle genomes for the cell and the organism as a whole.

The Genetic Systems of Mitochondria and Chloroplasts Resemble Those of Prokaryotes

As discussed in Chapter 12, it is thought that eukaryotic cells originated through a symbiotic relationship between an archaeon and an aerobic bacterium (an α-proteobacterium). The two organisms are postulated to have merged to form the ancestor of all nucleated cells, with the archaeon providing the nucleus and the proteobacterium serving as a respiring, ATP-producing endosymbiont—one that would eventually evolve into the mitochondrion (see Figure 12–3). This most likely occurred roughly 1.6 billion years ago, when oxygen had entered the atmosphere in substantial amounts (see Figure 14–54). The chloroplast was derived later, after the plant and animal lineages diverged, through endocytosis of an oxygen-producing cyanobacterium.

This *endosymbiont hypothesis* of organelle development receives strong support from the observation that the genetic systems of mitochondria and chloroplasts are similar to those of present-day bacteria. For example, bacterial, chloroplast, and mitochondrial genomes are typically circular and share many features of genome organization. In addition, chloroplast ribosomes are very similar to bacterial ribosomes, both in their structure and in their sensitivity to various antibiotics (such as chloramphenicol, streptomycin, erythromycin, and tetracycline). In addition, protein synthesis in chloroplasts starts with *N*-formylmethionine, as in bacteria, and not with methionine, as in the cytosol of eukaryotic cells. Although mitochondrial genetic systems are much less similar to those of present-day bacteria than are the genetic systems of chloroplasts, their ribosomes are also structurally similar and are sensitive to antibacterial antibiotics, and the transcriptional system of mitochondria is reminiscent of bacterial and bacteriophage systems.

The processes of organellar DNA transcription, protein synthesis, and DNA replication take place where the genome is located: in the matrix of mitochondria or the stroma of chloroplasts. The enzymes that mediate these genetic processes are unique to the organelle and resemble those of bacteria (or even of bacterial viruses) rather than their eukaryotic analogs. In spite of the complexity of the system required to maintain independent organellar genomes, the nuclear genome encodes the vast majority of mitochondrial proteins, including the enzymes required for genome expression in most species. The chloroplast genome is larger than that of mitochondria and encodes several enzymes involved in transcription and translation, but even so more than 95% of chloroplast proteins are typically encoded in the nuclear genome.

Over Time, Mitochondria and Chloroplasts Have Exported Most of Their Genes to the Nucleus by Gene Transfer

The nature of the organellar genes located in the nucleus of the cell reveals that, over the course of eukaryotic evolution, an extensive transfer of genes from organelle to nuclear DNA has occurred. Each such successful *gene transfer* is expected to be very difficult, because any gene moved from the organelle needs to adapt to both nuclear transcription and cytoplasmic translation requirements. In addition, the protein produced needs to acquire a signal sequence that directs it to the correct organelle after its synthesis in the cytosol (see Chapter 12).

By comparing the genes in the mitochondria from different organisms, we can infer that some of the gene transfers to the nucleus occurred relatively recently. The smallest and presumably most highly evolved mitochondrial genomes, for example, encode only a few hydrophobic inner-membrane proteins of the electron-transport chain, plus ribosomal RNAs (rRNAs) and transfer RNAs (tRNAs). Other mitochondrial genomes that have remained more complex tend to contain this same subset of genes along with others (**Figure 14–58**). The

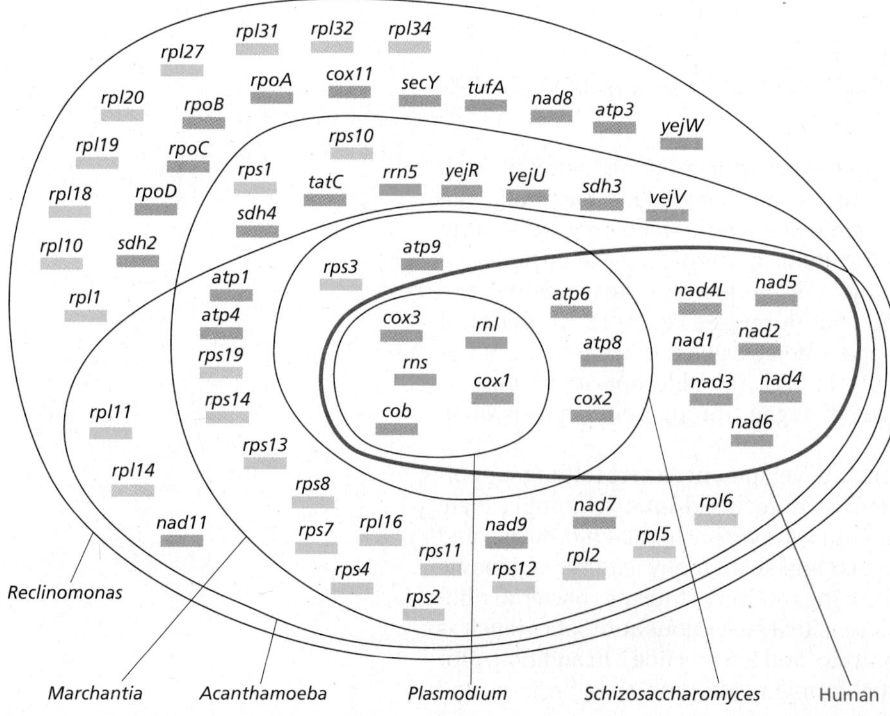

Figure 14–58 **Comparison of mitochondrial genomes.** Less complex mitochondrial genomes contain subsets of the genes found in larger mitochondrial genomes. Even the smallest mitochondrial genomes contain genes encoding respiratory-chain proteins and ribosomal RNAs. In the comparison shown, there are only five genes that are shared by all six mitochondrial genomes; these encode ribosomal RNAs (*rns* and *rnl*), cytochrome *b* (*cob*), and two cytochrome oxidase subunits (*cox1* and *cox3*). *Blue* indicates ribosomal RNAs; *green*, ribosomal proteins; and *brown*, components of the respiratory chain and other proteins. (Adapted from M.W. Gray et al., *Science* 283:1476–1481, 1999.)

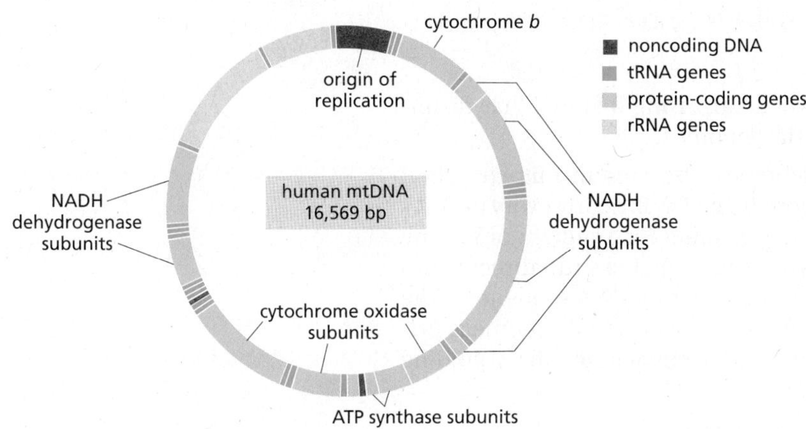

Figure 14–59 The organization of the human mitochondrial genome. The human mitochondrial genome of ~16,600 nucleotide pairs contains 2 rRNA genes, 22 tRNA genes, and 13 protein-coding sequences. There are two transcriptional promoters, one for each strand of the mitochondrial DNA (mtDNA). The DNAs of many other animal mitochondrial genomes have been completely sequenced. Most of these animal mitochondrial DNAs contain precisely the same genes as those contained in the mitochondrial DNA of humans, with the gene order being identical for animals ranging from fish to mammals.

most complex mitochondrial genomes include genes that encode components of the mitochondrial genetic system, such as RNA polymerase subunits and ribosomal proteins; these same genes are found in the cell nucleus in yeast and all animal cells. There are only 13 protein-coding genes in human mitochondrial DNA (**Figure 14–59**).

The proteins that are encoded by genes in the organellar DNA are synthesized on ribosomes within the organelle, using organelle-produced messenger RNA (mRNA) to specify their amino acid sequence (**Figure 14–60**). The protein traffic between the cytosol and these organelles seems to be almost exclusively unidirectional: protein export from mitochondria or chloroplasts to the cytosol is rare. An important exception occurs when a cell is about to undergo apoptosis. As will be discussed in detail in Chapter 18, during apoptosis the mitochondrion releases proteins (most notably cytochrome *c*) from the crista space through its outer mitochondrial membrane as part of an elaborate signaling pathway that is triggered to cause cells to undergo programmed cell death.

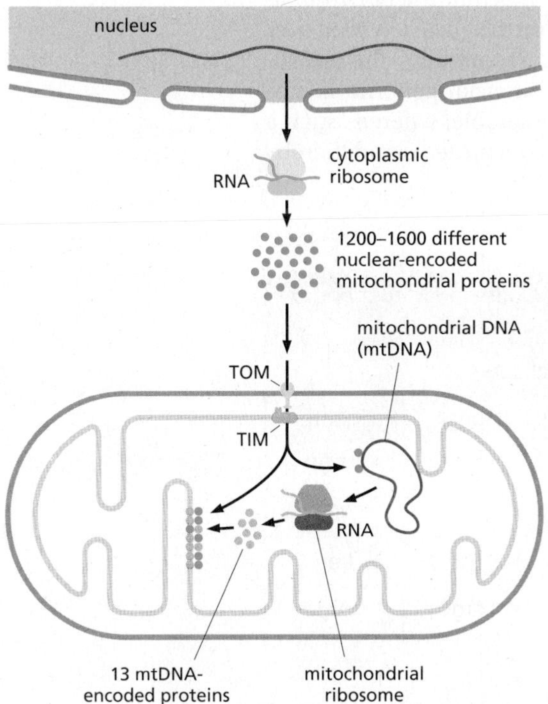

Figure 14–60 Biogenesis of the respiratory-chain proteins in human mitochondria. Most of the protein components of the mitochondrial respiratory chain are encoded by nuclear DNA *(dark green dots)*, with only a small number *(orange dots)* encoded by mitochondrial DNA (mtDNA). The mtDNA-encoded subunits assemble together with the nuclear-encoded subunits to form a functional oxidative phosphorylation system.

Transcription of mtDNA produces 13 mRNAs, all of which encode subunits of the oxidative phosphorylation system, and 24 of the RNAs (22 transfer RNAs and 2 ribosomal RNAs) needed for translation of these mRNAs on the mitochondrial ribosomes. The mRNAs produced by transcription of nuclear genes are translated on cytoplasmic ribosomes *(light green)*, which are distinct from the mitochondrial ribosomes *(brown)*. The nuclear-encoded mitochondrial proteins are imported into mitochondria through protein translocases called TOM and TIM (see Figure 12–49), and they constitute the vast majority of the approximately 1200–1600 different protein species present in mammalian mitochondria. These nuclear-encoded mitochondrial proteins in humans include the majority of the oxidative phosphorylation system subunits, all proteins needed for expression and maintenance of mtDNA, and the proteins of the mitochondrial ribosomes. (Adapted from N.G. Larsson, *Annu. Rev. Biochem.* 79:683–706, 2010.)

Mitochondria Have a Relaxed Codon Usage and Can Have a Variant Genetic Code

The human mitochondrial genome has several surprising features that distinguish it from nuclear, chloroplast, and bacterial genomes:

1. *Dense gene packing.* Unlike other genomes, the human mitochondrial genome seems to contain almost no noncoding DNA: nearly every nucleotide seems to be part of a coding sequence, either for a protein or for one of the rRNAs or tRNAs. Because these coding sequences run directly into each other, there is very little room left for regulatory DNA sequences. The little noncoding DNA present in the human mitochondrial genome is primarily at the origin of replication, which has essential functions in guiding replication and controlling transcription.
2. *Relaxed codon usage.* Whereas 30 or more tRNAs specify amino acids in the cytosol and in chloroplasts, only 22 tRNAs are required for mitochondrial protein synthesis. The normal codon–anticodon pairing rules are relaxed in mitochondria, so that many tRNA molecules recognize any one of the four nucleotides in the third (wobble) position. Such "2 out of 3" pairing allows one tRNA to pair with any one of four codons and permits protein synthesis with fewer tRNA molecules.
3. *Variant genetic code.* Perhaps most surprising, comparisons of mitochondrial gene sequences and the amino acid sequences of the corresponding proteins indicate that the genetic code is different: 4 of the 64 codons have different "meanings" from those of the same codons in other genomes (**Table 14–4**).

The close similarity of the genetic code in all organisms provides strong evidence that they all have evolved from a common ancestor. How, then, do we explain the differences in the genetic code in many mitochondria? A hint comes from the finding that the mitochondrial genetic code in different organisms is not the same. In the mitochondrion with the largest number of genes in Figure 14–58, that of the protozoan *Reclinomonas*, the genetic code is unchanged from the standard genetic code of the cell nucleus. Yet UGA, which is a stop codon elsewhere, is read as tryptophan in the mitochondria of mammals, fungi, and invertebrates. Similarly, the codon AGG normally codes for arginine, but it codes for *stop* in the mitochondria of mammals and codes for serine in the mitochondria of *Drosophila* (see Table 14–4). Such variation suggests that a random drift has occurred over evolutionary time in the genetic code in mitochondria. Presumably, the unusually small number of proteins encoded by the mitochondrial genome makes an occasional change in the meaning of a rare codon tolerable, whereas such a change in a larger genome would alter the function of many proteins and thereby destroy the cell.

TABLE 14–4 Some Differences Between the "Universal" Code and Mitochondrial Genetic Codes*

Codon	"Universal" code	Mitochondrial codes			
		Mammals	Invertebrates	Fungi	Plants
UGA	STOP	*Trp*	*Trp*	*Trp*	STOP
AUA	Ile	*Met*	*Met*	*Met*	Ile
CUA	Leu	Leu	Leu	*Thr*	Leu
AGA AGG	Arg	*STOP*	*Ser*	Arg	Arg

*Red italics indicate that the code differs from the "universal" code.

Notably, in many species, one or two tRNAs for mitochondrial protein synthesis are encoded in the nucleus. Some parasites, for example trypanosomes, have not retained any tRNA genes in their mitochondrial DNA. Instead, the required tRNAs are all produced in the cytosol and are thought to be imported into the mitochondrion by special tRNA translocases that are poorly characterized but are distinct from the mitochondrial protein import system.

Chloroplasts and Bacteria Share Many Striking Similarities

The chloroplast genomes of land plants range in size from 70,000 to 200,000 nucleotide pairs. Of the hundreds of chloroplast genomes that have now been sequenced, many are surprisingly similar, even in distantly related plants (such as tobacco and liverwort), and even those of green algae are closely related (**Figure 14–61**). Chloroplast genes are involved in three main processes: transcription, translation, and photosynthesis. Plant chloroplast genomes typically encode 80–90 proteins and around 45 RNAs, including 37 or more tRNAs. As in mitochondria, most of the organelle-encoded proteins are part of larger protein complexes that also contain one or more subunits encoded in the nucleus and imported from the cytosol.

The genomes of chloroplasts and bacteria have striking similarities. Basic regulatory sequences, such as transcription promoters and terminators, are virtually identical. The amino acid sequences of the proteins encoded in chloroplasts are clearly recognizable as bacterial, and several clusters of genes with related functions (such as those encoding ribosomal proteins) are organized in the same way in the genomes of chloroplasts, the bacterium *E. coli*, and cyanobacteria.

The mechanisms by which chloroplasts and bacteria divide are also similar. Both utilize *FtsZ* proteins, which are self-assembling GTPases related to tubulins (see Chapter 16). Bacterial FtsZ is a soluble protein that assembles into a dynamic ring of membrane-attached protofilaments beneath the plasma membrane in the middle of the dividing cell. The FtsZ ring acts as a scaffold for recruitment of other cell-division proteins and generates a contractile force that results in membrane constriction and eventually in cell division. Presumably, chloroplasts divide in very much the same way.

Although both employ membrane-interacting\GTPases, the mechanisms by which mitochondria and chloroplasts divide are fundamentally different. The machinery for chloroplast division acts from the inside, as in bacteria,

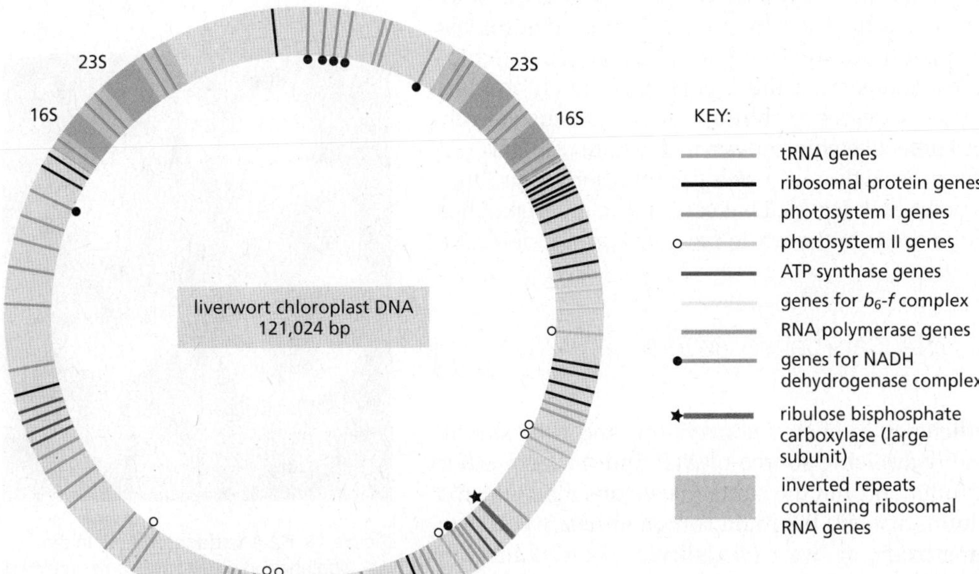

Figure 14–61 **The organization of the liverwort chloroplast genome.** The chloroplast genome organization is similar in all higher plants, although the size varies from species to species—depending on how much of the DNA surrounding the genes encoding the chloroplast's 16S and 23S ribosomal RNAs is present in two copies.

while the dynamin-like GTPases divide mitochondria from the outside (see Figure 14–8). The chloroplasts have remained closer to their bacterial origins than have mitochondria, inasmuch as the eukaryotic mechanisms of membrane constriction and vesicle formation have been adapted for mitochondrial fission.

The RNA editing and RNA processing that is prevalent in chloroplasts owes everything to their eukaryotic hosts. This RNA processing includes the generation of transcript 5′ and 3′ termini and the cleavage of polycistronic transcripts. In addition, an RNA editing process converts specific C residues to U and can change the amino acid specified by the edited codon. These and other RNA-based processes are catalyzed by protein families that are not found in prokaryotes.

Organellar Genes Are Maternally Inherited in Animals and Plants

In *Saccharomyces cerevisiae* (baker's yeast), when two haploid cells mate, they are equal in size and contribute equal amounts of mitochondrial DNA to the diploid zygote. Mitochondrial inheritance in yeasts is therefore *biparental*: both parents contribute equally to the mitochondrial gene pool of the progeny. However, during the course of the subsequent asexual, vegetative growth, the mitochondria become distributed more or less randomly to daughter cells. After a few generations, the mitochondria of any given cell contain only the DNA from one or the other parent cell, because only a small sample of the mitochondrial DNA passes from the mother cell to the bud of the daughter cell. This process is known as *mitotic segregation*, and it gives rise to a distinct form of inheritance that is called non-Mendelian, or *cytoplasmic inheritance*, in contrast to the Mendelian inheritance of nuclear genes.

The inheritance of mitochondria in animals and plants is quite different. In these organisms, the egg cell contributes much more cytoplasm to the zygote than does the male gamete (sperm in animals, pollen in plants). For example, a typical human oocyte contains about 100,000 copies of maternal mitochondrial DNA, whereas a sperm cell contains only a few. In addition, two active processes ensure that the sperm mitochondria do not compete with those in the egg. First, as sperm mature, the DNA in their mitochondria is degraded. Sperm mitochondria are also specifically recognized and eliminated from the fertilized egg cell by autophagy in very much the same way that damaged mitochondria are removed (by ubiquitylation followed by delivery to lysosomes, as discussed in Chapter 13). Because of both processes, the mitochondrial inheritance in both animals and plants is *uniparental*. More precisely, the mitochondrial DNA passes from one generation to the next by **maternal inheritance**.

In about two-thirds of higher plants, the chloroplast precursors from the male parent (contained in pollen grains) fail to enter the zygote, so that chloroplast as well as mitochondrial DNA is maternally inherited. In other plants, the chloroplast precursors from the pollen grains enter the zygote, making chloroplast inheritance biparental. In such plants, defective chloroplasts are a cause of variegation: a mixture of normal and defective chloroplasts in a zygote may sort out by mitotic segregation during plant growth and development, thereby producing alternating green and white patches in leaves. Leaf cells in the green patches contain normal chloroplasts, while those in the white patches contain defective chloroplasts (**Figure 14–62**).

Figure 14–62 **A variegated leaf.** In the white patches, the plant cells have inherited a defective chloroplast. (Courtesy of John Innes Foundation.)

Mutations in Mitochondrial DNA Can Cause Severe Inherited Diseases

Mitochondria are marvels of efficiency in energy conversion, and they supply the cells of our body with a readily available source of ATP and perform other critical metabolic and signaling functions. But the same marvelous mechanisms that enable these essential functions also are the main source of *reactive oxygen species* (*ROS*) such as H_2O_2, superoxide, or hydroxyl radicals. These damaging species contribute to the inevitable accumulation of deletions and point mutations in mitochondrial DNA, which are often observed.

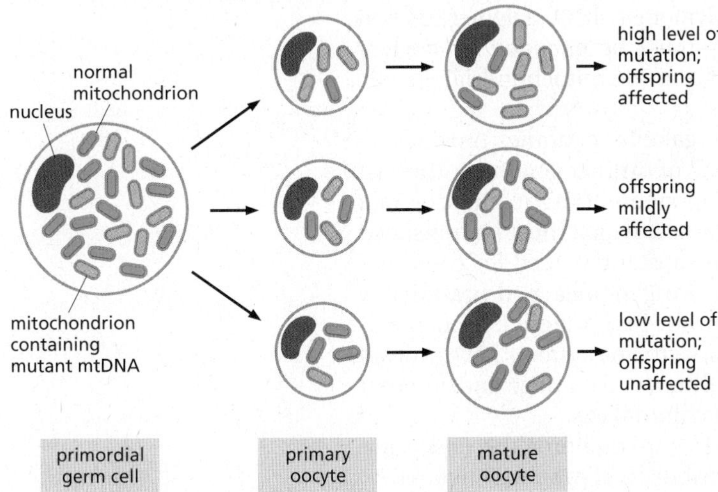

Figure 14–63 **Mitochondrial DNA mutations and inheritance.** When a primordial germ cell carries a mixed population of normal and mutated mtDNA, those genomes are distributed randomly to primary oocytes, where they are replicated during the oocyte maturation process. Those oocytes that inherit a higher load of mutant mtDNA lead to offspring that are more likely to be afflicted with mitochondrial disease. Those that inherit a lower fraction of mutated mtDNA are more likely to be unaffected. [Adapted from R.W. Taylor and D.M. Turnbull, *Nat. Rev. Genet.* 6(5):389–402, 2005. With permission from Nature.]

The less complex DNA replication and repair systems in mitochondria mean that such accidents are corrected less efficiently. This results in a 100-fold higher occurrence of deletions and point mutations than in nuclear DNA. Mathematical modeling suggests that most of these mutations and lesions are acquired in childhood or early adult life and then proliferate by *clonal expansion* in later life. Because of mitotic segregation, some cells will accumulate higher levels of faulty mitochondrial DNA than others and will exhibit impaired metabolic function. In rare cases, these mutations are passed to progeny.

In humans, as we have explained, all the mitochondrial DNA in a fertilized egg cell is inherited from the mother. Some mothers carry a mixed population of both mutant and normal mitochondrial genomes (**Figure 14–63**). Their daughters and sons typically inherit this mixture of normal and mutant mitochondrial DNAs, although oocytes (and therefore children) frequently have a higher or lower fraction of mutant mitochondrial DNA. If the child receives a low fraction of mutant DNA, she or he will be healthy unless the process of mitotic segregation results in enrichment of defective mitochondria in a particular tissue.

Diseases caused by mutations in mitochondrial DNA are clinically recognized by their passage from affected mothers to both their daughters and their sons, with the daughters but not the sons producing children with the disease. As expected from the random nature of mitotic segregation, the symptoms of these diseases vary greatly between different family members—including not only the severity and age of onset, but also which tissue is affected. Muscle and the nervous system are the most common body systems to be affected by mitochondrial disease, likely because of their particularly high demand for ATP. There are also mitochondrial diseases that are caused by mutations in nuclear-encoded mitochondrial proteins; these diseases are inherited in the regular, Mendelian fashion.

Why Do Mitochondria and Chloroplasts Maintain a Costly Separate System for DNA Transcription and Translation?

Why do mitochondria and chloroplasts require their own separate genetic systems, when other organelles that share the same cytoplasm, such as peroxisomes and lysosomes, do not? The question is not trivial, because maintaining a separate genetic system is costly: more than 90 proteins—including many ribosomal proteins, aminoacyl-tRNA synthetases, DNA polymerase, RNA polymerase, and RNA-processing and RNA-modifying enzymes—must be encoded by nuclear genes specifically for this purpose. As we have seen, the mitochondrial genetic system also entails the risk of disease.

The nature of the proteins encoded by the organellar genomes provides an additional layer of complexity in the maintenance of distinct genomes. The vast majority of organelle-encoded proteins act as components of large complexes that also contain proteins encoded in the nuclear genome, which must be translated in and imported from the cytosol. This creates a substantial need for coordination among the gene expression systems of the two genomes, and presumably also creates a substantial burden of quality control when protein abundance is not perfectly matched. For this and other reasons, signaling systems have evolved that regulate transcription of nuclear genes in response to damage and/or other perturbations in the function of mitochondria or chloroplasts.

Several possible reasons have been proposed for maintaining this costly and potentially hazardous arrangement. (1) The nonribosomal proteins encoded by the organellar genome are typically highly hydrophobic. This may make their production in and import from the cytoplasm simply too difficult and energy-consuming; however, several very hydrophobic proteins are synthesized in the cytosol and imported. (2) The current genome organization could be a remnant of an ancient endosymbiotic conflict. In the current arrangement, organelle function is completely dependent on the coordinated expression of both genomes. Neither the nucleus nor the organelle can control these critical functions without the concomitant response from the other. (3) It is also possible that the evolution (and eventual elimination) of the organellar genetic systems is still ongoing, but for now there is no alternative for the cell than to maintain separate genetic systems for its nuclear, mitochondrial, and chloroplast genes. As of now, the reason for the maintenance of distinct organellar genomes remains one of the most interesting and important unanswered questions in evolutionary biology.

Summary

Mitochondria are organelles that allow eukaryotes to carry out oxidative phosphorylation, while chloroplasts are organelles that allow plants to carry out photosynthesis. As a result of their prokaryotic origins, each organelle maintains and reproduces itself in a highly coordinated process that requires the contribution of two separate genetic systems—one in the organelle and the other in the cell nucleus. The vast majority of the proteins in these organelles are encoded by nuclear DNA, synthesized in the cytosol, and then imported individually into the organelle. Other organellar proteins, as well as organellar ribosomal and transfer RNAs, are encoded by the organellar DNA; these are synthesized in the organelle itself.

The ribosomes of chloroplasts closely resemble bacterial ribosomes, while the origin of mitochondrial ribosomes is more difficult to trace. Extensive protein similarities, however, suggest that both organelles originated from bacteria: the mitochondrion when a proto-eukaryotic archaeal cell entered into a stable endosymbiotic relationship with a respiratory bacterium, and the chloroplast when a distant descendant of this first eukaryotic cell entered into a stable endosymbiotic relationship with a cyanobacterium. Although some of the genes of these former bacteria still function to make organellar proteins and RNA, most of them have been transferred into the nuclear genome, where they encode bacteria-like enzymes that are synthesized on cytosolic ribosomes and then imported into the organelle. This situation creates substantial challenges for the inter-genome coordination of gene expression, as well as producing an increased risk of mutation and disease.

PROBLEMS

Which statements are true? Explain why or why not.

14–1 The three respiratory enzyme complexes in the mitochondrial inner membrane tend to associate with each other in ways that facilitate the correct transfer of electrons between appropriate complexes.

14–2 The number of *c* subunits in the rotor ring of ATP synthase defines how many protons need to pass through the turbine to make each molecule of ATP.

14–3 Mutations that are inherited according to Mendelian rules affect nuclear genes; mutations whose inheritance violates Mendelian rules are likely to affect organellar genes.

Discuss the following problems.

14–4 Heart muscle gets most of the ATP needed to power its continual contractions through oxidative phosphorylation. When oxidizing glucose to CO_2, heart muscle consumes O_2 at a rate of 10 μmol/min per gram of tissue, in order to replace the ATP used in contraction and give a steady-state ATP concentration of 5 μmol/g of tissue. At this rate, how many seconds would it take the heart to consume an amount of ATP equal to its steady-state levels? (Complete oxidation of one molecule of glucose to CO_2 yields 30 ATP, 26 of which are derived by oxidative phosphorylation using the 12 pairs of electrons captured in the electron carriers NADH and $FADH_2$. One pair of electrons is used to reduce each atom of oxygen.)

14–5 The transport of ions and small molecules across the inner mitochondrial membrane is often affected by the electrochemical gradient, whose components—the pH gradient and the membrane potential—may either oppose or drive the transport. How would the components of the electrochemical gradient affect the simultaneous transport of phosphate and protons into the matrix (**Figure Q14–1**)?

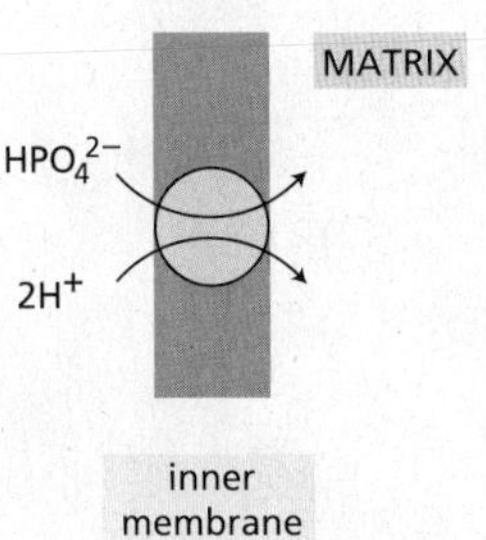

Figure Q14–1 Coupled transport of phosphate and protons across the inner mitochondrial membrane into the matrix (Problem 14–5).

14–6 If isolated mitochondria are incubated with a source of electrons such as succinate, but without oxygen, electrons enter the respiratory chain, reducing each of the electron carriers almost completely. When oxygen is then introduced, the carriers become oxidized at different rates (**Figure Q14–2**). How does this result allow you to order the electron carriers in the respiratory chain? What is their order?

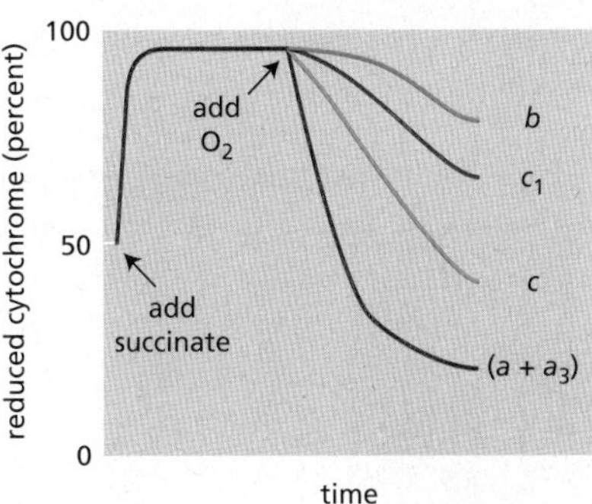

Figure Q14–2 Rapid spectrophotometric analysis of the rates of oxidation of electron carriers in the respiratory chain (Problem 14–6). Cytochromes *a* and a_3 cannot be distinguished and thus are listed as cytochrome ($a + a_3$).

14–7 Both H^+ and Ca^{2+} are ions that move through the cytosol. Why is the movement of H^+ ions so much faster than that of Ca^{2+} ions? How do you suppose the speed of these two ions would be affected by freezing the solution? Would you expect them to move faster or slower? Explain your answer.

14–8 In the 1860s, Louis Pasteur noticed that when he added O_2 to a culture of yeast growing anaerobically on glucose, the rate of glucose consumption declined dramatically. Explain the basis for this result, which is known as the Pasteur effect.

14–9 ATP synthase is the world's smallest rotary motor. Passage of H^+ ions through the membrane-embedded portion of ATP synthase (the F_o component) causes rotation of the single, central, axle-like γ subunit (the rotor stalk) inside the head group (the F_1 ATPase head). The tripartite head is composed of the three αβ dimers, the β subunits of which are responsible for synthesis of ATP. The rotation of the γ subunit induces conformational changes in the αβ dimers that allow ADP and phosphate to be converted into ATP. A variety of indirect evidence had suggested rotary catalysis by ATP synthase, but seeing is believing.

To demonstrate rotary motion, a modified form of the $\alpha_3\beta_3\gamma$ complex was used. The β subunits were modified so they could be firmly anchored to a solid support, and the γ subunit was modified (on the end that normally inserts into the F_o component in the inner membrane) so that a fluorescently tagged, readily visible filament of actin could be attached (**Figure Q14–3A**). This arrangement allows rotations of the γ subunit to be visualized as revolutions of the long actin filament. In these experiments, ATP synthase was studied in the reverse of its normal mechanism by allowing it to hydrolyze ATP. At low ATP concentrations, the actin filament was observed to revolve in steps of 120° and then pause for variable lengths of time, as shown in **Figure Q14–3B**.

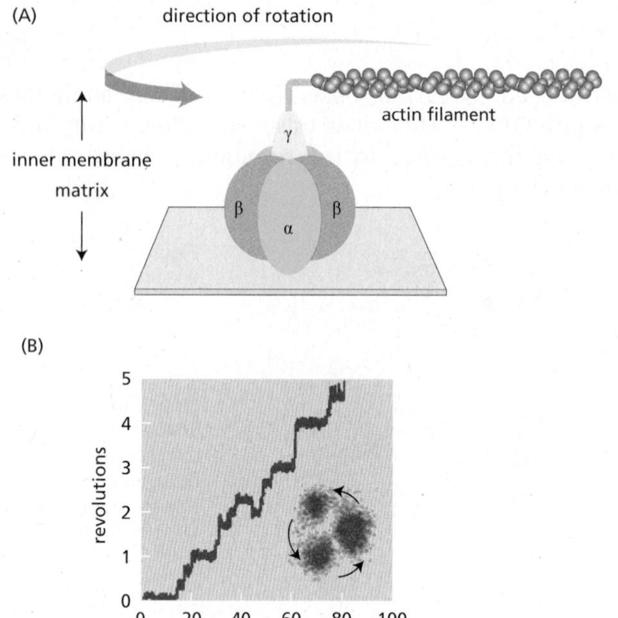

Figure Q14–3 Experimental setup for observing rotation of the γ subunit of ATP synthase (Problem 14–9). (A) The β subunits are anchored to a solid support, and a fluorescent actin filament is attached to the γ subunit. (B) The indicated trace is a typical example from one experiment. The inset shows the positions in the revolution at which the actin filament paused. (B, from R. Yasuda et al., *Cell* 93:1117–1124, 1998. With permission from Elsevier.)

A. Why does the actin filament revolve in steps with pauses in between? What does this rotation correspond to in terms of the structure of the $\alpha_3\beta_3\gamma$ complex?

B. In its normal mode of operation inside the cell, how many ATP molecules do you suppose would be synthesized for each complete 360° rotation of the γ subunit? Explain your answer.

14–10 In actively respiring liver mitochondria, the pH in the matrix is about half a pH unit higher than it is in the cytosol. Assuming that the cytosol is at pH 7 and the matrix is a sphere with a diameter of 1 μm [$V = (4/3)\pi r^3$], calculate the total number of protons in the matrix of a respiring liver mitochondrion. If the matrix began at pH 7 (equal to that in the cytosol), how many protons would have to be pumped out to establish a matrix pH of 7.5 (a difference of 0.5 pH units)?

14–11 Normally, the flow of electrons to O_2 is tightly linked to the production of ATP via the electrochemical gradient. If ATP synthase is inhibited, for example, electrons do not flow down the electron-transport chain and respiration ceases. Since the 1940s, several substances—such as 2,4-dinitrophenol—have been known to uncouple electron flow from ATP synthesis. Dinitrophenol was once prescribed as a diet drug to aid in weight loss. How would an uncoupler of oxidative phosphorylation promote weight loss? Why do you suppose dinitrophenol is no longer prescribed?

14–12 How much energy is available in visible light? How much energy does sunlight deliver to Earth? How efficient are plants at converting light energy into chemical energy? The answers to these questions provide an important backdrop to the subject of photosynthesis.

Each quantum or photon of light has energy $h\nu$, where h is Planck's constant (6.6×10^{-37} kJ sec/photon), and ν is the frequency in seconds^{-1}. The frequency of light is equal to c/λ, where c is the speed of light (3.0×10^{17} nm/sec), and λ is the wavelength in nanometers. Thus, the energy (E) of a photon is

$$E = h\nu = hc/\lambda$$

A. Calculate the energy of a mole of photons (6×10^{23} photons/mole) at 400 nm (violet light), at 680 nm (red light), and at 800 nm (near-infrared light).

B. Bright sunlight strikes Earth at the rate of about 1.3 kJ/sec per square meter. Assuming for the sake of calculation that sunlight consists of monochromatic light of wavelength 680 nm, how many seconds would it take for a mole of photons to strike a square meter?

C. Assuming that it takes eight photons to fix one molecule of CO_2 as carbohydrate under optimal conditions, calculate how long it would take a tomato plant with a leaf area of 1 square meter to make a mole of glucose from CO_2. Assume that photons strike the leaf at the rate calculated above and, furthermore, that all the photons are absorbed and used to fix CO_2.

D. If it takes 468 kJ/mole to fix a mole of CO_2 into carbohydrate, what is the efficiency of conversion of light energy into chemical energy after photon capture? Assume again that eight photons of red light (680 nm) are required to fix one molecule of CO_2.

14–13 Why are plants green? ("They contain chlorophyll" is not a sufficient answer.)

14–14 Examine the variegated leaf shown in Figure Q14–4. Yellow patches surrounded by green are common, but there are no green patches surrounded by yellow. Propose an explanation for this phenomenon.

Figure Q14–4 A variegated leaf of *Aucuba japonica* with green and yellow patches (Problem 14–14).

REFERENCES

General

Cramer WA & Knaff DB (1990) Energy Transduction in Biological Membranes: A Textbook of Bioenergetics. New York: Springer-Verlag.

Mathews CK, van Holde KE & Ahern K-G (2012) Biochemistry, 4th ed. San Francisco: Benjamin Cummings.

Nicholls DG & Ferguson SJ (2013) Bioenergetics, 4th ed. New York: Academic Press.

Schatz G (2012) The fires of life. *Annu. Rev. Biochem.* 81, 34–59.

The Mitochondrion

Ernster L & Schatz G (1981) Mitochondria: a historical review. *J. Cell Biol.* 91, 227s–255s.

Friedman JR & Nunnari J (2014) Mitochondrial form and function. *Nature* 505, 335–343.

Mitchell P (1961) Coupling of phosphorylation to electron and hydrogen transfer by a chemi-osmotic type of mechanism. *Nature* 191, 144–148.

Pebay-Peyroula E & Brandolin G (2004) Nucleotide exchange in mitochondria: insight at a molecular level. *Curr. Opin. Struct. Biol.* 14, 420–425.

Roger AJ, Muñoz-Gómez SA & Kamikawa R (2017) The origin and diversification of mitochondria. *Curr. Biol.* 27, R1177–R1192.

Scheffler IE (1999) Mitochondria. New York: Wiley-Liss.

The Proton Pumps of the Electron-Transport Chain

Althoff T, Mills DJ, Popot J-L & Kühlbrandt W (2011) Arrangement of electron transport chain components in bovine mitochondrial supercomplex $I_1III_2IV_1$. *EMBO J.* 30, 4652–4664.

Baradaran R, Berrisford JM, Minhas GS & Sazanov LA (2013) Crystal structure of the entire respiratory complex I. *Nature* 494, 443–448.

Beinert H, Holm RH & Münck E (1997) Iron-sulfur clusters: nature's modular, multipurpose structures. *Science* 277, 653–659.

Berry EA, Guergova-Kuras M, Huang LS & Crofts AR (2000) Structure and function of cytochrome *bc* complexes. *Annu. Rev. Biochem.* 69, 1005–1075.

Brandt U (2006) Energy converting NADH:quinone oxidoreductase (complex I). *Annu. Rev. Biochem.* 75, 69–92.

Chance B & Williams GR (1955) A method for the localization of sites for oxidative phosphorylation. *Nature* 176, 250–254.

Cooley JW (2013) Protein conformational changes involved in the cytochrome bc_1 complex catalytic cycle. *Biochim. Biophys. Acta* 1827, 1340–1345.

Efremov RG, Baradaran R & Sazanov LA (2010) The architecture of respiratory complex I. *Nature* 465, 441–445.

Gottschalk G (1997) Bacterial Metabolism, 2nd ed. New York: Springer.

Gray HB & Winkler JR (1996) Electron transfer in proteins. *Annu. Rev. Biochem.* 65, 537–561.

Hirst J (2013) Mitochondrial complex I. *Annu. Rev. Biochem.* 82, 551–575.

Hosler JP, Ferguson-Miller S & Mills DA (2006) Energy transduction: proton transfer through the respiratory complexes. *Annu. Rev. Biochem.* 75, 165–187.

Hongjun Y, Schut GJ, Haja D, Adams MWW & Huilin L (2021) Evolution of complex I-like respiratory complexes. *J. Biol. Chem.* 296, 100740, 1–10.

Hunte C, Zickermann V & Brandt U (2010) Functional modules and structural basis of conformational coupling in mitochondrial complex I. *Science* 329, 448–451.

Kampjut D & Sazanov LA (2020) The coupling mechanism of mammalian respiratory complex I. *Science* 370(6516), eabc4209.

Keilin D (1966) The History of Cell Respiration and Cytochromes. Cambridge, UK: Cambridge University Press.

Rouault TA, Tracey A & Tong WH (2008) Iron–sulfur cluster biogenesis and human disease. *Trends Genet.* 24, 398–407.

Trumpower BL (2002) A concerted, alternating sites mechanism of ubiquinol oxidation by the dimeric cytochrome bc_1 complex. *Biochim. Biophys. Acta* 1555, 166–173.

Tsukihara T, Aoyama H, Yamashita E . . . Yoshikawa S (1996) The whole structure of the 13-subunit oxidized cytochrome *c* oxidase at 2.8 Å. *Science* 272, 1136–1144.

ATP Production in Mitochondria

Abrahams JP, Leslie AG, Lutter R & Walker JE (1994) Structure at 2.8 Å resolution of F_1-ATPase from bovine heart mitochondria. *Nature* 370, 621–628.

Berg HC (2003) The rotary motor of bacterial flagella. *Annu. Rev. Biochem.* 72, 19–54.

Boyer PD (1997) The ATP synthase—a splendid molecular machine. *Annu. Rev. Biochem*. 66, 717–749.

Meier T, Polzer P, Diederichs K, Welte W & Dimroth P (2005) Structure of the rotor ring of F-type Na^+-ATPase from *Ilyobacter tartaricus*. *Science* 308, 659–662.

Stock D, Gibbons C, Arechaga I, Leslie AG & Walker JE (2000) The rotary mechanism of ATP synthase. *Curr. Opin. Struct. Biol.* 10, 672–679.

von Ballmoos C, Wiedenmann A & Dimroth P (2009) Essentials for ATP synthesis by F_1F_0 ATP synthases. *Annu. Rev. Biochem*. 78, 649–672.

Chloroplasts and Photosynthesis

Barber J (2013) Photosystem II: the water-splitting enzyme of photosynthesis. *Cold Spring Harb. Symp. Quant. Biol.* 77, 295–307.

Bassham JA (1962) The path of carbon in photosynthesis. *Sci. Am.* 206, 89–100.

Blankenship RE (2002) Molecular Mechanisms of Photosynthesis. Oxford, UK: Blackwell Scientific.

Blankenship RE & Bauer CE (eds.) (1995) Anoxygenic Photosynthetic Bacteria. Dordrecht: Kluwer.

Deisenhofer J & Michel H (1989) Nobel lecture. The photosynthetic reaction centre from the purple bacterium *Rhodopseudomonas viridis*. *EMBO J.* 8, 2149–2170.

De Las Rivas J, Balsera M & Barber J (2004) Evolution of oxygenic photosynthesis: genome-wide analysis of the OEC extrinsic proteins. *Trends Plant Sci*. 9, 18–25.

Hohmann-Marriott MF & Blankenship RE (2011) Evolution of photosynthesis. *Annu. Rev. Plant Biol.* 62, 515–548.

Jordan P, Fromme P, Witt HT, Klukas O, Saenger W & Krauss N (2001) Three-dimensional structure of cyanobacterial photosystem I at 2.5 Å resolution. *Nature* 411, 909–917.

Kobayashi K, Endo K & Wada H (2017) Specific distribution of phosphatidylglycerol to photosystem complexes in the thylakoid membrane. *Front. Plant Sci.* 8, 1991.

Kühlbrandt W, Wang DN & Fujiyoshi Y (1994) Atomic model of plant light-harvesting complex by electron crystallography. *Nature* 367, 614–621.

Lane N & Martin WF (2012) The origin of membrane bioenergetics. *Cell* 151, 1406–1416.

Lyons TW, Reinhard CT & Planavsky NJ (2014) The rise of oxygen in earth's early ocean and atmosphere. *Nature* 506, 307–315.

Nelson N & Ben-Shem A (2004) The complex architecture of oxygenic photosynthesis. *Nat. Rev. Mol. Cell Biol.* 5, 971–982.

Orgel LE (1998) The origin of life—a review of facts and speculations. *Trends Biochem. Sci.* 23, 491–495.

Pyc M, Cai Y, Greer MS, Yurchenko O, . . . Mullen RT (2017) Turning over a new leaf in lipid droplet biology. *Trends Plant Sci.* 22, 596–609.

Tang K-H, Tang YJ & Blankenship RE (2011) Carbon metabolic pathways in phototrophic bacteria and their broader evolutionary implications. *Front. Microbiol.* 2, 165.

Umena Y, Kawakami K, Shen J-R & Kamiya N (2011) Crystal structure of oxygen-evolving photosystem II at a resolution of 1.9 Å. *Nature* 473, 55–60.

Vinyard DJ, Ananyev GM & Dismukes GC (2013) Photosystem II: the reaction center of oxygenic photosynthesis. *Annu. Rev. Biochem.* 82, 577–606.

Yano J, Kern J, Sauer K . . . Yachandra VK (2006) Where water is oxidized to dioxygen: structure of the photosynthetic Mn_4Ca cluster. *Science* 314, 821–825.

Yoon HS (2004) A molecular timeline for the origin of photosynthetic eukaryotes. *Mol. Biol. Evol.* 21, 809–818.

The Genetic Systems of Mitochondria and Chloroplasts

Al Rawi S, Louvet-Vallee SS, Djeddi D . . . Galy V (2011) Postfertilization autophagy of sperm organelles prevents paternal mitochondrial DNA transmission. *Science* 334, 1144–1147.

Anderson S, Bankier AT, Barrell BG . . . Young IG (1981) Sequence and organization of the human mitochondrial genome. *Nature* 290, 457–465.

Bendich AJ (2004) Circular chloroplast chromosomes: the grand illusion. *Plant Cell* 16, 1661–1666.

Birky CW Jr (1995) Uniparental inheritance of mitochondrial and chloroplast genes: mechanisms and evolution. *Proc. Natl. Acad. Sci. USA* 92, 11331–11338.

Bullerwell CE & Gray MW (2004) Evolution of the mitochondrial genome: protist connections to animals, fungi and plants. *Curr. Opin. Microbiol.* 7, 528–534.

Chen XJ & Butow RA (2005) The organization and inheritance of the mitochondrial genome. *Nat. Rev. Genet.* 6, 815–825.

Clayton DA (2000) Vertebrate mitochondrial DNA—a circle of surprises. *Exp. Cell Res.* 255, 4–9.

Daley DO & Whelan J (2005) Why genes persist in organelle genomes. *Genome Biol.* 6, 110.

de Duve C (2007) The origin of eukaryotes: a reappraisal. *Nat. Rev. Genet.* 8, 395–403.

Dyall SD, Brown MT & Johnson PJ (2004) Ancient invasions: from endosymbionts to organelles. *Science* 304, 253–257.

Falkenberg M, Larsson NG & Gustafsson CM (2007) DNA replication and transcription in mammalian mitochondria. *Annu. Rev. Biochem.* 76, 679–699.

Harel A, Bromberg Y, Falkowski PG & Bhattacharya D (2014) Evolutionary history of redox metal-binding domains across the tree of life. *Proc. Natl. Acad. Sci. USA* 111, 7042–7047.

Hoppins S, Lackner L & Nunnari J (2007) The machines that divide and fuse mitochondria. *Annu. Rev. Biochem.* 76, 751–780.

Larsson NG (2010) Somatic mitochondrial DNA mutations in mammalian aging. *Annu. Rev. Biochem.* 79, 683–706.

Ma H, Xu H & O'Farrell PH (2014) Transmission of mitochondrial mutations and action of purifying selection in *Drosophila melanogaster. Nat. Genet.* 46, 393–397.

Neupert W & Herrmann JM (2007) Translocation of proteins into mitochondria. *Annu. Rev. Biochem.* 76, 723–749.

Taylor RW, Barron MJ, Borthwick GM . . . Turnbull DM (2003) Mitochondrial DNA mutations in human colonic crypt stem cells. *J. Clin. Invest.* 112, 1351–1360.

Wallace DC (1999) Mitochondrial diseases in man and mouse. *Science* 283, 1482–1488.

Williams TA, Foster PG, Cox CJ & Embley TM (2014) An archaeal origin of eukaryotes supports only two primary domains of life. *Nature* 504, 231–236.

CHAPTER

15

Cell Signaling

IN THIS CHAPTER

When things change, cells respond. Every cell, from the humble bacterium to the most sophisticated eukaryotic cell, monitors its intracellular and extracellular environment, processes the information it gathers, and responds accordingly. Unicellular organisms, for example, modify their behavior in response to changes in environmental nutrients or toxins. The cells of multicellular organisms detect and respond to countless internal and extracellular signals that control their growth, division, and differentiation during development, as well as their behavior in adult tissues. At the heart of all these communication systems are regulatory proteins that produce chemical signals, which are sent from one place to another in the body or within a cell, and other proteins that recognize the signals and respond to them, often integrating the signals and passing them on to produce an appropriate cell response.

The study of cell signaling has traditionally focused on the mechanisms by which eukaryotic cells communicate with one another using *extracellular signal molecules* such as hormones and growth factors. In this chapter, we describe the features of some of these cell–cell communication systems, and we use them to illustrate the general principles by which any regulatory system, inside or outside the cell, is able to generate, process, and respond to signals. Our main focus is on animal cells, but we end by considering the special features of cell signaling in plants.

PRINCIPLES OF CELL SIGNALING

Long before multicellular creatures roamed the Earth, unicellular organisms had developed mechanisms for responding to physical and chemical changes in their environment. These almost certainly included mechanisms for responding to the presence of other cells. Evidence comes from studies of present-day unicellular organisms such as bacteria and yeasts. Although these cells lead mostly independent lives, they can communicate and influence one another's behavior. Many bacteria, for example, respond to chemical signals that are secreted by their neighbors and accumulate at higher population density. This process, called *quorum sensing*, allows bacteria to coordinate their behavior, including their motility, antibiotic production, spore formation, and sexual conjugation. Similarly, yeast cells communicate with one another in preparation for mating. The budding yeast *Saccharomyces cerevisiae* provides a well-studied example: when a haploid individual is ready to mate, it secretes a peptide *mating factor* that signals cells of the opposite mating type to stop proliferating and prepare to mate. The subsequent fusion of two haploid cells of opposite mating type produces a diploid zygote.

Intercellular communication achieved an astonishing level of complexity during the evolution of multicellular organisms. These organisms are tightly knit societies of cells, in which the well-being of the individual cell is often set aside for the benefit of the organism as a whole. Complex systems of intercellular communication have evolved to allow the collaboration and coordination of different tissues and cell types. Bewildering arrays of signaling systems govern every conceivable feature of cell and tissue function during development and in the adult.

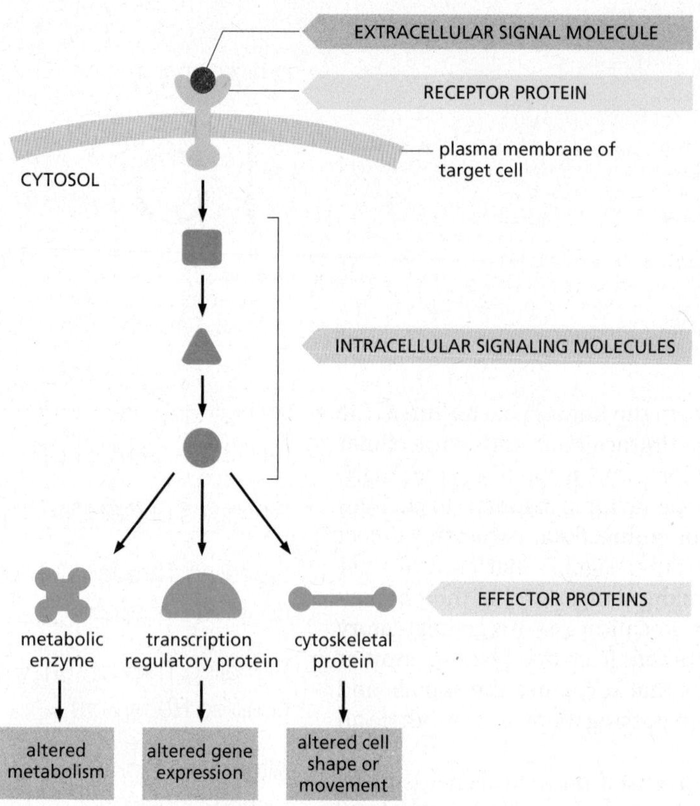

Figure 15–1 A simple intracellular signaling pathway activated by an extracellular signal molecule. The signal molecule usually binds to a receptor protein that is embedded in the plasma membrane of the target cell. The receptor activates one or more intracellular signaling pathways, involving a series of signaling proteins and small chemical messengers. Finally, one or more of the intracellular signaling molecules alters the activity of effector proteins and thereby the behavior of the cell.

Communication between cells in multicellular organisms is mediated mainly by **extracellular signal molecules**. Some of these operate over long distances, signaling to cells far away; others signal only to immediate neighbors. Most cells in multicellular organisms both emit and receive signals. Reception of the signals depends on *receptor proteins*, usually (but not always) at the cell surface, which bind the signal molecule. The binding activates the receptor, which in turn activates one or more *intracellular signaling pathways* or *systems*. These systems depend on *intracellular signaling proteins*, which process the signal inside the receiving cell and distribute it to the appropriate intracellular targets. Some of these proteins produce small chemical messengers called *second messengers*, which carry the signal to other signaling proteins. The targets that lie at the end of signaling pathways are generally called *effector proteins*, which are altered in some way by the incoming signal and implement the appropriate change in cell behavior. Depending on the signal and the type and state of the receiving cell, these effectors can be transcription regulators, ion channels, components of a metabolic pathway, or parts of the cytoskeleton (**Figure 15–1**).

The fundamental features of cell signaling have been conserved throughout the evolution of the eukaryotes. In budding yeast, for example, the response to mating factor depends on cell-surface receptor proteins, intracellular GTP-binding proteins, and protein kinases that are clearly related to functionally similar proteins in animal cells. Through gene duplication and divergence, however, the signaling systems in animals have become much more elaborate than those in yeasts; the human genome, for example, contains more than 1500 genes that encode receptor proteins, and the number of different receptor proteins is further increased by alternative RNA splicing and post-translational modifications.

Extracellular Signals Can Act Over Short or Long Distances

Many extracellular signal molecules remain bound to the surface of the signaling cell and influence only cells that contact it (**Figure 15–2A**). Such **contact-dependent signaling** is especially important during development and in immune responses.

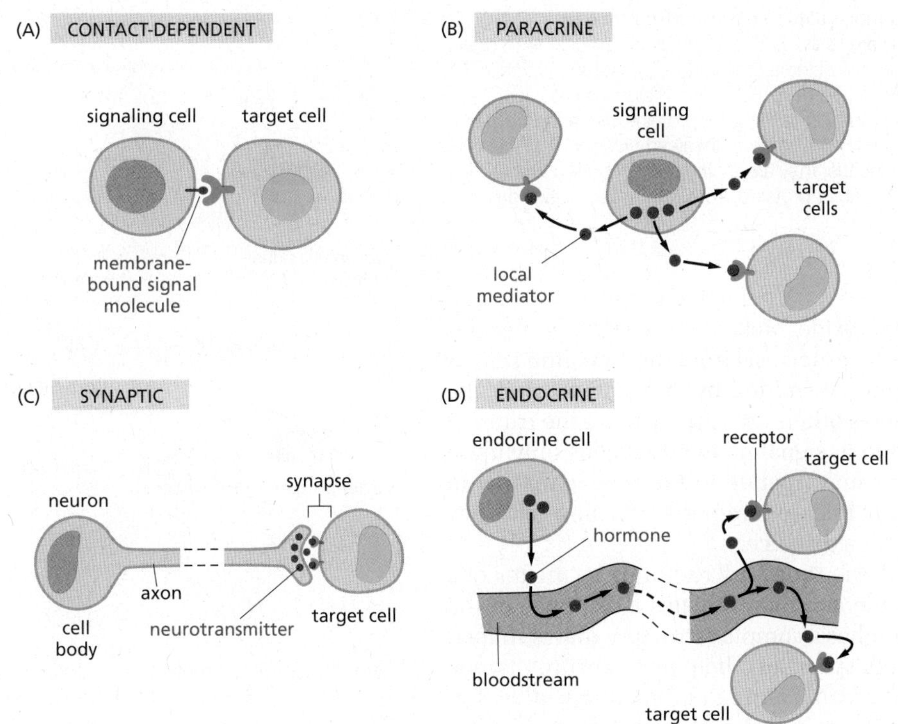

Figure 15–2 **Four forms of intercellular signaling.** (A) Contact-dependent signaling requires cells to be in direct membrane–membrane contact. (B) Paracrine signaling depends on local mediators that are released into the extracellular space and act on neighboring cells. (C) Synaptic signaling is performed by neurons that transmit signals electrically along their axons and release neurotransmitters at chemical synapses, which are often located far away from the neuronal cell body. (D) Endocrine signaling depends on endocrine cells, which secrete hormones into the bloodstream for distribution throughout the body. Many of the same types of signaling molecules are used in paracrine, synaptic, and endocrine signaling; the crucial differences lie in the speed and selectivity with which the signals are delivered to their targets.

Contact-dependent signaling during development can sometimes operate over relatively large distances if the communicating cells extend long, thin processes to make contact with one another.

In most cases, however, signaling cells secrete signal molecules into the extracellular fluid. Often, the secreted molecules are **local mediators**, which act only on cells in the local environment of the signaling cell. This is called **paracrine signaling** (Figure 15–2B). Usually, the signaling and target cells in paracrine signaling are of different cell types, but cells may also produce signals that they themselves respond to: this is referred to as *autocrine signaling.* Cancer cells, for example, often produce extracellular signals that stimulate their own survival and proliferation.

Large multicellular organisms like us also need long-range signaling mechanisms to coordinate the behavior of cells in remote parts of the body. Thus, they have evolved cell types specialized for intercellular communication over large distances. The most sophisticated of these are nerve cells, or neurons, which typically extend long, branching processes (axons) that enable them to contact target cells far away, where the processes terminate at the specialized sites of signal transmission known as *chemical synapses.* When a neuron is activated by stimuli from other nerve cells, it sends electrical impulses (action potentials) rapidly along its axon; when the impulse reaches the synapse at the end of the axon, it triggers secretion of a chemical signal that acts as a **neurotransmitter**. The tightly organized structure of the synapse ensures that the neurotransmitter is delivered specifically to receptors on the postsynaptic target cell (Figure 15–2C). The details of this **synaptic signaling** process are discussed in Chapter 11.

A quite different strategy for signaling over long distances makes use of **endocrine cells**, which secrete their signal molecules, called **hormones**, into the bloodstream. The blood carries the molecules far and wide, allowing them to act on target cells that may lie almost anywhere in the body (Figure 15–2D).

Extracellular Signal Molecules Bind to Specific Receptors

Cells in multicellular animals communicate by means of hundreds of kinds of extracellular signal molecules. These include proteins, small peptides, amino acids, nucleotides, steroids, retinoids, fatty acid derivatives, and even dissolved

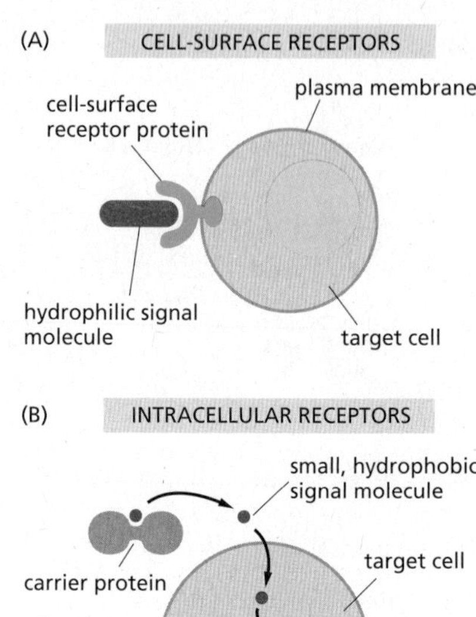

Figure 15–3 **The binding of extracellular signal molecules to either cell-surface or intracellular receptors.** (A) Most signal molecules are hydrophilic and are therefore unable to cross the target cell's plasma membrane directly; instead, they bind to cell-surface receptors, which in turn generate signals inside the target cell (see Figure 15–1). (B) Some small signal molecules, by contrast, diffuse across the plasma membrane and bind to receptor proteins inside the target cell—either in the cytosol or in the nucleus (as shown here). Many of these small signal molecules are hydrophobic and poorly soluble in aqueous solutions; they are therefore transported in the bloodstream and other extracellular fluids bound to carrier proteins, from which they dissociate before entering the target cell.

gases such as nitric oxide and carbon monoxide. Most of these signal molecules are released into the extracellular space by exocytosis from the signaling cell, as discussed in Chapter 13. Some, however, are emitted by diffusion through the signaling cell's plasma membrane, whereas others are displayed on the external surface of the cell and remain attached to it, signaling to target cells only upon contact. Transmembrane signal proteins may operate in this way, although in some cases their extracellular domains are released from the signaling cell's surface by proteolytic cleavage and then act at a distance.

Regardless of the nature of the signal, the *target cell* responds by means of a **receptor**, which binds the signal molecule and then initiates a response in the target cell. The binding site of the receptor has a complex structure that is shaped to recognize the signal molecule with high specificity, helping to ensure that the receptor responds only to the appropriate signal and not to the many other signaling molecules the cell is exposed to. Many signal molecules act at very low concentrations (typically $\leq 10^{-8}$ M), and their receptors usually bind them with high affinity (dissociation constant $K_d \leq 10^{-8}$ M; see Figure 3–42).

In most cases, receptors are transmembrane proteins on the target-cell surface. When these proteins bind an extracellular signal molecule (a *ligand*), they become activated and generate various intracellular signals that alter the behavior of the cell. In other cases, the receptor proteins are inside the target cell, and the signal molecule has to enter the cell to bind to them: this requires that the signal molecule be sufficiently small and hydrophobic to diffuse across the target cell's plasma membrane (**Figure 15–3**). This chapter focuses primarily on signaling through cell-surface receptors, but we will briefly describe signaling through intracellular receptors later in the chapter.

Each Cell Is Programmed to Respond to Specific Combinations of Extracellular Signals

A typical cell in a multicellular organism is exposed to hundreds of different signal molecules in its environment. The molecules can be soluble, bound to the extracellular matrix, or bound to the surface of a neighboring cell; they can be stimulatory or inhibitory; they can act in innumerable different combinations; and they can influence almost any aspect of cell behavior. The cell responds to this blizzard of signals selectively, in large part by expressing only those receptors and intracellular signaling systems that respond to the signals that are required for the regulation of that cell.

Most cells respond to many different signals in the environment, and some of these signals may influence the response to other signals. One of the key challenges in cell biology is to determine how a cell integrates all of this signaling information in order to make decisions—to divide, to move, to differentiate, and so on. Many cells, for example, require a specific combination of extracellular survival factors to allow the cell to continue living; when deprived of these signals, the cell activates a suicide program and kills itself—usually by *apoptosis*, as discussed in Chapter 18. Cell proliferation often depends on a combination of signals that promote both cell division and survival, as well as signals that stimulate cell growth (**Figure 15–4**). On the other hand, differentiation into a nondividing state (called *terminal differentiation*) frequently requires a different combination of survival and differentiation signals that must override any signal to divide.

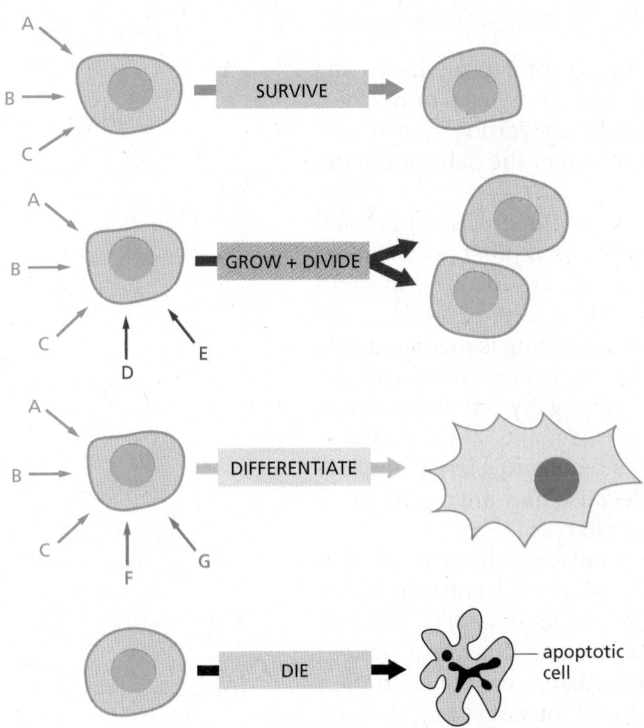

Figure 15–4 An animal cell's dependence on multiple extracellular signal molecules. Each cell type displays a set of receptors that enables it to respond to a corresponding set of signal molecules produced by other cells. These signal molecules work in various combinations to regulate the behavior of the cell. As shown here, an individual cell often requires multiple signals to survive *(blue arrows)* and additional signals to grow and divide *(red arrows)* or differentiate *(green arrows)*. If deprived of appropriate survival signals, a cell will undergo a form of cell suicide known as apoptosis. The actual situation is even more complex. Although not shown, some extracellular signal molecules act to inhibit these and other cell behaviors or even to induce apoptosis.

A signal molecule often has different effects on different types of target cells. The neurotransmitter acetylcholine (**Figure 15–5A**), for example, decreases the rate of action potential firing in heart pacemaker cells (**Figure 15–5B**) and stimulates the production of saliva by salivary gland cells (**Figure 15–5C**), even though the acetylcholine receptors are the same on both cell types. In skeletal muscle, acetylcholine causes the cells to contract by binding to a different type of acetylcholine receptor (**Figure 15–5D**). The different effects of acetylcholine in these cell types result from differences in the intracellular signaling proteins, effector proteins, and genes that are activated. Thus, an extracellular signal itself has little information content; it simply induces the cell to respond according to its predetermined state, which depends on the cell's developmental history and the specific genes it expresses.

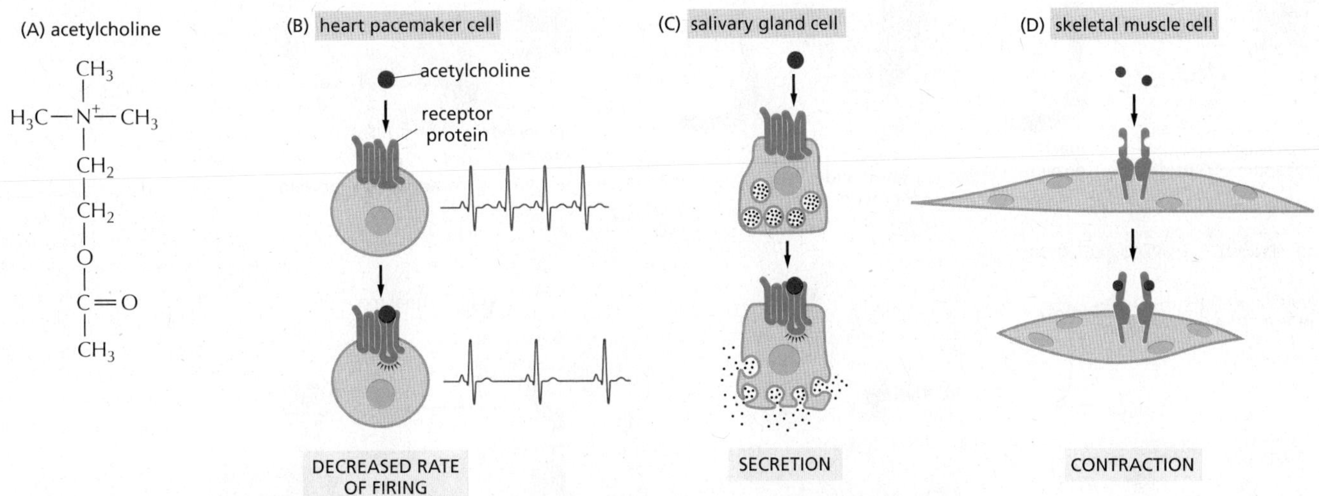

Figure 15–5 Various responses induced by the neurotransmitter acetylcholine. (A) The chemical structure of acetylcholine. (B–D) Different cell types are specialized to respond to acetylcholine in different ways. In some cases (B and C), acetylcholine binds to the same type of acetylcholine receptor (a G-protein-coupled receptor; see Figure 15–6), but the intracellular signals produced are interpreted differently in cells specialized for different functions. In other cases (D), the acetylcholine receptor protein is different (an ion-channel-coupled receptor; see Figure 15–6).

There Are Three Major Classes of Cell-Surface Receptor Proteins

Most extracellular signal molecules bind to specific receptor proteins on the surface of the target cells they influence and do not enter the cytosol or nucleus. These cell-surface receptors act as *signal transducers* by converting an extracellular ligand-binding event into intracellular signals that alter the behavior of the target cell.

Most cell-surface receptor proteins belong to one of three classes, defined by their transduction mechanism. *Ion-channel-coupled receptors,* also known as *transmitter-gated ion channels* or *ionotropic receptors,* are involved in rapid synaptic signaling between nerve cells and other electrically excitable target cells such as muscle cells (**Figure 15–6A**). This type of signaling is mediated by a small number of neurotransmitters that transiently open or close an ion channel formed by the protein to which they bind, briefly changing the ion permeability of the plasma membrane and thereby changing the excitability of the postsynaptic target cell. Most ion-channel-coupled receptors belong to a large family of homologous, multipass transmembrane proteins. Because they are discussed in detail in Chapter 11, we will not consider them further here.

G-protein-coupled receptors act by indirectly regulating the activity of a separate plasma-membrane-bound target protein, which is generally either an enzyme or an ion channel. A *heterotrimeric GTP-binding protein (G protein)* mediates the interaction between the activated receptor and this target protein (**Figure 15–6B**). The activation of the target protein can change the concentration of one or more small intracellular signaling molecules (if the target protein is an

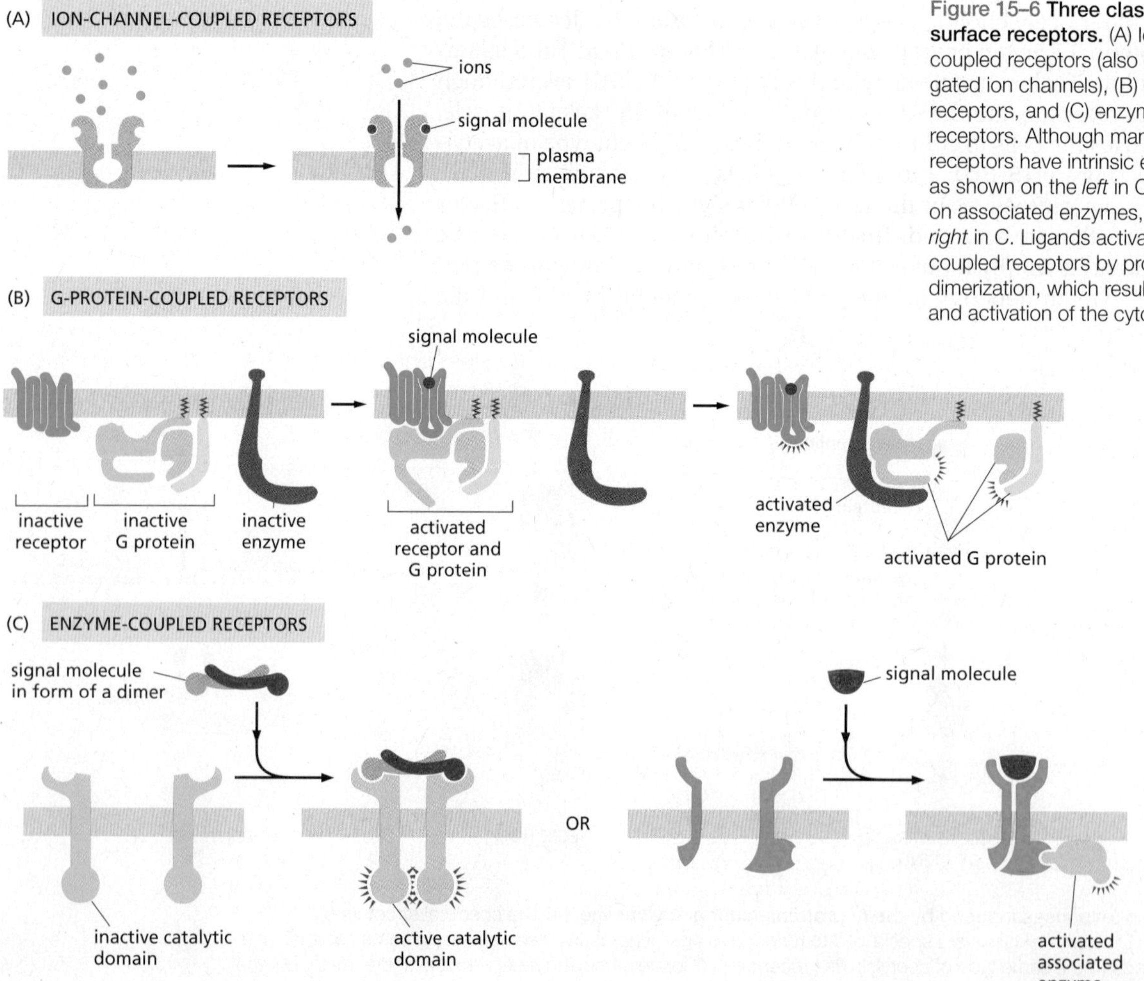

Figure 15–6 **Three classes of cell-surface receptors.** (A) Ion-channel-coupled receptors (also called transmitter-gated ion channels), (B) G-protein-coupled receptors, and (C) enzyme-coupled receptors. Although many enzyme-coupled receptors have intrinsic enzymatic activity, as shown on the *left* in C, many others rely on associated enzymes, as shown on the *right* in C. Ligands activate many enzyme-coupled receptors by promoting their dimerization, which results in the interaction and activation of the cytoplasmic domains.

enzyme) or it can change the ion permeability of the plasma membrane (if the target protein is an ion channel). The small intracellular signaling molecules act in turn to alter the behavior of yet other signaling proteins in the cell.

Enzyme-coupled receptors either function as enzymes or associate directly with enzymes that they activate (**Figure 15–6C**). They are usually single-pass transmembrane proteins that have their ligand-binding site outside the cell and their catalytic or enzyme-binding site inside. Enzyme-coupled receptors are heterogeneous in structure compared with the other two classes; the great majority, however, are either protein kinases or associate with protein kinases, which phosphorylate specific sets of proteins in the target cell when activated.

There are also some types of cell-surface receptors that do not fit easily into any of these classes but have important functions in controlling the specialization of different cell types during development and in tissue renewal and repair in adults. We discuss these in a later section, after we explain how G-protein-coupled receptors and enzyme-coupled receptors operate. First, we continue our general discussion of the principles of signaling via cell-surface receptors.

Cell-Surface Receptors Relay Signals Via Intracellular Signaling Molecules

Numerous intracellular signaling molecules relay signals received by cell-surface receptors into the cell interior. The resulting chain of intracellular signaling events ultimately alters effector proteins that are responsible for modifying the behavior of the cell (see Figure 15–1).

Some intracellular signaling molecules are small chemicals, which are often called **second messengers** (the “first messengers” being the extracellular signals). They are generated in large amounts in response to receptor activation and diffuse away from their source, spreading the signal to other parts of the cell. Some, such as *cyclic AMP* and Ca^{2+}, are water soluble and diffuse in the cytosol, while others, such as *diacylglycerol*, are lipid soluble and diffuse in the plane of the plasma membrane. In either case, they pass the signal on by binding to and altering the behavior of selected signaling or effector proteins.

Most intracellular signaling molecules are proteins, which help relay the signal into the cell by either generating second messengers or activating the next signaling or effector protein in the pathway. Many of these proteins behave like *molecular switches*. When they receive a signal, they switch from an inactive to an active state, until another process switches them off, returning them to their inactive state. The switching off can be just as important as the switching on. If a signaling pathway is to recover after transmitting a signal so that it can be ready to transmit another, every activated molecule in the pathway must return to its original, unactivated state.

The largest class of molecular switches consists of proteins that are activated or inactivated by **phosphorylation** (discussed in Chapter 3). For these proteins, the switch is thrown in one direction by a **protein kinase**, which covalently adds one or more phosphate groups to specific amino acids on the signaling protein, and in the other direction by a **protein phosphatase**, which removes the phosphate groups (**Figure 15–7A**). The activity of any protein regulated by phosphorylation depends on the balance between the activities of the kinases that phosphorylate it and of the phosphatases that dephosphorylate it. About 30–50% of human proteins contain covalently attached phosphate, and the human genome encodes about 520 protein kinases and about 150 protein phosphatases. A typical mammalian cell makes use of hundreds of distinct types of protein kinases at any moment.

Protein kinases attach phosphate to the hydroxyl group of specific amino acids on the target protein. There are two main types of protein kinase in eukaryotic cells. The great majority are **serine/threonine kinases**, which phosphorylate the hydroxyl groups of serines and threonines in their targets. Others are **tyrosine kinases**, which phosphorylate proteins on tyrosines. Tyrosine kinases are found

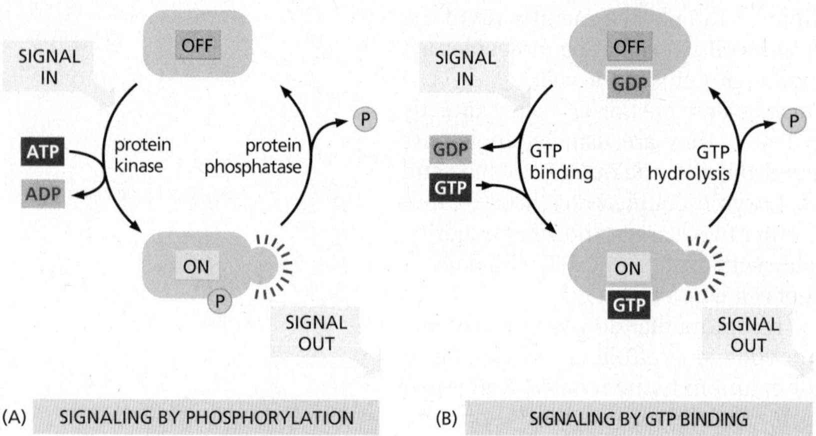

Figure 15–7 Two types of intracellular signaling proteins that act as molecular switches. (A) A protein kinase covalently adds a phosphate from ATP to the signaling protein, and a protein phosphatase removes the phosphate. Although not shown, many signaling proteins are activated by dephosphorylation rather than by phosphorylation. (B) A GTP-binding protein is induced to exchange its bound GDP for GTP, which activates the protein; the protein then inactivates itself by hydrolyzing its bound GTP to GDP.

primarily in multicellular animals; these kinases are not present, for example, in yeast.

Many intracellular signaling proteins controlled by phosphorylation are themselves protein kinases, and these are often organized into **kinase cascades**. In such a cascade, one protein kinase, activated by phosphorylation, phosphorylates the next protein kinase in the sequence, and so on, relaying the signal onward and, in some cases, amplifying it or spreading it to other signaling pathways.

Like the protein kinases, the protein phosphatases are categorized by their specificity for serine/threonine phosphate or tyrosine phosphate. There are about 100 **protein tyrosine phosphatases** encoded in the human genome, including some *dual-specificity phosphatases* that also dephosphorylate serines and threonines.

The other important class of molecular switches consists of **GTP-binding proteins** (discussed in Chapter 3). These proteins switch between two distinct structural conformations: an "on" state when GTP is bound and an "off" state when GDP is bound. In the "on" state, they bind and thereby activate specific signaling proteins. GTP-binding proteins usually have intrinsic GTPase activity and shut themselves off by hydrolyzing their bound GTP to GDP (**Figure 15–7B**). The inactive protein then returns to the "on" state when GDP dissociates, allowing a new GTP to bind. There are two major types of GTP-binding proteins. Large, *heterotrimeric GTP-binding proteins* (also called *G proteins*) help relay signals from G-protein-coupled receptors that activate them (see Figure 15–6B). Small **monomeric GTPases** (also called *monomeric GTP-binding proteins*) help relay signals from many classes of cell-surface receptors.

For most GTP-binding proteins, the inactivation process (GTP hydrolysis to GDP) and the activation process (GDP dissociation) are slow in the absence of other proteins. Inside the cell, regulatory proteins are used to accelerate one or the other process, thereby governing the activation state of the GTP-binding protein. **GTPase-activating proteins (GAPs)** drive the proteins into an "off" state by increasing the rate of hydrolysis of bound GTP. Conversely, **guanine nucleotide exchange factors (GEFs)** activate GTP-binding proteins by promoting the release of bound GDP, which allows a new GTP to bind. In the case of heterotrimeric G proteins, the activated receptor serves as the GEF. **Figure 15–8** illustrates the regulation of monomeric GTPases.

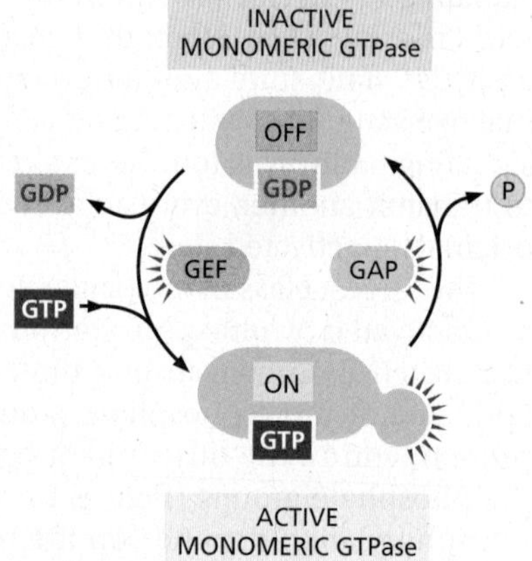

Figure 15–8 The regulation of a monomeric GTPase. GTPase-activating proteins (GAPs) inactivate the protein by stimulating it to hydrolyze its bound GTP to GDP, which remains tightly bound to the inactivated GTPase. Guanine nucleotide exchange factors (GEFs) activate the inactive protein by stimulating it to release its GDP; because the concentration of GTP in the cytosol is 10 times greater than the concentration of GDP, the protein rapidly binds GTP and is thereby activated.

Not all molecular switches in signaling systems depend on phosphorylation or GTP binding. We see later that some signaling proteins are switched on or off by the binding of another signaling protein or a second messenger, such as cyclic AMP or Ca^{2+}, or by covalent modifications other than phosphorylation or dephosphorylation, such as ubiquitylation (discussed in Chapter 3).

For simplicity, we often portray a signaling pathway as a series of activation steps (see Figure 15–1). It is important to note, however, that most signaling pathways contain inhibitory steps, and a sequence of two inhibitory steps can

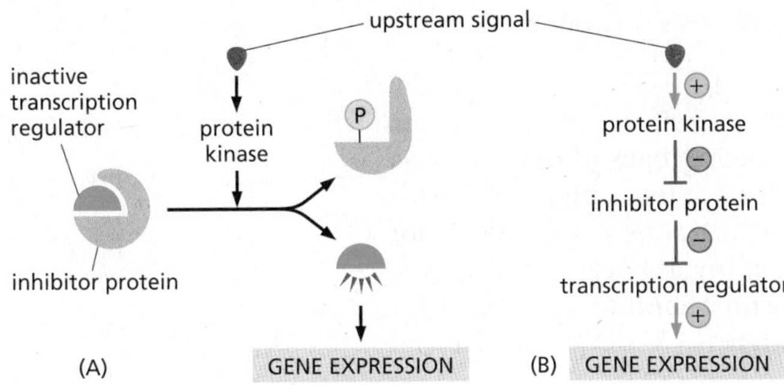

Figure 15–9 A sequence of two inhibitory signals produces a positive signal. (A) In this simple signaling system, a transcription regulator is kept in an inactive state by a bound inhibitor protein. In response to some upstream signal, a protein kinase is activated and phosphorylates the inhibitor, causing its dissociation from the transcription regulator, which can now activate gene expression. (B) The signaling pathway consists of a sequence of four steps, including two sequential inhibitory steps that are equivalent to a single activating step.

have the same effect as one activating step (**Figure 15–9**). This activation scheme is very common in signaling systems, as we will see when we describe specific pathways later in this chapter.

Intracellular Signals Must Be Specific and Robust in a Noisy Cytoplasm

In an idealized signaling pathway like that shown in Figure 15–1, each intracellular signaling molecule interacts only with the appropriate downstream target. Similarly, the target is activated only by the appropriate upstream signal. In reality, however, the cell is crowded with closely related signaling molecules that control a diverse array of cellular processes. It is inevitable that a signaling molecule will sometimes interact with molecules in other signaling pathways, potentially creating unwanted cross-talk and interference between signaling systems. How does a signal remain strong and specific under these noisy conditions?

A key to signaling specificity is the high affinity and specificity of the interactions between intracellular signaling molecules and their correct partners. The binding of a signaling molecule to its target is determined by precise and complex interactions between complementary surfaces on the two molecules. Some protein kinases, for example, contain active sites that recognize a specific amino acid sequence around the phosphorylation site on the correct target protein, and many signaling enzymes employ additional *docking sites*, outside their active site, that promote a specific, high-affinity interaction with a complementary site on the target. These and related mechanisms provide a strong and persistent interaction between the correct partners, thereby enhancing the likelihood that a signal is passed to the appropriate target.

The specificity of signaling systems also depends on noise filters that reduce or remove undesirable background signals. Consider a signaling pathway, for example, in which a response is triggered by phosphorylation of several sites on a target protein. Inside the cell, we can generally assume that a constant low level of phosphatase activity is present to remove these phosphorylations. As a result, a strong response is possible only if the appropriate protein kinase reaches a high and persistent level of activity that is sufficient to overcome the opposing phosphatase activity. If by some random accident another protein kinase interacts briefly with the target protein and catalyzes phosphorylation on one or two sites, these will be removed by the opposing phosphatase and no response will occur. The weak background signal is thereby ignored.

Cells in a population often exhibit random variations in the concentration or activity of their intracellular signaling molecules. Similarly, individual molecules in a large population of molecules vary in their activity or interactions with other molecules. This *signal variability* introduces another form of noise that can interfere with the precision and efficiency of signaling. Most signaling systems, however, generate remarkably robust and precise responses even when upstream signals are variable or some components of the system are disabled. In some cases, this *robustness* depends on the presence of parallel mechanisms; for example, a signal might employ two parallel pathways to activate a single common downstream target protein, allowing the response to occur even if one pathway is crippled.

Intracellular Signaling Complexes Form at Activated Cell-Surface Receptors

One simple and effective strategy for enhancing the specificity of interactions between intracellular signaling molecules and reducing background noise is to localize the molecules in the same part of the cell, often within large protein complexes, thereby promoting their interaction with one another and not with inappropriate partners. Such mechanisms often involve **scaffold proteins**, which bring together groups of interacting signaling proteins into *signaling complexes,* often before a signal has been received (**Figure 15–10A**). Because the

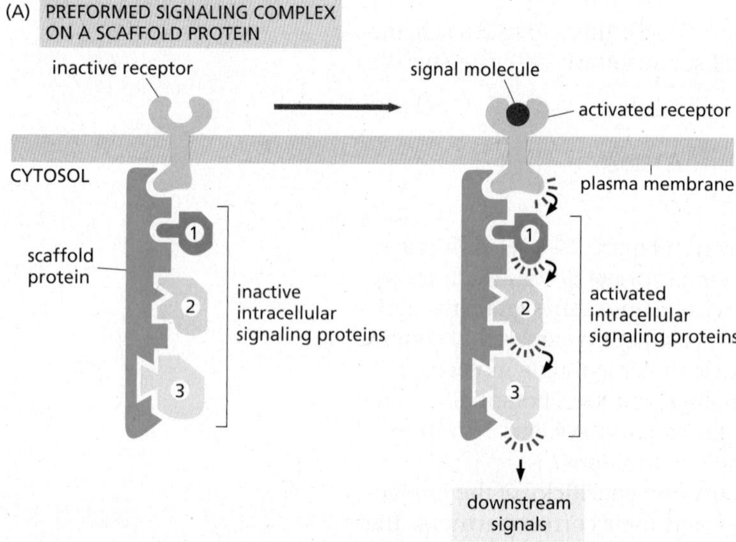

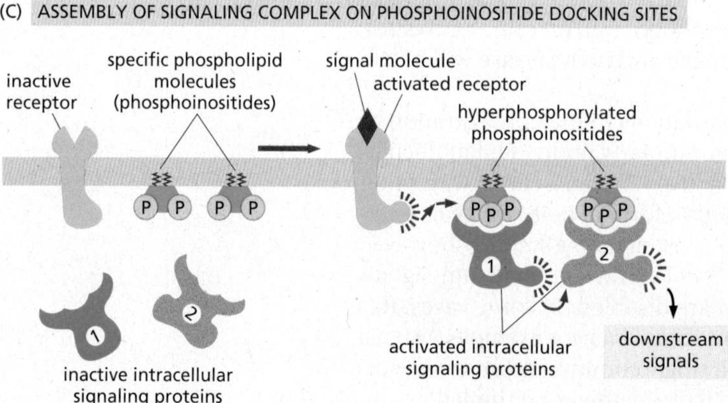

Figure 15–10 Three types of intracellular signaling complexes. (A) A receptor and some of the intracellular signaling proteins it activates in sequence are preassembled into a signaling complex on the inactive receptor by a large scaffold protein. (B) A signaling complex assembles transiently on a receptor only after the binding of an extracellular signal molecule has activated the receptor; here, the activated receptor phosphorylates itself at multiple sites, which then act as docking sites for intracellular signaling proteins. (C) Activation of a receptor leads to the increased phosphorylation of specific phospholipids (phosphoinositides) in the adjacent plasma membrane; these then serve as docking sites for specific intracellular signaling proteins, which can now interact with each other.

scaffold holds the proteins in close proximity, they can interact at high local concentrations and be activated rapidly, efficiently, and selectively in response to an appropriate extracellular signal, avoiding unwanted cross-talk with other signaling pathways.

In other cases, such signaling complexes form transiently in response to an extracellular signal and rapidly disassemble when the signal is gone. They often assemble around a cell-surface receptor after an extracellular signal molecule has activated it. In many of these cases, the cytoplasmic tail of an activated enzyme-coupled receptor is phosphorylated during the activation process, and the phosphorylated amino acids then serve as docking sites for the assembly of other signaling proteins (**Figure 15–10B**). In yet other cases, receptor activation leads to the production of modified phospholipid molecules (called phosphoinositides) in the adjacent plasma membrane, which then recruit specific intracellular signaling proteins to this region of membrane, where they are activated (**Figure 15–10C**).

Modular Interaction Domains Mediate Interactions Between Intracellular Signaling Proteins

Simply bringing intracellular signaling proteins together into close proximity is sometimes sufficient to activate them. Thus, *induced proximity*, where a signal triggers assembly of a signaling complex, is commonly used to relay signals from protein to protein along a signaling pathway. The assembly of such signaling complexes depends on various highly conserved, small **interaction domains**, which are found in many intracellular signaling proteins. Each of these compact protein modules binds to a particular structural motif in another protein or lipid. The recognized motif in the interacting protein can be a short peptide sequence, a covalent modification (such as a phosphorylated amino acid), or another protein domain. The use of modular interaction domains presumably facilitated the evolution of new signaling pathways. Because it can be inserted at many locations in a protein without disturbing the protein's folding or function, a new interaction domain can connect the protein to additional signaling pathways.

There are many types of interaction domains in signaling proteins. *Src homology 2* (*SH2*) *domains* and *phosphotyrosine-binding* (*PTB*) *domains*, for example, bind to phosphorylated tyrosines in a particular peptide sequence on activated receptors or intracellular signaling proteins. *Src homology 3* (*SH3*) domains bind to short, proline-rich amino acid sequences. Some *pleckstrin homology* (*PH*) domains bind to the charged head groups of specific phosphoinositides that are produced in the plasma membrane in response to an extracellular signal; they enable the protein they are part of to dock on the membrane and interact with other similarly recruited signaling proteins (see Figure 15–10C). Some signaling proteins consist solely of two or more interaction domains and function only as **adaptors** to link two other proteins together in a signaling pathway. Some adaptor proteins have multiple interaction domains as well as their own signal-propagation activity (see Figure 15–12).

Interaction domains enable signaling proteins to bind to one another in multiple specific combinations. Like Lego bricks, the proteins can form linear or branching chains or three-dimensional networks, which determine the route followed by the signaling pathway. As an example, **Figure 15–11** illustrates how some interaction domains mediate the formation of a large signaling complex around the receptor for the hormone *insulin*.

Modular interaction domains are generally located in flexible, unstructured regions of signaling proteins, arrayed along the polypeptide-like beads on a string. Proteins with multiple interaction domains can therefore nucleate the formation of large, cross-linked protein matrices around clusters of activated receptors (**Figure 15–12**). These protein matrices behave like gels or biomolecular condensates, thereby creating a local microenvironment that is distinct in composition from the surrounding cytosol (discussed in Chapter 3).

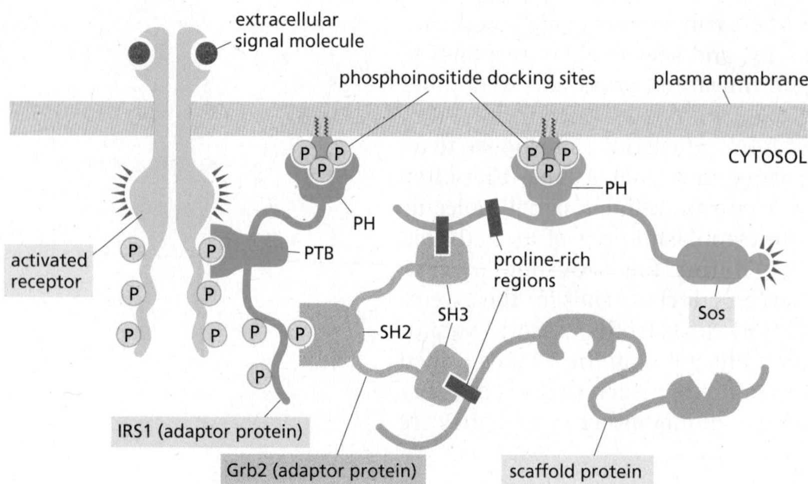

Figure 15–11 A specific signaling complex formed using modular interaction domains. This example is based on the insulin receptor, which is a dimeric enzyme-coupled receptor (a receptor tyrosine kinase, discussed later). First, the activated receptor phosphorylates itself on tyrosines, and one of the phosphotyrosines then recruits an adaptor protein called insulin receptor substrate-1 (IRS1) via a PTB domain of IRS1; the PH domain of IRS1 also binds to specific phosphoinositides on the inner surface of the plasma membrane. Then, the activated receptor phosphorylates IRS1 on tyrosines, and one of these phosphotyrosines binds the SH2 domain of the adaptor protein Grb2. Next, Grb2 uses one of its two SH3 domains to bind to a proline-rich region of a protein called Sos, which is thereby brought to the membrane to relay the signal downstream by acting as a GEF (see Figure 15–8) to activate a monomeric GTPase called Ras (not shown). Sos also binds to phosphoinositides in the plasma membrane via its PH domain. Grb2 uses its other SH3 domain to bind to a proline-rich sequence in a scaffold protein, which binds several other signaling proteins (not shown). The other phosphorylated tyrosines on IRS1 recruit additional signaling proteins that have SH2 domains (not shown).

Concentration of activated receptors and specific signaling proteins in these matrices is thought to enhance the strength and specificity of the receptor signal while reducing interference from other pathways.

Another way of bringing receptors and intracellular signaling proteins together is to locate them in a specific region of the cell. An important example is the *primary cilium* that projects like an antenna from the surface of most vertebrate cells (discussed in Chapter 16). A number of surface receptors and signaling proteins are concentrated there—particularly the components of the Hedgehog signaling system, as we discuss later. Light and smell receptors are also concentrated in specialized cilia.

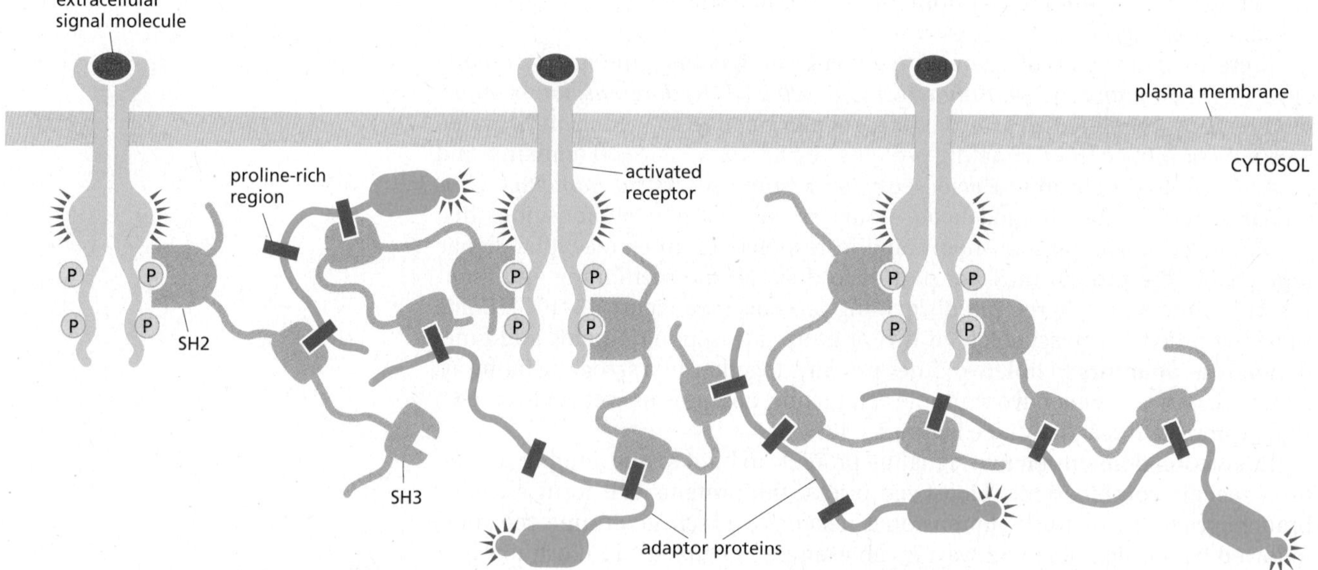

Figure 15–12 Formation of large receptor clusters by multivalent interactions among signaling proteins. The system pictured here contains activated receptor tyrosine kinases that are extensively phosphorylated on disordered regions in the receptor tails. The system also includes two adaptor proteins. One adaptor protein *(pink)* contains one SH2 domain, which binds phosphorylated tyrosines on the receptors, and two SH3 domains. The other adaptor protein *(blue)* contains three proline-rich regions that can bind to SH3 domains, plus a protein kinase domain. Numerous multivalent binding interactions can occur among the three components in this system, generating a cross-linked protein matrix or condensate in which the protein kinases of the receptor and adaptor protein are concentrated, potentially providing a more effective signal output. The cross-linking of the matrix can be enhanced further by including adaptor proteins with domains that interact with modified phospholipids in the membrane (see Figure 15–11).

The Relationship Between Signal and Response Varies in Different Signaling Pathways

The function of an intracellular signaling system is to detect and measure a specific stimulus in one location of a cell and then generate an appropriately timed and measured response, often at another location. The system accomplishes this task by sending information in the form of molecular "signals" from the receptor to the final effector proteins, often through a series of intermediaries that do not simply pass the signal along but also process it along the way. Signaling systems work in various ways: each has evolved to produce a response that is appropriate for the cell function the system controls. In the following paragraphs, we list some basic signaling properties and how they vary in different systems, as a foundation for more detailed discussions later.

1. *Response timing* varies dramatically in different signaling systems, according to the speed required for the response. In some cases, such as synaptic signaling (see Figure 15–2C), the response can occur within milliseconds. In other cases, as in the control of cell fate by morphogens during development, a full response can require hours or days.

2. *Sensitivity* to extracellular signals can vary greatly. Hormones tend to act at very low concentrations on their distant target cells, which are therefore highly sensitive to low concentrations of signal. Neurotransmitters, on the other hand, operate at much higher concentrations at a synapse, reducing the need for high sensitivity in postsynaptic receptors. Sensitivity is often controlled by changes in the number or affinity of the receptors on the target cell. A particularly important mechanism for increasing sensitivity is signal *amplification*, whereby a small number of activated cell-surface receptors evokes a large intracellular response by either producing large amounts of a second messenger or by activating many copies of a downstream signaling protein.

3. *Dynamic range* of a signaling system is related to its sensitivity. Some systems, like those involved in simple developmental decisions, are responsive over a narrow range of extracellular signal concentrations. Others, like those controlling vision or some metabolic responses to hormones, are highly responsive over a much broader range of signal strengths. We will see that a broad dynamic range is often achieved by *adaptation* mechanisms that adjust responsiveness according to the prevailing amount of signal.

4. *Persistence* of a response can vary greatly. A transient response of less than a second is appropriate in some synaptic responses, for example, while a prolonged or even permanent response is required in cell-fate decisions during development. Numerous mechanisms, including positive and negative feedback, can be used to alter the duration and reversibility of a response.

5. *Signal processing* can convert a simple signal into a complex response. In many systems, for example, a gradual increase in an extracellular signal is converted into an abrupt, switchlike response. In other cases, a simple input signal is converted into an oscillatory response, produced by a repeating series of transient intracellular signals. Feedback usually lies at the heart of biochemical switches and oscillators, as we describe later.

6. *Integration* allows a response to be governed by multiple inputs. As discussed earlier, for example, specific combinations of extracellular signals are generally required to stimulate complex cell behaviors such as cell growth, proliferation, and differentiation (see Figure 15–4). The cell therefore has to integrate information coming from multiple signals, which often depends on intracellular *coincidence detectors*; these proteins are equivalent to *AND gates* in the microprocessor of a computer, in that they are only activated if they receive multiple converging signals (**Figure 15–13**).

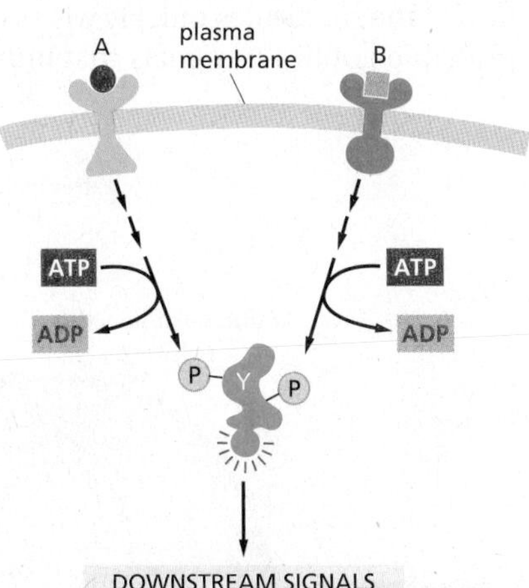

Figure 15–13 **An example of signal integration.** Extracellular signals A and B activate different intracellular signaling pathways, each of which leads to the phosphorylation of protein Y but at different sites on the protein. Protein Y is activated only when both of these sites are phosphorylated, and therefore it becomes active only when signals A and B are simultaneously present.

7. *Coordination* of multiple responses in one cell can be achieved by a single extracellular signal. Some extracellular signal molecules, for example, stimulate a cell to both grow and divide. This coordination generally depends on mechanisms for distributing a signal to multiple effectors, by creating branches in the signaling pathway. In some cases, the branching of signaling pathways can allow one extracellular signal to *modulate* the strength of a response to other extracellular signals.

Given the complexity that arises from behaviors like signal integration, coordination, and feedback, it is clear that signaling systems rarely depend on a simple linear sequence of steps but more often operate like a signaling network, in which information flows in multiple directions, including backwards. A major research challenge is to understand the nature of these networks and how they control complex cell behaviors.

The Speed of a Response Depends on the Turnover of Signaling Molecules

The speed of any signaling response depends on the nature of the intracellular signaling molecules that carry out the target cell's response. When the response requires only changes in proteins already present in the cell, it can occur very rapidly: an allosteric change in a neurotransmitter-gated ion channel (discussed in Chapter 11), for example, can alter the plasma membrane electrical potential in milliseconds, and responses that depend solely on protein phosphorylation can occur within seconds or minutes. When the response involves changes in gene expression and the synthesis of new proteins, however, it usually requires many minutes or hours, regardless of the mode of signal delivery (**Figure 15–14**).

When thinking about the speed of a response, it is natural to think of signaling systems in terms of the changes produced when the signal is delivered. But it can be just as important to consider what happens when the signal is withdrawn. In many signaling pathways, the response fades when the signal ceases. Often the effect is transient because the signal exerts its effects by increasing the concentrations of intracellular molecules that are short-lived (unstable), undergoing continual turnover. Thus, when the extracellular signal is removed, degradation of the molecules quickly wipes out all traces of the signal's action. The same principle applies to signals that induce protein phosphorylation by activating a

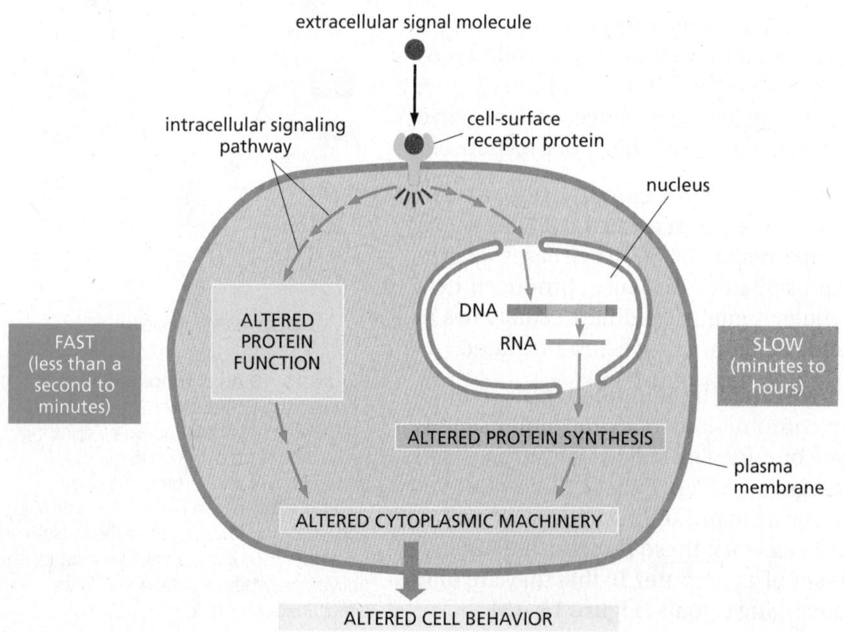

Figure 15–14 **Slow and rapid responses to an extracellular signal.** Certain types of signal-induced cellular responses, such as increased cell growth and division, involve changes in gene expression and the synthesis of new proteins; they therefore occur slowly, often starting an hour or more after the signal is received. Other responses—such as changes in cell movement, secretion, or metabolism—need not involve changes in gene transcription and therefore occur much more quickly, often starting in seconds or minutes; they may involve the rapid phosphorylation of effector proteins in the cytoplasm, for example. Synaptic responses mediated by changes in membrane potential are even quicker and can occur in milliseconds (not shown). Some signaling systems generate both rapid and slow responses as shown here, allowing the cell to respond quickly to a signal while simultaneously initiating a more long-term, persistent response.

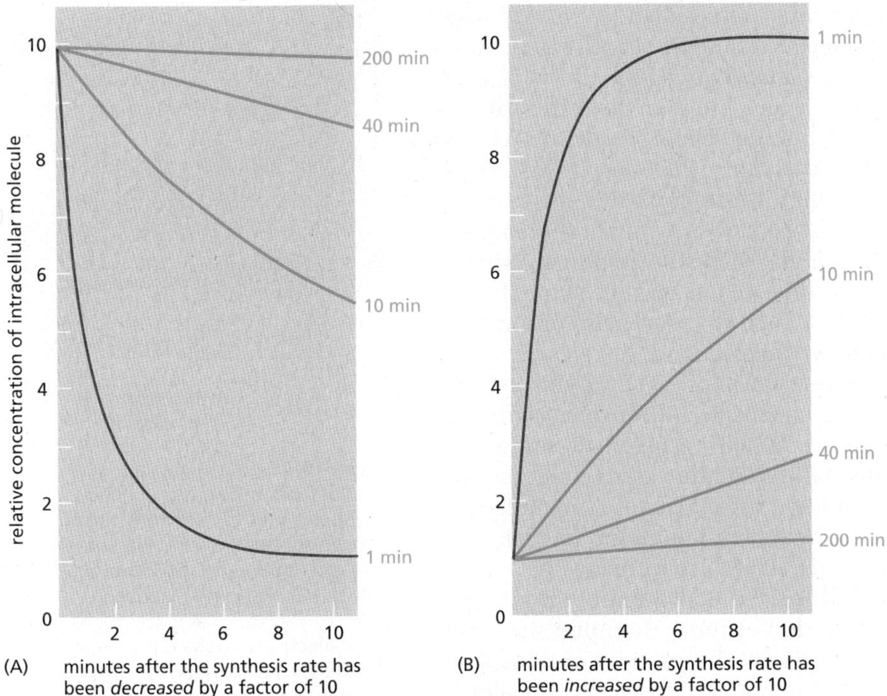

Figure 15–15 The importance of rapid turnover. The graphs show the predicted relative rates of change in the intracellular concentrations of molecules with differing turnover times when their synthesis rates are either (A) decreased or (B) increased suddenly by a factor of 10. In both cases, the concentrations of those molecules that are normally degraded rapidly in the cell *(red lines)* change quickly, whereas the concentrations of those that are normally degraded slowly *(green lines)* change proportionally more slowly. The numbers (in *blue*) on the right are the half-lives assumed for each of the different molecules.

protein kinase: because most phosphorylation is continually removed by phosphatases, the effects of the increased protein kinase activity are quickly reversed when kinase activity declines. It follows that the speed with which a cell responds to removal of a signal depends on the rate of destruction, or turnover, of the molecules or modifications that the signal affects.

It is also true, although much less obvious, that this turnover rate can determine the promptness of the response when an extracellular signal arrives. Consider, for example, two intracellular signaling molecules, X and Y, both of which are normally maintained at a steady-state concentration of 1000 molecules per cell. The cell synthesizes and degrades molecule Y at a rate of 100 molecules per second, with each molecule having an average lifetime of 10 seconds. Molecule X has a turnover rate that is 10 times slower than that of Y: it is both synthesized and degraded at a rate of 10 molecules per second, so that each molecule has an average lifetime in the cell of 100 seconds. If a signal acting on the cell causes a tenfold increase in the synthesis rates of both X and Y with no change in the molecular lifetimes, at the end of 1 second the concentration of Y will have increased by nearly 900 molecules per cell ($10 \times 100 - 100$), while the concentration of X will have increased by only 90 molecules per cell. In fact, after a molecule's synthesis rate has been either increased or decreased abruptly, the time required for the molecule to shift halfway from its old to its new equilibrium concentration is equal to its half-life; that is, equal to the time that would be required for its concentration to fall by half if all synthesis were stopped (**Figure 15–15**).

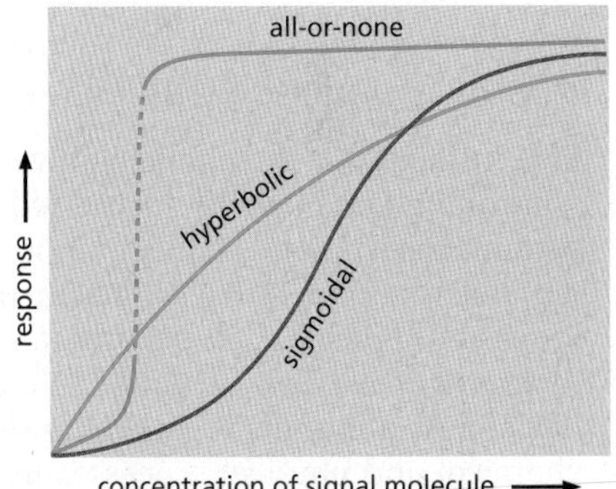

Figure 15–16 Signal processing can produce smoothly graded or switchlike responses. Some cell responses increase gradually as the concentration of extracellular signal molecule increases, eventually reaching a plateau as the signaling pathway is saturated, resulting in a *hyperbolic* response curve *(blue line)*. In other cases, the signaling system reduces the response at low signal concentrations and then produces a steeper response at some intermediate signal concentration—resulting in a *sigmoidal* response curve *(red line)*. In still other cases, the response is more abrupt and switchlike; the cell switches completely between a low and high response, without any stable intermediate response *(green line)*.

Cells Can Respond Abruptly to a Gradually Increasing Signal

Some signaling systems are capable of generating a smoothly graded response over a wide range of extracellular signal concentrations (**Figure 15–16**, ***blue line***); such systems are useful, for example, in the fine-tuning of metabolic processes by some hormones. Other signaling systems generate significant responses only when the signal concentration rises beyond some threshold value. These abrupt responses are of two types. One is a *sigmoidal* response, in which low concentrations of stimulus do not have much effect, but then the response rises steeply and continuously at intermediate stimulus levels (**Figure 15–16**, *red line*). Such systems provide a filter to reduce inappropriate responses to low-level background signals but respond with high sensitivity when the stimulus rises to physiological signal concentrations. A second type of abrupt response is the *discontinuous* or *all-or-none* response, in which the response switches on completely (and often irreversibly) when the signal reaches some threshold concentration (**Figure 15–16**, *green line*). Such responses are particularly useful for controlling the choice between two alternative cell states, and they generally involve positive feedback, as we describe in more detail shortly.

Cells use a variety of molecular mechanisms to produce a sigmoidal response to increasing signal concentrations. In one mechanism, more than one intracellular signaling molecule must bind to its downstream target protein to induce a response. As we discuss later, for example, four molecules of the second messenger cyclic AMP must bind simultaneously to each molecule of *cyclic-AMP-dependent protein kinase* (*PKA*) to activate the kinase. A similar sharpening of response is seen when the activation of an intracellular signaling protein requires phosphorylation at more than one site. Such responses become sharper as the number of required molecules or phosphate groups increases, and if the number is large enough, responses become almost all-or-none (**Figure 15–17**).

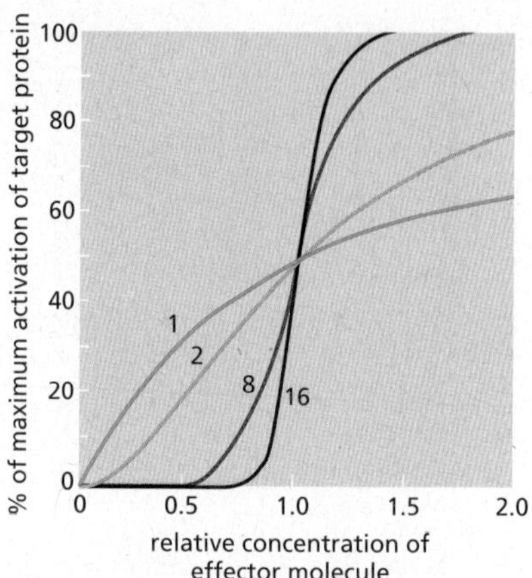

Figure 15–17 Activation curves for an allosteric protein as a function of effector molecule concentration. The curves show how the sharpness of the activation response increases with an increase in the number of effector molecules that must be bound simultaneously to activate the target protein. The curves shown are those expected, under certain conditions, if the activation requires the simultaneous binding of 1, 2, 8, or 16 effector molecules.

Responses are also sharpened when an intracellular signaling molecule activates one enzyme and also inhibits another enzyme that catalyzes the opposite reaction. A well-studied example of this common type of regulation is the stimulation of glycogen breakdown in skeletal muscle cells induced by the hormone epinephrine. Epinephrine's binding to a G-protein-coupled cell-surface receptor increases the intracellular concentration of cyclic AMP, which both activates an enzyme that promotes glycogen breakdown and inhibits an enzyme that promotes glycogen synthesis.

Positive Feedback Can Generate an All-or-None Response

Like intracellular metabolic pathways (discussed in Chapter 2) and the systems controlling gene activity (discussed in Chapter 7), most intracellular signaling systems incorporate feedback loops, in which the output of a process acts back to regulate that same process. We discussed the mathematical analysis of feedback loops in Chapter 8. In *positive feedback*, the output stimulates its own production; in *negative feedback*, the output inhibits its own production (**Figure 15–18**). Feedback loops are of great general importance in biology, and they regulate many chemical and physical processes in cells. Even the simplest of these loops can produce complex and interesting effects.

Positive feedback in a signaling pathway can transform the behavior of the responding cell. If the positive feedback is of only moderate strength, its effect will be simply to steepen the response to the signal, generating a sigmoidal response like those described earlier; but if the feedback is strong enough, it can produce an all-or-none response (see Figure 15–16). This response goes hand in hand with a further property: once the responding system has switched to the high level of activation, this condition is often self-sustaining and can persist even after the signal strength drops back below its critical value. In such a case, the system is said to be *bistable*: it can exist in either a "switched-off" or a "switched-on" state, and a transient stimulus can flip it from one state to the other (**Figure 15–19A and B**).

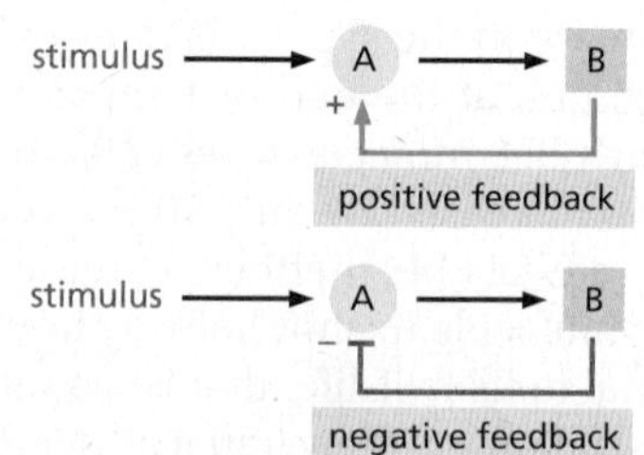

Figure 15–18 Positive and negative feedback. In these simple examples, a stimulus activates protein A, which, in turn, activates protein B. Protein B then acts back to either increase or decrease the activity of A.

Through positive feedback, a transient extracellular signal can induce long-term changes in cells and their progeny that can persist for the lifetime of the organism. The signals that trigger muscle-cell specification, for example, turn on the transcription of a series of genes that encode muscle-specific transcription regulatory proteins, which stimulate the transcription of their own genes, as well as genes encoding various other muscle-cell proteins; in this way, the decision to become a muscle cell is made permanent. This type of cell memory, which depends on positive feedback, is one of the basic ways in which a cell can undergo a lasting change of character without any alteration in its DNA sequence.

Studies of signaling responses in large populations of cells can give the false impression that a response is smoothly graded, even when strong positive feedback is causing an abrupt, discontinuous switch in the response in individual cells. Only by studying the response in single cells is it possible to see its all-or-none

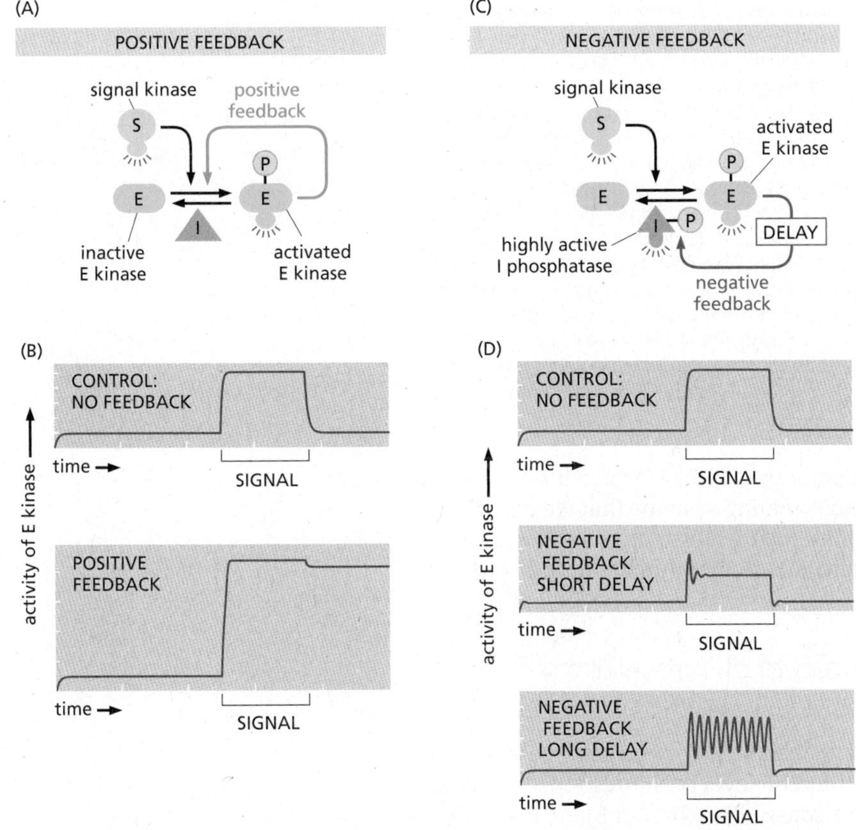

Figure 15–19 Some effects of simple feedback. The graphs show the computed effects of simple positive and negative feedback loops (discussed in Chapter 8). In each case, the input signal is an activated protein kinase (S) that phosphorylates and thereby activates another protein kinase (E); a protein phosphatase (I) dephosphorylates and inactivates the activated E kinase. In the graphs, the *red line* indicates the activity of the E kinase over time; the underlying *black bracket* indicates the time during which the input signal (activated S kinase) is present. (A) Diagram of a positive feedback loop, in which the activated E kinase acts back to promote its own phosphorylation and activation; the basal activity of the I phosphatase dephosphorylates activated E at a steady, low rate. (B) The top graph shows that, without feedback, the activity of the E kinase is simply proportional to the level of stimulation by the S kinase. The bottom graph shows that, with the positive feedback loop, the transient stimulation by S kinase switches the system from an "off" state to an "on" state, which then persists after the stimulus has been removed. (C) Diagram of a negative feedback loop, in which the activated E kinase phosphorylates and activates the I phosphatase, thereby increasing the rate at which the phosphatase dephosphorylates and inactivates the phosphorylated E kinase. (D) The top graph shows, again, the response in E kinase activity without feedback. The other graphs show the effects on E kinase activity of negative feedback operating after a short or long delay. With a short delay, the system shows a response when the signal is first increased, but the feedback quickly dampens the response—which then declines to some intermediate level at which the input signal and feedback are balanced. With a long delay, the response rises unopposed at first, allowing kinase activity to reach maximum levels before it feeds back to shut itself off. Then the sudden drop in activity removes the negative feedback, unleashing another pulse of kinase activity. If conditions are right, the result is sustained oscillations for as long as the stimulus is present.

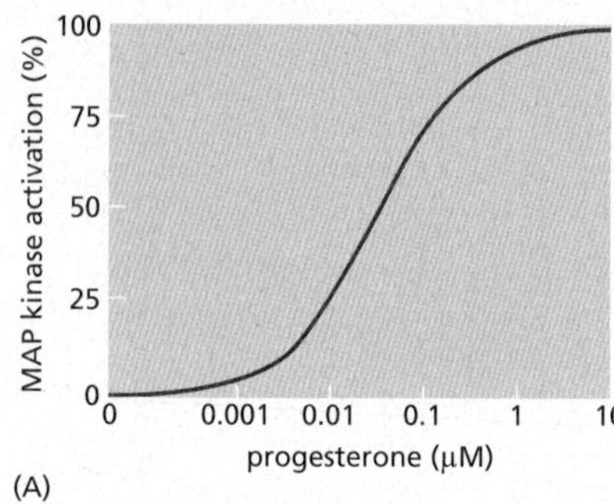

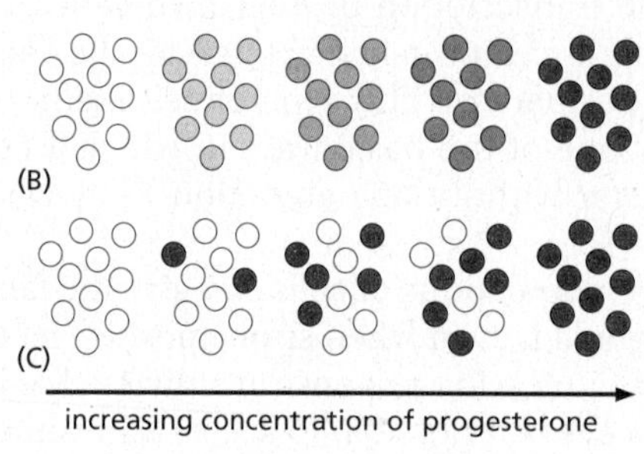

Figure 15–20 The importance of examining individual cells to detect all-or-none responses to increasing concentrations of an extracellular signal. In these experiments, immature frog eggs (oocytes) were stimulated with increasing concentrations of the hormone progesterone. The response was assessed by analyzing the activation of *MAP kinase* (discussed later), which is one of the protein kinases activated by phosphorylation in the response. The amount of phosphorylated (activated) MAP kinase in extracts of the oocytes was assessed biochemically. In (A), extracts of populations of stimulated oocytes were analyzed, and the activation of MAP kinase appeared to increase progressively with increasing progesterone concentration. There are two possible ways of explaining this result: (B) MAP kinase could have increased gradually in each individual cell with increasing progesterone concentration; or (C) individual cells could have responded in an all-or-none way, with the gradual increase in total MAP kinase activation reflecting the increasing number of cells responding with increasing progesterone concentration. When extracts of individual oocytes were analyzed, it was found that cells had either very low amounts or very high amounts, but not intermediate amounts, of the activated kinase, indicating that the response was essentially all-or-none at the level of individual cells, as diagrammed in C. Subsequent studies revealed that this all-or-none response is due in part to strong positive feedback in the progesterone signaling system. (Adapted from J.E. Ferrell and E.M. Machleder, *Science* 280:895–898, 1998. With permission from AAAS.)

character (**Figure 15–20**). The misleading smooth response in a cell population is due to the random, intrinsic variability in signaling systems that we described earlier: all cells in a population do not respond identically to the same concentration of extracellular signal, especially at intermediate signal concentrations where the receptors are only partially occupied.

Negative Feedback Is a Common Feature of Intracellular Signaling Systems

By contrast with positive feedback, negative feedback counteracts the effect of a stimulus and thereby abbreviates and limits the level of the response, making the system less sensitive to perturbations (discussed in Chapter 8). As with positive feedback, however, qualitatively different responses can be obtained when the feedback operates in different ways. Negative feedback with a long delay can produce responses that oscillate. The oscillations may persist for as long as the stimulus is present (**Figure 15–19C and D**) or they may even be generated spontaneously, without need of an external signal to drive them. Many such oscillators also contain positive feedback loops that generate sharper oscillations. Later in this chapter, we will encounter specific examples of oscillatory behavior in the intracellular responses to extracellular signals; all of them depend on negative feedback, generally accompanied by positive feedback.

If negative feedback operates with a short delay, the system generates a brief response to a stimulus, but the response decays rapidly even while the stimulus persists. If the stimulus is increased further, the system responds strongly again, but, again, the response soon decays. This is the phenomenon of *adaptation,* which we now discuss.

Cells Can Adjust Their Sensitivity to a Signal

We have seen that most signaling systems generate an output response that is proportional to the strength of the input signal. In many cases, the response remains constant as long as the input signal remains, and the response declines

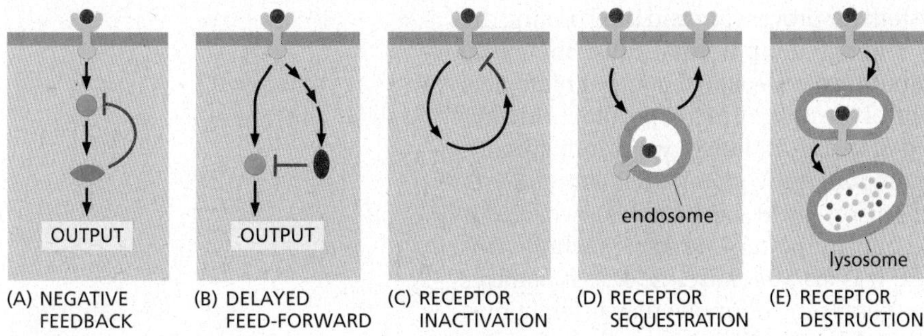

Figure 15–21 **Some ways in which target cells can become adapted (desensitized) to an extracellular signal molecule.** (A) Negative feedback with a short delay can dampen the initial response to receptor activation. (B) In some cases, the activated receptor rapidly activates a stimulatory pathway while also initiating a slower inhibitory pathway—resulting in a transient output response. This is called a delayed feed-forward loop. (C, D, E) Various mechanisms can inactivate a cell-surface receptor after a signal molecule binds, including mechanisms that depend on internalization of the receptor into endosomes, from which the receptor can be returned to the cell surface or destroyed in lysosomes.

when the input signal stops. In other systems, like those involving positive feedback, an input signal can generate a strong output signal that persists even after the input signal is removed. Finally, we have just discussed negative feedback systems in which an input signal triggers a response that rises and then falls even if the stimulus persists. A second, stronger input signal can then generate another transient response like the first one, and so on.

This phenomenon is called **adaptation**, or **desensitization**, and it allows cells to respond to *changes* in the strength of an input signal (rather than to the absolute amount of the signal) over a very wide range of signal levels. The visual system, discussed later in this chapter, provides the ideal illustration of this concept: it uses adaptation at varying signal strengths to allow us to see clearly over an astonishing range of light intensities, from starlight to bright sunshine.

Adaptation requires that some component of the signaling system generates a delayed inhibitory signal that reduces the strength of the output. There are several variations on this theme. One common mechanism, as just discussed, is negative feedback that operates with a short delay (**Figure 15–21A**). Many signaling pathways that begin with enzyme-coupled receptors, for example, include protein kinases that phosphorylate and thereby inhibit an upstream signaling protein in the pathway. A second mechanism of adaptation occurs when an extracellular signal rapidly activates a signaling response through one pathway while also triggering a parallel, slower signaling pathway that inhibits the response (**Figure 15–21B**). In some enzyme-coupled signaling pathways, for example, an activated receptor promotes recruitment to the membrane of a GEF called Sos, which stimulates the activity of a small GTP-binding protein called Ras (see Figure 15–11). Activation of the receptor also leads more slowly to recruitment of a GAP that inactivates Ras after a delay, thereby leading to a reduced output signal.

A particularly effective mechanism of adaptation, discussed in more detail later in the chapter, depends on receptor inactivation, whereby some activated receptors shut themselves off after some short period of activity. Ligand binding to some G-protein-coupled receptors, for example, leads to a change in the receptor that promotes its association with heterotrimeric G proteins and the resulting downstream response. Ligand binding also results in phosphorylation of the receptor and association with inhibitory molecules called *arrestins*, which interfere with G-protein activation and thereby reduce the response. The response is dampened further by endocytosis of the ligand-bound receptors, which can be sequestered inside the cell or simply destroyed (**Figure 15–21C, D, and E**).

Though bewildering in their complexity, the multiple cross-regulatory signaling pathways and feedback loops that we describe in this chapter are not just a haphazard tangle, but a highly evolved system for processing and interpreting the vast number of extracellular signals that impinge upon animal cells. The entire signaling network can be viewed as a computing device; like that other biological computing device, the brain, it presents one of the most difficult problems in biology. We can identify the components and discover how they work individually. We can understand how small subsets of components work together as regulatory modules, noise filters, or adaptation mechanisms, as we have seen. However, it is a much more daunting task to understand how the system works as a whole. This is not only because the system is complex; it is also because the way it behaves is strongly dependent on the quantitative details of the molecular interactions, and, for most animal cells, we have only rough qualitative information. A major challenge for the future of signaling research is to develop more sophisticated quantitative and computational methods for the analysis of signaling systems, as described in Chapter 8.

Summary

Each cell in a multicellular animal is programmed to respond to a specific set of extracellular signal molecules produced by other cells. The signal molecules act by binding to a complementary set of receptor proteins expressed by the target cells. Most extracellular signal molecules activate cell-surface receptor proteins, which act as signal transducers, converting the extracellular signal into intracellular ones that alter the behavior of the target cell. Activated receptors relay the signal into the cell interior by activating intracellular signaling proteins. Collectively, some of these signaling proteins help transduce, amplify, and spread the signal as they relay it, while others integrate signals from different signaling pathways. Some function as switches that are transiently activated by phosphorylation or GTP binding. Large signaling complexes form by means of modular interaction domains in the signaling proteins, which allow the proteins to form complex signaling networks.

Target cells use various mechanisms, including feedback loops, to adjust the ways in which they respond to extracellular signals. Positive feedback loops can help cells respond in an all-or-none fashion to a gradually increasing concentration of an extracellular signal or convert a short-lasting signal into a long-lasting or even irreversible response. Negative feedback is one way that allows cells to adapt to a signal molecule, which enables them to respond to small changes in the concentration of the signal molecule over a large concentration range.

SIGNALING THROUGH G-PROTEIN-COUPLED RECEPTORS

G-protein-coupled receptors (GPCRs) form the largest family of cell-surface receptors, and they mediate most responses to signals from the external world, as well as signals from other cells, including hormones, neurotransmitters, and local mediators. Our senses of sight, smell, and taste also depend on them. There are more than 800 GPCRs in humans, and in mice there are about 1000 concerned with the sense of smell alone. The signal molecules that act on GPCRs are as varied in structure as they are in function and include proteins and small peptides, derivatives of amino acids and fatty acids, photons of light, and all the molecules that we can smell or taste. The same signal molecule can activate many different GPCR family members; for example, epinephrine activates at least 9 distinct GPCRs, acetylcholine another 5, and the neurotransmitter serotonin at least 14. The different receptors for the same signal are usually expressed in different cell types and elicit different responses.

Despite the chemical and functional diversity of the signal molecules that activate them, all GPCRs have a similar structure. They consist of a single polypeptide chain that threads back and forth across the lipid bilayer seven times, forming a cylindrical structure, often with a deep ligand-binding site in its core

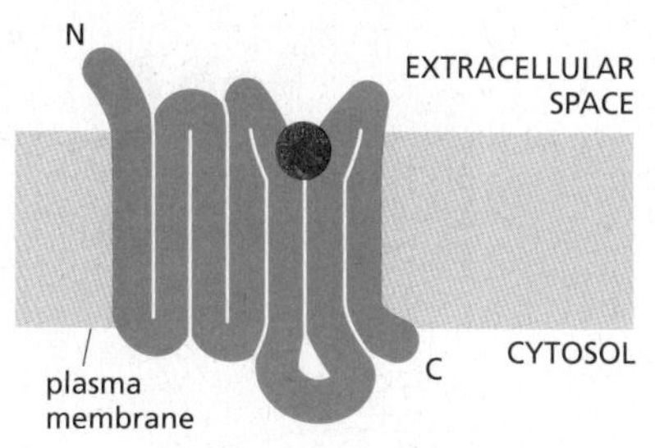

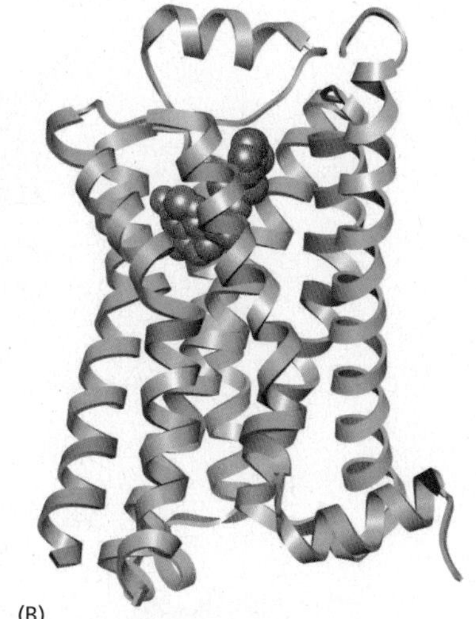

Figure 15-22 A G-protein-coupled receptor (GPCR). (A) GPCRs that bind small ligands such as epinephrine have small extracellular domains, and the ligand usually binds deep within the plane of the plasma membrane to a site that is formed by amino acids from several transmembrane segments. GPCRs that bind protein ligands have a large extracellular domain (not shown here) that contributes to ligand binding. (B) The structure of the β_2-adrenergic receptor, a receptor for the neurotransmitter epinephrine, illustrates the typical cylindrical arrangement of the seven transmembrane helices in a GPCR. The ligand *(orange)* binds in a pocket between the helices, resulting in conformational changes on the cytoplasmic surface of the receptor that promote G-protein activation (not shown). (PDB code: 3P0G.)

(**Figure 15-22**). In addition to their characteristic orientation in the plasma membrane, they all use G proteins to relay the signal into the cell interior.

The GPCR superfamily includes *rhodopsin,* the light-activated protein in the vertebrate eye, as well as the large number of olfactory receptors in the vertebrate nose. Other family members are found in unicellular organisms: the receptors in yeasts that recognize secreted mating factors are an example. It is likely that the GPCRs that mediate cell-cell signaling in multicellular organisms evolved from the sensory receptors in their unicellular eukaryotic ancestors.

It is remarkable that almost half of all known drugs work through GPCRs or GPCR-coupled signaling pathways. Of the many hundreds of genes in the human genome that encode GPCRs, about 150 encode orphan receptors, for which the ligand is unknown. Many of them are likely targets for new drugs that remain to be discovered.

Heterotrimeric G Proteins Relay Signals from GPCRs

When an extracellular signal molecule binds to a GPCR, the receptor undergoes a conformational change that enables it to activate a **heterotrimeric GTP-binding protein (G protein)**, which couples the receptor to enzymes or ion channels in the plasma membrane. In some cases the G protein is physically associated with the receptor before the receptor is activated, whereas in others it binds only after receptor activation. There are various types of G proteins, each specific for a particular set of GPCRs and for a particular set of target proteins in the plasma membrane. They all have a similar structure, however, and operate similarly.

G proteins are composed of three protein subunits: α, β, and γ. In the unstimulated state, the α subunit has GDP bound and the G protein is inactive (**Figure 15-23**). When a GPCR is activated, it acts like a guanine nucleotide

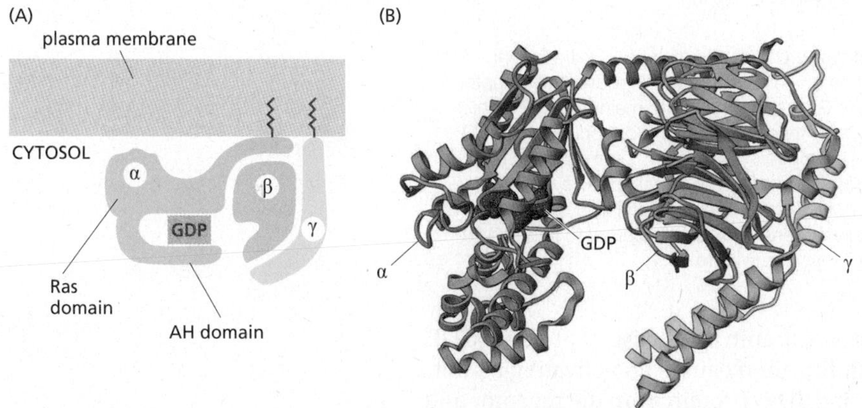

Figure 15-23 The structure of an inactive G protein. (A) Note that both the α and γ subunits have covalently attached lipid molecules *(red tails)* that help them bind to the plasma membrane, and the α subunit has GDP bound. (B) The three-dimensional structure of the inactive, GDP-bound form of a G protein called G_i, which interacts with numerous GPCRs, including the β_2-adrenergic receptor shown in Figure 15-22. The α subunit contains the GTPase domain and binds to one side of the β subunit. The γ subunit binds to the opposite side of the β subunit, and the β and γ subunits together form a single functional unit. The GTPase domain of the α subunit contains two major subdomains: the Ras domain, which is related to other GTPases and provides one face of the GTP-binding pocket; and the α-helical or AH domain, which clamps the GTP in place (see Figure 15-24). (PDB code: 1GG2.)

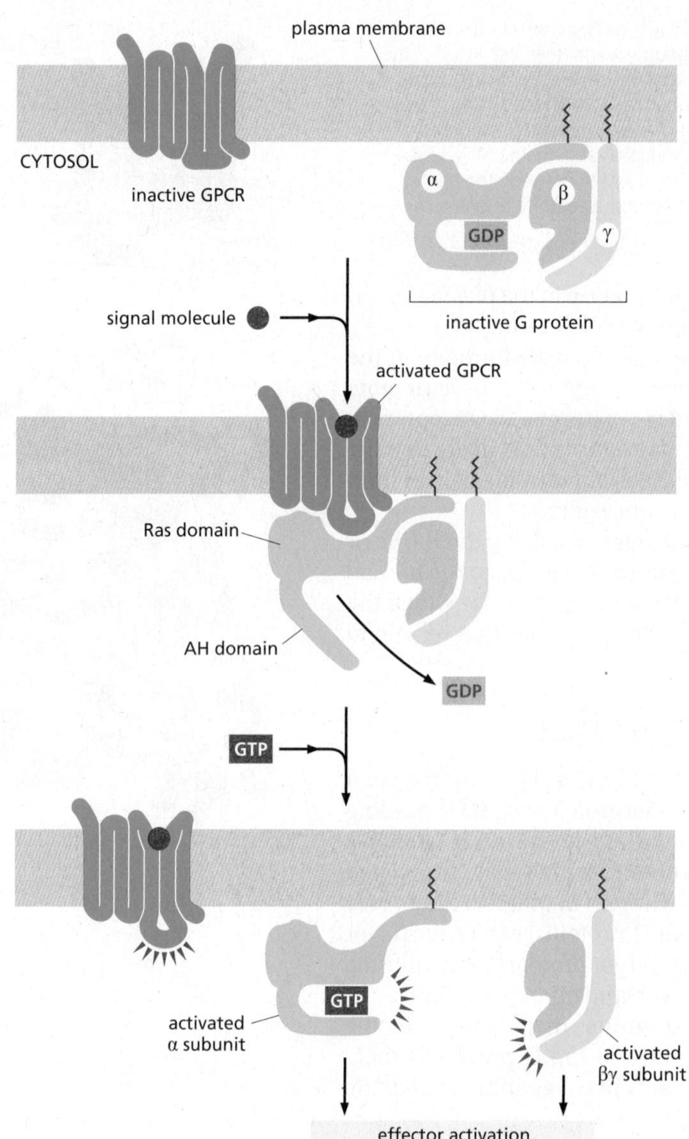

Figure 15–24 **Activation of a G protein by an activated GPCR.** Binding of an extracellular signal molecule to a GPCR changes the conformation of the receptor, which allows the receptor to bind and alter the conformation of a heterotrimeric G protein. The AH domain of the G protein α subunit moves outward to open the GTP-binding site, thereby promoting dissociation of GDP. GTP binding then promotes closure of the binding site, triggering conformational changes that cause dissociation of the α subunit from the receptor and from the βγ complex. The GTP-bound α subunit and the βγ complex each regulate the activities of downstream signaling molecules (not shown). The receptor stays active while the extracellular signal molecule is bound to it, and it can therefore catalyze the activation of many G-protein molecules (**Movie 15.1**).

exchange factor (GEF) and induces the α subunit to release its bound GDP, allowing GTP to bind in its place. GTP binding then causes an activating conformational change in the Gα subunit, releasing the G protein from the receptor and triggering dissociation of the GTP-bound Gα subunit from the Gβγ pair—both of which then interact with various targets, such as enzymes and ion channels in the plasma membrane, which relay the signal onward (**Figure 15–24**).

The α subunit is a GTPase and becomes inactive when it hydrolyzes its bound GTP to GDP. The time required for GTP hydrolysis is usually short because the GTPase activity is greatly enhanced by the binding of the α subunit to a second protein, which can be either the target protein or a specific **regulator of G protein signaling (RGS)**. RGS proteins act as α-subunit-specific GTPase-activating proteins (GAPs) (see Figure 15–8), and they help shut off G-protein-mediated

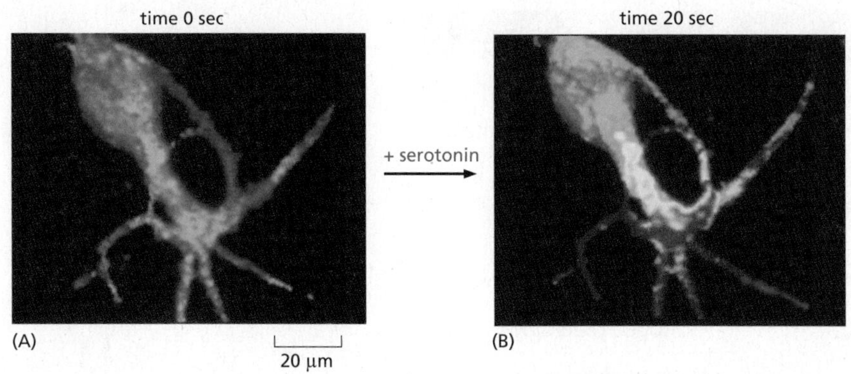

Figure 15–25 An increase in cyclic AMP in response to an extracellular signal. This *Aplysia* sensory nerve cell in culture is responding to the neurotransmitter serotonin, which acts through a GPCR to cause a rapid rise in the intracellular concentration of cAMP. To monitor the cAMP level, the cell has been loaded with a fluorescent protein that changes its fluorescence when it binds cAMP. *Blue* indicates a low level of cAMP, *yellow* an intermediate level, and *red* a high level. (A) In the resting cell, the cAMP level is about 5×10^{-8} M. (B) Twenty seconds after the addition of serotonin to the culture medium, the intracellular level of cAMP has increased to more than 10^{-6} M in the relevant parts of the cell, an increase of more than twentyfold. (From B.J. Bacskai et al., *Science* 260:222–226, 1993. With permission from AAAS.)

responses in all eukaryotes. There are about 25 RGS proteins encoded in the human genome, each of which interacts with a particular set of G proteins.

Some G Proteins Regulate the Production of Cyclic AMP

Cyclic AMP (cAMP) acts as a second messenger in some signaling pathways. An extracellular signal can increase cAMP concentration more than twentyfold in seconds (Figure 15–25). As explained earlier (see Figure 15–15), such a rapid response requires balancing a rapid synthesis of the molecule with its rapid breakdown or removal. Cyclic AMP is synthesized from ATP by an enzyme called **adenylyl cyclase**, and it is rapidly and continually destroyed by **cyclic AMP phosphodiesterases** (Figure 15–26). Adenylyl cyclase is a large, multipass transmembrane protein with its catalytic domain on the cytosolic side of the plasma membrane. There are at least eight isoforms in mammals, most of which are regulated by both G proteins and Ca^{2+}.

Many extracellular signals work by increasing cAMP concentrations inside the cell. These signals activate GPCRs that are coupled to a **stimulatory G protein (G_s)**. The activated α subunit of G_s binds and thereby activates adenylyl cyclase. Other extracellular signals, acting through different GPCRs, reduce cAMP levels by activating an **inhibitory G protein (G_i)**, which then inhibits adenylyl cyclase.

Both G_s and G_i are targets for medically important bacterial toxins. *Cholera toxin*, which is produced by the bacterium that causes cholera, is an enzyme that catalyzes the transfer of ADP ribose from intracellular NAD^+ to the α subunit of G_s. This ADP ribosylation alters the α subunit so that it can no longer hydrolyze its bound GTP, causing it to remain in an active state that stimulates adenylyl cyclase indefinitely. The resulting prolonged elevation in cAMP concentration within intestinal epithelial cells causes a large efflux of Cl^- and water into the gut, thereby causing the severe diarrhea that characterizes cholera. *Pertussis toxin*, which is made by the bacterium that causes pertussis (whooping cough), catalyzes the ADP ribosylation of the α subunit of G_i, preventing the protein from interacting with receptors; as a result, the G protein remains in the inactive GDP-bound state and is unable to regulate its target proteins. These two toxins are widely used in experiments to determine whether a cell's GPCR-dependent response to a signal is mediated by G_s or by G_i.

Some of the responses mediated by a G_s-stimulated increase in cAMP concentration are listed in Table 15–1. As the table shows, different cell types respond differently to an increase in cAMP concentration. Some cell types, such as fat cells, activate adenylyl cyclase in response to multiple hormones, all of which thereby

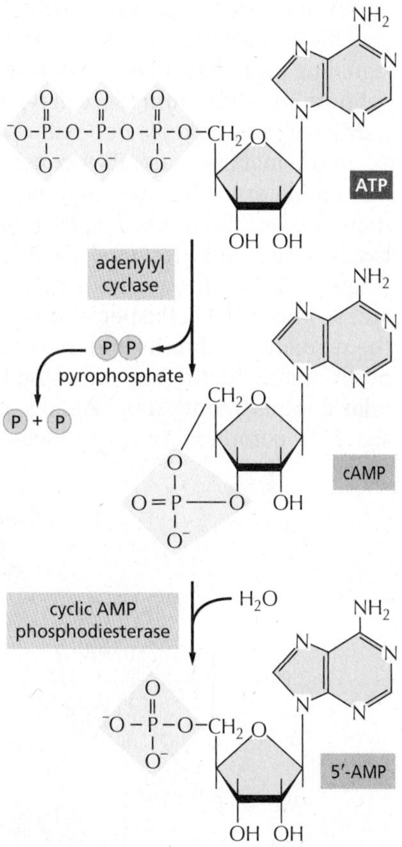

Figure 15–26 The synthesis and degradation of cyclic AMP. In a reaction catalyzed by the enzyme adenylyl cyclase, cyclic AMP (cAMP) is synthesized from ATP through a cyclization reaction that removes two phosphate groups as pyrophosphate (PP); a pyrophosphatase drives this synthesis by hydrolyzing the released pyrophosphate to phosphate. Cyclic AMP is short-lived (unstable) in the cell because it is hydrolyzed by specific phosphodiesterases to form 5′-AMP, as indicated.

TABLE 15–1 Some Hormone-induced Cell Responses Mediated by Cyclic AMP

Target tissue	Hormone	Major response
Thyroid gland	Thyroid-stimulating hormone (TSH)	Thyroid hormone synthesis and secretion
Adrenal cortex	Adrenocorticotrophic hormone (ACTH)	Cortisol secretion
Ovary	Luteinizing hormone (LH)	Progesterone secretion
Muscle	Epinephrine	Glycogen breakdown
Bone	Parathyroid hormone	Bone resorption
Heart	Epinephrine	Increase in heart rate and force of contraction
Liver	Glucagon	Glycogen breakdown
Kidney	Vasopressin	Water resorption
Fat	Epinephrine, ACTH, glucagon, TSH	Triglyceride breakdown

stimulate the breakdown of triglyceride (the storage form of fat) to fatty acids. Individuals with genetic defects in the G_S α subunit show decreased responses to certain hormones, resulting in metabolic abnormalities, abnormal bone development, and mental retardation.

Cyclic-AMP-dependent Protein Kinase (PKA) Mediates Most of the Effects of Cyclic AMP

In most animal cells, cAMP exerts its effects mainly by activating **cyclic-AMP-dependent protein kinase (protein kinase A; PKA)**. This kinase phosphorylates specific serines or threonines on selected target proteins, including intracellular signaling proteins and effector proteins, thereby regulating their activity. The target proteins differ from one cell type to another, which explains why the effects of cAMP vary so markedly depending on the cell type (see Table 15–1).

In the inactive state, PKA consists of a complex of two catalytic subunits and two regulatory subunits. The binding of cAMP to the regulatory subunits alters their conformation, causing them to dissociate from the complex. The released catalytic subunits are thereby activated to phosphorylate specific target proteins (**Figure 15–27**). The regulatory subunits of PKA are important for localizing the kinase inside the cell: special *A-kinase anchoring proteins* (*AKAPs*) bind both to the regulatory subunits and to a component of the cytoskeleton or a membrane of an organelle, thereby tethering the enzyme complex to a particular subcellular compartment. Some AKAPs also bind other signaling proteins, forming a signaling complex. An AKAP located around the nucleus of heart muscle cells,

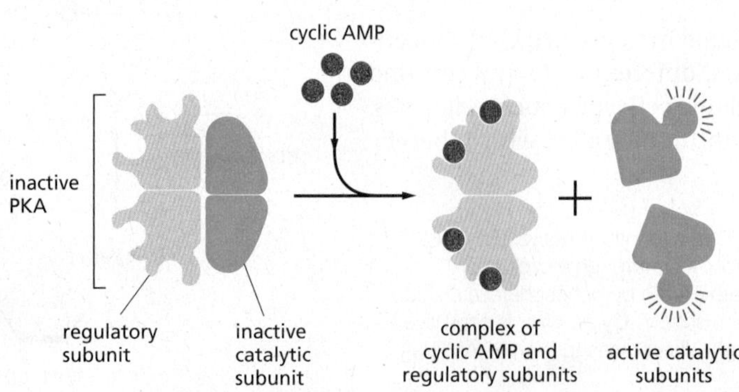

Figure 15–27 **The activation of cyclic-AMP-dependent protein kinase (PKA).** The binding of cAMP to the regulatory subunits of the PKA tetramer induces a conformational change, causing these subunits to dissociate from the catalytic subunits, thereby activating the kinase activity of the catalytic subunits. The release of the catalytic subunits requires the binding of more than two cAMP molecules to the regulatory subunits in the tetramer. This requirement greatly sharpens the response of the kinase to changes in cAMP concentration, as discussed earlier (see Figure 15–17).

for example, binds both PKA and a phosphodiesterase that hydrolyzes cAMP. In unstimulated cells, the phosphodiesterase keeps the local cAMP concentration low, so that the bound PKA is inactive; in stimulated cells, cAMP concentration rapidly rises, overwhelming the phosphodiesterase and activating the PKA. Among the target proteins that PKA phosphorylates and activates in these cells is the adjacent phosphodiesterase, which rapidly lowers the cAMP concentration again. This negative feedback arrangement converts what might otherwise be a prolonged PKA response into a brief, local pulse of PKA activity.

Whereas some responses mediated by cAMP occur within seconds (see Figure 15–25), others depend on changes in the transcription of specific genes and take hours to develop fully. In cells that secrete the peptide hormone *somatostatin*, for example, cAMP activates the gene that encodes this hormone. The regulatory region of the somatostatin gene contains a short *cis*-regulatory sequence, called the *cyclic AMP response element* (*CRE*), which is also found in the regulatory region of many other genes activated by cAMP. A specific transcription regulator called **CRE-binding (CREB) protein** recognizes this sequence. When PKA is activated by cAMP, it phosphorylates CREB on a single serine; phosphorylated CREB then recruits a transcription coactivator called *CREB-binding protein* (*CBP*), which stimulates the transcription of the target genes (**Figure 15–28**). Thus, CREB can transform a short cAMP signal into a long-term change in a cell, a process that, in the brain, is thought to play an important part in some forms of learning and memory.

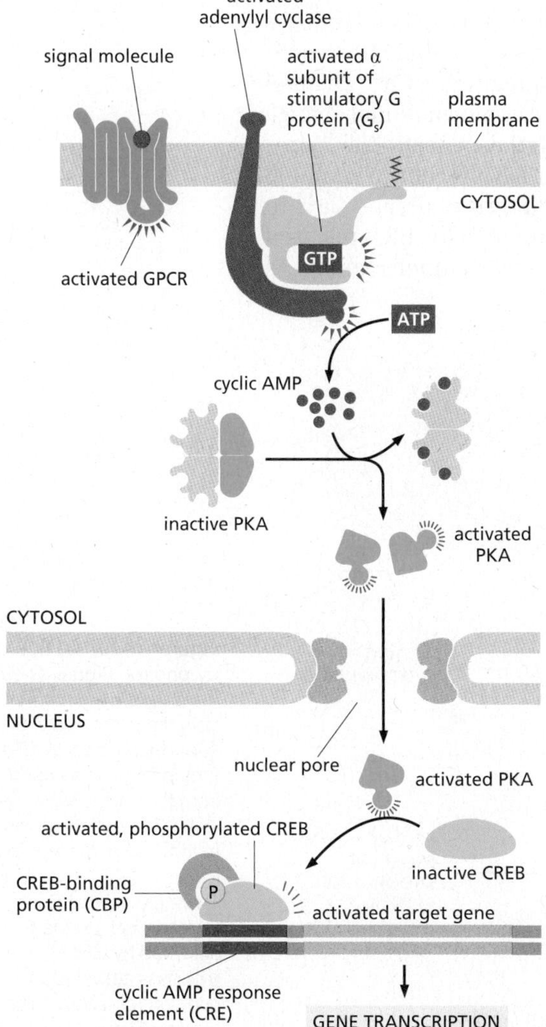

Figure 15–28 **How a rise in intracellular cyclic AMP concentration can alter gene transcription.** The binding of an extracellular signal molecule to its GPCR activates adenylyl cyclase via G_S and thereby increases cAMP concentration in the cytosol. This rise activates PKA, and the released catalytic subunits of PKA can then enter the nucleus, where they phosphorylate the transcription regulatory protein CREB. Once phosphorylated, CREB recruits the coactivator CBP, which stimulates gene transcription. In some cases, the inactive CREB protein is bound to the cyclic AMP response element (CRE) in DNA before it is phosphorylated (not shown). See **Movie 15.2**.

TABLE 15–2 Some Cell Responses in Which GPCRs Activate PLCβ

Target tissue	Signal molecule	Major response
Liver	Vasopressin	Glycogen breakdown
Pancreas	Acetylcholine	Amylase secretion
Smooth muscle	Acetylcholine	Muscle contraction
Blood platelets	Thrombin	Platelet aggregation

Some G Proteins Signal Via Phospholipids

Many GPCRs exert their effects through G proteins that activate the plasma-membrane-bound enzyme **phospholipase C-β (PLCβ)**. **Table 15–2** lists some examples of responses activated in this way. The phospholipase acts on a phosphorylated inositol phospholipid (a phosphoinositide) called **phosphatidylinositol 4,5-bisphosphate [PI(4,5)P$_2$]**, which is present in small amounts in the inner half of the plasma membrane lipid bilayer (**Figure 15–29**). Receptors that activate this **inositol phospholipid signaling pathway** do so primarily through a G protein called G_q, which activates phospholipase C-β in much the same way that G_s activates adenylyl cyclase. The activated phospholipase then cleaves the PI(4,5)P$_2$ to generate two products: **inositol 1,4,5-trisphosphate (IP$_3$)** and **diacylglycerol**. At this step, the signaling pathway splits into two branches.

IP$_3$ is a water-soluble molecule that leaves the plasma membrane and diffuses through the cytosol to the endoplasmic reticulum (ER), where it binds **IP$_3$ receptors** in the ER membrane. The IP$_3$ receptor is a large transmembrane Ca^{2+} channel that is closed in the absence of IP$_3$. IP$_3$ binding triggers a conformational change that exposes a high-affinity Ca^{2+}-binding site. Although the cytosolic Ca^{2+} concentration in the unstimulated cell is low (~10^{-7} M), it is sufficient to promote Ca^{2+} binding to some IP$_3$ receptors. The simultaneous binding of IP$_3$ and Ca^{2+} to an IP$_3$ receptor opens the receptor Ca^{2+} channel. Ca^{2+} stored in the ER is released and binds to other IP$_3$-bound receptors to cause widespread channel opening. As

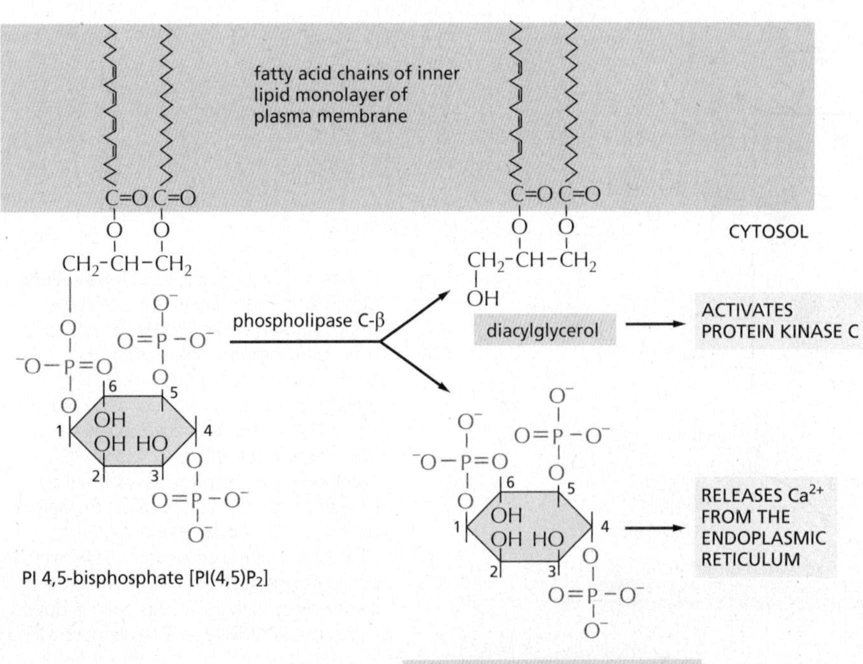

Figure 15–29 The hydrolysis of PI(4,5)P$_2$ by phospholipase C-β. Two second messengers are produced directly from the hydrolysis of PI(4,5)P$_2$: inositol 1,4,5-trisphosphate (IP$_3$), which diffuses through the cytosol and releases Ca^{2+} from the endoplasmic reticulum, and diacylglycerol, which remains in the membrane and helps to activate protein kinase C (PKC; see Figure 15–30). There are several classes of phospholipase C: these include the β class, which is activated by GPCRs; as we see later, the γ class is activated by a class of enzyme-coupled receptors called receptor tyrosine kinases (RTKs).

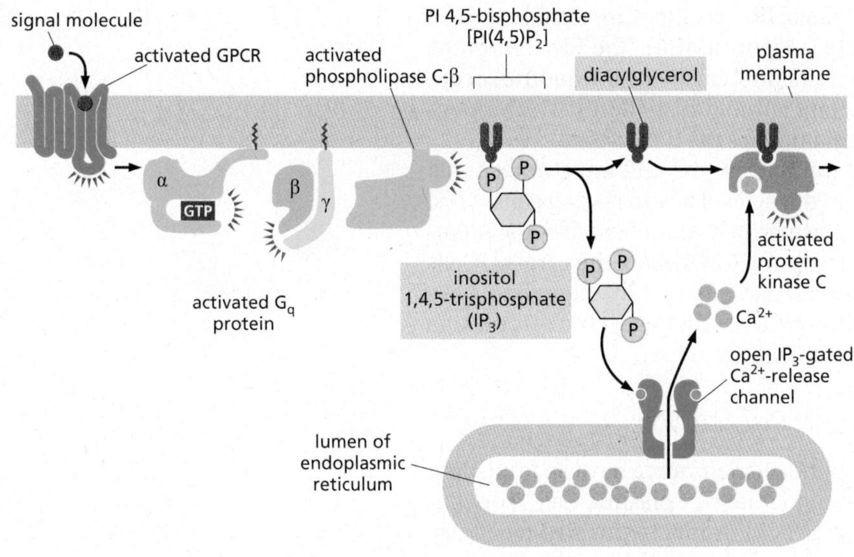

Figure 15–30 How GPCRs increase cytosolic Ca^{2+} and activate protein kinase C. The activated GPCR stimulates the plasma-membrane-bound phospholipase C-β (PLCβ) via a G protein called G_q. The α subunit and βγ complex of G_q are both involved in this activation. Two second messengers are produced when $PI(4,5)P_2$ is hydrolyzed by activated PLC. Inositol 1,4,5-trisphosphate (IP_3) diffuses through the cytosol and releases Ca^{2+} from the ER by binding to and opening IP_3-gated Ca^{2+}-release channels (IP_3 receptors) in the ER membrane (opening of these channels also requires binding of Ca^{2+}, not shown). The large electrochemical gradient for Ca^{2+} across this membrane causes Ca^{2+} to escape into the cytosol when the release channels are opened. Diacylglycerol remains in the plasma membrane and, together with phosphatidylserine (not shown) and Ca^{2+}, helps to activate protein kinase C (PKC), which is recruited from the cytosol to the cytosolic face of the plasma membrane. Of the 10 or more distinct isoforms of PKC in humans, at least 4 are activated by diacylglycerol (**Movie 15.3**).

a result, the concentration of cytosolic Ca^{2+} rises 10- to 20-fold (**Figure 15–30**). The increase in cytosolic Ca^{2+} propagates the signal by influencing the activity of Ca^{2+}-sensitive intracellular proteins, as we describe shortly.

At the same time that the IP_3 produced by the hydrolysis of $PI(4,5)P_2$ is increasing the concentration of Ca^{2+} in the cytosol, the other cleavage product of the $PI(4,5)P_2$, diacylglycerol, is exerting different effects. It also acts as a second messenger, but it remains embedded in the plasma membrane, where it has several potential signaling roles. One of its major functions is to activate a protein kinase called **protein kinase C (PKC)**, so named because it is Ca^{2+}-dependent. The initial rise in cytosolic Ca^{2+} induced by IP_3 alters the PKC so that it translocates from the cytosol to the cytoplasmic face of the plasma membrane. There it is activated by the combination of Ca^{2+}, diacylglycerol, and the negatively charged membrane phospholipid phosphatidylserine (see Figure 15–30). Once activated, PKC phosphorylates target proteins that vary depending on the cell type. The principles are the same as discussed earlier for PKA, although most of the target proteins are different.

Diacylglycerol can be further cleaved to release arachidonic acid, which can either act as a signal in its own right or be used in the synthesis of other small lipid signal molecules called *eicosanoids*. Most vertebrate cell types make eicosanoids, including *prostaglandins*, which have many biological activities. They participate in pain and inflammatory responses, for example, and many anti-inflammatory drugs (such as aspirin, ibuprofen, and cortisone) act in part by inhibiting their synthesis.

Ca^{2+} Functions as a Ubiquitous Intracellular Mediator

Many extracellular signals, and not just those that work via G proteins, trigger an increase in cytosolic Ca^{2+} concentration. In muscle cells, Ca^{2+} triggers contraction, and in many secretory cells, including nerve cells, it triggers secretion. Ca^{2+} has numerous other functions in a variety of cell types. Ca^{2+} is such an effective signaling mediator because its concentration in the cytosol is normally very low (~10^{-7} M), whereas its concentration in the extracellular fluid (~10^{-3} M) and in the lumen of the ER [and sarcoplasmic reticulum (SR) in muscle] is high. Thus, there is a large gradient tending to drive Ca^{2+} into the cytosol across both the plasma membrane and the ER or SR membrane. When a signal transiently opens Ca^{2+} channels in these membranes, Ca^{2+} rushes into the cytosol, and the resulting increase in the local Ca^{2+} concentration activates Ca^{2+}-responsive proteins in the cell.

Some stimuli, including membrane depolarization, membrane stretch, and certain extracellular signals, activate Ca^{2+} channels in the plasma membrane, resulting in Ca^{2+} influx from outside the cell. Other signals, including the GPCR-mediated signals described earlier, act primarily through IP_3 receptors to stimulate Ca^{2+} release from intracellular stores in the ER (see Figure 15–30). The ER membrane also contains a second type of regulated Ca^{2+} channel called the **ryanodine receptor** (so called because it is sensitive to the plant alkaloid ryanodine), which opens in response to rising Ca^{2+} levels and thereby amplifies the Ca^{2+} signal.

Several mechanisms rapidly terminate the Ca^{2+} signal and are also responsible for keeping the concentration of Ca^{2+} in the cytosol low in resting cells. Most important, there are Ca^{2+}-pumps in the plasma membrane and the ER membrane that use the energy of ATP hydrolysis to pump Ca^{2+} out of the cytosol. Cells such as muscle and nerve cells, which make extensive use of Ca^{2+} signaling, have an additional Ca^{2+} transporter (an Na^+-driven Ca^{2+} exchanger) in their plasma membrane that couples the efflux of Ca^{2+} to the influx of Na^+.

Feedback Generates Ca^{2+} Waves and Oscillations

The IP_3 receptors and ryanodine receptors of the ER membrane have an important feature: they are both stimulated by low to moderate cytoplasmic Ca^{2+} concentrations. This *Ca^{2+}-induced calcium release* (*CICR*) results in positive feedback, which has a major impact on the properties of the Ca^{2+} signal. The importance of this feedback is seen clearly in studies with Ca^{2+}-sensitive fluorescent indicators, such as *aequorin* or *fura-2*, which allow researchers to monitor cytosolic Ca^{2+} in individual cells under a microscope (**Figure 15–31** and **Movie 15.4**).

When cells carrying a Ca^{2+} indicator are treated with a small amount of an extracellular signal molecule that stimulates a small increase in the concentration of cytosolic IP_3, tiny bursts of Ca^{2+} are seen in one or more discrete regions of the cell. These Ca^{2+} puffs or sparks reflect the local opening of small numbers of IP_3 receptors in the ER membrane that have bound both IP_3 and Ca^{2+}, the concentrations of which are too low to bind to all of these receptors. Because various Ca^{2+}-binding proteins and Ca^{2+}-pumps restrict the diffusion of Ca^{2+}, the Ca^{2+} signal often remains localized to the site where the Ca^{2+} entered the cytosol. If the extracellular signal is stronger, however, IP_3 rises to a higher concentration and binds many of its receptors, although the low Ca^{2+} concentration still limits the activation of these receptors to some extent. Nevertheless, a local burst of Ca^{2+} release can now spread more easily to neighboring IP_3-bound receptors and activate them, resulting in a regenerative wave of Ca^{2+} release that moves through the cytosol (**Figure 15–32**), much like the spreading of an action potential along the membrane of an axon (see Figure 11–33). The presence of Ca^{2+}-stimulated ryanodine receptors in the ER membrane further enhances the positive feedback.

In addition to being regulated by positive feedback, IP_3 receptors and ryanodine receptors are also regulated by negative feedback, in that they are inhibited by high Ca^{2+} concentrations. Thus, the rise in Ca^{2+} in a stimulated cell leads eventually to inhibition of Ca^{2+} release. Because Ca^{2+}-pumps remove the Ca^{2+}, the Ca^{2+} concentration in the cytosol falls (see Figure 15–32). The decline in Ca^{2+} eventually relieves the negative feedback, allowing cytosolic Ca^{2+} to rise again. As in other cases of delayed negative feedback (see Figure 15–19), this sequence of events leads to oscillations in the Ca^{2+} concentration, which persist as long as cell-surface receptors are activated. The frequency of the oscillations reflects the

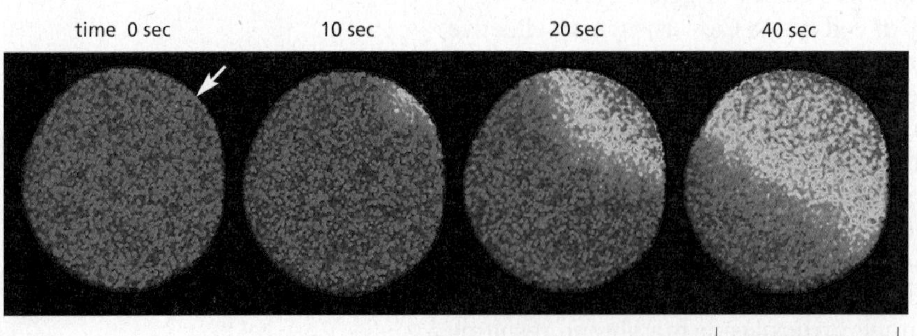

Figure 15–31 The fertilization of an egg by a sperm triggers a wave of cytosolic Ca^{2+}. This sea star egg was injected with a Ca^{2+}-sensitive fluorescent dye before it was fertilized. A wave of cytosolic Ca^{2+} *(red and yellow)*, caused by Ca^{2+} release from the ER, sweeps across the egg from the site of sperm entry *(arrow)*. This Ca^{2+} wave changes the egg cell surface, preventing the entry of other sperm, and it also initiates embryonic development (**Movie 15.5**). The initial increase in Ca^{2+} is thought to be caused by a sperm-specific form of PLC (PLCζ) that the sperm brings into the egg cytoplasm when it fuses with the egg; the PLCζ cleaves $PI(4,5)P_2$ to produce IP_3, which releases Ca^{2+} from the egg ER. The released Ca^{2+} stimulates further Ca^{2+} release from the ER, producing the spreading wave, as we explain in Figure 15–32. (Courtesy of Stephen Stricker.)

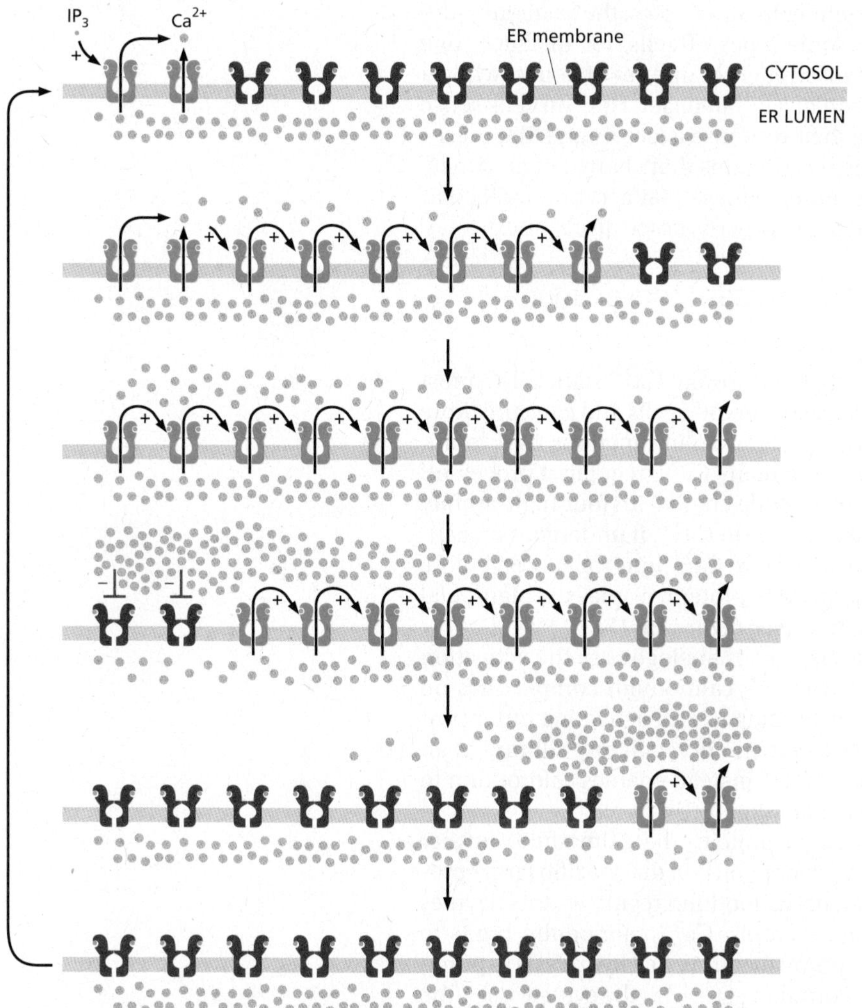

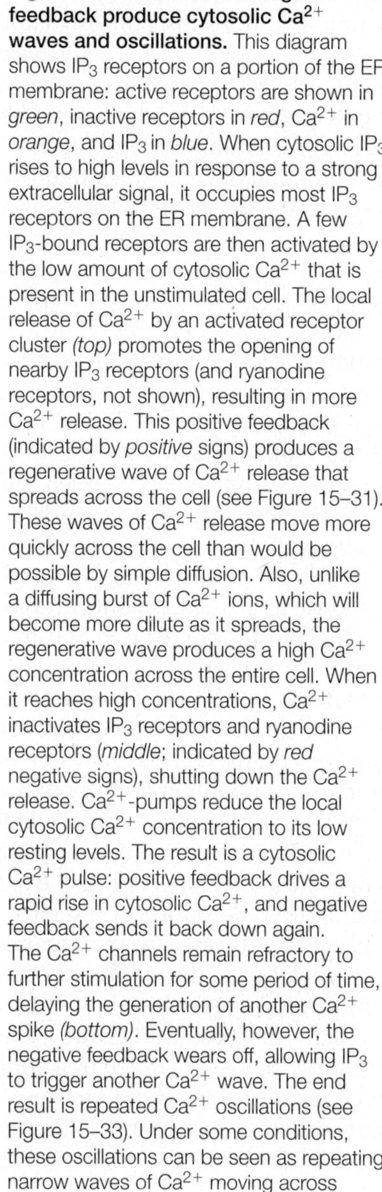
Figure 15–32 Positive and negative feedback produce cytosolic Ca^{2+} waves and oscillations. This diagram shows IP_3 receptors on a portion of the ER membrane: active receptors are shown in *green*, inactive receptors in *red*, Ca^{2+} in *orange*, and IP_3 in *blue*. When cytosolic IP_3 rises to high levels in response to a strong extracellular signal, it occupies most IP_3 receptors on the ER membrane. A few IP_3-bound receptors are then activated by the low amount of cytosolic Ca^{2+} that is present in the unstimulated cell. The local release of Ca^{2+} by an activated receptor cluster *(top)* promotes the opening of nearby IP_3 receptors (and ryanodine receptors, not shown), resulting in more Ca^{2+} release. This positive feedback (indicated by *positive* signs) produces a regenerative wave of Ca^{2+} release that spreads across the cell (see Figure 15–31). These waves of Ca^{2+} release move more quickly across the cell than would be possible by simple diffusion. Also, unlike a diffusing burst of Ca^{2+} ions, which will become more dilute as it spreads, the regenerative wave produces a high Ca^{2+} concentration across the entire cell. When it reaches high concentrations, Ca^{2+} inactivates IP_3 receptors and ryanodine receptors (*middle*; indicated by *red* negative signs), shutting down the Ca^{2+} release. Ca^{2+}-pumps reduce the local cytosolic Ca^{2+} concentration to its low resting levels. The result is a cytosolic Ca^{2+} pulse: positive feedback drives a rapid rise in cytosolic Ca^{2+}, and negative feedback sends it back down again. The Ca^{2+} channels remain refractory to further stimulation for some period of time, delaying the generation of another Ca^{2+} spike *(bottom)*. Eventually, however, the negative feedback wears off, allowing IP_3 to trigger another Ca^{2+} wave. The end result is repeated Ca^{2+} oscillations (see Figure 15–33). Under some conditions, these oscillations can be seen as repeating narrow waves of Ca^{2+} moving across the cell.

strength of the extracellular stimulus (**Figure 15–33**). The frequency and amplitude of oscillations can also be modulated by other signaling mechanisms, such as phosphorylation, which influence the Ca^{2+} sensitivity of Ca^{2+} channels or affect other components in the signaling system.

The frequency of Ca^{2+} oscillations can be translated into a frequency-dependent cell response. In some cases, the frequency-dependent response itself is also oscillatory: in hormone-secreting pituitary cells, for example, stimulation by an extracellular signal induces repeated Ca^{2+} spikes, each of which

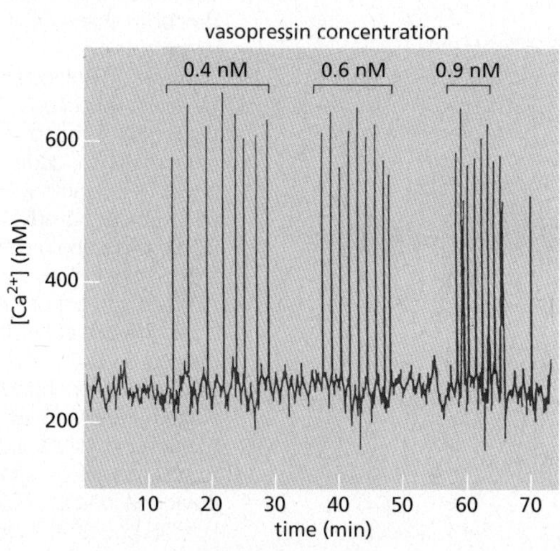

Figure 15–33 Vasopressin-induced cytosolic Ca^{2+} oscillations in a liver cell. The cell was loaded with the Ca^{2+}-sensitive protein aequorin and then exposed to increasing concentrations of the peptide signal molecule *vasopressin*, which activates a GPCR and thereby PLCβ (see Table 15–2). Note that the frequency of the Ca^{2+} spikes increases with an increasing concentration of vasopressin but that the amplitude of the spikes is not affected. Each spike lasts about 7 seconds. (Adapted from N.M. Woods et al., *Nature* 319:600–602, published 1986 by Nature Publishing Group. Reproduced with permission of SNCSC.)

is associated with a burst of hormone secretion. In other cases, the frequency-dependent response is non-oscillatory: in some types of cells, for instance, one frequency of Ca^{2+} spikes activates the transcription of one set of genes, while a higher frequency activates the transcription of a different set. How do cells sense the frequency of Ca^{2+} spikes and change their response accordingly? The mechanism presumably depends on Ca^{2+}-sensitive proteins that change their activity as a function of Ca^{2+}-spike frequency. A protein kinase that acts as a molecular memory device seems to have this remarkable property, as we discuss next.

Ca^{2+}/Calmodulin-dependent Protein Kinases Mediate Many Responses to Ca^{2+} Signals

Various Ca^{2+}-binding proteins help to relay the cytosolic Ca^{2+} signal. The most important is **calmodulin**, which is found in all eukaryotic cells and can constitute as much as 1% of a cell's total protein mass. Calmodulin functions as a multipurpose intracellular Ca^{2+} receptor, governing many Ca^{2+}-regulated processes. It consists of a highly conserved, single polypeptide chain with four high-affinity Ca^{2+}-binding sites (**Figure 15–34A**). When it binds to Ca^{2+}, it undergoes an activating conformational change. Because two or more Ca^{2+} ions must bind before calmodulin adopts its active conformation, the protein displays a sigmoidal response to increasing concentrations of Ca^{2+} (see Figure 15–17).

The allosteric activation of calmodulin by Ca^{2+} is analogous to the activation of PKA by cyclic AMP, except that the active Ca^{2+}/calmodulin complex has no enzymatic activity itself but instead acts by binding to and activating other proteins. In some cases, calmodulin serves as a permanent regulatory subunit of an enzyme complex, but usually the binding of Ca^{2+} instead enables calmodulin to bind to various target proteins in the cell to alter their activity.

When Ca^{2+}/calmodulin binds to its target protein, the calmodulin further changes its conformation, the nature of which depends on the specific target protein (**Figure 15–34B**). Among the many targets calmodulin regulates are enzymes and membrane transport proteins. As one example, Ca^{2+}/calmodulin binds to and activates the plasma membrane Ca^{2+}-pump that uses ATP hydrolysis to pump Ca^{2+} out of cells. Thus, whenever the concentration of Ca^{2+} in the cytosol rises, the pump is activated, which helps to return the cytosolic Ca^{2+} level to resting levels.

Many effects of Ca^{2+}, however, are more indirect and are mediated by protein phosphorylations catalyzed by a family of protein kinases called **Ca^{2+}/calmodulin-dependent kinases (CaM-kinases)**. Some CaM-kinases phosphorylate transcription regulators, such as the CREB protein (see Figure 15–28), and in this way activate or inhibit the transcription of specific genes.

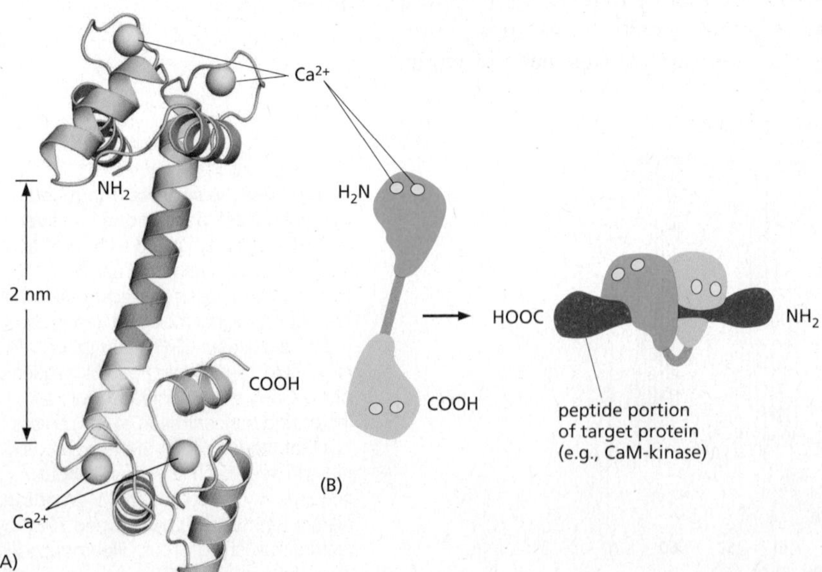

Figure 15–34 The structure of Ca^{2+}/calmodulin. (A) The molecule has a dumbbell shape, with two globular ends, which can bind to many different target proteins. The globular ends are connected by a long, exposed α helix, which allows the protein to adopt a number of different conformations, depending on the target protein it interacts with. Each globular head has two Ca^{2+}-binding sites (**Movie 15.6**). (B) Shown is the major structural change that occurs in Ca^{2+}/calmodulin when it binds to a target protein (in this example, a peptide that consists of the Ca^{2+}/calmodulin-binding domain of a Ca^{2+}/calmodulin-dependent protein kinase). Note that the Ca^{2+}/calmodulin has folded to surround the peptide. When it binds to other targets, it can adopt different conformations. (A, PDB code: 1CLL; B, PDB codes: 1CDL and 2BBM.)

One of the best-studied CaM-kinases is **CaM-kinase II**, which is found in most animal cells but is especially enriched in the nervous system. It constitutes up to 2% of the total protein mass in some regions of the brain, and it is highly concentrated in synapses. CaM-kinase II has several remarkable properties. To begin with, it has a spectacular quaternary structure: twelve copies of the enzyme are assembled into a stacked pair of rings, with kinase domains on the outside linked to a central hub (Figure 15–35). This structure helps the enzyme function as a

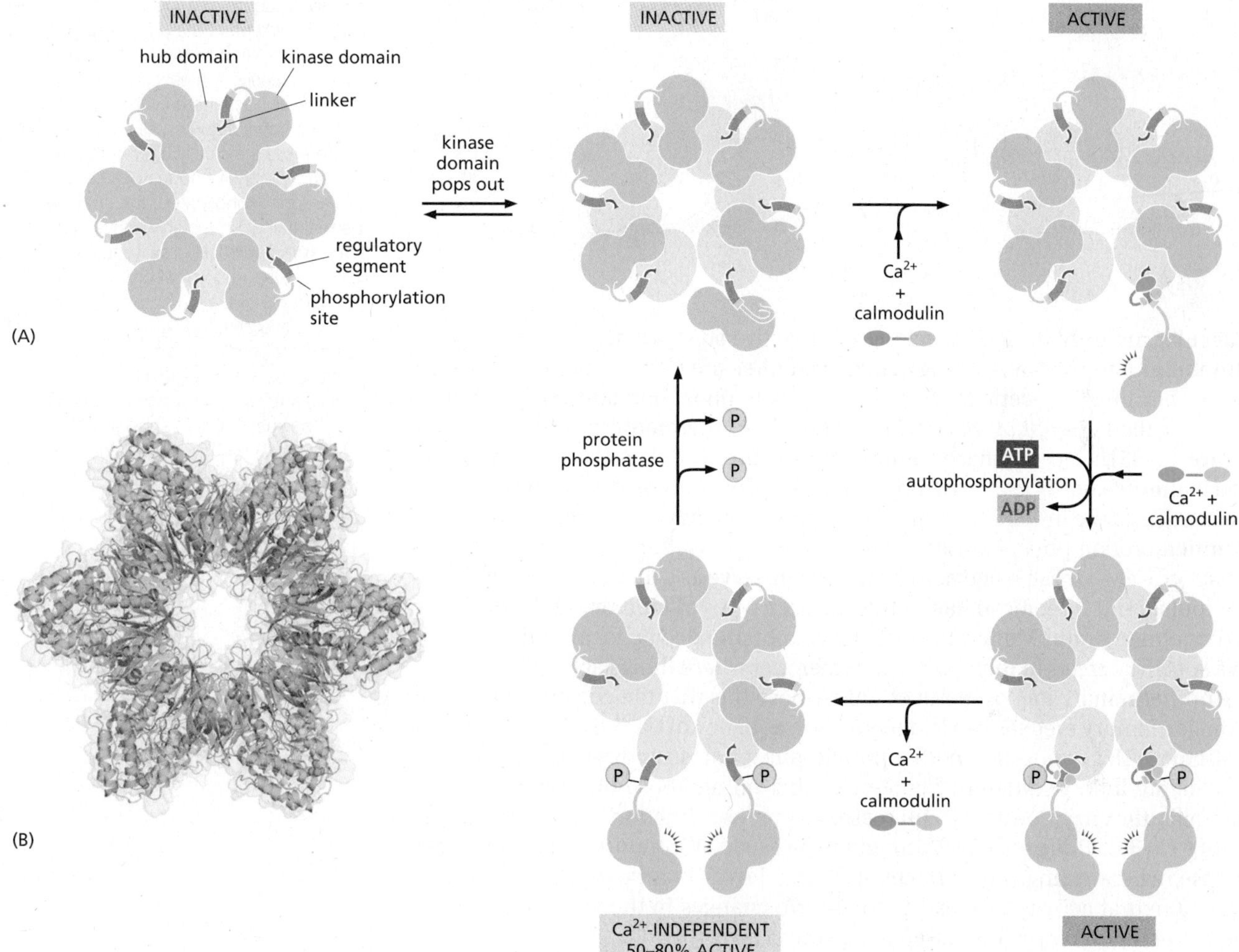

Figure 15–35 **The stepwise activation of CaM-kinase II.** (A) Each CaM-kinase II protein has two major domains: an amino-terminal kinase domain *(green)* and a carboxyl-terminal hub domain *(blue)*, linked by a regulatory segment. Six CaM-kinase II proteins are assembled into a giant ring in which the hub domains interact tightly to produce a central structure that is surrounded by kinase domains. The complete enzyme contains two stacked rings, for a total of 12 kinase proteins, but only one ring is shown here for clarity. When the enzyme is inactive, the ring exists in a dynamic equilibrium between two states. The first *(upper left)* is a compact state, in which the kinase domains interact with the hub domains, so that the regulatory segments are buried in the kinase active sites and thereby block catalytic activity. In the second inactive state *(upper middle)*, a kinase domain has popped out and is linked to its hub domain by its regulatory segment, which continues to inhibit the kinase domain but is now accessible to Ca^{2+}/calmodulin. If present, Ca^{2+}/calmodulin will bind the regulatory segment and prevent it from inhibiting the kinase, thereby locking the kinase in an active state *(upper right)*. If the adjacent kinase domain also pops out from the hub, it will also be activated by Ca^{2+}/calmodulin, and the two kinase domains will then phosphorylate each other on their regulatory segments *(lower right)*. This autophosphorylation further activates the enzyme. It also prolongs the activity of the enzyme in two ways. First, it traps the bound Ca^{2+}/calmodulin so that it does not dissociate from the enzyme until cytosolic Ca^{2+} levels return to basal values for at least 10 seconds (not shown). Second, it converts the enzyme to a Ca^{2+}-independent form, so that the kinase remains active even after the Ca^{2+}/calmodulin dissociates from it *(lower left)*. This activity continues until the action of a protein phosphatase overrides the autophosphorylation activity of CaM-kinase II. (B) This model of the enzyme is based on x-ray crystallography analysis of the CaM-kinase II dodecamer.

The remarkable dodecameric structure of the enzyme allows it to achieve a broad range of intermediate activity states in response to different Ca^{2+} oscillation frequencies: higher frequencies tend to cause more subunits in the enzyme to reach the phosphorylated active state (see Figure 15–36). The behavior of CaM-kinase II is also controlled by the length of the linker segment between the kinase and hub domains. The linker is longer in some isoforms of the enzyme; in these isoforms, the kinase domains tend to pop out of the ring more frequently, making it more sensitive to Ca^{2+}. These and other mechanisms allow the cell to tailor the responsiveness of the enzyme to the needs of different types of neurons. (A, adapted from L.H. Chao et al., *Cell* 146:732–745, 2011; B, PDB code: 3SOA.)

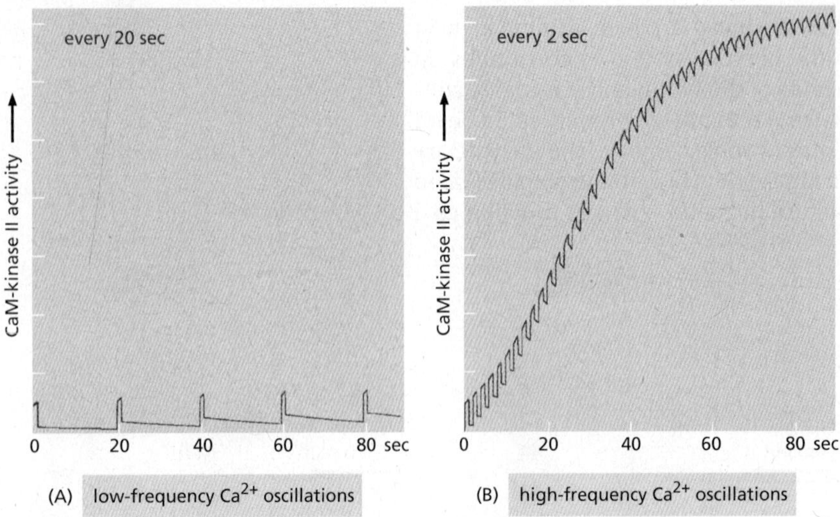

Figure 15–36 CaM-kinase II as a frequency decoder of Ca^{2+} oscillations. (A) At low frequencies of Ca^{2+} spikes, the enzyme becomes inactive after each spike, as the autophosphorylation induced by Ca^{2+}/calmodulin binding does not maintain the enzyme's activity long enough for the enzyme to remain active until the next Ca^{2+} spike arrives. (B) At higher spike frequencies, however, the enzyme fails to inactivate completely between Ca^{2+} spikes, so its activity ratchets up with each spike. If the spike frequency is high enough, this progressive increase in enzyme activity will continue until the enzyme is autophosphorylated on all subunits and is therefore maximally activated. Although not shown, once enough of its subunits are autophosphorylated, the enzyme can be maintained in a highly active state even with a relatively low frequency of Ca^{2+} spikes—a form of cell memory. The binding of Ca^{2+}/calmodulin to the enzyme is enhanced by the CaM-kinase II autophosphorylation (an additional form of positive feedback), helping to generate a more switchlike response to repeated Ca^{2+} spikes. (From P.I. Hanson et al., *Neuron* 12:943–956, 1994. With permission from Elsevier.)

molecular memory device, switching to an active state when exposed to Ca^{2+}/calmodulin and then remaining active even after the Ca^{2+} signal has decayed. This is because adjacent kinase subunits can phosphorylate each other (a process called *autophosphorylation*) when Ca^{2+}/calmodulin activates them (Figure 15–35). Once a kinase subunit is autophosphorylated, it remains active even in the absence of Ca^{2+}, thereby prolonging the duration of the kinase activity beyond that of the initial activating Ca^{2+} signal. The enzyme maintains this activity until a protein phosphatase removes the autophosphorylation and shuts the kinase off. CaM-kinase II activation can thereby serve as a memory trace of a prior Ca^{2+} pulse, and it seems to have a role in some types of memory and learning in the vertebrate brain. Mutant mice that lack a brain-specific form of the enzyme have specific defects in their ability to remember where things are.

Another remarkable property of CaM-kinase II is that the enzyme can use its intrinsic memory mechanism to decode the frequency of Ca^{2+} oscillations. When CaM-kinase II is exposed to both a protein phosphatase and repetitive pulses of Ca^{2+}/calmodulin at different frequencies that mimic those observed in stimulated cells, the enzyme's activity increases steeply as a function of pulse frequency (**Figure 15–36**). This property is thought to be especially important at a nerve cell synapse, where changes in intracellular Ca^{2+} levels in a postsynaptic cell as a result of neural activity can lead to long-term changes in the subsequent effectiveness of that synapse (discussed in Chapter 11).

Some G Proteins Directly Regulate Ion Channels

G proteins do not act exclusively by regulating the activity of membrane-bound enzymes that alter the concentration of cyclic AMP or Ca^{2+} in the cytosol. The α subunit of one type of G protein (called G_{12}), for example, activates a guanine nucleotide exchange factor that activates a monomeric GTPase of the *Rho family* (discussed later and in Chapter 16), which regulates the actin cytoskeleton.

In some other cases, G proteins directly activate or inactivate ion channels in the plasma membrane of the target cell, thereby altering the membrane's ion permeability, and hence its electrical excitability. As an example, acetylcholine released by the vagus nerve reduces the heart rate (see Figure 15–5B). This effect is mediated by a special class of acetylcholine receptors that activate the G_i protein discussed earlier. Once activated, the α subunit of G_i inhibits adenylyl cyclase (as described previously), while the $\beta\gamma$ subunits bind to K^+ channels in the plasma membrane of the heart muscle cell and open them. The opening of these K^+ channels makes it harder to depolarize the cell and thereby contributes to the inhibitory effect of acetylcholine on the heart. (These acetylcholine receptors, which can be activated by the fungal alkaloid muscarine, are called

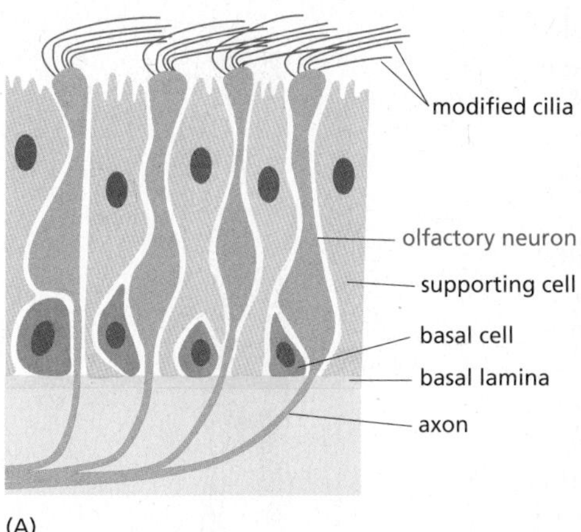

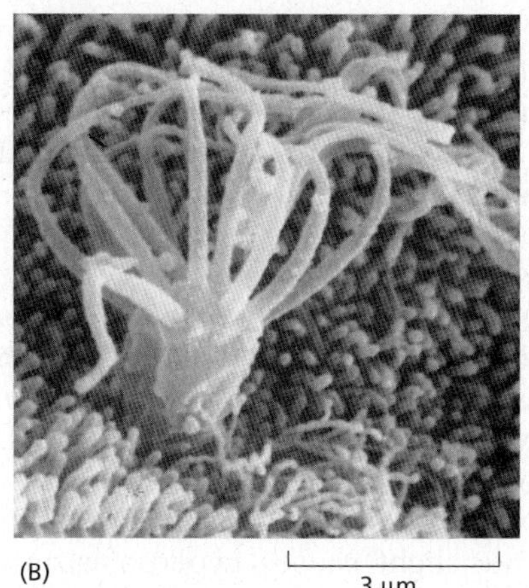

Figure 15–37 **Olfactory receptor neurons.** (A) A simplified drawing of a section of olfactory epithelium in the nose. Olfactory receptor neurons possess modified cilia, which project from the surface of the epithelium and contain the olfactory receptors, as well as the signal transduction machinery. The axon, which extends from the opposite end of the receptor neuron, conveys electrical signals to the brain when an odorant activates the cell to produce an action potential. In rodents, at least, the basal cells act as stem cells, producing new receptor neurons throughout life, to replace the neurons that die. (B) A scanning electron micrograph of the cilia on the surface of a human olfactory neuron. (B, from E.E. Morrison and R.M. Costanzo, *J. Comp. Neurol.* 297:1–13, 1990. With permission from Wiley-Liss.)

muscarinic acetylcholine receptors to distinguish them from the very different *nicotinic acetylcholine receptors*, which are ion-channel-coupled receptors on skeletal muscle and nerve cells that can be activated by the binding of nicotine, as well as by acetylcholine.)

Other G proteins regulate the activity of ion channels less directly, either by stimulating channel phosphorylation (by PKA, PKC, or CaM-kinase, for example) or by causing the production or destruction of cyclic nucleotides that directly activate or inactivate ion channels. These *cyclic-nucleotide-gated ion channels* have a crucial role in both smell (olfaction) and vision, as we now discuss.

Smell and Vision Depend on GPCRs That Regulate Ion Channels

Humans can distinguish more than 10,000 distinct smells, which they detect using specialized olfactory receptor neurons in the lining of the nose. These cells use specific GPCRs called **olfactory receptors** to recognize odors; the receptors are displayed on the surface of the modified cilia that extend from each olfactory neuron (Figure 15–37). The receptors act by increasing cAMP; when stimulated by odorant binding, they activate an olfactory-specific G protein (known as G_{olf}), which in turn activates adenylyl cyclase. The resulting increase in cAMP opens *cyclic-AMP-gated cation channels*, thereby allowing an influx of Na^+, which depolarizes the plasma membrane of the olfactory receptor neuron and initiates a nerve impulse that travels along its axon to the brain.

There are about 1000 different olfactory receptors in a mouse and about 350 in a human, each encoded by a different gene and each recognizing a different set of odorants. Each olfactory receptor neuron produces only one of these receptors; the neuron responds to a specific set of odorants by means of the specific receptor it displays, and each odorant activates its own characteristic set of olfactory receptor neurons. The same receptor also helps direct the elongating axon of each developing olfactory neuron to the specific target neurons that it will connect to in the brain. A different set of GPCRs acts in a similar way in some vertebrates to mediate responses to *pheromones*, chemical signals detected in a different part of the nose that are used in communication between members of the same species. Humans, however, are thought to lack functional pheromone receptors.

Vertebrate vision employs a similarly elaborate, highly sensitive, signal-detection mechanism that uses cyclic-nucleotide-gated cation channels, but the crucial cyclic nucleotide is **cyclic GMP** (Figure 15–38) rather than cAMP. As with cAMP, continual rapid synthesis (by *guanylyl cyclase*) and rapid degradation (by *cyclic GMP phosphodiesterase*) control the concentration of cyclic GMP. However, the light-activated GPCR in this system does not stimulate guanylyl cyclase and raise cyclic GMP levels; instead, it stimulates cyclic GMP phosphodiesterase, resulting in decreased cyclic GMP levels and thus decreased cation channel opening.

The visual signaling system has been especially well studied in **rod photoreceptors (rods)** in the vertebrate retina. Rods are responsible for noncolor vision

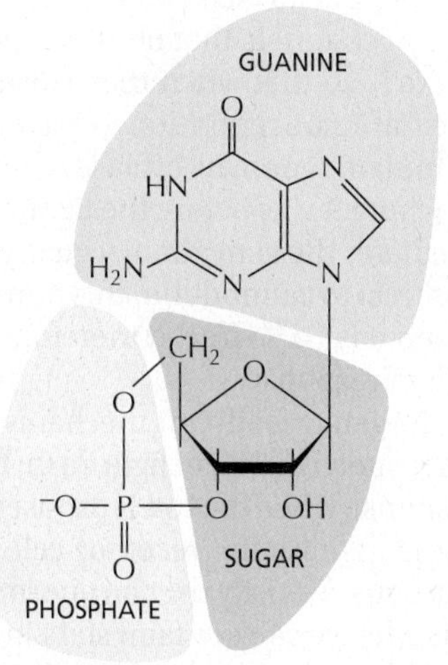

Figure 15-38 **Cyclic GMP.**

in dim light, whereas **cone photoreceptors (cones)** are responsible for color vision in bright light. Both types of photoreceptors are highly specialized cells with outer and inner segments, a cell body, and a synaptic region where the photoreceptor passes a chemical signal to a retinal neuron (**Figure 15–39**). After a network of retinal neurons processes the signals, the axons of a subset of the neurons transmit the signals to the brain.

The phototransduction apparatus is in the outer segment of the rod, which contains a stack of *discs*, each formed by a closed sac of membrane that is densely packed with a photosensitive GPCR called **rhodopsin**. The plasma membrane surrounding the outer segment contains *cyclic-GMP-gated cation channels.* Cyclic GMP bound to these channels keeps them open in the dark. Light-induced activation of rhodopsin molecules in the disc membrane decreases the cytosolic cyclic GMP concentration and closes the cation channels in the plasma membrane (**Figure 15–40**). Thus, light causes hyperpolarization (a more negative membrane potential—discussed in Chapter 11), which inhibits synaptic signaling.

Rhodopsin is a member of the GPCR family, but the extracellular signal that activates it is not a molecule but a photon of light. Each rhodopsin molecule contains a covalently attached chromophore, 11-*cis* retinal, which isomerizes almost instantaneously to all-*trans* retinal when it absorbs a single photon. The isomerization alters the shape of the retinal, forcing a conformational change in the rhodopsin protein. The activated rhodopsin molecule then alters the conformation of the G protein *transducin* (G_t), causing the transducin α subunit to activate **cyclic GMP phosphodiesterase**. The phosphodiesterase then hydrolyzes cytosolic cyclic GMP, causing its concentration to fall. As a result, the amount of cyclic GMP bound to the plasma membrane cation channels declines, allowing more of these channels to close. In this way, the signal passes quickly from the disc membrane to the plasma membrane, and a light signal is converted into an electrical one, through a hyperpolarization of the plasma membrane.

Rods use several negative feedback loops to allow the cells to revert quickly to a resting, dark state in the aftermath of a flash of light—a requirement for perceiving the shortness of the flash. A rhodopsin-specific protein kinase called *rhodopsin kinase* (*RK*) phosphorylates the cytosolic tail of activated rhodopsin on multiple serines, partially inhibiting the ability of the rhodopsin to activate transducin. An inhibitory protein called *arrestin* (discussed later) then binds to the phosphorylated rhodopsin, further inhibiting rhodopsin's activity. Mice or humans with a mutation that inactivates the gene encoding RK have a prolonged light response.

At the same time as arrestin shuts off rhodopsin, an RGS protein (discussed earlier) binds to activated transducin, stimulating the transducin to hydrolyze its bound GTP to GDP, which returns transducin to its inactive state. In addition, the cation channels that close in response to light are permeable to Ca^{2+}, as well as to Na^+, so that when they close, the normal influx of Ca^{2+} is inhibited, causing the Ca^{2+} concentration in the cytosol to fall. The decrease in Ca^{2+} concentration stimulates guanylyl cyclase to replenish the cyclic GMP, rapidly returning its level to where it was before the light was switched on. A specific Ca^{2+}-sensitive protein mediates the activation of guanylyl cyclase in response to the fall in Ca^{2+} levels. In contrast to calmodulin, this protein is inactive when Ca^{2+} is bound to it and active when it is Ca^{2+}-free. It therefore stimulates the cyclase when Ca^{2+} levels fall after a light response.

Negative feedback mechanisms do more than just return the rod to its resting state after a transient light flash; they also help the rod *adapt*, by stepping down the response when the rod is exposed to light continually. Adaptation, as we discussed earlier, allows the receptor cell to function as a sensitive detector of *changes* in stimulus intensity over an enormously wide range of baseline levels of stimulation. It is why we can see faint stars in a dark sky or a camera flash in bright sunlight.

The various heterotrimeric G proteins we have discussed in this chapter fall into four major families, as summarized in **Table 15–3**.

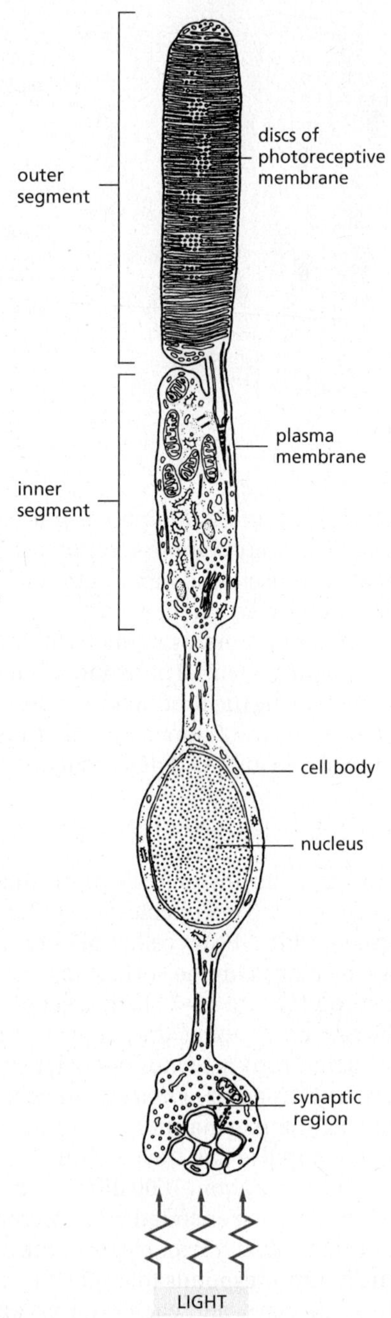

Figure 15–39 A rod photoreceptor cell. There are about 1000 discs in the outer segment. The disc membranes are not connected to the plasma membrane. The inner and outer segments are specialized parts of a *primary cilium* (discussed in Chapter 16). A primary cilium extends from the surface of most vertebrate cells, where it serves as a signaling compartment.

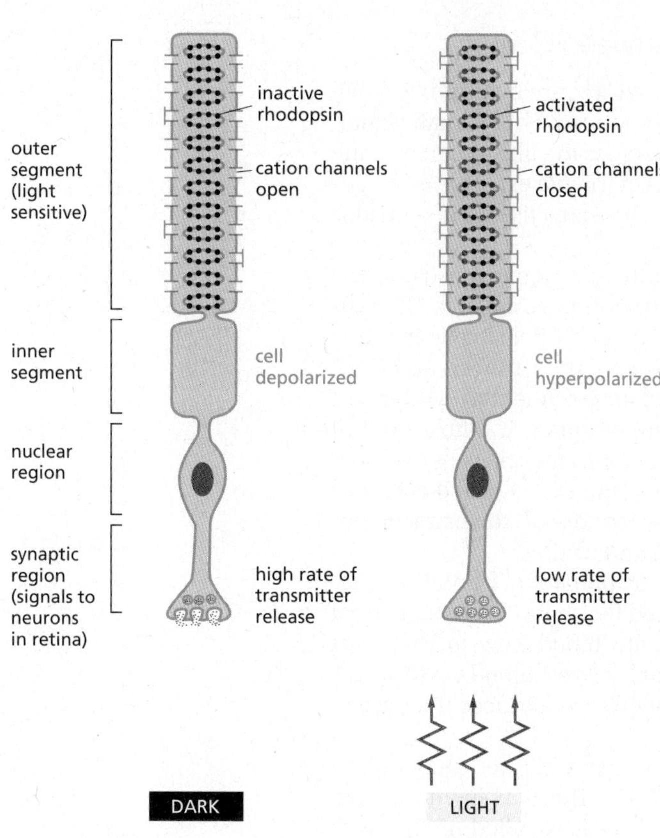

Figure 15–40 The response of a rod photoreceptor cell to light. Rhodopsin molecules in the outer-segment discs absorb photons. Photon absorption closes cation channels in the plasma membrane, which hyperpolarizes the membrane and reduces the rate of neurotransmitter release from the synaptic region. Because the neurotransmitter inhibits many of the postsynaptic retinal neurons in the absence of light, illumination serves to free the neurons from inhibition and thus, in effect, excites them. The neural connections of the retina lie between the light source and the outer segment, and so light must pass through the synapses and rod cell nucleus to reach the light sensors.

TABLE 15–3 Four Major Families of Heterotrimeric G Proteins*

Family	Some family members	Subunits that mediate action	Some functions
I	G_s	α	Activates adenylyl cyclase; activates Ca^{2+} channels
	G_{olf}	α	Activates adenylyl cyclase in olfactory sensory neurons
II	G_i	α	Inhibits adenylyl cyclase
		βγ	Activates K^+ channels
	G_o	βγ	Activates K^+ channels; inactivates Ca^{2+} channels
		α and βγ	Activates phospholipase C-β
	G_t (transducin)	α	Activates cyclic GMP phosphodiesterase in vertebrate rod photoreceptors
III	G_q	α	Activates phospholipase C-β
IV	$G_{12/13}$	α	Activates Rho family monomeric GTPases (via Rho GEF) to regulate the actin cytoskeleton

*Families are determined by amino acid sequence relatedness of the α subunits. Only selected examples are included. About 20 α subunits and at least 6 β subunits and 11 γ subunits have been described in humans.

Nitric Oxide Gas Can Mediate Signaling Between Cells

Small signaling molecules like cyclic nucleotides and Ca^{2+} are hydrophilic small molecules that act within the cell where they are produced. Some small signaling molecules, however, are hydrophobic enough to cross the plasma membrane and thereby affect nearby cells. An important and remarkable example is the gas **nitric oxide (NO)**, which acts as a signaling molecule in many tissues of both animals and plants.

In mammals, one of NO's many functions is to relax smooth muscle in the walls of blood vessels. The neurotransmitter acetylcholine stimulates NO synthesis by activating a GPCR on the membranes of the endothelial cells that line the interior of the vessel. The activated receptor triggers IP_3 synthesis and Ca^{2+} release (see Figure 15–30), leading to stimulation of an enzyme that synthesizes NO. Because dissolved NO passes readily across membranes, it diffuses out of the cell where it is produced and into neighboring smooth muscle cells, where it causes muscle relaxation and thereby vessel dilation (**Figure 15–41**). It acts only locally because it has a short half-life—about 5–10 seconds—in the extracellular space before oxygen and water convert it to nitrates and nitrites.

The effect of NO on blood vessels provides an explanation for the mechanism of action of nitroglycerine, which has been used for about 100 years to treat people with angina (pain resulting from inadequate blood flow to the heart muscle). The nitroglycerine is converted to NO, which relaxes blood vessels. This reduces the workload on the heart and, as a consequence, reduces the oxygen requirement of the heart muscle.

NO is made by the deamination of the amino acid arginine, catalyzed by enzymes called **NO synthases (NOS**; see Figure 15–41). The NOS in endothelial cells is called *eNOS*, while that in nerve and muscle cells is called *nNOS*. Both eNOS and nNOS are stimulated by an increase in cytosolic Ca^{2+}. Macrophages, by contrast, make yet another NOS, called inducible NOS (*iNOS*), that is constitutively active but synthesized only when the cells are activated, usually in response to an infection.

In some target cells, including smooth muscle cells, NO binds reversibly to iron in the active site of guanylyl cyclase, stimulating synthesis of cyclic GMP. NO

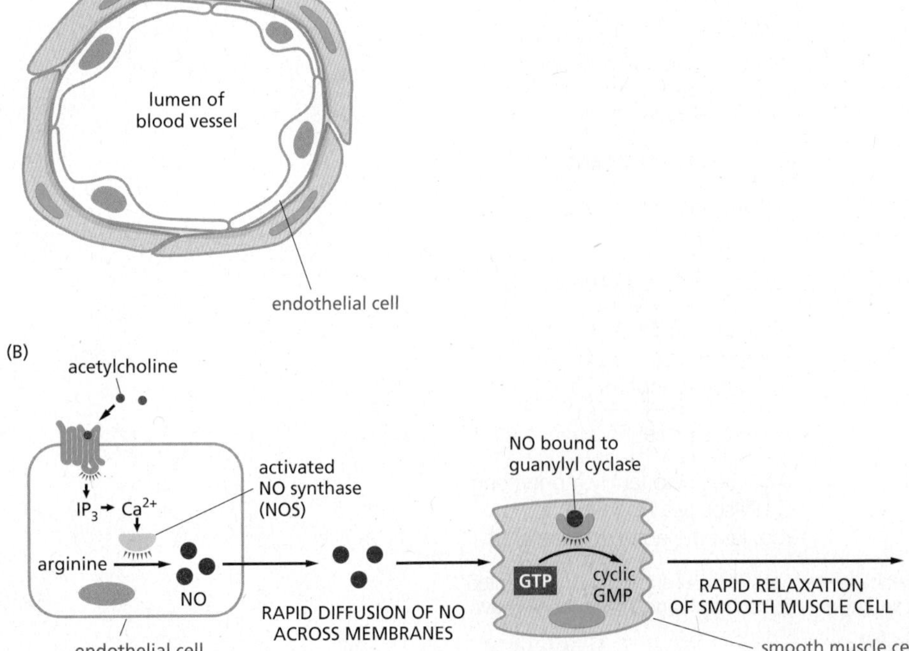

Figure 15–41 The role of nitric oxide (NO) in smooth muscle relaxation in a blood vessel wall. (A) Drawing of a cross section of a small blood vessel, showing the endothelial cells lining the lumen, the smooth muscle cells around them, and the basal lamina separating the two. (B) The neurotransmitter acetylcholine stimulates blood vessel dilation by activating a GPCR—the *muscarinic acetylcholine receptor*—on the surface of endothelial cells. This receptor activates a G protein, G_q, thereby stimulating IP_3 synthesis and Ca^{2+} release from the ER by the mechanisms illustrated in Figure 15–30. Increased Ca^{2+} activates nitric oxide synthase, causing the endothelial cells to produce NO from arginine. The NO diffuses out of the endothelial cells and into the neighboring smooth muscle cells, where it activates guanylyl cyclase to produce cyclic GMP. The cyclic GMP triggers a response that causes the smooth muscle cells to relax, increasing blood flow through the vessel.

can increase cyclic GMP in the cytosol within seconds, because the normal rate of turnover of cyclic GMP is high: rapid degradation to GMP by a phosphodiesterase constantly balances the production of cyclic GMP by guanylyl cyclase. The drug Viagra and its relatives inhibit the cyclic GMP phosphodiesterase in the penis, thereby increasing the amount of time that cyclic GMP levels remain elevated in the smooth muscle cells of penile blood vessels after NO production is induced by local nerve terminals. The cyclic GMP, in turn, keeps the blood vessels relaxed and thereby the penis erect. NO can also signal cells independently of cyclic GMP. It can, for example, alter the activity of an intracellular protein by covalently nitrosylating thiol (–SH) groups on specific cysteines in the protein.

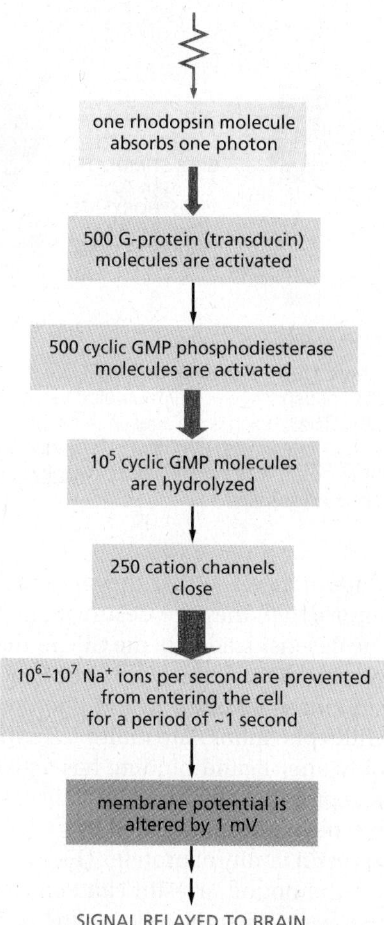
one rhodopsin molecule absorbs one photon
500 G-protein (transducin) molecules are activated
500 cyclic GMP phosphodiesterase molecules are activated
10^5 cyclic GMP molecules are hydrolyzed
250 cation channels close
10^6–10^7 Na^+ ions per second are prevented from entering the cell for a period of ~1 second
membrane potential is altered by 1 mV
SIGNAL RELAYED TO BRAIN

Figure 15–42 Amplification in the light-induced catalytic cascade in vertebrate rods. The *red arrows* indicate the steps where amplification occurs, with the thickness of the arrow roughly indicating the magnitude of the amplification.

Second Messengers and Enzymatic Cascades Amplify Signals

Despite the differences in molecular details, the different intracellular signaling pathways that GPCRs trigger share certain features and general principles. They depend on relay chains of intracellular signaling proteins and second messengers. These relay chains provide numerous opportunities for amplifying the responses to extracellular signals. In the rod visual transduction cascade, for example, a single activated rhodopsin molecule catalyzes the activation of about 500 transducin molecules per second. Each activated transducin molecule activates a molecule of cyclic GMP phosphodiesterase, resulting in the hydrolysis of more than 10^5 cyclic GMP molecules. The resulting drop in the concentration of cyclic GMP in turn transiently closes hundreds of cation channels in the plasma membrane (**Figure 15–42**). Thus, a rod cell can respond to even a single photon of light in a way that is highly reproducible in its timing and magnitude.

Likewise, when an extracellular signal molecule binds to a receptor that indirectly activates adenylyl cyclase via G_s, each receptor protein may activate many molecules of G_s protein, each of which can activate a cyclase molecule. Each cyclase molecule, in turn, can catalyze the conversion of a large number of ATP molecules to cAMP molecules. A similar amplification operates in the IP_3 signaling pathway. In these ways, a nanomolar (10^{-9} M) change in the concentration of an extracellular signal can induce micromolar (10^{-6} M) changes in the concentration of a second messenger such as cAMP or Ca^{2+}. Because these messengers function as allosteric effectors to activate specific enzymes or ion channels, a single extracellular signal molecule can alter many thousands of protein molecules within the target cell.

In signaling systems like those involved in smell and vision, any amplifying cascade of stimulatory signals requires counterbalancing mechanisms at every step of the cascade to restore the system to its resting state when stimulation ceases. As emphasized earlier, the response to stimulation in these systems can be rapid only if the inactivating mechanisms are also rapid. Cells therefore have efficient mechanisms for rapidly degrading (and resynthesizing) cyclic nucleotides and for buffering and removing cytosolic Ca^{2+}, as well as for inactivating the responding enzymes and ion channels once they have been activated. This is not only essential for turning a response off but is also important for establishing the resting state from which a response begins.

Each protein in the signaling relay chain can be a separate target for regulation, including the receptor itself, as we discuss next.

GPCR Desensitization Depends on Receptor Phosphorylation

As discussed earlier, when target cells are exposed to an extracellular signal for a prolonged period, they can become *desensitized*, or *adapted*, in several different ways. An important class of adaptation mechanisms depends on alteration of the quantity or condition of the receptor molecules themselves.

For GPCRs, there are three general modes of adaptation, all centered on inactivation of the receptor (see Figure 15–21C, D, and E): (1) In *receptor inactivation*, they become altered so that they can no longer interact with G proteins. (2) In *receptor sequestration*, they are temporarily moved to the interior of the cell

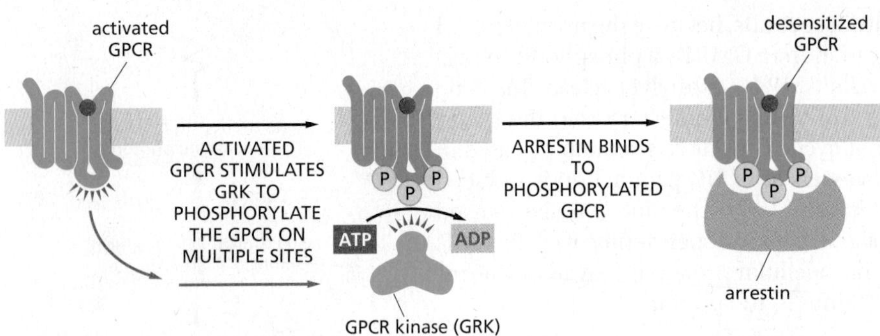

Figure 15–43 **The roles of GPCR kinases (GRKs) and arrestins in GPCR desensitization.** A GRK phosphorylates only activated receptors because it is the activated GPCR that turns on the GRK. The binding of an arrestin to the phosphorylated receptor prevents the receptor from binding to its G protein and also directs its endocytosis (not shown). Mice deficient in one form of arrestin fail to desensitize in response to morphine, for example, attesting to the importance of arrestins for desensitization.

(internalized) so that they no longer have access to their ligand. (3) In *receptor destruction*, they are destroyed in lysosomes after internalization. In each case, the desensitization of the GPCRs depends on their phosphorylation by PKA, PKC, or a member of the family of **GPCR kinases (GRKs)**, which includes the rhodopsin kinase involved in rod photoreceptor desensitization discussed earlier. The GRKs phosphorylate multiple serines and threonines on a GPCR, but they do so only after ligand binding has activated the receptor, because it is the activated receptor that allosterically activates the GRK. As with rhodopsin, once a receptor has been phosphorylated by a GRK, it binds with high affinity to a member of the **arrestin** family of proteins (Figure 15–43).

The bound arrestin can contribute to the desensitization process in at least two ways. First, it prevents the activated receptor from interacting with G proteins. Second, it serves as an adaptor protein to help couple the receptor to the clathrin-dependent endocytosis machinery (discussed in Chapter 13), inducing receptor-mediated endocytosis. The fate of the internalized GPCR–arrestin complex depends on other proteins in the complex. In some cases, the internalized receptor is dephosphorylated and recycled back to the plasma membrane for reuse; in others, it is degraded in lysosomes.

Receptor endocytosis does not necessarily stop the receptor from signaling. In some cases, the bound arrestin recruits other signaling proteins to relay the signal onward from the internalized GPCRs along new pathways.

Summary

GPCRs can indirectly activate or inactivate either plasma-membrane-bound enzymes or ion channels via G proteins. When an activated receptor stimulates a G protein, the G protein undergoes a conformational change that activates its α *subunit, thereby triggering release of a* $\beta\gamma$ *complex. Either component can then directly regulate the activity of target proteins in the plasma membrane. Some GPCRs either activate or inactivate adenylyl cyclase, thereby altering the intracellular concentration of the second messenger cyclic AMP. Others activate a phosphoinositide-specific phospholipase C (PLC*β*), which generates two second messengers. One is inositol 1,4,5-trisphosphate (*IP_3*), which releases* Ca^{2+} *from the ER and thereby increases the concentration of* Ca^{2+} *in the cytosol. The other is diacylglycerol, which remains in the plasma membrane and helps activate protein kinase C (PKC). An increase in cytosolic cyclic AMP or* Ca^{2+} *levels affects cells mainly by stimulating cAMP-dependent protein kinase (PKA) and* Ca^{2+}*/calmodulin-dependent kinases (CaM-kinases), respectively.*

PKC, PKA, and CaM-kinases phosphorylate specific target proteins and thereby alter the activity of the proteins. Each type of cell has its own characteristic set of target proteins that is regulated in these ways, enabling the cell to make its own distinctive response to the second messengers. The intracellular signaling cascades activated by GPCRs greatly amplify the responses, so that many thousands of target protein molecules are changed for each molecule of extracellular signaling ligand bound to its receptor. The responses mediated by GPCRs are rapidly turned off when the extracellular signal is removed, and activated GPCRs are inactivated by phosphorylation and association with arrestins.

SIGNALING THROUGH ENZYME-COUPLED RECEPTORS

Like GPCRs, **enzyme-coupled receptors** are transmembrane proteins with their ligand-binding domain on the outer surface of the plasma membrane. Instead of having a cytosolic domain that associates with a heterotrimeric G protein, however, their cytosolic domain either has intrinsic enzyme activity or associates directly with an enzyme. Whereas a GPCR has seven transmembrane segments, each subunit of an enzyme-coupled receptor typically has only one. GPCRs and enzyme-coupled receptors often activate some of the same signaling pathways. In this section, we describe some of the important features of signaling by enzyme-coupled receptors, with an emphasis on the most common class of these proteins, the *receptor tyrosine kinases*.

Activated Receptor Tyrosine Kinases (RTKs) Phosphorylate Themselves

Many extracellular signal proteins act through **receptor tyrosine kinases (RTKs)**. These include many secreted and cell-surface-bound proteins that control cell behavior in developing and adult animals. Some of these signal proteins and their RTKs are listed in **Table 15–4**.

There are about 60 human RTKs, which can be classified into about 20 structural subfamilies, each dedicated to its complementary family of protein ligands. **Figure 15–44** shows the basic structural features of a number of the families that operate in mammals. In all cases, the binding of the signal protein to the ligand-binding domain on the extracellular side of the receptor activates the tyrosine kinase domain on the cytosolic side. This leads to phosphorylation of tyrosine

TABLE 15–4 Some Extracellular Signal Proteins That Act Via RTKs

Signal protein family	Receptor family	Some representative responses
Epidermal growth factor (EGF)	EGF receptors	Stimulates cell survival, growth, proliferation, or differentiation of various cell types; acts as inductive signal in development
Insulin	Insulin receptor	Stimulates carbohydrate utilization and protein synthesis
Insulin-like growth factor (IGF1)	IGF receptor-1	Stimulates cell growth and survival in many cell types
Nerve growth factor (NGF)	Trk receptors	Stimulates survival and growth of some neurons
Platelet-derived growth factor (PDGF)	PDGF receptors	Stimulates survival, growth, proliferation, and migration of various cell types
Macrophage-colony-stimulating factor (MCSF)	MCSF receptor	Stimulates monocyte/macrophage proliferation and differentiation
Fibroblast growth factor (FGF)	FGF receptors	Stimulates proliferation of various cell types; inhibits differentiation of some precursor cells; acts as inductive signal in development
Vascular endothelial growth factor (VEGF)	VEGF receptors	Stimulates angiogenesis
Ephrin	Eph receptors	Stimulates angiogenesis; guides cell and axon migration

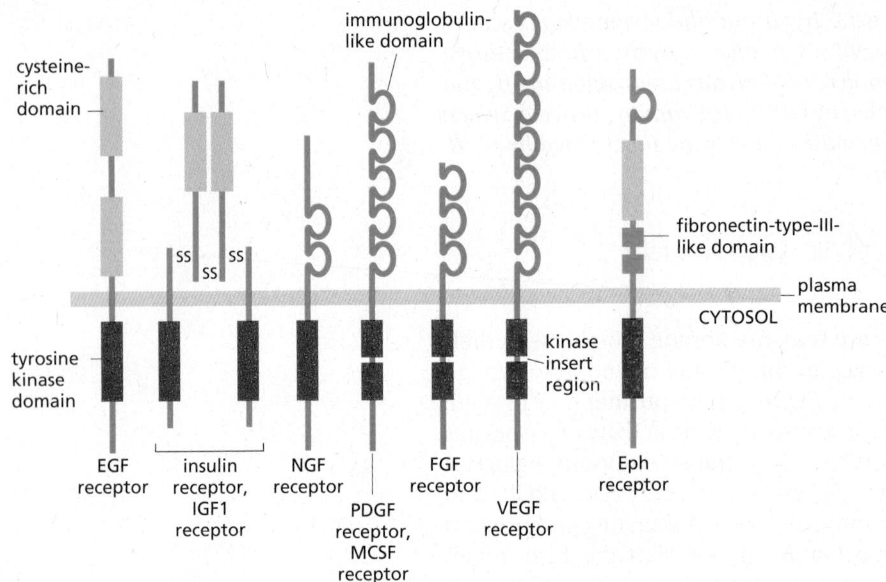

Figure 15–44 Some subfamilies of RTKs. Only one or two members of each subfamily are indicated. Note that in some cases, the tyrosine kinase domain is interrupted by a *kinase insert region* that is an extra segment emerging from the folded kinase domain. The functions of most of the cysteine-rich, immunoglobulin-like, and fibronectin-type-III-like domains are not known. Some of the ligands for the receptors shown are listed in Table 15–4, along with some representative responses that they mediate.

side chains on the cytosolic part of the receptor, creating phosphotyrosine docking sites for various intracellular signaling proteins that relay the signal.

How does the binding of an extracellular ligand activate the kinase domain on the other side of the plasma membrane? For a GPCR, ligand binding is thought to change the relative orientation of several of the transmembrane α helices, thereby shifting the position of the cytoplasmic loops relative to one another. It is unlikely, however, that a conformational change could propagate across the lipid bilayer through a single transmembrane α helix. Instead, the mechanism of RTK activation depends on ligand-stimulated changes in the interaction of two receptors, bringing the two cytoplasmic kinase domains together and thereby promoting their activation (**Figure 15–45**).

The cytoplasmic kinase domains of RTKs are dimerized and activated by a variety of mechanisms. In many cases, such as the receptor for *platelet-derived growth*

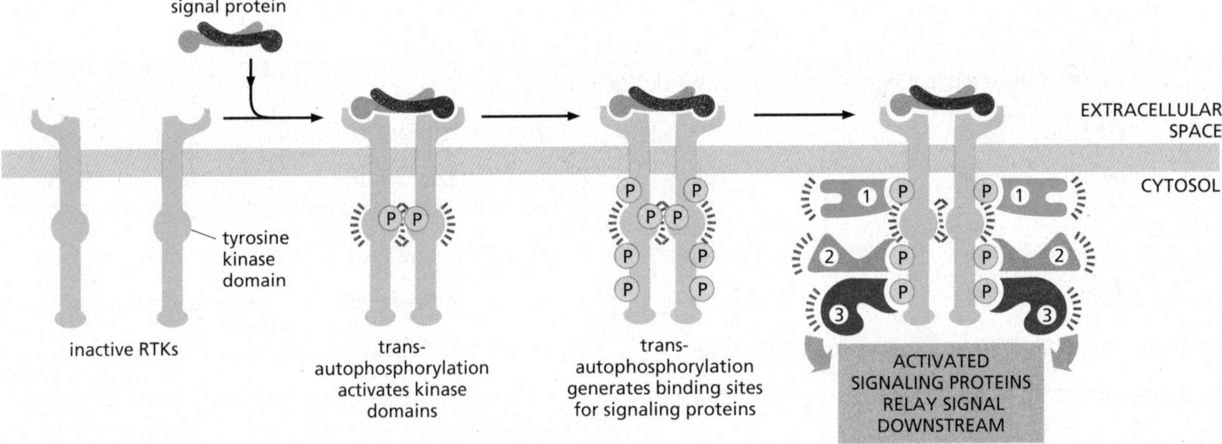

Figure 15–45 Activation of RTKs by dimerization. In the absence of extracellular signals, many RTKs exist as monomers in which the internal kinase domain is inactive. Binding of ligand brings two monomers together to form a dimer. The close proximity in the dimer leads the two kinase domains to phosphorylate each other, which has two effects. First, phosphorylation at some tyrosines in the kinase domains promotes the complete activation of the domains. Second, phosphorylation at tyrosines in other parts of the receptors generates docking sites for intracellular signaling proteins, resulting in the formation of large signaling complexes that can then broadcast signals along multiple signaling pathways.

Mechanisms of activation vary widely among different RTK family members. In some cases (as shown here), the ligand itself is a dimer and brings two receptors together by binding them simultaneously. In other cases, two ligands can bind independently on two receptors to promote receptor dimerization. Some RTKs exist normally as a dimer (see Figure 15–44), and ligand binding causes a conformational change that brings the two internal kinase domains closer together. Although many RTKs are activated by transautophosphorylation as shown here, there are some important exceptions, including the EGF receptor illustrated in Figure 15–46.

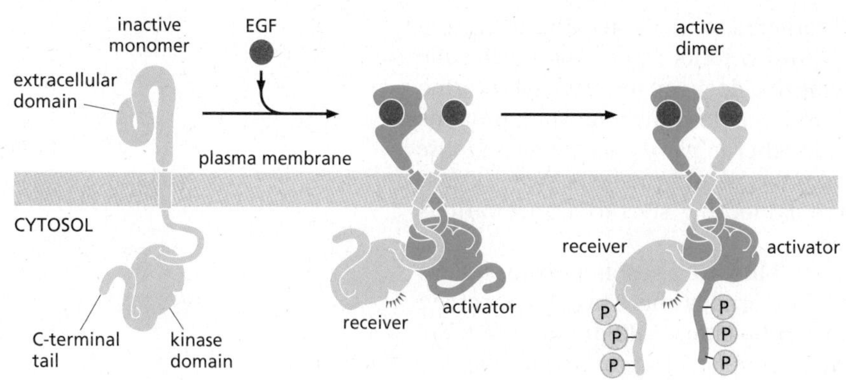

Figure 15–46 Activation of the EGF receptor kinase. In the absence of ligand, the EGF receptor exists primarily as an inactive monomer. EGF binding results in a conformational change that promotes dimerization of the external domains. The receptor kinase domain, unlike that of many RTKs, is not activated by transautophosphorylation. Instead, dimerization orients the internal kinase domains into an asymmetrical dimer, in which one kinase domain (the *activator*) pushes against the other kinase domain (the *receiver*), thereby causing an activating conformational change in the receiver. The active receiver domain then phosphorylates multiple tyrosines in the C-terminal tails of both receptors, generating docking sites for intracellular signaling proteins (see Figure 15–45).

Note that protein kinase activity is generated only in the receiver kinase domain; the activator kinase domain remains in an inactive conformation. In fact, some members of the EGF receptor family contain kinase domains that are missing key residues in their active site and are therefore inactive. These *pseudokinase* receptors can still dimerize with and activate EGF receptors containing an active kinase receiver domain.

factor (*PDGF*), dimerization of the external receptor domain simply brings two kinase domains close to each other in a stable orientation that allows them to phosphorylate each other on specific tyrosines in the kinase active sites, thereby promoting conformational changes that fully activate both kinase domains. Some RTKs, including the insulin receptor, exist as dimers in the absence of ligand (see Figure 15–44), and ligand binding re-orients the extracellular domains to bring the internal kinase domains into a closer position to promote kinase activation. In other cases, such as the receptor for *epidermal growth factor* (*EGF*), the kinase is not activated by phosphorylation but by conformational changes brought about by interactions between the two kinase domains outside their active sites (**Figure 15–46**).

Phosphorylated Tyrosines on RTKs Serve as Docking Sites for Intracellular Signaling Proteins

Once the kinase domains of an RTK dimer are activated, they phosphorylate multiple additional sites in the cytosolic parts of the receptors, typically in disordered regions outside the kinase domain (see Figure 15–45). This phosphorylation creates high-affinity docking sites for specific intracellular signaling proteins. Each signaling protein binds to a particular phosphorylated site on the activated receptors because it contains a specific phosphotyrosine-binding domain that recognizes surrounding features of the polypeptide chain in addition to the phosphotyrosine.

Once bound to the activated RTK, a signaling protein may become phosphorylated on tyrosines and thereby activated. In many cases, however, the binding alone may be sufficient to activate the docked signaling protein, by either inducing a conformational change in the protein or simply bringing it near the protein that is next in the signaling pathway. Thus, receptor phosphorylation serves as a switch to trigger the assembly of an intracellular signaling complex, which can then relay the signal onward, often along multiple routes, to various destinations in the cell. Because different RTKs bind different combinations of these signaling proteins, they activate different responses.

Some RTKs use additional docking proteins to enlarge the signaling complex at activated receptors. Insulin receptor signaling, for example, depends on a specialized adaptor protein called *insulin receptor substrate-1* (*IRS1*). IRS1 binds to specific phosphorylated tyrosines on the activated receptor and is then phosphorylated at multiple sites, thereby creating many more docking sites than could be accommodated on the receptor alone (see Figure 15–11).

Proteins with SH2 Domains Bind to Phosphorylated Tyrosines

A whole menagerie of intracellular signaling proteins can bind to the phosphotyrosines on activated RTKs (or on docking proteins such as IRS1). They help to relay the signal onward, mainly through chains of protein–protein interactions mediated by modular interaction domains, as discussed earlier (see Figure 15–11). Some of the docked proteins are enzymes, such as **phospholipase C-γ (PLCγ)**, which functions in the same way as phospholipase C-β—activating the inositol phospholipid signaling pathway discussed earlier in connection with GPCRs (see Figures 15–29 and 15–30). Through this pathway, RTKs can increase cytosolic Ca^{2+} levels and activate PKC. Another enzyme that docks on these receptors is the cytoplasmic tyrosine kinase *Src*, which phosphorylates other signaling proteins on tyrosines (discussed in Chapter 3). Yet another is *phosphoinositide 3-kinase* (*PI 3-kinase*), which phosphorylates lipids rather than proteins; as we discuss later, the phosphorylated lipids then serve as docking sites to attract various signaling proteins to the plasma membrane.

The intracellular signaling proteins that bind to phosphotyrosines have varied structures and functions. However, they usually share highly conserved phosphotyrosine-binding domains, which can be either **SH2 domains** (for *Src homology 2*) or, less commonly, *PTB domains* (for *phosphotyrosine-binding*). By recognizing specific phosphorylated tyrosines, these small interaction domains enable the proteins that contain them to bind to activated RTKs, as well as to many other intracellular signaling proteins that have been transiently phosphorylated on tyrosines (**Figure 15–47**). Many signaling proteins also contain other interaction domains that allow them to interact specifically with other proteins as part of the signaling process. These domains include the *SH3 domain*, which binds to proline-rich motifs in intracellular proteins (see Figures 15–11 and 15–12).

Not all proteins that bind to activated RTKs via SH2 domains help to relay the signal onward. Some act to decrease the signaling process, providing negative feedback. One example is the *c-Cbl protein*, which can dock on some activated receptors and catalyze their ubiquitylation, covalently adding one or more ubiquitin molecules to specific sites on the receptor. This promotes the endocytosis and degradation of the receptors in lysosomes—an example of receptor destruction (see Figure 15–21E). Endocytic proteins that contain *ubiquitin-interaction motifs*

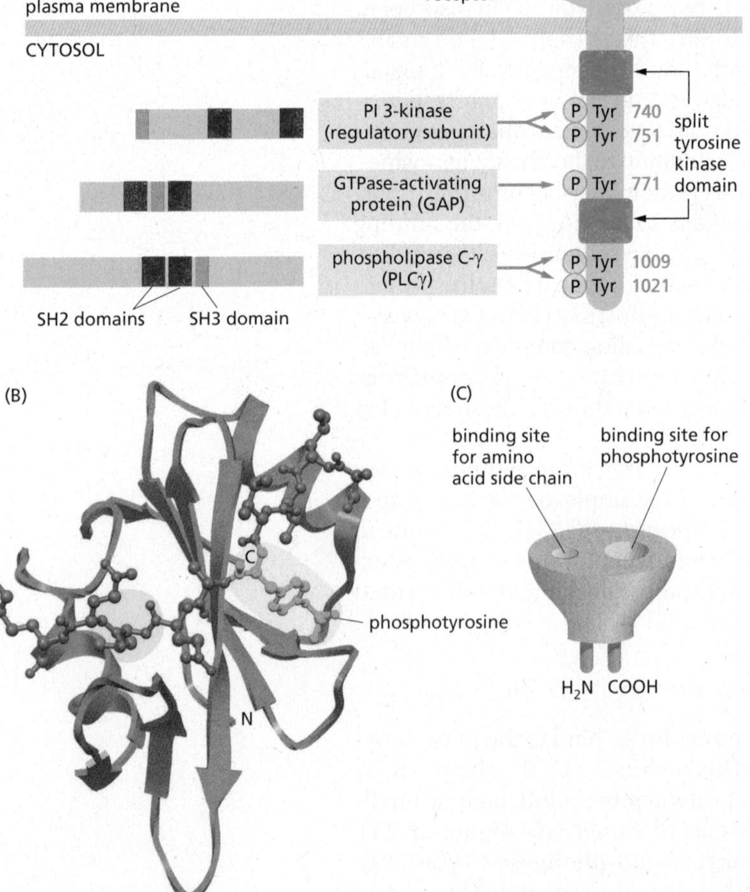

Figure 15–47 The binding of SH2-containing intracellular signaling proteins to an activated RTK. (A) Drawing based on an activated receptor for *platelet-derived growth factor* (*PDGF*); for simplicity, only one monomer of the dimer is shown, and the PDGF ligand is omitted. Five phosphotyrosines are shown, three in the kinase insert region and two on the C-terminal tail; these form three docking sites, each of which binds a different signaling protein as indicated. The numbers on the *right* indicate the positions of the tyrosines in the polypeptide chain. The functions of these phosphotyrosines were determined by mutation of specific tyrosines. Mutation of tyrosines 1009 and 1021, for example, prevents the binding and activation of PLCγ, so that receptor activation no longer stimulates the inositol phospholipid signaling pathway. The locations of the SH2 *(red)* and SH3 *(blue)* domains in the primary structures of the three signaling proteins are indicated. Additional phosphotyrosine docking sites on this receptor are not shown, including those that bind the cytoplasmic tyrosine kinase Src and two adaptor proteins. It is unclear how many signaling proteins can bind simultaneously to a single RTK. (B) The three-dimensional structure of an SH2 domain, as determined by x-ray crystallography. The binding pocket for phosphotyrosine is shaded in *yellow* on the right, and a pocket for binding a specific amino acid side chain (valine, in this case) is shaded in *green* on the *left*. The RTK polypeptide segment that binds the SH2 domain is shown in *green*. (C) The SH2 domain is a compact, "plug-in" module, which can be inserted in disordered regions of a protein without disturbing the protein's folding or function (discussed in Chapter 3). Because each domain has distinct sites for recognizing phosphotyrosine and for recognizing a particular amino acid side chain, different SH2 domains recognize phosphotyrosine in the context of different flanking amino acid sequences. (B, PDB code: 2SRC.)

(*UIMs*) recognize the ubiquitylated RTKs and direct them into clathrin-coated vesicles and, ultimately, into lysosomes (discussed in Chapter 13). Mutations that inactivate c-Cbl-dependent RTK down-regulation cause prolonged RTK signaling and thereby promote the development of cancer.

As is the case for GPCRs, ligand-induced endocytosis of RTKs does not always decrease signaling. In some cases, activated RTKs are endocytosed with their bound signaling proteins and continue to signal from endosomes or other intracellular compartments. This mechanism, for example, allows *nerve growth factor* (*NGF*) to bind to its specific RTK (called TrkA) at the end of a long nerve cell axon and signal to the cell body of the same cell a long distance away. The signaling endocytic vesicles containing activated TrkA, with NGF bound on the lumenal side and signaling proteins docked on the cytosolic side, are transported along the axon to the cell body, where they signal the cell to survive.

Some signaling proteins are composed almost entirely of SH2 and SH3 domains and function as *adaptors* to couple tyrosine-phosphorylated proteins to other proteins that do not have their own SH2 domains (see Figures 15–11 and 15–12). Adaptor proteins of this type help to couple activated RTKs to the important signaling protein *Ras*, a monomeric GTPase that, in turn, can activate various downstream signaling pathways, as we now discuss.

The Monomeric GTPase Ras Mediates Signaling by Most RTKs

The **Ras superfamily** consists of various families of monomeric GTPases, but only the Ras and Rho families relay signals from cell-surface receptors (**Table 15–5**). By interacting with different intracellular signaling proteins, a single Ras or Rho family member can coordinately spread the signal along several distinct downstream signaling pathways.

There are three major, closely related Ras proteins in humans: H-, K-, and N-Ras (see Table 15–5). Although they have subtly different functions, they are thought to work in the same way, and we will refer to them simply as **Ras**. Like many monomeric GTPases, Ras contains one or more covalently attached lipid groups that help anchor the protein to the cytoplasmic face of the plasma membrane, from where it relays signals to other parts of the cell. Ras is often required, for example, when RTKs signal to the nucleus to stimulate cell proliferation or differentiation, both of which require changes in gene expression. If Ras function is inhibited by various experimental approaches, the cell proliferation or differentiation responses normally induced by the activated RTKs do not occur. Conversely,

TABLE 15–5 The Ras Superfamily of Monomeric GTPases

Family	Some family members	Some functions
Ras	H-Ras, K-Ras, N-Ras	Relay signals from RTKs
	Rheb	Activates mTOR to stimulate cell growth
	Rap1	Activated by a cyclic-AMP-dependent GEF; influences cell adhesion by activating integrins
Rho*	Rho, Rac, Cdc42	Relay signals from surface receptors to the cytoskeleton and elsewhere
ARF*	ARF1–ARF6	Regulate assembly of protein coats on intracellular vesicles
Rab*	Rab1–60	Regulate intracellular vesicle traffic
Ran*	Ran	Regulates mitotic spindle assembly and nuclear transport of RNAs and proteins

*The Rho family is discussed in Chapter 16, the ARF and Rab proteins in Chapter 13, and Ran in Chapters 12 and 17. The three-dimensional structure of Ras is shown in Figure 3–64.

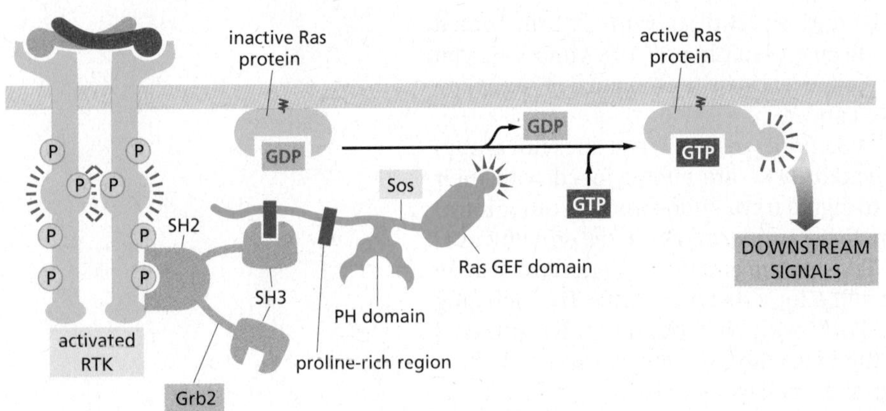

Figure 15–48 How an RTK activates Ras. The Grb2 adaptor protein recognizes a specific phosphorylated tyrosine on the activated receptor by means of an SH2 domain and recruits the Ras GEF Sos by means of an interaction between its SH3 domains and proline-rich regions in Sos. Sos also contains a PH domain that recognizes specific phospholipids but is not used here (see Figure 15–11). The Ras GEF domain of Sos stimulates the inactive Ras protein to replace its bound GDP by GTP, which activates Ras to relay the signal downstream.

30% of human tumors express hyperactive mutant forms of Ras, which contribute to the uncontrolled proliferation of the cancer cells.

Like other GTP-binding proteins, Ras functions as a molecular switch, cycling between two distinct conformational states—active when GTP is bound and inactive when GDP is bound (Movie 15.7). As discussed earlier for monomeric GTPases in general, two classes of signaling proteins regulate Ras activity by influencing its transition between active and inactive states (see Figure 15–8). *Ras guanine nucleotide exchange factors* (**Ras GEFs**) stimulate the dissociation of GDP and the subsequent binding of GTP from the cytosol, thereby activating Ras. *Ras GTPase-activating proteins* (**Ras GAPs**) increase the rate of hydrolysis of bound GTP by Ras, thereby inactivating Ras. Hyperactive mutant forms of Ras are resistant to Ras GAPs and are therefore locked in the GTP-bound active state, which is why they promote the development of cancer.

How do RTKs normally activate Ras? In principle, they could either activate a Ras GEF or inhibit a Ras GAP. Even though some GAPs bind directly (via their SH2 domains) to activated RTKs (see Figure 15–47A), it is the indirect coupling of the activated receptor to a Ras GEF that drives Ras into its active state. The loss of function of a Ras GEF has a similar effect to the loss of function of Ras itself. Activation of the other Ras superfamily proteins, including those of the Rho family, also occurs through the activation of GEFs. The particular GEF determines in which membrane the GTPase is activated, and, by acting as a scaffold, it can also determine which downstream proteins the GTPase activates.

The GEF that mediates Ras activation by RTKs was discovered by genetic studies of eye development in *Drosophila*, where an RTK called *Sevenless* (*Sev*) is required for the formation of a photoreceptor cell called R7. Genetic screens for components of this signaling pathway led to the discovery of a Ras GEF called *Son-of-sevenless* (*Sos*). Further genetic screens uncovered another protein, now called *Grb2*, which is an adaptor protein that links the Sev receptor to the Sos protein; the SH2 domain of the Grb2 adaptor binds to the activated receptor, while one or both of its SH3 domains bind to Sos. Sos then promotes Ras activation. Biochemical and cell biological studies have shown that Grb2 and Sos also link activated RTKs to Ras in mammalian cells, revealing that this mechanism in RTK signaling has been highly conserved in evolution (**Figure 15–48**). Once activated, Ras activates various other signaling proteins to relay the signal downstream.

Ras Activates a MAP Kinase Signaling Module

Both the tyrosine phosphorylations and the activation of Ras triggered by activated RTKs are usually short-lived (**Figure 15–49**). *Tyrosine-specific protein phosphatases* reverse the phosphorylations, and Ras GAPs induce activated Ras to inactivate itself by hydrolyzing its bound GTP to GDP. To stimulate cells to proliferate or differentiate, these short-lived signaling events must be converted into longer-lasting ones that can sustain the signal and relay it downstream to the nucleus to alter the pattern of gene expression. One of the key mechanisms used for this purpose is a system of proteins called the *mitogen-activated protein kinase*

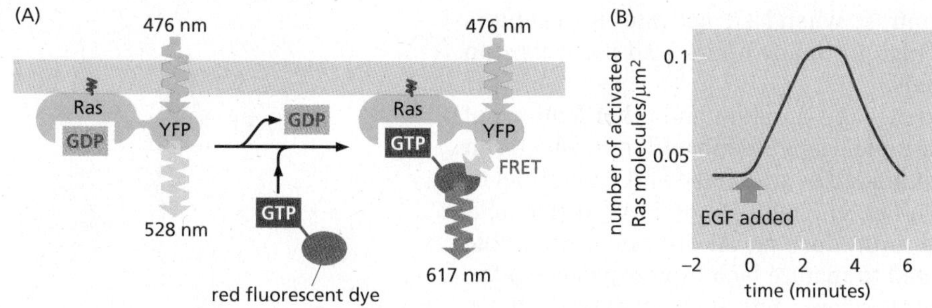

Figure 15–49 Transient activation of Ras revealed by single-molecule fluorescence resonance energy transfer (FRET). (A) Schematic drawing of the experimental strategy. Cells of a human cancer cell line were genetically engineered to express a Ras protein that was covalently linked to yellow fluorescent protein (YFP). GTP that was labeled with a red fluorescent dye was microinjected into some of the cells. The cells were then stimulated with the extracellular signal protein EGF, and single fluorescent molecules of Ras-YFP at the inner surface of the plasma membrane were measured by video fluorescence microscopy in individual cells. When a fluorescent Ras-YFP molecule becomes activated, it exchanges unlabeled GDP for fluorescently labeled GTP; the energy emitted by the YFP now activates the fluorescent GTP to emit red light (called fluorescence resonance energy transfer, or FRET; see Figure 9–22). Thus, the activation of single Ras molecules can be followed by the emission of red fluorescence from a previously yellow-green fluorescent spot at the plasma membrane. As shown in (B), activated Ras molecules could be detected after about 30 seconds of EGF stimulation. The red signal peaked at about 3 minutes and then decreased to baseline by 6 minutes. As Ras GAP was found to be recruited to the same spots at the plasma membrane as Ras, it presumably plays a major part in rapidly shutting off the Ras signal. (Modified from H. Murakoshi et al., *Proc. Natl. Acad. Sci. USA* 101:7317–7322, 2004. Copyright 2004 National Academy of Sciences, USA. With permission from National Academy of Sciences.)

module (**MAP kinase module**) (**Figure 15–50**). The three components of this system form a functional signaling module that has been remarkably well conserved during evolution and is used, with variations, in many different signaling contexts.

The three components are all protein kinases, which mainly phosphorylate serines and threonines. The final kinase in the series is called simply MAP kinase (MAPK). The next one upstream from this is MAP kinase kinase (MAPKK): it phosphorylates and thereby activates MAP kinase. Next above that, receiving an activating signal directly from Ras, is MAP kinase kinase kinase (MAPKKK): it phosphorylates and thereby activates MAPKK. In the mammalian **Ras–MAP-kinase signaling pathway**, these three kinases are known by shorter names: Raf (= MAPKKK), Mek (= MAPKK), and Erk (= MAPK).

Once activated, the MAP kinase relays the signal downstream by phosphorylating various proteins in the cell, including transcription regulators and other protein kinases (see Figure 15–50). Erk, for example, enters the nucleus and phosphorylates one or more components of a transcription regulatory complex. This activates the transcription of a set of *immediate early genes*, so named because they turn on within minutes after an RTK receives an extracellular signal, even if protein synthesis is experimentally blocked with drugs. Some of these genes encode other transcription regulators that turn on other genes, a process that requires both protein synthesis and more time. In this way, the Ras–MAP-kinase signaling pathway conveys signals from the cell surface to the nucleus and alters the pattern of gene expression. Among the genes activated by this pathway are some that stimulate cell proliferation, such as the genes encoding G_1 *cyclins* (discussed in Chapter 17).

MAP kinase activation lasts for different amounts of time, depending on the extracellular signal. When EGF activates its receptors in a neural precursor cell line, for example, Erk MAP kinase activity peaks at 5 minutes and rapidly declines,

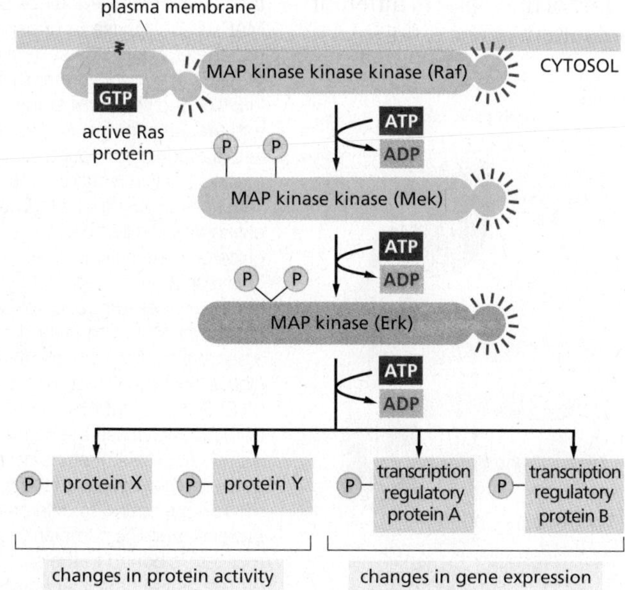

Figure 15–50 The MAP kinase module activated by Ras. The three-component module begins with a MAP kinase kinase kinase called *Raf.* Ras recruits Raf to the plasma membrane and helps activate it. Raf then activates the MAP kinase kinase *Mek*, which then activates the MAP kinase *Erk.* Erk in turn phosphorylates a variety of downstream proteins, including other protein kinases, as well as transcription regulators in the nucleus. The resulting changes in protein activities and gene expression cause complex changes in cell behavior (**Movie 15.8**).

and the cells later go on to divide. By contrast, when NGF activates its receptors on the same cells, Erk activity remains high for many hours, and the cells stop proliferating and differentiate into neurons.

Many intracellular mechanisms influence the duration and other features of the signaling response, including positive and negative feedback loops, which can combine to give responses that are either graded or switchlike and either brief or long lasting. In an example illustrated earlier, in Figure 15–20, MAP kinase activates a complex positive feedback loop to produce an all-or-none, irreversible response when frog oocytes are stimulated to mature by a brief exposure to the hormone progesterone. In many cells, MAP kinases activate a negative feedback loop by increasing the concentration of a protein phosphatase that removes the phosphates from MAP kinase. The increase in the phosphatase results from both an increase in the transcription of the phosphatase gene and the stabilization of the phosphatase protein against degradation. In the Ras–MAP-kinase pathway shown in Figure 15–50, Erk also phosphorylates and inactivates Raf, providing another negative feedback loop that helps shut off the MAP kinase module.

Scaffold Proteins Reduce Cross-Talk Between Different MAP Kinase Modules

Three-component MAP kinase signaling modules operate in all eukaryotic cells, with different modules mediating different responses in the same cell. In budding yeast, for example, one MAP kinase module mediates the response to mating pheromone, another the response to starvation, and yet another the response to osmotic shock. Some of these MAP kinase modules use one or more of the same kinases and yet manage to activate different downstream proteins and hence different responses. As discussed earlier, one way in which cells avoid cross-talk between the different parallel signaling pathways and ensure that each response is specific is to use scaffold proteins (see Figure 15–10A). In budding yeast cells, such scaffolds bind all or some of the kinases in each MAP kinase module to form a complex and thereby help to ensure response specificity (**Figure 15–51**).

Mammalian cells also use this scaffold strategy to prevent cross-talk between different MAP kinase modules. At least five parallel MAP kinase modules can operate in a mammalian cell. These modules make use of at least 12 MAP kinases, 7 MAPKKs, and 7 MAPKKKs. Two of these modules (terminating in MAP kinases called JNK and p38) are activated by different kinds of cell stresses, such as ultraviolet (UV) irradiation, heat shock, and osmotic stress; others mediate responses to signals from other cells.

Although the scaffold strategy provides precision and avoids cross-talk, it reduces the opportunities for amplification and spreading of the signal to different parts of the cell, which require at least some of the components to be diffusible. It

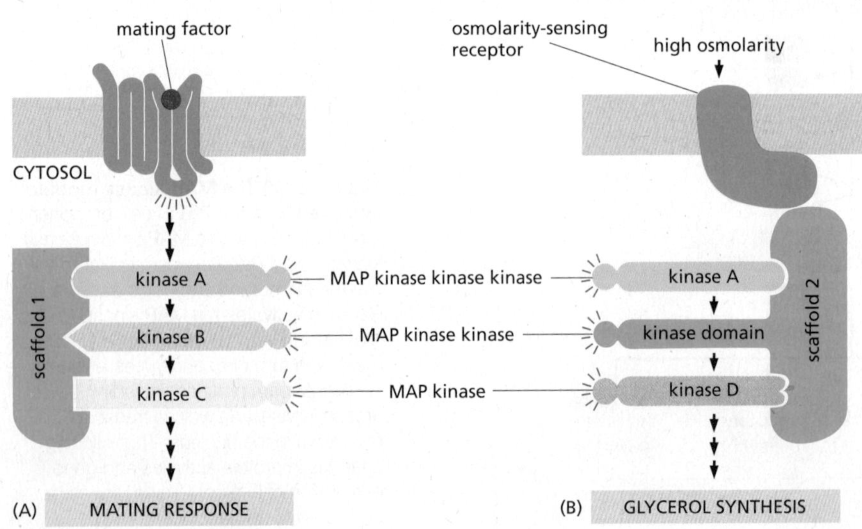

Figure 15–51 The organization of two MAP kinase modules by scaffold proteins in budding yeast. Budding yeast have at least six three-component MAP kinase modules involved in a variety of biological processes, including the two responses illustrated here—a mating response and the response to high osmolarity. (A) The mating response is triggered when a mating factor secreted by a yeast of opposite mating type binds to a GPCR. This activates a G protein, the $\beta\gamma$ complex of which indirectly activates the MAPKKK (kinase A), which then relays the response onward. Once activated, the MAP kinase (kinase C) phosphorylates and thereby activates several downstream proteins that mediate the mating response, in which the yeast cell stops dividing and prepares for fusion. The three kinases in this module are bound to scaffold protein 1. (B) In a second response, a yeast cell exposed to a high-osmolarity environment is induced to synthesize glycerol to increase its internal osmolarity. This response is mediated by an osmolarity-sensing receptor protein and a different MAP kinase module bound to a second scaffold protein. (Note that the kinase domain of scaffold 2 provides the MAPKK activity of this module.) Although both pathways use the same MAPKKK (kinase A, *green*), there is no cross-talk between them because the kinases in each module are bound to different scaffold proteins, and the osmolarity-sensing receptor is bound to the same scaffold protein as the particular kinase it activates.

is unclear to what extent the individual components of MAP kinase modules can dissociate from the scaffold during the activation process to permit amplification.

Rho Family GTPases Functionally Couple Cell-Surface Receptors to the Cytoskeleton

Besides the Ras proteins, the other class of Ras superfamily GTPases that relays signals from cell-surface receptors is the large **Rho family** (see Table 15–5). Rho family monomeric GTPases regulate both the actin and microtubule cytoskeletons, controlling cell shape, polarity, motility, and adhesion (discussed in Chapter 16); they also regulate cell-cycle progression, gene transcription, and membrane transport. They play a key part in the guidance of cell migration and nerve axon outgrowth, mediating cytoskeletal responses to the activation of a special class of guidance receptors. We focus on this aspect of Rho family function here.

The three best-characterized family members are **Rho** itself, **Rac**, and **Cdc42**, each of which affects multiple downstream target proteins. In the same way as for Ras, GEFs activate and GAPs inactivate the Rho family GTPases; there are more than 80 Rho GEFs and more than 70 Rho GAPs in humans. Some of the GEFs and GAPs are specific for one particular family member, whereas others are less specific. Unlike Ras, which is membrane-associated even when inactive (with GDP bound), inactive Rho family GTPases are often bound to *guanine nucleotide dissociation inhibitors* (*GDIs*) in the cytosol, which prevent the GTPases from interacting with their Rho GEFs at the plasma membrane.

Signaling by extracellular signaling proteins of the **ephrin** family provides an example of how RTKs can activate a Rho GTPase. Ephrins bind and thereby activate members of the *Eph* family of RTKs (see Figure 15–44). One member of the Eph family is found on the surface of motor neurons and helps guide the migrating tip of the axon (called a *growth cone*) to its muscle target. The binding of a cell-surface *ephrin* protein activates the Eph receptor, causing the growth cones to collapse, thereby repelling them from inappropriate paths and keeping them on track. The response depends on a Rho GEF called *ephexin*, which is stably associated with the cytosolic tail of the Eph receptor. When ephrin binding activates the Eph receptor, the receptor activates a cytoplasmic tyrosine kinase that phosphorylates ephexin on a tyrosine, enhancing the ability of ephexin to activate the Rho protein RhoA. The activated RhoA (RhoA-GTP) then regulates various downstream target proteins, including some effector proteins that control the actin cytoskeleton, causing the growth cone to collapse (**Figure 15–52**).

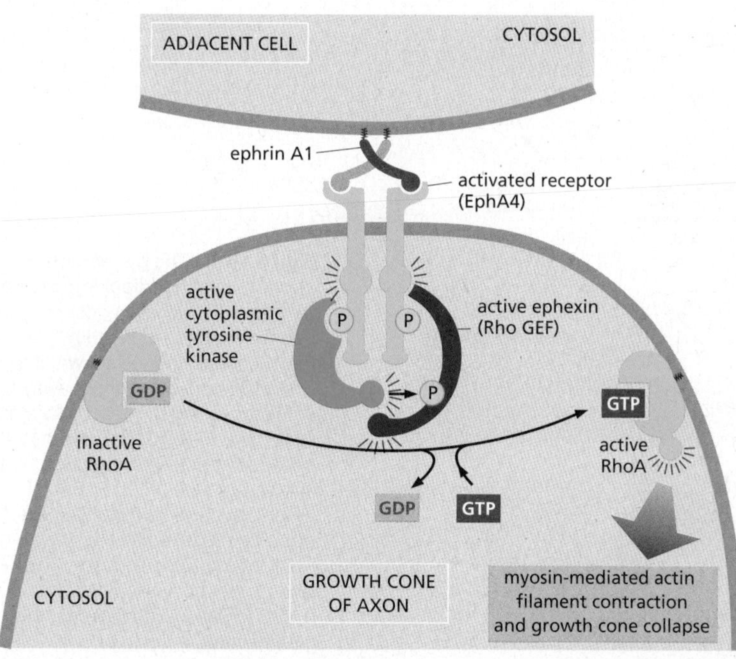

Figure 15–52 Growth cone collapse mediated by the Rho family GTPase RhoA. The binding of ephrin A1 proteins on an adjacent cell activates EphA4 RTKs on the growth cone. The resulting phosphotyrosines on the activated Eph receptors recruit and activate a cytoplasmic tyrosine kinase to phosphorylate the receptor-associated Rho GEF ephexin on a tyrosine. This enhances the ability of the ephexin to activate membrane-bound RhoA, which stimulates the myosin-dependent contraction of the actin cytoskeleton, thereby causing the growth cone to collapse.

Having considered how RTKs use GEFs and monomeric GTPases to relay signals into the cell, we now consider a second major RTK signaling strategy that depends on a different intracellular relay mechanism.

PI 3-Kinase Produces Lipid Docking Sites in the Plasma Membrane

As mentioned earlier (see Figure 15–47A), one of the proteins that binds to the intracellular tail of RTK molecules is the plasma-membrane-bound enzyme **phosphoinositide 3-kinase (PI 3-kinase)**. This kinase principally phosphorylates inositol phospholipids rather than proteins, and both RTKs and GPCRs can activate it. It plays a central part in promoting cell survival and growth.

Phosphatidylinositol (*PI*) is unusual among membrane lipids because it can undergo reversible phosphorylation at multiple sites on its inositol head group to generate a variety of phosphorylated PI lipids called **phosphoinositides**. When activated, PI 3-kinase catalyzes phosphorylation at the 3 position of the inositol ring to generate several phosphoinositides (**Figure 15–53**). The production of PI(3,4,5)P_3 matters most because it can serve as a docking site for various intracellular signaling proteins, which assemble into signaling complexes that relay the signal into the cell from the cytosolic face of the plasma membrane (see Figure 15–10C).

Notice the difference between this use of phosphoinositides and their use described earlier, in which PI(4,5)P_2 is cleaved by PLCβ (in the case of GPCRs) or PLCγ (in the case of RTKs) to generate soluble IP_3 and membrane-bound diacylglycerol (see Figures 15–29 and 15–30). By contrast, PI(3,4,5)P_3 is not cleaved by either PLC. It is made from PI(4,5)P_2 and then remains in the plasma membrane until specific *phosphoinositide phosphatases* dephosphorylate it. Prominent among these is the *PTEN* phosphatase, which dephosphorylates the 3 position of the inositol ring. Mutations in PTEN are found in many cancers: by prolonging signaling by PI 3-kinase, they promote uncontrolled cell growth.

There are various types of PI 3-kinases. Those activated by RTKs and GPCRs belong to class I. These are heterodimers composed of a common catalytic subunit and different regulatory subunits. RTKs activate *class Ia PI 3-kinases*, in which the regulatory subunit is an adaptor protein that binds to two phosphotyrosines

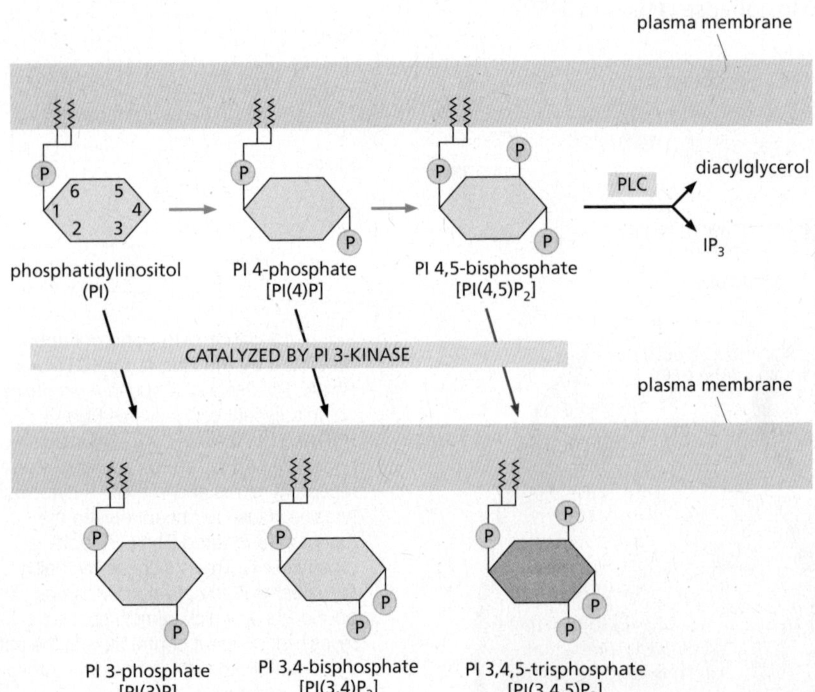

Figure 15–53 **The generation of phosphoinositide docking sites by PI 3-kinase.** PI 3-kinase phosphorylates the inositol ring on carbon atom 3 to generate the phosphoinositides shown at the bottom of the figure (diverting them away from the pathway leading to IP_3 and diacylglycerol; see Figure 15–29). The most important phosphorylation (indicated by the *red arrow*) is of PI(4,5)P_2 to PI(3,4,5)P_3, which can serve as a docking site for signaling proteins with PI(3,4,5)P_3-binding PH domains. Other inositol phospholipid kinases (not shown) catalyze the phosphorylations indicated by the *green arrows.*

on activated RTKs through its two SH2 domains (see Figure 15–47A). GPCRs activate *class Ib PI 3-kinases,* which have a regulatory subunit that binds to the $\beta\gamma$ complex of an activated heterotrimeric G protein (G_q) when GPCRs are activated by their extracellular ligand. The direct binding of activated Ras can also activate the common class I catalytic subunit.

Intracellular signaling proteins bind to PI(3,4,5)P_3 produced by activated PI 3-kinase via a specific interaction domain, such as a **pleckstrin homology (PH) domain**, first identified in the platelet protein pleckstrin. PH domains function mainly as protein–protein interaction domains, and it is only a small subset of them that bind to PI(3,4,5)P_3; at least some of these also recognize a specific membrane-bound protein as well as the PI(3,4,5)P_3, which greatly increases the specificity of the binding and helps explain why the signaling proteins with PI(3,4,5)P_3-binding PH domains do not all dock at all PI(3,4,5)P_3 sites. PH domains occur in about 200 human proteins, including the Ras GEF Sos discussed earlier (see Figure 15–48).

One especially important PH-domain-containing protein is the serine/threonine protein kinase *Akt.* The *PI-3-kinase–Akt signaling pathway* is the major pathway activated by the hormone *insulin.* It also plays a key part in promoting the survival and growth of many cell types in both invertebrates and vertebrates, as we now discuss.

The PI-3-Kinase–Akt Signaling Pathway Stimulates Animal Cells to Survive and Grow

As discussed earlier, extracellular signals are usually required for animal cells to grow and divide, as well as to survive (see Figure 15–4). Members of the *insulin-like growth factor* (*IGF*) family of signal proteins, for example, stimulate many types of animal cells to survive and grow. Like insulin, they bind to specific RTKs (see Figure 15–44), which activate PI 3-kinase to produce PI(3,4,5)P_3. The PI(3,4,5)P_3 recruits two protein kinases to the plasma membrane via their PH domains—**Akt** (also called *protein kinase B,* or *PKB*) and *phosphoinositide-dependent protein kinase 1* (*PDK1*)—and this leads to the activation of Akt (**Figure 15–54**). Once activated, Akt phosphorylates various target proteins at the plasma membrane, as well as in the cytosol and nucleus. The effect on most of the known targets is to inactivate them, but the targets are such that these actions of Akt all conspire

Figure 15–54 **One way in which signaling through PI 3-kinase can promote cell survival.** An extracellular survival signal activates an RTK, which recruits and activates PI 3-kinase. The PI 3-kinase produces PI(3,4,5)P_3, which serves as a docking site for two serine/threonine kinases with PH domains—Akt and the phosphoinositide-dependent kinase PDK1—and brings them into proximity at the plasma membrane. The Akt is phosphorylated on a serine by a third membrane-associated kinase (mTOR in complex 2, or mTORC2), which alters the conformation of the Akt so that it can be phosphorylated on a threonine by PDK1, which activates the Akt. The activated Akt now dissociates from the plasma membrane and phosphorylates various target proteins, including the Bad protein. When unphosphorylated, Bad holds the apoptosis-inhibitory Bcl2 in an inactive state (Bad and Bcl2 are both members of a family of proteins that regulates apoptosis, as discussed in Chapter 18). Once phosphorylated, Bad releases Bcl2, which now can block apoptosis and thereby promote cell survival. The phosphorylated Bad binds to a ubiquitous cytosolic protein called *14-3-3*, which keeps the protein out of action, as shown.

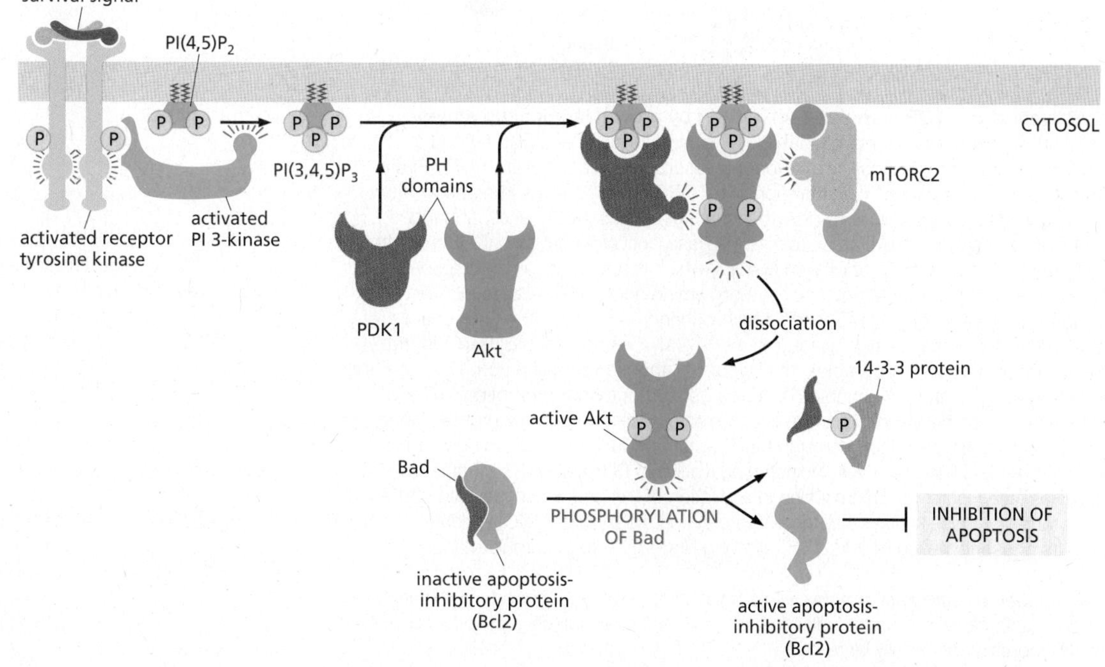

to enhance cell survival and growth, as illustrated for one cell survival pathway in Figure 15–54.

The control of cell growth by the **PI-3-kinase–Akt pathway** depends in part on a protein kinase called **TOR**, named as the target of *rapamycin*, a bacterial toxin that inactivates some forms of the kinase and is used clinically as both an immunosuppressant and anticancer drug. In mammalian cells, it is called **mTOR** and exists in cells in two functionally distinct multiprotein complexes. *mTOR complex 1* (*mTORC1*) contains the protein *raptor*; this complex is sensitive to rapamycin, and it stimulates cell growth. *mTOR complex 2* (*mTORC2*) contains the protein *rictor* and is insensitive to rapamycin; it helps to promote cell survival by activating Akt (see Figure 15–54) and also regulates the actin cytoskeleton via Rho family GTPases.

mTORC1 is a key regulator of cell physiology, which integrates inputs from multiple sources. The two best understood activators of mTORC1 are extracellular signal proteins referred to as *growth factors* and nutrients such as amino acids, both of which activate mTORC1 and thereby promote cell growth. The complex signaling network that governs mTORC1 activity includes examples of many of the classes of signaling molecules we have discussed in this chapter (**Figure 15–55**). We discuss in Chapter 17 how activated mTORC1 stimulates cell growth (see Figure 17–61).

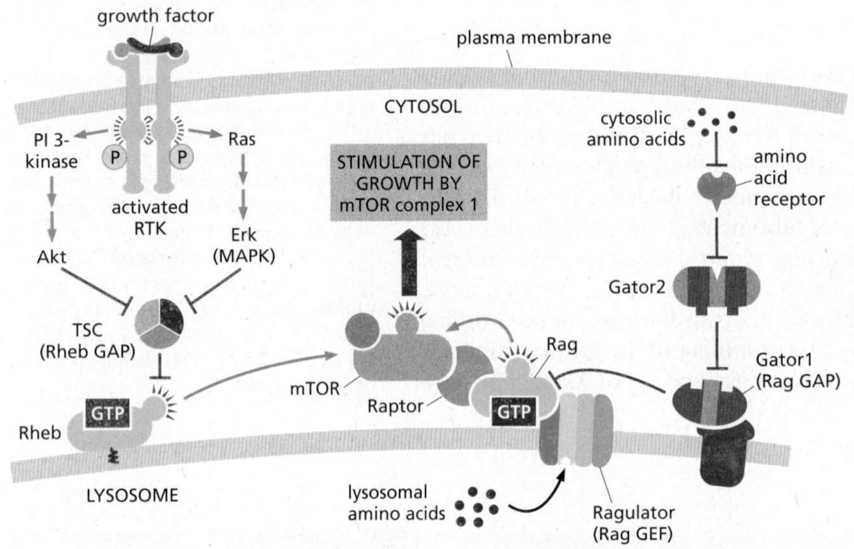

Figure 15–55 Activation of mTOR complex 1 (mTORC1) by growth factors and amino acids. mTORC1 activity depends on binding to the GTP-bound forms of two Ras-related GTPases, *Rag* and *Rheb*. In the presence of abundant cytosolic amino acids, Rag-GTP recruits mTORC1 to the surface of lysosomes. Full activation of mTORC1 then requires interaction with Rheb-GTP, which is activated in response to growth factor signaling. In both cases, enhanced GTP binding is driven primarily by inhibition of specific GTPase-activating proteins (GAPs). In the case of Rag, the GAP is a protein complex called *Gator1*, which is anchored on the lysosome and is regulated by a series of inhibitory interactions: cytosolic amino acids bind receptor proteins, thereby removing their inhibitory effect on *Gator2*, which is then free to inhibit the GAP activity of Gator1—resulting in Rag activation and binding to mTORC1. The interaction of Rag with the lysosome depends on a large protein complex, the *Ragulator*, that serves as an activating GEF for Rag; this GEF activity is stimulated by amino acids in the lysosome, further promoting mTORC1 activation. On the left side of the diagram are the mechanisms leading to activation of the Rheb GTPase by growth factors. As described earlier in the chapter, activation of RTKs by growth factor binding initiates various signaling pathways, including activation of PI 3-kinase (see Figure 15–54) and the GTPase Ras (see Figure 15–48), resulting in activation of the protein kinases Akt and Erk, respectively. One target of these kinases is a trimeric protein complex called *TSC*, which is a GAP for Rheb. Its phosphorylation inhibits TSC, allowing Rheb-GTP to accumulate and stimulate mTORC1.

TSC is short for *tuberous sclerosis complex*. Mutations in the genes that encode two of its three subunits, Tsc1 and Tsc2, cause the genetic disease *tuberous sclerosis,* which is associated with benign tumors that contain abnormally large cells.

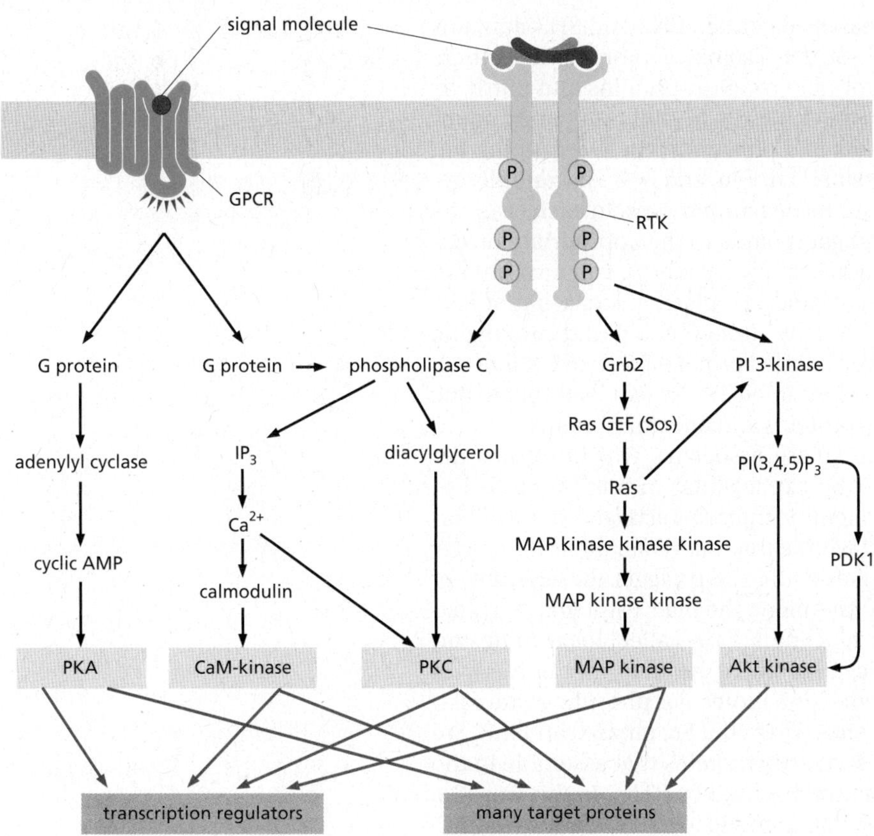

Figure 15–56 Five parallel intracellular signaling pathways activated by GPCRs, RTKs, or both. In this simplified example, the five kinases (shaded *gray*) at the end of each signaling pathway phosphorylate target proteins (shaded *red*), many of which are phosphorylated by more than one of the kinases. The phospholipase C activated by the two types of receptors is different: GPCRs activate PLCβ, whereas RTKs activate PLCγ (not shown). Although not shown, some GPCRs can also activate Ras, but they do so independently of Grb2, via a Ras GEF that is activated by Ca^{2+} and diacylglycerol.

RTKs and GPCRs Activate Overlapping Signaling Pathways

As mentioned earlier, RTKs and GPCRs activate some of the same intracellular signaling pathways. Both, for example, can activate the inositol phospholipid pathway triggered by phospholipase C. Moreover, even when they activate different pathways, the different pathways can converge on the same target proteins. **Figure 15–56** illustrates both of these types of signaling overlaps: it summarizes five parallel intracellular signaling pathways that we have discussed so far—one triggered by GPCRs, two triggered by RTKs, and two triggered by both kinds of receptors. Interactions among these pathways allow different extracellular signal molecules to modulate and coordinate each other's effects.

Some Enzyme-coupled Receptors Associate with Cytoplasmic Tyrosine Kinases

Many cell-surface receptors depend on tyrosine phosphorylation for their activity and yet lack a tyrosine kinase domain. These receptors act through **cytoplasmic tyrosine kinases**, which are associated with the receptors and phosphorylate various target proteins, often including the receptors themselves, when the receptors bind their ligand. These **tyrosine-kinase-associated receptors** thus function in much the same way as RTKs, except that their kinase domain is encoded by a separate gene and is noncovalently associated with the receptor polypeptide chain. A variety of receptor classes belong in this category, including the receptors for antigen and interleukins on lymphocytes (discussed in Chapter 24), integrins (discussed in Chapter 19), and receptors for various cytokines and some hormones. As with RTKs, many of these receptors depend on dimerization for their activation.

Some of these receptors work with members of the largest family of mammalian cytoplasmic tyrosine kinases, the **Src family**, which includes *Src, Yes, Fgr, Fyn,*

Lck, Lyn, Hck, and *Blk.* These protein kinases all contain SH2 and SH3 domains and are located on the cytoplasmic side of the plasma membrane, held there partly by their interaction with transmembrane receptor proteins and partly by covalently attached lipid chains (discussed in Chapter 3). Different family members are associated with different receptors and phosphorylate overlapping but distinct sets of intracellular signaling proteins. Lyn, Fyn, and Lck, for example, are each associated with different sets of receptors on lymphocytes. In each case, the kinase is activated when an extracellular ligand binds to the appropriate receptor protein. Src itself, as well as several other family members, can also bind to activated RTKs; in these cases, the receptor and cytoplasmic kinases mutually stimulate each other's catalytic activity, thereby strengthening and prolonging the signal (see Figure 15–52). There are even some G proteins (G_s and G_i) that can activate Src, which is one way that the activation of GPCRs can lead to tyrosine phosphorylation of intracellular signaling proteins and effector proteins.

Another type of cytoplasmic tyrosine kinase associates with *integrins,* the main receptors that cells use to bind to the extracellular matrix (discussed in Chapter 19). The binding of matrix components to integrins activates intracellular signaling pathways that influence the behavior of the cell. When integrins cluster at sites of contact with the extracellular matrix, they help trigger the assembly of cell–matrix junctions called *focal adhesions.* Among the many proteins recruited into these junctions is the cytoplasmic tyrosine kinase called **focal adhesion kinase (FAK)**, which binds to the cytosolic tail of one of the integrin subunits with the assistance of other proteins. The clustered FAK molecules phosphorylate each other, creating phosphotyrosine docking sites where the Src kinase can bind. Src and FAK then phosphorylate each other and other proteins that assemble in the junction, including many of the signaling proteins used by RTKs. In this way, the two tyrosine kinases signal to the cell that it has adhered to a suitable substratum, where the cell can now survive, grow, divide, migrate, and so on.

The largest and most diverse class of receptors that rely on cytoplasmic tyrosine kinases is the *cytokine receptors,* which we consider next.

Cytokine Receptors Activate the JAK–STAT Signaling Pathway

The large family of **cytokine receptors** includes receptors for many kinds of local mediators (collectively called *cytokines*), as well as receptors for some hormones, such as *growth hormone* and *prolactin* (Movie 15.9). These receptors are stably associated with cytoplasmic tyrosine kinases called **Janus kinases (JAKs**; after the two-faced Roman god), which phosphorylate and activate transcription regulators called STATs (signal transducers and activators of transcription). STAT proteins are located in the cytosol and are referred to as *latent transcription regulators* because they migrate into the nucleus and regulate gene transcription only after they are activated.

Although many intracellular signaling pathways lead from cell-surface receptors to the nucleus, where they alter gene transcription (see Figure 15–56), the **JAK–STAT signaling pathway** provides one of the more direct routes. Cytokine receptors are dimers or trimers and are stably associated with one or two of the four known JAKs (JAK1, JAK2, JAK3, and Tyk2). Cytokine binding alters the arrangement so as to bring two JAKs into close proximity so that they phosphorylate each other, thereby increasing the activity of their tyrosine kinase domains. The JAKs then phosphorylate tyrosines on the cytoplasmic tails of cytokine receptors, creating phosphotyrosine docking sites for STATs (**Figure 15–57**). Some adaptor proteins can also bind to some of these sites and couple cytokine receptors to the Ras–MAP-kinase signaling pathway discussed earlier, but these will not be discussed here.

There are at least six STATs in mammals. Each has an SH2 domain that performs two functions. First, it mediates the binding of the STAT protein to a phosphotyrosine docking site on an activated cytokine receptor. Once bound, the JAKs phosphorylate the STAT on tyrosines, causing the STAT to dissociate from the receptor. Second, the SH2 domain on the released STAT now mediates its binding to a phosphotyrosine on another STAT molecule, forming either a STAT homodimer

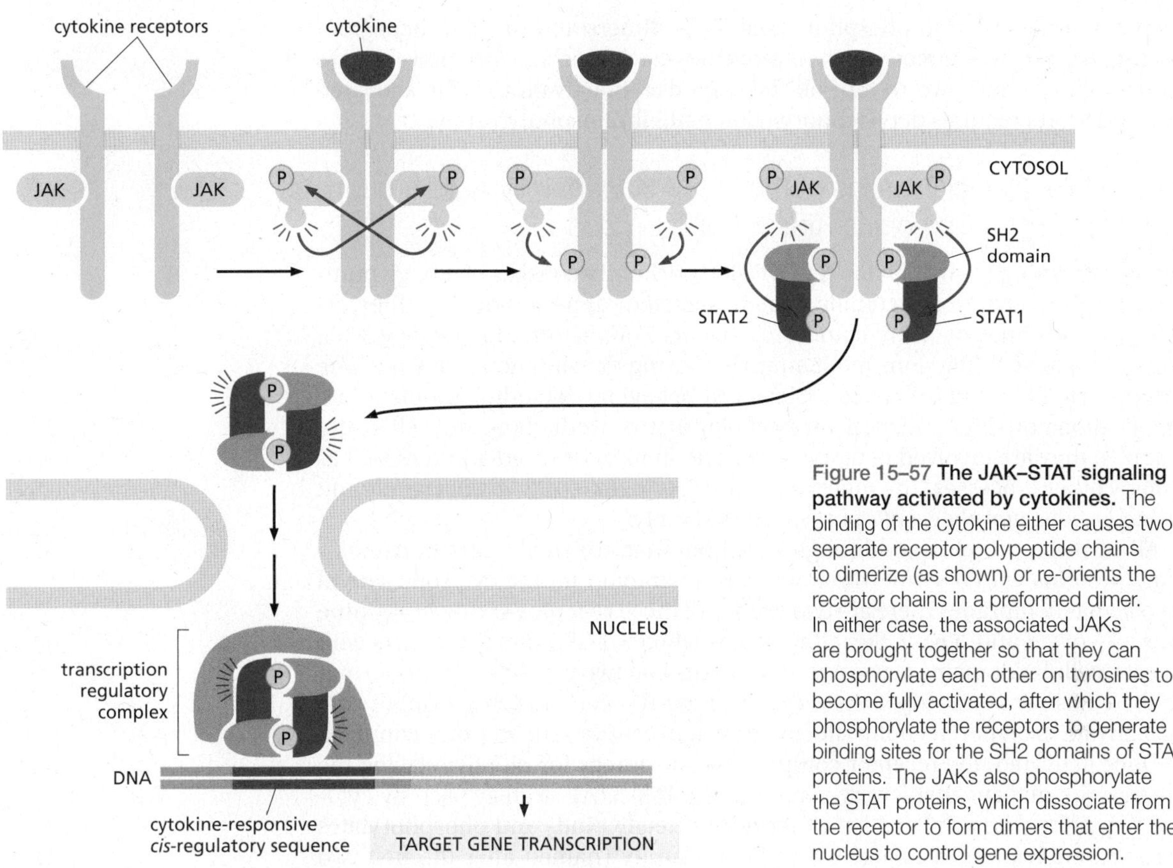

Figure 15–57 The JAK–STAT signaling pathway activated by cytokines. The binding of the cytokine either causes two separate receptor polypeptide chains to dimerize (as shown) or re-orients the receptor chains in a preformed dimer. In either case, the associated JAKs are brought together so that they can phosphorylate each other on tyrosines to become fully activated, after which they phosphorylate the receptors to generate binding sites for the SH2 domains of STAT proteins. The JAKs also phosphorylate the STAT proteins, which dissociate from the receptor to form dimers that enter the nucleus to control gene expression.

or a heterodimer. The STAT dimer then translocates to the nucleus, where, in combination with other transcription regulators, it binds to a specific *cis*-regulatory sequence in various genes and stimulates their transcription (see Figure 15–57). In response to the hormone prolactin, for example, which stimulates breast cells to produce milk, activated STAT5 stimulates the transcription of genes that encode milk proteins. **Table 15–6** lists some of the more than 30 cytokines and hormones that activate the JAK–STAT pathway by binding to cytokine receptors.

Negative feedback regulates the responses mediated by the JAK–STAT pathway. In addition to activating genes that encode proteins mediating the cytokine-induced response, the STAT dimers can also activate genes that encode inhibitory proteins that help shut off the response. Some of these proteins bind to and inactivate phosphorylated JAKs and their associated phosphorylated

TABLE 15–6 Some Extracellular Signal Proteins That Act Through Cytokine Receptors and the JAK–STAT Signaling Pathway

Signal protein	Receptor-associated JAKs	STATs activated	Some responses
Interferon-γ (IFNγ)	JAK1 and JAK2	STAT1	Activates macrophages
Interferon-α (IFNα)	Tyk2 and JAK2	STAT1 and STAT2	Increases cell resistance to viral infection
Erythropoietin	JAK2	STAT5	Stimulates production of erythrocytes
Prolactin	JAK1 and JAK2	STAT5	Stimulates milk production
Growth hormone	JAK2	STAT1 and STAT5	Stimulates growth by inducing IGF1 production
Granulocyte–macrophage-colony-stimulating factor (GMCSF)	JAK2	STAT5	Stimulates production of granulocytes and macrophages

receptors; others bind to phosphorylated STAT dimers and prevent them from binding to their DNA targets. Such negative feedback mechanisms, however, are not enough on their own to turn off the response. Inactivation of the activated JAKs and STATs requires dephosphorylation of their phosphotyrosines.

Extracellular Signal Proteins of the TGFβ Superfamily Act Through Receptor Serine/Threonine Kinases and Smads

The **transforming growth factor-β (TGFβ) superfamily** consists of a large number (33 in humans) of structurally related, secreted, dimeric proteins. They act either as hormones or, more commonly, as local mediators to regulate a wide range of biological functions in all animals. During development, they regulate pattern formation and influence various cell behaviors, including proliferation, specification and differentiation, extracellular matrix production, and cell death. In adults, they are involved in tissue repair and in immune regulation, as well as in many other processes. The superfamily consists of the TGFβ/*activin* family and the larger *bone morphogenetic protein* (*BMP*) family.

All of these proteins act through receptors that are single-pass transmembrane proteins with a serine/threonine kinase domain on the cytosolic side of the plasma membrane. There are two classes of these **receptor serine/threonine kinases**—*type I* and *type II*. Signaling begins when a TGFβ dimer interacts with the extracellular domains of two type-I receptors and two type-II receptors, bringing the kinase domains together so that the type-II receptors can phosphorylate and activate the type-I receptors, forming an active tetrameric receptor complex.

Once activated, the receptor complex uses a strategy for rapidly relaying the signal to the nucleus that is very similar to the JAK–STAT strategy used by cytokine receptors. The activated type-I receptor directly binds and phosphorylates a latent transcription regulator of the **Smad family** (named after the first two proteins identified, Sma in *Caenorhabditis elegans* and Mad in *Drosophila*). Activated TGFβ/activin receptors phosphorylate Smad2 or Smad3, while activated BMP receptors phosphorylate Smad1, Smad5, or Smad8. Once one of these *receptor-activated Smads* (*R-Smads*) has been phosphorylated, it binds to Smad4 (called a *co-Smad*), which can form a complex with any of the five R-Smads. The Smad complex then translocates into the nucleus, where it associates with other transcription regulators and controls the transcription of specific target genes (**Figure 15–58**). Because the partner proteins in the nucleus vary depending on the cell type and state of the cell, the genes affected vary.

Activated TGFβ receptors and their bound ligand are endocytosed by two distinct routes, one leading to further activation and the other leading to inactivation. The activation route depends on clathrin-coated vesicles and leads to early endosomes (discussed in Chapter 13), where most of the Smad activation occurs. An anchoring protein called *SARA* (for *Smad anchor for receptor activation*) has an important role in this pathway; it is concentrated in early endosomes and binds to both activated TGFβ receptors and Smads, increasing the efficiency of receptor-mediated Smad phosphorylation. The inactivation route depends on *caveolae* (discussed in Chapter 13) and leads to receptor ubiquitylation and degradation in proteasomes.

During the signaling response, the Smads shuttle continually between the cytoplasm and the nucleus: they are dephosphorylated in the nucleus and exported to the cytoplasm, where they can be rephosphorylated by activated receptors. In this way, the effect exerted on the target genes reflects both the concentration of the extracellular signal and the time the signal continues to act on the cell-surface receptors (often several hours). Cells exposed to a morphogen at high concentration, or for a long time, or both, will switch on one set of genes, whereas cells receiving a lower or more transient exposure will switch on another set.

As in other signaling systems, negative feedback regulates the Smad pathway. Among the target genes activated by Smad complexes are those that encode *inhibitory Smads*, either Smad6 or Smad7. Smad7 (and possibly Smad6) binds to the cytosolic tail of the activated receptor and inhibits its signaling ability in at least three ways: (1) it competes with R-Smads for binding sites on the receptor,

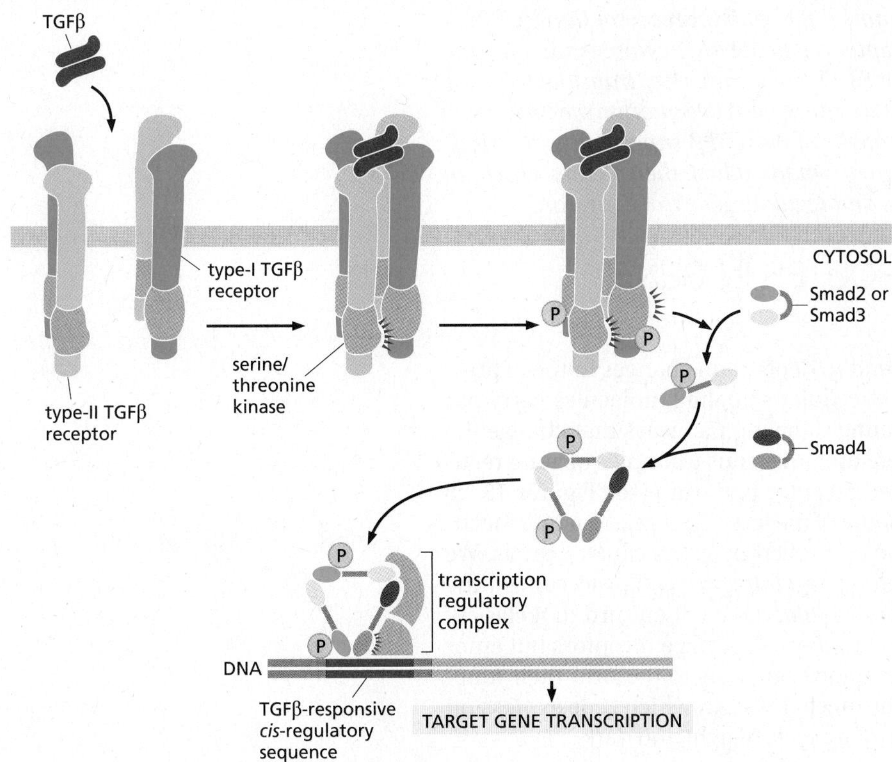

Figure 15–58 The Smad-dependent signaling pathway activated by TGFβ. The TGFβ dimer promotes the assembly of a tetrameric receptor complex containing two copies each of the type-I and type-II receptors. The type-II receptors phosphorylate specific sites on the type-I receptors, thereby activating their kinase domains and leading to phosphorylation of R-Smads such as Smad2 and Smad3. Smads open up to expose a binding surface when they are phosphorylated, leading to the formation of a trimeric Smad complex containing two R-Smads and the co-Smad, Smad4. The phosphorylated Smad complex enters the nucleus and collaborates with other transcription regulators to control the transcription of specific target genes.

decreasing R-Smad phosphorylation; (2) it recruits a ubiquitin ligase called *Smurf*, which ubiquitylates the receptor, leading to receptor internalization and degradation (it is because Smurfs also ubiquitylate and promote the degradation of Smads that they are called *Smad ubiquitylation regulatory factors*, or Smurfs); and (3) it recruits a protein phosphatase that dephosphorylates and inactivates the receptor. In addition, the inhibitory Smads bind to the co-Smad, Smad4, and inhibit it, either by preventing its binding to R-Smads or by promoting its ubiquitylation and degradation.

Summary

There are various classes of enzyme-coupled receptors, the most common of which are receptor tyrosine kinases (RTKs), tyrosine-kinase-associated receptors, and receptor serine/threonine kinases.

Ligand binding to RTKs causes their dimerization, which leads to activation of their kinase domains. These activated kinase domains phosphorylate multiple tyrosines on the receptors, producing a set of phosphotyrosines that serve as docking sites for a set of intracellular signaling proteins, which bind via their SH2 (or PTB) domains. One such signaling protein serves as an adaptor to couple some activated receptors to a Ras GEF (Sos), which activates the monomeric GTPase Ras; Ras, in turn, activates a three-component MAP kinase signaling module, which relays the signal to the nucleus by phosphorylating transcription regulators. Another important signaling protein that can dock on activated RTKs is PI 3-kinase, which phosphorylates specific phosphoinositides to produce lipid docking sites in the plasma membrane for signaling proteins with phosphoinositide-binding PH domains, including the serine/threonine protein kinase Akt, which plays a key part in the control of cell survival and cell growth. Many receptor classes, including some RTKs, activate Rho family monomeric GTPases, which functionally couple the receptors to the cytoskeleton.

Tyrosine-kinase-associated receptors depend on various cytoplasmic tyrosine kinases for their action. These kinases include members of the Src family, which associate with many kinds of receptors, and the focal adhesion kinase (FAK), which associates with integrins at focal adhesions. The cytoplasmic tyrosine kinases then phosphorylate a variety of signaling proteins to relay the signal onward. The largest family of tyrosine-kinase-associated receptors is the cytokine receptor family. When stimulated by ligand binding, these receptors activate JAK cytoplasmic tyrosine kinases, which phosphorylate STATs. The STATs then dimerize, translocate to the nucleus, and activate the transcription of specific genes. Receptor serine/threonine kinases, which are activated by signal proteins of the TGFβ superfamily, act similarly: they directly phosphorylate and activate Smads, which then oligomerize with another Smad, translocate to the nucleus, and regulate gene transcription.

ALTERNATIVE SIGNALING ROUTES IN GENE REGULATION

Major changes in the behavior of a cell tend to depend on changes in the expression of numerous genes. Thus, many extracellular signaling molecules carry out their effects, in whole or in part, by initiating signaling pathways that change the activities of transcription regulators. There are numerous examples of gene regulation in both GPCR and enzyme-coupled receptor pathways (see Figures 15–28 and 15–50). In this section, we describe some of the less common signaling mechanisms by which gene expression can be controlled by extracellular signals. We begin with several pathways that depend on *regulated proteolysis* to control the activity and location of latent *transcription regulators*. We then turn to a class of extracellular signal molecules that do not employ cell-surface receptors but enter the cell and interact directly with transcription regulators to perform their functions. Finally, we briefly discuss some of the mechanisms by which gene expression is controlled by the *circadian rhythm*: the daily cycle of light and dark.

The Receptor Notch Is a Latent Transcription Regulator

Signaling through the **Notch** protein is used widely in animal development. As discussed in Chapter 21, it has a general role in controlling cell-fate choices and regulating pattern formation during the development of most tissues, as well as in the continual cell renewal in tissues such as the lining of the gut. It is best known, however, for its role in the production of *Drosophila* neural cells, which usually arise as isolated single cells within an epithelial sheet of precursor cells. During this process, when a precursor cell commits to becoming a neural cell, it signals to its immediate neighbors not to do the same; the inhibited cells develop into epidermal cells instead. This process, called *lateral inhibition*, depends on a contact-dependent signaling mechanism that is activated by a single-pass transmembrane signal protein called **Delta**, displayed on the surface of the future neural cell. By binding to its receptor, the Notch protein, on a neighboring cell, Delta signals to the neighbor not to become neural (**Figure 15–59**). When this signaling process is defective, a huge excess of neural cells is produced at the expense of epidermal cells, which is lethal.

Notch is a single-pass transmembrane protein that requires proteolytic processing to function. It acts as a latent transcription regulator and provides the

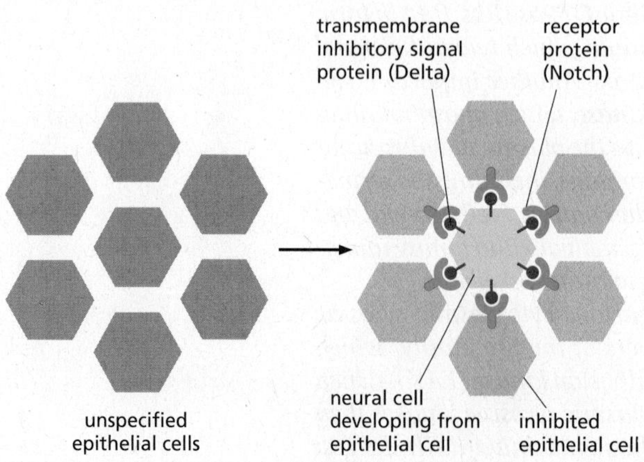

Figure 15–59 Lateral inhibition mediated by Notch and Delta during neural cell development in *Drosophila*. When individual cells in the epithelium begin to develop as neural cells, they signal to their neighbors not to do the same. This inhibitory, contact-dependent signaling is mediated by the ligand Delta, which appears on the surface of the future neural cell and binds to Notch protein on the neighboring cells. In many tissues, all the cells in a cluster initially express both Delta and Notch, and a competition occurs, with one cell emerging as winner, expressing Delta strongly and inhibiting its neighbors from doing likewise. In other cases, additional factors interact with Delta or Notch to make some cells susceptible to the lateral inhibition signal and others unresponsive to it.

simplest and most direct signaling pathway known from a cell-surface receptor to the nucleus. When activated by the binding of Delta on another cell, a plasma-membrane-bound protease cleaves off the cytoplasmic tail of Notch, and the released tail translocates into the nucleus to activate the transcription of a set of Notch-response genes. The Notch tail fragment acts by binding to a DNA-binding protein, converting it from a transcription repressor into a transcription activator.

Notch undergoes three successive proteolytic cleavage steps, but only the last two depend on Delta binding. As part of its normal biosynthesis, it is cleaved in the Golgi apparatus to form a heterodimer, which is then transported to the cell surface as the mature receptor. The binding of Delta to Notch induces a second cleavage in the extracellular domain, mediated by an extracellular protease. A final cleavage quickly follows, cutting free the cytoplasmic tail of the activated Notch (**Figure 15–60**). Note that, unlike most receptors, the activation of Notch is irreversible; once activated by ligand binding, the protein cannot be used again.

This final cleavage of the Notch tail occurs just within the transmembrane segment, and it is mediated by a protease complex called *γ-secretase*, which is also

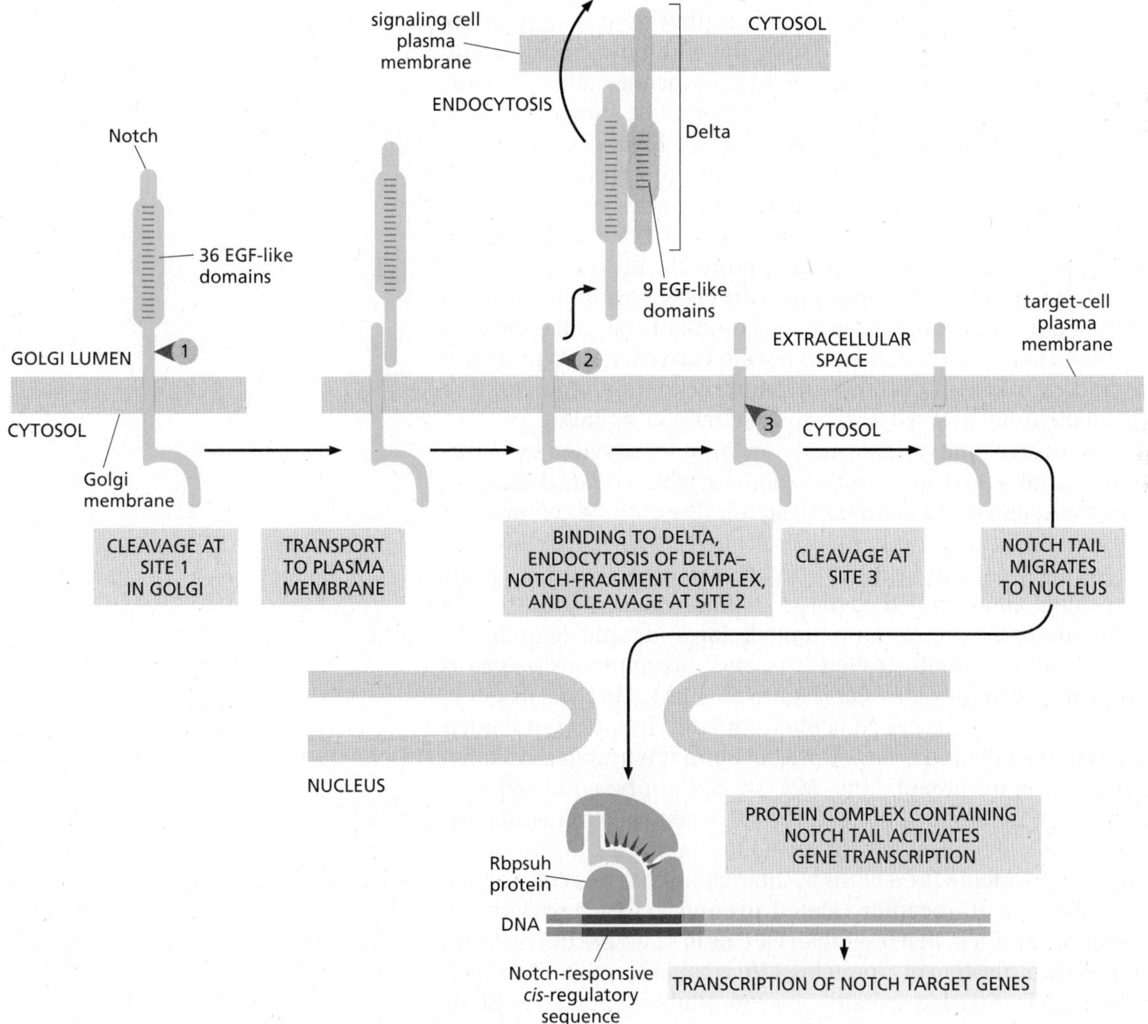

Figure 15–60 **The processing and activation of Notch by proteolytic cleavage.** The *numbered red arrowheads* indicate the sites of proteolytic cleavage. The first proteolytic processing step occurs within the *trans* Golgi network to generate mature heterodimeric Notch, which is then displayed on the cell surface. The binding to Delta on a neighboring cell triggers the next two proteolytic steps: the complex of Delta and the Notch fragment to which it is bound is endocytosed by the Delta-expressing cell, exposing the extracellular cleavage site in the transmembrane Notch subunit. Note that Notch and Delta interact through their repeated EGF-like domains. The released Notch tail migrates into the nucleus, where it binds to the Rbpsuh protein, which it converts from a transcription repressor to a transcription activator.

responsible for the intramembrane cleavage of various other single-pass transmembrane proteins. One of its essential subunits is *Presenilin*, so called because mutations in the gene encoding it are a frequent cause of early-onset familial Alzheimer's disease, a form of presenile dementia. The protease complex is thought to contribute to this and other forms of Alzheimer's disease by generating extracellular peptide fragments from a transmembrane neuronal protein; the fragments accumulate in excessive amounts and form aggregates of misfolded protein called amyloid plaques, which may injure nerve cells and contribute to their degeneration and loss.

Wnt Proteins Activate Frizzled and Thereby Inhibit β-Catenin Degradation

Wnt proteins are secreted signal molecules that control various aspects of animal development. They were discovered independently in flies and in mice: in *Drosophila*, the *Wingless* (*Wg*) gene originally came to light because of its role in wing development, while in mice, the *Int1* gene was found because it promoted the formation of breast tumors when activated by the integration of a virus next to it. Both of these genes encode Wnt proteins. There are 19 Wnts in humans, each having distinct, but often overlapping, functions.

Wnts are unusual extracellular signaling proteins in that they are covalently attached to a fatty acid chain after their synthesis in the endoplasmic reticulum. Wnts are therefore hydrophobic molecules that tend to associate with cell membranes and do not diffuse rapidly in the extracellular environment. They are thought to act primarily as local (paracrine) signaling molecules (see Figure 15–2B).

Wnts have been implicated in several signaling pathways. The best understood, and our primary focus here, is the **Wnt/β-catenin pathway** (also known as the *canonical Wnt pathway*), which centers on a protein called **β-catenin**. β-catenin is an intriguing example of a signaling protein with two seemingly unconnected functions in the same cell. A portion of the cell's β-catenin is located at cell–cell junctions and thereby contributes to the control of cell–cell adhesion; this function is discussed in Chapter 19 and does not concern us here. Wnt signaling acts primarily on cytoplasmic β-catenin, a latent transcription regulator that is degraded rapidly in unstimulated cells but stabilized by Wnt signaling.

In the absence of Wnt signaling, cytoplasmic β-catenin is degraded by a process that depends on a large protein *degradation complex*, which binds β-catenin, keeps it out of the nucleus, and promotes its destruction. The complex contains at least four other proteins, including a protein kinase called *casein kinase 1* (*CK1*), which phosphorylates the β-catenin on a serine, priming it for further phosphorylation by another protein kinase called *glycogen synthase kinase 3* (*GSK3*). This final phosphorylation marks the protein for ubiquitylation and rapid degradation in proteasomes. Two scaffold proteins called *axin* and *Adenomatous polyposis coli* (*APC*) hold the protein complex together (**Figure 15–61A**). APC gets its name from the finding that the gene encoding it is often mutated in a type of benign tumor (an adenoma) of the colon; the tumor projects into the lumen as a polyp and can eventually become malignant. (This APC should not be confused with the anaphase-promoting complex, or APC/C, that plays a central part in selective protein degradation during the cell cycle—discussed in Chapter 17.)

Wnt proteins regulate β-catenin proteolysis by interacting with two cell-surface co-receptors, **Frizzled** and **LDL-receptor-related protein (LRP)**. Frizzled is a seven-pass transmembrane protein that resembles GPCRs in structure but does not generally work through the activation of G proteins. LRP is a relatively simple single-pass transmembrane protein. Frizzled contains an extracellular domain that binds with high affinity to Wnt, due in part to a hydrophobic pocket on the receptor domain that interacts with the lipid modification on Wnt; this lipid is therefore required for Wnt signaling. LRP interacts with a different site on Wnt. Thus, as in so many other signaling pathways, the Wnt ligand drives formation of a co-receptor dimer.

Formation of the activated receptor complex promotes phosphorylation of the LRP receptor by the two protein kinases, CK1 and GSK3. Receptor activation also leads to recruitment of the scaffold protein **Dishevelled**. Axin is brought to the

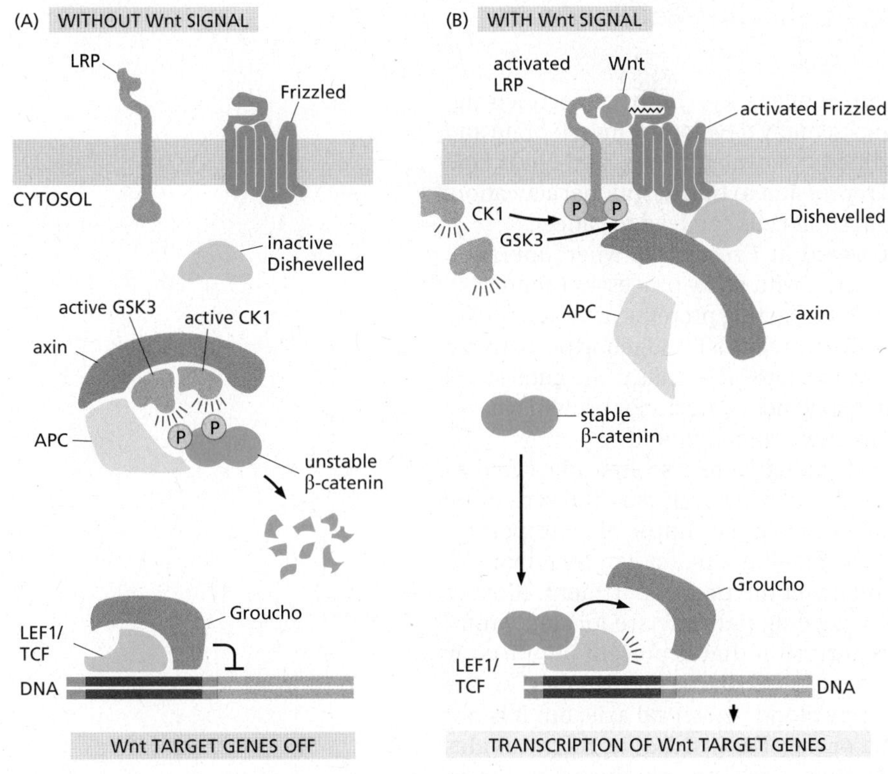

Figure 15–61 **The Wnt/β-catenin signaling pathway.** (A) In the absence of a Wnt signal, β-catenin that is not bound to cell–cell adherens junctions (not shown—discussed in Chapter 19) interacts with a degradation complex containing APC, axin, GSK3, and CK1. In this complex, β-catenin is phosphorylated by CK1 and then by GSK3, triggering its ubiquitylation and degradation in proteasomes. Wnt-responsive genes are kept inactive by the Groucho co-repressor protein bound to the transcription regulator LEF1/TCF. (B) Wnt binding to Frizzled and LRP brings the two co-receptors together, and the cytosolic tail of LRP is phosphorylated by CK1 and GSK3. The scaffold protein Dishevelled is recruited to the activated Frizzled protein. Axin binds to Dishevelled and the phosphorylated LRP and is inactivated, resulting in disassembly of the degradation complex. The phosphorylation of β-catenin is thereby prevented, and unphosphorylated β-catenin accumulates and translocates to the nucleus, where it binds to LEF1/TCF, displaces the co-repressor Groucho, and acts as a coactivator to stimulate the transcription of Wnt target genes. The scaffold protein Dishevelled is required for the signaling pathway to operate; it binds to Frizzled and becomes phosphorylated (not shown), but its precise role is unknown.

receptor complex and inactivated, thereby disrupting the β-catenin degradation complex in the cytoplasm. In this way, the phosphorylation and degradation of β-catenin are prevented, enabling β-catenin to accumulate and translocate to the nucleus, where it alters the pattern of gene transcription (**Figure 15–61B**).

In the absence of Wnt signaling, Wnt-responsive genes are kept silent by an inhibitory complex of transcription regulators. The complex includes proteins of the *LEF1/TCF* family bound to a co-repressor protein of the *Groucho* family (see Figure 15–61A). In response to a Wnt signal, β-catenin enters the nucleus and binds to the LEF1/TCF proteins, displacing Groucho. The β-catenin now functions as a coactivator, inducing the transcription of the Wnt target genes (see Figure 15–61B). Thus, as in the case of Notch signaling, Wnt/β-catenin signaling triggers a switch from transcriptional repression to transcriptional activation.

Among the genes activated by β-catenin is *Myc*, which encodes a protein (Myc) that is an important regulator of cell growth and proliferation (discussed in Chapter 17). Mutations of the *Apc* gene occur in 80% of human colon cancers (discussed in Chapter 20). These mutations inhibit the protein's ability to bind β-catenin, so that β-catenin accumulates in the nucleus and stimulates the transcription of *c-Myc* and other Wnt target genes, even in the absence of Wnt signaling. The resulting uncontrolled cell growth and proliferation promote the development of cancer.

The Wnt signaling pathway shown in Figure 15–61 is supplemented with additional regulatory inputs that fine-tune the strength of the Wnt signal. Some Wnt-stimulated genes, for example, suppress the Wnt signal, resulting in negative feedback. One of these genes encodes a secreted enzyme called Notum, which removes the fatty acid modification from Wnt, thereby inactivating it. Another Wnt-stimulated gene encodes a cell-surface transmembrane protein, Rnf43, which is a ubiquitin ligase that targets the Frizzled protein for degradation. This negative feedback mechanism can be suppressed by extracellular signaling molecules from other cells: a signaling protein called R-spondin, for example, binds to a GPCR called Lgr, which inactivates Rnf43, thereby enhancing the Wnt signal. Through these and a variety of other mechanisms, the localization and strength of the Wnt signal can be fine-tuned in different tissues.

Hedgehog Proteins Initiate a Complex Signaling Pathway in the Primary Cilium

Hedgehog proteins and Wnt proteins act in similar ways. Both are secreted signal molecules, which act as local mediators in many developing invertebrate and vertebrate tissues. Both proteins are modified by covalently attached lipids, and both trigger a switch from transcriptional repression to transcriptional activation. Excessive signaling along either pathway in adult cells can lead to cancer.

The first **Hedgehog protein** was discovered in *Drosophila*, where mutation of the *Hedgehog* gene produces a larva covered with spiky processes (denticles), like a hedgehog. At least three genes encode Hedgehog proteins in vertebrates—*Sonic*, *Desert*, and *Indian hedgehog*. The active forms of all Hedgehog proteins are covalently coupled to cholesterol, as well as to a fatty acid chain. The cholesterol is added during an unusual processing step in which a precursor protein cleaves itself to produce a smaller, cholesterol-containing signal protein.

The signaling proteins activated by Hedgehog were also first identified in *Drosophila* and are conserved in vertebrates and other animals. The vertebrate pathway, which we focus on here, provides a striking example of an important concept: the sensitivity and efficiency of a signaling system can be enhanced by concentrating its components in a small volume or compartment. Most of the signaling proteins of the vertebrate Hedgehog pathway are located within the **primary cilium**, a small membrane protrusion that is present in one copy on the surface of most vertebrate cell types. As we discuss in Chapter 16, the primary cilium contains a microtubule array along its central axis, but it is not motile like other microtubule-based cilia or flagella; instead, its microtubules are used as tracks for the transport of various signaling proteins to and from the tip. All the early signaling steps in the Hedgehog pathway occur in the cilium, and signaling is lost in cells with defects in primary cilium formation or function. Thus, the primary cilium serves as an antenna for the extracellular Hedgehog signal.

Hedgehog signaling begins at a cell-surface receptor called **Patched**, which employs a convoluted mechanism to activate a group of transcription regulators called *Gli proteins*, thereby increasing the expression of genes that drive changes in the target cell's proliferation or developmental fate. In the absence of Hedgehog ligand, the unoccupied Patched protein resides in the ciliary membrane and inhibits the activity of another transmembrane protein called **Smoothened**, which is located outside the cilium (**Figure 15–62A**). Smoothened is a GPCR-like protein that resembles the Wnt receptor Frizzled; like Frizzled, it has an external domain that serves as a lipid-binding activation domain, and its activation requires binding of this domain to cholesterol in the cell membrane. Patched, in contrast, is a large transmembrane protein that transports cholesterol out of the membrane. It has been proposed that Patched inhibits Smoothened by reducing the amount of cholesterol in the ciliary membrane, and Hedgehog inhibits Patched by blocking its cholesterol transport channel.

In cells lacking the Hedgehog signal, expression of Hedgehog-responsive genes is blocked by two mechanisms. First, an inhibitory protein called *SuFu* holds the Gli transcription regulators in an inactive state within the cilium. Second, one member of the family, Gli3, is proteolytically processed to form a smaller fragment that acts as a transcription repressor, helping to keep Hedgehog-responsive genes silent. This processing of the Gli3 protein depends on a signaling pathway that begins with a GPCR called Gpr161, which is found in the cilium membrane and stimulates adenylyl cyclase to produce cyclic AMP; cyclic AMP then stimulates PKA, which phosphorylates Gli3 to promote its partial processing into a transcription repressor (see Figure 15–62A).

The binding of Hedgehog to the Patched protein promotes the transport of Patched out of the cilium and removes its inhibitory effects on Smoothened, which moves into the cilium. Gpr161 is removed from the cilium, thereby reducing formation of the Gli3 transcription repressor. Inactive complexes of Gli proteins and SuFu are transported to the tip of the cilium, where the activated Smoothened

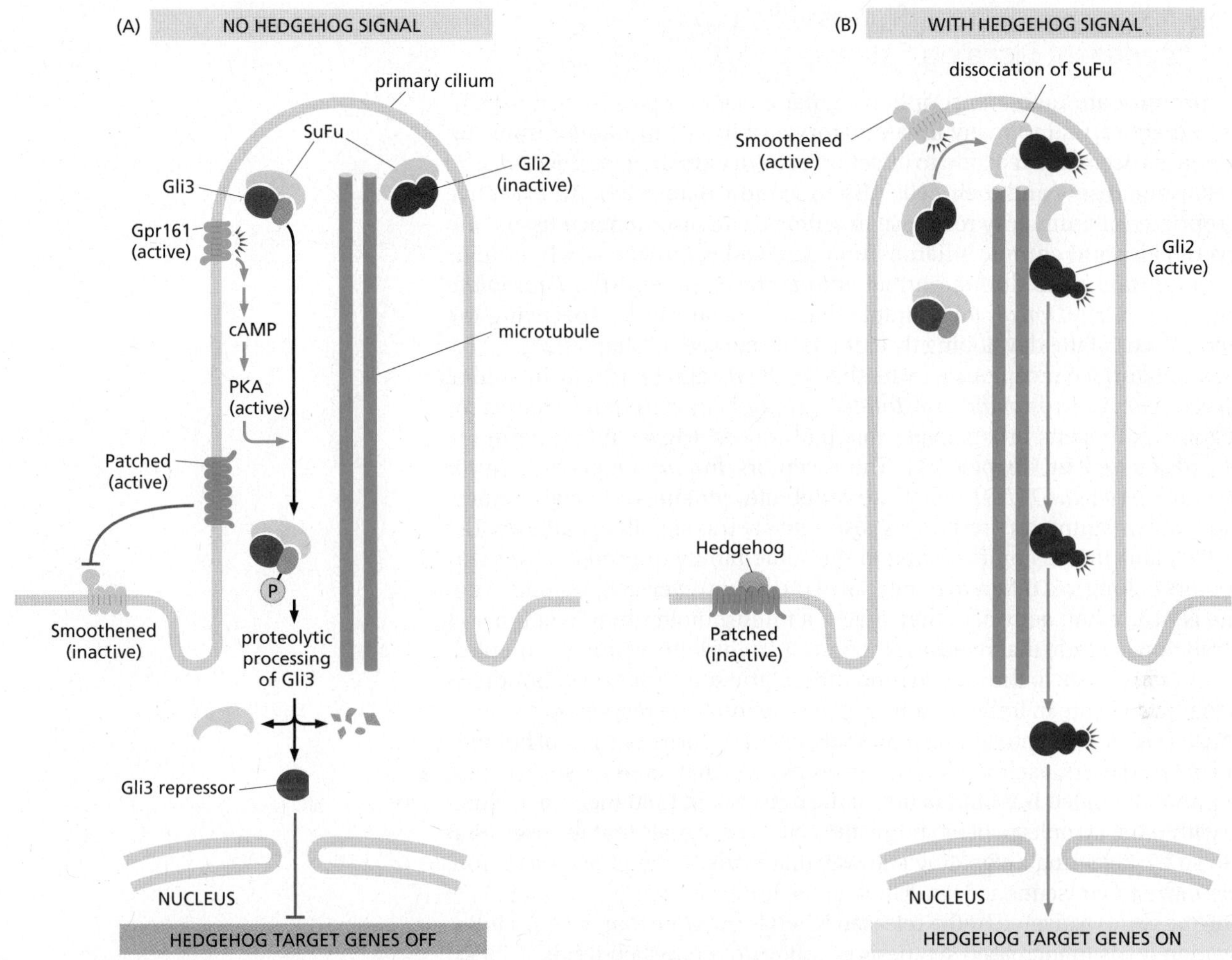

Figure 15–62 Vertebrate Hedgehog signaling in the primary cilium. (A) In the absence of Hedgehog, its receptor, Patched, is active in the cilium membrane and inhibits Smoothened, which is in the membrane adjacent to the cilium. Gli transcription regulators (primarily Gli2 and Gli3) are held in an inactive state by SuFu. In addition, an active GPCR (Gpr161) stimulates adenylyl cyclase, generating cyclic AMP that leads to PKA-dependent phosphorylation of the Gli3 protein. Phosphorylated Gli3 is processed to a transcription repressor, which accumulates in the nucleus to help keep Hedgehog target genes inactive. (B) Hedgehog binding to Patched removes the inhibition of Smoothened. Smoothened translocates to the cilium, where it triggers the dissociation of SuFu–Gli2 complexes and the conversion of Gli2 to an active transcription regulator; activated Gli2 is then transported to the cytoplasm, from where it moves to the nucleus to stimulate expression of Hedgehog-responsive genes. Hedgehog also promotes removal of Gpr161 from the cilium (not shown), thereby reducing the processing of Gli3 to a transcription repressor.

promotes their dissociation and triggers modifications of Gli2 proteins, thereby converting them into active transcription regulators. Activated Gli2 proteins are transported out of the cilium along microtubule tracks to the cytoplasm, from where they diffuse to the nucleus to promote gene expression (**Figure 15–62B**). The result is a rise in the expression of numerous Hedgehog target genes.

Hedgehog signaling can promote cell proliferation, and excessive Hedgehog signaling can lead to cancer. Inactivating mutations in one of the two human *Patched* genes, for example, which lead to excessive Hedgehog signaling, occur frequently in *basal cell carcinoma* of the skin, the most common form of cancer in Caucasians. A small molecule called *cyclopamine,* made by a meadow lily, is being used to treat cancers associated with excessive Hedgehog signaling. It blocks Hedgehog signaling by binding tightly to Smoothened and inhibiting its activity. It was originally identified because it causes severe developmental defects in the progeny of sheep grazing on such lilies; these include the presence of a single central eye (a condition called *cyclopia*), which is also seen in mice that are deficient in Hedgehog signaling.

Many Inflammatory and Stress Signals Act Through an NFκB-dependent Signaling Pathway

The **NFκB proteins** are latent transcription regulators that are present in most animal cells and are central to many inflammatory and innate immune responses. These responses occur as a reaction to infection or injury and help protect stressed multicellular organisms and their cells (discussed in Chapter 24). An excessive or inappropriate inflammatory response in animals can also damage tissue and cause severe pain, and chronic inflammation can lead to cancer. NFκB proteins also have important roles during normal animal development: the *Drosophila* NFκB family member *Dorsal,* for example, has a crucial role in specifying the dorsal–ventral axis of the developing fly embryo (discussed in Chapter 22).

Various cell-surface receptors activate the NFκB signaling pathway in animal cells. *Toll receptors* in *Drosophila* and *Toll-like receptors* in vertebrates, for example, recognize pathogens and activate this pathway to trigger innate immune responses (discussed in Chapter 24). The receptors for *tumor necrosis factor* α (*TNF*α) and *interleukin-1* (*IL1*), which are vertebrate cytokines especially important in inducing inflammatory responses, also activate this signaling pathway. The Toll, Toll-like, and IL1 receptors belong to the same family of proteins, whereas TNF receptors belong to a different family; all of them, however, act in similar ways to activate NFκB. When activated, they trigger a multiprotein ubiquitylation and phosphorylation cascade that releases NFκB from an inhibitory protein complex, so that it can translocate to the nucleus and turn on the transcription of hundreds of genes that participate in inflammatory and innate immune responses.

There are several NFκB proteins in mammals, and they form a variety of homodimers and heterodimers, each of which activates its own characteristic set of genes. Inhibitory proteins called **IκB** bind tightly to the dimers and hold them in an inactive state within the cytoplasm of unstimulated cells. The signals that release NFκB dimers do so by triggering a signaling pathway that leads to the phosphorylation, ubiquitylation, and consequent degradation of the IκB proteins (Figure 15–63).

Among the genes activated by the released NFκB is the gene that encodes IκBα. This activation leads to increased synthesis of IκBα protein, which binds to NFκB

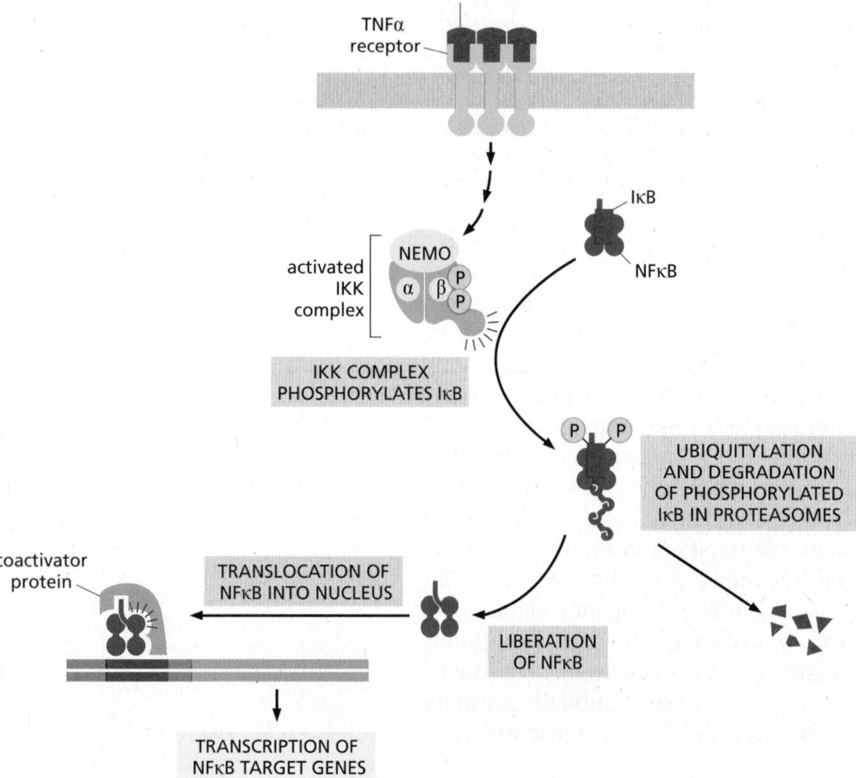

Figure 15–63 **The activation of the NFκB pathway by TNFα.** Both TNFα and its receptors are trimers. The binding of TNFα causes a rearrangement of the clustered cytosolic tails of the receptors, which now recruit various signaling proteins, resulting in the activation of a protein kinase that phosphorylates and activates IκB kinase kinase (IKK). IKK is a heterotrimer composed of two kinase subunits (IKKα and IKKβ) and a regulatory subunit called NEMO. IKKβ then phosphorylates IκB on two serines, which marks the protein for ubiquitylation and degradation in proteasomes. The released NFκB translocates into the nucleus, where, in collaboration with a coactivator protein, it stimulates the transcription of its target genes.

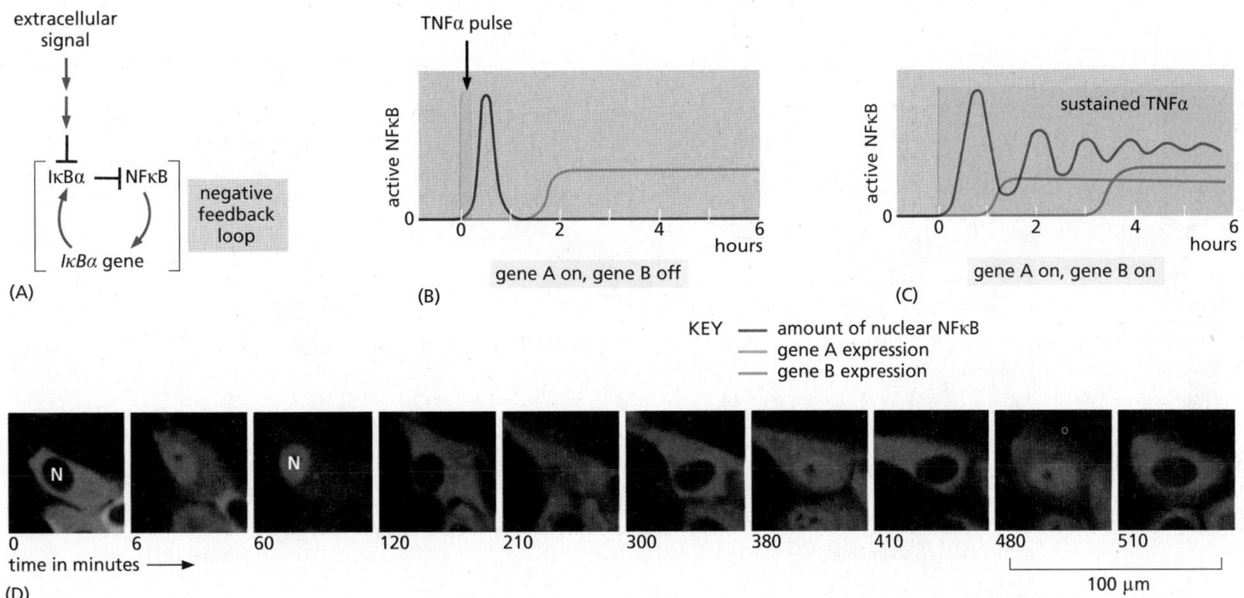

Figure 15–64 Negative feedback in the NFκB signaling pathway induces oscillations in NFκB activation. (A) Diagram showing how activated NFκB stimulates the transcription of the *IκBα* gene, the protein product of which acts back in the cytoplasm to sequester and inhibit NFκB there. If the stimulus is persistent, the newly made IκBα protein will then be ubiquitylated and degraded, liberating active NFκB again so that it can return to the nucleus and activate transcription (see Figure 15–63). (B) A short exposure to TNFα produces a single, short pulse of NFκB activation, beginning within minutes and ending by 1 hour. This response turns on the transcription of gene A but not gene B. (C) A sustained exposure to TNFα for the entire 6 hours of the experiment produces oscillations in NFκB activation that damp down over time. This response turns on the transcription of both genes A and B; gene B turns on only after several hours, indicating that gene B transcription requires prolonged activation of NFκB, for reasons that are not understood. (D) These time-lapse confocal fluorescence micrographs from a different study of TNFα stimulation show the oscillations of NFκB in a cultured cell, as indicated by the periodic movement into the nucleus (N) of a fusion protein composed of NFκB fused to a red fluorescent protein. In the cell at the center of the micrographs, NFκB is active and in the nucleus at 6, 60, 210, 380, and 480 minutes, but it is exclusively in the cytoplasm at 0, 120, 300, 410, and 510 minutes. (A–C, based on data from A. Hoffmann et al., *Science* 298:1241–1245, 2002, and adapted from A.Y. Ting and D. Endy, *Science* 298:1189–1190, 2002; D, from D.E. Nelson et al., *Science* 306: 704–708, 2004. All with permission from AAAS.)

and inactivates it, creating a negative feedback loop (**Figure 15–64A**). Experiments on TNFα-induced responses, as well as computer modeling studies of the responses, indicate that the negative feedback produces two types of NFκB responses, depending on the duration of the TNFα stimulus. Importantly, the two types of responses induce different patterns of gene expression (**Figure 15–64B, C, and D**).

Thus far, we have focused on the mechanisms by which extracellular signal molecules use cell-surface receptors to initiate changes in gene expression. We now turn to a class of extracellular signals that bypasses the plasma membrane entirely and controls, in the most direct way possible, transcription regulators inside the cell.

Nuclear Receptors Are Ligand-modulated Transcription Regulators

Various small, hydrophobic signal molecules diffuse directly across the plasma membrane of target cells and bind to intracellular receptors that are transcription regulators. These signal molecules include steroid hormones, thyroid hormones, retinoids, and vitamin D. Although they differ greatly from one another in both chemical structure (**Figure 15–65**) and function, they all act by a similar mechanism. They bind to their respective intracellular receptor proteins and alter the ability of these proteins to control the transcription of specific genes. Thus, these proteins serve both as intracellular receptors and as intracellular effectors for the signal.

The receptors are all structurally related, being part of the very large **nuclear receptor superfamily**. Many family members have been identified by DNA sequencing only, and their ligands are not yet known; they are therefore referred

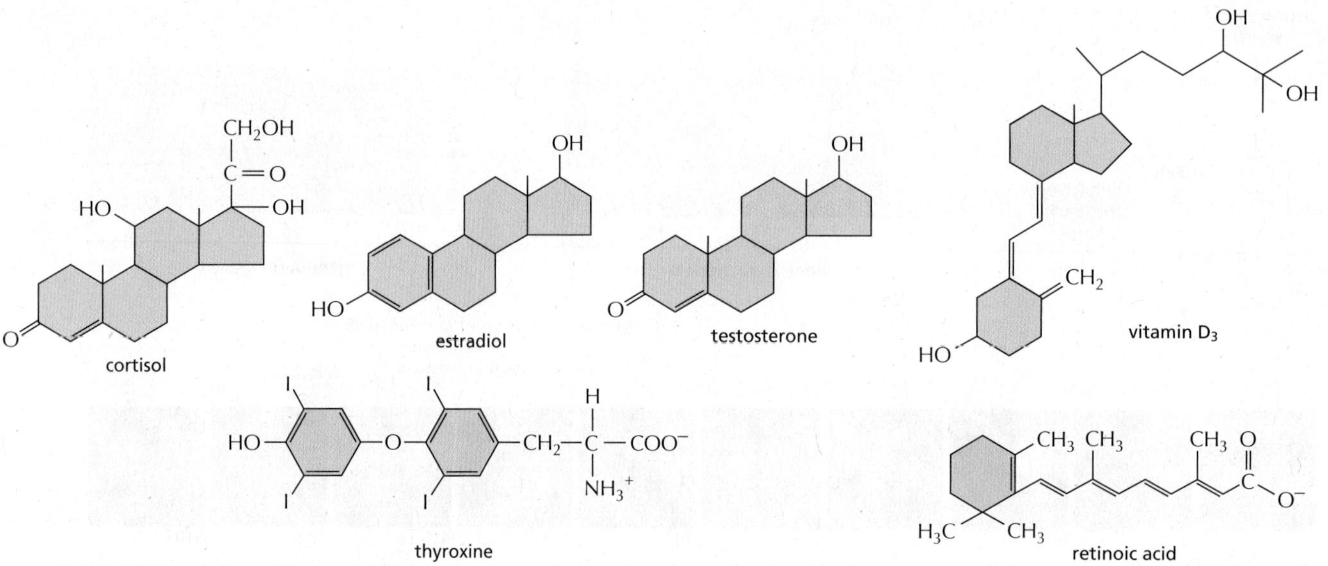

cortisol

estradiol

testosterone

vitamin D_3

thyroxine

retinoic acid

Figure 15–65 Some extracellular signal molecules that bind to intracellular receptors. Note that all of them are small and hydrophobic. The active, hydroxylated form of vitamin D_3 is shown. Estradiol and testosterone are steroid sex hormones.

to as *orphan nuclear receptors*, and they make up large fractions of the nuclear receptors encoded in the genomes of humans, *Drosophila*, and the nematode *C. elegans*. Some mammalian nuclear receptors are regulated by intracellular metabolites rather than by secreted signal molecules; the *peroxisome proliferation-activated receptors* (*PPARs*), for example, bind intracellular lipid metabolites and regulate the transcription of genes involved in lipid metabolism and fat-cell differentiation. It seems likely that the nuclear receptors for hormones evolved from such receptors for intracellular metabolites, which might help explain their intracellular location.

Steroid hormones—which include cortisol, the steroid sex hormones, vitamin D (in vertebrates), and the molting hormone *ecdysone* (in insects)—are all made from cholesterol. *Cortisol* is produced in the cortex of the adrenal glands and influences the metabolism of many types of cells. The *steroid sex hormones* are made in the testes and ovaries and are responsible for the secondary sex characteristics that distinguish males from females. *Vitamin D* is synthesized in the skin in response to sunlight; after it has been converted to its active form in the liver or kidneys, it regulates Ca^{2+} metabolism, promoting Ca^{2+} uptake in the gut and reducing its excretion in the kidneys. The *thyroid hormones*, which are made from the amino acid tyrosine, act to increase the metabolic rate of many cell types, while the *retinoids*, such as retinoic acid, are made from vitamin A and have important roles as local mediators in vertebrate development. Although all of these signal molecules are relatively insoluble in water, they are made soluble for transport in the bloodstream and other extracellular fluids by binding to specific carrier proteins, from which they dissociate before entering a target cell (see Figure 15–3B).

The nuclear receptors bind to specific DNA sequences adjacent to the genes that the ligand regulates. Some of the receptors, such as those for cortisol, are located primarily in the cytosol and enter the nucleus only after ligand binding; others, such as the thyroid and retinoid receptors, are bound to DNA in the nucleus even in the absence of ligand. In either case, the inactive receptors are usually bound to inhibitory protein complexes. Ligand binding alters the conformation of the receptor protein, causing the inhibitory complex to dissociate and the receptor to bind coactivator proteins that stimulate gene transcription (**Figure 15–66**). In other cases, however, ligand binding to a nuclear receptor inhibits transcription: some thyroid hormone receptors, for example, act as transcription activators in the absence of their hormone and become transcription repressors when hormone binds.

Thus far, we have focused on the control of gene expression by extracellular signal molecules produced by other cells. We now turn to gene regulation by a more global environmental signal: the cycle of light and dark that results from Earth's rotation.

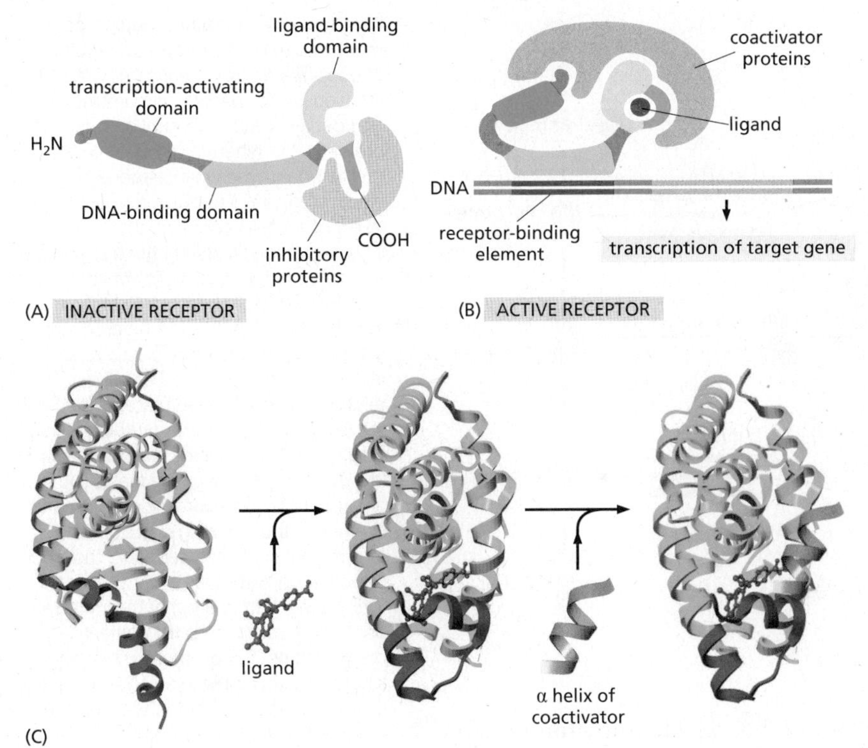

Figure 15–66 **The activation of nuclear receptors.** All nuclear receptors bind to DNA as either homodimers or heterodimers, but for simplicity we show them as monomers. (A) The receptors all have a related structure, which includes three major domains. An inactive receptor is bound to inhibitory proteins. (B) Typically, the binding of ligand to the receptor causes the ligand-binding domain of the receptor to clamp shut around the ligand, the inhibitory proteins to dissociate, and coactivator proteins to bind to the receptor's transcription-activating domain, thereby increasing gene transcription. In other cases, ligand binding has the opposite effect, causing co-repressor proteins to bind to the receptor, thereby decreasing transcription (not shown). (C) The structure of the ligand-binding domain of the retinoic acid receptor is shown in the absence *(left)* and presence *(middle)* of ligand (shown in *red*). When ligand binds, the *blue* α helix acts as a lid that snaps shut, trapping the ligand in place. The shift in the conformation of the receptor upon ligand binding also creates a binding site for a small α helix *(orange)* on the surface of coactivator proteins. (PDB codes: 6HN6, 2ZY0, and 2ZXZ.)

Circadian Clocks Use Negative Feedback Loops to Control Gene Expression

Life on Earth evolved in the presence of a daily cycle of day and night, and many present-day organisms possess an internal rhythm that regulates different behaviors at different times of day and night. These behaviors range from the cyclical change in metabolic enzyme activities of a bacterium to the elaborate sleep–wake cycles of humans. The intracellular oscillators that control such diurnal rhythms are called **circadian clocks**.

Having a circadian clock enables an organism to anticipate the regular daily changes in its environment and take appropriate action in advance. Of course, the internal clock cannot be perfectly accurate, and so it must be capable of being reset by external cues such as the light of day. Nonetheless, circadian clocks keep running even when the environmental cues (changes in light and dark) are removed, but the period of this free-running rhythm is generally a little less or more than 24 hours. External signals indicating the time of day cause small adjustments in the running of the clock, so as to keep the organism in synchrony with its environment. After more drastic shifts, circadian cycles become gradually reset (entrained) by the new cycle of light and dark, as anyone who has experienced jet lag can attest.

We might expect that the circadian clock would be a complex multicellular device, with different groups of cells responsible for different parts of the oscillation mechanism. Remarkably, however, in almost all multicellular organisms, including humans, the timekeepers are individual cells. Thus, our diurnal cycles of sleeping and waking, body temperature, and hormone release are controlled by a clock that operates in each member of a specialized group of brain cells in the suprachiasmatic nucleus (SCN) of the hypothalamus. Even if these cells are removed from the brain and dispersed in a culture dish, they will continue to oscillate individually, showing a cyclic pattern of gene expression with a period of approximately 24 hours. In the intact body, the SCN cells receive neural cues from the retina, entraining the SCN cells to the daily cycle of light and dark. SCN cells send information about the time of day to other brain areas, as well as the pineal gland, which relays the time signal to the rest of the body by releasing the hormone melatonin in time with the clock.

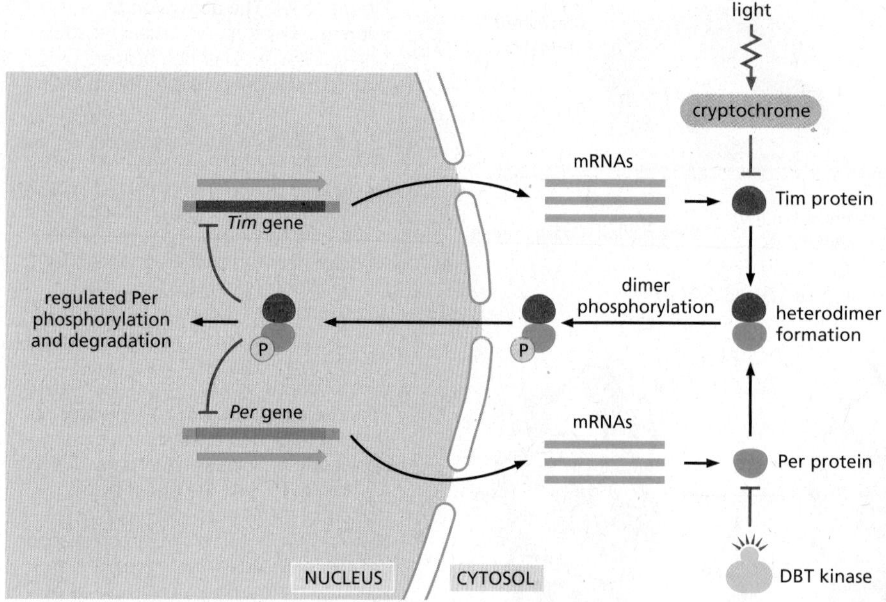

Figure 15–67 Simplified outline of the mechanism of the circadian clock in *Drosophila* cells. A central feature of the clock is the periodic accumulation and decay of two transcription regulatory proteins, Tim (short for timeless, based on the phenotype when its gene is inactivated) and Per (short for period). The levels of mRNAs encoding these proteins increase gradually during the day and are translated in the cytoplasm, where the two proteins slowly associate to form a Tim–Per heterodimer. The heterodimer is transported into the nucleus, where it represses the *Tim* and *Per* genes, resulting in negative feedback that causes the levels of Tim and Per proteins to fall during the night. Decreased Per and Tim levels remove the negative feedback, allowing renewed expression of *Tim* and *Per* genes the following day. In addition to this transcriptional feedback, the clock depends on numerous other proteins that influence the period of the circadian oscillator through post-translational modifications. For example, a protein kinase called Doubletime (DBT) promotes the degradation of Per in the cytoplasm, thereby delaying the formation of the Tim–Per dimer. Once formed, the nuclear import of the dimer depends on phosphorylation and glycosylation of one or both proteins, introducing additional delays that further influence the timing of the clock.

Entrainment (or resetting) of the clock occurs in response to new light–dark cycles. Although most *Drosophila* cells do not have true photoreceptors, light is sensed by intracellular flavoproteins called cryptochromes, which, in the presence of light, associate with the Tim protein and cause its degradation, thereby resetting the clock. (Adapted from J.C. Dunlap, *Science* 311:184–186, 2006.)

Although the SCN is the central regulator of the circadian rhythm in mammals, most of the other cells in the mammalian body also have circadian clocks, which have the ability to reset in response to light. Similarly, in *Drosophila*, many different types of cells have a similar circadian clock, which continues to cycle when isolated from the rest of the fly and can be reset by externally imposed light and dark cycles.

Circadian clocks are therefore a fundamental feature of many cells. Although we do not yet understand how these clocks work in detail, studies in a wide variety of organisms have revealed the basic principles and molecular components of some of them. A key principle is that circadian clocks generally depend on *negative feedback loops*. As discussed earlier, oscillations in the activity of an intracellular signaling protein can occur if that protein inhibits its own activity with a long delay (see Figure 15–19C and D). In *Drosophila* and many other animals, including humans, the heart of the circadian clock is a delayed negative feedback loop based on transcription regulators: accumulation of certain gene products switches off the transcription of their own genes, but with a delay, so that the cell oscillates between a state in which the products are present and transcription is switched off, and one in which the products are absent and transcription is switched on (**Figure 15–67**). The negative feedback underlying circadian rhythms is not always based on transcription regulators, however. In some cell types, the circadian clock is constructed of proteins that govern their own activities through post-translational mechanisms, as we discuss next.

Three Purified Proteins Can Reconstitute a Cyanobacterial Circadian Clock in a Test Tube

The best understood circadian clock is found in the photosynthetic cyanobacterium *Synechococcus elongatus*. The core oscillator in this organism is remarkably simple, being composed of just three proteins—*KaiA*, *KaiB*, and *KaiC*. The central player is KaiC, a multifunctional enzyme that catalyzes its own phosphorylation and dephosphorylation in a 24-hour cycle: it gradually phosphorylates itself sequentially at two sites during the day and dephosphorylates itself during the night. This timing depends on interactions with the two other Kai proteins: KaiA binds to unphosphorylated KaiC in the morning and stimulates KaiC autophosphorylation, first at one site and then, with a lengthy delay, at the other. At nightfall, the second phosphorylation promotes the gradual binding of the third

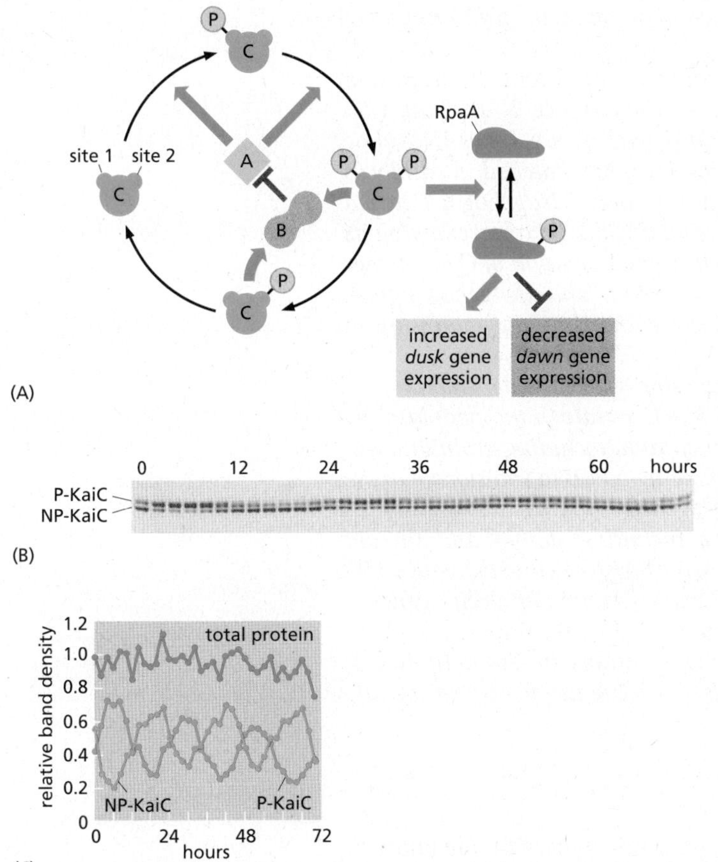

Figure 15–68 The core circadian oscillator of cyanobacteria. (A) KaiC protein *(C, blue)* is a combined kinase and phosphatase that phosphorylates and dephosphorylates itself on two adjacent sites. In the absence of other proteins, the phosphatase activity is dominant, and the protein is mostly unphosphorylated. During the day, the binding of KaiA *(A, green)* to KaiC suppresses the phosphatase activity and promotes the kinase activity, leading to KaiC phosphorylation, first at site 1 and then, slowly, at site 2, resulting in diphosphorylated KaiC. This form of KaiC interacts with KaiB *(B, orange)*, which blocks the stimulatory effects of KaiA, thereby reducing the rate of KaiC phosphorylation and allowing dephosphorylation at both sites to occur by the end of the night. KaiB then dissociates, allowing KaiA to promote KaiC phosphorylation again. Diphosphorylated KaiC increases in abundance during the day and peaks around dusk. It activates other proteins that phosphorylate a transcription regulator called RpaA *(dark green)*, which then stimulates expression of certain genes that peak in early evening (the *dusk* genes) and inhibits expression of other genes that peak in the morning (the *dawn* genes). When KaiC dephosphorylation gradually occurs during the night, these effects are reversed: *dusk* genes are turned off and *dawn* genes are turned on. (B) In this experiment, the three Kai proteins were purified and mixed in a test tube with ATP (which is required for KaiC kinase activity). Every 2 hours over the next 3 days, the KaiC protein was analyzed by polyacrylamide-gel electrophoresis, in which the phosphorylated form of KaiC migrates more slowly (*upper band*, P-KaiC) than the nonphosphorylated form (*lower band*, NP-KaiC). The three different phosphorylated forms of KaiC are not distinguished by this method. The phosphorylation of KaiC oscillates with a roughly 24-hour period. (C) The amount of phosphorylated and unphosphorylated KaiC in the experiment in B is plotted on this graph, along with the amount of total protein. (B and C, from M. Nakajima et al., *Science* 308:414–415, 2005. With permission from AAAS.)

protein, KaiB, which blocks the stimulatory effect of KaiA and thereby allows KaiC to dephosphorylate itself, slowly bringing KaiC back to its dephosphorylated state during the night. This clock depends on a negative feedback loop: KaiC drives its own phosphorylation until, after a delay, it recruits an inhibitor, KaiB, that stimulates KaiC to dephosphorylate itself. Amazingly, when the three Kai proteins are purified and incubated in a test tube with ATP, the KaiC phosphorylation cycle occurs with roughly 24-hour timing for several days (**Figure 15–68**).

Circadian oscillations in KaiC phosphorylation lead to parallel rhythms in the expression of a large number of genes involved in controlling metabolic activities and cell division (see Figure 15–68). Many aspects of cell behavior are thereby synchronized with the circadian cycle.

Even in continual darkness, cyanobacterial cells generate free-running oscillations of KaiC phosphorylation with roughly 24-hour periods. Like other circadian clocks, the cyanobacterial clock is entrained by the environmental light–dark cycle. Light is thought to affect the clock indirectly: the activities of Kai proteins are influenced by changes in intracellular redox potential, which occur as a result of increased photosynthetic activity during the day.

Summary

Some signaling pathways that are especially important in animal development depend on proteolysis to control the activity and location of latent transcription regulators. Notch proteins are latent transcription regulators that are activated by cleavage when Delta proteins on another cell binds to them; the cleaved cytosolic tail of Notch migrates into the nucleus, where it stimulates the transcription of Notch-responsive genes. In the Wnt/β-catenin signaling pathway, by contrast, the proteolysis of the latent transcription regulatory protein β-catenin is inhibited when a secreted Wnt protein binds to both a Frizzled and an LRP receptor protein;

as a result, β-catenin accumulates in the nucleus and activates the transcription of Wnt target genes.

Hedgehog signaling also depends on activation of latent transcription regulators—members of the Gli protein family. In the absence of a signal, Gli3 is proteolytically cleaved to form a transcription repressor that keeps Hedgehog target genes silenced. The unoccupied Hedgehog receptor, Patched, also blocks activation of another Gli family member, Gli2. The binding of Hedgehog to Patched has two effects: it inhibits the proteolytic processing of Gli3, thereby removing its inhibitory effect on gene expression, and it also triggers activation of Gli2, further promoting target gene expression. Thus, in Notch, Wnt, and Hedgehog signaling, the extracellular signal triggers a switch from transcriptional repression to transcriptional activation.

Signaling through the latent transcription regulator NFκB also depends on proteolysis. NFκB proteins are normally held in an inactive state in the cytoplasm by inhibitory IκB proteins. A variety of extracellular stimuli, including proinflammatory cytokines, trigger the degradation of IκB, allowing NFκB to translocate to the nucleus and activate the transcription of its target genes.

Some small, hydrophobic signal molecules, including steroid and thyroid hormones, diffuse across the plasma membrane of the target cell and activate intracellular receptor proteins that directly regulate the transcription of specific genes.

In many cell types, gene expression is governed by circadian clocks, in which delayed negative feedback produces 24-hour oscillations in the activities of transcription regulators, anticipating the cell's changing needs during the day and night.

SIGNALING IN PLANTS

In plants, as in animals, cells are in constant communication with one another. Plant cells communicate to coordinate their activities in response to the changing conditions of light, dark, and temperature, which guide the plant's cycle of growth, flowering, and fruiting. Plant cells also communicate to coordinate activities in their roots, stems, and leaves. In this final section, we consider how plant cells signal to one another and how they respond to light. Less is known about the receptors and intracellular signaling mechanisms involved in cell communication in plants than is known about that in animals, and we will concentrate mainly on how the receptors and intracellular signaling mechanisms differ from those used by animals.

Multicellularity and Cell Communication Evolved Independently in Plants and Animals

Although plants and animals are both eukaryotes, they have evolved separately for more than a billion years. Their last common ancestor is thought to have been a unicellular eukaryote that had mitochondria but no chloroplasts; the plant lineage acquired chloroplasts after plants and animals diverged. The earliest fossils of multicellular animals and plants date from almost 600 million years ago. Thus, it seems that plants and animals evolved multicellularity independently, each starting from a different unicellular eukaryote, sometime between 1.6 and 0.6 billion years ago (**Figure 15–69**).

If multicellularity evolved independently in plants and animals, some of the molecules and mechanisms used for cell communication will have evolved separately and would be expected to be different. Indeed, whereas both plants and animals use nitric oxide, cyclic GMP, Ca^{2+}, protein kinases, and small GTPases for signaling, modern flowering plants do not appear to contain homologs of the nuclear receptor family, JAK, STAT, TGFβ, Notch, Wnt, or Hedgehog signaling pathways. Similarly, plants do not seem to use cyclic AMP for intracellular signaling. Nevertheless, the general strategies underlying signaling are similar in plants and animals. Both, for example, use enzyme-coupled cell-surface receptors, as we now discuss.

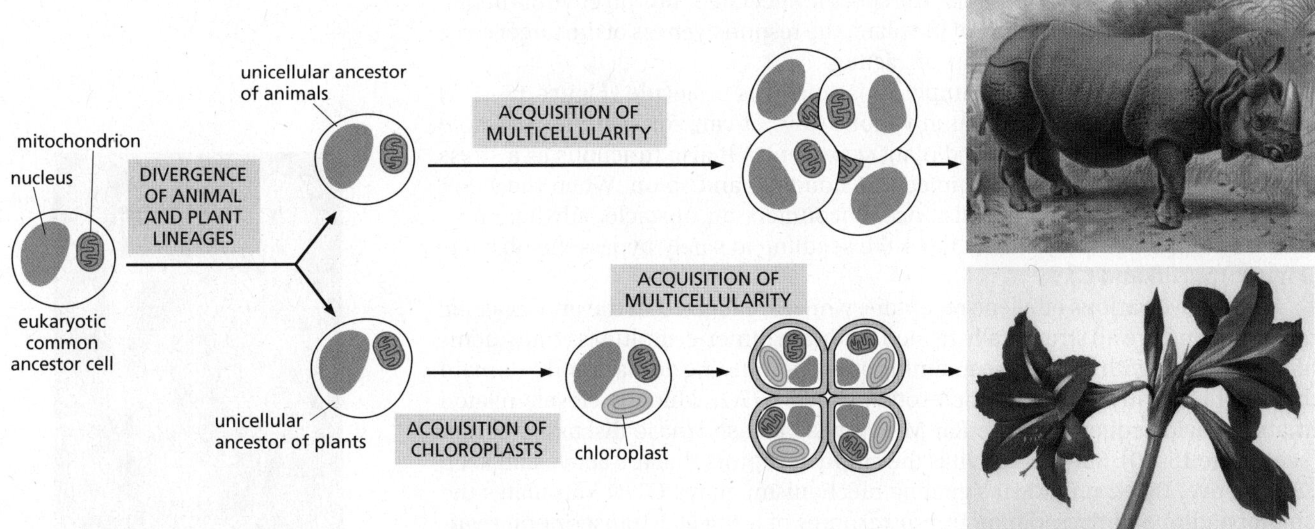

Figure 15–69 The proposed divergence of plant and animal lineages from a common unicellular eukaryotic ancestor. The plant lineage acquired chloroplasts after the two lineages diverged. Both lineages independently gave rise to multicellular organisms—plants and animals. (Paintings courtesy of John Innes Foundation.)

Receptor Serine/Threonine Kinases Are the Largest Class of Cell-Surface Receptors in Plants

Most cell-surface receptors in plants are enzyme-coupled. However, whereas the largest class of enzyme-coupled receptors in animals is the receptor tyrosine kinase (RTK) class, this type of receptor is extremely rare in plants. Instead, plants rely largely on a great diversity of transmembrane *receptor serine/threonine kinases,* which have a typical serine/threonine kinase cytoplasmic domain and an extracellular ligand-binding domain. The most abundant types of these receptors have a tandem array of extracellular leucine-rich repeat structures and are therefore called **leucine-rich repeat (LRR) receptor kinases.**

There are about 175 LRR receptor kinases encoded in the genome of the small flowering plant *Arabidopsis thaliana.* These include a protein called *Bri1,* which forms part of a cell-surface steroid hormone receptor. Plants synthesize a class of steroids that are called **brassinosteroids** because they were originally identified in the mustard family *Brassicaceae,* which includes *Arabidopsis.* These signal molecules regulate the growth and differentiation of plants throughout their life cycle. Binding of a brassinosteroid to a Bri1 cell-surface receptor kinase initiates an intracellular signaling cascade that uses a GSK3 protein kinase and a protein phosphatase to regulate the phosphorylation and degradation of specific transcription regulatory proteins in the nucleus, and thereby specific gene transcription. Mutant plants that are deficient in the Bri1 receptor kinase are insensitive to brassinosteroids and are therefore dwarfs.

The LRR receptor kinases are only one of many classes of transmembrane receptor serine/threonine kinases in plants. There are at least six additional families, each with its own characteristic set of extracellular domains. The *lectin receptor kinases,* for example, have extracellular domains that bind carbohydrate signal molecules. The *Arabidopsis* genome encodes more than 300 receptor serine/threonine kinases, which makes them the largest family of receptors known in plants. Many are involved in defense responses against pathogens.

Ethylene Blocks the Degradation of Specific Transcription Regulatory Proteins in the Nucleus

Various **plant growth regulators** (also called **plant hormones**) help to coordinate plant development. They include *ethylene, auxin, cytokinins, gibberellins,* and *abscisic acid,* as well as brassinosteroids. Growth regulators are all small molecules made by most plant cells. They diffuse readily through cell walls and can either act locally or be transported to influence cells further away. Each growth regulator can have multiple effects. The specific effect depends on environmental conditions, the nutritional state of the plant, the responsiveness of the target cells, and which other growth regulators are acting.

Ethylene is an important example. This small gas molecule (**Figure 15–70A**) can influence plant development in various ways; it can, for example, promote fruit ripening, leaf abscission, and plant senescence. It also functions as a stress signal in response to wounding, infection, flooding, and so on. When the shoot of a germinating seedling, for instance, encounters an obstacle, ethylene promotes a complex response that allows the seedling to safely bypass the obstacle (**Figure 15–70B and C**).

Plants have various ethylene receptors, which are located in the endoplasmic reticulum and are all structurally related. They are dimeric, multipass transmembrane proteins, with a copper-containing ethylene-binding domain and a domain that interacts with a cytoplasmic protein called *CTR1*, which is closely related in amino acid sequence to the Raf MAP kinase kinase kinase discussed earlier (see Figure 15–50). Surprisingly, it is the empty receptors that are active and keep CTR1 active. By an unknown signaling mechanism, active CTR1 stimulates the ubiquitylation and degradation in proteasomes of a nuclear transcription regulator called *EIN3*, which is required for the transcription of ethylene-responsive genes. In this way, the empty but active receptors keep ethylene-response genes off. Ethylene binding inactivates the receptors, altering their conformation so that they no longer activate CTR1. The EIN3 protein is no longer ubiquitylated and degraded and can now activate the transcription of the large number of ethylene-responsive genes (**Figure 15–71**).

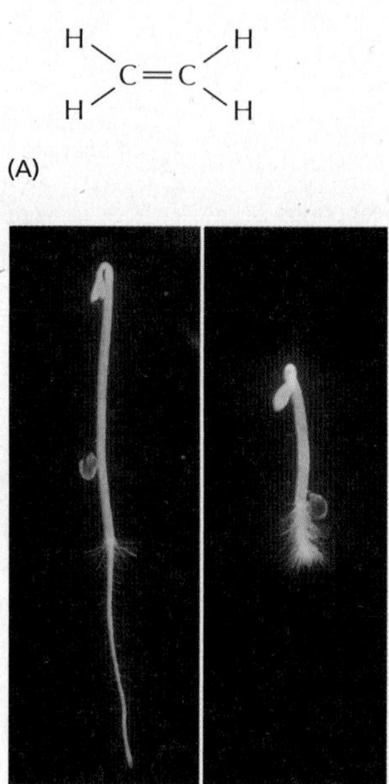

Figure 15–70 **The ethylene-mediated triple response that occurs when the growing shoot of a germinating seedling encounters an obstacle underground.** (A) The structure of ethylene. (B) In the absence of obstacles, the shoot grows upward and is long and thin. (C) If the shoot encounters an obstacle, such as a piece of gravel in the soil, the seedling responds to the encounter in three ways. First, it thickens its stem, which can then exert more force on the obstacle. Second, it shields the tip of the shoot (at *top*) by increasing the curvature of a specialized hook structure. Third, it reduces the shoot's tendency to grow away from the direction of gravity, so as to avoid the obstacle. (Courtesy of Melanie Webb.)

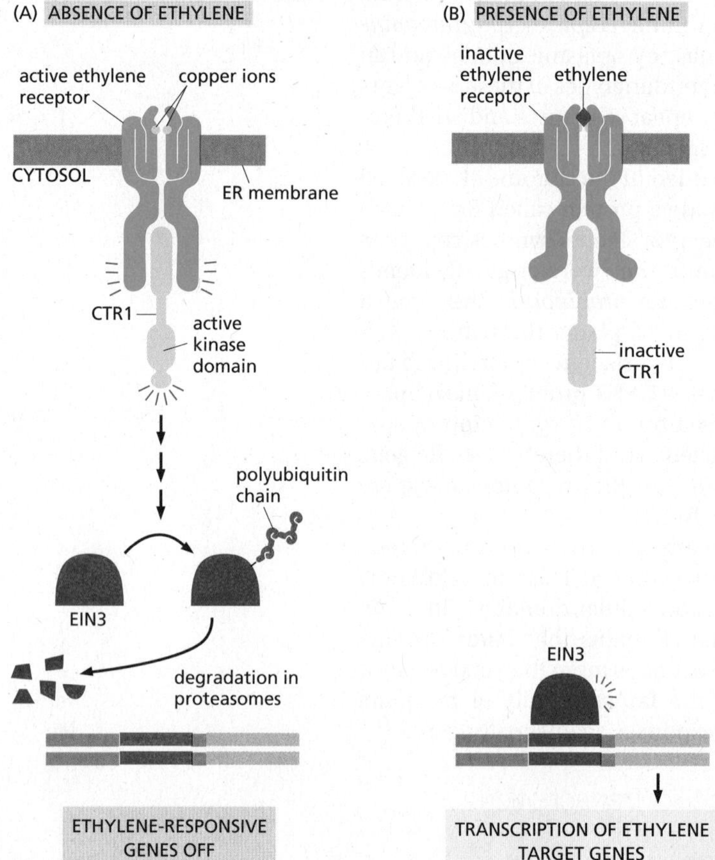

Figure 15–71 **The ethylene signaling pathway.** (A) In the absence of ethylene, the receptors and CTR1 are active, causing the ubiquitylation and destruction of EIN3, the transcription regulatory protein in the nucleus that is responsible for the transcription of ethylene-responsive genes. (B) The binding of ethylene inactivates the receptors and disrupts the activation of CTR1. The EIN3 protein is not degraded and can therefore activate the transcription of ethylene-responsive genes.

Regulated Positioning of Auxin Transporters Patterns Plant Growth

The plant hormone **auxin** (Figure 15–72A) binds to receptor proteins in the nucleus. It helps plants grow toward light, grow upward rather than branch out, and grow their roots downward. It also regulates organ initiation and positioning and helps plants flower and bear fruit. Like ethylene (and some of the animal signal molecules we have described in this chapter), auxin influences gene expression by controlling the degradation of transcription regulators. It works by stimulating the ubiquitylation and degradation of repressor proteins that block the transcription of auxin target genes in unstimulated cells (Figure 15–72B and C).

Auxin is unique in the way that it is transported. Unlike animal hormones, which are usually secreted by a specific endocrine gland and transported to target cells via the circulatory system, auxin has its own transport system. Specific plasma-membrane-bound *influx transporter proteins* and *efflux transporter proteins* move auxin into and out of plant cells, respectively. The efflux transporters can be distributed asymmetrically in the plasma membrane to make the efflux of auxin directional. A row of cells with their auxin efflux transporters confined to the basal plasma membrane, for example, will transport auxin from the top of the plant to the bottom.

In some regions of the plant, the localization of the auxin transporters, and therefore the direction of auxin flow, is highly dynamic and regulated. A cell can rapidly redistribute transporters by controlling the traffic of vesicles containing them. The auxin efflux transporters, for example, normally recycle between intracellular vesicles and the plasma membrane. A cell can redistribute these transporters on its surface by inhibiting their endocytosis in one domain of the plasma membrane, causing the transporters to accumulate there. One example occurs in the root, where gravity influences the direction of growth. The auxin efflux transporters are normally distributed symmetrically in the cap cells of the root. Within minutes of a change in the direction of the gravity vector, however, the efflux transporters redistribute to one side of the cells, so that auxin is pumped

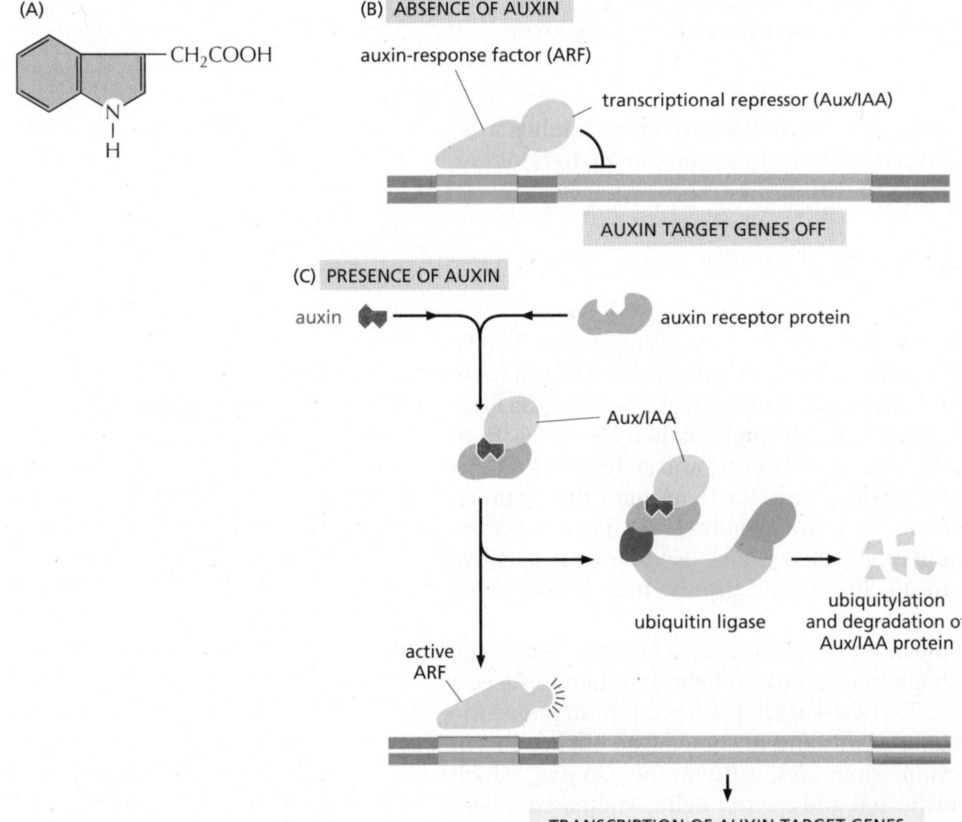

Figure 15–72 **The auxin signaling pathway.** (A) The structure of auxin (indole-3-acetic acid). (B) In the absence of auxin, a transcription repressor protein (called Aux/IAA) binds and suppresses a transcription regulator (called auxin-response factor; ARF), which is required for the transcription of auxin-responsive genes. (C) The auxin receptor proteins are located mainly in the nucleus. When activated by auxin binding, the receptor–auxin complex recruits a ubiquitin ligase, which ubiquitylates the Aux/IAA protein, marking it for degradation in proteasomes. ARF is now free to activate the transcription of auxin-responsive genes. There are many ARFs, Aux/IAA proteins, and auxin receptors that work as illustrated.

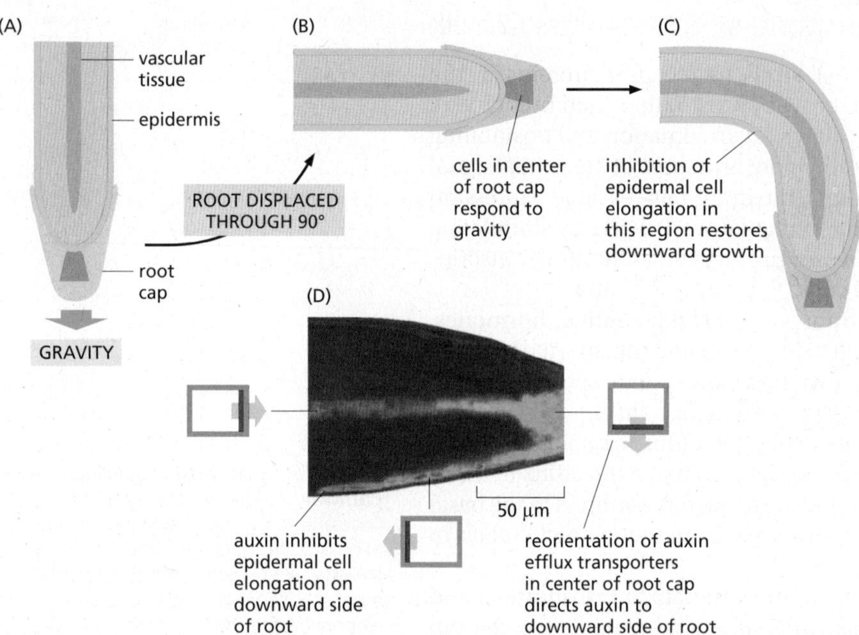

Figure 15–73 **Auxin transport and root gravitropism.** (A–C) Roots respond to a 90° change in the gravity vector and adjust their direction of growth so that they grow downward again. The cells that respond to gravity are in the center of the root cap, while it is the epidermal cells further back (on the lower side) that decrease their rate of elongation to restore downward growth. (D) The gravity-responsive cells in the root cap redistribute their auxin efflux transporters in response to the displacement of the root. This redirects the auxin flux mainly to the lower part of the displaced root, where it inhibits the elongation of the epidermal cells. The resulting asymmetrical distribution of auxin in the *Arabidopsis* root tip shown here is assessed indirectly, using an auxin-responsive reporter gene that encodes a protein fused to green fluorescent protein (GFP); the epidermal cells on the downward side of the root are green, whereas those on the upper side are not, reflecting the asymmetrical distribution of auxin. The distribution of auxin efflux transporters in the plasma membrane of cells in different regions of the root (shown as *gray rectangles*) is indicated in *red*, and the direction of auxin efflux is indicated by a *green arrow.* (D, photograph from T. Paciorek et al., *Nature* 435:1251–1256, published 2005 by Nature Publishing Group. Reproduced with permission of SNCSC.)

out toward the side of the root pointing downward. Because auxin inhibits root-cell elongation, this redirection of auxin transport causes the root tip to re-orient, so that it grows downward again (**Figure 15–73**).

Phytochromes Detect Red Light, and Cryptochromes Detect Blue Light

Plant development is greatly influenced by environmental conditions. Unlike animals, plants cannot move when conditions become unfavorable; they have to adapt or they die. The most important environmental influence on plants is light, which is their energy source and has a major role throughout their life cycle—from germination, through seedling development, to flowering and senescence. Plants have thus evolved a large set of light-sensitive proteins to monitor the quantity, quality, direction, and duration of light. These are usually referred to as *photoreceptors*. However, because the term photoreceptor is also used for light-sensitive cells in the animal retina (see Figure 15–39), we will use the term *photoprotein* instead.

All photoproteins sense light by means of a covalently attached light-absorbing chromophore, which changes its shape in response to light and then induces a change in the protein's conformation. The best-known plant photoproteins are the **phytochromes**, which are present in all plants and in some algae but are absent in animals. These are dimeric, cytoplasmic serine/threonine kinases, which respond differentially and reversibly to red and far-red light: whereas red light usually activates the kinase activity of the phytochrome, far-red light inactivates it. When activated by red light, the phytochrome is thought to phosphorylate itself

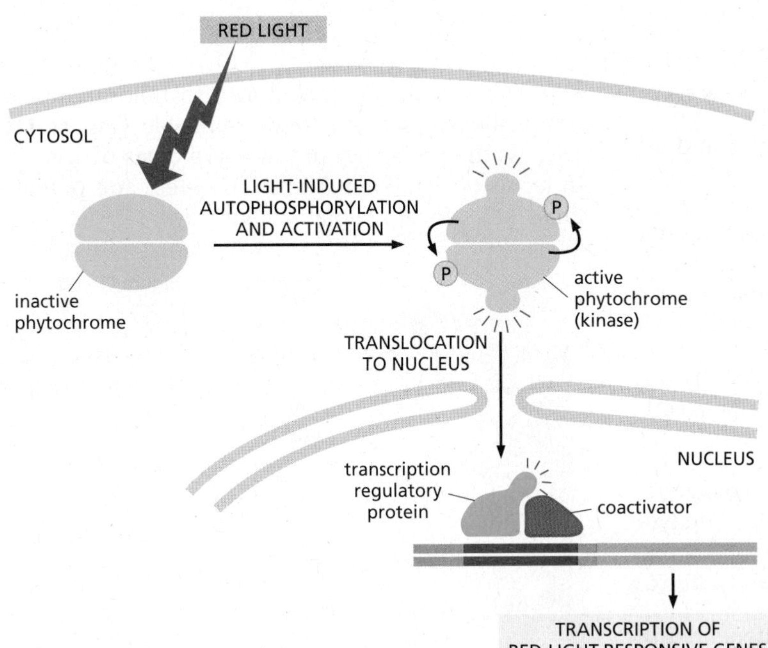

Figure 15–74 **One way in which phytochromes mediate a light response in plant cells.** When activated by red light, the phytochrome, which is a dimeric protein kinase, phosphorylates itself and then moves into the nucleus, where it activates transcription regulatory proteins to stimulate the transcription of red-light-responsive genes.

and then to phosphorylate one or more other proteins in the cell. In some light responses, the activated phytochrome translocates into the nucleus, where it activates transcription regulators to alter gene transcription (**Figure 15–74**). In other cases, the activated phytochrome activates a latent transcription regulator in the cytoplasm, which then translocates into the nucleus to regulate gene transcription. In still other cases, the activated phytochrome triggers signaling pathways in the cytosol that alter the cell's behavior without involving the nucleus.

Plants sense blue light using photoproteins of two other sorts, phototropin and cryptochromes. **Phototropin** is associated with the plasma membrane and is partly responsible for *phototropism*, the tendency of plants to grow toward light. Phototropism occurs by directional cell elongation, which is stimulated by auxin, but the links between phototropin and auxin are unknown.

Cryptochromes are flavoproteins that are sensitive to blue light. They are structurally related to blue-light-sensitive enzymes called *photolyases*, which are involved in the repair of ultraviolet-induced DNA damage in all organisms, except most mammals. Unlike phytochromes, cryptochromes are also found in animals, where they have an important role in circadian clocks (see Figure 15–67). Although cryptochromes are thought to have evolved from the photolyases, they do not have a role in DNA repair.

Summary

The mechanisms used to signal between cells in animals and in plants have both similarities and differences. Whereas animals rely heavily on GPCRs and RTKs, plants rely mainly on enzyme-coupled receptors of the receptor serine/threonine kinase type, especially those with extracellular leucine-rich repeats. Various plant hormones, or growth regulators, including ethylene and auxin, help coordinate plant development. Ethylene acts through intracellular receptors to stop the degradation of specific nuclear transcription regulators, which can then activate the transcription of ethylene-responsive genes. The receptors for some other plant hormones, including auxin, also regulate the degradation of specific transcription regulators, although the details vary. Auxin signaling is unusual in that it has its own highly regulated transport system, in which the dynamic positioning of plasma-membrane-bound auxin transporters controls the direction of auxin flow and thereby the direction of plant growth. Light has an important role in regulating plant development. These light responses are mediated by a variety of light-sensitive photoproteins, including phytochromes, which are responsive to red light, and cryptochromes and phototropin, which are sensitive to blue light.

PROBLEMS

Which statements are true? Explain why or why not.

15–1 All second messengers are water soluble and diffuse freely through the cytosol.

15–2 In the regulation of molecular switches, protein kinases and guanine nucleotide exchange factors (GEFs) always turn proteins on, whereas protein phosphatases and GTPase-activating proteins (GAPs) always turn proteins off.

15–3 Most intracellular signaling pathways provide numerous opportunities for amplifying the responses to extracellular signals.

15–4 Binding of extracellular ligands to receptor tyrosine kinases (RTKs) activates the intracellular catalytic domain by propagating a conformational change across the lipid bilayer through a single transmembrane α helix.

15–5 Even though plants and animals independently evolved multicellularity, they use virtually all the same signaling proteins and second messengers for cell–cell communication.

Discuss the following problems.

15–6 Cells communicate in ways that resemble human communication. Decide which of the following forms of human communication are analogous to autocrine, paracrine, endocrine, and synaptic signaling by cells.

A. A telephone conversation

B. Talking to people at a cocktail party

C. A radio announcement

D. Talking to yourself

15–7 Suppose that the circulating concentration of hormone is 10^{-10} M and the K_d for binding to its receptor is 10^{-8} M. What fraction of the receptors will have hormone bound? If a meaningful physiological response occurs when 50% of the receptors have bound a hormone molecule, how much will the concentration of hormone have to rise to elicit a response? The fraction of receptors (R) bound to hormone (H) to form a receptor–hormone complex (R–H) is $[R–H]/([R] + [R–H]) = [R–H]/[R]_{TOT} = [H]/([H] + K_d)$.

15–8 How is it that different cells can respond in different ways to exactly the same signaling molecule even when they have identical receptors?

15–9 Why do you suppose that phosphorylation/dephosphorylation, as opposed to allosteric binding of small molecules, for example, has evolved to play such a prominent role in switching proteins on and off in signaling pathways?

15–10 Consider a signaling pathway that proceeds through three protein kinases that are sequentially activated by phosphorylation. In one case, the kinases are held in a signaling complex by a scaffold protein; in the other, the kinases are freely diffusible (Figure Q15–1). Discuss the properties of these two types of organization in terms of signal amplification, speed, and potential for cross-talk between signaling pathways.

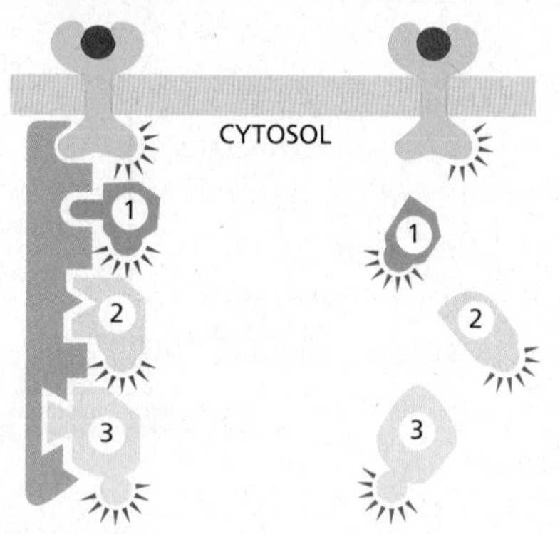

Figure Q15–1 A kinase cascade organized by a scaffold protein or composed of freely diffusing components (Problem 15–10).

15–11 Describe three ways in which a gradual increase in an extracellular signal can be sharpened by the target cell to produce an abrupt or nearly all-or-none response.

15–12 Why do signaling responses that involve changes in proteins already present in the cell occur in milliseconds to seconds, whereas responses that require changes in gene expression require minutes to hours?

15–13 Propose a specific type of mutation in the gene for the regulatory subunit of cyclic-AMP-dependent protein kinase (PKA) that could lead to a permanently active PKA.

15–14 Phosphorylase kinase integrates signals from the cyclic-AMP-dependent and Ca^{2+}-dependent signaling pathways that control glycogen breakdown in liver and muscle cells (Figure Q15–2). Phosphorylase kinase is composed of four subunits. One subunit is the protein kinase that catalyzes the addition of phosphate to glycogen phosphorylase to activate it for glycogen breakdown. The other three subunits are regulatory proteins that control the activity of the catalytic subunit. Two contain sites for phosphorylation by PKA, which is activated by cyclic AMP. The remaining subunit is calmodulin, which binds Ca^{2+} when the cytosolic Ca^{2+} concentration rises. The

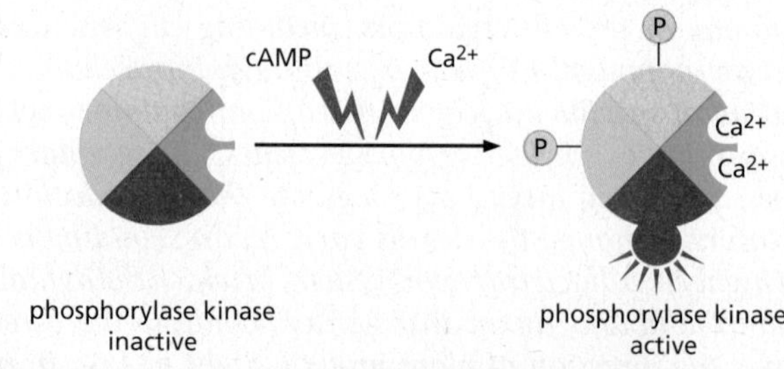

Figure Q15–2 Integration of cyclic-AMP-dependent and Ca^{2+}-dependent signaling pathways by phosphorylase kinase in liver and muscle cells (Problem 15–14).

regulatory subunits control the equilibrium between the active and inactive conformations of the catalytic subunit, with each phosphate and Ca^{2+} nudging the equilibrium toward the active conformation. How does this arrangement allow phosphorylase kinase to serve its role as an integrator protein for the two pathways that stimulate glycogen breakdown?

15–15 Activation ("maturation") of frog oocytes is signaled through a MAP kinase signaling module. An increase in the hormone progesterone triggers the module by stimulating the translation of mRNA encoding Mos, which is the frog's MAP kinase kinase kinase (**Figure Q15–3**). Maturation is easy to score visually by the presence of a white spot in the middle of the brown surface of the oocyte (see Figure Q15–3). To determine the dose–response curve for progesterone-induced activation of MAP kinase, you place 16 oocytes in each of six plastic dishes and add various concentrations of progesterone. After an overnight incubation, you crush the oocytes, prepare an extract, and determine the state of MAP kinase phosphorylation (hence, activation) by SDS polyacrylamide-gel electrophoresis (**Figure Q15–4A**). This analysis shows a graded increase in MAP kinase activity with increasing concentrations of progesterone.

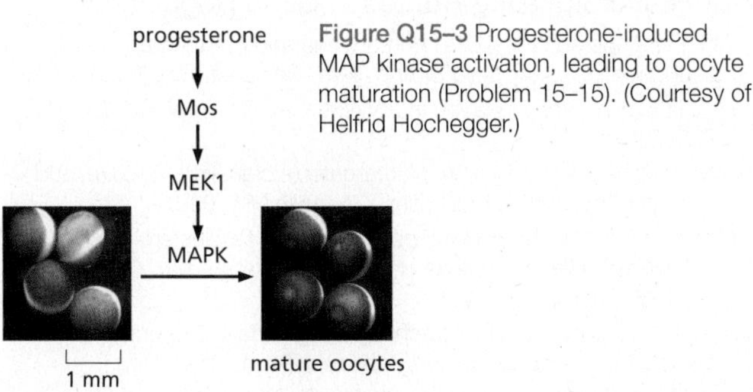

Figure Q15–3 Progesterone-induced MAP kinase activation, leading to oocyte maturation (Problem 15–15). (Courtesy of Helfrid Hochegger.)

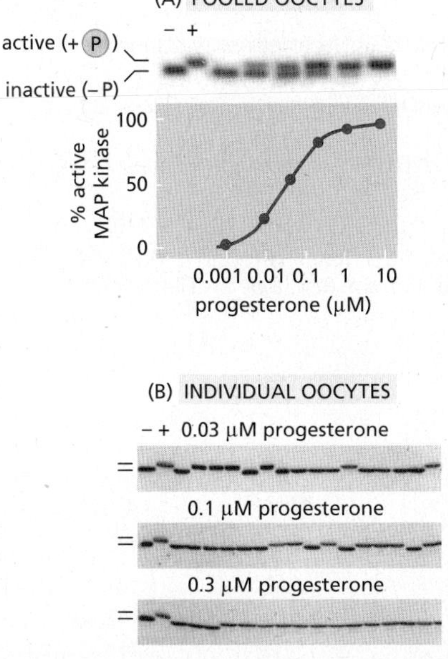

Figure Q15–4 Activation of frog oocytes (Problem 15–15). (A) Phosphorylation of MAP kinase in pooled oocytes. (B) Phosphorylation of MAP kinase in individual oocytes. MAP kinase was detected by immunoblotting using a MAP-kinase-specific antibody. The first two lanes in each gel contain nonphosphorylated, inactive MAP kinase (–) and phosphorylated, active MAP kinase (+), respectively. (From J.E. Ferrell, Jr., and E.M. Machleder, *Science* 280:895–898, 1998. With permission from AAAS.)

Before you crushed the oocytes, you noticed that not all the oocytes in individual dishes had white spots. Had some oocytes undergone partial activation and not yet reached the white-spot stage? To answer this question, you repeat the experiment, but this time you analyze MAP kinase activation in individual oocytes. You are surprised to find that each oocyte has either a fully activated or a completely inactive MAP kinase (**Figure Q15–4B**). How can an all-or-none response in individual oocytes give rise to a graded response in the population?

15–16 The Wnt planar polarity signaling pathway normally ensures that each wing cell in *Drosophila* has a single hair. Overexpression of the *Frizzled* gene from a heat-shock promoter (hs-*Fz*) causes multiple hairs to grow from many cells (**Figure Q15–5A**). This phenotype is suppressed if hs-*Fz* is combined with a heterozygous deletion (Dsh^{Δ}) of the *Dishevelled* gene (**Figure Q15–5B**). Do these results allow you to order the action of Frizzled and Dishevelled in the signaling pathway? If so, what is the order? Explain your reasoning.

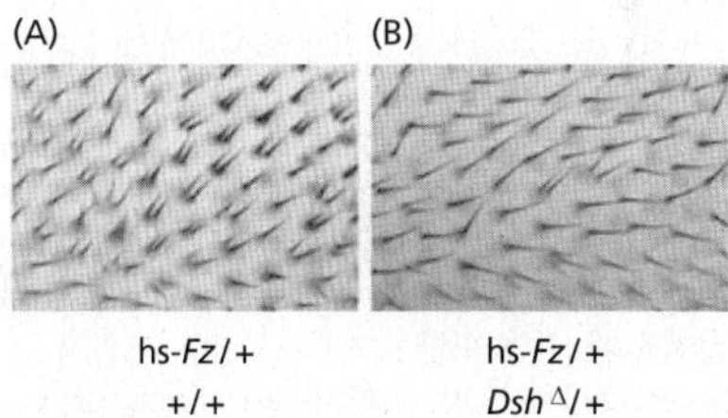

Figure Q15–5 Pattern of hair growth on wing cells in genetically different *Drosophila* (Problem 15–16). (From C.G. Winter et al., *Cell* 105:81–91, 2001. With permission from Elsevier.)

15–17 The last common ancestor to plants and animals was a unicellular eukaryote. Thus, it is thought that multicellularity and the attendant demands for cell communication arose independently in these two lineages. This evolutionary viewpoint accounts nicely for the vastly different mechanisms that plants and animals use for cell communication. Fungi use signaling mechanisms and components that are very similar to those used in animals. Which of the phylogenetic trees shown in **Figure Q15–6** do these observations support?

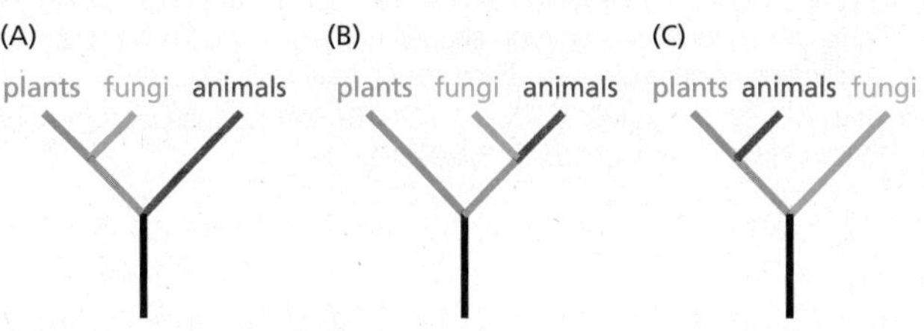

Figure Q15–6 Three possible phylogenetic relationships among plants, animals, and fungi (Problem 15–17).

REFERENCES

General

Ferrell, JE (2022) Systems Biology of Cell Signaling: Recurring Themes and Quantitative Models. Boca Raton, FL: Garland Science.

Lim W, Mayer B & Pawson T (2015) Cell Signaling: Principles and Mechanisms. New York: Garland Science.

Marks F, Klingmüller U & Müller-Decker K (2017) Cellular Signal Processing: An Introduction to the Molecular Mechanisms of Signal Transduction, 2nd ed. New York: Garland Science.

Principles of Cell Signaling

Alon U (2007) Network motifs: theory and experimental approaches. *Nat. Rev. Genet.* 8, 450–461.

Case LB, Ditlev JA & Rosen MK (2019) Regulation of transmembrane signaling by phase separation. *Annu. Rev. Biophys.* 48, 465–494.

Ferrell JE, Jr (2016) Perfect and near-perfect adaptation in cell signaling. *Cell Syst.* 2, 62–67.

Ferrell JE, Jr & Ha SH (2014) Ultrasensitivity part III: cascades, bistable switches, and oscillators. *Trends Biochem. Sci.* 39, 612–618.

Ladbury JE & Arold ST (2012) Noise in cellular signaling pathways: causes and effects. *Trends Biochem. Sci.* 37, 173–178.

Langeberg LK & Scott JD (2015) Signalling scaffolds and local organization of cellular behaviour. *Nat. Rev. Mol. Cell Biol.* 16, 232–244.

Needham EJ, Parker BL, Burykin T, James DE & Humphrey SJ (2019) Illuminating the dark phosphoproteome. *Sci. Signal.* 12, eaau8645.

Rosse C, Linch M, Kermorgant S, Cameron AJ, Boeckeler K & Parker PJ (2010) PKC and the control of localized signal dynamics. *Nat. Rev. Mol. Cell Biol.* 11, 103–112.

Shi Y (2009) Serine/threonine phosphatases: mechanism through structure. *Cell* 139, 468–484.

Taylor SS, Meharena HS & Kornev AP (2019) Evolution of a dynamic molecular switch. *IUBMB Life* 71, 672–684.

Tenner B, Mehta S & Zhang J (2016) Optical sensors to gain mechanistic insights into signaling assemblies. *Curr. Opin. Struct. Biol.* 41, 203–210.

Signaling Through G-Protein-coupled Receptors

Barnes CL, Malhotra H & Calvert PD (2021) Compartmentalization of photoreceptor sensory cilia. *Front. Cell Dev. Biol.* 9, 636737.

Bhattacharyya M, Karandur D & Kuriyan J (2020) Structural insights into the regulation of Ca(2+)/calmodulin-dependent protein kinase II (CaMKII). *Cold Spring Harb. Perspect. Biol.* 12, a035147.

Carafoli E & Krebs J (2016) Why calcium? How calcium became the best communicator. *J. Biol. Chem.* 291, 20849–20857.

Hilger D, Masureel M & Kobilka BK (2018) Structure and dynamics of GPCR signaling complexes. *Nat. Struct. Mol. Biol.* 25, 4–12.

Kadamur G & Ross EM (2013) Mammalian phospholipase C. *Annu. Rev. Physiol.* 75, 127–154.

Lobingier BT & von Zastrow M (2019) When trafficking and signaling mix: how subcellular location shapes G protein-coupled receptor activation of heterotrimeric G proteins. *Traffic* 20, 130–136.

Newton AC, Bootman MD & Scott JD (2016) Second messengers. *Cold Spring Harb. Perspect. Biol.* 8, a005926.

Prole DL & Taylor CW (2019) Structure and function of IP3 receptors. *Cold Spring Harb. Perspect. Biol.* 11, a035063.

Rajagopal S & Shenoy SK (2018) GPCR desensitization: acute and prolonged phases. *Cell. Signal.* 41, 9–16.

Sung CH & Chuang JZ (2010) The cell biology of vision. *J. Cell Biol.* 190, 953–963.

Signaling Through Enzyme-coupled Receptors

Balla T (2013) Phosphoinositides: tiny lipids with giant impact on cell regulation. *Physiol. Rev.* 93, 1019–1137.

Burke JE (2018) Structural basis for regulation of phosphoinositide kinases and their involvement in human disease. *Mol. Cell* 71, 653–673.

Kovacs E, Zorn JA, Huang Y, Barros T & Kuriyan J (2015) A structural perspective on the regulation of the epidermal growth factor receptor. *Annu. Rev. Biochem.* 84, 739–764.

Madsen RR & Vanhaesebroeck B (2020) Cracking the context-specific PI3K signaling code. *Sci. Signal.* 13, eaay2940.

Manning BD & Toker A (2017) AKT/PKB signaling: navigating the network. *Cell* 169, 381–405.

Saxton RA & Sabatini DM (2017) mTOR signaling in growth, metabolism, and disease. *Cell* 168, 960–976.

Terrell EM & Morrison DK (2019) Ras-mediated activation of the Raf family kinases. *Cold Spring Harb. Perspect. Med.* 9, a033746.

Villarino AV, Kanno Y & O'Shea JJ (2017) Mechanisms and consequences of Jak-STAT signaling in the immune system. *Nat. Immunol.* 18, 374–384.

Alternative Signaling Routes in Gene Regulation

Bangs F & Anderson KV (2017) Primary cilia and mammalian hedgehog signaling. *Cold Spring Harb. Perspect. Biol.* 9, a028175.

Bray SJ (2016) Notch signalling in context. *Nat. Rev. Mol. Cell Biol.* 17, 722–735.

Nusse R & Clevers H (2017) Wnt/beta-catenin signaling, disease, and emerging therapeutic modalities. *Cell* 169, 985–999.

Radhakrishnan A, Rohatgi R & Siebold C (2020) Cholesterol access in cellular membranes controls Hedgehog signaling. *Nat. Chem. Biol.* 16, 1303–1313.

Siebel C & Lendahl U (2017) Notch signaling in development, tissue homeostasis, and disease. *Physiol. Rev.* 97, 1235–1294.

Swan JA, Golden SS, LiWang A & Partch CL (2018) Structure, function, and mechanism of the core circadian clock in cyanobacteria. *J. Biol. Chem.* 293, 5026–5034.

Signaling in Plants

Belkhadir Y & Jaillais Y (2015) The molecular circuitry of brassinosteroid signaling. *New Phytol.* 206, 522–540.

Binder BM (2020) Ethylene signaling in plants. *J. Biol. Chem.* 295, 7710–7725.

Fiorucci AS & Fankhauser C (2017) Plant strategies for enhancing access to sunlight. *Curr. Biol.* 27, R931–R940.

Leyser O (2018) Auxin signaling. *Plant Physiol.* 176, 465–479.

Su SH, Gibbs NM, Jancewicz AL & Masson PH (2017) Molecular mechanisms of root gravitropism. *Curr. Biol.* 27, R964–R972.

CHAPTER

16

The Cytoskeleton

For cells to function properly, they must organize themselves in space and interact mechanically with each other and with their environment. They have to be correctly shaped, physically robust, and properly structured internally. Many have to change their shape and move from place to place. All cells have to be able to rearrange their internal components as they grow, divide, and adapt to changing circumstances. These spatial and mechanical functions depend on a remarkable system of filaments called the **cytoskeleton** (**Figure 16–1**).

The cytoskeleton's varied functions depend on the behavior of three main families of protein filaments—*actin filaments, microtubules,* and *intermediate filaments.* Each type of filament has distinct mechanical properties, dynamics, and biological roles, but all share certain fundamental features. Just as we require our ligaments, bones, and muscles to work together, so all three cytoskeletal filament systems must normally function collectively to give a cell its strength, its shape, and its ability to divide and move around.

In this chapter, we describe the function and evolutionary conservation of cellular filament systems. We explain the basic principles underlying filament assembly and disassembly and how other proteins interact with the filaments to alter their dynamics and direct their organization. Finally, we discuss how cytoskeletal systems work together with other cellular components to generate *cell polarity,* which is essential for many aspects of cell behavior and function.

IN THIS CHAPTER

FUNCTION AND DYNAMICS OF THE CYTOSKELETON

The three major cytoskeletal filaments are responsible for different aspects of the cell's spatial organization and mechanical properties. Actin filaments determine the shape of the cell's surface and are necessary for whole-cell locomotion; they also drive the pinching of one cell into two. Microtubules determine the positions of organelles, direct intracellular transport, and form the mitotic spindle that segregates chromosomes during cell division. Intermediate filaments provide mechanical strength. All of these cytoskeletal filaments interact with hundreds of accessory proteins that regulate and link the filaments to other cell components, as well as to each other. The accessory proteins are essential for the controlled assembly of the cytoskeletal filaments in particular locations, and they include the *motor proteins,* remarkable molecular machines that convert the energy of ATP hydrolysis into mechanical force that can either move organelles along the filaments or move the filaments themselves.

In this part of the chapter, we discuss the general features of the proteins that make up the filaments of the cytoskeleton. We focus on their ability to form intrinsically polarized and self-organized structures that are highly dynamic, allowing the cell to rapidly modify cytoskeletal structure and function when conditions change.

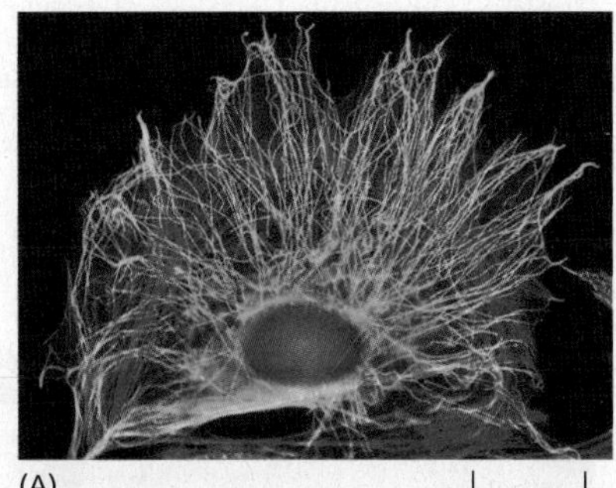

Figure 16–1 The cytoskeleton. (A) A cell in culture has been fixed and labeled to show its cytoplasmic arrays of microtubules *(green)* and actin filaments *(red)*. (B) This dividing cell has been labeled to show its spindle microtubules *(green)* and surrounding cage of intermediate filaments *(red)*. The DNA in both cells is labeled in *blue*. (A, courtesy of Albert Tousson, High Resolution Imaging Facility, University of Alabama at Birmingham; B, courtesy of Conly Rieder.)

ACTIN FILAMENTS

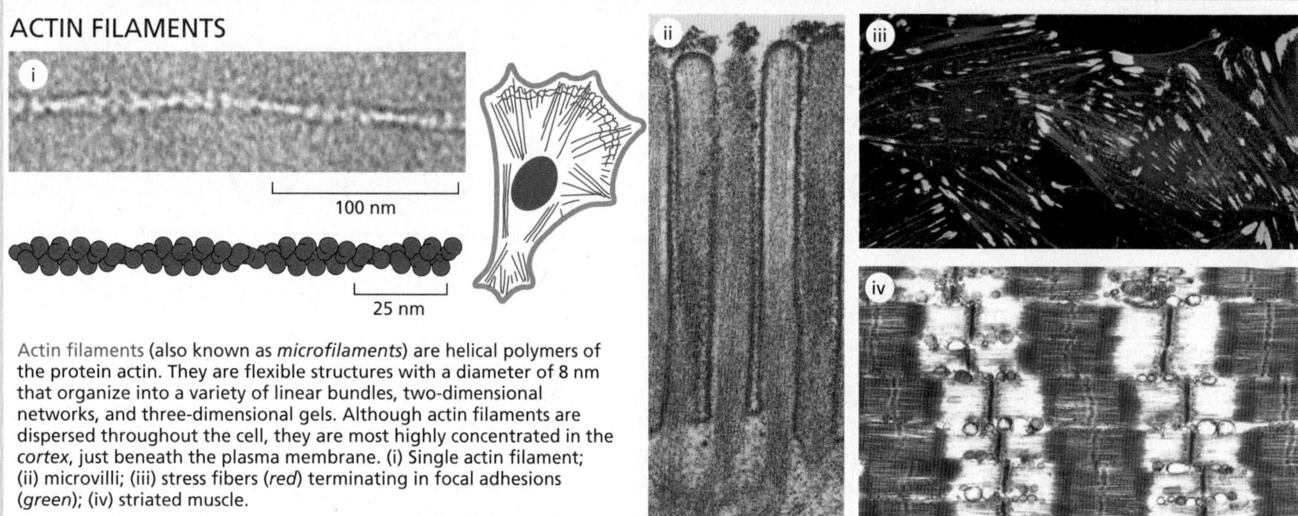

Actin filaments (also known as *microfilaments*) are helical polymers of the protein actin. They are flexible structures with a diameter of 8 nm that organize into a variety of linear bundles, two-dimensional networks, and three-dimensional gels. Although actin filaments are dispersed throughout the cell, they are most highly concentrated in the *cortex*, just beneath the plasma membrane. (i) Single actin filament; (ii) microvilli; (iii) stress fibers (*red*) terminating in focal adhesions (*green*); (iv) striated muscle.

Micrographs courtesy of R. Craig (i and iv); P.T. Matsudaira and D.R. Burgess (ii); K. Burridge (iii).

MICROTUBULES

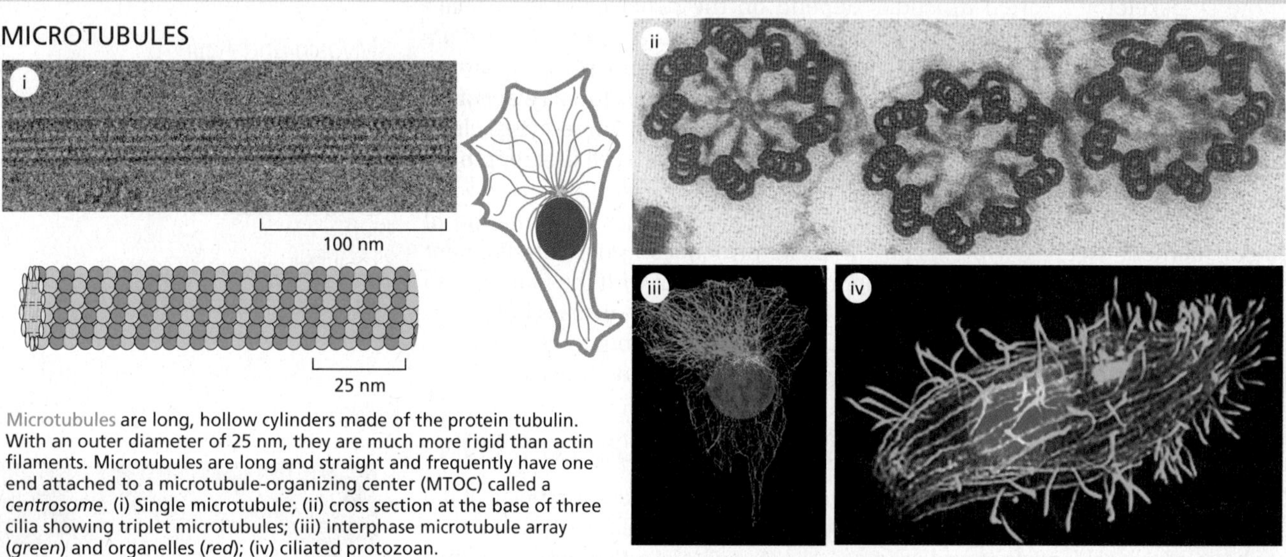

Microtubules are long, hollow cylinders made of the protein tubulin. With an outer diameter of 25 nm, they are much more rigid than actin filaments. Microtubules are long and straight and frequently have one end attached to a microtubule-organizing center (MTOC) called a *centrosome*. (i) Single microtubule; (ii) cross section at the base of three cilia showing triplet microtubules; (iii) interphase microtubule array (*green*) and organelles (*red*); (iv) ciliated protozoan.

Micrographs courtesy of R. Wade (i); D.T. Woodrow and R.W. Linck (ii); J. Seemann (iii); D. Burnette (iv).

INTERMEDIATE FILAMENTS

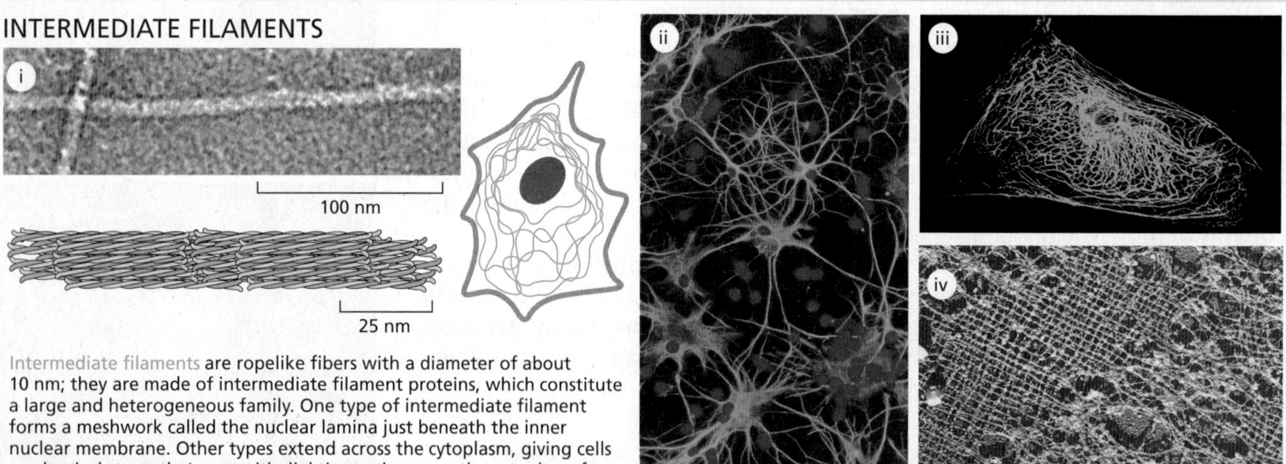

Intermediate filaments are ropelike fibers with a diameter of about 10 nm; they are made of intermediate filament proteins, which constitute a large and heterogeneous family. One type of intermediate filament forms a meshwork called the nuclear lamina just beneath the inner nuclear membrane. Other types extend across the cytoplasm, giving cells mechanical strength. In an epithelial tissue, they span the cytoplasm from one cell–cell junction to another, thereby strengthening the entire epithelium. (i) Individual intermediate filaments; (ii) intermediate filaments (*blue*) in neurons and (iii) epithelial cell; (iv) nuclear lamina.

(i) From K.N. Goldie et al., *J. Struct. Biol.* 158:378–385, 2007; (ii) N. Kedersha; (iii) Alvin Tesler/Science Source; (iv) from U. Aebi et al., *Nature* 323:560–564, published 1986 by Nature Publishing Group. Reproduced with permission of SNCSC.

Cytoskeletal Filaments Are Dynamic, but Can Nevertheless Form Stable Structures

Cytoskeletal systems are dynamic and adaptable, organized more like ant trails than interstate highways. A single trail of ants may persist for many hours, extending from the ant nest to a delectable picnic site, but the individual ants within the trail are anything but static. If the ant scouts find a new and better source of food or if the picnickers clean up and leave, the dynamic structure adapts with astonishing rapidity. In a similar way, large-scale cytoskeletal structures can change or persist, according to need, lasting for lengths of time ranging from less than a minute up to the cell's lifetime. But the individual macromolecular components that make up these structures are in a constant state of flux. As a result, like the alteration of an ant trail, a structural rearrangement in a cell requires little extra energy when conditions change.

Regulation of the dynamic behavior and assembly of cytoskeletal filaments allows eukaryotic cells to build an enormous range of structures from the three basic filament systems. The micrographs in **Panel 16–1** illustrate some of these structures. Actin filaments underlie the plasma membrane of animal cells in a layer called the *cell cortex*, providing strength and shape to the thin lipid bilayer. They also form many types of cell-surface projections. Some of these are dynamic structures, such as the *filopodia*, *lamellipodia*, and *pseudopodia* that cells use to explore territory and move around. More stable arrays allow cells to brace themselves against an underlying substratum and enable muscle to contract. The regular bundles of *stereocilia* on the surface of hair cells in the inner ear contain stable bundles of actin filaments that tilt as rigid rods in response to sound, and the *microvilli* on the surface of intestinal epithelial cells vastly increase the apical cell-surface area to enhance nutrient absorption. In plants, actin filaments drive rapid streaming of the cytoplasm inside cells.

Microtubules, which are frequently found in a cytoplasmic array that extends to the cell periphery, can quickly rearrange themselves to form a bipolar *mitotic spindle* during cell division. They can also form *cilia*, which function as motile whips or sensory devices on the surface of the cell, or tightly aligned bundles that serve as tracks for the transport of materials down long neuronal axons. In plant cells, organized arrays of microtubules help to direct the pattern of cell-wall synthesis, and in many protozoans they form the framework upon which the entire cell is built.

Intermediate filaments line the inner face of the nuclear envelope, forming a protective cage for the cell's DNA; in the cytosol, they are twisted into strong cables that can hold epithelial cell sheets together or help nerve cells to extend long and robust axons, and they allow us to form tough appendages such as hair and fingernails.

An important and dramatic example of rapid reorganization of the cytoskeleton occurs during cell division, as shown in **Figure 16–2** for a fibroblast growing in a tissue-culture dish. After the chromosomes have replicated, the interphase microtubule array that spreads throughout the cytoplasm is reconfigured into the bipolar *mitotic spindle*, which transfers the two copies of each chromosome into separate daughter nuclei. At the same time, the specialized actin structures that enable the fibroblast to crawl across the surface of the dish rearrange so that the cell stops moving and assumes a more spherical shape. Actin and its associated motor protein myosin then form a belt around the middle of the cell, the *contractile ring*, which constricts like a tiny muscle to pinch the cell in two. When division is complete, the cytoskeletons of the two daughter fibroblasts recreate their interphase structures, thereby converting the two rounded-up daughter cells into smaller versions of the flattened, crawling mother cell.

Many cells require rapid cytoskeletal rearrangements for their normal functioning during interphase as well. For example, the *neutrophil*, a type of white blood cell, chases and engulfs bacterial and fungal cells that accidentally gain

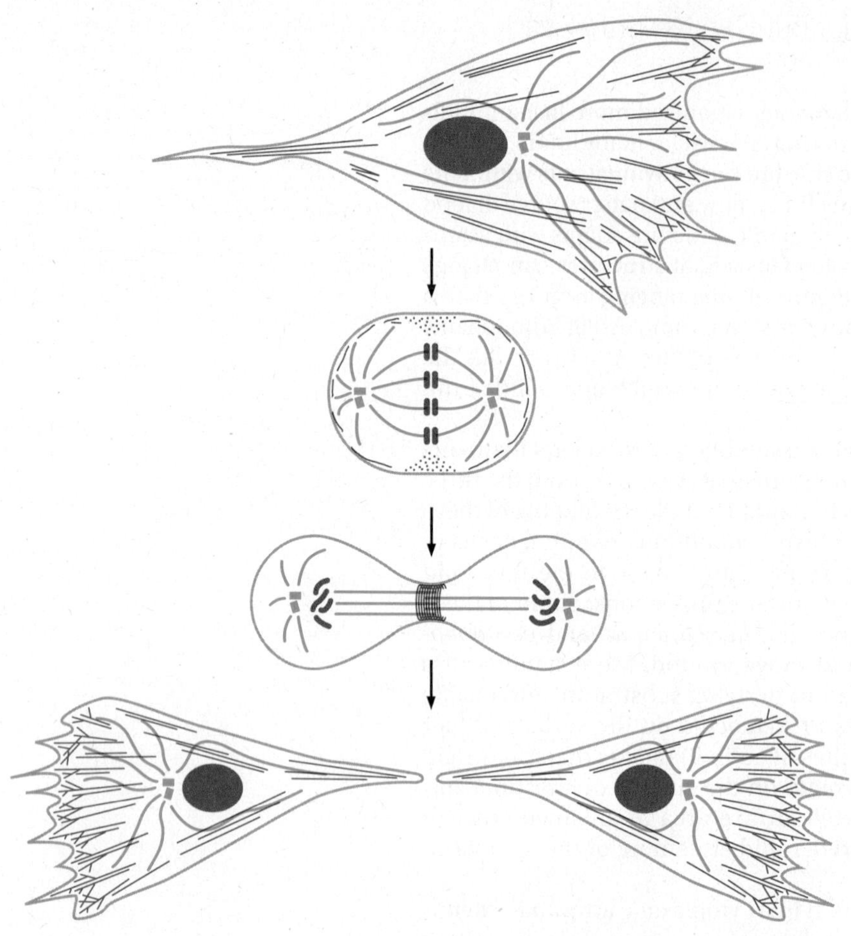

Figure 16–2 **Diagram of changes in cytoskeletal organization associated with cell division.** The crawling fibroblast drawn here has a polarized, dynamic actin cytoskeleton (shown in *red*) that assembles lamellipodia and filopodia to push its leading edge toward the right. The polarization of the actin cytoskeleton is assisted by the microtubule cytoskeleton *(green)*, consisting of long microtubules that emanate from a microtubule-organizing center located in front of the nucleus. When the cell divides, actin filaments are reorganized and the cell assumes a spherical shape. The polarized microtubule array rearranges to form a bipolar mitotic spindle, which is responsible for aligning and then segregating the duplicated chromosomes *(brown)*. After chromosome segregation, the actin filaments form a contractile ring at the center of the cell that pinches the cell in two. Then the two daughter cells reorganize their microtubule and actin cytoskeletons into smaller versions of those that were present in the mother cell, enabling them to crawl their separate ways.

access to the normally sterile parts of the body, as through a cut in the skin. Like most crawling cells, neutrophils advance by extending a protrusive structure filled with newly polymerized actin filaments. When the elusive bacterial prey moves in a different direction, the neutrophil can reorganize its polarized protrusive structures within seconds (**Figure 16–3**).

The Cytoskeleton Determines Cellular Organization and Polarity

In cells that have achieved a stable, differentiated morphology—such as mature neurons or epithelial cells—the dynamic elements of the cytoskeleton produce stable, large-scale structures for cellular organization. For example, on the specialized epithelial cells that line organs such as the intestine and the lung, cytoskeletal-based cell-surface protrusions including microvilli and cilia are able to maintain a constant location, length, and diameter over the entire lifetime of the cell. For the actin bundles at the cores of microvilli on intestinal epithelial cells, this is only a few days. But the actin bundles at the cores of

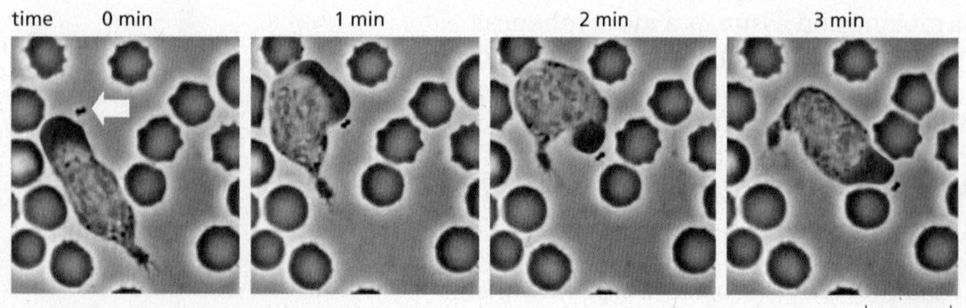

Figure 16–3 **A neutrophil in pursuit of bacteria.** In this preparation of human blood, a small clump of bacteria (*white arrow*) is about to be captured by a neutrophil. As the bacteria move, the neutrophil quickly reassembles the dense actin network within a *pseudopod* ("false foot") at its leading edge (highlighted in *red*) to push toward the location of the bacteria (**Movie 16.1**). Rapid disassembly and reassembly of the actin cytoskeleton in this cell enable it to change its orientation and direction of movement within a few minutes. (From a video recorded by David Rogers.)

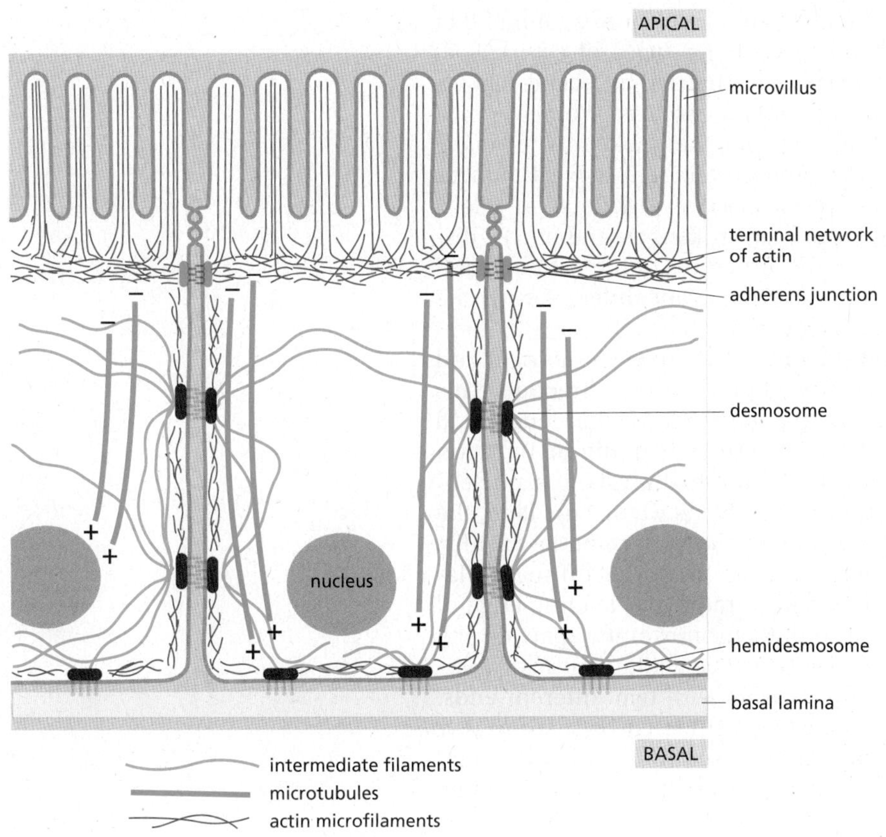

Figure 16–4 Organization of the cytoskeleton in polarized epithelial cells. All the components of the cytoskeleton cooperate to produce the characteristic shapes of specialized cells, including the epithelial cells that line the small intestine, diagrammed here. At the apical (upper) surface, facing the intestinal lumen, bundled actin filaments *(red)* form microvilli that increase the cell-surface area available for absorbing nutrients from food. Below the microvilli, a circumferential band of actin filaments is connected to cell–cell adherens junctions that anchor the cells to each other. Intermediate filaments *(blue)* are anchored to other kinds of adhesive structures, including desmosomes and hemidesmosomes, that connect the epithelial cells into a sturdy sheet and attach them to the underlying extracellular matrix; these structures are discussed in Chapter 19. Microtubules *(green)* run vertically from the top of the cell to the bottom and provide a global coordinate system that enables the cell to direct newly synthesized components to their proper locations.

stereocilia on the hair cells of the inner ear must maintain their stable organization for the entire lifetime of the animal, because these cells do not turn over. Remarkably, the actin filaments in stereocilia are very stable, and polymerization and depolymerization have been observed only at their tips. We do not understand how these actin structures are maintained at a constant length for decades.

Besides forming stable, specialized cell-surface protrusions, the cytoskeleton is also responsible for large-scale cellular polarity, enabling cells to tell the difference between top and bottom or front and back. The polarity information conveyed by cytoskeletal organization is often maintained over the lifetime of the cell. For example, polarized epithelial cells use organized arrays of microtubules, actin filaments, and intermediate filaments to maintain the critical differences between the *apical surface* and the *basolateral surface*. They also must maintain strong adhesive contacts with one another to enable this single layer of cells to serve as an effective physical barrier (**Figure 16–4**). How cell polarity is established is discussed in the last part of this chapter.

Filaments Assemble from Protein Subunits That Impart Specific Physical and Dynamic Properties

Cytoskeletal filaments can reach from one end of the cell to the other, spanning tens or even hundreds of micrometers. Yet the individual protein molecules that form the filaments are only a few nanometers in size. The cell builds the filaments by assembling large numbers of the small subunits, like building a skyscraper out of bricks. Because these subunits are small, they can diffuse rapidly in the cytosol, whereas the assembled filaments cannot. In this way, cells can undergo rapid structural reorganizations, disassembling filaments at one site and reassembling them at another site far away.

Actin filaments and microtubules are built from subunits that are compact and globular—*actin subunits* for actin filaments and *tubulin subunits* for microtubules—whereas intermediate filaments are made up of subunits that are themselves elongated and fibrous. All three major types of cytoskeletal filaments form as polymeric assemblies of subunits that self-associate, using a combination of end-to-end and side-to-side protein contacts. Whereas covalent linkages between their subunits hold together the backbones of many biological polymers—including DNA, RNA, and proteins—it is weak noncovalent interactions that hold together the three types of cytoskeletal polymers. Because covalent bonds are not being formed or broken, assembly and disassembly can occur rapidly. Differences in the structures of the subunits and how they associate with one another produce important differences in the stability and mechanical properties of each type of filament.

The subunits of actin filaments and microtubules are asymmetrical and bind to one another head-to-tail so that they all point in one direction. This subunit polarity gives the filaments structural polarity along their length and makes the two ends of each polymer behave differently. In addition, the actin and tubulin subunits are enzymes that catalyze the hydrolysis of a nucleoside triphosphate (NTP)—ATP and GTP, respectively. As we discuss later, the energy derived from NTP hydrolysis helps the filaments to remodel rapidly. It allows the cell to harness the polar and dynamic properties of these filaments to generate force in a specific direction to move the leading edge of a migrating cell forward, for example, or to pull chromosomes apart during cell division. In contrast, the subunits of intermediate filaments are symmetrical and thus do not form polarized filaments with two different ends. Intermediate filament subunits also do not catalyze the hydrolysis of ATP or GTP. Nevertheless, intermediate filaments can be disassembled rapidly when required. In mitosis, for example, kinases phosphorylate the subunits, leading to their dissociation.

Cytoskeletal filaments in living cells are not built by simply stringing subunits together in single file. A thousand tubulin subunits lined up end-to-end, for example, would span the diameter of a small eukaryotic cell, but a filament formed in this way would lack the strength to avoid breakage by ambient thermal energy, unless each subunit in the filament was bound extremely tightly to its two neighbors. Such tight binding would limit the rate at which the filaments could disassemble, making the cytoskeleton a static and less useful

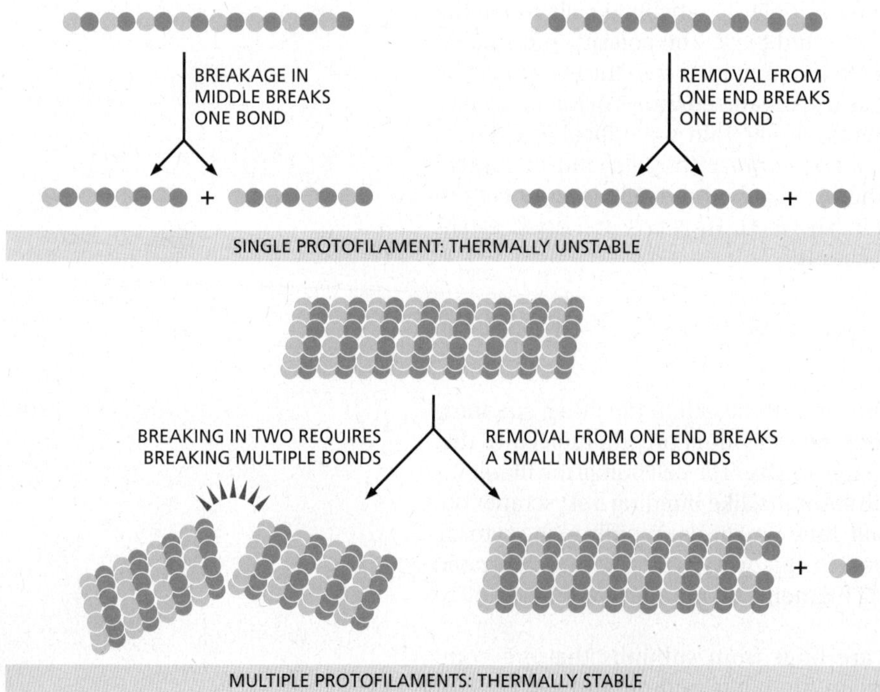

Figure 16–5 **The thermal stability of cytoskeletal filaments with dynamic ends.** A protofilament consisting of a single strand of subunits is thermally unstable, because breakage of a single bond between subunits is sufficient to break the filament. In contrast, formation of a cytoskeletal filament from more than one protofilament allows the ends to be dynamic, while enabling the filaments themselves to resist thermal breakage. In a microtubule, for example, removing a single subunit dimer from the end of the filament requires breaking noncovalent bonds with a maximum of three other subunits, whereas fracturing the filament in the middle requires breaking noncovalent bonds in all 13 protofilaments.

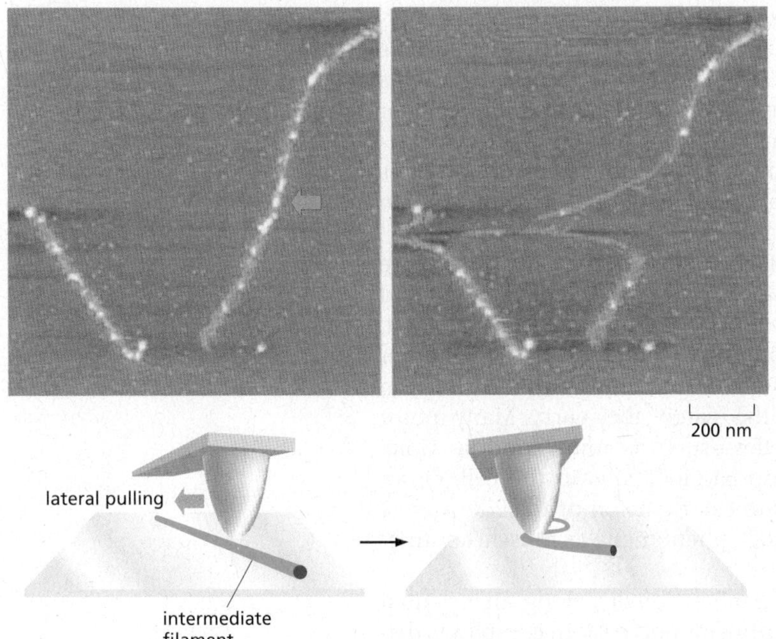

Figure 16–6 **Flexibility and stretch in an intermediate filament.** Intermediate filaments are formed from elongated fibrous subunits with strong lateral contacts, resulting in resistance to stretching forces. In this experiment, a tiny mechanical probe was used to rapidly scan across a surface to which intermediate filaments were adhered, providing a measurement of the height and contour of the filaments as shown in shades of *gray* to *white*. Along a single scan line *(arrow)* the applied force of the probe was increased from less than 1 nN to 30–40 nN. At this position one of the filaments stretched to more than three times its length before breaking, as illustrated in the micrographs. This technique is called atomic force microscopy. (Adapted from L. Kreplak et al., *J. Mol. Biol.* 354:569–577, 2005. With permission from Elsevier.)

structure. To provide both strength and adaptability, microtubules are built of 13 **protofilaments**—linear strings of subunits joined end to end—that associate with one another laterally to form a hollow cylinder. The addition or loss of a subunit at the end of one protofilament makes or breaks a small number of bonds. In contrast, loss of a subunit from the middle of a protofilament requires breaking many more bonds, while breaking it in two requires breaking bonds in multiple protofilaments all at the same time (**Figure 16–5**). The greater energy required to break multiple noncovalent bonds simultaneously allows microtubules to resist thermal breakage, while allowing rapid subunit addition and loss at the filament ends. Although made up of only two strands instead of 13, helical actin filament subunits also make both end-to-end and side-to-side contacts, allowing for rapid dynamics at filament ends while providing sufficient strength along the length of the filament. However, the tubular geometry of a microtubule makes it much stiffer than a two-stranded actin filament.

As with other specific protein–protein interactions, many hydrophobic interactions and noncovalent bonds hold the subunits in a cytoskeletal filament together (see Figure 3–4). The locations and types of subunit-subunit contacts differ for the different filaments. Intermediate filaments, for example, assemble by forming strong lateral contacts between α-helical coiled-coils, which extend over most of the length of each elongated fibrous subunit. Because the individual subunits are staggered in the filament, intermediate filaments form strong, ropelike structures that tolerate stretching and bending to a greater extent than do either actin filaments or microtubules (**Figure 16–6**).

Accessory Proteins and Motors Act on Cytoskeletal Filaments

The cell regulates the length and stability of its cytoskeletal filaments, their number and geometry, and their attachments to one another and to other components of the cell. Filaments can thereby form a wide variety of higher-order structures. Direct covalent modification of the filament subunits regulates some filament properties, but most of the regulation is performed by hundreds of accessory proteins that determine the spatial distribution and the dynamic behavior of the filaments, converting information received through signaling pathways into cytoskeletal action. These accessory proteins bind to the filaments or their subunits to determine the sites of assembly of new filaments, to regulate the partitioning of polymer proteins between filament and subunit forms, to change the kinetics of filament assembly and disassembly, to harness energy to generate force, and to link filaments to one another or to other cell structures such as organelles and the plasma membrane. In these processes, the accessory proteins bring cytoskeletal structure under the control of extracellular and intracellular signals, including those that trigger the dramatic transformations of the cytoskeleton that occur during each cell cycle. Acting in groups, the accessory proteins enable cells to maintain a highly organized but flexible internal structure and, in many cases, to move.

Among the most fascinating proteins that associate with the cytoskeleton are the motor proteins. These proteins bind to a polarized cytoskeletal filament and use the energy derived from repeated cycles of ATP hydrolysis to move along it. Dozens of different motor proteins coexist in every eukaryotic cell. They differ in the type of filament they bind to (either actin or microtubules), the direction in which they move along the filament, and the "cargo" they carry. Many motor proteins carry membrane-enclosed organelles—such as mitochondria, Golgi stacks, or secretory vesicles—to their appropriate locations in the cell. Other motor proteins cause cytoskeletal filaments to exert tension or to slide against each other, generating the force that drives such phenomena as muscle contraction, ciliary beating, and cell division.

Cytoskeletal motor proteins that move unidirectionally along an oriented polymer track are reminiscent of some other proteins and protein complexes discussed elsewhere in this book, such as DNA and RNA polymerases, helicases, and ribosomes. All of these proteins have the ability to use chemical energy to propel themselves along a linear track, with the direction of sliding dependent on the structural polarity of the track. All of them generate motion by coupling nucleoside triphosphate hydrolysis to a large-scale conformational change (see Figure 3–71).

Molecular Motors Operate in a Cellular Environment Dominated by Brownian Motion

Because of the microscopic size of cells, motor-based movements occur in an environment that is highly dynamic due to random thermal fluctuations called *Brownian motion*. As introduced in Chapter 1 (see p. 9), Brownian motion drives the diffusion of all of the molecules inside cells, and it is thereby critical for the rate of biochemical reactions, which require molecular collisions. Another important effect is its generation of viscous drag forces at small scales. For example, a motor protein moving along a cytoskeletal filament is constantly buffeted by random collisions with water and other molecules. As soon as its motor activity ceases, the motor protein will stop dead in its tracks due to the viscous drag forces generated by these collisions.

The stopping distance for any object moving through a fluid is determined by the **Reynolds number**, which is the ratio of inertial to viscous forces acting on that object. This ratio depends on the size of the object and the velocity of its movement through the fluid. For example, whereas both a bacterium and a fish can propel themselves through water, the size and speed of the fish provide it with significant inertia, such that when it stops swimming it continues to glide through the water for some distance. In contrast, with its much smaller size and slow speed of movement, a bacterium that stops propelling itself will come to an immediate halt (**Figure 16–7**). For a fish to behave similarly, it would need to be placed in an extremely viscous medium, such as roofing tar. Inside the cell, the tiny size and slow speeds of moving molecules result in extremely low Reynolds numbers, because viscous forces are far greater than inertial forces. Thus, there is no "gliding" inside a cell.

Random Brownian motion can also be harnessed to control movement at small scales, even in the absence of motor proteins. As we describe later in this chapter, some intracellular pathogens such as the bacterium *Listeria monocytogenes* use the force of actin polymerization to move around inside a cell. This movement does not involve motor proteins. Instead, actin filaments are induced to polymerize adjacent to the bacterial surface at one end. When the bacterium randomly

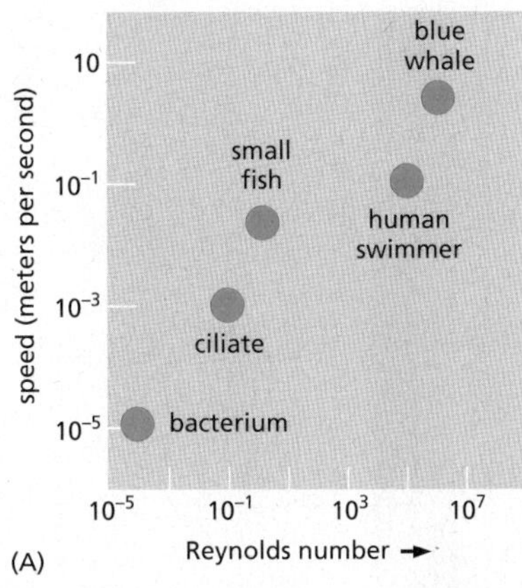

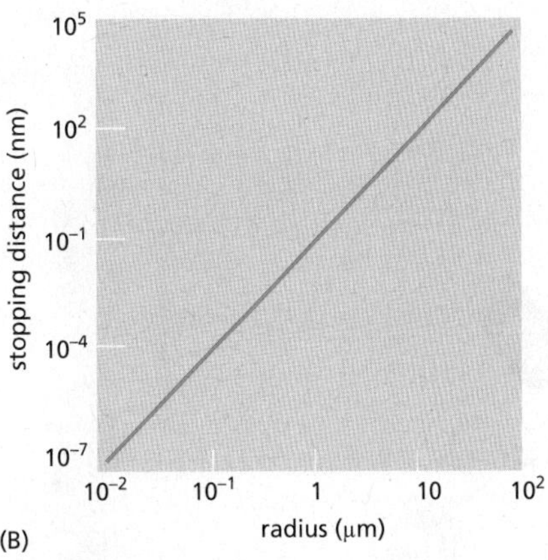

Figure 16–7 Cellular life occurs at a low Reynolds number. (A) The inertia of a swimming organism is counteracted by the viscous drag of water, and the ratio of these two forces is the *Reynolds number*, which is dimensionless. As illustrated in the graph, the Reynolds number is in the range of 10^6 for a blue whale traveling at about 10 meters per second. For a tiny swimming bacterium moving at 10^{-5} meters per second, the viscosity of the water dominates and the Reynolds number decreases to about 10^{-4}. (B) The Reynolds number indicates the extent to which an object will continue to coast or glide through a fluid after the motive force has stopped. The stopping distance in water for a sphere that has an initial velocity equal to its diameter per second is plotted. Thus, a one micron–diameter organelle would come to a halt within just a few angstroms once its forward motion ceases.

moves in the forward direction due to Brownian motion, actin quickly fills in the gap so that the bacterium cannot slip back to its original position. In this way, actin polymerization generates a force that pushes the bacterium through the cytoplasm (see Figure 16–17). A similar process drives forward movement of the plasma membrane at the leading edge of a migrating animal cell (see Figure 1–7). This phenomenon, in which random thermal motions are harnessed in a directed way, creates a *Brownian ratchet.*

Summary

The cytoplasm of eukaryotic cells is spatially organized by a network of protein filaments known as the cytoskeleton. This network contains three principal types of filaments: actin filaments, microtubules, and intermediate filaments. All three types of filaments form from assemblies of subunits that self-associate using a combination of end-to-end and side-to-side protein contacts. Differences in the structure of the subunits and the manner of their self-assembly give the filaments different mechanical properties. Subunit assembly and disassembly constantly remodel all three types of cytoskeletal filaments. Actin and tubulin (the subunits of actin filaments and microtubules, respectively) bind and hydrolyze nucleoside triphosphates (ATP and GTP, respectively) and assemble head-to-tail to generate polarized filaments capable of generating force. In living cells, accessory proteins including molecular motors modulate the dynamics and organization of cytoskeletal filaments, resulting in complex events such as cell division, migration, or muscle contraction, and generating elaborate cellular architecture to form polarized tissues such as epithelia.

ACTIN

The actin cytoskeleton performs a wide range of functions in diverse cell types. Each actin subunit, sometimes called globular or G-actin, is a 375-amino-acid polypeptide carrying a tightly associated molecule of ATP or ADP (**Figure 16–8A**). Actin is extraordinarily well conserved among eukaryotes. The amino acid sequences of actins from different eukaryotic species are usually about 90% identical. Small variations in actin amino acid sequence can cause significant functional differences: In vertebrates, for example, there are three isoforms of

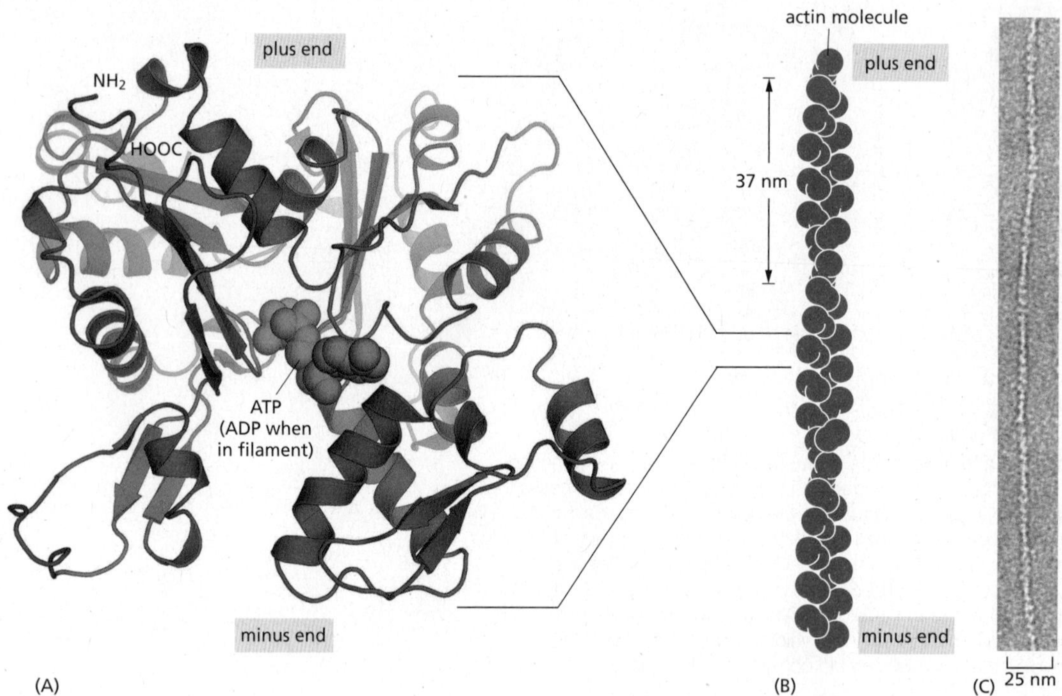

Figure 16–8 **The structures of an actin monomer and actin filament.** (A) The actin monomer has either ATP or ADP bound in a deep cleft in the center of the molecule. (B) Arrangement of monomers in a filament consisting of two protofilaments, held together by lateral contacts, which wind around each other as two parallel strands of a helix, with a twist repeating every 37 nm. All the subunits within the filament have the same orientation. (C) Electron micrograph of negatively stained actin filament. (C, courtesy of Roger Craig.)

actin, termed α, β, and γ, that differ slightly in their amino acid sequences and have distinct functions. α-Actin is expressed only in muscle cells, while β- and γ-actins are found together in almost all non-muscle cells.

Actin Subunits Assemble Head-to-Tail to Create Flexible, Polar Filaments

Actin subunits assemble head-to-tail to form a tight, right-handed helix, forming a structure about 8 nm wide called filamentous or F-actin (**Figure 16–8B and C**). Because the asymmetrical actin subunits of a filament all point in the same direction, filaments are polar and have structurally different ends: a slower-growing *minus end* and a faster-growing *plus end*. The minus end is also referred to as the *pointed end* and the plus end as the *barbed end* because of the arrowhead appearance of the complex formed between actin filaments and the motor protein myosin visible in electron micrographs (**Figure 16–9**). Within the filament, the subunits are positioned with their ATP-binding cleft directed toward the minus end.

Individual actin filaments are quite flexible. The stiffness of a filament can be characterized by its *persistence length*, the minimum filament length at which random thermal fluctuations are likely to cause it to bend. The persistence length of an actin filament is only a few tens of micrometers. In a living cell, however, accessory proteins frequently cross-link and bundle the filaments together, making large-scale actin structures that are much more rigid than an individual actin filament.

Nucleation Is the Rate-limiting Step in the Formation of Actin Filaments

The regulation of actin filament formation is an important mechanism by which cells control their shape and movement. Actin subunits can spontaneously bind one another, but the association is unstable until subunits assemble into an initial

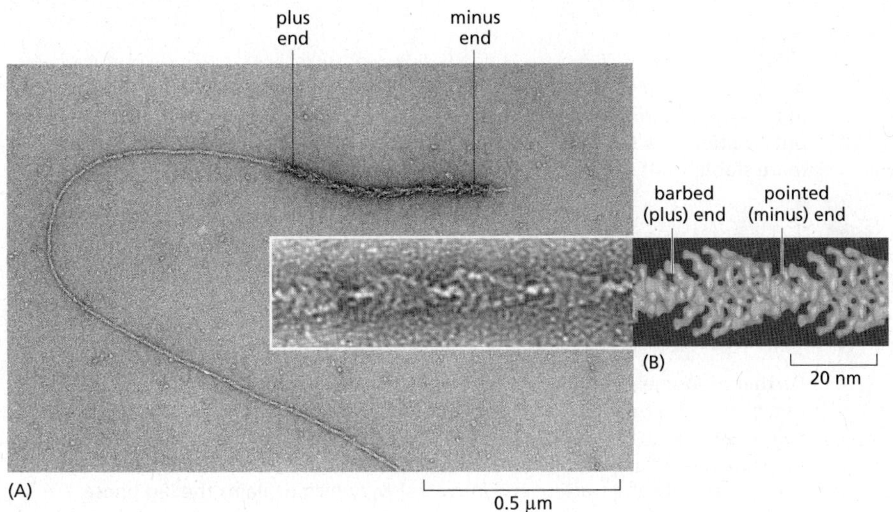

Figure 16–9 **Structural polarity of the actin filament.** (A) This electron micrograph shows an actin filament polymerized from a short actin filament seed that was decorated with myosin motor domains, resulting in an arrowhead pattern. The filament has grown much faster at the barbed (plus) end than at the pointed (minus) end. (B) Enlarged image and model showing the arrowhead pattern. (A, courtesy of Tom Pollard; B, adapted from M. Whittaker et al., *Ultramicroscopy* 58:245–259, 1995. With permission from Elsevier.)

oligomer, or nucleus, that is stabilized by multiple subunit–subunit contacts and can then elongate rapidly by addition of more subunits. This process is called filament *nucleation*.

Many features of actin nucleation and polymerization have been studied with purified actin in a test tube (**Figure 16–10**). The instability of smaller actin oligomers makes nucleation inefficient. When polymerization is initiated, this results in a lag phase during which no filaments are observed. During this lag phase, however, the small, unstable oligomers gradually succeed in making the transition to a more stable form that resembles an actin filament. This leads to a phase of rapid filament elongation during which subunits are added quickly to the ends of the nucleated filaments (Figure 16–10A). Finally, as the concentration of actin monomers declines, the system approaches a steady state at which the rate of addition of new subunits to the filament ends exactly balances the rate of subunit dissociation. The concentration of free subunits left in solution at this point is called the *critical concentration*, C_C. As explained in **Panel 16–2**, the

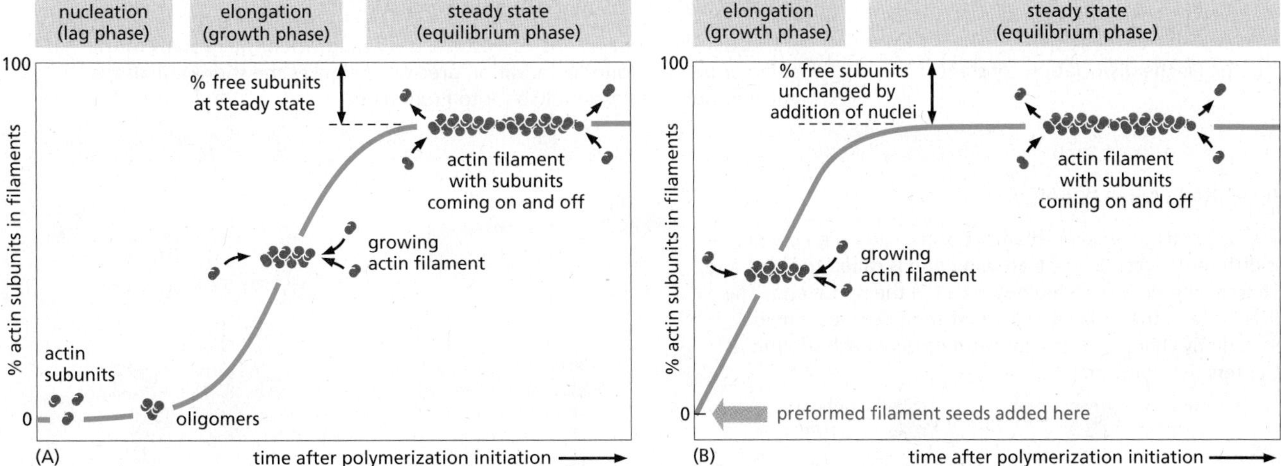

Figure 16–10 **The time course of actin polymerization.** Purified actin subunits were induced to polymerize in a test tube. (A) Polymerization of actin subunits into filaments occurs after a lag phase. (B) Polymerization occurs more rapidly in the presence of preformed fragments of actin filaments, which act as nuclei for filament growth. The percent (%) free subunits after polymerization reflects the critical concentration (C_C), at which there is no net change in polymer. The C_C has a constant value of ~0.1 μM, but the percent of actin subunits in filaments varies depending on how much actin is present in the reaction. Actin polymerization is often studied by observing the change in the light emission from a fluorescent probe, called pyrene, that has been covalently attached to the actin. Pyrene-actin fluoresces more brightly when it is incorporated into actin filaments.

ON RATES AND OFF RATES

A linear polymer of protein molecules, such as an actin filament or a microtubule, assembles (polymerizes) and disassembles (depolymerizes) by the addition and removal of subunits at the ends of the polymer. The rate of addition of these subunits (called monomers) is given by the rate constant k_{on}, which has units of M^{-1} sec^{-1}. The rate of loss is given by k_{off} (units of sec^{-1}).

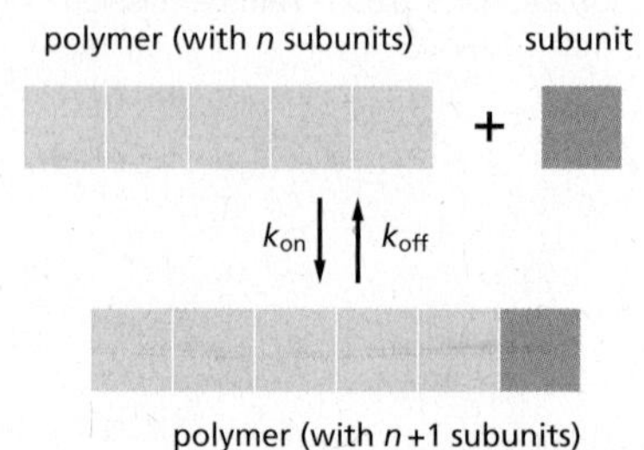

NUCLEATION

A helical polymer is stabilized by multiple contacts between adjacent subunits. In the case of actin, two actin molecules bind relatively weakly to each other, but addition of a third actin monomer to form a trimer makes the entire group more stable.

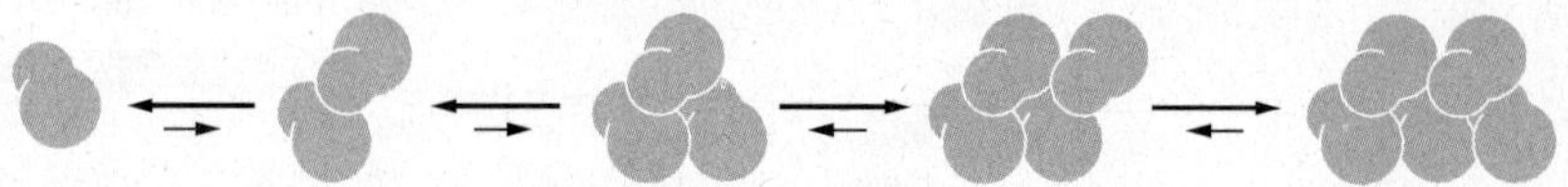

Further monomer addition can take place onto this trimer, which therefore acts as a nucleus for polymerization. For tubulin, the nucleus is larger and has a more complicated structure (possibly a ring of 13 or more tubulin molecules)—but the principle is the same.

The assembly of a nucleus is relatively slow, which explains the lag phase seen during polymerization. The lag phase can be reduced or abolished entirely by adding premade nuclei, such as fragments of already polymerized microtubules or actin filaments.

THE CRITICAL CONCENTRATION

The number of monomers that add to the polymer (actin filament or microtubule) per second will be proportional to the concentration of the free subunit ($k_{on}C$), but the subunits will leave the polymer end at a constant rate (k_{off}) that does not depend on C. As the polymer grows, subunits are used up, and C is observed to drop until it reaches a constant value, called the critical concentration (C_c). At this concentration, the rate of subunit addition equals the rate of subunit loss.

At this equilibrium,

$$k_{on}C = k_{off}$$

so that

$$C_c = \frac{k_{off}}{k_{on}} = K_d$$

where K_d is the dissociation constant.

TIME COURSE OF POLYMERIZATION

The *in vitro* assembly of a protein into a long polymer such as a cytoskeletal filament typically shows the following time course:

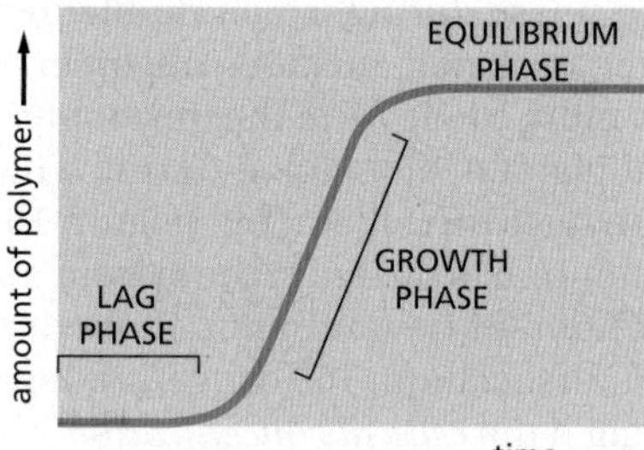

The lag phase corresponds to time taken for nucleation.

The growth phase occurs as monomers add to the exposed ends of the growing filament, causing filament elongation.

The equilibrium phase, or steady state, is reached when the growth of the polymer due to monomer addition precisely balances the shrinkage of the polymer due to disassembly back to monomers.

PLUS AND MINUS ENDS

The two ends of an actin filament or microtubule polymerize at different rates. The fast-growing end is called the plus end, whereas the slow-growing end is called the minus end. The difference in the rates of growth at the two ends is made possible by changes in the conformation of each subunit as it enters the polymer.

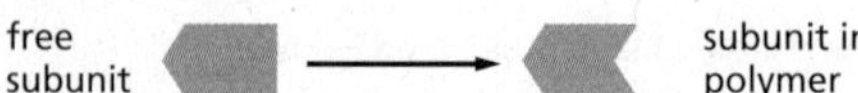

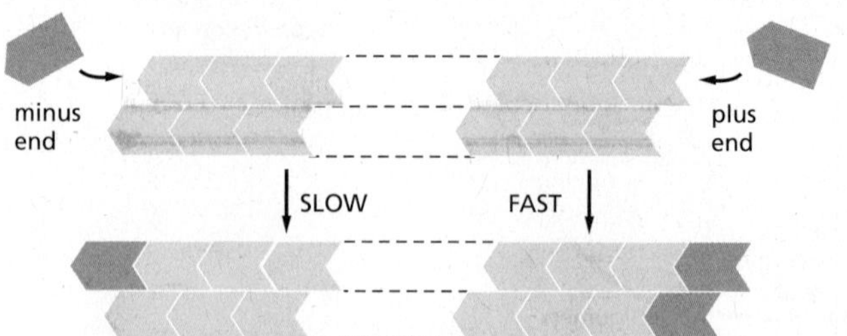

This conformational change affects the rates at which subunits add to the two ends.

Even though k_{on} and k_{off} will have different values for the plus and minus ends of the polymer, their ratio k_{off}/k_{on}—and hence C_c—must be the same at both ends for a simple polymerization reaction (no ATP or GTP hydrolysis). This is because exactly the same subunit interactions are broken when a subunit is lost at either end, and the final state of the subunit after dissociation is identical. Therefore, the ΔG for subunit loss, which determines the equilibrium constant for its association with the end, is identical at both ends: if the plus end grows four times faster than the minus end, it must also shrink four times faster. Thus, for $C > C_c$, both ends grow; for $C < C_c$, both ends shrink.

The nucleoside triphosphate hydrolysis that accompanies actin and tubulin polymerization removes this constraint.

NUCLEOTIDE HYDROLYSIS

Each actin molecule carries a tightly bound ATP molecule that is hydrolyzed to a tightly bound ADP molecule soon after its assembly into the polymer. Similarly, each tubulin molecule carries a tightly bound GTP molecule that is converted to a tightly bound GDP molecule soon after the molecule assembles into the polymer.

T → T → D (T = monomer carrying ATP or GTP) (D = monomer carrying ADP or GDP)

free monomer subunit in polymer

Hydrolysis of the bound nucleotide reduces the binding affinity of the subunit for neighboring subunits and makes it more likely to dissociate from each end of the filament. It is usually the T form that adds to the filament and the D form that leaves.

Considering events at the plus end only:

D D D D D T / D D D D D D; T k^T_{on}; k^D_{off} D

As before, the polymer will grow until $C = C_c$. For illustrative purposes, we can ignore k^D_{on} and k^T_{off} since they are usually very small, so that polymer growth ceases when

$$k^T_{on}C = k^D_{off} \quad \text{or} \quad C_c = \frac{k^D_{off}}{k^T_{on}}$$

This is a steady state and not a true equilibrium, because the ATP or GTP that is hydrolyzed must be replenished by a nucleotide exchange reaction of the free subunit (D → T).

ATP CAPS AND GTP CAPS

The rate of addition of subunits to a growing actin filament or microtubule can be faster than the rate at which their bound nucleotide is hydrolyzed. Under such conditions, the end has a "cap" of subunits containing the nucleoside triphosphate—an ATP cap on an actin filament or a GTP cap on a microtubule.

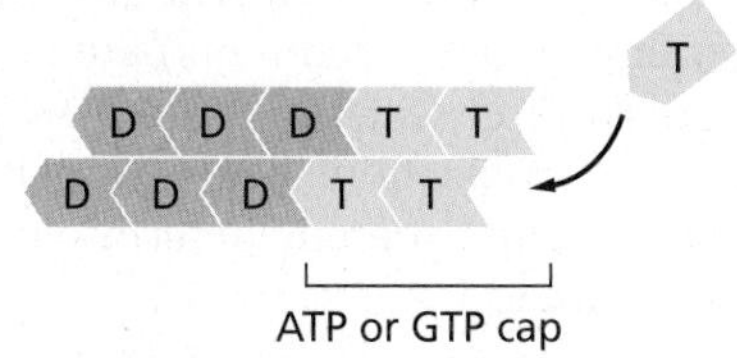

DYNAMIC INSTABILITY and TREADMILLING are two behaviors observed in cytoskeletal polymers. Both are associated with nucleoside triphosphate hydrolysis. Dynamic instability is believed to predominate in microtubules, whereas treadmilling may predominate in actin filaments.

TREADMILLING

One consequence of the nucleotide hydrolysis that accompanies polymer formation is to change the critical concentration at the two ends of the polymer. Because k^D_{off} and k^T_{on} refer to different reactions, their ratio k^D_{off}/k^T_{on} need not be the same at both ends of the polymer, so that:

$$C_c \text{ (minus end)} > C_c \text{ (plus end)}$$

Thus, if both ends of a polymer are exposed, polymerization proceeds until the concentration of free monomer reaches a value that is above C_c for the plus end but below C_c for the minus end. At this steady state, subunits undergo a net assembly at the plus end and a net disassembly at the minus end at an identical rate. The polymer maintains a constant length, even though there is a net flux of subunits through the polymer, known as treadmilling.

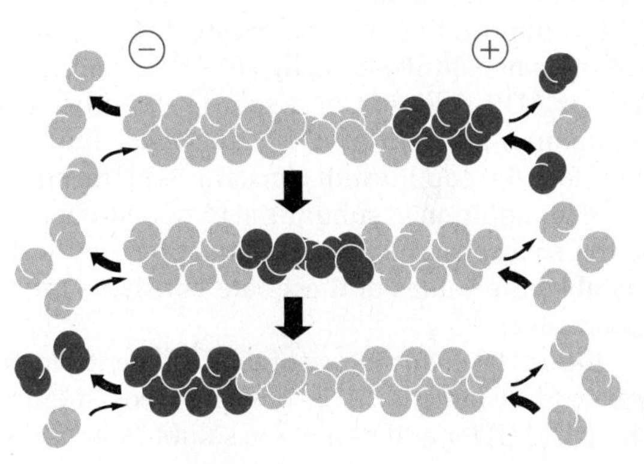

DYNAMIC INSTABILITY

Microtubules depolymerize about 100 times faster from an end containing GDP-tubulin than from one containing GTP-tubulin. A GTP cap favors growth, but if it is lost, then depolymerization ensues.

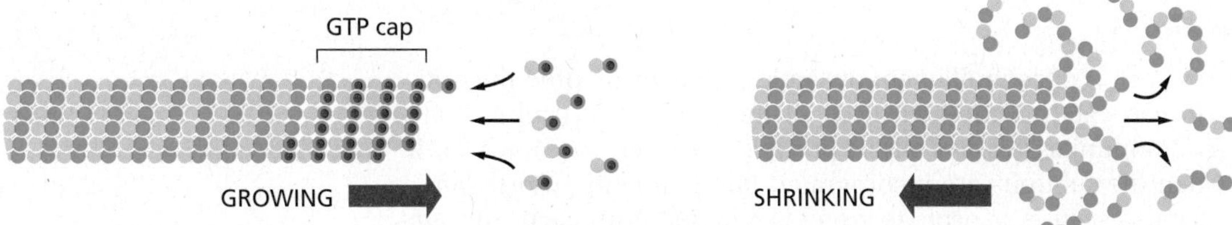

Individual microtubules can therefore alternate between a period of growth and a period of rapid disassembly, a phenomenon called dynamic instability.

value of the critical concentration is equal to the rate constant for subunit loss divided by the rate constant for subunit addition; that is, $C_c = k_{off}/k_{on}$, which is equal to the dissociation constant, K_d, and the inverse of the equilibrium constant, K (see Figure 3–42). In a test tube, the C_c for actin polymerization—that is, the free actin monomer concentration at which the fraction of actin in the polymer stops increasing—is about 0.1 μM. Inside the cell, the concentration of unpolymerized actin is much higher than this, and the cell has evolved mechanisms to prevent most of its monomeric actin from assembling into filaments, as we discuss later.

The lag phase in filament growth is eliminated if preexisting seeds (such as fragments of actin filaments) are added to the solution at the beginning of the polymerization reaction (Figure 16–10B). The cell takes great advantage of the barrier to nucleation: it uses special proteins to catalyze filament nucleation at specific sites, thereby determining the location at which new actin filaments are assembled.

Actin Filaments Have Two Distinct Ends That Grow at Different Rates

Because of the uniform orientation of asymmetrical actin subunits in the filament, the structures at its two ends are different. This orientation makes the two ends of each polymer different in ways that have a profound effect on filament growth rates. The kinetic rate constants for actin subunit association and dissociation—k_{on} and k_{off}, respectively—are much greater at the plus end than at the minus end. This can be seen when an excess of purified actin monomers is allowed to assemble onto polarity-marked filaments—the plus end of the filament elongates up to 10 times faster (see Figure 16–9). If filaments are rapidly diluted so that the free subunit concentration drops below the critical concentration, the plus end also depolymerizes faster.

It is important to note, however, that the two ends of an actin filament have the same net affinity for actin subunits, if all of the subunits are bound to either ATP or ADP. Addition of a subunit to either end of a filament of n subunits results in a filament of $n + 1$ subunits. Thus, the free-energy difference, and therefore the equilibrium constant (and the critical concentration), must be the same for addition of subunits at either end of the polymer. In this case, the ratio of the rate constants, k_{off}/k_{on}, must be identical at the two ends, even though the absolute values of these rate constants are very different at each end (see Panel 16–2).

The cell takes advantage of actin filament dynamics and polarity to do mechanical work. Filament elongation proceeds spontaneously when the free-energy change (ΔG) for addition of the soluble subunit is less than zero. This is the case when the concentration of subunits in solution exceeds the critical concentration. A cell can couple an energetically unfavorable process to this spontaneous process; thus, the cell can use free energy released during spontaneous filament polymerization to move an attached load. For example, by orienting the fast-growing plus ends of actin filaments toward its leading edge, a motile cell can push its plasma membrane forward, as we discuss later.

ATP Hydrolysis Within Actin Filaments Leads to Treadmilling at Steady State

Thus far in our discussion of actin filament dynamics, we have ignored the critical fact that actin can catalyze the hydrolysis of the nucleoside triphosphate ATP. For free actin subunits, this hydrolysis proceeds very slowly; however, it is accelerated when the subunits are incorporated into filaments. Shortly after ATP hydrolysis occurs, the free phosphate group is released from each subunit, but the ADP remains trapped in the filament structure. Thus, two different types of filament structures can exist, one in the ATP-bound *T form* and one in the ADP-bound *D form*.

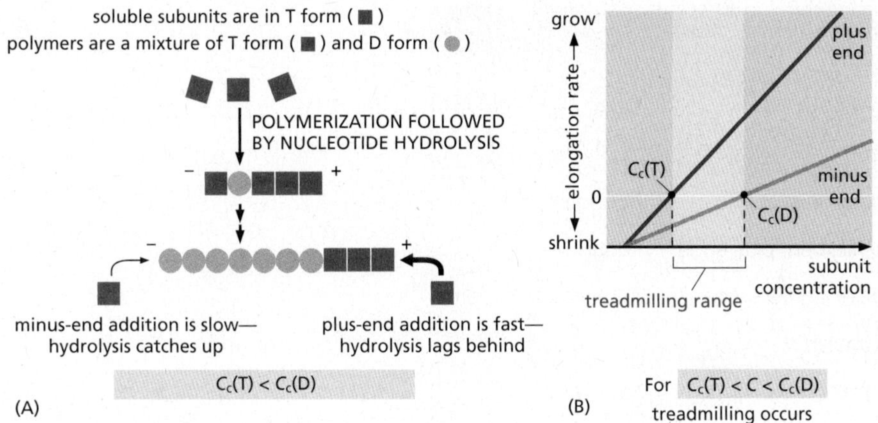

Figure 16–11 Treadmilling of an actin filament, made possible by the ATP hydrolysis that follows subunit addition. (A) Explanation for the different critical concentrations (C_c) at the plus and minus ends. Subunits with bound ATP (T-form subunits) polymerize at both ends of a growing filament and then undergo ATP hydrolysis within the filament. As the filament grows, elongation is faster than hydrolysis at the plus end in this example, and the terminal subunits at this end are therefore always in the T form. However, hydrolysis is faster than elongation at the minus end, and so terminal subunits at this end are in the D form. (B) Treadmilling occurs at intermediate concentrations of free subunits. The critical concentration for polymerization on a filament end in the T form, C_c(T), is lower than that for a filament end in the D form, C_c(D). If the actual subunit concentration is somewhere between these two values, the plus end grows while the minus end shrinks, resulting in treadmilling.

When the ATP is hydrolyzed, much of the free energy released by cleavage of the phosphate-phosphate bond is stored in the polymer. This makes the free-energy change for dissociation of a subunit from the D-form polymer more negative than the free-energy change for dissociation of a subunit from the T-form polymer. Consequently, the ratio of k_{off}/k_{on} for the D-form polymer, which is numerically equal to its critical concentration [C_c(D)], is larger than the corresponding ratio for the T-form polymer. Thus, C_c(D) is greater than C_c(T). At certain concentrations of free subunits, D-form polymers will therefore shrink while T-form polymers grow.

In living cells, most soluble actin subunits are in the T form, as the free concentration of ATP is about tenfold higher than that of ADP. However, the longer the time that subunits have been in the actin filament, the more likely they are to have hydrolyzed their ATP. Whether the subunit at each end of a filament is in the T or the D form depends on the rate of this hydrolysis compared with the rate of subunit addition. If the concentration of actin monomers is greater than the critical concentration for both the T-form and D-form polymers, then subunits will add to the polymers at both ends before the ATP in the previously added subunits is hydrolyzed; as a result, the tips of the actin filament will remain in the T form. On the other hand, if the subunit concentration is less than the critical concentrations for both the T-form and D-form polymers, then hydrolysis may occur before the next subunit is added, and both ends of the filament will be in the D form and will shrink. At intermediate concentrations of actin subunits, it is possible for the rate of subunit addition to be faster than ATP hydrolysis at the plus end but slower than ATP hydrolysis at the minus end. In this case, the plus end of the filament remains in the T conformation, while the minus end adopts the D conformation. The filament then undergoes a net addition of subunits at the plus end while simultaneously losing subunits from the minus end. This leads to the remarkable property of filament **treadmilling** (Figure 16–11; see Panel 16–2).

At a particular intermediate subunit concentration, the filament growth at the plus end exactly balances the filament shrinkage at the minus end. Under these conditions, the subunits cycle rapidly between the free and filamentous states, while the total length of the filament remains unchanged. This *steady-state treadmilling* requires a constant consumption of energy in the form of ATP hydrolysis.

The Functions of Actin Filaments Are Inhibited by Both Polymer-stabilizing and Polymer-destabilizing Chemicals

Chemical compounds that stabilize or destabilize actin filaments are important tools in studies of the filaments' dynamic behavior and function in cells. The *cytochalasins* are fungal products that prevent actin polymerization by binding to the plus end of actin filaments. *Latrunculin* prevents actin polymerization by binding to actin subunits. The *phalloidins* are toxins isolated from the *Amanita* mushroom that bind tightly all along the side of actin filaments and stabilize them

TABLE 16–1 Chemical Inhibitors of Actin and Microtubules

Chemical	Effect on filaments	Mechanism	Original source
		Actin	
Latrunculin	Depolymerizes	Binds actin subunits	Sponges
Cytochalasin B	Depolymerizes	Caps filament plus ends	Fungi
Phalloidin	Stabilizes	Binds along filaments	*Amanita* mushroom
		Microtubules	
Taxol (paclitaxel)	Stabilizes	Binds along filaments	Yew tree
Nocodazole	Depolymerizes	Binds tubulin subunits	Synthetic
Colchicine	Depolymerizes	Caps both filament ends	Autumn crocus

against depolymerization. All of these compounds cause dramatic changes in the actin cytoskeleton and are toxic to cells, indicating that the function of actin filaments depends on a dynamic equilibrium between filaments and actin monomers (**Table 16–1**).

Actin-binding Proteins Influence Filament Dynamics and Organization

In a test tube, polymerization of actin is controlled simply by its concentration, as described earlier, and by pH and the concentrations of salts and ATP. Within a cell, however, actin behavior is also regulated by numerous accessory proteins that bind actin monomers or filaments (summarized in **Panel 16–3**). At steady state *in vitro*, when the monomer concentration is 0.1 μM, filament half-life, a measure of how long an individual actin monomer spends in a filament as it treadmills, is approximately 30 minutes. In a non-muscle vertebrate cell, the actin half-life in filaments is only 30 seconds, demonstrating that cellular factors modify the dynamic behavior of actin filaments. Actin-binding proteins dramatically alter actin filament dynamics through spatial and temporal control of filament nucleation, elongation, and depolymerization. They also regulate the association of actin with membranes as well as how filaments are organized. In the following sections, we describe the ways in which these accessory proteins modify actin function in the cell, enabling actin polymerization to generate the forces required to support, shape, and move cellular membranes.

Actin Nucleation Is Tightly Regulated and Generates Branched or Straight Filaments

In most non-muscle vertebrate cells, approximately 50% of the actin is in filaments and 50% is soluble—and yet the soluble monomer concentration is 50–200 μM, well above the critical concentration. Why does so little of the actin polymerize into filaments? A major reason is that actin filament polymerization is tightly regulated in cells by a large number of proteins that control actin filament nucleation, almost always adjacent to a membrane surface. Proteins that contain actin monomer binding motifs linked in tandem mediate the simplest mechanism of filament nucleation. These actin-nucleating proteins bring several actin subunits together to form a seed. In most cases, actin nucleation is catalyzed by one of two different factors: the Arp2/3 complex or the formins. The first of these is a complex of proteins that includes two *actin-related proteins*, or *ARPs*, each of which is about 45% identical to actin. The **Arp2/3 complex** nucleates actin filament growth and remains bound to the

ACTIN FILAMENTS

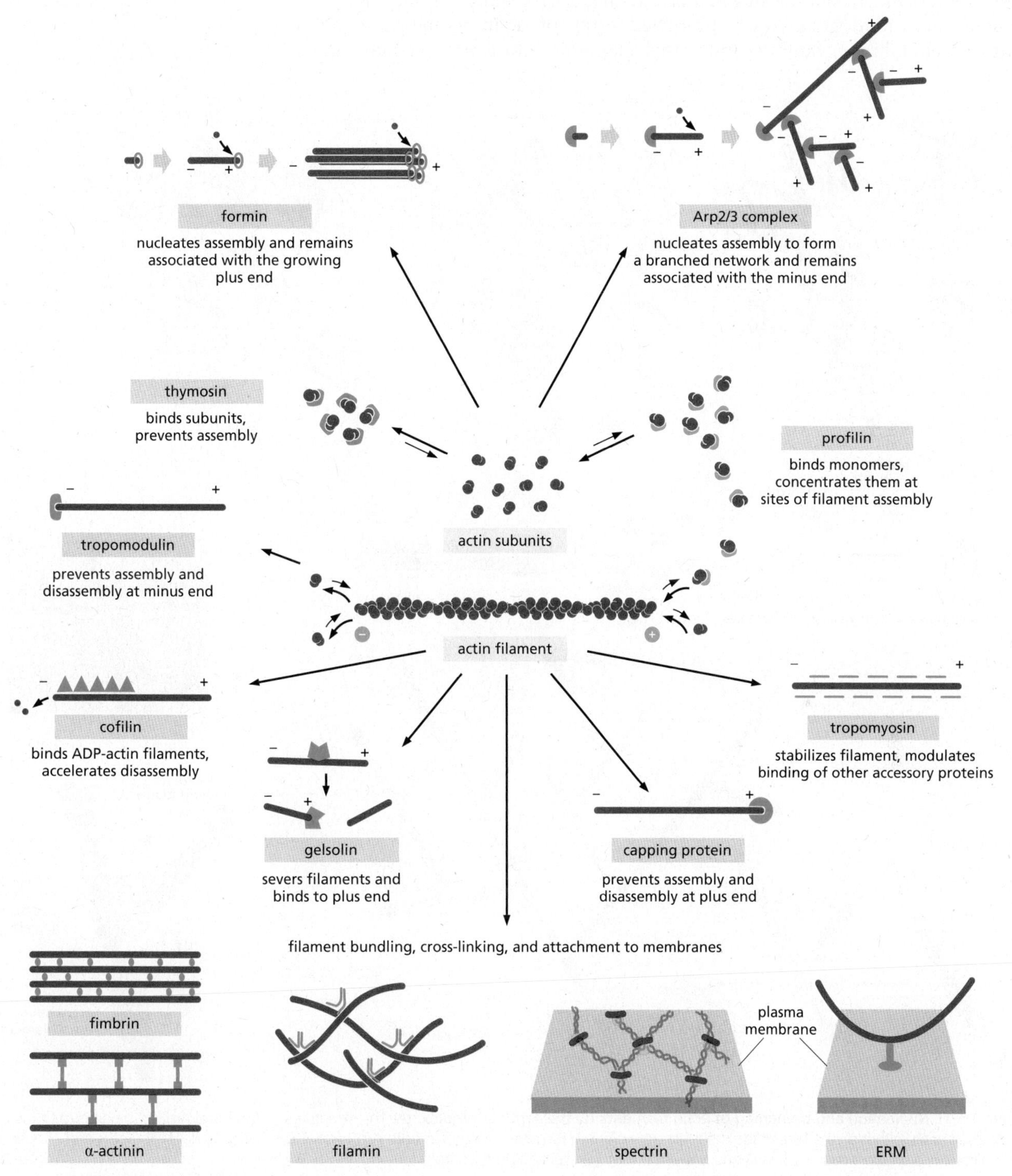

Some of the major accessory proteins of the actin cytoskeleton. Except for the myosin motor proteins, an example of each major type is shown. Each of these is discussed in the text. However, most cells contain more than a hundred different actin-binding proteins, and it is likely that there are important types of actin-associated proteins that are not yet recognized.

minus end, allowing rapid elongation at the plus end (**Figure 16–12A and B**). Arp2/3 complex-mediated actin filament nucleation requires the activity of a nucleation-promoting factor (NPF). The Arp2/3 complex is further stimulated when it attaches to the side of a preexisting actin filament. Thus, Arp2/3 complex activation generates a branched array of actin filaments adjacent to a membrane, building individual filaments into a treelike network

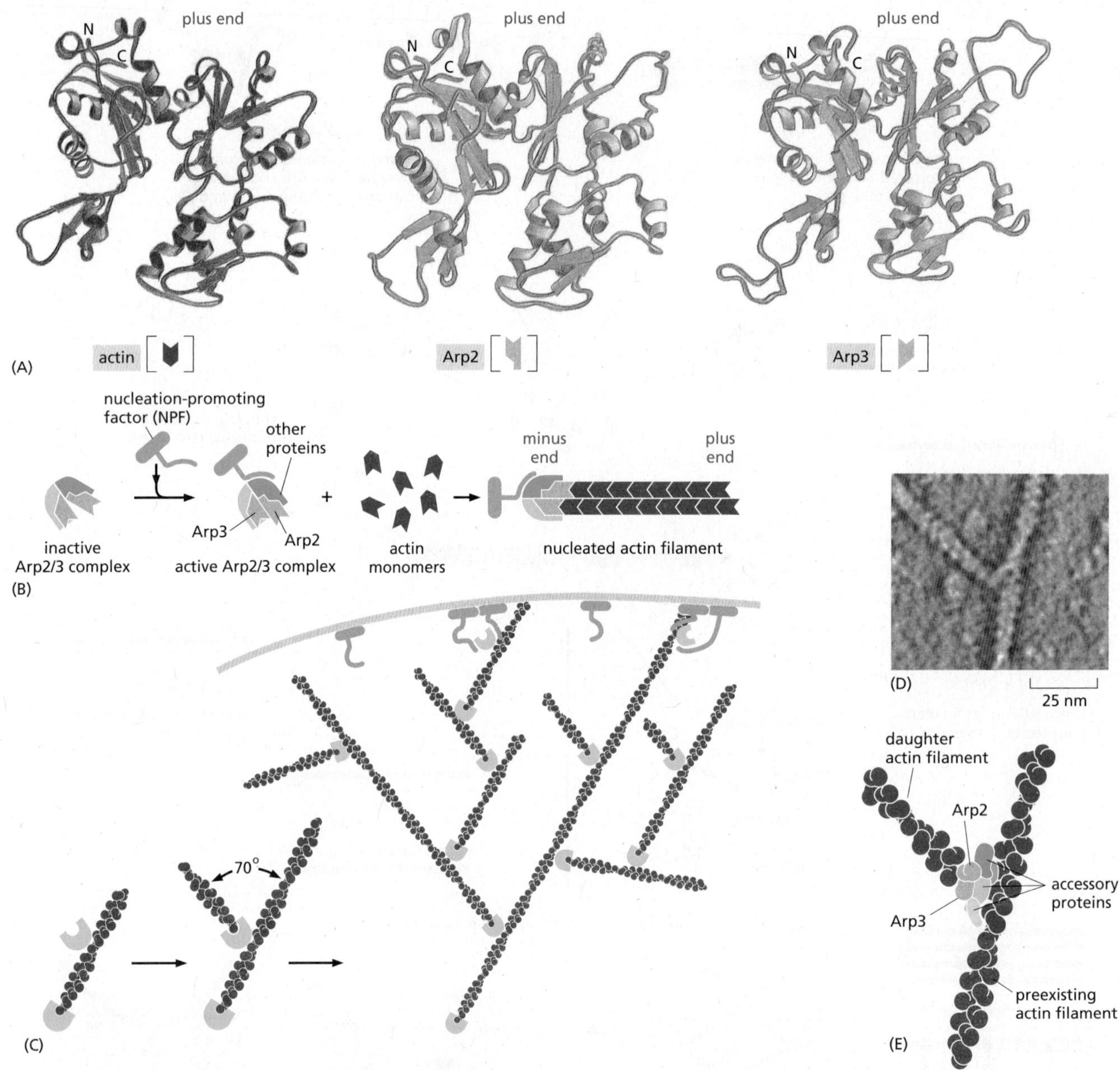

Figure 16–12 Nucleation and branching of actin filaments by the Arp2/3 complex. (A) The structures of Arp2 and Arp3 compared to the structure of actin. Although the face of the molecule equivalent to the plus end *(top)* in both Arp2 and Arp3 is very similar to the plus end of actin itself, differences on the sides and minus end prevent these actin-related proteins from forming filaments on their own or coassembling into filaments with actin. (B) A model for actin filament nucleation by the Arp2/3 complex. In the absence of an activating factor, Arp2 and Arp3 are held by their accessory proteins in an orientation that prevents them from nucleating a new actin filament. When a nucleation-promoting factor (NPF; *brown*) binds the complex, Arp2 and Arp3 are brought together into a new configuration that resembles the plus end of an actin filament. Actin subunits can then assemble onto this structure, bypassing the rate-limiting step of filament nucleation. In cells, NPF activation required to induce Arp2/3-dependent nucleation occurs at membrane surfaces. (C) The Arp2/3 complex nucleates filaments most efficiently when it is bound to the side of a preexisting actin filament. The result is a filament branch that grows at a 70° angle relative to the original filament. Repeated rounds of branching nucleation result in a treelike web of actin filaments. (D) Electron micrograph of a branched actin filament formed by mixing purified actin subunits with purified Arp2/3 complexes. (E) Diagram showing the position of the Arp2/3 complex at the branch between the preexisting mother filament and the daughter filament. (D, from I. Rouiller et al., *J. Cell Biol.* 180:887–895, 2008. © 2008A. Rouiller et al. Originally published in *J. Cell Biol.* 180:887–895. https://doi.org/10.1083/jcb.200709092. With permission from Rockefeller University Press.)

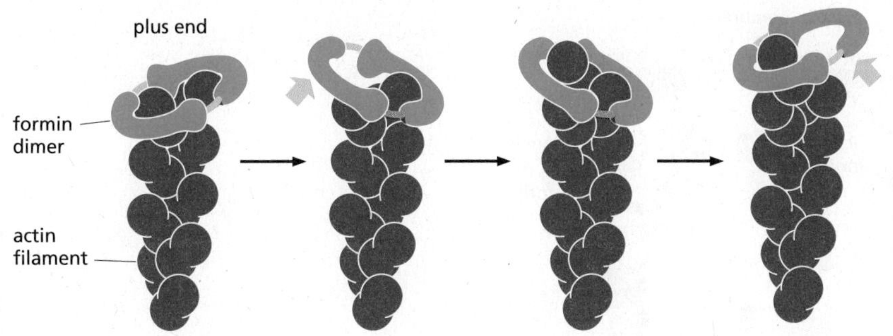

Figure 16–13 **Actin filament elongation mediated by formins.** Formin proteins *(green)* form a dimeric complex that can nucleate the formation of a new actin filament *(red)* and remain associated with the rapidly growing plus end as it elongates. The formin protein maintains its binding to one of the two actin subunits exposed at the plus end as it allows each new subunit to assemble. Only part of the large dimeric formin molecule is shown here. Other regions regulate its activity and link it to particular structures in the cell. Many formins are indirectly connected to the cell plasma membrane and aid the insertional polymerization of the actin filament directly beneath the membrane surface.

(Figure 16–12C, D, and E). Cells express multiple NPFs, each of which makes branched actin networks that are used to initiate various actin polymerization-dependent processes, such as transport of membrane-bound vesicles across short distances within the cell, the formation of *adherens junctions* (see Chapter 19), and *phagocytosis* (see Figure 13–70). As discussed below, NPF-driven actin polymerization frequently plays a direct role in protrusion of the leading edge of the cell during cell migration.

Formins are dimeric proteins that nucleate the growth of unbranched filaments that can be cross-linked by other proteins to form parallel bundles. Each formin subunit has a binding site for monomeric actin, and the formin dimer appears to nucleate actin filament polymerization by capturing two monomers. As the newly nucleated filament grows, the formin dimer remains associated with the rapidly growing plus end while still allowing the addition of new subunits at that end (Figure 16–13). This mechanism of filament assembly is clearly different from that used by the Arp2/3 complex, which remains stably bound to the filament minus end, preventing subunit addition or loss at that end. In addition to their role in actin nucleation, formin proteins also dramatically accelerate actin filament growth. Formins mediate assembly of a variety of cellular structures that contain polarized actin cables, including filopodia, stress fibers, and the contractile ring.

Actin filament nucleation by the Arp2/3 complex and formins occurs primarily at the plasma membrane, and therefore the highest density of actin filaments in most cells is at the cell periphery within the **cell cortex**. Actin filaments in this region determine the shape and movement of the cell surface, allowing the cell to change its shape and stiffness rapidly in response to changes in its external environment.

Actin Filament Elongation Is Regulated by Monomer-binding Proteins

Once nucleated, rapid polymerization of actin depends on the further addition of actin monomers at the plus end of each filament. A key factor is the monomer-binding protein *profilin.* Profilin binds to the face of the actin monomer opposite the ATP-binding cleft, blocking the side of the monomer that would normally associate with the filament minus end, while leaving exposed the site on the monomer that binds to the plus end (see Figure 16–8). When the profilin–actin complex binds a free plus end, a conformational change in actin reduces its affinity for profilin and the profilin falls off, leaving the actin filament one subunit longer. In some cell types, a small protein called *thymosin* is highly abundant and competes with profilin for actin monomer binding, thereby further regulating the pool of actin monomers available for polymerization.

Growth of actin filaments is greatly enhanced by profilin for two reasons. First, profilin maintains a large pool of actin monomers poised for polymerization at filament plus ends. Second, binding sites for profilin are present in many formin proteins, as well as in many NPFs that activate the Arp2/3 complex. By binding to the factors that stimulate filament nucleation, profilin-bound actin monomers are recruited directly to sites of filament elongation, rapidly accelerating actin filament assembly (Figure 16–14).

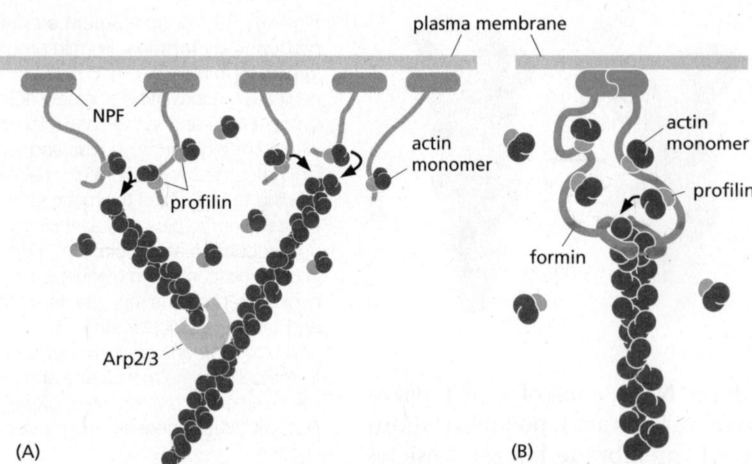

Figure 16–14 Profilin stimulates actin filament elongation. (A) Many nucleation-promoting factors (NPFs; *brown*) contain binding sites for profilin *(blue)*, which is bound to actin monomers. As a result, NPF activation leads not only to nucleation of branched actin filaments by Arp2/3 (*green*; see Figure 16–12), but also to rapid elongation of the new filaments. (B) Some members of the formin protein family possess whisker-like unstructured domains that contain several binding sites for profilin–actin complexes. This serves to recruit actin monomers to the growing plus end of the actin filament when formin is bound. Under some conditions, profilin can enhance the rate of actin filament elongation so that filament growth is faster than that expected for a diffusion-controlled reaction (see also Figure 3–76). Like NPFs, formin proteins inside the cell are activated to promote actin filament polymerization at membrane surfaces.

Actin Filament–binding Proteins Alter Filament Dynamics and Organization

Actin filament behavior is regulated by two major classes of binding proteins: those that bind along the side of a filament and those that bind to the ends (see Panel 16–3). Side-binding proteins include *tropomyosin,* an elongated protein unrelated to the myosin motor that binds simultaneously to six or seven adjacent actin subunits along each of the two grooves of the helical actin filament. In addition to stabilizing and stiffening the filament, the binding of tropomyosin can prevent the actin filament from interacting with other proteins; this aspect of tropomyosin function is important in the control of muscle contraction, as we discuss later.

An actin filament that stops growing and is not specifically stabilized in the cell will depolymerize rapidly, particularly at its plus end, once the actin molecules have hydrolyzed their ATP. The binding of plus-end *capping protein* (also called *CapZ* for its location in the muscle Z band) stabilizes an actin filament at its plus end by rendering it inactive, greatly reducing the rates of filament growth and depolymerization (**Figure 16–15**). At the minus end, an actin filament may be capped by the Arp2/3 complex that was responsible for its nucleation,

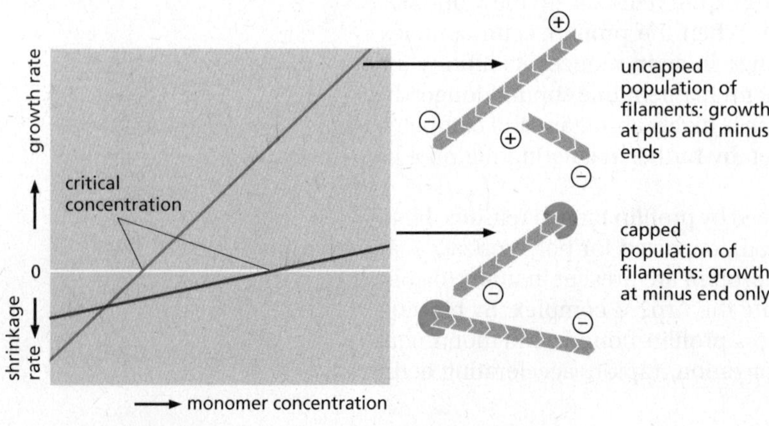

Figure 16–15 Filament capping and its effects on filament dynamics. A population of uncapped filaments adds and loses subunits at both the plus and minus ends, resulting in rapid growth or shrinkage, depending on the concentration of available free monomers *(green line)*. In the presence of a protein that caps the plus end *(red line)*, only the minus end is able to add or lose subunits; consequently, filament growth will be slower at all monomer concentrations above the critical concentration, and filament shrinkage will be slower at all monomer concentrations below the critical concentration. In addition, the critical concentration for the population shifts to that of the filament minus end.

although many minus ends in a typical cell are released from the Arp2/3 complex and are uncapped.

Tropomodulin, best known for its function in the capping of actin filaments in muscle, binds tightly to the minus ends of actin filaments that have been coated and thereby stabilized by tropomyosin. A large family of tropomodulin proteins regulates actin filament length and stability in many cell types.

End-binding proteins can affect filament dynamics even when they are present at very low levels. Because subunit addition and loss occur exclusively at filament ends, one molecule of an end-binding protein per actin filament (roughly one molecule per 200–500 actin subunits) can be enough to transform the architecture of an actin filament network. For example, rapid capping of actin filament plus ends after their nucleation at membranes focuses actin polymerization at new sites of nucleation, so that plus ends grow only where a pushing force is needed (see Figure 16–17).

The three-dimensional organization of cellular actin networks also depends on the activity of actin cross-linking proteins. Some of these cross-linking proteins bundle actin filaments into a parallel array, while others hold two actin filaments together at a large angle to each other, thereby creating a looser meshwork. Actin cross-linking proteins generally have two similar actin filament–binding sites, which can either be part of a single polypeptide chain or contributed by each of two polypeptide chains held together in a dimer (see Panel 16–3). The spacing and arrangement of these two filament-binding domains influence the type of actin structure that a given cross-linking protein forms. For example, the small monomeric protein *fimbrin* contributes to the tight packing of parallel bundles of actin filaments found in long cellular protrusions such as microvilli and the stereocilia of the inner ear's sensory hair cells. Other actin cross-linking proteins have either a flexible or a stiff, bent connection between their two binding domains, allowing them to form actin filament webs or three-dimensional networks, rather than actin bundles. The protein *filamin* achieves this by clamping together pairs of actin filaments roughly at right angles, thereby promoting the formation of a loose and highly viscous gel.

Each type of bundling protein also influences which other molecules can interact with the cross-linked actin filaments. Myosin II is the motor protein that enables stress fibers and other contractile arrays to exert tension. The very close packing of actin filaments caused by fimbrin apparently excludes myosin, and thus the parallel actin filaments held together by fimbrin are not contractile. On the other hand, α-actinin cross-links oppositely polarized actin filaments into loose bundles, allowing the binding of myosin and formation of contractile actin bundles. Because of the very different spacing and orientation of the actin filaments, bundling by fimbrin automatically discourages bundling by α-actinin, and vice versa, so that the two types of bundling protein are mutually exclusive.

As discussed previously, much of the actin polymerization in animal cells initiates at the plasma membrane where nucleation proteins are activated. Dozens of other actin-binding proteins associate with the cortical actin network. For example, a netlike meshwork of the protein **spectrin** connects to actin in the cortex of red blood cells, where it maintains cell shape and structural integrity of the plasma membrane (see pp. 630–631). More widely expressed are members of the *ERM* family (named for its first three members, ezrin, radixin, and moesin), which help organize membrane domains through their ability to interact with transmembrane proteins and the underlying actin cytoskeleton in many cell types. In so doing, they not only provide structural links to strengthen the cell cortex but also regulate the activities of signal transduction pathways. Moesin also increases cortical stiffness to promote cell rounding during mitosis. ERM proteins are thought to bind to and organize the cortical actin cytoskeleton in a variety of contexts, thereby affecting the shape and stiffness of the membrane as well as the localization and activity of signaling molecules.

Severing Proteins Regulate Actin Filament Depolymerization

Another important mechanism of actin filament regulation depends on proteins that break an actin filament into many smaller filaments, thereby generating a large number of new filament ends. The fate of these new ends depends on the presence of other accessory proteins. Under some conditions, newly formed ends nucleate filament elongation, thereby accelerating the assembly of new filament structures. Under other conditions, severing promotes the depolymerization of old filaments, speeding up the depolymerization rate by tenfold or more. In addition, severing changes the physical and mechanical properties of the cytoplasm: stiff, large bundles and gels become more fluid.

One class of actin-severing proteins is the *gelsolin superfamily*. These proteins are activated by high levels of cytosolic Ca^{2+}. Gelsolin interacts with the side of the actin filament and contains subdomains that bind to two different sites: one that is exposed on the surface of the filament and one that is hidden between adjacent subunits. Activated gelsolin is thought to sever an actin filament when a thermal fluctuation creates a small gap between neighboring subunits, at which point gelsolin inserts itself into the gap to break the filament. After the severing event, gelsolin remains attached to the actin filament and caps the new plus end.

A second actin filament–destabilizing protein, found in all eukaryotic cells, is *cofilin*. Also called *actin-depolymerizing factor*, cofilin binds along the length of the actin filament, forcing the filament to twist a little more tightly (**Figure 16–16**). Mechanical stress induced by cofilin binding weakens the contacts between actin subunits in the filament, making the filament less stable and more easily severed by thermal motions, generating filament ends that undergo rapid disassembly. As a result, most of the actin filaments inside cells are shorter lived than are filaments formed from pure actin in a test tube.

Cofilin binds preferentially to ADP-containing actin filaments rather than to ATP-containing filaments. Because ATP hydrolysis is usually slower than filament assembly, the newest actin filaments in the cell still contain mostly ATP and are resistant to depolymerization by cofilin. Cofilin therefore tends to dismantle the older filaments in the cell. As discussed later, the cofilin-mediated disassembly of old but not new actin filaments is critical for the polarized, directed growth of the actin network that drives the intracellular motility of pathogens as well as cell migration. Actin filaments can be protected from cofilin by tropomyosin binding. Thus, the dynamics of actin in different subcellular locations depend on the balance of stabilizing and destabilizing accessory proteins.

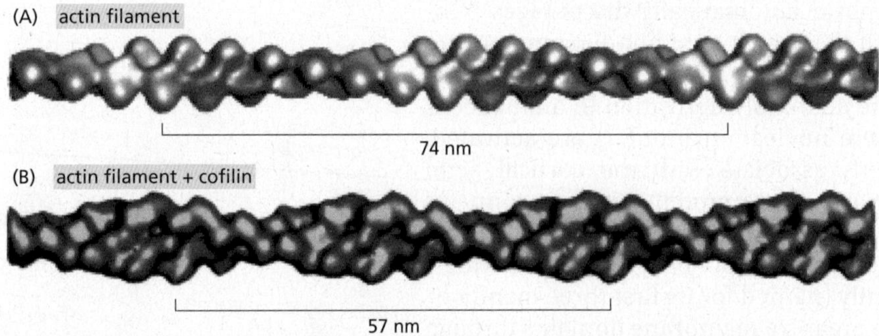

Figure 16–16 Twisting of an actin filament induced by cofilin. (A) Three-dimensional reconstruction from cryo-electron micrographs of filaments made of pure actin. The bracket shows the span of two twists of the actin helix. (B) Reconstruction of an actin filament coated with cofilin, which binds in a 1:1 stoichiometry to actin subunits all along the filament. Cofilin is a small protein (14 kilodaltons) compared to actin (43 kilodaltons), and so the filament appears only slightly thicker. The energy of cofilin binding serves to deform the actin filament, twisting it more tightly and reducing the distance spanned by each twist of the helix. (© 1997 A. McGough et al. Originally published in *J. Cell Biol.* https://doi.org.10.1083/jcb.138.4.771. With permission from Rockefeller University Press.)

Bacteria Can Hijack the Host Actin Cytoskeleton

The importance of accessory proteins in actin-based motility and force production is illustrated beautifully by studies of certain bacteria and viruses that use components of the host-cell actin cytoskeleton to move through the cytoplasm. The cytoplasm of mammalian cells is extremely viscous, containing organelles and cytoskeletal elements that inhibit diffusion of large particles such as bacteria or viruses. To move around in a cell and invade neighboring cells, several pathogens, including *Listeria monocytogenes* (which causes a rare but serious form of food poisoning), overcome this problem by recruiting and activating the Arp2/3 complex at their surface. The Arp2/3 complex nucleates the assembly of actin filaments that generate a substantial force and push the bacterium through the cytoplasm at rates of up to 15 μm/min, leaving behind a long actin "comet tail" (**Figure 16–17**; see also Figures 23–29 and 23–30). This motility can be reconstituted in a test tube by adding the bacteria to a mixture of pure actin, Arp2/3 complex, cofilin, and capping protein, illustrating how actin polymerization dynamics generates movement through spatial regulation of filament assembly and disassembly. As we shall see, actin-based movement of this sort also underlies membrane protrusion at the leading edge of motile cells.

Actin at the Cell Cortex Determines Cell Shape

Although actin is found throughout the cytoplasm of a eukaryotic cell, dynamic actin filament behavior occurs primarily at the cell cortex. Here, through interaction with many different proteins, actin filaments are organized into several types of arrays, including branched networks, parallel bundles, and combinations of straight and branched filaments (**Figure 16–18**). Different structures are initiated by the action of distinct nucleating proteins: the actin filaments of branched networks are nucleated by the Arp2/3 complex, while bundles are made of the long, straight filaments produced by formins. Dynamics here generate distinct

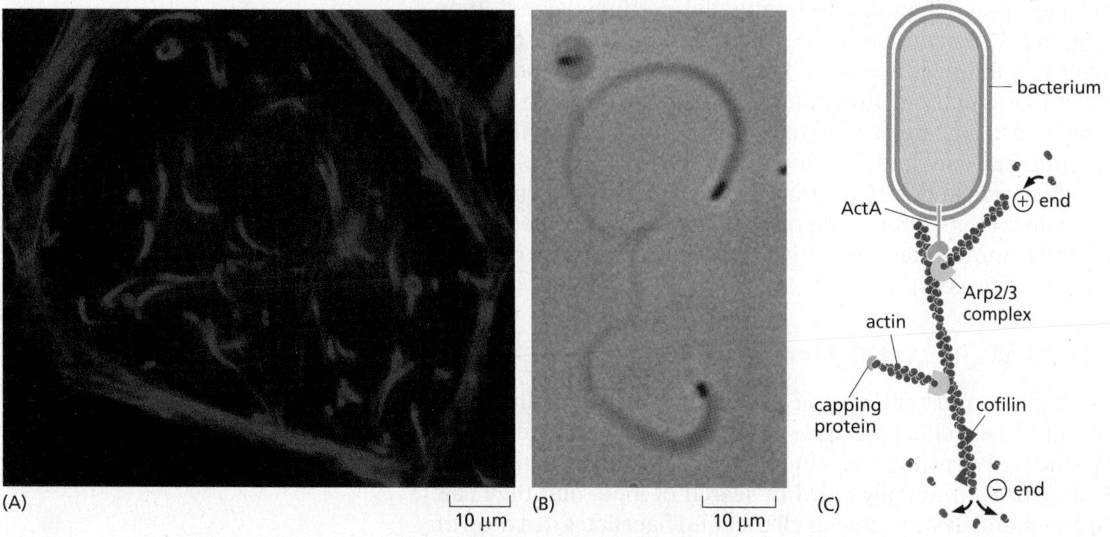

Figure 16–17 The actin-based movement of *Listeria monocytogenes*. (A) Fluorescence micrograph of an infected cell that has been stained to reveal bacteria in *blue* and actin filaments in *red*. Note the cometlike tail of actin filaments behind each moving bacterium. (B) *Listeria* motility can be reconstituted in a test tube with ATP and just four purified proteins: actin, Arp2/3 complex, capping protein, and cofilin. This micrograph shows the dense actin tails behind bacteria *(black)*. (C) The ActA protein on the bacterial surface activates the Arp2/3 complex to nucleate new filament assembly along the sides of existing filaments. Filaments grow at their plus end until capped by capping protein. Actin is recycled through the action of cofilin, which enhances depolymerization at the minus ends of the filaments. By this mechanism, polymerization is focused at the rear surface of the bacterium, propelling it forward (see Movie 23.7). (A, courtesy of Julie Theriot and Tim Mitchison; B, from T.P. Loisel et al., *Nature* 401:613–616, published 1999 by Nature Publishing Group. Reproduced with permission of SNCSC.)

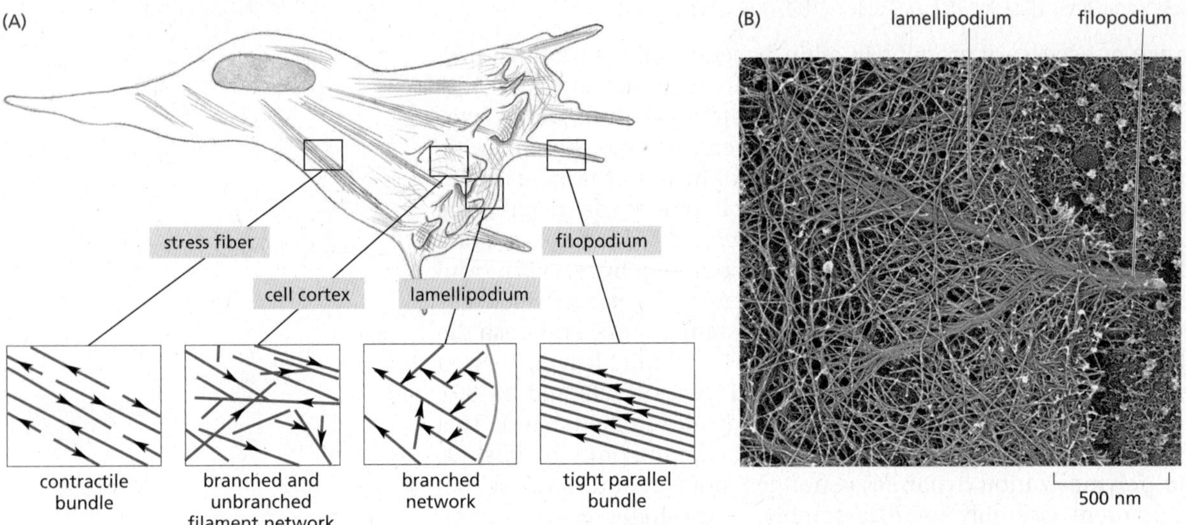

Figure 16–18 **Actin arrays in a cell moving along a flat surface.** (A) A schematic of a fibroblast migrating in a tissue-culture dish is shown with four areas enlarged to show the arrangement of actin filaments. The actin filaments are shown in *red*, with arrowheads pointing toward the minus end. Stress fibers are contractile and exert tension. The actin cortex underlies the plasma membrane and consists of actin networks that enable membrane protrusion at lamellipodia. Filopodia are spike-like projections of the plasma membrane that sense extracellular signals and allow a cell to explore its environment. (B) A migrating cell that was fixed, dried, and shadowed with platinum reveals a dense network of actin filaments at the leading edge (right side of image). The fast-growing plus ends are oriented toward the cell edge. Accessory proteins organize the filaments. Pseudo-colored in *red* are actin filaments in the process of becoming a filopodium. (B, courtesy of Tatyana Svitkina.)

cellular shapes and properties, as well as protrusive structures termed **filopodia** and **lamellipodia** observed in migrating fibroblasts. These are filled with dense cores of filamentous actin, which excludes membrane-enclosed organelles. The structures differ primarily in the way in which the actin is organized by actin cross-linking proteins (see Panel 16–3). Filopodia are essentially one-dimensional. They contain a core of long, bundled actin filaments, which are reminiscent of those in microvilli but longer and thinner, as well as more dynamic. Lamellipodia are two-dimensional, sheet-like structures. They contain a cross-linked mesh of actin filaments, most of which lie in a plane parallel to the solid substratum. Filopodia help cells sense environmental cues and function in cell migration. For example, filopodia extend from the leading edge of the growth cone in developing neurons and help guide them to their target. Protrusion of lamellipodia and related actin-rich structures, discussed in more detail later, drives the forward movement of the leading edge of many migrating cells.

Distinct Modes of Cell Migration Rely on the Actin Cytoskeleton

Cell movement is one of the most striking of all cell behaviors. While some cells employ microtubule-based cilia or flagella to swim (discussed later in this chapter), many cells undergo crawling movements that rely on the actin cytoskeleton. Predatory amoebae continually crawl in search of food, and they can easily be observed to attack and devour smaller ciliates and flagellates in a drop of pond water (see Movie 1.4). During embryogenesis, the structure of an animal is created by the migrations of individual cells to specific target locations and by the coordinated movements of whole epithelial sheets (discussed in Chapter 21). In vertebrates, *neural crest cells* migrate long distances from their site of origin in the neural tube to a variety of sites throughout the embryo (see Movie 21.7). Similarly, actin-rich growth cones at the advancing tips of developing axons travel to distant synaptic targets.

The adult animal seethes with crawling cells. Macrophages and neutrophils crawl to sites of infection and engulf foreign invaders as a critical part of the innate immune response. Osteoclasts tunnel into bone, forming channels that are filled in by the osteoblasts that follow them, in a continual process of bone remodeling and renewal. Similarly, fibroblasts migrate through connective tissues, remodeling them where necessary and helping to rebuild damaged structures at sites of injury. In an ordered procession, the cells in the epithelial lining of the intestine travel up the sides of the intestinal villi, replacing absorptive cells lost at the tip of the villus. Cell migration also has a role in many cancers, when cells in a primary tumor invade neighboring tissues and crawl into blood vessels or lymph vessels and then emerge at other sites in the body to form metastases.

Cell migration depends on the actin-rich cortex that lies beneath the plasma membrane. Figure 16–19 illustrates distinct modes of cell migration on the basis of how the actin is organized. All migration is characterized by *protrusion,* in which the plasma membrane is pushed out at the front of the cell. In **mesenchymal cell migration**, characteristic of fibroblasts or epithelial cells grown on glass surfaces, the Arp2/3 complex mediates actin filament

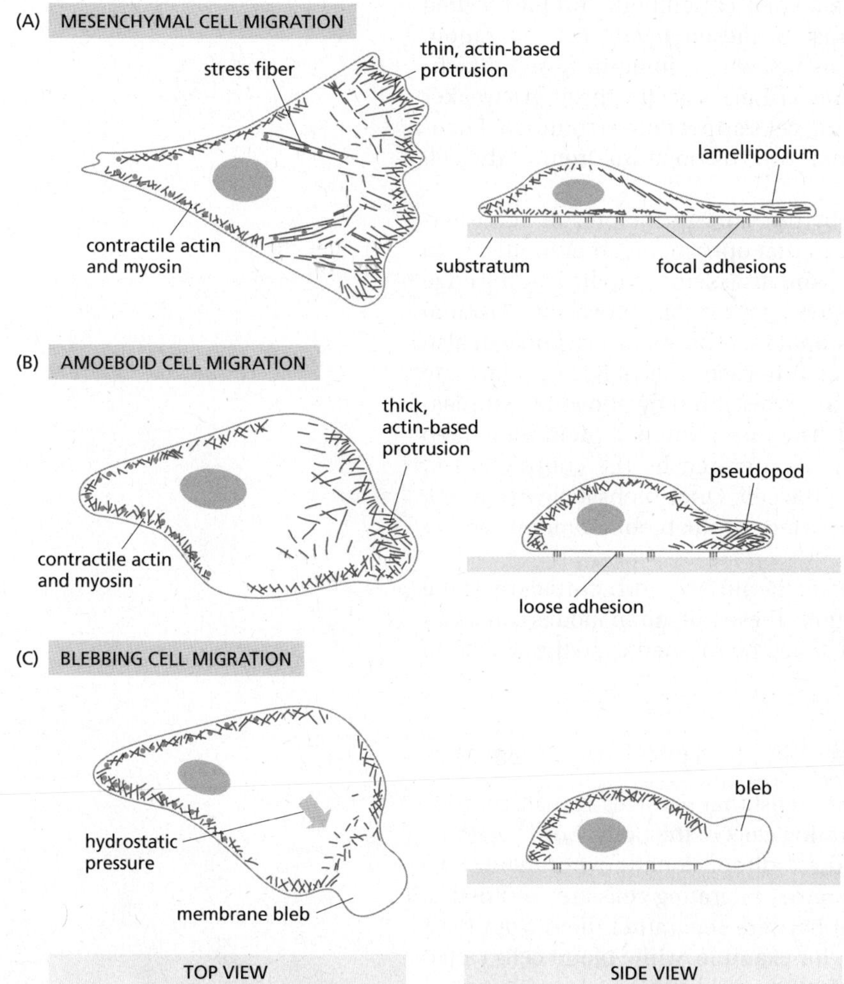

Figure 16–19 Modes of cell migration. (A) In the mesenchymal mode of cell migration, focal adhesions form behind protruding lamellipodia that connect to contractile stress fibers, bracing the cell against the surface (substratum) across which it is migrating. (B) In amoeboid cell migration, a three-dimensional pseudopod is formed by explosive actin polymerization at the leading edge. Cells maintain traction even though adhesion to the substratum is reduced, enabling much more rapid cell movement. (C) Leading edge protrusion can also occur through the formation of a bleb if the plasma membrane detaches from the underlying cortex and is pushed out by hydrostatic pressure. In all three modes of migration, contraction of cortical actin and myosin II at the rear of the cell drives locomotion of the cell body.

nucleation at the leading edge of the cell, thereby generating a *lamellipodium* (see Figure 16–18). Behind the leading edge, *cofilin* (see Figure 16–16) disassembles the older (ADP-bound) actin filaments. The actin network is therefore assembling at the front and disassembling at the back, reminiscent of the treadmilling that occurs in individual actin filaments as discussed previously (see Figure 16–11). Mesenchymal cell migration requires firm attachment to the underlying substratum, which allows the cell body to generate traction and propel itself forward. The actin cytoskeleton braces itself using stress fibers connected to integrin-based focal adhesions on the bottom surface of the cell (see Figure 19–56). Contraction of bundles of actin filaments and myosin II motors at the rear of the cell, coordinated with disassembly of attachment sites, enables the cell body to translocate.

Mesenchymal cell migration is slow, with movement rates of less than 1 μm per minute. In contrast, a second mode of cell migration, **amoeboid cell migration**, resembles the rapid movement and shape changes typical of amoebae and can be hundreds of times faster (Movie 16.2). This mode of movement is typical of white blood cells such as neutrophils and is characterized by the explosive extension of cell protrusions by local activation of the Arp2/3 complex. The protrusions are much thicker than lamellipodia and are called **pseudopodia** (or pseudopods). In contrast to mesenchymal cell migration, amoeboid-type movement does not rely as heavily on integrin-based attachments to the underlying surface. Traction nevertheless occurs through a weaker association to the substratum, and migrating cells appear more rounded. Locomotion is accompanied by a combination of protrusion at the front of the cell and contraction at the rear.

Protrusion of pseudopodia is driven by the same principles of actin filament polymerization and depolymerization that operate in lamellipodia, with cycles of actin nucleation, polymerization, and disassembly at the leading edge of the cell, but generating a three-dimensional rather than a two-dimensional structure. A distinct type of three-dimensional membrane protrusion can also result from a process called **blebbing**. In this case, the plasma membrane detaches locally from the underlying actin cortex, thereby allowing cytoplasmic flow to push the membrane outward. The formation of a membrane bleb depends on hydrostatic pressure, which is generated by the contraction of actin and myosin assemblies in the rear of the cell. Once blebs have extended, actin filaments reassemble on the inside surface of the blebbed membrane to form a new actin cortex.

Thus, leading-edge protrusion, adhesion to the surface, and contraction at the cell cortex underlie all modes of cell migration. These migration modes can interconvert depending on cell state, the extracellular environment, and the activation of different signaling pathways.

Cells Migrating in Three Dimensions Can Navigate Around Barriers

Studying the movement of cells on two-dimensional surfaces reveals important principles of cell migration. Most migrating cells in the body, however, must negotiate their way through a complex, three-dimensional environment. Surrounded by other cells and extracellular matrix, migrating cells are confined in space and must navigate around physical barriers separating them from their destination. During immune surveillance, for example, white blood cells called *dendritic cells* migrate between sites of infection and lymph nodes to initiate a systemic immune response (see Chapter 24). Dendritic cell migration can be mimicked by embedding them in a three-dimensional collagen gel matrix, in which they undergo rapid amoeboid migration toward a chemotactic signal, extending a large number of actin-rich protrusions (**Figure 16–20A and B**). When introduced into a microchamber containing different-sized channels, a cell will choose the widest path (**Figure 16–20C**). Later in this chapter, we discuss how chemotactic cues induce cell polarization and guide a migrating cell in the right direction.

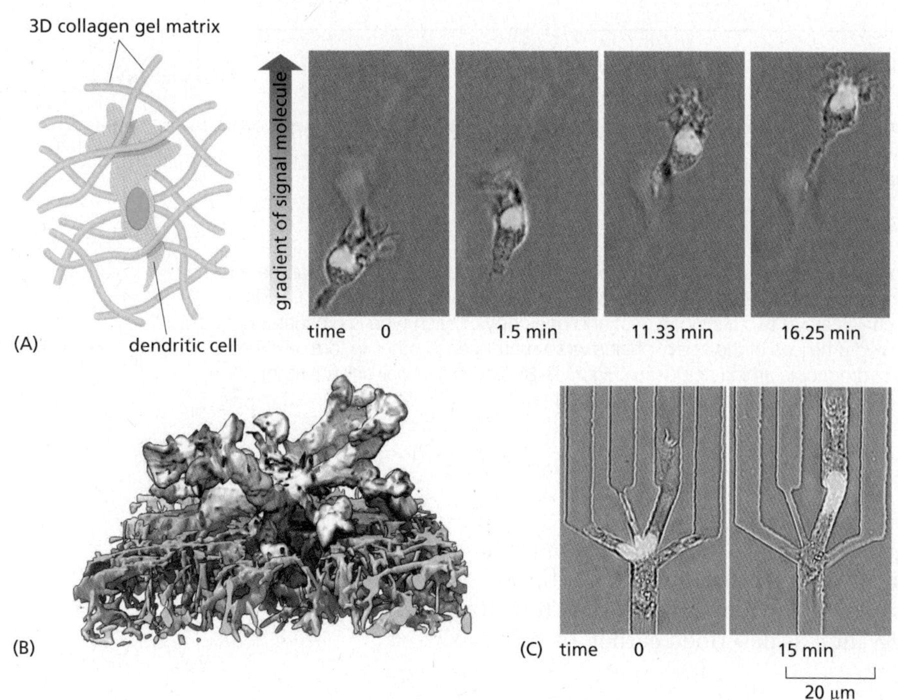

Figure 16–20 A dendritic cell migrating in three dimensions. (A) Schematic of a cell introduced into a collagen gel matrix. Images at right show that a chemotactic signal gradient induces polarized extension of pseudopodia that explore the environment, seeking out a path through gaps in the gel matrix that are typically smaller than the cell diameter. The nucleus is stained with a *blue* dye. (B) Three-dimensional image of a dendritic cell and its many pseudopodia (outlined in *red*) that is based on fluorescence microscopy. A portion of the collagen matrix surrounding the cell is shown in *gray*. (C) In this experiment, a dendritic cell was introduced into a microchannel that contains a decision point with four different pore sizes. Using the nucleus as a gauge, the cell ultimately chooses a path through the largest pore (Movie 16.3). (Courtesy of Michael Sixt.)

Some cells use specialized adhesion structures called **podosomes** that are important for cells to cross tissue barriers, for example during embryonic development. **Invadopodia** are related structures that contribute to the migration of metastatic cancer cells invading a tissue. Podosomes and invadopodia contain many of the same actin-regulatory components as other protrusions. They also secrete proteases that degrade the extracellular matrix, allowing them to carve through matrix barriers in their path.

Summary

Actin is a highly conserved cytoskeletal protein that is present in high concentrations in nearly all eukaryotic cells. Nucleation presents a kinetic barrier to actin polymerization, but once formed, actin filaments undergo dynamic behavior due to hydrolysis of the bound ATP. Actin filaments are polarized and can undergo treadmilling when a filament assembles at the plus end while simultaneously depolymerizing at the minus end. In cells, actin filament dynamics are regulated at every step, and the varied forms and functions of actin depend on a versatile repertoire of accessory proteins. Approximately half of the actin is kept in a monomeric form through association with proteins such as profilin. Nucleation factors such as the Arp2/3 complex and formins promote formation of branched and parallel filaments, respectively. Interplay between proteins that bind or cap actin filaments and those that promote filament severing or depolymerization can slow or accelerate the kinetics of filament assembly and disassembly. Another class of accessory proteins assembles the filaments into larger ordered structures by cross-linking them to one another in geometrically defined ways. Connections between these actin arrays and the plasma membrane of cells give an animal cell mechanical strength and permit the elaboration of cortical cellular structures such as lamellipodia, filopodia, pseudopodia, and microvilli. By inducing actin filament polymerization at their surface, intracellular pathogens can hijack the host-cell cytoskeleton and move around inside the cell. Cell migration depends on the formation of protrusions at the leading edge by assembly of new actin filaments or membrane blebbing, traction against the underlying substratum, and contractile forces generated by myosin motors to bring the cell body forward.

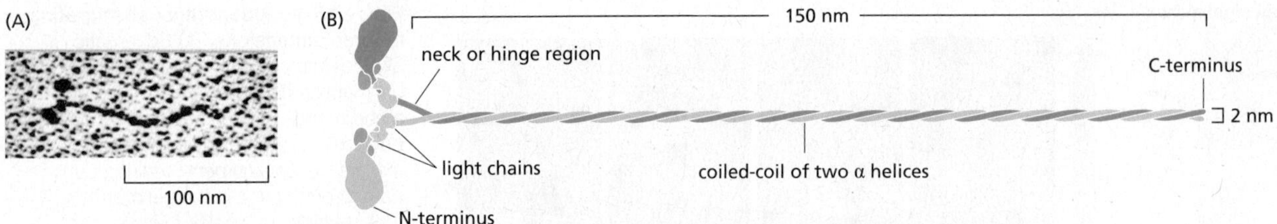

Figure 16–21 Myosin II. (A) The two globular heads and long tail of a myosin II molecule shadowed with platinum can be seen in this electron micrograph. (B) A myosin II molecule is composed of two heavy chains (each about 2000 amino acids long; *green*) and four light chains *(blue)*. The light chains are of two distinct types, and one copy of each type is present on each myosin head. Dimerization occurs when the two α helices of the heavy chains wrap around each other to form a coiled-coil, driven by the association of regularly spaced hydrophobic amino acids (see Figure 3–8). The coiled-coil arrangement makes an extended rod in solution, and this part of the molecule forms the tail. (A, courtesy of David Shotton.)

MYOSIN AND ACTIN

A crucial feature of the actin cytoskeleton is that it can form contractile structures that cross-link and slide actin filaments relative to one another through the action of **myosin** motor proteins. Actin–myosin assemblies perform important functions in all eukaryotic cells, such as in cell migration as described earlier. In addition, actin and myosin drive muscle contraction.

Actin-based Motor Proteins Are Members of the Myosin Superfamily

The first motor protein to be identified was skeletal muscle myosin, which generates the force for muscle contraction. This protein, now called *myosin II*, is an elongated protein formed from two heavy chains and two copies of each of two light chains. Each heavy chain has a globular head domain at its N-terminus that contains the force-generating machinery, followed by a very long α-helical amino acid sequence that forms an extended coiled-coil to mediate heavy-chain dimerization (**Figure 16–21**). The two light chains bind close to the N-terminal head domain, while the long coiled-coil tail bundles itself with the tails of other myosin molecules. In skeletal muscle, these tail–tail interactions form large, bipolar *thick filaments* that have several hundred myosin heads, oriented in opposite directions at the two ends of the thick filament (**Figure 16–22**).

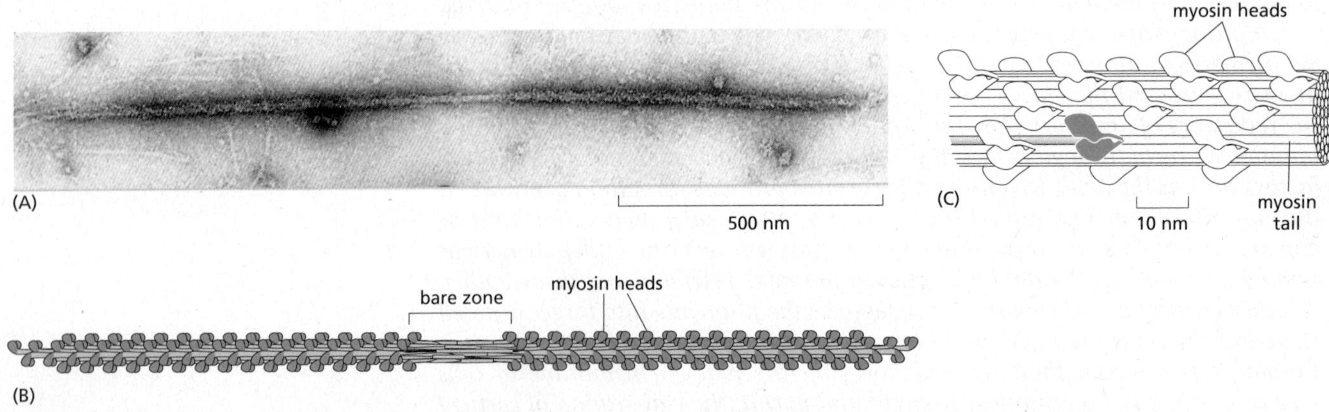

Figure 16–22 The myosin II bipolar thick filament in muscle. (A) Electron micrograph of a myosin II thick filament isolated from frog muscle. Note the central bare zone, which is free of head domains. (B) Schematic diagram, not drawn to scale. The myosin II molecules aggregate by means of their tail regions, with their heads projecting to the outside of the filament. The bare zone in the center of the filament consists entirely of myosin II tails. (C) A small section of a myosin II filament as reconstructed from electron micrographs. In the relaxed (noncontracting) state, the two heads of a myosin molecule are bent backward and sterically interfere with each other to switch off their activity. An individual myosin molecule in the inactive state is highlighted in *green*. The cytoplasmic myosin II filaments in non-muscle cells are much smaller, although similarly organized (see Figure 16–34). (A, from M. Stewart and R.W. Kensler, *J. Mol. Biol.* 192:831–851, 1986. With permission from Elsevier; C, based on R.A. Crowther et al., *J. Mol. Biol.* 184:429–439, 1985.)

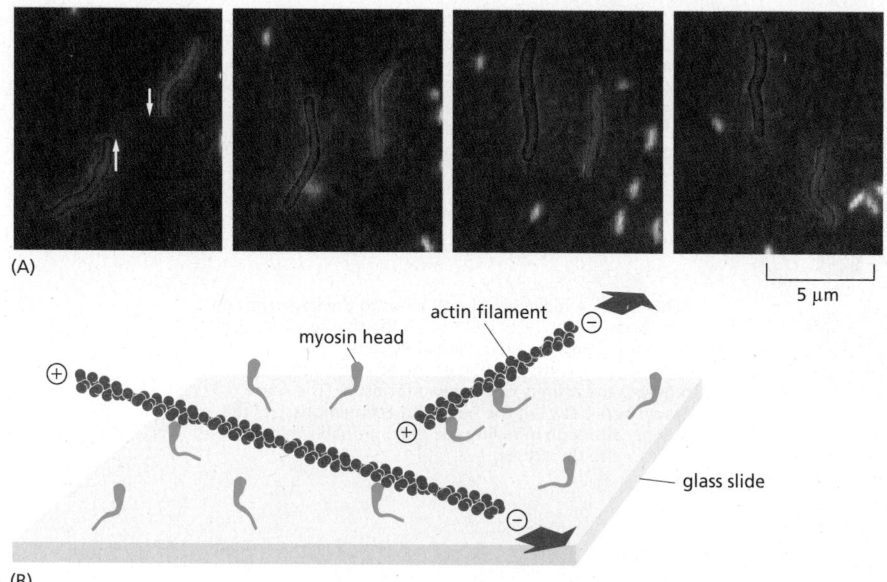

Figure 16–23 **Direct evidence for the motor activity of the myosin head.** In this experiment, purified myosin heads were attached to a glass slide, and then actin filaments labeled with fluorescent phalloidin were added and allowed to bind to the myosin heads. (A) When ATP was added, the actin filaments began to glide along the surface, owing to the many individual steps taken by each of the dozens of myosin heads bound to each filament. The video frames shown in this sequence were recorded about 0.6 second apart; the two actin filaments shown *(red)* were moving in opposite directions at a rate of about 4 μm/sec. (B) Diagram of the experiment. The large *red arrows* indicate the direction of actin filament movement (**Movie 16.4**). (A, courtesy of James Spudich.)

Each myosin head binds and hydrolyzes ATP, using the energy of ATP hydrolysis to walk toward the plus end of an actin filament (**Figure 16–23**). The opposing orientation of the heads in the thick filament makes the filament efficient at sliding pairs of oppositely oriented actin filaments toward each other, shortening the muscle. In skeletal muscle, in which carefully arranged actin filaments are aligned in *thin filament* arrays surrounding the myosin thick filaments, the ATP-driven sliding of actin filaments results in a powerful contraction. Cardiac and smooth muscle contain myosin II molecules that are similarly arranged, although different genes encode them.

Myosin Generates Force by Coupling ATP Hydrolysis to Conformational Changes

Motor proteins use structural changes in their ATP-binding sites to produce cyclic interactions with a cytoskeletal filament. Each cycle of ATP binding, hydrolysis, and release propels them forward in a single direction to a new binding site along the filament. For myosin II, each step of the movement along actin is generated by the swinging of an 8.5-nm-long α helix, or *lever arm*, which is structurally stabilized by the binding of light chains. At the base of this lever arm next to the head, there is a pistonlike helix that connects movements at the ATP-binding cleft in the head to small rotations of the so-called converter domain. A small change at this point can swing the helix like a long lever, causing the far end of the helix to move by about 5.0 nm.

These changes in the conformation of the myosin are coupled to changes in its binding affinity for actin, allowing the myosin head to release its grip on the actin filament at one point and snatch hold of it again at another. The full mechanochemical cycle of ATP binding, ATP hydrolysis, and phosphate release (which causes the *power stroke*) produces a single step of movement (**Figure 16–24**). At low ATP concentrations, the interval between the force-producing step and the binding of the next ATP is long enough that single steps can be observed (**Figure 16–25**).

Sliding of Myosin II Along Actin Filaments Causes Muscles to Contract

Muscle contraction is the most familiar and best-understood form of movement in animals. In vertebrates, running, walking, swimming, and flying all depend on the rapid contraction of skeletal muscle on its scaffolding of bone, while

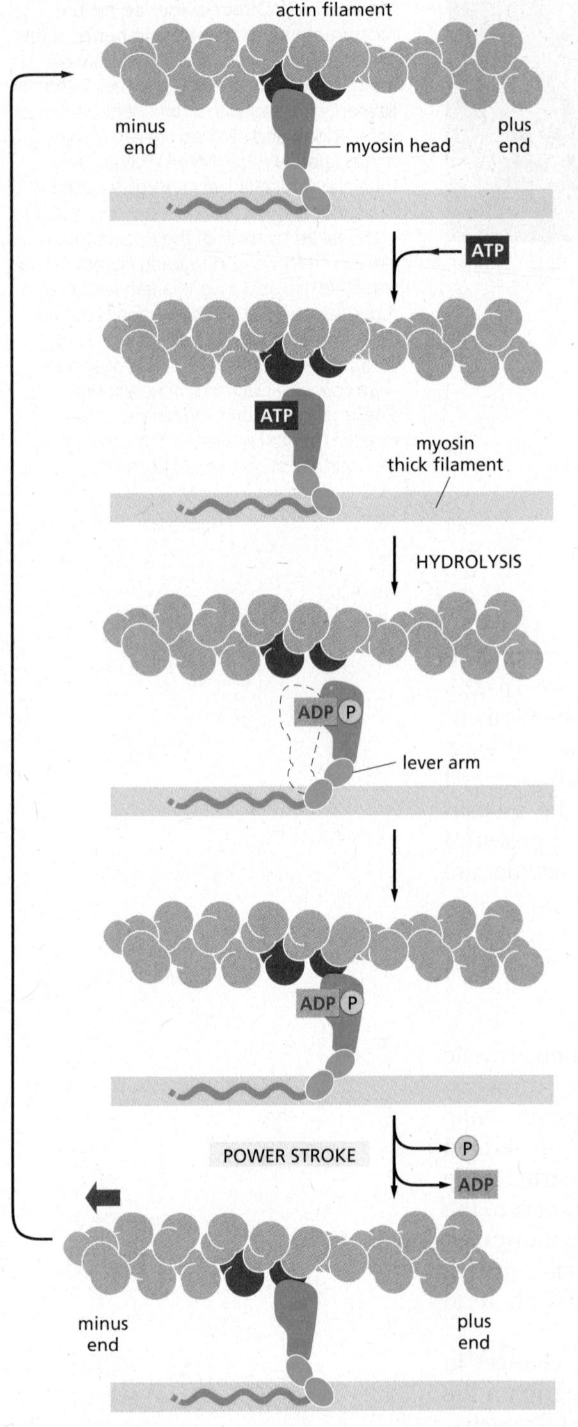

ATTACHED At the start of the cycle shown in this figure, a myosin head lacking a bound nucleotide is locked tightly onto an actin filament in a *rigor* configuration (so named because it is responsible for *rigor mortis*, the rigidity of death). In an actively contracting muscle, this state is very short-lived, being rapidly terminated by the binding of a molecule of ATP.

RELEASED A molecule of ATP binds to the large cleft on the "back" of the head (that is, on the side furthest from the actin filament) and immediately causes a slight change in the conformation of the actin-binding site, reducing the affinity of the head for actin. (The space drawn here between the head and actin emphasizes this change, although in reality the head probably remains very close to the actin.)

COCKED ATP binding triggers a conformational change in the cleft that leads to a rotation in the converter domain, causing the lever arm to swing out and the head to be displaced along the filament by a distance of about 5 nm. Hydrolysis of ATP occurs, but the ADP and inorganic phosphate (P) remain tightly bound to the protein.

RE-BINDING AND POWER STROKE The myosin head binds weakly to a new site on the actin filament, causing release of the inorganic phosphate produced by ATP hydrolysis, concomitantly with the tight binding of the head to actin. This release triggers the power stroke—the force-generating change in shape during which the head regains its original conformation. In the course of the power stroke, the head loses its bound ADP, thereby returning to the start of a new cycle.

FORCE GENERATING At the end of the cycle, the myosin head is again locked tightly to the actin filament in a rigor configuration. Note that the head has moved to a new position on the actin filament.

Figure 16–24 **The cycle of structural changes used by myosin II to walk along an actin filament.** In the myosin II cycle, the head remains bound to the actin filament for only about 5% of the entire cycle time, allowing many myosins to work together to move a single actin filament (**Movie 16.5**). (Based on I. Rayment et al., *Science* 261:58–65, 1993.)

involuntary movements such as heart pumping and gut peristalsis depend on the contraction of cardiac muscle and smooth muscle, respectively. All these forms of muscle contraction depend on the ATP-driven sliding of highly organized arrays of actin filaments against arrays of myosin II filaments.

Skeletal muscle was a relatively late evolutionary development, and muscle cells are highly specialized for rapid and efficient contraction. The long, thin muscle fibers of skeletal muscle are actually huge single cells that form during

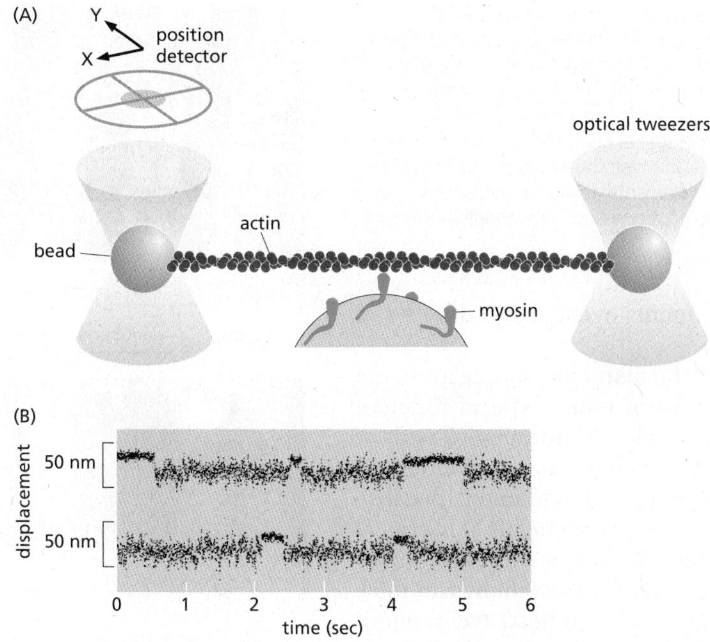

Figure 16–25 **The force of a single myosin molecule moving along an actin filament measured using an optical trap.** (A) Schematic of the experiment, showing an actin filament with beads attached at both ends and held in place by focused beams of light called optical tweezers (**Movie 16.6**). The tweezers trap and move the bead and can also be used to measure the force exerted on the bead through the filament. In this experiment, the filament was positioned over another bead to which myosin II motors were attached, and the optical tweezers were used to determine the effects of myosin binding on movement of the actin filament. (B) These traces show filament movement in two separate experiments. Initially, when the actin filament is unattached to myosin, thermal motion of the filament produces noisy fluctuations in filament position. When a single myosin binds to the actin filament, thermal motion decreases abruptly, and a roughly 10-nm displacement results from movement of the filament by the motor. The motor then releases the filament. Because the ATP concentration is very low in this experiment, the myosin remains attached to the actin filament for much longer than it would in a muscle cell. (Adapted from C. Rüegg et al., *News Physiol. Sci.* 17:213–218, 2002. With permission from the American Physiological Society.)

development by the fusion of many separate cells. The large muscle cell retains the many nuclei of the contributing cells. These nuclei lie just beneath the plasma membrane (**Figure 16–26**). The bulk of the cytoplasm inside is made up of myofibrils, which is the name given to the basic contractile elements of the muscle cell. A **myofibril** is a cylindrical structure 1–2 μm in diameter that is often as long as the muscle cell itself. It consists of a long, repeated chain of tiny contractile units—called *sarcomeres,* each about 2.2 μm long—which give the vertebrate myofibril its striated appearance (**Figure 16–27**).

Each sarcomere is formed from a miniature, precisely ordered array of parallel and partly overlapping thin and thick filaments. The *thin filaments* are composed of actin and associated proteins, and they are attached at their plus ends to a *Z disc* at each end of the sarcomere. The capped minus ends of the actin filaments extend in toward the middle of the sarcomere, where they overlap with *thick filaments,* the bipolar assemblies formed from specific muscle isoforms of myosin II (see Figure 16–22). When this region of overlap is examined in cross section by electron microscopy, the myosin filaments are arranged

Figure 16–26 **Skeletal muscle cells (also called muscle fibers).** (A) These huge multinucleated cells form by the fusion of many muscle cell precursors, called myoblasts. Here, a single muscle cell is depicted. In an adult human, a muscle cell is typically 50 μm in diameter and can be up to several centimeters long. (B) Fluorescence micrograph of rat muscle, showing the peripherally located nuclei *(blue)* in these giant cells. Myofibrils are stained *red*. (B, courtesy of Nancy L. Kedersha.)

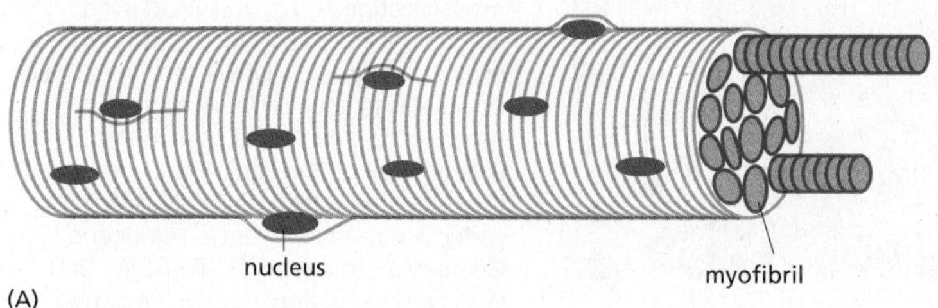

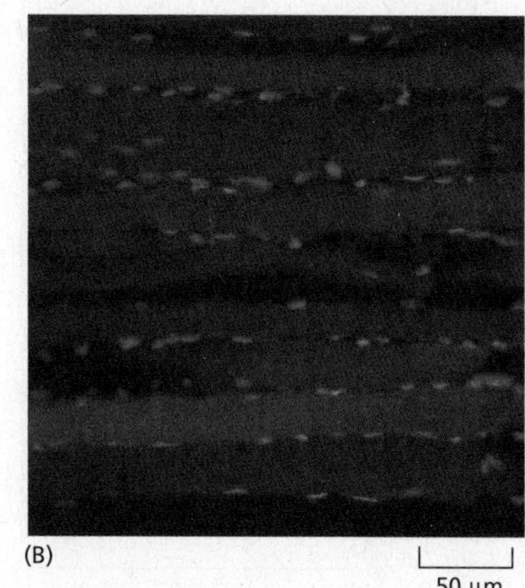

Figure 16–27 **Skeletal muscle myofibrils.** (A) Low-magnification electron micrograph of a longitudinal section through a skeletal muscle cell of a rabbit, showing the regular pattern of cross-striations. The cell contains many myofibrils aligned in parallel (see Figure 16–26). (B) Detail of the skeletal muscle shown in A, showing portions of two adjacent myofibrils and the definition of a sarcomere *(black arrow)*. (C) Schematic diagram of a single sarcomere, showing the origin of the dark and light bands seen in the electron micrographs. The Z discs, at each end of the sarcomere, are attachment sites for the plus ends of actin filaments (thin filaments); the M line, or midline, is the location of proteins that link adjacent myosin II filaments (thick filaments) to one another. (D) When the sarcomere contracts, the actin and myosin filaments slide past one another without shortening. (A and B, courtesy of Roger Craig.)

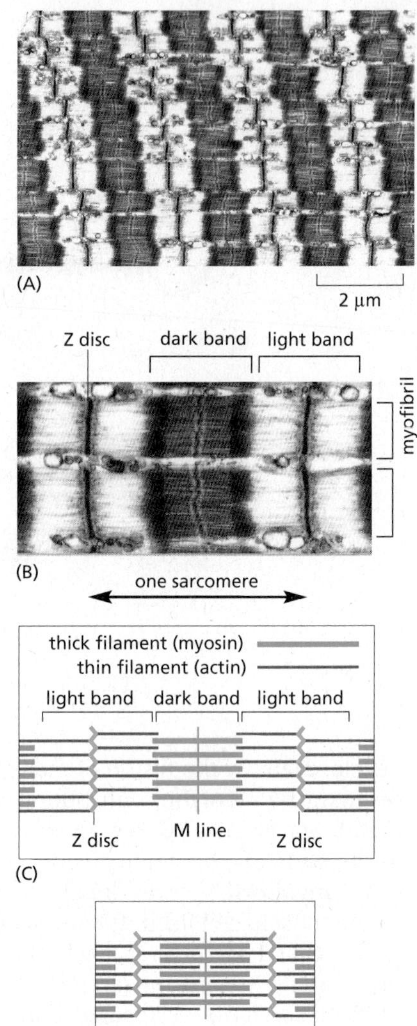

in a regular hexagonal lattice, with the actin filaments evenly spaced between them (Figure 16–28).

Sarcomere shortening is caused by the myosin filaments sliding past the actin thin filaments, with no change in the length of either type of filament (see Figure 16–27C and D). Bipolar thick filaments walk toward the plus ends of two sets of thin filaments of opposite orientations, driven by dozens of independent myosin heads that are positioned to interact with each thin filament. Because there is no coordination among the movements of the myosin heads, it is critical that they remain tightly bound to the actin filament for only a small fraction of each ATPase cycle so that they do not hold one another back. Each myosin thick filament has about 300 heads (294 in frog muscle), and each head cycles about five times per second in the course of a rapid contraction—sliding the myosin and actin filaments past one another at rates of up to 15 μm/sec and enabling the sarcomere to shorten by 10% of its length in less than one-fiftieth of a second. The rapid synchronized shortening of the thousands of sarcomeres lying end-to-end in each myofibril enables skeletal muscle to contract rapidly enough for running and flying or for playing the piano.

Accessory proteins produce the remarkable uniformity in filament organization, length, and spacing in the sarcomere and enable it to withstand the constant wear-and-tear of contraction (Figure 16–29). The actin filament plus ends are anchored in the Z disc, which is built from CapZ and α-actinin; CapZ in the Z disc caps the filaments (preventing depolymerization), while α-actinin holds them together in a regularly spaced bundle. Actin filaments are stabilized along their length by tropomyosin and also by a protein of enormous size, called *nebulin*, which consists almost entirely of a repeating 35-amino-acid actin-binding motif. Nebulin stretches from the Z disc toward the minus end of each thin filament, which is capped and stabilized by tropomodulin. Although there is some slow exchange of actin subunits at both ends of the muscle thin filament, such that the components of the thin filament turn over with a half-life of several days, the actin filaments in sarcomeres are remarkably stable compared with those found

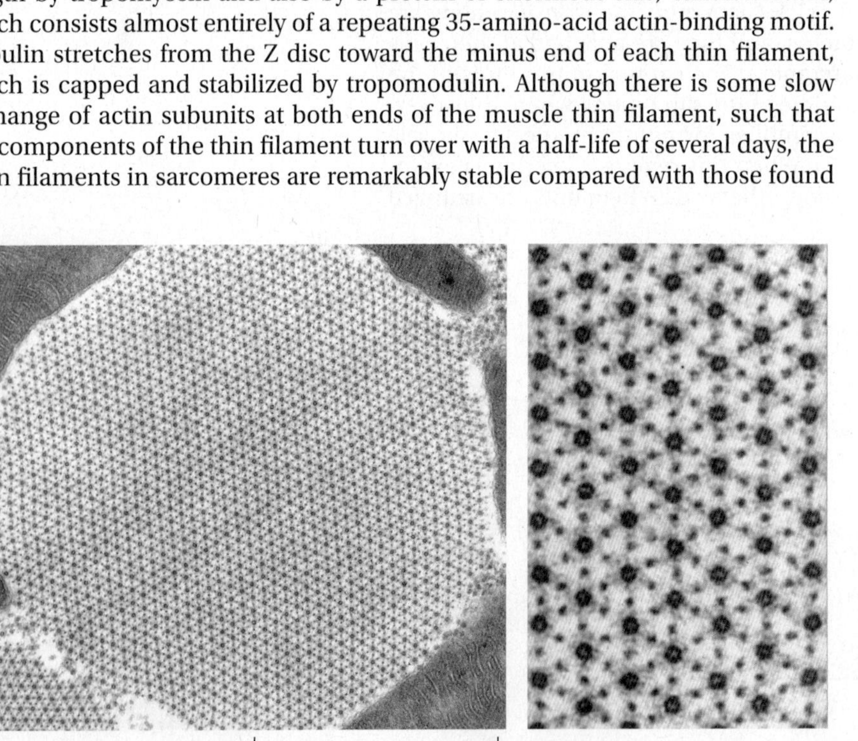

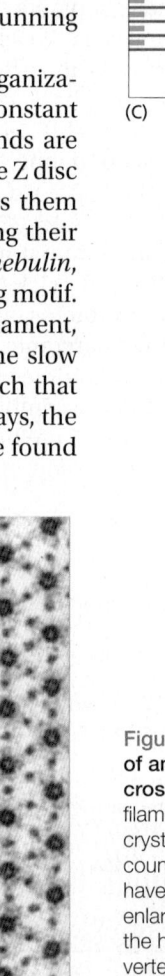

Figure 16–28 **Electron micrographs of an insect flight muscle viewed in cross section.** The myosin and actin filaments are packed together with almost crystalline regularity. Unlike their vertebrate counterparts, these myosin filaments have a hollow center, as seen in the enlargement on the right. The geometry of the hexagonal lattice is slightly different in vertebrate muscle. (From J. Auber and R. Couteaux, *J. Microsc*. 2:309–324, 1963. With permission from Société Française de Microscopie Électronique.)

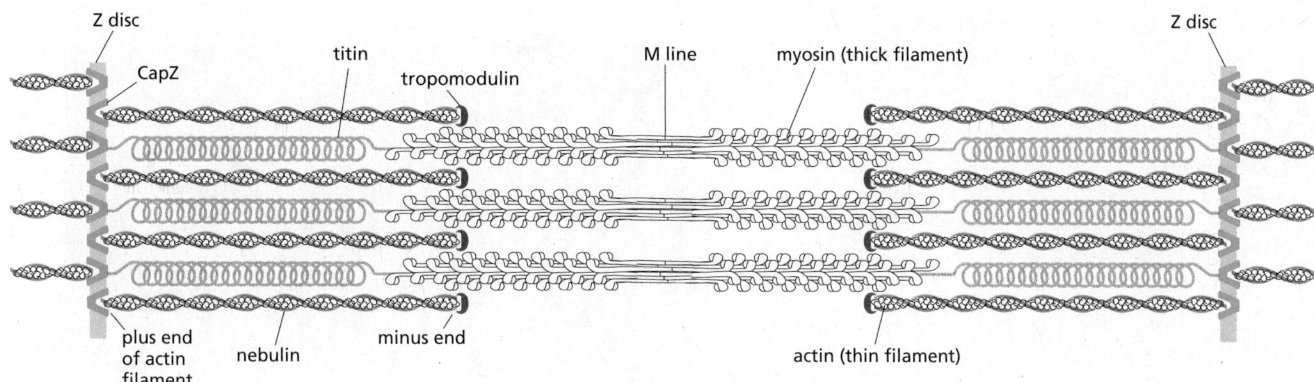

Figure 16–29 Organization of accessory proteins in a sarcomere. Each giant titin molecule extends from the Z disc to the M line—a distance of more than 1 μm. Part of each titin molecule is closely associated with a myosin thick filament (which switches polarity at the M line); the rest of the titin molecule is elastic and changes length as the sarcomere contracts and relaxes. Each nebulin molecule is exactly the length of a thin filament. The actin filaments are also coated with tropomyosin and bound intermittently by troponin (not shown; see Figure 16–31) and are capped at both ends. Tropomodulin caps the minus end of the actin filaments, and CapZ anchors the plus end at the Z disc, which also contains α-actinin (not shown).

in most other cell types, whose dynamic actin filaments turn over with half-lives of a few minutes or less.

Opposing pairs of an even longer protein, called *titin*, position the thick filaments midway between the Z discs. Titin acts as a molecular spring, with a long series of immunoglobulin-like domains that can unfold one by one as stress is applied to the protein. A springlike unfolding and refolding of these domains keeps the thick filaments poised in the middle of the sarcomere and allows the muscle fiber to recover after being overstretched. In *Caenorhabditis elegans*, whose sarcomeres are longer than those in vertebrates, titin is longer as well, suggesting that it serves also as a molecular ruler, determining in this case the overall length of each sarcomere.

A Sudden Rise in Cytosolic Ca^{2+} Concentration Initiates Muscle Contraction

The force-generating molecular interaction between myosin thick filaments and actin thin filaments takes place only when a signal passes to the skeletal muscle from the nerve that stimulates it. Immediately upon arrival of the signal, the muscle cell needs to be able to contract very rapidly, with all the sarcomeres shortening simultaneously. Two major features of the muscle cell make extremely rapid contraction possible. First, as previously discussed, the individual myosin motor heads in each thick filament spend only a small fraction of the ATP cycle time bound to the filament and actively generating force, so many myosin heads can act in rapid succession on the same thin filament without interfering with one another. Second, a specialized membrane system relays the incoming signal rapidly throughout the entire cell. The signal from the nerve triggers an action potential in the muscle cell plasma membrane (discussed in Chapter 11), and this electrical excitation spreads swiftly into a series of membranous folds—the transverse tubules, or *T tubules*—that extend inward from the plasma membrane around each myofibril. The signal is then relayed across a small gap to the *sarcoplasmic reticulum*, an adjacent weblike sheath of modified endoplasmic reticulum that surrounds each myofibril like a net stocking (**Figure 16–30A and B**).

When the incoming action potential activates a Ca^{2+} channel in the T-tubule membrane, it triggers the opening of a Ca^{2+}-release channel in the closely associated sarcoplasmic reticulum membrane (**Figure 16–30C**). Ca^{2+} flooding into the cytosol then initiates the contraction of each myofibril. Because the signal from the muscle cell plasma membrane is passed within milliseconds (via the T tubules and sarcoplasmic reticulum) to every sarcomere in the cell, all of the myofibrils in the cell contract at once. The increase in Ca^{2+} concentration is transient because the Ca^{2+} is rapidly pumped back into the sarcoplasmic reticulum by an abundant, ATP-dependent Ca^{2+}-pump (also called a Ca^{2+}-ATPase) in its membrane (see Figure 11–14). Typically, the cytoplasmic Ca^{2+} concentration is restored to resting levels within 30 milliseconds, allowing the myofibrils to relax. Thus, muscle contraction depends

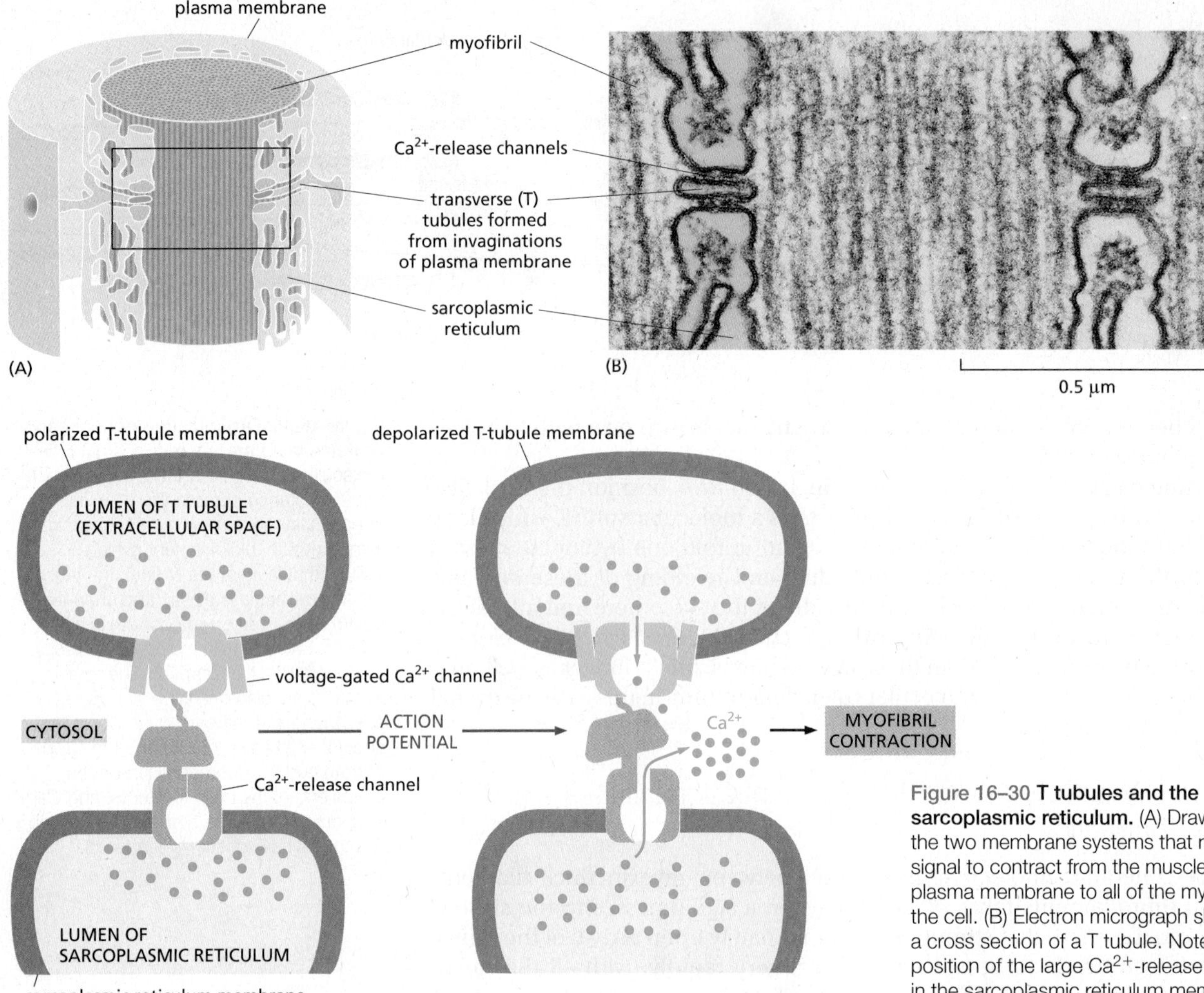

Figure 16–30 T tubules and the sarcoplasmic reticulum. (A) Drawing of the two membrane systems that relay the signal to contract from the muscle cell plasma membrane to all of the myofibrils in the cell. (B) Electron micrograph showing a cross section of a T tubule. Note the position of the large Ca^{2+}-release channels in the sarcoplasmic reticulum membrane that connect to the adjacent T-tubule membrane. (C) Schematic diagram showing how a Ca^{2+}-release channel in the sarcoplasmic reticulum membrane is thought to be opened by the activation of a voltage-gated Ca^{2+} channel in the membrane of the T tubule (**Movie 16.7**). (B, courtesy of Clara Franzini-Armstrong.)

on two processes that consume enormous amounts of ATP: filament sliding, driven by the ATPase of the myosin motor domain, and Ca^{2+} pumping, driven by the Ca^{2+}-pump.

The Ca^{2+} dependence of vertebrate skeletal muscle contraction, and hence its dependence on commands transmitted via nerves, is due entirely to a set of specialized accessory proteins that are closely associated with the actin thin filaments. One of these accessory proteins is a muscle form of *tropomyosin*, the elongated protein that binds along the groove of the actin filament helix. The other is *troponin*, a complex of three polypeptides, troponins T, I, and C (named for their tropomyosin-binding, inhibitory, and Ca^{2+}-binding activities, respectively). Troponin I binds to actin as well as to troponin T. In a resting muscle, the troponin I–T complex pulls the tropomyosin out of its normal binding groove into a position along the actin filament that interferes with the binding of myosin heads, thereby preventing any force-generating interaction. When the level of Ca^{2+} is raised, troponin C—which binds up to four molecules of Ca^{2+}—causes troponin I to release its hold on actin. This allows the tropomyosin molecules to slip back into their normal position so that the myosin heads can walk along the actin filaments (**Figure 16–31**). Troponin C is closely related to the ubiquitous Ca^{2+}-binding protein calmodulin (see Figure 15–34); it can be thought of as a specialized form of calmodulin that has acquired binding sites for troponin I and troponin T, thereby ensuring that the myofibril responds extremely rapidly to an increase in Ca^{2+} concentration.

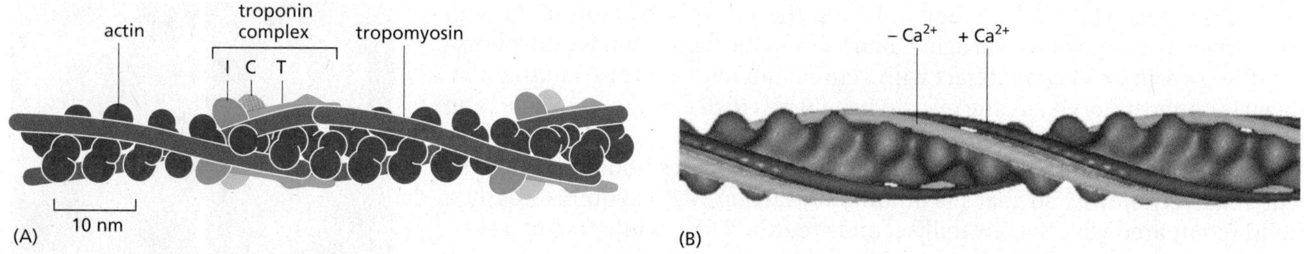

Figure 16–31 The control of skeletal muscle contraction by troponin. (A) A thin filament of a skeletal muscle cell, showing the positions of tropomyosin and troponin along the actin filament. Each tropomyosin molecule has seven evenly spaced regions with similar amino acid sequences, each of which is thought to bind to an actin subunit in the filament. (B) Reconstructed cryo-electron microscopy image of an actin filament showing the relative position of a superimposed tropomyosin strand in the presence *(dark purple)* or absence *(light purple)* of calcium. (A, adapted from G.N. Phillips et al., *J. Mol. Biol.* 192:111–131, 1986; B, adapted from C. Xu et al., *Biophys. J.* 77:985–992, 1999. With permission from the Biophysical Society.)

In smooth muscle cells, so called because they lack the regular striations of skeletal muscle, contraction is also triggered by an influx of calcium ions, but the regulatory mechanism is different. Smooth muscle forms the contractile portion of the stomach, intestine, and uterus, as well as the walls of arteries and many other structures requiring slow and sustained contractions. Smooth muscle is composed of sheets of highly elongated spindle-shaped cells, each with a single nucleus. Smooth muscle cells do not express the troponins. Instead, elevated intracellular Ca^{2+} levels regulate contraction by a mechanism that depends on calmodulin (**Figure 16–32**). Ca^{2+}-bound calmodulin activates myosin

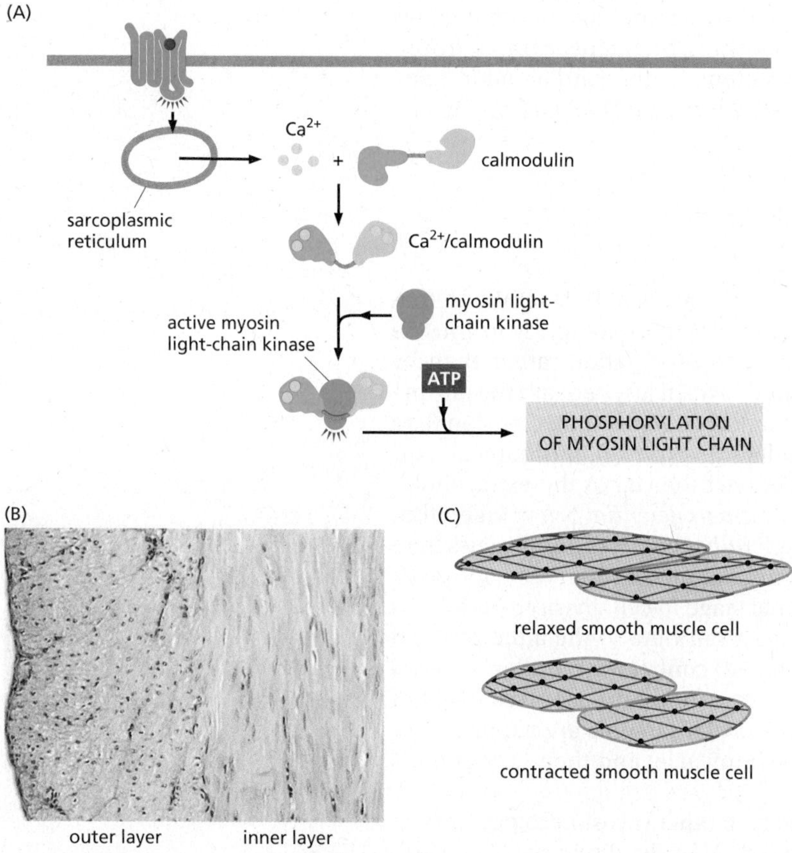

Figure 16–32 Smooth muscle contraction. (A) Upon muscle stimulation by activation of cell-surface receptors, Ca^{2+} released into the cytoplasm from the sarcoplasmic reticulum (SR) binds to calmodulin (see Figure 15–34). Ca^{2+}-bound calmodulin then binds myosin light-chain kinase (MLCK), which phosphorylates myosin light chain, stimulating myosin activity. Non-muscle myosin is regulated by the same mechanism (see Figure 16–34). (B) Smooth muscle cells in a cross section of cat intestinal wall. The outer layer of smooth muscle is oriented with the long axis of its cells extending parallel along the length of the intestine, and upon contraction will shorten the intestine. The inner layer is oriented circularly around the intestine and when contracted will cause the intestine to become narrower. Contraction of both layers squeezes material through the intestine, much like squeezing toothpaste out of a tube. (C) A model for the contractile apparatus in a smooth muscle cell, with bundles of contractile filaments containing actin and myosin *(red)* oriented obliquely to the long axis of the cell. Their contraction greatly shortens the cell. In this diagram, the bundle angles are exaggerated to schematically illustrate the effect of contraction. In addition, only a few of the many bundles are shown. (B, courtesy of Gwen V. Childs.)

light-chain kinase (MLCK), thereby inducing the phosphorylation of smooth muscle myosin on one of its two light chains. When the light chain is phosphorylated, the myosin head can interact with actin filaments and cause contraction; when it is dephosphorylated, the myosin head tends to dissociate from actin and becomes inactive.

The phosphorylation events that regulate contraction in smooth muscle cells occur relatively slowly, so that maximum contraction often requires nearly a second (compared with the few milliseconds required for contraction of a skeletal muscle cell). But rapid activation of contraction is not important in smooth muscle: its myosin II hydrolyzes ATP about 10 times more slowly than does skeletal muscle myosin, producing a slow cycle of myosin conformational changes that results in slow contraction.

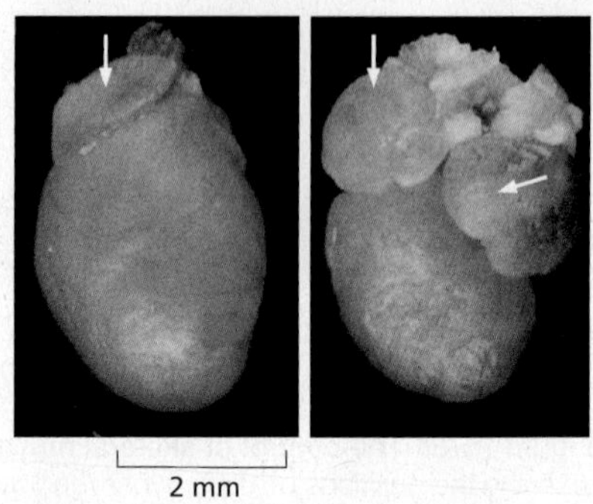

Figure 16–33 **Effect on the heart of a subtle mutation in cardiac myosin.** *Left*, normal heart from a 6-day-old mouse pup. *Right*, heart from a pup with a point mutation in both copies of its cardiac myosin gene, changing Arg403 to Gln. The arrows indicate the atria. In the heart from the pup with the cardiac myosin mutation, both atria are greatly enlarged (hypertrophic), and the mice die within a few weeks of birth. (From D. Fatkin et al., *J. Clin. Invest.* 103:147–153, 1999. With permission from the American Society for Clinical Investigation.)

Heart Muscle Is a Precisely Engineered Machine

The heart is the most heavily worked muscle in the body, contracting about 3 billion (3×10^9) times during the course of a human lifetime (**Movie 16.8**). Heart cells express several specific isoforms of cardiac muscle myosin and cardiac muscle actin. Even subtle changes in these cardiac-specific contractile proteins—changes that would not cause any noticeable consequences in other tissues—can cause serious heart disease (**Figure 16–33**).

The normal cardiac contractile apparatus is such a highly tuned machine that a tiny abnormality anywhere in the works can be enough to gradually wear it down over years of repetitive motion. *Familial hypertrophic cardiomyopathy* is a common cause of sudden death in young athletes. It is a genetically dominant inherited condition that affects about two out of every thousand people, and it is associated with heart enlargement, abnormally small coronary vessels, and disturbances in heart rhythm (cardiac arrhythmias). The cause of this condition is either any one of more than 40 subtle point mutations in the genes encoding cardiac β-myosin heavy chain (almost all causing changes in or near the motor domain) or one of about a dozen mutations in other genes encoding contractile proteins—including myosin light chains, cardiac troponin, and tropomyosin. Minor missense mutations in the cardiac actin gene cause another type of heart condition, called *dilated cardiomyopathy*, which can also result in early heart failure.

Actin and Myosin Perform a Variety of Functions in Non-Muscle Cells

Most non-muscle cells contain contractile actin–myosin II assemblies that form transiently, enabling dynamic changes in cell morphology. Non-muscle contractile bundles are regulated by myosin phosphorylation rather than by troponin (**Figure 16–34A**). Contractile actin and myosin just beneath the plasma membrane in the cell cortex creates tension, and gradients in this tension lead to cell-shape changes. Actin–myosin II bundles also provide mechanical support by assembling into **stress fibers** that connect the cell to the extracellular matrix through *focal adhesions* or by forming a *circumferential belt* in an epithelial cell, connecting it to adjacent cells through *adherens junctions* (discussed in Chapter 19). As described in Chapter 17, actin and myosin II in the *contractile ring* generate the force for cytokinesis, the final stage in cell division. Finally, as discussed previously, contractile bundles also contribute to the adhesion and forward motion of migrating cells. Organization of contractile bundles is somewhat similar to the periodic organization of sarcomeres and involves many of the same proteins. However, non-muscle myosin II filaments are about sevenfold shorter than the thick filaments of skeletal muscle, and their formation is highly dynamic (**Figure 16–34B and C**).

Non-muscle cells also express a large family of other myosin proteins, which have diverse structures and functions in the cell. After the discovery of conventional muscle myosin, a second member of the family was found in the freshwater

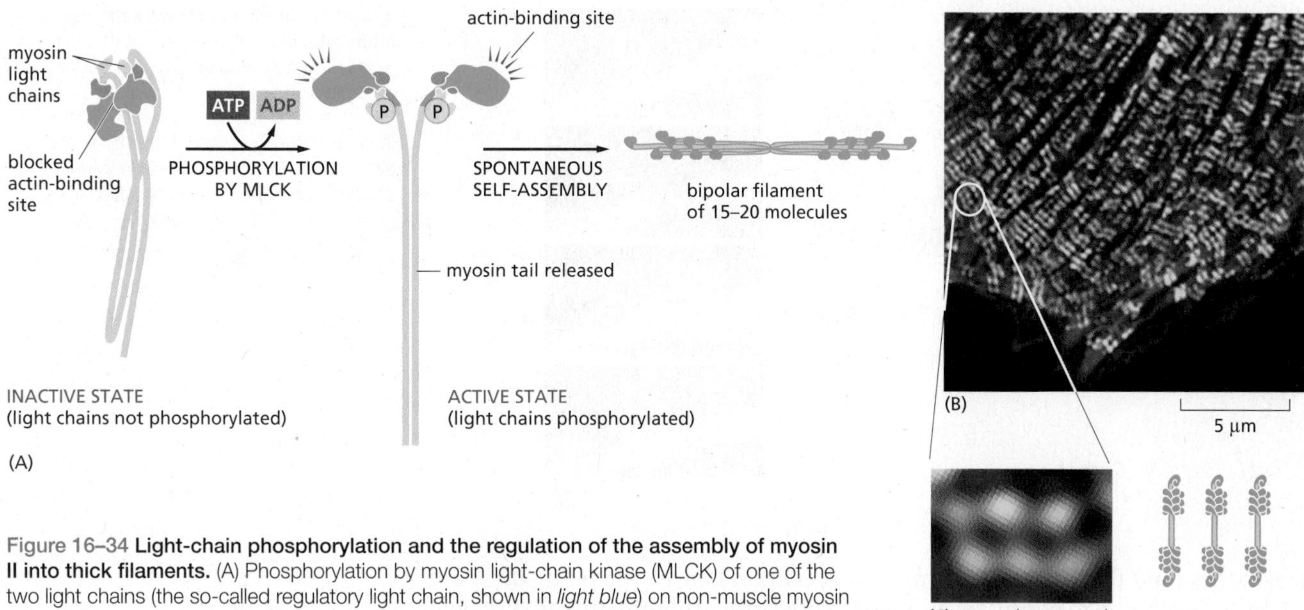

Figure 16–34 **Light-chain phosphorylation and the regulation of the assembly of myosin II into thick filaments.** (A) Phosphorylation by myosin light-chain kinase (MLCK) of one of the two light chains (the so-called regulatory light chain, shown in *light blue*) on non-muscle myosin II in a test tube has at least two effects: it causes a change in the conformation of the myosin heads, relieving steric inhibition to permit actin binding, and it releases the myosin tail from a "sticky patch" on the myosin head, thereby allowing the myosin molecules to assemble into short, bipolar, thick filaments. Smooth muscle is regulated by the same mechanism (see Figure 16–32). (B) Fluorescence micrograph showing the distribution of myosin light chain *(yellow)* and the actin filament cross-linker α-actinin *(blue)* in the contractile fibers of a rat embryo fibroblast. (C) A parallel stack of myosin filaments at higher magnification. These myosin II filaments are much smaller than those found in skeletal muscle cells (see Figure 16–22), but nevertheless form sarcomere-like units. (B and C, courtesy of Shiqiong Hu and Alexander Bershadsky.)

amoeba *Acanthamoeba castellanii.* This protein had a different tail structure and seemed to function as a monomer, and so it was named *myosin I* (for one-headed). Conventional muscle myosin was renamed *myosin II* (for two-headed). Subsequently, many other myosin types were discovered. The heavy chains generally start with a recognizable myosin motor domain at the N-terminus and then diverge widely with a variety of C-terminal tail domains (**Figure 16–35**). The myosin family includes a number of one-headed and two-headed varieties that are about equally related to myosin I and myosin II, and the nomenclature now reflects their approximate order of discovery (myosin III through at least myosin XVIII). Sequence comparisons among diverse eukaryotes indicate that there are at least 37 distinct myosin families in the superfamily. All of the myosins except one move toward the plus end of an actin filament, although they do so at different speeds. The exception is myosin VI, which moves toward the minus end. The

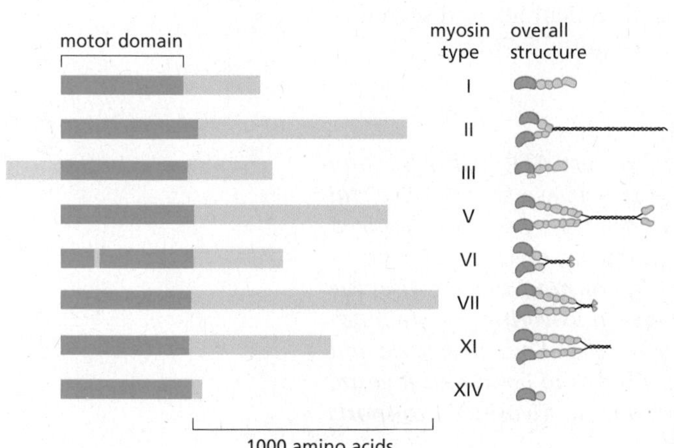

Figure 16–35 **Myosin superfamily members.** Comparison of the domain structure of the heavy chains of some myosin types. All myosins share similar motor domains (shown in *dark green*), but their C-terminal tails *(light green)* and N-terminal extensions *(light blue)* are very diverse. On the right are depictions of the molecular structures for these family members. Many myosins form dimers, with two motor domains per molecule, but a few (such as I, III, and XIV) seem to function as monomers, with just one motor domain. Myosin VI, despite its overall structural similarity to other family members, is unique in moving toward the minus end (instead of the plus end) of an actin filament. The small insertion within its motor head domain, not found in other myosins, is probably responsible for this change in direction.

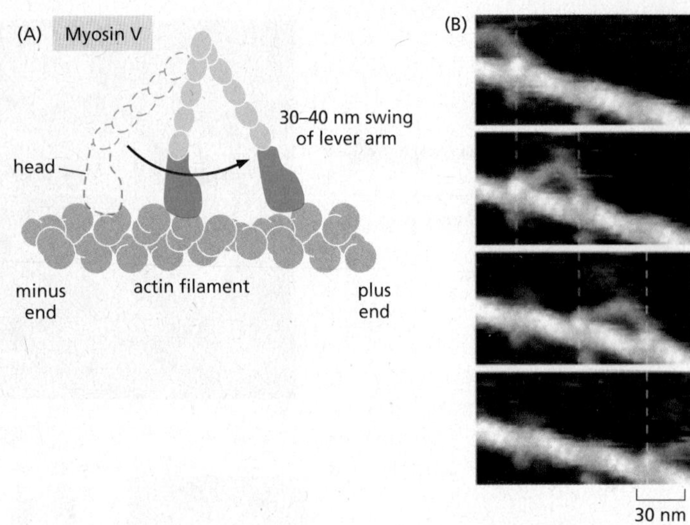

Figure 16–36 **Myosin V walks along actin filaments.** (A) The lever arm of myosin V is long, allowing it to take a bigger step along an actin filament than can myosin II (see Figure 16–24). (B) Atomic force microscopy images showing myosin V *(green)* walking along an actin filament. Myosin V functions to carry cargo in cells. (B, adapted from N. Kodera and T. Ando, *Biophys. Rev.* 6:237–260, 2014. Reproduced with permission of SNCSC.)

myosin tails (and the tails of motor proteins generally) have apparently diversified during evolution to permit the proteins to bind other subunits and to interact with different cargoes.

Some myosins are found only in plants, and some are found only in vertebrates. Most, however, are found in all eukaryotes, suggesting that myosins arose early in eukaryotic evolution. The human genome includes about 40 myosin genes. Nine of the human myosins are expressed primarily or exclusively in the hair cells of the inner ear, and mutations in five of them are known to cause hereditary deafness. These extremely specialized myosins are important for the construction and function of the complex and beautiful bundles of actin found in stereocilia that project from the apical surface of these cells; these cellular protrusions tilt in response to sound and convert sound waves into electrical signals.

The functions of most of the myosins remain to be determined, but several are well characterized. The myosin I proteins often contain either a second actin-binding site or a membrane-binding site in their tails, and they are generally involved in intracellular organization—including the protrusion of actin-rich structures at the cell surface, such as microvilli (see Panel 16–1 and Figure 16–4), and endocytosis. Myosin V is a two-headed myosin with a large step size (**Figure 16–36**) and is involved in organelle transport along actin filaments. In contrast to myosin II motors, which work in ensembles and are attached only transiently to actin filaments so as not to interfere with one another, myosin V moves continuously, or *processively*, along actin filaments without letting go. Myosin V functions are well studied in the yeast *Saccharomyces cerevisiae*, which undergoes a stereotypical pattern of growth and division called budding. Actin cables in the mother cell point toward the bud, where actin is found in patches that concentrate where cell-wall growth is taking place. Myosin V motors carry a wide range of cargoes—including mRNA, endoplasmic reticulum, and secretory vesicles—along the actin cables and into the bud (see Figure 16–76).

Summary

Using their neck domain as a lever arm, myosins convert ATP hydrolysis into mechanical work to move along actin filaments in a stepwise fashion. Skeletal muscle is made up of myofibrils containing thousands of sarcomeres assembled from highly ordered arrays of actin and myosin II filaments, together with many accessory proteins. Muscle contraction is stimulated by calcium, which causes the actin filament–associated protein tropomyosin to move, uncovering myosin binding sites and allowing the filaments to slide past one another. Smooth muscle and non-muscle cells have less well-ordered contractile bundles of actin and myosin, which are regulated by myosin light-chain phosphorylation. Myosin V transports cargo by walking along actin filaments.

MICROTUBULES

Microtubules are structurally more complex than actin filaments, but they are also highly dynamic and play similarly diverse and important roles in the cell. Microtubules are polymers of the protein **tubulin**. The tubulin subunit is a heterodimer formed from two closely related globular proteins called *α-tubulin* and *β-tubulin*, each comprising 445–450 amino acids, which are tightly held together by noncovalent bonds (**Figure 16–37A**). These two tubulin proteins are found only in this heterodimer, and each α or β monomer has a binding site for one molecule of GTP. The GTP that is bound to α-tubulin is physically trapped at the dimer interface and is never hydrolyzed or exchanged; it can therefore be considered to be an integral part of the tubulin heterodimer structure. In contrast, β-tubulin may be bound to either GTP or GDP and—as we shall see—this difference is important for microtubule dynamics.

Tubulin is found in all eukaryotic cells, and it exists in multiple isoforms. Its amino acid sequence has been highly conserved during evolution: thus, yeast and human tubulins are 75% identical. In mammals, there are at least six forms of α-tubulin and a similar number of β-tubulins, each encoded by a different gene. Although the different forms of tubulin are very similar and can copolymerize into mixed microtubules in a test tube, they can have distinct locations in cells and tissues and perform subtly different functions.

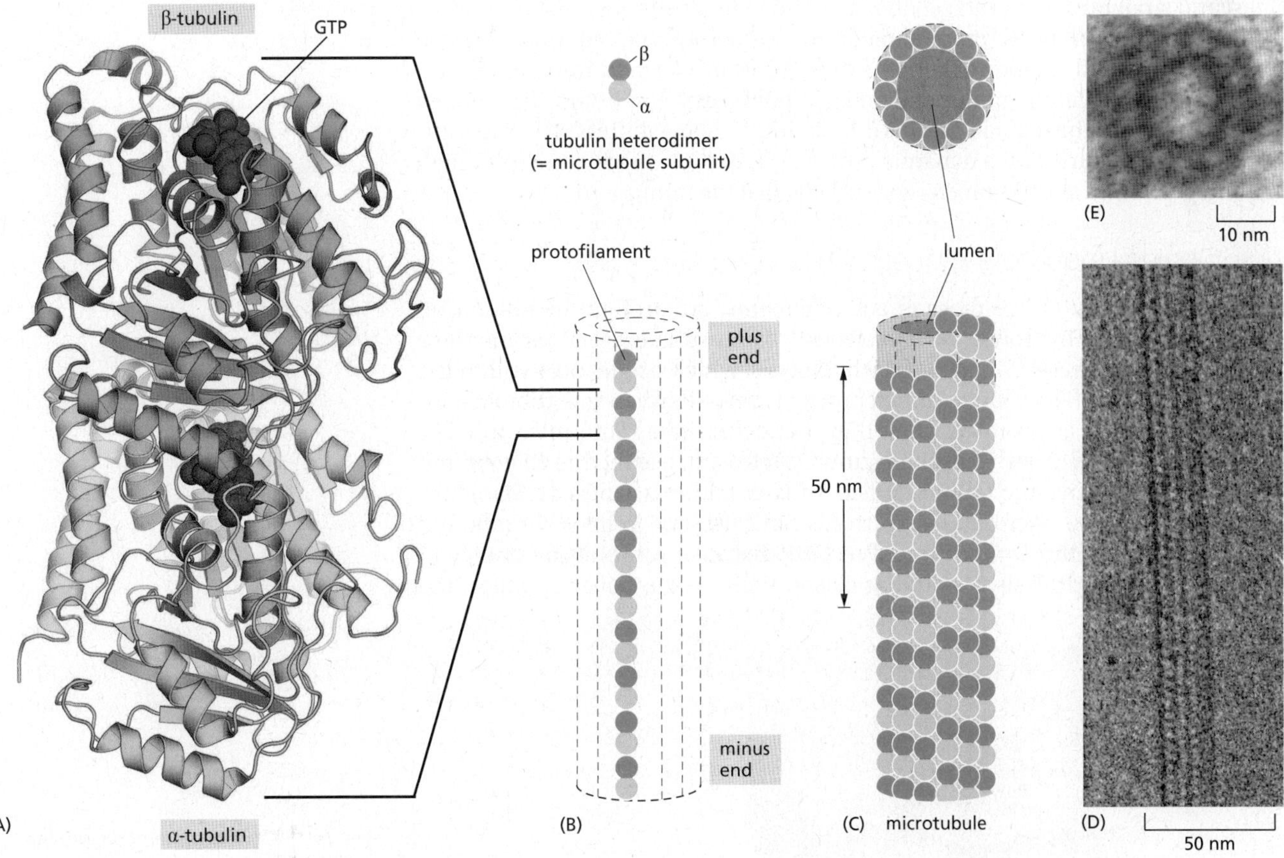

Figure 16–37 The structure of a microtubule and its subunit. (A) The subunit of each protofilament is a tubulin heterodimer, formed from a tightly linked pair of α- and β-tubulin monomers. The GTP molecule in the α-tubulin monomer is so tightly bound that it can be considered an integral part of the protein. The GTP molecule in the β-tubulin monomer, however, is less tightly bound and has an important role in filament dynamics. GTP is shown in *red*. (B) One tubulin subunit (αβ-heterodimer) and one protofilament are shown schematically. Each protofilament consists of many adjacent subunits with the same orientation. (C) The microtubule is a stiff hollow tube formed from 13 protofilaments aligned in parallel. (D) A short segment of a microtubule viewed in an electron microscope. (E) Electron micrograph of a cross section of a microtubule showing a ring of 13 protofilaments. (A, PDB code: 1JFF; D, courtesy of Richard Wade; E, courtesy of Richard Linck.)

As a striking example, mutations in a particular human β-tubulin gene give rise to a paralytic eye-movement disorder because of loss of ocular nerve function. Some human neurological diseases have also been linked to mutations in a specific tubulin gene.

Microtubules Are Hollow Tubes Made of Protofilaments

A microtubule is a hollow cylindrical structure built from 13 parallel protofilaments, each composed of αβ-tubulin heterodimers stacked head to tail and then folded into a tube (**Figure 16–37B, C, and D**). The assembly of a microtubule generates two new types of protein–protein contacts. Along the longitudinal axis of a protofilament, the "top" of one β-tubulin molecule forms an interface with the "bottom" of the α-tubulin molecule in the adjacent heterodimer. This interface, which is very similar to the interface holding the α and β monomers together in the dimer subunit, has a high binding energy. Perpendicular to these interactions, neighboring protofilaments form lateral contacts, with the main lateral contacts occurring between monomers of the same type (α–α and β–β). A slight stagger in lateral contacts gives rise to the helical microtubule lattice, which holds most of the subunits tightly in place. As a result, the addition and loss of subunits occur almost exclusively at the microtubule ends (see Figure 16–5).

The multiple contacts among subunits make microtubules stiff and difficult to bend. The average length over which microtubules stay straight (persistence length) is more than 10 times that of actin filaments, making microtubules the stiffest and straightest structural elements found in most animal cells.

The subunits in each protofilament in a microtubule all point in the same direction, and the protofilaments themselves are aligned in parallel, with α-tubulins exposed at the minus end and β-tubulins exposed at the plus end. As for actin filaments, the regular, parallel orientation of their subunits gives microtubules both a structural and a dynamic polarity (**Figure 16–38**): the microtubule's plus end grows and shrinks much more rapidly than its minus end.

Microtubules Undergo a Process Called Dynamic Instability

Microtubule dynamics, like those of actin filaments, are profoundly influenced by the binding and hydrolysis of a nucleoside triphosphate—GTP for microtubules, as opposed to ATP for actin. GTP hydrolysis, which occurs only within the β-tubulin subunit of the tubulin dimer, proceeds very slowly in free tubulin subunits and is greatly accelerated when they are incorporated into microtubules. After GTP hydrolysis, a free phosphate group is released, leaving the GDP bound to β-tubulin within the microtubule lattice. Thus, as in the case of actin filaments, two different types of microtubule structures can exist, one in the *T form* bound to GTP and one in the *D form* bound to GDP. Because some of the energy of phosphate bond hydrolysis is stored as elastic strain in the polymer lattice, the

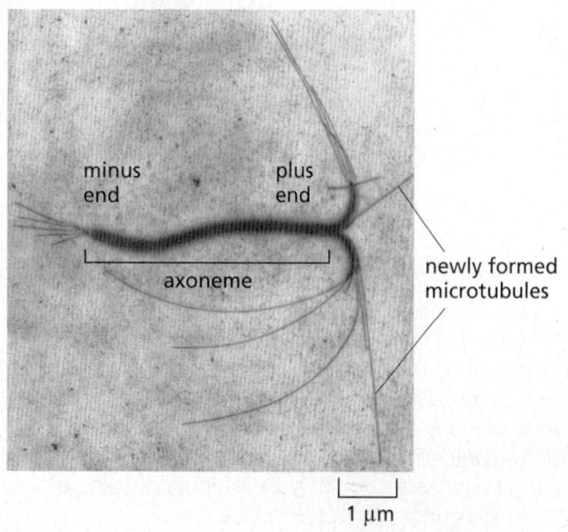

Figure 16–38 The preferential growth of microtubules at the plus end. Microtubules grow faster at one end than at the other. A stable bundle of microtubules obtained from the core of a cilium (called an axoneme) was incubated for a short time with tubulin subunits under polymerizing conditions. Microtubules grew fastest from the plus end of the microtubule bundle, the end on the right in this micrograph. (Courtesy of Gary Borisy.)

free-energy change for dissociation of a subunit from the D-form polymer is more favorable (negative) than the free-energy change for dissociation of a subunit from the T-form polymer. This makes the ratio of k_{off}/k_{on} for GDP-tubulin [which is equal to its critical concentration, C_c(D)] much higher than that of GTP-tubulin. As a result, under physiological conditions, the T form tends to polymerize and the D form tends to depolymerize.

Whether the tubulin subunits at the very end of a microtubule are in the T or the D form depends on the relative rates of tubulin addition and GTP hydrolysis. If the rate of subunit addition is low, hydrolysis is likely to occur before the next subunit is added, and the tip of the filament will then be in the D form. If the rate of subunit addition is high—and thus the filament is growing rapidly—then it is likely that a new subunit will be added to the end of the polymer before the GTP in the previously added subunit has been hydrolyzed. In this case, the tip of the polymer remains in the T form, forming a *GTP cap*. But this T form may not persist. Often GTP-tubulin subunits will assemble at the end of the microtubule at a rate similar to the rate of GTP hydrolysis, and in this case hydrolysis will sometimes "catch up" with the rate of subunit addition and transform the end to a D form. This transformation will be sudden and random, with a certain probability per unit time that depends on the concentration of free GTP-tubulin subunits, and it produces the microtubule's *dynamic instability*.

Suppose that the concentration of free tubulin is intermediate between the critical concentration for a T-form end and the critical concentration for a D-form end (that is, above the concentration necessary for T-form assembly, but below that for the D form). Now, any end that happens to be in the T form will grow, whereas any end that happens to be in the D form will shrink. On a single microtubule, an end might grow for a certain length of time in a T form, but then suddenly change to the D form and begin to shrink rapidly. At some later time, it might regain a T-form end and begin to grow again. This rapid interconversion between a growing and shrinking state, at a uniform free tubulin concentration, is called **dynamic instability** (**Figure 16–39** and **Figure 16–40A**; see Panel 16–2). The change from growth to shrinkage is called a *catastrophe*, while the change from shrinkage to growth is called a *rescue*.

The structural basis for dynamic instability is uncertain. On the basis of observations of the ends of growing and shrinking microtubules *in vitro*, one model proposed that tubulin subunits in the T form, with GTP bound to the β monomer, produce straight protofilaments that make strong and regular lateral contacts with one another, and that the hydrolysis of GTP to GDP makes these protofilaments curve (**Figure 16–40B**). More recent studies indicate that free tubulin subunits possess a similar bent conformation in both the T form and the D form. Growing microtubules with curved protofilaments at their tips have now been observed both *in vitro* and *in vivo*, and a straightening of the T form–containing protofilaments may occur subsequent to subunit incorporation as favorable lateral interactions zip them into the microtubule lattice. Regardless of the mechanism of microtubule assembly, the loss of a GTP cap and subsequent catastrophe causes protofilaments containing D-form subunits to spring apart and depolymerize (**Figure 16–40C**).

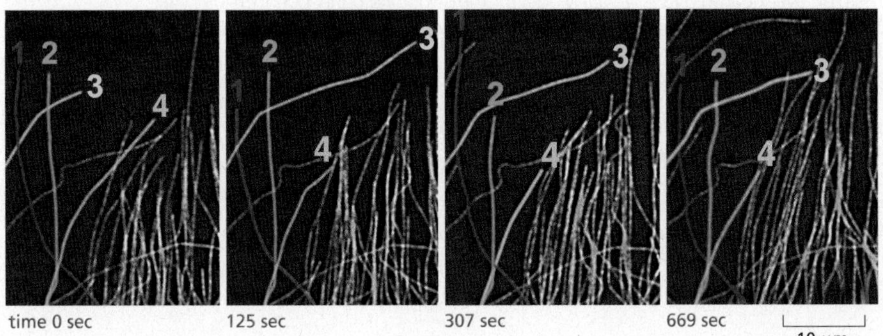

Figure 16–39 **Direct observation of the dynamic instability of microtubules in a living cell.** Microtubules in a newt lung epithelial cell were observed after the cell was injected with a small amount of fluorescently labeled tubulin. Notice the dynamic instability of microtubules at the edge of the cell. Four individual microtubules are highlighted for clarity; each of these shows alternating shrinkage and growth (**Movie 16.9**). (Courtesy of Wendy C. Salmon and Clare Waterman.)

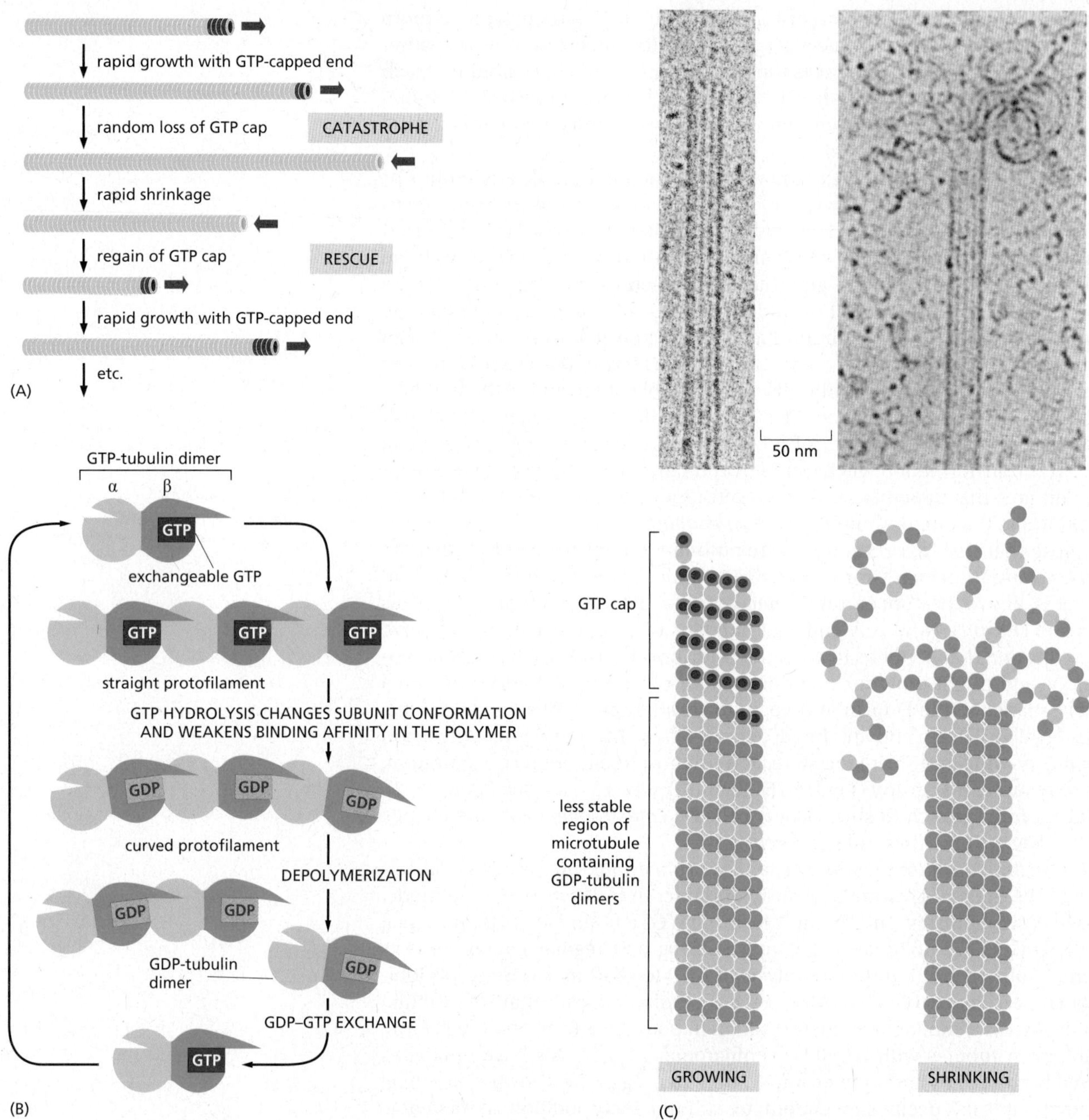

Figure 16–40 Dynamic instability due to the structural differences between a growing and a shrinking microtubule end. (A) If the free tubulin concentration in solution is near the critical concentration, a single microtubule end may undergo transitions between a growing state and a shrinking state. A growing microtubule has GTP-containing subunits at its end, forming a GTP cap. If GTP hydrolysis proceeds more rapidly than subunit addition, the cap is lost and the microtubule begins to shrink, an event called a *catastrophe*. But GTP-containing subunits may still add to the shrinking end, and if enough add to form a new cap, then microtubule growth resumes, an event called a *rescue*. (B) Model for the structural consequences of GTP hydrolysis in the microtubule lattice. The addition of GTP-containing tubulin subunits to the end of a protofilament causes the end to grow in a linear conformation that can readily pack into the cylindrical wall of the microtubule. Hydrolysis of GTP after assembly changes the conformation of the subunits and tends to force the protofilament into a curved shape that is less able to pack into the microtubule wall. (C) In an intact microtubule with a stable cap of GTP-tubulin, protofilaments made from GDP-containing subunits are forced into a linear conformation by the many lateral bonds within the microtubule wall. Loss of the GTP cap, however, allows the GDP-containing protofilaments to relax into their more curved conformation. This leads to a progressive disruption of the microtubule. Above the drawings of a growing and a shrinking microtubule, electron micrographs show actual microtubules in each of these two states. Note particularly the curling, disintegrating GDP-containing protofilaments at the end of the shrinking microtubule. (C, © 1991 E.M. Mandelkow, E. Mandelkow, and R.A. Milligan. Originally published in *J. Cell Biol.* https://doi.org/10.1083/jcb.114.5.977. With permission from Rockefeller University Press.)

Why does nucleoside triphosphate hydrolysis lead to treadmilling of actin filaments, while microtubules undergo dynamic instability? One explanation is that the tubulin subunits with GDP bound have a very low affinity for one another, with a very high k_{off}/k_{on}. Therefore, as soon as GTP hydrolysis occurs at the tip of a microtubule growing *in vitro*, it undergoes a catastrophe and depolymerizes completely. In contrast, ADP-actin has a lower k_{off}/k_{on} ratio and it depolymerizes more slowly. A second reason is thought to be differences in critical concentrations (C_c) for the two ends of the filaments. The minus end of an actin filament has a much higher C_c than the plus end, whereas measurements of microtubules indicate that the plus and minus ends possess a similar C_c. Therefore, although the basic principles of actin and microtubule dynamics are very similar, with nucleoside triphosphate binding and hydrolysis leading to dynamic behaviors, quantitative differences in affinities lead to dramatic differences in their intrinsic behavior.

Microtubule Functions Are Inhibited by Both Polymer-stabilizing and Polymer-destabilizing Drugs

Chemical compounds that impair polymerization or depolymerization of microtubules are powerful tools for investigating the roles of these polymers in cells. Whereas *colchicine* and *nocodazole* interact with tubulin subunits and lead to microtubule depolymerization, *Taxol* binds to and stabilizes microtubules, causing a net increase in tubulin polymerization (see Table 16–1). Drugs like these have a rapid and profound effect on the organization of the microtubules in living cells. Both microtubule-depolymerizing drugs (such as nocodazole) and microtubule-polymerizing drugs (such as Taxol) preferentially kill dividing cells, because microtubule dynamics are crucial for correct function of the mitotic spindle (discussed in Chapter 17). Some of these drugs kill certain types of tumor cells in a human patient, although not without toxicity to rapidly dividing normal cells, including those in the bone marrow, intestine, and hair follicles. Taxol in particular has been widely used to treat cancers of the breast and lung, and it is frequently successful in treatment of tumors that are resistant to other chemotherapeutic agents.

A Protein Complex Containing γ-Tubulin Nucleates Microtubules

Because formation of a microtubule requires the interaction of many tubulin heterodimers, the concentration of tubulin subunits required for spontaneous nucleation of microtubules is very high. Microtubule nucleation therefore requires help from other factors. While α- and β-tubulins are the regular building blocks of microtubules, another type of tubulin, called *γ-tubulin*, is present in much smaller amounts than α- and β-tubulin and is involved in the nucleation of microtubule growth in organisms ranging from yeasts to humans. Microtubules are generally nucleated from a specific intracellular location known as a **microtubule-organizing center (MTOC)** where γ-tubulin is most enriched. Nucleation in many cases depends on the **γ-tubulin ring complex (γ-TuRC)**. Within this complex, two accessory proteins bind directly to the γ-tubulin, along with several other proteins that help create a spiral ring of γ-tubulin molecules, which serves as a template that creates a microtubule with 13 protofilaments (**Figure 16–41**).

The Centrosome Is a Prominent Microtubule Nucleation Site

Many animal cells possess a single, well-defined MTOC called the centrosome, which is located adjacent to the nucleus and from which microtubules are nucleated at their minus ends, so the plus ends point outward and continually grow and shrink, probing the entire three-dimensional volume of the cell. A centrosome typically recruits more than 50 copies of γ-TuRC. However, most animal cells contain many hundreds of microtubules, most of which could not be stably anchored at the centrosome simply because they would not fit. Thus,

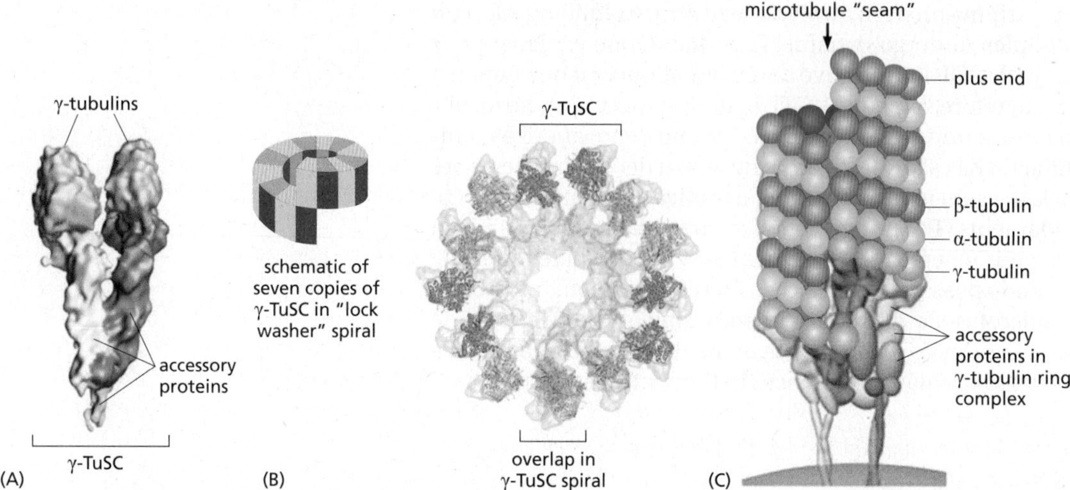

Figure 16–41 Microtubule nucleation by the γ-tubulin ring complex. (A) Two copies of γ-tubulin associate with a pair of accessory proteins to form the γ-tubulin small complex (γ-TuSC). This image was generated by high-resolution electron microscopy of individual purified complexes. (B) Seven copies of the γ-TuSC associate to form a spiral structure in which the last γ-tubulin lies beneath the first, resulting in 13 exposed γ-tubulin subunits in a circular orientation that matches the orientation of the 13 protofilaments in a microtubule. (C) In many cell types, the γ-TuSC spiral associates with additional accessory proteins to form the γ-tubulin ring complex (γ-TuRC), which is likely to nucleate the minus end of a microtubule as shown here. Note the longitudinal discontinuity between two protofilaments, which results from the spiral orientation of the γ-tubulin subunits. Microtubules often have one such "seam" breaking the otherwise uniform helical packing of the protofilaments. (A and B, from J.M. Kollman et al., *Nature* 466:879–883, published 2010 by Macmillan Publishers Ltd. Reproduced with permission of SNCSC.)

the majority of γ-TuRC is found in the cytoplasm, and centrosomes are not absolutely required for microtubule nucleation, as destroying them with a laser pulse does not prevent microtubule nucleation elsewhere in the cell. A variety of proteins have been identified that anchor γ-TuRC to the centrosome, but mechanisms that activate microtubule nucleation at MTOCs and at other sites in the cell are poorly understood.

Embedded in the centrosome are the **centrioles**, a pair of cylindrical structures arranged at right angles to each other in an L-shaped configuration (**Figure 16–42**).

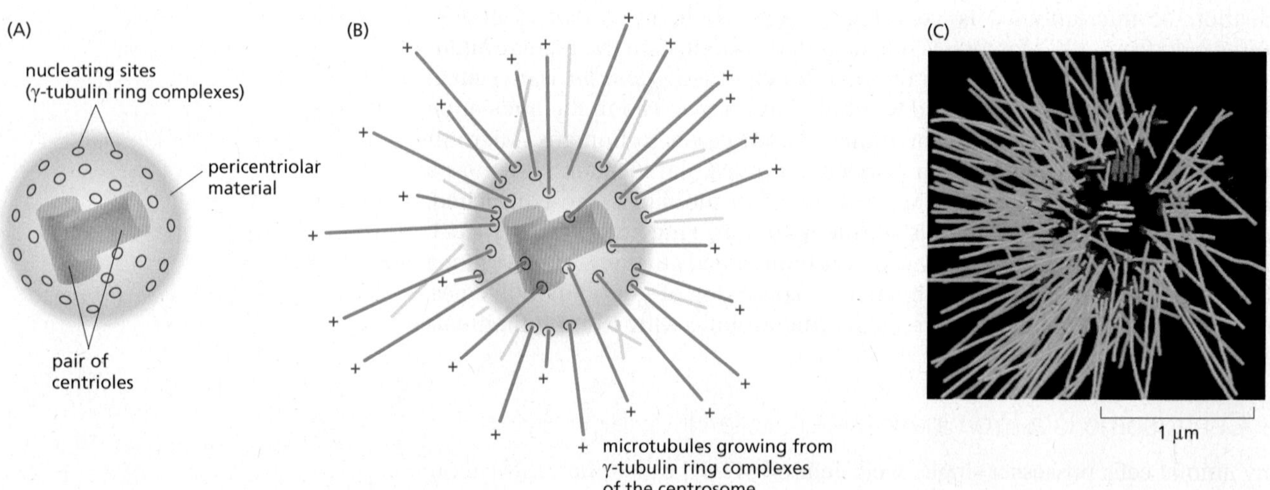

Figure 16–42 The centrosome. (A) The centrosome is a major MTOC in animal cells. Located in the cytoplasm next to the nucleus, it consists of a pair of centrioles surrounded by an amorphous matrix of fibrous proteins, the pericentriolar material, in which the γ-tubulin ring complexes that nucleate microtubule growth are embedded. (B) A centrosome with attached microtubules. The minus end of each microtubule is embedded in the centrosome, having grown from a γ-tubulin ring complex, whereas the plus end of each microtubule is free in the cytoplasm. (C) In a reconstructed image of the MTOC from a *C. elegans* cell, a dense thicket of microtubules can be seen emanating from the centrosome. (C, from E.T. O'Toole et al., *J. Cell Biol.* 163:451–456, 2003. With permission from the authors.)

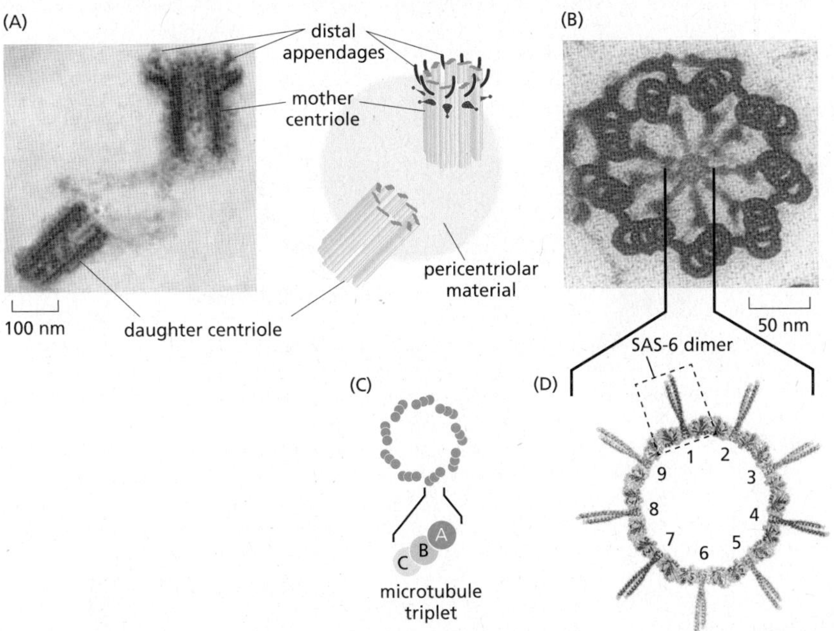

Figure 16–43 A pair of centrioles in the centrosome. (A) An electron micrograph of a thin section of an isolated centrosome from an interphase cell showing the mother centriole with its distal appendages and the adjacent daughter centriole, which formed through a duplication event during S phase (see Figure 17–29). In the centrosome, the centriole pair is surrounded by a dense matrix of pericentriolar material from which microtubules nucleate. Centrioles also function as basal bodies to nucleate the formation of ciliary axonemes (see Figure 16–58). (B) Electron micrograph of a cross section through a centriole in the cortex of a protozoan. Each centriole is composed of nine sets of microtubule triplets arranged to form a cylinder. (C) Each triplet contains one complete microtubule (the A microtubule) fused to two incomplete microtubules (the B and C microtubules). (D) The centriolar protein SAS-6 forms a coiled-coil dimer. Nine SAS-6 dimers can self-associate to form a ring. Located at the hub of the centriole cartwheel-like structure, the SAS-6 ring is thought to generate the ninefold symmetry of the centriole. (A, from M. Paintrand et al., *J. Struct. Biol.* 108:107, 1992. With permission from Elsevier; B, courtesy of Richard Linck; D, courtesy of Michel Steinmetz.)

A centriole consists of a cylindrical array of short, modified microtubules arranged into a barrel shape with striking ninefold symmetry (**Figure 16–43**). Together with a large number of accessory proteins, the centrioles recruit the *pericentriolar material,* where microtubule nucleation takes place. The pericentriolar material consists of a dense spherical matrix that is thought to form through a process of biomolecular condensation (see Figure 12–6). As described in Chapter 17 (see Figure 17–29), the centrosome duplicates before mitosis, forming a pair of centrosomes that each contain a centriole pair. When mitosis begins, the two centrosomes move apart to form the poles of the mitotic spindle (see Panel 17–1).

Microtubule Organization Varies Widely Among Cell Types

The arrangement of microtubules in the cytoplasm varies in different cell types (**Figure 16–44**). In budding yeast, microtubules are nucleated at an MTOC that is embedded in the nuclear envelope as a small, multilayered structure called the *spindle pole body,* also found in other fungi and diatoms. Higher-plant cells lack centrosomes and nucleate microtubules at sites distributed all around the nuclear envelope and at the cell cortex. Neither fungi nor most plant cells contain centrioles. Despite these differences, all these cells seem to use γ-TuRC to nucleate their microtubules.

Cultured fibroblasts contain an aster-like configuration of microtubules, with dynamic plus ends pointing outward toward the cell periphery and stable

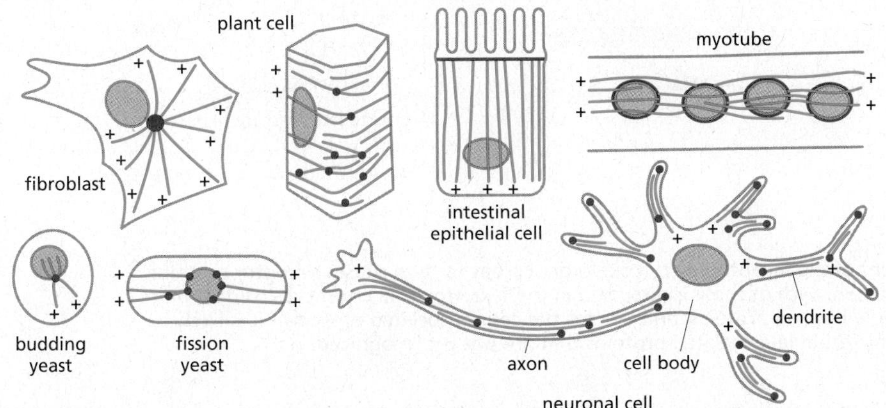

Figure 16–44 Microtubule organization in different cell types. Microtubules *(green)* are organized by MTOCs *(red)*, which nucleate, anchor, or stabilize the microtubule minus ends. A single focal MTOC in a fibroblast or yeast cell is the primary nucleation site and leads to microtubules organized with their plus ends extending out toward the cell periphery. A more complex distribution of microtubule plus and minus ends is observed in other cell types.

MICROTUBULES

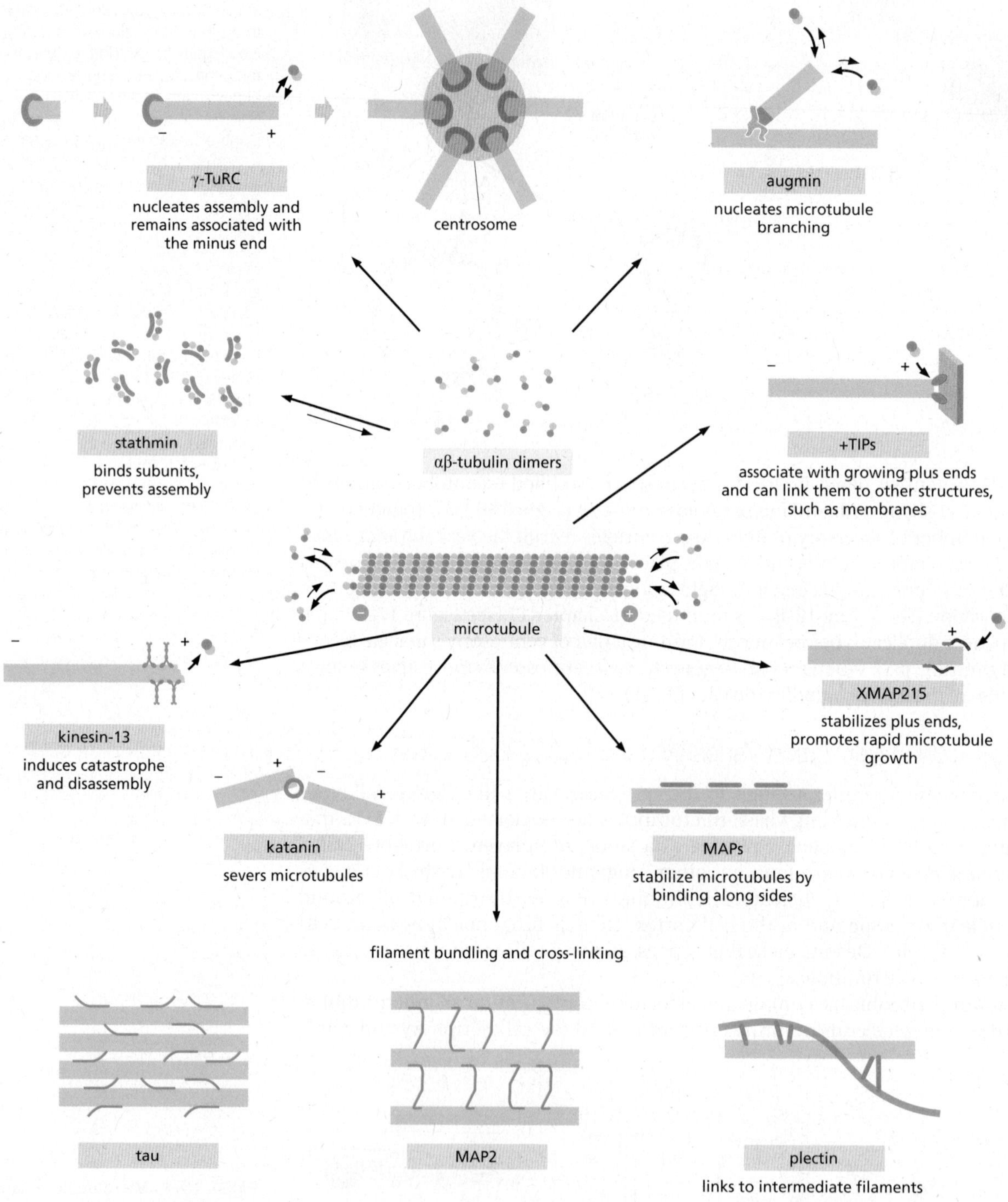

Some of the major accessory proteins of the microtubule cytoskeleton. Except for two classes of motor proteins, an example of each major type is shown. Each of these is discussed in the text. However, most cells contain more than a hundred different microtubule-binding proteins, and—as for the actin-associated proteins—it is likely that there are important types of microtubule-associated proteins that are not yet recognized.

minus ends gathered near the nucleus. Most microtubule minus ends detach from the centrosome and can be organized by other cellular structures, such as the Golgi apparatus, which functions as an MTOC in mesenchymal cells. The organization of microtubules in the cell establishes a general coordinate system, which is then used to position many organelles within the cell. Highly differentiated cells with complex morphologies, such as neurons, muscles, and epithelial cells, use additional mechanisms to establish their more elaborate internal organization. Thus, for example, when an epithelial cell forms cell–cell junctions and becomes highly polarized, the microtubule minus ends relocalize to a region near the apical plasma membrane. From this asymmetrical location, a microtubule array extends along the long axis of the cell, with plus ends directed toward the basal surface (see Figure 16–4). In many cells, the minus ends of microtubules are stabilized by association with γ-TuRC or other capping proteins, or else they serve as microtubule-depolymerization sites. One class of capping proteins, called CAMSAPs in vertebrates, acts to protect minus ends from depolymerization and anchor them to cellular structures, such as the Golgi apparatus or the cell cortex.

Neurons contain complex cytoskeletal structures. As they differentiate, neurons send out specialized processes that will either receive electrical signals (*dendrites*) or transmit electrical signals (*axons*) (see Figure 16–45). Axons and dendrites (collectively called *neurites*) are filled with bundles of microtubules that are critical to their structure and their function. In axons, all the microtubules are oriented with their minus ends pointing back toward the cell body and their plus ends pointing toward the axon terminals (see Figure 16–44). Microtubules do not reach from the cell body all the way to the axon terminals; each is typically only a few micrometers in length, but large numbers are staggered in an overlapping array. These aligned microtubule tracks act as a highway to transport specific proteins, protein-containing vesicles, and mRNAs to the axon terminals, where synapses are constructed and maintained. The longest axon in the human body reaches from the base of the spinal cord to the foot and is up to a meter in length. Dendrites are generally much shorter. The microtubules in dendrites lie parallel to one another but their polarities are mixed, with some pointing their plus ends toward the dendrite tip, and others pointing in the opposite direction.

Microtubule-binding Proteins Modulate Filament Dynamics and Organization

Microtubule polymerization dynamics are very different in cells than in solutions of pure tubulin. Microtubules in cells exhibit a much higher polymerization rate (typically 10–15 μm/min, relative to about 1.5 μm/min with purified tubulin at similar concentrations), a greater catastrophe frequency, and extended pauses in microtubule growth, a dynamic behavior rarely observed in pure tubulin solutions. These and other differences arise because microtubule dynamics inside the cell are governed by a variety of proteins that bind tubulin dimers or microtubules, as summarized in Panel 16–4.

Proteins that bind to microtubules are collectively called **microtubule-associated proteins**, or **MAPs**. Because the short (~20 amino acid) C-terminal tails of both α- and β-tubulin that protrude from the microtubule are enriched in glutamic and aspartic acids, the surface of the microtubule possesses a net negative charge. Many MAPs are positively charged and bind to microtubules through electrostatic interactions. Some MAPs can stabilize microtubules against disassembly. A subset of MAPs can also mediate the interaction of microtubules with other cell components. This subset is prominent in neurons, where stabilized microtubule bundles form the core of the axons and dendrites that extend from the cell body (**Figure 16–45**). These MAPs have at least one domain that binds to the microtubule surface and another that projects outward. The length of the projecting domain can determine how closely MAP-coated microtubules pack together, as demonstrated in cells engineered to overproduce different MAPs.

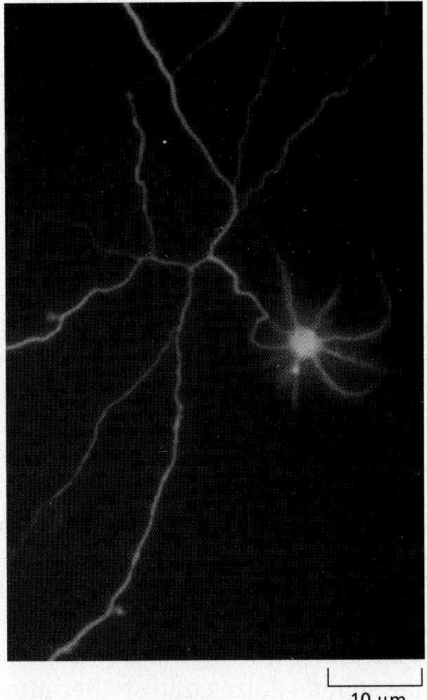

Figure 16–45 **Localization of MAPs in the axon and dendrites of a neuron.** This immunofluorescence micrograph shows the distribution of the proteins tau *(green)* and MAP2 *(orange)* in a hippocampal neuron in culture. Whereas tau staining is confined to the axon (long and branched in this neuron), MAP2 staining is confined to the cell body and its dendrites. The antibody used here to detect tau binds only to unphosphorylated tau; phosphorylated tau is also present in dendrites. (Courtesy of James W. Mandell and Gary A. Banker.)

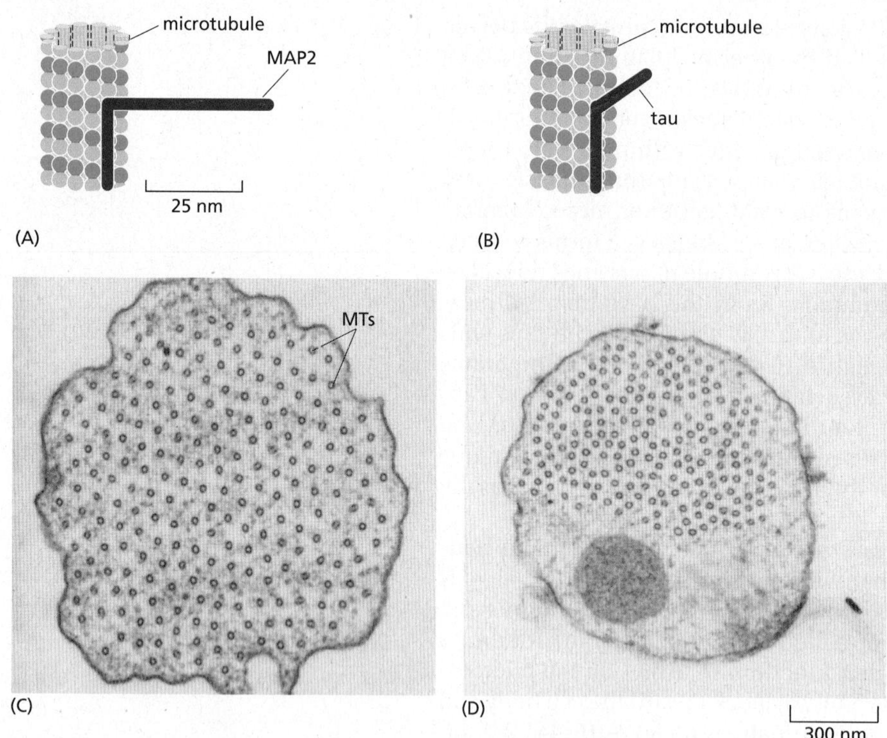

Figure 16–46 Organization of microtubule bundles by MAPs. (A) One end of MAP2 binds along the microtubule lattice and extends a long projecting arm. (B) Tau possesses a shorter projection arm. (C) Electron micrograph showing a cross section through a microtubule bundle in a cell overexpressing MAP2. The regular spacing of the microtubules (MTs) in this bundle results from the constant length of the projecting arms of MAP2. (D) Similar cross section through a microtubule bundle in a cell overexpressing tau. Here the microtubules are spaced more closely together than they are in C because of tau's relatively short projecting arm. (C and D, from J. Chen et al., *Nature* 360:674–677, published 1992 by Nature Publishing Group. Reproduced with permission of SNCSC.)

Cells overexpressing MAP2, which has a long projecting domain, form bundles of stable microtubules that are kept widely spaced, while cells overexpressing tau, a MAP with a much shorter projecting domain, form bundles of more closely packed microtubules (**Figure 16–46**). MAPs are the targets of several protein kinases, and phosphorylation of a MAP can control both its activity and localization inside cells by disrupting its electrostatic interaction with microtubules.

MAPs can also recruit other proteins that organize the microtubule cytoskeleton. An important example is augmin, an 8-subunit protein complex that binds to sites along the microtubule and recruits γ-TuRC, which nucleates a new microtubule to form a microtubule branch (**Figure 16–47A**). Thus, similar to the activity of the Arp2/3 complex on actin filaments, augmin causes branching nucleation on preexisting microtubules (**Movie 16.10**). Augmin-induced branches help build the spindle during mitosis. Because they lack centrosomes, plant cells rely extensively on augmin-dependent microtubule branching nucleation to organize the microtubule cytoskeleton (**Figure 16–47B and C**).

Microtubule Plus End–binding Proteins Modulate Microtubule Dynamics and Attachments

While the minus ends of microtubules in many cells are usually stabilized and inert, plus ends, in contrast, efficiently explore and probe the entire volume of the cell. This process is facilitated by numerous proteins that bind to microtubule plus ends and thereby influence microtubule dynamics. These proteins can influence the rate at which a microtubule switches from a growing to a shrinking state (the frequency of catastrophes) or from a shrinking to a growing state

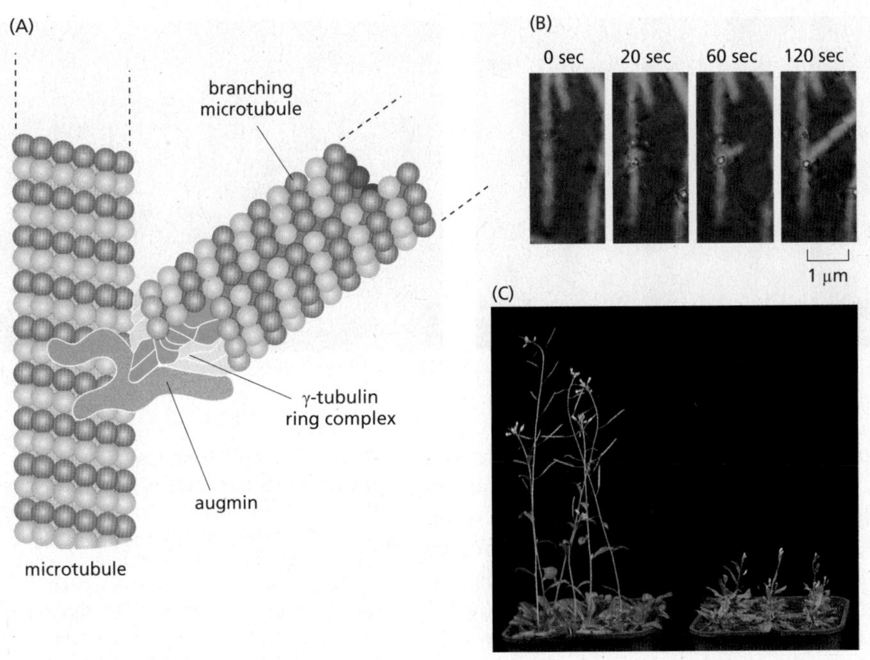

Figure 16–47 **Microtubule branching by augmin.** (A) Augmin binds along the side of an existing microtubule and recruits a γ-tubulin ring complex that nucleates a new microtubule with a low branching angle. (B) Fluorescence micrographs showing augmin *(orange)* nucleating a microtubule branch in the cortex of an epidermal cell in the plant *Arabidopsis thaliana*. (C) Depletion of augmin severely stunts plant growth. (B, from W. Liu et al., *J. Integr. Plant Biol.* 61:388–393, 2019; C, from T. Liu et al., *Curr. Biol.* 24:2708–2713, 2014. With permission from Elsevier.)

(the frequency of rescues). For example, members of a family of kinesin proteins known as *catastrophe factors* (or kinesin-13s) bind to microtubule ends and appear to pry protofilaments apart, lowering the normal activation-energy barrier that prevents a microtubule from springing apart into the curved protofilaments that are characteristic of the shrinking state (**Figure 16–48**). Other plus end–associated proteins act to promote rapid microtubule growth. A particularly ubiquitous example is *XMAP215*, which has close homologs in organisms that range from yeast to humans. Like formin proteins that concentrate actin subunits at the plus end of a growing actin filament, XMAP215 binds free tubulin subunits and delivers them to the plus end of a microtubule, dramatically accelerating polymerization (see Figure 16–48).

A large subset of MAPs is enriched at microtubule plus ends. Called *plus-end tracking proteins* (*+TIPs*), these MAPs bind an actively growing plus end and dissociate when the microtubule begins to shrink (**Figure 16–49**). The kinesin-13 catastrophe factors and XMAP215 mentioned above behave as +TIPs and act to modulate the growth and shrinkage of the microtubule end to which they are

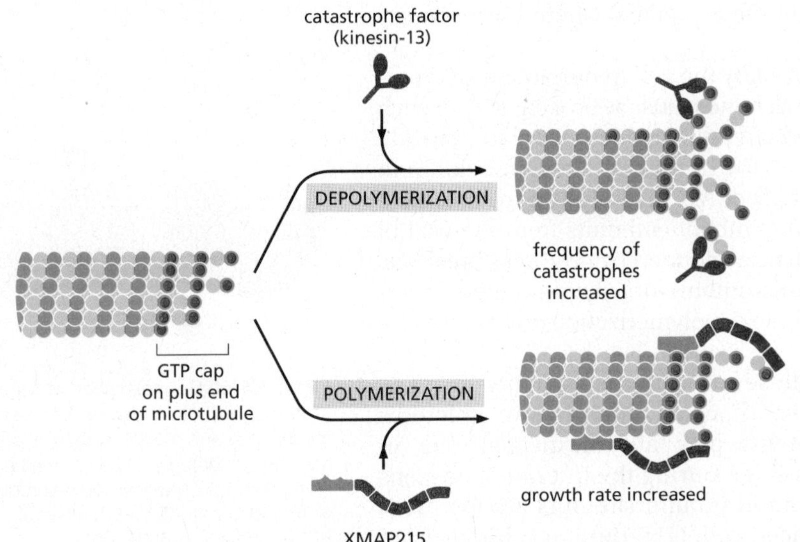

Figure 16–48 **The effects of proteins that bind to microtubule ends.** The transition between microtubule growth and shrinkage is controlled in cells by a variety of proteins. Catastrophe factors such as kinesin-13, a member of the kinesin motor protein superfamily, bind to microtubule ends and pry them apart, thereby promoting depolymerization. On the other hand, a MAP such as XMAP215 promotes rapid microtubule polymerization (XMAP stands for *Xenopus* microtubule-associated protein, and the number refers to its molecular mass in kilodaltons). XMAP215 binds tubulin dimers and delivers them to the microtubule plus end, thereby increasing the microtubule growth rate.

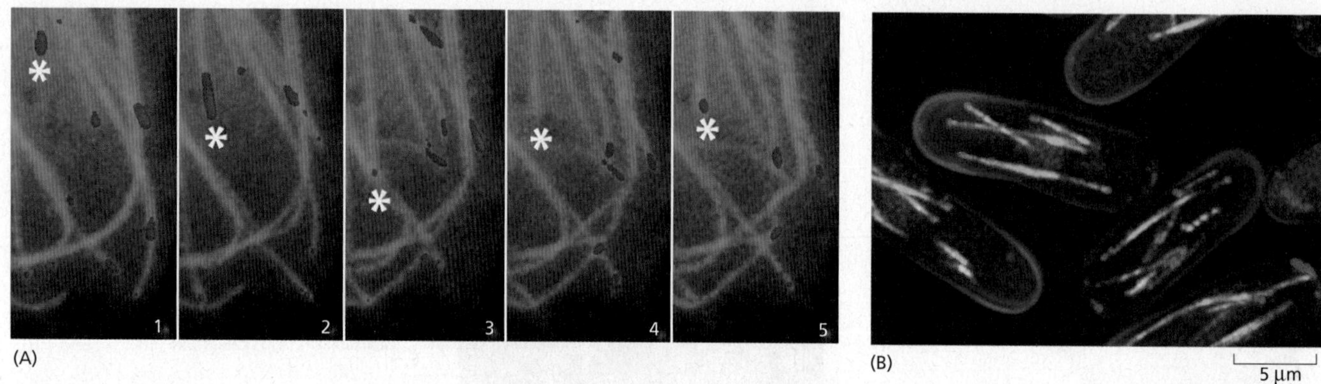

Figure 16–49 **+TIP proteins found at the growing plus ends of microtubules.** (A) Frames from a fluorescence time-lapse movie of the edge of a cell expressing fluorescently labeled tubulin that incorporates into microtubules *(green)* as well as the +TIP protein EB1 tagged with a different color *(red)*. The same microtubule is marked *(asterisk)* in successive movie frames. When the microtubule is growing (frames 1, 2), EB1 is associated with the tip. When the microtubule undergoes a catastrophe and begins shrinking, EB1 is lost (frames 3, 4). The labeled EB1 is regained when growth of the microtubule is rescued (frame 5). See **Movie 16.11**. (B) In the fission yeast *Schizosaccharomyces pombe*, microtubules *(green)* are bound at their plus ends by the homolog of EB1 *(red)* as they grow toward the two poles of the rod-shaped cells. (A, courtesy of Anna Akhmanova and Ilya Grigoriev; B, courtesy of Takeshi Toda.)

attached. Other +TIPs control microtubule positioning by helping to capture and stabilize the growing microtubule end at specific cellular targets, such as the cell cortex or the kinetochore of a mitotic chromosome. EB1 and its relatives, small dimeric proteins that are highly conserved in animals, plants, and fungi, are key players in this process. EB1 proteins do not actively move toward plus ends, but rather recognize a structural feature of the growing plus end (see Figure 16–49). Several +TIPs depend on EB1 proteins for their plus-end accumulation and also interact with each other and with the microtubule lattice. By attaching to the plus end, these factors control microtubule dynamics and also allow the cell to harness the energy of microtubule polymerization to generate pushing forces that can be used for positioning the spindle, chromosomes, or organelles.

Tubulin-sequestering and Microtubule-severing Proteins Modulate Microtubule Dynamics

As it does with actin monomers, the cell sequesters unpolymerized tubulin subunits to maintain a pool of active subunits at a level near the critical concentration. One molecule of the small protein *stathmin* (also called Op18) binds two tubulin heterodimers and prevents their addition to the ends of microtubules (**Figure 16–50**). Stathmin thus decreases the effective concentration of tubulin subunits that are available for polymerization (an action analogous to that of the drug colchicine) and enhances the likelihood that a growing microtubule will switch to the shrinking state. Phosphorylation of stathmin inhibits its binding to tubulin, and signals that cause stathmin phosphorylation can increase the rate of microtubule elongation and suppress dynamic instability. Stathmin has been implicated in the regulation of both cell proliferation and cell death. Notably, mice lacking stathmin develop normally but are less fearful than wild-type mice, reflecting a role for stathmin in neurons of the amygdala, where it is normally expressed at high levels.

Severing is another mechanism employed by the cell to destabilize microtubules. To sever a microtubule, 13 longitudinal bonds must be broken, one for each protofilament. The protein *katanin*, named after the Japanese word for "sword," accomplishes this demanding task (**Figure 16–51A and B**). Katanin belongs to a large family of proteins that use the energy of ATP hydrolysis to disassemble or remodel protein complexes. By extracting tubulin subunits from the wall of the microtubule, katanin weakens the structure and thereby promotes breakage. Katanin also releases microtubules from microtubule-organizing centers and is thought to contribute to the rapid microtubule depolymerization observed at the poles of spindles during mitosis.

Paradoxically, the loss of microtubule-severing protein activity in many cell types leads to a decrease rather than an increase in microtubules. Thus, microtubule-severing proteins play an unexpected role in stabilizing microtubules. How is this possible? During the intermediate steps of a microtubule-severing event, GDP-bound tubulin subunits are lost from the wall of the microtubule and are replaced with GTP-tubulin subunits from

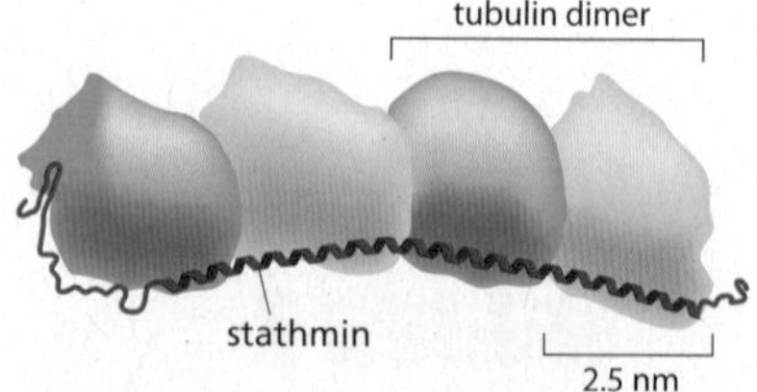

Figure 16–50 **Sequestration of tubulin by stathmin.** Structural studies with electron microscopy and crystallography suggest that the elongated stathmin protein binds along the side of two tubulin heterodimers. (Adapted from M.O. Steinmetz et al., *EMBO J.* 19:572–580, 2000.)

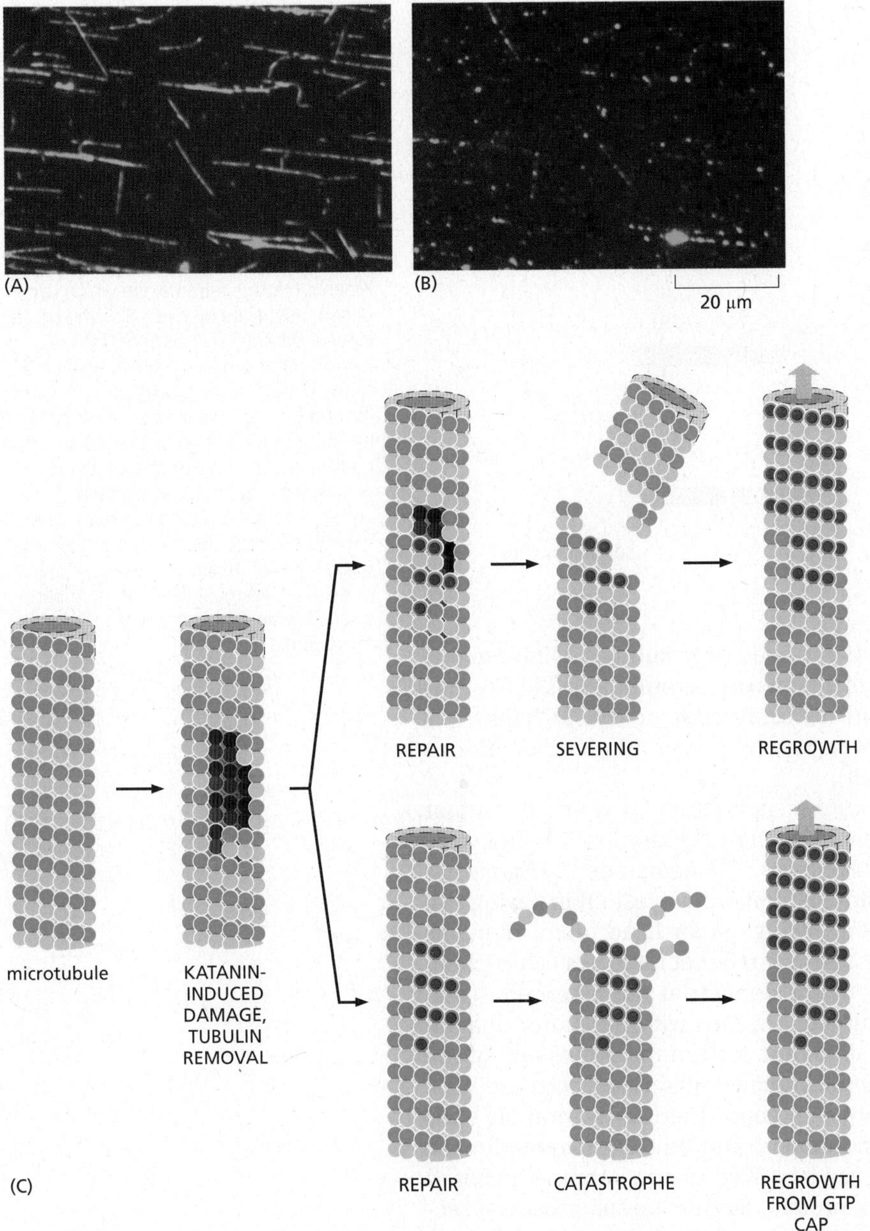

Figure 16–51 Microtubule severing by katanin can destabilize or amplify microtubules. (A) Taxol-stabilized, fluorescently labeled microtubules were adsorbed on the surface of a glass slide, to which purified katanin was added along with ATP. There are a few breaks in the microtubules 30 seconds after the addition of katanin. (B) Three minutes after the addition of katanin, the filaments have been severed in many places, leaving a series of small fragments at the previous locations of the long microtubules. (C) Incorporation of GTP-tubulin subunits from the soluble pool into sites of katanin-induced damage in the microtubule lattice stabilizes the severed end or generates an island of GTP-tubulin that promotes rescue. (A and B, from J.J. Hartman et al., *Cell* 93:277–287, 1998. With permission from Elsevier. C, adapted from A. Vemu et al., *Science* 361: eaau1504, 2018. With permission from AAAS.)

the soluble pool. If a sufficient number of GTP-tubulin subunits are incorporated before the severing is complete, the new plus end of the cut microtubule will possess a stabilizing GTP-tubulin cap and will therefore polymerize. Thus microtubule severing can generate plus ends that promote growth of more polymer. Alternatively, incomplete severing could lead to an island of GTP-tubulin in the microtubule lattice that could promote a future rescue event when this site is exposed after catastrophe (**Figure 16–51C**). Although insertion of GTP-tubulin into the lattice has been observed *in vitro* with pure microtubules, the importance of this activity in cells has not yet been fully established.

Two Types of Motor Proteins Move Along Microtubules

Like actin filaments, microtubules also work together with motor proteins in a variety of cellular processes. There are two major classes of microtubule-based motors, **kinesins** and **dyneins**, which perform three major functions. First, they move cargo such as organelles and macromolecules within the cell. Unlike actin-based transport, however, microtubule-based motors are used to transport

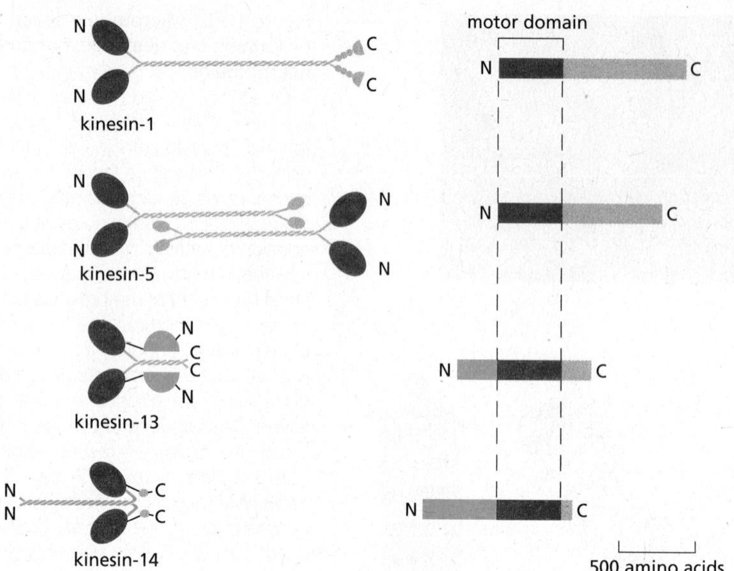

Figure 16–52 **Kinesins.** Structures of four kinesin superfamily members. As in the myosin superfamily, only the motor domains are conserved. Kinesin-1 has the motor domain at the N-terminus of the heavy chain and moves toward the microtubule plus end. The middle domain forms a long coiled-coil, mediating dimerization. The C-terminal domain forms a tail that attaches to cargo, such as a membrane-enclosed organelle. Kinesin-5 forms tetramers in which two dimers associate by their tails. The bipolar kinesin-5 tetramer is able to slide two microtubules past each other, analogous to the activity of the bipolar thick filaments formed by myosin II. Kinesin-13 has its motor domain located in the middle of the heavy chain. It is a member of a family of kinesins that have lost typical motor activity and instead bind to microtubule ends to promote depolymerization (see Figure 16–48). Kinesin-14 is a C-terminal kinesin. Unlike most kinesins, members of the kinesin-14 family travel toward the microtubule minus end.

cargo over long distances. Second, motors can slide microtubules relative to one another, thereby generating specific arrangements of microtubules, as in neurons and epithelial cells (see Figure 16–44), and in the mitotic spindle (see Chapter 17). Third, a subset of microtubule-based motors regulates microtubule dynamics, as illustrated by kinesin-13 (see Figure 16–48).

Like myosins, kinesins are a large protein superfamily in which the motor domain of the heavy chain is the common element (**Figure 16–52**). The yeast *Saccharomyces cerevisiae* has six distinct kinesins. The nematode *C. elegans* has 20 kinesins, and humans have 45. **Kinesin-1** is similar to myosin II in having two heavy chains per active motor; these form two globular head motor domains that are held together by an elongated coiled-coil tail that mediates heavy-chain dimerization. Most kinesins have their motor domain at the N-terminus and walk toward the plus end of the microtubule. Kinesins with the motor domain at the C-terminus walk in the opposite direction, toward the minus end of the microtubule, while kinesin-13 has a central motor domain and does not walk at all, but uses the energy of ATP hydrolysis to depolymerize microtubule ends (see Figure 16–48). Some kinesins are monomers, and others are homodimers, heterodimers, or tetramers. The motor may be linked to a membrane-enclosed organelle via a light chain or an adaptor protein. Some kinesins possess a second microtubule-binding domain that increases its affinity for the microtubule or mediates cross-linking and sliding of two microtubules.

In kinesin-1, small movements at the ATP-binding site regulate the docking and undocking of the motor head domain to a long linker region. This acts to throw the second head forward along the protofilament to a binding site 8 nm closer to the microtubule plus end, which is the distance between tubulin dimers of a protofilament. The ATP-hydrolysis cycles in the two heads are closely coordinated, so that this cycle of linker docking and undocking allows the two-headed motor to move in a hand-over-hand (or head-over-head) stepwise manner (**Figure 16–53**).

The dyneins are a family of minus end–directed microtubule motors unrelated to the kinesins. They are composed of one, two, or three heavy chains (that include the motor domain) and a large and variable number of associated intermediate, light-intermediate, and light chains. The dynein family has two major branches. The first contains the *cytoplasmic dyneins*, which are homodimers of two heavy chains (**Figure 16–54**). Cytoplasmic dynein 1 is encoded by a single gene in almost all eukaryotic cells but is missing from flowering plants and some algae. It is used for organelle and mRNA trafficking, for positioning the centrosome and nucleus during cell migration, and for construction of the microtubule spindle in mitosis and meiosis. Cytoplasmic dynein 2 is found only in eukaryotic

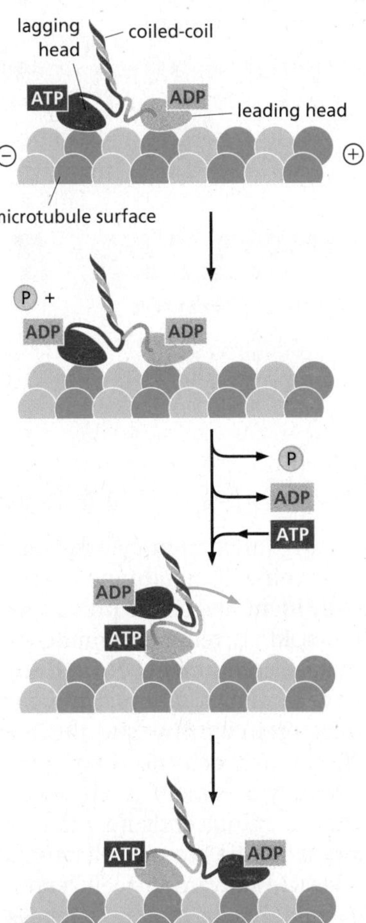

Figure 16–53 The mechanochemical cycle of kinesin. Kinesin-1 is a dimer of two ATP-binding motor domains (heads) that are connected through a long coiled-coil tail (see Figure 16–52). The two kinesin motor domains work in a coordinated manner; during a kinesin "step," the rear head detaches from its tubulin binding site on the microtubule, passes the partner motor domain, and then rebinds to the next available binding site. Using this "hand-over-hand" motion, the kinesin dimer can move for long distances on the microtubule without completely letting go of its track.

At the start of each step, one of the two kinesin motor domain heads, the rear or lagging head *(dark red)*, is tightly bound to the microtubule and to ATP, while the front or leading head is loosely bound to the microtubule with ADP in its binding site. The forward displacement of the rear motor domain is driven by the dissociation of ADP and binding of ATP in the leading head (between panels 2 and 3 in this drawing). The binding of ATP to this motor domain causes a small peptide called the *neck linker* to shift from a rearward-pointing to a forward-pointing conformation (the neck linker is drawn here as a *purple* connecting line between the leading motor domain and the intertwined coiled-coil). This shift pulls the rear head forward, once it has detached from the microtubule with ADP bound [detachment requires ATP hydrolysis and phosphate (P) release]. The kinesin molecule is now poised for the next step, which proceeds by an exact repeat of the process shown (**Movie 16.12**).

organisms that have cilia and is used to transport material from the tip to the base of the cilia—a process called *intraflagellar transport* (IFT). *Axonemal dyneins* comprise the second branch and include monomers, heterodimers, and heterotrimers, with one, two, or three motor-containing heavy chains, respectively. They are highly specialized for the rapid and efficient microtubule sliding movements that drive the beating of cilia and flagella (discussed later).

Dyneins are the largest of the known molecular motors. Although structurally unrelated to myosins and kinesins, dyneins follow the general rule of coupling ATP hydrolysis to microtubule binding and unbinding as well as to a force-generating conformational change (**Figure 16–55**).

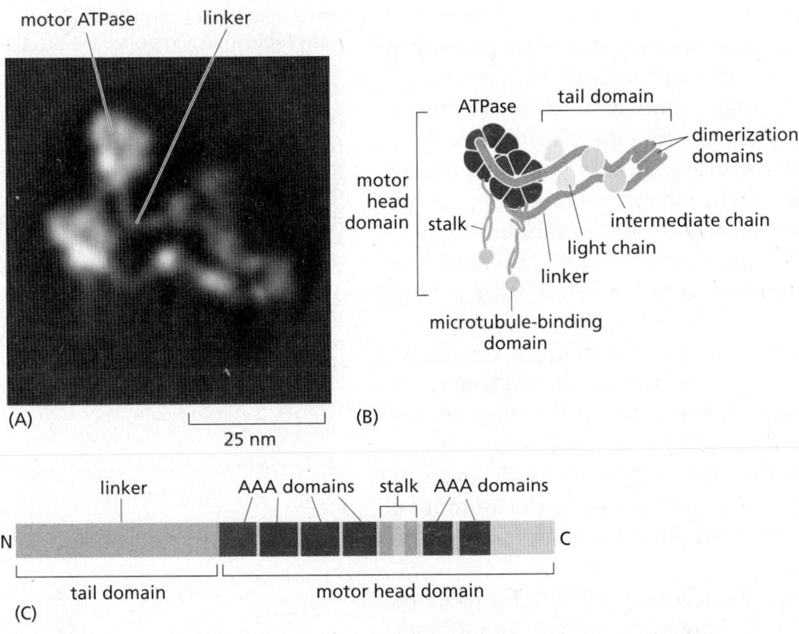

Figure 16–54 Cytoplasmic dynein. (A) Cryo-electron microscopy (cryoEM) reconstruction of a molecule of cytoplasmic dynein. Like myosin II and kinesin-1, cytoplasmic dynein is a two-headed molecule. The dynein head is very large compared with the head of either myosin or kinesin. (B) Schematic depiction of cytoplasmic dynein showing the two heavy chains that contain a motor head with domains for microtubule binding and ATP hydrolysis, connected by a long stalk. The tail domain consists of a linker that connects the motor heads to a dimerization domain. Bound to the linker domain are multiple intermediate chains and light chains *(blue)* that help to mediate many of dynein's functions. (C) The organization of domains in a dynein heavy chain. This is a huge polypeptide, containing more than 4000 amino acids. The conserved dynein motor head domain contains six AAA domains, four of which retain ATP-binding sequences, but only one of which has the major ATPase activity *(brown)*. The tail domain is not as highly conserved as the head domain and varies among different dynein subtypes. (A, courtesy of Andrew Carter.)

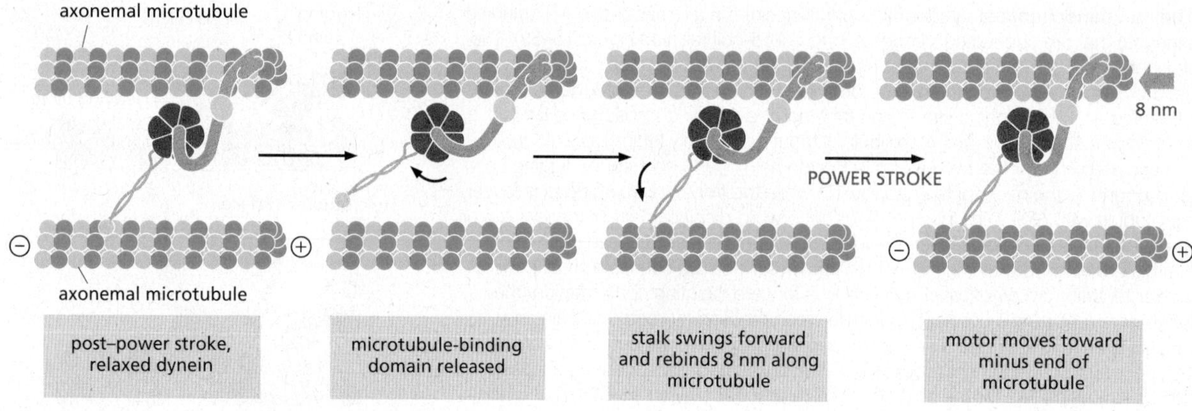

Figure 16–55 The power stroke of dynein. Illustration of the movement of a monomeric axonemal dynein found in the flagellum of the unicellular green alga *Chlamydomonas reinhardtii*. As in cytoplasmic dynein, the motor-containing head domain of axonemal dynein connects to a long, coiled-coil stalk with the microtubule-binding site at the tip. The tail attaches to an adjacent microtubule in the axoneme. Movement is thought to occur through a "linker-swing, dynein-winch" mechanism. ATP binding and hydrolysis cause the linker to throw the head domain toward the microtubule minus end like a fishing hook. The microtubule-binding domain reattaches 8 nm along the microtubule. Release of ATP and phosphate then leads to a large conformational power stroke in the linker domain, pulling the tail and its attached microtubule toward the minus end. Each cycle generates a step of about 8 nm, thereby contributing to flagellar beating (see Figure 16–60). In the case of cytoplasmic dynein, the tail is attached to a cargo such as a vesicle, and a single power stroke transports the cargo about 8 nm along the microtubule toward its minus end (see Figure 16–56).

Microtubules and Motors Move Organelles and Vesicles

A major function of cytoskeletal motors in interphase cells is the transport and positioning of membrane-enclosed organelles (**Movie 16.13**). Kinesin was originally identified as the protein responsible for fast *anterograde axonal transport*, the rapid movement of mitochondria, secretory vesicle precursors, and various synapse components down the microtubule highways of the axon to the distant nerve terminals. Cytoplasmic dynein 1 was identified as the motor responsible for transport in the opposite direction, *retrograde axonal transport*. Although organelles in most cells need not cover such long distances, their polarized transport is equally necessary. A typical microtubule array in an interphase cell is oriented with the minus ends near the center of the cell at the centrosome and the plus ends extending to the cell periphery. Thus, centripetal movements of organelles or vesicles toward the cell center require the action of minus end–directed cytoplasmic dynein motors, whereas centrifugal movements toward the periphery require plus end–directed kinesin motors. Notably, in animal cells, nearly all minus end–directed transport is driven by the single cytoplasmic dynein 1 motor, whereas at least 15 different kinesins are used for plus end–directed transport.

A clear example of the effect of microtubules and microtubule motors on the behavior of intracellular membranes is their role in organizing the endoplasmic reticulum (ER) and the Golgi apparatus. The network of ER membrane tubules aligns with microtubules and extends almost to the edge of the cell (**Movie 16.14**), whereas the Golgi apparatus is located near the centrosome. When cells are treated with a drug that depolymerizes microtubules, such as colchicine or nocodazole, the ER collapses to the center of the cell, while the Golgi apparatus fragments and disperses throughout the cytoplasm. *In vitro*, kinesins can tether ER-derived membranes to preformed microtubule tracks and walk toward the microtubule plus ends, dragging the ER membranes out into tubular protrusions and forming a membranous web that looks very much like the ER in cells. Conversely, dyneins are required for positioning the Golgi apparatus near the cell center of animal cells; they do this by moving Golgi vesicles along microtubule tracks toward the microtubules' minus ends at the centrosome.

The different tails and their associated light chains on specific motor proteins allow the motors to attach to their appropriate organelle cargo. Membrane-associated motor receptors that are sorted to specific membrane-enclosed compartments interact directly or indirectly with the tails of the appropriate kinesin family members. Many viruses take advantage of microtubule motor-based transport during infection and use kinesin to move from their site of replication and assembly to the plasma membrane, from which they are poised to infect neighboring cells.

For dynein, a large macromolecular assembly mediates attachment to cargoes. To translocate organelles effectively, cytoplasmic dynein, itself a huge protein complex, requires association with a second large protein complex called *dynactin* as well as with an adaptor protein that mediates their interaction and links to a cargo such as a vesicle. The dynactin complex includes a short, actin-like filament that forms from the actin-related protein Arp1 (distinct from Arp2 and Arp3, the components of the Arp2/3 complex involved in the nucleation of conventional actin filaments) (**Figure 16–56**). A number of other proteins also contribute to dynein cargo binding and motor regulation, and their function is especially important in neurons, where defects in microtubule-based transport have been linked to neurological diseases. A striking example is smooth brain, or lissencephaly, a human disorder in which cells fail to migrate to the cerebral cortex of the developing brain. One type of lissencephaly is caused by defects in Lis1, a dynein-binding protein required for nuclear migration in several species. In the normal brain, migration of the nucleus directs the developing neural cell body toward its correct position in the cortex. In the absence of Lis1, however, this process fails, and affected children suffer from developmental delays as well as a variety of neurological defects. Dynein is required continually for neuronal function, as mutations in a dynactin subunit or in the tail region of cytoplasmic dynein lead to neuronal degeneration in humans and mice. These effects are associated with decreased retrograde axonal transport and provide strong evidence for the importance of robust axonal transport in neuronal viability.

The cell can regulate the activity of motor proteins and thereby cause either a change in the positioning of its membrane-enclosed organelles or whole-cell movements. Fish melanophores provide one of the most dramatic examples. These giant cells, which are responsible for rapid changes in skin coloration in several species of fish, contain large pigment granules that can alter their location in response to neuronal or hormonal stimulation (**Figure 16–57**). The pigment granules aggregate or disperse by moving along an extensive network of microtubules that are anchored at the centrosome by their minus ends. The tracking of individual pigment granules reveals that the inward movement is rapid and smooth, while the outward movement is jerky, with frequent backward steps. Both dynein and kinesin microtubule motors are associated with the pigment granules. The jerky outward movements may result from a tug-of-war between the two opposing

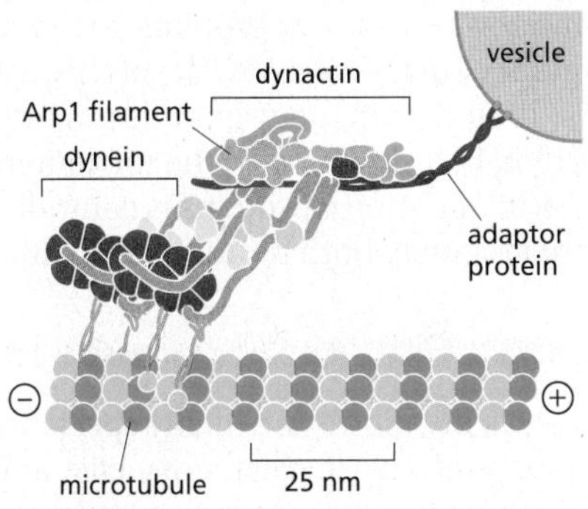

Figure 16–56 **Dynactin and an adaptor protein mediate the attachment of dynein to a membrane-enclosed organelle.** Dynein requires the presence of a large number of accessory proteins to associate with membrane-enclosed organelles. Dynactin is a large complex that includes components that bind weakly to microtubules, components that bind to dynein itself, and components that form a small, actin-like filament made of the actin-related protein Arp1. Dynactin associates with two molecules of cytoplasmic dynein as well as with an adaptor protein that mediates the connection to a cargo.

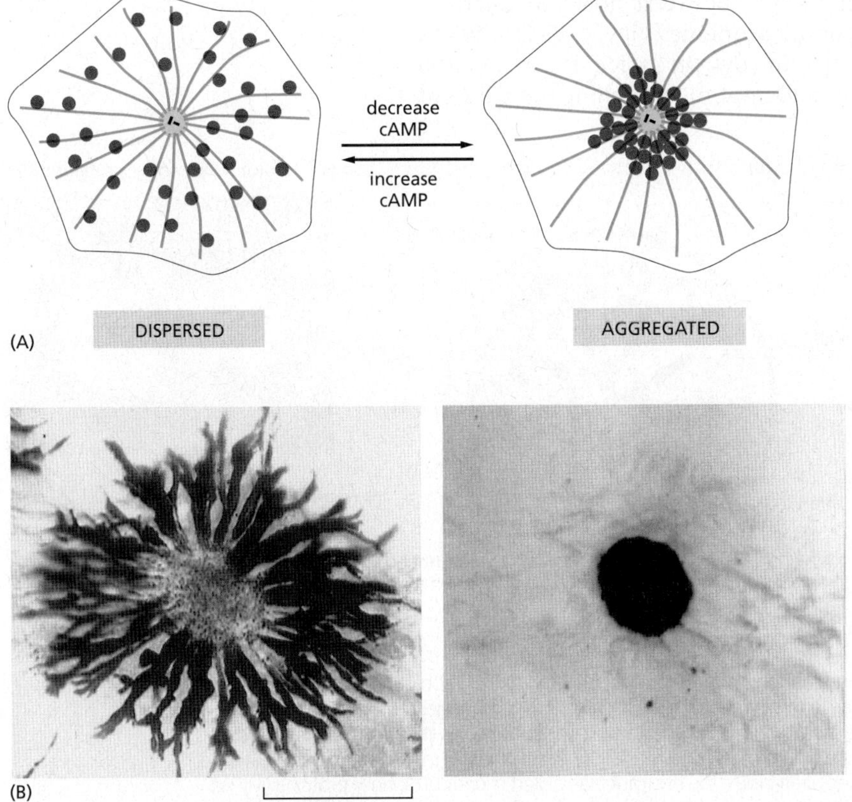

Figure 16–57 **Regulated melanosome movements in fish pigment cells.** These giant cells, which are responsible for changes in skin coloration in several species of fish, contain large pigment granules called melanosomes. The melanosomes can change their location in the cell in response to a hormonal or neuronal stimulus. (A) Schematic view of a pigment cell, showing the dispersal and aggregation of melanosomes *(brown)* in response to an increase or decrease in intracellular cyclic AMP (cAMP), respectively. Both redistributions of melanosomes occur along microtubules. (B) Bright-field images of a single cell in a scale of an African cichlid fish, showing its melanosomes either dispersed throughout the cytoplasm *(left)* or aggregated in the center of the cell *(right)*. (B, courtesy of Leah Haimo.)

microtubule motor proteins, with the stronger kinesin winning out overall. When intracellular cyclic AMP levels decrease, kinesin is inactivated, leaving dynein free to drag the pigment granules rapidly toward the cell center, changing the fish's color. In a similar way, the movement of other membrane organelles coated with particular motor proteins is controlled by a complex balance of competing signals that regulate both motor protein attachment and activity.

Motile Cilia and Flagella Are Built from Microtubules and Dyneins

Just as myofibrils are highly specialized and efficient motility machines built from actin and myosin filaments, cilia and flagella are highly specialized and efficient motility structures built from microtubules and dynein. Both cilia and flagella are hairlike cell appendages that have a bundle of microtubules at their core. **Flagella** are found on sperm and many protozoa. By their undulating motion, they enable the cells from which they emerge to swim through liquid media. **Cilia** are organized in a similar fashion, but they beat with a whiplike motion that resembles the breast-stroke in swimming. Ciliary beating can either propel single cells through a fluid (as in the swimming of the protozoan *Paramecium*) or can move fluid over the surface of a group of cells in a tissue. In the human body, huge numbers of cilia (10^9/cm^2 or more) line our respiratory tract, sweeping layers of mucus, trapped particles of dust, and bacteria up to the mouth where they are swallowed and ultimately eliminated. Likewise, cilia along the oviduct help to sweep eggs toward the uterus.

The movement of a cilium or a flagellum is produced by the bending of its core, which is called the **axoneme**. The axoneme is composed of microtubules and their associated proteins, arranged in a distinctive and regular pattern. Nine special microtubule doublets (comprising one complete and one partial microtubule fused together so that they share a common tubule wall) are arranged in a ring around a pair of single microtubules (**Figure 16–58**). Almost all forms of motile eukaryotic flagella and cilia (from protozoans to humans) have this characteristic arrangement. The microtubules extend continuously for the length of the axoneme, which can be 10–200 μm. At regular positions along the length of the microtubules, accessory proteins cross-link the microtubules together.

Molecules of *axonemal dynein* form bridges between adjacent microtubule doublets around the circumference of the axoneme (**Figure 16–59**). When the motor domain of this dynein is activated, the dynein molecules attached to one microtubule doublet (see Figure 16–60) attempt to walk along the adjacent

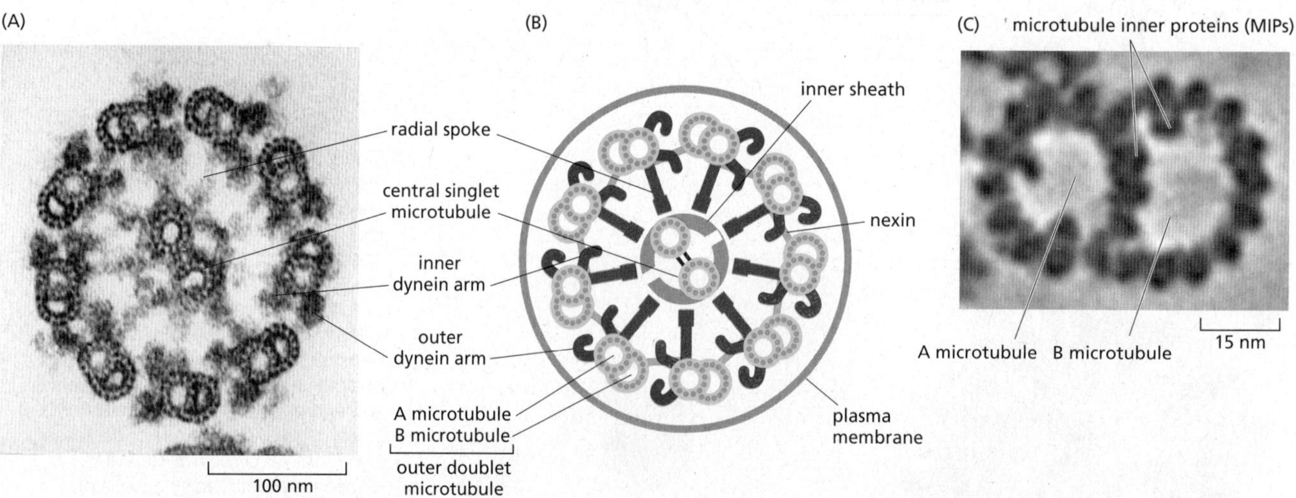

Figure 16–58 The arrangement of microtubules in a flagellum or cilium. (A) Electron micrograph of the flagellum of a green-alga cell (*Chlamydomonas*) shown in cross section, illustrating the distinctive "9 + 2" arrangement of microtubules. (B) Diagram of the parts of a flagellum or cilium. The various projections from the microtubules link the microtubules together and occur at regular intervals along the length of the axoneme. (C) High-resolution electron tomography image of an outer microtubule doublet showing structural details and features inside the microtubules called microtubule inner proteins (MIPs). (A, courtesy of Lewis Tilney; C, courtesy of Daniela Nicastro.)

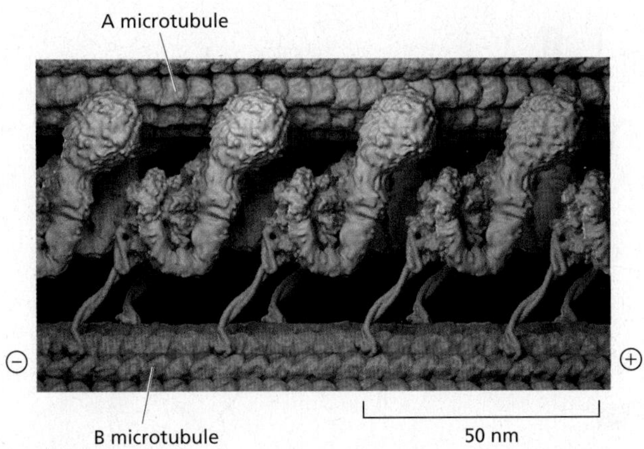

Figure 16–59 Axonemal dynein. CryoEM reconstruction of a sea urchin sperm flagellum showing dynein arms connecting the A microtubule of one doublet with the B microtubule of an adjacent doublet at regular intervals. Sperm axonemal dynein is dimeric. The tail of the molecule binds tightly to an A microtubule, while the two globular heads each have a stalk that connects to an ATP-dependent binding site on a B microtubule (see Figure 16–58). When the heads hydrolyze their bound ATP, they move toward the minus end of the B microtubule, thereby producing a sliding force between the adjacent microtubule doublets in a cilium or flagellum (see Figure 16–60). (Courtesy of Daniela Nicastro.)

microtubule doublet, tending to force the adjacent doublets to slide relative to one another, much as actin thin filaments slide during muscle contraction. However, the presence of other links between the microtubule doublets prevents this sliding, and the dynein force is instead converted into a bending motion (**Figure 16–60**). Not all dyneins in the axoneme are active at the same time, which results in the characteristic wave-like motion of the cilium or flagellum (**Movie 16.15**).

In humans, hereditary defects in axonemal dynein cause a condition called primary ciliary dyskinesia, or Kartagener's syndrome. This syndrome is characterized by inversion of the normal asymmetry of internal organs (situs inversus) due to disruption of fluid flow in the developing embryo, male sterility due to immotile sperm, and a high susceptibility to lung infections due to paralyzed cilia being unable to clear the respiratory tract of debris and bacteria.

Bacteria also swim using cell-surface structures called flagella, but these do not contain microtubules or dynein and do not wave or beat. Instead, *bacterial flagella* are long, rigid helical filaments, made up of repeating subunits of the protein flagellin. The flagella rotate like propellers, driven by a special rotary motor embedded in the bacterial cell wall. The use of the same name to denote these two very different types of swimming apparatus is an unfortunate historical accident.

Primary Cilia Perform Important Signaling Functions in Animal Cells

Many cells possess a shorter, nonmotile counterpart of cilia and flagella called the *primary cilium*. Primary cilia can be viewed as specialized compartments or organelles that perform a wide range of cellular functions but share many

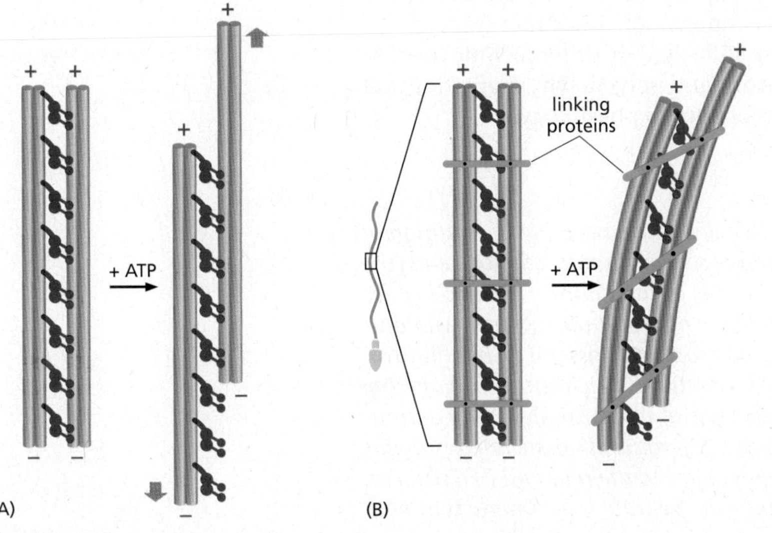

Figure 16–60 The bending of an axoneme. (A) When axonemes are exposed to the proteolytic enzyme trypsin, the flexible protein links holding adjacent microtubule doublets together are broken. In this case, the addition of ATP allows the motor action of the dynein heads to slide one microtubule doublet against the adjacent doublet. (B) In an intact axoneme (such as in a spermatozoon), the flexible protein links prevent the sliding of the doublet. The motor action therefore causes a bending motion, creating waves or beating motions.

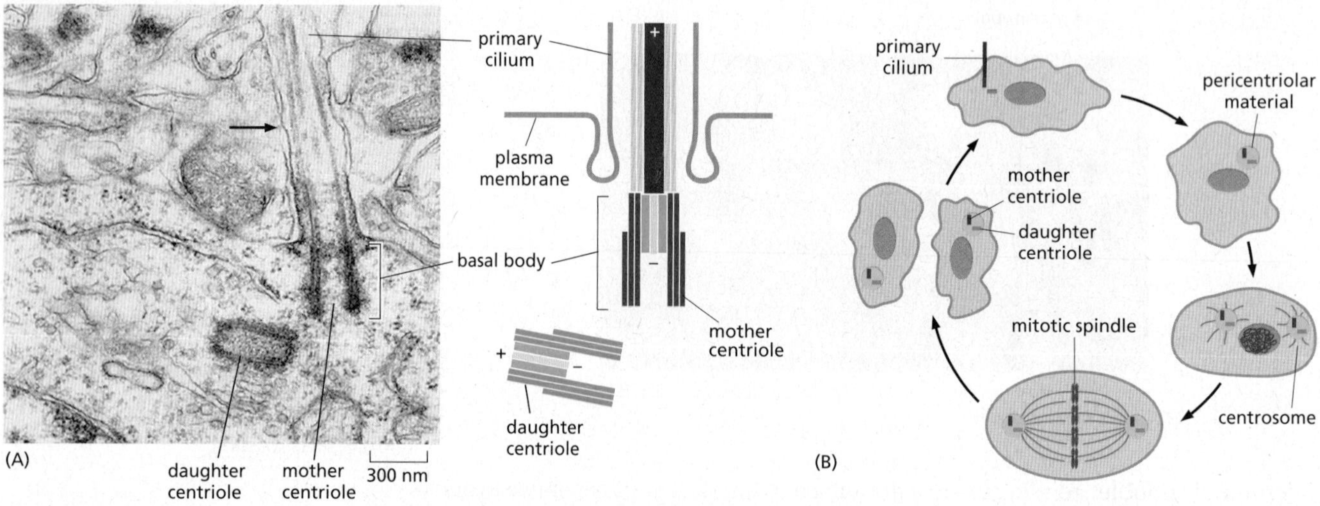

Figure 16–61 Primary cilia. (A) Electron micrograph and diagram of the basal body of a mouse neuron primary cilium. The axoneme of the primary cilium *(black arrow)* is nucleated by the mother centriole at the basal body, which localizes at the plasma membrane near the cell surface. (B) Centrioles function alternately as basal bodies and as the core of centrosomes. Before a cell enters the cell-division cycle, the primary cilium is shed or resorbed. The centrioles recruit pericentriolar material and duplicate during S phase, generating two centrosomes, each of which contains a pair of centrioles. The centrosomes nucleate microtubules and localize to the poles of the mitotic spindle. Upon exit from mitosis, a primary cilium again grows from the mother centriole. (A, courtesy of Josef Spacek.)

structural features with motile cilia. Both motile and nonmotile cilia are generated during interphase at plasma membrane–associated structures called *basal bodies*, which anchor them at the cell surface. At the core of each basal body is a single centriole, the same structure found in pairs embedded at the center of animal centrosomes, with nine groups of fused microtubule triplets arranged in a cartwheel (see Figure 16–43). Centrioles are multifunctional, contributing to assembly of the mitotic spindle in dividing cells but migrating to the plasma membrane of interphase cells to template the nucleation of the axoneme (**Figure 16–61**). Because no protein translation occurs in cilia, construction of the axoneme requires intraflagellar transport (IFT), a transport system discovered in the green algae *Chlamydomonas*. Analogous to the axon, motors move cargoes in both anterograde and retrograde directions, in this case driven by kinesin-2 and cytoplasmic dynein 2, respectively.

Primary cilia are found on the surface of almost all cell types, where they sense and respond to the exterior environment, functions best understood in the context of smell and sight. In the nasal epithelium, cilia protruding from dendrites of olfactory neurons are the site of both odorant reception and signal amplification. Similarly, the rod and cone cells of the vertebrate retina possess a specialized primary cilium called the outer segment, which is specialized for converting light into a neural signal (see Figure 15–40). Maintenance of the outer segment requires continual IFT-mediated transport of large quantities of lipids and proteins into the cilium, at rates of up to 1000 molecules per second. The links between cilia function and the senses of sight and smell are underscored by the ciliopathies, a set of disorders associated with defects in IFT, the cilium, or the basal body. In the ciliopathy Bardet–Biedl syndrome, patients cannot smell and suffer from retinal degeneration. Other characteristics of this multifaceted disorder include hearing loss, polycystic kidney disease, diabetes, obesity, and polydactyly, suggesting that primary cilia have functions in many aspects of human physiology.

Summary

Microtubules are stiff polymers of tubulin molecules. They assemble by addition of GTP-containing tubulin subunits to the free end of a microtubule, with one end (the plus end) growing faster than the other. Hydrolysis of the bound GTP takes place after assembly and weakens the bonds that hold the microtubule together. Microtubules are dynamically unstable and liable to catastrophic disassembly, but they can be stabilized in cells by association with other structures. Microtubule-organizing centers such as centrosomes protect the minus ends of microtubules and continually nucleate the formation of new microtubules. Microtubule-associated proteins (MAPs) stabilize microtubules, and those that localize to the plus end (+TIPs) can alter the dynamic properties of the microtubule or mediate their interaction with other structures. Counteracting the stabilizing activity of MAPs are catastrophe factors, such as kinesin-13 proteins, that act to peel apart microtubule ends. Other kinesin family members as well as dynein use the energy of ATP hydrolysis to move unidirectionally along a microtubule. The motor dynein moves toward the minus end of microtubules, and its sliding of axonemal microtubules underlies the beating of cilia and flagella. Primary cilia are nonmotile sensory organelles found on many cell types.

INTERMEDIATE FILAMENTS AND OTHER CYTOSKELETAL POLYMERS

All eukaryotic cells contain actin and tubulin. But the third major type of cytoskeletal protein, the *intermediate filament*, forms a cytoplasmic filament only in some metazoans—including vertebrates, nematodes, and mollusks. Intermediate filaments are particularly prominent in the cytoplasm of cells that are subject to mechanical stress and are generally not found in animals that have rigid exoskeletons, such as arthropods and echinoderms. It seems that intermediate filaments impart mechanical strength to tissues for the squishier animals.

Cytoplasmic intermediate filaments are closely related to their ancestors, the much more prevalent *nuclear lamins*, which are found in many eukaryotes but missing from unicellular organisms. The nuclear lamins form a meshwork lining the inner membrane of the nuclear envelope, where they provide anchorage sites for chromosomes and nuclear pores. Several times during metazoan evolution, lamin genes have apparently duplicated, and the duplicates have evolved to produce rope-like, cytoplasmic intermediate filaments. In contrast to the highly conserved actins and tubulin isoforms that are encoded by a handful of genes, different families of intermediate filaments are much more diverse and are encoded by 70 different human genes with distinct, cell type–specific functions (**Table 16–2**).

Intermediate Filament Structure Depends on the Lateral Bundling and Twisting of Coiled-Coils

Although their amino- and carboxyl-terminal domains differ, all intermediate filament family members are elongated proteins with a conserved central α-helical domain containing 40 or so heptad repeat motifs that form an extended

TABLE 16–2 Major Types of Intermediate Filament Proteins in Vertebrate Cells

Types of intermediate filament	Component polypeptides	Location
Nuclear	Lamins A, B, and C	Nuclear lamina (inner lining of nuclear envelope)
Vimentin-like	Vimentin	Many cells of mesenchymal origin
	Desmin	Muscle
	Glial fibrillary acidic protein	Glial cells (astrocytes and some Schwann cells)
	Peripherin	Some neurons
Epithelial	Type I keratins (acidic)	Epithelial cells and their derivatives (e.g., hair and nails)
Epithelial	Type II keratins (neutral/basic)	
Axonal	Neurofilament proteins (NF-L, NF-M, and NF-H)	Neurons

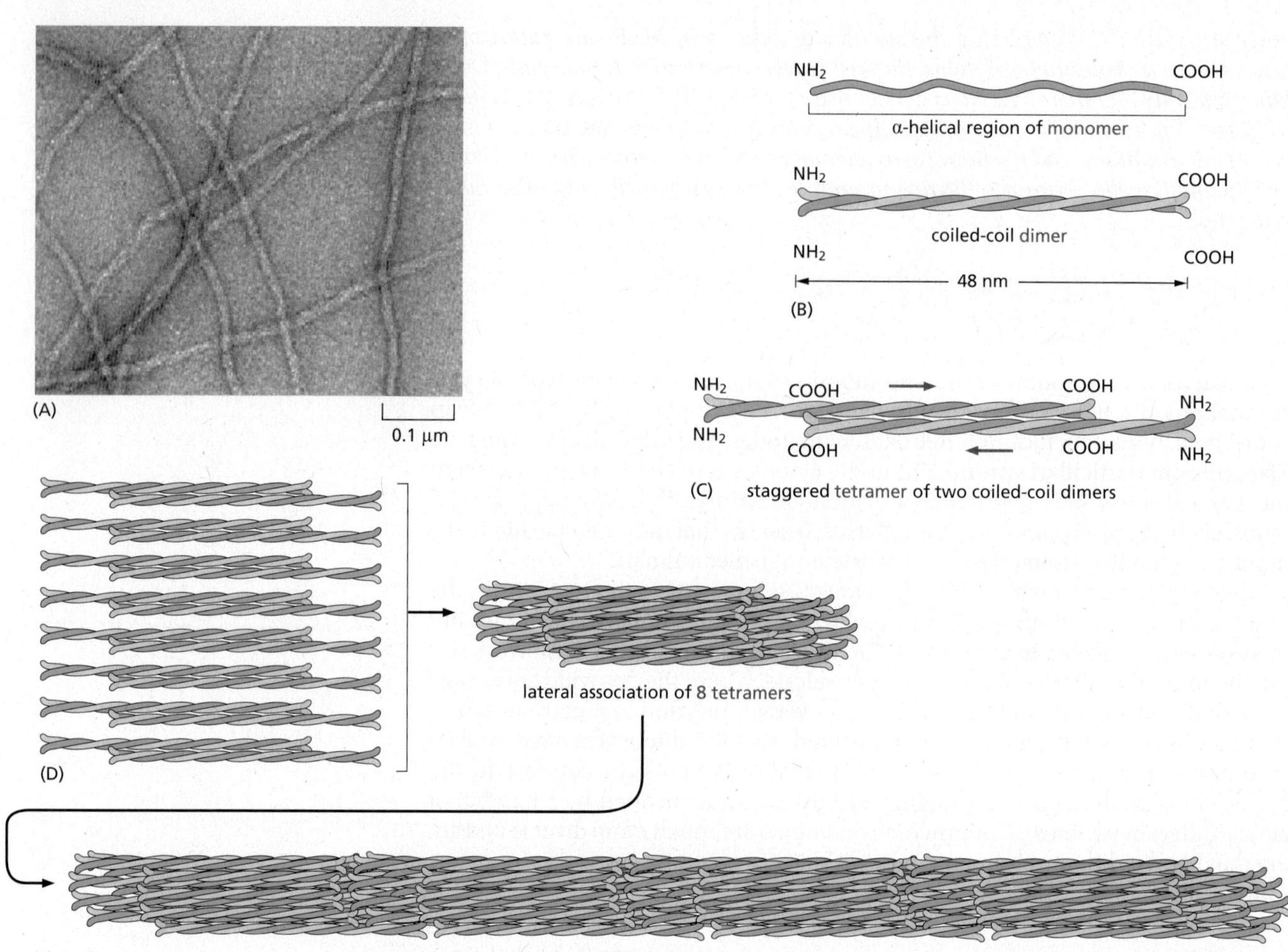

Figure 16–62 A model of intermediate filament construction. An electron micrograph of intermediate filaments is shown in (A). The monomer shown in (B) pairs with another monomer to form a dimer, in which the conserved central rod domains are aligned in parallel and wound together into a coiled-coil. (C) Two dimers then line up side by side to form an antiparallel tetramer of four polypeptide chains. Dimers and tetramers are the soluble subunits of intermediate filaments. (D) Within each tetramer, the two dimers are offset with respect to one another, thereby allowing it to associate with another tetramer. (E) In the final 10-nm-diameter filament, tetramers are packed together in a rope-like array, which has 16 dimers (32 coiled-coils) in cross section. Half of these dimers are pointing in each direction. An electron micrograph of intermediate filaments is shown on the upper left (**Movie 16.16**). (A, from L. Norlén et al., *Exp. Cell Res.* 313:2217–2227, 2007. With permission from Elsevier.)

coiled-coil structure with another monomer (see Figure 3–8). A pair of parallel dimers then associates in an antiparallel fashion to form a staggered tetramer (**Figure 16–62**). Unlike actin or tubulin subunits, intermediate filament subunits do not contain a binding site for ATP or GTP. Furthermore, because the tetrameric subunit is made up of two dimers pointing in opposite directions, its two ends are the same. The assembled intermediate filament therefore lacks the overall structural polarity that is critical for actin filaments and microtubules. The tetramers pack together laterally to form the filament, which includes eight parallel protofilaments made up of tetramers. Each individual intermediate filament therefore has a cross section of 32 individual α-helical coils. This large number of polypeptides all lined up together, with the strong lateral hydrophobic interactions typical of coiled-coil proteins, gives intermediate filaments a rope-like character. They can be easily bent, with a persistence length of less than 1 μm (compared to several millimeters for microtubules and about 10 μm for actin), but they are extremely difficult to break and can be stretched to more than three times their length (see Figure 16–6).

Less is understood about the mechanisms of assembly and disassembly of intermediate filaments than of actin filaments and microtubules. In pure protein solutions, intermediate filaments are extremely stable due to tight association of subunits, but some types of intermediate filaments, including *vimentin*, form highly dynamic structures in cells such as fibroblasts. Protein phosphorylation probably regulates their disassembly, in much the same way that phosphorylation regulates the disassembly of nuclear lamins in mitosis (see Figure 12–65). As evidence for rapid turnover, labeled subunits microinjected into tissue-culture cells incorporate into intermediate filaments within a few minutes. Remodeling of the intermediate filament network accompanies events requiring dynamic cellular reorganization, such as division, migration, and differentiation.

Intermediate Filaments Impart Mechanical Stability to Animal Cells

Keratins are the most diverse intermediate filament family: there are about 20 found in different types of human epithelial cells and about 10 more that are specific to hair and nails; analysis of the human genome sequence has revealed that there are 54 distinct keratins. Every keratin filament is made up of an equal mixture of type I (acidic) and type II (neutral/basic) keratin proteins; these form a heterodimer filament subunit (see Figure 16–62). Cross-linked keratin networks held together by disulfide bonds can survive even the death of their cells, forming tough coverings for animals, as in the outer layer of skin and in hair, nails, claws, and scales. The diversity in keratins is clinically useful in the diagnosis of epithelial cancers (carcinomas), as the particular set of keratins expressed gives an indication of the epithelial tissue in which the cancer originated and thus can help to guide the choice of treatment.

A single epithelial cell may produce multiple types of keratins, and these copolymerize into a single network (**Figure 16–63**). Keratin filaments impart mechanical strength to epithelial tissues in part by anchoring the intermediate filaments at sites of cell–cell contact, called *desmosomes*, or cell–matrix contact, called *hemidesmosomes* (see Figure 16–4). We discuss these important adhesive structures in Chapter 19. Accessory proteins, such as *filaggrin*, bundle keratin filaments in differentiating cells of the epidermis to give the outermost layers of the skin their special toughness. Individuals with mutations in the gene encoding filaggrin are strongly predisposed to dry skin diseases such as eczema.

Mutations in keratin genes cause several human genetic diseases. For example, when defective keratins are expressed in the basal cell layer of the epidermis, they produce a disorder called *epidermolysis bullosa simplex*, in which the skin blisters in response to even very slight mechanical stress, which ruptures the basal

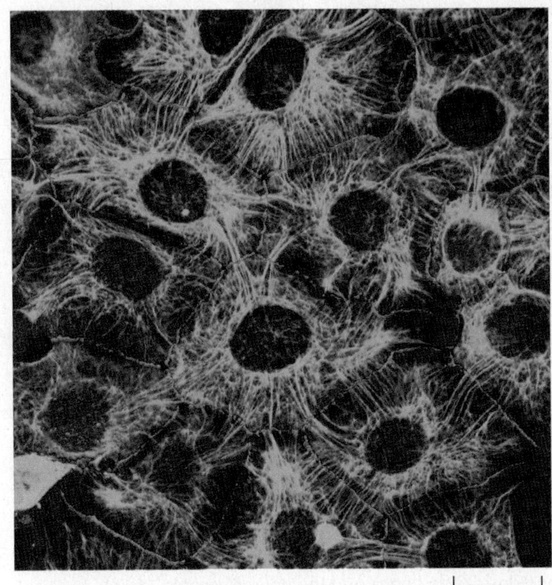

Figure 16–63 **Keratin filaments in epithelial cells.** Immunofluorescence micrograph of the network of keratin filaments *(blue)* in a sheet of epithelial cells in culture. The filaments in each cell are indirectly connected to those of its neighbors by desmosomes (discussed in Chapter 19). A second protein *(red)* has been stained to reveal the location of the cell boundaries. (From K.J. Green and C.A. Gaudry, *Nat. Rev. Mol. Cell Biol.* 1:208–216, published 2000 by Nature Publishing Group. Reproduced with permission of SNCSC.)

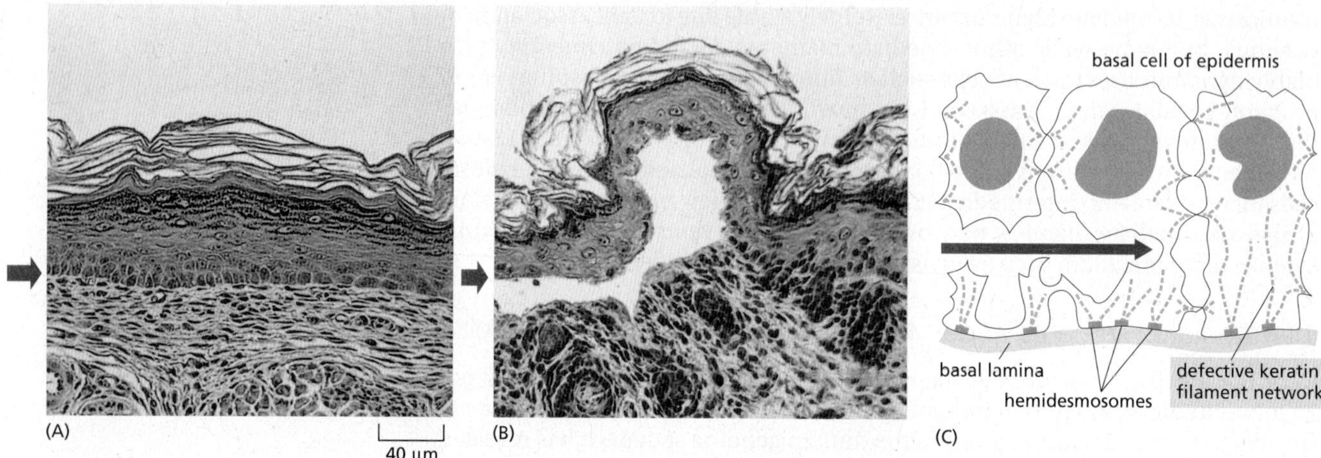

Figure 16–64 **Blistering of the skin caused by a mutant keratin gene.** A mutant gene encoding a truncated keratin protein (lacking both the N- and C-terminal domains) was expressed in a transgenic mouse. The defective protein assembles with the normal keratins and thereby disrupts the keratin filament network in the basal cells of the skin. Light micrographs of cross sections of (A) normal and (B) mutant skin show that the blistering results from the rupturing of cells in the basal layer of the mutant epidermis *(short red arrows)*. (C) A sketch of three cells in the basal layer of the mutant epidermis, as observed by electron microscopy. As indicated by the *red arrow*, the cells rupture between the nucleus and the hemidesmosomes (discussed in Chapter 19), which connect the keratin filaments to the underlying basal lamina. (A and B, © 1991 P.A. Coulombe et al. Originally published in *J. Cell Biol.* https://doi.org/10.1083/jcb.115.6.1661. With permission from Rockefeller University Press.)

cells (**Figure 16–64**). Other types of blistering diseases, including disorders of the mouth, esophageal lining, and the cornea of the eye, are caused by mutations in the different keratins whose expression is specific to those tissues. All of these maladies are typified by cell rupture as a consequence of mechanical trauma and a disorganization or clumping of the keratin filament cytoskeleton. Many of the specific mutations that cause these diseases alter the ends of the central rod domain, demonstrating the importance of this particular part of the protein for correct filament assembly.

Members of another family of intermediate filaments, called **neurofilaments**, are found in high concentrations along the axons of vertebrate neurons (**Figure 16–65**). Three types of neurofilament proteins (NF-L, NF-M, and NF-H)

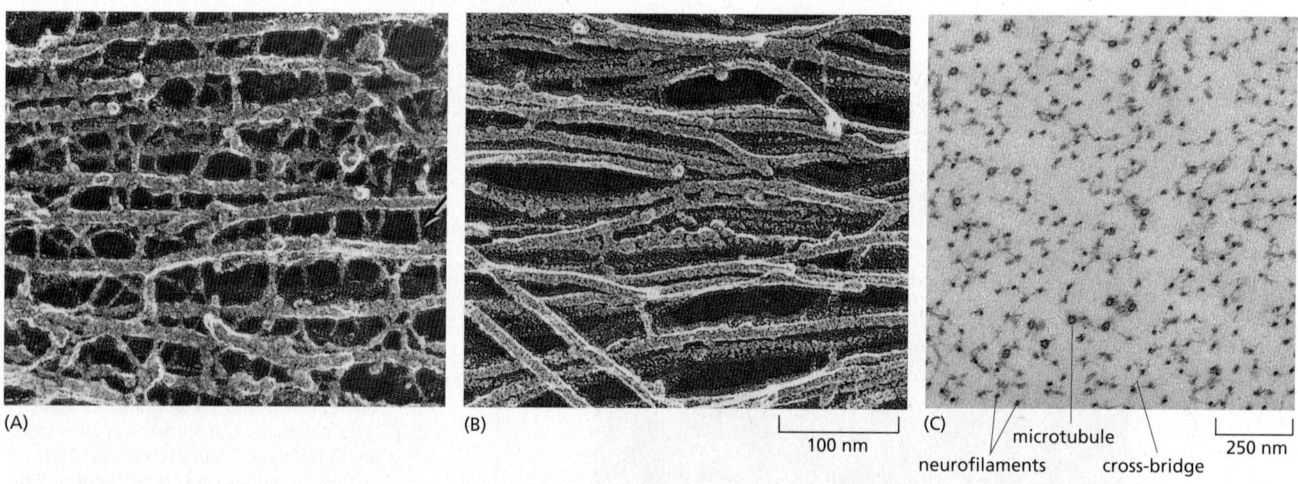

Figure 16–65 **Two types of intermediate filaments in cells of the nervous system.** (A) Freeze-etch electron microscopy image of neurofilaments in a nerve cell axon, showing the extensive cross-linking through protein cross-bridges—an arrangement believed to give this long cell process great tensile strength. The cross-bridges are formed by the long, nonhelical extensions at the C-terminus of the largest neurofilament protein (NF-H). (B) Freeze-etch image of glial filaments in glial cells, showing that these intermediate filaments are smooth and have few cross-bridges. (C) Conventional transmission electron micrograph of a cross section of an axon showing the regular side-to-side spacing of the neurofilaments, which greatly outnumber the microtubules. (A and B, courtesy of Nobutaka Hirokawa; C, courtesy of Anthony Brown.)

coassemble *in vivo*, forming heteropolymers. The NF-H and NF-M proteins have lengthy C-terminal tail domains that bind to neighboring filaments, generating aligned arrays with a uniform interfilament spacing. During axonal growth, new neurofilament subunits are incorporated all along the axon in a dynamic process that involves the addition of subunits along the filament length and at the ends. After an axon has grown and connected with its target cell, the diameter of the axon may increase as much as fivefold. The level of neurofilament gene expression seems to directly control axonal diameter, which in turn influences how fast electrical signals travel down the axon. In addition, neurofilaments provide strength and stability to the long cell processes of neurons.

The neurodegenerative disease amyotrophic lateral sclerosis (ALS, or Lou Gehrig's disease) is associated with an accumulation and abnormal assembly of neurofilaments in motor neuron cell bodies and in the axon, aberrations that may interfere with normal axonal transport. The degeneration of the axons leads to muscle weakness and atrophy, which is usually fatal. The overexpression of human NF-L or NF-H in mice results in mice that have an ALS-like disease. However, a causative link between neurofilament pathology and ALS has not been firmly established.

The vimentin-like filaments are a third family of intermediate filaments. *Desmin*, a member of this family, is expressed in skeletal, cardiac, and smooth muscle, where it forms a scaffold around the Z disc of the sarcomere (see Figure 16–29). Mice lacking desmin show normal initial muscle development, but adults have various muscle cell abnormalities, including misaligned muscle fibers. In humans, mutations in desmin are associated with various forms of muscular dystrophy and cardiac myopathy, illustrating the important role of desmin in stabilizing muscle fibers.

Inside the nucleus, nuclear lamins maintain the mechanical stability of the nucleus. In addition, it is becoming increasingly evident that one class of lamins, the A-type, together with many proteins of the nuclear envelope, are scaffolds for proteins that control myriad cellular processes including transcription, chromatin organization, and signal transduction. The majority of *laminopathies* is associated with mutant versions of lamin A and include tissue-specific diseases. Skeletal and cardiac abnormalities might be explained by a weakened nuclear envelope leading to cell damage and death, but laminopathies are also thought to arise from pathogenic and tissue-specific alterations in gene expression.

Linker Proteins Connect Cytoskeletal Filaments and Bridge the Nuclear Envelope

The intermediate filament network is linked to the rest of the cytoskeleton by members of a family of proteins called *plakins*. Plakins are large and modular, containing multiple domains that connect cytoskeletal filaments to each other and to junctional complexes. *Plectin* is a particularly interesting example. In addition to bundling intermediate filaments, it links the intermediate filaments to microtubules, actin filament bundles, and filaments of the motor protein myosin II; it also helps attach intermediate filament bundles to adhesive structures at the plasma membrane (**Figure 16–66**).

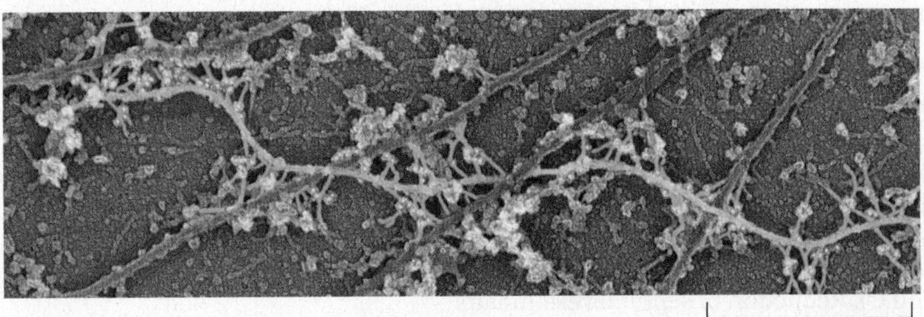

Figure 16–66 **Plectin cross-linking of diverse cytoskeletal elements.** Plectin *(green)* is seen here making cross-links from intermediate filaments *(blue)* to microtubules *(red)*. In this electron micrograph, the dots *(yellow)* are gold particles linked to anti-plectin antibodies. The entire actin filament network was removed to reveal these proteins. (© 1996 T.M. Svitkina et al. Originally published in *J. Cell Biol.* http://doi.org/10.1083/jcb.135.4.991. With permission from Rockefeller University Press.)

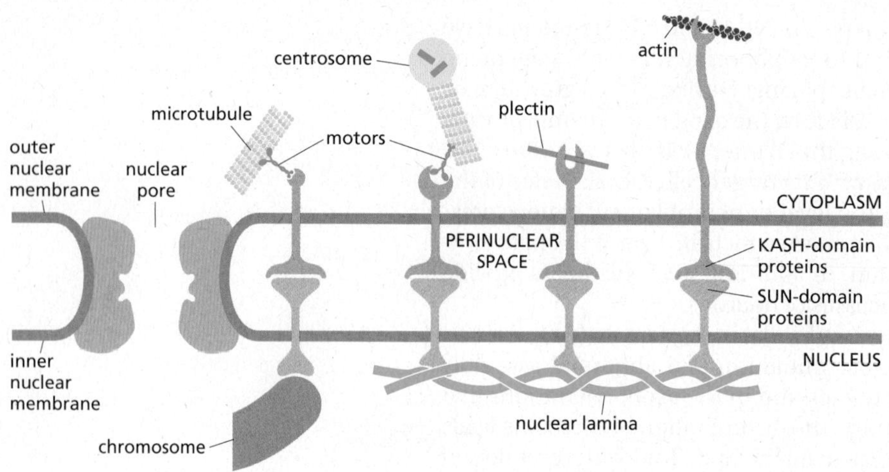

Figure 16–67 SUN–KASH protein complexes connect the nucleus and cytoplasm through the nuclear envelope. The cytoplasmic cytoskeleton is linked across the nuclear envelope to the nuclear lamina or chromosomes through SUN and KASH proteins *(orange* and *purple*, respectively). The SUN and KASH domains of these proteins bind within the lumen of the nuclear envelope. From the inner nuclear envelope, SUN proteins connect to the nuclear lamina or chromosomes. KASH proteins in the outer nuclear envelope connect to the cytoplasmic cytoskeleton by binding microtubule motor proteins, actin filaments, or plectin.

Mutations in the gene for plectin cause a devastating human disease that combines epidermolysis bullosa (caused by disruption of skin keratin filaments), muscular dystrophy (caused by disruption of desmin filaments), and neurodegeneration (caused by disruption of neurofilaments). Mice lacking a functional plectin gene die within a few days of birth, with blistered skin and abnormal skeletal and heart muscles. Thus, although plectin may not be necessary for the initial formation and assembly of intermediate filaments, its cross-linking action is required to provide cells with the strength they need to withstand the mechanical stresses inherent to vertebrate life.

Plectin and other plakins can interact with protein complexes that connect the cytoskeleton to the nuclear interior. These complexes consist of SUN proteins of the inner nuclear membrane and KASH proteins of the outer nuclear membrane (**Figure 16–67**). SUN and KASH proteins bind to each other within the lumen of the nuclear envelope, forming a bridge that connects the nuclear and cytoplasmic cytoskeletons. Inside the nucleus, the SUN proteins bind to the nuclear lamina or chromosomes, whereas in the cytoplasm, KASH proteins can bind directly to actin filaments and indirectly to microtubules and intermediate filaments through association with motor proteins and plakins, respectively. This linkage serves to mechanically couple the nucleus to the cytoskeleton and is involved in many cellular functions, including chromosome movements inside the nucleus during meiosis, nuclear and centrosome positioning, nuclear migration, and global cytoskeletal organization.

Septins Form Filaments That Contribute to Subcellular Organization

GTP-binding proteins called *septins* serve as an additional filament system in all eukaryotes except terrestrial plants. Septins assemble into nonpolar filaments that form rings and cage-like structures, which act as scaffolds to compartmentalize membranes into distinct domains or to recruit and organize the actin and microtubule cytoskeletons. First identified in budding yeast, septin filaments localize to the neck between a dividing yeast mother cell and its growing bud (**Figure 16–68A**). At this location, septins form a barrier that restricts lateral diffusion of proteins embedded in the plasma membrane, enabling cell growth to be concentrated preferentially within the bud. Septins also recruit the actin–myosin machinery that forms the contractile ring required for cytokinesis. In animal cells, septins function in cell division, migration, vesicle trafficking, and cell signaling. In primary cilia, for example, a ring of septin filaments assembles at the base of the cilium and serves as a diffusion barrier at the plasma membrane, restricting the movement of membrane proteins and establishing a specific composition in the ciliary membrane (**Figure 16–68B and C**). Reduction of septin levels impairs primary cilium formation and signaling.

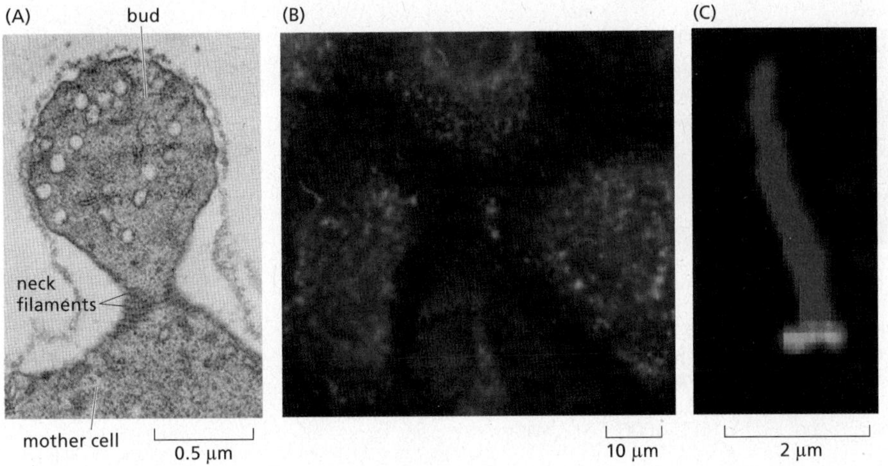

Figure 16–68 Cell compartmentalization by septins. (A) Septins form filaments in the neck region between a mother yeast cell and bud. (B) In this photomicrograph of human cultured cells, the DNA is stained *blue* and septins are labeled in *green*. The microtubules of primary cilia are labeled with an antibody that recognizes a modified (acetylated) form of tubulin *(red)* that is enriched in the axoneme. (C) A magnified image reveals a collar of septin at the base of the cilium. (A, © 1976 B. Byers and L. Goetsch. Originally published in *J. Cell Biol.* https://doi.org/10.1083/jcb.69.3.717. With permission from Rockefeller University Press; B and C, from Q. Hu et al., *Science* 329:436–439, 2010. With permission from AAAS.)

There are 7 septin genes in yeast and 13 in humans, and septin proteins fall into four groups on the basis of sequence relationships. In a test tube, purified septins assemble into symmetrical hetero-hexamers or hetero-octamers that form nonpolar paired filaments (**Figure 16–69**). GTP binding is required for the folding of septin polypeptides, but the role of GTP hydrolysis in septin function is not understood. Septin structures assemble and disassemble inside cells, but they are not as dynamic as actin filaments and microtubules.

Bacterial Cell Shape and Division Depend on Homologs of Eukaryotic Cytoskeletal Proteins

Although they are much smaller than a typical eukaryotic cell, bacterial cells of different species assume a variety of shapes, from spheres or rods to more elaborate morphologies including stars, spirals, and branched filaments (see Figure 23–3).

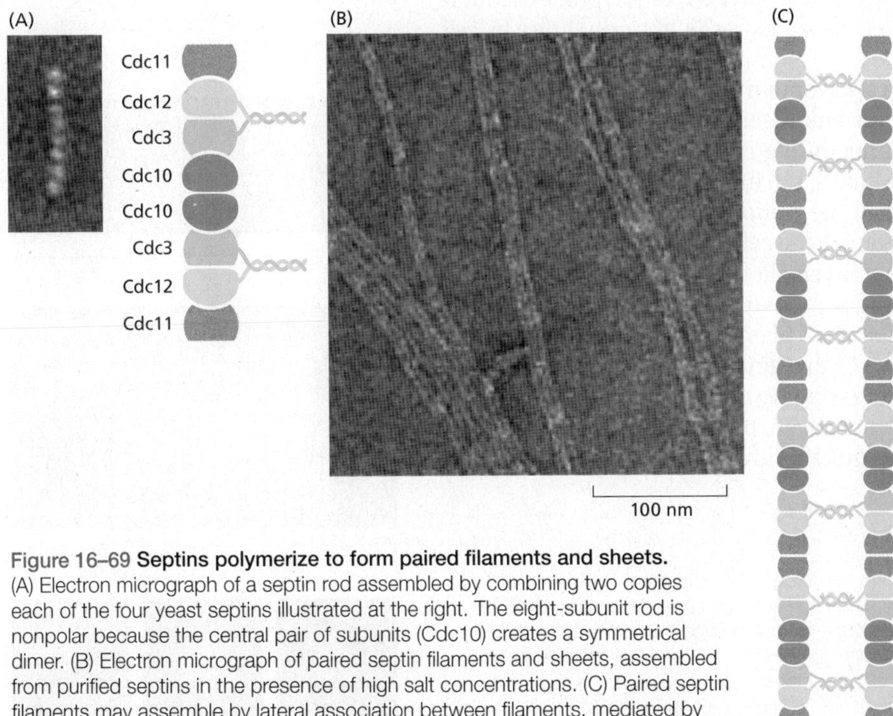

Figure 16–69 Septins polymerize to form paired filaments and sheets. (A) Electron micrograph of a septin rod assembled by combining two copies each of the four yeast septins illustrated at the right. The eight-subunit rod is nonpolar because the central pair of subunits (Cdc10) creates a symmetrical dimer. (B) Electron micrograph of paired septin filaments and sheets, assembled from purified septins in the presence of high salt concentrations. (C) Paired septin filaments may assemble by lateral association between filaments, mediated by coiled-coils formed between the paired C-terminal extensions of Cdc3 and Cdc12 that project from each filament. (From A. Bertin et al., *Proc. Natl. Acad. Sci. USA* 105:8274–8279, 2008. With permission from the National Academy of Sciences.)

Figure 16–70 Actin homologs in bacteria determine cell shape. The common soil bacterium *Bacillus subtilis* normally forms cells with a regular rodlike shape when viewed by scanning electron microscopy *(left)*. In contrast, *B. subtilis* cells lacking the actin homolog MreB or Mbl grow in distorted or twisted shapes and eventually die *(center* and *right)*. WT = wild type. [From A. Chastanet and R. Carballido-Lopez, *Front. Biosci. (Schol. Ed.)* 4:1582–1606, 2012. With permission from Frontiers in Bioscience.]

Historically, biologists assumed that a cytoskeleton was not necessary in such simple cells that also lack extensive networks of intracellular membrane-enclosed organelles. We now know, however, that the outwardly simple morphology of bacterial cells is deceptive. Bacterial cells are highly organized and contain homologs of actin, tubulin, and intermediate filaments. These filament systems are essential to regulate the synthesis and remodeling of the peptidoglycan cell wall to define cell shape and mediate cell division. Furthermore, the bacterial cytoskeleton can also play important roles in DNA segregation and intracellular organization.

Many bacteria contain homologs of actin. Two of these, MreB and Mbl, are found primarily in rod-shaped or spiral-shaped cells where they assemble to form dynamic patches that move circumferentially along the length of the cell. These proteins contribute to cell shape by serving as a scaffold to direct the synthesis of the peptidoglycan cell wall, in much the same way that microtubules help organize the synthesis of the cellulose cell wall in higher plant cells (see Figure 19–66). MreB and Mbl filaments are highly dynamic, with half-lives of a few minutes, and ATP hydrolysis accompanies the polymerization process. Mutations disrupting MreB or Mbl expression cause extreme abnormalities in cell shape (**Figure 16–70**).

Nearly all bacteria and many archaea also contain a homolog of tubulin called FtsZ, which can polymerize into filaments and assemble into a ring (called the Z-ring) at the site where the **septum** forms during cell division (**Figure 16–71**). Although the Z-ring persists for many minutes, the individual filaments within it are highly dynamic and have an average half-life of about 30 seconds, as they treadmill along the cell circumference and organize the cell-wall synthesis machinery. As the bacterium divides, the Z-ring becomes smaller until it has completely disassembled. However, because of high turgor pressure within bacterial cells, the energy derived from GTP hydrolysis within FtsZ filaments is unlikely to generate enough bending force to drive the membrane invagination necessary to complete cell division. Instead, FtsZ dynamics in the Z-ring are thought to spatially organize the cell-wall synthesis machinery to promote even, processive constriction. This hypothesis may explain why FtsZ GTPase mutants are viable, but yield deformed septa in *Escherichia coli* cells.

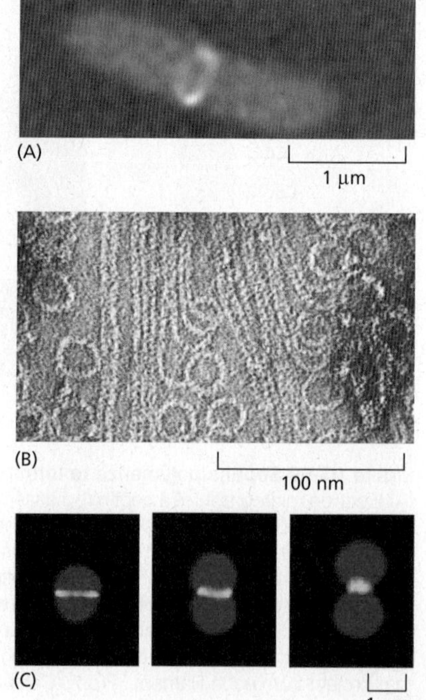

Figure 16–71 The bacterial FtsZ protein, a tubulin homolog in prokaryotes. (A) A band of FtsZ protein forms a ring in a dividing bacterial cell. This ring has been labeled by fusing the FtsZ protein to green fluorescent protein (GFP), which allows it to be observed in living *E. coli* cells with a fluorescence microscope. (B) FtsZ filaments and circles, formed *in vitro*, as visualized using electron microscopy. (C) Dividing chloroplasts *(red)* from a red alga also cleave using a protein ring made from FtsZ *(yellow)*. (A, from X. Ma et al., *Proc. Natl. Acad. Sci. USA* 93:12998–13003, 1996; B, from H.P. Erickson et al., *Proc. Natl. Acad. Sci. USA* 93:519–523, 1996. Both with permission from National Academy of Sciences; C, from S. Miyagishima et al., *Plant Cell* 13:2257–2268, 2001, with permission from American Society of Plant Biologists.)

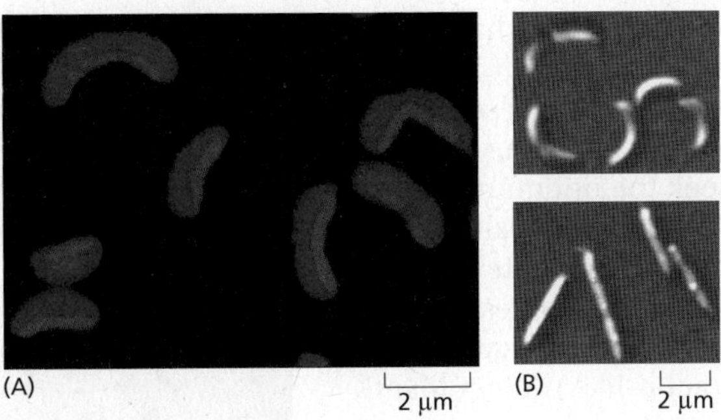

Figure 16–72 ***Caulobacter*** **and crescentin.** The sickle-shaped bacterium *Caulobacter crescentus* expresses a protein, crescentin, with a series of coiled-coil domains similar in size and organization to the domains of eukaryotic intermediate filaments. (A) The crescentin protein forms a fiber (labeled in *red*) that runs down the inner side of the curving bacterial cell wall. (B) When the gene is disrupted, the bacteria grow as straight rods *(bottom)*. (From N. Ausmees et al., *Cell* 115:705–713, 2003. With permission from Elsevier.)

In addition to actin and tubulin homologs, at least one bacterial species, *Caulobacter crescentus,* also harbors a protein with significant structural similarity to intermediate filaments. A protein called crescentin forms a filamentous structure that influences the unusual crescent shape of this species; when the gene encoding crescentin is deleted, the *Caulobacter* cells grow as straight rods (**Figure 16–72**). Attached to the membrane along the side of inner curvature, crescentin is thought to exert a compressive force that locally decreases peptidoglycan insertion and therefore biases cell-wall synthesis to the opposite side.

Other bacterial cytoskeletal proteins function in DNA segregation during cell division. A particularly intriguing bacterial actin homolog is ParM, which is encoded by a gene on certain bacterial plasmids that also carry genes responsible for antibiotic resistance and cause the spread of multidrug resistance in epidemics. Bacterial plasmids typically encode all the gene products that are necessary for their own segregation, presumably as a strategy to ensure their inheritance and propagation in bacterial hosts after plasmid replication. ParM assembles into filaments that associate at each end with a copy of the plasmid, and growth of the ParM filament pushes the replicated plasmid copies apart (**Figure 16–73**). This spindle-like structure apparently arises from the selective stabilization of filaments that bind to specialized proteins recruited to the origins of replication on

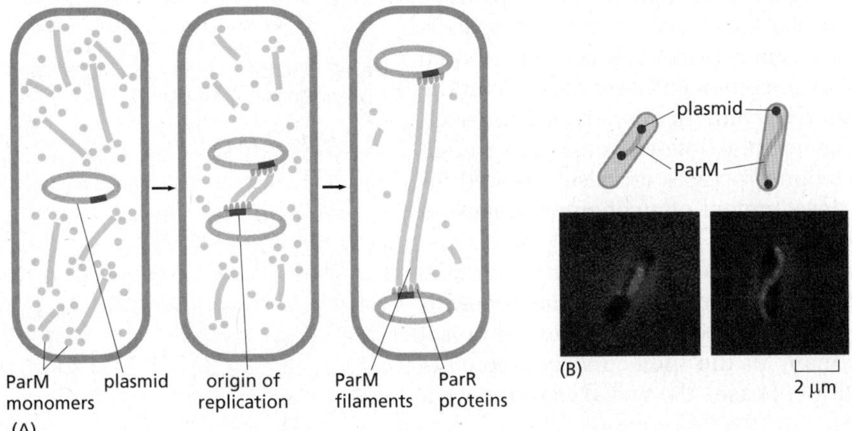

Figure 16–73 **Role of the actin homolog ParM in plasmid segregation in bacteria.** (A) Some bacterial drug-resistance plasmids *(orange)* encode an actin homolog, ParM, that will spontaneously nucleate to form small, dynamic filaments *(green)* throughout the bacterial cytoplasm. A second plasmid-encoded protein called ParR *(blue)* binds to specific DNA sequences in the plasmid and also stabilizes the dynamic ends of the ParM filaments. When the plasmid duplicates, both ends of the ParM filaments become stabilized, and the growing ParM filaments push the duplicated plasmids to opposite ends of the cell. (B) In these bacterial cells harboring a drug-resistance plasmid, the plasmids are labeled in *red* and the ParM protein in *green*. *Left*, a short ParM filament bundle connects the two daughter plasmids shortly after their duplication. *Right*, the fully assembled ParM filament has pushed the duplicated plasmids to the cell poles. (A, adapted from E.C. Garner et al., *Science* 306:1021–1025, 2004; B, from J. Møller-Jensen et al., *Mol. Cell* 12:1477–1487, 2003. With permission from Elsevier.)

the plasmids. A distant relative of both tubulin and FtsZ, called TubZ, has a similar function in other bacterial species.

Finally, the cytoskeleton plays a role in the internal organization of some bacterial cells. The best-known example applies to magnetotactic bacteria, which are capable of swimming along Earth's magnetic field to seek the optimum aquatic environment. These bacteria contain *magnetosomes*, which are small vesicles invaginated from the plasma membrane that surround magnetite crystals. Arranged in a straight line along the length of the cell, magnetosomes produce a dipole, reminiscent of a compass needle. The organization of magnetosome chains requires their association with filaments of the actin-like protein MamK (**Figure 16–74**).

Thus, all cells contain cytoskeletal proteins that perform a wide variety of functions, and the actin and tubulin families are very ancient, predating the split between the eukaryotic and bacterial kingdoms.

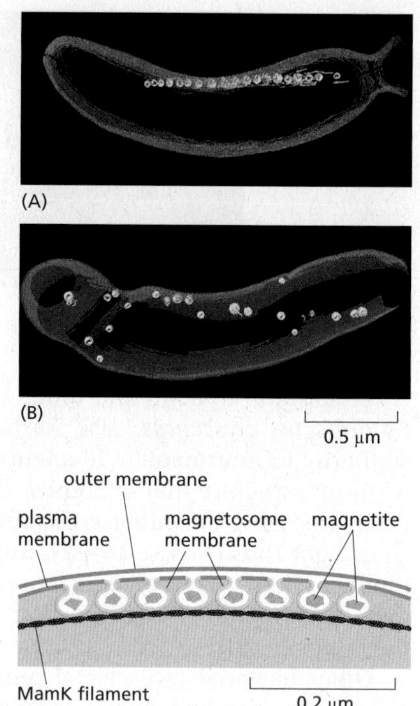

Figure 16–74 MamK organizes chains of magnetosomes. (A) Three-dimensional reconstruction of a wild-type *Magnetospirillum magneticum* cell showing the cell membrane *(gray)*, magnetosomes *(yellow)*, and magnetosome-associated MamK filaments *(green)*. (B) Magnetosomes appear disordered in a *mamk* deletion mutant, and no filaments are observed. (C) Magnetosomes are formed from invaginations of the plasma membrane and form chains along MamK filaments. (A and B, adapted from A. Komeili et al., *Science* 311:242–245, 2006. With permission from AAAS.)

Summary

Whereas tubulin and actin have been highly conserved in evolution, intermediate filament proteins are very diverse. There are many tissue-specific forms of intermediate filaments in the cytoplasm of animal cells, including keratin filaments in epithelial cells, neurofilaments in nerve cells, and desmin filaments in muscle cells. The primary function of these filaments is to provide mechanical strength. Septins comprise an additional system of filaments that organize compartments inside cells. Bacterial cells also contain homologs of actin, tubulin, and intermediate filaments that form dynamic structures essential for cell shape and division.

CELL POLARITY AND COORDINATION OF THE CYTOSKELETON

A central challenge in cell biology is to understand how multiple individual molecular components collaborate to produce complex cell behaviors. Cell polarity, which we describe in this final part of the chapter, controls many aspects of cell function, such as the direction of protein secretion and signaling, the orientation of cell division, and the path a migrating cell will take. Cells polarize in response to extracellular cues or intracellular landmarks to establish specific domains on their surface. Coordination with the cytoskeleton is then required for a cell to build different structures with distinct molecular components at the front versus the back, or at the top versus the bottom. In this way, the cytoskeleton acts to transduce polarity signals to generate whole-cell organization and behavior. Carefully controlled cell-polarization processes are also required for oriented cell divisions in tissues and for development of a coherent, organized multicellular organism.

Genetic studies in yeast, flies, and worms have provided most of our current understanding of the molecular basis of cell polarity. As we shall see, with increasing complexity of a cellular system comes greater elaboration of polarity-determining mechanisms. However, many of the molecular components have been evolutionarily conserved, and in all cases the cytoskeleton plays a central role.

Cell Polarity Is Governed by Small GTPases in Budding Yeast

The establishment of cell polarity often begins with local regulation of the actin cytoskeleton by external or internal signals. Many polarity signals converge just beneath the plasma membrane through activation of a group of closely related small monomeric GTPases that are members of the **Rho family**—*Cdc42, Rac*, and *Rho*. Like other monomeric GTPases, the Rho proteins act as molecular switches that cycle between an active GTP-bound state and an inactive GDP-bound state (see Figure 3–63). The state of each GTPase depends on dedicated regulatory proteins. Guanine nucleotide exchange factors (GEFs) are required to activate

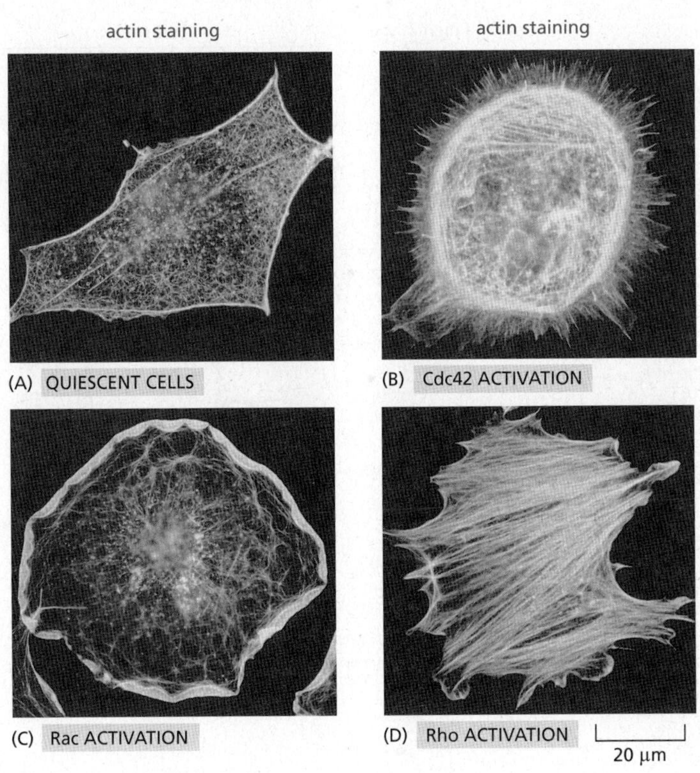

Figure 16–75 **The dramatic effects of Cdc42, Rac, and Rho on actin organization in fibroblasts.** In each case, the actin filaments have been labeled with fluorescent phalloidin. (A) Serum-starved fibroblasts have actin filaments primarily in the cortex and relatively few stress fibers. (B) Microinjection of a constitutively activated form of Cdc42 results in many long filopodia at the cell periphery. (C) Microinjection of a constitutively activated form of Rac causes the formation of an enormous lamellipodium that extends from the entire circumference of the cell. (D) Microinjection of a constitutively activated form of Rho causes the rapid assembly of many prominent stress fibers. (From A. Hall, *Science* 279:509–514, 1998. With permission from Catherine Nobes.)

the GTPase by replacing tightly bound GDP with GTP, whereas GTPase activating proteins (GAPs) inactivate the GTPase by promoting GTP hydrolysis, which is otherwise very slow. In addition, guanine nucleotide dissociation inhibitors (GDIs) can bind to the GDP-bound form of a GTPase and inhibit GTP exchange by a GEF.

When the GTP-bound form of each Rho family GTPase is introduced into a fibroblast cell, dramatic changes in cell shape are observed. Active Cdc42 leads to the formation of many long filopodia at the cell surface. Activation of Rac promotes actin polymerization at the cell periphery, leading to the formation of sheet-like lamellipodial protrusions. Activation of Rho promotes both the bundling of actin filaments with myosin II filaments into stress fibers and the clustering of integrins and associated proteins to form focal adhesions (**Figure 16–75**). These dramatic and complex structural changes occur because each of these three molecular switches has numerous downstream target proteins that affect actin organization and dynamics. Normally, however, these GTPase pathways are not activated uniformly throughout the cell as in this experiment, but are deployed with precise spatial and temporal regulation to generate polarized subcellular structures, which in turn give rise to changes in cell shape and behavior.

Cdc42 is the most highly conserved of all the Rho family GTPases and a master regulator of cell polarity in many cell types. Its importance in the establishment of cell polarity is illustrated by its role in the budding yeast *Saccharomyces cerevisiae*, which undergoes a highly polarized cell division. The formation of a new bud begins with the selection of a single bud site on the cell surface. It is crucial that only a single site is selected, because producing more than one bud would be detrimental to cell division. Before bud site selection, inactive GDP-bound Cdc42 is uniformly distributed on the cell membrane. Occasionally, a Cdc42 molecule will release its GDP and bind GTP, leading to the formation of multiple GTP-Cdc42 foci at random locations in the membrane. Eventually, one of these Cdc42 molecules recruits a protein kinase called PAK, which in turn recruits a scaffold protein together with the Cdc42 GEF. The GEF-containing complex promotes activation of neighboring GDP-bound Cdc42 molecules, resulting in positive feedback. The clustering of Cdc42-GTP molecules at a single site depletes the cytoplasmic pool of

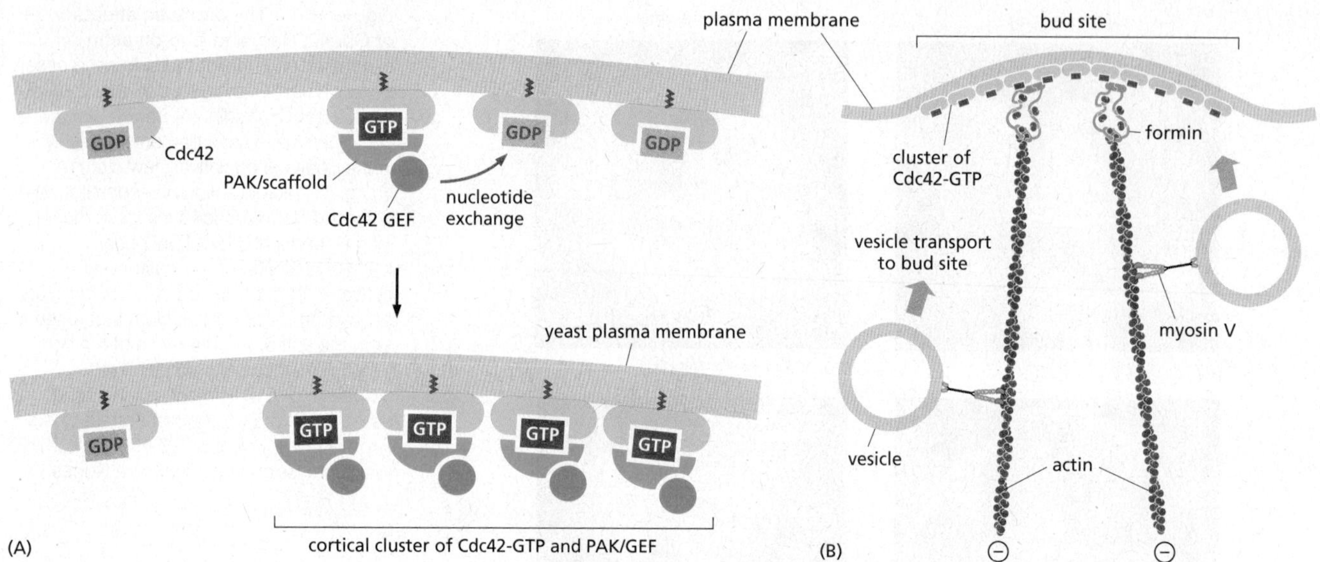

Figure 16–76 **Cdc42 establishes yeast-cell polarity.** (A) A positive feedback loop by which Cdc42-GTP recruits its own GEF to the plasma membrane to generate a focal site of Cdc42 activity. (B) Local activation of a formin protein by Cdc42-GTP nucleates actin filament assembly. Transport of vesicles along these actin filaments toward their plus ends by myosin V delivers cargoes necessary for growth of the bud.

the Cdc42 GEF, thereby ensuring the formation of just one localized site of Cdc42 activation (**Figure 16–76A**). This cluster of Cdc42-GTP then transmits a signal that polarizes the cytoskeleton by recruiting and activating a formin protein. Recall that formins stimulate rapid assembly of long, straight actin filaments and remain tethered at their plus ends (see Figure 16–13). The resulting actin filaments enable bud growth through the delivery of secretory vesicles and other cargoes to the polarity site by type V myosins (**Figure 16–76B**; see also Figure 16–36). In addition, both PAK and formin contribute to assembly of septin filaments at the bud neck (see Figure 16–68A). Thus, a low level of Cdc42 activity is locally amplified by positive feedback to initiate the polarized assembly of actin and septins at a single site on the mother cell. Downstream effectors then contribute to bud growth and polarized cell division to produce a daughter cell.

PAR Proteins Generate Anterior–Posterior Polarity in Embryos

Most animal cells do not polarize toward a single membrane site like budding yeast, but instead form complementary cortical domains that mark opposite ends of the cell. This form of polarity has been studied extensively in the early stages of embryonic development of the nematode *Caenorhabditis elegans*. The unfertilized egg is symmetrical, but the fertilized egg, or zygote, rapidly establishes the anterior–posterior axis. The first cell division then occurs asymmetrically along this axis, giving rise to two daughter cells that are of different sizes, compositions, and fates (see Figure 17–50). Genetic screens for mutants defective in this asymmetrical cell division identified *partitioning defective* (*par*) genes that encode the so-called *PAR proteins*. Subsequent studies revealed that other factors are also involved, and the precise molecular mechanisms by which anterior–posterior polarity is established in this system remain under active investigation.

The initial events leading to polarization of the zygote depend on regulation of the cortical actin cytoskeleton by the GTPase Rho. Egg symmetry is first broken when the sperm enters the egg, which marks the location of the posterior end of the embryo. The centrosome associated with the sperm nucleates a microtubule aster, which through an unknown mechanism depletes the Rho GEF from the acto-myosin cortex in that region of the cell. Local loss of Rho activity decreases myosin II–dependent cortical contractility, resulting in greater tension toward the

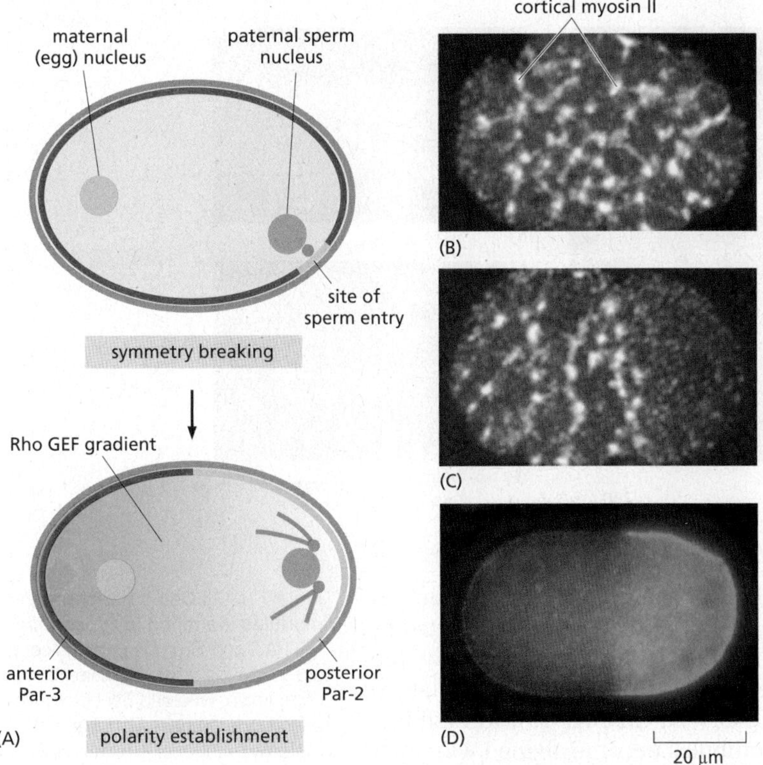

Figure 16–77 **PAR proteins establish two distinct cortical domains in *C. elegans*.** (A) Symmetry breaking and polarity establishment occur in the fertilized egg before the maternal and paternal nuclei meet. After sperm entry, its centrosome duplicates and nucleates microtubules that decrease Rho GEF activity at what will become the posterior end of the embryo. This leads to accumulation of the anterior PAR proteins (including Par-3), allowing posterior PAR proteins (including Par-2) to bind the cortex at the posterior end. (B) Prior to fertilization, bundles of fluorescently labeled myosin II *(white)* are distributed throughout the cortex of the unpolarized egg because of the uniform distribution of activated Rho all along the plasma membrane, resulting in uniform cortical acto-myosin contractility. (C) After fertilization, local depletion of the Rho GEF near the sperm entry site (at right in this image) reduces myosin levels and contractility in the posterior cortex of the cell. (D) Par complex localization after polarization with Par-3 *(red)* at the anterior and Par-2 *(green)* at the posterior of the zygote. Multiple mechanisms operate to maintain this asymmetry through mutual antagonism between anterior and posterior PAR proteins. (B and C, from L. Rose and P. Gönczy, The *C. elegans* Research Community, WormBook, 2014, doi 10.1895/wormbook.1.30.2; D, from J. Nance and J.A. Zallen, *Development* 138:799–809, 2011. With permission from the Company of Biologists.)

other end of the cell. This asymmetry then sets up localization of the PAR proteins so that they occupy separate cortical domains.

One set of PAR proteins, the anterior PAR proteins, includes Par-3, Par-6, Cdc42, and atypical protein kinase C (aPKC). Initially unpolarized at the cortex, these proteins move toward the anterior because of the change in contractility that causes flow within the membrane. Enrichment of the anterior PAR proteins actively displaces another set of proteins (the posterior PAR proteins, including Par-1 and Par-2), which then bind to the posterior cortex (**Figure 16–77**). These complementary cortical domains are maintained by mutual antagonism between the anterior and posterior components, which occurs through multiple mechanisms. For example, phosphorylation by aPKC in the anterior region excludes the posterior PAR proteins, while the posterior PAR proteins suppress aPKC activity in their domain. It remains unclear how PAR proteins then direct the trafficking and localization of other proteins to define the cortical domains. As in yeast, the master polarity regulator Cdc42, which is activated exclusively at the anterior cortex, likely plays a key role.

In all animal species, mechanisms that establish polarity in the embryo determine the overall body plan (see Chapter 21). Subsequent polarized cell divisions produce daughter cells that are destined to become different tissues in the organism. In this way, the establishment of intracellular polarity early in development sets the stage for polarity throughout the adult animal.

Conserved Complexes Polarize Epithelial Cells and Control Their Growth

All animal cells are polarized in some way, perhaps most prominently in the epithelial cells that make up many tissues. Epithelia are sheets of tightly connected cells that line the surface of organs and act as highly selective barriers, such as the barriers between the skin and the outside environment or between the lumen of the gut and its surrounding tissues. Epithelial cells have two distinct domains: the *apical* domain at the upper surface, which faces the outside environment, and

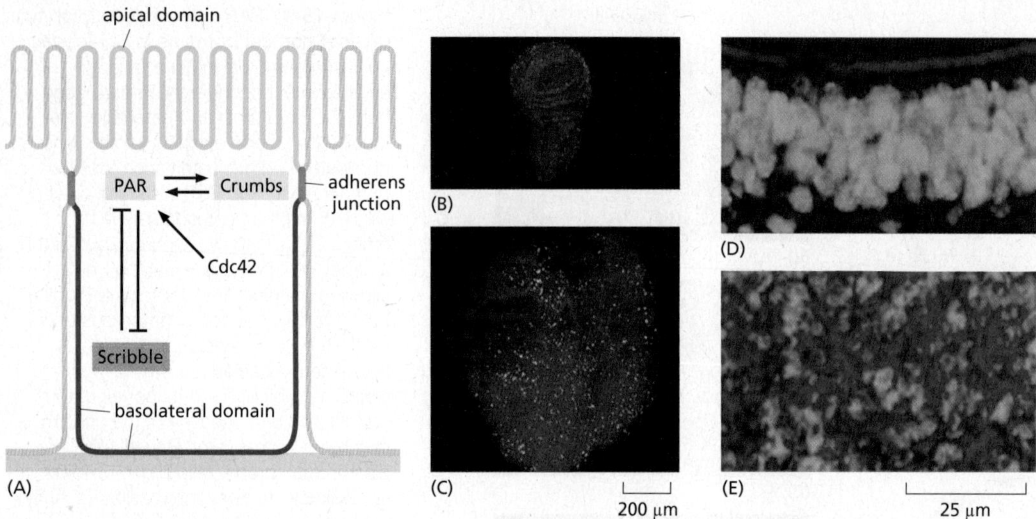

Figure 16–78 Cell-polarity protein modules identified in *Drosophila*. (A) PAR and Crumbs proteins cooperate to assemble the apical domain and junctional complexes, whereas Scribble defines the basolateral domain. Scribble and PAR are mutually antagonistic, whereas PAR and Crumbs reinforce each other. Cdc42 helps to recruit PAR proteins. (B) Fluorescence micrograph of a wild-type *Drosophila* larval epithelial tissue called the imaginal disc stained for actin *(red)* shows the well-organized, folded monolayer epithelium. (C) In a Scribble mutant, the imaginal disc has lost its normal morphology and grown dramatically larger because of hyperproliferation. Actively dividing cells are labeled *yellow*. (D) Higher-magnification image of a wild-type disc shows the polarized localization of actin *(red)* to the apical domain. (E) Actin organization is disrupted in the absence of Scribble. Nuclei are stained *blue*. (B and C, courtesy of David Bilder; D and E, from D. Bilder et al., *Science* 289:113–116, 2000. With permission from AAAS.)

the *basolateral* domain at the bottom and side surfaces, which face the underlying matrix and adjacent cells (see Figure 16–4). Cell polarization is initiated by cues from the adjacent cells or the extracellular matrix. An important early step is the formation of cell-cell junctions that separate the apical and basolateral domains and hold the epithelial sheet together (see Chapter 19). In addition, as discussed earlier (see Figure 16–44), microtubules become aligned with their minus ends anchored at the apical surface and their plus ends pointing basally. This arrangement helps to polarize the secretory system so that nascent secreted and membrane proteins are transported to the appropriate domain. For example, digestive enzymes must be secreted exclusively into the lumen of the gut to avoid extensive damage to the surrounding tissues.

Much of what we know about the regulators of epithelial polarity comes from genetic screens that identified polarity-deficient mutants in *Drosophila*. These screens revealed some familiar proteins, including Cdc42 and the anterior PAR proteins, which in this system localize to the apical domain. In addition, two other protein groups were identified, termed *Crumbs* and *Scribble*, which are located in the apical and basolateral domains, respectively. Cells cannot form an apical domain in the absence of PAR or Crumbs proteins. In contrast, the apical domain is greatly expanded in the absence of Scribble proteins. This and other evidence indicate that mutual antagonism between apical and basolateral regulators helps maintain cell polarity and junctional contacts (**Figure 16–78A**). Downstream of the PAR, Crumbs, and Scribble modules, the organization of epithelial cell junctions, cytoskeleton, and secretory pathway is mediated by Rho family GTPases; the mechanisms underlying this regulation are not well understood.

Disruption of these polarity modules can lead to cellular growth defects, often through misregulation of a signaling system called the Hippo pathway (see Figure 21–67). For example, mutation of Scribble in epithelia of the developing fly larva not only leads to the loss of polarized cell organization but also promotes massive and invasive overgrowth, all characteristics of malignant tumors (**Figure 16–78B, C, D, and E**). Because Scribble proteins also help maintain polarity and normal growth in human epithelial tissues, they act as tumor suppressors (see Chapter 20).

Cell Migration Requires Dynamic Cell Polarity

Whereas epithelial cells maintain a polarized state throughout their lifetime to perform an essential barrier function within tissues, migrating cells display a dynamic polarity that requires continual long-distance communication and coordination between one end of a cell and the other. In addition to driving local mechanical processes such as protrusion at the front and retraction at the rear,

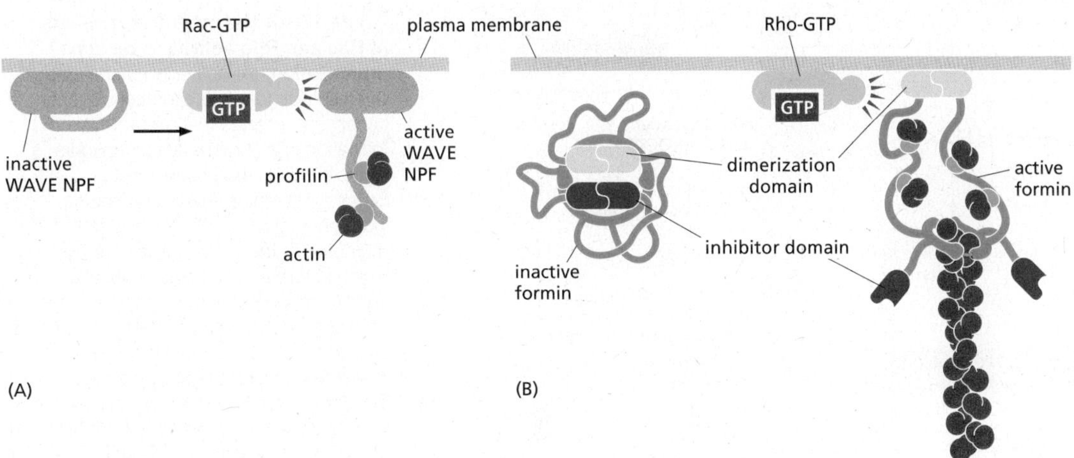

Figure 16–79 **Regulation of nucleation-promoting factors and formins by Rho family GTPases.** (A) Members of the WAVE family of nucleation-promoting factors are activated upon binding to Rac-GTP at the plasma membrane. A conformational change in WAVE opens up the protein, allowing domains that were previously inaccessible to interact with both the Arp2/3 complex and profilin-bound actin subunits. For clarity, the WAVE regulatory complex is not shown. (B) The activity of some formin proteins is inhibited by an autoinhibitory interaction in the absence of an active GTPase. Binding to Rho-GTP exposes a binding site for the plus end of an actin filament, as well as profilin–actin binding domains.

the cytoskeleton is responsible for coordinating cell shape, organization, and mechanical properties along its entire length, a distance that is typically tens of micrometers for animal cells. As we shall see, the Rho family GTPases and several cell-polarity proteins that we have already introduced are central to this process.

Cell locomotion requires an initial polarization of the cell to propel it in a particular direction. Once again, Cdc42 appears to be critical, as it sets up the overall polarity of a migrating cell. The downstream effectors of Cdc42-GTP include the polarity protein Par-3, and Cdc42 is also thought to stimulate extension of filopodia (see Figure 16–75) that help sense and respond to extracellular cues. Once polarized, protrusion of the leading edge of the cell is driven through nucleation of branched actin filaments by the Arp2/3 complex (see Figure 16–12). This process depends on Rac-GTP, which stimulates members of the WAVE family of actin nucleation-promoting factors (NPFs). A WAVE protein, present in a large regulatory complex, can exist in an inactive folded conformation and an activated open conformation. Association with Rac-GTP recruits WAVE to the plasma membrane and stabilizes the open conformation (**Figure 16–79A**). Active WAVE then stimulates nucleation of new actin filaments by Arp2/3 along the sides of existing filaments (see Figures 16–12 and 16–14A). In this way, Rac-GTP–dependent WAVE promotes the formation of branched actin networks that drive lamellipod formation at the leading edge of migrating cells (**Figure 16–80A**).

Rho-GTP is equally important for cell migration and has a very different set of targets. Instead of activating the Arp2/3 complex to build branched actin networks, Rho-GTP induces formin proteins to construct parallel actin bundles. Similar to the effects of Rac-GTP on WAVE, association of formins with Rho-GTP stabilizes an open, active conformation (**Figure 16–79B**). As discussed previously, formins stimulate both the nucleation and elongation of straight actin filaments (see Figures 16–13 and 16–14B). At the same time, Rho-GTP activates the Rho-associated kinase (ROCK), which stimulates another kinase called LIM, which phosphorylates and inhibits the activity of the actin-destabilizing protein cofilin. The resulting stable, unbranched actin filaments are ideal for interacting with myosin II. Furthermore, ROCK activates myosin II by inhibiting a phosphatase acting on myosin light chains (see Figure 16–34). The consequent increase in the net amount of myosin light-chain phosphorylation increases the level of contractile myosin motor-protein activity in the cell, enhancing the formation of tension-dependent structures such as stress fibers and focal adhesions (**Figure 16–80B**).

Spatial and temporal separation between the Rac and Rho pathways is thought to facilitate maintenance of the large-scale differences between the cell front and the cell rear during migration. Whereas Rac is activated exclusively at the leading edge of the cell, Rho activation predominates at the rear. Furthermore, the two pathways are mutually antagonistic. For example, Rac inhibits Rho activity through one of its effectors, the kinase PAK, which inhibits myosin II and therefore

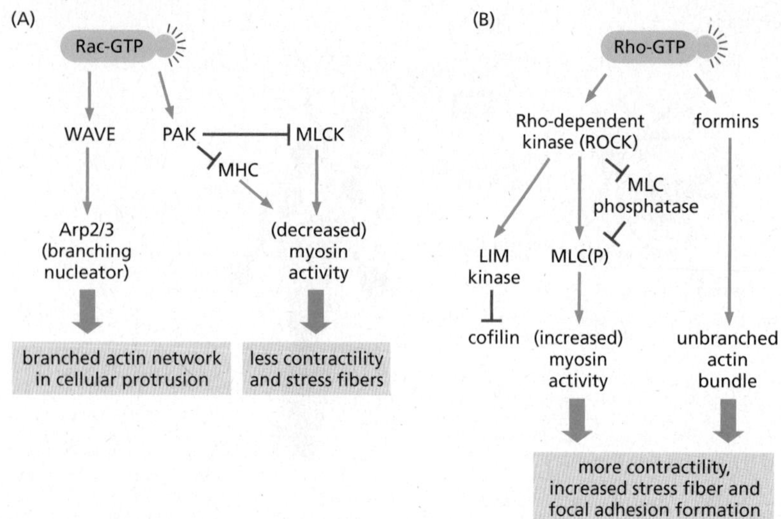

Figure 16–80 **The contrasting effects of Rac and Rho activation on actin organization.** (A) Activation of the small GTPase Rac leads to alterations in actin accessory proteins that promote the formation of protrusive actin networks in lamellipodia and pseudopodia. Several different pathways contribute independently. Rac-GTP activates members of the WAVE protein family, which in turn activate actin nucleation and branched network formation by the Arp2/3 complex. In a parallel pathway, Rac-GTP activates the protein kinase PAK, which has several targets including the myosin light-chain kinase (MLCK), which is inhibited by phosphorylation. Inhibition of MLCK results in decreased phosphorylation of the myosin regulatory light chain and leads to myosin II filament disassembly and a decrease in contractile activity. In some cells, PAK also directly inhibits myosin II activity by phosphorylation of the myosin heavy chain (MHC). (B) Activation of the related GTPase Rho leads to nucleation of actin filaments by formins and increases contraction by myosin II, promoting the formation of contractile actin bundles at the rear of the cell and assembly of stress fibers. Activation of myosin II by Rho requires a Rho-associated kinase called ROCK. This kinase inhibits the phosphatase that removes the activating phosphate groups from myosin II light chains (MLC); it may also directly phosphorylate the myosin light chains in some cell types. ROCK also activates other protein kinases, such as LIM kinase, which in turn contributes to the formation of stable contractile actin filament bundles by inhibiting the actin-depolymerizing factor cofilin. A similar signaling pathway is important for forming the contractile ring necessary for cytokinesis (see Figure 17–45).

contractility. Rac and Rho also modulate the activities of the other's GEFs, GAPs, or GDIs to reinforce spatial separation of their activities.

External Signals Can Dictate the Direction of Cell Migration

Chemotaxis is the movement of a cell toward or away from a source of some diffusible chemical. These external signals act through cell surface receptors that trigger Rho family proteins to polarize and orient the cell motility apparatus, enabling directed cell migration. One well-studied example is the chemotactic movement of a class of white blood cells, called *neutrophils*, toward a source of bacterial infection. Receptor proteins on the surface of neutrophils enable them to detect very low concentrations of *N*-formylated peptides that are derived from bacterial proteins (only prokaryotes begin protein synthesis with *N*-formylmethionine). Using these receptors, neutrophils are guided to bacterial targets by their ability to detect a difference of only 1% in the concentration of these diffusible peptides on one side of the cell versus the other (**Figure 16–81A**).

In this case, and in the chemotaxis of *Dictyostelium* amoebae toward a source of cyclic AMP, binding of the chemoattractant to its G-protein-coupled receptor activates phosphoinositide 3-kinases (PI3Ks) (see Figure 15–53), which generate a signaling molecule [PI(3,4,5)P_3] that in turn activates the Rac GTPase. Rac then activates the Arp2/3 complex leading to lamellipodial protrusion. Through an unknown mechanism, accumulation of the polarized actin network at the leading edge causes further local enhancement of PI3K activity in a positive feedback loop, strengthening the induction of protrusion. The PI(3,4,5)P_3 that activates Rac cannot diffuse far from its site of synthesis, because it is rapidly converted back into PI(4,5)P_2 by a constitutively active lipid phosphatase. At the same time, binding of the chemoattractant ligand to its receptor activates another signaling pathway that turns on Rho and enhances myosin-based contractility. As described in the previous section, these two pathways directly inhibit each other, such that Rac activation dominates in the front of the cell and Rho activation dominates in the rear (**Figure 16–81B**). This enables the cell to maintain its functional polarity with protrusion at the leading edge and contraction at the back.

Nondiffusible chemical cues attached to the extracellular matrix or to the surface of cells can also influence the direction of cell migration. When these signals activate receptors, they can cause increased cell adhesion and directed actin polymerization. Most long-distance cell migrations in animals, including neural-crest-cell migration and the travels of neuronal growth cones, depend on a combination of diffusible and nondiffusible signals to steer the locomoting cells or growth cones to their proper destinations.

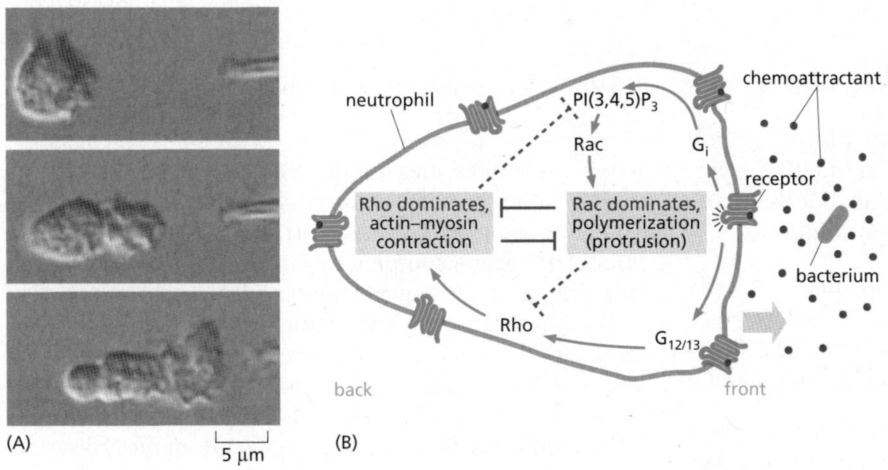

Figure 16–81 Neutrophil polarization and chemotaxis. (A) The pipette tip at the right is leaking a small amount of the bacterial peptide formyl-Met-Leu-Phe, which is recognized by the human neutrophil as the product of a foreign invader. The neutrophil quickly extends a new lamellipodium toward the source of the chemoattractant peptide *(top)*. It then extends this lamellipodium and polarizes its cytoskeleton so that contractile myosin II is located primarily at the rear, opposite the position of the lamellipodium *(middle)*. Finally, the cell crawls toward the source of the peptide *(bottom)*. If a real bacterium were the source of the peptide, rather than an investigator's pipette, the neutrophil would engulf the bacterium and destroy it (see also Figure 16–3 and **Movie 16.17**). (B) Binding of bacterial molecules to G-protein-coupled receptors on the neutrophil stimulates directed motility. These receptors are found all over the surface of the cell, but are more likely to be bound to the bacterial ligand at the front. Two distinct signaling pathways contribute to the cell's polarization. At the front of the cell, stimulation of the Rac pathway leads, via the trimeric G protein G_i, to growth of protrusive actin networks. Second messengers within this pathway are short-lived, so protrusion is limited to the region of the cell closest to the stimulant. The same receptor also stimulates a second signaling pathway, via the trimeric G proteins G_{12} and G_{13}, that triggers the activation of Rho. The two pathways are mutually antagonistic. Because Rac-based protrusion is active at the front of the cell, Rho is activated only at the rear of the cell, stimulating contraction of the cell rear and assisting directed movement. (A, from O.D. Weiner et al., *Nat. Cell Biol.* 1:75–81, published 1999 by Macmillan Magazines Ltd. Reproduced with permission of SNCSC.)

Communication Among Cytoskeletal Elements Supports Whole-Cell Polarity and Locomotion

The interconnected cytoskeleton is crucial for cell polarity and migration. Although polarity signals are frequently transduced by Rho family GTPases that act primarily on the actin cytoskeleton and myosin contractility, microtubules, septins, and intermediate filaments also participate. For example, vimentin intermediate filament networks associate with integrins at focal adhesions, and vimentin-deficient fibroblasts display impaired mechanical stability, migration, and contractile capacity. Furthermore, disruption of linker proteins that connect different cytoskeletal elements, including several plakins and KASH proteins, leads to defects in cell polarization and migration. Thus, interactions among cytoplasmic filament systems, as well as mechanical linkage to the nucleus, are required for complex, whole-cell behaviors such as migration.

Cells also use microtubules to support cell polarity and to organize persistent movement in a specific direction. Cross-linking proteins connect microtubule minus and plus ends to actin at the apical and basal cortex of epithelial cells, respectively. They also link the plus ends of microtubules and actin at the front of migrating cells. An example of such a cross-linker is the formin proteins, a subset of which binds to microtubules in addition to regulating actin filament assembly. These interactions enable microtubules to influence actin rearrangements and cell adhesion. By extending from the centrosome into the protrusive region of a migrating cell, microtubules can also serve as a compass to aid in directed cell migration. Microtubules also influence actin and focal adhesions by serving as tracks for motor-dependent transport of cargoes to and from the cell periphery. They can also deliver regulatory proteins, such as Rac GEFs, which bind to the +TIPs traveling on growing microtubule ends. Thus, microtubules reinforce the polarity information that the actin cytoskeleton receives from the outside world, allowing a sensitive response to weak signals and enabling motility to persist in the same direction for a prolonged period.

Summary

Cell polarity and migration require large-scale shaping and structuring of cells. This involves the coordinated activities of all three basic filament systems along with a large variety of regulatory and motor proteins. Rho family proteins work together with cell-polarity proteins to establish stable cytoskeletal structures necessary to generate higher levels of polarity within an organism or to maintain epithelial tissues. These same factors also operate during the dynamic polarization required for directed cell migration—a widespread behavior important in embryonic development and also in wound healing, tissue maintenance, and immune system function in the adult animal—providing a prime example of complex, coordinated cytoskeletal action influenced by external cues.

PROBLEMS

Which statements are true? Explain why or why not.

16–1 The role of ATP hydrolysis in actin polymerization is similar to the role of GTP hydrolysis in tubulin polymerization: both serve to weaken subunit bonds in the polymer and thereby promote depolymerization.

16–2 Motor neurons trigger action potentials in muscle cell membranes that open voltage-gated Ca^{2+} channels in T tubules, allowing extracellular Ca^{2+} to enter the cytosol, bind to troponin C, and initiate rapid muscle contraction.

16–3 In most animal cells, minus end–directed microtubule motors deliver their cargo to the periphery of the cell, whereas plus end–directed microtubule motors deliver their cargo to the interior of the cell.

16–4 Because bacteria are very small and lack the elaborate networks of intracellular membrane-enclosed organelles typical of eukaryotic cells, they do not require cytoskeletal filaments.

Discuss the following problems.

16–5 A scallop is a hinged bivalve that swims by slowly opening its two-part shell and then rapidly closing it, forcing a jet of water out the back and propelling itself forward. Imagine a bacterium constructed analogously. Could it swim in its low-Reynolds-number environment using the same mechanism as the scallop? Why or why not?

16–6 The plus and minus ends of actin filaments grow at different rates and have different critical concentrations (C_c). Between these critical concentrations, the filaments grow at their plus ends but shrink at their minus ends, a property termed treadmilling (**Figure Q16–1**). Is there any concentration of actin subunits at which the filament length does not change? If so, describe the concentration at which it occurs.

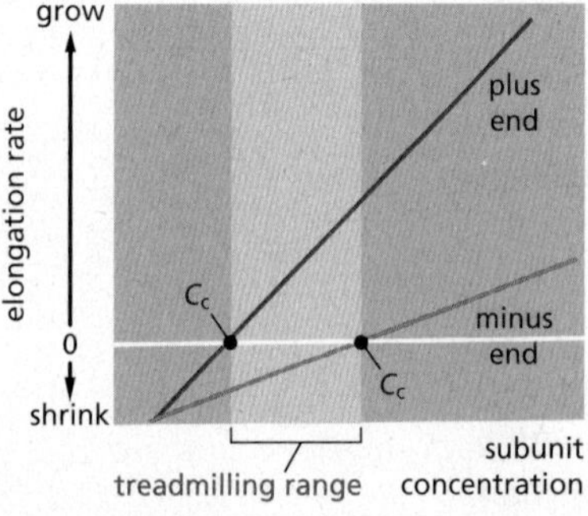

Figure Q16–1 Treadmilling at intermediate concentrations of free actin subunits (Problem 16–6).

16–7 Cofilin preferentially binds to older actin filaments and promotes their disassembly. How does cofilin distinguish old filaments from new ones?

16–8 How is the unidirectional motion of a lamellipodium maintained?

16–9 Detailed measurements of sarcomere length and tension during isometric contraction in striated muscle provided crucial early support for the sliding-filament model of muscle contraction. On the basis of your understanding of the sliding-filament model and the structure of a sarcomere, propose a molecular explanation for the relationship of tension to sarcomere length in the portions of **Figure Q16–2** marked I, II, III, and IV. (In this muscle, the length of the myosin filament is 1.6 μm, and the lengths of the actin thin filaments that project from the Z discs are 1.0 μm.)

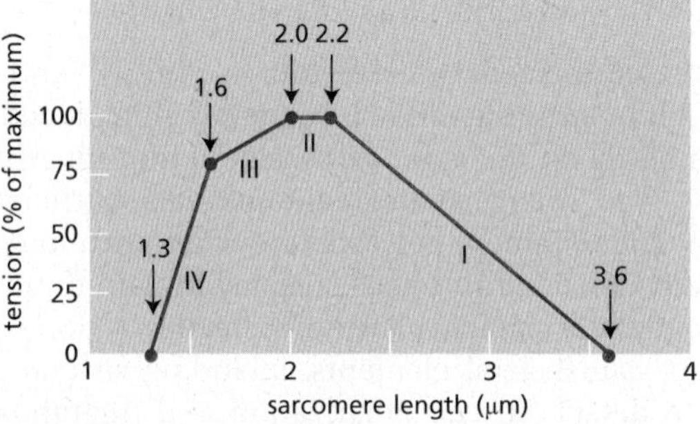

Figure Q16–2 Tension as a function of sarcomere length during isometric contraction (Problem 16–9).

16–10 At 1.4 mg/mL pure tubulin, microtubules grow at a rate of about 2 μm/min. At this growth rate, how many αβ-tubulin dimers (8 nm in length) are added to the ends of a microtubule each second?

16–11 The movements of single motor-protein molecules can be analyzed directly. Using polarized laser light, it is possible to create interference patterns that exert a centrally directed force, ranging from zero at the center to a few piconewtons at the periphery (about 200 nm from the center). Individual molecules that enter the interference pattern are rapidly pushed to the center, allowing them to be captured and moved at the experimenter's discretion.

These so-called optical tweezers can be used to position single kinesin molecules on a microtubule that is fixed to a coverslip. Although a single kinesin molecule cannot be seen optically, it can be tagged with a silica bead and tracked indirectly by following the bead (**Figure Q16–3A**). In the absence of ATP, the kinesin molecule remains at the center of the interference pattern, but with ATP it moves toward the plus end of the microtubule. As kinesin moves along the microtubule, it encounters the force of the interference pattern, which simulates the load kinesin carries during its actual function in the cell. Moreover, the pressure against the silica bead counters the effects of Brownian (thermal) motion, so that the position of the bead more accurately reflects the position of the kinesin molecule on the microtubule.

A trace of the movements of a kinesin molecule along a microtubule is shown in **Figure Q16–3B**.

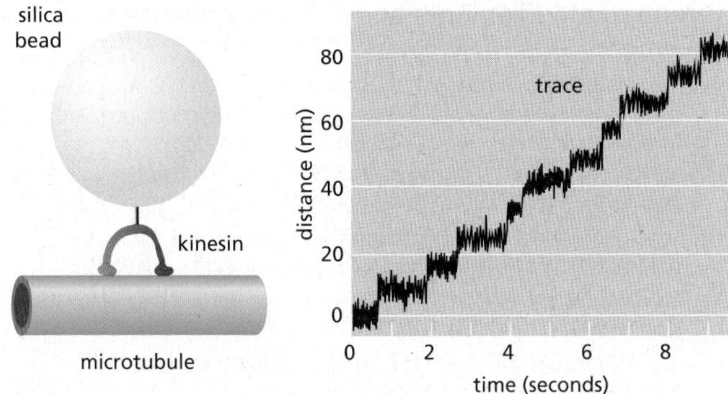

Figure Q16–3 Movement of kinesin along a microtubule (Problem 16–11). (A) Experimental setup, with kinesin linked to a silica bead, moving along a microtubule. (B) Position of kinesin (as visualized by the position of the silica bead) relative to the center of the interference pattern, as a function of time of movement along the microtubule. The jagged nature of the trace results from Brownian motion of the bead.

A. As shown in Figure Q16–3B, all movement of kinesin is in one direction (toward the plus end of the microtubule). What supplies the free energy needed to ensure a unidirectional movement along the microtubule?

B. What is the average rate of movement of kinesin along the microtubule?

C. What is the length of each step that kinesin takes as it moves along the microtubule?

D. Kinesin has two globular domains that can each bind to β-tubulin, and it moves along a single protofilament in a microtubule. In each protofilament, the β-tubulin subunit repeats at 8-nm intervals. Given the step length and the interval between β-tubulin subunits, how do you suppose a kinesin molecule moves along a microtubule?

E. Is there anything in the data in Figure Q16–3 that tells you how many ATP molecules are hydrolyzed per step?

16–12 A mitochondrion 1 μm long can travel the 1-meter length of the axon from the spinal cord to the big toe in a day. The Olympic men's freestyle swimming record for 200 meters is 1.72 minutes. In terms of body lengths per day, who is moving faster: the mitochondrion or the Olympic record holder? (Assume that the swimmer is 2 meters tall.)

16–13 Why is it that intermediate filaments have identical ends and lack polarity, whereas actin filaments and microtubules each have two distinct ends with a defined polarity?

16–14 Once cell polarity has been established in the fertilized egg of *C. elegans* (**Figure Q16–4**), the female pronucleus migrates along the microtubules emanating from the centrosomes adjacent to the paternal nucleus, which leads to fusion of the haploid genomes. What motor is likely involved in nuclear migration?

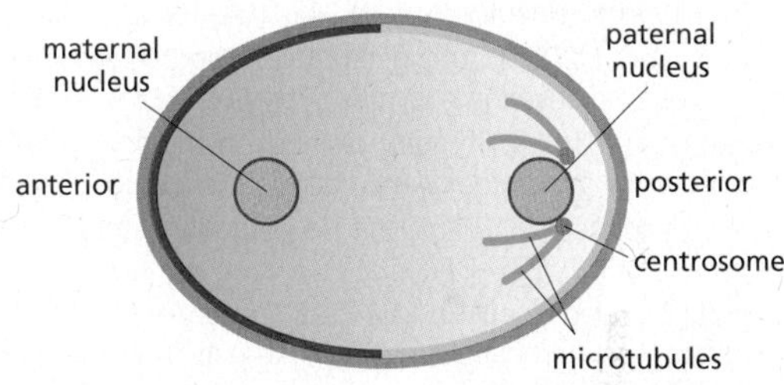

Figure Q16–4 Polarity establishment in a fertilized *C. elegans* egg prior to nuclear fusion (Problem 16–14).

REFERENCES

General

Pollard TD & Goldman RD (2017) The Cytoskeleton. Cold Spring Harbor, NY: Cold Spring Harbor Laboratory Press.

Function and Dynamics of the Cytoskeleton

Hill TL & Kirschner MW (1982) Bioenergetics and kinetics of microtubule and actin filament assembly-disassembly. *Int. Rev. Cytol.* 78, 1–125.

Luby-Phelps K (2000) Cytoarchitecture and physical properties of cytoplasm: volume, viscosity, diffusion, intracellular surface area. *Int. Rev. Cytol.* 192, 189–221.

Oosawa F & Asakura S (1975) Thermodynamics of the Polymerization of Protein, pp. 41–55, 90–108. New York: Academic Press.

Pauling L (1953) Aggregation of globular proteins. *Discuss. Faraday Soc.* 13, 170–176.

Purcell EM (1977) Life at low Reynolds number. *Am. J. Phys.* 45, 3–11.

Actin

Loisel TP, Boujemaa R, Pantaloni D & Carlier MF (1999) Reconstitution of actin-based motility of *Listeria* and *Shigella* using pure proteins. *Nature* 401, 613–616.

Mullins RD, Heuser JA & Pollard TD (1998) The interaction of Arp2/3 complex with actin: nucleation, high affinity pointed end capping, and formation of branching networks of filaments. *Proc. Natl. Acad. Sci. USA* 95, 6181–6186.

Pelaseyed T & Bretscher A (2018) Regulation of actin-based apical structures on epithelial cells. *J. Cell Sci.* 131, jcs221853.

Renkawitz J, Kopf A, Stopp J . . . Sixt M (2019) Nuclear positioning facilitates amoeboid migration along the path of least resistance. *Nature* 568, 546–550.

Rottner K & Schaks M (2019) Assembling actin filaments for protrusion. *Curr. Opin. Cell Biol.* 56, 53–63.

Skau CT & Waterman CW (2015) Specification of architecture and function of actin structures by actin nucleation factors. *Annu. Rev. Biophys*. 44, 285–310.

Svitkina T (2018) The actin cytoskeleton and actin-based motility. *Cold Spring Harb. Perspect. Biol.* 10(1), a018267.

Yamada KM & Sixt M (2019) Mechanisms of 3D cell migration. *Nat. Rev. Mol. Cell Biol.* 20, 738–752.

Myosin and Actin

Cooke R (2004) The sliding filament model: 1972–2004. *J. Gen. Physiol.* 123, 643–656.

Hammer JA III & Sellers JR (2011) Walking to work: roles for class V myosins as cargo transporters. *Nat. Rev. Mol. Cell Biol.* 13, 13–26.

Hartman MA, Finan D, Sivaramakrishnan S & Spudich JA (2011) Principles of unconventional myosin function and targeting. *Annu. Rev. Cell Dev. Biol.* 27, 133–155.

Hu S, Dasbiswas K, Guo Z . . . Bershadsky AD (2017) Long-range self-organization of cytoskeletal myosin II filament stacks. *Nat. Cell Biol.* 19, 133–141.

Walklate J, Ujfalusi Z & Greeves MA (2016) Myosin isoforms and the mechanochemical cross-bridge cycle. *J. Exp. Biol.* 219, 168–174.

Yildiz A, Forkey JN, McKinney SA . . . Selvin PR (2003) Myosin V walks hand-over-hand: single fluorophore imaging with 1.5-nm localization. *Science* 300, 2061–2065.

Microtubules

Brouhard GJ, Stear JH, Noetzel TL . . . Hyman AA (2008) XMAP215 is a processive microtubule polymerase. *Cell* 132, 79–88.

Dogterom M & Yurke B (1997) Measurement of the force-velocity relation for growing microtubules. *Science* 278, 856–860.

Goodson HV & Jonasson EM (2018) Microtubules and microtubule-associated proteins. *Cold Spring Harb. Perspect. Biol.* 10(6), a022608.

Hotani H & Horio T (1988) Dynamics of microtubules visualized by darkfield microscopy: treadmilling and dynamic instability. *Cell Motil. Cytoskeleton* 10, 229–236.

Howard J, Hudspeth AJ & Vale RD (1989) Movement of microtubules by single kinesin molecules. *Nature* 342, 154–158.

Kuo Y-W & Howard J (2020) Cutting, amplifying, and aligning microtubules with severing enzymes. *Trends Cell Biol.* 31, 50–61.

Lin J & Nicastro D (2018) Asymmetric distribution and spatial switching of dynein activity generates ciliary motility. *Science* 360, eaar1968.

Mitchison T & Kirschner M (1984) Dynamic instability of microtubule growth. *Nature* 312, 237–242.

Reck-Peterson SL, Redwine WB, Vale RD & Carter AP (2018) The cytoplasmic dynein transport machinery and its many cargoes. *Nat. Rev. Mol. Cell Biol.* 19, 382–398.

Reiter JF & Leroux MR (2017) Genes and molecular pathways underpinning ciliopathies. *Nat. Rev. Mol. Cell Biol.* 18, 533–547.

Rice S, Lin AW, Safer D . . . Milligan RA & Vale RD (1999) A structural change in the kinesin motor protein that drives motility. *Nature* 402, 778–784.

Stearns T & Kirschner M (1994) *In vitro* reconstitution of centrosome assembly and function: the central role of gamma-tubulin. *Cell* 76, 623–637.

Svoboda K, Schmidt CF, Schnapp BJ & Block SM (1993) Direct observation of kinesin stepping by optical trapping interferometry. *Nature* 365, 721–727.

Théry M & Blanchoin L (2021) Microtubule self-repair. *Curr. Opin. Cell Biol.* 68, 144–154.

Woodruff JB, Ferreira Gomes B, Widlund PO . . . Hyman AA (2017) The centrosome is a selective condensate that nucleates microtubules by concentrating tubulin. *Cell* 169, 1066–1077.

Wu J & Akhmanova A (2017) Microtubule-organizing centers. *Annu. Rev. Cell Dev. Biol.* 33, 51–75.

Intermediate Filaments and Other Cytoskeletal Polymers

Garner EC, Bernard R, Wang W . . . Mitchison T (2011) Coupled, circumferential motions of the cell wall synthesis machinery and MreB filaments in *B. subtilis*. *Science* 333, 222–225.

Garner EC, Campbell CS & Mullins RD (2004) Dynamic instability in a DNA-segregating prokaryotic actin homolog. *Science* 306, 1021–1025.

Herrmann H & Aebi U (2016) Intermediate filaments: structure and assembly. *Cold Spring Harb. Perspect. Biol.* 8(11), a018242.

Isermann P & Lammerding J (2013) Nuclear mechanics and mechanotransduction in health and disease. *Curr. Biol.* 23, R1113–R1121.

Köster S, Weitz DA, Goldman RD, Aebi U & Herrmann H (2015) Intermediate filament networks in vitro and in vivo: from coiled-coils to filaments, fibers and networks. *Curr. Opin. Cell Biol.* 32, 82–91.

Wagstaff J & Lowe J (2018) Prokaryotic cytoskeletons: protein filaments organizing small cells. *Nat. Rev. Microbiol.* 16, 187–201.

Woods BL & Gladfelter AS (2021) The state of the septin cytoskeleton from assembly to function. *Curr. Opin. Cell Biol.* 68, 105–112.

Cell Polarity and Coordination of the Cytoskeleton

Bilder D, Li M & Perrimon N (2000) Cooperative regulation of cell polarity and growth by *Drosophila* tumour suppressors. *Science* 289, 113–116.

Devreotes P & Horwitz AR (2015) Signaling networks that regulate cell migration. *Cold Spring Harb. Perspect. Biol.* 7(8), a005959.

Dogterom M & Koenderink GH (2019) Actin-microtubule crosstalk in cell biology. *Nat. Rev. Mol. Cell Biol.* 20, 38–54.

Chiou J, Balasubramanian MK & Lew D (2017) Cell polarity in yeast. *Annu. Rev. Cell Dev. Biol.* 33, 77–101.

Lawson CD & Ridley AJ (2017) Rho GTPase signalling complexes in cell migration and invasion. *J. Cell Biol.* 217, 447–457.

Munro E, Nance J & Priess JR (2004) Cortical flows powered by asymmetrical contraction transport PAR proteins to establish and maintain anterior-posterior polarity in the early *C. elegans* embryo. *Dev. Cell* 7, 413–424.

Schaks M, Giannone G & Rottner K (2019) Actin dynamics in cell migration. *Essays Biochem.* 63, 483–495.

St. Johnston D (2018) Establishing and transducing cell polarity: common themes and variations. *Curr. Opin. Cell Biol.* 51, 33–41.

CHAPTER 17

The Cell Cycle

IN THIS CHAPTER

The only way to make a new cell is to duplicate a cell that already exists. This simple fact, first established in the middle of the nineteenth century, carries with it a profound message for the continuity of life. All living organisms, from the unicellular bacterium to the multicellular mammal, are products of repeated rounds of cell growth and division extending back in time to the beginnings of life on Earth more than 3 billion years ago.

A cell reproduces by performing an orderly sequence of events in which it duplicates its contents and then divides in two. This cycle of duplication and division, known as the **cell cycle**, is the essential mechanism by which all living things reproduce. In unicellular species, such as bacteria and yeasts, each cell division produces a complete new organism. In multicellular species, long and complex sequences of cell divisions are required to produce a functioning organism. Even in the adult body, cell division is usually needed to replace cells that die. In fact, each of us must manufacture many millions of cells every second simply to survive: if all cell division were stopped—by exposure to a very large dose of x-rays, for example—we would die within a few days.

The details of the cell cycle vary from organism to organism and at different times in an organism's life. Certain characteristics, however, are universal. At a minimum, the cell must accomplish its most fundamental task: the passing on of its genetic information to the next generation of cells. To produce two genetically identical daughter cells, the DNA in each chromosome is replicated faithfully to produce two complete copies. The replicated chromosomes are then distributed (*segregated*) to the two daughter cells, so that each receives a copy of the entire genome (**Figure 17–1**). In addition to duplicating their genome, most cells also duplicate their other organelles and macromolecules; otherwise, daughter cells would get smaller with each division. To maintain their size, dividing cells coordinate their growth (that is, their increase in cell mass) with their division.

This chapter describes the events of the eukaryotic cell cycle and how they are controlled and coordinated. We begin with a brief overview of the cell cycle. We then describe the *cell-cycle control system*, a complex network of regulatory proteins that triggers the different events of the cycle. We next consider in detail the major stages of the cell cycle, in which the chromosomes are duplicated and then segregated into the two daughter cells. Finally, we consider how extracellular signals govern the rates of cell growth and division and how these two processes are coordinated.

OVERVIEW OF THE CELL CYCLE

The most basic function of the cell cycle is to duplicate the vast amount of DNA in the chromosomes and then segregate the copies into two genetically identical daughter cells. These processes define the two major phases of the cell cycle. Chromosome duplication occurs during *S phase* (S for DNA synthesis), which requires 10–12 hours and occupies about half of the cell-cycle time in a typical mammalian cell. After S phase, chromosome segregation and cell division occur

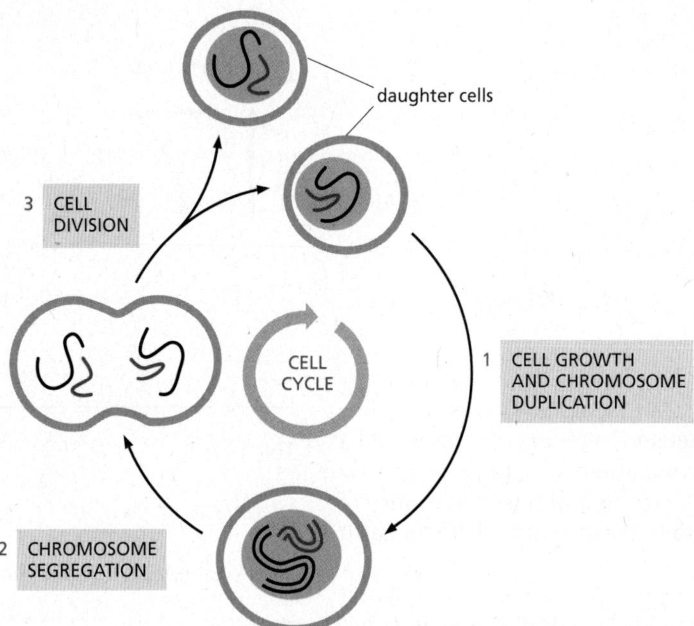

Figure 17–1 **The cell cycle.** The division of a hypothetical eukaryotic cell with two chromosomes (one *red*, and one *black*) is shown to illustrate how two genetically identical daughter cells are produced in each cycle. Each of the daughter cells will often continue to divide by going through additional cell cycles.

in *M phase* (M for *m*itosis), which requires much less time (less than an hour in a mammalian cell). M phase comprises two major events: nuclear division, or *mitosis*, during which the copied chromosomes are distributed into a pair of daughter nuclei; and cytoplasmic division, or *cytokinesis*, when the cell itself divides in two (**Figure 17–2**).

At the end of S phase, the DNA molecules in each pair of duplicated chromosomes remain intertwined and held tightly together by specialized protein linkages. Early in mitosis, at a stage called *prophase*, the two DNA molecules are disentangled and condensed into pairs of rigid, compact rods called **sister chromatids**, which remain linked by *sister-chromatid cohesion*. When the nuclear envelope then disassembles, the sister-chromatid pairs become attached to the *mitotic spindle*, a giant bipolar array of microtubules (discussed in Chapter 16). Sister chromatids are attached to opposite poles of the spindle and, eventually, align at the spindle equator in a stage called *metaphase*. The destruction of sister-chromatid cohesion at the start of *anaphase* separates the sister chromatids, which are pulled to opposite poles of the spindle. The spindle is then disassembled, and the segregated chromosomes are packaged into separate nuclei at *telophase*. Cytokinesis then cleaves the cell in two, so that each daughter cell inherits one of the two nuclei (**Figure 17–3**).

The Eukaryotic Cell Cycle Usually Consists of Four Phases

Most cells require much more time to grow and double their mass of proteins and organelles than they require to duplicate their chromosomes and divide. Partly to allow time for growth, most cell cycles have *gap phases*—a G_1 **phase** between M phase and S phase and a G_2 **phase** between S phase and mitosis. Thus, the eukaryotic cell cycle is traditionally divided into four sequential phases: G_1, S, G_2, and M. G_1, S, and G_2 together are called **interphase** (**Figure 17–4**, and see Figure 17–3). In a typical human cell proliferating in culture, interphase might

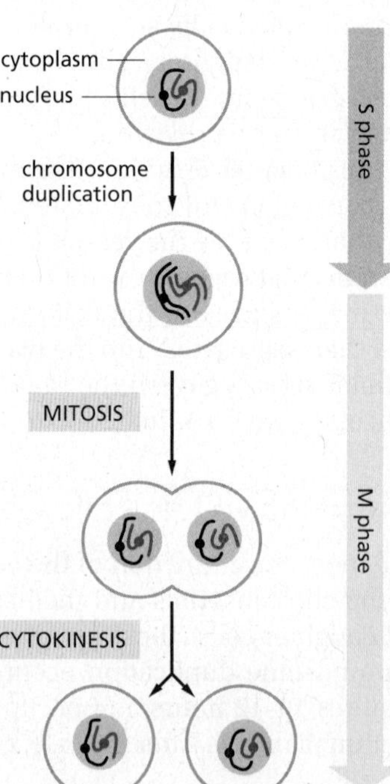

Figure 17–2 **The major events of the cell cycle.** The major chromosomal events of the cell cycle occur in S phase, when the chromosomes are duplicated, and M phase, when the duplicated chromosomes are segregated into a pair of daughter nuclei (in mitosis), after which the cell itself divides into two (cytokinesis).

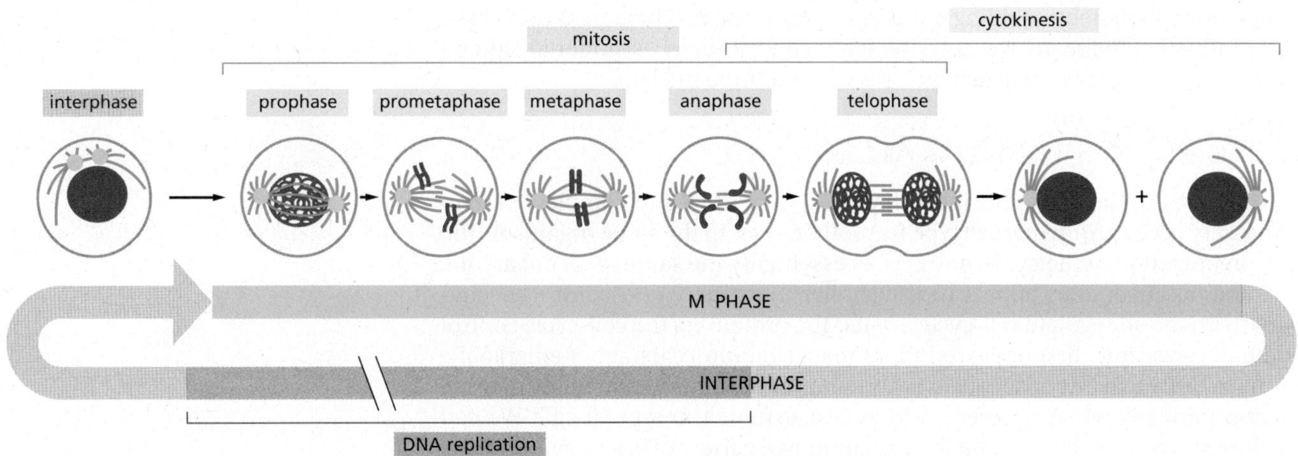

Figure 17–3 **The events of eukaryotic cell division as seen under a microscope.** The easily visible processes of nuclear division (mitosis) and cell division (cytokinesis), collectively called M phase, typically occupy only a small fraction of the cell cycle. The other, much longer, part of the cycle is known as interphase, which includes S phase and the gap phases.

occupy 23 hours of a 24-hour cycle, with 1 hour for M phase. Cell growth occurs throughout the cell cycle.

The two gap phases are more than simple time delays to allow cell growth. They also provide time for the cell to monitor the internal and external environment to ensure that conditions are suitable and preparations are complete before the cell commits itself to the complex events of S phase and mitosis. The G_1 phase is especially important in this respect. Its length can vary greatly depending on external conditions and extracellular signals from other cells. If extracellular conditions are unfavorable, for example, cells delay progress through G_1 and may even enter a specialized resting state known as *G_0* (G zero), in which they can remain for days, weeks, or even years. Indeed, many cells remain permanently in G_0 until the organism dies. If extracellular conditions become favorable or signals to grow and divide are introduced, cells in G_0 progress through a commitment point in G_1 known as **Start** (in yeasts) or the **restriction point** (in mammalian cells). We will use the term "Start" for both yeast and animal cells. After passing this point, cells are committed to DNA replication, even if the extracellular signals that stimulate cell growth and division are removed.

Not all cells undergo the conventional four-phase cell cycle. The early cleavage divisions of vertebrate embryos, for example, are not accompanied by cell growth, and these rapid divisions simply include alternating S and M phases without intervening gaps. Another important and common variation is the **endocycle**, also known as *endoreduplication,* in which multiple rounds of S phase occur without intervening M phases, resulting in cells with many copies

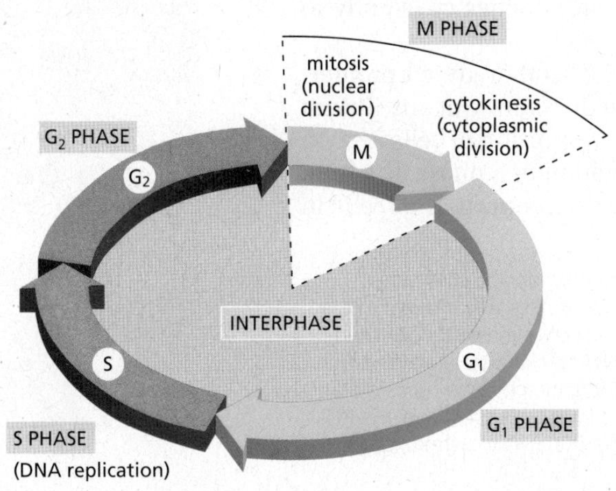

Figure 17–4 **The four phases of the cell cycle.** In most cells, gap phases separate the major events of S phase and M phase. G_1 is the gap between M phase and S phase, while G_2 is the gap between S phase and M phase.

of the genome—thereby enabling rapid increases in the production of numerous gene products. Finally, as we describe later, some cell types undergo mitosis without cytokinesis, resulting in large cells with multiple nuclei.

Cell-Cycle Control Is Similar in All Eukaryotes

Some features of the cell cycle, including the time required to complete certain events, vary greatly from one cell type to another, even in the same organism. The basic organization of the cycle, however, is essentially the same in all eukaryotic cells, and all eukaryotes appear to use similar machinery and control mechanisms to drive and regulate cell-cycle events. The proteins of the cell-cycle control system, for example, first appeared more than a billion years ago. Remarkably, they have been so well conserved over the course of evolution that many of them function perfectly when transferred from a human cell to a yeast cell. We can therefore study the cell cycle and its regulation in a variety of organisms and use the findings from all of them to assemble a unified picture of how eukaryotic cells divide.

Several model organisms are used in the analysis of the eukaryotic cell cycle. The budding yeast *Saccharomyces cerevisiae* and the fission yeast *Schizosaccharomyces pombe* are simple eukaryotes in which powerful molecular and genetic approaches can be used to identify and characterize the genes and proteins that govern the fundamental features of cell division. The early embryos of certain animals, particularly those of the frog *Xenopus laevis*, are excellent tools for biochemical dissection of cell-cycle control mechanisms, while the fruit fly *Drosophila melanogaster* is useful for the genetic analysis of mechanisms underlying the control and coordination of cell growth and division in multicellular organisms. Cultured human cells provide an excellent system for the molecular and microscopic exploration of the complex processes by which our own cells divide.

Cell-Cycle Progression Can Be Studied in Various Ways

How can we tell what stage a cell has reached in the cell cycle? One way is simply to look at living cells with a microscope. A glance at a population of mammalian cells proliferating in culture reveals that a fraction of the cells have rounded up and are in mitosis (cell rounding allows the mitotic spindle to function more effectively). Other cells can be observed in the process of cytokinesis. We can gain additional clues about cell-cycle position by staining cells with DNA-binding fluorescent dyes (which reveal the condensation of chromosomes in mitosis) or with antibodies that recognize specific cell components such as the microtubules (revealing the mitotic spindle). S-phase cells can be identified in the microscope by supplying them with visualizable molecules that are incorporated into newly synthesized DNA, such as the artificial thymidine analog 5-ethynyl-2′-deoxyuridine (EdU); cell nuclei that have incorporated EdU are then revealed by treatment with a fluorescent dye that attaches covalently to EdU (**Figure 17–5**).

Typically, in a population of cultured mammalian cells that are all proliferating rapidly but asynchronously, about 30–40% will be in S phase at any instant and become labeled by a brief pulse of EdU. From the proportion of cells in such a population that are labeled, we can estimate the duration of S phase as a fraction of the whole cell-cycle duration. Similarly, from the proportion of cells in

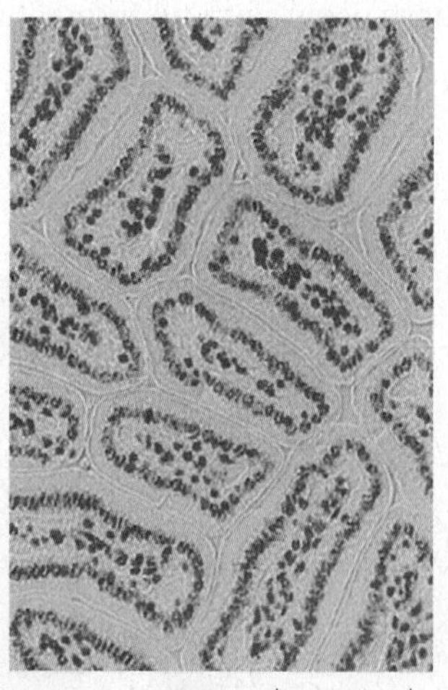

Figure 17–5 Labeling S-phase cells. A fluorescence micrograph of EdU-labeled cells of the mouse small intestine, showing intestinal villi in transverse section. The mouse was injected with a single brief dose of EdU, which then became incorporated into the newly synthesized DNA of any cell that was progressing through S phase at the time of injection. Ninety-six hours later, the tissue was fixed and labeled with a fluorescent dye that attaches to EdU *(red)*, thereby labeling cells that were in S phase 96 hours earlier. All the cell nuclei are stained with a *blue* fluorescent dye. (From A. Salic and T.J. Mitchison, *Proc. Natl. Acad. Sci. USA* 105:2415–2420, 2008. With permission from National Academy of Sciences.)

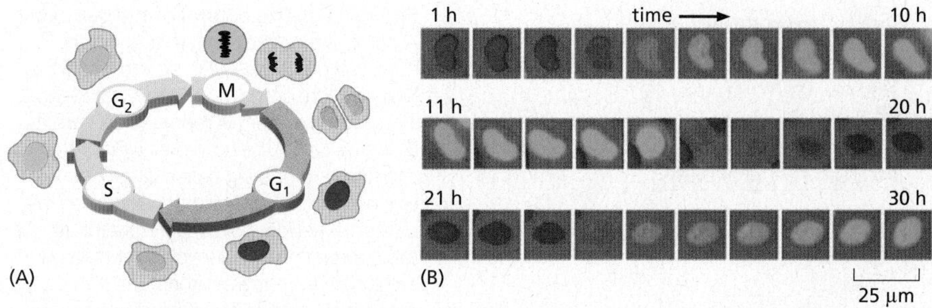

Figure 17–6 **Measuring cell-cycle timing in live cells.** (A) The method shown here depends on fluorescent proteins that are present only at specific cell-cycle stages, as illustrated in the diagram. First, a protein called geminin is labeled with a green fluorescent protein. This protein is targeted for degradation by the APC/C, a ubiquitin ligase that is active from metaphase to the end of G_1 (as discussed later). Thus, the green fluorescence of this protein is seen from early S phase to mid-mitosis. A second protein, called Cdt1, is tagged with a red fluorescent protein. This protein is targeted for ubiquitylation and destruction from late G_1 to telophase of mitosis. Cells therefore glow red from the end of mitosis to the end of G_1. (B) Fluorescence microscopy of a single mammalian cell expressing these two proteins reveals alternating red and green fluorescence as the cell progresses through the cell cycle. These images were obtained every hour over a 30-hour period. The cell was in late mitosis and cytokinesis at the 16- and 17-hour time points, and only one of the daughter cells is shown in subsequent images. This method is called Fucci (fluorescent ubiquitylation-based cell-cycle indicator). (From A. Sakaue-Sawano et al., *Cell* 132:487–498, 2008. With permission from Elsevier.)

mitosis (the *mitotic index*), we can estimate the duration of M phase. The timing of cell-cycle phases can also be measured in living cells using fluorescently labeled proteins that appear and disappear at specific stages (**Figure 17–6**).

Another way to assess the stage that a cell has reached in the cell cycle is by measuring its DNA content, which doubles during S phase. This approach is greatly facilitated by the use of fluorescent DNA-binding dyes and a *flow cytometer,* which allows the rapid and automatic analysis of large numbers of cells (**Figure 17–7**). We can use flow cytometry to determine the fraction of cells in G_1, S, and G_2 + M phases by measuring DNA content in a cell population.

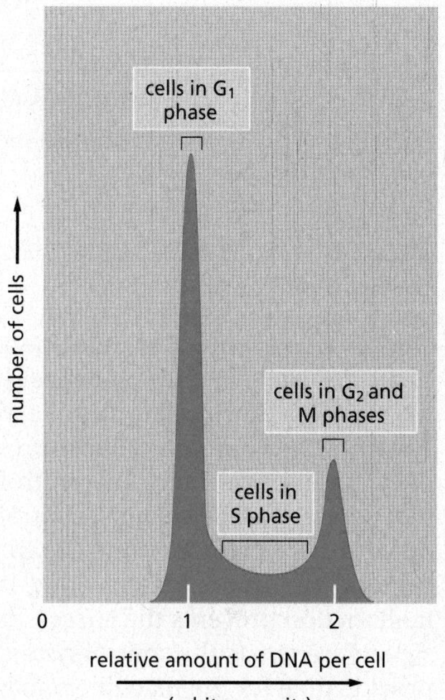

Figure 17–7 **Analysis of DNA content with a flow cytometer.** This graph shows typical results obtained for a proliferating cell population when the DNA content of its individual cells is determined in a flow cytometer. (A flow cytometer, also called a fluorescence-activated cell sorter, or FACS, can also be used to sort cells according to their fluorescence.) The cells analyzed here were stained with a dye that becomes fluorescent when it binds to DNA, so that the amount of fluorescence is directly proportional to the amount of DNA in each cell. The cells fall into three categories: those that have an unreplicated complement of DNA and are therefore in G_1, those that have a fully replicated complement of DNA (twice the G_1 DNA content) and are in G_2 or M phase, and those that have an intermediate amount of DNA and are in S phase. The distribution of cells indicates that there are greater numbers of cells in G_1 than in G_2 + M phase, showing that G_1 is longer than G_2 + M in this population.

Summary

Cell division usually begins with duplication of the cell's contents, followed by distribution of those contents into two daughter cells. Chromosome duplication occurs during S phase of the cell cycle, whereas most other cell components are duplicated continually throughout the cycle. During M phase, the replicated chromosomes are segregated into individual nuclei (mitosis), and the cell then splits in two (cytokinesis). S phase and M phase are usually separated by gap phases called G_1 and G_2, when various intracellular and extracellular signals regulate cell-cycle progression. Cell-cycle organization and control have been highly conserved during evolution, and studies in a wide range of organisms have led to a unified view of eukaryotic cell-cycle control.

THE CELL-CYCLE CONTROL SYSTEM

For many years, cell biologists watched the puppet show of DNA synthesis, mitosis, and cytokinesis but had no idea of what lay behind the curtain controlling these events. It was not even clear whether there was a separate control system or whether the processes of DNA synthesis, mitosis, and cytokinesis somehow controlled themselves. A major breakthrough came in the late 1980s with the identification of the key proteins of the control system, along with the realization that they are distinct from the proteins that perform the processes of DNA replication, chromosome segregation, and so on.

In this section, we first consider the basic principles upon which the cell-cycle control system operates. We then discuss the protein components of the system and how they work together to time and coordinate the events of the cell cycle.

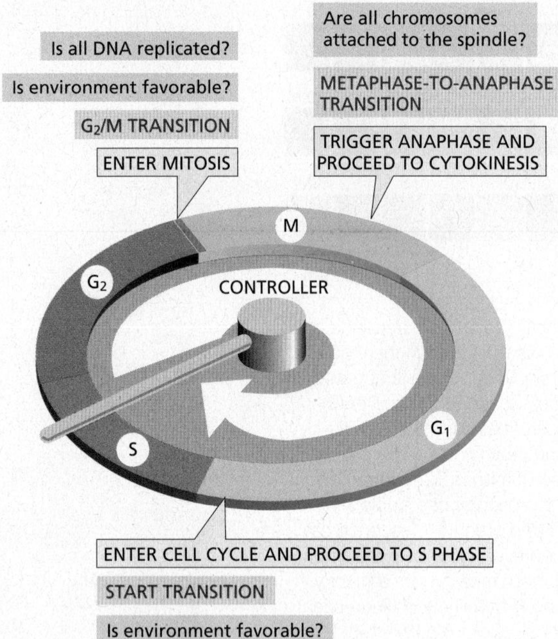

Figure 17–8 The control of the cell cycle. A cell-cycle control system triggers the essential processes of the cycle—such as DNA replication, mitosis, and cytokinesis. The control system is represented here as a central arm—the controller—that rotates clockwise, triggering essential processes when it reaches specific transitions on the outer dial *(yellow boxes)*. Information about the completion of cell-cycle events, as well as signals from the environment, can cause the control system to arrest the cycle at these transitions.

The Cell-Cycle Control System Triggers the Major Events of the Cell Cycle

The **cell-cycle control system** operates much like a timer that triggers the events of the cell cycle in a set sequence (**Figure 17–8**). In its simplest form—as seen in the stripped-down cell cycles of early animal embryos, for example—the control system is rigidly programmed to provide a fixed amount of time for the completion of each cell-cycle event. The control system in these early embryonic divisions is independent of the events it controls, so that its timing mechanisms continue to operate even if those events fail. In most cells, however, the control system does respond to information received back from the processes it controls. If some malfunction prevents the successful completion of DNA synthesis, for example, signals are sent to the control system to delay progression to M phase. Such delays provide time for the machinery to be repaired and also prevent the disaster that might result if the cycle progressed prematurely to the next stage—and segregated incompletely replicated chromosomes, for example.

The cell-cycle control system is based on a connected series of biochemical switches, each of which initiates a specific cell-cycle event. This system of switches possesses many important features that increase the accuracy and reliability of cell-cycle progression. First, the switches are generally *binary* (on/off) and launch events in a complete, irreversible fashion. It would clearly be disastrous, for example, if events such as chromosome condensation or nuclear-envelope breakdown were only partially begun but not completed. Second, the cell-cycle control system is remarkably robust and reliable, allowing the system to operate effectively under a variety of conditions and even if some components fail. Finally, the control system is highly adaptable and can be modified to suit specific cell types or to respond to specific intracellular or extracellular signals.

In most eukaryotic cells, the cell-cycle control system governs cell-cycle progression at three major regulatory transitions (see Figure 17–8). The first is Start (or the restriction point) in late G_1, when the cell commits to cell-cycle entry and chromosome duplication. The second is the **G_2/M transition**, when the control system triggers the early mitotic events that lead to chromosome alignment on the mitotic spindle in metaphase. The third is the **metaphase-to-anaphase transition**, when the control system stimulates sister-chromatid separation, leading to the completion of mitosis and cytokinesis. The control system blocks progression through each of these transitions if it detects problems inside or outside the cell. If the control system senses problems in the completion of DNA replication, for example, it will hold the cell at the G_2/M transition until those problems are solved. Similarly, if extracellular conditions are not appropriate for cell proliferation, the control system blocks progression through Start, thereby preventing cell division until conditions become favorable.

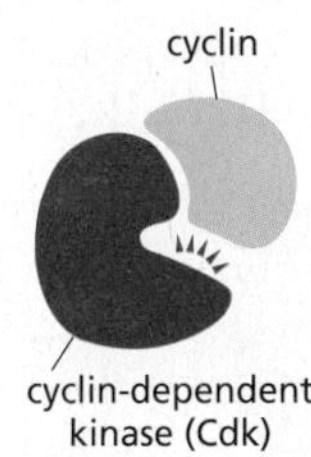

Figure 17–9 **Two key components of the cell-cycle control system.** When cyclin forms a complex with Cdk, the protein kinase is activated to trigger specific cell-cycle events. Without cyclin, Cdk is inactive.

The Cell-Cycle Control System Depends on Cyclically Activated Cyclin-dependent Protein Kinases

Central components of the cell-cycle control system are members of a family of protein kinases known as **cyclin-dependent kinases (Cdks)**. The activities of these kinases rise and fall as the cell progresses through the cycle, leading to cyclical changes in the phosphorylation of intracellular proteins that initiate or regulate the major events of the cell cycle. An increase in Cdk activity at the G_2/M transition, for example, increases the phosphorylation of proteins that control chromosome condensation, nuclear-envelope breakdown, spindle assembly, and other events that occur in early mitosis.

Cyclical changes in Cdk activity are controlled by a complex array of other proteins. The most important of these Cdk regulators are proteins known as **cyclins**. Cdks, as their name implies, depend on cyclins for their activity: unless they are bound tightly to a cyclin, they have no protein kinase activity (**Figure 17–9**). Cyclins were originally named because they undergo a cycle of synthesis and degradation in each cell cycle. The levels of the Cdk proteins, by contrast, are constant. Cyclical changes in cyclin protein levels result in the cyclic assembly and activation of **cyclin–Cdk complexes** at specific stages of the cell cycle.

There are three major classes of cyclins, each defined by the stage of the cell cycle at which they bind Cdks and carry out their functions (**Figure 17–10**):

1. **G_1/S-cyclins** activate Cdks in late G_1 and thereby help trigger progression through Start, resulting in a commitment to cell-cycle entry. Their levels fall in S phase.

2. **S-cyclins** bind Cdks soon after progression through Start and help stimulate chromosome duplication. S-cyclin levels remain elevated until mitosis, and these cyclins also contribute to the control of some early mitotic events.

3. **M-cyclins** activate Cdks that stimulate entry into mitosis at the G_2/M transition. M-cyclin levels fall in mid-mitosis.

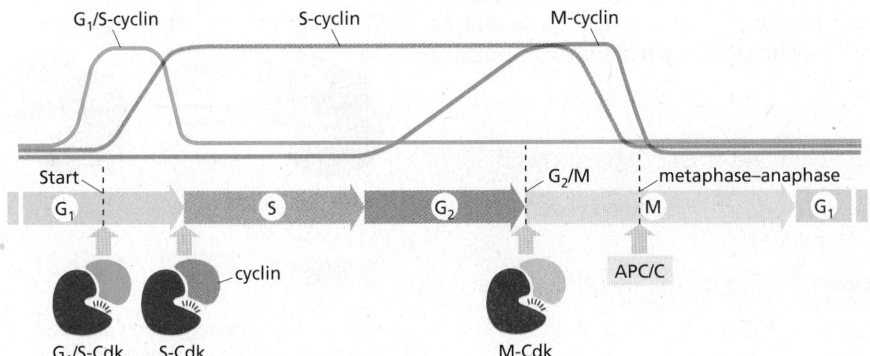

Figure 17–10 **Cyclin–Cdk complexes of the cell-cycle control system.** The concentrations of the three major cyclin types oscillate during the cell cycle, while the concentrations of Cdks (not shown) exceed cyclin amounts and do not change. In late G_1, rising G_1/S-cyclin levels lead to the formation of G_1/S-Cdk complexes that trigger progression through the Start transition. S-Cdk complexes form later in G_1 and trigger DNA replication, as well as some early mitotic events. M-Cdk complexes form during G_2 but are held in an inactive state; they are activated at the end of G_2 and trigger entry into mitosis at the G_2/M transition. A separate regulatory protein complex, the APC/C, initiates the metaphase-to-anaphase transition, as we discuss later.

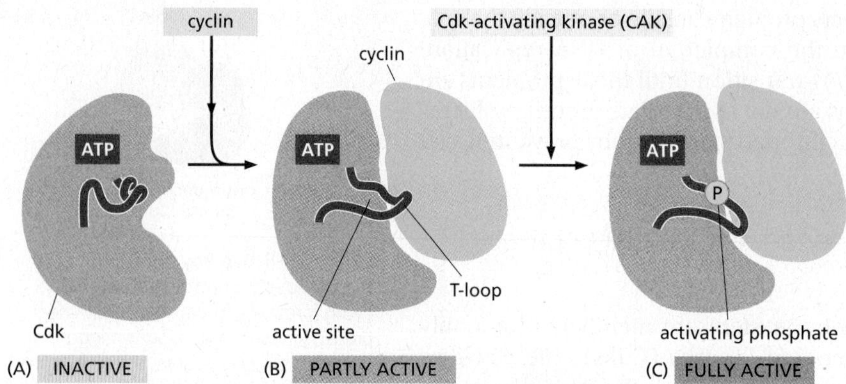

Figure 17–11 The structural basis of Cdk activation. These drawings are based on three-dimensional structures of human Cdk2 and cyclin A, as determined by x-ray crystallography. The location of the bound ATP is indicated. The enzyme is shown in three states. (A) In the inactive state, without cyclin bound, the active site is blocked by a region of the protein called the T-loop *(red)*. (B) The binding of cyclin causes the T-loop to move out of the active site, resulting in partial activation of Cdk2. (C) Phosphorylation of Cdk2 (by CAK) at a threonine residue in the T-loop further activates the enzyme by changing the shape of the T-loop, improving the ability of the enzyme to bind its protein substrates (**Movie 17.1**).

In most cells, a fourth class of cyclins, the **G_1-cyclins**, helps govern the activities of the G_1/S-cyclins, thereby controlling progression through Start in late G_1. Extracellular signals that stimulate cell proliferation act in part by increasing the production of G_1-cyclins, as we discuss later in the chapter.

In yeast cells, a single Cdk protein binds all classes of cyclins and triggers different cell-cycle events by changing cyclin partners at different stages of the cycle. In vertebrate cells, by contrast, there are four Cdks. Two interact with G_1-cyclins, one with G_1/S- and S-cyclins, and one with S- and M-cyclins. In this chapter, we simply refer to the different cyclin–Cdk complexes as **G_1-Cdk**, **G_1/S-Cdk**, **S-Cdk**, and **M-Cdk**. **Table 17–1** lists the names of the individual Cdks and cyclins.

Cyclin binding alone does not fully activate the associated Cdk. Complete activation requires a separate kinase, the Cdk-activating kinase (CAK), which phosphorylates an amino acid near the entrance of the Cdk active site. This causes a conformational change that greatly increases the activity of the Cdk subunit (**Figure 17–11**). CAK activity is constant through the cell cycle, and this modification therefore occurs constitutively throughout the cycle.

At certain cell-cycle stages, phosphorylation at a pair of amino acids near the kinase active site, by a protein kinase known as **Wee1**, inhibits Cdk activity. Dephosphorylation of these sites by a phosphatase known as **Cdc25** increases Cdk activity (**Figure 17–12**). This regulatory mechanism is particularly important

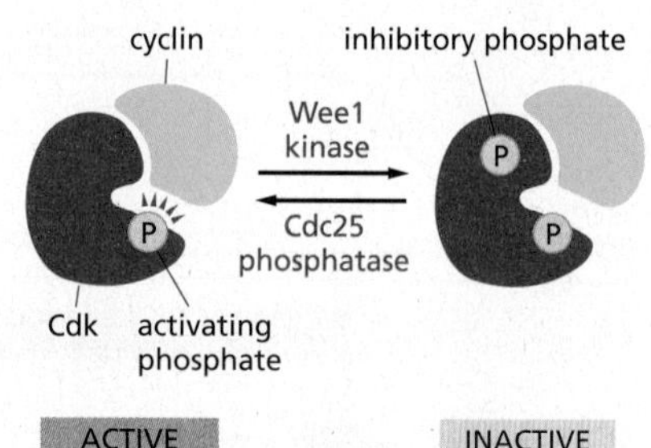

Figure 17–12 The regulation of Cdk activity by inhibitory phosphorylation. The active cyclin–Cdk complex is turned off when the kinase Wee1 phosphorylates two closely spaced sites above the active site. Removal of these phosphates by the phosphatase Cdc25 activates the cyclin–Cdk complex. For simplicity, only one inhibitory phosphate is shown. CAK adds the activating phosphate, as shown in Figure 17–11.

TABLE 17–1 The Major Cyclins and Cdks of Vertebrates and Budding Yeast

	Vertebrates		Budding yeast	
Cyclin–Cdk complex	Cyclin	Cdk partner	Cyclin	Cdk partner
G_1-Cdk	Cyclin D*	Cdk4, Cdk6	Cln3	Cdk1**
G_1/S-Cdk	Cyclin E	Cdk2	Cln1, 2	Cdk1
S-Cdk	Cyclin A	Cdk2, Cdk1**	Clb5, 6	Cdk1
M-Cdk	Cyclin B	Cdk1	Clb1, 2, 3, 4	Cdk1

*There are three D cyclins in mammals (cyclins D1, D2, and D3).
**The original name of Cdk1 was Cdc2 in vertebrates and fission yeast and Cdc28 in budding yeast.

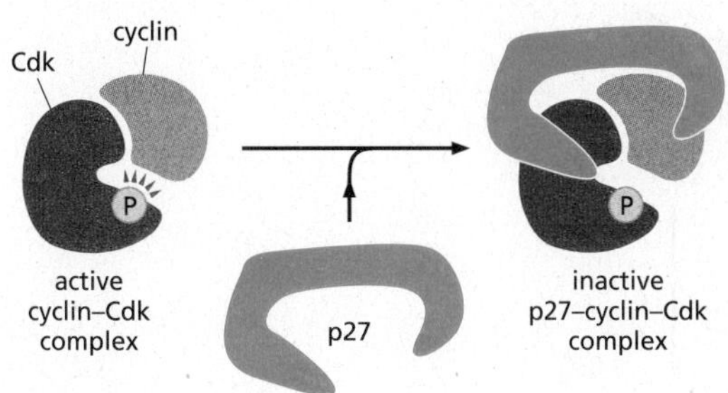

Figure 17–13 The inhibition of a cyclin–Cdk complex by a CKI. This drawing is based on the three-dimensional structure of the human cyclin A–Cdk2 complex bound to the CKI p27, as determined by x-ray crystallography. The p27 binds to both the cyclin and Cdk in the complex, distorting the active site of the Cdk. It also inserts into the ATP-binding site, further inhibiting the enzyme activity.

for generating the rapid activation of M-Cdk activity at the onset of mitosis, as we discuss later.

The activities of G_1/S- and S-Cdks early in the cell cycle are governed in part by **Cdk inhibitor proteins (CKIs)**. These small proteins wrap around the cyclin–Cdk complex, promoting a rearrangement in the Cdk active site that renders it inactive (**Figure 17–13**).

Protein Phosphatases Reverse the Effects of Cdks

As we learned in Chapters 3 and 15, protein phosphorylation is not controlled simply by the protein kinases that attach phosphate but also by the protein phosphatases that remove it. Phosphatases that reverse the effects of Cdks and other kinases are therefore key players in the cell-cycle control system.

We can think about the level of protein phosphorylation like the level of water in a sink, which depends on the rate of water flow in through the tap (protein kinase activity) and the rate of flow out of the drain (phosphatase activity). The fastest way to raise the water level is to plug the drain at the same time as you increase the flow through the tap. Indeed, we will see that the activity of phosphatases tends to decline when Cdk activity increases, resulting in a more robust increase in the protein phosphorylation state.

Protein phosphatase 2A (PP2A) is a particularly critical regulator of Cdk substrates during the cell cycle. This three-subunit enzyme comes in multiple forms, depending on the identity of a subunit called the regulatory subunit, or B subunit (**Figure 17–14**). The B subunit influences the substrate selectivity, localization, and regulation of the enzyme. Two B subunits, B55 and B56, are the most important.

The cell-cycle regulation of PP2A associated with the B55 subunit is particularly well understood and illustrates how opposing Cdk and phosphatase activities are coordinated during the cell cycle. PP2A-B55 activity is high during interphase but inhibited during early mitosis when M-Cdk activity rises. The underlying mechanism is conceptually simple: M-Cdk turns off PP2A-B55 via the phosphorylation of an intermediary protein kinase called *Greatwall* (**Figure 17–15**). As a result, the kinase activity of M-Cdk goes relatively unopposed in early mitosis, contributing to the rapid phosphorylation of M-Cdk substrates. When anaphase is initiated and M-Cdk activity declines after cyclin destruction, the system works in reverse: PP2A-B55 is reactivated to promote rapid dephosphorylation of Cdk substrates during anaphase and telophase.

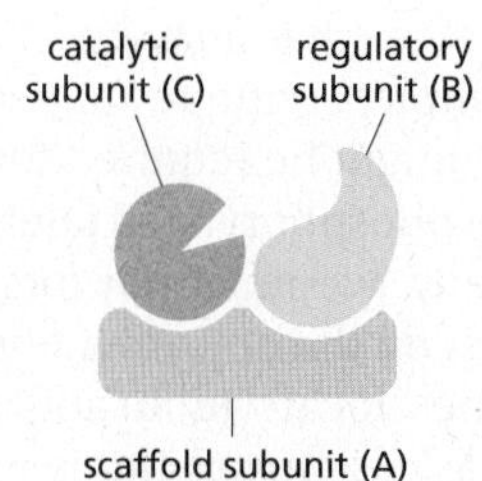

Figure 17–14 Structure of the phosphatase PP2A. PP2A is composed of three subunits. The two core subunits include the small catalytic subunit and a large structural subunit called the scaffold. This dimer associates with one of several different regulatory subunits, which are positioned next to the active site of the catalytic subunit and can influence its interaction with substrates.

Hundreds of Cdk Substrates Are Phosphorylated in a Defined Order

Cdks catalyze the phosphorylation of hundreds of different proteins in the cell. Clearly, however, these proteins are not all phosphorylated at the same time: proteins that trigger DNA replication in early S phase, for example, are phosphorylated much earlier than proteins that promote spindle assembly in early mitosis. How is the correct ordering of substrate phosphorylation achieved?

The answer is only partly understood. First, it is clear that cyclins do not simply activate the Cdk partner but also direct it to specific target proteins. The surface of each cyclin contains a binding site for short amino acid sequences that

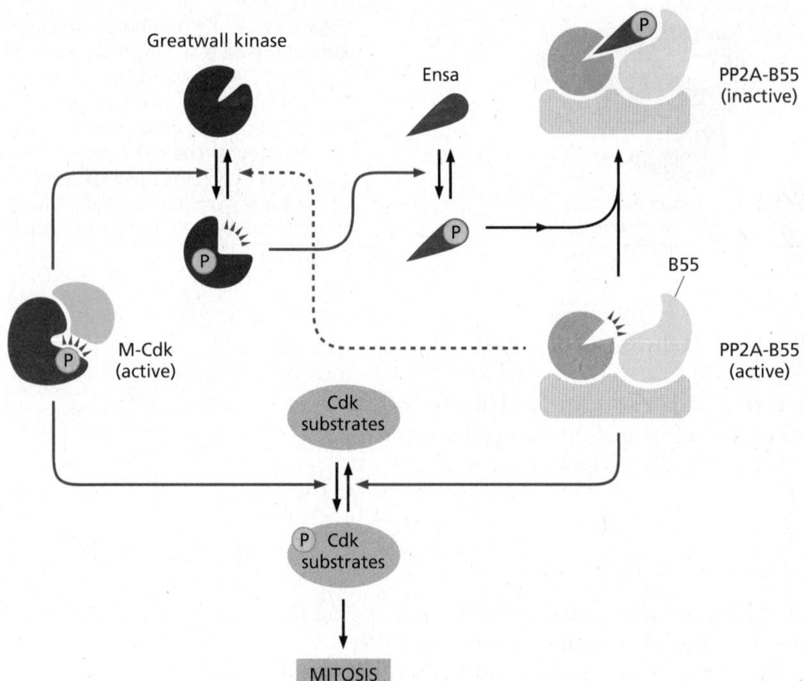

Figure 17–15 **Control of PP2A-B55 activity in mitosis.** Prior to mitosis, PP2A-B55 is active, helping to reduce the phosphorylation of M-Cdk targets. When M-Cdk activity begins to rise at the beginning of mitosis, it phosphorylates and thereby activates another protein kinase called Greatwall. This kinase in turn phosphorylates a small protein called Ensa, which binds tightly to PP2A-B55 and inhibits phosphatase activity. M-Cdk thereby inactivates its opponent. As an added twist, PP2A-B55 can dephosphorylate Greatwall *(dashed line)*, thereby inhibiting its own inhibitor—a form of positive feedback. We discuss the implications of this feedback in a later section.

are found on certain Cdk substrates. As a result, each cyclin–Cdk complex interacts more tightly with specific targets: S-Cdk, for example, has a high affinity for DNA replication proteins and therefore phosphorylates those proteins at a high rate while essentially ignoring low-affinity mitotic targets. It is likely, therefore, that the ordering of substrate phosphorylation depends in part on the activation timing of each cyclin–Cdk complex.

Cyclin specificity is not the whole story, however. Even the same cyclin–Cdk complex can induce different effects at different times in the cycle, indicating that a single enzyme phosphorylates different targets in a specific order. This ordering is likely to result from differences in the affinities of the interactions between the Cdk active site and the substrate: high-affinity substrates are phosphorylated earlier. The total amount of enzyme activity is also important: M-Cdk activity, for example, continues to rise as the cell progresses through mitosis, and higher activity might be required for the phosphorylation of certain low-affinity targets that are phosphorylated later in mitosis.

Finally, we must not forget that the timing of protein phosphorylation also depends on the opposing phosphatases, each of which will have differing activation times, localization, and affinities for specific targets. With all these factors at play, it is easy to see that a combination of mechanisms is likely to generate the perfectly timed choreography of protein phosphorylation during the cell cycle. Disentangling these mechanisms is a major goal of current research.

Positive Feedback Generates the Switchlike Behavior of Cell-Cycle Transitions

As mentioned earlier, a key feature of the cell-cycle control system is its ability to generate switchlike, binary decisions: progression through each major cell-cycle transition is a complete, irreversible commitment. The cell-cycle control system achieves this behavior through the use of positive feedback. As we discussed in Chapters 8 and 15, positive feedback is used frequently in cell regulation to generate robust, all-or-none regulatory effects, and this mechanism is well suited for generating the switchlike behavior that is so critical in cell-cycle progression.

The activation of M-Cdk at the G_2/M transition provides the best-understood example of positive feedback in cell-cycle control. M-Cdk activation begins with the accumulation of M-cyclin during G_2, which leads to a corresponding accumulation of M-Cdk complexes as the cell approaches mitosis. Although the Cdk subunit in these complexes is phosphorylated at an activating site by the Cdk-activating kinase (CAK), as discussed earlier, the protein kinase Wee1 holds it in an inactive state by inhibitory phosphorylation at two neighboring sites (see Figure 17–12). Thus, by the time the cell reaches the end of G_2, it contains an abundant stockpile of M-Cdk that is primed and ready to act but is suppressed by phosphates that block the kinase active site.

What, then, causes the activation of the M-Cdk stockpile? The crucial event is activation of the protein phosphatase Cdc25, which removes the inhibitory phosphates that restrain M-Cdk (**Figure 17–16**). At the same time, the inhibitory activity of the kinase Wee1 is suppressed, further ensuring that M-Cdk activity increases. Notably, Cdc25 is activated, at least in part, by its target, M-Cdk. M-Cdk also inhibits the inhibitory kinase Wee1. The ability of M-Cdk to activate its own activator (Cdc25) and inhibit its own inhibitor (Wee1) results in positive feedback (see Figure 17–16). The result is that all M-Cdk complexes in the cell are rapidly and irreversibly activated, leading to rapid phosphorylation of the many proteins that drive the early events of mitosis.

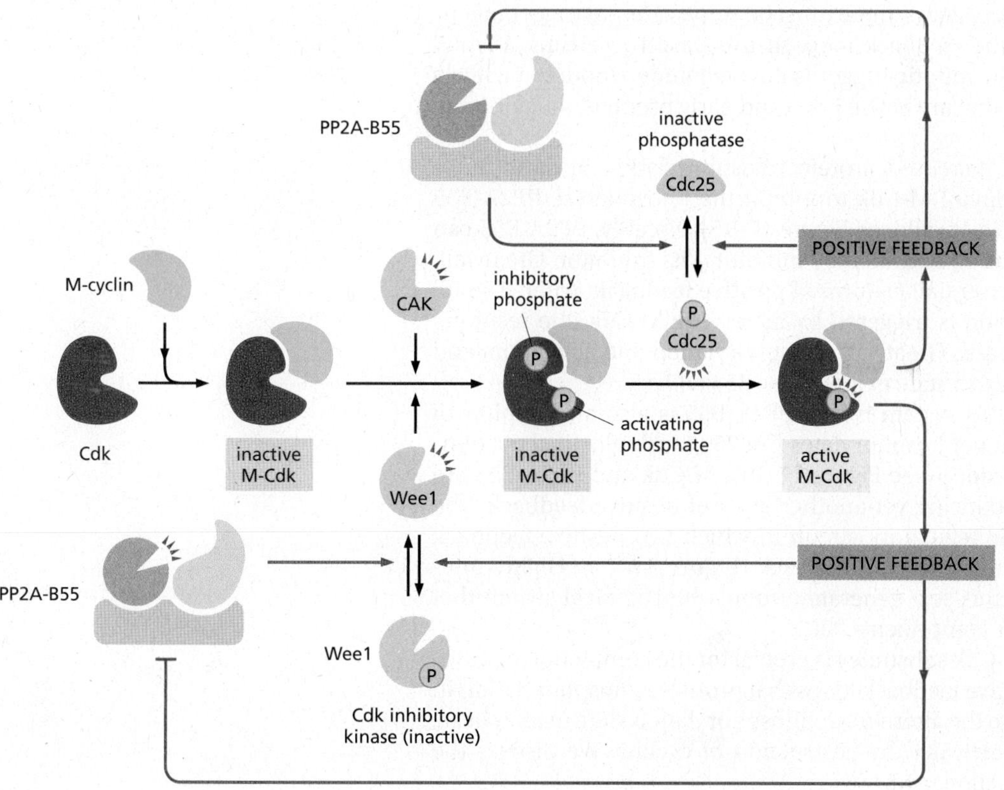

Figure 17–16 Positive feedback in the activation of M-Cdk. Cdk1 associates with M-cyclin as the levels of M-cyclin gradually rise. The resulting M-Cdk complex is phosphorylated on an activating site by Cdk-activating kinase (CAK) and on a pair of inhibitory sites by Wee1 kinase (for simplicity, only one inhibitory phosphate is shown). The resulting inactive M-Cdk complex is then activated at the end of G_2 by the phosphatase Cdc25. Cdc25 is further stimulated by active M-Cdk, resulting in positive feedback. This feedback is enhanced by the ability of M-Cdk to inhibit Wee1. The phosphorylation of both Cdc25 and Wee1 is reversed by the phosphatase PP2A-B55. As described earlier, this phosphatase is inactivated by M-Cdk (see Figure 17–15), providing another mechanism by which M-Cdk activates itself.

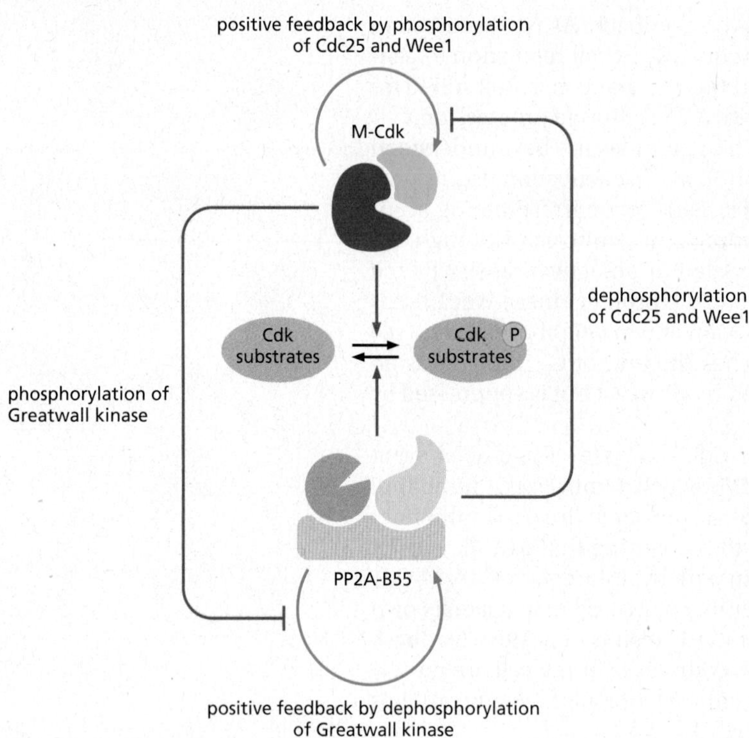

Figure 17–17 The mitotic regulatory circuit. This diagram summarizes the interactions described in Figures 17–15 and 17–16. M-Cdk and PP2A-B55 are each activated by positive feedback *(green)*. They also inhibit each other *(red)*. The result is an integrated system of multiple positive feedback loops. Much remains to be learned about other mechanisms and molecules that feed into this system in early and late mitosis.

As in all positive feedback systems, there must be some external trigger mechanism that first unleashes the feedback loop, in this case by causing a small increase in Cdc25 activity. The mitotic trigger is not well understood, but a likely possibility is that the S-Cdks that are active in G_2 and early prophase help initiate Cdc25 activation.

As we discussed earlier, increased protein phosphorylation in early mitosis also depends on the ability of M-Cdk to inhibit the phosphatase PP2A-B55 through the protein kinase Greatwall (see Figure 17–15). Notably, PP2A-B55 can fight back by dephosphorylating and thereby inhibiting its inhibitor, Greatwall. This *mutual antagonism* is essentially a form of positive feedback: when a small amount of Greatwall activation is triggered by increasing M-Cdk, the resulting inhibition of PP2A-B55 increases Greatwall phosphorylation and activation, and the system switches abruptly to a state of low PP2A-B55 activity.

A fascinating feature of this system is that PP2A-B55 is also an inhibitor of M-Cdk activation, because it dephosphorylates Cdc25 (causing its inactivation) and Wee1 (causing its activation) (see Figure 17–16). M-Cdk and PP2A-B55 are thus mutual antagonists, producing yet another layer of positive feedback. The overall result is a remarkable regulatory circuit in which two positive feedback loops are themselves linked by positive feedback (**Figure 17–17**). Highly integrated feedback systems of this sort generate a robust biochemical switch that can operate even when some components fail.

Dephosphorylation of M-Cdk substrates is crucial for the completion of mitosis and cytokinesis. The positive feedback loops that promote phosphorylation in early mitosis can be flipped to the alternate dephosphorylation state in anaphase, when M-Cdk is abruptly inactivated by destruction of cyclins. We discuss the mechanisms of cyclin destruction next.

The Anaphase-promoting Complex/Cyclosome (APC/C) Triggers the Metaphase-to-Anaphase Transition

Whereas activation of specific cyclin–Cdk complexes drives progression through the Start and G_2/M transitions (see Figure 17–10), progression through the metaphase-to-anaphase transition is triggered not by protein phosphorylation but by protein destruction, leading to the final stages of cell division.

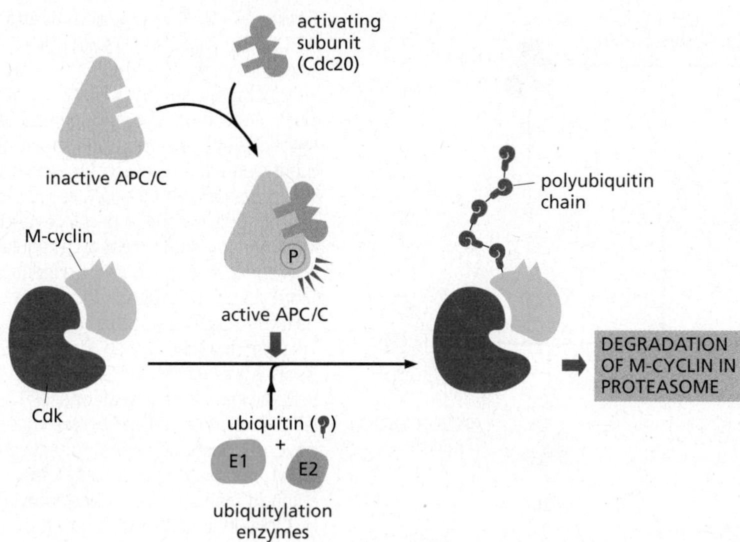

Figure 17–18 **The control of proteolysis by the APC/C.** The APC/C is activated in metaphase by association with Cdc20, which recognizes specific amino acid sequences on M-cyclin and other target proteins. With the help of two additional proteins called E1 and E2 (see Chapter 3), the APC/C assembles polyubiquitin chains on the target protein. The polyubiquitylated target is then recognized and degraded in a proteasome.

The key regulator of the metaphase-to-anaphase transition is the **anaphase-promoting complex, or cyclosome (APC/C)**, a member of the ubiquitin ligase family of enzymes. As discussed in Chapter 3, ubiquitin ligases are used in numerous cell processes to stimulate the proteolytic destruction of specific regulatory proteins. They polyubiquitylate specific target proteins, resulting in their destruction in proteasomes. Other ubiquitin ligases mark proteins for purposes other than destruction (discussed in Chapter 3).

The APC/C catalyzes the ubiquitylation and destruction of two major proteins. The first is *securin*; its destruction in metaphase activates a protease that separates the sister-chromatid pairs and unleashes anaphase, as described later. The S- and M-cyclins are the second major targets of the APC/C. Destroying these cyclins inactivates most Cdks in the cell (see Figure 17–10). As a result, the many proteins phosphorylated by Cdks from S phase to early mitosis are dephosphorylated by PP2A and other phosphatases in the anaphase cell. This dephosphorylation of Cdk targets is required for the completion of M phase, including the final steps in mitosis and then cytokinesis.

APC/C activity increases in mid-mitosis and remains high through G_1. Activation depends primarily on association with one of two activating subunits, **Cdc20** or **Cdh1**. These subunits are essential for APC/C activity for two reasons. First, their binding causes a conformational change that enhances enzyme activity. Second, they provide the main binding site for the enzyme's protein substrates. Activators interact with short amino acid sequences on APC/C substrates, holding them in place while the APC/C builds polyubiquitin chains on the target (**Figure 17–18**).

The two APC/C activator subunits interact sequentially with the APC/C. Cdc20 acts first, in metaphase, to trigger the destruction of securin and cyclins, resulting in chromosome segregation in anaphase. Cdc20 is then replaced by Cdh1, which maintains APC/C activity through late mitosis and G_1, ensuring that cyclins and other proteins are kept at low levels until the following cell cycle.

Sequential activation of the APC/C by Cdc20 and Cdh1 is based on the opposing effects of M-Cdk: phosphorylation activates APC/C–Cdc20 and inhibits APC/C–Cdh1. As a result, APC/C–Cdc20 is turned on when M-Cdk levels rise in early mitosis, whereas APC/C–Cdh1 activation occurs only after M-Cdk activity declines due to cyclin destruction in anaphase (**Figure 17–19**).

The cell-cycle control system also employs another ubiquitin ligase called **SCF** (see Figure 3–67). Its major role in the cell cycle is to ubiquitylate certain CKI proteins in late G_1, thereby helping to control the activation of S-Cdks and DNA replication. SCF is also responsible for the destruction of G_1/S-cyclins in early S phase. SCF activity depends on substrate-binding subunits called F-box proteins. Unlike APC/C activity, however, SCF activity is constant during the cell

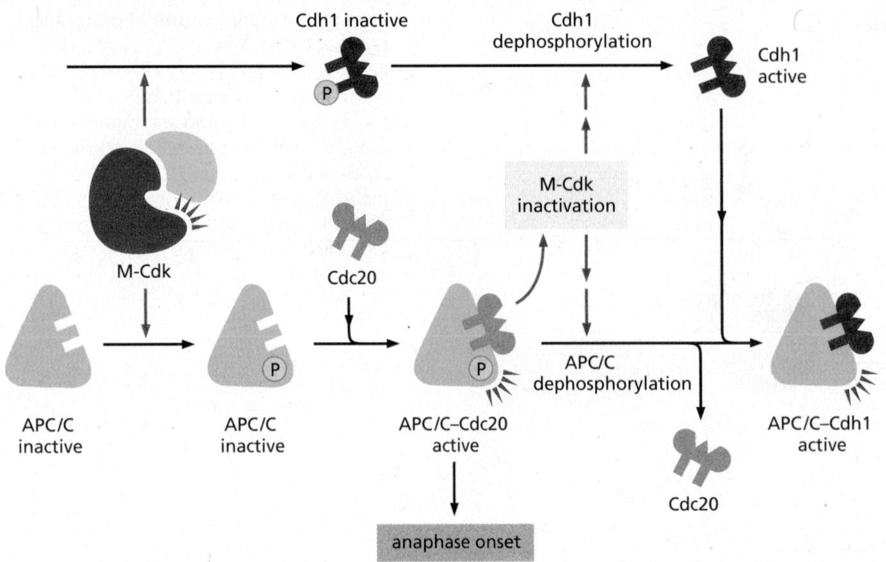

Figure 17–19 Sequential activation of APC/C by Cdc20 and Cdh1. M-Cdk phosphorylates the APC/C, thereby enhancing its association with Cdc20. Thus, M-Cdk not only triggers the early mitotic events leading up to metaphase, but it also sets the stage for progression into anaphase and the destruction of cyclins. This creates a negative feedback loop: M-Cdk sets in motion a regulatory process that leads to its own inactivation. More negative feedback then follows: APC/C–Cdc20 also turns itself off by inactivating Cdks, which allows APC/C dephosphorylation. APC/C–Cdc20 is thereby inactivated in anaphase. The APC/C remains active through mitosis and beyond because the second activator, Cdh1, is regulated in a completely different fashion: Cdh1 phosphorylation by Cdks inhibits its binding to the APC/C, keeping it inactive from late G_1 to anaphase. When cyclins are destroyed in anaphase, Cdk inactivation allows Cdh1 dephosphorylation, which activates Cdh1 and thereby stimulates formation of APC/C–Cdh1. APC/C–Cdh1 remains active until Cdh1 is phosphorylated and thereby inactivated by G_1/S- and S-Cdks at the beginning of the next cell cycle.

cycle. Ubiquitylation by SCF is controlled instead by changes in the phosphorylation state of its target proteins, as F-box subunits recognize only specifically phosphorylated proteins.

The G_1 Phase Is a Stable State of Cdk Inactivity

A key regulatory event in late M phase is the inactivation of Cdks, which resets the cell-cycle control system as the cell prepares to enter a new cell cycle. In most cells, this state of Cdk inactivity generates a stable G_1 gap phase, during which the cell grows and monitors its environment before committing to a new cell cycle.

Cells employ several mechanisms to suppress Cdk activity after mitosis. One mechanism, as we have just seen, depends on the late mitotic activation of APC/C–Cdh1, which ensures that cyclin destruction continues throughout G_1. A second mechanism for Cdk suppression depends on the increased production of CKIs, the Cdk inhibitor proteins discussed earlier. Budding yeast cells, in which this mechanism is best understood, contain a CKI protein called *Sic1*, which binds to and inactivates M-Cdk in late mitosis and G_1. Like Cdh1, Sic1 is inhibited by M-Cdk, which phosphorylates Sic1 during mitosis and thereby promotes its ubiquitylation by SCF. Thus, Sic1 and M-Cdk, like Cdh1 and M-Cdk, inhibit each other, resulting in more examples of positive feedback. As a result, the decline in M-Cdk activity that occurs in late mitosis causes the Sic1 protein to accumulate, and this CKI helps keep M-Cdk activity low after mitosis. A CKI protein called *p27* may serve similar functions in animal cells.

In many cells, decreased transcription of M-cyclin genes helps reduce M-Cdk activity in late mitosis. In budding yeast, for example, M-Cdk promotes the expression of these genes, resulting in another positive feedback loop. This loop is turned off as cells exit from mitosis: the inactivation of M-Cdk by Cdh1 and Sic1 leads to decreased M-cyclin gene transcription and thus decreased M-cyclin synthesis. Gene regulatory proteins that promote the expression of G_1/S- and S-cyclins are also inhibited during G_1.

Thus, APC/C–Cdh1 activity, CKI accumulation, and decreased cyclin gene expression act together to ensure that G_1 phase is a time when most Cdk activity is suppressed. As in many other aspects of cell-cycle control, the use of multiple regulatory mechanisms allows the system to operate with reasonable efficiency even if one mechanism is defective. So how does the cell escape from this stable G_1 state to initiate a new cell cycle? The answer is that G_1/S-Cdk activity, which rises in late G_1, releases all of the braking mechanisms that suppress Cdk activity, as we describe later in this chapter when we discuss the control of cell proliferation.

The Cell-Cycle Control System Functions as a Linked Series of Biochemical Switches

Table 17–2 summarizes some of the major components of the cell-cycle control system. These proteins are functionally linked to form a robust network, which operates essentially autonomously to activate a series of biochemical switches, each of which triggers specific cell-cycle events.

When conditions for cell proliferation are right, various external and internal signals stimulate the activation of G_1-Cdk, which in turn stimulates the expression of genes encoding G_1/S- and S-cyclins (**Figure 17–20**). The resulting activation of G_1/S-Cdk then drives progression through the Start transition, in part by releasing many of the inhibitory mechanisms, discussed above, that restrain Cdk activity in G_1. By mechanisms we discuss later, G_1/S-Cdks also unleash a wave of S-Cdk

TABLE 17–2 Summary of the Major Cell-Cycle Regulatory Proteins

General name	Functions and comments
	Protein kinases and protein phosphatases that modify Cdks
Cdk-activating kinase (CAK)	Phosphorylates an activating site in Cdks
Wee1 kinase	Phosphorylates inhibitory sites in Cdks; primarily involved in suppressing Cdk1 activity before mitosis; animals also contain a related kinase, Myt1, with similar functions
Cdc25 phosphatase	Removes inhibitory phosphates from Cdks; three family members (Cdc25A, B, C) in mammals; primarily involved in controlling Cdk1 activation at the onset of mitosis
	Protein phosphatases that dephosphorylate substrates of Cdks and other cell-cycle kinases
PP2A-B55	One of two major forms of PP2A involved in Cdk substrate regulation
PP2A-B56	Second of two major forms of PP2A involved in Cdk substrate regulation
PP1	Second major class of phosphatase involved in Cdk substrate regulation
	Cdk inhibitor proteins (CKIs)
Sic1 (budding yeast)	Suppresses Cdk1 activity in G_1; phosphorylation by Cdk1 at the end of G_1 triggers its destruction
p27 (mammals)	Suppresses G_1/S-Cdk and S-Cdk activities in G_1; helps cells withdraw from cell cycle when they terminally differentiate; phosphorylation by Cdk2 triggers its ubiquitylation by SCF
p21 (mammals)	Suppresses G_1/S-Cdk and S-Cdk activities after DNA damage
p16 (mammals)	Suppresses G_1-Cdk activity in G_1; frequently inactivated in cancer
	Ubiquitin ligases and their activators
APC/C	Catalyzes ubiquitylation of regulatory proteins involved primarily in exit from mitosis, including securin and S-cyclins and M-cyclins; regulated by association with activating subunits Cdc20 or Cdh1
Cdc20	APC/C-activating subunit in all cells; triggers initial activation of APC/C at metaphase-to-anaphase transition; stimulated by M-Cdk activity
Cdh1	APC/C-activating subunit that maintains APC/C activity after anaphase and throughout G_1; inhibited by Cdk activity
SCF	Employs various substrate-binding (F-box) subunits to catalyze ubiquitylation of regulatory proteins involved in G_1 control, including some CKIs (Sic1 in budding yeast, p27 in mammals); phosphorylation of target protein usually required for this activity
CRL4–Cdt2	Related to SCF: catalyzes ubiquitylation of the replication regulator Cdt1 during S phase, thereby suppressing further loading of Mcm helicase at origin; target recognition depends on association with proteins at the replication fork

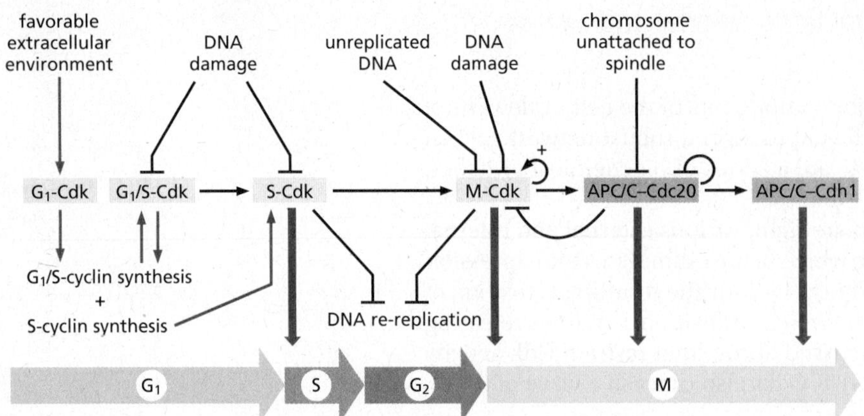

Figure 17–20 Sequential activation of Cdks during the cell cycle. The core of the cell-cycle control system consists of a series of cyclin–Cdk complexes *(yellow)*. The activity of each complex is also influenced by various inhibitory mechanisms, which provide information about the extracellular environment, DNA damage, and spindle assembly *(top)*. We will discuss the mechanisms underlying these inhibitory effects later in this chapter.

activity, which initiates chromosome duplication in S phase and also contributes to M-Cdk activation and some early events of mitosis. M-Cdk then triggers progression through the G_2/M transition and the events of early mitosis, leading to the alignment of sister-chromatid pairs on the mitotic spindle. Finally, M-Cdk also activates APC/C–Cdc20, thereby triggering the destruction of securin and cyclins—leading to sister-chromatid segregation and late mitotic events. Cdk inactivation also leads to activation of APC/C–Cdh1 and other mechanisms that suppress Cdk activity, resulting in a stable G_1 phase. We are now ready to discuss these cell-cycle stages in more detail, starting with S phase.

Summary

The cell-cycle control system triggers the events of the cell cycle and ensures that they are properly timed and coordinated with each other. Central components of the control system are the cyclin-dependent protein kinases (Cdks), which depend on cyclin subunits for their activity. Oscillations in the activities of different cyclin–Cdk complexes control various cell-cycle events. Thus, activation of S-phase cyclin–Cdk complexes (S-Cdk) initiates S phase, whereas activation of M-phase cyclin–Cdk complexes (M-Cdk) triggers mitosis. The mechanisms that control the activities of cyclin–Cdk complexes include phosphorylation of the Cdk subunit and binding of Cdk inhibitor proteins (CKIs). Protein phosphatases, including PP2A, oppose the actions of Cdks and thereby help control Cdk substrate phosphorylation and thus cell-cycle progression. The cell-cycle control system also depends crucially on a ubiquitin ligase called the APC/C, which catalyzes the ubiquitylation and consequent destruction of cyclins and other regulatory proteins that control progression through late mitosis. Together, the many components of the cell-cycle control system are assembled into a complex regulatory system containing a linked series of biochemical switches that drive stepwise progression through the phases of the cell cycle.

S PHASE

The linear chromosomes of eukaryotic cells are vast and dynamic assemblies of DNA and protein, and their duplication is a complex process that takes up a major fraction of the cell cycle. Not only must the long DNA molecule of each chromosome be duplicated accurately—a remarkable feat in itself—but the chromatin proteins in each region of that DNA must also be reproduced, ensuring that the daughter cells inherit all features of chromosome structure.

The central event of chromosome duplication—DNA replication—poses two problems for the cell. First, replication must occur with extreme accuracy to minimize the risk of mutations in the next cell generation. Second, every nucleotide in the genome must be copied once, but only once, to prevent the damaging effects of gene amplification. In Chapter 5, we discuss the sophisticated protein machinery that performs DNA replication with astonishing speed and accuracy. In this section, we consider the elegant mechanisms by which the cell-cycle control system initiates the replication process and, at the same time, prevents it from happening more than once per cycle.

S-Cdk Initiates DNA Replication Once Per Cell Cycle

As we discussed in Chapter 5, DNA replication in eukaryotic cells begins at *origins of replication*, which are scattered at numerous locations in every chromosome. During S phase, DNA replication is initiated at these origins when a *DNA helicase* unwinds the double helix and DNA replication enzymes are loaded onto the two single-stranded templates. This leads to the *elongation* phase of replication, when the replication machinery moves outward from the origin at two *replication forks*.

To ensure that chromosome duplication occurs only once per cell cycle, the initiation of DNA replication is divided into two distinct steps that occur at different times in the cell cycle (Figure 17–21). The first step occurs only in late mitosis or early G_1, when two inactive DNA helicases, called **Mcm helicases**, are loaded onto the DNA at the replication origin. This step is sometimes called *licensing* of replication origins because initiation of DNA synthesis is permitted only at origins that are preloaded with Mcm helicases. The second step occurs in S phase, when the Mcm helicases are activated, primarily by S-Cdks, resulting in DNA unwinding and the initiation of DNA synthesis. Once a replication origin has been activated in this way, it cannot be reused until new Mcm helicases are loaded at that origin, which can occur only when the cell reaches late mitosis or G_1. As a result, origins can be activated only once per cell cycle, ensuring that the DNA is replicated once and only once.

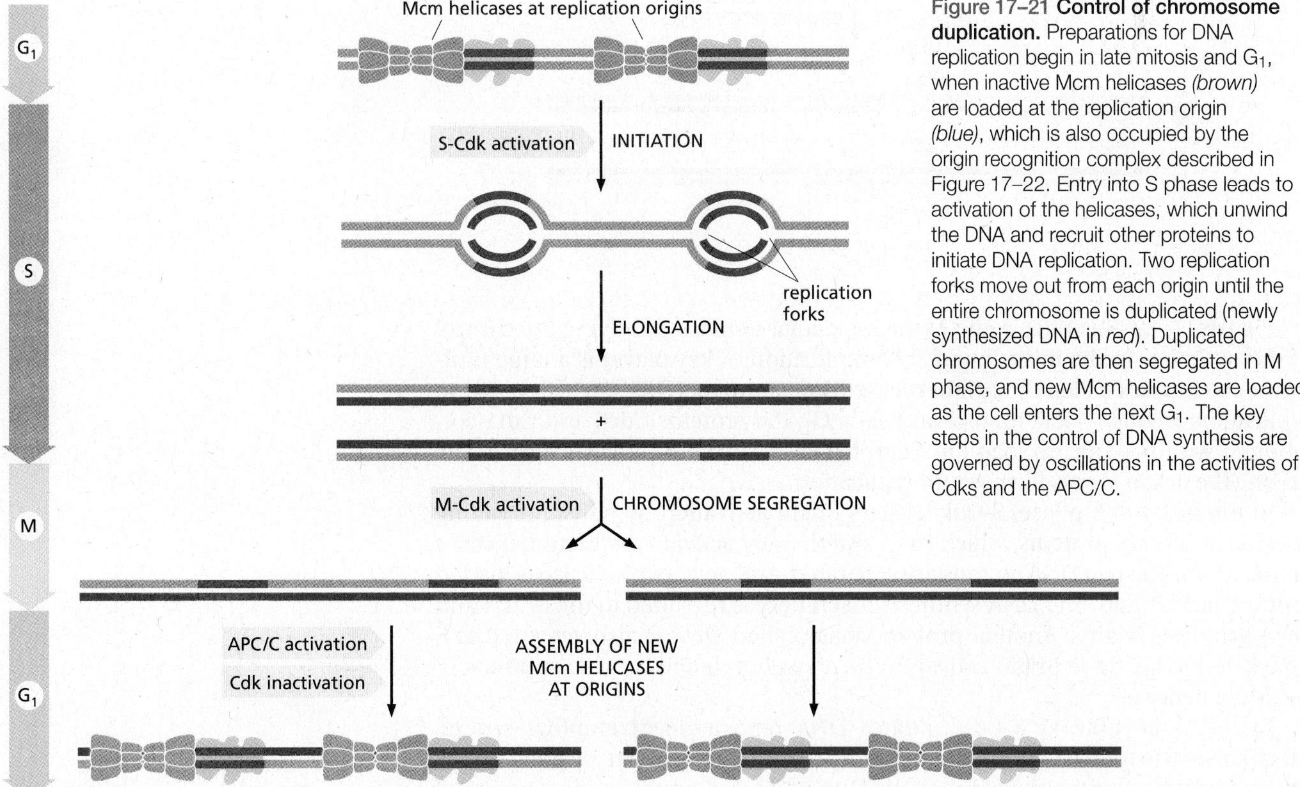

Figure 17–21 Control of chromosome duplication. Preparations for DNA replication begin in late mitosis and G_1, when inactive Mcm helicases *(brown)* are loaded at the replication origin *(blue)*, which is also occupied by the origin recognition complex described in Figure 17–22. Entry into S phase leads to activation of the helicases, which unwind the DNA and recruit other proteins to initiate DNA replication. Two replication forks move out from each origin until the entire chromosome is duplicated (newly synthesized DNA in *red*). Duplicated chromosomes are then segregated in M phase, and new Mcm helicases are loaded as the cell enters the next G_1. The key steps in the control of DNA synthesis are governed by oscillations in the activities of Cdks and the APC/C.

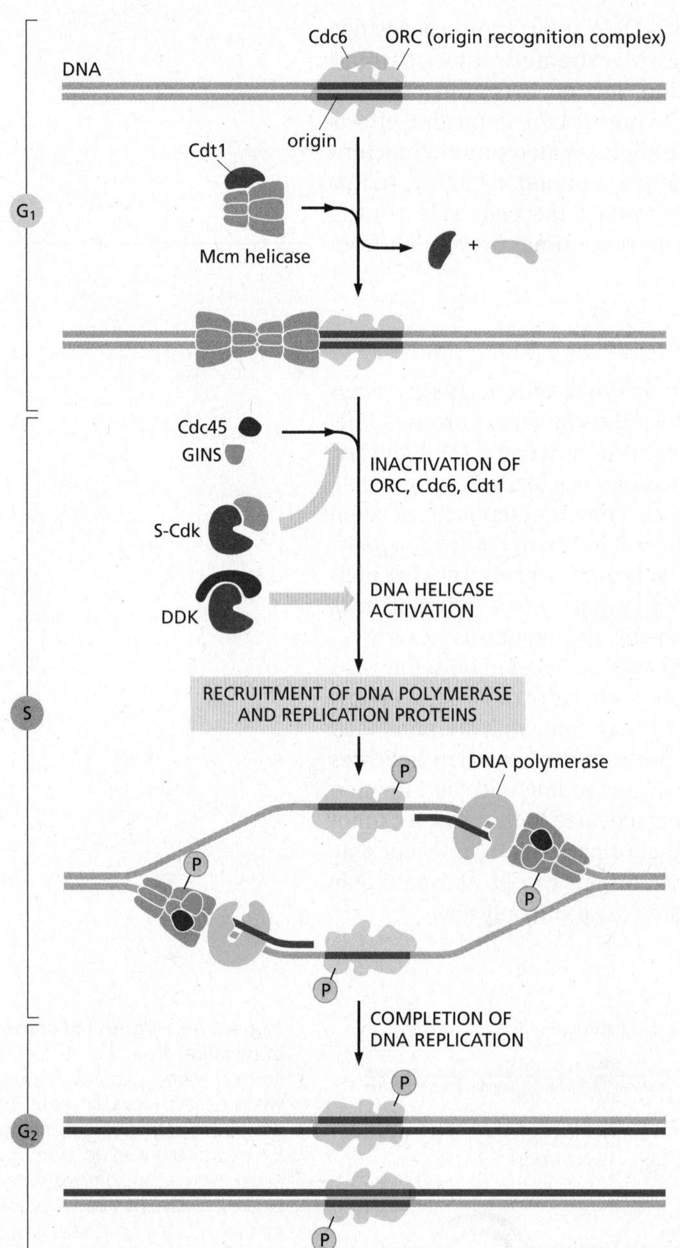

Figure 17–22 Control of the initiation of DNA replication. The replication origin is bound by the origin recognition complex (ORC) throughout the cell cycle, but ORC functions only in late mitosis and early G_1 when it associates with Cdc6. ORC–Cdc6 binds the Mcm helicase, which contains six closely related subunits arranged in a barrel shape. The helicase also associates with a protein called Cdt1. Using energy provided by ATP hydrolysis, the ORC and Cdc6 proteins load two copies of the Mcm helicase around the DNA next to the origin. At the onset of S phase, S-Cdk stimulates the assembly of several accessory proteins, including Cdc45 and GINS, on each Mcm helicase. Another protein kinase, DDK, phosphorylates subunits of the Mcm helicase. The result is a large protein complex called the CMG helicase (for Cdc45–Mcm–GINS), which unwinds the DNA at the origin. DNA polymerase and other replication proteins arrive at the origin, and DNA replication begins. For clarity, this diagram does not show synthesis of the lagging strand (discussed in Chapter 5). The ORC is displaced by the replication machinery, but new ORCs bind to both replication origins after their replication. S-Cdk and other mechanisms also inactivate the loading factors ORC, Cdc6, and Cdt1, thereby preventing loading of new Mcm helicases at the origins until the end of mitosis.

Figure 17–22 illustrates some of the molecular details underlying the control of the two steps in the initiation of DNA replication. A key player is a large multiprotein complex called the **origin recognition complex (ORC)**, which binds to replication origins. In late mitosis and early G_1, the proteins **Cdc6** and **Cdt1** collaborate with the ORC to load the Mcm helicases around the DNA next to the origin. The origin is now licensed for replication.

At the onset of S phase, S-Cdk triggers origin activation by phosphorylating specific accessory proteins, which bind and thereby activate the Mcm helicases at the origin. The two DNA strands are separated, and an active helicase is loaded around each strand. The DNA synthesis machinery is recruited to the origin and DNA synthesis begins. Another protein kinase called *DDK* is also activated in S phase and helps drive origin activation by phosphorylating specific subunits of the Mcm helicase.

At the same time as S-Cdk initiates DNA replication, it employs several mechanisms to prevent the loading of new Mcm helicases at origins. S-Cdk phosphorylates and thereby inhibits the ORC and Cdc6 proteins. Inactivation of

APC/C–Cdh1 in late G_1 also helps prevent Mcm helicase loading, as follows. In late mitosis and early G_1, APC/C–Cdh1 triggers the destruction of a Cdt1 inhibitor called **geminin**, thereby allowing Cdt1 to be active. When APC/C–Cdh1 is turned off in late G_1, geminin accumulates and inhibits the Cdt1 that is not associated with DNA. Also, the association of Cdt1 with a protein at active replication forks stimulates Cdt1 ubiquitylation by a ubiquitin ligase called CRL4–Cdt2, leading to Cdt1 degradation. In these various ways, Mcm complex loading is blocked from S phase to mid-mitosis. Thus, once an origin is used, it cannot be reloaded with a new Mcm complex in the same cell cycle.

How, then, is the cell-cycle control system reset to allow replication in the next cell cycle? In late mitosis, APC/C activation leads to the inactivation of Cdks and the destruction of geminin. ORC and Cdc6 are dephosphorylated and Cdt1 is activated, allowing Mcm helicase loading to prepare the cell for the next S phase.

Chromosome Duplication Requires Duplication of Chromatin Structure

The DNA of the chromosomes is complexed with a variety of protein components, including histones and various regulatory proteins involved in the control of gene expression (discussed in Chapter 4). Thus, duplication of a chromosome is not simply a matter of replicating the DNA at its core but also requires the duplication of these chromatin proteins and their proper assembly on the DNA.

The production of chromatin proteins increases during S phase to provide the raw materials needed to package the newly synthesized DNA. Most important, S-Cdks stimulate a large increase in the synthesis of the four histone subunits that form the histone octamers at the core of each nucleosome. These subunits are assembled into nucleosomes on the DNA by nucleosome assembly factors, which typically associate with the replication fork and distribute nucleosomes on both strands of the DNA as they emerge from the DNA synthesis machinery.

Chromatin packaging helps to control gene expression. In some parts of the chromosome, the chromatin is highly condensed and is called *heterochromatin*, whereas in other regions it has a more open structure called *euchromatin* (discussed in Chapter 4). These differences in chromatin structure depend on a variety of mechanisms, including modification of histone tails and the presence of non-histone proteins. Because these differences are important in gene regulation, it is crucial that chromatin structure, like the DNA within, is reproduced accurately during S phase. How chromatin structure is reproduced is not well understood, however. During DNA synthesis, histone-modifying enzymes and various non-histone proteins are probably deposited onto the two new DNA strands as they emerge from the replication fork, and these proteins are thought to reproduce the local chromatin structure of the parent chromosome (see Figure 4–44).

Cohesins Hold Sister Chromatids Together

At the end of S phase, each replicated chromosome consists of a pair of identical sister chromatids glued together along their length. This sister-chromatid cohesion sets the stage for a successful mitosis because it greatly facilitates the attachment of the two sister chromatids to opposite poles of the mitotic spindle. Imagine how difficult it would be to achieve bipolar spindle attachment if sister chromatids were allowed to drift apart after S phase. Indeed, defects in sister-chromatid cohesion—in yeast mutants, for example—lead inevitably to major errors in chromosome segregation.

Sister-chromatid cohesion depends on a large protein complex called **cohesin**, which forms a ring structure that surrounds the two chromatids (**Figure 17–23**). Cohesin is first loaded around unduplicated chromosomes before S phase, with assistance from a specialized loading complex. During S

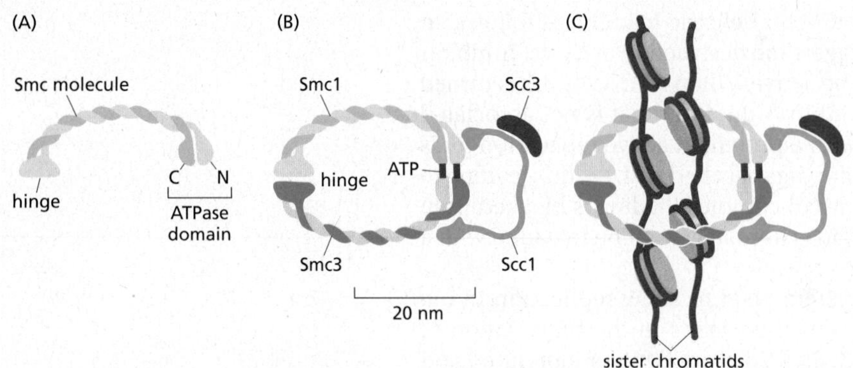

Figure 17–23 Cohesin. Cohesin is a protein complex with four subunits. (A) Two subunits of cohesin are members of a large family of proteins called *SMC proteins* (for structural maintenance of chromosomes). These subunits, Smc1 and Smc3, are coiled-coil proteins whose N- and C-terminal regions fold together to form a globular ATPase domain. (B) The Scc1 subunit contains two globular domains separated by a long disordered region that binds an additional subunit called Scc3. Binding of the globular domains of Scc1 to the Smc1 and Smc3 subunits results in a ring structure that encircles the sister chromatids as shown in (C). ATP binding promotes the interaction of the ATPase domains as shown here, while ATP hydrolysis dissociates the ATPase domains. Loading and unloading of cohesin on DNA requires ATP hydrolysis and is expected to involve the opening of one of the three intersubunit gates, but this process remains poorly understood. During S phase, acetylation of the ATPase domain of Smc3 (not shown) prevents the opening of the ring, thereby locking the cohesin ring around the sister chromatids. In metaphase, proteolytic cleavage of sites in the disordered region of Scc1 triggers sister-chromatid separation, as discussed later.

phase, by mechanisms that remain obscure, the cohesin ring is held in place during passage of the replication fork, such that the cohesin ring encircles the pair of new sister chromatids as they are being synthesized. Also during S phase, an acetyltransferase modifies a subpopulation of cohesins, locking them around the sisters to provide the stable sister-chromatid cohesion that is required to hold the sisters together until mitosis.

Sister-chromatid cohesion also results, at least in part, from the intertwining of sister DNA molecules that occurs when two replication forks meet during DNA synthesis. The enzyme topoisomerase II gradually disentangles the sister DNAs between S phase and early mitosis by cutting one DNA molecule, passing the other through the break, and then resealing the cut DNA (see Figure 5–23). Once the intertwining has been removed, sister-chromatid cohesion depends on cohesin. The loss of sister cohesion at the metaphase-to-anaphase transition therefore depends primarily on disruption of these complexes, as we describe later.

Summary

Duplication of the chromosomes in S phase involves the accurate replication of the entire DNA molecule in each chromosome, as well as the duplication of the chromatin proteins that associate with the DNA and govern various aspects of chromosome function. Chromosome duplication is triggered by the activation of S-Cdk, which activates proteins that unwind the DNA and initiate its replication at replication origins. Once a replication origin is activated, S-Cdk also inhibits proteins that are required to allow that origin to initiate DNA replication again. Thus, each origin is fired once and only once in each S phase and cannot be reused until the next cell cycle. During S phase, the duplicated chromosomes are linked together by cohesin, which provides the sister-chromatid cohesion that is required for alignment of the sister-chromatid pairs on the bipolar spindle in mitosis.

MITOSIS

Following the completion of S phase and transition through G_2, the cell undergoes the dramatic changes of M phase. This begins with mitosis, during which the sister chromatids are separated and distributed (*segregated*) to a pair of identical daughter nuclei, each with its own copy of the genome. Mitosis is traditionally divided into five stages—*prophase, prometaphase, metaphase, anaphase,* and *telophase*—defined primarily on the basis of chromosome behavior as seen in a microscope. As mitosis is completed, the second major event of M phase—cytokinesis—divides the cell into two halves, each with an identical nucleus. **Panel 17–1** summarizes the major events of M phase (**Movie 17.2**, **Movie 17.3**, **Movie 17.4**, and **Movie 17.5**).

From a regulatory point of view, mitosis can be divided into two major parts, each governed by distinct components of the cell-cycle control system. First, an increase in M-Cdk activity at the G_2/M transition triggers the events of early mitosis (prophase, prometaphase, and metaphase). M-Cdk and other mitotic protein kinases phosphorylate a variety of proteins, leading to the assembly of the mitotic spindle and its attachment to the sister-chromatid pairs. The second major part of mitosis begins at the metaphase-to-anaphase transition, when the APC/C triggers the destruction of securin, liberating a protease that cleaves cohesin and thereby initiates separation of the sister chromatids. The APC/C also promotes the destruction of cyclins, which leads to Cdk inactivation and the dephosphorylation of Cdk targets, which is required for the events of late M phase, including the completion of anaphase, the disassembly of the mitotic spindle, and the division of the cell by cytokinesis.

M-Cdk and Other Protein Kinases Drive Entry into Mitosis

One of the most remarkable features of cell-cycle control is that a single protein kinase, M-Cdk, brings about so many of the diverse and complex cell rearrangements that occur in the early stages of mitosis. At a minimum, M-Cdk must induce the assembly of the mitotic spindle and ensure that each sister chromatid in a pair is attached to the opposite pole of the spindle. It also triggers *chromosome condensation*, the large-scale reorganization of the intertwined sister chromatids into compact, rodlike structures. In animal cells, M-Cdk also promotes the breakdown of the nuclear envelope and rearrangements of the actin cytoskeleton and the Golgi apparatus. Each of these processes is initiated when M-Cdk phosphorylates specific proteins involved in the process. Many of these Cdk substrates have been identified, and in many cases we understand in considerable detail how their phosphorylation alters their function.

M-Cdk directly phosphorylates many of the proteins involved in mitotic processes, but it also acts indirectly by phosphorylating and thereby activating other protein kinases that carry out some mitotic functions. Two families of protein kinases, the *Polo-like kinases* and the *Aurora kinases*, make particularly important contributions to the control of early mitotic events. The Polo-like kinase Plk1, for example, is required for the normal assembly of a bipolar mitotic spindle, in part because it phosphorylates proteins involved in separation of the spindle poles early in mitosis. The Aurora kinase Aurora-A also helps control proteins that govern the assembly and stability of the spindle, whereas Aurora-B controls attachment of sister chromatids to the spindle, as we discuss later.

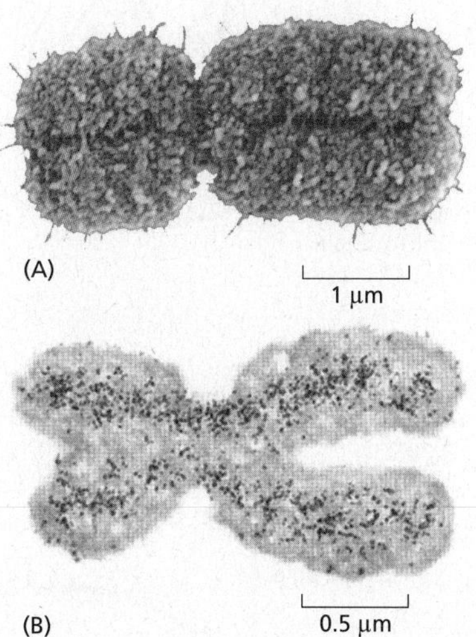

Figure 17–24 **The mitotic chromosome.** (A) Scanning electron micrograph of a human mitotic chromosome, consisting of two sister chromatids joined along their length. The constricted regions are the centromeres. (B) An electron micrograph of a duplicated mitotic chromosome in which condensin is labeled with antibodies attached to tiny gold particles *(dark dots)*, showing that condensin is found mainly in the central core of the chromosome. (A, courtesy of Terry D. Allen; B, from N. Kireeva et al., *J. Cell Biol.* 166:775–785, 2004. With permission from Rockefeller University Press.)

Condensin Helps Configure Duplicated Chromosomes for Separation

At the end of S phase, the immensely long DNA molecules of the sister chromatids are tangled in a mass of partially intertwined DNA and proteins. Any attempt to pull the sisters apart in this state would undoubtedly lead to breaks in the chromosomes. To avoid this disaster, the cell devotes a great deal of time and energy in early mitosis to reorganizing the sister chromatids into relatively short, distinct structures that can be pulled apart more easily in anaphase. These chromosomal changes involve two overlapping processes: *chromosome condensation*, in which the chromatids are dramatically compacted; and *sister-chromatid resolution*, whereby the two sisters are resolved into distinct, separable units (**Figure 17–24A**). Resolution results from the disentangling of the sister DNAs, accompanied by the partial removal of cohesin molecules along the chromosome arms. As a result, when the cell reaches metaphase, the sister chromatids appear in the microscope as compact, rodlike structures that are joined tightly at their centromeric regions and only loosely along their arms.

The condensation and resolution of sister chromatids depend, at least in part, on a five-subunit protein complex called **condensin**, which is concentrated along the central axes of mitotic chromosomes (**Figure 17–24B**). Condensin structure is related to that of the cohesin complex that holds sister chromatids together (see Figure 17–23). It contains two SMC subunits like those of cohesin, plus

1 PROPHASE

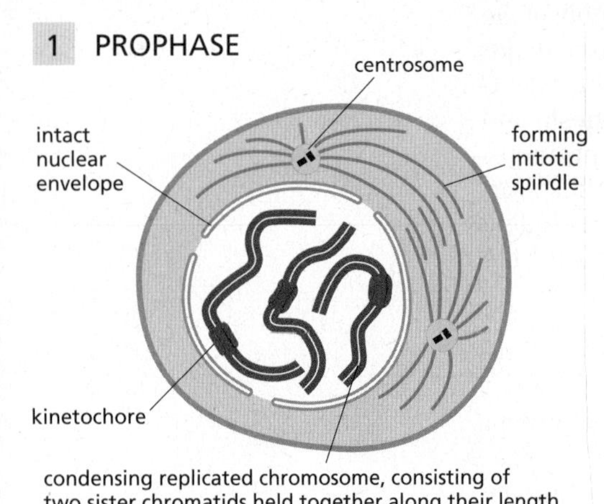

At prophase, the replicated chromosomes, each consisting of two closely associated sister chromatids, condense. Outside the nucleus, the mitotic spindle assembles between the two centrosomes, which have replicated and moved apart. For simplicity, only three chromosomes are shown. In diploid cells, there would be two copies of each chromosome present. In the fluorescence micrograph, chromosomes are stained *orange* and microtubules are *green*.

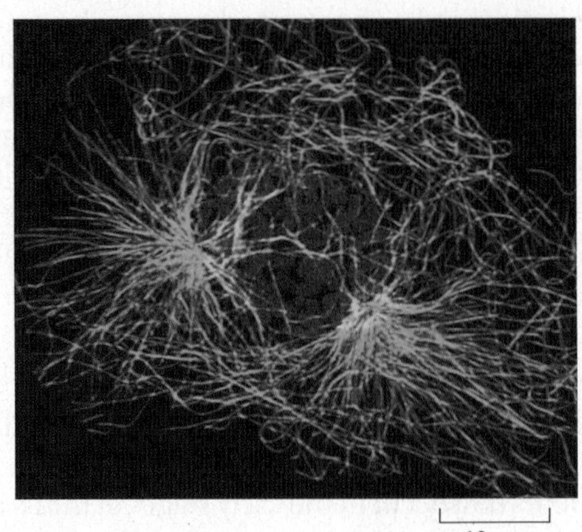

2 PROMETAPHASE

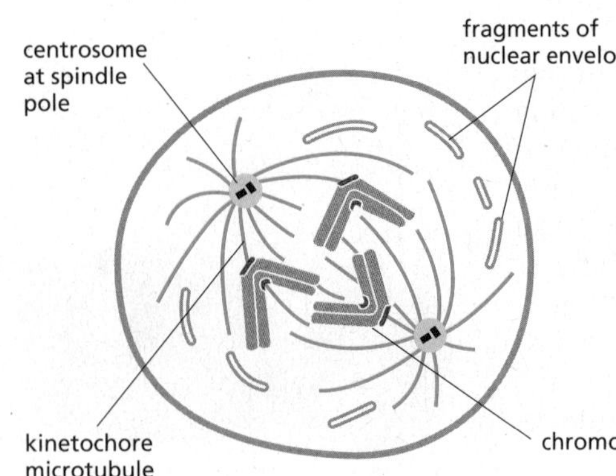

Prometaphase starts abruptly with the breakdown of the nuclear envelope. Chromosomes can now attach to spindle microtubules via their kinetochores and undergo active movement.

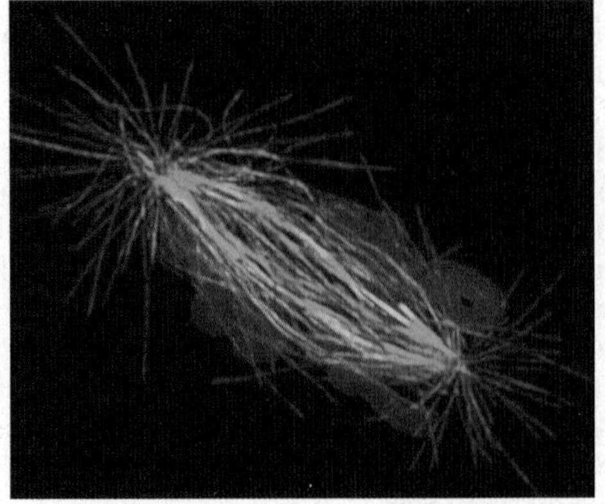

3 METAPHASE

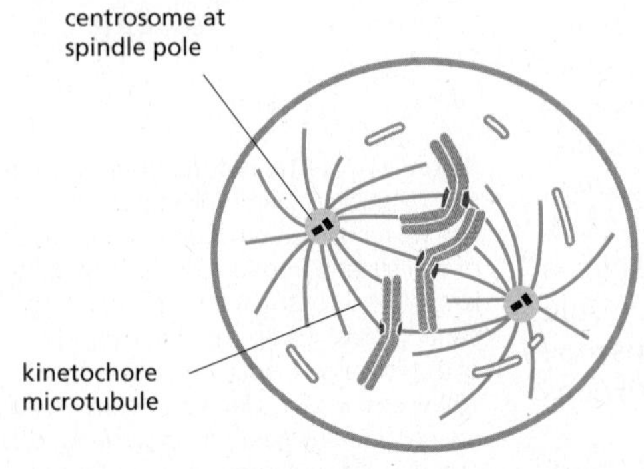

At metaphase, the chromosomes are aligned at the equator of the spindle, midway between the spindle poles. The kinetochore microtubules attach sister chromatids to opposite poles of the spindle.

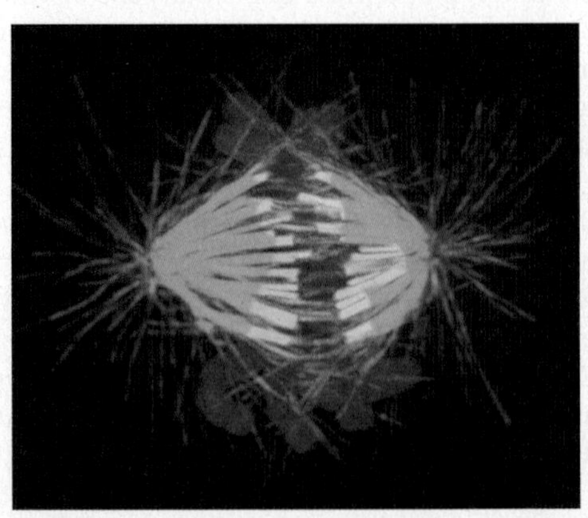

4 ANAPHASE

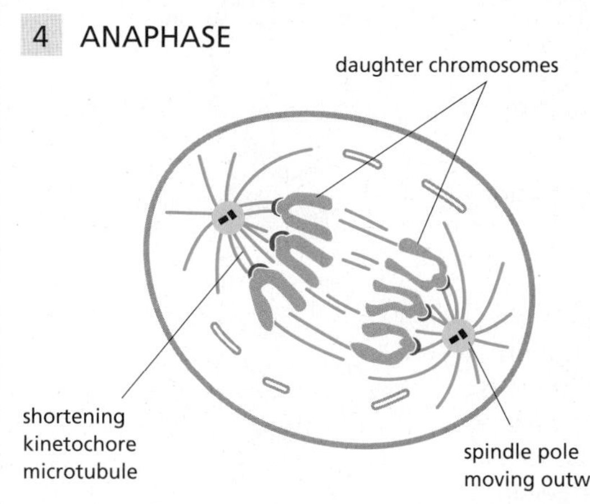

At anaphase, the sister chromatids synchronously separate to form two daughter chromosomes, and each is pulled slowly toward the spindle pole it faces. The kinetochore microtubules get shorter, and the spindle poles also move apart; both processes contribute to chromosome segregation.

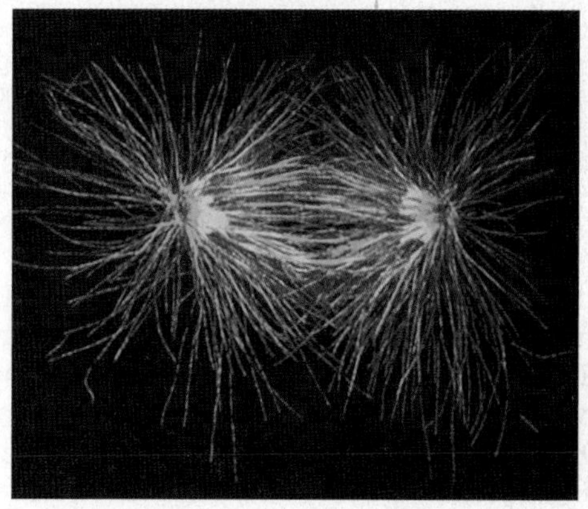

5 TELOPHASE

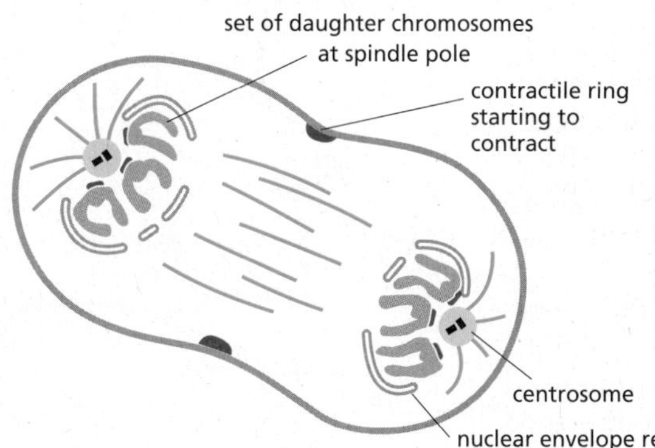

During telophase, the two sets of daughter chromosomes arrive at the poles of the spindle and decondense. A new nuclear envelope reassembles around each set, completing the formation of two nuclei and marking the end of mitosis. The division of the cytoplasm begins with contraction of the contractile ring.

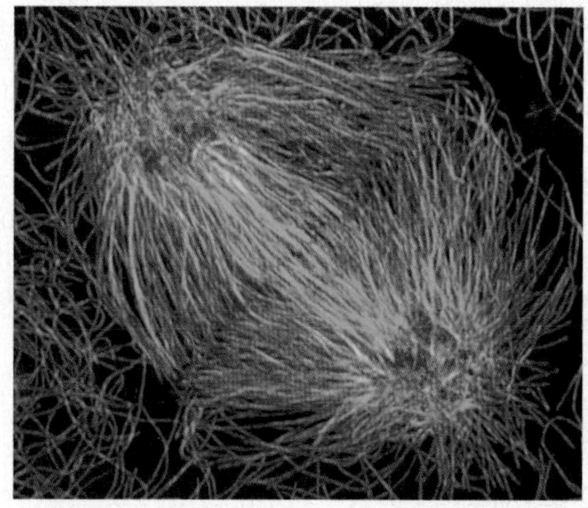

6 CYTOKINESIS

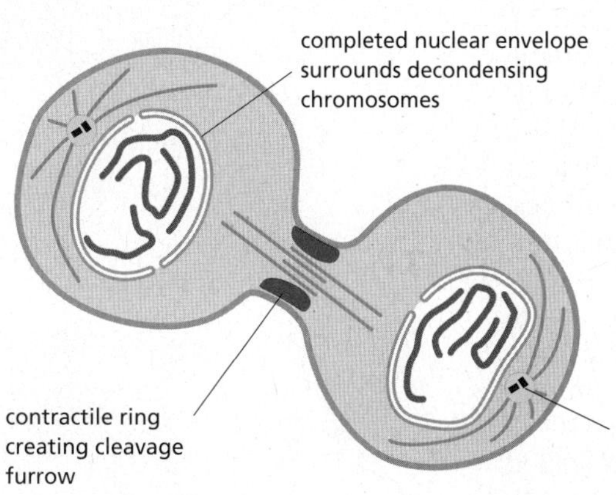

During cytokinesis, the cytoplasm is divided in two by a contractile ring of actin and myosin filaments, which pinches the cell in two to create two daughters, each with one nucleus.

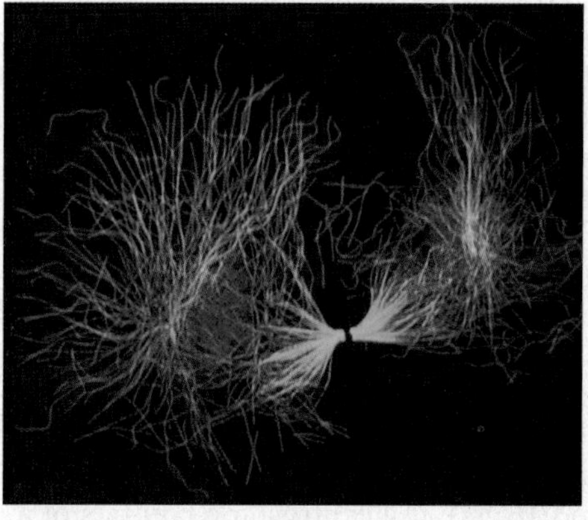

(Micrographs courtesy of Julie Canman and Ted Salmon.)

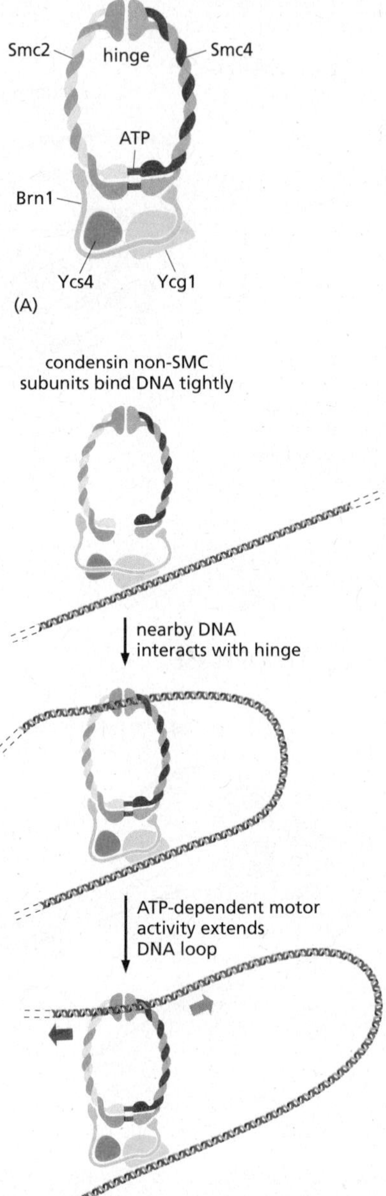

Figure 17–25 Condensin. (A) Condensin is a five-subunit protein complex that resembles cohesin (see Figure 17–23). The ATPase head domains of its two major subunits, Smc2 and Smc4, are held together by a linker protein called Brn1, which is associated with two additional proteins called Ycs4 and Ycg1. (B) These diagrams illustrate a mechanism by which condensin might generate DNA loops, thereby promoting the compaction of a chromosome. For an alternative model, see Figure 4-56. In the model shown here, the process begins when the Ycg1 subunit interacts tightly with a strand of chromosome DNA, anchoring the condensin firmly in place on the DNA. The nearby DNA curls around to interact with the hinge domain. By mechanisms that remain unclear, the hinge and ATPase domains work together, using energy from ATP hydrolysis, to move leftward along the top DNA strand (*red* arrow), thereby generating a chromosome loop. Such loops are common structural elements in chromosome packaging (discussed in Chapter 4).

three non-SMC subunits (Figure 17–25A). Like cohesin, condensin forms a ring that encircles DNA. In addition, condensin has the ability to use energy provided by ATP hydrolysis to promote the compaction of sister chromatids. There is evidence that the ATPase domains of condensin act as a motor that moves DNA through the ring, thereby allowing it to create chromosome loops (Figure 17–25B). Much remains to be learned, however, before we understand how the local coordination of condensin, cohesin, and histone proteins is choreographed by M-Cdk to generate the efficient packaging and resolution of the sister chromatids.

The Mitotic Spindle Is a Dynamic Microtubule-based Machine

The central event of mitosis—chromosome segregation—depends in all eukaryotes on a complex and beautiful machine called the **mitotic spindle** (see Panel 17–1). The spindle is a bipolar array of microtubules, which pulls sister chromatids apart in anaphase, thereby segregating the two sets of chromosomes to opposite ends of the cell, where they are packaged into daughter nuclei (Movie 17.6). M-Cdk triggers the assembly of the spindle early in mitosis, in parallel with the chromosome restructuring just described. Before we consider how the spindle assembles and how its microtubules attach to sister chromatids, we briefly review the key features of metaphase spindle structure.

As we discussed in Chapter 16, microtubules are dynamic polar polymers with plus and minus ends that display distinct behaviors. The metaphase spindle of an animal cell contains many thousands of microtubules radiating from two spindle poles, with minus ends oriented toward the pole and plus ends directed outward (Figure 17–26). The **kinetochore microtubules** are the central players in chromosome segregation: their plus ends are attached to sister-chromatid pairs at large protein structures called *kinetochores,* which are located at the *centromere* of each sister chromatid. Each kinetochore binds large numbers of microtubules that are cross-linked to form thick microtubule bundles called *K-fibers.* These fibers are responsible for moving separated sister chromatids toward the poles in anaphase.

The second major type of spindle microtubule, and by far the most numerous, are the **non-kinetochore microtubules** (also known as interpolar microtubules). These relatively short and unstable microtubules are densely packed between the poles, cross-linked by various proteins to form a dynamic and adaptable scaffolding network that provides structural stability to the spindle. Some of these microtubules are embedded in the spindle poles, but many are found away from the poles, sometimes with minus ends attached to the sides of other microtubules. Near the spindle equator, they are often cross-linked with antiparallel microtubules oriented with minus ends directed toward the opposite spindle pole.

In most animal cells, spindles also contain **astral microtubules** that radiate outward from the poles and contact the cell cortex, helping to position the spindle in the cell. Each of the two poles in these spindles is focused at a large protein organelle called the **centrosome.** As described in Chapter 16 (see Figures 16–42 and 16–43), the centrosome consists of a cloud of amorphous material (called the *pericentriolar material*) that surrounds a pair of *centrioles* (Figure 17–27). The pericentriolar material nucleates a radial array of microtubules, with their dynamic plus ends projecting outward and their minus ends associated with

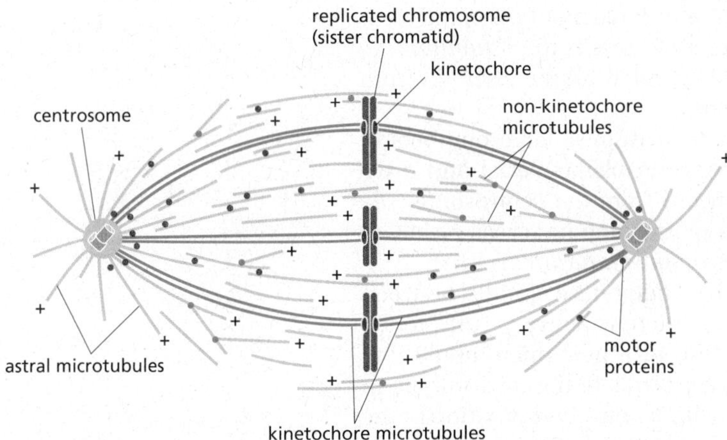

Figure 17–26 **The metaphase mitotic spindle in an animal cell.** Bundles of parallel kinetochore microtubules (K-fibers) connect the spindle poles with the kinetochores of sister chromatids. Non-kinetochore microtubules are scattered throughout the spindle with their minus ends directed toward the nearest pole; these microtubules are cross-linked by various microtubule-associated proteins to form a dynamic, interconnected meshwork. Near the spindle equator, antiparallel microtubules are cross-linked by specific motor and other proteins *(light purple dots)*. In other regions of the spindle, parallel microtubules are cross-linked *(red dots)*. Some microtubules grow from nucleating factors in the centrosome, while others originate in nucleating factors on the sides of other microtubules *(green dots)*, as we discuss later. Astral microtubules radiate out from the poles into the cytoplasm and are not present in spindles lacking centrosomes.

the centrosome. Some cells—notably the cells of higher plants and the oocytes of many vertebrates—do not have centrosomes and therefore do not have astral microtubules. Thus, as we discuss later, centrosomes are not essential for spindle assembly.

The assembly and function of a bipolar spindle depend on hundreds of different microtubule-associated proteins that can be placed in three general categories: nucleating factors that govern the formation of new microtubules, regulatory proteins that control rates of polymerization and depolymerization at both ends of the microtubules, and motor proteins that cross-link and move microtubules relative to each other. We briefly describe each of these important microtubule regulators in the following sections, after which we discuss how these components collaborate in the assembly of the spindle.

Microtubules Are Nucleated in Multiple Regions of the Spindle

Microtubules are the building blocks of the spindle, and spindle assembly requires the synthesis of enormous numbers of new microtubules. Mature metaphase spindles contain tens or hundreds of thousands of microtubules, most of which are turning over rapidly, resulting in the need for a constant supply of new

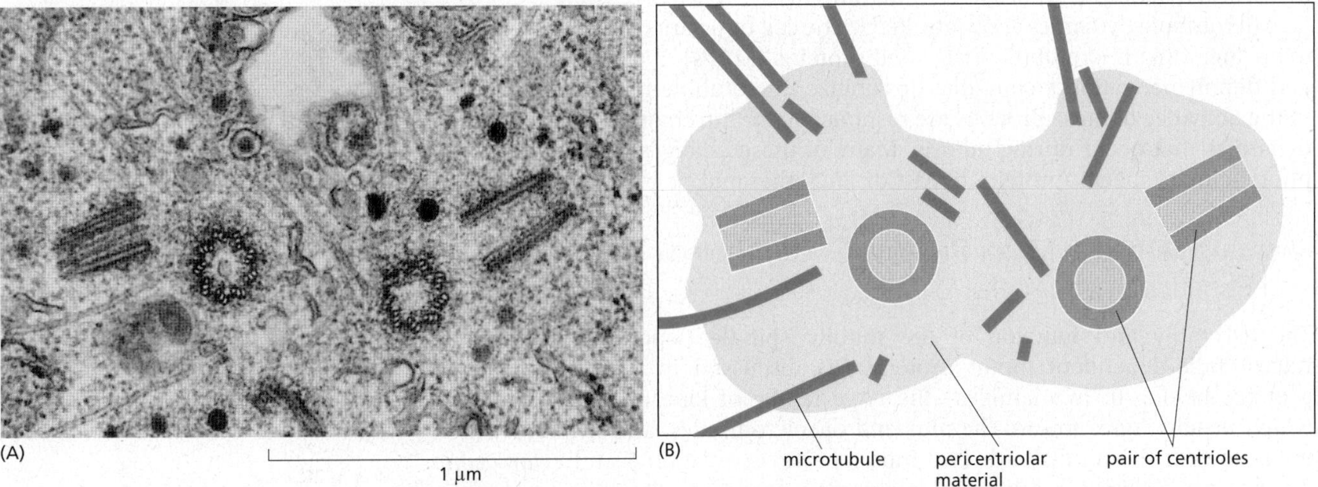

Figure 17–27 **The centrosome.** (A) Electron micrograph of an S-phase mammalian cell in culture, showing a duplicated centrosome. Each centrosome contains a pair of centrioles; although the centrioles have duplicated, they remain together in a single complex, as shown in the drawing of the micrograph in (B). One centriole of each centriole pair has been cut in cross section, while the other is cut in longitudinal section, indicating that the two members of each pair are aligned at right angles to each other. The two halves of the replicated centrosome, each consisting of a centriole pair surrounded by pericentriolar material, will split and migrate apart to initiate the formation of the two poles of the mitotic spindle when the cell enters M phase (see Figures 16–42 and 16–43). (A, from M. McGill et al., *J. Ultrastruct. Res.* 57:43–53, 1976. With permission from Academic Press.)

microtubules. To meet this need, the spindle contains large numbers of microtubule nucleating factors. By far the most important of these is the *γ-tubulin ring complex* (*γ-TuRC*; see Figure 16–41), which nucleates microtubule assembly from the minus end, leaving the plus end to grow rapidly.

In the many animal cell types that contain centrosomes, large numbers of γ-TuRCs are anchored and activated in the pericentriolar material and drive the formation of microtubules that radiate outward from the centrosome (see Figure 16-42). The number of γ-TuRCs in each centrosome increases greatly at the beginning of mitosis, in a process called *centrosome maturation*.

Microtubule formation also occurs within the metaphase spindle between the poles. A large protein complex called *augmin* anchors active γ-TuRC to the side of a microtubule, resulting in the nucleation of a new microtubule that branches off the other (see Figure 16-47). Augmin binds to the microtubule in such a way that it orients the new microtubule with its plus end pointing in the same direction as the microtubule to which it binds; as a result, the new microtubule has the correct orientation when it is released and repositioned in the spindle microtubule network.

Spindle assembly also depends on increased microtubule synthesis in the vicinity of the chromosomes. Mitotic chromosomes generate local signals that activate γ-TuRC and thereby promote microtubule formation. As we describe later, this mechanism, together with microtubule sorting by various motor proteins, allows chromosomes to make a major contribution to the formation of a bipolar spindle—particularly in the absence of centrosomes.

Microtubule Instability Increases Greatly in Mitosis

As discussed in Chapter 16, microtubules are in a state of *dynamic instability*, in which individual microtubules are either growing or shrinking and stochastically switch between the two states. New microtubules are continually being created to balance the loss of those that disappear completely by depolymerization.

Entry into mitosis signals an abrupt change in the cell's microtubules. During prophase, and particularly in prometaphase and metaphase (see Panel 17–1), the average lifetime of a microtubule decreases dramatically—particularly the lifetime of non-kinetochore microtubules, which exist for only 15–30 seconds. This increase in microtubule instability, coupled with the increased ability of the spindle to nucleate microtubules as mentioned above, results in a remarkably dense and dynamic array of spindle microtubules.

Microtubule dynamics are controlled in the cell by a variety of regulatory proteins, including microtubule-associated proteins (MAPs) that promote stability and depolymerization factors that destabilize microtubule plus ends. Changes in the activities of these proteins are responsible for the changes in microtubule dynamics that occur during mitosis. Many of these changes result from phosphorylation of specific proteins by M-Cdk and other mitotic protein kinases.

Microtubule-based Motor Proteins Govern Spindle Assembly and Function

The assembly and function of the mitotic spindle depend on numerous microtubule-dependent motor proteins. As discussed in Chapter 16, these proteins belong to two families—the large family of kinesin-related proteins, which usually move toward the plus end of microtubules, and dyneins, which move toward the minus end. Two motor proteins—*kinesin-5* and *cytoplasmic dynein*—are particularly important in spindle assembly and function, and many others, including *kinesin-14* and *kinesins-4/10*, are also involved (**Figure 17–28**).

Kinesin-5 is a large tetramer containing two dimeric motor domains at each end. The motor domains both move toward the plus end of a microtubule but can be oriented in opposite directions; as a result, they can associate with two antiparallel microtubules and slide them in opposite directions. When this occurs near the center of the spindle, the result is that the minus ends of the microtubule

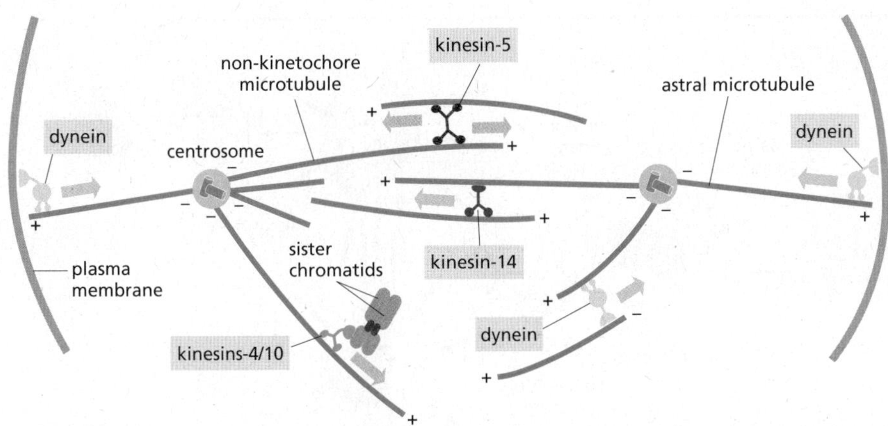

Figure 17–28 Major motor proteins of the spindle. Microtubule-dependent motor proteins contribute to spindle assembly and function (see text). The colored arrows indicate the direction of motor protein movement along a microtubule—*blue* toward the minus end and *light red* toward the plus end. Because non-kinetochore microtubules are extensively cross-linked to each other in the spindle and at the poles, it is thought that microtubule sliding like that shown here can alter spindle length. (From D.O. Morgan, The Cell Cycle: Principles of Control, p. 117. London: New Science Press, 2007. With permission from Oxford University Press.)

are pushed toward the poles. This process is fundamentally important in generating bipolarity in the spindle.

Cytoplasmic dynein is a minus end–directed motor that, together with associated proteins, organizes microtubules at various locations in the cell. By attaching to the minus end of one microtubule and transporting it toward the minus end of a second microtubule, dynein functions to connect new microtubules formed in the body of the spindle to microtubules nucleated at a centrosome, and to focus the spindle poles. Dynein is also present on the plus ends of astral microtubules at the cell cortex. By moving toward the minus end of astral microtubules, dynein motors pull the spindle poles toward the cell cortex and away from each other.

Kinesin-14, unlike most kinesins, is a minus end–directed motor. It contains a dimeric motor domain and a second domain that can bind to a neighboring microtubule in a specific orientation, enabling it to cross-link antiparallel microtubules at the center of the spindle and pull the poles together. Kinesin-4 and kinesin-10 proteins, also called *chromokinesins*, are plus end–directed motors that associate with chromosome arms and push the attached chromosome away from the pole (or the pole away from the chromosome).

Bipolar Spindle Assembly in Most Animal Cells Begins with Centrosome Duplication

The mitotic spindle must have two poles if it is to pull the two sets of sister chromatids to opposite ends of the cell in anaphase. In most animal cells, several mechanisms ensure the bipolarity of the spindle. One depends on centrosomes. A typical animal cell enters mitosis with a pair of centrosomes, each of which nucleates a radial array of microtubules. The two centrosomes provide prefabricated spindle poles that greatly facilitate bipolar spindle assembly. The other mechanisms depend on the ability of mitotic chromosomes to nucleate and stabilize microtubules and on the ability of motor proteins to organize microtubules into a bipolar array. We will discuss these "self-organization" mechanisms later in this section.

In interphase, most animal cells contain a single centrosome that nucleates most of the cell's cytoplasmic microtubules. The centrosome duplicates when the cell enters the cell cycle, so that by the time the cell reaches mitosis there are two centrosomes. Centrosome duplication begins at about the same time as the cell enters S phase. The G_1/S-Cdk (a complex of cyclin E and Cdk2 in animal cells; see Table 17–1) that triggers cell-cycle entry also helps initiate centrosome duplication, together with another protein kinase called Plk4. Each of the two centrioles in the centrosome nucleates the formation of a new centriole, which is gradually constructed in S phase to generate a closely linked pair of centrosomes (**Figure 17–29**). This centrosome pair remains together on one side of the nucleus until the cell enters mitosis, when the two centrosomes undergo numerous changes to form the poles of the spindle.

There are interesting parallels between centrosome duplication and chromosome duplication. Both use a semiconservative mechanism of duplication, in

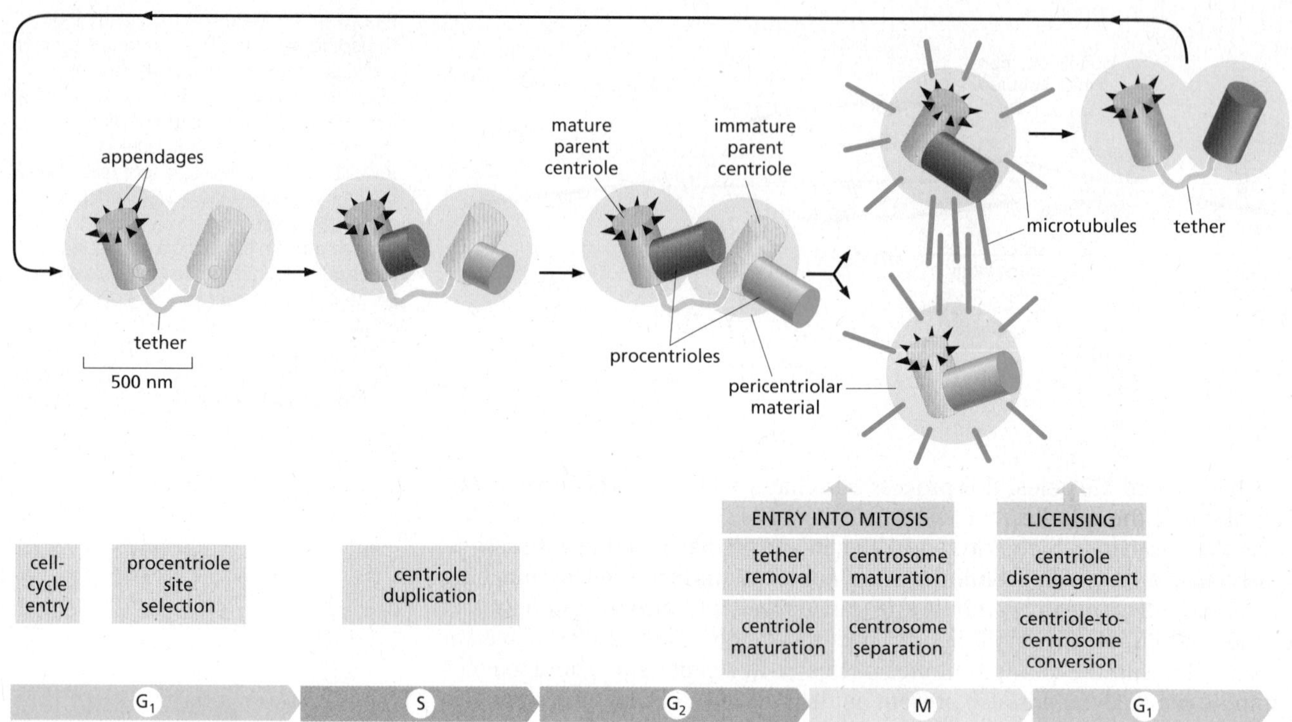

Figure 17–29 The centrosome cycle. In early G_1, the centrosome consists of two centrioles linked by a protein tether, as well as associated pericentriolar material for nucleation of microtubules *(light green)*. One centriole, the mature parent, carries protein appendages required for certain centrosome functions. Upon entry into the cell cycle, G_1/S-Cdk and Plk4 initiate centriole duplication, whereby a new centriole (procentriole) is assembled at a single site on the side of each parent centriole *(yellow dots)*. The elongation of the procentrioles is usually completed in G_2. The two centriole pairs remain close together in a single centrosomal complex until entry into mitosis, when the protein kinases M-Cdk, Plk1, and other regulators trigger numerous changes: the tether between centrosomes is removed, the immature parent centriole acquires appendages (centriole maturation), the pericentriolar material expands to enable more microtubule nucleation (centrosome maturation), and the centrosomes separate, forming a new spindle between them. After the completion of mitosis, the two centrioles in each centrosome detach (centriole disengagement), and the new centriole acquires pericentriolar material (centriole-to-centrosome conversion). These two processes are required for centriole duplication in the subsequent cell cycle; thus, progression through mitosis is required for duplication, helping to ensure that centrioles (and centrosomes) duplicate only once per cell cycle.

which the two halves separate and serve as templates for construction of a new half. Centrosomes, like chromosomes, must replicate once and only once per cell cycle, to ensure that the cell enters mitosis with precisely two copies: an incorrect number of centrosomes can lead to defects in spindle assembly and thus errors in chromosome segregation.

We are beginning to unravel the complex mechanisms that limit centrosome duplication to once and only once per cell cycle. These mechanisms are reminiscent of those that restrict DNA replication to once per cell cycle: after duplication has occurred in S phase, passage through mitosis is required for the "licensing" of centrosome duplication in the next cell cycle (see Figure 17–29).

Spindle Assembly in Animal Cells Requires Nuclear-Envelope Breakdown

In cells that contain centrosomes, spindle assembly begins when the two centrosomes move apart along the nuclear envelope in prophase, pulled by dynein motor proteins that link astral microtubules to the cell cortex (see Figure 17–28). The plus ends of the microtubules between the centrosomes interdigitate to form an array of antiparallel microtubules, and kinesin-5 motor proteins cross-link these microtubules and push the centrosomes apart.

Centrosomes and microtubules are located in the cytoplasm, where they have no access to the sister-chromatid pairs inside the nucleus. Thus, the attachment of the nascent spindle to sister-chromatid pairs requires the removal of the nuclear envelope. In addition, many of the motor proteins and microtubule regulators that promote spindle assembly are associated with the chromosomes inside the nucleus, and they require nuclear-envelope breakdown to carry out their functions.

Nuclear-envelope breakdown is a complex, multistep process, which is thought to begin when M-Cdk phosphorylates several subunits of the nuclear pore complexes in the nuclear envelope. This phosphorylation initiates the disassembly of nuclear pore complexes and their dissociation from the envelope. M-Cdk also phosphorylates components of the nuclear lamina, the structural framework beneath the envelope, resulting in nuclear lamina disassembly. In parallel, phosphorylation of several inner-nuclear-envelope proteins leads to detachment of

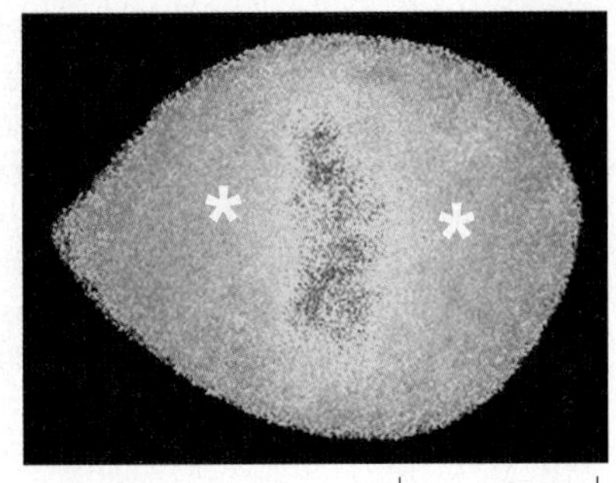

Figure 17–30 **Activation of the GTPase Ran around mitotic chromosomes.** The Ran protein, like other members of the small GTPase family (discussed in Chapter 15), can exist in two conformations depending on whether it is bound to GDP (inactive state) or GTP (active state). The localization of active Ran in mitosis was determined using a protein that emits fluorescence at a specific wavelength when it is activated by Ran-GTP. In the metaphase human cell shown here, Ran activity (*yellow* and *red*) is highest around the chromosomes, between the poles of the mitotic spindle (indicated by *asterisks*). (From P. Kaláb et al., *Nature* 440:697–701, published 2006 by Nature Publishing Group. Reproduced with permission from SNCSC.)

lamin proteins and chromosomes from the nuclear envelope, which is then incorporated into the membranes of the endoplasmic reticulum.

Mitotic Chromosomes Promote Bipolar Spindle Assembly

When they are present, centrosomes drive spindle assembly. Centrosomes are very efficient microtubule nucleators that also act to organize the spindle poles, rapidly generating a bipolar microtubule array that rains down on the chromosomes when the nuclear envelope is removed. However, chromosomes are not just passive passengers in the process of spindle assembly. By creating a local environment that favors both microtubule nucleation and microtubule stabilization, they play an active part in spindle formation. The influence of the chromosomes can be demonstrated by using a fine glass needle to reposition them after the spindle has formed. For some cells in metaphase, if a single chromosome is tugged out of alignment, a mass of new spindle microtubules rapidly appears around the newly positioned chromosome, while the spindle microtubules at the chromosome's former position depolymerize. This property of the chromosomes seems to depend, at least in part, on a guanine nucleotide exchange factor (GEF) that is bound to chromatin; the GEF stimulates a small GTPase in the cytosol called *Ran* to bind GTP in place of GDP. The activated Ran-GTP, which is also involved in nuclear transport (discussed in Chapter 12), releases microtubule-regulatory proteins from protein complexes in the cytosol, thereby stimulating the local nucleation and stabilization of microtubules around chromosomes (**Figure 17–30**). The best-understood mechanism depends on the release of a protein called TPX2, which then binds and activates the protein kinase Aurora-A. Aurora-A phosphorylates regulatory proteins that activate γ-TuRCs, leading to an increase in local formation of new microtubules. TPX2 might also activate augmin to promote the formation of new microtubule branches on the sides of existing microtubules. Local microtubule stabilization is also promoted by the protein kinase Aurora-B, which associates with mitotic chromosomes.

In the absence of centrosomes, spindle assembly is thought to begin with the formation of microtubules around the chromosomes. Various microtubule-associated proteins then organize the microtubules into a bipolar spindle (**Figure 17–31**). Two motor proteins that we discussed earlier are particularly

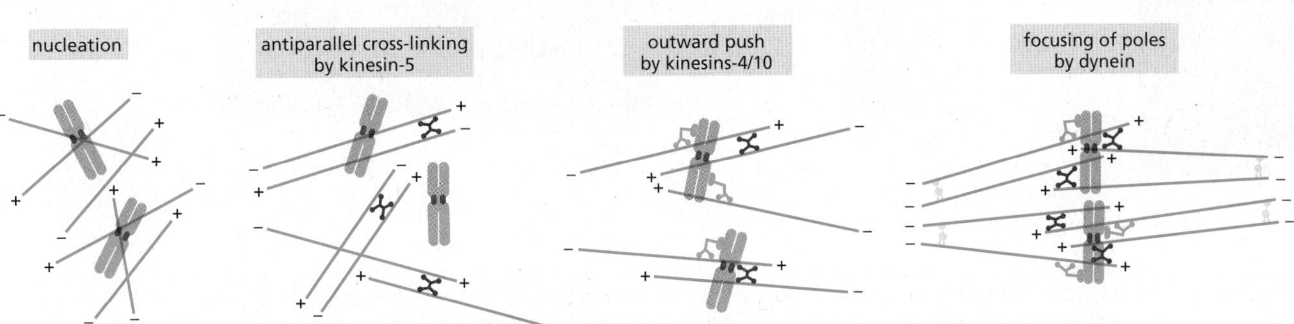

Figure 17–31 **Spindle self-organization by motor proteins.** Mitotic chromosomes stimulate the local activation of proteins that promote the formation of microtubules in the vicinity of the chromosomes. Kinesin-5 motor proteins (see Figure 17–28) organize these microtubules into antiparallel bundles, while plus end–directed kinesin-4 and kinesin-10 link the microtubules to chromosome arms and push minus ends away from the chromosomes. Dynein, together with numerous other proteins, focuses these minus ends into a pair of spindle poles.

important. Kinesin-5, by cross-linking antiparallel microtubules and pushing their minus ends outward, is essential for the bipolar character of the spindle. Dynein, by transporting one microtubule minus end toward the minus end of another, is required for the focusing of minus ends in a discrete pole at each end of the bipolar array. Many other microtubule regulators, including other motor proteins and augmin, also contribute to spindle assembly.

Cells that normally lack centrosomes, such as those of higher plants and many animal oocytes, use chromosome-based self-organization mechanisms to form spindles. These mechanisms also assemble spindles in experimental systems where centrosomes are removed—such as in certain animal embryos that have been induced to develop from eggs without fertilization (called *parthenogenesis*). As the sperm normally provides the centrosome when it fertilizes an egg, the mitotic spindles in these parthenogenetic embryos develop without centrosomes (**Figure 17–32**). Although the resulting acentrosomal spindle can segregate chromosomes normally, it lacks astral microtubules, which are responsible for positioning the spindle in animal cells; as a result, the spindle can be mispositioned in the cell.

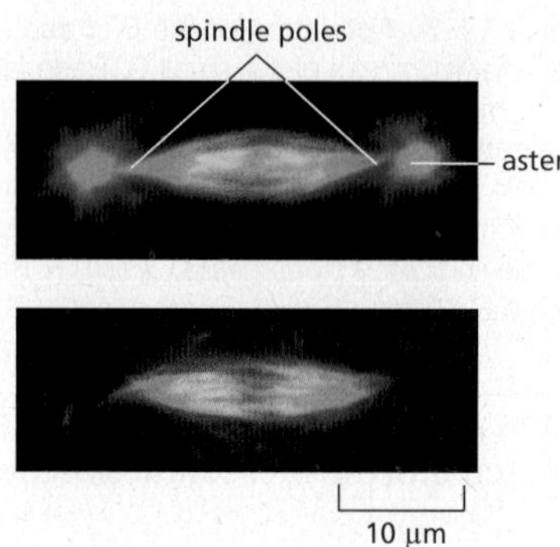

Figure 17–32 Bipolar spindle assembly without centrosomes in parthenogenetic embryos of the insect *Sciara* (or fungus gnat). The microtubules are stained *green*, the chromosomes *red*. The top fluorescence micrograph shows a normal spindle formed with centrosomes in a normally fertilized *Sciara* embryo. The bottom micrograph shows a spindle formed without centrosomes in an embryo that initiated development without fertilization. Note that the spindle with centrosomes has an aster at each pole of the spindle, whereas the spindle formed without centrosomes does not. Both types of spindles are able to segregate the replicated chromosomes. (© 1998 B. de Saint Phalle and W. Sullivan. Originally published in *J. Cell Biol.* https://doi.org/10.1083/jcb.141.6.1383. With permission from Rockefeller University Press.)

Kinetochores Attach Sister Chromatids to the Spindle

After the assembly of a bipolar microtubule array, the second major step in spindle formation is the attachment of the array to the sister-chromatid pairs. Spindle microtubules become attached to each chromatid at its **kinetochore**, a giant, multilayered protein structure that is built at the centromeric region of the chromatid (**Figure 17–33**; also see Chapter 4). In metaphase, the plus ends of kinetochore microtubules are embedded head-on in specialized microtubule-attachment sites within the outer region of the kinetochore, furthest from the DNA. The kinetochore of an animal cell can bind 10–40 microtubules, whereas a budding yeast kinetochore can bind only one. Attachment of each microtubule depends on multiple copies of a rod-shaped protein complex called the Ndc80 complex, which is anchored in the kinetochore at one end and interacts with the

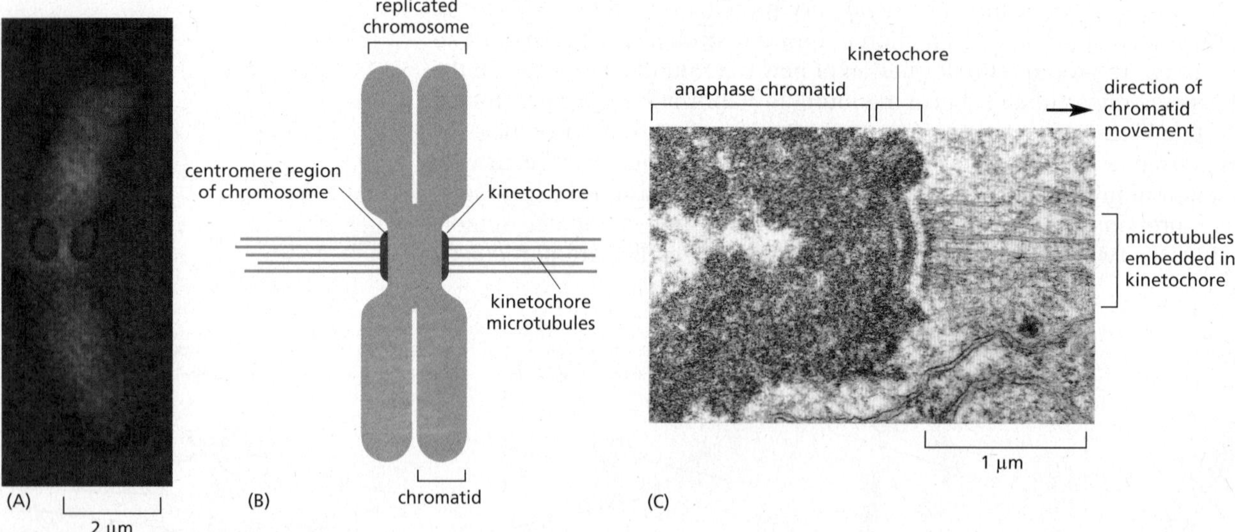

Figure 17–33 The kinetochore. (A) A fluorescence micrograph of a metaphase chromosome stained with a DNA-binding fluorescent dye and with human autoantibodies that react with specific kinetochore proteins. The two kinetochores, one associated with each sister chromatid, are stained *red.* (B) A drawing of a metaphase chromosome showing its two sister chromatids attached to the plus ends of kinetochore microtubules. Each kinetochore forms a plaque on the surface of the centromere. (C) Electron micrograph of an anaphase chromatid with microtubules attached to its kinetochore. While most kinetochores have a trilaminar structure, the one shown here (from a green alga) has an unusually complex structure with additional layers. (A, © 1991 R.P. Zinkowski et al. Originally published in *J. Cell Biol.* https://doi.org/10.1083/jcb.113.5.1091. With permission from Rockefeller University Press. C, from J.D. Pickett-Heaps and L.C. Fowke, *Aust. J. Biol. Sci.* 23:71–92, 1970. With permission from CSIRO Publishing.)

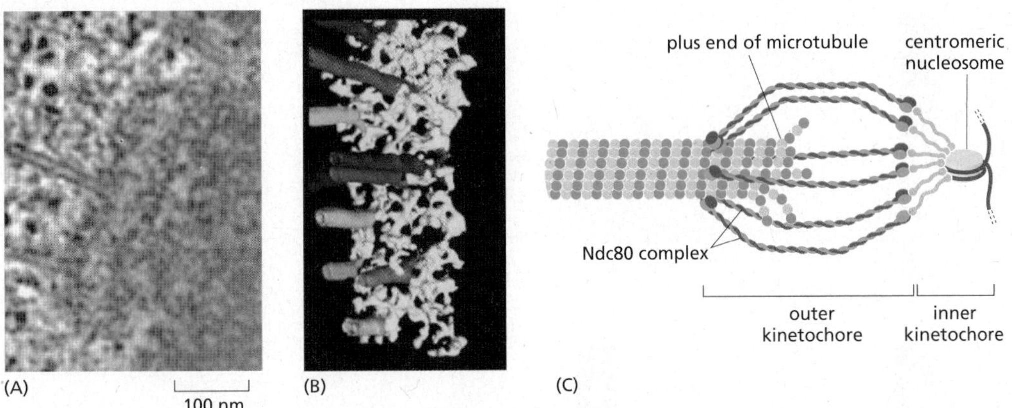

Figure 17–34 Microtubule attachment sites in the kinetochore. (A) In this electron micrograph of a mammalian kinetochore, the chromosome is on the right, and the plus ends of multiple microtubules are embedded in the outer kinetochore on the left. (B) Electron tomography (discussed in Chapter 9) was used to construct a low-resolution three-dimensional image of the outer kinetochore in A. Several microtubules *(in multiple colors)* are embedded in fibrous material of the kinetochore, which is thought to be composed of the Ndc80 complex and other proteins. (C) Each microtubule is attached to the kinetochore by interactions with multiple copies of the Ndc80 complex *(blue)*. This complex binds to the sides of the microtubule near its plus end, allowing polymerization and depolymerization to occur while the microtubule remains attached to the kinetochore. The opposite end of the Ndc80 complex is attached through intermediary proteins to centromeric nucleosomes containing a specialized version of histone H3 called CENP-A (see Chapter 4). Budding yeast kinetochores contain a single centromeric nucleosome and thus a single microtubule-binding site like that shown here, while animal kinetochores (like those in B) contain large arrays of these binding sites distributed over large numbers of centromeric nucleosomes. (A and B, from Y. Dong et al., *Nat. Cell Biol.* 9:516–522, published 2007 by Nature Publishing Group. Reproduced with permission of SNCSC.)

sides of the microtubule at the other, thereby linking the microtubule to the kinetochore while still allowing the addition or removal of tubulin subunits at this end (**Figure 17–34**). Regulation of plus-end polymerization and depolymerization at the kinetochore is critical for the control of chromosome movement on the spindle, as we discuss later.

Kinetochore attachment to the spindle occurs by a complex sequence of events. At the end of prophase in animal cells, the centrosomes of the growing spindle generally lie on opposite sides of the nuclear envelope. Thus, when the envelope breaks down, the sister-chromatid pairs are bombarded by dynamic microtubule plus ends coming from two directions. However, the kinetochores do not instantly achieve the correct "end-on" microtubule attachment to both spindle poles. Instead, detailed studies with light and electron microscopy show that most initial attachments are unstable *lateral* attachments, in which a kinetochore attaches to the side of a passing microtubule, with assistance from dynein motor proteins in the outer kinetochore. Eventually, microtubule plus ends are captured by one and then the other kinetochore in the correct end-on orientation.

Another attachment mechanism also plays a part, particularly in the absence of centrosomes. Short microtubules in the vicinity of the chromosomes become embedded in the plus end-binding sites of the kinetochore. Polymerization at these plus ends then results in growth of the microtubules away from the kinetochore. The minus ends of these kinetochore microtubules are eventually cross-linked to other minus ends and focused by dynein motor proteins at the spindle pole (see Figure 17–31).

Bi-orientation Is Achieved by Trial and Error

The success of mitosis demands that sister chromatids in a pair attach to opposite poles of the mitotic spindle, so that they move to opposite ends of the cell when they separate in anaphase. How is this mode of attachment, called **bi-orientation**, achieved? What prevents the attachment of both kinetochores to the same spindle pole or the attachment of one kinetochore to both spindle poles? Part of the answer is that sister kinetochores are constructed in a back-to-back orientation

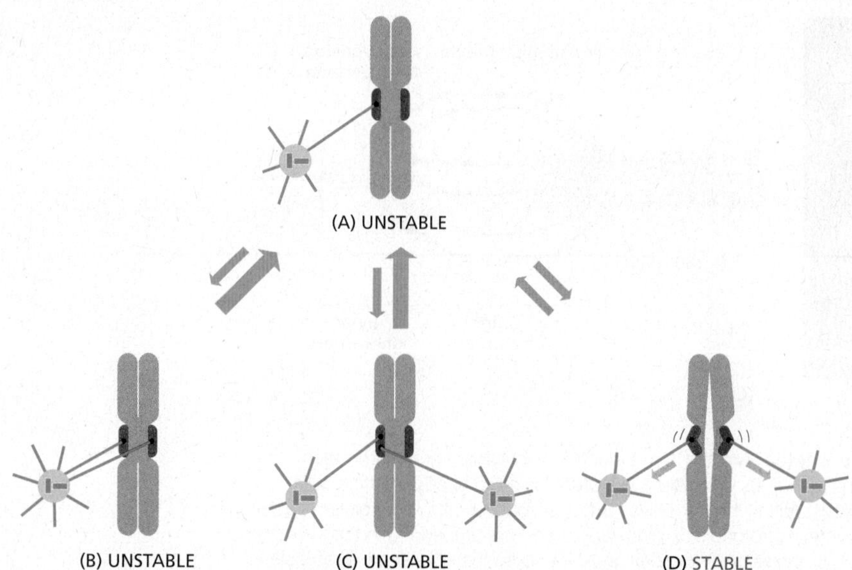

Figure 17–35 Alternative forms of kinetochore attachment to the spindle poles. (A) Initially, a single microtubule from a spindle pole binds to one kinetochore in a sister-chromatid pair. Additional microtubules can then bind to the chromosome in various ways. (B) A microtubule from the same spindle pole can attach to the other sister kinetochore, or (C) microtubules from both spindle poles can attach to the same kinetochore. These incorrect attachments are unstable, however, so that one of the two microtubules tends to dissociate. (D) When a microtubule from the opposite pole binds to the second kinetochore, the sister kinetochores are thought to sense tension across their microtubule-binding sites. This triggers an increase in microtubule binding affinity, thereby locking the correct attachment in place. Occasionally (not shown), attachments can form that are a combination of C and D; that is, the two sister kinetochores are attached to opposite poles and are under tension, but there is also an inappropriate microtubule link between one kinetochore and both spindle poles. This combination is stable, and cells that progress to anaphase with such an attachment risk moving both sister chromatids to the same daughter, which is usually lethal for the cell.

that reduces the likelihood that both kinetochores can face the same spindle pole. Nevertheless, incorrect attachments do occur, and elegant regulatory mechanisms have evolved to correct them.

Incorrect attachments are corrected by a system of trial and error that is based on a simple principle: most incorrect attachments are highly unstable and do not last, whereas correct attachments are held in place. How does the kinetochore sense a correct attachment? The answer appears to be tension (**Figure 17–35**). When a sister-chromatid pair is properly bi-oriented on the spindle, the two kinetochores are pulled in opposite directions by strong poleward forces. Sister-chromatid cohesion resists these poleward forces, creating high levels of tension within the kinetochores. When chromosomes are incorrectly attached—when both sister chromatids are attached to the same spindle pole, for example—tension is low and the kinetochore generates an inhibitory signal that loosens the grip of its microtubule attachment site, allowing detachment to occur. When bi-orientation occurs, the high tension at the kinetochore shuts off the inhibitory signal, strengthening microtubule attachment. In animal cells, tension not only increases the affinity of the attachment site but also leads to the attachment of additional microtubules to the kinetochore. This results in the formation of a thick *kinetochore fiber* composed of multiple microtubules.

The tension-sensing mechanism depends on the protein kinase Aurora-B, which is associated with the kinetochore and is thought to generate the inhibitory signal that reduces the strength of microtubule attachment in the absence of tension. It phosphorylates several components of the microtubule attachment site, including the Ndc80 complex, decreasing the site's affinity for a microtubule plus end. When bi-orientation occurs, the resulting tension reduces

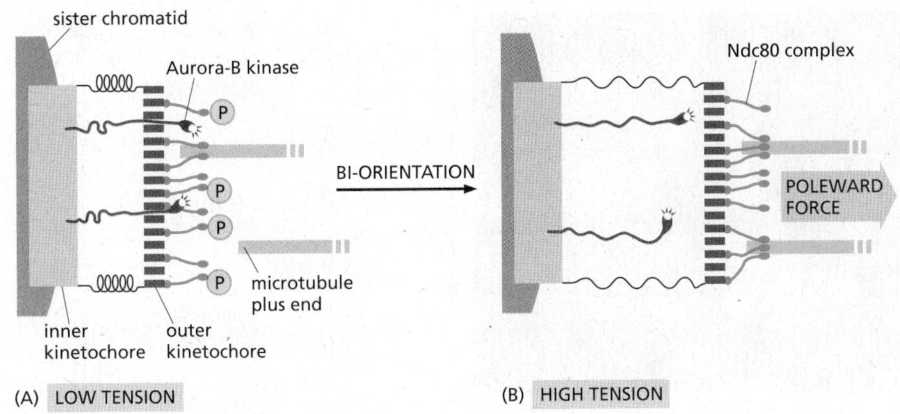

Figure 17–36 How tension might increase microtubule attachment to the kinetochore. These diagrams illustrate one speculative mechanism by which bi-orientation might increase microtubule attachment to the kinetochore. A single kinetochore is shown for clarity; the spindle pole is on the right. (A) When a sister-chromatid pair is unattached to the spindle or attached to just one spindle pole, there is little tension between the outer and inner kinetochores. The protein kinase Aurora-B is tethered to the inner kinetochore and phosphorylates the microtubule attachment sites, including the Ndc80 complex *(blue)*, in the outer kinetochore as shown, thereby reducing the affinity of microtubule binding. Microtubules therefore associate and dissociate rapidly, and attachment is unstable. (B) When bi-orientation is achieved, the forces pulling the kinetochore toward the spindle pole are resisted by forces pulling the other sister kinetochore toward the opposite pole, and the resulting tension pulls the outer kinetochore away from the inner kinetochore. As a result, Aurora-B is unable to reach the outer kinetochore, and microtubule attachment sites are not phosphorylated. Microtubule binding affinity is therefore increased, resulting in the stable attachment of multiple microtubules to both kinetochores. The dephosphorylation of outer kinetochore proteins depends on a phosphatase that is not shown here.

phosphorylation by Aurora-B, thereby increasing the affinity of the attachment site (**Figure 17–36**).

After their attachment to the two spindle poles, the chromosomes are tugged back and forth, eventually assuming a position equidistant between the two poles, a position called the **metaphase plate**. In vertebrate cells, the chromosomes then oscillate gently at the metaphase plate, awaiting the signal for the sister chromatids to separate. The signal is produced, with a predictable lag time, after the bi-oriented attachment of the last sister-chromatid pair.

Multiple Forces Act on Chromosomes in the Spindle

The metaphase spindle, with the sister chromatids aligned at the metaphase plate, appears as a stable bipolar structure of a fixed length, but appearances can be deceiving. The spindle is a highly dynamic assembly that exists at a steady state that depends on the precise balance of numerous forces generated by various motor proteins and by microtubule polymerization and depolymerization. These forces move chromosomes once they are attached to the spindle and produce the tension that is so important for the stabilization of correct attachments. In anaphase, similar forces pull the separated chromatids to opposite ends of the spindle. Three major spindle forces are particularly critical, although their strength and importance vary at different stages of mitosis.

The first major force pulls the kinetochore and its associated chromatid along the kinetochore microtubule toward the spindle pole. This force is generated by depolymerization at the plus end of the microtubule. It pulls on chromosomes during prometaphase and metaphase but is particularly important for moving sister chromatids toward the poles after they separate in anaphase. Notably, this kinetochore-generated poleward force does not require ATP or motor proteins. This might seem implausible at first, but it has been shown that purified kinetochores in a test tube, with no ATP present, can remain attached to depolymerizing microtubules and thereby move. The energy that drives the movement is stored in the microtubule and is released when the microtubule depolymerizes; it ultimately comes from the hydrolysis of GTP that occurs after a tubulin subunit adds to the end of a microtubule (discussed in Chapter 16).

How does plus-end depolymerization drive the kinetochore toward the pole? As we discussed earlier (see Figure 17–34C), Ndc80 complexes in the kinetochore make multiple low-affinity attachments along the side of the microtubule. Because the attachments are constantly breaking and re-forming at new sites, the kinetochore remains attached to a microtubule even as the microtubule depolymerizes. In principle, this could move the kinetochore toward the spindle pole.

A second poleward force is provided in some cell types by **microtubule flux**, whereby the microtubules themselves are pulled toward the spindle poles and dismantled at their minus ends. The mechanism underlying this poleward movement is not clear, although it is likely to depend on forces generated by motor proteins and minus-end depolymerization at the spindle pole. In metaphase, the addition

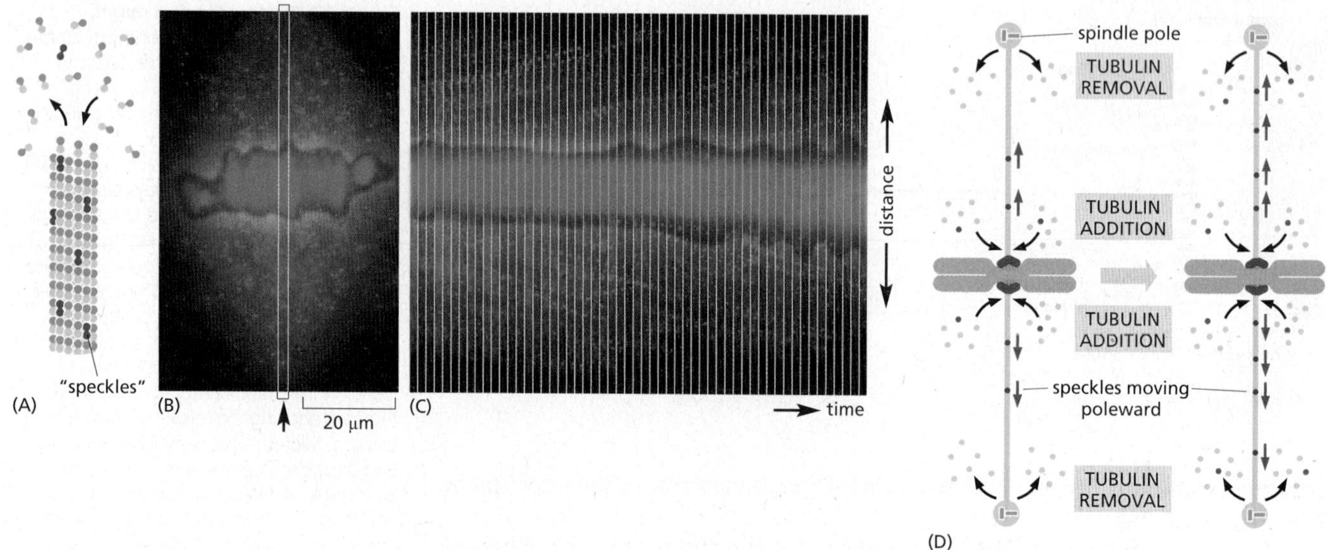

Figure 17–37 Microtubule flux in the metaphase spindle. (A) To observe microtubule flux, a small amount of fluorescent tubulin is injected into living cells or cell extracts so that individual microtubules form with a small proportion of fluorescent tubulin. Such microtubules have a speckled appearance when viewed by fluorescence microscopy. (B) Much of what we know about mitotic spindle behavior comes from studies in extracts of the eggs of the frog *X. laevis*. These extracts carry out many cell-cycle processes in a test tube, including mitotic spindle assembly. This image shows a fluorescence micrograph of a mitotic spindle in an extract mixed with a small amount of fluorescent tubulin. Because there are no centrosomes in these extracts, the spindle does not have astral microtubules. The chromosomes are colored *brown*, and the tubulin speckles are *red*. (C) The movement of individual speckles can be followed by time-lapse video microscopy. Images of the thin vertical boxed region *(arrow)* in B, taken every 10 seconds, are aligned here to show that individual speckles move toward the poles at a rate of about 0.75 μm/min, indicating that the microtubules are moving poleward. (D) The length of a kinetochore microtubule does not change significantly during this experiment because new tubulin subunits are added at the microtubule plus end at the same rate as tubulin subunits are removed from the minus end. (B and C, from T.J. Mitchison and E.D. Salmon, *Nat. Cell Biol.* 3:E17–21, published 2001 by Nature Publishing Group. Reproduced with permission of SNCSC.)

of new tubulin at the plus end of a microtubule compensates for the loss of tubulin at the minus end, so that microtubule length remains constant despite the movement of microtubules toward the spindle pole (**Figure 17–37**). Any kinetochore that is attached to a microtubule undergoing such flux experiences a poleward force, which contributes to the generation of tension at the kinetochore in metaphase. Together with the kinetochore-based forces discussed above, flux also contributes to the poleward forces that move sister chromatids after they separate in anaphase.

A third force acting on chromosomes is the *polar ejection force*, or *polar wind*. Plus end–directed kinesin-4 and kinesin-10 motors on chromosome arms interact with microtubules and transport the chromosomes away from the spindle poles (see Figure 17–28). This force is particularly important in prometaphase and metaphase, when it helps push chromosome arms out from the spindle. This force might also help align the sister-chromatid pairs at the metaphase plate.

The APC/C Triggers Sister-Chromatid Separation and the Completion of Mitosis

The cell cycle reaches its most dramatic moment with the separation of the sister chromatids at the metaphase-to-anaphase transition (**Figure 17–38**). Although M-Cdk activity sets the stage for this event, the anaphase-promoting complex, or cyclosome (APC/C), discussed earlier throws the switch that initiates sister-chromatid separation by ubiquitylating several mitotic regulatory proteins and thereby triggering their destruction (see Figure 17–18).

As we discussed earlier, cohesins hold sister-chromatid pairs together after S phase. In early mitosis, the resolution of the sister chromatids is accompanied by the removal of most cohesin from the chromosome arms via a mechanism that depends on a protein that pulls open the cohesin ring at the junction of its Smc3 and Scc1 subunits (see Figure 17–23). When the cell reaches metaphase, cohesins

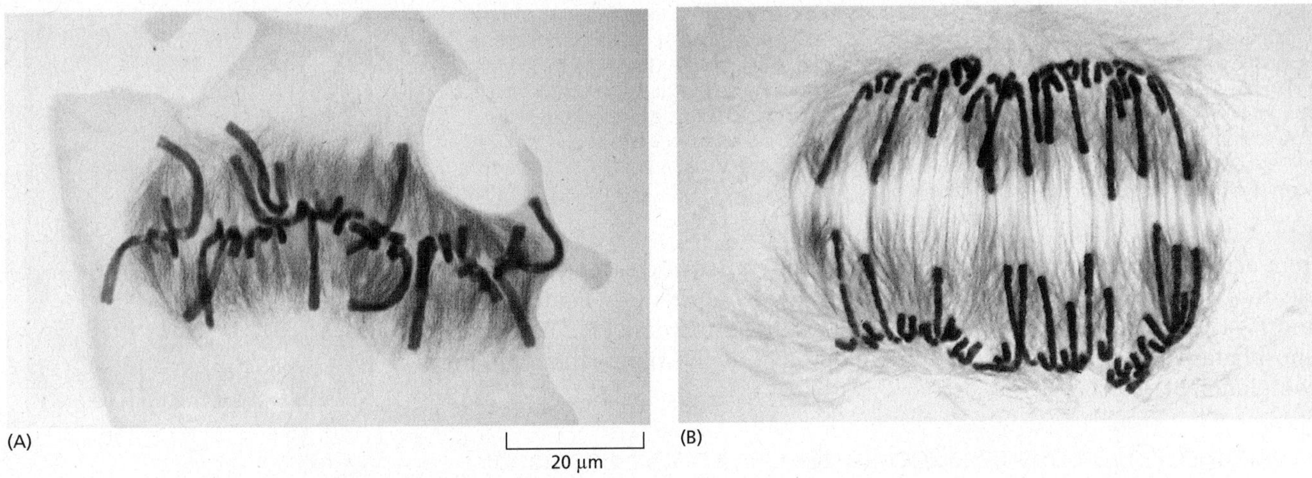

Figure 17–38 **Sister-chromatid separation at anaphase.** In the transition from metaphase (A) to anaphase (B), sister chromatids suddenly and synchronously separate and move toward opposite poles of the mitotic spindle—as shown in these light micrographs of *Haemanthus* (lily) endosperm cells that were stained with gold-labeled antibodies against tubulin *(pink)*. (Courtesy of Andrew Bajer.)

remain primarily at the centromeric regions of the chromosomes, adjacent to the kinetochores, where they serve to resist the poleward forces that pull the sister chromatids apart. Anaphase begins with the abrupt removal of the remaining cohesin, which allows the sisters to separate and move to opposite poles of the spindle. The APC/C initiates the process by targeting the inhibitory protein **securin** for destruction. Before anaphase, securin binds to and inhibits the activity of a protease called **separase**. The destruction of securin in metaphase releases separase, which is then free to cleave the Scc1 subunit of cohesin (see Figure 17–23). The cohesins fall away, and the sister chromatids separate (**Figure 17–39**).

We saw earlier that phosphorylation of various proteins by M-Cdk promotes spindle assembly, chromosome condensation, and nuclear-envelope breakdown in early mitosis. It is thus not surprising that the dephosphorylation of these same proteins is required for spindle disassembly and the re-formation of daughter nuclei in telophase. Dephosphorylation of Cdk targets depends in part on the inactivation of most Cdks in the cell, which results when the APC/C targets S- and M-cyclins for destruction. Protein dephosphorylation also results

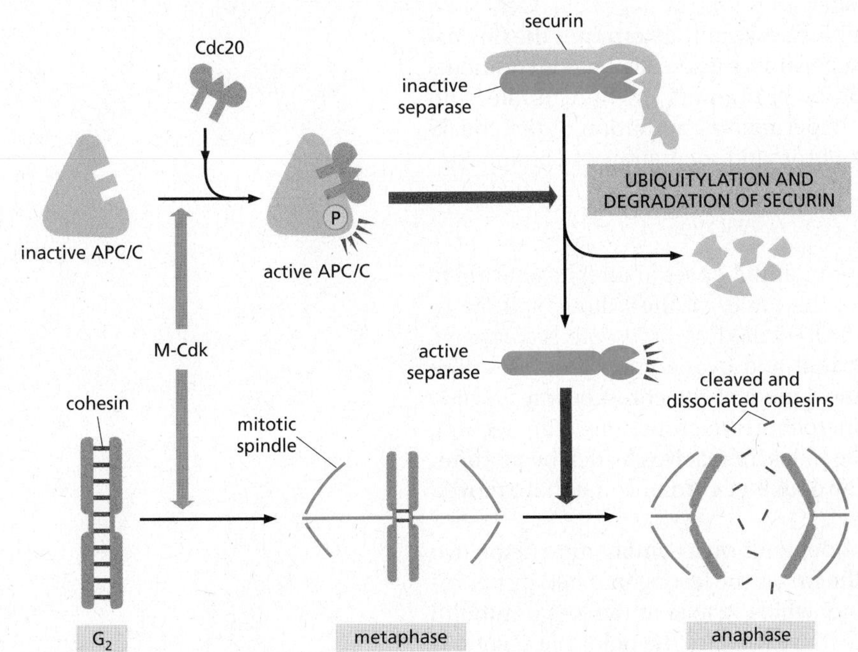

Figure 17–39 **The initiation of sister-chromatid separation by the APC/C.** The activation of APC/C by Cdc20 leads to the ubiquitylation and destruction of securin, which normally holds separase in an inactive state. The destruction of securin allows separase to cleave Scc1, a subunit of the cohesin complex holding the sister chromatids together (see Figure 17–23). The forces of the mitotic spindle then pull the sister chromatids apart. In animal cells, phosphorylation by Cdks also inhibits separase (not shown). Thus, Cdk inactivation in anaphase (resulting from cyclin destruction) also promotes separase activation by allowing its dephosphorylation.

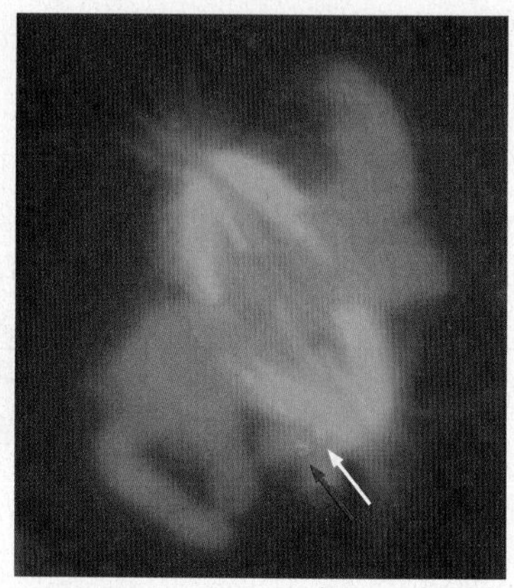

Figure 17–40 **Mad2 protein on unattached kinetochores.** This fluorescence micrograph shows a mammalian cell in prometaphase, with the mitotic spindle in *green* and the sister chromatids in *blue*. One sister-chromatid pair is attached to only one pole of the spindle. Staining with anti-Mad2 antibodies indicates that Mad2 is bound to the kinetochore of the unattached sister chromatid (*red dot,* indicated by *red arrow*). A small amount of Mad2 is associated with the kinetochore of the sister chromatid that is attached to the spindle pole (*pale dot,* indicated by *white arrow*). (© 1998 J. Waters et al. Originally published in *J. Cell Biol.* https://doi.org/10.1083/jcb.141.5.1181. With permission from Rockefeller University Press.)

from activation of phosphatases. Recall from our earlier discussions, for example, that the phosphatase PP2A-B55 is inactivated by M-Cdk (see Figures 17–15 and 17–17). When Cdk activity declines after cyclin destruction, PP2A and other phosphatases are activated, further driving protein dephosphorylation and the completion of mitosis.

Unattached Chromosomes Block Sister-Chromatid Separation: The Spindle Assembly Checkpoint

Drugs that destabilize microtubules, such as colchicine or vinblastine (discussed in Chapter 16), arrest cells in mitosis for hours or even days. This observation led to the identification of a **spindle assembly checkpoint** mechanism that is activated by the drug treatment and blocks progression through the metaphase-to-anaphase transition. The checkpoint mechanism ensures that cells do not enter anaphase until all chromosomes are correctly bi-oriented on the mitotic spindle.

The spindle assembly checkpoint depends on a sensor mechanism that monitors microtubule attachment at the kinetochore. Any kinetochore that is not properly attached to microtubules sends out a diffusible negative signal that blocks APC/C–Cdc20 activation throughout the cell and thus blocks the metaphase-to-anaphase transition. When the last sister-chromatid pair is properly attached and bi-oriented, this block is removed, allowing sister-chromatid separation to occur.

The negative checkpoint signal depends on several proteins, including *Mad2*, which are recruited to unattached kinetochores (**Figure 17–40**). The unattached kinetochore acts as an enzyme that catalyzes a change in the conformation of Mad2, so that Mad2 then interacts with other proteins to form a large multiprotein complex that binds and thereby inhibits APC/C–Cdc20. When proper microtubule attachment is achieved, these inhibitory complexes are disassembled, and APC/C–Cdc20 inhibition is thereby relieved.

In mammalian cells, the spindle assembly checkpoint determines the normal timing of anaphase. The destruction of securin in these cells begins moments after the last sister-chromatid pair becomes bi-oriented on the spindle, and anaphase begins about 20 minutes later. Experimental inhibition of the checkpoint mechanism causes premature sister-chromatid separation and anaphase.

Chromosomes Segregate in Anaphase A and B

The sudden loss of sister-chromatid cohesion at the onset of anaphase leads to sister-chromatid separation, which allows the forces of the mitotic spindle to pull the sisters to opposite poles of the cell—called *chromosome segregation.* The chromosomes move by two independent and overlapping processes. The first, **anaphase A**, is the initial poleward movement of the chromosomes, which is accompanied by shortening of the kinetochore microtubules. The second, **anaphase B**, is the separation of the spindle poles themselves, which begins after the sister chromatids have separated and the daughter chromosomes have moved some distance apart (**Figure 17–41**).

Chromosome movement in anaphase A depends on a combination of the two major poleward forces described earlier. The first is the force generated by microtubule depolymerization at the kinetochore, which results in the loss of tubulin subunits at the plus end as the kinetochore moves toward the pole. The second is

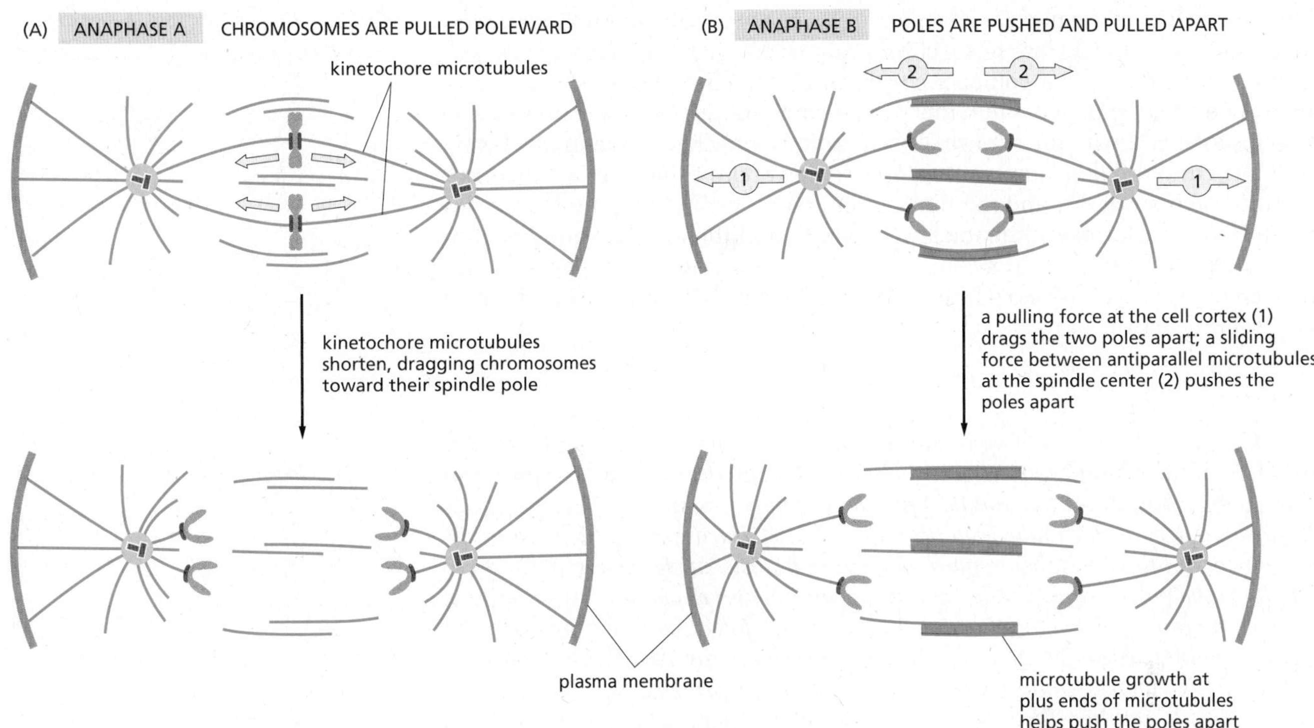

Figure 17–41 **The two processes of anaphase in mammalian cells.** (A) Separated sister chromatids move toward the poles in anaphase A. (B) In anaphase B, the two spindle poles move apart.

provided by microtubule flux, which is the poleward movement of the microtubules toward the spindle pole (see Figure 17–37). The relative importance of these two forces during anaphase varies in different cell types: in embryonic cells, chromosome movement depends mainly on microtubule flux, for example, whereas movement in yeast and vertebrate somatic cells results primarily from forces generated at the kinetochore.

Spindle-pole separation during anaphase B depends on motor-driven mechanisms similar to those that separate the two centrosomes in early mitosis. Dynein motor proteins that anchor astral microtubule plus ends to the cell cortex pull the poles apart. Kinesin-5, which cross-links antiparallel microtubules at the center of the spindle, pushes the poles apart (see Figure 17–28).

Although sister-chromatid separation initiates the chromosome movements of anaphase A, other mechanisms also ensure correct chromosome movements in anaphase A and spindle elongation in anaphase B. Most important, the completion of a normal anaphase depends on the dephosphorylation of Cdk substrates, which in most cells results from the APC/C-dependent destruction of cyclins. If M-cyclin destruction is prevented—by the production of a mutant form that is not recognized by the APC/C, for example—sister-chromatid separation generally occurs, but the chromosome movements and microtubule behavior of anaphase are abnormal.

The relative contributions of anaphase A and anaphase B to chromosome segregation vary greatly, depending on the cell type. In mammalian cells, anaphase B begins shortly after anaphase A and stops when the spindle is about twice its metaphase length; in contrast, the spindles of yeasts and certain protozoa primarily use anaphase B to separate the chromosomes at anaphase, and their spindles elongate to up to 15 times their metaphase length.

Segregated Chromosomes Are Packaged in Daughter Nuclei at Telophase

By the end of anaphase, the daughter chromosomes have segregated into two equal groups at opposite ends of the cell. In **telophase**, the final stage of mitosis, the two sets of chromosomes are packaged into a pair of daughter nuclei. The first major event of telophase is the disassembly of the mitotic spindle, followed by the re-formation of the nuclear envelope. This process occurs in multiple stages. First, proteins on the surface of the chromosomes promote their interaction with each other, resulting in a compact cluster of all the chromosomes. Next, fragments of endoplasmic reticulum membrane containing inner nuclear-envelope proteins associate with the surface of the chromosome cluster, eventually fusing to re-form the complete nuclear envelope. Nuclear pore complexes are incorporated into the envelope, and the nuclear lamina re-forms. The pore complexes pump in nuclear proteins, the nucleus expands, and the mitotic chromosomes are reorganized into their less-condensed interphase state. A new nucleus has been created, and mitosis is complete. All that remains is for the cell to complete its division into two.

Summary

M-Cdk triggers the events of early mitosis, including chromosome condensation, assembly of the mitotic spindle, and bipolar attachment of the sister-chromatid pairs to microtubules of the spindle. Spindle assembly in animal cells depends on the nucleation of microtubules at multiple locations. Centrosomes, which are duplicated before mitosis and then separated in early mitosis, nucleate microtubules to help form the poles of the spindle. Spindle formation also depends on the ability of mitotic chromosomes to stimulate local microtubule formation and the ability of motor proteins to organize microtubules into a bipolar array. Anaphase is triggered by the APC/C, which stimulates the destruction of the proteins that hold the sister chromatids together. The APC/C also promotes cyclin destruction and thus the inactivation of M-Cdk. The resulting dephosphorylation of Cdk targets is required for the events that complete mitosis, including the disassembly of the spindle and the re-formation of the nuclear envelope.

CYTOKINESIS

The final step in the cell cycle is **cytokinesis**, the division of the cytoplasm in two. In most cells, cytokinesis follows every mitosis, although some cells, such as early *Drosophila* embryos and some mammalian hepatocytes and heart muscle cells, undergo mitosis without cytokinesis and thereby acquire multiple nuclei. In most animal cells, cytokinesis begins in anaphase and ends shortly after the completion of mitosis in telophase.

Cytokinesis begins in an animal cell with the appearance of a *cleavage furrow* on the cell surface. The furrow rapidly deepens and spreads around the cell until it completely divides the cell in two. The structure underlying this process is the *contractile ring*—a dynamic assembly composed of actin filaments, myosin II filaments, and many structural and regulatory proteins. During anaphase, the ring assembles just beneath the plasma membrane (**Figure 17–42**; see also Panel 17–1). The ring gradually contracts, and, at the same time, fusion of intracellular vesicles

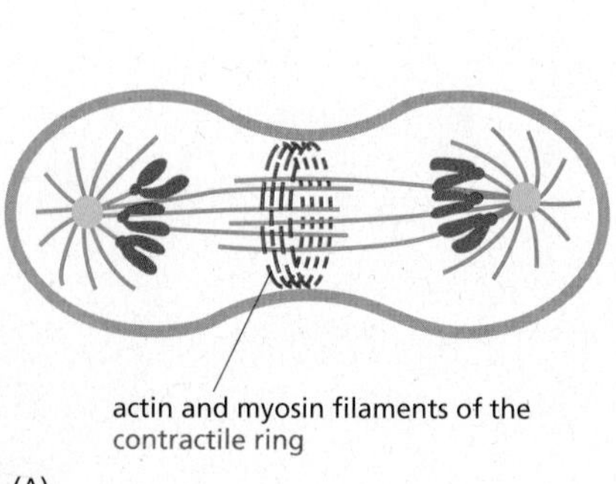

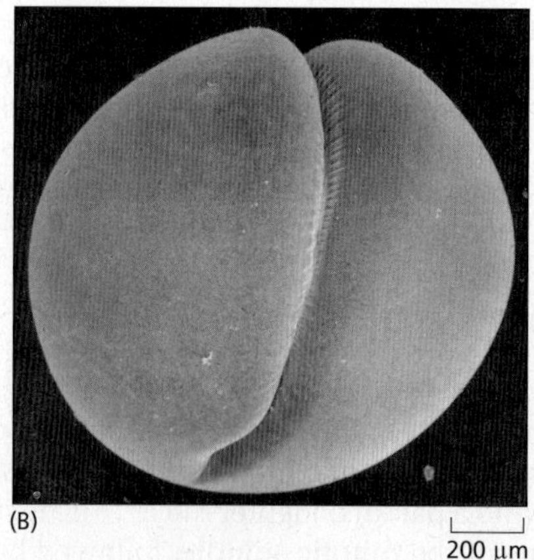

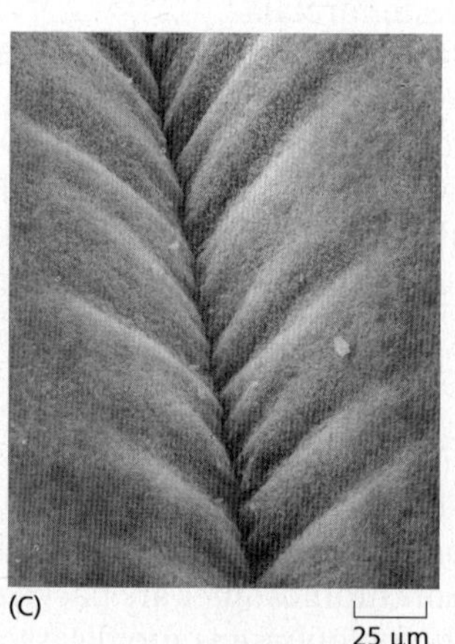

Figure 17–42 Cytokinesis. (A) The actin–myosin bundles of the contractile ring are oriented as shown, so that their contraction pulls the membrane inward. (B) In this low-magnification scanning electron micrograph of a cleaving frog egg, the cleavage furrow is especially prominent, as the cell is unusually large. The furrowing of the cell membrane is caused by the activity of the contractile ring underneath it. (C) The surface of a furrow at higher magnification. (B and C, from H.W. Beams and R.G. Kessel, *Am. Sci.* 64:279–290, 1976. With permission from Sigma Xi.)

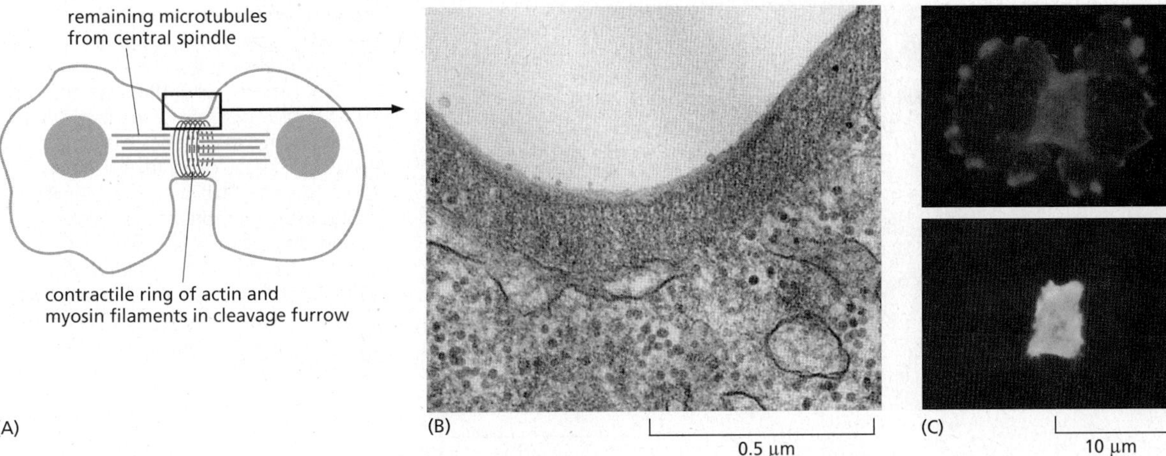

Figure 17–43 **The contractile ring.** (A) A drawing of the cleavage furrow in a dividing cell. (B) An electron micrograph of the ingrowing edge of a cleavage furrow of a dividing animal cell. (C) Fluorescence micrographs of a dividing slime mold amoeba stained for actin *(red)* and myosin II *(green)*. Whereas all of the visible myosin II has redistributed to the contractile ring, only some of the actin has done so; the rest remains in the cortex of the nascent daughter cells. (B, from H.W. Beams and R.G. Kessel, *Am. Sci.* 64:279–290, 1976. With permission from Sigma Xi. C, courtesy of Yoshio Fukui.)

with the plasma membrane inserts new membrane. This addition of membrane compensates for the increase in surface area that accompanies cytoplasmic division. When ring contraction is completed, the contractile ring is disassembled, and the narrow membrane bridge between the daughter cells is severed.

Actin and Myosin II in the Contractile Ring Guide the Process of Cytokinesis

In interphase cells, actin and myosin II filaments form a cortical network underlying the plasma membrane. In some cells, they also form large cytoplasmic bundles called *stress fibers* (discussed in Chapter 16). As cells enter mitosis, these arrays of actin and myosin disassemble; much of the actin reorganizes, and myosin II filaments are released. As the sister chromatids separate in anaphase, actin and myosin II begin to accumulate in the rapidly assembling **contractile ring** (Figure 17–43), which also contains numerous other proteins that provide structural support or assist in ring assembly. Assembly of the contractile ring results in part from the local formation of new actin filaments, which depends on *formin* proteins that nucleate the assembly of parallel arrays of linear, unbranched actin filaments (discussed in Chapter 16). After anaphase, the overlapping arrays of actin and myosin II filaments contract to generate the force that divides the cytoplasm in two. Once contraction begins, the ring exerts a force large enough to bend a fine glass needle that is inserted in its path. As the ring constricts, it maintains the same thickness, suggesting that its total volume and the number of filaments it contains decrease steadily. Moreover, unlike actin in muscle, the actin filaments in the ring are highly dynamic, and their arrangement changes continually during cytokinesis.

The contractile ring is finally dispensed with altogether when cleavage ends and the plasma membrane of the cleavage furrow narrows to form the **midbody**. The midbody persists as a tether between the two daughter cells and contains the remains of the *central spindle*, a large protein structure derived from the antiparallel microtubules of the spindle midzone, packed tightly together within a dense matrix material (Figure 17–44). Cytokinesis is completed by a process called *abscission*: the membranes on both sides of the midbody are constricted and severed by filaments formed from a polymeric protein called ESCRT-III.

Local Activation of RhoA Triggers Assembly and Contraction of the Contractile Ring

RhoA, a small GTPase of the Ras superfamily (see Table 15–5), controls the assembly and function of the contractile ring at the site of cleavage. RhoA is attached to the inner surface of the cell membrane at the future division site, where it promotes actin filament formation, myosin II assembly, and ring contraction. It

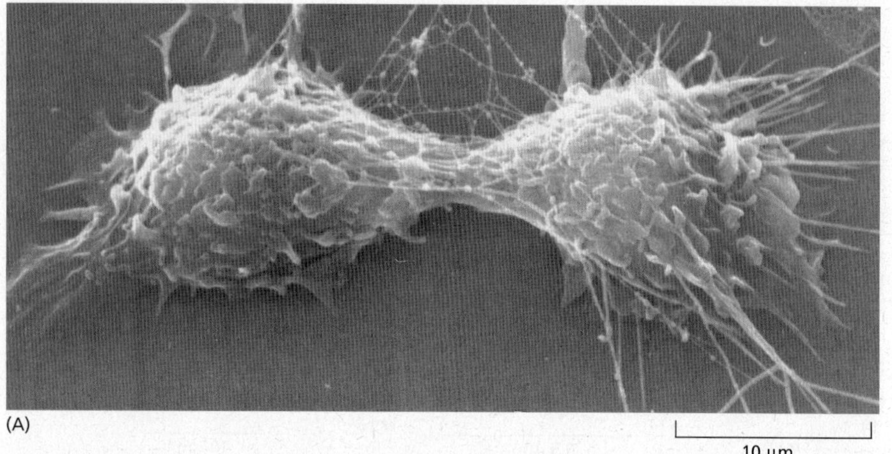

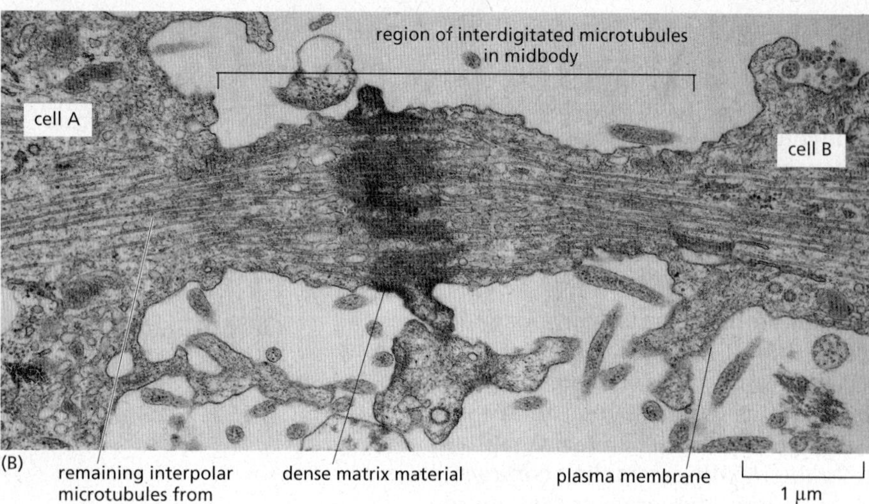

Figure 17–44 The midbody. (A) A scanning electron micrograph of a cultured animal cell dividing; the midbody still joins the two daughter cells. (B) A conventional electron micrograph of the midbody of a dividing animal cell. Cleavage is almost complete, but the daughter cells remain attached by this thin strand of cytoplasm containing the remains of the central spindle. Abscission results when the membranes on both sides of the dense midbody are constricted and severed. (A, courtesy of Guenter Albrecht-Buehler; B, courtesy of J.M. Mullins.)

stimulates actin filament formation by activating formins, and it promotes myosin II assembly and contractions by activating multiple protein kinases, including the Rho-associated kinase (ROCK) (**Figure 17–45**). These kinases phosphorylate the regulatory myosin light chain, a subunit of myosin II, thereby stimulating bipolar myosin II filament formation and motor activity.

RhoA is activated by a guanine nucleotide exchange factor (GEF) called Ect2, which stimulates the release of GDP and binding of GTP to RhoA (see Figure 17–45). Ect2 is localized to the division site and activated by complex mechanisms involving spindle microtubules, as we discuss next.

The Microtubules of the Mitotic Spindle Determine the Plane of Animal Cell Division

The central problem in cytokinesis is how to ensure that division occurs at the right time and in the right place. Cytokinesis must occur only after the two sets of chromosomes are fully segregated from each other, and the site of division must

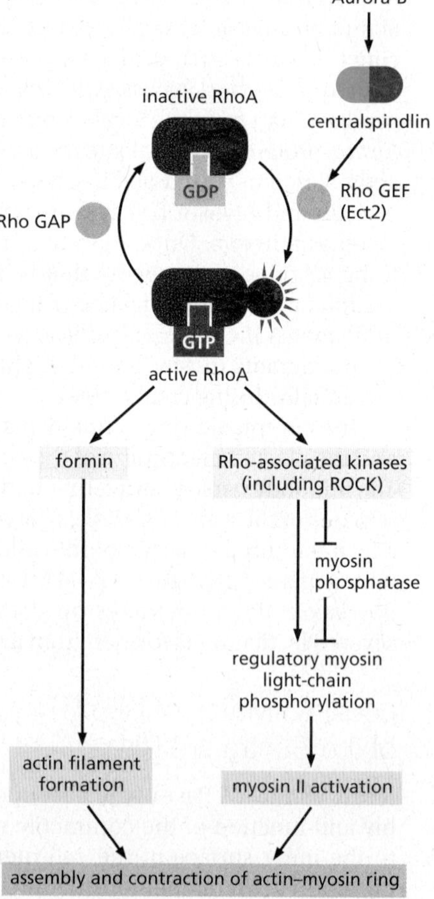

Figure 17–45 Regulation of the contractile ring by the GTPase RhoA. Like other Rho family GTPases, RhoA is activated by a Rho GEF protein (called Ect2) and inactivated by a Rho GTPase-activating protein (Rho GAP). By binding formins, activated RhoA promotes the assembly of actin filaments in the contractile ring. By activating Rho-associated protein kinases, such as ROCK, it stimulates myosin II filament formation and activity, thereby promoting contraction of the ring. Activation of RhoA at the future cleavage site depends on Ect2 activation by another protein complex called centralspindlin, discussed in the next section.

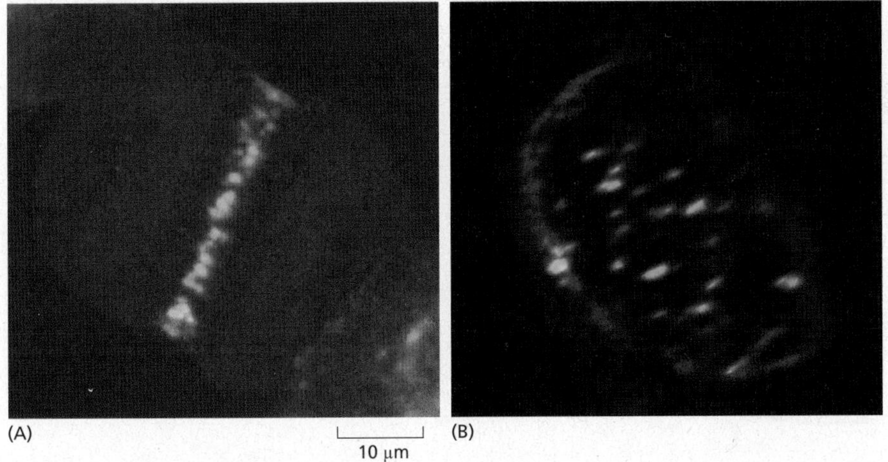

Figure 17–46 **Localization of cytokinesis regulators at the central spindle of the human cell.** (A) At center is a cultured human cell at the beginning of cytokinesis, showing the locations of the GTPase RhoA *(red)* and a protein called Cyk4 *(green)*, which is one of two subunits of centralspindlin, a protein complex that is concentrated at the overlapping plus ends of antiparallel microtubules. (B) When the same three-dimensional image is viewed in the plane of the contractile ring, as shown here, RhoA *(red)* is seen as a ring beneath the cell surface, while the centralspindlin subunit Cyk4 *(green)* is associated with microtubule bundles scattered throughout the equatorial plane of the cell. Note that the small population of centralspindlin at the cell cortex is not readily detected in these images. (Courtesy of Alisa Piekny and Michael Glotzer.)

be placed between the two sets of daughter chromosomes, thereby ensuring that each daughter cell receives a complete set. The correct timing and positioning of cytokinesis in animal cells are achieved by mechanisms that depend on the mitotic spindle. During anaphase, the spindle generates signals that initiate furrow formation at a position midway between the spindle poles, thereby ensuring that division occurs between the two sets of separated chromosomes. Because these signals originate in the anaphase spindle, this mechanism also contributes to the correct timing of cytokinesis in late mitosis.

Studies of the fertilized eggs of marine invertebrates first revealed the importance of spindle microtubules in determining the placement of the contractile ring. After fertilization, these embryos cleave rapidly without intervening periods of growth. In this way, the original egg is progressively divided into smaller and smaller cells. Because the cytoplasm is clear, the spindle can be observed in real time with a microscope. If the spindle is tugged into a new position with a fine glass needle in early anaphase, the incipient cleavage furrow disappears, and a new one develops in accord with the new spindle site—supporting the idea that signals generated by the spindle induce local furrow formation.

How does the mitotic spindle specify the site of division? The key mechanism appears to be that the midzone of the anaphase spindle generates a signal that promotes furrow formation at the cell cortex. The central component of this regulatory system is a two-subunit protein complex called *centralspindlin*, which forms oligomeric assemblies that are concentrated primarily on the antiparallel microtubules at the spindle midzone (**Figure 17–46**). Centralspindlin assembly at the midzone is stimulated by the protein kinase Aurora-B, which also localizes to the spindle midzone in anaphase (see Figure 17–45). Aurora-B localization to the central spindle depends on dephosphorylation of Cdk substrates, providing one mechanism that delays cytokinesis until anaphase.

Centralspindlin interacts with the RhoA GEF, Ect2, to activate RhoA at the equatorial cell cortex, halfway between the spindle poles (**Figure 17–47**). In some cell types, small subpopulations of centralspindlin and Ect2 can be seen to migrate from the spindle midzone to the cell cortex, where Ect2 then interacts with RhoA to trigger furrow formation (see Figure 17–45). The focusing of centralspindlin and Ect2 at the equator depends in part on the ability of astral microtubules to somehow inhibit centralspindlin localization outside the equatorial region.

In some cell types, the site of ring assembly is chosen before mitosis. In budding yeasts, for example, a ring of proteins called *septins* assembles in late G_1 at the future division site. The septins are thought to form a scaffold onto which other components of the contractile ring, including myosin II, assemble. In plant cells, an organized band of microtubules and actin filaments, called the **preprophase band**, assembles just before mitosis and marks the site where the cell wall will assemble and divide the cell in two, as we now discuss.

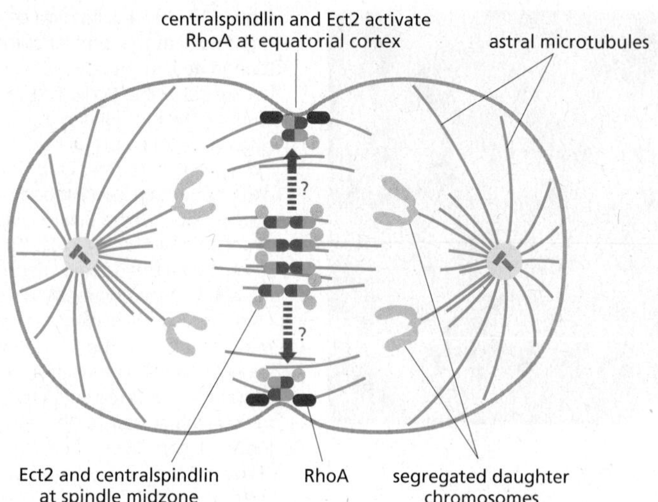

Figure 17–47 **Activation of RhoA at the site of cleavage furrow formation.** Centralspindlin multimers *(blue)* associate with the RhoA GEF, Ect2 *(orange)*, at the spindle midzone. By uncertain mechanisms *(dashed lines)*, some of these proteins move to the cortex at the equator of the cell, where they activate RhoA *(purple)* to trigger furrow formation. To focus the signal at the equator, the plus ends of astral microtubules use unknown mechanisms to inhibit the centralspindlin–Ect2 complex at other regions of the cortex.

The Phragmoplast Guides Cytokinesis in Higher Plants

In most animal cells, the inward movement of the cleavage furrow depends on an increase in the surface area of the plasma membrane. New membrane is added primarily at the cleavage furrow and is generally provided by small membrane vesicles that are transported on microtubules from the Golgi apparatus to the furrow.

Membrane deposition is particularly important for cytokinesis in higher-plant cells. These cells are enclosed by a semirigid *cell wall*. Rather than a contractile ring dividing the cytoplasm from the outside in, the cytoplasm of the plant cell is partitioned from the inside out by the construction of a new cell wall, called the **cell plate**, between the two daughter nuclei (Figure 17–48; Figure 17–49). The assembly of the cell plate begins in late anaphase and is guided by a structure called the **phragmoplast**, which contains microtubules derived from the mitotic

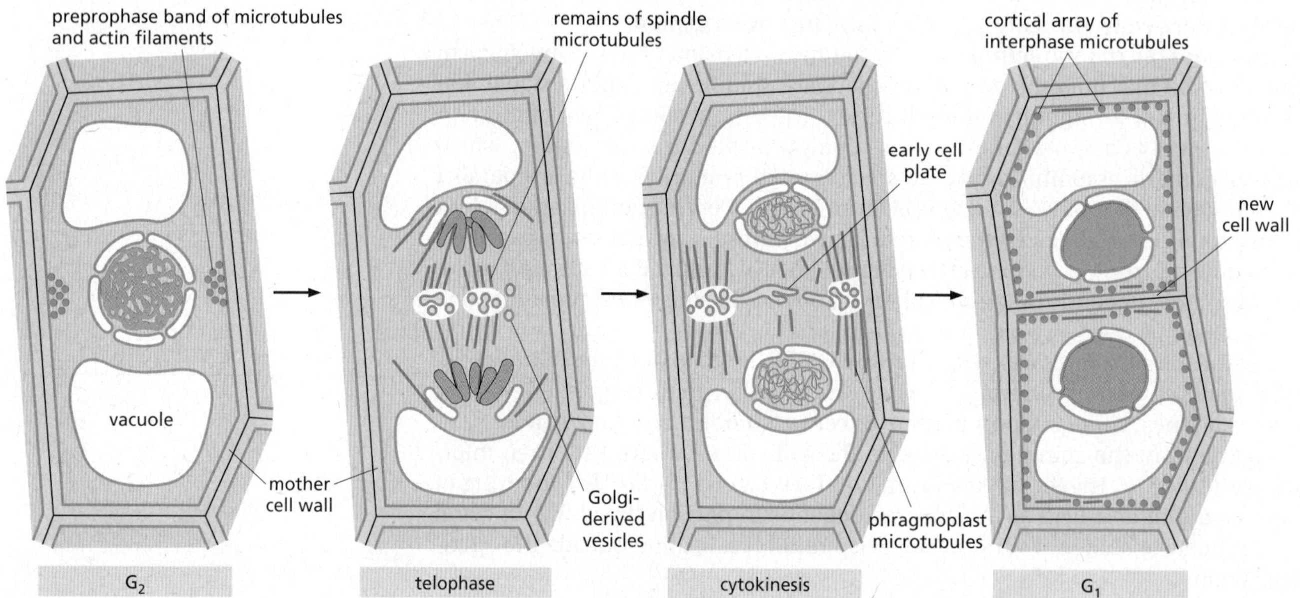

Figure 17–48 **The special features of cytokinesis in a higher-plant cell.** The division plane is established before M phase by a band of microtubules and actin filaments (the preprophase band) at the cell cortex. At the beginning of telophase, after the chromosomes have segregated, a new cell wall starts to assemble inside the cell at the equator of the old spindle. The microtubules of the mitotic spindle remaining at telophase form the phragmoplast. The plus ends of these microtubules no longer overlap but end at the cell equator. Golgi-derived vesicles, filled with cell-wall material, are transported along these microtubules and fuse to form the new cell wall, which grows outward to reach the plasma membrane and original cell wall. The plasma membrane and the membrane surrounding the new cell wall fuse, separating the two daughter cells.

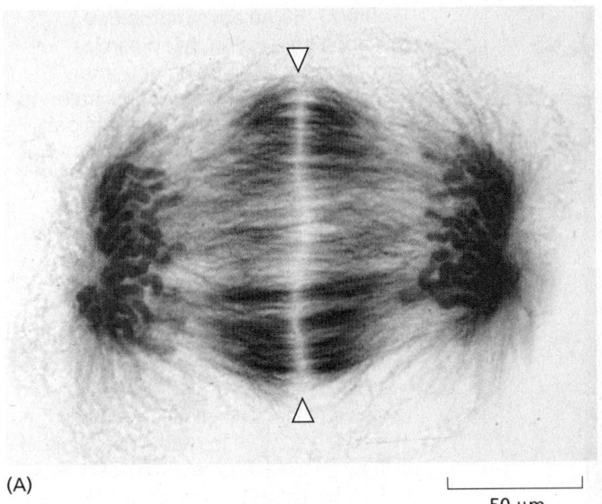

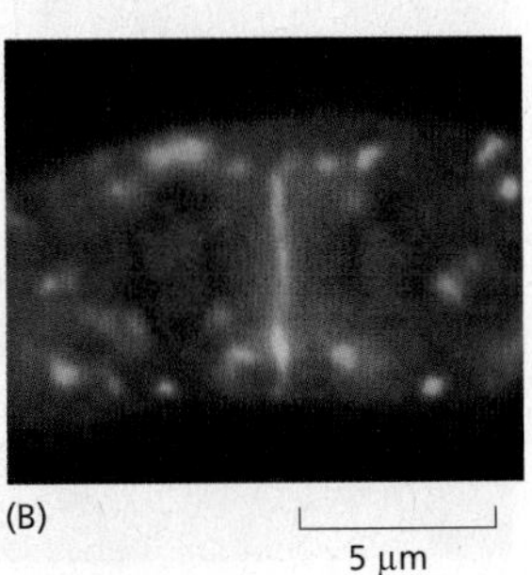

Figure 17–49 Cytokinesis in plant cells during telophase. (A) In this light micrograph, the early cell plate (between the two *arrowheads*) has formed in a plane perpendicular to the plane of the page. The microtubules of the spindle are stained with gold-labeled antibodies against tubulin, and the DNA in the two sets of daughter chromosomes is stained with a dark blue dye. Note that there are no astral microtubules, because there are no centrosomes in higher-plant cells. (B) In this fluorescence micrograph of a plant cell, DNA is stained *blue* and microtubules are *red*. A protein called Syntaxin *(green)* lies along the cell plate at the cell equator, where it is responsible for stimulating the fusion of Golgi vesicles delivering cell-wall materials. (A, courtesy of Andrew Bajer; B, from C.-M.K. Ho et al., *Plant Cell* 23:2909–2923, 2011. Republished with permission of American Society of Plant Biologists.)

spindle. Motor proteins transport small vesicles along these microtubules from the Golgi apparatus to the cell center. These vesicles, filled with polysaccharide and glycoproteins required for the synthesis of the new cell wall, fuse to form a disc-like, membrane-enclosed structure called the *early cell plate*. The plate expands outward by further vesicle fusion until it reaches the plasma membrane and the original cell wall and divides the cell in two. Later, cellulose microfibrils are laid down within the matrix of the cell plate to complete the construction of the new cell wall.

Membrane-enclosed Organelles Must Be Distributed to Daughter Cells During Cytokinesis

The process of mitosis ensures that each daughter cell receives a full complement of chromosomes. When a eukaryotic cell divides, however, each daughter cell must also inherit all of the other essential cell components, including the membrane-enclosed organelles. As discussed in Chapter 12, organelles such as mitochondria and chloroplasts cannot be assembled *de novo* from their individual components; they can arise only by the growth and division of the preexisting organelles. Similarly, cells cannot make a new endoplasmic reticulum (ER) unless some part of it is already present.

How, then, do the various membrane-enclosed organelles segregate when a cell divides? Organelles such as mitochondria and chloroplasts are usually present in large enough numbers to be safely inherited if, on average, their numbers roughly double once each cycle. The ER in interphase cells is continuous with the nuclear membrane and is organized by the microtubule cytoskeleton. Upon entry into M phase, the reorganization of the microtubules and breakdown of the nuclear envelope releases the ER. In most cells, the ER remains largely intact and is cut in two during cytokinesis. The Golgi apparatus is reorganized and fragmented during mitosis. Golgi fragments associate with the spindle poles and are thereby distributed to opposite ends of the spindle, ensuring that each daughter cell inherits the materials needed to reconstruct the Golgi in telophase.

Some Cells Reposition Their Spindle to Divide Asymmetrically

Most animal cells divide symmetrically: the contractile ring forms around the equator of the parent cell, producing two daughter cells of equal size and with the same components. This symmetry results from the placement of the mitotic spindle, which in most cases tends to center itself in the cytoplasm. Astral microtubules and motor proteins that either push or pull on these microtubules contribute to the centering process.

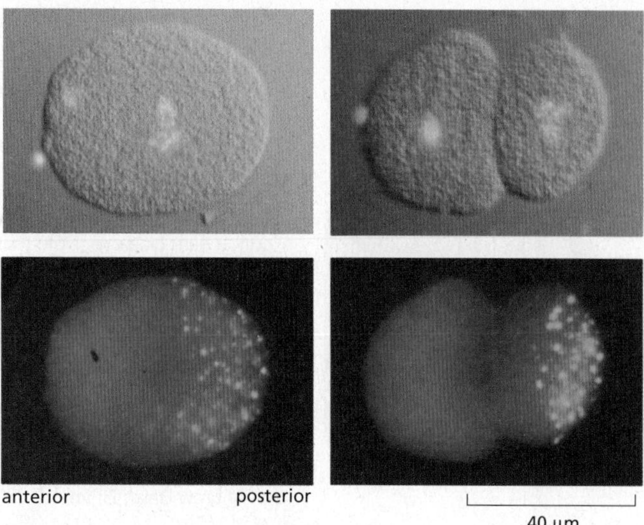

Figure 17–50 **An asymmetric cell division segregating cytoplasmic components to only one daughter cell.** These light micrographs illustrate the controlled asymmetric segregation of specific cytoplasmic components to one daughter cell during the first division of a fertilized egg of the nematode *C. elegans.* The fertilized egg is shown in the left micrographs and the two daughter cells in the right micrographs. The cells above have been stained with a *blue*, DNA-binding, fluorescent dye to show the nucleus (and polar bodies); they are viewed by both differential-interference-contrast microscopy and fluorescence microscopy. The cells below are the same cells stained with an antibody against P-granules and viewed by fluorescence microscopy. These small granules are made of RNA and proteins and determine which cells become germ cells. They are distributed randomly throughout the cytoplasm of the unfertilized egg (not shown) but become segregated to the posterior pole of the fertilized egg. The cleavage plane is oriented to ensure that only the posterior daughter cell receives the P-granules when the egg divides. The same segregation process is repeated in several subsequent cell divisions, so that the P-granules end up only in cells that give rise to eggs and sperm. (Courtesy of Susan Strome.)

There are many instances in development, however, when cells divide asymmetrically to produce two cells that differ in size, in the cytoplasmic contents they inherit, or in both. Usually, the two different daughter cells are destined to develop along different pathways. To create daughter cells with different fates in this way, the mother cell must first segregate certain components (called *cell-fate determinants*) to one side of the cell and then position the plane of division so that the appropriate daughter cell inherits these components (**Figure 17–50**). To position the plane of division asymmetrically, the spindle has to be moved in a controlled manner within the dividing cell. It seems likely that changes in local regions of the cell cortex direct such spindle movements and that motor proteins localized there pull one of the spindle poles, via its astral microtubules, to the appropriate region. Genetic analyses in *Caenorhabditis elegans* and *Drosophila* have identified some of the proteins required for such asymmetric divisions, and some of these proteins seem to have a similar role in vertebrates.

Mitosis Can Occur Without Cytokinesis

Although nuclear division is usually followed by cytoplasmic division, there are exceptions. Some cells undergo multiple rounds of nuclear division without intervening cytoplasmic division. In the early *Drosophila* embryo, for example, the first 13 rounds of nuclear division occur without cytoplasmic division, resulting in the formation of a single large cell containing several thousand nuclei that are arranged in a monolayer near the surface. A cell in which multiple nuclei share the same cytoplasm is called a **syncytium**. This arrangement greatly speeds up early development, as the cells do not have to take the time to go through all the steps of cytokinesis for each division. After these rapid nuclear divisions, membranes are created around each nucleus in one round of coordinated cytokinesis called *cellularization*. The plasma membrane extends inward and, with the help of an actin–myosin ring, pinches off to enclose each nucleus (**Figure 17–51**).

Nuclear division without cytokinesis also occurs in some types of mammalian cells. Megakaryocytes, which produce blood platelets, and some hepatocytes and muscle cells, for example, become multinucleated in this way.

Summary

After mitosis completes the formation of a pair of daughter nuclei, cytokinesis finishes the cell cycle by dividing the cell itself. Cytokinesis depends on a ring of actin and myosin filaments that contracts in late mitosis at a site midway between the segregated chromosomes. In animal cells, the positioning of the

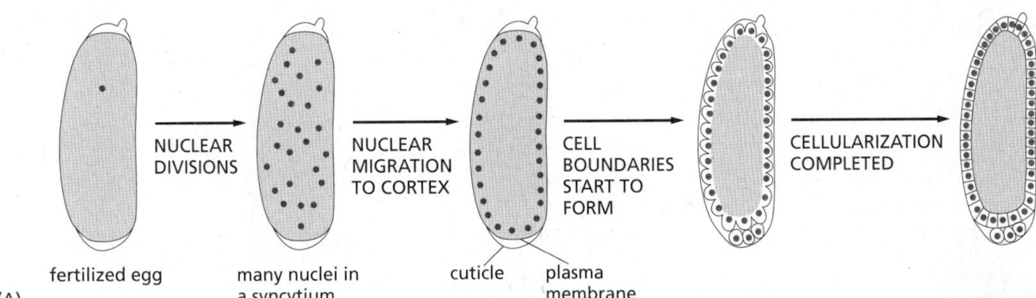

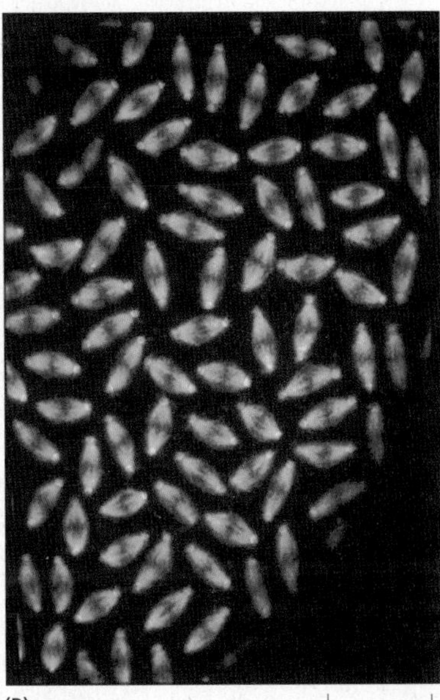

Figure 17–51 **Mitosis without cytokinesis in the early *Drosophila* embryo.** (A) The first 13 nuclear divisions occur synchronously and without cytoplasmic division to create a large syncytium. Most of the nuclei migrate to the cortex, and the plasma membrane extends inward and pinches off to surround each nucleus to form individual cells in a process called cellularization. (B) Fluorescence micrograph of multiple mitotic spindles in a *Drosophila* embryo before cellularization. The microtubules are stained *green* and the centrosomes *red*. Note that all the nuclei go through the cycle synchronously; here, they are all in metaphase, with the unlabeled chromosomes seen as a dark band at the spindle equator. (B, courtesy of Kristina Yu and William Sullivan.)

contractile ring is determined by proteins associated with the midzone microtubules of the anaphase spindle. Dephosphorylation of Cdk targets, which results from Cdk inactivation in anaphase, triggers cytokinesis at the correct time after anaphase.

MEIOSIS

Most eukaryotic organisms reproduce sexually: the genomes of two parents mix to generate offspring that are genetically distinct from either parent. The cells of these organisms are generally *diploid*; that is, they contain two slightly different copies, or *homologs*, of each chromosome, one from each parent. Sexual reproduction depends on a specialized nuclear division process called *meiosis*, which produces *haploid* cells carrying only a single copy of each chromosome. In many organisms, the haploid cells differentiate into specialized reproductive cells called *gametes*—eggs and sperm in most species. In these species, the reproductive cycle ends when a sperm and egg fuse to form a diploid *zygote*, which has the potential to form a new individual. In this section, we consider the basic mechanisms and regulation of meiosis, with an emphasis on how they compare with those of mitosis.

Meiosis Includes Two Rounds of Chromosome Segregation

Meiosis reduces the chromosome number by half using many of the same molecular machines and control systems that operate in mitosis. As in the mitotic cell cycle, the cell begins the meiotic program by duplicating its chromosomes in meiotic S phase, resulting in pairs of sister chromatids that are tightly linked along their entire lengths by cohesin complexes. Unlike mitosis, however, two successive rounds of chromosome segregation then occur (**Figure 17–52**). The first of these divisions (**meiosis I**) solves the problem, unique to meiosis, of segregating the homologs. The duplicated paternal and maternal homologs pair up alongside each other and become physically linked by the process of genetic recombination. These pairs of homologs, each containing a pair of sister chromatids, then line up on the first meiotic spindle. In the first meiotic anaphase, duplicated homologs rather than sister chromatids are pulled apart and segregated into the two daughter nuclei. Only in the second division (**meiosis II**), which occurs without further DNA replication, are the sister chromatids pulled apart and segregated (as in mitosis) to produce haploid daughter nuclei. In this way, each diploid nucleus that enters meiosis produces four haploid nuclei, each of which contains either the maternal or paternal copy of each chromosome, but not both (**Movie 17.7**).

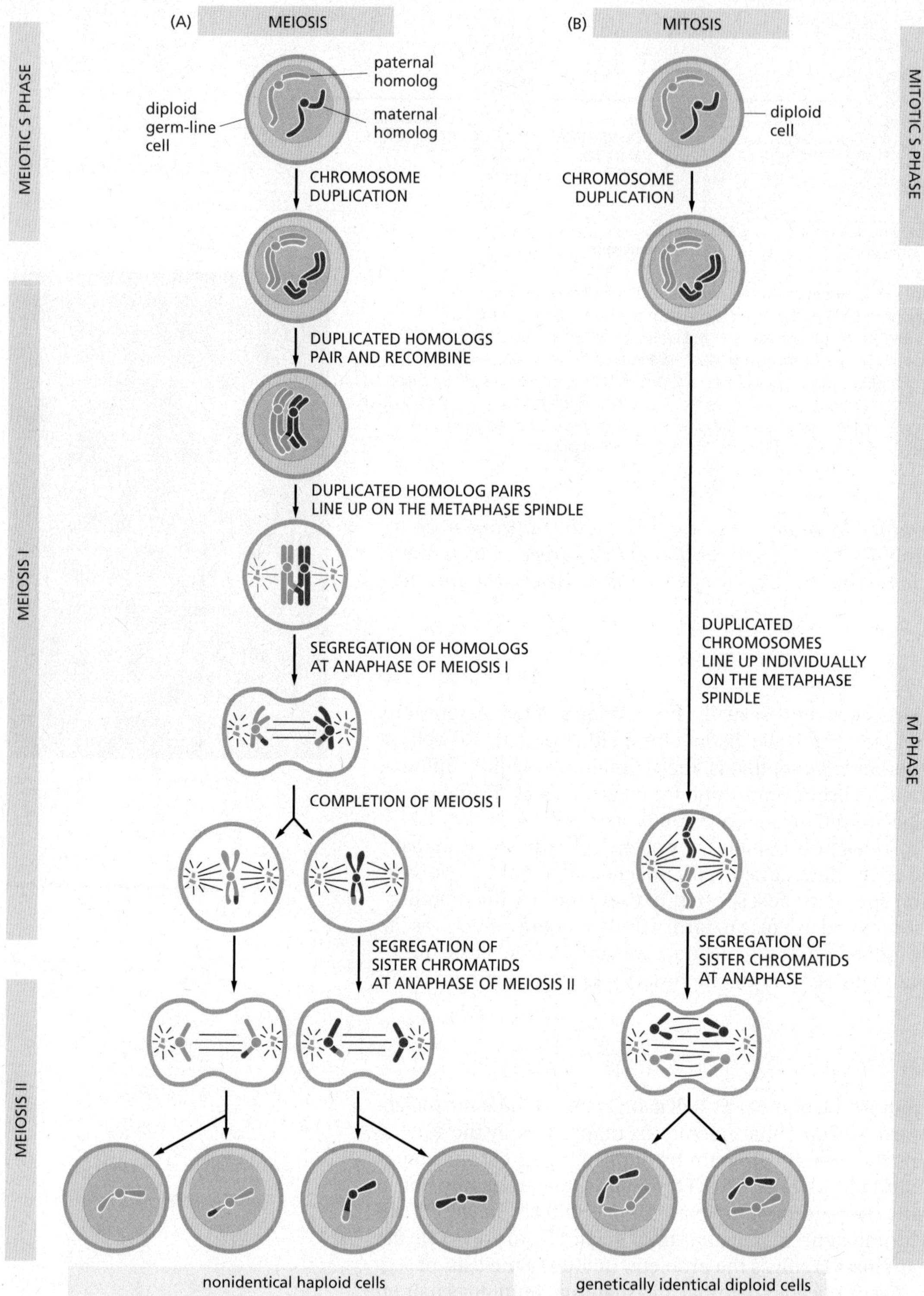

Figure 17–52 Comparison of meiosis and mitosis. For clarity, only one pair of homologous chromosomes (homologs) is shown. (A) Meiosis is a form of nuclear division in which a single round of chromosome duplication (meiotic S phase) is followed by two rounds of chromosome segregation. The duplicated homologs, each consisting of tightly bound sister chromatids, pair up and are segregated into different daughter nuclei in meiosis I; the sister chromatids are segregated in meiosis II. As indicated by the formation of chromosomes that are partly *red* and partly *blue*, homolog pairing in meiosis leads to genetic recombination (crossing-over) during meiosis I. Each diploid cell that enters meiosis therefore produces four genetically different haploid nuclei, which are distributed by cytokinesis into haploid cells that differentiate into gametes. (B) In mitosis, by contrast, homologs do not pair up, and the sister chromatids are segregated during the single division. Thus, each diploid cell that divides by mitosis produces two genetically identical diploid daughter nuclei, which are distributed by cytokinesis into a pair of daughter cells.

Duplicated Homologs Pair During Meiotic Prophase

During mitosis in most organisms, homologous chromosomes behave independently of each other. During meiosis I, however, it is crucial that homologs recognize each other and associate physically in order for the maternal and paternal homologs to be bi-oriented on the first meiotic spindle. Special mechanisms mediate these interactions.

The gradual juxtaposition of homologs occurs during a prolonged period called meiotic prophase (or prophase I), which can take hours in yeasts, days in mice, and weeks in higher plants. Like their mitotic counterparts, duplicated meiotic prophase chromosomes first appear as long threadlike structures, in which the sister chromatids are so tightly glued together that they appear as one. It is during early prophase I that the homologs begin to associate along their length in a process called **pairing**, which, in some organisms at least, begins with interactions between complementary DNA sequences (called *pairing sites*) in the two homologs. As prophase progresses, the homologs become more closely juxtaposed, forming a four-chromatid structure called a **bivalent** (Figure 17–53A). In most species, homolog pairs are then locked together by homologous recombination: DNA double-strand breaks are formed at several locations in each sister chromatid, resulting in large numbers of DNA recombination events between the homologs (as described in Chapter 5). Some of these events lead to reciprocal DNA exchanges called *crossovers*, where the DNA of a chromatid crosses over to become continuous with the DNA of a homologous chromatid (Figure 17–53B; also see Figure 5–53).

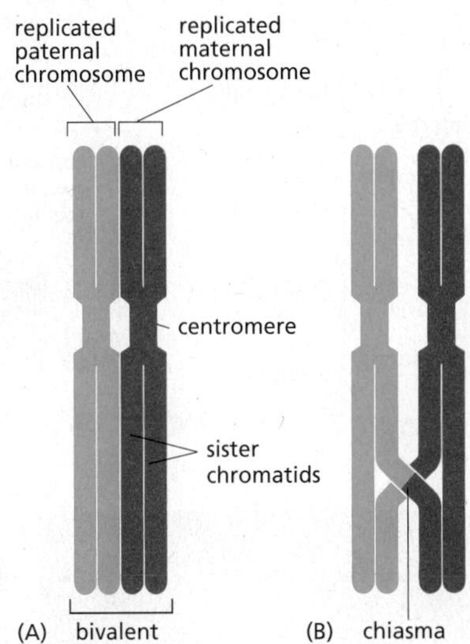

Figure 17–53 **Homolog pairing and crossing-over.** (A) The structure formed by two closely aligned duplicated homologs is called a *bivalent*. As in mitosis, the sister chromatids in each homolog are tightly connected along their entire lengths and at their centromeres. At this stage, the homologs are usually joined by a protein complex called the *synaptonemal complex* (not shown; see Figure 17–54). (B) A later-stage bivalent in which a single crossover has occurred between nonsister chromatids. It is only when the synaptonemal complex disassembles and the paired homologs separate a little at the end of prophase I, as shown, that the crossover is seen microscopically as a thin connection between the homologs called a *chiasma*.

Homolog Pairing Culminates in the Formation of a Synaptonemal Complex

The paired homologs are brought into close juxtaposition, with their structural axes (*axial cores*) about 400 nm apart, by a mechanism that depends in most species on the double-strand DNA breaks that occur in sister chromatids. What pulls the axes together? One possibility is that the large protein machine, called a *recombination complex*, which assembles on a double-strand break in a chromatid, binds the matching DNA sequence in the nearby homolog and helps reel in this partner. This so-called *presynaptic alignment* of the homologs is followed by *synapsis*, in which the axial core of a homolog becomes tightly linked to the axial core of its partner by a closely packed array of *transverse filaments* to create a **synaptonemal complex**, which bridges the gap, now only 100 nm, between the homologs (Figure 17–54). Although crossing-over begins

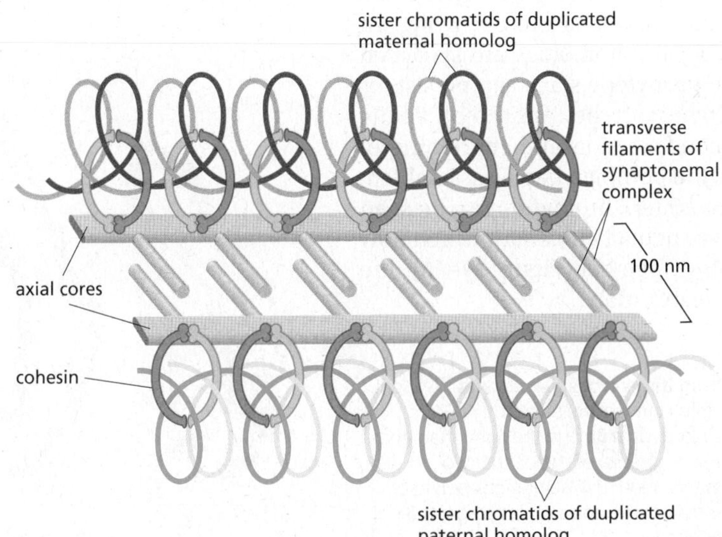

Figure 17–54 **Simplified schematic drawing of a synaptonemal complex.** Each homolog is organized around a protein axial core, and the synaptonemal complex forms when these homolog axes are linked by rod-shaped transverse filaments. The axial core of each homolog also interacts with the cohesin complexes that hold the sister chromatids together (see Figure 9–28). (Modified from K. Nasmyth, *Annu. Rev. Genet.* 35:673–745, 2001.)

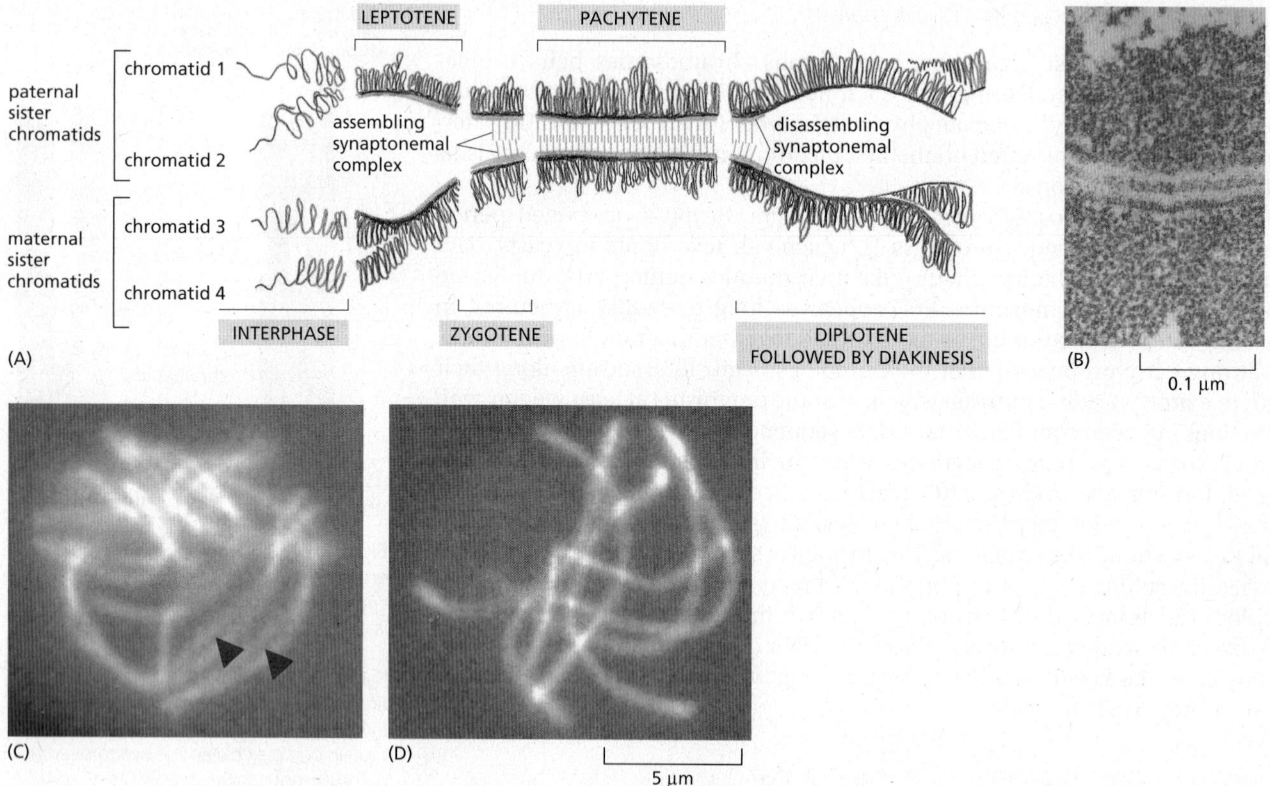

Figure 17–55 **Homolog synapsis and desynapsis during the different stages of prophase I.** (A) A single bivalent is shown schematically. At leptotene, the two sister chromatids coalesce, and their chromatid loops extend out from a common axial core. Assembly of the synaptonemal complex begins in early zygotene and is complete in pachytene. The complex disassembles in diplotene. (B) An electron micrograph of a synaptonemal complex from a meiotic cell at pachytene in a lily flower. (C and D) Immunofluorescence micrographs of prophase I cells of the fungus *Sordaria*. Partially synapsed bivalents at zygotene are shown in C and fully synapsed bivalents are shown in D. *Red arrowheads* in C point to regions where synapsis is still incomplete. (B, courtesy of Brian Wells; C and D, from A. Storlazzi et al., *Genes Dev.* 17:2675–2687, 2003. With permission from Cold Spring Harbor Laboratory Press.)

before the synaptonemal complex assembles, the final steps occur while the DNA is held in the complex.

The morphological changes that occur during homolog pairing are the basis for dividing meiotic prophase into five sequential stages—leptotene, zygotene, pachytene, diplotene, and diakinesis (**Figure 17–55**). Prophase starts with *leptotene*, when homologs condense and pair and genetic recombination begins. At *zygotene*, the synaptonemal complex begins to assemble at sites where the homologs are closely associated and recombination events are occurring. At *pachytene*, the assembly process is complete, and the homologs are synapsed along their entire lengths (see Figure 9–28). The pachytene stage can persist for days or longer, until desynapsis begins at *diplotene* with the disassembly of the synaptonemal complex and the concomitant condensation and shortening of the chromosomes. It is only at this stage, after the complex has disassembled, that the individual crossover events between nonsister chromatids can be seen as inter-homolog connections called **chiasmata** (singular, **chiasma**), which now play a crucial part in holding the compact homologs together (**Figure 17–56**). The homologs are now ready to begin the process of segregation.

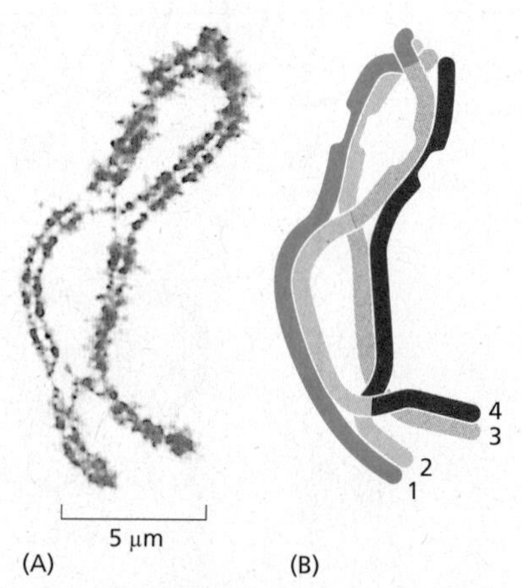

Figure 17–56 **A bivalent with three chiasmata resulting from three crossover events.** (A) Light micrograph of a grasshopper bivalent. (B) Schematic of the three crossovers shown in A. Each sister chromatid is numbered. Note how the combination of the chiasmata and the tight attachment of the sister chromatid arms to each other (mediated by cohesin complexes) holds the two homologs together after the synaptonemal complex has disassembled; if either the chiasmata or the sister-chromatid cohesion failed to form, the homologs would come apart at this stage and not be segregated properly in meiosis I. (A, courtesy of Bernard John.)

Homolog Segregation Depends on Several Unique Features of Meiosis I

A fundamental difference between meiosis I and mitosis (and meiosis II) is that in meiosis I, homologs rather than sister chromatids separate and then segregate (see Figure 17–52). This difference depends on three features of meiosis I that distinguish it from mitosis (**Figure 17–57**).

First, both sister kinetochores in a homolog must attach stably to the same spindle pole. This type of attachment is normally avoided during mitosis (see Figure 17–35). In meiosis I, however, the two sister kinetochores are fused into a single microtubule-binding unit that attaches to just one pole (see Figure 17–57A). The fusion of sister kinetochores is achieved by a complex of proteins that is localized at the kinetochores in meiosis I, but we do not know in any detail how these proteins work. They are removed from kinetochores after meiosis I, so that in meiosis II the sister-chromatid pairs can be bi-oriented on the spindle as they are in mitosis.

Second, crossovers generate a strong physical linkage between homologs, allowing their bi-orientation at the equator of the spindle—much like cohesion between sister chromatids is important for their bi-orientation in mitosis (and meiosis II). Crossovers hold homolog pairs together only because the

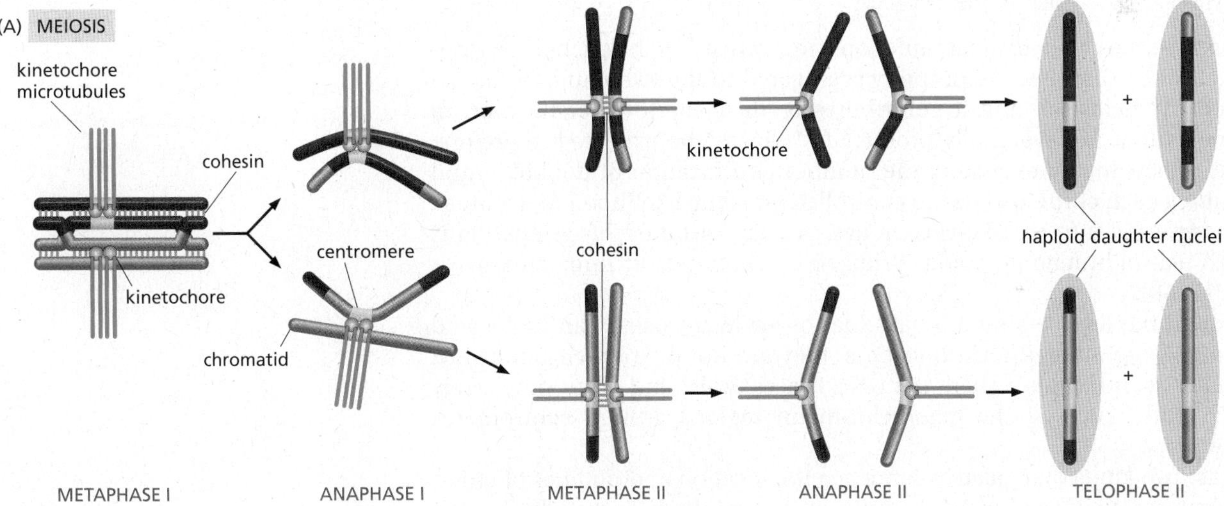

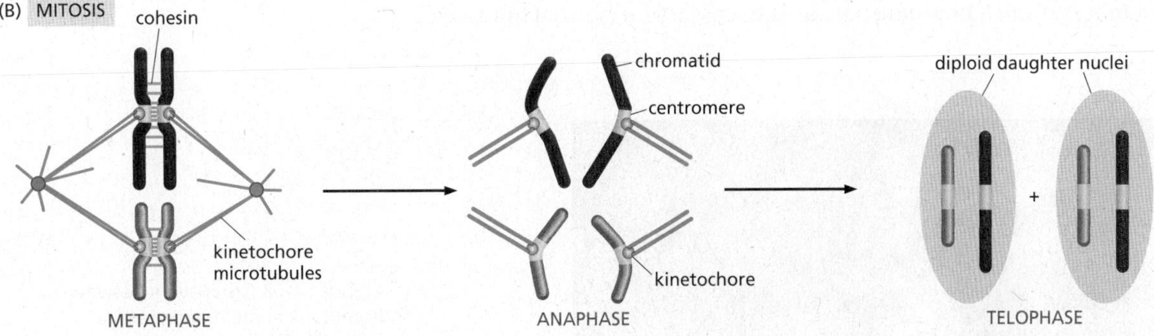

Figure 17–57 Comparison of chromosome behavior in meiosis I, meiosis II, and mitosis. Chromosomes behave similarly in mitosis and meiosis II, but they behave very differently in meiosis I. (A) In meiosis I, the two sister kinetochores are located side-by-side on each homolog and attach to microtubules from the same spindle pole. The proteolytic cleavage of cohesin along the sister-chromatid arms unglues the arms and resolves the crossovers, allowing the duplicated homologs to separate at anaphase I, while the residual cohesin at the centromeres keeps the sisters together. Cleavage of centromeric cohesin allows the sister chromatids to separate at anaphase II. (B) In mitosis, by contrast, the two sister kinetochores attach to microtubules from different spindle poles, and the two sister chromatids come apart at the start of anaphase and segregate into separate daughter nuclei.

arms of the sister chromatids are connected by sister-chromatid cohesion (see Figure 17–57A).

Third, cohesin is removed in anaphase I only from chromosome arms and not from the regions near the centromeres, where the kinetochores are located. The loss of arm cohesion triggers homolog separation at the onset of anaphase I. This process depends on APC/C activation, which leads to securin destruction, separase activation, and cohesin cleavage along the arms (see Figure 17–39).

Cohesins near the centromeres are protected from separase in meiosis I by a kinetochore-associated protein called *shugoshin* (from the Japanese word for "guardian spirit"). Shugoshin acts by recruiting a protein phosphatase that removes phosphates from centromeric cohesins. Cohesin phosphorylation is normally required for separase to cleave cohesin; thus, removal of this phosphorylation near the centromere prevents cohesin cleavage. Sister-chromatid pairs therefore remain linked through meiosis I, allowing their correct bi-orientation on the spindle in meiosis II. Shugoshin is inactivated after meiosis I. At the onset of anaphase II, APC/C activation triggers centromeric cohesin cleavage and sister-chromatid separation—much as it does in mitosis. After anaphase II, nuclear envelopes form around the chromosomes to produce four haploid nuclei, after which cytokinesis and other differentiation processes lead to the production of haploid gametes.

Crossing-Over Is Highly Regulated

Crossing-over has two distinct functions in meiosis: it helps hold homologs together so that they are properly segregated to the two daughter nuclei produced by meiosis I, and it contributes to the genetic diversification of the gametes that are eventually produced. As might be expected, therefore, crossing-over is highly regulated: the number and location of double-strand breaks along each chromosome are controlled, as is the likelihood that a break will be converted into a crossover. On average, the result of this regulation is that each pair of human homologs is linked by about two or three crossovers (**Figure 17–58**).

Although the double-strand breaks that occur in meiosis I can be located almost anywhere along the chromosome, they are not distributed uniformly: they cluster at "hot spots," where the DNA is accessible, and occur only rarely in "cold spots," such as the heterochromatin regions around centromeres and telomeres.

At least two kinds of regulation influence the location and number of crossovers that form, neither of which is well understood. Both operate before the synaptonemal complex assembles. One ensures that at least one crossover forms between the members of each homolog pair, as is necessary for normal homolog

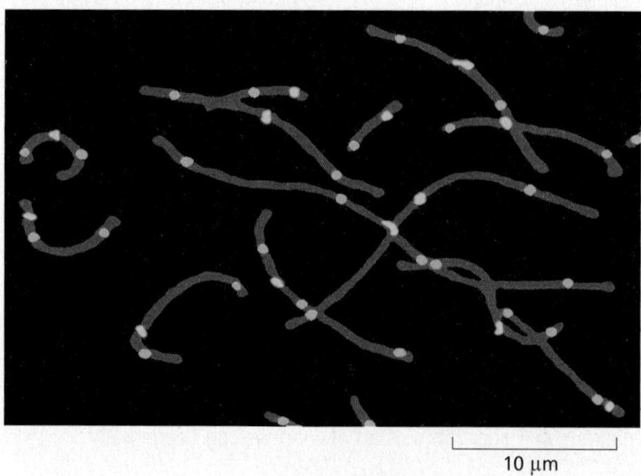

Figure 17–58 **Crossovers between homologs in the human testis.** In these immunofluorescence micrographs, antibodies were used to stain the synaptonemal complexes *(red)*, the centromeres *(blue)*, and the sites of crossing-over *(green)*. Note that all of the bivalents have at least one crossover and none have more than four. (Modified from A. Lynn et al., *Science* 296:2222–2225, 2002. With permission from AAAS.)

segregation in meiosis I. In the other, called *crossover interference*, the presence of one crossover event inhibits another from forming close by, perhaps by locally depleting proteins required for converting a double-strand DNA break into a stable crossover.

Meiosis Frequently Goes Wrong

The sorting of chromosomes that takes place during meiosis is a remarkable feat of intracellular bookkeeping. In humans, each meiosis requires that the starting cell keep track of 92 chromatids (46 chromosomes, each of which has duplicated), distributing one complete set of each type of autosome to each of the four haploid progeny. Not surprisingly, mistakes can occur in allocating the chromosomes during this elaborate process. Mistakes are especially common in human female meiosis, which arrests for years after diplotene: meiosis I is completed only at *ovulation*, and meiosis II only after the egg is fertilized. Indeed, such chromosome segregation errors during egg development are the most common cause of spontaneous abortion (miscarriage) and intellectual disability in humans.

When homologs fail to separate properly—a phenomenon called **nondisjunction**—the result is that some of the resulting haploid gametes lack a particular chromosome, while others have more than one copy of it. Upon fertilization, these gametes form abnormal embryos, most of which die. Some survive, however. *Down syndrome* in humans, for example, which is the leading cause of intellectual disability, is caused by an extra copy of chromosome 21, usually resulting from nondisjunction during meiosis I in the female ovary. Segregation errors during meiosis I increase greatly with advancing maternal age.

Summary

Haploid gametes are produced by meiosis, in which a diploid nucleus undergoes two successive cell divisions after one round of DNA replication. Meiosis is dominated by a prolonged prophase. At the start of prophase, the chromosomes have replicated and consist of two tightly joined sister chromatids. Homologous chromosomes then pair up and become progressively more closely juxtaposed as prophase proceeds. The tightly aligned homologs undergo genetic recombination, forming crossovers that help hold each pair of homologs together during metaphase I. Meiosis-specific, kinetochore-associated proteins help ensure that both sister chromatids in a homolog attach to the same spindle pole; other kinetochore-associated proteins ensure that the homologs remain connected at their centromeres during anaphase I, so that homologs rather than sister chromatids are segregated in meiosis I. After meiosis I, meiosis II follows rapidly, without DNA replication, in a process that resembles mitosis, in that sister chromatids are pulled apart at anaphase.

CONTROL OF CELL DIVISION AND CELL GROWTH

A fertilized mouse egg and a fertilized human egg are similar in size, yet they produce animals of very different sizes. What factors in the control of cell behavior in humans and mice are responsible for these size differences? The same fundamental question can be asked for each organ and tissue in an animal's body. What factors determine the length of an elephant's trunk or the size of its brain or its liver? These questions are largely unanswered, but it is nevertheless possible to say what the ingredients of an answer must be.

The size of an organ or organism depends on its total cell mass, which depends on both the total number of cells and their size. Cell number, in turn, depends on the rates of cell division and cell death. Organ and body size are therefore determined by three fundamental processes: cell growth, cell division, and cell survival. Each is tightly regulated—both by intracellular programs and by extracellular signal molecules that control these programs.

The extracellular signal molecules that regulate cell growth, division, and survival are generally soluble secreted proteins, proteins bound to the surface of cells, or components of the extracellular matrix. They can be divided operationally into three major classes:

1. *Mitogens*, which stimulate cell division, primarily by triggering a wave of G_1/S-Cdk activity that relieves intracellular negative controls that otherwise block progress through the cell cycle.

2. *Growth factors*, which stimulate cell growth (an increase in cell mass) by promoting the synthesis of proteins and other macromolecules and by inhibiting their degradation.

3. *Survival factors*, which promote cell survival by suppressing the form of programmed cell death known as *apoptosis*.

Many extracellular signal molecules promote all of these processes, while others promote one or two of them. Indeed, the term "growth factor" is often used inappropriately to describe a factor that has any of these activities. Even worse, the term "cell growth" is often used to mean an increase in cell number, or *cell proliferation*.

In addition to these three classes of stimulating signals, there are extracellular signal molecules that suppress cell proliferation, cell growth, or both. There are also extracellular signal molecules that activate apoptosis.

In this section, we focus primarily on how mitogens and other factors, such as DNA damage, control the rate of cell division. We then turn to the important but poorly understood problem of how a proliferating cell coordinates its growth with cell division so as to maintain its appropriate size. We discuss the control of cell survival and cell death by apoptosis in Chapter 18.

Mitogens Stimulate Cell Division

Unicellular organisms tend to grow and divide as fast as they can, and their rate of proliferation depends largely on the availability of nutrients in the environment. The cells of a multicellular organism, however, divide only when the organism needs more cells. Thus, for an animal cell to proliferate, it must receive stimulatory extracellular signals, in the form of **mitogens**, from other cells, usually its neighbors. Mitogens overcome intracellular braking mechanisms that block progress through the cell cycle.

More than 50 animal proteins are known to act as mitogens. Most of these proteins have a broad specificity. *Platelet-derived growth factor* (*PDGF*), for example, can stimulate many types of cells to divide, including fibroblasts, smooth muscle cells, and neuroglial cells. Similarly, *epidermal growth factor* (*EGF*) acts not only on epidermal cells but also on many other cell types, including both epithelial and nonepithelial cells. Some mitogens, however, have a narrow specificity: *erythropoietin*, for example, only induces the proliferation of red blood cell precursors. Many mitogens, including PDGF, also have actions other than the stimulation of cell division: they can stimulate cell growth, survival, differentiation, or migration, depending on the circumstances and the cell type.

In some tissues, inhibitory extracellular signal proteins oppose the positive regulators and thereby inhibit organ growth. The best-understood inhibitory signal proteins are transforming growth factor-β (TGFβ) and its relatives. TGFβ inhibits the proliferation of several cell types, mainly by blocking cell-cycle progression in G_1.

Cells Can Enter a Specialized Nondividing State

In the absence of a mitogenic signal to proliferate, Cdk inhibition in G_1 is maintained by the multiple mechanisms discussed earlier, and progression into a new cell cycle is blocked. In some cases, cells partly disassemble their cell-cycle control system and withdraw from the cycle to a specialized nondividing state called $\mathbf{G_0}$.

Most cells in our body are in G_0, but the molecular basis and reversibility of this state vary in different cell types. Most of our neurons and skeletal muscle cells, for example, are in a *terminally differentiated* G_0 state, in which their cell-cycle control system is completely dismantled: the expression of the genes encoding various Cdks and cyclins is permanently turned off, and cell division rarely occurs. Some cell types withdraw from the cell cycle only transiently and retain the ability to reassemble the cell-cycle control system quickly and reenter the cycle. Most liver cells, for example, are in G_0, but they can be stimulated to divide if the liver is damaged. Still other types of cells, including fibroblasts and some lymphocytes, withdraw from and reenter the cell cycle repeatedly throughout their lifetime.

Almost all the variation in cell-cycle length in the adult body occurs during the time the cell spends in G_1 or G_0. By contrast, the time a cell takes to progress from the beginning of S phase through mitosis is usually brief (typically 12–24 hours in mammals) and relatively constant, regardless of the interval from one division to the next.

Mitogens Stimulate G_1-Cdk and G_1/S-Cdk Activities

For the vast majority of animal cells, mitogens control the rate of cell division by acting in the G_1 phase of the cell cycle. As discussed earlier, multiple mechanisms act during G_1 to suppress Cdk activity. Mitogens release these brakes on Cdk activity, thereby allowing entry into a new cell cycle.

As we discuss in Chapter 15, mitogens interact with cell-surface receptors to trigger multiple intracellular signaling pathways. One major pathway acts through the monomeric GTPase **Ras**, which leads to the activation of a *mitogen-activated protein kinase (MAP kinase) cascade* (see Figure 15–50). This leads to an increase in the production of transcription regulatory proteins, including **Myc**. Myc is thought to promote cell-cycle entry by several mechanisms, one of which is to increase the expression of genes encoding G_1-cyclins (D cyclins), thereby increasing G_1-Cdk (cyclin D–Cdk4) activity. Myc also has a major role in stimulating the transcription of genes that increase cell growth.

The key function of G_1-Cdk complexes in animal cells is to activate a group of gene regulatory factors called the **E2F proteins**, which bind to specific DNA sequences in the promoters of a wide variety of genes that encode proteins required for S-phase entry, including G_1/S-cyclins, S-cyclins, and proteins involved in DNA synthesis and chromosome duplication. In the absence of mitogenic stimulation, E2F-dependent gene expression is inhibited by an interaction between E2F and members of the **retinoblastoma protein (Rb)** family. When cells are stimulated to divide by mitogens, active G_1-Cdk accumulates and phosphorylates Rb family members, reducing their binding to E2F. The liberated E2F proteins then activate expression of their target genes (**Figure 17–59**).

This transcriptional control system, like so many other control systems that regulate the cell cycle, includes feedback loops that ensure that entry into the cell cycle is complete and irreversible. The liberated E2F proteins, for example, increase the transcription of their own genes. In addition, E2F-dependent transcription of G_1/S-cyclin (cyclin E) and S-cyclin (cyclin A) genes leads to increased G_1/S-Cdk and S-Cdk activities, which in turn increase Rb protein phosphorylation and promote further E2F release (see Figure 17–59).

The central member of the Rb family, the Rb protein itself, was identified originally through studies of an inherited form of eye cancer in children, known as *retinoblastoma* (discussed in Chapter 20). The loss of both copies of the *Rb* gene leads to excessive proliferation of some cells in the developing retina, suggesting that the Rb protein is particularly important for restraining cell division in this tissue. The complete loss of Rb does not immediately cause increased proliferation of retinal or other types of cells, in part because Cdh1 and CKIs also help inhibit progression through G_1 and in part because other cell types contain Rb-related proteins that provide backup support in the absence of Rb. It is also likely that other proteins, unrelated to Rb, help to regulate the activity of E2F.

Additional layers of control promote an overwhelming increase in S-Cdk activity at the beginning of S phase. We mentioned earlier that the APC/C

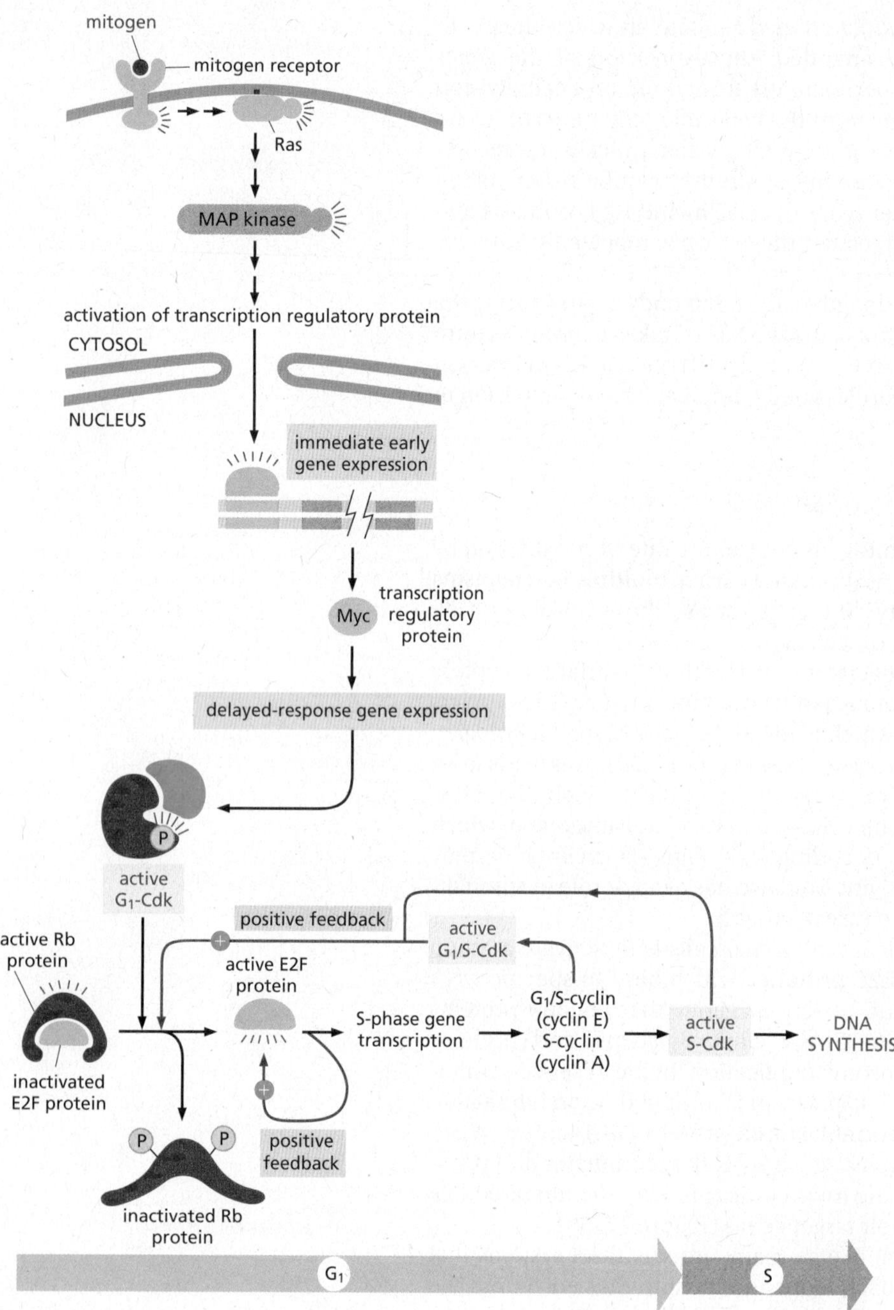

Figure 17–59 Mitogen stimulation of cell-cycle entry. As discussed in Chapter 15, mitogens bind to cell-surface receptors to initiate intracellular signaling pathways. One of the major pathways involves activation of the small GTPase Ras, which activates a MAP kinase cascade, leading to increased expression of numerous *immediate early* genes, including the gene encoding the transcription regulatory protein Myc. Myc increases the expression of many *delayed-response* genes, including some that lead to increased G_1-Cdk activity (cyclin D–Cdk4), which triggers the phosphorylation of members of the Rb family of proteins. This inactivates the Rb proteins, freeing the gene regulatory protein E2F to activate the transcription of G_1/S genes, including the genes for a G_1/S-cyclin (cyclin E) and S-cyclin (cyclin A). The resulting G_1/S-Cdk and S-Cdk activities further enhance Rb protein phosphorylation, forming a positive feedback loop. E2F proteins also stimulate the transcription of their own genes, forming another positive feedback loop.

activator Cdh1 suppresses cyclin levels after mitosis. In animal cells, however, G_1- and G_1/S-cyclins are resistant to APC/C–Cdh1 and can therefore act unopposed by the APC/C to promote Rb protein phosphorylation and E2F-dependent gene expression. S-cyclin, by contrast, is not resistant, and its level is initially restrained by APC/C–Cdh1 activity. However, G_1/S-Cdk also phosphorylates and inactivates APC/C–Cdh1, thereby allowing the accumulation of S-cyclin, further promoting S-Cdk activation. G_1/S-Cdk also inactivates CKI proteins that suppress S-Cdk activity. The overall effect of all these interactions is the rapid and complete activation of the S-Cdk complexes required for S-phase initiation.

DNA Damage Blocks Cell Division

Progression through the cell cycle, and thus the rate of cell proliferation, is controlled not only by extracellular mitogens but also by other extracellular and intracellular signals. One of the most important influences is DNA damage,

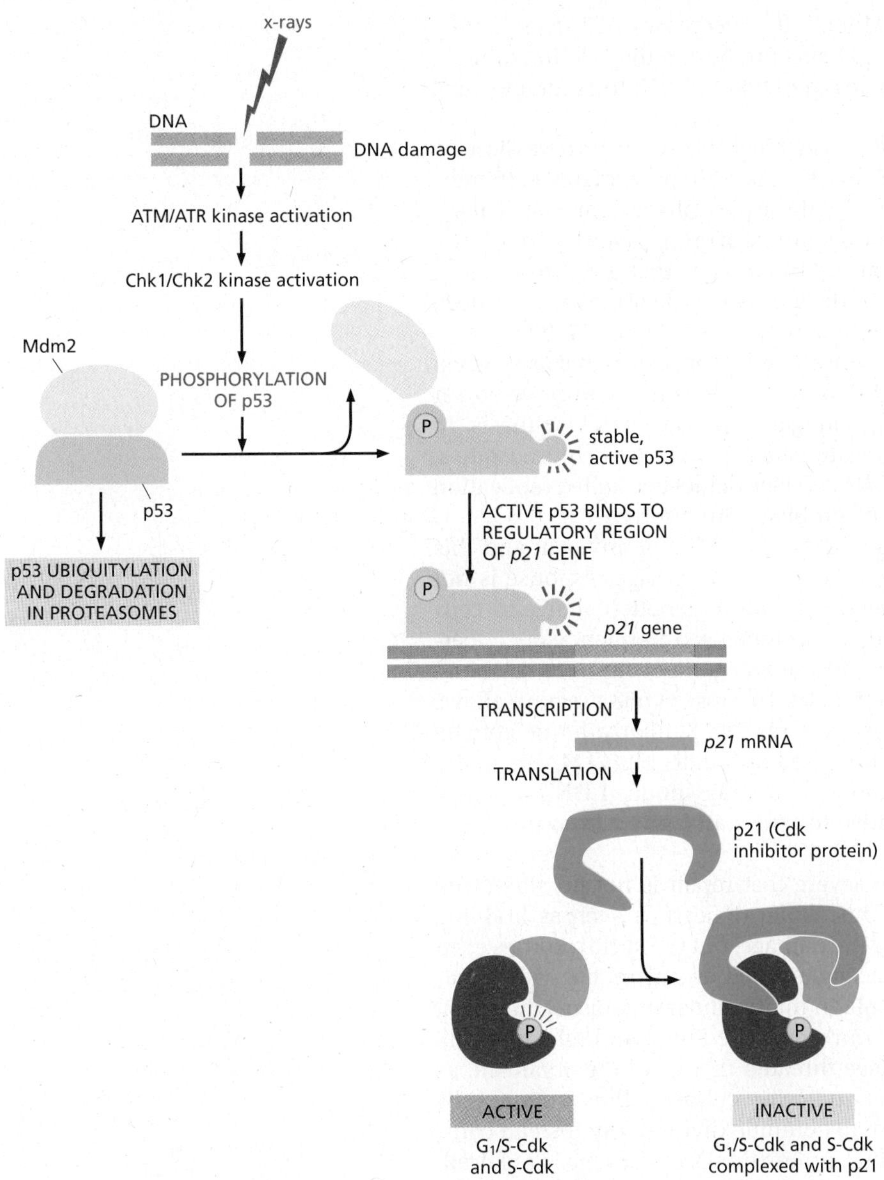

Figure 17–60 **How DNA damage arrests the cell cycle in G_1.** When DNA is damaged, various protein kinases are recruited to the site of damage and initiate a signaling pathway that causes cell-cycle arrest. The first kinase at the damage site is either ATM or ATR, depending on the type of damage. Additional protein kinases, called Chk1 and Chk2, are then recruited and activated, resulting in the phosphorylation of the transcription regulatory protein p53. Mdm2 normally binds to p53 and promotes its ubiquitylation and destruction in proteasomes. Phosphorylation of p53 blocks its binding to Mdm2; as a result, p53 accumulates to high levels and stimulates transcription of numerous genes, including the gene that encodes the CKI protein p21. The p21 binds and inactivates G_1/S-Cdk and S-Cdk complexes, arresting the cell in G_1. In some cases, DNA damage also induces either the phosphorylation of Mdm2 or a decrease in Mdm2 production, which causes a further increase in p53 (not shown).

which can occur as a result of spontaneous chemical reactions in DNA, errors in DNA replication, or exposure to radiation or certain chemicals (discussed in Chapter 5). It is essential that damaged chromosomes are repaired before attempting to duplicate or segregate them. The cell-cycle control system can readily detect DNA damage and arrest the cycle at either of two transitions—one at Start, which prevents entry into the cell cycle and into S phase, and one at the G_2/M transition, which prevents entry into mitosis (see Figure 17–20).

DNA damage initiates a signaling pathway by activating one of a pair of related protein kinases called **ATM** and **ATR**, which associate with the site of damage and phosphorylate various target proteins, including two other protein kinases called *Chk1* and *Chk2*. These various kinases phosphorylate other target proteins that lead to cell-cycle arrest. A major target is the gene regulatory protein **p53**, which stimulates transcription of the gene encoding *p21*, a CKI protein; p21 binds to G_1/S-Cdk and S-Cdk complexes and inhibits their activities, thereby helping to block entry into the cell cycle (**Figure 17–60** and **Movie 17.8**).

DNA damage activates p53 by an indirect mechanism. In undamaged cells, p53 is highly unstable and is present at very low concentrations. This is largely because it interacts with another protein, *Mdm2*, which acts as a ubiquitin ligase that targets p53 for destruction by proteasomes. Phosphorylation of p53 after

DNA damage reduces its binding to Mdm2. This decreases p53 degradation, which results in a marked increase in p53 concentration in the cell. In addition, the decreased binding to Mdm2 enhances the ability of p53 to stimulate gene transcription (see Figure 17–60).

The protein kinases Chk1 and Chk2 also block cell-cycle progression by phosphorylating members of the Cdc25 family of protein phosphatases, thereby inhibiting their function. As described earlier, these phosphatases are particularly important in the activation of M-Cdk at the beginning of mitosis (see Figure 17–16). Chk1 and Chk2 phosphorylate Cdc25 at inhibitory sites that are distinct from the phosphorylation sites that stimulate Cdc25 activity. The inhibition of Cdc25 activity by DNA damage helps block entry into mitosis (see Figure 17–20).

The DNA damage response can also be activated by problems that arise when a replication fork fails during DNA replication. When nucleotides are depleted, for example, replication forks stall during the elongation phase of DNA synthesis. To prevent the cell from attempting to segregate partially replicated chromosomes, the same mechanisms that respond to DNA damage detect the stalled replication forks and block entry into mitosis until the problems are resolved.

A low level of DNA damage occurs in the normal life of any cell, and this damage accumulates in the cell's progeny if the DNA damage response is not functioning. Over the long term, the accumulation of genetic damage in cells lacking the DNA damage response leads to an increased frequency of cancer-promoting mutations. Indeed, mutations in the *p53* gene occur in at least half of all human cancers (discussed in Chapter 20). This loss of p53 function allows the cancer cell to accumulate mutations more readily. Similarly, a rare genetic disease known as *ataxia telangiectasia* is caused by a defect in ATM, one of the protein kinases that are activated in response to x-ray–induced DNA damage; people with this disease are very sensitive to x-rays and suffer from increased rates of cancer.

What happens if DNA damage is so severe that repair is not possible? The answer differs in different organisms. Unicellular organisms such as budding yeast arrest their cell cycle to try to repair the damage, but the cycle resumes even if the repair cannot be completed. For a single-celled organism, life with mutations is apparently better than no life at all. In multicellular organisms, however, the health of the organism takes precedence over the life of an individual cell. Cells that divide with severe DNA damage threaten the life of the organism, as genetic damage can often lead to cancer and other diseases. Thus, animal cells with severe DNA damage do not attempt to continue division, but instead commit suicide by undergoing apoptosis. Thus, unless the DNA damage is repaired, the DNA damage response can lead to either cell-cycle arrest or cell death. DNA damage–induced apoptosis often depends on the activation of p53.

Many Human Cells Have a Built-In Limitation on the Number of Times They Can Divide

Many human cells divide a limited number of times before they stop and undergo a permanent cell-cycle arrest. Fibroblasts taken from normal human tissue, for example, go through only about 25–50 population doublings when cultured in a standard mitogenic medium. Toward the end of this time, proliferation slows down and finally halts, and the cells enter a nondividing state from which they never recover. This phenomenon is called **replicative cell senescence**.

Replicative cell senescence in human fibroblasts seems to be caused by changes in the structure of the **telomeres**, the repetitive DNA sequences and associated proteins at the ends of chromosomes. As discussed in Chapter 5, when a cell divides, telomeric DNA sequences are not replicated in the same manner as the rest of the genome but instead are synthesized by the enzyme **telomerase**. Telomerase also promotes the formation of protein cap structures that protect the chromosome ends. Because human fibroblasts, and many other human somatic cells, do not produce telomerase, their telomeres become shorter with every cell division, and their protective protein caps progressively deteriorate. Eventually, the exposed chromosome ends are sensed as DNA damage, which activates a p53-dependent cell-cycle arrest (see Figure 17–60). Rodent cells, by contrast, maintain telomerase activity when they proliferate in culture and therefore do not have such a telomere-dependent mechanism for limiting proliferation. The forced expression of telomerase in normal human fibroblasts, using genetic engineering techniques, blocks this form of senescence. Unfortunately, most cancer cells have regained the ability to produce telomerase and therefore maintain telomere function as they proliferate; as a result, they do not undergo replicative cell senescence.

Cell Proliferation Is Accompanied by Cell Growth

If cells proliferated without growing, they would get progressively smaller and there would be no net increase in total cell mass. In most proliferating cell populations, therefore, cell growth accompanies cell division. In single-celled organisms such as yeasts, both cell growth and cell division require only nutrients. In animals, by contrast, both cell growth and cell proliferation depend on extracellular signal molecules, produced by other cells, which we call **growth factors** and mitogens, respectively.

Like mitogens, the extracellular growth factors that stimulate animal cell growth bind to receptors on the cell surface and activate intracellular signaling pathways. These pathways stimulate the accumulation of proteins and other macromolecules, and they do so by both increasing their rate of synthesis and decreasing their rate of degradation. They also trigger increased uptake of nutrients and production of the ATP required to fuel the increased protein synthesis. One of the most important intracellular signaling pathways activated by growth factor receptors involves the enzyme phosphoinositide 3-kinase (*PI 3-kinase*), which adds a phosphate from ATP to the 3′ position of inositol phospholipids in the plasma membrane (discussed in Chapter 15). The activation of PI 3-kinase leads to the activation of a protein kinase called *mTORC1*, which lies at the heart of cell growth regulatory pathways in all eukaryotes (see Figure 15–55). mTORC1 activates many targets in the cell that stimulate metabolic processes, including protein and lipid synthesis, or reduce protein turnover (**Figure 17–61**).

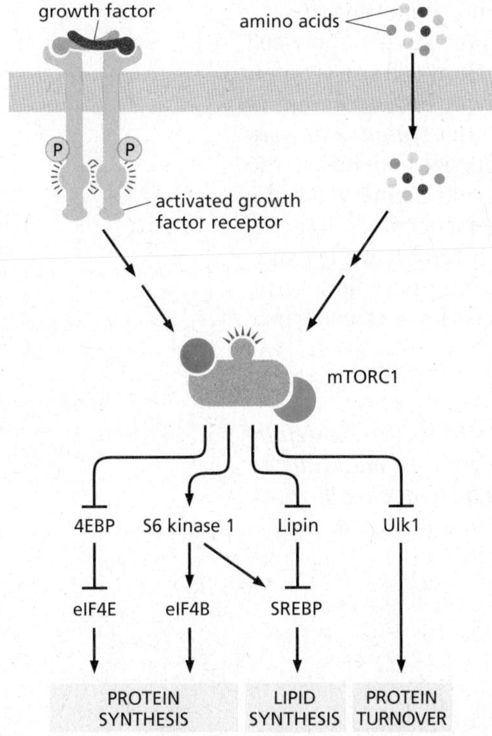

Figure 17–61 Stimulation of cell growth by extracellular growth factors and nutrients. The occupation of cell-surface receptors by growth factors leads to the activation of a complex signaling pathway that results in the activation of the multisubunit protein kinase mTORC1 (see Figure 15–55). Cytosolic amino acids also help activate mTORC1. mTORC1 phosphorylates multiple proteins, including 4EBP and the protein kinase S6 kinase 1 (S6K1), to stimulate the activity of the translation initiation factors eIF4E and eIF4B, thereby stimulating protein synthesis. mTORC1 also acts through S6K1, as well as another protein called Lipin, to activate a transcription regulator called SREBP, which increases the expression of genes involved in lipid synthesis. Finally, mTORC1 phosphorylates another protein kinase, Ulk1, reducing its ability to promote protein turnover.

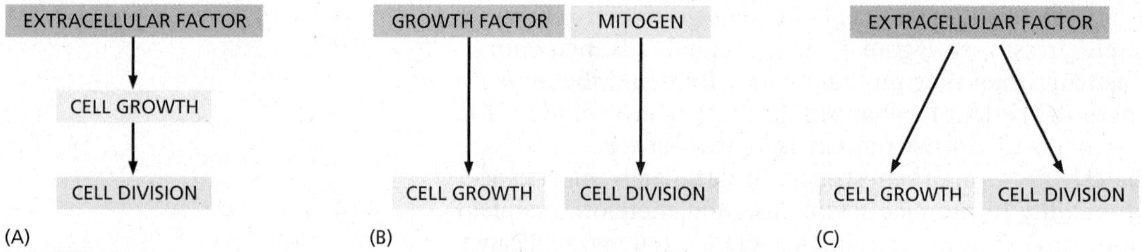

(A) EXTRACELLULAR FACTOR → CELL GROWTH → CELL DIVISION

(B) GROWTH FACTOR → CELL GROWTH; MITOGEN → CELL DIVISION

(C) EXTRACELLULAR FACTOR → CELL GROWTH, CELL DIVISION

Figure 17–62 **Potential mechanisms for coordinating cell growth and division.** In proliferating cells, cell size is maintained by mechanisms that coordinate rates of cell division and cell growth. Numerous alternative coupling mechanisms are thought to exist, and different cell types appear to employ different combinations of these mechanisms. (A) In many cell types—particularly yeast—the rate of cell division is governed by the rate of cell growth, so that division occurs only when growth rate achieves some minimal threshold; in yeasts, it is mainly the levels of extracellular nutrients that regulate the rate of cell growth and thereby the rate of cell division. (B) In some animal cell types, growth and division can each be controlled by separate extracellular factors (growth factors and mitogens, respectively), and cell size depends on the relative levels of the two types of factors. (C) Some extracellular factors can stimulate both cell growth and cell division by simultaneously activating signaling pathways that promote growth and other pathways that promote cell-cycle progression.

Proliferating Cells Usually Coordinate Their Growth and Division

For proliferating cells to maintain a constant size, they must coordinate their growth with cell division to ensure that cell size doubles with each division: if cells grow too slowly, they will get smaller with each division, and if they grow too fast, they will get larger with each division. It is not clear how cells achieve this coordination, but it is likely to involve multiple mechanisms that vary in different organisms and even in different cell types of the same organism (**Figure 17–62**).

Animal cell growth and division are not always coordinated, however. In many cases, they are completely uncoupled to allow growth without division or division without growth. Muscle cells and nerve cells, for example, can grow dramatically after they have permanently withdrawn from the cell cycle. Similarly, the eggs of many animals grow to an extremely large size without dividing; after fertilization, however, this relationship is reversed, and many rounds of division occur without growth.

Compared to cell division, there has been surprisingly little study of how cell size is controlled in animals. As a result, it remains a mystery how cell size is determined and why different cell types in the same animal grow to be so different in size. One of the best-understood cases in mammals is the adult *sympathetic neuron*, which has permanently withdrawn from the cell cycle. Its size depends on the amount of *nerve growth factor* (NGF) secreted by the target cells it innervates; the greater the amount of NGF the neuron has access to, the larger it becomes. It seems likely that the genes a cell expresses set limits on the size it can be, while extracellular signal molecules and nutrients regulate the size within these limits. The challenge is to identify the relevant genes and signal molecules for each cell type.

Summary

In multicellular animals, cell size, cell division, and cell survival are carefully controlled to ensure that the organism and its organs achieve and maintain an appropriate size. Mitogens stimulate the rate of cell division by removing intracellular molecular brakes that restrain cell-cycle progression in G_1. *Growth factors promote cell growth (an increase in cell mass) by stimulating the synthesis and inhibiting the degradation of macromolecules. To maintain a constant cell size, proliferating cells employ multiple mechanisms to ensure that cell growth is coordinated with cell division.*

PROBLEMS

Which statements are true? Explain why or why not.

17–1 As there are about 10^{13} cells in an adult human, and about 10^{10} cells die and are replaced each day, we become new people every 3 years.

17–2 All three of the major cell-cycle transitions—Start, G_2/M, and metaphase-to-anaphase—depend on the activity of Cdks.

17–3 Initiation of DNA synthesis is permitted only at origins of replication that are licensed by being loaded with Mcm complexes.

17–4 Chromosomes are positioned on the metaphase plate by equal and opposite forces that pull them toward the two poles of the spindle.

17–5 Meiosis segregates the paternal homologs into sperm and the maternal homologs into eggs.

17–6 If we could turn on telomerase activity in all our cells, we could prevent aging.

Discuss the following problems.

17–7 Some cell-cycle genes from human cells function perfectly well when expressed in yeast cells. Why do you suppose that is considered remarkable? After all, many human genes encoding enzymes for metabolic reactions also function in yeast, and no one thinks that is remarkable.

17–8 Hoechst 33342 is a membrane-permeant dye that fluoresces when it binds to DNA. When a population of cells is incubated briefly with the Hoechst dye and then sorted in a flow cytometer, which measures the fluorescence of each cell, the cells display various levels of fluorescence as shown in **Figure Q17–1**.

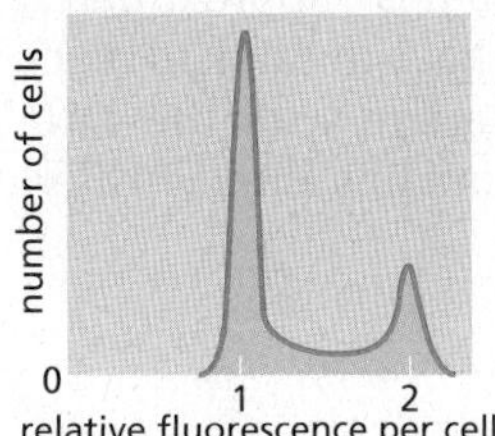

Figure Q17–1 Analysis of Hoechst 33342 fluorescence in a population of cells sorted in a flow cytometer (Problem 17–8).

A. Which cells in Figure Q17–1 are in the G_1, S, G_2, and M phases of the cell cycle? Explain the basis for your answer.

B. Sketch the sorting distributions you would expect for cells that were treated with inhibitors that block the cell cycle in the G_1, S, or M phase. Explain your reasoning.

17–9 A two-component Fucci (fluorescent ubiquitylation-based cell-cycle indicator) used in fruit flies gives different-colored cells at different points in the cell cycle, as shown in **Figure Q17–2**. If this result was obtained by tagging protein A with GFP (green) and protein B with RFP (red), when during the cell cycle are these proteins expressed?

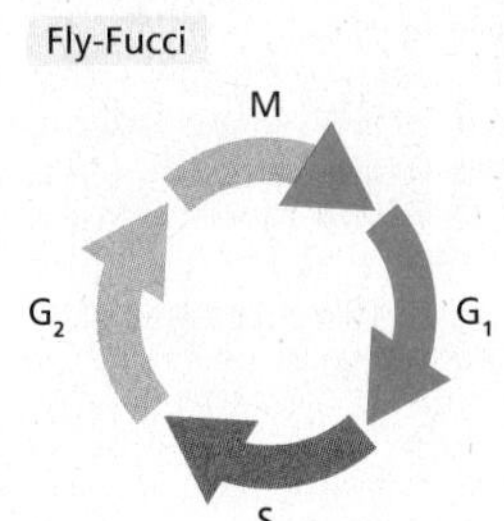

Figure Q17–2 A two-component Fucci system in fruit flies (Problem 17–9). (From Figure 5a of N. Zielke and B.A. Edgar, *Wiley Interdiscip. Rev. Dev. Biol.* 4:469–487, 2015. With permission from Wiley.)

A. Protein A in G_1, S, and G_2; protein B in G_1, S, and G_2

B. Protein A in G_2, M, and G_1; protein B in S, G_2, and early M

C. Protein A in late M and G_1; protein B in S

D. Protein A in late M, G_1, and S; protein B in G_2 and early M

17–10 What specific event does the cell-cycle control system stimulate at the metaphase-to-anaphase transition?

17–11 Which one of the following combinations of activities of a protein kinase and a protein phosphatase would give the highest activity for a target protein that is most active when phosphorylated?

A. Kinase OFF; phosphatase OFF

B. Kinase OFF; phosphatase ON

C. Kinase ON; phosphatase OFF

D. Kinase ON; phosphatase ON

17–12 The yeast cohesin subunit Scc1, which is essential for sister-chromatid cohesion, can be artificially regulated for expression at any point in the cell cycle. If expression is turned on at the beginning of S phase, all the cells divide satisfactorily and survive. By contrast, if Scc1 expression is turned on only after S phase is completed, the cells fail to divide and they die, even though Scc1 accumulates in the nucleus and interacts efficiently with chromosomes. Why do you suppose that cohesin must be present during S phase for cells to divide normally?

17–13 A living cell from the lung epithelium of a newt is shown at different stages in M phase in **Figure Q17–3**. Order these light micrographs into the correct sequence and identify the stage in M phase that each represents.

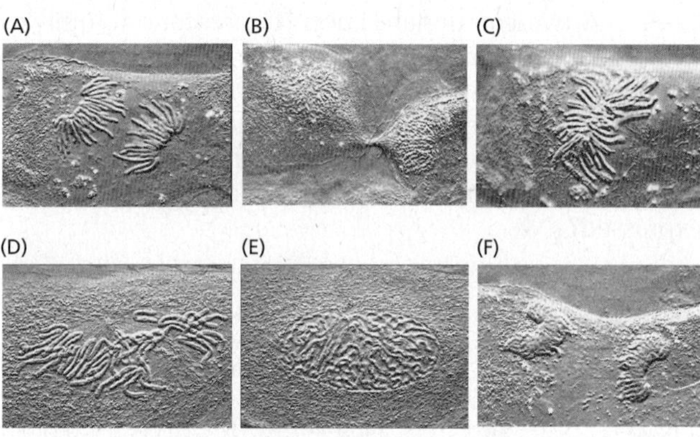

Figure Q17–3 Light micrographs of a single cell at different stages of M phase (Problem 17–13). (Courtesy of Conly L. Rieder.)

17–14 How many kinetochores are there in a human cell at mitosis?

17–15 If a cell just entering mitosis is treated with nocodazole, which destabilizes microtubules, the nuclear envelope breaks down and chromosomes condense, but no spindle forms and the cell cycle arrests in mitosis. In contrast, if such a cell is treated with cytochalasin D, which destabilizes actin filaments, mitosis proceeds normally but generates a binucleate cell that proceeds into G_1 phase. Explain the basis for the different outcomes of these treatments with cytoskeleton inhibitors. What do these results tell you about cell-cycle checkpoints in M phase?

17–16 Early on in the study of recombination during meiosis, geneticists concluded that there was about one crossover per chromosome arm. Using fluorescent tags that stain the synaptonemal complex red, the centromeres blue, and the crossovers green, it is now possible to observe directly the distribution of crossovers in chromosomes. For the 13 complete bivalents shown in **Figure Q17–4**, how many have more than one crossover in the arm of a chromosome?

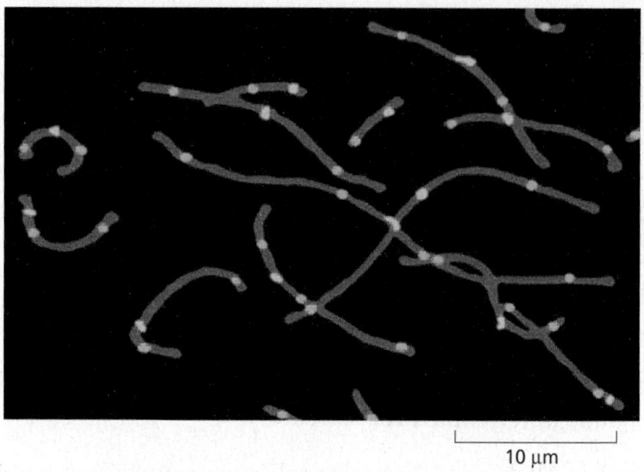

Figure Q17–4 Crossovers between homologs in human cells (Problem 17–16). (Modified from A. Lynn et al., *Science* 296:2222–2225, 2002. With permission from AAAS.)

17–17 Down syndrome (trisomy 21) and Edwards syndrome (trisomy 18) are the most common autosomal trisomies seen in human infants. Does this fact mean that these chromosomes are the most difficult to segregate properly during meiosis?

17–18 The human genome consists of 23 pairs of chromosomes (22 pairs of autosomes and one pair of sex chromosomes). During meiosis, the maternal and paternal sets of homologs pair, and then are separated into gametes, so that each contains 23 chromosomes. If you assume that the chromosomes in the paired homologs are randomly assorted to daughter cells, how many potential combinations of paternal and maternal homologs can be generated during meiosis? (For the purposes of this calculation, assume that no recombination occurs.)

17–19 High doses of caffeine interfere with the DNA damage response in mammalian cells. Why then do you suppose the Surgeon General has not yet issued an appropriate warning to heavy coffee and cola drinkers? A typical cup of coffee (150 mL) contains 100 mg of caffeine (196 g/mole). Assuming that the caffeine is not metabolized or excreted (but that all the liquid is), approximately how many cups of coffee would you have to drink to reach the dose (10 mM) required to interfere with the DNA damage response? (A typical adult contains about 40 liters of water.)

REFERENCES

General

Morgan DO (2007) The Cell Cycle: Principles of Control. London: New Science Press.

The Cell-Cycle Control System

Alfieri C, Zhang S & Barford D (2017) Visualizing the complex functions and mechanisms of the anaphase promoting complex/cyclosome (APC/C). *Open Biol.* 7, 170204.

Echalier A, Endicott JA & Noble ME (2010) Recent developments in cyclin-dependent kinase biochemical and structural studies. *Biochim. Biophys. Acta* 1804, 511–519.

Evans T, Rosenthal ET, Youngblom J, Distel D & Hunt T (1983) Cyclin: a protein specified by maternal mRNA in sea urchin eggs that is destroyed at each cleavage division. *Cell* 33, 389–396.

Hartwell LH, Culotti J, Pringle JR & Reid BJ (1974) Genetic control of the cell division cycle in yeast. *Science* 183, 46–51.

Holder J, Poser E & Barr FA (2019) Getting out of mitosis: spatial and temporal control of mitotic exit and cytokinesis by PP1 and PP2A. *FEBS Lett.* 593, 2908–2924.

Holt LJ, Tuch BB, Villen J, Johnson AD, Gygi SP & Morgan DO (2009) Global analysis of Cdk1 substrate phosphorylation sites provides insights into evolution. *Science* 325, 1682–1686.

Kamenz J, Gelens L & Ferrell JE, Jr. (2021) Bistable, biphasic regulation of PP2A-B55 accounts for the dynamics of mitotic substrate phosphorylation. *Curr. Biol.* 31, 794–808.

Novak B & Tyson JJ (2021) Mechanisms of signalling-memory governing progression through the eukaryotic cell cycle. *Curr. Opin. Cell Biol.* 69, 7–16.

Nurse P, Thuriaux P & Nasmyth K (1976) Genetic control of the cell division cycle in the fission yeast *Schizosaccharomyces pombe*. *Mol. Gen. Genet.* 146, 167–178.

Örd M & Loog M (2019) How the cell cycle clock ticks. *Mol. Biol. Cell* 30, 169-172.

S Phase

Arias EE & Walter JC (2007) Strength in numbers: preventing rereplication via multiple mechanisms in eukaryotic cells. *Genes Dev.* 21, 497–518.

Bell SP & Labib K (2016) Chromosome duplication in *Saccharomyces cerevisiae*. *Genetics* 203, 1027–1067.

Li H & O'Donnell ME (2018) The eukaryotic CMG helicase at the replication fork: emerging architecture reveals an unexpected mechanism. *Bioessays* 40(3).

Siddiqui K, On KF & Diffley JF (2013) Regulating DNA replication in eukarya. *Cold Spring Harb. Perspect. Biol.* 5, a012930.

Tanaka S & Araki H (2013) Helicase activation and establishment of replication forks at chromosomal origins of replication. *Cold Spring Harb. Perspect. Biol.* 5, a010371.

Mitosis

Banterle N & Gonczy P (2017) Centriole biogenesis: from identifying the characters to understanding the plot. *Annu. Rev. Cell Dev. Biol.* 33, 23–49.

Champion L, Linder MI & Kutay U (2017) Cellular reorganization during mitotic entry. *Trends Cell Biol.* 27, 26–41.

Davidson IF & Peters JM (2021) Genome folding through loop extrusion by SMC complexes. *Nat. Rev. Mol. Cell Biol.* 22, 445–464.

Dey G & Baum B (2021) Nuclear envelope remodelling during mitosis. *Curr. Opin. Cell Biol.* 70, 67–74.

Hinshaw SM & Harrison SC (2018) Kinetochore function from the bottom up. *Trends Cell Biol.* 28, 22–33.

Lampson MA & Grishchuk EL (2017) Mechanisms to avoid and correct erroneous kinetochore-microtubule attachments. *Biology (Basel)* 6, 1.

McIntosh JR (2016) Mitosis. *Cold Spring Harb. Perspect. Biol.* 8, a023218.

Musacchio A (2015) The molecular biology of spindle assembly checkpoint signaling dynamics. *Curr. Biol.* 25, R1002–R1018.

Musacchio A & Desai A (2017) A molecular view of kinetochore assembly and function. *Biology (Basel)* 6, 5.

Nigg EA & Holland AJ (2018) Once and only once: mechanisms of centriole duplication and their deregulation in disease. *Nat. Rev. Mol. Cell Biol.* 19, 297–312.

Oriola D, Needleman DJ & Brugues J (2018) The physics of the metaphase spindle. *Annu. Rev. Biophys.* 47, 655–673.

Paulson JR, Hudson DF, Cisneros-Soberanis F & Earnshaw WC (2021) Mitotic chromosomes. *Semin. Cell Dev. Biol.* 117, 7–29.

Petry S (2016) Mechanisms of mitotic spindle assembly. *Annu. Rev. Biochem.* 85, 659–683.

Prosser SL & Pelletier L (2017) Mitotic spindle assembly in animal cells: a fine balancing act. *Nat. Rev. Mol. Cell Biol.* 18, 187–201.

Yatskevich S, Rhodes J & Nasmyth K (2019) Organization of chromosomal DNA by SMC complexes. *Annu. Rev. Genet.* 53, 445–482.

Yi P & Goshima G (2018) Microtubule nucleation and organization without centrosomes. *Curr. Opin. Plant Biol.* 46, 1–7.

Cytokinesis

Basant A & Glotzer M (2018) Spatiotemporal regulation of RhoA during cytokinesis. *Curr. Biol.* 28, R570–R580.

Fremont S & Echard A (2018) Membrane traffic in the late steps of cytokinesis. *Curr. Biol.* 28, R458–R470.

Glotzer M (2017) Cytokinesis in metazoa and fungi. *Cold Spring Harb. Perspect. Biol.* 9, a022343.

Livanos P & Muller S (2019) Division plane establishment and cytokinesis. *Annu. Rev. Plant Biol.* 70, 239–267.

Pollard TD & O'Shaughnessy B (2019) Molecular mechanism of cytokinesis. *Annu. Rev. Biochem.* 88, 661–689.

Rappaport R (1986) Establishment of the mechanism of cytokinesis in animal cells. *Int. Rev. Cytol.* 105, 245–281.

Meiosis

Gray S & Cohen PE (2016) Control of meiotic crossovers: from double-strand break formation to designation. *Annu. Rev. Genet.* 50, 175–210.

Hughes SE & Hawley RS (2020) Alternative synaptonemal complex structures: too much of a good thing? *Trends Genet.* 36, 833–844.

Hunter N (2015) Meiotic recombination: the essence of heredity. *Cold Spring Harb. Perspect. Biol.* 7, a016618.

Petronczki M, Siomos MF & Nasmyth K (2003) Un ménage à quatre: the molecular biology of chromosome segregation in meiosis. *Cell* 112, 423–440.

Subramanian VV & Hochwagen A (2014) The meiotic checkpoint network: step-by-step through meiotic prophase. *Cold Spring Harb. Perspect. Biol.* 6, a016675.

Webster A & Schuh M (2017) Mechanisms of aneuploidy in human eggs. *Trends Cell Biol.* 27, 55–68.

Zickler D & Kleckner N (2015) Recombination, pairing, and synapsis of homologs during meiosis. *Cold Spring Harb. Perspect. Biol.* 7, a016626.

Zickler D & Kleckner N (2016) A few of our favorite things: pairing, the bouquet, crossover interference and evolution of meiosis. *Semin. Cell Dev. Biol.* 54, 135–148.

Control of Cell Division and Cell Growth

Bertoli C, Skotheim JM & de Bruin RA (2013) Control of cell cycle transcription during G1 and S phases. *Nat. Rev. Mol. Cell Biol.* 14, 518–528.

Blackford AN & Jackson SP (2017) ATM, ATR, and DNA-PK: the trinity at the heart of the DNA damage response. *Mol. Cell* 66, 801–817.

Boutelle AM & Attardi LD (2021) p53 and tumor suppression: it takes a network. *Trends Cell Biol.* 31, 298–310.

Dick FA & Rubin SM (2013) Molecular mechanisms underlying RB protein function. *Nat. Rev. Mol. Cell Biol.* 14, 297–306.

Rubin SM, Sage J & Skotheim JM (2020) Integrating old and new paradigms of G1/S control. *Mol. Cell* 80, 183–192.

Saxton RA & Sabatini DM (2017) mTOR signaling in growth, metabolism, and disease. *Cell* 168, 960–976.

Sharpless NE & Sherr CJ (2015) Forging a signature of in vivo senescence. *Nat. Rev. Cancer* 15, 397–408.

Shimobayashi M & Hall MN (2014) Making new contacts: the mTOR network in metabolism and signalling crosstalk. *Nat. Rev. Mol. Cell Biol.* 15, 155–162.

Vollmer J, Casares F & Iber D (2017) Growth and size control during development. *Open Biol.* 7, 170190.

Zatulovskiy E & Skotheim JM (2020) On the molecular mechanisms regulating animal cell size homeostasis. *Trends Genet.* 36, 360–372.

CHAPTER 18

Cell Death

The development and maintenance of multicellular organisms depend not only on cell growth and cell division but also on cell death. During animal development, for example, carefully orchestrated patterns of cell growth, division, and death help determine the size and shape of limbs and organs. The maintenance of tissue size in adult animals often requires that cells die at the same rate as they are produced—a process called cell turnover. Such "normal" cell death also occurs in plants during development and in the senescence of flowers and leaves, and it can also occur in unicellular organisms, including yeasts and bacteria. Most of these normal cell deaths are suicides, in which a sequence of molecular events destroys the cells from within, but the molecular mechanisms can differ widely. In this chapter, we focus on a molecularly distinct form of cell suicide called **apoptosis** (from the Greek word meaning "falling off," as leaves from a tree), even though it occurs only in animal cells. Apoptosis is thought to occur in all animals, and it is by far the most common way for our cells to die. It has also been the most intensely studied form of cell death.

A cell dying by apoptosis undergoes characteristic morphological and biochemical changes, in part, to package itself to be eaten and digested quickly by a nearby cell. **Figure 18–1** shows a normal rat cell (panel A) and rat cells of the same type undergoing two different forms of cell death (panels B–D). In apoptosis, the cell shrinks and condenses, the cytoskeleton collapses, the nuclear envelope disassembles, and the nuclear chromatin condenses and fragments (Figure 18–1B). The apoptotic cell surface often forms multiple large protrusions called blebs (**Movie 18.1**), and, if the cell is large, it breaks up into membrane-enclosed fragments called apoptotic bodies. Importantly, the surface of the cell and its fragments become chemically altered, so that a neighboring cell, often a macrophage in vertebrates, rapidly engulfs the cell and fragments before they can spill their contents (Figure 18–1C). In this way, the cell dies neatly, and the cell and its fragments are rapidly cleared away, without causing a damaging inflammatory response. Because apoptotic cells are eaten and digested so quickly, there are often few dead cells to be seen, even in a tissue where large numbers of cells have died by apoptosis.

Damaged or infected cells can also die by apoptosis, ensuring that they are eliminated before they can threaten the health of the animal. There are, however, many non-apoptotic ways for severely damaged or stressed animal cells to die. These are usually lumped together under the umbrella term *cell necrosis*. A common, largely passive form of cell necrosis occurs when cells lyse in response to an acute tissue insult, such as trauma or a blocked blood supply: the cells swell and burst (Figure 18–1D), spilling their contents over their neighbors and eliciting an inflammatory response. In other forms of cell necrosis, the cell takes an active part in the death process, and they are given special names such as necroptosis or pyroptosis, depending on the nature of the inducing stress and the molecular mechanisms involved.

In this chapter, we discuss the major functions of apoptosis in vertebrates, its molecular mechanism and regulation, and how excessive or insufficient apoptosis can contribute to human disease.

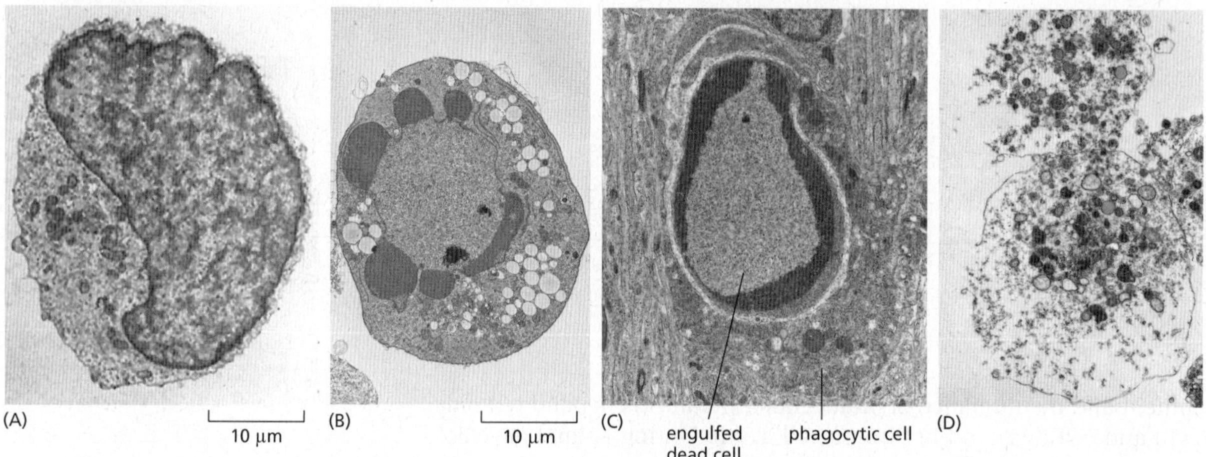

Figure 18–1 Two distinct forms of cell death. These electron micrographs show developing rat oligodendrocyte precursor cells in different states: (A) a normal cell in culture; (B) a cell in culture that has died by apoptosis because it was deprived of extracellular survival signals (discussed later); (C) a cell in a normal developing optic nerve that has died by apoptosis and has been engulfed by a neighboring phagocytic cell; (D) a cell in culture that died by necrosis. Note that the apoptotic cells in B and C have an intact plasma membrane, but the chromatin has become condensed and distorted and concentrated at the margin of the nucleus, whereas the necrotic cell in D seems to have exploded. The large vacuoles visible in the cytoplasm of the cell in B are a variable feature of apoptosis. (Courtesy of Julia Burne and Martin Raff.)

Apoptosis Eliminates Unwanted Cells

The amount of apoptotic cell death that occurs in many developing and adult vertebrate tissues is astonishing: at least a million cells die this way each second in a healthy adult human (and are replaced by cell division). It seems remarkably wasteful for so many cells to die, especially as the vast majority are perfectly healthy at the time they kill themselves. What is the benefit of such massive cell death?

In some cases, especially in animal development, the function of cell death is clear. Apoptosis helps sculpt our hands and feet during embryonic development: these appendages start out as spade-like structures, and the individual digits separate only as the cells between them die, as illustrated for a mouse paw in **Figure 18–2**. In other cases, cells die by apoptosis when the structure they form is no longer needed. When a tadpole changes into a frog at metamorphosis, for example, the cells in the tail die by apoptosis, and the tail, which is not needed in the frog, disappears. Apoptosis also functions as a quality-control process in development, eliminating cells that are abnormal, misplaced, nonfunctional, or potentially dangerous to the animal. Striking examples occur in the vertebrate adaptive immune system, where apoptosis eliminates developing T and B lymphocytes that either fail to produce potentially useful antigen-specific receptors or produce self-reactive receptors that make the cells potentially dangerous (discussed in Chapter 24); it also eliminates most of the lymphocytes activated to proliferate by an infection, after they have helped destroy the responsible microbes.

In adult tissues that are neither growing nor shrinking, cell death and cell division must be tightly regulated to ensure they are in balance. If part of the liver is removed in an adult rat, for example, liver cell proliferation increases to make up the loss. Conversely, if a rat is treated with the drug phenobarbital—which stimulates liver cell growth and division (and thereby liver enlargement)—and then the phenobarbital treatment is stopped, apoptosis in the liver greatly increases until the liver cell number has returned to normal, usually within a week or so. Thus, liver cell number is kept constant through the regulation of both the cell death rate and the cell birthrate. The control mechanisms responsible for such remarkable regulation are largely unknown.

Animal cells can recognize damage in their various organelles and, if the damage is great enough, they can kill themselves by undergoing apoptosis. An important example is DNA damage, which can produce cancer-promoting

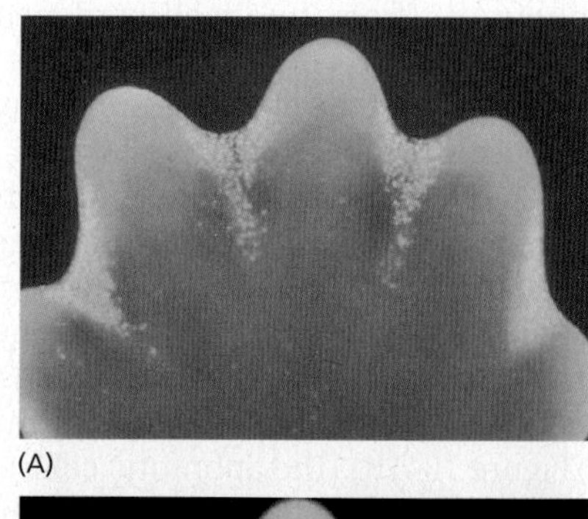

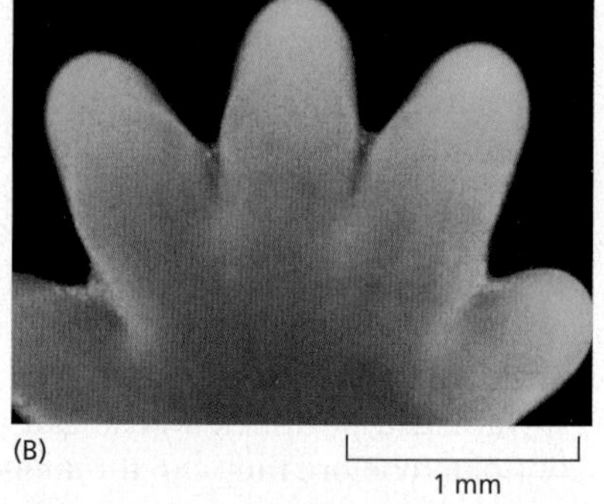

Figure 18–2 Sculpting the digits in the developing mouse paw by apoptosis. (A) The paw in this mouse fetus has been stained with the dye acridine orange, which enters apoptotic cells and thereby brightly labels them in the normal developing paw. The apoptotic cells appear as *bright green dots* concentrated between the developing digits. (B) The interdigital cell death has eliminated much of the tissue between the developing digits, as seen one day later, when there are very few apoptotic cells. (From W. Wood et al., *Development* 127:5245–5252, 2000. With permission from the Company of Biologists.)

mutations if not repaired. Cells have various ways of detecting DNA damage (see Figure 17–60) and can die by apoptosis if they cannot repair it.

Apoptosis Depends on an Intracellular Proteolytic Cascade Mediated by Caspases

A family of specialized intracellular proteases triggers apoptosis by cleaving numerous, but specific, intracellular proteins at specific amino acid sequences, thereby bringing about the dramatic changes that occur during apoptosis. Because these proteases have a cysteine at their active site and cleave their target proteins at specific aspartic acids, they are called **caspases** (*c* for cysteine and *asp* for aspartic acid). Not all caspases are involved in apoptosis: indeed, the first human caspase identified, caspase-1, mainly helps stimulate inflammatory responses by cleaving the precursors of two pro-inflammatory, extracellular signal molecules (cytokines, discussed in Chapter 24). The caspases involved in apoptosis preexist in the cytosol of nearly all our cells as inactive precursors (often called procaspases), which are activated during apoptosis. There are two major classes of these apoptotic caspases: *initiator* caspases and *executioner* caspases.

Initiator caspases, as their name implies, begin the apoptotic program. In mammals, they are mainly *caspase-8* and *caspase-9*. They are made as inactive soluble monomers and are activated only when the monomers dimerize. The dimerization occurs when an apoptotic signal triggers the assembly of a specific *adaptor-protein* complex, which then recruits multiple copies of identical initiator caspase monomers to form larger *activation complexes*, within which the monomers dimerize and become activated. Each monomer in the activated caspase dimer then cleaves its partner at specific sites to form the mature, activated, initiator caspase dimer (**Figure 18–3**).

The major function of initiator caspases is to activate the **executioner caspases**, which orchestrate the apoptosis program. There are three executioner caspases in vertebrates—*caspase-3, caspase-6,* and *caspase-7*. Unlike initiator caspases, executioner caspases normally exist as inactive soluble dimers, which are activated by cleavage, almost always mediated by an initiator caspase (**Figure 18–4**). Each initiator caspase can activate many copies of one or more executioner caspases, resulting in an amplifying *caspase cascade*. Once activated, executioner caspases catalyze the widespread protein cleavage events that are responsible for killing the cell in a characteristic way and preparing it for rapid engulfment and digestion

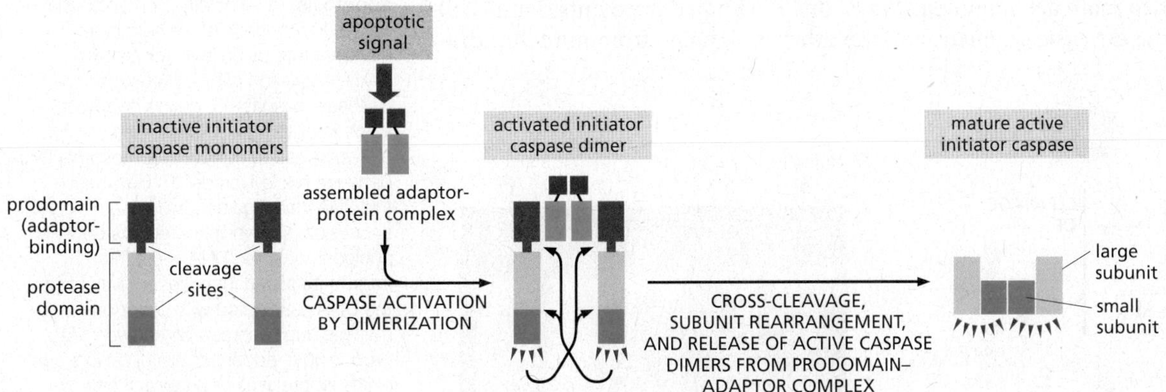

Figure 18–3 Activation of an initiator caspase at the start of apoptosis. An initiator caspase contains a protease domain in its large carboxyl-terminal region and a smaller, adaptor-protein-binding prodomain in its amino-terminal region. The caspase is made as an inactive monomer, which is activated by dimerization. The dimerization and activation only occurs when apoptotic signals trigger the assembly of specific adaptor-protein complexes (shown here as a hypothetical simplest form—a homodimer). Each type of adaptor complex then recruits multiple copies of one type of initiator caspase monomer, allowing them to dimerize and thereby become active. In the case shown, once activated, each monomer in the activated caspase dimer cross-cleaves its partner at specific sites; the cleavage in the protease domain enables the large and small caspase subunits to rearrange, and cleavage between the large protease domain and the prodomain releases the mature, active caspase dimer from the prodomain–adaptor complex.

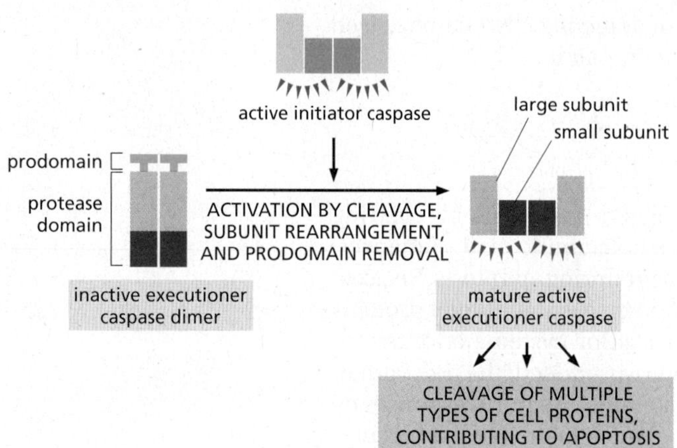

Figure 18–4 Executioner caspase activation during apoptosis. Executioner caspases have very short prodomains, which lack sites for interacting with other proteins. They are initially formed as inactive dimers, which are activated by cleavage at a site in each protease domain, almost always by an initiator caspase. The cleavages allow the large and small subunits to rearrange to form two active protease sites, each of which then cleaves off the prodomain of its partner monomer, to produce the mature, active executioner caspase, as shown. The mature activated caspase then cleaves a variety of cell target proteins, leading to the controlled apoptotic death of the cell.

by neighboring cells. The caspase-initiated proteolytic cascade is not only self-amplifying and destructive but also irreversible, so that once a cell starts out along the path to apoptotic death, it cannot turn back.

Executioner caspases cleave hundreds of different cell proteins during apoptosis, but the roles of these target proteins in apoptosis are known in only a minority of cases, several of which we mention here. The cleavage of nuclear lamins by caspase-6, for example, causes the irreversible breakdown of the nuclear lamina (discussed in Chapter 12). The cleavage by caspase-3 of an inhibitor protein that normally holds a particular DNA-degrading endonuclease in an inactive form frees the endonuclease to cut up the DNA in the cell nucleus during apoptosis (**Figure 18–5**). The cleavage of certain proteins that regulate the actin cytoskeleton results in the actin polymerization in the cell cortex that is responsible for the surface blebbing in apoptosis, mentioned earlier (and see Movie 18.1). The cleavage of other actin regulators and some cell–cell adhesion proteins that attach cells to their neighbors helps an apoptotic cell round up and detach from its neighbors, making it easier for a neighboring cell to engulf it. As we discuss later, the cleavage of two phospholipid transfer proteins in the plasma membrane results in the exposure of phosphatidylserine on the surface of apoptotic cells, where it serves as an "eat me" signal to neighboring phagocytic cells. Importantly, preventing any one of these individual protein-cleavage steps does not stop apoptotic cell death, although it changes its characteristics as expected in each case.

How is an initiator caspase first activated in response to an apoptotic signal? Our cells use two main activation pathways: one is signaled from outside the cell and is called the *extrinsic pathway*, and the other is signaled from mitochondria

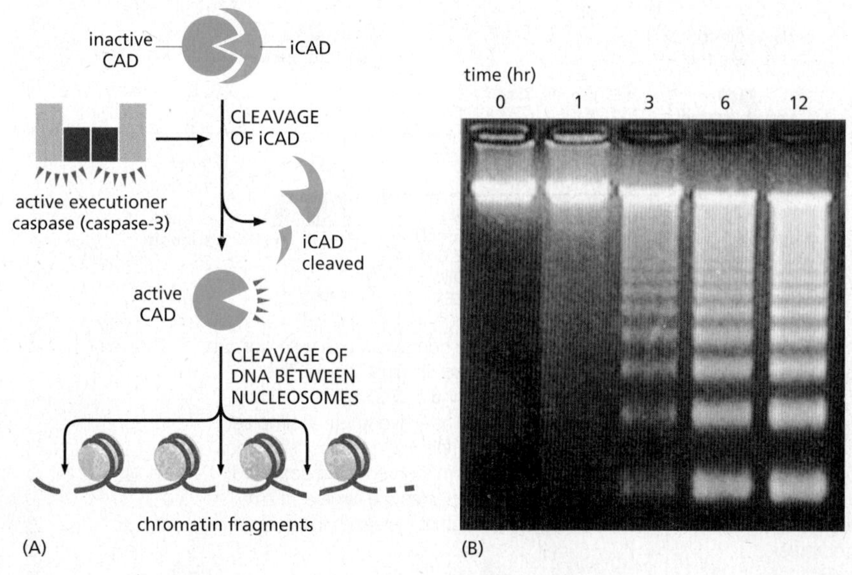

Figure 18–5 DNA fragmentation during apoptosis. (A) In healthy cells, a caspase-activated DNase (CAD) is held in an inactive state by an inhibitor protein, iCAD. Activation of an executioner caspase, caspase-3, during apoptosis leads to cleavage of iCAD, releasing the active DNase to cut the chromosomal DNA between nucleosomes. (B) Because the DNA cleavage occurs only at accessible sites in linker regions between nucleosomes, the DNA is cut into fragments of variable size, equivalent to the DNA associated with either one or multiple nucleosomes (as shown in A), producing a ladder pattern upon DNA gel electrophoresis. The pattern shown was obtained by inducing apoptosis in mouse thymus lymphocytes with dexamethasone, extracting DNA at the times indicated at the top of the gel, separating the fragments by size by electrophoresis in an agarose gel, and staining the DNA in the gel with ethidium bromide. (B, from D. McIlroy et al., *Genes Dev.* 14:549–558, 2000. With permission from Cold Spring Harbor Laboratory Press.)

inside the cell and is called the *intrinsic*, or *mitochondrial, pathway*. Each uses its own initiator caspase and adaptor proteins, as we now discuss.

Activation of Cell-Surface Death Receptors Initiates the Extrinsic Pathway of Apoptosis

Extracellular signal proteins binding to cell-surface **death receptors** trigger the **extrinsic pathway** of apoptosis. Death receptors are transmembrane proteins containing an extracellular ligand-binding domain, a single transmembrane domain, and an intracellular *death domain* that is required for the receptors to activate the apoptotic program. The receptors are homotrimers and belong to the *tumor necrosis factor (TNF) receptor* family, which has eight members, including a receptor for TNF itself and the *Fas* death receptor. The ligands that activate the death receptors are also homotrimers, belonging to the *TNF family* of signal proteins.

A relatively well-understood example of how death receptors trigger the extrinsic pathway of apoptosis is the activation of the **Fas receptor** on the surface of a target cell by **Fas ligand** on the surface of a killer (cytotoxic) lymphocyte (discussed in Chapter 24). Fas signaling has a role in regulating the numbers of T and B lymphocytes, as indicated by the finding that inactivation of Fas signaling results in an abnormal increase in these cells. As shown in **Figure 18–6**, the binding of

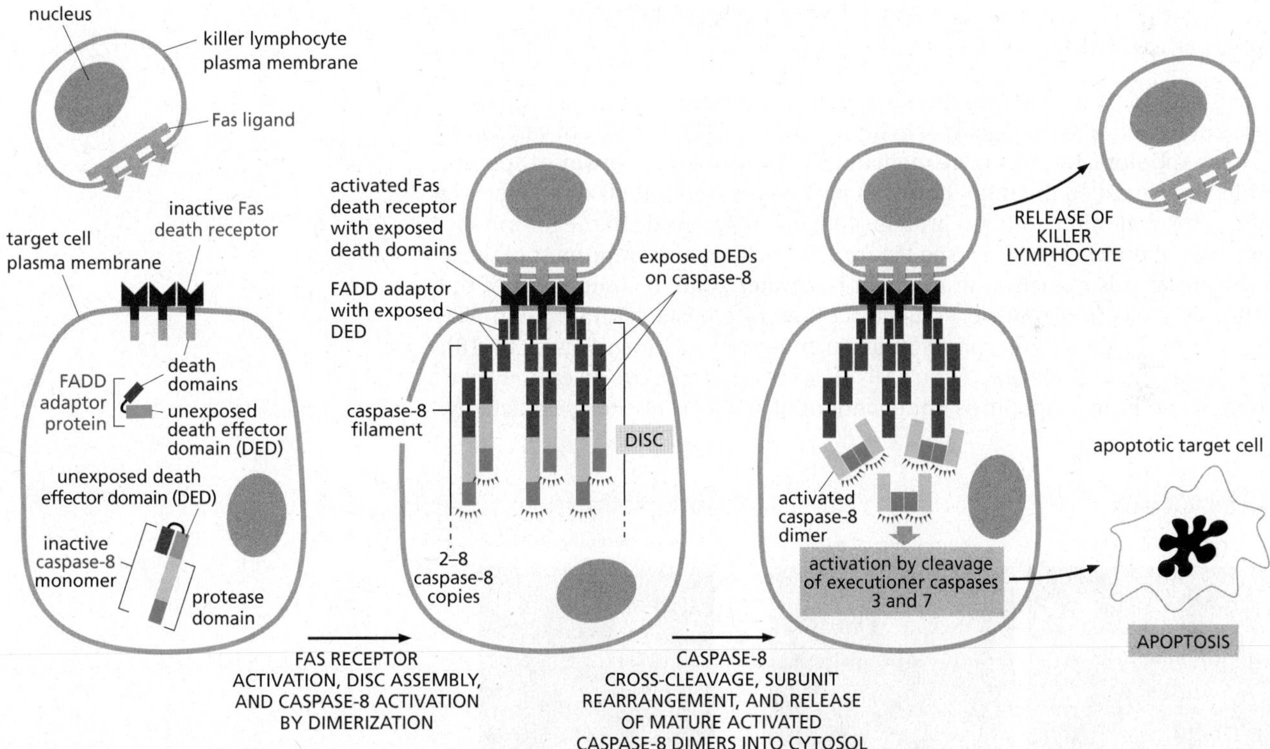

Figure 18–6 The extrinsic pathway of apoptosis activated through Fas death receptors. Trimeric Fas ligands on the surface of a killer lymphocyte bind to trimeric Fas death receptors on the surface of a target cell, inducing the target cell to kill itself by undergoing apoptosis by the extrinsic pathway. Although not shown, at least two trimeric Fas ligands have to bind to and cluster at least two trimeric Fas receptors to activate the pathway; for clarity, only a single copy of the ligand and receptor is shown. The ligand-induced receptor clustering (not shown) exposes a *death domain* on the receptor tails, as indicated here by the change in the color of the domain from *light red* to *dark red* upon exposure. Each exposed death domain binds to a similar exposed death domain on the cytosolic adaptor protein FADD (for Fas-associated death domain). The bound FADD protein then exposes a *death effector domain* (DED; *dark blue*), enabling FADD to recruit an inactive, monomeric initiator caspase (mainly caspase-8) by binding to an exposed DED *(dark blue)* on the prodomain of the caspase. Each caspase-8 monomer has two DEDs, and, when one binds to an exposed DED, the other becomes exposed (as indicated by the change in the DED color from *light blue* to *dark blue*) and can recruit another caspase-8 monomer; this results in a chain reaction in which the caspase-8 monomers oligomerize into a three-dimensional helical filament (not shown), with each FADD protein attached to up to eight caspase-8 molecules. The end result is the assembly of a large death-inducing signaling complex (DISC) composed of multiple copies of Fas, FADD, and caspase-8. Within the DISC, neighboring caspase-8 monomers interact to form activated dimers, which can now cross-cleave their partner monomers (not shown, but see Figure 18–3), cleaving off the prodomain and forming the mature activated dimers, which are released into the cytosol, where they can cleave and activate executioner caspases to induce apoptosis. In human cells, the caspase-10 initiator caspase can also be incorporated into the DISC, but, unlike caspase-8, it is not essential for Fas-induced apoptosis.

trimeric Fas ligands to the trimeric Fas receptors clusters the receptors, exposing death domains on the receptor tails, which then bind and cluster a small intracellular adaptor protein called *FADD*. The clustered FADD proteins, in turn, recruit multiple copies of an inactive, monomeric initiator caspase (mainly caspase-8), which oligomerize, completing the formation of a large **death-inducing signaling complex (DISC)** on the cytoplasmic face of the target-cell plasma membrane. The oligomerization of caspase-8 allows the dimerization and activation of the caspase, which then cleaves itself to form mature, active caspase-8 dimers that can cleave and activate downstream executioner caspases to induce apoptosis.

The sensitivity of different cell types to Fas-induced apoptosis varies. Many cells produce inhibitory proteins that act to restrain the extrinsic apoptotic pathway. Some cells, for example, produce a protein called *FLIP*, which resembles caspase-8 but lacks protease activity because it is missing the key cysteine required in the active site. FLIP dimerizes with caspase-8 in the DISC and prevents it from activating executioner caspases to initiate apoptosis. In this way, FLIP sets an inhibitory threshold that the extrinsic pathway, through activated caspase-8, must overcome to trigger apoptosis. As we discuss later, for the extrinsic pathway to kill some cell types, it has to overcome another caspase inhibitor, which it does by recruiting the intrinsic apoptotic pathway—the pathway we now discuss.

The Intrinsic Pathway of Apoptosis Depends on Proteins Released from Mitochondria

Cells can also activate their apoptosis program from inside the cell, often in response to developmental signals or to injury such as DNA damage. In vertebrate cells, these apoptotic responses are mediated by the **intrinsic pathway** of apoptosis, which is also called the **mitochondrial pathway**, as it depends on the release into the cytosol of mitochondrial proteins that normally reside in the intermembrane space of these organelles (see Figure 12–47). The most important of these released proteins is **cytochrome c**, which is a water-soluble component of the mitochondrial electron-transport chain and therefore has a central role in ATP production by oxidative phosphorylation in mitochondria (see Figure 14–18). When released into the cytosol (**Figure 18–7**), however, it takes on an entirely new function: it can induce apoptosis, independent of its electron-transport activity.

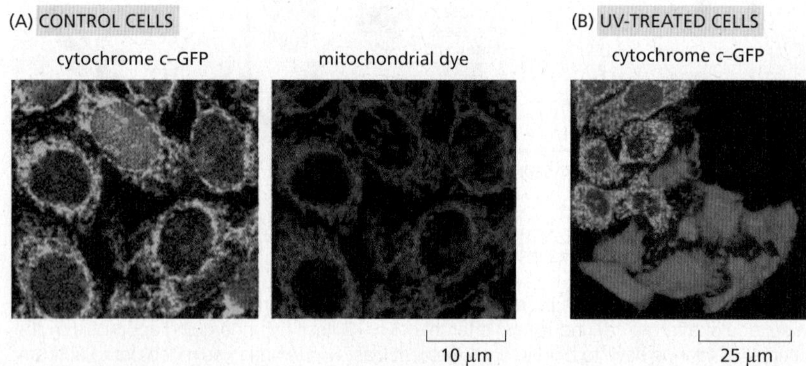

Figure 18–7 **Release of cytochrome *c* from mitochondria into the cytosol during the intrinsic pathway of apoptosis.** These fluorescence micrographs show human cancer cells in culture. The cells were transfected with a gene encoding a fusion protein consisting of cytochrome *c* linked to green fluorescent protein (cytochrome *c*–GFP), and they were also treated with a red dye that accumulates in mitochondria. (A) Unstimulated control cells: the overlapping distribution of the *green* and *red* confirms that the cytochrome *c*–GFP is located in mitochondria. (B) Cells were irradiated with ultraviolet (UV) light to induce the intrinsic pathway of apoptosis and were photographed 5 hours later. The seven cells in the bottom right of this micrograph have released their cytochrome *c* from mitochondria into the cytosol, whereas the other cells in the micrograph have not yet done so. (From J.C. Goldstein et al., *Nat. Cell Biol.* 2:156–162, published 2000 by Nature Publishing Group. Reprinted with permission of SNCSC.)

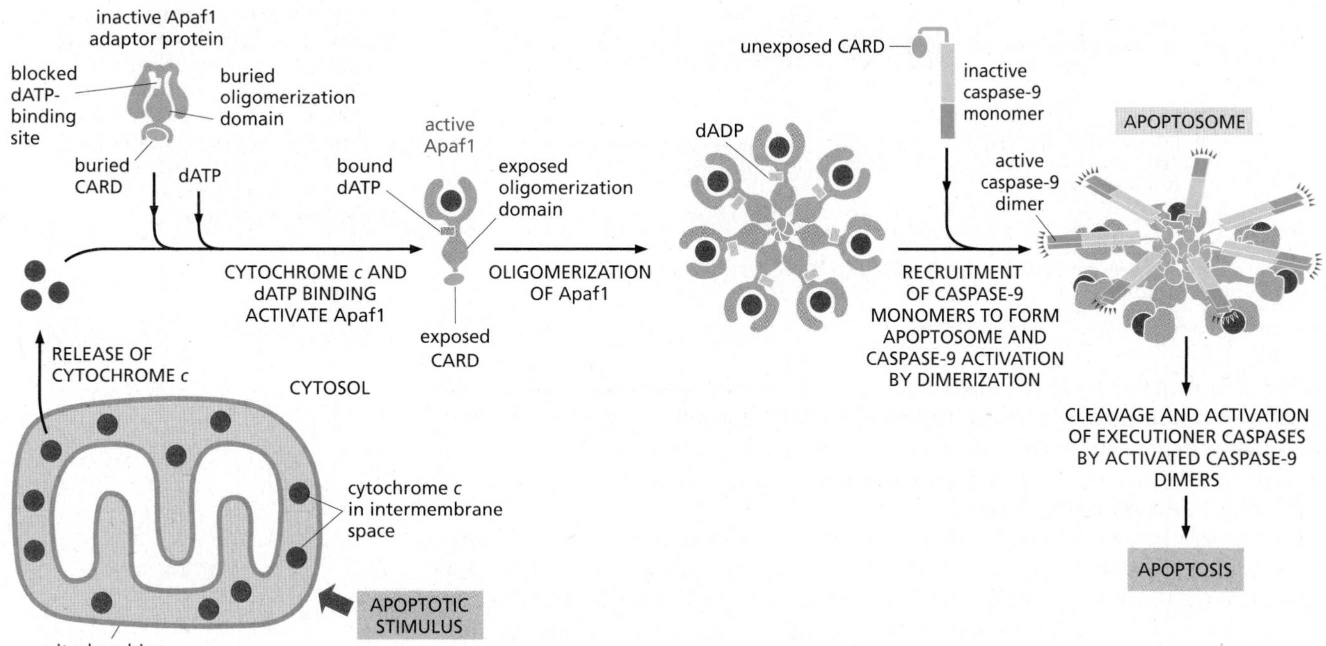

Figure 18–8 **The intrinsic pathway of apoptosis.** Intracellular apoptotic stimuli cause mitochondria to release cytochrome *c*. The binding of cytochrome *c* to the cytosolic adaptor protein Apaf1 induces a conformational change in Apaf1, which activates it, exposing a binding site for deoxy-ATP (dATP), an oligomerization domain, and a caspase recruitment domain (CARD). The exposed oligomerization domain mediates the assembly of Apaf1 into a wheel-like heptamer. Each CARD in the heptamer then recruits an inactive caspase-9 monomer through its own CARD, forming a large apoptosome, with the interacting CARDs clustered above the central hub of the apoptosome. The dATP is hydrolyzed to dADP during the assembly process. Within the apoptosome, the caspase-9 monomers are activated by dimerization. Activated caspase-9 dimers then cleave and activate downstream executioner caspases, leading to apoptosis. Note that the CARD domain is related in structure and function to the death effector domain (DED) of caspase-8 (see Figure 18–6).

(It remains an important mystery how, during evolution, mitochondria and cytochrome *c* came to acquire their surprising role in apoptosis.)

Once released, into the cytosol, cytochrome *c* binds to an adaptor protein called **Apaf1** (*apoptotic protease activating factor 1*), causing the adaptor to bind deoxy-ATP and oligomerize into a large, wheel-like heptamer. The heptamer then recruits inactive initiator caspase-9 monomers into the complex, forming an even larger structure called an **apoptosome** (**Figure 18–8**). Caspase-9 is activated by dimerization within the apoptosome, just as the other major initiator caspase, caspase-8, is activated by dimerization within the DISC (see Figure 18–6). Once activated, caspase-9 cleaves and activates the downstream executioner caspases that mediate apoptosis.

The intrinsic pathway is responsible for the great majority of apoptotic cell deaths in vertebrates. For upstream signals to activate the intrinsic pathway, they have to alter the outer mitochondrial membrane so that the soluble proteins in the intermembrane space such as cytochrome *c* can diffuse into the cytosol. This crucial outer membrane permeabilization step is controlled by interactions between members of the *Bcl2 family* of proteins, as we now discuss.

Bcl2 Proteins Are the Critical Controllers of the Intrinsic Pathway of Apoptosis

The intrinsic pathway is tightly regulated to ensure that cells kill themselves by apoptosis only when it is appropriate. This regulation is largely the function of the **Bcl2 family** of proteins, which are named after the first family member described (B cell lymphoma-2), as explained later. In mammalian cells, complex interactions between these proteins control the permeabilization of the outer mitochondrial membrane and thereby govern the release into the cytosol of cytochrome *c* and

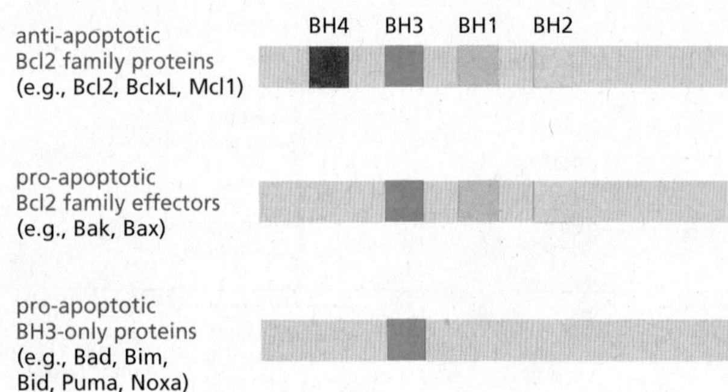

Figure 18–9 Schematic drawing of the BH domains in the three classes of Bcl2 family proteins. Note that the BH3 domain is the only one of the four BH domains shared by all Bcl2 family members; it mediates the direct interactions between pro-apoptotic and anti-apoptotic family members (see Figure 18–10C and D).

other soluble proteins in the intermembrane space, a process called **mitochondrial outer membrane permeabilization (MOMP)**. Like the caspase family, the Bcl2 family of proteins is found in all animals and has been remarkably conserved; a human Bcl2 protein, for example, can suppress apoptosis when expressed in the worm *Caenorhabditis elegans.*

There are three structural and functional classes of mammalian Bcl2 family proteins: (1) *Anti-apoptotic Bcl2 family proteins*, including Bcl2 itself, inhibit apoptosis by preventing MOMP. (2) *Pro-apoptotic Bcl2 family effectors* can directly induce MOMP by creating openings in the outer mitochondrial membrane. (3) A second class of pro-apoptotic Bcl2 family proteins, called *BH3-only proteins* (for reasons explained shortly), promotes apoptosis by regulating the other two classes. The balance between the activities of these three classes largely determines whether MOMP occurs or not and, therefore, whether a mammalian cell lives or dies by the intrinsic pathway of apoptosis.

As illustrated in **Figure 18–9**, the anti-apoptotic Bcl2 family proteins, including Bcl2 itself and *Bcl extra-large* (*BclxL*), share four distinctive *Bcl2 homology (BH) domains* (BH1–4). The two main pro-apoptotic Bcl2 family effectors, *Bak* and *Bax*, are structurally similar to Bcl2 but lack the BH4 domain. The members of the second class of pro-apoptotic Bcl2 family proteins share sequence homology with Bcl2 in only the BH3 domain and are therefore called *BH3-only proteins*; they are by far the largest class of Bcl2 proteins.

When an apoptotic stimulus triggers the intrinsic pathway, the **pro-apoptotic Bcl2 family effectors**, **Bak** and **Bax**, become activated and trigger MOMP by aggregating into oligomers of various sizes in the mitochondrial outer membrane, producing openings in the membrane by an uncertain mechanism, allowing cytochrome *c* and other intermembrane proteins to escape into the cytosol (**Figure 18–10A and B**). At least one of these pro-apoptotic effectors is required for the intrinsic pathway of apoptosis to operate in mammalian cells: mutant mouse cells that lack both proteins do not undergo MOMP or engage the intrinsic apoptotic pathway. Whereas Bak is bound to the mitochondrial outer membrane even in the absence of an apoptotic signal (see Figure 18–10A and B), Bax is mainly located in the cytosol until an apoptotic signal activates it, causing it to relocate to the outer membrane, where it oligomerizes. As we discuss below, the activation of Bak and Bax usually depends on activated BH3-only proteins.

The **anti-apoptotic Bcl2 family proteins** such as **Bcl2** and **BclxL** are also located on the cytosolic surface of the outer mitochondrial membrane, where they help prevent inappropriate MOMP. They do so by binding to the BH3 domain of active pro-apoptotic effectors Bak and Bax, thereby preventing their oligomerization (**Figure 18–10C**). There are at least five mammalian anti-apoptotic Bcl2 family proteins, and every mammalian cell requires at least one to avoid apoptosis and therefore to survive. Moreover, a number of these anti-apoptotic proteins must be inhibited for the intrinsic pathway to induce apoptosis, and BH3-only proteins mediate this inhibition.

The **BH3-only proteins**, the largest subclass of Bcl2 family proteins, promote MOMP and thereby apoptosis when they are either produced or activated in cells in response to an apoptotic stimulus. They do so in at least two ways. (1) Some BH3-only proteins, including Bad (see Figure 18–9), inhibit certain

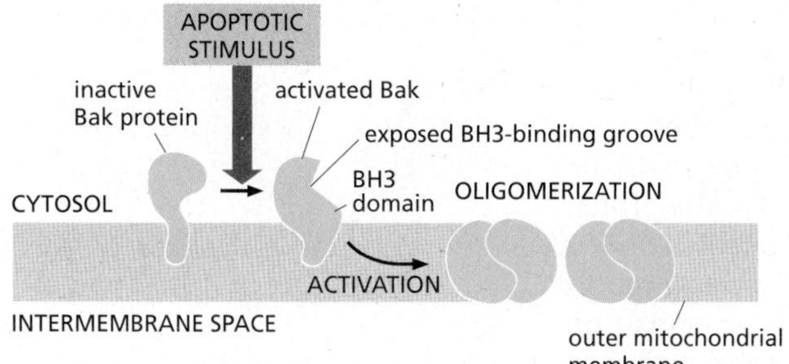

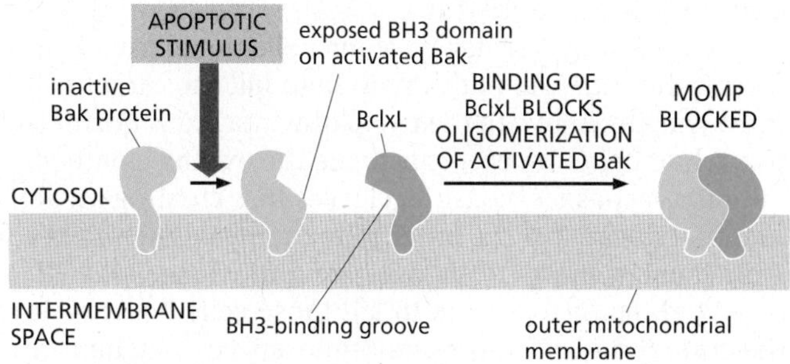

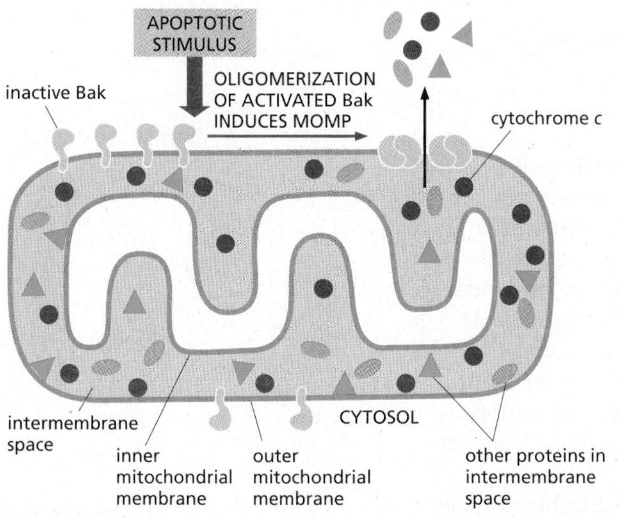

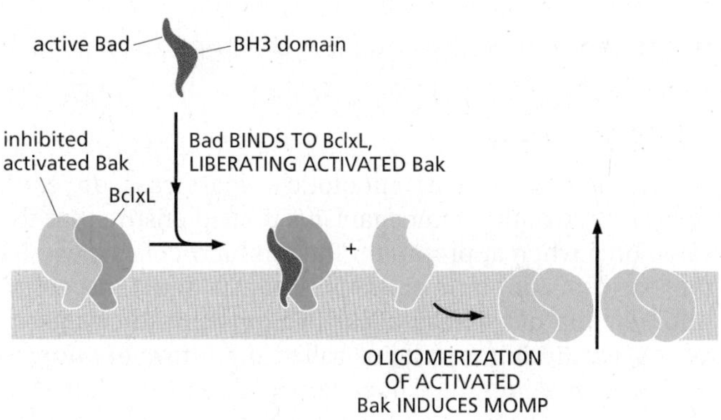

Figure 18–10 How pro-apoptotic Bcl2 family effectors induce MOMP and how anti-apoptotic Bcl2 family proteins block it. (A) Most of the pro-apoptotic effector Bak is already attached to the outer mitochondrial membrane before the protein is activated. When activated by an apoptotic stimulus, the protein undergoes a conformational change that both exposes a BH3 domain and creates a BH3-binding groove, allowing Bak–Bak oligomerization in the outer membrane. (B) The Bak oligomers induce MOMP by creating openings in the outer membrane that allow cytochrome *c* and other soluble proteins in the intermembrane space to diffuse into the cytosol; although the details of how Bak (and Bax) oligomers produce the openings are uncertain, it is thought that they form large ring structures that disrupt the integrity of the outer membrane. Once released into the cytosol, cytochrome *c* stimulates the assembly of apoptosomes (see Figure 18–8). (C) The anti-apoptotic Bcl2 family protein BclxL, like Bak, is normally bound to the outer mitochondrial membrane, where it can interact via its BH3-binding groove to the exposed BH3 domain on activated Bak, thereby blocking Bak–Bak oligomerization, MOMP, and apoptosis. (D) One way BH3-only proteins such as Bad are thought to indirectly induce MOMP and apoptosis is by inhibiting certain anti-apoptotic Bcl2 family proteins such as BclxL.

anti-apoptotic Bcl2 family proteins by binding via their BH3 domain to the BH3-binding groove on the anti-apoptotic protein. This binding blocks the anti-apoptotic activity of the Bcl2 family protein, thereby allowing Bax and/or Bak to oligomerize in the outer mitochondrial membrane to trigger MOMP (**Figure 18–10D**). (2) Some BH3-only proteins, including Bim and Bid (see Figure 18–9), can directly bind to and activate Bak and Bax, stimulating them to oligomerize and trigger MOMP.

In these ways, BH3-only proteins provide the crucial link between apoptotic stimuli and the intrinsic pathway of apoptosis, with different stimuli activating or inducing the production of different BH3-only proteins. When a cell suffers DNA damage that it cannot repair, for example, the tumor suppressor protein p53 (discussed in Chapters 17 and 20) accumulates in the nucleus and activates the transcription of genes that encode the BH3-only proteins *Puma* and *Noxa* (see Figure 18–9), which then trigger MOMP and apoptosis, thereby eliminating a potentially dangerous cell that could become cancerous. As we will see, some extracellular survival signals promote a cell's survival by preventing it from undergoing apoptosis by inhibiting the synthesis or activity of certain BH3-only proteins (see Figure 18–13B).

As mentioned earlier, in some cells the extrinsic apoptotic pathway has to recruit the intrinsic pathway to help kill the target cell; the BH3-only protein *Bid* is the link between the two pathways. Bid is normally inactive in these cells, but, when death receptors activate the extrinsic pathway (see Figure 18–6), the initiator caspase, caspase-8, cleaves Bid, creating an active form of the protein that translocates to the outer mitochondrial membrane to activate Bax and/or Bak, causing them to oligomerize and trigger MOMP. MOMP is required for activated death receptors to kill these cells because it releases proteins from the mitochondrial intermembrane space that neutralize an inhibitory protein present in the cytosol of these cells that normally blocks apoptosis, as we explain next.

An Inhibitor of Apoptosis (an IAP) and Two Anti-IAP Proteins Help Control Caspase Activation in the Cytosol of Some Mammalian Cells

Because activation of an apoptotic caspase cascade leads to certain cell death, cells employ multiple mechanisms to help ensure that these proteases are activated only when appropriate. The inhibition of caspase-8 by the FLIP protein in the extrinsic apoptotic pathway is one example discussed earlier (see p. 1095). Another line of defense against inappropriate caspase activation is provided by caspase inhibitor proteins called **inhibitors of apoptosis (IAPs)**. These proteins were first identified in certain insect viruses (baculoviruses), which encode IAP proteins to prevent a host cell that is infected by the virus from activating caspases and killing itself by apoptosis, which would curtail the virus's replication. Most animal cells also make IAP proteins, although most of these proteins do not regulate caspases and apoptosis.

Mammalian cells seem to have only one IAP that can directly inhibit caspase activity. It is called **XIAP** (because it is encoded on the X chromosome), and it resides in the cytosol of many of our cells. It binds to and inhibits caspase-9, an initiator caspase, and the executioner caspases, caspase-3 and caspase-7, thereby setting an inhibitory threshold that these caspases must overcome to trigger apoptosis. In these cells, XIAP also helps regulate the levels of the three caspase proteins: it has a ubiquitin-ligase domain that polyubiquitylates the caspases that XIAP binds to, marking them for destruction in proteasomes (see Figure 3–67).

When the intrinsic pathway of apoptosis is activated, among the proteins released from the mitochondrial intermembrane space by MOMP are two **anti-IAP proteins** called *Smac* and *Omi*. In cells with XIAP in their cytosol, these anti-IAP proteins bind to XIAP and prevent it from inhibiting caspases, thereby promoting caspase activation and apoptosis (**Figure 18–11**). This is why the extrinsic apoptotic pathway needs to recruit the intrinsic pathway (involving MOMP and caspases 9, 3, and 7) to induce apoptosis in some cells, as mentioned earlier; these are cells that express XIAP in their cytosol.

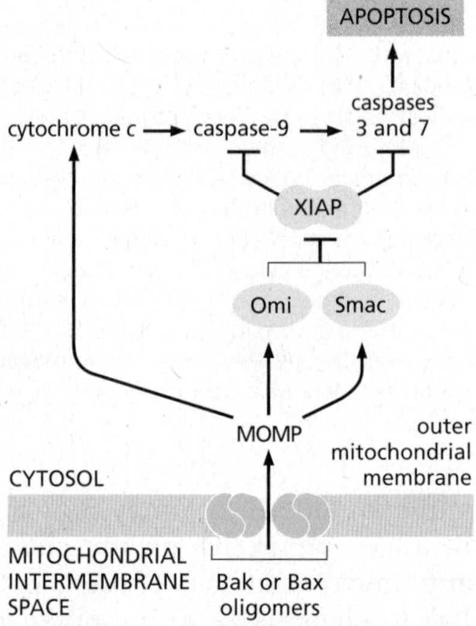

Figure 18–11 How MOMP overcomes XIAP inhibition. In some mammalian cells, the presence of XIAP in the cytosol inhibits initiator caspase-9 and executioner caspases 3 and 7; the XIAP binds to the active site of the three enzymes, blocking their activities. However, as well as releasing cytochrome *c*, MOMP releases two anti-IAP proteins, Smac and Omi, which inhibit the anti-caspase activity of XIAP, thereby allowing the activation of these caspases in the cytosol during the intrinsic pathway of apoptosis (see Figure 18–8).

Extracellular Survival Factors Inhibit Apoptosis in Various Ways

As discussed in Chapters 15 and 21, extracellular signals regulate most activities of animal cells, including apoptosis. They are part of the normal "social" controls that ensure individual cells behave for the good of the organism as a whole—in this case, by surviving when the cells are needed and killing themselves when they are not. Some extracellular signal molecules stimulate apoptosis, whereas others inhibit it. We have discussed signal proteins such as Fas ligand that activate death receptors to trigger the extrinsic pathway of apoptosis. Other extracellular signal molecules that stimulate apoptosis are especially important during vertebrate development: a surge of thyroid hormone in the bloodstream, for example, signals cells in the tadpole tail to undergo apoptosis at metamorphosis. In mice, locally produced signal proteins stimulate cells between developing fingers and

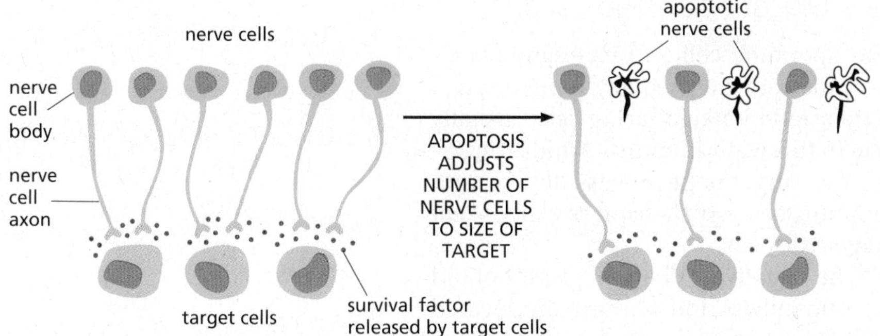

Figure 18–12 How survival factors and apoptosis can help adjust the number of developing nerve cells to the size of the target tissue the nerve cells innervate. In this example, more nerve cells are produced than can be supported by the limited amount of survival factors produced by the cells in a target tissue. As a result, some nerve cells receive an insufficient amount of survival factor to avoid apoptosis. This strategy of overproduction followed by culling during development helps ensure that all the appropriate target cells are contacted by appropriate nerve cells and that surplus nerve cells are automatically eliminated.

toes to kill themselves, thereby sculpting these digits (see Figure 18–2). Here, however, we focus on extracellular signal molecules that promote cell survival by inhibiting apoptosis and are collectively called **survival factors**.

Most animal cells require such signals from other cells to avoid undergoing apoptosis, usually by the intrinsic pathway. This surprising arrangement apparently helps ensure that cells survive only when and where they are needed. Some nerve cells, for example, are produced in excess in the developing nervous system and then compete for limited amounts of survival factors that are secreted by the target cells they normally connect to. Nerve cells that receive enough survival signals live, while the others die by apoptosis. In this way, the number of surviving neurons is automatically adjusted so that it is appropriate for the number of target cells they connect with (**Figure 18–12**). A similar competition for limited amounts of survival factors produced locally or systemically might help control cell numbers in other tissues, both during development and in adulthood.

Survival factors usually bind to cell-surface receptors, which activate intracellular signaling pathways (discussed in Chapter 15) that suppress the apoptotic program, usually by regulating the expression or activity of members of the Bcl2 family of proteins. Some survival factors, for example, stimulate the synthesis of anti-apoptotic Bcl2 family proteins such as Bcl2 itself or BclxL (**Figure 18–13A**). Some others act by inhibiting the function of pro-apoptotic BH3-only proteins such as Bad (**Figure 18–13B**). Some developing neurons, like those illustrated in Figure 18–12, use a counterintuitive alternative strategy, in which survival-factor receptors stimulate apoptosis when they are unoccupied and stop doing so when survival factors bind. The end result in all these cases is the same: cell survival depends on survival-factor binding to cell-surface receptors.

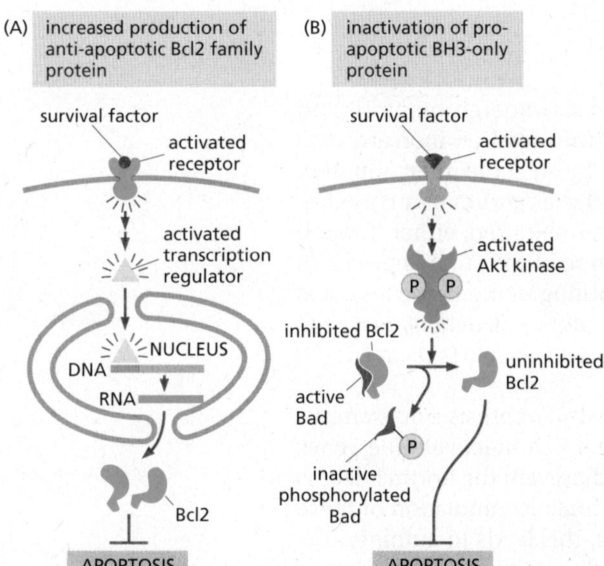

Figure 18–13 Two of the various ways that extracellular survival factors can inhibit apoptosis. (A) Some survival factors suppress apoptosis by stimulating the transcription of genes that encode anti-apoptotic Bcl2 family proteins such as Bcl2 (as shown here) or BclxL. (B) Many others activate the serine/threonine protein kinase Akt, which, among many other targets, phosphorylates and inactivates the pro-apoptotic BH3-only protein Bad (see Figure 18–9). When not phosphorylated, Bad promotes apoptosis by binding to and inhibiting an anti-apoptotic Bcl2 family protein such as Bcl2 itself. Once phosphorylated, Bad dissociates, freeing Bcl2 to suppress apoptosis (see Figure 18–10C and Figure 15–54). Akt can also suppress apoptosis by phosphorylating and inactivating transcription regulatory proteins that stimulate the transcription of genes encoding proteins that promote apoptosis, such as the BH3-only protein Bim (not shown). There are also many other ways that survival factors can inhibit apoptosis that are not illustrated here.

Healthy Neighbors Phagocytose and Digest Apoptotic Cells

Apoptotic cell death is a tidy process. The apoptotic cell and its fragments do not break open and release their contents. Instead, they usually remain intact until they are rapidly phagocytosed and digested by neighboring cells (usually macrophages in humans), leaving no trace. In this way, apoptosis avoids triggering a destructive inflammatory response. The engulfment process depends on chemical changes on the surface of the apoptotic cell, which displays various "eat me" signals that are recognized by the phagocytic cells.

The most important of these signals is the negatively charged phospholipid **phosphatidylserine**. In healthy cells, phosphatidylserine is normally located exclusively in the inner leaflet of the lipid bilayer of the plasma membrane (see Figure 10–15); it is kept there by a specific *phospholipid flippase* that uses ATP hydrolysis to flip phosphatidylserine (PS) and phosphatidylethanolamine (PE) from the outer to the inner leaflet. In cells undergoing apoptosis, however, PS accumulates on the cell surface by two mechanisms. First, executioner caspases cleave the phospholipid flippase, thereby inactivating it, preventing PS and PE from being transferred from the outer to the inner leaflet. Second, executioner caspases cleave and thereby activate a *phospholipid scramblase* that transfers plasma membrane phospholipids nonspecifically between the inner and outer lipid leaflets, thereby scrambling them between the two leaflets. These two mechanisms responsible for the exposure of PS on the surface of apoptotic cells are illustrated in **Figure 18–14** and **Movie 18.2**. The appearance of PS on the cell surface also occurs in some forms of necrotic cell death, even in the absence of caspase activation; disruption of the plasma membrane is sufficient to do it.

The PS on apoptotic and necrotic cells is recognized by a variety of soluble "bridging" proteins that interact with both the exposed PS and specific receptors on the surface of a neighboring phagocytic cell, triggering the cytoskeletal and other changes that initiate the engulfment process.

Macrophages are professional phagocytes and will phagocytose dead cells, microbes, cell debris, latex beads, and almost any other particles, but they do not phagocytose healthy cells within the organism's own body—even those healthy cells that transiently expose phosphatidylserine on their surface when activated, such as platelets and some T lymphocytes. One reason is that almost all of our cells display signal proteins on their surface that bind to inhibitory receptors on the surface of macrophages, stimulating the receptors to block phagocytosis. Thus, in addition to expressing cell-surface "eat me" signals such as phosphatidylserine that stimulate phagocytosis, apoptotic cells must also remove or inactivate the "don't eat me" signals that block phagocytosis.

Either Excessive or Insufficient Apoptosis Can Contribute to Disease

There are many human disorders in which too many cells undergo apoptosis and thereby contribute to pathological tissue loss. Among the most common and dramatic examples are heart attacks and strokes caused by an acute interruption of the blood supply to the heart or brain, respectively. In these conditions, many cells initially die by necrosis. If the blocked blood vessel is unblocked, either through an arterial catheter or with clot-disrupting drugs, some of the surviving cells in the oxygen-deprived tissue die by apoptosis, contributing to the tissue loss. It is hoped that, in the future, drugs that can block the apoptotic cell deaths, as well as drugs that can block some forms of necrotic cell death, will be able to decrease the tissue loss and its debilitating consequences.

There are other disorders in which too few cells die by apoptosis. For example, as mentioned earlier, mutations in mice and humans that inactivate the genes that encode either the Fas death receptor or its ligand prevent the normal deaths of some types of lymphocytes, resulting in the abnormal accumulation of these cells in the spleen and lymph glands. In many cases, this leads to autoimmune disorders, because some self-reactive lymphocytes fail to be eliminated and react

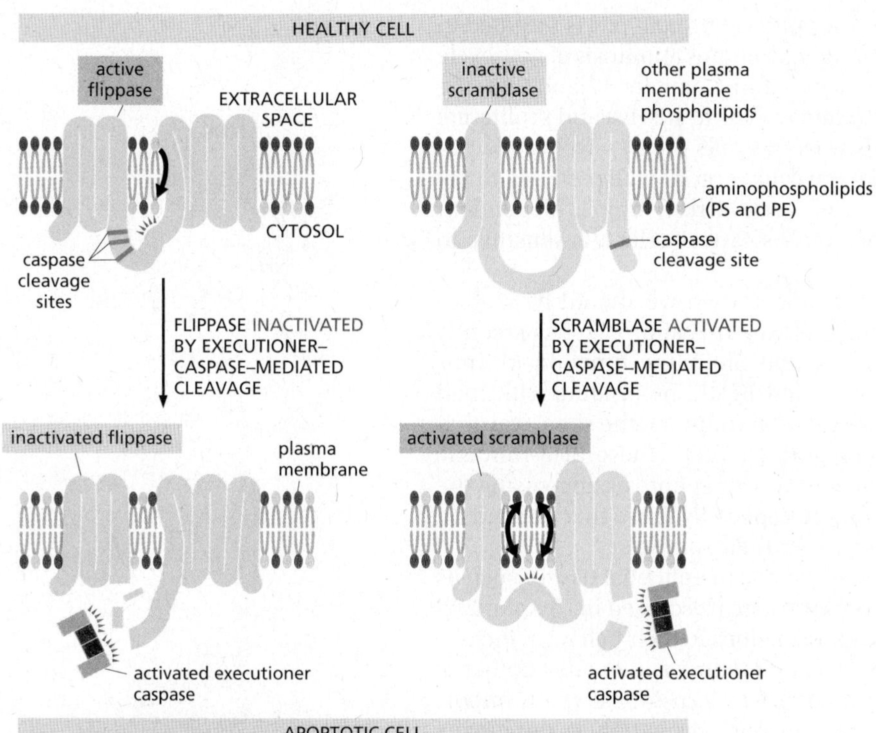

Figure 18–14 The two caspase-dependent mechanisms responsible for the accumulation of phosphatidylserine on the surface of apoptotic cells. In healthy cells *(top half of figure)*, an ATP-dependent aminophospholipid flippase (ATP11C) actively flips phosphatidylserine (PS) and phosphatidylethanolamine (PE) from the outer to the inner leaflet of the plasma membrane lipid bilayer, keeping these lipids mainly confined to the inner leaflet. During apoptosis *(bottom half of figure)*, activated executioner caspases (caspase-3, caspase-7, or both) cleave and thereby *inactivate* the ATP11C flippase, preventing its lipid translocation activity; at the same time, the activated executioner caspases cleave and thereby *activate* a phospholipid scramblase (Xkr8) in the plasma membrane that nonspecifically flips phospholipids between the two lipid leaflets of the membrane, scrambling them between the leaflets. (Although not shown, the ATP11C flippase is tightly associated with a smaller transmembrane protein, CDC50, which is required to chaperone the flippase to the plasma membrane and possibly to assist in the lipid translocation process.)

With the flippase permanently inactivated and the scramblase permanently activated, PS and PE become rapidly and irreversibly exposed on the apoptotic cell surface, where the PS serves as an "eat me" signal to neighboring phagocytic cells. Both of these caspase-dependent mechanisms are required to create this rapid and effective "eat me" signal: if only the flippase were inactivated, it could take a very long time for PS to accumulate in effective amounts on the apoptotic cell surface, because phospholipids rarely flip spontaneously between the two leaflets of a lipid bilayer without an enzyme (such as the scramblase) catalyzing it (see Figure 10–10B). Similarly, if only the scramblase were activated, the active flippase would rapidly return any scrambled PS from the external leaflet to the internal one, thereby removing the PS from the cell surface. Figure 12–39 illustrates the actions of phospholipid scramblases in lipid bilayer synthesis in the ER membrane and of phospholipid scramblases in generating lipid bilayer asymmetry in some other intraceullar membranes. (Modified from S. Nagata et al., *Cell Death Differ.* 23:952–961, 2016.)

against the individual's own tissues. The increase in lymphocytes can also lead to lymphocyte cancers called lymphomas.

Indeed, decreased apoptosis makes an important contribution to many types of cancer, because the normal inhibitory controls on apoptosis are often defective in cancer cells. The *Bcl2* gene, for example, was first identified in a common form of human lymphoma, in which a chromosome translocation causes excessive production of the anti-apoptotic Bcl2 protein (as mentioned earlier, Bcl2 gets its name from this *B cell lymphoma*). The increased amount of Bcl2 protein in the lymphocytes that carry the translocation promotes the development of cancer by inhibiting apoptosis, thereby abnormally prolonging the cells' survival and increasing their number; the increase in Bcl2 also decreases the cells' sensitivity to anticancer drugs, which often work by causing cancer cells to undergo apoptosis (discussed in Chapter 20).

Similarly, the gene encoding the tumor suppressor protein p53 is mutated in about 50% of human cancers so that it no longer promotes apoptosis or cell-cycle arrest in response to DNA damage (as discussed in Chapters 17 and 20). The lack of p53 function, therefore, enables the cancer cells to survive and proliferate even when their DNA is damaged; in this way, the cells progressively accumulate more mutations, some of which make the cancer more malignant. As many anticancer drugs induce apoptosis (and cell-cycle arrest) by p53-dependent mechanisms, the loss of p53 function also makes cancer cells less sensitive to these drugs.

If decreased apoptosis contributes to a cancer, then we should be able to treat the cancer with drugs that promote apoptosis. This approach has recently led to the development of small chemicals that block the function of anti-apoptotic Bcl2 family proteins such as Bcl2 and BclxL, by binding with high affinity to the BH3-binding groove of these proteins, in much the same way that pro-apoptotic BH3-only proteins do (see Figure 18–10D). These BH3-mimetic drugs (**Figure 18–15**) activate the intrinsic pathway of apoptosis, increasing the amount of tumor cell death in certain cancers, especially those that depend on particular anti-apoptotic Bcl2 family proteins for their survival.

Most human cancers are carcinomas, which arise in epithelial tissues such as those in the lung, intestinal tract, breast, and prostate (discussed in Chapter 20). Such epithelial cancer cells display many abnormalities in their behavior, including a decreased ability to adhere to the extracellular matrix and to one another at specialized cell–cell junctions. In the next chapter, we discuss these vitally important structures, which are responsible for holding our cells in tissues and organs in their proper place.

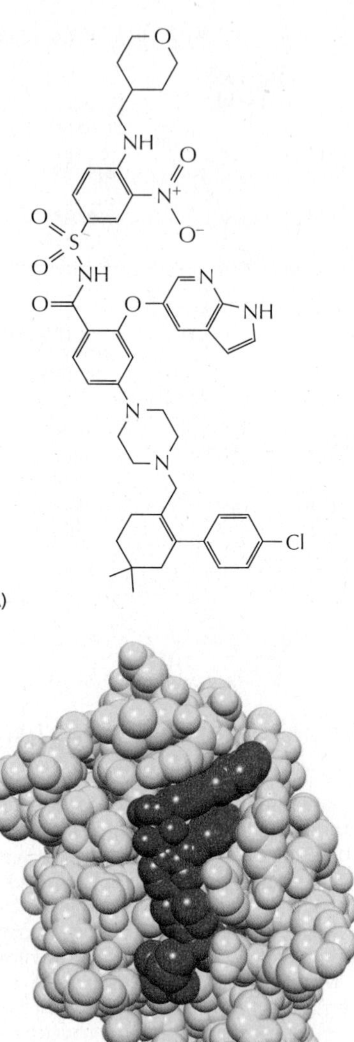

Figure 18–15 A BH3-mimetic drug that specifically inhibits the Bcl2 anti-apoptotic Bcl2 family protein. As shown in Figure 18–10D, one way BH3-only proteins can promote apoptosis is by binding to the long BH3-binding groove in anti-apoptotic Bcl2 family proteins such as Bcl2 itself, thereby preventing the protein from blocking apoptosis. (A) The chemical structure of *venetoclax*, which was designed and synthesized to bind tightly and specifically in the BH3-binding groove of Bcl2. (B) Crystal structure of venetoclax *(red)* bound to the human Bcl2 protein *(yellow)*. By inhibiting the activity of Bcl2, the drug promotes apoptosis in any cell that depends on this protein for survival, as is the case for cells of human chronic lymphocytic leukemia, for which this drug is in clinical use. (PDB code: 6O0K.)

Summary

Animal cells can activate various intracellular death programs and kill themselves when they are seriously damaged or stressed, no longer needed, or are a threat to the organism. In most cases, these deaths occur by apoptosis, in which the cells shrink, the nucleus and cells condense and often fragment, and neighboring phagocytic cells rapidly engulf the cells or fragments before there is any leakage of cytoplasmic contents. Apoptosis is mediated by proteolytic enzymes called caspases, which cleave specific intracellular proteins to kill the cell quickly and neatly.

Apoptotic caspases are present as inactive precursors in almost all nucleated animal cells. The activation of initiator caspases occurs when an apoptotic stimulus activates adaptor proteins, which bring inactive initiator caspase monomers into proximity within large activation complexes, in which the monomers are activated by dimerization. The activated initiator caspase dimers cleave themselves and then activate downstream executioner caspase dimers by cleaving them; the activated executioner caspases then cleave hundreds of target proteins in the cell. The amplifying, irreversible caspase cascade is responsible for all the events of apoptosis, including those that collectively kill the cell and prepare it for being phagocytosed and rapidly digested by a neighboring cell.

Cells use two distinct pathways to activate initiator caspases to trigger apoptosis: the extrinsic pathway is activated by extracellular ligands binding to cell-surface death receptors; the intrinsic pathway is activated from within the cell by developmental signals or stress signals. Each pathway uses its own initiator caspases, adaptor proteins, and activation complexes. In the extrinsic pathway, the death receptors recruit caspase-8 via adaptor proteins to form the activation complex called the DISC. In the intrinsic pathway, intracellular signals induce mitochondrial outer membrane permeabilization (MOMP), which releases soluble proteins from the mitochondrial intermembrane space into the cytosol; released cytochrome c activates the adaptor protein Apaf1, which recruits caspase-9 monomers to form a large activation complex called an apoptosome. Anti-apoptotic and pro-apoptotic intracellular Bcl2 family proteins interact with one another to tightly control MOMP to ensure that the intrinsic pathway of apoptosis is normally only activated when the death of the cell benefits the animal.

PROBLEMS

Which statements are true? Explain why or why not.

18–1 In normal adult tissues, cell death usually balances cell division.

18–2 Mammalian cells that do not have cytochrome *c* should be resistant to apoptosis induced by DNA damage.

Discuss the following problems.

18–3 Fas ligand is a homotrimeric plasma membrane protein on killer lymphocytes that binds to a homotrimeric death receptor, Fas, on the surface of target cells, including some lymphocytes (**Figure Q18–1**). The clustering of trimeric Fas by the binding of clusters of Fas ligand alters the conformation of Fas so that it binds an adaptor protein, which then recruits and activates caspase-8, triggering a caspase cascade that leads to apoptotic cell death.

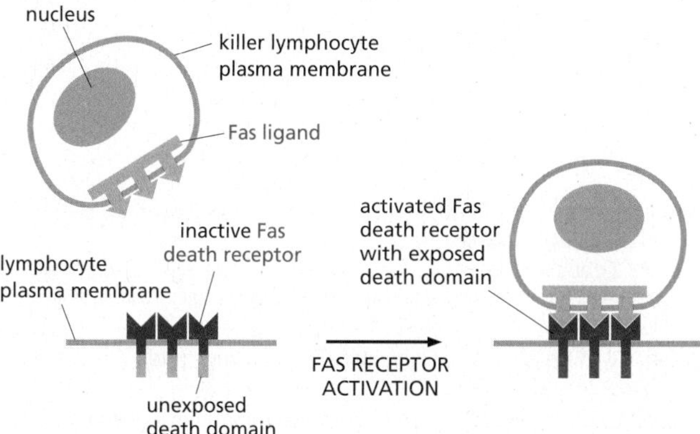

Figure Q18–1 The binding of trimeric Fas ligand to Fas (Problem 18–3). Although not shown, at least two trimeric Fas ligands have to bind to and cluster at least two trimeric Fas receptors to activate the pathway; for clarity, only a single copy of the ligand and receptor is shown.

In humans, the autoimmune lymphoproliferative syndrome (ALPS) is associated with dominant mutations in Fas that include point mutations and C-terminal truncations. In individuals that are heterozygous for such mutations, some lymphocytes do not die at their normal rate and accumulate in abnormally large numbers, causing a variety of clinical problems. In contrast to these patients, individuals heterozygous for mutations that eliminate Fas expression entirely do not have these clinical problems.

Assuming that the normal and dominant forms of Fas are expressed to the same level and assemble randomly into trimers, what fraction of Fas–Fas ligand complexes on a lymphocyte from a heterozygous ALPS patient would be expected to be composed entirely of normal Fas subunits? Does your calculation suggest an explanation for why individuals heterozygous for expressed Fas mutants have clinical problems, whereas heterozygous individuals with unexpressed Fas mutants have no clinical problems?

18–4 In contrast to their similar brain abnormalities, newborn mice deficient in Apaf1 or caspase-9 have distinctive abnormalities in their paws. Apaf1-deficient mice fail to eliminate the webs between their developing digits, whereas caspase-9–deficient mice have normally formed digits (**Figure Q18–2**). If Apaf1 and caspase-9 function in the same apoptotic pathway, how is it possible for these deficient mice to differ in web-cell apoptosis?

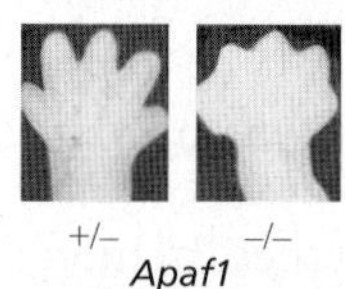

Figure Q18–2 Appearance of paws in $Apaf1^{-/-}$ and $Casp9^{-/-}$ newborn mice relative to normal newborn mice (Problem 18–4). (From H. Yoshida et al., *Cell* 94:739–750, 1998. With permission from Elsevier.)

18–5 When human cancer cells are exposed to ultraviolet (UV) light at 90 mJ/cm^2, most of the cells undergo apoptosis within 24 hours. Release of cytochrome *c* from mitochondria can be detected as early as 6 hours after exposure of a population of such cells to UV light, and it continues to increase for more than 10 hours thereafter. Does this mean that individual cells slowly release their cytochrome *c* over this time period? Or, alternatively, do individual cells release their cytochrome *c* rapidly but with different cells being triggered at variable times?

To answer this fundamental question, you have fused the gene for green fluorescent protein (GFP) to the gene for cytochrome *c*, so that you can observe the behavior of individual cells by confocal fluorescence microscopy. In cells that are expressing the cytochrome *c*–GFP fusion protein, fluorescence shows the punctate pattern typical of mitochondrial proteins. You then irradiate these cells with UV light and observe individual cells for changes in the punctate pattern. Two such cells (outlined in white) are shown in **Figure Q18–3A and B**. Release of

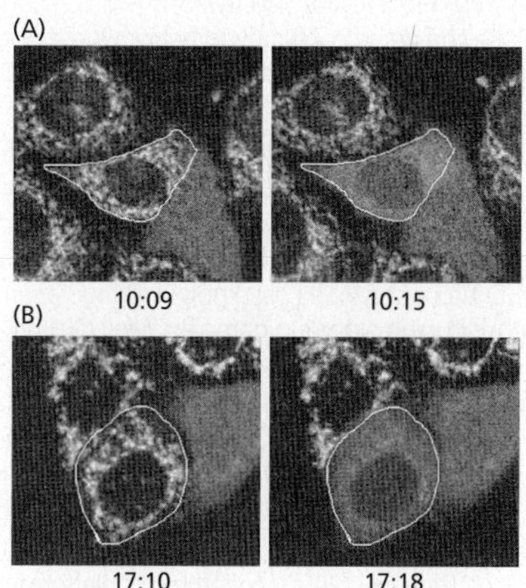

Figure Q18–3 Analysis of cytochrome *c*–GFP release from mitochondria of individual cells by time-lapse video fluorescence microscopy (Problem 18–5). (A) Cells observed for 6 minutes, 10 hours after UV irradiation. (B) Cells observed for 8 minutes, 17 hours after UV irradiation. One cell in A and one in B, each *outlined in white*, have released their cytochrome *c*–GFP during the time frame of the observation, which is shown as hours:minutes below each panel. (From J.C. Goldstein et al., *Nat. Cell Biol.* 2:156–162, published 2000 by Nature Publishing Group. Reproduced with permission from SNCSC.)

cytochrome *c*–GFP is detected as a change from a punctate to a diffuse pattern of fluorescence characteristic of a cytosolic protein. Times after UV exposure are indicated as hours:minutes below the individual panels.

Which model for cytochrome *c* release do these observations support? Explain your reasoning.

18–6 Imagine that you could microinject cytochrome *c* into the cytosol of wild-type mammalian cells, or into the cytosol of cells that were defective for the process of mitochondrial outer membrane permeabilization (MOMP). Would you expect the injected wild-type cells or the injected MOMP-defective cells to undergo apoptosis? Explain your reasoning.

18–7 Which one of the following statements about Bcl2 family members is correct?

A. Bak is pro-apoptotic and Bax is anti-apoptotic.
B. Bax is pro-apoptotic and BclxL is anti-apoptotic.
C. Bcl2 is pro-apoptotic and Bak is anti-apoptotic.
D. BclxL is pro-apoptotic and Bcl2 is anti-apoptotic.

18–8 One important role of Fas and Fas ligand is to mediate the elimination of tumor cells by killer lymphocytes. In a study of 35 primary lung and colon tumors, half the tumors were found to have amplified and overexpressed a gene for a secreted protein that binds to Fas ligand. How do you suppose that overexpression of this protein might contribute to the survival of these tumor cells? Explain your reasoning.

REFERENCES

Adams JM & Cory S (2017) The BCL-2 arbiters of apoptosis and their growing role as cancer targets. *Cell Death Differ.* 25, 27–36.

Birkinshaw RW & Czabotar PE (2017) The BCL-2 family of proteins and mitochondrial outer membrane permeabilization. *Semin. Cell Dev. Biol.* 72, 152–162.

Czabotar PE, Lessene G, Strasser A & Adams JM (2014) Control of apoptosis by the BCL-2 protein family: implications for physiology and therapy. *Nat. Rev. Mol. Cell Biol.* 5, 49–63.

Danial NN & Korsmeyer SJ (2004) Cell death: critical control points. *Cell* 116, 205–219.

Datta SR, Ranger AM, Lin MZ . . . Greenberg ME (2002) Survival factor-mediated BAD phosphorylation raises the mitochondrial threshold for apoptosis. *Dev. Cell* 3, 631–643.

Davies AM (1996) The neurotrophic hypothesis: where does it stand? *Philos. Trans. R. Soc. Lond. B Biol. Sci.* 351, 389–394.

Dorstyn L, Akey WA & Kumar S (2018) New insights into apoptosome structure and function. *Cell Death Differ.* 25, 1194–1208.

Ellis RE, Yuan JY & Horvitz HR (1991) Mechanisms and functions of cell death. *Annu. Rev. Cell Biol.* 7, 663–698.

Fu T-M, Li Y, Lu A . . . Wu H (2016) Cryo-EM structure of caspase-8 tandem DED filament reveals assembly and regulation mechanisms of the death-inducing signaling complex. *Mol. Cell* 64, 236–250.

Galluzzi L, Vitale I, Aaronson SA . . . Kroemer G (2018) Molecular mechanisms of cell death: recommendations of the Nomenclature Committee on Cell Death. *Cell Death Differ.* 25, 486–541.

Green DR (2018) Cell Death: Apoptosis and Other Means to an End, 2nd ed. Cold Spring Harbor, NY: Cold Spring Harbor Laboratory Press.

Ha HJ, Chun HL & Park HH (2020) Assembly of platforms for signal transduction in the new era: dimerization, helical filament assembly, and beyond. *Exp. Mol. Med.* 52, 356–366.

Jiang X & Wang X (2004) Cytochrome C-mediated apoptosis. *Annu. Rev. Biochem.* 73, 87–108.

Jost PJ, Grabow S, Gray D . . . Kaufmann T (2009) XIAP discriminates between type I and type II FAS-induced apoptosis. *Nature* 460, 1935–1039.

Julien O & Wells JA (2017) Caspases and their substrates. *Cell Death Differ.* 24, 1380–1389.

Jullien M, Gomez-Bougie, Chiron D & Touzeau C (2020) Restoring apoptosis with BH3 mimetics in mature B-cell malignancies. *Cells* 9, 717.

Kale J, Osterlund EJ & Andrews DW (2018) BCL-2 family proteins: changing partners in the dance towards death. *Cell Death Differ.* 25, 65–80.

Kalkavan H & Green DR (2018) MOMP, cell suicide as a BCL-2 family business. *Cell Death Differ.* 25, 46–55.

Kerr JF, Wylie AH & Currie AR (1972) Apoptosis: a basic biological phenomenon with wide-ranging implications in tissue kinetics. *Br. J. Cancer* 26, 239–257.

Mandal R, Barron JC, Kostova I . . . Strebhardt K (2020) Caspase-8: the double-edged sword. *Biochim. Biophys. Acta Rev. Cancer* 1873, 1–18.

Nagata S (2005) DNA degradation in development and programmed cell death. *Annu. Rev. Immunol.* 23, 853–875.

Nagata S, Suzuki J & Fujii T (2016) Exposure of phosphatidylserine on the cell surface. *Cell Death Differ.* 23, 952–961.

Silke J & Vince J (2017) IAPs and cell death. *Curr. Top. Microbiol. Immunol.* 403, 96–117.

Voss AK & Strasser A (2020) The essentials of developmental apoptosis. *F1000Res.* 9, F1000 Faculty Rev-148.

I II III IV PART V

CELLS IN THEIR SOCIAL CONTEXT

Cell Junctions and the Extracellular Matrix

CHAPTER 19

IN THIS CHAPTER

Of all the social interactions between cells in a multicellular organism, the most fundamental are those that hold the cells together. Cells may be linked by direct interactions or they may be held together within the *extracellular matrix*, a complex network of proteins and polysaccharide chains that the cells secrete. By one means or another, cells must cohere if they are to form an organized multicellular structure that can withstand and respond to the various external forces that try to pull it apart.

Mechanisms of cell cohesion govern the architecture of tissues and organs—their shape, strength, and the arrangement of different cell types. The making and breaking of the attachments between cells and the modeling of the extracellular matrix govern the way cells move within the organism, guiding them as the body grows, develops, and repairs itself. Attachments to other cells and to the extracellular matrix control the orientation and behavior of the cell's cytoskeleton, thereby allowing cells to sense and respond to changes in the mechanical features of their environment. Thus, the apparatus of cell junctions and the extracellular matrix is critical for every aspect of the organization, function, and dynamics of multicellular structures. Defects in this apparatus underlie an enormous variety of diseases.

The key features of cell junctions and the extracellular matrix are best illustrated by considering two broad categories of tissues that are found in all animals (Figure 19–1). **Connective tissues**, such as bone or tendon, are formed from an extracellular matrix produced by cells that are distributed sparsely in the matrix. It is the matrix—rather than the cells—that bears most of the mechanical stress to which the tissue is subjected. *Cell-matrix junctions* link connective tissue cells to the matrix, allowing the cells to move through the matrix and monitor changes in its mechanical properties.

In **epithelial tissues**, such as the lining of the gut or the epidermal covering of the skin, cells are tightly bound together into sheets called **epithelia**. The extracellular matrix is less pronounced, consisting mainly of a thin mat called the *basal lamina* (or *basement membrane*) underlying the sheet. Within the epithelium, cells are attached to each other directly by *cell-cell junctions*, where cytoskeletal filaments are anchored, transmitting stresses across the interiors of the cells, from

epithelial tissue
basal lamina
connective tissue
collagen fibers
mechanical stresses are transmitted from cell to cell by cytoskeletal filaments anchored to cell–matrix and cell–cell adhesion sites
extracellular matrix directly bears mechanical stresses of tension and compression

adhesion site to adhesion site. The cytoskeletons of epithelial cells are also linked to the basal lamina through cell-matrix junctions.

Figure 19–2 provides a closer view of epithelial cells to illustrate the major types of cell-cell and cell-matrix junctions that we will discuss in this chapter. The diagram shows the typical arrangement of junctions in a *simple columnar* epithelium such as the lining of the small intestine of a vertebrate. Here, a single layer of tall cells stands on a basal lamina, with the cells' uppermost surface, or **apical** surface, free and exposed to the extracellular medium. On their sides, or *lateral* surfaces, the cells make junctions with one another. Two types of cell-cell junctions link the cytoskeletons of adjacent cells: **adherens junctions** are connected to actin filaments, and **desmosomes** are linked to intermediate filaments. Because these junctions anchor the cells strongly to each other and thus help the

Figure 19–1 **How animal cells are bound together in two major tissue types.** In connective tissue, the extracellular matrix is the main stress-bearing component. In epithelial tissue, the cytoskeleton of the cell is the main stress-bearing component: the cytoskeletons of cells are linked from cell to cell by adhesive junctions and transmit mechanical stresses across the interiors of the cells. Cell–matrix attachments bond epithelial tissue to the connective tissue beneath it.

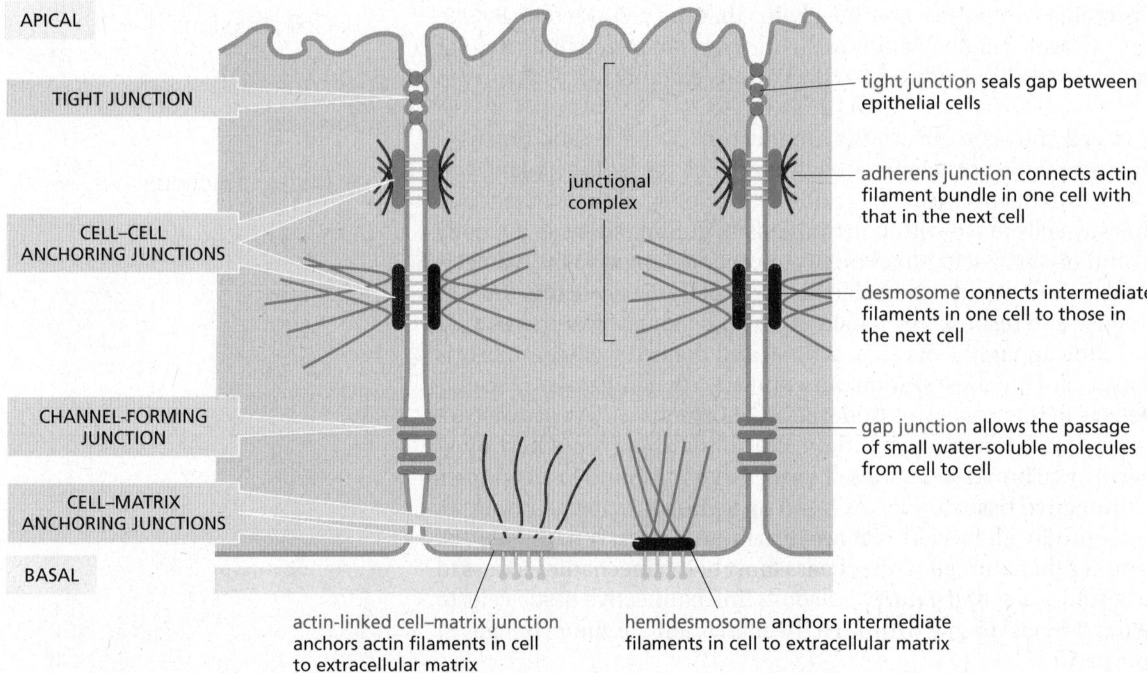
APICAL
TIGHT JUNCTION
CELL–CELL ANCHORING JUNCTIONS
CHANNEL-FORMING JUNCTION
CELL–MATRIX ANCHORING JUNCTIONS
BASAL
junctional complex
tight junction seals gap between epithelial cells
adherens junction connects actin filament bundle in one cell with that in the next cell
desmosome connects intermediate filaments in one cell to those in the next cell
gap junction allows the passage of small water-soluble molecules from cell to cell
actin-linked cell–matrix junction anchors actin filaments in cell to extracellular matrix
hemidesmosome anchors intermediate filaments in cell to extracellular matrix

Figure 19–2 **A summary of the various cell junctions found in a vertebrate epithelial cell.** In the apical region of the cell, the relative positions of the junctions are the same in nearly all vertebrate epithelia. The tight junction occupies the most apical position, followed by the adherens junction (adhesion belt) and then by a special parallel row of desmosomes; together, these three junctions form a structure called a junctional complex. Gap junctions and additional desmosomes are less regularly organized. Two types of cell–matrix anchoring junctions tether the basal surface of the cell to the basal lamina. The drawing is based on epithelial cells of the small intestine.

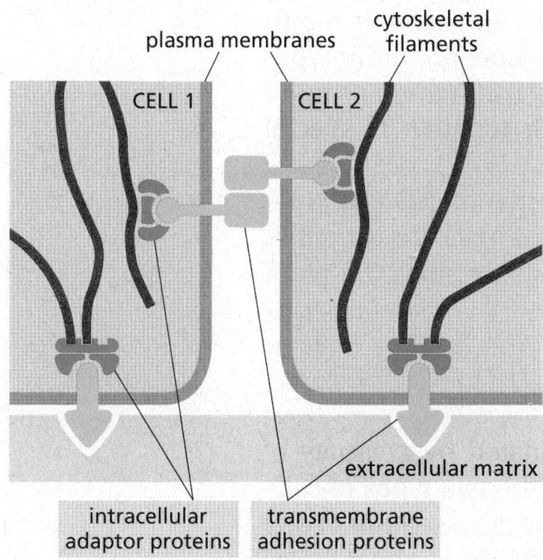

Figure 19–3 Transmembrane adhesion proteins link the cytoskeleton to extracellular structures. The external linkage may be either to other cells (cell–cell junctions, mediated typically by cadherins) or to extracellular matrix (cell–matrix junctions, mediated typically by integrins). The internal linkage to the cytoskeleton is generally indirect, via intracellular adaptor proteins, to be discussed later.

tissue withstand mechanical stress, they are sometimes called **anchoring junctions**. At the **basal** surface of the cells, two additional types of anchoring junctions link the cytoskeleton of the epithelial cell to the basal lamina: *actin-linked cell–matrix junctions* anchor actin filaments to the matrix, while *hemidesmosomes* anchor intermediate filaments to it.

Two other types of cell–cell junctions are shown in Figure 19–2. *Tight junctions* hold the cells closely together near the apical surface, sealing the gap between the cells and thereby preventing molecules from leaking across the epithelium. Near the basal end of the cells are channel-forming junctions, called *gap junctions,* that create passageways linking the cytoplasms of adjacent cells.

Each of the four major anchoring junction types depends on **transmembrane adhesion proteins** that span the plasma membrane, with one end linking to the cytoskeleton inside the cell and the other end linking to other structures outside it (**Figure 19–3**). These cytoskeleton-linked transmembrane proteins fall neatly into two superfamilies, corresponding to the two basic kinds of external attachment. Proteins of the **cadherin** superfamily chiefly mediate attachment of cell to cell (**Movie 19.1**). Proteins of the **integrin** superfamily chiefly mediate attachment of cells to matrix. There is specialization within each family: some cadherins link to actin and form adherens junctions, while others link to intermediate filaments and form desmosomes; likewise, some integrins link to actin and form actin-linked cell–matrix junctions, while others link to intermediate filaments and form hemidesmosomes (**Table 19–1**).

TABLE 19–1 **Anchoring Junctions**

Junction	Transmembrane adhesion protein	Extracellular ligand	Intracellular cytoskeletal attachment	Intracellular adaptor proteins
		Cell–cell		
Adherens junction	Classical cadherins	Classical cadherin on neighboring cell	Actin filaments	α-Catenin, β-catenin, p120-catenin, vinculin
Desmosome	Nonclassical cadherins (desmoglein, desmocollin)	Desmoglein and desmocollin on neighboring cell	Intermediate filaments	Plakoglobin, plakophilin, desmoplakin
		Cell–matrix		
Actin-linked cell–matrix junction	Integrin	Extracellular matrix proteins	Actin filaments	Talin, kindlin, vinculin, paxillin, focal adhesion kinase (FAK), numerous others
Hemidesmosome	$\alpha_6\beta_4$ integrin, type XVII collagen	Extracellular matrix proteins	Intermediate filaments	Plectin, BP230

There are some exceptions to these rules. Some integrins, for example, mediate cell–cell rather than cell–matrix attachment. Moreover, there are other types of cell adhesion molecules that can provide transient cell–cell attachments more flimsy than anchoring junctions but sufficient to stick cells together in special circumstances.

We begin the chapter with a discussion of the major forms of cell–cell junctions. We then consider in turn the extracellular matrix of animals, the structure and function of integrin-mediated cell–matrix junctions, and, finally, the plant cell wall, a special form of extracellular matrix.

CELL–CELL JUNCTIONS

Cell–cell junctions come in many forms and can be regulated by a variety of mechanisms. The best understood and most common are the two types of cell–cell anchoring junctions, which employ cadherins to link the cytoskeleton of one cell with that of its neighbor. Their primary function is to resist the external forces that pull cells apart. The epithelial cells of your skin, for example, must remain tightly linked when they are stretched, pinched, or poked. Cell–cell anchoring junctions must also be dynamic and adaptable, so that they can be altered or rearranged when tissues are remodeled or repaired or when there are changes in the forces acting on them.

In this section, we focus primarily on the cadherin-based anchoring junctions. We then briefly describe tight junctions and gap junctions. Finally, we consider the more transient cell–cell adhesion mechanisms employed by some cells in the bloodstream.

Cadherins Form a Diverse Family of Adhesion Molecules

Cadherins are present in all multicellular animals. They are also present in the choanoflagellates, which are closely related to animals but can exist either as free-living unicellular organisms or as multicellular colonies. Other eukaryotes, including fungi and plants, lack cadherins, and they are also absent from bacteria and archaea. Cadherins therefore seem to be part of the essence of what it is to be an animal.

The cadherins take their name from their dependence on Ca^{2+} ions: removing Ca^{2+} from the extracellular medium causes adhesions mediated by cadherins to come apart. The first three cadherins to be discovered were named according to the main tissues in which they were found: *E-cadherin* is present on many types of epithelial cells; *N-cadherin* on nerve, muscle, and lens cells; and *P-cadherin* on cells in the placenta and epidermis. All are also found in other tissues. These and other **classical cadherins** are closely related in sequence throughout their extracellular and intracellular domains.

There are also a large number of **nonclassical cadherins** that are more distantly related in sequence, with more than 50 expressed in the brain alone. The nonclassical cadherins include proteins with known adhesive function, such as the diverse *protocadherins* found in the brain, and the *desmocollins* and *desmogleins* that form desmosomes (see Table 19–1). Together, the classical and nonclassical cadherin proteins constitute the **cadherin superfamily** (Figure 19–4), with more than 180 members in humans.

Cadherins Mediate Homophilic Adhesion

Anchoring junctions between cells are usually symmetrical: if the linkage is to actin in the cell on one side of the junction, it will be to actin in the cell on the other side. In fact, the binding between cadherins is generally **homophilic** (like-to-like; Figure 19–5): cadherin molecules of a specific subtype on one cell bind to cadherin molecules of the same or closely related subtype on adjacent cells.

The spacing between the cell membranes at an anchoring junction is precisely defined and depends on the structure of the participating cadherin molecules. All

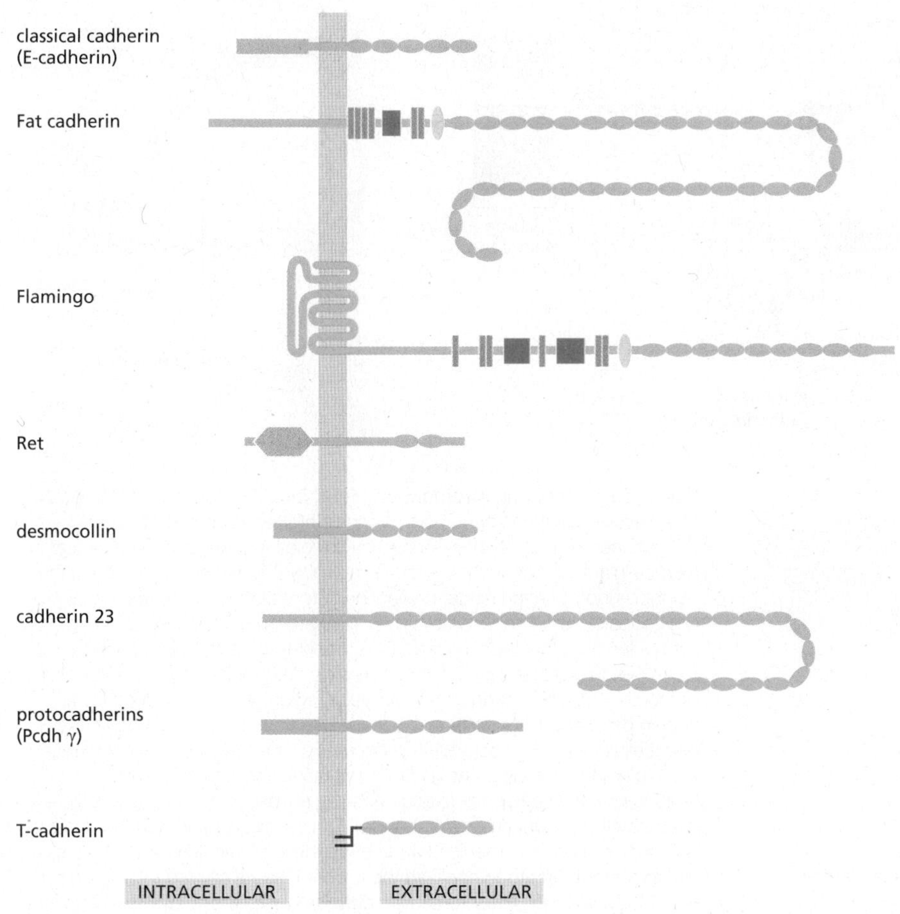

Figure 19–4 **The cadherin superfamily.** The diagram shows some of the diversity among cadherin superfamily members. These proteins all have extracellular portions containing multiple copies of the extracellular cadherin domain *(green ovals)*. In the classical cadherins of vertebrates there are 5 of these domains, and in desmogleins and desmocollins there are 4 or 5, but some nonclassical cadherins have more than 30. The intracellular portions are more varied, reflecting interactions with a wide variety of intracellular ligands, including signaling molecules and adaptor proteins that connect the cadherin to the cytoskeleton. In some cases, such as T-cadherin, a transmembrane domain is not present, and the protein is attached to the plasma membrane by a glycosylphosphatidylinositol (GPI) anchor. The differently colored motifs in Fat, Flamingo, and Ret represent conserved domains that are also found in other protein families.

the members of the superfamily, by definition, have an extracellular portion consisting of several copies of the *extracellular cadherin* (*EC*) *domain*. Homophilic binding occurs at the N-terminal tips of the cadherin molecules—the cadherin domains that lie farthest from the membrane. These terminal domains each form a knob and a nearby pocket, and the cadherin molecules protruding from opposite cell membranes bind by insertion of the knob of one domain into the pocket of the other (**Figure 19–6A**).

Each cadherin domain forms a more-or-less rigid unit, joined to the next cadherin domain by a hinge. Ca^{2+} ions bind to sites near each hinge and prevent it from flexing, so that the whole string of cadherin domains behaves as a rigid and slightly curved rod. When Ca^{2+} is removed, the hinges can flex, and the structure becomes floppy (**Figure 19–6B**). At the same time, the conformation at the N-terminus is thought to change slightly, weakening the binding affinity for the matching cadherin molecule on the opposite cell.

Unlike receptors for soluble signaling molecules, which bind their specific ligand with high affinity, cadherins (and most other cell–cell adhesion proteins) typically bind to their partners with relatively low affinity. Strong attachments result from the formation of many such weak bonds in parallel. When binding to oppositely oriented partners on another cell, cadherin molecules are often clustered side-to-side with many other cadherin molecules on the same cell (**Figure 19–6C**). The strength of this junction is far greater than that of any individual intermolecular bond, and yet regulatory mechanisms can easily disassemble the junction by separating the molecules sequentially, just as two pieces of fabric can be joined strongly by Velcro and yet easily peeled apart from the sides. A similar "Velcro principle" also operates at cell–cell and cell–matrix adhesions formed by other types of transmembrane adhesion proteins.

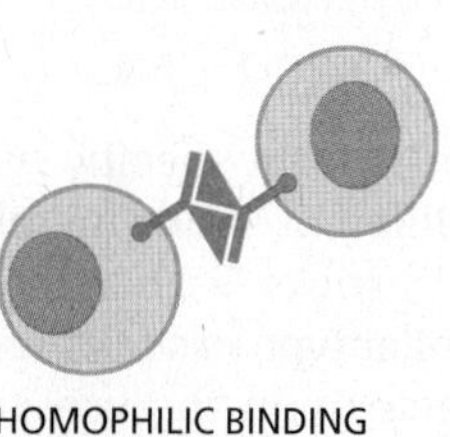

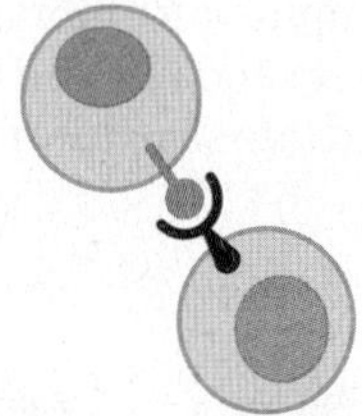

Figure 19–5 **Homophilic versus heterophilic binding.** Cadherins in general bind homophilically; some other cell adhesion molecules, discussed later, bind heterophilically.

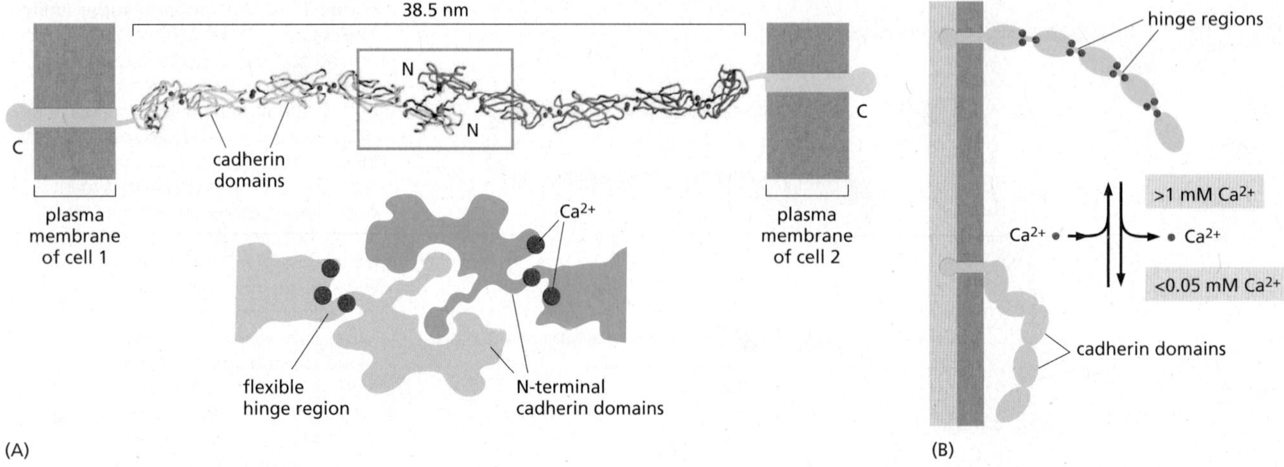

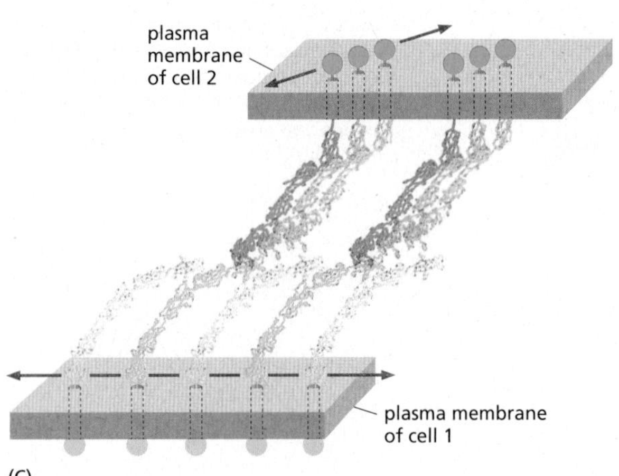

Figure 19–6 Cadherin structure and function. (A) The extracellular region of a classical cadherin contains five copies of the extracellular cadherin domain (see Figure 19–4) separated by flexible hinge regions. At a typical extracellular Ca^{2+} concentration (>1 mM), Ca^{2+} ions *(red dots)* bind in the neighborhood of each hinge, preventing it from flexing. To generate cell–cell adhesion, the cadherin domain at the N-terminal tip of one cadherin molecule binds the cadherin domain at the N-terminal tip of a cadherin molecule on another cell. The structure was determined by x-ray diffraction of the crystallized C-cadherin extracellular region. The two cadherins shown here, although identical, are colored differently for clarity. (B) If the extracellular Ca^{2+} concentration is decreased artificially in an experiment, Ca^{2+} binding decreases. As a result, increased flexibility in the hinge regions results in a floppier molecule that is no longer oriented correctly to interact with a cadherin on another cell—and adhesion fails. (C) At a typical cell–cell junction, an organized array of cadherin molecules functions like Velcro to hold cells together. Cadherins on the same cell are thought to be coupled by side-to-side interactions between their N-terminal head regions, resulting in a linear array like the alternating *green* and *light green* cadherins on the lower cell shown here. These arrays are thought to interact with perpendicular arrays on an adjacent cell (*blue* cadherin molecules, top cell). Multiple perpendicular arrays on both cells interact to form a tight-knit mat of cadherin proteins. (A, based on T.J. Boggon et al., *Science* 296:1308–1313, 2002; C, based on O.J. Harrison et al., *Structure* 19:244–256, 2011.)

Cadherin-dependent Cell–Cell Adhesion Guides the Organization of Developing Tissues

Cadherins form specific homophilic attachments, explaining why there are so many different family members. Cadherins are not like glue, making cell surfaces generally sticky. Rather, they mediate highly selective recognition, enabling cells of a similar type to stick together and to stay segregated from other types of cells.

Selectivity in the way that animal cells consort with one another was first demonstrated in the 1950s, long before the discovery of cadherins, in experiments in which amphibian embryos were dissociated into single cells. These cells were then mixed up and allowed to reassociate. Remarkably, the dissociated cells often reassembled into structures resembling those of the original embryo (**Figure 19–7**).

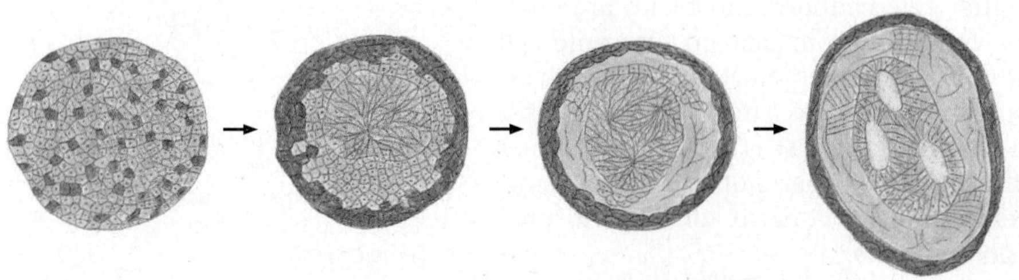

Figure 19–7 Sorting out. Cells from different layers of an early amphibian embryo will sort out according to their origins. In the classical experiment shown here, mesoderm cells *(green)*, neural plate cells *(blue)*, and epidermal cells *(red)* have been disaggregated and then reaggregated in a random mixture. They sort out into an arrangement reminiscent of a normal embryo, with a "neural tube" internally, epidermis externally, and mesoderm in between. (Modified from P.L. Townes and J. Holtfreter, *J. Exp. Zool.* 128:53–120, 1955. With permission from John Wiley & Sons.)

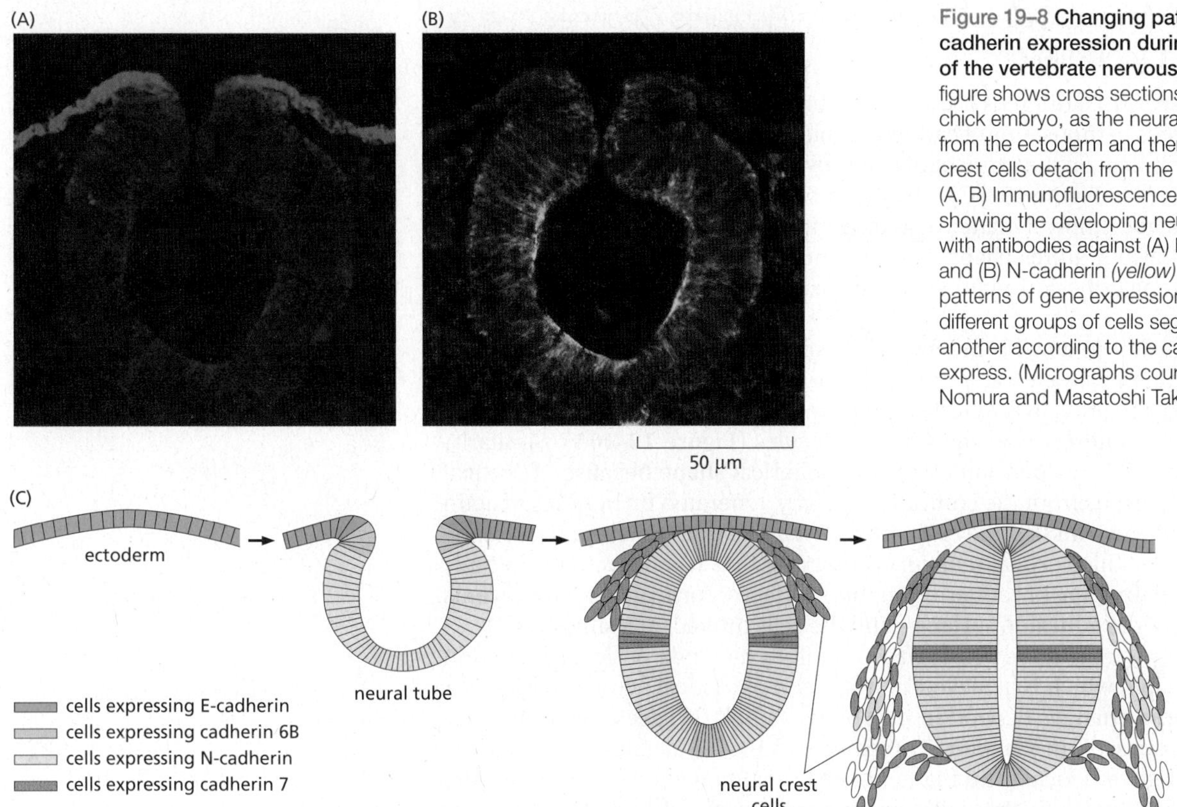

Figure 19–8 Changing patterns of cadherin expression during construction of the vertebrate nervous system. The figure shows cross sections of the early chick embryo, as the neural tube detaches from the ectoderm and then as neural crest cells detach from the neural tube. (A, B) Immunofluorescence micrographs showing the developing neural tube labeled with antibodies against (A) E-cadherin *(blue)* and (B) N-cadherin *(yellow)*. (C) As the patterns of gene expression change, the different groups of cells segregate from one another according to the cadherins they express. (Micrographs courtesy of Miwako Nomura and Masatoshi Takeichi.)

These experiments, together with numerous more recent experiments, reveal that selective cell–cell recognition systems make cells of the same differentiated tissue preferentially adhere to one another.

Cadherins play a crucial part in these cell-sorting processes. The appearance and disappearance of specific cadherins correlate with steps in embryonic development where cells regroup and change their contacts to create new tissue structures. In the vertebrate embryo, for example, changes in cadherin expression are seen when the neural tube forms and pinches off from the overlying ectoderm: neural tube cells lose E-cadherin and acquire other cadherins, including N-cadherin, while the cells in the overlying ectoderm continue to express E-cadherin (**Figure 19–8A and B**). Then, when the neural crest cells migrate away from the neural tube, these cadherins become scarcely detectable, and another cadherin (cadherin 7) appears that helps hold the migrating cells together as loosely associated cell groups (**Figure 19–8C**). Finally, when some of the neural crest cells aggregate to form a ganglion, they switch on expression of N-cadherin again. If N-cadherin is artificially overexpressed in the emerging neural crest cells, the cells fail to escape from the neural tube.

Studies with cultured cells further support the importance of homophilic cadherin binding in tissue segregation. In a line of cultured fibroblasts called *L cells*, for example, cadherins are not expressed and the cells do not adhere to one another. When these cells are transfected with DNA encoding E-cadherin, E-cadherins on one cell bind to E-cadherins on another, resulting in cell–cell adhesion. If L cells expressing different cadherins are mixed together, they sort out and aggregate separately, indicating that different cadherins preferentially bind to their own type (**Figure 19–9A**), mimicking what happens when cells derived from tissues that express different cadherins are mixed together. A similar segregation of cells occurs if L cells expressing different amounts of the same cadherin are mixed together (**Figure 19–9B**). It therefore seems likely that both qualitative and quantitative differences in the expression of cadherins have a role in organizing tissues.

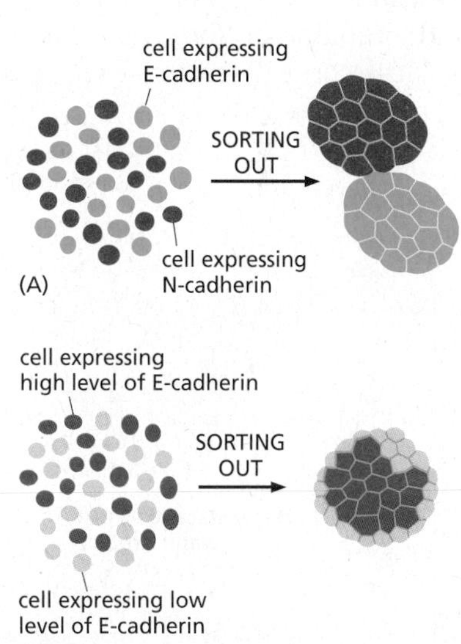

Figure 19–9 Cadherin-dependent cell sorting. Cells in culture can sort themselves out according to the type and level of cadherins they express. This can be visualized by labeling different populations of cells with dyes of different colors. (A) Cells expressing N-cadherin sort out from cells expressing E-cadherin. (B) Cells expressing high levels of E-cadherin sort out from cells expressing low levels of E-cadherin. The cells expressing high levels adhere more strongly and congregate internally.

Assembly of Strong Cell–Cell Adhesions Requires Changes in the Actin Cytoskeleton

In their mature form, adherens junctions are enormous protein complexes containing hundreds to thousands of cadherin molecules, packed into dense, regular arrays that are linked on the extracellular side by lateral interactions between cadherin domains (see Figure 19–6C). The assembly of these junctions over a large surface area is not simply a matter of adding more cadherins to an initial attachment site, but also requires changes in the underlying actin cytoskeleton. Most important, strong adhesion requires a decrease in *cortical tension* at the site of the adhesion.

Cortical tension in a cell is much like the surface tension of a water droplet. In water, binding between water molecules at the air-water interface pulls the surface inward, resulting in a spherical shape that resists disruption by other surfaces that do not interact with the water molecules (**Figure 19–10A**). Similarly, an unattached cell in suspension assumes a spherical shape because of cortical tension, which results from the contractile activity generated by bundles of actin and non-muscle myosin II at the cell cortex, just beneath the plasma membrane (**Figure 19–10B**). This cortical tension is so strong that when two cells initially interact through binding of cadherins on their surfaces, cortical tension prevents the spreading of the adhesion surface—and the cells interact at a single point.

Assembly of a large adhesion surface therefore depends on local reduction of cortical tension, which is achieved by inhibition of cortical actin-myosin fiber formation. These changes depend in part on two small GTPases called Rac and Rho. As we discussed in Chapter 16 (see Figure 16–75), signals generated by these GTPases govern local actin filament behavior: Rho generally promotes the formation of actin-myosin stress fibers at the cell cortex, while Rac inhibits the formation of these fibers and instead promotes branched actin networks. When two epithelial precursor cells first interact at a small cluster of cadherin linkages, the cadherins generate intracellular signals that promote local activation of Rac and inhibition of Rho. The result is disassembly of actin-myosin fibers and loss of local cortical tension—which then allows further cadherin recruitment to

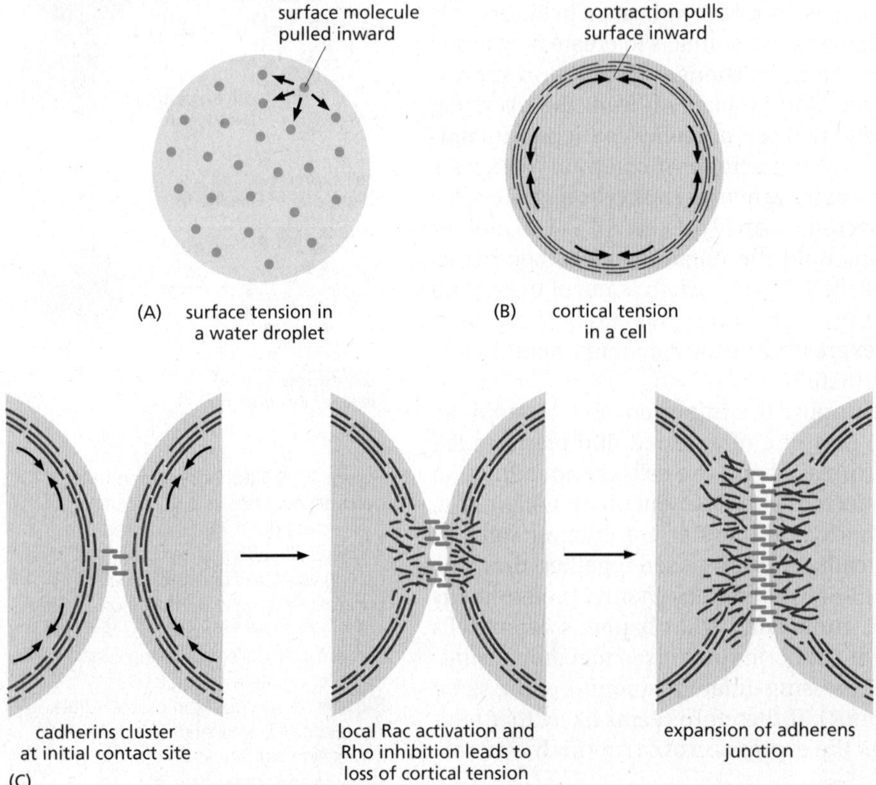

Figure 19–10 Local changes in cortical tension help promote the initial formation of an adherens junction. (A) Surface tension of a water droplet results from binding among water molecules at the air–water interface, pulling the surface inward. (B) In an unattached cell, the contraction of actin–myosin bundles at the cell cortex *(red)* creates cortical tension, drawing the surface inward. (C) When two epithelial cell precursors first interact, small cadherin clusters *(green)* assemble at the contact site. Cortical tension prevents this initial interaction site from spreading. The cadherins generate local signals to inhibit the GTPase Rho and activate the GTPase Rac, leading to localized disassembly of actin–myosin fibers, loss of cortical tension, and formation of branched actin networks and cell protrusions—all of which allow the recruitment of more cadherins and the spreading of the cell–cell junction over a greater surface area. In the long term (not shown), the large adherens junction inhibits Rac and stimulates Rho, thereby promoting formation of local actin–myosin fibers that interact with the cadherins (see Figure 19–11).

spread the adhesion to a larger surface area (**Figure 19–10C**). Rac activation has the added benefit of stimulating local actin protrusions from the cell, which also contributes to expansion of the junction. The implications of this mechanism are clear: the development of strong adhesions depends not only on the cell adhesion molecules themselves but also on the associated regulatory systems that govern actin behavior.

Eventually, after large numbers of cadherin molecules are aligned at a nascent adherens junction, Rac is inhibited and Rho is activated, which promotes the assembly of linear, contractile actin filament bundles that link the adherens junction to the actin cytoskeleton. These new linkages pull the junction inward, generating tension that stimulates further actin recruitment and expansion of the junction, as we describe in the next sections.

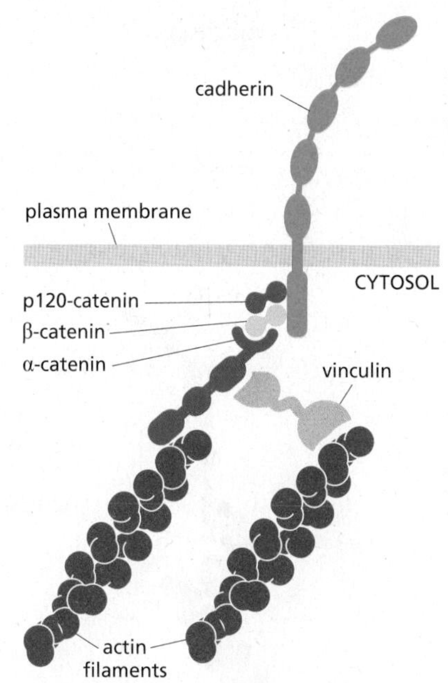

Figure 19–11 **The linkage of classical cadherins to actin filaments.** The cadherins are coupled indirectly to actin filaments through an adaptor protein complex containing p120-catenin, β-catenin, and α-catenin. Other proteins, including vinculin, associate with α-catenin and help provide the linkage to actin. β-Catenin has a second, and very important, function in intracellular signaling, as we discuss in Chapter 15 (see Figure 15–61). For clarity, this diagram does not show the cadherin of the adjacent cell in the junction.

Catenins Link Classical Cadherins to the Actin Cytoskeleton

The extracellular domains of cadherins mediate homophilic binding at adherens junctions. The intracellular domains of typical cadherins, including all classical and some nonclassical ones, interact with filaments of the cytoskeleton: actin at adherens junctions and intermediate filaments at desmosomes (see Table 19–1). These cytoskeletal linkages are essential for efficient cell–cell adhesion, as cadherins that lack their cytoplasmic domains cannot stably hold cells together.

The linkage of cadherins to the cytoskeleton depends on adaptor proteins that assemble on the cytoplasmic tail of the cadherin. At adherens junctions, the cadherin tail binds two such proteins, *β-catenin* and a distant relative called *p120-catenin*; a third protein called *α-catenin* interacts with β-catenin and recruits other proteins to provide a dynamic linkage to actin filaments (**Figure 19–11**). At desmosomes, cadherins are linked to intermediate filaments through other adaptor proteins, including a β-catenin-related protein called *plakoglobin*, as we discuss later.

Adherens Junctions Respond to Tension from Inside and Outside the Tissue

The protein complexes that mediate junctions either between cells or between cells and the extracellular matrix are dynamic machines that have the remarkable ability to sense mechanical stresses and generate biochemical signals that lead to an appropriate response. We call this *mechanotransduction*.

As we discussed in the previous sections, adherens junctions are linked through catenins to contractile bundles of actin and myosin II. These junctions are therefore subjected to pulling forces generated by the attached actin. The pulling forces are important for junction assembly and maintenance: inhibition of myosin activity, for example, results in the disassembly of many adherens junctions. Furthermore, the contractile forces acting on a junction in one cell are balanced by contractile forces at the junction of the neighboring cell, so that no cell pulls others toward it.

Adherens junctions sense the forces acting on them and modify local actin and myosin behavior to balance the forces on both sides of the junction. Evidence for these mechanisms comes from studies of pairs of cultured mammalian cells connected by adherens junctions. If contractile activity in one cell is increased experimentally, the adherens junctions linking the two cells increase in size, and the contractile activity of the second cell increases to match that of the first—resulting in a balance of forces across the junction. These and other experiments reveal that adherens junctions are not simply passive sites of protein–protein binding but are tension sensors that regulate their behavior in response to changing mechanical conditions.

Mechanotransduction at cell–cell junctions is thought to depend, at least in part, on proteins in the cadherin complex that alter their shape when stretched by tension. The protein α-catenin, for example, is stretched from a folded to an extended conformation when contractile activity increases at the junction. The

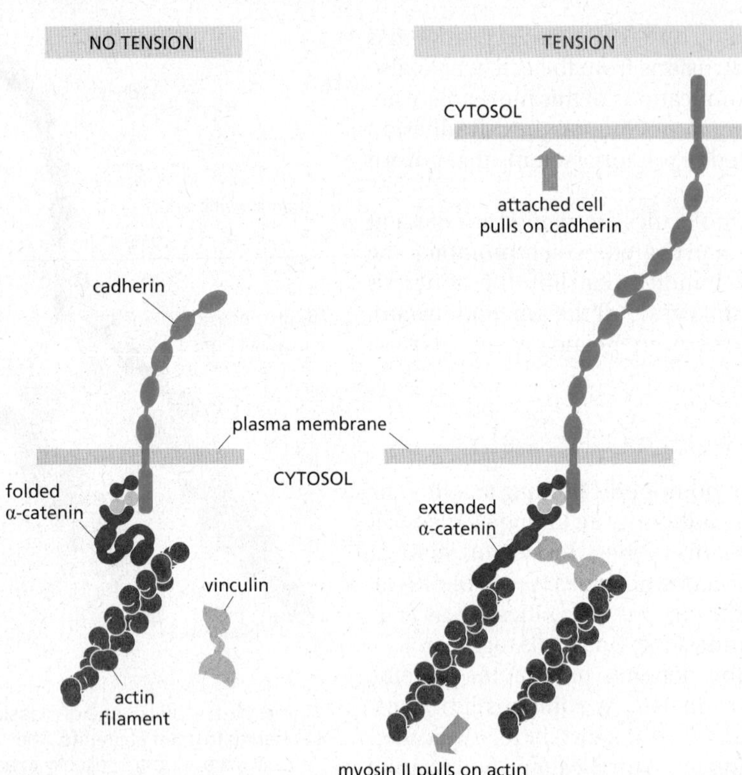

Figure 19–12 Mechanotransduction in an adherens junction. Cell–cell junctions are able to sense increased tension and respond by strengthening their actin linkages. When actin filaments are pulled from within the cell by non-muscle myosin II, the resulting force unfolds a domain in α-catenin, thereby exposing an otherwise hidden binding site for the adaptor protein vinculin. Vinculin then promotes additional actin recruitment, strengthening the linkages between the junction and the cytoskeleton.

unfolding exposes a cryptic binding site for another protein, vinculin, which promotes the recruitment of more actin to the junction (**Figure 19–12**). It is likely that these changes also lead to local regulation of the small GTPases that control actin and myosin behavior. By mechanisms such as this, pulling on a junction alters local actin behavior to make the junction stronger. Furthermore, as noted above, pulling on a junction in one cell will increase the contractile force generated in the attached cell.

Because all the cells in an epithelium are mechanically linked through their cell–cell junctions and cytoskeletons, mechanotransduction can also work over long distances. In the developing wing of *Drosophila*, for example, contraction of cells in the hinge of the wing results in mechanical stresses that spread throughout the wing, triggering changes in cell movement and orientation that are important for wing formation. It is likely that tension sensors throughout the tissue detect shifts in mechanical stress and trigger a molecular response. Because tension spreads quickly through all connected cells in a tissue, it provides an unusually rapid and effective signal for modifying cell behavior throughout the tissue.

Tissue Remodeling Depends on the Coordination of Actin-mediated Contraction with Cell–Cell Adhesion

Adherens junctions are an essential part of the machinery for modeling the shapes of multicellular structures in the animal body. By indirectly linking the actin filaments in one cell to those in its neighbors, they enable the cells in the tissue to use their actin cytoskeletons in a coordinated way.

Adherens junctions occur in various forms. In many nonepithelial tissues, they appear as small punctate or linear attachments that connect the cortical actin filaments beneath the plasma membranes of two interacting cells. In heart muscle, they anchor the actin bundles of the contractile apparatus and act in parallel with desmosomes to link the contractile cells end-to-end. But the prototypical examples of adherens junctions occur in epithelia, where they often form a continuous **adhesion belt** (or *zonula adherens*) that encircles each of the cells

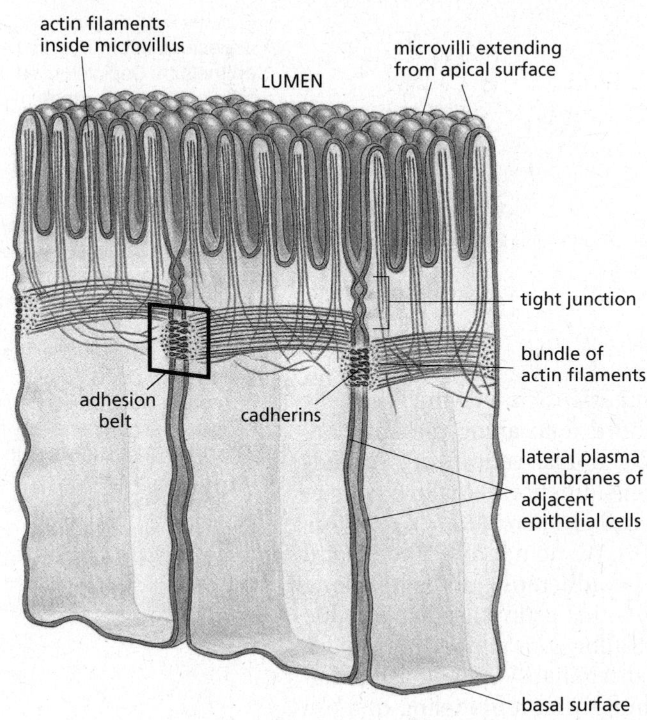

Figure 19–13 **Adherens junctions between epithelial cells in the small intestine.** These cells are specialized for absorption of nutrients; at their apical surfaces, facing the lumen of the gut, they have many microvilli (protrusions that increase the absorptive surface area). The adherens junction takes the form of an *adhesion belt*, encircling each of the interacting cells. Its most obvious feature is a contractile bundle of actin filaments running along the cytoplasmic surface of the junctional plasma membrane. The actin filament bundles are tethered by adaptor proteins to cadherins, which bind to cadherins on the adjacent cell. In this way, the actin filament bundles in adjacent cells are tied together. For clarity, this drawing does not show most of the other cell–cell and cell–matrix junctions of epithelial cells (see Figure 19–2).

just beneath the apical surface of the epithelium (**Figure 19–13**). Within each cell, a contractile bundle of actin filaments and myosin II lies adjacent to the adhesion belt, oriented parallel to the plasma membrane and tethered to it by the cadherins and their associated intracellular adaptor proteins. The actin-myosin bundles are thus linked, via the cadherins, into an extensive transcellular network. Coordinated contraction of this network provides the motile force for a fundamental process in animal morphogenesis—the folding of epithelial-cell sheets into tubes, spheres, and related structures (**Figure 19–14**).

In some forms of tissue remodeling, actin-myosin contractility is coordinated with major changes in local cell-cell adhesion patterns. An example can be found in cellular rearrangements that occur early in the development of the fruit fly *Drosophila melanogaster*. Soon after gastrulation, the outer epithelium of the

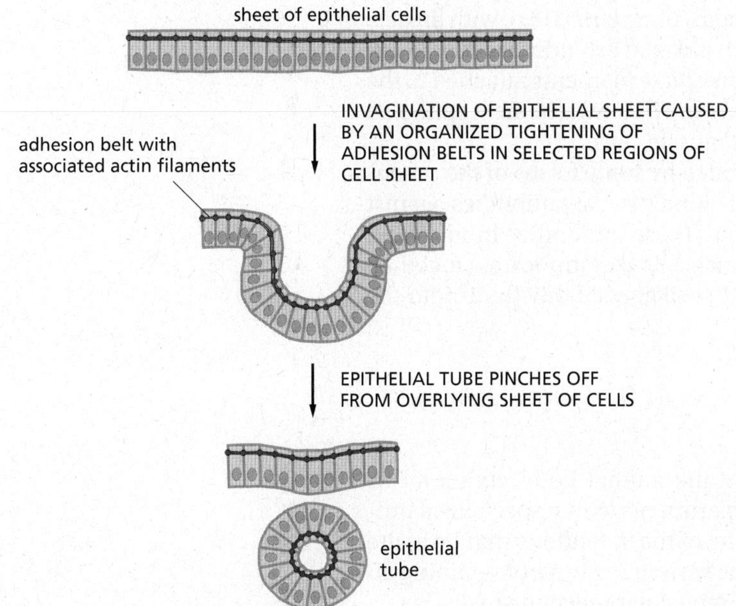

Figure 19–14 **The folding of an epithelial sheet to form an epithelial tube.** The oriented contraction of the bundles of actin and myosin filaments running along adhesion belts causes the epithelial cells to narrow at their apical surfaces, thereby helping the epithelial sheet to roll up into a tube. An example is the formation of the neural tube in early vertebrate development (see Figure 19–8).

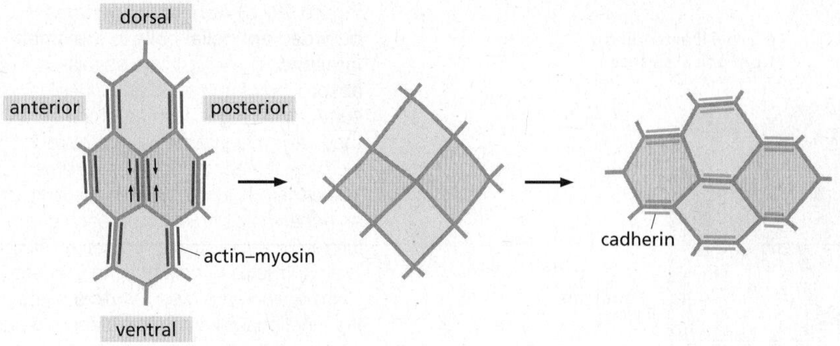

Figure 19–15 **Remodeling of cell–cell adhesions in embryonic *Drosophila* epithelium.** Depicted at *left* is a group of cells in the outer epithelium of a *Drosophila* embryo. During germ-band extension, cells converge toward each other *(middle)* on the dorsal–ventral axis and then extend *(right)* along the anterior–posterior axis. The result is intercalation: cells that were originally far apart along the dorsal–ventral axis *(green)* are inserted between the cells *(gray)* that separated them. These rearrangements depend on the spatial regulation of actin–myosin contractile bundles, which are localized primarily at the vertical cell boundaries *(red, left)*. Contraction of these bundles is accompanied by removal of E-cadherin (not shown) at the same cell boundaries, resulting in shrinkage and loss of adhesion along the vertical axis *(middle)*. New cadherin-based adhesions *(blue, right)* then form and expand along horizontal boundaries, resulting in extension of the cells in the anterior–posterior dimension.

embryo is elongated by a process called *germ-band extension,* in which the cells converge inward along the dorsal-ventral axis and extend along the anterior-posterior axis. Actin-dependent contraction along dorsal-ventral cell boundaries is accompanied by a loss of specific adherens junctions to allow cells to insert themselves between other cells (a process called *intercalation*), resulting in a longer and narrower epithelium (**Figure 19–15**). We do not fully understand the mechanisms underlying the disassembly of adherens junctions along dorsal-ventral cell boundaries, but one possibility is that actin-based contractile forces pull sufficiently hard on the edges of cell-cell adhesions to peel them apart, particularly if contraction is coupled to additional regulatory mechanisms that weaken the adhesion. There is evidence, for example, that remodeling of these and other adhesions during development depends on removal of cadherins from the cell surface by clathrin-mediated endocytosis.

Desmosomes Give Epithelia Mechanical Strength

Desmosomes are structurally similar to adherens junctions but contain specialized cadherins that link to intermediate filaments instead of actin filaments. Their main function is to provide mechanical strength. Desmosomes are important in vertebrates but are not found, for example, in *Drosophila.* They are present in most mature vertebrate epithelia and are particularly plentiful in tissues that are subject to high levels of mechanical stress, such as heart muscle and the epidermis, the epithelium that forms the outer layer of the skin.

Figure 19–16A shows the general structure of a desmosome, and **Figure 19–16B** shows some of the proteins that form it. Desmosomes typically appear as buttonlike spots of adhesion, riveting the cells together (**Figure 19–16C**). Inside the cell, the bundles of rope-like intermediate filaments that are anchored to the desmosomes form a structural framework of great tensile strength (**Figure 19–16D**), with linkage to similar bundles in adjacent cells, creating a network that extends throughout the tissue (**Figure 19–17**). The particular type of intermediate filaments attached to the desmosomes depends on the cell type: they are *keratin filaments* in most epithelial cells, for example, and *desmin filaments* in heart muscle cells.

The importance of desmosomes is demonstrated by some forms of the potentially fatal skin disease *pemphigus.* Affected individuals make antibodies against one of their own desmosomal cadherin proteins. These antibodies bind to and disrupt the desmosomes that hold their epidermal cells (keratinocytes) together. This results in a severe blistering of the skin, with leakage of body fluids into the loosened epithelium.

Tight Junctions Form a Seal Between Cells and a Fence Between Plasma Membrane Domains

Sheets of epithelial cells enclose and partition the animal body, lining all its surfaces and cavities, and creating internal compartments where specialized processes occur. The epithelial sheet seems to be one of the inventions that lie at the origin of animal evolution, diversifying in a huge variety of ways but retaining an organization that is based on a set of conserved molecular mechanisms.

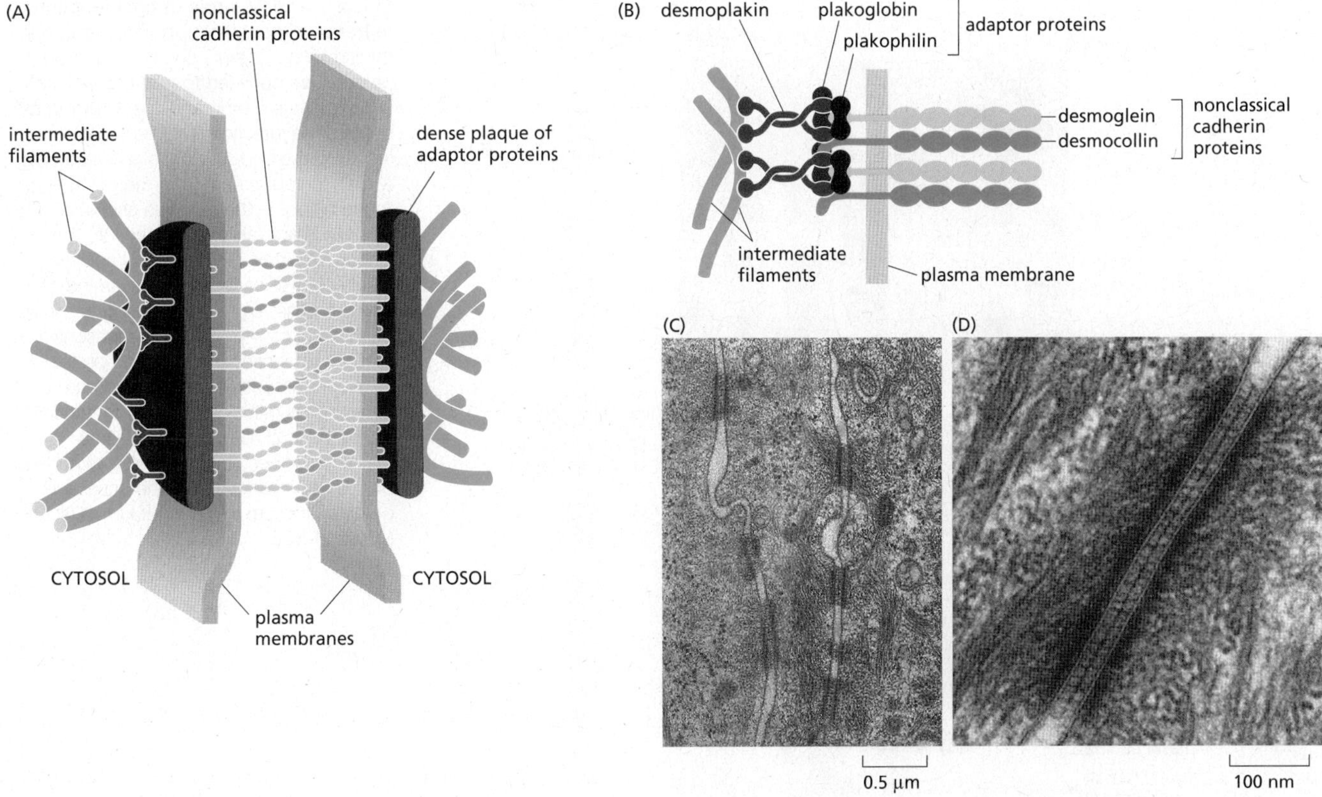

Figure 19–16 **Desmosomes.** (A) The structural components of a desmosome. On the cytoplasmic surface of each interacting plasma membrane is a dense plaque composed of a mixture of intracellular adaptor proteins. A bundle of intermediate filaments is attached to the surface of each plaque. Transmembrane nonclassical cadherins bind to the plaques and interact through their extracellular domains to hold the adjacent membranes together. (B) Some of the molecular components of a desmosome. Desmoglein and desmocollin are nonclassical cadherins. Their cytoplasmic tails bind *plakoglobin* (γ-catenin) and *plakophilin* (a distant relative of p120-catenin), which in turn bind to *desmoplakin*. Desmoplakin binds to the sides of intermediate filaments, thereby tying the desmosome to these filaments. (C) An electron micrograph of desmosome junctions between three epidermal cells in the skin of a baby mouse. (D) Part of the same tissue at higher magnification, showing a single desmosome, with intermediate filaments attached to it. (C and D, from W. He et al., *Science* 302:109–113, 2003. With permission from AAAS.)

Essentially all epithelia are anchored to other tissue on one side (the basal side) and free of such attachment on their opposite side (the apical side). A basal lamina lies at the interface with the underlying tissue, mediating the attachment, while the apical surface of the epithelium is generally bathed in extracellular fluid. Thus, as discussed in Chapter 16, all epithelia are structurally **polarized**, and so are their individual cells: the basal end of a cell, adherent to the basal lamina below, differs from the apical end, exposed to the medium above.

Correspondingly, all epithelia have at least one function in common: they serve as selective permeability barriers, separating the fluid that permeates the tissue on their basal side from fluid with a different chemical composition on their apical side. This barrier function requires that the adjacent cells be sealed together by **tight junctions**, so that molecules cannot leak freely across the cell sheet.

The epithelium of the small intestine provides a good illustration of tight-junction structure and function (see Figure 19–2). This epithelium has a *simple columnar* structure; that is, it consists of a single layer of tall (columnar) cells. These are of several differentiated types, but the majority are absorptive cells, specialized for uptake of nutrients from the internal cavity, or *lumen*, of the gut.

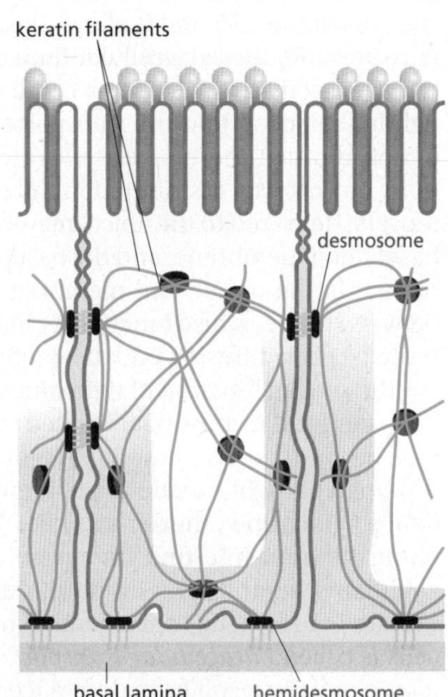

Figure 19–17 **Desmosomes, hemidesmosomes, and the intermediate filament network.** The keratin intermediate filament networks of adjacent cells—in this example, epithelial cells of the small intestine—are indirectly connected to one another through desmosomes, and to the basal lamina through hemidesmosomes.

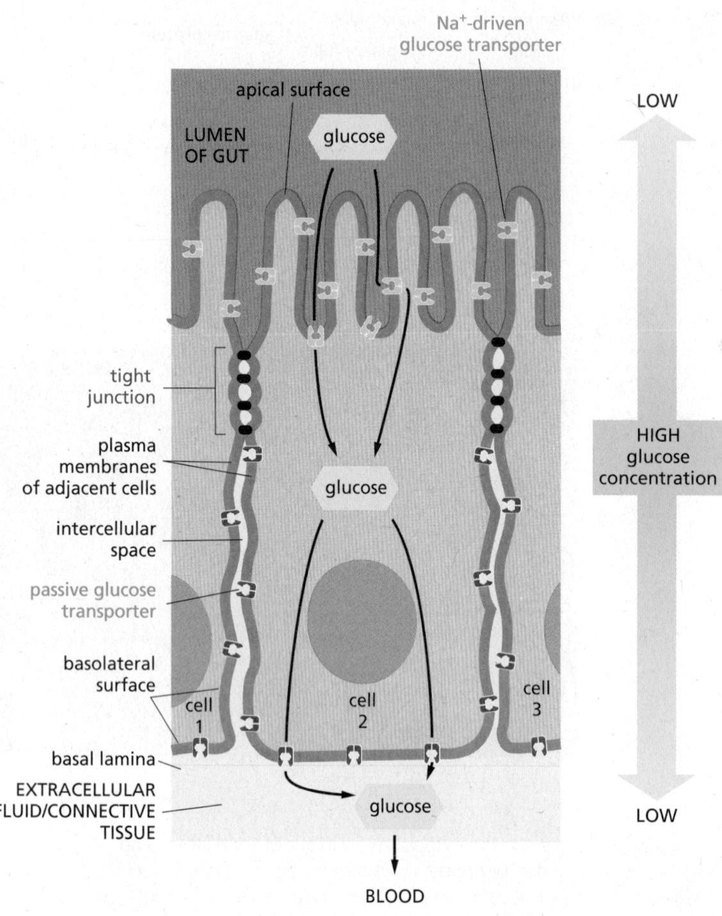

Figure 19–18 **The role of tight junctions in transcellular transport.** For clarity, only the tight junctions are shown. Transport proteins are confined to different regions of the plasma membrane in epithelial cells of the small intestine. This segregation results in the one-way transfer of nutrients across the epithelium from the gut lumen to the blood. In the example shown, glucose is actively transported into the cell by Na^+-driven glucose transporters at its apical surface, and it leaves the cell through passive glucose transporters in its basolateral membrane. Tight junctions are thought to confine the transport proteins to their appropriate membrane domains by acting as diffusion barriers, or "fences," within the lipid bilayer of the plasma membrane; these junctions also block the backflow of glucose from the basal side of the epithelium into the gut lumen (see Movie 11.2).

The absorptive cells have to transport selected nutrients across the epithelium from the lumen into the extracellular fluid on the other side. From there, these nutrients diffuse into small blood vessels to provide nourishment to the organism. This *transcellular transport* depends on two sets of transport proteins in the plasma membrane of the absorptive cell. One set is confined to the apical surface of the cell (facing the lumen) and actively transports selected molecules into the cell from the gut. The other set is confined to the *basolateral* (basal and lateral) surfaces of the cell, and it allows the same molecules to leave the cell by passive transport into the extracellular fluid on the other side of the epithelium. For this transport activity to be effective, the spaces between the epithelial cells must be tightly sealed, so that the transported molecules cannot leak back into the gut lumen through these spaces (**Figure 19–18**). Moreover, the transport proteins must be correctly distributed in the plasma membranes: the apical transporters must be delivered to the apical membrane and must not be allowed to drift to the basolateral membrane, and the basolateral transporters must be delivered to and remain in the basolateral membrane. Tight junctions, besides sealing the gaps between the cells, also function as "fences" that help prevent apical or basolateral proteins from diffusing into the wrong region.

The sealing function of tight junctions is easy to demonstrate experimentally: a low-molecular-mass tracer added to one side of an epithelium will generally not pass beyond the tight junction (**Figure 19–19**). This seal is not absolute, however. Although all tight junctions are impermeable to macromolecules, their permeability to ions and other small molecules varies. Tight junctions in the epithelium lining the small intestine, for example, are 10,000 times more permeable to inorganic ions, such as Na^+, than the tight junctions in the epithelium lining the urinary bladder. The movement of ions and other molecules between epithelial cells is called *paracellular transport*, and tissue-specific differences in transport rates generally result from differences in the proteins that form tight junctions.

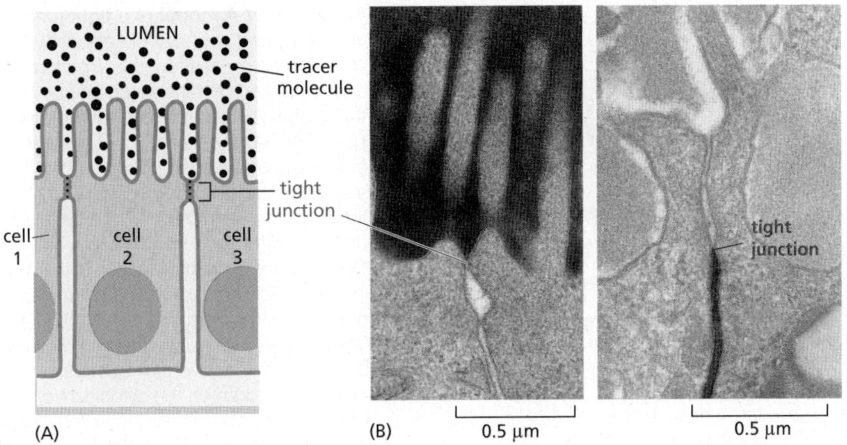

Figure 19–19 **The role of tight junctions in allowing epithelia to serve as barriers to solute diffusion.** (A) The drawing shows how a small extracellular tracer molecule added on one side of an epithelium is prevented from crossing the epithelium by the tight junctions that seal adjacent cells together. Adherens junctions and other cell junctions are not shown for clarity. (B) Electron micrographs of cells in an epithelium in which a small, extracellular, electron-dense tracer molecule has been added to either the apical side (on the *left*) or the basolateral side (on the *right*). The tight junction blocks passage of the tracer in both directions. (B, courtesy of Daniel Friend, by permission of E.L. Bearer.)

Tight Junctions Contain Strands of Transmembrane Adhesion Proteins

When tight junctions are visualized by freeze-fracture electron microscopy, they are seen as a branching network of *sealing strands* that completely encircles the apical end of each cell in the epithelial sheet (**Figure 19–20A and B**). In conventional electron micrographs, the outer leaflets of the two interacting plasma membranes are tightly apposed where sealing strands are present (**Figure 19–20C**). Each sealing strand is composed of a long row of transmembrane homophilic adhesion proteins embedded in each of the two interacting plasma membranes. The extracellular domains of these proteins adhere directly to one another to occlude the intercellular space (**Figure 19–21**).

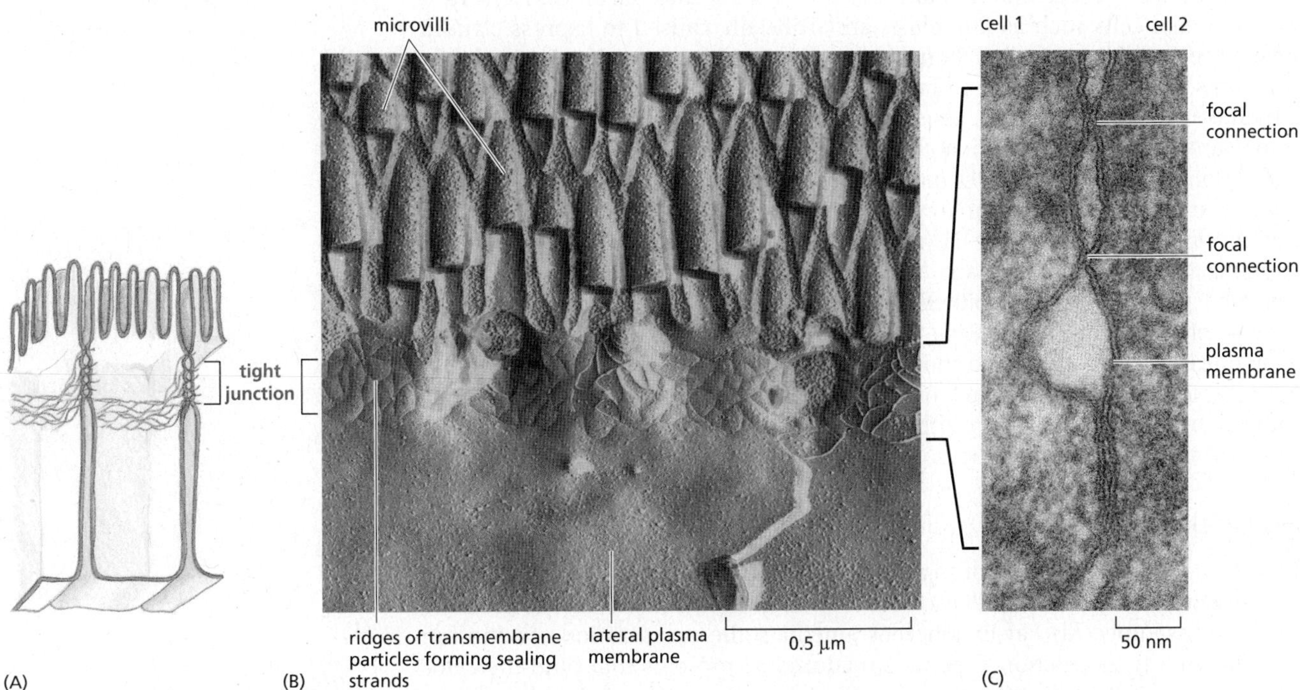

Figure 19–20 **The structure of a tight junction between epithelial cells of the small intestine.** The junctions are shown (A) schematically, (B) in a freeze-fracture electron micrograph, and (C) in a conventional electron micrograph. In B, the plane of the micrograph is parallel to the plane of the membrane, and the tight junction appears as a band of branching sealing strands that encircle each cell in the epithelium (see Figure 19–21A). In C, the junction is seen in cross section as a series of focal connections between the outer leaflets of the two interacting plasma membranes, each connection corresponding to a sealing strand in cross section. [B and C, from N.B. Gilula, in Cell Communication (R.P. Cox, ed.), pp. 1–29. New York: Wiley, 1974. With permission from John Wiley & Sons.]

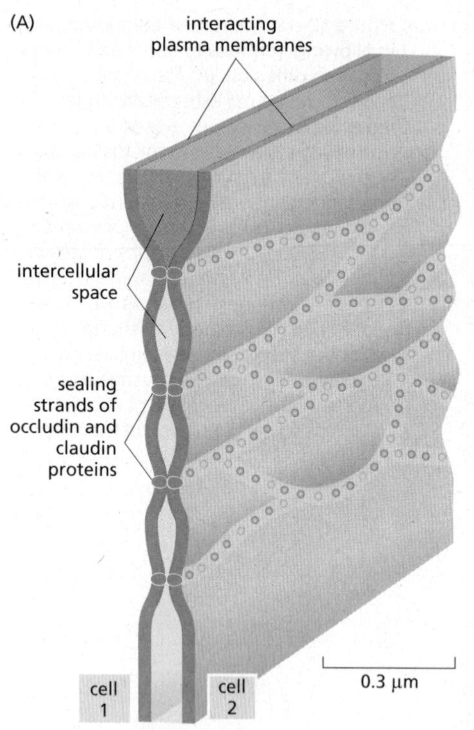

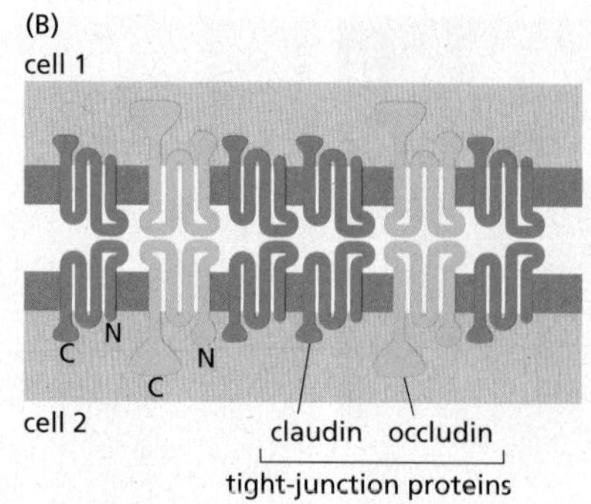

Figure 19–21 A model of a tight junction. (A) The sealing strands hold adjacent plasma membranes together. The strands are composed of transmembrane proteins that make contact across the intercellular space and create a seal. (B) The molecular composition of a sealing strand. The major extracellular components of the tight junction are members of a family of proteins with four transmembrane domains. One of these proteins, claudin, is the most important for the assembly and structure of the sealing strands, whereas the related protein occludin governs junction permeability. The two termini of these proteins are both on the cytoplasmic side of the membrane, where they interact with large scaffolding proteins that organize the sealing strands and link the tight junction to the actin cytoskeleton (not shown here, but see Figure 19–22).

The main transmembrane proteins forming these strands are the *claudins*, which are essential for tight-junction formation and function. Mice that lack the *claudin-1* gene, for example, fail to make tight junctions between the cells in the epidermal layer of the skin; as a result, the baby mice lose water rapidly by evaporation through the skin and die within a day after birth. Conversely, if nonepithelial cells such as fibroblasts are artificially caused to express claudin genes, they will form tight-junctional connections with one another. Normal tight junctions also contain a second major transmembrane protein called *occludin*, which is not essential for the assembly or structure of the tight junction but is important for limiting junctional permeability. A third transmembrane protein, *tricellulin*, is required to seal cell membranes together and prevent transepithelial leakage at the points where three cells meet.

The claudin protein family has many members (24 in humans), and these are expressed in different combinations in different epithelia to confer particular permeability properties on the epithelial sheet. They are thought to form *paracellular pores*—selective channels allowing specific ions to cross the tight-junctional barrier, from one extracellular space to another. A specific claudin found in kidney epithelial cells, for example, is needed to let Mg^{2+} pass between the cells of the kidney tubules so that this ion can be resorbed from the urine into the blood. A mutation in the gene encoding this claudin results in excessive loss of Mg^{2+} in the urine.

Scaffold Proteins Organize Junctional Protein Complexes

Like the cadherin molecules of an adherens junction, the claudins and occludins of a tight junction interact with each other on their extracellular sides to promote junction assembly. Also as in adherens junctions, the organization of adhesion proteins in a tight junction depends on additional proteins that bind the cytoplasmic side of the adhesion proteins. The key organizational proteins at tight junctions are the *zonula occludens* (*ZO*) proteins. The three major members of the ZO family—ZO-1, ZO-2, and ZO-3—are large scaffold proteins that provide a structural support on which the tight junction is built. These intracellular molecules consist of strings of protein-binding domains, typically including several **PDZ domains** that can recognize and bind the C-terminal tails of specific partner proteins (**Figure 19–22**). One domain of these scaffold proteins can attach to

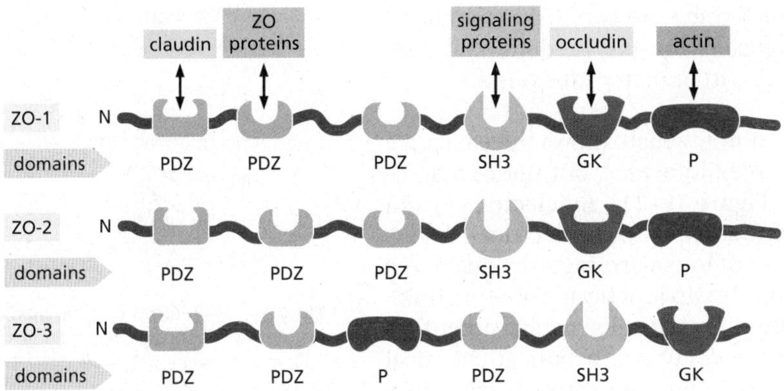

Figure 19–22 **Scaffold proteins at the tight junction.** The scaffold proteins ZO-1, ZO-2, and ZO-3 are concentrated beneath the plasma membrane at tight junctions. Each of the proteins contains multiple protein-binding domains, including three PDZ domains, an SH3 domain, a GK domain, and a proline-rich domain (P), linked together like beads on a flexible string. These domains enable the proteins to interact with each other and with numerous other partners, as indicated here by arrows, to generate a tightly woven protein network that organizes the sealing strands of the tight junction and links them to the actin cytoskeleton. Scaffold proteins with similar structure help organize other junctional complexes, including those at neural synapses.

a claudin protein, while others can attach to occludin or the actin cytoskeleton. Moreover, one molecule of scaffold protein can bind to another. In this way, the cell assembles a meshwork of intracellular proteins that organizes and positions the sealing strands of the tight junction.

The tight-junctional network of sealing strands usually lies just apical to adherens and desmosome junctions that bond the cells together mechanically; the whole assembly is called a *junctional complex* (see Figure 19–2). The parts of this junctional complex depend on each other for their formation. For example, anti-cadherin antibodies that block the formation of adherens junctions also inhibit the formation of tight junctions.

Gap Junctions Couple Cells Both Electrically and Metabolically

Tight junctions block the passageways through the gaps between epithelial cells, preventing extracellular molecules from leaking from one side of an epithelium to the other. Another type of junctional structure has a radically different function: it bridges gaps between adjacent cells so as to create direct channels from the cytoplasm of one to that of the other. These channels are called **gap junctions**.

Gap junctions are present in most animal tissues, including connective tissues as well as epithelia and heart muscle. Each gap junction appears in conventional electron micrographs as a patch where the membranes of two adjacent cells are separated by a uniform narrow gap of about 2–4 nm (**Figure 19–23**). The gap is

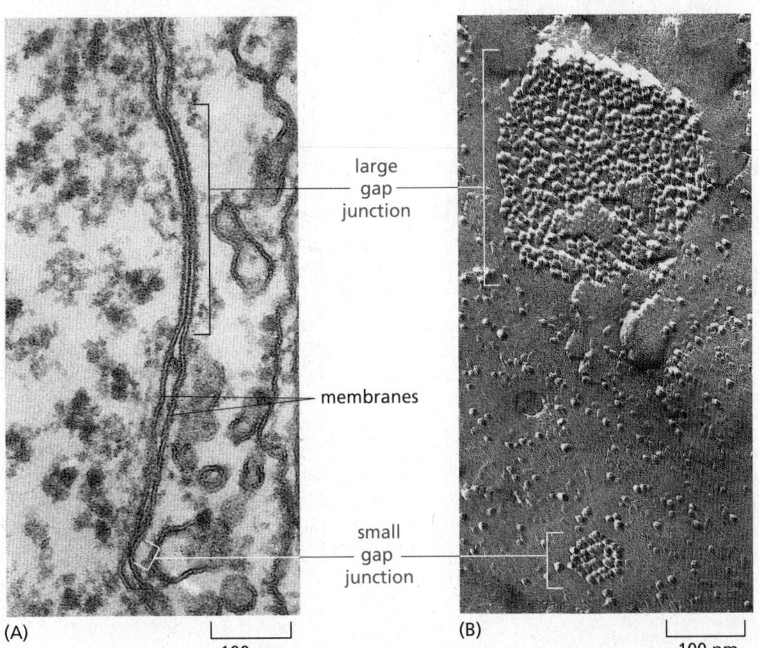

Figure 19–23 **Gap junctions as seen in the electron microscope.** (A) Thin-section and (B) freeze-fracture electron micrographs of a large and a small gap-junction plaque between fibroblasts in culture. In B, each gap junction is seen as a cluster of homogeneous intramembrane particles. Each intramembrane particle corresponds to a connexon (see Figure 19–25). [From N.B. Gilula, in Cell Communication (R.P. Cox, ed.), pp. 1–29. New York: Wiley, 1974. With permission from John WIley and Sons.]

spanned by channel-forming proteins, of which there are two distinct families, called the *connexins* and the *innexins*. Connexins are the predominant gap-junction proteins in vertebrates, with 21 isoforms in humans. Innexins are found in the gap junctions of invertebrates.

Gap junctions have a pore size of about 1.4 nm, which allows the exchange of inorganic ions and other small water-soluble molecules, but not of macromolecules such as proteins or nucleic acids (**Figure 19–24**). An electric current injected into one cell through a microelectrode causes an electrical disturbance in the neighboring cell, due to the flow of ions carrying electric charge through gap junctions. This electrical coupling via gap junctions serves an obvious purpose in tissues containing electrically excitable cells: action potentials can spread rapidly from cell to cell, without the delay that occurs at chemical synapses. In vertebrates, for example, electrical coupling through gap junctions synchronizes the contractions of heart muscle cells as well as those of the smooth muscle cells responsible for the peristaltic movements of the intestine. Gap junctions also occur in many tissues whose cells are not electrically excitable. In principle, the sharing of small metabolites and ions provides a mechanism for coordinating the activities of individual cells in such tissues and for smoothing out random fluctuations in small-molecule concentrations in different cells.

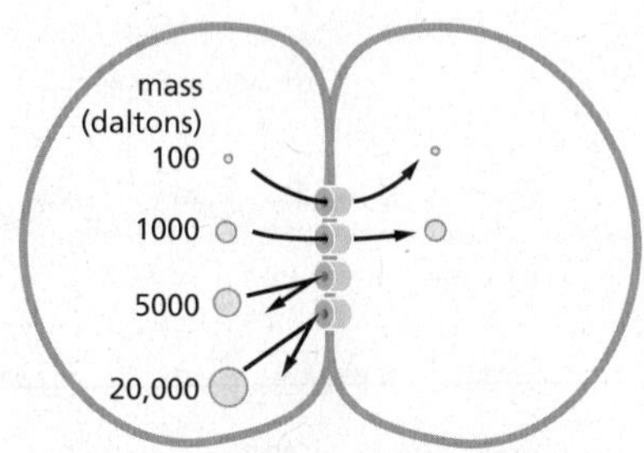

Figure 19–24 **Determining the size of a gap-junction channel.** When fluorescent molecules of various sizes are injected into one of two cells coupled by gap junctions, molecules with a mass of less than about 1000 daltons can pass into the other cell, but larger molecules cannot. Thus, the coupled cells share their small molecules (such as inorganic ions, sugars, amino acids, nucleotides, vitamins, and the intracellular signaling molecules cyclic AMP and inositol trisphosphate) but not their macromolecules (proteins, nucleic acids, and polysaccharides).

A Gap-Junction Connexon Is Made of Six Transmembrane Connexin Subunits

Connexins are four-pass transmembrane proteins, six of which assemble to form a *hemichannel*, or **connexon**. When the connexons in the plasma membranes of two cells in contact are aligned, they form a continuous aqueous channel that connects the two cell interiors (**Figure 19–25**). A gap junction consists of many such connexon pairs, forming a sort of molecular sieve. Not only does this sieve provide a communication channel between cells, but it also provides a form of cell-cell adhesion that supplements the cadherin- and claudin-mediated adhesions we discussed earlier.

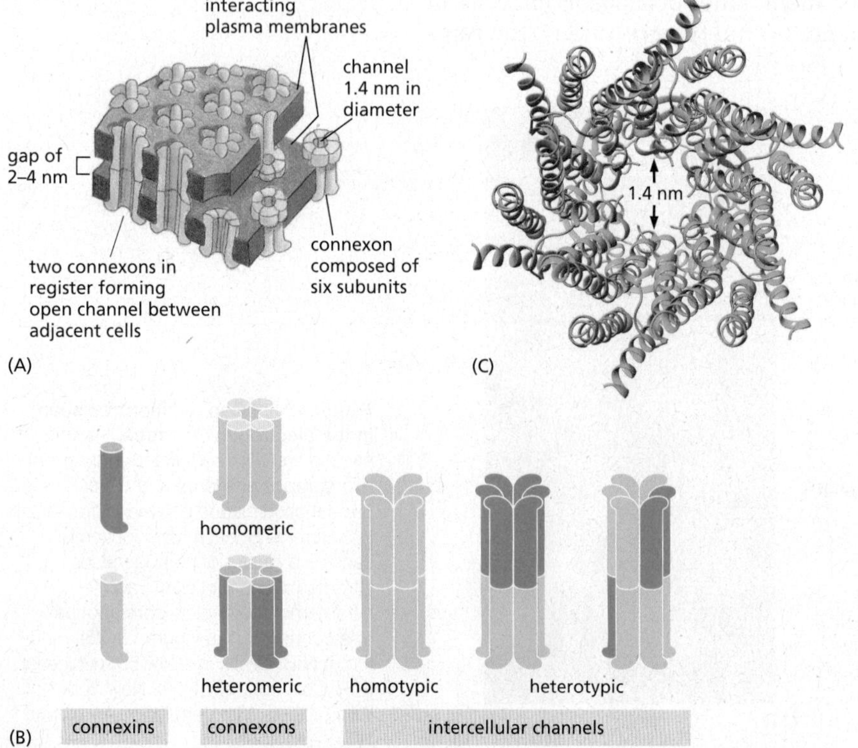

Figure 19–25 **Gap junctions.** (A) A drawing of the interacting plasma membranes of two adjacent cells connected by gap junctions. Each lipid bilayer is shown as a *gray* sheet. Protein assemblies called connexons *(green)*, each of which is formed by six connexin subunits, penetrate the apposed lipid bilayers. Two connexons join across the intercellular gap to form a continuous aqueous channel connecting the two cells. (B) The organization of connexins into connexons, and connexons into intercellular channels. The connexons can be homomeric or heteromeric, and the intercellular channels can be homotypic or heterotypic. (C) The high-resolution structure of a homomeric gap-junction channel, determined by x-ray crystallography of human connexin 26. In this view, we are looking down on the pore, formed from six connexin subunits. The structure illustrates the general features of the channel and suggests a pore size of about 1.4 nm, as predicted from studies of gap-junction permeability with molecules of various sizes (see Figure 19–24). (PDB code: 2ZW3.)

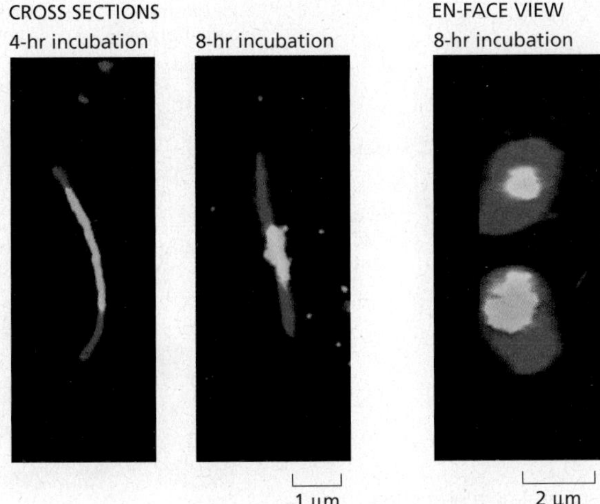

Figure 19–26 Connexin turnover at a gap junction. Cells were transfected with a slightly modified connexin gene, coding for a connexin with a short amino acid tag containing four cysteines in the sequence Cys-Cys-X-X-Cys-Cys (where X denotes an arbitrary amino acid). This *tetracysteine tag* can bind strongly to certain small fluorescent dye molecules, which can be added to the culture medium and will readily enter cells by diffusing across the plasma membrane. In the experiment shown, a green dye was added first to label all the connexin molecules in the cells, and the cells were then washed and incubated for 4 or 8 hours. At the end of this time, a red dye was added to the medium and the cells were washed again and fixed. Connexin molecules already present at the beginning of the experiment are labeled green (and take up no red dye because their tetracysteine tags are already saturated with green dye), while connexins synthesized subsequently, during the 4- or 8-hour incubation, are labeled red. The fluorescence images show gap junctions between pairs of cells treated in this way. The central part of the gap-junction plaque is *green*, indicating that it consists of old connexin molecules, while the periphery is *red*, indicating that it consists of connexins synthesized during the previous 4 or 8 hours. The longer the time of incubation, the smaller the green central patch of old molecules, and the larger the peripheral ring of new molecules that have been recruited to replace the old ones. (From G. Gaietta et al., *Science* 296:503–507, 2002. With permission from AAAS.)

Gap junctions in different tissues can have different properties because they are formed from different combinations of connexins, creating channels that differ in permeability and regulation. Most cell types express more than one type of connexin, and two different connexin proteins can assemble into a heteromeric connexon, with its own distinct properties. Moreover, adjacent cells expressing different connexins can form intercellular channels in which the two aligned half-channels are different (see Figure 19–25B).

Like conventional ion channels (discussed in Chapter 11), individual gap-junction channels do not remain open all the time; instead, they flip between open and closed states. These changes are triggered by a variety of stimuli, including the voltage difference between the two connected cells, the membrane potential of each cell, and various chemical properties of the cytoplasm, including the pH and concentration of free Ca^{2+}. Some subtypes of gap junctions can also be regulated by extracellular signals such as neurotransmitters. We are only just beginning to understand the physiological functions and structural basis of these various gating mechanisms.

Each gap-junctional plaque is a dynamic structure that can readily assemble, disassemble, or be remodeled, and it can contain a cluster of a few to many thousands of connexons (see Figure 19–23B). Studies with fluorescently labeled connexins in living cells show that new connexons are continually added around the periphery of an existing junctional plaque, while old connexons are removed from the middle of it and destroyed (**Figure 19–26**). This turnover is rapid: the connexin molecules have a half-life of only a few hours.

The mechanism of removal of old connexons from the middle of the plaque is not known, but the route of delivery of new connexons to its periphery seems clear: they are inserted into the plasma membrane by exocytosis, like other integral membrane proteins, and then diffuse in the plane of the membrane until they bump into the periphery of a connexon plaque and become trapped. This has a corollary: the plasma membrane away from the gap junction should contain connexons—hemichannels—that have not yet paired with their counterparts on another cell. It is thought that these unpaired hemichannels are normally held in a closed conformation, preventing the cell from losing its small molecules by leakage through them. But there is also evidence that in some circumstances they can open and serve as channels for the release of small signal molecules.

In Plants, Plasmodesmata Perform Many of the Same Functions as Gap Junctions

The tissues of a plant are organized on different principles from those of an animal. Plant cells are imprisoned within tough *cell walls* composed of an extracellular matrix rich in cellulose and other polysaccharides, as we discuss later.

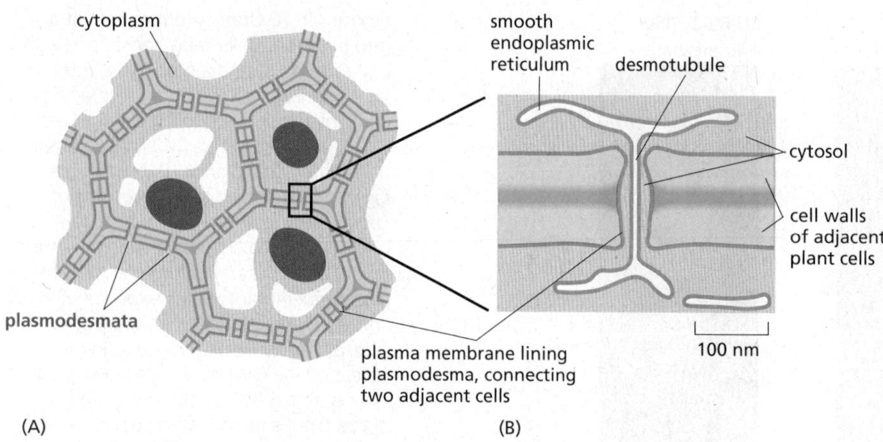

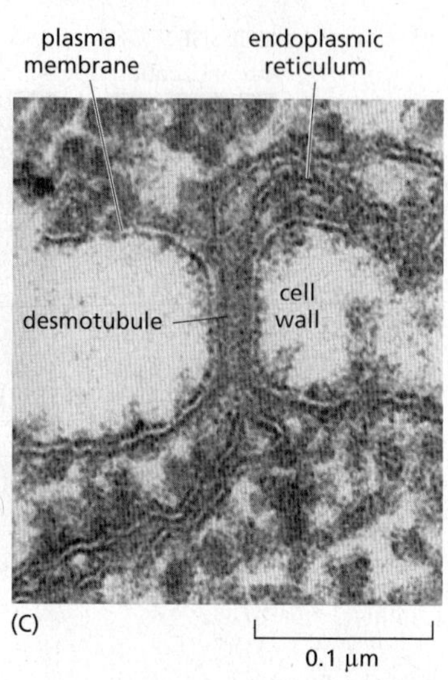

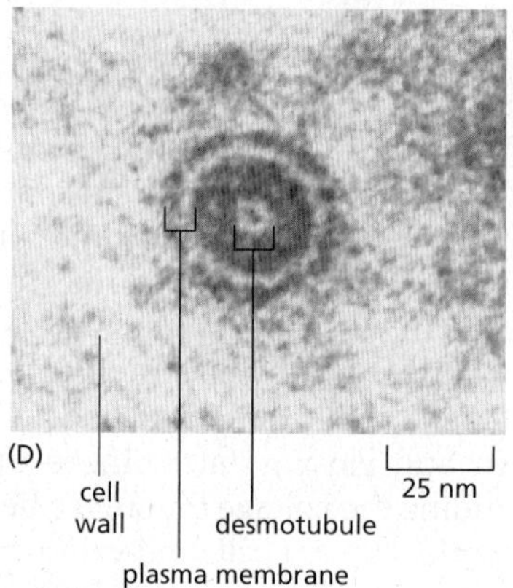

Figure 19–27 Plasmodesmata. (A) The cytoplasmic channels of plasmodesmata pierce the plant cell wall and connect cells in a plant together. (B) Each plasmodesma is lined with plasma membrane that is common to the two connected cells. It usually also contains a fine tubular structure, the desmotubule, derived from smooth endoplasmic reticulum. (C) Electron micrograph of a longitudinal section of a plasmodesma from a water fern. The plasma membrane lines the pore and is continuous from one cell to the next. Endoplasmic reticulum and its association with the central desmotubule can also be seen. (D) A similar plasmodesma seen in cross section. (C and D, from R. Overall et al., *Protoplasma* 111:134–150, 1982.)

The cell walls of adjacent cells are firmly cemented to one another, which eliminates the need for anchoring junctions to hold the cells in place. But a need for direct cell–cell communication remains. Thus, plant cells have only one class of intercellular junctions, **plasmodesmata.** Like gap junctions, they directly connect the cytoplasms of adjacent cells.

In plants, the cell wall between a typical pair of adjacent cells is at least 0.1 μm thick, and so a structure very different from a gap junction is required to mediate communication across it. Plasmodesmata solve the problem. With a few specialized exceptions, every cell in a higher plant is connected to its neighbors by these structures, which form fine cytoplasmic channels through the intervening cell walls. As shown in **Figure 19–27A**, the plasma membrane of one cell is continuous with that of its neighbor at each plasmodesma, which connects the cytoplasms of the two cells by a roughly cylindrical channel with a diameter of 20–40 nm.

Running through the center of the channel in most plasmodesmata is a narrower cylindrical structure, the *desmotubule,* which is continuous with elements of the smooth endoplasmic reticulum (ER) in each of the connected cells (**Figure 19–27B, C, and D**). Between the outside of the desmotubule and the inner face of the cylindrical channel formed by plasma membrane is an annulus of cytosol through which small molecules can pass from cell to cell. As each new cell wall is assembled during the cytokinesis phase of cell division, plasmodesmata are created within it. They form around elements of smooth ER that become trapped across the developing cell plate (discussed in Chapter 17). They can also be inserted *de novo* through preexisting cell walls, where they are commonly found in dense clusters called *pit fields.* When no longer required, plasmodesmata can be removed.

In spite of the radical difference in structure between plasmodesmata and gap junctions, they seem to function in remarkably similar ways. Evidence obtained by injecting tracer molecules of different sizes suggests that plasmodesmata allow the passage of molecules with a mass of less than about 800 daltons, which is similar to the size cutoff for gap junctions. As with gap junctions, transport through plasmodesmata is regulated. Dye-injection experiments, for example, show that there can be barriers to the movement of even low-molecular-mass molecules between certain cells or groups of cells that are connected by apparently normal plasmodesmata; the mechanisms that restrict communication in these cases are not understood.

Selectins Mediate Transient Cell–Cell Adhesions in the Bloodstream

We now complete our overview of cell-cell junctions and adhesion by briefly describing some of the more specialized adhesion mechanisms used in some tissues. In addition to those we have already discussed, at least three other superfamilies of cell-cell adhesion proteins are important: the *integrins*, the *selectins*, and the adhesive *immunoglobulin (Ig) superfamily* members. We shall discuss integrins in more detail later: their main function is in cell-matrix adhesion, but a few of them mediate cell-cell adhesion in specialized circumstances. Ca^{2+} dependence provides one simple way to distinguish among these classes of adhesion proteins experimentally. Selectins, like cadherins and integrins, require Ca^{2+} for their adhesive function; Ig superfamily members do not.

Selectins are cell-surface carbohydrate-binding proteins (*lectins*) that mediate a variety of transient cell-cell adhesion interactions in the bloodstream. Their main role, in vertebrates at least, is in governing the traffic of white blood cells into lymphoid organs and inflamed tissues. White blood cells lead a nomadic life, roving between the bloodstream and the tissues, and this necessitates special adhesive behavior. The selectins control the binding of white blood cells to the endothelial cells lining blood vessels, thereby enabling the blood cells to migrate out of the bloodstream into a tissue.

Each selectin is a transmembrane protein with a conserved lectin domain that binds to a specific oligosaccharide on another cell (**Figure 19–28A**). There are at least three types: *L-selectin* on white blood cells, *P-selectin* on blood platelets and on endothelial cells that have been locally activated by an inflammatory response, and *E-selectin* on endothelial cells later in the inflammatory response. In a lymphoid organ, such as a lymph node or the spleen, the endothelial cells express oligosaccharides that are recognized by L-selectin on lymphocytes, causing the lymphocytes to loiter and become trapped. At sites of inflammation, the roles are reversed: the endothelial cells switch on expression of selectins that recognize the oligosaccharides on white blood cells and platelets, flagging the cells down to help deal with the local emergency. Selectins do not act alone, however; they collaborate with integrins, which strengthen the binding of the blood cells to the endothelium. The cell-cell adhesions mediated by both selectins and integrins are *heterophilic*; that is, the binding is to a molecule of a different type. Selectins bind to specific oligosaccharides on glycoproteins and glycolipids, while integrins bind to specific Ig-family proteins.

Selectins and integrins act in sequence to let white blood cells leave the bloodstream and enter tissues (**Figure 19–28B**). The selectins mediate a weak adhesion because the binding of the lectin domain of the selectin to its carbohydrate ligand

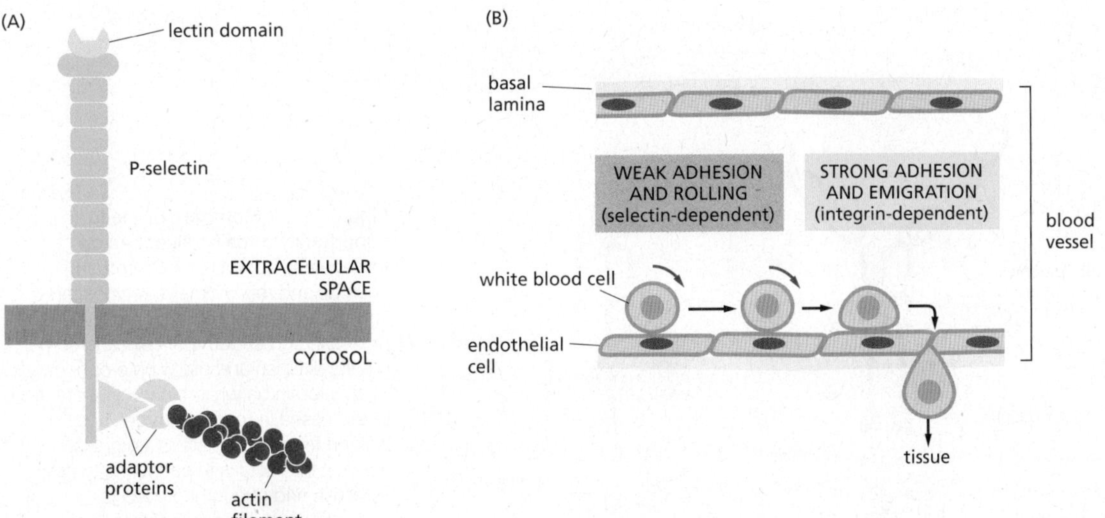

Figure 19–28 **The structure and function of selectins.** (A) Diagram of P-selectin, which attaches to the actin cytoskeleton through adaptor proteins. (B) How selectins and integrins mediate the cell–cell adhesions required for a white blood cell to migrate out of the bloodstream into a tissue. Selectins on endothelial cells bind weakly to oligosaccharides on the white blood cell, so that it becomes loosely attached and rolls along the vessel wall. The white blood cell then activates a cell-surface integrin called LFA1, which binds to a protein called ICAM1 (belonging to the Ig superfamily) on the membrane of the endothelial cell. The white blood cell adheres to the vessel wall and then crawls out of the vessel (**Movie 19.2**).

is of low affinity. This allows the white blood cell to adhere weakly and reversibly to the endothelium, rolling along the surface of the blood vessel, propelled by the flow of blood. The rolling continues until the blood cell activates its integrins. As we discuss later, these transmembrane molecules can be switched into an adhesive conformation that enables them to latch onto specific macromolecules external to the cell—in the present case, proteins on the surfaces of the endothelial cells. Once it has attached in this way, the white blood cell escapes from the bloodstream into the tissue by crawling out of the blood vessel between adjacent endothelial cells.

Members of the Immunoglobulin Superfamily Mediate Ca^{2+}-independent Cell–Cell Adhesion

The chief endothelial-cell proteins that are recognized by the white blood cell integrins are called *ICAMs* (*intercellular cell adhesion molecules*) or *VCAMs* (*vascular cell adhesion molecules*). They are members of another large and ancient family of cell-surface molecules—the **immunoglobulin (Ig) superfamily**. These contain one or more extracellular Ig-like domains that are characteristic of antibody molecules. They have many functions outside the immune system that are unrelated to immune defenses.

While ICAMs and VCAMs on endothelial cells both mediate heterophilic binding to integrins, many other Ig superfamily members appear to mediate homophilic binding. An example is the *neural cell adhesion molecule* (*NCAM*), which is expressed by various cell types, including most nerve cells (**Figure 19–29**). NCAM can take different forms, generated by alternative splicing of an RNA transcript produced from a single gene. Some forms of NCAM carry an unusually large quantity of sialic acid (with chains containing hundreds of repeating sialic acid units). By virtue of their negative charge, the long polysialic acid chains can interfere with cell adhesion (because like charges repel one another); thus, these forms of NCAM can serve to inhibit adhesion rather than cause it.

Another group of Ig superfamily members, the *nectins*, collaborates with cadherins to help build and strengthen adherens junctions in many tissues. Members of the nectin family contain three Ig-like domains (see Figure 19–29) and interact with other nectins, sometimes with the same family member (homophilic) and sometimes with a different family member (heterophilic). Their short intracellular tail binds to an adaptor protein that links the nectin to the actin cytoskeleton and to cadherin at adherens junctions. Nectins help set up cell-cell interactions during embryonic development. In auditory and olfactory epithelia, for example, sensory cells are distributed as single cells in a field of larger support cells. The distribution of these cell types in the epithelium depends on specific interactions

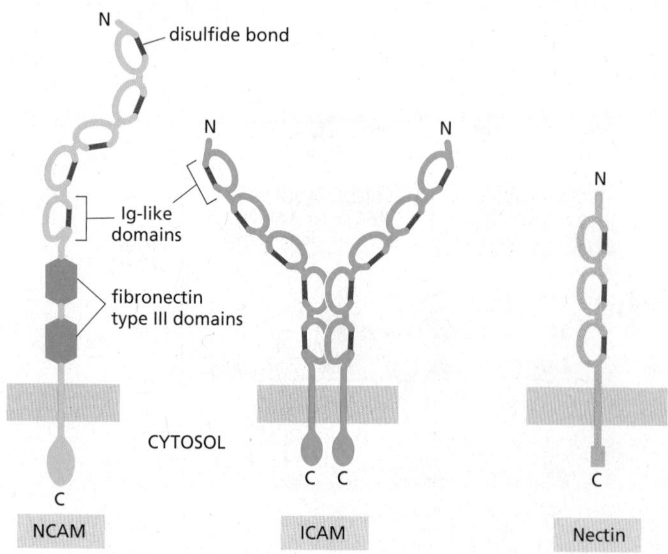

Figure 19–29 **Members of the Ig superfamily of cell–cell adhesion molecules.** NCAM is expressed on neurons and many other cell types, and it mediates homophilic binding. ICAM is expressed on endothelial cells and some other cell types and binds heterophilically to an integrin on white blood cells. Nectin is expressed in many cell types and is often found at adherens junctions, where it interacts with cadherins to help establish and strengthen specific cell–cell interactions during tissue formation.

between nectin family members on the surfaces of the sensory and support cells. The formation of these epithelia is defective when nectin expression is experimentally reduced.

A cell of a given type generally uses an assortment of different adhesion proteins to interact with other cells, just as each cell uses an assortment of different receptors to respond to the many soluble extracellular signal molecules in its environment. Although cadherins and Ig superfamily members are frequently expressed on the same cells, the adhesions mediated by cadherins are much stronger, and they are largely responsible for holding cells together, segregating cell collectives into discrete tissues, and maintaining tissue integrity. Members of the Ig superfamily seem to contribute more to the fine-tuning of these adhesive interactions during development and regeneration, playing a part in various specialized adhesive phenomena, such as that discussed for blood cells and specialized epithelia. Thus, while mutant mice that lack N-cadherin die early in development, those that lack Ig superfamily members develop relatively normally but show moderate abnormalities in the development of certain tissues.

Summary

In epithelia, as well as in some other tissues, cells are directly attached to one another through strong cell–cell adhesions, mediated by transmembrane proteins called cadherins, which are anchored intracellularly to the cytoskeleton. Cadherins generally bind to one another homophilically: the head of one cadherin molecule binds to the head of a similar cadherin on an opposite cell. This selectivity enables mixed populations of cells of different types to sort out from one another according to the specific cadherins they express, and it helps to control cell rearrangements during development.

The "classical" cadherins at adherens junctions are linked to the actin cytoskeleton by intracellular adaptor proteins called catenins. These form an anchoring complex on the intracellular tail of the cadherin molecule and are involved not only in physical anchorage but also in the detection of and response to tension and other regulatory signals at the junction.

Tight junctions seal the gaps between cells in epithelia, creating a barrier to the diffusion of molecules across the cell sheet and also helping to separate the populations of proteins in the apical and basolateral plasma membrane domains of the epithelial cell. Claudins are the major transmembrane proteins forming tight junctions. Intracellular scaffold proteins organize the claudins and other junctional proteins into a complex protein network that is linked to the actin cytoskeleton.

The cells of many animal tissues are coupled by gap junctions, which take the form of plaques of clustered connexons, which usually allow molecules smaller than about 1000 daltons to pass directly from the inside of one cell to the inside of the next. Cells connected by gap junctions share many of their inorganic ions and other small molecules and are therefore chemically and electrically coupled.

Three additional classes of transmembrane adhesion proteins mediate more transient cell–cell adhesion: selectins, immunoglobulin (Ig) superfamily members, and integrins. Selectins are expressed on white blood cells, blood platelets, and endothelial cells; they bind heterophilically to carbohydrate groups on cell surfaces, helping to mediate the adhesive interactions between these cells. Ig superfamily proteins also play a part in these interactions, as well as in many other adhesive processes; some of them bind homophilically, some heterophilically. Integrins, though they mainly serve to attach cells to the extracellular matrix, can also mediate cell–cell adhesion by binding to specific Ig superfamily proteins.

THE EXTRACELLULAR MATRIX OF ANIMALS

Tissues are not made up solely of cells. They also contain a remarkably complex and intricate network of macromolecules constituting the *extracellular matrix*. This matrix is composed of many different proteins and polysaccharides that are

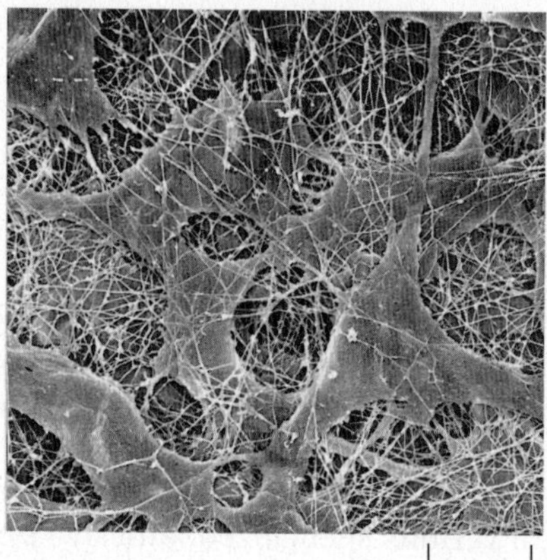

Figure 19–30 **Fibroblasts in connective tissue.** This scanning electron micrograph shows tissue from the cornea of a rat. The extracellular matrix surrounding the fibroblasts is here composed largely of collagen fibrils. The glycoproteins, hyaluronan, and proteoglycans, which normally form a hydrated gel filling the interstices of the fibrous network, have been removed by enzyme and acid treatment. (Courtesy of T. Nishida.)

secreted locally and assembled into an organized meshwork in close association with the surfaces of the cells that produce them.

The classes of macromolecules constituting the extracellular matrix in different animal tissues are broadly similar, but variations in the relative amounts of these different classes of molecules and in the ways in which they are organized give rise to an amazing diversity of materials. The matrix can become calcified to form the rock-hard structures of bone or teeth, or it can form the transparent substance of the cornea, or it can adopt the rope-like organization that gives tendons their enormous tensile strength. It forms the jelly in a jellyfish. Covering the body of a beetle or a lobster, it forms a rigid carapace. Moreover, the extracellular matrix is more than a passive scaffold to provide physical support. It has an active and complex role in regulating the behavior of the cells that touch it, inhabit it, or crawl through its meshes, influencing their survival, development, migration, proliferation, shape, and function.

In this section, we describe the major features of the extracellular matrix in animal tissues, with an emphasis on vertebrates. We begin with an overview of the major classes of macromolecules in the matrix, after which we turn to the structure and function of the *basal lamina*, the thin layer of specialized extracellular matrix that lies beneath all epithelial cells. In the next section, we then describe the varied types of junctions that connect cells to the matrix.

The Extracellular Matrix Is Made and Oriented by the Cells Within It

The macromolecules that constitute the extracellular matrix are mainly produced locally by cells in the matrix. These cells also help to organize the matrix: the orientation of the cytoskeleton inside the cell can control the orientation of the matrix produced outside. In most connective tissues, the matrix macromolecules are secreted by cells called **fibroblasts** (**Figure 19–30**). In certain specialized types of connective tissues, such as cartilage and bone, however, they are secreted by cells of the fibroblast family that have more specific names: *chondrocytes*, for example, form cartilage, and *osteoblasts* form bone.

The extracellular matrix is constructed from three major classes of macromolecules: (1) *glycosaminoglycans* (*GAGs*), which are large and highly charged polysaccharides that are usually covalently linked to protein in the form of *proteoglycans*; (2) fibrous proteins, which are primarily members of the *collagen* family; and (3) a large class of noncollagen *glycoproteins*, which carry conventional asparagine-linked oligosaccharides (described in Chapter 12). All three classes of macromolecule have many members and come in a great variety of shapes and sizes (**Figure 19–31**). Mammals are thought to have almost 300 matrix proteins, including about 36 proteoglycans, about 40 collagens, and more than 200 glycoproteins, which usually contain multiple subdomains and self-associate to form multimers. Add to this the large number of matrix-associated proteins and enzymes that can modify matrix behavior by cross-linking, degradation, or other mechanisms, and one begins to see that the matrix is an almost infinitely variable material. Each tissue contains its own unique blend of matrix components, resulting in an extracellular matrix that is specialized for the needs of that tissue.

The proteoglycan molecules in connective tissue typically form a highly hydrated, gel-like "ground substance" in which collagens and glycoproteins are embedded. The polysaccharide gel resists compressive forces on the matrix while permitting the rapid diffusion of nutrients, metabolites, and hormones between the blood and the tissue cells. The collagen fibers strengthen and help organize

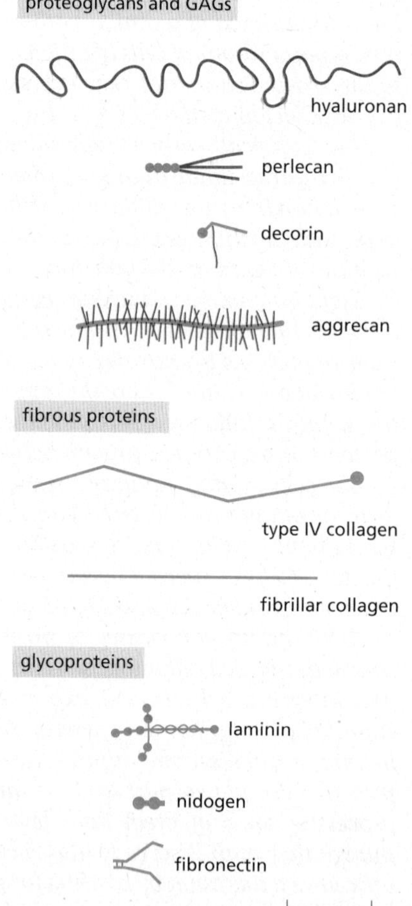

Figure 19–31 **The comparative shapes and sizes of some of the major extracellular matrix macromolecules.** Protein is shown in *green*, and glycosaminoglycan (GAG) in *red*.

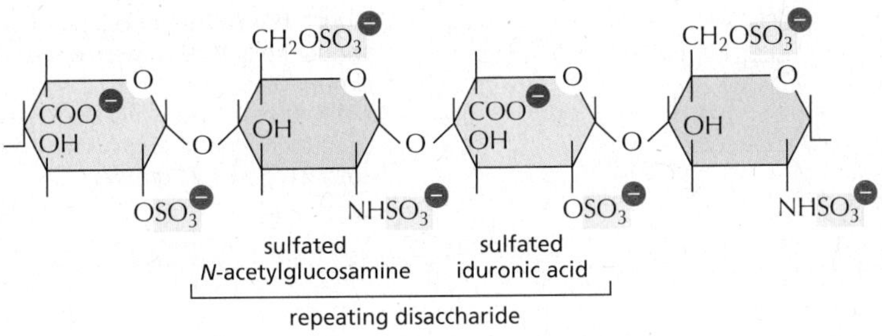

$CH_2OSO_3^-$ $CH_2OSO_3^-$ COO^- OH COO^- OH OH OH OSO_3^- $NHSO_3^-$ OSO_3^- $NHSO_3^-$

sulfated *N*-acetylglucosamine

sulfated iduronic acid

repeating disaccharide

Figure 19–32 The repeating disaccharide sequence of a heparin glycosaminoglycan (GAG) chain. These chains can consist of as many as 75 disaccharide units but are typically less than half that size. There is a high density of negative charges along the chain because of the presence of both carboxyl and sulfate groups; indeed, heparin is the most densely charged biological molecule known. The most common form of heparin carries three sulfate groups in each disaccharide, as shown here. *In vivo*, the proportion of sulfated and nonsulfated groups is highly variable. Heparin has an average of about 2.7 sulfates per disaccharide. Heparan sulfate is a closely related GAG that is generally about twice the length of heparin and less charged, with an average of about 1 sulfate per disaccharide.

the matrix, while other fibrous proteins, such as the rubberlike *elastin*, give it resilience. Finally, the many matrix glycoproteins help cells migrate, settle, and differentiate in the appropriate locations.

Glycosaminoglycan (GAG) Chains Occupy Large Amounts of Space and Form Hydrated Gels

Glycosaminoglycans (GAGs) are unbranched polysaccharide chains composed of repeating disaccharide units. One of the two sugars in the repeating disaccharide is always an amino sugar (*N*-acetylglucosamine or *N*-acetylgalactosamine), which in most cases is sulfated. The second sugar is usually a uronic acid (glucuronic or iduronic). Because there are sulfate or carboxyl groups on most of their sugars, GAGs are highly negatively charged (**Figure 19–32**). Indeed, they are the most anionic molecules produced by animal cells. Four main groups of GAGs are distinguished by their sugars, the type of linkage between the sugars, and the number and location of sulfate groups: (1) *hyaluronan*, (2) *chondroitin sulfate* and *dermatan sulfate*, (3) *heparin* and *heparan sulfate*, and (4) *keratan sulfate*.

Polysaccharide chains are too stiff to fold into compact globular structures, and they are strongly hydrophilic. Thus, GAGs tend to adopt highly extended conformations that occupy a huge volume relative to their mass (**Figure 19–33**), and they form hydrated gels even at very low concentrations. The weight of GAGs in connective tissue is usually less than 10% of the weight of proteins, but GAG chains fill most of the extracellular space. Their high density of negative charges attracts a cloud of cations, especially Na^+, that are osmotically active, causing large amounts of water to be sucked into the matrix. This creates a swelling pressure, or turgor, that enables the matrix to withstand compressive forces (in contrast to collagen fibrils, which resist stretching forces). The cartilage matrix that lines the knee joint, for example, can support pressures of hundreds of atmospheres in this way.

Defects in the production of GAGs can affect many different body systems. In one rare human genetic disease, for example, there is a severe deficiency in the synthesis of dermatan sulfate disaccharide. The affected individuals have a short stature, a prematurely aged appearance, and generalized defects in their skin, joints, muscles, and bones.

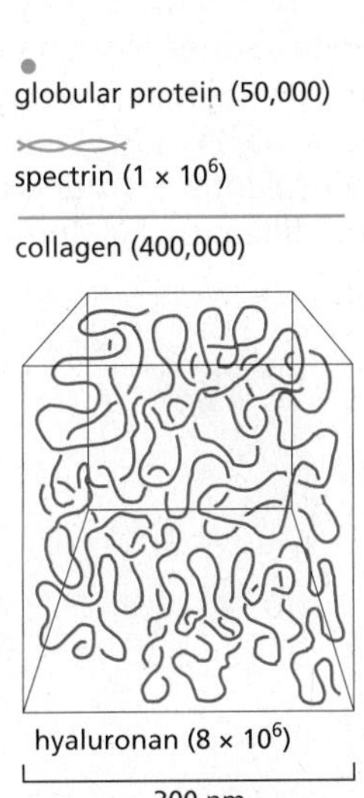

Figure 19–33 The relative dimensions and volumes occupied by various macromolecules. Several proteins and a single hydrated molecule of hyaluronan are shown, with molecular mass in daltons.

Hyaluronan Acts as a Space Filler During Tissue Morphogenesis and Repair

Hyaluronan (also called *hyaluronic acid* or *hyaluronate*) is the simplest of the GAGs (**Figure 19–34**). It consists of a regular repeating sequence of up to 25,000 disaccharide units, is found in variable amounts in all tissues and fluids in adult animals, and is especially abundant in early embryos. Hyaluronan is not a typical GAG because it contains no sulfated sugars, all its disaccharide units are identical, its chain length is enormous, and it is not generally linked covalently to any core protein. Moreover, whereas other GAGs are synthesized inside the cell and released by exocytosis, hyaluronan is spun out directly from the cell surface by an enzyme complex embedded in the plasma membrane.

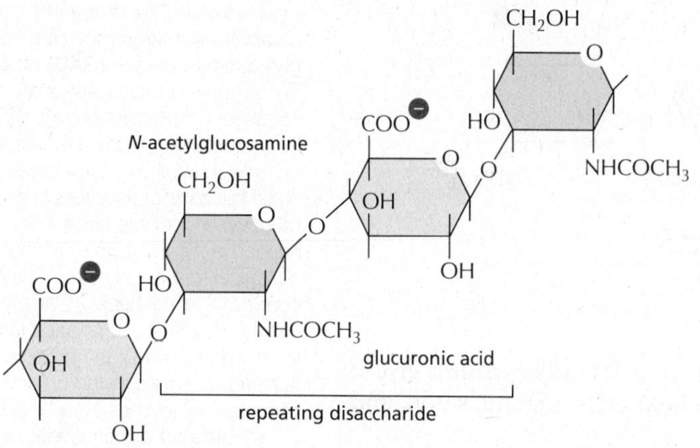

N-acetylglucosamine
glucuronic acid
repeating disaccharide

Figure 19–34 The repeating disaccharide sequence in hyaluronan, a relatively simple GAG. This ubiquitous molecule in vertebrates consists of a single long chain of up to 25,000 disaccharides. Note the absence of sulfate groups.

Hyaluronan is thought to have a role in resisting compressive forces in tissues and joints. It is also important as a space filler during embryonic development, where it can be used to force a change in the shape of a structure, as a small quantity expands with water to occupy a large volume. Hyaluronan synthesized locally from the basal side of an epithelium can deform the epithelium by creating a cell-free space beneath it, into which cells subsequently migrate. In the developing heart, for example, hyaluronan synthesis helps in this way to drive formation of the valves and septa that separate the heart's chambers. Similar processes occur in several other organs. When cell migration ends, the excess hyaluronan is generally degraded by the enzyme *hyaluronidase*. Hyaluronan is also produced in large quantities during wound healing, and it is an important constituent of joint fluid, in which it serves as a lubricant.

Proteoglycans Are Composed of GAG Chains Covalently Linked to a Core Protein

Except for hyaluronan, all GAGs are covalently attached to protein as **proteoglycans**, which are produced by most animal cells. Membrane-bound ribosomes make the polypeptide chain, or *core protein*, of a proteoglycan, which is then threaded into the lumen of the endoplasmic reticulum. The polysaccharide chains are mainly assembled on this core protein in the Golgi apparatus before delivery to the exterior of the cell by exocytosis. First, a special *linkage tetrasaccharide* is attached to a serine side chain on the core protein to serve as a primer for polysaccharide growth; then, one sugar at a time is added by specific glycosyl transferases (**Figure 19–35**). While still in the Golgi apparatus, many of the polymerized sugars are covalently modified by a sequential and coordinated series of reactions. These modifications include sulfation, which increases the negative charge, and

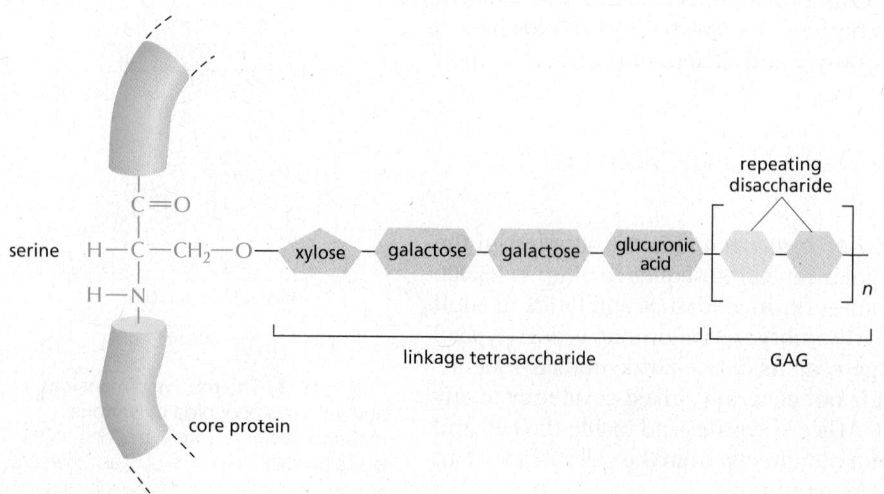

Figure 19–35 The linkage between a GAG chain and its core protein in a proteoglycan molecule. A specific linkage tetrasaccharide is first assembled on a serine side chain. The rest of the GAG chain, consisting mainly of a repeating disaccharide unit, is then synthesized, with one sugar added at a time. In chondroitin sulfate, the disaccharide is composed of D-glucuronic acid and *N*-acetyl-D-galactosamine; in heparan sulfate, it is either D-glucuronic acid or L-iduronic acid and *N*-acetyl-D-glucosamine.

epimerization, which alters the configuration of the substituents around individual carbon atoms in the sugar molecule.

Proteoglycans are clearly distinguished from other glycoproteins by the nature, quantity, and arrangement of their sugar side chains. By definition, at least one of the sugar side chains of a proteoglycan must be a GAG. Whereas glycoproteins generally contain relatively short, branched oligosaccharide chains that contribute only a small fraction of their mass, proteoglycans can contain as much as 95% carbohydrate by mass, mostly in the form of long, unbranched GAG chains, each typically about 80 sugars long.

In principle, proteoglycans have the potential for almost limitless heterogeneity. Even a single type of core protein can carry highly variable numbers and types of attached GAG chains. Moreover, the underlying repeating sequence of disaccharides in each GAG can be modified by a complex pattern of sulfate groups. The core proteins, too, are diverse, though many of them belong to structurally related families that share specific domains involved in binding to GAGs or other proteins.

Proteoglycans can be huge. The proteoglycan *aggrecan*, for example, which is a major component of cartilage, has a mass of about 3×10^6 daltons with more than 100 GAG chains. Other proteoglycans are much smaller and have only 1–10 GAG chains; an example is *decorin*, which is secreted by fibroblasts and has a single GAG chain (see Figure 19–31). Decorin binds to collagen fibrils and regulates fibril assembly and fibril diameter; mice that cannot make decorin have fragile skin that has reduced tensile strength. The GAGs and proteoglycans of these various types can associate to form even larger polymeric complexes in the extracellular matrix. Molecules of aggrecan, for example, assemble with hyaluronan in cartilage matrix to form aggregates that are as big as a bacterium (**Figure 19–36**). Moreover, besides associating with one another, GAGs and proteoglycans associate with fibrous matrix proteins such as collagen and with protein meshworks such as the basal lamina, creating extremely complex composites (**Figure 19–37**).

Not all proteoglycans are secreted components of the extracellular matrix. Some are integral components of plasma membranes and have their core protein

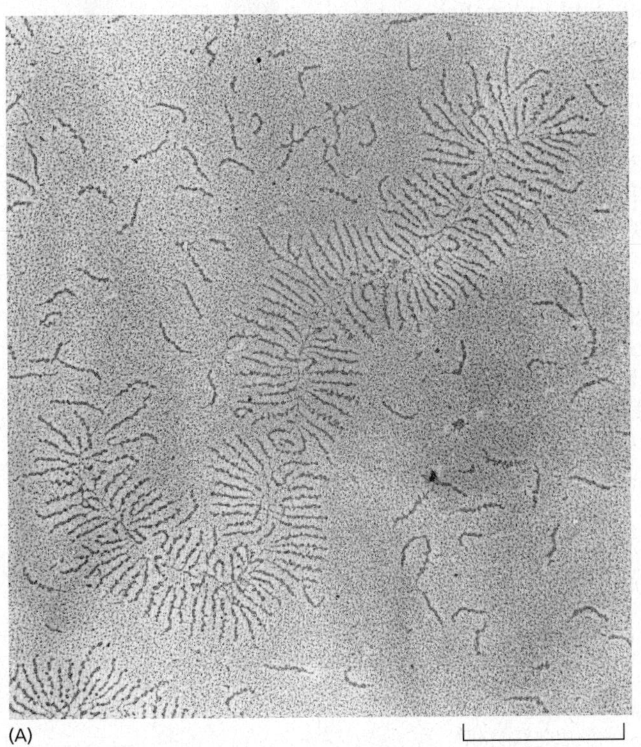

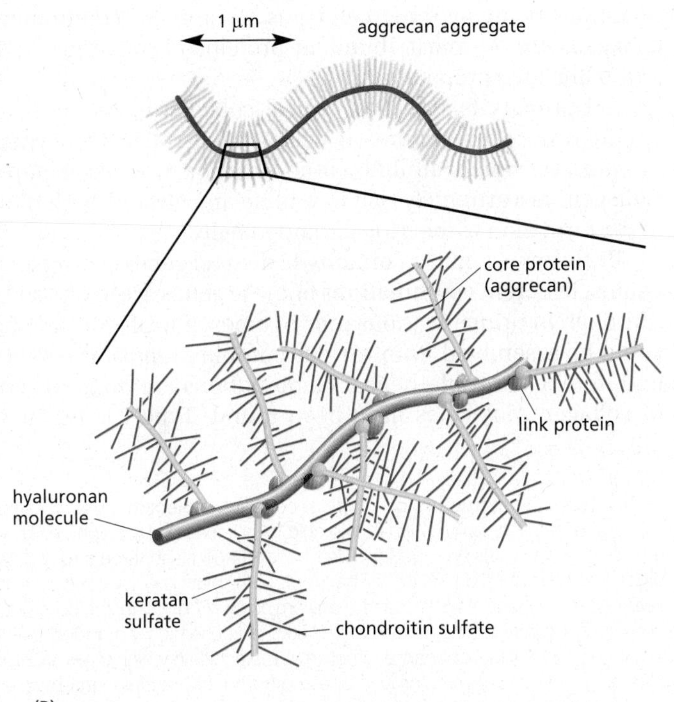

Figure 19–36 **An aggrecan aggregate from fetal bovine cartilage.** (A) An electron micrograph of an aggrecan aggregate shadowed with platinum. Many free aggrecan molecules are also visible. (B) A drawing of the giant aggrecan aggregate shown in A. It consists of about 100 aggrecan monomers (see Figure 19–31) noncovalently bound through the N-terminal domain of the core protein to a single hyaluronan molecule. A link protein binds both to the core protein of the proteoglycan and to the hyaluronan molecule, thereby stabilizing the aggregate. The molecular mass of such a complex can be 10^8 daltons or more, and it occupies a volume equivalent to that of a bacterium, which is about 2 μm^3. (A, courtesy of Lawrence Rosenberg.)

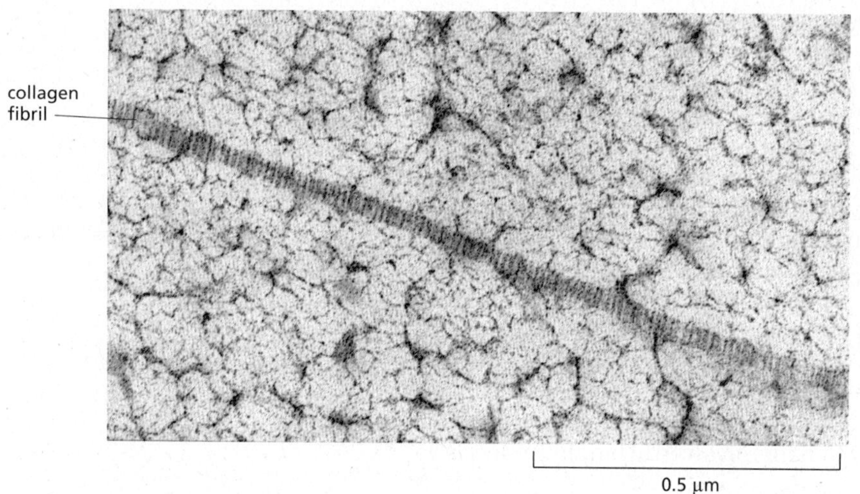

Figure 19–37 **Proteoglycans in the extracellular matrix of rat cartilage.** The tissue was rapidly frozen at −196°C and fixed and stained while still frozen (a process called freeze substitution) to prevent the GAG chains from collapsing. In this electron micrograph, the proteoglycan molecules are seen to form a fine filamentous network in which a single striated collagen fibril is embedded. The more darkly stained parts of the proteoglycan molecules are the core proteins; the faintly stained threads are the GAG chains. (© 1984 E.B. Hunziker and R.K. Schenk. Originally published in *J. Cell Biol.* https://doi.org/10.1083/jcb.98.1.277. With permission from Rockefeller University Press.)

either inserted across the lipid bilayer or attached to the lipid bilayer by a glycosylphosphatidylinositol (GPI) anchor. Among the best-characterized plasma membrane proteoglycans are the *syndecans*, which have a membrane-spanning core protein whose intracellular domain interacts with the actin cytoskeleton and with signaling molecules in the cell cortex. The extracellular domain is linked to multiple GAG chains (primarily heparan sulfate). Syndecans are located on the surface of many types of cells, including fibroblasts and epithelial cells. In fibroblasts, syndecans can be found in cell–matrix adhesions, where they modulate integrin function by interacting with fibronectin on the cell surface and with cytoskeletal and signaling proteins inside the cell. As we discuss later, syndecans and related proteoglycans called *glypicans* also interact with soluble peptide growth factors, influencing their effects on cell growth and proliferation.

Collagens Are the Major Proteins of the Extracellular Matrix

The **collagens** are a family of fibrous proteins found in all multicellular animals. They are secreted in large quantities by connective-tissue cells and in smaller quantities by many other cell types. As a major component of skin and bone, collagens are the most abundant proteins in mammals, where they constitute 25% of the total protein mass.

The primary feature of a typical collagen molecule is its long, stiff, triple-stranded helical structure, in which three collagen polypeptide chains, called *α chains*, are wound around one another in a rope-like superhelix (**Figure 19–38**). Collagens are extremely rich in proline and glycine, both of which are important in the formation of the triple-stranded helix.

The human genome contains 42 distinct genes coding for different collagen α chains. Different combinations of these genes are expressed in different tissues. Although in principle thousands of types of triple-stranded collagen molecules could be assembled from various combinations of the 42 α chains, only a limited number of triple-helical combinations are possible, and roughly 40 types of collagen molecules have been found. Type I is by far the most common,

Figure 19–38 **The structure of a typical collagen molecule.** (A) A model of part of a single collagen α chain, in which each amino acid is represented by a sphere. The chain is about 1000 amino acids long. It is arranged as a left-handed helix, with three amino acids per turn and with glycine as every third amino acid. Therefore, an α chain is composed of a series of triplet Gly-X-Y sequences, in which X and Y can be any amino acid (although X is commonly proline and Y is commonly hydroxyproline, a form of proline that is chemically modified during collagen synthesis in the cell). (B) A model of part of a collagen molecule, in which three α chains, each shown in a different color, are wrapped around one another to form a triple-stranded helical rod. Glycine is the only amino acid small enough to occupy the crowded interior of the triple helix. Only a short length of the molecule is shown; the entire molecule is 300 nm long. (From a model by B.L. Trus.)

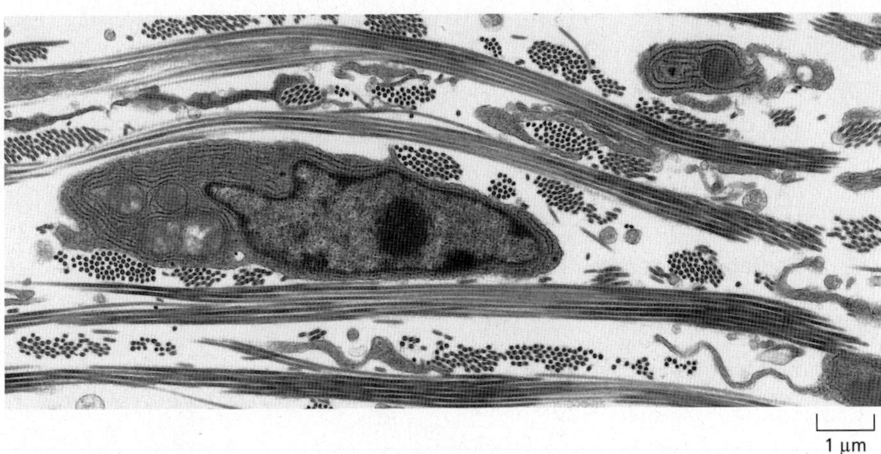

Figure 19–39 **A fibroblast surrounded by collagen fibrils in the connective tissue of embryonic chick skin.** In this electron micrograph, the fibrils are organized into bundles that run approximately at right angles to one another. Therefore, some bundles are oriented longitudinally, whereas others are seen in cross section. The collagen fibrils are produced by fibroblasts. (From C. Ploetz et al., *J. Struct. Biol.* 106:73–81, 1991. With permission from Elsevier.)

being the principal collagen of skin and bone. It belongs to the class of **fibrillar collagens**, or fibril-forming collagens: after being secreted into the extracellular space, they assemble into higher-order polymers called **collagen fibrils**, which are thin structures (10–300 nm in diameter) many hundreds of micrometers long in mature tissues, where they are clearly visible in electron micrographs (**Figure 19–39**; see also Figure 19–37). Collagen fibrils often aggregate into larger, cablelike bundles, several micrometers in diameter, that are visible in the light microscope as *collagen fibers*.

Collagen types IX and XII are called *fibril-associated collagens* because they decorate the surface of collagen fibrils. They are thought to link these fibrils to one another and to other components in the extracellular matrix. Types IV and VII are *network-forming collagens*: type IV forms a major part of the basal lamina, while type VII molecules form dimers that assemble into specialized structures called *anchoring fibrils*. Anchoring fibrils help attach the basal lamina of multilayered epithelia to the underlying connective tissue and therefore are especially abundant in the skin. There are also a number of "collagen-like" proteins containing short collagen-like segments. These include collagen type XVII, which has a transmembrane domain and is found in hemidesmosomes, and type XVIII, the core protein of a proteoglycan in the basal lamina.

Many proteins appear to have evolved by repeated duplications of an original DNA sequence, giving rise to a repetitive pattern of amino acids. The genes that encode the α chains of most of the fibrillar collagens provide a good example: they are very large (up to 44 kilobases in length) and contain about 50 exons. Most of the exons are 54 or multiples of 54 nucleotides long, suggesting that these collagens originated through multiple duplications of a primordial gene containing 54 nucleotides and encoding exactly six Gly-X-Y repeats (see Figure 19–38).

Table 19–2 provides additional details for some of the collagen types discussed in this chapter.

Collagen Chains Undergo a Series of Post-translational Modifications

Individual collagen polypeptide chains are synthesized on membrane-bound ribosomes and injected into the lumen of the endoplasmic reticulum (ER) as larger precursors, called *pro-α chains.* These precursors not only have the short amino-terminal signal peptide required to direct the nascent polypeptide to the

TABLE 19–2 Some Types of Collagen and Their Properties

	Type	Polymerized form	Tissue distribution	Mutant phenotype
Fibril-forming (fibrillar)	I	Fibril	Bone, skin, tendons, ligaments, cornea, internal organs (accounts for 90% of body collagen)	Severe bone defects, fractures (*osteogenesis imperfecta*)
	II	Fibril	Cartilage, intervertebral disc, notochord, vitreous humor of the eye	Cartilage deficiency, dwarfism (*chondrodysplasia*)
	III	Fibril	Skin, blood vessels, internal organs	Fragile skin, loose joints, blood vessels prone to rupture (*vascular Ehlers–Danlos syndrome*)
	V	Fibril (with type I)	As for type I	Fragile skin, loose joints (*classical Ehlers–Danlos syndrome*)
	XI	Fibril (with type II)	As for type II	Myopia, blindness
Fibril-associated	IX	Lateral association with type II fibrils	Cartilage	Osteoarthritis
	XII	Lateral association with type I fibrils	Tendons	Skeletal and muscle abnormalities
Network-forming	IV	Sheetlike network	Basal lamina	Kidney disease (glomerulonephritis), deafness
	VII	Anchoring fibrils	Beneath stratified squamous epithelia	Skin blistering
Transmembrane	XVII	Nonfibrillar	Hemidesmosomes	Skin blistering
Proteoglycan core protein	XVIII	Nonfibrillar	Basal lamina	Myopia, detached retina, hydrocephalus

Note that types I, IV, V, IX, and XI are each composed of two or three types of α chains (distinct, nonoverlapping sets in each case), whereas types II, III, VII, XII, XVII, and XVIII are composed of only one type of α chain each.

ER but also have, at both their N- and C-terminal ends, additional amino acids, called *propeptides,* that are clipped off at a later step of collagen assembly. Moreover, in the lumen of the ER, selected prolines and lysines are hydroxylated to form *hydroxyproline* and *hydroxylysine*, respectively, and some hydroxylysines are then glycosylated.

Each pro-α chain combines with two others to form a triple-stranded, helical molecule known as *procollagen*. The hydroxyl groups of hydroxyprolines and hydroxylysines (**Figure 19–40**) form interchain hydrogen bonds that help stabilize the triple-stranded helix. The enzyme that catalyzes proline hydroxylation requires ascorbic acid (vitamin C). In *scurvy*, the disease caused by a dietary deficiency of vitamin C that was common in sailors until the nineteenth century, defective pro-α chains fail to form a stable triple helix and are degraded, thereby inhibiting the production of new collagen fibrils. In healthy tissues, collagen is continually degraded and replaced (with a turnover time of months or years, depending on the tissue). In scurvy, replacement fails, and within a few months, with the gradual loss of the preexisting normal collagen in the matrix, blood vessels become fragile, teeth become loose in their sockets, and wounds cease to heal.

After secretion, the propeptides of the fibrillar procollagen molecules are removed by specific proteolytic enzymes outside the cell. This converts the procollagen molecules to collagen, which assemble in the extracellular space to form much larger collagen fibrils. The propeptides have at least two functions. First,

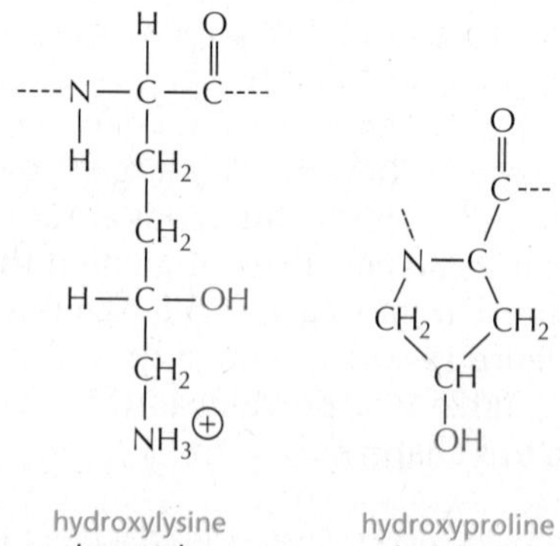

hydroxylysine in protein

hydroxyproline in protein

Figure 19–40 **Hydroxylysine and hydroxyproline.** These modified amino acids are common in collagen. They are formed by enzymes that act after the lysine and proline have been incorporated into procollagen molecules.

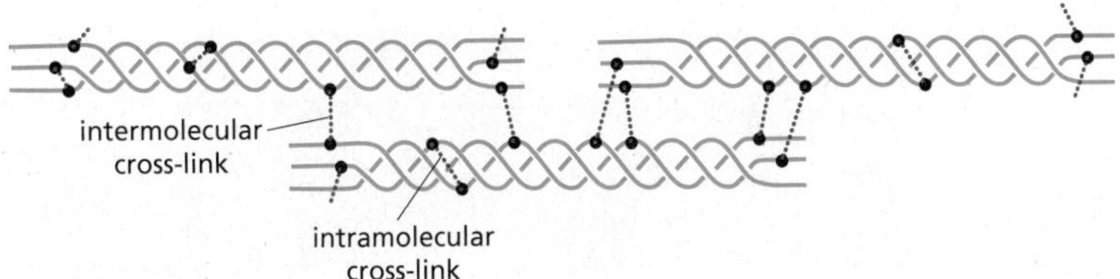

Figure 19–41 **Cross-links formed between modified lysine side chains within a collagen fibril.** Covalent intramolecular and intermolecular cross-links are formed in several steps. First, the extracellular enzyme lysyl oxidase deaminates certain lysines and hydroxylysines to yield highly reactive aldehyde groups. The aldehydes then react spontaneously to form covalent bonds with each other or with other lysines or hydroxylysines. Most of the cross-links form between the short nonhelical segments at each end of the collagen molecules.

they guide the intracellular formation of the triple-stranded collagen molecules. Second, because they are retained until after secretion, they prevent the intracellular formation of large collagen fibrils, which could be catastrophic for the cell.

After the fibrils have formed in the extracellular space, they are greatly strengthened by the formation of covalent cross-links between lysine residues of the constituent collagen molecules (**Figure 19–41**). The types of covalent bonds involved are found only in collagen and elastin. If cross-linking is inhibited, the tensile strength of the fibrils is drastically reduced: collagenous tissues become fragile, and structures such as skin, tendons, and blood vessels tend to tear. The extent and type of cross-linking vary from tissue to tissue. Collagen is especially highly cross-linked in the Achilles tendon, for example, where tensile strength is crucial.

Secreted Fibril-associated Collagens Help Organize the Fibrils

In contrast to GAGs, which resist compressive forces, collagen fibrils form structures that resist tensile forces. The fibrils have various diameters and are organized in different ways in different tissues. In mammalian skin, for example, they are woven in a wickerwork pattern so that they resist tensile stress in multiple directions; leather consists of this material, suitably preserved. In tendons, collagen fibrils are organized in parallel bundles aligned along the major axis of tension. In mature bone and in the cornea, they are arranged in orderly plywoodlike layers, with the fibrils in each layer lying parallel to one another but nearly at right angles to the fibrils in the layers on either side. The same arrangement occurs in tadpole skin (**Figure 19–42**).

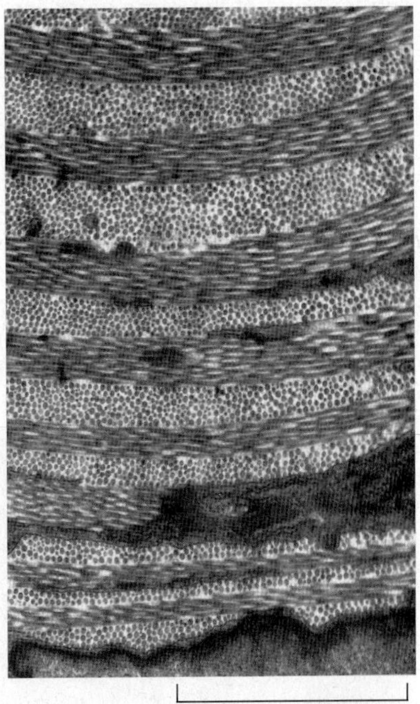

Figure 19–42 **Collagen fibrils in the tadpole skin.** This electron micrograph shows the plywoodlike arrangement of the fibrils: successive layers of fibrils are laid down nearly at right angles to each other. This organization is also found in mature bone and in the cornea. (Courtesy of Jerome Gross.)

The connective-tissue cells themselves determine the size and arrangement of the collagen fibrils. The cells can express one or more genes for the different types of fibrillar collagen molecules. But even fibrils composed of the same mixture of collagens have different arrangements in different tissues. How is this achieved? Part of the answer is that cells can regulate the disposition of the collagen molecules after secretion by guiding collagen fibril formation near the plasma membrane. In addition, cells can influence this organization by secreting, along with their fibrillar collagens, different kinds and amounts of other matrix macromolecules. In particular, they secrete the fibrous protein *fibronectin*, as we discuss later, and this precedes the formation of collagen fibrils and helps guide their organization.

Fibril-associated collagens, such as types IX and XII collagens, are thought to be especially important in organizing collagen fibrils. They differ from fibrillar collagens in the following ways. First, their triple-stranded helical structure is interrupted by one or two short nonhelical domains, which makes the molecules more flexible than fibrillar collagen molecules. Second, they do not aggregate with one another to form fibrils in the extracellular space. Instead, they bind in a periodic manner to the surface of fibrils formed by the fibrillar collagens. Type IX molecules bind to type II collagen-containing fibrils in cartilage, the cornea, and the vitreous of the eye (**Figure 19–43**), whereas type XII molecules bind to type I collagen-containing fibrils in tendons and various other tissues.

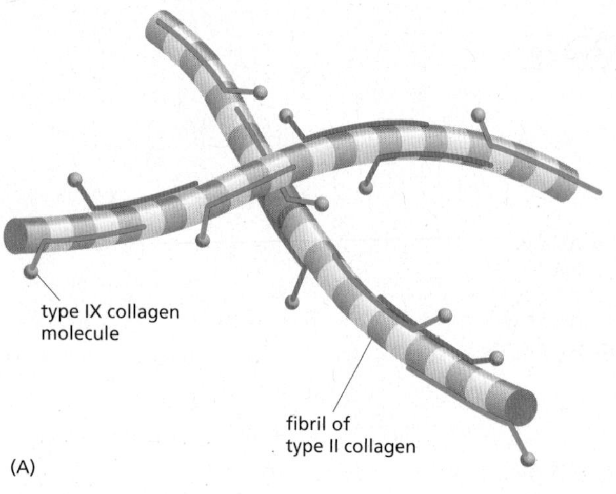

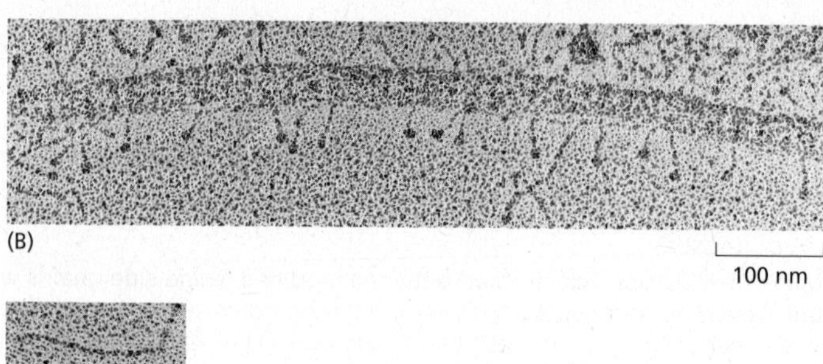

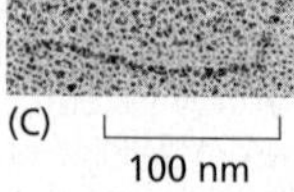

Figure 19–43 **Type IX collagen.** (A) Type IX collagen molecules binding in a periodic pattern to the surface of a fibril containing type II collagen. (B) Electron micrograph of a rotary-shadowed type II collagen–containing fibril in cartilage, decorated by type IX collagen molecules. (C) An individual type IX collagen molecule. (B and C, © 1988, L. Vaughan et al. Originally published in *J. Cell Biol.* https://doi.org/10.1083/jcb.106.3.991. With permission from Rockefeller University Press.)

Fibril-associated collagens are thought to mediate the interactions of collagen fibrils with one another and with other matrix macromolecules to help determine the organization of the fibrils in the matrix.

Elastin Gives Tissues Their Elasticity

Many vertebrate tissues, such as skin, blood vessels, and lungs, need to be both strong and elastic in order to function. A network of **elastic fibers** in the extracellular matrix of these tissues gives them the resilience to recoil after transient stretch (**Figure 19–44**). Elastic fibers are at least five times more extensible than a rubber band of the same cross-sectional area. Long, inelastic collagen fibrils are interwoven with the elastic fibers to limit the extent of stretching and prevent the tissue from tearing.

The main component of elastic fibers is **elastin**, a highly hydrophobic protein (about 750 amino acids long), which, like collagen, is unusually rich in proline and glycine but, unlike collagen, is not glycosylated. Soluble *tropoelastin* (the biosynthetic precursor of elastin) is secreted into the extracellular space and assembled into elastic fibers close to the plasma membrane, generally in cell-surface infoldings. After secretion, the tropoelastin molecules become highly cross-linked to one another, generating an extensive network of elastin fibers and sheets. A mechanism similar to the one that operates in cross-linking collagen molecules forms cross-links between lysines in elastin molecules.

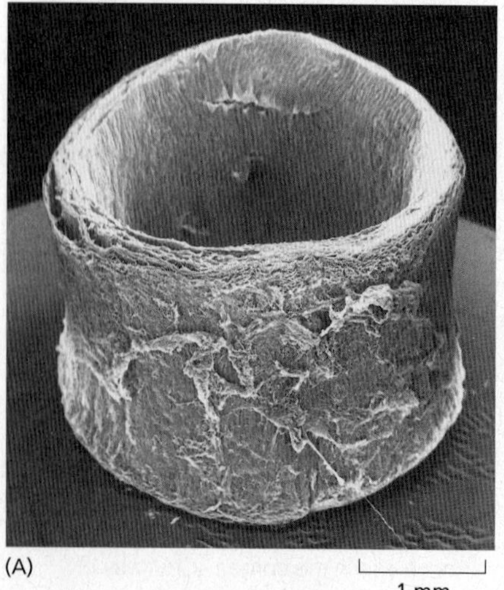

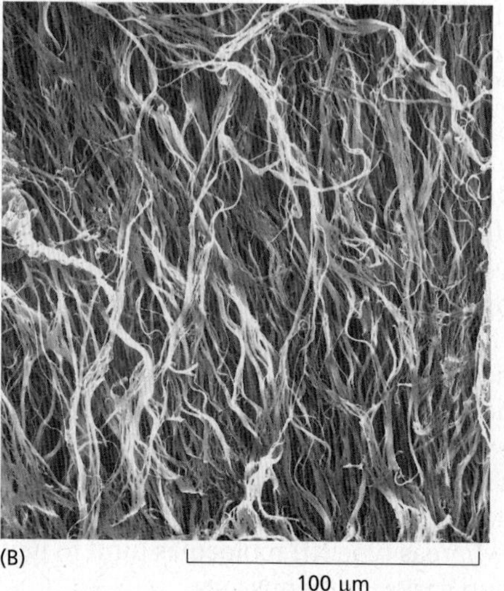

Figure 19–44 **Elastic fibers.** These scanning electron micrographs show (A) a low-power view of a segment of a dog's aorta and (B) a high-power view of the dense network of longitudinally oriented elastic fibers in the outer layer of the same blood vessel. All the other components have been digested away with enzymes and formic acid. (From K.S. Haas et al., *Anat. Rec.* 230:86–96, 1991. With permission from Wiley-Liss.)

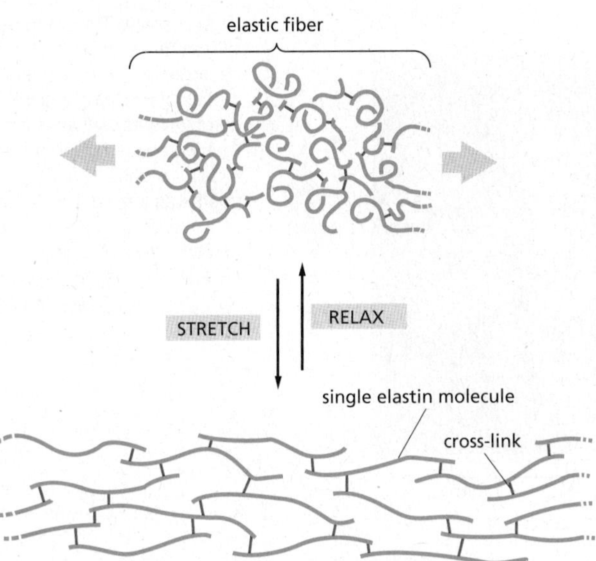

Figure 19–45 Stretching a network of elastin molecules. The molecules are joined together by covalent bonds *(red)* to generate a cross-linked network. In this model, each elastin molecule in the network can extend and contract in a manner resembling a random coil, so that the entire assembly can stretch and recoil like a rubber band.

The elastin protein is composed largely of two types of short segments that alternate along the polypeptide chain: hydrophobic segments, which are responsible for the elastic properties of the molecule; and alanine- and lysine-rich α-helical segments, which are cross-linked to adjacent molecules. Each segment is encoded by a separate exon. There is still uncertainty concerning the conformation of elastin molecules in elastic fibers and how the structure of these fibers accounts for their rubberlike properties. However, it seems that parts of the elastin polypeptide chain, like the polymer chains in ordinary rubber, adopt a loose "random-coil" conformation, and it is the random-coil nature of the component molecules cross-linked into the elastic fiber network that allows the network to stretch and recoil like a rubber band (**Figure 19–45**).

Elastin is the dominant extracellular matrix protein in arteries, comprising 50% of the dry weight of the largest artery—the aorta (see Figure 19–44). Mutations in the elastin gene causing a deficiency of the protein in mice or humans result in narrowing of the aorta and other arteries and excessive proliferation of smooth muscle cells in the arterial wall. Apparently, the normal elasticity of an artery is required to restrain the proliferation of these cells.

Elastic fibers do not consist solely of elastin. The elastin core is covered with a sheath of *microfibrils*, each of which has a diameter of about 10 nm. The microfibrils appear before elastin in developing tissues and seem to provide scaffolding to guide elastin deposition. Arrays of microfibrils are elastic in their own right, and in some places they persist in the absence of elastin: they help to hold the lens in its place in the eye, for example. Microfibrils are composed of a number of distinct glycoproteins, including the large glycoprotein *fibrillin*, which binds to elastin and is essential for the integrity of elastic fibers. Mutations in the fibrillin gene result in *Marfan syndrome*, a relatively common human disorder. In the most severely affected individuals, the aorta is prone to rupture; other common effects include displacement of the lens and abnormalities of the skeleton and joints. Affected individuals are often unusually tall and lanky: Abraham Lincoln is suspected to have had the condition.

Cells Govern and Respond to the Mechanical Properties of the Matrix

Cells interact with the extracellular matrix mechanically as well as chemically, and studies in culture suggest that the mechanical interaction can have dramatic effects on the architecture of connective tissue. Thus, when fibroblasts are mixed with a meshwork of randomly oriented collagen fibrils that form a gel in a culture dish, the fibroblasts tug on the meshwork, drawing in collagen from their

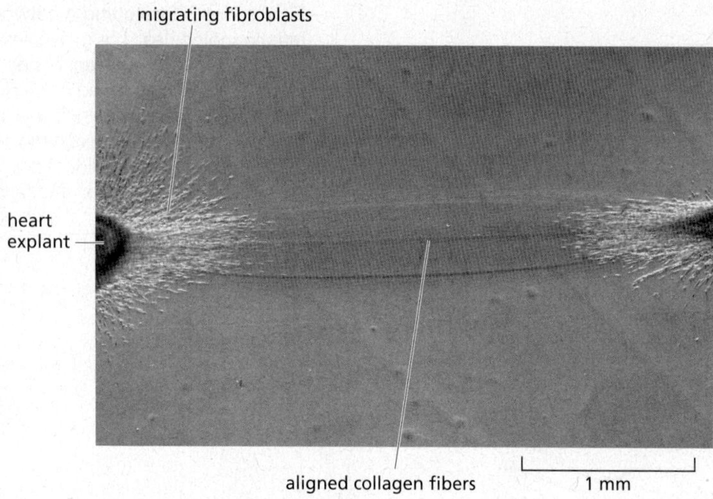

Figure 19–46 The shaping of the extracellular matrix by cells. This micrograph shows a region between two pieces of embryonic chick heart (rich in fibroblasts as well as heart muscle cells) that were cultured on a collagen gel for 4 days. A dense tract of aligned collagen fibers has formed between the explants, presumably as a result of the fibroblasts in the explants tugging on the collagen. (From D. Stopak and A.K. Harris, *Dev. Biol.* 90:383–398, 1982. With permission from Elsevier.)

surroundings and thereby causing the gel to contract to a small fraction of its initial volume. By similar activities, a cluster of fibroblasts surrounds itself with a capsule of densely packed and circumferentially oriented collagen fibers.

If two small pieces of embryonic tissue containing fibroblasts are placed far apart on a collagen gel, the intervening collagen becomes organized into a compact band of aligned fibers that connect the two explants (**Figure 19–46**). The fibroblasts subsequently migrate out from the explants along the aligned collagen fibers. Thus, the fibroblasts influence the alignment of the collagen fibers, and the collagen fibers in turn affect the distribution of the fibroblasts.

Fibroblasts may have a similar role in organizing the extracellular matrix inside the body. First they synthesize the collagen fibrils and deposit them in the correct orientation. Then they work on the matrix they have secreted, crawling over it and tugging on it so as to create tendons and ligaments and the tough, dense layers of connective tissue that surround and bind together most organs.

In addition to determining the orientation of the collagen fibrils they produce, fibroblasts control the overall density and composition of the extracellular matrix, which varies dramatically in different tissues. Some tissues, such as tendons and cartilage, are composed of dense matrix that is far more rigid and resistant to deformation than the soft, elastic matrix of tissues like fat and the brain. These differences depend on the ability of cells in these tissues to regulate the types of collagen and other proteins produced, the relative rates of matrix protein synthesis and degradation, and the amount of collagen and elastin cross-linking. The density of the matrix, in turn, regulates the behavior of the fibroblasts and other cells that travel through it; for example, the proliferation, migration, and developmental fate of stem cells are influenced by matrix composition and density. Abnormally high matrix density is associated with certain fibrotic diseases and appears to be a risk factor in some forms of cancer.

Fibronectin and Other Multidomain Glycoproteins Help Organize the Matrix

In addition to proteoglycans, collagens, and elastic fibers, the extracellular matrix contains a large and varied assortment of glycoproteins that typically have multiple domains, each with specific binding sites for other matrix macromolecules and for receptors on the surface of cells (**Figure 19–47**). These proteins therefore contribute to both organizing the matrix and helping cells attach to it. Like the proteoglycans, they also guide cell movements in developing tissues by serving as tracks along which cells can migrate or as repellents that keep cells out of forbidden areas. They can also bind and thereby influence the function of peptide growth factors and other small molecules produced by nearby cells.

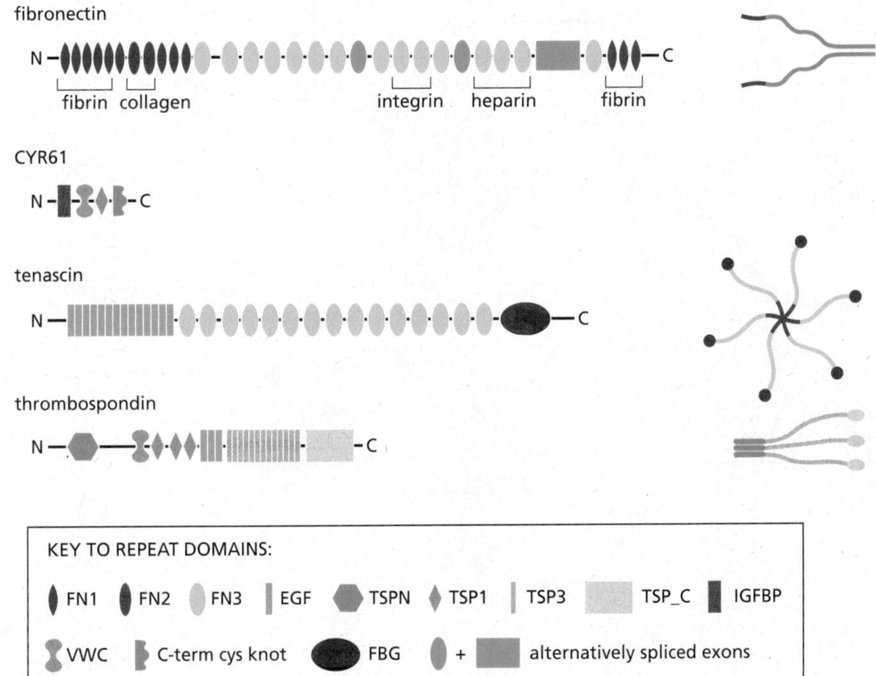

Figure 19–47 Complex glycoproteins of the extracellular matrix. Many matrix glycoproteins are large scaffold proteins containing multiple copies of specific protein-interaction domains. Each domain is folded into a discrete globular structure, and many such domains are arrayed along the protein like beads on a string. This diagram shows four representative proteins among the roughly 200 matrix glycoproteins that are found in mammals. Each protein contains multiple repeat domains, with the names listed in the key at the bottom. Fibronectin, for example, contains numerous copies of three different *fibronectin repeats* (types I–III, labeled here as FN1, FN2, and FN3). Two type III repeats near the center of the protein contain important binding sites for cell-surface integrins, whereas three nearby type III repeats form a binding site for heparin or heparan sulfate proteoglycans. FN repeats at the N-terminus are involved in binding fibrin or collagen. Other matrix proteins contain repeated sequences resembling those of epidermal growth factor (EGF), a major regulator of cell growth and proliferation; these repeats might serve a similar signaling function in matrix proteins. Other proteins contain domains, such as the insulin-like growth factor–binding protein (IGFBP) repeat, that bind and regulate the function of soluble growth factors. To add more structural diversity, many of these proteins are encoded by RNA transcripts that can be spliced in different ways, adding or removing exons, such as those in fibronectin. Finally, the scaffolding and regulatory functions of many matrix proteins are further expanded by assembly into multimeric forms, as shown at the right: fibronectin forms dimers linked at the C-termini, whereas tenascin and thrombospondin form N-terminally linked hexamers and trimers, respectively. Other domains include four repeats from thrombospondin (TSPN, TSP1, TSP3, TSP_C). VWC, von Willebrand type C; FBG, fibrinogen-like. (Adapted from R.O. Hynes and A. Naba, *Cold Spring Harb. Perspect. Biol.* 4:a004903, 2012. With permission from Cold Spring Harbor Laboratory Press.)

The best-understood member of this class of matrix proteins is **fibronectin,** a large glycoprotein found in all vertebrates and important for many cell-matrix interactions. Mutant mice that are unable to make fibronectin die early in embryogenesis because their endothelial cells fail to form proper blood vessels. The defect is thought to result from abnormalities in the interactions of these cells with the surrounding extracellular matrix, which normally contains fibronectin.

Fibronectin is a dimer composed of two very large subunits joined by disulfide bonds at their C-terminal ends. Each subunit contains a series of small repeated domains, or modules, separated by short stretches of flexible polypeptide chain (Figure 19–48). Each domain is usually encoded by a separate exon, suggesting that the fibronectin gene, like the genes encoding many matrix proteins, evolved by multiple exon duplications. In the human genome, there is only one fibronectin gene, containing about 50 exons of similar size, but the transcripts can be spliced in different ways to produce multiple fibronectin isoforms (see Figure 19–48B). The major repeat domain in fibronectin is called the **type III fibronectin repeat,** which is about 90 amino acids long and occurs at least 15 times in each subunit. This repeat is among the most common of all protein domains in vertebrates.

Fibronectin Binds to Integrins

One way to analyze a complex multifunctional protein molecule such as fibronectin is to synthesize individual regions of the protein and test their ability to bind other proteins. By these and other methods, it was possible to show that one region of fibronectin binds to collagen, another to proteoglycans, and another to specific integrins on the surface of various types of cells (see Figure 19–48B). Synthetic peptides corresponding to different segments of the integrin-binding domain were then used to show that binding depends on a specific tripeptide sequence (*Arg-Gly-Asp,* or *RGD*) that is found in one of the type III repeats (see Figure 19–48C). Even very short peptides containing this **RGD sequence** can compete with fibronectin for the binding site on cells, thereby inhibiting the attachment of the cells to a fibronectin matrix.

Several extracellular proteins besides fibronectin also have an RGD sequence that mediates cell-surface binding. Many of these proteins are components of the

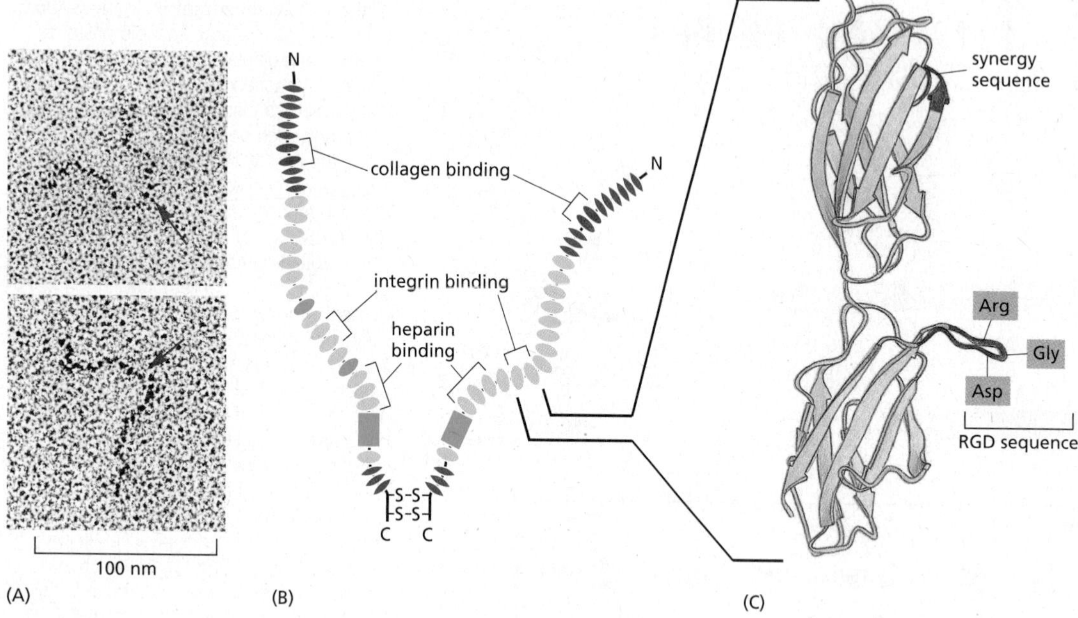

Figure 19–48 **The structure of a fibronectin dimer.** (A) Electron micrographs of individual fibronectin dimer molecules shadowed with platinum; *red arrows* mark the joined C-termini. (B) The two polypeptide chains are similar but generally not identical (being made from the same gene but from differently spliced mRNAs). They are joined by two disulfide bonds near the C-termini. Each chain is almost 2500 amino acids long and is folded into multiple domains (see Figure 19–47). As indicated, some domains are specialized for binding to a particular molecule. For simplicity, not all of the known binding sites are shown. (C) The three-dimensional structure of the ninth and tenth type III fibronectin repeats, as determined by x-ray crystallography. Both the Arg-Gly-Asp (RGD) and the "synergy" sequences shown in *red* are important for binding to integrins on cell surfaces. (A, from J. Engel et al., *J. Mol. Biol.* 150:97–120, 1981. With permission from Elsevier; C, from D.J. Leahy, *Annu. Rev. Cell Dev. Biol.* 13:363–393, 1997. With permission from Annual Reviews.)

extracellular matrix, while others are involved in blood clotting. Peptides containing the RGD sequence have been useful in the development of anti-clotting drugs. Some snakes use a similar strategy to cause their victims to bleed: they secrete RGD-containing anti-clotting proteins called *disintegrins* into their venom.

The cell-surface receptors that bind RGD-containing proteins are members of the integrin family, which we describe in detail later. Each integrin specifically recognizes its own small set of matrix molecules, indicating that tight binding requires more than just the RGD sequence. Moreover, RGD sequences are not the only sequence motifs used for binding to integrins: many integrins recognize and bind to other motifs instead.

Tension Exerted by Cells Regulates the Assembly of Fibronectin Fibrils

Fibronectin can exist both in a soluble form, circulating in the blood and other body fluids, and as insoluble *fibronectin fibrils*, in which fibronectin dimers are cross-linked to one another by additional disulfide bonds and form part of the extracellular matrix. Unlike fibrillar collagen molecules, however, which can self-assemble into fibrils in a test tube, fibronectin molecules assemble into fibrils only on the surface of cells, and only where those cells possess appropriate fibronectin-binding proteins—in particular, integrins. The integrins provide a linkage from the fibronectin outside the cell to the actin cytoskeleton inside it. The linkage transmits tension to the fibronectin molecules—provided that they also have an attachment to some other structure—and stretches them, exposing cryptic binding sites in the fibronectin molecules (**Figure 19–49**). This allows them

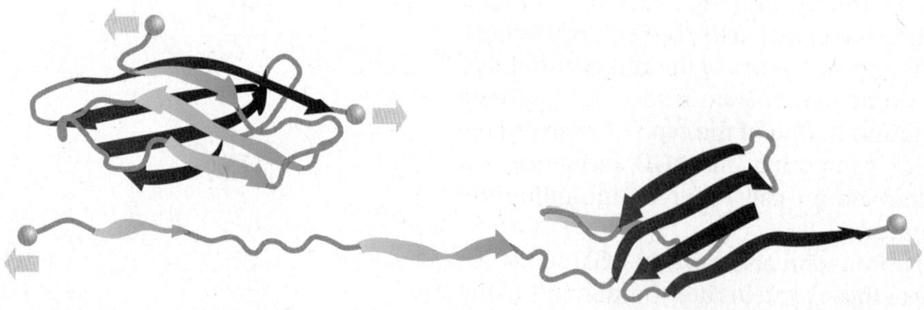

Figure 19–49 **Tension-sensing by fibronectin.** Some type III fibronectin repeats are thought to unfold when fibronectin is stretched. The unfolding exposes cryptic binding sites that interact with other fibronectin molecules resulting in the formation of fibronectin filaments like those shown in Figure 19–50. (From V. Vogel and M. Sheetz, *Nat. Rev. Mol. Cell Biol.* 7:265–275, 2006. With permission from Springer Nature.)

to bind directly to one another and to recruit additional fibronectin molecules to form a fibril (Figure 19–50). This dependence on tension and interaction with cell surfaces ensures that fibronectin fibrils assemble where there is a mechanical need for them and not in inappropriate locations such as the bloodstream.

Many other extracellular matrix proteins contain multiple copies of the type III fibronectin repeat (see Figure 19–47), and it is possible that tension exerted on these proteins also uncovers cryptic binding sites and thereby influences their behavior.

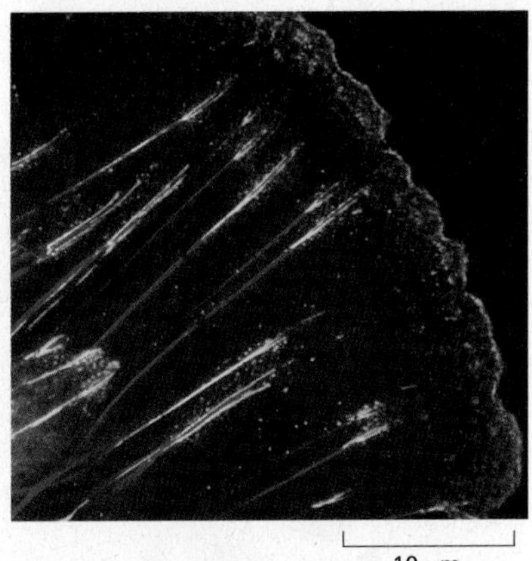

Figure 19–50 **Organization of fibronectin into fibrils at the cell surface.** This fluorescence micrograph shows the front end of a migrating mouse fibroblast. Extracellular fibronectin is stained *green*, and intracellular actin filaments are stained *red*. The fibronectin is initially present as small dotlike aggregates near the leading edge of the cell. It accumulates at focal adhesions (sites of anchorage of actin filaments, discussed later) and becomes organized into fibrils parallel to the actin filaments. Integrin molecules spanning the cell membrane link the fibronectin outside the cell to the actin filaments inside it (see Figure 19–56). Tension exerted on the fibronectin molecules through this linkage is thought to stretch them, exposing binding sites that promote fibril formation. (Courtesy of Roumen Pankov and Kenneth Yamada.)

The Basal Lamina Is a Specialized Form of Extracellular Matrix

Thus far in this section, we have reviewed the general principles underlying the structure and function of the major classes of extracellular matrix components. We now describe how some of these components are assembled into a specialized type of extracellular matrix called the **basal lamina** (also known as the **basement membrane**). This exceedingly thin, tough, flexible sheet of matrix molecules is an essential underpinning of all epithelia. Although small in volume, it has a critical role in the architecture of the body. Like the cadherins, it seems to be one of the defining features common to all multicellular animals, and it seems to have appeared very early in their evolution. The major molecular components of the basal lamina are among the most ancient extracellular matrix macromolecules.

The basal lamina is typically 40–120 nm thick. It lies beneath epithelial cells and also surrounds individual muscle cells, fat cells, and Schwann cells (which wrap around peripheral nerve cell axons to form myelin). The basal lamina thus separates these cells and epithelia from the underlying or surrounding connective tissue and forms the mechanical connection between them. In other locations, such as the kidney glomerulus, a basal lamina lies between two cell sheets and functions as a selective filter (Figure 19–51). Basal laminae have more than simple structural and filtering roles, however. They are able to determine cell polarity; influence cell metabolism; organize the proteins in adjacent plasma membranes; promote cell survival, proliferation, or differentiation; and serve as highways for cell migration.

Laminin and Type IV Collagen Are Major Components of the Basal Lamina

The basal lamina is synthesized by the cells on each side of it: the epithelial cells contribute one set of basal lamina components, while cells of the underlying bed of connective tissue (called the *stroma*, Greek for "bedding") contribute another set (Figure 19–52). Although the precise composition of the mature basal lamina varies from tissue to tissue and even from region to region in the same lamina, it typically contains the glycoproteins *laminin*, *type IV collagen*, and *nidogen*, along with the proteoglycan *perlecan*. Other common basal lamina components are fibronectin and *type XVIII collagen* (an atypical member of the collagen family, forming the core protein of a proteoglycan).

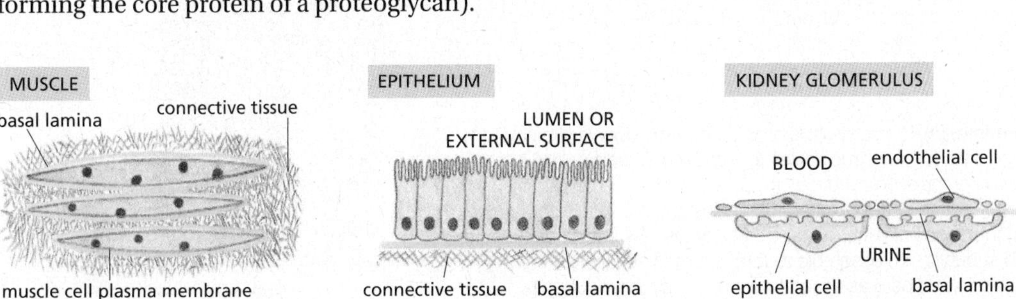

Figure 19–51 **Three ways in which the basal lamina is organized.** Sheets of basal lamina *(yellow)* surround certain cells (such as skeletal muscle cells), underlie epithelia, and are interposed between two cell sheets (as in the kidney glomerulus). Note that, in the kidney glomerulus, both cell sheets have gaps in them, and the basal lamina has a filtering as well as a supportive function, helping to determine which molecules will pass into the urine from the blood. The filtration also depends on other protein-based structures, called *slit diaphragms*, that span the intercellular gaps in the epithelial sheet.

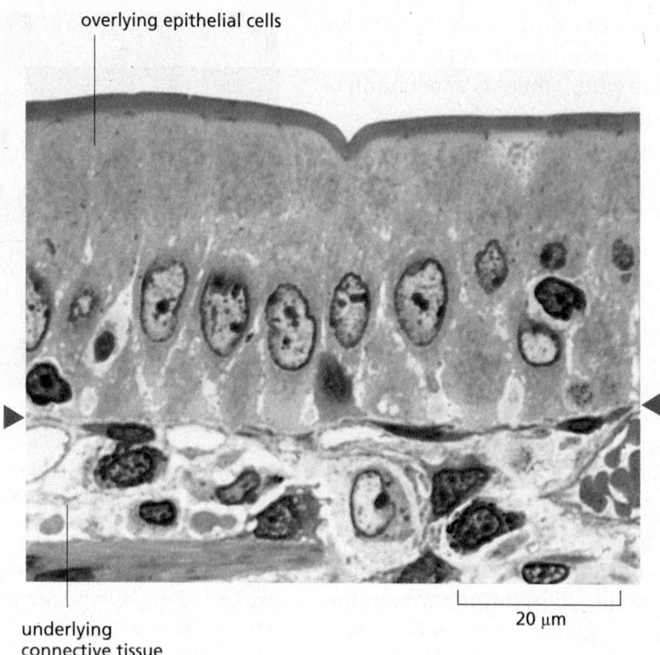

Figure 19–52 **The basal lamina supports a sheet of epithelial cells.** In this light micrograph of a cross section of the small intestine, the sheet of columnar epithelial cells rests on the basal lamina *(red arrowheads)*. A network of collagen fibrils and other fibers in the underlying connective tissue interacts with the lower face of the basal lamina. (Jose Luis Calvo/ Shutterstock.)

Laminin is the primary organizer of the sheet structure, and, early in development, the basal lamina consists mainly of laminin molecules. Laminins constitute a large family of proteins, each composed of three long polypeptide chains (α, β, and γ) held together by disulfide bonds and arranged in the shape of an asymmetric bouquet, like a bunch of three flowers whose stems are twisted together at the foot but whose heads remain separate (**Figure 19–53**). These heterotrimers can self-assemble *in vitro* into a network, largely through interactions between their heads, although interaction with cells is needed to organize the network into an orderly sheet. Because there are several isoforms of each type of chain, and

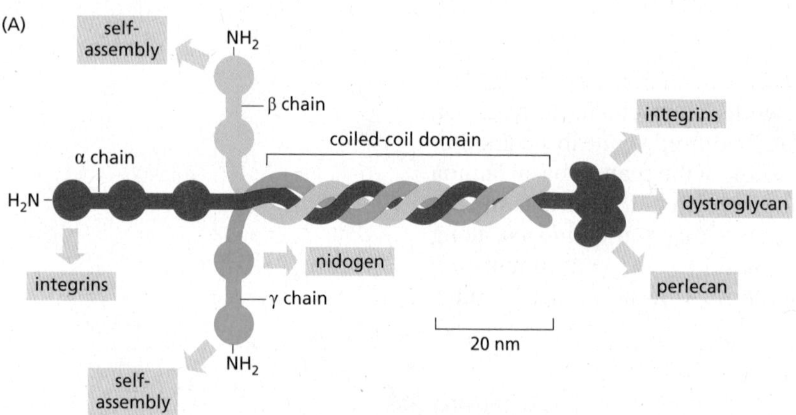

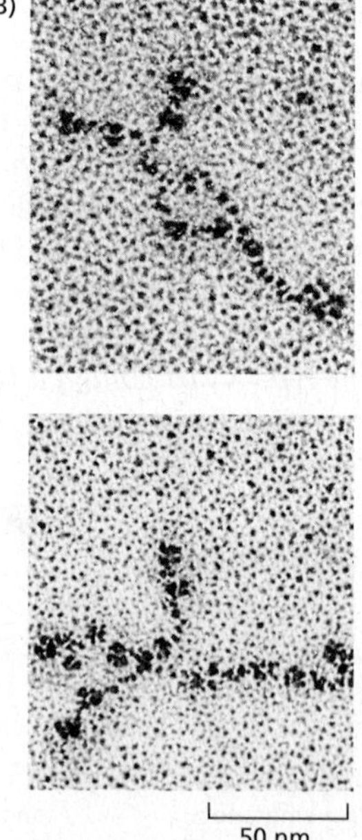

Figure 19–53 **The structure of laminin.** (A) The best-understood family member is laminin-111, shown here with some of its binding sites for other molecules *(gray boxes)*. Laminins are multidomain glycoproteins composed of three polypeptides (α, β, and γ) that are disulfide-bonded into an asymmetric crosslike structure. Each of the polypeptide chains is more than 1500 amino acids long. Five types of α chains, four types of β chains, and three types of γ chains are known, and various combinations of these subunits can assemble to form a large variety of different laminins, which are named according to numbers assigned to each of their three subunits: laminin-111, for example, contains α1, β1, and γ1 subunits. Each isoform tends to have a specific tissue distribution: laminin-332 is found in skin, laminin-211 in muscle, and laminin-411 in endothelial cells of blood vessels. Through their binding sites for other proteins, laminin molecules play a central part in organizing the basal lamina and anchoring it to cells. (B) Electron micrographs of laminin molecules shadowed with platinum. (B, from J. Engel et al., *J. Mol. Biol.* 150:97–120, 1981. With permission from Elsevier.)

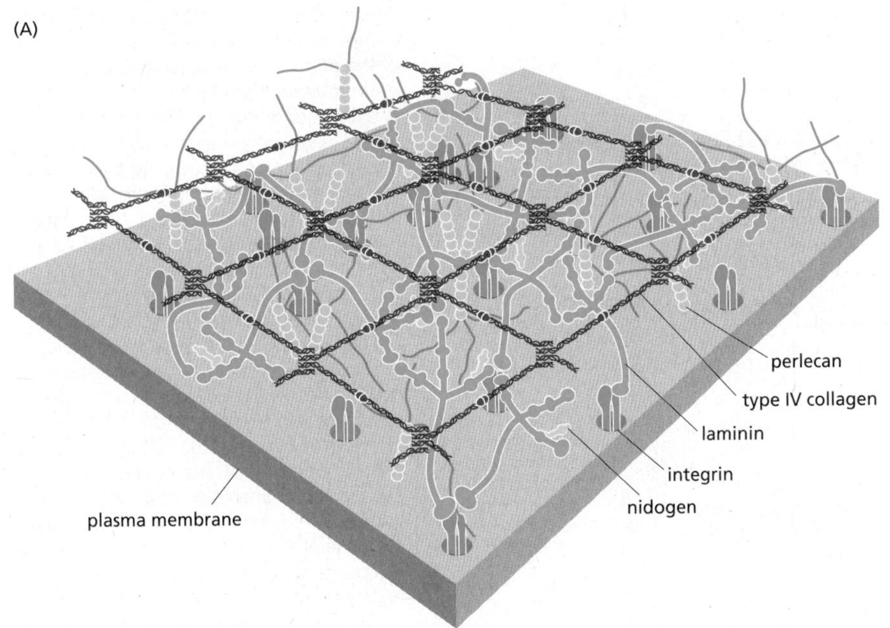

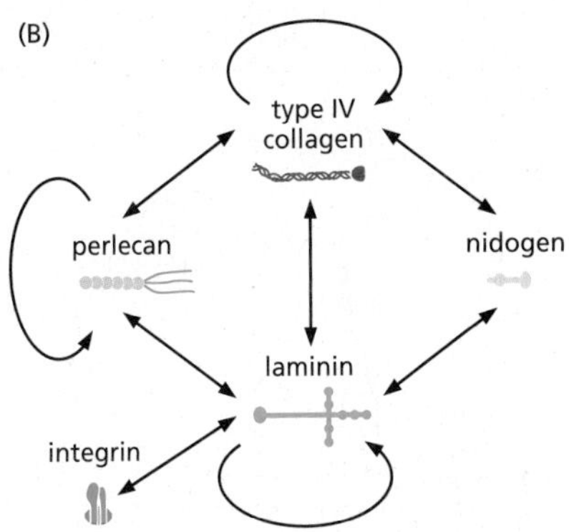

Figure 19–54 **A model of the molecular structure of a basal lamina.** (A) The basal lamina is formed by specific interactions (B) between the proteins laminin, type IV collagen, and nidogen and the proteoglycan perlecan. *Arrows* in B connect molecules that can bind directly to each other. There are various isoforms of type IV collagen and laminin, each with a distinctive tissue distribution. Transmembrane laminin receptors (integrins and dystroglycan) in the plasma membrane are thought to organize the assembly of the basal lamina; only the integrins are shown. (Based on H. Colognato and P.D. Yurchenco, *Dev. Dyn.* 218:213–234, 2000. With permission from John Wiley & Sons.)

these can associate in various combinations, many different laminins can be produced, creating basal laminae with distinctive properties. The laminin $\gamma 1$ chain is, however, a component of most laminin heterotrimers; mice lacking it die during embryogenesis because they are unable to make a basal lamina.

Type IV collagen is a second essential component of a mature basal lamina, and it, too, exists in several isoforms. Like the *fibrillar collagens* that constitute the bulk of the protein in connective tissues such as bone or tendon, type IV collagen molecules consist of three separately synthesized long protein chains that twist together to form a rope-like superhelix; however, they differ from the fibrillar collagens in that the triple-stranded helical structure is interrupted in more than 20 regions, allowing multiple bends. Type IV collagen molecules interact via their terminal domains to assemble extracellularly into a flexible, felt-like network that gives the basal lamina tensile strength.

Laminin and type IV collagen interact with other basal lamina components, such as the glycoprotein nidogen and the proteoglycan perlecan, resulting in a highly cross-linked network of proteins and proteoglycans (**Figure 19–54**). The laminin molecules that generate the initial sheet structure first join to each other while bound to receptors on the surface of the cells. The cell-surface receptors are primarily members of the integrin family, but another important type of laminin receptor is *dystroglycan*, a highly glycosylated transmembrane protein. Together, these receptors organize basal lamina assembly: they hold the laminin molecules by their feet, leaving the laminin heads positioned to interact so as to form a two-dimensional network. This laminin network then coordinates the assembly of the other basal lamina components.

Interactions between laminin and cell-surface receptors are critical for the adhesion of epithelia to the underlying connective tissue. In the skin, for example, the epithelial outer layer—the epidermis—depends on tight interactions with the basal lamina to keep it attached to the underlying dermis. In people with genetic defects in laminin-332, a key component of the skin basal lamina, the epidermis is poorly attached to the dermis. This causes a blistering disease called *junctional epidermolysis bullosa*, a severe and sometimes lethal condition.

Basal Laminae Have Diverse Functions

The basal lamina can act as a selective barrier to the movement of cells, as well as a filter for molecules. The lamina beneath an epithelium, for example, usually prevents fibroblasts in the underlying connective tissue from making contact

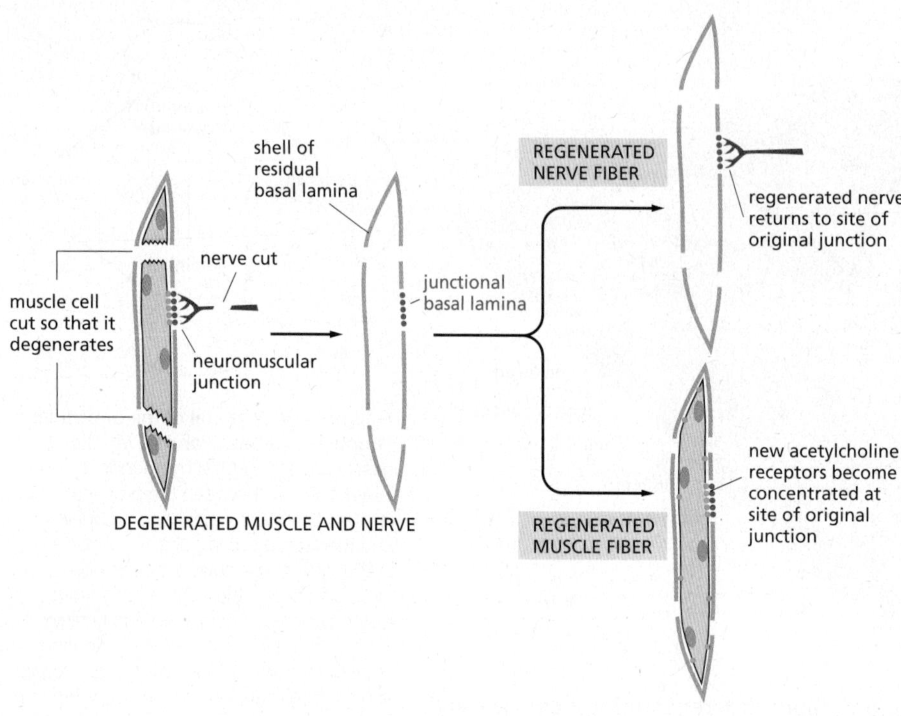

Figure 19–55 Regeneration experiments demonstrating the special character of the junctional basal lamina at a neuromuscular junction. If a frog muscle and its motor nerve are destroyed, the basal lamina around each muscle cell remains intact, and the sites of the old neuromuscular junctions are still recognizable. When the nerve, but not the muscle, is allowed to regenerate *(upper right)*, the junctional basal lamina directs the regenerating nerve to the original synaptic site. When the muscle, but not the nerve, is allowed to regenerate *(lower right)*, the junctional basal lamina causes newly made acetylcholine receptors *(blue)* to accumulate at the original synaptic site. These experiments show that the junctional basal lamina controls the localization of synaptic components on both sides of the lamina. Some of the molecules responsible for these effects have been identified. Motor neuron axons, for example, deposit agrin in the junctional basal lamina, where it regulates the assembly of acetylcholine receptors and other proteins in the junctional plasma membrane of the muscle cell. Reciprocally, muscle cells deposit a particular isoform of laminin in the junctional basal lamina, and this molecule is likely to interact with specific ion channels on the presynaptic membrane of the neuron.

with the epithelial cells. It does not, however, stop macrophages, lymphocytes, or nerve processes from passing through it, using specialized protease enzymes to cut a hole for their transit. The basal lamina is also important in tissue regeneration after injury. When cells in tissues such as muscles, nerves, and epithelia are damaged or killed, the basal lamina often survives and provides a scaffold along which regenerating cells can migrate. In this way, the original tissue architecture is readily reconstructed.

A particularly striking example of the role of the basal lamina in regeneration comes from studies of the *neuromuscular junction,* the site where the nerve terminals of a motor neuron form a chemical synapse with a skeletal muscle cell (discussed in Chapter 11). In vertebrates, the basal lamina that surrounds the muscle cell separates the nerve-cell and muscle-cell plasma membranes at the synapse, and the synaptic region of the lamina has a distinctive chemical character, with special isoforms of type IV collagen and laminin and a proteoglycan called *agrin*. After a nerve or muscle injury, the basal lamina at the synapse has a central role in reconstructing the synapse at the correct location (**Figure 19–55**). Defects in components of the basal lamina at the synapse are responsible for some forms of muscular dystrophy, in which muscles develop normally but then degenerate later in life.

Cells Have to Be Able to Degrade Matrix, as Well as Make It

The ability of cells to degrade and destroy extracellular matrix is as important as their ability to make it and bind to it. Rapid matrix degradation is required in processes such as tissue repair, and even in the seemingly static extracellular matrix of adult animals there is a slow, continual turnover, with matrix macromolecules being degraded and resynthesized. This allows bone, for example, to be remodeled so as to adapt to changes in the stresses on it.

From the point of view of individual cells, the ability to cut through matrix is crucial in two ways: it enables them to divide while embedded in matrix, and it enables them to travel through it. Cells in connective tissues generally need to be able to stretch out in order to divide. If a cell lacks the enzymes needed to degrade the surrounding matrix, it is strongly inhibited from dividing and hindered from migrating.

Localized degradation of matrix components is also required wherever cells have to escape from confinement by a basal lamina. It is needed during normal branching growth of epithelial structures such as glands, for example, to allow the population of epithelial cells to increase, and needed also when white blood cells migrate across the basal lamina of a blood vessel into tissues in response to infection or injury. Matrix degradation is important both for the spread of cancer cells through the body and for their ability to proliferate in the tissues that they invade (discussed in Chapter 20).

In general, matrix components are degraded by extracellular proteolytic enzymes (proteases) that act close to the cells that produce them. Many of these proteases belong to one of two general classes. The largest group, with about 50 members in vertebrates, is the **matrix metalloproteases**, which depend on bound Ca^{2+} or Zn^{2+} for activity. The second group is the **serine proteases**, which have a reactive serine in their active site. Together, metalloproteases and serine proteases cooperate to degrade matrix proteins such as collagen, laminin, and fibronectin. Some metalloproteases, such as the *collagenases*, are highly specific, cleaving particular proteins at a small number of sites. In this way, the structural integrity of the matrix is largely retained, while the limited amount of proteolysis that occurs is sufficient for cell migration. Other metalloproteases may be less specific, but, because they are anchored to the plasma membrane, act just where they are needed; it is this type of matrix metalloprotease that is crucial for a cell's ability to divide when embedded in matrix.

Clearly, the activities of the proteases that degrade the matrix must be tightly controlled if the fabric of the body is not to collapse in a heap. Numerous mechanisms are therefore employed to ensure that matrix proteases are activated only at the correct time and place. Protease activity is generally confined to the cell surface by specific anchoring proteins, by membrane-associated activators, and by the production of specific protease inhibitors in regions where protease activity is not needed.

The proteolytic cleavage of matrix proteins does not always lead simply to their destruction, but sometimes generates protein fragments with specific biological activities. For example, cleavage of type IV collagen by matrix metalloproteases leads to the release of protein fragments that inhibit the local formation of blood vessels. Similarly, cleavage of laminin in certain tissues can generate protein fragments that help govern local cell proliferation.

Matrix Proteoglycans and Glycoproteins Regulate the Activities of Secreted Proteins

The physical properties of extracellular matrix are important for its fundamental roles as a scaffold for tissue structure and as a substrate for cell anchorage and migration. The matrix also has an important impact on cell signaling. Cells communicate with each other by secreting peptide signal molecules, like growth factors and morphogens, that diffuse through the extracellular fluid to influence other cells (discussed in Chapter 15). On the way to their targets, the signal molecules encounter the tightly woven meshwork of the extracellular matrix, which contains a high density of negative charges and protein-interaction domains that can interact with the signal molecules, thereby altering their function in a variety of ways.

The highly charged heparan sulfate chains of proteoglycans, for example, interact with numerous secreted signal molecules, including *fibroblast growth factors* (*FGFs*) and *vascular endothelial growth factor* (*VEGF*), which (among other effects) stimulate a variety of cell types to proliferate. By providing a dense array of growth factor–binding sites, proteoglycans are thought to generate large local reservoirs of these factors, limiting their diffusion and focusing their actions on nearby cells. Similarly, proteoglycans might help generate steep morphogen gradients in an embryo, which can be important in the patterning of tissues during development. FGF activity can also be enhanced by proteoglycans, which oligomerize the FGF molecules and also interact with cell-surface FGF receptors, enabling the FGF to cross-link and activate its receptors more effectively.

The importance of proteoglycans as regulators of the distribution and activity of signal molecules is illustrated by the severe developmental defects that can occur when specific proteoglycans are inactivated by mutation. In *Drosophila*, for example, the function of several signal proteins during development is governed by interactions with the membrane-associated proteoglycans *Dally* and *Dally-like*. These members of the *glypican* family, like the syndecan proteoglycans described earlier, are membrane-associated proteins linked to multiple heparan sulfate molecules. They are thought to concentrate signal proteins in specific locations and act as co-receptors that collaborate with conventional cell-surface receptor proteins; as a result, they promote signaling in the correct location and prevent it in the wrong locations. In the *Drosophila* ovary, Dally is partly responsible for the restricted localization and function of a signaling protein called Dpp, which blocks differentiation of the germ-line stem cells: when the gene encoding Dally is mutated, Dpp activity is greatly reduced and oocyte development is abnormal.

Several other matrix molecules interact with signal proteins. The type IV collagen of the basal lamina interacts with Dpp in *Drosophila*, for example. Fibronectin contains a type III fibronectin repeat that interacts with VEGF and another domain that interacts with hepatocyte growth factor (HGF), thereby promoting the activities of these factors. As discussed earlier, many matrix glycoproteins contain extensive arrays of binding domains, and the arrangement of these domains is likely to influence the presentation of signal proteins to their target cells (see Figure 19–47).

Finally, many matrix glycoproteins contain domains that bind directly to specific cell-surface receptors, thereby generating signals that influence the behavior of the cells, as we describe in the next section.

Summary

Cells are embedded in an intricate extracellular matrix, which not only binds the cells together but also influences their survival, development, shape, polarity, and migratory behavior. The matrix contains various protein fibers interwoven in a network of glycosaminoglycan (GAG) chains. GAGs are negatively charged polysaccharide chains that (except for hyaluronan) are covalently linked to protein to form proteoglycan molecules. GAGs attract water and occupy a large volume of extracellular space. Proteoglycans are also found on the surface of cells, where they often function as co-receptors to help cells respond to secreted signal proteins. Fiber-forming proteins give the matrix strength and resilience. The fibrillar collagens are rope-like, triple-stranded helical molecules that aggregate into long fibrils in the extracellular space, thereby providing tensile strength. They also form structures to which cells can be anchored, often via large multidomain glycoproteins, such as laminin and fibronectin, that bind to integrins on the cell surface. Elasticity is provided by elastin molecules, which form an extensive cross-linked network of fibers and sheets that can stretch and recoil.

The basal lamina is a specialized form of extracellular matrix that underlies epithelial cells or is wrapped around certain other cell types, such as muscle cells. The basal lamina is organized on a framework of laminin molecules, which are linked together by their side-arms and bind to integrins and other receptors in the basal plasma membrane of overlying epithelial cells. Type IV collagen molecules, together with the protein nidogen and the large proteoglycan perlecan, assemble into a sheetlike mesh that is an essential component of the mature basal lamina. Basal laminae provide mechanical support for epithelia; they form the interface and attachment between epithelia and connective tissue; they serve as filters in the kidney; they act as barriers to keep cells in their proper compartments; they influence cell polarity and cell differentiation; and they guide cell migration during development and tissue regeneration.

CELL–MATRIX JUNCTIONS

Cells make extracellular matrix, organize it, and degrade it. The matrix in turn exerts powerful influences on the cells. The influences are exerted chiefly through transmembrane cell adhesion proteins that act as *matrix receptors*. These proteins tie the matrix outside the cell to the cytoskeleton inside it, but their role goes far beyond simple passive mechanical attachment. Through them, components of the matrix can affect almost any aspect of a cell's behavior. The matrix receptors have a crucial role in epithelial cells, mediating their interactions with the basal lamina beneath them. They are no less important in connective-tissue cells, mediating the cells' interactions with the matrix that surrounds them.

Several types of molecules can function as matrix receptors or co-receptors, including the transmembrane proteoglycans. But the principal receptors on animal cells for binding most extracellular matrix proteins are the integrins. Like the cadherins and the key components of the basal lamina, integrins are part of the fundamental architectural tool kit that is characteristic of multicellular animals. The members of this large family of transmembrane adhesion molecules have a remarkable ability to transmit signals in both directions across the plasma membrane. The binding of a matrix component to an integrin can send a message into the interior of the cell, and conditions in the cell interior can send a signal outward to control binding of the integrin to the matrix. Tension applied to an integrin can cause it to tighten its grip on intracellular and extracellular structures, and loss of tension can loosen its hold, so that molecular signaling complexes fall apart on either side of the membrane. In this way, integrins can serve not only to transmit mechanical and molecular signals but also to convert one type of signal into the other.

Integrins Are Transmembrane Heterodimers That Link the Extracellular Matrix to the Cytoskeleton

There are many varieties of integrins, but they all operate in a similar fashion. An integrin molecule is composed of two noncovalently associated glycoprotein subunits called α and β. Both subunits span the cell membrane and have short intracellular C-terminal tails and large N-terminal extracellular domains (**Figure 19–56**). The extracellular domains bind to specific amino acid sequence motifs in extracellular matrix proteins or, in some cases, in proteins on the surfaces

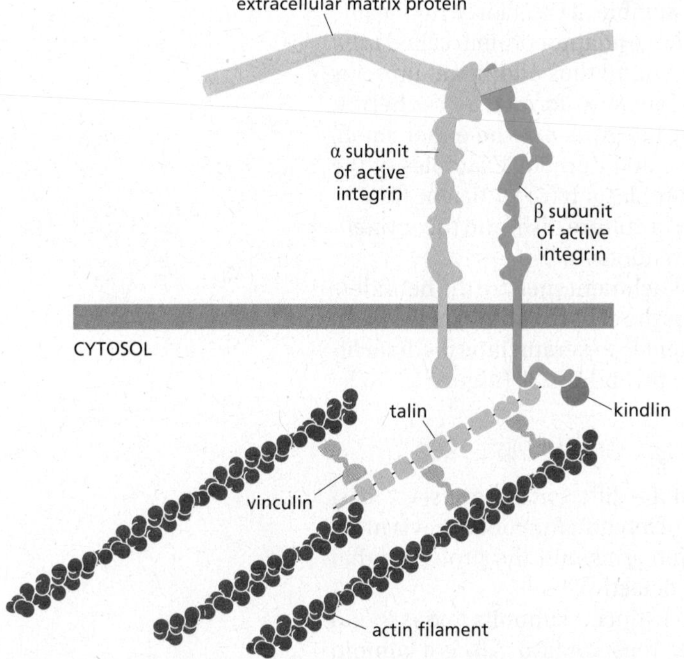

Figure 19–56 The subunit structure of an active integrin molecule, linking extracellular matrix to the actin cytoskeleton. The N-terminal heads of the integrin chains attach directly to an extracellular protein such as fibronectin; the C-terminal intracellular tail of the integrin β subunit binds to adaptor proteins that interact with filamentous actin. The best-understood adaptor is a giant protein called talin, which contains a string of multiple domains for binding actin and other proteins, such as vinculin, that help reinforce and regulate the linkage to actin filaments. One end of talin binds to a specific site on the integrin β-subunit cytoplasmic tail; other regulatory proteins, such as kindlin, bind at another site on the tail.

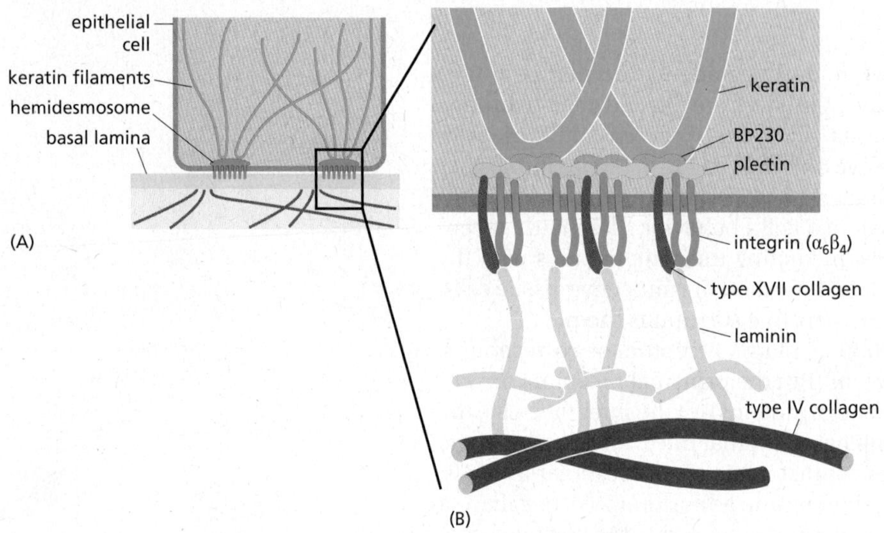

Figure 19–57 **Hemidesmosomes.** (A) Hemidesmosomes spot-weld epithelial cells to the basal lamina, linking laminin outside the cell to keratin filaments inside it. (B) Molecular components of a hemidesmosome. A specialized integrin ($\alpha_6\beta_4$ integrin) spans the membrane, attaching to keratin filaments intracellularly via adaptor proteins called plectin and BP230, and attaching to laminin extracellularly. The adhesive complex also contains, in parallel with the integrin, an unusual collagen family member known as collagen type XVII; this has a membrane-spanning domain attached to its extracellular collagen region. Defects in any of these components can give rise to a blistering disease of the skin. One such disease, called *bullous pemphigoid*, is an autoimmune disease in which the immune system develops antibodies against collagen XVII or BP230.

of other cells. The best-understood binding site for integrins is the RGD sequence mentioned earlier (see Figure 19–48), which is found in fibronectin and other extracellular matrix proteins. Some integrins bind a Leu-Asp-Val (LDV) sequence in fibronectin and other proteins. Additional integrin-binding sequences exist in laminins and collagens.

Humans contain 24 types of integrins, which are formed from the products of 8 different β-chain genes and 18 different α-chain genes that are dimerized in different combinations. Each integrin dimer has specific properties and functions. Moreover, because the same integrin molecule in different cell types can have different ligand-binding specificities, it seems that additional cell type–specific factors can interact with integrins to modulate their binding activity. The binding of integrins to their matrix ligands is also affected by the concentration of Ca^{2+} and Mg^{2+} in the extracellular medium, reflecting the presence of divalent cation-binding domains in the α and β subunits. The divalent cations influence both the affinity and the specificity of the binding of an integrin to its extracellular ligands.

The intracellular portion of an integrin dimer binds to a complex of several different proteins, which together form a linkage to the cytoskeleton. For all but one of the 24 varieties of human integrins, this intracellular linkage is to actin filaments. These linkages depend on proteins that assemble at the short cytoplasmic tails of the integrin subunits (see Figure 19–56). A large adaptor protein called *talin* is a component of the linkage in many cases, but numerous additional proteins are also involved. Like the actin-linked cell-cell junctions formed by cadherins, the actin-linked cell-matrix junctions formed by integrins may be either small, inconspicuous, and transient or large, prominent, and durable. Examples of the latter are the *focal adhesions* that form when fibroblasts have sufficient time to establish strong attachments to the rigid surface of a culture dish, and the *myotendinous junctions* that attach muscle cells to their tendons.

In epithelia, the most prominent cell-matrix attachment sites are the hemidesmosomes, where a specific type of integrin anchors the cells to laminin in the basal lamina. Here, uniquely, the intracellular attachment is to keratin intermediate filaments, via the intracellular adaptor proteins plectin and BP230 (**Figure 19–57**).

Integrin Defects Are Responsible for Many Genetic Diseases

Although there is some overlap in the activities of the different integrins—at least five bind laminin, for example—it is the diversity of integrin functions that is more remarkable. **Table 19–3** lists some varieties of integrins and the problems that result when individual integrin α or β chains are defective.

The β1 subunit forms dimers with at least 12 distinct α subunits and is found on almost all vertebrate cells: $\alpha_5\beta_1$ is a fibronectin receptor and $\alpha_6\beta_1$ is a laminin

TABLE 19–3 Some Types of Integrins

Integrin	Ligand*	Distribution	Phenotype when α subunit is mutated	Phenotype when β subunit is mutated
$\alpha_5\beta_1$	Fibronectin	Ubiquitous	Death of embryo; defects in blood vessels, somites, neural crest	Early death of embryo (at implantation)
$\alpha_6\beta_1$	Laminin	Ubiquitous	Severe skin blistering; defects in other epithelia also	Early death of embryo (at implantation)
$\alpha_7\beta_1$	Laminin	Muscle	Muscular dystrophy; defective myotendinous junctions	Early death of embryo (at implantation)
$\alpha_L\beta_2$ (LFA1)	Ig superfamily counterreceptors (ICAM1)	White blood cells	Impaired recruitment of leukocytes	Leukocyte adhesion deficiency (LAD); impaired inflammatory responses; recurrent life-threatening infections
$\alpha_{IIb}\beta_3$	Fibrinogen	Platelets	Bleeding; no platelet aggregation (Glanzmann disease)	Bleeding; no platelet aggregation (Glanzmann disease); mild osteopetrosis
$\alpha_6\beta_4$	Laminin	Hemidesmosomes in epithelia	Severe skin blistering; defects in other epithelia also	Severe skin blistering; defects in other epithelia also

*Not all ligands are listed.

receptor on many types of cells. Mutant mice that cannot make any β_1 integrins die early in embryonic development. Mice that are only unable to make the α_7 subunit (the partner for β_1 in muscle) survive but develop muscular dystrophy (as do mice that cannot make the laminin ligand for the $\alpha_7\beta_1$ integrin).

The β_2 subunit forms dimers with at least four types of α subunit and is expressed exclusively on the surface of white blood cells, where it has an essential role in enabling these cells to fight infection. The β_2 integrins mainly mediate cell–cell rather than cell–matrix interactions, binding to specific ligands on another cell, such as an endothelial cell. The ligands are members of the Ig superfamily of cell–cell adhesion molecules. We have already described an example earlier in the chapter: an integrin of this class ($\alpha_L\beta_2$, also known as LFA1) on white blood cells enables them to attach firmly to the Ig family protein ICAM1 on vascular endothelial cells at sites of infection (see Figure 19–28B). People with the genetic disease *leukocyte adhesion deficiency* fail to synthesize functional β_2 subunits. As a consequence, their white blood cells lack the entire family of β_2 receptors, and they suffer repeated bacterial infections.

The β_3 integrins are found on blood platelets (as well as various other cells), and they bind several matrix proteins, including the blood clotting factor *fibrinogen.* Platelets have to interact with fibrinogen to mediate normal blood clotting, and humans with *Glanzmann disease*, who are genetically deficient in β_3 integrins, suffer from defective clotting and bleed excessively.

Integrins Can Switch Between an Active and an Inactive Conformation

A cell crawling through a tissue—a fibroblast or a macrophage, for example, or an epithelial cell migrating along a basal lamina—has to be able both to make and to break attachments to the matrix, and to do so rapidly if it is to travel quickly. Similarly, a circulating white blood cell has to be able to switch on or off its tendency to bind to endothelial cells in order to crawl out of a blood vessel at a site of inflammation. Furthermore, the making and breaking of the extracellular attachments in all these cases has to be coupled to the prompt assembly and disassembly of

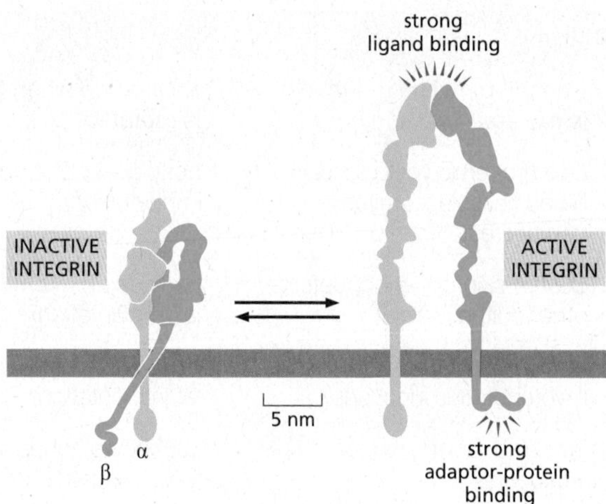

Figure 19–58 Integrins exist in two major activity states. Inactive (folded) and active (extended) structures of an integrin molecule, based on data from x-ray crystallography and other methods.

cytoskeletal attachments inside the cell. The integrin molecules that span the membrane and mediate the attachments cannot simply be passive, rigid objects with sticky patches at their two ends. They must be able to switch between an active state, where they readily form attachments, and an inactive state, where they do not.

Structural studies, using a combination of electron microscopy and x-ray crystallography, suggest that integrins exist in multiple structural conformations that reflect different states of activity (**Figure 19–58**). In the inactive state, the external segments of the integrin dimer are folded together into a compact structure that binds poorly to matrix proteins. In this state, the cytoplasmic tails of the dimer are hooked together, preventing their interaction with cytoskeletal linker proteins. In the active state, the two integrin subunits are unhooked at the membrane to expose the intracellular binding sites for cytoplasmic adaptor proteins, and the external domains unfold and extend, like a pair of legs, to expose a high-affinity matrix-binding site at the tips of the subunits. Thus, the switch from inactive to active states depends on a major conformational change that simultaneously exposes the external and internal ligand-binding sites at the ends of the integrin molecule. External matrix binding and internal cytoskeleton linkages are thereby coupled.

Switching between the inactive and active states is regulated by a variety of mechanisms, depending on the needs of the cell. In some cases, activation occurs by an "outside-in" mechanism: the binding of an external matrix protein, such as the RGD sequence of fibronectin, can drive some integrins to switch from the low-affinity inactive state to the high-affinity active state. As a result, binding sites for talin and other cytoplasmic adaptor proteins are exposed on the tail of the β chain. The binding of these adaptor proteins then leads to attachment of actin filaments to the intracellular end of the integrin molecule (see Figure 19–56). In this way, when the integrin catches hold of its ligand outside the cell, the cell reacts by tying the integrin molecule to the cytoskeleton, so that force can be applied at the point of cell attachment.

The chain of cause and effect can also operate in reverse. This "inside-out" integrin-activation process generally depends on intracellular regulatory signals that stimulate the ability of talin and other proteins to interact with the β chain of the integrin. Talin competes with the integrin α chain for its binding site on the tail of the β chain. Thus, when talin binds to the β chain, it blocks the intracellular α–β linkage, allowing the two legs of the integrin molecule to spring apart.

The regulation of "inside-out" integrin activation is particularly well understood in platelets, where an extracellular signal protein called thrombin binds to a specific G-protein-coupled receptor (GPCR) on the cell surface and thereby activates an intracellular signaling pathway that leads to integrin activation (**Figure 19–59**). It is likely that similar signaling pathways govern integrin activation in numerous other cell types.

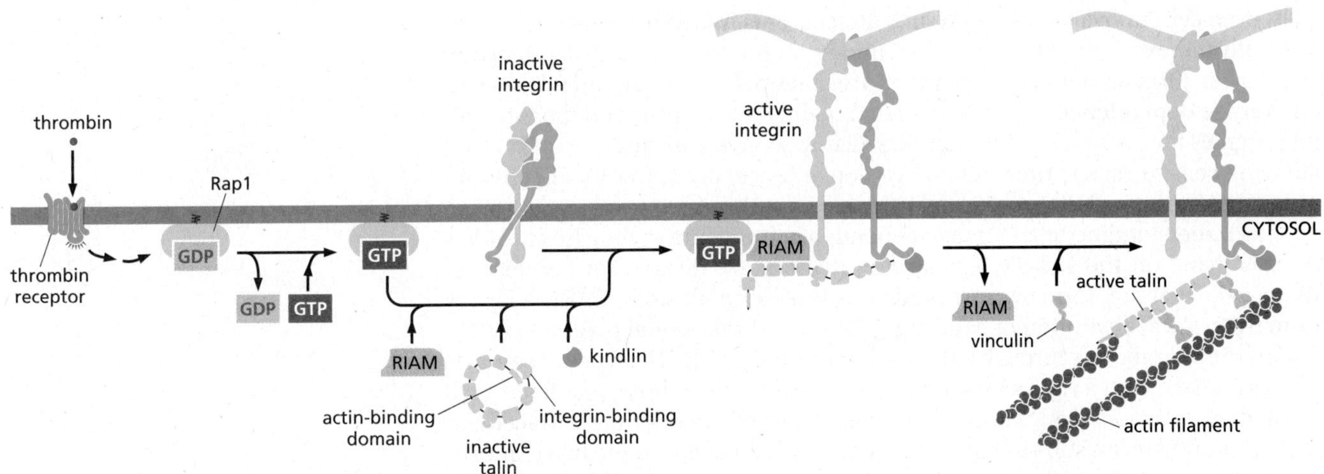

Figure 19–59 **Activation of integrins by intracellular signaling.** Signals received from outside the cell can act through intracellular signaling proteins to stimulate integrin activation. In platelets, as illustrated here, the extracellular signal protein thrombin activates a G-protein–coupled receptor on the cell surface, thereby initiating a signaling pathway that leads to activation of Rap1, a member of the monomeric GTPase family. Activated Rap1 interacts with the protein RIAM, which then recruits talin to the plasma membrane. Prior to this recruitment, talin is held in an inactive state by an interaction between its C-terminal actin-binding domain and its N-terminal integrin-binding domain. When it is recruited by RIAM to the plasma membrane, talin unfolds to expose its binding sites for integrin and actin. Together with another protein called kindlin, talin interacts with the integrin β chain to trigger integrin activation. Talin then interacts with actin and with adaptor proteins such as vinculin, resulting in the formation of multiple actin linkages (see Figure 19–56 and Figure 19–61).

Integrins Cluster to Form Strong Adhesions

Integrins, like other cell adhesion molecules, differ from cell-surface receptors for hormones and for other extracellular soluble signal molecules in that they usually bind their ligand with lower affinity and are present at a 10- to 100-fold higher concentration on the cell surface. The Velcro principle, mentioned earlier in the context of cadherin adhesion (see Figure 19–6C), operates here too. After their activation, integrins cluster together to create a dense plaque in which many integrin molecules are anchored to cytoskeletal filaments. The resulting protein structure can be remarkably large and complex, as seen in the focal adhesion made by a fibroblast on the fibronectin-coated surface of a culture dish.

The assembly of mature cell–matrix junctional complexes depends on the recruitment of dozens of different scaffolding and signaling proteins. Talin is a major component of many cell–matrix complexes, but numerous other proteins also make important contributions. These include the *integrin-linked kinase* (*ILK*) and its binding partners *pinch* and *parvin*, which together form a trimeric complex that serves as an organizing hub at many junctions. Cell–matrix junctions also employ several actin-binding proteins, such as vinculin, *zyxin*, *VASP*, and *α-actinin*, to promote the assembly and organization of actin filaments. Another critical component of many cell–matrix junctions is the *focal adhesion kinase* (*FAK*), which interacts with multiple components in the junction and serves an important function in signaling, as we describe shortly.

Extracellular Matrix Attachments Act Through Integrins to Control Cell Proliferation and Survival

Like other transmembrane cell adhesion proteins, integrins do more than just create attachments. They also activate intracellular signaling pathways and thereby allow control of almost any aspect of the cell's behavior according to the nature of the surrounding matrix and the state of the cell's attachments to it.

Many cells will not grow or proliferate in culture unless they are attached to extracellular matrix; nutrients and soluble mitogens in the culture medium are not enough. For some cell types, including epithelial, endothelial, and muscle cells, even cell survival depends on such attachments. When these cells lose contact with the extracellular matrix, they undergo apoptosis. This dependence of cell growth, proliferation, and survival on attachment to a substratum is known as **anchorage dependence**, and it is mediated mainly by integrins and the intracellular signals they generate. Mutations that disrupt or override this form of control, allowing cells to escape from anchorage dependence, occur in cancer cells and play a major part in their invasive behavior.

Our understanding of anchorage dependence has come mainly from studies of cells living on the surface of matrix-coated culture dishes. For connective-tissue cells that are normally surrounded by matrix on all sides, this is a far cry from the natural environment. Walking over a two-dimensional plain is very different from clambering through a three-dimensional jungle. The types of contacts that cells make with a rigid substratum are not the same as those, much less well studied, that they make with the deformable web of fibers of the extracellular matrix, and there are substantial differences in cell behavior in the two contexts. Nevertheless, it is likely that the same basic principles apply. Both *in vitro* and *in vivo*, intracellular signals generated at cell–matrix adhesion sites are crucial for cell proliferation and survival.

Integrins Recruit Intracellular Signaling Proteins at Sites of Cell–Matrix Adhesion

The mechanisms by which integrins signal into the cell interior are complex, involving several pathways, and integrins and conventional signaling receptors often influence one another and work together to regulate cell behavior, as we have already emphasized. The Ras-MAP kinase pathway (see Figure 15–50), for example, can be activated both by conventional signaling receptors and by integrins, but cells often need both kinds of stimulation of this pathway at the same time to give sufficient activation to induce cell proliferation. Integrins and conventional signaling receptors also cooperate to promote cell survival (discussed in Chapters 15 and 18).

One of the best-studied modes of integrin signaling depends on a cytoplasmic protein tyrosine kinase called focal adhesion kinase (FAK). In studies of cells cultured on plastic dishes, focal adhesions are often prominent sites of tyrosine phosphorylation (**Figure 19–60**), and FAK is one of the major tyrosine-phosphorylated proteins found at these sites. When integrins cluster at cell–matrix contacts, FAK is recruited to the integrin β subunit by intracellular adaptor proteins such as talin or *paxillin* (which binds to one type of integrin α subunit). The clustered FAK molecules phosphorylate each other on a specific tyrosine, creating a

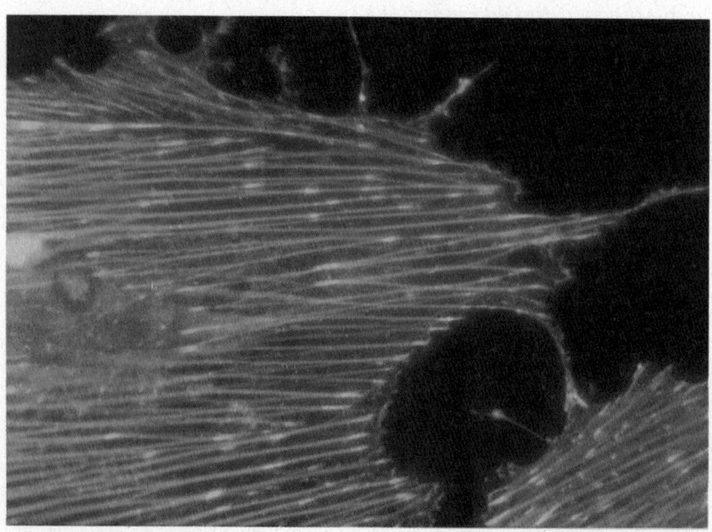

Figure 19–60 Tyrosine phosphorylation at focal adhesions. A fibroblast cultured on a fibronectin-coated substratum and stained with fluorescent antibodies: actin filaments are stained *green* and proteins that contain phosphotyrosine are *red*, giving *orange* where the two components overlap. The actin filaments terminate at focal adhesions, where the cell attaches to the substratum by means of integrins. Proteins containing phosphotyrosine are also concentrated at these sites, reflecting the local activation of FAK and other protein kinases. Signals generated at such adhesion sites help regulate cell division, growth, and survival. (Courtesy of Keith Burridge.)

phosphotyrosine-docking site for members of the Src family of cytoplasmic tyrosine kinases. In addition to phosphorylating other proteins at the adhesion sites, these kinases then phosphorylate FAK on additional tyrosines, creating docking sites for a variety of additional intracellular signaling proteins. In this way, outside-in signaling from integrins, via FAK and Src family kinases, is relayed into the cell in much the same way as receptor tyrosine kinases generate signals (as discussed in Chapter 15).

Cell–Matrix Adhesions Respond to Mechanical Forces

Like the cell-cell junctions we described earlier, cell-matrix junctions can sense and respond to the mechanical forces that act on them. Most cell-matrix junctions, for example, are connected to a contractile actin network that tends to pull the junctions inward. When cells are attached to a rigid matrix that strongly resists such pulling forces, the cell-matrix junction is able to sense the resulting high tension and trigger a response in which it recruits additional integrins and other proteins to increase the junction's ability to withstand that tension. Cell attachment to a relatively soft matrix generates less tension and therefore a less robust response. These mechanisms allow cells to sense and respond to differences in the rigidity of extracellular matrices in different tissues.

We saw earlier that mechanotransduction at cadherin-based cell-cell junctions likely depends on junctional proteins that change their structure when the junction is stretched by tension (see Figure 19–12). The same is true for cell-matrix junctions. Talin, for example, includes a large number of binding sites for the actin-regulatory protein vinculin. Many of these sites are hidden inside folded protein domains but are exposed when those domains are unfolded by stretching the protein (**Figure 19–61**). The N-terminal end of talin binds integrin and the C-terminal end binds actin (see Figure 19–56); thus, when actin filaments are pulled by myosin motors inside the cell, the resulting tension stretches the talin rod, thereby exposing vinculin-binding sites. The vinculin molecules then recruit

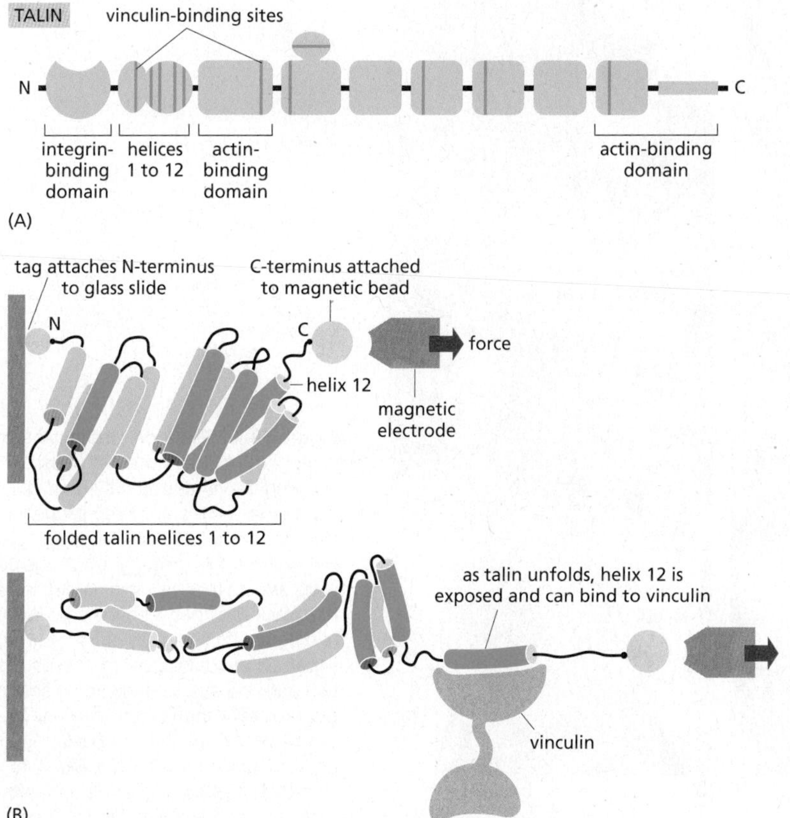

Figure 19–61 Talin is a tension sensor at cell–matrix junctions. Tension across cell–matrix junctions stimulates the local recruitment of vinculin and other actin-regulatory proteins, thereby strengthening the junction's attachment to the cytoskeleton. The experiment presented here tested the hypothesis that tension is sensed by the talin adaptor protein that links integrins to actin filaments (see Figure 19–56). (A) The long, flexible talin protein is divided into a series of folded domains, some of which contain vinculin-binding sites *(dark green lines)* that are thought to be hidden and therefore inaccessible. One domain near the N-terminus, for example, comprises a folded bundle of 12 α helices containing five vinculin-binding sites. (B) This experiment tested the hypothesis that tension stretches the 12-helix domain, thereby exposing vinculin-binding sites. A fragment of talin containing this domain was attached to an apparatus in which the domain could be stretched, as shown here. The fragment was labeled at its N-terminus with a tag that sticks to the surface of a glass slide on a microscope stage. The C-terminal end of the fragment was bound to a tiny magnetic bead, so the talin fragment could be stretched using a small magnetic electrode. The solution around the protein contained fluorescently tagged vinculin proteins. After the talin protein was stretched, excess vinculin solution was washed away, and the microscope was used to determine if any fluorescent vinculin proteins were bound to the talin protein. In the absence of stretching *(top)*, most talin molecules did not bind vinculin. When the protein was stretched *(bottom)*, two or three vinculin molecules were bound (only one is shown here for clarity). (Adapted from A. del Rio et al., *Science* 323:638–641, 2009.)

and organize additional actin filaments. Tension thereby increases the strength of the junction.

Summary

Integrins are the principal cell-surface receptors used by animal cells to bind to the extracellular matrix: they function as transmembrane linkers between the extracellular matrix and the cytoskeleton. Most integrins connect to actin filaments, while those at hemidesmosomes bind to intermediate filaments. Integrin molecules are heterodimers, and the binding of extracellular matrix ligands or intracellular activator proteins such as talin results in a dramatic conformational switch from an inactive to an active state. This creates an allosteric coupling between binding to matrix outside the cell and binding to the cytoskeleton inside it, allowing the integrin to convey signals in both directions across the plasma membrane. Complex assemblies of proteins become organized around the intracellular tails of activated integrins, producing intracellular signals that can influence almost any aspect of cell behavior, from proliferation and survival, as in the phenomenon of anchorage dependence, to cell polarity and guidance of migration. Integrin-based cell–matrix junctions are also capable of mechanotransduction: they can sense and respond to mechanical forces acting across the junction.

THE PLANT CELL WALL

Each cell in a plant deposits, and is in turn completely enclosed by, an elaborate extracellular matrix called the *plant cell wall*. It was the thick cell walls of cork, visible in a primitive microscope, that in 1665 enabled Robert Hooke to observe and name cells for the first time. The walls of neighboring plant cells, cemented together to form the intact plant (**Figure 19–62**), are generally thicker, stronger,

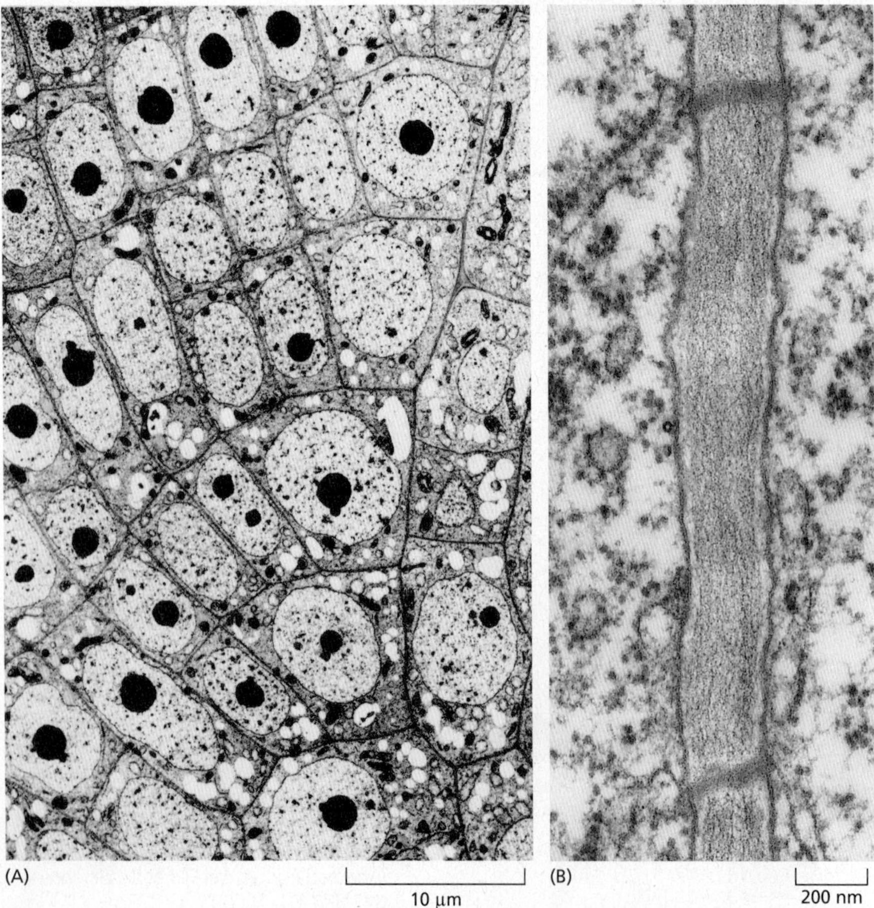

Figure 19–62 Plant cell walls. (A) Electron micrograph of the root tip of a rush, showing the organized pattern of cells that results from an ordered sequence of cell divisions in cells with relatively rigid cell walls. In this growing tissue, the cell walls are still relatively thin, appearing as fine black lines between the cells in the micrograph. (B) Section of a typical cell wall separating two adjacent plant cells. The two dark transverse bands correspond to plasmodesmata that span the wall (see Figure 19–27). (A, courtesy of C. Busby and B. Gunning, *Eur. J. Cell Biol.* 21: 214–223, 1980. With permission from Elsevier. B, courtesy of Jeremy Burgess.)

and, most important of all, more rigid than the extracellular matrix produced by animal cells. In evolving relatively rigid walls, which can be up to many micrometers thick, early plant cells forfeited the ability to crawl about and adopted a sedentary lifestyle that has persisted in all present-day plants.

The Composition of the Cell Wall Depends on the Cell Type

All cell walls in plants have their origin in dividing cells, as the cell plate forms during cytokinesis to create a new partition wall between the daughter cells (discussed in Chapter 17). The new cells are usually produced in special regions called *meristems*, and they are generally small in comparison with their final size. To accommodate subsequent cell growth, the walls of the newborn cells, called **primary cell walls**, are thin and extensible, although tough. Once cell growth stops, the primary wall is sometimes retained without major modification, but, more commonly, a rigid **secondary cell wall** is produced by depositing new layers of matrix inside the old ones. These new layers generally have a composition that is significantly different from that of the primary wall. The most common additional polymer in secondary walls is **lignin**, a complex network of covalently linked phenolic compounds found in the walls of the xylem vessels and fiber cells of woody tissues.

Although the cell walls of higher plants vary in both composition and organization, they are all constructed, like animal extracellular matrices, using a structural principle common to all fiber-composites, including fiberglass and reinforced concrete. One component provides tensile strength, while another, in which the first is embedded, provides resistance to compression. While the principle is the same in plants and animals, the chemistry is different. Unlike the animal extracellular matrix, which is rich in protein and other nitrogen-containing polymers, the plant cell wall is made almost entirely of polymers that contain no nitrogen, including *cellulose* and lignin. For a sedentary organism that depends on CO_2, H_2O, and sunlight, these two abundant biopolymers represent "cheap," carbon-based structural materials, helping to conserve the scarce fixed nitrogen available in the soil that generally limits plant growth. Thus trees, for example, make a huge investment in the cellulose and lignin that compose the bulk of their biomass.

In the cell walls of higher plants, the tensile fibers are made from the polysaccharide cellulose, the most abundant organic macromolecule on Earth, tightly linked into a network by *cross-linking glycans.* In primary cell walls, the matrix in which the cross-linked cellulose network is embedded is composed of *pectin*, a highly hydrated network of polysaccharides rich in galacturonic acid. Secondary cell walls contain additional molecules to make them rigid and permanent; lignin, in particular, forms a hard, waterproof filler in the interstices between the other components. All of these molecules are held together by a combination of covalent and noncovalent bonds to form a highly complex structure, whose composition, thickness, and architecture depend on the cell type.

The plant cell wall thus has a "skeletal" role in supporting the structure of the plant as a whole, a protective role as an enclosure for each cell individually, and a transport role, helping to form channels for the movement of fluid in the plant. When plant cells become specialized, they generally adopt a specific shape and produce specially adapted types of walls, according to which the different types of cells in a plant can be recognized and classified. We focus here, however, on the primary cell wall and the molecular architecture that underlies its remarkable combination of strength, resilience, and plasticity, as seen in the growing parts of a plant.

The Tensile Strength of the Cell Wall Allows Plant Cells to Develop Turgor Pressure

The aqueous extracellular environment of a plant cell consists of the fluid contained in the walls that surround the cell. Although the fluid in the plant cell wall contains more solutes than does the water in the plant's external milieu (for example, soil), it is still hypotonic in comparison with the cell interior. This osmotic

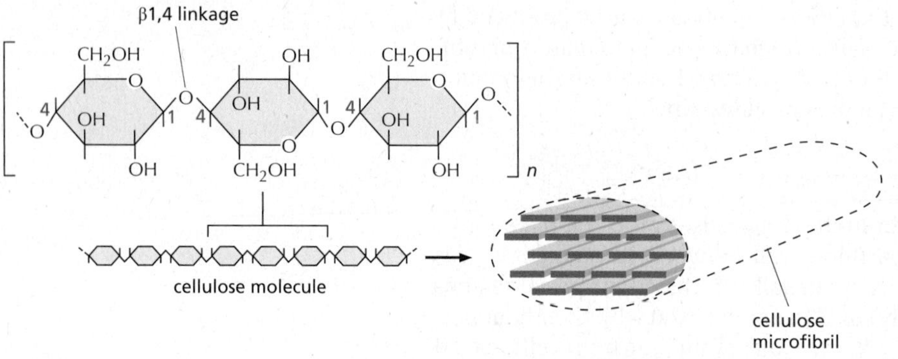

Figure 19–63 **Cellulose.** Cellulose molecules are long, unbranched chains of β1,4-linked glucose units. Each glucose residue is inverted with respect to its neighbors, and the resulting disaccharide repeat occurs hundreds of times in a single cellulose molecule. In most higher plant cells, about 18 individual cellulose molecules assemble in parallel to form a strong, hydrogen-bonded cellulose microfibril.

imbalance causes the cell to develop a large internal hydrostatic pressure, or **turgor pressure**, which pushes outward on the cell wall, just as an inner tube pushes outward on a bicycle tire. The turgor pressure increases just to the point where the cell is in osmotic equilibrium, with no net influx of water despite the salt imbalance. The turgor pressure generated in this way may reach 10 or more atmospheres, about five times that in the average car tire. This pressure is vital to plants because it is the main driving force for cell expansion during growth, and it provides much of the mechanical rigidity of living plant tissues. Compare the wilted leaf of a dehydrated plant, for example, with the turgid leaf of a well-watered one. It is the mechanical strength of the cell wall that allows plant cells to sustain this internal pressure.

The Primary Cell Wall Is Built from Cellulose Microfibrils Interwoven with a Network of Pectic Polysaccharides

Cellulose gives the primary cell wall tensile strength. Each cellulose molecule consists of a linear chain of at least 500 glucose residues that are covalently linked to one another to form a ribbonlike structure, which is stabilized by hydrogen bonds within the chain (Figure 19–63). In addition, hydrogen bonds between adjacent cellulose molecules cause them to stick together in overlapping parallel arrays, forming bundles of about 18 cellulose chains, all of which have the same polarity. These highly ordered crystalline aggregates, many micrometers long, are called **cellulose microfibrils**, and they have a tensile strength comparable to that of steel. Sets of microfibrils are arranged in layers, or lamellae, with each microfibril about 20–40 nm from its neighbors and connected to them by long cross-linking glycan molecules, which are attached by hydrogen bonds to the surface of the microfibrils. The primary cell wall consists of several such lamellae arranged in a plywoodlike network (Figure 19–64).

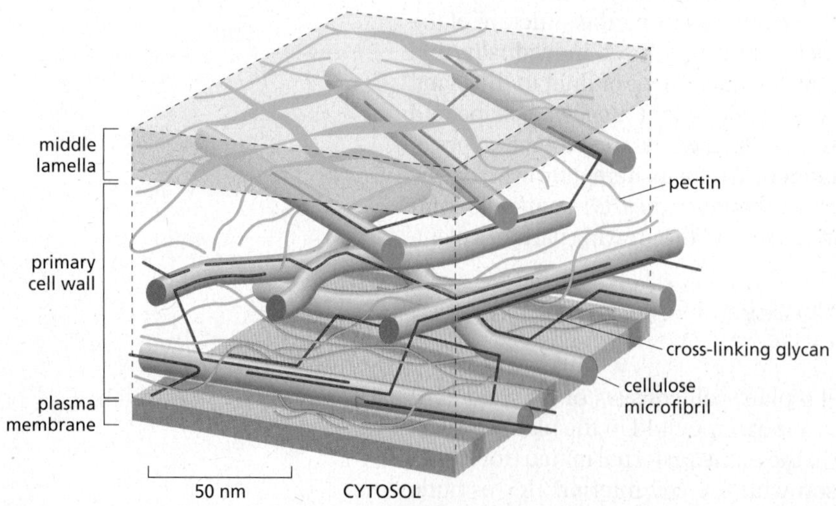

Figure 19–64 **Scale model of a portion of a primary plant cell wall showing the two major polysaccharide networks.** The orthogonally arranged layers of cellulose microfibrils *(blue)* are tied into a network by the cross-linking glycans *(red)* that form hydrogen bonds with the microfibrils. This network is accompanied by a network of pectin polysaccharides *(green)*. The network of cellulose and cross-linking glycans provides tensile strength, while the pectin network resists compression. Cellulose, cross-linking glycans, and pectin are typically present in roughly equal amounts in a primary cell wall. The middle lamella is especially rich in pectin, and it cements adjacent cells together.

The **cross-linking glycans** are a heterogeneous group of branched polysaccharides that bind tightly to the surface of each cellulose microfibril and thereby help to cross-link the microfibrils into a complex network. There are many classes of cross-linking glycans, but they all have a long linear backbone composed of one type of sugar (glucose, xylose, or mannose) from which short side chains of other sugars protrude. The backbone sugar molecules form hydrogen bonds with the surface of cellulose microfibrils, cross-linking them in the process. Both the backbone and the side-chain sugars vary according to the plant species and its stage of development.

The network of cellulose microfibrils and cross-linking glycans includes another cross-linked polysaccharide network that is based on **pectins** (see Figure 19–64). Pectins are a heterogeneous group of branched polysaccharides that contain many negatively charged galacturonic acid units. Because of their negative charge, pectins are highly hydrated and associated with a cloud of cations, resembling the glycosaminoglycans of animal cells in the large amount of space they occupy (see Figure 19–33). When Ca^{2+} is added to a solution of pectin molecules, it cross-links them to produce a semirigid gel (it is pectin that is added to fruit juice to make jam or jelly). Certain pectins are particularly abundant in the *middle lamella*, the specialized region that cements together the walls of adjacent cells (see Figure 19–64); here, Ca^{2+} cross-links are thought to help hold cell wall components together. Although covalent bonds also play a part in linking the components, very little is known about their nature. Regulated separation of cells at the middle lamella underlies such processes as the ripening of tomatoes and the abscission (detachment) of leaves in the fall.

In addition to the two polysaccharide-based networks that form the bulk of all plant primary cell walls, proteins are present, contributing up to about 5% of the wall's dry mass. Many of these proteins are enzymes, responsible for wall turnover and remodeling, particularly during growth. Another class of wall proteins, like collagen, contains high levels of hydroxyproline. These proteins are thought to strengthen the wall, and they are produced in greatly increased amounts as a local response to attack by pathogens. From the genome sequence of *Arabidopsis*, it has been estimated that more than 700 genes are required to synthesize, assemble, and remodel the plant cell wall.

Oriented Cell Wall Deposition Controls Plant Cell Growth

Once a plant cell has left the meristem where it is generated, it can grow dramatically, commonly by more than a thousand times in volume. The manner of this expansion determines the final shape of each cell, and hence the final form of the plant as a whole. Turgor pressure inside the cell drives the expansion, but it is the behavior of the cell wall that governs its direction and extent. Complex wall-remodeling activities are required, as well as the deposition of new wall materials. Because of their crystalline structure, the individual cellulose microfibrils in the wall are unable to stretch, and this gives them a crucial role in the process. For the cell wall to stretch or deform, the microfibrils must either slide past one another or become more widely separated, or both. The orientation of the microfibrils in the innermost layers of the wall governs the direction in which the cell expands. Cells in plants therefore anticipate their future morphology by controlling the orientation of the cellulose microfibrils that they deposit in the wall (**Figure 19–65**).

Unlike most other matrix macromolecules, which are made in the endoplasmic reticulum and Golgi apparatus and are secreted, cellulose is spun out from the surface of the cell by a plasma membrane–bound enzyme complex (*cellulose synthase*), which uses as its substrate the sugar nucleotide UDP-glucose supplied from the cytosol. Each enzyme complex, or *rosette*, is a radial array of six trimers, each containing the protein products of three separate cellulose synthase (*CESA*) genes (see Figure 19–66). Three *CESA* genes are required for primary cell wall synthesis and a different three for secondary cell wall synthesis.

As they are being synthesized, the nascent cellulose chains assemble into microfibrils. These are spun out on the extracellular surface of the plasma

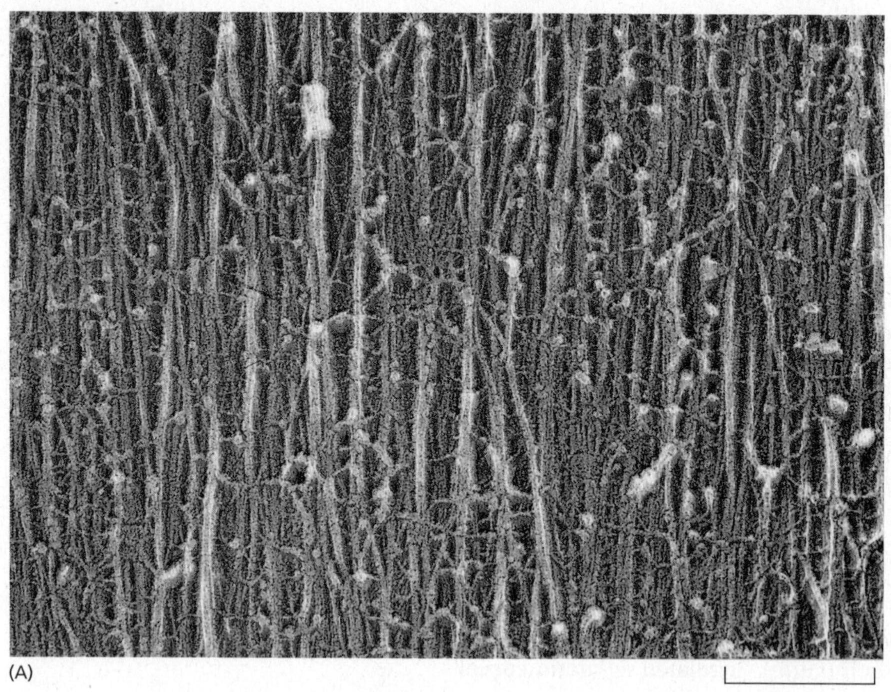

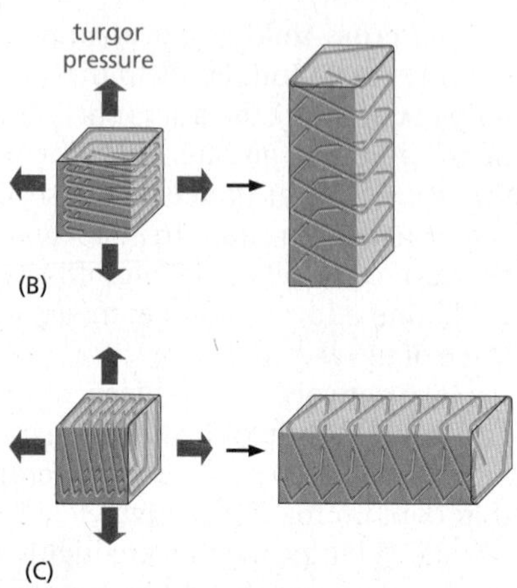

Figure 19–65 **Cellulose microfibrils influence the direction of cell elongation.** (A) The orientation of cellulose microfibrils in the primary cell wall of an elongating carrot cell is shown in this electron micrograph of a shadowed replica from a rapidly frozen and deep-etched cell wall. The cellulose microfibrils are aligned parallel to one another and perpendicular to the axis of cell elongation. The microfibrils are cross-linked by, and interwoven with, a complex web of matrix molecules (compare with Figure 19–64). (B, C) The cells in B and C start off with identical shapes (shown here as cubes) but with different net orientations of cellulose microfibrils in their walls. Although turgor pressure is uniform in all directions, cell wall loosening allows each cell to elongate only in a direction perpendicular to the orientation of the innermost layer of microfibrils, which have great tensile strength. Cell expansion occurs in concert with the insertion of new wall material. The final shape of an organ, such as a shoot, is determined in part by the direction in which its component cells can expand. (A, courtesy of Brian Wells and Keith Roberts.)

membrane, forming a layer, or lamella, in which all the microfibrils have more or less the same alignment (see Figure 19–64). Each new lamella is deposited internally to the previous one, so that the wall consists of concentrically arranged lamellae, with the oldest on the outside. The most recently deposited microfibrils in elongating cells commonly lie perpendicular to the axis of cell elongation, although the orientation of the microfibrils in the outer lamellae that were laid down earlier may be different (see Figure 19–65B and C).

Microtubules Orient Cell Wall Deposition

An important clue to the mechanism that dictates microfibril orientation came from observations of the microtubules in plant cells. These are frequently arranged in the cortical cytoplasm with the same orientation as the cellulose microfibrils that are currently being deposited in the cell wall in that region. These cortical microtubules form a *cortical array* close to the cytosolic face of the plasma membrane, held there by poorly characterized proteins. The congruent orientation of the cortical array of microtubules (lying just inside the plasma membrane) and cellulose microfibrils (lying just outside) is seen in many types and shapes of plant cells and is present during both primary and secondary cell wall deposition, suggesting a causal relationship.

This suggestion can be tested by treating a plant tissue with a microtubule-depolymerizing drug so as to disassemble the entire system of cortical microtubules. The consequences for subsequent cellulose deposition, however, are not as straightforward as might be expected. The drug treatment does not disrupt the production of new cellulose microfibrils, and in some cases cells can continue to deposit new microfibrils in the preexisting orientation. Any developmental switch in the orientation of the microfibril pattern that would normally occur between successive lamellae, however, is invariably blocked. It seems that a preexisting orientation of microfibrils can be propagated even in the absence of microtubules, but any change in the deposition of cellulose microfibrils requires that intact microtubules be present to determine the new orientation.

These observations are consistent with the following model. The cellulose-synthesizing rosettes embedded in the plasma membrane spin out long cellulose

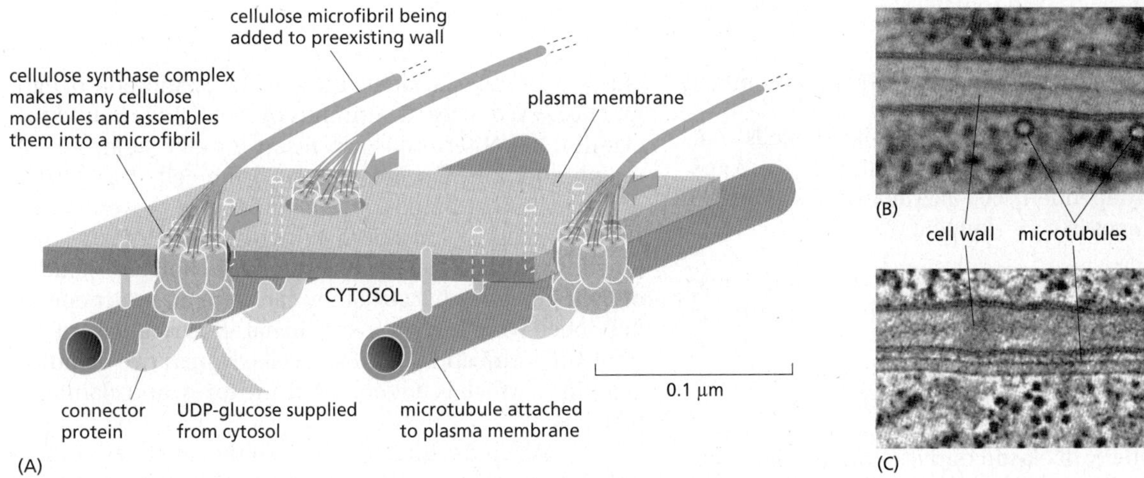

Figure 19–66 **One model of how the orientation of newly deposited cellulose microfibrils might be determined by the orientation of cortical microtubules.** (A) The large cellulose synthase complexes, or *rosettes*, are integral membrane proteins that synthesize cellulose microfibrils on the outer face of the plasma membrane. Each rosette contains six enzyme trimers, resulting in the synthesis of the 18 cellulose chains that make up a cellulose microfibril in many plant cells (see Figure 19–63). The distal ends of the stiff microfibrils become integrated into the texture of the wall, and their elongation at the proximal end pushes the synthase complex along in the plane of the membrane. Because the cortical array of microtubules is attached to the plasma membrane in a way that confines this complex to defined membrane channels, the orientation of these microtubules—when they are present—determines the axis along which the new microfibrils are laid down. (B, C) Two electron micrographs show the tight association of the cortical microtubules with the plasma membrane. One shows the microtubules in cross section while the other shows a microtubule in longitudinal section. Both emphasize the constant gap of about 20 nm between membrane and microtubule. (B and C, courtesy of Andrew Staehelin.)

molecules. As the synthesis of cellulose molecules and their self-assembly into microfibrils proceeds, the distal end of each microfibril presumably forms indirect cross-links to the previous layer of wall material, orienting the new microfibril in parallel with the old ones as it becomes integrated into the texture of the wall. Because the microfibril is stiff, the rosette at its growing, proximal end has to move as it deposits the new material. Traveling in the plane of the membrane, the rosette moves in the direction defined by the way in which the far end of the microfibril is anchored in the existing wall. In this way, each layer of microfibrils would tend to be spun out from the membrane in the same orientation as the layer laid down previously, with the rosettes following the direction of the preexisting oriented microfibrils outside the cell. Oriented microtubules inside the cell, however, can force a change in the direction in which the rosettes move: they can create boundaries in the plasma membrane that act like the banks of a canal to constrain rosette movement (Figure 19–66). In this view, cellulose synthesis can occur independently of microtubules; but it is constrained spatially when cortical microtubules are present to define membrane microdomains within which the enzyme complex can move.

In this way, plant cells can change their direction of expansion by a sudden change in the orientation of their cortical array of microtubules. Because plant cells cannot move (being constrained by their walls), the entire morphology of a multicellular plant presumably depends on a coordinated, highly patterned deployment of cortical microtubule orientations during plant development. It is not known how these orientations are controlled, although it has been shown that the microtubules can reorient rapidly in response to extracellular stimuli, including plant growth regulators such as ethylene and auxins (discussed in Chapter 15).

Microtubules are not, however, the only cytoskeletal elements that influence wall deposition. Local foci of cortical actin filaments can also direct the deposition of new wall material at specific sites on the cell surface, contributing to the elaborate final shaping of many differentiated plant cells.

Summary

Plant cells are surrounded by a tough extracellular matrix, or cell wall, which is responsible for many of the unique features of a plant's lifestyle. The wall is composed of a network of cellulose microfibrils and cross-linking glycans, embedded in a highly cross-linked matrix of pectin polysaccharides. In secondary cell walls, lignin may be deposited to make them waterproof, hard, and woody. A cortical array of microtubules can control the orientation of newly deposited cellulose microfibrils, which in turn determine the direction of cell expansion and therefore the final shape of the cell and, ultimately, of the plant as a whole.

PROBLEMS

Which statements are true? Explain why or why not.

19–1 Given the numerous processes inside cells that are regulated by changes in Ca^{2+} concentration, it seems likely that Ca^{2+}-dependent cell–cell adhesions are also regulated by changes in Ca^{2+} concentration.

19–2 Tight junctions perform two distinct functions: they seal the space between cells to restrict paracellular flow, and they fence off plasma membrane domains to prevent the mixing of apical and basolateral membrane proteins.

19–3 The elasticity of elastin derives from its high content of α helices, which act as molecular springs.

19–4 Integrins can convert mechanical signals into intracellular molecular signals.

19–5 If the entire cortical array of microtubules were disassembled by drug treatment, new cellulose microfibrils would be laid down in random orientations.

Discuss the following problems.

19–6 Comment on the following (1922) quote from Warren Lewis, who was one of the pioneers of cell biology. "Were the various types of cells to lose their stickiness for one another and for the supporting extracellular matrix, our bodies would at once disintegrate and flow off into the ground in a mixed stream of cells."

19–7 Cell adhesion molecules were originally identified using antibodies raised against cell-surface components to block cell aggregation. In the adhesion-blocking assays, the researchers found it necessary to use antibody fragments, each with a single binding site (so-called Fab fragments), rather than intact IgG antibodies, which are Y-shaped molecules with two identical binding sites. The Fab fragments were generated by digesting the IgG antibodies with papain, a protease, to separate the two binding sites (**Figure Q19–1**). Why do you suppose it was necessary to use Fab fragments to block cell aggregation?

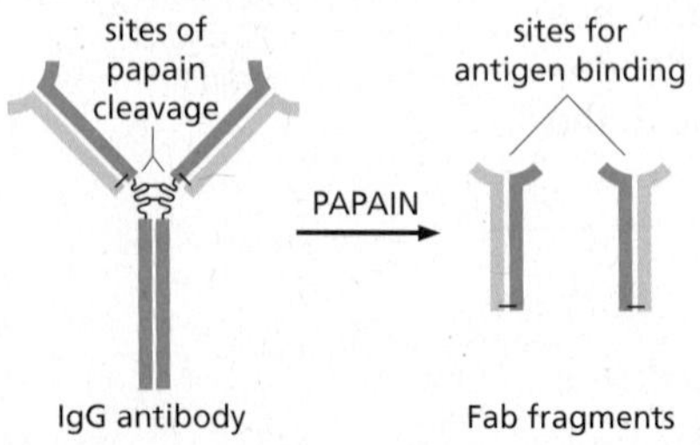

Figure Q19–1 Production of Fab fragments from IgG antibodies by digestion with papain (Problem 19–7).

19–8 The food-poisoning bacterium *Clostridium perfringens* makes a toxin that binds to members of the claudin family of proteins, which are the main constituents of tight junctions. When the C-terminus of the toxin is bound to a claudin, the N-terminus can insert into the adjacent cell membrane, forming holes that kill the cell. The portion of the toxin that binds to the claudins has proven to be a valuable reagent for investigating the properties of tight junctions. MDCK cells are a common choice for studies of tight junctions because they can form an intact epithelial sheet with high transepithelial electrical resistance (low ion permeability). MDCK cells express two claudins: claudin-1, which is not bound by the toxin, and claudin-4, which is.

When an intact MDCK epithelial sheet is incubated with the C-terminal toxin fragment, claudin-4 disappears, becoming undetectable within 24 hours. In the absence of claudin-4, the cells remain healthy and the epithelial sheet appears intact. The mean number of strands in the tight junctions that link the cells also decreases over 24 hours from about four to about two, and they are less highly branched. A functional assay for the integrity of the tight junctions shows that transepithelial resistance decreases dramatically in the presence of the toxin fragment, but the resistance can be restored by washing out the toxin fragment (**Figure Q19–2A**). Curiously, the toxin fragment produces these effects only when it is added to the basolateral side of the sheet; it has no effect when added to the apical surface (**Figure Q19–2B**).

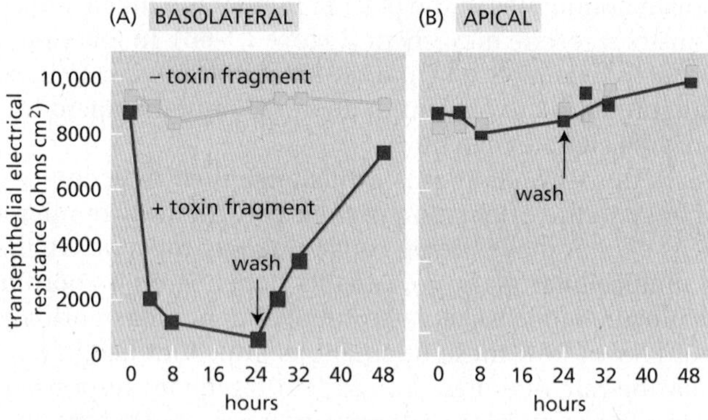

Figure Q19–2 Effects of *Clostridium* toxin fragment on the barrier function of MDCK cells (Problem 19–8). (A) Addition of toxin fragment from the basolateral side of the epithelial sheet. (B) Addition of toxin fragment from the apical side of the epithelial sheet. For a given voltage, a higher resistance (ohms cm^2) gives less paracellular current.

A. How can it be that two tight-junction strands remain, even though all of the claudin-4 has disappeared?

B. Why do you suppose the toxin fragment works when it is added to the basolateral side of the epithelial sheet but not when added to the apical side?

19–9 The glycosaminoglycan polysaccharide chains that are linked to specific core proteins to form the proteoglycan components of the extracellular space are highly negatively charged. How do you suppose these negatively charged polysaccharide chains help to establish a hydrated gel-like environment around the cell? How would the properties of these molecules differ if the polysaccharide chains were uncharged?

19–10 At body temperature, L-aspartate in proteins is converted to its optimal isomer D-aspartate at an appreciable rate. Most proteins in the body have a very low level of D-aspartate, if it can be detected at all. Elastin, however, has a fairly high level of D-aspartate. Moreover, the amount of D-aspartate increases in direct proportion to the age of the person from whom the sample was taken. Why do you suppose that most proteins have little if any D-aspartate, while elastin has levels of D-aspartate that increase steadily with age?

19–11 It is not an easy matter to assign particular functions to specific components of the basal lamina, because the overall structure is a complicated composite material with both mechanical and signaling properties. Nidogen, for example, cross-links two central components of the basal lamina by binding to the laminin γ chain and to type IV collagen. Given such a key role, it was surprising that mice with a homozygous knockout of the gene for nidogen-1 were entirely healthy, with no abnormal phenotype. Similarly, mice homozygous for a knockout of the gene for nidogen-2 also appeared completely normal. By contrast, mice that were homozygous for a defined mutation in the gene for the laminin γ chain, which eliminated just the binding site for nidogen, died at birth with severe defects in lung and kidney formation. The mutant portion of the laminin γ chain is thought to have no other function than to bind nidogen and does not affect laminin structure or its ability to assemble into the basal lamina. How would you explain these genetic observations, which are summarized in **Table Q19–1**? What would you predict would be the phenotype of a mouse that was homozygous for knockouts of both nidogen genes?

TABLE Q19–1 Phenotypes of mice with genetic defects in components of the basal lamina (Problem 19–11)

Protein	Genetic defect	Phenotype
Nidogen-1	Gene knockout (–/–)	None
Nidogen-2	Gene knockout (–/–)	None
Laminin γ chain	Nidogen binding-site deletion (+/–)	None
Laminin γ chain	Nidogen binding-site deletion (–/–)	Dead at birth

+/– stands for heterozygous, –/– stands for homozygous.

19–12 Discuss the following statement: "The basal lamina of muscle fibers serves as a molecular bulletin board, in which adjoining cells can post messages that direct the differentiation and function of the underlying cells."

19–13 Platelets are flat, disc-like cells with a surface area of about 20 μm^2. They have about 80,000 integrin molecules on their surface. If the transmembrane portion of an integrin approximates a cylinder with a 10-nm diameter, how tightly packed are integrins on the surface of a platelet? Imagine that the surface area of the platelet is represented as a grid containing 80,000 squares, each containing one integrin. What is the average distance from one integrin to its neighbor? (Assume each integrin is at the center of its square.)

19–14 The affinity of integrins for matrix components can be modulated by changes to their cytoplasmic domains: a process known as inside-out signaling. You have identified a key region in the cytoplasmic domains of $\alpha_{IIb}\beta_3$ integrin that seems to be required for inside-out signaling (**Figure Q19–3**). Substitution of alanine for either D723 in the β chain or R995 in the α chain leads to a high level of spontaneous activation, under conditions where the wild-type chains are inactive. Your advisor suggests that you convert the aspartate in the β chain to an arginine (D723R) and the arginine in the α chain to an aspartate (R995D). You compare all three α chains (R995, R995A, and R995D) against all three β chains (D723, D723A, and D723R). You find that all pairs have a high level of spontaneous activation, except D723 versus R995 (the wild type) and D723R versus R995D, which have low levels. On the basis of these results, how do you think the $\alpha_{IIb}\beta_3$ integrin is held in its inactive state?

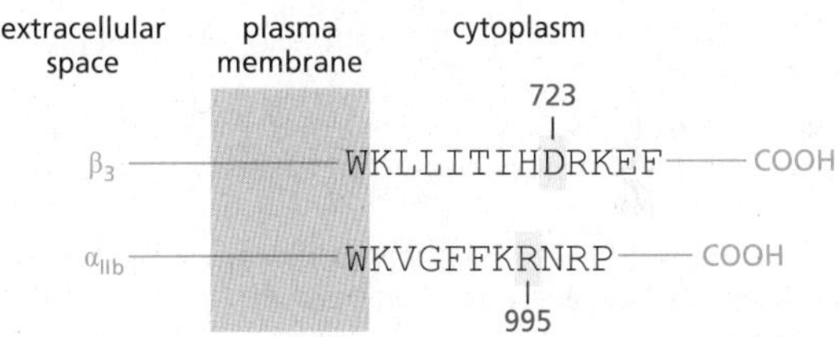

Figure Q19–3 Schematic representation of $\alpha_{IIb}\beta_3$ integrin (Problem 19–14). (From P.E. Hughes et al., *J. Biol. Chem.* 271:6571–6574, 1996. With permission from American Society for Biochemistry and Molecular Biology.)

19–15 Your boss is coming to dinner! All you have for a salad is some wilted, day-old lettuce. You vaguely recall that there is a trick to rejuvenating wilted lettuce, but you cannot remember what it is. Should you soak the lettuce in saltwater, soak it in tap water, or soak it in sugar water, or maybe just shine a bright light on it and hope that photosynthesis will perk it up?

19–16 A plant must be able to respond to changes in the water status of its surroundings. It does so by the flow of water molecules through water channels called aquaporins. The hydraulic conductivity of a single aquaporin is 4.4×10^{-22} m^3 per second per MPa (megapascal) of pressure. What does this correspond to in terms of water molecules per second at atmospheric pressure? [Atmospheric pressure is 0.1 MPa (1 bar) and the concentration of water is 55.5 M.]

REFERENCES

General

Beckerle M (ed.) (2002) Cell Adhesion. Oxford: Oxford University Press.

Hynes RO & Yamada KM (eds.) (2011) Extracellular Matrix Biology (Cold Spring Harbor Perspectives in Biology). Cold Spring Harbor, NY: Cold Spring Harbor Laboratory Press.

Cell–Cell Junctions

Brasch J, Harrison OJ, Honig B & Shapiro L (2012) Thinking outside the cell: how cadherins drive adhesion. *Trends Cell Biol.* 22, 299–310.

Bruser L & Bogdan S (2017) Adherens junctions on the move—membrane trafficking of E-cadherin. *Cold Spring Harb. Perspect. Biol.* 9, a0219140.

Garcia MA, Nelson WJ & Chavez N (2018) Cell-cell junctions organize structural and signaling networks. *Cold Spring Harb. Perspect. Biol.* 10, a029181.

Goodenough DA & Paul DL (2009) Gap junctions. *Cold Spring Harb. Perspect. Biol.* 1, a002576.

Harris TJ & Tepass U (2010) Adherens junctions: from molecules to morphogenesis. *Nat. Rev. Mol. Cell Biol.* 11, 502–514.

Honig B & Shapiro L (2020) Adhesion protein structure, molecular affinities, and principles of cell-cell recognition. *Cell* 181, 520–535.

Lecuit T, Lenne PF & Munro E (2011) Force generation, transmission, and integration during cell and tissue morphogenesis. *Annu. Rev. Cell Dev. Biol.* 27, 157–184.

Lu KJ, Danila FR, Cho Y & Faulkner C (2018) Peeking at a plant through the holes in the wall—exploring the roles of plasmodesmata. *New Phytol.* 218, 1310–1314.

Maule AJ, Benitez-Alfonso Y & Faulkner C (2011) Plasmodesmata—membrane tunnels with attitude. *Curr. Opin. Plant Biol.* 14, 683–690.

McEver RP & Zhu C (2010) Rolling cell adhesion. *Annu. Rev. Cell Dev. Biol.* 26, 363–396.

Najor NA (2018) Desmosomes in human disease. *Annu. Rev. Pathol.* 13, 51–70.

Nakagawa S, Maeda S & Tsukihara T (2010) Structural and functional studies of gap junction channels. *Curr. Opin. Struct. Biol.* 20, 423–430.

Pinheiro D & Bellaïche Y (2018) Mechanical force-driven adherens junction remodeling and epithelial dynamics. *Dev. Cell* 47, 3–19.

Takeichi M (2014) Dynamic contacts: rearranging adherens junctions to drive epithelial remodelling. *Nat. Rev. Mol. Cell Biol.* 15, 397–410.

Yap AS, Duszyc K & Viasnoff V (2018) Mechanosensing and mechanotransduction at cell-cell junctions. *Cold Spring Harb. Perspect. Biol.* 10, a028761.

The Extracellular Matrix of Animals

Couchman JR (2010) Transmembrane signaling proteoglycans. *Annu. Rev. Cell Dev. Biol.* 26, 89–114.

Domogatskaya A, Rodin S & Tryggvason K (2012) Functional diversity of laminins. *Annu. Rev. Cell Dev. Biol.* 28, 523–553.

Fidler AL, Boudko SP, Rokas A & Hudson BG (2018) The triple helix of collagens—an ancient protein structure that enabled animal multicellularity and tissue evolution. *J. Cell Sci.* 131, jcs203950.

Fidler AL, Darris CE, Chetyrkin SV . . . Hudson BG (2017) Collagen IV and basement membrane at the evolutionary dawn of metazoan tissues. *eLife* 6, e24176.

Hynes RO & Naba A (2012) Overview of the matrisome—an inventory of extracellular matrix constituents and functions. *Cold Spring Harb. Perspect. Biol.* 4, a004903.

Jayadev R & Sherwood DR (2017) Basement membranes. *Curr. Biol.* 27, R207–R211.

Karamanos NK, Theocharis AD, Piperigkou Z . . . Onisto M (2021) A guide to the composition and functions of the extracellular matrix. *FEBS J.* (in press). Available at https://febs.onlinelibrary.wiley.com/doi/epdf/10.1111/febs.15776

Lu P, Takai K, Weaver VM & Werb Z (2011) Extracellular matrix degradation and remodeling in development and disease. *Cold Spring Harb. Perspect. Biol.* 3, a005058.

Mouw JK, Ou G & Weaver VM (2014) Extracellular matrix assembly: a multiscale deconstruction. *Nat. Rev. Mol. Cell Biol.* 15, 771–785.

Muncie JM & Weaver VM (2018) The physical and biochemical properties of the extracellular matrix regulate cell fate. *Curr. Top. Dev. Biol.* 130, 1–37.

Pozzi A, Yurchenco PD & Iozzo RV (2017) The nature and biology of basement membranes. *Matrix Biol.* 57–58, 1–11.

Ricard-Blum S (2011) The collagen family. *Cold Spring Harb. Perspect. Biol.* 3, a004978.

Schmelzer CEH & Duca L (2021) Elastic fibers: formation, function and fate during aging and disease. *FEBS J.* (in press). Available at https://febs.onlinelibrary.wiley.com/doi/epdf/10.1111/febs.15899

Townley RA & Bülow HE (2018) Deciphering functional glycosaminoglycan motifs in development. *Curr. Opin. Struct. Biol.* 50, 144–154.

Cell–Matrix Junctions

Bachmann M, Kukkurainen S, Hytonen VP & Wehrle-Haller B (2019) Cell adhesion by integrins. *Physiol. Rev.* 99, 1655–1699.

Calderwood DA, Campbell ID & Critchley DR (2013) Talins and kindlins: partners in integrin-mediated adhesion. *Nat. Rev. Mol. Cell Biol.* 14, 503–517.

Goult BT, Yan J & Schwartz MA (2018) Talin as a mechanosensitive signaling hub. *J. Cell Biol.* 217, 3776–3784.

Hogg N, Patzak I & Willenbrock F (2011) The insider's guide to leukocyte integrin signalling and function. *Nat. Rev. Immunol.* 11, 416–426.

Humphrey JD, Dufresne ER & Schwartz MA (2014) Mechanotransduction and extracellular matrix homeostasis. *Nat. Rev. Mol. Cell Biol.* 15, 802–812.

Jansen KA, Atherton P & Ballestrem C (2017) Mechanotransduction at the cell-matrix interface. *Semin. Cell Dev. Biol.* 71, 75–83.

Matellan C & Del Rio Hernandez AE (2019) Engineering the cellular mechanical microenvironment—from bulk mechanics to the nanoscale. *J. Cell Sci.* 132, jcs229013.

Ross TD, Coon BG, Yun S, . . . Schwartz MA (2013) Integrins in mechanotransduction. *Curr. Opin. Cell Biol.* 25, 613–618.

Shattil SJ, Kim C & Ginsberg MH (2010) The final steps of integrin activation: the end game. *Nat. Rev. Mol. Cell Biol.* 11, 288–300.

Sun Z, Costell M & Fässler R (2019) Integrin activation by talin, kindlin and mechanical forces. *Nat. Cell Biol.* 21, 25–31.

The Plant Cell Wall

Albersheim P, Darvill A, Roberts K et al. (2011) Plant Cell Walls: From Chemistry to Biology. New York: Garland Science.

Braidwood L, Breuer C & Sugimoto K (2014) My body is a cage: mechanisms and modulation of plant cell growth. *New Phytol.* 201, 388–402.

Keegstra K (2010) Plant cell walls. *Plant Physiol.* 154, 483–486.

Lampugnani ER, Khan GA, Somssich M & Persson S (2018) Building a plant cell wall at a glance. *J. Cell Sci.* 131, jcs207373.

Lloyd C (2011) Dynamic microtubules and the texture of plant cell walls. *Int. Rev. Cell Mol. Biol.* 287, 287–329.

McFarlane HE, Doring A & Persson S (2014) The cell biology of cellulose synthesis. *Annu. Rev. Plant Biol.* 65, 69–94.

Meents MJ, Watanabe Y & Samuels AL (2018) The cell biology of secondary cell wall biosynthesis. *Ann. Bot.* 121, 1107–1125.

Polko JK & Kieber JJ (2019) The regulation of cellulose biosynthesis in plants. *Plant Cell* 31, 282–296.

Szymanski DB & Cosgrove DJ (2009) Dynamic coordination of cytoskeletal and cell wall systems during plant cell morphogenesis. *Curr. Biol.* 19, R800–R811.

Wolf S, Hematy K & Hofte H (2012) Growth control and cell wall signaling in plants. *Annu. Rev. Plant Biol.* 63, 381–407.

CHAPTER 20

Cancer

About one in five of us will die of cancer, but that is not why we devote a chapter to this disease. Cancer cells break the most basic rules of cell behavior by which multicellular organisms are built and maintained, and they exploit every kind of opportunity to do so. These transgressions, while often tragic, help to reveal what the normal rules are and how they are enforced. As a result, cancer research helps to illuminate the fundamentals of cell biology—especially cell signaling (Chapter 15), the cell cycle and cell growth (Chapter 17), apoptosis (Chapter 18), and the control of tissue architecture (Chapters 19 and 22). Of course, with a deeper understanding of these normal processes, we also gain a deeper understanding of the disease and better tools to treat it.

In this chapter, we first consider what cancer is and describe the natural history of the disease from a cellular standpoint. We then discuss the molecular changes that make a cell cancerous. And we end the chapter by considering how our enhanced understanding of the molecular basis of cancer is leading to improved methods for its prevention and treatment.

IN THIS CHAPTER

CANCER AS A MICROEVOLUTIONARY PROCESS

The body of an animal operates as a society or ecosystem, whose individual members are cells that reproduce by cell division and organize themselves into collaborative assemblies called *tissues*. This ecosystem is very peculiar, however, because self-sacrifice—as opposed to survival of the fittest—is the rule. Ultimately, all of the somatic cell lineages in animals dedicate their existence to the support of the germ cells, which alone have a chance of continued survival (discussed in Chapter 21). Because the genome of the somatic cells is the same as that of the germ-cell lineage that gives rise to sperm or eggs, through their self-sacrifice the somatic cells help to propagate copies of their own genes.

Unlike free-living cells such as bacteria or yeast, which compete to survive, the cells of a multicellular organism must be committed to collaboration. To coordinate their behavior, the cells send, receive, and interpret an elaborate set of extracellular signals that serve as *social controls*, directing cells how to act (discussed in Chapter 15). As a result, each cell normally behaves in a socially responsible manner—resting, growing, dividing, differentiating, or dying—as needed for the good of the organism.

Molecular disturbances that upset this harmony mean trouble for a multicellular society. In a human body with more than 10^{13} cells, billions of cells experience mutations every day, potentially disrupting the social controls. Most dangerously, a mutation may give one cell a selective advantage, allowing it to grow and divide slightly more vigorously and survive more readily than its neighbors. In this way, a mutated cell can become a founder of a growing mutant clone. Over time, repeated rounds of mutation, competition, and natural selection operating within the population of somatic cells can cause matters to go from bad to worse, jeopardizing the future of the multicellular organism. These are the basic ingredients of cancer: it is a disease in which an individual mutant clone of cells begins by prospering at the expense of its neighbors. In the end—as the

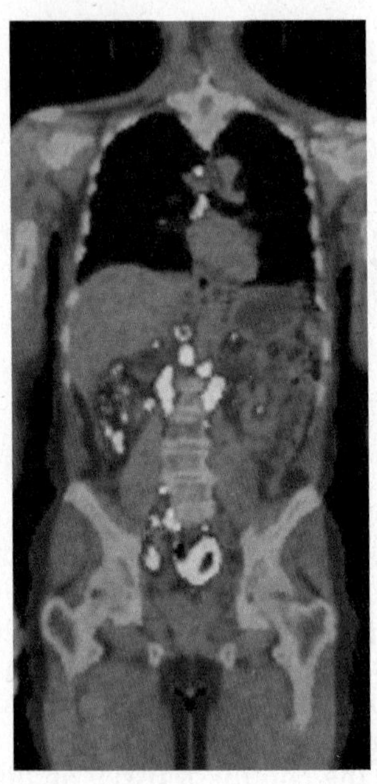

Figure 20–1 **Metastasis.** Malignant tumors typically give rise to metastases, making the cancer hard to eradicate. Shown in this fusion image is a whole-body scan of a patient with metastatic non-Hodgkin's lymphoma (NHL). The background image of the body's tissues was obtained by CT (computed x-ray tomography) scanning. Overlaid on this image, a PET (positron emission tomography) scan reveals the tumor tissue (yellow), detected by its unusually high uptake of radioactively labeled fluorodeoxyglucose (FDG). High FDG uptake occurs in cells with unusually active glucose uptake and metabolism, a characteristic of cancer cells (see Figure 20–18). The yellow spots in the abdominal region reveal multiple metastases. (Courtesy of S.S. Gambhir.)

clone evolves over time and spreads—it can destroy the entire cellular society (**Movie 20.1**).

In this section, we discuss the development of cancer as a microevolutionary process that takes place within the course of a human life span in a subpopulation of cells in the body. As we shall see, the process depends on the same principles of mutation and natural selection that have driven the evolution of living organisms on Earth for billions of years.

Cancer Cells Bypass Normal Proliferation Controls and Colonize Other Tissues

Cancer cells are defined by two heritable properties: (1) they reproduce in defiance of the normal restraints on cell growth and division, and (2) they invade and colonize territories normally reserved for other cells. It is the combination of these properties that makes cancers particularly dangerous. An abnormal cell that goes through successive and inappropriate rounds of growth and division to proliferate out of control will give rise to a tumor, or *neoplasm*—literally, a new growth. As long as the *neoplastic* cells have not yet become invasive, however, the tumor is said to be **benign**. For most types of such neoplasms, removing or destroying the mass locally usually achieves a complete cure.

A tumor is considered a true cancer only if it is **malignant**; that is, when its cells have acquired the ability to invade surrounding tissue. Invasiveness is an essential characteristic of cancer cells. It allows them to break loose, enter blood or lymphatic vessels, and form secondary tumors called **metastases** at other sites in the body (**Figure 20–1**). In general, the more widely a cancer spreads, the harder it becomes to eradicate. It is metastases that usually kill the cancer patient by causing failure of a vital organ.

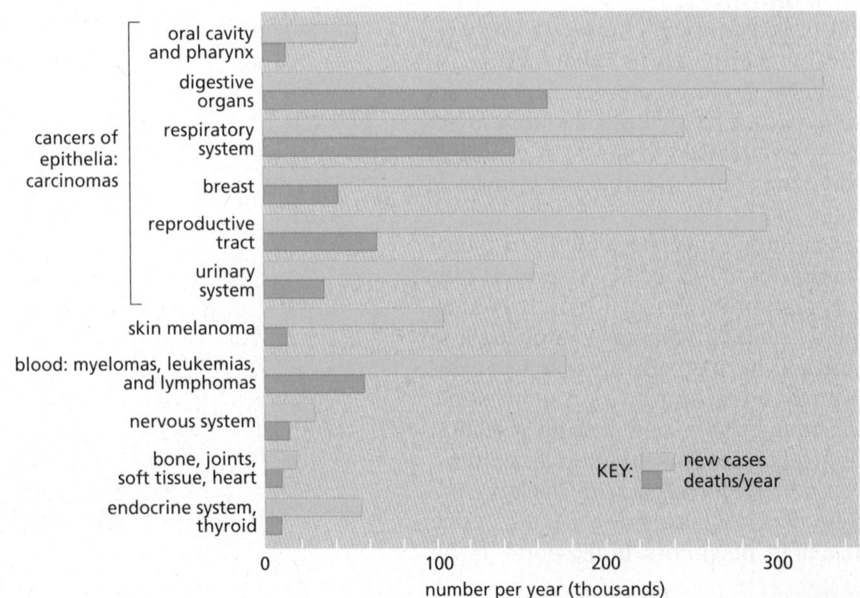

Figure 20–2 **Cancer incidence and mortality in the United States.** The estimated total number of new cases diagnosed in 2018 in the United States was 1,762,450, and total cancer deaths were 606,880. Note that deaths reflect cases diagnosed at many different stages and that well under half of the people who develop cancer die of it. The most common cancers are those of the digestive organs (including colon, pancreas, and liver), respiratory system (primarily lung and bronchus), reproductive tract (prostate and uterine), and breast. Skin cancers other than melanomas are not included in these figures, as almost all are cured easily and many are unrecorded. Each broad category has many subdivisions according to the specific cell type, the location in the body, and the microscopic appearance of the tumor. The data for the United Kingdom are similar. However, incidences are different in some other parts of the world, reflecting widespread exposures to different infectious agents and environmental toxins. (Data from American Cancer Society, Cancer Facts & Figures 2019.)

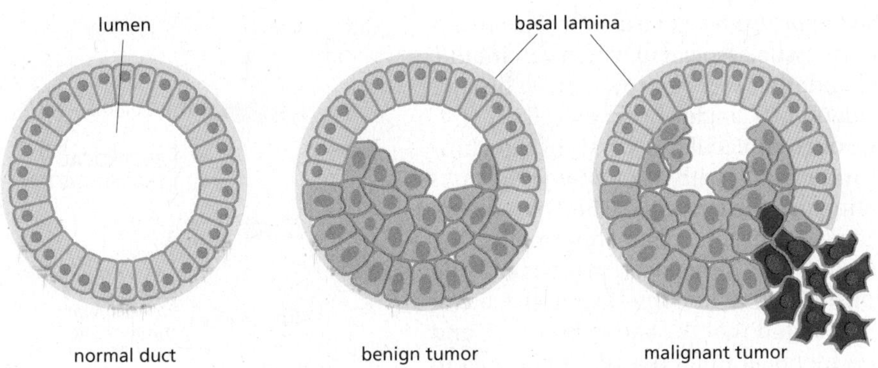

Figure 20–3 Benign versus malignant tumors. A benign glandular tumor (*pink cells*; an adenoma) remains inside the basal lamina (*yellow*) that marks the boundary of the normal structure (a duct, in this example). In contrast, a malignant glandular tumor (*red cells*; an adenocarcinoma) can develop from a benign tumor cell, and it destroys the integrity of the tissue, as shown. There are many different forms that such tumors may take.

Cancers are traditionally classified according to the tissue and cell type from which they arise. **Carcinoma** is the name given to cancers arising from epithelial cells, and they are by far the most common cancers in humans, accounting for about 85% of cancers. Most of the normal cell turnover by proliferation and death in adults occurs in epithelia, and the cell types that undergo the most cell division cycles have the greatest probability of accumulating the multiple mutations needed to become cancerous. In addition, epithelial tissues are the most likely to be exposed to the various forms of physical and chemical damage that favor the development of cancer. Figure 20–2 shows the types of cancers diagnosed in the United States, together with their incidence and death rates. After carcinomas, the next most common cancer types include the various **myelomas**, **leukemias**, and **lymphomas**, derived from white blood cells and their precursors (hemopoietic cells). **Sarcomas** that arise from connective tissue or muscle cells and cancers derived from cells of the nervous system are much less common.

In parallel with the set of names for malignant tumors, there is a related set of names for benign tumors: an *adenoma*, for example, is a benign epithelial tumor with a glandular organization; the corresponding type of malignant tumor is an *adenocarcinoma* (Figure 20–3).

Most cancers have characteristics that reflect their origin. Thus, for example, the cells of a *basal-cell carcinoma*, derived from a keratinocyte stem cell in the skin, generally continue to synthesize cytokeratin intermediate filaments, whereas the cells of a *melanoma*, derived from a pigment cell in the skin, will often (but not always) continue to make pigment granules. Cancers originating from different cell types are, in general, very different diseases. Basal-cell carcinomas of the skin, for example, are only locally invasive and rarely metastasize, whereas melanomas can become much more malignant and often form metastases. Basal-cell carcinomas are readily cured by surgery or local irradiation, whereas malignant melanomas, once they have metastasized widely, are frequently fatal.

Later, we shall see that there is also a different, newer way to classify cancers, one that cuts across the traditional classification by site of origin: we can now classify many of them in terms of the mutations that make the particular tumor cells cancerous. The final section of the chapter will show how this information can be crucial to the design and choice of treatments.

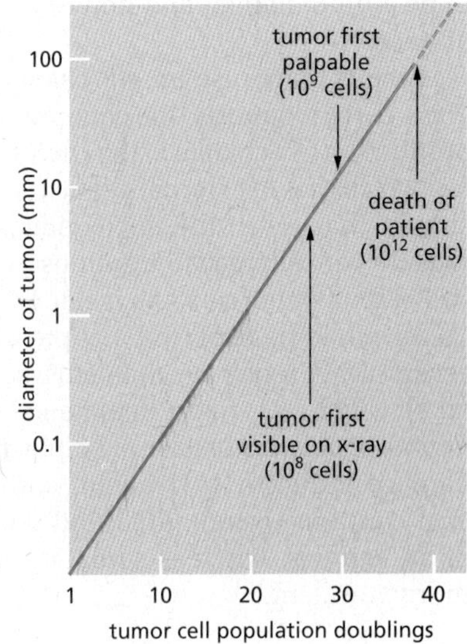

Figure 20–4 The growth of a typical human tumor, such as a tumor of the breast. The diameter of the tumor is plotted on a logarithmic scale. Years may elapse before the tumor becomes noticeable. The doubling time for a typical breast tumor, for example, is about 100 days. However, particularly aggressive tumors may grow much more rapidly.

Most Cancers Derive from a Single Abnormal Cell

Even after a cancer has metastasized, we can usually trace its origins to a single **primary tumor**, arising in a specific organ. The primary tumor is thought to arise by cell division from a single cell that initially experienced some heritable change. Subsequently, additional changes have accumulated in some of the descendants of this cell, allowing them to outgrow, out-divide, and often outlive their neighbors.

By the time it is first detected, a typical human cancer will have been developing for many years and will already contain a billion cancer cells or more (Figure 20–4). Tumors will usually also contain a variety of other cell types and associated extracellular matrix, termed the *tumor stroma*. For example, fibroblasts will be present in the supporting connective tissue associated with a carcinoma, in addition to immune cells and vascular endothelial cells. How can we be sure that the cancer cells are the clonal descendants of a single abnormal cell?

One way of proving clonal origin is through molecular analysis of the chromosomes in tumor cells. In almost all individuals with *chronic myelogenous leukemia* (*CML*), for example, we can distinguish the leukemic white blood cells from the individual's normal cells by a specific chromosomal abnormality: the so-called *Philadelphia chromosome*, created by a translocation between the long arms of chromosomes 9 and 22 (Figure 20–5). When the DNA at the site of translocation is cloned and sequenced, it is found that the site of breakage and rejoining of the translocated fragments is identical in all the leukemic cells in any given individual, but that this site differs slightly (by a few hundred or thousand base pairs) from one individual to another. This is the expected result if, and only if, the cancer in each individual arises from a unique accident occurring in a single cell. (We will see later how this particular translocation promotes the development of CML by creating a novel hybrid gene encoding a protein that promotes cell proliferation.)

Many other lines of evidence, from a variety of different cancers, point to the same conclusion: most cancers originate from a single aberrant cell.

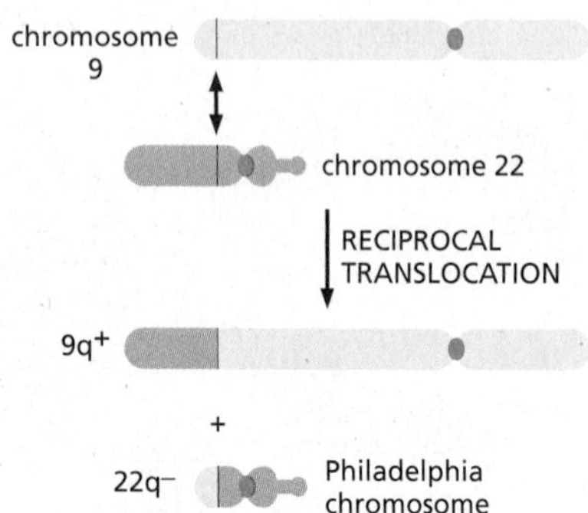

Figure 20–5 The translocation between chromosomes 9 and 22 responsible for chronic myelogenous leukemia. The normal structures of chromosomes 9 and 22 are shown at the top. When a reciprocal translocation occurs between them at the indicated site, the result is the abnormal pair at the bottom. The smaller of the two resulting abnormal chromosomes (22q−) is called the Philadelphia chromosome, after the city where the abnormality was first recorded.

Cancer Cells Contain Somatic Mutations

If a single abnormal cell is to give rise to a tumor, it must pass on its abnormality to its progeny: the aberration has to be heritable. Thus, the development of a clone of cancer cells depends on genetic changes. The tumor cells contain **somatic mutations**: they have one or more shared detectable abnormalities in their DNA sequence that distinguish them from the normal cells surrounding the tumor, as in the example of CML just described. (The mutations are called *somatic* because they occur in the soma, or body cells, not in the germ line.) Cancers are also driven by *epigenetic changes*—persistent, heritable changes in gene expression that result from modifications of chromatin structure without alteration of the cell's DNA sequence. But somatic mutations that alter a DNA sequence appear to be a fundamental and universal feature, and cancer is in this sense a genetic disease.

Factors that cause genetic changes tend to provoke the development of cancer. Thus, **carcinogenesis** (the generation of cancer) can be linked to *mutagenesis* (the production of a change in the DNA sequence). This correlation is particularly clear for two classes of external agents: (1) *chemical carcinogens* (which typically cause simple changes in the nucleotide sequence), and (2) *radiation,* such as x-rays (which typically cause chromosome breaks and translocations) or ultraviolet (UV) light (which causes specific DNA base alterations).

As would be expected, people who have inherited a genetic defect in one of several DNA repair mechanisms, causing their cells to accumulate mutations at an elevated rate, run a heightened risk of cancer. Those with the disease *xeroderma pigmentosum*, for example, have defects in the system that repairs DNA damage induced by UV light, and they have a greatly increased incidence of skin cancers. Overall, inherited mutations are thought to play a role in 5–10% of all cancers, whereas somatic mutations and epigenetic changes are much more prevalent.

A Single Mutation Is Not Enough to Change a Normal Cell into a Cancer Cell

It is estimated that there are 3.7×10^{13} cells in an average-sized human (not counting bacteria), and that 10^{16} cell divisions occur within the body over the course of a typical lifetime. This cell proliferation occurs primarily in epithelia and the hemopoietic system and is balanced by cell death to maintain normal tissue homeostasis. Even in an environment that is free of mutagens, mutations would occur spontaneously at an estimated rate of three per cell division, corresponding to about 10^{-6} mutations per gene per cell division. This unavoidable error rate is set by fundamental limitations on the accuracy of DNA replication and repair (see pp. 253–254). This means that, over the course of a typical lifetime, every single human gene will have undergone mutation somewhere in the body on roughly 10^{10} separate occasions. Among the resulting mutant cells, there will be many that have sustained deleterious mutations in genes that regulate cell growth and division, causing the cells to disobey the normal rules controlling cell turnover. Given the huge number of unavoidable mutations in a large organism like ourselves, the problem of cancer might seem to be not why it occurs, but why it occurs so infrequently.

If a mutation in a single gene were enough to convert a typical healthy cell into a cancer cell, we would not be viable organisms. Many lines of evidence indicate that the development of a cancer typically requires that a substantial number of independent, rare genetic and epigenetic accidents occur in the lineage that emanates from a single cell. One such indication comes from epidemiological studies of the incidence of cancer as a function of age (**Figure 20–6**). If a single mutation were responsible for cancer, occurring with a fixed probability per year, the chance of developing cancer in any given year of life should be independent of age. In fact, for most types of cancer, the incidence rises steeply with age—as would be expected if cancer is caused by a progressive, random accumulation of a set of mutations in a single lineage of cells. Notably, however, cancer incidence declines markedly among the very elderly. One interpretation of this phenomenon is that decreased cell proliferation, which is characteristic of declining stem cell function in octogenarians, provides fewer opportunities for mutation.

As discussed later, these indirect arguments linking the number of accumulated mutations to cancer development have now been confirmed by systematically sequencing the genomes of the tumor cells from individual cancer patients and cataloging the mutations that they contain.

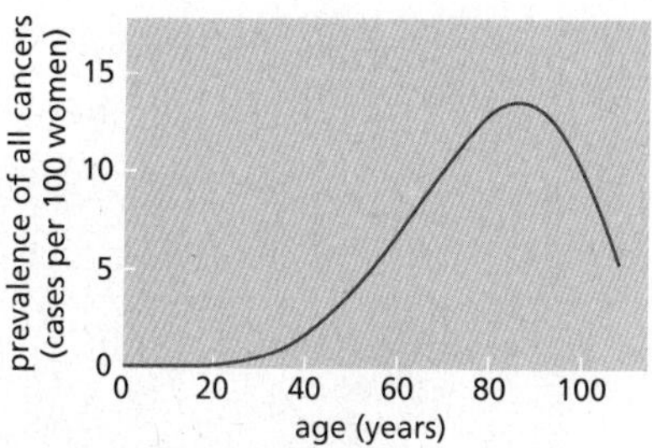

Figure 20–6 **Cancer prevalence as a function of age.** Age versus malignant cancer prevalence is plotted for women in the Surveillance, Epidemiology, and End Results (SEER) 9 cancer registries for the year 2000. The prevalence of cancer rises steeply as a function of age. If only a single mutation were required to trigger a cancer and the mutation had an equal chance of occurring at any time, the prevalence of cancer would be the same at all ages. (Data from C. Harding et al., *Cancer* 118:1371–1386, 2012, doi 10.1002/cncr.26376.)

Many Cancers Develop Gradually Through Successive Rounds of Random Inherited Change Followed by Natural Selection

For those cancers known to have a specific external cause, the disease does not usually become apparent until long after exposure to the causal agent. The incidence of lung cancer, for example, does not begin to rise steeply until after decades of heavy smoking (**Figure 20–7**). Similarly, the incidence of leukemias in those exposed to intense radiation in Hiroshima and Nagasaki did not show a marked rise until about 5 years after the explosion of the atomic bombs. And industrial workers exposed for a limited period to chemical carcinogens do not usually develop the cancers characteristic of their occupation until 10, 20, or even more years after the exposure. During this long incubation period, the prospective cancer cells undergo a succession of changes, and the same presumably applies to cancers where the initial genetic lesion has no such obvious external cause.

The fact that the development of a cancer requires a gradual accumulation of mutations in a number of different genes within a cell helps to explain the well-known phenomenon of **tumor progression**, whereby an initial mild disorder of cell behavior evolves gradually into a full-blown cancer (**Figure 20–8**).

At each stage of progression, some individual cell acquires an additional mutation or epigenetic change that gives it a selective advantage over its neighbors, making it better able to thrive in its environment—an environment that, inside a tumor, may be harsh, with low levels of oxygen, scarce nutrients, and the natural barriers to growth presented by the surrounding normal tissues. The larger the number of tumor cells, the higher the chance that at least one of them will undergo a change that favors it over its neighbors. Thus, as the tumor grows, progression accelerates. The offspring of the best-adapted cells continue to divide, eventually producing the dominant clones in a developing lesion

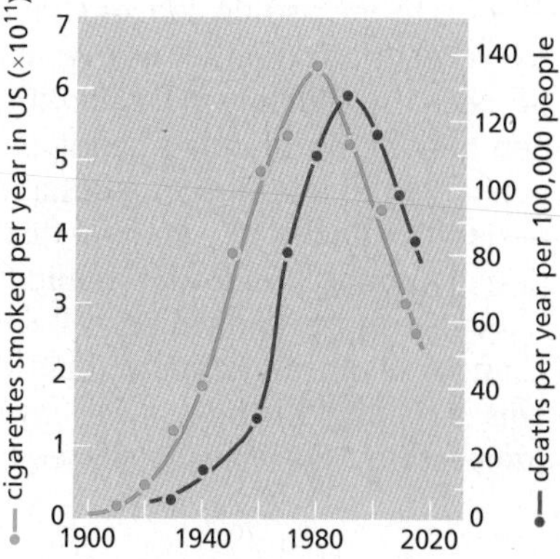

Figure 20–7 **Smoking and lung cancer.** A major increase in cigarette smoking that started in the early 1900s *(blue line)* caused a dramatic rise in lung cancer deaths *(red line)* after a lag time of about 20 years. Because cigarette smoking peaked in 1980, lung cancer deaths are now declining after a similar lag. (Data from National Center for Health Statistics, Centers for Disease Control and Prevention, 2017.)

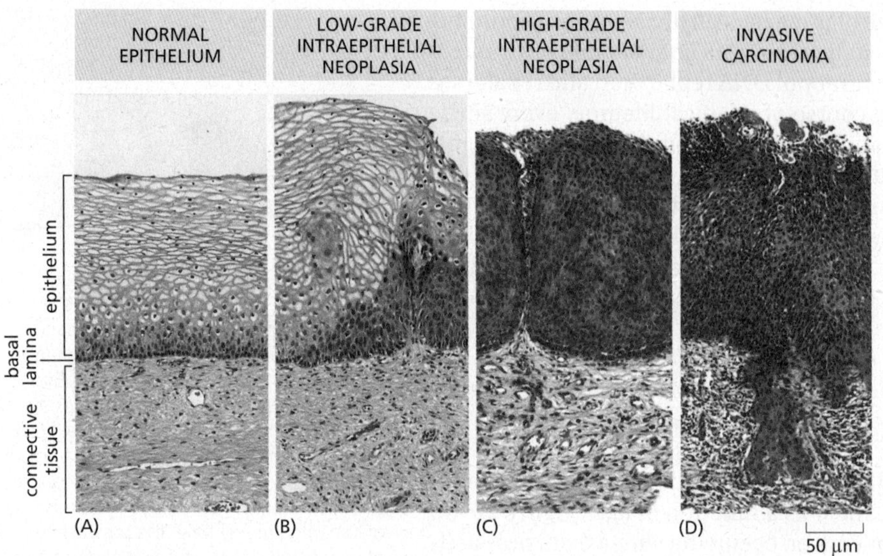

Figure 20–8 **Stages of progression in the development of cancer of the epithelium of the uterine cervix.** Pathologists use standardized terminology to classify the types of disorders they see in tissue slices like those shown here. (A) In a stratified squamous epithelium, dividing cells are confined to the basal layer. (B) In this low-grade intraepithelial neoplasia (right half of image), dividing cells can be found throughout the lower third of the epithelium; the superficial cells are still flattened and show signs of differentiation. (C) In high-grade intraepithelial neoplasia, cells in all the epithelial layers are proliferating and exhibit defective differentiation. (D) True malignancy begins when the cells move through or destroy the basal lamina that underlies the epithelium and invade the underlying connective tissue. (Courtesy of Andrew J. Connolly.)

(**Figure 20–9**). Thus, tumor progression involves a large element of chance and usually takes many years, which may be why the majority of us will die of causes other than cancer.

Just as in the evolution of plants and animals, a kind of speciation often occurs: the original cancer cell lineage can diversify to give many genetically different subclones of cells. These may coexist in the same mass of tumor tissue; or they may migrate and colonize separate environments suited to their individual quirks, where they settle, thrive, and progress as independently evolving metastases. As new mutations arise within each tumor mass, different subclones may gain an advantage and come to predominate, only to be overtaken by others or outgrown by their own sub-subclones. The large amount of genetic diversity in most tumors is one of the chief factors that make cancer cures difficult; it also increases the importance of detecting a tumor as early as possible.

Cancers Can Evolve Abruptly Due to Genetic Instability

If cancer cells evolved exclusively through the gradual accumulation of single deleterious mutations, then the time scale of transformation from premalignant disease to metastatic cancer should be predictable for a particular cancer type. However, this is not the case. As in the evolution of species, cancer evolution may proceed through long periods with no discernible change punctuated by the sudden generation of new phenotypes. The "Big Bang" theory of cancer evolution posits that in addition to gradual mutagenesis and selection for the fittest cells, periodic, cataclysmic genome disruption can promote rapid steps in the evolution of a cancer cell toward malignancy.

Unlike normal dividing cells, most human cancer cells accumulate genetic changes at an abnormally rapid rate and are said to be **genetically unstable.** This instability provides a selective advantage by speeding the process of tumor progression—allowing the subsequent accumulation of many additional mutations

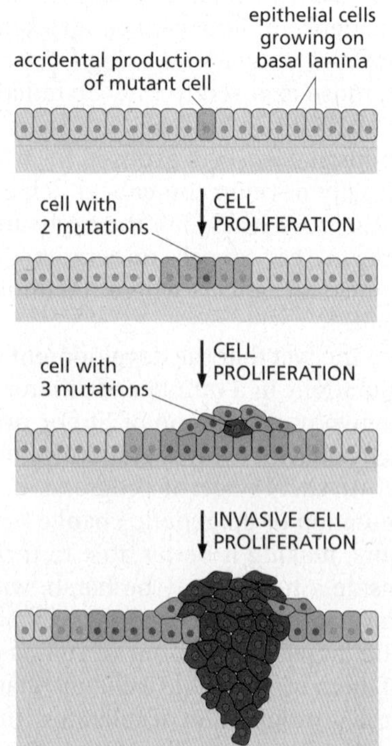

Figure 20–9 **Clonal evolution during tumor progression.** In this schematic diagram, a tumor develops through repeated rounds of mutation and proliferation, giving rise to a clone of fully malignant cancer cells. At each step, a single cell undergoes a mutation that either enhances cell proliferation or decreases cell death, so that its progeny become the dominant clone in the tumor. Proliferation of each clone hastens the occurrence of the next step of tumor progression by increasing the size of the cell population that is at risk of undergoing an additional mutation. The final step depicted here is invasion through the basal lamina, an initial step in metastasis. In reality, there may be more than the three steps shown here, depending on the tumor type, and a combination of genetic and epigenetic changes is involved. Not shown here is the fact that, over time, a variety of competing subclones will often arise in a tumor. As we will discuss later, this heterogeneity complicates cancer therapies.

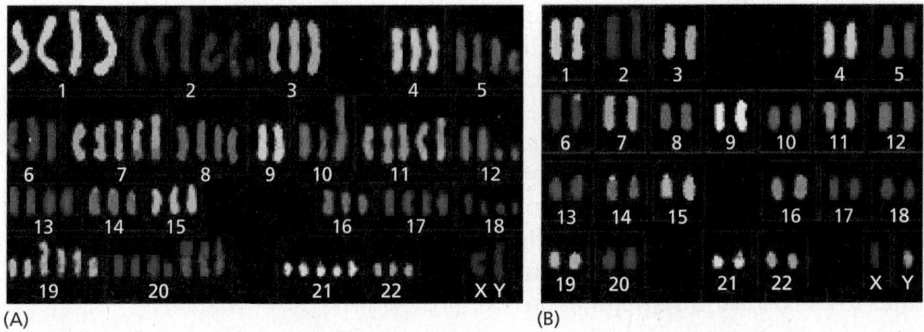

Figure 20–10 **Chromosome complements (karyotypes) of colon cancers showing different kinds of genetic instability.** (A) The karyotype of a typical cancer shows many gross abnormalities in chromosome number and structure. Considerable variation can also exist from cell to cell (not shown). (B) The karyotype of a tumor that has a stable chromosome complement with few chromosomal anomalies; the genetic abnormalities in these tumors are mostly invisible, having been created by defects in DNA repair. All of the chromosomes in this figure were stained as in Figure 4–11, the DNA of each human chromosome being marked with a different combination of fluorescent dyes. (Courtesy of Wael Abdel-Rahman and Paul Edwards.)

required to produce a cancer. But the extent of the instability and its molecular origins differ from cancer to cancer and from individual to individual, both in severity and in character. In some cases, the karyotype—the set of chromosomes as they appear at mitosis—is normal or nearly so, but many point mutations are detected in individual genes, suggesting a failure of the repair mechanisms that normally correct errors in the replication or maintenance of DNA sequences. Often, however, the cancer cell karyotype is severely disordered, with many chromosomal breaks and rearrangements, resulting in many deletions, duplications, and amplifications of parts of the genome (**Figure 20–10**). Such highly disrupted genomes indicate that catastrophic events have occurred, likely due to defects in chromosome duplication or segregation during mitosis. For example, a common feature of many cancer cells is a failure to correctly attach all of the chromosomes to the mitotic spindle, which can result in chromosome breakage or aneuploidy, the gain or loss of individual chromosomes. More dramatically, mitotic defects can also lead to the isolation of a single chromosome in one of the daughter cells within a "micronucleus," where it is prone to massive DNA damage and chromosomal rearrangement, termed "chromothripsis" (**Figure 20–11**).

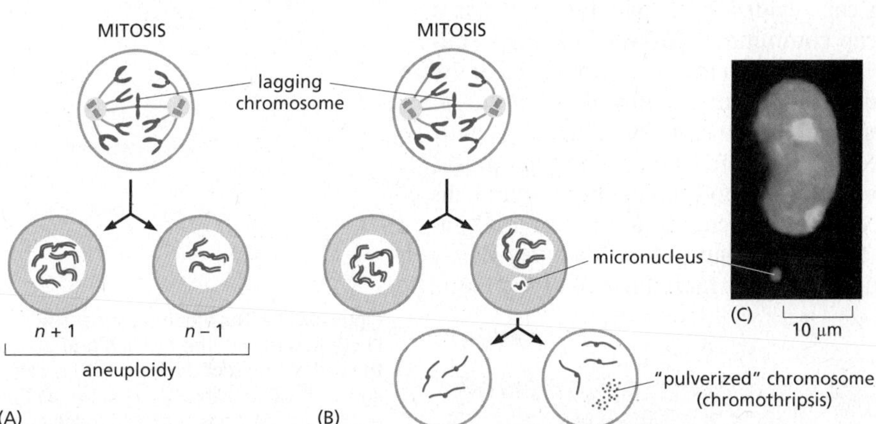

Figure 20–11 **Chromosome segregation defects can give rise to aneuploidy and/or chromothripsis.** Correct chromosome segregation requires that replicated copies of each chromosome, termed sister chromatids, are attached to opposite spindle poles. A mistake in this process is depicted in which one sister of a chromatid pair is attached to both spindle poles and is therefore pulled in opposite directions during anaphase, causing it to lag behind in the center of the cell, resulting in a *lagging chromosome*. (A) A lagging chromosome may end up in the same daughter cell as its sister, resulting in aneuploidy with $n + 1$ and $n - 1$ karyotypes (n = chromosome number). (B) Another possible outcome is that the lagging chromosome remains separated from the rest of the chromosomes in the daughter cell, forming its own "micronucleus" in the following interphase. DNA replication in micronuclei is frequently incomplete, and DNA damage accumulates, but this does not delay the cell cycle. In the next mitosis, the isolated chromosome is prone to undergo fragmentation during chromosome condensation, in a process called chromothripsis. (C) Micronuclei can persist over several generations of cell division, undergoing rounds of chromosome fragmentation and reassembly, or they can be reincorporated into a daughter cell nucleus (**Movie 20.2**). The fluorescence micrograph shows a primary nucleus and a micronucleus present in the same cell. Nuclear DNA staining *(blue)* appears uneven due to normal nuclear substructures. (C, from M.L. Leibowitz et al., *Annu. Rev. Genet.* 49:183–211, 2015. With permission from Annual Reviews.)

From an evolutionary perspective, none of this should be a surprise: anything that increases the probability of random changes in gene function that are heritable from one cell generation to the next—and that are not too deleterious—will likely speed the evolution of a clone of cells toward malignancy, thereby causing this property to be selected for during tumor progression. Genome instability likely also contributes to the heterogeneity of cancer cells frequently observed within an individual tumor. By leading to multiple aberrant karyotypes, chromosome mis-segregation, aneuploidy—and more rarely chromothripsis—all act to shuffle the genetic cards, allowing cancer cells to sample a variety of different phenotypes and for different clones to coexist within a population.

Some Cancers May Harbor a Small Population of Stem Cells

Self-renewing tissues, where cell division continues throughout life, are the breeding ground for the great majority of human cancers. They include the epidermis (the outer epithelial layer of the skin), the epithelial lining of the digestive and reproductive tracts, and the bone marrow, where blood cells are generated (see Chapter 22). With each cell division cycle comes the chance of mutation, which can be amplified dramatically by environmental factors. As a result, there is a strong correlation between the frequency of cell division and the incidence of cancer in a particular tissue (**Figure 20–12**).

In almost all proliferating tissues, renewal depends on the presence of stem cells, which divide to give rise to terminally differentiated cells, which do not divide. This creates a mixture of cells that are genetically identical and closely related by lineage but are in different states of differentiation. Many tumors similarly appear to consist of populations of cells in various states of differentiation. A comparison of tumor development to the normal homeostasis of stem cell-derived tissues may help us better understand the origin of some cancers and also why some tumors are so resistant to treatment.

To consider the implications, it is helpful to consider how normal stem-cell systems operate. When a normal stem cell divides, each daughter cell has a choice—it can remain a stem cell or it can commit to a pathway leading to differentiation. A stem-cell daughter remains in place to generate more cells in the future. A committed daughter typically undergoes some rounds of cell proliferation (as a so-called *transit amplifying cell*), but it then stops dividing, terminally differentiates, and eventually is discarded and replaced (it may die by apoptosis, with recycling of its materials, or be shed from the body). Therefore, stem cells tend to be vastly outnumbered by the cells that are committed to terminal differentiation. However, though few and far between and often relatively slowly dividing, stem cells carry the entire responsibility for maintenance of the tissue

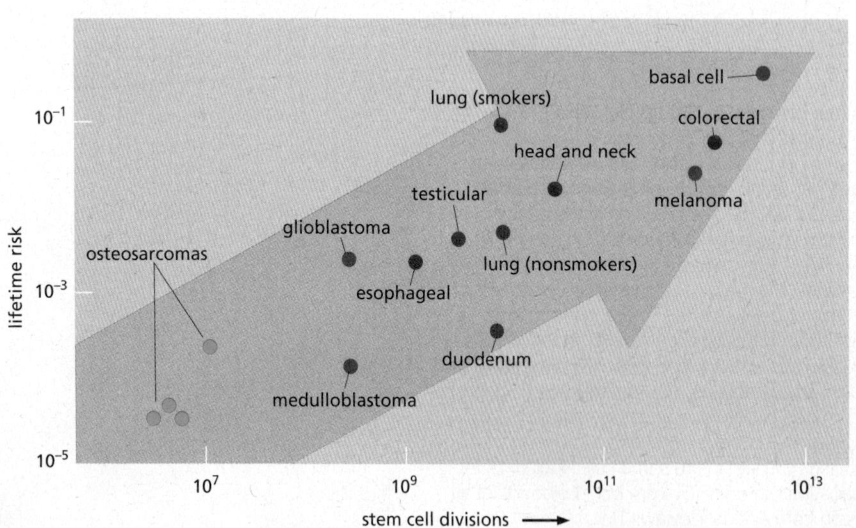

Figure 20–12 The lifetime cancer risk is correlated with the division rate of the cell of origin of the cancer. This plot shows the relationship between the number of stem cell divisions in a given tissue and the lifetime risk of cancer in that tissue. At the lower extreme are osteosarcomas (cancers that originate in bone), which are derived from mesenchymal cells that divide infrequently and rarely give rise to cancer. In contrast, epithelial cells of the skin and digestive tract are highly proliferative and give rise to malignancies much more frequently. Dots represent specific cancer types and are colored to indicate their classification: sarcoma *(light blue)*; neuronal *(dark blue)*; carcinoma *(purple)*; and skin *(red)*. Note that environmental factors strongly amplify the risk of many cancers, such as for lung cancer between smokers and nonsmokers. (Data from C. Tomasetti and B. Vogelstein, *Science* 347:78–81, 2015, doi 10.1126/science.1260825.)

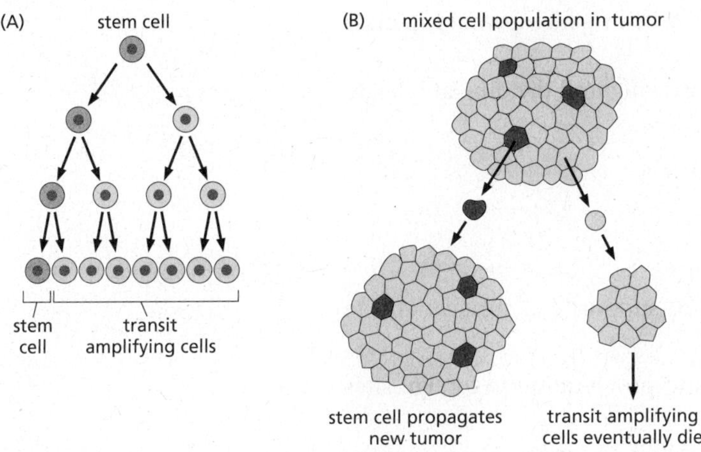

Figure 20–13 Cancer stem cells can be responsible for a tumor's growth and yet remain only a small part of the tumor-cell population. (A) Lineage of stem cells producing transit amplifying cells. (B) How a small proportion of cancer stem cells can maintain a tumor. Although the much more abundant transit amplifying cells will eventually die, the number of cancer stem cells will increase slowly but steadily to produce a growing tumor.

over the long term. In a healthy body, feedback controls regulate the process, adjusting the balance of cell-fate choices, cell proliferation, and cell death in a way that corrects for any departure from proper cell population numbers.

During the development of cancer, mutations have the potential to subvert the normal cellular differentiation program in multiple ways; for example, by leading to an overproliferation of transit amplifying cells or to an inhibition of their terminal differentiation or death. More insidious, however, are mutations that lead to the generation of cancer stem cells. Some tumors appear to exhibit this etiology: they consist of rare **cancer stem cells** capable of self-renewal, together with much larger numbers of dividing transit amplifying cells that are derived from the cancer stem cells but have a limited capacity for self-renewal (**Figure 20–13**).

Evidence for the existence of cancer stem cells comes from experiments in which individual cells from a cancer are tested for their ability to give rise to fresh tumors when implanted into a mouse. It has been known for more than half a century that there is usually only a small chance—typically much less than 1%—that a tumor cell chosen at random and tested in this way will generate a new tumor. This by itself does not prove that the tumor cells are heterogeneous: like seeds scattered on difficult ground, each of them may have only a small chance of finding a spot where it can survive and grow. Modern technologies for sorting cells have shown, however, that subpopulations of cancer cells expressing markers typically found on the surface of stem cells have a greatly enhanced ability to found new tumors. Moreover, the new tumors consist of mixtures of cells that express the stem cell markers and cells that do not, all generated from the same founder cell that expressed the markers. The cancer stem-cell phenomenon, whatever its basis, implies that even when the tumor cells are genetically similar, they may be phenotypically diverse. A treatment that wipes out tumor cells in one state will often allow survival of others that remain a danger. Radiotherapy or a cytotoxic drug, for example, may selectively kill off the rapidly dividing cells, reducing the tumor volume to almost nothing, and yet spare a few slowly dividing cells that go on to resurrect the disease. This greatly adds to the difficulty of cancer therapy, and it is part of the reason why treatments that seem at first to succeed often end in relapse and disappointment.

A Common Set of Hallmarks Typically Characterizes Cancerous Growth

Clearly, to produce a cancer, a cell must acquire a range of aberrant properties—a collection of subversive new skills—as it evolves. Different cancers require different combinations of these properties. Nevertheless, cancers all share some common hallmarks. These defining properties are commonly combined with other features, such as genetic instability, that help the miscreants to arise and thrive. A list of the key attributes of cancer cells in general would include the following:

1. Altered homeostasis that results in cells growing and dividing at a faster rate than they die
2. Bypass of normal limits to cell proliferation
3. Evasion of cell-death signals
4. Altered cellular metabolism
5. Manipulation of the tissue environment to support cell survival and to evade a deleterious immune response
6. Escape of cells from their home tissues and proliferation in foreign sites (metastasis)

Below we discuss these key features in more detail. In the next section of the chapter, we examine the mutations and molecular mechanisms that underlie these and other properties of cancer cells.

Cancer Cells Display an Altered Control of Growth and Homeostasis

Mutability and large cell population numbers create opportunities for mutations to occur, but the driving force for development of a cancer has to come from some sort of *selective advantage* possessed by the mutant cells. Most obviously, a mutation or epigenetic change can confer such an advantage by increasing the rate at which a clone of cells proliferates or by enabling it to continue proliferating when normal cells would stop or die. One of the most important properties of many types of cancer cells is that they fail to undergo apoptosis when a normal cell would do so (Figure 20–14).

Cancer cells that can be grown in culture or cultured cells artificially engineered to contain the types of mutations encountered in cancers typically show a **transformed** phenotype. They are abnormal in their shape, their motility, their responses to growth factors in the culture medium, and, most characteristically, in the way they react to contact with the culture dish and with one another. Whereas most normal cells will not divide unless they are attached to the surface, transformed cells will often divide even if held in suspension. More generally, transformed cells no longer require all of the positive signals from their surroundings that normal cells require. In addition, transformed cells fail to recognize some negative influences. Thus, for example, normal cells become inhibited from moving and dividing when the culture reaches confluence (where the cells are touching one another), while transformed cells continue moving and dividing even after confluence, and so pile up in layer upon layer in the culture dish (Figure 20–15).

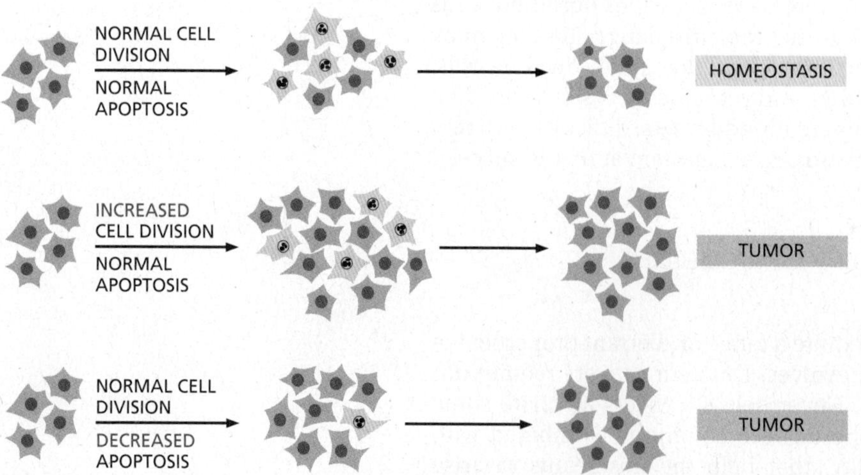

Figure 20–14 **Both increased cell division and decreased apoptosis can contribute to tumorigenesis.** In normal tissues, apoptosis balances cell division to maintain homeostasis (see Movie 18.1). During the development of cancer, either an increase in cell division or an inhibition of apoptosis can lead to the increased cell numbers important for tumorigenesis. The cells fated to undergo apoptosis are gray in this diagram. Both an increase in cell division and a decrease in apoptosis commonly contribute to tumor growth.

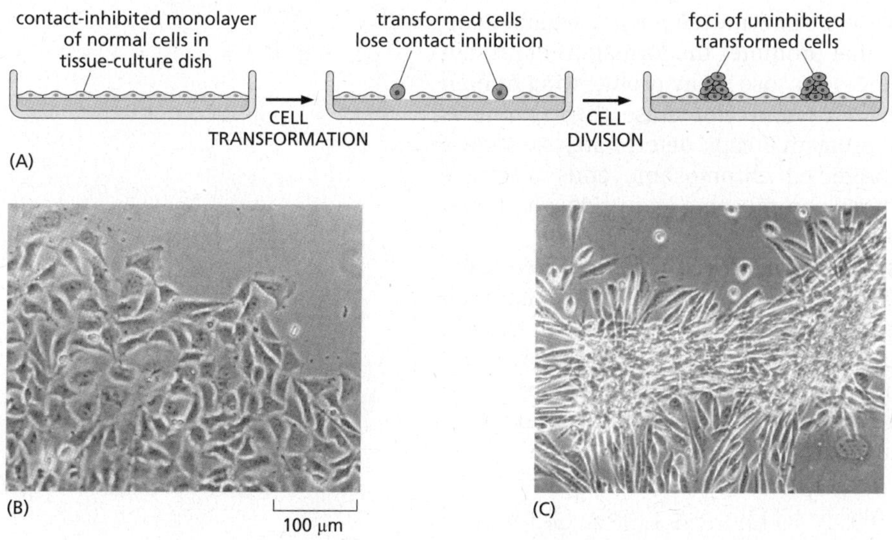

Figure 20–15 Loss of contact inhibition by cancer cells in cell culture. Most normal cells stop proliferating once they have carpeted the dish with a single layer of cells: proliferation seems to depend on contact with the dish and to be inhibited by contacts with other cells—a phenomenon known as *contact inhibition*. Cancer cells, in contrast, usually disregard these restraints and continue to grow, so that they pile on top of one another (**Movie 20.3**). (A) Schematic drawing. (B and C) Light micrographs of normal (B) and transformed (C) fibroblasts. (B and C, courtesy of Lan Bo Chen.)

Cancer cells also misbehave in their natural environment, embedded in a tissue. For example, normal cells of the gut epithelium are constantly turning over through rounds of cell division and death, but maintain their barrier function by seamlessly extruding old or damaged cells into the gut lumen. Once detached from the matrix and matrix-associated survival signals, these cells die through a form of apoptosis. By overriding normal death signals, transformed cells may be able to survive after their ejection from the epithelium. However, because the normal direction of cell extrusion is into the lumen of the gut, they would nevertheless be swept away through the waste canal. However, some of the mutations selected for during cancer progression can change the direction of extrusion, thereby enabling the tumor cells to cross the basement membrane and invade the surrounding tissue, potentially leading to metastasis (**Figure 20–16**).

In conclusion, by disobeying the normal etiquette of where to live and when to die, cancer cells both evade growth suppression and explore new opportunities to prosper and multiply.

Human Cancer Cells Escape a Built-in Limit to Cell Proliferation

Many normal human cells have a built-in limit to the number of times that they can divide when stimulated to proliferate in culture: they permanently stop dividing after a certain number of population doublings (25–50 for human fibroblasts, for example). This cell-division-counting mechanism is termed **replicative cell senescence**, and it generally depends on the progressive shortening of the telomeres at the ends of chromosomes, a process that eventually changes their structure (discussed in Chapter 17). As discussed in Chapter 5, the maintenance

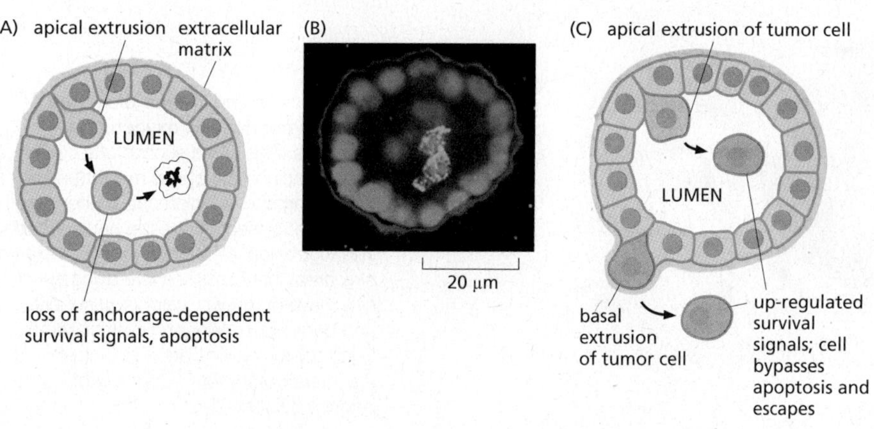

Figure 20–16 The direction in which a cell extrudes from an epithelium has important consequences for its fate. (A) Under normal conditions, as cells proliferate in a tissue such as the gut or mammary epithelium and become more crowded, they are extruded into the lumen to maintain cell number. Extruded cells generally undergo apoptosis due to loss of survival signals. (B) Fluorescence micrograph of a cluster of mammary epithelial cells grown in three-dimensional (3D) culture and stained for its basal boundary *(red)* and nuclei *(blue)*. As the sac of tissue grows, the central cells occupying the lumen undergo apoptosis as shown by staining with caspase-3 *(green)*. (C) Tumor cells with up-regulated survival signals that have been extruded apically might survive but are likely to be eliminated nevertheless; for example, by excretion through the digestive tract or by secretion from the mammary gland. In contrast, tumor cells that have been extruded basally can more readily initiate invasion. (B, courtesy of J. Debnath.)

of telomere DNA during S phase depends on the enzyme *telomerase*, which maintains a special telomeric DNA sequence that promotes the formation of protein cap structures to protect chromosome ends. Because many proliferating human cells (stem cells being an exception) are deficient in telomerase, their telomeres shorten with every division, and their protective caps deteriorate, creating a DNA damage signal because the unprotected chromosome ends resemble double-strand breaks. Eventually, the altered chromosome ends trigger a permanent cell-cycle arrest or cause the cell to die.

Human cancer cells avoid replicative cell senescence in one of two ways. Most often, they reactivate the telomerase gene as they proliferate, so that their telomeres do not shorten or become uncapped; alternatively, they can evoke a mechanism based on a DNA repair process that depends on homologous recombination for elongating their chromosome ends (called ALT). Regardless of the strategy used, the result is that the cancer cells continue to proliferate under conditions when normal cells would stop due to telomere erosion.

Cancer Cells Have an Abnormal Ability to Bypass Death Signals

A large multicellular organism requires powerful safety mechanisms that guard against the trouble caused by damaged and deranged cells. These mechanisms are essential because, as previously explained, a large number of mutated cells will inevitably be produced. Normally, internal disorder gives rise to danger signals in the faulty cell, activating protective measures to reverse and cure this disorder or, failing that, activating decisions that lead to cell death by apoptosis (see Chapter 18). To survive, cancer cells require mutations to elude or break through these defenses designed to eliminate defective cells.

Cancer cells generally contain mutations that drive the cell into an abnormal state, where metabolic processes may be unbalanced and essential cell components may be produced in ill-matched proportions. States of this type, where the cell's homeostatic mechanisms are inadequate to cope with an imposed disturbance, are loosely referred to as states of *cell stress*. As one example, chromosome breakage and other forms of DNA damage are commonly observed during the development of cancer, reflecting the genetic instability that cancer cells display. Thus, to survive and divide without limit, a prospective cancer cell must accumulate changes that disable the normal safety mechanisms that would otherwise cause a cell that is stressed to commit suicide by apoptosis.

While cancer cells tend to avoid apoptosis, this does not mean that they rarely die. On the contrary, in the interior of a large solid tumor, cell death often occurs on a massive scale: living conditions are difficult, with severe competition among the cancer cells for oxygen and nutrients. Most of these cells die by an alternative cell-death mechanism termed necrosis (**Figure 20–17**).

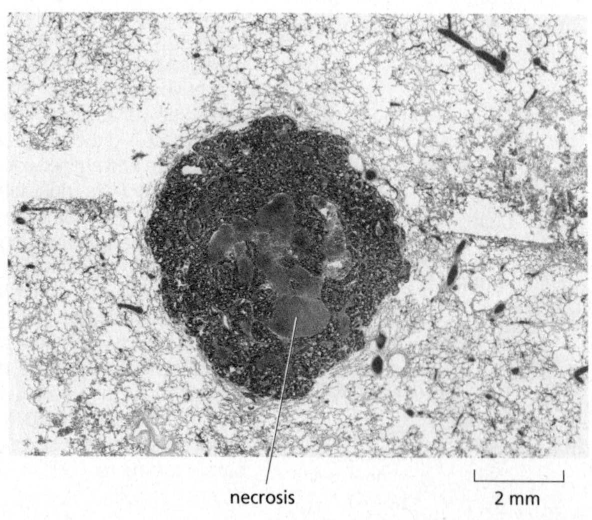

Figure 20–17 **The interior of a large tumor is deprived of oxygen and nutrients.** This cross section of a colon adenocarcinoma that has metastasized to the lung shows colorectal cancer cells that have formed a cohesive nodule *(dark-staining region)*. The metastasis has central *pink areas* of necrosis where dying cancer cells have outgrown their blood supply and burst open, releasing their contents. Such anoxic regions are common in the interior of tumors. (Courtesy of Andrew J. Connolly.)

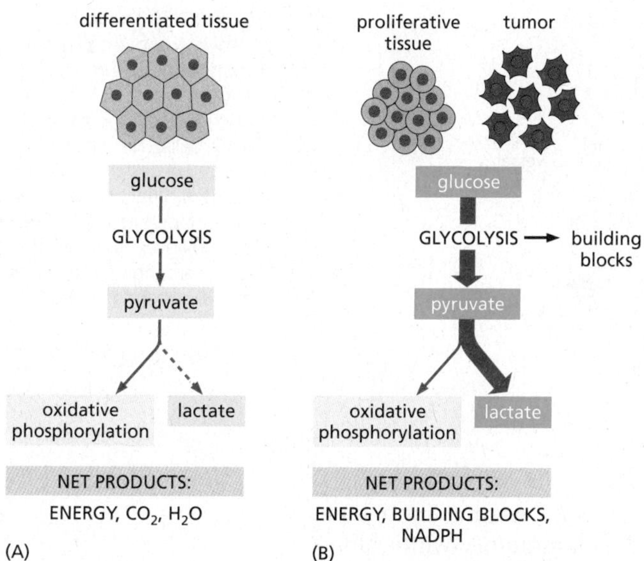

Figure 20–18 **The Warburg effect in tumor cells reflects a dramatic change in glucose uptake and sugar metabolism.** (A) Cells that are not proliferating will normally oxidize nearly all of the glucose that they import from the blood to produce ATP through oxidative phosphorylation that takes place in their mitochondria. Only when deprived of oxygen will these cells generate most of their ATP from glycolysis, converting the pyruvate produced to lactate in order to regenerate the NAD^+ that they need to keep glycolysis going (see Figure 2–50). (B) Tumor cells, by contrast, will generally produce abundant lactate even in the presence of oxygen. This results from a greatly increased rate of glycolysis that is fed by a very large increase in the rate of glucose import. In this way, tumor cells resemble the rapidly proliferating cells in embryos, which likewise require for biosynthesis a large supply of the small-molecule building blocks that can be produced from imported glucose (see also Figure 2–60).

The tumor grows because the cell birthrate outpaces the cell death rate, but often by only a small margin. For this reason, the time that a tumor takes to double in size can be far longer than the cell-cycle time of the tumor cells.

Cancer Cells Have Altered Sugar Metabolism

Given sufficient oxygen, differentiated cells in most adult tissues will fully oxidize almost all the carbon in the glucose they take up to CO_2, which is eventually exhaled by the lungs as a waste product. A growing tumor needs nutrients in abundance to provide the building blocks to make new macromolecules. Correspondingly, most tumors are metabolically more similar to a growing embryo than to normal adult tissue. Tumor cells consume glucose avidly, importing it from the blood at a rate that can be as much as 100 times higher than that for neighboring normal cells. Moreover, only a fraction of this imported glucose is fully oxidized to CO_2 by mitochondrial oxidative phosphorylation, which normally enables highly efficient production of ATP. Instead, the metabolism of carbon atoms from glucose is rewired to support the production of raw materials for the synthesis of the proteins, nucleic acids, and lipids that enable cellular proliferation (**Figure 20–18**). In other words, even though glycolysis is a much less efficient mode of ATP production than oxidative phosphorylation, it can continue unabated in cancer cells that find themselves in an oxygen-poor environment, and it has the critical added benefit of producing abundant cellular building blocks.

This tendency of tumor cells to de-emphasize oxidative phosphorylation even when oxygen is plentiful, while at the same time taking up large quantities of glucose, is necessary for the rapid proliferation of many cancer cells and is called the *Warburg effect*—so named because Otto Warburg first noticed the phenomenon in the early twentieth century. It is this abnormally high glucose uptake that allows tumors to be selectively imaged in whole-body scans (see Figure 20–1), thereby providing one way to monitor cancer progression and responses to treatment.

The Tumor Microenvironment Influences Cancer Development

While the cancer cells in a tumor are the bearers of dangerous mutations and are often grossly abnormal, the other cells in the tumor—especially those of the supporting connective tissue, or **stroma**—are far from passive bystanders. The development of a tumor relies on a two-way communication between the cancer cells and tumor stroma, just as the normal development of epithelial organs relies on communication between epithelial cells and mesenchymal cells (discussed in Chapter 22).

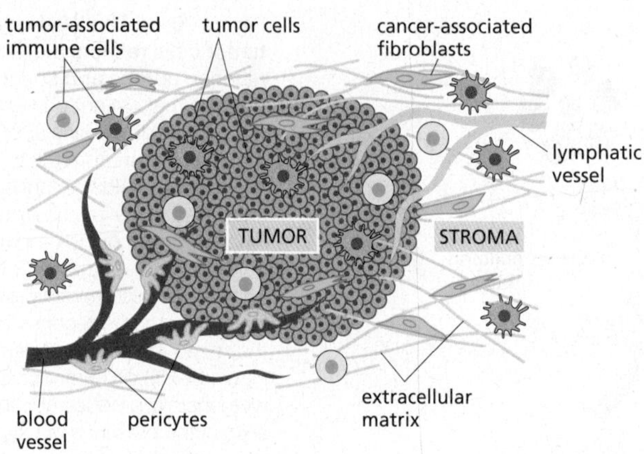

Figure 20–19 **The tumor microenvironment plays a role in tumorigenesis.** Tumors consist of many cell types, including cancer cells, endothelial cells, pericytes (vascular smooth muscle cells), fibroblasts, and inflammatory white blood cells. Communication among these and other cell types plays an important part in tumor development. Note, however, that only the cancer cells in a tumor are thought to contain mutations that make them genetically abnormal.

The stroma provides a framework for the tumor. As for normal connective tissue, the stroma is composed of fibroblasts and inflammatory white blood cells and of the endothelial cells that form blood and lymphatic vessels with their attendant smooth muscle cells (**Figure 20–19**). As a carcinoma progresses, the cancer cells induce changes in this stroma by secreting signal proteins that alter the behavior of the stromal cells and proteolytic enzymes that modify the extracellular matrix. The stromal cells in turn act back on the tumor cells by secreting signal proteins that stimulate cancer cell growth and division and proteases that further remodel the extracellular matrix. In these ways, the tumor and its stroma evolve together, like weeds and the ecosystem that they invade, and the tumor can become dependent on its particular stromal cells. Experiments using mice indicate that the growth of some transplanted carcinomas depends on the tumor-associated fibroblasts and that normal fibroblasts will not suffice.

Cancer cells have a complex interaction with the cells of the immune system that are present in the stroma, which have the potential to destroy the tumor if it is recognized as aberrant tissue but can also promote tumor growth by providing signals that stimulate cancer cell proliferation. The tumor can manipulate the immune system to its advantage in at least two ways. First, tumors may invoke an inflammatory reaction similar to what occurs when normal tissue is damaged, which helps them acquire the stroma they need for survival and growth. Tumors have therefore been likened to unhealed wounds, eliciting some of the same responses, including an increase in the permeability of nearby blood vessels, which allows the flow of signaling molecules in and out of the vessels, and the deposition of extracellular matrix. Tumors also stimulate the formation of new blood vessels, a process termed *angiogenesis*, which promotes survival of a tumor as it grows larger and becomes hypoxic in its interior. Second, and of equal importance, the tumor establishes an immunosuppressive microenvironment by blocking the activation of white blood cells that could lead to its destruction. Strategies that override this ability of a tumor to suppress an immune response have recently emerged as powerful tools in cancer therapy, and they are described in the last section of this chapter.

Cancer Cells Must Survive and Proliferate in a Foreign Environment

In order to kill us, cancer cells generally need to spread and multiply at new sites in the body through a process called metastasis. This is the most deadly—and least understood—aspect of cancer, being responsible for 90% of cancer-associated deaths. By spreading through the body, a cancer becomes almost impossible to eradicate by either surgery or local irradiation. **Metastasis** is itself a multistep process: the cancer cells first have to invade local tissues and vessels, move through the circulation, leave the vessels, and then establish new cellular colonies

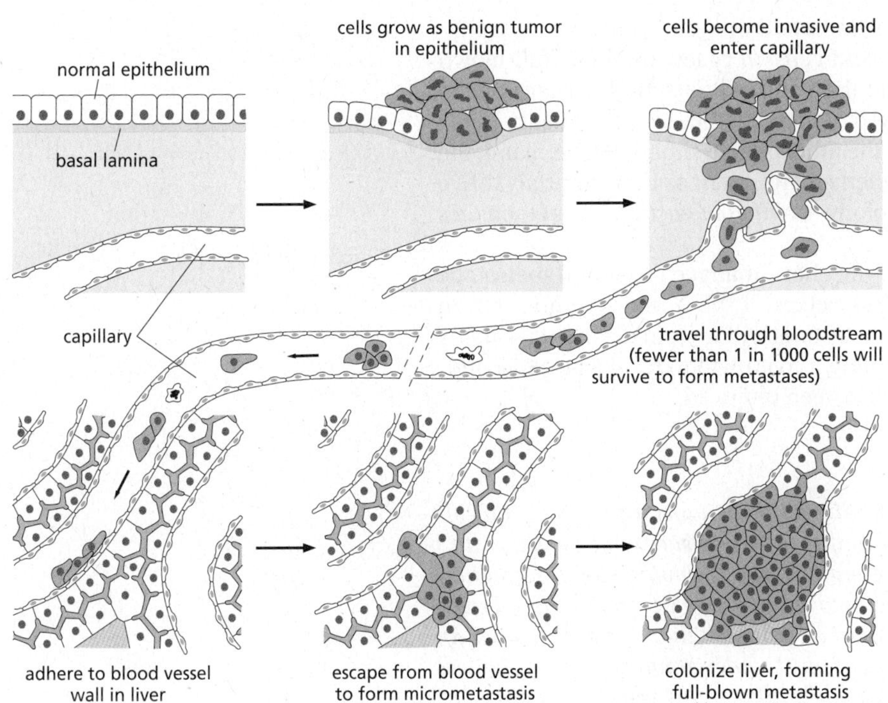

Figure 20–20 **Steps in the process of metastasis.** This example illustrates the spread of a tumor from its site of origin (such as the bladder) to another organ (such as the liver). Tumor cells may enter the bloodstream directly by crossing the wall of a blood vessel, as diagrammed here, or, more commonly perhaps, by crossing the wall of a lymphatic vessel that ultimately discharges its contents (lymph) into the bloodstream. Tumor cells that have entered a lymphatic vessel often become trapped in lymph nodes along the way, giving rise to lymph-node metastases. Studies in animals show that typically fewer than one in every thousand malignant tumor cells introduced into the bloodstream is viable after 24 hours, and that less than 0.1% of these surviving circulating tumor cells (CTCs) will colonize a new tissue so as to produce a detectable tumor at a new site.

at distant sites (Figure 20–20). Each of these events is complex, and most of the molecular mechanisms involved are not yet clear, but the last step—colonization of distant sites—is rate limiting.

For a cancer cell to become malignant, it must break free of constraints that keep normal cells in their proper places and prevent them from invading neighboring tissues. Invasiveness is thus one of the defining properties of malignant tumors, which show a disorganized pattern of growth and ragged borders, with extensions into the surrounding tissue (see, for example, Figure 20–8). Although the underlying molecular changes are not well understood, invasiveness almost certainly requires a disruption of the adhesive mechanisms that normally keep cells tethered to their proper neighbors and to the extracellular matrix.

The next step in metastasis—the establishment of colonies in distant organs—begins with entry into the circulatory system: the invasive cancer cells must penetrate the wall of a blood or lymphatic vessel. Lymphatic vessels, being larger and having more flimsy walls than blood vessels, allow cancer cells to enter in small clumps; such clumps may then become trapped in lymph nodes, giving rise to lymph-node metastases. The cancer cells that enter blood vessels, in contrast, do so singly or, more rarely, in small clusters. With modern techniques for sorting cells according to their surface properties, it has become possible to detect these *circulating tumor cells* (*CTCs*) in samples of blood from cancer patients, even though they are only a minute fraction of the total blood-cell population. Notably, in a mouse model, CTC clusters gave rise to metastases at a significantly higher rate than single CTCs did. Presumably the adhesion of epithelial-derived cancer cells to one another helps to override death signals and suppress apoptosis.

Of the cancer cells that enter the lymphatics or bloodstream, only a tiny proportion succeed in making their exit, settling in new sites, and surviving and proliferating there as founders of metastases. To discover which of the later steps in metastasis present cancer cells with the greatest difficulties, one can label the cells with a fluorescent dye or green fluorescent protein (GFP), inject them into the bloodstream of a mouse, and then monitor their fate (Movie 20.4). In such experiments, one observes that many cells survive in the circulation, lodge in small vessels, and exit into the surrounding tissue, regardless of whether they come from a tumor that metastasizes or one that does not. Some cells die immediately after they enter foreign tissue; others survive entry into the foreign tissue but fail to proliferate. Still others divide a few times and then stop, forming micrometastases containing ten to several thousand cells. Very few establish full-blown metastases. Experiments show that fewer than one in thousands, perhaps one in millions, manages this feat. The final step of colonization seems to be the most difficult: like the Vikings who landed on the inhospitable shores of Greenland, the migrant cells may fail to survive in the alien environment or they may only thrive there for a short while to found a little colony—a *micrometastasis*—that then dies out (**Movie 20.5**).

Many cancers are discovered before they have managed to found metastatic colonies, and these can be cured by destruction of the primary tumor. But on occasion, an undetected, distant micrometastasis will remain dormant for many years, only to reveal its presence by erupting into growth to form a large secondary tumor long after the primary tumor has been removed.

Summary

Cancer cells, by definition, grow and proliferate in defiance of normal controls (that is, they are neoplastic) and gain the ability to invade surrounding tissues and colonize distant organs (that is, they are malignant). By giving rise to secondary tumors, or metastases, they become difficult to eradicate by surgery or local irradiation. Cancers are thought to originate from a single cell that has experienced an initial mutation, but the progeny of this cell must undergo many further changes to become cancerous. Tumor progression usually takes many years and reflects the operation of a Darwinian-like process of evolution, in which somatic cells undergo mutation and epigenetic changes accompanied by natural selection and occasional bursts of genomic chaos that give rise to heterogeneity.

Cancer cells acquire a variety of special properties as they evolve, multiply, and spread. Their mutant genomes enable them to grow and divide in defiance of the signals that normally keep cell proliferation under tight control. As part of the evolutionary process of tumor progression, cancer cells acquire a collection of additional abnormalities, including defects in the controls that permanently stop cell division or induce apoptosis in response to cell stress or DNA damage and defects in the mechanisms that normally keep cells from straying from their proper place. All of these changes increase the ability of cancer cells to survive, grow, and divide in their original tissue and then to metastasize, founding new colonies in foreign environments. The evolution of a tumor also depends on other cells present in the tumor microenvironment, collectively called stromal cells, that the cancer attracts and manipulates.

Because many changes are needed to confer this collection of asocial behaviors, it is not surprising that most cancer cells are genetically and/or epigenetically unstable. This instability is thought to be selected for in the clones of aberrant cells that are able to produce tumors, because it greatly accelerates the accumulation of the further genetic and epigenetic changes that are required for tumor progression.

CANCER-CRITICAL GENES: HOW THEY ARE FOUND AND WHAT THEY DO

As we have seen, cancer depends on the accumulation of heritable changes in somatic cells; that is, changes that are transmitted by a cell to its progeny. To understand it at a molecular level we need to identify the mutations and epigenetic alterations involved and to discover how they give rise to cancerous cell behavior. Finding the relevant cells is often easy; they are favored by natural selection and call attention to themselves by giving rise to tumors. But how do we identify those genes that have undergone cancer-promoting changes among all the other genes in the cancerous cells? A typical cancer depends on a whole set of mutations and epigenetic changes that are never exactly the same in two different patients, although there are commonalities among certain tumor types. In addition, a given cancer cell will also contain a large number of somatic mutations that are accidental by-products—so-called *passengers* rather than *drivers*—of its genetic instability, and it can be difficult to distinguish these incidental changes from those changes that have a causative role in the disease. Despite these difficulties, many of the genes that are repeatedly altered in human cancers have been identified over the past 40 years. We will call such genes, for want of a better term, *cancer-critical genes*, meaning all genes whose alteration can contribute to the causation or evolution of cancer.

In this section, we shall first discuss how cancer-critical genes are identified. We shall then examine their functions and the parts they play in conferring on cancer cells the properties outlined in the first part of the chapter. We shall end the section by discussing colon cancer as an extended example, showing how a succession of changes in cancer-critical genes enables a tumor to evolve from one pattern of bad behavior to another that is worse.

The Identification of Gain-of-Function and Loss-of-Function Cancer Mutations Has Traditionally Required Different Methods

Cancer-critical genes are grouped into two broad classes, according to whether the cancer risk arises from too much activity of the gene product or too little. Genes of the first class, in which a gain-of-function mutation can drive a cell toward cancer, are called **proto-oncogenes**; their mutant, overactive or overexpressed forms are called **oncogenes**. Genes of the second class, in which a loss-of-function mutation can contribute to cancer, are called **tumor suppressor genes**. In either case, the mutation may lead toward cancer directly (by causing cells to proliferate when they should not) or indirectly; for example, by causing genetic or epigenetic instability and so hastening the occurrence of other inherited changes that directly stimulate tumor growth. Those genes whose alteration results in genomic instability represent a subclass of cancer-critical genes that are sometimes called *genome maintenance genes*.

As we shall see, mutations in oncogenes and tumor suppressor genes can have similar effects in promoting the development of cancer; overproduction of a signal for cell proliferation, for example, can result from either kind of mutation. Thus, from the point of view of a cancer cell, oncogenes and tumor suppressor genes—and the mutations that affect them—are flip sides of the same coin. The techniques that led to the discovery of these two categories of genes, however, are quite different.

The mutation of a single copy of a proto-oncogene that converts it to an oncogene has a dominant, growth-promoting effect on a cell (**Figure 20–21A**). Thus,

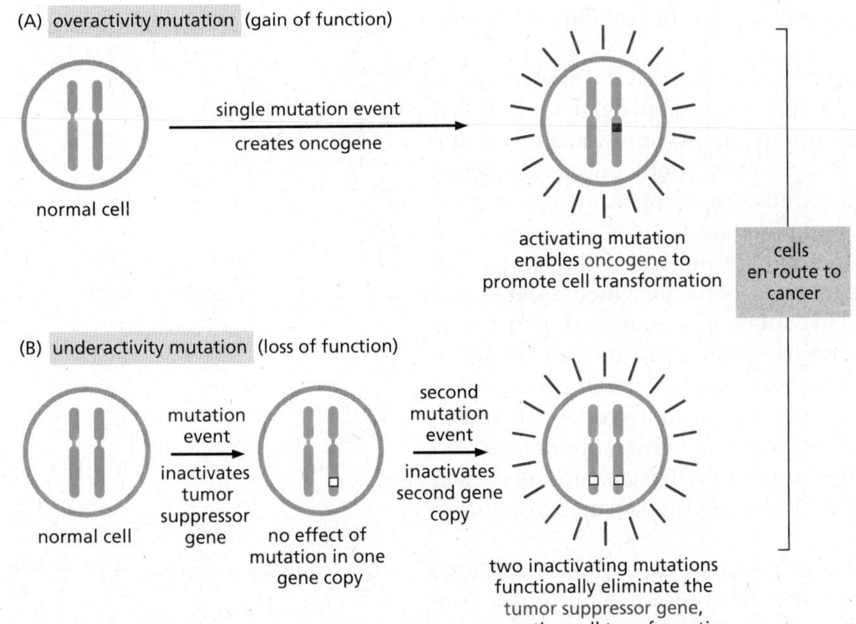

Figure 20–21 Cancer-critical mutations fall into two readily distinguishable categories, dominant and recessive. In this diagram, activating mutations are represented by *solid red boxes*, inactivating mutations by *hollow red boxes*. (A) Oncogenes act in a dominant manner: a gain-of-function mutation in a single copy of the cancer-critical gene can drive a cell toward cancer. (B) Mutations in tumor suppressor genes, on the other hand, generally act in a recessive manner: the function of both alleles of the cancer-critical gene must be lost to drive a cell toward cancer. Although in this diagram the second allele of the tumor suppressor gene is inactivated by mutation, it is often inactivated instead by loss of the second chromosome. Not shown is the fact that mutation of some tumor suppressor genes can have an effect even when only one of the two gene copies is damaged.

we can identify the oncogene by its effect when it is *added*—by DNA transfection, for example, or through transduction with a viral vector—to the genome of a suitable type of tester cell or experimental animal. In the case of the tumor suppressor gene, on the other hand, the cancer-causing alleles produced by the change are generally recessive: often (but not always) both copies of the normal gene must be removed or inactivated in the diploid somatic cell before an effect is seen (**Figure 20–21B**). This calls for a different experimental approach, one focusing on discovering what is *missing* in the cancer cell.

We begin by discussing a few examples of each class of cancer-critical genes to illustrate basic principles. These examples are chosen also for their historical importance: the experiments that led to their discovery—at different times and by different methods—marked turning points in the understanding of cancer.

Retroviruses Led to the Identification of Oncogenes

The search for the genetic causes of human cancer took a circuitous route, beginning with clues that came from the study of **tumor viruses**. Although viruses are involved only in a minority of human cancers, a set of viruses that infect animals provided critical early tools for studying cancer.

One of the first animal viruses to be implicated in cancer was discovered in chickens more than 100 years ago, when an infectious agent that causes connective-tissue tumors, or sarcomas, was identified as a virus—the *Rous sarcoma virus*. Like all the other *RNA tumor viruses* discovered since, it is a **retrovirus**. When it infects a cell, its RNA genome is copied into DNA by reverse transcription, and the DNA is inserted into the host genome, where it can persist and be inherited by subsequent generations of cells. Something in the DNA inserted by the Rous sarcoma virus made the host cells cancerous, but what was it? The answer was a surprise. It turned out to be a piece of DNA that was unnecessary for the virus's own survival or reproduction; instead, it was a passenger, a gene called *v-Src*, that the virus had picked up on its travels. *v-Src* was unmistakably similar, but not identical, to a gene—*c-Src*—that is found in all vertebrate genomes and encodes a protein tyrosine kinase. *c-Src* had evidently been taken up accidentally by the retrovirus from the genome of a previously infected host cell, and it had undergone mutation in the process to become an oncogene (*v-Src*).

This Nobel Prize–winning finding was followed by a flood of discoveries of other viral oncogenes carried by retroviruses that cause cancer in nonhuman animals. Each such oncogene turned out to have a counterpart proto-oncogene in the normal vertebrate genome. As was the case for *Src*, these other oncogenes generally differed from their normal counterparts, either in structure or in level of expression.

But how did this relate to typical human cancers, in which retroviruses were not known to play a role? In an assay to identify human oncogenes, DNA was extracted from human tumor cells, broken into fragments, and introduced into cultured mouse cells. Occasional colonies of abnormally proliferating cells began to appear in the culture dish that showed a transformed phenotype, outgrowing the untransformed cells in the culture and piling up in layer upon layer (see Figure 20–15). Each colony was a clone originating from a single cell that had incorporated a DNA fragment that drove cancerous behavior. Once isolated and sequenced, the DNA fragments were found to contain a human version of a gene already known from study of a retrovirus that caused tumors in rats—an oncogene called *v-Ras*. The newly discovered oncogene was clearly derived by mutation from a normal human gene, one of a small family of proto-oncogenes called ***Ras***. This discovery in the early 1980s of the same oncogene in human tumor cells and in an animal tumor virus was electrifying. The implication that cancers are caused by mutations in a limited number of cancer-critical genes transformed our understanding of the molecular biology of cancer.

As discussed in Chapter 15, normal Ras proteins are monomeric GTPases that help transmit signals from cell-surface receptors to the cell interior (see Movie 15.7). The *Ras* oncogenes isolated from human tumors contain point

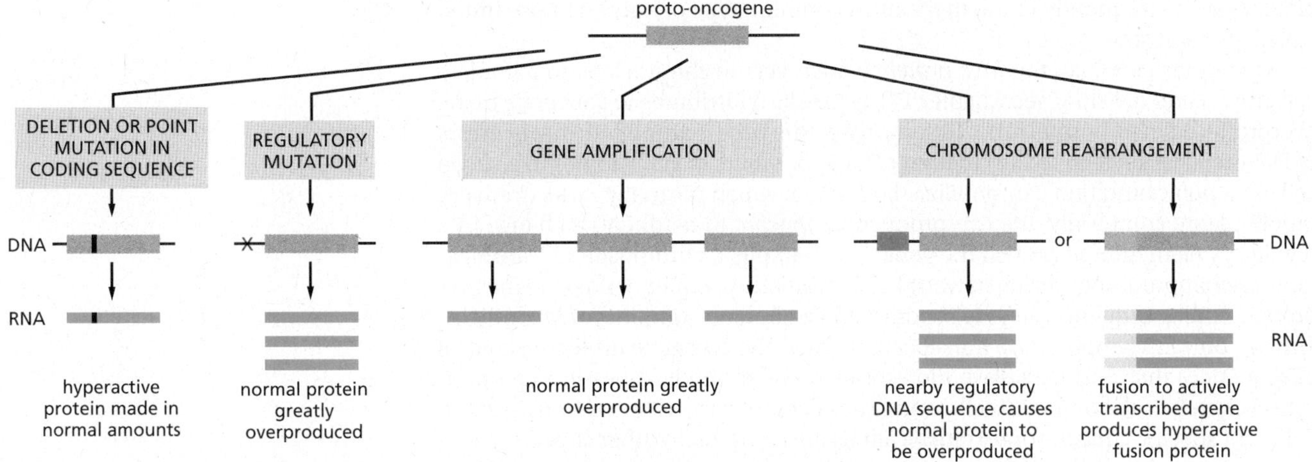

Figure 20–22 **The types of accidents that can convert a proto-oncogene into an oncogene.**

mutations that create a hyperactive Ras protein that cannot shut itself off by hydrolyzing its bound GTP to GDP. Because this makes the protein hyperactive, its effect is dominant; that is, only one of the cell's two gene copies needs to change to have an effect. One or another of the three human *Ras* family members is mutated in about 30% of all human cancers. *Ras* genes are thus among the most important of all cancer-critical genes.

Genes Mutated in Cancer Can Be Made Overactive in Many Ways

Figure 20–22 summarizes the types of accidents that can convert a proto-oncogene into an oncogene. (1) A small change in DNA sequence such as a point mutation or deletion may produce a hyperactive protein when it occurs within a protein-coding sequence or lead to protein overproduction when it occurs within a regulatory region for that gene. (2) Gene amplification events, such as those that can be caused by errors in DNA replication, may produce extra gene copies; this can lead to overproduction of the protein. (3) A chromosomal rearrangement—involving the breakage and rejoining of the DNA helix—may either change the protein-coding region, resulting in a hyperactive fusion protein, or alter the control regions for a gene so that a normal protein is overproduced.

As one example, the receptor for the extracellular signal protein *epidermal growth factor* (*EGF*) can be activated by a deletion that removes part of its extracellular domain, causing it to be active even in the absence of EGF (**Figure 20–23**). The mutant EGF receptor thus produces an inappropriate stimulatory signal, like a faulty doorbell that rings even when nobody is pressing the button. Mutations

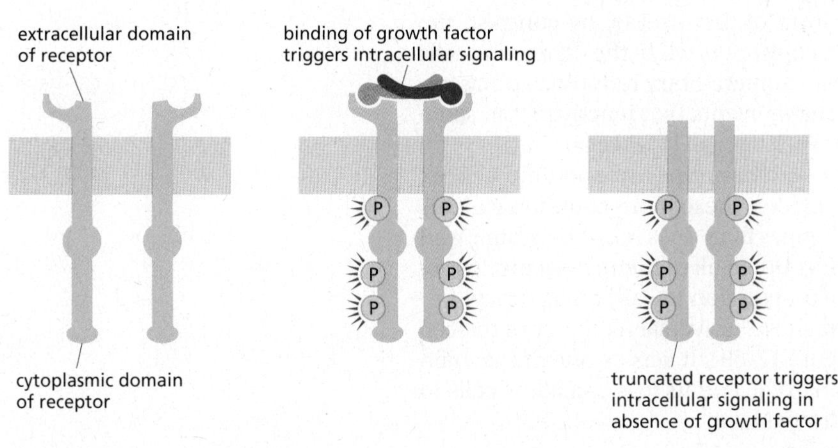

Figure 20–23 **Mutation of the epidermal growth factor (EGF) receptor can make it active even in the absence of EGF, and consequently oncogenic.** Normally, binding of EGF to the receptor's extracellular domain leads to phosphorylation of the intracellular domain, which activates signaling. Truncation of the extracellular domain leads to hyperphosphorylation and inappropriate activation. Other types of activating mutations are observed in different cancers.

of this type are frequently found in the most common type of human brain tumor, called glioblastoma.

As another example, the *Myc protein*, which acts in the nucleus to stimulate cell growth and division (see Chapter 17), generally contributes to cancer by being overproduced in its normal form. In some cases, the gene is amplified; that is, errors of DNA replication lead to the creation of large numbers of gene copies in a single cell. Or a point mutation can stabilize the protein, which normally turns over very rapidly. More commonly, the overproduction appears to be due to a change in a regulatory element that acts on the gene. For example, a chromosomal translocation can inappropriately bring powerful gene regulatory sequences next to the *Myc* protein-coding sequence, so as to produce unusually large amounts of *Myc* mRNA. Thus, in Burkitt's lymphoma, a translocation brings the *Myc* gene under the control of sequences that normally drive the expression of antibody genes in B lymphocytes. As a result, the mutant B cells tend to proliferate excessively and form a tumor. Different specific chromosome translocations are common in other cancers.

Studies of Rare Hereditary Cancer Syndromes First Identified Tumor Suppressor Genes

Identifying a gene that has been inactivated in the genome of a cancer cell requires a different strategy from finding a gene that has become hyperactive: one cannot, for example, use a cell transformation assay to identify something that simply is not there. The key insight that led to the discovery of the first tumor suppressor gene came from studies of a rare type of human cancer, **retinoblastoma**, which arises from cells in the retina of the eye that are converted to a cancerous state by an unusually small number of mutations. As often happens in biology, the discovery arose from examination of a special case, but it turned out to reveal a gene of widespread importance.

Retinoblastoma occurs in childhood, and tumors develop from neural precursor cells in the immature retina. About one child in 20,000 is afflicted. One form of the disease is hereditary, and the other is not. In the hereditary form, multiple tumors usually arise independently, affecting both eyes; in the nonhereditary form, only one eye is affected, and by only one tumor. A few individuals with retinoblastoma have a visibly abnormal karyotype, with a deletion of a specific band on chromosome 13 that, if inherited, predisposes an individual to the disease. Deletions of this same region are also encountered in tumor cells from some patients with the nonhereditary disease, which suggested that the cancer was caused by loss of a critical gene in that location.

By mapping the location of this chromosomal deletion, it was possible to identify the *Rb* gene. It was then discovered that those who suffer from the hereditary form of the disease have a deletion or loss-of-function mutation present in one copy of the *Rb* gene in every somatic cell. These cells are predisposed to becoming cancerous but do not do so if they retain one good copy of the gene. The retinal cells that are cancerous are defective in both copies of *Rb* because of a somatic event that has eliminated the function of the previously good copy.

In individuals with the nonhereditary form of the disease, by contrast, the noncancerous cells show no defect in either copy of *Rb*, while the cancerous cells have become defective in both copies. These nonhereditary retinoblastomas are very rare because they require two independent events that inactivate the same gene on two chromosomes in a single retinal cell lineage (Figure 20–24).

The *Rb* gene is also missing in several common types of sporadic cancer, including carcinomas of lung, breast, and bladder. These more common cancers arise by a more complex series of genetic changes than does retinoblastoma, and they make their appearance much later in life. But in all of them, it seems, loss of *Rb* function is frequently a major step in the progression toward malignancy.

The *Rb* gene encodes the **Rb** protein, which is a universal regulator of the cell cycle in almost all cells of the body (see Figure 17–59). It acts as one of the main brakes on progress through the cell-division cycle, and its loss can allow cells to enter the cell cycle inappropriately, as we discuss later.

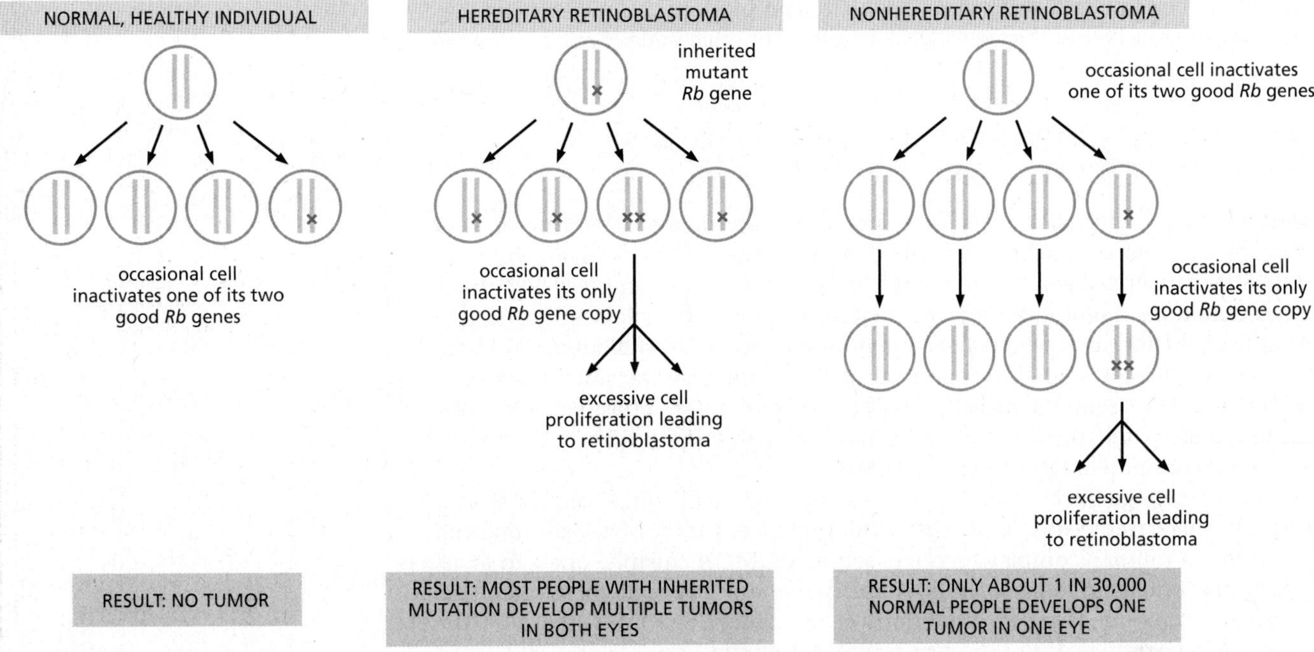

Figure 20–24 The genetic mechanisms that cause retinoblastoma. In the hereditary form, all cells in the body lack one of the normal two functional copies of the *Rb* tumor suppressor gene, and tumors arise from a clone of cells where the remaining copy is lost or inactivated by a somatic event (either mutation or epigenetic silencing). In the nonhereditary form, all cells initially contain two functional copies of the gene, and the tumor arises because both copies are lost or inactivated through the coincidence of two somatic events in a single line of cells.

Both Genetic and Epigenetic Mechanisms Can Inactivate Tumor Suppressor Genes

For tumor suppressor genes, it is their inactivation that is dangerous. This inactivation can occur in many ways, with different combinations of mishaps serving to eliminate or cripple both gene copies. The first copy may, for example, be lost by a small chromosomal deletion or inactivated by a point mutation due to a random error in DNA replication. The second copy is more commonly eliminated by a less specific mechanism that is likely to occur in cells progressing toward cancer that have become genetically unstable. For example, the chromosome carrying the remaining normal copy may be lost from the cell or damaged due to errors in chromosome segregation (see Figure 20–11) or the normal gene, along with neighboring genetic material, may be replaced by a mutant version through either a *mitotic recombination* event or a *gene conversion* that accompanies it (see pp. 305–306). Epigenetic changes provide another important way to permanently inactivate a tumor suppressor gene. Most commonly, the C nucleotides in CG sequences in its promoter may become methylated in a heritable manner, which can irreversibly silence the gene in a cell and in all of its progeny (see pp. 435–436). Figure 20–25

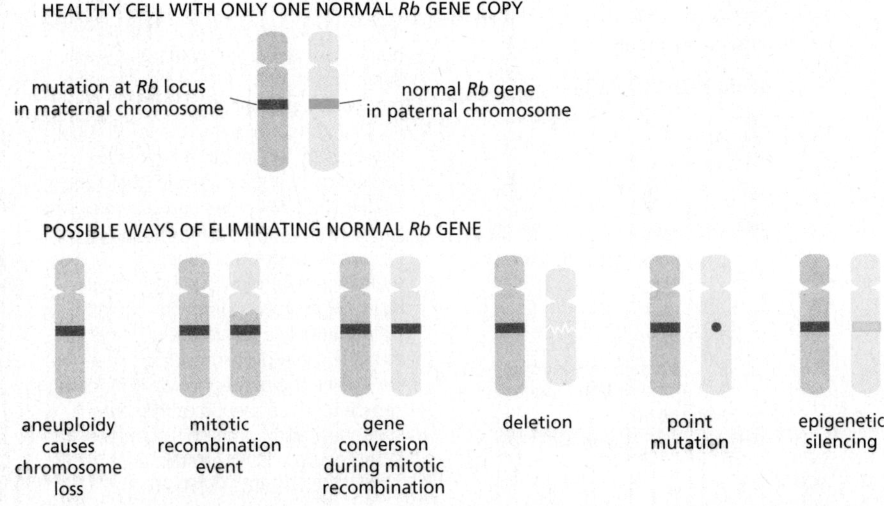

Figure 20–25 Six ways of inactivating the remaining good copy of a tumor suppressor gene through changes in DNA sequence or an epigenetic mechanism. A cell that is defective in only one of its two copies of a tumor suppressor gene—for example, the *Rb* gene—usually behaves as a normal, healthy cell; the diagrams show how this cell may lose the function of the other gene copy as well and thereby progress toward cancer.

summarizes the range of ways in which the remaining good copy of a tumor suppressor gene can be lost through a DNA sequence or epigenetic change, using the *Rb* gene as an example.

Systematic Sequencing of Cancer Cell Genomes Has Transformed Our Understanding of the Disease

Methods such as those we have described above shone a spotlight on a set of cancer-critical genes that were identified in a piecemeal fashion. Meanwhile, the rest of the cancer cell genome remained in darkness: it was a mystery how many other mutations might lurk there, of what types, in which varieties of cancer, at what frequencies, with what variations from individual to individual, and with what consequences. With the sequencing of the human genome and the dramatic advances in DNA sequencing technology (see Figure 8–44), it has become possible to see the whole picture—to view cancer cell genomes in their entirety. This transforms our understanding of the disease.

Cancer cell genomes can be scanned systematically in several different ways. At one extreme—the most costly, but no longer prohibitively so—one can determine a tumor's complete genome sequence. More cheaply, one can focus just on the 21,000 or so genes in the human genome that code for protein (the so-called *exome*), looking for mutations in the cancer cell DNA that alter the amino acid sequence of the product or prevent its synthesis (**Figure 20–26**). In addition, DNA sequencing enables a survey of the genome for regions that have undergone deletion or duplication to reveal copy number variations and the loss or gain of chromosomes (aneuploidy). The genome can also be scanned for epigenetic changes to reveal changes in methylation patterns of DNA or other changes that accompany transcriptional silencing or activation without affecting DNA sequence. Systematic sequencing of RNAs can reveal alterations in levels of gene expression by analysis of mRNAs (see Figure 7–5), as well as levels of regulatory noncoding RNAs. Finally, to measure the expression of known **cancer-critical genes** directly by detecting their protein products, tumor cell lysates can be surveyed quantitatively using mass spectrometry. These approaches involve comparing cancer cells with normal controls—ideally, noncancerous cells originating in the same tissue and from the same individual.

A combination of the high-throughput methods described above has been applied to more than 10,000 tumors spanning 33 cancer types through a consortium called The Cancer Genome Atlas, which is coordinated by the National Institutes of Health. Other international and freely available resources, including COSMIC, the Catalog of Somatic Mutations in Cancer, are continually compiling new cancer data. As we discuss below, these large-scale projects have shed important new light on what types of alterations are present in a cancer genome, what kinds of genes or pathways are altered across and within cancer types, how

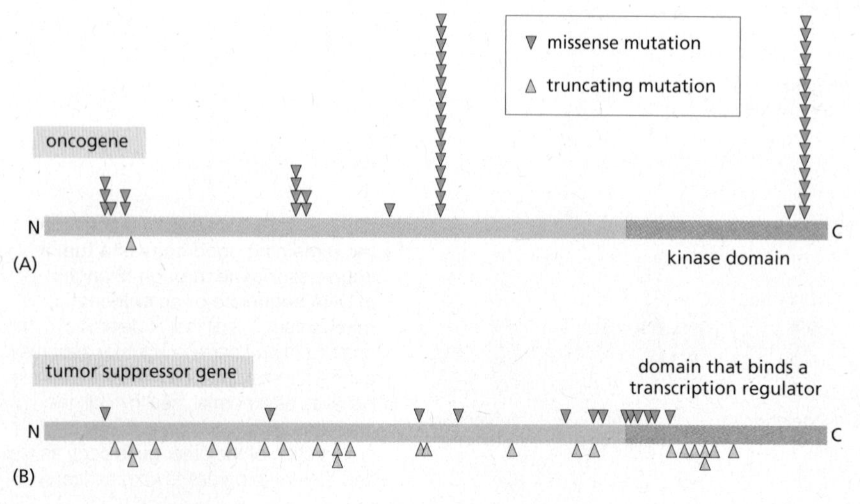

Figure 20–26 The distinct types of DNA sequence changes found in oncogenes compared to those in tumor suppressor genes. In this diagram, mutations that change an amino acid are denoted by *blue arrowheads*, whereas mutations that truncate the polypeptide chain are marked by *yellow arrowheads*. (A) As in this example (the *PIK3CA* gene, which produces a phosphoinositide 3-kinase subunit), oncogene mutations can be detected by the fact that the same nucleotide change is repeatedly found among the missense mutations in a gene. (B) As in this example (the *Rb* gene), for tumor suppressor genes missense mutations that abort protein synthesis by creating stop codons predominate. Note that only a few of the possible mutations in a protein-coding sequence are likely to be activating, while inactivation can be a consequence of missense, nonsense, and frameshift mutations. (Adapted from B. Vogelstein et al., *Science* 339:1546–1558, 2013.)

heterogeneous the cancer cells within a tumor are, and how the patterns of alterations evolve over time during disease progression and treatment.

Many Cancers Have an Extraordinarily Disrupted Genome

Cancer genome analysis reveals, first of all, the scale of gross genetic disruption in cancer cells. This varies greatly among cancer types and from one cancer patient to another, both in severity and in character. As discussed previously (see p. 1169) the chromosomal karyotype may appear normal, yet is riddled with numerous point mutations in individual genes due to a failure of the repair mechanisms that normally correct errors in the replication or maintenance of DNA sequences. Frequently, however, the karyotype is extremely disrupted, with many chromosomal breaks and rearrangements and chromosome sequences that are completely scrambled, indicating that extensive fragmentation of the DNA occurred followed by random re-assembly (as in chromothripsis; see Figure 20–11). From the pattern of changes, one can infer that disruptive events have occurred repeatedly during the evolution of the tumor, with a progressive increase in genetic disorder.

A common feature of many human cancers is aneuploidy—a change in chromosome karyotype from the normal number (46). A study of chromosome-level gain or loss across a panel of more than 10,000 cancer genomes revealed that most cancer types harbor a characteristic pattern of chromosomal abnormalities, with different arms or whole chromosomes altered at different frequencies. In many cases, the entire set of chromosomes was doubled or quadrupled because of complete failures in cell division, generating polyploid cells that subsequently experience additional chromosome gain and loss. Almost 90% of all the cancers surveyed showed some level of aneuploidy. Figure 20–27 depicts a subset of the data showing selected tumor types, each of which displays a unique pattern of aneuploidy. Thus, for example, whereas 74% of thyroid carcinomas displayed a normal karyotype, less than 5% of glioblastoma and cervical carcinoma tumors possessed the normal chromosome number. Although the degree of aneuploidy is often characteristic of a cancer class, heterogeneity is also apparent. For example, one subset of glioblastomas is characterized by gain of chromosome 7 and loss of chromosome 10, but this combination of defects is not present in all cases, indicating that there are multiple pathways of genome disorder that can lead to different cancer subtypes.

Cancer genome studies also illuminate the underlying defects that bring about genome disruption. In ovarian cancers, for example, chromosome breaks, translocations, and deletions are very common, and these aberrations correlate with a high frequency of mutations and epigenetic silencing in the genes needed for repair of DNA double-strand breaks by homologous recombination, especially *Brca1* and *Brca2* (see pp. 300–301). In a subset of endometrial cancers, on the other hand, one instead finds many point mutations scattered throughout the genome, which could for example be caused by mutation of enzymes required for proofreading during DNA replication (see pp. 267–269). Thus, different cancers and cancer subtypes possess highly variable mutation rates, ranging from one base substitution per exome in some pediatric cancers to thousands of mutations per exome in mutagen-induced malignancies such as lung cancer and melanoma.

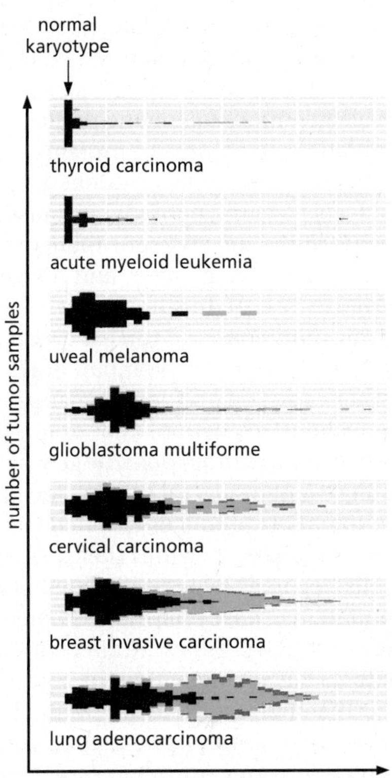

Figure 20–27 **The prevalence of aneuploidy among different tumor types.** For each tumor type, the total number of chromosome arms detected is plotted on the X axis. The number of chromosome arms in a normal karyotype is indicated. Genome doubling status is also shown (*black*, not double; *blue*, one genome doubling; *red*, two or more genome doublings). The number of tumor samples possessing the respective arm numbers is represented by the length of the bar on the Y axis for each tumor type. Note that samples from some tumor types predominantly possess a normal karyotype, while samples from other tumor types display extreme heterogeneity with a dramatic increase in chromosome arm numbers because of extensive aneuploidy. (Adapted from A.M. Taylor et al., *Cancer Cell* 33:676–689, 2018.)

Epigenetic and Chromatin Changes Contribute to Most Cancers

So far we have focused primarily on the mutations in cancer-critical genes that contribute directly to the hallmarks of cancer. Equally important to the genesis of cancer, however, are changes that modify gene expression epigenetically, without altering the DNA sequence. As discussed previously, increased methylation of DNA in the promoter region of a gene such as *Rb* can permanently silence it. In addition, reversible covalent modifications such as methylation and acetylation also occur on the histone proteins that package DNA into chromatin. These histone marks modulate gene expression by altering the conformation and accessibility of chromatin to regulatory factors such as transcription factors and chromatin remodeling complexes (see Chapter 7). An important finding of large-scale genome sequencing projects is that roughly 50% of human cancers harbor mutations in chromatin proteins, which includes mutations in enzymes that add, remove, or recognize covalent marks on histones and DNA. Many of the resulting defects have the potential to cause heritable, epigenetic changes that modulate gene expression and contribute to tumorigenesis.

Frequently, genetic and epigenetic changes cooperate to cause cancer (**Figure 20–28**). In some cases, epigenetic changes precede characteristic oncogenic mutations and can even lead to them, such as when epigenetic silencing of genes required for the repair of DNA damage acts to increase mutation rates. In other cases, an accidental change in DNA sequence can disrupt chromatin and epigenetic regulation. One example is a gain-of-function mutation in the gene encoding isocitrate dehydrogenase (IDH), which is known to be a frequent initiating event in several cancers including glioma, leukemia, and other tumors. The mutant IDH produces elevated levels of a metabolite, 2-hydroxyglutarate, which acts to inhibit enzymes that demethylate DNA. Therefore, DNA in IDH-mutant cells becomes hypermethylated at random sites in the genome. Just as for DNA mutations, those epigenetic changes that provide a selective advantage in growth or survival will allow the affected cell and its descendants to persist and accumulate further changes, thereby contributing to cancer progression.

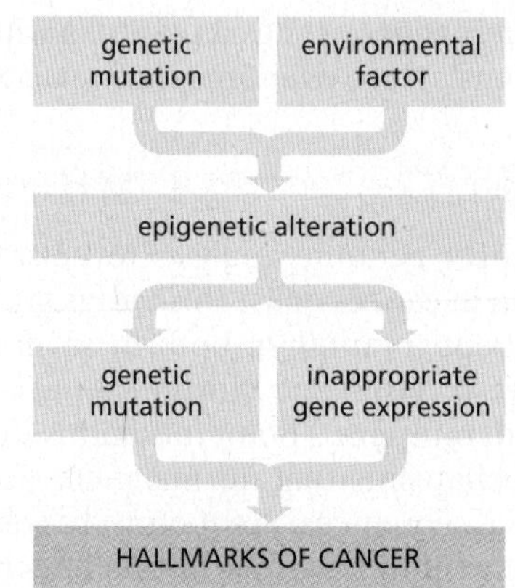

Figure 20–28 Epigenetic and genetic mechanisms can cooperate to promote the evolution of cancer. Both genetic mutations and environmental factors (such as metabolic state) can lead to epigenetic changes. As indicated, these epigenetic changes can in turn increase the rate at which genetic as well as further epigenetic changes accumulate—thereby speeding the process of cancer progression by altering gene expression in ways that promote cancer cell proliferation and are inherited in clones of cells.

The indirect effect of mutant IDH on DNA hypermethylation illustrates another important point: metabolic conditions affect the status of chromatin, which in turn can influence gene expression. Notably, only a minority of cancers with aberrant DNA methylation can be explained by an underlying genetic event. These observations suggest that environmental conditions, which may not themselves be mutagenic, can provide a source of epigenetic changes. Many DNA- and histone-modifying enzymes require metabolites as cofactors, providing a potential link between known risk factors for human cancer, such as diet and inflammation, and cancer development.

In addition to modulating the expression of individual genes, epigenetic changes can also have more global effects on the chromatin state, thereby influencing gene networks that underlie programs of cellular behavior. For example, different patterns of gene expression determined by epigenetic marks could restrict the ability of a cell to activate apoptosis or prevent it from exiting the cell cycle and undergoing differentiation. Evidence for such changes in cancer comes from characterization of rare pediatric brain tumors that lack somatic mutations and instead display aberrant DNA methylation profiles.

It is crucial to remember that altered patterns of gene activation and silencing due to epigenetic regulation are a normal feature of cellular differentiation, enabling the same genome to produce and maintain a multitude of different cell types through precise developmental programs (see Chapter 21). Thus, as for other properties of cancer cells, epigenetic changes do not represent new biology, but rather the subversion of existing cellular mechanisms.

Hundreds of Human Genes Contribute to Cancer

Among the billions of somatic mutations now identified in cancer cells, how can we discover which of them are **drivers** of cancer; that is, causal factors in the development of the disease? Clearly, most will be merely **passengers**—mutations that happen to have occurred in the same cell as the driver mutations, thanks to genetic instability, but are irrelevant to the development of the cancer. One criterion is frequency of occurrence. Driver mutations affecting a gene that play a part in the disease will be seen repeatedly, in many different individuals with a particular type of cancer. In contrast, a passenger mutation that confers no selective advantage on the cancer cell is likely to be encountered only rarely.

By compiling the genome sequence data for different types of cancer, each with its own set of identified driver mutations, we can develop a comprehensive catalog of those genes that are strongly suspected to be cancer-critical for at least one type of tumor. Current estimates put the total number of such genes at around 300, about 1% of the genes in the human genome. These cancer-critical genes are amazingly diverse, revealing an unexpected breadth of mechanisms. Whereas some of the new cancer genes encode classical signaling and cell-cycle proteins that might be anticipated to become mutated in cancer cells, others populate new and sometimes surprising categories. These include functions as diverse as metabolism, chromatin biology, RNA splicing, protein homeostasis, and cell differentiation. It is clear that alterations in any of these processes can contribute, in one tissue or another, to the evolution of cells with the cancerous properties that were listed on page 1172.

Clearly, the molecular changes that cause cancer are complex. As we now explain, however, the complexity is not quite as daunting as it might initially seem.

Disruptions in a Handful of Key Pathways Are Common to Many Cancers

Some genes, like *Rb* and *Ras*, are mutated in many cases of cancer and in cancers of many different types. The involvement of genes such as *Rb* and *Ras* in cancer is no surprise, now that we understand their normal functions: they control fundamental processes of cell division and growth. But even these common culprits feature in considerably less than half of individual cases. What is happening to the control of these processes in the many cases of cancer where, for example, *Rb* is intact or *Ras* is not mutated? What part do mutations in the hundreds of other cancer-critical genes play in the development of the disease? With our increasing knowledge of the normal functions of the genes in the human genome, it is becoming easier to detect patterns in the cataloged driver mutations and to give some simplifying answers to these questions.

In order to clarify the type of patterns revealed, consider the deadly disease *glioblastoma*—the most common type of human brain cancer. Analysis of the genomes of tumor cells from 91 individuals identified a total of at least 79 genes that were mutated in more than one individual. The normal functions of most of these genes were known or could be guessed, allowing them to be assigned to specific biochemical or regulatory pathways. Three functional groupings stood out, accounting for 21 of the recurrently mutated genes. One group consists of genes in the *Rb pathway* (that is, *Rb* itself, along with genes that directly regulate *Rb*); this pathway governs initiation of the cell-division cycle. Another consists of genes in the same regulatory subnetwork as *Ras*—referred to as the *RTK/Ras/PI 3-kinase pathway*, after three of its core components; this pathway serves to transmit signals for cell growth and cell division from outside the cell to the cell interior. A third grouping consists of genes in a pathway regulating responses to stress and DNA damage—the *p53 pathway*. We shall have more to say about these pathways below.

For the 91 glioblastomas, 74% had identifiable mutations in all three pathways. If one were to trace these three pathways further upstream and include all the components, known and unknown, on which they depend, this percentage would almost certainly be even higher. In other words, in almost every case of glioblastoma, there are mutations that disrupt each of three fundamental controls: the control of cell growth, the control of cell division, and the control of responses to stress and DNA damage.

Strikingly, in any given tumor-cell clone, there is a strong tendency for no more than one gene to be mutated in each pathway. Evidently, what matters for tumor evolution is the disruption of the control mechanism, and not the genetic means by which that is achieved. Thus, for example, in an individual whose tumor cells have no mutation in *Rb* itself, there is generally a mutation in some other component of the Rb pathway, producing a similar biological effect. This indicates that there is no further selection for a mutant in a pathway if the pathway has been disabled by a mutation elsewhere.

Similar patterns are seen in other types of cancers. A survey of many specimens of the major variety of ovarian cancer, for example, identified 67% of individuals

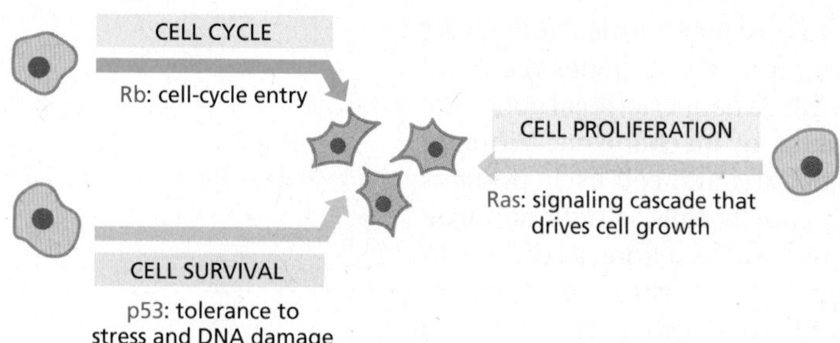

Figure 20–29 The three major cellular pathways that contribute to tumorigenesis. The specific examples listed here are described in this chapter. Cell-cycle events controlled by Rb are described in detail in Chapter 17.

as having mutations in the Rb pathway, 45% in the Ras/PI 3-kinase pathway (defined more narrowly than in the glioblastoma study), and more than 96% in the p53 pathway. Allowing for additional pathway components not included in the analysis, it seems that most cases of this type of cancer, too, have mutations disrupting the same three controls, leading to misregulated cell growth, misregulated cell proliferation, and an abnormal disregard of stress and DNA damage.

It seems that these three fundamental controls are subverted in one way or another in virtually every type of cancer. However, because specialized tissues can depend on different mechanisms to relay environmental signals to the core control machinery, these controls are open to subversion in a different set of ways in different types of cancers. In fact, one can find examples of driver mutations in practically all the major signaling pathways through which cells communicate during development and tissue maintenance (discussed in Chapters 15, 21, and 22).

Figure 20–29 outlines the three central pathways just described, abbreviating them as cell cycle, cell proliferation, and cell survival. We have previously discussed Rb (see pp. 1182–1183) and devoted an entire chapter to cell-cycle controls (Chapter 17). Some important details of the other two control pathways in Figure 20–29 are reviewed next.

Mutations in the PI 3-kinase/Akt/mTOR Pathway Drive Cancer Cells to Grow

Cell proliferation is not simply a matter of progression through the cell cycle; it also requires coordinated cell growth, which involves complex anabolic processes through which the cell synthesizes all the necessary macromolecules from small-molecule precursors (**Figure 20–30**). Cancer depends, therefore, not only on a loss of restraints on cell-cycle progression but also on a disrupted control of cell growth.

Downstream of RTK/Ras activation, the PI 3-kinase/Akt/mTOR intracellular signaling pathway is critical for cell growth control. As described in Chapter 15, various extracellular signal proteins, including insulin and insulin-like growth factors, normally activate this pathway. In cancer cells, however, the pathway is activated by mutation so that the cell can grow in the absence of such signals. The resulting abnormal activation of the protein kinases Akt and mTOR not only stimulates protein synthesis but also greatly increases both glucose uptake and the production of the acetyl CoA in the cytosol required for cell lipid synthesis, as outlined in Figure 20–30B.

The abnormal activation of the PI 3-kinase/Akt/mTOR pathway, which normally occurs early in the process of tumor progression, helps to explain the excessive rate of glycolysis that is observed in tumor cells, known as the Warburg effect, as discussed earlier (see Figure 20–18). As expected from our previous discussion, cancers can activate this pathway in many different ways. Thus, for example, a growth factor receptor can become abnormally activated, as in Figure 20–23. Also very common in cancers is the loss of the phosphatase and tensin homolog (PTEN) phosphatase that normally functions to counteract this pathway. PTEN suppresses the PI 3-kinase/Akt/mTOR pathway by dephosphorylating the phosphatidylinositol 3,4,5-trisphosphate [PI(3,4,5)P_3]

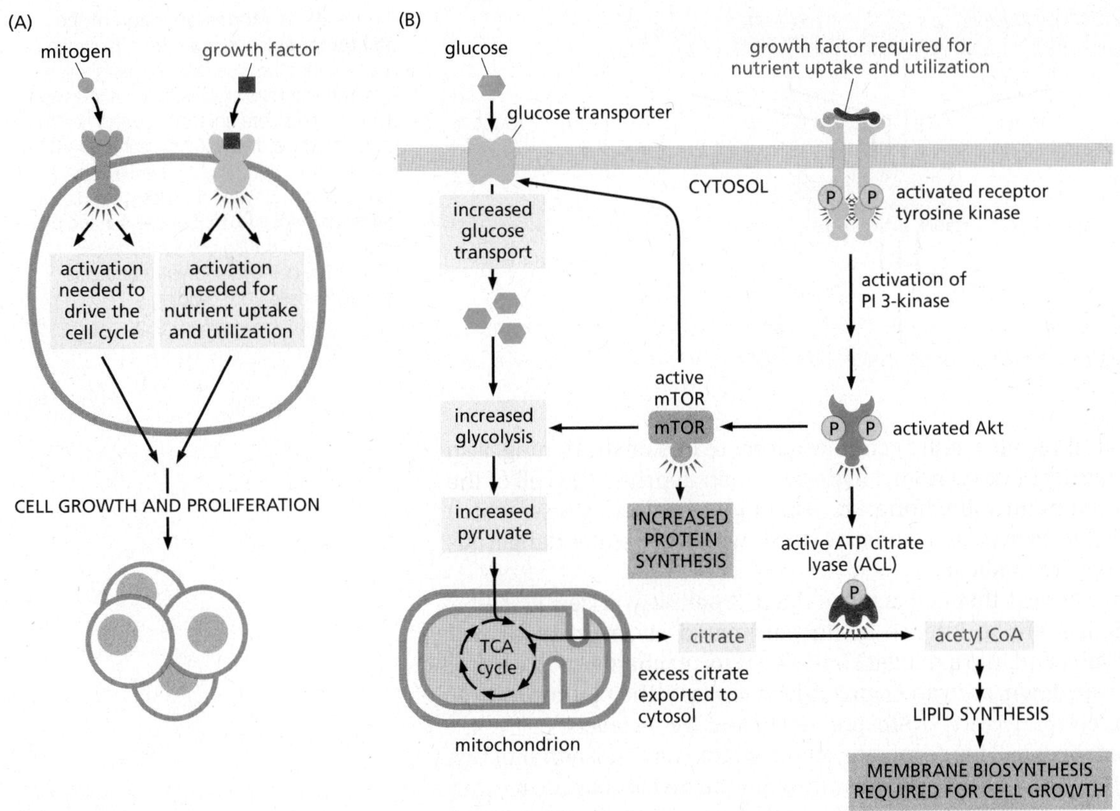

Figure 20–30 **Cells seem to require two types of signals to proliferate.** (A) In order to multiply successfully, most normal cells require both extracellular signals that drive cell-cycle progression (shown here as *blue* mitogen) and extracellular signals that drive cell growth (shown here as *red* growth factor). How mitogens activate signaling through the Rb pathway to drive entry into the cell cycle is described in Figure 17–59. (B) Diagram of the signaling system involving activation of Akt and mTOR that drives cell growth through greatly stimulating glucose uptake and utilization, including a conversion of the excess citric acid produced from sugar intermediates in mitochondria into the acetyl CoA that is needed in the cytosol for lipid synthesis and new membrane production. As indicated, protein synthesis is also increased. This system becomes abnormally activated early in tumor progression. TCA cycle indicates the tricarboxylic acid cycle (citric acid cycle).

molecules that the PI 3-kinase generates (see p. 920). *PTEN* is commonly mutated in tumors.

Mutations in the p53 Pathway Enable Cancer Cells to Survive and Proliferate Despite Stress and DNA Damage

That cancer cells must break the normal rules governing cell growth and cell division is obvious: that is part of the definition of cancer. It is not so obvious why cancer cells should also be abnormal in their response to stress and DNA damage, and yet this too is an almost universal feature. The gene that lies at the center of this response, the *p53* gene, is mutated in about 50% of all cases of cancer—a higher proportion than for any other known cancer-critical gene. When we include with *p53* the other genes that are closely involved in its function, we find that most cases of cancer harbor mutations in the p53 pathway. Why should this be? To answer, we must first consider the normal function of this pathway.

In contrast to Rb, most cells in the body contain very little **p53** protein under normal conditions: although the protein is synthesized, it is rapidly degraded. Mice in which both copies of the gene have been deleted or inactivated typically appear normal in all respects except one—they universally develop cancer before 10 months of age. These observations suggest that p53 has a function that is required only in special circumstances. In fact, cells raise their concentration of p53 protein in response to a whole range of conditions that have only one obvious thing in common: they are, from the cell's point of view, pathological, putting the cell in danger of death or serious injury. These conditions include DNA damage,

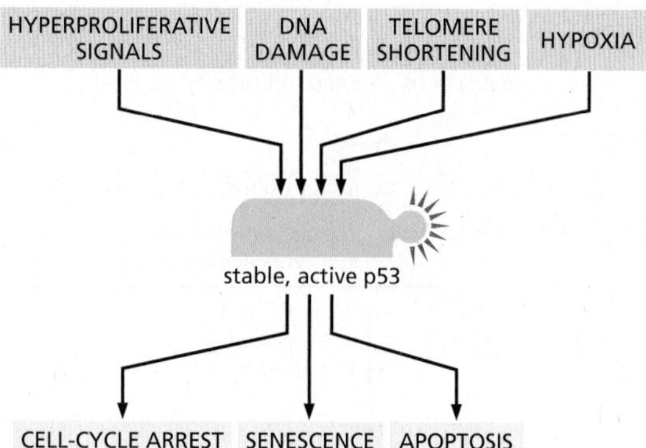

Figure 20–31 Modes of action of the p53 tumor suppressor. The p53 protein is a cellular stress sensor. In response to hyperproliferative signals, DNA damage, hypoxia, telomere shortening, and various other stresses, the p53 levels in the cell rise. As indicated, this may either arrest cell cycling in a way that allows the cell to adjust and survive, trigger cell suicide by apoptosis, or cause cell *senescence*—an irreversible cell-cycle arrest that stops damaged cells from dividing.

which puts the cell at risk from a faulty genome; telomere loss or shortening, also dangerous to the integrity of the genome; hypoxia, which deprives the cell of the oxygen it needs to maintain mitochondrial respiration; osmotic stress, which causes the cell to swell or shrivel; and oxidative stress, which generates dangerous levels of highly reactive free radicals.

Yet another form of stress that can activate the p53 pathway arises, it seems, when regulatory signals are so intense or uncoordinated as to drive the cell beyond its normal limits and into a danger zone where its mechanisms of control and coordination break down, as in an engine driven too fast. The p53 concentration rises, for example, when *Myc* is overexpressed to oncogenic levels.

All these circumstances call for desperate action, which may take either of two forms: the cell can block any further progress through the division cycle in order to take time out to repair or recover from the pathological condition or it can accept that it must die, and do so in a way that minimizes damage to the organism. A good death, from this point of view, is a death by apoptosis. In apoptosis, the cell is phagocytosed by its neighbors and its contents are efficiently recycled. A bad death is a death by necrosis. In necrosis, the cell bursts or disintegrates and its contents are spilled into the extracellular space, inducing inflammation.

The p53 pathway, therefore, behaves as a sort of antenna, sensing the presence of a wide range of dangerous conditions and, when any are detected, triggering appropriate action—either a temporary or permanent arrest of cell cycling or suicide by apoptosis (Figure 20–31). These responses serve to prevent deranged cells from proliferating. Cancer cells are indeed generally deranged, and their survival and proliferation thus depend on inactivation of the p53 pathway. If the p53 pathway were active in them, they would be halted in their tracks or die (Movie 20.6). For example, if the p53 pathway is functional, a cell with unrepaired DNA damage will stop dividing or die; it cannot proliferate.

The p53 protein performs its job mainly by acting as a transcription regulator (see Movie 17.8). Indeed, the most common mutations observed in p53 in human tumors are in its DNA-binding domain, where they cripple the ability of p53 to bind to its DNA target sequences. As discussed in Chapter 17, the p53 protein exerts its inhibitory effects on the cell cycle, in part at least, by inducing the transcription of *p21*, which encodes a protein that binds to and inhibits the cyclin-dependent kinase (Cdk) complexes required for progression through the cell cycle. By blocking the kinase activity of these Cdk complexes, the p21 protein prevents the cell from progressing through S phase and replicating its DNA.

The mechanism by which p53 induces apoptosis includes stimulation of the expression of many pro-apoptotic genes, as described in Chapter 18.

Studies Using Mice Help to Define the Functions of Cancer-critical Genes

The ultimate test of a gene's role in cancer has to come from investigations in the intact, mature organism. The most favored organism for experimental studies is the mouse. To explore the function of a candidate oncogene or tumor suppressor

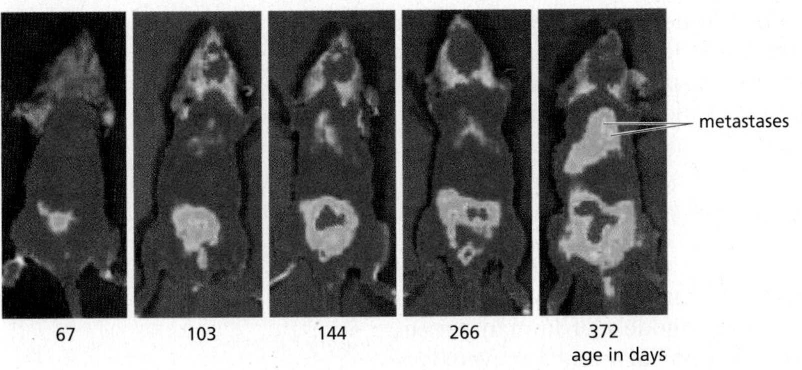

Figure 20–32 **Monitoring tumor growth and metastasis in a mouse with a luminescent reporter.** A mouse was genetically engineered in a way that allows both copies of its *PTEN* tumor suppressor gene to be inactivated in the prostate gland, simultaneously with the prostate-specific activation of a gene engineered to produce the enzyme luciferase (derived from fireflies). After an injection of luciferin (the substrate molecule for luciferase) into the mouse's bloodstream, the cells in the prostate emit light and can be detected by their bioluminescence in a live mouse, as seen in the 67-day-old animal at the left. Cells lacking the PTEN phosphatase enzyme contain elevated amounts of the Akt activator, $PI(3,4,5)P_3$, and this causes the prostate cells to proliferate abnormally, progressing over time to form a cancer. In this way, the process of metastasis could be followed in the same animal over the course of a year. The light intensity in these experiments is proportional to the number of prostate-cell descendants, increasing from *light blue* to *green*, to *yellow*, to *red* in this representation. (Adapted from C.-P. Liao et al., *Cancer Res*. 67:7525–7533, 2007. With permission from the American Association for Cancer Research.)

gene, one can make a transgenic mouse that overexpresses it or a knockout mouse that lacks it. Using the techniques described in Chapter 8, one can engineer mice in which the misexpression or deletion of the gene is restricted to a specific set of cells, or in which expression of the gene can be switched on at will at a chosen point in time, or both, to see whether and how tumors develop. Moreover, to follow the growth of tumors from day to day in the living mouse, the cells of interest can be genetically marked and made visible by expression of a fluorescent or luminescent reporter (Figure 20–32). In these ways, one can begin to clarify the part that each cancer-critical gene plays in cancer initiation or progression.

Transgenic mouse studies confirm, for example, that a single oncogene is generally not enough to turn a normal cell into a cancer cell. Thus, in mice engineered to express a *Myc* or *Ras* oncogenic transgene, some of the tissues that express the oncogene may show enhanced cell proliferation, and, over time, occasional cells will undergo further changes to give rise to cancers. Most cells expressing the oncogene, however, do not give rise to cancers. Nevertheless, from the point of view of the whole animal, the inherited oncogene is a serious menace because it creates a high risk that a cancer will arise somewhere in the body. Mice that express both *Myc* and *Ras* oncogenes develop cancers earlier and at a much higher rate than mice that express either gene alone (Figure 20–33); but, again, the cancers originate as scattered, isolated tumors among noncancerous

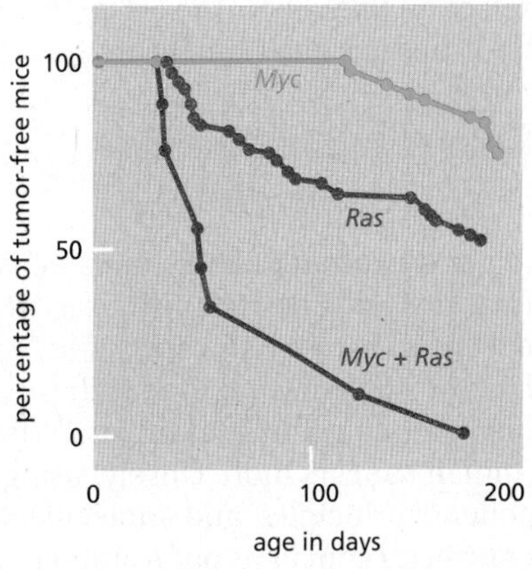

Figure 20–33 **Oncogene collaboration in transgenic mice.** The graphs show the incidence of tumors in three types of transgenic mouse strains, one carrying a *Myc* oncogene, one carrying a *Ras* oncogene, and one carrying both oncogenes. For these experiments, two lines of transgenic mice were first generated. One carries an inserted copy of an oncogene created by fusing the proto-oncogene *Myc* with the mouse mammary tumor virus regulatory DNA (which then drives *Myc* overexpression in the mammary gland). The other line carries an inserted copy of the *Ras* oncogene under the control of the same regulatory element. Both strains of mice develop tumors much more frequently than normal, most often in the mammary or salivary glands. Mice that carry both oncogenes together are obtained by crossing the two strains. These hybrids develop tumors at a far higher rate still, much greater than the sum of the rates for the two oncogenes separately. Nevertheless, the tumors arise only after a delay and only from a small proportion of the cells in the tissues where the two genes are expressed. Further accidental changes, in addition to the two oncogenes, are apparently required for the development of cancer. (After E. Sinn et al., *Cell* 49:465–475, 1987.)

cells. Thus, even cells expressing these two oncogenes must undergo further, randomly generated changes to become cancerous. This strongly suggests that multiple mutations are required for tumorigenesis, as supported by a great deal of other evidence discussed earlier. Experiments using mice with deletions of tumor suppressor genes lead to similar conclusions.

Cancers Become More and More Heterogeneous as They Progress

Although useful to study the function and interaction of cancer driver genes *in vivo*, there are important limitations to mouse models of human cancer. Whereas a mouse tumor is small and arises over a period of months or even days, a human tumor may be the size of a mouse or larger and will often have been growing for decades. From simple histology looking at stained tissue sections, it is clear that some tumors contain distinct sectors, all clearly cancerous, but differing in appearance because they differ genetically or epigenetically: the cancer cell population is heterogeneous. Evidently, within the initial clone of cancerous cells, additional mutations have arisen and conferred selective advantages, such as increased growth rates, creating diverse subclones. Today, the ability to analyze cancer genomes lets us look much deeper into the process. Comparison of samples from different regions of a tumor and the metastases it has spawned reveal a classic picture of Darwinian evolution, occurring on a time scale of months or years rather than millions of years, but governed by the same rules of natural selection.

One approach to examine the development of tumor heterogeneity takes advantage of human cells grown outside the body, termed "organoids" (described in Chapter 22). Unlike the immortalized cancer cell lines described previously that have adapted to an artificial, two-dimensional lifestyle (see Figure 20–15), organoids are grown from adult stem cells under more physiological conditions within a three-dimensional matrix. In this environment, a single stem cell can proliferate and differentiate into a self-organizing structure that resembles a miniature, simplified version of the organ from which it was taken, with realistic microanatomy. One study applied this organoid system to examine the genetic heterogeneity within individual colorectal tumors (**Figure 20–34A**). Organoids were grown from single cells isolated from different regions of a tumor as well as from neighboring, normal tissue. Genome analysis revealed detailed patterns of mutations in each organoid that indicated how closely it was related to the others, and from these data one could draw up a family tree. Those organoids derived from the same tumor site bore the most similarities, and all possessed the same set of cancer-driving mutations that arose in their common ancestor within the trunk of the tree, corresponding to the early stages of tumor growth.

Clearly, cancer cells are constantly mutating, multiplying, competing, evolving, and diversifying as they exploit new ecological niches and react to the treatments that are used against them (**Figure 20–34B**). Diversification accelerates as they metastasize and colonize new territories, where they encounter new selection pressures. The longer the evolutionary process continues, the harder it becomes to catch them all in the same net and kill them.

Colorectal Cancers Evolve Slowly Via a Succession of Visible Changes

At the beginning of this chapter, we saw that most cancers develop gradually from a single aberrant cell, progressing from benign to malignant tumors by the accumulation of a number of independent genetic and epigenetic changes. We have discussed what some of these changes are in molecular terms and have seen how they contribute to cancerous behavior. We now examine one of the common human cancers more closely, using it to illustrate and enlarge upon some of the general principles and molecular mechanisms we have introduced. We take **colorectal cancer** as our example.

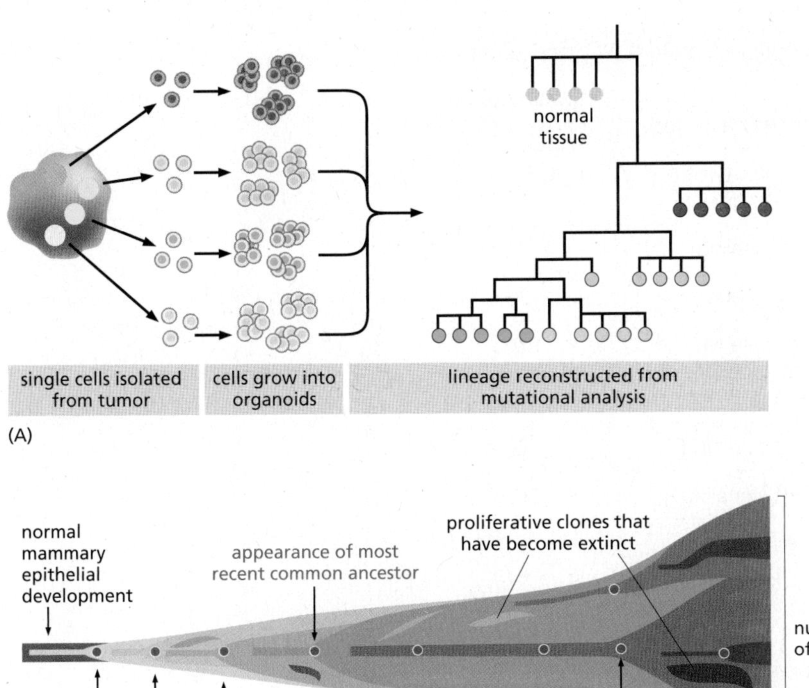

Figure 20–34 **How cancers progress as a series of subclones.** (A) Cancer analysis using organoids. Single cells were isolated from four different regions of a human colorectal tumor (color coded) as well as from nearby healthy tissue (not shown). These cells were then used to produce organoids, allowing large amounts of material to be obtained for analysis. DNA sequencing was used to determine the mutations in each organoid. These data then enabled construction of a phylogenetic tree that reveals the order in which the mutations emerged in the original tumor. (B) A depiction of how driver mutations are thought to cause cancer progression over long periods of time, before producing a large enough clone of proliferating cells to be detected as a tumor. The data indicate that driver mutations occur only rarely in a background of long-lived subclones of cells that continually accumulate passenger mutations without gaining a growth advantage. (A, adapted from C.J. Kuo and C. Curtis, *Nature* 556:441–442, 2016. With permission from Springer Nature. B, adapted from S. Nik-Zainal et al., *Cell* 149:994–1007, 2012. This article is distributed under the terms of the Creative Commons Attribution License.)

Colorectal cancers arise from the epithelium lining the colon (the large intestine) and rectum (the terminal segment of the gut). The organization of this tissue is broadly similar to that of the small intestine, discussed in Chapter 22 (pp. 1281–1282). For both the small and large intestine, the epithelium is renewed at an extraordinarily rapid rate, taking about a week to completely replace most of the epithelial sheet. In both regions, the renewal depends on stem cells that lie in deep pockets of the epithelium, called intestinal crypts. The signals that maintain the stem cells and control the normal organization and renewal of the epithelium are beginning to be quite well understood, as explained in Chapter 22. Mutations that disrupt these signals begin the process of tumor progression for most colorectal cancers (Movie 20.7).

Colorectal cancers are common, currently causing nearly 60,000 deaths a year in the United States, or about 10% of total deaths from cancer. Like most cancers, they are not usually diagnosed until late in life (90% occur after the age of 55). However, routine examination of normal adults with a colonoscope (a fiber-optic device for viewing the interior of the colon and rectum) often reveals a small benign tumor, or adenoma, of the gut epithelium in the form of a protruding mass of tissue called a *polyp*. These adenomatous polyps are believed to be the

TABLE 20–1 Some Genetic Abnormalities Detected in Colorectal Cancer Cells

Gene	Class	Pathway affected	Human colon cancers (%)
K-Ras	Oncogene	Receptor tyrosine kinase signaling	40
β-Catenin[1]	Oncogene	Wnt signaling	5–10
Apc[1]	Tumor suppressor	Wnt signaling	>80
p53	Tumor suppressor	Response to stress and DNA damage	60
TGFβ receptor II[2]	Tumor suppressor	TGFβ signaling	10
Smad4[2]	Tumor suppressor	TGFβ signaling	30
MLH1 and other DNA mismatch repair genes (often silenced by DNA methylation)	Tumor suppressor (genetic stability)	DNA mismatch repair	15

[1,2]The genes with the same superscript numeral act in the same pathway, and therefore only one of the components is mutated in an individual cancer.

precursors of a large proportion of colorectal cancers. Because the progression of the disease is usually very slow, there is typically a period of about 10 years in which the slowly growing tumor is detectable but has not yet turned malignant. Thus, when people are screened by colonoscopy in their fifties and the polyps are removed through the colonoscope—a quick and easy procedure—the subsequent incidence of colorectal cancer is much lower: according to some studies, less than a quarter of what it would be otherwise.

In microscopic sections of polyps smaller than 1 cm in diameter, the cells and their arrangement in the epithelium usually appear almost normal. The larger the polyp, the more likely it is to contain cells that look aberrant and are abnormally organized. Sometimes, two or more distinct areas can be distinguished within a single polyp, with the cells in one area appearing relatively normal and those in the other appearing clearly cancerous, as though they have arisen as a mutant subclone within the original clone of adenomatous cells. At later stages in the disease, some tumor cells become invasive in a small fraction of the polyps, first breaking through the epithelial basal lamina, then spreading through the layer of muscle that surrounds the gut, and finally metastasizing to lymph nodes via lymphatic vessels and to liver, lung, and other organs via blood vessels.

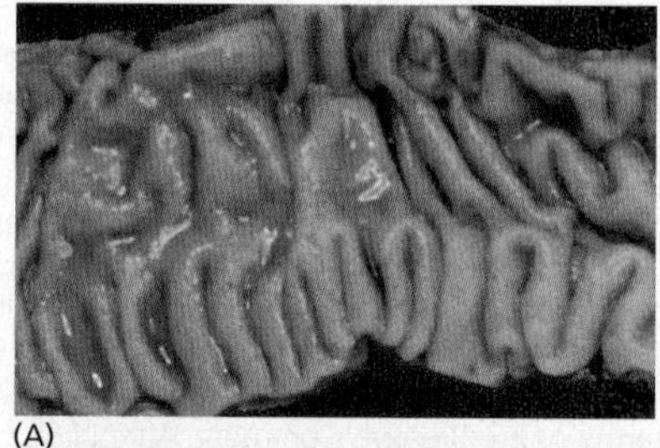

(A)

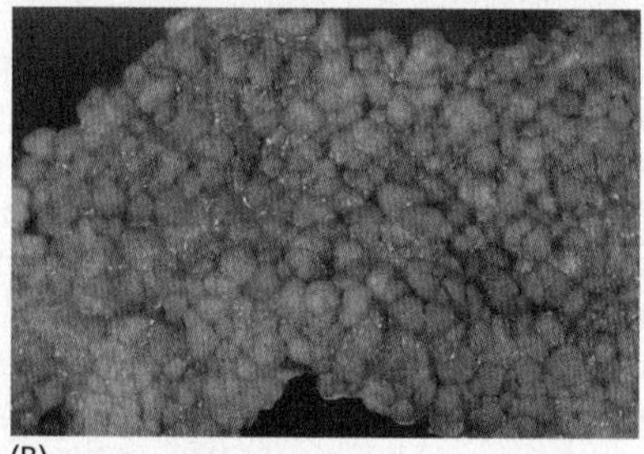

(B)

Figure 20–35 Colon of individual with familial adenomatous polyposis coli compared to normal colon. (A) The normal colon wall is a gently undulating but smooth surface. (B) The polyposis colon is completely covered by hundreds of projecting polyps, each resembling a tiny cauliflower when viewed with the naked eye. (Courtesy of Mark Arends.)

A Few Key Genetic Lesions Are Common to a Large Fraction of Colorectal Cancers

What are the mutations that accumulate with time to produce this chain of events? Of those genes so far discovered to be involved in colorectal cancer, three stand out as most frequently mutated: the proto-oncogene *K-Ras* (a member of the *Ras* gene family), in about 40% of cases; *p53*, in about 60% of cases; and the tumor suppressor gene *Apc* (discussed below), in more than 80% of cases. Others are involved in smaller numbers of colon cancers, and some of these are listed in **Table 20–1**.

The role of *Apc* first came to light through study of certain families showing a rare type of hereditary predisposition to colorectal cancer, called *familial adenomatous polyposis coli* (*FAP*). In this syndrome, hundreds or thousands of polyps develop along the length of the colon (**Figure 20–35**). These polyps start to appear in early adult life, and if they are not removed, one or more will almost always progress to become malignant; the average time from the first detection of polyps to the diagnosis of cancer is 12 years. The disease can be traced to a deletion or inactivation of the tumor suppressor gene *Apc*, named after the syndrome. Individuals with FAP have inactivating mutations or deletions of one copy of the *Apc* gene in all their cells and show loss of heterozygosity in tumors, meaning that both copies have been lost or inactivated, even in the benign polyps. Most individuals with colorectal cancer do not have the hereditary condition. Nevertheless, in more than 80% of the cases, their cancer cells (but not their normal cells) have inactivated both copies of the *Apc* gene through mutations acquired during the individual's lifetime. Thus, by a route similar to what we discussed for retinoblastoma, mutation of the *Apc* gene was identified as one of the central ingredients of colorectal cancer.

The APC protein is an inhibitory component of the *Wnt signaling pathway* (discussed in Chapter 15). It binds to the *β-catenin protein*, another component of the Wnt pathway, and helps to induce the protein's degradation. By inhibiting β-catenin in this way, APC prevents it from localizing to the nucleus, where it would act as a transcriptional regulator to drive cell proliferation and maintain the stem-cell state (see Figure 15–61). Loss of APC results in an excess of free β-catenin and thus leads to an uncontrolled expansion of the stem-cell population. This causes massive increase in the number and size of the intestinal crypts (see Figure 22–4).

When the *β-catenin* gene was sequenced in a collection of colorectal tumors, it was discovered that many of the tumors that did not have *Apc* mutations had activating mutations in the β-catenin protein instead. Thus, it is excessive activity in the Wnt signaling pathway that is critical for the initiation of this cancer, rather than any single oncogene or tumor suppressor gene that the pathway contains.

This being so, why is the *Apc* gene in particular so often the most common culprit in colorectal cancer? The APC protein is large and it interacts not only with β-catenin but also with various other cell components, including microtubules. Loss of APC appears to increase the frequency of mitotic spindle defects, leading to chromosome abnormalities when cells divide. This additional, independent cancer-promoting effect could explain why *Apc* mutations feature so prominently in the causation of colorectal cancer.

Some Colorectal Cancers Have Defects in DNA Mismatch Repair

In addition to the hereditary disease (FAP) associated with *Apc* mutations, there is a second, more common kind of hereditary predisposition to colon carcinoma in which the course of events differs from the one we have described for FAP. In this more common condition, called *hereditary nonpolyposis colorectal cancer (HNPCC)*, or *Lynch syndrome*, the probability of colon cancer is increased without any increase in the number of colorectal polyps (adenomas). Moreover, the cancer cells are unusual, in that they have a normal (or almost normal) karyotype. The majority of colorectal tumors in non-HNPCC individuals, in contrast, have gross chromosomal abnormalities, with multiple translocations, deletions, and other aberrations, and have many more chromosomes than normal (see Figure 20–10).

The mutations that predispose HNPCC individuals to colorectal cancer occur in one of several genes that code for central components of the DNA *mismatch repair system*. These genes are homologous in structure and function to the *MutL* and *MutS* genes in bacteria and yeast (see Figure 5–20). Only one of the two copies of the involved gene is defective, so the repair system is still able to remove the inevitable DNA replication errors that occur in the individual's cells. However, as discussed previously, these individuals are at risk because the accidental loss or inactivation of the remaining good gene copy will immediately elevate the spontaneous mutation rate by a hundredfold or more (discussed in

Chapter 5). These genetically unstable cells can then speed through the standard processes of mutation and natural selection that allow clones of cells to progress to malignancy.

This particular type of genetic instability produces invisible changes in the chromosomes—most notably changes in individual nucleotides and short expansions and contractions of mononucleotide and dinucleotide repeats such as AAAA... or CACACA.... Once the defect in HNPCC individuals was recognized, the epigenetic silencing or mutation of mismatch repair genes was found in about 15% of the colorectal cancers occurring in people with no inherited predisposing mutation.

Thus, the genetic instability found in many colorectal cancers can be acquired in at least two ways. The majority of the cancers display a form of chromosomal instability that leads to visibly altered chromosomes, whereas in the others the instability occurs on a much smaller scale and reflects a defect in DNA mismatch repair. Indeed, many carcinomas show either chromosomal instability or defective mismatch repair—but rarely both. These findings clearly demonstrate that genetic instability is not an accidental by-product of malignant behavior but a contributory cause—and that cancer cells can acquire this instability in multiple ways.

The Steps of Tumor Progression Can Often Be Correlated with Specific Mutations

In what order do *K-Ras*, *p53*, *Apc*, and the other identified colorectal cancer-critical genes mutate, and what contribution does each of them make to the asocial behavior of the cancer cell? There is no single answer because colorectal cancer can arise by more than one route: thus, we know that in some cases, the first mutation can be in a DNA mismatch repair gene; in others, it can be in a gene regulating cell proliferation. Moreover, as previously discussed, a general feature such as genetic instability or a tendency to proliferate abnormally can arise in a variety of ways, through mutations in different genes.

Nevertheless, certain sets of mutations are particularly common in colorectal cancer, and they occur in a characteristic order. Thus, in most cases, mutations inactivating the *Apc* gene appear to be the first or, at least, a very early step, as they are detected at the same high frequency in small benign polyps as in large malignant tumors. Changes that lead to genetic and epigenetic instability are likely also to arise early in tumor progression, as they are needed to drive the later steps.

Activating mutations in the *K-Ras* gene occur later, as they are rare in small polyps but common in larger ones that show disturbances in cell differentiation and histological pattern.

Inactivating mutations in *p53* are thought to come later still, as they are rare in polyps but common in carcinomas (**Figure 20–36**). We have seen that loss of *p53* function allows cancer cells to endure stress and to avoid apoptosis and cell-cycle arrest. Additionally, loss of *p53* is related to the heightened activation of oncogenes such as *Ras*. Experiments in mice show that an initial low level of oncogene

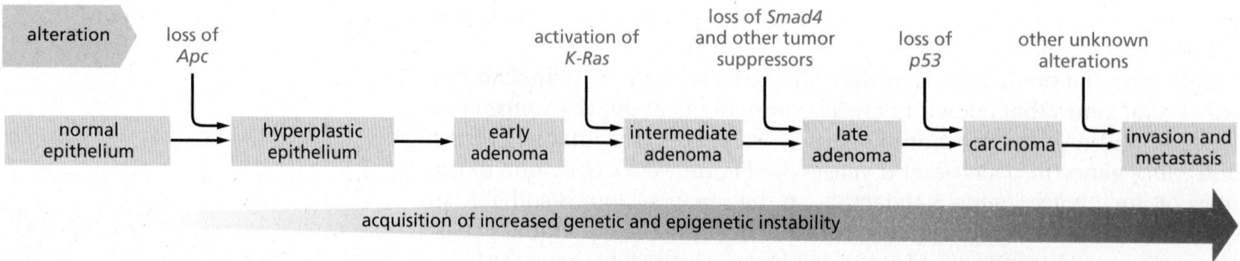

Figure 20–36 Suggested typical sequence of genetic changes underlying the development of a colorectal carcinoma. This oversimplified diagram provides a general idea of the way mutation and tumor development are related. But many other mutations are generally involved, and different colon cancers can progress through different sequences of mutations (and/or epigenetic changes).

activation can give rise to a slowly growing tumor even while *p53* is functional: genes such as *Ras* are, after all, part of the normal machinery of growth control, and moderate activation is not stressful for a cell and does not call the p53 protein into play. Progression of a tumor from slow to rapid, malignant growth, however, involves activation of oncogenes beyond normal physiological limits to a higher, stressful level. If the p53 protein is present and functional, this should lead to cell-cycle arrest or death. Only by losing p53 function can the cancer cells with hyperactive oncogenes survive and progress.

The steps we have just described are only part of the picture. It is important to emphasize that each case of colorectal cancer is different, with its own detailed combination of mutations, and that even for the mutations that are commonly shared, the sequence of occurrence may vary.

The Changes in Tumor Cells That Lead to Metastasis Are Still Largely a Mystery

Perhaps the most significant gap in our understanding of cancer concerns invasiveness and metastasis (see Figure 20–20). More than 25 years after the multistep progression of colorectal cancer was first described, no genetic mutations have been identified that are characteristically associated with metastatic disease. To date, even large-scale genomic sequencing efforts have yet to uncover recurrent mutations that adequately explain the escape of cancer cells from a primary tumor and their dissemination and colonization of a distant tissue. One possibility is that metastasis can be initiated by epigenetic changes that alter patterns of gene expression. These changes could allow cancer cells to take on new traits in the absence of further genetic changes, thereby reprogramming their behavior to promote invasiveness and motility or to generate cancer stem cells. Importantly, changes in cellular programming and behavior operate normally in a variety of contexts during cellular differentiation. The example most relevant to metastasis is the *epithelial-to-mesenchymal transition* (*EMT*), a highly regulated and poorly understood process that allows epithelial cells to lose their characteristic polarity and adhesiveness and take on a mesenchymal phenotype, which includes enhanced migratory behavior. An EMT program operates at several stages of embryogenesis as well as during wound healing, and its activation in cancer cells could explain how they escape from a primary tumor and invade a new tissue. However, EMT is unlikely to explain the major puzzle of how disseminated cells acquire the ability to survive in the microenvironment of that new tissue, with its unfamiliar mix of growth factors and extracellular matrix components. Because the EMT program (like almost all cellular differentiation programs) involves changes in gene expression, it does not alter DNA sequences and would not be detected by sequencing the genomes of metastasized cancer cells.

Observations of the many stages and varied features of tumorigenesis highlight that cancer cells do not invent new biological phenomena, but instead deploy existing mechanisms and pathways at inappropriate times and places. A better understanding of the normal, underlying cell biology, and of the molecular changes that act to subvert it in cancer, will provide promising leads for cancer treatment, as we discuss in the final section of this chapter.

Summary

The molecular analysis of cancer cells reveals two classes of cancer-critical genes: oncogenes and tumor suppressor genes. A set of these genes becomes altered by a combination of genetic and epigenetic accidents to drive tumor progression. Many cancer-critical genes code for components of the social control pathways that regulate when cells grow, divide, differentiate, or die. In addition, a subclass of tumor suppressors can be categorized as genome maintenance genes, because their normal role is to help maintain genome integrity.

The inactivation of the p53 pathway, which occurs in nearly all human cancers, allows genetically damaged cells to escape apoptosis and continue to proliferate.

Inactivation of the Rb pathway also occurs in most human cancers, illustrating how fundamental each of these pathways is for protecting us against cancer. More generally, the extensive sequencing of cancer cell genomes reveals that the development of a cancer requires the acquisition of heritable disturbances in each of three types of normal controls—those in the cell cycle, cell proliferation, and cell survival pathways.

The sequencing of cancer cell genomes indicates that—except for the cancers of childhood—many cancers acquire multiple driver mutations over the long course of tumor progression, along with a considerably larger number of passenger mutations of no consequence. The same methods reveal how subclones of cells arise and die out as a tumor ages. Tumors thus contain a heterogeneous mixture of cells, some—the so-called cancer stem cells—being much more dangerous than others.

We can often correlate the steps of tumor progression with mutations that activate specific oncogenes and inactivate specific tumor suppressor genes, and colon cancer provides a good example. But different combinations of mutations and epigenetic changes are found in different types of cancer, and even in different individuals with the same type of cancer, reflecting the random way in which these inherited changes arise. Nevertheless, many of the same changes are encountered repeatedly, suggesting that there are a limited number of ways to breach our defenses against cancer. However, the molecular basis of the final and most deadly step in cancer, metastasis, remains poorly understood.

CANCER PREVENTION AND TREATMENT: PRESENT AND FUTURE

We can apply the growing understanding of the molecular biology of cancer to sharpen our attack on the disease at three levels: prevention, diagnosis, and treatment. Prevention is always better than cure, and indeed many cancers can be prevented, especially by avoiding smoking. Highly sensitive molecular assays promise new opportunities for earlier and more precise diagnosis, with the aim of detecting primary tumors while they are still small and have not yet metastasized. Cancers caught at these early stages can often be nipped in the bud by surgery or radiotherapy, as we saw for colorectal polyps. Nevertheless, full-blown malignant disease will continue to be common for many years to come, and cancer treatments will continue to be needed.

In this section, we first examine the preventable causes of cancer and then consider how advances in our understanding at a molecular level are beginning to transform the treatment of the disease.

Epidemiology Reveals That Many Cases of Cancer Are Preventable

A certain irreducible background incidence of cancer is to be expected regardless of circumstances. As discussed in Chapter 5, mutations can never be absolutely avoided because they are an inescapable consequence of fundamental limitations on the accuracy of DNA replication and repair. If a person could live long enough, it is inevitable that at least one of his or her cells would eventually accumulate a set of mutations sufficient for cancer to develop.

Nevertheless, environmental factors seem to play a large part in determining the risk for cancer. This is demonstrated most clearly by a comparison of cancer incidence in different countries: for almost every cancer that is common in one country, there is another country where the incidence is much lower. Because migrant populations tend to adopt the pattern of cancer incidence typical of their new host country, the differences are thought to be due mostly to environmental, not genetic, factors (**Figure 20–37A**). From epidemiologic studies compiled by the American Cancer Society, it is estimated that at least 45% of cancer deaths can be attributed to modifiable risk factors, indicating that more than half of all cancers should be avoidable (**Figure 20–37B**).

(A)

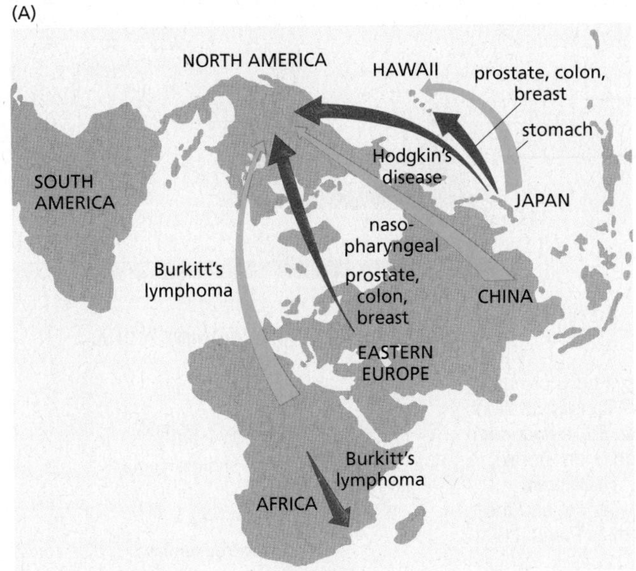

(B)

cause	deaths (percent of total)	number of deaths in US (2018)	estimated reduction possible (percent)
all risk factors	45%	265,150	50%
cigarette smoke	29%	169,180	75%
excess body weight	6.6%	38,230	50%
diet	4.9%	28,630	50%
alcohol	4.0%	23,510	50%
viruses	2.7%	16,100	100%
physical inactivity	2.2%	12,800	85%
UV radiation	1.5%	8,750	50%

Figure 20–37 **Cancer incidence is related to environmental influences.** (A) This map of the world shows the rates of cancer increasing *(red arrows)* or decreasing *(blue arrows)* when specific populations move from one location to another. Such observations suggest the importance of environmental factors, including diet, in dictating cancer risk. (B) Some estimated effects of environment and lifestyle on cancer in the United States (US). The table shows both the yearly deaths in the United States attributable to each factor and the estimated percentage of deaths that could be eliminated through prevention. (B, data from F. Islami et al., *CA Cancer J. Clin.* 68:31–54, 2018; and G.A. Colditz et al., *Sci. Transl. Med.* 4:127rv4, 2012.)

Unfortunately, different cancers have different environmental risk factors, and a population that escapes one such danger is usually exposed to another. This is not, however, inevitable. There are some human subgroups whose way of life substantially reduces the total cancer death rate among individuals of a given age. Under the current conditions in the United States and Europe, approximately one in five people will die of cancer. But the incidence of cancer among strict Mormons in Utah—who avoid alcohol, coffee, cigarettes, drugs, and unsafe sex—is only about half the incidence for nonpracticing members of the same family or for Americans in general.

Although such observations on human populations indicate that cancer can often be avoided, it has been difficult in most cases—with cigarette smoking as a striking exception—to pinpoint the specific environmental factors responsible for these large population differences or to establish how they act. Nevertheless, several important classes of environmental cancer risk factors have been identified (Figure 20–37B). But there are also many other influences—including the chemicals in our environment, the hormones that circulate in our bodies, and the irritations, infections, and damage to which we expose our tissues—that are no less important and favor development of the disease in other ways.

Sensitive Assays Can Detect Those Cancer-causing Agents That Damage DNA

Many quite disparate chemicals are carcinogenic when they are fed to experimental animals or painted repeatedly on their skin. Examples include a range of aromatic hydrocarbons and derivatives of them such as aromatic amines, nitrosamines, and alkylating agents such as mustard gas. Although these **chemical carcinogens** are diverse in structure, a large proportion of them have at least one shared property—they cause mutations. In one common test for mutagenicity (the *Ames test*), the carcinogen is mixed with an activating extract prepared from rat liver cells (to mimic the biochemical processing that occurs in an intact animal). The mixture is then added to a culture of specially designed test bacteria and the bacterial mutation rate measured. Most of the compounds scored as mutagenic by this rapid and convenient assay in bacteria also cause mutations or chromosome aberrations when tested on mammalian cells.

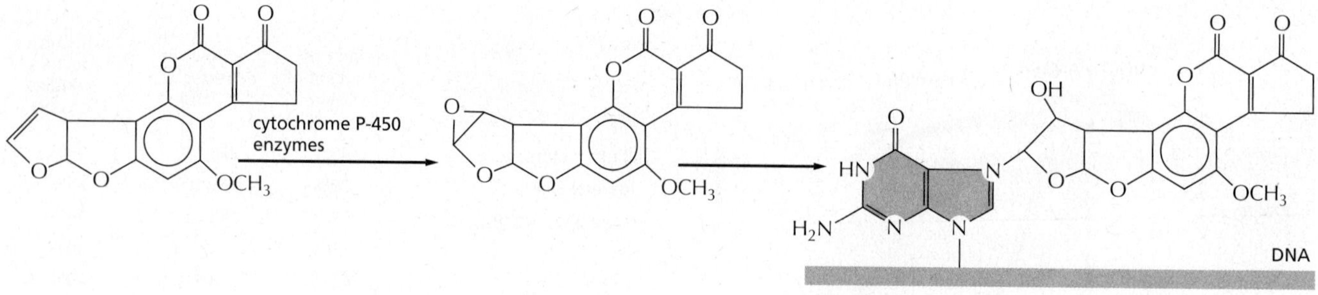

(A) AFLATOXIN B1 AFLATOXIN-2,3-EPOXIDE CARCINOGEN BOUND TO GUANINE IN DNA

- VINYL CHLORIDE: liver angiosarcoma
- BENZENE: acute leukemias
- ARSENIC: skin carcinomas, bladder cancer
- ASBESTOS: mesothelioma
- RADIUM: osteosarcoma

(B)

Figure 20–38 Some known carcinogens. (A) Carcinogen activation. A metabolic transformation that occurs in the liver must activate many chemical carcinogens before they will cause mutations by reacting with DNA (labeled *orange*). The compound illustrated here is *aflatoxin B1*, a toxin from a mold (*Aspergillus flavus*) that grows on grain and peanuts when they are stored under humid tropical conditions. Aflatoxin is an important cause of liver cancer in the tropics. (B) Different carcinogens cause different types of cancer. (B, data from Institute of Medicine, Cancer and the Environment: Gene-Environment Interactions. Washington, DC: National Academies Press, 2002.)

A few of these carcinogens act directly on DNA. But generally the more potent ones are relatively inert chemically; these chemicals become damaging only after they have been converted to a more reactive molecule by metabolic processes in the liver, catalyzed by a set of intracellular enzymes known as the *cytochrome P-450 oxidases.* These enzymes normally help to convert ingested toxins into harmless and easily excreted compounds. Unhappily, their activity on certain chemicals generates products that are highly mutagenic. Examples of carcinogens activated in this way include *benzo[a]pyrene,* a cancer-causing polycyclic aromatic hydrocarbon present in coal tar, tobacco smoke, and the fungal toxin *aflatoxin B1* (**Figure 20–38**).

Fifty Percent of Cancers Could Be Prevented by Changes in Lifestyle

Tobacco smoke is the most important carcinogen in the world today. Even though many other risk factors have been identified, none of these appear to be responsible for anything like the same numbers of human cancer deaths attributable to tobacco smoke. It is sometimes thought that the main environmental causes of cancer are the products of a highly industrialized way of life—the rise in pollution, the enhanced use of food additives, and so on—but there is little evidence to support this view. The idea may have come in part from the identification of some highly carcinogenic materials used in industry, such as 2-naphthylamine and asbestos. Except for the increase in cancers caused by smoking, however, age-adjusted death rates for most common human cancers have stayed much the same over the past half-century or, in some cases, have declined significantly (**Figure 20–39**).

Most of the carcinogenic factors that are known to be significant are by no means specific to the modern world. The most potent known carcinogen, by certain assays at least, is aflatoxin B1 (see Figure 20–38). It is produced by fungi that naturally contaminate foods such as tropical peanuts and is an important cause of liver cancer in Africa and Asia.

Except for tobacco, chemical toxins and mutagens are of lesser importance as contributory causes of cancer than other factors that are a matter of human behavior. For example, currently nearly three-fourths of adults and one-third of children and adolescents in the United States are overweight or obese. The combination of four risk factors—excess body weight, alcohol intake, poor diet, and physical inactivity—now accounts for the highest proportion of all cancer cases in women and is second only to smoking in men. Thus, it is estimated that as many

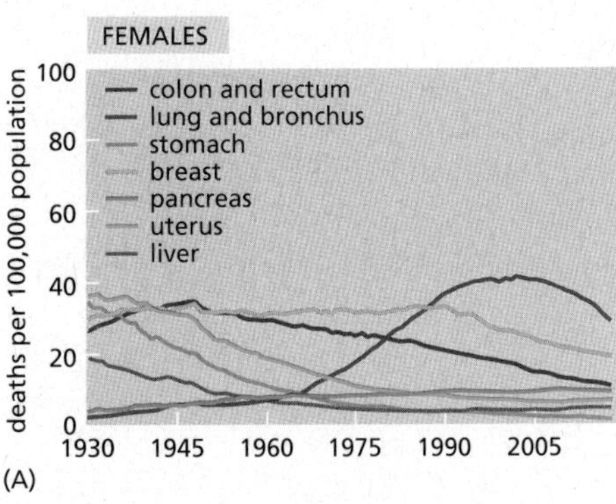

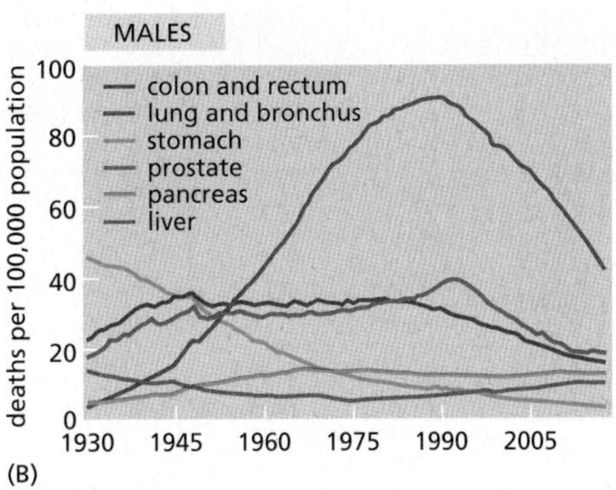

Figure 20–39 **Age-adjusted cancer death rates, United States, 1930–2018.** Selected death rates, adjusted to the age distribution of the US population, are plotted for (A) females and (B) males. Note the dramatic rise in lung cancer for both sexes, following the pattern of tobacco smoking, and the fall in deaths from stomach cancer, thought to be related to a fall in rates of infection with *Helicobacter pylori*. Recent reductions in other cancer death rates may correspond to improvements in detection and treatment. Adjustment of the data for the age of individuals is needed to compensate for the inevitable increase in cancer as people live longer, on average. (Adapted from American Cancer Society. Cancer Facts and Figures, 2021. Atlanta: American Cancer Society, Inc. Data from US Mortality Volumes 1930 to 1959, US Mortality Data 1960 to 2016, National Center for Health Statistics, Centers for Disease Control and Prevention. Note that uterus includes uterine cervix and uterine corpus combined due to limitations in coding prior to 1970. With permission from American Cancer Society.)

as 50% of all cancers could be avoided by identifiable changes in lifestyle (see Figure 20–37B).

Viruses and Other Infections Contribute to a Significant Proportion of Human Cancers

Human cancers are not contagious, but a significant minority of cancers are related directly or indirectly to infections with viruses and, less frequently, bacteria or parasites. Such infections account for approximately 15% of cancers worldwide and are more common in developing countries. Evidence for their involvement comes partly from the detection of viruses in individuals with cancer and partly from epidemiology. Thus, cancer of the uterine cervix is associated with infection with a papillomavirus, while liver cancer is very common in parts of the world (Africa and Southeast Asia) where hepatitis-B viral infections are common. Chronic infection with hepatitis-C virus, which has infected 170 million people worldwide, is also clearly associated with the development of liver cancer.

The main culprits, as shown in Table 20–2, are the DNA viruses. The **DNA tumor viruses** cause cancer by the most direct route—by interfering with controls of the cell cycle and apoptosis. To understand this type of viral carcinogenesis, it is important to review the life history of viruses. Many DNA viruses use the host cell's DNA replication machinery to replicate their own genomes. However, to produce a large number of infectious virus particles within a single host cell, the DNA virus has to commandeer this machinery and drive it hard, breaking through the normal constraints on DNA replication and usually killing the host cell in the process. Many DNA viruses reproduce only in this way. But some have a second option: they can propagate their genome as a quiet, well-behaved passenger in the host cell, replicating in parallel with the host cell's DNA (either integrated into the host genome or as an extrachromosomal plasmid) in the course of ordinary cell-division cycles. These viruses will switch between two modes of existence according to circumstances, remaining latent and harmless for a long time, but then proliferating in occasional cells in a process that kills the host cell and generates large numbers of infectious particles.

Neither of these conditions converts the host cell to a cancerous character, nor is it in the interest of the virus to do so. But for viruses with a latent phase, accidents can occur that prematurely activate some of the viral proteins that the

TABLE 20–2 Viruses Associated with Human Cancers

Virus	Associated cancer	Areas of high incidence
	DNA viruses	
Papovavirus family		
Papillomavirus (many distinct strains)	Warts (benign)	Worldwide
Papillomavirus (many distinct strains)	Carcinoma of the uterine cervix	Worldwide
Hepadnavirus family		
Hepatitis-B virus	Liver cancer (hepatocellular carcinoma)	Southeast Asia, tropical Africa
Herpesvirus family		
Epstein–Barr virus	Burkitt's lymphoma (cancer of B lymphocytes)	West Africa, Papua New Guinea
Epstein–Barr virus	Nasopharyngeal carcinoma	Southern China, Greenland
Human herpesvirus 8	Kaposi's sarcoma	Central and southern Africa
	RNA viruses	
Retrovirus family		
Human T-cell leukemia virus type I (HTLV-1)	Adult T-cell leukemia/lymphoma	Japan, West Indies
Human immunodeficiency virus (HIV, the AIDS virus)	Kaposi's sarcoma (via human herpesvirus 8)	Central and southern Africa
Flavivirus family		
Hepatitis-C virus	Liver cancer (hepatocellular carcinoma)	Worldwide

For all these viruses, the number of people infected is much larger than the number who develop cancer: the viruses must act in conjunction with other factors. As described in the text, different viruses contribute to cancer in different ways.

virus would normally use in its replicative phase to allow the viral DNA to replicate independently of the cell cycle. As in the case of papillomavirus described below, this type of accident can switch on the persistent proliferation of the host cell itself, leading to cancer.

Cancers of the Uterine Cervix Can Be Prevented by Vaccination Against Human Papillomavirus

The **papillomaviruses** are a prime example of DNA tumor viruses. They are responsible for human warts and are especially important as a cause of carcinoma of the uterine cervix: this is the second most common cancer of women in the world as a whole, representing about 6% of all human cancers. Human papillomaviruses (**HPVs**) infect the cervical epithelium and maintain themselves in a latent phase in the basal layer of cells as extrachromosomal plasmids, which replicate in step with the chromosomes. Infectious virus particles are generated through a switch to a replicative phase in the outer epithelial layers, as progeny of these cells begin to differentiate before being sloughed from the surface. Here, cell division should normally stop, but the virus interferes with this cell-cycle arrest so

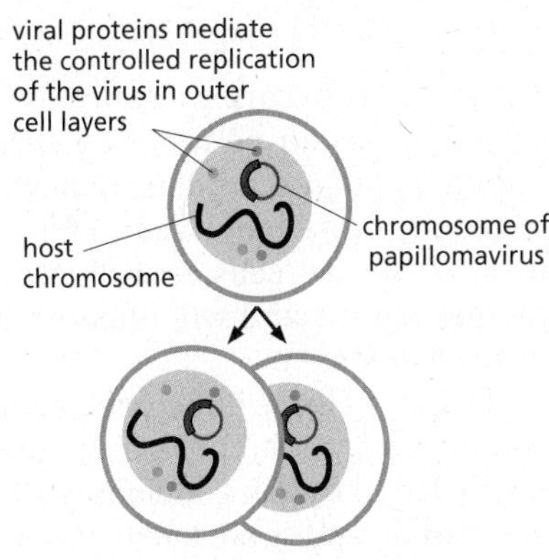

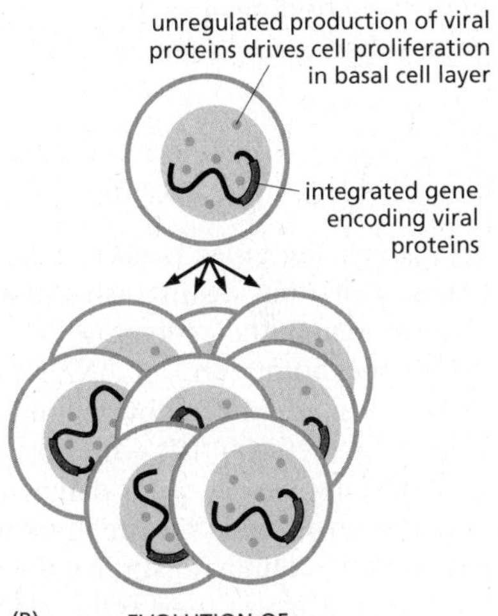

Figure 20–40 **How certain papillomaviruses are thought to give rise to cancer of the uterine cervix.** Papillomaviruses have double-stranded circular DNA chromosomes of about 8000 nucleotide pairs. These chromosomes are normally stably maintained in the basal cells of the epithelium as plasmids *(red circles)*, whose replication is regulated so as to keep step with the chromosomes of the host. (A) Normally, the virus perturbs the host cell cycle only when the virus is programmed to produce infectious progeny, in the outer layers of an epithelium. This is relatively harmless. (B) Rare accidents can cause the integration of a fragment of such a plasmid into a chromosome of the host, altering the environment of the viral genes in the basal cells of an epithelium. This can disrupt the normal control of viral gene expression. The unregulated production of certain viral proteins (named E6 and E7) interferes with the control of cell division in the basal cells, thereby helping to generate a cancer *(bottom)*.

as to allow replication of its own genome. Usually, the effect is restricted to the outer layers of cells and is relatively harmless, as in a wart. Occasionally, however, a genetic accident causes the viral genes that encode the proteins that prevent cell-cycle arrest to integrate into the host chromosome and become active in the basal layer, where the stem cells of the epithelium reside (see Figure 22–6). This can lead to cancer, with the viral genes acting as oncogenes (**Figure 20–40**).

The whole process, from initial infection to invasive cancer, is slow, taking many years. It involves a long intermediate stage when the affected patch of cervical epithelium is visibly disordered but the cells have not yet begun to invade the underlying connective tissue—a phenomenon called *intraepithelial neoplasia.* Many such lesions regress spontaneously. Moreover, at this stage, it is still easy to cure the condition by destroying or surgically removing the abnormal tissue. Fortunately, the presence of such lesions can be detected by scraping off a sample of cells from the surface of the cervix and viewing it under the microscope (the *Pap smear* technique).

Better still, a vaccine has now been developed that protects against infection with the relevant strains of human papillomavirus. This vaccine, if given to girls before puberty and thus before they become sexually active, has been shown to greatly reduce their risk of ever developing cervical cancer. Because the virus spreads through sexual activity, it is now recommended that both young males and young females be routinely vaccinated. Mass immunization programs have begun in several countries.

Infectious Agents Can Cause Cancer in a Variety of Ways

In papillomaviruses, the viral genes that are mainly to blame are called *E6* and *E7.* The protein products of these viral oncogenes interact with many host-cell proteins, but, in particular, they bind to two key tumor suppressor proteins of the host cell, putting them both out of action and so permitting the cell to replicate its DNA and divide in an uncontrolled way. One of these host proteins is Rb; the other is p53. Other DNA tumor viruses use similar mechanisms to inhibit Rb and p53, underlining the central importance of altering both the cell-cycle control pathway and the cell survival pathway if a cell is to escape the normal constraints on proliferation, as we have previously discussed.

In other cancers, viruses have indirect tumor-promoting actions. The hepatitis-B and hepatitis-C viruses, for example, favor the development of liver cancer by causing chronic inflammation (hepatitis), which stimulates extensive cell division in the liver that promotes the eventual evolution of tumor cells. In AIDS, the human immunodeficiency virus (HIV) promotes development of an otherwise rare cancer called Kaposi's sarcoma by destroying the immune system, thereby permitting a secondary infection with a human herpesvirus (HHV-8) that has a direct carcinogenic action. By causing severe inflammation, chronic infection with parasites and bacteria can also promote the development of some cancers. For example, chronic infection of the stomach with the bacterium *Helicobacter pylori,* which causes ulcers, appears to be a major cause of stomach cancer; dramatic falls in the incidence of stomach cancer over the past half-century (see Figure 20–39) correlate with a decline in the incidence of *Helicobacter* infections.

The Search for Cancer Cures Is Difficult but Not Hopeless

The difficulty of curing a cancer is similar to the difficulty of getting rid of weeds. Cancer cells can be removed surgically or destroyed with toxic chemicals or radiation, but it is hard to eradicate every single one of them. Surgery can rarely ferret out every metastasis, and treatments that kill cancer cells are generally toxic to normal cells as well. Moreover, unlike normal cells, cancer cells can mutate rapidly and will often evolve resistance to the poisons and irradiation used against them.

In spite of these difficulties, effective cures using anticancer drugs (alone or in combination with other treatments) have already been found for some formerly highly lethal cancers, including Hodgkin's lymphoma, testicular cancer, choriocarcinoma, and some leukemias and other cancers of childhood. Even for types of cancer where a cure at present seems beyond our reach, there are treatments that will prolong life or at least relieve distress. But what prospect is there of doing better and finding cures for the most common forms of cancer, which still cause great suffering and so many deaths?

Traditional Therapies Exploit the Genetic Instability and Loss of Cell-Cycle Checkpoint Responses in Cancer Cells

Anticancer therapies need to take advantage of some molecular peculiarity of cancer cells that distinguishes them from normal cells. One such property is genetic instability, reflecting deficiencies in chromosome maintenance, cell-cycle checkpoints, and/or DNA repair. Remarkably, the most widely used cancer therapies seem to work by exploiting these abnormalities, although this was not known by the scientists who first developed the treatments. Ionizing radiation and most anticancer drugs damage DNA or interfere with chromosome segregation at mitosis, and they preferentially kill cancer cells because cancer cells have a diminished ability to survive the damage. Normal cells treated with radiation, for example, arrest their cell cycle until they have repaired the damage to their DNA, thanks to the cell-cycle checkpoint responses discussed in Chapter 17. Because cancer cells generally have defects in their checkpoint responses, they may continue to divide after irradiation, only to die after a few days because the genetic damage remains unrepaired. More generally, most cancer cells are physiologically deranged to a stressful degree: they live dangerously. Even though the cells in a tumor have evolved to be unusually tolerant of minor DNA damage, they are hypersensitive to the much greater amount of damage that can be created by radiation and by DNA-damaging drugs. A small increase of genetic damage can be enough to tip the balance between proliferation and death. Unfortunately, however, DNA-damaging therapies typically cause terrible side effects on normal, rapidly dividing cells in the digestive system, bone marrow, and mucous membranes. Furthermore, these treatments are themselves carcinogenic and can lead to second cancers.

Furthermore, while the molecular abnormalities present in cancer cells often enhance their sensitivity to cytotoxic agents, they can also increase their resistance. For example, where a normal cell might die by apoptosis in response to DNA damage, thanks to the stress response mediated by p53, a cancer cell may escape apoptosis because it lacks p53. Cancers vary widely in their sensitivity to cytotoxic treatments, some responding to one drug, some to another, probably reflecting the particular kinds of defects that a particular cancer has in its DNA repair processes, cell-cycle checkpoints, and control of apoptosis.

New Drugs Can Kill Cancer Cells Selectively by Targeting Specific Mutations

Radiotherapy and traditional cytotoxic drugs are rather weakly selective: they hurt normal cells as well as the cancer cells, and the safety margin is narrow. The dose often cannot be raised high enough to kill all the cancer cells, because this would kill the patient, and curative treatments, where achievable, generally require a combination of several cytotoxic agents. The side effects can be harsh and hard to endure. How can we do better?

An ideal treatment is one that is cell-lethal in combination with some lesion that is present in the cancer cells, but harmless to cells where this lesion is absent. Such a treatment is said to be *synthetic-lethal* (from the original sense of the word *synthesis*, meaning "putting together"): it kills only in partnership with the cancer-specific mutation. As we become increasingly able to pinpoint the specific molecular alterations in cancer cells that make them different from their normal neighbors, new opportunities for such precisely targeted treatments are coming into view. We end this chapter with some examples of new treatments that are already being put into practice.

PARP Inhibitors Kill Cancer Cells That Have Defects in *Brca1* or *Brca2* Genes

As we have emphasized, the genetic instability of cancer cells makes the cells both dangerous and vulnerable—dangerous because of their enhanced ability to evolve and proliferate, and vulnerable because treatment that leads to still more extreme genetic disruption can take them over the brink and kill them. In some cancers, genetic instability results from an identified fault in one of the many devices on which normal cells depend for DNA repair and maintenance. In this case, a drug tailored to block a complementary part of the DNA repair machinery can lead to such severe genetic damage that the cancer cells die.

Detailed studies of the mechanisms for DNA maintenance discussed in Chapter 5 reveal a surprising amount of apparent redundancy. Thus, knocking out a particular pathway for DNA repair is generally less disastrous than one might expect, because alternate repair pathways exist. For example, stalled DNA replication forks can arise when the fork encounters a single-strand break in a template strand, but cells can avoid the disaster that would otherwise result either by directly repairing these single-strand breaks or, if that fails, by using homologous recombination to repair the broken fork (see Figure 5–49). Suppose that the cells in a particular cancer have become genetically unstable by acquiring a mutation that reduces their ability to repair broken replication forks by homologous recombination. Might it be possible to eradicate that cancer by treating it with a drug that inhibits the repair of single-strand breaks, thereby greatly increasing the number of forks that break? The consequences of such drug treatment might be expected to be relatively harmless for normal cells, but lethal for the cancer.

This strategy appears to work to kill the cells in at least one class of cancers—those that have inactivated both copies of either their *Brca1* or their *Brca2* tumor suppressor genes. As described in Chapter 5, Brca2 is an accessory protein that interacts with the Rad51 protein (the RecA analog in humans) in the repair of DNA double-strand breaks by homologous recombination. Brca1 is another protein that is also required for this repair process. Like *Rb*, the *Brca1* and *Brca2* genes were discovered as mutations that predispose humans to cancer—in this case, chiefly cancers of the breast and ovaries (though unlike *Rb*, they seem to be involved in only a small proportion of such cancers). Individuals who inherit one mutant copy of *Brca1* or *Brca2* develop tumors that have inactivated the second copy of the same gene, presumably because this change makes the cells genetically unstable and speeds tumor progression.

While Brca1 and Brca2 are needed for the repair of DNA double-strand breaks, single-strand breaks are repaired by other machinery, involving an enzyme called PARP (polyADP-ribose polymerase). This understanding of the basic mechanisms of DNA repair led to a striking discovery: drugs that block PARP activity kill *Brca*-deficient cells with extraordinary selectivity. At the same time, PARP inhibition has very little effect on normal cells; in fact, mice that have been engineered to lack PARP1—the major PARP family member involved in DNA repair—remain healthy under laboratory conditions. This result suggests that,

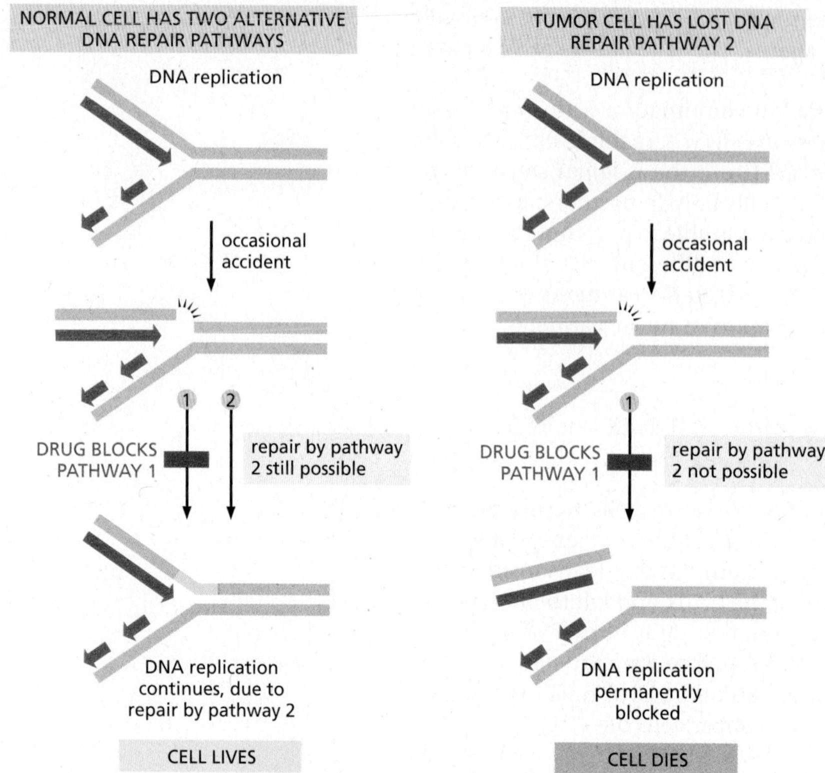

Figure 20–41 **How a tumor's genetic instability can be exploited for cancer therapy.** As explained in Chapter 5, the maintenance of DNA sequences is so critical for life that cells have evolved multiple pathways for repairing DNA damage and reducing DNA replication errors. As illustrated, a DNA replication fork will stall whenever it encounters a break in a DNA template strand. In this example, normal cells have two different repair pathways that help them to avoid the problem, pathways 1 and 2. They are therefore not harmed by treatment with a drug that blocks repair pathway 1. But, because the inactivation of repair pathway 2 was selected for during the evolution of the tumor cell, the tumor cells are killed by the same drug treatment.

In the actual case that underlies this example, the function of repair pathway 1 (requiring the PARP protein discussed in the text) is to remove persistent, accidental breaks in a DNA single strand before they are encountered by a moving replication fork. Pathway 2 is the recombination-dependent process (requiring the Brca2 and Brca1 proteins) for repairing stalled replication forks illustrated in Figure 5–49. PARP inhibitors are often used for treating cancers with defective *Brca2* or *Brca1* tumor suppressor genes.

while the repair pathway requiring PARP provides a first line of defense against persistent breaks in a DNA strand, these breaks can be repaired efficiently by a genetic recombination pathway in normal cells. In contrast, tumor cells that have acquired their genetic instability by the loss of Brca1 or Brca2 have lost this second line of defense, and they are therefore uniquely sensitive to PARP inhibitors (Figure 20–41).

PARP inhibitors have produced some striking results, causing tumors to regress in many Brca-deficient patients and delaying progression of their disease, with relatively few disagreeable side effects. These drugs also appear to be applicable to cancers with other mutations that cause defects in the cell's homologous recombination machinery—a small, though significant, proportion of cancer cases.

PARP inhibition provides an example of the type of rational, highly selective approach to cancer therapy that is beginning to be possible. Along with other new treatments to be discussed below, it raises high hopes for treating many other cancers.

Small Molecules Can Be Designed to Inhibit Specific Oncogenic Proteins

An obvious tactic for treating cancer is to attack a tumor expressing an oncogene with a drug designed to specifically block the function of the protein that the oncogene produces. But how can such a treatment avoid hurting the normal cells that depend on the function of the proto-oncogene from which the oncogene has evolved, and why should the drug kill the cancer cells, rather than simply calm them down? One answer may lie in the phenomenon of *oncogene dependence.* Once a cancer cell has undergone an oncogenic mutation, it will often undergo further mutations, epigenetic changes, or physiological adaptations that make it reliant on the hyperactivity of the initial oncogene, just as drug addicts become reliant on high doses of their drug. Blocking the activity of the oncogenic protein may then kill the cancer cell without significantly harming its normal neighbors. Some remarkable successes have been achieved in this way.

As we saw earlier, chronic myelogenous leukemia (CML) is usually associated with a particular chromosomal translocation, visible as the Philadelphia chromosome (see Figure 20–5). This results from chromosome breakage and rejoining at the sites of two specific genes, *Abl* and *Bcr*. The fusion of these genes creates a hybrid gene, called *Bcr-Abl*, that codes for a chimeric protein consisting of the N-terminal fragment of Bcr fused to the C-terminal portion of Abl (**Figure 20–42**). Abl is a tyrosine kinase involved in cell signaling. The substitution of the Bcr fragment for the normal N-terminus of Abl makes it hyperactive, so that it stimulates inappropriate proliferation of the hemopoietic precursor cells that contain it and prevents these cells from dying by apoptosis—which many of them would normally do. As a result, excessive numbers of white blood cells accumulate in the bloodstream, producing CML.

The chimeric Bcr-Abl protein is an obvious target for therapeutic attack. Searches for synthetic drug molecules that can inhibit the activity of tyrosine kinases discovered one, called *imatinib* (trade name Gleevec), that blocks Bcr-Abl (**Figure 20–43**). When the drug was first given to patients with CML, nearly

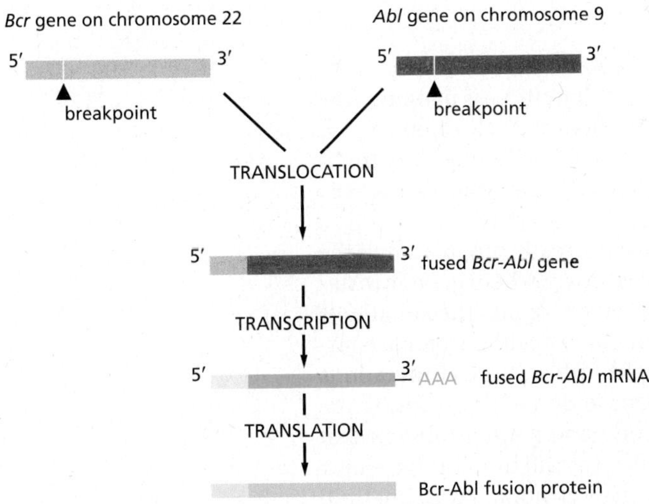

Figure 20–42 **The conversion of the *Abl* proto-oncogene into an oncogene in individuals with chronic myelogenous leukemia.** The chromosome translocation responsible joins the *Bcr* gene on chromosome 22 to the *Abl* gene from chromosome 9, thereby generating a Philadelphia chromosome (see Figure 20–5). The resulting fusion protein has the N-terminus of the Bcr protein joined to the C-terminus of the Abl tyrosine protein kinase; in consequence, the Abl kinase domain becomes inappropriately active, driving excessive proliferation of a clone of hemopoietic cells in the bone marrow.

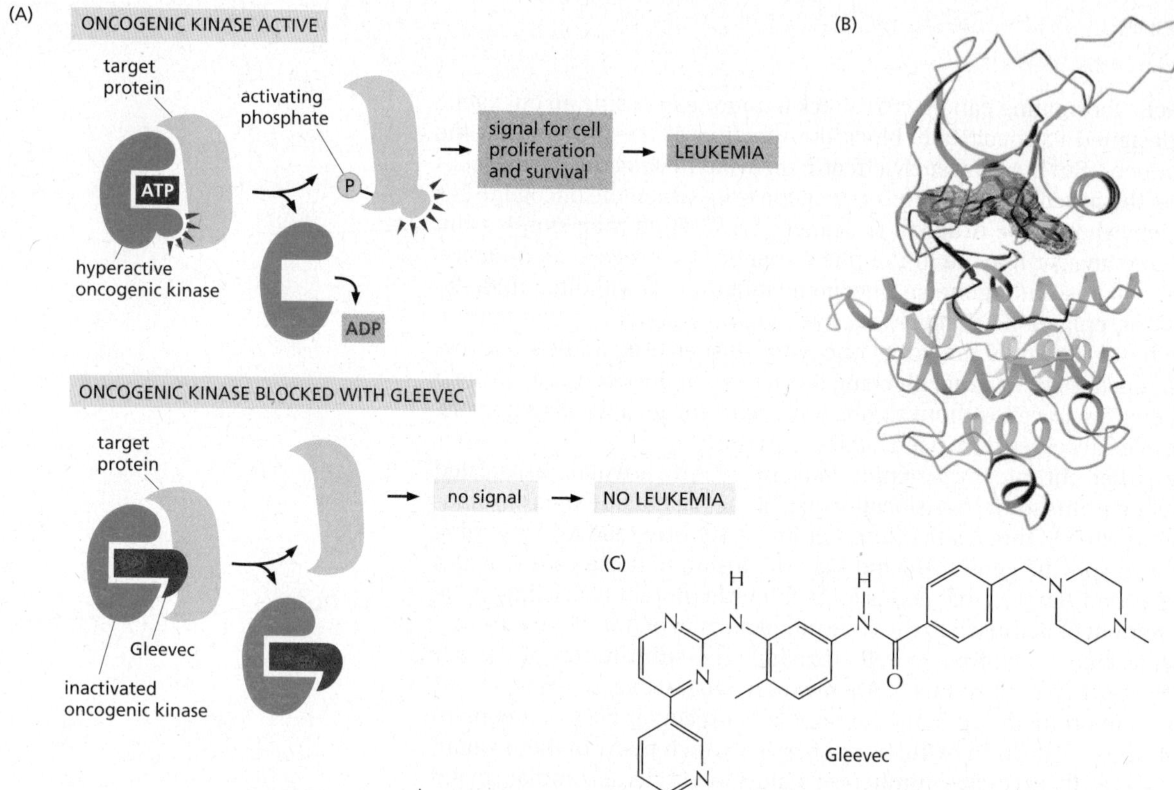

Figure 20–43 **How imatinib (Gleevec) blocks the activity of Bcr-Abl protein and halts chronic myelogenous leukemia.** (A) Imatinib sits in the ATP-binding pocket of the tyrosine kinase domain of Bcr-Abl and thereby prevents Bcr-Abl from transferring a phosphate group from ATP onto a tyrosine residue in a substrate protein. This blocks transmission of a signal for cell proliferation and survival. (B) The structure of the complex of imatinib (solid *blue* object) with the tyrosine kinase domain of the Abl protein (ribbon diagram), as determined by x-ray crystallography. (C) The chemical structure of the drug. It can be administered orally; it has side effects, but they are usually quite tolerable. (B, from T. Schindler et al., *Science* 289:1938–1942, 2000. With permission from AAAS. PDB code: 3K5V.)

all of them showed a dramatic response, with an apparent disappearance of the cells carrying the Philadelphia chromosome in more than 80% of patients. The response appears relatively durable: after years of continual treatment, many patients have not progressed to later stages of the disease—although imatinib-resistant cancers emerge with a probability of about 5% per year during the early years.

Results are not so good for those patients who have already progressed to the more acute phase of myeloid leukemia, known as blast crisis, where genetic instability has set in and the march of the disease is far more rapid. These patients show a response at first and then relapse because the cancer cells develop a resistance to imatinib. This resistance is usually associated with secondary mutations in the part of the *Bcr-Abl* gene that encodes the kinase domain, disrupting the ability of imatinib to bind to Bcr-Abl kinase. Second-generation inhibitors that function effectively against a whole range of imatinib-resistant mutants have now been developed. By combining one or more of these new inhibitors with imatinib as the initial therapy, it seems that CML—at least in the chronic (early) stage—may be on its way to becoming a curable disease.

Despite the complications with resistance, the extraordinary success of imatinib is enough to drive home an important principle: once we understand precisely what genetic lesions have occurred in a cancer, we can begin to design effective rational methods to treat it. This success story has fueled efforts to identify small-molecule inhibitors for other oncogenic protein kinases and to use them to attack the appropriate cancer cells. Increasing numbers are being developed. These include molecules that target the EGF receptor and are currently approved

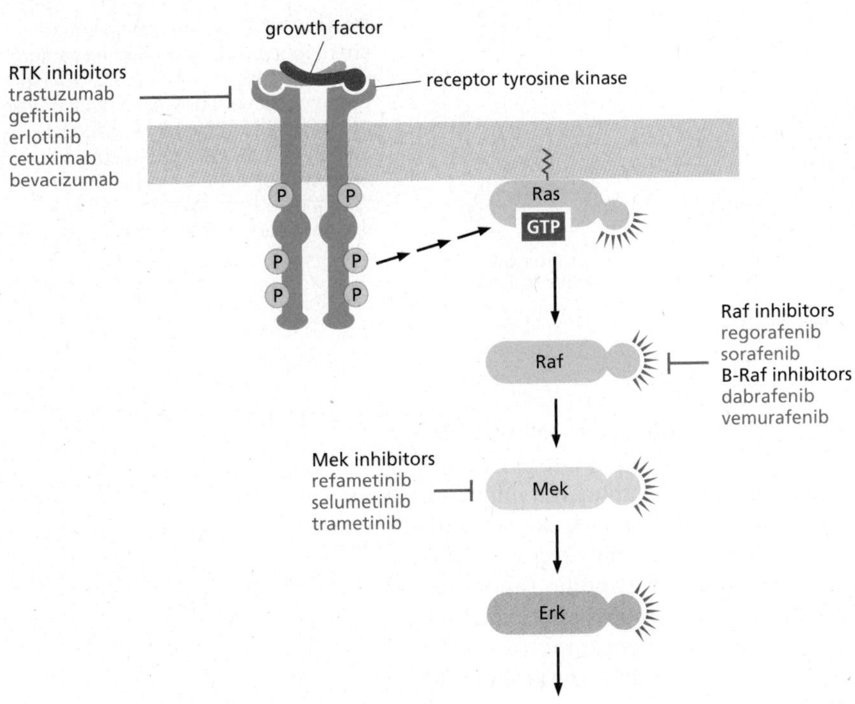

Figure 20–44 **Some anticancer drugs and drug targets in the Ras–MAP-kinase signaling pathway.** This Ras–MAP-kinase signaling pathway is triggered by a variety of receptor tyrosine kinases (RTKs), including the EGF receptor (see Figures 15–48 and 15–50. Raf kinase inhibitors include several that specifically target the oncoprotein B-Raf. By convention, those drugs that are antibodies end in "mab," while those that are small molecules end in "nib." (Adapted from B. Vogelstein et al., *Science* 339:1546–1558, 2013.)

for the treatment of some lung cancers, as well as drugs that specifically target the B-Raf oncoprotein in melanomas.

Protein kinases have been relatively easy to inhibit with small molecules such as imatinib, and many kinase inhibitors are being produced by pharmaceutical companies in the hope that they can be effective as drugs for some forms of cancer. Many cancers lack an oncogenic mutation in a protein kinase. But most tumors contain inappropriately activated signaling pathways, for which a target somewhere in the pathway can hopefully be found (Movie 20.8). As an example, Figure 20–44 displays some of the anticancer drugs and drug targets that are currently being tested for a pathway frequently activated in cancers.

Many Cancers May Be Treatable by Enhancing Immune Responses

Cancers have complex interactions with the immune system, and its various components may sometimes help as well as hinder tumor progression. But for more than a century it has been a dream of cancer researchers to somehow harness our immune systems in a controlled and efficient way to exterminate cancer cells, just as it exterminates pathogenic microorganisms. There are finally signs that this dream may one day be realized, at least for some forms of cancer.

The simplest type of immunological therapy, conceptually at least, is to inject the patient with antibodies that target the cancer cells. This approach has had some successes. About 25% of breast cancers, for example, express unusually high levels of the Her2 (human epidermal growth factor receptor 2) protein, a receptor tyrosine kinase related to the EGF receptor that plays a part in the normal development of mammary epithelium. A monoclonal antibody called *trastuzumab* (trade name Herceptin), which binds to Her2 and inhibits its function, slows the growth of human breast cancers that overexpress Her2, and it is now a standard therapy for these cancers (see Figure 20–44). A related approach uses antibodies to deliver poisons to the cancer cells. Antibodies against proteins that are abundant on the surface of a particular type of cancer cell but rare on normal cells can be coupled to a toxin that kills the cells that the antibody binds to.

A great deal of current excitement centers around a different type of immunological approach, which is based on a class of lymphocytes called T cells. As

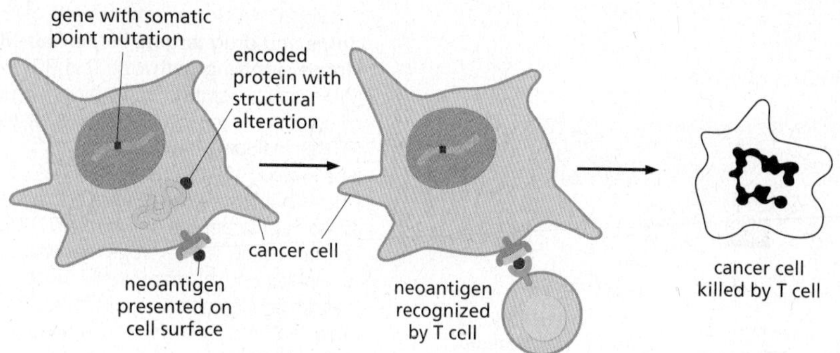

Figure 20–45 Tumor-specific antigens are recognized by the immune system. Due to somatic mutation, cells in tumors will produce many different mutant proteins. As described in Chapter 24, peptides from these proteins will be displayed as neoantigens on the tumor cell surface and have the potential to activate a T cell response that kills the cell (see Figure 24–42).

described in detail in Chapter 24, cytotoxic T cells can kill a host cell that displays foreign peptide antigens on its surface. Because the peptide is usually derived from an invading pathogen, the response helps terminate the infection. The challenge is to find ways of recruiting cytotoxic T cells to attack cancers with similar efficiency and specificity, presuming that the cancer cells express tumor-specific antigens. Recall that, from the thousands of tumor genome sequences thus far determined, we know that a typical cancer cell contains approximately 50 proteins with a mutation that alters an amino acid sequence, most of these being passenger mutations, as previously explained (see p. 1186). The amino acid changes resulting from either passenger or driver mutations have the potential to produce cell-surface *neoantigens* that can be recognized as foreign by T cells, resulting in cancer cell death (**Figure 20–45**). However, because cytotoxic T cells kill infected host cells, they are under tight control to keep their activity within safe bounds. Furthermore, cancer cells typically lack some of the components present in pathogen-infected cells that are necessary to initiate a robust T cell response. Thus, very few tumor-reactive T cells exist, and therapies to boost their numbers are at the forefront of immunotherapy.

One approach to augment the T cell response to cancer cells is to collect a patient's own T cells from his or her blood, expand the cancer-reactive population, and then re-infuse them into the patient. This method of introducing a large number of T cells that already recognize the tumor cells bearing cancer-specific antigens has been very effective in metastatic melanoma patients, achieving complete remission rates as high as 20%. In an even more elaborate variation of this strategy, the patient's T cells can be genetically engineered before their numbers are increased, in a procedure called chimeric antigen receptor T cell (CAR T) therapy (**Movie 20.9**). The engineered CAR T cells not only recognize the patient's tumor-specific antigen but also possess co-stimulatory activities that boost the response of the T cells. This therapy is extremely effective against certain blood cancers, but, unfortunately, it also has severe side effects due to immune system hyperactivation.

Immunosuppression Is a Major Hurdle for Cancer Immunotherapy

If cancer cell neoantigens are recognized as foreign, why doesn't the immune system work better to eliminate the cancer in the first place? As discussed previously (see pp. 1175–1176), cancer cells manipulate their surrounding stroma, including immune cells, to create a microenvironment that simultaneously promotes tumor growth and suppresses immune responses. Investigation of how tumors escape immune destruction led to the discovery of crucial immunosuppressive mechanisms—the expression on the cancer cell surface of one or more proteins that bind to inhibitory receptors on various immune cells, including activated T cells. As a result, even if cytotoxic T cells recognize tumor antigens, they are prevented from killing the tumor cells. The inhibitory receptors expressed on the surface of T cells are part of the *immune checkpoint* that plays an important normal function in preventing excessive, tissue-damaging immune responses during infections. But, in the context of cancer, immune

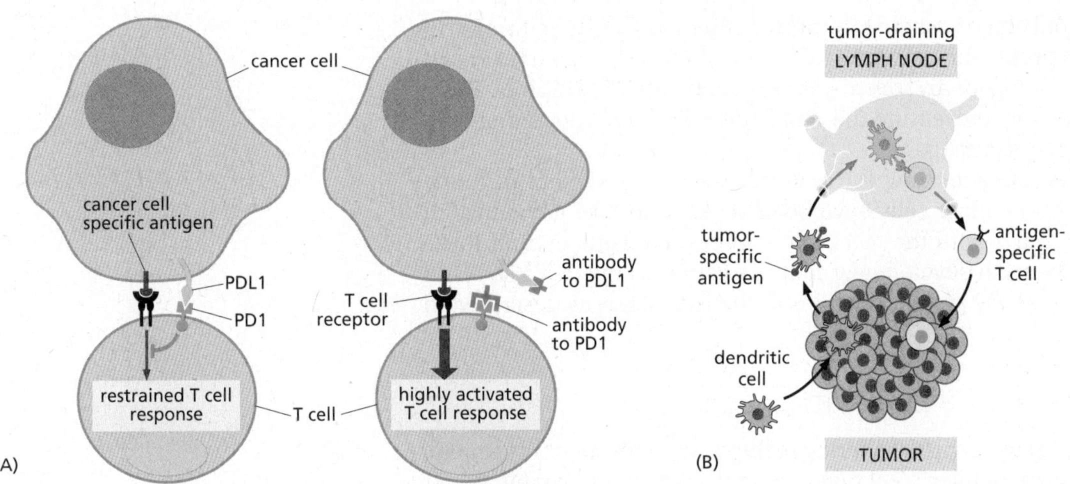

Figure 20–46 **How the immune system interacts with cancer.** (A) An antibody therapy for overcoming the immunosuppressive microenvironment in tumors. Cancer cells often protect themselves from immune attack by expressing proteins on their surface that bind to inhibitory receptors on T cells. In the example shown, the cancer cell expresses PDL1, which binds the PD1 receptor on a T cell and interferes with T cell activation. This makes the tumor susceptible to antibodies that unleash T cell attack. (B) Dendritic cells play a key role in activating T cells by acquiring tumor cell antigens from tumors and presenting them in lymph nodes, as illustrated, and some cancers have evolved mechanisms that interfere with this process (not shown).

checkpoint mechanisms can prevent an individual from killing the cancer cells that are threatening his or her survival.

To overcome the immunosuppressive environment and provide a strong T cell activating signal, a promising new type of anticancer therapy focuses on developing *immune checkpoint inhibitors* such as antibodies that prevent the tumor cells from engaging with the inhibitory receptors on T cells. As illustrated in **Figure 20–46A**, targeting inhibitory receptors or their ligands with antibodies can unleash an immune attack on the cancer cells. Importantly, multiple neoantigens are recognized as foreign, so that cancer cells cannot escape an immune attack through the mutational loss of a single neoantigen, making it difficult for the tumor to become resistant to the antibody treatment. A substantial fraction of metastatic melanoma patients injected repeatedly with monoclonal antibodies that bind to either one such receptor, PD1 (programmed cell death 1), or its ligand on the surface of cancer cells, PDL1, respond in a dramatic way, with their cancer being driven into remission for years. Unfortunately, the treatment fails to help others with the same type of cancer.

An important clue as to why immunotherapies are effective in some patients but not others comes from the extensive analysis of cancer genomes described in the second part of this chapter. A fortuitous property of tumor samples is that they contain not only the cancer cells but also the cells of the associated stroma, including fibroblasts and endothelial and various immune cells. Because each cell type expresses a signature set of RNAs, the presence and proportion of each cell type in the tumor can be assessed, including the T cells. A comparison of the immune cell profiles of thousands of different tumors has revealed interesting correlations. For example, as one might predict, cancers with genomes that contain large numbers of point mutations are frequently found to possess a higher proportion of T cells compared to cancers with genomes that contain fewer mutations, presumably due to an increase in neoantigens. In turn, this tumor profile correlates with better treatment outcomes after treatment with immune checkpoint inhibitors.

Some cancers appear to avoid immune destruction by preventing T cells from infiltrating the tumor in the first place. The cellular basis of this phenomenon is poorly understood and is an important ongoing area of investigation. It is known that T cell activation requires that T cells physically interact with antigen-presenting cells called dendritic cells. Dendritic cells are highly migratory and pick up pathogens or their products at sites of infection and deliver them to lymph nodes or other lymphoid organs, where they present them to T cells. In the case of cancer, dendritic cells present neoantigens to T cells, so that they can track down the cells expressing the antigens and destroy them (**Figure 20–46B**). Some cancer cells inhibit this process by preventing the dendritic cells from interacting with them and acquiring their neoantigens.

These observations illustrate how the immune system is subject to many levels of control and how cancer cells have been selected to take advantage of them. Importantly, research into the cell biological mechanisms behind these phenomena not only helps to develop new cancer treatments but will also teach us basic principles of how the system operates in the first place, as described in Chapter 24.

Cancers Evolve Resistance to Therapies

High hopes have to be tempered with sobering realities. We have seen that genetic instability can provide an Achilles' heel that cancer therapies can exploit, but at the same time it can make eradicating the disease more difficult by allowing the cancer cells to evolve resistance to therapeutic drugs, often at an alarming rate. This applies even to the drugs that target genetic instability itself. Thus, PARP inhibitors give valuable remission of illness, but in the long term the disease generally comes back. For example, *Brca*-deficient cancers can sometimes develop resistance to PARP inhibitors by undergoing a second mutation in an affected *Brca* gene that restores its function. By then, the cancer is already out of control and it may be too late to affect the course of the disease with additional treatments.

There are many different strategies by which cancers can evolve resistance to anticancer drugs. Often, a cancer will be dramatically reduced in size by an initial drug treatment, with all of the detectable tumor cells seeming to disappear. But months or years later the cancer will reappear in an altered form that is resistant to the drug that was at first so successful. In such cases, the initial drug treatment has evidently failed to destroy some tiny fraction of cells in the original tumor-cell population. These cells may have escaped death because they carry a protective mutation or epigenetic change or perhaps simply because they were lurking in a protected environment. Alternatively, and most insidiously, a phenotypically distinct subpopulation of tumor cells may function as cancer stem cells. The surviving cells eventually regenerate the cancer by continuing to proliferate, mutating and evolving still further as they do so. Combination therapies, in which the right two agents are used simultaneously to target the same cancer cells, should in principle help greatly with such problems.

However, in some cases, cells that are exposed to one anticancer drug evolve a resistance not only to that drug but also to other drugs to which they have never been exposed. This phenomenon of **multidrug resistance** frequently correlates with amplification of a part of the genome that contains a gene called *Mdr1* or *Abcb1*. This gene encodes a plasma membrane–bound transport ATPase of the ABC transporter superfamily (discussed in Chapter 11), which pumps lipophilic drugs out of the cell (see Movie 11.5). The overproduction of this protein (or some of its other family members) by a cancer cell can prevent the intracellular accumulation of many cytotoxic drugs, making the cell insensitive to them.

In the to-and-fro struggle between advanced metastatic cancer and the therapist, as current practice stands, the cancer usually wins in the end. Does it have to be so? As we discuss below, there is reason to think that by attacking a cancer with many weapons at once—instead of using them one after another, each until it fails—it may be possible to do much better.

We Now Have the Tools to Devise Combination Therapies Tailored to the Individual

Nowadays, cancers caught at an early stage can often be cured by surgery, radiation, or drugs. For most cancers that have progressed and metastasized widely, however, cure is still frequently beyond us. Treatments such as those described earlier can give valuable remissions, but sooner or later these are typically followed by relapse.

Nevertheless, for some forms of advanced cancer, curative therapies are being developed that utilize multiple strategies; for example, combining chemotherapy and radiation therapy with immunotherapy. Ideally, the choice of drugs to be given in combination should be tailored to the individual patient. Cancers evolve by a fundamentally random process, and each individual is different, but modern methods of genome analysis now let us characterize the cells from a tumor biopsy in exhaustive detail so as to discover which cancer-critical genes are affected in a particular case. Admittedly, this is not straightforward: the tumor cells in an individual are heterogeneous and do not all contain the same genetic lesions. With increased understanding of the pathways of cancer evolution, however, and with the experience gained from many different cases, it should become possible to make informed decisions about the optimal therapies to use.

Improvements in tissue sampling, genomics, and biostatistics have enabled the direct characterization of primary human tumors. However, these analyses do not take into account the contributions of the entire body. Advances in cancer therapy are also coming from the development of more sophisticated models of the disease, which can be used to investigate the factors involved in tumor growth and metastasis, as well as to examine response to therapy. In one approach, mice can be genetically engineered to introduce particular combinations of human mutations (see Figure 20–33). A second approach is the human tumor xenograft. In this model, human cancer cells are transplanted—either under the skin or into the organ type in which the cancer originated—into immunocompromised mice that do not reject human cells. In an approach that bypasses the use of mice, organoids can be grown in 3D culture from patient-derived healthy and tumor tissues (see Figure 20–34A). These systems facilitate patient-specific drug testing and the development of individualized treatment regimens.

From the perspective of the patient, the pace of advance in cancer research can seem frustratingly slow. Each new drug has to be tested in the clinic, first for safety and then for efficacy, before it can be released for general use. And if the drug is to be used in combination with others, the combination therapy must then go through the same long process. Strict ethical rules constrain the conduct of trials, which means that they take time—typically several years. But slow and cautious steps, taken systematically in the right direction, can lead to great advances. There is still far to go, but the examples that we have discussed provide proof of principle and grounds for optimism.

From the cancer research effort, we have learned a great deal of what we know about the molecular biology of the normal cell. Now, more and more, we are discovering how to put that knowledge to use in the battle with cancer itself.

Summary

Our growing understanding of the cell biology of cancers has already begun to lead to better ways of preventing, diagnosing, and treating these diseases. Anticancer therapies can be designed to destroy cancer cells preferentially by exploiting the properties that distinguish cancer cells from normal cells, including the cancer cells' dependence on oncogenic proteins and the defects they harbor in their DNA repair mechanisms. We now have good evidence that, by increasing our understanding of normal cell control mechanisms and exactly how they are subverted in specific cancers, we can eventually devise drugs to kill cancers precisely by attacking specific molecules critical for the proliferation and survival of the cancer cells. In addition, great progress has recently been made through sophisticated immunological approaches to cancer therapy. And, as we become better able to determine which genes are altered in the cells of any given tumor, we can begin to tailor treatments more accurately to each individual.

PROBLEMS

Which statements are true? Explain why or why not.

20–1 Cancer therapies directed solely at killing the rapidly dividing cells that make up the bulk of a tumor are unlikely to eliminate the cancer from many patients.

20–2 In the regulatory pathways that control cell growth and proliferation, the products of oncogenes are stimulatory components and the products of tumor suppressor genes are inhibitory components.

20–3 The chemical carcinogen dimethylbenz[*a*]anthracene (DMBA) must be an extraordinarily specific mutagen because 90% of the skin tumors it causes have an A-to-T alteration at exactly the same site in the mutant *Ras* gene.

20–4 The main environmental causes of cancer are the products of our highly industrialized way of life such as pollution and food additives.

Discuss the following problems.

20–5 What is the term for a cancer arising from epithelial cells?

20–6 In contrast to colon cancer, the incidence of which increases dramatically with age, the incidence of osteosarcoma—a tumor that occurs most commonly in the long bones—peaks during adolescence. Osteosarcomas are relatively rare in young children (up to age 9) and in adults (over age 20). Why do you suppose that the incidence of osteosarcoma does not show the same sort of age dependence as colon cancer?

20–7 Mortality due to lung cancer was followed in groups of males in the United Kingdom for 50 years. Figure Q20–1 shows the cumulative risk of dying from lung cancer as a function of age and smoking habits for four groups of males: those who never smoked, those who stopped at age 30, those who stopped at age 50, and those who continued to smoke. These data show clearly that individuals can substantially reduce their cumulative risk of dying from lung cancer by stopping smoking. What do you suppose is the biological basis for this observation?

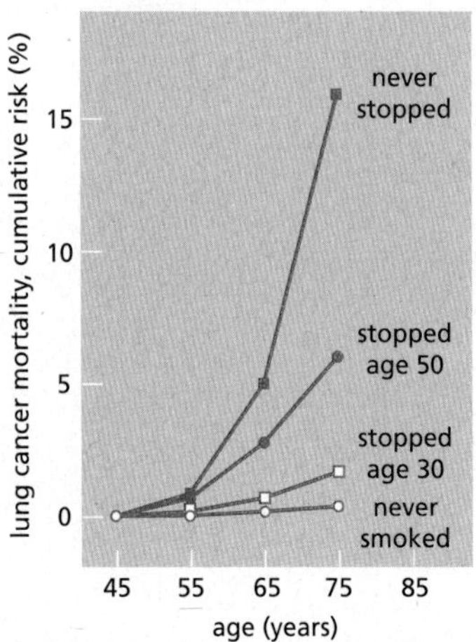

Figure Q20–1 Cumulative risk of lung cancer mortality for nonsmokers, smokers, and former smokers (Problem 20–7). Cumulative risk is the running total of deaths, as a percentage, for each group. Thus, for continuing smokers, 1% died of lung cancer between ages 45 and 55; an additional 4% died between 55 and 65 (giving a cumulative risk of 5%); and 11% more died between 65 and 75 (for a cumulative risk of 16%).

20–8 A small fraction—2 to 3%—of all cancers, across many subtypes, displays a quite remarkable phenomenon: tens to hundreds of rearrangements that primarily involve a single chromosome or chromosomal region. The breakpoints can be tightly clustered, with several in a few kilobases, and the junctions of the rearrangements often involve segments of DNA that were not originally close together on the chromosome. The copy number of various segments within the rearranged chromosome is found to be 0, indicating deletion, or 1, indicating retention.

You can imagine two ways in which such multiple, localized rearrangements might happen: a progressive rearrangements model with ongoing inversions, deletions, and duplications involving a localized area, or a catastrophic model in which the chromosome is shattered into fragments that are stitched back together in random order by a double-strand break repair process such as nonhomologous end joining (Figure Q20–2).

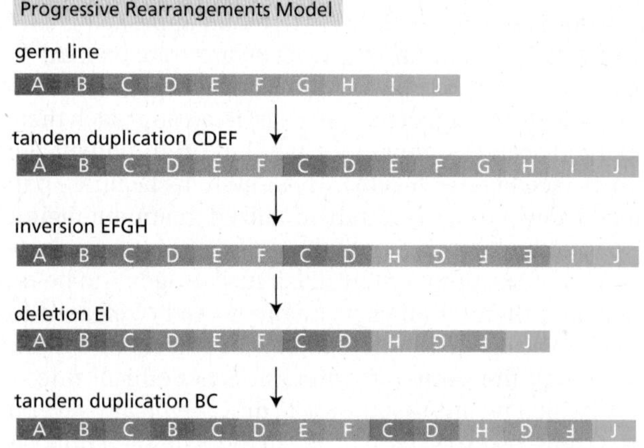

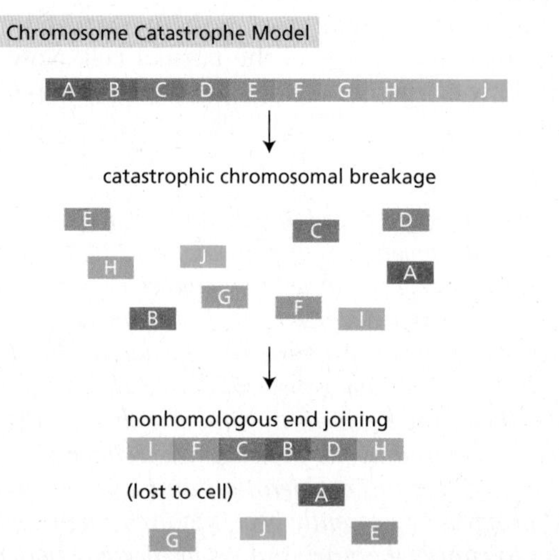

Figure Q20–2 Two models to explain the multiple, localized chromosome rearrangements found in some cancers (Problem 20–8). The progressive rearrangements model shows a sequence of rearrangements that disrupts the chromosome, generating increasingly complex chromosomal configurations. The chromosome catastrophe model shows the chromosome being fragmented and then reassembled randomly, with some pieces left out.

Which of the two models in Figure Q20–2 accounts more readily for the features of these highly rearranged chromosomes? Explain your reasoning.

20–9 The Tasmanian devil (Figure Q20–3A) is a carnivorous Australian marsupial that is threatened with extinction by the spread of a fatal disease in which a malignant oral-facial tumor interferes with the animal's ability to feed. You have been called in to analyze the source of this unusual cancer. It seems clear to you that the cancer is somehow spread from animal to animal, very likely by their frequent fighting, which is accompanied by biting around the face and mouth. To uncover the source of the cancer, you isolate tumors from 11 Tasmanian devils captured in widely separated regions and examine them. The karyotypes of nontumor cells from these Tasmanian devils are normal (Figure Q20–3B). As expected, the tumor cells are highly rearranged relative to the normal karyotype, but, surprisingly, the karyotypes from all 11 tumor samples are very similar (Figure Q20–3C). Moreover, one of the Tasmanian devils has an inversion on chromosome 5 that is not present in its facial tumor. How do you suppose this cancer is transmitted from animal to animal? Is it likely to arise as a consequence of an infection by a virus or microorganism? Explain your reasoning.

(A) Tasmanian devil (*Sarcophilus harrisii*)

(B) normal karyotype

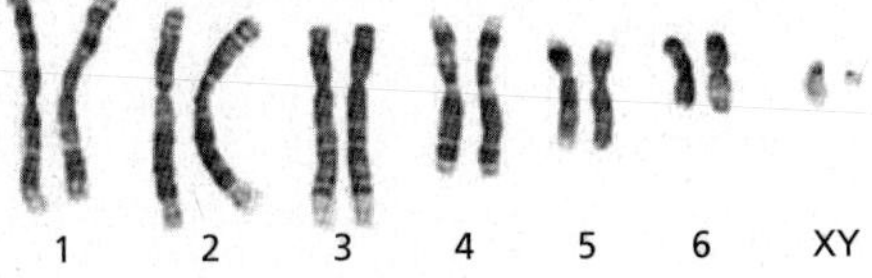

(C) tumor karyotype

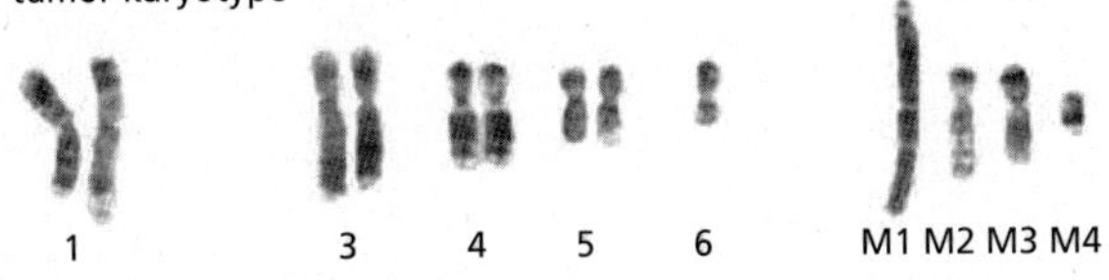

Figure Q20–3 Karyotypes of cells from Tasmanian devils (Problem 20–9). (A) A Tasmanian devil. (B) Normal karyotype for a male Tasmanian devil. The karyotype has 14 chromosomes, including XY. (C) Karyotype of cancer cells found in each of the 11 facial tumors studied. The karyotype has 13 chromosomes, no sex chromosomes, no chromosome 2 pair, one chromosome 6, two chromosomes 1 with deleted long arms, and four highly rearranged marker chromosomes (M1–M4). (A, reproduced courtesy of Museum Victoria; B and C, from A.-M. Pearse and K. Swift, *Nature* 439:549, 2006. Reproduced with permission from SNCSC.)

20–10 Virtually all cancer treatments are designed to kill cancer cells. However, one particular cancer—acute promyelocytic leukemia (APL), which is caused by too many immature blood-forming cells (promyelocytes) in the blood—has been successfully treated with all-*trans*-retinoic acid, which causes the promyelocytes to differentiate into neutrophils. How might a change in the state of differentiation of APL cancer cells help the patient?

20–11 PolyADP-ribose polymerase (PARP) plays a key role in the repair of DNA single-strand breaks. In the presence of the PARP inhibitor olaparib, single-strand breaks accumulate. When a replication fork encounters a single-strand break, it converts it to a double-strand break, which in normal cells is then repaired by homologous recombination. In cells defective for homologous recombination, however, inhibition of PARP triggers cell death.

Individuals who have only one functional copy of the *Brca1* gene, which is required for homologous recombination, are at much higher risk for cancer of the breast and ovary. Cancers that arise in these tissues in these individuals can be treated successfully with olaparib. Explain how it is that treatment with olaparib kills the cancer cells in these individuals but does not harm their normal cells.

20–12 One major goal of modern cancer therapy is to identify small molecules—anticancer drugs—that can be used to inhibit the products of specific cancer-critical genes. If you were searching for such molecules, would you design inhibitors for the products of oncogenes or the products of tumor suppressor genes? Explain why you would (or would not) select each type of gene.

REFERENCES

General

Bishop JM (2004) How to Win the Nobel Prize: An Unexpected Life in Science. Cambridge, MA: Harvard University Press.

Hanahan D & Weinberg RA (2011) Hallmarks of cancer: the next generation. *Cell* 144, 646–674.

Vogelstein B, Papadopoulos N, Velculescu VE . . . Kinzler KW (2013) Cancer genome landscapes. *Science* 339, 1546–1558.

Weinberg RA (2013) The Biology of Cancer, 2nd ed. New York: Garland Science.

Cancer as a Microevolutionary Process

Batlle E & Clevers H (2017) Cancer stem cells revisited. *Nat. Med.* 23, 1124–1134.

Castro-Giner F, Scheidmann MC & Aceto N (2018) Beyond enumeration: functional and computational analysis of circulating tumor cells to investigate cancer metastasis. *Front. Med.* 5, article 34.

Cross WCH, Graham TA & Wright NA (2016) New paradigms in clonal evolution: punctuated equilibrium in cancer. *J. Pathol.* 240, 125–136.

Greaves M (2018) Nothing in cancer makes sense except . . . *BMC Biol.* 16, 22.

Nowell PC (1976) The clonal evolution of tumor cell populations. *Science* 194, 23–28.

Slattum GM & Rosenblatt J (2014) Tumour cell invasion: an emerging role for basal epithelium cell extrusion. *Nat. Rev. Cancer* 14, 495–501.

Targa A & Rancati G (2018) Cancer: a CINful evolution. *Curr. Opin. Cell Biol.* 52, 136–144.

Vander Heiden MG, Cantley LC & Thompson CB (2009) Understanding the Warburg effect: the metabolic requirements of cell proliferation. *Science* 324, 1029–1033.

Wu S, Powers S, Chu W & Hannum YA (2016) Substantial contribution of extrinsic risk factors to cancer development. *Nature* 529, 43–47.

Cancer-critical Genes: How They Are Found and What They Do

Bailey MH, Tokheim C, Porta-Pardo E . . . Ding L (2018) Comprehensive characterization of cancer driver genes and mutations. *Cell* 173, 371–385.

Brabletz T, Kalluri R, Nieto MA & Weinberg RA (2018) EMT in cancer. *Nat. Rev. Cancer* 18, 128–134.

Brognard J & Hunter T (2011) Protein kinase signaling networks in cancer. *Curr. Opin. Genet. Dev.* 21, 4–11.

Drost J & Clevers H (2018) Organoids in cancer research. *Nat. Rev. Cancer* 18, 407–418.

Eilers M & Eisenman R (2008) Myc's broad reach. *Genes Dev.* 22, 2755–2766.

Flavahan WA, Gaskell E & Bernstein BE (2017) Epigenetic plasticity and the hallmarks of cancer. *Science* 357, eaal2380.

Junttila MR & Evan GI (2009) p53—a Jack of all trades but master of none. *Nat. Rev. Cancer* 9, 821–829.

Levine AJ (2009) The common mechanisms of transformation by the small DNA tumor viruses: the inactivation of tumor suppressor gene products: *p53*. *Virology* 384, 285–293.

Levine AJ, Martincorena I & Cambell PJ (2015) Somatic mutation in cancer and normal cells. *Science* 349, 1483–1489.

Shaw RJ & Cantley LC (2006) Ras, PI(3)K and mTOR signalling controls tumour cell growth. *Nature* 441, 424–430.

Shen H & Laird PW (2013) Interplay between the cancer genome and epigenome. *Cell* 153, 38–55.

Tomasetti C, Marchionni L, Nowak MA, Parmigiani G & Vogelstein B (2014) Only three driver gene mutations are required for the development of lung and colorectal cancers. *Proc. Natl. Acad. Sci. USA* 112, 118–123.

Cancer Prevention and Treatment: Present and Future

Ames B, Durston WE, Yamasaki E & Lee FD (1973) Carcinogens are mutagens: a simple test system combining liver homogenates for activation and bacteria for detection. *Proc. Natl. Acad. Sci. USA* 70, 2281–2285.

Chen DS & Mellman I (2017) Elements of cancer immunity and the cancer-immune set point. *Nature* 541, 321–330.

Druker BJ & Lydon NB (2000) Lessons learned from the development of an Abl tyrosine kinase inhibitor for chronic myelogenous leukemia. *J. Clin. Invest.* 105, 3–7.

Hyman DM, Taylor BS & Baselga J (2017) Implementing genome-driven oncology. *Cell* 168, 584–599.

Jonkers J & Berns A (2004) Oncogene addiction: sometimes a temporary slavery. *Cancer Cell* 6, 535–538.

Lim WA & June CH (2017) The principles of engineering immune cells to treat cancer. *Cell* 168, 724–740.

Lord CJ & Ashworth A (2012) The DNA damage response and cancer therapy. *Nature* 481, 287–294.

Sliwkowski MX & Mellman I (2013) Antibody therapeutics in cancer. *Science* 341, 1192–1198.

Varmus H, Pao W, Politi K, Podsypanina K & Du Y-CN (2005) Oncogenes come of age. *Cold Spring Harb. Symp. Quant. Biol.* 70, 1–9.

Development of Multicellular Organisms

CHAPTER 21

IN THIS CHAPTER

Every multicellular organism, be it animal or plant, starts its life as a single cell—a fertilized egg, or **zygote**. During development, this cell divides repeatedly to produce many different kinds of cells, arranged in a final pattern of spectacular complexity and precision. The goal of developmental cell biology is to understand the cellular and molecular mechanisms that direct this amazing transformation (Movie 21.1).

Plants and animals have very different ways of life, and they use different developmental strategies. In this chapter, we focus mainly on animals. Three processes are fundamental to animal development: (1) cell proliferation, which produces many cells from one; (2) cell specialization, or **differentiation**, by which cells take on different characteristics and functions; and (3) morphogenesis, in which cells rearrange themselves to form structured tissues and organs (Figure 21–1).

Development of the zygote begins with multiple rounds of cell division, generating a large population of cells that can then be specialized for different functions. At each subsequent stage in its development, a cell is presented with a limited set of options, so that its developmental pathway branches repeatedly, reflecting a large set of sequential choices. Like the decisions we make in our own lives, the choices made by the cell are based on its internal state—which largely reflects its history—and on current influences from other cells, especially its close neighbors. To understand development, we need to know how each choice is controlled and how it depends on previous choices. Beyond that, we need to understand how the choices, once made, influence the cell's chemistry and behavior, and how cell behaviors act synergistically to determine the structure and function of the body.

As cells become specialized, they change not only their biochemistry but also their shape and their attachments to other cells and to the extracellular matrix. They move and rearrange themselves to create the complex architecture of the body, with all its tissues and organs, each structured precisely and defined in size. To understand this process of form generation, or **morphogenesis**, we will need to take account of the mechanical, as well as the biochemical, interactions between the cells.

At first glance, one would no more expect the worm, the flea, the eagle, and the giant squid to be generated by the same developmental mechanisms than one would suppose that the same methods were used to make a shoe and an airplane. Remarkably, however, research has revealed that much of the basic machinery of development is essentially the same in all animals—not just in all vertebrates, but in all invertebrates too. Recognizably similar, evolutionarily related molecules

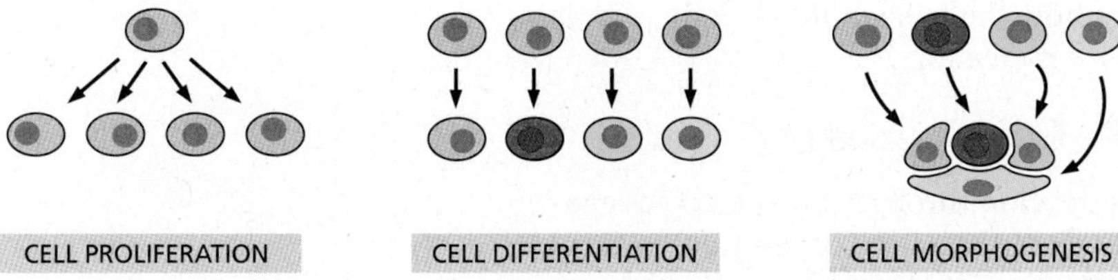

Figure 21–1 **The three essential cell processes that allow a multicellular organism to develop.**

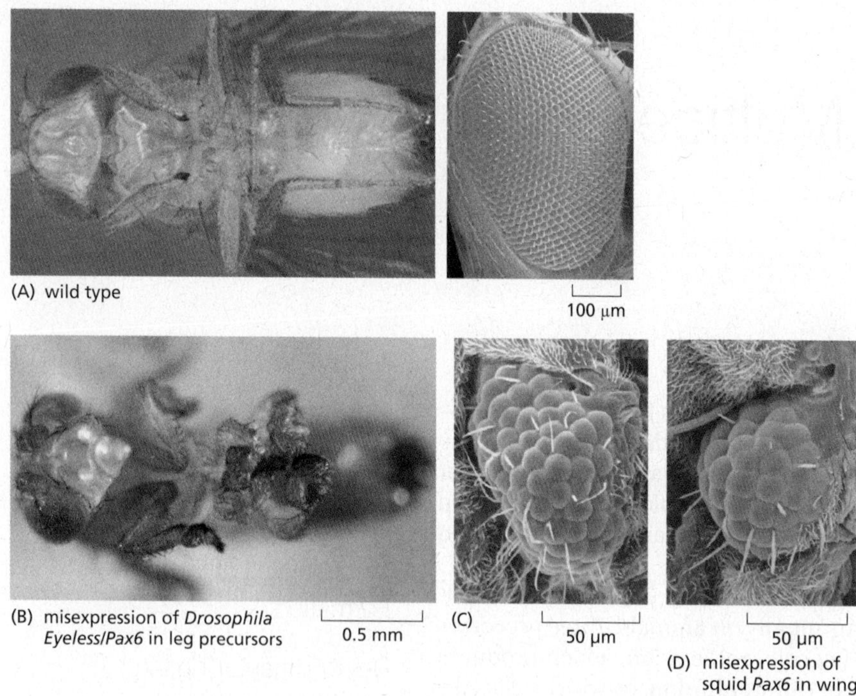

Figure 21–2 **Homologous proteins often have conserved functions.** (A) The Eyeless protein (also called Pax6) controls eye development in *Drosophila*. In a wild-type fly *(top row)*, the *Eyeless/Pax6* gene is expressed in the developing eye and directs formation of the eye structure shown in the scanning electron micrograph at right. Misexpression of *Eyeless/Pax6* in the legs *(bottom row)* results in the formation of eye tissue on leg precursor (B) (see Figure 7–38) or wing precursor (C). The homologous *Pax6* from a squid, when misexpressed in developing *Drosophila* legs, has the same effect (D). (Fly images courtesy of Katy Ong and Justin Kumar; electron micrographs from S.I. Tomarev et al., *Proc. Natl. Acad. Sci. USA* 94:2421–2426. Copyright 1997 National Academy of Sciences, USA. With permission from National Academy of Sciences.)

define the specialized animal cell types, mark the differences between body regions, and help create the animal body pattern. Homologous proteins often perform the same role in different species and are functionally interchangeable. Thus, a mouse or squid protein produced artificially in a fly, for example, can perform the same function as the fly's own version of that protein (**Figure 21–2**). Thanks to an underlying unity of mechanisms, developmental biologists have been making great strides toward a coherent understanding of animal development.

We begin this chapter with an overview of some of the basic mechanisms that operate in animal development, focusing on those that are best understood and generate differences between cells. We then discuss, in sequence, how cells in the embryo diversify to form patterns in space, how the timing of developmental events is controlled, how changing cell behaviors drive morphogenesis, and how the size of an animal is regulated.

OVERVIEW OF DEVELOPMENT

Animals live by eating other organisms. Thus, despite their remarkable diversity, animals as different as worms, mollusks, insects, and vertebrates share anatomical features that are fundamental to this way of life. Epidermal cells form a protective outer layer; gut cells absorb nutrients from ingested food; muscle cells allow movement toward food sources; and neurons and sensory cells control behavior. These various cell types are organized into tissues and organs, forming a sheet of skin covering the exterior, a mouth for feeding, and an internal gut tube for digestion—with muscles, nerves, and other tissues arranged in the space between the skin and the gut. Many animals have clearly defined axes—an anteroposterior axis, with mouth and brain anterior and anus posterior; a dorsoventral axis, with back dorsal and belly ventral; and a left-right axis. In this section, we discuss some fundamental mechanisms underlying how the basic animal body plan is established and how this diversity of cell types is created.

Conserved Mechanisms Establish the Core Tissues of Animals

The shared anatomical features of animals develop through conserved mechanisms. After fertilization, the zygote usually divides rapidly, or **cleaves**, to form many smaller cells called **blastomeres**. During this cleavage phase, the embryo, which cannot yet feed, does not grow. This step of development is initially driven

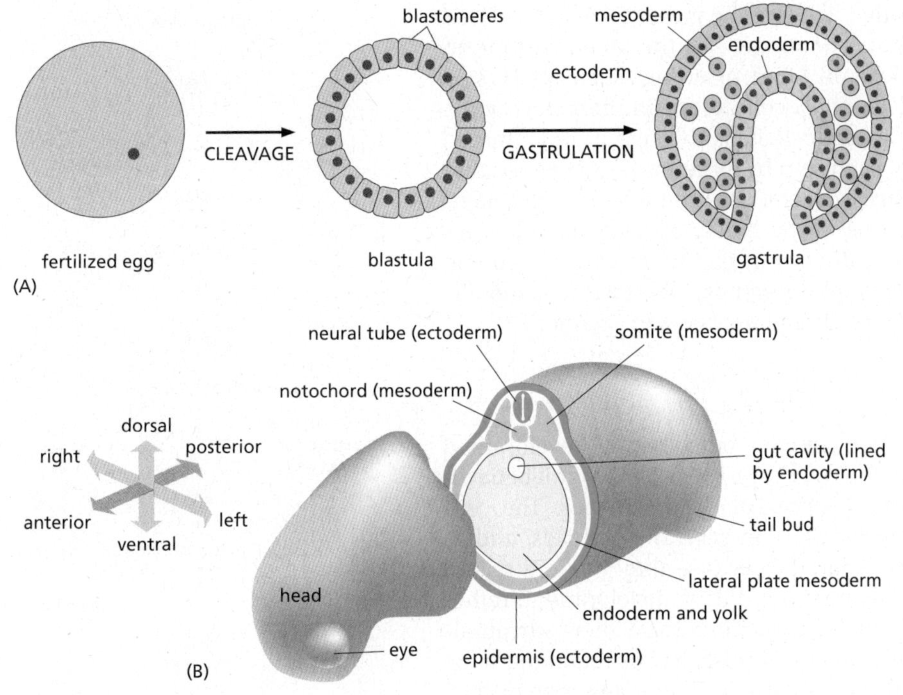

Figure 21–3 The early stages of development, as exemplified by a frog. (A) A fertilized egg divides to produce many blastomere cells that form an epithelial sheet often surrounding a cavity. During gastrulation, some of the cells tuck into the interior to form the mesoderm *(green)* and endoderm *(yellow)*. Ectodermal cells *(blue)* remain on the outside. (B) A cross section through the trunk of an amphibian embryo shows the basic animal body plan, with a sheet of ectoderm on the outside, a tube of endoderm on the inside, and mesoderm sandwiched between them. The endoderm forms the epithelial lining of the gut, from the mouth to the anus. It gives rise not only to the pharynx, esophagus, stomach, and intestines, but also to many associated structures. The salivary glands, liver, pancreas, trachea, and lungs, for example, all develop from the wall of the digestive tract and grow to become systems of branching tubes that open into the gut or pharynx. The endoderm forms only the epithelial components of these structures—the lining of the gut and the secretory cells of the pancreas, for example. The supporting muscular and fibrous elements arise from the mesoderm.

The mesoderm gives rise to the connective tissues—at first, to the loose mesh of cells in the embryo known as *mesenchyme*, and ultimately to cartilage, bone, and fibrous tissue, including the dermis (the inner layer of the skin). The mesoderm also forms the muscles, the entire vascular system—including the heart, blood vessels, and blood cells—and the tubules, ducts, and supporting tissues of the kidneys and gonads. The notochord forms from the mesoderm and serves as the core of the future backbone and the source of signals that coordinate the development of surrounding tissues.

The ectoderm will form the epidermis (the outer, epithelial layer of the skin) and epidermal appendages such as hair, sweat glands, and mammary glands. It will also give rise to the whole of the nervous system, central and peripheral, including not only neurons and glia but also the sensory cells of the nose, the ear, the eye, and other sense organs. (B, after T. Mohun et al., *Cell* 22:9–15, 1980.)

and controlled entirely by the material deposited in the egg by the mother. The embryonic genome remains largely inactive until a point is reached when maternal mRNAs and proteins are used up or abruptly degraded. The embryo's genome is then activated—the maternal-zygotic transition that we will discuss later—and the cells cohere to form a **blastula**, typically a solid or a hollow fluid-filled ball of cells.

Even in these early stages of embryogenesis, genetic programs are under way that give rise to the basic tissue types. Complex cell rearrangements called **gastrulation** (from the Greek *gaster*, meaning "belly") soon transform the blastula into a multilayered structure containing a rudimentary internal gut (**Figure 21–3**). Some cells of the blastula remain external, constituting the **ectoderm**, which will give rise to the epidermis and the nervous system; other cells invaginate, forming the **endoderm**, which will give rise to the gut tube and its appendages, such as lung, pancreas, and liver. In most animals, another group of cells moves into the space between ectoderm and endoderm and forms the **mesoderm**, which will give rise to muscles, connective tissues, blood, kidneys, and various other components. Further cell movements and accompanying cell differentiation create and refine the embryo's architecture.

The ectoderm, mesoderm, and endoderm formed during gastrulation constitute the three **germ layers** of the early embryo. This initial subdivision is the first step in defining the multitude of cell fates that will emerge in development. Many later developmental transformations will produce the elaborately structured organs. But the basic body plan and axes set up in miniature during gastrulation are preserved into adult life, when the organism may be billions of times larger (**Movie 21.2**).

The Developmental Potential of Cells Becomes Progressively Restricted

Concomitant with the refinement of the body plan, the individual cells within a *lineage*—that is, the progeny of a particular proliferating mother cell—become more and more restricted in their developmental potential. During the blastula stages, cells are often **totipotent** or **pluripotent**—they have the potential to give rise to all or almost all of the cell types of the adult body. The pluripotency is lost as gastrulation proceeds: a cell located in the endodermal germ layer, for example, can give rise to the cell types that will line the gut or form gut-derived organs such as the liver or pancreas, but it no longer has the potential to form mesoderm-derived structures such as skeleton, heart, or kidney. Such a cell is said to be *determined* for an endodermal fate. Thus, **cell determination** starts early and progressively narrows the options as the cell steps through a programmed series of intermediate states—guided at each step by its genome, its history, and its interactions with neighbors. The process reaches its limit when a cell undergoes **terminal differentiation** to form one of the highly specialized cell types of the adult body (**Figure 21–4**). Some cell types in the adult, which maintain the ability to divide, also retain a degree of pluripotency, albeit with a generally narrow range of options. These adult stem cells are discussed in Chapter 22.

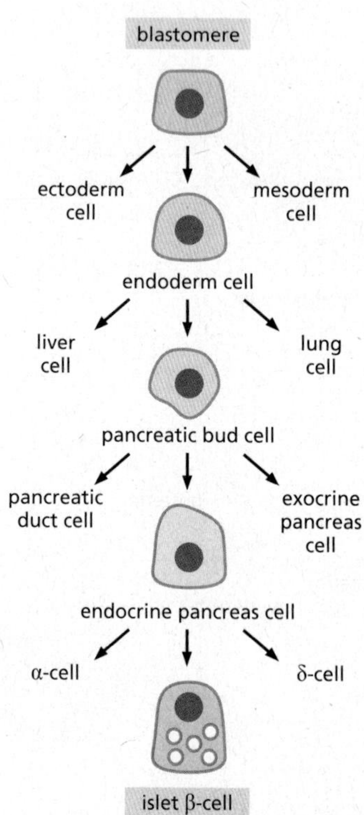

Figure 21–4 The lineage from blastomere to a terminally differentiated cell type. As development proceeds, cells become more and more specialized. Blastomeres have the potential to give rise to most or all cell types. Under the influence of signaling molecules and gene regulatory factors, cells acquire more restricted fates until they differentiate into highly specialized cell types, such as the pancreatic β cells that secrete the hormone insulin.

Cell Memory Underlies Cell Decision-Making

Underlying the richness and astonishingly complex outcomes of development is **cell memory** (see p. 435). Both the genes a cell expresses and the way it behaves depend on the cell's past, as well as on its present circumstances. The cells of our body—the muscle cells, the neurons, the skin cells, the gut cells, and so on—maintain their specialized characters largely because they retain a record of the extracellular signals their ancestors received during development, rather than because they continually receive such instructions from their surroundings. Despite their radically different phenotypes, almost all cells retain the same complete genome that was present in the zygote. Their differences arise instead from differential gene expression, which can lead to the stable inheritance of a particular cellular program. We have discussed the molecular mechanisms of gene regulation, cell memory, cell division, cell signaling, and cell movement in previous chapters. In this chapter, we shall see how these basic processes are collectively deployed as the developing animal self-assembles.

Several Model Organisms Have Been Crucial for Understanding Development

The anatomical features that animals share have undergone many extreme modifications in the course of evolution. As a result, the differences between species are usually more striking to our human eye than the similarities. But at the level of the underlying molecular mechanisms and the particular macromolecules involved, the reverse is true: the similarities among all animals are profound and extensive. Through more than half a billion years of evolutionary divergence, all animals have retained unmistakably similar sets of genes and proteins that are responsible for generating their body plans and for forming their specialized cells and organs.

This astonishing degree of evolutionary conservation was discovered not by broad surveys of animal diversity, but through intensive study of a small number of experimentally convenient species—the model organisms discussed in Chapter 1. For animal developmental biology, the most important have been the fly *Drosophila melanogaster*, the frog *Xenopus laevis*, the roundworm *Caenorhabditis elegans*, the mouse *Mus musculus*, and the zebrafish *Danio rerio*. In discussing the mechanisms of development, we shall draw our examples mainly from these few species, keeping in mind that they represent only a portion of the tree of animal life.

Regulatory DNA Seems Largely Responsible for the Differences Between Animal Species

Although many developmental mechanisms are conserved, they nevertheless produce animals that can be quite different. These differences arise primarily from variation in the activity of key development-controlling genes. As discussed in Chapter 7, each gene in a multicellular organism is associated with many thousands of nucleotides of noncoding DNA that contains regulatory elements. These

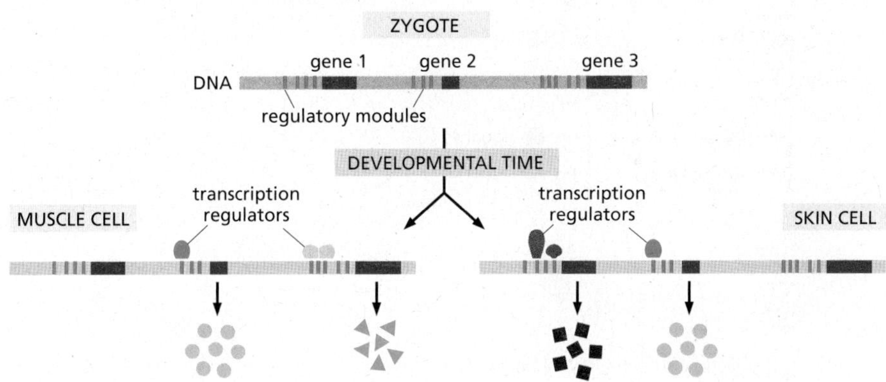

Figure 21–5 **Regulatory DNA defines the gene expression patterns in development.** The genome is the same in a muscle cell as in a skin cell, but different genes are active because these cells express different transcription regulators that bind to gene regulatory elements. For example, transcription regulators in skin cells recognize a regulatory element in gene 1, leading to its activation, whereas a different set of regulators is present in muscle cells, binding to and activating gene 3. Transcriptional regulators that activate the expression of gene 2 are present in both cell types.

regulatory elements determine when, where, and how strongly the gene product is expressed, according to the transcription regulators and chromatin structures that are present in the particular cell (**Figure 21–5**). Consequently, a change in sequence of the regulatory DNA that arises in evolution, even without any change in the coding DNA, can alter the logic of the gene regulatory network and change the outcome of development.

As discussed in Chapter 4, when we compare the genomes of different animal species, we find that evolution has altered the coding and regulatory DNA to different extents. The coding DNA can be quite conserved, but the noncoding regulatory DNA is usually much less so. It seems that changes in regulatory DNA are largely responsible for the dramatic differences between one class of animals and another (see p. 239). We can view the protein products of the conserved coding sequences as a kit of common molecular parts and the regulatory DNA as instructions for assembly: with different instructions, the same kit of parts can be used to make a whole variety of different body structures. We will return to this important concept later.

Small Numbers of Conserved Cell–Cell Signaling Pathways Coordinate Spatial Patterning

Spatial patterning of a developing animal requires that cells become different according to their positions in the embryo, which means that cells must respond to extracellular signals produced by other cells, especially their neighbors. In the most common mode of spatial patterning, a group of cells starts out with the same developmental potential, and a signal from cells outside the group then induces one or more members of the group to change their character. This process is called *inductive signaling*. Generally, the inductive signal is limited in time and space so that only a subset of the cells capable of responding—the cells close to the source of the signal—take on the induced character (**Figure 21–6**). Some inductive signals depend on cell-cell contact; others act over a longer range and are mediated by molecules that diffuse through the extracellular medium or are transported in the bloodstream (see Figure 15–2).

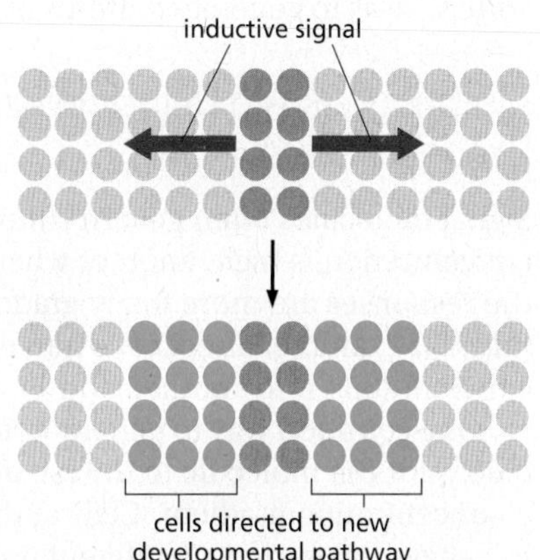

Figure 21–6 **Inductive signaling.** Cells *(gray)* expressing an extracellular signaling molecule direct a new cell fate in nearby neighbors *(blue)*.

Most of the known inductive events in animal development are governed by a small number of highly conserved signaling pathways, including transforming growth factor-β (TGFβ), Wnt, Hedgehog, Notch, and receptor tyrosine kinase (RTK) pathways (discussed in Chapter 15). The discovery of the limited vocabulary that developing cells use for intercellular communication has emerged as one of the great simplifying features of developmental biology.

Through Combinatorial Control and Cell Memory, Simple Signals Can Generate Complex Patterns

How can a small number of signaling pathways generate the huge diversity of cells and patterns? Several kinds of mechanisms are responsible. First, the effect of activating a signaling pathway depends on the previous experiences of the

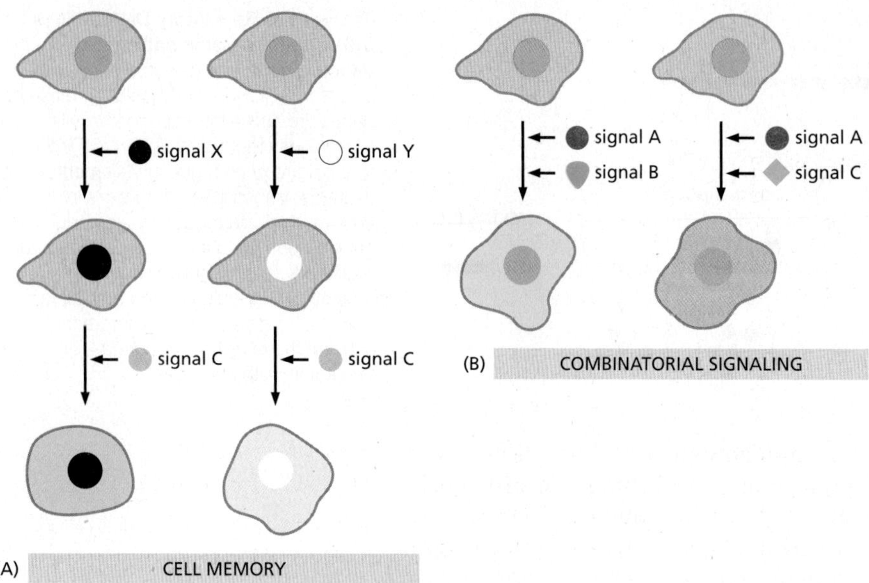

Figure 21–7 **Two mechanisms for generating different responses to the same inductive signal.** (A) Through cell memory, previous signals (or other events) can leave a lasting trace that alters the response to the current signal (see Figure 7–56). The memory trace is represented here in the coloring of the cell nucleus. (B) In combinatorial signaling, the effect of a signal depends on the presence of other signals received at the same time.

responding cell: past influences leave a lasting mark, registered in the state of the cell's chromatin and the selection of transcription regulators and RNA molecules that the cell contains. This cell memory enables cells with different histories to respond to the same signals differently (Figure 21–7A). Second, the response of a cell to a given signal depends on the other signals that the cell is receiving concurrently (Figure 21–7B). As a result, different combinations of signals can generate a large variety of different responses.

Alongside these two mechanisms, some components of a signaling pathway, such as ligands or receptors, are encoded by genes that have undergone duplication followed by functional divergence during evolution. These closely homologous genes can then be expressed in distinct cell types, where they can direct different signaling outcomes. Notch signaling, for example, may be mediated by Notch1 in one tissue, but by Notch4 in another, with each homolog inducing transcription of different target genes. Thus, the same few signaling pathways can be used repeatedly at different times and places with different outcomes, so as to generate patterns of unlimited complexity.

Morphogens Are Diffusible Inductive Signals That Exert Graded Effects

Signal molecules often govern simple yes–no choices—one outcome when their concentration is high, another when it is low or absent. In other cases, however, the responses are more finely graded: a high concentration of a signal molecule may, for example, direct cells into one developmental pathway, an intermediate concentration into another, and a low concentration into yet another.

One common way to generate such different concentrations of a signal molecule is for the molecule to diffuse out from a localized signaling source, creating a concentration gradient. Cells at different distances from the source are driven to behave in a variety of different ways, according to the signal concentration that they experience (Figure 21–8). A signal molecule that imposes a pattern on a whole field of cells in this way is called a **morphogen**. In the simplest case, a specialized group of cells produces a morphogen at a steady rate, and the morphogen is then degraded as it diffuses away from this source. The speed of diffusion and

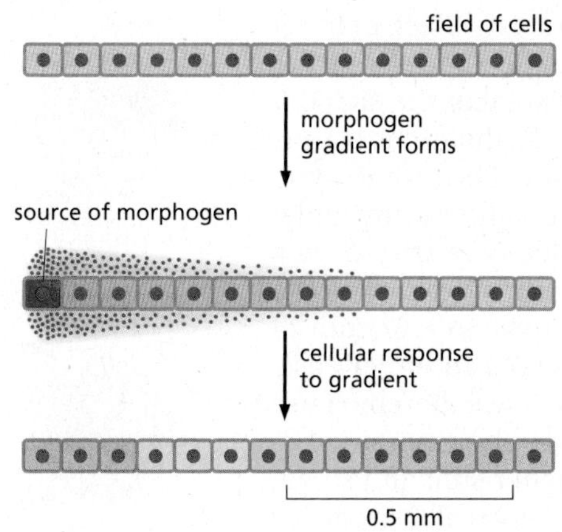

Figure 21–8 **Gradient formation and interpretation.** A gradient forms by the localized production of an inducer—a morphogen—that diffuses away from its source. Different concentrations of morphogen (or different durations of exposure) induce different gene expression patterns and cell fates in responding cells. Diffusive transport can generate steep gradients only over short distances, and morphogens generally act over distances of 1 mm or less.

the half-life of the morphogen will together determine the range and steepness of its resulting gradient (Figure 21–9).

Nearby cells then interpret their distance from the morphogen source according to how much signal they are exposed to, detected by binding to cell-surface receptors. After signal transduction from these receptors, target genes will become transcribed only if the morphogen concentration exceeds a specific threshold; low concentrations that fall beneath this threshold will not activate the target. In this manner, a graded signal can be converted into multiple discrete, on-or-off changes in gene activity. Thus, a single secreted protein, in combination with the physical properties of diffusion and the ability of cells to interpret the information, can generate several distinct fates within a field of cells.

This simple mechanism can be modified in various ways. For example, cell-surface receptors may trap the diffusing morphogen and cause it to be endocytosed and degraded, shortening its effective half-life. Alternatively, the morphogen may bind to molecules in the extracellular matrix such as heparan sulfate proteoglycan (discussed in Chapter 19), thereby greatly reducing its diffusion rate.

Lateral Inhibition Can Generate Patterns of Different Cell Types

Morphogen gradients, and other types of inductive signals, exploit an existing asymmetry in the embryo to create further asymmetries and differences between cells: already, at the outset, some cells are specialized to produce the morphogen and thereby impose a pattern on another class of cells that are sensitive to it. But what if there is no clear initial asymmetry? Can a regular pattern arise spontaneously within a set of cells that are initially all alike?

The answer is yes. The fundamental principle underlying such *de novo* pattern formation is positive feedback: cells can exchange signals in such a way that any small initial discrepancy between cells at different sites becomes self-amplifying, driving the cells toward different fates. This is most clearly illustrated in the phenomenon of *lateral inhibition*, a form of cell–cell interaction that forces close neighbors to become different and thereby generates fine-grained patterns of different cell types.

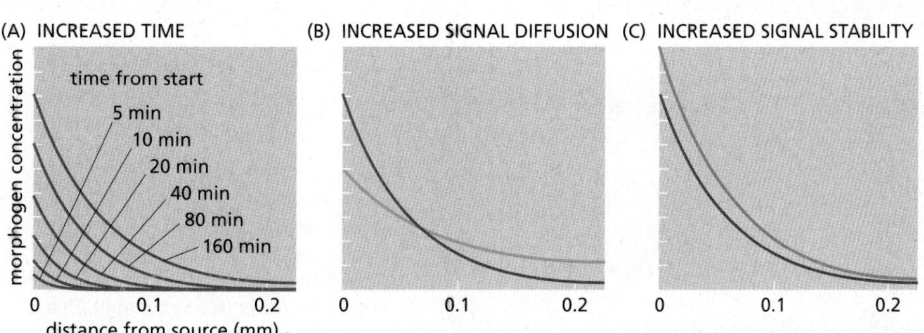

Figure 21–9 **Setting up a signal gradient by diffusion.** Each graph shows the concentration of a morphogen signal molecule that is produced at a steady rate at the origin. In all cases, the molecule undergoes degradation as it diffuses away from the source, and the graphs are calculated on the assumption that diffusion is occurring along two axes in space (for example, radially from a source in an epithelial sheet). (A) The pattern of the morphogen assuming that production starts at time 0, the molecule has a half-life of 170 minutes, and that it diffuses with an effective diffusion constant of $D = 1\ \mu m^2\ sec^{-1}$, typical of a small protein molecule in extracellular tissues. *Red lines* show six successive stages in the buildup of the morphogen, which falls off exponentially with distance from the source. Panels B and C show how simple changes in the properties of the system alter the gradient at the 160-minute time point. (B) A threefold increase in the diffusion constant of the morphogen extends its range *(blue line)* but lowers its concentration next to the source. (C) A threefold increase in morphogen half-life *(green line)* increases its concentration throughout the tissue. Effects of the morphogen will depend not just on its concentration at some critical moment, but also on how each target cell integrates its response over time. (Courtesy of Patrick Müller.)

Consider a pair of adjacent cells that start off in a similar state. Each of these cells can both produce and respond to a certain signal molecule X, with the added rule that the stronger the signal a cell receives, the weaker the signal it generates itself (Figure 21–10). If one cell produces more X, the other is thus forced to produce less, for instance by reducing transcription of the X-encoding gene. This gives rise to a positive feedback loop that tends to amplify any initial difference between the two adjacent cells. Such a difference may arise from a bias imposed by some present or past external factor or it may simply originate from spontaneous random fluctuations, or "noise"—an inevitable feature of the genetic control circuitry in cells (discussed in Chapter 7). In either case, lateral inhibition means that if cell 1 makes a little more of X, it will thereby cause cell 2 to make less; and because cell 2 makes less X, it delivers less inhibition to cell 1 and so allows the production of X in cell 1 to rise higher still; and so on, until a stable state is reached where cell 1 produces a lot of X and cell 2 produces very little. The result is that the two cells are driven along different pathways of differentiation.

In almost all tissues, a balanced mixture and distribution of different cell types are required. Lateral inhibition provides a common way to generate the mixture. As we shall see, lateral inhibition is very often mediated by exchange of signals at cell-cell contacts via the Notch signaling pathway, driving cell diversification by enabling individual cells that express one set of genes to direct their immediate neighbors to express a different set, in exactly the way we have described (see also Figure 15–60).

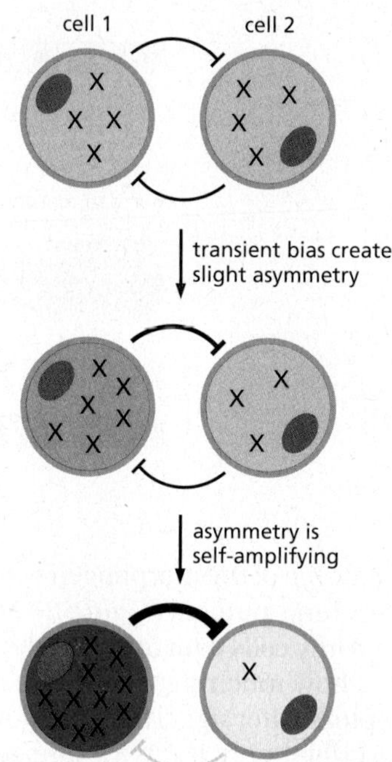

Figure 21–10 **Genesis of asymmetry through lateral inhibition and positive feedback.** In this example, two cells interact, each producing a substance X that acts on the other cell to inhibit its production of X, an effect known as lateral inhibition. An increase of X in one of the cells leads to a positive feedback that tends to increase X in that cell still further, while decreasing X in its neighbor. This can create an accelerating instability, making the two cells become radically different. Ultimately, the system comes to rest in one or the other of two opposite stable states. The final choice of state represents a form of memory: the small influence that initially directed the choice is no longer required to maintain it.

Asymmetric Cell Division Can Also Generate Diversity

Cell diversification does not always depend on extracellular signals: in some cases, cells are born different as a result of an asymmetric cell division, in which some important molecule or molecules present in the mother cell are distributed unequally between the two daughters. This asymmetric inheritance during mitosis ensures that the two daughter cells develop differently (Figure 21–11). The mechanism here is intrinsic to the dividing cell, although daughter cells can also become asymmetric after division as a result of the extrinsic, inductive signals discussed earlier. Intrinsically asymmetric division is a common feature of early development. The fertilized egg may already possess an internal pattern, and cleavage of this large cell segregates different fate determinants into separate blastomeres. We shall see later that asymmetric division also plays a part in later developmental processes, as well as in stem cells (see Chapter 22).

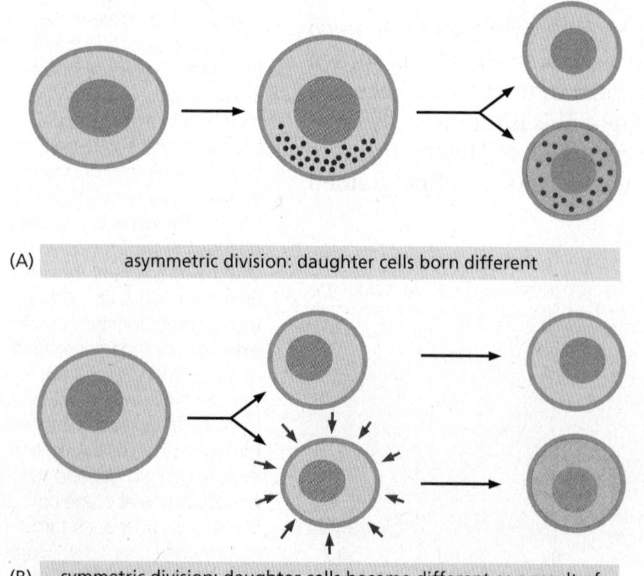

Figure 21–11 **Two ways of making daughter cells different.** Daughters can assume different fates either through (A) an intrinsically asymmetric division in which there is differential inheritance of cytoplasmic molecules or through (B) a symmetric division followed by signaling to only one daughter.

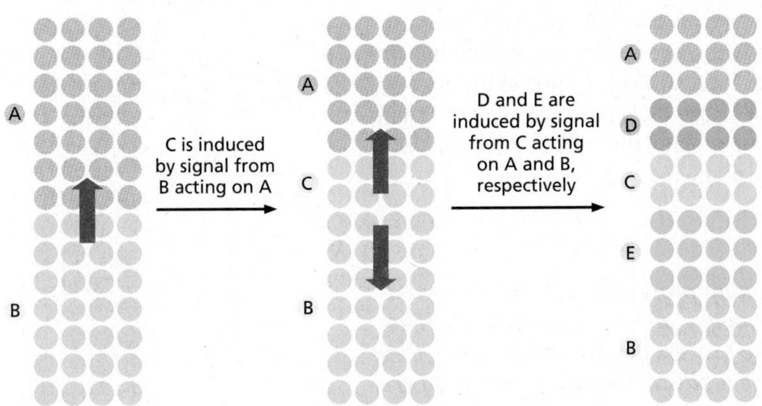

Figure 21–12 **Patterning by sequential induction.** A series of inductive interactions can generate many types of cells, starting from only a few.

Initial Patterns Are Established in Small Fields of Cells and Refined by Sequential Induction as the Embryo Grows

The signals that organize the spatial pattern of cells in an embryo generally act over short distances and govern relatively simple choices. A morphogen, for example, typically acts over a distance of less than 1 mm—an effective range for diffusion—and directs choices between several developmental options for the cells on which it acts. Yet the organs that eventually develop are much larger and more complex than this.

As the organ grows via cell proliferation, the refinement of its initial pattern is explained by a series of successive interactions that add increasing levels of detail to an initially simple sketch. For example, as soon as two types of cells are present in a developing tissue, one of them can produce a signal that induces a subset of neighboring cells to specialize in a third way. The third cell type can in turn signal back to the other two cell types nearby, generating a fourth and a fifth cell type, and so on (Figure 21–12).

This strategy for generating a progressively more complicated pattern is called **sequential induction**. It is chiefly through sequential inductions that the body plan of a developing animal, after being first roughed out in miniature, becomes elaborated with finer and finer details as development proceeds.

Developmental Biology Provides Insights into Disease and Tissue Maintenance

The rapid progress in understanding animal development has been one of the great success stories in biology, and it has important practical implications. Some 2–5% of all human babies are born with anatomical abnormalities, such as heart malformations, truncated limbs, cleft palate, or spina bifida. Advances in developmental biology help us understand how these defects arise, even if we cannot yet prevent or cure most of them.

Less obvious, but even more important from a practical point of view, is that developmental biology also provides insights into the workings of cells and tissues in the adult body. Developmental processes do not halt at birth; they continue throughout life, as tissues are maintained and repaired. The fundamental mechanisms of cell growth and division, cell–cell signaling, cell memory, cell adhesion, and cell movement are all involved in adult tissue maintenance and repair—just as they are in embryo development. These are also the main mechanisms perturbed in tumor cells, as we saw in Chapter 20.

Embryos are simpler than adults, and they allow us to analyze such basic processes more easily. Studies of the early *Drosophila* embryo, for example, were crucial to the discovery of several conserved signaling pathways, including the Wnt, Hedgehog, and Notch pathways. They also provided the key to understanding the central role of these pathways in the maintenance of normal adult human tissues and laid the foundation for manipulating the pathways in the cause of regenerative medicine. Finally, they have identified both targets and rational mechanisms for therapies to fight cancer and other diseases.

In Chapter 22, we shall consider how these and other core developmental mechanisms operate in the normal adult body, especially in tissues that are continually renewed by means of stem cells—including the gut, skin, and the hematopoietic system. But now, we must look more closely at the way in which an early embryo generates its spatial pattern of specialized cells, beginning with the transformations that create the adult body plan.

Summary

Animal development is an amazing self-assembly process, in which the initially similar cells of the embryo become different from one another and organize themselves into increasingly complex structures. The process begins with a single large cell, the fertilized egg, which cleaves to produce a blastula that undergoes gastrulation to generate the three germ layers of the embryo—ectoderm, mesoderm, and endoderm. As development continues, the cells become more and more narrowly specialized according to their locations and their interactions with one another, eventually forming one of the differentiated cell types of the adult body.

Differences between developing cells arise in various ways and have to be properly coordinated in space. In one common strategy, initially similar cells within a group become different by exposure to different levels of an inductive signal or morphogen emanating from a source outside the group. Neighboring cells can also become different by lateral inhibition, in which a cell signals to its neighbors not to follow the same fate. These cell–cell interactions are mediated by a small number of highly conserved signaling pathways, which are used repeatedly in different organisms and at different times during development. Not all cell diversification arises by cell–cell interactions, however: daughter cells can be born different as a result of asymmetric cell division.

Regulators of transcription and chromatin structure bind to regulatory DNA and determine the fate of each cell. Differences of body plan seem to arise to a large extent from differences in the regulatory DNA associated with each gene. This DNA has a central role in defining the sequential program of development, calling genes into action at specific times and places according to the pattern of gene expression that was present in each cell at the previous developmental stage.

Development has been most thoroughly studied in a handful of model organisms. But most of the genes and mechanisms thereby identified are used in all animals and repeatedly at different stages of development. Thus, insights from worms, flies, fish, frogs, and mice deeply inform our understanding of embryology and adult tissue maintenance in humans, as well as how aberrations in these processes cause birth defects and cancer.

MECHANISMS OF PATTERN FORMATION

A developing multicellular organism has to create different cell fates in fields of cells that were nearly indistinguishable and accomplish this task in a spatially ordered manner so that functional tissues are formed. Some of the early microscopists imagined the entire shape and structure of the human body to be already present in the sperm as a "homunculus," a miniature human; after fertilization, the homunculus would simply grow and generate a full-sized person. We now know that this view is incorrect, and that development is a progression from simple to complex, through a gradual refinement of an animal's anatomy. To see how the whole sequence of events of spatial patterning and cell determination is set in motion, we must return to the egg and the early embryo.

Different Animals Use Different Mechanisms to Establish Their Primary Axes of Polarization

Surprisingly, the earliest steps of animal development are among the most variable, even within a phylum. A frog, a chicken, and a mammal, for example, even though they develop in similar ways later, make eggs that differ radically in size and structure, and they begin their development with different sequences of cell divisions and cell specializations. Likewise, there is great variation in the time and manner in which the primary axes of the body become marked out. However, this *polarization* of the embryo usually becomes discernible very early, before gastrulation begins—it is the first step of spatial patterning.

Two axes have to be established. The *anteroposterior* (*A-P*) axis specifies the locations of future head and tail. The *dorsoventral* (*D-V*) axis specifies the future back and belly. For species that show bilateral asymmetry, creation of an additional *left-right (L-R)* axis is also important. Finally, the eggs of many animals have an *animal-vegetal (A-V)* axis that is not evident in the adult, but defines which parts are to become internal (through the movements of gastrulation) and which are to remain external. (The bizarre name dates from a century ago and has nothing to do with vegetables.)

At one extreme within animal diversity, the egg is spherically symmetrical, and the axes only become defined during embryogenesis. The mouse is one example, with little obvious sign of polarity in the egg. Correspondingly, the blastomeres produced by the first few cell divisions all seem to be alike and are remarkably adaptable. If the early mouse embryo is split in two, a pair of identical twins can be produced—two complete, normal individuals from a single cell. Similarly, if one of the cells in a two-cell mouse embryo is destroyed by pricking it with a needle and the resulting "half-embryo" is placed in the uterus of a foster mother to develop, in many cases a perfectly normal mouse will emerge.

At the opposite extreme, an egg is built with an asymmetric structure that itself defines the future axes of the body. This is the case for most species, including insects such as *Drosophila*, as we shall see shortly. Other organisms lie somewhere in between. The egg of the frog *Xenopus*, for example, has a clearly defined A-V axis before fertilization: the pronucleus near the top defines the animal pole, while the mass of yolk (the embryo's food supply, destined to be incorporated in the gut) toward the bottom defines the vegetal pole. Several types of mRNA molecules are already localized in the vegetal cytoplasm of the egg, where they produce their protein products. After fertilization, these mRNAs and proteins act in and on the cells in the lower and middle part of the embryo, giving the cells their specialized characters.

Formation of the D-V axis of the *Xenopus* embryo is triggered by fertilization. At the site where the sperm enters, which is only permitted in the region of the animal pole, the sperm centrosome nucleates a microtubule aster. This reorganization of the microtubule cytoskeleton causes the outer cortex of the egg to rotate relative to the central core of the egg cytoplasm, so that the animal pole of the cortex becomes slightly shifted to one side (**Figure 21–13**). Cortical rotation

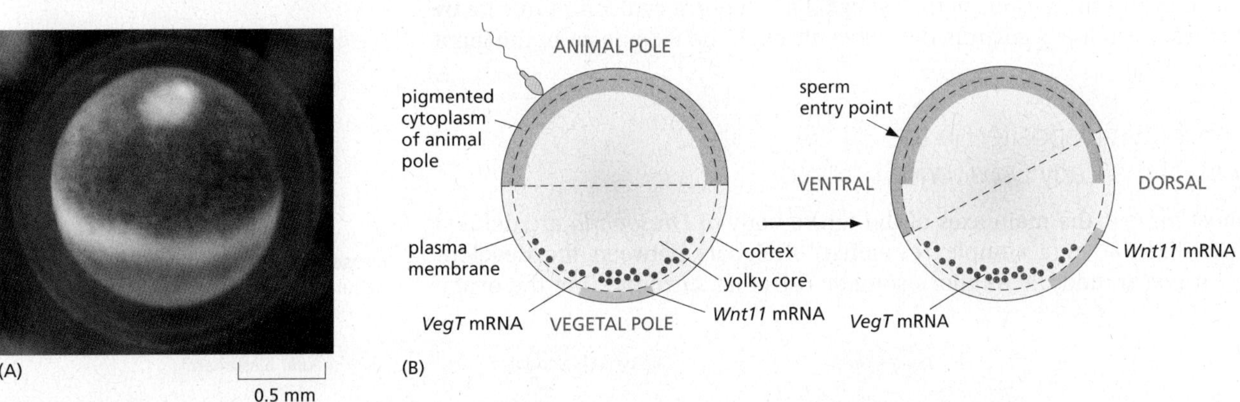

Figure 21–13 The frog egg and its asymmetries. (A) Side view of a *Xenopus* egg photographed just before fertilization. (B) The asymmetric distribution of molecules inside the egg along the animal–vegetal axis, and how fertilization activates dorsoventral asymmetry. Vegetally localized *VegT* in the unfertilized egg defines the vegetal source of signals that will induce endoderm and mesoderm. Sperm entry initiates a reorganization of the microtubule cytoskeleton that triggers a rotation of the egg cortex (a layer a few micrometers deep) through about 30° relative to the core of the egg. Cortical rotation relocalizes Wnt signaling components including *Wnt11* mRNA to the future dorsal side, which sets up the dorsoventral axis of the embryo. (A, courtesy of Tony Mills.)

relocalizes signaling molecules and initiates a cascade of events that will organize the dorsoventral axis of the body. (The A-P axis of the embryo will only become clear later, in the process of gastrulation.) This general process, by which a cue creates a new axis in the embryo, is known as *symmetry breaking*.

Although different animal species use a variety of different mechanisms to specify their overall organization, the outcome has been relatively well conserved in evolution: head is distinguished from tail, back from belly, and gut from skin. It seems that it does not much matter what tricks the embryo uses, it always manages to break the initial symmetry and set up its basic body plan.

Studies in *Drosophila* Have Revealed Many Genetic Control Mechanisms Underlying Development

It is the fly *Drosophila*, more than any other organism, that has provided the key to our present understanding of how genes govern early development. Decades of study culminated in a large-scale genetic screen, focusing on the early embryo and searching for mutations that disrupt its pattern. This revealed a group of key developmental genes that act in a relatively small set of regulatory pathways. The discovery of these genes and the subsequent analysis of their functions is a famous *tour de force* that has had a revolutionary impact on all of biology, earning its discoverers a Nobel Prize. Some parts of the machinery revealed in this way are conserved between flies and vertebrates, some parts not. But the logic of the experimental approach and the general strategies of genetic control that it revealed have transformed our understanding of multicellular development in general.

To understand how the early developmental machinery operates in *Drosophila*, it is important to note a peculiarity of fly development. Like the eggs of other insects, but unlike most vertebrates, the *Drosophila* egg—shaped like a cucumber—begins its development with an extraordinarily rapid series of nuclear divisions without cell division, producing multiple nuclei in a common cytoplasm—a **syncytium**. The nuclei then migrate to the cell cortex, forming a structure called the *syncytial blastoderm*. After about 6000 nuclei have been produced, the plasma membrane folds inward between them and partitions them into separate cells, converting the syncytial blastoderm into the *cellular blastoderm* (Figure 21–14).

We shall see that the initial patterning of the *Drosophila* embryo depends on molecules that diffuse through the cytoplasm at the syncytial stage and exert their actions on genes in the rapidly dividing nuclei, before the partitioning of the egg into separate cells. Here, there is no need for the usual forms of cell-cell signaling; neighboring regions of the syncytial blastoderm can communicate by means of transcription regulators that move through the cytoplasm of the giant multinuclear cell.

Gene Products Deposited in the Egg Organize the Axes of the Early *Drosophila* Embryo

As in most insects, the main axes of the future body of *Drosophila* are defined before fertilization by a complex exchange of signals between the developing egg, or *oocyte*, and the mother's somatic cells that surround it in the ovary.

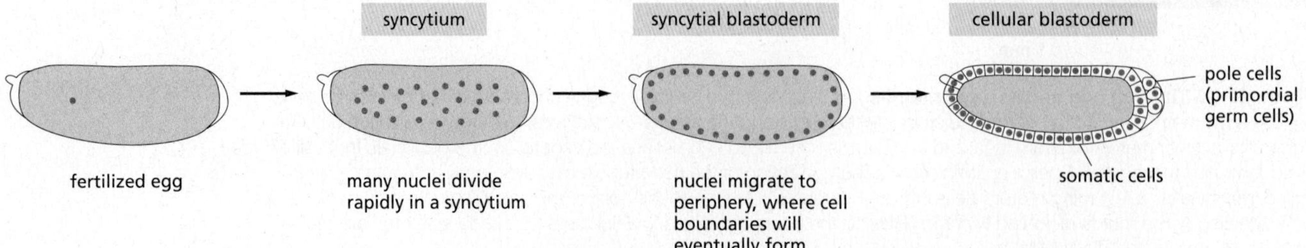

Figure 21–14 **Development of the *Drosophila* egg from fertilization to the cellular blastoderm stage.**

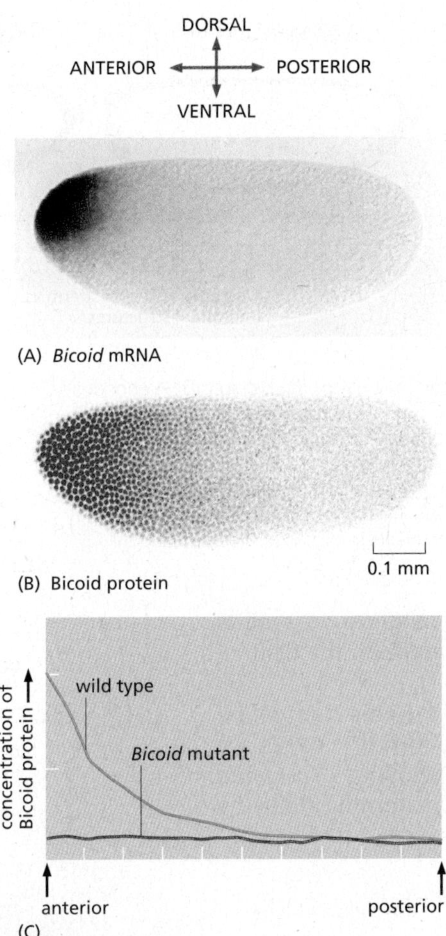

Figure 21–15 **The Bicoid protein gradient.** (A) *Bicoid* mRNA is deposited at the anterior pole during oogenesis. (B) Local translation followed by diffusion generates the Bicoid protein gradient. (C) Absence of the Bicoid protein gradient in embryos from *Bicoid* homozygous mutant mothers. (A and B, courtesy of Stephen Small.)

These signals can be described as **maternal effect**, because it is the genome of the mother rather than the zygote that produces them. Before fertilization, the anteroposterior and dorsoventral axes of the future embryo become defined by systems of **egg-polarity genes** that create landmarks—either mRNA or protein—in the oocyte. After fertilization, each landmark serves as a beacon, providing a signal that organizes the developmental process in its neighborhood.

The nature of the egg-polarity genes emerged from studies of mutants in which the patterning of the embryo was altered. Some of these mutations gave embryos with disrupted polarity; for example, one caused tail-end structures at both ends of the body, with no head-end structures. This particular mutation allowed the identification of the landmark that organizes the anterior end of the embryo, called *Bicoid*. A deposit of *Bicoid* mRNA molecules is localized, before fertilization, at the anterior end of the egg. Upon fertilization, the mRNA is translated to produce Bicoid protein. This protein is an intracellular morphogen and transcription regulator that diffuses away from its source to form a concentration gradient within the syncytial cytoplasm, with its maximum at the head end of the embryo (**Figure 21–15**). The different concentrations of Bicoid along the A-P axis help determine different cell fates by directly regulating the transcription of genes in the nuclei of the syncytial blastoderm (discussed in Chapter 7).

There are three other egg-polarity gene systems that pattern the syncytial nuclei; two act along the A-P axis and one acts along the D-V axis. Together with the *Bicoid* group of genes, and acting in a broadly similar way, their gene products mark out three fundamental partitions of body regions—head versus rear,

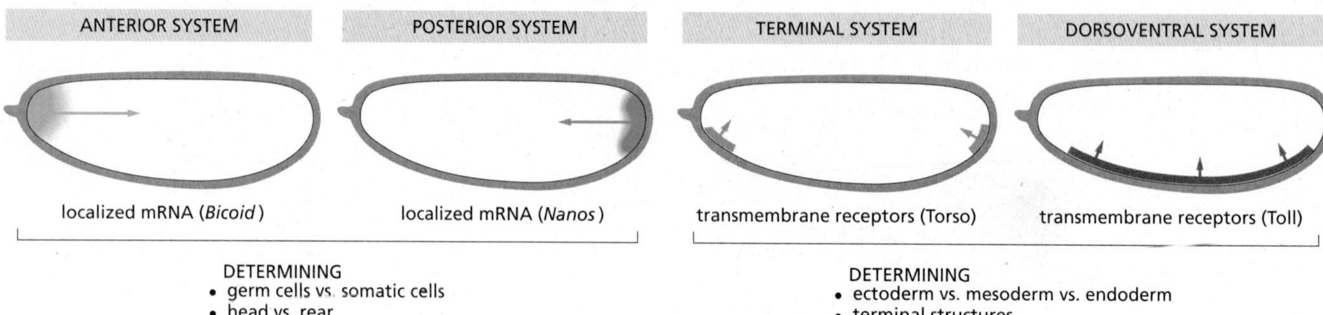

Figure 21–16 **The organization of the four egg-polarity gradient systems in *Drosophila*.** *Bicoid* mRNA encodes a transcriptional activator that determines the head and thoracic regions. Nanos is a translational repressor that governs the formation of the abdomen. Localized *Nanos* mRNA is also incorporated into the germ cells as they form at the posterior of the embryo, and Nanos protein is necessary for germ-line development. Toll and Torso are receptor proteins that are distributed all over the membrane but are activated only at the sites indicated by the coloring, through localized exposure to the extracellular ligands Spaetzle (the ligand for Toll) and Trunk (the ligand for Torso). Toll activity determines the mesoderm and Torso activity determines the formation of terminal structures at the head and tail.

dorsal versus ventral, and endoderm versus mesoderm and ectoderm—as well as a fourth partition, no less fundamental to the body plan of animals: the distinction between germ cells and somatic cells (Figure 21–16).

The egg-polarity genes act first in a hierarchy of gene systems that define a progressively more detailed pattern of body parts. In the next few pages, we begin with the molecular mechanisms that pattern the developing *Drosophila* embryo and larva along the A-P axis, before considering the patterning along the D-V axis.

Three Groups of Genes Control *Drosophila* Segmentation Along the A-P Axis

The body of an insect is divided along its A-P axis into a series of **segments**. The segments are repetitions of a theme with variations: each segment forms highly specialized structures, all built according to a similar fundamental plan (Figure 21–17). The gradients of transcription regulators set up along the A-P axis in the early embryo by the egg-polarity genes are the prelude to the creation of the segments. These regulators initiate the orderly transcription of *segmentation genes*, which refine the pattern of gene expression to define the boundaries and ground plan of the individual segments. Segmentation genes are expressed by subsets of cells in the embryo, and their products are among the first components that the embryo's own genome contributes to embryonic development; they are therefore called *zygotic-effect genes*, to distinguish them from the earlier-acting maternal-effect genes. Mutations in segmentation genes can alter either the number of segments or their basic internal organization.

The **segmentation genes** fall into three groups according to their mutant phenotypes (Figure 21–18). It is convenient to think of the three groups as acting in sequence, although in reality their functions overlap in time. First to be expressed is a set of at least six **gap genes**, whose products mark out coarse A-P subdivisions of the embryo. Mutations in a gap gene eliminate one or more groups of adjacent segments: in the mutant *Krüppel*, for example, the larva lacks eight segments. Next comes a set of eight **pair-rule genes**. Mutations in these genes

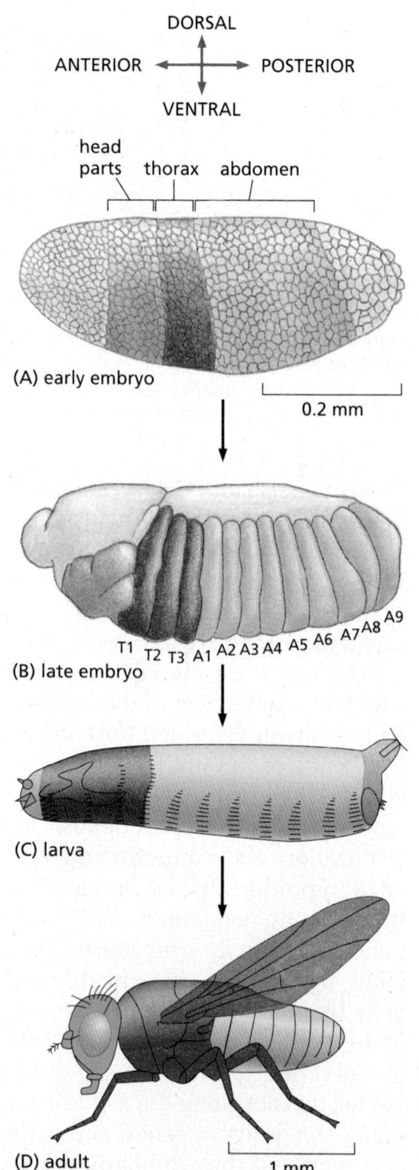

Figure 21–17 **The origins of the *Drosophila* body segments.** (A) At 3 hours, the embryo (shown in side view) is at the blastoderm stage and no segmentation is visible, although a fate map can be drawn showing the future segmented regions (color). (B) At 10 hours, all the segments are clearly defined (T1: first thoracic segment; A1: first abdominal segment). See Movie 21.3. (C) The segments of the *Drosophila* larva and their correspondence with regions in the embryo. (D) The segments of the *Drosophila* adult and their correspondence with regions in the embryo.

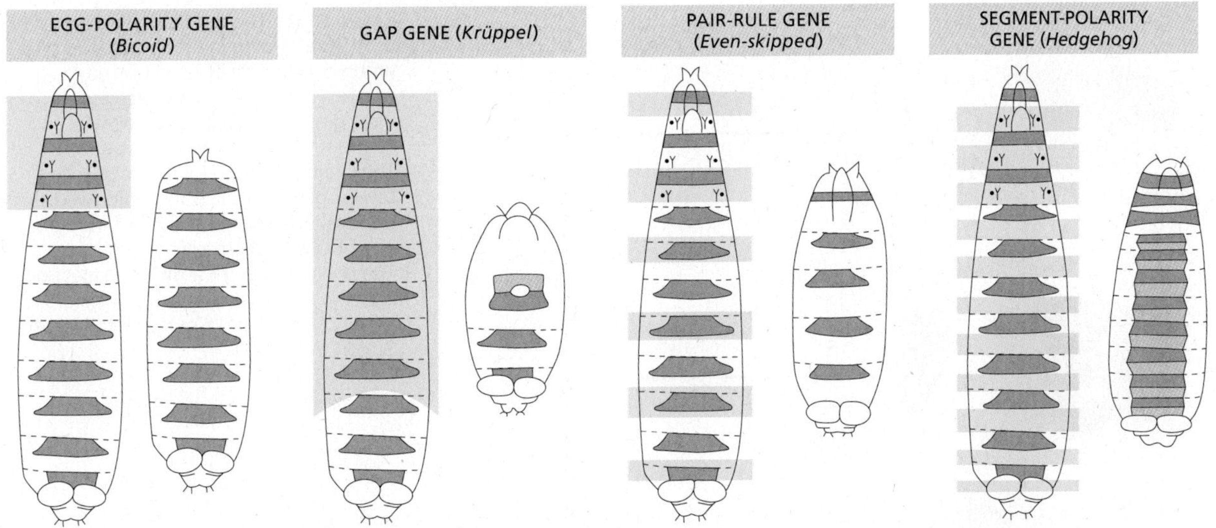

Figure 21–18 **Examples of the phenotypes of mutations affecting egg-polarity genes and the three types of segmentation genes.** In each case, the areas shaded in *green* on the normal larva *(left)* are deleted in the mutant *(right)* or are replaced by mirror-image duplicates of the unaffected regions. (Modified from C. Nüsslein-Volhard and E. Wieschaus, *Nature* 287:795–801, 1980.)

cause a series of deletions affecting alternate segments, leaving the embryo with only half as many segments as usual; although all the mutants display this two-segment periodicity, they differ in the precise pattern. Finally, there are at least 10 **segment-polarity genes**, in which mutations produce a normal number of segments but with a part of each segment deleted and replaced by a mirror-image duplicate of all or part of the rest of the segment.

The phenotypes of the various segmentation mutants suggest that the segmentation genes form a coordinated system that subdivides the embryo progressively into smaller and smaller domains along the A-P axis, each distinguished by a different pattern of gene expression. Molecular genetics has helped to reveal how this system works.

A Hierarchy of Gene Regulatory Interactions Subdivides the *Drosophila* Embryo

Like *Bicoid*, most of the segmentation genes encode transcription regulators. Their control by the egg-polarity genes and their actions on one another and on still other genes can be deciphered by comparing gene expression in normal and mutant embryos. By using appropriate probes to detect RNA transcripts or their protein products, one can observe genes switch on and off in changing patterns. These patterns reveal the wealth of spatial information created within the morphologically uniform embryo by the egg-polarity gene network. By comparing these patterns in different mutants, one can begin to discern the logic of the entire gene control system.

The products of the egg-polarity genes provide the global positional signals in the early embryo (see Figure 21–16). The Bicoid protein, as we have seen, acts as a morphogen and activates different sets of genes at different positions along the A-P axis: some gap genes are only activated in regions with high levels of Bicoid, others only where levels of Bicoid are lower. There are only six gap genes, but a combination of overlapping expression as well as different levels within their domains provides each cell along the A-P axis with a rich variety of positional identities. After the gap-gene products refine their positions by repressing each other's expression, they provide a second tier of positional signals that act more locally to regulate finer details of patterning. They control the expression of the pair-rule genes, through combinatorial effects as discussed in Chapter 7 for the pair-rule gene *Even-skipped* (see pp. 423–424). The pair-rule genes demarcate

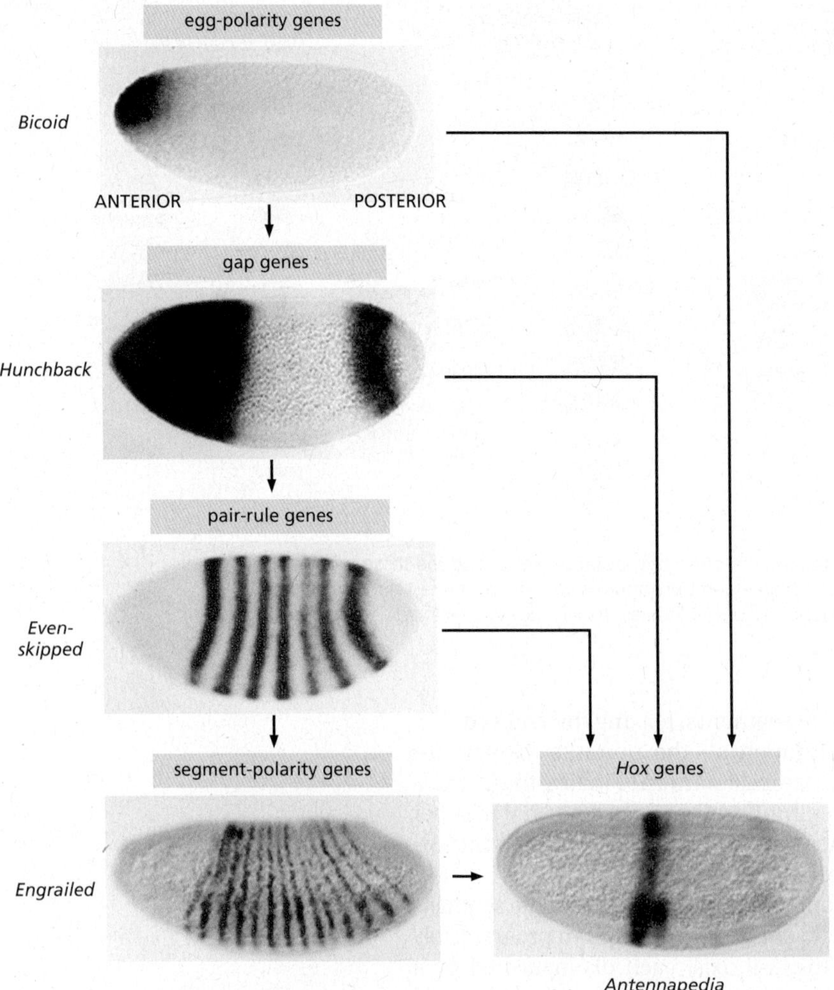

Figure 21–19 **The regulatory hierarchy of A-P patterning in the *Drosophila* embryo.** Egg-polarity genes define the A-P axis and also initiate expression of three groups of genes (gap, pair-rule, and segment polarity) that create segments. The identity of each segment is specified by *Hox* genes (discussed shortly), whose expression is controlled by input from both egg-polarity and segmentation genes. The photographs show mRNA expression patterns of representative examples of genes of each type. (Courtesy of Stephen Small.)

the repeated groups of cells that will later become segments and, in turn, collaborate with one another and with the gap genes to set up a regular, periodic pattern of expression of the segment-polarity genes, which define the internal pattern of each individual segment (**Figure 21–19**).

A large subset of the segment-polarity genes codes for components of two signaling pathways—the Wnt pathway and the Hedgehog pathway, including the secreted signal proteins Wingless (the first-named member of the Wnt family) and Hedgehog. (The Hedgehog pathway was first discovered through study of *Drosophila* segmentation, and it takes its name from the prickly appearance of the surface of the *Hedgehog* mutant embryo.) Wingless and Hedgehog are synthesized in different bands of cells that serve as signaling centers within each segment. The two proteins mutually maintain each other's expression while regulating the expression of genes such as *Engrailed* in neighboring cells (**Figure 21–20**). In such a manner, a series of sequential inductions creates a fine-grained pattern of gene expression within each segment.

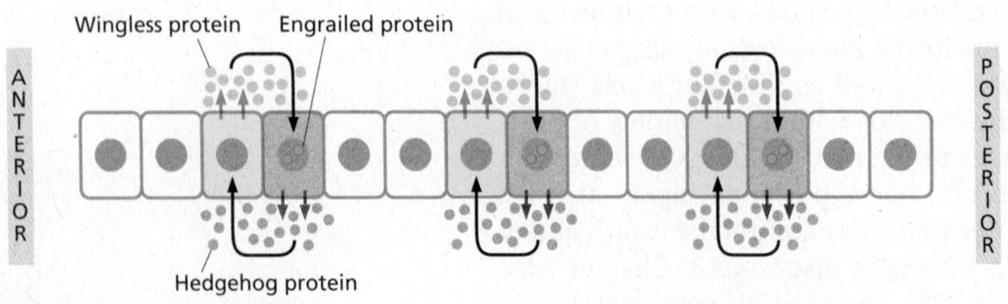

Figure 21–20 **Mutual maintenance of *Hedgehog* and *Wingless* expression.** Engrailed is a transcription regulator *(blue)* that drives the expression of *Hedgehog*. *Hedgehog* encodes a secreted protein *(red)* that activates a signaling pathway in neighboring cells and thereby drives them to express the *Wingless* gene. In turn, *Wingless* encodes a secreted protein *(green)* that acts back on neighbors of the *Wingless*-expressing cell to maintain their expression of *Engrailed*. Engrailed then maintains *Hedgehog* expression to complete the loop. As indicated, the same network repeats along the A-P axis of the fly. (Based on S. DiNardo et al., *Curr. Opin. Genet. Dev.* 4:529–534, 1994.)

Egg-Polarity, Gap, and Pair-Rule Genes Create a Transient Pattern That Is Remembered by Segment-Polarity and *Hox* Genes

The gap genes and pair-rule genes are activated within the first few hours after fertilization. Their mRNA products initially appear in patterns that only approximate the final picture; then, within a short time, this fuzzy initial pattern resolves itself into a regular, crisply defined system of stripes. But this pattern itself is unstable and transient: as the embryo proceeds through gastrulation and beyond, the pattern disintegrates. The genes' actions, however, have passed on an enduring memory of their patterns of expression by inducing the expression of certain segment-polarity genes along with another class of genes called *Hox* genes (discussed shortly). After a period of pattern refinement mediated by cell–cell interactions, the expression patterns of these new groups of patterning genes are stabilized to provide positional labels that serve to maintain the segmental organization of the larva and adult fly.

The segment-polarity gene *Engrailed* provides a good example. Its RNA transcripts form a series of 14 bands in the cellular blastoderm, each approximately one cell wide. These stripes lie immediately anterior to similar stripes of expression of another segment-polarity gene, *Wingless.* As the cells in the developing embryo continue to divide and move, signaling between the Wingless-expressing cells and the Engrailed-expressing cells maintains narrow stripes of their expression (see Figure 21–20). This interaction triggers a stable Engrailed expression pattern that will last throughout the life of the fly, long after the signals that induced and refined it have disappeared. The segment borders in embryo, larva, and adult will all form at the posterior edge of each such Engrailed stripe (**Figure 21–21**).

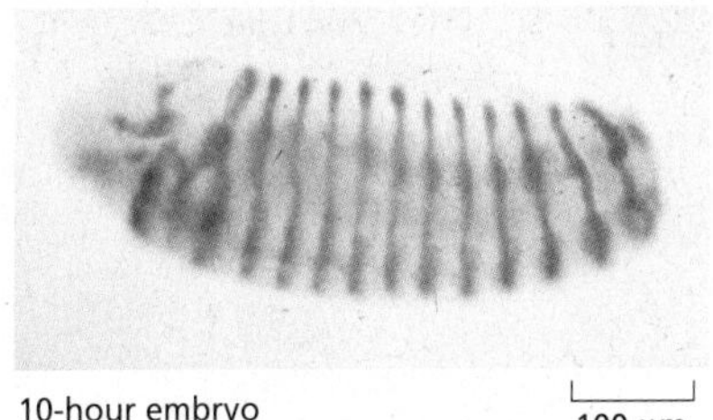

Figure 21–21 **The pattern of expression of *Engrailed*, a segment-polarity gene.** The *Engrailed* pattern is shown in a 10-hour embryo and an adult (whose wings have been removed in this preparation). The pattern is revealed by constructing a strain of *Drosophila* containing the control sequences of the *Engrailed* gene coupled to the coding sequence of the reporter *LacZ*, whose product is detected histochemically through the brown product generated by immunohistochemistry against LacZ itself (10-hour embryo) or through the blue product generated by a reaction that LacZ catalyzes (adult). Note that the *Engrailed* pattern marks segment boundaries and, once established, is preserved throughout the animal's life. (Courtesy of Tom Kornberg.)

In addition to regulating the segment-polarity genes, the products of pair-rule genes collaborate with those of gap genes to induce the precisely localized activation of a further set of genes—the *Hox genes* (see Figure 21–19). It is the *Hox* genes that first define and then permanently distinguish one segment from another. In the next section, we examine these important genes in detail; we shall see that this role is critical in a wide range of animals, including ourselves.

Hox Genes Permanently Pattern the A-P Axis

As animal development proceeds, the body becomes more and more complex. But again and again, in every species and at every level of organization, we find that complex structures are made by repeating a few elementary themes, with variations. Thus, a subset of basic differentiated cell types, such as muscle cells or fibroblasts, recur at different sites and are organized into tissues such as muscle or tendon. Subtle variations in how and where patterning mechanisms are deployed determines how structures such as teeth or digits are built, giving rise to molars and incisors, fingers and thumbs and toes.

Wherever we find this phenomenon of *modulated repetition,* we can break down the developmental biologist's problem into two kinds of questions: What is the basic construction mechanism common to all the objects of the given class, and how is this mechanism modified to give the observed variations in different animals? The segments of the insect body provide a good example. We have thus far sketched the way in which the rudiment of a single body segment is constructed and how cells within each segment become different from one another. We now consider how one segment becomes determined, or *specified,* to be different from another.

The first glimpse of the answer to this problem came more than 80 years ago, with the discovery of a set of mutations in *Drosophila* that cause bizarre disturbances in the organization of the adult fly. In the *Antennapedia* mutant, for example, legs sprout from the head in place of antennae, whereas in the *Bithorax* mutant, portions of an extra pair of wings appear where normally there should be the much smaller appendages called halteres (**Figure 21–22**). These mutations transform parts of the body into structures appropriate to other positions, and

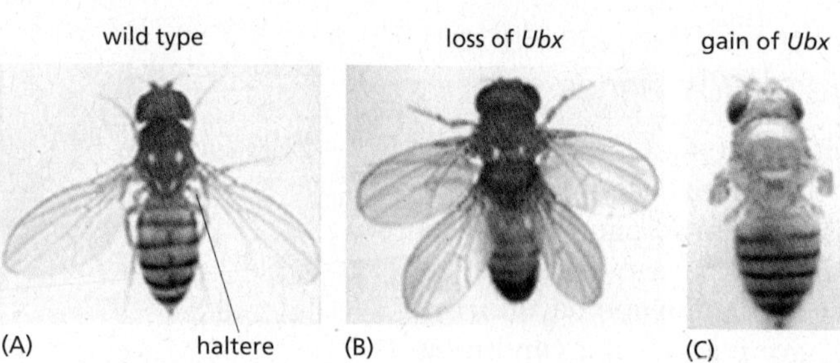

Figure 21–22 Homeotic mutations. *Ultrabithorax*, or *Ubx*, is one of three genes in the *Bithorax* gene complex (a *Hox* gene cluster). *Ubx* is responsible for all of the differences between the second (wing-bearing) and third (haltere-bearing) thoracic segments. (A and B) *Ubx* loss-of-function mutations transform the haltere-bearing segment (A) into a wing-bearing segment, resulting in four-winged flies (B). (C) *Ubx* gain-of-function in the second thoracic segment transforms this wing-bearing segment into a haltere-bearing segment, resulting in wingless flies. (A, courtesy of the Archives, California Institute of Technology; C, courtesy of L.S. Shashidhara.)

they are called *homeotic* mutations (from the Greek *homoios*, meaning "similar") because the transformation is between structures of a recognizably similar general type, changing one kind of limb or one kind of segment into another. It was eventually discovered that a whole set of genes, the **homeotic selector genes** or **Hox genes**, serve to permanently specify the A-P characters of the whole set of animal segments. These genes are all related to one another as members of a multigene family.

There are eight *Hox* genes in the fly, and they all lie in one or the other of two gene clusters known as the ***Bithorax* complex** and the ***Antennapedia* complex**. The genes in the *Bithorax* complex control the differences among the abdominal and thoracic segments of the body, while those in the *Antennapedia* complex control the differences among thoracic and head segments. Comparisons with other species show that the same genes are present in essentially all animals, including humans. These comparisons also reveal that the *Antennapedia* and *Bithorax* complexes are two halves of a single entity, called the ***Hox* complex**, that has become split in the course of the fly's evolution, and whose members operate in a coordinated way to exert their control over the head-to-tail pattern of the body.

The products of the *Hox* genes, the **Hox proteins**, are transcription regulators, all of which possess a highly conserved, 60-amino-acid-long DNA-binding *homeodomain* (see p. 404). The homeodomain-encoding DNA sequence is called a "homeobox," from which, by abbreviation, the *Hox* complex takes its name. There are many homeobox-containing genes, but only those located in a *Hox* complex are *Hox* genes.

Hox Proteins Give Each Segment Its Individuality

The Hox proteins can be viewed as molecular address labels possessed by the cells of each segment: these labels give the cells in each region a **positional value**; that is, an intrinsic character that differs according to a cell's location. If the address labels in a developing *Drosophila* segment are changed, the segment behaves as though it were located somewhere else; if all the *Hox* genes in an embryo are deleted, the body segments in the larva will all be alike.

To a first approximation, each *Hox* gene is normally expressed in those regions that develop abnormally when that gene is mutated or absent. How does each Hox protein give a segment its permanent identity? Recall that the Hox proteins are transcription regulators, which can bind to gene regulatory DNA; each Hox protein targets a different set of genes for activation or repression. Hundreds of genes are under this type of Hox-modulated control, including genes that control cell–cell signaling, transcriptional regulation, cell polarity, cell adhesion, cytoskeletal function, cell growth, and cell death, all conspiring to give each segment its distinctive Hox-dependent character.

Hox Genes Are Expressed According to Their Order in the *Hox* Complex

How, then, is the expression of the *Hox* genes themselves regulated? The coding sequences of the eight *Hox* genes in *Drosophila* are interspersed amid a much larger quantity of regulatory DNA. This DNA includes binding sites for the

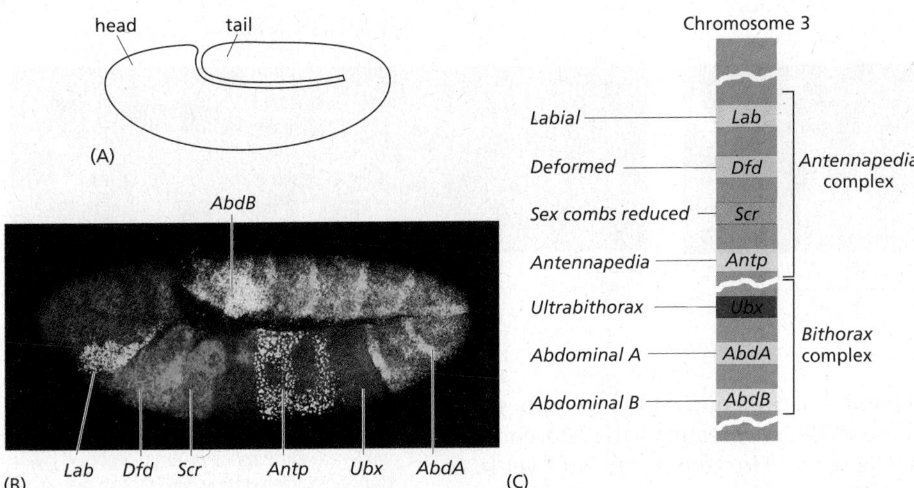

Figure 21–23 The patterns of expression compared to the chromosomal locations of the genes of the *Hox* complex. (A) Diagram of a *Drosophila* embryo at the so-called germ band retraction stage, about 10 hours after fertilization when the developing body axis is folded over on itself. (B) An embryo at this stage has been stained by *in situ* hybridization using differently labeled probes to detect the mRNA products of different *Hox* genes in different colors. (C) The spatial pattern in the photograph corresponds, with minor deviations, to the sequence of genes in each of the two subdivisions of the chromosomal complex. (B, courtesy of William McGinnis, adapted from D. Kosman et al., *Science* 305:846, 2004. With permission from AAAS.)

products of the egg-polarity and segmentation genes, thereby serving as an interpreter of the detailed spatial information supplied to it by all of these transcription regulators. The net result is that a particular set of *Hox* genes is transcribed in a specific region along the A-P body axis.

The pattern of *Hox* gene expression exhibits a remarkable regularity that suggests an additional form of control. The sequence in which the genes are ordered along the chromosome, in both the *Antennapedia* and the *Bithorax* complexes, corresponds almost exactly to the order in which they are expressed along the A-P axis of the body (**Figure 21–23**). This hints at some process of gene activation, perhaps dependent on chromatin structures that propagate along the *Hox* complexes, switching on one *Hox* gene after another according to their order along the chromosome. The most "posterior" of the *Hox* genes that are expressed in a cell generally dominates, driving down expression and activity of the "anterior" genes and dictating the character of the segment. The gene regulatory mechanisms underlying these phenomena are still not well understood, but their consequences are profound. We shall see that the serial organization of gene expression in the *Hox* complex is a fundamental feature that has been highly conserved in the course of animal evolution.

Trithorax and Polycomb Group Proteins Regulate *Hox* Expression to Maintain a Permanent Record of Positional Information

The spatial pattern of expression of the genes in the *Hox* complex is set up by signals acting early in development, but the effects are long lasting. Although the pattern of expression undergoes complex adjustments as development proceeds, the *Hox* pattern stamps each cell and all of its progeny with a permanent record of the A-P position that the cell occupied in the early embryo. In this way, the cells of each segment maintain a memory of their location along the A-P axis of the body, which governs the segment-specific identity not only of the larval segments but also of the structures of the adult fly.

Two molecular mechanisms ensure that a cell remembers its positional information. One is from the *Hox* genes themselves: many of the Hox proteins autoactivate the transcription of their own genes, thereby helping to keep the genes on indefinitely. Another crucial input is from two large, complementary sets of proteins, called the **Trithorax group** and the **Polycomb group**, which imprint the chromatin of the *Hox* complex with a heritable record of its embryonic state of activation or repression. These are key general regulators of chromatin structure that are critical for cell memory: if genes of the *Trithorax* or *Polycomb* group are defective, the pattern of expression of the *Hox* genes is set up correctly at first, but it is not correctly maintained as cells divide and the embryo grows older.

The two sets of regulators act in opposite ways. Trithorax group proteins are needed to maintain the transcription of *Hox* genes in cells where their

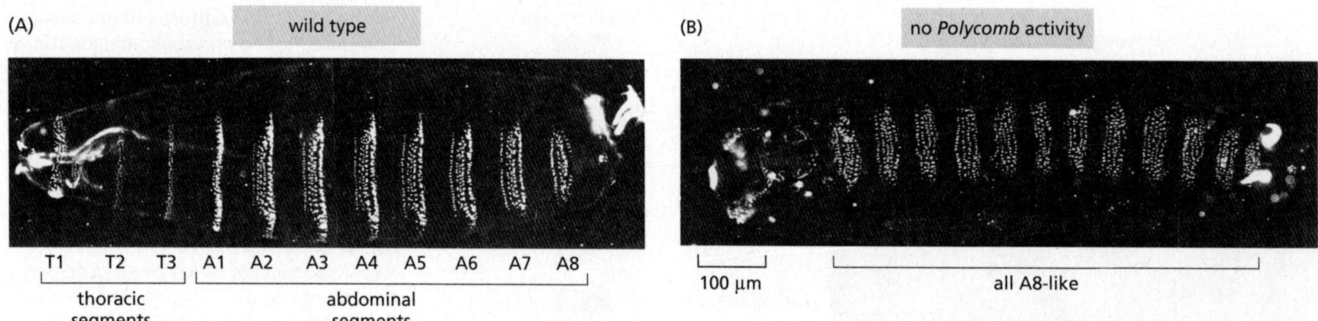

Figure 21–24 **The role of genes of the *Polycomb* group.** (A) Photograph of a wild-type *Drosophila* embryo, imaged by dark-field microscopy. (B) Photograph of a mutant embryo defective for the gene *Extra sex combs* (*Esc*) and derived from a mother also lacking this gene. The gene belongs to the *Polycomb* group. Essentially all segments have been transformed to resemble the most posterior abdominal segment, A8. In the mutant, the pattern of expression of the homeotic selector genes, which is roughly normal initially, is unstable in such a way that all these genes soon become switched on all along the body axis. (From G. Struhl, *Nature* 293:36–41, published 1981 by Nature Publishing Group. Reproduced with permission of SNCSC.)

transcription has already been switched on. In contrast, Polycomb group proteins form stable complexes that bind to the chromatin of the *Hox* complex and maintain the repressed state in cells where *Hox* genes have not yet been activated (Figure 21–24). Although first discovered because of their influence on *Hox* genes in flies, Polycomb and Trithorax group proteins are general regulators of chromatin structure that control many genes in plants as well as animals. How such changes in chromatin can store developmental cell memory is discussed in Chapters 4 and 7.

The D-V Signaling Genes Create a Gradient of the Transcription Regulator Dorsal

We now turn to patterning of the second major axis of the *Drosophila* embryo. As with the patterning along the A-P axis just discussed, the patterning along the dorsoventral (D-V) axis begins with maternal gene products that define this axis in the egg (see Figure 21–16) and then progresses through zygotic gene products that further subdivide the D-V axis in the embryo.

Initially, a protein that is produced by the mother's somatic cells underneath the future ventral region of the embryo leads to the localized activation of a transmembrane receptor called **Toll** on the ventral side of the egg membrane. (Curiously, *Drosophila* Toll and vertebrate Toll-like proteins also operate in innate immune responses, as discussed in Chapter 24.) The localized activation of Toll controls the distribution of **Dorsal**, a transcription regulator of the NFκB family discussed in Chapter 15. The Toll-regulated activity of Dorsal, like that of NFκB, depends on the translocation of Dorsal protein from the cytosol, where it is held in an inactive form, to the nucleus, where it regulates gene expression (see Figure 15–63). In the newly laid egg, both Dorsal mRNA and protein are distributed uniformly in the cytosol. After the nuclei in the syncytial blastoderm have migrated to the surface of the embryo, but before cellularization (see Figure 21–14), Toll receptor activation on the ventral side induces a remarkable redistribution of the Dorsal protein. On the dorsal side, the protein remains in the cytosol, but ventrally it becomes concentrated in the nuclei, with a smooth gradient of nuclear localization between these two extremes (Figure 21–25).

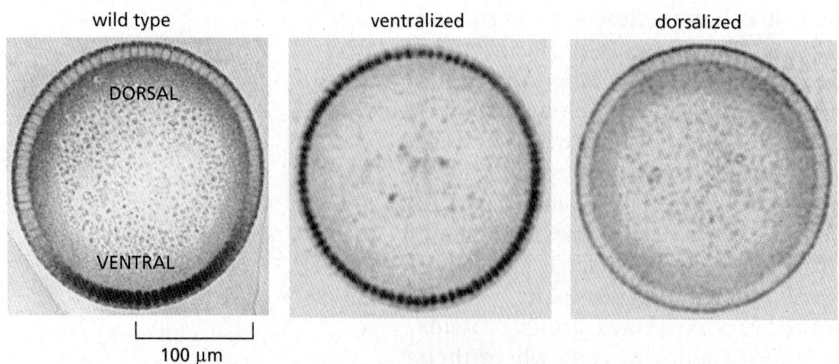

Figure 21–25 **The concentration gradient of Dorsal protein in the nuclei of the blastoderm.** In wild-type *Drosophila* embryos, the protein is present in the dorsal cytoplasm and absent from the dorsal nuclei; ventrally, it is depleted in the cytoplasm and concentrated in the nuclei. In a mutant in which the Toll pathway is activated everywhere and not just ventrally, Dorsal protein is everywhere concentrated in the nuclei; the result is a ventralized embryo. Conversely, in a mutant in which the Toll signaling pathway is inactivated, Dorsal protein everywhere remains in the cytoplasm and is absent from the nuclei; the result is a dorsalized embryo. (From S. Roth et al., *Cell* 59:1189–1202, 1989. With permission from Elsevier.)

Similar to Bicoid along the A-P axis, Dorsal acts as a morphogen along the D-V axis. Once inside the nucleus, the Dorsal protein turns on or off the expression of different sets of genes depending on Dorsal's concentration. The expression of each responding gene depends on its regulatory DNA—specifically, on the number and affinity of the binding sites that this DNA contains for Dorsal and other transcription regulators. In this way, the regulatory DNA interprets the positional signal provided by the nuclear Dorsal protein gradient, so as to define a distinct D-V series of territories—complementary bands of cells that run the length of the embryo. Most ventrally—where the nuclear concentration of Dorsal protein is highest—it switches on, for example, the expression of a gene called *Twist*, which directs mesodermal fate. Most dorsally, where the nuclear concentration of Dorsal protein is lowest, the cells switch on a gene called *Decapentaplegic* (*Dpp*). And in an intermediate region, where the nuclear concentration of Dorsal protein is high enough to repress *Dpp* but too low to activate *Twist*, the cells switch on another set of genes, including one called *Short gastrulation* (*Sog*) (**Figure 21–26A**).

Products of the genes directly regulated by the Dorsal protein generate in turn more local signals, which define finer subdivisions along the D-V axis. These signals act after cellularization and take the form of conventional extracellular diffusible proteins. In particular, *Dpp* codes for a secreted TGFβ family protein, which forms a local morphogen gradient in the dorsal part of the embryo. *Sog*, produced ventrally to *Dpp*, encodes another secreted protein that acts as an antagonist of Dpp protein, by binding to it and preventing Dpp from activating its receptor. The opposing diffusion gradients of these two signal proteins create a steep gradient of Dpp activity: the highest Dpp activity levels, in combination with certain other factors, cause development of the most dorsal tissue of all—an

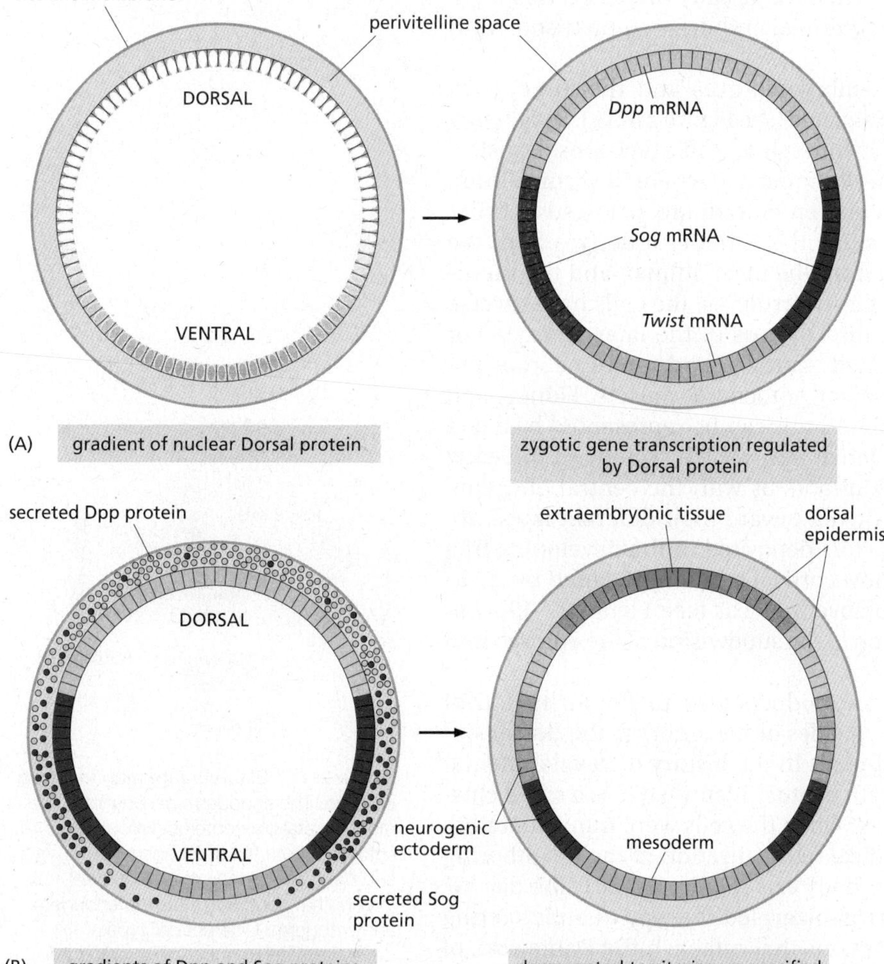

Figure 21–26 How morphogen gradients guide a patterning process along the dorsoventral axis of the *Drosophila* embryo. (A) Initially, a gradient of Dorsal protein defines three broad territories of gene expression, marked here by the expression of three representative genes: *Dpp*, *Sog*, and *Twist*. (B) Slightly later, the cells expressing *Dpp* and *Sog* secrete, respectively, the signal proteins Dpp (a TGFβ family member) and Sog (an antagonist of Dpp). These two proteins then diffuse and interact with one another (and with certain other factors) to create the dorsoventral (D-V) territories shown.

extraembryonic membrane. Intermediate levels cause development of dorsal epidermis; and the absence of Dpp activity in cells expressing Sog allows the development of neurogenic ectoderm, which will give rise to the nervous system (**Figure 21–26B**).

A Hierarchy of Inductive Interactions Subdivides the Vertebrate Embryo

The molecular genetic analysis of *Drosophila* development has uncovered how a cascade of transcription regulators and signaling pathways sequentially subdivides the embryo. The same principle of progressive pattern refinement is used during the development of all animal embryos, including vertebrates. Remarkably, conservation is not restricted to the general strategy of pattern formation, but also extends to many of the molecules involved.

As mentioned previously, the earliest phases of vertebrate development are surprisingly variable, even between closely related species, and it is even hard to say precisely how the A-P and D-V axes of an early fly embryo correspond to those of an early frog or mouse embryo. Nevertheless, we shall see that amid this display of evolutionary plasticity, some features of early development turn out to be highly conserved. The same is true of later developmental stages also, often to an astonishing degree. From our own anatomy, it is obvious that we are cousins to birds and fish. But looking at molecular mechanisms, we see that we are related to flies and worms too.

In the following pages, we discuss how vertebrate embryos are patterned by the interplay of signaling molecules and transcription regulators. We begin by discussing the formation and patterning of the embryonic axes in amphibians, taking the frog *Xenopus* as our example. We have already broached this topic earlier in the chapter. Here, we pick up the thread and draw comparisons with the fly.

As noted earlier, the origins of the embryonic axes and the three germ layers in the frog can be traced back to the blastula (see Figure 21–3A). By labeling individual blastomeres, we can track cells through all their divisions, transformations, and migrations and see what they become and where they come from. The precursors of ectoderm, mesoderm, and endoderm are arranged in order along the animal-vegetal axis of the blastula: the endoderm derives from the most vegetal blastomeres, the ectoderm from the most animal, and the mesoderm from a middle set. Within each of these territories, the cells have diverse fates according to their positions along the D-V axis of the later embryo. For ectoderm, epidermal precursors are located ventrally, and future neurons are found dorsally; for mesoderm, precursors for notochord, muscle, kidney, and blood are arranged from dorsal to ventral. All this can be represented by a **fate map** that shows which later cell types derive from which regions of the early embryo (**Figure 21–27**). The fate map confronts us with the central question: How are the cells in different positions driven toward their different fates? We have already explained how maternal factors deposited in the developing frog egg define its animal-vegetal axis, and how cortical rotation triggered by fertilization defines the orientation of the dorsoventral axis (see Figure 21–13). But how does the establishment of axes lead on to the subdivision of the embryo into the future body parts?

The answer is that the maternal gene products lead to the formation of *signaling centers* on the vegetal and dorsal sides of the embryo. The dorsal signaling center in particular has a special place in the history of developmental biology. Experiments in the early twentieth century identified it as a small cluster of cells with an extraordinary property: when the cells were transplanted to an opposite site, they could trigger a radical reorganization of the neighboring tissue, causing it to form a second whole-body axis (**Figure 21–28**). The discovery of this signaling center, called the **Organizer**, led the way to a pioneering analysis of the chain of inductive interactions that establish the framework of the vertebrate body.

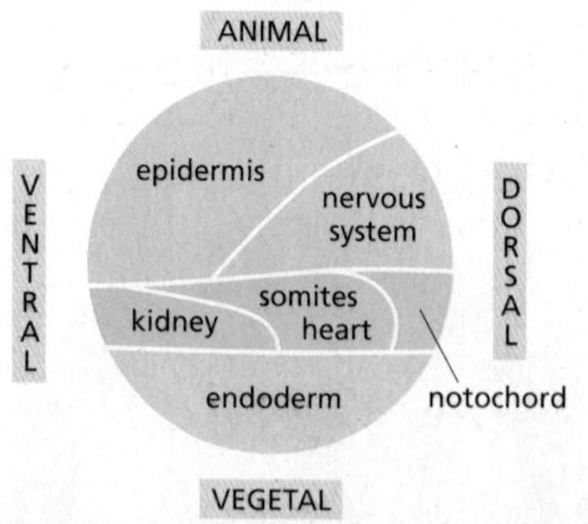

Figure 21–27 **Blastula fate map in a frog embryo.** The endoderm derives from the most vegetal blastomeres *(yellow)*, the ectoderm from the most animal *(blue)*, and the mesoderm from a middle set *(green)* that contributes also to endoderm and ectoderm. Different cell types derive from different positions along the dorsoventral axis.

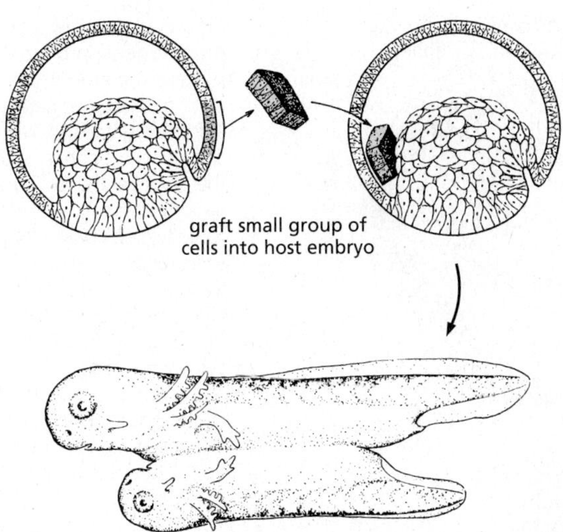

Figure 21–28 **Induction of a secondary axis by the Organizer.** An amphibian embryo receives a graft of a small cluster of cells taken from a specific site, called the Organizer region, on the dorsal side of another embryo at the same stage. Signals from the graft organize the behavior of neighboring cells of the host embryo, causing development of a pair of conjoined (Siamese) twins. See Movie 21.4. [After J. Holtfreter and V. Hamburger, in Analysis of Development (B.H. Willier, P.A. Weiss, and V. Hamburger, eds.), pp. 230–296. Philadelphia: Saunders, 1955.]

In contrast to the *Drosophila* syncytial embryo, the fertilized frog egg undergoes conventional cleavage divisions that result in an embryo consisting of thousands of cells. Patterning must therefore be mediated by extracellular signal molecules that diffuse through the embryo from cell to cell, not by transcription regulators that move through the cytoplasm of a syncytium. Not surprisingly, the Organizer is now known to be a major source of secreted signals. As we shall see, this includes not only ligands that bind and activate transmembrane receptors (see Chapter 15), but also secreted proteins that inhibit the activity of these ligands.

A Competition Between Secreted Signaling Proteins Patterns the Vertebrate Embryonic Axes

The signal molecules that pattern the frog embryo along the animal–vegetal (A-V) axis belong to the TGFβ family: they are secreted by a signaling center at the vegetal pole and form concentration gradients along the A-V axis. These *Nodal* proteins act over a relatively short range: cells closest to the vegetal pole are exposed to high levels and respond by switching on genes that promote the development of endoderm; cells farther away are exposed to lower levels and activate genes that promote the formation of mesoderm. The cells at the vegetal pole that produce Nodal also produce a second, more rapidly diffusing protein called *Lefty*, which antagonizes Nodal. The high ratio of Lefty to Nodal at the animal pole allows Lefty to block Nodal signaling; this causes the cells there to develop as ectoderm (Figure 21–29A). Thus, a mid-range activation by Nodal, combined with a long-range inhibition by Lefty, sets up the pattern of progenitors along the A-V axis for the three germ layers—endoderm, mesoderm, and ectoderm.

The frog uses a somewhat related strategy to subdivide the germ-layer territories along the D-V axis of the embryo. It relies on patterned inhibition of otherwise uniform signaling by the *bone morphogenetic proteins* (*BMPs*; members of yet another subclass of the TGFβ family), which are secreted throughout the embryo. The dorsal signaling system exerts its influence by secreting several proteins, including *Chordin* and *Noggin*, that block BMP signaling when their own concentrations are high. In this way, Chordin and Noggin create a dorsal-to-ventral gradient of BMP, with low activity on the dorsal side and high activity on the ventral side (Figure 21–29B). Ectodermal cells that experience high levels of BMP signaling are driven to epidermal fates, whereas cells that experience little or no BMP signaling remain neural. We can note that this strategy for patterning the D-V axis by opposing gradients of BMP family signals and diffusible inhibitors is similar to that used in *Drosophila*, and indeed the particular molecules used are homologous.

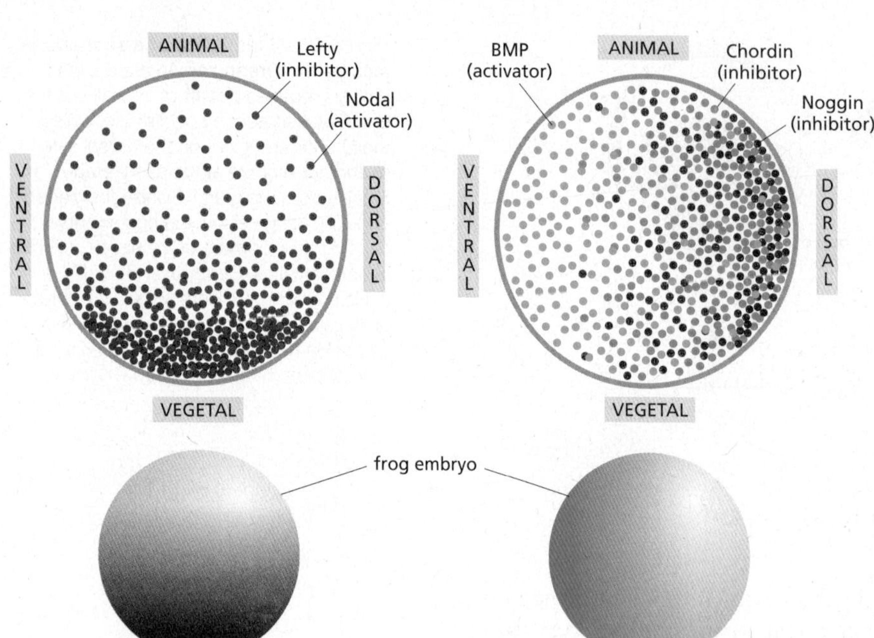

Figure 21–29 How Nodal and bone morphogenic protein (BMP) signaling pattern the embryonic axes. Nodal and its antagonist Lefty pattern the animal–vegetal axis, while BMP and its antagonists Chordin and Noggin pattern the dorsoventral axis. (A) In the animal-pole region, where Nodal levels are low relative to Lefty, Lefty blocks Nodal from binding to its receptors. In the vegetal region, there is an excess of Nodal, resulting in Nodal pathway activation. (B) Along the dorsoventral axis, BMP is widely present, but Chordin and Noggin are concentrated at the dorsal side: there, they bind to BMP and block its binding to receptors. The resulting patterns of Nodal and BMP activity are illustrated at the bottom of the figure.

Knowing the signals that specify the three germ layers and various tissue types of the vertebrate body, one can reproduce this specification in a culture dish. Frog cells taken from the animal-pole region of the embryo, for example, will differentiate into blood (a ventral mesodermal tissue) when diverted from their original fate by exposure to intermediate concentrations of Nodal and high concentrations of BMP. Similarly, mouse or human embryonic stem cells can be coaxed into generating specific cell types by exposing them in culture to appropriate combinations of signal molecules. In this way, the insights gained through studies of animal development can be used to generate the cell types needed for regenerative medicine, as we discuss in the next chapter.

Hox Genes Control the Vertebrate A-P Axis

The conservation of developmental mechanisms between *Drosophila* and vertebrates extends far beyond the D-V signaling system. *Hox* genes are found in almost every animal species studied, where they are often grouped in complexes similar to the insect *Hox* complex. In mice and humans, for example, there are four such complexes—called the *HoxA, HoxB, HoxC,* and *HoxD* complexes—each on a different chromosome. Individual genes in each complex can be recognized by their sequences as counterparts of specific members of the *Drosophila* set. Indeed, mammalian *Hox* genes can function in *Drosophila* as partial replacements for the corresponding *Drosophila Hox* genes. It appears that each of the four mammalian *Hox* complexes is, roughly speaking, the equivalent of one complete insect *Hox* complex (that is, an *Antennapedia* complex plus a *Bithorax* complex) (**Figure 21–30**).

The ordering of the genes within each vertebrate *Hox* complex is essentially the same as in the insect *Hox* complex, suggesting that all four vertebrate complexes originated by duplications of a single primordial complex present in the common ancestor of vertebrates and insects, and have preserved its basic organization. Most tellingly, the members of each vertebrate *Hox* complex are expressed in a head-to-tail series along the axis of the embryo, just as they are in *Drosophila*. As in *Drosophila*, vertebrate *Hox* gene expression patterns are often aligned with vertebrate segments. This alignment is especially clear in the hindbrain (see Figure 21–30), where the segments are called *rhombomeres*.

The products of the vertebrate *Hox* genes, the Hox proteins, specify positional values that control the A-P pattern of parts in the hindbrain, neck, and trunk (as well as some other parts of the body). As in *Drosophila*, when a posterior *Hox*

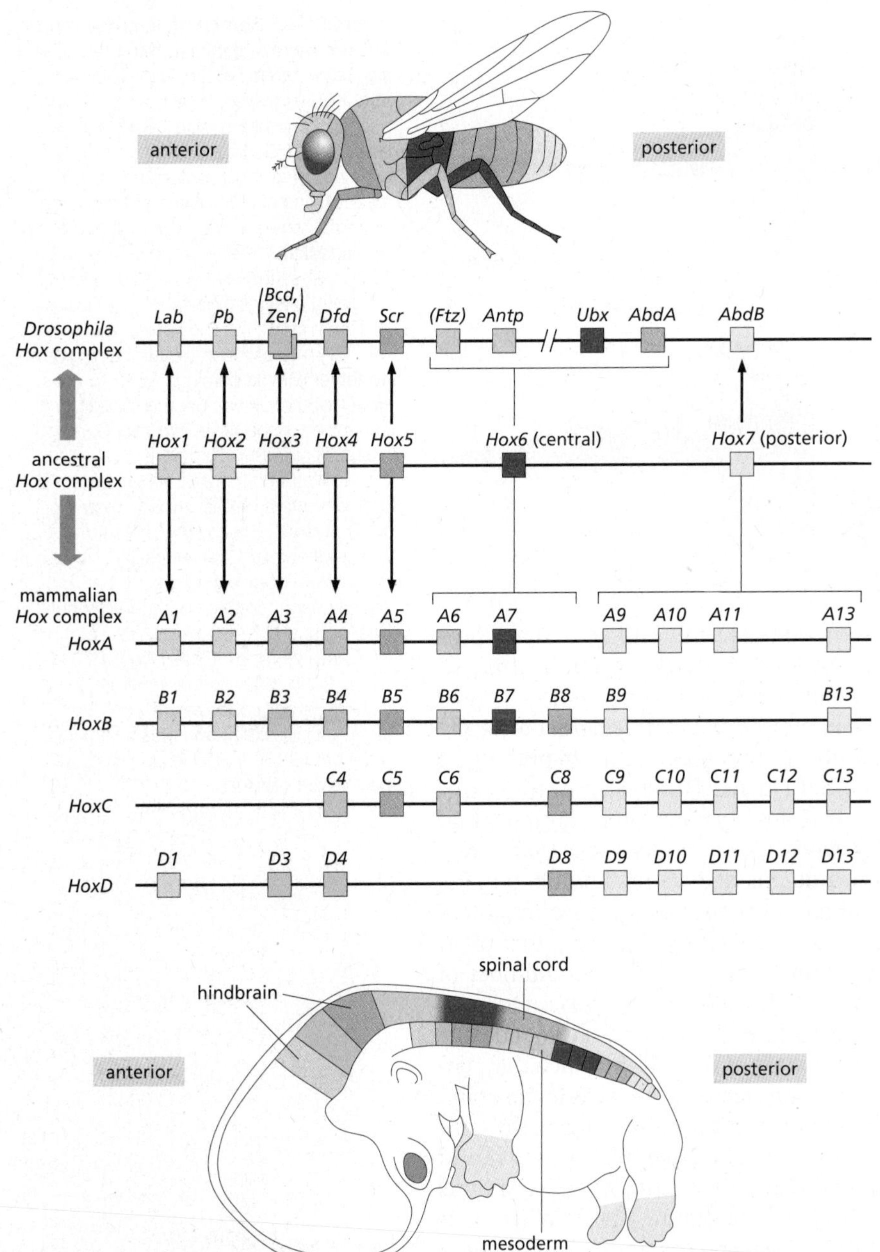

Figure 21–30 The *Hox* complexes of an insect and a mammal, compared and related to body regions. The genes of the *Antennapedia* and *Bithorax* complexes of *Drosophila* are shown in their chromosomal order in the top line (see Figure 21–23). The corresponding genes of the four mammalian *Hox* complexes are shown below, also in chromosomal order. The gene expression domains in fly and mammal are indicated in a simplified form by color in the diagrams of animals above and below. There is a remarkable parallelism. However, the details of the patterns depend on developmental stage and vary somewhat from one mammalian *Hox* complex to another. Also, in many cases, genes shown here as expressed in an anterior domain are also expressed more posteriorly, overlapping the domains of more posterior *Hox* genes.

The complexes are thought to have evolved as follows: first, in some common ancestor of worms, flies, and vertebrates, a single primordial homeotic selector gene underwent repeated duplication to form a series of such genes in tandem—the ancestral *Hox* complex. In the *Drosophila* sublineage, this single complex became split into separate *Antennapedia* and *Bithorax* complexes. Meanwhile, in the lineage leading to the mammals, the whole complex was repeatedly duplicated to give four *Hox* complexes. The parallelism is not perfect because apparently some individual genes have been duplicated and others lost. Still others have been co-opted for different purposes (genes in parentheses in the top line) over the time that has elapsed since the complexes diverged. (Based on a diagram courtesy of William McGinnis.)

gene is artificially expressed in an anterior region, it can convert the anterior tissue to a posterior character. Conversely, loss of posterior *Hox* genes allows the posterior tissue where they are normally expressed to adopt an anterior character (**Figure 21–31**). Because of overlapping functions between genes in the four *Hox* gene clusters, the transformations observed in mouse *Hox* mutants are not always so straightforward as those in the fly, and they are often incomplete. Nonetheless, it seems clear that the fly and the mouse use essentially the same molecular machinery to impart individual characteristics to successive regions along at least a part of the A-P axis.

Some Transcription Regulators Can Activate a Program That Defines a Cell Type or Creates an Entire Organ

Just as there are single genes that specify the identity of a particular region of the body, there are also genes whose products act as triggers for the development of a specific cell type or even a specific organ, initiating and coordinating the whole complex program of gene expression that is required. An example is the *MyoD*

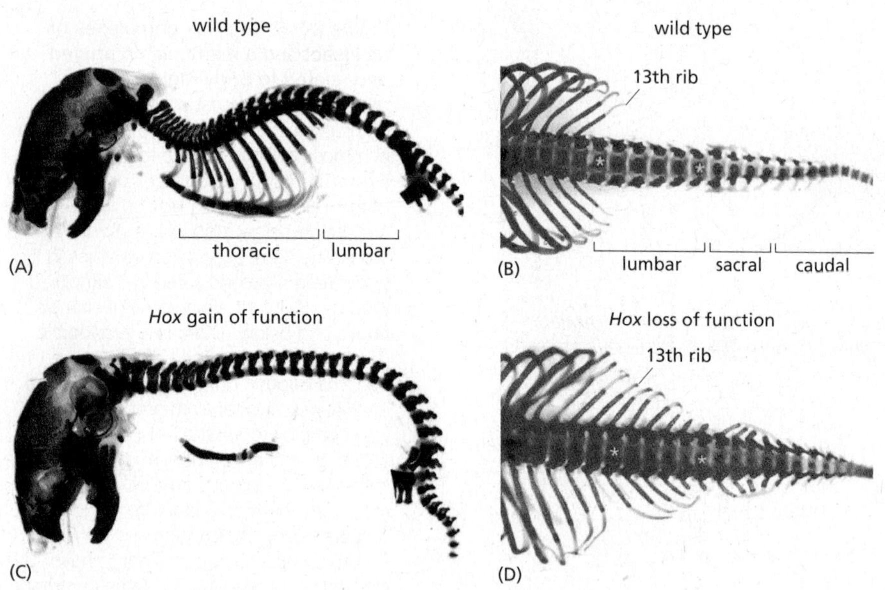

Figure 21–31 Control of anteroposterior pattern by *Hox* genes in the mouse. (A and B) A normal mouse (wild type) has about 65 vertebrae, differing in structure according to their position along the body axis: 7 cervical (neck), 13 thoracic (with ribs), 6 lumbar (bracketed by *yellow asterisks* in panel B), 4 sacral (bracketed by *red asterisks* in panel B), and about 35 caudal (tail). Panel A shows a side view, and panel B shows a dorsal view; for clarity, the limbs have been removed in each picture. (C) The *HoxA10* gene is normally expressed in the lumbar region (together with its paralogs *HoxC10* and *HoxD10*); here it has been artificially expressed in the developing vertebral tissue all along the body axis. As a result, the cervical and thoracic vertebrae are all converted to a ribless, lumbar character. (D) Conversely, when *HoxA10* is removed along with *HoxC10* and *HoxD10*, vertebrae that should normally have a lumbar or sacral character take on a thoracic, rib-bearing character instead. (A and C, from M. Carapuço et al., *Genes Dev*. 19:2116–2121, 2005. With permission from Cold Spring Harbor Laboratory Press; B and D, from D.M. Wellik and M.R. Capecchi, *Science* 301:363–367, 2003. With permission of AAAS.)

family of transcription regulators that we encountered in Chapter 7. These proteins drive cells to differentiate into muscle, expressing muscle-specific actins and myosins and all the other specialized cytoskeletal, metabolic, and membrane proteins that a muscle cell needs. Analogously, members of the Achaete/Scute family of transcription regulators drive cells to become neural progenitors. In both these examples, the proteins belong to the helix–loop–helix (HLH) class of transcription regulators (see p. 405), and the same is true for several other proteins that induce the differentiation of particular cell types. These *master transcription regulators* exert their powerful differentiation-inducing activity by binding to many different regulatory sites in the genome and thereby controlling the expression of large numbers of downstream target genes. In one well-studied case, that of an Achaete/Scute family member called Atonal homolog 1 (Atoh1), the number of direct target genes in the mouse genome is more than 600. It is important to note, however, that even such powerful drivers of cell differentiation can have radically different effects according to the context and history of the cells in which they act: Atoh1, for example, drives differentiation of certain classes of neurons in the brain, of sensory hair cells in the inner ear, and of secretory cells in the lining of the gut.

Other genes encoding transcription regulators can drive the formation and assembly of the multiple cell types that constitute an entire organ. A famous example is the transcription regulator Eyeless (see Figure 21–2). When it is artificially expressed in a patch of cells in the leg precursors of *Drosophila*, a well-organized eye-like organ develops on the leg, with the various eye cell types correctly arranged; conversely, loss of the *Eyeless* gene results in flies that lack eyes. Moreover, loss of the *Eyeless* homolog *Pax6* in vertebrates likewise leads to loss of eye structures; humans with *Pax6* mutations display a condition called congenital aniridia (**Figure 21–32**). Similar organ-selector proteins are known for foregut, heart, pancreas, and other organs. They are all master transcription regulators that directly regulate hundreds of target genes, the products of which then specify and construct the different elements of the appropriate organ. However, as in the example of Atoh1, they usually exert their specific effect only in combination with the right partners, which are only expressed in cells that were appropriately primed during their earlier development.

Notch-mediated Lateral Inhibition Refines Cellular Spacing Patterns

After the establishment of the basic body plan and the generation of organ precursors, many further steps of pattern refinement are required to achieve the adult pattern of terminally differentiated cells in tissues and organs. Lateral

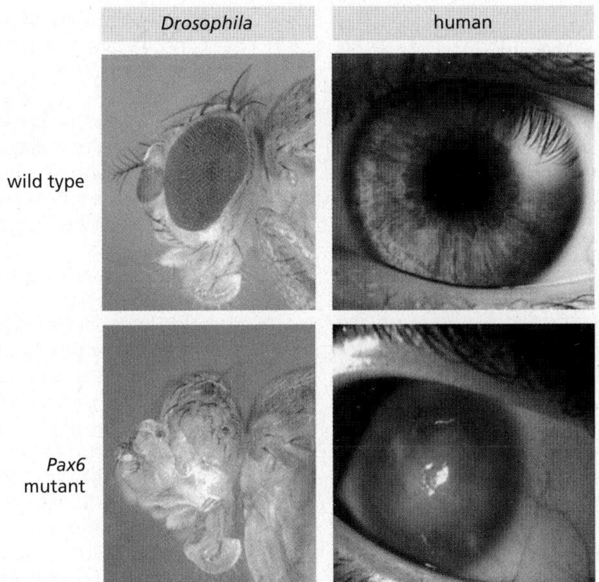

Figure 21–32 **The Pax6 protein (also called Eyeless in *Drosophila*) controls development of light-sensing organs in many animal species.** Flies or humans carrying mutations in the *Pax6* gene lack certain eye structures present in normal animals. [Fly pictures courtesy of Katy Ong; human pictures from N.L. Washington et al., *PLoS Biol.* 7(11):e1000247, 2009.]

inhibition (see Figure 21–10) mediated by Notch signaling is crucial for both cell diversification and fine-grained patterning in an enormous variety of tissues in all animals.

Certain tissues require the evenly spaced distribution of a specialized cell type, such as sensory cells, secretory cells, or stem cells, across their expanse. A good example is the development of **sensory bristles** in *Drosophila*, most easily seen on the fly's back, but also present on most of its other exposed surfaces. Each of these is a miniature sense organ, consisting of a sensory neuron and a small set of supporting cells. Some bristles respond to chemical stimuli and others to mechanical stimuli, but they are all constructed in a similar way (**Figure 21–33**).

The proneural genes *Achaete* and *Scute* mentioned earlier mark the clusters of epidermal cells within which bristles will form, but exactly which cells form a bristle depends on competitive interactions among them. Signaling within each cluster selects a single cell—called the *sensory organ precursor cell*—to serve as the bristle progenitor. The progenitor becomes distinct from surrounding cells through lateral inhibition (see Figure 21–10), which is mediated by the Notch signaling pathway. Binding of the transmembrane receptor *Notch* on the surface of a cell in the cluster by its transmembrane ligand *Delta* on a neighboring cell generates signals that inhibit differentiation to form a sensory organ precursor and also inhibit production of Delta in that same cell. At first, all cells in the cluster express both Notch and Delta and inhibit one another from differentiating.

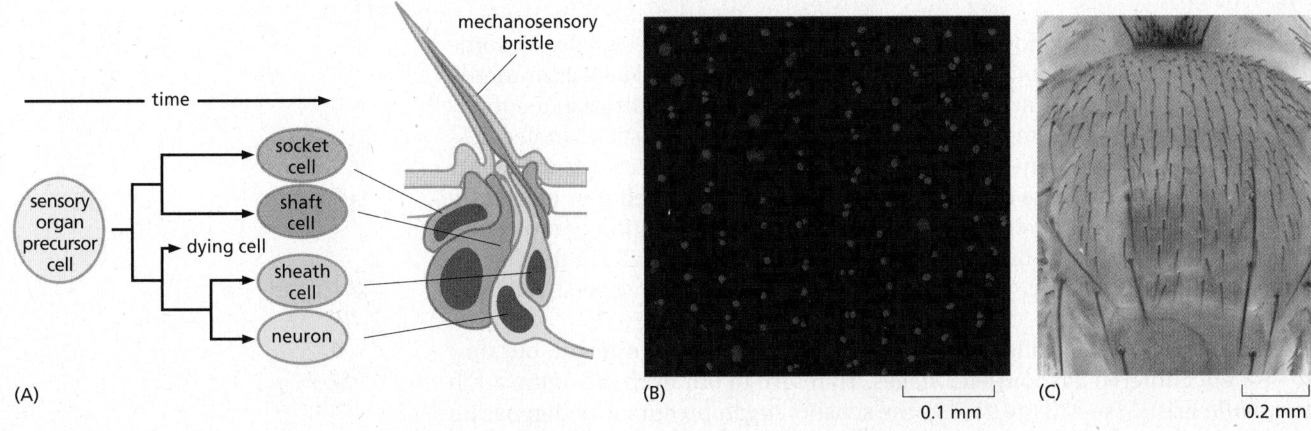

Figure 21–33 ***Drosophila* mechanosensory bristles.** (A) The lineage of the four cells of the bristle—all descendants of a single sensory organ precursor cell—is shown on the left. The sensory organ precursor, once it is specified, generates this set of cells through a short program of division cycles. In each generation of the progeny, lateral inhibition operates again to drive the newborn cells toward different fates: one of the ultimate progeny cells will become the neuron; another, the shaft of the bristle; others, supporting cells of various sorts. (B) The distribution of sensory organ precursors in the pupal epidermis, seen using a fluorescent reporter for the *senseless* gene *(pink)*. (C) The pattern of bristles on the thorax of an adult fly. (B and C, courtesy of François Schweisguth.)

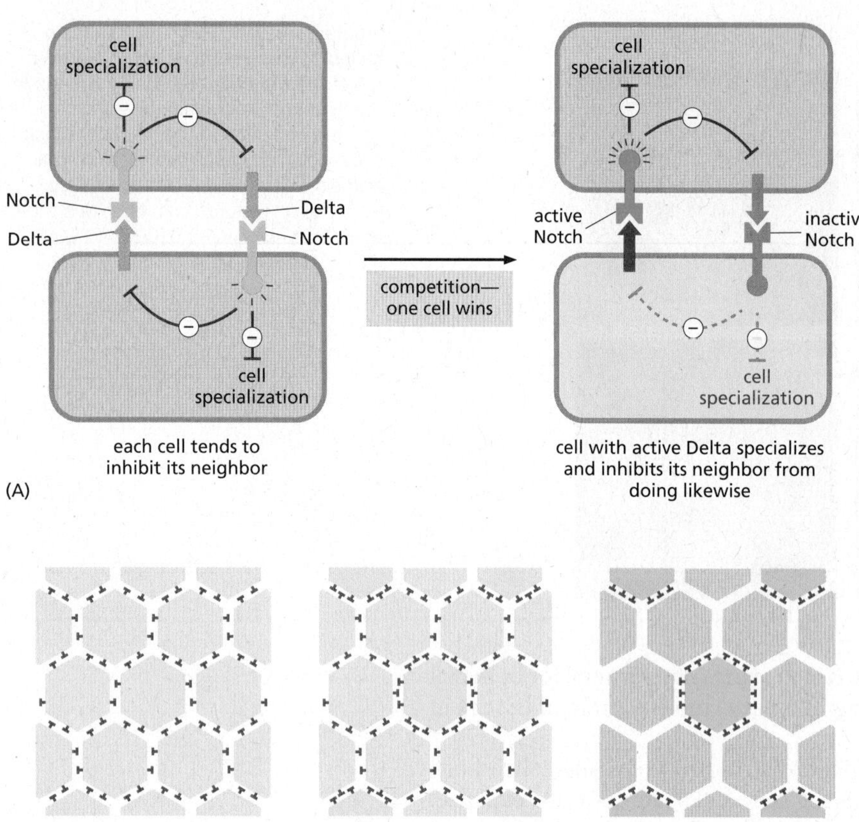

Figure 21–34 Lateral inhibition. (A) The basic mechanism of Notch-mediated competitive lateral inhibition, illustrated for just two interacting cells. Proteins or effector lines shown in *gray* indicate inactivity. (B) The outcome of the same process operating in a larger patch of cells. Initially, all cells in the patch are equivalent and signal through Delta–Notch interactions to discourage their neighbors from specializing as a sensory organ precursor. This balanced inhibition is disrupted when an individual cell generates a slightly stronger Delta signal than that of its neighbors: it then inhibits its neighbors more strongly while itself becoming more committed to the sensory organ precursor fate. As neighboring cells lose their capacity to differentiate, they also lose their capacity to inhibit other cells from doing so. Lateral inhibition thus makes adjacent cells follow different fates. Although the interaction is thought to be dependent primarily on cell–cell contacts, the sensory organ precursor can also deliver an inhibitory signal to cells that are more than one cell diameter away; for example, by sending out long protrusions to touch them.

Gradually, however, small signaling differences lead to a positive feedback loop in which a cell producing a slightly stronger Delta signal both discourages its neighbors from differentiating and reduces their production of Delta. The result is a competition, from which a single sensory organ precursor cell eventually emerges as winner, sending a strong inhibitory signal to its immediate neighbors but receiving no such signal in return (Figure 21–34).

The sensory organ precursor goes through a short program of further divisions to generate the set of cells that form the final bristle (see Figure 21-33). Notch signaling acts repeatedly at successive stages in this program to drive the descendants of the sensory organ precursor along different pathways and assign them to their various specialized fates. However, it does so in conjunction with additional mechanisms that bias the outcome of the competition mediated by lateral inhibition. Determinants that are asymmetrically localized inside the dividing cells have this role in sensory bristle development, as we now discuss.

Cell-fate Determinants Can Be Asymmetrically Inherited

Cell diversification does not always depend on extracellular signals: in some cases, daughter cells are born different as a result of an intrinsically asymmetric cell division, during which some significant set of molecules is divided unequally between them. This asymmetrically segregated molecule (or set of molecules) then acts as a determinant for one of the cell fates by directly or indirectly altering the pattern of gene expression within the daughter cell that receives it (see Figure 21-11). A good example of the asymmetric segregation of molecules occurs in the early frog embryo: *VegT* RNA is localized specifically in the vegetal region of the fertilized egg. After the cleavage divisions, only vegetal cells will inherit *VegT* RNA.

Asymmetric divisions often occur at the beginning of development, but they are also encountered at some later stages. To return to our example of the adult *Drosophila* bristle (see Figure 21-33), the sensory organ precursor undergoes an

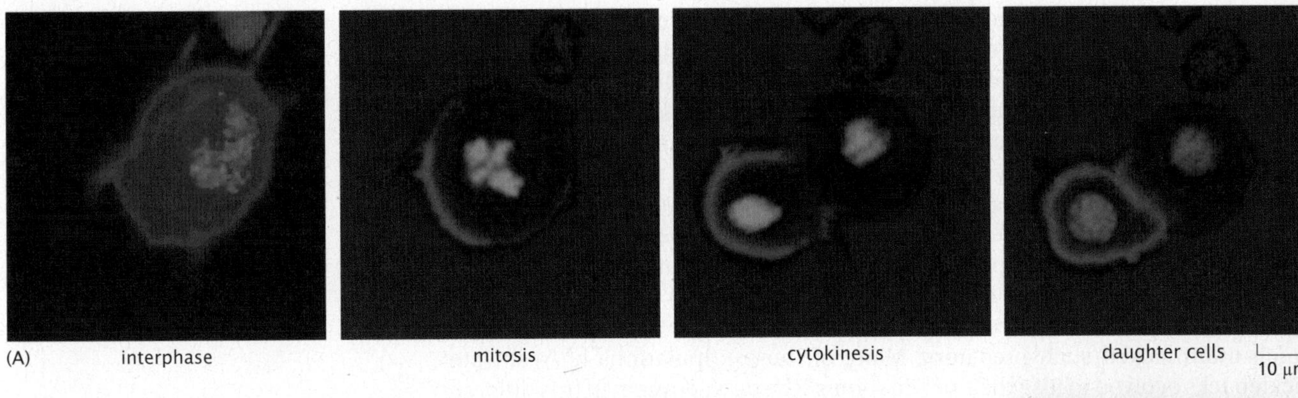

asymmetric cell division to generate one daughter that will produce the outer shaft and socket cells and one daughter that will produce the inner neuron and sheath cell. Each fate is dictated by the differential inheritance of a protein called Numb, which influences a Notch-mediated lateral inhibition decision between the daughter cells. Numb is a cytosolic protein that inhibits Notch signaling, and its presence in a sensory organ precursor cell specifies the "inner" precursor fate. Prior to the division of this cell, Numb is associated uniformly with the cell cortex. As the mitotic spindle forms, Numb is displaced from the cortical region close to one spindle pole, restricting its localization to the cortex adjacent to the other pole (**Figure 21–35**). This restriction depends on the polarizing activity of the Par protein complex described in Chapter 16 (and see Figure 16–76). When cytokinesis is completed, Numb is present in only one of the two daughters, where it inhibits Notch signaling within that cell—thereby acting as an asymmetrically segregated fate determinant.

Numb operates in an analogous manner during asymmetric divisions of the *Drosophila* neural stem cell lineage to generate the large numbers and balanced proportions of neurons and glia. Asymmetric division of neural progenitors also drives neurogenesis in the cortex of the vertebrate brain. In both cases, Notch signaling determines the difference between a more progenitor-like and a more differentiated fate.

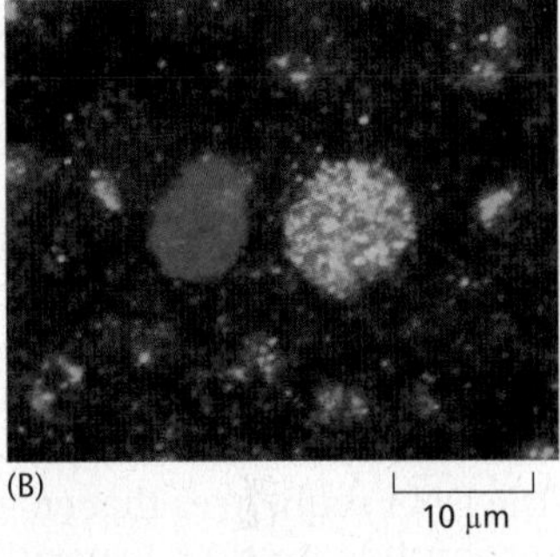

Figure 21–35 Asymmetric cell division creates different daughter-cell fates in the *Drosophila* bristle lineage. (A) Images of a dividing sensory organ precursor cell in the *Drosophila* pupa. Initially, the Numb protein *(pink)* is distributed uniformly around the cell cortex, but it partitions to one side of the cell during metaphase. Nuclear DNA is stained *blue*. Division results in the segregation of Numb into only one of the two daughter cells. (B) In this pair of daughter cells, asymmetrically inherited Numb (not shown) prevents Notch signaling in the daughter cell on the left; Notch activity is restricted to the daughter cell on the right, as shown by a reporter of the Notch target gene *Enhancer of Split m8 (green)*, which appears as speckles in the cell nucleus (stained *blue*). (A, from J.A. Knoblich, *Nat. Rev. Mol. Cell Biol.* 11:849–860, 2010; B, courtesy of François Schweisguth.)

Evolution of Regulatory DNA Explains Many Morphological Differences

In the preceding sections, we have seen that animals contain the same essential cell types, have a similar collection of genes, and share many of the molecular mechanisms of pattern formation. But how can we reconcile these ideas with the radical differences that we see in the body structures of animals as diverse as a worm, a fly, a frog, and a mouse? We asserted earlier, in a general way, that these differences usually seem to reflect differences in the regulatory DNA that controls the conserved basic set of developmental regulatory proteins. We now examine the evidence more closely.

When we compare animal species with similar basic body plans—different vertebrates, for example, such as fish, birds, and mammals—we find that corresponding genes can have similar sets of regulatory elements that are conserved and recognizably homologous in the different animals. The same is true if we compare different species of nematode worms or insects. But, when we compare vertebrate regulatory regions with those of worms or flies, it is hard to see any such resemblance. The protein-coding sequences are unmistakably similar, but the corresponding regulatory DNA sequences appear very different, suggesting that the differences in body plans mainly reflect differences in regulatory DNA. Although variations in the proteins themselves also contribute, differences in regulatory DNA would be enough to generate extremely different tissues and body structures even if the proteins were the same.

It is not yet possible to trace all of the genetic steps that have led to the spectacular diversity of animals. Their lineages have diverged over hundreds of millions of years, and in most cases too many changes have occurred for us to be able to say that this or that feature results from this or that mutation. The picture is clearer, however, for more recent evolutionary events. Studies of closely related animal populations and plant populations whose members have different morphologies have revealed that dramatic developmental effects can result from specific changes in regulatory DNA.

A well-studied example is the morphological diversity found in stickleback fish. Marine sticklebacks extend sharp spines from their pelvic skeleton. These spines are thought to help protect the fish from soft-mouthed fish predators that live in the ocean. After the last ice age ended about 10,000 years ago, some marine sticklebacks colonized newly formed freshwater streams and lakes, which often lacked such predators. Many of these populations of freshwater sticklebacks evolved to lose their pelvic spines. The development of this different morphology reflects differences in control of the expression of a transcription regulator called Pitx1. Whereas marine sticklebacks express the *Pitx1* gene in the hindlimb bud cells that will give rise to pelvic spikes, freshwater sticklebacks have lost this expression as a result of a change at the *Pitx1* locus. These changes do not lie in the coding sequence. Instead, each is a small deletion of a block of regulatory DNA that controls *Pitx1* expression specifically in the pelvic cells (**Figure 21–36**).

The Pitx1 protein has important functions elsewhere in the fish body, for example in the jaw and pituitary gland, and these functions must be retained. The DNA sequences that encode the Pitx1 protein, as well as the regulatory DNA responsible for tissue-specific *Pitx1* expression at these other body sites, is preserved in both marine and freshwater fish. The evolution of pelvic development

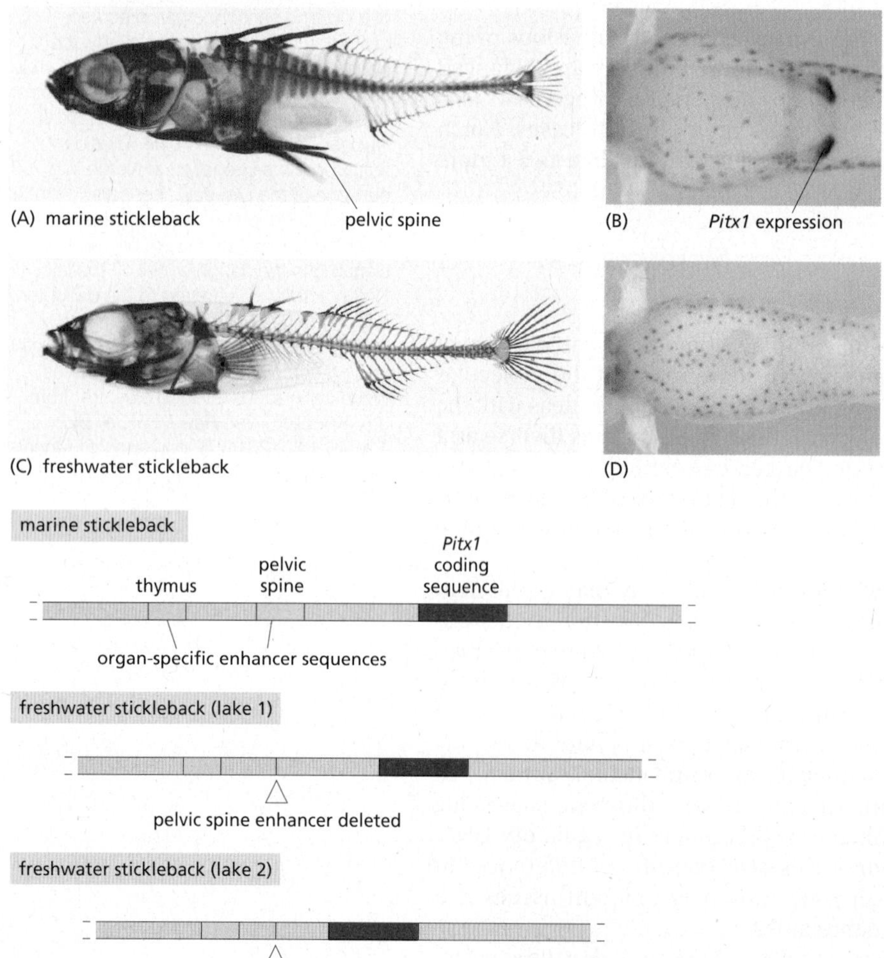

Figure 21–36 Morphological diversity in stickleback fish is caused by changes in regulatory elements. (A–D) Pelvic spines are present in marine (A) but missing in some freshwater (C) populations. Correspondingly, *Pitx1* is expressed in the pelvic area in marine (B) but not in freshwater (D) fish. (E) The lack of expression in the pelvic area of freshwater populations is caused by regulatory deletions that removed pelvic-specific enhancer sequences; these deletions occurred independently in different populations and are of different sizes. The coding sequence and other enhancers and sites of expression for *Pitx1* are conserved in marine and freshwater sticklebacks. (A–D, courtesy of Michael D. Shapiro.)

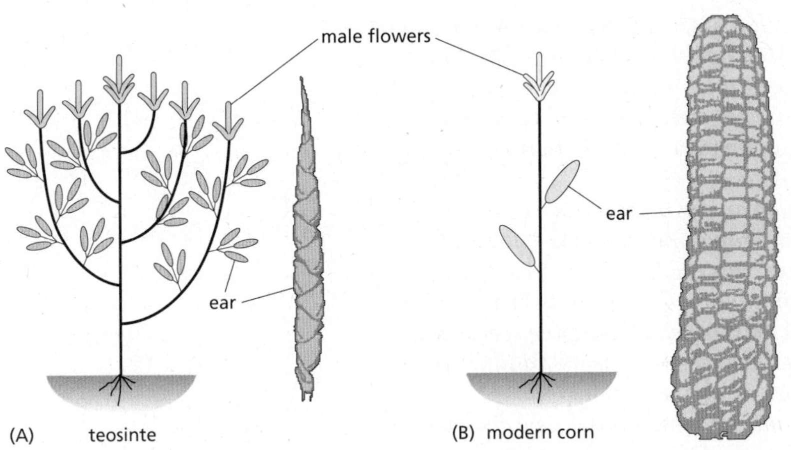

Figure 21–37 **The insertion of a mobile genetic element helped produce modern corn.** Today's corn plants were originally bred from a wild plant called teosinte (A). This wild ancestor produced numerous ears that contained small, hard seeds. (B) Modern corn, by contrast, produces fewer ears—but they contain numerous, plump, sweet seeds (kernels). The insertion of a mobile genetic element in the regulatory region of a gene involved in seed development helped drive the change. Here, the two plants are drawn to the same scale; for simplicity, the leaves are not shown.

in sticklebacks shows how the modular nature of regulatory DNA elements that we encountered in Chapter 7 (see Figure 7–31) allows independent modification of the different parts of the body, even when formation of multiple body parts depends on the same proteins.

In the recent evolution of plants, changes in their structure can be traced in a similar way to changes in regulatory DNA. In an important example for agriculture, such changes account for a large part of the dramatic difference between the wild teosinte plant and its modern descendant, maize, through some 10,000 years of selection upon natural genetic variation by Native Americans (**Figure 21–37**).

Summary

Drosophila has been the foremost model organism for the study of the genetics of animal development. Its embryonic pattern is initiated by the products of maternal-effect genes called egg-polarity genes, which operate by setting up graded distributions of transcription regulators in the egg and early embryo. The gradient of Bicoid protein along the anteroposterior (A-P) axis, for example, helps initiate the orderly expression of gap genes, pair-rule genes, and segment-polarity genes. These three classes of segmentation genes, through a hierarchy of interactions, become expressed in some regions of the embryo and not others, progressively subdividing an initially uniform field of cells along the A-P axis into a regular series of repeating modular units called segments.

Superimposed on the pattern of gene expression that repeats itself in every segment, there is a serial pattern of expression of Hox genes that confers on each segment a different identity. These genes are grouped in complexes and are arranged in a sequence that matches their sequence of expression along the A-P axis of the body.

Although Hox gene expression is initiated in the embryo, it is subsequently maintained by the action of chromatin-binding proteins of the Polycomb and Trithorax group, which stamp the chromatin of the Hox complex with a heritable record of its embryonic state of repression or activation, respectively. Hox complexes homologous to that of Drosophila are found in virtually every type of animal, where they help pattern the A-P axis of the body.

Signaling gradients are also set up along the dorsoventral (D-V) axis. Initially, Toll signaling generates a nuclear gradient of Dorsal protein, which induces an extracellular signaling gradient of the TGFβ family protein Dpp and its antagonist, Sog. This creates a gradient of Dpp activity that helps refine the assignment of different characters to cells at different positions along the D-V axis.

In Xenopus, the polarity of the egg and the site of sperm entry set up the embryonic axes. A gradient generated by the TGFβ family protein Nodal induces different fates along the animal-vegetal axis, whereas BMP and Chordin—proteins homologous to Drosophila Dpp and Sog, respectively—control the patterning of the D-V axis.

Transcription regulators control the formation of specific cell types. Members of the MyoD family drive the process of muscle-cell determination, coordinating the many components required, whereas Achaete/Scute transcription regulators control neural fate. Other genes encoding such master transcriptional regulators can regulate the formation of entire organs. Eyeless, for example, is both necessary and sufficient to generate eye structures in Drosophila.

To refine the anatomical pattern within such an organ, the cells interact locally, both by diffusible inductive signals and by short-range mechanisms. Often, the cells compete with one another by lateral inhibition. This process results in activation of the Notch signaling pathway in one cell and inhibition in its neighbors, generating two different cell types. Asymmetric cell divisions, in which daughter cells inherit different molecular determinants from the mother cell, provide an additional way to organize a fine-grained diversity of cell types.

Evidence from recent evolutionary events indicates that anatomical changes are mostly driven by changes in regulatory DNA sequences that determine when and where developmental genes are expressed. How the striking diversity in body structures has evolved over longer times remains largely unknown, although it seems likely that similar principles apply.

DEVELOPMENTAL TIMING

Developmental events unfold over minutes, hours, days, weeks, months, or even years, with each organism following its own strict timetable. The cascades of inductive interactions and transcriptional regulatory events described earlier take time, as signals are transmitted and transcription regulators are synthesized and then bind to DNA to activate or repress their target genes. So too do the changes in cell size, shape, and organization that result from changes in gene expression. All of these events require correct parameters of timing, including order (the sequence of events), interval (the time between events), and, in some cases, rhythm (repeated cycles of an event). Each developmental process must thus occur at an appropriate rate, tuned by evolution to fit with the timing of other processes in the embryo or in the environment. The control of timing is a compelling problem in developmental biology, but one of the least understood.

Molecular Lifetimes Play a Critical Part in Developmental Timing

Developmental processes are complex, but they are built up from simple steps. A first challenge is to understand the timing of these steps. How long does it take, for example, to switch the expression of a gene on or off? This is not like throwing a light switch: it involves delays. First, it takes time to make an mRNA molecule: the RNA polymerase must travel the length of the gene, the primary RNA transcript must be spliced and otherwise processed, and the resulting mRNA must be exported from the nucleus and delivered to the site where it will be translated. This adds up to what one might call the *gestation time* of the individual molecule. Second, it takes time for the individual mRNA molecules to accumulate to their fully effective concentration; as explained in Chapter 15, this *accumulation time* is dictated by the average lifetime of the molecules—the longer they last, the higher their ultimate concentration, and the longer the time taken to attain it. Similar delays occur at the next step, where the mRNA is translated into protein: synthesis of each individual protein molecule involves a gestation delay, and attainment of an effective concentration of protein molecules involves an accumulation delay that depends on the time it takes for each protein to fold into its three-dimensional structure and its lifetime once produced. The time for the whole gene switching process is the sum of the gestation delays and the accumulation delays (basically, the molecular lifetimes) for both the mRNA and the protein molecules. Somewhat counterintuitively, it is the combined length of these delays, rather than the rate of molecular synthesis (the number of molecules synthesized per second), that chiefly determines the switching time.

The same additive principle applies to long cascades of gene switching, where gene *A* activates gene *B*, and gene *B* activates gene *C*, and so on. It also applies in other circumstances, such as in signaling pathways where one protein directly regulates the activation of the next. In all these cases, molecular lifetimes, along with gestation delays, play a key part in determining the pace of development. The lifetimes of mRNA and protein molecules are enormously variable, from a few minutes or hours to days or more, explaining much of the tempo of developmental events.

Gene switching delays, however, are not the be-all and end-all of developmental timing. Development involves many other kinds of delay that contribute to timing. Chromatin structure takes time to remodel. Inductive signals take time to diffuse across a field of cells (see Figure 21–9). Cells take time to move and rearrange themselves in space. Nevertheless, the timing of gene switching plays a fundamental part in developmental timing, as illustrated in an especially clear and striking way by a gene expression oscillator that controls the segmentation of the vertebrate body axis, as we now explain.

A Gene Expression Oscillator Acts as a Clock to Control Vertebrate Segmentation

The main body axis of all vertebrates has a repetitive, periodic structure, seen in the series of vertebrae, ribs, and segmental muscles of the neck, trunk, and tail. These segmental structures originate from the mesoderm that lies as a long slab on either side of the embryonic midline. This slab becomes broken up into a regular repetitive series of separate blocks, or **somites**—cohesive groups of cells, separated by clefts (**Figure 21–38A**). In contrast to the segments of a *Drosophila* embryo that appear simultaneously, as we discussed previously, somites in the vertebrate are formed sequentially. The somites form (as bilateral pairs) one after another, in a regular rhythm, starting in the region of the head and ending in the tail. Depending on the species, the final number of somites ranges from fewer than 40 (in a frog or a zebrafish) to more than 300 (in a snake).

The cells that will form somites are generated in the posterior, most immature part of the mesodermal slab, called the *presomitic mesoderm*. This is a proliferative zone, and as new cells are generated, the zone is pushed tailward, extending

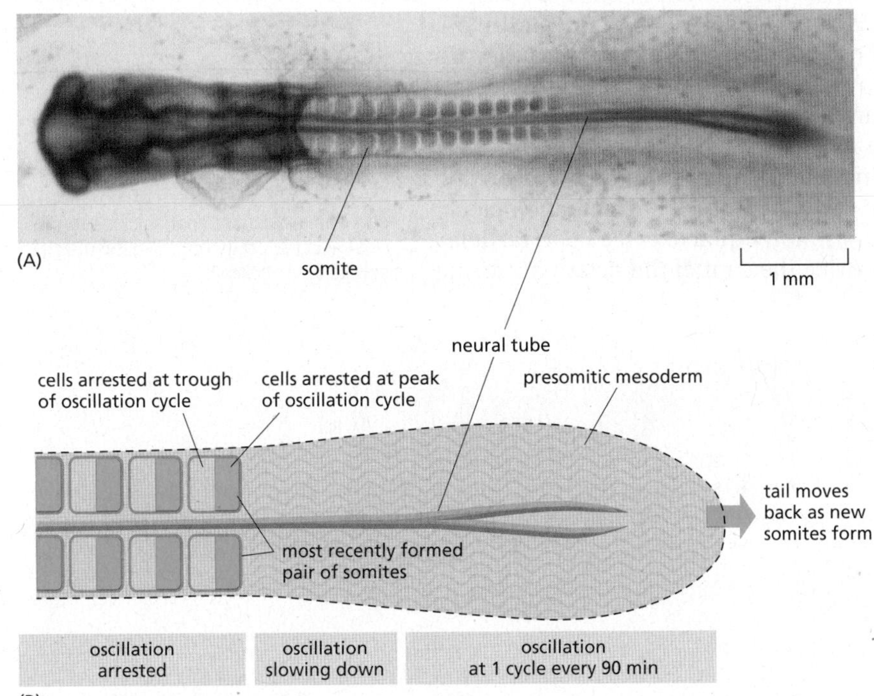

Figure 21–38 Somite formation in the chick embryo. (A) Brightfield image of a chick embryo at 40 hours of incubation. (B) How the temporal oscillation of gene expression in the presomitic mesoderm becomes converted into a spatial alternating pattern of gene expression in the formed somites. In the posterior part of the presomitic mesoderm, each cell oscillates with a cycle time of 90 minutes. As cells mature and emerge from the presomitic region, their oscillation is gradually slowed down and finally brought to a halt, leaving them in a state that depends on the phase of the cycle they happen to be in at the critical moment. In this way, a temporal oscillation of gene expression traces out an alternating spatial pattern. (**Movie 21.5**.) (A, from Y.J. Jiang et al., *Curr. Biol.* 8:R868–R871, 1998. With permission from Elsevier.)

the embryo (Figure 21–38B). As the presomitic mesoderm moves tailward, it deposits a trail of somites formed from cells that group together into blocks as they emerge from the anterior end of the presomitic region. The special character of the presomitic mesoderm is maintained by a combination of fibroblast growth factor (FGF) and Wnt signals, produced by a signaling center at the tail end of the embryo, and the range of these signals seems to define the region from which somites can be formed. The somites emerge with clocklike timing, but what determines the rhythm of the process?

In the posterior part of the presomitic mesoderm that is destined to be segmented, the expression of certain genes oscillates in time. Such oscillations can be observed in time-lapse movies of embryos containing fluorescent reporters of individual genes. One new somite pair is formed in each oscillation cycle, and, in mutants where the oscillations fail to occur, somite segmentation is disrupted: the cells may still break up, belatedly, into separate clusters, but they do so in a haphazard, irregular way. The gene expression oscillator controlling regular segmentation is called the **segmentation clock**. The length of one complete oscillation cycle depends on the species: it is 30 minutes in a zebrafish, 90 minutes in a chick, 120 minutes in a mouse, and 300 minutes in a human.

As cells emerge from the presomitic mesoderm to form somites—in other words, as they escape from the influence of the FGF and Wnt signals—their oscillating gene expression stops. Some become arrested in one state, some in another, according to the phase of the oscillation cycle at the time they leave the presomitic region. In this way, the temporal oscillation of gene expression in the presomitic mesoderm leaves its trace in a spatially periodic pattern of gene expression in the maturing mesoderm. This in turn dictates how the tissue will break up into physically separate blocks, through effects on the pattern of cell–cell adhesion.

How does the segmentation clock work? The first somite oscillator genes to be discovered were *Hes* genes, which code for inhibitory transcription regulators. As well as regulating other genes, the products of *Hes* genes can directly regulate their own expression, creating a simple negative feedback loop. Autoregulation of *Hes* genes is thought to be the basic generator of the oscillations of the somite clock. Although the machinery has been modified in various ways in different species, the underlying principle seems to be conserved. When the key *Hes* gene is transcribed, the amount of Hes protein builds up until it is sufficient to block *Hes* gene transcription. Synthesis of the protein then ceases, and as the protein decays, transcription is permitted to begin again; and so on, cyclically (Figure 21–39). The period of oscillation, which determines the size of each somite, depends on the delay in the feedback loop. This delay equals the sum of the gestation delays and accumulation delays (that is, the molecular lifetimes) of the *Hes* mRNA and protein molecules, according to the principle discussed earlier. Mathematical modeling (see Chapter 8) allows us to relate these basic molecular parameters to the cycle time of the segmentation clock: to a first approximation, the cycle period is simply equal to twice the total delay in the negative feedback loop, and thus twice the sum of the delays occurring

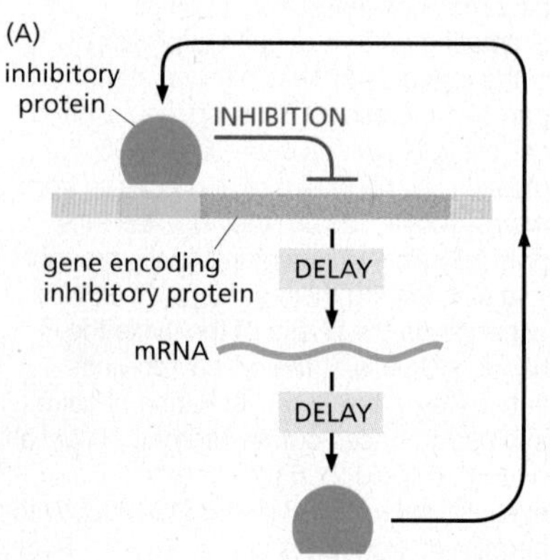

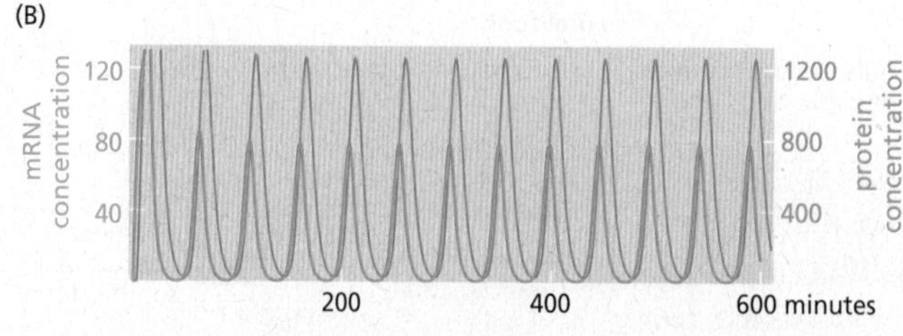

Figure 21–39 Delayed negative feedback giving rise to oscillating gene expression. (A) A single gene, coding for a transcription regulator that inhibits its own expression, can behave as an oscillator. For oscillation to occur, there must be a delay (or several delays) in the feedback circuit, and the lifetimes of the mRNA and protein (which contribute to the delay) must be short compared with the total delay. The total delay determines the period of oscillation. It is thought that a feedback circuit like this, based on a pair of redundantly acting genes called *Her1* and *Her7* in the zebrafish—or their counterpart, *Hes7*, in the mouse—is the pacemaker of the segmentation clock governing somite formation. (B) The predicted oscillation of *Her1* and *Her7* mRNA and protein, computed using rough estimates of the feedback circuit parameters appropriate to this gene in the zebrafish. Concentrations are measured as numbers of molecules per cell. The predicted period is close to the observed period, which is 30 minutes per somite in the zebrafish.

at each step of the loop. Moreover, an experimental manipulation that changes one of these parameters (for example, altering the size of a *Hes* intron to perturb mRNA-processing time) results in the predicted effect on the period of somite formation.

The feedback loop just described is intracellular, and each cell in the presomitic mesoderm can generate oscillations on its own, even when cultured in a dish. But these oscillations at the single-cell level are somewhat erratic and imprecise, reflecting the fundamentally noisy, stochastic nature of the control of gene expression, as discussed in Chapter 7. A mechanism is needed to keep all the cells in the presomitic mesoderm that will form a particular somite oscillating in synchrony. This is achieved in part through cell-cell communication via the Notch signaling pathway. In this context, Notch signaling does not drive neighboring cells to be different, as in lateral inhibition, but does just the opposite: it keeps them in unison. In mutants where Notch signaling fails, including mutants defective in Delta or Notch itself, the cells drift out of synchrony, and somite segmentation is again disrupted. This leads to gross deformity of the vertebral column—an extraordinary display of the consequences of the noisy temporal control of gene expression at the single-cell level, writ large in the structure of the vertebrate body as a whole.

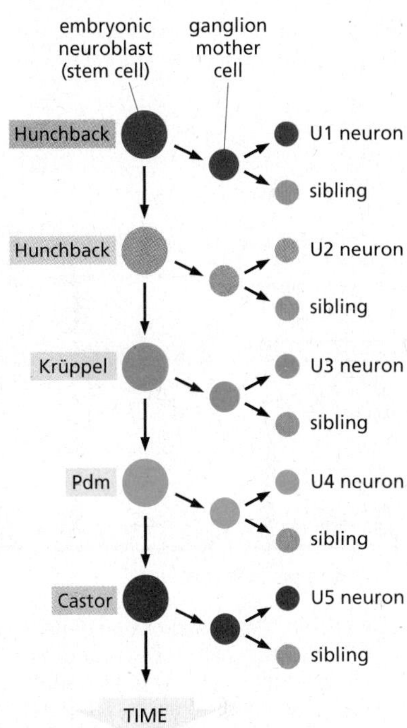

Figure 21–40 Temporal patterning of neural fate in *Drosophila* embryos. Each neuroblast undergoes an asymmetric cell division to renew itself and to produce a daughter called a ganglion mother cell. The ganglion mother cell subsequently divides to produce a pair of terminally differentiated neurons or glia. The types of neurons and glia are determined by the expression of a transcription regulator in the parent neuroblast that is then inherited by its ganglion-mother-cell daughter. Each neuroblast will sequentially express the transcription regulators Hunchback, Krüppel, Pdm, and Castor as it progresses through its lineage. After dividing several times while expressing one of these regulators, the neuroblast switches its pattern of gene expression, expressing an updated set of genes that will be passed along to its daughter cells, which differentiate into specific cell types accordingly. (After B.J. Pearson and C.Q. Doe, *Nature* 425:624–628, 2003.)

Cell-intrinsic Timing Mechanisms Can Lead to Different Cell Fates

Although signaling between cells plays an essential part in driving the progress of development, this does not mean that cells always require signals from other cells to prod them into changing their character as development proceeds. Some of these changes are intrinsic to the cell (like the ticking of a circadian clock, discussed in Chapter 15) and depend on *intracellular developmental programs* that can operate even when the cell is removed from its normal environment.

The best-understood example is the development of neuroblasts, the stem cells of the *Drosophila* central nervous system. These cells go through a succession of asymmetric cell divisions to generate distinct types of neurons and glial cells, always in the same sequence and with the same timing. Each neuroblast produces this wide range of daughter-cell fates using a time-dependent developmental program. As the neuroblast goes through its set schedule of divisions, it successively changes its internal state by expressing a series of different transcription regulators. For example, most embryonic neuroblasts sequentially express the transcription regulators Hunchback, Krüppel, Pdm, and Castor in a fixed order (**Figure 21–40**). When a neuroblast divides, the transcription regulators expressed at that moment are inherited by and maintained in its progeny; thus, the differentiated neural cells are endowed with different characters according to their time of birth.

When neuroblasts are taken from an embryo and maintained in culture, isolated from their normal surroundings, they step through much the same stereotyped developmental program as if they had been left in the embryo. Moreover, many of the neuroblast transitions occur even when cell division is blocked. The neuroblasts seem to have a built-in timer that determines when each of the transcription regulators is expressed, and this timer can continue to run in the absence of cell-cycle progression. The molecular basis of the timing is not well understood. Because cross-regulatory transcriptional cascades are involved, it likely depends on the time taken for gene switching, as described earlier, but it might also depend on slow progressive changes in chromatin structure and genome organization, which also can serve to measure the passage of time in the embryo.

Timing mechanisms can coordinate cell fates with the spatial organization of tissues. For example, during development of the mammalian cerebral cortex, the ordered expression of different transcription factors in the progenitor cells that divide to produce neurons and glial cells leads to sequential production of specific neural cell types. The newborn neurons migrate to their appropriate position

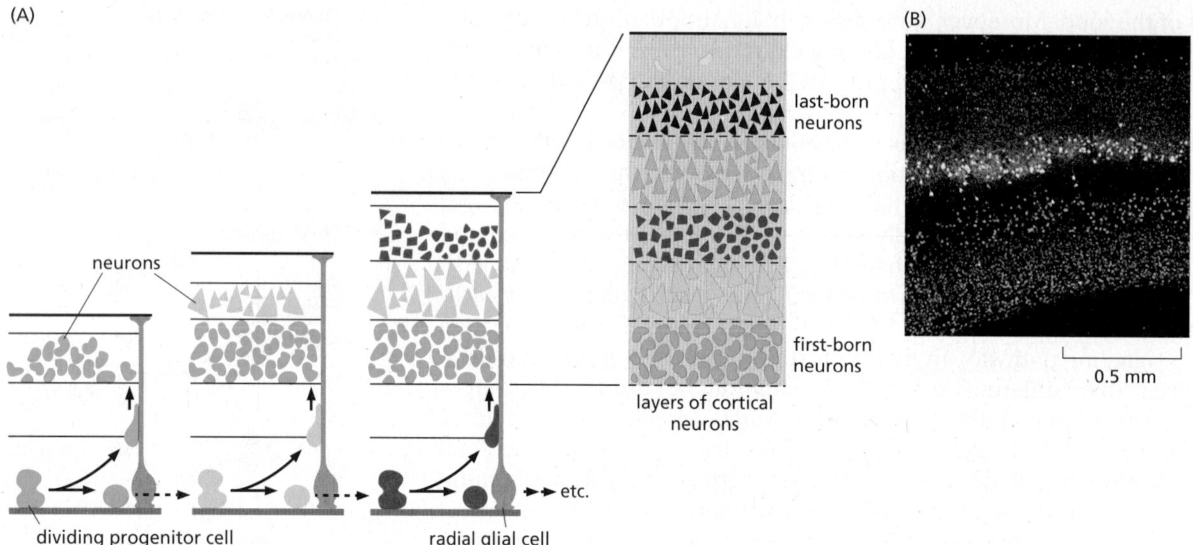

Figure 21–41 **Timed expression of neural genes drives laminar organization of the brain.** (A) Many neurons and glial cells in the mammalian cerebral cortex develop from a common progenitor, which proliferates on the inner surface of the cortical neuroepithelium to produce successive generations of neurons (colored here *blue*, *green*, *red*, *orange*, and *black*). The developing neurons migrate outward, crawling along processes that extend from radial glial cells, which also develop from the same progenitor cells. The first-born neurons settle closest to their birthplace, while neurons born later crawl past them to settle farther out. Successive generations of neurons thus occupy different layers in the cortex and have different intrinsic characters according to their birth dates. (B) Fluorescence micrograph showing the laminar organization of excitatory neurons labeled with antibodies to different transcription factors (*blue*, *green*, *white*, and *red*) in the mammalian cerebral cortex. Neural cell types are specified by the temporal sequence of transcription factor expression in their progenitors, so that birth order specifies both the fate and location of differentiated neurons. (From I. Holguera and C. Desplan, *Science* 362: 176–180, 2018.)

in the developing cortex, ultimately occupying different layers of the adult brain. Thus, all neurons that populate a layer form in the same time window and have common fates (**Figure 21–41**).

Cells Rarely Count Cell Divisions to Time Their Development

Many specialized cells develop from proliferating progenitor cells that stop dividing and terminally differentiate after a limited number of cell divisions. In these cases, it is tempting to speculate that the cell-division cycle serves as an intracellular timer to control the timing of cell differentiation. The cell cycle would be the ticking clock that sets the tempo of other developmental processes, with maturational changes in gene expression being dependent on cell-cycle progression. Most of the evidence, however, indicates that this appealing idea is wrong. Although there are examples where cells change their maturation state with each division and the change depends on cell division, this is not the general rule. As we just saw for neuroblasts in the *Drosophila* embryo, cells in developing animals often carry on with their normal timetable of maturation and differentiation even when cell division is artificially blocked; necessarily, some abnormalities occur, if only because a single undivided cell cannot differentiate in two ways at once. But it seems that most developing cells can change their developmental state without a requirement for cell division. Developmental control genes can switch the cell-division-cycle machinery on or off, and it is the dynamics of these genes, rather than the cell cycle, that sets the tempo of development.

MicroRNAs Can Regulate Developmental Transitions

Because genetic screens are useful for tracking down the genes involved in almost any biological process, they have been used to search for mutations that alter developmental timing. Such screens were performed in the nematode worm

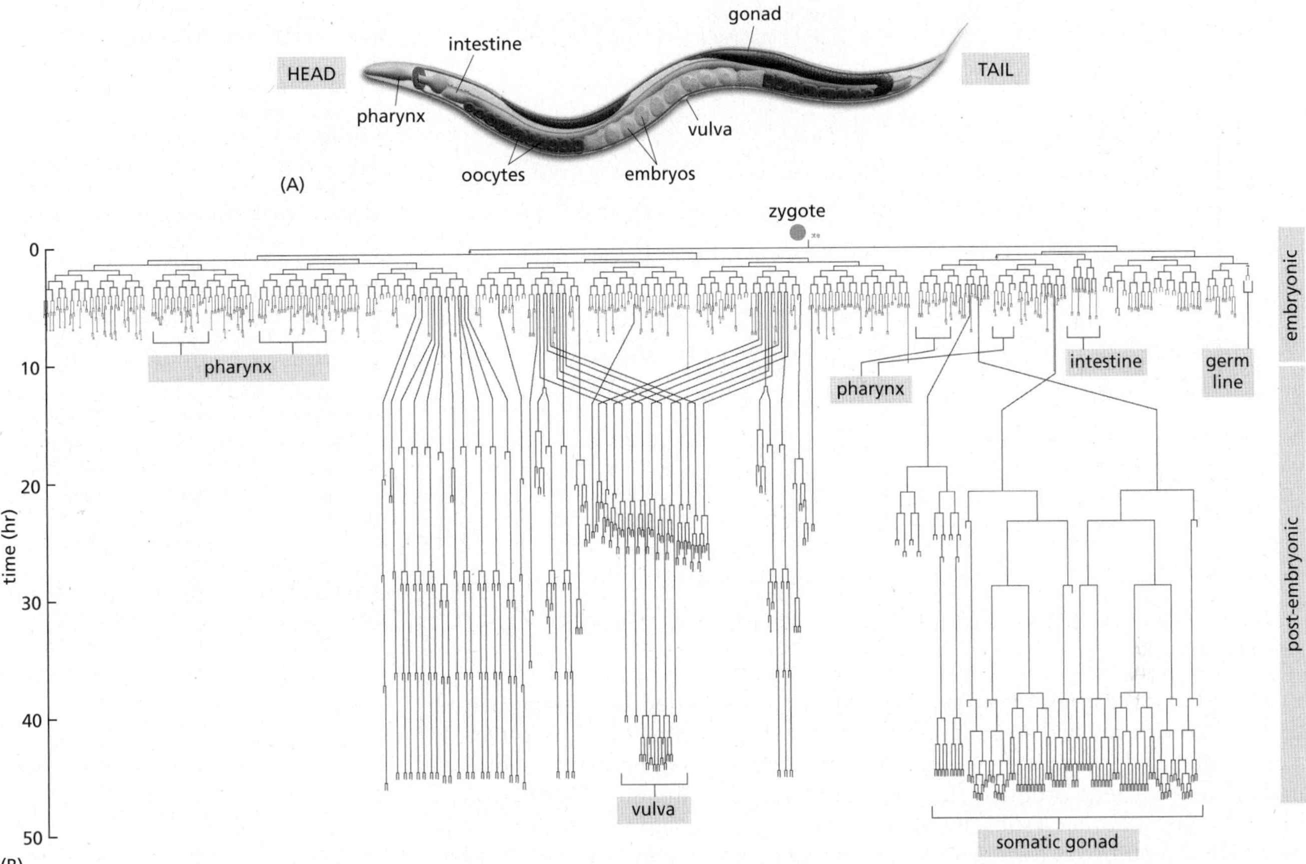

Figure 21–42 Cell lineage in *Caenorhabditis elegans*. (A) Some of the major tissues in the adult worm (see Figure 1–42). (B) The complete cell lineage of the worm was determined by meticulous analysis of the fate of every daughter cell after each cell division, beginning with the single-celled zygote and ending in the 959 cells of the adult. Each vertical line represents a single cell, with branch points reflecting a cell division. Labels indicate the cells in the lineage that make up the tissues shown in panel A. (A and B, courtesy of D.H. Hall, Worm Atlas from https://www.wormatlas.org/celllineages.html.)

Caenorhabditis elegans (**Figure 21–42**). This small creature follows an astonishingly precise and predictable developmental program that has been described in extraordinary detail, so that one can map out the exact lineage of every cell in the body. The time and place of each cell division, as well as the fate of each of its daughters, are invariant from one individual to the next.

A great advantage of the lineage diagram describing the normal development of *C. elegans* is that one can see exactly how the developmental program is altered in a mutant. Genetic screens in *C. elegans* revealed mutations that disrupt developmental timing in a particularly striking way: in these so-called **heterochronic** mutants, certain cells in a larva at one stage of development behave as though they were in a larva at a different stage of development or cells in the adult carry on dividing as though they belonged to a larva (**Figure 21–43**).

Genetic analyses showed that the products of the heterochronic genes act in series, forming regulatory cascades. Unexpectedly, two genes at the top of their respective cascades, called *Lin4* and *Let7*, were found to code not for protein but instead for **microRNAs (miRNAs)**—short, untranslated, regulatory RNA molecules, 21 or 22 nucleotides long. These act by binding to complementary sequences in the noncoding regions of mRNA molecules transcribed from other heterochronic genes, thereby repressing their translation and promoting their degradation, as discussed in Chapter 7. The *Lin4* miRNA binds to the 3′ untranslated region of mRNA produced by the heterochronic gene *Lin14*, which itself promotes early-stage larva-cell behaviors. A developmentally regulated increase of *Lin4* levels thus governs the progression from early to late-stage larva-cell behaviors, by gradually reducing levels of Lin14 protein. Increasing levels of *Let7* miRNA govern the progression from late larva to adult in an analogous way by regulating a different target gene called *Lin41*. In fact, *Lin4* and *Let7* were the first miRNAs to be described in any animal: it was through developmental genetic studies in *C. elegans* that the importance of this whole class of molecules for gene regulation in animals was discovered.

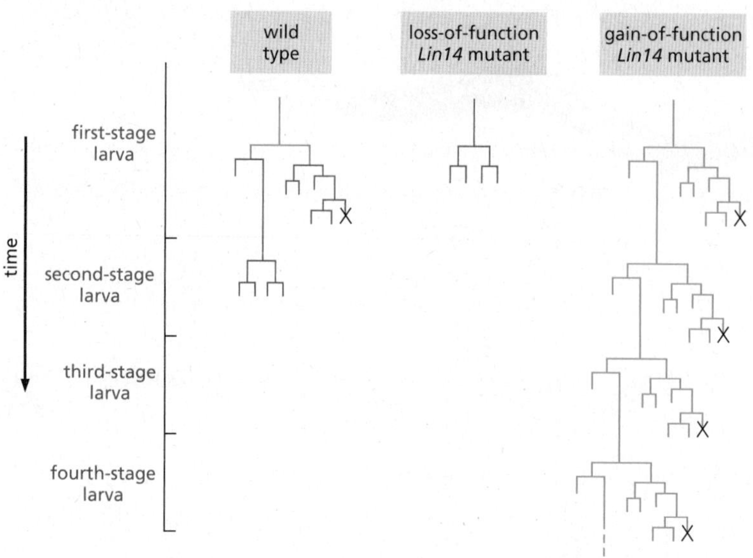

Figure 21–43 Heterochronic mutations in the *Lin14* gene of *C. elegans*. Only the effects on one of the many altered lineages are shown. A loss-of-function (recessive) mutation in *Lin14* causes premature occurrence of the pattern of cell division and differentiation characteristic of a late larva, so that the animal reaches its final state prematurely and with an abnormally small number of cells. The gain-of-function (dominant) mutation has the opposite effect, causing cells to reiterate patterns of cell divisions characteristic of the first larval stage, continuing through as many as five or six molt cycles. The *Lin14* gain-of-function mutations can result from altering the binding site for the *Lin4* miRNA, rendering Lin14 protein immune from down-regulation. The *black X* denotes a programmed cell death. *Green lines* represent cells that contain Lin14 (which binds to DNA), *red lines* those that do not. (Adapted from V. Ambros and H.R. Horvitz, *Science* 226:409–416, 1984; and P. Arasu et al., *Genes Dev.* 5:1825–1833, 1991.)

Cell and Nuclear Size Relationships Schedule the Onset of Zygotic Gene Expression

As we discussed earlier, many gene products are deposited in the egg by the mother, allowing the first steps of development to begin immediately upon fertilization. Maternal supplies eventually run out, however, and the zygote begins to produce its own gene products. This **maternal–zygotic transition (MZT)** occurs in most organisms after a substantial delay and marks the temporal window during which the embryo's own genome largely takes over control of development from maternal macromolecules (**Figure 21–44**).

The MZT occurs in most species shortly before gastrulation, when rapid and synchronous cleavage divisions give way to slower and more conventional cell cycles. During this transition, thousands of genes initiate transcription from the previously silent zygotic genome, and many maternal gene products are cleared from the embryo. In several species, including frogs, fish, and flies, some of the earliest zygotic genes to be transcribed include specific miRNAs that target maternal mRNAs for translational repression and degradation. Here again, miRNAs function to sharpen developmental transitions by blocking and removing mRNAs that define an earlier developmental stage.

How is the timing of the MZT controlled? One trigger appears to be the nuclear-to-cytoplasmic ratio. Many animal embryos do not grow or change volume during the earliest stages of development. Rather, rapid rounds of DNA replication and mitosis transform the fertilized egg into thousands of smaller cells. During these cleavage divisions, the total amount of cytoplasm in the embryo remains constant, but the number of cell nuclei in the embryo increases exponentially, thereby increasing the ratio of DNA to cytoplasm. Strikingly, haploid embryos containing

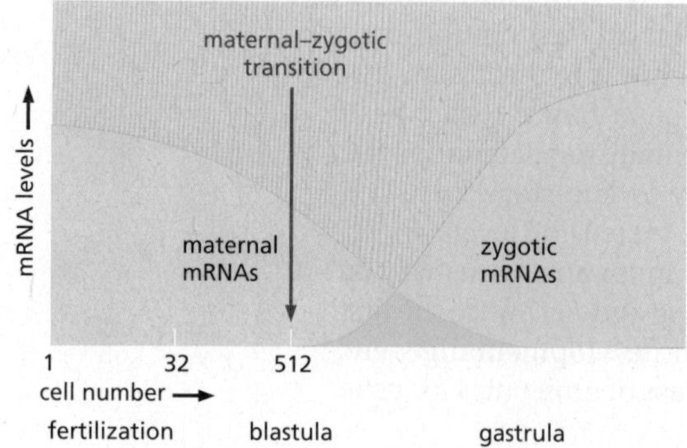

Figure 21–44 The maternal–zygotic transition in a zebrafish embryo. Maternal mRNAs are deposited by the mother into the egg and drive early development. These mRNAs are degraded during different stages of embryogenesis, including blastula and gastrula stages, but a relatively abrupt change occurs at the maternal–zygotic transition (MZT). Before this, the embryonic (zygotic) genome is transcriptionally inactive; afterward, zygotic genes start to be transcribed. In zebrafish embryos, the zygotic genome begins to be activated at the 512-cell stage.

half the amount of DNA per cell undergo the MZT one cell cycle later than normal diploid embryos. According to one model, the nuclear-to-cytoplasmic ratio might be measured through the titration of a maternally provided transcription repressor against the increasing amount of nuclear DNA. The total amount of repressor would stay constant during cleavage divisions, but the amount of repressor per genome would decrease by half with each round of DNA synthesis until a threshold is reached that allows the zygotic genome to become transcriptionally active. One candidate for the repressor is maternal histone proteins, whose concentration regulates MZT timing in several organisms. Other evidence suggests that certain maternally provided transcription regulators trigger the MZT after they accumulate to some threshold sufficient to activate their gene targets. Different species may rely on either or both mechanisms to time activation of the zygotic genome appropriately.

Hormonal Signals Coordinate the Timing of Developmental Transitions

We have so far emphasized timing mechanisms that operate locally and separately in the different parts of the embryo or in specific subsystems of the molecular control machinery. Evolution has tuned each of these largely independent processes to run at an appropriate rate, matched to the needs of the organism as a whole. For some purposes, however, this is not enough: a global coordinating signal is required. This is especially true where changes have to occur throughout the body in response to a cue that depends on the environment. For example, when an insect or amphibian undergoes *metamorphosis*—the transition from larva to adult—almost every part of the body is transformed. The timing of metamorphosis depends on external factors such as the supply of food, which determines when the animal reaches an appropriate size. All the bodily changes have to be triggered together at the right time, even though they are occurring in widely separated sites. The coordination in such cases is provided by **hormones**—signal molecules that spread throughout the body.

The metamorphosis of amphibians provides a spectacular example. During this developmental transition, amphibians switch from an aquatic to a terrestrial life. Larva-specific organs such as gills and tail disappear, and adult-specific organs such as legs form. This dramatic transformation is triggered by thyroid hormone, produced in the thyroid gland. If the gland is removed or if thyroid hormone action is blocked, metamorphosis does not occur, although growth continues, producing a giant tadpole. Conversely, a dose of thyroid hormone given to a tadpole by an experimenter can trigger metamorphosis prematurely.

Thyroid hormone is distributed throughout the body by the vascular system and induces changes in receiving cells by binding to intracellular nuclear hormone receptors, which regulate hundreds of genes. This does not mean, however, that target tissues all respond in the same way to the hormone: organs differ not only in their levels of thyroid hormone receptors and levels of extracellular proteins that locally regulate the amount of active hormone, but also in the sets of genes that respond. Thyroid hormone induces muscle in the limbs to grow and muscle in the tail to die. The timing of the responses also differs; for example, the legs form early in response to a very low concentration of circulating hormone, but it requires a high level of the hormone to induce resorption of the tail.

A surge of thyroid hormone triggers metamorphosis, but how is the timing of the surge controlled? One mechanism depends on coupling hormone synthesis to the size of the thyroid gland, which reflects the size of the tadpole. Only when the gland attains a certain size does it produce enough thyroid hormone to initiate metamorphosis. However, environmental cues other than nutrition also play a part: conditions such as temperature and light are sensed by the nervous system, which regulates the secretion of another tier of hormones (neurohormones) that stimulate the secretion of thyroid hormone. Thus, tadpole-intrinsic factors such as size combine with environmental factors to determine when metamorphosis begins.

Environmental Cues Determine the Time of Flowering

Another striking example of environmentally controlled developmental timing is the flowering of plants. Flowering involves a transformation of the behavior of the cells at the growing apex of the plant shoot—the *apical meristem*. During ordinary vegetative growth, these cells behave as stem cells, generating a steady succession of new leaves and new segments of stalk. In flowering, the meristem cells switch to making the components of a flower, with its sepals and petals, its stamens carrying pollen, and its ovary containing the female gametes.

To time the switch correctly, the plant has to take account of both past and present conditions. One important cue, for many plants, is day length. To sense this, the plant uses its circadian clock—an endogenous 24-hour rhythm of gene expression—to generate a signal for flowering only when there is light for the appropriate part of the day. The clock itself is influenced by light, and the plant in effect uses the clock to compare past to present lighting conditions. Important parts of the genetic circuitry underlying these phenomena have been identified, including the phytochromes and cryptochromes that act as light receptors (discussed in Chapter 15). The flowering signal that is carried from the leaves to the stem cells via the vasculature depends on the product of a gene called *Flowering locus T* (*Ft*).

But this signal will trigger flowering only if the plant is in a receptive condition from prior long-term cold exposure. Many plants need winter before they will flower—a process called *vernalization*. Cold over a period of weeks or months progressively reduces the level of expression of a remarkable gene called *Flowering locus C* (*Flc*). *Flc* encodes a transcriptional repressor that suppresses expression of the *Ft* flowering promoter.

How does vernalization shut down *Flc* so as to lift the block to flowering? The effect involves at least three long noncoding RNAs, including an antisense transcript called *Coolair* that overlaps with the *Flc* gene and is produced when the temperature is low (**Figure 21–45**). Together with cold-induced chromatin modifiers, including Polycomb-group proteins, *Coolair* coordinates the switching of *Flc* chromatin to a silent state (discussed in Chapters 4 and 7). The degree of silencing depends on the length of cold exposure, enabling the plants to distinguish the odd chilly night from the whole of winter.

The effect on the chromatin is long lasting, persisting through many rounds of cell division even as the weather grows warmer. Thus vernalization creates a

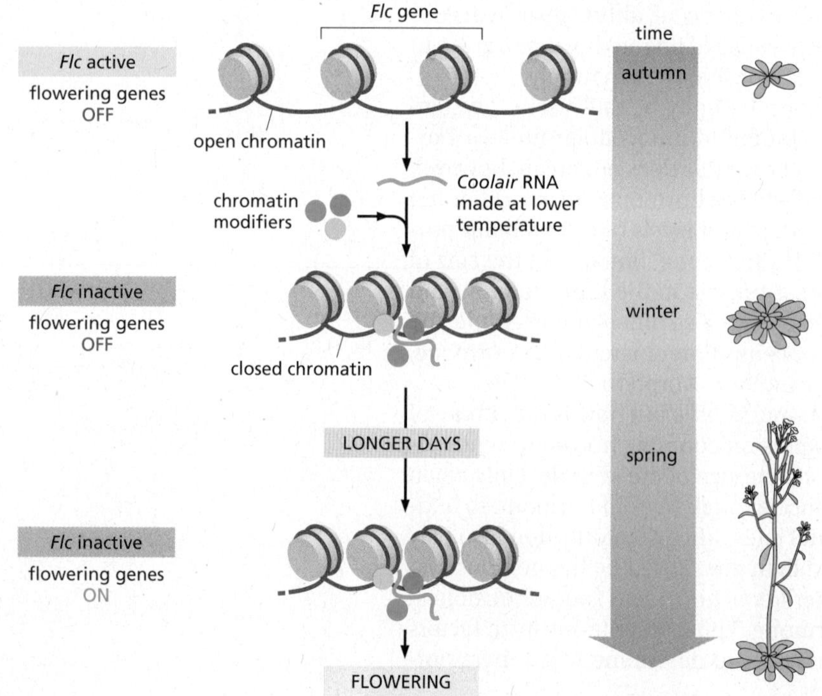

Figure 21–45 Temporal control of flowering in *Arabidopsis*. The *Flc* gene is active and blocks flowering when plants have been grown without exposure to winterlike temperatures. Exposure to a prolonged period of cold leads to the production of the noncoding RNA *Coolair*, which overlaps with the *Flc* gene. *Coolair* induces long-term chromatin changes that turn off *Flc*. These changes persist after the end of the cold period and allow the plant to flower when longer days promote flowering.

persistent block in production of Flc, enabling the Ft signal to be generated when day length is sufficiently long.

Mutations affecting the regulation of *Flc* expression alter the time of flowering and thus the ability of a plant to flourish in a given climate. The whole control system governing the switch to flowering is thus of vital importance for agriculture, especially in an era of rapid climate change.

The example of vernalization suggests a general point about the role of chromatin modification in developmental timing. The plant uses changes in chromatin to record its experience of prolonged cold. It may be that in other organisms—animals as well as plants—slow, progressive changes in chromatin structure provide long-term timers for those mysterious developmental processes that unfold slowly, over a period of days, weeks, months, or years. Such chromatin timers may be among the most important clocks in the embryo, but as yet we understand very little about them.

Summary

Developmental timing is controlled at many levels. It takes time to switch a gene on or off, and this time delay depends on the lifetimes of the molecules involved, which can vary widely. Cascades of gene regulation involve cascades of delays. Feedback loops can give rise to temporal oscillations in gene expression, and these may serve to generate spatially periodic structures. During vertebrate segmentation, for example, expression of the Hes genes oscillates, and one new pair of somites is formed during each oscillation cycle. Hes genes encode transcription repressor proteins that can act back on expression of the Hes genes themselves. This negative feedback generates oscillations with a period that reflects the delay in the autoregulatory gene switching loop. The period of oscillation of this "segmentation clock" controls the sizes of the somites. Notch signaling between neighboring cells synchronizes their oscillations: when Notch signaling fails, the cells drift out of synchrony because of genetic noise in their individual clocks, and the segmental organization of the vertebral column is disrupted.

Timing does not always depend on cell-cell interactions; many developing animal cells have intrinsic developmental programs that play out even in isolated cells in culture. Neuroblasts in Drosophila embryos, for example, go through set programs of asymmetric divisions, generating different neural-cell types at each division with a predictable sequence and timing and through a cascade of gene switching events. Studies in both vertebrates and invertebrates show that such programs are rarely governed by the timing of cell division and can unfold even when cell division is blocked. MicroRNAs produced at critical moments sharpen developmental transitions by blocking the translation and promoting the degradation of specific sets of mRNAs. One such event is the maternal-zygotic transition, when transcription from the embryonic genome initiates. The onset of this event involves both titration of a maternal repressor by multiplying zygotic nuclei and the time-dependent accumulation of maternal transcriptional activators. Global coordination of developmental timing is achieved by hormones that disseminate and act on cells throughout the organism: as a tadpole grows, for example, thyroid hormone levels surge and trigger its metamorphosis into a frog. Environmental control of developmental timing is especially striking in plants and reveals the presence of molecular timers that act over the long term. In vernalization, for example, prolonged cold induces changes in chromatin that chart the passage through winter so as to allow flowering only in the spring. Slow, progressive changes in chromatin structure are likely to be important timers in the long-term programming of development in animals too.

MORPHOGENESIS

The specialization of cells into distinct types at specific times is important, but it is only one aspect of animal development. Equally important are the movements and deformations that cells go through to assemble into tissues and organs with specific shapes and sizes. These physical forms, which are a central output of cell-fate programs, are essential for each organ's physiological function. Like developmental timing, the process of *morphogenesis* ("form generation") is less well understood than the processes of differential gene expression and inductive signaling that lead to cell-type specialization. The cell movements can be readily described, and the underlying molecular mechanisms that coordinate the movements are now being deciphered.

In Chapter 19, we saw how cells cohere to form epithelial sheets or surround themselves with extracellular matrix to create connective tissues. We also discussed how the basic features of tissues, such as the polarity of epithelia, arise from the properties of individual cells. In this section, we consider how the dynamic rearrangements of cells during animal development give shape to the embryo and all the individual organs and appendages of the body.

A small number of cell behaviors are basic to morphogenesis. Cells can change shape, and coordinated shape changes can bend an epithelial sheet into a tube or a hollow ball. Cells can rearrange with respect to their neighbors, causing elongation, constriction, or thickening of a tissue. By stretching out while holding onto their companions, specialized sets of cells can form growing tubular networks such as the system of blood or lymph vessels. Individual cells can also extract themselves from their neighbors and form physically separate groups that then migrate through the embryo along defined tracks. Group migrations, as occur in gastrulation, can transform the entire topology of the embryo. Underlying all these processes are changes in physical forces that alter cell shape and cell contacts—either with other cells or with extracellular matrix. We begin by considering how these forces are generated.

Imbalance in Physical Forces Acting on Cells Drives Morphogenesis

Although it is alive, a cell is governed by the same physical principles that shape inanimate objects. Cells and tissues possess specific mechanical properties, and their shapes reflect the forces exerted upon them. At rest, a cell is in a mechanical equilibrium in which all forces are balanced. Morphogenesis entails a shift in that balance, as new forces are generated in an oriented way. These organized force changes then play out across increasing distances. For example, the activity of myosin motor proteins on the actin cytoskeleton may change the shape of a cell, which in turn affects its interactions with neighboring cells and thus the shape of a tissue. Such coordinated changes act to sculpt entire organs and animal bodies.

While some proteins, such as actin and myosin, generate forces within cells, others act to sense the applied forces. Such proteins, called *mechanotransducers*, initiate intracellular signaling in response to mechanical stimuli, just as the molecules described in Chapter 15 trigger signaling pathways in response to chemical cues. α-Catenin and Talin are mechanotransducers that bind to the cytosolic tail of cell adhesion molecules (see Chapter 19); these unfold in response to tension exerted by neighboring cells, and this conformational change sends biochemical signals that can elicit rapid responses in cellular mechanics as well as long-term effects on gene expression. Piezo (see p. 660) is a mechanosensitive ion channel that spans the cell membrane; it opens in response to shear stress or stretching of the plasma membrane, passing ions that then initiate signaling cascades. Through such mechanosensitive molecules, cells and tissues can read the physical environment within which they exist and respond accordingly.

Tension and Adhesion Determine Cell Packing Within Epithelial Sheets

The two-dimensional epithelial sheet found in many tissues is a simple system to illustrate principles of morphogenesis. Across animal species, stable epithelial sheets display a remarkably consistent pattern of cell shapes, in which most

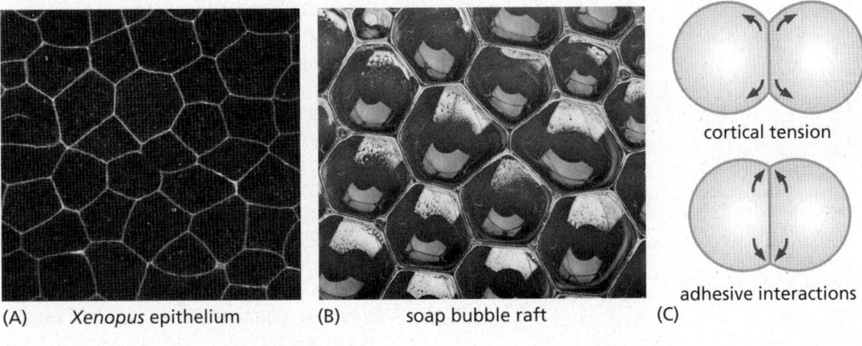

Figure 21–46 **Regular distribution of cell shapes in epithelia.** (A) Fluorescence micrograph of cell membranes in an early *Xenopus* embryo shows that most epithelial cells have hexagonal profiles, with six neighbors forming three-way junctions where cells meet. (B) A two-dimensional "raft" of soap bubbles on the surface of water shows a similar organization. (C) The length and geometry of cell–cell contacts are set by opposing activities. The contractile network of the cortical actomyosin cytoskeleton within each cell tends to pull cells apart, while adhesive interactions mediated by cadherins and other molecules tend to hold cell surfaces together. (A, courtesy of Saranyaraajan Varadarajan and Ann Miller; B, Shebeko/Shutterstock.)

cells possess a hexagonal profile at their apical surface (**Figure 21–46A**). This arrangement is determined by the mechanical properties of the cells and their connections with neighbors, which are mediated primarily by *adherens junctions* (see Chapter 19). Contractility of an actomyosin ring underlying the plasma membrane adjacent to the junctions generates cortical tension within each cell, which tends to pull it away from its neighbors. However, this force is opposed by cell–cell adhesion, mediated by molecules such as cadherins within the junctions. The balance between these inward and outward forces generates a predictable configuration of cells, which can be mathematically modeled and is analogous to the packing arrangement of inorganic entities such as soap bubbles (**Figure 21–46B and C**). Although the physical mechanisms that balance surface-tension minimization and contact interfaces in bubbles are very different from those that shape cells, the mathematical solution representing a minimal energy of arrangement is similar.

Changing Patterns of Cell Adhesion Molecules Force Cells into New Arrangements

The shapes of tissues are ultimately determined by patterns of gene expression, which produce molecules that change the physical properties of cells in which they are expressed. One important class of genes that contributes to these properties codes for the adhesion molecules that cells display on their surface. Through changes in its surface molecules, a cell can break old attachments and make new ones. For example, cells in one region of a tissue may develop surface properties that make them cohere with one another and become segregated from a neighboring group of cells that possess different surface molecules. Cadherins and other cell–cell adhesion proteins that are differentially expressed in the various tissues of a developing animal play important roles. In Chapter 19, we saw how selective cell–cell adhesion is sufficient to drive the sorting of cells dissociated from an embryo into groups of a similar type (see Figure 19–7); treatment with antibodies that bind cadherins interferes with this process. Changes in expression of the various cadherin genes correlate closely with the different modes of association among cells during embryogenesis, as occurs, for example, during gastrulation, neural tube formation, and somitogenesis. Cadherins also control the formation and dissolution of epithelial sheets and clusters of cells (see Movie 19.1). Finally, because adhesion molecules not only glue one cell to another but also provide anchorage for intracellular actin filaments, the sites of cell–cell adhesion influence tension and cell movements in the developing tissue.

Repulsive Interactions Help Maintain Tissue Boundaries

The different types of cadherins enable different types of cells to cohere selectively: cells expressing one type of cadherin will maximize their contact with cells expressing the same cadherin and thereby segregate from other cells, creating specific tissue boundaries. Cell mixing can be inhibited and boundaries created and maintained in another way as well: cells of different types

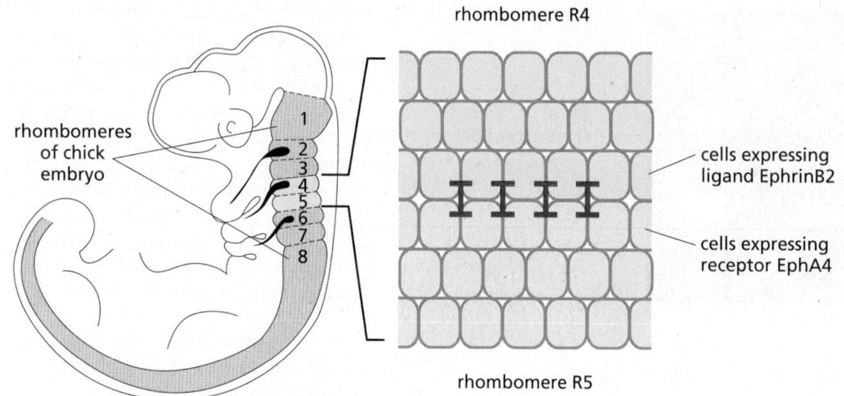

Figure 21–47 Sorting out by repulsion. Ephrin–Eph signaling in hindbrain segmentation in a chick embryo. Each pair of rhombomeres (segments in the hindbrain) is associated with a branchial arch (a modified gill rudiment) to which the pair sends innervation. Rhombomeres are distinguished from one another by expression of different *Hox* genes (see Figure 21–30). Differences in surface tension at the interface between cells that express EphrinB2 in rhombomere R4 and EphA4 in rhombomere R5 lead to mutual repulsion *(red bars)*, creating a sharp boundary.

can sometimes actively repel one another. The bidirectional activation of Eph receptors and ephrins discussed in Chapter 15 often mediates such repulsion, acting at interfaces between different groups of cells to keep the groups from mixing, and repelling invasion by inappropriate visitors. Ephrin–Eph signaling operates at the boundaries of the somites in the vertebrate embryo discussed earlier (see Figure 21–38). By modulating the cortical actomyosin cytoskeleton, ephrin–Eph signaling is thought to induce differences in cortical tension between cells, thereby decreasing the force of the cell–cell contact. Another example is in rhombomeres, which are segmented regions of the developing vertebrate brain. Neighboring rhombomeres express complementary combinations of ephrins and Eph receptors, and this keeps the cells in adjacent rhombomeres strictly segregated, with a boundary between them that is sharply defined (**Figure 21–47**).

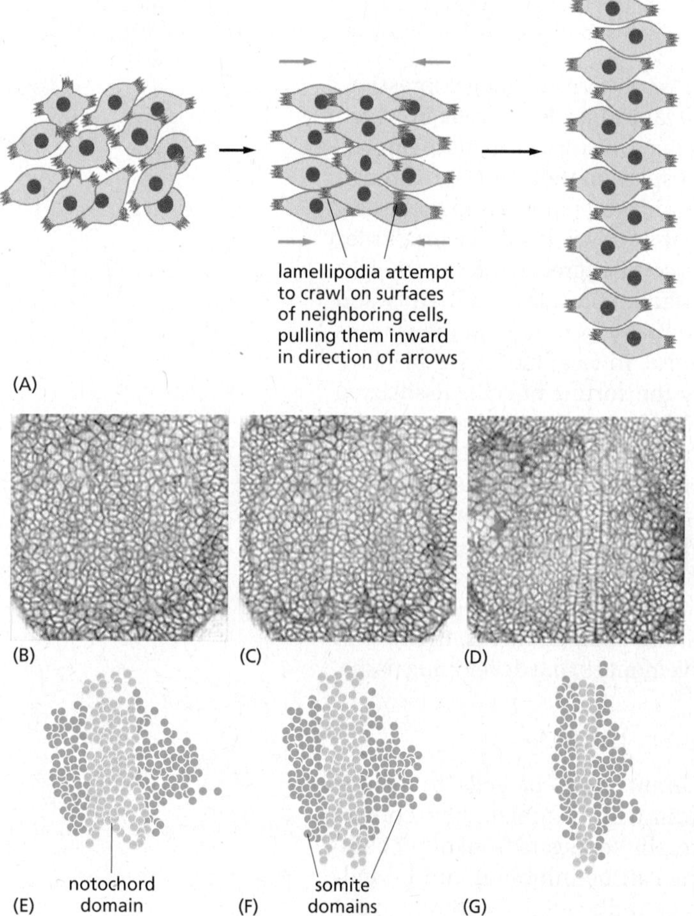

Figure 21–48 Convergent extension by mesenchymal cell migration. (A) A population of cells form lamellipodia, with which they attempt to crawl over one another. Alignment of the lamellipodial movements along a common axis, controlled by planar-polarity signaling (discussed shortly), leads to convergent extension. (B–G) The pattern of convergent extension of dorsal mesoderm during zebrafish gastrulation at 8.8 (B, E), 9.3 (C, F), and 11.3 (D, G) hours after fertilization. Cells that will give rise to the notochord are labeled in *green*, and cells that will give rise to somites and muscle are labeled in *blue*. The notochord and somite domains are spatially separate from the start of the recording (B, E), but their boundaries are at first barely visible and only a little later become obvious. Convergence narrows the notochord domain to a width of about two cells at the last time point (D, G). (A, after J. Shih and R. Keller, *Development* 116:901–914, 1992. With permission from the Company of Biologists; B–G, after N.S. Glickman et al., *Development* 130:873–887, 2003. With permission from the Company of Biologists.)

Groups of Similar Cells Can Perform Dramatic Collective Rearrangements

Cell sorting mediated by cadherin and repulsion mediated by ephrin–Eph exemplify how differences in cell-surface properties can drive tissue arrangements, causing cells that express different sets of genes to separate from one another. However, groups of cells that are all similar can also undergo dramatic rearrangements. Many animal embryos extend their major body axis through such rearrangements, in a process called **convergent extension**. Convergent extension occurs when a population of cells moves coordinately toward a specific boundary, intercalating between each other as they do so. This causes the tissue to narrow along one axis (converge) and elongate along another (extend). The overall change in tissue shape results from cell displacement, rather than a change in shape of individual cells, and is largely autonomous to the group of cells involved. A striking demonstration is seen when small, square fragments of tissue from the appropriate region of a frog embryo are isolated in culture. These will spontaneously narrow and elongate, just as they would in the developing animal.

Convergent extension is a common movement that elongates many organs, although the cell behaviors that underlie it can differ. During frog gastrulation, for example, cells in one region of the surface epithelium lose their epithelial character as they migrate into the interior of the embryo. The loosely connected cells, called *mesenchymal cells*, form actin-based protrusions called lamellipodia that are oriented so that they crawl over one another and interdigitate as they converge toward the embryonic midline (**Figure 21–48**). In contrast, when the *Drosophila* embryo extends its anteroposterior axis, the cells involved remain epithelial. These cells maintain tight adhesive junctions but rearrange the relationships to their neighbors by shrinking cell junctions along one axis while extending them in another axis to accomplish cell intercalation (**Figure 21–49**; see also Figure 19–15). Remodeling of the junctions involves differential contractile tension on them that is generated by localized myosin activity. In both cases, whether mesenchymal migration or epithelial junction remodeling, the coordination of cell behaviors depends on a signaling pathway that can generate *planar cell polarity*, as we discuss next.

Planar Cell Polarity Orients Cell Behaviors Within an Embryo

Just as a compass allows us to orient ourselves in a landscape, cells undergoing morphogenetic movements must have a system to orient themselves with respect to the different body axes of the animal. We discussed in Chapter 19 how all

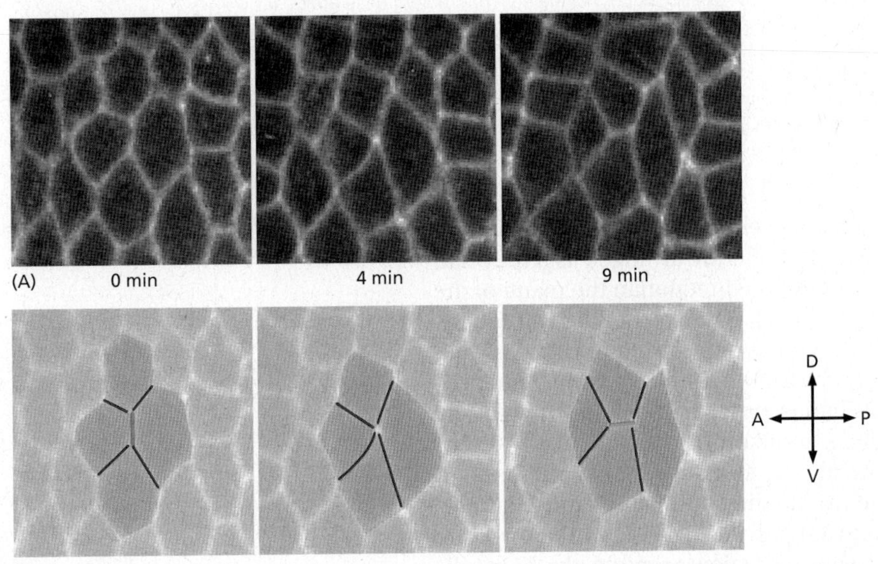

Figure 21–49 **Convergent extension by rearrangement of epithelial junctions.** Shown in (A) are fluorescence images of the *Drosophila* embryonic epidermis over a 9-minute period, with drawings in (B) that colorize specific cells and cell junctions (**Movie 21.6**). As in other epithelia, most cells are initially hexagonal, and their junctions are shared between three neighbors. To elongate the tissue along the anteroposterior axis, some of the dorsoventral-oriented junctions *(green)* shrink, forming a transient arrangement in which cells are pentagonal or quadrilateral, and a junction is shared between four neighbors. A new junction *(blue)* then forms along the anteroposterior axis, now shared between two cells that did not previously contact each other and separating two cells that previously did contact each other. This process of shrinking and elongating junctions in a planar-polarized fashion, occurring all along the epidermis, narrows and extends the tissue while maintaining epithelial integrity throughout. (Images courtesy of Huapeng Yu and Jennifer Zallen.)

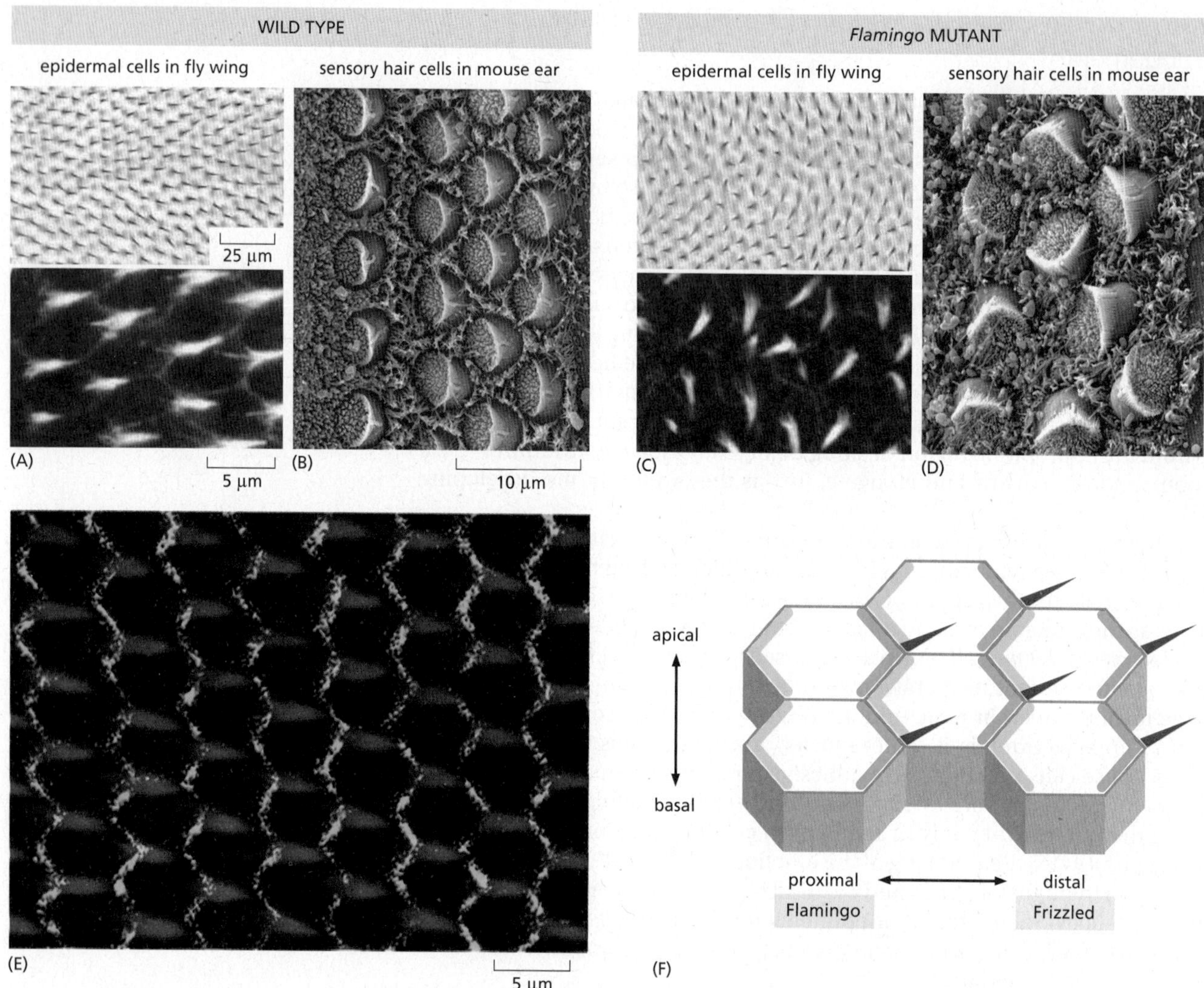

Figure 21–50 Planar cell polarity. (A) Wing hairs on the wing of a fly. Each cell in the wing epithelium forms a small, spiky protrusion or "hair" at its apex, and all the hairs point the same way, toward the tip of the wing. This reflects a planar polarity in the structure of each cell. (B) Sensory hair cells in the inner ear of a mouse similarly have a well-defined planar polarity, manifest in the oriented pattern of stereocilia (actin-filled protrusions) on their surface. The detection of sound depends on the correct, coordinated orientation of the hair cells. (C) A mutation in the gene *Flamingo* in the fly, coding for a nonclassical cadherin, disrupts the pattern of planar cell polarity in the wing. (D) A mutation in a homologous *Flamingo* gene in the mouse randomizes the orientation of the planar cell polarity vector of the hair cells in the ear. The mutant mice are deaf. (E) Micrograph showing localization of Flamingo *(green)* in a pupal fly wing, in relation to the hair growing from each cell *(red)*. (F) Diagram illustrating planar-polarized localization of Flamingo to one side of each epithelial cell, and localization of another planar polarity–regulating protein called Frizzled to the opposite side, from which the hair will form. (A and C, from J. Chae et al., *Development* 126:5421–5429, 1999. With permission from the Company of Biologists; B and D, from J.A. Curtin et al., *Curr. Biol.* 13:1129–1133, 2003. With permission from Elsevier; E, courtesy of Paul Adler.)

epithelial cells have apical–basal polarity, but the cells of many epithelia show an additional polarity at right angles to this axis: the cells are arranged as if they had an arrow written on them, all pointing in a common direction in the plane of the epithelium. This type of polarity is called **planar cell polarity**. In the wing of a fly, for example, each epithelial cell has a tiny asymmetric projection, called a wing hair, on its surface, and the hairs all point toward the tip of the wing. Similarly, in the inner ear of a vertebrate, each mechanosensory hair cell has a precisely oriented asymmetric bundle of actin-filled, rodlike protrusions called stereocilia sticking up from its apical plasma membrane as a detector of sound and of forces such as gravity. Tilting the bundle in one direction causes ion channels in the membrane to open, electrically activating the cell; tilting in the opposite direction has the opposite effect. For the ear to function properly, the hair cells must be oriented correctly. Planar cell polarity is also important in the respiratory tract, where every ciliated cell must orient the beating of its cilia so as to sweep mucus upward, away from the lungs.

Screens for mutants with misoriented wing hairs in *Drosophila* have identified a set of genes that control planar cell polarity in many tissues. Some of these genes code for components of the Wnt signaling pathway, and others code for transmembrane proteins such as specialized members of the cadherin superfamily that interact across neighboring cell–cell junctions. Together with intracellular proteins that influence the localization of the transmembrane proteins, two complementary, mutually exclusive domains are defined along the apical surface in such a way as to exert a polarizing influence that propagates from cell to cell, leading to polarization of the entire epithelial tissue. Essentially the same system of proteins controls planar cell polarity in vertebrates; mice deficient in homologs of the *Drosophila* planar-polarity genes have a variety of defects, including incorrectly oriented hair cells in the inner ear, making them deaf (**Figure 21–50**).

An Epithelium Can Bend During Development to Form a Tube

Convergent extension is an example of morphogenesis in two dimensions. When cells need to leave the plane of a tissue and produce three-dimensional structures, they engage in additional cell behaviors. The universal process of gastrulation involves future endodermal and/or mesodermal cells migrating into the interior of the embryo, and this is often initiated by a localized bending of the blastula epithelium. Similar bending can generate biological tubes from flat epithelial sheets, including the neural tube characteristic of chordates, although there are other routes to tube formation as sketched in **Figure 21–51**.

Tissues can be bent by a widely used cell behavior called *apical constriction*, which occurs when cells use myosin motor proteins to contract bundles of actin filaments anchored at adherens junctions near their apical surface. Shortening of these filament bundles causes the epithelial cells to narrow at their apex, transforming the cell from a columnar shape into a distinctive wedge or bottle shape. Because the cells are connected to each other by adherens junctions, coordinated contraction in a group can result in the entire sheet bending or even rolling up into a tube (**Figure 21–52**). The cell groups that undergo apical constriction are selected by localized expression of patterning genes that trigger the activity of the contractile actomyosin networks. In *Drosophila* embryos, the ventrally restricted transcription regulator Twist (Figure 21–26) controls myosin behavior in presumptive mesodermal cells, which then gastrulate by creating a transient tube-like invagination. In vertebrate embryos, patterns of Sonic hedgehog (Shh) and BMP

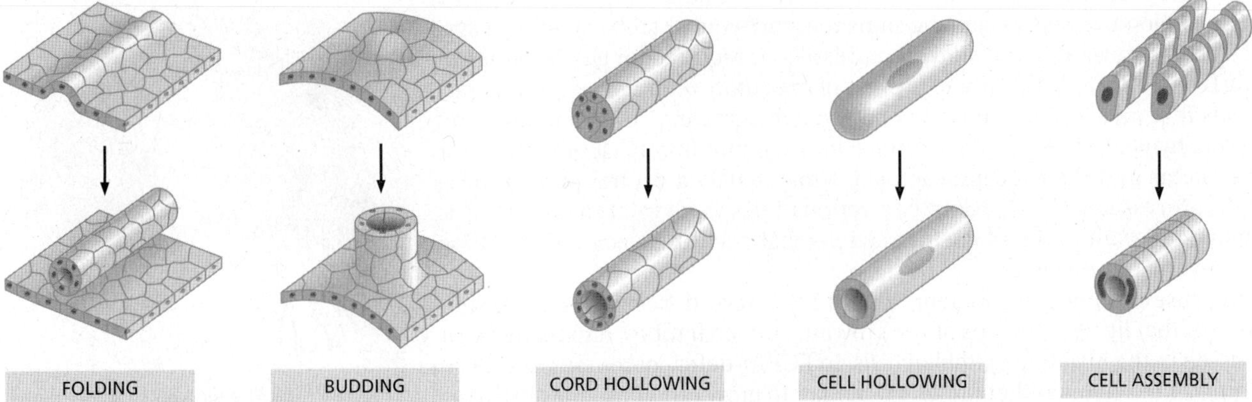

Figure 21–51 Cell behaviors involved in tube formation. Folding generates the neural tube, budding underlies the formation of lungs and trachea, cord hollowing occurs during the formation of mammalian salivary glands, cell hollowing is involved in the formation of tracheal terminal cell tubes, and cell assembly generates the heart tube that forms at the earliest stage of heart development.

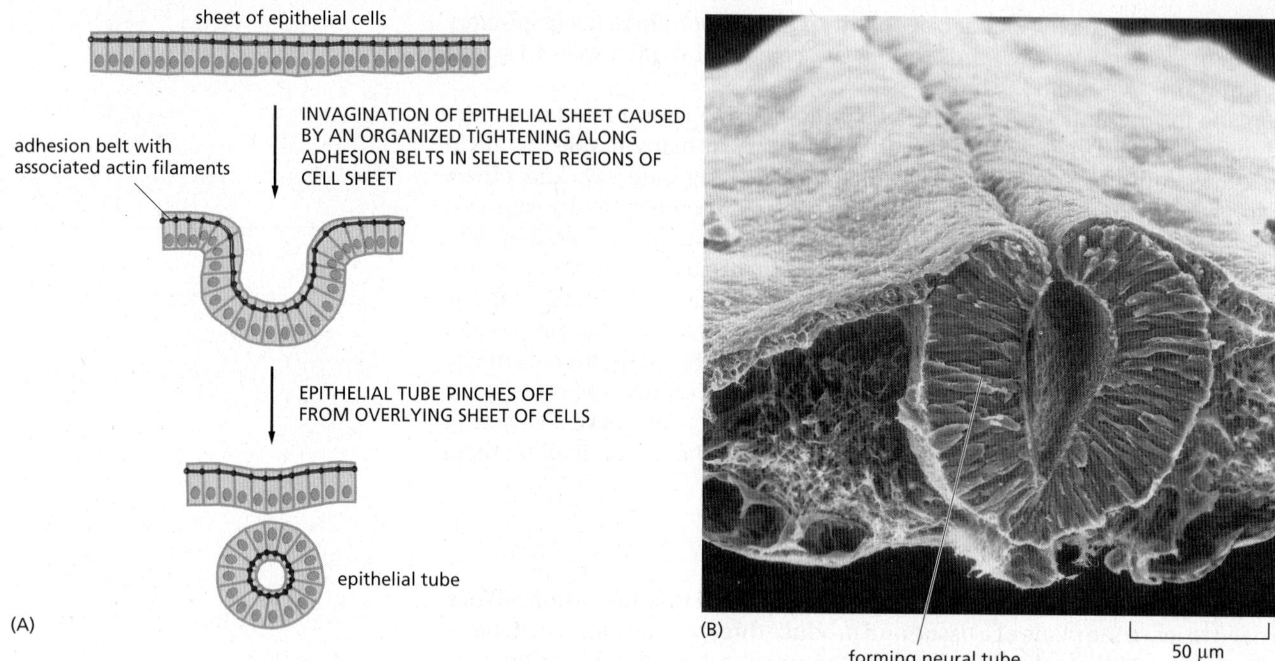

Figure 21–52 Bending of an epithelial sheet to form a tube. Contraction of apical bundles of actin filaments linked from cell to cell by adherens junctions causes the epithelial cells to narrow at their apex. Depending on the number and arrangement of cells in which the contraction occurs, the invaginated epithelium can remain connected to the original sheet or alternatively can roll up into a separated tube or a hollow sphere. (A) Diagram showing how an apical contraction along one axis of an epithelial sheet can cause the sheet to form a tube. (B) Scanning electron micrograph of a cross section through the trunk of a 2-day chick embryo, showing the formation of the neural tube by the process diagrammed in panel A. The vertebrate neural tube gives rise to the brain and spinal cord. (B, courtesy of Jean-Paul Revel.)

signaling induce apical constriction at three hinge points that transform the neural plate into a closed neural tube.

Interactions Between an Epithelium and Mesenchyme Generate Branching Tubular Structures

Embryos often develop branching tubular structures when large surfaces are required for functions such as excretion, absorption of nutrients, and gas exchange. The lungs are an example. They originate from epithelial buds that grow out from the floor of the foregut and penetrate neighboring mesenchyme to form the bronchial trees, systems of tubes that branch repeatedly as they extend, eventually making an estimated 17 million branches in humans. Endothelial cells that form the lining of blood vessels invade the same mesenchyme, thereby creating a system of closely apposed airways and blood vessels, as required for gas exchange in the lung (**Figure 21–53**). This whole process of *branching morphogenesis* depends on signals that pass in both directions between the growing epithelial buds and the mesenchyme. Genetic studies in mice indicate that fibroblast growth factor (FGF) proteins and their receptor tyrosine kinases play a central part in these signaling processes. FGF signaling has various roles in development, but it is especially important in the many interactions that occur between a developing epithelium and mesenchyme.

In the case of lung development, FGF10 is expressed in clusters of mesenchyme cells that lie near the tips of the growing epithelial tubes, and its receptor is expressed in the invading epithelial cells. In FGF10-deficient mutant mice, primary buds of the lung epithelium form, but fail to grow out of the mesenchyme to create branching bronchial trees. Conversely, a microscopic bead soaked in FGF10 and placed near embryonic lung epithelium in culture will induce a bud to form and grow out from the epithelium toward the bead. Evidently, the epithelium enters the mesenchyme only by invitation, in response to FGF10.

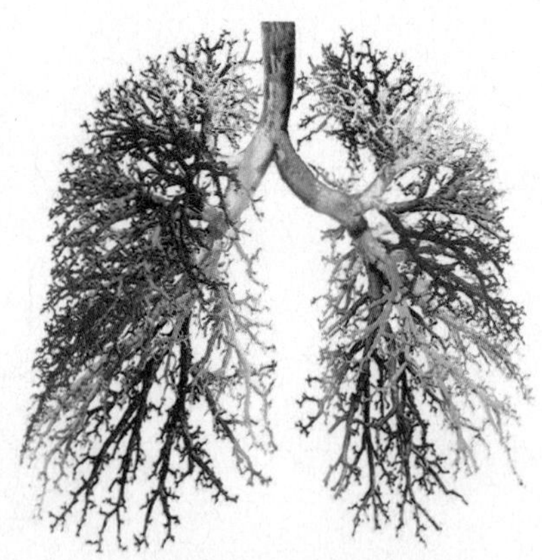

Figure 21–53 The airways of the lungs, shown in a cast of the adult human bronchial trees. Resins of different colors have been injected into different large branches of the trees. (James Cavallini/ Science Source.)

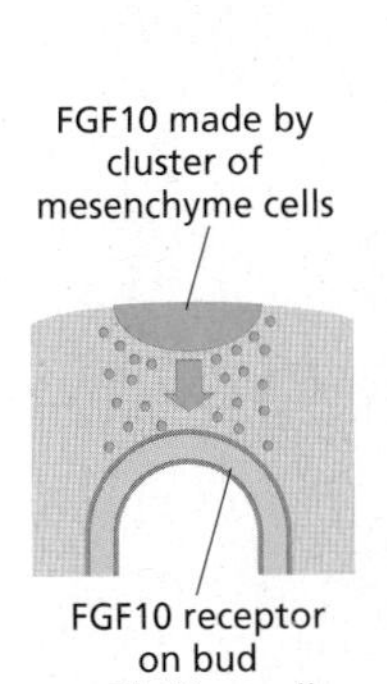

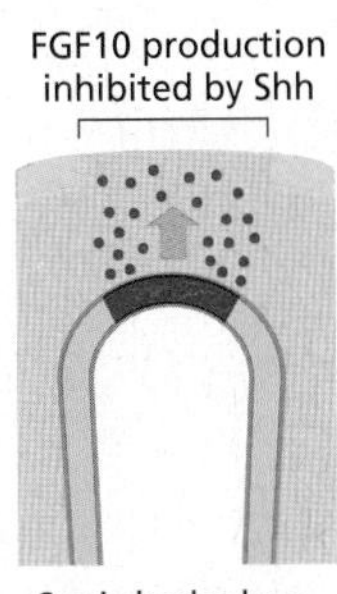

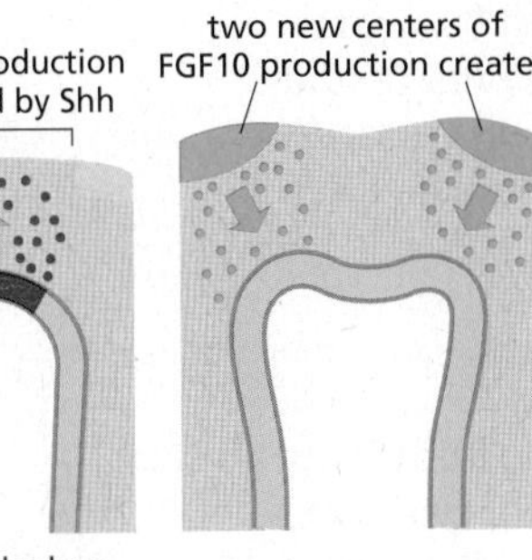

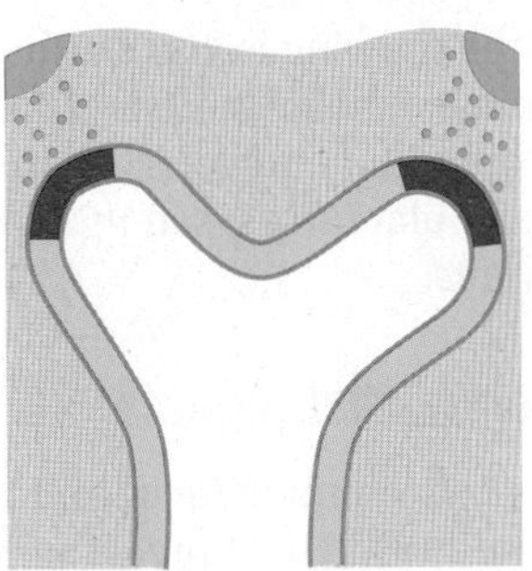

Figure 21–54 **Branching morphogenesis of the lung.** How FGF10 and Sonic hedgehog are thought to induce the growth and branching of the buds of the bronchial tree. Many other signal molecules, such as BMP4, are also expressed in this system, and the illustrated branching mechanism is only one of several possibilities. As indicated, FGF10 protein is expressed in clusters of mesenchyme cells near the tips of the growing epithelial tubes, and its receptor is expressed in the epithelial cells themselves. The Sonic hedgehog signal is sent in the opposite direction, from the epithelial cells at the tips of the buds back to the mesenchyme. The patterns of gene expression and their timing suggest that the Sonic hedgehog signal may serve to shut off FGF10 expression in the mesenchyme cells closest to the growing tip of a bud, splitting the FGF10-secreting cluster into two separate clusters, which in turn cause the bud to branch into two.

But what makes the growing epithelial tubes of the lung branch repeatedly as they invade the mesenchyme? This depends on a Sonic hedgehog signal that is sent in the opposite direction, from the epithelial cells at the tips of the buds back to the mesenchyme, where it inhibits FGF10 production to promote bifurcation as shown in Figure 21–54. In mice lacking Sonic hedgehog, the lung epithelium grows and differentiates, but it forms a sac instead of a branching tree of tubules.

FGF is also the major signal in formation of the air-exchange system of insects, which consists of a pattern of fine, air-filled channels called *tracheae* and *tracheoles*. These originate from the epidermis covering the surface of the body and extend inward to invade the underlying tissues, branching and narrowing as they go (Figure 21–55). The FGF acts on cells at the tips of the advancing tracheae, causing them to extend filopodia and migrate toward the source of the FGF signal. Because the tip cells remain connected to the remainder of the tracheal epithelium, the pulling force that they generate elongates the tracheal tube.

Initially, the pattern of FGF production in fly embryos is defined by the D-V and A-P patterning systems discussed earlier. In later stages of development, however, FGF expression is induced by transcription regulators called *hypoxia-inducible factors* (*HIFs*) that are activated by hypoxia (low oxygen levels). In this way, hypoxia stimulates the formation of finer and finer and more extensively branched trachea, until the oxygen supply is sufficient to stop the process.

The Extracellular Matrix Also Influences Tissue Shape

So far, we have considered only forces generated by cells during morphogenesis. However, we must not ignore another important constituent of organs: the extracellular matrix. As detailed in Chapter 19, this network of insoluble molecules is found in most animal tissues, and a specialized matrix called the basal lamina underlies all epithelial sheets. Extracellular matrices provide substrates for morphogenetic cell-migration events, as we will discuss shortly. Furthermore, the mechanical stiffness of the matrix can provide instructive signals for shaping organs by modifying the resistance against which cellular forces are generated. During branching morphogenesis of the mammalian salivary gland, for

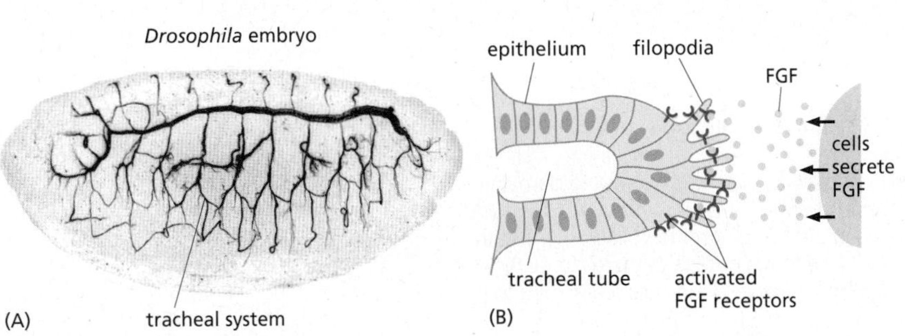

Figure 21–55 **Branching morphogenesis of airways in a fly.** (A) *Drosophila* embryonic tracheal system. (B) FGF (produced in *Drosophila* by the *Branchless* gene) signals from surrounding cells to the tracheal epithelium and activates its FGF receptors, leading to filopodia formation and tube elongation. [A, from G. Manning and M.A. Krasnow, in The Development of *Drosophila melanogaster* (A. Martinez-Arias and M. Bate, eds.), Vol. 1, pp. 609–685. New York: Cold Spring Harbor Laboratory Press, 1993. With permission from Cold Spring Harbor Laboratory Press.]

example, the basal lamina surrounding the growing bud is perforated by matrix-degrading enzymes. This localized softening of the matrix encourages preferential outgrowth of the epithelium into the more pliable regions. Changes in extracellular resistance, as well as dynamic intracellular forces, can therefore create force imbalances that drive morphogenetic events.

Cell Migration Is Guided by Environmental Signals

The birthplace of cells is often far from their ultimate location in the body. Our skeletal muscles, for example, derive from muscle-cell precursors, or *myoblasts*, in somites from which they migrate into the limbs and other regions. The routes that the migrant cells follow and the selection of sites that they colonize determine the eventual pattern of muscles in the body. The embryonic connective tissues form the framework through which the myoblasts travel, and these tissues provide the cues that guide myoblast distribution. No matter which somite they come from, the myoblasts that migrate into a forelimb bud will form the pattern of muscles appropriate to a forelimb, and those that migrate into a hindlimb bud will form the pattern appropriate to a hindlimb. It is the connective tissue that provides the patterning information.

As a migrant cell travels through the embryonic tissues, it repeatedly extends surface projections that probe its immediate surroundings, testing for cues to which it is particularly sensitive by virtue of its specific assortment of cell-surface receptor proteins. Inside the cell, these receptors are connected to the cortical actin and myosin cytoskeleton, which moves the cell along (see Chapter 16). Some extracellular matrix molecules, such as the protein fibronectin, provide adhesive sites that help the cell advance; others, such as chondroitin sulfate proteoglycan, inhibit locomotion and repel immigration. The nonmigrant cells along the migration pathway may likewise have inviting or repellent macromolecules on their surface; some may even extend filopodia to make their presence known.

In addition to general adhesive molecules, migratory cells can also be guided by specific chemical or mechanical signals that are sensed by dedicated receptors. Among many such influences, a few stand out as especially important. In particular, many types of migrating cells are guided by chemotaxis that depends on a G-protein-coupled receptor, called CXCR4, which is activated by an extracellular ligand, CXCL12. Cells expressing this receptor can shuffle their way along tracks marked out by CXCL12 (**Figure 21–56**). Chemotaxis toward sources of CXCL12 plays a major part in guiding the migrations of lymphocytes and various other white blood cells; of neurons in the developing brain; of myoblasts entering limb

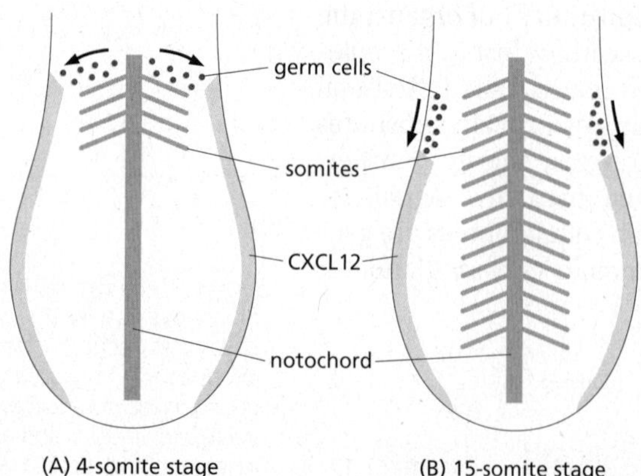

Figure 21–56 CXCL12 guides migrating germ cells. Zebrafish germ cells migrate to domains that express CXCL12. As the sites of CXCL12 expression change, cells follow the CXCL12 track and are guided to the region where the gonad develops at a later developmental stage. (A) At the 4-somite stage, germ cells move from a position that is close to the midline to more lateral regions where CXCL12 is expressed. (B) As the CXCL12 expression retracts, germ cells are guided to more posterior positions, where the gonads are developing.

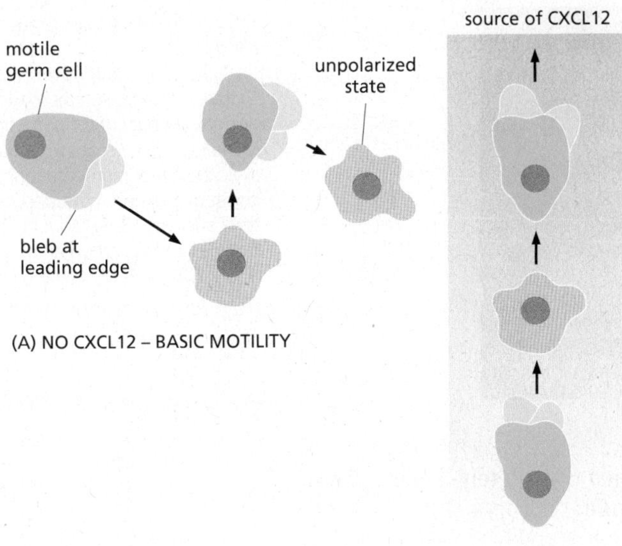

Figure 21–57 Directional migration by local blebbing. (A) Germ cells migrate via protrusions at the leading edge of the cell called blebs, where the plasma membrane detaches locally from the underlying actin cortex and is pushed outward (see Figure 16–18). (B) The persistence and site of the protrusions are biased toward higher levels of CXCL12. Thus, germ cells migrate up the CXCL12 gradient. Note that this form of cell migration is distinct from that based on actin-rich lamellipodia (see Figures 21–48 and 21–59).

buds; of primordial germ cells as they travel toward the developing gonads; and of cancer cells when they metastasize.

Detailed studies of primordial-germ-cell migration have shown that CXCL12 signaling does not induce cell migration per se but rather serves to control its direction. In the absence of CXCL12 signaling, germ cells still display the membrane blebbing associated with cell migration, but the position of the cell front where blebs form is randomly chosen (**Figure 21–57**), and migration is more of a random walk; if CXCL12 signaling is intact, blebbing is more frequent on the side of the cell that faces the source of CXCL12, resulting in directional migration.

The Distribution of Migrant Cells Depends on Survival Factors

The final distribution of migrant cells depends not only on the routes they take, but also on whether they survive the journey and thrive in the environment they find at their journey's end. Specific sites provide survival factors needed for specific types of migrant cells to survive.

Among the most important sets of migrant cells in the vertebrate embryo is that of the **neural crest**. They arise from the border region between the part of the ectoderm that will form epidermis and the part that will form the central nervous system. As the neural ectoderm rolls up to form the neural tube, the neural crest cells break loose from the epithelial sheet along this border region and set out on their long migrations (see Figure 19–8 and **Movie 21.7**). They settle ultimately in many sites and give rise to a surprising diversity of cell types. Some will differentiate into the neurons and glial cells of the peripheral nervous system—not only in the sensory ganglia that lie close to the spinal cord, but also, following a much longer migration, in the wall of the gut. Others will form skeletal tissue in the face; still others will lodge in the skin and specialize as pigment cells.

A signaling pathway required for neural crest cell survival was identified by studying animals with defective pigmentation. Many naturally occurring mouse mutants that show white spots on their normally black coat of hair are defective in a gene that encodes a secreted peptide called Stem cell factor. Stem cell factor is produced by tissues along the migration pathways and acts as a survival factor for the migrating neural crest cells. In animals defective for Stem cell factor or its receptor, a transmembrane tyrosine kinase called Kit, many of the migrating neural crest cells die by apoptosis. As a result, the mutant individuals have nonpigmented (albino) patches of skin (**Figure 21–58**). Stem cell factor is an important survival signal for other types of migratory cells as well, including

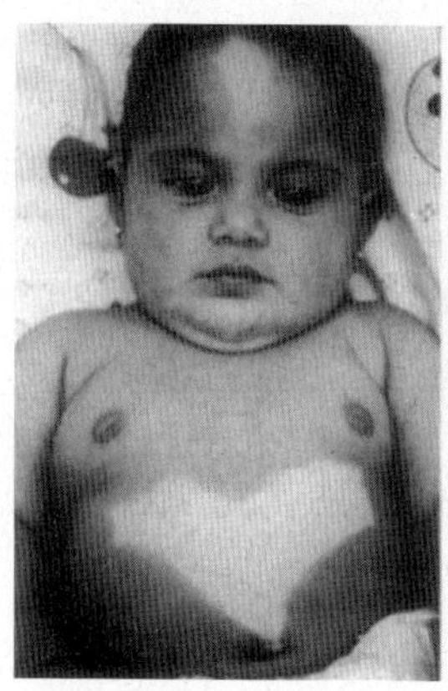

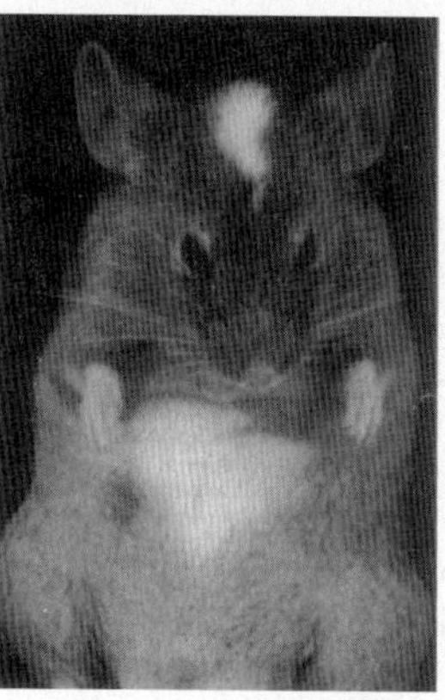

Figure 21–58 **Effect of mutations in the *Kit* gene.** Both the baby and the mouse are heterozygous for a loss-of-function mutation that leaves them with only half the normal quantity of *Kit* gene product, which is the receptor for stem cell factor (SCF). In both cases, pigmentation is defective because pigment cells depend on SCF for their survival. (Courtesy of R.A. Fleischman, from R.A. Fleischman et al., *Proc. Natl. Acad. Sci. USA* 88:10885–10889, 1991.)

primordial germ cells and blood-cell precursors. Mice completely lacking Stem cell factor or Kit are sterile and eventually die of anemia.

Cells Migrate in Groups to Achieve Large-Scale Morphogenetic Movements

In addition to traveling solo, cells can migrate in groups, maintaining contact with one another and exhibiting coordinated behaviors that differ from those of independently migrating cells. These movements are termed *collective cell migration,* and they come in several forms distinguished by how the cells are organized. For example, neural crest cells undergo *chain migration,* in which cells follow one another in streamlike files (Figure 21–59A). Although their association is loose, contact between the cells is required for directed migration along paths determined by chemotactic factors such as CXCL12. Whereas individual neural crest cells respond poorly to this signal, head-to-tail cell contact induces polarity within the group that drives protrusion at the leading edge of each cell in the migrating chain.

Zebrafish development provides a good example of a second form of collective cell movement, called *clustered migration* (Figure 21–59B). Such migration occurs in groups of cohesive cells that are usually mesenchymal in origin, with leader cells of the group exhibiting the most protrusive behavior. In the zebrafish embryo, a cell cluster called the lateral line primordium migrates from a position just behind the ear and crosses the trunk musculature toward the tail. Cells in the migrating primordium continue to divide, and some are left behind to form sensory structures in its wake. The specific migratory route is laid out by a prepatterned stripe of CXCL12, but interestingly this ligand is not expressed in a gradient. Instead, the cell cluster itself generates a local gradient. Leader and follower cells both express the CXCR4 receptor, but in addition the followers express a nonfunctional CXCL12 receptor called CXCL7, which acts as a "sink" to diminish the effect of the CXCL12 signal at the rear of the cluster. The result is differential signaling within the cell cluster that orients the direction of travel.

Entire epithelia can move in a third form of collective migration called sheet migration (see Figure 21–59C). Here, strong adhesion and apical–basal polarity are maintained in the group; cells at the leading edge of the sheet extend robust protrusions that explore the environment, while follower cells create short basal protrusions that are also oriented in the direction of migration. Collective sheet migration can fuse flanking epithelia together, sealing gaps that have formed during development, as seen at the end of *Drosophila* and *C. elegans* embryogenesis or during wound healing. It can also create single organs from bilateral primordia, such as through the fusion that forms the mammalian palate. Interestingly, epithelial migrations can occur in the absence of a free leading edge, as seen in the spherical epithelium of the developing *Drosophila* egg chamber. In this case, collective epithelial migration driven by oriented basal protrusions drives repeated rounds of rotation of the entire tissue, a process required to shape the mature egg. Migration of cells to the top of mammalian gut villi, which

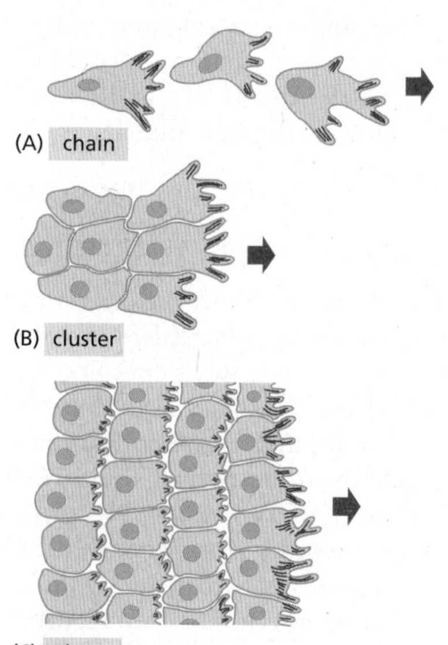

Figure 21–59 **Collective cell migration.** Cell protrusions are shown in *red*. (A) Cells undergoing chain migration are connected only loosely, but transient contact with the cell ahead polarizes the protrusions *(red)* of the cell behind, promoting a shared trajectory. (B) Clustered migration involves groups of adherent cells in which a few "leader" cells extend protrusions in the direction of travel. (C) In sheet migration, large groups of tightly adherent epithelial cells move coherently together, usually with a free edge where cells take on a partially mesenchymal morphology with robust protrusions that lead to migration. Smaller oriented protrusions are also seen in cells behind the leading edge.

preserves intestinal homeostasis (see Figure 22–4), appears to involve a similar movement of a continuous epithelial sheet.

Summary

Animal tissues are sculpted by dramatic changes in cell shape, arrangement, or position. These changes result from shifts in the balance of forces on cells, often triggered by the expression of genes that regulate actomyosin contractility or cell adhesion. Cells that have similar adhesion molecules on their surfaces cohere and tend to segregate from other cell groups with different surface properties. Selective cell-cell adhesion is often mediated by cadherins; repulsion is often driven by ephrin–Eph signaling. Within a tissue, cells can rearrange themselves to drive convergence and extension movements that result in elongation of a tissue along a body axis. Many movements are coordinated through a planar-polarity signaling pathway that is also responsible for orienting cells correctly in various types of epithelium. Epithelial tubes can originate in various ways, most simply by the rolling up and pinching off of a segment of epithelium due to constriction of the apical surface. Elaborate branched tubular structures, such as the airways of the lung, are generated through bidirectional signaling between an epithelial bud and the mesenchyme that it invades in a process called branching morphogenesis. Branching and other morphogenetic events are regulated by changes both in the extracellular matrix and within the cells themselves. Long-range migrating cells, such as those of the neural crest, break loose from their original neighbors and travel through the embryo to colonize new sites. Many migrant cells, including primordial germ cells, are guided by chemotaxis dependent on the receptor CXCR4 and its ligand CXCL12. Interconnected groups of cells can also migrate, with leader and follower cells showing distinctive behaviors. Such collective cell migrations accomplish large-scale tissue shaping.

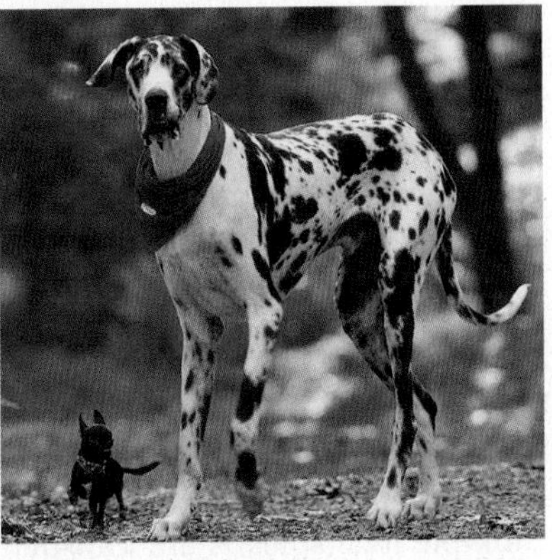

Figure 21–60 **Members of the same species can have dramatically different sizes.** The Chihuahua weighs 2–5 kg, whereas a Great Dane weighs 45–90 kg. (Courtesy of Deanne Fitzmaurice.)

GROWTH

One of the most fundamental aspects of development is one we know surprisingly little about—how the size of an animal or an organ is determined. Why, for example, do we grow to be so much larger than a mouse? All mammals develop from eggs of a similar size, but there is a 100-million-fold difference in adult weight between the smallest (a shrew) and the largest (a blue whale). Even within a species, size can vary greatly; a Great Dane, for instance, can weigh over 40 times more than a Chihuahua (**Figure 21–60**). In addition to hereditary factors, the environment also plays a role—a goldfish released into the wild may grow 10 times the size of one kept in a home aquarium.

Three variables define the size of an organ or organism: the number of cells, the size of the cells, and the quantity of extracellular material per cell. Size differences can arise from changes in any of these factors (**Figure 21–61**). If we compare a mouse with a human, for example, we find that the difference lies chiefly in the number of cells, there being roughly 3000 times more cells in a human, corresponding to a body that is roughly 3000 times more massive. Wild and cultivated species of food plants, on the other hand, often differ in body size chiefly because of differences in cell size.

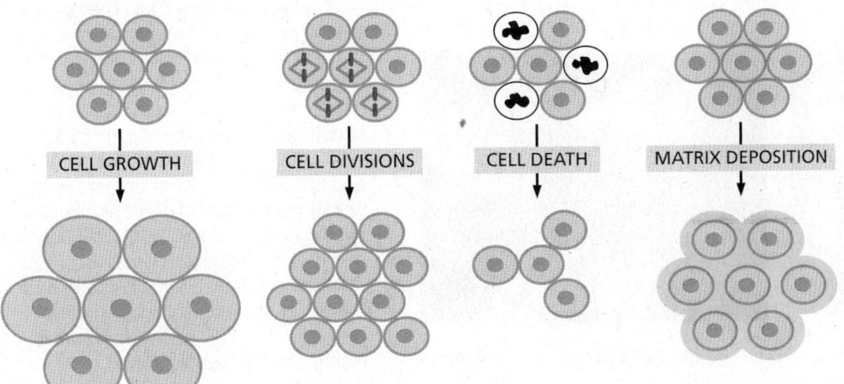

Figure 21–61 **Determinants of organ size.** A tissue may tune cell growth, proliferation, death, or a combination of these to regulate size. Extracellular matrix deposition also contributes, especially in animal tissues such as bone and cartilage, and in plants.

The challenge, therefore, is to understand how cell numbers, cell size, and extracellular matrix production are regulated. First, we need to identify the signals that drive or inhibit growth. Then we need to discover how the signals themselves are regulated. In some cases, organ growth is dictated by developmental programs that do not monitor the size the structure has attained. In other cases, the growth of an organ or of the body as a whole is controlled homeostatically, so that a "target size" is reached even in the face of drastic disturbances. This suggests the developing structure somehow senses its own size, through local or systemic signals, and uses this information to regulate its growth or shrinkage.

The variation in control strategies is nicely illustrated by some classic transplantation experiments. If several fetal thymus glands are transplanted into a developing mouse, each grows to its characteristic adult size. In contrast, if multiple fetal spleens are transplanted, each ends up smaller than normal, but collectively they grow to the size of one adult spleen. Thus, thymus growth is regulated by local mechanisms intrinsic to the individual organ, whereas spleen growth is controlled by a systemic feedback mechanism that senses the quantity of spleen tissue in the body as a whole; in neither case is the mechanism known. Very often, the sizes and proportions of body parts depend on combinations of size-measuring feedback controls and intracellular programs, as well as on environmental influences such as nutrition and habitat.

The Proliferation, Death, and Size of Cells Determine Organ and Organism Size

The worm *C. elegans* demonstrates the ways in which size differences can arise. As we discussed earlier, each individual of a given sex is generated by almost exactly the same sequences of cell divisions and cell deaths, and consequently has precisely the same number of somatic cells—959 in the adult hermaphrodite (see Figure 21–42). More than 1000 cell divisions generate 1090 somatic cells during hermaphrodite development, but 131 of these cells undergo apoptotic cell death. Thus, precise regulation of both cell division and cell death determines the final numbers of somatic cells in the worm.

It takes only 3 days for *C. elegans* to generate all of its adult somatic cells, after which cell divisions occur only in the germ line. Yet the worm continues to grow, doubling in size between sexual maturity and death 2–3 weeks later. This doubling results from somatic-cell growth: although the cells no longer divide, they continue to go through rounds of DNA synthesis; this *endoreplication* of the genome makes the cells *polyploid*. As in all organisms, the size of a cell is proportional to its ploidy (that is, the number of genome copies that it contains): a doubling of ploidy roughly doubles cell volume. Thus the worm's final size is set by a combination of programmed cell divisions and cell deaths, along with regulation of the sizes of individual cells through changes in ploidy.

Endoreplication of the genome as a means to increase cell size is not peculiar to *C. elegans* and occurs in a developmentally regulated manner in specific tissues of many animals. In one extreme case, neurons that innervate large areas of the mollusk *Aplysia* may contain hundreds of thousands more copies of the genome than does a diploid cell and grow up to 1 mm in diameter (**Figure 21–62**).

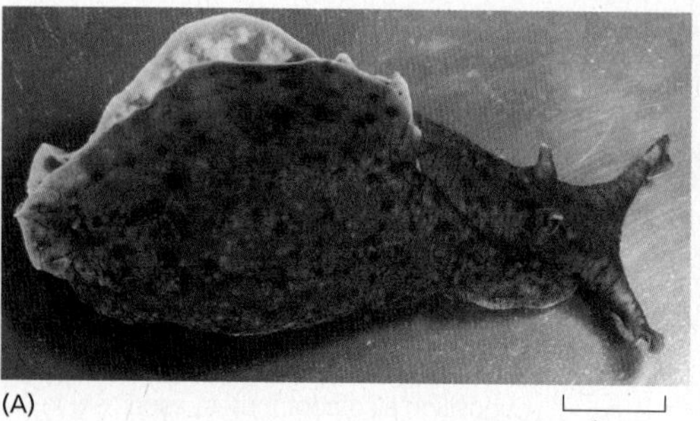

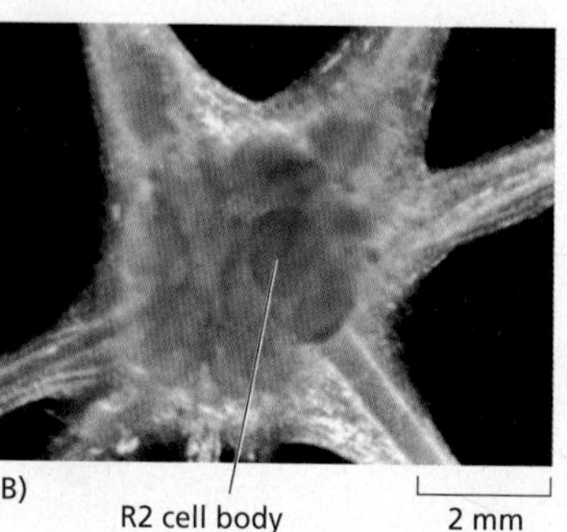

Figure 21–62 **Giant neurons of the sea slug are highly polyploid.** (A) The mollusk species *Aplysia californica*, commonly called a sea hare because its chemosensory tentacles stick up like ears, can weigh more than 2 kg. The *Aplysia* central nervous system contains groups of neural cell bodies called ganglia. (B) A dissected abdominal ganglion contains massive neurons such as R2, which has undergone endoreplication to produce ~300,000 copies of the diploid genome and is thought to be the largest somatic cell found in nature. *Aplysia* and its giant neurons enable physiological, biochemical, and genomic studies at the level of single cells, providing an accessible model for studies of how neural circuits control behavior. (A and B, courtesy of Lynne Fieber and Michael Schmale.)

Figure 21–63 **Two species of the frog genus *Xenopus*.** *X. tropicalis (top)* has an ordinary diploid genome. *X. laevis (bottom)* has roughly twice as much DNA per cell and evolved through hybridization of two *X. tropicalis*–like species followed by a whole-genome duplication event roughly 30 million years ago. (From E. Amaya, M. Offield, and R. Grainger, *Trends Genet.*14:253–255, 1998. With permission from Elsevier.)

Changes in ploidy can also arise during the evolution of new species, where they can drive size changes of the entire animal. The frog *Xenopus laevis* possesses a tetraploid genome and has a mass of about twice that of diploid relatives such as *Xenopus tropicalis* (**Figure 21–63**). These two species show robust size-scaling relationships at the organism, cellular, and subcellular levels, with the smaller frog possessing smaller cells, and the smaller cells, in turn, containing smaller nuclei. In plants, as in animals, cell size increases as ploidy increases (**Figure 21–64**). This effect has been exploited in the agricultural breeding of plants for large size: most of the major fruits and vegetables that we consume are polyploids.

Changes in Cell Size Usually Result from Modified Cell Cycles

Although fundamental to understanding how organ and organism size are determined, how cell size itself is set and maintained remains an open question. As discussed in Chapter 17, increases in cell size (growth) are usually coordinated with progression through the cell cycle (division) to maintain cell-size homeostasis. However, cells can uncouple the two processes, for example during growth of an egg, which enlarges dramatically without replicating its genome or dividing. After fertilization, this phenomenon is reversed when rounds of DNA replication and cell division occur in the absence of growth to form many small cells.

Endoreplication, which reduplicates the genome to enlarge a somatic cell, results from another variation on the cell cycle. Although S phase proceeds,

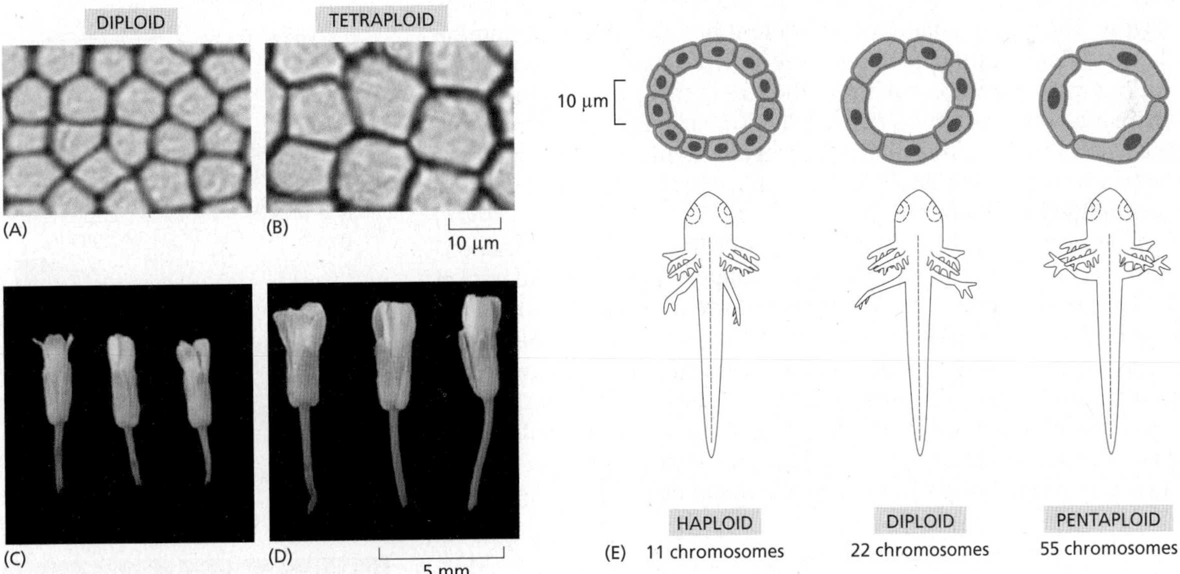

Figure 21–64 **Effects of ploidy on cell size and organ size.** In all organisms, from bacteria to humans, cell size is proportional to ploidy—the number of copies of the genome per cell. This is illustrated for (A–D) *Arabidopsis* flowers and (E) for salamanders. In each case, the upper panels show cells in a specific tissue [a petal for *Arabidopsis*, a pronephric (kidney) tubule for the salamander]; the lower panels show the gross anatomy—flowers for *Arabidopsis*, the whole body for the salamander. In the case of *Arabidopsis* flowers, increase in cell size increases organ size. By contrast, the salamander and its individual organs attain their normal standard size regardless of ploidy, because large cell size is compensated for by fewer cells. This indicates that the size of an organism or organ in this species is not controlled simply by counting cell divisions or cell numbers; size must somehow be regulated at the level of total cell mass. [A–D, from C. Breuer et al., *Plant Cell* 19:3655–3668, 2007. With permission from the American Society of Plant Biologists; E, adapted from G. Fankhauser, in Analysis of Development (B.H. Willier et al., eds.), pp. 126–150. Philadelphia: Saunders, 1955.]

mitotic cyclin levels are too low to induce M phase, and the polyploid cells usually cease dividing completely, entering a *post-mitotic* state. Additional rounds of DNA replication in the absence of cell division lead to an increase in cell size. In some tissues, such as the liver and cardiac muscle, DNA replication is followed by nuclear division, but not cytokinesis, resulting in polyploid cells with multiple nuclei, through a process called *endomitosis*. Whether or not nuclear division occurs, ploidy increases are accompanied by cell growth and contribute to tissue function. For example, the trophoblast cells of the mammalian placenta undergo eight rounds of endoreplication, producing giant cells crucial for formation of the tissue barrier between mother and fetus.

How do increases in ploidy increase the size of cells? One idea is that elevated gene-copy numbers increase the biosynthetic capacity by allowing production of more gene products. However, a strong positive correlation between genome size and cell size is observed across all species, independent of the fraction of the genome that codes for proteins. For example, although all vertebrates have a similar number of coding genes, axolotl salamanders possess a diploid genome that is 10 times the size of the human genome (containing 10 times more noncoding DNA). Remarkably, axolotl cells are also 10 times larger than human cells. The molecular basis of the scaling relationship between genome size and cell size is one of the many outstanding mysteries in biology.

Animals and Organs Can Assess and Regulate Total Cell Mass

The size of an animal or organ depends on both cell size and cell number; that is, on total cell mass. Remarkably, many animals and organs can somehow assess their total cell mass and regulate it, providing evidence for feedback controls of the sort highlighted earlier in our introductory account of the general principles of growth control. In contrast with *C. elegans*, if cell size is artificially increased or decreased in these cases, cell numbers adjust to maintain the same total cell mass. This has been beautifully illustrated by experiments done long ago in salamanders, where cell size can be manipulated by altering the animal's ploidy. As shown in Figure 21–64E, salamanders of different ploidies end up being the same size with very different numbers of cells. The individual cells in a pentaploid salamander, for example, are about five times the size of those in a haploid salamander, but there are only one-fifth as many cells. This scaling operates not only in the body as a whole, but also in its individual organs. Pentaploid salamanders possess the same size brain as their diploid counterparts, but with many fewer neurons and a simplified structure.

The **imaginal discs** of *Drosophila* provide another striking example of homeostatic size control. The discs are epithelial pouches that grow by cell proliferation during the larval period and, during the pupal stage, form the organs and extremities of the adult fly (**Figure 21–65**). Experiments have been chiefly done on the wing imaginal disc. Mutations in components of the cell-cycle control machinery can be used to speed up or slow down the rate of cell division in the disc. Remarkably, such mutations result in an excessive number of abnormally small cells or a reduced number of abnormally large cells, respectively, leaving the size (area) and patterning of the adult wing practically unchanged. Thus, the size of

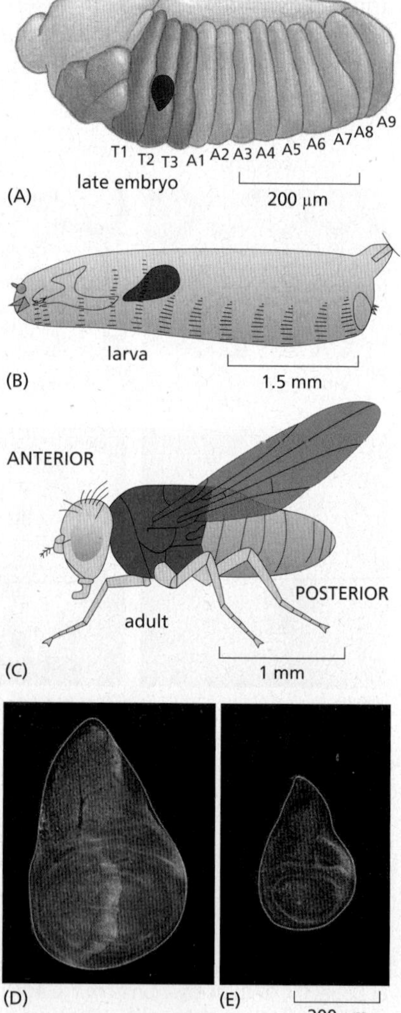

Figure 21–65 **Growth of the *Drosophila* wing imaginal disc.** Drawings in panels A–C show the locations of cells that give rise to the adult wing and dorsal thorax after metamorphosis. (A) Initially, ~30 cells are specified in the second thoracic segment of the embryo as precursors to the wing and dorsal thorax. These cells divide rapidly during the 4 days of larval life. (B) In the larva before metamorphosis, these cells are organized into a simple pouch-like epithelial organ called the wing imaginal disc, which contains ~30,000 cells and is ~400 μm in diameter. Other imaginal discs (not shown) give rise to most of the other external structures of the fly. (C) The final adult wing is ~2.25 mm long. (D, E) The morphogen Dpp is expressed in a stripe down the center of the wing imaginal disc *(green)*, diffusing out toward the edges and influencing disc growth as well as cell fates. When Dpp expression is removed from the disc late during larval development, the wing disc and resultant wing is proportionally smaller. (D and E, from S. Matsuda and M. Affolter, *eLife* 6:e22319, 2017, doi 10.7554/eLife.22319.)

the disc is not determined by counting the number of cell divisions so as to produce a set number of cells. Instead, there must be a regulatory mechanism that halts growth when the disc's total cell mass reaches the appropriate value, so that the adult wing that subsequently develops is normal. Moreover, if young discs or even disc fragments are taken out of their normal context and transplanted into the growth-permissive abdominal cavity of an adult female, they will grow until they reach their normal size and then stop. Clearly, the mechanisms that regulate organ size are intrinsic to the disc.

We know little about how organisms or organs assess total cell mass or monitor their own growth. Nevertheless, we are beginning to understand some of the signal molecules that drive or halt growth in response to the mysterious cues that convey information about the size attained.

Various Extracellular Signals Stimulate or Inhibit Growth

To produce an organ with the correct morphology and proportions, cell proliferation and cell growth must be coordinated within a tissue. While some control occurs systemically through growth-regulatory hormones, discussed later, there are also interesting examples in which tissues regulate their own growth. One example of how this may be achieved has been investigated in the *Drosophila* wing imaginal disc (see Figure 21–65), where the TGFβ family member Dpp plays a role. We encountered Dpp before as a morphogen involved in patterning the dorsoventral axis of the fly embryo; in the disc, it works analogously, diffusing from a narrow stripe down the middle of the disc to dictate different cell fates at different distances. Some experiments indicate that the gradient of Dpp emanating from this central source also influences tissue size, perhaps because growth of the disc leads cells farther from the source to experience lower levels of Dpp signaling (see Figure 21–65D and E). Linking tissue growth to pattern formation would ensure that changes in organ size are accompanied by appropriate changes in cell differentiation and distribution.

Not all growth-regulating extracellular signals stimulate growth; some inhibit it by promoting cell death or inhibiting cell growth, cell division, or both. *Myostatin* is another TGFβ family member that specifically inhibits the growth and proliferation of myoblasts—the precursor cells that fuse to form the huge, multinucleated cells of skeletal muscle. In this case, the tissue itself produces the signal, so as more muscle forms, myostatin levels increase and eventually turn off muscle growth. The importance of this simple organ-size feedback mechanism can be seen when the *Myostatin* gene is deleted: the muscles of mutant mice grow to be several times larger than normal. Remarkably, two breeds of cattle that were bred long ago for large muscles both have mutations in the *Myostatin* gene, as does a breed of dogs (**Figure 21–66**).

The Hippo Pathway Relays Mechanical Signals Regulating Growth

Organ growth is governed in many tissues by an intracellular signaling system called the Hippo pathway. It was discovered in *Drosophila*, but it operates in vertebrates as well. Named after the overgrown, buckled tissues that result when

(A) wild type

(B) myostatin mutant

Figure 21–66 **Myostatin limits muscle growth.** (A) A standard whippet and (B) a bully whippet that lacks myostatin. (A, Bianca Grueneberg/Getty Images; B, ©2020 Stuart Isett. All rights reserved.)

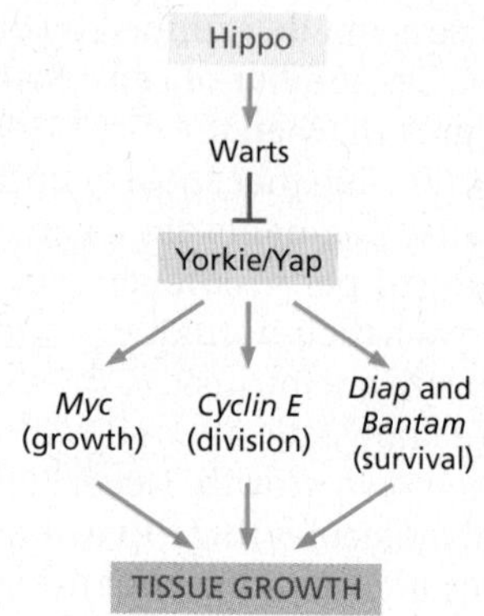

Figure 21–67 **The Hippo pathway in *Drosophila*.** Hippo, a protein kinase, limits growth by phosphorylation and activation of the kinase Warts, which in turn phosphorylates and inactivates the transcriptional coactivator Yorkie (Yap and Taz in vertebrates). When unphosphorylated, Yorkie/Yap drives tissue growth: it activates the transcription of the growth-promoting gene *Myc*, the cell-cycle progression gene *Cyclin E*, the anti-apoptotic gene *Diap*, and the microRNA *Bantam*. Hippo-induced phosphorylation of Yorkie/Yap blocks this effect.

the pathway is inactive, Hippo signaling in wild-type animals inhibits growth both by promoting cell death (by blocking an apoptosis inhibitor) and by inhibiting cell-cycle progression (by inhibiting the expression of the cell-cycle gene *Cyclin E*). Some components of the pathway in *Drosophila* are shown in **Figure 21–67**. The organs of animals that are abnormally resistant to Hippo repression can grow to a monstrous size (**Figure 21–68**).

Unlike many signaling pathways that involve binding of a signal molecule to a dedicated receptor, the Hippo pathway instead appears to be controlled primarily by mechanical forces and cell architecture. Cell-cell and cell-matrix contact, cytoskeletal tension, and the apical-basal polarity of the cell can each influence Hippo activation. Through this mechanism, tissues can sense physical as well as chemical cues in their environment and regulate their growth accordingly.

Hormones Coordinate Growth Throughout the Body

We have already seen how some signals act systemically as hormones to regulate the development of the animal as a whole. Some of these serve to regulate growth. In mammals, for instance, growth hormone (GH) is secreted by the pituitary gland into the bloodstream and stimulates growth throughout the body: excessive production of GH leads to pituitary gigantism, and too little leads to dwarfism (**Figure 21–69**). Pituitary dwarfs with lower GH levels have bodies and organs that are proportionately small, and stand in contrast to achondroplastic dwarfs, whose limbs are disproportionately short usually because of a mutation in a gene encoding an FGF receptor that disrupts normal cartilage and bone development (**Figure 21–70**).

Growth hormone stimulates growth primarily by inducing the liver and other organs to produce insulin-like growth factor 1 (IGF1), which acts mainly as a local signal within many tissues. IGF1 and related factors are potent growth promoters that increase cell survival, cell growth, cell proliferation, or some combination of these, depending on the cell type. Large breeds of dogs such as Great Danes owe their great size to high levels of IGF1, while miniature breeds such as Chihuahuas have low levels (see Figure 21–60).

It is important to note that in all species, nutritional conditions also play a fundamental part in regulating the pace and extent of growth, and in animals they do so through hormonal signal networks that are highly conserved between

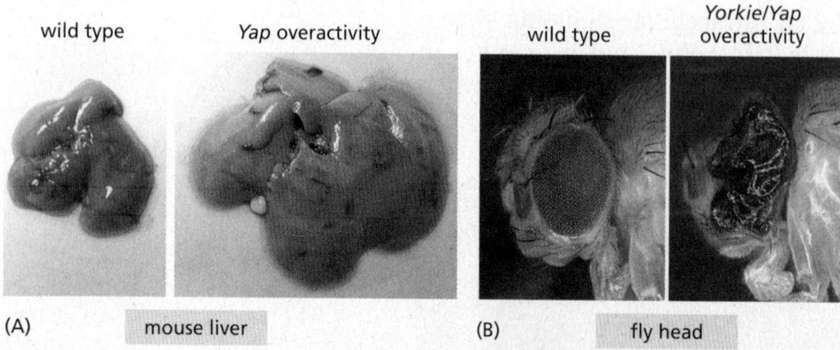

Figure 21–68 **Overcoming Hippo repression increases organ size.** (A) When *Yap* is overexpressed in the mouse liver, Hippo signaling cannot repress its activity and excess growth results. (B) Similar overgrowth is seen when fly eyes express a form of Yorkie/Yap that cannot be phosphorylated and inhibited by Hippo signaling. (A, from J. Dong et al., *Cell* 130:1120–1133, 2007. With permission from Elsevier; B, courtesy of Jung Kim.)

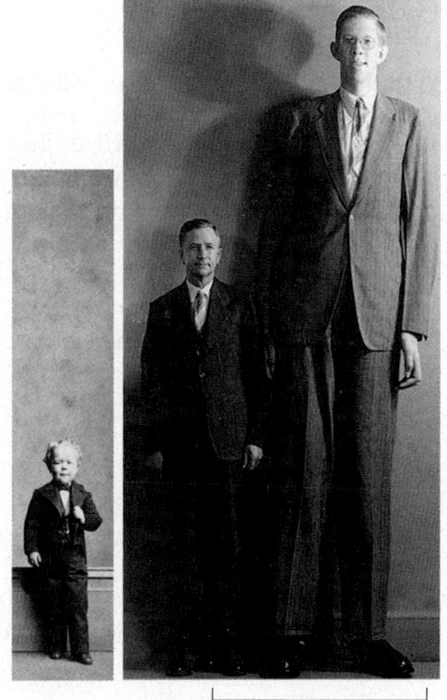

Figure 21–69 **Pituitary dwarf and pituitary giant.** The "giant" on the right is Robert Wadlow (1914–1940), the tallest recorded man at 8 feet 11 inches (2.72 m), together with his father, who was almost 6 feet tall (1.82 m). The dwarf on the left is General Tom Thumb, which was the stage name of Charles Sherwood Stratton (1838–1883). On his 18th birthday, he was measured at 2 feet 8.5 inches (82.6 cm) tall, and at his death, he was 3 feet 4 inches (102 cm). (Left, Bettman/Getty; right, History and Art Collection/Alamy Stock Photo.)

vertebrates and invertebrates. Genetic experiments, especially in *Drosophila*, have begun to unravel the logic of these controls and to indicate how they may operate alongside other machinery, such as the Hippo pathway, to determine final size.

The Duration of Growth Influences Organism Size

In addition to rate of growth, the final size of an organism will be determined by the time over which growth occurs. Some animals have a defined period of the life span during which growth is permitted; these species are said to show *determinate growth.* In humans, the growth period terminates at the end of puberty, while in many insects it terminates with the onset of metamorphosis. In both cases the change is triggered by the release of hormones that transition the organism into sexual maturity. Thereafter, cell proliferation is restricted to supporting tissue homeostasis via the action of adult stem cells (see Chapter 22), although significant cell growth can continue in cells such as fat cells. The timing of hormone release is set by many influences, among which both nutritional status and juvenile body size provide important inputs to set final body size.

Other species show *indeterminate growth* and are capable of size increases throughout their life as long as environmental conditions are amenable. Lobsters and many fish are familiar examples, but indeterminate growth is found along numerous branches of the tree of life. Plants, in which cell proliferation takes place in specialized regions called meristems (see Chapter 19), also show both determinate and indeterminate growth patterns. Giant sequoias and other redwood trees are an example of the latter, whereas the former include agricultural crops selected for maximal and synchronous fruit production. Some species, including tomatoes, exist in both determinate and indeterminate varieties. The mechanisms that assign a terminal growth point to one organism while enabling a closely related species to continue growth throughout life remain unknown.

Summary

The sizes of animal species and their organs vary widely and depend mainly on the size and number of cells, which are increased through cell growth and cell division, respectively. Cell numbers are reduced by cell death—mainly by apoptosis. The mystery is how decisions about growth, division, and death within individual cells are regulated and coordinated to produce the characteristic final size of the adult animal.

Some signals such as growth factors, mitogens, and survival factors act in organs to stimulate growth, while other signal molecules do the opposite. These signals are deployed under the control of developmental patterning programs and influence the cell cycle or apoptotic machinery to produce tissue-specific growth outcomes. Many animals and organs can assess their own total cell mass and regulate it, reaching a consistent target size in the face of artificial changes to cell growth or cell numbers. The compensatory mechanisms are not known, but signaling pathways that ensure that organs do not grow beyond a certain size have been identified.

Although most of these signals operate locally to help sculpt the size of organs and appendages, others act as hormones to regulate the growth of the animal as a whole. Nutrients can regulate growth through stimulating the release of hormonal signals throughout the body. The period of life during which growth is permitted can also be controlled by hormones. Final organism size reflects the integrated output of all the above mechanisms.

Figure 21–70 **Achondroplasia.** This type of dwarfism, shown in Velázquez's painting, occurs in one of 10,000–100,000 births; in more than 99% of cases it results from a mutation at an identical site in the genome, corresponding to amino acid 380 in the FGF receptor FGFR3 (a glycine in the transmembrane domain). The mutation is dominant, and almost all cases are due to new, independently occurring mutations, implying an extraordinarily high mutation rate at this particular site in the genome. The defect in FGF signaling causes dwarfism by interfering with the growth of cartilage in developing long bones. (The Picture Art Collection/Alamy Stock Photo.)

PROBLEMS

Which statements are true? Explain why or why not.

21–1 In the early cleavage stages, when the embryo cannot yet feed, the developmental program is driven and controlled entirely by the material deposited in the egg by the mother.

21–2 Because of the many later developmental transformations that produce the elaborately structured organs, the body plan set up during gastrulation bears little resemblance to the body plan in the adult.

21–3 As development progresses, individual cells within a lineage become more and more restricted in the range of cell types they can give rise to.

21–4 At different stages of embryonic development, the same signals are used over and over again by different cells, but with different biological outcomes.

21–5 Changes in the coding regions of genes involved in development are primarily responsible for the differences between species.

21–6 The cell cycle is the ticking clock that sets the tempo of developmental processes, with maturational changes in gene expression being dependent on cell-cycle progression.

Discuss the following problems.

21–7 Name the three processes that are fundamental to animal development, and describe each of them in a single sentence.

21–8 Name the three germ layers of the early embryo that are formed during gastrulation, and list the principal structures each gives rise to in the adult.

21–9 In the early *Drosophila* embryo, there seems to be no requirement for the usual forms of cell-cell signaling; instead, transcriptional regulators and mRNA molecules move freely between nuclei. How can that be?

21–10 Morphogens play a key role in development, creating concentration gradients that inform cells of where they are and how to behave. Examine the simple patterns represented by the flags in Figure Q21–1. Which do you suppose could be created by a gradient of a single morphogen? Which would require gradients of two morphogens? Assuming that such patterns were present in a sheet of cells, explain how they could be created by morphogens.

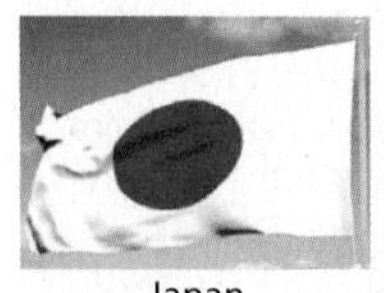

Japan

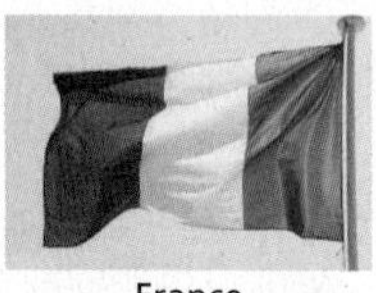

France

Norway

Figure Q21–1 National flags from three countries (Problem 21–10). (Left, railway fx/Shutterstock; center, Creative Photo Corner/Shutterstock; right, Derek Brumby/Shutterstock.)

21–11 Two adjacent cells in the nematode worm normally differentiate into an anchor cell (AC) and a ventral uterine precursor (VU) cell, but which of the two becomes the AC and which becomes the VU cell is completely random. The cells have an equal chance of adopting either fate, but they always adopt opposite fates. Mutations of the *Lin12* gene alter these fates. In hyperactive *Lin12* mutants, both cells become VU cells, while in inactive *Lin12* mutants, both cells become ACs. Thus, *Lin12* is central to the decision-making process. In genetic mosaics in which one precursor cell has the hyperactive *Lin12* and the other precursor has the inactive *Lin12*, the cell with the hyperactive *Lin12* gene always becomes the VU cell, and the cell with the inactive *Lin12* gene always becomes the AC. Assuming that one cell sends a signal and the other cell receives it, explain how these results suggest that the *Lin12* gene encodes a protein required to receive the signal. Offer a suggestion for how the fates of these two precursor cells are normally decided in wild-type worms.

21–12 It was clear from the early days of studying development that certain "morphogenetic" substances were present in the egg and segregated asymmetrically into cells of the developing embryo. One such investigation in ascidian (sea squirt) embryos examined endodermal alkaline phosphatase, which could be visualized using a histochemical stain. Treatment of embryos with cytochalasin B stopped cell division but did not block expression of alkaline phosphatase at the appropriate time. Treatment with actinomycin D, which blocks transcription, did not interfere with expression of alkaline phosphatase. Treatment with puromycin, which blocks translation, eliminated expression of alkaline phosphatase. What is the likely nature of the morphogenetic substance that gives rise to alkaline phosphatase?

21–13 The mouse *HoxA3* and *HoxD3* genes are paralogs that occupy equivalent positions in their respective *Hox* gene clusters and share roughly 50% identity in their protein-coding sequences. Mice with defects in *HoxA3* have deficiencies in pharyngeal tissues, whereas mice with defects in *HoxD3* have deficiencies in the axial skeleton, suggesting quite different functions for the paralogs. Thus, it came as a surprise when it was found that replacing a defective *HoxD3* gene with the normal *HoxA3* gene corrected the deficiency, as did the reciprocal experiment of replacing a mutant *HoxA3* gene with a normal *HoxD3* gene. Neither transplaced gene, however, could supply its normal function; that is, a normal *HoxA3* gene at the *HoxD3* locus could not correct the deficiency caused by a mutant *HoxA3* gene at the *HoxA3* locus. The same was true for the *HoxD3* gene. If the *HoxA3* and *HoxD3* genes are equivalent, how do you suppose they can play such distinct roles in development? Why do you suppose they cannot perform their normal function in a new location?

21–14 The segmentation of somites in vertebrate embryos is thought to depend on oscillations in the expression of the *Hes7* gene. Mathematical modeling explains these oscillations in terms of the delays in production of the unstable Hes7 protein, which acts as a transcription regulator to shut off its own expression. Once Hes7 decays, with a half-life of about 20 minutes, its transcription resumes. To test this model, you decide to reduce the total delay by removing one, two, or all three of the introns from the *Hes7* gene in mice. Why do you expect that intron removal would reduce the delay? What would you predict would happen to the oscillation time, and somite formation, if the model were correct?

21–15 The oscillatory clock that drives somite formation in vertebrates involves three essential components: Her7 (an unstable repressor of its own synthesis), Delta (a transmembrane signaling molecule), and Notch (a transmembrane receptor for Delta). Notch is bound by Delta on neighboring cells, activating the Notch signaling pathway, which then activates *Her7* transcription. Normally, this system works flawlessly to create sharply defined somites (**Figure Q21–2A**). In the absence of Delta, however, only the first five somites form normally, and the rest are poorly defined (**Figure Q21–2B**). If a pulse of Delta is supplied later, somite formation returns to normal in the regions where Delta was present (**Figure Q21–2C**). A diagram of the connections between the components of the clock and how they interact in adjacent cells is shown in **Figure Q21–2D**. In the absence of Delta, why do the cells become unsynchronized? What is it about the presence of Delta that keeps adjacent cells oscillating in synchrony?

(A) control

(B) Delta-minus

(C) Delta pulse

(D)

delay

Her7 ← Notch

Delta

delay

Notch → Her7

Delta

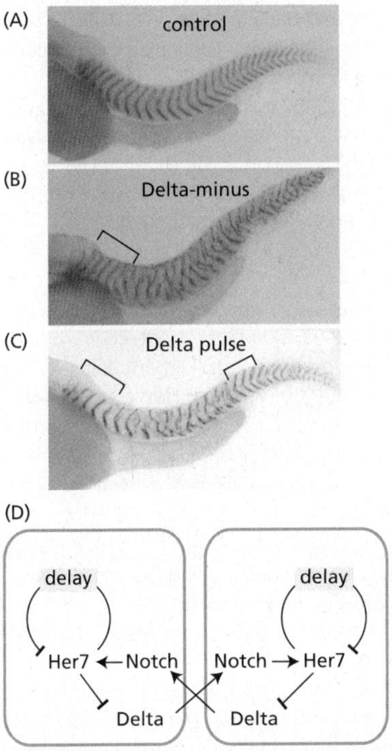

Figure Q21–2 Somite formation in zebrafish embryos (Problem 21–15). (A) Wild-type embryos with normal somites. (B) Somite formation in embryos lacking Delta. The bracket indicates normal-looking somites where they initially form. (C) Somite formation in embryos lacking Delta but receiving a pulse of Delta expression at the time indicated by the right-hand bracket. (D) Interactions among components of the oscillatory clock in adjacent cells. (Adapted from C. Soza-Ried et al., *Development* 141:1780–1788, 2014. With permission from the Company of Biologists.)

21–16 The extracellular protein factor Decapentaplegic (Dpp) is critical for proper wing development in *Drosophila* (**Figure Q21–3A**). It is normally expressed in a narrow stripe in the middle of the wing, along the anterior–posterior boundary. Flies that are defective for Dpp form stunted "wings" (**Figure Q21–3B**). If an additional copy of the gene is placed under control of a promoter that is active in the anterior part of the wing or in the posterior part of the wing, a large mass of wing tissue composed of normal-looking cells is produced at the site of Dpp expression (**Figure Q21–3C and D**). Does Dpp stimulate cell division, cell growth, or both? How can you tell?

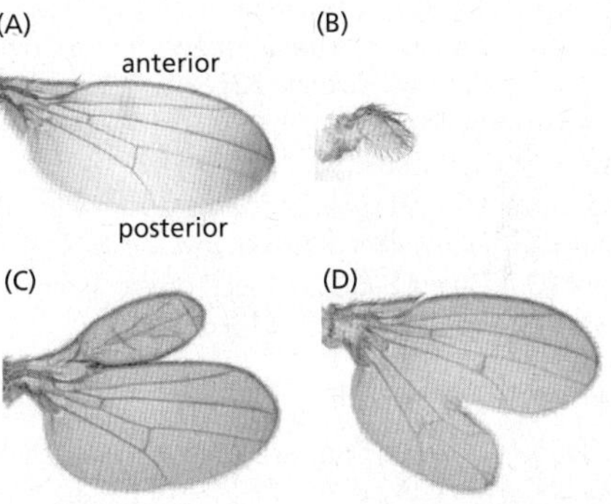

Figure Q21–3 Effects of Dpp expression on wing development in *Drosophila* (Problem 21–16). (A) Normal Dpp expression. (B) Absence of Dpp expression. (C) Additional anterior Dpp expression. (D) Additional posterior Dpp expression. (From M. Zecca et al., *Development* 121:2265–2278, 1995. With permission from the Company of Biologists.)

REFERENCES

General

Baressi MJ & Gilbert SF (2019) Developmental Biology, 12th ed. Sunderland, MA: Sinauer Associates.

Carroll SB (2006) Endless Forms Most Beautiful: The New Science of Evo Devo. New York: Norton.

Wallingford JD (2019) We Are All Developmental Biologists. *Dev. Cell* 50(2), 132–137.

Wolpert L, Tickle C & Martinez-Arias A (2019) Principles of Development, 6th ed. Oxford: Oxford University Press.

Overview of Development

Gurdon JB (2013) The egg and the nucleus: a battle for supremacy (Nobel Lecture). *Angew. Chem. Int. Ed. Engl.* 52, 13890–13899.

Istrail S & Davidson EH (2005) Logic functions of the genomic cis-regulatory code. *Proc. Natl. Acad. Sci. USA* 102, 4954–4959.

Levine M (2010) Transcriptional enhancers in animal development and evolution. *Curr. Biol.* 20, R754–R763.

Lewis J (2008) From signals to patterns: space, time, and mathematics in developmental biology. *Science* 322, 399–403.

Meinhardt H & Gierer A (2000) Pattern formation by local self-activation and lateral inhibition. *Bioessays* 22, 753–760.

Rogers KW & Schier AF (2011) Morphogen gradients: from generation to interpretation. *Annu. Rev. Cell Dev. Biol.* 27, 377–407.

Shubin N, Tabin C & Carroll S (2009) Deep homology and the origins of evolutionary novelty. *Nature* 457, 818–823.

Mechanisms of Pattern Formation

Andrey G & Duboule D (2014) SnapShot: Hox gene regulation. *Cell* 156, 856–856.e1.

Chan YF, Marks ME, Jones FC . . . Kingsley DM (2010) Adaptive evolution of pelvic reduction in sticklebacks by recurrent deletion of a Pitx1 enhancer. *Science* 327, 302–305.

Davis RL, Weintraub H & Lassar AB (1987) Expression of a single transfected cDNA converts fibroblasts to myoblasts. *Cell* 51, 987–1000.

De Robertis EM (2006) Spemann's organizer and self-regulation in amphibian embryos. *Nat. Rev. Mol. Cell Biol.* 4, 296–302.

Driever W & Nüsslein-Volhard C (1988) A gradient of bicoid protein in *Drosophila* embryos. *Cell* 54, 83–93.

Halder G, Callaerts P & Gehring WJ (1995) Induction of ectopic eyes by targeted expression of the eyeless gene in *Drosophila*. *Science* 267, 1788–1792.

Jaeger J (2011) The gap gene network. *Coll. Mol. Life Sci.* 68, 243–274.

Knoblich JA (2010) Asymmetric cell division: recent developments and their implications for tumour biology. *Nat. Rev. Mol. Cell Biol.* 11, 849–860.

Lander AD (2013) How cells know where they are. *Science* 339, 923–927.

Lawrence PA (1992) The Making of a Fly: The Genetics of Animal Design. Oxford: Wiley-Blackwell

Lewis EB (1978) A gene complex controlling segmentation in *Drosophila*. *Nature* 276, 565–570.

Müller P, Rogers KW, Jordan BM . . . Schier AF (2012) Differential diffusivity of Nodal and Lefty underlies a reaction-diffusion patterning system. *Science* 336, 721–724.

Nüsslein-Volhard C & Wieschaus E (1980) Mutations affecting segment number and polarity in *Drosophila*. *Nature* 287, 795–801.

Schuettengruber B, Bourbon HM, Di Croce L & Cavalli G (2017) Genome regulation by polycomb and trithorax: 70 years and counting. *Cell* 171, 34–57.

Sjöqvist M & Andersson ER (2019) Do as I say, Not(ch) as I do: lateral control of cell fate. *Dev. Biol.* 447(1), 58–70.

von Dassow G, Meir E, Munro EM & Odell GM (2000) The segment polarity network is a robust developmental module. *Nature* 406, 188–192.

Developmental Timing

Brown DD & Cai L (2007) Amphibian metamorphosis. *Dev. Biol.* 306, 20–33.

Giraldez AJ, Mishima Y, Rihel J . . . Schier AF (2006) Zebrafish MiR-430 promotes deadenylation and clearance of maternal mRNAs. *Science* 312, 75–79.

Holguera I & Desplan C (2018) Neuronal specification in space and time. *Science* 362, 176–180.

Hubaud A & Pourquié O (2014) Signalling dynamics in vertebrate segmentation. *Nat. Rev. Mol. Cell Biol.* 15, 709–721.

Isshiki T, Pearson B, Holbrook S & Doe CQ (2001) *Drosophila* neuroblasts sequentially express transcription factors which specify the temporal identity of their neuronal progeny. *Cell* 106, 511–521.

Jukam D, Shariati SAM & Skotheim JM (2017) Zygotic genome activation in vertebrates. *Dev. Cell* 42, 316–332.

Lee RC, Feinbaum RL & Ambros V (1993) The *C. elegans* heterochronic gene lin-4 encodes small RNAs with antisense complementarity to lin-14. *Cell* 75, 843–854.

Oates AC (2020) Waiting on the Fringe: cell autonomy and signaling delays in segmentation clocks. *Curr. Opin. Genet. Dev.* 63, 61–70.

Song J, Irwin J & Dean C (2013) Remembering the prolonged cold of winter. *Curr. Biol.* 23, R807–R811.

Wightman B, Ha I & Ruvkun G (1993) Posttranscriptional regulation of the heterochronic gene lin-14 by lin-4 mediates temporal pattern formation in *C. elegans*. *Cell* 75, 855–862.

Morphogenesis

Bussmann J & Raz E (2015) Chemokine-guided cell migration and motility in zebrafish development. *EMBO J.* 34, 1309–1318.

Butler MT & Wallingford JB (2017) Planar cell polarity in development and disease. *Nat. Rev. Mol. Cell Biol.* 18, 375–388.

Glimour D, Rembold M & Leptin M (2017) From morphogen to morphogenesis and back. *Nature* 541, 311–320.

Goodwin K & Nelson CM (2020) Branching morphogenesis. *Development* 147, dev184499.

Lecuit T & Lenne PF (2007) Cell surface mechanics and the control of cell shape, tissue patterns and morphogenesis. *Nat. Rev. Mol. Cell Biol.* 8, 633–644.

Le Douarin NM & Kalcheim C (1999) The Neural Crest, 2nd ed. Cambridge: Cambridge University Press.

Solnica-Krezel L & Sepich DS (2012) Gastrulation: making and shaping germ layers. *Annu. Rev. Cell Dev. Biol.* 28, 687–717.

Takeichi M (2011) Self-organization of animal tissues: cadherin-mediated processes. *Dev. Cell* 21, 24–26.

Te Boekhorst V, Preziosi L & Friedl P (2016) Plasticity of cell migration in vivo and in silico. *Annu. Rev. Cell Dev. Biol.* 32, 491–526.

Walck-Shannon E & Hardin J (2014) Cell intercalation from top to bottom. *Nature* 15, 34–48.

Growth

Andersen DS, Colombani J & Léopold P (2013) Coordination of organ growth: principles and outstanding questions from the world of insects. *Trends Cell Biol.* 23, 336–344.

Conlon I & Raff M (1999) Size control in animal development. *Cell* 96, 235–244.

Hariharan IK & Bilder D (2006) Regulation of imaginal disc growth by tumor-suppressor genes in *Drosophila*. *Annu. Rev. Genet.* 40, 335–361.

Øvrebø JI & Edgar BA (2018) Polyploidy in tissue homeostasis and regeneration. *Development* 145, dev156034.

Penzo-Méndez AI & Stanger BZ (2015) Organ-size regulation in mammals. *Cold Spring Harb. Perspect. Biol.* 7, a019240.

Restrepo S, Zartman JJ & Basler K (2014) Coordination of patterning and growth by the morphogen DPP. *Curr. Biol.* 24, R245–R255.

Zheng Y & Pan D (2019) The hippo signaling pathway in development and disease. *Dev. Cell* 50, 264–282.

Stem Cells in Tissue Homeostasis and Regeneration

CHAPTER

22

IN THIS CHAPTER

Cells evolved originally as free-living individuals, and such cells still dominate Earth and its oceans. But the cells that matter most to us, as humans, are specialized members of a multicellular community. These cells have lost features needed for independent survival and acquired peculiarities that serve the needs of the body as a whole. Although they share the same genome, they are spectacularly diverse in structure, chemistry, and behavior. There are more than 200 different named cell types in the human body that collaborate with one another to form many different tissues, which are arranged into organs performing widely varied functions. To understand them, it is not enough to analyze cells in a culture dish: we need also to know how they live, work, and die in their natural habitat, the intact body.

In Chapters 7 and 21, we saw how the various cell types become different in the embryo and how cell memory and signals from their neighbors enable them to remain different thereafter. In Chapter 19, we discussed the mechanisms that cells use to self-assemble into multicellular tissues, including the use of molecular devices that bind cells together and the extracellular materials that give tissues and organs support. But the adult body is not static: it is a structure in dynamic equilibrium, where new cells are continually being born, differentiating, and dying. Homeostatic mechanisms maintain a proper balance, so that tissue architecture is preserved despite the constant replacement of old cells by new.

In this chapter, we focus on these homeostatic processes that continue throughout life. In doing so, we illustrate some of the diversity of differentiated cell types, examining in particular the part played in many adult tissues by **stem cells**—undifferentiated cells that are specialized to provide a fresh supply of differentiated cells where these need to be continually replaced or when they are required in large number for tissue repair and regeneration. We will see that many adult tissues constantly renew and repair themselves, but others do not, in which case, lost cells are lost forever, causing deafness, blindness, dementia, and other disorders. We discuss how stem cells are maintained within a self-renewing tissue and how the fate of their daughter cells is determined. We then describe the remarkable ability of some animal species to regenerate entire limbs or organs, in one extreme example, reproducing all of the tissues of the organism from a single stem cell. In the final section of the chapter, we discuss how stem cells can be generated and manipulated artificially, raising the practical question that underlies a current storm of interest in stem-cell technology: How can we use our understanding of the processes of cell differentiation and tissue renewal to improve upon nature and make good those injuries and degenerations associated with disease and aging of the human body that have hitherto seemed beyond repair?

STEM CELLS AND TISSUE HOMEOSTASIS

In self-renewing adult tissues, cells are continually being born, differentiating, and dying. This "flow" of cells can be compared with the flow of water in a river: the river may look the same from day to day, but, as it flows downstream, the water

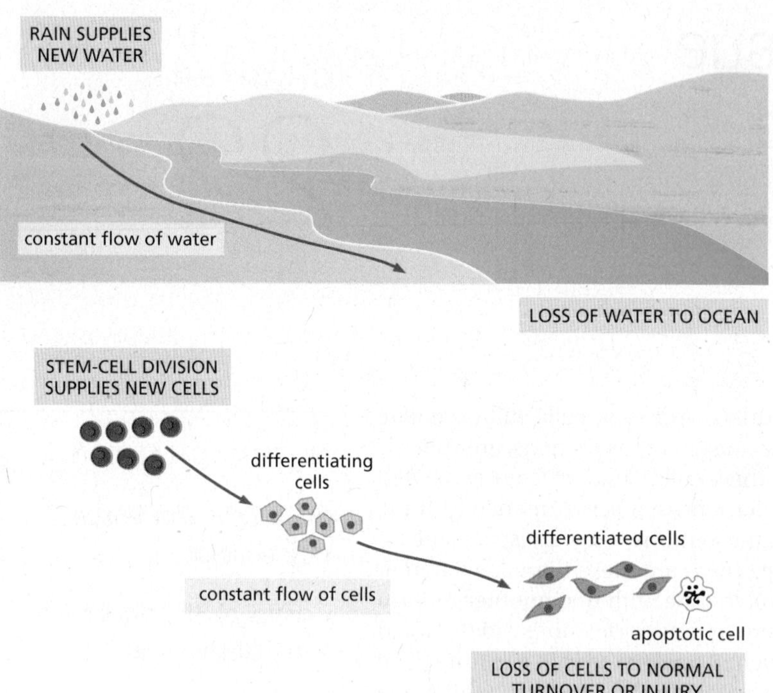

Figure 22–1 Tissue homeostasis compared to a river. In a self-renewing adult tissue, a constant flow of new cells produced by cell division "upstream" and a constant loss of differentiated cells "downstream" maintain the tissue in a dynamic equilibrium.

in the river is never the same (**Figure 22–1**). Similarly, the characteristic architecture of a self-renewing tissue is maintained even though the cell population is always changing, with differentiated cells lost "downstream" and stem cells producing new cells "upstream." Thus, the stem cells in such tissues must be able to both replicate themselves and produce differentiated cells for the life span of the organism. Without stem cells, these tissues and organs would fail rapidly, being unable to keep up with the natural turnover of cells. Indeed, defects in stem-cell functions can contribute to disease and aging for this reason.

In this first section, we define the essential characteristics of stem cells, using two epithelial tissues as examples, the lining of the gut and the outer surface of the skin. We then consider how we can identify stem cells in tissues—by cell lineage tracing or, in the case of blood cells, by cell transplantation studies. We finally discuss tissues that can maintain themselves in the absence of stem cells.

Stem Cells Are Defined by Their Ability to Self-renew and Produce Differentiated Cells

Many adult tissues, especially those with high cell turnover rates, contain **tissue-specific stem cells** (also called **adult stem cells**). Each of these tissues contains its own unique stem-cell population, capable of producing the differentiated cells characteristic of that tissue and not those of other tissues. The stem cells of each tissue possess their own distinct developmental history, without necessarily sharing molecular characteristics with the stem cells of other tissues.

All stem cells, however, share two fundamental properties that define them: (1) they are able to replenish themselves as stem cells, generally throughout the lifetime of the organism—a process called *self-renewal*; and (2) they can also produce differentiated cells. Thus, when a stem cell divides, each daughter has a choice: it can either remain a stem cell or it can embark on a course that commits it to differentiation (**Figure 22–2**).

Stem cells usually do not produce differentiated cells directly; instead, they make an intermediate cell type that is committed to a differentiation pathway but continues to proliferate, thereby generating greater numbers of differentiated cells. These cells are called *progenitor cells*; they are also called *transit-amplifying cells* because their divisions serve to amplify the number of differentiated cells that

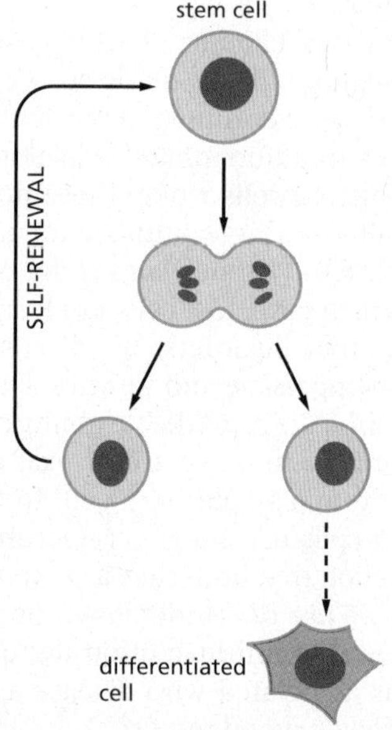

Figure 22–2 The defining characteristics of a stem cell. Each daughter cell produced when a stem cell divides can either remain a stem cell, in the process of self-renewal, or commit to differentiation, usually after a number of cell divisions. The self-renewal process maintains the pool of stem cells in the tissue.

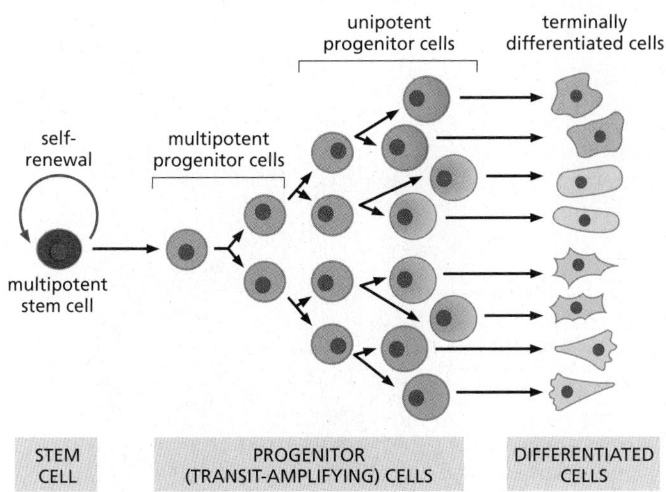

Figure 22–3 **A hierarchy of stem cells, progenitor cells, and differentiated cells.** In addition to self-renewal, tissue-specific stem cells generally produce progenitor (transit-amplifying) cells that divide a limited number of times before they terminally differentiate. Stem cells and progenitor cells can be unipotent or multipotent, depending on whether they produce only one type or multiple types of differentiated cells.

ultimately result from each stem-cell division (Figure 22–3). Unlike stem cells, these intermediate cells only go through a limited number of divisions before they differentiate. When a cell reaches the end of its differentiation pathway and does not divide again, it is *terminally differentiated.* When an adult stem cell or progenitor cell generates multiple differentiated cell types, it is *multipotent*; when it generates only one differentiated cell type, it is *unipotent* (see Figure 22–3).

We now consider two epithelial tissues to illustrate how these different categories of cells are organized in a self-renewing adult tissue—the lining of the gut and the outer layer of the skin. In both cases, the stem cells continually produce transit-amplifying progenitor cells at one surface of the tissue, while terminally differentiated cells are lost from the opposite surface—upstream and downstream, respectively, in our river analogy.

The Epithelial Lining of the Small Intestine Is Continually Renewed Through Cell Proliferation in Crypts

The lining of the small intestine (and of most other regions of the gut) is a single-layered epithelium, only one cell thick. The **intestinal epithelium** covers the surfaces of the *villi* that project into the gut lumen, and it lines the *crypts* that descend into the underlying connective tissue (Figure 22–4). Dividing cells, including the stem cells and progenitor cells, are restricted to the crypts, and terminally differentiated cells pour out of the crypts in a steady stream onto the villi. There are four main types of differentiated cells—one absorptive and three secretory (Figure 22–5):

1. *Absorptive cells* (also called *brush-border cells* or *enterocytes*) are the majority cell type in the epithelium and have densely packed microvilli on their exposed surfaces. Their job is to take up nutrients from the gut lumen. To this end, they also produce hydrolytic enzymes that perform some of the final steps of extracellular digestion of food.

2. *Goblet cells* secrete mucus into the gut lumen; this mucus covers the epithelium with a protective coat.

3. *Paneth cells* form part of the innate immune defense system (discussed in Chapter 24) and secrete proteins that kill bacteria; they also secrete Wnt signal proteins (discussed in Chapter 15) required to maintain the stem-cell population.

4. *Enteroendocrine cells*, of more than 15 different subtypes, secrete serotonin and peptide hormones that act on neurons and other cell types in the gut wall and regulate the growth, proliferation, and digestive activities of cells of the gut and other tissues.

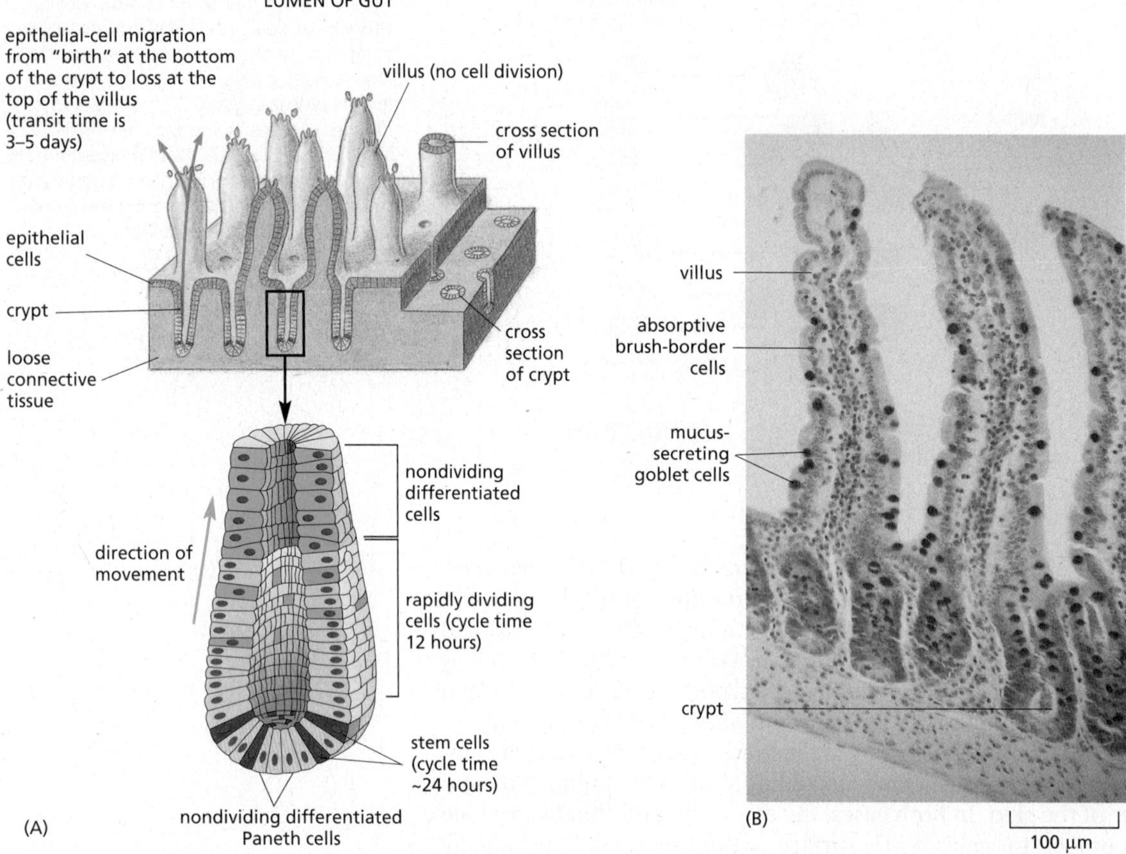

Figure 22–4 Renewal of the gut epithelial lining. (A) The pattern of cell turnover and proliferation in the epithelium that forms the lining of the small intestine. Stem cells *(red)* lie at the crypt base, interspersed among nondividing differentiated cells (Paneth cells). Progeny of the stem cells move mainly upward from the crypts onto the villi; after a few quick divisions, they cease dividing and differentiate—some of them while still in the crypt, most of them as they emerge from the crypt. The Paneth cells, like the other nondividing differentiated cells, are continually replaced by the progeny of the stem cells, but they migrate downward to the crypt base and survive there for many weeks. (B) Micrograph of a section of part of the lining of the small intestine, showing the crypts and villi. Note the mixture of differentiated cell types, all generated from the stem cells; these are primarily absorptive cells, with mucus-secreting goblet cells (stained *red*) interspersed among them.

As if on a conveyor belt, the absorptive, goblet, and enteroendocrine cells travel upward from their site of birth in the crypt, by a sliding movement in the plane of the epithelial sheet, to cover the surfaces of the villi. Within 3–4 days (in the mouse) after emerging from the crypts, the differentiated cells reach the tips of the villi, where they are discarded into the gut lumen (see Movie 20.7). The Paneth cells in the crypts are produced in much smaller numbers and have a different migration pattern. They live at the bottom of the crypts, where they too are continually replaced, although not so rapidly, persisting for several weeks before undergoing apoptosis and being phagocytosed by their neighbors.

The stem cells that give rise to the intestinal epithelium are located just above the base of the crypt, interspersed among the Paneth cells. The stem cells can be identified because only they express a particular G-protein-coupled receptor called *Lgr5*, which serves as a specific marker for the stem cell population. When the stem cells divide, which they do every 24 hours or so in the mouse intestine, some of the progeny commit to differentiation, becoming transit-amplifying progenitor cells that migrate upward, while others remain stem cells in the process of self-renewal. The stem cells are multipotent, producing all the differentiated cell types in the epithelium.

Epidermal Stem Cells Maintain a Self-renewing, Waterproof, Epithelial Barrier on the Body Surface

Stem-cell systems are organized in a variety of different ways depending on the tissue. For example, the outer epithelial covering of the body, the **epidermis**, undergoes continual renewal, but, unlike the intestinal epithelium, it is multilayered, or *stratified*. Stem cells are located in the basal layer, as are the dividing transit-amplifying progenitor cells. Once the progenitor cells stop dividing, they leave the basal layer and move outward toward the exposed surface, undergoing

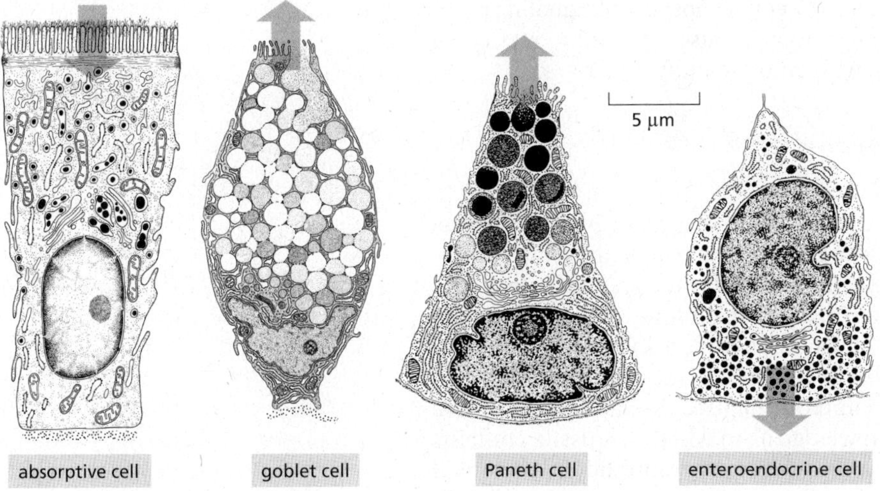

Figure 22–5 **The four main differentiated cell types found in the epithelial lining of the small intestine.** All cells are oriented with the gut lumen at *top*. Broad *orange arrows* indicate direction of secretion or uptake of materials for each type of cell. All of these cells are generated from undifferentiated multipotent stem cells living near the bottoms of the crypts (see Figure 22–4). *Absorptive (brush-border) cells* outnumber the other cell types in the epithelium by about 10:1 or more. The microvilli on their apical surface provide a 30-fold increase in surface area, not only for the import of nutrients but also for the anchorage of enzymes that perform the final stages of extracellular digestion, breaking down small peptides and disaccharides into monomers that can be transported across the cell membrane. *Goblet cells* secrete mucus; these are the most common of the secretory cell types. *Paneth cells* secrete (along with some growth factors) *cryptdins*—proteins of the defensin family that kill bacteria. Different subtypes of *enteroendocrine cells* secrete serotonin and peptide hormones into the gut wall (and thence the blood). Cholecystokinin is a hormone released from enteroendocrine cells in response to the presence of nutrients in the gut. It binds to receptors on nearby sensory nerve endings and causes the release of digestive enzymes from the pancreas and bile from the gall bladder; it also signals to the brain to stop the feeling of hunger once one has eaten enough. (Absorptive and goblet cells, Don W. Fawcett/Science Source; Paneth and enteroendocrine cells, from R.V. Krstić, Illustrated Encyclopedia of Human Histology. Berlin: Springer-Verlag, 1964. With permission from Springer Nature.)

terminal differentiation as they go. They end up as lifeless scales, or *squames*, which are eventually shed from the surface of the skin (**Figure 22–6**).

Even though the architecture of the epidermis is very different from that of the intestinal epithelium, many of the same basic principles apply. The stem cells are maintained by signals within a specific region of the tissue, which in the epidermis is the basal lamina and underlying connective tissue. The daughters of stem cells that are committed to differentiation undergo several divisions as transit-amplifying

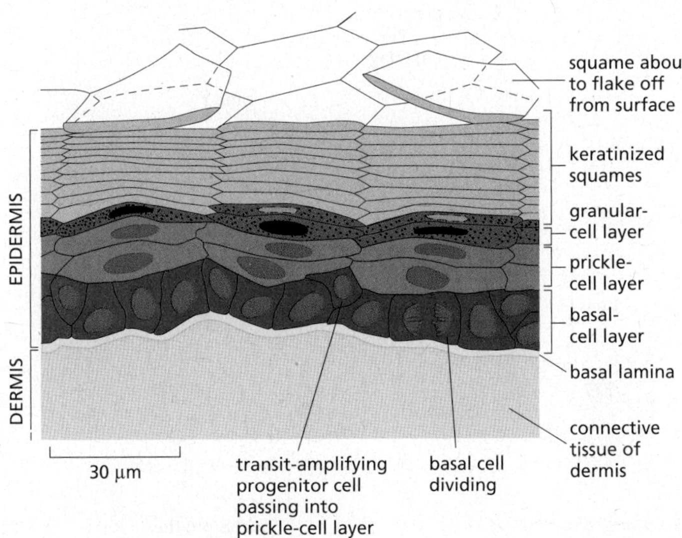

Figure 22–6 **The multilayered structure of the epidermis.** The epidermis forms the outer covering of the skin, creating a waterproof barrier that is self-repairing and continually renewed. Beneath this lies a relatively thick layer of connective tissue, which includes the tough, collagen-rich dermis (from which leather is made). The cells of the epidermis are called keratinocytes, because their characteristic differentiated activity is the synthesis of keratin intermediate filament proteins, which give the epidermis its toughness. These cells change their appearance and properties from one layer to the next, progressing through a regular program of terminal differentiation. Those in the innermost layer, attached to an underlying basal lamina, are termed basal cells, and it is normally only these that divide: the basal-cell population includes relatively small numbers of stem cells along with larger numbers of transit-amplifying progenitor cells derived from them. Above the basal cells are several layers of larger prickle cells. Beyond the prickle cells lies the thin, darkly staining granular-cell layer, where the cells are sealed together to form a waterproof barrier; this marks the boundary between the inner, metabolically active strata and the outermost layer of the epidermis, consisting of dead cells whose intracellular organelles have disappeared. These outermost cells are reduced to flattened scales, or squames, filled with densely packed keratin, which are eventually shed from the surface of the skin. The time from exit of a cell from the basal layer to its loss by shedding at the surface is a week or two, depending on body region and species.

cells in the basal layer before differentiating. Moreover, most of the signaling pathways that organize the intestinal stem-cell system are also involved in regulating the epidermal stem-cell system, although with different individual roles.

Cell Lineage Tracing Reveals the Location of Stem Cells and Their Progeny

Stem cells in adult tissues are usually rare and difficult to identify in conventional tissue sections, unless a stem cell-specific marker like Lgr5 is available. Recombinant DNA technology provides a general and powerful way to identify stem cells and their progeny in any renewing tissues using a technique called *cell lineage tracing*. The method uses transgenic animals to create a visible genetic mark in just a few cells, which, over time, give rise to widely separated and easily distinguished clones of progeny cells, as explained in **Figure 22–7** and **Figure 22–8**. This approach does not require prior knowledge as to whether a tissue contains stem cells or not. If they exist, the stem cells will be marked randomly and will lead to a persistent clonal lineage that contains stem cells as well as differentiated cells (see Figure 22–7B); dividing progenitor cells will also be randomly marked and produce labeled clones, but all of these will eventually disappear (see Figure 22–7C). The analysis of the clones not only indicates whether stem cells are

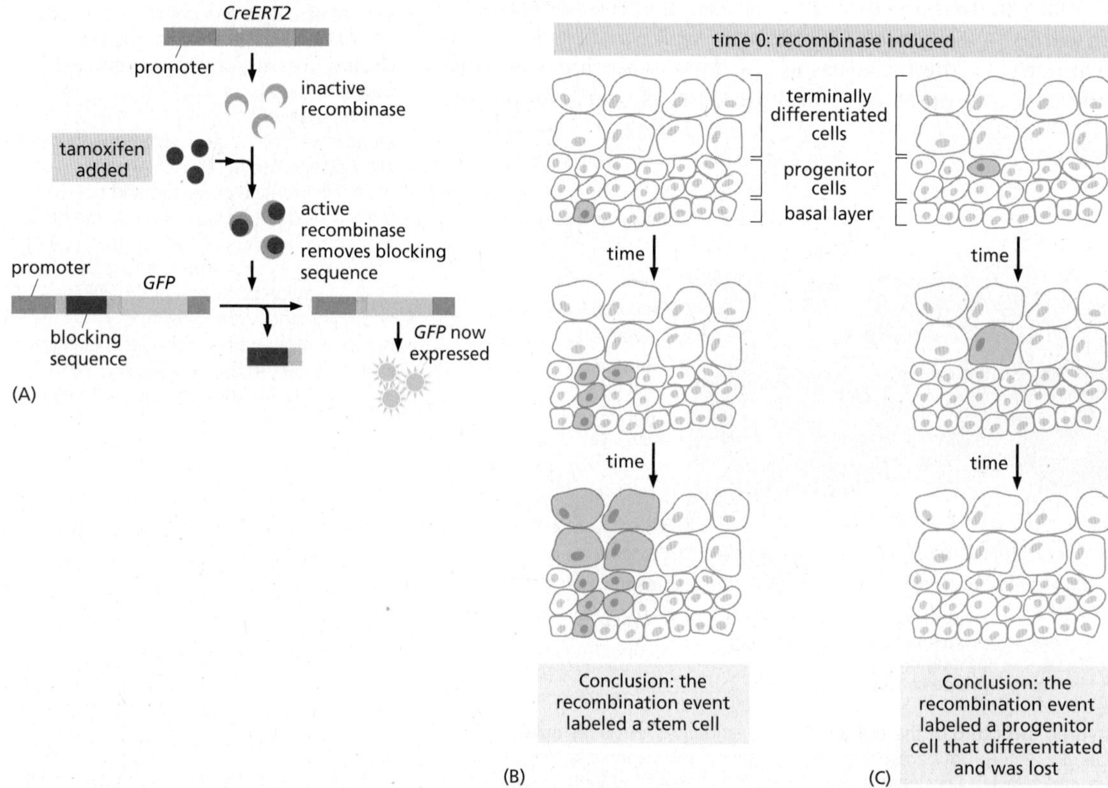

Figure 22–7 Clonal analysis using a genetic marker. (A) Transgenic animals containing two transgenes can be used to drive expression of a readily detected and heritable marker in a small set of cells. The first transgene encodes a marker gene, such as one that encodes green fluorescent protein (GFP). However, the expression of the *GFP* transgene, shown here in *green*, is prevented by a blocking sequence *(red)* that is flanked by LoxP sites (*pink*; see Figure 5–66). The second transgene, *CreERT2 (brown),* encodes a chimeric form of the Cre recombinase called CreERT, which consists of the Cre recombinase linked to the estrogen receptor protein; this enzyme becomes active as a recombinase only when it binds the artificial estrogen analog tamoxifen *(red spheres)*. Addition of tamoxifen leads to a recombination event that removes the blocking DNA sequence. As a result, the GFP marker is expressed. Because the blocking DNA has been permanently removed from the genome, the marker continues to be expressed in all the descendants of a cell in which the recombination event has occurred. With a low dose of the inducer molecule tamoxifen, it is possible to activate the marker at random in just a few widely spaced cells, giving rise to distinguishable clones. (B) If the recombination event occurs in a stem cell, a clonal lineage will be marked, and the labeling will persist over time as the marked stem cell self-renews and produces differentiated cells. (C) If the recombination event occurs in a cell that is not a stem cell, the label will disappear over time as the marked cell differentiates and is eventually lost.

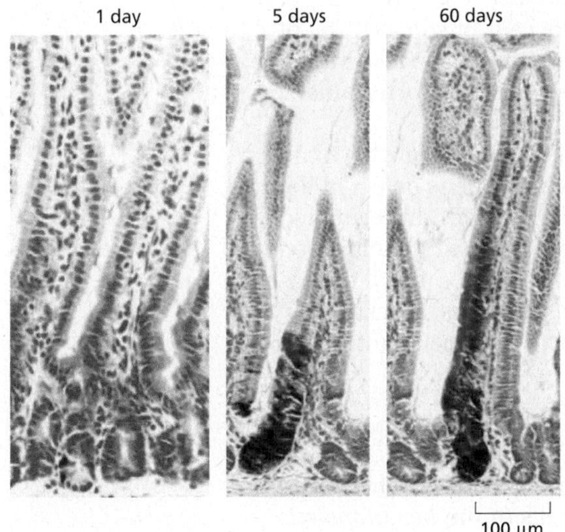

Figure 22–8 **Lgr5-expressing stem cells and their progeny in the small intestine.** The basic method shown in Figure 22–7 was modified here to mark single intestinal stem cells and trace the fates of their progeny. The *Lgr5* gene encodes a member of the family of G-protein-linked transmembrane receptors, and it is expressed specifically in stem cells near the crypt base. In this case, the *Lgr5* promoter was used to drive expression of *CreERT2*, and treatment with a low dose of tamoxifen resulted in occasional stem cells expressing the marker protein LacZ (rather than GFP). These cells and all of their progeny could subsequently be detected with a *blue* histochemical stain. All of the *blue* cells in these images derive from a single Lgr5-expressing stem cell. After 60 days, the *blue* progeny of this cell are seen to extend almost all the way up a villus. These progeny can be shown to include all types of differentiated cells, as well as persistent Lgr5-expressing cells at the crypt base. This proves that Lgr5-expressing cells are multipotent stem cells. (From N. Barker et al., *Nature* 449:1003–1007, published 2007 by Nature Publishing Group. Reproduced with permission of SNCSC.)

present, but also where the stem cells are located and whether they are unipotent or multipotent.

A more directed approach can be used for cell lineage tracing if a gene that is expressed specifically in the stem cells in a tissue is known, as is the case for the *Lgr5* gene in the mouse intestine. In this case, one can use the gene's promoter to express the genetic marker specifically in the stem cells. It was this type of experiment that initially established that Lgr5-expressing cells in the mouse intestinal epithelium are stem cells and that they are multipotent (see Figure 22–8).

Quiescent Stem Cells Are Difficult to Identify by Lineage Tracing

The lineage tracing method just described assumes that the stem cells in a tissue are actively dividing to generate daughter cells that self-renew or differentiate. Some adult stem cells, however, reside in a quiescent state, serving as a "reservoir" for when they are needed: they divide only rarely or not at all, unless they are induced to do so by a stimulus, such as tissue injury. In these cases, it can require extra time or stimulation to reveal the stem cells by lineage tracing.

Human **skeletal muscle** provides an example. It consists of multinucleated muscle cells (muscle fibers) that form during development by the fusion of terminally differentiated *myoblasts*. Humans do not normally generate new skeletal muscle in adult life, but they still have the capacity to do so when there is a need for muscle growth or repair. Cells capable of serving as myoblasts are retained as small, flattened, and nondividing cells lying in close contact with the mature muscle fiber and contained within its sheath of basal lamina (**Figure 22–9**). If the muscle is damaged or stimulated to grow, these *satellite cells* are activated to proliferate, and their progeny can fuse with the existing muscle fiber to repair

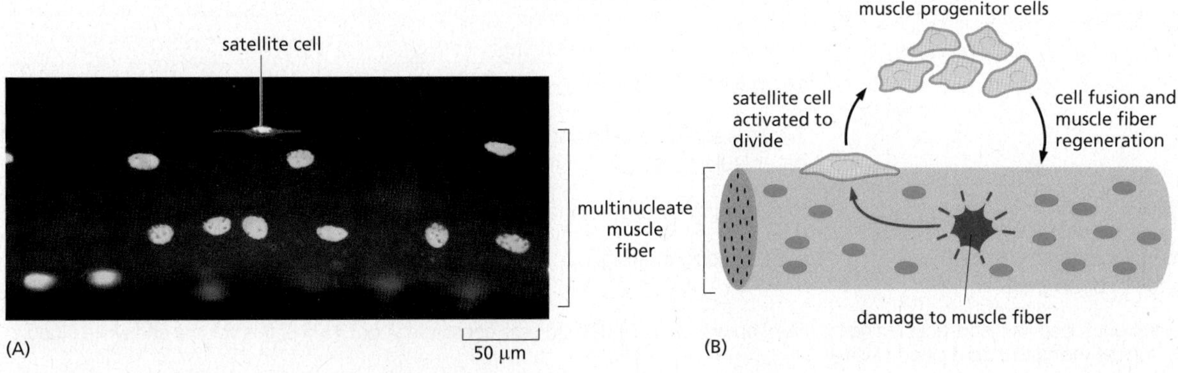

Figure 22–9 **The repair of skeletal muscle fibers by satellite cells.** (A) The specimen is stained with an antibody *(red)* against a muscle cadherin, M-cadherin, which is present on both the satellite cell and the muscle fiber and is concentrated at the site where their membranes are in contact. The nuclei of the muscle fiber are stained *green*, and the nucleus of the satellite cell is stained *blue*. (B) Schematic drawing of the repair of a damaged muscle fiber by the proliferation and fusion of satellite cells. (A, courtesy of Terence Partridge.)

the damaged muscle or to allow muscle growth. Satellite cells or some subset of them are thus the stem cells of adult skeletal muscle, normally held in reserve in a quiescent state but available when needed as a self-renewing source of terminally differentiated myoblasts.

The process of muscle repair by means of satellite cells is limited in what it can achieve. In one form of *muscular dystrophy*, for example, a genetic defect in the cytoskeletal protein dystrophin slowly but progressively damages differentiated skeletal muscle cells. As a result, satellite cells proliferate to repair the damaged muscle fibers. But this regenerative response is unable to keep pace with the damage, and connective tissue eventually replaces the muscle fibers, blocking any further possibility of repair. A similar decline in the capacity for repair contributes to the progressive muscle weakening that occurs in the elderly.

Hematopoietic Stem Cells Can Be Identified by Transplantation

A different method to identify stem cells in adult tissues is by cell transplantation. The method was first used to identify the stem cells of the *hematopoietic* (blood-making) *system*, the most complex stem-cell system in the adult mammalian body. The **hematopoietic stem cells** that give rise to both the red blood cells (**erythrocytes**) and white blood cells (**leukocytes**) are located in the adult bone marrow, where they also produce blood *platelets*. There are many different white blood cell types, including monocytes that can exit the bloodstream and develop into *macrophages*, which are found in most organs. When an animal is exposed to a large dose of x-rays, most of the hematopoietic cells in the bone marrow are destroyed, and, as a result, the animal dies within days because of its inability to produce new blood cells. The animal can be saved, however, by a transfusion of cells taken from the bone marrow of a healthy donor mouse of the same inbred

TABLE 22–1 **Blood Cells**

Type of cell	Main functions	Typical concentration in human blood (cells/liter)
Red blood cells (erythrocytes)	Transport O_2 to and CO_2 from tissues	5×10^{12}
White blood cells (leukocytes)		
Granulocytes		
Neutrophils (polymorphonuclear leukocytes)	Phagocytose and kill invading bacteria	5×10^9
Eosinophils	Destroy larger parasites and modulate allergic inflammatory responses	2×10^8
Basophils	Release histamine (and in some species serotonin) in certain immune reactions	4×10^7
Monocytes	Become tissue macrophages, which phagocytose and digest invading microorganisms and foreign bodies as well as damaged senescent cells; some also differentiate into dendritic cells	4×10^8
Lymphocytes		
B cells	Make and secrete antibodies	$\sim 0.3 \times 10^9$
T cells	Kill virus-infected cells and regulate activities of other leukocytes	$\sim 2 \times 10^9$
Natural killer (NK) cells	Kill virus-infected cells and some tumor cells	1×10^8
Platelets (cell fragments arising from megakaryocytes in bone marrow)	Initiate blood clotting	3×10^{11}

Humans contain about 5 liters of blood, accounting for 7% of body weight. Red blood cells constitute about 45% of this volume and white blood cells about 1%, the rest being the liquid blood plasma.

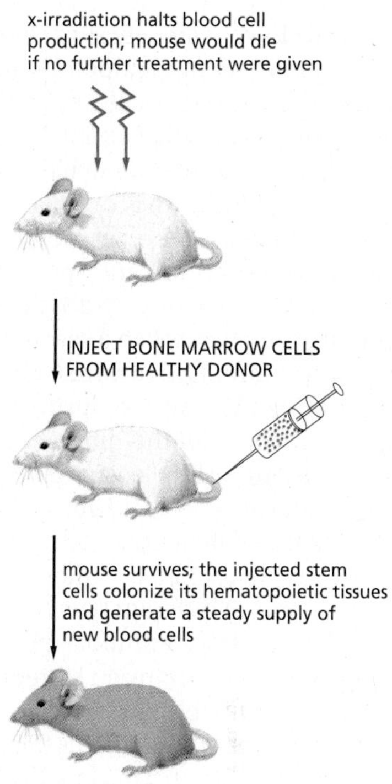

Figure 22–10 Rescue of an irradiated mouse by a transfusion of bone marrow cells. An essentially similar procedure is used in the treatment of leukemia in humans by bone marrow transplantation after irradiation or chemotherapy.

x-irradiation halts blood cell production; mouse would die if no further treatment were given

INJECT BONE MARROW CELLS FROM HEALTHY DONOR

mouse survives; the injected stem cells colonize its hematopoietic tissues and generate a steady supply of new blood cells

strain. Among these cells there are some that can colonize the irradiated host and permanently re-equip it with hematopoietic tissue (Figure 22–10). Such experiments prove that the bone marrow contains an entire hematopoietic stem-cell system, and they have allowed scientists to isolate the relevant stem cells and discover the molecular features that distinguish them.

For this purpose, cells taken from mouse bone marrow are sorted (using a fluorescence-activated cell sorter) according to their cell-surface antigens, and the different fractions are transfused into irradiated mice. If a fraction rescues the irradiated mice, it must contain hematopoietic stem cells. In this way, it was shown that the hematopoietic stem cells display a specific combination of cell-surface proteins and that, by appropriate cell sorting, one can obtain virtually pure stem-cell preparations. The stem cells turn out to be a tiny fraction of the mouse bone marrow population—about 1 cell in 50,000–100,000; but this is enough. Remarkably, a single such cell injected into a host mouse with defective hematopoiesis is sufficient to reconstitute its entire hematopoietic system, generating a complete set of blood-cell types, as well as fresh stem cells. This and lineage tracing experiments have established that an individual hematopoietic stem cell is *multipotent* and can self-renew and give rise to the complete range of blood-cell types.

Blood contains large numbers of many types of differentiated cells (Table 22–1), many of which can be seen in a standard, stained smear of human blood (Figure 22–11). Erythrocytes are homogeneous and remain in the blood vessels, where they transport O_2 and CO_2 bound to hemoglobin. By contrast, leukocytes are heterogeneous in morphology and function and must crawl across the walls of small blood vessels into tissues to function. Terminally differentiated

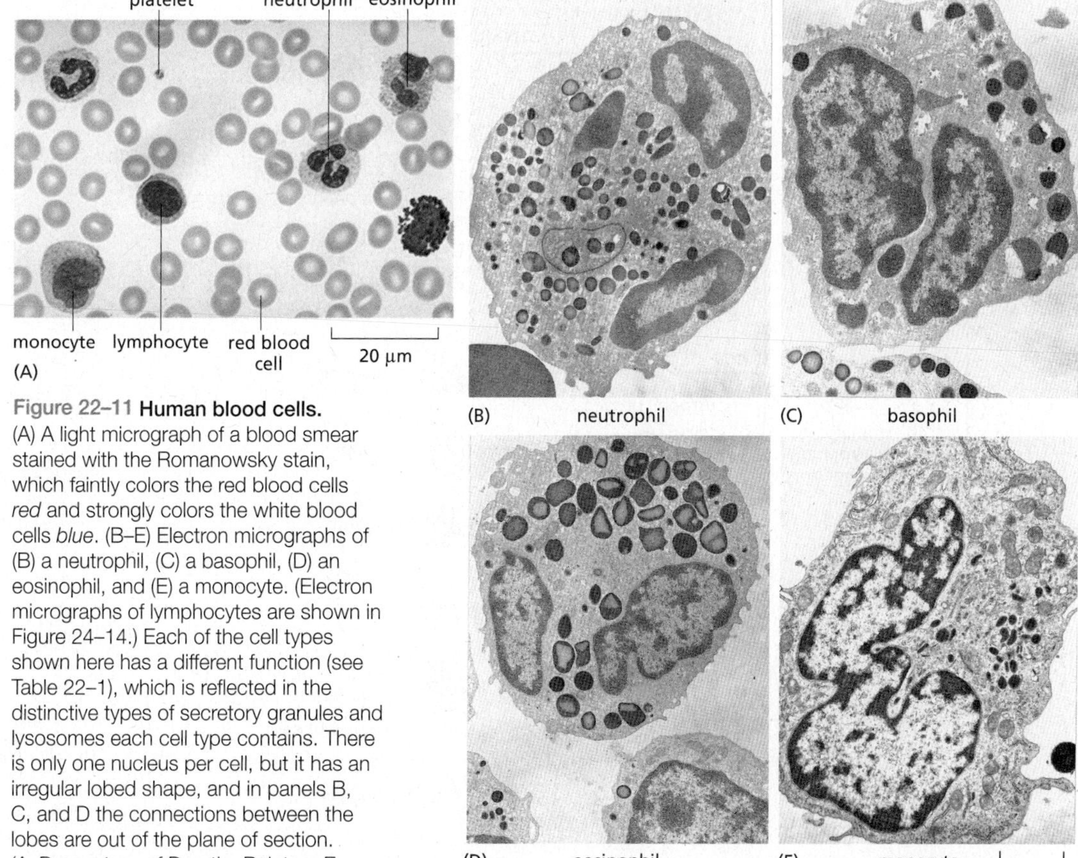

Figure 22–11 Human blood cells. (A) A light micrograph of a blood smear stained with the Romanowsky stain, which faintly colors the red blood cells *red* and strongly colors the white blood cells *blue*. (B–E) Electron micrographs of (B) a neutrophil, (C) a basophil, (D) an eosinophil, and (E) a monocyte. (Electron micrographs of lymphocytes are shown in Figure 24–14.) Each of the cell types shown here has a different function (see Table 22–1), which is reflected in the distinctive types of secretory granules and lysosomes each cell type contains. There is only one nucleus per cell, but it has an irregular lobed shape, and in panels B, C, and D the connections between the lobes are out of the plane of section. (A–D, courtesy of Dorothy Bainton; E, courtesy of David Mason.)

blood cells have relatively short life spans and are produced throughout the life of the animal, so hematopoietic stem cells must generate enormous numbers of these differentiated cells each day. But they do not produce them directly. Instead, the stem cells continually produce large numbers of transit-amplifying progenitor cells in the bone marrow, which are committed to differentiation but go through multiple divisions before they terminally differentiate.

The stem cells, however, do not jump directly from a multipotent stem-cell state into a committed and specific pathway of differentiation; instead, they go through a number of cell divisions, in which they progressively restrict their developmental options in a series of steps (Figure 22–12). The first step is usually to become committed to either a myeloid or a lymphoid fate, by way of two kinds of multipotent transit-amplifying progenitor cells. One is capable of generating large numbers of all the different types of **myeloid cells**, including blood **granulocytes** (neutrophils, eosinophils, and basophils), **monocytes** (the precursors of macrophages and dendritic cells), erythrocytes, and **megakaryocytes** (which remain in the bone marrow and produce blood platelets by pinching off cell fragments) (Movie 22.1). The other type of multipotent progenitor cell gives rise to large numbers of different types of **lymphoid cells**, including the B and T lymphocytes of the adaptive immune system and the lymphocyte-like *natural killer* (*NK*) cells of the innate immune system (discussed in Chapter 24). Further commitment steps ultimately give rise to progenitor cells committed to the production of just one cell type, although this final commitment step occurs well before the cells cease proliferating and terminally differentiate. Many different signal molecules, produced inside and outside the bone marrow, control the survival, proliferation, and pathway commitment of hematopoietic cells and their committed progeny in the bone marrow and thereby regulate how many of each type of differentiated blood cell is eventually produced.

Figure 22–12 A simplified scheme of mouse and human hematopoiesis. A multipotent hematopoietic stem cell normally divides infrequently to generate either more multipotent stem cells, which are self-renewing, or multipotent progenitor cells, which give rise to all the cells of the blood and immune system. The progenitor cells divide a limited number of times and go through multiple stepwise intermediates before they develop into fully differentiated cells. As they go through their divisions, the progenitors become progressively more specialized in the range of cell types that they can give rise to, as indicated by the branching of this cell lineage diagram. In adult mammals, all of the cells shown develop mainly in the bone marrow—except for T lymphocytes, which as indicated develop in the thymus, and macrophages and some dendritic cells, which develop from monocytes that are circulating in the blood. Note that not all stem cells generate the identical patterns of progeny via precisely the same sequence.

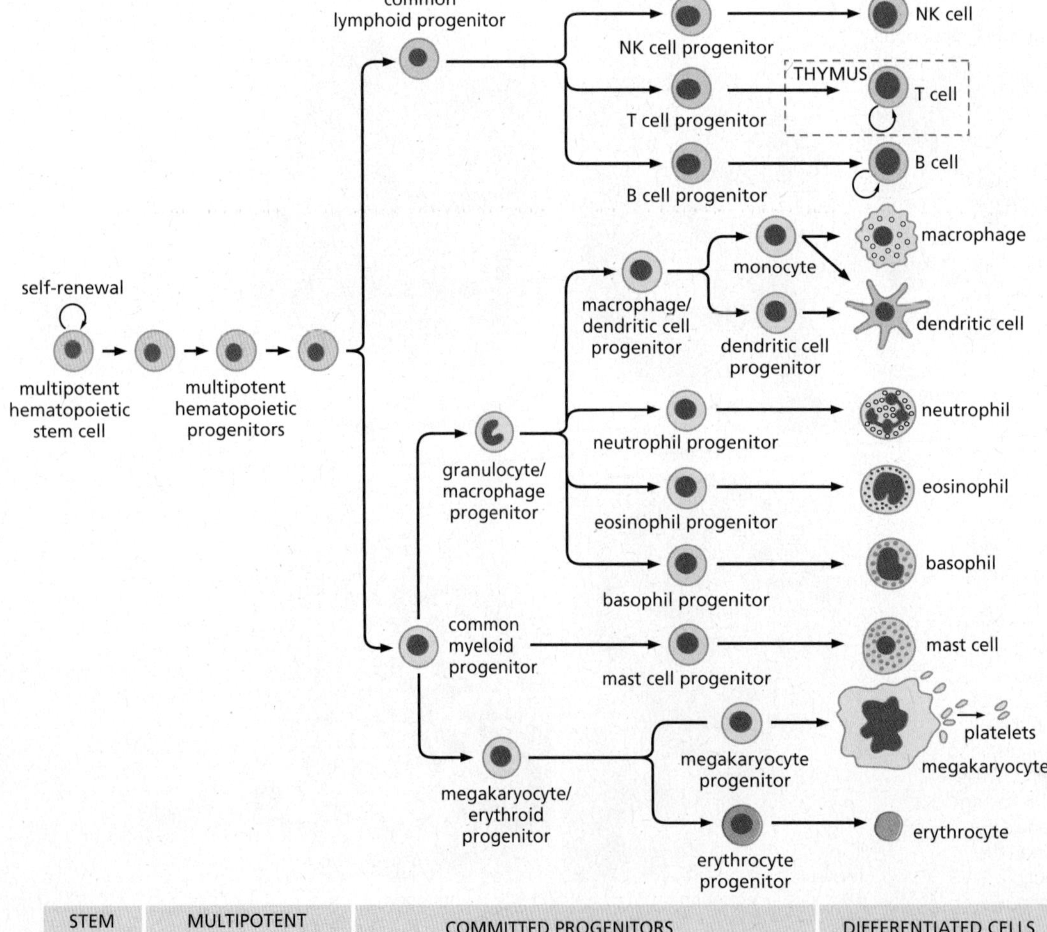

In this hierarchical hematopoietic system, only the stem cells can self-renew for the life of the individual, and a single stem-cell division can lead to the production of thousands to millions of differentiated progeny. This explains why the number of stem cells is such a tiny fraction of the total population of hematopoietic cells in the bone marrow. Keeping the number of stem-cell divisions low has important advantages. If these divisions on their own had to keep up with the high demand for terminally differentiated blood cells, they would result in rapid replicative cell senescence and exhaustion of the stem-cell pool, with dire consequences. Moreover, the lower the number of stem-cell divisions, the lower the risk of the cells accumulating dangerous mutations, which would persist in mutant clones; as discussed in Chapter 20, such clones are a particular danger in the hematopoietic system, where a relatively small number of mutations can be sufficient to cause blood-cell cancers.

Some Tissues Do Not Require Stem Cells for Their Maintenance

Some types of cells can divide even though fully differentiated, allowing for renewal and regeneration without the use of stem cells. The insulin-secreting **pancreatic β cells** are one example. Their mode of renewal has a special importance, because it is their loss through autoimmune attack that is responsible for type 1 (juvenile-onset) diabetes, and their functional decline with age and obesity is also a significant factor in type 2 (adult-onset) diabetes. The β cells are normally sequestered in cell clusters called *islets of Langerhans*. The islets seem not to contain stem cells, yet new β cells are continually generated within the islets. Lineage tracing studies, similar to those described earlier, show that the renewal of this population normally occurs by the simple division of differentiated insulin-producing β cells.

Another tissue that can renew by simple division of fully differentiated cells is the liver. The main cell type in the liver is the **hepatocyte**, a large cell that performs the liver's many metabolic functions. Hepatocytes normally live for a year or more and divide at a very slow rate. Powerful homeostatic mechanisms operate to adjust both their rate of cell proliferation and their rate of cell death, to keep the liver at its normal size and to restore that size in the case of damage. A dramatic effect is seen if large numbers of hepatocytes are removed surgically or killed by poisoning with carbon tetrachloride. Within a day or so after either sort of damage, a surge of cell division occurs among the surviving hepatocytes, quickly replacing the lost tissue. If two-thirds of a rat's liver is removed, for example, a liver of nearly normal size can regenerate by hepatocyte proliferation within about 2 weeks.

Both the pancreas and the liver also contain small populations of stem cells that can be called into play as a backup mechanism to produce the differentiated cell types in more extreme circumstances.

In Response to Injury, Some Differentiated Cells Can Revert to Progenitor Cells and Some Progenitor Cells Can Revert to Stem Cells

Although the pathways from stem cell to progenitor cell to differentiated cell are normally unidirectional, there are some cases where injury can reverse the direction. One striking example occurs when a myelinated mammalian nerve is cut: the axon distal to the cut degenerates, and the differentiated myelinating Schwann cells (see Figure 11–35) *dedifferentiate* to form proliferating Schwann-cell progenitor cells. These progenitor cells help guide the regenerating axons back to their original targets and then remyelinate the axons to complete the regeneration process.

Similarly, in some tissues, when stem cells are lost, progenitor cells that have committed to differentiation can reprogram to revert to stem cells. In both the mouse and *Drosophila* testis, for example, spermatogonial cells typically follow a unidirectional development pathway from stem cell to proliferating progenitor cells, which undergo meiosis and finally differentiate into sperm. If stem cells are lost, either naturally or experimentally, the mitotically proliferating progenitor cells can reprogram and revert to stem cells. More generally, this process might contribute to the long-term maintenance of stem-cell populations, allowing the lifetime of individual stem cells to be shorter than the lifetime of the organism. Unfortunately, cancer cells often also acquire stem cell–like properties, allowing them to self-renew indefinitely (discussed in Chapter 20).

Some Tissues Lack Stem Cells and Are Not Renewable

Some adult tissues that lack stem cells are not able to regenerate. The remarkable variation in the ability of different tissues to regenerate is illustrated by comparing the olfactory epithelium in the nose, the auditory epithelium of the inner ear, and the photoreceptive epithelium of the retina, which exhibit striking differences in their renewal capacity. The *olfactory epithelium* contains a population of stem cells that give rise to differentiated cells that have a limited life span and are continually replaced. But unlike the epidermis discussed earlier, these differentiated olfactory cells are neurons; they have their cell bodies in the olfactory epithelium and extend their axons back to the olfactory bulbs in the brain. The renewal of this epithelium therefore involves the continual production of new axons that have to navigate back to specific sites in the brain, where they form new synapses.

In contrast, in mammals at least, the *auditory epithelium* and *retinal photoreceptive epithelium* lack stem cells, and their sensory receptor cells—the sensory hair cells in the ear and the photoreceptors in the retina—are irreplaceable. If they are destroyed—whether by too much exposure to loud noise, by looking into the beam of a laser, or through degenerative processes occurring in disease or in old age—the loss is permanent.

We will return to tissue regeneration later in the chapter.

Summary

Many adult tissues, particularly those with a high cell turnover rate such as the intestinal lining, skin epidermis, and blood, are continually renewed by stem cells to maintain tissue homeostasis throughout the lifetime of the organism. Stem cells are defined by their ability to both self-renew and to generate terminally differentiated cells, usually by way of rapidly dividing, transit-amplifying progenitor cells. These properties can be revealed experimentally through lineage-tracing or transplantation experiments. Tissue-specific, or adult, stem cells are restricted in their differentiation potential, only generating one or more of the specific cell types of a particular tissue.

In the single-layer lining of the small intestine, multipotent stem cells are located near the base of each crypt, where they self-renew and produce dividing committed progenitor cells, most of which flow upward and terminally differentiate into one of three main types of gut cells when they reach the villus; other progenitor cells move in the opposite direction and become Paneth cells, which remain at the base of the crypt and help maintain the stem cells. Other self-renewing epithelia, such as the epidermis, have a multilayered (stratified) architecture, with stem cells and their differentiating progeny arranged in different ways, but are governed by similar basic principles.

The hematopoietic system is the most complex mammalian stem-cell system; all the red blood cells and the many types of white blood cells derive from a common, multipotent, hematopoietic stem cell in the adult bone marrow, where it divides slowly and produces multipotent and unipotent progenitor (transit-amplifying) cells, which divide rapidly and differentiate into a large number and variety of terminally differentiated cell types every day. In other tissues, such as skeletal muscle, stem cells are quiescent and only divide and differentiate when tissue growth or repair is required. Adult tissue renewal and repair do not always depend on stem cells; in the pancreas and liver, for example, differentiated cells can divide throughout life to replace lost cells and maintain tissue homeostasis. In some cases, progenitor cells can be generated from differentiated cells, and stem cells can be generated from progenitor cells. At the opposite extreme, some sensory epithelia of the adult mammalian ear and eye are not renewable; their sensory cells do not undergo turnover—once lost, they are lost forever.

CONTROL OF STEM-CELL FATE AND SELF-RENEWAL

Adult stem cells maintain tissue homeostasis in renewing tissues, owing to their ability both to self-renew and to produce differentiated cells for the lifetime of the organism. How do these cells maintain their stem-cell identity for so long, and how do their daughter cells choose between self-renewal and a commitment to differentiation? While maintaining a stem-cell population is critically important for self-renewing tissues, stem-cell proliferation must be kept in check, as uncontrolled production of undifferentiated cells is a hallmark of cancer (discussed in Chapter 20). In this section, we discuss the mechanisms that allow stem cells to maintain their identity while producing differentiated cells and keeping their own proliferation under control.

The Stem-Cell Niche Maintains Stem-Cell Self-Renewal

The identity and self-renewal of stem cells depend on extracellular signals from their environment that both promote their proliferation and inhibit their commitment to differentiation. In many tissues, stem cells inhabit a special microenvironment called the **stem-cell niche**, where these signal molecules are provided in high concentration (**Figure 22–13**). The stem cells remain in close physical proximity to the niche-supporting cells, which produce the signals and provide a local special environment. Outside this environment, the concentration of the signal molecules is insufficient to maintain the self-renewal of the stem cells, and their daughter cells therefore commit to differentiation.

The existence of the stem-cell niche was first inferred from the experiment in which bone marrow cells were transplanted into a mouse that had been subjected to a high dose of x-rays (see Figure 22–10). Without the prior irradiation, the transplanted cells were unable to reestablish hematopoiesis, suggesting that the resident hematopoietic stem cells first had to be removed before the transplanted stem cells could successfully seed the host's bone marrow. The hypothetical special "place" occupied by either the host or donor hematopoietic stem cells was referred to as the *stem-cell niche.* Because of the complexity of bone marrow and the rarity of the stem cells there, the location of the niche remained unclear until many years later, when immunofluorescence microscopy and the use of cell markers provided evidence that the surface of sinusoids in the bone marrow is the likely site of the niche.

The molecular nature of stem cell–renewal signals provided within the niche varies across species and tissues, but they frequently are secreted signal proteins of the Wingless (Wnt), Hedgehog (Hh), or transforming growth factor (TGF) families (discussed in Chapter 15). Other niche signals are cell-surface proteins that depend on direct contact between the stem cells and the niche-supporting cells. The mammalian intestinal crypt provides a well-studied example (see

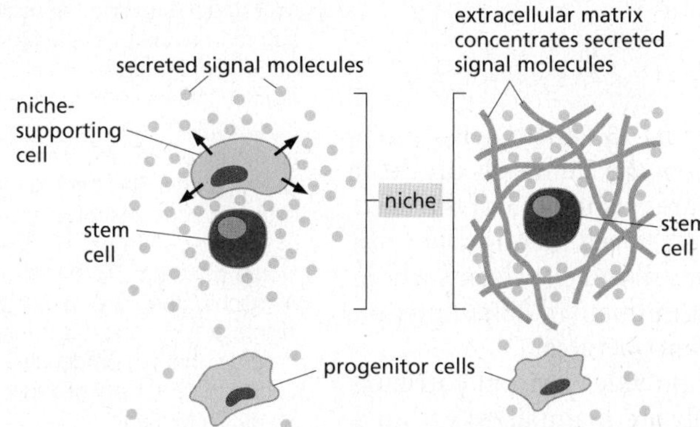

Figure 22–13 **The stem-cell niche.** Stem cells undergo self-renewal only in a specialized microenvironment where they are exposed to the necessary signal molecules. These signal molecules can be provided by niche-supporting cells or by a specialized extracellular matrix that serves to concentrate them. The niche environment can be very small. In some cases, a stem cell directly adjacent to the niche is capable of self-renewal, while a cell just one cell-diameter away from the niche is not.

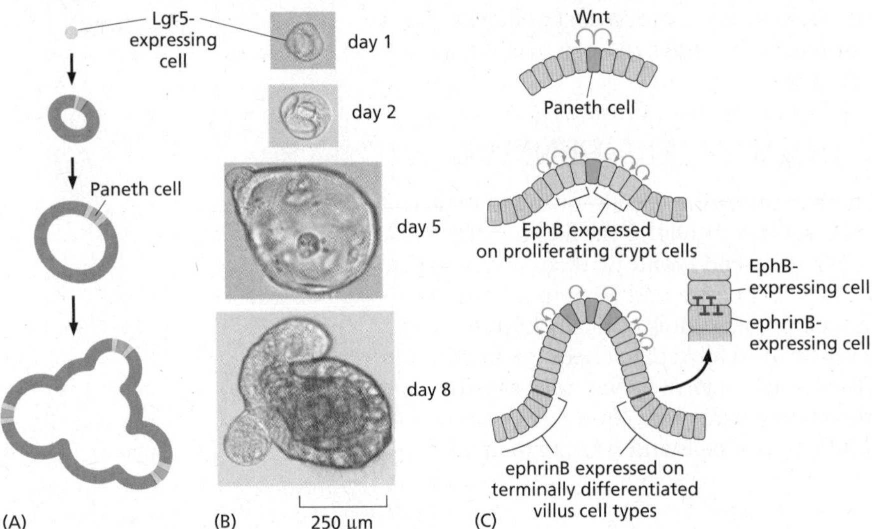

Figure 22–14 **Genesis of a minigut from a single Lgr5-expressing cell cultured in a cell-free matrix.** (A) Schematic drawing and (B) phase-contrast micrographs of the developing organoid. The founder cell first divides to form a small sack or vesicle. At random, one or more of the cells in this vesicle differentiates as a Paneth cell *(blue)*. The Paneth cells secrete Wnt proteins that stimulate stem-cell self-renewal and maintain Lgr5 expression *(yellow)* in their immediate neighbors; besides generating more stem cells, the stem cells produce progenitor cells that differentiate into the full range of intestinal epithelial-cell types. (C) Schematic diagram showing how Paneth cell–derived Wnt signals help organize the developing crypt: besides keeping neighboring cells in the crypt proliferating in a stem-cell state, these signals activate repulsive interactions mediated by membrane-bound ephrin and Eph proteins on contacting cells (discussed in Chapter 15). These interactions cause dividing crypt cell types (which express EphB, induced by Wnt) to segregate from the nondividing, terminally differentiated villus cell types (which express ephrinB). In many tissues, cells that interact via ephrin–Eph binding repel each other when they touch (see Figure 15–52 or Figure 21–47). (Adapted from T. Sato and H. Clevers, *Science* 340:1190–1194, 2013. With permission from AAAS.)

Figure 22–4). We saw earlier that the stem cells that give rise to all the terminally differentiated cells in the intestinal epithelium express the cell-surface receptor protein Lgr5 and are located just above the base of the crypt, interspersed among the Paneth cells. The Paneth cells help create the stem-cell niche in two ways. They secrete Wnt proteins that act over a short range to stimulate stem-cell self-renewal. They also express *Delta*, a ligand for the signaling receptor Notch, on their surface, which activates Notch on the stem cells they directly contact, thereby inhibiting the cells from differentiating—an example of lateral inhibition (see Figure 15–59).

Remarkably, a single Lgr5-expressing stem cell embedded in a cell-free extracellular matrix can proliferate in a culture dish and form a tiny gut-like structure, an *organoid*, containing all the cell types normally found in gut tissue, including stem cells. Figure 22–14 shows this process and outlines the key signaling events that establish the crypt stem-cell niche, where different cell types are produced that arrange themselves into miniature villi to form the three-dimensional organoid structure.

The Size of the Niche Can Determine the Number of Stem Cells

In each intestinal crypt, the stem-cell niche is created by 15 Paneth cells, and it only has space for a limited number of stem cells. When the stem cells divide, it is a random matter as to which of the daughter cells are pushed out of the nest; failing to get the signals they need to maintain their stem-cell identity, they are condemned to commit to differentiation. In most other stem-cell systems where this question of balance between self-renewal and commitment to differentiation has been examined, it appears that a similar mechanism operates.

The *Caenorhabditis elegans* germ-line cell lineage provides a visually striking example (Figure 22–15). In this system, the stem cells are maintained within a

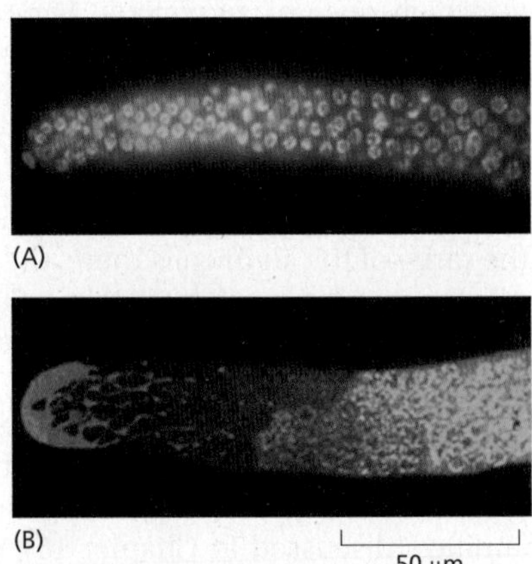

Figure 22–15 **Stem-cell niche of the *C. elegans* gonad.** (A) Fluorescence micrograph showing the nuclei *(blue)* in a portion of the *C. elegans* gonad. The nucleus of the distal tip cell is shown in *red*. (B) The same tissue is stained to show the cytoplasm and processes of the distal tip cell in *red* and the differentiated germ cells in *green*. The distal tip cell processes extend across multiple stem-cell diameters, maintaining stem-cell identity through Notch signaling. Those cells moved out of reach of the distal tip cell processes initiate meiosis and begin the differentiation program that will produce either eggs or sperm. (Courtesy of Judith Kimble and Sarah Crittenden.)

niche that is formed by a large somatic cell called a *distal tip cell*, which forms long, thin cellular processes. Notch ligand located on the surface of the distal tip cell signals to Notch on the stem cells to maintain their stem-cell identity. As in the intestinal crypt, the stem cells undergo self-renewal proliferation in the niche, until some of their progeny are pushed out of range of the distal tip cell processes, and, as a result, they begin to differentiate. In this way, a steady stream of cells at different stages of differentiation move away from the distal tip cell and the stem cells, as if on a conveyor belt, on their way to becoming terminally differentiated germ cells—either eggs or sperm.

Niche size, however, is not the only way to control stem-cell numbers—asymmetric stem-cell division is another, as we now discuss.

Asymmetric Stem-Cell Division Can Maintain Stem-Cell Number

As discussed earlier, when a stem cell divides, the two daughter cells must make a choice between remaining a stem cell or committing to differentiation. The balance between these two fates is critically important for tissue homeostasis, because excessive self-renewal would lead to an excess of stem cells, whereas excessive differentiation would deplete the stem-cell pool. We have just discussed how the size of the stem-cell niche can help control stem-cell numbers. Asymmetric stem-cell division is another mechanism for such control.

In an asymmetric stem-cell division, a process internal to the dividing stem cell causes a biased inheritance of one or more important molecules that influences the fate of the two daughter cells. For example, only one of the two daughter cells might inherit a *cell-fate determinant* that is required to maintain stem-cell identity, causing the other daughter to commit to differentiation (**Figure 22–16A**). Each such asymmetric stem-cell division produces one stem cell and one cell committed to differentiation; as a result, the number of stem cells is preserved irrespective of how many cell divisions the stem cells undergo. In symmetric stem-cell divisions, both daughters inherit such cell-fate determinants and remain stem cells, thereby increasing the pool of stem cells (**Figure 22–16B**).

As shown in Figure 22–16, prior to an asymmetric stem-cell division, the cell becomes polarized, with the cell-fate determinant localized on one side of the cell. The cell then positions the mitotic spindle to orient the plane of cell division so that only one daughter inherits the fate determinant.

Another way the orientation of the division plane in a stem cell can influence the fate of the daughter cells is to determine their position relative to the stem-cell

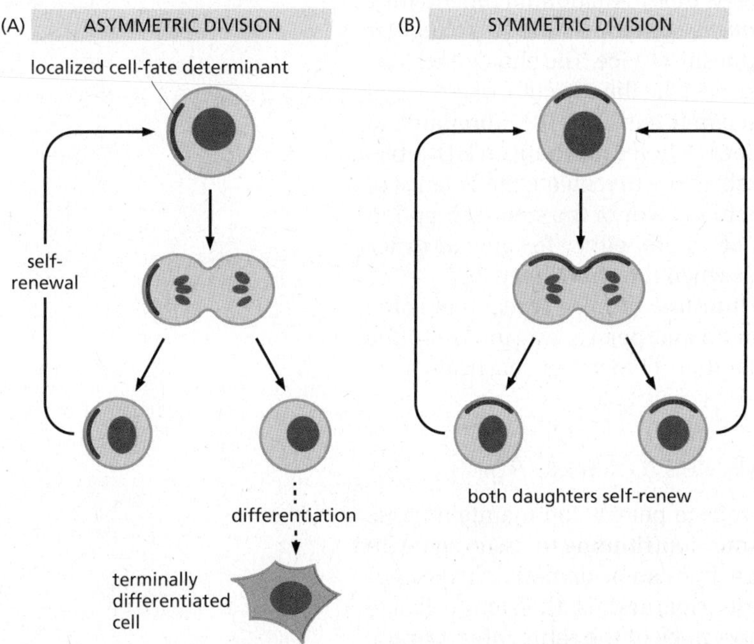

Figure 22–16 Asymmetric and symmetric stem-cell divisions. (A) In the asymmetric stem-cell division schematized here, a cell-fate determinant *(red)* that maintains stem-cell identity is localized to the cell cortex at one side of the cell. When the cell divides, only one daughter cell inherits the determinant and remains a stem cell, while the other daughter commits to differentiation, differentiating after a number of progenitor-cell divisions (not shown). (B) In the symmetric stem-cell division shown, both daughter cells inherit such determinants and remain stem cells.

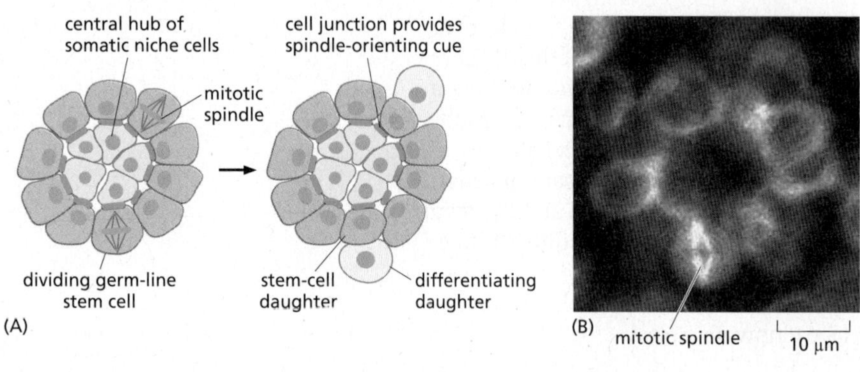

Figure 22–17 How the plane of division in a dividing, germ-line stem cell in the *Drosophila* testis determines which daughter cell maintains contact with the niche. (A) The niche consists of a hub of somatic cells, and the germ-line stem cells are arranged in a ring around the hub. When a stem cell divides, its mitotic spindle is oriented perpendicular to the niche, so that only one daughter cell maintains contact with the niche and remains a stem cell; the other daughter loses contact with the niche cells and is therefore deprived of self-renewal signals and commits to differentiation. (B) The light micrograph shows germ-line stem cells surrounding the niche and a dividing stem cell with its mitotic spindle oriented perpendicular to the niche. The stem cells have been genetically engineered to express fluorescently labeled tubulin. (B, from Y.M. Yamashita et al., *Science* 301: 1547–1550, 2003. With permission from AAAS.)

niche, such that the niche is asymmetrically inherited. In the *Drosophila* testis, for example, the germ-line stem cells orient their spindles perpendicular to the niche-supporting cells: as a result, when the stem cell divides, one daughter cell remains attached to the niche and adopts a stem-cell fate, while the other daughter is displaced away from the niche, where it is deprived of self-renewal signals and therefore commits to differentiation (**Figure 22–17**).

In Many Symmetric Stem-Cell Divisions, Daughter Cells Choose Their Fates Independently and Stochastically

As we have seen, asymmetric stem-cell division provides a straightforward way to maintain stem-cell numbers while generating differentiated cells in a self-renewing adult tissue. However, it leaves little room for a tissue to adapt to changing conditions, such as fluctuating nutrient availability or in response to tissue injury. In many self-renewing adult tissues, the mode of stem-cell division changes to adapt to such altered conditions. And similar adaptive changes occur during tissue development: for example, both the neural and epidermal stem cells in early developing mice mainly divide symmetrically to expand the stem-cell pool to support the rapidly growing organs; later in development, however, the stem cells shift to more asymmetric cell divisions to produce more differentiated cells (see Figure 22–16).

Stem cells in many renewing tissues, including the intestinal epithelium, adopt an even more flexible strategy. In these cases, when the stem cells divide, they do so symmetrically, and each daughter cell makes the choice between remaining a stem cell (self-renewal) and committing to differentiation, independently of its sister. In this *independent-choice mechanism,* sometimes the two daughter cells make the same choice, other times the opposite choice. The choice each cell makes might either be stochastic (probabilistic), like the flip of a coin, or governed by the environment the daughter cell finds itself in (**Figure 22–18**). Compared to the asymmetric-division strategy, the independent-choice mechanism is flexible, allowing local and more general environmental factors to regulate the balance of probabilities according to need—adjusting them in favor of the stem-cell option where more stem cells are required, as they often are, either for growth or for damage repair, or in favor of differentiated cells when they have been lost.

Lineage tracing experiments in the small intestine of mice reveal that some stem-cell lineages disappear while others expand and persist, as expected if the daughter cells of the stem-cell divisions make their choices stochastically (see Figure 22–18).

A Decline in Stem-Cell Function Contributes to Tissue Aging

Despite the remarkable ability of adult stem cells to persist and maintain tissue homeostasis, this ability often declines over time, contributing to tissue aging. In the case of the hematopoietic stem cells, this decline can be demonstrated experimentally using *serial transplantation.* In these experiments, the stem cells are transplanted from donor mice to irradiated host mice of the same inbred strain.

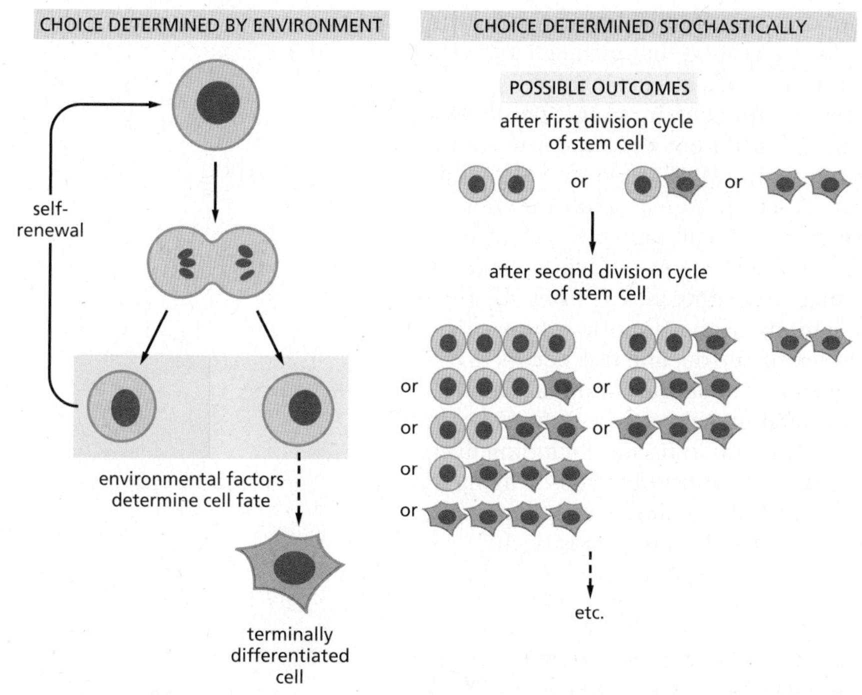

Figure 22–18 The independent-choice mechanism for how the two daughters of a stem-cell division choose their fates. The outcomes of this mechanism are more variable than that of symmetric stem-cell divisions. The choice might be determined by the local environment a daughter finds itself in (left side of the drawing) or it might be entirely stochastic (right side of drawing). With a choice made stochastically by each daughter and with a 50% probability for each one to remain a stem cell or commit to differentiation, there is, for example, a 25% chance at the first division that both daughters will commit to differentiation, so that the clone will eventually disappear. Or, at this division or later, a preponderance of daughters might choose to remain stem cells, creating a clone that persists and increases in size. With the help of some mathematics, the probability distribution of clone sizes generated from a single stem cell at any given time can be predicted on this stochastic assumption. The observations in the gut and often elsewhere fit this stochastic independent-choice strategy.

After the donor cells have repopulated the hematopoietic system of the host mice (see Figure 22–10), the hosts' hematopoietic stem cells are then transplanted into new irradiated hosts, and this sequence is repeated again and again. After multiple rounds of transplantation, the stem cells gradually lose their ability to repopulate the hematopoietic system of the new hosts, demonstrating that the stem cells change with age, apparently undergoing replicative cell senescence after repeated divisions.

To investigate whether the supportive functions of the hematopoietic stem-cell niche also decline with age, hematopoietic stem cells from young or old donor mice are transplanted into irradiated young or old host mice. Comparisons of the results from these four types of experiments (young-to-young, young-to-old, old-to-young, old-to-old—**Figure 22–19A**) indicate that a young stem-cell

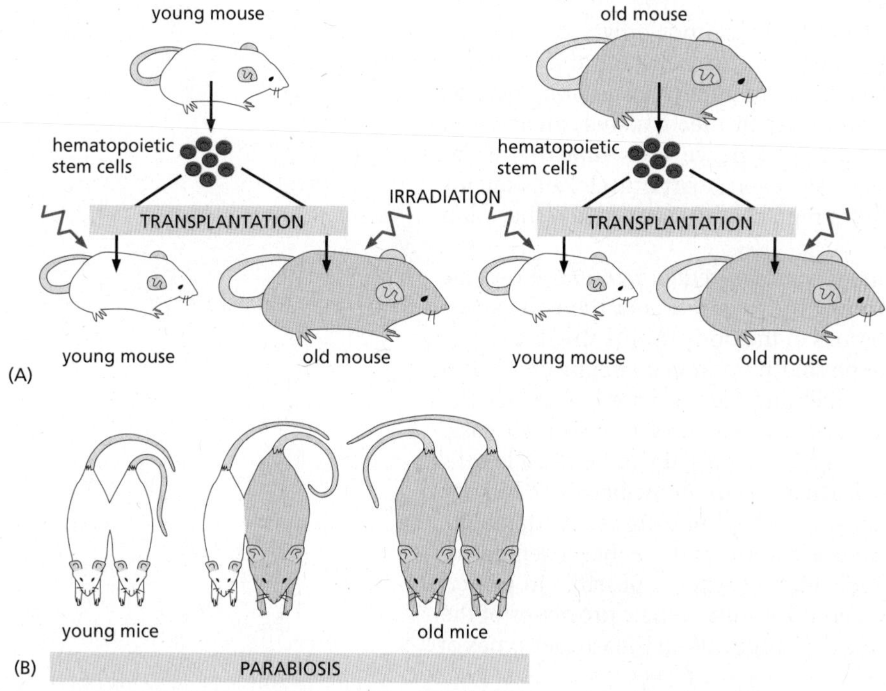

Figure 22–19 Methods to distinguish the contribution of cell-autonomous versus environmental factors to declines in hematopoietic stem-cell function during aging. (A) Hematopoietic stem cells are transplanted from young or old donor to young or old host, so that the function of stem cells can be assessed. A four-way comparison will reveal the contribution of the autonomous aging of stem cells versus the aging of the environment or niche in the decline of stem-cell function. (B) In parabiosis, animals are joined to develop a shared circulatory system. Cells in one animal will be exposed to the systemic environment of the other animal. The cellular functions in young or old partners in the heterochronic pairs (young-to-old) are compared with those in isochronic pairs (young-to-young and old-to-old). Such comparisons reveal the role of autonomous stem-cell aging compared to the role of circulating factors or the stem-cell niche in the decline of stem-cell function. [Based on M.A. Goodell and T.A. Rando, *Science* 350(6265):1199–1204, 2015, doi 10.1126/science.aab3388.]

niche cannot rescue old hematopoietic stem cells and that an old stem-cell niche can support young hematopoietic stem cells, suggesting that niche changes do not contribute to the decline of hematopoiesis with age, at least in mice.

A more general method for studying stem-cell function in aging is *parabiosis*, in which the circulatory systems of two mice of different age are connected—young-with-young, young-with-old, or old-with-old (**Figure 22–19B**). This three-way comparison provides a way of determining the contributions of different factors to a decline in stem-cell function with aging: the stem cells themselves, the stem-cell niche, and circulating factors in the blood. There is evidence from such studies, for example, that circulating factors from old mice can decrease neurogenesis in the hippocampus of the brain in young mice, impairing their learning and memory, mimicking the decline that occurs in the normal mouse brain during aging. Conversely, other studies find that circulating factors from young mice can restore skeletal muscle stem-cell function in old mice and can counteract age-related dysfunction in tissues including liver, pancreas, bone, and heart. Blood components that contribute to tissue aging or rejuvenation are likely to include molecules with direct actions on stem cells, and identification of these circulating factors is an active area of research.

Summary

Multiple extracellular signals operate in each stem-cell niche, enabling tissue-specific stem cells to maintain adult tissue homeostasis throughout an organism's life. These signals regulate when and where the stem cells self-renew and their daughter cells commit to differentiation. An individual stem cell's decision to self-renew or commit to differentiation is controlled to ensure the tissue maintains an adequate number of stem cells and produces an appropriate number of terminally differentiated cells. In some cases, asymmetric stem-cell divisions that produce one stem cell and one differentiating cell at each division achieve this balance. In other cases, the stem-cell divisions are symmetric, and the two daughter cells independently choose between self-renewal and commitment to differentiation, with the choice being either stochastic or in response to local environmental signals. The ability of adult stem cells to self-renew and produce terminally differentiated cells declines over time, contributing to tissue aging.

REGENERATION AND REPAIR

As we have seen, many of the tissues of the body are not only self-renewing but also self-repairing, and this is largely thanks to stem cells and the feedback controls that regulate the stem cells' behavior to maintain tissue homeostasis. There are, however, limits to what these natural repair mechanisms can achieve. In most parts of the human brain, for example, nerve cells that die, as in Alzheimer's disease, are not replaced. Likewise, when heart muscle cells die for lack of oxygen during a heart attack, they are replaced by scar tissue rather than by new heart muscle cells.

Some animals do far better than humans and can regenerate entire organs, such as whole limbs after amputation. Among the invertebrates, there are some species that can even regenerate all the tissues of the body from a single somatic cell. These phenomena encourage the hope that human cells might be coaxed by artificial measures into similar feats of repair and regeneration, to replace the nerve cells that die in individuals with Alzheimer's disease or Parkinson's disease, the insulin-secreting β cells that are lost in type 1 diabetes, the heart muscle cells that die in a heart attack, and so on. As we learn more about the basic cell biology of regeneration, these goals, once only a dream, are beginning to seem attainable.

In this section, we start with some examples of the remarkable regenerative abilities of some animal species, as an indication of what is possible in principle. We then discuss how we can improve on the natural repair processes of the human body and treat disease by exploiting the properties of the various types of stem cells found in adult human tissues.

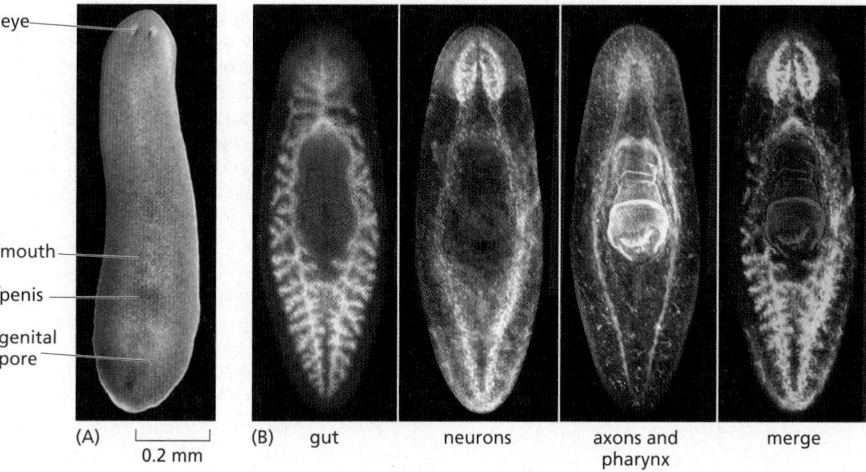

Figure 22–20 **The planarian worm, *Schmidtea mediterranea*.** (A) External view. (B) Immunostaining with three different antibodies, revealing the internal anatomy. (A, courtesy of A. Sánchez Alvarado; B, from A. Sánchez Alvarado, *BMC Biol.* 10:88, 2012. With permission from the author.)

Planarian Flatworms Contain Stem Cells That Can Regenerate a Whole New Body

Schmidtea mediterranea is a small freshwater flatworm, or *planarian*, a centimeter or so long when grown to full size (Figure 22–20). It has an epidermis, a gut, a brain, a pair of primitive eyes, a peripheral nervous system, musculature, and excretory and reproductive organs—most of the basic body parts familiar in other animals, although all relatively simple by vertebrate standards and built from about 20-25 distinct differentiated cell types. For more than a century, planarians such as *Schmidtea* have intrigued biologists because of their extraordinary capacity for regeneration: a small tissue fragment taken from almost any part of the body will grow and regenerate a complete new animal. This property goes with another: when the animal is starved, it gets smaller and smaller, by reducing its cell numbers while maintaining essentially normal body proportions. This behavior is called *degrowth*, and it can continue until the animal is as little as one-twentieth of its full size. Supplied with food, it will grow back to full size again. Cycles of degrowth and growth can be repeated indefinitely, without impairing survival or fertility.

Underlying this behavior is a process of continual cell turnover. Along with the differentiated cells, which do not divide, there is a population of small, undifferentiated dividing cells called *neoblasts*. The neoblasts constitute about 20% of the cells in the body and are widely distributed within it; by cell division, they serve as stem cells for the production of new differentiated cells. The differentiated cells, meanwhile, are continually dying by apoptosis, allowing their corpses to be phagocytosed and digested by neighboring cells (discussed in Chapter 18). Through this cell cannibalism, the constituents of the dying cells can be efficiently recycled. Cell division continues in a dynamic balance with cell death and cell cannibalism, no matter whether the animal is fed or starved. In conditions of starvation, the balance is tilted toward cell death and cannibalism, and in conditions of plenty, toward cell growth and division.

A high dose of x-rays halts all cell division, putting a stop to cell turnover and destroying the capacity for regeneration. The result is death after a delay of several weeks. The animal can be rescued, however, by an injection into it of a single neoblast isolated from an unirradiated donor (Figure 22–21). In a certain proportion of cases, the injected cell divides to form a clone of progeny that eventually repopulates the entire body, creating a healthy regenerative individual with an apparently complete set of terminally differentiated cell types, as well as dividing neoblasts. A gradient of positional information regarding head-to-tail identity is continually expressed by the muscle cells along the worm's body and instructs the neuroblasts' appropriate development. Genetic markers prove that the differentiated cells are all derived from the single neoblast that was injected, suggesting that at least some neoblasts are *totipotent* stem cells, in that they are able to give

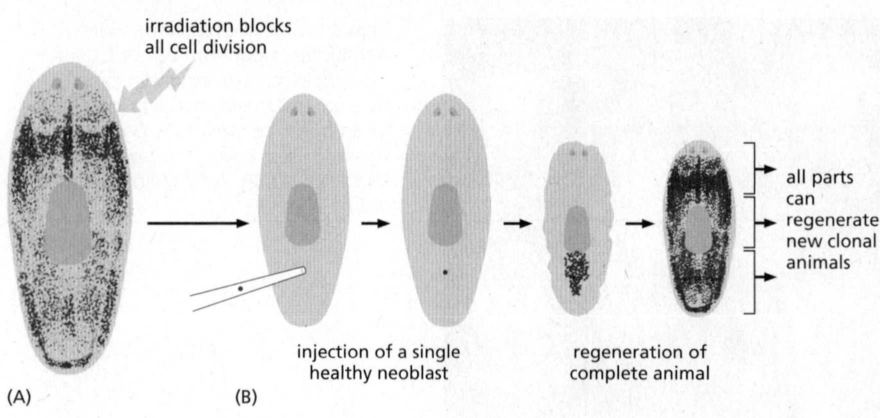

Figure 22–21 Regeneration of a planarian from a single somatic cell. (A) The distribution of dividing cells (neoblasts; *blue*) in the adult body. Irradiation blocks all cell division and prevents regeneration. (B) A single unirradiated neoblast cell injected into the irradiated animal is able to proliferate and reconstitute all tissues. This eventually produces a complete animal that consists entirely of the progeny of this one cell and can regenerate normally. A gradient of information concerning head-to-tail identity is continually expressed in muscle cells along the body and instructs the neuroblasts. (Adapted from E.M. Tanaka and P.W. Reddien, *Dev. Cell* 21:172–185, 2011.)

rise to all the cell types that make up the body of the flatworm, including more neoblasts like themselves.

Some Vertebrates Can Regenerate Entire Limbs and Organs

One might think that such powers of regeneration would be a prerogative of only small, simple, primitive animals. But some vertebrates, too, especially fish and amphibians, show remarkable regenerative abilities. Salamanders, including newts and axolotls, for example, can regenerate a whole or a part of an amputated limb, as well as many other body parts, including brain tissue and spinal cord (**Figure 22–22**). In the process of limb regeneration, a *blastema*—a small bud resembling an embryonic limb bud—forms at the site of amputation. The rapid lateral migration of adjacent epidermal cells serves to seal the wound site. The underlying blastema cells are mostly derived from activated stem cells and progenitor cells in the stump that are highly proliferative. Although the cells all look alike, they retain a memory of their tissue of origin and enough positional information to enable them to differentiate into the appropriate cell types and to form a correctly patterned replacement for the limb or the part of the limb that was amputated. The regenerative process looks like a recapitulation of embryonic limb development. As in the developing limb, all of the required cell behaviors are orchestrated by intercellular signals that are produced by the overlying epidermis, the ingrowing nerves, the proliferating lineage-restricted stem cells, and the transit-amplifying progenitor cells and their differentiated

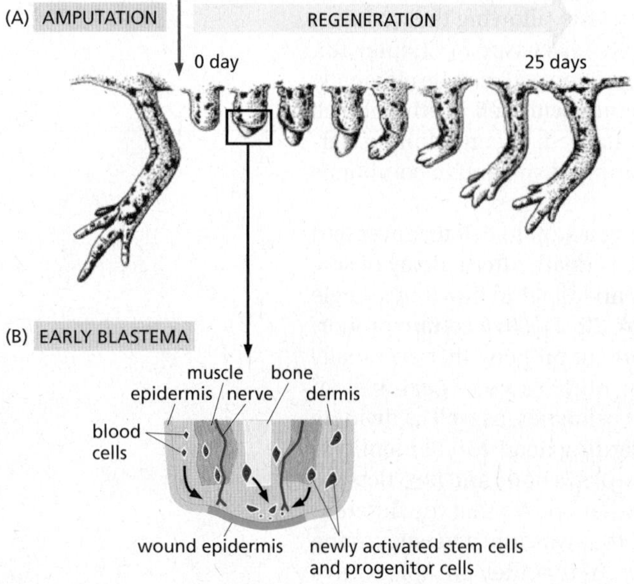

Figure 22–22 Newt limb regeneration. (A) The time-lapse sequence shows the stages of regeneration after amputation at the mid-humerus level. The sequence spans the events of wound healing, activation and proliferation of stump stem cells and progenitor cells in the process of blastema formation, and differentiation of various limb-cell types. (B) Schematic diagram of the early blastema. (A, courtesy of Susan Bryant and David Gardiner.)

progeny. Most of the signaling pathways involved are the same as those used in limb development, including those activated by Hedgehog, Wnt, and FGF family members.

Why a salamander can regenerate so many body parts whereas a mammal cannot remains a profound mystery.

Stem Cells Can Be Used Clinically to Replace Lost Hematopoietic or Skin Cells

Earlier in this chapter, we saw how mice can be irradiated to kill off their hematopoietic stem and progenitor cells and then rescued by a transfusion of new stem cells, which repopulate the bone marrow and restore blood-cell production (see Figure 22–10). In the same way, human individuals with some forms of leukemia or lymphoma can be irradiated or chemically treated to destroy their cancer cells, along with the rest of their hematopoietic tissue, and then can be rescued by a transfusion of healthy, noncancerous hematopoietic stem cells. In favorable cases, the stem cells can be sorted out from samples of an individual's own hematopoietic tissue before it is ablated; the stem cells are then transfused back afterward, thereby avoiding problems of immune rejection.

Another example of the use of stem cells is in the repair of the skin after extensive burns. By culturing cells from undamaged regions of the burned person's skin, it is possible to generate enough epidermal stem cells by long and complicated procedures to repopulate the damaged body surface.

Neural Stem Cells Can Be Manipulated in Culture and Used to Repopulate a Diseased Central Nervous System

The vertebrate central nervous system (the CNS) is the most complex tissue in the body, at an opposite extreme from the epidermis. And yet fish and amphibians can regenerate large parts of their brain, spinal cord, and eyes after these have been cut away. In adult mammals, however, these tissues have very little capacity for self-repair, and stem cells capable of generating new neurons are hard to find—so hard to find that for many years they were thought to be absent.

We now know, however, that neural stem cells that generate neurons, glial cells, or both, do persist in certain parts of the adult mammalian brain, as shown in **Figure 22–23** for stem cells that give rise to neurons in the mouse olfactory bulb. In both the adult mouse and human brain, there is also a continual turnover of neurons in the hippocampus, a region specifically concerned with learning and memory. Here, plasticity of adult brain function is associated with turnover of a particular subset of neurons: about 1400 new neurons in this class are generated every day, which is a turnover rate of 1.75% of the cells per year.

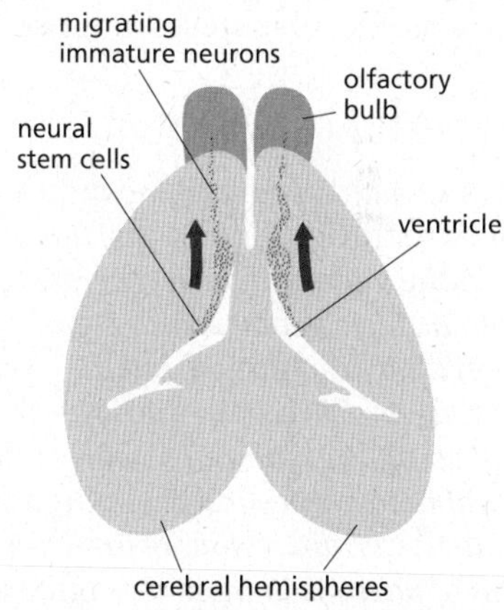

Figure 22–23 The continuing production of neurons in an adult mouse brain. The brain is viewed from above, in a cutaway section, to show the region lining the ventricles of the forebrain where neural stem cells are found. These cells continually produce progeny that migrate to the olfactory bulb, where they terminally differentiate into neurons. The constant turnover of neurons in the olfactory bulb is presumably linked in some way to the turnover of the olfactory receptor neurons that project to it from the olfactory epithelium in the nose. In adult mice and humans, there is also a continuing turnover of neurons in the hippocampus (not shown), a region specially concerned with learning and memory. (Adapted from B. Barres, *Cell* 97:667–670, 1999.)

A more dramatic neuronal turnover is observed in the brains of certain songbirds. In these birds, large numbers of neurons die each year and are replaced by new neurons, as part of a process by which the birds refine their song for each new breeding season.

Brain neural stem cells can be studied in culture. Fragments taken from self-renewing regions of the adult or fetal mammalian brain, for example, can be dissociated and cultured under conditions where they form floating "neurospheres"—clusters consisting of a mixture of neural stem cells and their neuronal and glial progeny cells. The neurospheres can be propagated through many cell generations, and their cells can be taken at any time and implanted back into the brain of an intact animal, where they will produce differentiated neurons and glial cells.

Using slightly different culture conditions, with the right combination of growth factors in the culture medium, the neural stem cells can be grown as dissociated cells in a culture dish and induced to proliferate as an almost pure stem-cell population without attendant differentiated progeny. By a further change in the culture conditions, these cells can be induced at any time to differentiate to give

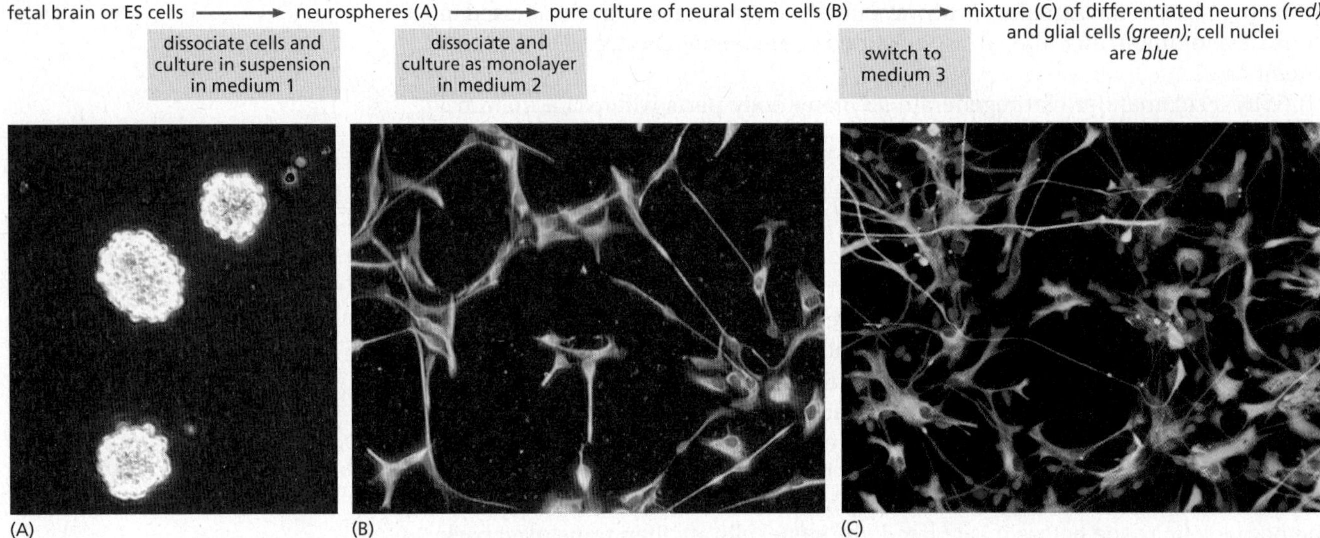

Figure 22–24 **Neural stem cells.** Shown are the steps leading from fetal brain tissue or embryonic stem cells, via neurospheres (A), to a pure culture of neural stem cells (B). These stem cells can be kept proliferating as such indefinitely or, through a change of medium, can be caused to differentiate (C) into neurons *(red)* and glial cells *(green)*. Neural stem cells with the same properties can also be derived, via a similar series of steps, from embryonic stem (ES) or induced pluripotent stem (iPS) cells (discussed later in this chapter). (Micrographs from L. Conti et al., *PLoS Biol.* 3:e283, 2005. With permission from the authors.)

either a mixture of neurons and glial cells (**Figure 22–24**) or just one of these two classes of cells, according to the composition of the medium.

Neural stem cells, whether derived as above or from the pluripotent stem cells that we describe in the final section of this chapter, can be grafted into an adult brain. Once there, they show a remarkable ability to adjust their behavior to match their new location. Stem cells from the mouse hippocampus, for example, when implanted into the migration pathway of the mouse olfactory-bulb precursor cells (see Figure 22–23), give rise to neurons that become correctly incorporated into the olfactory bulb. This capacity of neural stem cells and their progeny to adapt to a new environment in animals suggests applications in the treatment for central nervous system diseases or injury.

Summary

Animals vary in their capacity for regeneration. At one extreme, planarian flatworms contain stem cells (neoblasts) that support the continual turnover of all cell types, and an entire worm can be regenerated from practically any small body fragment or even from a single neoblast cell. Salamanders can regenerate limbs and other large body parts after amputation, but the regenerating cells remain restricted according to their origins: muscle cells in the regenerate derive from muscle, epidermis from epidermis, and so on. In mammals, regeneration is more limited. Nevertheless, it is becoming possible to go beyond the natural limits of wound healing by exploiting stem-cell biology. Certain regions of the nervous system contain stem cells that support production of neurons in these sites throughout life. Neural stem cells can be obtained from these sites or from fetal brains, grown in culture, and then grafted back into other sites in the brain, where they are able to generate neurons appropriate to the new location.

CELL REPROGRAMMING AND PLURIPOTENT STEM CELLS

When cells are transplanted from one site in the mammalian body to another or are removed from the body and maintained in culture, they remain largely faithful to their origins. Each type of specialized cell has a memory of its developmental history and seems fixed in its fate. Some limited transformations can certainly occur, and, as we have seen, some stem cells can generate a variety of differentiated cell types, but the possibilities are restricted. Each type of adult stem cell serves for the renewal of one particular type of tissue, and the whole pattern of self-renewing stem cells to differentiated cells in the adult body is amazingly stable. What, at a fundamental molecular level, is the nature of these stable differences between cell types and cell states? Is there any way to override the cell-memory mechanisms and force a switch from one state to another that is radically different?

We have already discussed these fundamental questions from a general standpoint in Chapter 7. Here we consider them more closely in the context of stem-cell biology, where there has been a recent revolution in our understanding and in our ability to manipulate states of cell differentiation. With further research, these advances will likely have many important medical applications.

Nuclei Can Be Reprogrammed by Transplantation into Foreign Cytoplasm

If we cannot switch the basic character of a specialized cell by changing its environment, can we do so by interfering with its inner workings in a more direct and drastic way? An extreme treatment of this sort is to take the nucleus of the cell and transplant it into the cytoplasm of a large cell of a different type. If the factors that define and maintain a particular cell type are in the cytoplasm, the transplanted nucleus should switch its pattern of gene expression to conform with that of the host cell. In Chapter 7, we described a famous experiment of this sort, using the frog *Xenopus*. In this experiment, the nucleus of a differentiated cell (a cell from the lining of a tadpole's gut) was used to replace the nucleus of an oocyte (an egg-cell precursor arrested in prophase of the first meiotic division, in readiness for fertilization). The resulting hybrid cell went on, in a certain fraction of cases, to develop into a completely normal frog (see Figure 7–2A). This was crucial evidence for what is now a central principle of developmental biology: the cell nucleus, even that of a differentiated cell, contains a complete genome, capable of supporting development of all the normal cell types of the organism. At the same time, the experiment showed that cytoplasmic factors can indeed reprogram a nucleus: the oocyte cytoplasm can drive the gut-cell nucleus back to an early embryonic state, from which it can then step through the changing patterns of gene expression that lead all the way to a complete adult organism.

The full story, however, is not quite so simple. First, the reprogramming in such experiments is not perfect. When the transplanted nucleus is taken from a gut cell, for example, a gene that is normally specific to the gut is found to be expressed persistently, even in the muscle cells of the final animal. Second, the experiment succeeds in only a limited proportion of cases, and this success rate becomes lower and lower, the more mature the animal from which the transplanted nucleus is taken: very large numbers of transplantations must be performed to achieve a single success if the nucleus comes from a differentiated cell of an adult frog.

Nuclear transplantation can be done in mammals too, with basically similar results. For example, a nucleus taken from a differentiated cell in the mammary gland of an adult sheep and transplanted into an enucleated sheep's egg was able to support development of an apparently normal sheep—the famous Dolly. Again, the success rate is low: many transplantations have to be done to obtain one such individual.

Reprogramming of a Transplanted Nucleus Involves Drastic Changes in Chromatin

In a typical fully differentiated cell, there seem to be mechanisms maintaining the pattern of gene expression that cytoplasmic factors cannot easily override. An obvious possibility is that the stability of the pattern of gene expression in an adult cell may depend, in part at least, on self-perpetuating modifications of chromatin, as discussed in Chapter 4 (see Figure 4–44). As explained in Chapter 7, the phenomenon of X-inactivation in mammals provides a clear example of such epigenetic control. Two X chromosomes are present in each female cell, exposed to the same chemical environment, but while one remains active, the other persists from one cell generation to the next in a condensed inactive state; cytoplasmic factors cannot be responsible for the difference, which must instead reflect mechanisms intrinsic to the individual chromosome. Elsewhere in the genome also, controls at the level of chromatin act in combination with other forms of regulation to govern the expression of each gene. Genes can be shut down completely, or switched on constitutively, or maintained in a labile state where they can be readily switched on or off according to changing circumstances.

The reprogramming of a nucleus transplanted into an oocyte involves dramatic changes in chromatin. The nucleus swells, increasing its volume 50-fold as the chromosomes decondense; there is a wholesale alteration in patterns of methylation of DNA and histones; the linker histone H1 (the histone that links adjacent nucleosomes) is replaced by a variant form that is peculiar to the oocyte and early embryo; and the preexisting type of histone H3 is also replaced at many sites by a distinct isoform. Evidently, the egg contains factors that reset the state of the chromatin in the nucleus, wiping out old histone modifications on chromatin and imposing new ones. Reprogrammed in this way, the genome becomes competent once again to initiate embryonic development and to give rise to the full range of differentiated cell types.

Embryonic Stem (ES) Cells Can Generate Any Part of the Body

A fertilized egg, or an equivalent cell produced by nuclear transplantation, is a remarkable thing: it can generate a whole new multicellular individual, which means that it can give rise to every normal type of specialized cell, including even egg or sperm cells for production of the next generation. A cell in such a state is said to be **totipotent**; a cell that can give rise to most cell types but not absolutely all is said to be **pluripotent**. Nevertheless, such a totipotent or pluripotent cell is not a stem cell as it is not self-renewing, but is instead dedicated to a program of progressive differentiation. If it were the only available starting point for study and exploitation of pluripotent cells, the enterprise would require a continual supply of fresh fertilized eggs or fresh nuclear transplantation procedures—an awkward requirement for studies in experimental animals, and unacceptable for practical applications in humans.

Here, however, nature has been unexpectedly kind to scientists. It is possible to take an early mouse embryo, at the blastocyst stage, and through cell culture to derive from it a class of stem cells called **embryonic stem cells**, or **ES cells**. ES cells originate from the inner cell mass of the early embryo (the cluster of cells that give rise to the body of the embryo proper, as opposed to extraembryonic structures), and they have an extraordinary property: given suitable culture conditions, they will continue proliferating indefinitely and yet retain an unrestricted developmental potential. Their only limitation is that they do not give rise to extraembryonic tissues such as those of the placenta. Thus they are classified as pluripotent, rather than totipotent. But this is a minor restriction. If ES cells are put back into a blastocyst, they become incorporated into the embryo and can give rise to all the tissues and cell types in the body, integrating perfectly into whatever site they may come to occupy, and adopting the character and behavior that normal cells would show at that site (Figure 22–25). They can even give rise to germ cells, from which a new generation of animals can be derived.

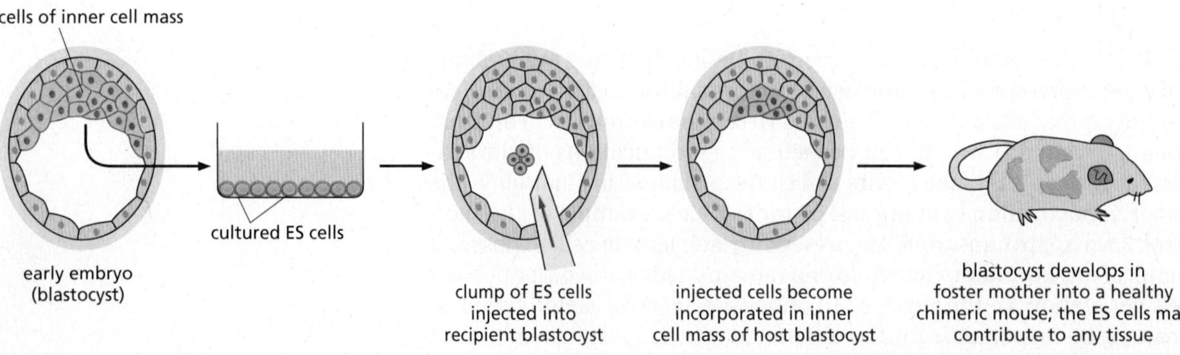

Figure 22–25 **Production and pluripotency of ES cells.** ES cells are derived from the inner cell mass (ICM) of the early embryo. The ICM cells are transferred to a culture dish containing an appropriate medium, where they become converted to ES cells and can be kept proliferating indefinitely without differentiating. The ES cells can be taken at any time—after genetic manipulation, if desired—and injected back into a developing blastocyst. There they incorporate into the inner cell mass and take part in formation of a well-formed chimeric animal that is a mixture of ordinary and ES-derived cells. The ES-derived cells can differentiate into any of the cell types in the body, including germ cells from which a new generation of mice can be produced, which are no longer chimeric, but consist of cells that all inherit half their genes from the cultured ES cell line.

ES cells let us move between cell culture, where we can use powerful techniques for genetic transformation and selection, and the intact organism, where we can discover how such genetic manipulations affect development and physiology. Thus, ES cells opened the way to efficient genetic engineering in mammals, leading to a revolution in our understanding of mammalian molecular and developmental biology.

Cells with properties similar to those of mouse ES cells can also be derived from early human embryos and from human fetal germ cells, and even, as we explain shortly, from differentiated cells taken from adult mammalian tissues. In this way, one can obtain a potentially inexhaustible supply of pluripotent cells. Grown in culture, these cells can be manipulated, by suitable choice of culture conditions, to give rise to large quantities of almost any type of differentiated cell, opening the way to many practical applications. Before discussing them, however, we consider the underlying biology.

A Core Set of Transcription Regulators Defines and Maintains the ES-Cell State

What is it that gives ES cells and related types of pluripotent stem cells their extraordinary developmental potential? And what can they tell us about the fundamental mechanisms underlying stemness, commitment to differentiation, and the stability of the differentiated state?

For some ES-cell attributes, the answer is simple. For example, an essential feature of ES cells is that they must avoid *replicative cell senescence.* As discussed in Chapter 17, this is the fate of fibroblasts and many other types of proliferating somatic cells: such cells are limited in the number of times they will divide, in part at least because they lack telomerase activity, with the result that their telomeres become shorter with each division cycle, leading eventually to a permanent cell-cycle arrest. ES cells, by contrast, express high levels of active telomerase, allowing them to escape replicative cell senescence and continue to divide indefinitely. This is a property shared with other, more developmentally restricted types of stem cells, such as those of the adult intestine, which similarly can carry on dividing for hundreds or thousands of cell cycles.

The deeper problem is to explain how the whole complex pattern of gene expression in an ES cell is organized and maintained. As a first step, one can look for genes expressed specifically in ES cells or in the corresponding pluripotent cells of the early embryo. This approach identifies a relatively small number of candidate *ES-critical genes*; that is, genes that seem to be essential in one way or another for the peculiar character of ES cells. A gene called *Oct4*, for example, is exclusively expressed in ES cells and in related classes of cells in the intact organism—specifically, in the germ-cell lineage and in the inner cell mass and its precursors. *Oct4* codes for a transcription regulator. When it is lost from ES cells, they lose their ES-cell character, and when it is missing in an embryo, the cells that should specialize as inner cell mass are diverted into an extraembryonic pathway of differentiation, and the embryo's development is aborted.

Fibroblasts Can Be Reprogrammed to Create Induced Pluripotent Stem (iPS) Cells

In Chapter 7, we saw that fibroblasts and some other cell types can be driven to switch their character and differentiate as muscle cells if the master muscle-specific transcription regulator MyoD is artificially expressed in them. Could the same technique be used to convert fibroblasts and other cell types into ES cells, through forced expression of ES-critical genes such as *Oct4*? This question was tackled by transfecting mouse fibroblasts with retroviral vectors carrying genes that one might hope to have such an effect. A total of 24 candidate ES-critical genes were tested in this way. None of them was able by itself to cause the conversion, but in certain combinations they could do so. In 2006, the first breakthrough experiments whittled down the requirement to a core set of four genes, all of them encoding transcription regulators—Oct4, Sox2, Klf4, and Myc, known as the OSKM factors for short. When co-expressed, these could reprogram mouse fibroblasts, permanently converting them into cells very similar to ES cells (**Figure 22–26**). ES-like cells created in this way are called **induced pluripotent stem cells**, or **iPS cells**. Like ES cells, iPS cells can continue dividing indefinitely in culture, and when incorporated into a mouse blastocyst they can participate in creation of a perfectly formed chimeric animal. In this animal, they can contribute to the development of any tissue and can turn into any differentiated cell type, including functional germ cells from which a new generation of mice can be raised (see Figure 22–25).

iPS cells can now be derived from adult human cells, including from various differentiated cell types besides fibroblasts. Numerous methods can be used to drive expression of the transforming OSKM factors, including methods that leave no trace of foreign DNA in the reprogrammed cells. Variations of the original cocktail of transcription regulators can drive the conversion, with different specialized cell types having somewhat different requirements. Myc overexpression, for example, turns out not to be absolutely necessary, although it enhances the efficiency of the process. And differentiated cell types may express some of the required factors as part of their normal phenotype. For example, certain cells of hair follicles already express Sox2, Klf4, and Myc; to convert them into iPS cells, it is enough to force them artificially to express Oct4.

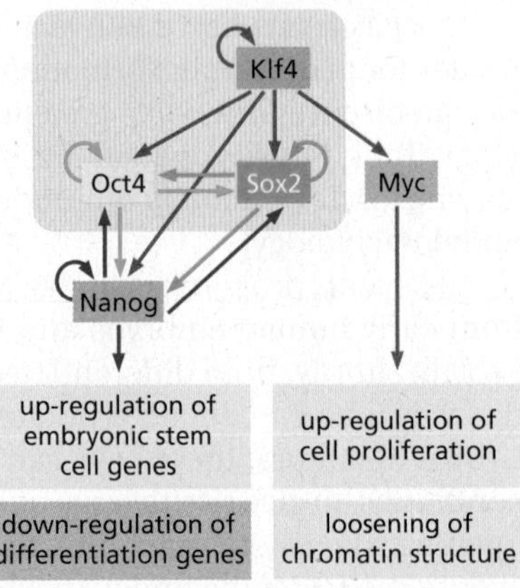

Figure 22–26 Reprogramming fibroblasts to iPS cells with the OSKM factors. As indicated, the transcription regulatory proteins Oct4, Sox2, and Klf4 (OSK factors) induce both their own and each other's synthesis *(gray shading)*. This generates a self-sustaining feedback loop that helps to maintain cells in an ES cell–like state, even after all of the experimentally added OSKM initiators have been removed. Myc overexpression speeds up early stages of the reprogramming process through the mechanisms shown (see Figure 17–59). Stable reprogramming also involves the permanently induced expression of the *Nanog* gene, which encodes an additional transcription regulator (see Figure 7–10). (Adapted from J. Kim et al., *Cell* 132: 1049–1061, 2008.)

Reprogramming Involves a Massive Upheaval of the Gene Control System

Converting a differentiated cell into an iPS cell is not like flicking a switch on some predictable, precisely engineered piece of machinery. Only a few of the cells that receive the OSKM factors will actually become iPS cells—one in several thousand in the original experiments, and still only a small minority with more recent, improved techniques. In fact, the success of the original experiments depended on clever selection strategies to pick out those few cells where the conversion had occurred (**Figure 22–27**).

Conversion to an iPS state by the OSKM factors is not only inefficient but also slow: fibroblasts take 10 days or more from introduction of the conversion factors before they begin to express markers characteristic of iPS cells. This suggests the transformation involves a long cascade of changes. These changes have been extensively studied, and they affect both the expression of individual genes and the state of the chromatin. The time course is outlined in **Figure 22–28**. The process begins with a Myc-induced cell proliferation and loosening of chromatin structure that promotes the binding of the other three transcription regulators to many hundreds of different sites in the genome. At a large proportion of these sites, Oct4, Sox2, and Klf4 all bind in concert. The binding sites include the endogenous *Oct4, Sox2,* and *Klf4* genes themselves, which eventually creates positive feedback loops like those just described, making expression of these genes self-sustaining (see Figure 22–26). But self-induction

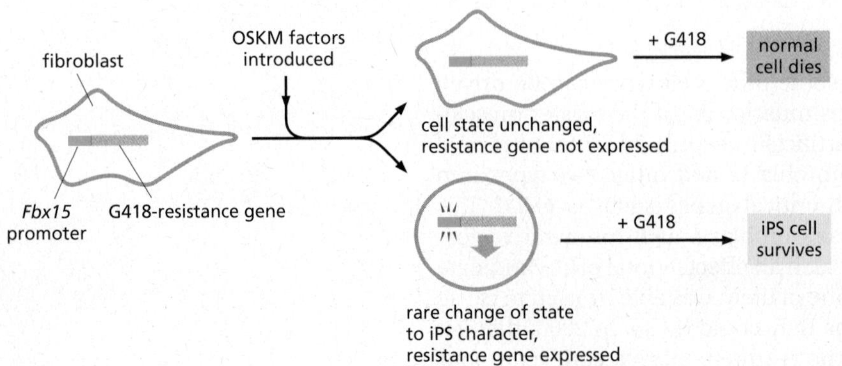

Figure 22–27 A strategy used to select cells that have converted into iPS cells. The experiment makes use of a gene (*Fbx15*) that is present in all cells but is normally expressed only in ES and early embryonic cells (although not required for their survival). G418 is an aminoglycoside antibiotic that blocks protein synthesis in both bacteria and eukaryotic cells. A fibroblast cell line is genetically engineered to contain a gene that produces an enzyme that degrades G418 under the control of the *Fbx15* regulatory sequence. When the OSKM factors are artificially expressed in this cell line, a small proportion of the cells undergo a change of state and activate the *Fbx15* regulatory sequence, driving expression of the G418-resistance gene. When G418 is added to the culture medium, these are the only cells that survive and proliferate. When tested, they turn out to have iPS-cell characteristics.

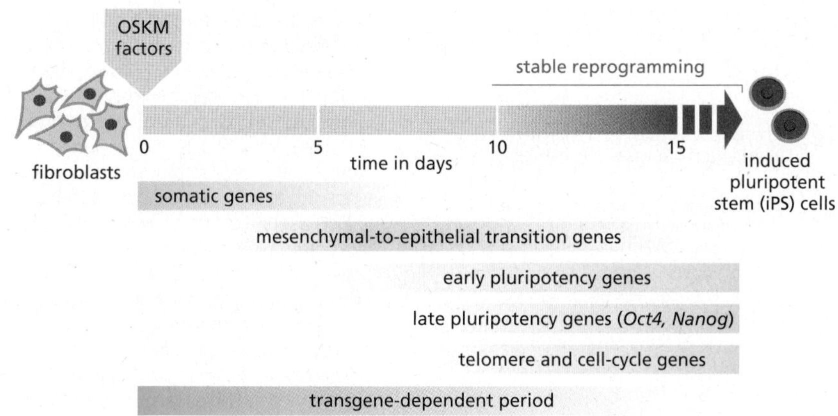

Figure 22–28 **A summary of some of the major events that accompany the reprogramming of fibroblasts to iPS cells.** Expression of the OSKM factors induces a series of events over a period of days to weeks. Reprogramming begins with down-regulation of somatic-cell markers. Morphological changes characteristic of a *mesenchymal-to-epithelial transition* ensue (see p. 1197), driven in part by changes in the expression of cell adhesion and signal proteins. Induction of early pluripotency markers is followed by the expression of pluripotency genes such as *Nanog* and *Oct4*. With similar timing, cells become immortalized as telomerase is induced and cell-cycle genes are regulated to enable stem-cell self-renewal. Stable reprogramming occurs in the time window when cells activate endogenous pluripotency genes and become independent of the OSKM factors. Notably, the vast majority of fibroblasts expressing the OSKM factors fail to down-regulate somatic markers and activate pluripotency genes and do not convert to iPS cells. (Adapted from M. Stadtfeld et al., *Cell Stem Cell* 2:230–240, 2008. With permission from Elsevier.)

of *Oct4*, *Sox2*, and *Klf4* is only a small part of the transformation that occurs. These three core factors activate some target genes and repress others, producing a cascade of effects that reorganize the gene control system globally and at every level, changing the patterns of histone modification, DNA methylation, and chromatin compaction, as well as the expression of innumerable proteins and noncoding RNAs. By the end of this complex process, the resulting iPS cell is no longer dependent on the artificially generated OSKM factors that triggered the change: it has settled into a stable, self-sustaining state of coordinated gene expression, making its own OSKM factors (and all the other essential ingredients of a pluripotent stem cell) from its own endogenous copies of the genes.

An Experimental Manipulation of Factors That Modify Chromatin Can Increase Reprogramming Efficiencies

The low efficiency and slow rate of conversion in early iPS-cell studies suggested that there are barriers that normally block the switch from the differentiated state to the iPS state in these experiments, and that overcoming them can be a difficult process that involves a large element of chance. This probably helps explain why the outcome is often variable, with significant differences between the individual iPS cell lines generated, even when the initial differentiated cells are genetically and phenotypically identical. Only some of the candidate iPS lines produced have passed all the tests of pluripotency. Moreover, at a molecular level, there are differences even among the fully validated iPS lines: although they share many features, they vary in details of their gene expression patterns and, for example, in their patterns of DNA methylation.

Overcoming these difficulties is critical for improving our understanding of how cell specialization is controlled and organized in multicellular organisms; it should also facilitate many medical advances. Thus, intensive research is being carried out on the reprogramming process. One approach aims at obtaining a much clearer picture of the role that chromatin structures play in gene regulation in eukaryotes.

From our discussion of nuclear transplantation, one might expect that any reprogramming of a differentiated cell would require a radical and widespread change in the chromatin structure associated with selected genes. Not only are such changes observed, but a large number of different experiments reveal that the efficiency of the reprogramming process can be substantially increased by altering the activity of proteins that affect chromatin structure. Figure 22–29 categorizes some of the factors that when manipulated can enhance the transformation of fibroblasts to iPS cells; those in the top three rows—chromatin remodelers, histone modifiers, and histone variants—are especially well known to have profound effects on the organization of nucleosomes in chromatin (discussed in Chapter 4).

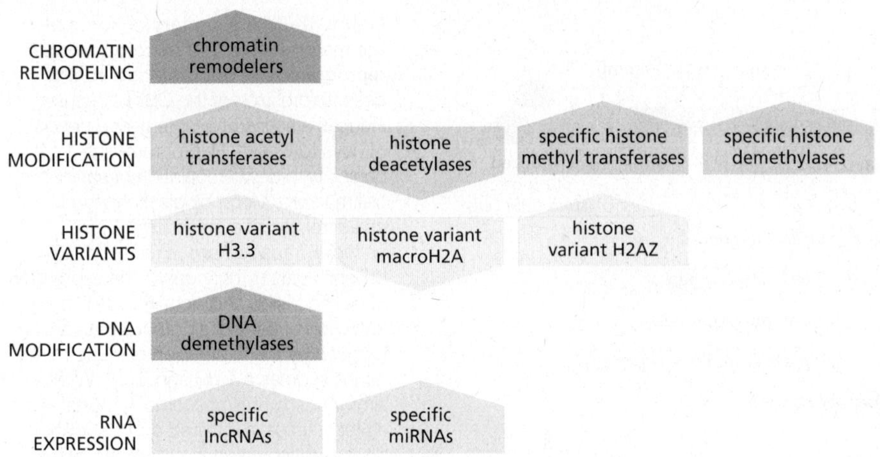

Figure 22–29 **Factors that have been observed to enhance reprogramming efficiency.** Emphasized here are those factors that can alter chromatin states, with those in the top three rows having the most direct effects. An *up arrow* indicates that reprogramming is increased when the activity of the indicated factor is increased; a *down arrow* indicates that reprogramming is increased when the activity of the indicated factor is decreased. Thus, for example, increased activity of histone acetyl transferases and increased activity of histone deacetylases have opposite effects, as expected from their biochemical activities (see p. 206). Note that histone chaperones also contribute to programming efficiency and may do so either positively or negatively depending on the histone variants assembled into chromatin.

ES and iPS Cells Can Be Guided to Generate Specific Adult Cell Types and Even Organoids

We can think of embryonic development in terms of a series of choices presented to cells as they follow a road that leads from the fertilized egg to terminal differentiation. After their long sojourn in culture, the ES cells or iPS cells and their progeny can still read the signs at each branch in the highway and respond as normal early embryonic cells would. If ES or iPS cells are implanted directly into an embryo at a later stage of development or into an adult tissue, however, they fail to receive the appropriate sequence of cues; their differentiation then is not properly controlled, and they will often give rise to a tumor called a *teratoma*, containing a mixture of cell types inappropriate to the site in the body (**Figure 22–30**).

By exposing ES or iPS cells in culture to an appropriate sequence of growth factors and other signal proteins, delivered with the right timing, it is possible to guide the cells along a pathway that approximates a normal developmental pathway and convert them into standard, differentiated, adult cell types (**Figure 22–31** and **Movie 22.2**). Success requires trial and error but has now been achieved for many differentiated mouse and human cell types, including human dopaminergic neurons and human insulin-producing β cells, the cells lost in Parkinson's disease and type 1 diabetes, respectively.

Remarkably, under appropriate conditions, mouse or human ES and iPS cells and their progeny can proliferate, differentiate, and self-assemble in culture to form miniature, three-dimensional organs called **organoids**, which closely resemble the normal organ in its organization. An early striking example is shown in **Figure 22–32**, where a developing eye organoid is formed from human ES cells.

Mouse and human ES and iPS cells, and progenitor cells derived from them, have been used to form organoids that resemble a large variety of developing organs, including important parts of the human brain, arguably the most complex and sophisticated structure on Earth. Such organoids provide powerful models for studying organ development in a culture dish, where one can identify and study the genes involved and explore the roles of cell–cell interactions in ways not possible in the intact organism. We discussed earlier the remarkable demonstration that a single multipotent intestinal stem cell can form a complex intestinal organoid under appropriate conditions in culture (see Figure 22–14).

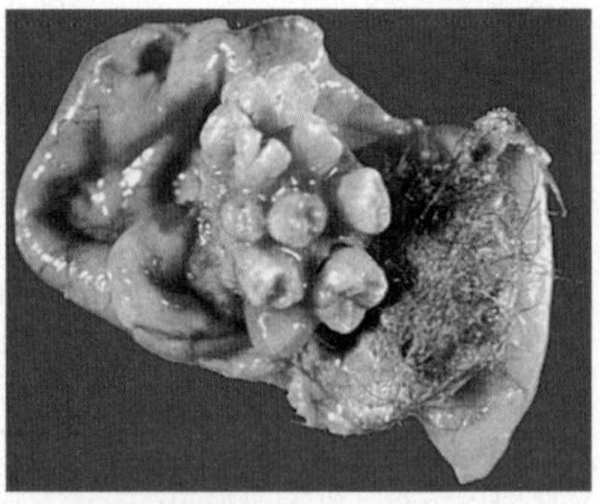

Figure 22–30 **A teratoma.** These tumors contain tissues derived from each of the embryonic germ layers: endoderm, mesoderm, and ectoderm (including hair and teeth, as seen here). The tumors are usually benign and can develop when ES or iPS cells are injected into an immunocompatible adult animal, demonstrating the pluripotency of these stem cells. (Courtesy of Cao Xuan Cu.)

Cells of One Specialized Type Can Be Forced to Transdifferentiate Directly into Another

The route we have just described, from one type of differentiated cell to another type of differentiated cell via conversion to an iPS cell, seems needlessly roundabout. Could we not convert differentiated cell type A into differentiated cell type B directly, without backtracking to a pluripotent stem cell? For many years, it has been known that such *transdifferentiation* can be achieved in a few special cases,

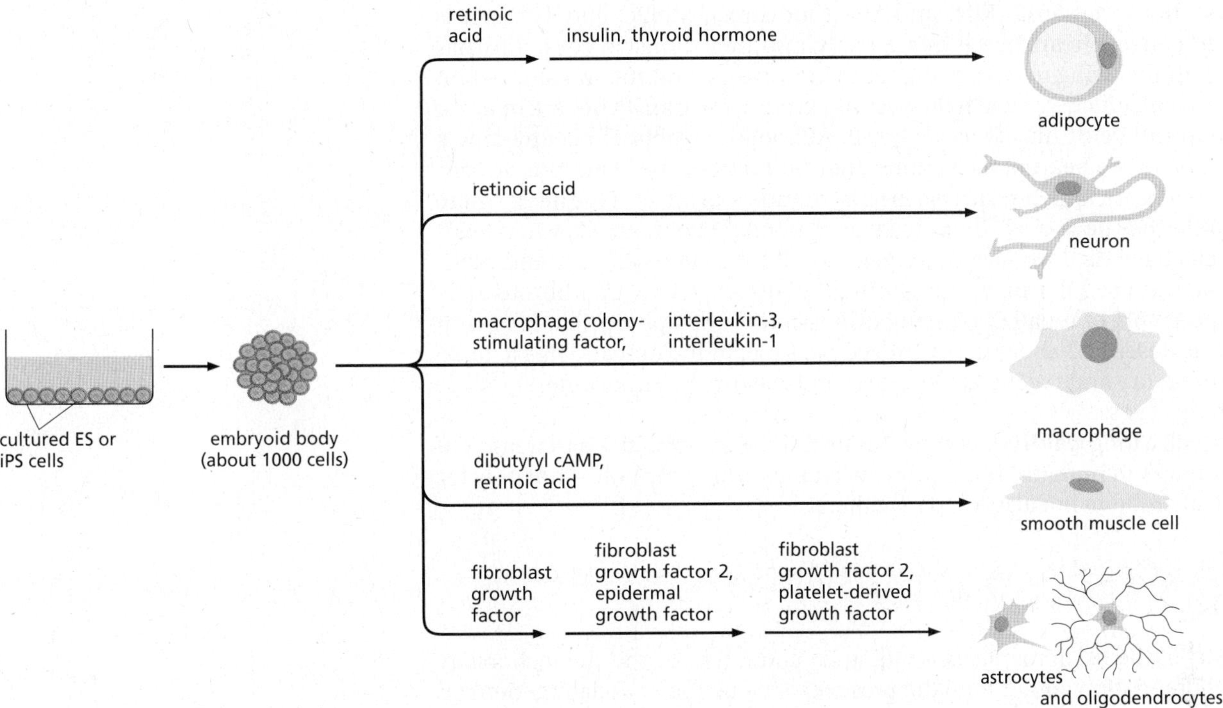

Figure 22–31 **Production of differentiated cells from mouse or human ES or iPS cells in culture.** ES and iPS cells can be cultured indefinitely as pluripotent cells when attached as a monolayer to a dish. Alternatively they can be detached and allowed to form aggregates called *embryoid bodies*, which causes the cells to begin to specialize. Cells from embryoid bodies, cultured in media with different factors added, can then be driven to differentiate in various ways. Notably, it takes very much longer to generate the differentiated cells from human compared to mouse pluripotent stem cells, reflecting the much slower rate of human development compared to a mouse. (Based on E. Fuchs and J.A. Segre, *Cell* 100:143–155, 2000.)

such as the conversion of fibroblasts into skeletal muscle cells by forced expression of MyoD (see p. 428). But now, with the insights that have come from the study of ES and iPS cells, ways have been found to bring about such interconversions in a much wider range of cases, including conversion of fibroblasts into neurons, hepatocytes, and intestinal epithelial cells.

An elegant example with special medical relevance comes from studies of the heart. By forcing expression of an appropriate combination of transcription

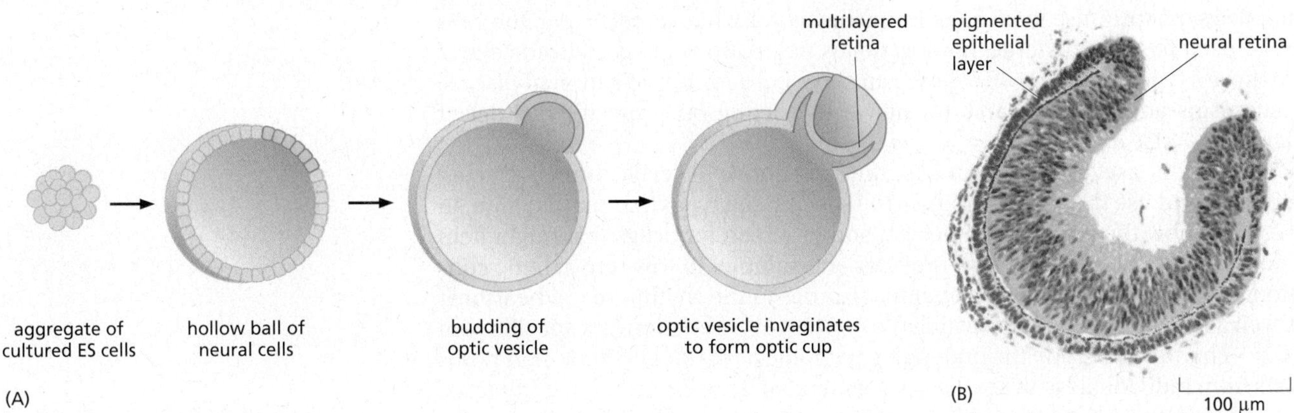

Figure 22–32 **Cultured human ES cells can give rise to a three-dimensional organoid.** (A) Schematic drawing shows how, under appropriate conditions, mouse or human pluripotent stem cells and their progeny in culture can proliferate, differentiate, and self-assemble into a three-dimensional eye-like structure (an optic cup), which includes a multilayered retina similar in organization to the one that forms during normal eye development *in vivo*. (B) Fluorescence micrograph of an optic cup formed by human ES cells in culture. The structure includes a developing retina containing multiple layers of neural cells (stained *green*) and an underlying layer of pigmented epithelium, the apical surface of which is stained *red*. All nuclei are stained *blue*. (A, adapted from M. Eiraku and Y. Sasai, *Curr. Opin. Neurobiol.* 22:768–777, 2012; B, from T. Nakano et al., *Cell Stem Cell* 10:771–785, 2012. With permission from Elsevier.)

regulators—not Oct4, Sox2, Klf4, and Myc, but Gata4, Mef2c, and Tbx5—it is possible to convert heart fibroblasts directly into heart muscle cells. This has been done in the living mouse, using retroviral vectors, and the transformation occurs with high efficiency when the vectors carrying the transgenes are injected directly into the heart muscle tissue itself. Although fibroblasts occupy only a small fraction of the heart tissue volume, they outnumber the heart muscle cells in the normal heart, and they survive in large numbers in areas of the heart where heart muscle cells have died. Thus, in a typical nonfatal heart attack, when heart muscle cells have died for lack of oxygen, the fibroblasts proliferate and make collagenous extracellular matrix, replacing the lost muscle with a fibrous scar. This is a poor sort of repair. By forcing expression of the appropriate factors in the heart, as described above, it has proved possible, in the mouse at least, to do better than nature and regenerate lost heart muscle through transdifferentiation of heart fibroblasts.

We are still a long way from putting this technique into practice as a treatment for heart attacks in humans, but it shows what the future may hold—not only for this medical problem but also for many others.

ES and iPS Cells Are Also Useful for Drug Discovery and Analysis of Disease

A large part of the excitement surrounding ES and iPS cells and the technology of transdifferentiation comes from the prospect of using the artificially generated cells for tissue repair. It begins to seem that virtually any type of tissue might be replaceable, allowing treatment of degenerative diseases that have previously had no cure, other than by organ transplantation. Research in this area is moving rapidly, but there are many difficulties to be overcome.

With the advent of iPS cells and direct transdifferentiation, at least one major hurdle that has challenged organ transplantation has been surmounted, in principle at least: the problem of immune rejection. ES cells, because they are created from early embryos that generally come from unrelated donors, will never be genetically identical to the cells of the individual receiving the cell transplant. The transplanted cells and their progeny are therefore prone to rejection by the immune system. Both iPS and transdifferentiated cells, in contrast, can be generated from a small sample of the individual's own tissue and so should escape immune attack when transplanted back into the same individual.

Tissue repair by cell and organoid transplantation, however, is not the only application for which ES, iPS, and transdifferentiated cells can be used: there are other ways that promise to be more immediately valuable. In particular, the cells can be used to generate large, homogeneous populations of specialized cells of any chosen type in culture, and these can serve both for investigation of disease mechanisms and for the search for new drugs acting on a specific cell defect (**Figure 22–33**).

Where a disease has a genetic cause, one can derive iPS cells from an affected individual and use the cells to produce the specific cell types that malfunction, to investigate how the malfunction occurs, and to screen for drugs that might help to put it right. *Timothy syndrome* provides an example. In this rare genetic condition, there is a severe, life-threatening disorder in the rhythm of the heartbeat (as well as several other abnormalities) as a result of a mutation in a specific type of Ca^{2+} channel. To study the underlying pathology, researchers took skin fibroblasts from individuals with the disorder, generated iPS cells from the fibroblasts, and induced the iPS cells to differentiate into heart muscle cells. These cells, when compared with heart muscle cells prepared similarly from normal control individuals, showed irregular heart muscle contractions and abnormal patterns of Ca^{2+} influx and electrical activity that could be characterized in detail. From these findings, it is a small step to development of an *in vitro* assay for drugs that might correct the misbehavior of the heart muscle cells.

Even in human disorders without a known genetic cause, iPS cells can be useful in understanding the condition. In one example, brain organoids were

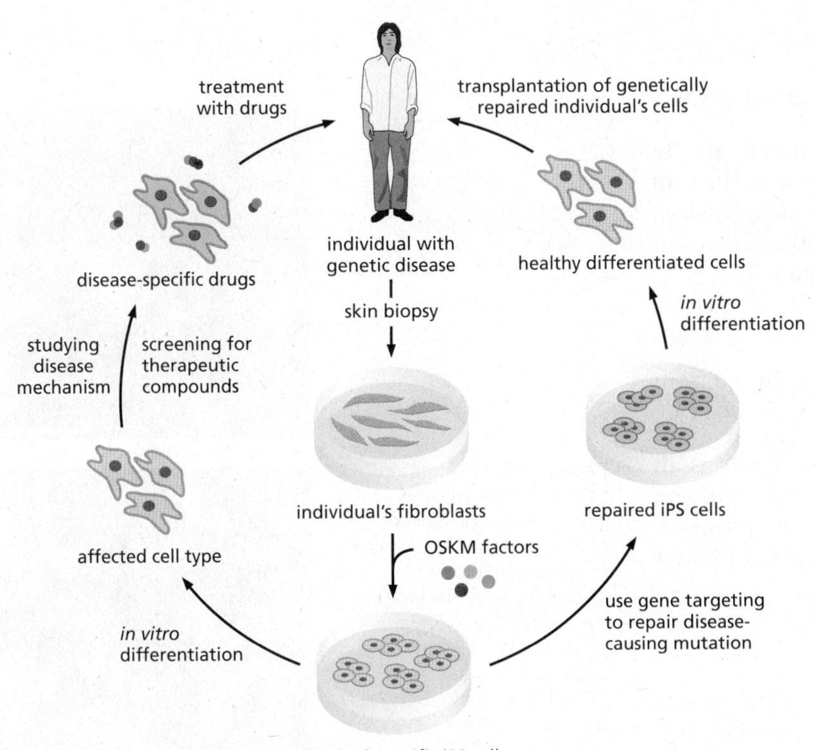

Figure 22–33 **Use of iPS cells for drug discovery and for analysis and treatment of genetic disease.** The left side of the diagram shows how differentiated cells that are generated from iPS cells derived from an individual with a genetic disease can be used for analysis of the disease mechanism and for discovery of therapeutic drugs. The right side of the diagram shows how the genetic defect might be repaired in the iPS cells, which could then be induced to differentiate in an appropriate way and grafted back into the individual without danger of immune rejection. (Adapted from D.A. Robinton and G.Q. Daley, *Nature* 481:295–305. Reproduced with permission of SNCSC.)

produced using iPS cells from an individual with microcephaly, a condition characterized by severely stunted brain growth and development. Careful analysis of these developing brain organoids revealed that the microcephaly in this case seemed to be caused by the premature cessation of proliferation and differentiation of brain progenitor cells, resulting in abnormally small numbers of differentiated brain cells.

Summary

In the adult mammalian body, the various types of tissue-specific stem cells are highly specialized, each giving rise to a limited range of differentiated cell types. Cells become restricted to specific pathways of differentiation during embryonic development. One way to force a return to a pluripotent or totipotent state is by nuclear transplantation: the nucleus of a differentiated cell can be injected into an enucleated oocyte, whose cytoplasm reprograms the genome of the injected nucleus back to an approximation of an early embryonic state. This allows the injected oocyte to develop into an entire new individual. The reversion of the genome to this state involves radical, genome-wide changes in chromatin structure and DNA methylation.

Remarkably, cells taken from the inner cell mass of an early mammalian embryo can be propagated in culture indefinitely in a pluripotent state. When transplanted back into a host early embryo, these embryonic stem (ES) cells can contribute cells to any tissue, including the germ line. ES cells have been invaluable for genetic engineering in mice. Cells with similar properties, called induced pluripotent stem (iPS) cells, can be generated from adult differentiated cells such as fibroblasts by forced expression of a cocktail of key transcription regulators. A similar method can be used to reprogram differentiated adult cells directly from one specialized cell type to another. In principle, iPS cells generated from cells taken from an adult human individual could be used for tissue repair in that same individual, avoiding the problem of immune rejection. More immediately, iPS cells provide a source of specialized cells that can be used to analyze in vitro the effects of mutations affecting human cells and to screen for drugs for treatment of genetic diseases. Both ES and iPS cells and their progeny can form tiny organs (organoids) in culture, which can serve as powerful models for studies of human development and disease.

PROBLEMS

Which statements are true? Explain why or why not.

22–1 In the mouse small intestine, stem cells in the crypts divide asymmetrically to maintain the population of cells that make up the crypts and villi; after each division, one daughter remains a stem cell and the other begins to divide rapidly to produce differentiated progeny.

22–2 Stem cells are the same in all tissues.

22–3 Every tissue that can be renewed is renewed from a tissue-specific population of stem cells.

22–4 Although mice continually replace neurons in their olfactory bulbs and songbirds replace large numbers of neurons to refine their songs for each mating season, humans do not have the ability to replace neurons in their brains.

Discuss the following problems.

22–5 In the 1950s, scientists fed ^{3}H-thymidine to rats to label cells that were synthesizing DNA, and then followed the fates of labeled cells for periods of up to a year. They found three patterns of cell labeling in different tissues. Cells in some tissues such as neurons in the central nervous system and the retina did not get labeled. Muscle, kidney, and liver, by contrast, each showed a small number of labeled cells that retained their label, apparently without further division or loss. Finally, cells such as those in the squamous epithelia of the tongue and esophagus were labeled in fairly large numbers, with radioactive pairs of nuclei visible in 12 hours; however, the labeled cells disappeared over time. Which of these three patterns of labeling would you expect to see if the labeled cells were generated by stem cells? Explain your answer.

22–6 At any given time, a single intestinal crypt of mice comprises about 15 stem cells and 10 Paneth cells. After cell division, which occurs about once a day, the daughter cells remain stem cells only if they maintain contact with a Paneth cell. This constant competition for Paneth-cell contact raises the possibility that crypts might become monoclonal over time; that is, the crypt cells at one point in time might derive from only one of the 15 stem cells that existed at some earlier time. To test this possibility, you use the so-called confetti marker that upon activation expresses one of three fluorescent proteins in the stem cells of the crypt. You then examine crypts at various times to determine whether they contain cells with multiple colors or only one color (Figure Q22–1). Do the crypts become monoclonal over time or not? How can you tell?

22–7 The origin of new β cells of the pancreas—from stem cells or from preexisting β cells—was not resolved until recently, when the technique of lineage tracing was used to decide the issue. Transgenic mice were engineered to express a tamoxifen-activated form of Cre recombinase under the control of the insulin promoter, which is active only in β cells. In these mice, investigators could remove an inhibitory segment of DNA by adding tamoxifen and thereby allow expression of human placental alkaline phosphatase (HPAP), which can be detected by histochemical staining. After a pulse of tamoxifen that converted about 30% of β cells in young mice to cells that express HPAP, the investigators followed the percentage of labeled β cells for a year, during which time the total number of β cells in the pancreas increased by 6.5-fold. How do you suppose the percentage of β cells would change over time if new β cells were derived from stem cells? What if new β cells were derived from preexisting β cells? Which hypothesis do the results in Figure Q22–2 support?

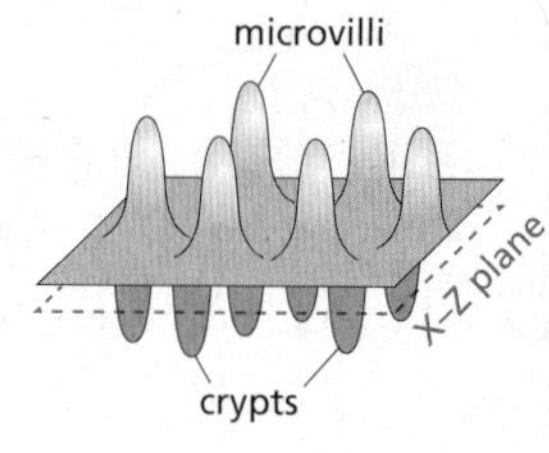

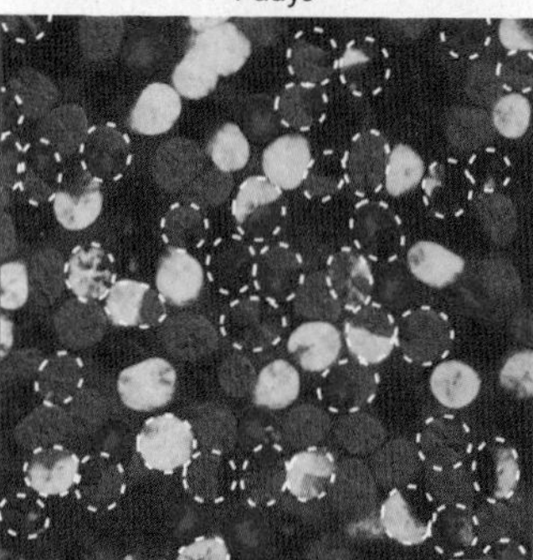

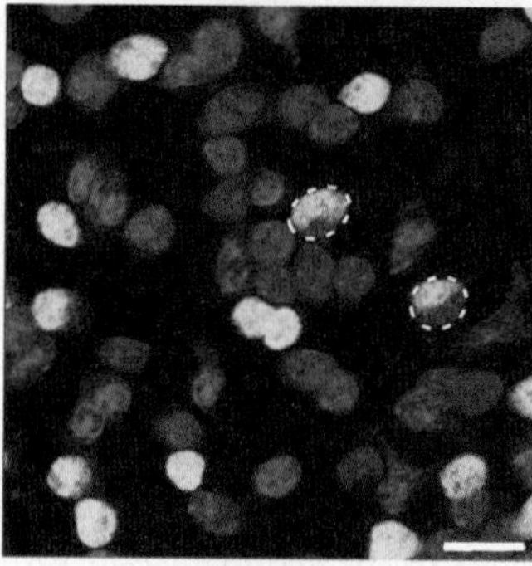

Figure Q22–1 Fluorescent cells in crypts in mouse intestines at various times after activation of expression of fluorescent proteins (Problem 22–6). The images are taken in the X–Z plane, which cuts through multiple crypts, as indicated in the schematic drawing. Roughly 50 crypts are visible in each section. *Dotted white circles* identify some individual crypts. Scale bars are 100 μm. (Adapted from H.J. Snippert et al., *Cell* 143:134–144, 2010. With permission from Elsevier.)

22–8 One of the earliest assays for hematopoietic stem cells made use of their ability to form colonies in the spleens of heavily irradiated mice. By varying the amounts of transplanted bone marrow cells, investigators showed that the number of spleen colonies varied linearly with dose and that the curve passed through the origin,

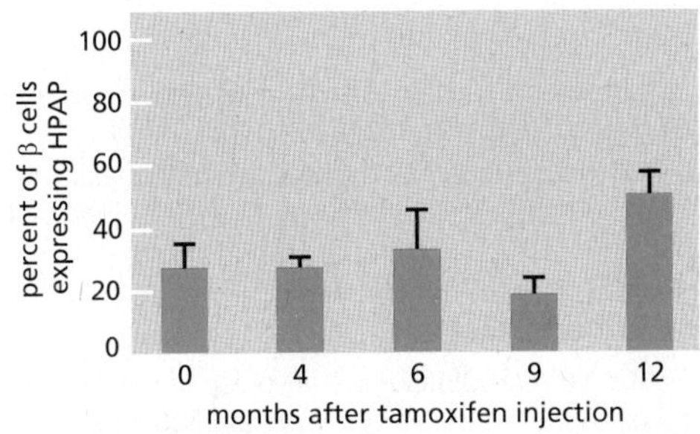

Figure Q22–2 Percentage of labeled β cells in pancreatic islets of mice at different ages (Problem 22–7). All mice were injected with a pulse of tamoxifen at 6–8 weeks of age, and then their pancreatic cells were stained for HPAP at various times afterward. Error bars represent standard deviations for all animals analyzed for that time point. (Adapted from Figure 2 of Y. Dor et al., *Nature* 429:41–46, 2004.)

suggesting that single cells were capable of forming individual colonies. However, because colony formation was rare relative to the numbers of transplanted cells, it was possible that undispersed clumps of two or more cells were the actual initiators.

A classic paper resolved this issue by exploiting rare, cytologically visible genome rearrangements generated by irradiation. Recipient mice were first irradiated to deplete bone marrow cells, and then they were irradiated a second time after transplantation to generate rare genome rearrangements in the transplanted cell population. Spleen colonies were then screened to find ones that carried genome rearrangements. How do you suppose this experiment distinguishes between colonization by single cells versus cellular aggregates?

22–9 It is possible to purify hematopoietic stem cells using a combination of antibodies directed against cell-surface targets. By removing cells that expressed surface markers characteristic of specific lineages such as B cells, granulocytes, myelomonocytic cells, and T cells, investigators generated a population of cells enriched for stem cells. They further enriched this population for putative stem cells by positively selecting for cells that expressed suspected stem-cell surface markers. Spleen colony formation in irradiated mice by these putative stem cells and the unfractionated bone marrow cells is shown in Figure Q22–3. Given that only about 1 in 10 cells lodges in the spleen, do these results support the idea that the enriched population consists mostly of hematopoietic stem cells? What additional information would you need to have to feel confident that the enriched cells are true stem cells? What proportion of bone marrow cells are hematopoietic stem cells?

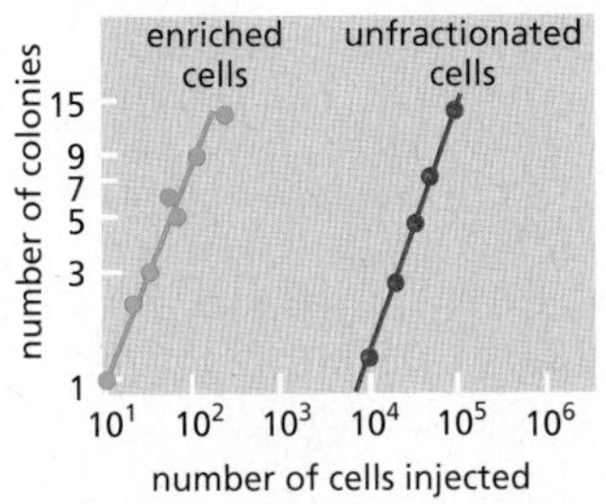

Figure Q22–3 Spleen colony formation by cells enriched for stem cells and by unfractionated bone marrow cells (Problem 22–9).

22–10 Generation of induced pluripotent stem (iPS) cells was first accomplished using retroviral vectors to carry the OSKM (Oct4, Sox2, Klf4, and Myc) set of transcription regulators into cells. The efficiency of fibroblast reprogramming was typically low (0.01%), in part because large numbers of retroviruses must integrate to bring about reprogramming, and each integration event carries with it the risk of inappropriately disrupting or activating a critical gene. In what other ways, or other forms, do you suppose you might deliver the OSKM transcription regulators so as to avoid these problems?

22–11 To test whether blood-borne factors alter neurogenesis in the central nervous system in mice, you use the technique of parabiosis—linking of circulatory systems in two individuals—to measure effects on the neurogenic niche in the dentate gyrus of the hippocampus. As indicated in Figure Q22–4A, you link the circulatory systems of two young mice, or two old mice, or a young mouse with an old mouse. Five weeks later, you stain slices of the dentate gyrus with antibodies to doublecortin (Dcx), a marker for newly born neurons, and count the new neurons, as summarized in Figure Q22–4B and C. Results in young-young parabionts and old-old parabionts were no different than in individual young or old mice. Do these results support the idea that blood-borne factors affect neurogenesis in the dentate gyrus of the hippocampus? Why or why not? Which, if either, mouse benefits in the parabiosis of a young mouse with an old mouse?

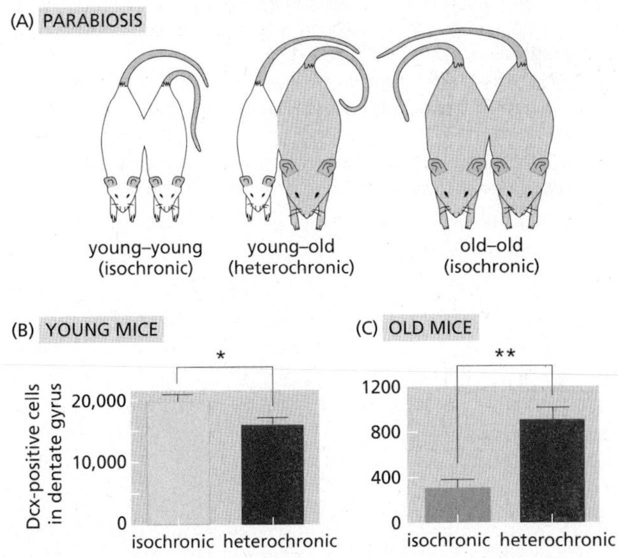

Figure Q22–4 Effects of parabiosis on the number of newly born neurons in the dentate gyrus of the mouse hippocampus (Problem 22–11). (A) Linked circulatory systems (parabiosis) between pairs of mice. (B) Dcx-positive cells (newly born neurons) in a young mouse paired with a young mouse (isochronic) or in a young mouse paired with an old mouse (heterochronic). (C) Dcx-positive cells in an old mouse paired with an old mouse (isochronic) or in an old mouse paired with a young mouse (heterochronic). Isochronic refers to parabiosis between mice of the same age; heterochronic refers to parabiosis between mice of different ages. Asterisks refer to statistical significance of the results: * is for $P < 0.05$, which means the result would be expected to occur by chance in less than 1 in 20 repeats; ** is for < 0.01, which means the result would be expected to occur by chance in less than 1 in 100 repeats.

REFERENCES

General

Gurdon JB & Melton DA (2008) Nuclear reprogramming in cells. *Science* 322, 1811–1815.

He S, Nakada D & Morrison SJ (2009) Mechanisms of stem-cell renewal. *Annu. Rev. Cell Dev. Biol.* 25, 377–406.

Karagiannis P, Takahashi K, Saito M . . . Osafune K (2019) Induced pluripotent stem cells and their use in human models of disease and development. *Physiol. Rev.* 99, 79–114.

Morrison SJ & Spradling AC (2008) Stem cells and niches: mechanisms that promote stem cell maintenance throughout life. *Cell* 132, 598–611.

Stem Cells and Tissue Homeostasis

Barker N, van Es JH, Kuipers J . . . Clevers H (2007) Identification of stem cells in small intestine and colon by marker gene *Lgr5*. *Nature* 449, 1003–1007.

Baron CS & van Oudenaarden A (2019) Unravelling cellular relationships during development and regeneration using genetic lineage tracing. *Nat. Rev. Mol. Cell Biol.* 20, 753–765.

Blanpain C & Fuchs E (2014) Plasticity of epithelial stem cells in tissue regeneration. *Science* 344, 1242281.

Morrison SJ, Uchida N & Weissman IL (1995) The biology of hematopoietic stem cells. *Annu. Rev. Cell Dev. Biol.* 11, 35–70.

Taub R (2004) Liver regeneration: from myth to mechanism. *Nat. Rev. Mol. Cell Biol.* 5, 836–847.

Control of Stem-Cell Fate and Self-Renewal

Crane GM, Jeffery E & Morrison SJ (2017) Adult haematopoeitic stem cell niches. *Nat. Rev. Immunol.* 17(9), 573–590.

Goodell MA & Rando TA (2015) Stem cells and healthy aging. *Science* 350, 1199–1204.

Losick VP, Morris LX, Fox DT & Spradling A (2011) *Drosophila* stem cell niches: a decade of discovery suggests a unified view of stem cell regulation. *Dev. Cell* 21, 159–171.

Sato T, Van Es JH, Snippert HJ . . . Clevers H (2011) Paneth cells constitute the niche for Lgr5 stem cells in intestinal crypts. *Nature* 469, 415–418.

Sunchu B & Cabernard C (2020) Principles and mechanism of asymmetric cell division. *Development* 147(13), dev167650.

Venkei ZG & Yamashita Y (2018) Emerging mechanisms of asymmetric stem cell division. *J. Cell Biol.* 217(11), 3785–3795.

Yoshida S (2019) Heterogeneous, dynamic, and stochastic nature of mammalian spermatogenic stem cells. *Curr. Top. Dev. Biol.* 135, 245–285.

Regeneration and Repair

Brockes JP & Kumar A (2008) Comparative aspects of animal regeneration. *Annu. Rev. Cell Dev. Biol.* 24, 525–549.

Tanaka EM (2016) The molecular and cellular choreography of appendage regeneration. *Cell* 165(7), 1598–1608.

Tanaka EM & Reddien PW (2011) The cellular basis for animal regeneration. *Dev. Cell* 21, 172–185.

Wagner DE, Wang IE & Reddien PW (2011) Clonogenic neoblasts are pluripotent adult stem cells that underlie planarian regeneration. *Science* 332, 811–816.

Cell Reprogramming and Pluripotent Stem Cells

Apostolou E & Hochedlinger K (2013) Chromatin dynamics during cellular reprogramming. *Nature* 502, 462–471.

Fox IJ, Daley GQ, Goldman SA . . . Trucco M (2014) Stem cell therapy. Use of differentiated pluripotent stem cells as replacement therapy for treating disease. *Science* 345, 1247391.

Hoechedlinger K & Jaenisch R (2015) Induced pluripotency and epigenetic reprogramming. *Cold Spring Harb. Perspect. Biol.* 7, a019448.

Inoue H, Nagata N, Kurokawa H & Yamanaka S (2014) IPS cells: a game changer for future medicine. *EMBO J.* 33, 409–417.

Radzisheuskaya A & Silver JCR (2014) Do all roads lead to Oct4? The emerging concepts of induced pluripotency. *Trends Cell Biol.* 24, 275–284.

Sasai Y, Eiraku M & Suga H (2012) *In vitro* organogenesis in three dimensions: self-organising stem cells. *Development* 139, 4111–4121.

Takahashi K & Yamanaka S (2006) Induction of pluripotent stem cells from mouse embryonic and adult fibroblast cultures by defined factors. *Cell* 126, 663–676.

Theunissen TW & Jaenisch R (2014) Molecular control of induced pluripotency. *Cell Stem Cell* 14, 720–734.

CHAPTER 23

Pathogens and Infection

IN THIS CHAPTER

Infectious diseases currently cause about one-quarter of all human deaths worldwide, more than all forms of cancer combined and second only to cardiovascular diseases. There is a continuing heavy burden of ancient diseases such as tuberculosis and malaria, and these are increasingly difficult to treat because of rising drug resistance. Newer infectious diseases continually emerge, including the current pandemic (worldwide epidemic) of COVID-19 (coronavirus disease 2019), which was first clinically observed in 2019 and has since infected hundreds of millions of people and caused millions of deaths worldwide. Moreover, some diseases long thought to result from other causes are now recognized to be associated with infections. Most gastric ulcers, for example, are caused not by stress or spicy food, but by infection of the stomach lining by the bacterium *Helicobacter pylori*.

The burden of infectious diseases is not spread equally across the planet. Poorer countries and communities suffer disproportionately, often due to poor public sanitation and overburdened health systems, as occurred in Haiti in 2010 with a severe cholera outbreak after a devastating earthquake. Some infectious diseases, however, occur primarily or exclusively in industrialized communities: Legionnaires' disease, for example, a bacterial infection of the lungs, commonly spreads through air-conditioning systems.

Since the mid-1800s, physicians and scientists have struggled to identify the microbes—collectively called **pathogens**—that are capable of causing infectious diseases. More recently, the advent of microbial genetics and molecular cell biology has greatly enhanced our understanding of the causes and mechanisms of infectious diseases. We now know that pathogens frequently exploit the attributes of their host's cells in order to infect them. This understanding can give us new insights into normal cell biology, as well as strategies for treating and preventing infections.

Although pathogens are understandably a focus of attention, only a relatively small fraction of the microbial species we encounter are pathogens. Much of the biomass of Earth is made up of microbes. They produce everything from the oxygen we breathe to the soil nutrients we use to grow food and represent the bottom of the ecological pyramid. Even those species of microbes that colonize the human body do not generally cause disease. Many of these microbes have a beneficial effect on the health of the organism, assisting its normal development and physiology.

In this chapter, we give an overview of the different kinds of pathogens. We then discuss the cell biology of infection—the molecular interactions between pathogens and their hosts. Finally, we discuss an emerging concept that the many microbes that normally colonize our body, the so-called human **microbiota**, may actually benefit our health. In Chapter 24, we consider how our innate and adaptive immune systems collaborate to defend us against pathogens while leaving our microbiota intact.

INTRODUCTION TO PATHOGENS

We normally think of pathogens as hostile invaders, but a pathogen, like any other organism, is simply exploiting an available niche in which to live and procreate. Living on or in a host organism is a very effective strategy, and it is possible that

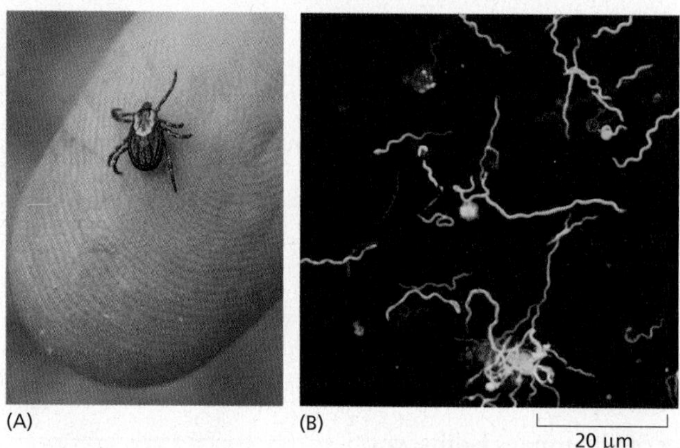

Figure 23–1 **Parasitism at many levels.** (A) Most animals harbor parasites, an example being the blacklegged tick, or deer tick (*Ixodes scapularis*), shown here on a human finger. Although ticks of this species thrive on white-tailed deer and other wild mammals, they can also live on humans. (B) Ticks themselves harbor their own parasites including the bacterium *Borrelia burgdorferi*, stained here with a vital dye that labels living bacteria *green* and dead bacteria *red*. These long spiral-shaped bacteria live in deer ticks and can be transmitted to humans during a tick's blood meal. *B. burgdorferi* causes Lyme disease, which is characterized by a bull's-eye–shaped skin rash and fever; if the infection is left untreated, various complications can result, including arthritis and neurological abnormalities. The idea that parasites have their own parasites was noted by Jonathan Swift in 1733:
"So, naturalists observe, a flea
Has smaller fleas that on him prey;
And these have smaller still to bite 'em;
And so proceed ad infinitum." (A, National Geographic Image Collection/Alamy Stock Photo; B, courtesy of M. Embers.)

every organism on Earth is subject to some type of infection (**Figure 23–1**). Even bacteria can be infected by viruses called bacteriophages (see Figure 1–37). It is not surprising then that many microorganisms have evolved the ability to survive and reproduce in the human body, a nutrient-rich, warm, and moist environment that remains at a uniform temperature and constantly renews itself. In this section, we discuss some of the common features that microorganisms must have in order to colonize the human body or cause disease, and we explore the wide variety of organisms that are known to cause disease.

Pathogens Can Be Viruses, Bacteria, or Eukaryotes

Many types of microbes cause disease in humans, and others reside on or in our bodies without causing harm. The most familiar are viruses and bacteria. Viruses cause diseases ranging from COVID-19 and Ebola virus disease to the common cold. Viruses are essentially fragments of nucleic acid (DNA or RNA) that generally encode a relatively small number of gene products and are wrapped in a protective shell of proteins (**Figure 23–2A**) and (in some cases) an outer membrane envelope. Much larger and more complex than viruses, bacteria are prokaryotic cells that perform most of their basic metabolic functions themselves, relying on the host primarily for nutrition (**Figure 23–2B**). Bacteria cause illnesses ranging from tuberculosis, which causes roughly 1.4 million deaths per year, to pneumonia, as well as diarrheal and sexually transmitted diseases.

Some other infectious agents are eukaryotic organisms. These range from single-celled protozoa (**Figure 23–2C**) and fungi (**Figure 23–2D**) to large, complex metazoa such as parasitic worms. One of the most common human parasites, shared by about a billion people at present, is the nematode worm *Ascaris lumbricoides*, which infects the gut (**Figure 23–2E**). It resembles its harmless nematode cousin *Caenorhabditis elegans*, which is used as a model organism for genetic and developmental biological research (see Figure 1–42). *C. elegans*, however, is only about 1 mm in length, whereas *Ascaris* can reach 30 cm.

Pathogens Interact with Their Hosts in Different Ways

Although the ability of a particular microorganism to cause disease depends on many factors, it requires that the pathogen possess specialized pathogenic characteristics that allow it to live in humans. One way we distinguish between pathogens is by classifying them as primary versus opportunistic on the basis of the circumstances under which they cause human disease. **Primary pathogens** can cause overt disease in most healthy, nonimmune people. **Opportunistic pathogens** do not cause disease in healthy people but can cause illness in individuals suffering from other conditions. Some primary pathogens cause acute, life-threatening epidemic infections and spread rapidly from one sick or dying host to another; historically important examples include the bacterium *Vibrio cholerae*, which causes cholera, and the sudden acute respiratory syndrome coronavirus 2 (SARS-CoV-2) and influenza viruses, which cause COVID-19

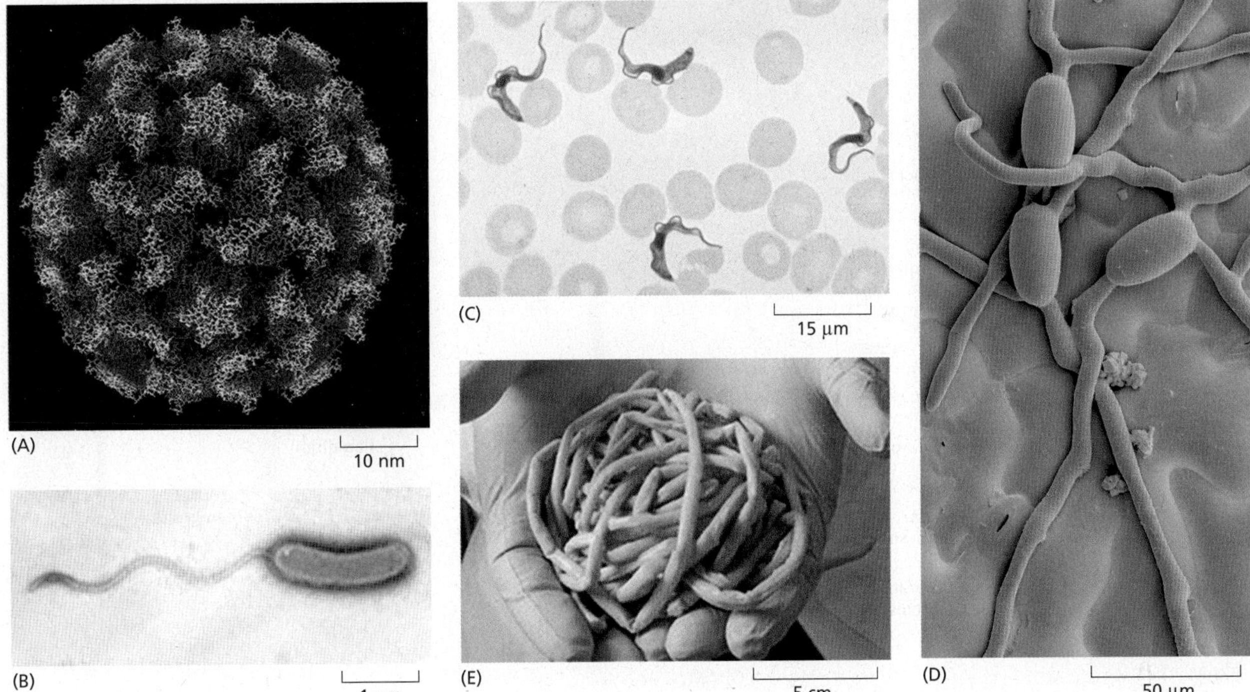

Figure 23–2 **Pathogens in many forms.** (A) The structure of the protein coat, or *capsid*, of poliovirus. This virus was once a common cause of paralysis, but the disease (poliomyelitis) has been greatly reduced by widespread vaccination. (B) The bacterium *Vibrio cholerae*, the causative agent of the epidemic diarrheal disease cholera, shown here with its attached flagellum. (C) The protozoan parasite *Trypanosoma brucei (purple)* in a field of erythrocytes (red blood cells; *pink*). This parasite causes African sleeping sickness, a potentially fatal disease of the central nervous system. (D) Powdery mildew spores germinating on an *Arabidopsis* leaf. (E) This clump of *Ascaris* nematodes was passed rectally by a child in Africa. (A, courtesy of Robert Grant, Stephan Crainic, and James M. Hogle; B, photograph kindly provided by John Mekalanos; C, CDC, Department of Health and Human Services; D, courtesy of Kim Findlay and John Innes Centre; E, CDC/James Gathany.)

and flu, respectively. Others may persistently infect a single individual for years without causing overt disease; examples include the bacterium *Mycobacterium tuberculosis*, which causes tuberculosis, and the intestinal worm *Ascaris* (see Figure 23–2E). Although these potential primary pathogens can make some people critically ill, billions of people carry these foreign organisms in an asymptomatic way, often unaware that they are infected.

In order to survive and multiply, a successful pathogen must be able to (1) enter the host (usually by breaking an epithelial barrier); (2) find a nutritionally compatible niche in the host's body; (3) avoid, subvert, or circumvent the host's innate and adaptive immune responses; (4) replicate, using host resources; and (5) exit one host and spread to another. Pathogens have evolved various mechanisms that exploit the biology of their host organisms to help accomplish these five tasks. For some pathogens, these mechanisms are adapted to a unique host species, whereas for others the mechanisms are sufficiently general to permit invasion, survival, and replication in a wide variety of hosts. Because pathogens have evolved the ability to interface directly with the molecular machinery of host cells, we have learned a great deal about cell biological principles by studying them.

Our constant exposure to pathogens has strongly influenced human evolution. In modern times, humans have learned how to limit the ability of pathogens to infect us through improvements in public health measures and childhood nutrition, vaccines, antimicrobial drugs, and routine testing of blood used for transfusions. As we learn more about the mechanisms by which pathogens cause disease (called *pathogenesis*), we will devise new ways to supplement or augment our immune systems in fighting infections.

We now introduce the basic features of each of the three major types of pathogens—bacteria, eukaryotic parasites, and viruses—before we examine the detailed mechanisms used by these pathogens to infect their hosts.

Bacteria Are Diverse and Occupy a Remarkable Variety of Ecological Niches

Bacteria are small but highly sophisticated cells whose organization and behaviors have attracted the attention of many scientists for well over a century. Bacteria are classified broadly on the basis of their cellular and molecular features. One

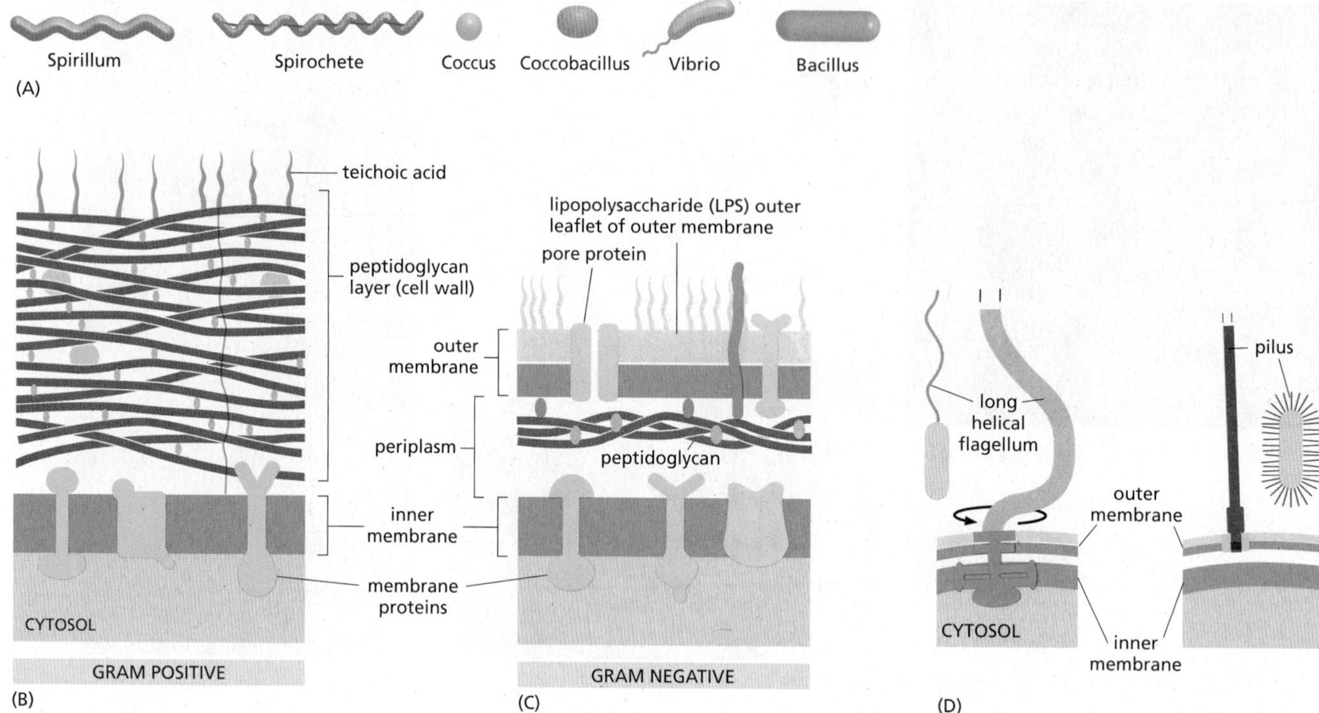

Figure 23–3 **Bacterial shapes and cell-surface structures.** (A) Bacteria are traditionally classified by shape. For example, the bacterium *Borrelia burgdorferi*, shown in Figure 23–1B, is spiral shaped and is thus classified as a spirochete. (B and C) Bacteria are also classified as *Gram positive* or *Gram negative* on the basis of a staining procedure first described in 1884. (B) Gram-positive bacteria such as *Streptococcus* and *Staphylococcus* have a single membrane and a thick cell wall made of cross-linked *peptidoglycan*. They are called Gram positive because they retain the violet dye used in the Gram-staining procedure. (C) Gram-negative bacteria such as *Escherichia coli* (*E. coli*) and *Salmonella enterica* have two membranes, separated by the *periplasm* (see Figure 11–17). The peptidoglycan cell wall of these organisms is located in the periplasm and is thinner than in Gram-positive bacteria; they therefore fail to retain the dye in the Gram-staining procedure. The inner membrane of both Gram-positive and Gram-negative bacteria is a phospholipid bilayer. The inner leaflet of the outer membrane of Gram-negative bacteria is also made primarily of phospholipids, whereas the outer leaflet of the outer membrane is composed of the unique glycosylated lipid *LPS*. (D) Cell-surface appendages are important for bacterial behavior. Both Gram-positive and Gram-negative bacteria swim using the rotation of helical flagella. An example is *V. cholerae*, depicted in panel A and shown in Figure 23–2B. Although the bacterium illustrated has only a single flagellum at one pole, many have multiple flagella. *Pili* (also called *fimbriae*) are used to adhere to various surfaces in the host and to facilitate genetic exchange between bacteria. Some kinds of pili can retract to generate force and thereby help bacteria move along surfaces.

such feature is their shape—rods, spheres (*cocci*), or spirals (**Figure 23–3A**). They are also traditionally classified by their so-called **Gram-staining** properties, which is an older staining method that is still informative. This staining reflects differences in the structure of the bacterial surface, a layer that is important for pathogens because it directly contacts host cells and the immune system. **Gram-positive** bacteria have a thick layer of peptidoglycan cell wall outside their inner (plasma) membrane (**Figure 23–3B**), whereas **Gram-negative** bacteria have a thinner peptidoglycan cell wall. In both cases, the cell wall protects against lysis by osmotic swelling, and it is a target of host antibacterial proteins such as lysozyme and antibiotics such as penicillin. Gram-negative bacteria are also covered outside the cell wall by an outer membrane containing *lipopolysaccharide* (*LPS*) (**Figure 23–3C**). Both peptidoglycan and LPS are unique to bacteria and are recognized as *pathogen-associated molecular patterns* (*PAMPs*) by the host innate immune system, as discussed in Chapter 24. The surface of bacterial cells can also display an array of appendages, including flagella and pili, which enable bacteria to swim or adhere to desirable surfaces, respectively (**Figure 23–3D**). Apart from cell shape and structure, differences in ribosomal RNA and genomic DNA sequence are also used for phylogenetic classification. Because bacterial genomes are small—typically between 1,000,000 and 5,000,000 nucleotide pairs (compared to more than 3,000,000,000 for humans)—they are now simple to sequence, making this an important new classification tool.

Bacteria also exhibit extraordinary molecular, metabolic, and ecological diversity. At the molecular level, bacteria are far more diverse than eukaryotes, and they can occupy ecological niches having extremes of temperature, salt concentrations, and nutrient limitation. To describe how flexible bacterial pathogens are with respect to their growth niche, we discriminate between **facultative pathogens** and **obligate pathogens**. Facultative bacteria can replicate in an environmental reservoir such as water or soil and only cause disease if they happen to encounter a susceptible host. Obligate pathogens can only replicate inside the body of their host. Bacteria also differ in the range of hosts they will infect. A champion generalist is the opportunistic pathogen *Pseudomonas aeruginosa*, which can cause disease in a wide variety of plants and animals. In contrast, *Salmonella enterica* serovar Typhi only infects humans, causing typhoid fever, an ancient and life-threatening human disease that was sensationalized by the story of the cook "Typhoid Mary," who carried this bacterium asymptomatically and spread typhoid fever in New York City in the early 1900s.

Bacterial Pathogens Carry Specialized Virulence Genes

Genes that contribute to the ability of an organism to cause disease (as opposed to genes needed solely to grow and replicate) are called **virulence genes**, and the proteins they encode are called **virulence factors**. In general, the presence of a relatively small number of such virulence genes distinguishes pathogenic bacteria from their closest nonpathogenic relatives. Virulence genes are often clustered together on the bacterial chromosome; large clusters are called *pathogenicity islands*. Virulence genes can also be carried on *bacteriophages* (bacterial viruses) or *transposons* (see Table 5–4), both of which integrate into the bacterial chromosome, or on extrachromosomal *virulence plasmids* (**Figure 23–4A**).

Pathogenic bacteria are thought to emerge when groups of virulence genes are transferred together into a previously avirulent bacterium by a process called **horizontal gene transfer** (to distinguish it from vertical gene transfer from parent to offspring). Horizontal transfer can occur by one of three mechanisms: natural *transformation* by released naked DNA, *transduction* (infection) by bacteriophages, or sexual exchange by *conjugation* (**Figure 23–4B** and **Movie 23.1**). Sequencing the genomes of pathogenic and nonpathogenic bacteria has revealed that horizontal gene transfer has made important contributions to bacterial evolution, enabling species to inhabit new ecological and nutritional niches and to cause disease. Even within a single bacterial species, the amount of chromosomal

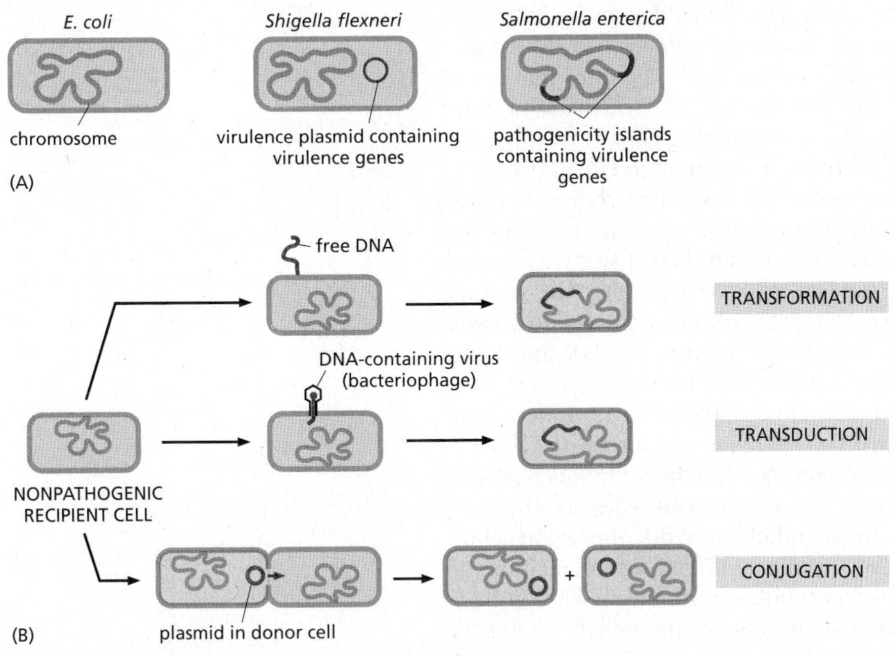

Figure 23–4 **Genetic differences between pathogenic and nonpathogenic bacteria.** (A) Genetic differences between nonpathogenic *E. coli* and two closely related food-borne pathogens—*Shigella flexneri*, which causes dysentery, and *Salmonella enterica* serovar Typhimurium, a common cause of food poisoning. Nonpathogenic *E. coli* has a single circular chromosome. The chromosome of *S. flexneri* differs from that of *E. coli* in a limited number of locations; most of the genes required for pathogenesis (virulence genes) are carried on an extrachromosomal virulence plasmid. The chromosome of *S. enterica* serovar Typhimurium carries two large inserts (pathogenicity islands) not found in the *E. coli* chromosome; these inserts each contain many virulence genes. (B) Bacterial pathogens evolve by horizontal gene transfer. This can occur by three mechanisms: natural *transformation*, in which naked DNA is taken in by competent bacteria; *transduction*, in which bacterial viruses (*bacteriophages*) transfer DNA from one bacterium (in the example shown, a pathogen) into another; and *conjugation* (see Movie 23.1), during which plasmid DNA, and even chromosomal DNA, is transferred from a donor to a recipient bacterium.

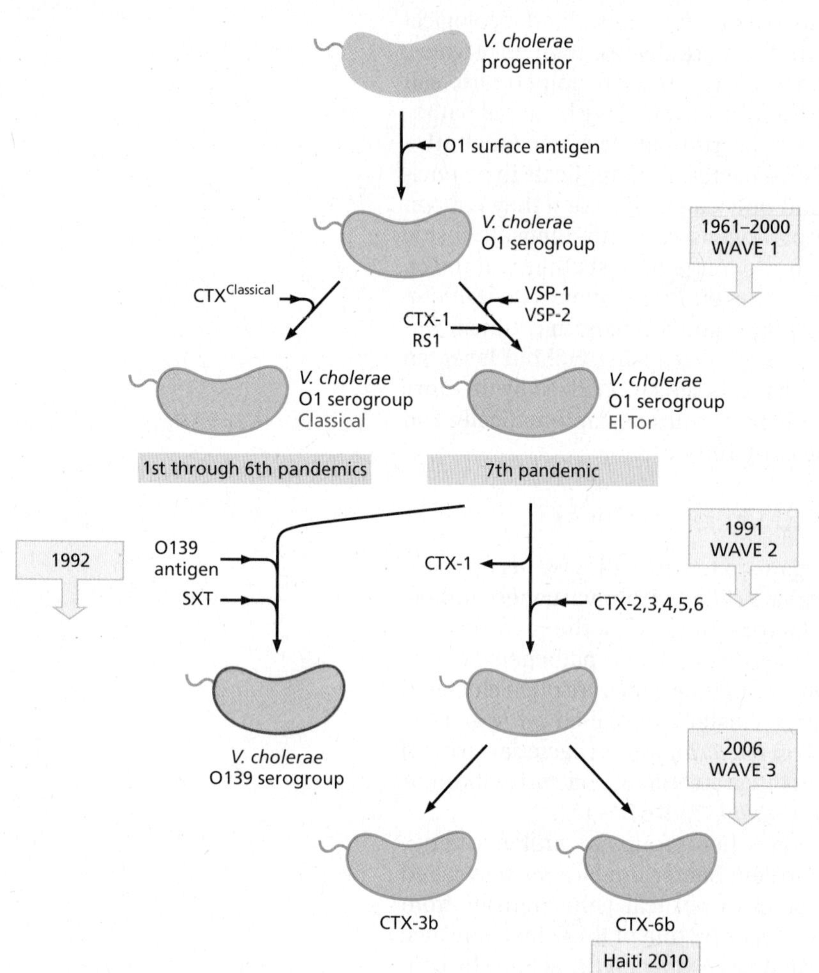

Figure 23–5 Model for the evolution of pathogenic *V. cholerae* strains. Progenitor strains in the wild first acquired the biosynthetic pathway necessary to make the O1 antigen type of carbohydrate chain *(blue outline)* on the outer-membrane LPS (see Figure 23–3C), a feature that is associated with the ability to cause epidemic cholera disease. Incorporation of the CTX bacteriophage created the classical pathogenic strains responsible for the first six worldwide epidemics of cholera between 1817 and 1923. Sometime in the twentieth century, an O1 strain in the environment picked up the CTX bacteriophage again, along with an associated bacteriophage RS1 and two pathogenicity islands (VSP-1 and VSP-2), creating the El Tor (wave 1) strain that emerged as the seventh worldwide pandemic in 1961. In the 1990s, an El Tor strain was isolated that had picked up a new DNA cassette, enabling it to produce the O139 antigen type of carbohydrate chain *(purple outline)* rather than the O1 type. This altered the bacterium's interaction with the human immune system without diminishing its virulence; this bacterium also picked up a new pathogenicity island (SXT). Separately, wave 2 strains emerged, in which CTX-1 was replaced with CTX-2–6 bacteriophages. In 2006, wave 3 strains emerged with variant CTX-3 and CTX-6 bacteriophages. The history of the *V. cholerae* strains shown in this diagram was deduced by comparing the DNA sequences of the genome of each strain to one another and noting the key differences. An electron micrograph of *V. cholerae* is shown in Figure 23–2B.

variation is astonishing; the genomes of different strains of *Escherichia coli* can differ by as much as 25%. Such variation has led to the concept that a bacterial species has both a *core genome* common to all isolates within the species and a larger *pangenome* consisting of all genes present in the full spectrum of isolates.

Acquisition of genes and gene clusters can drive the rapid evolution of pathogens and turn nonpathogens into pathogens. Consider, for example, *V. cholerae*—the Gram-negative bacterium that causes the epidemic diarrheal disease cholera. Of the hundreds of strains of *V. cholerae*, the only ones that cause pandemic human disease are those infected (see Figure 23–4B) with a mobile bacteriophage (CTXϕ; hereafter referred to as CTX) containing genes encoding the two subunits of the toxin that causes the diarrhea. As summarized in **Figure 23–5**, seven pandemics of *V. cholerae* have arisen since 1817. The first six were caused by the periodic reemergence of so-called *classical* strains. In addition to the toxin-encoding bacteriophage, these strains shared a similar O1 surface antigen, part of the LPS in the outer membrane (see Figure 23–3C). In 1961, the seventh pandemic began, caused by a new strain named *El Tor*, now also called a *wave 1* strain, which arose when an O1-expressing strain acquired CTX-1 and RS1ϕ (hereafter referred to as RS1) bacteriophages and at least two new pathogenicity islands. El Tor strains have now displaced the classical strains. In 1991, new *wave 2* and *early wave 3* strains emerged, in which CTX-1 was replaced with CTX-2–6 bacteriophages. In 1992, another new strain emerged in which O1 was replaced with another variant called O139, which differs in a surface antigen, and thus was not recognized by antibodies present in the blood of survivors of previous cholera epidemics. Then, in 2006, current wave 3 strains emerged, which include the strain containing CTX-6b that caused the severe cholera outbreak in Haiti in 2010, infecting more than 700,000 people and leading to approximately 8500 deaths.

As this example makes clear, the rapid evolution of bacterial pathogens can be likened to an arms race that pits the survival of a bacterium against our immune systems and the tools of modern medicine. Similar struggles for survival take place between all pathogens and humans, and understanding these conflicts provides key insights into the evolution of pathogens and greatly informs how we treat new outbreaks of infectious diseases.

Bacterial Virulence Genes Encode Toxins and Secretion Systems That Deliver Effector Proteins to Host Cells

What are the gene products that enable a bacterium to cause disease in a healthy host? The precise nature of these products differs between classes of pathogens that are largely distinguished by whether they live outside of host cells, so-called *extracellular bacterial pathogens* (discussed in this section), or inside host cells, so-called *intracellular bacterial pathogens* (discussed in a later section). For extracellular bacterial pathogens, virulence genes often encode secreted toxic proteins (*toxins*) that are released by bacteria. These toxins diffuse to target host cells, where they bind to specific receptors on the cell surface and enter the cell, interfering with host-cell proteins to elicit a response that is beneficial to the pathogen. Several of these bacterial toxins are among the most potent of known human poisons. Bacterial toxins are often composed of two protein components—an A subunit with enzymatic activity, and a B subunit that binds to host-cell receptors and directs the trafficking of the A subunit to the cytosol by various routes (**Figure 23–6**). The

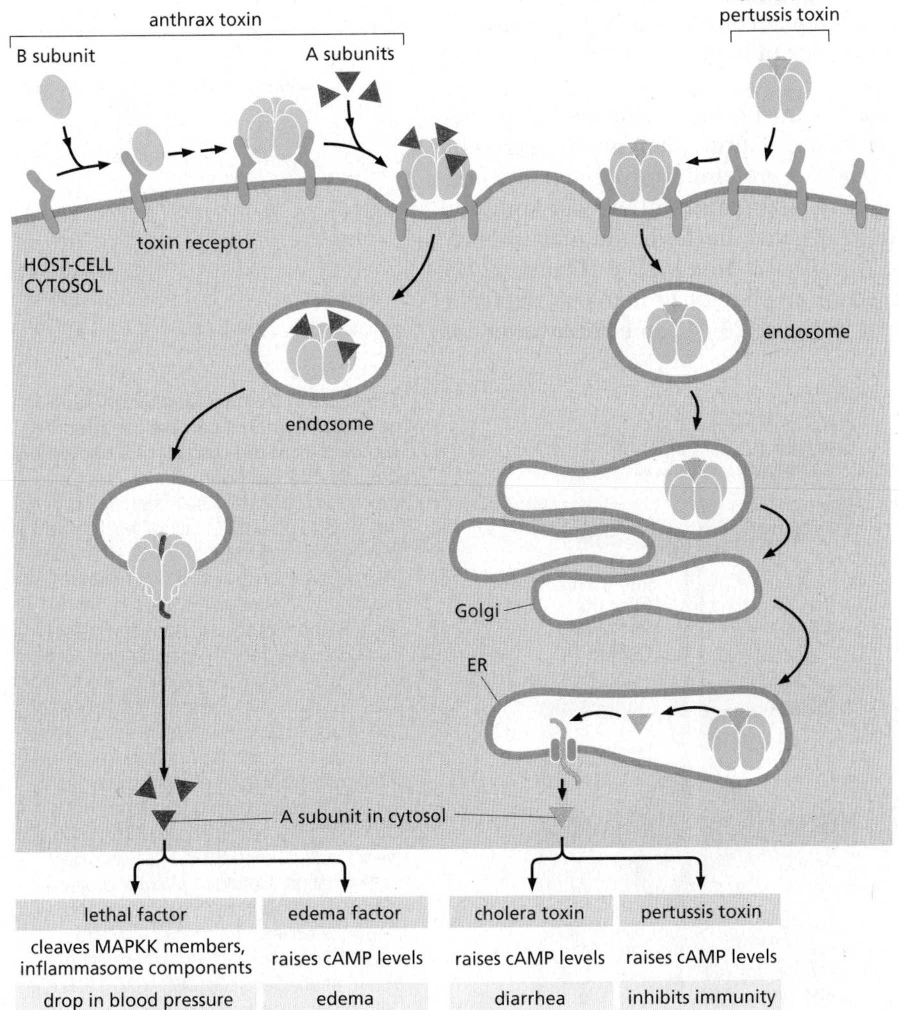

Figure 23–6 Toxins released by bacteria bind to and enter into host cells, disrupting host-cell processes. Bacterial toxins are often composed of A and B protein subunits. The B (binding) subunit of the toxin, which interacts with host-cell toxin receptors, can assemble after binding to toxin receptors (as with anthrax toxin) or prior to receptor binding (as with cholera and pertussis toxins). Binding to toxin receptors enables endocytosis and intracellular trafficking of the B subunits and the associated and enzymatically active A subunit(s). In the case of *Bacillus anthracis*, the B subunit changes conformation in the low-pH environment of the endosome to form a pore through which two different A subunits, lethal factor and edema factor, are transported across the membrane of the endosome in an unfolded conformation. In the cases of *V. cholerae* toxin and *Bordetella pertussis* toxin, the B and A subunits are transported to the Golgi apparatus and then to the endoplasmic reticulum (ER), where the A subunits are then translocated into the cytosol in an unfolded conformation through a protein-translocation channel (see Movie 23.2).

two subunits of **cholera toxin**, for example, are encoded by the *V. cholerae* CTX phage (see Figure 23–5), which is incorporated into the bacterial genome (Movie 23.2). The A subunit catalyzes the transfer of an ADP-ribose moiety from NAD^+ to the trimeric G protein G_S. This ADP ribosylation alters the G-protein α subunit so that it can no longer hydrolyze its bound GTP, causing it to remain in an active state that stimulates adenylyl cyclase indefinitely (see Chapter 15). The resulting prolonged elevation in cAMP concentration within intestinal epithelial cells causes a large efflux of Cl^- and water into the gut, thereby causing the severe diarrhea that characterizes cholera. Released bacteria then contaminate food and water, spreading infection to new hosts.

Some pathogenic bacteria secrete multiple toxins, each of which targets a different signaling pathway in host cells. Anthrax, for example, is an acute infectious disease of sheep, cattle, and occasionally humans. It is caused by contact with spores of the Gram-positive bacterium *Bacillus anthracis.* Dormant spores can survive in soil for long periods. If inhaled, ingested, or rubbed into breaks in the skin, spores can germinate and the bacteria replicate. The bacteria secrete two toxins with identical B subunits but different A subunits. The B subunits bind to a host-cell surface receptor protein to transfer the two different A subunits into host cells (see Figure 23–6). The A subunits are called **lethal factor** and **edema factor**. The latter is a calmodulin-dependent adenylyl cyclase that catalyzes the production of cyclic AMP (see Figure 15–26). Injection of edema toxin into the bloodstream of an animal leads to an ion imbalance that can cause an accumulation of extracellular fluid (*edema*) in the intestine and liver. Lethal toxin is a protease that cleaves several activated members of the mitogen-activated protein kinase kinase (MAP kinase kinase; MAPKK) family (see Figure 15–50) and components of the inflammasome (see Chapter 24), disrupting intracellular signaling and leading to immune-cell dysfunction. In animals, lethal toxin targets cells of the cardiovascular system, causing shock (a large fall in blood pressure) and death.

Whereas toxins are released and diffuse to target cells, pathogens that physically contact host cells often possess specialized **contact-dependent secretion systems** that directly inject so-called *effector proteins* from the bacterium into host cells. There are several such secretion systems, and we will limit our discussion to two that are present in Gram-negative bacteria and play important roles in bacterial pathogenesis: the *type III secretion system* (**Figure 23–7**) and the *type IV secretion system.* The injection of effector proteins, as depicted to occur via the **type III secretion system** in Figure 23–7, can elicit a variety of

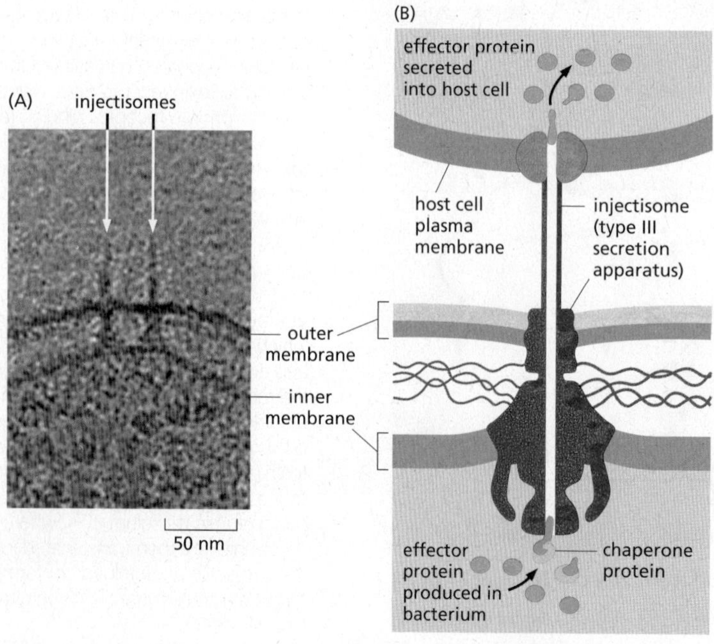

Figure 23–7 Contact-dependent type III secretion systems can deliver effector proteins from the cytosol of a bacterium directly into the host cell. (A) Electron micrograph of type III secretion systems on the surface of a bacterial cell, each of which consists of more than two dozen proteins. (B) The large lower ring is embedded in the bacterial inner membrane, and the smaller upper ring is embedded in the bacterial outer membrane. During infection, docking of the tip of the hollow needle at a host-cell plasma membrane results in the secretion of bacterial translocator proteins *(dark green)*, which form a pore in the host membrane. Bacterial effector proteins *(blue)* are initially bound to chaperone proteins in the bacterial cytosol *(light green)*, which keep them in an unfolded conformation. Unfolded effector proteins then pass through the injectisome and are secreted into the host cell, where they fold into their active conformation. (A, Hu et al., *Cell* 168:1065–1074, 2017. With permission from Elsevier.)

host-cell responses. For extracellular pathogens, it can enable the bacterium to block phagocytosis by immune cells. For intracellular pathogens, it can promote phagocytosis by nonimmune cells, or survival inside cells. The type III and type IV secretion systems appear to have evolved independently. There is a remarkable degree of structural similarity between type III secretion systems and the bacterial flagellum (see Figures 23–2B and 23–3D), suggesting they have a common evolutionary origin. Similarly, type IV secretion systems are closely related to the conjugation apparatus that many bacteria use to exchange genetic material (see Figure 23–4B). Although the mechanism of effector protein translocation through the type III and type IV secretion systems is not well understood, the inner diameter of the type III secretion system needle is only approximately 2 nm, so proteins must be moved through in an unfolded conformation.

Fungal and Protozoan Parasites Have Complex Life Cycles Involving Multiple Forms

Pathogenic fungi and protozoan parasites are eukaryotes, as are their hosts. Because antifungal and antiparasitic drugs often target core molecules that are similar between parasite and host, they can be less effective and more toxic to the host than are antibiotics that target bacteria. A second characteristic of fungal and parasitic infections that makes them difficult to treat is the tendency of the pathogens to switch among several different forms during their life cycles. A drug that is effective at killing one form can be ineffective at killing another form; therefore, the population of the pathogen can survive the treatment.

Fungi include both unicellular *yeasts* (such as *Saccharomyces cerevisiae* and *Schizosaccharomyces pombe*, which are used to bake bread and brew beer, and as model organisms for cell biology research) and filamentous, multicellular *molds* (like those found on moldy fruit or bread; see Figure 23–2D). Most of the important pathogenic fungi exhibit *dimorphism*—the ability to grow in either yeast or mold form. The yeast-to-mold or mold-to-yeast transition is frequently associated with infection. *Histoplasma capsulatum*, for example, grows as a mold at low temperature in the soil, but it switches to a yeast form when inhaled into the lung, where it can cause the disease histoplasmosis (**Figure 23–8**). The most common human fungal pathogen, *Candida albicans*, which often exists as a resident of the human oral and gastrointestinal microbiota and is also an opportunistic pathogen that can cause serious disease in immunocompromised people by entering the bloodstream and rapidly dividing.

Protozoan parasites are single-celled eukaryotes with more elaborate life cycles than fungi, and they frequently require more than one host. **Malaria** is the most devastating protozoal disease, infecting more than 200 million people every year and killing upward of 400,000. It is caused by four species of *Plasmodium*, which are transmitted to humans by the bite of the female *Anopheles* mosquito. *Plasmodium falciparum* causes the most serious form of malaria and is the most intensively studied of the malaria-causing parasites. It exists in many distinct

Figure 23–8 **Dimorphism in the pathogenic fungus *Histoplasma capsulatum*.** (A) At low temperature in the soil, *H. capsulatum* grows as a multicellular filamentous mold consisting of many individual cells connected together. (B) After it is inhaled into the lung of a mammal, the increase in temperature to 37°C causes a switch to a yeast form consisting of small clumps of round cells. (C) A stained histologic section of a mouse lung infected with *H. capsulatum*, showing a macrophage containing yeast forms of the pathogen. (A and B, courtesy of Sinem Beyhan and Anita Sil; C, courtesy of Davina Hocking Murray and Anita Sil.)

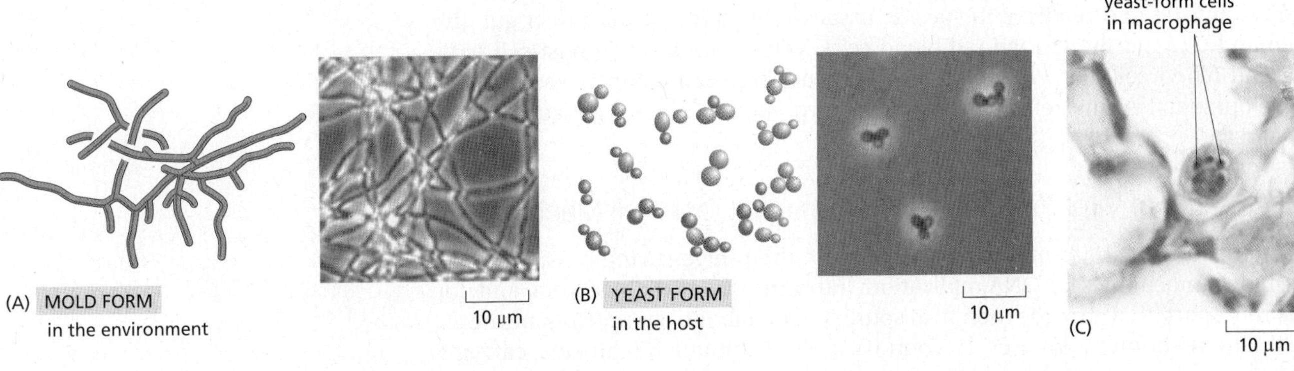

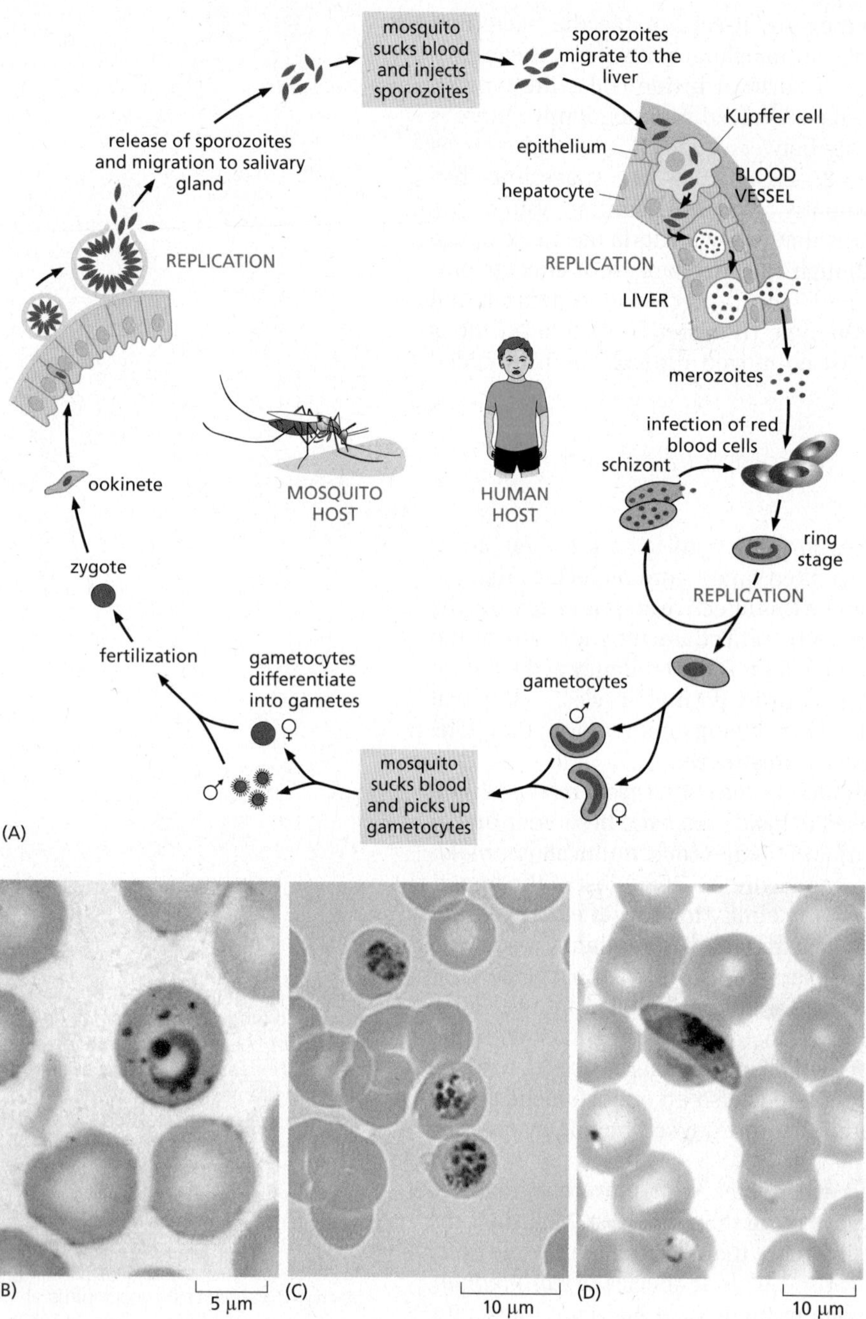

Figure 23–9 The complex life cycle of malaria parasites. (A) The sexual cycle of *Plasmodium falciparum* requires passage between a human host and an insect host (see Movie 23.3). (B–D) Blood smears from people with malaria, showing three different forms of the parasite that appear in red blood cells: B, ring stage; C, schizont; and D, gametocyte. (B–D, courtesy of the Centers for Disease Control, Division of Parasitic Diseases, DPDx.)

forms, and it requires both the human and mosquito hosts to complete its sexual cycle (**Figure 23–9A** and **Movie 23.3**). Several of these forms are highly specialized to invade and replicate in specific tissues—the lining of the insect gut, the human liver, and the human red blood cell. Even within a single host-cell type, the red blood cell, the *Plasmodium* parasite undergoes a complex sequence of developmental events, reflected in striking morphological changes (**Figure 23–9B, C, and D**).

All Aspects of Viral Propagation Depend on Host-Cell Machinery

Bacteria, fungal, and protozoan pathogens are themselves living cells. They use their own machinery for DNA replication, transcription, and translation, and, for the most part, they provide their own sources of metabolic energy from nutrients in their environment. **Viruses**, by contrast, are the ultimate hitchhikers, carrying little more than genetic information in the form of nucleic acid. Many clinically

TABLE 23–1 Viruses That Cause Human Disease

Virus	Genome type	Disease
Adenovirus	Double-stranded DNA	Respiratory disease
Epstein–Barr virus (EBV)	Double-stranded DNA	Infectious mononucleosis
Herpes simplex virus 1	Double-stranded DNA	Recurrent cold sores
Human papillomavirus	Double-stranded DNA	Warts, cancer
Varicella-zoster virus	Double-stranded DNA	Chickenpox and shingles
Smallpox virus (Variola)	Double-stranded DNA	Smallpox
Hepatitis-B virus	Part single-, part double-stranded DNA	Hepatitis B
Human immunodeficiency virus (HIV)	Single-stranded RNA [+] strand	Acquired immune deficiency syndrome (AIDS)
Coronavirus	Single-stranded RNA [+] strand	Common cold, respiratory disease, COVID-19
Hepatitis-A virus	Single-stranded RNA [+] strand	Hepatitis A
Hepatitis-C virus	Single-stranded RNA [+] strand	Hepatitis C
Poliovirus	Single-stranded RNA [+] strand	Poliomyelitis
Rhinovirus	Single-stranded RNA [+] strand	Common cold
Yellow fever virus	Single-stranded RNA [+] strand	Yellow fever
Zika virus	Single-stranded RNA [+] strand	Zika virus disease
Ebola virus	Single-stranded RNA [–] strand	Ebola virus disease
Influenza virus type A	Single-stranded RNA [–] strand	Respiratory disease (flu)
Measles virus	Single-stranded RNA [–] strand	Measles
Mumps virus	Single-stranded RNA [–] strand	Mumps
Rabies virus	Single-stranded RNA [–] strand	Rabies

important human viruses have relatively small genomes consisting of double-stranded DNA or single-stranded RNA [either positive (+) sense RNA that can be directly translated into proteins or negative (–) sense RNA that must be replicated to produce (+) sense RNA] (**Table 23–1**). We now have complete genome sequences of almost all of them.

Viral genomes typically encode three types of protein: proteins for replicating the genome, proteins for packaging the genome and delivering it to more host cells, and proteins for modifying the structure or function of the host cell to enhance the replication of the virus. In general, viral replication involves the following sequence of steps: (1) entry into the host cell, (2) disassembly of the infectious virus particle, (3) replication of the viral genome, (4) transcription of viral genes and synthesis of viral proteins, (5) assembly of these viral components into progeny virus particles, and (6) release of progeny virions (**Figure 23–10**). A single virus particle (a *virion*) that infects a single host cell can produce thousands of progeny.

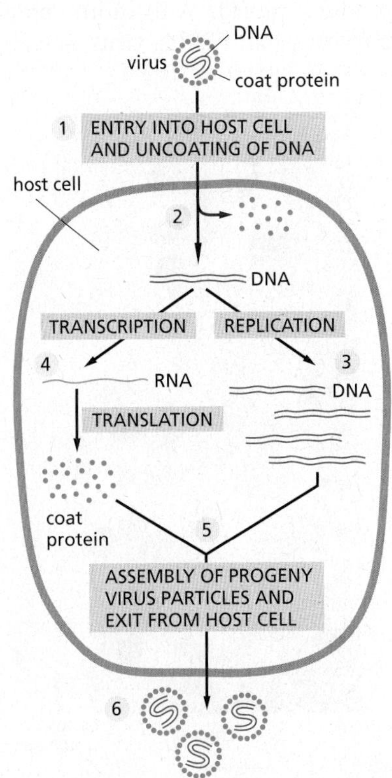

Figure 23–10 A simple viral life cycle. The hypothetical simple virus shown here consists of a small double-stranded DNA molecule that codes for only a single viral capsid protein. To reproduce, the viral genome must (1) enter a host cell, where it (2) disassembles and releases its genetic material. The genome is (3) replicated to produce multiple copies, and (4) transcribed and translated to produce the viral coat protein. The viral genomes can then (5) assemble spontaneously with the coat protein to form a new virus particle, which (6) escapes from the host cell. No known virus is this simple.

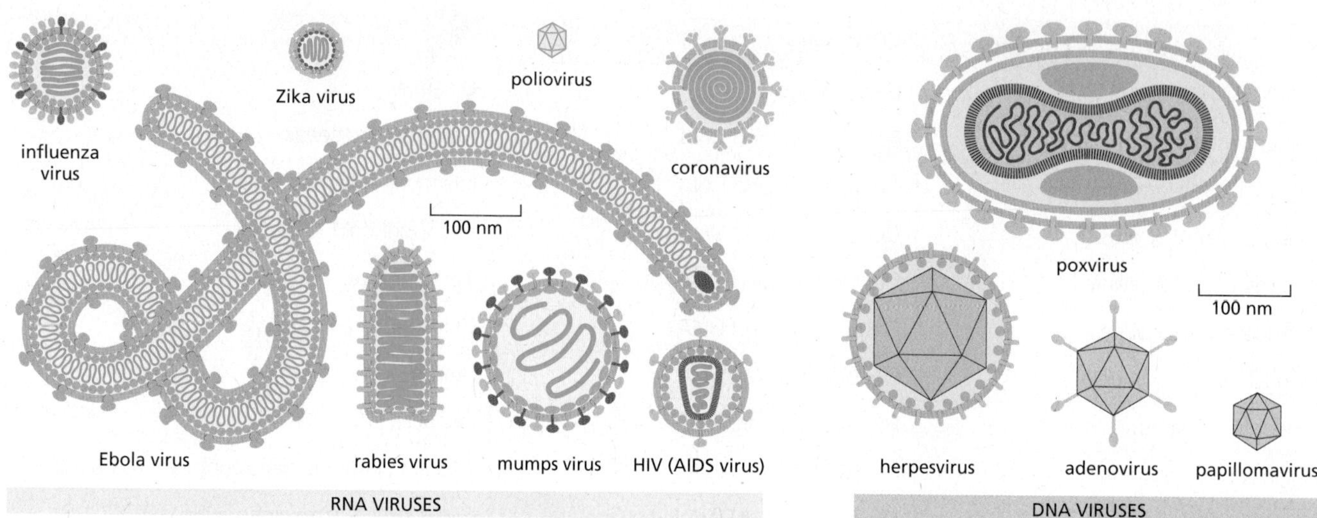

Figure 23–11 Examples of viral morphology. As shown, both DNA and RNA viruses vary greatly in both size and shape. Although there is a general correlation between virus physical size and genome size, there are likely to be outliers such as Ebola virus, which has a large physical size but a smaller genome size suggesting a lower degree of genome compaction.

Virions come in a wide variety of shapes and sizes (Figure 23–11), and although most have relatively small genomes, genome size can vary considerably. A giant virus of amoebae, called *pithovirus*, was recently revived from a 30,000-year-old ice core harvested from permafrost in Siberia and is the largest known virus by physical size, with 1.5-μm-long particles and a double-stranded DNA genome of 610,000 nucleotides. The virions of *poxviruses* are also large: they are 250–350 nm long and enclose a genome of double-stranded DNA of about 270,000 nucleotides. At the other end of the size scale are the virions of *parvovirus*, which are less than 30 nm in diameter and have a single-stranded DNA genome of fewer than 5000 nucleotides.

Viral genomes are packaged in a protein coat, called a **capsid**, which in some viruses is further enclosed by a lipid bilayer membrane, or envelope. The capsid is made of one or several proteins, arranged in regular arrays that often produce structures with either helical symmetry, which results in a cylindrical structure (for example, influenza, measles, and bunyavirus), or icosahedral symmetry (for example, poliovirus and herpesvirus; see Figure 23–11). Other viruses instead produce capsids with more complicated or irregular structures (for example, poxviruses and Ebola virus; see Figure 23–11). A capsid packaged with the viral

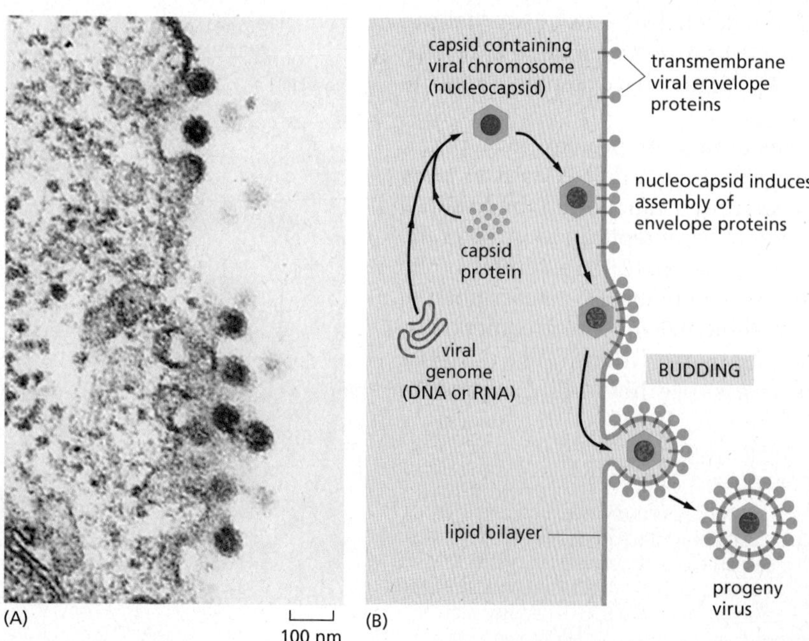

Figure 23–12 Acquisition of a viral envelope. (A) Electron micrograph of an animal cell from which several copies of an enveloped virus (*Semliki Forest virus*) are budding. (B) Schematic drawing of the envelope assembly and budding processes. The lipid bilayer that surrounds the viral capsid is derived directly from the plasma membrane of the host cell. In contrast, the proteins in this lipid bilayer (shown in *green*) are encoded by the viral genome. (A, from A. Loewy et al., *J. Virol.* 69:469–475, 1995. Reproduced with permission from the American Society for Microbiology.)

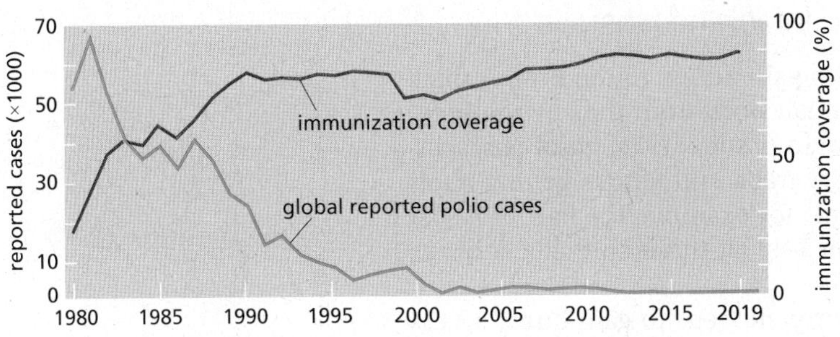

Figure 23–13 Effective control of a viral disease through vaccination. The graph shows the number of cases of poliomyelitis reported per year worldwide *(blue line)* and the global immunization coverage as a percentage of the human population *(red line)*. As immunization coverage increased, disease incidence decreased. (Figure based on information found at https://polioeradication.org.)

genome is called a *nucleocapsid*. The nucleocapsids of *nonenveloped viruses* usually leave an infected cell by lysing it. For *enveloped viruses*, by contrast, the nucleocapsid is enclosed within a lipid bilayer membrane that the virus acquires in the process of budding from the host-cell plasma membrane, which it does without disrupting the membrane or killing the cell (**Figure 23–12**).

Because the host cell performs most of the critical steps in viral replication, the identification of effective antiviral drugs that do not harm the host can be difficult. The most effective strategy for containing viral diseases is through vaccination of potential hosts. For example, a highly successful vaccination program eradicated smallpox infection from the planet by 1980. Vaccines effective against *poliovirus*, which causes poliomyelitis and paralysis, have existed since the 1950s, and a more recent global vaccination initiative, begun in 1988, seeks to completely eradicate poliovirus infection. The number of poliomyelitis cases worldwide has declined from an estimated 350,000 in 1988, to 176 in 2019, and there is hope of eradication in the coming years (**Figure 23–13**).

Summary

Infectious diseases are caused by pathogens, which include viruses, bacteria, and fungi, as well as protozoan and metazoan parasites. All pathogens must have mechanisms for entering their host and for evading immediate destruction by the host. Pathogenic bacteria produce specific virulence factors that mediate their interactions with the host; these proteins change the behavior of host cells in ways that promote the replication and spread of the bacteria. Eukaryotic pathogens such as fungi and protozoan parasites typically pass through several different forms during the course of infection; the ability to switch among these forms is usually required for these pathogens to survive in a host and cause disease. In some cases, such as malaria, parasites must pass sequentially through several host species to complete their life cycles. Unlike bacteria and eukaryotic parasites, viruses have no metabolism of their own and no intrinsic ability to produce the proteins encoded by their DNA or RNA genomes; they rely on subverting the machinery of their host cell.

CELL BIOLOGY OF PATHOGEN INFECTION

The mechanisms through which pathogens cause disease are extremely diverse. Nonetheless, all pathogens must carry out certain common tasks: they must gain access to the host, reach an appropriate growth niche, avoid host defenses, replicate, and exit from the infected host to spread to an uninfected one. In this section, we examine the cell-biological strategies that many pathogens use to accomplish these tasks. We describe how pathogens initially gain access to the host by overcoming epithelial barriers and colonizing epithelia. We then recount how extracellular pathogens disturb host cells without entering them to establish a niche and avoid host defenses. We next cover the myriad ways in which intracellular pathogens, the master manipulators of host-cell biology, enter host cells, reach an appropriate niche in the cytosol or in a membrane-bound compartment, avoid or alter membrane traffic, mobilize the cytoskeleton, manipulate autophagy, and take over metabolism. Finally, we cover how pathogen evolution shapes the immune avoidance and drug resistance.

Pathogens Breach Epithelial Barriers to Infect the Host

The first step in infection is for the pathogen to gain access to the host. A thick covering of skin protects most parts of the human body from the environment. The protective boundaries of some other human tissues (eyes, nasal passages, respiratory tract, mouth, digestive tract, urinary tract, and female genital tract) are less robust. In the lungs and small intestine, for example, the barrier is just a single monolayer of epithelial cells. Nonetheless, all these epithelia serve as barriers to infection.

Wounds in barrier epithelia are one way for pathogens to gain direct access to unoccupied niches within otherwise sterile host tissues. This avenue of entry requires little in the way of pathogen specialization, and some pathogens can cause serious illness if they enter through such wounds. Staphylococci from the skin and nose and streptococci from the throat and mouth are two examples of opportunistic bacterial pathogens that can be residents of the normal microbiota (described in more detail in the final section of this chapter) in certain individuals, yet are also responsible for many serious infections resulting from breaches in epithelial barriers. The emergence of bacterial strains of *Staphylococcus* that are resistant to the antibiotics commonly used for treatment (for example, methicillin-resistant *Staphylococcus aureus*, or MRSA, which can cause serious skin and tissue infections) is of particular concern. Papillomaviruses, which cause warts and cervical cancer, also take advantage of breaches in epithelial barriers.

Another efficient way for a pathogen to cross the skin is to catch a ride in the saliva of a biting arthropod. A diverse group of bacteria, viruses, and protozoa has developed the ability to survive in arthropods that they use as *vectors* for transmission to a mammalian host. Many *zoonoses*, a term that refers to diseases spread to humans from other animals, are spread in this way. As discussed earlier, the *Plasmodium* protozoan that causes malaria develops through several forms in its life cycle, including some that are specialized for survival in a human and others that are specialized for survival in a mosquito vector (see Figure 23–9). Viruses that are spread by mosquito bites also cause yellow fever, dengue fever, as well as Zika virus disease—the latter garnered worldwide attention during an epidemic in 2015–2016, and although Zika infection is often asymptomatic in adults, in pregnant women it can cause birth defects in the fetus. These three viruses replicate in both insect cells and mammalian cells, as required for their transmission by an insect vector.

The efficient spread of a pathogen via an insect vector requires that an individual insect consumes a blood meal from an infected host and transfers the pathogen to a nonimmune host. In a few striking cases, the behavior of the insect is altered by the pathogen so that its transmission to a new host is more likely. An example is the bacterium *Yersinia pestis*, which causes bubonic plague. It multiplies in the flea's foregut to form aggregated masses that physically block the digestive tract; during each repeated, but futile, attempt at feeding, some of the bacteria in the foregut are flushed into the bite site, thus transmitting plague to a new host (**Figure 23–14**).

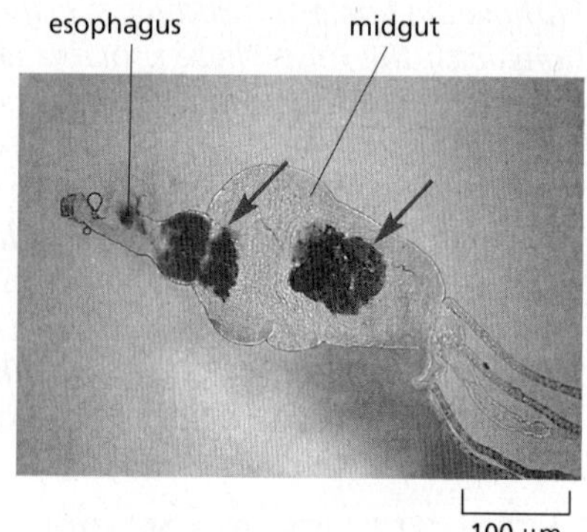

Figure 23–14 Plague bacteria within a flea. This light micrograph shows the digestive tract dissected from a flea that had dined about 2 weeks previously on the blood of an animal infected with the plague bacterium, *Yersinia pestis*. The bacteria multiplied in the flea gut to produce large cohesive aggregates *(red arrows)*; the bacterial mass on the left is occluding the passage between the esophagus and the midgut. This type of blockage prevents a flea from digesting its blood meals, so that hunger causes it to bite repeatedly, disseminating the infection. (From B.J. Hinnebusch, E.R. Fischer, and T.G. Schwan, *J. Infect. Dis.* 178:1406–1415, 1998.)

Pathogens That Colonize an Epithelium Must Overcome Its Protective Mechanisms

Epithelial barriers such as the skin lining of the mouth and large intestine are densely populated by the microbiota, whereas others, including the lining of the lower lung and the bladder, are more sparsely populated. Nevertheless, these epithelial barriers have mechanisms to prevent excessive microbial colonization. A layer of protective mucus covers the respiratory epithelium, and the coordinated beating of motile cilia sweeps the mucus and trapped bacteria up and out of the lung. The epithelial lining of the bladder and the upper gastrointestinal tract also has a thick layer of mucus, and these organs are periodically flushed by urination and by peristalsis, respectively, which washes away microbes.

Pathogens that infect these epithelial surfaces, including pathogenic bacteria and eukaryotic parasites, have evolved specific features to overcome these protective mechanisms. Those that infect the urinary tract, for example, adhere

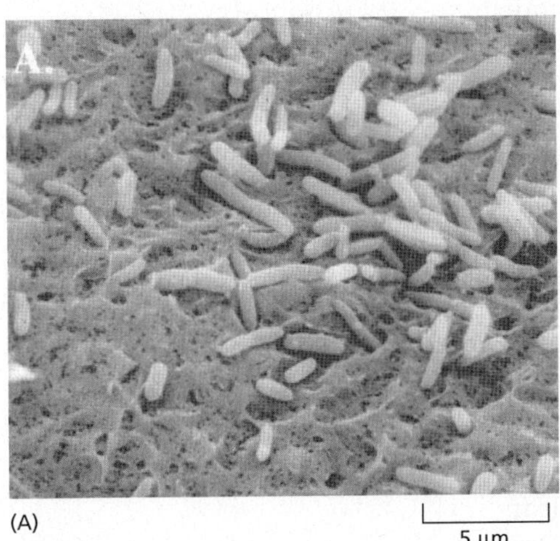

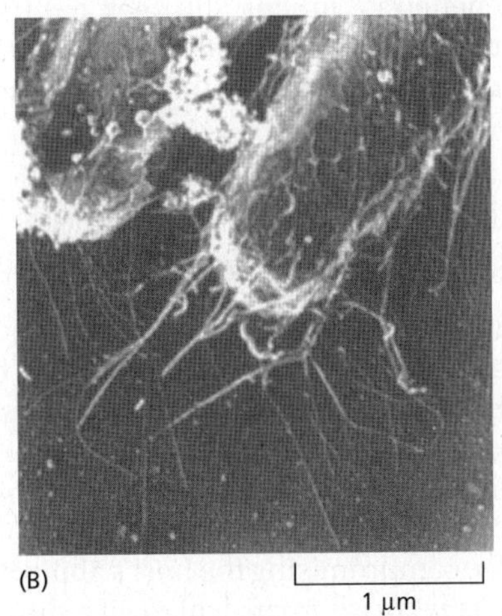

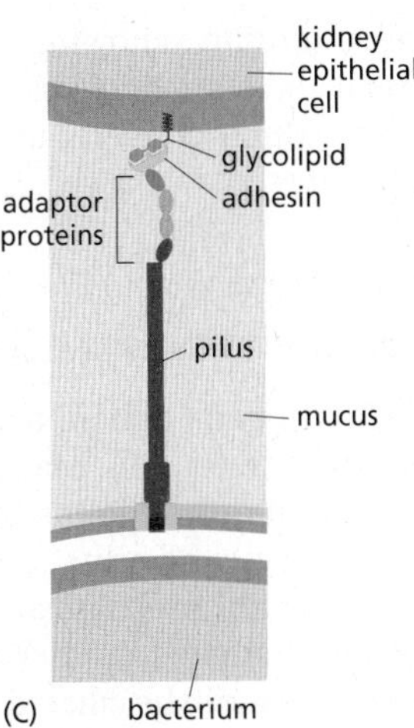

Figure 23–15 Pathogenic *E. coli* in the infected bladder of a mouse. (A) Scanning electron micrograph of uropathogenic *E. coli*, a common cause of bladder and kidney infections. The bacteria are attached to the surface of epithelial cells lining the infected bladder. (B) A close-up view of one of the bacteria showing the pili on its surface. (C) An *E. coli* pilus spans the mucus layer and has adaptor proteins on its tip that bind to glycolipids on the surface of kidney cells. (A, from G.E. Soto and S.J. Hultgren, *J. Bacteriol.* 181:1059–1071, 1999. With permission from the American Society for Microbiology; B, from D.G. Thanassi and S.J. Hultgren, *Methods* 20:111–126, 2000. With permission from Elsevier.)

tightly to the epithelial lining via specific **adhesins**, which are proteins or protein complexes that recognize and bind to cell-surface molecules on the epithelium. An important group of adhesins in *E. coli* strains that infect the kidney are components of *pili*—surface projections that can be several micrometers long and thus able to span the thickness of the protective mucus layer; at the tip of each pilus is an adhesin protein that binds tightly to the D-galactose–D-galactose disaccharide on glycolipids on the surface of kidney cells (Figure 23–15). Strains of *E. coli* that infect the bladder rather than the kidney express a second kind of pilus with a different adhesin protein that binds to D-mannose-decorated proteins on bladder epithelial cells. It is the specificity of the adhesin proteins on the tips of the two types of pili that is responsible for the bacteria colonizing different parts of the urinary tract.

The epithelial lining of the stomach is an especially hostile environment for microbes. Besides the thick layer of mucus and peristaltic washing, the acidic pH (average pH ≈ 2) is lethal to almost all bacteria ingested in food. Yet, the stomach is home to resident microbial species including *H. pylori*, which persists for life as part of the stomach microbiota of approximately half of all humans on Earth. Although it does not cause disease in most individuals, *H. pylori* can cause stomach ulcers and cancers. The hypothesis that a persistent bacterial infection could cause stomach ulcers was initially met with skepticism. The young Australian doctor who made the initial discovery finally proved the point: he drank a pure culture of bacteria and developed inflammation of the stomach, which often precedes the development of ulcers. Antibiotics can now effectively cure a patient of recurrent ulcers. *H. pylori* has several adaptations that allow it to colonize the harsh environment in the stomach (Figure 23–16). One is to use its flagella

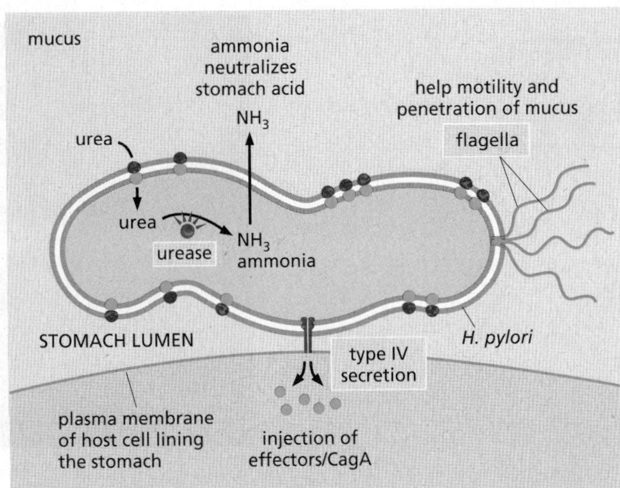

Figure 23–16 *H. pylori* interaction with epithelial cells in the stomach lining. These bacteria colonize the gastric epithelium by using their flagella to swim through the mucus layer lining the stomach and attach to the surface of epithelial cells. *H. pylori* produces the enzyme urease, which converts urea obtained from the environment into ammonia that is released and neutralizes acid surrounding the bacterium, raising the pH in its local environment. *H. pylori* also uses its type IV secretion system to secrete the effector protein CagA, which assists in colonization but also causes chronic disruption of host-cell pathways and prolonged inflammation, which predisposes infected individuals to gastric ulcers and cancer.

for chemotactic motility, allowing it to penetrate mucus and seek out the more neutral pH near the surface of gastric epithelial cells. Another is to produce the enzyme *urease*, which converts urea to ammonia to neutralize the acid in its immediate vicinity.

Extracellular Pathogens Use Toxins and Contact-dependent Secretion Systems to Disturb Host Cells Without Entering Them

Extracellular pathogens adhere to epithelial and other surfaces without entering their underlying host cells. Many such pathogens exert effects on the host by secreting toxins that diffuse to target cells. For example, *Bordetella pertussis*, the bacterium that causes whooping cough, colonizes the respiratory epithelium and circumvents the normal mechanism that clears the respiratory tract by expressing adhesins that bind ciliated epithelial cells. The adherent bacteria also produce toxins that eventually kill the ciliated cells, compromising the host's ability to clear the infection. The most familiar of these is *pertussis toxin*, which, like the cholera toxin discussed earlier (both are shown in Figure 23–6), has an A subunit that ADP-ribosylates the α subunit of the G protein G_i, causing the G protein to remain in the inactive GDP-bound state and preventing it from inhibiting the activity of the host cell's adenylyl cyclase, thereby increasing the production of cyclic AMP. This toxin also interferes with the chemotactic pathway that neutrophils use to seek out and destroy bacteria (see Figures 16–3 and 16–81). *B. pertussis* colonization of the respiratory tract causes severe coughing, which helps spread the infection.

Other extracellular pathogens use contact-dependent secretion systems to directly inject bacterial effector proteins into the host cells to which they adhere. An example is *H. pylori*, which infects the stomach epithelium as described above. The *H. pylori* genome contains a pathogenicity island (see Figure 23–4A) that encodes a type IV secretion system and the effector protein CagA, which is injected by the secretion system into the host cell (Figure 23–16). CagA helps the bacterium persist in the stomach by affecting cell movement, inducing inflammation, altering host gene expression, changing cell proliferation and apoptosis, and disrupting cell–cell junctions. However, chronic alteration of these pathways is also a predisposing factor for gastric ulcers and cancer.

Another example is enteropathogenic *E. coli* (EPEC), which causes diarrhea and can be lethal to young children. EPEC uses a type III secretion system (see Figure 23–7) to deliver its own special receptor protein (called *Tir*) into the plasma membrane of a host intestinal epithelial cell (**Figure 23–17** and **Movie 23.4**). The extracellular domain of Tir binds to the bacterial surface protein *intimin*, triggering actin polymerization in the host cell that results in the formation of a unique cell-surface protrusion called a *pedestal*; this pushes the tightly adherent bacteria up about 1–5 μm from the host-cell membrane, thereby promoting bacterial

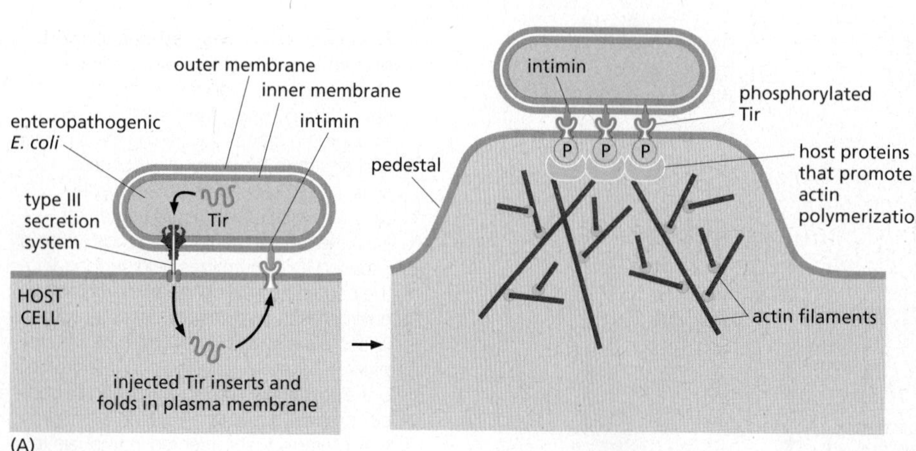

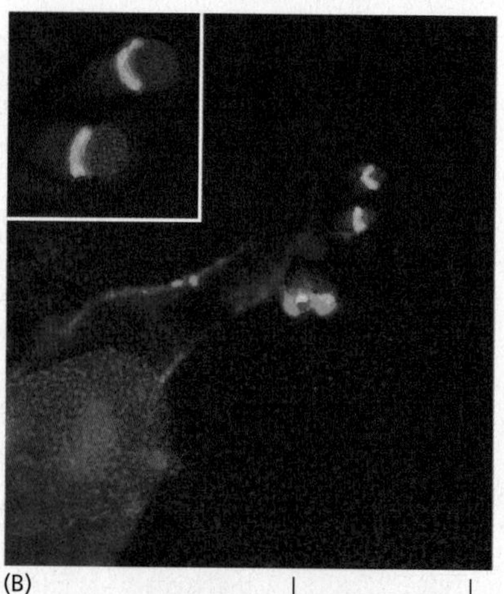

Figure 23–17 Interaction of enteropathogenic *E. coli* (EPEC) with host intestinal epithelial cells. (A) When EPEC contacts an epithelial cell in the lining of the human gut, it delivers a bacterial protein called Tir into the host cell through a type III secretion system. Tir then inserts into the plasma membrane of the host cell, where it functions as a receptor for the bacterial adhesin protein intimin. Next, a host-cell protein tyrosine kinase phosphorylates the intracellular domain of Tir on tyrosines. Phosphorylated Tir recruits host-cell proteins [including an adaptor protein, a nucleation-promoting factor (NPF), and the Arp2/3 complex] that trigger actin polymerization (see Figure 16–12). Consequently, a branched network of actin filaments assembles underneath the bacterium, forming an actin pedestal (see Movie 23.4). (B) EPEC on a pedestal. In this fluorescence micrograph, the DNA of the EPEC and host cell is labeled in *blue*, Tir protein is labeled in *green*, and host-cell actin filaments are labeled in *red*. The inset shows a close-up view of the two upper bacteria on pedestals. (B, from D. Goosney et al., *Annu. Rev. Cell Dev. Biol.* 16:173–189, 2000. With permission from Annual Reviews.)

movement along the cell surface by a mechanism described later for the actin-based motility of intracellular pathogens. A similar strategy is used by vaccinia virus (the virus that was used as a vaccine to eradicate smallpox) to form mobile actin-rich membrane extensions, which promote spread of the virus from cell to cell. The study of how EPEC and vaccinia virus promote actin polymerization has been of major importance in understanding how intracellular signaling pathways regulate the cytoskeleton in normal, uninfected cells (discussed in Chapter 16). Although actin mobilization promotes the spread of these pathogens, the symptoms of EPEC infection (severe diarrhea) are caused by the loss of absorptive microvilli and disruption of signaling pathways in epithelial cells, which are triggered by Tir and other secreted bacterial effector proteins.

Intracellular Pathogens Have Mechanisms for Both Entering and Leaving Host Cells

Intracellular pathogens have to cross barriers, adhere to host cells, and also enter host cells to cause disease. These include all viruses and many bacteria and protozoa. Each of these has a preferred niche for replication and survival within host cells. Bacteria and protozoa replicate either in the cytosol or within a membrane-enclosed compartment. While most RNA viruses replicate within the cytosol, most DNA viruses replicate in the nucleus (poxviruses are a notable exception). Life inside a host cell has several advantages. The pathogens are not accessible to *antibodies*, nor are they easy targets for phagocytic cells (discussed in Chapter 24); furthermore, intracellular bacteria and protozoa are bathed in a rich source of nutrients, and viruses have access to the host cell's biosynthetic machinery for their reproduction. This lifestyle, however, requires that the pathogen have mechanisms for entering host cells, for finding a suitable subcellular niche where it can replicate, and for exiting from the infected cell to spread the infection. Below we consider some of the myriad ways that individual intracellular pathogens exploit and modify host-cell biology to satisfy these requirements.

Viruses Bind to Virus Receptors at the Host-Cell Surface

The first step in infection for any intracellular pathogen is to bind to the surface of the host target cell. Viruses accomplish this by the binding of viral surface proteins to **virus receptors**, which are cell-surface proteins that perform various functions in uninfected cells and have been co-opted as binding sites for viral proteins. The first virus receptor identified was an *E. coli* surface protein that is recognized by the bacteriophage lambda; the protein normally functions to transport the sugar maltose from outside the bacterium to the inside where it is used as an energy source. Receptors need not be proteins, however: an envelope protein of herpes simplex virus, for example, binds to heparan sulfate proteoglycans (discussed in Chapter 19) on the surface of certain vertebrate host cells, and simian virus 40 (SV40) binds to a glycolipid. The specificity of virus–receptor interactions often serves as a barrier preventing the spread of a virus from one species to another. Acquiring the ability to bind to a new receptor often requires multiple changes in a virus, but it can be crucial in allowing the cross-species transmission that can result in new disease outbreaks.

Viruses that infect animal cells generally exploit cell-surface receptor molecules that are either ubiquitous [such as angiotensin-converting enzyme 2 (ACE2) used by the coronavirus SARS-CoV-2] or are found uniquely on those cell types in which the virus replicates (such as the neuron-specific proteins used by rabies virus). Although a virus usually uses a single type of host-cell receptor, some viruses use more than one type. An important example is HIV, which requires two types of receptors to enter a host cell. Its primary receptor is CD4, a cell-surface protein on helper T cells and macrophages that is involved in immune recognition (discussed in Chapter 24). It also requires a co-receptor, which is either CCR5 (a receptor for β-chemokines) or CXCR4 (a receptor for α-chemokines), depending on the particular variant of the virus; macrophages are susceptible

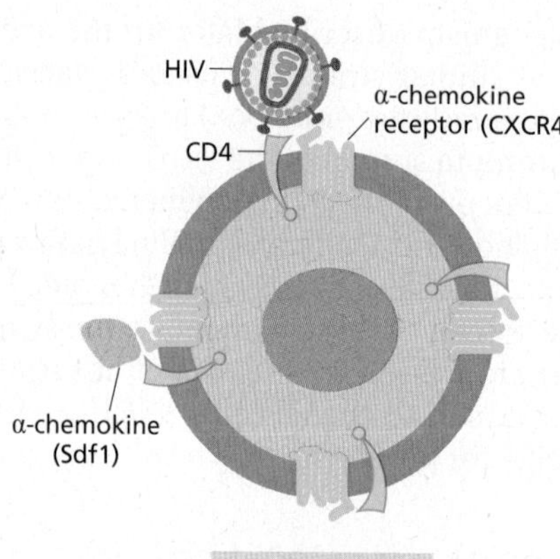

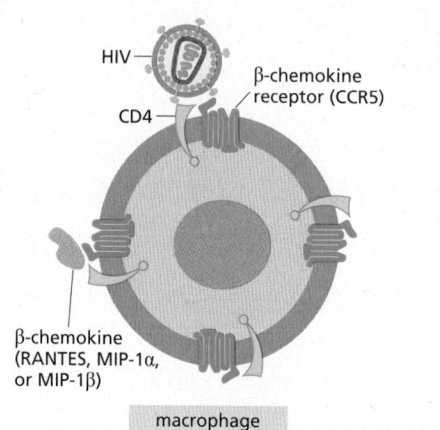

Figure 23–18 Receptor and co-receptors for HIV. All strains of HIV require the CD4 protein as a primary receptor. Early in an infection, most of the viruses use CCR5 as a co-receptor, allowing them to infect macrophages and their precursors, monocytes. As the infection progresses, mutant variants of the virus arise that now use CXCR4 as a co-receptor, enabling them to infect helper T cells efficiently. The natural ligand for the chemokine receptors (Sdf1 for CXCR4; RANTES, MIP-1α, or MIP-1β for CCR5) blocks co-receptor function and prevents viral invasion.

only to HIV variants that use CCR5 for entry, whereas helper T cells are most efficiently infected by variants that use CXCR4 (**Figure 23–18**). The viruses that are found within the first few months after HIV infection almost invariably use CCR5, which explains why individuals that carry an altered *CCR5* gene are less susceptible to HIV infection. In the later stages of infection, viruses often either switch to use CXCR4 or adapt to use both co-receptors through the accumulation of viral mutations. In this way, the virus can change the cell types it infects as the disease progresses. It may seem paradoxical that viruses would infect immune cells, as we might expect that virus binding would trigger an immune response; but invasion of an immune cell can be a useful way for a virus to weaken the immune response and travel around the body to infect other immune cells.

Viruses Enter Host Cells by Membrane Fusion, Pore Formation, or Membrane Disruption

After recognition and attachment to the host-cell surface, viruses must enter into the cell for viral replication to proceed. Entry into host cells requires that viruses overcome different challenges that depend on virus size and structure (see Figure 23–11). **Enveloped viruses** must regulate membrane fusion processes, both to ensure that their membrane envelopes fuse only with the appropriate host-cell membrane and to prevent fusion with one another. Such regulation is achieved by the coronavirus SARS-CoV-2, for example, by requiring both binding of the virus spike protein to the ACE2 receptor and cleaving of the spike protein by a host-cell protease to enable fusion with the host-cell membrane.

Some enveloped viruses, such as HIV, enter the host cell by fusing their envelope membrane at neutral pH with the plasma membrane (**Figure 23–19A**). In this scenario, binding to receptors or co-receptors usually triggers a conformational change in a viral envelope protein that exposes a normally buried fusion peptide (discussed in Chapter 13).

Most enveloped (and nonenveloped) viruses enter cells by activating signaling pathways in the cell that induce endocytosis, commonly via clathrin-coated pits (see Figure 13–7), leading to internalization into endosomes. Large viruses that do not fit into clathrin-coated vesicles, such as poxviruses, often enter cells by *macropinocytosis*, a process by which membrane ruffles fold over and entrap fluid into macropinosomes (see Figure 13–68).

Once inside endosomes, enveloped viruses such as influenza A virus must fuse their envelope with the endosomal membrane from the luminal side, and they frequently sense the acid environment in the late endosome as a cue to trigger a conformational change in a viral surface protein that exposes a fusion peptide (**Figure 23–19B** and **Movie 23.5**). The mechanism of membrane fusion mediated by viral spike glycoproteins has similarities with SNARE-mediated membrane fusion during normal vesicular trafficking (discussed in Chapter 13). The H^+ pumped into the early endosome also has another effect; it enters the influenza virion through an ion channel in the viral envelope and triggers changes in the

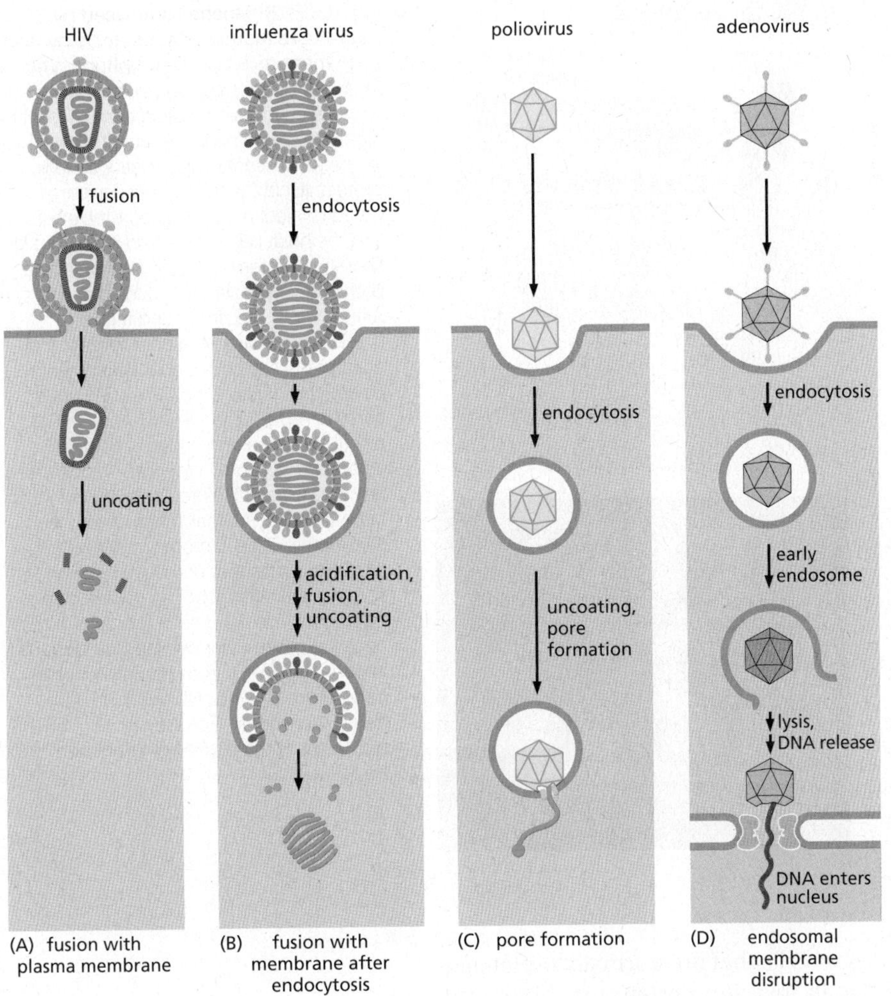

Figure 23–19 Four virus entry strategies that follow virus binding to virus receptors. (A) Some enveloped viruses, such as HIV, fuse directly with the host-cell plasma membrane to release their RNA genome *(blue)* and capsid proteins *(brown)* into the cytosol. (B) Other enveloped viruses, such as influenza virus, first bind to cell-surface receptors, triggering receptor-mediated endocytosis; when the endosome acidifies, the virus envelope fuses with the endosomal membrane, releasing the viral RNA genome *(blue)* and capsid proteins *(tan)* into the cytosol. (C) Poliovirus, a nonenveloped virus, induces receptor-mediated endocytosis, and then forms a pore in the endosomal membrane to extrude its RNA genome *(blue)* into the cytosol. (D) Adenovirus, another nonenveloped virus, uses a more complicated strategy: it induces receptor-mediated endocytosis and then disrupts the endosomal membrane, releasing the capsid including its DNA genome into the cytosol. The trimmed-down virus eventually docks onto a nuclear pore and releases its DNA *(red)* directly into the nucleus where it is transcribed and replicated (see Movie 23.5).

viral capsid. These priming steps allow the capsids to disassemble once released into the cytosol after virus fusion with the late endosomal membrane.

Because they lack a surrounding lipid bilayer, **nonenveloped viruses** enter host cells in a fundamentally different way. Poliovirus binds to a cell-surface receptor, triggering both receptor-mediated endocytosis (see Figure 13–54) and a conformational change in the viral particle. The conformational change exposes a hydrophobic projection on one of the capsid proteins, which inserts into the endosomal membrane to form a pore. The viral RNA genome then enters the cytosol through the pore, leaving the capsid in the endosome (**Figure 23–19C**). *Adenovirus* disrupts the endosomal membrane after it is taken up by receptor-mediated endocytosis, releasing the remainder of the virus into the cytosol. During endosomal trafficking and subsequent transport within the cytosol, adenoviruses undergo multiple uncoating steps, which sequentially remove structural proteins and ready the virus particles to release their DNA into the nucleus through nuclear pore complexes (**Figure 23–19D**).

Bacteria Enter Host Cells by Phagocytosis

Bacteria are much larger than viruses—too large to be taken up either through pores or by receptor-mediated endocytosis. Instead, they enter host cells by phagocytosis, which is a normal function of phagocytes such as neutrophils, macrophages, and dendritic cells (discussed in Chapter 24). These phagocytes patrol the tissues of the body and ingest and destroy microbes; however, some intracellular bacterial pathogens such as *M. tuberculosis* use this to their advantage and have evolved to survive and multiply inside macrophages.

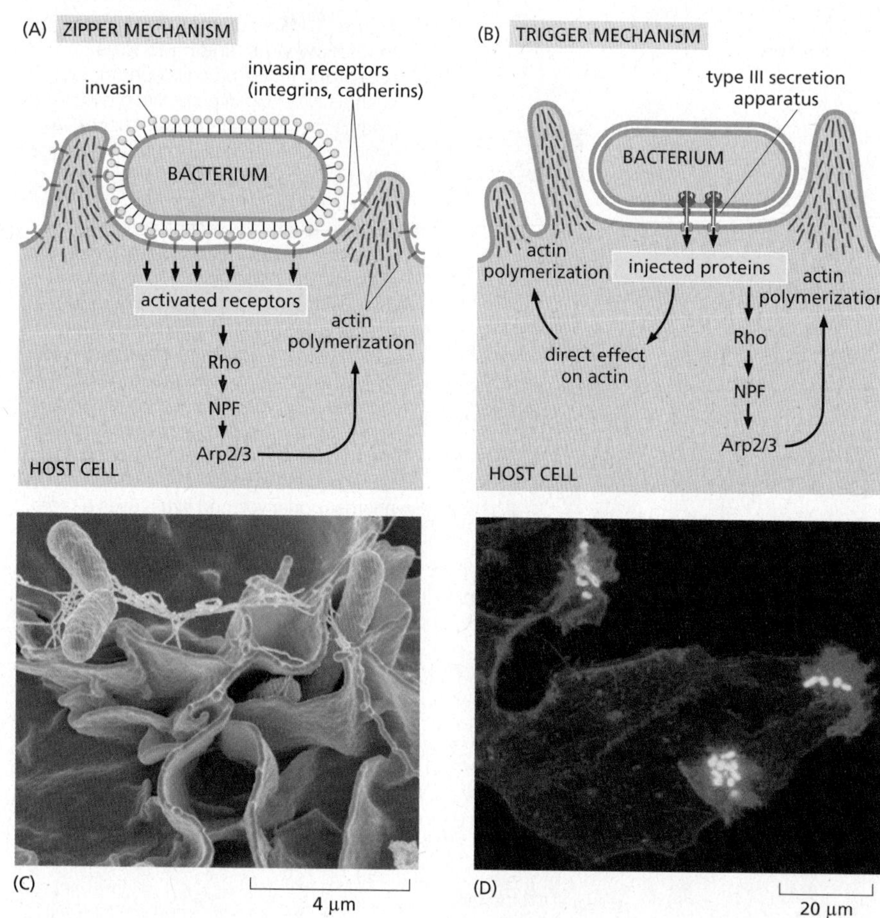

Figure 23–20 Mechanisms used by bacteria to induce phagocytosis by host cells that are normally nonphagocytic. (A) In the zipper mechanism, bacteria express an invasion protein that binds with high affinity to a host-cell receptor, which is often a cell–cell or cell–matrix adhesion protein. (B) In the trigger mechanism, bacteria inject a set of effector molecules into the host-cell cytosol through a type III secretion system called SPI1 (*Salmonella* pathogenicity island 1), inducing membrane ruffling. Both the zipper and trigger mechanisms cause the polymerization of actin at the site of bacterial attachment by activating Rho family small GTPases and the Arp2/3 complex. (C) A scanning electron micrograph showing *Salmonella enterica* invasion by the trigger mechanism. Bacteria (pseudocolored *yellow*) are shown surrounded by a small membrane ruffle. (D) Fluorescence micrograph showing that the large ruffles that engulf the *Salmonella* bacteria are actin rich. The bacteria are labeled in *green* and actin filaments in *red*; because of the color overlap, the bacteria appear *yellow*. (C, from Rocky Mountain Laboratories, NIAID, NIH; D, from J.E. Galán, *Annu. Rev. Cell Dev. Biol.* 17:53–86, 2001. With permission from Annual Reviews.)

Some bacterial pathogens can invade host cells that are normally nonphagocytic. One way they do so is by expressing an invasion protein that binds with high affinity to a host-cell receptor, which is often a cell–cell or cell–matrix adhesion protein (discussed in Chapter 19). For example, *Yersinia pseudotuberculosis* (a bacterium that causes diarrhea and is a close relative of the plague bacterium *Y. pestis*) expresses a protein called invasin that has an RGD motif that is similar to fibronectin's and which likewise is recognized by host-cell β_1-integrins (see Figure 19–48). *Listeria monocytogenes*, which causes a rare but serious form of food poisoning, invades host cells by expressing a protein that binds to the cell–cell adhesion protein E-cadherin (see Figures 19–4 and 19–5). For both these bacterial species, binding of the bacterial invasion proteins to the host-cell adhesion proteins stimulates signaling through members of the Rho family of small GTPases (discussed in Chapter 16). This in turn activates NPFs and the Arp2/3 complex, leading to actin polymerization at the site of bacterial attachment. Actin polymerization, sometimes accompanied by the assembly of a clathrin coat, drives the advancement of the host cell's plasma membrane over the adhesive surface of the microbe, resulting in the phagocytosis of the bacterium—a process known as the *zipper mechanism* of invasion (**Figure 23–20A**).

A second pathway by which bacteria can invade nonphagocytic cells is known as the *trigger mechanism* (**Figure 23–20B**). It is used by various pathogens, including the food-borne pathogen *Salmonella enterica* serovar Typhimurium, and it is initiated when the bacterium injects a set of effector molecules into the host-cell cytosol through a type III secretion system (see Figure 23–7). Some of these effector molecules activate Rho family proteins, which in turn stimulate actin polymerization, as just discussed. Other bacterial effector proteins directly interact with host-cell cytoskeletal elements, nucleating and stabilizing actin filaments and causing the rearrangement of actin cross-linking proteins. The overall effect is to cause the formation of ruffles on the surface of the host cell (**Figure 23–20C and D**),

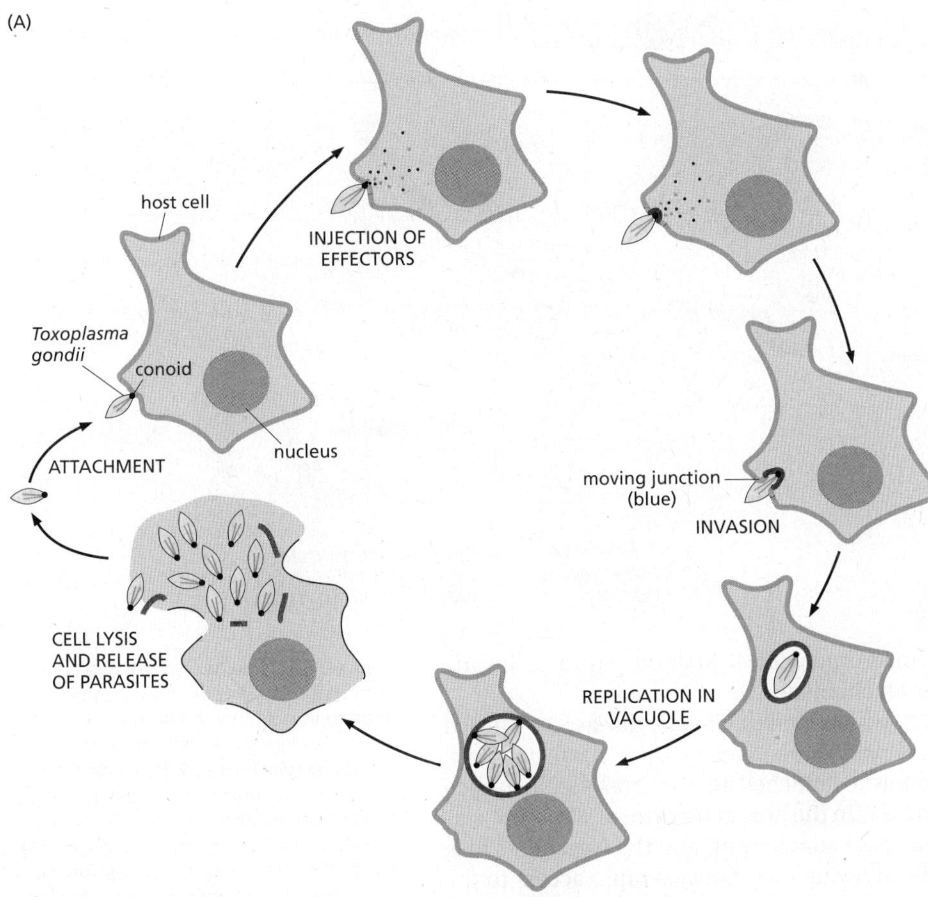

(B)
parasite in vacuole
host-cell nucleus
14 μm

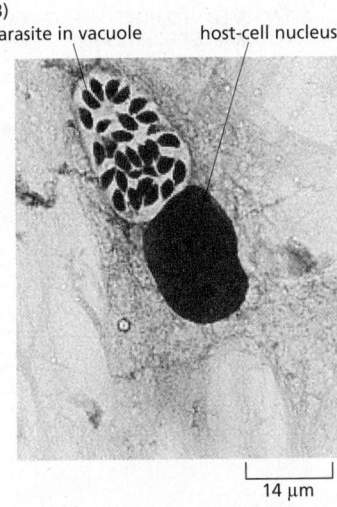

Figure 23–21 The life cycle of the intracellular parasite *Toxoplasma gondii*. (A) After attachment to a host cell, *T. gondii* uses its conoid to inject effector proteins that facilitate invasion. These effector proteins include a receptor that binds a parasite surface protein, as well as components of the moving junction (shown in *blue*) through which the parasite squeezes as it enters the host cell. As the host cell's plasma membrane invaginates, the parasite somehow removes the normal host-cell membrane proteins, so that the compartment (shown in *red*) does not fuse with lysosomes. After several rounds of replication, the parasite causes the compartment to break down and the host cell to lyse, releasing the progeny parasites to infect other host cells (see Movie 23.6). (B) Light micrograph of *T. gondii* replicating within a membrane-enclosed compartment (a parasitophorous vacuole) in a cultured cell. (B, courtesy of Manuel Camps and John Boothroyd.)

which fold over and engulf the bacteria by a process that resembles macropinocytosis (see Figure 13–68). The appearance of cells being invaded by use of the trigger mechanism is similar to the ruffling induced by some extracellular growth factors, suggesting that the bacteria hijack normal intracellular signaling pathways.

Intracellular Eukaryotic Parasites Actively Invade Host Cells

The uptake of viruses and bacteria into host cells is carried out largely by the host, with the pathogen being a relatively passive participant. In contrast, intracellular eukaryotic parasites, which are typically much larger than other types of intracellular pathogens, invade host cells through a variety of complex pathways that usually require energy expenditure by the parasite.

Toxoplasma gondii, a cat parasite that also causes occasional serious human infections in pregnant and immunocompromised individuals, is an example. *T. gondii* invasion involves the secretion of parasite effector proteins that target host-cell components, as well as the activities of the parasite's own cytoskeleton (**Figure 23–21** and **Movie 23.6**). When this protozoan contacts a host cell, it protrudes an unusual microtubule-based structure called a *conoid*, which facilitates host-cell entry. At the point of contact, the parasite discharges effector proteins into the host cell from specialized secretory organelles. One of these effector proteins is a receptor that inserts into the host-cell plasma membrane and binds to a parasite surface protein. Other effector proteins form a ring-like *moving junction* through which the parasite squeezes using forces generated by its own unusual actin and myosin cytoskeleton. Remarkably, as the parasite invades it removes host transmembrane proteins from the surrounding membrane, so that it is eventually protected in a membrane-enclosed compartment, or *parasitophorous vacuole*, that does not fuse with lysosomes and does not participate in host-cell membrane trafficking processes. The specialized membrane is selectively porous: it allows the parasite to take up small metabolic intermediates and nutrients from

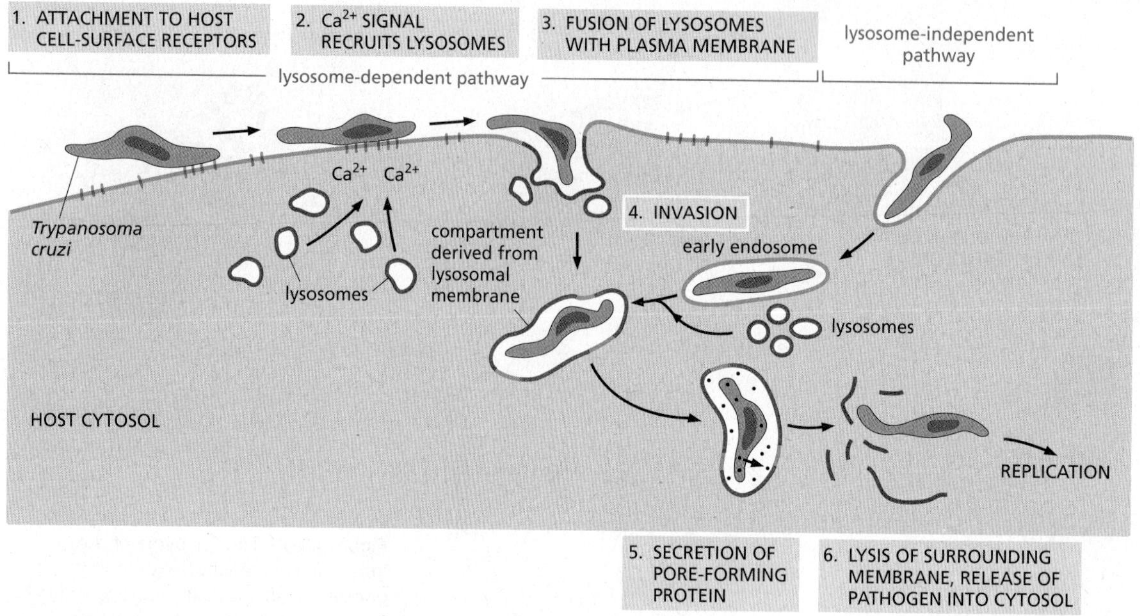

Figure 23–22 **The two alternative strategies that *Trypanosoma cruzi* uses to invade host cells.** In the lysosome-dependent pathway *(left)*, *T. cruzi* recruits host-cell lysosomes to its site of attachment to the host cell. The lysosomes fuse with the invaginating plasma membrane to create an intracellular compartment constructed almost entirely of lysosomal membrane. After a brief stay in the compartment, the parasite secretes a pore-forming protein that disrupts the surrounding membrane, thereby allowing the parasite to escape into the host-cell cytosol and proliferate. In the lysosome-independent pathway *(right)*, the parasite induces the host plasma membrane to invaginate and pinch off without recruiting lysosomes; then, lysosomes fuse with the endosome prior to the parasite's escape into the cytosol.

the host cell's cytosol but excludes macromolecules. Malaria parasites invade human red blood cells using a similar mechanism.

The protozoan *Trypanosoma cruzi*, which causes Chagas disease in Mexico and Central and South America, uses two alternative invasion strategies. In a *lysosome-dependent pathway*, the parasite attaches to the host's cell-surface receptors, inducing a local increase in Ca^{2+} in the host cell's cytosol. The Ca^{2+} signal recruits lysosomes to the site of parasite attachment, and the lysosomes fuse with the host cell's plasma membrane, allowing the parasites rapid access to the lysosomal compartment (**Figure 23–22**). In a *lysosome-independent pathway*, the parasite penetrates the host-cell plasma membrane by inducing the membrane to invaginate, without lysosome recruitment.

Some Intracellular Pathogens Escape from the Phagosome into the Cytosol

All intracellular pathogens, including viruses, bacteria, and eukaryotic parasites, face a similar problem: they must find a compartment within the host cell where they can replicate themselves. After their endocytosis by a host cell, they usually find themselves in an endosomal compartment, which normally would fuse with lysosomes (see Figure 13–67)—a dangerous place for pathogens because of the presence of many antimicrobial factors. To survive, pathogens use a variety of strategies (**Figure 23–23**). Some escape from the endosomal compartment before such fusion. Others remain in the endosomal compartment but modify it so that

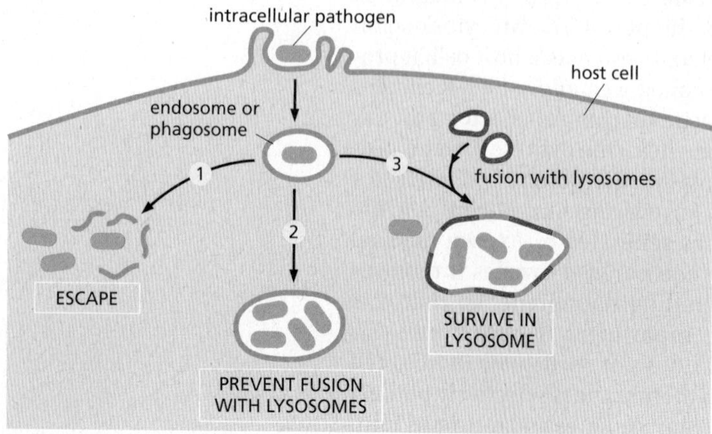

Figure 23–23 **Choices that an intracellular pathogen faces.** After entry into a host cell, generally through phagocytosis into a membrane-enclosed compartment, intracellular pathogens can use one of three strategies to survive and replicate. Pathogens that escape into the cytosol (1) include all viruses, *Trypanosoma cruzi*, *Listeria monocytogenes*, and *Shigella flexneri*. Those that prevent fusion with lysosomes (2) include *Mycobacterium tuberculosis*, *Legionella pneumophila*, and *Toxoplasma gondii*. Those that survive in the lysosome (3) include *Salmonella enterica*, *Coxiella burnetii*, and *Leishmania*.

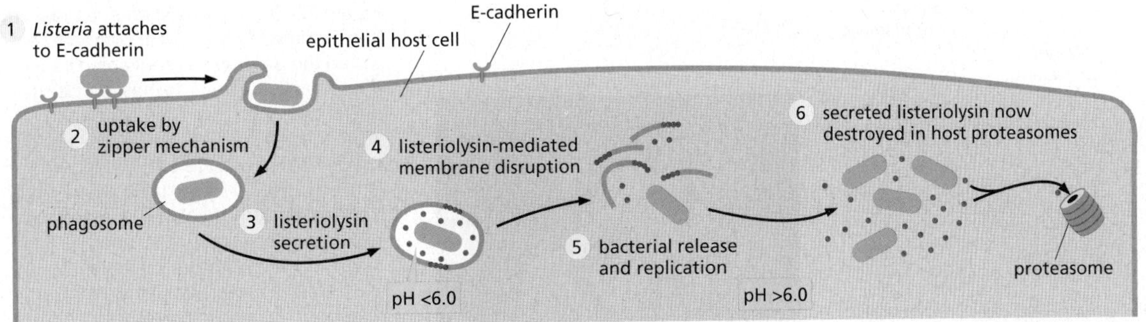

Figure 23–24 **Escape of *Listeria monocytogenes* by selective destruction of the phagosomal membrane.** The bacterium attaches to E-cadherin or other receptors on the surface of host epithelial cells and induces its own uptake by the zipper mechanism (see Figure 23–20A). Within the phagosome, the bacterium secretes the protein listeriolysin O, which is activated at pH <6 and forms oligomers in the phagosome membrane, thereby creating large pores and eventually disrupting the membrane. Once in the host-cell cytosol, the bacteria begin to replicate and continue to secrete listeriolysin O; because the pH in the cytosol is >6, however, the listeriolysin O there is less active and is also rapidly degraded by proteasomes. Thus, the host cell's plasma membrane remains intact.

it no longer fuses with lysosomes. Still others have evolved to weather the harsh conditions in the lysosome.

Trypanosoma cruzi uses the escape strategy by secreting a pore-forming toxin that lyses the lysosome membrane, releasing the parasite into the host cell's cytosol (see Figure 23–22). The bacterium *Listeria monocytogenes* uses a similar strategy. After phagocytosis by the zipper mechanism, it secretes a protein called *listeriolysin O*, which disrupts the phagosomal membrane, releasing the bacteria into the cytosol (**Figure 23–24**).

Many Pathogens Alter Membrane Traffic in the Host Cell to Survive and Replicate

Intracellular pathogens that remain in membrane-bound compartments of the host cell after invasion (see Figure 23–23) often alter these compartments to reduce their exposure to antimicrobial molecules and allow for pathogen survival and reproduction. To do so, these pathogens must alter membrane (vesicular) traffic, for example by slowing or preventing normal fusing of endosomes with lysosomes (**Figure 23–25**).

Different pathogens have distinct strategies for altering host-cell membrane traffic. *M. tuberculosis* prevents the maturation of the early endosome that contains the bacteria, so the endosome never acidifies or acquires the other characteristics of a late endosome or lysosome. Phagosomes containing *Salmonella enterica*, in contrast, acidify and acquire markers of late endosomes and lysosomes, but the bacteria slow the process of phagosomal maturation. They do so by injecting effector proteins through a second type III secretion system, distinct from that

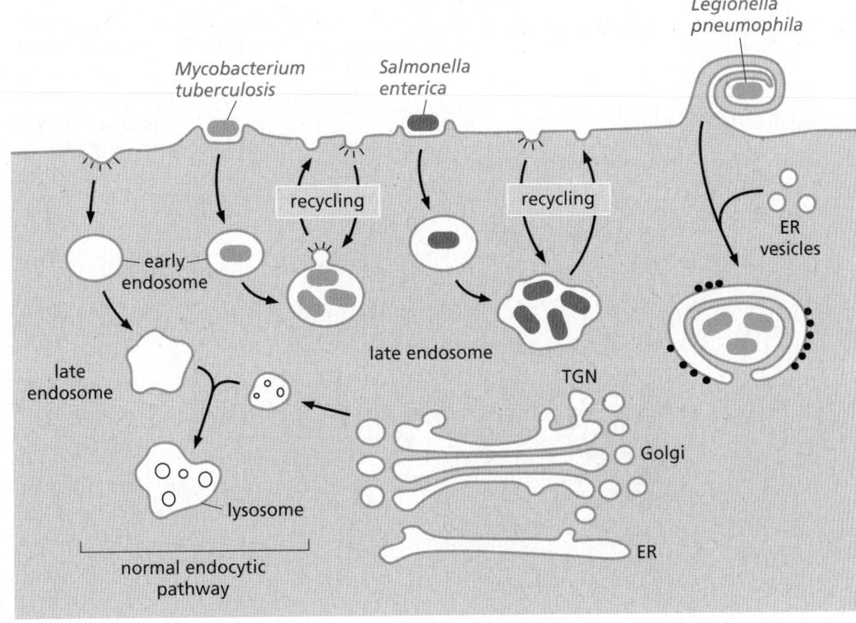

Figure 23–25 **Modifications of membrane traffic in host cells by bacterial pathogens to slow or prevent normal fusing of endosomes with lysosomes.** Intracellular bacterial pathogens, including *Mycobacterium tuberculosis*, *Salmonella enterica*, and *Legionella pneumophila*, all replicate in membrane-enclosed compartments, but the compartments differ. *M. tuberculosis* remains in a compartment that has early endosomal markers and continues to communicate with the plasma membrane via transport vesicles. *S. enterica* replicates in a compartment that has late endosomal markers. *L. pneumophila* replicates in an unusual compartment that is wrapped in rough endoplasmic reticulum (ER) membrane and communicates with the ER via transport vesicles. TGN = *trans* Golgi network.

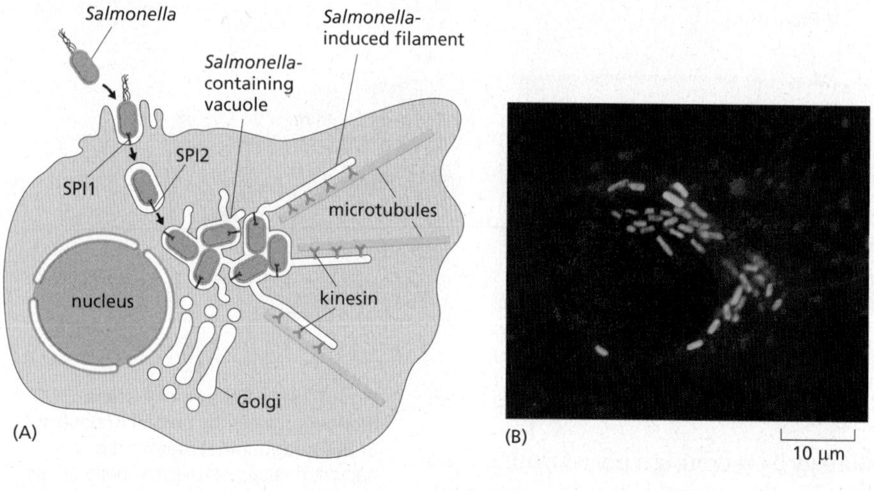

Figure 23–26 ***Salmonella enterica*** **residing in a modified phagosomal compartment called the *Salmonella*-containing vacuole.** These bacteria invade the host cell using one of two type III secretion systems to inject effector proteins that induce the trigger mechanism of microbe entry illustrated in Figure 23–20B. (A) After its engulfment into a phagosome, the bacterium inactivates the first type III secretion system and activates the second type III secretion system to inject different effector proteins, which remodel the phagosome into the specialized *Salmonella*-containing vacuole. One of the injected effector proteins activates host kinesin motor proteins to pull membrane tubules outward toward the plus ends of the microtubules (see Figure 16–53), forming a structure with an unusual tubulated shape that is of unknown functional significance. (B) Fluorescence micrograph showing *S. enterica* in a *Salmonella*-containing vacuole. The bacteria are stained *green*, the microtubules *red*, and the nucleus *blue*. (B, courtesy of Stephane Meresse.)

involved in invasion by the trigger mechanism (see Figure 23–20). Some of these bacterial effectors activate host kinesin motor proteins to pull membrane tubules outward from the phagosome along cytoplasmic microtubules, forming a specialized compartment called the *Salmonella*-containing vacuole (**Figure 23–26**).

Other bacteria seem to find shelter in intracellular compartments that are distinct from those of the usual endocytic system. One example is *Legionella pneumophila,* which was first recognized as a human pathogen in 1976, when it was found to be the cause of a type of pneumonia known as **Legionnaires' disease.** *L. pneumophila* is normally a parasite of freshwater amoebae but can be spread to humans by central air-conditioning systems, which harbor infected amoebae and produce microdroplets of water that are easily inhaled. Once in the lung, the bacteria are engulfed by macrophages by an unusual process called coiling phagocytosis (**Figure 23–27A**). *L. pneumophila* uses a type IV secretion system to inject effector proteins into the phagocyte that modulate the accumulation of phosphoinositides and the activity of proteins that regulate vesicular traffic, including SNARE proteins and Rab and Arf family small GTPases (discussed in Chapter 13). The effector proteins thereby prevent the phagosome from fusing with endosomes and promote its fusion with the endoplasmic reticulum, converting the phagosome into a compartment that resembles the rough endoplasmic reticulum (**Figure 23–27B**).

Viruses can also alter membrane traffic in the host cell. Many enveloped viruses, including the coronavirus SARS-CoV-2, make use of host-cell membranes to acquire their own envelope membrane. In the simplest cases, virally encoded glycoproteins are inserted into the endoplasmic reticulum membrane and follow

human phagocyte
Legionella bacteria
(A)
0.5 μm

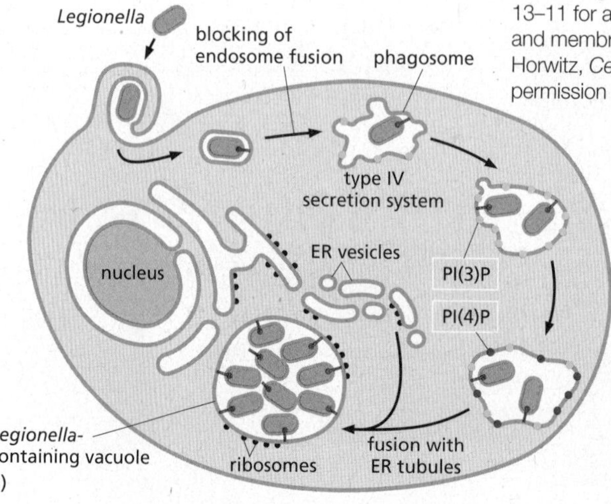

Figure 23–27 ***Legionella pneumophila*** **residing in a compartment with characteristics similar to those of the rough endoplasmic reticulum (ER).** (A) Electron micrograph showing the unusual coiled structure that the *Legionella pneumophila* bacterium induces on the surface of a phagocyte during the invasion process. Some other pathogens, including the bacterium *Borrelia burgdorferi* (which causes Lyme disease), the eukaryotic pathogen *Leishmania*, and the yeast *Candida albicans*, can also invade cells using this type of coiling phagocytosis. (B) After invasion, *L. pneumophila* uses its type IV secretion system to secrete effector proteins that block phagosome–endosome fusion and phagosome maturation. It also secretes effector proteins that promote phosphoinositide conversion from PI(3)P to PI(4)P and the fusion of the phagosome with ER-derived vesicles, thereby converting the characteristics of the *Legionella*-containing vacuole from those of an endosome to those of the rough ER. See Figures 13–10 and 13–11 for a discussion of phosphoinositides and membrane targeting. (A, from M.A. Horwitz, *Cell* 36:27–33, 1984. With permission from Elsevier.)

the secretory pathway through the Golgi apparatus to the plasma membrane; the viral capsid proteins and genome assemble into nucleocapsids, which acquire their envelope as they bud off from the plasma membrane (see Figure 23–12). The process for SARS-CoV-2 is more complicated, as this virus induces the formation of convoluted double-membrane vesicles derived from the endoplasmic reticulum, which house its genome replication factories (**Figure 23–28A**). These

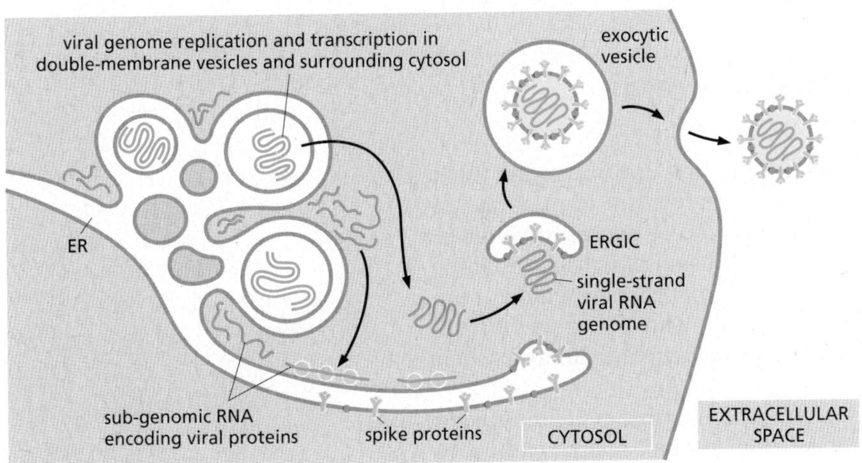

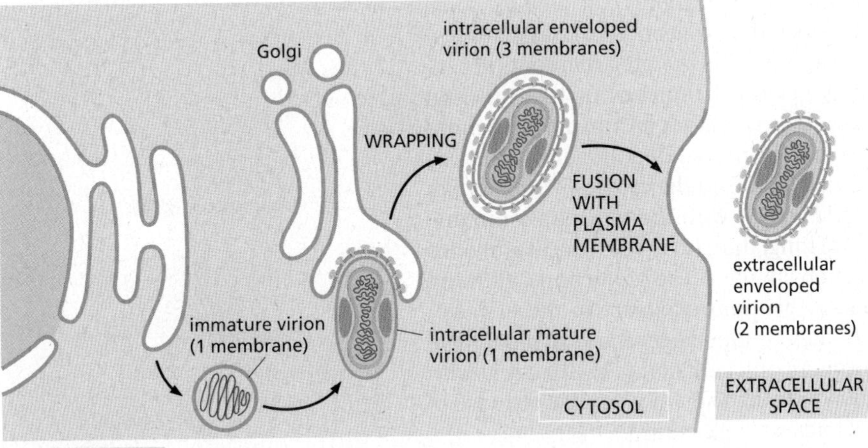

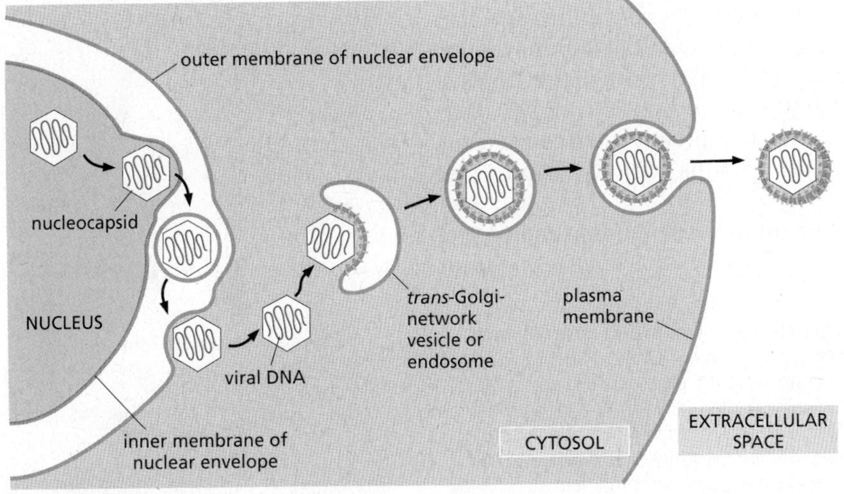

Figure 23–28 **Complex strategies for viral envelope acquisition.** (A) The coronavirus SARS-CoV-2 replicates its positive sense RNA genome in double membrane vesicles associated with the endoplasmic reticulum (ER). Viral genomes are released into the cytoplasm by a poorly understood mechanism that may involve passage through pores in the double membrane. Transcription of sub-genomic RNAs encoding viral structural proteins occurs in the surrounding cytosol. Translation of transmembrane viral proteins, including the spike protein, occurs on the ER. These proteins are then trafficked from the ER to the ER Golgi intermediate compartment (ERGIC), where the single-stranded viral RNA genome associates with viral proteins to drive formation of virus particles in the lumen of exocytic vesicles. When these vesicles fuse with the plasma membrane, viruses are released from the cell surface. (B) Vaccinia virus assembles in "replication factories" in the cytosol, far away from the plasma membrane. The immature virion, surrounded by a single membrane, then acquires two additional membranes from the Golgi apparatus by a poorly understood wrapping mechanism, to form the intracellular enveloped virion. After fusion of the outermost membrane with the host-cell plasma membrane, the extracellular enveloped virion is released from the host cell. (C) Herpesvirus nucleocapsids assemble in the nucleus and then bud through the inner nuclear membrane into the space between the inner and outer nuclear membranes, acquiring a membrane coat. The virus particles then apparently lose this coat when they fuse with the endoplasmic reticulum membrane to escape into the cytosol. Subsequently, the nucleocapsids travel through the Golgi apparatus, acquiring two new membrane coats in the process. The virus then buds from the cell surface with a single membrane when its outer membrane fuses with the plasma membrane.

may concentrate replication intermediates and sequester such intermediates from innate immune sensor molecules (see Chapter 24). Viral genomes are then released into the cytoplasm by an unknown mechanism and interact with viral structural proteins in the endoplasmic reticulum Golgi intermediate compartment (ERGIC), acquiring two new lipid bilayer membrane coats. Subsequent fusion of virus-containing exocytic vesicles with the plasma membrane leads to the release of viruses with a single membrane envelope. DNA viruses such as herpesviruses and vaccinia virus also alter membrane traffic and acquire their lipid envelopes in complex ways (**Figure 23–28B and C**).

Bacteria and Viruses Use the Host-Cell Cytoskeleton for Intracellular Movement

As mentioned earlier, many pathogens escape into the cytosol rather than remaining in a membrane-enclosed compartment. The cytosol of mammalian cells is extremely viscous, as it is crowded with protein complexes, organelles, and cytoskeletal filaments, all of which inhibit the diffusion of particles the size of a bacterium or a viral nucleocapsid. Thus, to reach a particular region of the host cell, a pathogen must be actively moved there. As with transport of intracellular organelles, pathogens generally use the host cell's cytoskeleton for their active movement.

Several pathogens have adopted a remarkable mechanism that depends on actin polymerization for their movement. These include the human bacterial pathogens *Listeria monocytogenes, Shigella flexneri, Rickettsia rickettsii* (which causes Rocky Mountain spotted fever), and *Burkholderia pseudomallei* (which causes melioidosis, a disease characterized by severe respiratory symptoms), as well as Ebola virus and the insect virus baculovirus. All induce the nucleation and assembly of host-cell actin filaments at one pole of the microbe. The growing filaments generate force and push the pathogens through the cytosol at rates of up to 1 μm/min (**Figure 23–29**). New filaments form at the rear of each pathogen and are left behind like a rocket trail as the microbe advances; the filaments depolymerize within a minute or so as they encounter depolymerizing factors in the cytosol. For *L. monocytogenes* and *S. flexneri*, the moving bacteria collide with the plasma membrane and move outward, inducing the formation of long, thin, host-cell protrusions with the bacteria at their tip. A neighboring cell often engulfs these projections, allowing the bacteria to enter the neighbor's cytoplasm without exposure to the extracellular environment, thereby avoiding antibodies produced by the host's adaptive immune system. For *B. pseudomallei*, movement and collision of the bacteria with the plasma membrane promote cell–cell fusion, which serves a similar purpose of immune avoidance while aiding continued bacterial replication.

The molecular mechanisms of pathogen-induced actin assembly differ for the different pathogens, suggesting that they evolved independently (**Figure 23–30**). *L. monocytogenes* and baculovirus produce proteins that directly bind to and activate the Arp2/3 complex to initiate the formation of an actin tail and movement (see Figure 16–17). *S. flexneri* produces an unrelated surface protein that binds to and activates the NPF N-WASp, which then activates the Arp2/3 complex. *Rickettsia* and *Burkholderia* species produce proteins that directly polymerize actin, for example by mimicking the function of host proteins such as formins (see Figure 16–13).

Many viral pathogens rely primarily on microtubule-dependent motor proteins, rather than actin polymerization, to move within the host-cell cytosol,

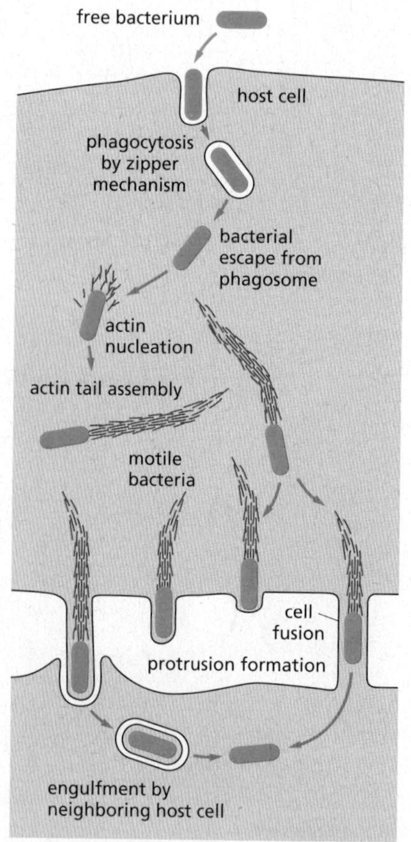

Figure 23–29 The actin-based movement of bacterial pathogens within and between host cells. After invasion, bacterial pathogens such as *L. monocytogenes*, *S. flexneri*, *R. rickettsii*, and *B. pseudomallei* induce the assembly of actin-rich tails in the host-cell cytoplasm, which drives rapid bacterial movement. For most of these pathogens, the moving bacteria collide with the host-cell plasma membrane to form membrane-covered protrusions, which are engulfed by neighboring cells—spreading the infection from cell to cell. In contrast, for *B. pseudomallei*, collision with the plasma membrane promotes cell–cell fusion, creating a conduit through which bacteria can invade neighboring cells (**Movie 23.7**).

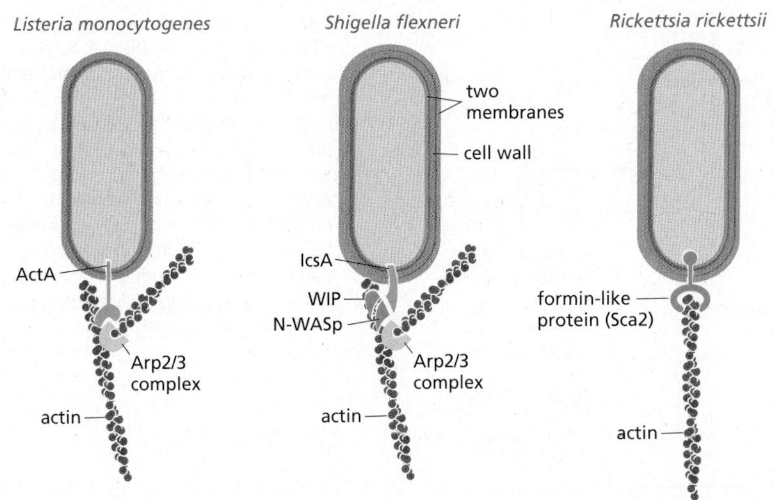

Figure 23–30 Molecular mechanisms for actin nucleation by various bacterial pathogens. *L. monocytogenes* and *S. flexneri* induce actin nucleation by recruiting and activating the host Arp2/3 complex (see Figure 16–12), although each uses a different recruitment strategy: *L. monocytogenes* expresses a surface protein, ActA, that directly binds to and activates the Arp2/3 complex; *S. flexneri* expresses a surface protein, IcsA (unrelated to ActA), that recruits the host NPF N-WASp, which in turn recruits the Arp2/3 complex, along with other host proteins, including WIP (WASp-interacting protein). *R. rickettsii* uses an entirely different strategy: it expresses a surface protein, Sca2, that directly nucleates actin polymerization by mimicking the activity of host formin proteins.

taking advantage of the inherent polarity of microtubules to enable directed long-range movements. Important examples are viruses that infect neurons, such as the *neurotropic alphaherpesviruses*, which include the virus that causes chickenpox (**Figure 23–31**). These virions enter sensory neurons at the tips of their axons and move by retrograde ("backward") axonal transport along the axon toward the microtubule minus end. The transport is mediated by attachment of viral capsid proteins to the motor protein dynein. These viruses then establish either productive or latent infection in the nuclei of neurons of the peripheral nervous system. After replication or reactivation, virions are then carried by anterograde ("forward") axonal transport along microtubules to the axon tips, with the transport being mediated by the attachment of a different viral capsid protein to a kinesin motor protein. Many other viruses associate with either dynein or kinesin motor proteins to move along microtubules at some stage in their replication, and some even alter the dynamic assembly and disassembly of microtubules. As microtubules serve as oriented tracks for vesicular transport in all cell types, not just neurons, it is not surprising that many viruses have independently evolved the ability to exploit them for their own transport.

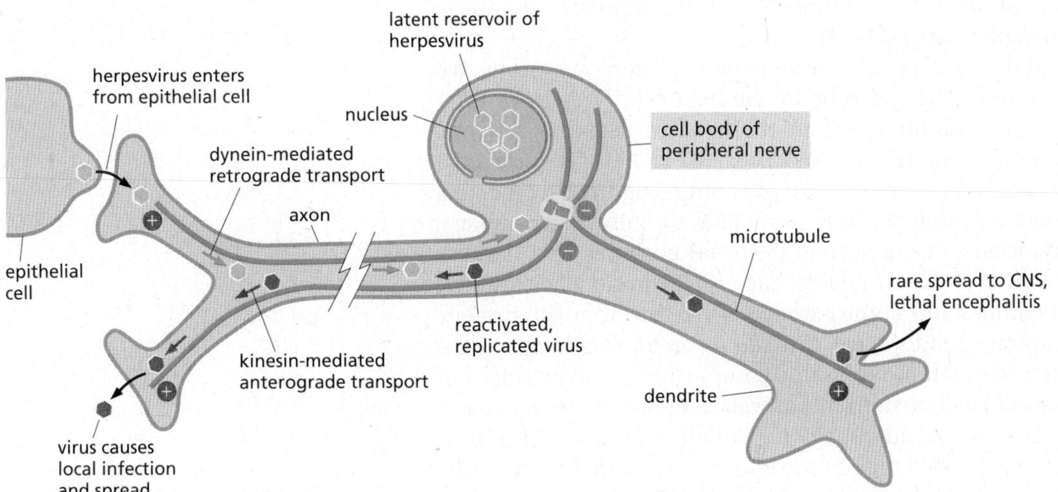

Figure 23-31 Microtubule-based movement of viruses. Neurotropic alphaherpesviruses infect peripheral nerve cells from neighboring epithelial cells in the skin. Virus particles are then transported along microtubules in the nerve-cell axon toward the microtubule minus ends in the cell body (retrograde transport), a process that is driven by dynein motor proteins. Upon reaching the cell body, viruses can establish a latent reservoir in the nucleus. After reactivation (for example, following stress or immune suppression) or upon replication, virus particles are again transported along microtubules in the axon, this time toward the microtubule plus ends at the nerve terminal (anterograde transport), where they are released to enable local reinfection and spread. In rare instances, viruses can instead spread to the central nervous system (CNS), which can cause encephalitis.

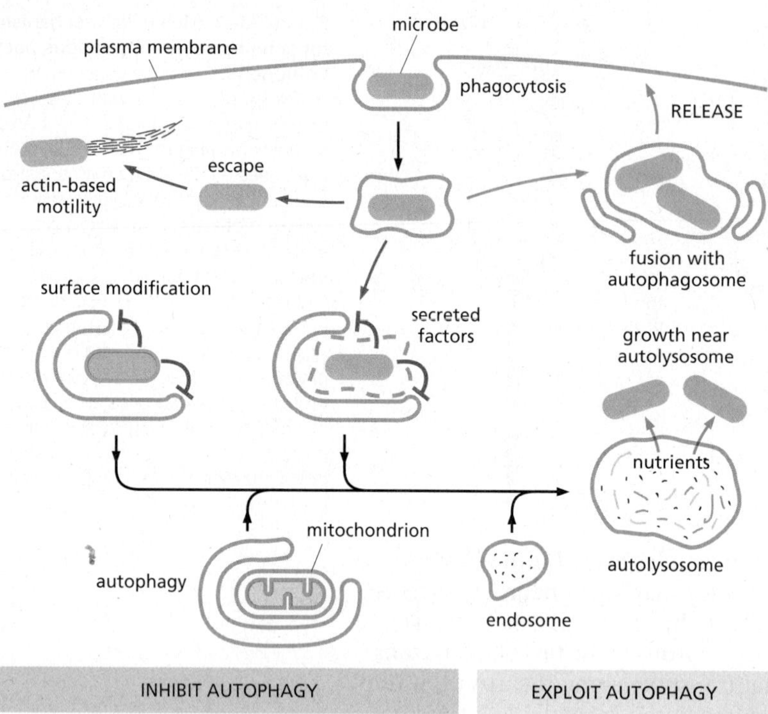

Figure 23–32 **Microbial manipulation of autophagy.** Microbial pathogens have various mechanisms for avoiding antimicrobial autophagy *(red lines)*, which include modifying their surface to avoid autophagy initiation or secreting factors and initiating actin-based motility to avoid enclosure in autophagosomes. Microbes can also exploit autophagy *(green arrows)*, for example by recruiting and fusing with autophagosomes to gain nutrients and lipids or by growing near autolysosomes for nutrient acquisition.

Many Microbes Manipulate Autophagy

Whether a microbe lives within the cytosol or inside a membrane-bound compartment, it must contend with the host cell's defenses, which include the *autophagy* pathway (see Figure 13–67). Autophagy is the process through which organelles or other cytoplasmic cargoes become surrounded by a double-membrane autophagosome that fuses with lysosomes to promote degradation (see Figure 13–71). It has recently become apparent that targeting of intracellular microbes for destruction by autophagy is an important pathway in the innate immune defense against infection (see Chapter 24). When this occurs, the process is called *antimicrobial autophagy*, or *xenophagy*. Not surprisingly, pathogens have evolved various strategies either to avoid antimicrobial autophagy or to manipulate autophagy pathways for their own benefit (**Figure 23–32**).

One strategy for avoiding antimicrobial autophagy is to deploy a protective shield that prevents detection of the microbe by the host cell. *Francisella tularensis*, which causes the zoonosis tularemia, or rabbit fever, uses this scheme. Once it escapes from the phagosome into the cytosol, it shields its surface by producing a variant of its outer membrane LPS that is not recognized by the host cell's autophagy initiation machinery. Other strategies to subvert autophagy, employed by *L. monocytogenes*, are to secrete bacterial enzymes and harness actin-based movement, both of which enable bacteria to avoid enclosure by autophagosomes. On the other hand, some pathogens take the opposite strategy of activating and then exploiting autophagy. *Coxiella burnetii*, which causes the zoonosis Q fever, replicates in a membrane-bound compartment that recruits and fuses with autophagosomes to deliver nutrients as well as lipids for compartment expansion. Poliovirus makes use of autophagy for the alternative effect of promoting virus trafficking to the plasma membrane and release from the host cell.

Viruses Can Take Over the Metabolism of the Host Cell

Viruses use basic host-cell machinery for most aspects of their reproduction: they depend on host-cell ribosomes to produce their proteins, and many use host-cell DNA and RNA polymerases for their own replication and transcription. Many viruses encode proteins that modify the host transcription or translation apparatus to favor the synthesis of viral RNAs and proteins over those of the host cell, shifting the synthetic capacity of the cell toward the production of new virus particles. Poliovirus, for example, encodes a protease that specifically cleaves the TATA-binding component of TFIID (see Figure 6–15), shutting off transcription of most of the host cell's protein-coding genes. Influenza virus produces a protein that blocks both the splicing and the polyadenylation of host-cell RNA transcripts, preventing their export into the cytosol (see Figure 6–39), but does not affect viral RNA transcripts.

Viruses also alter translation by the host. Translation initiation for most host-cell mRNAs depends on recognition of their 5′ cap by translation initiation factors (see Figure 6–40). This initiation process is often inhibited during viral infection, so that the host-cell ribosomes can be used more efficiently for the synthesis of viral proteins. Some viral genomes encode endonucleases that cleave off the 5′ cap from host-cell mRNAs; some go even further by using the liberated 5′ caps as primers to synthesize viral mRNAs, a process called *cap snatching.* Several other viral RNA genomes encode proteases that cleave certain translation initiation factors; these viruses rely on 5′ cap–independent translation of their own RNA, using internal ribosome entry sites (IRESs; see Figure 7–72).

DNA viruses that replicate in the nucleus use host-cell DNA polymerase to replicate their genome. Because DNA polymerase is expressed at high levels only during S phase of the cell cycle, adenovirus has evolved a mechanism to drive the host cell into S phase, so that the cell produces large amounts of active DNA polymerase, which then replicates the viral genome. To accomplish this, the adenovirus genome also encodes proteins that inactivate both Rb (see Figure 17–59) and p53 (see Figure 17–60), two key suppressors of cell-cycle progression. As might be expected for any mechanism that encourages unregulated DNA replication, these viruses can promote, under some circumstances, the development of cancer.

Animal RNA viruses encode their own replication proteins because most animals lack polymerase enzymes that use RNA as a template (such as the RNA-dependent RNA polymerases used in RNA interference; see Chapter 7). For RNA viruses with a single-stranded genome, the replication strategy depends on whether the RNA is a positive [+] strand, which contains translatable information like mRNA does, or a complementary negative [-] strand. When the RNA genome is a positive [+] strand (such as for the coronavirus SARS-CoV-2), the incoming viral genome is translated to produce the viral RNA polymerase and viral proteins; the viral polymerase is then used to replicate the viral RNA and to generate mRNAs for the production of more viral proteins. For viruses with a negative [-] strand RNA genome (such as influenza and measles virus), an RNA polymerase enzyme is packaged as a structural protein of the incoming viral capsids.

Retroviruses such as HIV, which have a positive [+] strand RNA genome, are a special class of RNA virus because they carry with them a viral *reverse transcriptase* enzyme. After entry to the host cell, the reverse transcriptase uses the viral RNA genome as a template to synthesize a double-stranded DNA copy of the viral genome, which enters into the nucleus and integrates into the host cell's chromosomes (see Figure 5–61). It is later transcribed by the cell's DNA-dependent RNA polymerase to produce viral genomes and proteins.

Pathogens Can Evolve Rapidly by Antigenic Variation

The complexity and specificity of the interplay between pathogens and their host cells might suggest that virulence would be difficult to acquire by random mutation. Yet, new pathogens are constantly emerging, and old pathogens are constantly changing in ways that make familiar infections more difficult to prevent or treat. Pathogens have two advantages that enable them to evolve rapidly. First, they replicate very quickly, providing a great deal of mutational variation for natural selection to work with. Whereas humans and chimpanzees have acquired a 2% difference in genome sequences over about 8 million years of divergent evolution, poliovirus manages a 2% change in its genome in 5 days—about the time it takes the virus to pass from the human mouth to the gut. Second, selective pressures act rapidly on this genetic variation. The host's adaptive immune

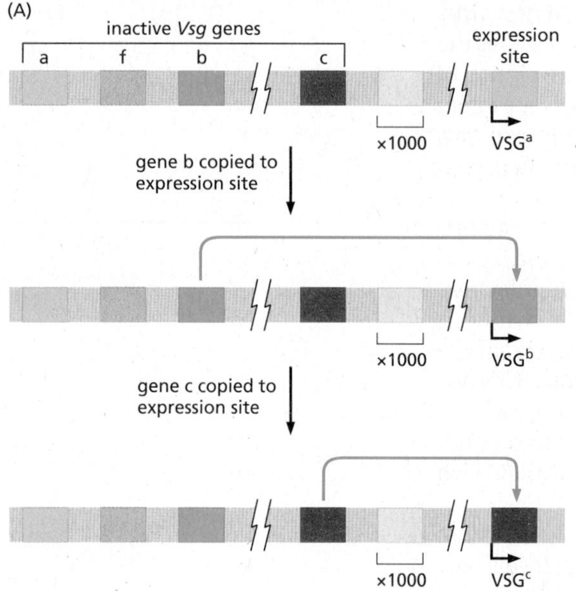

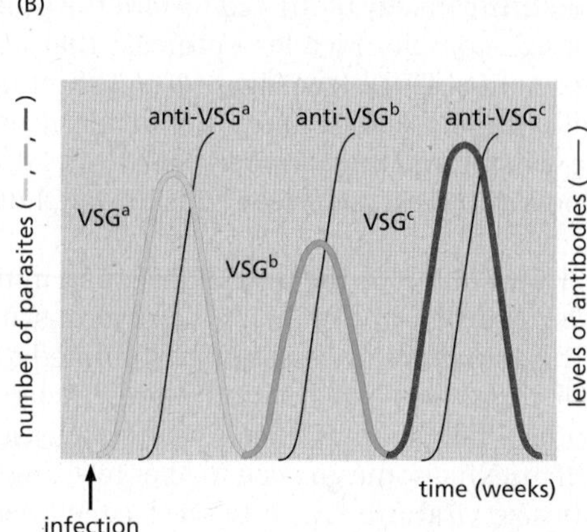

Figure 23-33 Antigenic variation in trypanosomes. (A) There are about 1000 distinct *Vsg* genes in *Trypanosoma brucei*, and they are expressed one at a time from approximately 20 expression sites in the genome. To be expressed, an inactive gene is copied and the copy is moved into an expression site through DNA recombination. Each *Vsg* gene encodes a different surface protein (antigen). These switching events allow the trypanosome to repeatedly change the surface antigen it expresses. (B) A person infected with trypanosomes expressing VSG^a mounts an antibody response against this particular antigen, which clears most of the VSG^a-expressing parasites. However, a few of the trypanosomes will have spontaneously switched to expression of VSG^b, which can now proliferate until anti-VSG^b antibodies are made. By that time, however, some parasites will have switched to VSG^c, and so the cycle continues.

system and modern microbicidal drugs, both of which destroy pathogens that fail to change, are the main sources of these selective pressures.

An example of an adaptation to the selective pressure imposed by the adaptive immune system is the phenomenon of **antigenic variation**. An important adaptive immune response against many pathogens is the host's production of antibodies that recognize specific molecules (*antigens*) on the pathogen's surface (discussed in Chapter 24). Many pathogens have evolved mechanisms that change these antigens during the course of an infection, enabling them to stay one step ahead of the antibody response. Some eukaryotic parasites, for example, undergo programmed rearrangements of the genes encoding their surface antigens (Figure 23–33). A striking example occurs in *Trypanosoma brucei*, a protozoan parasite that causes African sleeping sickness and is spread by tsetse flies. (*T. brucei* is a relative of *T. cruzi*—see Figure 23–22—but it replicates extracellularly rather than intracellularly.) *T. brucei* is covered with a single type of glycoprotein, called *variant-specific glycoprotein* (*VSG*), which elicits in the host an antibody response that rapidly clears most of the parasites. The trypanosome genome, however, contains about 1000 different inactive *Vsg* genes (or pseudogenes), each encoding a VSG with a distinct amino acid sequence. Only one *Vsg* gene is expressed at any one time, from one of approximately 20 possible expression sites in the genome. Gene rearrangements that copy different inactive *Vsg* genes into expression sites repeatedly change the VSG protein displayed on the surface of the pathogen. In this way, a few trypanosomes with an altered VSG escape the initial antibody-mediated clearance, replicate, and cause the disease to recur, leading to a chronic cyclic infection.

Bacterial pathogens can also rapidly change their surface antigens. Species of the genus *Neisseria* are champions at this. These Gram-negative cocci can cause sexually transmitted disease, in the case of *Neisseria gonorrhoeae*, or meningitis, in the case of *Neisseria meningitidis*. They undergo genetic recombination very similar to that just described for eukaryotic pathogens, which enables them to vary the pilin protein they use to attach to host cells. By inserting one of the multiple silent copies of variant *pilin* genes into a single expression locus, they can express many slightly different versions of the protein and repeatedly change the amino acid sequence over time. *Neisseria* bacteria are also extremely adept at taking up DNA from their environment by natural transformation and incorporating it into their genomes, further contributing to their extraordinary variability. The end result of this considerable variation is a plethora of different surface protein compositions with which to bewilder the host adaptive immune system. It is therefore not surprising that it has been difficult to develop an effective vaccine against *N. gonorrhoeae* infections, although there are now several that protect against *N. meningitidis*, due to the much smaller number of variants of its surface polysaccharide capsule.

Error-prone Replication Dominates Viral Evolution

In contrast to the DNA rearrangements in bacteria and parasites, viruses rely on an error-prone replication mechanism for antigenic variation. Retroviral genomes, for example, acquire on average one point mutation every replication cycle, because the viral reverse transcriptase (see Figure 5–61) needed to produce DNA from the viral RNA genome lacks the proofreading activity of DNA polymerases. A typical, untreated HIV infection may eventually produce HIV genomes with every possible point mutation. By a process of mutation and selection within each host, most HIV viruses change over time—from a form that is most efficient at infecting macrophages to one more efficient at infecting T cells, as described earlier (see Figure 23–18). Similarly, once a patient is treated with an antiviral drug, the viral genome can quickly mutate and be selected for its resistance to the drug. Remarkably, only about one-third of the nucleotide positions in the coding sequence of the viral genome are invariant (because mutations at these positions would be detrimental to the virus), and nucleotide sequences in some parts of the genome, such as the *Env* gene (see Figure 7–66), can differ by as much as 30% from one HIV isolate to another. This extraordinary genomic plasticity greatly complicates attempts to develop vaccines against HIV.

The rapid evolution of HIV by error-prone replication has also led to the swift emergence and spread of new HIV strains. Nucleotide sequence comparisons between various strains of HIV and the very similar simian immunodeficiency virus (SIV) isolated from a variety of monkey species suggest that the most virulent type of HIV, HIV-1, may have jumped from primates to humans multiple independent times, starting as long ago as 1908 (**Figure 23–34A**). Sequence

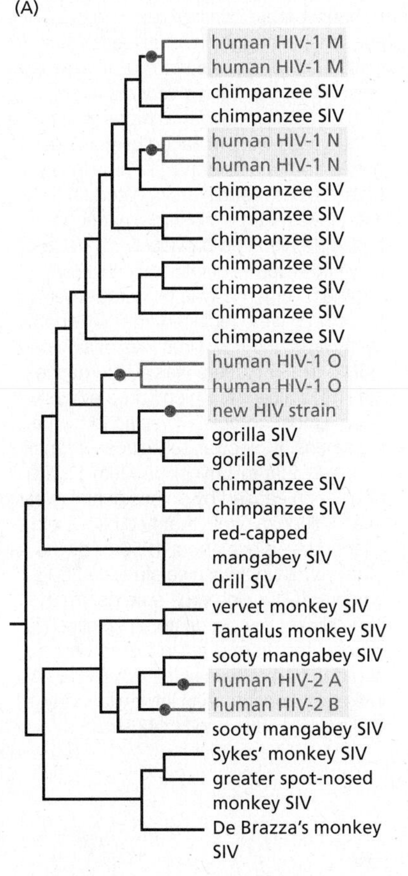

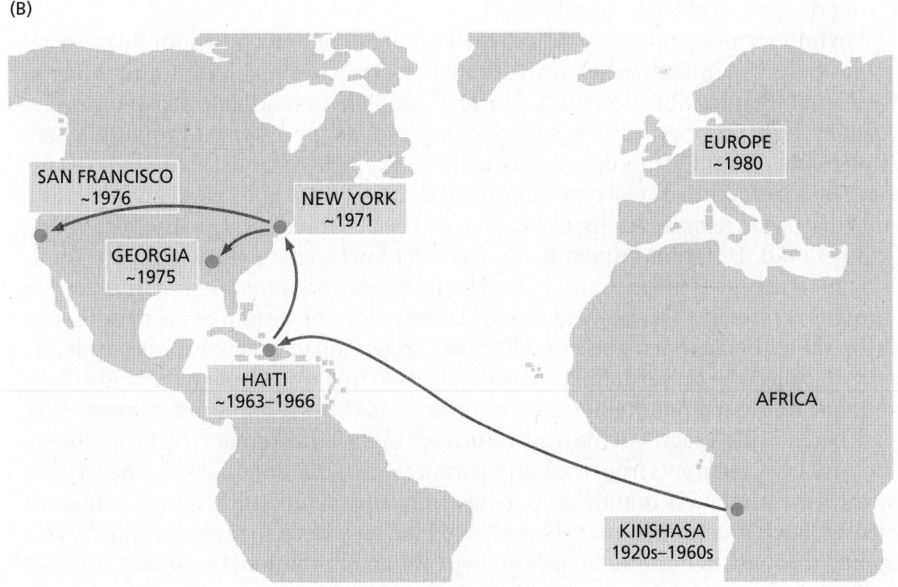

Figure 23–34 **Diversification of HIV-1, HIV-2, and related strains of SIV.** (A) HIV comprises different viral families, all descended from SIV (simian immunodeficiency virus). On three separate occasions, SIV was passed from a chimpanzee to a human, resulting in three HIV-1 groups: major (M), outlier (O), and non-M non-O (N). HIV-1 M is the most common and is primarily responsible for the global AIDS epidemic. On two separate occasions, SIV was passed from a sooty mangabey monkey to a human, resulting in the two pandemic HIV-2 groups A and B. In 2009, a new strain of HIV was discovered that appears to have resulted from SIV passage from a gorilla to a human. (B) Geography and timing of HIV-1 spread from Africa to other parts of the world.

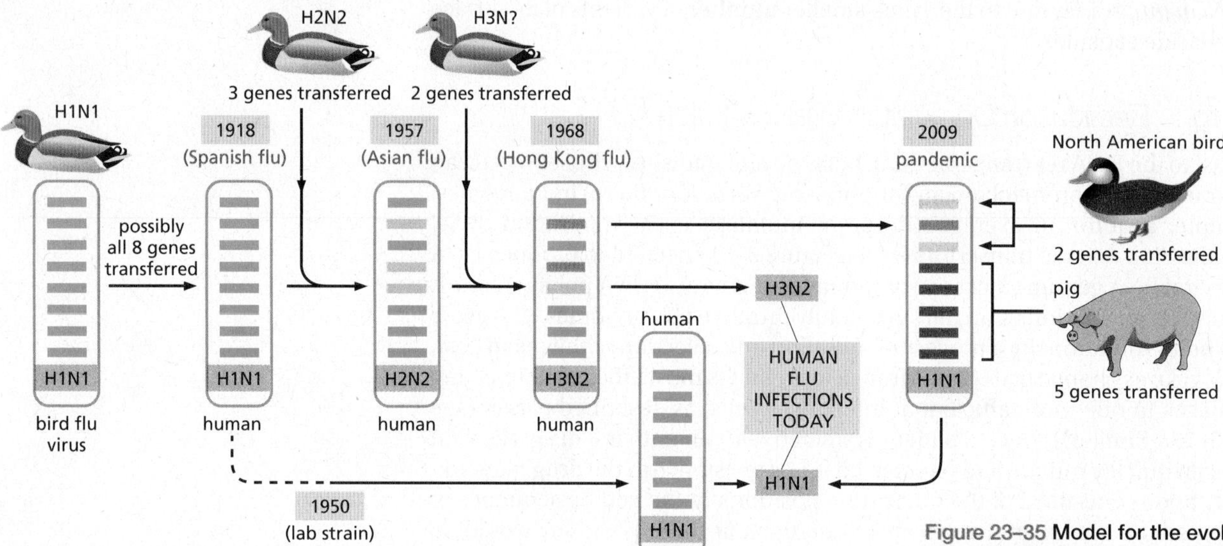

Figure 23–35 Model for the evolution of pandemic strains of influenza virus by recombination. Influenza A virus is a natural pathogen of birds, particularly waterfowl, and it is always present in wild bird populations. In 1918, a particularly virulent form of the virus crossed the species barrier from birds to humans and caused a devastating worldwide epidemic. This strain was designated H1N1, referring to the specific forms of its main antigens, hemagglutinin (H) and neuraminidase (N). Changes in the virus, rendering it less virulent, and the rise of adaptive immunity in the human population prevented the pandemic from continuing in subsequent seasons, although H1N1 influenza strains continued to cause serious disease every year in very young and very old people. In 1957, a new pandemic arose when three genes were replaced by equivalent genes from a different avian virus *(green bars)*; the new strain (designated H2N2) was not effectively cleared by antibodies in people who had previously contracted only H1N1 forms of influenza. In 1968, another pandemic was triggered when two genes were replaced from yet another avian virus; the new virus was designated H3N2. In 1977, there was a resurgence of H1N1 influenza, which had previously been almost completely replaced by the N2 strains. Molecular sequence information suggests that this minor pandemic may have been caused by an accidental release of an influenza strain that had been held in a laboratory since about 1950 or by the use of this strain in a vaccine study. In 2009, a new H1N1 swine virus emerged that had derived five genes from pig influenza viruses, two from avian influenza viruses, and one from a human influenza virus. As indicated, most human influenza today is caused by H1N1 and H3N2 strains.

comparisons between human HIV strains has further enabled a reconstruction of the evolutionary and geographic history of the current pandemic of HIV-1 group M (major). The pandemic strain is thought to have originated in Kinshasa, Democratic Republic of the Congo, in the 1920s. Virus spread increased in Africa in the 1960s due to changing sexual behaviors and emerging transportation networks (**Figure 23–34B**). It then crossed the Atlantic to Haiti in the late 1960s and spread to New York and other locations in the United States in the 1970s, eventually giving rise to a worldwide pandemic.

An important exception to the rule that error-prone replication dominates viral evolution is the influenza viruses. Although they accumulate point mutations as they replicate, they differ from other viruses in that their genome consists of several (usually eight) strands of RNA, each of which codes for different proteins. When two strains of influenza infect the same host, the RNA strands of the two strains can reassort to form a new type of influenza virus. In normal years, influenza is a mild disease in healthy adults, although it can be life-threatening in the very young and very old. Different influenza strains infect fowl such as ducks and chickens, but only a subset of these strains can infect humans, and transmission from fowl to humans is rare. In 1918, however, a particularly virulent variant of avian influenza crossed the species barrier to infect humans, triggering the catastrophic pandemic of 1918 called the Spanish flu, which killed 20–50 million people worldwide. Subsequent influenza pandemics have been triggered by genome reassortment, in which a new RNA segment from an avian form of the virus replaced one or more of the viral RNA segments from the human form (**Figure 23–35**). In 2009, a new H1N1 swine virus emerged that derived genes from pig, avian, and human influenza viruses. Such recombination events allowed the new virus to replicate rapidly and spread through an immunologically naïve human population. Generally, within 2 or 3 years, the human population develops immunity to a new recombinant strain of virus, and the infection rate drops to a steady-state level. Because the recombination events are unpredictable, it is not possible to know when the next influenza pandemic will occur or how severe it might be.

Drug-resistant Pathogens Are a Growing Problem

The development of drugs that cure rather than prevent infections has had a major impact on human health. **Antibiotics**, which are either bactericidal (they kill bacteria) or bacteriostatic (they inhibit bacterial growth without killing), are the most successful class of such drugs. Penicillin was one of the first antibiotics

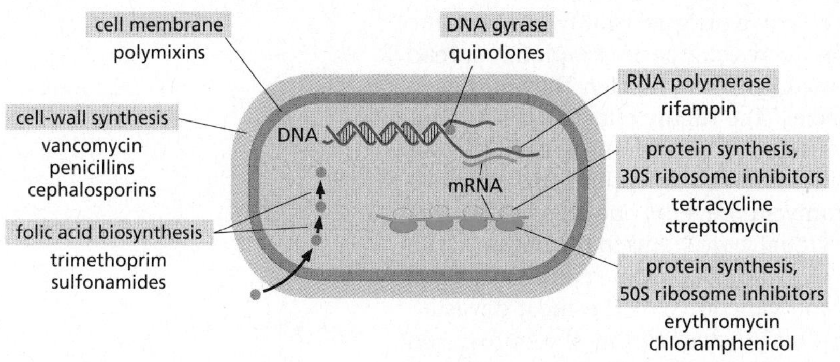

Figure 23–36 **Antibiotic targets.** Although there are many antibiotics in clinical use, they have a narrow range of targets, which are highlighted in *blue*. A few representative antibiotics in each class are listed. Nearly all antibiotics used to treat human infections fall into one of these categories. The vast majority inhibit either bacterial protein synthesis or bacterial cell-wall synthesis. The illustration example is a Gram-positive bacterium.

used to treat infections in humans, just in time to prevent tens of thousands of deaths from infected battlefield wounds in World War II. Because bacteria (see Figure 1–9) are not closely related evolutionarily to the eukaryotes they infect, much of their basic machinery for cell-wall synthesis, DNA replication and transcription, RNA translation, and metabolism differs from that of their host. These differences enable us to develop antibacterial drugs that exhibit *selective toxicity*, in that they specifically inhibit these processes in bacteria without disrupting them in the host. Most of the antibiotics that we use to treat bacterial infections are small molecules that inhibit macromolecular synthesis in bacteria by targeting bacterial enzymes that either are distinct from their eukaryotic counterparts or are involved in pathways such as cell-wall biosynthesis that are absent in animals (**Figure 23–36**; see also Table 6–4).

However, bacteria continually evolve and strains resistant to antibiotics rapidly develop, often within a few years of the introduction of a new drug. Similar drug resistance also arises rapidly when treating viral infections with antiviral drugs. The virus population in an HIV-infected person treated with the reverse transcriptase inhibitor azidothymidine (AZT), for example, will acquire complete resistance to the drug within a few months. The current protocol for treatment of HIV infections involves the simultaneous use of three drugs, which helps to minimize the acquisition of resistance for any one of them. Even eukaryotic pathogens rapidly evolve resistance. The malaria parasite *Plasmodium falciparum* is now generally resistant to the heavily used drug chloroquine, which was introduced in the 1930s, and resistance to the newer drug artemisinin is emerging.

There are three general strategies by which a pathogen can develop drug resistance: (1) it can alter the molecular target of the drug so that it is no longer sensitive to the drug; (2) it can produce an enzyme that modifies or destroys the drug; or (3) it can prevent the drug's access to the drug target by, for example, actively pumping the drug out of the pathogen (**Figure 23–37**).

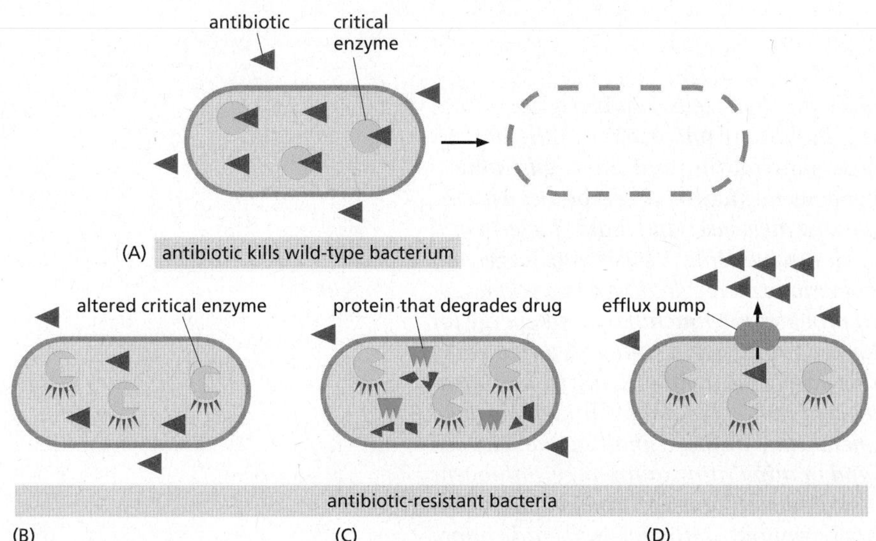

Figure 23–37 **Three general mechanisms of antibiotic resistance.** (A) A nonresistant wild-type bacterial cell bathed in a drug *(red triangles)* that binds to and inhibits an essential enzyme *(light green)* will be killed because of enzyme inhibition. (B) A bacterium that has altered the drug's target enzyme so that the drug no longer binds to the enzyme will survive and proliferate. In many cases, a single point mutation in the gene encoding the target protein can generate resistance. (C) A bacterium that expresses an enzyme *(dark green)* that either degrades or covalently modifies the drug will survive and proliferate. Some resistant bacteria, for example, make β-lactamase enzymes, which cleave penicillin and similar molecules. (D) A bacterium that expresses or up-regulates an efflux pump that ejects the drug from the bacterial cytoplasm (using energy derived from either ATP hydrolysis or the electrochemical gradient across the bacterial plasma membrane) will survive and proliferate. Some efflux pumps, such as the TetR efflux pump, are specific for a single drug (in this case, tetracycline), whereas others, called multidrug resistance (MDR) efflux pumps, are capable of exporting a wide variety of structurally dissimilar drugs. Up-regulation of an MDR pump can render a bacterium resistant to a very large number of different antibiotics in a single step.

Once a pathogen has chanced upon an effective drug-resistance strategy, the newly acquired or mutated genes that confer the resistance are frequently spread throughout the pathogen population by horizontal gene transfer. They may even spread between pathogens of different species. The highly effective but expensive antibiotic *vancomycin*, for example, is used as a treatment of last resort for many severe, hospital-acquired, Gram-positive bacterial infections that are resistant to most other known antibiotics. Vancomycin prevents one step in bacterial cell-wall synthesis—the cross-linking of peptidoglycan chains in the bacterial cell wall (see Figure 23–3B). Resistance can arise if the bacterium synthesizes a cell wall using different subunits that do not bind vancomycin. The most devastating form of vancomycin resistance depends on the acquisition of a transposon (see Figure 5–58) containing seven genes, the products of which work together to sense the presence of vancomycin, shut down the normal pathway for bacterial cell-wall synthesis, and produce a different type of cell wall.

Drug-resistance genes acquired by horizontal transfer frequently come from environmental microbial reservoirs. Nearly all antibiotics used to treat bacterial infections today are based on natural products produced by fungi or bacteria. Penicillin, for example, is made by the mold *Penicillium*, and more than 50% of the antibiotics currently used in the clinic are made by Gram-positive bacteria of the genus *Streptomyces*, which reside in the soil. It is believed that microorganisms produce antimicrobial compounds, many of which have probably existed on Earth for hundreds of millions of years, as weapons in their competition with other microorganisms in the environment. Surveys of bacteria taken from soil samples that have never been exposed to antibiotic drugs used in modern medicine reveal that there are bacteria already resistant to about seven or eight of the antibiotics widely used in clinical practice. When pathogenic microorganisms are faced with the selective pressure provided by antibiotic treatments, they can apparently draw upon the immense source of genetic material in environmental microbial reservoirs to acquire resistance.

Like most other aspects of infectious disease, human behavior has exacerbated the problem of drug resistance. Many patients take antibacterial antibiotics for symptoms that are typically caused by viruses (flu-like illnesses, colds, and sore throats), and these drugs have no effects. Persistent and chronic misuse of antibiotics can eventually result in antibiotic-resistant microbes, which can then transfer the resistance to pathogens. Antibiotics are also misused in the livestock industry, where they are commonly employed as food additives to promote the growth and health of farm animals. An antibiotic closely related to vancomycin was commonly added to cattle feed in Europe; the resulting resistance in the microbiota of these animals is widely believed to be one of the original sources for vancomycin-resistant bacteria that now threaten the lives of hospitalized patients.

Summary

All pathogens share the ability to interact with host cells in diverse ways that promote pathogen replication and spread. Pathogens often colonize the host by adhering to or invading the epithelial surfaces that line the respiratory, gastrointestinal, and urinary tracts, as well as the other body surfaces in direct contact with the environment. Intracellular pathogens, including all viruses and many bacteria and protozoa, invade host cells by one of several mechanisms. Viruses rely largely on receptor-mediated endocytosis, whereas bacteria exploit cell adhesion and phagocytic pathways; in both cases, the host cell provides the machinery and energy for the invasion. Protozoa, by contrast, employ unique invasion strategies that usually require significant energy expenditure on the part of the invader. Once inside, intracellular pathogens seek out a cell compartment that is favorable for their survival and replication, frequently altering host membrane traffic, exploiting the host-cell cytoskeleton for intracellular movement, and manipulating autophagy. Pathogens evolve rapidly, so that new infectious diseases frequently emerge, and old pathogens acquire new ways to evade our attempts at prevention, treatment, and eradication.

Because the acquisition of drug resistance is almost inevitable, it is crucial that we take measures to delay the onset of resistance and develop innovative new drug treatments.

THE HUMAN MICROBIOTA

Although pathogens can cause disease in otherwise healthy humans, it is now appreciated that our bodies are colonized by many microbes that normally cause no harm. These microbes compose the so-called microbiota. Recent estimates suggest that the human body contains about 3×10^{13} human cells, as well as approximately 4×10^{13} resident microbial cells. As detailed below, we now appreciate that the microbiota is a complex microbial community that makes important contributions to human biology.

The Human Microbiota Is a Complex Ecological System

The human microbiota is usually confined to specific locations of the body (**Figure 23–38**), including the skin, mouth, digestive tract, and vagina, with distinct communities of species inhabiting each body part. Of the resident microbes, bacterial cells make up the vast majority by sheer numbers (>99%), whereas there are smaller numbers of archaeal, fungal, and protozoan cells. The concentrations and total numbers of microbial cells differ vastly between body locations (Figure 23–38), with the large intestine containing by far the most microbial residents.

The microbiota of an individual person consists of thousands of different microbial species. The species composition varies considerably between individual humans, even between close relatives or identical twins. Different body sites also have different degrees of species diversity. The digestive tract, for example, contains between 500 and 1000 species, most belonging to only a few phyla of bacteria. The microbiota of the digestive tract also provides an interesting illustration of how the diversity and composition of microbial species can change over time, from birth through adulthood. The digestive tract of human infants is colonized by environmental microbes during birth, and the mode of birth (vaginal delivery versus cesarean section) can influence species composition. During the first year of life, the microbiota consists of fewer species that vary considerably over time and between individuals, whereas species diversity increases and composition stabilizes at 1–2 years of age. Thereafter, the microbiota of an individual is generally consistent over time but is influenced by a variety of factors, including age, pregnancy, diet, health status, hygiene, and antibiotic use.

To appreciate the contributions of the microbiota to human biology, it is helpful to consider the various ecological terms that are used to classify relationships between microbes and their host. For the pathogens discussed earlier in this chapter, the microbe benefits to the detriment of the host, a situation referred to as **parasitism**. In contrast, most constituents of the microbiota exhibit less sinister ecological relationships with the host. Some exhibit **commensalism**, which describes the circumstance in which the microbe benefits but has no known beneficial or harmful effects on the host. In **mutualism**, both the microbe and host benefit. It is sometimes difficult to draw a line between these categories.

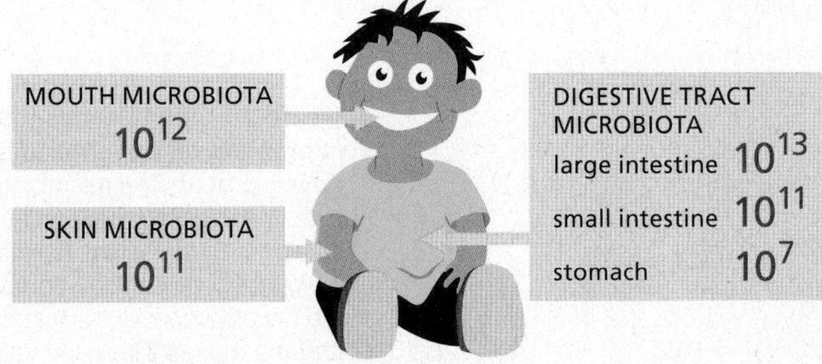

Figure 23–38 **Sites in the human body that harbor the microbiota include the skin, mouth, and digestive tract.** An estimate of the total number of microbes in each location is indicated.

Some constituents of the microbiota that exhibit commensalism or mutualism under normal circumstances can also act as parasites or *opportunistic pathogens* that can cause disease if our immune systems are weakened or if they gain access to a normally sterile part of the body.

The Microbiota Influences Our Development and Health

It is increasingly recognized that many constituents of the human microbiota exhibit a mutualistic relationship with our bodies by supporting metabolism, development, and immunity (**Figure 23–39**). The anaerobic bacteria that inhabit our intestines, for example, gain shelter and a nutrient supply but also contribute to the digestion of our food and produce important metabolites. One reason for the beneficial influence of the microbiota on our metabolism is that the combined genomes of the various microbial species, called the **microbiome**, contain 100 times more genes than the number in the human genome itself. A consequence of this genomic diversity is that the microbiota expands the range of biochemical activities available to humans by producing small-molecule metabolites of different chemical composition or in greater abundance than that produced by our own cells. An example is the production of abundant short-chain fatty acids by the microbiota in the digestive tract, which may contribute to metabolism and may also signal to our cells to influence human physiology.

The microbiota is also important for normal development, most notably of the epithelial tissue and immune system in the digestive tract. Studies comparing mice with and without a gastrointestinal microbiota (so-called *germ-free* mice) show that the microbiota affects properties of the intestinal lining including villus geometry, stem-cell proliferation, blood vessel density, and mucus thickness (Figure 23–39). The *mucosal immune system* that is associated with the mucus-containing surface of the intestinal epithelium is also strongly influenced by the microbiota. The mucosal immune system must be fine-tuned to be nonresponsive to beneficial microbes in the microbiota, yet responsive to pathogens and microbes that inappropriately penetrate into the epithelial layer. The microbiota plays an important role in the development of lymphoid tissues and in the appropriate differentiation of various immune cell types to control the overall number, species composition, and location of microbes in the digestive tract (Figure 23–39).

There is increasing evidence that an imbalance in the community of microbes that constitute the microbiota, referred to as *dysbiosis*, is correlated with various human diseases. These include autoimmune and allergic diseases, obesity,

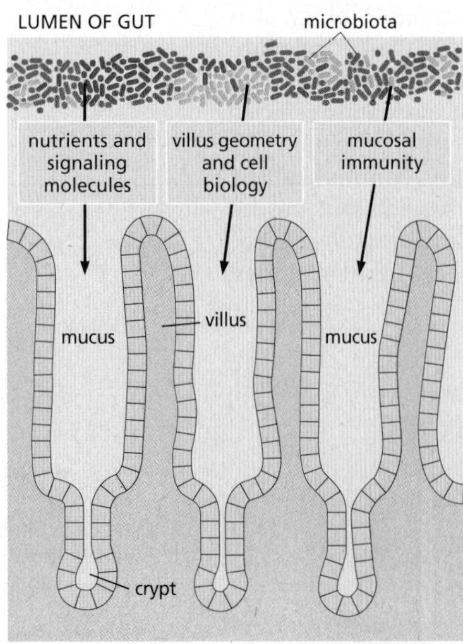

Figure 23–39 **Influences of the microbiota on metabolism and development.** The microbiota (not drawn exactly to scale) produces nutrients as well as signaling molecules that affect host-cell biology and physiology and also influences the development of the intestinal epithelium and the mucosal immune system.

inflammatory bowel disease, and diabetes, and the list is expanding rapidly. A common example of a disease caused by dysbiosis is colitis, which occurs in response to antibiotic treatment that kills off beneficial microbes in the digestive tract, resulting in overgrowth of the bacterial pathogen *Clostridium difficile*. Although first-line treatment for colitis is an additional course of antibiotics, recurrent colitis has been successfully treated by the transfer of the microbiota from a healthy individual to someone suffering from the disease.

As research intensifies into the influences of the microbiota on human health and disease, a future challenge will be to shift scientific inquiry from investigation of correlations and associations between the presence of certain microbes and health status to investigation of whether there are causal relationships between the composition of the microbiota and its molecular products and health or disease. An impediment to making this shift is the sheer number of variables that must be considered when studying a community consisting of hundreds of microbial species that produce thousands of proteins and small molecules. One path forward may be to search for clear instances in which individual microbes or molecules exert strong effects on human physiology. As microbiomes with defined composition are developed and genetic tools are improved, we may begin to understand how individual microbial species or even individual microbial genes influence human development and health, and we may be able to harness the microbiome to treat human disease.

Summary

Humans are colonized by a community of microbes, collectively called the microbiota, that consists of roughly the same number of microbial cells as there are human cells in the body and is located primarily on the skin, in the mouth, and in the digestive tract. The microbiota as a whole is thought to exhibit a mutualistic ecological relationship with the human host, meaning that both microbes and humans benefit from their interactions. The microbiota produces nutrients that aid in metabolism, as well as molecules that influence tissue and immune system development and function. However, it has been difficult to establish causal relationships between individual microbes or their molecular products and human health. Knowledge of the microbiota is expanding rapidly, and advances promise to enhance our understanding of human biology and result in new disease treatments.

PROBLEMS

Which statements are true? Explain why or why not.

23–1 Pathogens must enter host cells to cause disease.

23–2 Viruses replicate their genomes in the nucleus of the host cell.

23–3 You should not take antibiotics for diseases caused by viruses.

23–4 Our adult bodies harbor about 10 times more resident microbial cells than human cells.

23–5 The microbiomes from healthy humans are all very similar.

Discuss the following problems.

23–6 In order to survive and multiply, a successful pathogen must accomplish five tasks. Name them.

23–7 What are the three general mechanisms for horizontal gene transfer?

23–8 John Snow is widely regarded as the father of modern epidemiology. Most famously, he investigated an outbreak of cholera in London in 1854 that killed more than 600 people before it was finished. Snow recorded where the victims lived and plotted the data on a map, along with the locations of the water pumps that served as the source of water for the public (Figure Q23–1). He concluded that the disease was most likely spread in the water, although he could find nothing suspicious-looking in it. His conclusion ran counter to the then-current belief that cholera was from "miasmas" in bad air. Very few believed his theory during the next 50 years, with the "bad air" theory persisting until at least 1901. What do you suppose Snow saw in the data that led him to his conclusion? Why do you think most scientists remained skeptical for so long?

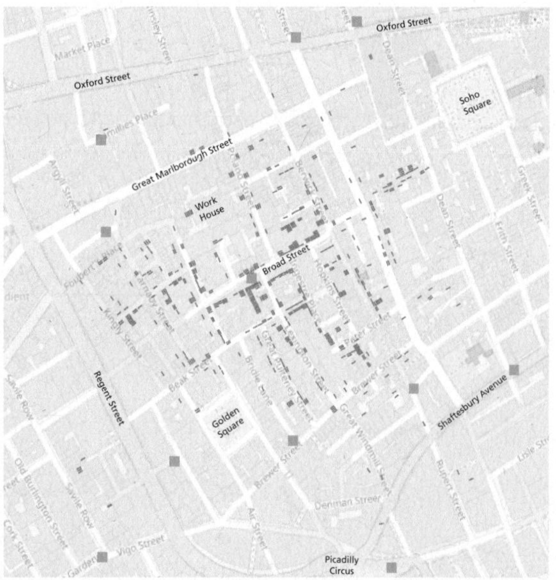

Figure Q23–1 A map of where the victims of the 1854 cholera outbreak lived, superimposed on a modern map of the area (Problem 23–8). The locations of the victims' houses are indicated by the small *red rectangles*. Stacks of rectangles indicate multiple cases occurring in the same house. Public water pumps are shown as *blue squares*. (Adapted from Wellcome Library, London/Google Maps.)

23–9 The Gram-negative bacterium *Yersinia pestis*, the causative agent of the plague, is extremely virulent. Upon infection, *Y. pestis* injects a set of effector proteins into macrophages that suppresses their phagocytic behavior and also interferes with their innate immune responses. One of the effector proteins, YopJ, acetylates serines and threonines on various MAP kinases, including the MAP kinase kinase kinase (MAPKKK) TAK1, which controls a key signaling step in the innate immune response pathway. To determine how YopJ interferes with TAK1, you transfect human cells with catalytically active YopJ (YopJ^{WT}) or inactive YopJ (YopJ^{C172A}) and with FLAG-tagged active TAK1 (TAK1^{WT}) or inactive TAK1 (TAK1^{K63W}) and assay for total TAK1 and for phosphorylated TAK1, using antibodies against the FLAG tag or against phosphorylated TAK1 (Figure Q23–2). How does YopJ block the TAK1 signaling pathway? How do you suppose the serine/threonine acetylase activity of YopJ might interfere with TAK1 activation?

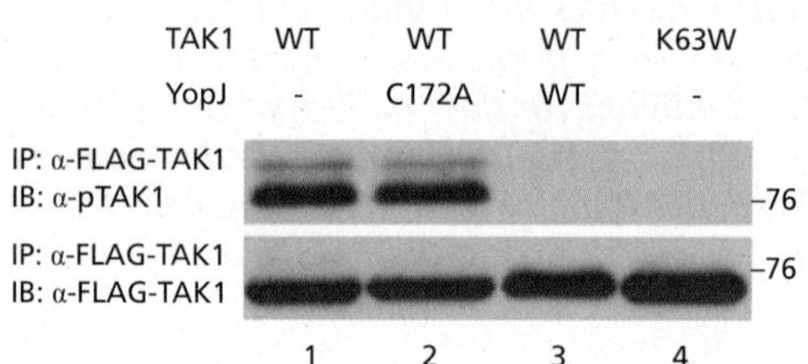

Figure Q23–2 Effects of YopJ on TAK1 phosphorylation (Problem 23–9). TAK1 was immunoprecipitated (IP) using antibodies against the FLAG tag (α-FLAG-TAK1). Total TAK1 in the immunoprecipitation was assayed by immunoblot (IB) using the same antibody. Phosphorylated TAK1 was assayed by IB using antibodies specific for phospho-TAK1 (α-pTAK1). A marker of protein molecular mass is shown at *right* in kilodaltons. (From N. Paquette et al., *Proc. Natl. Acad. Sci. USA* 109:12710–12715, 2012. With permission from National Academy of Sciences.)

23–10 The intracellular bacterial pathogen *Salmonella enterica* serovar Typhimurium, which causes gastroenteritis, injects effector proteins to promote its invasion into nonphagocytic host cells by the trigger mechanism. *S. enterica* serovar Typhimurium first stimulates membrane ruffling to promote invasion, and then suppresses membrane ruffling once invasion is complete. This behavior is mediated in part by injection of two effector proteins: SopE, which promotes membrane ruffling and invasion, and SptP, which blocks the effects of SopE. Both effector proteins target the monomeric GTPase, Rac, which in its active form promotes membrane ruffling. How do you suppose SopE and SptP affect Rac activity? How do you suppose the effects of SopE and SptP are staggered in time if they are injected simultaneously?

23–11 Several negative-strand viruses carry their genome as a set of discrete RNA segments. Examples include influenza virus (eight segments), Rift Valley fever virus (three segments), hantavirus (three segments), and Lassa virus (two segments), to name a few. Why does segmentation of the genome provide a strong evolutionary advantage for these viruses?

23–12 Influenza epidemics account for 250,000–500,000 deaths globally each year. These epidemics are markedly seasonal, occurring in temperate climates in the Northern and Southern Hemispheres during their respective winters. By contrast, in the tropics, there is significant influenza activity year round, with a peak in the rainy season (**Figure Q23–3**). Can you suggest some possible explanations for the patterns of influenza epidemics in temperate zones and the tropics?

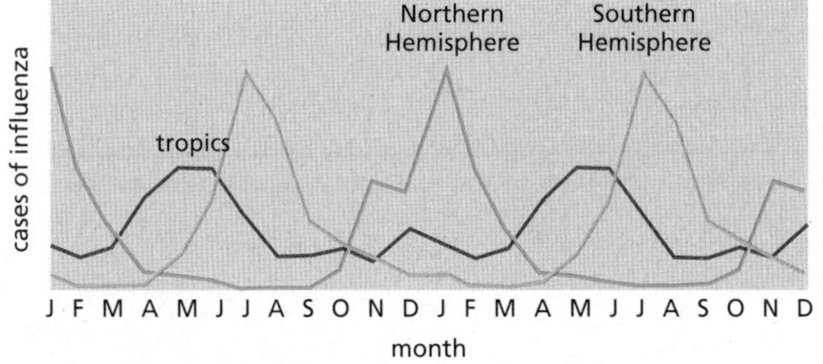

Figure Q23–3 Seasonal patterns of influenza epidemics (Problem 23–12). Cases of influenza at different times of the year are shown for the Northern Hemisphere *(blue)*, the Southern Hemisphere *(orange)*, and the tropics *(red)*.

23–13 Avian influenza viruses readily infect birds but are transmitted to humans very rarely. Similarly, human influenza viruses spread readily to other humans but have never been detected in birds. The key to this specificity lies in the viral capsid protein, hemagglutinin, which binds to sialic acid residues on cell-surface glycoproteins, triggering virus entry into the cell (**Movie 23.8**). Hemagglutinin on human viruses recognizes sialic acid in a 2,6 linkage with galactose, whereas avian hemagglutinin recognizes sialic acid in a 2,3 linkage with galactose. Humans make carbohydrate chains that have only the 2,6 linkage between sialic acid and galactose; birds make only the 2,3 linkage; but pigs make carbohydrate chains with both linkages. How does this situation make pigs ideal hosts for generating new strains of human influenza viruses?

23–14 The majority of antibiotics used in the clinic are made as natural products by bacteria. Why do you suppose bacteria make the very agents we use to kill them?

23–15 In the early days of penicillin research, it was discovered that bacteria in the air could destroy the penicillin, a big problem for large-scale production of the drug. How do you suppose this occurs?

23–16 When the Oxford team of Ernst Chain and Norman Heatley had laboriously collected their first two grams of penicillin (probably no more than 2% pure!), Chain injected two normal mice with 1 g each of this preparation and waited to see what would happen. The mice survived with no apparent ill effects. His boss, Howard Florey, was furious at what he saw as a waste of good antibiotic. Why was this experiment important?

REFERENCES

General

Acheson NH (2021) Fundamentals of Molecular Virology, 3rd ed. Hoboken, NJ: Wiley.

Cossart P, Roy CR & Sansonetti P (eds.) (2019) Bacteria and Intracellularity. Washington, DC: ASM Press.

Fauci A & Morens DM (2012) The perpetual challenge of infectious diseases. *N. Engl. J. Med.* 366, 454–461.

Gunn A & Pitt SJ (2012) Parasitology: An Integrated Approach. Hoboken, NJ: Wiley.

Heitman J, Filler SG, Edwards JE & Mitchell AP (2006) Molecular Principles of Fungal Pathogenesis. Washington, DC: ASM Press.

Wilson BA, Winkler ME & Ho BT (2019) Bacterial Pathogenesis: A Molecular Approach, 4th ed. Washington, DC: ASM Press.

Introduction to Pathogens

Baltimore D (1971) Expression of animal virus genomes. *Bacteriol. Rev.* 35, 235–241.

Coffin JM (2021) 50th anniversary of the discovery of reverse transcriptase *Mol. Biol. Cell* 32, 91–97.

Cowman AF, Healer J, Marapana D & Marsh K (2016) Malaria: biology and disease. *Cell* 167, 610–624.

Crick FH & Watson JD (1956) Structure of small viruses. *Nature* 177, 473–475.

Faruque SM & Mekalanos JJ (2012) Phage-bacterial interactions in the evolution of toxigenic *Vibrio cholerae*. *Virulence* 3, 556–565.

Galán JE & Gabriel Waksman G (2018) Protein-injection machines in bacteria. *Cell* 172, 1306–1318.

Rappleye CA & Goldman WE (2006) Defining virulence genes in the dimorphic fungi. *Annu. Rev. Microbiol.* 60, 281–303.

Soucy S, Huang J & Gogarten J (2015) Horizontal gene transfer: building the web of life. *Nat. Rev. Genet.* 16, 472–482.

Young JAT & Collier RJ (2007) Anthrax toxin: receptor binding, internalization, pore formation, and translocation. *Annu. Rev. Biochem.* 76, 243–265.

Cell Biology of Pathogen Infection

Asrat S, de Jesús DA, Hempstead AD . . . Isberg RR (2014) Bacterial pathogen manipulation of host membrane trafficking. *Annu. Rev. Cell Dev. Biol.* 30, 79–109.

Davies J & Davies D (2010) Origins and evolution of antibiotic resistance. *Microbiol. Mol. Biol. Rev.* 74, 417–433.

Faria NR, Rambaut A, Suchard MA . . . Lemey P (2014) HIV epidemiology. The early spread and epidemic ignition of HIV-1 in human populations. *Science* 346, 56–61.

Forsberg KJ, Reyes A, Wang B . . . Dantas G (2012) The shared antibiotic resistome of soil bacteria and human pathogens. *Science* 337, 1107–1111.

Goldberg DE, Siliciano RF & Jacobs WR, Jr (2012) Outwitting evolution: fighting drug-resistant TB, malaria, and HIV. *Cell* 148, 1271–1283.

Grove J & Marsh M (2011) The cell biology of receptor-mediated virus entry. *J. Cell Biol.* 195, 1071–1082.

Hartenian E, Nandakumar D, Lari A . . . Glaunsinger BA (2020) The molecular virology of coronaviruses. *J. Biol. Chem.* 295, 12910–12934.

Hernandez-Gonzalez M, Larocque G & Way M (2021) Viral use and subversion of membrane organization and trafficking. *J. Cell Sci.* 134, jcs252676.

Huang J & Brumell J (2014) Bacteria–autophagy interplay: a battle for survival. *Nat. Rev. Microbiol.* 12, 101–114.

Mengaud J, Ohayon H, Gounon P . . . Cossart P (1996) E-cadherin is the receptor for internalin, a surface protein required for entry of *L. monocytogenes* into epithelial cells. *Cell* 84, 923–932.

Petrova V & Russell C (2018) The evolution of seasonal influenza viruses. *Nat. Rev. Microbiol*. 16, 47–60.

Polk DB & Peek RM, Jr (2010) *Helicobacter pylori*: gastric cancer and beyond. *Nat. Rev. Cancer* 10, 403–414.

Ray K, Marteyn B, Sansonetti PJ & Tang CM (2009) Life on the inside: the intracellular lifestyle of cytosolic bacteria. *Nat. Rev. Microbiol.* 7, 333–340.

Ribet D & Cossart P (2015) How bacterial pathogens colonize their hosts and invade deeper tissues. *Microbes Infect.* 17, 173–183.

Sibley LD (2011) Invasion and intracellular survival by protozoan parasites. *Immunol. Rev.* 240, 72–91.

Sun E, He J & Zhuang X (2013) Live cell imaging of viral entry. *Curr. Opin. Virol.* 3, 34–43.

Tilney LG & Portnoy DA (1989) Actin filaments and the growth, movement, and spread of the intracellular bacterial parasite, *Listeria monocytogenes*. *J. Cell Biol.* 109, 1597–1608.

Vink C, Rudenko G & Seifert HS (2012) Microbial antigenic variation mediated by homologous DNA recombination. *FEMS Microbiol. Rev.* 36, 917–948.

Walsh D, Mathews MB & Mohr I (2013) Tinkering with translation: protein synthesis in virus-infected cells. *Cold Spring Harb. Perspect. Biol.* 5, a012351.

Welch MD & Way M (2013) Arp2/3-mediated actin-based motility: a tail of pathogen abuse. *Cell Host Microbe* 14, 242–255.

Worobey M, Watts TD, McKay RA . . . Jaffe HW (2016) 1970s and 'Patient 0' HIV-1 genomes illuminate early HIV/AIDS history in North America. *Nature* 39, 98–101.

The Human Microbiota

Fan Y & Pedersen O (2021) Gut microbiota in human metabolic health and disease. *Nat. Rev. Microbiol.* 19, 55–71.

Fischbach MA (2019) Microbiome: focus on causation and mechanism. *Cell* 174, 785–790.

Sender R, Fuchs S & Milo M (2016) Are we really vastly outnumbered? Revisiting the ratio of bacterial to host cells in humans. *Cell* 164, 337–340.

CHAPTER 24

The Innate and Adaptive Immune Systems

IN THIS CHAPTER

As we discussed in Chapter 23, all living organisms serve as hosts for other species, usually in relationships that are benign or even mutually helpful. But all organisms, and all cells in a multicellular organism, need to defend themselves against infection by harmful invaders, collectively called **pathogens**, which can be microbes (bacteria, viruses, or fungi) or larger parasites. The first line of defense against pathogens is provided by the **innate immune responses**, which can include protective barriers, toxic molecules, and phagocytic cells that ingest and destroy the invading pathogen. These and other innate immune defenses are not pathogen specific, but they usually can prevent or halt an infection early; if they fail to do so, some organisms, including all vertebrates and some bacteria and archaea (see Figure 7–81), can activate more sophisticated, powerful, and pathogen-specific *adaptive immune responses*. In this chapter, we mainly discuss the innate and adaptive immune systems of humans.

The diverse cells of our innate immune system can respond directly to a pathogen, and some of them can then help activate adaptive immune responses. The innate and adaptive immune responses then work together to help eliminate the pathogen (**Figure 24–1**). Unlike the innate responses, our adaptive responses are highly specific to the particular pathogen that induced them, and they depend on white blood cells called B and T lymphocytes. *B lymphocytes* (*B cells*) secrete *antibodies* that bind specifically to the pathogen. *T lymphocytes* (*T cells*) can either directly kill cells infected with the pathogen (**Figure 24–2**) or produce secreted or cell-surface signal proteins that stimulate other host cells to help destroy the pathogen. Whereas innate immune responses are generally brief, the adaptive responses provide long-lasting protection: a person who recovers from measles or is vaccinated specifically against it, for example, is protected for life against measles by the adaptive immune system, although not against other common viruses, such as those that cause mumps or chickenpox.

Both the innate and adaptive immune systems have evolved sensing mechanisms that enable the systems to recognize pathogens and their harmful products, and distinguish them from both the host's own cells and molecules and from harmless or beneficial foreign organisms and their molecules. The innate system relies on various sensor proteins to distinguish self from nonself by recognizing particular types or patterns of molecules that are common to microbes but are absent or sequestered in the host. Our adaptive system, by contrast, uses unique genetic mechanisms to produce a virtually limitless diversity of related proteins—receptors on T and B cells and secreted antibodies—that, among them, can bind almost any foreign molecule. This remarkable strategy enables our adaptive immune system to react specifically against any pathogen, even if we never encountered it before. However, it also requires that the system learn not to react against self molecules or harmless foreign ones; if these learning mechanisms fail, harmful autoimmune or allergic responses result.

In this chapter, we focus mainly on features of our immune responses that distinguish them from other kinds of human cell and tissue responses. We begin with innate immune defenses and then discuss the highly specialized properties of our adaptive immune system.

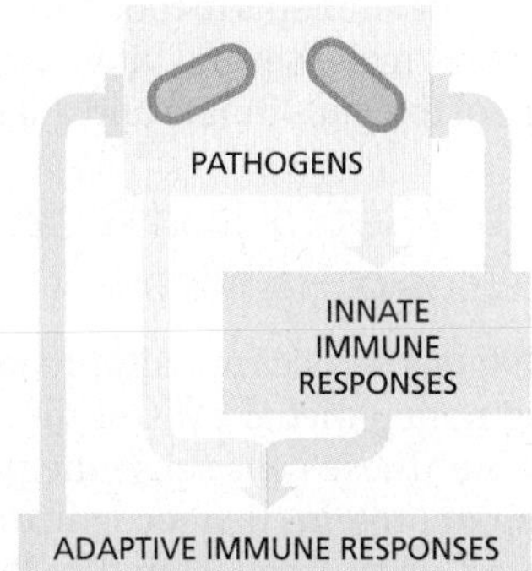

Figure 24–1 **Innate and adaptive immune responses.** Innate immune responses are activated directly by pathogens and defend all multicellular organisms against infection. In vertebrates, pathogens, together with the innate immune responses they activate, also stimulate adaptive immune responses, which then work together with innate immune responses to help fight the infection. Whereas adaptive responses are specific to a particular pathogen, innate responses are not.

THE INNATE IMMUNE SYSTEM

Our adaptive immune responses are slow to develop when we first encounter a new pathogen. This is because the specific B cells and T cells that can respond to a particular pathogen are initially few in number and must be stimulated to proliferate and differentiate before they can mount effective adaptive immune responses, which can take days. By contrast, a single bacterium that divides every hour can generate almost 20 million progeny in a single day, producing a full-blown infection. We therefore rely on our **innate immune system** to defend us against infection during the first critical hours and days of exposure to a new pathogen.

In this section, we consider some of the strategies our innate immune system uses to recognize pathogens and to provide a first line of defense against them.

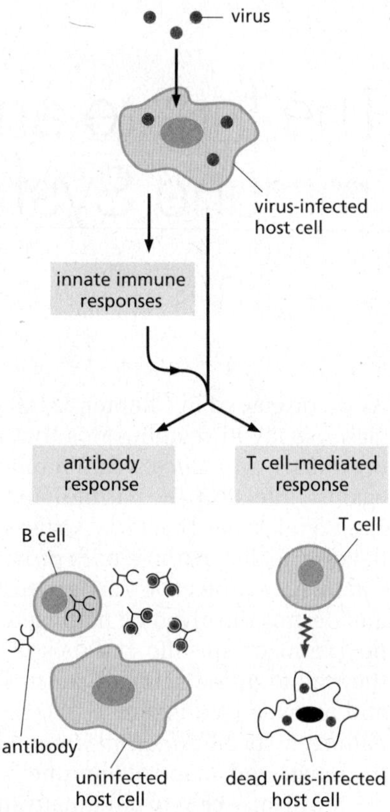

Figure 24–2 Two classes of vertebrate adaptive immune responses. Lymphocytes carry out both classes, shown here as responses to a viral infection. In one class, B cells secrete antibodies that specifically bind to and neutralize an extracellular virus, thereby preventing the virus from infecting host cells. In the other, T cells mediate the response; in this example, they kill the virus-infected host cells. In both cases, innate immune responses help activate the adaptive immune responses through pathways that we discuss later.

Epithelial Surfaces Serve as Barriers to Infection

Our first encounters with infectious organisms are typically at the epithelial surfaces that form our skin and line our respiratory, digestive, and genitourinary tracts. These epithelia provide both physical and chemical barriers to invasion by pathogens: tight junctions between epithelial cells bar entry between the cells, and a variety of substances secreted by the cells discourage the attachment and entry of pathogens. The keratinized epithelial cells of the skin, for example, form a thick physical barrier, and the sebaceous glands in the skin secrete fatty acids and lactic acid, which inhibit bacterial growth. In addition, epithelial cells in all tissues, including those in plants and invertebrates, secrete antimicrobial molecules called **defensins**. Defensins are positively charged, amphipathic peptides that bind to and disrupt the membranes of many pathogens, including enveloped viruses, bacteria, fungi, and parasites.

The epithelial cells that line internal organs such as the respiratory and digestive tracts also secrete slimy mucus, which sticks to the epithelial surface and makes it difficult for pathogens to adhere. The beating of cilia on the surface of the epithelial cells lining the respiratory tract and the peristaltic action of the intestine also discourage the adherence of pathogens. Moreover, as we discuss in Chapter 23, healthy skin and gut are normally populated by enormous numbers of harmless (and often helpful) *commensal* microbes, collectively called the *flora*, which compete for nutrients with pathogens; some also produce antimicrobial peptides that actively inhibit pathogen proliferation. Commensal microbes also bring other benefits to their host: some of those in the gut, for example, help digest food and make several vitamins; some are also needed for the normal development of the gut's innate and adaptive immune systems.

Pattern Recognition Receptors (PRRs) Recognize Conserved Features of Pathogens

Pathogens do occasionally breach the epithelial barricades, in which case underlying nonepithelial cells of the innate immune system provide the next line of defense. These cells sense the presence of pathogens largely through the use of receptor proteins that recognize microbe-associated molecules that are either not present or are sequestered in the host organism. Because these microbial molecules often occur in repeating patterns, they are called **pathogen-associated molecular patterns**, or **PAMPs** (because the molecular patterns are shared with commensal microbes, they are also called microbe-associated molecular patterns, or MAMPs). PAMPs are present in various microbial macromolecules, including nucleic acids, lipids, polysaccharides, and proteins.

The diverse receptor proteins that recognize PAMPs are collectively called **pattern recognition receptors (PRRs)**, which not only bind to PAMPs but can also activate intracellular signaling pathways that lead to the production and secretion of various signal molecules that help fight the pathogen, as we discuss shortly. Some PRRs are transmembrane proteins on the surface of many types of host cells, where they recognize extracellular pathogens. On specialized phagocytic cells (phagocytes) such as *macrophages* and *neutrophils*, for example, they

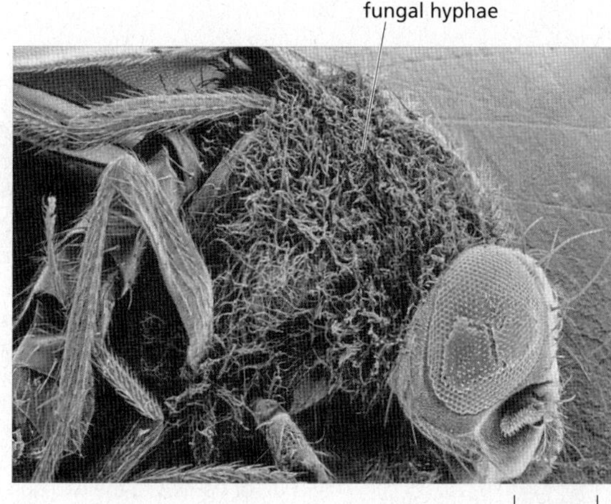

Figure 24–3 **A scanning electron micrograph of a mutant fruit fly that died from a fungal infection.** The fly is covered with fungal hyphae, as it lacked Toll receptors, which help protect *Drosophila* from fungal infections. (From B. Lemaitre et al., *Cell* 86:973–983, 1996. With permission from Elsevier.)

can help mediate the uptake of the pathogens into phagosomes, which then fuse with lysosomes to form *phagolysosomes*, where the pathogens are destroyed. Other PRRs are located intracellularly, where they can detect intracellular pathogens such as viruses; these PRRs are either free in the cytosol or associated with the membranes of the endolysosomal system (discussed in Chapter 13). Still other PRRs are secreted and bind to the surface of extracellular pathogens, marking them for destruction by either phagocytes or blood proteins that are part of the *complement system* (discussed later).

There Are Multiple Families of PRRs

The first PRR identified was the *Toll receptor* in *Drosophila*, which was already well known for its role in fly development (see Figure 21–16). It was later discovered to be required also for the production of antimicrobial peptides that protect the fly against fungal infections (**Figure 24–3**). Toll is a transmembrane glycoprotein with a large extracellular domain that contains a series of leucine-rich repeats. Soon it was discovered that both plants and animals have a variety of **Toll-like receptors (TLRs)** that function as PRRs in innate immune responses against various pathogens. A human makes at least 10 different TLRs, each recognizing distinct ligands: TLR3, for example, recognizes double-stranded viral RNA in the endosomal lumen (**Figure 24–4**); TLR4 recognizes lipopolysaccharide (LPS) on the outer membrane of Gram-negative bacteria; TLR5 recognizes the protein that forms the bacterial flagellum; TLR7 and TLR8 recognize single-stranded viral RNA; and TLR9 recognizes short, unmethylated sequences of bacterial, viral, or protozoan DNA, called CpG motifs, which are uncommon in vertebrate DNA.

In addition to TLRs, humans use several other families of PRRs to detect pathogens. One is the large family of **NOD-like receptors (NLRs)**. Like TLRs, NLRs have leucine-rich repeat motifs, but they are exclusively cytoplasmic and recognize a distinct set of bacterial molecules. Individuals who are homozygous for an inactivating mutation in the NLR gene *NOD2* have a greatly increased risk of developing Crohn's disease, a chronic inflammatory disease of the small intestine, thought to involve chronic immune responses against harmless commensal gut microbes. Another family of PRRs consists of *RIG-like receptors* (*RLRs*), which are members of the RNA helicase family of proteins. They are also exclusively cytoplasmic and detect viral pathogens. A fourth family of PRRs consists of *C-type lectin receptors* (*CLRs*), which are transmembrane cell-surface proteins that recognize carbohydrates (which is why they are called lectins) on various microbes; they are called C-type because the binding to carbohydrate is dependent on Ca^{2+}. **Table 24–1** summarizes some PRRs and their ligands and locations in cells. Collectively, these and other PRRs act as an alarm system to alert the innate and adaptive immune systems that an infection is brewing (**Movie 24.1**).

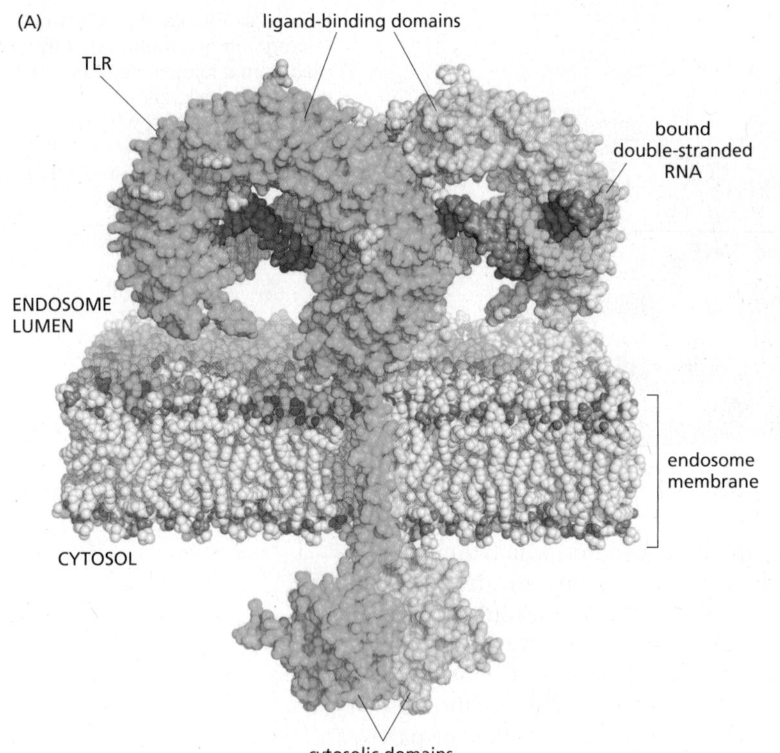

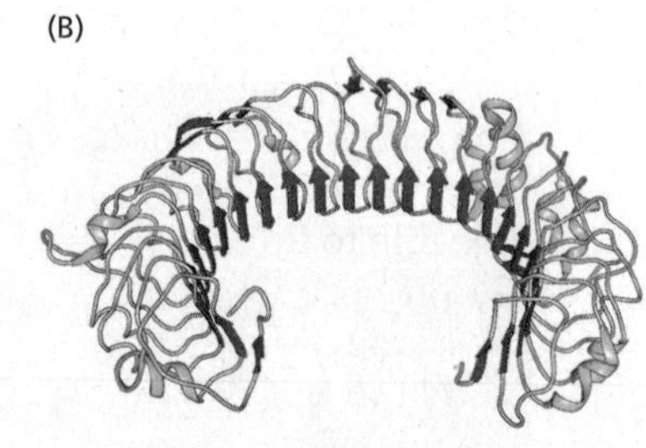

Figure 24–4 **A Toll-like receptor.** (A) The structure of human TLR3 is shown *(green)*, bound to a double-stranded RNA molecule (dsRNA; *blue*). The receptor is a transmembrane homodimer in the membrane of endosomes. The binding of dsRNA to the two horseshoe-shaped domains on the lumenal side of the endosome brings the two cytosolic domains together, allowing adaptor proteins in the cytosol to assemble into a large signaling complex, leading to the production of antivirus cytokines (not shown, but discussed later). (B) The crystal structure of a lumenal domain of the transmembrane receptor, which contains 23 conventional leucine-rich repeats, each of which contributes a β strand to the continuous β sheet *(red)* that lines the concave surface of the structure. (A, adapted from L. Liu et al., *Science* 320:379–381, 2008; B, adapted from J. Choe et al., *Science* 309:581–585, 2005. Both with permission from AAAS. PDB code: 1ZIW.)

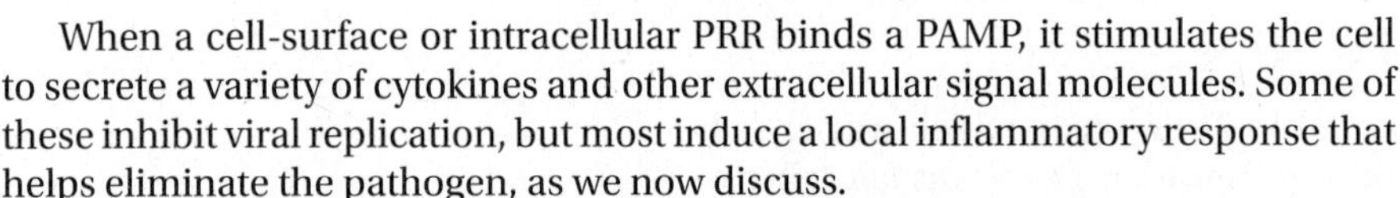

When a cell-surface or intracellular PRR binds a PAMP, it stimulates the cell to secrete a variety of cytokines and other extracellular signal molecules. Some of these inhibit viral replication, but most induce a local inflammatory response that helps eliminate the pathogen, as we now discuss.

Activated PRRs Trigger an Inflammatory Response at Sites of Infection

When a pathogen invades a tissue, it activates PRRs on or in various cells of the innate immune system, resulting in an **inflammatory response** at the site of infection. Resident macrophages are usually the first cells to respond, and

TABLE 24–1 **Some Pattern Recognition Receptors (PRRs)**

Receptor	Location	Ligand	Origin of ligand
		Toll-like receptors (TLRs)	
TLR3	Endolysosomal system	Double-stranded RNA	Viruses
TLR4	Plasma membrane	Bacterial lipopolysaccharide (LPS); viral coat proteins	Bacteria, viruses
TLR5	Plasma membrane	Flagellin	Bacteria
TLR9	Endolysosomal system	Unmethylated CpG dinucleotides in DNA	Bacteria, viruses, protozoa
		NOD-like receptors (NLRs)	
NOD2	Cytoplasm	Degradation products of peptidoglycans	Bacteria
		Retinoic acid–inducible gene 1 (RIG)-like receptors (RLRs)	
RIG1	Cytoplasm	Double-stranded RNA	Viruses
		C-type lectin receptors (CLRs)	
Dectin1	Plasma membrane	β-Glucan	Fungi

they orchestrate the subsequent responses by secreting short-range signal molecules that recruit other cells of the innate immune system. The inflammatory response involves changes in local blood vessels and is characterized clinically by local pain, redness, heat, and swelling. The blood vessels dilate and become permeable to fluid and proteins, leading to local swelling and an accumulation of blood proteins, including some that aid in defense against pathogens. At the same time, the endothelial cells lining the local blood vessels are stimulated to express cell adhesion proteins, which promote the attachment and escape of white blood cells, or *leukocytes* (see Figure 19–28), adding to the local swelling; initially neutrophils escape, followed later by lymphocytes and monocytes (the blood-borne precursors of macrophages—see Figure 22–12).

The activation of PRRs results in the production of a large variety of extracellular signal molecules that mediate the inflammatory response at the site of an infection. These include both lipid signal molecules, such as prostaglandins, and protein (or peptide) signal molecules called **cytokines**, which mainly influence nearby cells. Some of the most important **pro-inflammatory cytokines** are *tumor necrosis factor-α* (*TNFα*), *interferon-γ* (*IFNγ*), a variety of *chemokines* that recruit leukocytes, and various *interleukins* (*ILs*) that we discuss later, including IL1β, IL6, IL12, IL17, and IL18. In addition, a secreted PRR (mannose-binding lectin) activates the complement system when the PRR binds to a pathogen; fragments of complement proteins released during complement activation stimulate an inflammatory response (discussed shortly; see Figure 24–7).

When activated by PAMPs, most cell-surface and intracellular PRRs stimulate the production of multiple pro-inflammatory cytokines by activating intracellular signaling pathways that switch on transcription regulators, including NFκB, to induce the transcription of the relevant cytokine genes (see Figure 15–63). Some PRRs, however, can also stimulate pro-inflammatory cytokine production by a different mechanism: when activated, several cytoplasmic NLRs assemble with adaptor proteins and specific proteases of the caspase family (discussed in Chapter 18) to form **inflammasomes**, in which the pro-inflammatory cytokines such as IL1β and IL18 are cleaved from their inactive precursor proteins by caspase-1. These cytokines are then released from the cell by unconventional secretion pathways. Inflammasomes closely resemble *apoptosomes* in their assembly and structure, but, in apoptosomes, caspases are activated to initiate an intracellular, proteolytic, caspase cascade that leads to apoptotic cell death (see Figure 18–8).

NLR-dependent inflammasome assembly can also be triggered in the absence of infection if cells are damaged or stressed. Such cells produce *damage-associated molecular patterns* (*DAMPs*), including those on altered or misplaced self molecules, which can activate the relevant NLRs: the arthritis caused by uric acid crystals formed in the joints of individuals with gout, who have abnormally high uric acid levels in their blood, is a painful example.

The inflammatory response is amplified by various positive feedback loops. Activated macrophages, for example, secrete IL1β, which acts back on macrophages to increase their production of more precursor of IL1β; at the same time, other cytokines increase the assembly of inflammasomes that produce yet more IL1β by cleaving its precursor. As another example, activated macrophages secrete chemokines that recruit leukocytes that also secrete chemokines that recruit more leukocytes, some of which are monocytes that mature into macrophages, which can be become activated to drive more rounds of this positive feedback cycle.

Besides their local effects, pro-inflammatory cytokines can produce widespread changes in the body. IL1β, IL6, and TNFα, for example, can act on the brain hypothalamus, muscle cells, and fat cells to increase body temperature, producing a fever that helps some immune cells fight infection. These cytokines can also stimulate the liver to secrete acute-phase proteins, such as C-reactive protein, which binds to the surface of various pathogens, where it recruits complement components that stimulate phagocytosis of the pathogen (discussed shortly). Because it increases several hundredfold, the increase in C-reactive protein is widely used clinically as a test for infection, inflammation, and tissue damage.

Phagocytic Cells Seek, Engulf, and Destroy Pathogens

In all animals, the recognition of a microbial invader is usually quickly followed by its engulfment by a phagocytic cell. In humans, these are usually macrophages, which are long-lived cells that are resident in most tissues and are therefore the first phagocytes to respond. Neutrophils, by contrast, are short-lived and, although they are the most numerous leukocytes in blood, they are not present in other healthy tissues. They are rapidly recruited from the blood to sites of infection by various attractive molecules, including formylmethionine-containing peptides (which are released by microbes but are not made by mammalian cells), chemokines secreted by activated macrophages, and peptide fragments produced from cleaved, activated complement proteins. The recruited neutrophils phagocytose the pathogens and secrete their own pro-inflammatory cytokines, thereby amplifying the local inflammatory response.

In addition to their PRRs, macrophages and neutrophils display a variety of cell-surface receptors that recognize antibodies or fragments of complement proteins bound to the surface of a pathogen. The binding of such a coated pathogen to these receptors leads to its rapid phagocytosis (**Figure 24–5**) and the mounting of a ferocious attack on the pathogen once it is inside a phagolysosome. Both macrophages and neutrophils possess an impressive armory of weapons to kill ingested invaders, including enzymes such as lysozyme and acid hydrolases that can degrade the pathogen's cell wall. The cells assemble *NADPH oxidase complexes* on the phagolysosomal membrane, where the complexes catalyze the production of highly toxic oxygen-derived compounds, including superoxide (O_2^-), hydrogen peroxide, and hydroxyl radicals. A transient increase in oxygen consumption by the phagocytic cells, called the *respiratory burst*, helps power the production of these toxic compounds. Whereas macrophages generally survive this killing frenzy and live to kill again, neutrophils do not: they are programmed to die by apoptosis (discussed in Chapter 18) after they have destroyed their prey and are then phagocytosed by macrophages; some neutrophils die by a form of cell necrosis, releasing decondensed chromatin that forms extracellular nets, which are thought to trap and kill pathogens. Dead and dying neutrophils are a major component of the pus that forms in acute wounds infected with bacteria.

If a pathogen is too large to be successfully phagocytosed (if it is a large parasite such as a worm, for example), a group of macrophages, neutrophils, or eosinophils (another type of leukocyte—see Figure 22–11) will gather around the invader. They secrete defensins and other damaging agents and release the oxygen-derived toxic products of the respiratory burst. This barrage is often sufficient to destroy the pathogen (**Figure 24–6**).

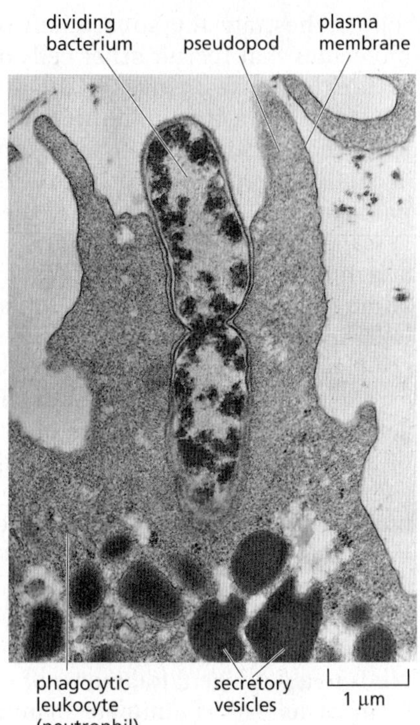

Figure 24–5 **Phagocytosis of an antibody-coated pathogen.** Electron micrograph of a neutrophil phagocytosing an antibody-coated bacterium, which is in the process of dividing. The process in which antibody (or complement) coating of a pathogen increases the efficiency with which the pathogen is phagocytosed is called *opsonization*. (Courtesy of Dorothy F. Bainton, from R.C. Williams, Jr., and H.H. Fudenberg, Phagocytic Mechanisms in Health and Disease. New York: Intercontinental Medical Book Corporation, 1971.)

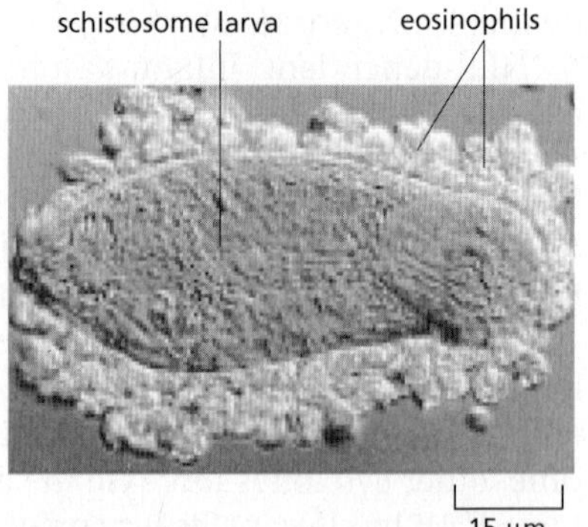

Figure 24–6 **Eosinophils attacking a parasite.** Phagocytes cannot ingest large parasites such as the schistosome larva shown here. When such a parasite is coated with antibody or complement components, however, eosinophils (and other leukocytes) can recognize it and collectively kill it by secreting a large variety of toxic molecules. (Courtesy of Anthony Butterworth.)

Complement Activation Targets Pathogens for Phagocytosis or Lysis

The blood and other extracellular fluids contain numerous proteins with antimicrobial activity, some of which are produced in response to an infection, while others are produced constitutively. The most important of these are components of the **complement system**, which consists of more than 30 interacting soluble proteins that are mainly made continually by the liver and are inactive until an infection or another trigger activates them. They were originally identified by their ability to amplify and thereby "complement" the action of antibodies made by B cells, but some are also secreted PRRs, which directly recognize PAMPs on microbes.

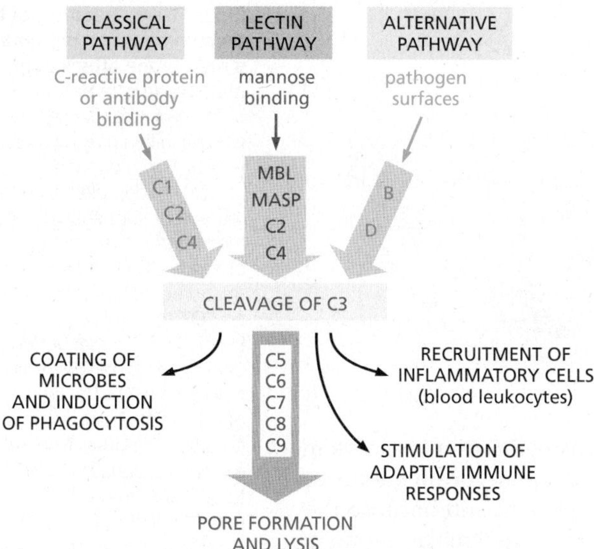

Figure 24–7 **The principal stages in complement activation by the classical, lectin, and alternative pathways.** In all three pathways, the reactions of complement activation usually take place on the surface of an invading microbe, such as a bacterium, and lead to the cleavage of C3 and the various consequences shown. As indicated, the complement proteins C1 to C9, mannose-binding lectin (MBL), MBL-associated serine protease (MASP), and factors B and D are the central components of the complement system. The early components are shown within *gray arrows*, while the late components are shown within a *brown arrow*. The *black arrows* indicate the functions of the protein fragments produced during complement activation. The various complement proteins that regulate the system are omitted. C-reactive protein is a secreted PRR protein that is made by the liver; it increases in the blood during an infection (and other inflammatory conditions) and binds to the surface of some bacteria.

The *early complement components* consist of three sets of proteins, belonging to three distinct pathways of complement activation—the *classical pathway*, the *lectin pathway*, and the *alternative pathway*. The early components of all three pathways act locally to cleave and activate C3, which is the pivotal complement component (**Figure 24–7**); individuals with a C3 deficiency are subject to repeated severe infections. The early components are proenzymes, which are activated sequentially by proteolytic cleavage. The cleavage of each proenzyme in the series activates the next component to generate a serine protease, which cleaves the next proenzyme in the series, and so on. Because each activated enzyme cleaves many molecules of the next proenzyme in the chain, the activation of the early components consists of an amplifying *proteolytic cascade*.

Many of these protein cleavages liberate a biologically active small fragment, which can attract neutrophils, plus a membrane-binding larger fragment. The binding of the large fragment to a cell membrane, usually the surface of a pathogen, helps stimulate the next reaction in the sequence. In this way, complement activation is largely kept confined to the cell surface where it began. In particular, the large fragment of C3, called C3b, binds covalently to the surface of the pathogen. Here, it recruits protein fragments produced by cleavage of other early complement components to form proteolytic complexes that catalyze the subsequent steps in the complement cascade. The early events in complement activation have diverse functions: C3b-binding receptors on phagocytic cells enhance the ability of these cells to phagocytose the pathogen, and similar receptors on B cells enhance the ability of these cells to make antibodies against various microbial molecules on C3b-coated pathogens. The smaller fragment of C3 (called C3a), as well as small fragments of C4 and C5, act independently as diffusible signals to promote an inflammatory response by recruiting leukocytes to the site of infection.

As indicated in Figure 24–7, C-reactive protein or antibodies bound to the surface of a pathogen activate the *classical pathway*. *Mannose-binding lectin*, mentioned earlier, is a secreted PRR that initiates the *lectin pathway* of complement activation when it recognizes bacterial or fungal glycolipids and glycoproteins bearing terminal mannose and fucose sugars in a particular spatial conformation. These initial binding events in the classical and lectin pathways cause the recruitment and activation of the early complement components. Because molecules on the surface of pathogens can directly activate the *alternative pathway*, it is usually the first complement pathway activated at the start of an infection.

Membrane-immobilized C3b, produced by any of the three pathways, triggers a further cascade of reactions that leads to the assembly of the *late complement*

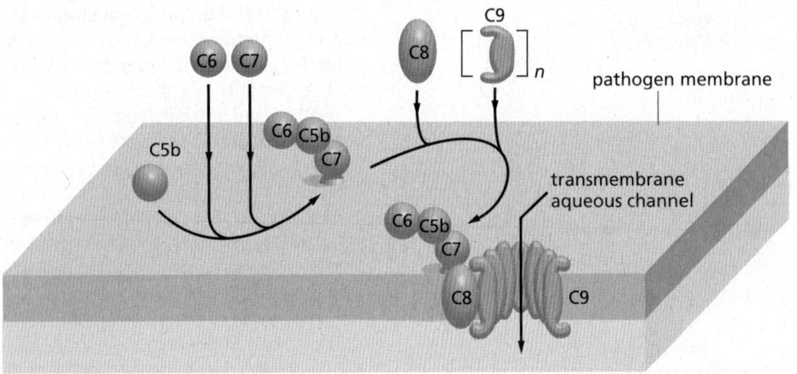

Figure 24–8 Assembly of the late complement components to form a membrane attack complex in the membrane of a pathogen. The cleavage of the early complement components (shown within *gray arrows* in Figure 24–7) results in the formation of C3b-containing proteolytic complexes on the pathogen membrane (not shown). These then cleave the first of the late components, C5, to produce C5a (not shown) and C5b. As illustrated, C5b rapidly assembles with C6 and C7 to form C567, which then binds firmly via C7 to the pathogen membrane. One molecule of C8 binds to the complex to form C5678. The binding of a molecule of C9 to C5678 induces a conformational change in C9 that exposes a hydrophobic region and causes C9 to insert into the target membrane. This starts a chain reaction in which the altered C9 binds a second molecule of C9, which can then bind another molecule of C9, and so on. In this way, a ring of C9 molecules forms a large, transmembrane aqueous channel in the pathogen membrane.

components to form *membrane attack complexes*. These protein complexes assemble in the pathogen membrane near the site of C3 activation, forming aqueous pores through the membrane (**Figure 24–8**). For this reason, and because they perturb the structure of the lipid bilayer in their vicinity, they make the membrane leaky and can, in some cases, cause the microbe to lyse.

The self-amplifying, inflammatory, and destructive properties of the complement cascade make it essential that the cascade is tightly controlled, which is achieved in various ways. One way is that key activated components are unstable and rapidly inactivate after they are generated, unless they bind immediately to either the next component in the cascade or to a nearby membrane. In addition, specific inhibitor proteins in the blood or on the surface of host cells abort the cascade by inactivating certain complement components once the components have been activated by proteolytic cleavage. One such inhibitor protein in the blood is recruited to sialic acid on host-cell glycoproteins and glycolipids (see Figure 10–16); because pathogens generally lack sialic acid, they are singled out for complement-mediated phagocytosis and destruction, while host cells are spared. Some pathogens, including the bacterium *Neisseria gonorrhoeae* that causes the sexually transmitted disease gonorrhea, coat themselves with a layer of sialic acid to effectively hide from the complement system.

Virus-infected Cells Take Drastic Measures to Prevent Viral Replication

A common way for a host-cell PRR to recognize the presence of an infecting virus is to detect unusual elements of the viral genome, such as the double-stranded RNA (dsRNA) that is an intermediate in the life cycle of many viruses and is recognized by several PRRs, including the Toll-like receptor TLR3 (see Figure 24–4A). In addition, DNA virus genomes frequently contain significant amounts of the CpG motifs mentioned earlier, which can be recognized by TLR9 (see Table 24–1, p. 1356).

Mammalian cells are particularly adept at recognizing the presence of dsRNA, which activates intracellular PRRs that induce the host cell to produce and secrete two antiviral cytokines: **interferon-α (IFNα)** and **interferon-β (IFNβ)**. These interferons are referred to as *type I interferons* to distinguish them from IFNγ, which is a type II interferon and has different functions, as we discuss later. A type I interferon acts in both an autocrine fashion on the infected cells that produced it and a paracrine fashion on uninfected neighbors. Type I interferons bind to a common cell-surface receptor, which activates the JAK–STAT intracellular signaling pathway (see Figure 15–57) to stimulate the transcription of many specific genes and thereby promote the production of hundreds of proteins, including many cytokines, reflecting the complexity of the cell's acute response to a viral infection.

The production of type I interferons appears to be a general response of our cells to a viral infection, and viral components other than dsRNA and CpG motifs in viral DNA can trigger it. The type I interferons help block viral replication in multiple ways. They activate a latent ribonuclease, for example, that nonspecifically degrades single-stranded RNAs of both the virus and host cell. They also indirectly activate a protein kinase that phosphorylates and inactivates the protein synthesis initiation factor eIF2 (discussed in Chapter 6), thereby shutting down most protein synthesis in the infected host cell. Apparently, by destroying most of its own RNA and transiently halting most of its protein synthesis, the host cell inhibits viral replication without killing itself. If these measures fail, the cell takes an even more extreme step to prevent the virus from replicating: it kills itself by undergoing apoptosis, often with the help of immune killer cells that are activated by type I interferons, as we discuss next.

Natural Killer Cells Induce Virus-infected Cells to Kill Themselves

An indirect way that type I interferons block viral replication is by enhancing the activity of **natural killer cells (NK cells)**. These lymphocyte-like leukocytes are part of the innate immune system. They are recruited early to sites of inflammation by cytokines secreted by activated resident macrophages; once there, they secrete cytokines that further activate macrophages to increase their ability to ingest and destroy pathogens and secrete cytokines—yet another positive feedback loop that amplifies the inflammatory response. Like *cytotoxic T cells* of the adaptive immune system (discussed later), NK cells also directly destroy virus-infected cells by inducing the infected cells to kill themselves by undergoing apoptosis. Thus NK cells help defend us against both extracellular and intracellular pathogens. We consider later how NK cells induce apoptosis when we discuss how cytotoxic T cells do it (see Figure 24–43). Although the two types of killer cells kill in the same ways, the means by which they distinguish the surface of virus-infected cells from that of uninfected cells are different (Movie 24.2).

Both cytotoxic T cells and NK cells recognize the same special class of cell-surface proteins on a host cell to help determine if the cell is virus-infected, but they use distinct receptors to do so. The special cell-surface proteins recognized are called *class I MHC proteins*. As we discuss in detail later, *MHC proteins* are so called because they are encoded by a cluster of genes in the *major histocompatibility complex*. Class I MHC proteins are present on almost all nucleated cells in vertebrates, and cytotoxic T cells use specific *T cell receptors* (*TCRs*) to recognize peptide fragments of viral proteins bound to these MHC proteins on the surface of virus-infected host cells to induce the cells to undergo apoptosis (discussed later). By contrast, NK cells have a variety of cell-surface *inhibitory receptors* that monitor the level of class I MHC proteins on the surface of other host cells: the high levels of these MHC proteins normally present on healthy host cells engage these receptors and thereby inhibit the killing activity of the NK cells. The NK cells thus focus primarily on host cells expressing abnormally low levels of class I MHC proteins and induce the cells to kill themselves; these are mainly virus-infected cells and some cancer cells (Figure 24–9). NK-cell killing activity is stimulated when various *activating receptors* on the NK cell surface recognize specific proteins that are greatly increased on the surface of virus-infected cells and some cancer cells.

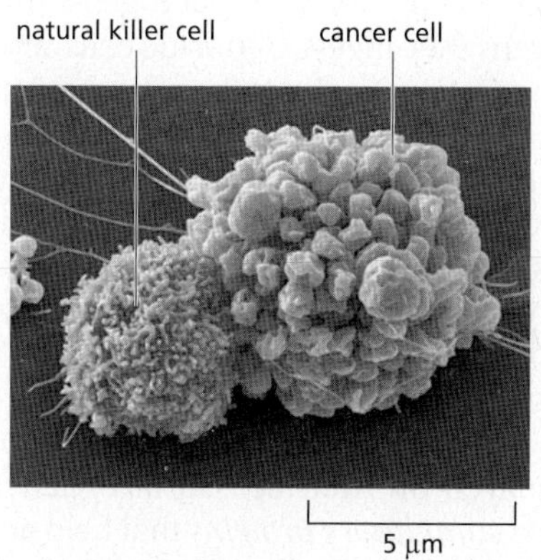

Figure 24–9 **A natural killer (NK) cell attacking a cancer cell.** This scanning electron micrograph was taken shortly after the NK cell attached to the cancer cell, causing the cancer cell to undergo apoptosis. The blebbing of the cancer cell's plasma membrane is characteristic of cells dying in this way (discussed in Chapter 18; see Movie 18.1). (From Eye of Science/Science Source.)

The reason that class I MHC protein levels are often low on virus-infected cells is that many viruses have developed a variety of mechanisms to inhibit the expression of these proteins on the surface of the host cells they infect, in order to avoid detection by cytotoxic T cells: some viruses encode proteins that block class I MHC gene transcription; others block the intracellular assembly of peptide–MHC complexes; still others block the transport of these complexes to the cell surface. By evading recognition by cytotoxic T cells in these ways, however, a virus incurs the wrath of NK cells, which recognize the infected cells as being different—both because the infected cells express little class I MHC protein and because they express large amounts of other

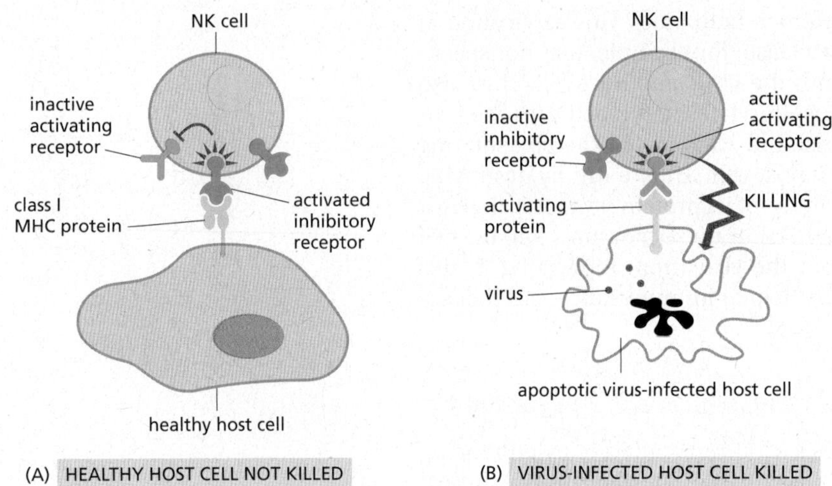

Figure 24–10 How an NK cell recognizes its target. An NK cell displays a variety of activating and inhibitory receptors on its surface, and the decision to kill or not kill a host cell depends on the sum of interactions between these receptors and the molecules they recognize on the host cell. One simplified example is shown here. (A) The high levels of class I MHC proteins found on healthy host cells activate inhibitory receptors on the NK cell, suppressing the NK cell's killing activity. (B) In contrast, the high levels of activating proteins and abnormally low level of class I MHC proteins on infected cells stimulate the NK cell to kill the virus-infected host cell by inducing the host cell to kill itself by undergoing apoptosis.

cell-surface proteins that are recognized by the activating receptors on the NK cells (**Figure 24–10**).

NK cells belong to a large class of lymphocyte-like cells of the innate immune system, collectively called *innate lymphoid cells* (*ILCs*); although these cells share some characteristics with T cells, they lack TCRs. Besides NK cells, the ILCs include more recently discovered cell types with diverse distributions and functions: some promote the early development of lymphoid tissues; some secrete various cytokines during innate immune responses to a wide variety of pathogens; others promote the repair of damaged tissues; and some help prevent adaptive immune responses against commensal microbes in the gut. And new ILCs and functions are still being discovered.

Dendritic Cells Provide the Link Between the Innate and Adaptive Immune Systems

Dendritic cells are crucially important components of the innate immune system. Like macrophages, they are made in the bone marrow, are resident in most of our tissues, express a large variety of PRRs that enable them to recognize and phagocytose invading pathogens or their products, and they become activated during the encounter with pathogens. But, unlike macrophages, which kill the pathogens they ingest, dendritic cells act indirectly to fight the pathogens they ingest by activating T cells of the adaptive immune system to join the fight.

As discussed later, an activated dendritic cell cleaves the proteins of the ingested pathogen or its products into peptide fragments, which bind to newly synthesized MHC proteins that then carry the fragments to the dendritic-cell surface. The activated cells then migrate to a nearby lymphoid organ such as a lymph node (also called a lymph gland), where they present the peptide–MHC complexes to T cells, activating the T cells to proliferate and help fight the specific pathogen (**Figure 24–11**).

In addition to the complexes of MHC proteins and microbial peptides displayed on their cell surface, activated dendritic cells also display cell-surface *co-stimulatory proteins* that help activate T cells (see Figure 24–11). The activated dendritic cells also secrete a variety of cytokines that influence the type of T cell response induced, ensuring that it is appropriate to fight the particular pathogen. In these ways, dendritic cells serve as crucial links between the innate immune system, which provides a rapid first line of defense against invading pathogens, and the adaptive immune system, which mounts slower but more powerful and highly specific responses to attack a particular invader, as we discuss next.

Summary

All multicellular organisms possess innate immune defenses against invading pathogens; these defenses include physical and chemical barriers and various defensive cell responses that are not specific to a particular pathogen. In vertebrates,

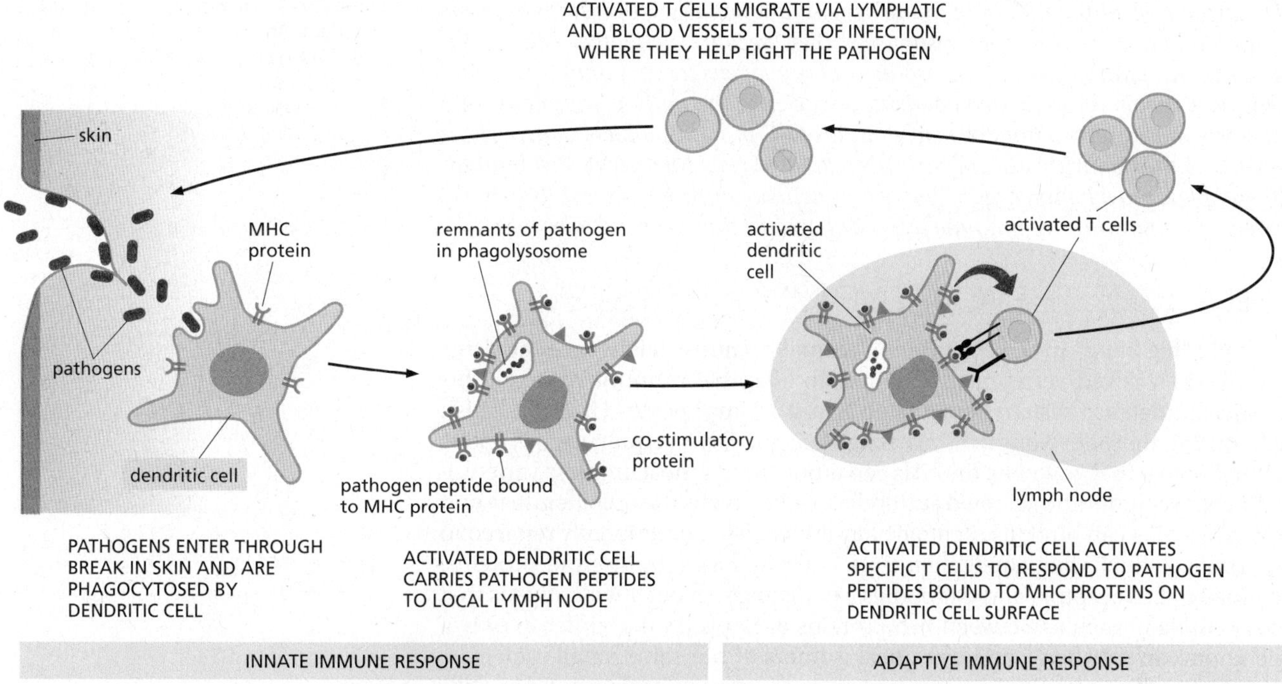
ACTIVATED T CELLS MIGRATE VIA LYMPHATIC AND BLOOD VESSELS TO SITE OF INFECTION, WHERE THEY HELP FIGHT THE PATHOGEN
skin
MHC protein
remnants of pathogen in phagolysosome
activated dendritic cell
activated T cells
pathogens
dendritic cell
co-stimulatory protein
pathogen peptide bound to MHC protein
lymph node
PATHOGENS ENTER THROUGH BREAK IN SKIN AND ARE PHAGOCYTOSED BY DENDRITIC CELL
ACTIVATED DENDRITIC CELL CARRIES PATHOGEN PEPTIDES TO LOCAL LYMPH NODE
ACTIVATED DENDRITIC CELL ACTIVATES SPECIFIC T CELLS TO RESPOND TO PATHOGEN PEPTIDES BOUND TO MHC PROTEINS ON DENDRITIC CELL SURFACE
INNATE IMMUNE RESPONSE
ADAPTIVE IMMUNE RESPONSE

Figure 24–11 Dendritic cells as functional links between the innate and adaptive immune systems. Dendritic cells pick up invading pathogens or their products at the site of an infection. The pathogen PAMPs activate the dendritic cells to express *co-stimulatory proteins* and increased amounts of MHC proteins on their surface and to migrate via lymphatic vessels to a nearby lymph node. In the lymph node, the activated dendritic cells activate T cells that express appropriate receptors for the co-stimulatory proteins and the pathogen peptides bound to MHC proteins on the dendritic-cell surface. The activated T cells proliferate, and some of their progeny migrate via lymphatic and blood vessels to the original site of infection, where they help eliminate the pathogen, either by activating local macrophages to engulf and kill the pathogen or by directly killing infected host cells (not shown). In addition, some of the activated T cells help stimulate specific B cells in the lymph node to secrete antibodies against the pathogen (not shown).

A crucial feature of dendritic-cell activation is that the pathogen provides an individual dendritic cell with both the peptides for presentation to T cells and the PAMP signals that activate the dendritic cell to express co-stimulatory proteins. In this way, the individual dendritic cell has all it needs to activate specific T cells that recognize the peptide–MHC complexes on its surface (**Movie 24.3**).

these innate immune responses can also recruit more powerful adaptive immune responses, which are pathogen-specific and help fight the infection. Innate immune responses rely on the ability of host cells to recognize characteristic features of microbial molecules called pathogen-associated molecular patterns, or PAMPs, which can be associated with a pathogen's proteins, lipids, sugars, or nucleic acids. PAMPs are mainly recognized by a variety of pattern recognition receptors (PRRs), including the Toll-like receptors (TLRs) found on or in both plant and animal cells. In vertebrates, some PRRs are secreted and can activate complement when they bind to PAMPs on the pathogen surface. The complement system, which can also be activated by antimicrobial antibodies bound to pathogens, consists of a group of blood proteins that are activated in sequence to help fight infections by disrupting the pathogen's membrane, stimulating an inflammatory response, or, most important, by targeting the microbe for phagocytosis—mainly by macrophages and neutrophils. The phagocytes use a combination of hydrolytic enzymes, antimicrobial peptides, and oxygen-derived toxic molecules to kill invading pathogens; in addition, they secrete various signal molecules that help trigger an inflammatory response.

Cells infected by a virus produce and secrete type I interferons (IFNα and IFNβ), which induce a complex set of host-cell responses that inhibit viral replication. The interferons also enhance the killing activity of natural killer (NK) cells. An NK cell kills infected host cells because they express large amounts of surface proteins that activate the NK cell; the killing is especially efficient when infected cells express reduced amounts of class I MHC proteins, which, when present in normal amounts on the surface of a healthy host cell, inhibit the killing activity of NK cells.

Dendritic cells of the innate immune system functionally link innate immune responses to adaptive immune responses. They become activated when their PRRs recognize pathogens and the products of pathogens at sites of infection and phagocytose them. The activated dendritic cells cleave the pathogen proteins into peptide fragments, which bind to newly made MHC proteins, which transport the fragments to the dendritic-cell surface. The activated cells then carry the peptide–MHC complexes to a lymph organ, where they activate appropriate T cells to make pathogen-specific adaptive immune responses against the invading microbes.

OVERVIEW OF THE ADAPTIVE IMMUNE SYSTEM

A dramatic "big bang" in the evolution of animal immune defense mechanisms occurred when jawed vertebrates acquired an **adaptive immune system.** This sophisticated defense system depends on B and T lymphocytes (B and T cells), which, during their development, rearrange specific DNA sequences in various combinations so that, together, the cells can produce an almost limitless variety of B and T cell receptors and secreted antibodies. Collectively, these cell-surface and secreted proteins can bind to essentially any molecule—operationally referred to as an **antigen**—including small chemicals, carbohydrates, lipids, and proteins. Individually, the receptors and antibodies can distinguish between antigens that are very similar—such as between two proteins or peptides that differ in only a single amino acid or between two optical isomers of the same small molecule. By this strategy, the adaptive immune system can recognize and respond specifically to any pathogen, including new mutant forms. However, because the genetic rearrangement processes involved produce receptors that can bind to self molecules as well as receptors that can bind to foreign molecules, vertebrates have had to evolve special mechanisms to ensure that B and T cells do not react against the host's own molecules and cells—a process called *immunological self-tolerance*.

Moreover, many harmless foreign substances enter the body, for example, as food or inhaled material, and it would be pointless and potentially dangerous to mount adaptive immune responses against them. Such inappropriate responses are normally avoided because innate immune responses are required to call adaptive immune responses into play and do so only when the innate cells' PRRs recognize microbial PAMPs, as we discussed earlier. One can trick the adaptive immune system into responding to a harmless foreign molecule, such as a foreign protein, by co-injecting a molecule (often of microbial origin) called an *adjuvant*, which activates PRRs. This trick is called **immunization**, and it can be exploited in vaccination (discussed later).

There are two broad classes of adaptive immune responses—*antibody responses* and *T cell–mediated immune responses*—and most pathogens induce both classes of responses. In **antibody responses**, B cells are activated to secrete antibodies, which are proteins that circulate in the bloodstream and permeate other body fluids, where they can bind specifically to the foreign antigen that stimulated their production. Binding of antibody can neutralize extracellular viruses (see Figure 24–2) and microbial toxins (such as tetanus toxin or cholera toxin) by blocking their ability to bind to receptors on host cells. Antibody binding can also mark invading pathogens for destruction, both by making it easier for phagocytes of the innate immune system to ingest and destroy them and by activating the complement system by the classical pathway (see Figure 24–7).

In **T cell–mediated immune responses**, T cells recognize foreign antigens that are bound to MHC proteins on the surface of host cells such as dendritic cells, which are specialized for presenting antigen to T cells and are therefore often referred to as "professional" *antigen-presenting cells* (*APCs*). Because MHC proteins carry fragments of pathogen proteins from inside a host cell to the cell surface, T cells can detect pathogens hiding inside a host cell and either kill the infected cell (see Figure 24–2) or stimulate phagocytes or B cells to help eliminate the pathogens.

In this section, we discuss the origins and general properties of B and T cells. In later sections, we consider the specific properties and functions of these cells.

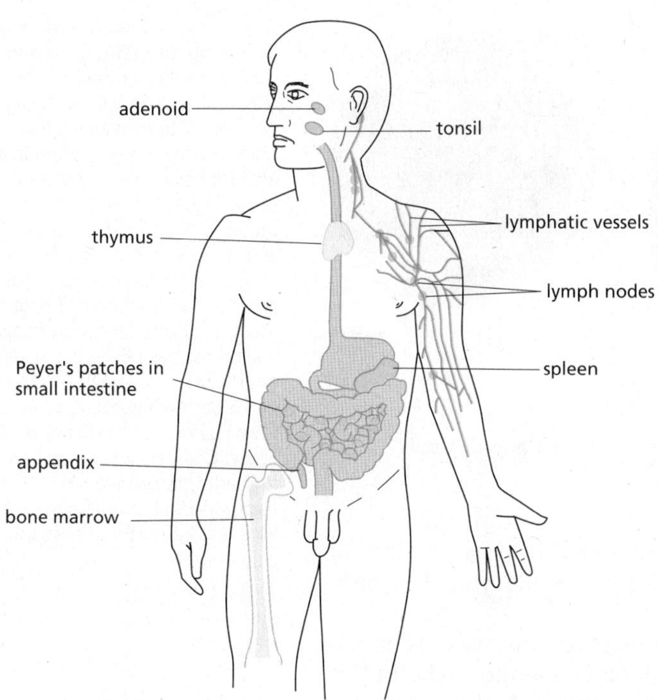

Figure 24–12 Human lymphoid organs. Lymphocytes develop from lymphoid progenitor cells in the thymus and bone marrow *(yellow)*, which are therefore called *central* (or *primary*) *lymphoid organs*. Newly formed lymphocytes migrate from these primary organs to *peripheral* (or *secondary*) *lymphoid organs (blue)*, where B and T cells respond to foreign antigen. Only some of the peripheral lymphoid organs and lymphatic vessels *(green)* are shown; many lymphocytes, for example, are found in the skin and respiratory tract. As we discuss later, the lymphatic vessels ultimately empty into the bloodstream (not shown).

B Cells Develop in the Bone Marrow, T Cells in the Thymus

There are about 2×10^{12} lymphocytes in the human body, making the immune system comparable in cell mass to the liver or the brain. They occur in large numbers in the blood and lymph (the colorless fluid in the lymphatic vessels, which connect the lymph nodes in the body to each other and to the bloodstream). They are also concentrated in **lymphoid organs**, such as the thymus, lymph nodes, and spleen (**Figure 24–12**), and many are also found in other organs, including skin, lung, and gut.

T cells and B cells derive their names from the organs in which they develop: T cells develop in the *thymus*, and B cells, in adult mammals, develop in the *bone marrow*. Both types of cells develop from *lymphoid progenitor cells* that are produced from multipotent *hematopoietic stem cells*, which, in adults, are found mainly in the bone marrow (**Figure 24–13**). The hematopoietic stem cells give rise

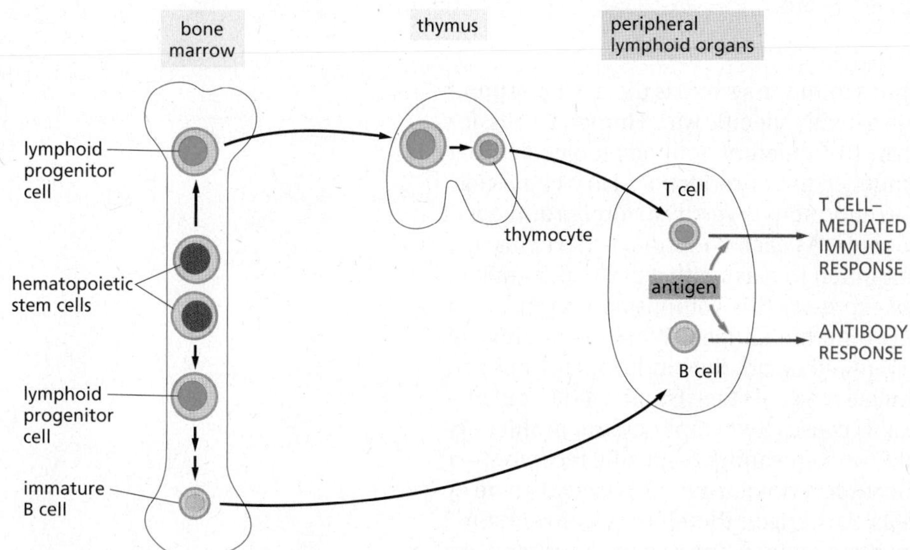

Figure 24–13 The development of B and T cells in adult humans. The central lymphoid organs, where lymphocytes develop from lymphoid progenitor cells, are labeled in *yellow boxes*. The lymphoid progenitor cells develop from multipotent hematopoietic stem cells in the bone marrow. Some lymphoid progenitor cells develop locally in the bone marrow into immature B cells, while others migrate via the bloodstream to the thymus where they develop into thymocytes (developing T cells). Foreign antigens activate B cells and T cells mainly in peripheral lymphoid organs, such as lymph nodes and the spleen.

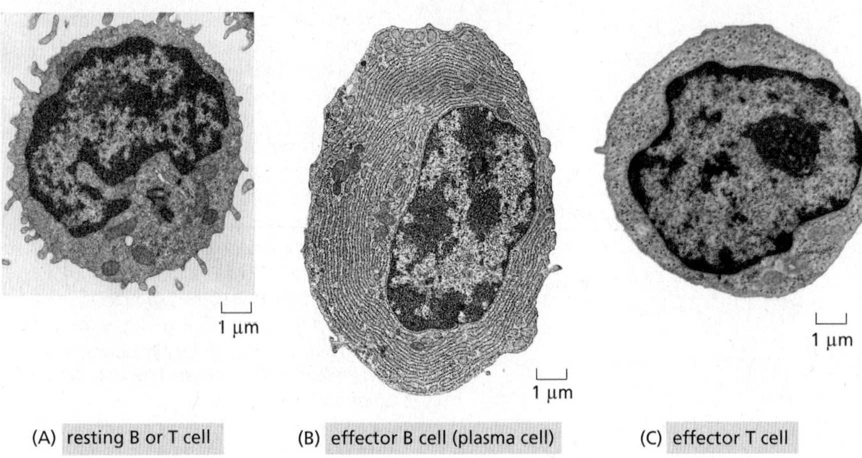

Figure 24–14 **Electron micrographs of resting and effector B and T cells.** (A) This resting lymphocyte could be either a B cell or a T cell, as these cells are difficult to distinguish morphologically until antigen activates them to become effector cells. (B) An effector B cell (a plasma cell). It is filled with an extensive rough endoplasmic reticulum (ER), which is distended with antibody molecules that are secreted in large amounts. (C) An effector T cell, which has relatively little rough ER but is filled with free ribosomes; it secretes cytokines, but in relatively small amounts. The three cells are shown at the same magnification. (A and B, from D. Zucker-Franklin et al., Atlas of Blood Cells: Function and Pathology, 2nd ed. Philadelphia: Lea & Febiger, 1988. Reprinted with permission of Wolters Kluwer; C, David M. Phillips/Science Source.)

to more than just lymphocytes: as discussed in Chapter 22, they produce all of the cells of the hematopoietic system, including erythrocytes, leukocytes, and platelets (see Figure 22–12).

Because they are sites where B and T lymphocytes (as well as innate lymphoid cells, such as NK cells) develop from lymphoid progenitor cells, the thymus and bone marrow are referred to as **central, or primary, lymphoid organs** (see Figure 24–12). Here, lymphocytes differentiate and then migrate via the blood to the **peripheral, or secondary, lymphoid organs**—mainly the lymph nodes, spleen, and epithelium-associated lymphoid tissues in the gastrointestinal tract, respiratory tract, and skin. It is in these peripheral lymphoid organs that foreign antigens activate B and T cells (see Figure 24–13).

B and T cells become morphologically distinguishable from each other only after antigen has activated them: resting B and T cells look very similar, even in an electron microscope (**Figure 24–14A**). After activation by an antigen, both B and T cells proliferate and mature into *effector cells*. Effector B cells secrete antibodies; in their most mature form, called *plasma cells*, they are filled with an extensive rough endoplasmic reticulum that is busily making antibodies (**Figure 24–14B**). In contrast, effector T cells (**Figure 24–14C**) contain very little endoplasmic reticulum and secrete a variety of cytokines rather than antibodies. Whereas B cell–derived antibodies are widely distributed by the bloodstream, T cell–derived cytokines mainly act locally, on cells the T cell contacts or on noncontacted neighboring cells, although some are carried via the blood and act on distant host cells.

Immunological Memory Depends on Both Clonal Expansion and Lymphocyte Differentiation

The most remarkable feature of the adaptive immune system is that it can respond to millions of different foreign antigens in a highly specific way. Human B cells, for example, collectively can make more than 10^{12} different antibody molecules that react specifically with the antigen that induced their production. How can B cells and T cells respond specifically to such an enormous diversity of foreign antigens? The answer for both B and T cells is the same. As each lymphocyte develops in a central lymphoid organ, it becomes committed to react with a particular antigen before ever being exposed to it. The cell expresses this commitment in the form of cell-surface receptors that specifically bind the antigen. When a lymphocyte encounters its antigen in a peripheral lymphoid organ, the binding of the antigen to the receptors, with help from co-stimulatory signals (see Figure 24–11, and discussed later), activates the lymphocyte; this causes the lymphocyte to proliferate, thereby producing many more cells with the same antigen-specific receptors—a process called *clonal expansion*. The encounter with antigen also causes some of the cells to differentiate into effector cells. An antigen therefore selectively stimulates those cells that express complementary antigen-specific receptors and are

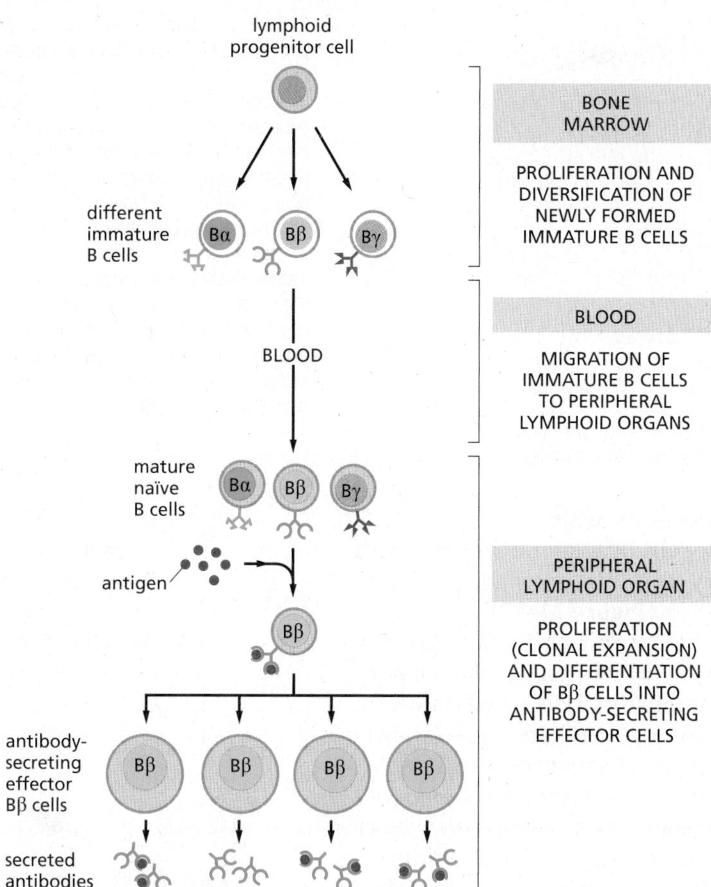

Figure 24–15 Clonal selection. An antigen activates only those B and T cells that are already committed to respond to it. The committed cell expresses cell-surface receptors that specifically recognize the antigen. The human adaptive immune system consists of many millions of different B and T cell clones, with cells within a clone expressing the same unique antigen receptor. Before its first encounter with antigen, a clone would usually contain only one or a small number of cells. A particular antigen may activate hundreds of different clones, each expressing a different antigen receptor that binds either a different part of the antigen or the same part with a different binding affinity. Although only B cells are shown here, T cells are selected in a similar way. Note that the antigen receptors on the B cells labeled Bβ in this diagram have the same antigen-binding site as the antibodies secreted by the effector Bβ cells. As we discuss later, B cells require co-stimulatory signals from T cells to become activated by antigen to proliferate and differentiate into antibody-secreting cells (not shown).

thus already committed to respond to it (Figure 24–15). This arrangement, called **clonal selection**, provides an explanation for **immunological memory**, whereby we develop longlasting immunity to many common infectious diseases after our initial exposure to the pathogen—either through natural infection or vaccination.

It is easy to demonstrate such immunological memory in experimental animals. If an animal is immunized once with antigen A, an adaptive immune response (antibody, T cell–mediated, or both) can be detected after several days; the response rises rapidly and exponentially, and then, more gradually, declines. This is the characteristic course of a **primary immune response**, occurring on an animal's first exposure to an antigen. If, after some weeks, months, or even years have elapsed, the animal is immunized again with antigen A, it will usually produce a **secondary immune response** that differs from the primary response: the lag period is shorter, because there are now many more preexisting B or T cells (or both) with specificity for antigen A, and the response is greater and more efficient. These differences indicate that the animal has "remembered" its first exposure to antigen A. If the animal is given a different antigen (for example, antigen B) instead of a second immunization with antigen A, the response is typical of a primary, and not a secondary, immune response. The secondary response therefore reflects antigen-specific immunological memory for antigen A (Figure 24–16).

Immunological memory depends on both lymphocyte proliferation and differentiation. In an adult animal, the peripheral lymphoid organs contain a mixture of B and T cells in at least three stages of maturation: *naïve cells, effector cells,* and *memory cells.* When **naïve cells** encounter their specific foreign antigen for the first time, the antigen stimulates some of them to proliferate and differentiate into **effector cells**, which then carry out an immune response: effector B cells secrete antibody, whereas effector T cells either kill infected cells (see Figure 24–2) or influence the response of other immune cells—by secreting cytokines, for example. Whereas most of the effector cells die after the pathogen has been eliminated,

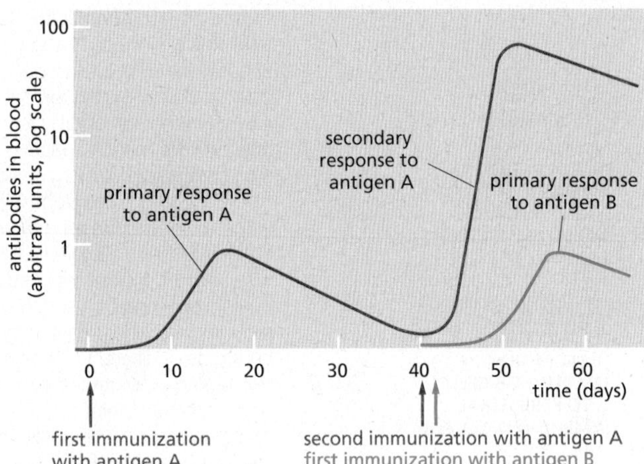

Figure 24–16 **Immunological memory: primary and secondary antibody responses.** The secondary response induced by a second exposure to antigen A is faster and greater than the primary response and is specific for A, indicating that the adaptive immune system has specifically remembered its previous encounter with antigen A. The same type of immunological memory is observed in T cell–mediated responses (not shown). As we discuss later, the types of antibodies produced in the secondary response are different from those produced in the primary response, and these antibodies bind the antigen more tightly.

a long-lived population of **memory cells** usually persists, which can more easily and more quickly be induced to become effector cells by a later encounter with the same antigen: like naïve cells, when memory cells encounter their antigen, they give rise to effector cells and more memory cells (**Figure 24–17**).

Thus, during the primary response, clonal expansion and differentiation creates many antigen-specific memory cells, some of which can persist as a population for the lifetime of the animal, even in the absence of their specific antigen, thereby providing lifelong protection against the pathogen. Whereas many newly formed memory T and B cells join the pool of T and B cells that continually recirculate through peripheral lymphoid organs (discussed shortly), some memory T cells remain permanently at the site of pathogen entry as *tissue-resident memory T cells,* which provide long-term local protection against reinfection. In addition, a small proportion of the plasma cells produced in a primary B cell response in a peripheral lymphoid organ migrate to the bone marrow, where they can survive for years and continue to secrete their specific antibodies into the bloodstream, contributing to long-term immunological memory.

Most B and T Cells Continually Recirculate Through Peripheral Lymphoid Organs

Pathogens generally enter the body through an epithelial surface, usually through the skin, gut, or respiratory tract. To induce an adaptive immune response, pathogenic microbes or their products must travel from these entry points to a peripheral lymphoid organ such as a lymph node, where B and T cells become activated (see Figure 24–11). The route and destination depend on the site of entry. Lymphatic vessels carry antigens that enter through the skin or respiratory tract to local lymph nodes; antigens that enter through the gut end up in gut-associated peripheral lymphoid organs such as Peyer's patches; and the spleen filters out antigens that enter the blood (see Figure 24–12). As discussed earlier (see Figure 24–11), in many cases, activated dendritic cells will carry the antigen from the site of infection to the peripheral lymphoid organ, where they play a crucial part in activating T cells, as we discuss later.

But only a tiny fraction of naïve B and T cells can recognize a particular microbial antigen in a peripheral lymphoid organ, a reasonable estimate being between 1/10,000 and 1/1,000,000 of each class of lymphocyte, depending on the antigen. How do these rare cells find an antigen-presenting cell displaying their specific antigen? The answer is that the lymphocytes continually recirculate between one peripheral lymphoid organ and another via the lymph and blood. In a lymph node, for example, lymphocytes continually leave the bloodstream by squeezing out between specialized endothelial cells lining small veins called *postcapillary venules.* After percolating through the node, they accumulate in small lymphatic vessels that leave the node and connect with other lymphatic vessels that pass through other lymph nodes downstream (see Figure 24–12). Passing into larger

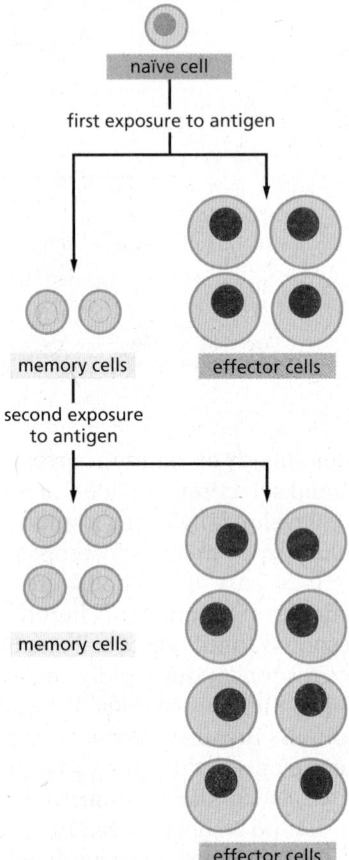

Figure 24–17 **Clonal expansion and production of memory cells as the cellular basis of immunological memory.** When stimulated by their specific antigen and co-stimulatory signals, both naïve B and T cells proliferate and differentiate, producing both effector cells and memory cells. In principle, memory cells could form either directly from naïve cells, as shown here, or from "retired" effector cells (not shown); for some types of T cells, at least, effector cells can become memory cells. When exposed to the same antigen in the presence of co-stimulatory signals, memory cells respond more readily, rapidly, and efficiently than do naïve cells.

and larger vessels, the lymphocytes eventually enter the main lymphatic vessel (the *thoracic duct*), which carries them back into the blood (Figure 24–18).

The continual recirculation of a lymphocyte between the blood and lymph ends only if its specific antigen activates it in a peripheral lymphoid organ. In that case, the lymphocyte remains in the peripheral lymphoid organ, where it proliferates and differentiates to produce effector and memory T and B cells. Many of the effector T cells leave the lymphoid organ via the lymph and migrate through the blood to the site of infection (see Figure 24–11), whereas others stay in the lymphoid organ and help activate B cells to differentiate into antibody-secreting cells and undergo further maturation (discussed later). Some effector B cells (including some of the most mature, plasma cells) remain in the peripheral lymphoid organ and secrete antibodies into the blood and undergo further maturation (discussed later). Many of the memory B and T cells produced in a peripheral lymphoid organ join the recirculating pool of naïve and memory lymphocytes.

Lymphocyte recirculation depends on specific interactions between the lymphocyte cell surface and the surface of the endothelial cells lining the blood vessels in the peripheral lymphoid organs. Lymphocytes that enter a lymph node via the blood, for example, adhere weakly to specialized endothelial cells lining the postcapillary venules via *homing receptors* that belong to the *selectin* family of cell-surface lectins that bind to specific sugar groups on the endothelial cell surface (see Figure 19–28). The lymphocytes roll slowly along the surface of the endothelial cells until another, much stronger adhesion system, dependent on an integrin protein, is called into play by chemokines secreted by the endothelial cells. Now, the lymphocytes stop rolling, and they crawl out of the blood vessel into the lymph node by using yet another cell adhesion protein called CD31 (Figure 24–19). Although B and T cells initially enter the same region of a lymph node, different chemokines guide them to separate regions of the node—B cells to *lymphoid follicles* and T cells to the *paracortex* (Figure 24–20).

Unless they encounter their antigen, both B and T cells soon leave the lymph node via efferent lymphatic vessels. If they encounter their antigen, however, they are stimulated to display adhesion receptors that trap the cells in the node; the cells accumulate at the junction between the B cell and T cell areas, where the

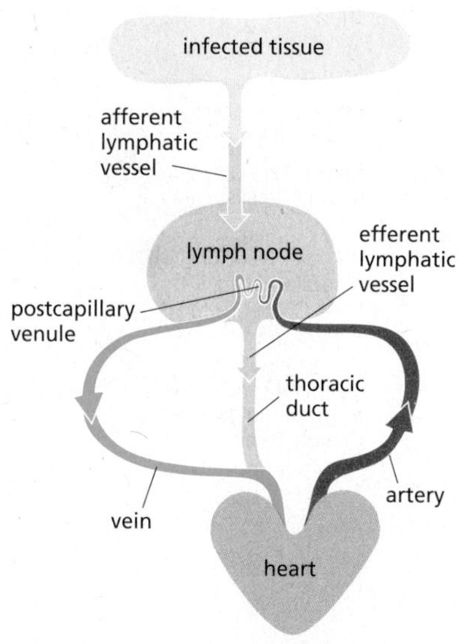

Figure 24–18 **The path followed by lymphocytes that continually recirculate between the lymph and blood.** The circulation through a lymph node *(yellow)* is shown here. Microbial antigens are usually carried into the lymph node by activated dendritic cells (not shown), which enter the node via afferent lymphatic vessels draining an infected tissue *(green)*. B and T cells, by contrast, enter via the blood, migrating out of the bloodstream into the lymph node through postcapillary venules. Unless they encounter their antigen, the B and T cells leave the lymph node via efferent lymphatic vessels, which eventually join the thoracic duct. The thoracic duct empties into a large vein carrying blood to the heart, completing the recirculation cycle for T and B cells. A typical recirculation cycle for these lymphocytes takes about 12–24 hours.

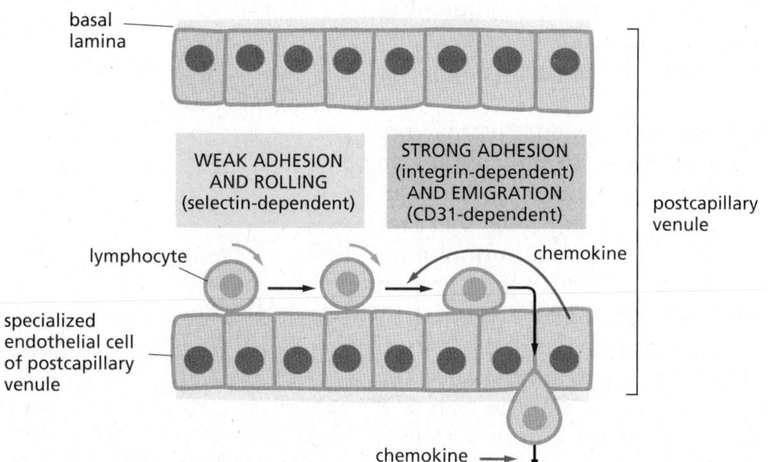

Figure 24–19 **Migration of a lymphocyte out of the bloodstream into a lymph node.** A recirculating lymphocyte adheres weakly to the surface of the specialized endothelial cells lining a postcapillary venule in a lymph node (see Figure 24–18). This initial adhesion is mediated by L-selectin (discussed in Chapter 19) on the lymphocyte surface. The adhesion is sufficiently weak to enable the lymphocyte, pushed by the flow of blood, to roll along the surface of the endothelial cells. Stimulated by chemokines secreted by specialized endothelial cells in the node *(curved red arrow)*, the lymphocyte rapidly activates a stronger adhesion system, mediated by an integrin protein. This strong adhesion enables the cell to stop rolling. The lymphocyte then uses an immunoglobulin-like cell adhesion protein (CD31) to bind to the junctions between adjacent endothelial cells and migrate out of the venule. The subsequent migration of the lymphocyte in the lymph node is directed by chemokines produced within the node *(straight red arrow)*. The migration of other types of leukocytes out of the bloodstream into sites of infection occurs in a similar way (see Figure 19–28 and Movie 19.2).

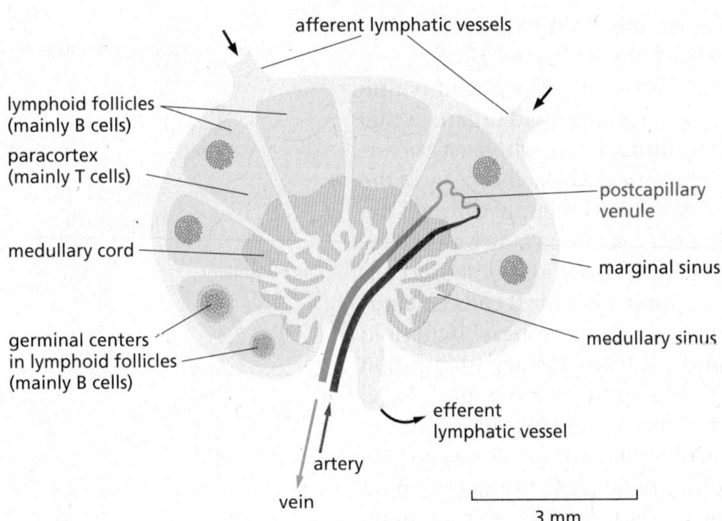

Figure 24–20 A simplified drawing of a human lymph node. B cells are mainly found in lymphoid follicles, whereas T cells are found mainly in the paracortex. Some of the lymphoid follicles contain germinal centers, where B cells, activated by their specific antigen (with the help of activated T cells), proliferate rapidly and differentiate into memory and effector cells (as discussed later). Chemokines attract resting B and T cells into the lymph node from the blood via postcapillary venules (see Figure 24–19), after which the two cell types migrate to their respective areas, attracted by different chemokines. If they do not encounter their specific antigen, both B and T cells then migrate to the medullary sinus and leave the node via the efferent lymphatic vessel, which carries the lymph away from the node to a downstream lymph node (not shown). After traveling from node to node, the lymph enters a large lymphatic vessel, the thoracic duct, that empties into the bloodstream to begin another cycle of B and T cell circulation through peripheral lymphoid organs (see Figure 24–18).

rare antigen-specific B and T cells can interact, leading to their proliferation and differentiation into either effector cells or memory cells. Many of the effector cells leave the node, expressing different chemokine receptors that help guide them to their new destinations—effector plasma cells to the bone marrow, for example, and effector T cells to sites of infection.

Immunological Self-tolerance Ensures That B and T Cells Do Not Attack Normal Host Cells and Molecules

As discussed earlier, cells of the innate immune system use PRRs to distinguish microbial molecules from self molecules made by the host. The adaptive immune system has the far more difficult recognition task of responding specifically to an almost unlimited number of foreign molecules while not responding to the large number of self molecules. How does it accomplish this feat? It helps that self molecules normally do not induce the innate immune reactions required to activate adaptive immune responses. But even when an infection or tissue injury triggers innate reactions, the vast majority of self molecules normally still fail to induce an adaptive immune response. Why?

One important reason is that the adaptive immune system "learns" not to respond to self molecules. Normal mice, for example, cannot mount an immune response against one of their own protein components of the complement system called C5 (see Figure 24–7). However, mutant mice that lack the gene encoding C5 but are otherwise genetically identical to normal mice of the same strain can make a strong immune response to this blood protein when immunized with it. The **immunological self-tolerance** exhibited by normal mice persists only for as long as the self molecule remains in the body: if a self molecule such as C5 is experimentally removed from an adult mouse, the animal gains the ability to respond to it after a few weeks or months, as new B and T cells develop in the absence of C5. Thus, the adaptive immune system is genetically capable of responding to self molecules but deploys multiple strategies to avoid doing so.

Self-tolerance depends on a number of distinct mechanisms, including the following (**Figure 24–21**):

1. In *receptor editing,* developing B cells that recognize self molecules change their antigen receptors so that the cells no longer do so.

2. In *clonal deletion,* large numbers of developing, potentially self-reactive B and T cells die by apoptosis when they encounter their particular self molecule.

3. In *clonal inactivation* (also called clonal anergy), self-reactive B and T cells become functionally inactivated when they encounter their self molecule.

4. In *clonal suppression*, self-reactive *regulatory T cells* (discussed later) suppress the activity of other types of potentially self-reactive lymphocytes.

As we discuss later, some of these mechanisms—especially the first two, receptor editing in B cells and clonal deletion of B and T cells—operate in central lymphoid organs when self-reactive developing B and T cells first encounter their self molecules, and they are largely responsible for the process called *central tolerance*. Clonal inactivation and clonal suppression, by contrast, operate mainly when mature B and T cells encounter their self molecules in peripheral lymphoid organs, and they are largely responsible for the process called *peripheral tolerance*. Clonal deletion, however, can also operate peripherally, and clonal inactivation can also operate centrally.

Why does the binding of a self molecule lead to tolerance rather than activation? The answer is still not completely known. As we discuss later, the activation of a B or T cell by its antigen in a peripheral lymphoid organ requires more than just antigen binding: it requires co-stimulatory signals, which are provided by a *helper T cell* in the case of a B cell and by an activated dendritic cell in the case of a naïve T cell. The production of such signals is usually triggered by exposure to a pathogen, but a self-reactive lymphocyte normally encounters its self antigen in the absence of such signals. Under these conditions, the lymphocyte will not only fail to be activated, it will often be rendered tolerant—being killed (deleted), or inactivated, or suppressed by a regulatory T cell (see Figure 24–21). In peripheral lymphoid organs, both T cell tolerance and activation usually occur through interactions with a dendritic cell, although the type or state of the dendritic cell is different in the two cases.

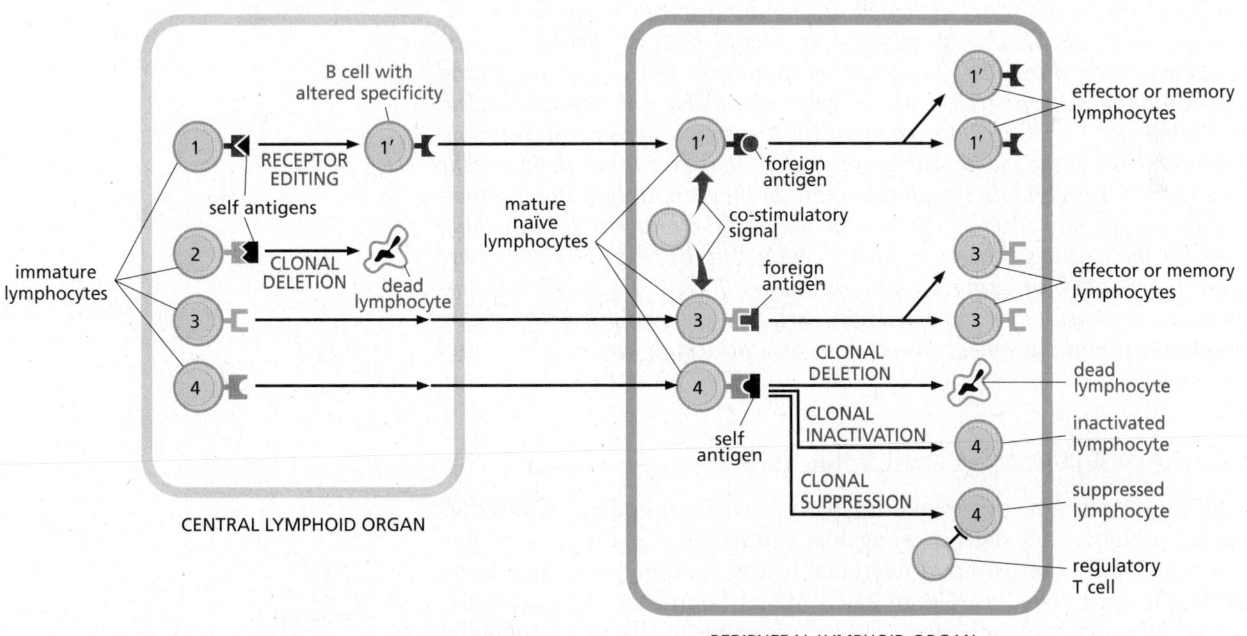

Figure 24–21 **Mechanisms of immunological self-tolerance.** When a potentially self-reactive immature B cell recognizes a specific self antigen in the central lymphoid organ (the bone marrow) where the cell is produced, it may alter its antigen receptor so that it no longer recognizes the self antigen (cell 1); this process is called receptor editing. Alternatively, when either a potentially self-reactive developing B or T cell recognizes a specific self antigen in a central lymphoid organ, it may die by apoptosis, a process called clonal deletion (cell 2). Because these two forms of tolerance (shown on the left) occur in central lymphoid organs, they are called *central tolerance*.

When a potentially self-reactive developing B or T cell escapes tolerance in the central lymphoid organ and binds its specific self antigen in a peripheral lymphoid organ (cell 4) or in a nonlymphoid peripheral tissue (not shown), it will generally not be activated, because the binding usually occurs in the absence of sufficient co-stimulatory signals; instead, the cell may die by apoptosis (often after a period of proliferation), be inactivated, or be suppressed by a regulatory T cell. These forms of tolerance (shown on the right) are called *peripheral tolerance*. As discussed later, the cells providing the co-stimulatory signals are T lymphocytes for B cells and usually dendritic cells for T cells (not shown).

For reasons that are usually unknown, self-tolerance mechanisms sometimes fail, causing T or B cells (or both) to react against the animal's own molecules. *Myasthenia gravis* is an example of such an **autoimmune disease**. Most of the affected individuals make antibodies against the acetylcholine receptors on their own skeletal muscle cells; these receptors are required for the muscle to contract normally in response to nerve stimulation, which releases acetylcholine (see Figure 11–39). The antibodies interfere with the normal functioning of the receptors so that the individuals become weak and may die because they cannot breathe. Similarly, in *juvenile (type 1) diabetes*, adaptive immune reactions against insulin-secreting β cells in the pancreas kill these cells, leading to severe insulin deficiency.

Summary

The human adaptive immune system is composed of many millions of B and T cell clones, with the cells in each clone sharing a unique cell-surface receptor that enables them to bind a particular pathogen antigen. The binding of antigen to these receptors, with the help of membrane-bound co-stimulatory signals, activates the lymphocyte to proliferate and differentiate into an effector cell that helps eliminate the pathogen. Effector B cells secrete antibodies, which can act over long distances to help eliminate extracellular pathogens and to neutralize their toxins. Effector T cells, by contrast, produce cell-surface co-stimulatory molecules and secreted cytokines, which act locally to help other immune cells eliminate the pathogen; in addition, some T cells induce infected host cells to kill themselves.

During a primary adaptive immune response to an antigen, B and T cells that recognize the antigen proliferate, so there are more of them to respond the next time, during a secondary response to the same antigen. Moreover, although most effector cells produced during a primary response die after the pathogen is eliminated, some of the lymphocytes activated during the primary response become long-lived memory cells, which can respond faster and more efficiently the next time the same pathogen invades. These two mechanisms—clonal expansion and differentiation into memory cells—are largely responsible for immunological memory. Both B and T cells circulate continually between one peripheral lymphoid organ and another via the blood and lymph; only if they encounter their specific foreign antigen in a peripheral lymphoid organ do they stop migrating, proliferate, and differentiate into effector cells and memory cells. B and T cells that would react against self molecules either alter their receptors (in the case of B cells) or are eliminated, inactivated, or suppressed by regulatory T cells. These mechanisms collectively are responsible for immunological self-tolerance, which helps ensure that the adaptive immune system normally avoids attacking the molecules and cells of the host.

B CELLS AND IMMUNOGLOBULINS

We would die of infection if we were unable to make antibodies. **Antibodies** are secreted proteins that defend us against extracellular pathogens in several ways. They bind to viruses and microbial toxins, thereby preventing them from binding to host cells (see Figure 24–2). When bound to an extracellular pathogen or its products, antibodies also recruit some of the components of the innate immune system, including various types of leukocytes and components of the complement system, which work together to inactivate or eliminate the invaders.

Antibodies are synthesized exclusively by B cells. They are produced in billions of different varieties, each with a unique antigen-binding site formed by one or more unique amino acid sequences. They belong to the class of proteins called **immunoglobulins** (abbreviated as **Igs**) and are among the most abundant protein components in the blood. In this section, we discuss the structure and function of immunoglobulins and how each of us can make them with so many different antigen-binding sites.

B Cells Make Immunoglobulins (Igs) as both Cell-Surface Antigen Receptors and Secreted Antibodies

The first Igs made by a developing B cell are not secreted but are instead inserted into the plasma membrane, where they serve as receptors for antigen. They are called **B cell receptors (BCRs)**, and each B cell has approximately 10^5 of them on its surface. Each BCR is stably associated with invariant transmembrane proteins that activate intracellular signaling pathways when antigen binds to the BCR; we discuss these invariant proteins later, when we consider how B cells are activated with the assistance of *helper T cells*.

Each B cell clone produces a single species of BCR, with a unique antigen-binding site. When an antigen and a helper T cell activate a naïve or a memory B cell, the B cell proliferates and differentiates into an effector cell, which then produces and secretes large amounts of soluble (rather than membrane-bound) Ig. The secreted Ig is now called an antibody, and it has the same unique antigen-binding site as the BCR (**Figure 24–22**; and see Figure 24–15).

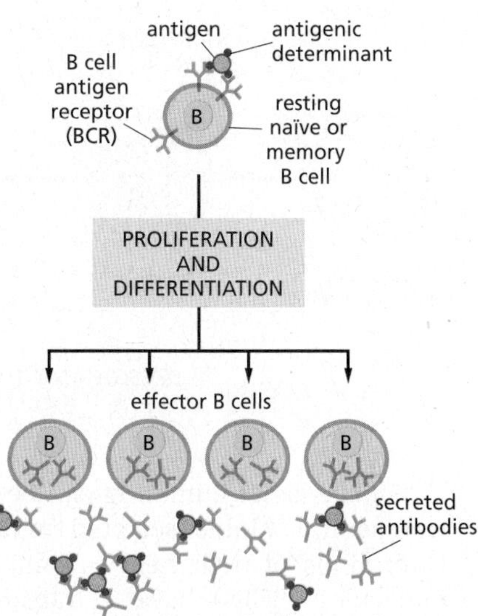

Figure 24–22 **Activation of a B cell by antigen.** The binding of an antigen to the B cell receptors (BCRs) on either a naïve or a memory B cell (together with co-stimulatory signals provided by helper T cells—not shown) activates the cell to proliferate and differentiate into antibody-secreting, effector B cells. The effector cells produce and secrete antibodies with a unique antigen-binding site, which is the same as that of the cell-surface BCRs. Because typical antibodies have two identical antigen-binding sites, they can cross-link antigens, as shown for an antigen with multiple, identical, *antigenic determinants*—the parts of an antigen an antibody or antigen receptor binds to.

A typical Ig molecule is bivalent, with two identical antigen-binding sites. It consists of four polypeptide chains—two identical *light chains* and two identical *heavy chains.* The N-terminal parts of both light and heavy chains usually cooperate to form the antigen-binding surface, while the more C-terminal parts of the heavy chains form the tail of the Y-shaped protein (**Figure 24–23**). The tail mediates many of the activities of antibodies, and antibodies with the same antigen-binding sites can have any one of a number of different tail regions, each of which gives the antibody different functional properties, such as the ability to activate complement or to bind to receptor proteins on various immune cells that bind a specific type of antibody tail.

Mammals Make Five Classes of Igs

We can make five major *classes* of Igs, each of which mediates a characteristic biological response after antigen binding to an antibody: IgA, IgD, IgE, IgG, and IgM, each with its own class of heavy chain (α, δ, ε, γ, and μ, respectively). IgA molecules have α chains, IgG molecules have γ chains, and so on. Moreover, there are four IgG subclasses (IgG1, IgG2, IgG3, and IgG4), with γ_1, γ_2, γ_3, and γ_4 heavy chains, respectively, and there are two IgA subclasses. In addition to the various classes and subclasses of heavy chains, we make two types of light chains, κ and λ, which seem to be functionally indistinguishable. Either type of light chain can be associated with any of the heavy chains, but an individual Ig molecule always contains identical light chains and identical heavy chains: an IgG molecule, for instance, can have either κ or λ light chains, but not one of each. As a result, an Ig's antigen-binding sites are always identical (see Figures 24–22 and 24–23).

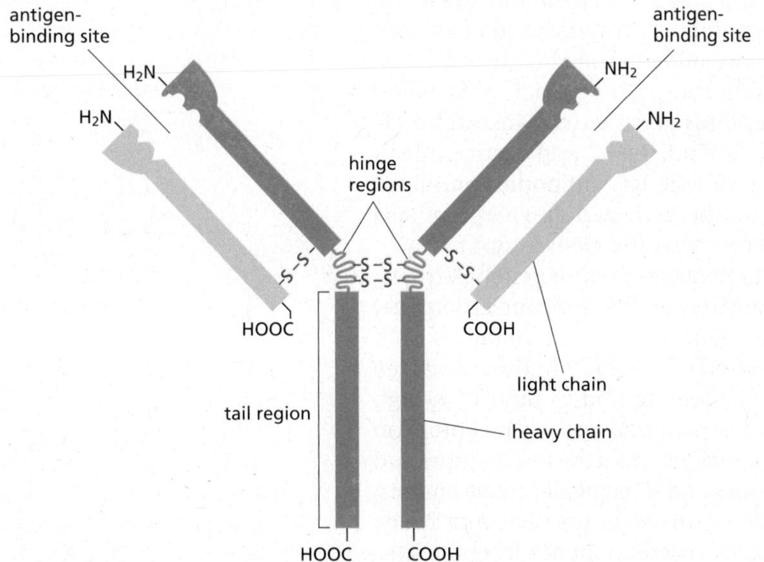

Figure 24–23 **A schematic drawing of a bivalent antibody molecule.** The distal ends of the light and heavy chains together form the two antigen-binding sites. The two heavy chains each have a hinge region (not present in all classes of heavy chains), which, because of its flexibility, improves the efficiency with which the antibody can cross-link antigens (see Figure 24–22). The two heavy chains also form the tail of the antibody, which determines the functional properties of the antibody. The heavy and light chains are held together by a combination of covalent S–S bonds *(red)* and noncovalent bonds (not shown).

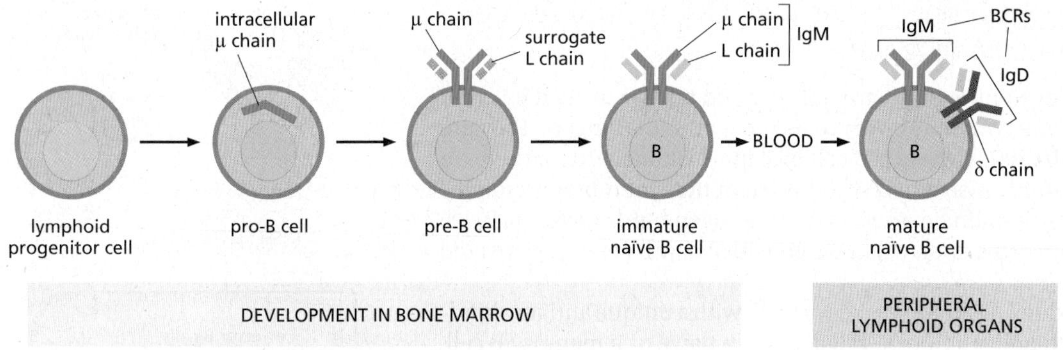

Figure 24–24 Stages of B cell development. All of the stages shown occur before the B cells bind their specific foreign antigen. The first cells in the B cell lineage that make Ig are called *pro-B cells*; they make μ heavy chains, which remain in the membrane of the endoplasmic reticulum until a special type of light (L) chain is made, called a surrogate light chain. The surrogate light chains substitute for genuine light chains and assemble with μ chains to form a temporary receptor molecule that is delivered to the plasma membrane. The cells are now called *pre-B cells*. Signaling from this pre-B cell receptor allows the cells to make bona fide light chains, which combine with μ chains to form four-chain IgM molecules that serve as antigen-specific, cell-surface BCRs on *immature naïve B cells*. After these cells leave the bone marrow (labeled in *yellow shading*), they start to express IgD BCRs as well, which have the same antigen-binding sites as the IgM BCRs; it is this *mature naïve B cell* that reacts with its specific foreign antigen in peripheral lymphoid organs (labeled in *blue shading*).

All classes of human Ig can be made in a membrane-bound BCR form, as well as in a soluble, secreted antibody form. The two forms differ only in the C-terminus of their heavy chain. The heavy chains of membrane-bound Ig molecules (BCRs) have a transmembrane hydrophobic C-terminus, which anchors them in the lipid bilayer of the B cell's plasma membrane. The heavy chains of secreted antibody molecules, by contrast, have instead a hydrophilic C-terminus, which allows them to escape from the cell. The switch in the character of the Ig molecules made occurs because the activation of B cells by antigen and helper T cells induces a change in the way in which the heavy-chain RNA transcripts are made and processed in the nucleus (see Figure 7–62).

The various Ig heavy chains give a distinctive conformation to the tail region of secreted antibodies, so that each class (and subclass) has characteristic properties of its own. **IgM** is always the first class of Ig that a developing B cell in the bone marrow makes. It forms the BCRs on the surface of *immature naïve B cells*. After these cells leave the bone marrow, they start to produce **IgD** BCRs as well, with the same antigen-binding site as the IgM BCRs. These cells are now called *mature naïve B cells*, as they can now respond to their specific foreign antigen if they encounter it in a peripheral lymphoid organ (**Figure 24–24**). IgM is also the major class of antibody secreted into the blood in the early stages of a primary antibody response on first exposure to an antigen. In its secreted form, IgM is a wheel-like pentamer composed of five four-chain units, giving it a total of 10 antigen-binding sites that allow it to bind strongly to pathogens; in its antigen-bound form, IgM is highly efficient at activating complement, which is important in early antibody responses to pathogens.

IgM and IgD are the only classes of Igs made before a B cell is activated by antigen, at which stage they are made only as BCRs. All the other Ig classes are made only after antigen stimulation, as both BCRs and antibodies. **IgG** antibodies are four-chain monomers (see Figure 24–23), and they are secreted into the blood in especially large quantities during secondary antibody responses to most antigens. The tail region of some subclasses of IgG antibodies that are bound to antigen can activate complement and also bind to specific receptors on macrophages and neutrophils. Largely by means of such **Fc receptors** (so named because antibody tails are called Fc regions), these phagocytic cells bind, ingest, and destroy infecting microorganisms that have become coated with the IgG antibodies produced in response to the infection (see Figure 24–5); the activated Fc receptors also signal the phagocyte to secrete pro-inflammatory cytokines (**Movie 24.4**).

IgE antibodies are produced in response to parasites and, in genetically susceptible individuals, to otherwise harmless environmental antigens (allergens) such as pollen, foods, and drugs. The IgE tail region binds to another class of Fc receptors on the surface of *mast cells* in tissues and of *basophils* in the blood (see Figures 22–11 and 22–12). Because antigen-free IgE antibodies bind with high affinity to such Fc receptors, the antibodies act as acquired antigen receptors on these cells. Antigen binding to the bound antibodies activates the Fc receptors and stimulates the cells to secrete a variety of cytokines and biologically active amines, especially *histamine*, which causes blood vessels to dilate and become leaky; this helps leukocytes, antibodies, and complement components to enter sites

TABLE 24–2 **Properties of the Major Classes of Antibodies in Humans**

	Class of antibody				
Properties	IgM	IgD	IgG (1–4)	IgA (1, 2)	IgE
Heavy chains	μ	δ	γ (1–4)	α (1, 2)	ε
Light chains	κ or λ	κ or λ	κ or λ	κ or λ	κ or λ
Number of four-chain units	5	1	1	1 or 2	1
Mean blood serum level (mg/mL)	1.5	0.04	IgG1 = 9 IgG2 = 3 IgG3 = 1 IG4 = 0.5	2.1	3×10^{-5}
Activates classical complement pathway	+	–	+ (IgG1–3)	–	–
Crosses from mother to fetus	–	–	+ (all subclasses)	–	–
Binds to macrophages and neutrophils	+ (macrophages only)	–	+ (all subclasses)	+	–
Binds to mast cells and basophils	–	–	+ (IgG1 and IgG3)	–	+

where mast cells have been activated. The release of amines from mast cells and basophils is largely responsible for the symptoms of such *allergic reactions* as hay fever, asthma, and hives. In addition, mast cells secrete factors that attract and activate *eosinophils* (see Figures 22–11 and 22–12), which also have Fc receptors that bind IgE molecules and can kill extracellular parasitic worms, especially if the worms are coated with IgE antibodies (see Figure 24–6).

IgA is the principal antibody class in secretions, including saliva, tears, milk, and respiratory and intestinal tract secretions. Yet another class of Fc receptors, located on the relevant epithelial cells, guides the secretion by binding antigen-free IgA dimers and transporting them from the extracellular fluid across the epithelium by the process of transcytosis (see Figure 13–56). Some of this IgA is directed against commensal microbes and keeps them in check by preventing them from binding to mucosal epithelial cells. The properties of the various classes of antibodies in humans are summarized in **Table 24–2**.

Ig Light and Heavy Chains of Antibodies Consist of Constant and Variable Regions

Both the light and heavy chains of antibodies have a variable amino acid sequence at their N-terminal ends but a constant sequence at their C-terminal ends. Whereas the **constant region** and **variable region** of a light chain are the same size, the constant region of a heavy chain is about three or four times longer than its variable region, depending on the class (**Figure 24–25**).

The variable regions of the light and heavy chains come together to form the antigen-binding sites, and the variability of their amino acid sequences provides the structural basis for the diversity of these binding sites. The greatest diversity

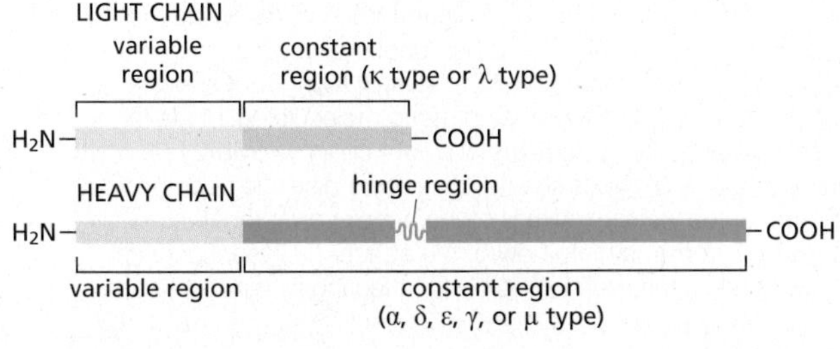

Figure 24–25 **Constant and variable regions of antibody chains.** The variable regions of both the light and heavy chains form the antigen-binding sites, while the constant regions of the heavy chains determine the other functional properties of an antibody molecule. The different subclasses of IgG antibodies have different γ-chain constant regions. An individual Ig protein has either κ or λ light chains but not both. Note that the heavy chain of the BCR form of each class of Ig has an additional, transmembrane domain at its C-terminus (not shown), and the constant regions of the μ and ε heavy chains do not have a hinge region (not shown).

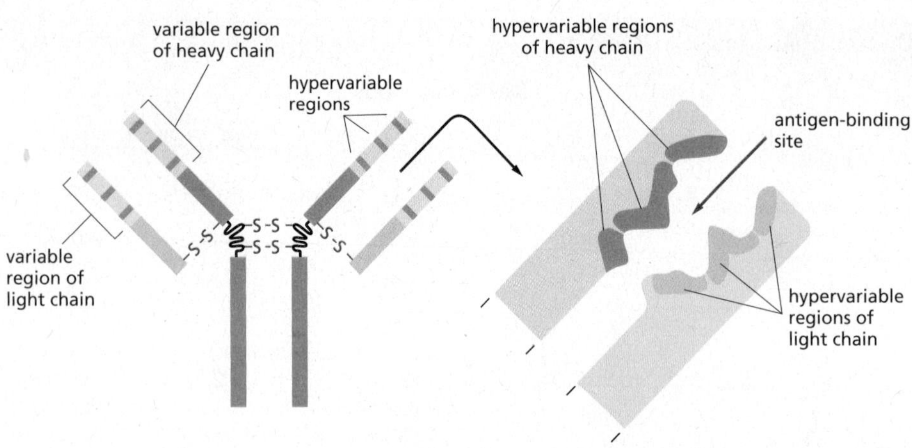

Figure 24–26 Ig hypervariable regions. Highly schematized drawing of how the three hypervariable regions in each light and heavy chain together form each antigen-binding site of an Ig protein.

occurs in three small **hypervariable regions** in the variable regions of both light and heavy chains. Only about 5–10 amino acids in each hypervariable region form the actual antigen-binding site (**Figure 24–26**). As a result, the size of the **antigenic determinant** that an Ig molecule recognizes is generally small: it can consist of fewer than 10 amino acids on the surface of a globular protein, for example (see Figure 24–22).

Both light and heavy chains are made up of repeating segments—each about 110 amino acids long and each containing one intrachain disulfide bond. Each repeating segment folds independently to form a compact functional unit called an **immunoglobulin (Ig) domain.** As shown in **Figure 24–27A**, a light chain consists of one variable (V_L) and one constant (C_L) domain, whereas a heavy chain

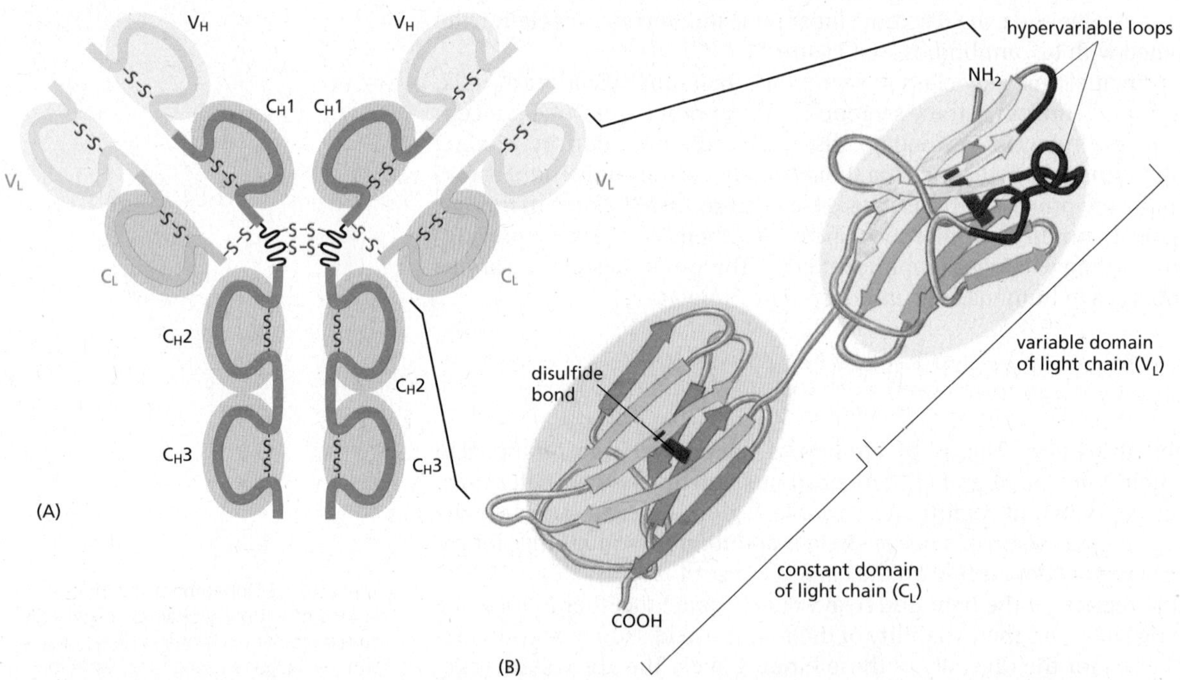

Figure 24–27 Ig domains. (A) The light and heavy chains in an Ig protein are each folded into similar repeating domains. The variable domains (shaded in *blue*) of the light and heavy chains (V_L and V_H) make up the antigen-binding sites, while the constant domains (shaded in *gray*) of the heavy chains (mainly C_H2 and C_H3) determine the other functional properties of the protein. The heavy chains of IgM and IgE do not have a hinge region and have an extra constant domain (C_H4—not shown). Hydrophobic interactions between domains on adjacent chains help hold the chains together in the Ig molecule: V_L binds to V_H, C_L binds to C_H1, and so on. (B) X-ray crystallography–based structures of the Ig domains of a light chain (**Movie 24.5**). The variable and constant domains have a similar overall structure, consisting of two β sheets joined by a disulfide bond *(red)*. Note that all the hypervariable regions *(black)* form loops at the far end of the variable domain, where they come together to form part of the antigen-binding site. All Igs are glycosylated on their C_H2 domains (not shown); the attached oligosaccharide chains vary from Ig to Ig and can greatly influence the functional properties of the protein, mainly by affecting its binding to Fc receptors on immune cells.

has one variable and three or four constant domains: the variable domains of the light and heavy chains pair to form the antigen-binding region. Each Ig domain has a very similar three-dimensional structure, consisting of a sandwich of two β sheets held together by a disulfide bond; the variable domains are unique in that each has its particular set of hypervariable regions, which are arranged in three *hypervariable loops* that cluster together at the ends of the variable domains to form the antigen-binding site (Figure 24–27B).

Ig Genes Are Assembled from Separate Gene Segments During B Cell Development

Prior to antigen stimulation, the human **primary Ig repertoire** is probably composed of more than 10^{12} different BCRs. This repertoire consists of IgM and IgD proteins and is apparently large enough to ensure that there will be an antigen-binding site to fit almost any potential antigenic determinant, albeit with low affinity ($K_a \approx 10^5$–10^7 liters/mole). After stimulation by antigen and helper T cells, B cells can switch from making IgM and IgD to making other classes of Ig—a process called *class switching*. In addition, the binding affinity of these Igs for their antigen progressively increases over time—a process called *affinity maturation*. Thus, antigen stimulation generates a **secondary Ig repertoire**, with a greatly increased affinity (K_a up to 10^{11} liters/mole) and a greater diversity of both Ig classes and antigen-binding sites.

How can each of us make so many different Igs? The problem is not quite as formidable as it might first appear. Recall that the variable regions of the Ig light and heavy chains usually combine to form the antigen-binding site. Thus, if we had 1000 genes encoding light chains and 1000 genes encoding heavy chains, we could, in principle, combine their products in 1000 × 1000 different ways to make 10^6 different antigen-binding sites. Nonetheless, we have evolved special genetic mechanisms to enable our B cells to generate an almost unlimited number of different light and heavy chains in a remarkably economical way. We do so in two steps. First, before antigen stimulation, developing B cells join together separate *gene segments* in DNA to create the genes that encode the primary repertoire of low-affinity IgM and IgD proteins. Second, after antigen stimulation, the assembled *Ig genes* can undergo two further changes—point mutations that can increase the affinity of their antigen-binding site and DNA rearrangements that switch the class of Ig made. Together, these changes produce the secondary repertoire of high-affinity IgG, IgE, and IgA proteins.

We produce our primary Ig repertoire by joining separate **Ig gene segments** together during B cell development. Each type of Ig chain—κ light chains, λ light chains, and heavy chains—is encoded by a separate locus on a separate chromosome. Each locus contains a large number of gene segments encoding the V region of an Ig chain and one or more gene segments encoding the C region. During the development of a B cell in the bone marrow, a complete coding sequence for each of the two Ig chains to be synthesized is assembled by site-specific genetic recombination (discussed in Chapter 5). Once a V-region coding sequence is assembled next to a C-region sequence, it can then be co-transcribed and the resulting RNA transcript processed to produce an mRNA molecule that codes for the complete Ig polypeptide chain.

Each light-chain V region, for example, is encoded by a DNA sequence assembled from two gene segments—a long ***V* gene segment** and a short *joining segment*, or ***J* gene segment** (Figure 24–28). Each heavy-chain V region is similarly constructed by combining gene segments, but here an additional *diversity segment*, or ***D* gene segment**, is also required (Figure 24–29). In addition to bringing together the separate gene segments of the Ig gene, these rearrangements also activate transcription from the gene promoter through changes in the relative positions of the *cis*-regulatory DNA sequences acting on the gene. Thus, a complete Ig chain can be synthesized only after the DNA has been rearranged.

The large number of inherited *V*, *J*, and *D* gene segments available for encoding Ig chains contributes substantially to Ig diversity, and the combinatorial joining of

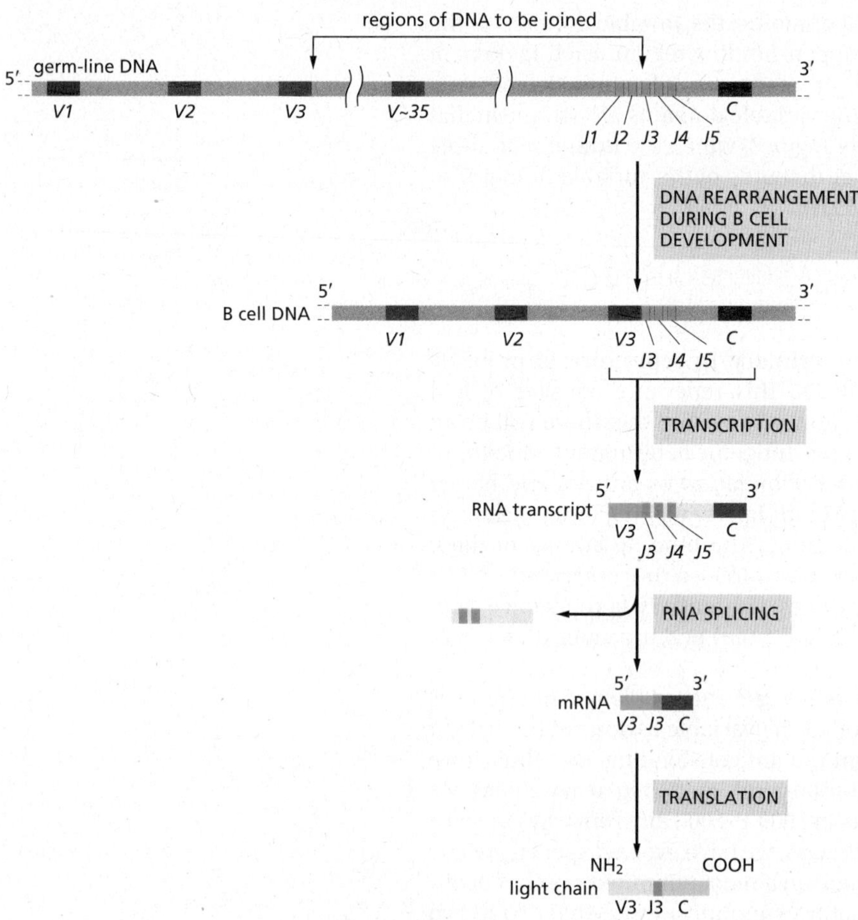

Figure 24–28 **The V–J joining process involved in making a human κ light chain.** In the "germ-line" DNA (where the Ig gene segments are not rearranged and are therefore not being expressed), the cluster of five *J* gene segments *(green)* is separated from the C-region coding sequence *(purple)* by a short intron and from the 35 or so functional *V* gene segments *(red)* by thousands of nucleotide pairs. During the development of a B cell, a randomly chosen *V* gene segment (*V3* in this case) is moved to lie precisely next to one of the *J* gene segments (*J3* in this case). The "extra" *J* gene segments (*J4* and *J5*) and the intron sequence are transcribed (along with the joined *V3* and *J3* gene segments and the C-region coding sequence) and then removed by RNA splicing to generate mRNA molecules with contiguous *V3*, *J3*, and *C* sequences, as shown. These mRNAs are then translated into κ light chains. A *J* gene segment encodes the 15 or so C-terminal amino acids of the V region, and a short sequence containing the *V–J* segment junction encodes the third hypervariable region, which is the most variable part of the light-chain V region.

these segments (called *combinatorial diversification*) greatly increases this contribution. Any of the 35 or so functional *V* segments in our κ light-chain locus, for example, can be joined to any of the 5 *J* segments (see Figure 24–28), so that this locus can encode at least 175 (35 × 5) different κ-chain V regions. Similarly, any of the 40 *V* segments in the human heavy-chain locus can be joined to any of the 23 or so *D* segments and to any of the 6 *J* segments to encode at least 5520 (40 × 23 × 6) different heavy-chain V regions. By this mechanism alone, called **V(D)J recombination**, a human can produce 295 different V_L regions (175 κ and 120 λ) and 5520 different V_H regions. In principle, these could then be combined to make more than 1.6×10^6 (295 × 5520) different antigen-binding sites.

V(D)J recombination is mediated by an enzyme complex called *V(D)J recombinase*, which recognizes recombination signal sequences in the DNA that flanks each gene segment to be joined. Although the process ensures that only appropriate gene segments recombine, a variable number of nucleotides are often lost from the ends of the recombining gene segments, and one or more randomly chosen nucleotides are also inserted. This random loss and gain of nucleotides at joining sites is called **junctional diversification**, and it greatly increases the diversity of V-region coding sequences created by V(D)J recombination (between 10^7-fold and 10^8-fold), specifically in the third hypervariable region. This increased diversification comes at a price, however. In many cases, it shifts the reading frame to produce a nonfunctional gene, in which case the developing B cell fails to make a functional Ig molecule and consequently dies in the bone marrow by apoptosis. Once a B cell makes a functional heavy chain and light chain that form an

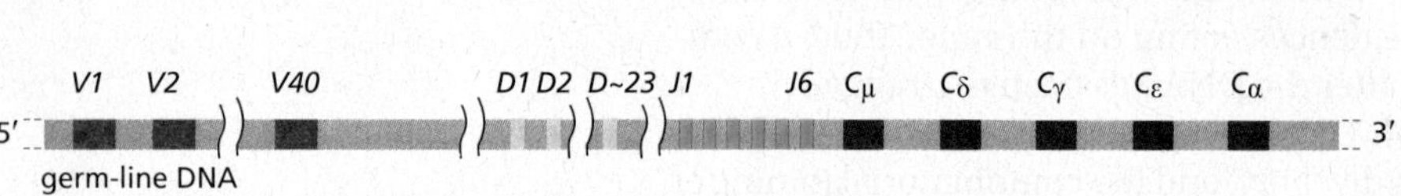

Figure 24–29 **The human heavy-chain locus.** There are 40 *V* segments, about 23 *D* segments, 6 *J* segments, and an ordered cluster of C-region coding sequences, each cluster encoding a different class of heavy chain. The *D* segment (and part of the *J* segment) encodes amino acids in the third hypervariable region, which is the most variable part of the heavy-chain V region. The genetic mechanisms involved in producing a heavy chain are the same as those shown in Figure 24–28 for light chains, except that two DNA rearrangement steps are required instead of one: first a *D* segment joins to a *J* segment, and then a *V* segment joins to the rearranged *DJ* segment. The rearrangements lead to the production of a *VDJC* mRNA that encodes a complete Ig heavy chain. The figure is not drawn to scale: the total length of the heavy-chain locus is more than 2 megabases. Moreover, a number of details are omitted; for example, the exons encoding each C-region Ig domain and the hinge region (see Figure 24–27) and the different subclasses of C_γ-coding segments are not shown.

antigen-binding site, it turns off the V(D)J recombination process, thereby ensuring that the cell makes Ig of only one antigen-binding specificity.

B cells making BCRs that bind strongly to self antigens in the bone marrow would be dangerous. Such B cells are signaled by the self-binding to maintain expression of an active V(D)J recombinase and undergo a second round of V(D)J recombination in a light-chain locus, thereby changing the specificity of its BCR—in the process of **receptor editing** discussed earlier; self-reactive B cells that fail to change their specificity die by apoptosis, in the process of clonal deletion (see Figure 24–21).

Antigen-driven Somatic Hypermutation Fine-Tunes Antibody Responses

As mentioned earlier, with the passage of time after an infection or vaccination, there is usually a progressive increase in the affinity of the antibodies produced against the pathogen. This phenomenon of **affinity maturation** is due to the accumulation of point mutations in both heavy-chain and light-chain V-region coding sequences. The mutations occur long after the coding regions have been assembled. After B cells have been stimulated by both antigen and helper T cells in a peripheral lymphoid organ, some of the activated B cells proliferate rapidly in some lymphoid follicles and form *germinal centers* (see Figure 24–20). Here, the B cells mutate at the rate of about one mutation per V-region coding sequence per cell generation. Because this is about a million times greater than the spontaneous mutation rate in other genes and occurs in somatic cells rather than germ cells, the process is called **somatic hypermutation.**

Very few of the altered Igs generated by hypermutation will have an increased affinity for the antigen. However, the antigen will stimulate preferentially those few B cells that do make BCRs with increased affinity for the antigen. Clones of these altered B cells will preferentially survive and proliferate, especially as the amount of antigen decreases to very low levels late in the response. Most other B cells in the germinal center will die by apoptosis. Thus, as a result of repeated cycles of somatic hypermutation followed by antigen-driven proliferation of selected clones of effector and memory B cells, BCRs and antibodies of increasingly higher affinity become abundant during an adaptive immune response, providing progressively better protection against the pathogen (**Movie 24.6**).

A breakthrough in understanding the molecular mechanism of somatic hypermutation came with the identification of an enzyme that is required for the process. It is called **activation-induced deaminase (AID)** because it is expressed specifically in activated B cells and deaminates cytosine (C) to uracil (U) during transcription of V-region coding DNA. The deamination produces U:G mismatches in the DNA double helix, and the repair of these mismatches produces various types of mutations, depending on the repair pathway used. Somatic hypermutation affects only actively transcribed DNA, both because AID works only on single-stranded DNA (which is transiently exposed during transcription) and because proteins involved in the transcription of V-region coding sequences are required to recruit the AID enzyme. AID is also required for activated B cells to switch from IgM and IgD production to the production of the other classes of Ig, as we now discuss.

B Cells Can Switch the Class of Ig They Make

After a developing B cell leaves the bone marrow, but before it interacts with antigen, it expresses both IgM and IgD BCRs on its surface, both with the same antigen-binding sites (see Figure 24–24). Stimulation by antigen and helper T cells in a peripheral lymphoid organ activates many of these mature naïve B cells to become IgM-secreting effector cells, so that IgM antibodies dominate the primary antibody response. Later in the immune response, however, when the activated B cells are undergoing somatic hypermutation in germinal centers, the combination of antigen and cytokines derived from helper T cells (discussed later) stimulates many of the B cells to switch from making membrane-bound IgM and IgD to making IgG, IgE, or IgA, in the process of **class switching**. Some of these cells become memory cells that express the corresponding class of Ig as BCRs on their surface, while others become effector cells that secrete the Ig molecules as antibodies. The IgG, IgE, and IgA molecules initially retain their original antigen-binding sites, which can then undergo affinity maturation in the germinal centers. These class-switched Igs are collectively referred to as *secondary classes* of Igs, because they are produced only after antigen stimulation, dominate secondary antibody responses, and make up the secondary Ig repertoire.

As discussed earlier, the constant region of an Ig heavy chain determines the class of the Ig and hence the Ig's functional properties. Thus, the ability of B cells to switch the class of Ig they make without changing their antigen specificity implies that the same assembled V_H-region coding sequence (which specifies the antigen-binding part of the heavy chain) can sequentially associate with different C_H-coding sequences. This has important functional implications. It means that, in an individual animal, a particular antigen-binding site that has been selected by environmental antigens can be distributed among the various classes of antibodies, thereby acquiring the different functional properties of each class. In this way, antipathogen antibodies harness a variety of effector mechanisms to help clear the pathogen, including activation of the complement system and Fc receptors on phagocytes and other innate immune cells (see Table 24–2).

When a B cell switches from making IgM and IgD to one of the secondary classes of Ig, an irreversible change occurs in the DNA—a process called **class switch recombination**. It entails the deletion of all the C_H-coding sequences between the assembled VDJ-coding sequence and the particular C_H-coding sequence that the cell is destined to express. Class switch recombination differs from V(D)J recombination in several ways. (1) It happens after antigen stimulation, mainly in germinal centers, and depends on helper T cells. (2) It uses different recombination signal sequences, called *switch sequences*, which flank the different C_H-coding segments. (3) It involves cutting and joining the switch sequences, which are noncoding sequences, and leaves the assembled V_H-region coding sequence unchanged (**Figure 24–30**). (4) Most important, the molecular mechanism is different. It depends on AID, which is also involved in somatic hypermutation, rather than on the V(D)J recombinase. The cytokines that activate class switching induce the production of transcription regulators that activate transcription from the relevant switch sequences, allowing the recruitment of

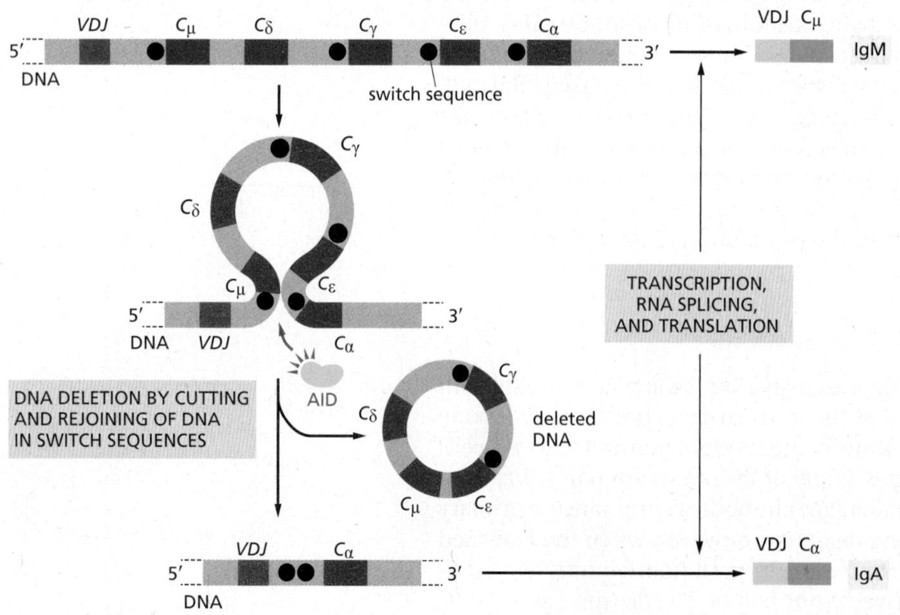

Figure 24–30 An example of the DNA rearrangement that occurs in class switch recombination. A B cell making IgM (and IgD—not shown) molecules with the same heavy-chain V region encoded by a particular assembled *VDJ* DNA sequence is stimulated to switch to making IgA molecules with the same heavy-chain V region. In the process, it deletes the DNA between the *VDJ* sequence and the C_α-coding sequence. Specific DNA sequences (*switch sequences*) located upstream of each C_H-coding sequence (except C_δ, as B cells don't switch from C_μ to C_δ) can recombine with one another, with the deletion of the intervening DNA, as shown here. As discussed in the text, the recombination process depends on AID, the same enzyme that is involved in somatic hypermutation. When switching from IgM and IgD to IgG or IgE, the C-region coding sequences downstream of C_γ or C_ε, which remain after the DNA deletion, are removed during RNA splicing.

AID to these sites. Once bound, AID initiates switch recombination by deaminating some cytosines to uracil in the vicinity of these switch sequences. Excision of these uracils is thought to lead to double-strand breaks in the participating switch regions, which are then joined by a form of nonhomologous end joining (discussed in Chapter 5).

Thus, whereas the primary Ig repertoire in humans (and mice) is generated by V(D)J joining mediated by V(D)J recombinase, the secondary antibody repertoire is generated by somatic hypermutation and class switch recombination, both of which are mediated by AID. Figure 24–31 lists the main mechanisms for diversifying Igs that we have discussed in this chapter.

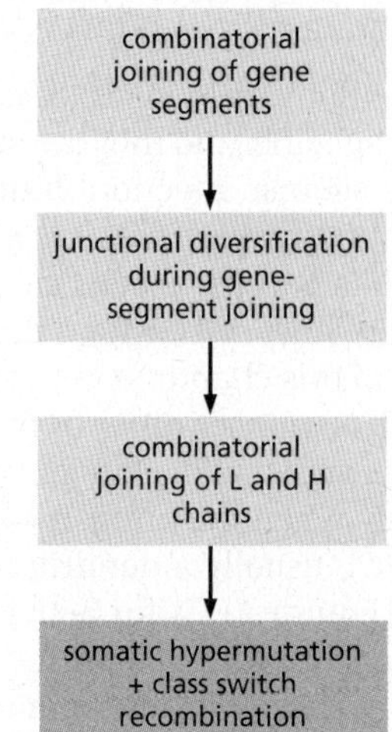

Figure 24–31 The main mechanisms of Ig diversification in mice and humans. Those shaded in *gray* occur during B cell development in the bone marrow, whereas the two mechanisms shaded in *red* occur when B cells are stimulated by foreign antigen and helper T cells in germinal centers in peripheral lymphoid organs, either late in a primary response or in a secondary response.

Summary

Each B cell clone makes Ig molecules with a unique antigen-binding site. Initially, the Ig molecules are inserted into the plasma membrane and serve as B cell receptors (BCRs) for antigen. Antigen binding to the BCRs, together with co-stimulatory signals from helper T cells, activate the B cells to proliferate and differentiate into antibody-secreting effector cells or memory cells. The effector cells secrete large amounts of antibodies with the same antigen-binding site as the BCRs.

A typical Ig molecule is composed of four polypeptide chains—two identical heavy chains and two identical light chains. Parts of both the heavy and light chains form the two identical antigen-binding sites. There are multiple classes of Ig (IgA, IgD, IgE, IgG, and IgM), each with a distinctive heavy chain, which determines the functional properties of the Ig class. Each light and heavy chain is composed of a number of similarly folded Ig domains. The amino acid sequence variation in the variable domains of both light and heavy chains is concentrated in several small hypervariable regions, which form loops at one end of these domains to form the antigen-binding site.

Igs are encoded by loci on three different chromosomes, each of which is responsible for producing a different polypeptide chain—a κ light chain, a λ light chain, or a heavy chain. Each locus contains separate gene segments that encode different parts of the variable region of the particular Ig chain. Each light-chain locus contains one or more constant (C)-region coding sequences and sets of variable (V) and joining (J) gene segments. The heavy-chain locus contains sets of C-region coding sequences and sets of V, diversity (D), and J gene segments.

During B cell development in the bone marrow, before foreign antigen stimulation, separate gene segments are brought together by site-specific recombination mediated by V(D)J recombinase. A V_L gene segment recombines with a J_L gene segment to produce a DNA sequence coding for the V region of a light chain, and a V_H gene segment recombines with a D and a J_H gene segment to produce a DNA sequence coding for the V region of a heavy chain. Each of the newly assembled V-region coding sequences is then co-transcribed with the appropriate C-region sequence to produce an RNA molecule that codes for the complete Ig polypeptide chain.

By randomly combining inherited gene segments that code for the variable regions during B cell development, we can make hundreds of different light chains and thousands of different heavy chains. Because the antigen-binding site is formed where the hypervariable loops of the V_L and V_H domains come together in the final Ig molecule, the heavy and light chains can potentially pair to form Igs with more than a million different antigen-binding sites. A loss or gain of nucleotides at the site of gene-segment joining increases this number enormously. The Igs made by such V(D)J recombination before antigen stimulation are IgMs and IgDs with low affinity for binding antigen, and they constitute the primary Ig repertoire.

Igs are further diversified after antigen stimulation in peripheral lymphoid organs by the AID-mediated and helper T cell–dependent processes of somatic hypermutation and class switch recombination, which together produce the high-affinity IgG, IgE, and IgA Igs that constitute the secondary Ig repertoire. The process of class switching allows the same antigen-binding site to be incorporated into antibodies that have different tails and therefore different functional properties.

T CELLS AND MHC PROTEINS

Like antibody responses, T cell–mediated immune responses are exquisitely antigen specific, and they are at least as important as antibodies in defending vertebrates against infection. Indeed, most adaptive immune responses, including most antibody responses, require helper T cells for their initiation. Most important, unlike B cells, T cells can help eliminate pathogens that have entered the interior of host cells, where they are invisible to B cells and antibodies. Much of the rest of this chapter is concerned with how T cells accomplish this feat.

T cell responses differ from B cell responses in at least two crucial ways. First, a T cell is activated by foreign antigen to proliferate and differentiate into effector cells only when the antigen is displayed on the surface of an *antigen-presenting cell* (*APC*), usually a dendritic cell in a peripheral lymphoid organ. One reason T cells require APCs for activation is that the form of antigen they recognize is different from that recognized by the Igs produced by B cells. Whereas Igs can recognize antigenic determinants on the surface of pathogens and soluble folded proteins, for example, T cells can only recognize fragments of protein antigens that have been produced by partial proteolysis inside a host cell. As mentioned earlier, newly formed *MHC proteins* capture these peptide fragments and carry them to the surface of the host cell, where T cells can recognize them.

The second difference is that, once activated, effector T cells act mainly at short range, usually contacting the cells they influence, either within a secondary lymphoid organ or after they have migrated to a site of infection. Effector B cells, by contrast, secrete antibodies that can act far away. Effector T cells interact directly with another host cell in the body, which they either kill (if it is an infected host cell, for example) or signal in some way (if it is a B cell or macrophage, for example). We will refer to such host cells as *target cells*. As is the case with APCs, target cells must display an antigen bound to an MHC protein on their surface for a T cell to recognize them.

There are three main classes of T cells—cytotoxic T cells, helper T cells, and regulatory T cells. When activated, they develop into effector cells, each class having its own distinct activities. Effector *cytotoxic T cells* directly kill cells that are infected with an intracellular pathogen. Effector *helper T cells* help stimulate the responses of other immune cells—mainly macrophages, dendritic cells, and B cells; as we will see, there are various functionally distinct subtypes of effector helper T cells. Effector *regulatory T cells* suppress the activity of other immune cells. After functioning, some effector T cells become memory T cells.

In this section, we discuss how T cells recognize foreign antigens on the surface of APCs or target cells and the crucial part played by MHC proteins in the recognition process. We describe the different classes and subclasses of T cells and their respective functions and how T cells start their development in the thymus. We begin by considering the cell-surface receptors that T cells use to recognize antigen.

T Cell Receptors (TCRs) Are Ig-like Heterodimers

T cell receptors (TCRs), unlike Igs made by B cells, exist only in membrane-bound form. They are composed of two transmembrane, disulfide-linked polypeptide chains, each of which contains two Ig-like domains—one variable and one constant. On most T cells, the TCRs have one α chain and one β chain (**Figure 24–32**).

The genetic loci that encode the α and β chains are located on different chromosomes. Like an Ig heavy-chain locus (see Figure 24–29), the TCR loci contain separate *V*, *D*, and *J* gene segments (or just *V* and *J* gene segments in the case of the α-chain locus), which are brought together by site-specific recombination during T cell development in the thymus. With one exception, T cells use the same mechanisms to generate antigen-binding-site diversity of their TCRs as B cells use to generate antigen-binding-site diversity of their Igs, and they use the same V(D)J recombinase; thus, humans or mice deficient in this recombinase cannot make functional B or T cells. The mechanism that does not operate in TCR diversification is antigen-driven somatic hypermutation. Therefore, the affinities

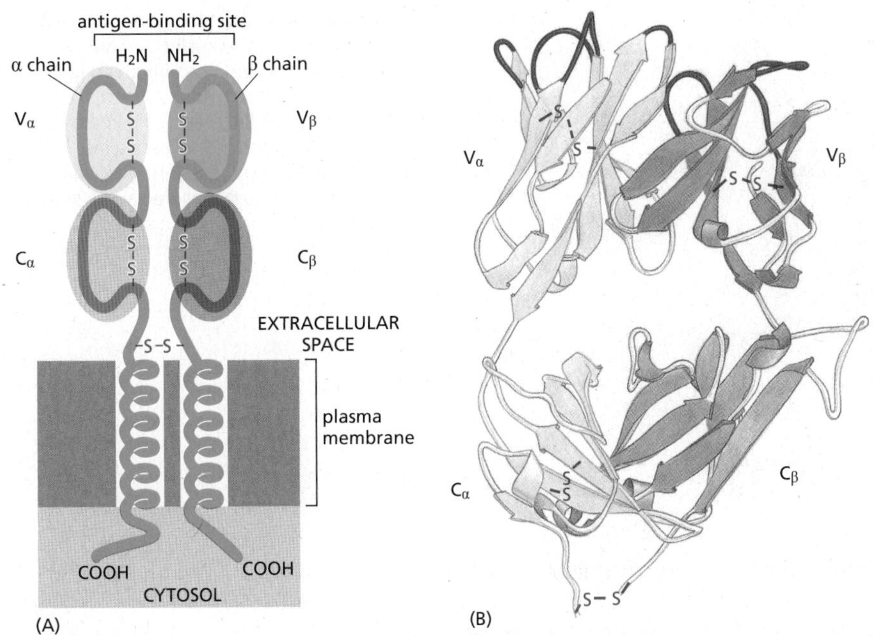

Figure 24–32 **A T cell receptor (TCR) heterodimer.** (A) Schematic drawing of a TCR composed of an α and a β polypeptide chain. Each chain has a large extracellular part that is folded into two Ig-like domains—one variable and one constant. A V_α and a V_β domain (shaded in *blue*) form the antigen-binding site. Unlike Igs, which have two binding sites for antigen, TCRs have only one. Although not shown, the αβ-heterodimer is noncovalently associated with a large set of invariant membrane-bound proteins that help activate the T cell when the TCRs bind their specific antigen (see Figure 24–45B). A typical T cell has about 30,000 TCRs on its surface. (B) The three-dimensional structure of the extracellular part of a TCR. The antigen-binding site is formed by the hypervariable loops of both the V_α and V_β domains *(black)*, located at the distal end of the domains, and it is similar in its overall dimensions and geometry to the antigen-binding site of an Ig molecule. (B, PDB code: 1TCR.)

of TCRs tend to be low ($K_a \approx 10^5$–10^7 liters/mole). Various co-receptors and cell-cell adhesion proteins, however, greatly strengthen the binding of a T cell to an APC or target cell. TCRs also do not diversify by class switching.

Instead of making TCR α and β chains, a minority of T cells makes a different but related type of TCR heterodimer, composed of γ chains and δ chains. Although these *γδ T cells* normally make up only 5–10% of the T cells in human blood, they can be the dominant T cell population in epithelia such as the skin. They differ from conventional αβ T cells in other important ways: their surface receptors have restricted diversity and generally do not recognize antigens as peptides presented by MHC proteins. We will not discuss them further, so that, from here on, when we refer to T cells, we mean T cells with αβ TCRs.

As with BCRs, TCRs are tightly associated in the plasma membrane with a number of invariant membrane-bound proteins that are involved in passing the signal from an antigen-activated receptor to the cell interior. We will discuss these proteins in more detail later, when we consider some of the molecular events involved in T and B cell activation. First, we consider the special ways in which T cells recognize foreign antigen on the surface of an APC or target cell.

Activated Dendritic Cells Activate Naïve T Cells

Generally, naïve T cells, including naïve helper and cytotoxic T cells, proliferate and differentiate into effector cells and memory cells only when they see their specific antigen on the surface of an activated **dendritic cell** in a peripheral lymphoid organ (Figure 24–33). The activated dendritic cell displays the antigen in a complex with MHC proteins on its surface, along with co-stimulatory proteins (see Figure 24–11). The memory T cells that develop, however, can be re-activated by the same antigen–MHC complex on the surface of other types of APCs (target cells), including macrophages and B cells—as well as by dendritic cells.

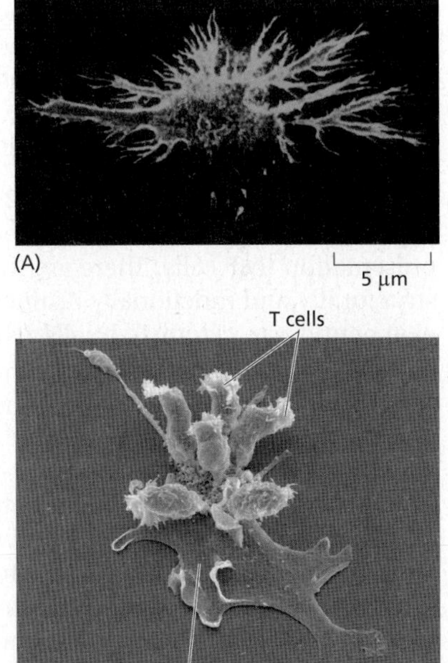

Figure 24–33 **Dendritic cells.** (A) Immunofluorescence micrograph of a mouse dendritic cell in culture. These APCs derive their name from their long processes, or "dendrites." The cell has been labeled with a monoclonal antibody that recognizes a surface antigen on these cells. (B) Scanning electron micrograph of T cells bound to the surface of an activated dendritic cell in a mouse lymph node. (A, courtesy of David Katz; B, courtesy of I. Mellman, P. Pierres, and S. Turley.)

Immature dendritic cells are located in most tissues—underlying epithelial layers of the skin and gut, for example—where they are constantly sampling and processing proteins in their environment. They become activated to mature when their pattern recognition receptors (PRRs) encounter pathogen-associated molecular patterns (PAMPs) on an invading pathogen or its products. The pathogen or products are ingested, and the microbial proteins are cleaved into peptide fragments, which are loaded onto MHC proteins, as we discuss later. The activated dendritic cells then migrate via the lymph from the site of infection to local lymph nodes or gut-associated lymphoid organs, a process aided by the expression of the chemokine receptor CCR7. Here, they present the foreign antigens, displayed

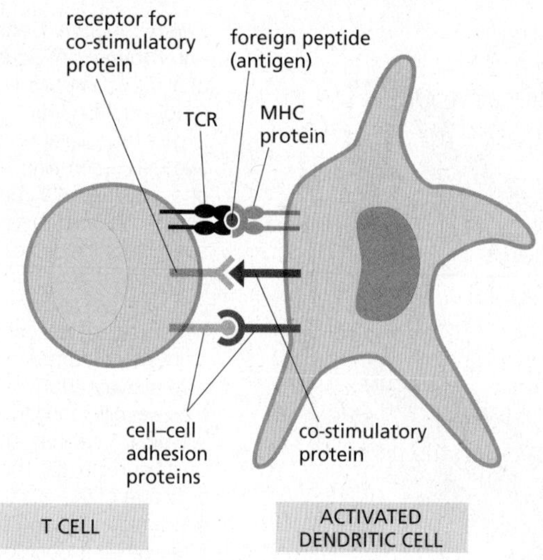

Figure 24–34 The three general types of proteins on the surface of an activated dendritic cell involved in activating a T cell. Although not shown, activated dendritic cells also secrete soluble cytokines that influence the T cell activation process. The invariant polypeptide chains that are always stably associated with the TCR are also not shown; they are discussed later and illustrated in Figure 24–45B and **Movie 24.7**.

as peptide–MHC complexes on the dendritic cell surface, for recognition by the relevant T cells (see Figure 24–11).

Activated dendritic cells display three types of protein molecules on their surface that have a role in activating a T cell to become an effector cell or memory cell (**Figure 24–34**): (1) *MHC proteins,* which present foreign peptides to the TCRs; (2) *co-stimulatory proteins,* which bind to complementary receptors on the T cell surface; and (3) *cell–cell adhesion molecules,* which enable a T cell to bind to the dendritic cell for long enough to become activated, typically several hours or more. In addition, activated dendritic cells secrete a variety of cytokines that influence the type of effector helper T cell that develops, and different types of dendritic cells promote different outcomes (discussed later).

T Cells Recognize Foreign Peptides Bound to MHC Proteins

As discussed earlier (see Figure 24–11), **MHC proteins** capture intracellular peptide fragments of foreign proteins and display them on a cell surface for presentation to T cells. There are two main classes of MHC proteins, which are structurally and functionally distinct. **Class I MHC proteins** mainly present foreign peptides to cytotoxic T cells, whereas **class II MHC proteins** mainly present foreign peptides to helper and regulatory T cells (**Figure 24–35**). Some class I–like MHC proteins present microbial lipid and glycolipid antigens to T cells, but these proteins are not encoded within the MHC region of the genome, and we will not consider them further.

Both class I and class II MHC proteins are heterodimers, in which two extracellular domains form a *peptide-binding groove,* which always has a variable small peptide noncovalently bound in it. In class I MHC proteins, the two domains that form the peptide-binding groove are provided by the transmembrane α chain, which is noncovalently associated with a small subunit called *β_2-microglobulin;*

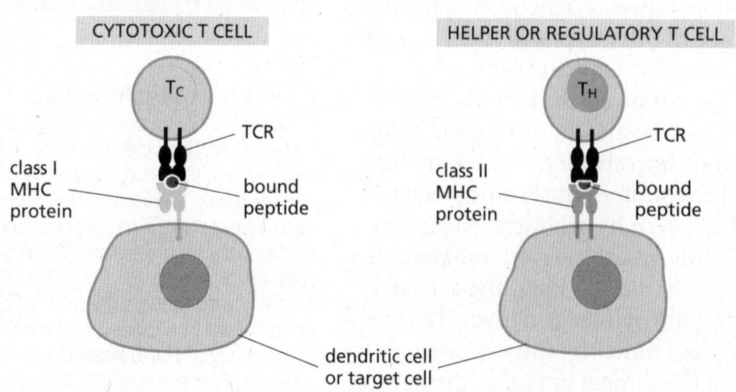

Figure 24–35 Recognition by T cells of peptides bound to MHC proteins. Cytotoxic T (T_C) cells recognize foreign peptides in association with class I MHC proteins, whereas helper T (T_H) cells recognize foreign peptides in association with class II MHC proteins; regulatory T cells also recognize self or foreign peptides in association with class II MHC proteins. In all cases, the T cell recognizes the peptide–MHC complexes on the surface of an APC—either a dendritic cell or a target cell. Different types of dendritic cells activate naïve cytotoxic and helper T cells (not shown).

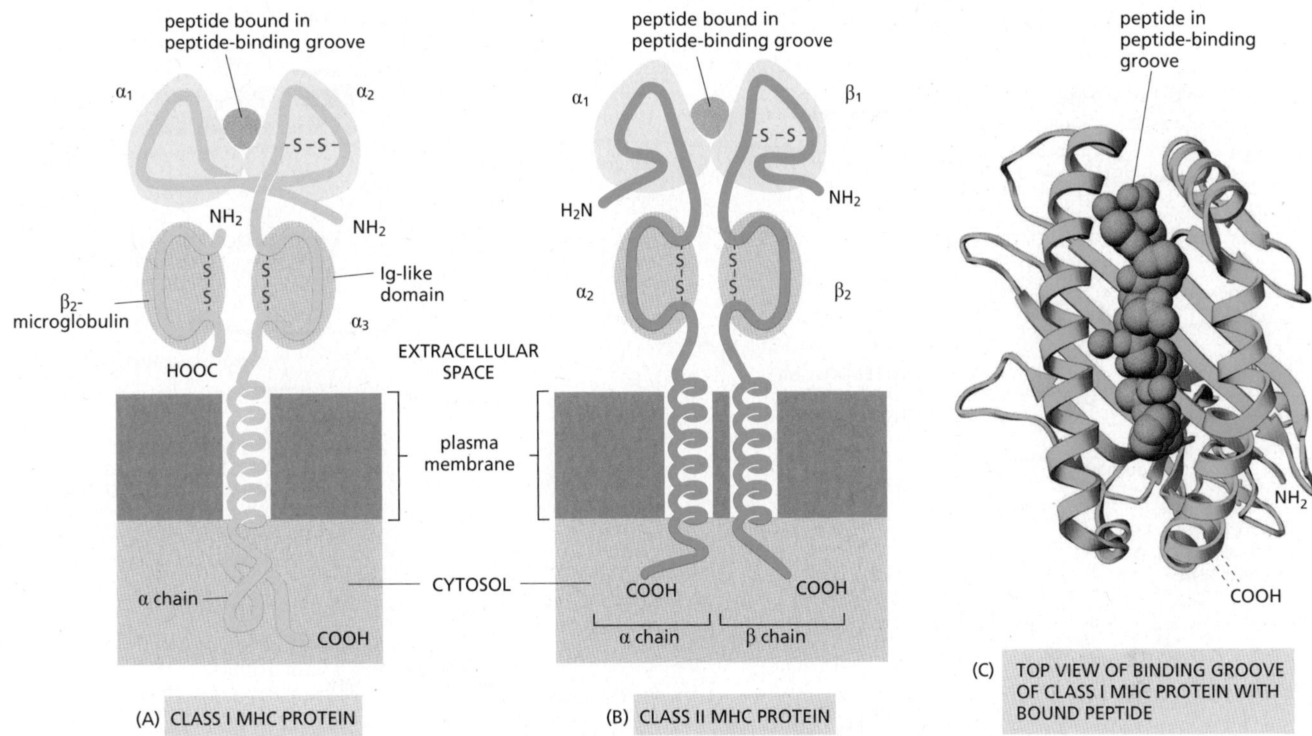

Figure 24–36 **Class I and class II MHC proteins.** (A) The transmembrane α chain of the class I molecule has three extracellular domains, α_1, α_2, and α_3, each encoded by a separate exon. The α chain is noncovalently associated with a smaller, nontransmembrane polypeptide chain, β_2-microglobulin, which is not encoded within the MHC region of the genome. The α_3 domain and β_2-microglobulin are Ig-like. While β_2-microglobulin is invariant, the α chain is extremely polymorphic, mainly in the α_1 and α_2 domains. (B) In class II MHC proteins, both the α chain and the β chain are transmembrane polypeptides encoded within the MHC and are polymorphic, mainly in the α_1 and β_1 domains; the α_2 and β_2 domains are Ig-like. Thus, there are striking similarities between class I and class II MHC proteins: in both, the two outermost domains (shaded in *blue*) are polymorphic and interact to form a groove that binds peptide fragments. The S–S disulfide bonds in A and B are shown in *red*. (C) Top view of the three-dimensional structure of the peptide-binding groove of a human class I MHC protein *(green)* with a bound peptide *(red)*, as would be seen by a TCR on a cytotoxic T cell. Two α helices form the sides of the groove, and a β pleated sheet forms the groove's floor (Movie 24.8 and Movie 24.9). (C, PDB code: 1FZK.)

in class II MHC proteins, a different α chain and a large, noncovalently associated β chain each contribute an extracellular domain to form the peptide-binding groove (Figure 24–36). A TCR binds to both the peptide and the ridges of the binding groove. Humans have three major class I proteins, called *HLA-A, HLA-B,* and *HLA-C,* and three class II proteins, called *HLA-DP, HLA-DQ,* and *HLA-DR* (HLA stands for human leukocyte antigen, as these proteins were first demonstrated on human leukocytes). Figure 24–37 shows how the genes that encode these proteins are arranged in the MHC region of human chromosome 6.

There are important differences between the class I and class II MHC proteins with regard to the cell types that express them and the origin of the peptides in their peptide-binding grooves. Almost all of our nucleated cells express class I proteins. Their peptide-binding groove displays one of a diverse collection of peptides (typically 8–10 amino acids in length). In a healthy cell, the peptides originate from the

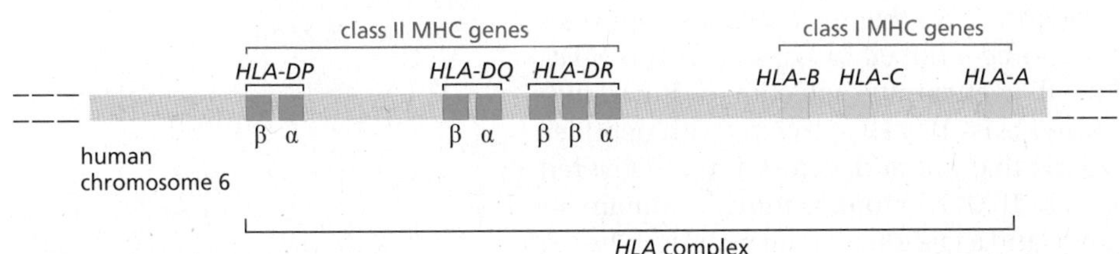

Figure 24–37 **Human MHC genes.** This simplified schematic drawing shows the location of the genes that encode the transmembrane subunits of class I *(light green)* and class II *(dark green)* MHC proteins. The genes shown encode the three main types of class I MHC proteins (HLA-A, HLA-B, and HLA-C) and three types of class II MHC proteins (HLA-DP, HLA-DQ, and HLA-DR). An individual can therefore make six types of class I MHC proteins (three encoded by maternal genes and three by paternal genes) and more than six types of class II MHC proteins. Because of the extreme polymorphism of the MHC genes, the chances are very low that the maternal and paternal alleles will be the same. The number of class II MHC proteins that an individual can make is greater than six because there are two *DRβ* genes and because maternally encoded and paternally encoded polypeptide chains can sometimes pair. The entire region shown spans about 7 million base pairs and contains other genes that are not shown.

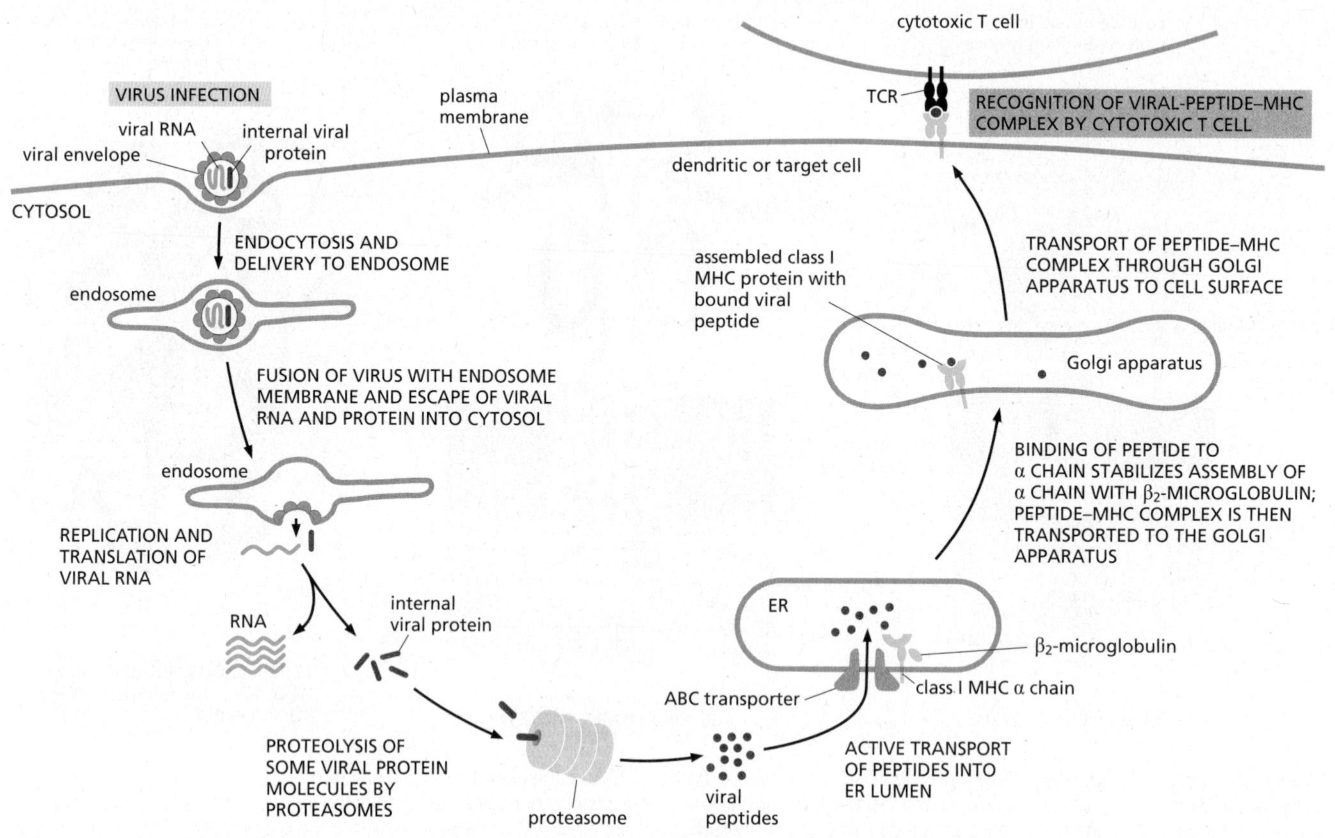

Figure 24–38 A simplified drawing of antigen processing of an internal viral protein for presentation by class I MHC proteins to a cytotoxic T cell. An effector cytotoxic T cell kills a virus-infected cell when it recognizes fragments of an internal viral protein bound to class I MHC proteins on the surface of the infected cell. Not all viruses enter the cell in the way that this enveloped RNA virus does, but fragments of internal viral proteins always follow the pathway shown. Only a small proportion of the viral proteins synthesized in the cytosol are degraded and transported to the cell surface, but this is sufficient to attract an attack by a cytotoxic T cell. Several chaperone and other proteins in the ER lumen combine with the ABC transporter and MHC protein to form a *peptide-loading complex* that aids peptide selection (editing) and the assembly of peptide–class I MHC protein complexes (not shown). Both the final assembly of a class I MHC protein and its transport in vesicles to the cell surface require the presence of either a self or foreign peptide in the peptide-binding groove of the MHC protein (**Movie 24.10**).

cell's own cytosolic and nuclear proteins that have undergone partial degradation in proteasomes in the processes of normal protein turnover and quality control mechanisms. Some of the peptide fragments produced in this way are actively transported into the lumen of the endoplasmic reticulum (ER) by a specialized transporter in the ER membrane; there, they are loaded onto newly synthesized class I MHC α chains. This process depends on a peptide-loading complex that assembles transiently in the ER membrane; the complex consists of the transporter, chaperone proteins, and other proteins that assist in peptide selection by the class I MHC α chain and its partner β_2-microglobulin. The self-peptide–MHC complex produced is then transported through the Golgi apparatus to the cell surface. Such complexes are not dangerous, however, because the cytotoxic T cells that could recognize them have been eliminated or inactivated or suppressed by regulatory T cells in the process of self-tolerance (see Figure 24–21). By contrast, in a cell infected by a pathogen such as a virus, the pathogen proteins will be processed in the same way, and peptides derived from them will be displayed on the infected cell surface bound to class I MHC proteins; there, they are recognized by effector cytotoxic T cells expressing the appropriate TCRs, thereby targeting the infected cell for destruction (**Figure 24–38**).

In general, **antigen-presenting cells (APCs)** are the main cells that express class II MHC proteins. Dendritic cells are the major APCs, as they are specialized for this function, and only they can activate naïve T cells. Other immune cells that are targets of effector T cell regulation, including B cells and macrophages, are also APCs and express class II MHC proteins, as do thymus epithelial cells (discussed later). Other epithelial cells can also be induced to express class II MHC proteins and act as APCs, but only when they encounter infection or inflammation. All APCs load their newly synthesized class II MHC proteins with peptides derived mainly from extracellular proteins that are endocytosed and delivered to endosomes. The newly synthesized class II MHC proteins initially contain an *invariant chain*, which acts as both a guide and a guardian of the peptide-binding

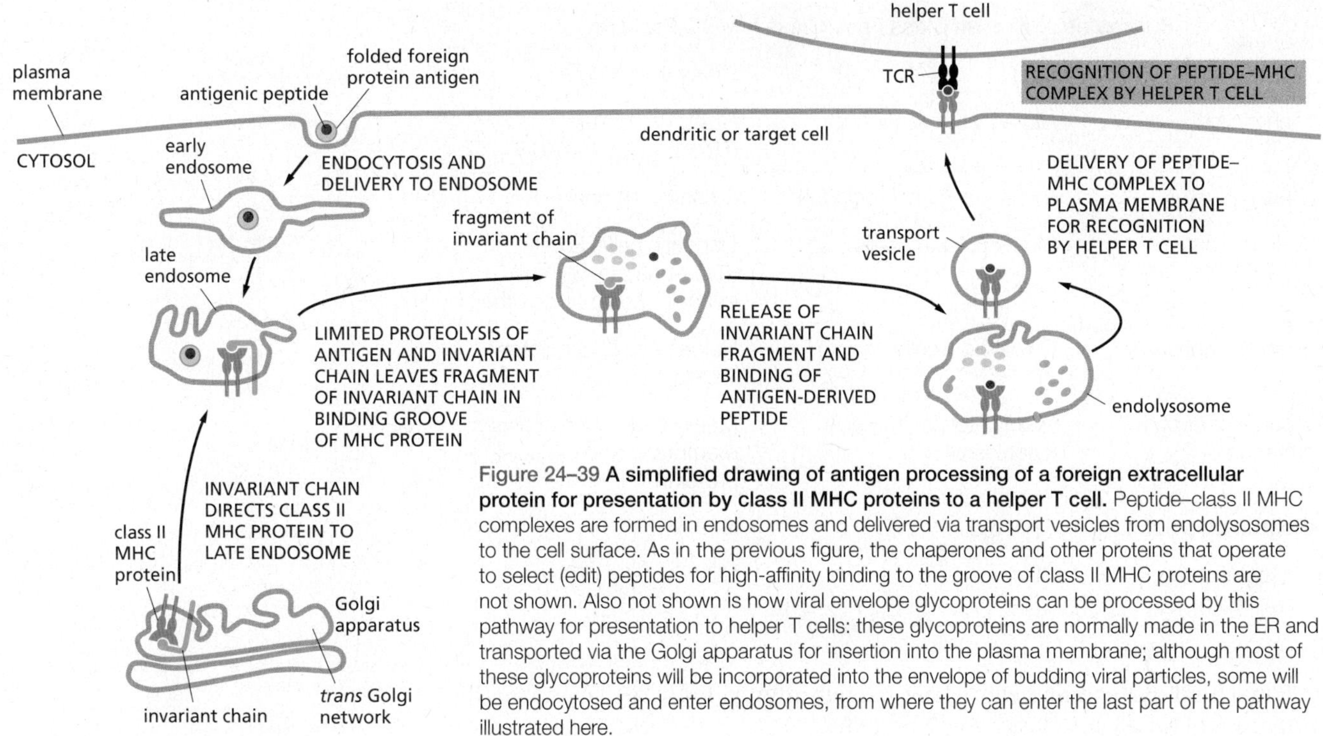

Figure 24–39 **A simplified drawing of antigen processing of a foreign extracellular protein for presentation by class II MHC proteins to a helper T cell.** Peptide–class II MHC complexes are formed in endosomes and delivered via transport vesicles from endolysosomes to the cell surface. As in the previous figure, the chaperones and other proteins that operate to select (edit) peptides for high-affinity binding to the groove of class II MHC proteins are not shown. Also not shown is how viral envelope glycoproteins can be processed by this pathway for presentation to helper T cells: these glycoproteins are normally made in the ER and transported via the Golgi apparatus for insertion into the plasma membrane; although most of these glycoproteins will be incorporated into the envelope of budding viral particles, some will be endocytosed and enter endosomes, from where they can enter the last part of the pathway illustrated here.

groove, preventing the groove from prematurely binding a peptide until the class II MHC protein reaches the endocytic pathway. Here, invariant chain removal is initiated by proteolysis and completed by the action of a chaperone protein, allowing peptide fragments (typically 12–20 amino acids long) produced from endocytosed proteins to bind to the groove of the class II MHC proteins. These peptide–MHC complexes are then transported to the plasma membrane for display on the surface of the APC. In a healthy host cell, class II MHC protein grooves are loaded with self peptides derived from normal proteins and will be ignored by most T cells because of self-tolerance mechanisms (although they will be recognized by regulatory T cells as part of these self-tolerance mechanisms—see Figure 24–21). During an infection, however, pathogen proteins are also endocytosed and processed in the same way, enabling APCs to present pathogen peptides bound to class II MHC proteins to T cells expressing an appropriate TCR (**Figure 24–39**).

The distinction just discussed between the antigen-processing pathways for loading peptides onto class I and class II MHC proteins is not absolute. Dendritic cells, for example, need to be able to activate cytotoxic T cells to kill virus-infected cells even when the virus does not infect dendritic cells themselves. To do so, a specialized subset of dendritic cells uses a process called **cross-presentation**, which begins when these noninfected dendritic cells phagocytose virus-infected host cells or their fragments. In one pathway, at least, the ingested viral proteins are transported by a special mechanism from phagolysosomes into the cytosol, where they are degraded in proteasomes; the resulting protein fragments are then transported into the ER lumen and loaded onto assembling class I MHC proteins, as described earlier (see Figure 24–38). Cross-presentation by dendritic cells is not confined to endocytosed pathogens and their products: it also operates to activate cytotoxic T cells against tumor antigens of cancer cells and the foreign MHC proteins on the cells of foreign organ grafts.

During an infection, only a small fraction of the many thousands of MHC proteins on the surface of an APC or target cell will have pathogen peptides bound to them. This is sufficient, however: fewer than 50 copies of the same peptide–MHC complex on a dendritic cell, for example, can activate a helper T cell that has a

TABLE 24–3 Properties of Human Class I and Class II MHC Proteins

	Class I	Class II
Genetic loci	*HLA-A, HLA-B, HLA-C*	*HLA-DM, HLA-DO, HLA-DP, HLA-DQ, HLA-DR*
Chain structure	α chain + β_2-microglobulin	α chain + β chain
Cell distribution	Most nucleated cells	Dendritic cells, B cells, macrophages, thymus epithelial cells, some others
Presents antigen to	Cytotoxic T cells	Helper T cells, regulatory T cells
Source of peptide fragments	Mainly proteins made in cytoplasm	Mainly endocytosed plasma membrane and extracellular proteins
Polymorphic domains	$\alpha_1 + \alpha_2$	$\alpha_1 + \beta_1$
Recognition by co-receptor	CD8	CD4

TCR that binds the complex with a high-enough affinity. Table 24–3 compares the properties of class I and class II MHC proteins.

MHC Proteins Are the Most Polymorphic Human Proteins Known

Although any individual can make only a small number of different class I and class II MHC proteins, together these proteins must be able to present peptide fragments from almost any foreign protein to T cells. Thus, unlike the antigen-binding site of an Ig protein, the peptide-binding groove of each MHC protein must be able to bind a very large number of different peptides. The genes encoding class I and class II MHC proteins (see Figure 24–37) are the most *polymorphic* known in higher vertebrates: in the human population, for example, there are more than 2000 allelic variants of these genes. The corresponding variations in the MHC proteins are functionally important, as they are concentrated in the floor and walls of the peptide-binding grooves and allow MHC molecules in different individuals to bind different arrays of peptides.

It is thought that infectious diseases have been an important driving force for generating this remarkable MHC polymorphism. In the evolutionary war between pathogens and the adaptive immune system, pathogens will tend to change their proteins through mutation so that the peptides derived from them will not fit in the MHC peptide-binding grooves. When a pathogen succeeds, it can sweep through a population as an epidemic. In such circumstances, the few individuals who produce a new allelic form of MHC protein that can bind peptides derived from the altered pathogen will have a large selective advantage. This type of selection will tend to promote and maintain a large diversity of MHC proteins in the population. In West Africa, for example, individuals with a specific MHC allele (*HLA-B53*) have a reduced susceptibility to a severe form of malaria that is endemic there; although this allele is rare elsewhere, it is found in 25% of the West African population.

The extensive diversity of human MHC proteins is the main reason that individuals who receive a foreign organ transplant must be treated with strong immunosuppressive drugs to prevent the immunological rejection of the grafted organ. Of all the foreign proteins that the graft expresses, the MHC proteins are by far the most powerful stimulators of the recipient's T cells, which would rapidly destroy the graft if they were not prevented from doing so by such drugs. Foreign MHC proteins are powerful T cell stimulants because T cells respond to them in the same way they respond to self MHC proteins that have foreign peptides bound to them; for this reason, the proportion of a person's T cells that can specifically recognize any foreign MHC protein is relatively high.

CD4 and CD8 Co-receptors on T Cells Bind to Invariant Parts of MHC Proteins

The affinity of TCRs for peptide–MHC complexes on an APC is usually too low by itself to mediate a functional interaction between the two cells. T cells normally require *accessory receptors* to help stabilize the interaction by increasing the overall strength of the cell–cell adhesion. Unlike TCRs or MHC proteins, the accessory receptors are invariant and do not bind to foreign peptides. Once bound to the surface of a dendritic cell, for example, a T cell increases the strength of the binding by activating an integrin adhesion protein, which then binds more strongly to an Ig-like protein on the surface of the dendritic cell. This increased adhesion enables the T cell to remain bound long enough to become activated.

When an accessory receptor has a direct role in activating the T cell by generating its own intracellular signals, it is called a **co-receptor**. The most important and best understood of the co-receptors on T cells are the CD4 and CD8 proteins, both of which are single-pass transmembrane proteins with extracellular Ig-like domains. Like TCRs, they recognize MHC proteins, but, unlike TCRs, they bind to invariant parts of the MHC protein, far away from the peptide-binding groove. **CD4** is expressed on both helper T cells and regulatory T cells and binds to class II MHC proteins, whereas **CD8** is expressed on cytotoxic T cells and binds to class I MHC proteins (**Figure 24–40**).

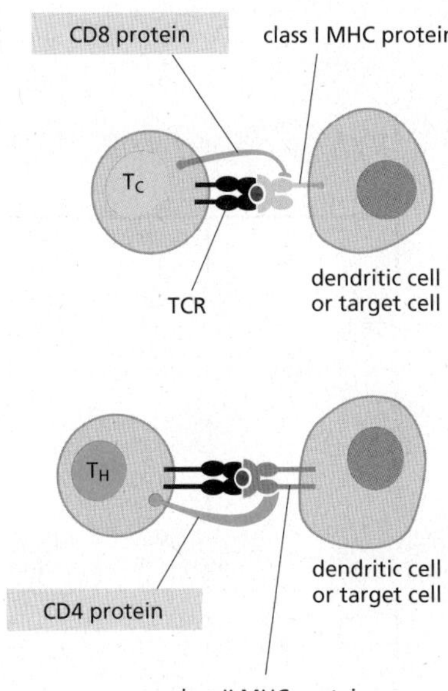

Figure 24–40 **CD4 and CD8 co-receptors on the surface of T cells.** Cytotoxic T (T_C) cells express CD8, which recognizes class I MHC proteins, whereas helper T (T_H) cells and regulatory T cells (not shown) express CD4, which recognizes class II MHC proteins. Note that the co-receptors bind to the same MHC protein that the TCR has engaged, so that they are brought together with TCRs during the antigen recognition process. Whereas the TCR binds to the variable (polymorphic) parts of the MHC protein that form the peptide-binding groove, the co-receptor binds to the invariant part, well away from the binding groove.

CD4 and CD8 contribute to T cell recognition by helping the T cell to focus on particular MHC proteins, and thereby on particular types of target cells. Thus, the recognition of class I MHC proteins by CD8 allows cytotoxic T cells to focus on any type of infected host cell, while the recognition of class II MHC proteins by CD4 allows helper and regulatory T cells to focus on the target immune cells that they help or suppress, respectively. The cytoplasmic tail of the CD4 and CD8 proteins is associated with a member of the Src family of cytoplasmic tyrosine kinases (discussed in Chapter 15) called *Lck*, which phosphorylates various intracellular signaling proteins on tyrosines and thereby participates in the activation of the T cell (discussed later).

The AIDS virus (HIV) uses CD4 molecules (as well as chemokine receptors) to gain entry to helper T cells (see Figure 23–18). Individuals with AIDS are susceptible to infection by microbes that are not normally dangerous because HIV depletes helper T cells. As a result, most individuals with AIDS die of infection within several years of the onset of symptoms, unless they are treated with a combination of anti-HIV drugs. HIV also uses CD4 and chemokine receptors to enter macrophages, which also have both of these types of receptors on their surface.

Developing Thymocytes Undergo Positive and Negative Selection

The development of αβ T cells begins when bone marrow–derived lymphoid progenitor cells enter the outer part of the thymus, the *cortex*, from the bloodstream. There, the progenitor cells develop into thymus lymphocytes, **thymocytes**, which undergo stepwise development under the influence of a variety of signals from thymus epithelial cells, dendritic cells, macrophages, fibroblasts, and other stromal cells.

In an early step, the thymocytes are induced to express V(D)J recombinase, which enables them to begin to rearrange their TCR gene segments and make αβ TCRs. If the cells fail to express such a cell-surface TCR, they will not receive the signals they need to survive and continue to develop, and they will die by apoptosis by default. Because peripheral αβ T cells can only see pathogen-derived peptides in the context of self MHC proteins, developing αβ thymocytes need to express TCRs that have some affinity for self MHC proteins to be potentially useful; those expressing TCRs unable to bind to any self-peptide–self-MHC complex would generally also fail to receive survival signals and therefore also die by apoptosis.

Soon after producing αβ TCRs, the developing thymocytes express both CD4 and CD8 co-receptors. These so-called *double-positive thymocytes* interact with

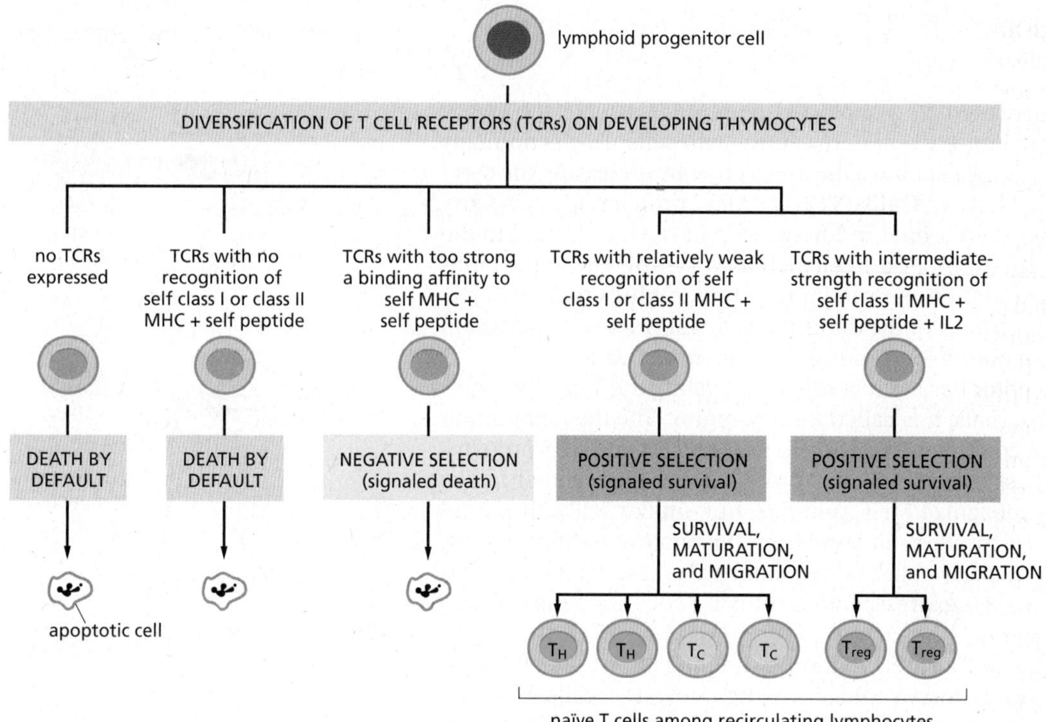

Figure 24–41 **Negative and positive selection in the thymus.** Developing thymocytes with no TCRs or with TCRs that have no recognition of MHC proteins with a peptide bound would be of no use and fail to receive survival signals and die by default ("from neglect") by apoptosis. In the process of *negative selection*, thymocytes with TCRs with such strong binding affinity to self-peptide–self-MHC complexes that the cells would be potentially dangerous are actively signaled to die by apoptosis. In the process of *positive selection*, thymocytes with TCRs with appropriate binding affinity for self-peptide–self-MHC complexes are signaled to survive, mature, and migrate to peripheral lymphoid organs, where they function as helper (T_H), cytotoxic (T_C), or regulatory T (T_{reg}) cells, depending on the class of MHC they recognize and the binding affinity of their TCRs: T_C cells recognize peptides in association with class I MHC proteins, whereas T_H and T_{reg} cells recognize peptides in association with class II MHC proteins. As indicated, whereas thymocytes with TCRs that bind weakly to self peptides bound to self class II MHC proteins end up as T_H cells, those with TCRs that bind with intermediate strength to self peptides bound to self class II MHC proteins (in the presence of IL2) end up as T_{reg} cells.

epithelial cells in the thymus cortex that express self peptides bound to class I and class II MHC proteins. In a process called **positive selection**, a thymocyte expressing a TCR that binds with an appropriate affinity to a self peptide bound to either a class I MHC protein (using CD8 as a co-receptor) or a class II MHC protein (using CD4 as a co-receptor) on cortical epithelial cells is signaled to survive and continue to mature. As part of this process, depending on the TCR's preference for class I or class II MHC proteins, the CD4 or CD8 co-receptor that will not be needed is silenced by DNA methylation of the respective gene. The resulting CD4 or CD8 *single-positive thymocytes* ultimately leave the inner part of the thymus (the *medulla*) to become *naïve T cells*: the CD4 cells become either helper T cells or regulatory T cells (discussed later), while the CD8 cells become cytotoxic T cells.

Before that, however, after the developing single-positive thymocytes move from the cortex into the medulla, they undergo two more selection processes (on the surface of medullary epithelial and dendritic cells) that have crucial roles in immunological self-tolerance. Thymocytes with TCRs that bind too strongly to self peptides bound to class I or class II MHC proteins could be dangerous if they were to continue to mature and leave the thymus, as they could then potentially attack similar self complexes on normal cells in peripheral tissues and thereby cause an autoimmune disease. Such strong binding (via TCRs and either CD4 or CD8 co-receptors) signals such thymocytes to undergo apoptosis in a process called **negative selection**, an example of *clonal deletion* acting in central immunological self-tolerance (see Figure 24–21). In a second form of positive selection, some CD4-positive thymocytes that bind strongly to self peptides bound to class II MHC proteins (but not strongly enough for negative selection to operate) in the presence of secreted *IL2* develop into *regulatory T cells* (*T_{reg} cells*), which mediate *clonal suppression* in peripheral lymphoid tissues, an important mechanism in peripheral self-tolerance (see Figure 24–21). The various positive and negative selection processes that operate during thymocyte development are summarized in **Figure 24–41**.

For negative selection in the thymus to be an effective mechanism of peripheral T cell self-tolerance, APCs in the thymus must display an array of self peptides bound to self MHC molecules that reflect the self peptides derived from self proteins in peripheral tissues, as well as in the thymus. The thymus, however, would not be expected to produce many of the proteins that are specifically expressed in other organs. As an example, it would not be expected to produce insulin, and yet it is crucial to delete thymocytes with TCRs that could recognize insulin-derived peptides bound to self MHC proteins on the surface of insulin-secreting β cells in the pancreas; a failure to do so would result in the T cell–dependent destruction of the β cells and, as a consequence, *type 1 (or juvenile) diabetes*.

The mechanism that enables the deletion of such cells in the thymus depends on a subpopulation of epithelial cells in the thymus medulla that expresses a promiscuous transcriptional regulator called **AIRE** (autoimmune regulator), which is specific to these epithelial cells. The AIRE protein acts as part of a multiprotein complex to induce the production of small amounts of mRNA from 75 to 90% of all genes, including those that encode tissue-restricted proteins such as insulin. When the peptides derived from the proteins encoded by these genes are bound to MHC proteins and displayed on the surface of these epithelial cells, this is sufficient to ensure the deletion of most of the potentially self-reactive thymocytes. Mutations that inactivate the *AIRE* gene cause a severe multi-organ autoimmune disease in both mice and humans, indicating the importance of AIRE in self-tolerance.

Having left the thymus, naïve helper and cytotoxic T cells continually receive survival signals as a result of weak binding to self-peptide–self-MHC-protein complexes. These naïve T cells, however, are normally only activated to proliferate and differentiate into effector and memory T cells in a peripheral lymphoid organ and only when their TCRs bind with high affinity to a pathogen-derived peptide in the binding groove of an MHC protein on the surface of an activated dendritic cell that expresses co-stimulatory signals. We now consider the functions of the different subclasses of effector T cells, beginning with cytotoxic T cells.

Cytotoxic T Cells Induce Infected Target Cells to Undergo Apoptosis

Cytotoxic T cells (T_C cells), like the NK cells discussed earlier, protect us against intracellular pathogens (including viruses, bacteria, and parasites) that multiply in the cytoplasm of a host cell. Both types of cytotoxic cells kill infected host cells before the pathogen can escape to infect neighboring host cells. Before a naïve T_C cell can kill, however, it has to become an effector cell by activation on an APC, usually an activated dendritic cell that has pathogen-derived peptides bound to class I MHC proteins on its surface. The effector T_C cell can then recognize any target cell harboring the same pathogen and expressing some of the same peptide–MHC complexes on its surface: its TCRs cluster, along with CD8 co-receptors, adhesion molecules, and intracellular signaling proteins (discussed later), at the interface between the two cells, forming an **immunological synapse**. In this process, the effector T_C cell reorganizes its cytoskeleton to focus its killing apparatus on the target cell, secreting its toxic proteins into a confined space (**Figure 24–42**); in this way, it avoids killing neighboring cells. A similar synapse forms when an effector helper T cell interacts with its target cell, except that the co-receptor is CD4 (**Movie 24.11**).

An effector T_C cell (or an NK cell) can employ one of two strategies to kill the target, both of which operate by inducing the target cell to activate caspases and kill itself by undergoing apoptosis. One mechanism uses a protein called *Fas ligand* on the killer-cell surface, which binds to a transmembrane receptor protein called *Fas* on the target cell; this mechanism is discussed in Chapter 18 (see Figure 18–6). The other mechanism is the main one used by both NK cells and T_C cells to kill an infected target cell. The killer cell stores various toxic proteins within secretory vesicles in its cytoplasm that it releases into the synaptic space by exocytosis. The toxic proteins include *perforin* and proteases called *granzymes*. The perforin is homologous to complement component C9 and polymerizes in the target-cell plasma membrane (see Figure 24–8), forming a transmembrane pore that locally disrupts the membrane and allows the granzymes to enter the target cell. Once in the cytosol, one of the granzymes,

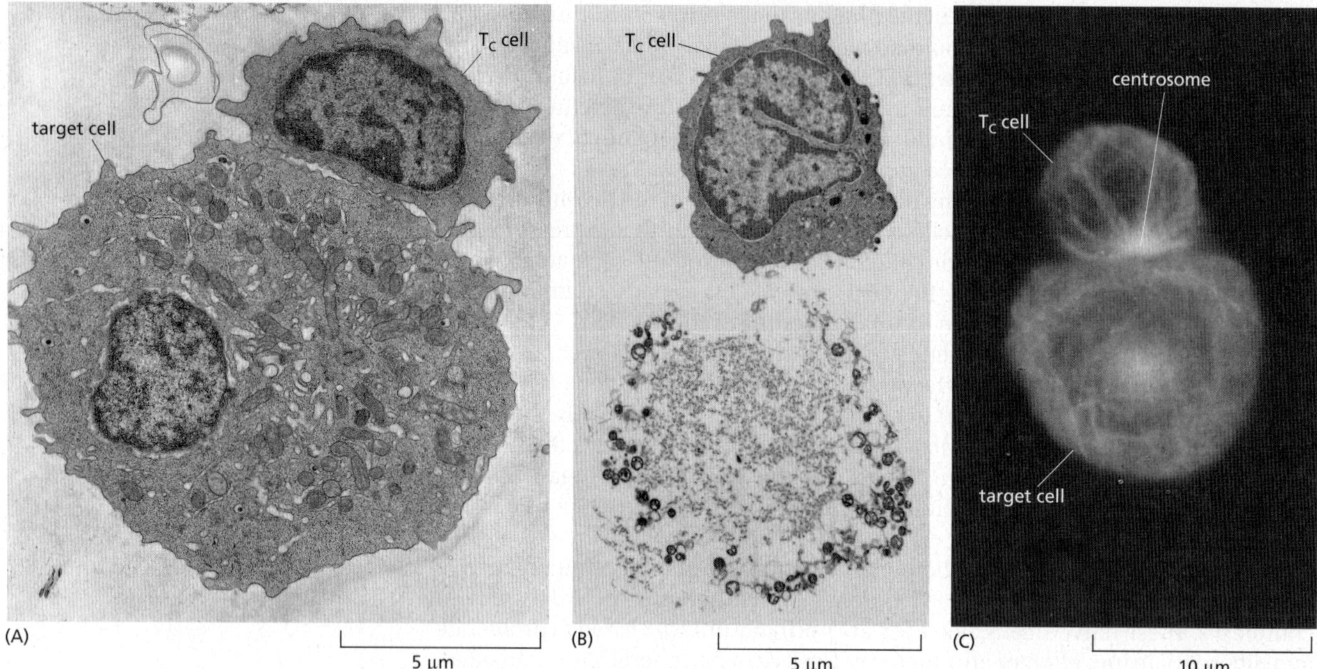

Figure 24–42 Effector cytotoxic T cells killing target cells in culture. (A) Electron micrograph showing an effector cytotoxic T cell (T_C cell) binding to a target cell in culture. The T_C cells were obtained from mice immunized with the target cells, which are foreign tumor cells. (B) Electron micrograph showing a T_C cell and a tumor cell that the T_C cell has killed. In an animal, as opposed to in a culture dish, the killed target cell would be phagocytosed by a neighboring cell (often a macrophage) long before it disintegrated in the way that it has here. (C) Micrograph of a T_C cell and a tumor cell after immunofluorescence staining with anti-tubulin antibodies. Note that the centrosome in the T_C cell is located at the point of cell–cell contact with the target cell—called an immunological synapse. The secretory granules (not visible) in the T_C cell are initially transported along microtubules to the centrosome, which then moves to the synapse, delivering the granules to where they can release their contents directly onto the target-cell surface. (A and B, from D. Zagury et al., *Eur. J. Immunol.* 5:818–822, 1975. With permission from John Wiley & Sons; C, © 1982 B. Geiger et al. Originally published in *J. Cell Biol.* https://doi.org/10.1083/jcb.95.1.137. With permission from Rockefeller University Press.)

granzyme B, cleaves and activates executioner caspases, thereby inducing apoptosis (**Figure 24–43**).

Effector Helper T Cells Help Activate Other Cells of the Innate and Adaptive Immune Systems

In contrast to T_C cells, **helper T cells (T_H cells)** are crucial for defense against both extracellular and intracellular pathogens, and they express CD4 rather than CD8 co-receptors and recognize foreign peptides bound to class II rather than class I MHC proteins. Once naïve T_H cells are induced on activated dendritic cells to become effector cells, they can help activate other immune cells: they help activate B cells to become antibody-secreting cells and later to undergo Ig class switching and somatic hypermutation; they help activate macrophages to destroy any intracellular pathogens multiplying within the macrophage's phagosomes; and they further stimulate the activated dendritic cells that activated either naïve

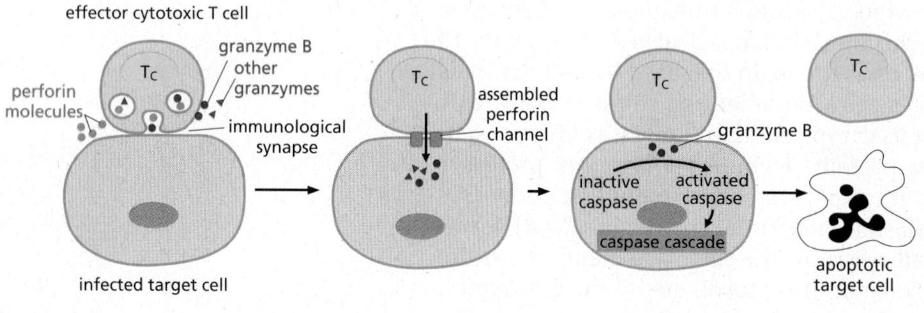

Figure 24–43 The main way that an effector T_C cell (or NK cell) kills an infected target cell. This simplified drawing shows how the killer cell releases perforin and granzymes onto the surface of an infected target cell by localized exocytosis at an immunological synapse. The high concentration of Ca^{2+} in the extracellular fluid causes the perforin to assemble into transmembrane channels that disrupt the target-cell plasma membrane, allowing the granzymes to enter the target-cell cytosol. One of the granzymes, granzyme B, cleaves and thereby activates executioner caspases (caspase-3 and caspase-7) to initiate a caspase cascade, leading to apoptosis (see Figure 18–6). A single cytotoxic cell can kill multiple target cells in sequence. It remains a mystery why the released perforins do not disrupt the plasma membrane of the killer cell itself (**Movie 24.12** and **Movie 24.13**).

T_C cells or naïve T_H cells—to maintain or increase the dendritic cells' activated state. In each case, the effector T_H cell recognizes the same complex of foreign peptide and class II MHC protein on the target-cell surface that it initially recognized on the activated dendritic cell. As discussed later, the T_H cell stimulates the target cell both by secreting a variety of cytokines (mainly interleukins) and by displaying co-stimulatory proteins on the T_H cell surface. As we discuss now, different subtypes of T_H effector cells have different functions.

Naïve Helper T Cells Can Differentiate into Different Types of Effector T Cells

In response to infection, a naïve T_H cell can differentiate into several distinct types of effector T cells, depending on the nature of the pathogen and on the cytokines the T_H cell encounters while being activated, which mainly depends on the activating dendritic cell. These different effector T cell types include four subtypes of helper T cells—*T_{FH}, T_H1, T_H2,* and *T_H17* cells—and regulatory T cells (*T_{reg} cells*). Initially, the naïve T_H cells differentiate into two subpopulations of helper T cells: *follicular helper T cells* (*T_{FH} cells*), which support B cell antibody production in lymphoid follicles (see Figure 24–20) in peripheral lymphoid organs, and non-T_{FH} helper T cells that will develop into T_H1, T_H2, or T_H17 effector T_H cells; most of the latter three cell types migrate from the peripheral lymphoid organ where they differentiated to the site of infection, where they help innate immune cells fight the pathogen.

Figure 24–44 summarizes both the cytokines that help induce these different effector T cells and some of the cytokines the effector cells then secrete, as well as the main transcription regulators that control the development of the effector cells. Naïve T_H cells activated in the presence of *interleukin-6* (*IL6*) develop into effector **T_{FH} cells**, which are located in lymphoid follicles, where they secrete a variety of cytokines, including IL21, to help specific B cells to proliferate and differentiate into antibody-secreting effector cells and undergo somatic hypermutation and Ig class switching within germinal centers (see Figure 24–20). Naïve T_H cells activated by dendritic cells secreting *IL12* develop into effector **T_H1** cells, which

Figure 24–44 **Differentiation of naïve helper T cells into different types of effector helper cells or regulatory T cells in a peripheral lymphoid organ.** The nature of the pathogen and the cytokines produced by the activating dendritic cell (and by other innate immune cells in the environment) mainly determine which type of effector T cell develops, as indicated. Some of the main cytokines produced by each type of effector cell are also shown, and the master transcription regulator for each subset is indicated in the nucleus. Some of the differentiated effector cells are "plastic," in that they are able to change the cytokines they produce in response to changes in their environment (not shown).

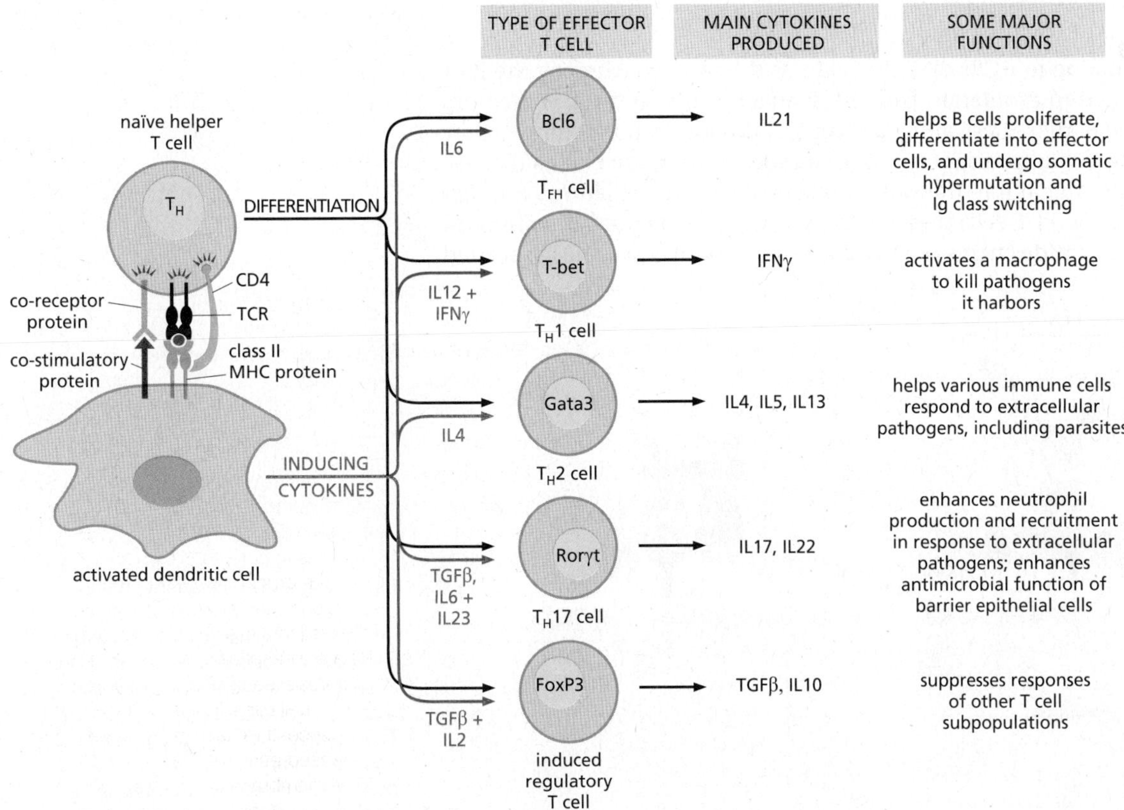

produce *interferon-γ* (*IFNγ*) to help activate macrophages to destroy pathogens that either invaded the macrophage or were ingested by it; the IFNγ can also help activate naïve cytotoxic T cells in a peripheral lymphoid organ to become effector T_C cells. Naïve T_H cells activated in the presence of *IL4* develop into effector **T_H2 cells**, which secrete *IL4, IL5*, and *IL13* to help fight extracellular pathogens, including parasites; they stimulate B cells to switch from making IgM and IgD to making IgE antibodies, which can bind to mast cells and basophils, as discussed earlier. Naïve T_H cells activated in the presence of *IL6, IL23*, and *TGFβ* develop into effector **T_H17 cells**; these cells secrete *IL17* and *IL22* to promote production and recruitment of neutrophils and to stimulate barrier epithelial cells to secrete cytokines and antimicrobial peptides.

In some cases, naïve T_H cells that encounter their antigen in a peripheral lymphoid organ in the presence of TGFβ and in the absence of IL6 develop into **regulatory T cells (T_{reg})**, which suppress rather than help immune cells. They are called **induced T_{reg} cells** to distinguish them from **natural T_{reg} cells**, which develop in the thymus by positive selection during thymocyte development (see Figure 24–41). Both types of T_{reg} cells suppress the activation or function of various other kinds of innate and adaptive immune cells, by means of both secreted suppressive cytokines such as *IL10* and TGFβ and inhibitory proteins on the T_{reg} cell surface. Induced T_{reg} cells are mainly concerned with suppressing immune responses to foreign antigens (preventing responses to harmless ingested or inhaled antigens and commensal microbes, as well as limiting responses against pathogens to avoid excessive reactions that cause unwanted tissue injury); natural T_{reg} cells are concerned with preventing immune responses to self molecules (see Figure 24–21). Both induced and natural T_{reg} cells express the transcription regulator *FoxP3*, which serves as both a marker of these cells and a master controller of their development: if the gene encoding this protein is inactivated in mice or humans, the individuals fail to produce any T_{reg} cells and develop a fatal autoimmune disease involving multiple organs—findings that establish the crucial importance of T_{reg} cells in self-tolerance.

Both T and B Cells Require Multiple Extracellular Signals for Activation

Foreign antigen binding to BCRs or TCRs initiates the process whereby the B or T cells are stimulated to proliferate and differentiate into effector and memory cells. As mentioned earlier, these antigen receptors do not act on their own: they are stably associated with invariant transmembrane polypeptide chains that are required to relay the signal into the cell. In B cells, these are called *Igα* and *Igβ* (**Figure 24–45A**), while in T cells they exist in a complex called *CD3*, composed of four types of polypeptide chains (**Figure 24–45B**). In both cases, the associated

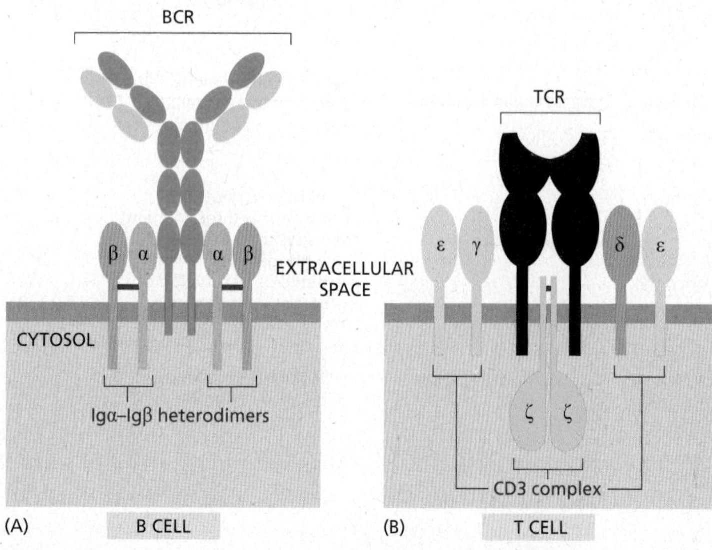

Figure 24–45 The invariant chains associated with BCRs and TCRs. (A) Each BCR is associated with two invariant heterodimers, each composed of an Igα and an Igβ polypeptide chain linked by a disulfide bond *(red)*. (B) Each TCR is associated with an invariant CD3 complex composed of two disulfide-bonded ζ chains, two ε chains, and one δ and one γ chain; these chains form homodimers or heterodimers, as shown.

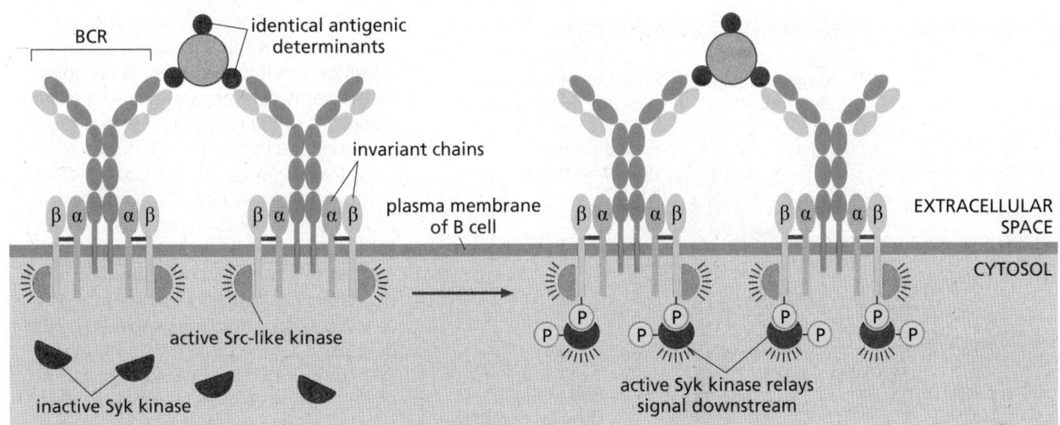

Figure 24–46 **Early signaling events in a B cell activated by the binding of specific foreign antigen to its BCRs.** If the antigen is on the surface of a pathogen or is a soluble macromolecule with two or more identical antigenic determinants (as shown), it cross-links adjacent BCRs, causing them and their associated invariant chains to cluster, as shown. A Src-like cytoplasmic tyrosine kinase (which can be either *Fyn* or *Lyn*) is associated with the cytosolic tail of Igβ; it joins the cluster and phosphorylates both the Igα and Igβ invariant chains (for simplicity, only the phosphorylation on Igβ is shown). An important transmembrane protein tyrosine phosphatase called *CD45* is also required to remove inactivating phosphates from these Src-like kinases for the signaling process to occur (not shown). The resulting phosphotyrosines on Igα and Igβ serve as docking sites for another Src-like tyrosine kinase called *Syk*, which becomes phosphorylated and thereby activated to relay the signal onward.
The pathway from TCRs is similar (including a requirement for CD45), except that the first Src-like kinase is *Lck* (instead of Fyn or Lyn), and it is associated with a CD4 or CD8 co-receptor and phosphorylates tyrosines on all the CD3 polypeptide chains shown in Figure 24–45B; the second Src-like kinase is *ZAP70*, which is homologous to the Syk kinase in B cells (**Movie 24.14**).

proteins help convert extracellular antigen binding to the BCR or TCR into intracellular signals, and they do so in similar ways.

Antigen binding to BCRs or TCRs clusters these receptors and their associated invariant chains (and CD4 or CD8 co-receptors in the case of TCRs). This clustering activates a Src family cytoplasmic tyrosine kinase (discussed in Chapter 15) to phosphorylate tyrosines on the cytoplasmic tails of some of the invariant chains. The phosphotyrosines then serve as docking sites for a second cytoplasmic tyrosine kinase, which becomes phosphorylated and activated by the first kinase; the second kinase then relays the signal downstream by phosphorylating other intracellular signaling proteins on tyrosines. Some of these early events in the signaling pathway activated by BCRs are shown in **Figure 24–46**.

Signaling through BCRs or TCRs and their associated proteins alone is not sufficient to activate a lymphocyte to proliferate and differentiate. Extracellular **co-stimulatory signals** produced by another cell are also required, and they are provided by both membrane-bound proteins (see Figure 24–34) and secreted cytokines. Indeed, signaling through the BCR or TCR with insufficient co-stimulation can either eliminate the lymphocyte (clonal deletion) or inactivate it, with both of these mechanisms contributing to peripheral self-tolerance (see Figure 24–21). For a naïve T cell, an activated dendritic cell provides the co-stimulatory signals; these include the transmembrane *B7 proteins*, which are recognized by the co-receptor protein *CD28* on the surface of the T cell (**Figure 24–47A**). For a B cell, an effector T_{FH} cell provides the co-stimulatory signals; these include the transmembrane *CD40 ligand*, which binds to *CD40 receptors* on the B cell (**Figure 24–47B**). The CD40 ligand on effector T_H cells acts in two other situations: (1) it acts back on CD40 receptors on the dendritic cell surface to increase and sustain the activation of the dendritic cell, creating a positive feedback loop; and (2) it acts as a co-stimulatory signal on the surface of an effector T_H1 cell, allowing the T cell to help activate an infected macrophage to destroy the pathogens it harbors.

In addition to receptors for co-stimulatory proteins, both B and T cells have inhibitory proteins on their surface that negatively regulate the cell's activity, preventing excessive or inappropriate responses. Two such proteins, *CTLA4* and *PD1*, expressed by T cells have attracted great attention because of their roles in suppressing the ability of T cells to inhibit cancer progression. They inhibit T cell activity in different ways: CTLA4 inhibits T cell activation by competing with the

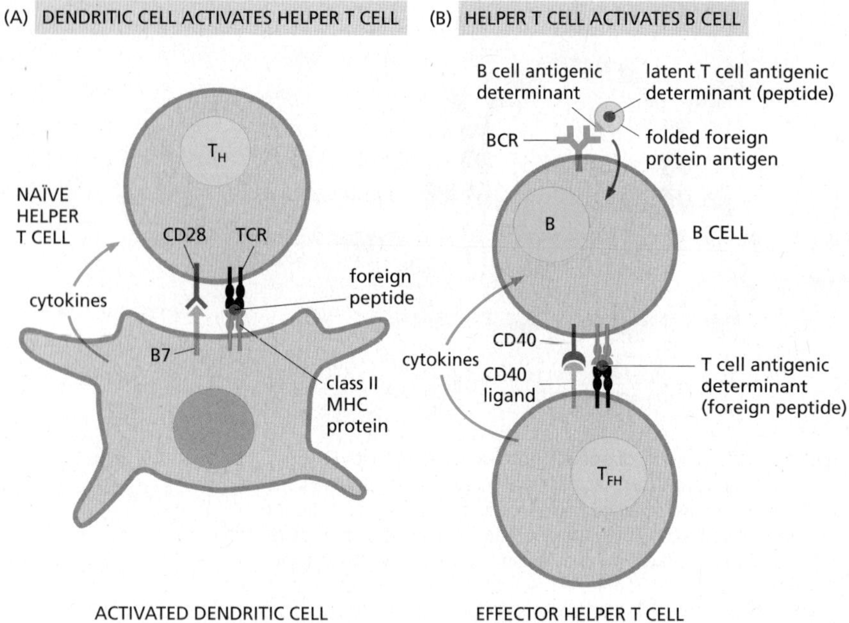

Figure 24–47 **Comparison of the co-stimulatory proteins required to activate a helper T cell and a B cell in response to the same foreign protein.** (A) A naïve helper T cell (T_H) is activated by a peptide fragment of a foreign protein bound to a class II MHC protein on the surface of an activated dendritic cell. The co-stimulatory protein on the dendritic cell (a B7 protein—either CD80 or CD86) binds to the CD28 co-receptor on the T_H cell, providing a necessary co-stimulatory signal to the T_H cell; in addition, cytokines secreted by the dendritic cell (and by other nearby innate immune cells) influence what subtype of effector helper cell the T_H cell becomes (see Figure 24–44). (B) Once activated to become an effector cell, the T_H cell—most commonly a follicular T_H cell (T_{FH}; see Figure 24–44)—can help activate B cells that have the same peptide–MHC protein complexes on their surface as the dendritic cell that activated the naïve T_H cell. These B cells have BCRs that bind an antigenic determinant on the surface of a folded foreign protein, and the B cells endocytose the protein *(red arrow)*; the protein is then cleaved into peptides, some of which are carried to the B cell surface by class II MHC proteins. There, some of the peptide–MHC complexes are recognized by the TCRs on the T_{FH} cell (see Figure 24–39). Note that the BCRs and TCRs recognize different antigenic determinants of the protein. As indicated, the co-stimulatory protein used by the effector T_{FH} cell is CD40 ligand, which binds to the CD40 co-receptor on the B cell; the T_{FH} cell also secretes cytokines *(green arrow)* including IL21 (see Figure 24–44) to help stimulate the B cell to undergo somatic hypermutation and class switching (not shown). The CD4 co-receptor on T_H cells and the invariant chains associated with the TCRs and BCRs are omitted in both A and B for simplicity.

transmembrane co-receptor protein CD28 (see Figure 24–47A), whereas PD1 is expressed on activated T cells, and its prolonged binding to its cell-surface ligand (PDL1 or PDL2) on various cell types, including cancer cells, inhibits the activity of the T cells. Monoclonal antibodies against either CTLA4 or PD1 (or PDL1), or especially against both of these inhibitory pathways, can relieve the inhibition and allow T_C cells to destroy the tumor cells in many patients with metastatic cancer (see Figure 20–46).

Many Cell-Surface Proteins Belong to the Ig Superfamily

Most of the proteins that mediate antigen recognition and cell–cell recognition in the adaptive immune system contain one or more Ig or Ig-like domains, suggesting that the proteins have a common evolutionary history. Included in this very large **Ig superfamily** are antibodies, TCRs, MHC proteins, the CD4, CD8, and CD28 co-receptors, the B7 co-stimulatory proteins, and most of the invariant polypeptide chains associated with TCRs and BCRs, as well as the various Fc receptors on lymphocytes and other leukocytes. Many of these proteins are dimers or higher oligomers, in which Ig or Ig-like domains of one chain interact with those in another (**Figure 24–48**).

In both vertebrates and invertebrates, many proteins in the Ig superfamily are also found outside immune systems, where they often function in cell–cell recognition and adhesion processes, both during development and in adult tissues. It seems likely that the entire gene superfamily evolved from a primordial gene coding for a single Ig-like domain, similar to that encoding β_2-microglobulin (see Figure 24–36). In present-day family members, a separate exon usually encodes the amino acids in each Ig-like domain, consistent with the likelihood that new family members arose during evolution by exon and gene duplications.

Vaccination Against Pathogens Has Been Immunology's Greatest Contribution to Human Health

As all organisms do, we continually battle with our pathogens. Despite the sophistication of our immune defenses described in this chapter, this battle has been remarkably evenly matched. As discussed in Chapter 23, all pathogens have developed multiple ways of overcoming their host's defenses, at least for long enough to replicate and establish an infection. Most pathogens have the important advantage of multiplying and evolving much more rapidly than humans; in particular, viruses, as a collection, have evolved mechanisms for blocking essentially every

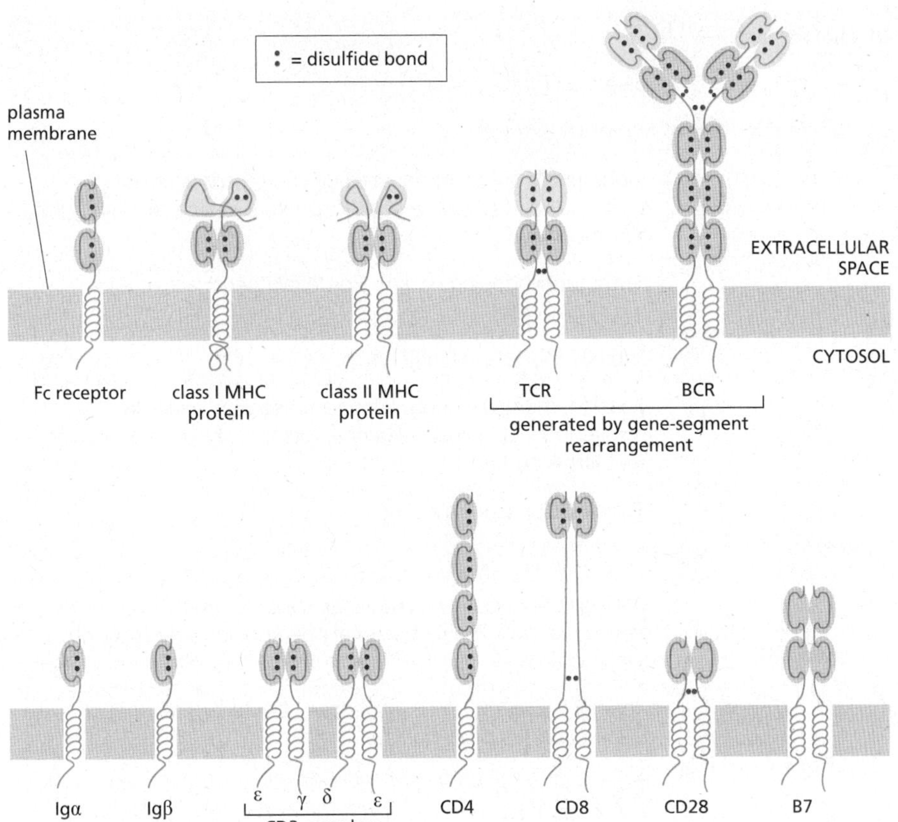

Figure 24–48 **Some of the cell-surface proteins discussed in this chapter that belong to the Ig superfamily.** The Ig and Ig-like domains are shaded in *gray*, except for the antigen-binding Ig and Ig-like domains of BCRs and TCRs, respectively, which are shaded in *blue*. Note that the antigen-binding domains of class I and class II MHC proteins (also shaded in *blue*) are not Ig-like. The Ig superfamily also includes many cell-surface proteins involved in cell–cell interactions outside the immune system, such as the neural cell adhesion molecules (NCAMs; see Figure 19–29) and the receptors for various protein growth factors discussed in Chapter 15 (not shown). There are more than 750 members of the Ig superfamily in humans, making it the most populous family of proteins encoded in the human genome.

step of our innate and adaptive immune responses. In response, humans have developed public health measures and powerful antipathogen drugs and vaccines that reinforce our natural immunological defenses.

The modern era of **vaccination** began in 1796, when Edward Jenner reported that material from skin lesions in cattle suffering from cowpox protects humans against smallpox, which at that time was a devastating epidemic disease. Although the cowpox virus is closely related to the smallpox virus, it does not cause disease in humans. Smallpox was officially eradicated worldwide by 1980 through widespread and coordinated vaccination with intact *vaccinia* virus (a relative of the cowpox virus), after the disease had killed roughly 500 million people in the preceding 100 years. Today, vaccination (from the Latin word *vaccus*, meaning "cow") is widely used to control the spread of many different pathogens, saving millions of human lives a year worldwide. In the best cases, a vaccine induces strong, long-lasting, adaptive immunological memory, mimicking and often bettering the protection induced by the natural infection, but without causing disease; that is, it is strongly immunogenic but not pathogenic.

There are currently three classes of vaccines approved for use against different pathogens in humans (**Table 24–4**). In the first class are *whole microbe vaccines*. These are generally strongly immunogenic but require specific manipulations to avoid pathogenicity. A common method is *attenuation*, in which the pathogen is passaged repeatedly through either a foreign host or cultured foreign cells until it accumulates mutations that render it no longer pathogenic in humans. To eliminate the risk of a genetic reversion to a pathogenic form, some vaccines of this class consist of either *extracts* of the pathogen or whole pathogens that have been *inactivated* by heating, chemical treatment, or, now commonly, by genetic manipulation.

The second class, *subunit vaccines*, are even safer, consisting of one or more individual components of the pathogen—either purified from the pathogen or, more commonly, produced by recombinant DNA technology. Most subunit vaccines are composed of viral proteins: common examples consist of viral coat proteins self-assembled *in vitro* to form *virus-like particles*. Other subunit vaccines are composed of bacterial macromolecules: examples are inactivated forms

TABLE 24–4 Some Vaccines Approved for Human Use

Class of vaccine	Diseases (pathogens)*
Whole microbe (or microbial extract)	
Attenuated bacteria	Tuberculosis (BCG, attenuated bovine *Mycobacterium tuberculosis*), typhoid fever (*Salmonella enterica* serovar Typhi, oral vaccine)
Killed bacteria	Pertussis (whooping cough; *Bordetella pertussis*), cholera (*Vibrio cholerae*)
Bacterial extract	Cholera (*Vibrio cholerae*)
Attenuated virus	Measles, mumps, rubella (German measles), varicella (chickenpox, shingles), influenza, rotavirus, polio (oral vaccine), hepatitis A and B
Inactivated virus	Rabies, influenza, polio, hepatitis A and B
Subunit	
Bacterial capsular polysaccharide	Meningitis (*Neisseria meningitidis*), typhoid fever (*Salmonella enterica* serovar Typhi), bacterial pneumonia (*Streptococcus pneumoniae*)
Bacterial polysaccharide–protein conjugate	Meningitis (*Haemophilus influenzae*), bacterial pneumonia (*Streptococcus pneumoniae*)
Bacterial protein toxids	Tetanus (*Clostridium tetani*), diphtheria (*Corynebacterium diphtheriae*), pertussis (toxoid + other bacterial proteins)
Viral protein (usually recombinant)	Hepatitis A and B, human influenza viruses
Virus-like particles composed of viral coat proteins	Cervical cancer (human papillomavirus types 16 and 18)
Nucleic acid**	
Recombinant RNA virus vector, based on vesicular stomatitis virus, engineered to express an Ebola virus surface protein***	Ebola hemorrhagic fever (Ebola virus)
Recombinant DNA virus vector, based on adenovirus, engineered to express a form of SARS-CoV-2 spike protein (see Figure 24–50)****	COVID-19 (SARS-CoV-2 coronavirus)
Lipid nanoparticles containing modified mRNA that encodes a form of SARS-CoV-2 spike protein (see Figure 24–50)****	COVID-19 (SARS-CoV-2 coronavirus)

*There are often different types of vaccines available for the same disease.
**Some (such as the Ebola vaccine) can replicate in recipient cells, others (including those shown in Figure 24–50) cannot.
***This was the first nucleic acid vaccine approved (in 2019) for human use.
****These vaccines were given Emergency Use Authorization at the end of 2020 or in 2021 and were approved in 2021.

of secreted protein toxins (*toxoids*), which in their active forms cause diseases such as tetanus and diphtheria, and the capsular polysaccharides that surround various bacteria (see Table 24–4).

To be maximally effective, a vaccine needs to stimulate both B cell and T cell immune responses. Subunit vaccines composed solely of bacterial capsular polysaccharides, for example, are relatively ineffective because they only activate B cells. Because helper T cells are required to induce B cell memory, antibody affinity maturation, and Ig class switching, such polysaccharide vaccines only induce short-lived, low-affinity, IgM anti-polysaccharide antibodies, providing weak and brief protection. Chemically coupling (conjugating) the polysaccharide to a protein such as tetanus or diphtheria toxoid to produce a *conjugate vaccine* solves this problem: the protein component provides the foreign peptides that

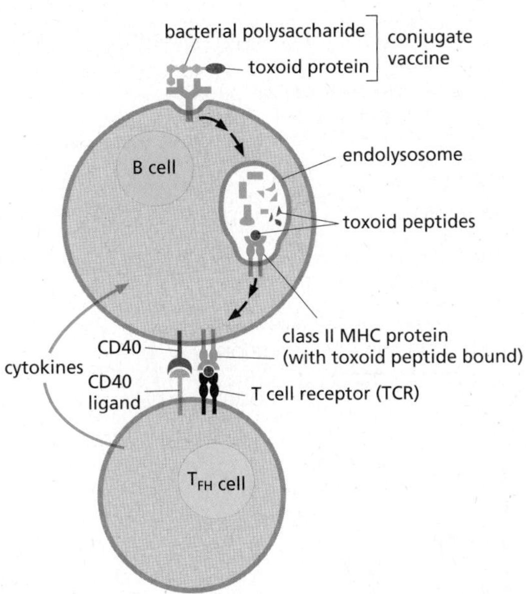

Figure 24–49 **How conjugate vaccines against bacterial polysaccharide antigens activate both B cells and helper T cells.** In this case, the polysaccharide is chemically coupled to a protein toxoid. The B cells recognize the polysaccharide part of the conjugate, endocytose the conjugate, and cleave the protein part into peptide fragments in endolysosomes. The peptide fragments are then transported by class II MHC proteins to the B cell surface, where the peptide–MHC complexes are recognized by helper T cells (see Figure 24–39). The helper T cells (T_{FH} cells) help activate the B cells to produce both high-affinity IgG anti-polysaccharide antibodies (see Figure 24–47B) and memory B cells (not shown).

activate the helper T cells required for optimal stimulation of the appropriate polysaccharide-specific B cells, as illustrated in **Figure 24–49**.

As discussed earlier in the chapter, for a vaccine to activate adaptive immune responses, it first must activate dendritic cells. A vaccine composed of a whole microbe or a microbial extract fulfills this requirement because it contains pathogen-associated molecular patterns (PAMPs) that activate pattern recognition receptors (PRRs) on innate immune cells, especially dendritic cells. Subunit vaccines that lack PAMPs require the addition of an *adjuvant*, a substance that enhances the immunogenicity of antigens, usually by activating PRRs.

The third and most recent type of vaccines contains genetically engineered DNA or RNA molecules that encode pathogen proteins or parts of such proteins. The development of these *nucleic acid vaccines* has required the knowledge and tools produced by decades of fundamental research to make the vaccines tolerable, safe, and effective. The first such vaccine approved for human use (in 2019) was a recombinant RNA virus vector vaccine encoding an Ebola virus surface protein (see Table 24–4). But the use of nucleic acid vaccines only took off in 2021, in response to the COVID-19 pandemic, which has infected hundreds of millions of people and caused millions of deaths. Because only a pathogen's genome sequence is needed to develop a nucleic acid vaccine against the pathogen, it was possible to produce the first safe and effective DNA- and RNA-based vaccines for large-scale human use within a year from the publication (in January 2020) of the RNA genome sequence of the causative SARS-CoV-2 virus.

Figure 24–50 illustrates how two different kinds of nucleic acid vaccines against COVID-19 work: both encode forms of the protein "spike" found on the surface of the SARS-CoV-2 virus (see Figure 1–49). A number of whole microbe and subunit vaccines are also in wide use or in development around the world to fight the COVID-19 pandemic. But, given the recent success of nucleic acid vaccines, it seems likely that they will revolutionize how most new vaccines are produced in the future.

A successful vaccination program against a pathogen usually requires that most of the population becomes immune to it, either naturally by infection or by vaccination. Such *herd immunity* slows the spread of an infection by decreasing the number of susceptible individuals in the population. To produce herd immunity, a vaccine needs to be well tolerated, safe and effective, and widely perceived to be so. The importance of both herd immunity and the public acceptance of a vaccine was inadvertently demonstrated by the response to a false and fraudulent report of a link between a measles vaccine and autism. After publication of the report in 1998, the uptake of the vaccine in Britain decreased substantially, leading to an increase in both the size and frequency of measles outbreaks (**Figure 24–51**).

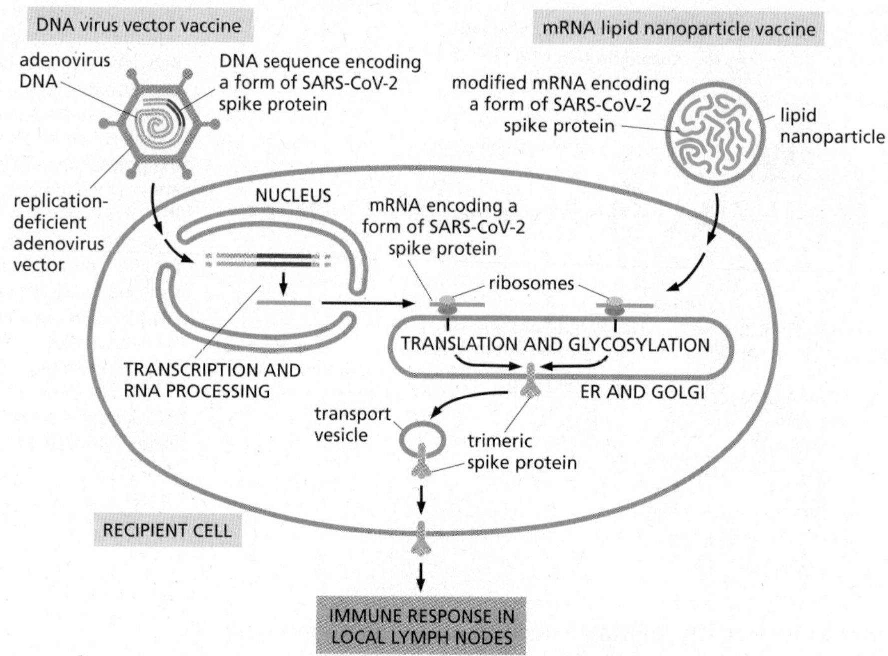

Figure 24–50 How two kinds of widely used nucleic acid COVID-19 vaccines work. In both, the nucleic acids encode forms of the spike (S) protein, which normally protrudes from the surface of the SARS-CoV-2 coronavirus (see Figure 1–49 and Figure 9–50); these spikes on the virus bind to receptors on various human cells and help the virus enter the cells. The *DNA virus vector vaccine* shown on the left is representative of two COVID-19 vaccines that received early authorization for emergency use in humans—known as the Oxford–AstraZeneca vaccine and the Janssen–Johnson & Johnson vaccine—both of which are based on adenovirus vectors (chimpanzee and human, respectively). These vectors, which have been genetically engineered to prevent their replication in human cells, transfer their DNA into the nucleus of these cells, where it produces mRNAs. As indicated, the vector-encoded mRNAs move to the cytoplasm, where they are translated on endoplasmic reticulum (ER)-bound ribosomes to produce the spike proteins. After glycosylation in the ER and Golgi apparatus, the spike proteins are transported to the cell surface as trimeric spikes.

The *mRNA lipid nanoparticle vaccine* shown on the right is representative of the first COVID-19 vaccines of this kind authorized for emergency use in humans—known as the Pfizer–BioNTech and the Moderna vaccines. The mRNA is synthesized *in vitro* with nucleoside modifications (uridines replaced by methylpseudouridines) to prevent the mRNA from binding to PRRs and triggering a destructive inflammatory response, which would otherwise decrease the vaccine's tolerability and the mRNA's translation. In addition, the mRNA is designed with a nucleotide sequence that maximizes its translation efficiency. Finally, the mRNA molecules are encapsulated in tiny, complex lipid nanoparticles: this coating protects the mRNAs, enhances the vaccine's uptake into host-cell endosomes, and mediates the release of the mRNA from the endosomes into the cytosol.

In all cases the vaccine is injected into an arm muscle, from where it is carried via lymphatic vessels to local lymph nodes. Both in the muscle and in lymph nodes, antigen-presenting cells (most important, dendritic cells) take up the vaccine and are activated in the process. In these cells, the mRNAs are translated into spike protein, which is then glycosylated and transported to the cell surface as indicated. In the lymph nodes, the activated dendritic cells present peptide fragments of the spike protein, bound to MHC proteins on the dendritic-cell surface, to spike-specific helper and cytotoxic T cells, which are thereby activated to proliferate and differentiate into effector cells. The effective helper T cells then help stimulate spike-specific B cells to proliferate and differentiate into effector cells that secrete neutralizing anti-spike antibodies, while the cytotoxic T cells kill any host cells that subsequently become infected with the SARS-CoV-2 virus. At the same time, the activated dendritic cells stimulate the production of spike-specific memory B and T cells, which a second immunization can further increase in number and effectiveness. Later, a natural encounter with the SARS-CoV-2 virus will call these memory cells into action to protect vaccinated individuals against COVID-19 disease.

It remains to be seen how long the protection lasts and how effective it will be against new mutant SARS-CoV-2 virus variants, as they arise and are selected to resist our collective adaptive immune responses to either immunization or natural infection. Fortunately, the nucleic acid vaccines, especially mRNA vaccines, can be resynthesized rapidly to increase their effectiveness against any such genetic variants.

Summary

There are three main functionally distinct classes of T cells. Cytotoxic T cells (T_C cells) directly kill infected host cells by the targeted secretion of perforins and granzymes onto the surface of the infected cells, inducing them to kill themselves by undergoing apoptosis. Helper T cells (T_H cells) help B cells to make antibody responses, macrophages to destroy the microorganisms they harbor, and dendritic cells to activate T_C cells. By contrast, regulatory T cells (T_{reg} cells) produce suppressive proteins (such as the cytokines IL10 and TGFβ) to inhibit other immune cells.

All T cells express cell-surface antigen receptors (TCRs), which are encoded by genes that are assembled from multiple gene segments during T cell development in the thymus. αβ TCRs recognize peptide fragments of foreign proteins that are displayed in association with MHC proteins on the surface of antigen-presenting cells (APCs), including T cell targets. Naïve T cells are activated in peripheral

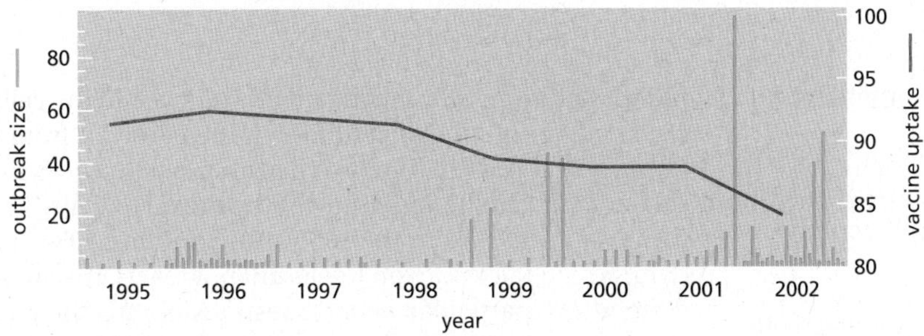

Figure 24–51 **The importance of herd immunity and public acceptance of a vaccine.** After the publication in 1998 of a false link between measles vaccination and autism, fewer children in Britain were vaccinated against the measles virus *(red line)*, and outbreaks of measles *(blue bars)* increased in both frequency and size. About 95% of a population needs to be immune to measles to achieve herd immunity (the comparable value for COVID-19 is estimated to be between 65 and 90%, depending on the virus variant). The vaccine uptake shown is the percentage of children completing a primary course of the measles, mumps, and rubella (MMR) vaccine at their second birthday. (Data courtesy of V.A.A. Jansen and M.E. Ramsey. Adapted from V.A.A. Jansen et al., *Science* 301:804, 2003. With permission from AAAS.)

lymphoid organs by activated dendritic cells, which secrete cytokines and express peptide–MHC complexes, co-stimulatory proteins, and various cell–cell adhesion molecules on their surface.

Class I MHC proteins present foreign peptides to T_C cells, whereas class II MHC proteins present foreign peptides to T_H cells and self and foreign peptides to T_{reg} cells. Whereas class I MHC proteins are expressed on almost all nucleated vertebrate cells, class II MHC proteins are normally restricted to APCs, including dendritic cells, macrophages, and B lymphocytes. Both classes of MHC proteins have a single peptide-binding groove, which binds a large set of small peptide fragments produced intracellularly by normal protein-degradation processes: class I MHC proteins mainly bind fragments produced in the cytosol, whereas class II MHC proteins mainly bind fragments produced in endocytic compartments. The peptide–MHC complexes are transported to the cell surface, where complexes that contain a peptide derived from a foreign protein are recognized by TCRs, which interact with both the peptide and the walls of the peptide-binding groove. T cells also express CD4 or CD8 co-receptors, which recognize invariant regions of MHC proteins: T_H cells and T_{reg} cells express CD4, which recognizes class II MHC proteins; T_C cells express CD8, which recognizes class I MHC proteins.

A combination of positive and negative selection processes operate during thymocyte development to help ensure that only T cells with potentially useful TCRs survive, mature, and emigrate from the thymus, while all other thymocytes die by apoptosis. After leaving the thymus, the naïve T_H and T_C cells continually receive survival signals when their TCRs recognize self-peptide–self-MHC complexes, but they can only be activated to become effector cells when their TCRs encounter foreign peptides in the grooves of MHC proteins on an activated dendritic cell. The cells that leave the thymus as natural T_{reg} cells help maintain self-tolerance by suppressing self-reactive T cells that escape negative selection in the thymus.

The production of an effector T cell from a naïve T cell requires multiple signals from an activated dendritic cell. MHC–peptide complexes on the dendritic cell surface provide one signal, by binding to both TCRs and to either a CD4 co-receptor on a T_H or T_{reg} cell or to a CD8 co-receptor on a T_C cell; co-stimulatory proteins on the dendritic cell surface and secreted cytokines provide the other signals. When naïve T_H cells are initially activated on a dendritic cell in a peripheral lymphoid organ, they can differentiate into various types of effector T cells, depending on the invading pathogen and the cytokines in their environment. They can become effector T_{FH} cells, which remain in the lymphoid organ to support B cell antibody production in lymph follicles, or they can develop into effector T_H1, T_H2, or T_H17 cells, which mainly migrate to sites of infection, where they help other immune cells fight the pathogen; alternatively, they can develop into induced T_{reg} cells, which suppress other immune cells. The various effector T_H cells recognize the same complex of foreign peptide and class II MHC protein on the surface of the target cell they are helping as they initially recognized on the dendritic cell that activated them; they activate their target cells by a combination of membrane-bound co-stimulatory proteins and secreted cytokines. T_{reg} cells use cell-surface and secreted inhibitory proteins to suppress their target cells.

Both T cells and B cells require multiple signals for activation. Antigen binding to the TCRs or BCRs provides one signal, while co-stimulatory proteins binding to co-receptors and cytokines binding to their complementary cell-surface receptors provide the others. Effector T_H cells provide the co-stimulatory signals and secreted cytokines for B cells, whereas APCs provide them for T cells.

PROBLEMS

Which statements are true? Explain why or why not.

24–1 T cells whose receptors strongly bind a self-peptide–self-MHC complex are killed off in peripheral lymphoid organs when they encounter the self peptide on an antigen-presenting dendritic cell.

24–2 To guarantee that the antigen-presenting cells in the thymus will display a complete repertoire of self peptides to allow elimination of self-reactive T cells, the thymus recruits dendritic cells from all over the body.

24–3 The antibody diversity created by the combinatorial joining of V, D, and J segments by V(D)J recombination pales in comparison to the enormous diversity created by the random gain and loss of nucleotides at V, D, and J joining sites.

Discuss the following problems.

24–4 Why do living trees not rot? Redwood trees, for example, can live for centuries, but once they die they decay fairly quickly. What might this suggest?

24–5 It would be disastrous if a complement lytic attack were not confined to the surface of the pathogen that is the target of the attack. Yet, the proteolytic cascade involved in the attack liberates biologically active molecules at several steps: one that diffuses away and one that remains bound to the target surface. How does the complement attack remain localized to the pathogen when active products leave the pathogen surface?

24–6 On the basis of its sequence similarity to Apobec1, which deaminates C to U in RNA, activation-induced deaminase (AID) was originally proposed to work on RNA. But definitive experiments in *Escherichia coli* demonstrated that AID deaminates C to U in DNA. The authors of the paper expressed AID in bacteria and followed mutations in a selectable gene. They found that AID expression increased mutations about fivefold above the background level in the absence of AID expression. More important, they found that 80% of the induced mutations were G → A or C → T. Does this fit with your expectation if AID-induced mutations arose by deamination of C to U in the DNA?

[Hint: Imagine what would happen if the G:U mismatch created by AID was replicated several times; how would the sequences of the final mutations relate to the original G-C base pair?]

24–7 For many years it was a complete mystery how cytotoxic T cells could recognize and respond to a viral protein in a virus-infected cell when the protein seemed to be present only in the nucleus of the cell. The answer was revealed in a classic paper that took advantage of a clone of cytotoxic T cells whose T cell receptor (TCR) was directed against an antigen associated with the nuclear protein of the 1968 strain of influenza virus. The authors of the paper found that when they incubated noninfected living cells in high concentrations of certain peptides derived from the viral nuclear protein, the cells became sensitive to lysis by subsequent incubation with the cytotoxic T cells. Using various peptides from the 1968 strain and the 1934 strain (with which the cytotoxic T cells did not react), the authors defined the particular peptide responsible for the killing response of the T cell (Figure Q24–1).

A. Which part of the viral protein gives rise to the peptide that is recognized by the clone of cytotoxic T cells? Why do not all of the viral peptides sensitize the target cells for lysis by the cytotoxic T cells?

B. It is known that the MHC proteins come to the cell surface with peptides already bound. How then do you imagine that these experiments worked when the peptides were added to the outside of the living target cells?

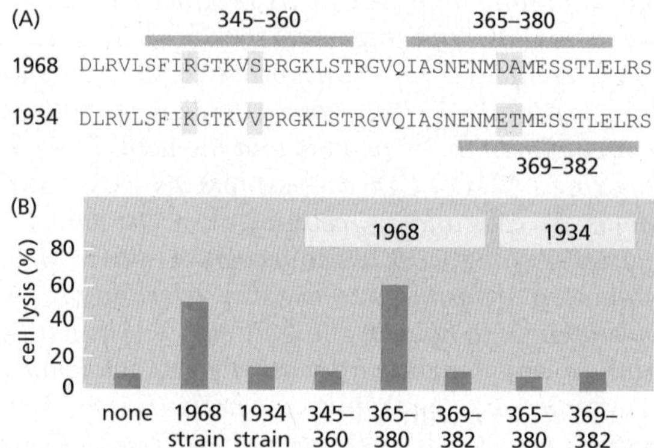

Figure Q24–1 Viral nuclear protein recognition by cytotoxic T cells (Problem 24–7). (A) Sequences of a segment of the nuclear protein from the 1968 and 1934 strains of influenza virus. Peptides used in the experiments in B are highlighted by *pink bars*. The amino acid differences between the viral proteins are highlighted in *blue*. (B) Cytotoxic T cell–mediated lysis of target cells. The target cells were untreated (none), transfected with nuclear protein genes (1968 or 1934 strain), or preincubated with high concentrations of nuclear protein peptides from the 1968 or 1934 strain, as indicated by the *yellow* highlights.

24–8 Working out the roles of MHC proteins in T cell antigen recognition was complicated. One of the key observations came from studying how different kinds of class I MHC proteins influence the way cytotoxic T cells killed cells infected with lymphocytic choriomeningitis virus (LCMV). Cytotoxic T cells derived from mice expressing "k-type" class I MHC proteins lysed LCMV-infected cells expressing the same k-type MHC protein, but they did not lyse LCMV-infected cells from mice expressing "d-type" class I MHC proteins (Figure Q24–2). Similarly, cytotoxic T cells from d-type mice infected with LCMV lysed infected d-type cells, but not infected k-type cells. LCMV can kill both k-type and d-type mice.

A. If homozygous d-type mice were bred to homozygous k-type mice to generate d-type/k-type heterozygous progeny, would you expect that cytotoxic T cells derived

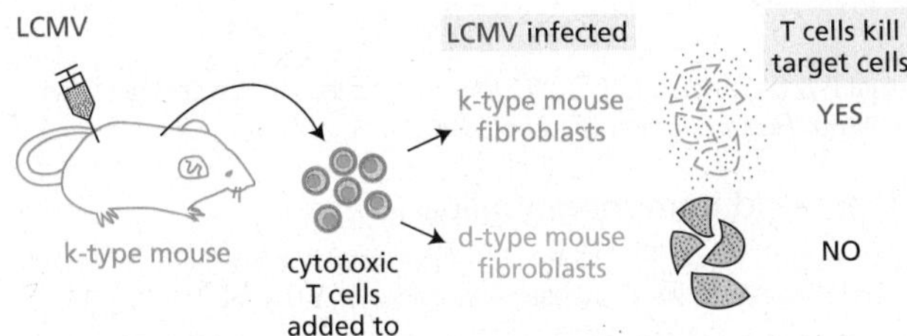

Figure Q24–2 Pattern of killing of LCMV-infected fibroblasts by cytotoxic T cells from an LCMV-infected k-type mouse (Problem 24–8).

from these LCMV-infected heterozygotes would be able to lyse LCMV-infected d-type cells? How about LCMV-infected k-type cells? Explain your answers.

B. Oddly enough, LCMV infection does not kill mice that lack a thymus (and therefore lack T cells)—such as "nude" mice, so called because they also lack hair. If a thymus is transplanted back into a nude mouse, the mouse will then die when infected with LCMV. Suppose that a developing d-type/k-type heterozygous nude mouse was given a thymus from a d-type donor and was later infected with LCMV. Would you expect its cytotoxic T cells to be able to lyse LCMV-infected d-type cells? How about infected k-type cells? Explain your answers.

24–9 Before exposure to a foreign antigen, T cells with receptors specific for the antigen are a tiny fraction of the T cells—of the order 1 in 10^5 or 1 in 10^6 T cells. After exposure to the antigen, only a small number of dendritic cells typically display the antigen on their surface. How long does it take for such antigen-presenting dendritic cells to interact with the antigen-specific T cells, which is the key first step in T cell activation and clonal expansion? To begin to address this question, scientists studied T-cell–dendritic-cell interactions in the absence of specific antigen. They isolated T cells and dendritic cells from unimmunized mice, labeled the T cells *green* and the dendritic cells *red*, and injected them into an unimmunized mouse. After 20 hours, they isolated a local lymph node and scored the contacts between the two cell types visually using two-photon fluorescence microscopy (**Figure Q24–3A**). The frequency of contacts between the two types of cells is given in **Figure Q24–3B**. Assuming that 100 dendritic cells present a specific antigen in an immunized mouse, how long would it take them to scan 10^5 T cells to find one or more antigen-specific T cells? How long to scan 10^6 T cells?

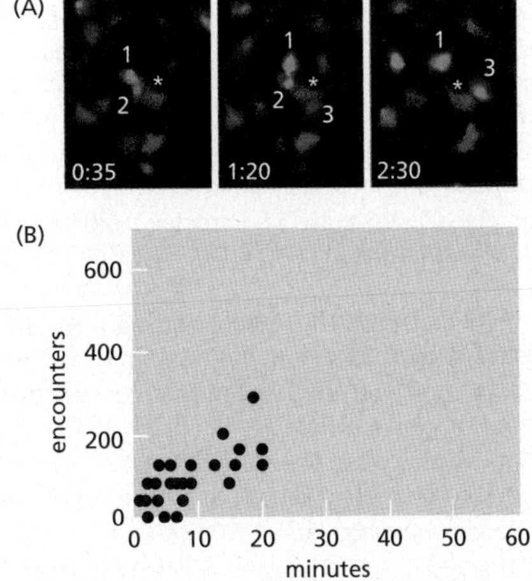

Figure Q24–3 Scanning of the T cell repertoire by dendritic cells in the absence of antigen (Problem 24–9). (A) Contacts between different T cells and one dendritic cell. T cells are *green* and dendritic cells are *red*. The dendritic cell labeled with an asterisk contacts a total of three T cells (numbered) over time in this sequence of images. Times are shown as hours:minutes. (B) Plot of T cell contacts for individual dendritic cells over time. (A, from P. Bousso and E. Robey, *Nat. Immunol.* 4:579–585, published 2003 by Nature Publishing Group. Reproduced with permission of SNCSC.)

24–10 At first glance, it would seem a dangerous strategy for the thymus to actively promote the survival, maturation, and emigration of developing T cells that bind weakly to self peptides bound to self MHC proteins. Would it not be safer to get rid of these T cells, along with those that bind strongly to such self-peptide–self-MHC complexes, as this would seem a more secure way to avoid autoimmune reactions?

24–11 CD4 proteins on helper and regulatory T cells serve as co-receptors that bind to invariant parts of class II MHC proteins. CD4 is thought to increase the adhesion between T cells and antigen-presenting cells (APCs) that are initially connected only weakly by the TCR bound to its specific peptide–MHC complex. To test this possibility, you label cell-surface MHC proteins with a fluorescently labeled peptide so that you can detect individual peptide–MHC complexes at the interface between the APCs and the T cells in a culture dish. To detect T cell responses—the sign of a productive contact—you load them with a Ca^{2+} indicator dye to detect the increase in cytosolic Ca^{2+} that occurs when lymphocytes are activated. You now count the peptide–MHC complexes at a large number of interfaces (immunological synapses) and measure the Ca^{2+} signal in the adherent T cells (**Figure Q24–4**, *red dots*). When you repeat the experiment in the presence of blocking antibodies against CD4, you get a different result (*blue dots*). Do these results support or refute the notion that CD4 augments T-cell–APC binding? Explain your answer.

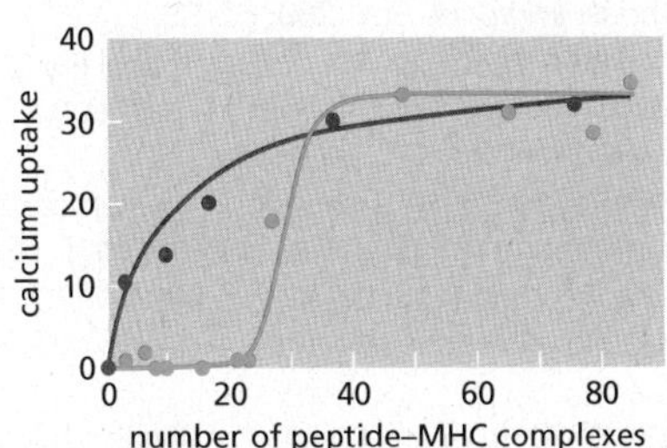

Figure Q24–4 Role of CD4 in the T cell response (Problem 24–11). The uptake of Ca^{2+} in cells with different numbers of fluorescently labeled peptide–MHC complexes at the interface between the T cells and the antigen-presenting cells. The results in the absence of CD4-blocking antibodies are shown by the *red curve*; results in the presence of CD4 antibodies are shown by the *blue curve*.

REFERENCES

General

Abbas AK, Lichtman AH & Pillai S (2019) Basic Immunology: Functions and Disorders of the Immune System, 6th ed. Philadelphia: Elsevier.

Murphy K, Weaver C & Berg L (2022) Janeway's Immunobiology, 10th ed. New York: Norton.

Parham P (2021) The Immune System, 5th ed. New York: Norton.

The Innate Immune System

Chan AH & Schroder K (2020) Inflammasome signaling and regulation of interleukin-1 cytokines. *J. Exp. Med.* 217, e20190314.

Chao S & Pamer EG (2011) Monocyte recruitment during infection and inflammation. *Nat. Rev. Immunol.* 11, 762–774.

Gallo RL & Hooper LV (2012) Epithelial antimicrobial defence of the skin and intestine. *Nat. Rev. Immunol.* 12, 503–516.

Iwasaki A & Medzhitov R (2015) Control of adaptive immunity by the innate immune system. *Nat. Immunol.* 16, 343–353.

Janeway CA, Jr & Medzhitov R (2002) Innate immune recognition. *Adv. Immunol.* 109, 87–124.

Kumar H, Kawai T & Akira S (2011) Pathogen recognition by the innate immune system. *Int. Rev. Immunol.* 30, 16–34.

Kumar S (2018) Natural killer cell toxicity and its regulation by inhibitory receptors. *Immunology* 154, 383–393.

Liew PX & Kubes P (2019) The neutrophil's role during health and disease. *Physiol. Rev.* 99, 1223–1248.

Mandal A & Viswanathan C (2015) Natural killer cells: in health and disease. *Hematol. Oncol. Stem Cell Ther.* 8, 47–55.

McNab F, Mayer-Barber K, Sher A . . . O'Garra A (2015) Type I interferons in infectious disease. *Nat. Rev. Immunol.* 15, 87–103.

Meffre E & Iwasaki A (2020) Interferon deficiency can lead to severe COVID. *Nature* 587, 374–376.

Newton K & Dixit VM (2012) Signaling in innate immunity and inflammation. *Cold Spring Harb. Perspect. Biol.* 4, a006049.

Reis ES, Mastellos DC, Hajishengallis G & Lambris JD (2019) New insights into the immune functions of complement. *Nat. Rev. Immunol.* 19, 503–516.

Vijay K (2018) Toll-like receptors in immunity and inflammatory diseases: past, present, and future. *Int. Immunopharmacol.* 59, 391–412.

Vivier E, Artis D, Colonna M . . . Spits H (2018) Innate lymphoid cells: 10 years on. *Cell* 174, 1054–1066.

Zhao L & Lu W (2014) Defensins in innate immunity. *Curr. Opin. Hematol.* 2, 37–42.

Overview of the Adaptive Immune System

Clark RA (2015) Resident memory T cells in human health and disease. *Sci. Transl. Med.* 7, 269rv1.

Eckert N, Permanyer M, Yu K, Werth K & Förster R (2019) Chemokines and other mediators in the development and functional organization of lymph nodes. *Immunol. Rev.* 289, 62–83.

Farber DL, Netea MG, Radbruch A . . . Zinkernagel RM (2016) Immunological memory: lessons from the past and a look to the future. *Nat. Rev. Immunol.* 16, 124–128.

Mueller SN & Mackay LK (2016) Tissue-resident memory T cells: local specialists in immune defence. *Nat. Rev. Immunol.* 16, 79–89.

Nguyen QP, Deng TZ, Witherden DA & Goldrath AW (2019) Origins of $CD4^+$ circulating and tissue-resident memory T-cells. *Immunology* 157, 3–12.

Palm AE & Henry C (2019) Remembrance of things past: long-term B cell memory after infection and vaccination. *Front. Immunol.* 10, 1787.

Pradeu T & Du Pasquier L (2018) Immunological memory: what's in a name? *Immunol. Rev.* 283, 7–20.

Schwartz RH (2012) Historical overview of immunological tolerance. *Cold Spring Harb. Perspect. Biol.* 4, a006908.

Weisel F & Shlomchik M (2017) Memory B cells of mice and humans. *Annu. Rev. Immunol.* 35, 255–284.

B Cells and Immunoglobulins

Bournazos S, Wang TT, Dahan R . . . Ravetch JV (2017) Signaling by antibodies: recent progress. *Annu. Rev. Immunol.* 35, 285–311.

Carmona LM & Schatz DG (2017) New insights into the evolutionary origins of the recombination-activating gene proteins and V(D)J recombination. *FEBS J.* 284, 1590–1605.

Ebert A, Hill L & Busslinger M (2015) Spatial regulation of V-(D)J recombination at antigen receptor loci. *Adv. Immunol.* 128, 93–121.

Eibel H, Kraus H, Sic H . . . Rizzi M (2014) B cell biology: an overview. *Curr. Allergy Asthma Rep.* 14, 434–444.

Gitlin AD, Shulman Z & Nussenzweig MC (2014) Clonal selection in the germinal centre by regulated proliferation and hypermutation. *Nature* 29, 509, 637–640.

Imkeller K & Wardemann H (2018) Assessing human B cell repertoire diversity and convergence. *Immunol. Rev.* 284, 51–66.

Kwak K, Akkaya M & Pierce SK (2020) B cell signaling in context. *Nat. Immunol.* 20, 963–969.

Lu LL, Suscovich TJ, Fortune SM & Alter G (2018) Beyond binding: antibody effector functions in infectious diseases. *Nat. Rev. Immunol.* 18, 46–61.

Methot SP & Di Noia JM (2017) Molecular mechanisms of somatic hypermutation and class switch recombination. *Adv. Immunol.* 133, 37–87.

Wang L & Müschen M (2018) Portending death in germinal centers—when B cells know their time is up. *Cell Res.* 28, 5–6.

T Cells and MHC Proteins

Alcover A, Alarcón B & Di Bartolo V (2018) Cell biology of T cell receptor expression and regulation. *Annu. Rev. Immunol.* 36, 103–125.

Blum JS, Wearsch PA & Cresswell P (2013) Pathways of antigen processing. *Annu. Rev. Immunol.* 31, 443–473.

Chudnovskiy A, Pasqual G & Victora GD (2019) Studying interactions between dendritic cells and T cells in vivo. *Curr. Opin. Immunol.* 58, 24–30.

DiToro D, Winstead CJ, Pham D . . . Weaver CT (2018) Differential IL-2 expression defines developmental fates of follicular versus nonfollicular helper T cells. *Science* 361(6407), eaao2933.

Dustin ML & Choudhuri K (2016) Signaling and polarized communication across the T cell immunological synapse. *Annu. Rev. Cell Dev. Biol.* 32, 303–325.

Eisenbarth SC (2019) Dendritic cell subsets in T cell programming: location dictates function. *Nat. Rev. Immunol.* 19, 89–103.

Gascoigne NR, Rybakin V, Acuto O & Brzostek J (2016) TCR signal strength and T cell development. *Annu. Rev. Cell Dev. Biol.* 32, 327–348.

Halle S, Halle O & Förster R (2017) Mechanisms and dynamics of T cell-mediated cytotoxicity In vivo. *Trends Immunol.* 38, 432–443.

Kotsias F, Cebrian I & Alloatti A (2019) Antigen processing and presentation. *Int. Rev. Cell Mol. Biol.* 348, 69–121.

Perniola R (2018) Twenty years of AIRE. *Front. Immunol.* 9, 98.

Plitas G & Rudensky AY (2016) Regulatory T cells: differentiation and function. *Cancer Immunol. Res.* 4, 721–725.

Pollard AJ & Bijker EM (2021) A guide to vaccinology: from basic principles to new developments. *Nat. Rev. Immunol.* 21, 83–100.

Rossjohn J, Gras S, Miles JJ . . . McCluskey J (2015) T cell antigen receptor recognition of antigen-presenting molecules. *Annu. Rev. Immunol.* 33, 169–200.

Sakaguchi S, Mikami N, Wing JB. . . Ohkura N (2020) Regulatory T cells and human disease. *Annu. Rev. Immunol.* 38, 541–566.

Steinman RM (2012) Decisions about dendritic cells: past, present, and future. *Annu. Rev. Immunol.* 30, 1–22.

Glossary

ABC transporters A large class of membrane transport proteins that use the energy of ATP hydrolysis to transfer peptides or small molecules across membranes. (Chapter 11) (Figure 11–16)

acetyl CoA Small water-soluble activated carrier molecule. Consists of an acetyl group linked to coenzyme A (CoA) by an easily hydrolyzable, energy-rich thioester bond. (Chapter 2) (Figure 2–38)

acetylcholine receptor (AChR) Membrane protein that responds to binding of acetylcholine (ACh). The nicotinic AChR is a transmitter-gated ion channel that opens in response to ACh. The muscarinic AChR is not an ion channel, but a G-protein-coupled cell-surface receptor. (Chapter 11)

acid A proton donor. Substance that releases protons (H^+) when dissolved in water, forming hydronium ions (H_3O^+) and lowering the pH. (Chapter 2) (Panel 2–2, pp. 96–97)

acid hydrolases Hydrolytic enzymes—including proteases, nucleases, glycosidases, lipases, phospholipases, phosphatases, and sulfatases—that work best at acidic pH; these enzymes are found within the lysosome. (Chapter 13)

action potential Rapid, transient, self-propagating electrical excitation in the plasma membrane of a cell such as a neuron or muscle cell. Action potentials, or nerve impulses, make possible long-distance signaling in the nervous system. (Chapter 11) (Figure 11–33)

activated carrier Small diffusible molecule that stores easily exchangeable energy in the form of one or more energy-rich covalent bonds. Examples are ATP, acetyl CoA, $FADH_2$, NADH, and NADPH. (Chapter 2) (Figure 2–31)

activation energy The extra energy that must be acquired by atoms or molecules in addition to their ground-state energy in order to reach the transition state required for them to undergo a particular chemical reaction. (Figure 2–21)

activation-induced deaminase (AID) The enzyme catalyzing the processes of somatic hypermutation and immunoglobulin class switching in activated B cells. (Chapter 24)

active site Region of an enzyme surface to which a substrate molecule binds in order to undergo a catalyzed reaction. (Chapter 3)

active transport Movement of a molecule across a membrane or other barrier driven by energy other than that stored in the electrochemical or concentration gradient of the transported molecule itself. (Chapter 11)

adaptation (1) Adaptation (desensitization): adjustment of sensitivity following repeated stimulation. The mechanism that allows a cell to react to small changes in stimuli even against a high background level of stimulation. (2) Evolutionary adaptation: an evolved trait. (Chapters 11, 15)

adaptive immune system System of lymphocytes providing highly specific and long-lasting defense against pathogens in vertebrates. It consists of two major classes of lymphocytes: B lymphocytes (B cells), which secrete antibodies that bind specifically to the pathogen or its products, and T lymphocytes (T cells), which can either directly kill cells infected with the pathogen or produce secreted or cell-surface signal proteins that stimulate other host cells to help eliminate the pathogen. (Chapter 24) (Figure 24–2)

adaptor protein, adaptor General term for a protein that functions solely to link two or more different proteins together in an intracellular signaling pathway or protein complex. (Chapters 13, 15) (Figure 15–11)

adenylyl cyclase (adenylate cyclase) Membrane-bound enzyme that catalyzes the formation of cyclic AMP from ATP. An important component of some intracellular signaling pathways. (Chapter 15)

adherens junction Cell junction in which the cytoplasmic face of the plasma membrane is attached to actin filaments. Examples include adhesion belts linking adjacent epithelial cells and focal contacts on the lower surface of cultured fibroblasts. (Chapter 19)

adhesins Specific proteins or protein complexes of pathogenic bacteria that recognize and bind cell-surface molecules on the host cells to enable tight adhesion and colonization of tissues. (Chapter 23)

adhesion belt Adherens junctions in epithelia that form a continuous belt (*zonula adherens*) just beneath the apical face of the epithelium, encircling each of the interacting cells in the sheet. (Chapter 19)

ADP (adenosine 5'-diphosphate) Nucleotide derivative (a nucleoside diphosphate) produced by hydrolysis of the terminal phosphate of ATP. Regenerates ATP when phosphorylated by an energy-generating process such as oxidative phosphorylation. (Chapter 2) (Figure 2–33)

adult stem cells Undifferentiated cells found throughout the body that undergo cell divisions to maintain homeostasis of adult tissues and replenish damaged tissues; also known as somatic stem cells. (Chapter 22)

aerobic respiration Process by which a cell obtains energy from sugars or other organic molecules by allowing their carbon and hydrogen atoms to combine with the oxygen in air to produce CO_2 and H_2O, respectively. (Chapters 2 and 14)

affinity constant (K_a) The equilibrium constant for a simple binding interaction, when expressed as [AB]/[A][B]. Also known as the *association constant*. (Chapter 3)

affinity maturation Progressive increase in the affinity of antibodies for the immunizing antigen with the passage of time after immunization. (Chapter 24)

AIRE (**autoimmune regulator**) A protein expressed by a subpopulation of epithelial cells in the thymus that stimulates the production of small amounts of self proteins characteristic of other organs, exposing developing thymocytes to these proteins for the purpose of self-tolerance. (Chapter 24)

Akt Serine/threonine protein kinase that acts in the PI-3-kinase–Akt intracellular signaling pathway; involved especially in signaling cells to grow and survive. Also called protein kinase B (PKB). (Chapter 15)

allele One of several alternative forms of a gene. In a diploid cell, each gene will typically have two alleles, occupying the same corresponding position (locus) on homologous chromosomes. (Chapters 5, 8)

allosteric protein A protein that can adopt at least two distinct conformations, and for which the binding of a ligand at one site causes a conformational change that alters the activity of the protein at a second site; this allows one type of molecule in a cell to alter the fate of a molecule of another type, a feature widely exploited in enzyme regulation. (Chapter 3)

allostery (adjective **allosteric**) Change in a protein's conformation brought about by the binding of a regulatory ligand (at a site other than the protein's catalytic site) or by covalent modification. The change in conformation alters the activity of the protein; it can also form the basis of directed movement. (Chapter 3) (Figures 3–71 and 16–53)

alpha helix (**α helix**) Common folding pattern in proteins, in which a linear sequence of amino acids folds into a right-handed helix stabilized by internal hydrogen-bonding between backbone atoms. (Chapter 3) (Figure 3–6)

alternative RNA splicing Production of different RNAs from the same gene by splicing the transcript in different ways. (Chapter 7) (Figure 7–59)

amino acid Organic molecule containing both an amino group and a carboxyl group. Those that serve as building blocks of proteins are alpha amino acids, having both the amino and carboxyl groups linked to the same carbon atom. (Chapter 1) (Figure 3–1 and Panel 3–1, pp. 118–119)

aminoacyl-tRNA synthetase Enzyme that attaches the correct amino acid to a tRNA molecule to form an aminoacyl-tRNA. (Chapter 6) (Figure 6–58)

amoeboid cell migration A rapid mode of cell locomotion typical of amoebae and white blood cells that is characterized by protrusion of actin-rich pseudopodia at the leading edge, formation of weak attachments to the substratum, and acto-myosin–based contraction at the rear of the cell. (Chapter 16)

AMPA receptor Glutamate-gated ion channel in the mammalian central nervous system that carries most of the depolarizing current responsible for excitatory postsynaptic potentials. (Chapter 11)

amphiphilic Having both hydrophobic and hydrophilic regions, as in a phospholipid or a detergent molecule. (Chapter 10)

amyloid fibrils Self-propagating, stable β-sheet aggregates built from hundreds of identical polypeptide chains that become layered one over the other to create a continuous stack of β sheets. The unbranched fibrous structure can contribute to human diseases when not controlled. (Chapter 3)

anaphase (1) Stage of mitosis during which sister chromatids separate and move away from each other. (2) In meiosis, anaphase I and II are the stages during which chromosome homolog pairs separate (I), and then sister chromatids separate (II). (Panel 17–1, pp. 1048–1049)

anaphase A The stage of mitosis during which chromosome segregation occurs as chromosomes move toward the two spindle poles. (Chapter 17)

anaphase B The stage of mitosis during which chromosome segregation occurs as spindle poles separate and move apart. (Chapter 17)

anaphase-promoting complex, or cyclosome (**APC/C**) Ubiquitin ligase that catalyzes the ubiquitylation and destruction of securin and M- and S-cyclins, initiating the separation of sister chromatids in the metaphase-to-anaphase transition during mitosis. (Chapter 17)

anchorage dependence Dependence of cell growth, proliferation, and survival on attachment to a substratum. (Chapter 19)

anchoring junction Cell junction that attaches cells to neighboring cells or to the extracellular matrix. (Chapter 19) (Table 19–1, p. 1107)

angiogenesis Growth of new blood vessels by sprouting from existing ones. (Chapter 20)

antenna complex Part of a photosystem that captures light energy and channels it into the photochemical reaction center. It consists of protein complexes that bind large numbers of chlorophyll molecules and other pigments. (Chapter 14)

***Antennapedia* complex** One of two gene clusters in *Drosophila* that contain *Hox* genes; genes in the *Antennapedia* complex control the differences among the thoracic and head segments of the body. (Chapter 21)

anti-apoptotic Bcl2 family proteins Proteins (for example, Bcl2, BclxL) on the cytosolic surface of the outer mitochondrial membrane that bind and inhibit pro-apoptotic Bcl2 family proteins and thereby help prevent inappropriate activation of the intrinsic pathway of apoptosis. (Chapter 18)

anti-IAP proteins Produced in response to various apoptotic stimuli and, by binding to IAPs (inhibitory antiapoptotic proteins), prevent their binding to a caspase—thereby blocking the inhibition of apoptosis provided by IAPs. (Chapter 18)

antibiotic Substance such as penicillin or streptomycin that is toxic to microorganisms. Often a natural product of a particular microorganism or plant. (Chapter 23)

antibody (or **immunoglobulin**) Protein secreted by activated B cells in response to a pathogen or foreign molecule. Binds tightly to the pathogen or foreign molecule, inactivating it or marking it for destruction by phagocytosis or complement-induced lysis. (Chapters 3, 24) (Figure 24–23)

antibody response Adaptive immune response in which B cells are activated to secrete antibodies that circulate in the bloodstream or enter other body fluids, where they can bind specifically to the foreign antigen that stimulated their production. (Chapter 24)

anticodon Sequence of three nucleotides in a transfer RNA (tRNA) molecule that is complementary to a three-nucleotide codon in a messenger RNA (mRNA) molecule. (Chapter 6)

antigen Any molecule that can bind specifically to an antibody or B cell receptor, or any protein fragment bound to an MHC protein that can bind specifically to a T cell receptor. (Chapters 3, 24)

antigen-presenting cell (**APC**) Cell that displays foreign antigen complexed with an MHC protein on its surface for presentation to T lymphocytes. (Chapter 24)

antigenic determinant Specific region of an antigen that binds to an antibody or a complementary receptor on the surface of a B cell (BCR) or T cell (TCR). (Chapter 24)

antigenic variation Ability to change the antigens displayed on the cell surface; a property of some pathogenic microorganisms that enables them to evade attack by the adaptive immune system. (Chapter 23)

antiparallel Describes the relative orientation of the two strands in a DNA double helix or two paired regions of a polypeptide chain; the chemical polarity of one strand is opposite to that of the other. (Chapter 4)

antiporter Carrier protein that transports two different ions or small molecules across a membrane in opposite directions, either simultaneously or in sequence. (Chapter 11) (Figure 11–8)

Apaf1 Adaptor protein of the intrinsic apoptotic pathway; on binding cytochrome *c*, oligomerizes to form an apoptosome. (Chapter 18)

apical Referring to the tip of a cell, a structure, or an organ. The apical surface of an epithelial cell is the exposed free surface, opposite to the basal surface. The basal surface rests on the basal lamina that separates the epithelium from other tissue. (Chapter 19)

apoptosis Form of programmed cell death, in which a "suicide" program is activated within an animal cell, leading to rapid cell death mediated by intracellular proteolytic enzymes called caspases. (Chapter 18)

apoptosome Heptamer of Apaf1 proteins that forms on activation of the intrinsic apoptotic pathway; it recruits and activates initiator caspases that subsequently activate downstream executioner caspases to induce apoptosis. (Chapter 18)

aquaporin (**water channel**) Channel protein embedded in the plasma membrane that greatly increases the cell's permeability to water, allowing transport of water, but not ions, at a high rate across the membrane. (Chapter 11)

archaea (**archaebacteria**) Single-celled organisms without a nucleus, superficially similar to bacteria. At a molecular level, more closely related to eukaryotes in genetic machinery than are bacteria. Archaea and bacteria together make up the prokaryotes. (Figure 1–9)

ARF proteins Monomeric GTPase in the Ras superfamily responsible for regulating both COPI coat assembly and clathrin coat assembly. (Chapter 13) (Table 15–5, p. 915)

Arp2/3 complex Complex containing two actin-related proteins that binds to an actin filament and then nucleates actin filament growth from the minus end. (Chapter 16)

arrestin Member of a family of proteins that contributes to GPCR desensitization by preventing the activated receptor from interacting with G proteins; also serves as an adaptor to couple the receptor to clathrin-dependent endocytosis. (Chapter 15) (Figure 15–43)

association constant (**K_a**) The equilibrium constant for a simple binding interaction, when expressed as [AB]/[A][B]. Also known as the *affinity constant*. (Chapter 3)

astral microtubule In the mitotic spindle, any of the microtubules radiating from the aster that are not attached to a kinetochore of a chromosome. (Chapter 17)

asymmetric cell division Cell division in which some important molecule or molecules are distributed unequally between the two daughter cells, causing these cells to become different from each other. (Figures 21–35 and 22–16)

ATM (**ataxia telangiectasia mutated protein**) Protein kinase activated by double-strand DNA breaks. If breaks are not repaired, ATM initiates a signal cascade that culminates in cell-cycle arrest. Related to ATR. (Chapter 17)

ATP (**adenosine 5'-triphosphate**) Nucleoside triphosphate composed of adenine, ribose, and three phosphate groups. The principal carrier of chemical energy in cells. The terminal phosphate groups are highly reactive in the sense that their hydrolysis, or transfer to another molecule, takes place with the release of a large amount of free energy. (Chapter 2) (Figure 2–33)

ATP synthase (**F_1F_o ATP synthase**) Abundant transmembrane enzyme complex in the inner membrane of mitochondria and the thylakoid membrane of chloroplasts. Driven by an electrochemical proton gradient, it catalyzes the formation of ATP from ADP and phosphate during oxidative phosphorylation and photosynthesis. Also present in the plasma membrane of bacteria. (Chapter 14)

ATR (**ataxia telangiectasia and Rad3 related protein**) Protein kinase activated by DNA damage. If damage remains unrepaired, ATR helps initiate a signal cascade that culminates in cell-cycle arrest. Related to ATM. (Chapter 17)

autoimmune disease Pathological state in which the body mounts a disabling adaptive immune response against one or more of its own molecules. (Chapter 24)

autophagosome Organelle surrounded by a double membrane that contains engulfed cytoplasmic cargo in the initial stages of autophagy. (Chapter 13)

autophagy Digestion of cytoplasm and worn-out organelles by the cell's own lysosomes. (Chapter 13)

auxin Plant hormone, indole-3-acetic acid, with numerous roles in plant growth and development. (Chapter 15)

axon Long nerve-cell projection that can rapidly conduct nerve impulses over long distances so as to deliver signals to other cells. (Chapter 11)

axoneme Bundle of microtubules and associated proteins that forms the core of a cilium or a flagellum in eukaryotic cells and is responsible for their movements. (Chapter 16)

B cell receptor (**BCR**) The transmembrane immunoglobulin protein on the surface of a B cell that serves as its receptor for antigen. (Chapter 24)

bacterial artificial chromosome (**BAC**) Cloning vector that can accommodate large pieces of DNA, typically up to 1 million base pairs. (Chapter 8)

bacteriorhodopsin Pigmented protein found in the plasma membrane of a salt-loving archaeon, *Halobacterium salinarum* (*Halobacterium halobium*). Pumps protons out of the cell in response to light. (Chapter 10)

bacterium (plural **bacteria**) (**eubacterium**) Member of the domain bacteria, one of the three main branches of the tree of life (archaea, bacteria, and eukaryotes). Bacteria and archaea both lack a distinct nuclear compartment and together comprise the prokaryotes. (Chapter 23) (Figure 1–9)

Bak A main effector Bcl2 family protein of the intrinsic pathway of apoptosis in mammalian cells that is bound to the mitochondrial outer membrane even in the absence of an apoptotic signal; activation is usually by activated pro-apoptotic BH3-only proteins. (Chapter 18)

basal Situated near the base. Opposite the apical surface. (Chapter 19)

basal lamina (plural **basal laminae**) Thin mat of extracellular matrix that separates epithelial sheets, and many other types of cells such as muscle or fat cells, from connective tissue. Sometimes called basement membrane. (Chapter 19) (Figure 19–51)

base (1) A substance that can reduce the number of protons in solution, either by accepting H^+ ions directly or by releasing OH^- ions, which then combine with H^+ to form H_2O. (Chapter 2) (2) The purines and pyrimidines in DNA and RNA are organic nitrogenous bases and are commonly referred to simply as bases. (Panel 2–2, pp. 96–97)

base excision repair DNA repair pathway in which single faulty bases are removed from the DNA helix and replaced. *Compare* **nucleotide excision repair**. (Chapter 5) (Figure 5–41)

base pair Two nucleotides in an RNA or DNA molecule that are held together by hydrogen bonds; for example, G paired with C, and A paired with T or U. (Chapter 4)

basement membrane Thin mat of extracellular matrix that separates epithelial sheets, and many other types of cells such as muscle or fat cells, from connective tissue. Also called basal lamina. (Chapter 19) (Figure 19–51)

Bax A main effector Bcl2 family protein of the intrinsic pathway of apoptosis in mammalian cells; located mainly in the cytosol, it translocates to the mitochondria only after activation, usually by activated pro-apoptotic BH3-only proteins. (Chapter 18)

Bcl2 Anti-apoptotic Bcl2 family protein of the outer mitochondrial membrane that binds and inhibits pro-apoptotic Bcl2 family proteins and prevents inappropriate activation of the intrinsic pathway of apoptosis. (Chapter 18)

Bcl2 family Family of intracellular proteins that either promote or inhibit apoptosis by regulating the release of cytochrome *c* and other mitochondrial proteins from the intermembrane space into the cytosol. (Chapter 18)

BclxL Anti-apoptotic Bcl2 family protein of the outer mitochondrial membrane that binds and inhibits pro-apoptotic Bcl2 family proteins and prevents inappropriate activation of the intrinsic pathway of apoptosis. (Chapter 18)

benign Of tumors: self-limiting in growth, and noninvasive. (Chapter 20)

beta sheet (β sheet) Common structural motif in proteins in which different sections of the polypeptide chain run alongside each other, joined together by hydrogen-bonding between atoms of the polypeptide backbone. Also known as a β pleated sheet. (Chapter 3) (Figure 3–6)

beta-catenin (β-catenin) Multifunctional cytoplasmic protein involved in cadherin-mediated cell–cell adhesion, linking cadherins to the actin cytoskeleton. Can also act independently as a transcription regulatory protein. Has an important role in animal development as part of a Wnt signaling pathway. (Chapter 15)

BH3-only proteins The largest subclass of Bcl2 family proteins. Produced or activated in response to an apoptotic stimulus, these proteins promote apoptosis mainly by inhibiting anti-apoptotic proteins in the Bcl2 family. (Chapter 18)

bi-orientation In mitosis, the attachment of sister chromatids to opposite poles of the mitotic spindle, so that they move to opposite ends of the cell when they separate in anaphase. (Chapter 17)

binding site Region on the surface of one molecule (usually a protein or nucleic acid) that can interact with another molecule through noncovalent bonding. (Chapter 3)

biomolecular condensate An aggregate inside cells, formed by a process analogous to liquid–liquid phase separation and based on fluctuating weak interactions between scaffold proteins; concentrates selected protein and RNA molecules in a membraneless compartment. (Chapters 1, 3, 12) (Figures 3–77 and 12–5)

BiP Endoplasmic reticulum (ER)-resident chaperone protein. Member in the family of hsp70-type chaperones. (Chapter 12)

***Bithorax* complex** One of two gene clusters in *Drosophila* that contain *Hox* genes; genes in the *Bithorax* complex control the differences among the abdominal and thoracic segments of the body. (Chapter 21)

bivalent A four-chromatid structure formed during meiosis, consisting of a duplicated chromosome tightly paired with its homologous duplicated chromosome. (Chapter 17)

blastomere One of the many similar cells formed by the early cleavages of a fertilized egg. (Chapter 21)

blastula Early stage of an animal embryo, usually consisting of a hollow ball of epithelial cells surrounding a fluid-filled cavity, before gastrulation begins. (Chapter 21)

blebbing Membrane protrusion formed when the plasma membrane detaches locally from the underlying actin cortex, allowing cytoplasmic flow and hydrostatic pressure within the cell to push the membrane outward. (Chapter 16)

bone Dense and rigid connective tissue comprising a mixture of tough fibers (type I collagen fibrils), which resist pulling forces, plus solid particles (calcium phosphate as hydroxylapatite crystals), which resist compression. (Chapter 21)

brassinosteroids Class of steroid signal molecules in plants that regulate the growth and differentiation of plants throughout their life cycle via binding to a cell-surface receptor kinase to initiate a signaling cascade. (Chapter 15)

bright-field microscope Normal light microscope in which the image is obtained by simple transmission of light through the object being viewed. (Chapter 9)

Brownian motion The random movement of particles or molecules suspended in a liquid or gas, caused by molecular collisions. (Chapter 1)

buffer Solution of weak acid or weak base that resists the pH change that would otherwise occur when small quantities of acid or base are added. (Chapter 2)

C3 The pivotal complement protein that is activated by the early components of all three complement pathways (the classical pathway, the lectin pathway, and the alternative pathway). (Figure 24–7)

Ca^{2+} pump (calcium pump, Ca^{2+} ATPase) Transport protein in the membrane of sarcoplasmic reticulum of muscle cells (and elsewhere). Pumps Ca^{2+} out of the cytoplasm into the sarcoplasmic reticulum using the energy of ATP hydrolysis. (Chapter 11)

Ca^{2+}-activated K^+ channel Opens in response to the raised concentration of Ca^{2+} in nerve cells that occurs in response to an action potential. Increased K^+ permeability makes the membrane harder to depolarize, increasing the delay between action potentials and decreasing the response of the cell to constant, prolonged stimulation (adaptation). (Chapter 11)

Ca^{2+}/calmodulin-dependent kinase (CaM-kinase) Serine/threonine protein kinase that is activated by Ca^{2+}/calmodulin. Indirectly mediates the effects of an increase in cytosolic Ca^{2+} by phosphorylating specific target proteins. (Chapter 15) (Figure 15–34)

cadherin Member of the large cadherin superfamily of transmembrane adhesion proteins. Mediates homophilic Ca^{2+}-dependent cell–cell adhesion in animal tissues. (Chapter 19) (Figure 19–3 and Table 19–1, p. 1107)

cadherin superfamily Family of classical and nonclassical cadherin proteins with more than 180 members in humans. (Chapter 19)

calmodulin Ubiquitous intracellular Ca^{2+}-binding protein that undergoes a large conformation change when it binds Ca^{2+}, allowing it to regulate the activity of many target proteins. In its activated (Ca^{2+}-bound) form, it is called Ca^{2+}/calmodulin. (Chapter 15) (Figure 15–34)

calnexin Carbohydrate-binding chaperone protein in the endoplasmic reticulum (ER) membrane; binds to oligosaccharides on incompletely folded proteins and retains them in the ER. (Chapter 12)

calreticulin Carbohydrate-binding chaperone protein in the endoplasmic reticulum (ER) lumen; binds to oligosaccharides on incompletely folded proteins and retains them in the ER. (Chapter 12)

CaM-kinase II Multifunctional Ca^{2+}/calmodulin-dependent protein kinase that phosphorylates itself and various target proteins when activated. Found in most animal cells but is especially abundant at synapses in the brain; it is involved in some forms of synaptic plasticity in vertebrates. (Chapter 15) (Figure 15–35)

cancer stem cells Rare cancer cells capable of dividing indefinitely. (Chapter 20)

cancer-critical genes Genes whose alteration contributes to the causation or evolution of cancer by driving tumorigenesis. (Chapter 20)

capsid Protein coat of a virus, formed by the self-assembly of one or more types of protein subunit into a geometrically regular structure. (Chapter 23) (Figure 3–27)

carbohydrate layer The carbohydrate-rich zone on the eukaryotic cell surface attributable to glycoproteins, glycolipids, and proteoglycans of the plasma membrane. (Chapter 10)

carbon-fixation reaction Process by which inorganic carbon (as atmospheric CO_2) is incorporated into organic molecules. The second stage of photosynthesis. (Chapter 14) (Figure 14–40)

carcinogenesis The generation of cancer. (Chapter 20)

carcinoma Cancer of epithelial cells. The most common form of human cancer. (Chapter 20)

cargo The membrane components and soluble molecules carried by transport vesicles. (Chapter 13)

cartilage Form of connective tissue composed of cells (chondrocytes) embedded in a matrix rich in type II collagen and chondroitin sulfate proteoglycan. (Chapters 19, 21)

caspase Intracellular protease that is involved in mediating the intracellular events of apoptosis. (Figures 18–3 and 18–4)

catalyst Substance that can lower the activation energy of a reaction (thus increasing its rate), without itself being consumed by the reaction. (Chapters 1, 2, 3)

caveola (plural **caveolae**) Invaginations at the cell surface that bud off internally to form pinocytic vesicles. Thought to form from lipid rafts, regions of membrane rich in certain lipids. (Chapter 13)

caveolins Family of unusual integral membrane proteins that are the major structural proteins in caveolae. (Chapter 13)

CD4 Co-receptor protein on helper T cells and regulatory T cells that binds to a nonvariable part of class II MHC proteins (on antigen-presenting cells) outside the peptide-binding groove. (Chapter 24) (Figure 24–40)

CD8 Co-receptor protein on cytotoxic T cells that binds to a nonvariable part of class I MHC proteins (on antigen-presenting cells and infected target cells) outside the peptide-binding groove. (Chapter 24) (Figure 24–40)

Cdc20 Activating subunit of the anaphase-promoting complex/cyclosome (APC/C). (Chapter 17)

Cdc25 Protein phosphatase that dephosphorylates Cdks and increases their activity. (Chapter 17)

Cdc42 Member of the Rho family of monomeric GTPases that regulate the actin and microtubule cytoskeletons, cell-cycle progression, gene transcription, and membrane transport. (Chapter 15)

Cdc6 Protein essential in the preparation of DNA for replication. With Cdt1 it binds to an origin recognition complex on chromosomal DNA and helps load the Mcm proteins onto the DNA. (Figure 5–31)

Cdh1 Activating subunit of the anaphase-promoting complex/cyclosome (APC/C). (Chapter 17)

Cdk inhibitor protein (CKI) Protein that binds to and inhibits cyclin–Cdk complexes, primarily involved in the control of G_1 and S phases. (Chapter 17)

Cdt1 Protein essential in the preparation of DNA for replication. With Cdc6 it binds to origin recognition complexes on chromosomes and helps load the Mcm proteins on to the DNA. (Chapter 17) (Figure 5–31)

cDNA clone Clone containing double-stranded cDNA molecules derived from the protein-coding mRNA molecules present in a cell. (Chapter 8)

cDNA library Collection of cloned DNA molecules representing complementary DNA copies of the mRNA produced by a cell. (Chapter 8)

cell cortex Specialized layer of cytoplasm on the inner face of the plasma membrane. In animal cells it is an actin-rich layer responsible for movements of the cell surface. (Chapter 16)

cell cycle (cell-division cycle) Reproductive cycle of a cell: the orderly sequence of events by which a cell duplicates its chromosomes and other cell contents, and then divides into two. (Chapters 4, 17) (Figure 17–4)

cell-cycle control system Network of regulatory proteins that governs progression of a eukaryotic cell through the cell cycle. (Chapter 17)

cell determination Process whereby a cell progressively loses the potential to form other cell types, as development proceeds. (Chapter 21)

cell doctrine The nineteenth-century proposal that all living organisms are composed of one or more cells and that all cells arise from the division of other living cells. (Chapter 9)

cell memory Retention by cells and their descendants of persistently altered patterns of gene expression, without any change in DNA sequence. *See also* **epigenetic inheritance**. (Chapters 7, 21)

cell plate Flattened membrane-bounded structure that forms by fusing vesicles in the cytoplasm of a dividing plant cell and is the precursor of the new cell wall. (Chapter 17)

cellulose Long, unbranched chains of glucose; major constituent of plant cell walls. (Chapter 19)

cellulose microfibril Highly ordered crystalline aggregate formed from bundles of about 18 cellulose chains, arranged with the same polarity and stuck together in overlapping parallel arrays by hydrogen bonds between adjacent cellulose molecules. (Chapter 19)

central (primary) lymphoid organ Organ in which T or B lymphocytes are produced from precursor cells. In adult mammals, these are the thymus and bone marrow, respectively. (Chapter 24) (Figure 24–12)

centriole Short cylindrical array of microtubules, closely similar in structure to a basal body. A pair of centrioles is usually found at the center of a centrosome in animal cells. (Chapter 16) (Figure 16–42)

centromere Constricted region of a mitotic chromosome that holds sister chromatids together. This is also the site on the DNA where the kinetochore forms so as to capture microtubules from the mitotic spindle. (Chapter 4) (Figure 4–43)

centrosome Centrally located organelle of animal cells that is the primary microtubule-organizing center (MTOC) and acts as the spindle pole during mitosis. In most animal cells it contains a pair of centrioles. (Chapter 17) (Figures 16–41 and 17–27)

cerebral cortex Outermost layer of the hemispheres of the brain; the most complex structure in the human body. (Figure 21–41)

CG island Region of DNA in vertebrate genomes with a greater than average density of CG sequences; the C nucleotides in these regions generally remain unmethylated. (Chapter 7)

channel (membrane channel) Transmembrane protein complex that allows inorganic ions or other small molecules to diffuse passively across the lipid bilayer. (Chapter 11) (Figure 11–3)

channelrhodopsin Photosensitive protein forming a cation channel across the membrane that opens in response to light. (Chapter 11)

charge separation In photosynthesis, the light-induced transfer of a high-energy electron from chlorophyll to an acceptor molecule resulting in the formation of a positive charge on the chlorophyll and a negative charge on a mobile electron carrier. (Figure 14-45)

chemical carcinogens Disparate chemicals that are carcinogenic—due to the ability to cause mutations—when fed to experimental animals or painted repeatedly on their skin. (Chapter 20)

chemical group Certain common combinations of atoms—such as methyl ($-CH_3$), hydroxyl ($-OH$), carboxyl ($-COOH$), carbonyl ($-C{=}O$), phosphate ($-PO_3^{2-}$), sulfhydryl ($-SH$), and amino ($-NH_2$) groups—that have distinct chemical and physical properties and influence the behavior of the molecule in which the group occurs. (Chapter 2)

chemiosmotic coupling (chemiosmosis) Mechanism in which an electrochemical proton gradient across a membrane (composed of a pH gradient plus a membrane potential) is used to drive an energy-requiring process, such as ATP production or the rotation of bacterial flagella. (Chapter 14)

chemotaxis Movement of a cell toward or away from some diffusible chemical. (Chapter 16)

chiasma (plural **chiasmata**) X-shaped connection visible between paired homologous chromosomes during meiosis. Represents a site of chromosomal crossing-over, a form of genetic recombination. (Chapter 17)

chlorophyll Light-absorbing green pigment that plays a central part in photosynthesis in bacteria, plants, and algae. (Chapter 14)

chloroplast Membrane-bounded organelle in green algae and plants that contains chlorophyll and carries out photosynthesis. (Chapters 12, 14)

cholera toxin Secreted toxic protein of *Vibrio cholerae* responsible for causing the watery diarrhea associated with cholera. Comprises an A subunit with enzymatic activity and a B subunit that binds to host-cell receptors to direct subunit A to the host-cell cytosol. (Chapter 23)

cholesterol An abundant lipid molecule with a characteristic four-ring steroid structure. An important component of the plasma membranes of animal cells. (Chapter 10) (Figure 10–4)

chromatin Complex of DNA, histones, and non-histone proteins found in the nucleus of a eukaryotic cell. The material of which chromosomes are made. (Chapter 4)

chromatin immunoprecipitation Technique by which chromosomal DNA bound by a particular protein can be isolated and identified by precipitating it by means of an antibody against the protein. (Chapter 8) (Figures 8–67 and 8–68)

chromosome Structure composed of a very long DNA molecule and associated proteins that carries part (or all) of the hereditary information of an organism. Especially evident in plant and animal cells undergoing mitosis or meiosis, during which each chromosome becomes condensed into a compact rodlike structure visible in the light microscope. (Chapter 4)

cilium (plural **cilia**) Hairlike extension of a eukaryotic cell containing a core bundle of microtubules. Many cells contain a single nonmotile cilium, while others contain large numbers that perform repeated beating movements. *Compare* **flagellum**. (Chapter 16)

circadian clock Internal cyclical process that produces a particular change in a cell or organism with a period of around 24 hours; for example, the sleep–wakefulness cycle in humans. (Chapter 15)

***cis* face** Face on the same or near side. (Chapter 13)

***cis* Golgi network (CGN)** Network of fused vesicular tubular clusters that is closely associated with the *cis* face of the Golgi apparatus and is the compartment at which proteins and lipids enter the Golgi from the ER. (Chapter 13)

***cis*-regulatory sequences** DNA sequences to which transcription regulators bind to control the rate of gene transcription. In nearly all cases, these sequences must be on the same chromosome (that is, *in cis*) to the genes they control. (Chapter 7) (Figure 7–18)

cisternal maturation mechanism One hypothesis for how the Golgi apparatus achieves and maintains its polarized structure and how molecules move from one cisterna to another. This model views the cisternae as dynamic structures that mature from early to late by acquiring and then losing specific Golgi-resident proteins as they move through the Golgi stack with cargo. (Chapter 13)

citric acid cycle [tricarboxylic acid (TCA) cycle; Krebs cycle] Central metabolic pathway found in aerobic organisms. Oxidizes acetyl groups derived from food molecules, generating the activated carriers NADH and $FADH_2$, some GTP, and waste CO_2. In eukaryotic cells, it occurs in the mitochondria. (Chapter 2) (Panel 2–9, pp. 110–111)

clamp loader Protein complex that utilizes ATP hydrolysis to load the sliding clamp onto a primer–template junction in the process of DNA replication. (Chapter 5)

class I MHC protein One of two classes of major histocompatibility complex (MHC) protein. Found on the surface of almost all vertebrate cell types, where it can present foreign peptides derived from a pathogen such as a virus to cytotoxic T cells. (Chapter 24) (Figures 24–35 and 24–36A)

class II MHC protein One of two classes of major histocompatibility complex (MHC) protein. Found on the surface of various antigen-presenting cells, where it presents peptides to helper and regulatory T cells. (Chapter 24) (Figures 24–35 and 24–36B)

class switch recombination An irreversible change at the DNA level when a B cell switches from making IgM and IgD to making one of the secondary classes of immunoglobulin. (Chapter 24)

class switching Change from making one class of immunoglobulin (for example, IgM) to making another class (for example, IgG) that many B cells undergo during the course of an adaptive immune response. Involves DNA rearrangements called class switch recombination. (Chapter 24) (Figure 24–30)

classical cadherins Family of cadherin proteins, including E-cadherin, N-cadherin, and P-cadherin, that are closely related in sequence throughout their extracellular and intracellular domains. (Chapter 19)

clathrin Protein that assembles into a polyhedral cage on the cytosolic side of a membrane so as to form a clathrin-coated pit, which buds off by endocytosis to form an intracellular clathrin-coated vesicle. (Chapter 13) (Figure 13–6)

clathrin-coated pits Specialized regions typically occupying about 2% of the total plasma membrane area at which the endocytic pathway often begins. (Chapter 13)

clathrin-coated vesicles Coated vesicles inside the cell that transport material from the plasma membrane and between endosomal and Golgi compartments. (Chapter 13)

cleave, cleavage (1) Physical splitting of a cell into two. (2) Specialized type of cell division seen in many early embryos whereby a large cell becomes subdivided into many smaller cells without growth. (Chapter 21)

clonal selection The process whereby, from a population of T and B lymphocytes with a vast repertoire of randomly generated antigen-specific receptors, a given foreign antigen activates (selects) only those lymphocyte clones that display a receptor that fits the antigen. Explains how the adaptive immune system can respond to millions of different antigens in a highly specific way. (Chapter 24) (Figure 24–15)

co-receptor In immunology: an accessory receptor on B cells or T cells that does not bind antigen but binds to a co-stimulatory signal and helps activate the lymphocyte, by helping to activate an intracellular signaling pathway. (Chapter 24)

co-stimulatory signal In immunology: a secreted or membrane-bound signal protein that helps activate an antigen-responding B cell or T cell. (Chapter 24)

co-translational Occurring as translation proceeds. Examples include the import of a protein into the endoplasmic reticulum before the polypeptide chain is completely synthesized (co-translational translocation; Figure 12–24), and the folding of a nascent protein into its secondary and tertiary structure as it emerges from a ribosome (Figure 6–83C). (Chapter 12)

coat-recruitment GTPases Members of a family of monomeric GTPases that have important roles in vesicle transport, being responsible for coat assembly at the membrane. (Chapter 13)

coated vesicle Small membrane-enclosed organelle with a cage of proteins (the coat) on its cytosolic surface. Formed by the pinching off of a coated region of membrane (coated pit). Some coats are made of clathrin, others are made from other proteins. (Chapter 13)

codon Sequence of three nucleotides in a DNA or mRNA molecule that represents the instruction for incorporation of a specific amino acid into a growing polypeptide chain. (Chapter 6)

coenzyme Small molecule tightly associated with an enzyme that participates in the reaction that the enzyme catalyzes, often by forming a covalent bond to the substrate. Examples include biotin, NAD^+, and coenzyme A. (Chapter 3)

cohesin, cohesin complex Complex of proteins that uses ATP hydrolysis energy to organize an interphase chromosome into a series of looped domains; during mitosis, cohesins also hold sister chromatids together along their length before their separation. (Chapters 4, 17) (Figures 4–57 and 17–23)

coiled-coil Especially stable rodlike protein structure formed by two or more α helices coiled around each other. (Chapter 3) (Figure 3–8)

collagen Fibrous protein rich in glycine and proline that is a major component of the extracellular matrix in animals, conferring tensile strength. Exists in many forms: type I, the most common, is found in skin, tendon, and bone; type II is found in cartilage; type IV is present in basal laminae. (Chapter 19) (Figures 3–24 and 19–38)

collagen fibril A higher-order collagen polymer of fibrillar collagens that assemble into thin structures (10–300 nm in diameter) many hundreds of micrometers long in mature tissues. (Chapter 19)

colony-stimulating factor (CSF) General name for numerous signal molecules that control differentiation of blood cells. (Chapter 15)

colorectal cancer Cancer arising from the epithelium lining the colon (the large intestine) and rectum (the terminal segment of the gut). (Chapter 20)

column chromatography Technique for separation of a mixture of substances in solution by passage through a column containing a porous solid matrix. Substances are retarded to different extents by their interaction with the matrix and can be collected separately from the column. Depending on the matrix, separation can be according to charge, hydrophobicity, size, or the ability to bind to other molecules. (Chapter 8)

commensalism Ecologic relationship between microbes and their host in which the microbe benefits but offers no benefit and causes no harm. (Chapter 23)

committed progenitor Cell derived from a stem cell that divides for a limited number of times before terminally differentiating; also known as a transit amplifying cell.

complement system System of blood proteins that can be activated by antibody–antigen complexes or pathogens to help eliminate the pathogens by directly causing their lysis, by promoting their phagocytosis, or by activating an inflammatory response. (Chapter 24) (Figure 24–7)

complementary (1) Of nucleic acid sequences: capable of forming a perfect base-paired duplex with each other (Figure 4–5). (2) Of other interacting molecules, such as an enzyme and its substrate: having biochemical or structural features that marry up, so that noncovalent bonding is facilitated. (Chapter 4) (Figure 2–4)

complementation test Test to determine whether two mutations that produce similar phenotypes are in the same or different genes. (Chapter 8) (Panel 8–1, pp. 520–521)

complex oligosaccharides Broad class of *N*-linked oligosaccharides, attached to mammalian glycoproteins in the endoplasmic reticulum and modified in the Golgi apparatus, containing *N*-acetylglucosamine, galactose, sialic acid, and fucose residues. (Chapter 13)

condensin, condensin complex Complex of proteins involved in chromosome condensation prior to mitosis. Target for M-Cdk. (Chapters 4, 17) (Figure 17–25)

conditional mutation Mutation that changes a protein or RNA molecule so that its function is altered only under some conditions, such as at an unusually high or unusually low temperature. (Chapter 8)

cone photoreceptor (cone) Photoreceptor cell in the vertebrate retina that is responsible for color vision in bright light. (Chapter 15)

confocal microscope Type of light microscope that produces a clear image of a given plane within a solid object. It uses a laser beam as a pinpoint source of illumination and scans across the plane to produce a two-dimensional "optical section." (Chapter 9) (Figure 9–24)

conformation The folded, three-dimensional structure of a polypeptide chain. (Chapter 3)

connective tissue Any supporting tissue that lies between other tissues and consists of cells embedded in a relatively large amount of extracellular matrix. Includes bone, cartilage, and loose connective tissue. (Chapter 19)

connexin Protein component of gap junctions, a four-pass transmembrane protein. Six connexins assemble in the plasma membrane to form a connexon, or "hemichannel." (Chapter 19) (Figure 19–25)

connexon Water-filled pore in the plasma membrane formed by a ring of six connexin protein subunits. Half of a gap junction: connexons from two adjoining cells join to form a continuous channel through which ions and small molecules can pass. (Chapter 19) (Figure 19–25)

consensus nucleotide sequence A summary or "average" of a large number of individual nucleotide sequences derived by comparing many sequences with the same basic function and tallying up the most common nucleotides found at each position. (Chapter 6) (Figure 6–12)

conservative site-specific recombination A type of DNA recombination that takes place between short, specific sequences of DNA and occurs without the gain or loss of nucleotides. Unlike homologous recombination, it does not require an extensive region of homology between the two recombining DNA molecules. (Chapter 5)

constant region In immunology: region of an immunoglobulin or T cell receptor chain that has a constant amino acid sequence. (Chapter 24)

constitutive secretory pathway Pathway present in all cells by which molecules such as plasma membrane proteins are continually delivered to the plasma membrane from the Golgi apparatus in vesicles that fuse with the plasma membrane. The default route to the plasma membrane if no other sorting signals are present. (Chapter 13) (Figure 13–38)

contact-dependent secretion system Specialized bacterial systems that secrete effector proteins directly into host cell targets. (Chapter 23) (Figure 23–7)

contact-dependent signaling Form of intercellular signaling in which signal molecules remain bound to the surface of the signaling cell and influence only cells that contact it. (Chapter 15)

contractile ring Ring containing actin and myosin that forms under the surface of animal cells undergoing cell division. It contracts to pinch the two daughter cells apart. (Chapter 17) (Figure 17–43)

convergent extension Rearrangement of cells within a tissue that causes it to extend in one dimension and shrink in another. (Chapter 21) (Figure 21–48)

COPI-coated vesicles Coated vesicles that transport material in the secretory pathway, budding from Golgi compartments. (Chapter 13)

COPII-coated vesicles Coated vesicles that transport material early in the secretory pathway, budding from the endoplasmic reticulum. (Chapter 13)

copy number variations (CNVs) A difference between two individuals in the same population in the number of copies of a particular block of DNA sequence. This variation arises from occasional duplications and deletions of these sequences.

cortex The cytoskeletal network in the cortical region of the cytosol just beneath the plasma membrane. (Chapter 10)

coupled reaction Linked pair of chemical reactions in which the free energy released by one reaction serves to drive the other. (Chapter 2) (Figure 2–29)

covalent bond Stable chemical link between two atoms produced by sharing one or more pairs of electrons. (Chapter 2) (Panel 2–1, pp. 94–95)

CRE-binding (CREB) protein Transcription regulator that recognizes the cyclic AMP response element (CRE) in the regulatory region of genes activated by cAMP. On activation by PKA, phosphorylated CREB recruits a transcriptional coactivator (CREB-binding protein; CBP) to stimulate transcription of target genes. (Chapter 15) (Figure 15–28)

CRISPR A defense mechanism in bacteria using small noncoding RNA molecules (crRNAs) to seek out and destroy invading viral genomes through complementary base-pairing and targeted nuclease digestion. (Chapter 7)

crista (plural **cristae**) A specialized invagination of the inner mitochondrial membrane. (Chapter 14)

cross-linking glycan One of a heterogeneous group of branched polysaccharides that help to cross-link cellulose microfibrils into a complex network. Has a long linear backbone of one sugar type (glucose, xylose, or mannose) with short side chains of other sugars. (Chapter 19)

cross-presentation A process in which extracellular proteins taken up by specialized dendritic cells can give rise to peptides that can be presented by their class I MHC proteins to cytotoxic T cells. (Chapter 24)

crRNAs Small noncoding RNAs (~30 nucleotides) that are the effectors of CRISPR-mediated immunity in bacteria. (Chapter 7)

cryo-electron microscopy Technique for examining a thin film of an aqueous suspension of biological material that has been frozen rapidly enough to create vitreous ice. The specimen is then kept frozen and transferred to the electron microscope. Image contrast is low, but this type of microscopy permits the determination of atomic-level structures because the image is generated solely by the macromolecular structures present. (Chapters 9, 12)

cryptochrome Plant flavoprotein sensitive to blue light. Structurally related to blue-light-sensitive enzymes called photolyases (involved in the repair of ultraviolet-induced DNA damage). Also found in animals, where they have an important role in circadian clocks. (Chapter 15)

cyclic AMP (cAMP) Nucleotide that is generated from ATP by adenylyl cyclase in response to various extracellular signals. It acts as a small intracellular signaling molecule, mainly by activating cAMP-dependent protein kinase (PKA). It is hydrolyzed to AMP by a phosphodiesterase. (Chapter 15) (Figure 15–25)

cyclic-AMP-dependent protein kinase (protein kinase A; PKA) Enzyme that phosphorylates target proteins in response to a rise in intracellular cyclic AMP. (Chapter 15) (Figure 15–27)

cyclic AMP phosphodiesterase Specific enzyme that rapidly and continually destroys cyclic AMP, forming 5′-AMP. (Chapter 15) (Figure 15–26)

cyclic GMP (cGMP) Nucleotide that is generated from GTP by guanylyl cyclase in response to various extracellular signals. (Chapter 15)

cyclic GMP phosphodiesterase Specific enzyme that rapidly hydrolyzes and degrades cyclic GMP. (Chapter 15)

cyclin Protein that periodically rises and falls in concentration in step with the eukaryotic cell cycle. Cyclins activate crucial protein kinases (called cyclin-dependent protein kinases, or Cdks) and thereby help control progression from one stage of the cell cycle to the next. (Chapter 17)

cyclin–Cdk complex Protein complex formed periodically during the eukaryotic cell cycle as the level of a particular cyclin increases. A cyclin-dependent kinase (Cdk) thereby becomes activated. (Chapter 17) (Figures 17–9 and 17–10, and Table 17–1, p. 1034)

cyclin-dependent kinase (Cdk) Protein kinase that has to be complexed with a cyclin protein in order to act. Different cyclin–Cdk complexes trigger different steps in the cell-division cycle by phosphorylating specific target proteins. (Chapter 17) (Figure 17–9)

cyclosome *See* **anaphase-promoting complex**. (Chapter 17)

cytochrome Family of colored heme-containing proteins that transfer electrons during respiration and photosynthesis. (Chapter 14)

cytochrome *c* Soluble component of the mitochondrial electron-transport chain. As a second type of function, its release into the cytosol from the mitochondrial intermembrane space initiates apoptosis. (Chapters 14, 18)

cytochrome *c* oxidase complex Third of the three electron-driven proton pumps in the respiratory chain. It accepts electrons from cytochrome *c* and generates water using molecular oxygen as an electron acceptor. (Chapter 14) (Figure 14–18)

cytochrome *c* reductase Second of the three electron-driven proton pumps in the respiratory chain. Accepts electrons from ubiquinone and passes them to cytochrome *c*. (Chapter 14) (Figure 14–18)

cytokine Extracellular signal protein or peptide that acts as a local mediator in cell–cell communication. (Chapter 24)

cytokine receptor Cell-surface receptor that binds a specific cytokine or hormone and acts through the JAK–STAT signaling pathway. (Chapter 15) (Figure 15–57)

cytokinesis Division of the cytoplasm of a plant or animal cell into two, as distinct from the associated division of its nucleus (which is mitosis). Part of M phase. (Chapter 17) (Panel 17–1, pp. 1048–1049)

cytoplasm Contents of a cell that are contained within its plasma membrane but, in the case of eukaryotic cells, outside the nucleus. (Chapter 12)

cytoplasmic tyrosine kinase Enzyme activated by certain cell-surface receptors (tyrosine-kinase-associated receptors) that transmits the receptor signal onward by phosphorylating target cytoplasmic proteins on tyrosine side chains. (Chapter 15)

cytoskeleton System of protein filaments in the cytoplasm of a eukaryotic cell that gives the cell shape and the capacity for directed movement. Its most abundant components are actin filaments, microtubules, and intermediate filaments. (Chapter 16)

cytosol Contents of the main compartment of the cytoplasm, defined as excluding membrane-bounded organelles such as endoplasmic reticulum and mitochondria. (Chapter 12)

cytotoxic T cell (T_C cell) Type of T cell responsible for killing host cells infected with a virus or another type of intracellular pathogen. (Chapter 24) (Figure 24–42)

***D* gene segment** A short DNA sequence that encodes a part of the variable region of an immunoglobulin heavy chain or the β chain of a T cell receptor (TCR). (Chapter 24)

dark-field microscopy Type of light microscopy in which oblique rays of light focused on the specimen do not enter the objective lens, but light that is scattered by components in the living cell can be collected to produce a bright image on a dark background. (Chapter 9) (Figure 9–6)

death-inducing signaling complex (DISC) Complex in which initiator caspases interact and are activated after extracellular ligands bind to cell-surface death receptors in the extrinsic pathway of apoptosis. (Chapter 18)

death receptor Transmembrane receptor protein that can signal the cell to undergo apoptosis when it binds its extracellular ligand. (Chapter 18) (Figure 18–6)

default pathway The transport pathway of proteins directly to the cell surface via the nonselective constitutive secretory pathway, entry into which does not require a particular signal. (Chapter 13)

defensin Positively charged, amphipathic, antimicrobial peptide—secreted by epithelial cells—that binds to and disrupts the membranes of many pathogens. (Chapter 24)

delayed K^+ channel Neuronal voltage-gated K^+ channel that opens after membrane depolarization during the falling phase of an action potential, being delayed because of its slower activation kinetics than that of Na^+ channels. Its opening causes a K^+ efflux that drives the membrane potential back toward its original negative value, enabling it to transmit a second impulse. (Chapter 11)

Delta Single-pass transmembrane signal protein displayed on the surface of cells that binds to the Notch receptor protein on a neighboring cell, activating a contact-dependent signaling mechanism. (Chapter 15)

dendrite Extension of a nerve cell, often elaborately branched, that receives stimuli from other nerve cells. (Chapter 11)

dendritic cell The most potent type of antigen-presenting cell, which takes up antigen and processes it for presentation to T cells. It is required for activating naïve T cells. (Chapter 24) (Figure 24–11)

deoxyribonucleic acid (DNA) Polynucleotide formed from covalently linked deoxyribonucleotide units. The store of hereditary information within a cell and the carrier of this information from generation to generation. (Figure 4–3 and Panel 2–6, pp. 104–105)

depolarization Deviation in the electrical potential across the plasma membrane toward a positive value. A depolarized cell has a potential that is positive outside and negative inside. (Chapter 11)

desensitization *See* **adaptation**. (Chapter 15)

desmosome Anchoring cell–cell junction, usually formed between two epithelial cells. Characterized by dense plaques of protein into which intermediate filaments in the two adjoining cells insert. (Chapter 19) (Figure 19–2)

detergent Small amphiphilic molecule, more soluble in water than in lipids, that disrupts hydrophobic associations and destroys the lipid bilayer thereby solubilizing membrane proteins. (Chapter 10)

diacylglycerol (DAG) Lipid produced by the cleavage of inositol phospholipids in response to extracellular signals. Composed of two fatty acid chains linked to glycerol, it serves as a small signaling molecule to help activate protein kinase C (PKC). (Chapter 15) (Figure 15–29)

dideoxy sequencing The standard enzymatic method of DNA sequencing. (Chapter 8)

differential-interference-contrast microscope Type of light microscope that exploits the interference effects that occur when light passes through parts of a cell of different refractive indices. Used to view unstained living cells. (Chapter 9)

differentiation Process by which a cell undergoes a change to an overtly specialized cell type. (Chapter 21)

diffusion The net drift of molecules through space due to random thermal movements. (Chapter 2)

Dishevelled Scaffold protein recruited to the Frizzled family of cell-surface receptors upon their activation by Wnt binding that helps relay the signal to other signaling molecules. (Chapter 15)

dissociation constant The reciprocal of the association constant, with units of moles/liter. (Chapter 3)

DNA cloning (1) The act of making many identical copies (typically billions) of a DNA molecule—the amplification of a particular DNA sequence. (2) Also, the isolation of a particular stretch of DNA (often a particular gene) from the rest of the cell's genome. (Chapter 8)

DNA helicase Enzyme that harnesses ATP hydrolysis energy to open a region of the DNA helix into its single strands as an aid to DNA replication or DNA repair. (Chapter 5)

DNA library Collection of cloned DNA molecules, representing either an entire genome (genomic library) or complementary DNA copies of the mRNA produced by a cell (cDNA library). (Chapter 8)

DNA ligase Enzyme that joins the ends of two strands of DNA together with a covalent bond to produce one continuous DNA strand. (Chapter 5)

DNA methylation Addition of methyl groups to DNA. Extensive methylation of the cytosine base in CG sequences is used in plants and animals to help keep genes in an inactive state. (Chapter 7)

DNA-only transposon Transposable element that exists as DNA throughout its life cycle. Many of these elements move by cut-and-paste transposition. *See also* **transposon**. (Chapter 5)

DNA polymerase Enzyme that synthesizes DNA by joining nucleotides together using a DNA template as a guide; its substrates are the four nucleoside triphosphates: A, G, C, and T. (Chapter 5)

DNA primase Enzyme that synthesizes a short strand of RNA on a DNA template, producing an RNA primer for DNA synthesis. (Chapter 5) (Figure 5–10)

DNA repair A set of different enzymatic processes for repairing the many accidental lesions that occur continually in DNA. (Chapter 5)

DNA replication Process by which a copy of a DNA molecule is made. (Chapter 1)

DNA supercoiling A conformation with loops or coils that DNA adopts in response to superhelical tension; conversely, creating various loops or coils in the helix can create such tension. (Chapter 6)

DNA topoisomerase (topoisomerase) Enzyme that binds to DNA and reversibly breaks a phosphodiester bond in one or both strands. Topoisomerase I creates transient single-strand breaks, allowing the double helix to swivel and relieving superhelical tension. Topoisomerase II creates transient double-strand breaks, allowing one double helix to pass through another and thus resolving knots and tangles. (Chapter 5) (Figures 5–22 and 5–23)

DNA transcription *See* **transcription**. (Chapter 6)

DNA tumor virus General term for a variety of different DNA viruses that can cause tumors. (Chapter 20)

dolichol Isoprenoid lipid molecule that anchors the precursor oligosaccharide in the endoplasmic reticulum membrane during protein glycosylation. (Chapter 12)

domain (protein domain) Portion of a protein that has a tertiary structure of its own. Larger proteins are generally composed of several domains, each connected to the next by short flexible regions of polypeptide chain. Homologous domains are recognized in many different proteins. (Chapter 3)

Dorsal Transcription regulator of the NFκB family regulating gene expression and involved in establishing the dorsoventral axis in an embryo. (Chapter 21)

double helix The three-dimensional structure of DNA, in which two antiparallel DNA chains, held together by hydrogen-bonding between the bases, are wound into a helix. (Chapter 4) (Figure 4–5)

drivers Mutations that are causal factors in the development of cancer. (Chapter 20)

dynamic instability Sudden conversion from growth to shrinkage, and vice versa, in a protein filament such as a microtubule or actin filament. (Chapter 16) (Panel 16–2, pp. 960–961, and Figure 16–40)

dynamin Cytosolic GTPase that binds to the neck of a clathrin-coated vesicle in the process of budding from the membrane and which is involved in completing vesicle formation. (Chapter 13)

dynein Large motor protein that undergoes ATP-dependent movement along microtubules. (Chapter 16)

E2F protein Transcription regulatory protein that switches on many genes that encode proteins required for entry into the S phase of the cell cycle. (Chapter 17)

early endosome Common receiving compartment with which most endocytic vesicles fuse and where internalized cargo is sorted either for return to the plasma membrane or for degradation by inclusion in a late endosome. (Chapter 13)

ectoderm Embryonic epithelial tissue that is the precursor of the epidermis and nervous system. (Chapter 21)

edema factor One of the two A subunits of anthrax toxin; an adenylyl cyclase that catalyzes production of cAMP, leading to ion imbalance and consequent edema in the skin or lung. (Chapter 23)

effector cell Cell that carries out the final response or function in a particular process. The main effector cells of the immune system, for example, are activated lymphocytes and phagocytes that help eliminate pathogens. (Chapter 24)

egg-polarity genes Genes in the *Drosophila* egg that define the anteroposterior and dorsoventral axes of the future embryo through the creation of landmarks (mRNA or protein) in the egg that provide signals organizing the developmental process. (Chapter 21)

elastic fiber Extensible fiber formed by the protein elastin in many animal connective tissues, such as in skin, blood vessels, and lungs, which gives them their stretchability and resilience. (Chapter 19)

elastin Extracellular protein that forms extensible fibers (elastic fibers) in connective tissues. (Chapters 3, 19)

electrochemical gradient Combined influence of a difference in the concentration of an ion on two sides of a membrane and the electrical charge difference across the membrane (membrane potential). Ions or charged molecules can move passively only down their own electrochemical gradient. (Chapter 11)

electrochemical proton gradient The driving force for the production of ATP by ATP synthase. *See also* **electrochemical gradient**. (Chapter 14)

electron microscope Microscope that uses a beam of electrons to create the image. (Chapter 9)

electron microscope (EM) tomography Technique for viewing three-dimensional specimens in the electron microscope in which multiple views are taken from different directions by tilting the specimen holder. The views are combined computationally to give a three-dimensional image. (Chapter 9)

electron-transport chain Series of reactions in which electron carrier molecules pass electrons "down the chain" from higher to successively lower energy levels. The energy released during such electron movement can be used to power various processes. Electron-transport chains present in the inner mitochondrial membrane (called the respiratory chain) and in the thylakoid membrane of chloroplasts generate an electrochemical proton gradient across the membrane that is used to drive ATP synthesis. See especially Figures 14–18 and 14–52. (Chapters 2, 14)

electrostatic attraction A noncovalent, ionic attraction between two molecules carrying groups of opposite charge. (Chapter 2) (Panel 2–3, pp. 98–99)

embryonic stem cells (ES cells) Cells derived from the inner cell mass of the early mammalian embryo. Capable of giving rise to all the cells in the body. Can be grown in culture, genetically modified, and inserted into a blastocyst to develop a transgenic animal. (Chapter 22)

endocrine cell Specialized animal cell that secretes a hormone into the blood. Usually part of a gland, such as the thyroid or pituitary gland. (Chapter 15)

endocycle Variation of the cell cycle in which multiple rounds of DNA replication occur without intervening M phases. (Chapter 17)

endocytic vesicle Vesicle formed as material ingested by the cell during endocytosis is enclosed by a small portion of the plasma membrane, which first invaginates and then pinches off to form the vesicle. (Chapter 13)

endocytosis Uptake of material into a cell by an invagination of the plasma membrane and its internalization in a membrane-enclosed vesicle. *See also* **pinocytosis** and **phagocytosis**. (Chapter 13)

endoderm Embryonic tissue that is the precursor of the gut and associated organs. (Chapter 21)

endoplasmic reticulum (ER) Extensive, net-like membrane-bounded compartment in the cytoplasm of eukaryotic cells, where lipids are synthesized and membrane-bound proteins and secretory proteins are made. (Chapter 12) (Figure 12–15)

endosome maturation Process by which early endosomes mature to late endosomes and endolysosomes; in the conversion process, the endosome membrane protein composition changes, the endosome moves away from the cell periphery and close to the nucleus, and the endosome ceases to recycle material to the plasma membrane and irreversibly commits its remaining contents to degradation. (Chapter 13)

endothelial cell Flattened cell type that forms a sheet (the endothelium) lining all blood and lymphatic vessels. (Chapter 21)

engulfment Process by which a portion of the cytoplasm becomes enclosed by a double membrane, such as during autophagy. (Chapter 12)

entropy (*S*) Thermodynamic quantity that measures the degree of disorder or randomness in a system; the higher the entropy, the greater the disorder. (Chapter 2) (Panel 2–7, pp. 106–107)

enveloped virus Virus with a capsid surrounded by a lipid bilayer membrane (the envelope), which is often derived from the host-cell plasma membrane when the virus buds from the cell. (Chapter 23) (Figure 23–12)

enzyme Protein that catalyzes a specific chemical reaction. (Chapters 2, 3)

enzyme-coupled receptor A major type of cell-surface receptor that has a cytoplasmic domain that either has enzymatic activity or is associated with an intracellular enzyme. In either case, the enzymatic activity is stimulated by an extracellular ligand binding to the receptor. (Chapter 15) (Figure 15–6)

ephrin One of a family of membrane-bound protein ligands for the Eph receptor tyrosine kinases (RTKs) that, among many other functions, stimulate repulsion or attraction responses that guide the migration of cells and nerve cell axons during animal development. (Chapter 15)

epidermis Epithelial layer covering the outer surface of the body. Has different structures in different animal groups. The outer layer of plant tissue is also called the epidermis. (Chapter 22)

epigenetic inheritance Inheritance of phenotypic changes in a cell or organism that do not result from changes in the nucleotide sequence of DNA. Can be due to positive feedback loops of transcription regulators or to heritable modifications in chromatin such as DNA methylation or histone modifications. (Chapters 4, 7) (Figure 7–56)

epistasis analysis Analysis to discover the order in which genes act, by investigating if a mutation in one gene can mask the effect of a mutation in another gene when both mutations are present in the same organism or cell. (Chapter 8)

epithelial tissue A tissue, such as the lining of the gut or the epidermal covering of the skin, in which cells are closely bound together into sheets called epithelia. (Chapter 19)

epithelium (plural **epithelia**) Sheet of cells covering the outer surface of a structure or lining a cavity. (Chapter 19)

equilibrium State in a chemical reaction where there is no net change in free energy to drive the reaction in either direction. The ratio of product to substrate reaches a constant value at chemical equilibrium. (Chapters 2, 3) (Figure 2–30)

equilibrium constant (*K*) The ratio of forward and reverse rate constants for a reaction. Equal to the association or affinity constant (K_a) for a simple binding reaction (A + B → AB). *See also* **affinity constant**, **association constant**, **dissociation constant**. (Chapters 2, 3) (See p. 146.)

ER lumen The space enclosed by the membrane of the endoplasmic reticulum (ER). (Chapter 12)

ER resident protein Protein that remains in the lumen of the endoplasmic reticulum (ER) or its membranes and carries out its function there, as opposed to the many proteins that are present in the ER only in transit. (Chapter 12)

ER retention signal Short amino acid sequence on a protein that prevents it from moving out of the endoplasmic reticulum (ER). Found on those proteins that are resident in the ER and function there. (Chapter 12)

ER signal sequence N-terminal signal sequence that directs proteins to enter the endoplasmic reticulum (ER). Cleaved off by signal peptidase after entry. (Chapter 12)

erythrocyte Small hemoglobin-containing blood cell of vertebrates that transports oxygen to, and carbon dioxide from, tissues. Also called a red blood cell. (Chapter 22)

erythropoietin A hormone produced by the kidney that stimulates the production of red blood cells in bone marrow. (Chapter 15)

ESCRT protein complexes Four protein complexes (ESCRT-0, ESCRT-I, ESCRT-II, and ESCRT-III) that act sequentially to shepherd mono-ubiquitylated membrane proteins on endosomal membranes into intralumenal vesicles. The ESCRT-III complex catalyzes the pinching-off reaction. (Chapter 13)

ethylene Small gas molecule that is a plant growth regulator influencing plant development in various ways including promoting fruit ripening, leaf abscission, and plant senescence and functioning as a stress signal in response to wounding, infection, and flooding. (Chapter 15)

euchromatin Region of an interphase chromosome that stains diffusely; "normal" chromatin, as opposed to the more condensed heterochromatin. (Chapter 4)

eukaryote Organism composed of one or more cells that have a distinct nucleus. Member of one of the three main divisions of the living world, the other two being bacteria and archaea. (Chapter 1) (Figures 1–9 and 1–21)

eukaryotic initiation factor (eIF) Protein that helps load initiator tRNA on to the ribosome, thus initiating translation. (Chapter 6)

excitatory neurotransmitter Neurotransmitter that opens cation channels in the postsynaptic membrane, causing an influx of Na^+, and in many cases Ca^{2+}, that depolarizes the postsynaptic membrane toward the threshold potential for firing an action potential. (Chapter 11)

executioner caspases Apoptotic caspases that catalyze the widespread cleavage events during apoptosis that kill the cell. (Chapter 18)

exocytosis Excretion of material from the cell by vesicle fusion with the plasma membrane; can occur constitutively or be regulated. (Chapter 13)

exon Segment of a eukaryotic gene that consists of a sequence of nucleotides that will be represented in mRNA or in a final transfer, ribosomal, or other mature RNA molecule. In protein-coding genes, exons encode the amino acids in the protein. An exon is usually adjacent to a noncoding DNA segment called an intron. (Chapters 4, 6) (Figure 4–15)

expansion microscopy (ExM) A microscopy technique in which superresolution is achieved by physically enlarging the specimen. Biological material is labeled with fluorescent probes before being embedded in a polymer gel. The sample is then gently swollen before examination in a light microscope. (Chapter 9)

extracellular pathogens Pathogens that disturb host cells and can cause serious disease without replicating in host cells. (Chapter 23)

extracellular signal molecule Any secreted or cell-surface chemical signal that binds to receptors and regulates the activity of the cell expressing the receptor. (Chapter 15)

extrinsic pathway Pathway of apoptosis triggered by extracellular signal proteins binding to cell-surface death receptors. (Chapter 18)

facultative pathogens Bacteria that replicate in an environmental reservoir such as water or soil and only cause disease if they happen to encounter a susceptible host. (Chapter 23)

FAD/FADH$_2$ (flavin adenine dinucleotide/reduced flavin adenine dinucleotide) Electron carrier pair that functions in the citric acid cycle and fatty acid oxidation. One molecule of FAD gains two electrons plus two protons in becoming the activated carrier $FADH_2$. (Chapter 2) (Figure 2–39)

Fas (**Fas protein**, **Fas death receptor**) Transmembrane death receptor that initiates apoptosis when it binds its extracellular ligand (Fas ligand). (Chapter 18) (Figure 18–6)

Fas ligand Ligand that activates the cell-surface death receptor, Fas, triggering the extrinsic pathway of apoptosis. (Chapter 18)

fat Energy-storage lipid in cells. Composed of triglycerides—fatty acids esterified with glycerol. (Chapter 2)

fate map Representation showing which cell types will later derive from which regions of a developing tissue; for example, from the blastula. (Chapter 21) (Figure 21–27)

Fc receptor One of a family of cell-surface receptors that bind the tail region (Fc region) of an antibody molecule. Different Fc receptors are specific for different classes of antibodies, such as IgG, IgA, or IgE. (Chapter 24)

feedback inhibition The process in which a product of a reaction feeds back to inhibit a previous reaction in the same pathway. (Chapter 3) (Figures 3–52 and 3–53)

fermentation Anaerobic energy-yielding metabolic pathway involving the oxidation of organic molecules. Anaerobic glycolysis refers to the process whereby pyruvate is converted into lactate or ethanol, with the conversion of NADH to NAD^+. (Chapter 2) (Figure 2–50)

fibril-associated collagen A collagen (including types IX and XII) that has a flexible triple-stranded helical structure and binds to the surface of collagen fibrils. Mediates the interactions of these collagen fibrils with one another and with other matrix macromolecules to help determine their organization in the matrix. (Chapter 19)

fibrillar collagen Class of fibril-forming collagens (including type I collagen, the most common type and the principal collagen of skin and bone) that have long rope-like structures with few or no interruptions and which assemble into collagen fibrils. (Chapter 19)

fibroblast Common cell type found in connective tissue. Secretes an extracellular matrix rich in collagen and other extracellular matrix macromolecules. Migrates and proliferates readily in wounded tissue and in tissue culture. (Chapter 19)

fibronectin Extracellular matrix protein involved in adhesion of cells to the matrix and guidance of migrating cells during embryogenesis. Integrins on the cell surface are receptors for fibronectin. (Chapter 19)

filopodium (plural **filopodia**) (**microspike**) Thin, spike-like protrusion with an actin filament core, generated on the leading edge of a crawling animal cell. (Chapter 16) (Figure 16–17)

flagellum (plural **flagella**) Long, whiplike protrusion whose undulations drive a cell through a fluid medium. Eukaryotic flagella are longer versions of cilia. Bacterial flagella are smaller and completely different in construction and mechanism of action. *Compare* **cilium**. (Chapter 16)

fluorescence microscope Microscope designed to view material stained with fluorescent dyes or proteins. Similar to a light microscope but the illuminating light is passed through one set of filters before the specimen, to select those wavelengths that excite the dye, and through another set of filters before the light reaches the eye, to select only those wavelengths emitted when the dye fluoresces. (Chapter 9) (Figure 9–10)

fluorescence recovery after photobleaching (**FRAP**) Technique for monitoring the kinetic parameters of a protein by analyzing how fluorescent protein molecules move into an area of the cell bleached by a beam of laser light. (Chapter 9) (Figure 9–20)

fluorescence resonance energy transfer (**FRET**) Technique for monitoring the closeness of two fluorescently labeled molecules (and thus their interaction) in cells. Also known as Förster resonance energy transfer. (Chapter 9) (Figure 9–19)

focal adhesion kinase (**FAK**) Cytoplasmic tyrosine kinase present at cell–matrix junctions (focal adhesions) in association with the cytoplasmic tails of integrins. (Chapter 15)

follicular helper T cell (**T_{FH} cell**) Type of T cell located in lymphoid follicles that secretes various cytokines to stimulate B cells to undergo antibody class switching and somatic hypermutation. (Chapter 24)

formin Dimeric protein that nucleates the growth of straight, unbranched actin filaments that can be cross-linked by other proteins to form parallel bundles. (Chapter 16)

Förster resonance energy transfer *See* **fluorescence resonance energy transfer** (**FRET**). (Chapter 9)

FRAP *See* **fluorescence recovery after photobleaching**. (Chapter 9)

free energy (**G**) (**Gibbs free energy**) The energy that can be extracted from a system to drive reactions. Takes into account changes in both energy and entropy. (Chapter 2) (Panel 2–7, pp. 106–107)

free-energy change (**ΔG**) Change in the free energy during a reaction: the free energy of the product molecules minus the free energy of the starting molecules. A large negative value of ΔG indicates that the reaction has a strong tendency to occur. (Chapter 2) (Panel 2–7, pp. 106–107)

free ribosome Ribosome that is free in the cytosol, unattached to any membrane. (Chapter 12)

FRET *See* **fluorescence resonance energy transfer**. (Chapter 9)

Frizzled Family of cell-surface receptors that are seven-pass transmembrane proteins that resemble GPCRs in structure but do not generally work through the activation of G proteins. Activated by Wnt binding to recruit the scaffold protein Dishevelled, which helps relay the signal to other signaling molecules. (Chapter 15)

fungus (plural **fungi**) Kingdom of eukaryotic organisms that includes the yeasts, molds, and mushrooms. Many plant diseases and a relatively small number of animal diseases are caused by fungi. (Chapter 23)

fusion protein Engineered protein that combines two or more normally separate polypeptides. Produced from a recombinant gene. (Chapter 8)

ΔG Change in the free energy during a reaction: the free energy of the product molecules minus the free energy of the starting molecules. A large negative value of ΔG indicates that the reaction has a strong tendency to occur. (Chapter 2) (Panel 2–7, pp. 106–107)

G_0 State of withdrawal from the eukaryotic cell-division cycle by entry into a quiescent digression from the G_1 phase. A common, sometimes permanent, state for differentiated cells. (Chapter 17)

G_1 phase Gap 1 phase of the eukaryotic cell-division cycle, between the end of mitosis and the start of DNA synthesis. (Chapter 17) (Figure 17–4)

G_1-Cdk Cyclin–Cdk complex formed in vertebrate cells by a G_1-cyclin and the corresponding cyclin-dependent kinase (Cdk). (Chapter 17) (Table 17–1, p. 1034)

G_1-cyclin Cyclin present in the G_1 phase of the eukaryotic cell cycle. Forms complexes with Cdks that help govern the activity of the G_1/S-cyclins, which control progression to S phase. (Chapter 17)

G_1/S-Cdk Cyclin–Cdk complex formed in vertebrate cells by a G_1/S-cyclin and the corresponding cyclin-dependent kinase (Cdk). (Chapter 17) (Figure 17–10 and Table 17–1, p. 1034)

G_1/S-cyclin Cyclin that activates Cdks in late G_1 of the eukaryotic cell cycle and thereby helps trigger progression through Start, resulting in a commitment to cell-cycle entry. Its level falls at the start of S phase. (Chapter 17) (Figure 17–10)

G_2 phase Gap 2 phase of the eukaryotic cell-division cycle, between the end of DNA synthesis and the beginning of mitosis. (Chapter 17) (Figure 17–4)

G_2/M transition Point in the eukaryotic cell cycle at which the cell checks for completion of DNA replication before triggering the early mitotic events that lead to chromosome alignment on the spindle. (Chapter 17) (Figure 17–8)

ganglioside Any glycolipid having one or more sialic acid residues in its structure. Found in the plasma membrane of eukaryotic cells and especially abundant in nerve cells. (Chapter 10) (Figure 10–16)

gap gene In *Drosophila* development, a gene that is expressed in specific broad regions along the anteroposterior axis of the early embryo, and which helps designate the main divisions of the insect body. (Chapter 21) (Figure 21–18)

gap junction Communicating channel-forming cell–cell junction present in most animal tissues that allows ions and small molecules to pass from the cytoplasm of one cell to the cytoplasm of the next. (Chapter 19)

gastrulation Important stage in animal embryogenesis during which the embryo is transformed from a ball of cells to a structure with a gut (a gastrula). (Chapter 21)

gated transport Movement of proteins between the cytosol and the nucleus through the nuclear pore complexes in the nuclear envelope; these complexes function as selective gates. (Chapter 12)

geminin Protein that prevents the formation of the protein–DNA complexes required to initiate DNA replication forks during S phase and mitosis, thus ensuring that the chromosomes are replicated only once in each cell cycle. (Chapter 17)

gene Region of DNA that is transcribed as a single unit and carries information for a discrete hereditary characteristic, usually corresponding to (1) a single protein (or set of related proteins generated by variant post-transcriptional processing) or (2) a single RNA (or set of closely related RNAs). (Chapters 1, 7)

gene control region The set of linked DNA sequences regulating expression of a particular gene. Includes promoter and *cis*-regulatory sequences required to initiate transcription of the gene and control the rate of transcription. (Chapter 7) (Figure 7–20)

gene conversion Process by which DNA sequence information can be transferred from one DNA helix (which remains unchanged) to another DNA helix whose sequence is altered. It often accompanies general recombination events. (Chapter 5) (Figure 5–57)

gene family The set of genes in an organism related in DNA sequence because of their derivation from the same ancestor. (Chapter 1)

general transcription factor Any of the proteins whose assembly at all promoters of a given type is required for the binding and activation of RNA polymerase and the initiation of transcription. (Chapter 6) (Table 6–3, p. 333)

genetic code The set of rules specifying the correspondence between nucleotide triplets (codons) in DNA or RNA and amino acids in proteins. (Chapter 6) (Figure 6–52)

genetic instability Abnormally increased spontaneous mutation rate, such as occurs in cancer cells. (Chapter 20)

genetic screen Procedure for discovery of genes affecting a specific phenotype by surveying large numbers of mutagenized individuals. (Chapter 8)

genetics The study of the genes of an organism on the basis of heredity and variation. (Chapter 8)

genome The totality of genetic information belonging to a cell or an organism; in particular, the DNA that carries this information. (Chapters 1, 4)

genome annotation Process attempting to mark out all the genes (protein-coding and noncoding) in a genome and ascribing functions to each. (Chapter 8)

genomic imprinting Phenomenon in which a gene is either expressed or not expressed in the offspring depending on which parent it is inherited from. (Chapter 7) (Figure 7–52)

genomic library Collection of cloned DNA molecules representing an entire genome. (Chapter 8)

genotype Genetic constitution of an individual cell or organism. The particular combination of alleles found in a specific individual. (Chapter 8) (Panel 8–1, pp. 520–521)

germ cell A cell in the germ line of an organism, which includes the haploid gametes and their specified diploid precursor cells. Germ cells contribute to the formation of a new generation of organisms and are distinct from somatic cells, which form the body and leave no descendants. (Chapter 5)

germ layer One of the three primary tissue layers (endoderm, mesoderm, and ectoderm) of an animal embryo. (Chapter 21) (Figure 21–3)

glial cell Supporting non-neural cell of the nervous system. Includes oligodendrocytes and astrocytes in the vertebrate central nervous system and Schwann cells in the peripheral nervous system. (Chapter 11)

glycerophospholipid Phospholipid derived from glycerol, abundant in biomembranes. (Chapter 10) (Figures 10–2 and 10–3)

glycogen Polysaccharide composed exclusively of glucose units. Used to store energy in animal cells. Large granules of glycogen are especially abundant in liver and muscle cells. (Chapter 2) (Figure 2–52 and Panel 2–4, pp. 100–101)

glycolipid Lipid molecule with a sugar residue or oligosaccharide attached. (Chapter 10) (Panel 2–5, pp. 102–103)

glycolysis Ubiquitous metabolic pathway in the cytosol in which sugars are incompletely degraded with production of ATP. Literally, "sugar splitting." (Chapter 2) (Figure 2–46 and Panel 2–8, pp. 108–109)

glycoprotein Any protein with one or more sugars or oligosaccharide chains covalently linked to amino acid side chains. Most secreted proteins and most proteins exposed on the outer surface of the plasma membrane are glycoproteins. (Chapter 12)

glycosaminoglycan (GAG) Long, linear, highly charged polysaccharide composed of a repeating pair of sugars, one of which is always an amino sugar. Mainly found covalently linked to a protein core in extracellular matrix proteoglycans. Examples include chondroitin sulfate, hyaluronan, and heparin. (Chapter 19) (Figure 19–32)

glycosylphosphatidylinositol anchor (GPI anchor) Lipid linkage by which some membrane proteins are bound to the membrane. The protein is joined, via an oligosaccharide linker, to a phosphatidylinositol anchor during its travel through the endoplasmic reticulum. (Chapters 10, 12) (Figure 12–30)

Golgi apparatus (Golgi complex) Complex organelle in eukaryotic cells, centered on a stack of flattened, membrane-enclosed spaces, in which proteins and lipids transferred from the endoplasmic reticulum are modified and sorted. It is also the site of synthesis of many cell-wall polysaccharides in plants and extracellular matrix glycosaminoglycans in animal cells. (Chapter 13) (Figure 13–27)

GPCR kinase (GRK) Member of a family of enzymes that phosphorylates multiple serines and threonines on a GPCR to produce receptor desensitization. (Chapter 15) (Figure 15–43)

G protein (heterotrimeric GTP-binding protein) A heterotrimeric GTP-binding protein with intrinsic GTPase activity that couples GPCRs to enzymes or ion channels in the plasma membrane. (Chapter 15) (Table 15–3, p. 907)

G-protein-coupled receptor (GPCR) A seven-pass cell-surface receptor that, when activated by its extracellular ligand, activates a G protein, which in turn activates either an enzyme or ion channel in the plasma membrane. (Chapter 15) (Figures 15–6 and 15–22)

Gram negative Description for bacteria that do not stain with Gram stain as a result of having a thin peptidoglycan cell wall outside their inner (plasma) membrane that is covered by a second lipid-containing outer membrane. (Chapter 23) (Figure 23–3)

Gram positive Description for bacteria that stain positive with Gram stain because of a thick layer of peptidoglycan cell wall outside their inner (plasma) membrane. (Chapter 23)

Gram staining A technique for classifying bacteria that is based on differences in the structure of the bacterial cell wall and outer surface. (Chapter 23)

granulocyte Category of white blood cell distinguished by conspicuous cytoplasmic granules. Includes neutrophils, basophils, and eosinophils. Arises from a granulocyte/macrophage (GM) progenitor cell. (Chapter 22) (Figure 22–11)

granulocyte/macrophage (GM) progenitor cell Committed progenitor cell in the bone marrow that gives rise to neutrophils and macrophages. (Figure 22–12)

green fluorescent protein (GFP) Fluorescent protein isolated from a jellyfish. Widely used as a marker in cell biology. (Chapters 8, 9) (Figure 9–16)

growth cone Migrating motile tip of a growing nerve cell axon or dendrite. (Chapter 16)

growth factor Extracellular signal protein that can stimulate a cell to grow. Growth factors often have other functions as well, including stimulating cells to survive or proliferate. Examples include epidermal growth factor (EGF) and platelet-derived growth factor (PDGF). (Chapter 17)

growth hormone (GH) Mammalian hormone secreted by the pituitary gland into the bloodstream that stimulates growth throughout the body. (Chapter 15)

GTP (guanosine 5'-triphosphate) Nucleoside triphosphate produced by the phosphorylation of GDP (guanosine diphosphate). Like ATP, it releases a large amount of free energy on hydrolysis of its terminal phosphate group. Has a special role in microtubule assembly, protein synthesis, and cell signaling. (Figure 2–59)

GTP-binding protein Also called GTPase; an enzyme that converts GTP to GDP. (Chapters 3, 15)

GTPase An enzyme that converts GTP to GDP. GTPases fall into two large families. Large *G proteins* (*heterotrimeric G proteins*) are composed of three different subunits and mainly couple GPCRs to enzymes or ion channels in the plasma membrane. Small monomeric *GTP-binding proteins* (also called *monomeric GTPases*) consist of a single subunit and help relay signals from many types of cell-surface receptors and have roles in intracellular signaling pathways, regulating intracellular vesicle trafficking, and signaling to the cytoskeleton. Both heterotrimeric G proteins and monomeric GTPases cycle between an active GTP-bound form and an inactive GDP-bound form and frequently act as molecular switches in intracellular signaling pathways. (Chapter 3) (Figure 3–63)

GTPase-activating protein (GAP) Protein that binds to a GTPase and inhibits it by stimulating its GTPase activity, causing the enzyme to hydrolyze its bound GTP to GDP. (Chapter 15) (Figure 15–8)

guanine nucleotide exchange factor (GEF) Protein that binds to a GTPase and activates it by stimulating it to release its tightly bound GDP, thereby allowing it to bind GTP in its place. (Chapter 15) (Figure 15–8)

haplotype block Combination of alleles and DNA markers that has been inherited in a large, linked block on one chromosome of a homologous pair—undisturbed by genetic recombination—across many generations. (Chapter 8)

Hedgehog protein Secreted extracellular signal molecule that has many different roles controlling cell differentiation and gene expression in animal embryos and adult tissues. Excessive Hedgehog signaling can lead to cancer. (Chapter 15)

helper T cell (T_H cell) Type of T cell that helps activate B cells to make antibodies, cytotoxic T cells to become effector cells, and macrophages to kill ingested pathogens. They can also help activate dendritic cells. (Chapter 24)

hematopoietic stem cells Stem cells located in the bone marrow that give rise to almost all blood-cell types. (Chapter 22)

hepatocyte The main liver cell type that carries out a variety of functions including blood protein synthesis, carbohydrate and lipid metabolism, and detoxification of harmful substances. (Chapter 22)

heterochromatin Chromatin that is highly condensed even in interphase; generally transcriptionally inactive. (*Compare* with **euchromatin**.) (Chapter 4)

heterochronic Describes genes involved in developmental timing; mutation results in cells of a specific fate behaving as cells at a different stage of development. (Chapter 21)

heterotrimeric GTP-binding protein *See* **G protein**. (Chapter 15)

high-mannose oligosaccharides Broad class of *N*-linked oligosaccharides, attached to mammalian glycoproteins in the endoplasmic reticulum, containing two *N*-acetylglucosamine residues and many mannose residues. (Chapter 13)

high-performance liquid chromatography (HPLC) Type of chromatography that uses columns packed with tiny beads of matrix; the solution to be separated is pushed through under high pressure. (Chapter 8)

histone One of a group of small abundant proteins, rich in arginine and lysine, that combine to form the nucleosome cores around which DNA is wrapped in eukaryotic chromosomes. (Chapter 4) (Figure 4–24)

histone chaperone (**chromatin assembly factor**) Protein that binds free histones, releasing them as they are incorporated into newly replicated chromatin. (Chapter 5) (Figure 4–27)

histone H1 "Linker" (as opposed to "core") histone protein that binds to DNA where it exits from a nucleosome and helps to compact nucleosomes. (Chapter 4) (Figure 4–30)

Holliday junction (**cross-strand exchange**) X-shaped structure formed in DNA molecules undergoing recombination, in which the two DNA molecules are held together by exchanging one of their two strands; also called a cross-strand exchange. (Chapter 5) (Figure 5–54)

homeotic selector gene In *Drosophila* development, a gene that defines and preserves the differences between body segments. (Chapter 21)

homolog One of two or more genes that are similar in sequence as a result of derivation from the same ancestral gene. The term covers both orthologs and paralogs. (Chapter 1) (Figure 1–20) *See* **homologous chromosomes**.

homologous chromosomes (**homologs**) The maternal and paternal copies of a particular chromosome in a diploid cell. (Chapter 4)

homologous recombination (**general recombination**) Genetic exchange between a pair of identical or very similar DNA sequences, often those located on two copies of the same chromosome. Provides an error-free mechanism for repairing DNA double-strand breaks. (Chapter 5) (Figures 5–47, 5–49, and 5–53)

homophilic Binding between molecules of the same kind, especially those involved in cell–cell adhesion. (Chapter 19) (Figure 19–5)

horizontal gene transfer Gene transfer between bacteria or archaea via natural transformation by released naked DNA, transduction by bacteriophages, or sexual exchange by conjugation. (Chapter 23)

hormone Signal molecule secreted by an endocrine cell into the bloodstream, which can then carry the signal to distant target cells. (Chapters 15, 21)

***Hox* complex** A gene complex consisting of a series of *Hox* genes. (Chapter 21)

***Hox* genes** Genes coding for transcription regulators, each gene containing a homeodomain, and specifying body-region differences. *Hox* mutations typically cause homeotic transformations. (Chapter 21)

Hox proteins Transcription regulatory proteins encoded by *Hox* genes; possess a highly conserved, 60-amino-acid-long DNA-binding homeodomain. (Chapter 21)

HPV Human papillomavirus; infects the cervical epithelium and is important as a cause of carcinoma of the uterine cervix. (Chapter 20)

hyaluronan (**hyaluronic acid**) Type of nonsulfated glycosaminoglycan with a regular repeating sequence of up to 25,000 identical disaccharide units and not linked to a core protein. Found in the fluid lubricating joints and in many other tissues. (Chapter 19) (Figures 19–33 and 19–34)

hybridization In molecular biology, the process whereby two complementary nucleic acid strands form a base-paired duplex DNA–DNA, DNA–RNA, or RNA–RNA molecule. Forms the basis of a powerful technique for detecting specific nucleotide sequences. (Chapter 8) (Figure 8–31)

hybridoma Hybrid cell line generated by fusion of a tumor cell and another cell type. Monoclonal antibodies are produced by hybridoma lines obtained by fusing antibody-secreting B cells with cells of a B-lymphocyte tumor. (Chapter 8) (Figure 8–3)

hydrogen bond Noncovalent bond in which an electropositive hydrogen atom is partially shared by two electronegative atoms. (Chapter 2) (Panel 2–3, pp. 98–99)

hydronium ion (**H_3O^+**) Water molecule associated with an additional proton. The form generally taken by protons in aqueous solution. (Chapter 2)

hydrophilic Dissolving readily in water. Literally, "water loving." (Chapters 2, 10)

hydrophobic (**lipophilic**) Not dissolving readily in water. Literally, "water-fearing." (Chapters 2, 10)

hydrophobic force Force exerted by the hydrogen-bonded network of water molecules that brings two nonpolar surfaces together by excluding water between them. (Panel 2–3, pp. 98–99)

hyperpolarization Deviation in the electrical potential across the plasma membrane towards a more negative value. (Chapter 11)

hypervariable region In immunology: any of the three small parts of the variable region of an immunoglobulin or T cell receptor chain that show the highest variability from molecule to molecule and contribute to the antigen-binding site. (Chapter 24) (Figure 24–26)

IκB Inhibitory proteins that bind tightly to NFκB dimers and hold them in an inactive state within the cytoplasm of unstimulated cells. (Chapter 15)

Ig gene segments In immunology: short DNA sequences that are joined together during B cell and T cell development to produce the coding sequences for immunoglobulins and T cell receptors, respectively. (Chapter 24) (Figure 24–28)

Ig superfamily Large and diverse family of proteins that contain immunoglobulin or immunoglobulin-like domains. Most are involved in cell–cell interactions or antigen recognition. (Chapter 24) (Figure 24–48)

IgA Immunoglobulin A; the principal class of antibody in secretions, including saliva, tears, milk, and respiratory and intestinal secretions. (Chapter 24)

IgD Immunoglobulin D; produced by immature naïve B cells after leaving the bone marrow. Transmembrane IgD and IgM proteins, with the same antigen-binding site, form the B cell receptors (BCRs) on these cells. (Chapter 24)

IgE Immunoglobulin E; binds with high affinity via its tail region to a class of Fc receptors on the surface of mast cells (tissues) or basophils (blood), where it acts as an antigen receptor; antigen binding stimulates the secretion of cytokines and biologically active amines, which help attract white blood cells, antibodies, and complement proteins to the site of activation. (Chapter 24)

IgG Immunoglobulin G; the major antibody class in the blood, produced in especially large quantities during secondary antibody responses. The tail region of some IgG subclasses can bind to specific Fc receptors on macrophages and neutrophils. Antigen–IgG complexes can activate complement. (Chapter 24)

IgM Immunoglobulin M; the first class of immunoglobulin that a developing B cell in the bone marrow makes, forming B-cell receptors on its surface. IgM antibodies are the major class of antibody secreted into the blood in the early stages of a primary antibody response on first exposure to an antigen, where their pentameric structure (with 10 antigen-binding sites) allows strong binding to pathogens. When bound to antigen, it is highly efficient at activation of complement. (Chapter 24)

image processing Computer-based techniques in microscopy that process digital images in order to extract latent information. Enables compensation for some optical faults in microscopes, enhanced contrast to improve detection of small differences in light intensity, and subtraction of background irregularities in the optical system. (Chapter 9)

imaginal disc Group of cells that are set aside, apparently undifferentiated, in the *Drosophila* embryo and which will develop into an adult structure (for example, eye, leg, wing). Overt differentiation occurs at metamorphosis. (Chapter 21) (Figure 21–65)

immunization Method of inducing adaptive immune responses to pathogens or foreign molecules, usually involving the co-injection of an adjuvant, a molecule (often of microbial origin) that helps activate innate immune responses required for the adaptive responses. (Chapter 24)

immunoblotting *See* **western blotting**. (Chapter 8)

immunoglobulin (Ig) *See* **antibody**. (Chapters 3, 24)

immunoglobulin (Ig) domain Characteristic protein domain of about 110 amino acids that is found in immunoglobulin light and heavy chains. Similar domains, known as immunoglobulin-like (Ig-like) domains, are present in many other proteins, which, together with Igs, constitute the Ig superfamily. (Chapter 24) (Figure 24–27)

immunoglobulin (Ig) superfamily Large and diverse family of proteins that contain immunoglobulin domains or immunoglobulin-like domains. Most are involved in cell–cell interactions or antigen recognition. (Chapter 19) (Figure 24–48)

immunogold electron microscopy Method to localize specific macromolecules using a primary antibody that binds to the molecule of interest and is then detected with a secondary antibody to which a colloidal gold particle has been attached. The gold particle is electron-dense and can be seen as a black dot in the electron microscope. (Figure 9–43)

immunological memory Long-lived property of the adaptive immune system that follows a primary immune response to many antigens, such that a subsequent encounter with the same antigen will provoke a more rapid and stronger secondary immune response. (Chapter 24) (Figure 24–16)

immunological self-tolerance The lack of response of the adaptive immune system to an antigen. Tolerance to self molecules is crucial to avoid autoimmune diseases. (Chapter 24) (Figure 24–21)

immunological synapse The highly organized interface that develops between a T cell and an antigen-presenting cell (APC) or target cell it is in contact with, formed by T-cell receptors binding to antigen–MHC complexes on the APC and cell adhesion proteins binding to their counterparts on the APCs. (Chapter 24)

induced fit A principle for increasing the specificity of substrate recognition by proteins and RNAs. In protein synthesis, the ribosome folds around a codon–anticodon interaction, and only when the match is correct does the subsequent reaction proceed efficiently. (Chapter 6)

induced pluripotent stem cells (iPS cells) Cells that are induced by artificial expression of specific transcription regulators to look and behave like the pluripotent embryonic stem cells that are derived from embryos. (Chapters 7, 22)

induced regulatory T cell (induced T_{reg} cell) A regulatory T cell (T_{reg} cell) that develops from naïve helper T cells when they are activated in the presence of TGFβ in the absence of IL6. (Chapter 24)

inflammasome Intracellular protein complex formed after activation of cytoplasmic NOD-like receptors with adaptor proteins. It contains a caspase enzyme that cleaves pro-inflammatory cytokines from their precursor proteins. (Chapter 24)

inflammatory response Local response of a tissue to injury or infection—characterized clinically by redness, swelling, heat, and pain. Caused by invasion of white blood cells, which are attracted by and secrete various cytokines. (Chapter 24)

inhibitors of apoptosis (IAPs) Intracellular protein inhibitors of apoptosis. (Chapter 18)

inhibitory G protein (G_i) Heterotrimeric G protein that can regulate ion channels and inhibit the enzyme adenylyl cyclase in the plasma membrane. *See also* **G protein**. (Chapter 15) (Table 15–3, p. 907)

inhibitory neurotransmitter Neurotransmitter that opens transmitter-gated Cl^- or K^+ channels in the postsynaptic membrane of a nerve or muscle cell and thus tends to inhibit the generation of an action potential. (Chapter 11)

initial segment Specialized membrane region at the base of a nerve axon (adjacent to the cell body) that is rich in voltage-gated Na^+ channels plus other classes of ion channels that all contribute to the encoding of membrane depolarization into action potential frequency. (Chapter 11)

initiator caspases Apoptotic caspases that begin the apoptotic process, activating the executioner caspases. (Chapter 18)

initiator tRNA Special tRNA that initiates translation. It always carries the amino acid methionine, forming the complex Met–tRNAi. (Chapter 6) (Figure 6–74)

innate immune response An early immune response in all organisms to a pathogen, which includes the production of antimicrobial molecules and the activation of phagocytic cells. Such a response is not specific for the pathogen, in contrast to an adaptive immune response. (Chapter 24)

innate immune system A variety of defense mechanisms that help prevent and fight infection. Unlike adaptive immune mechanisms, innate immune mechanisms act from the start of an infection, do not adapt to a specific pathogen, and do not generate immunological memory. (Chapter 24)

inner mitochondrial membrane Mitochondrial membrane that encloses the matrix space and forms extensive invaginations called cristae. (Chapters 12, 14)

inner nuclear membrane One of two concentric membranes comprising the nuclear envelope; contains specific proteins as anchoring sites for chromatin and the nuclear lamina. (Chapter 12)

inositol 1,4,5-trisphosphate (IP$_3$) Small intracellular signaling molecule produced during activation of the inositol phospholipid signaling pathway. Acts to release Ca^{2+} from the endoplasmic reticulum. (Chapter 15) (Figures 15–29 and 15–30)

inositol phospholipid signaling pathway Intracellular signaling pathway that starts with the activation of phospholipase C and the generation of IP$_3$ and diacylglycerol (DAG) from inositol phospholipids in the plasma membrane. The DAG helps to activate protein kinase C. (Chapter 15) (Figures 15–29 and 15–30)

integrin Transmembrane adhesion protein that is involved in the attachment of cells to the extracellular matrix and to each other. (Chapter 19) (Figure 19–3 and Table 19–1, p. 1107)

interaction domain Compact protein module, found in many intracellular signaling proteins, that binds to a particular structural motif (for example, a short peptide sequence, a covalent modification, or another protein domain) in another protein or lipid. (Chapter 15)

interferon-α (**IFNα**) and **interferon-β** (**IFNβ**) Cytokines (type I interferons) produced by mammalian cells as a general response to a viral infection. (Chapter 24)

intermembrane space The compartment in a mitochondrion between the outer and inner mitochondrial membranes. (Chapters 12, 14)

internal ribosome entry site (**IRES**) Specific site in a eukaryotic mRNA, other than at the 5′ end, at which translation can be initiated. (Chapter 7) (Figure 7–72)

interphase Long period of the cell cycle between one mitosis and the next. Includes G_1 phase, S phase, and G_2 phase. (Chapter 17) (Figure 17–4)

intestinal epithelium The single cell layer of polarized epithelial cells that makes up the lining of the small and large intestine, where it acts as a barrier to the lumen of the gut and mediates absorption of nutrients. (Chapter 22)

intracellular pathogens Pathogens, including all viruses and many bacteria and protozoa, that enter and replicate inside host cells to cause disease. (Chapter 23)

intrinsic pathway (**mitochondrial pathway**) Pathway of apoptosis activated from inside the cell in response to stress or developmental signals; depends on the release into the cytosol of mitochondrial proteins normally resident in the mitochondrial intermembrane space. (Chapter 18)

intron Noncoding region of a eukaryotic gene that is transcribed into an RNA molecule but is then excised by RNA splicing during production of the mRNA or other functional RNA. (Chapters 4, 6) (Figure 4–15)

invadopodia Actin-rich protrusions extending in three dimensions that are important for cells to cross tissue barriers by degrading the extracellular matrix. (Chapter 16)

ion channel Transmembrane protein complex that forms a water-filled channel across the lipid bilayer through which specific inorganic ions can diffuse down their electrochemical gradients. (Chapter 11) (Figure 11–22)

ion-sensitive indicators Molecules whose light emission reflects the local concentration of a particular ion; some are luminescent (emitting light spontaneously) while others are fluorescent (emitting light on exposure to light). (Figure 9–21)

ionotropic receptor (**transmitter-gated ion channel**) Ion channel found at chemical synapses in the postsynaptic plasma membranes of nerve and muscle cells. Opens only in response to the binding of a specific extracellular neurotransmitter. The resulting inflow of ions leads to the generation of a local electrical signal in the postsynaptic cell. (Figures 11–38 and 15–6)

IP$_3$ receptor (**IP$_3$-gated Ca^{2+}-release channel**) Gated Ca^{2+} channel in the ER membrane that opens on binding cytosolic IP$_3$, releasing stored Ca^{2+} into the cytosol. (Chapter 15) (Figure 15–30)

iron–sulfur cluster Electron-transporting group consisting of either two or four iron atoms bound to an equal number of sulfur atoms, found in a class of electron-transport proteins. (Figure 14–16)

iron–sulfur protein A protein that contains one or more iron–sulfur clusters, typically using those cofactors for electron transport. (Chapter 14)

***J* gene segment** Short DNA sequence that encodes part of the variable region of light and heavy immunoglobulin chains and of α and β chains of T cell receptors. (Chapter 24) (Figures 24–28 and 24–29)

JAK–STAT signaling pathway Signaling pathway activated by cytokines and some hormones, providing a rapid route from the plasma membrane to the nucleus to alter gene transcription. Involves cytoplasmic Janus kinases (JAKs), and signal transducers and activators of transcription (STATs). (Chapter 15)

Janus kinases (**JAKs**) Cytoplasmic tyrosine kinases associated with cytokine receptors, which phosphorylate and activate transcription regulators called STATs. (Chapter 15)

junctional diversification The random loss and gain of nucleotides at joining sites during V(D)J recombination that occurs during B and T cell development when the cells are assembling the gene segments that encode their antigen receptors. It enormously increases the diversity of V-region coding sequences. (Chapter 24)

K^+ leak channel K^+-transporting ion channel in the plasma membrane of animal cells that remains open even in a "resting" cell. (Chapter 11)

K_m The Michaelis constant, equal to the concentration of an enzyme's substrate that allows that enzyme to produce product at one-half of its maximum rate. (Chapter 3)

karyotype Display of the full set of chromosomes of a cell, arranged with respect to size, shape, and number. (Chapter 4)

keratin Type of intermediate filament, commonly produced by epithelial cells. (Chapter 16)

kinase cascade Intracellular signaling pathway in which one protein kinase, activated by phosphorylation, phosphorylates the next protein kinase in the sequence, and so on, relaying the signal onward. (Chapter 15)

kinesin Member of one of the two main classes of motor proteins that use the energy of ATP hydrolysis to move along microtubules. (Chapter 16) (Figure 16–51)

kinesin-1 Motor protein associated with microtubules that transports cargo within the cell; also called "conventional kinesin." (Chapter 16)

kinetic proofreading A principle for increasing the specificity of catalysis. In the synthesis of DNA, RNA, and proteins, it refers to a time delay that begins with an irreversible step (such as ATP or GTP hydrolysis) and during which incorrect base pairs are more likely to dissociate than correct pairs. (Chapter 6)

kinetochore Large protein complex that connects the centromere of a chromosome to microtubules of the mitotic spindle. (Chapter 17) (Figure 17–33)

kinetochore microtubule In the mitotic or meiotic spindle, a microtubule that connects the spindle pole to the kinetochore of a chromosome. (Chapter 17)

lagging strand One of the two newly synthesized strands of DNA found at a replication fork. The lagging strand is made in discontinuous lengths that are later joined covalently. (Chapter 5) (Figure 5–7)

lamellipodium (plural **lamellipodia**) Flattened, sheetlike protrusion supported by a meshwork of actin filaments, which is extended at the leading edge of a crawling animal cell. (Chapter 16) (Figures 16–18 and 16–19)

laminin Extracellular matrix fibrous protein found in basal laminae, where it forms a sheetlike network. (Chapter 19) (Figures 19–53 and 19–54)

lampbrush chromosome Huge chromosome paired in preparation for meiosis, found in immature amphibian eggs; consisting of large loops of chromatin extending out from a linear central axis. (Chapter 4) (Figure 4–45)

late endosome Compartment formed from a bulbous, vacuolar portion of early endosomes by a process called endosome maturation; late endosomes fuse with one another and with lysosomes to form endolysosomes that degrade their contents. (Chapter 13)

LDL-receptor-related protein (**LRP**) Co-receptor bound by Wnt proteins in the regulation of β-catenin proteolysis. (Chapter 15)

leading strand One of the two newly synthesized strands of DNA found at a replication fork. The leading strand is made by continuous synthesis in the 5′-to-3′ direction. (Chapter 5) (Figure 5–7)

lectin Protein that binds tightly to a specific sugar. Abundant lectins from plant seeds are used as affinity reagents to purify glycoproteins or to detect them on the surface of cells. (Chapter 10)

Legionnaires' disease Type of pneumonia resulting from infection with *Legionella pneumophila*, a bacterial parasite of freshwater amoebae that is spread to humans by air-conditioning systems that harbor infected amoebae. (Chapter 23)

lethal factor One of the subunits of anthrax toxin; a protease that cleaves several activated members of the MAP kinase kinase family and causes a large fall in blood pressure and death on entry into the bloodstream of an animal. (Chapter 23)

leucine-rich repeat (**LRR**) **receptor kinases** Common type of receptor serine/threonine kinase in plants that contains a tandem array of leucine-rich repeat sequences in its extracellular portion. (Chapter 15)

leukemia Cancer of white blood cells. (Chapter 20)

leukocyte General name for all the nucleated blood cells lacking hemoglobin. Also called white blood cells. Includes lymphocytes, granulocytes, and monocytes. (Chapter 22) (Figure 22–11)

ligand Any molecule that binds to a specific site on a protein or other molecule. From Latin *ligare*, "to bind." (Chapter 3)

light microscope One of a class of microscopes that uses visible light to create the image. (Chapter 9)

lignin Network of cross-linked phenolic compounds that forms a supporting network throughout the cell walls of xylem and woody tissue in plants. (Chapter 19)

limit of resolution In microscopy, the smallest distance apart at which two point objects can be resolved as separate. Just under 0.2 μm for conventional light microscopy, a limit determined by the wavelength of light. (Chapter 9)

linkage In ligand binding, the conformational coupling between two separate ligand-binding sites on a protein, such that a conformational change in the protein induced by binding of one ligand affects the binding of a second ligand. (Chapter 3)

lipid bilayer (**phospholipid bilayer**) Thin double sheet of phospholipid molecules that forms the core structure of all cell membranes. The two layers of lipid molecules are packed with their hydrophobic tails pointing inward and their hydrophilic heads outward, exposed to water. (Chapter 10) (Figure 10–1 and Panel 2–5, pp. 102–103)

lipid droplets Storage form in cells for excess lipids; composed of a single monolayer of phospholipids and proteins that surrounds neutral lipids that can be retrieved from droplets as required by the cell. (Chapter 10)

lipid raft Small region of a membrane enriched in sphingolipids and cholesterol. (Chapter 10) (Figure 10–13)

liposome Artificial phospholipid bilayer vesicle formed from an aqueous suspension of phospholipid molecules. (Chapter 10) (Figure 10–9)

local mediator Extracellular signal molecule that acts on neighboring cells. (Chapter 15)

long noncoding RNA (**lncRNA**) One of a large group (~5000 in humans) of RNAs longer than 200 nucleotides and not coding for protein. The functions, if any, of most lncRNAs are unknown, but individual lncRNAs are known to play important roles in the cell; for example, in telomerase function and genomic imprinting. In a general sense, lncRNAs are believed to act as scaffolds, holding together proteins and nucleic acids to speed up a wide variety of reactions in the cell. (Chapter 7)

long-term depression (**LTD**) A long-lasting (hours or more) decrease in the sensitivity of certain synapses in the brain triggered by NMDA-receptor activation. As the opposing process to long-term potentiation, it is thought to be involved in learning and memory. (Chapter 11)

long-term potentiation (**LTP**) Long-lasting increase (days to weeks) in the sensitivity of certain synapses in the brain, induced by a short burst of repetitive firing in the presynaptic neurons. (Chapter 11) (Figure 11–46)

loss of heterozygosity The result of errant homologous recombination that uses the homolog from the other parent instead of the sister chromatid as the template, converting the sequence of the repaired DNA to that of the other homolog. (Chapter 5)

low-density lipoprotein (LDL) Large complex composed of a single protein molecule and many esterified cholesterol molecules, together with other lipids. The form in which cholesterol is transported in the blood and taken up into cells. (Chapter 13) (Figure 13–53)

lumen The space inside a hollow structure. In cells: the cavity enclosed by an organelle membrane. In tissues: the cavity enclosed by a sheet of cells. (Chapters 10, 12)

lymphocyte White blood cell responsible for the specificity of adaptive immune responses. Two main types: B cells, which produce antibody, and T cells, which interact directly with other effector cells of the immune system and with infected cells. T cells develop in the thymus and are responsible for cell-mediated immunity. B cells develop in the bone marrow in mammals and are responsible for the production of circulating antibodies. (Chapter 24)

lymphoid cells Immune cells including B and T lymphocytes and natural killer cells. (Chapter 22)

lymphoid organ An organ containing large numbers of lymphocytes. Lymphocytes are produced in central (or primary) lymphoid organs and respond to antigen in peripheral (or secondary) lymphoid organs. (Chapter 24) (Figure 24–12)

lymphoma Cancer of lymphocytes, in which the cancer cells are mainly found in lymphoid organs (rather than in the blood, as in leukemias). (Chapter 20)

lysosomal storage diseases Genetic diseases resulting from defects in or a lack of one or more functional hydrolases in lysosomes of some cells, leading to accumulation of undigested substrates in lysosomes and consequent cell pathology. (Chapter 13)

lysosome Membrane-enclosed organelle in eukaryotic cells containing digestive enzymes, which are typically most active at the acid pH found in the lumen of lysosomes. (Chapter 13) (Figure 13–62)

lysozyme Enzyme that catalyzes the cutting of polysaccharide chains in the cell walls of bacteria. (Chapter 3)

M-Cdk (M-phase Cdk) Cyclin–Cdk complex formed in vertebrate cells by an M-cyclin and the corresponding cyclin-dependent kinase (Cdk). (Chapter 17) (Figure 17–10 and Table 17–1, p. 1034)

M-cyclin A cyclin found in all eukaryotic cells that promotes the events of mitosis. (Chapter 17) (Figure 17–10)

M6P receptor proteins Transmembrane receptor proteins present in the *trans* Golgi network that recognize the mannose 6-phosphate (M6P) groups added exclusively to lysosomal enzymes, marking the enzymes for packaging and delivery to early endosomes. (Chapter 13)

macromolecule Polymers constructed of long chains of covalently linked, small organic (carbon-containing) molecules. The principal building blocks from which a cell is constructed and the components that confer the most distinctive properties of living things. (Chapter 2)

macrophage Phagocytic cell derived from blood monocytes, resident in most tissues but able to roam. It has both scavenger and antigen-presenting functions in immune responses. (Chapter 13)

macropinocytosis Clathrin-independent, dedicated degradative endocytic pathway induced in most cell types by cell-surface receptor activation by specific cargoes. (Chapter 13)

malaria Protozoal disease caused by any one of four species of *Plasmodium*, which are transmitted to humans by the bite of the female *Anopheles* mosquito. (Chapter 23)

malignant Of tumors and tumor cells: invasive and/or able to undergo metastasis. A malignant tumor is a cancer. (Chapter 20) (Figure 20–3)

MAP kinase module (mitogen-activated protein kinase module) An intracellular signaling module composed of three protein kinases, acting in sequence, with MAP kinase as the third. Typically activated by a Ras protein in response to extracellular signals. (Chapter 15) (Figure 15–50)

master transcription regulator A transcription regulator specifically required for formation of a particular cell type. Artificial expression of master transcription regulators (alone or in combination with others) can often convert one cell type into another. (Chapter 7)

maternal inheritance A form of inheritance observed in animals and plants, caused by the fact that mitochondrial DNA is inherited only through the female germ line. (Chapter 14)

maternal effect Describes a gene that acts in the mother to specify maternal mRNAs and proteins in the egg (that is, maternal-effect gene). Maternal-effect mutations affect the development of the embryo even if the embryo itself has not inherited the mutated gene. (Chapter 21)

maternal–zygotic transition (MZT) Event in animal development where the embryo's own genome largely takes over control of development from maternally deposited macromolecules. (Chapter 21)

matrix metalloprotease Ca^{2+}- or Zn^{2+}-dependent proteolytic enzyme present in the extracellular matrix that degrades matrix proteins. Includes the collagenases. (Chapter 19)

matrix space The large internal compartment of the mitochondrion. (Chapter 12)

Mcm helicase Six-subunit protein complex that serves as the replicative helicase in eukaryotic DNA replication, unwinding the DNA to enable DNA synthesis. (Chapters 5, 17)

mechanosensitive channels Transmembrane ion channels that open in response to a mechanical stress on the lipid bilayer in which they are embedded. (Chapter 11)

megakaryocyte Large myeloid cell with a multilobed nucleus that remains in the bone marrow when mature. Buds off platelets from long cytoplasmic processes. (Chapter 22)

meiosis I The first of two rounds of chromosome segregation after meiotic chromosome duplication; segregates the homologs, each composed of a tightly linked pair of sister chromatids. (Chapter 17)

meiosis II The second of two rounds of chromosome segregation after meiotic chromosome duplication; segregates the sister chromatids of each homolog. (Chapter 17)

membrane-associated protein Membrane protein not extending into the hydrophobic interior of the lipid bilayer but bound to either face of the membrane by noncovalent interactions with other membrane proteins. (Chapter 10) (Figure 10–17)

membrane-bending proteins Attach to specific membrane regions as needed, where they act to control local membrane curvature and thus confer on these regions their three-dimensional shapes. (Chapter 10)

membrane-bound ribosome Ribosome attached to the cytosolic face of the endoplasmic reticulum. The site of synthesis of proteins that enter the endoplasmic reticulum. (Chapter 12) (Figure 12–21)

membrane potential Voltage difference across a membrane due to a slight excess of positive ions on one side and of negative ions on the other. A typical membrane potential for an animal cell plasma membrane is –60 mV (inside negative relative to the surrounding fluid). (Chapter 11) (Figure 11–23)

membrane protein Amphiphilic protein of diverse structure and function that associates with the lipid bilayer of cell membranes. (Chapter 10) (Figure 10–17)

membrane transport protein Membrane protein that mediates the passage of ions or molecules across a membrane. The two main classes are transporters (also called carriers or permeases) and channels. (Chapter 11) (Figure 11–3)

membraneless organelle An assembly of specific proteins held together by multivalent, low-affinity interactions. Also referred to as a *biomolecular condensate*. (Chapter 12)

memory cell In immunology: a T or B lymphocyte generated after antigen stimulation that is more easily and more quickly induced to become an effector cell or another memory cell by a later encounter with the same antigen. (Chapter 24) (Figure 24–17)

mesenchymal cell migration A mode of cell locomotion typical of fibroblasts that is characterized by protrusion of actin-rich lamellipodia at the leading edge, formation of integrin-based attachments to the underlying substratum, and acto-myosin–based contraction at the rear of the cell. (Chapter 16)

mesoderm Embryonic tissue that is the precursor to muscle, connective tissue, skeleton, and many of the internal organs. (Chapter 21) (Figure 21–3)

messenger RNA (mRNA) RNA molecule that specifies the amino acid sequence of a protein. Produced in eukaryotes by processing of an RNA molecule made by RNA polymerase as a complementary copy of DNA. It is translated into protein in a process catalyzed by ribosomes. (Chapter 6) (Figure 6–21)

metabolism The sum total of the chemical processes that take place in living cells. All of catabolism plus anabolism. (Chapter 2) (Figure 2–14)

metabotropic receptors Neurotransmitter receptors that regulate ion channels indirectly through the activation of second-messenger molecules. (Chapter 11)

metaphase The stage of mitosis when all the chromosomes have been moved to the center of the cell, having been positioned there by the mitotic spindle. Metaphase follows prophase and prometaphase; it is succeeded by anaphase. (Chapter 17)

metaphase plate Imaginary plane at right angles to the mitotic spindle and midway between the spindle poles; the plane in which chromosomes are positioned at metaphase. (Chapter 17) (Panel 17–1, pp. 1048–1049)

metaphase-to-anaphase transition Transition in the eukaryotic cell cycle preceding sister-chromatid separation at anaphase. If the cell is not ready to proceed to anaphase, the cell cycle is halted at this point. (Chapter 17) (Figure 17–8 and Panel 17–1, pp. 1048–1049)

metastases Secondary tumors, at sites in the body additional to that of the primary tumor; they result from cancer cells breaking loose, entering blood or lymphatic vessels, and colonizing separate environments. (Chapter 20)

metastasis The spread of cancer cells from their site of origin to other sites in the body. (Chapter 20) (Figures 20–1 and 20–20)

MHC genes Cluster of genes in one vertebrate chromosome (chromosome 6 in humans) that code for a set of highly polymorphic cell-surface glycoproteins (MHC proteins). (Figure 24–37)

MHC protein Cell-surface glycoprotein encoded within the major histocompatibility complex (MHC) of genes. The proteins are highly polymorphic and exist in two main classes—class I and class II MHC proteins, both of which function to present fragments of foreign proteins on the surface of antigen-presenting cells to T cells. (Chapter 24)

microbiome The combined genomes of the various species of a defined microbiota. (Chapter 23)

microbiota The collective of microorganisms that reside in or on an organism. (Chapter 23)

microelectrode A piece of fine glass tubing, pulled to an even finer tip, that is used to inject electric current into cells or to study the intracellular concentrations of common inorganic ions (such as H^+, Na^+, K^+, Cl^-, and Ca^{2+}) in a single living cell by insertion of its tip directly into the cell interior through the plasma membrane. (Figure 11–36)

microRNAs (miRNAs) Short (~21 nucleotide) eukaryotic RNAs, produced by the processing of specialized RNA transcripts coded in the genome, that regulate gene expression through base-pairing with mRNA. (Chapters 7, 21) (Figure 7–78)

microsome Small vesicle derived from endoplasmic reticulum that is produced by fragmentation when cells are homogenized. (Chapter 12) (Figure 12–17)

microtubule flux Movement of individual tubulin molecules in the microtubules of the spindle toward the poles by loss of tubulin at their minus ends. Helps to generate the poleward movement of sister chromatids after they separate in anaphase. (Chapter 17) (Figure 17–37)

microtubule-associated protein (MAP) Any protein that binds to microtubules and modifies their properties. Many different kinds have been found, including structural proteins, such as MAP2, and motor proteins, such as dynein. [Not to be confused with the "MAP" (mitogen-activated protein) of "MAP kinase."] (Chapter 16)

microtubule-organizing center (MTOC) Region in a cell, such as a centrosome or a basal body, from which microtubules grow. (Chapter 16)

midbody Structure formed at the end of cleavage that can persist for some time as a tether between the two daughter cells in animals. (Chapter 17) (Figure 17–44)

mitochondrial hsp70 Part of a multisubunit protein assembly bound to the matrix side of the TIM23 complex that acts as a motor to pull mitochondrial precursor proteins into the matrix space. (Chapter 12)

mitochondrial matrix Large internal compartment of the mitochondrion. The corresponding compartment in a chloroplast is known as the stroma. (Chapter 14)

mitochondrial outer membrane permeabilization (MOMP) The change in the outer mitochondrial membrane that releases cytochrome *c* and other soluble proteins from the intermembrane space into the cytosol—a critical step in the intrinsic pathway of apoptosis. (Chapter 18)

mitochondrial precursor proteins Proteins that are first fully synthesized in the cytosol and then translocated into mitochondrial subcompartments, as directed by one or more signal sequences. (Chapter 12)

mitochondrion (plural **mitochondria**) Membrane-bounded organelle, about the size of a bacterium, that carries out oxidative phosphorylation and produces most of the ATP in eukaryotic cells. (Chapters 12, 14) (Figure 1–25)

mitogen Extracellular signal molecule that stimulates cells to proliferate. (Chapter 17)

mitotic chromosome Highly condensed duplicated chromosome as seen at mitosis, consisting of two sister chromatids held together at the centromere. (Chapter 4)

mitotic spindle Bipolar array of microtubules and associated molecules that forms in a eukaryotic cell during mitosis; it serves to position duplicated chromosomes and then segregate their sister chromatids to opposite spindle poles. (Chapter 17) (Figure 17–26 and Panel 17–1, pp. 1048–1049)

model organism A species that has been studied intensively over a long period and thus serves as a "model" for deriving fundamental biological principles. (Chapter 1)

molecular chaperone (**chaperone**) Protein that helps guide the proper folding of other proteins or helps them avoid misfolding. Includes heat-shock proteins. (Chapter 6)

monoallelic gene expression Expression of only one of the two copies of a gene in a diploid genome, occurring, for example, as a result of imprinting or X-inactivation. (Chapter 7)

monoclonal antibody Antibody secreted by a hybridoma cell line. Because the hybridoma is generated by the fusion of a single B cell with a single tumor cell, each hybridoma produces antibodies that are all identical. (Chapters 4, 8)

monocyte Type of white blood cell that leaves the bloodstream and matures into a macrophage in tissues. (Chapter 22) (Figure 22–11)

monomeric GTPase A single-subunit enzyme that converts GTP to GDP (also called monomeric GTP-binding protein). Cycles between an active GTP-bound form and an inactive GDP-bound form and frequently acts as a molecular switch in intracellular signaling pathways. (Chapter 15)

morphogen Diffusible signal molecule that can impose a pattern on a field of cells by causing cells in different places to adopt different fates. (Chapter 21) (Figure 21–8)

morphogenesis Developmental process in which cells undergo movements and deformations in order to assemble into tissues and organs with specific shapes and sizes. (Chapter 21)

motor protein Protein that uses energy derived from nucleoside triphosphate hydrolysis to propel itself along a linear track (protein filament or other polymeric molecule). (Chapter 16)

mRNA degradation control Cell regulation of gene expression by selectively preserving or destroying certain mRNA molecules in the cytoplasm. (Chapter 7)

mTOR The mammalian version of the large protein kinase called TOR, involved in cell signaling; mTOR exists in two functionally distinct multiprotein complexes. (Chapter 15)

multidrug resistance An observed phenomenon in which cells exposed to one anticancer drug evolve a resistance not only to that drug but also to other drugs to which they have never been exposed. (Chapter 20)

multidrug resistance (**MDR**) **protein** Type of ABC transporter protein that can pump hydrophobic drugs (such as some anticancer drugs) out of the cytoplasm of eukaryotic cells. (Chapter 11)

multipass transmembrane protein Membrane protein in which the polypeptide chain crosses the lipid bilayer more than once. (Chapter 10) (Figure 10–17)

multivesicular bodies Intermediates in the endosome maturation process; early endosomes that are on their way to becoming late endosomes. (Chapter 13)

mutation Heritable change in the nucleotide sequence of a chromosome. (Chapter 1) (Panel 8–1, pp. 520–521)

mutation rate The rate at which changes (mutations) occur in DNA sequences. (Chapter 5)

mutualism Ecologic relationship between microbes and their host in which both the microbe and host benefit. (Chapter 23)

Myc Transcription regulatory protein that is activated when a cell is stimulated to grow and divide by extracellular signals. It activates the transcription of many genes, including those that stimulate cell growth. (Chapter 17) (Figure 17–59)

myelin sheath Insulating layer of specialized cell membrane wrapped around vertebrate axons. Produced by oligodendrocytes in the central nervous system and by Schwann cells in the peripheral nervous system. (Chapter 11) (Figure 11–35)

myeloid cell Any white blood cell other than a lymphocyte. (Chapter 22) (Figure 22–12)

myeloma Cancer of antibody-producing (B) cells found in bone marrow. (Chapter 20)

myoblast Mononucleated, undifferentiated muscle precursor cell. A skeletal muscle cell is formed by the fusion of multiple myoblasts. (Chapter 22)

myofibril Long, highly organized bundle of actin, myosin, and other proteins in the cytoplasm of muscle cells that contracts by a sliding filament mechanism. (Chapter 16)

myosin Type of motor protein that uses the energy of ATP hydrolysis to move along actin filaments. (Chapter 16)

Na^+-K^+ pump (**Na^+-K^+ ATPase**) Transmembrane carrier protein found in the plasma membrane of most animal cells that pumps Na^+ out of and K^+ into the cell, using energy derived from ATP hydrolysis. (Chapter 11) (Figure 11–15)

NAD^+/NADH (**nicotinamide adenine dinucleotide/reduced nicotinamide adenine dinucleotide**) Electron carrier system that participates in oxidation–reduction reactions, such as the oxidation of food molecules. NAD^+ accepts the equivalent of a hydride ion (H^-, a proton plus two electrons) to become the activated carrier NADH. The NADH formed donates its high-energy electrons to the ATP-generating process of oxidative phosphorylation. (Chapter 2) (Figure 2–36)

NADH dehydrogenase complex First of the three electron-driven proton pumps in the mitochondrial respiratory chain, also known as Complex I. It accepts electrons from NADH and passes them to a quinone. (Chapter 14) (Figure 14–18)

$NADP^+$/NADPH (**nicotinamide adenine dinucleotide phosphate/reduced nicotinamide adenine dinucleotide phosphate**) Electron carrier system closely related to

NAD^+/NADH, but used almost exclusively in reductive biosynthetic, rather than catabolic, pathways. (Chapter 2) (Figure 2–36)

naïve cell In immunology: a T or B lymphocyte that proliferates and differentiates into an effector cell or memory cell when it encounters its specific foreign antigen for the first time. (Chapter 24) (Figure 24–17)

natural killer cell (NK cell) Cytotoxic cell of the innate immune system that can kill virus-infected cells and some cancer cells. (Chapter 24)

natural regulatory T cell (natural T_{reg} cell) A regulatory T cell (T_{reg} cell) that develops in the thymus and helps maintain self-tolerance. (Chapter 24)

negative selection Process by which thymocytes expressing a T cell receptor with high affinity for a self peptide bound to a self MHC protein are eliminated by undergoing apoptosis. (Chapter 24)

negative staining A technique in electron microscopy enabling fine detail of isolated macromolecules to be seen. Samples are prepared such that a very thin film of heavy-metal salt covers everywhere except where excluded by the presence of macromolecules, which allow electrons to pass through, creating a reverse or negative image of the molecule. (Chapter 9)

Nernst equation Equation that relates the equilibrium electrical potential (membrane potential in volts) to differences in ion concentrations across a membrane. (Chapter 11)

neural crest Collection of cells located along the line where the neural tube pinches off from the surrounding epidermis in the vertebrate embryo. Neural crest cells migrate to give rise to a variety of tissues, including neurons and glia of the peripheral nervous system, pigment cells of the skin, and the bones of the face and jaws. (Chapter 21) (Figure 19–8)

neural tube Tube of ectoderm that will form the brain and spinal cord in a vertebrate embryo. (Chapter 21)

neurofilament Type of intermediate filament found in nerve cells. (Chapter 16) (Figure 16–64)

neuromuscular junction Specialized chemical synapse between an axon terminal of a motor neuron and a skeletal muscle cell. (Chapter 11) (Figure 11–39)

neuron (nerve cell) Impulse-conducting cell of the nervous system, with extensive processes specialized to receive, conduct, and transmit signals. (Chapter 11) (Figure 11–29)

neurotransmitter Small signal molecule secreted by the presynaptic nerve cell at a chemical synapse to relay the signal to the postsynaptic cell. Examples include acetylcholine, glutamate, GABA, glycine, and many neuropeptides. (Chapters 11, 15)

neutrophil White blood cell that is specialized for the uptake of particulate material by phagocytosis. Enters tissues that become infected or inflamed. (Chapter 13) (Figure 24–5)

NFκB protein Latent transcription regulator that is activated by various intracellular signaling pathways when cells are stimulated during immune, inflammatory, or stress responses. Also has important roles in animal development. (Chapter 15) (Figure 15–63)

nitric oxide (NO) Gaseous signal molecule that is widely used in cell–cell communication in both animals and plants. (Chapter 15) (Figure 15–41)

nitrogen fixation Biochemical process carried out by certain bacteria that reduces atmospheric nitrogen (N_2) to ammonia, leading eventually to various nitrogen-containing metabolites. (Chapter 2)

NMDA receptor Subclass of glutamate-gated ion channel in the mammalian central nervous system critical for long-term potentiation and long-term depression. NMDA-receptor channels are doubly gated, opening only when glutamate is bound to the receptor and, simultaneously, the membrane is strongly depolarized. (Chapter 11)

NO synthase (NOS) Enzyme that synthesizes nitric oxide (NO) by the deamination of arginine. (Chapter 15) (Figure 15–41B)

NOD-like receptors (NLRs) Large family of pattern recognition receptors (PRRs) with leucine-rich repeat motifs; they are exclusively cytoplasmic and recognize a distinct set of microbial molecules. (Chapter 24)

nonclassical cadherins Large family of cadherins that are more distantly related in sequence than classical cadherins and include proteins involved in adhesion (including protocadherins, desmocollins, and desmogleins) and signaling. (Chapter 19)

noncoding RNA An RNA molecule that is the final product of a gene and does not code for protein. These RNAs serve as enzymatic, structural, and regulatory components for a wide variety of processes in the cell. (Chapter 6)

noncovalent bond Weak bonds that require multiple sets to hold two molecules together. Includes hydrogen bonds, electrostatic attractions, and van der Waals attractions. (Chapter 2)

nondisjunction Event occurring occasionally during meiosis in which a pair of homologous chromosomes fails to separate so that the resulting germ cell has either too many or too few chromosomes. (Chapter 17)

nonenveloped virus Virus consisting of a nucleic acid core and a protein capsid only. (Chapter 23) (Figure 23–19C and D)

nonhomologous end joining A DNA repair mechanism for rejoining the ends at double-strand breaks in which the two broken ends of DNA are brought together and rejoined by DNA ligation, generally with the loss of one or more nucleotides at the site of joining. (Chapter 5)

non-kinetochore microtubule In the mitotic or meiotic spindle, a microtubule that forms between the spindle poles and is not attached directly to kinetochores; it can cross-link at the spindle equator with an antiparallel microtubule from the other pole. Also called interpolar microtubule. (Chapter 17)

nonretroviral retrotransposon A type of transposable element that moves by being first transcribed into an RNA copy that is converted to DNA by reverse transcriptase and then inserted elsewhere in the genome. Its mechanism of insertion differs from that of the retroviral-like transposons. (Chapter 5) (Table 5–4, p. 308)

nonsense-mediated mRNA decay Mechanism for degrading aberrant mRNAs containing in-frame internal stop codons before they can be translated into protein. (Chapter 6) (Figure 6–80)

Notch Transmembrane receptor protein (and latent transcription regulator) involved in many cell-fate choices in animal development, for example in the specification of nerve cells from ectodermal epithelium. Its ligands are cell-surface proteins such as Delta and Serrate. (Chapter 15) (Figure 15–60)

NSF Hexameric ATPase that disassembles a complex of a v-SNARE and a t-SNARE. (Chapter 13) (Figure 13–21)

nuclear envelope The double membrane (two bilayers) surrounding the nucleus. Consists of an outer and inner membrane and is perforated by nuclear pores. The outer membrane is continuous with the endoplasmic reticulum. (Chapter 12) (Figures 4–10 and 12–54)

nuclear export receptors Proteins that bind to both the export signal and nuclear pore complex proteins, so as to guide their cargo through the nuclear pore complex to the cytosol. (Chapter 12)

nuclear export signal Sorting signal contained in the structure of those protein molecules and RNA–protein complexes that are to be transported from the nucleus to the cytosol through nuclear pore complexes; includes nuclear RNPs and new ribosomal subunits. (Chapter 12) (Figure 12–61)

nuclear import receptors Proteins that recognize nuclear localization signals to initiate the nuclear import of proteins containing the appropriate nuclear localization signal. (Chapter 12)

nuclear lamin Protein subunit of the intermediate filaments that form the nuclear lamina. (Chapter 12)

nuclear lamina Fibrous meshwork of proteins on the inner surface of the inner nuclear membrane. It is made up of a network of intermediate filaments formed from nuclear lamins. (Chapter 12)

nuclear localization signal (NLS) Signal sequence or signal patch found in proteins destined for the nucleus that enables their selective transport into the nucleus from the cytosol through the nuclear pore complexes. (Chapter 12) (Figures 12–56 and 12–61)

nuclear magnetic resonance (NMR) spectroscopy NMR is the resonant absorption of electromagnetic radiation at a specific frequency by atomic nuclei in a magnetic field, due to flipping of the orientation of their magnetic dipole moments. The NMR spectrum provides information about the chemical environment of the nuclei. NMR is used widely to determine the three-dimensional structure of small proteins and other small molecules. The principles of NMR are also used for medical diagnostic purposes in magnetic resonance imaging (MRI). (Chapter 8) (Figure 8–21)

nuclear pore complex (NPC) Large multiprotein structure forming an aqueous channel (the nuclear pore) through the nuclear envelope; it allows selected molecules to move between nucleus and cytoplasm. (Chapters 6, 12) (Figure 12–55)

nuclear receptor superfamily Intracellular receptors for hydrophobic signal molecules such as steroid and thyroid hormones and retinoic acid. The receptor–ligand complex acts as a transcription factor in the nucleus. (Chapter 15) (Figure 15–66)

nuclear transport The process by which cargo molecules are moved into and out of the nucleus through nuclear pore complexes. (Chapter 12)

nuclear transport receptor (karyopherin) Protein that escorts macromolecules either into or out of the nucleus: nuclear import receptor or nuclear export receptor. (Chapter 12) (Figure 12–61)

nucleolus A prominent structure in the nucleus where rRNA is transcribed and ribosomal subunits are assembled. (Figures 6–44 and 6–46)

nucleoporin Any of a number of different proteins that make up nuclear pore complexes. (Chapter 12)

nucleosome Beadlike structure in eukaryotic chromatin, composed of a short length of DNA wrapped around an octameric core of histone proteins. The fundamental structural unit of chromatin. (Chapter 4) (Figures 4–22 and 4–23)

nucleotide Nucleoside with a phosphate group joined in ester linkage to the sugar moiety. DNA and RNA are polymers of nucleotides. (Chapter 1) (Panel 2–6, pp. 104–105)

nucleotide excision repair Type of DNA repair that corrects irreversible damage of the DNA double helix, such as that caused by certain chemicals or UV light, by cutting out the damaged region on one strand and resynthesizing it using the undamaged strand as template. *Compare* **base excision repair**. (Chapter 5) (Figure 5–41)

***O*-linked glycosylation** Addition of one or more sugars to a hydroxyl group on a protein. (Chapter 13)

obligate pathogens Bacteria that can only replicate inside their host. (Chapter 23)

olfactory receptors G-protein-coupled receptors on the modified cilia of olfactory receptor neurons that recognize odors. The receptors activate adenylyl cyclase via an olfactory-specific G protein (G_{olf}), and resultant increases in cAMP open cyclic-AMP-gated cation channels, allowing Na^+ influx and depolarization and initiation of a nerve impulse. (Chapter 15)

oligodendrocyte Glial cell in the vertebrate central nervous system that forms a myelin sheath around axons. *Compare* **Schwann cell**. (Chapter 11)

oligosaccharyl transferase Endoplasmic reticulum (ER) enzyme complex that transfers core oligosaccharides from dolichol lipid anchors to selected asparagine side chains in newly synthesized proteins as they enter the ER lumen. (Chapter 12)

oncogene An altered gene whose product can act in a dominant fashion to help make a cell cancerous. Typically, an oncogene is a mutant form of a normal gene (proto-oncogene) involved in the control of cell growth or division. (Chapter 20) (Figure 20–21)

open reading frame (ORF) A continuous nucleotide sequence free from stop codons in at least one of the three reading frames (and thus with the potential to code for protein). (Chapters 7, 8)

opportunistic pathogens Microbes of the normal flora that can cause disease only if the immune systems are weakened or if they gain access to a normally sterile part of the body. (Chapter 23)

optogenetics Use of genetically engineered channelrhodopsin and other light-responsive ion channels and transporters to modulate neuron function and hence analyze the neurons and circuits underlying complex functions, including behaviors in whole animals. (Chapter 11) (Figure 11–47)

organelle Subcellular compartment or large macromolecular complex, often but not always membrane-enclosed, that has a distinct structure, composition, and function. Examples of membrane-enclosed organelles are the nucleus, mitochondrion, ER, and Golgi apparatus; examples of organelles that form as biomolecular condensates and lack a membrane are the nucleolus and centrosomes. (Chapters 1, 12) (Figure 1–21)

organelle contact site Region of contact between two organelles stabilized by specific tethering proteins. (Chapter 12)

Organizer Specialized tissue at the dorsal lip of the blastopore in an amphibian embryo; a source of signals that help to orchestrate formation of the embryonic body axis. (Chapter 21)

organoid A miniaturized and simplified organ produced in three-dimensional cell culture that possesses realistic microanatomy. (Chapter 22)

origin recognition complex (**ORC**) Large protein complex that is bound to the DNA at origins of replication in eukaryotic chromosomes throughout the cell cycle. (Chapters 5, 17) (Figure 5–31)

orthologs Genes or proteins from different species that are similar in sequence because they are descendants of the same gene in the last common ancestor of those species; orthologs often have the same or a very similar function in each organism. *Compare* **paralogs**. (Chapter 1) (Figure 1–20)

outer mitochondrial membrane The mitochondrial membrane that separates the organelle from the cytosol; surrounds the inner mitochondrial membrane. (Chapters 12, 14)

outer nuclear membrane One of two concentric membranes comprising the nuclear envelope; surrounds the inner nuclear membrane and is continuous with the inner nuclear membrane and the membrane of the endoplasmic reticulum. (Chapter 12)

OXA complex Protein translocator in the inner mitochondrial membrane that mediates insertion of inner membrane proteins. (Chapter 12)

oxidation (verb **oxidize**) Loss of electrons from an atom, as occurs during the addition of oxygen to a molecule or when a hydrogen is removed. Opposite of reduction. (Chapter 2) (Figure 2–20)

oxidative phosphorylation Process in bacteria and mitochondria in which ATP formation is driven by the transfer of electrons through an electron-transport chain to molecular oxygen. Involves the intermediate generation of an electrochemical proton gradient across a membrane and a chemiosmotic coupling of that gradient to drive ATP production by the ATP synthase. (Chapters 2, 14) (Figure 14–12)

P-type pumps A class of ATP-driven pumps comprising structurally and functionally related multipass transmembrane proteins that phosphorylate themselves during the pumping cycle. The class includes many of the ion pumps responsible for setting up and maintaining gradients of Na^+, K^+, H^+, and Ca^{2+} across cell membranes. (Chapter 11) (Figure 11–12)

p53 A transcription regulatory protein that is activated by damage to DNA and is involved in blocking further progression through the cell cycle until the damage can be repaired. The *p53* gene is a tumor suppressor gene that is mutated in about half of human cancers. (Chapters 17, 20) (Figure 20–31)

pair-rule gene In *Drosophila* development, a gene expressed in a series of regular transverse stripes along the body of the embryo and which helps to determine its segments. (Chapter 21) (Figure 21–18)

pairing In meiosis, the lining up of the two homologous chromosomes along their length. (Chapter 17) (Figure 17–53)

pancreatic β cell Insulin-secreting cells of the pancreas. (Chapter 22)

papillomaviruses Class of viruses responsible for human warts and a prime example of DNA tumor viruses, being a cause of cancer of the uterine cervix. (Chapter 20)

paracrine signaling Short-range cell–cell communication via secreted signal molecules that act on neighboring cells. (Chapter 15) (Figure 15–2)

paralogs Genes or proteins that are similar in sequence because they are the result of a gene duplication event occurring in an ancestral organism. Those in two different organisms are less likely to have the same function than are orthologs. *Compare* **orthologs**. (Chapter 1) (Figure 1–20)

parasitism Ecologic relationship between microbes and their host in which the microbe benefits to the detriment of the host, as is often the case for pathogens. (Chapter 23)

passenger mutations Mutations that have occurred in the same cell as driver mutations, but which are irrelevant to the development of the cancer. (Chapter 20)

passive transport (**facilitated diffusion**) Transport of a solute across a membrane down its concentration gradient or its electrochemical gradient, using only the energy stored in the gradient. (Chapter 11) (Figure 11–4)

patch-clamp recording Electrophysiological technique in which a tiny electrode tip is sealed onto a patch of cell membrane, thereby making it possible to record the flow of current through individual ion channels in the patch. (Chapter 11) (Figure 11–36)

Patched Transmembrane protein predicted to cross the plasma membrane 12 times; located both in intracellular vesicles and on the cell surface where it binds the Hedgehog protein. (Chapter 15)

pathogen (adjective **pathogenic**) An organism, cell, virus, or prion that causes disease. (Chapters 23, 24)

pathogen-associated molecular patterns (**PAMPs**) Microbe-associated molecules, either not present or kept sequestered in the host organism, that often occur in a repeating pattern that is recognized by pattern recognition receptors (PRRs) of the innate immune system. PAMPs can be present in various microbial molecules, including nucleic acids, lipids, polysaccharides, and proteins. (Chapter 24)

pattern recognition receptor (**PRR**) Receptor present on or in cells of the innate immune system that recognizes and is activated by microbial pathogen-associated molecular patterns (PAMPs). (Chapter 24)

PDZ domain Protein-binding domain present in many scaffold proteins and often used as a docking site for the intracellular tails of transmembrane proteins. (Chapter 19) (Figure 19–22)

pectin Mixture of polysaccharides rich in galacturonic acid that forms a highly hydrated matrix in which cellulose is embedded in plant cell walls. (Chapter 19) (Figure 19–64)

peripheral (**secondary**) **lymphoid organ** Lymphoid organ in which T cells and B cells interact and respond to foreign antigens. Examples are spleen, lymph nodes, and mucosal-associated lymphoid organs. (Chapter 24) (Figure 24–12)

peroxins The proteins that form a protein translocator that participates in the import of proteins into peroxisomes. (Chapter 12)

peroxisome Small membrane-bounded organelle that uses molecular oxygen to oxidize organic molecules. Contains some enzymes that produce and others that degrade hydrogen peroxide (H_2O_2). (Chapter 12) (Figure 12–43)

pH scale Common measure of the acidity of a solution: "p" refers to power of 10, "H" to hydrogen. Defined as the negative logarithm of the hydrogen ion concentration in moles per liter (M). $pH = -\log[H^+]$. Thus a solution of pH 3 will contain 10^{-3} M hydrogen ions. pH less than 7 is acidic and pH greater than 7 is alkaline. (Chapter 2)

phagocytosis Process by which unwanted cells, debris, and other bulky particulate material is endocytosed ("eaten") by a cell. Prominent in carnivorous cells, such as *Amoeba proteus*, and in vertebrate macrophages and neutrophils. From Greek *phagein*, "to eat." (Chapter 13)

phagosome Large intracellular membrane-enclosed vesicle that is formed as a result of phagocytosis. Contains ingested extracellular material. (Chapter 13) (Figure 13–70)

phase-contrast microscope Type of light microscope that exploits the interference effects that occur when light passes through material of different refractive indices. Used to view living cells. (Chapter 9) (Figure 9–6)

phase variation The random switching of phenotype of an infectious agent that is caused by changes in expression of proteins at frequencies much higher than mutation rates. (Chapter 5)

phenotype The observable character (including both physical appearance and behavior) of a cell or organism. (Chapter 8) (Panel 8–1, pp. 520–521)

phosphatidylinositol 4,5-bisphosphate [PI(4,5)P_2, PIP_2] Membrane inositol phospholipid (a phosphoinositide) that is cleaved by phospholipase C into IP_3 and diacylglycerol at the beginning of the inositol phospholipid signaling pathway. It can also be phosphorylated by PI 3-kinase to produce PIP_3 docking sites for signaling proteins in the PI-3-kinase–Akt signaling pathway. (Chapter 15) (Figures 15–28 and 15–54)

phosphatidylserine Negatively charged phospholipid that is normally confined to the cytosolic leaflet of the lipid bilayer of the plasma membrane; in apoptotic cells, it accumulates in the outer leaflet, where it serves as an "eat me" signal to neighboring phagocytic cells. (Chapter 18)

phosphoinositide A lipid containing a phosphorylated inositol derivative. Minor component of the plasma membrane, but important in demarking different membranes and for intracellular signal transduction in eukaryotic cells. (Chapter 15) (Figure 15–53)

phosphoinositide 3-kinase (**PI 3-kinase**) Membrane-bound enzyme that is a component of the PI-3-kinase–Akt intracellular signaling pathway. It phosphorylates phosphatidylinositol 4,5-bisphosphate at the 3 position on the inositol ring to produce PIP_3 docking sites in the membrane for other intracellular signaling proteins. (Chapter 15) (Figure 15–54)

phosphoinositides (**phosphatidylinositol phosphates**, or **PIPs**) A lipid containing a phosphorylated inositol derivative. Minor component of the plasma membrane, but important in demarking different membranes and for intracellular signal transduction in eukaryotic cells. (Chapter 13) (Figure 13–10)

phospholipase C (**PLC**) Membrane-bound enzyme that cleaves inositol phospholipids to produce IP_3 and diacylglycerol in the inositol phospholipid signaling pathway. PLCβ is activated by GPCRs via specific G proteins, while PLCγ is activated by RTKs. (Chapter 15) (Figure 15–56)

phospholipid The main category of lipids used to construct biomembranes. Generally composed of two fatty acids linked through glycerol (or sphingosine) phosphate to one of a variety of polar groups. (Chapter 10) (Figure 10–3 and Panel 2–5, pp. 102–103)

phosphorylation Reaction in which a phosphate group is covalently coupled to another molecule. (Chapters 2, 3, 15)

photoactivation Technique for studying intracellular processes in which an inactive form of a molecule of interest is introduced into the cell and is then activated by a focused beam of light at a precise spot in the cell. (Chapter 9)

photochemical reaction center The part of a photosystem that converts light energy into chemical energy in photosynthesis. (Chapter 14) (Figure 14–46)

photorespiration A wasteful form of metabolism conducted by photosynthetic plants in low CO_2 environments that consumes O_2 and liberates CO_2, but does not result in the production of an energy carrier useful to the plant. (Chapter 14)

photosynthetic electron-transfer reactions Light-driven reactions in photosynthesis in which electrons move along an electron-transport chain in a membrane, generating ATP and NADPH. (Chapter 14)

photosystem Multiprotein complex involved in photosynthesis that captures the energy of sunlight and converts it to useful forms of energy: a reaction center plus an antenna (Chapter 14) (Figure 14–46)

phototropin Photoprotein associated with the plant plasma membrane that senses blue light and is partly responsible for phototropism. (Chapter 15)

phragmoplast Structure made of microtubules and actin filaments that forms in the prospective plane of division of a plant cell and guides formation of the cell plate. (Chapter 17) (Figure 17–48)

phytochrome Plant photoprotein that senses light via a covalently attached light-absorbing chromophore, which changes its shape in response to light and then induces a change in the protein's conformation. Plant phytochromes are cytoplasmic serine/threonine kinases, which respond differentially and reversibly to red and far-red light to alter cell behavior. (Chapter 15)

PI-3-kinase–Akt pathway Intracellular signaling pathway that stimulates animal cells to survive and grow. (Chapter 15) (Figure 15–54)

pinocytosis Literally, "cell drinking." Type of endocytosis in which soluble materials are continually taken up from the environment in small vesicles and moved into endosomes along with the membrane-bound molecules. *Compare* **phagocytosis**. (Chapter 13) (Figure 13–51)

piRNAs (**piwi-interacting RNAs**) A class of small noncoding RNAs made in the germ line that, in complex with Piwi proteins, keep the movement of transposable elements in check by transcriptionally silencing transposon genes and destroying RNAs produced by them. (Chapter 7)

planar cell polarity Type of cellular asymmetry seen in some epithelia, such that each cell has a polarity vector oriented in the plane of the epithelium. (Chapter 21) (Figure 21–50)

plant growth regulator (**plant hormone**) Signal molecule that helps coordinate growth and development. Examples are ethylene, auxins, gibberellins, cytokinins, abscisic acid, and the brassinosteroids. (Chapter 15)

plasma membrane The membrane that surrounds a living cell. (Chapters 1, 10) (Figure 10–1)

plasmid vector Small, circular molecules of double-stranded DNA that have been derived from plasmids that occur naturally in bacterial cells; widely used for gene cloning. (Chapter 8)

plasmodesma (plural **plasmodesmata**) Plant equivalent of a gap junction. Communicating cell–cell junction in plants in which a channel of cytoplasm lined by plasma membrane connects two adjacent cells through a small pore in their cell walls. (Chapter 19)

platelet Cell fragment, lacking a nucleus, that breaks off from a megakaryocyte in the bone marrow and is found in large numbers in the bloodstream. Helps initiate blood clotting when blood vessels are injured. (Figure 22–12)

pleckstrin homology domain (PH domain) Protein domain found in some intracellular signaling proteins. Some PH domains in intracellular signaling proteins bind to phosphatidylinositol 3,4,5-trisphosphate produced by PI 3-kinase, bringing the signaling protein to the plasma membrane when PI 3-kinase is activated. (Chapter 15)

pluripotent Describes a cell that has the potential to give rise to all or almost all of the cell types of the adult body. (Chapters 21, 22)

podosomes Actin-rich protrusive and adhesive structures on the cell surface important for crossing tissue barriers. (Chapter 16)

point spread function The distribution of light intensity within the three-dimensional, blurred image that is formed when a single point source of light is brought to a focus with a lens. (Chapter 9)

polarized In epithelia, the fact that the basal end of a cell, adherent to the basal lamina below, differs from the apical end, exposed to the medium above; thus, all epithelia and their individual cells are structurally polarized. (Chapter 19)

Polycomb group Set of proteins critical for cell memory for some genes. They form complexes as part of the chromatin of the *Hox* complex, where they maintain a repressed state in cells where *Hox* genes have not been activated. (Chapters 4, 21) (Figure 4–40)

polymerase chain reaction (PCR) Technique for amplifying specific regions of DNA by the use of sequence-specific primers and multiple cycles of DNA synthesis, each cycle being followed by a brief heat treatment to separate complementary strands. (Chapter 8) (Figure 8–34)

polymorphisms Describes genome sequences that coexist as two or more sequence variants at high frequency in a population. (Chapter 8)

polypeptide Linear polymer of amino acids. Proteins are large polypeptides, and the two terms can be used interchangeably. (Figure 3–1)

polypeptide backbone Repeating sequence of atoms along the core of the polypeptide chain. (Figure 3–1)

polyribosome mRNA engaged with multiple ribosomes in the act of translation. (Figure 6–77)

polytene chromosome Giant chromosome in which the DNA has undergone repeated replication and the many copies have stayed together in precise alignment. (Chapter 4) (Figures 4–47 and 4–48)

porins Channel-forming proteins of the outer membranes of bacteria, mitochondria, and chloroplasts. (Chapter 12)

position effect variegation Alteration in gene expression resulting from a change in the position of the gene on a chromosome, caused by the spread of neighboring heterochromatic domains. When an active gene is placed next to heterochromatin, the inactivating influence of the heterochromatin can spread to affect the gene to a variable degree, giving rise to a variegated phenotype. (Figure 4–32)

positional value A cell's internal record of its positional information in a multicellular organism; an intrinsic character that differs according to a cell's location. (Chapter 21)

positive feedback Control mechanism whereby the end product of a reaction or pathway stimulates its own production or activation. (Figure 7–42)

positive selection In immunology: a process of thymocyte maturation in which thymocytes expressing a T cell receptor with appropriate affinity for a self peptide bound to a self MHC protein is signaled to survive and continue development. (Chapter 24)

post-transcriptional controls Any control on gene expression that is exerted at a stage after transcription has begun. (Chapter 7) (Figure 7–57)

post-translational Occurring after completion of translation, thus after the completion of protein synthesis. (Chapter 12)

preprophase band Circumferential band of microtubules and actin filaments that forms around a plant cell under the plasma membrane prior to mitosis and cell division. (Chapter 17) (Figure 17–48)

primary cell wall The first cell wall produced by a developing plant cell; it is thin and flexible, allowing room for cell growth. (Chapter 19) (Figure 19–64)

primary cilium Short, single, nonmotile cilium lacking dynein that arises from a centriole and projects from the surface of many animal cell types. Involved in cell signaling, with some signaling proteins being concentrated in these (Chapter 15) (Figure 15–39)

primary Ig repertoire The billions of IgM and IgD immunoglobulin molecules made by the B cells of an adaptive immune system in the absence of antigen stimulation. (Chapter 24)

primary immune response Adaptive immune response to an antigen that is made on first encounter with that antigen. (Chapter 24) (Figure 24–16)

primary pathogens Pathogens that can cause overt disease in most healthy people. Some cause acute, life-threatening epidemic infections and spread rapidly between hosts; other potential primary pathogens may persistently infect a single individual for years without causing overt disease, the host often being unaware of being infected. (Chapter 23)

primary structure Linear sequence of monomer units in a polymer, such as the amino acid sequence of a protein. (Chapter 3)

primary tumor Tumor at the original site at which a cancer first arose. Secondary tumors develop elsewhere by metastasis. (Chapter 20)

prion disease Transmissible spongiform encephalopathy—such as kuru and Creutzfeldt–Jakob disease (CJD) in humans, scrapie in sheep, and bovine spongiform encephalopathy (BSE, or "mad cow disease") in cows—that is caused and transmitted by an infectious, abnormally folded protein (prion). (Figure 3–33)

pro-apoptotic Bcl2 family effectors Pro-apoptotic proteins of the intrinsic pathway of apoptosis that in response to an apoptotic stimulus become activated and aggregate to form oligomers in the mitochondrial outer membrane, inducing the release of cytochrome *c* and other intermembrane proteins. Bax and Bak are the main effector Bcl2 family proteins in mammalian cells. (Chapter 18)

pro-inflammatory cytokine Any cytokine that stimulates an inflammatory response. (Chapter 24)

prokaryote Single-celled microorganism whose cells lack a well-defined, membrane-enclosed nucleus. Either a bacterium or an archaeon. (Chapter 1) (Figure 1–9)

promoter Nucleotide sequence in DNA to which RNA polymerase binds to begin transcription. (Chapters 6, 7) (Figure 6–11)

prophase The initial stage of mitosis, during which the chromosomes condense and the mitotic spindle begins to assemble outside of the nucleus. It is followed by prometaphase—where the nuclear envelope disassembles, and the microtubules from each spindle pole engage with the chromosomes. (Panel 17–1, pp. 1048–1049)

proteasome Large protein complex in the cytosol with proteolytic activity that is responsible for degrading proteins that have been marked for destruction by ubiquitylation or by some other means. (Chapter 6) (Figures 6–86 and 6–87)

protein The major macromolecular constituent of cells. A linear polymer of amino acids linked together by peptide bonds in a specific sequence. (Chapters 1, 3) (Figure 3–1)

protein activity control The selective activation, inactivation, degradation, or compartmentalization of specific proteins after they have been made. One of the means by which a cell controls which proteins are active at a given time or location in the cell. (Chapter 7)

protein degradation control The means by which the concentration of a protein in the cell can be reduced by selectively degrading it in response to external signals or to stages of the cell division cycle. (Chapter 7)

protein domain *See* **domain**. (Chapter 3)

protein glycosylation The process of transferring either a single saccharide or a preformed precursor oligosaccharide to proteins. (Chapter 12)

protein kinase Enzyme that transfers the terminal phosphate group of ATP to one or more specific amino acids (serine, threonine, or tyrosine) of a target protein. (Chapters 3, 15)

protein kinase C (PKC) Ca^{2+}-dependent protein kinase that, when activated by diacylglycerol and an increase in the concentration of cytosolic Ca^{2+}, phosphorylates target proteins on specific serine and threonine residues. (Chapter 15) (Figure 15–30)

protein phosphatase Enzyme that catalyzes phosphate removal from amino acids of a target protein. (Chapters 3, 15)

protein phosphatase 2A (PP2A) Protein phosphatase, composed of three subunits, that dephosphorylates Cdk substrates and thereby helps to govern cell-cycle progression. (Chapter 17)

protein phosphorylation The covalent addition of a phosphate group to a serine, threonine, or tyrosine side chain of a protein. (Chapter 3)

protein subunit An individual protein chain in a protein composed of more than one chain. (Chapter 3)

protein translocation The process of moving a protein across a membrane. (Chapter 12)

protein translocator Any membrane-bound protein that mediates the transport of another protein across a membrane. (Chapter 12) (Figure 12–48)

protein tyrosine phosphatase Enzyme that removes phosphate groups from phosphorylated tyrosine residues on proteins. (Chapter 15)

proteoglycan Molecule consisting of one or more glycosaminoglycan chains attached to a core protein. (Chapters 13, 19) (Figure 19–36)

proto-oncogene Normal gene, usually concerned with the regulation of cell proliferation, that can be converted into a cancer-promoting oncogene by mutation. (Chapter 20)

protofilament Linear string of subunits joined end to end; multiple protofilaments associate with one another laterally to construct and provide strength and adaptability to microtubules. (Chapter 16)

proton (H^+) Positively charged subatomic particle that forms part of an atomic nucleus. Hydrogen has a nucleus composed of a single proton (H^+). (Chapter 2)

proton-motive force The force exerted by the electrochemical proton gradient that moves protons across a membrane. (Chapter 14)

protozoan parasite Parasitic, nonphotosynthetic, single-celled, motile eukaryotic organism; for example, *Plasmodium*. (Chapter 23)

pseudogene Nucleotide sequence of DNA that has accumulated multiple mutations that have rendered an ancestral gene inactive and nonfunctional. (Chapter 4)

pseudopodia Three-dimensional actin-based protrusions at the leading edge of rapidly migrating cells. (Chapter 16)

purified cell-free system Fractionated cell homogenate that retains a particular biological function of the intact cell, and in which biochemical reactions and cell processes can be most easily studied. (Chapter 8)

purifying selection Natural selection operating in a population to slow genome changes and reduce divergence by eliminating individuals carrying deleterious mutations. (Chapter 4)

Q cycle The cyclic series of reactions that allow the quinone (Q) electron carrier to transfer an electron to the cytochrome *c* reductase complex and then recapture it, in order to pump two rather than one proton per electron across the membrane in a respiratory system; a similar Q cycle occurs as part of the electron transfers critical for photosynthesis. (Figure 14–23)

quantitative RT-PCR (reverse transcription–polymerase chain reaction) Technique in which a population of mRNAs is converted into cDNAs via reverse transcription, and the cDNAs are then amplified by PCR. The quantitative part relies on a direct relationship between the rate at which the PCR product is generated and the original concentration of the mRNA species of interest. Widely used for detecting viral infection. (Chapter 8)

quaternary structure Three-dimensional relationship of the different polypeptide chains in a multisubunit protein or protein complex. (Chapter 3)

quinone (Q) Small, lipid-soluble, mobile electron carrier molecule found in the respiratory and photosynthetic electron-transport chains. (Chapter 14) (Figure 14–17)

Rab cascade An ordered recruitment of sequentially acting Rab proteins into Rab domains on membranes, which changes the identity of an organelle and determines membrane dynamics. (Chapter 13)

Rab effectors Molecules that bind activated, membrane-bound Rab proteins and act as downstream mediators of vesicle transport, membrane tethering, and fusion. (Chapter 13)

Rab proteins Monomeric GTPases in the Ras superfamily present in plasma and organelle membranes in their

GTP-bound state and as soluble cytosolic proteins in their GDP-bound state. Involved in conferring specificity on vesicle docking. (Chapter 13) (Table 15–5, p. 915)

Rac (**Rac protein**) Member of the Rho family of monomeric GTPases that regulate the actin and microtubule cytoskeletons, cell-cycle progression, gene transcription, and membrane transport. (Chapter 15)

Rad51 protein Eukaryotic protein that catalyzes the pairing of homologous DNA strands during recombination and repair processes . Analogous to the RecA protein in *E. coli* and other bacteria. (Chapter 5)

Ran (**Ran protein**) Monomeric GTPase of the Ras superfamily present in both cytosol and nucleus. It is required for the active transport of macromolecules into and out of the nucleus through nuclear pore complexes. (Chapter 12) (Table 15–5, p. 915)

rapidly inactivating K^+ channel Neuronal voltage-gated K^+ channel, open when the membrane is depolarized, with a specific voltage sensitivity and kinetics of inactivation that induce a reduced rate of action potential firing at levels of stimulation only just above the threshold required, thereby resulting in a firing rate proportional to the strength of the depolarizing stimulus. (Chapter 11)

Ras A small family of proto-oncogenes that are frequently mutated in cancers, each of which produces a Ras monomeric GTPase. (Chapter 20)

Ras (**Ras protein**) Monomeric GTPase of the Ras superfamily that helps to relay signals from cell-surface receptor tyrosine kinase receptors to the nucleus, frequently in response to signals that stimulate cell division. Named for the *Ras* gene, first identified in viruses that cause rat sarcomas. (Chapters 3, 15, 17) (Figure 3–64)

Ras GAPs Ras GTPase-activating proteins; increase the rate of hydrolysis of bound GTP by Ras, thereby inactivating Ras. (Chapter 15)

Ras GEFs Ras guanine nucleotide exchange factors; stimulate the dissociation of GDP and the subsequent uptake of GTP from the cytosol, thereby activating Ras. (Chapter 15)

Ras–MAP-kinase signaling pathway Intracellular signaling pathway that relays signals from activated receptor tyrosine kinases to effector proteins in the cell, including transcription regulators in the nucleus. (Chapter 15)

Ras superfamily Large superfamily of monomeric GTPases (also called small GTP-binding proteins) of which Ras is the prototypical member. (Chapter 15) (Table 15–5, p. 915)

Rb A cell-cycle control protein that inhibits E2F transcription regulatory proteins and entry into S phase of the cell cycle. The *Rb* gene is a tumor suppressor that is defective in both copies in individuals with retinoblastoma. (Chapter 20) (Figures 17–59 and 20–24)

reading frame The phase in which nucleotides are read in sets of three to encode a protein. An mRNA molecule can be read in any one of three reading frames, only one of which will give the required protein. (Chapter 6) (Figure 6–53)

RecA protein Prototype for a ubiquitous class of DNA-binding proteins that catalyze synapsis of DNA strands during genetic recombination in bacteria; analogous to Rad51 protein in eukaryotes. (Chapter 5) (Figure 5–48)

receptor Any protein that binds a specific signal molecule (ligand) and initiates a response in the cell. Some are on the cell surface, while others are inside the cell. (Chapter 15) (Figure 15–3)

receptor editing Process by which a developing B cell that recognizes a self molecule changes its antigen receptors so that the cell no longer does so. (Chapter 24)

receptor-mediated endocytosis Internalization of receptor–ligand complexes from the plasma membrane by endocytosis, a type of receptor down-regulation. (Chapter 13) (Figure 13–54)

receptor serine/threonine kinase Cell-surface receptor with an extracellular ligand-binding domain and an intracellular kinase domain that phosphorylates signaling proteins on serine or threonine residues in response to ligand binding. The TGF receptor is an example. (Chapter 15) (Figure 15–58)

receptor tyrosine kinase (**RTK**) Cell-surface receptor with an extracellular ligand-binding domain and an intracellular kinase domain that phosphorylates signaling proteins on tyrosine residues in response to ligand binding. (Chapter 15) (Figure 15–44 and Table 15–4, p. 911)

recombinant DNA technology Collection of techniques by which DNA segments from different sources are combined to make a new DNA, often called a recombinant DNA. Recombinant DNAs are widely used in the cloning of genes, in the genetic modification of organisms, and in the production of large amounts of rare proteins. (Chapter 8)

recycling endosome Organelle that provides an intermediate stage on the passage of recycled receptors back to the cell membrane. Regulates plasma membrane insertion of some proteins. (Chapter 13) (Figure 13–56)

red blood cell Small hemoglobin-containing blood cell of vertebrates that transports oxygen to, and carbon dioxide from, tissues. Also called an erythrocyte. (Figures 22–11 and 22–12)

redox pair Pair of molecules in which one acts as an electron donor and one as an electron acceptor in an oxidation–reduction reaction; for example, NADH (electron donor) and NAD^+ (electron acceptor). (Panel 14–1, p. 825)

redox potential The affinity of a redox pair for electrons, generally measured as the voltage difference between an equimolar mixture of the pair and a standard reference. NADH/NAD^+ has a low redox potential, and O_2/H_2 has a high redox potential (high affinity for electrons). (Panel 14–1, p. 825)

redox reaction Reaction in which one component becomes oxidized and the other reduced; an oxidation–reduction reaction. (Chapter 14) (Panel 14–1, p. 825)

reduction (verb **reduce**) Addition of electrons to an atom, as occurs during the addition of hydrogen to a biological molecule or the removal of oxygen from it. Opposite of oxidation. (Chapter 2) (Figure 2–20)

regulated secretory pathway A secretory pathway found mainly in cells specialized for secreting products rapidly on demand—such as hormones, neurotransmitters, or digestive enzymes—in which soluble proteins and other substances are initially stored in secretory vesicles for later release. (Chapter 13) (Figure 13–38)

regulator of G protein signaling (**RGS**) A type of GAP protein that binds to a trimeric G protein and enhances its GTPase activity, thus helping to limit G-protein-mediated signaling. (Chapter 15) (Figure 15–8)

regulatory site Region of an enzyme surface to which a regulatory molecule binds and thereby influences the catalytic events at the separate active site. (Chapter 3)

regulatory T cell (T_{reg}) A type of T cell that suppresses the development, activation, or function of other immune cells via secreted cytokines or cell-surface inhibitory proteins. (Chapter 24)

replication fork Y-shaped region of a replicating DNA molecule; the point at which the two strands of the parent DNA helix are being separated and the daughter strands are being formed. (Chapter 5) (Figures 5–7 and 5–18)

replication origin A location on a DNA molecule at which duplication of the DNA begins by the formation of replication forks. (Chapters 4, 5) (Figures 4–19 and 5–24)

replicative cell senescence Phenomenon observed in primary cell cultures in which cell proliferation slows down and finally irreversibly halts. (Chapters 17, 20)

respiratory chain (electron-transport chain) Electron-transport chain present in the inner mitochondrial membrane that generates an electrochemical proton gradient across the membrane: this gradient is used to drive ATP synthesis by ATP synthase. (Chapter 14) (Figures 14–4 and 14–10)

resting membrane potential Electrical potential across the plasma membrane of a cell at rest; that is, a cell that has not been stimulated to open additional ion channels beyond those that are normally open. (Chapter 11)

restriction nuclease One of a large number of nucleases that can cleave a DNA molecule at any site where a specific short sequence of nucleotides occurs. Extensively used in recombinant DNA technology. (Chapter 8) (Figure 8–23)

restriction point Important transition at the end of G_1 in the eukaryotic cell cycle; passage past this point commits the cell to enter S phase. The term was originally used for this transition in the mammalian cell cycle; in this book we use the term "Start." (Chapter 17) (Figure 17–8)

retinoblastoma A rare type of human cancer arising from cells in the retina of the eye that are converted to a cancerous state by an unusually small number of mutations. Studies of retinoblastoma led to the discovery of the first tumor suppressor gene. (Chapter 20)

retinoblastoma protein (Rb protein) Tumor suppressor protein involved in the regulation of cell division. Mutated in the cancer retinoblastoma, as well as in many other tumors. Its normal activity is to regulate the eukaryotic cell cycle by binding to and inhibiting the E2F proteins, thus blocking progression to DNA replication and cell division. (Chapter 17) (Figure 17–59)

retroviral-like retrotransposons A large family of transposons that move themselves in and out of chromosomes by a mechanism similar to that used by retroviruses, being first transcribed into an RNA copy that is converted to DNA by reverse transcriptase then inserted elsewhere in the genome. (Chapter 5) (Table 5–4, p. 308)

retrovirus RNA-containing virus that replicates in a cell by first making an RNA–DNA intermediate and then a double-strand DNA molecule that becomes integrated into the cell's DNA. (Chapters 5, 20) (Figure 5–61)

reverse genetics Approach to discovering gene function that starts from the DNA (gene) and its protein product and then creates mutants to analyze the gene's function. (Chapter 8)

reverse transcriptase Enzyme first discovered in retroviruses that makes a double-strand DNA copy from a single-strand RNA template molecule. (Chapter 5)

Reynolds number A dimensionless number that represents the ratio of inertial forces to viscous forces acting on an object moving through a fluid. The inertial force or momentum relates to the size of an object and its speed, while viscous forces depend on the properties of the fluid. (Chapter 16)

RGD sequence Tripeptide sequence of arginine-glycine-aspartic acid that forms a binding site for integrins; present in fibronectin and some other extracellular proteins. (Chapter 19) (Figure 19–48C)

Rho family Family of monomeric GTPases within the Ras superfamily involved in signaling the rearrangement of the cytoskeleton. Includes Rho, Rac, and Cdc42. (Chapters 15, 16) (Table 15–5, p. 915)

Rho protein Member of the Rho family of monomeric GTPases that regulate the actin and microtubule cytoskeletons, cell-cycle progression, gene transcription, and membrane transport. (Chapter 15)

rhodopsin Seven-span membrane protein of the GPCR family that acts as a light sensor in rod photoreceptor cells in the vertebrate retina. Contains the light-sensitive prosthetic group retinol. (Chapter 15) (Figure 15–40)

ribonucleic acid *See* **RNA**. (Chapter 1)

ribosomal RNA (rRNA) Any one of a number of specific RNA molecules that form part of the structure of a ribosome and participate in the synthesis of proteins. Often distinguished by their sedimentation coefficient (for example, 28S rRNA, 5S rRNA). (Chapter 6)

ribosome Particle composed of rRNAs and ribosomal proteins that catalyzes the synthesis of protein using information provided by mRNA. (Chapter 6) (Figure 6–68)

ribozyme An RNA molecule with catalytic activity. (Chapter 6)

RNA (ribonucleic acid) Polymer formed from covalently linked ribonucleotide monomers. *See also* **messenger RNA**, **ribosomal RNA**, **transfer RNA**. (Chapter 1) (Figures 6–4 and 6–7)

RNA editing Type of RNA processing that alters the nucleotide sequence of an RNA transcript after it is synthesized by inserting, deleting, or altering individual nucleotides. (Chapter 7)

RNA exosome Large protein complex with an interior rich in 3′-to-5′ RNA exonucleases; degrades RNA molecules to produce ribonucleotides. (Chapter 6)

RNA interference (RNAi) As originally described, mechanism by which an experimentally introduced double-stranded RNA induces sequence-specific destruction of complementary mRNAs. The term RNAi is often used to include the inhibition of gene expression by microRNAs (miRNAs) and piwi-interacting RNAs (piRNAs), which are encoded in the cell's own genome. (Chapter 7)

RNA polymerase Enzyme that catalyzes the synthesis of an RNA molecule on a DNA template from ribonucleoside triphosphate precursors. (Chapter 6) (Figure 6–9)

RNA primer Short stretch of RNA synthesized on a DNA template. It is required by DNA polymerases to start their DNA synthesis. (Chapter 5)

RNA-processing control Regulation by a cell of gene expression by controlling the processing of RNA transcripts, which includes their splicing. (Chapter 7)

RNA splicing Process in which intron sequences are excised from RNA transcripts. A major process in the nucleus of eukaryotic cells leading to formation of messenger RNAs (mRNAs). (Chapter 6)

proliferation, and gene expression in animal embryos and adult tissues. (Chapter 15)

Wnt/β-catenin pathway Signaling pathway activated by binding of a Wnt protein to its cell-surface receptors. The pathway has several branches. In the major (canonical) branch, activation causes increased amounts of β-catenin to enter the nucleus, where it regulates the transcription of genes controlling cell differentiation and proliferation. Overactivation of the Wnt/β-catenin pathway can lead to cancer. (Chapter 15) (Figure 15–61)

XIAP An inhibitor of apoptosis protein (IAP) that is encoded on the X chromosome. (Chapter 18)

X-inactivation Inactivation of one copy of the X chromosome in the somatic cells of female mammals. (Chapter 7)

X-inactivation center (XIC) Site in an X chromosome at which inactivation is initiated and spreads outward. (Figure 7–55)

Xist A long (20,000 nucleotide) noncoding RNA responsible for inactivating one of the two X chromosomes in female mammals. (Chapter 7)

x-ray crystallography Technique for determining the three-dimensional arrangement of atoms in a molecule on the basis of the diffraction pattern of x-rays passing through a crystal of the molecule. (Chapter 8) (Figure 8–20)

zygote Diploid cell produced by fusion of a male and female gamete. A fertilized egg. (Chapter 21)

Index

Note

F indicates coverage in a Figure only, on a page separated from the indexed text treatment, T indicates inclusion in a Table, and P indicates a Panel.

Acronyms have generally been preferred to their expansions.

A

B

C

D

E

G

J

K

M

N

O

Q

R

S

T

U

Z

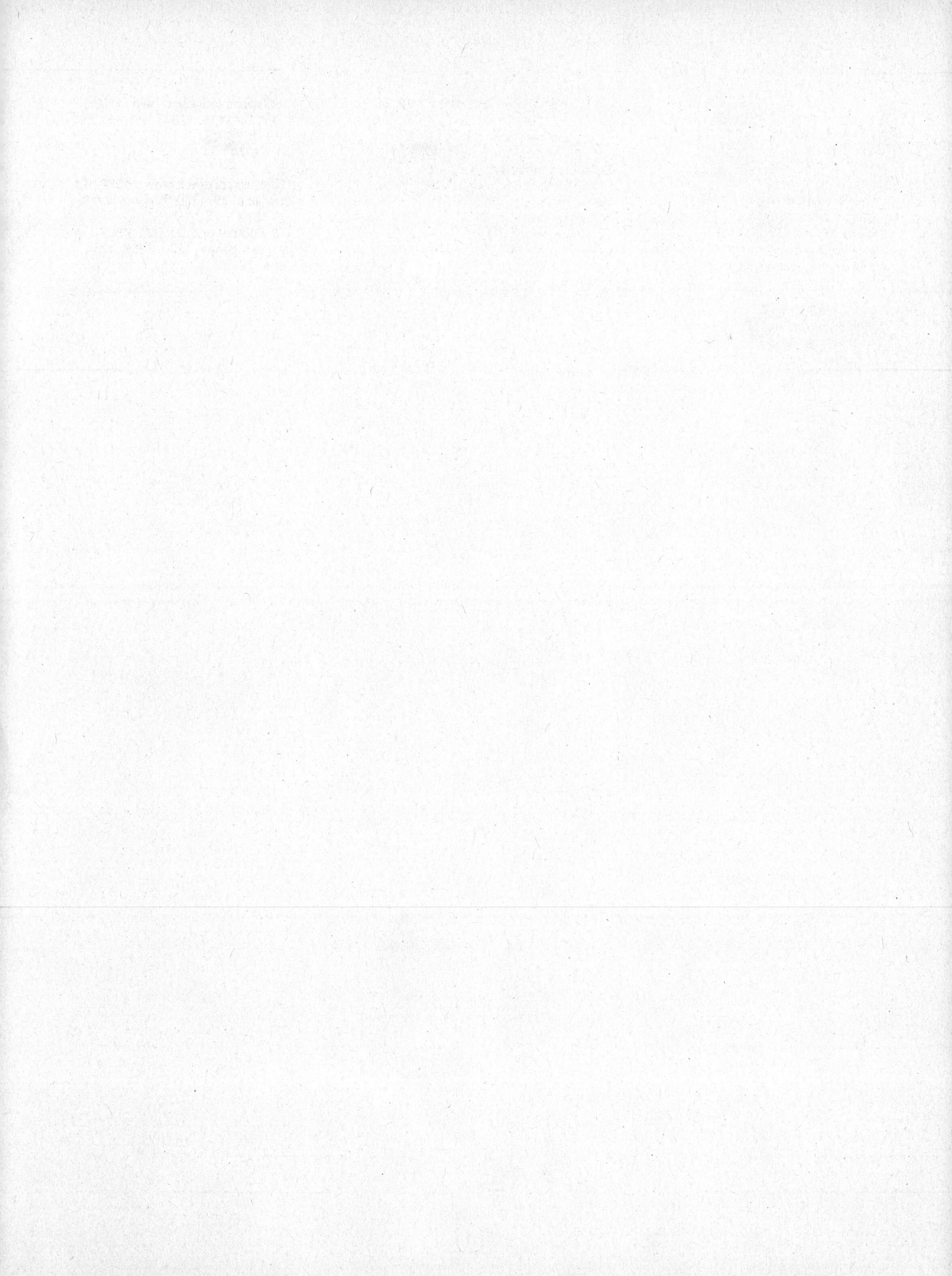